HANDBOOK OF HUMAN FACTORS AND ERGONOMICS

HANDBOOK OF HUMAN FACTORS AND ERGONOMICS

Third Edition

Edited by
Gavriel Salvendy
Purdue University
West Lafayette, Indiana
and
Tsinghua University
Beijing, People's Republic of China

WILEY
JOHN WILEY & SONS, INC.

This book is printed on acid-free paper. ♾

Published by John Wiley & Sons, Inc., Hoboken, New Jersey
Published simultaneously in Canada

Disclaimer

The editor, authors, and the publisher have made every effort to provide accurate and complete information in the Handbook but the Handbook is not intended to serve as a replacement for professional advice. Any use of the information in this Handbook is at the reader's discretion. The editor, authors, and the publisher specifically disclaim any and all liability arising directly or indirectly from the use or application of any information contained in this Handbook. An appropriate professional should be consulted regarding your specific situation.

For general information about our other products and services, please contact our Customer Care Department within the United States at (800) 762-2974, outside the United States at (317) 572-3993 or fax (317) 572-4002.

Wiley also publishes its books in a variety of electronic formats. Some content that appears in print may not be available in electronic books. For more information about Wiley products, visit our web site at www.wiley.com.

Library of Congress Cataloging-in-Publication Data:

Handbook of human factors and ergonomics / edited by Gavriel Salvendy.–3rd ed.

p. cm.
ISBN-13 978-0-471-44917-1 (Cloth)
ISBN-10 0-471-44917-2 (Cloth)
1. Human engineering–Handbooks, manuals, etc. I. Salvendy, Gavriel, 1938-
TA166.H275 2005
620.8′2–dc22

2005003111

Printed in the United States of America

10 9 8 7 6 5 4 3 2

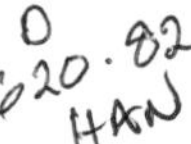

ABOUT THE EDITOR

Gavriel Salvendy is a professor of industrial engineering at Purdue University and Chair, Professor, and Head of the Department of Industrial Engineering at Tsinghua University, Beijing, People's Republic of China. He is the author and co-author of over 430 research publications, including over 230 journal papers, and is the author or editor of 28 books. His publications have appeared in seven languages. Gavriel Salvendy is the founding editor of the *International Journal on Human–Computer Interaction* and *Human Factors and Ergonomics in Manufacturing*. He was the founding chair of the International Commission on Human Aspects in Computing, headquartered in Geneva, Switzerland. In 1990 he became the first member of either the Human Factors and Ergonomics Society or the International Ergonomics Association to be elected to the National Academy of Engineering. He was elected "for fundamental contributions to and professional leadership in human, physical, and cognitive aspects of engineering systems." In 1995 he received an honorary doctorate from the Chinese Academy of Science "for great contributions to the development of science and technology and for the great influence upon the development of science and technology in China." He is the fourth person in all fields of science and engineering in the 45 years of the Academy ever to receive this award. He is an honorary fellow and life member of the Ergonomics Society and fellow of Human Factors and Ergonomics Society, Institute of Industrial Engineers, and the American Psychological Association. He has advised organizations in 30 countries on the human side of effective design, implementation, and management of advanced technologies in the workplace. He earned his Ph.D. in engineering production at the University of Birmingham, United Kingdom.

ADVISORY BOARD

CONTRIBUTORS

Chadia Abras
Director, Educational Technology & Distance Learning
Graduate & Professional Studies
Goucher College
1021 Dulaney Valley Road
Baltimore, Maryland
cabras@goucher.edu

Margherita Antona
Research Associate
Foundation for Research and Technology–Hellas (FORTH)
Institute of Computer Science
Vassilika Vouton
Heraklion, Crete, Greece
antona@ics.forth.gr

Susan Archer
Director of Operations
Micro Analysis and Design, Inc.
4949 Pearl East Circle
Boulder, Colorado
sarcher@maad.com

Nuray Aykin
Director of Special Projects
Office of the President
The New School
55 West 13th Street, Room 211
New York, New York
aykinn@newschool.edu

Kevin B. Bennett
Professor and Graduate Program Director
Department of Psychology
Wright State University
3640 Colonel Glenn Highway
Dayton, Ohio
kevin.bennett@wright.edu

Carolyn K. Bensel
Senior Research Psychologist
Supporting Science and Technology Directorate
U.S. Army Natick Soldier Center
Kansas Street
Natick, Massachusetts
Carolyn.Bensel@Natick.Army.Mil

Kenneth R. Boff
Chief Scientist
Human Effectiveness Division
Air Force Research Laboratory
Wright-Patterson Air Force Base, Ohio
Ken.Boff@wpafb.af.mil

Walter C. Borman
Professor
Department of Psychology
University of South Florida
4202 Fowler Avenue
Tampa, Florida
wally.borman@pdri.com

Peter R. Boyce
Consultant
60, Riverside Close,
Bridge, Canterbury, Kent, Great Britain
peter@boycepeter.freeserve.co.uk

Martin Braun
Senior Scientist
Fraunhofer-Institut für Arbeitswirtschaft und Organisation (IAO)
Nobelstrasse 12
Stuttgart, Germany
martin.braun@iao.fraunhofer.de

Michael A. Campion
Professor of Management
Krannert Graduate School of Management
Purdue University
403 West State Street
West Lafayette, Indiana
campion@mgmt.purdue.edu

Pascale Carayon
Proctor & Gamble Bascom Professor in Total Quality
Department of Industrial and Systems Engineering
Director of the Center for Quality and Productivity Improvement
University of Wisconsin–Madison
610 Walnut Street, 575 WARF
Madison, Wisconsin
carayon@engr.wisc.edu

C. Melody Carswell
Associate Professor
Department of Psychology
University of Kentucky
205 Kastle Hall
Lexington, Kentucky
cmcars00@pop.uky.edu

John G. Casali
John Grado Professor
Grado Department of Industrial and Systems Engineering
Virginia Polytechnic Institute and State University
Blacksburg, Virginia,
jcasali@vt.edu

Joseph Cohn
Naval Research Laboratory
4555 Overlook Avenue SW
Washington, D.C.
cohn@itd.nrl.navy.mil

Sara J. Czaja
Professor and Co-Director
Department of Psychiatry and Behavioral Center on Aging Sciences
University of Miami Miller School of Medicine
1695 NW 9th Avenue, Suite 3208G
Miami, Florida
sczaja@med.miami.edu

Patrick G. Dempsey
Director of Experimental Programs
Liberty Mutual Research Institute for Safety
71 Frankland Road
Hopkinton, Massachusetts
Patrick.Dempsey@LibertyMutual.com

Karen M. Dettinger
University of Wisconsin Hospital and Clinics
Madison, Wisconsin
km.dettinger@hosp.wisc.edu

Colin G. Drury
UB Distinguished Professor and Chair
Department of Industrial Engineering
University at Buffalo: State University of New York
438 Bell Hall
Buffalo, New York
drury@buffalo.edu

David W. Eby
Associate Professor and Head
Social and Behavioral Division Research
University of Michigan Transportation Research Institute
2901 Baxter Road
Ann Arbor, Michigan
eby@umich.edu

Paula J. Edwards
Research Assistant
Institute for Health Systems Engineering
Georgia Institute of Technology
Atlanta, Georgia
pedwards@isye.gatech.edu

Gabriele Elke
Department of Psychology
Ruhr University–Bochum
Postfach 102148
Bochum, Germany
ge@auo.psy.ruhr-uni-bochum.de

Mica R. Endsley
President
SA Technologies, Inc.
3750 Palladian Village Drive
Building 600
Marietta, Georgia
Mica@SATechnologies.com

Donald L. Fisher
Professor
Department of Mechanical and Industrial Engineering
University of Massachusetts Amherst
Marston Hall, 130 Natural Resources Road
Amherst, Massachusetts
fisher@ecs.umass.edu

Arthur D. Fisk
Professor
School of Psychology
Georgia Institute of Technology
654 Cherry Street
Atlanta, Georgia
af7@mail.gatech.edu

John M. Flach
Professor and Chair
Department of Psychology
Wright State University
3640 Colonel Glenn Highway
Dayton, Ohio
john.flach@wright.edu

Wolfgang Friesdorf
Professor
Department of Human Factors and Product Ergonomics
Technische Universitat Berlin
Fasanenstrasse 1/1
Berlin, Germany
wolfgang.friesdorf@awb.tu-berlin.de

Michael J. Griffin
Professor
Institute of Sound and Vibration Research
University of Southampton
Southampton, Hampshire, England
mjg@isvr.soton.ac.uk

Joseph W. Guthrie
Institute of Simulation and Training
University of Central Florida
3100 Technology Parkway
Orlando, Florida
jguthrie@ist.ucf.edu

Lorenz Hagenmeyer
Senior Scientist
Fraunhofer-Institut für Arbeitswirtschaft und Organisation (IAO)
Nobelstrasse 12
Stuttgart, Germany
lorenz.hagenmeyer@iao.fraunhofer.de

Jerry W. Hedge
President
Organizational Solutions Group
449 Gum Street
Holly Hill, South Carolina
jerwhedge@aol.com

Martin G. Helander
Professor
School of Mechanical and Aerospace Engineering
Nanyang Technological University
Singapore
martin@ntu.edu.sg

Erik Hollnagel
Professor
Department of Computer and Information Science
LIU/IDA/CSELAB
University of Linköping
Linköping, Sweden
eriho@ida.liu.se

Pia Honold Quaet-Faslem
User Interface Designer
Strategy and Marketing: User Experience
Siemens AG Information and Communication Mobile
Haidenauplatz 1
Munich, Germany
pia.quaet-faslem@siemens.com

Juan Pablo Hourcade
Assistant Professor
Department of Computer Science
14 MacLean Hall
University of Iowa
Iowa City, Iowa
jpablo@acm.org

Heidi D. Howarth
Engineering Psychologist
Volpe National Transportation Systems Center
U.S. Department of Transportation
55 Broadway, Kendall Square
Cambridge, Massachusetts
heidi.howarth@volpe.dot.gov

Jeffrey A. Hudson
General Dynamics AIS
5200 Springfield Pike, Suite 200
Dayton, Ohio
jeff.hudson@wpafb.af.mil

Julie A. Jacko
Professor
Institute for Health Systems Engineering
Wallace H. Coulter Department of Biomedical Engineering
Georgia Institute of Technology and Emory University School of Medicine
313 Ferst Drive
Atlanta, Georgia
jacko@bme.gatech.edu

Elżbieta Jankowska
Head of the Laboratory of Filtration and Ventilation
Central Institute for Labour Protection
National Research Institute
Czerniakowska 16
Warsaw, Poland

Tanja Kabel
Research Assistant
Institute of Industrial Engineering and Ergonomics
RWTH Aachen University
Bergdriesch 27
Aachen, Germany
tanja.kabel@t-online.de

Barry H. Kantowitz
Professor of Industrial and Operations Engineering
Professor of Psychology
University of Michigan Transportation Research Institute, 2901 Baxter Road
Ann Arbor, Michigan
barrykan@umich.edu

Jolanta Karpowicz
Head of the Laboratory of Electromagnetic Hazards
Central Institute for Labour Protection
National Research Institute
Czerniakowska 16
Warsaw, Poland
jokar@ciop.pl

Waldemar Karwowski
Professor and Director
Center for Industrial Ergonomics
University of Louisville
Lutz Hall, Room 445
Warnock Street
Louisville, Kentucky
karwowski@louisville.edu

Halimahtun M. Khalid
Director
Damai Sciences
A-31-3 Suasana Sentral
Jalan Stesen Sentral 5
50470 Kuala Lumpur
Malaysia
mahtun@damai-sciences.com

Mika Kivimäki
Professor of Occupational Health Psychology
Department of Psychology
Finnish Institute of Occupational Health
Topeliuksenkatu 41 aA
Helsinki, Finland
Mika.Kivimaki@ttl.fi

Thitima Kongnakorn
Graduate Assistant
School of Industrial and Systems Engineering
Georgia Institute of Technology
765 Ferst Drive
Atlanta, Georgia

Danuta Koradecka
Director
Central Institute for Labour Protection
National Research Institute
Czerniakowska 16
Warsaw, Poland
dakor@ciop.pl

Kenneth R. Laughery
Professor
Psychology Department
Rice University
6100 Main Street
Houston, Texas
laugher@ruf.rice.edu

K. Ronald Laughery, Jr.
President
Micro Analysis and Design, Inc.
4949 Pearl East Circle
Boulder, Colorado
rlaughery@maad.com

Jonathan Lazar
Associate Professor
Department of Computer and Information Sciences
Towson University
8000 York Road
Towson, Maryland
jlazar@towson.edu

Christian Lebiere
Principal Research Scientist
Micro Analysis and Design, Inc.
600 Thomas Boulevard
Pittsburgh, PA

John D. Lee
Associate Professor
Department of Mechanical and Industrial Engineering
University of Iowa
Iowa City, Iowa
jdlee@engineering.uiowa.edu

Mark Lehto
Associate Professor
School of Industrial Engineering
Purdue University
315 North Grant Street
West Lafayette, Indiana
lehto@purdue.edu

V. Kathlene Leonard
Postdoctoral Fellow
Institute for Health Systems Engineering
Wallace H. Coulter Department of Biomedical Engineering
Georgia Institute of Technology and Emory University School of Medicine
313 Ferst Drive
vkleonard@gmail.com

James R. Lewis
Senior Human Factors Engineer
International Business Machines Corporation Software Group
8051 Congress Avenue Suite 2227
Boca Raton, Florida
jimlewis@us.ibm.com

Torsten Licht
Research Assistant
Institute of Industrial Engineering and Ergonomics
RWTH Aachen University
Bergdriesch 27
Aachen, Germany
t.licht@iaw.rwth-aachen.de

Kari Lindström
Professor and Director of the Department of Psychology
Finnish Institute of Occupational Health
Topeliuksenkatu 41 aA
Helsinki, Finland
Kari.Lindstrom@ttl.fi

Holger Luczak
Director and Chair
Institute of Industrial Engineering and Ergonomics
Research Institute for Rationalization and Operations Management
Pontdriesch 14/16
Aachen, Germany
lcz@fir.rwth-aachen.de

Nicolas Marmaras
Associate Professor
School of Mechanical Engineering
National Technical University of Athens
GR 15780 Zografos
Athens, Greece
marmaras@central.ntua.gr

W. S. Marras
Honda Professor and Director
Biodynamics Laboratory
The Ohio State University
1971 Neil Avenue
Columbus, Ohio
marras.1@osu.edu

Gina J. Medsker
Manager
Strategic Human Capital Management Program
Human Resources Research Organization
66 Canal Center Plaza, Suite 400
Alexandria, Virginia
gmedsker@humrro.org

Allen Milewski
Associate Professor
Software Engineering Department
Monmouth University
West Long Branch, New Jersey
amilewsk@monmouth.edu

Frederick P. Morgeson
Associate Professor of Management
Eli Broad Graduate School of Management
Michigan State University
475 North Business Complex
East Lansing, Michigan
morgeson@msu.edu

Frances Mount
Senior Human Factors Specialist
National Space Biomedical Research Institute, Mail Code SF
NASA Johnson Space Center
Houston, Texas
frances.e.mount1@jsc.nasa.gov

Allen L. Nagy
Professor of Psychology
Department of Psychology
Wright State University
3640 Colonel Glenn Highway
Dayton, Ohio
allen.nagy@wright.edu

Fiona Fui-Hoon Nah
Associate Professor of Management Information Systems
Department of Management
University of Nebraska–Lincoln
209 College of Business Administration
Lincoln, Nebraska
fnah@unlnotes.unl.edu

Sankaran N. Nair
Director, Data Management
Center on Aging
University of Miami Miller School of Medicine
1695 NW 9th Avenue, Suite 3204
Miami, Florida
SNair@med.miami.edu

Dimitris Nathanael
Researcher
School of Mechanical Engineering
National Technical University of Athens
GR 15780 Zografos
Athens, Greece
dnathan@central.ntua.gr

Timothy A. Nichols
Graduate Student
School of Psychology
Georgia Institute of Technology
654 Cherry Street
Atlanta, Georgia
gte966q@mail.gatech.edu

Chris North
Assistant Professor
Department of Computer Science
Virginia Polytechnic Institute and State University
660 McBryde Hall
Blacksburg, Virginia
north@vt.edu

Roland Örtengren
Professor
Human Factors Engineering
Department of Product and Production Development
Chalmers University of Technology
Göteborg, Sweden
roland.ortengren@chalmers.se

Stephen M. Popkin
Chief, Human Factors Division
Volpe National Transportation Systems Center
U.S. Department of Transportation
55 Broadway, Kendall Square
Cambridge, Massachusetts
stephen.popkin@volpe.dot.gov

Małgorzata Pośniak
Head of the Department of Chemical and Aerosol Hazards
Central Institute for Labour Protection
National Research Institute
Czerniakowska 16
Warsaw, Poland
mapos@ciop.pl

Heather A. Priest
Institute of Simulation and Training
University of Central Florida
3100 Technology Parkway
Orlando, Florida
hpriest@ist.ucf.edu

Janet D. Proctor
Lead Academic Advisor, Psychology
Liberal Arts
Purdue University
100 North University Street
West Lafayette, Indiana
jproctor@sla.purdue.edu

Robert W. Proctor
Professor
Department of Psychological Sciences
Purdue University
703 Third Street
West Lafayette, Indiana
proctor@psych.purdue.edu

Ronald E. Rice
Arthur N. Rupe Endowed Professor
Department of Communication
University of California–Santa Barbara
4840 Ellison Hall
Santa Barbara, California
rrice@comm.ucsb.edu

Kathleen M. Robinette
Principal Research Anthropologist
Air Force Research Laboratory
2800 Q Street
Wright-Patterson Air Force Base, Ohio
kathleen.robinette@wpafb.af.mil

David Rodrick
Associate in Research
Learning Systems Institute
Florida State University
Tallahassee, Florida
drodrick@lsi.fsu.edu

Wendy A. Rogers
Professor
School of Psychology
Georgia Institute of Technology
654 Cherry Street
Atlanta, Georgia
wr43@prism.gatech.edu

William B. Rouse
Executive Director & Professor
Tennenbaum Institute
Georgia Institute of Technology
Atlanta, Georgia
bill.rouse@ti.gatech.edu

François Sainfort
William W. George Professor of Health Systems
College of Engineering
Georgia Institute of Technology
225 North Building
Atlanta, Georgia
sainfort@isye.gatech.edu

Eduardo Salas
Professor and Trustee Chair
Department of Psychology
Institute of Simulation and Training
University of Central Florida
3100 Technology Parkway
Orlando, Florida
esalas@ist.ucf.edu

William R. Santee
Research Physical Scientist
Biophysics and Biomedical Modeling Division
U.S. Army Institute of Environmental Medicine
Kansas Street
Natick, Massachusetts
william.santee@na.amedd.army.mil

Anthony Savidis
Research Associate
Institute of Computer Science
Foundation for Research and Technology–Hellas (FORTH)
Vassilika Vouton
Heraklion, Crete Greece
as@ics.forth.gr

Dylan Schmorrow
Program Officer
Office of Naval Research
875 N. Randolph Street, Room 1425
Arlington, Virginia
schmord@onr.navy.mil

E. Eugene Schultz
High Tower Software
Aliso Viejo, California
eeschultz@sbcglobal.net

Richard Schweickert
Professor of Psychological Sciences
Department of Psychology
Purdue University
703 Third Street
West Lafayette, Indiana
swike@psych.purdue.edu

Andrew Sears
Professor and Chair of Information Systems
Interactive Systems Research Center
UMBC
1000 Hilltop Circle
Baltimore, Maryland
asears@umbc.edu

Joseph Sharit
Research Professor
Department of Industrial Engineering
University of Miami
Coral Gables, Florida
jsharit@miami.edu

Bohdana Sherehiy
Graduate Fellow
Center for Industrial Ergonomics
University of Louisville
Lutz Hall, Room 445
Warnock Street
Louisville, Kentucky

Thomas B. Sheridan
Ford Professor of Engineering and Applied Psychology Emeritus
Department of Mechanical Engineering and Department of Aeronautics and Astronautics
Massachusetts Institute of Technology
77 Massachusetts Avenue, No. 3-435
Cambridge, Massachusetts
sheridan@mit.edu

Jolanta Skowroń
Head, Laboratory of Toxicology
Department of Chemistry and Aerosol Hazards
Central Institute for Labour Protection
National Research Institute
Czerniakowska 16
Warsaw, Poland

Michael J. Smith
The Duane H. and Dorothy M. Bluemke Professor of Engineering
Department of Industrial and Systems Engineering
University of Wisconsin–Madison
1513 University Avenue, Room 360
Madison, Wisconsin
mjsmith@engr.wisc.edu

Carolyn M. Sommerich
Associate Professor
Department of Industrial, Welding and Systems Engineering
The Ohio State University
1971 Neil Avenue
Columbus, Ohio
sommerich.1@osu.edu

Dieter Spath
Professor and Head of Institute
Fraunhofer-Institut für Arbeitswirtschaft und Organisation (IAO)
Nobelstrasse 12
Stuttgart, Germany
dieter.spath@iao.fraunhofer.de

Kay M. Stanney
Professor and Trustee Chair
Industrial Engineering and Management Systems
University of Central Florida
4000 Central Florida Boulevard
Orlando, Florida
stanney@mail.ucf.edu

Constantine Stephanidis
Professor and Director
Foundation for Research and Technology–Hellas (FORTH)
Institute of Computer Science
Vassilika Vouton
Heraklion, Crete, Greece
cs@ics.forth.gr

Cynthia Stohl
Professor of Communication
Department of Communication
University of California–Santa Barbara
4840 Ellison Hall
Santa Barbara, California
cstohl@comm.ucsb.edu

Anders Sundin
WM-data Caran AB
Göteborg, Sweden
anders.sundin@caran.com.

Alvaro D. Taveira
Professor and Department Chair
Department of Occupational and Environmental Safety and Health
University of Wisconsin–Whitewater
800 West Main Street
Whitewater, Wisconsin
taveiraa@uww.edu

Donald I. Tepas
Connecticut Transportation Institute
University of Connecticut
Storrs, Connecticut

Pamela S. Tsang
Associate Professor
Department of Psychology
Wright State University
3640 Colonel Glenn Highway
Dayton, Ohio
pamela.tsang@wright.edu

Gregg C. Vanderheiden
Professor of IE and BioMed Engineering, and Director
Trace Center
University of Wisconsin–Madison
1550 Engineering Drive
Madison, Wisconsin
gv@trace.wisc.edu

Michael A. Vidulich
Senior Research Psychologist
Air Force Research Laboratory (AFRL/HECP)
2255 H Street
Wright-Patterson Air Force Base, Ohio
Michael.Vidulich@wpafb.af.mil

Kim-Phuong L. Vu
Assistant Professor
Department of Psychology
California State University–Long Beach
1250 Bellflower Boulevard
Long Beach, California
kvu8@csulb.edu

Christopher Wickens
Professor Emeritus, Aviation and Psychology
Human Factors Division
Institute of Aviation
University of Illinois
1 Airport Road
Savoy, Illinois
cwickens@uiuc.edu

Glenn Wilson
AFRL/HECP
Air Force Research Laboratory
2255 H Street
Wright-Patterson Air Force Base, Ohio
Glenn.Wilson@wpafb.af.mil

Katherine A. Wilson
Institute of Simulation and Training
University of Central Florida
3100 Technology Parkway
Orlando, Florida
kwilson@ist.ucf.edu

Michael S. Wogalter
Professor
Psychology Department
North Carolina State University
640 Poe Hall, CB 7801
Raleigh, North Carolina
WogalterM@aol.com

Barbara Woolford
Human Factors Research Manager
NASA Johnson Space Center
2101 NASA Parkway
Houston, Texas
Barbara.j.woolford@nasa.gov

Ji Soo Yi
Research Assistant
School of Industrial and Systems Engineering
Georgia Institute of Technology
765 Ferst Drive NW
Atlanta, Georgia
jyi@isye.gatech.edu

Peter Young
Associate Professor
Department of Electrical and Computer Engineering
Colorado State University
Fort Collins, Colorado
pmy@engr.colostate.edu

Bernhard M. Zimolong
Professor
Department of Psychology
Ruhr University–Bochum
Bochum, Germany
bz@auo.psy.ruhr-uni-bochum.de

FOREWORD TO THE SECOND EDITION

With the rapid introduction of highly sophisticated computer, communication, and manufacturing systems, we are seeing dramatic changes in the ways people work and use technology. That is why every industry today should recognize the importance of human factors and ergonomics.

The practice of this science has done much to improve the interaction of people with their environment. Ergonomics is a valuable tool in the design of products that are safe, convenient, and user friendly. It is equally important in the creation of jobs and workplaces that increase worker safety and satisfaction. Ergonomics allows designers to comprehend the capabilities, limitations, and motivations of workers in order to improve efficiency and cut costs—especially those associated with human error and occupational injury or illness.

By applying human factors and ergonomics to our products, our office technology, and our manufacturing processes, we can enhance the satisfaction and enthusiasm of both consumers and producers.

Thus, the publication of this Third Edition of the *Handbook of Human Factors and Ergonomics* is very timely. It is a comprehensive guide that contains practical knowledge and technical background on virtually all aspects of physical, cognitive, and social ergonomics. As such, it can be a valuable source of information for any individual or organization committed to providing competitive, high-quality products and safe, productive work environments.

JOHN F. SMITH, JR.
Chairman of the Board
Chief Executive Officer and President
General Motors Corporation

PREFACE

This Handbook is concerned with the role of humans in complex systems, the design of equipment and facilities for human use, and the development of environments for comfort and safety. The first and second editions of the Handbook were a major success and profoundly influenced the human factors profession. It was translated and published in Japanese and Russian and won the Institute of Industrial Engineers Joint Publishers Book of the Year Award. It has received strong endorsement from top management; the late Elliott Estes, retired president of General Motors Corporation, who wrote the Foreword to the first edition of the Handbook, indicated that "regardless of what phase of the economy a person is involved in, this Handbook is a very useful tool. Every area of human factors from environmental conditions and motivation to use of new communication systems ... is well covered in the Handbook by experts in every field."

In a literal sense, human factors and ergonomics is as old as the machine and the environment, for it was aimed at designing them for human use. However, it was not until World War II that human factors emerged as a separate discipline.

The field of human factors and ergonomics has developed and broadened considerably since its inception more than 60 years ago and has generated a body of knowledge in the following areas of specialization:

- The human factors function
- Human factors fundamentals
- Design of tasks and jobs
- Equipment, workplace, and environmental design
- Design for health, safety, and comfort
- Performance modeling
- Evaluation
- Human–computer interaction
- Design for individual differences
- Selected applications

The foregoing list shows how broad the field has become. As such, this Handbook should be of value to all human factors and ergonomics specialists, engineers, industrial hygienists, safety engineers, and human–computer interaction specialists.

Such a breadth of subject matter presents a serious challenge to represent successfully the entire field of human factors and ergonomics in a single Handbook. I did not believe in 2002, when this all began, that any one person could properly select the subjects to be included in the Handbook without serious distortions to fit his or her own particular area of knowledge and bias. Accordingly, an advisory board composed of experts in the more important areas of human factors and ergonomics was invited to advise the editor in planning the contents of the Handbook. The advisory board members are listed on pages vii and viii. I sincerely appreciate their excellent counsel and advice during the preparation of this Handbook. Nevertheless, any sampling deficiencies that remain are of course my own responsibility.

The 61 chapters constituting the third edition of the Handbook were written by 108 people. In creating this Handbook, the authors gathered information from over 2700 references and presented over 500 figures and 250 tables to provide theoretically based and practically oriented material for use by both practioners and researchers. In the third edition of the *Handbook of Human Factors and Ergonomics*, all of the 61 chapters have been completely newly written. This third edition of the Handbook covers totally new subject areas that were not included in the second edition. These include the following subjects:

- Communications
- Cultural ergonomics
- Human factors and ergonomics methods
- Situation awareness
- Affective and emotional design
- Virtual environments
- Human factors and ergonomics Inspection and Audits
- Multimodal user interface
- Online communities
- Human factors and information security
- Usability evaluation and testing
- Design of e-business Web sites
- Augmented cognition in HCI
- Design for disability
- Design for children
- Design for all
- Human factors and ergonomics standards
- Human factors and ergonomics in transportation

The main purpose of this Handbook is to serve the needs of the human factors practitioner. Each chapter has a strong theory and science base and is heavily tilted toward application orientation. As such, a significant number of case studies, examples, figures, and tables are utilized to facilitate the usability of the material presented.

The many contributing authors came through magnificently. I thank them all most sincerely for agreeing so willingly to create the Handbook with me.

I had the privilege of working with Robert L. Argentieri, our Wiley Executive Editor, who significantly facilitated my editorial work. I was truly fortunate to have during the preparation of this Handbook the most able contribution of Kim Gilbert, Editorial Manager of the Handbook, who has done a truly outstanding job with the cooperation of all the authors and compilation of the Handbook for production.

This Handbook would not have been possible without the excellent value judgment and support of Dr. Dennis Engi, Professor and Head of the School of Industrial Engineering at Purdue University.

GAVRIEL SALVENDY

West Lafayette, Indiana
October 2004

CONTENTS

PART 1

THE HUMAN FACTORS FUNCTION

CHAPTER 1

THE DISCIPLINE OF ERGONOMICS AND HUMAN FACTORS

Waldemar Karwowski
University of Louisville
Louisville, Kentucky

The purpose of science is mastery over nature.

F. Bacon (*Novum Organum, 1620*)

1 INTRODUCTION

Ergonomics (Gr. *ergon* + *nomos*), the study of work, was originally defined and proposed by the Polish scientist B. W. Jastrzebowski (1857a–d), as a scientific discipline with a very broad scope and a wide range of interests and applications, encompassing all aspects of human activity, including labor, entertainment, reasoning, and dedication (Karwowski, (1991, 2001). In his paper published in the journal *Nature and Industry*, Jastrzebowski (1857) divided work into two main categories: *useful work*, which brings improvements for the common good, and *harmful work*, which brings deterioration (discreditable work). Useful work, which aims to improve things and people, is classified into physical, aesthetic, rational, and moral work. According to Jastrzebowski, such work requires utilization of motor forces, sensory forces, forces of reason (thinking and reasoning), and spiritual forces. He lists the four main benefits of useful work as being exemplified by property, ability, perfection, and felicity.

The contemporary ergonomics discipline, introduced independently by Murrell in 1949 (Edholm and Murrell, 1973), was viewed at that time as an applied science, a technology, and sometimes both. British scientists founded the Ergonomics Research Society in 1949. According to Kuorinka (2000), the development of ergonomics internationally can be linked to a project initiated by the European Productivity Agency (EPA), a branch of the Organization for European Economic Cooperation, which established a Human Factors Section in 1955. Under the EPA project, in 1956 specialists from European countries visited the United States to observe human factors research. In 1957 the EPA organized a technical seminar, "Fitting the Job to the Worker," at the University of Leiden, The Netherlands, during which a set of proposals was presented to form an international association of work scientists. A Steering Committee consisting of H. S. Belding, G. C. E. Burger, S. Forssman, E. Grandjean, G. Lehman, B. Metz, K. U. Smith, and R. G. Stansfield, was charged with developing specific proposals for such an association. The committee decided to adopt the name International Ergonomics Association. At a meeting in Paris in 1958, it was decided to proceed with forming the new association. The Steering Committee designated itself the Committee for the International Association of Ergonomic Scientists and elected G. C. E. Burger as its first president,

K. U. Smith as treasurer, and E. Grandjean as secretary. The Committee for the International Association of Ergonomic Scientists met in Zurich in 1959 during a conference organized by EPA, and decided to retain the name International Ergonomics Association. On April 6, 1959, at a meeting in Oxford, England, Grandjean declared the founding of the International Ergonomics Association (IEA). The committee met again in Oxford later in 1959 and agreed on a set of bylaws for the IEA. These were formally approved by the IEA General Assembly at the first International Congress of Ergonomics, held in Stockholm in 1961.

Over the last 50 years, *ergonomics*, a term that is used here synonymously with *human factors* (denoted HFE), has been evolving as *a unique and independent discipline that focuses on the nature of human–artifact interactions, viewed from the unified perspective of the science, engineering, design, technology, and management of human-compatible systems, including a variety of natural and artificial products, processes, and living environments* (Karwowski, 2005). The various dimensions of the ergonomics discipline are shown in Figure 1.

The International Ergonomics Association (IEA, 2003) defines *ergonomics* (human factors) as *the scientific discipline concerned with the understanding of the interactions among humans and other elements of a system, and the profession that applies theory, principles, data, and methods to design in order to optimize human well-being and overall system performance*. Ergonomists contribute to the design and evaluation of tasks, jobs, products, environments, and systems to make them compatible with the needs, abilities, and limitations of people. Ergonomics discipline promotes a holistic, human-centered approach to work systems design that considers physical, cognitive, social, organizational, environmental, and other relevant factors (Grandjean, 1986; Wilson and Corlett, 1990; Sanders and McCormick, 1993; Chapanis, 1999; Salvendy, 1997; Karwowski, 2001; Vicente, 2004; Stanton et al., 2004).

Traditionally, the domains of specialization within HFE cited most often are physical, cognitive, and organizational ergonomics. *Physical ergonomics* is concerned primarily with human anatomical, anthropometric, physiological, and biomechanical characteristics as they relate to physical activity (Chaffin and Anderson, 1993; Pheasant, 1986; Kroemer et al., 1994; Karwowski and Marras, 1999; NRC, 2001). *Cognitive ergonomics* focuses on mental processes such as perception, memory, information processing, reasoning, and motor response as they affect interactions among humans and other elements of a system (Vicente, 1999; Hollnagel, 2003; Diaper and Stanton, 2004). *Organizational ergonomics* (also known as *macroergonomics*) is concerned with the optimization of sociotechnical systems, including their organizational structures, policies, and processes (Reason, 1999; Holman et al., 2003; Nemeth,

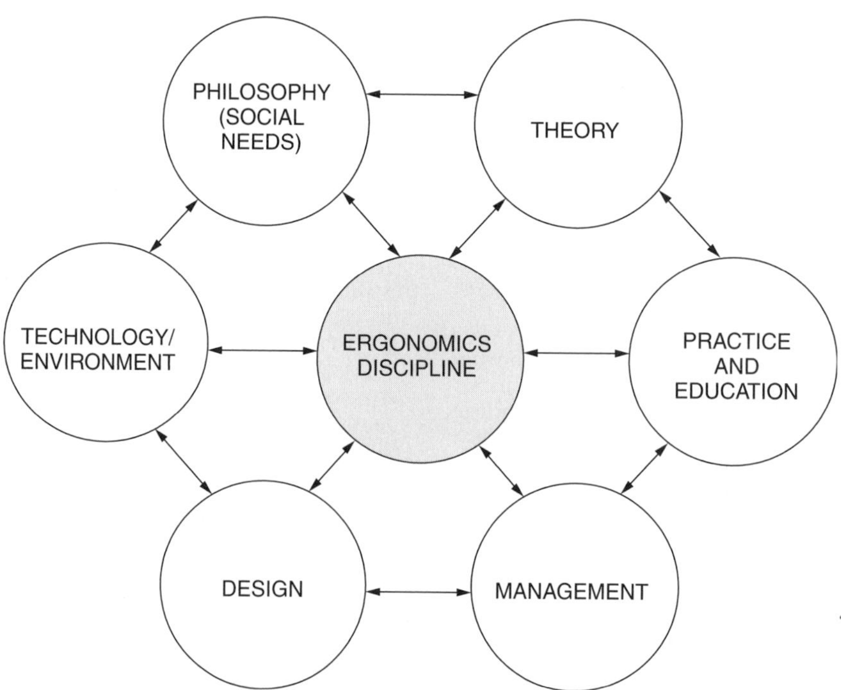

Figure 1 General dimensions of the HFE discipline. (After Karwowski, 2005.)

Table 1 Exemplary Domains of the Disciplines of Medicine, Psychology, and Ergonomics

Medicine	Psychology	Ergonomics
Cardiology	Applied psychology	Affective ergonomics
Community medicine	Child psychology	Cognitive ergonomics
Dermatology	Clinical psychology	Community ergonomics
Endocrinology	Cognitive psychology	Consumer ergonomics
Gastroenterology	Community psychology	Ecological ergonomics
Gerontology	Counseling psychology	Ergonomics of aging
Internal medicine	Developmental psychology	Forensic ergonomics
Nephrology	Educational psychology	Human–computer interaction
Neurology	Environmental psychology	Human–system integration
Neuroscience	Experimental psychology	Information ergonomics
Oncology	Forensic psychology	Knowledge ergonomics
Ophthalmology	Health psychology	Macroergonomics
Physical medicine	Organizational psychology	Nanoergonomics
Psychiatry	Positive psychology	Neuroergonomics
Pulmonology		Participatory ergonomics
Radiology	Quantitative psychology	Physical ergonomics
Urology	Social psychology	Rehabilitation ergonomics

Source: Karwowski (2005).

Table 2 Objectives of the HFE Discipline

- Basic operational objectives
  - Reduce errors
  - Increase safety
  - Improve system performance
- Objectives bearing on reliability, maintainability, availability and integrated logistic support
  - Increase reliability
  - Improve maintainability
  - Reduce personnel requirements
  - Reduce training requirements
- Objectives affecting users and operators
  - Improve the working environment
  - Reduce fatigue and physical stress
  - Increase ease of use
  - Increase user acceptance
  - Increase aesthetic appearance
- Other objectives
  - Reduce losses of time and equipment
  - Increase economy of production

Source: Chapanis (1995).

2004). Examples of relevant topics include communication, crew resource management, design of working times, teamwork, participatory work design, community ergonomics, computer-supported cooperative work, new work paradigms, virtual organizations, telework, and quality management. The traditional domains noted above, together with new domains, are listed in Table 1.

According to the discussion above, the paramount objective of HFE is to understand interactions between people and everything that surrounds us, and based on such knowledge to optimize human well-being and overall system performance. Table 2 provides a summary of specific HFE objectives as discussed by Chapanis (1995). As pointed out by the National Academy of Engineering in the United States (NAE, 2004), in the future, ongoing developments in engineering will *expand toward tighter connections between technology and the human experience, including new products customized to the physical dimensions and capabilities of the user, and the ergonomic design of engineered products*.

2 HUMAN–TECHNOLOGY INTERACTIONS

Whereas in the past, ergonomics has been driven by technology (reactive design approach), in the future, ergonomics should drive technology (proactive design approach). Technology can be defined as the entire system of people and organizations, knowledge, processes, and devices that go into creating and operating technological artifacts, as well as the artifacts themselves (NRC, 2001). Technology is a product and a process involving both science and engineering. Science aims to understand the "why" and "how" of nature (through a process of scientific inquiry that generates knowledge about the natural world). Engineering represents "design under constraints" of cost, reliability, safety, environmental impact, ease of use, available human and material resources, manufacturability, government regulations, laws, and politics (Wulf, 1988). Engineering seeks to shape the natural world to meet human needs and wants: a body of knowledge of design and creation of human-made products and a process for solving problems.

Contemporary HFE discovers and applies information about human behavior, abilities, limitations, and other characteristics to the design of tools, machines, systems, tasks, jobs, and environments for productive, safe, comfortable, and effective human use (Sanders and McCormick, 1993; Helander, 1997b). In this context, HFE deals with a broad scope of problems relevant to the design and evaluation of work systems, consumer products, and working environments, in which human–machine interactions affect human

Table 3 Classification Scheme for Human Factors/Ergonomics

1. General

Human Characteristics

2. Psychological aspects
3. Physiological and anatomical aspects
4. Group factors
5. Individual differences
6. Psychophysiological state variables
7. Task-related factors

Information Presentation and Communication

8. Visual communication
9. Auditory and other communication modalities
10. Choice of communication media
11. Person–machine dialogue mode
12. System feedback
13. Error prevention and recovery
14. Design of documents and procedures
15. User control features
16. Language design
17. Database organization and data retrieval
18. Programming, debugging, editing, and programming aids
19. Software performance and evaluation
20. Software design, maintenance, and reliability

Display and Control Design

21. Input devices and controls
22. Visual displays
23. Auditory displays
24. Other modality displays
25. Display and control characteristics

Workplace and Equipment Design

26. General workplace design and buildings
27. Workstation design
28. Equipment design

Environment

29. Illumination
30. Noise
31. Vibration
32. Whole body movement
33. Climate
34. Altitude, depth, and space
35. Other environmental issues

System Characteristics

36. General system features

Table 3 *(continued)*

Work Design and Organization

37. Total system design and evaluation
38. Hours of work
39. Job attitudes and job satisfaction
40. Job design
41. Payment systems
42. Selection and screening
43. Training
44. Supervision
45. Use of support
46. Technological and ergonomic change

Health and Safety

47. General health and safety
48. Etiology
49. Injuries and illnesses
50. Prevention

Social and Economic Impact of the System

51. Trade unions
52. Employment, job security, and job sharing
53. Productivity
54. Women and work
55. Organizational design
56. Education
57. Law
58. Privacy
59. Family and home life
60. Quality of working life
61. Political comment and ethical considerations

Methods and Techniques

62. Approaches and methods
63. Techniques
64. Measures

Source: EIAC (2004).

performance and product usability. The wide scope of issues addressed by the contemporary HFE discipline is presented in Table 3. Figure 2 illustrates the evolution of the scope of HFE with respect to the nature of human–system interactions. Originally, HFE focused on local human–machine interactions, whereas today, the primary focus is on broadly defined human–technology interactions. In this view the HFE can also be called the discipline of *technological ecology*. Tables 4 and 5 present the taxonomy of human- and technology-related components, respectively, which are of great importance to HFE discipline.

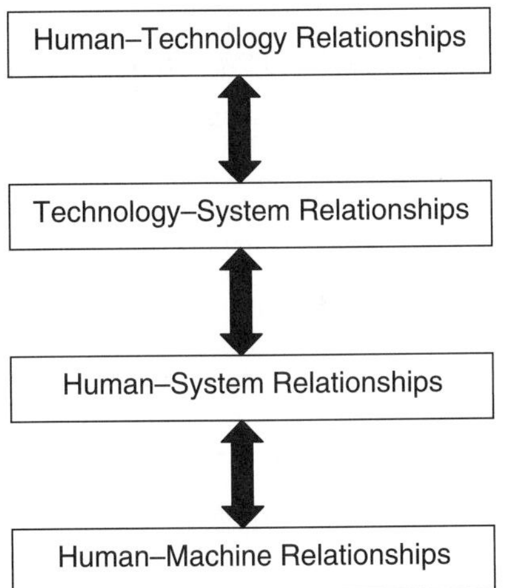

Figure 2 Expanded view of the human–technology relationships. (Modified from Meister, 1999.)

According to Meister (1987), the traditional concept of a *human–machine system* is an organization of people and the machines they operate and maintain in order to perform assigned jobs that implement the purpose for which the system was developed. In this context, system is a *construct* whose characteristics are manifested in physical and behavioral phenomena (Meister, 1991). The system is critical to HFE theorizing because it describes the substance of the human–technology relationship. General system variables of interest to HFE discipline are shown in Table 6.

The human functioning in human–machine systems can be described in terms of perception, information processing, decision making, memory, attention, feedback, and human response processes. Furthermore, the human work taxonomy can be used to describe five distinct levels of human functioning, ranging from primarily physical tasks to cognitive tasks (Karwowski, 1992a). These basic but universal human activities are (1) tasks that produce force (primarily, muscular work), (2) tasks of continuously coordinating sensory-monitoring functions (e.g., assembling or tracking tasks), (3) tasks of converting information into motor actions (e.g., inspection tasks), (4) tasks of converting information into output information (e.g., required control tasks), and (5) tasks of producing information (primarily creative work) (Luczak et al., 1999). Any task in a human–machine system requires processing of information that is gathered based on perceived and interpreted relationships between system elements. The information processed may need to be stored by either a human or a machine for later use.

Table 5 Taxonomy of HFE Elements: Technology

Technology elements	Effects of technology on the human
Components	Changes in human role
Tools	Changes in human behavior
Equipments	Organization–technology relationships
Systems	Definition of organization
Degree of automation	Organizational variables
Mechanization	
Computerization	
Artificial intelligence	
System characteristics	
Dimensions	
Attributes	
Variables	

Source: Meister (1999).

Table 4 Taxonomy of HFE Elements: The Human Factor

Human elements	Effects of the human on technology
Physical/sensory	Improvement in technology effectiveness
Cognitive	Absence of effect
Motivational/emotional	Reduction in technological effectiveness
Human conceptualization	Human–technological relationships
Stimulus–response orientation (limited)	Controller relationship
Stimulus–conceptual–response orientation (major)	Partnership relationship
Stimulus–conceptual–motivational–response orientation (major)	Client relationship
Effects of technology on the human	Human operations in technology
Performance effects	Equipment operation
Goal accomplishment	Equipment maintenance
Goal nonaccomplishment	System management
Error/time discrepancies	Type/degree of human involvement
Feeling effect	Direct (operation)
Technology acceptance	Indirect (recipient)
Technology indifference	Extensive
Technology rejection	Minimal
Demand effects	None
Resource mobilization	
Stress/trauma	

Source: Meister (1999).

Table 6 General System Variables

1. Requirements constraints imposed on the system
2. Resources required by the system
3. Nature of its internal components and processes
4. Functions and missions performed by the system
5. Nature, number, and specificity of goals
6. Structural and organizational characteristics of the system (e.g., its size, number of subsystems and units, communication channels, hierarchical levels, and amount of feedback)
7. Degree of automation
8. Nature of the environment in which the system functions
9. System attributes (e.g., complexity, sensitivity, flexibility, vulnerability, reliability, and determinancy)
10. Number and type of interdependencies (human–machine interactions) within the system and type of interaction (degree of dependency)
11. Nature of the system's terminal output(s) or mission effects

Source: Meister (1999).

The scope of HFE factors that need to be considered in the design, testing, and evaluation of any human–system interactions is shown in Table 7 in the form of an exemplary ergonomics checklist. It should be noted that such checklists also reflect practical application of the discipline. According to the Board of Certification in Professional Ergonomics (BCPE), a practitioner of ergonomics is a person who (1) has a mastery of a body of ergonomics knowledge, (2) has a command of the methodologies used by ergonomists in applying that knowledge to the design of a product, system, job, or environment, and (3) has applied his or her knowledge to the analysis, design testing, and evaluation of products, systems, and environments. The areas of current practice in the field can best be described by examining the focus of the Technical Groups of the Human Factors and Ergonomics Society, as illustrated in Table 8.

3 HFE AND ECOLOGICAL COMPATIBILITY

HFE discipline advocates *systematic use of knowledge concerning relevant human characteristics to achieve compatibility in the design of interactive systems of people, machines, environments, and devices of all kinds to ensure specific goals* [Human Factors and Ergonomics Society (HFES), 2003]. Typically, such goals include improved (system) effectiveness, productivity, safety, ease of performance, and the

Table 7 Examples of Factors to Be Used in Ergonomics Checklists

I. Anthropometric, Biomechanical, and Physiological Factors

1. Are the differences in human body size accounted for by the design?
2. Have the right anthropometric tables been used for specific populations?
3. Are the body joints close to neutral positions?
4. Is the manual work performed close to the body?
5. Are any forward-bending or twisted trunk postures involved?
6. Are sudden movements and force exertion present?
7. Is there a variation in worker postures and movements?
8. Is the duration of any continuous muscular effort limited?
9. Are the breaks of sufficient length and spread over the duration of the task?
10. Is the energy consumption for each manual task limited?

II. Factors Related to Posture (Sitting and Standing)

1. Is sitting/standing alternated with standing/sitting and walking?
2. Is the work height dependent on the task?
3. Is the height of the worktable adjustable?
4. Are the height of the seat and backrest of the chair adjustable?
5. Is the number of chair adjustment possibilities limited?
6. Have good seating instructions been provided?
7. Is a footrest used where the work height is fixed?
8. Has work above the shoulder or with hands behind the body been avoided?
9. Are excessive reaches avoided?

Table 7 *(continued)*

10. Is there enough room for the legs and feet?
11. Is there a sloping work surface for reading tasks?
12. Have combined sit–stand workplaces been introduced?
13. Are handles of tools bent to allow for working with the straight wrists?

III. Factors Related to Manual Materials Handling (Lifting, Carrying, Pushing and Pulling Loads)

1. Have tasks involving manual displacement of loads been limited?
2. Have optimum lifting conditions been achieved?
3. Is anybody required to lift more than 23 kg?
4. Have lifting tasks been assessed using the NIOSH (Waters et al., 1993) method?
5. Are handgrips fitted to the loads to be lifted?
6. Is more than one person involved in lifting or carrying tasks?
7. Are there mechanical aids for lifting or carrying available and used?
8. Is the weight of the load carried limited according to recognized guidelines?
9. Is the load held as close to the body as possible?
10. Are pulling and pushing forces limited?
11. Are trolleys fitted with appropriate handles and handgrips?

IV. Factors Related to the Design of Tasks and Jobs

1. Does the job consist of more than one task?
2. Has a decision been made about allocating tasks between people and machines?
3. Do workers performing the tasks contribute to problem solving?
4. Are difficult and easy tasks performed interchangeably?
5. Can workers decide independently on how the tasks are carried out?
6. Are there sufficient possibilities for communication between workers?
7. Is sufficient information provided to control the tasks assigned?
8. Can the group take part in management decisions?
9. Are shift workers given enough opportunities to recover?

V. Factors Related to Information and Control Tasks

Information

l. Has an appropriate method of displaying information been selected?
2. Is the information presentation as simple as possible?
3. Has the potential confusion between characters been avoided?
4. Has the correct character/letter size been chosen?
5. Have texts with capital letters only been avoided?
6. Have familiar typefaces been chosen?
7. Is the text/background contrast good?
8. Are the diagrams easy to understand?
9. Have the pictograms been used properly?
10. Are sound signals reserved for warning purposes?

Control

1. Is the sense of touch used for feedback from controls?
2. Are differences between controls distinguishable by touch?
3. Is the location of controls consistent, and is sufficient spacing provided?
4. Have the requirements for control–display compatibility been considered?
5. Is the type of cursor control suitable for the intended task?
6. Is the direction of control movements consistent with human expectations?
7. Are the control objectives clear from the position of the controls?
8. Are controls within easy reach of female workers?

(continued overleaf)

Table 7 ***(continued)***

9. Are labels or symbols identifying controls used properly?
10. Is the use of color in controls design limited?

Human–computer interaction

1. Is the human–computer dialogue suitable for the intended task?
2. Is the dialogue self-descriptive and easy to control by the user?
3. Does the dialogue conform to the expectations on the part of the user?
4. Is the dialogue error-tolerant and suitable for user learning?
5. Has command language been restricted to experienced users?
6. Have detailed menus been used for users with little knowledge and experience?
7. Is the type of help menu fitted to the level of the user's ability?
8. Has the QWERTY layout been selected for the keyboard?
9. Has a logical layout been chosen for the numerical keypad?
10. Is the number of function keys limited?
11. Have the limitations of speech in human–computer dialogue been considered?
12. Are touch screens used to facilitate operation by inexperienced users?

VI. Environmental Factors

Noise and vibration

1. Is the noise level at work below 80 dBA?
2. Is there an adequate separation between workers and source of noise?
3. Is the ceiling used for noise absorption?
4. Are acoustic screens used?
5. Are hearing conservation measures fitted to the user?
6. Is personal monitoring to noise/vibration used?
7. Are the sources of uncomfortable and damaging body vibration recognized?
8. Is the vibration problem being solved at the source?
9. Are machines regularly maintained?
10. Is the transmission of vibration prevented?

Illumination

1. Is the light intensity for normal activities in the range 200 to 800 lux?
2. Are large brightness differences in the visual field avoided?
3. Are the brightness differences between task area, close surroundings, and wider surroundings limited?
4. Is the information easily legible?
5. Is ambient lighting combined with localized lighting?
6. Are light sources properly screened?
7. Can light reflections, shadows, or flicker from the fluorescent tubes be prevented?

Climate

1. Are workers able to control the climate themselves?
2. Is the air temperature suited to the physical demands of the task?
3. Is the air prevented from becoming either too dry to too humid?
4. Are drafts prevented?
5. Are the materials/surfaces that have to be touched neither too cold nor too hot?
6. Are the physical demands of the task adjusted to the external climate?
7. Are undesirable hot and cold radiation prevented?
8. Is the time spent in hot or cold environments limited?
9. Is special clothing used when spending long periods in hot or cold environments?

Source: Based on Dul and Weerdmeester (1993).

Table 8 Subject Interests of Technical Groups of the Human Factors and Ergonomics Society

Technical Group	Description/Areas of Concern
I. Aerospace systems	Applications of human factors to the development, design, operation, and maintenance of human–machine systems in aviation and space environments (both civilian and military).
II. Aging	Human factors applications appropriate to meeting the emerging needs of older people and special populations in a wide variety of life settings.
III. Cognitive engineering and decision making	Research on human cognition and decision making and the application of this knowledge to the design of systems and training programs. Emphasis is on considerations of descriptive models, processes, and characteristics of human decision making, alone or in conjunction with other people or with intelligent systems; factors that affect decision making and cognition in naturalistic task settings; technologies for assisting, modifying, or supplementing human decision making; and training strategies for assisting or influencing decision making.
IV. Communications	All aspects of human-to-human communication, with an emphasis on communication mediated by telecommunications technology, including multimedia and collaborative communications, information services, and interactive broadband applications. Design and evaluation of enabling technologies and infrastructure technologies in education, medicine, business productivity, and personal quality of life.
V. Computer systems	Human factors aspects of (1) interactive computer systems, especially user-interface design issues; (2) the data-processing environment, including personnel selection, training, and procedures; and (3) software development.
VI. Consumer products	Development of consumer products that are useful, usable, safe, and desirable. Application of the principles and methods of human factors, consumer research, and industrial design to ensure market success.
VII. Education	Design of educational systems, environments, interfaces, and technologies, as well as human factors education. Improvement in educational design and addressing educational needs of those seeking to increase their knowledge and skills in the human factors/ergonomics field.
VIII. Environmental design	Ergonomic and macroergonomic aspects of the constructed physical environment, including architectural and interior design aspects of home, office, and industrial settings. Promotion of the use of human factors principles in environmental design.
IX. Forensics professional	Application of human factors knowledge and technique to "standards of care" and accountability established within legislative, regulatory, and judicial systems. Emphasis on providing a scientific basis to issues being interpreted by legal theory.
X. Industrial ergonomics	Application of ergonomics data and principles for improving safety, productivity, and quality of work in industry. Concentration on service and manufacturing processes, operations, and environments.
XI. Internet	Human factor aspects of user-interface design of Web content, Web-based applications, Web browsers, Webtops, Web-based user assistance, and Internet devices; behavioral and sociological phenomena associated with distributed network communication; human reliability in administration and maintenance of data networks; and accessibility of Web-based products.
XII. Macroergonomics	Improving productivity and quality of work life and integrating psychosocial, cultural, and technological factors with human–machine performance interface factors in the design of jobs, workstations, organizations, and related management systems.
XIII. Medical systems and functionally impaired populations	All aspects of the application of human factors principles and techniques toward the improvement of medical systems, medical devices, and the quality of life for functionally impaired user populations.
XIV. Perception and performance	The relationship between vision and human performance, including (1) the nature, content, and quantification of visual information and the context in which it is displayed; (2) the physics and psychophysics of information display; (3) perceptual and cognitive representation and interpretation of displayed information; (4) workload assessment using visual tasks; and (5) actions and behaviors that are consequences of visually displayed information.
XV. Safety	Research and applications concerning human factors in safety and injury control in all settings and attendant populations, including transportation, industry, military, office, public building, recreation, and home improvements.

(continued overleaf)

Table 8 ***(continued)***

Technical Group	Description/Areas of Concern
XVI. System development	Concerned with research and exchange of information for integrating human factors into the development of systems. Integration of human factors activities into system development processes in order to provide systems that meet user requirements.
XVII. Surface transportation	Human factor aspects of mechanisms for conveying humans and resources: (1) passenger, commercial, and military vehicles, on- and off-road; (2) mass transit; maritime transportation; (3) rail transit, including vessel traffic services; (4) pedestrian and bicycle traffic; (5) and highway and infrastructure systems, including intelligent transportation systems.
XVIII. Test and evaluation	A forum for test and evaluation practitioners and developers from all areas of human factors and ergonomics. Concerned with methodologies and techniques that have been developed in their respective areas.
XIX. Training	Fosters information and interchange among people interested in the fields of training and training research.
XX. Virtual environment	Human factors issues associated with human–virtual environment interaction, including (1) maximizing human performance efficiency in virtual environments; (2) ensuring health and safety; and (3) circumventing potential social problems through proactive assessment.

contribution to overall human well-being and quality of life. Although the term *compatibility* is a key word in the definition above, it has been used primarily in a narrow sense only, often in the context of the design of displays and controls, including studies of spatial (location) compatibility or the intention–response–stimulus compatibility related to the movement of controls (Wickens and Carswell, 1997). Karwowski and his co-workers (Karwowski et al., 1988; Karwowski, 1985, 1991) advocated the use of *compatibility* in a greater context of the ergonomics system. For example, Karwowski (1997) introduced the term *human-compatible systems* to focus on the need for comprehensive treatment of compatibility in the human factors discipline.

The *American Heritage Dictionary of the English Language* (Morris, 1978) defines *compatible* as (1) capable of living or performing in harmonious, agreeable, or congenial combination with another or others; and (2) capable of orderly, efficient integration and operation with other elements in a system. From the beginning of contemporary ergonomics, measurements of compatibility between the system and the human, and evaluation of the results of ergonomics interventions, were based on the measures that best suited specific purposes (Karwowski, 2001). Such measures included the specific psychophysiological responses of the human body (e.g., heart rate, electromyography, perceived human exertion, satisfaction, comfort or discomfort), as well as a number of indirect measures, such as the incidence of injury, economic losses or gains, system acceptance, or operational effectiveness, quality, or productivity. The lack of a universal matrix to quantify and measure human-system compatibility is an important obstacle in demonstrating the value of ergonomics science and profession (Karwowski, 1998). However, even though 20 years ago ergonomics was perceived by some (see, e.g., Howell, 1986) as a highly unpredictable area of human scientific endeavor, today HFE has positioned itself as a unique, *design-oriented* discipline, independent of engineering and medicine (Moray, 1994; Sanders and McCormick, 1993; Helander, 1997a; Karwowski, 1991, 2003).

Figure 3 illustrates the human-system compatibility approach to ergonomics in the context of quality of working life and system (an enterprise or business entity) performance. This approach reflects the nature of complex compatibility relationships among a human operator (human capacities and limitations), technology (in terms of products, machines, devices, processes, and computer-based systems), and broadly defined environment (business processes, organizational structure, the nature of work systems, and the effects of work-related multiple stressors). The operator's performance is an outcome of the compatibility matching between individual human characteristics (capacities and limitations) and the requirements and affordances of both the technology and environment. The quality of working life and system (enterprise) performance is affected by matching of the positive and negative outcomes of the complex compatibility relationships among the human operator, technology, and the environment. Positive outcomes include such measures as work productivity, performance times, product quality, and subjective psychological (desirable) behavioral outcomes such as job satisfaction, employee morale, human well-being, and commitment. The negative outcomes include both human and system-related errors, loss of productivity, low quality, accidents, injuries, physiological stresses, and subjective psychological (undesirable) behavioral outcomes such as job dissatisfaction, job/occupational stress, and discomfort.

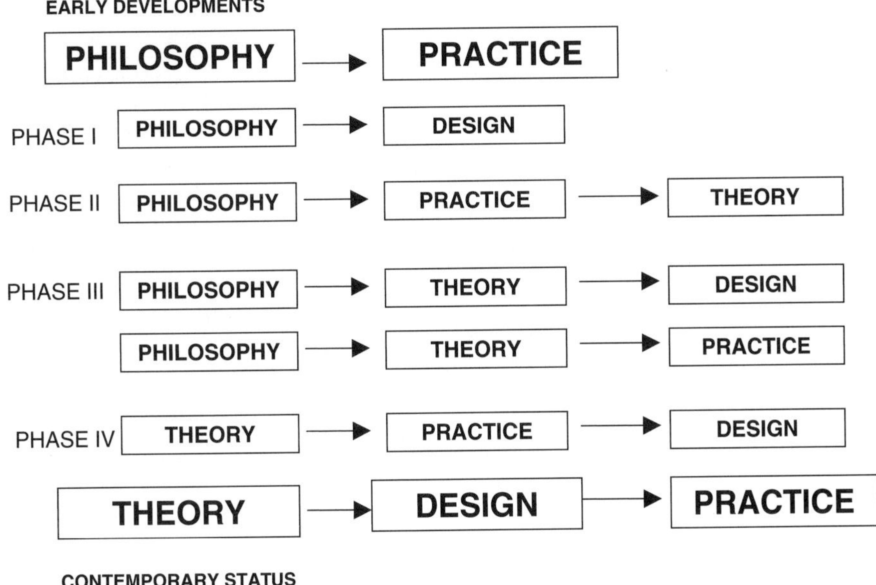

Figure 3 Evolution in development of the HFE discipline. (After Karwowski, 2005.)

4 DISTINGUISHING FEATURES OF THE CONTEMPORARY HFE DISCIPLINE AND PROFESSION

The main focus of the HFE discipline in the twenty-first century is on the design and management of systems that satisfy customer demands in terms of human compatibility requirements. Karwowski (2005) has discussed 10 characteristics of contemporary HFE discipline and profession. These distinguishing features are as follows:

1. HFE is very ambitious in its goals, but poorly funded compared to other contemporary disciplines.
2. HFE experiences continuing evolution of its "fit" philosophy, including diverse and ever-expanding human-centered design criteria (from safety to comfort, productivity, usability, or affective needs such as job satisfaction or life happiness).
3. HFE has yet to establish its unique disciplinary identity and credibility among other sciences, engineering, and technology.
4. HFE covers extremely diverse subject matters in a manner similar to medicine, engineering, and psychology (see Table 1).
5. HFE deals with very complex phenomena that are not easily understood and cannot be simplified by making nondefendable assumptions about their nature.
6. Historically, HFE has been developing from the "philosophy of fit" toward practice. Today, HFE is developing a sound theoretical basis for design and practical applications (Figure 4).
7. HFE attempts to "by-step" the need for fundamental understanding of human–system interactions, without separation from a consideration of knowledge utility for practical applications, in the quest for immediate and useful solutions (Figure 5).
8. HFE enjoys limited recognition by decision makers, the general public, and politicians as to the value that it can bring to a global society at large, especially in the context of facilitating socioeconomic development.
9. HFE has relatively weak and limited professional educational base.
10. HFE is adversely affected by the ergonomics illiteracy of students and professionals in other disciplines, the mass media, and the public at large.

Theoretical ergonomics is interested in the fundamental understanding of the interactions between people and their environments. Also central to HFE interests is an understanding of how human–system interactions should be designed. On the other hand, HFE also falls under the category of applied research. Taxonomy of research efforts with respect to the quest for fundamental understanding and the consideration of use, originally proposed by Stokes (1997), allows for differentiation of main categories of research dimensions as follows: (1) pure basic research, (2) use-inspired basic research, and (3) pure applied research. Figure 5 illustrates interpretation of these categories for HFE theory, design, and applications. Table 9 presents relevant specialties and subspecialties in HFE research as outlined by Meister (1999), who classified them into three main categories: (1) system/technology-oriented specialties,

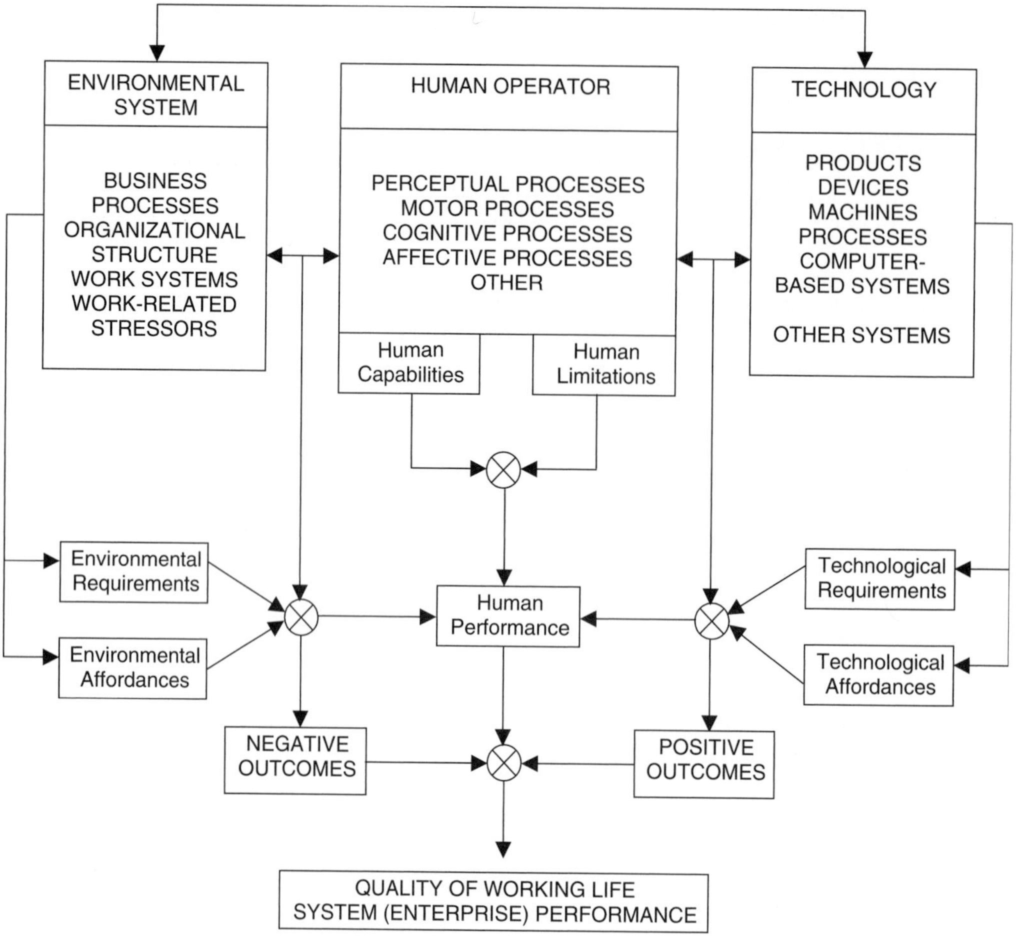

ENVIRONMENTAL SYSTEM
BUSINESS PROCESSES
ORGANIZATIONAL STRUCTURE
WORK SYSTEMS
WORK-RELATED STRESSORS

HUMAN OPERATOR
PERCEPTUAL PROCESSES
MOTOR PROCESSES
COGNITIVE PROCESSES
AFFECTIVE PROCESSES
OTHER
Human Capabilities
Human Limitations

TECHNOLOGY
PRODUCTS
DEVICES
MACHINES
PROCESSES
COMPUTER-BASED SYSTEMS
OTHER SYSTEMS

Environmental Requirements
Environmental Affordances
Human Performance
Technological Requirements
Technological Affordances
NEGATIVE OUTCOMES
POSITIVE OUTCOMES
QUALITY OF WORKING LIFE
SYSTEM (ENTERPRISE) PERFORMANCE

Note: ⊗ – Matching of compatibility relationships

Figure 4 Human–system compatibility approach to ergonomics. (After Karwowski, 2005.)

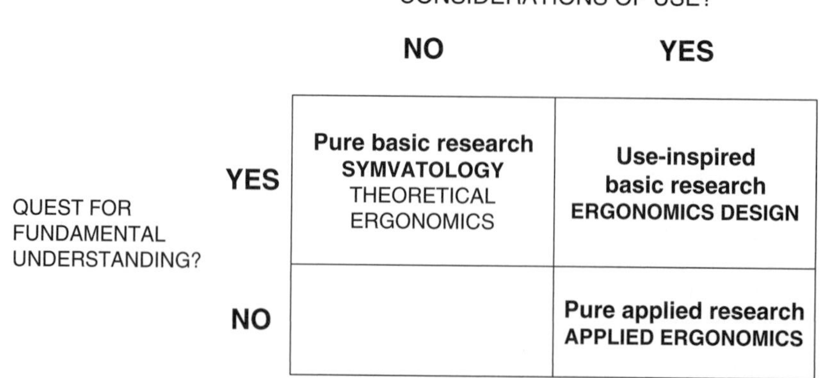

CONSIDERATIONS OF USE?
NO
YES
QUEST FOR FUNDAMENTAL UNDERSTANDING?
YES
NO
Pure basic research
SYMVATOLOGY
THEORETICAL ERGONOMICS
Use-inspired basic research
ERGONOMICS DESIGN
Pure applied research
APPLIED ERGONOMICS

Figure 5 Considerations of fundamental understanding and use in HFE research. (After Karwowski, 2005.)

Table 9 Specialties and Subspecialties in HFE Research

System/Technology-Oriented Specialties

1. *Aerospace*: civilian and military aviation and outer-space activities.
2. *Automotive*: automobiles, buses, railroads, transportation functions (e.g., highway design, traffic signs, ships).
3. *Communication*: telephone, telegraph, radio, direct personnel communication in a technological context.
4. *Computers*: anything associated with the hardware and software of computers.
5. *Consumer products*: other than computers and automobiles, any commercial product sold to the general public (e.g., pens, watches, TV sets).
6. *Displays*: equipment used to present information to operators (e.g., HMO, HUD, meters, scales).
7. *Environmental factors/design*: the environment in which human–machine system functions are performed (e.g., offices, noise, lighting).
8. *Special environment*: this turns out to be underwater.

Process-Oriented Specialties

1. *Biomechanics*: human physical strength as it is manifested in such activities as lifting and pulling.
2. *Industrial ergonomics (IE)*: related primarily to manufacturing; processes and resultant problems (e.g., carpal tunnel syndrome).
3. *Methodology/measurement*: ways of answering HFE questions or solving HFE problems.
4. *Safety*: closely related to IE but with a major emphasis on analysis and prevention of accidents.
5. *System design/development*: processes of analyzing, creating, and developing systems.
6. *Training*: how personnel are taught to perform functions/tasks in a human–machine system.

Behaviorally Oriented Specialties

1. *Aging*: the effect of this process on technological performance.
2. *Human functions*: emphasizes perceptual-motor and cognitive functions. The latter differs from training in the sense that training also involves cognition but is the process of implementing cognitive capabilities. (The HFES specialty called *cognitive ergonomics/decision making* has been categorized.)
3. *Visual performance*: how people see; differs from displays in that the latter relate to equipment for seeing, whereas the former deals with the human capability and function of seeing.

Source: Adapted from Meister (1999).

(2) process-oriented specialties, and (3) behaviorally oriented specialties. In addition, Table 10 presents a list of contemporary HFE research methods that can be used to advance knowledge, discovery, and utilization through its practical applications.

5 PARADIGMS FOR THE ERGONOMICS DISCIPLINE

The paradigms for any scientific discipline include theory, abstraction, and design (Pearson and Young, 2002). Theory is a foundation of the mathematical sciences. Abstraction (modeling) is a foundation of the natural sciences, where progress is achieved by formulating hypotheses and following the modeling process systematically to verify and validate them. Design as the basis for engineering, where progress is achieved primarily by posing problems and systematically following the design process to construct systems that solve them.

In view of the above, Karwowski (2005) discussed the following paradigms for HFE discipline: (1) ergonomics theory, (2) ergonomics abstraction, and (3) ergonomics design. Ergonomics theory is concerned with the ability to identify, describe, and evaluate human–system interactions. Ergonomics abstraction is concerned with the ability to use those interactions to make predictions that can be compared with the real world. Ergonomics design is concerned with the ability to implement knowledge about those interactions and use them to develop systems that satisfy customer needs and relevant human compatibility requirements.

Furthermore, the pillars for any scientific discipline include a definition, a teaching paradigm, and an educational base (NRC, 2001). A definition of ergonomics discipline and profession adopted by IEA (2003) emphasizes fundamental questions and significant accomplishments, recognizing that the HFE field is constantly changing. A teaching paradigm for ergonomics should conform to established scientific

Table 10 Contemporary HFE Research Methods

Physical Methods
PLIBEL: method assigned for identification of ergonomic hazards Musculoskeletal discomfort surveys used at NIOSH
Dutch musculoskeletal questionnaire
Quick exposure checklist for the assessment of workplace risks for work-related musculoskeletal disorders
Rapid upper limb assessment
Rapid entire body assessment
Strain index
Posture checklist using personal digital assistant technology
Scaling experiences during work: perceived exertion and difficulty
Muscle fatigue assessment: functional job analysis technique
Psychophysical tables: lifting, lowering, pushing, pulling, and carrying
Lumbar motion monitor
Occupational repetitive action (OCRA) methods: OCRA index and OCRA checklist
Assessment of exposure to manual patient handling in hospital wards: MAPO (movement and assistance of hospital patients) index
Psychophysiological Methods
Electrodermal measurement
Electromyography
Estimating mental effort using heart rate and heart rate variability
Ambulatory methods and sleepiness
Assessing brain function and mental chronometry with event-related potentials
MEG and fMRI Magnetoencephalography and magnetic resonance imaging.
Ambulatory assessment of blood pressure to evaluate workload
Monitoring alertness by eyelid closure
Measurement of respiration in applied human factors and ergonomics research
Behavioral and Cognitive Methods
Observation
Heuristics
Applying interviews to usability assessment
Verbal protocol analysis
Repertory grid for product evaluation
Focus groups
Hierarchical task analysis
Allocation of functions
Critical decision method
Applied cognitive work analysis
Systematic human error reduction and prediction approach
Predictive human error analysis
Hierarchical task analysis
Mental workload
Multiple resource time sharing
Critical path analysis for multimodal activity
Situation awareness measurement and the situation awareness
keystroke level model
GOMS (Goals, operators, methods, and selection rules)
Link analysis
Global assessment technique
Team Methods
Team training
Distributed simulation training for teams
Synthetic task environments for teams
Event-based approach to training
Team building
Measuring team knowledge
Team communications analysis
Questionnaires for distributed assessment of team mutual awareness

Table 10 ***(continued)***

Team decision requirement exercise: making team decision requirements explicit
Targeted acceptable responses to generated events or tasks
Behavioral observation scales
Team situation assessment training for adaptive coordination
Team task analysis
Team workload
Social network analysis
Environmental Methods
Thermal conditions measurement
Cold stress indices
Heat stress indices
Thermal comfort indices
Indoor air quality: chemical exposures
Indoor air quality: biological/particulate-phase contaminant
Exposure assessment methods
Olfactometry: human nose as a detection instrument
Context and foundation of lighting practice
Photometric characterization of the luminous environment
Evaluating office lighting
Rapid sound-quality assessment of background noise
Noise reaction indices and assessment
Noise and human behavior
Occupational vibration: a concise perspective
Habitability measurement in space vehicles and earth analogs
Macroergonomic Methods
Macroergonomic organizational questionnaire survey
Interview method
Focus groups
Laboratory experiment
Field study and field experiment
Participatory ergonomics
Cognitive walk-through method
Kansei engineering
HITOP analysis TM
TOP-Modeler C
CIMOP system C
Anthropotechnology
Systems analysis tool
Macroergonomic analysis of structure
Macroergonomic analysis and design

Source: Based on Stanton et al. (2004).

standards, emphasize the development of competence in the field, and integrate theory, experimentation, design, and practice. Finally, an introductory course sequence in ergonomics should be based on the curriculum model and the disciplinary description.

6 ERGONOMICS COMPETENCY AND LITERACY

As pointed out by the National Academy of Engineering (Pearson and Young, 2002), many consumer products and services promise to make people's lives easier, more enjoyable, more efficient, or healthier, but very often do not deliver on this promise. Design of interactions with technological artifacts and work systems requires involvement of ergonomically competent people: people with ergonomics proficiency in a certain area, although not generally in other areas of application, similar to medicine or engineering.

One of the critical issues in this context is the ability of users to understand the utility and limitations of technological artifacts. Ergonomics literacy prepares people to perform their roles in the workplace and outside the working environment. Ergonomically literate people can learn enough about how technological systems operate to protect themselves by making informed choices and making use of beneficial affordances of the artifacts and environment. People trained in ergonomics typically possess a high level of knowledge and skill related to one or more specific area of ergonomics application. Ergonomics literacy is a prerequisite to ergonomics competency. The following can be proposed as dimensions for ergonomics literacy (Figure 6):

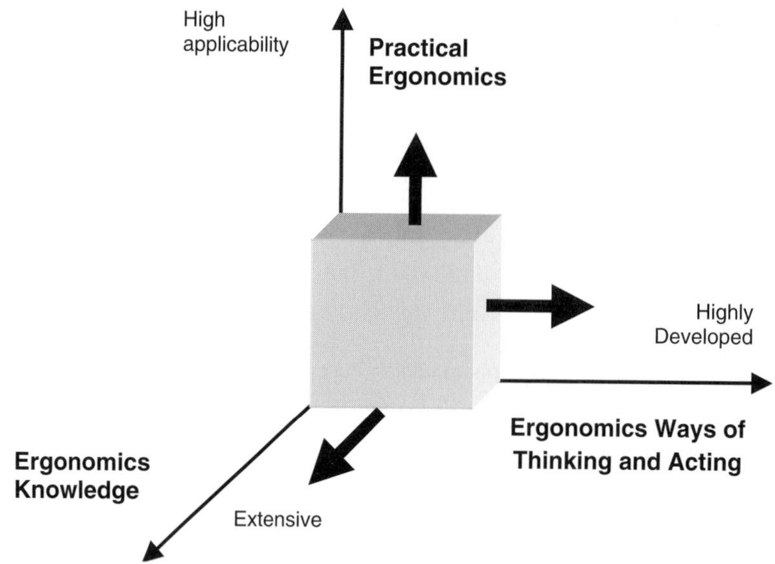

Figure 6 Desired goals for ergonomics literacy. (After Karwowski, 2005.)

1. *Ergonomics knowledge and skills.* A person has a basic knowledge of the philosophy of human-centered design and principles for accommodating human limitations.

2. *Ways of thinking and acting.* A person seeks information about benefits and risks of artifacts and systems (consumer products, services, etc.) and participates in decisions about purchasing and use and/or development of artifacts/ systems.

3. *Practical ergonomics capabilities.* A person can identify and solve simple task (job)-related design problems at work or home and can apply basic concepts of ergonomics to make informed judgments about usability of artifacts and the related risks and benefits of their use.

Finally, Table 11 presents a list of 10 standards for ergonomics literacy, which were proposed by Karwowski (2005) in parallel to a model of technological literacy reported by National Academy of Engineering (Pearson and Young, 2002). Eight of these standards are related to developing an understanding of the nature, scope, attributes, and the role of HFE discipline in modern society; two standards refer to the need for developing the abilities to apply the ergonomics design process and evaluate the impact of artifacts on human safety and well-being.

Table 11 Standards for Ergonomics Literacy: Ergonomics and Technology

An understanding of:	
Standard 1:	characteristics and scope of ergonomics
Standard 2:	core concepts of ergonomics
Standard 3:	connections between ergonomics and other fields of study, and relationships among technology, environment, industry, and society
Standard 4:	cultural, social, economic, and political effects of ergonomics
Standard 5:	role of society in the development and use of technology
Standard 6:	effects of technology on the environment
Standard 7:	attributes of ergonomics design
Standard 8:	role of ergonomics research, development, invention, and experimentation
Abilities to:	
Standard 9:	apply the ergonomics design process
Standard 10:	assess the impact of products and systems on human health, well-being, system performance, and safety

Source: Karwowski (2005).

7 ERGONOMICS DESIGN

Ergonomics is a design-oriented discipline. However, as discussed by Karwowski (2003), ergonomists do not design systems; rather, HFE professionals design the interactions between artifact systems and humans. A fundamental problem involved in such a design is that typically there are multiple functional system–human compatibility requirements that must be satisfied at the same time. To address this issue, structured design methods for complex human–artifact systems are needed. In such a perspective, ergonomics design can be defined in general as mapping from the human capabilities and limitations to system (technology–environment) requirements and affordances (Figure 7), or, more specifically, from the system–human compatibility needs to the relevant human–system interactions.

Suh (1990, 2001) proposed a framework for axiomatic design, which utilizes four different domains that reflect mapping between the identified needs ("what one wants to achieve") and the ways to achieve them ("how to satisfy the stated needs"): (1) customer requirements (customer needs or desired attributes), (2) functional domain (functional requirements and constraints), (3) physical domain (physical design parameters), and (4) processes domain (processes and resources). Karwowski (2005) conceptualized the foregoing domains for ergonomics design purposes, as illustrated in Figure 8, using the concept of compatibility requirements and compatibility mappings between the domains of (1) HFE requirements (goals in terms of human needs and system performance), (2) functional requirements and constraints expressed in terms of human capabilities and limitations, (3) physical domain in terms of design of compatibility, expressed through human–system interactions and specific work system design solutions, and (4) processes domain, defined as management of compatibility (see Figure 9).

7.1 Axiomatic Design

Axiomatic design process is described by the mapping process from functional requirements (FRs) to design parameters (DPs). The relationship between the two vectors FR and DP is as follows:

$$\{\text{FR}\} = [\mathbf{A}]\{\text{DP}\}$$

where [**A**] is the design matrix that characterizes the product design. The design matrix [**A**] for three FRs and three DPs is

$$[\mathbf{A}] = \begin{bmatrix} A_{11} & A_{12} & A_{13} \\ A_{21} & A_{22} & A_{23} \\ A_{31} & A_{32} & A_{33} \end{bmatrix}$$

The following two design axioms, proposed by Suh (1990), are the basis for a formal methodology of design.

Axiom 1: Independence Axiom This axiom stipulates a need for independence of the FRs, which are defined as the minimum set of independent requirements that characterize the design goals (defined by DPs).

Axiom 2: Information Axiom This axiom stipulates minimizing the information content of the design. Among those designs that satisfy the independence axiom, the design that has the smallest information content is the best design.

According to the second design axiom, the information content of the design should be minimized. The information content I_i for a given functional requirement (FR_i) is defined in terms of the probability P_i of satisfying FR_i:

$$I_i = \log_2(1/P_i) = -\log_2 P_i \qquad \text{bits}$$

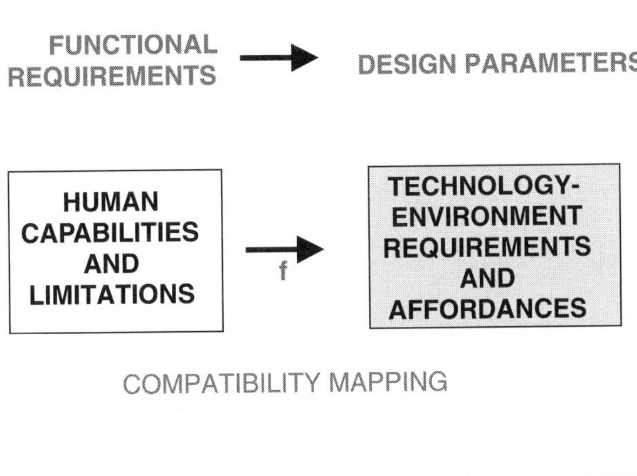

Figure 7 Ergonomics design process: compatibility mapping. (After Karwowski, 2005.)

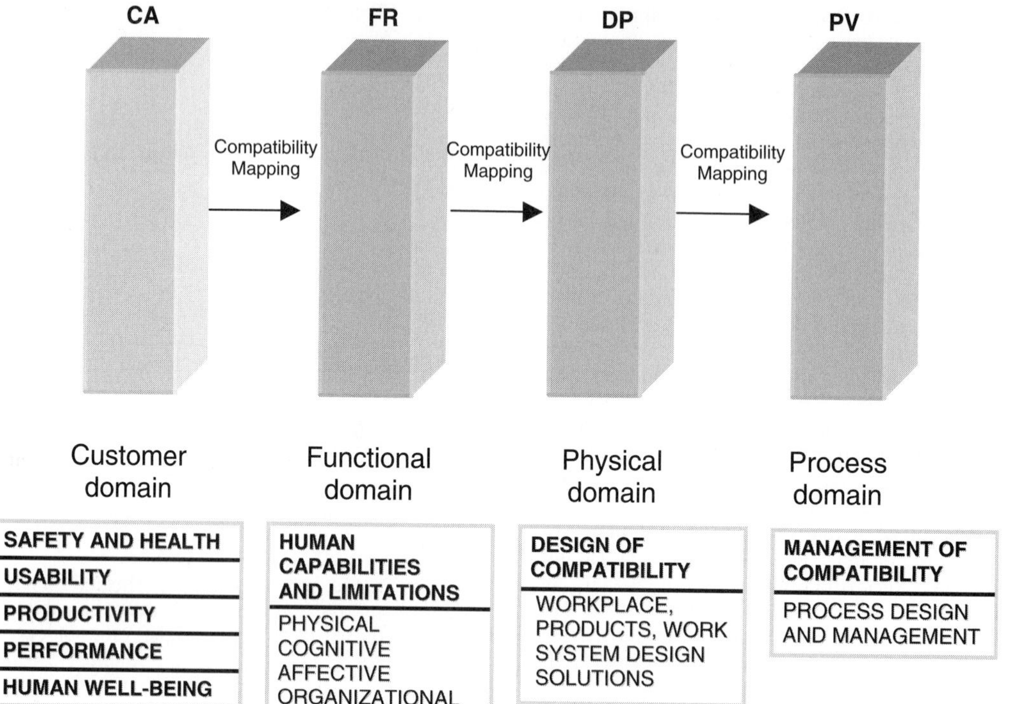

Figure 8 Four domains of axiomatic HFE design. (After Karwowski, 2005.)

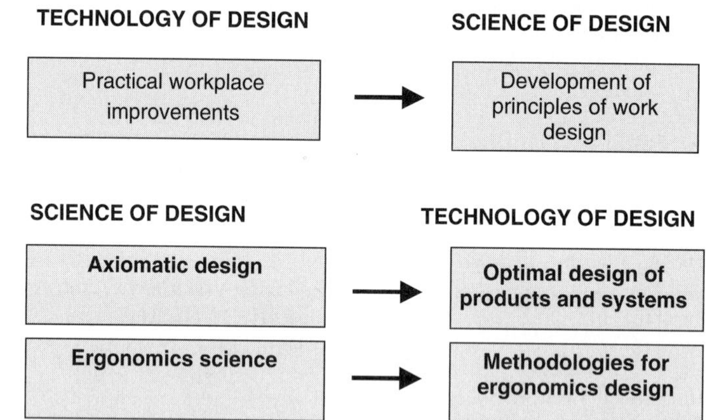

Figure 9 Axiomatic approach to ergonomics design. (After Karwowski, 2005.)

The information content will be additive when there are many functional requirements that must be satisfied simultaneously. In the general case of m FRs, the information content for the entire system,

$$I_{\text{sys}} = -\log_2 P_{\{m\}}$$

where $P_{\{m\}}$ is the joint probability that all m FRs are satisfied.

The axioms above can be adapted for ergonomics design purposes as follows.

Axiom 1: Independence Axiom This axiom stipulates a need for independence of the functional compatibility requirements (FCRs), which are defined as the minimum set of independent compatibility requirements that characterize the design goals [defined by ergonomics design parameters (EDPs)].

Axiom 2: Human Incompatibility Axiom This axiom stipulates a need to minimize the incompatibility content of the design. Among those designs that satisfy the independence axiom, the design that has the smallest incompatibility content is the best design.

As discussed by Karwowski (2001, 2003), in ergonomics design, axiom 2 can be interpreted as follows. The human incompatibility content of the design I_i for a given functional requirement (FR_i) is defined in terms of the compatibility C_i index satisfying FR_i:

$$I_i = \log_2(1/C_i) = -\log_2 C_i \qquad \text{ints}$$

where I denotes the incompatibility content of a design.

7.2 Theory of Axiomatic Design in Ergonomics

As discussed by Karwowski (1985, 1991, 1999, 2001, 2005), a need to remove the system–human incompatibility (or ergonomics entropy) plays the central role in ergonomics design. In view of this, the second axiomatic design axiom can be adopted for the purpose of ergonomics theory as follows.

The incompatibility content of the design I_i for a given functional compatibility requirement (FCR_i) is defined in terms of the compatibility C_i index that satisfies this FCR_i:

$$I_i = \log_2(1/C_i) = -\log_2 C_i \qquad \text{ints}$$

where I denotes the incompatibility content of a design, and the compatibility index $C_i (0 < C < 1)$ is defined depending on the specific design goals (i.e., the applicable or relevant ergonomics design criterion used for system design or evaluation).

To minimize system–human incompatibility, one can either (1) minimize exposure to the negative (undesirable) influence of a given design parameter on the system–human compatibility, or (2) maximize the positive influence of the desirable design parameter (adaptability) on system–human compatibility. The first design scenario [i.e., a need to minimize exposure to the negative (undesirable) influence of a given design parameter (A_i)] typically occurs when A_i exceeds some maximum exposure value of R_i: for example, when the compressive force on the human spine (lumbosacral joint) due to manual lifting of loads exceeds the accepted (maximum) reference value. It should be noted that if $A_i < R_i$, then C can be set to 1, and the related incompatibility due to considered design variable will be zero.

The second design scenario [i.e., a need to maximize positive influence (adaptability) of the desirable feature (design parameter A_i) on system human compatibility], typically occurs when A_i is less than or below some desired or required value of R_i (i.e., minimum reference value): for example, when the range of chair height adjustability is less than the recommended (reference) range of adjustability to accommodate 90% of the mixed (male/female) population. It should be noted that if $A_i > R_i$, then C can be set to 1 and the related incompatibility due to considered design variable will be zero. In both of the cases described above, the human–system incompatibility content can be assessed as discussed below.

1. *Ergonomics design criterion: minimize exposure when $A_i > R_i$*. The compatibility index C_i is defined by the ratio R_i/A_i, where R_i is the maximum exposure (standard) for design parameter i and A_i is the actual value of a given design parameter i:

$$C_i = R_i/A_i$$

and hence

$$\begin{aligned} I_i &= -\log_2 C_i \\ &= -\log_2(R_i/A_i) = \log_2(A_i/R_i) \qquad \text{ints} \end{aligned}$$

Note that if $A_i < R_i$, then C can be set to 1 and the incompatibility content I_i is zero.

2. *Ergonomics design criterion: maximize adaptability when $A_i < R_i$*. The compatibility index C_i is defined by the ratio A_i/R_i, where A_i is the actual value of a given design parameter i, and R_i is the desired reference or required (ideal) design parameter standard: i:

$$C_i = A_i/R_i$$

and hence

$$\begin{aligned} I_i &= -\log_2 C_i \\ &= -\log_2(A_i/R_i) = \log_2(R_i/A_i) \qquad \text{ints} \end{aligned}$$

Note that if $A_i > R_i$, then C can be set to 1 and the incompatibility content I_i is zero.

As discussed by Karwowski (2004), the proposed units of measurement for system–human incompatibility (ints) are parallel and numerically identical to the measure of information (bits). The information content of the design in expressed in terms of the (ergonomics) incompatibility of design parameters with the optimal, ideal, or desired reference values, expressed in terms of ergonomics design parameters, such as range of table height or chair height adjustability, maximum acceptable load of lift, maximum compression on the spins, optimal number of choices, maximum number of hand repetitions per cycle time on a production line, minimum required decision time, and maximum heat load exposure per unit of time.

The general relationships between technology of design and science of design are illustrated in Figure 9. Furthermore, Figure 10 depicts such relationships for the HFE discipline. In the context of axiomatic design in ergonomics, the functional requirements are the human–system compatibility requirements,

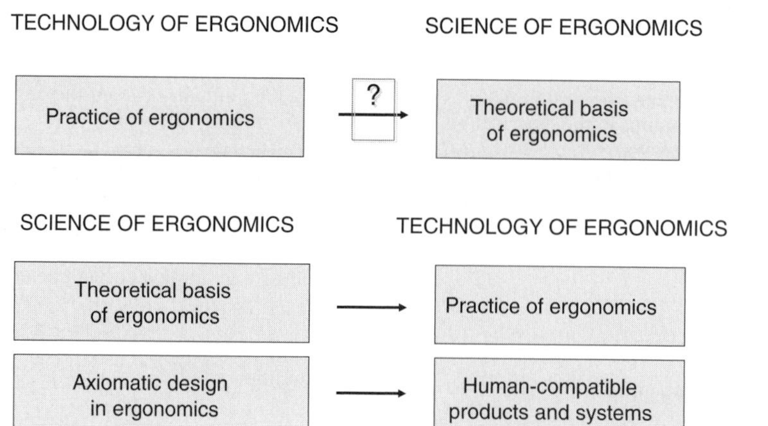

Figure 10 Science, technology, and design of human-compatible systems. (After Karwowski, 2005.)

while the design parameters are the human–system interactions. Therefore, ergonomics design can be defined as mapping from human–system compatibility requirements to human–system interactions. More generally, HFE can be defined as the science of design, testing, evaluation, and management of human system interactions according to the human–system compatibility requirements.

7.3 Applications of Axiomatic Design

Helander (1995) was first to provide a conceptualization of the second design axiom in ergonomics by considering selection of a chair based on the information content of specific chair design parameters. Recently, Karwowski (2003) introduced the concept of system incompatibility measurements and the measure of incompatibility for ergonomics design and evaluation. Furthermore, Karwowski (2003) has also illustrated an application of the first design axiom adapted to the needs of ergonomics design, using an example of the design of the rear lighting system utilized to provide information about application of brakes in a passenger car. The rear lighting system is illustrated in Figure 11. In this highway safety–related example, the FRs of the rear lighting (braking display) system were defined in terms of FRS and DPs as follows:

FR_1 = provides early warning to maximize the lead response time(MLRT)| (information about the car in front that is applying brakes)

FR_2 = assures safe braking (ASB)

The traditional (old) design solution is based on two design parameters (DPs):

DP_1 = two rear brake lights on the sides (TRLS)

DP_2 = efficient braking mechanism (EBM)

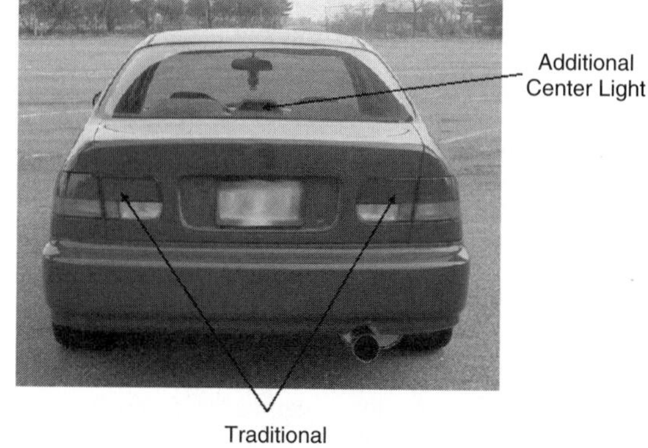

Figure 11 Redesigned rear light system of an automobile. (After Karwowski, 2005.)

The design matrix of the traditional rear lighting system (TRLS) is as follows:

$$\begin{Bmatrix} FR_1 \\ FR_2 \end{Bmatrix} = \begin{bmatrix} X & 0 \\ X & X \end{bmatrix} \begin{Bmatrix} DP_1 \\ DP_2 \end{Bmatrix}$$

MLRT	X	0	TRLS
ASB	X	X	EBM

This rear lighting warning system (old solution) can be classified as a *decoupled design* and is not an optimal design. The reason for such classification is that even with the efficient braking mechanism, one cannot compensate for the lack of time in the driver's response to braking of the car in front due to a sudden traffic slowdown. In other words, this rear lighting system does not provide early warning that would allow the driver to maximize his or her lead response time (MLRT) to braking.

The solution that was implemented two decades ago utilizes a new concept for the rear lighting of the braking system (NRLS). The new design is based on addition of the third brake light, positioned in the center and at a height that allows this light to be seen through the windshields of the car preceding the car immediately in front. This new design solution has two design parameters:

DP1 = a new rear lighting system (NRLS)

DP2 = efficient braking mechanism (EBM)(the same as before)

The formal design classification of the new solution is uncoupled design. The design matrix for this new design is as follows:

$$\begin{Bmatrix} FR_1 \\ FR_2 \end{Bmatrix} = \begin{bmatrix} X & 0 \\ 0 & X \end{bmatrix} \begin{Bmatrix} DP_1 \\ DP_2 \end{Bmatrix}$$

MLRT	X	0	**NRLS**
ASB	0	X	EBM

It should be noted that the original (traditional) rear lighting system (TRLS) can be classified as decoupled design. This old design ($DP_{1,O}$) does not compensate for the lack of early warning that would make it possible to maximize the driver's lead response time (MLRT) whenever braking is needed, and therefore violates the second functional requirement (FR_2) for a safe braking requirement. The design matrix for new system (NRLS) is an uncoupled design that satisfies the independence of functional requirements (independence axiom). This uncoupled design ($DP_{1,N}$) fulfills the requirement of maximizing lead response time (MLRT) whenever braking is needed and does not violate the FR_2 (safe braking requirement).

8 THEORETICAL ERGONOMICS: SYMVATOLOGY

It should be noted that the system–human interactions often represent complex phenomena with dynamic compatibility requirements. The are often nonlinear and can be unstable (chaotic) phenomena, modeling of which requires a specialized approach. Karwowski (2001) indicated a need for symvatology as a corroborative science to ergonomics that can help in developing solid foundations for the ergonomics science. The proposed subdiscipline is *symvatology*, the science of the artifact–human (system) compatibility. Symvatology aims to discover the laws of artifact–human compatibility, propose theories of artifact–human compatibility, and develop a quantitative matrix for measurement of such compatibility. Karwowski (2001) coined the term *symvatology* by joining two Greek words: *symvatotis* (compatibility) and *logos* (logic, or reasoning about). Symvatology is the systematic study (which includes theory, analysis, design, implementation, and application) of interaction processes that define, transform, and control compatibility relationships between artifacts (systems) and people. An *artifact system* is defined as a set of all artifacts (meaning objects made by human work) as well as natural elements of the environment and their interactions occurring in time and space afforded by nature. A human system is defined as a human (or humans) with all the characteristics (physical, perceptual, cognitive, emotional, etc.) that are relevant to an interaction with the artifact system.

To optimize both the human and system well-being and performance, system–human compatibility should be considered at all levels, including the physical, perceptual, cognitive, emotional, social, organizational, managerial, environmental, and political. This requires a way to measure the inputs and outputs that characterize the set of system–human interactions (Karwowski, 1991). The goal of quantifying artifact–human compatibility can be realized only if we understand its nature. Symvatology aims to observe, identify, describe, perform empirical investigations, and produce theoretical explanations of the natural phenomena of artifact–human compatibility. As such, symvatology should help to advance the progress of ergonomics discipline by providing methodology for design for compatibility, as well as design of compatibility between the artificial systems (technology) and humans. In the perspective described above, the goal of ergonomics should be to optimize both human and system well-being and their mutually dependent performance. As pointed out by Hancock (1997), it is not enough to assure the well-being of humans; one must also optimize the well-being of systems (i.e., artifact-based technology and nature) to make proper uses of life.

Due to the nature of the interactions, an artifact system is often a dynamic system with a high level of complexity, that exhibits nonlinear behavior. The *American Heritage Dictionary of the English Language* (1978) defines *complex* as consisting of interconnected or interwoven parts. Karwowski et al. (1988) and Karwowski and Jamaldin (1995) proposed representing an artifact–human system as a construct that contains a human subsystem, an artifact subsystem, an environmental subsystem, and a set of interactions occurring between the various elements of these subsystems over time. In the framework above, compatibility is a dynamic, natural phenomenon that is affected by the artifact–human system structure, its inherent complexity, and its entropy or the level of incompatibility between the system's elements. Since the structure of system interactions determines the complexity and related compatibility relationships in a given system, compatibility should be considered in relation to the system's complexity.

The system space, denoted here as an ordered set [(complexity, compatibility)], is defined by four pairs: (high, high), (high, low), (low, high), (low, low). Under the best scenario (i.e., the most optimal state of system design), the artifact–human system exhibits high compatibility and low complexity levels. It should be noted that the transition from a high to a low level of system complexity does not necessarily lead to an improved (higher) level of system compatibility. Also, it is often the case in most of artifact–human systems that improved (higher) system compatibility can be achieved only at the expense of increasing system complexity.

As discussed by Karwowski and Jamaldin (1995) lack of compatibility, or ergonomics incompatibility (EI), defined as degradation (disintegration) of an artifact–human system, is reflected in the system's measurable inefficiency and associated human losses. To express the innate relationship between the systems's complexity and compatibility, Karwowski et al. (1988, 1994a) proposed the *complexity–incompatibility principle*, which can be stated as follows: *As artifact–human system complexity increases, the incompatibility between system elements, as expressed through their ergonomic interactions at all system levels, also increases, leading to greater ergonomic (nonreducible) entropy of the system and decreasing the potential for effective ergonomic intervention.*

The foregoing principle was illustrated by Karwowski (1995) using as examples design of a chair (see Figure 12) and design of a computer display, two common problems in the area of human–computer interaction. In addition, Karwowski and Jamaldin (1996) discussed the complexity–compatibility paradigm in the context of organizational design. It should be noted that the principle reflects the natural phenomena that others in the field have described in terms of difficulties encountered when humans interact with consumer products and technology in general. For example, according to Norman (1989), the paradox of technology is that adding functionality to an artifact typically comes with the trade-off of increased complexity. These added complexities often lead to increased human difficulty and frustration when interacting with these artifacts. One reason for the above is that technology, which has more features, also has less feedback. Moreover, Norman noted that added complexity cannot be avoided when functions are added, and can be minimized only with good design, which follows natural mapping between system elements (i.e., control–display compatibility). Following Ashby's (1964) law of requisite variety, Karwowski and Jamaldin (1995) proposed a corresponding law, called the *law of requisite complexity*, which states that only design complexity can reduce system complexity. This means that only added complexity of the regulator, expressed by system compatibility requirements, can be used to reduce ergonomics system entropy (i.e., reduce overall artifact–human system incompatibility).

9 CONGRUENCE BETWEEN MANAGEMENT AND ERGONOMICS

Advanced technologies with which humans interact toady constitute complex systems that require a high level of integration from both the *design and management* perspectives. *Design integration*

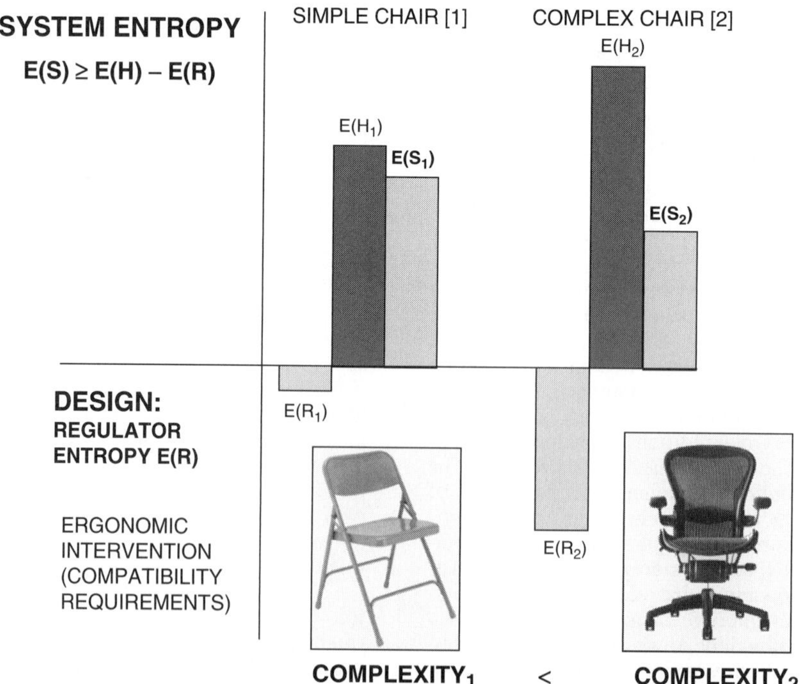

Figure 12 System entropy determination: example of chair design. (After Karwowski, 2002.)

typically focuses on the interactions between hardware (computer-based technology), organization (organizational structure), information system, and people (human skills, training, and expertise). *Management integration* refers to interactions among various system elements across process and product quality, workplace and work system design, occupational safety and health programs, and corporate environmental protection polices.

Scientific management originated with the work of Frederick W. Taylor (1911), who studied, among other problems, *how jobs were designed* and how workers could be trained to perform these jobs. The natural congruence between contemporary management and HFE can be described in the context of the respective definitions of these two disciplines. Management is defined today as a set of activities, including (1) planning and decision making, (2) organizing, (3) leading, and (4) controlling, directed at an organization's resources (human, financial, physical, and information) with the aim of achieving organizational goals in an efficient and effective manner (Griffin, 2001). The main elements of the management definition presented above and central to ergonomics are (1) *organizing*, (2) *human resource planning*, and (3) *effective and efficient achievement of organizational goals*. In the description of these elements, the original terms proposed by Griffin (2001) are applied to ensure precision of the concepts and terminology used. *Organizing* is deciding which method of organizational element grouping is best. *Job design* is the basic building block of organization structure. Job design focuses on identification and determination of the tasks and activities for which the particular worker is responsible.

It should be noted that the basic ideas of management (i.e., planning and decision making, organizing, leading, and controlling) are also essential to HFE. Specifically, common to management and ergonomics are the issues of job design and job analysis. Job design is widely considered to be the first building block of an organizational structure. Job analysis as a systematic analysis of jobs within an organization allows determination of a person's work-related responsibilities. *Human resource planning* is an integral part of human resource management. The starting point for this business function is a *job analysis*: a systematic analysis of the workplaces in an organization. *Job analysis* consists of job description and job specification. *Job description* should include description of task demands and work environment conditions, such as work tools, materials, and machines needed to perform specific tasks. *Job specification* determines abilities, skills, and other worker characteristics necessary for *effective and efficient* task performance in particular jobs.

The discipline of management also considers important human factors that play a role in *achieving organizational goals in an effective and efficient way*. Such factors include work stress in the context of individual worker behavior and human resource management in the context of safety and heath management. Work stress may be caused by the four categories of organizational and individual factors: (1) decision related to *task demands*; (2) *work environment demands*, including physical, perceptional, and cognitive task demands, as well as quality of the work environment (i.e., adjustment of tools and machines to human characteristics and capabilities); (3) *role demands* related to relations with supervisor and co-workers; and (4) *interpersonal demands*, which can cause conflict between workers (e.g., management style, group pressure). Human resource management includes the provision of safe work conditions and environment at each workstation, workplace, and in the entire organization.

It should also be noted that the elements of management discipline described above, such as *job design, human resource planning* (job analysis and job specification), *work stress management, and safety and health management*, are essential components of an HFE subdiscipline often called *industrial ergonomics*. Industrial ergonomics investigates human–system relationships at the individual workplace (workstation) level or at the work system level, embracing knowledge that is also of central interest to management. From this point of view, industrial ergonomics, in congruence with management, is focusing on organization and management at the workplace level (work system level) through the design and assessment (testing and evaluation) of job tasks, tools, machines, and work environments in order to adapt these to the capabilities and needs of workers.

Another important subdiscipline of HFE with respect to the central focus of management discipline is *macroergonomics*. According to Hendrick and Kleiner (2001), macroergonomics is concerned with the analysis, design, and evaluation of work systems. *Work* denotes any form of human effort or activity. *System* refers to *sociotechnical* systems, which range from a single person to a complex multinational organization. A *work system* consists of people interacting with some form of (1) job design (work modules, tasks, knowledge, and skill requirements), (2) hardware (machines or tools) and/or software, (3) internal environment (physical parameters and psychosocial factors), (4) external environment (political, cultural, and economic factors), and (5) organizational design (i.e., the work system's structure and processes used to accomplish desired functions).

The unique technology of human factors/ergonomics is the *human–system interface technology*. Human–system interface technology can be classified into five subparts, each with a related design focus (Hendrick, 1997; Hendrick and Kleiner, 2001):

1. *Human–machine* interface technology: hardware ergonomics
2. *Human–environment* interface technology: environmental ergonomics
3. *Human–software* interface technology: cognitive ergonomics
4. *Human–job* interface technology: work design ergonomics
5. *Human–organization* interface technology: macroergonomics

In this context, as discussed by Hendrick and Kleiner (2001), the HFE discipline discovers knowledge of human performance capabilities, limitations, and other human characteristics in order to develop human–system interface (HSI) technology, which includes interface design principles, methods, and guidelines. Finally, the HFE profession applies HSI technology to the design, analysis, test and evaluation, standardization, and control of systems.

10 INTERNATIONAL ERGONOMICS ASSOCIATION

Over the past 20 years, ergonomics as a scientific discipline and as a profession has grown rapidly by expanding the scope and breadth of theoretical inquiries, methodological basis, and practical applications (Meister 1997, 1999; Chapanis, 1999; Stanton and Young, 1999; Kuorinka, 2000; Karwowski, 2001; IEA, 2003). As a profession, the field of ergonomics has seen the development of formal organizational structures (i.e., national and cross-national ergonomics societies and networks) in support of HFE professionals internationally. As of 2004, the International Ergonomics Association (www.iea.cc), a federation of 42 ergonomics and human factors societies around the world, accounted for over 14,000 HFE members worldwide (see Table 12). The main goals of the IEA are to elaborate and advance the science and practice of ergonomics at the international level and to improve the quality of life by expanding the scope

Table 12 Membership by Federated Societies

Federated society	Name	Initials	Members
Australia	Ergonomics Society of Australia	ESA	536
Austria	Österreichische Arbeitsgemeinschaft für Ergonomie	OAE	32
Belgium	Belgian Ergonomics Society	BES	176
Brazil	Brazilian Ergonomics Society	ABERGO	140
Canada	Association of Canadian Ergonomists	ACE	545
Chile	Sociedad Chileana de Ergonomia	SOCHERGO	30
China	Chinese Ergonomics Society	ChES	450
Colombia	Sociedad Colombiana de Ergonomia	SCE	30
Croatia	Croatian Ergonomics Society	CrES	40
Czech Republic	Czech Ergonomics Society	CzES	33
Francophone Society	Societé d'Ergonomie Langue Française	SELF	680
Germany	Gesellschaft für Arbeitswissenschaft	GfA	578
Greece	Hellenic Ergonomics Society	HES	34
Hong Kong	Hong Kong Ergonomics Society	HKES	33
Hungary	Hungarian Ergonomics Society	MES	70
India	Indian Society of Ergonomics	ISE	53
Iran	Iranian Ergonomics Society	IES	30
Ireland	Irish Ergonomics Society	IrES	35
Israel	Israeli Ergonomics Society	IES	38
Italy	Societá Italiana di Ergonomia	SIA	191
Japan	Japan Ergonomics Society	JES	2,155
Mexico	Sociedad de Ergonomistas de Mexico	SEM	30
Netherlands	Nederlandse Vereniging Voor Ergonomie	NVVE	444
New Zealand	New Zealand Ergonomics Society	NZES	115
Nordic countries	Nordic Ergonomics Society	NES	1,510
Poland	Polish Ergonomics Society	PES	373
Portugal	Associação Portuguesa de Ergonomia	APERGO	101
Russia	Inter-Regional Ergonomics Association	IREA	207
Slovakia	Slovak Ergonomics Association	SEA	27
South Africa	Ergonomics Society of South Africa	ESSA	60
South Korea	Ergonomics Society of Korea	ESK	520
Southeast Asia	Southeast Asian Ergonomics Society	SEAES	64
Spain	Association Espanola de Ergonomia	AEE	151
Switzerland	Swiss Society for Ergonomics	SSE	128
Taiwan	Ergonomics Society of Taiwan	EST	98
Turkey	Turkish Ergonomics Society	TES	50
Ukraine	All-Ukrainian Ergonomics Association	AUEA	107
United Kingdom	Ergonomics Society	ES	1,024
United States	Human Factors and Ergonomics Society	HFES	3,655
Yugoslavia	Ergonomics Society of F. R. of Yugoslavia	ESFRY	50
Affiliated society	Human Ergology Society (Japan)	HES(J)	222
Total			14,845

Source: IEA, http://www.iea.cc/newsletter/nov2003.cfm.

of ergonomics applications and contributions to global society (Table 13).

Past and current IEA activities focus on the development of programs and guidelines to facilitate the discipline and profession of ergonomics worldwide. Examples of such activities include:

- International directory of ergonomics programs
- Core competencies in ergonomics
- Criteria for IEA endorsement of certifying bodies in professional ergonomics
- Guidelines for the process of endorsing a certification body in professional ergonomics
- Guidelines on standards for accreditation of ergonomics education programs at the tertiary (university) level
- Ergonomics quality in design (EQUID) program

More information about these programs can be found on the IEA Web site: www.ie.cc. In addition to the above, the IEA endorses scientific journals in the field. A list of the core HFE journals is shown in Table 14. A complete classification of the core and related HFE journals was proposed by Dul and Karwowski (2004).

IEA has also developed several actions for stimulating development of HFE in industrially developing countries (IDCs). Such actions include the following elements:

- Cooperating with international agencies such as the ILO (International Labour Organisation), WHO (World Health Organisation), and professional scientific associations with which the IEA has signed formal agreements
- Working with major publishers of ergonomics journals and texts to extend their access to federated societies, with particular focus on developing countries
- Development of support programs for developing countries to promote ergonomics and extend ergonomics training programs
- Promotion of workshops and training programs in developing countries through educational kits and visiting ergonomists
- Extending regional ergonomics networks of countries to countries with no ergonomics programs in their region
- Support to non-IEA member countries in applying for affiliation to IEA in conjunction with the IEA Development Committee

Table 13 IEA Technical Committees

Aging	Human–Computer Interaction
Agriculture	Human Reliability
Auditory Ergonomics	Musculoskeletal Disorders
Building and Architecture	Organizational Design and Management
Building and Construction	Process Control
Consumer Products	Psychophysiology in Ergonomics
Cost-Effective Ergonomics	Quality Management
Ergonomics for Children and Educational Environments	Rehabilitation Ergonomics
Hospital Ergonomics	Safety and Health
Human Aspects of Advanced Manufacturing	Standards

Table 14 Core HFE Journals

Official IEA journal	Ergonomics[a]
IEA-endorsed journals	Applied Ergonomics[a]
	Ergonomia: An International Journal of Ergonomics and Human Factors
	Human Factors and Ergonomics in Manufacturing[a]
	International Journal of Industrial Ergonomics[a]
	International Journal of Human–Computer Interaction[a]
	International Journal of Occupational Safety and Ergonomics
	Theoretical Issues in Ergonomics Science
Other core journals	Human Factors[a]
	Le Travail Human[a]
Non-ISI journals	Asian Journal of Ergonomics
	Japanese Journal of Ergonomics
	Occupational Ergonomics
	Tijdschrift voor Ergonomie
	Zeitschrift für Arbeitswissenschaft
	Zentralblatt für Arbeirsmedizin, Arbeitsschurz und Ergonomie

Source: Based on Dul and Karwowski (2004).
[a]ISI-ranked journal.

11 FUTURE CHALLENGES

Contemporary HFE discipline exhibits rapidly expanding application areas, continuing improvements in research methodologies, and increased contributions to fundamental knowledge as well as important applications to the needs of the society at large. For example, the subfield of neuroergonomics focuses on the neural control and brain manifestations of the perceptual, physical, cognitive, emotional, etc. interrelationships in human work activities (Parasuraman, 2003). As the *science of brain and work environment, neuroergonomics* aims to explore the premise of design of work to match the neural capacities and limitations of people. The potential benefits of this emerging branch of HFE are improvements in medical therapies and applications of more sophisticated workplace design principles. The near future will also see development of an entirely new HFE domain that could be called *nanoergonomics*. The idea of building machines at the molecular scale, once fulfilled, will affect every facet of our lives: medicine, health care, computer, information, communication, environment, economy, and many more (Henry T. Yang, Chancellor, University of California–Santa Barbara). Nanoergonomics will address issues of humans interacting with devices and machines of extremely small dimensions, and in general with nanotechnology.

Developments in technology and the socioeconomic dilemmas of the twenty-first century pose significant challenges for the discipline and profession of HFE. According to the report "Major Predictions for Science and Technology in the Twenty-First *Century*, published by the Japan Ministry of Education, Science and Technology (MITI, 2001), the following issues will affect the future of our civilization:

- Developments in genetics (DNA, human evolution, creation of an artificial life, extensive outer-space exploration, living outside Earth)
- Developments in cognitive sciences (human cognitive processes through artificial systems)
- Revolution in medicine (cell and organ regeneration, nanorobotics for diagnostics and therapy, super-prostheses, artificial photosynthesis of foods)
- Elimination of starvation and malnutrition (artificial photosynthesis of foods, safe genetic foods manipulation)
- Full recycling of resources and reusable energy (biomass and nanotechnology)
- Changes in human habitat (outer-space cities, 100% underground industrial manufacturing, separation of human habitat from natural environments, protection of diversity of life-forms on Earth)
- Cleanup of the negative effects of the twentieth century on the environment (organisms for environmental cleaning, regeneration of the ozone)
- Communication (nonverbal communication technology, new three-dimensional projections systems)
- Politics (computerized democracy)
- Transport and travel (natural sources of clean energy, automated transport systems, revolution in supersonic small aircraft and supersonic travel, underwater ocean travel)
- Safety and control over one's life (prevention of crime by brain intervention, human error avoidance technology, control of the forces of nature, intelligent systems for safety in all forms of transport)

The issues listed above will also affect future directions in development of the science, engineering, design, technology, and management of human-compatible systems.

REFERENCES

Ashby, W. R. (1964), *An Introduction to Cybernetics*, Methuen, London.

Awad, E., and Ghaziri, H. M. (2004), *Knowledge Management*, Prentice-Hall, Upper Saddle River, NJ.

Baber, C. (1996), "Repertory Grid Theory and Its Application to Product Evaluation," in *Usability Evaluation in Industry*, P. W. Jordan, B. Thomas, B. A. Weerdmeester, and I. L. McClelland, Eds., Taylor & Francis, London, pp. 157–166.

Card, S., Moran, T., and Newell, A. (1983), *The Psychology of Human–Computer Interaction*, Lawrence Erlbaum Associates, Mahwah, NJ.

Chaffin, D. B., and Anderson, G. B. J. (1993), *Occupational Biomechanics*, 2nd ed., Wiley, New York.

Chapanis, A. (1995), *Human Factors in System Engineering*, Wiley, New York.

Chapanis, A. (1999), *The Chapanis Chronicles: 50 Years of Human Factors Research, Education, and Design*, Aegean, Santa Barbara, CA.

Conrad, M. (1983), *Adaptability*, Plenum Press, New York.

Diaper, D., and Stanton, N. A. (2004), *The Handbook of Task Analysis for Human–Computer Interaction*, Lawrence Erlbaum Associates, Mahwah, NJ.

Dix, A., Finlay, J., Abowd, G., and Beale R. (1993), *Human–Computer Interaction*, Prentice-Hall, Englewood Cliffs, NJ.

Dul, J., and Karwowski, W. (2004), "An Assessment System for Rating Scientific Journals in the Field of Ergonomics and Human Factors," *Applied Ergonomics*, Vol. 35 No. 4, pp. 301–310.

Dul, J., and Weerdmeester, B. (1993), *Ergonomics for Beginners: A Quick Reference Guide*, Taylor & Francis, London.

Dzissah, J., Karwowski, W., and Yang, Y. N. (2001), "Integration of Quality, Ergonomics, and Safety Management Systems," in *International Encyclopedia of Ergonomics and Human Factors*, W. Karwowski, Ed., Taylor & Francis, London, pp. 1129–1135.

Edholm, O. G., and Murrell, K. F. H. (1973), *The Ergonomics Society: A History, 1949–1970*, Ergonomics Research Society, London.

EIAC (2000), *Ergonomics Abstracts*, Ergonomics Information Analysis Centre, School of Manufacturing and

Mechanical Engineering, University of Birmingham, Birmingham England.

Goldman, S. L., Nagel, R. N., and Preiss, K. (1995), *Agile Competitors and Virtual Organizations*, Van Nostrand Reinhold, New York.

Gould, J. D., and Lewis, C. (1983), "Designing for Usability: Key Principles and What Designers Think," in *Proceedings of the CHI '83 Conference on Human Factors in Computing Systems*, ACM, New York, pp. 50–53.

Grandjean, E. (1986), *Fitting the Task to the Man*, Taylor & Francis, London.

Griffin, R. W. (2001), *Management*, 7th ed., Houghton Mifflin, Boston.

Hancock, P. (1997), *Essays on the Future of Human–Machine Systems*, BANTA Information Services Group, Eden Prairie, MN.

Harre, R. (1972), *The Philosophies of Science*, Oxford University Press, London.

Helander, M. G. (1995), "Conceptualizing the Use of Axiomatic Design Procedures in Ergonomics," in *Proceedings of the IEA World Conference*, Associação Brasileira de Ergonomia, Rio de Janeiro, Brazil, pp. 38–41.

Helander, M. G. (1997a), "Forty Years of IEA: Some Reflections on the Evolution of Ergonomics," *Ergonomics*, Vol. 40, pp. 952–961.

Helander, M. G. (1997b), "The Human Factors Profession," in *Handbook of Human Factors and Ergonomics*, 2nd ed., G. Salvendy, Ed., Wiley, New York, pp. 3–16.

Helander, M. G., and Lin, L. (2002), "Axiomatic Design in Ergonomics and Extension of Information Axiom," *Journal of Engineering Design*, Vol. 13, No. 4, pp. 321–339.

Helander, M. G., Landaur, T. K., and Prabhu, P. V., Eds. (1997), *Handbook of Human–Computer Interaction*, Elsevier, Amsterdam.

Hendrick, H. W. (1997), "Organizational Design and Macroergonomics," in *Handbook of Human Factors and Ergonomics*, G. Salvendy, Ed., Wiley, New York, pp. 594–636.

Hendrick, H. W. and Kleiner, B. M. (2001), *Macroergonomics—An Introduction to Work System Design* The Human Factors and Ergonomics Society, Santa Monica, CA.

Hendrick, H. W., and Kleiner, B. W. (2002a), *Macroergonomics: An Introduction to Work Systems Design*, Human Factors and Ergonomics Society, Santa Monica, CA.

Hendrick, H. W., and Kleiner, B. M., Eds. (2002b), *Macroergonomics: Theory, Methods, and Applications*, Lawrence Erlbaum Associates, Mahwah, NJ.

Hollnagel, E., Ed. (2003), *Handbook of Cognitive Task Design*, Lawrence Erlbaum Associates, Mahwah, NJ.

Holman, D., Wall, T. D., Clegg, C. W., Sparrow, P., and Howard, A. (2003), *New Workplace: A Guide to the Human Impact of Modern Working Practices*, Wiley, Chichester, West Sussex, England.

HFES (2003), *Directory and Yearbook*, Human Factors and Ergonomics Society, Santa Monica, CA.

Iacocca Institute (1991), *21st Century Manufacturing Enterprise Strategy: An Industry-Led View*, Vol. 1 and 2, Iacocca Institute, Bethlehem, PA.

IEA (2003), *IEA Triennial Report, 2000–2003*, IEA Press, Santa Monica, CA.

Jamaldin, B., and Karwowski, W. (1997), "Quantification of Human–System Compatibility (HUSYC): An Application to Analysis of the Bhopal Accident," in *From Experience to Innovation: Proceedings of the 13th Triennial Congress of the International Ergonomics Association*, P. Seppala, T. Luopajarvi, C.-H. Nygard, and M. Mattila, Eds., Tampere, Finland, Vol. 3, pp. 46–48.

Jastrzebowski, W. B. (1857a), "An Outline of Ergonomics, or the Science of Work Based upon the Truths Drawn from the Science of Nature, Part I," *Nature and Industry*, Vol. 29, pp. 227–231.

Jastrzebowski, W. B. (1857b), "An Outline of Ergonomics, or the Science of Work Based upon the Truths Drawn from the Science of Nature, Part II," *Nature and Industry*, Vol. 30, pp. 236–244.

Jastrzebowski, W. B. (1857c), "An Outline of Ergonomics, or the Science of Work Based upon the Truths Drawn from the Science of Nature, Part III," *Nature and Industry*, Vol. 31, pp. 244–251.

Jastrzebowski, W. B. (1857d), "An Outline of Ergonomics, or the Science of Work Based upon the Truths Drawn from the Science of Nature, Part IV," *Nature and Industry*, Vol. 32, pp. 253–258.

Karwowski, W. (1985), "Why Do Ergonomists Need Fuzzy Sets?" in *Ergonomics International 85, Proceedings of the 9th Congress of the International Ergonomics Association, Bernemouth, England*, I.D. Brown, R. Goldsmith, K. Coombes and M.A. Sinclair, Eds., London, Taylor & Francis, pp. 409–411.

Karwowski, W. (1991), "Complexity, Fuzziness and Ergonomic Incompatibility Issues: The Control of Dynamic Work Environments," *Ergonomics*, Vol. 34, No. 6, pp. 671–686.

Karwowski, W. (1992a), "The Complexity–Compatibility Paradigm in the Context of Organizational Design of Human–Machine Systems," in *Human Factors in Organizational Design and Management*, V. O. Brown and H. Hendrick, Eds., Elsevier, Amsterdam, pp. 469–474.

Karwowski, W. (1992b), "The Human World of Fuzziness, Human Entropy, and the Need for General Fuzzy Systems Theory," *Journal of the Japan Society for Fuzzy Theory and Systems*, Vol. 4, No. 5, pp. 591–609.

Karwowski, W., (1994), "A General Modelling Framework for the Human-Computer Interaction Based on the Principles of Ergonomic Compatibility Requirements and Human Entropy," in *Fourth International Scientific Conference Book of Short Papers*, Molteni, G., Occhipinti, E. and Piccoli, B., Eds., Institute of Occupational Health, University of Milan, October 2–5, Vol. 1, pp. A12–A19.

Karwowski, W. (1995), "A General Modeling Framework for the Human–Computer Interaction Based on the Principles of Ergonomic Compatibility Requirements and Human Entropy," in *Work with Display Units, Vol. 94*, A. Grieco, G. Molteni, E. Occhipinti, and B. Piccoli, Eds., North-Holland, Amsterdam, pp. 473–478.

Karwowski, W. (1997), "Ancient Wisdom and Future Technology: The Old Tradition and the New Science of Human Factors/Ergonomics," in *Proceedings of the Human Factors and Ergonomics Society 4th Annual Meeting*, Albuquerque, NM, Human Factors and Ergonomics Society, Santa Monica, CA, pp. 875–877.

Karwowski, W., (1998), "Selected Directions and Trends in Development of Ergonomics in USA" (in Polish), *Ergonomia*, Vol. 21, No. 1–2, pp. 141–155.

Karwowski, W. (2000), "Symvatology: The Science of an Artifact–Human Compatibility," *Theoretical Issues in Ergonomics Science*, Vol. 1, No. 1, pp. 76–91.

Karwowski, W., Ed. (2001), *International Encyclopedia of Ergonomics and Human Factors*, Taylor & Francis, London.

Karwowski, W., (2005), "Ergonomics and Human Factors: The Paradigms for Science, Engineering, Design, Technology, and Management of Human-Compatible Systems," *Ergonomics*, (in press).

Karwowski, W. (2006), "On Measure of the Human–System Compatibility," *Theoretical Issues in Ergonomics Science* (in press).

Karwowski, W., and Jamaldin, B. (1995), "The Science of Ergonomics: System Interactions, Entropy, and Ergonomic Compatibility Measures," in *Advances in Industrial Ergonomics and Safety*, Vol. VII, A. C. Bittner, Jr., and P. C. Champney, Eds., Taylor & Francis, London, pp. 121–126.

Karwowski, W., and Jamaldin, B. (1996), "New Methodological Framework for Quantifying Compatibility of Complex Ergonomics Systems," *Proceedings of the 6th Pan Pacific Conference on Occupational Ergonomics*, Taipei, Taiwan, Vol. 10, pp. 676–679.

Karwowski, W., and Marras, W. S., Eds. (1999), *The Occupational Ergonomics Handbook*, CRC Press, Boca Raton, FL.

Karwowski, W., and Mital, A., Eds. (1986), *Applications of Fuzzy Set Theory in Human Factors*, Elsevier Science, Amsterdam.

Karwowski, W., and Rodrick, D. (2001), "Physical Tasks: Analysis, Design and Operation," in *Handbook of Industrial Engineering*, 3rd ed., G. Salvendy, Ed., Wiley, New York, pp. 1041–1110.

Karwowski, W., and Salvendy, G., Eds. (1994), *Organization and Management of Advanced Manufacturing*, Wiley, New York.

Karwowski, W., Marek, T., and Noworol, C. (1988), "Theoretical Basis of the Science of Ergonomics," in *Proceedings of the 10th Congress of the International Ergonomics Association*, Sydney, Australia, August, Taylor & Francis, London, pp. 756–758.

Karwowski, W., Marek, T., and Noworol, C. (1994a), "The Complexity–Incompatibility Principle in the Science of Ergonomics," in *Advances in Industrial Ergonomics and Safety*, Vol. VI, F. Aghazadeh, Ed., Taylor & Francis, London, pp. 37–40.

Karwowski, W., Salvendy, G., Badham, R., Brodner, P., Clegg, C., Hwang, L., Iwasawa, J., Kidd, P. T., Kobayashi, N., Koubek, R., Lamarsh, J., Nagamachi, M., Naniwada, M., Salzman, H., Seppälä, P., Schallock, B., Sheridan, T., and Warschat, J. (1994b), "Integrating People, Organization and Technology in Advance Manufacturing," *Human Factors and Ergonomics in Manufacturing*, Vol. 4, pp. 1–19.

Karwowski, W., Kantola, J., Rodrick, D., and Salvendy, G. (2002a), "Macroergonomics Aspects of Manufacturing," in *Macroergonomics: An Introduction to Work System Design*, H. W. Hendrick and B. M. Kleiner, Eds., Lawrence Erlbaum Associates, Mahwah, NJ, pp. 223–248.

Karwowski, W., Rizzo, F., and Rodrick, D. (2002b), "Ergonomics in Information Systems," in *Encyclopedia of Information Systems*, H. Bidgoli Ed., Academic Press, San Diego, CA, pp. 185–201.

Karwowski, W., Siemionow, W., and Gielo-Perczak, K. (2003), "Physical Neuroergonomics: The Human Brain in Control of Physical Work Activities," *Theoretical Issues in Ergonomics Science*, Vol. 4, No. 1–2, pp. 175–199.

Kroemer, K., and Grandjean, E. (1997), *Fitting the Task to the Human*, 5th ed., Taylor & Francis, London.

Kroemer, K., Kroemer, H., and Kroemer-Elbert, K. (1994), *Ergonomics: How to Design for Ease and Efficiency*, Prentice-Hall, Englewood Cliffs, NJ.

Kuorinka, I., Ed. (2000), *History of the Ergonomics Association: The First Quarter of Century*, IEA Press, Santa Monica, CA.

Luczak, H., Schlick, C., and Springer, J. (1999), "A Guide to Scientific Sources of Ergonomics Knowledge," *The Occupational Ergonomics Handbook*, W. Karwowski and W. S. Marras, Eds., CRC Press, London, S. 27–50.

Meister, D. (1987), "Systems Design, Development and Testing," in *Handbook of Human Factors*, G. Salvendy, Ed., Wiley, New York, pp. 17–42.

Meister, D. (1991), *The Psychology of System Design*, Elsevier, Amsterdam.

Meister, D. (1997), *The Practice of Ergonomics*, Board of Certification in Professional Ergonomics, Bellingham, WA.

Meister, D. (1999), *The History of Human Factors and Ergonomics*, Lawrence Erlbaum Associates, London.

MITI (2001), "Major Predictions for Science and Technology in the Twenty First Century," Japan Ministry of Education, Science and Technology, Tokyo, unpublished report.

Moray, M. (1994), "The Future of Ergonomics: The Need for Interdisciplinary Integration," in *Proceedings of the IEA Congress*, Human Factors and Ergonomics Society, Santa Monica, CA, pp. 1791–1793.

Morris, W., Ed. (1978), *The American Heritage Dictionary of the English Language*, Houghton Mifflin, Boston.

NAE (National Academy of Engineering) (2004), *The Engineer of 2020: Visions of Engineering in the New Century*, National Academy Press, Washington, DC.

Nemeth, C. (2004), *Human Factors Methods for Design.*, CRC Press, Boca Raton, FL.

Nielsen, J. (1997), "Usability Engineering," in *The Computer Science and Engineering Handbook*, A. B. Tucker, Jr., Ed., CRC Press, Boca Raton, FL, pp. 1440–1460.

Nielsen, J. (2000), *Designing Web Usability: The Practice of Simplicity*, New Readers, Indianapolis, IN.

Norman, D. (1988), *The Design of Everyday Things*, Doubleday, New York.

Norman, D. A. (1993), *Things That Make Us Smart*, Addison Wesley, Reading, MA.

NRC (National Research Council) (2001), *Musculoskeletal Disorders and the Workplace: Low Back and Upper Extremities*, National Academy Press, Washington, DC.

Opperman, R. (1994), "Adaptively Supported Adaptability," *International Journal of Human–Computer Studies*, Vol. 40, pp. 455–472.

Parasuraman, R. (2003), "Neuroergonomics: Research and Practice," *Theoretical Issues in Ergonomics Science*, Vol. 4, No. 1–2, pp. 5–20.

Pearson, G., and Young, T., Ed. (2002), *Technically Speaking: Why All Americans Need to Know More About Technology*, National Academy Press, Washington, DC.

Pheasant, S. (1986), *Bodyspace: Anthropometry, Ergonomics and Design*, Taylor & Francis, London.

Putz-Anderson, V., Ed. (1988), *Cumulative Trauma Disorders: A Manual for Musculoskeletal Diseases of the Upper Limbs*, Taylor & Francis, London.

Reason, J. (1999), *Managing the Risk of Organizational Accidents*, Ashgate, Aldershot, Hampshire, England.

Rouse, W., Kober, N., and Mavor, A., Eds. (1997), *The Case of Human Factors in Industry and Government: Report of a Workshop*, National Academy Press, Washington, DC.

Salvendy, G., Ed. (1997), *Handbook of Human Factors and Ergonomics*, 2nd ed., Wiley, New York.

Sanders, M. M., and McCormick, E. J. (1993), *Human Factors in Engineering and Design*, 7th ed., McGraw-Hill, New York.

Silver, B. L. (1998), *The Ascent of Science*, Oxford University Press, Oxford.

Stanton, N. A., and Young, M. (1999), *A Guide to Methodology in Ergonomics: Designing for Human Use*, Taylor & Francis, London.

Stanton, N., Hedge, A., Brookhuis, K., Salas, E., and Hendrick, H. W. (2004), *Handbook of Human Factors and Ergonomics Methods*, CRC Press, Boca Raton, FL.

Stokes, D. E. (1997), *Pasteur's Quadrant: Basic Science and Technological Innovation*, Brookings Institution Press, Washington, DC.

Suh, N. P. (1990), *The Principles of Design*, Oxford University Press, New York.

Suh, N. P. (2001), *Axiomatic Design: Advances and Applications*, Oxford University Press, New York.

Taylor, F. W. (1911), *The Principles of Scientific Management* Harper Bros., New York.

Vicente, K. J. (1999), *Cognitive Work Analysis: Towards Safe, Productive, and Healthy Computer-Based Work*, Lawrence Erlbaum Associates, Mahwah, NJ.

Vicente, K. J. (2004), *The Human Factor*, Routledge, New York.

Waters, T. R., Putz-Anderson, V., Garg, A., and Fine, L. J. (1993), "Revised NIOSH Equation for the Design and Evaluation of Manual Lifting Tasks," *Ergonomics*, Vol. 36, No. 7, pp. 749–776.

Wickens, C. D., and Carswell, C. M. (1997), "Information Processing," in *Handbook of Human Factors and Ergonomics*, 2nd ed., G. Salvendy, Ed., Wiley, New York, pp. 89–129.

Wilson, J. R., and Corlett, E. N., Eds. (1995), *Evaluation of Human Work: A Practical Ergonomics Methodology*, 2nd ed., Taylor & Francis, London.

Woodson, W. E., Tillman, B., and Tillman, P. (1992), *Human Factors Design Handbook*, 2nd ed., McGraw-Hill, New York.

Womack, J., Jones, D., and Roos, D. (1990), *The Machine That Changed the World*, Rawson Associates, New York.

Wulf, W. A. (1988), "Tech Literacy: Letter to the White House," National Academy of Engineering Website. Available online at: www.nae.edu/nae/naehome.nse/weblinks/NAEW-4NHM87.

CHAPTER 2

HUMAN FACTORS ENGINEERING AND SYSTEMS DESIGN

Sara J. Czaja and Sankaran N. Nair
University of Miami Miller School of Medicine
Miami, Florida

1 INTRODUCTION

1.1 Human Factors Engineering and the Systems Approach

Human factors is generally defined as the study of human beings and their interaction with products, environments, and equipment in performing tasks and activities. The focus of human factors is on the application of knowledge about human capabilities, limitations, and other characteristics to the design of human–machine systems. By definition, a human–machine system is a system in which an interaction occurs between people and other system components, such as hardware, software, tasks, environments, and work structures. The system may be simple, such as a human interacting with a tool, or it may be complex, such as a flexible manufacturing system. The general objectives of human factors are to maximize human and system efficiency and health, safety, comfort, and quality of life [Sanders and McCormick, 1993; Human Factors and Ergonomics Society (HFES), 1998]. In terms of research, this involves studying human performance to develop design principles, guidelines, methodologies, and tools for the design of the human–system interface. In terms of practice, this involves the application of these principles, guidelines, and tools to the actual design and evaluation of real-world systems and system components (Hendrick and Kleiner, 2001). In both instances, human factors is concerned with optimizing the interaction between the human and the other components of the human–system interface.

Given that the focus of human factors is on studying performance within the context of tasks and environments, systems theory and the systems approach is fundamental to human factors engineering. Generally, systems theory argues for a unified nature of reality and the belief that the components of a system are meaningful only in terms of the general goals and purposes of the entire system. A basic tenet among systems theorists is that all systems are synergistic and that the whole is greater than the sum of its parts. This is in contrast to a reductionist approach, which focuses on a particular system component or element in isolation. The *reductionist* approach has traditionally been the "popular" approach to system design, where the focus has been on the physical or technical components of a system, with little regard for the behavioral component. In recent years the increased incidence of human error in the medical, transportation, and nuclear power environments and the limited success of many technical developments has demonstrated the shortcomings of this approach and the need for a systems prospective. Implicit in the belief in systems theory is adoption of the systems approach. Generally, the systems approach considers the interaction among all of the components of a system relative to system goals when evaluating particular phenomena. Systems methodology represents a set of methods and tools applicable to (1) the analysis of systems and system problems; (2) the design, development, and implementation of complex systems; and (3) the management of systems and change in systems (Banathy and Jenlink, 2004).

Applied to the field of human factors, the systems concept implies that performance must be evaluated in

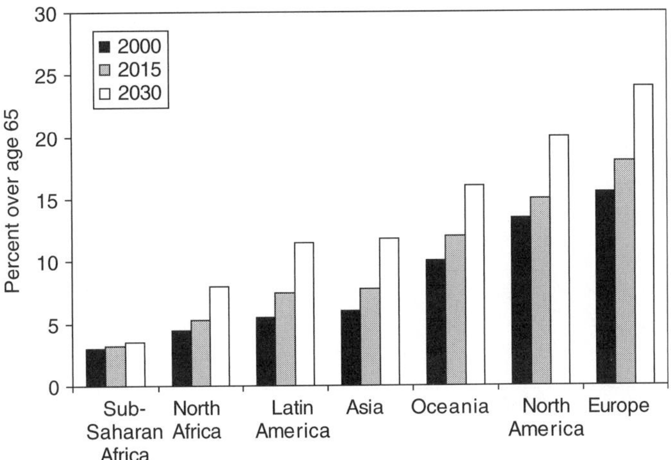

Figure 1 Current percentages and future estimates for persons over the age of 65. (From Kinsella and Velkoff, 2001.)

terms of the context of the human–machine system: equipment, environment, operating procedures, and goals. Human factors engineers generally agree that the overall efficiency of a system is determined by optimizing the performance of both the human and physical components. Traditionally, the design engineer focuses solely on the technical component of the system, and the behavioral scientist focuses solely on the performance component. Human factors is unique in that it is concerned with both the behavioral and physical domains. The systems concept provides a unifying framework for the study of these domains (Meister, 1989).

A basic tenet of human factors is that optimization of human and system efficiency requires adoption of the systems approach, where all major system components are given adequate consideration throughout the system design process. Designs that do not consider the human element will not achieve the maximum level of performance. Thus, a central activity of human factors is the application of information regarding human performance to all phases of system development and design. For this reason, a discussion of the role of human factors in system design and evaluation is central to a handbook on human factors engineering. This is especially true in today's era of computerization and automation where systems are becoming increasingly complex. Work and social environments have changed enormously over the past decade. The rate of technological change is unprecedented, affecting not only the way work is performed and communications patterns but working arrangements. For example, more people are telecommuting on a regular basis. In 1995, at least 3 million Americans were telecommuting to work, and this number is expected to increase by 20% per year (Fisk et al., 2003). The organization of work is also changing dramatically, due to the growth of new management practices and techniques, such as teamwork, supply chain partnering, process-based work, and just-in-time procedures. Unfortunately, many technical innovations and modern management practices have been less effective than intended (Clegg, 2000). Finally, the demographics of the population is changing and we are witnessing "graying of the population" not only in the United States but worldwide (Figure 1). In addition, the number of workers with more advanced levels of education has increased, and the number of women in the labor force has been increasing steadily. All of these issues underscore the need for a more human factors/systems–based approach.

In this chapter we discuss the role of human factors engineering in system design. The focus is on the approaches and methodologies used by human factors engineers to integrate knowledge regarding human performance into the design process. The topic of system design is vast and encompasses many areas of specialization within human factors. Thus, we introduce several concepts that are covered in depth in other chapters of the handbook. Prior to discussing the design process, a brief history of the systems approach is provided. Our overall intent in the chapter is to provide an overview of the system design process and to demonstrate the importance of human factors to systems design. Further, we introduce new approaches to system design that are being applied to complex, integrated systems.

1.2 Brief History of the Systems Approach and Human Factors Engineering

The systems concept was initially a philosophy associated with thinkers such as Hegel, who recognized that the whole is more than the sum of its parts. It was also a fundamental concept among Gestalt psychologists, who recognized the importance of "objectness" or wholeness to human perception. The idea of a general systems theory was developed by Bertalanffy in the late 1930s and developed further by

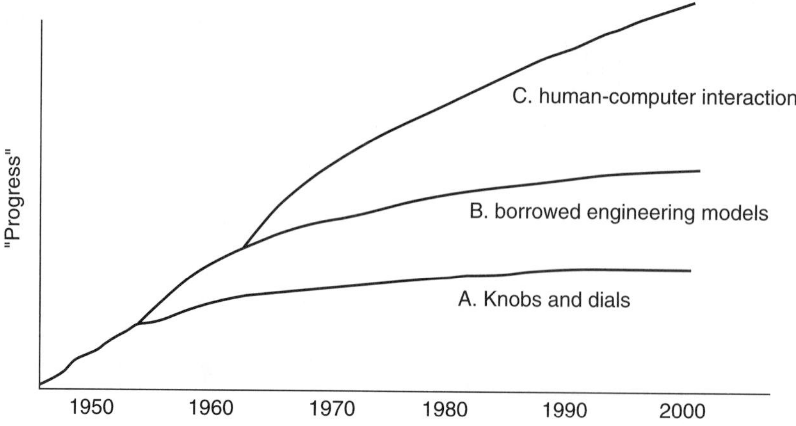

Figure 2 History of human–machine systems engineering. (From Sheridan, 2002.)

Ashby in the 1940s (Banathy and Jenlink, 2004). The systems approach, which evolved from systems thinking, was developed initially in the biological sciences and refined by communication engineers in the 1940s. Adoption of this approach was bolstered during World War II when it was recognized that military systems were becoming too complex for humans to operate successfully. This discovery gave rise to the emergence of the field of human factors engineering and its emphasis on human–machine systems.

Sheridan (2002) has classified the progress of human factors and the study of human–machine systems into three phases: phase A (knobs and dials), phase B (borrowed engineering models), and phase C (human–computer interaction) (Figure 2). The initial time period, phase A, gave birth to the concept of human–machine systems. The focus of human factors engineers was primarily on aircraft (civilian and military) and weapon systems, with limited applications in the automotive and communication industries. Following World War II there was an appreciation of the need to continue to develop human factors. The initial focus of this effort was on the design of displays and controls and workstations for defense systems. In this era, human factors study was often equated with the study of knobs and dials.

During phase B the field began to evolve beyond knobs and dials when human factors engineers recognized the applicability of system engineering models to the study of human performance. During the 1960s, systems theory became a dominant way of thinking within engineering, and human factors engineers began to use modeling techniques, such as control theory, to predict human–system performance. A number of investigators were concerned with developing models of human performance and applying these models to system design. For example, signal detection theory was applied to human detection performance, and the concept of the "ideal observer" evolved (Green and Swets, 1966). At the same time, the application of human factors expanded beyond the military, and many companies began to establish human factors groups. The concept of the human–machine system also expanded as human factors engineers became involved with the design of consumer products and workplaces. In 1960, a human factors group was formed at Eastman Kodak. The group specialized in problems related to workplace and job design and also became involved in the design and evaluation of products (Eastman Kodak Company, 1983).

Phase C refers to the era of human–computer interaction. Advances in computing power and automation have changed the nature of human–machine systems dramatically, resulting in new challenges for human factors engineers and system designers. In many work domains the deployment of computers and automation has changed the nature of the demands placed on the worker. In essence, people are doing less physical work and are interacting mentally with computers and automated systems, with an emphasis on perceiving, attending, thinking, decision making, and problem solving (Rasmussen et al., 1994; Sheridan 2002). To design these types of work systems effectively, we need to apply to the design process knowledge regarding human information-processing capabilities. The need for this type of knowledge has created a greater emphasis on issues related to human cognition within the field of human factors and has lead to the emergence of cognitive engineering (Woods, 1988). Cognitive engineering focuses on complex, cognitive thinking and knowledge-related aspects of human performance, whether carried out by humans or by machine agents (Wickens et al., 2004). It is closely aligned with the field of cognitive science and artificial intelligence. As shown in Figure 2, phase C is continuing to grow at a rapid pace and human factors engineers are confronted with many new types of technology and work systems, such as artificial intelligence agents, human supervisory control, and virtual reality. These types of systems present new challenges for system designers and human factors specialists.

Table 1 Design Process

Elements

1. Design specification
2. Design history (e.g., predecessor system data and analyses)
3. Design components transferred from a predecessor system
4. Design goals (technological and idiosyncratic)

Processes

1. Analysis of design goals (performed by both designers and HFE specialists)
2. Determination of design problem parameters (both)
3. Search for information to understand the design problem and parameters (both)
4. Behavioral analysis of functions and tasks (specialist only)
5. Transformation of behavioral information into physical surrogates (specialist only)
6. Development and evaluation of alternative solution to the design problem (both, mostly designers)
7. Selection of one design solution to be followed by detailed design (both, mostly designers)
8. Design of the human–machine interface and human–computer interface (both; either may be primary)
9. Evaluation and testing of design outputs (both)
10. Determination of system status and development progress (both)

Factors Affecting Design

1. Nature of the design problem and of the system, equipment, or product to be designed
2. Availability of needed relevant information
3. Strategies for solution of design problem (information-processing methods)
4. Idiosyncratic factors (designer/specialist intelligence, training, experience, skill, personality)
5. Project organization and management

Source: Meister (2000).

Further need for new approaches to system design comes from the changing nature of the design process. Developments in technology and automation have not only increased the complexities of the types of systems that are being designed but have also changed the design process itself and the way designers think, act, and communicate. As noted by Meister (2000), design is an extremely complex process that proceeds over relatively long time periods in an atmosphere of uncertainty. The process is influenced by many factors, some of which are behavioral and some of which are technical and organizational (Table 1). Design also involves interaction among many people with different types and levels of knowledge. At the most basic level, this interaction involves engineers from many different specialties; however, in reality it also involves the users of the system being designed and managerial/organizational representatives. Further, system design often takes place under time constraints in turbulent economic and social markets.

Overall, it has become apparent that we can not restrict the application of human factors to the design of specific jobs, workplaces, or human–machine interfaces and must broaden our view of system design and consider broader sociotechnical issues. In other words, design of today's systems requires the adoption of a more *macroergonomic approach*, a top-down sociotechnical system approach to design that is concerned with the human–organizational interface and represents a broad perspective to systems design. As noted by Hendrick and Kleiner (2001), a number of important trends are related to the organization and design of work systems that underscore the need for a macroergonomic approach, including (1) rapid developments in technology, (2) demographic shifts, (3) changes in the value system of the workforce, (4) world competition, (5) an increased concern for safety and the resulting increase in ergonomics-based litigations, and (6) the failure of traditional microergonomics.

In sum, the nature of human–machine systems has changed drastically since the era of knobs and dials, presenting new challenges and opportunities for human factors engineers. We are not only faced with designing and evaluating new types of systems and a wider variety of systems (e.g., health care systems, living environments) but also with many different types of user populations. Many people with limited technical background and of varying ages are operating computer systems, which raises many new issues for system designers. For example, older workers may require different types of training or different work schedules to interact effectively with new technology, or operators with a limited technical background may require a different type of interface than those who are more experienced. Emergence of these types of issues reinforces the need to include human factors in system design. As noted by Meister (2000), the influx of technology into most environments and tasks has called increased attention to the importance of system design within the field of human factors. As he eloquently states, one way in which humans can control technology and the effects of technology is to understand the effects of technology on human performance and to apply behavioral principles and data to the design of technology so that technology can become "more humane." In the following section we present a general model of a system that will serve as background to a discussion of the system design process.

2 DEFINITION OF A SYSTEM

2.1 General System Characteristics

A *system* is an aggregation of elements organized in some structure (usually, hierarchical) to accomplish system goals and objectives. All systems have the following characteristics: interaction of elements, structure, purpose, and goals, and inputs and outputs. A system is usually composed of humans and machines and has a definable structure and organization and external boundaries that separate it from elements outside the system. All the elements within a system interact and function to achieve system goals. Further, each system component has an effect on the other components. It is through the system inputs and outputs that the elements of a system interact and communicate. Systems also exist within an environment (physical and social), and the characteristics of this environment have an impact on the structure and the overall effectiveness of the system (Meister, 1989, 1991). For example, to be responsive to today's highly competitive and unstable environment, manufacturing systems have to be flexible and dynamic. Traditional models of manufacturing systems that are based on long-term planning for production where production takes place in a sequence of separate processes in separate departments are no longer adequate. Instead, production needs to be organized around simultaneous activities where there is decentralized decision making and quick and easy access to information (Drucker, 1998). This creates the need for a change in the organizational structure. Formal, hierarchical organizations do not effectively support distributed decision making and flexible production processes.

Generally, all systems have the following components: (1) elements (personnel, equipment, procedures); (2) conversion processes (processes that result in changes in system states); (3) inputs or resources (personnel abilities, technical data); (4) outputs (e.g., number of units produced); (5) an environment (physical and social and organizational); (6) purpose and functions (the starting point in system development); (7) attributes (e.g., reliability); (8) components and programs; (9) management, agents, and decision makers; and (10) structure. These components must be considered in the design and evaluation of every system. For example, the nature of the system inputs has a significant impact on the ability of a system to produce the desired outputs. Inputs that are complex, ambiguous, or unanticipated may lead to errors or time delays in information processing, which in turn may lead to inaccurate or inappropriate responses. If there is conflicting or confusing information on a patient's chart, a physician might have difficulty diagnosing the illness and prescribing the appropriate course of treatment.

Systems can also be characterized according to the nature of the system variables. Meister (1991) distinguishes between two types of system variables: behavioral variables and physical variables. *Behavioral variables* describe requirements for system operators and include factors such as task and function requirements, skill and training requirements, and number and type of interdependencies. *Physical variables* describe the physical and structural function of the system and include the number of subsystems, the size and complexity of the system, the number and specificity of goals and missions, the requirements placed on the system, and the nature of feedback mechanisms. The manner in which these variables are treated in system design determines the overall performance of the system (Meister, 1991). Decisions regarding function allocation or amount, type, and scheduling of feedback have an obvious impact on overall system efficiency.

2.2 System Classifications

There are various ways in which systems are classified. Systems can be distinguished according to degree of automation, functions and tasks, feedback mechanisms, system class, hierarchical levels, and combinations of system elements (Meister, 1991). A basic distinction between open- and closed-loop systems is usually made on the basis of the nature of a system's feedback mechanisms. *Closed-loop systems* perform a process that requires continuous control and feedback for error correction. Feedback mechanisms exist that provide continuous information regarding the difference between the actual and the desired states of the system. In contrast, *open-loop systems* do not use feedback for continuous control; when activated, no further control is executed. However, feedback can be used to improve future operations of the system (Sanders and McCormick, 1993). The distinction between open- and closed-loop systems is important, as they require different design strategies.

Systems are also distinguished according to their service orientation. In this regard, there are mission- and service-oriented systems. In *mission-oriented systems*, the needs of the personnel are subordinated to the goals of the system. Military and production systems are examples of this type of system. *Service-oriented systems* exist to meet the needs of clients or users.

A governmental agency is an example of this system. In reality, most systems contain components of both types of systems. It is important to understand the service orientation of a particular system because this will have an impact on the degree to which personnel needs and desires may be considered relative to system demands (Meister, 1989).

We are also able to describe different classes of systems. For example, we can distinguish at a very general level among educational systems, production systems, maintenance systems and health care systems, transportation systems, communication systems, and military systems. Within each of these systems we can also identify subsystems, such as the social system or the technical system. Complex systems generally contain a number of subsystems. Finally, we are able to distinguish systems according to components or elements. For example, we can distinguish among machine systems, human systems (biological systems), and human–machine systems. It is the latter type of system that is of interest to human factors engineers.

2.3 Human–Machine Systems

A human–machine system is some combination of humans and machines that interact to achieve the goals of a system. These systems are characterized by elements that interact, structure, goals, conversion processes, inputs, and outputs. Further, they exist in an environment and have internal and external boundaries. A simple model of a human–machine system is presented in Figure 3. This general systems model applies to human–machine systems; inputs are received and processed, and outputs are produced through the interaction of the system components.

With the emergence of computer and automation technologies, the nature of the human–machine system has changed dramatically. For example, computers have changed display technology, and information can be presented in a variety of formats. Control functions have also changed, and humans can even speak commands. Perhaps more important, machines have become more intelligent and capable of performing tasks formerly restricted to humans. Prior to the development of intelligent machines, the model of the human–machine interface was formed around a control relationship in which the machine was under human control. In current human–machine systems (which involve some form of computer technology), the machine is intelligent and capable of extending the capabilities of the human. Computer/automation systems can now perform routine, elementary tasks, complex computations, suggest ways to perform tasks, or engage in reasoning or decision making. In these instances, the human–machine interface can no longer be conceptualized in terms of a control relationship where the human controls the machine. A more accurate representation is a partnership where the human and the machine are engaged in two-way cognitive interaction (Eggleston, 1987).

For example, in aircraft piloting, the introduction of the flight management system (FMS) has dramatically changed the tasks of the pilot. The FMS is capable of providing the pilot with advice on navigation,

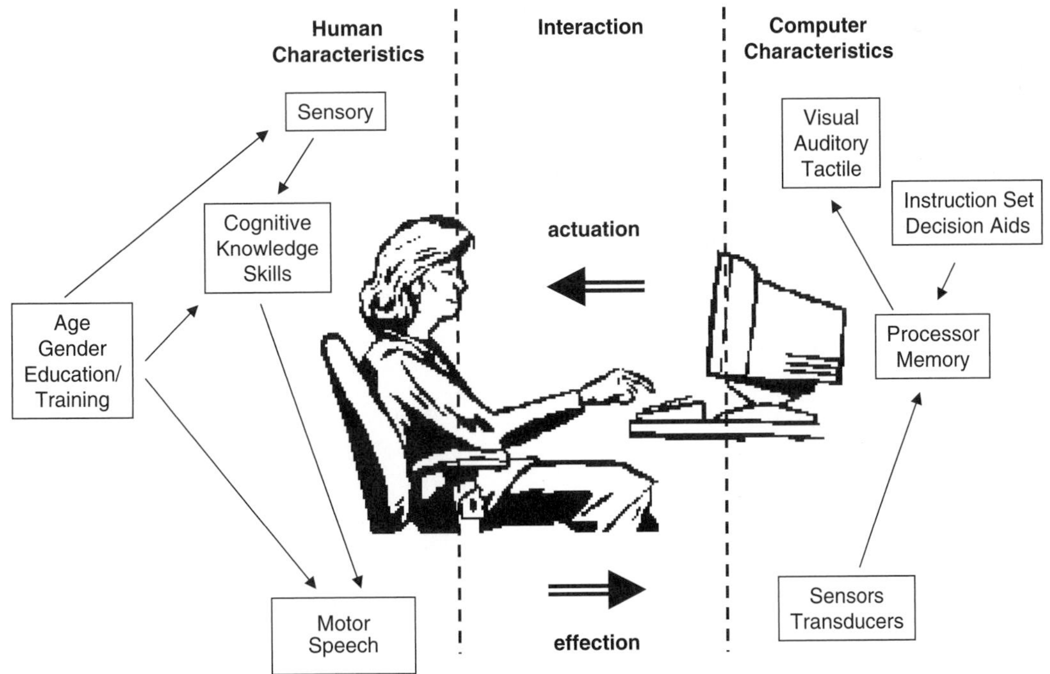

Figure 3 Example of a human–machine system.

weather patterns, airport traffic patterns, and other topics and is also capable of detecting and diagnosing abnormalities. The job of the pilot has become that of a process manager, and in essence the workspace of the pilot has become a desk; there is limited manual control of the flight system (Sheridan, 2002). Other types of transportation systems, such as trains, ships, and even automobiles, are also incorporating new computer, communication, and control technologies that change the way that operators interact with these systems and raise new design concerns. With respect to automobiles, a number of issues related to driver safety are emerging: For example, are maps and route information systems a decision aid or a distraction?

Similar issues are emerging in other domains. For example, flexible manufacturing systems represent some combination of automatic, computer-based, and human control. In these systems the operators largely assume the role of a supervisory controller and must plan and manage the manufacturing operation. Issues regarding function allocation are critical within these systems, as is the provision of adequate cognitive and technical support to the humans. Computers now offer the potential of assisting humans in the performance of cognitive activities, such as decision making, and a question arises as to what level of machine power should be deployed to assist human performance so that the overall performance of the system is maximized. As noted by several authors (e.g., Rasmussen and Goldstein, 1985; Woods, 1988), the answer to this question should not be machine driven but situation or problem driven, where the technology is used to augment human capabilities. This question has added complexity, as in most complex systems the problem is not restricted to one operator but to two or more operators who cooperate and have access to different databases. Today's automated systems are becoming even more complex with more decision elements, multiple controller set points, more rules, and more distributed objective functions and goals. Further, different parts of the system, both human and machine, may attempt to pursue different goals, and these goals may be in conflict. This is commonly referred to as the *mixed-initiative problem*, in which mixed human initiatives combine with mixed automation initiatives. Most systems of this type are supervised by teams of people in which the operator is part of a decision-making team of people who together with the automated system control the process (Sheridan, 2002). The mixed-initiative problem presents a particular challenge for system designers and human factors engineers.

Obviously, there are many different types of human–machine systems, and they vary greatly in size, structure, complexity, and so on. Although the emphasis in this chapter is on work systems where computerization is an integral system component, we should not restrict our conceptualization of systems to large, complex technological systems in production or process environments. We also need to consider other types of systems, such as a person using an appliance within a living environment, a physician interacting with a heart monitor in an intensive care unit, or an older person driving an automobile within a highway environment. In all cases, the overall performance of the system will be improved with the application of human factors engineering to system design.

New challenges for system design also arise from the evolution of virtual environments (VEs). Designers of these systems need to consider characteristics unique to VE systems, such as the design of navigational techniques, object selection and manipulation mechanisms, and the integration of visual, auditory, and haptic system outputs. Designers of these types of systems must enhance presence, immersion, and system comfort while minimizing consequences such as

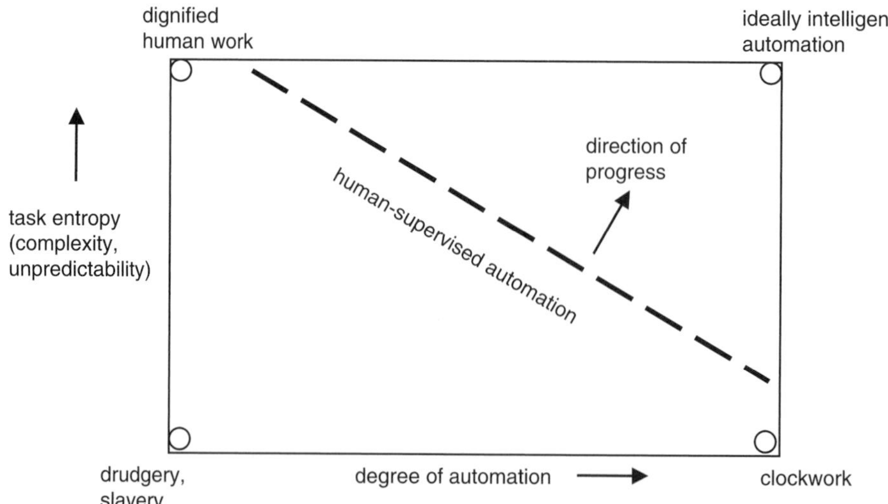

Figure 4 Progress of human-supervised automation. (From Sheridan, 2002.)

motion sickness. VE user interfaces are fundamentally different from traditional user interfaces with unique input–output devices, perspectives, and physiological interactions. Currently, very few principles exist for the design of VE interfaces (Stanney et al., 2003).

Thus, in today's world, human–machine systems, which increasingly involve machine intelligence, can take many forms, depending on the technology involved and the function allocation between human and machine. Figure 4 presents the extremes of various degrees of automation and the complexity of various task scenarios. The lower left represents a system in which the human is left to perform completely predictable and, in most cases, "leftover tasks." In contrast, the upper right represents ideally intelligent automation where automated systems are deployed to maximal efficiency—a state not attainable in the foreseeable future. The lower right also represents an effective use of full automation, and the upper left represents the most effective deployment of humans—working on undefined and unpredictable problems. As discussed by Sheridan (2002), few real situations occur at these extremes; most human-automated systems represent some trade-off of these options, which gradually progress toward the upper right—ideally, intelligent automation. Clearly, specification of the human–machine relationship is an important design decision. The relationship must be such that the abilities of both the human and machine components are maximized, as is cooperation among these components. Too often, technology is viewed as a panacea and implemented without sufficient attention to human and organizational issues.

The impact of the changing nature of human–machine systems on the system design process and current approaches to system design is discussed in a later section. However, before this topic is addressed, concepts of system and human reliability are introduced because these concepts are important to a discussion of system design and evaluation.

2.4 System Reliability

System reliability refers to the dependability of performance of system, subsystem, or system component in carrying out its intended function for a specified period of time. Reliability is usually expressed as the probability of successful performance; therefore, for the probability estimate to be meaningful, the criteria for successful performance must be specified (Proctor and Van Zandt, 1994). The overall reliability of a system depends on the reliability of the individual components and how they are combined within a system. Reliability of a component is the probability that it does not fail and is defined as r, where $r = 1 - p$; p represents the probability of failure.

Generally, components in a system are arranged in series, in parallel, or some combination of both. For total performance of the system to be satisfactory, if the components are arranged in series, they must all operate adequately. In this case, if the component failures are independent of each other, system reliability is the product of the reliability of the individual components. Further, as more components are added to a system, the reliability of the system decreases unless the reliability of these components is equal to 1.0. The reliability of the overall system can only be as great as that of the least reliable component.

In parallel systems, two or more components perform the same function such that successful performance of the system requires that only one component operate successfully. This is often referred to as *system redundancy*; the additional components provide redundancy to guard against system failure. For these types of systems, adding components in parallel increases the reliability of the system. If all of the components are equally reliable, system reliability is determined by calculating the probability that at least one component remains functional and considering the reliability of each of the parallel subsystems. Parallel redundancy is often provided for human functions because the human component within a system is the least reliable.

2.5 Human Reliability

Human reliability is the probability that each human component of the system will perform successfully for an extended period of time and is defined as 1 − (operator error probability) (Proctor and Van Zandt, 1994). The study of human error has become an increasingly important research concern because it has become apparent that the control of human error is necessary for the successful operation of complex, integrated systems. The incidence of human error has risen dramatically over the past few years, with many disastrous consequences. It has been estimated that human error is the primary cause of 60 to 90% of major accidents and incidents in complex systems such as process control, aviation, and health care environments (Wickens and Hollands, 2000). The topic of human error is discussed in detail elsewhere in this handbook. It is discussed briefly in this chapter because the analysis of human error has important implications for system design. It is generally recognized that many errors that people make are the result of poor system design or organizational structure, and the error is usually only one in a lengthy and complex chain of breakdowns.

Due to the prevalence of human error and the enormous and often costly consequences, the study of human error has become an important focus within human factors engineering and in fact has emerged as a well-defined discipline. In recent years a number of techniques have emerged to study human error. Generally, these techniques fall into two categories, quantitative techniques and qualitative techniques. *Quantitative techniques* attempt to predict the likelihood of human error for the development of risk assessment for the entire system. There are well-developed techniques for human reliability prediction such as THERP and SLIM MAUD. These techniques can provide useful insights into human factors deficiencies in system design and thus can be used to identify areas where human factors knowledge needs to be incorporated. However, there are shortcomings associated with these techniques, such as limitations in providing precise estimates of human performance abilities, especially for cognitive processes (Wickens, 1992). Further, designers are not able to identify all of the contingencies of the work process.

Qualitative techniques emphasize the causal element of human error and attempt to develop an understanding of the causal events and factors contributing to human error. Clearly, when using these approaches, the circumstances under which human error is observed and the resultant causal explanation for error occurrence have important implications for system design. If the causal explanation stops at the level of the operator, remedial measures might encompass better training or supervision; for example, a common solution for back injuries is to provide operators with training on "how to lift," overlooking opportunities for other, perhaps more effective, changes in the system, such as modifications in management, work procedures, work planning, or resources.

As noted, analyses of many major accident events indicate that the root cause of these events can be traced to latent failures and organizational errors. In other words, human errors and their resulting consequences usually result from inadequacies in system design. An example is the crash at Dryden Airport in Ontario. The analysis of this accident revealed that the accident was linked to organizational failings such as poor training, lack of management commitment to safety, and inadequate maintenance and regulatory procedures (Reason, 1995). These findings indicate that when analyzing human error it is important to look at the entire system and the organizational context in which the error occurred.

Several researchers have developed taxonomies for classifying human errors into categories. These taxonomies are useful, as they help identify the source of human error and strategies that might be effective in coping with error. Different taxonomies emphasize different aspects of human performance. For example, some taxonomies emphasize human actions, whereas others emphasize information-processing aspects of behavior. Rasmussen and colleagues (Rasmussen, 1982; Rasmussen et al., 1994) developed a taxonomy of human errors from analyses of human involvement in failures in complex processes. This schema is based on a decomposition of mental processes and states involved in erroneous behavior. For the analysis, the events of the causal chain are followed backward from the observed accidental event through mechanisms involved at each stage. The taxonomy is based on an analysis of the work system and considers the context in which the error occurred (e.g., workload, work procedures, shift requirements). This taxonomy has been applied to the analysis of work systems and has proven to be useful for understanding the nature of human involvement in accident events.

Reason (1990, 1995) has developed a similar scheme for examining the etiology of human error for the design and analysis of complex work systems. The model is based on a systems approach and describes a pathway for identifying the organizational causes of human error. The model includes two interrelated causal sequences for error events: (1) an active failure pathway where the failure originates in top management decisions and proceeds through error-producing conditions in various workplaces to unsafe acts committed by workers at the immediate human–machine interface, and (2) a latent failure pathway that runs directly from the organizational processes to deficiencies in the system's defenses. The model can be used to assess organizational safety health in order to develop proactive measures for remediating system difficulties and as an investigation technique for identifying the root causes of system breakdowns.

The implications of error analysis for system design depend on the nature of the error as well as the nature of the system. Errors and accidents have multiple causes, and different types of errors require different remedial measures. For example, if an error involves deviations from normal procedures in a well-structured technical system, it is possible to derive a corrective action for a particular aspect of an interface or task element. This might involve redesign of equipment or of some work procedure to minimize the potential for the error occurrence. However, in complex dynamic work systems it is often difficult or undesirable to eliminate the incidence of human error completely. In these types of systems, there are many possible strategies for achieving system goals; thus, it is not possible to specify precise procedures for performing tasks. Instead, operators must be creative and flexible and engage in exploratory behavior in order to respond to the changing demands of the system. Further, designers are not able to anticipate the entire set of possible events; thus, it is difficult to build in mechanisms to cope with these events. This makes inevitable a certain amount of error.

Several researchers (Rouse and Morris, 1987; Rasmussen et al., 1994) advocate the design of *error-tolerant systems*, where the system tolerates the occurrence of errors but avoids the consequences; there is a means to control the impact of error on system performance. Design of these interfaces requires an understanding of the work domain and the acceptable boundaries of behavior and modeling the cognitive activity of operators dealing with incidents in a dynamic environment. A simple example of this type of design would be a computer system which holds a record of a file so that it is not lost permanently if an operator mistakenly deletes the files. A more sophisticated example would be an intelligent monitoring system which is capable of varying levels of intervention.

Rouse and Morris (1987) describe an error-tolerant system that provides three levels of support. Two levels involve feedback (current state and future state) and rely on an operator's ability to perceive his or her own errors and act appropriately. The third level involves *intelligent monitoring*, that is, online identification and error control. They propose architecture for the development of this type of system that is based on an operator-centered design philosophy and involves incremental support and automation. Rasmussen and Vincente (1989) have developed a framework for an interface that supports recovery from human errors. The framework, called *ecological interface design*, is based on an analysis of the work system. This approach is described in more detail in a later section.

3 SYSTEM DESIGN PROCESS

3.1 Approaches to System Design

System design is usually depicted as a highly structured and formalized process characterized by stages in which various activities occur. These activities vary as a function of system requirements, but they generally involve planning, designing, testing, and evaluating. More details regarding these activities are given in a subsequent section. Generally, system design is characterized as a top-down process that proceeds, in an interactive fashion, from broad molar functions to progressively more molecular tasks and subtasks. It is also a time-driven process and is constrained by cost, resources, and organizational and environmental requirements. The overall goal of system design is to develop an entity that is capable of transforming inputs into outputs to accomplish specified goals and objectives.

Meister (1991) distinguishes among three levels within system design:

1. *Design process*: how the system is designed
2. *Design philosophy*: the conceptual framework of the design
3. *Design architecture*: the specification of the structure of the system and the human–machine interface

In recent years, within the realm of system design, a great deal of attention has been given to the design philosophy and the resulting design architecture as it has become apparent that new design approaches are required to design modern work systems. The design and analysis of such systems cannot be based on design models developed for work systems characterized by a stable environment and stable task procedures. Instead, the design approach is concerned with supplying resources to operators who operate in a dynamic work space and change their work patterns according to changing environmental conditions. In other words, a structural perspective whereby we describe the behavior of the system in terms of cause-and-effect patterns and arrange system elements in cause-and-effect chains is no longer adequate. Instead, we need a framework for system design that represents all aspects of work systems in a coordinated and compatible fashion (Rasmussen et al., 1994).

3.1.1 Traditional Model of System Design

The traditional view of the system design process is that it is a linear sequence of activities where the output of each stage serves as input to the next stage. The stages generally proceed from the conceptual level to physical design through implementation and evaluation. Human factors inputs are generally considered in the design and evaluation stages (Eason, 1991). The general characteristics of this approach are that it represents a reductionist approach where various components are designed in isolation and made to fit together, it is dominated by technological considerations where humans are considered secondary components. The focus is on fitting the person to the system, and different components of the system are developed on the basis of narrow functional perspectives (Kidd, 1992; Liker and Majchrzak, 1994). Generally, this approach has dominated the design of overall work systems, such as manufacturing systems, as well as the design of the human–machine interface. For example, the emphasis in the design of human–computer systems has largely been on the individual level of the human–computer interaction without much attention to task and environmental factors that may affect performance. Hendrick and Kleiner (2001) maintain that the primary emphasis of human factors engineers has been on the microergonomic aspects of design without sufficient attention to social and organizational issues.

The implementation of computers of automation into most work systems, coupled with the enhanced capabilities of technological systems, has created a need for new approaches to system design. As discussed, there are many instances where technology has failed to achieve its potential, resulting in failures in system performance with adverse and often disastrous consequences. These events have demonstrated that the traditional design approach is no longer adequate. In this vein, Liker and Majchrzak (1994) identify nine features that need to be incorporated into a design process in order to effectively design the human–technology infrastructure (Table 2). Features 1 to 4 are concerned with the content of the design requirements and mandate that the design process should produce an integrated sociotechnical design in accordance with open system principles at a level of specificity required to implement the design. Features 5 to 9 are concerned with the actual design process. Overall, these features incorporate many of the aspects of concurrent engineering methodologies that have proven to be effective within manufacturing systems. They provide a useful framework for the development and analysis of design strategies. Clearly, the traditional method of design fails to incorporate most of these features.

Liker and Majchrzak (1994) review four design approaches, including the sociotechnical systems approach, participatory ergonomics, human-centered human factors design, and computer models of integrated systems in terms of their potential effectiveness in designing human–technical systems. They conclude that these approaches hold promise, as they have many of the characteristics outlined in Table 2. Further, they suggest that each approach has unique strengths and that perhaps a combination of these approaches would be most effective. A brief overview of these approaches and some other design approaches will be presented to provide some

Table 2 Desired Features for System Design Processes

1. *Integrated human infrastructure design*. The design process should provide a means for considering the fit between the formal organization, informal organization, individual characteristics, and linking mechanism to the environment.
2. *Concurrent human infrastructure and technical system design*. The design process should provide not only analytic tools for understanding the likely implications of any particular technical solution for people and the organization, but also the likely implications of different organizational arrangements for technology design.
3. *Fit with environment*. The design process should consider the requirements and constraints imposed on the human infrastructure and technology by the larger environment of the system and its missions.
4. *Specificity*. The design process should provide precise information on implications so that engineers can alter specific technical designs and organizational planners can modify organizational scenarios as the design process proceeds.
5. *Multiple-scenario generation*. The design process should facilitate the development of multiple design scenarios that can then be compared.
6. *Design evaluation and refinement*. An interactive design procedure, which facilitates systematic comparisons among scenarios, should then be used to gradually fine-tune the design based on prototyping and testing design concepts.
7. *Life-cycle user involvement*. The design process should provide a process for involving potential users of the new technology throughout the design and implementation process.
8. *Facilitating design team learning*. The design process should increase planners' knowledge of the sociotechnical system about issues relevant to organizational–technology integration.
9. *Facilitating design modification*. The design process should enable redesign as circumstances, such as technology, change.

Source: Liker and Majchrzak (1994).

examples of alternative approaches to system design and demonstrate methodologies and concepts that can be applied to the design of current human–machine systems. This will be followed by a discussion of the specification application of human factors engineering to design activities.

3.1.2 Alternative Approaches to System Design

Sociotechnical Systems Approach The sociotechnical systems approach, which evolved from work conducted at the Tavistock Institute, represents a complete design process for the analysis, design, and implementation of systems. The approach is based on open systems theory and emphasizes the fit between social and technical systems and the environment. This approach includes methods for analyzing the environment, the social system, and the technical system. The overall design objective is the joint optimization of the social and technical systems (Pasmore, 1988). Some drawbacks associated with sociotechnical design are that the design principles are often vague and there is often an overemphasis on the social system without sufficient emphasis on the design of the technical system.

Clegg (2000) recently presented a set of sociotechnical principles to guide system design. The principles are intended for the design of new systems that involve new technologies and modern management practices. The principles are organized into three interrelated categories: meta-principles, content principles, and process principles. *Meta-principles* are intended to demonstrate a world view of design, *content principles* focus on more specific aspects of the content of the new designs, and *process principles* are concerned with the design process. The principles also provide a potential for evaluative purposes. They are based on a macroergonomic perspective.

The central focus of macroergonomics is on interfacing organizational design with the technology employed in the system to optimize human–system functioning. Macroergonomics considers the human–organization–environment–machine interface as opposed to microergonomics, which focuses on the human–machine interface. Macroergonomics is considered to be the driving force for microergonomics. Macroergonomics concepts have been applied successfully to manufacturing, service, and health care organizations as well as to the design, of computer-based information systems (Hendrick and Kleiner, 2001).

For example, the design of the Romeo Engine Plant of the Ford Motor Company was based largely on a sociotechnical approach. The plant proved to be one of the best engine plants within the Ford Motor Company. Berger (1994) presents data from three case studies aimed at testing the notion that balancing the technological, organizational, and human aspects is important to manufacturing development. The case studies involved medium-sized manufacturing companies in Sweden. The companies were concerned with reducing lead times and engaged in redesign efforts. The cases varied according to the degree of balance achieved. None of the cases achieved a balance on all three aspects. Two of the cases emphasized the balance between the human and organizational aspects, and the other case emphasized the technical aspects, largely ignoring the human and organizational aspects. All three of the cases showed considerable improvements in reducing lead times; therefore, few conclusions regarding the balancing aspects could be drawn in terms of an amount relationship. Berger concluded that a balanced approach needs to be integrated within the problem-solving sequence of the redesign effort and that a key feature in manufacturing development is the ability to rebalance according to the time course and the prevailing problem of the redesign process.

Participatory Ergonomics Participatory ergonomics is the application of ergonomic principles and concepts to the design process by people who are part of the work group and users of the system. These people are typically assisted by ergonomic experts who serve as trainers and resource centers. The overall goal of participatory design is to capitalize on the knowledge of users and to incorporate their needs and concerns into the design process. Methods, such as focus groups, quality circles, and inventories, have been developed to maximize the value of user participation. Participatory ergonomics has been applied to the design of jobs and workplaces and to the design of products. For example, the quality circle approach was adopted by a refrigerator manufacturing company that needed a system-wide method for assessing the issues of aging workers. The assembly line for medium-sized refrigerators was chosen as an area for job redesign. The project redesign team involved workers from the line as well as other staff members. The team was instructed with respect to the principles of ergonomics and design for older workers. The solution, proposed by the team, for improving the assembly line resulted in improved performance and also allowed older workers to continue to perform the task (Imada et al., 1986). The design of current personal computer systems also typically involves user participation. Representative users participate in usability studies. In general, participatory ergonomics does not represent a design process because it does not consider broader system design issues but rather, focuses on individual components. However, the benefits of user participation should not be overlooked and should be a fundamental aspect of system design.

User-Centered Design The user-centered design approach represents an approach where human factors are of central concern within the design process. It is based on an open-systems model and considers the human and technical subsystems within the context of the broader environment. User-centered approaches propose general specifications for system design, such as that the system must maximize user involvement at the task level, and that the system should be designed to support cooperative work and allow users to maintain control over operations (Liker and Majchrzak, 1994). Essentially, this design approach incorporates user requirements, user goals, and user tasks as early as possible into the design of a system, when the design is still relatively flexible and when changes can be made at least cost.

Eason (1989) has developed a detailed process for user-centered design in which a system is developed in an evolutionary incremental fashion and development of the social system complements development of the technical system. Eason maintains that the technical system should follow the design of jobs and that the design of the technical system must involve user participation and consider criteria for four factors: functionality, usability, user acceptance, and organizational acceptance. Once these criteria are identified, alternative design solutions are developed and evaluated. There are different philosophies with respect to the nature of user involvement. Eason emphasizes user involvement throughout the design process, whereas with other models the users are considered sources of data and the emphasis is on translating knowledge about users into practice. Advocates of the user participation approach argue that users should participate in the choice between alternatives because they have to live with the results. Advocates of the knowledge approach express concern about the ability of users to make informed judgments. Eason (1991) maintains that designers and users can form a partnership where both can play an effective role. A number of methods are used in user-centered design, including checklists and guidelines, observations, interviews, focus groups, and task analysis.

Computer-Supported Design The design of complex technical systems involves the interpretation and integration of vast amounts of technical information. Further, design activities are typically constrained by time and resources and involve the contributions of many persons with varying backgrounds and levels of technical expertise. In this regard, computer-based design support tools have emerged to aid designers and support the design of effective systems. These systems are capable of offering a variety of supports, including information retrieval, information management, and information transformation. The type of support warranted depends on the needs and expertise of the designer (Rouse, 1987). A common example of this type of support is a computer-aided design/computer-aided manufacturing (CAD/CAM) system.

Majchrzak and Gasser (1992) have developed a computer modeling system, Action, for manufacturing system designers. Action (which evolved from Hitop) is a knowledge-based design, decision support, and simulation software package that is to be used as a tool to aid in the design and planning of technology of discrete-parts manufacturing systems. Action represents a design tool rather than a design process; however, the tool has some design methodology features. The design of this tool was based on an open systems model, and the design methodology features support a concurrent organizational design approach and a supportive human infrastructure.

There are many issues surrounding the development and deployment of computer-based design support tools, including specification of the appropriate level of support, determination of optimal ways to characterize the design problem and the type of knowledge most useful to designers, and the identification of factors that influence the acceptance of these tools. A discussion of these issues is beyond the scope of this chapter. Refer to Rouse and Boff (1987a,b) for an excellent review of this topic.

Ecological Interface Design Ecological interface design (EID) is a theoretical framework for designing human–computer interfaces for complex sociotechnical systems (Rasmussen et al., 1994; Vincente, 2002). The primary aim of EID is to support knowledge workers who are required to engage in adaptive problem solving in order to respond to novelty and change in system demands. EID is based on a cognitive systems engineering approach and involves an analysis of the work domain and the cognitive characteristics and behavior tendencies of the individual. Analysis of the work domain is based on an abstraction hierarchy (means–end analysis) (Rasmussen, 1986) and relates to the specification of information content. The skills–rules–knowledge taxonomy (Rasmussen, 1983) is used to derive inferences for how information should be presented. The aims of EID are to support the entire range of activities that confront operators, including familiar, unfamiliar, and unanticipated events, without contributing to the difficulty of the task.

EID has been applied to a variety of domains, such as process control, aviation, software engineering, and medicine and has been shown to improve performance over that achieved by more traditional design approaches. However, there are still some challenges confronting the widespread use of EID in the industry. These challenges include the time and effort required to analyze the work domain, choice of the interface form, and the difficulty of integrating EID with the design of other components of a system (Vincente, 2002).

3.2 Incorporating Human Factors in System Design

One problem faced by human factors engineers in system design is convincing project managers, engineers, and designers of the value of incorporating human factors knowledge and expertise into the system design process. In many instances, human factors issues are ignored or human factors activities are restricted to the evaluation stage. This is referred to as the "too little too late" phenomenon (Lim et al., 1992). Restricting human factors inputs to the evaluation stage limits the utility and effectiveness of human factors contributions. Either the contributions are ignored because it would be too costly or time consuming to alter the design of the system ("too late") or minor alterations are made to the design to pay lip service to human issues ("too little"). In either case there is limited realization of human factors contributions. For human factors to be effective, human factors engineers need to be involved throughout the design process.

There are a variety of reasons why human factors engineers are not considered as equal partners in a design team. One reason is that other team members (e.g., designers, engineers) have misconceptions about the potential contributions of human factors and the importance of human issues. They perceive, for example, that humans are flexible and can adapt to system requirements, or that accommodating human issues will compromise the technical system. Another reason is that sometimes, human factors inputs are of limited value to designers (Meister, 1989; Chapanis, 1995). The inputs are either so specific that they apply to a particular design situation and not to the design process in question, or they are vague and overly general. For example, a design guideline which specifies that "older people need larger characters on computer screens" is of little value. How does one define "larger characters"? Obviously, the type of input required depends on the nature of the design problem. Design of a kitchen to accommodate people in wheelchairs requires precise information, such as counter height dimensions or required turning space. In contrast, guidelines for designing intelligent interfaces need to be expressed at the cognitive task level, independent of a particular technology (Woods and Roth, 1988). Thus, one important task for human factors engineers is to ensure that design inputs are in a form that is usable and useful to designers. Williges and colleagues (Williges et al., 1992) demonstrate how integrated empirical models can be used as quantitative design guidelines. Their approach involved integrating data from four sequential experiments and developing a model for the design of a telephone-based information system.

To ensure that human factors will be applied to system design systematically, we need to market the potential contributions of human factors to engineers, project managers, and designers. One approach is to use case studies, relevant to the design problem, that illustrate the benefits of human factors. Case studies of this nature can be found in technical journals (e.g., *Applied Ergonomics*, *Ergonomics in Design*) and technical reports. Another approach is to perform a cost–benefit analysis. Estimating the costs and benefits associated with human factors is difficult because it is difficult to isolate the contribution of human factors relative to other variables, baseline measures of performance are unavailable, or performance improvements are hard to quantify and link to system improvements. There are methods available to conduct this type of analysis.

3.3 Applications of Human Factors to the System Design Process

System design can be conceptualized as a problem-solving process that involves the formulation of the problem, the generation of solutions to the problem, analysis of these alternatives, and selection of the most effective alternative (Rouse, 1985). There are various ways to classify the various stages in system design. Meister (1989), on the basis of a military framework, distinguishes among four phases:

1. *System planning.* The need for the system is identified and system objectives are defined.
2. *Preliminary design.* Alternative system concepts are identified, and prototypes are developed and tested.

3. *Detail design.* Full-scale engineering is developed.
4. *Production and testing.* The system is built and undergoes testing and evaluation.

To maximize system effectiveness, human factors engineers need to be involved in all phases of the process. In addition to human factors engineers, a representative sample of operators (users) should also be included.

The basic role of human factors in system design is the application of behavioral principles, data, and methods to the design process. Within this role, human factors get involved in a number of activities. These activities include specifying inputs for job, equipment and interface design, human performance criteria, operator selection and training, and inputs regarding testing and evaluation. The nature of these activities is discussed at a general level in the next section. Most of these issues are discussed in detail in subsequent chapters. The intent of this discussion is to highlight the nature of human factors involvement in the design process.

3.3.1 System Planning

During system planning, the need for the system is established and the goals and objectives and performance specifications of the system are identified. Performance specifications define what a system must do to meet its objectives and the constraints under which the system will operate. These specifications determine the system's performance requirements. Human factors should be a part of the system planning process. The major role of human factors engineers during this phase is to ensure that human issues are considered in the specification of design requirements and the statement of system goals and objectives. This includes understanding personnel requirements, general performance requirements, the intended users of the system, user needs, and the relationship of system objectives relative to these needs.

3.3.2 System Design

System design encompasses both preliminary design and detailed design. During this phase of the process, alternative design concepts are identified and tested and a detailed model of the system is developed. To ensure adequate consideration of human issues during this phase, the involvement of human factors engineers is critical. The major human factors activities include (1) function allocation, (2) task analysis, (3) job design, (4) interface design, (5) design of support materials, and (6) workplace design. The primary role of the human factors engineer is to ensure joint optimization of the human and technical systems.

Function Allocation Function allocation is a critical step in work system design. This is especially true in today's work systems, as machines are becoming more and more capable of performing tasks once restricted to humans. A number of studies have shown (e.g., Morris et al., 1985; Sharit et al., 1987) that proper allocation of functions between humans and machines results in improvements in overall system performance.

Function allocation involves formulating a functional description of a system and subsequent allocation of functions among system components. A frequent approach to function allocation is to base allocation decisions on machine capabilities; automate wherever possible. Although this approach may appear expedient, there are several drawbacks. In most systems not all tasks can be automated, and thus some tasks must be performed by humans. These tasks are typically "leftover" tasks. Allocating them to humans generally leads to problems of underload, inattention, and job dissatisfaction. A related problem is that automated systems fail and humans have to take over. This can be problematic if the humans are out of the loop or if their skills have become rusty due to disuse. In essence the machine-based allocation strategy is inadequate. As discussed previously, there are numerous examples of technocentered design. It has become clear that a better approach is complementary where functions are allocated so that human operators are complemented by technical systems. This approach involves identifying how to couple humans and machines to maximize system performance. In this regard, there is much research aimed at developing methods to guide function allocation decisions. These methods include lists (e.g., Fitts's list), computer simulation packages, and general guidelines for function allocation (e.g., Price, 1985).

The traditional static approach (humans are better at . . .) to function allocation has been challenged and dynamic allocation approaches have been developed. With dynamic allocation, responsibility for a task at any particular instance is allocated to the component most capable at that point in time. Hou et al. (1993) developed a framework to allocate functions between humans and computers for inspection tasks. Their framework represents a dynamic allocation framework and provides *for* a quantitative evaluation of the allocation strategy chosen. Morris et al. (1985) investigated the use of a dynamic adaptive allocation approach within an aerial search environment. They found that the adaptive approach resulted in an overall improvement in system performance. Similar to this approach is the adaptive automation approach. This approach involves invoking some form of automation as a function of the person's momentary needs (e.g., transient increase in workload or fatigue). The intent of this approach is to optimize the control of human–machine systems in varying environments. To date, few studies have examined the benefit of this approach. However, several important issues have emerged in the design of these types of systems, such as what aspect of the task should be adapted and who should make the decision to implement or remove automation.

Task Analysis Task analysis is also a central activity in system design. Task analysis helps ensure that human performance requirements match operators' (users') needs and capabilities and that the system can be operated in a safe and efficient manner. The output of a task analysis is also essential to the design of the interface, workplaces, support materials, training programs, and test and evaluation procedures.

A task analysis is generally performed after function allocation decisions are made; however, sometimes the results of the task analysis alter function allocation decisions. A task analysis usually consists of two phases: a task description and a task analysis. A *task description* involves a detailed decomposition of functions into tasks which are further decomposed into subtasks or steps. A *task analysis* specifies the physical and cognitive demands associated with each of these subtasks.

A number of methods are available for conducting task analysis. Commonly used methods include flow process charts, critical task analysis, and hierarchical task analysis. Techniques for collecting task data include documentation review, surveys and questionnaires, interviews, observation, and verbal protocols.

As the demands of tasks have changed and become more cognitive in nature, methods have been developed for performing *cognitive task analysis*, which attempts to describe the knowledge and cognitive processes involved in human performance in particular task domains. The results of a cognitive task analysis are important to the design of interfaces for intelligent machines. A common approach used to carry out a cognitive task analysis is a goal–means decomposition. This approach involves an analysis of the work domain to identify the cognitive demands inherent in a particular situation and building a model that relates these cognitive demands to situational demands (Roth et al., 1992). Another approach involves the use of cognitive simulation.

Job Design The type of work that a person performs is largely a function of job design. Jobs involve more than tasks and include work content, distribution of work, and work roles. Essentially, a job represents a person's prescribed role within an organization. *Job design* involves determining how tasks will be grouped together, how work will be coordinated among individuals, and how people will be rewarded for their performance (Davis and Wacker, 1987). To design jobs effectively, consideration must be given to workload requirements and to the psychosocial aspects of work (people's needs and expectations). This consideration is especially important in automated work systems, where the skills and potential contributions of humans are often overlooked.

In terms of workload, the primary concern is that work requirements are commensurate with human abilities and that individuals are not placed in situations of underload or overload, as both situations can lead to performance decrements, job dissatisfaction, and stress. Both the physical and mental demands of a task need to be considered. There are well-established methods for evaluating the physical demands of tasks and for determination of work and rest schedules. The concept of mental workload is more esoteric. This issue has received a great deal of attention in the literature, and a variety of methods have been developed to evaluate the mental demands associated with a task.

Consideration of operator characteristics is also an essential element of job design, as the workforce is becoming more heterogeneous. For example, older workers may need different work/rest schedules than younger workers or may be unsuited to certain types of tasks. Those who are physically challenged may also require different job specifications.

In terms of psychosocial considerations, a number of studies have identified critical job dimensions. Generally, these dimensions include task variety, task identity, feedback, autonomy, task significance, opportunity to use skills, and challenge. As far as possible, these characteristics should be designed into jobs. Davis and Wacker (1987) have developed a quality of working life criteria checklist which lists job dimensions important to the satisfaction of individual needs. These dimensions relate to the physical environment, institutional rights and privileges, job content, internal social relations, external social relations, and career path.

A number of approaches to job design have been identified. These include work simplification, job enrichment, job enlargement, job rotation, and teamwork design. The method chosen should depend on the actual design problem, work conditions, and individuals. However, it is generally accepted that the work simplification approach does not lead to optimal job design.

Interface Design Interface design involves specification of the nature of the human–machine interaction, that is, the means by which the human is connected to the machine. During this stage of design, the human factors specialist typically works closely with engineers and designers. The role of human factors is to provide the design team with information regarding the human performance implications of design alternatives. This generally involves three major activities: (1) gathering and interpreting human performance data, (2) conducting attribute evaluations of suggested designs, and (3) human performance testing (Sanders and McCormick, 1993). Human performance testing typically involves building mock-ups and prototypes and testing them with a sample of users. This type of testing can be expensive and time consuming. Recently, the development of rapid prototyping tools has made it possible to speed up and compress this process. These tools have been used primarily in the testing of computer interfaces; however, they can be applied to a variety of situations.

Interface design encompasses the design of both the physical and cognitive components of the interface and includes the design and layout of controls and displays, information content, and information representation. Physical components include factors such as type of control or input device, size and shape of controls, control location, and visual and auditory specifications (e.g., character size, character contrast, labeling, signal rate, signal frequency).

Cognitive components refer to the information-processing aspects of the interface (e.g., information content, information layout). As machines have become more intelligent, much of the focus of interface design has been on the cognitive aspects of the interface: Issues of concern include determination of the optimal level of machine support, identification of the type of information that users need, determination of how this information should be presented, and identification of methodologies to analyze work domains and cognitive activities. The central concern is developing interfaces that best support human task performance. In this regard, a number of approaches have evolved for interface design. Ecological interface design (Rasmussen and Vincente, 1989) is an example of recent design method.

There are a variety of sources of data on the characteristics of human performance that can serve as inputs to the design process. These include handbooks, textbooks, standards (e.g., ANSI), and technical journals. There are also a variety of models of human performance including cognitive models (e.g., GOMS; Card et al., 1983), control theory models, and engineering models. These models can be useful in terms of predicting the effects of design parameters on human performance outcomes. As discussed previously, it is the responsibility of the human factors engineer to make sure that information regarding human performance is in a form that is useful to designers. It is also important when using these data to consider the nature of the task, the task environment, and the user population.

Design of Support Materials This phase of the design process includes identifying and developing materials that facilitate the user's interaction with the system. These materials include job aids, instructional materials, and training devices and programs. All too often this phase of the design process is neglected or given little attention. A common example is the cumbersome manuals that accompany software packages or VCRs.

Support materials should not be used as a substitute for good design, however; the design of effective support materials is an important part of the system design process. Users typically need training and support to interact successfully with new technologies and complex systems. To maximize their effectiveness, human factors principles need to be applied to the design of instructional materials, job aids, and training programs. Guidelines are available for the design of instructional materials and job aids. Bailey (1982) provides a thorough discussion of these issues. A great deal has also been written on the design of training programs.

Design of the Work Environment The design of the work environment is an important aspect of work system design. Systems exist within a context, and the characteristics of this context affect overall system performance. The primary concern of workplace design is to ensure that the work environment supports the operator and activity performance and allows the worker to perform tasks in an efficient, comfortable, and safe manner. Important issues include workplace and equipment layout, furnishings, reach dimensions, clearance dimensions, visual dimensions, and the design of the ambient environment. There are numerous sources of information related to workplace design and evaluation that can be used to guide this process. These issues are also covered in detail in other chapters of this handbook.

3.4 Test and Evaluation

Test and evaluation are critical aspects of system design and usually take place throughout the system design process. Test and evaluation provide a means for continuous improvement during system development. Human factors inputs are essential to the testing and evaluation of systems. The primary role of human factors is to assess the impact of system design features on human performance outputs, including objective outputs such as speed and accuracy of performance and workload, and subjective outputs such as comfort and user satisfaction. Human factors specialists are also interested in ascertaining the impact of human performance on overall system performance. Issues related to evaluation and the assessment of system effectiveness are covered in detail in Chapters 41–44.

Because the evaluation of systems and system components involves measurement of human performance in operational terms (relative to the system or subsystem in question), human factors engineers face a number of challenges when evaluating systems. Generally, the standards of generalizability are higher for human factors research, as the research results must be extended to real-world systems (Kantowitz, 1992). At the same time, it is often difficult to achieve an appropriate level of control. Unfortunately, in many instances the utility of test and evaluation results are limited because of deficiencies in test and evaluation procedures (Bitner, 1992).

In this regard, there are three key issues that need to be addressed when developing methods for evaluating system effectiveness: (1) subject representativeness, (2) variable representativeness, and (3) setting representativeness (Kantowitz, 1992). *Subject representativeness* refers to the extent to which subjects tested in the research study represent the population to which the research results apply. In most cases, the sample involved in system evaluation should represent the population of interest on relevant characteristics. *Variable representativeness* refers to the extent that the study variables are representative of the research question. It is important to select variables that capture the essential issues being assessed in the research study. *Setting representativeness* is the degree of congruence between the test situation in which the research is performed and the target situation in which the research must be applied. The important issue is the comparability of the psychological processes captured in these situations, not necessarily physical fidelity.

A variety of techniques are available for conducting human factors research, including experimental methods, observational methods, surveys and questionnaires, and audits. There is no single preferred method; each has its associated strengths and weaknesses. The method one chooses depends on the nature of the research question. It is generally desirable to use several methods in conjunction.

4 CONCLUSIONS

System design and development is an important area of application for human factors engineers. System performance will be improved by consideration of behavioral issues. Although much has been written on system design, our knowledge of this topic is far from complete. The changing nature of systems coupled with the increased diversity of users presents new challenges for human factors specialists and affords many research opportunities. The goals of this chapter were to summarize some of the current issues in system design and to illustrate the important role of human factors engineers within the system design process. Further, the chapter provides a framework for many of the topics addressed in this handbook.

REFERENCES

Bailey, R. W. (1982), *Human Performance Engineering: A Guide for System Design*, Prentice-Hall, Englewood Cliffs, NJ.

Banathy, B. H., and Jenlink, P. M. (2004), "Systems Inquiry and Its Application in Education," in *Handbook of Research on Educational Communications and Technology*, D. H. Jonassen, Ed., Lawrence Erlbaum Associates, Mahwah, NJ, pp. 37–55.

Berger, A. (1994), "Balancing Technological, Organizational, and Human Aspects in Manufacturing Development," *International Journal of Human Factors in Manufacturing*, Vol. 4, pp. 261–280.

Bitner, A. C. (1992), "Robust Testing and Evaluation of Systems," *Human Factors*, Vol. 34, pp. 477–484.

Card, S. K., Moran, T. P., and Newell, A. (1983), *The Psychology of Human–Computer Interaction*, Lawrence Erlbaum Associates, Hillsdale, NJ.

Chapanis, A. (1995). "Ergonomics in Product Development: A Personal View," *Ergonomics*, Vol. 38, pp. 1625–1638.

Clegg, C. W. (2000), "Sociotechnical Principles for System Design," *Applied Ergonomics*, Vol. 31, pp. 463–477.

Davis, L. E., and Wacker, G. J. (1987), "Job Design," in *Handbook of Human Factors*, G. Salvendy, Ed., Wiley, New York, pp. 431–452.

Drucker, P. F. (1998), "The Coming of a New Organization," *Harvard Business Review*, January–February, pp. 45–53.

Eason, K. D. (1989), *Information Technology and Organizational Change*, Taylor & Francis, London.

Eason, K. D. (1991), "Ergonomic Perspectives on Advances in Human–Computer Interaction," *Ergonomics*, Vol. 34, pp. 721–741.

Eastman Kodak Company (1983), *Ergonomic Design for People at Work*, Lifetime Learning Publications, Belmont, CA.

Eggleston, R. G. (1987), "The Changing Nature of the Human–Machine Design Problem: Implications for System Design and Development," in *System Design: Behavioral Perspectives on Designers, Tools and Organizations*, W. B. Rouse and K. R. Boff, Eds., North-Holland, New York, pp. 113–126.

Fisk, A. D., Rogers, W. A., Charness, N., Czaja, S. J., and Sharit, J. (2004), *Designing for Older Adults: Principles and Creative Human Factors Approaches*, CRC Press, Boca Raton, FL.

Green, D. M., and Swets, J. A. (1966), *Signal Detection Theory and Psychophysics*, Wiley, New York.

Hendrick, H. W., and Kleiner, B. M. (2001), *Macro ergonomics: An Introduction to Work System Design.* Human Factors and Ergonomics Society, Santa Monica, CA.

HFES (1998), "HFES Strategic Plan," in *Human Factors Directory and Yearbook*, Human Factors and Ergonomics Society, Santa Monica, CA.

Hou, T., Lin, L., and Drury, C. G. (1993), "An Empirical Study of Hybrid Inspection Systems and Allocation of Inspection Function," *International Journal of Human Factors in Manufacturing*, Vol. 3, pp. 351–367.

Imada, A. S., Nora, K., and Nagamachi, M. (1986), "Participatory Ergonomics: Methods for Improving Individual and Organizational Effectiveness," in *Human Factors in Organizational Design and Management*, Vol. II, O. Brown, Jr., and H. W. Hendrick, Eds., North-Holland, New York, pp. 403–406.

Kantowitz, B. H. (1992), "Selecting Measures for Human Factors Research," *Human Factors*, Vol. 34, pp. 387–398.

Kidd, P. T. (1992), "Interdisciplinary Design of Skill–based Computer-Aid Technologies: Interfacing in Depth." *International Journal of Human Factors in Manufacturing*, 209–228.

Kinsella, K., and Velkoff, V. A. (2001), *An Aging World, 2001*, U.S. Census Bureau, Series Vol. P95/01-1, U.S. Government Printing Office, Washington, DC.

Liker, J. K., and Majchrzak, A. (1994), "Designing the Human Infrastructure for Technology," in *Organization and Management of Advanced Manufacturing*, W. Karwowski and G. Salvendy, Eds., Wiley, New York, pp. 121–164.

Lim, K. Y., Long, J. B., and Hancock, N. (1992), "Integrating Human Factors with the Jackson System Development Method: An Illustrated Overview," *Ergonomics*, Vol. 35, pp. 1135–1161.

Majchrzak, A., and Gasser, L. (1992), "On Using Artificial Intelligence to Integrate the Design of Organizational Midprocess Changes in U.S. Manufacturing," *AI and Society*, Vol. 5, pp. 321–338.

Meister, D. (1989), *Conceptual Aspects of Human Factors*, John Hopkins University Press, Baltimore.

Meister, D. (1991), *Psychology of System Design*, Elsevier, New York.

Meister, D. (2000), "Cognitive Processes in System Design," *Theoretical Issues in Ergonomics Science*, Vol. 2, pp. 113–138.

Morris, N. M., Rouse, W. B., and Ward, S. L. (1985), "Experimental Evaluation of Adaptive Task Allocation in an Aerial Search Environment," in *Proceedings of the 2nd IFAC/IFIP/IFORS/IEA Conference: Analysis, Design and Evaluation of Man–Machine Systems*, G. Mancini, G. Johannsen, and L. Martensson, Eds., Varese, Italy, pp. 67–72.

Pasmore, W. A. (1988), *Designing Effective Organizations: The Sociotechnical Systems Perspective*, Wiley, New York.

Price, H. E. (1985), "The Allocation of Functions in Systems," *Human Factors*, Vol. 2, pp. 33–45.

Proctor, R. W., and Van Zandt, T. (1994), *Human Factors in Simple and Complex Systems*, Allyn and Bacon, Boston.

Rasmussen, J. (1982), "Human Errors: A Taxonomy for Describing Human Malfunctioning in Industrial Installations," *Journal of Occupational Accidents*, Vol. 4, pp. 311–333.

Rasmussen, J. (1983), "Skill, Rules and Knowledge: Signals, Signs and Symbols, and Other Distinctions in Human Performance Models," *IEEE Transactions of Systems, Man, and Cybernetics*, Vol. 13, pp. 257–266.

Rasmussen, J. (1986), *Information Processing and Human–Machine Interaction: An Approach to Cognitive Engineering*, North-Holland, Amsterdam.

Rasmussen, J., and Goldstein, L. P. (1985), "Decision Support in Supervisory Control," in *Proceedings of the 2nd IFAC/IFIP/IFORS/IEA Conference: Analysis, Design and Evaluation of Man–Machine Systems*, G., G. Johannsen, and L. Martensson, Eds., Varese, Italy pp. 79–90.

Rasmussen, J., and Vincente, K. J. (1989), "Coping with Human Errors Through System Design: Implications for Ecological Interface Design," *International Journal of Man–Machine Studies*, Vol. 31, pp. 517–534.

Rasmussen, J., Pejtersen, A. M., and Goodstein, L. P. (1994), *Cognitive Systems Engineering*, Wiley, New York.

Reason, J. (1990), *Human Error*, Cambridge University Press, New York.

Reason, J. (1995), "A Systems Approach to Organizational Error," *Ergonomics*, Vol. 38, pp. 1708–1721.

Roth, E. M., Woods, D. D., and Pople, H. E., Jr. (1992), "Cognitive Simulation as a Tool for Cognitive Task Analysis," *Ergonomics*, Vol. 35, pp. 1163–1198.

Rouse, W. B. (1985), "On the Value of Information in System Design: A Framework for Understanding and Aiding Designers," *Information Processing and Management*, Vol. 22, pp. 217–228.

Rouse, W. B. (1987), "Designers, Decision Making and Decision Support," in *System Design: Behavioral Perspectives on Designers, Tools and Organizations*, W. B. Rouse and K. R. Boff, Eds., North-Holland, New York, pp. 275–284.

Rouse, W. B., and Boff, K. R. (1987a), "Workshop Themes and Issues: The Psychology of Systems Design," in *System Design: Behavioral Perspectives on Designers, Tools and Organizations*, W. B. Rouse and K. R. Boff, Eds., North-Holland, New York, pp. 7–18.

Rouse, W. B. and Boff, K. R. (1987b), *System Design: Behavioral Perspectives on Designers, Tools and Organizations*, North-Holland, New York.

Rouse, W. B., and Morris, N. M. (1987), "Conceptual Design of Error Tolerant Interface for Complex Engineering Systems," *Automatica*, Vol. 23, pp. 231–235.

Sanders, M. S., and McCormick, E. J. (1993), *Human Factors in Engineering and Design*, 7th ed., McGraw-Hill, New York.

Sharit, J., Chang, J. C., and Salvendy, G. (1987), "Technical and Human Aspects of Computer-Aided Manufacturing," in *Handbook of Human Factors*, G. Salvendy, Ed., Wiley, New York, pp. 1694–1724.

Sheridan, T. B. (2002), *Humans and Automation: System Design and Research Issues*, Human Factors and Ergonomics Society, Santa Monica, CA.

Stanney, K. M., Mollaghasemi, M., Reeves, L., Breaux, R., and Graeber, D. A. (2003), "Usability Engineering of Virtual Environments (VEs): Identifying Multiple Criteria That Drive Effective VE System Design," *International Journal of Human–Computer Studies*, Vol. 58, pp. 447–481.

Vicente, K. J. (2002), "Ecological Interface Design: Progress and Challenges," *Human Factors*, Vol. 44, No. 1, pp. 62–78.

Wickens, C. D. (1992), *Engineering Psychology and Human Performance*, 2nd ed., Harper-Collins, New York.

Wickens, C. D., and Hollands, J. G. (2000), *Engineering Psychology and Human Performance*, 3rd ed., Prentice-Hall, Upper Saddle River, NJ.

Wickens, C. D., Lee, J., Liu, Y. D., and Gordon-Becker, S. (2004), *Introduction to Human Factors Engineering*, 2nd ed., Prentice-Hall, Upper Saddle River, NJ.

Williges, R. C., Williges, B. H., and Han, S. (1992), "Developing Quantitative Guidelines Using Integrated Data from Sequential Experiments," *Human Factors*, Vol. 34, pp. 399–408.

Woods, D. D. (1988), "Commentary: Cognitive Engineering in Complex and Dynamic Worlds," in *Complex Engineering in Complex Dynamic Worlds*, E. Hollnagel, G. Mancini, and D. D. Woods, Eds., Academic Press, New York, pp. 115–129.

Woods, D. D., and Roth, E. M. (1988), "Cognitive Systems Engineering," In *Handbook of Human–Computer Interaction*, M. E. Helander, Ed., North-Holland, New York, pp. 3–43.

PART 2

THE HUMAN FACTORS FUNDAMENTALS

CHAPTER 3

SENSATION AND PERCEPTION

Robert W. Proctor and Janet D. Proctor
Purdue University
West Lafayette, Indiana

1 INTRODUCTION

Human–machine interaction, as all other interactions of persons with their environment, involves a continuous exchange of information between the operator(s) and the machine. The operator provides input to the machine, which acts on this input and displays information back to the operator regarding its status and the consequences of the input. The operator must process this information, decide what, if any, controlling actions are needed, and then provide new input to the machine. One important facet of this exchange of information between the machine and the operator is the displaying of information from the machine as input to the operator. All such information must enter through the operator's senses and be organized and recognized accurately to ensure correct communication of the displayed information. Thus, an understanding of how people sense and perceive is essential for display design. An effective display is consistent with the characteristics and limitations of the human sensory and perceptual systems. These systems are also involved intimately in both the control of human interactions with the environment and of actions taken to operate machines. However, because selection and control of action is the topic of Chapter 4, in this chapter we focus primarily on the nature of sensation and perception. Similarly, because other chapters focus on the applied topics of motion and vibration (Chapter 23), noise (Chapter 24), illumination (Chapter 25), and displays (Chapter 45), we concentrate primarily on the nature of sensory and perceptual processes and the general implications for human factors and ergonomics.

Many classifications of sensory systems exist, but most commonly, distinctions are made between five sensory modalities: vision, audition, olfaction, gustation, and somasthesis. The vestibular system, which provides the sense of balance, is also of importance. All sensory systems extract information about four characteristics of stimulation: (1) the sensory modality and submodalities (e.g., touch as opposed to pain), (2) the stimulus intensity, (3) the duration of the stimulation, and (4) its location (Gardner and Martin, 2000). Each system has receptors that are sensitive to some aspect of the physical environment. These receptors are responsible for sensory transduction, or the conversion of physical stimulus energy into electrochemical energy in the nervous system. The sensory information for each sense is encoded in the activity of neurons and travels to the brain via specialized, structured pathways consisting of highly interconnected networks of neurons. For most modalities, two or more pathways operate in parallel to analyze and convey different types of information from the sensory signal. The pathways project to primary receiving areas in the cerebral cortex (see Figure 1), in most cases after passing through relay areas in the thalamus. From the primary receiving area, the pathways then project to many other areas within the brain.

Each neuron in the sensory pathways is composed of a cell body, dendrites at the input side, and an axon with branches at the output side. Most neurons fire in an all-or-none manner, sending spike, or action, potentials down the axon away from the cell body. The rate at which a neuron fires varies as a function of the input that the neuron is receiving from other neurons (or directly from sensory receptors) at its dendrites. Most neurons exhibit a baseline firing rate in the absence of stimulation, usually on the order

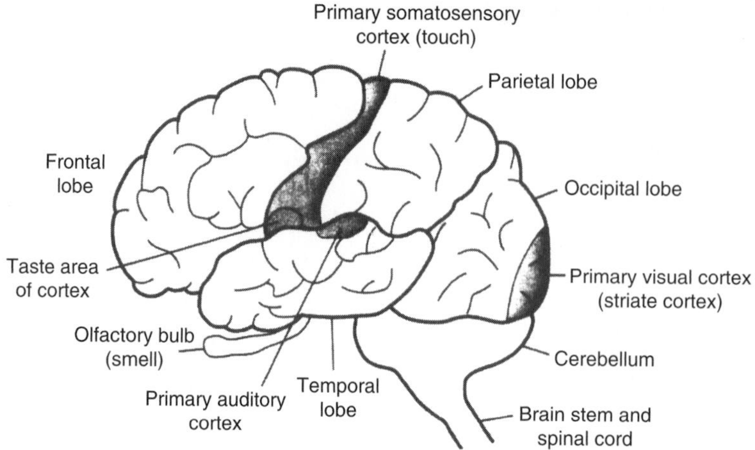

Figure 1 Primary sensory receiving areas (visual, auditory, and somatosensory) of the cerebral cortex and other important landmarks and areas. (From Schiffman, 1996.)

of 5 to 10 spikes/s, and information is signaled by deviations above or below this baseline rate. The speed of transmission of a spike along the fiber varies across different types of neurons, ranging from 20 to 100 m/s. Immediately after an action potential occurs, the neuron is in a refractory state in which another action potential cannot be generated. This sets an upper limit on the firing rate of about 1000 spikes/s.

Transmission between neurons occurs at small gaps, called *synapses*, between the axonal endings of one neuron and the dendrites of another. Communication at the synapse takes place by means of transmitter substances that have an excitatory effect of increasing the firing rate of the neuron or an inhibitory effect of decreasing the firing rate. Because as many as several hundred neurons may have synapses with the dendrites of a specific neuron, whether the firing rate will increase or decrease is a function of the sum of the excitatory and inhibitory inputs that the neuron is receiving. Which specific neurons provide excitatory and inhibitory inputs will determine the patterns of stimulation to which the neuron will be sensitive (i.e., to which the firing rate will increase or decrease from baseline). The patterns may be either rather general (e.g., an increase in illumination) or quite specific (e.g., a pair of lines at a particular angle, moving in a particular direction). In general, increases in stimulus intensity result in increased firing rates for individual neurons and in a larger population of neurons that respond to the stimulus. Thus, intensity is coded by frequency and population codes.

The study of sensation and perception involves not only the anatomy and physiology of the sensory systems, but also behavioral measures of perception. Psychophysical data obtained from tasks in which observers are asked to detect, discriminate, rate, or recognize stimuli provide information about how the properties of the sensory systems relate to what is perceived. Behavioral measures also provide considerable information about the functions of higher-level brain processes for which current knowledge of the physiological bases is rudimentary. The sensory information must be interpreted by these higher-level processes, which include mental representations, decision making, and inference. Thus, perceptual experiments provide evidence about how the sensory input is organized into a coherent percept. In the next section we review methods used to investigate sensory and perceptual processes.

2 METHODS FOR INVESTIGATING SENSATION AND PERCEPTION

Many methods have been, and can be, used to obtain data pertinent to understanding sensation and perception (see, e.g., Scharff, 2003). The most basic distinction is between methods that involve anatomy and physiology as opposed to methods that involve behavioral responses. Because the former are not of much direct use in human factors and ergonomics, we do not cover them in as much detail as we do the latter.

2.1 Anatomical and Physiological Methods

A wide variety of specific techniques exist for analyzing and mapping out the pathways associated with sensation and perception. These include injecting tracer substances into the neurons, classifying neurons in terms of the size of their cell bodies and characteristics of their dendritic trees, and lesioning areas of the brain (see Wandell, 1995). Such techniques have provided a relatively detailed understanding of the sensory pathways.

One particular technique that has produced a wealth of information about the functional properties of specific neurons in the sensory pathways and their associated regions in the brain is *single-cell recording*. Such recording is typically performed on a monkey, cat, or other nonhuman species; an electrode is inserted that is sufficiently small to record only the activity of a single neuron. The responsivity of this neuron to various features of stimulation can be examined to gain some understanding of the neuron's role in the sensory system. By systematic examinination of the responsivities of neurons in a given region, it has been possible to determine much about the way that sensory input is coded. In our discussion of sensory systems, we will have opportunity to refer to the results of single-cell recordings.

Neuropsychological and psychophysiological investigations of humans have been used increasingly in recent years to evaluate issues pertaining to information processing. Neuropsychological studies typically examine patients who have some specific neurological disorder, Several striking phenomena have been observed that enhance our understanding of higher-level vision (Farah, 2000). One example is visual neglect, in which a person with a lesion in the right cerebral hemisphere, often in a region called the *right posterior parietal lobe*, fails to detect or respond to stimuli in the left visual field (Mort et al., 2003). This is in contrast to people with damage to regions of the temporal lobe, who have difficulty recognizing stimuli (Milner and Goodale, 1995). These and other results have provided evidence that a dorsal system, also called the *parietal pathway*, determines where something is (and how to act on it), whereas a ventral system, also called the *temporal pathway*, determines what that something is (Merigan and Maunsell, 1993).

A widely used psychophysiological method involves the measurement of *event-related potentials* (ERPs) (Rugg and Coles, 1995). To record ERPs, electrodes attached to a person's scalp measure voltage variations in the electroencephalogram (EEG), which reflects the summed electrical activity of neuron populations as recorded at various sites on the scalp. An ERP is those changes that involve the brain's response to a particular event, usually onset of a stimulus. ERPs provide good temporal resolution, but the spatial resolution is not very high. Those ERP components occurring within 100 ms after onset of a stimulus are sensory components that reflect transmission of sensory information to, and its arrival at, the sensory cortex. The latencies for these components differ across sensory modalities. Later components reflect other aspects of information processing. For example, a negative component called *mismatch negativity* is evident in the ERP about 200 ms after presentation of a stimulus event other than the one that is most likely. It is present regardless of whether the stimulus is in an attended stream of stimuli or an unattended stream, suggesting that it reflects an automatic detection of physical deviance. The latency of a positive component called the *P*300 is thought to reflect stimulus evaluation time, that is, the time to update the perceiver's current model of the physical environment.

During the past decade, use of functional neuroimaging techniques such as *functional magnetic resonance imaging* (fMRI) and *positron emission tomography* (PET), which provide insight into the spatial organization of brain functions, has become prominent (e.g., Kanwisher and Duncan, 2004). Both fMRI and PET provide images of the neural activity in different areas of the brain by measuring the volume of blood flow, which increases as the activity in an area increases. They have good spatial resolution, but the temporal resolution is not as good as that of ERPs. By comparing measurements taken during a control period to those taken while certain stimuli are present or tasks performed, the brain imaging techniques can be used to identify which areas of the brain are involved in the processing of different types of stimuli and tasks.

Electrophysiological and functional imaging methods, as well as other psychophysiological techniques, provide tools that can be used to address many issues of concern in human factors. Among other things, these methods can be used to determine whether a particular experimental phenomenon has its locus in processes associated with sensation and perception or with those involving subsequent response selection and execution. Because of this diagnosticity, it has been suggested that psychophysiological measures may be applied to provide precise measurement of dynamic changes in mental workload (e.g., Wilson, 2002) and to other problems in human factors (Kramer and Weber, 2000). Parasuraman (2003) has recently coined the term *neuroergonomics* to refer to a neuroscience approach to ergonomics, which he advocates.

2.2 Psychophysical Methods

The more direct concern in human factors and ergonomics is with behavioral measures, because our interest is primarily with what people can and cannot perceive and with evaluating specific perceptual issues in applied settings. Because many of the methods used for obtaining behavioral measures can be applied to evaluating aspects of displays and other human factors concerns, we cover them in some detail. The reader is referred to textbooks on psychophysical methods by Gescheider (1997) and Baird and Noma (1978) and to a chapter by Schiffman (2003) for more thorough coverage.

2.2.1 Psychophysical Measures of Sensitivity

Classical Threshold Methods The goal of one class of psychophysical methods is to obtain some estimate of sensitivity to detecting either the presence of some stimulation or differences between stimuli. The classical methods were based on the concept of a threshold, with an *absolute threshold* representing the minimum amount of stimulation necessary for an observer to tell that a stimulus was presented on a trial, and a *difference threshold* representing the minimal amount of difference in stimulation along some dimension required to tell that a comparison stimulus differs from a standard stimulus. Fechner (1860) developed several techniques for finding absolute

Table 1 Determination of a Sensory Threshold by the Method of Limits Using Alternating Ascending (A) and Descending (D) Series

Stimulus Intensity (Arbitrary Units)	A	D	A	D	A	D	A	D	A	D
15						Y				
14				Y		Y				Y
13				Y		Y				Y
12		Y		Y		Y				Y
11		Y		Y	Y	Y		Y		Y
10	Y	Y		Y	N	Y	Y	Y		Y
9	N	Y	Y	N	N	N	N	Y	Y	N
8	N	N	N		N		N	Y	N	
7	N		N		N		N	N	N	
6	N		N		N		N		N	
5	N				N		N		N	
4	N				N				N	
3	N								N	
2	N								N	
1	N									
Transition points[a]:	9.5	8.5	8.5	9.5	10.5	9.5	9.5	7.5	8.5	9.5

[a]Mean threshold value = 9.1.

thresholds, with the methods of limits and constant stimuli being among the most widely used.

To find a threshold using the *method of limits*, equally spaced stimulus values along the dimension of interest (e.g., magnitude of stimulation) that bracket the threshold are selected (see Table 1). In alternating series, the stimuli are presented in ascending or descending order, beginning each time from a different, randomly chosen starting value below or above the threshold. For the ascending order, the first response typically would be, "No, I do not detect the stimulus." The procedure is repeated, incrementing the stimulus value each time, until the observer's response changes to "yes," and the average of that stimulus value and the last one to which a "no" response was given is taken as the threshold for that series. A descending series is conducted in the same manner, but from a stimulus above threshold, until the response changes from "yes" to "no." The thresholds for the individual series are then averaged to produce the final threshold estimate. A particularly efficient variation of the method of limits is the *staircase* method (Cornsweet, 1962). For this method, rather than having distinct ascending and descending series started from randomly selected values below and above threshold, only a single continuous series is conducted in which the direction of the stimulus sequence—ascending or descending—is reversed when the observer's response changes. The threshold is then taken to be the average of the stimulus values at which these transitions occur. The staircase method has the virtue of bracketing the threshold closely, thus minimizing the number of stimulus presentations that is needed to obtain a certain number of response transitions on which to base the threshold estimate.

The *method of constant stimuli* differs from the method of limits primarily in that the different stimulus values are presented randomly, with each stimulus value presented many different times. The basic data in this case are the percentage of "yes" responses for each stimulus value. These typically plot as an S-shaped psychophysical function (see Figure 2). The threshold is taken to be the estimated stimulus value for which the percentage of "yes" responses would have been 50%.

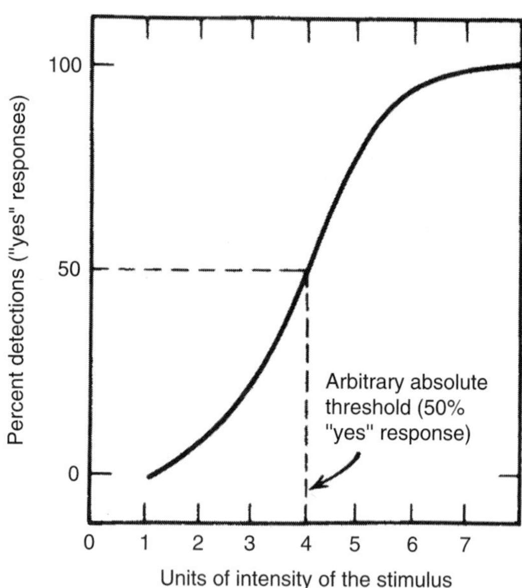

Figure 2 Typical S-shaped psychophysical function obtained with the method of constant stimuli. The absolute threshold is the stimulus intensity estimated to be detected 50% of the time. (From Schiffman, 1996.)

Both the methods of limits and constant stimuli can be extended to difference thresholds in a straightforward manner (see Gescheider, 1997). The most common extension is to use stimulus values for the comparison stimulus that range from being distinctly less than that of the standard stimulus to being distinctly greater. For the method of limits, ascending and descending series are conducted in which the observer responds "less," then "equal," and then "greater" as the magnitude of the comparison increases, or vice versa as it decreases, The average stimulus value for which the responses shift from "less" to "equal" is the lower threshold, and from "equal" to "greater" is the upper threshold, The difference between these two values is called the *interval of uncertainty*, and the difference threshold is found by dividing the interval of uncertainty by 2. The midpoint of this interval is the point of subjective equality, and the difference between this point and the true value of the standard stimulus reflects constant error, or the influence of any factors that cause the observer to overestimate or underestimate systematically the value of the comparison in relation to that of the standard.

When the method of constant stimuli is used to obtain difference thresholds, the order in which the standard and comparison are presented is varied, and the observer judges which stimulus is greater than the other. The basic data then are the percentages of "greater" responses for each value of the comparison stimulus. The stimulus value corresponding to the 50th percentile is taken as the point of subjective equality. The difference between that stimulus value and the one corresponding to the 25th percentile is taken as the lower difference threshold, and the difference between the subjectively equal value and the stimulus value corresponding to the 75th percentile is the upper threshold: The two values are averaged to get a single estimate of the difference threshold.

Although threshold methods are often used to investigate basic sensory processes, variants can be used to investigate applied problems as well. Shang and Bishop (2000) argued that the concept of visual threshold is of value for measuring and monitoring landscape attributes. They measured three types of different thresholds—detection, recognition, and visual impact (changes in visual quality as a consequence of landscape modification)—for two types of objects, a transmission tower and an oil refinery tank, as a function of size, contrast, and landscape type. Shang and Bishop were able to obtain thresholds of high reliability and concluded that a visual variable that combined the effects of contrast and size, which they called *contrast weighted visual size*, was the best predictor of all three thresholds.

Signal Detection Methods Although many variants of the classical methods are still used, they are not as popular as they once were. The primary reason is that the threshold measures confound perceptual sensitivity, which they are intended to measure, with response criterion or bias (e.g., willingness to say "yes"), which they are not intended to measure. The threshold estimates can also be influenced by numerous other extraneous factors, although the impact of most of these factors can be minimized with appropriate control procedures. Alternatives to the classical methods, signal detection methods, have come to be preferred in many situations because they contain the means for separating sensitivity and response bias. Authoritative references for signal detection methods and theory include Green and Swets (1966), Macmillan and Creelman (2005), and Wickens (2001). Macmillan (2002) provides a briefer introduction to its principles and assumptions.

The typical signal detection experiment differs from the typical threshold experiment in that only a single stimulus value is presented for a series of trials, and the observer must discriminate trials on which the stimulus was not presented (noise trials) from trials on which it was (signal-plus-noise, or signal, trials). Thus, the signal detection experiment is much like a true–false test in that it is objective; the accuracy of the observer's responses with respect to the state of the world can be determined. If the observer says "yes" most of the time on signal trials and "no" most of the time on noise trials, we know that the observer was able to discriminate between the two states of the world. If, on the other hand, the proportion of "yes" responses is equal on signal and noise trials, we know that the observer could not discriminate between them. Similarly, we can determine whether the observer has a bias to say one response or the other by considering the relative frequencies of "yes" and "no" responses regardless of the state of the world. If half of the trials included the signal and half did not, yet the observer said "yes" 70% of the time, we know that the observer had a bias to say "yes."

Signal detection methods allow two basic measures to be computed, one corresponding to discriminability (or *sensitivity)* and the other to *response bias*. Thus, the key advantage of the signal detection methods is that they allow the extraction of a pure measure of perceptual sensitivity separate from any response bias that exists, rather than combining the two in a single measure, as in the threshold techniques. There are many alternative measures of sensitivity (Swets, 1986) and bias (Macmillan and Creelman, 1990), based on a variety of psychophysical models and assumptions. We will base our discussion around signal detection theory and the two most widely used measures of sensitivity and bias, d' and β. Sorkin (1999) describes how signal detection measures can be calculated using spreadsheet application programs such as Excel.

Signal detection theory assumes that the sensory effect of a signal or noise presentation on any given trial can be characterized as a point along a continuum of evidence indicating that the signal was in fact presented. Across trials, the evidence will vary, such that for either type of trial it will sometimes be higher (or lower) than at other times. For computation of d' and β, it is assumed that the resulting distribution of values is normal (i.e., bell-shaped and symmetric), or Gaussian, for both the signal and noise trials and that the variances for the two distributions are

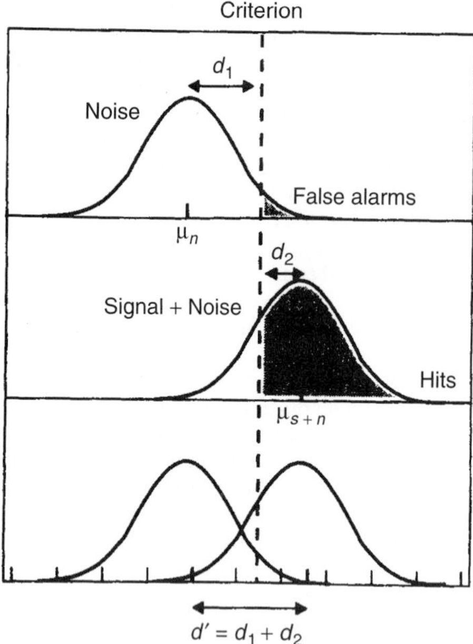

Figure 3 Equal variance, normal probability distributions for the noise, and signal-plus-noise distributions on the sensory continuum, with a depiction of the proportion of false alarms, the proportion of hits, and the computation of d'. Bottom panel shows both distributions on a single continuum. (From Proctor and Van Zandt, 1994.)

equal (see Figure 3). To the extent that the signal is discriminable from the noise, the distribution for the signal trials should be shifted to the right (i.e., higher on the continuum of evidence values) relative to that for the noise trials. The measure d' is therefore the distance between the means of the signal and noise distributions, in standard deviation units. That is,

$$d' = \frac{\mu_s - \mu_n}{\sigma}$$

where μ_s is the mean of the signal distribution, μ_n is the mean of the noise distribution, and σ is the standard deviation of both distributions. The assumption is that the observer will respond "yes" whenever the evidence value on any trial exceeds a criterion. The measure of β, which is expressed by the formula

$$\beta = \frac{f_s(C)}{f_n(C)}$$

where C is the criterion and f_s and f_n are the heights of the signal and noise distributions, respectively, is the likelihood ratio for the two distributions at the criterion. It indicates the placement of this criterion with respect to the distributions and thus reflects the relative bias to respond "yes" or "no."

Computation of d' and β is relatively straightforward. The placement of the distributions with respect to the criterion can be determined as follows. The *hit rate* is the proportion of signal trials on which the observer correctly said "yes"; this can be depicted graphically by placing the criterion with respect to the signal distribution so that the proportion of the distribution exceeding it corresponds to the hit rate. The *false-alarm rate* is the proportion of noise trials on which the observer incorrectly said "yes." This corresponds to the proportion of the noise distribution that exceeds the criterion; when the noise distribution is placed so that the proportion exceeding the criterion is the false-alarm rate, relative positions of the signal and noise distributions are depicted. Sensitivity, as measured by d', is the difference between the means of the signal and noise distributions, and this difference can be found by separately calculating the distance of the criterion from each of the respective means and then combining those two distances. Computationally, this involves converting the false-alarm rate and hit rate into standard normal z scores. If the criterion is located between the two means, d' is the sum of the two z scores. If the criterion is located outside that range, the smaller of the two z scores must be subtracted from the larger to obtain d'. The likelihood ratio measure of bias, β, can be found from the hit and false-alarm rates by using a z table that specifies the height of the distribution for each z value. When β is 1.0, no bias exists to give one or the other response. A value of β greater than 1.0 indicates a bias to respond "no," whereas a bias less than 1.0 indicates a bias to respond "yes."

Although β has been used most often as the measure of bias to accompany d', several investigations have indicated that an alternative bias measure, C, is better (Snodgrass and Corwin, 1988; Macmillan and Creelman, 1990; Corwin, 1994). C is a measure of criterion location rather than likelihood ratio. Specifically,

$$C = -0.5[z(H) + z(F)]$$

where H is the hit rate and F the false-alarm rate. C is superior to β on several grounds, including that it is less affected by the level of accuracy than is β and will yield a meaningful measure of bias when accuracy is near chance.

For a given d', the possible combinations of hit rates and false-alarm rates that the observer could produce through adopting different criteria can be depicted in a *receiver operating characteristic* (ROC) curve (see Figure 4). The farther an ROC is from the diagonal that extends from hit and false-alarm rates of 0 to 1, which represents chance performance (i.e., d' of 0), the greater the sensitivity. The procedure described above yields only a single point on the ROC, but in many cases it is advantageous to examine performance under several criteria settings, so that the form of the complete ROC is evident (Swets, 1986). One advantage is that the estimate of sensitivity will be more reliable when it is based on several points along

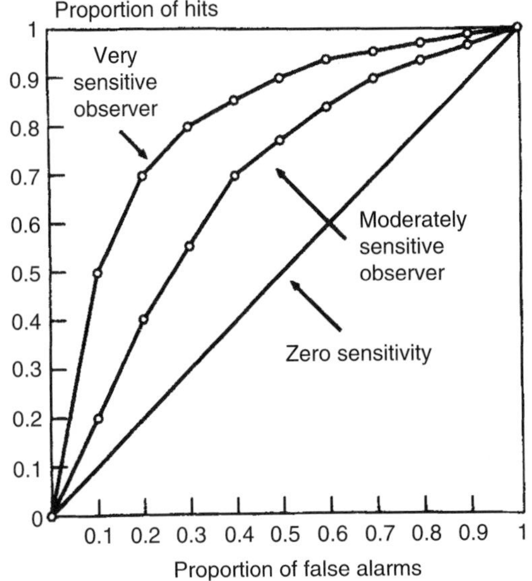

Figure 4 ROC curves showing the possible hit and false-alarm rates for different sensitivities. (From Proctor and Van Zandt, 1994.)

the ROC than when it is based on only one. Another is that the empirical ROC can be compared to the ROC implied by the psychophysical model that underlies a particular measure of sensitivity to determine whether serious deviations occur. For example, when enough points are obtained to estimate complete ROC curves, it is possible to evaluate the assumptions of equal variance, normal distributions on which the measures of d' and β are based. When plotted on z-score coordinates, the ROC curve will be linear with a slope of 1.0 if both assumptions are supported; deviations from a slope of 1.0 mean that one distribution is more variable than the other, whereas systematic deviations from linearity indicate that assumption of normality is violated. If either of these deviations is present, alternative measures of sensitivity and bias that do not rely on the assumptions of normality and equal variance should be used.

In cases in which a complete ROC curve is desired, several procedures exist for varying response criteria. The relative payoff structure may be varied across blocks of trials to make one or the other response more preferable; similarly, instructions may be varied regarding how the observer is to respond when uncertain. Another way to vary response criteria is to manipulate the relative probabilities of the signal and noise trials; the response criterion should be conservative when signal trials are rare and become increasingly more liberal as the signal trials become increasingly more likely. One of the most efficient techniques is to use rating scales (e.g., from 1, meaning very sure that the signal was not present, to 5, meaning very sure that it was present) rather than "yes"–"no" responses. The ratings are then treated as a series of criteria, ranging from high to low, and hit and false-alarm rates are calculated with respect to each.

Signal detection methods are powerful tools for investigating basic and applied problems pertaining not only to sensation and perception but also to many other areas in which an observer's response must be based on probabilistic information, such as distinguishing normal from abnormal x-rays (Manning and Leach, 2002) or detecting whether severe weather will occur within the next hour (Harvey, 2003). Although most work on signal detection theory has involved discriminations along a single psychological continuum, it has been extended also to situations in which multidimensional stimuli are presumed to produce values on multiple psychological continua such as color and shape (e.g., Macmillan, 2002). Kadlec (1999; http://web.uvic.ca/psyc/software/) provides a computer program for performing such analyses, which have the benefit of allowing evaluation of whether the stimulus dimensions are processed in perceptually separable and independent manners and whether the decisions for each dimension are also separable. As these examples illustrate, signal detection methods can be extremely effective when used with discretion.

2.2.2 Psychophysical Scaling

Another concern in psychophysics is to construct scales for the relation between physical intensity and perceived magnitude (see Marks and Gescheider, 2002, for a review). Scales can be constructed from three types of tasks that differ in the responses required of subjects (Gescheider, 1988, 1997). One way to build such scales is to do so from discriminative responses to stimuli that differ only slightly. Fechner (1860) established procedures for constructing psychophysical scales from difference thresholds. Later, Thurstone (1927) proposed a method for constructing a scale from paired comparison procedures in which each stimulus is compared to all others. Thurstonian scaling methods can even be used for complex stimuli for which physical values are not known. The second type of task involves having subjects divide the sensory continuum into two or more intervals that are subjectively equal. Such methods date back to the work of Plateau (1872).

Much recent work on scaling has followed the lead of Stevens (1975) in using direct methods that require some type of magnitude judgment (see, e.g., Bolanowski and Gescheider, 1991, for an overview). The technique of *magnitude estimation* is the most widely used. With this procedure, the observer is either presented a standard stimulus and told that its sensation is a particular numerical value (modulus) or is allowed to choose his or her own modulus. Stimuli of different magnitudes are then presented randomly, and the observer is to assign values to them proportional to their perceived magnitudes. These values then provide a direct scale relating physical magnitude to perceived magnitude. A technique called *magnitude production* can also be used, in which the observer is instructed to adjust the value of a stimulus to be a particular magnitude. Variations of magnitude estimation and production have been used to measure such things as emotional stress (Holmes and Rahe, 1967) and pleasantness of voice quality for normal speakers and persons with a range of vocal pathology (Eadie and Doyle, 2002). Furthermore, Walker (2002) provided evidence that magnitude estimation can be used as a design tool in the development of data sonifications, that is, representations of data by sound.

Baird and Berglund (1989) coined the term *environmental psychophysics* for the application of psychophysical methods such as magnitude estimation to applied problems of the type examined by Berglund (1991) that are associated with odorous air pollution and community noise. As Berglund puts it: "The method of ratio scaling developed by S. S. Stevens (1956) is a contribution to environmental science that ranks as good and important as most methods from physics or chemistry" (p. 141). When any measurement technique developed for laboratory research is applied to problems outside the laboratory, special measurement issues may arise. In the case of environmental psychophysics, the environmental stimulus of concern typically is complex and multisensory, diffuse, and naturally varying, presented against an uncontrollable background. The most serious measurement problem is that often it is not possible to obtain repeated measurements from a given observer under different magnitude concentrations, necessitating that a scale be derived from judgments of different observers at different points in time.

Because differences exist in the way that people assign magnitude numbers to stimuli, each person's scale must be calibrated properly. Berglund and her colleagues have developed what they call the *master scaling procedure* to accomplish this purpose. The procedure has observer's make magnitude judgments for several values of a referent stimulus, as well as for the environmental stimulus. Each observer's power function for the referent stimulus is transformed to a single master function (this is much like converting different normal distributions to the standard normal distribution for comparison). The appropriate transformation for each observer is then applied to her or his magnitude judgment for the environmental stimulus so that all such judgments are in terms of the master scale.

2.2.3 Other Techniques

Many other techniques have been used to investigate issues in sensation and perception. Most important are methods that use response times either instead of, or in conjunction with, response accuracy (see Welford, 1980; Van Zandt, 2002). Reaction-time methods have a history of use approximately as long as that of the classical psychophysical methods, dating back to Donders (1868), but they have been particularly widespread since about 1950. Simple reaction times require the observer to respond as quickly as possible with a single response (e.g., a keypress) whenever a stimulus event occurs. Alternative hypotheses of various factors that affect detection of the stimulus and the decision to respond, such as the locus of influence of visual masking and whether the detection of two signals presented simultaneously can be conceived of as an independent race, can be evaluated using simple reaction times.

Decision processes play an even larger role in go/no go tasks, where responses must be made to some stimuli but not to others, and in choice–reaction tasks, where there is more than one possible stimulus, more than one possible response, and the stimulus must be identified if the correct response is to be made. Methods such as the *additive factors logic* (Sternberg, 1969) can be used to isolate perceptual and decisional factors. This logic proposes that two variables whose effects are additive affect different processing stages, but two variables whose effects are interactive affect the same processing stage. Variables that interact with marker variables whose effects can be assumed to be in perceptual processes, but not with marker variables whose effects are on response selection or programming, can be assigned a perceptual locus. Analyses based on the distributions of reaction times have gained in popularity in recent years. Van Zandt (2002) provides MATLAB code for performing such analyses.

3 SENSORY SYSTEMS AND BASIC PERCEPTUAL PHENOMENA

The ways in which the sensory systems encode information have implications not only for the structure and function of the sensory pathways but ultimately, also for the nature of human perception. They also place restrictions on the design of displays. Displays must be designed to satisfy known properties of sensory encoding (e.g., visual information that would be legible if presented in central vision will not be legible if the display were presented in the visual periphery), but they do not need to exceed the capabilities of sensory encoding. The sensory information that is encoded also must be represented in the nervous system. The nature of this representation also has profound implications for perception.

3.1 Vision

3.1.1 Visual System

The sensory receptors in the eye are sensitive to energy within a limited range of the electromagnetic spectrum. One way of characterizing such energy is as continuous waves of different wavelengths. The visible spectrum ranges from wavelengths of approximately 370 nanometers (nm; billionths of a meter) to 730 nm. Any energy outside this range, such as ultraviolet rays, will not be detected because they have no effect on the receptors. Light can also be characterized in terms of small units of energy called *photons*. Describing light in terms of wavelength is important for some aspects of perception, such as color vision, whereas for others it is more useful to treat it in terms of photons. As with any system in which light energy is used to create a representation of the physical world, the light must be focused and a clear image created. In the case of the eye, the image is focused on the photoreceptors located on the retina, which lines the back wall of the eye.

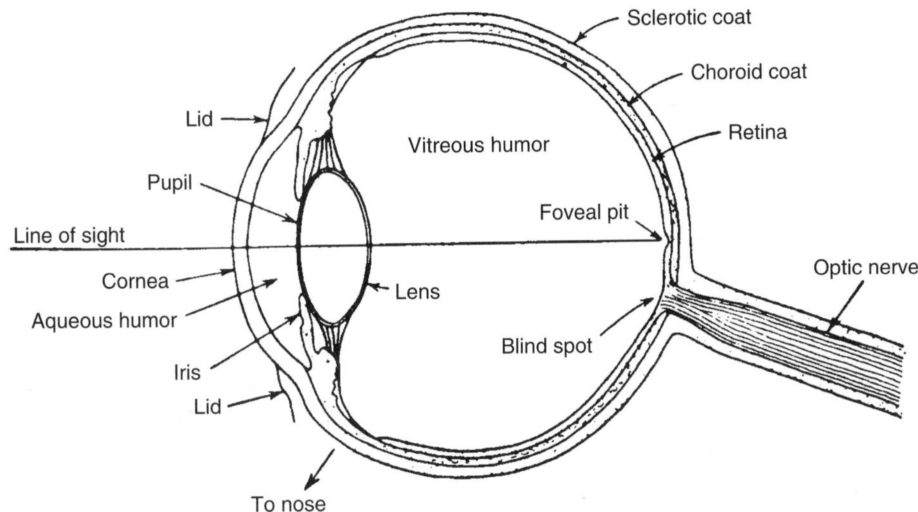

Figure 5 Structure of the human eye. (From Schiffman, 1996.)

Focusing System Light enters the eye (see Figure 5) through the cornea, which acts as a lens of fixed optical power and provides the majority of the focusing. The remainder of the focusing is accomplished by the lens, whose power varies automatically as a function of the distance from the observer of the object that is being fixated. Beyond a distance of approximately 6 m, the far point, the lens is relatively flat; for distances closer than the far point, muscles attached to the lens cause it to become progressively more spherical the closer the fixated object is to the observer, thus increasing its refractive power. The reason why this process, called *accommodation*, is needed is that without an increase in optical power for close objects, their images would be focused at a point beyond the retina and the retinal image would be out of focus. Accommodation is effective for distances as close as 20 cm (the near point), but the extent of accommodation, and the speed at which it occurs, decreases with increasing age, with the near point receding to approximately 100 cm by age 60. This decrease in accommodative capability, called *presbyopia*, can be corrected with reading glasses. Other imperfections of the lens system—*myopia*, where the focal point is in front of the receptors; *hyperopia*, where the focal point is behind the receptors; and *astigmatism*, where certain orientations are out of focus while others are not—also typically are treated with glasses.

Between the cornea and the lens, the light passes through the pupil, which can vary in size from 2 to 8 mm. The pupil size is large when the light level is low, to maximize the amount of light that gets into the eye, and small when the light level is high, to minimize the imperfections in imaging that arise when light passes through the extreme periphery of the lens system. One additional consequence of these changes in image quality as a function of pupil size is that the *depth of field*, or the distance in front of or behind a fixated object at which the images of other objects will be in focus also, will be greatest when the pupil size is 2 mm and decrease as pupils size increases, at least up to intermediate diameters (Marcos et al., 1999). In other words, under conditions of low illumination, accommodation must be more precise and work that requires high acuity, such as reading, can be fatiguing (Randle, 1988). The pupil size also varies from moment to moment as a function of arousal level and the amount of mental effort being devoted to a task (Kahneman, 1973), although the changes in size associated with these factors are less than those associated with the lighting level.

When the eyes fixate on an object at a distance of approximately 6 m or farther, the lines of sight are parallel. As the object is moved progressively closer, the eyes turn inward and the lines of sight converge. Thus, the degree of *vergence* of the eyes varies systematically as a function of the distance of the object being fixated. The near point for vergence is approximately 5 cm, and if an object closer than that is fixated, the images at the two eyes will not be fused and a double image will be seen.

The natural resting states for accommodation and vergence, called *dark focus* and *dark vergence*, respectively, are intermediate to the near and far points (Leibowitz and Owens, 1975; Owens and Leibowitz, 1983; Andre, 2003). One view for which there is considerable support is that dark focus and vergence provide zero reference points about which accommodative and vergence effort varies (Ebenholtz, 1992). A practical implication of this is that less eye fatigue will occur if a person working at a visual display screen for long periods of time is positioned at a distance that corresponds approximately to the dark focus and vergence points. As with most other human characteristics of concern in human factors and ergonomics, considerable individual differences in dark focus and vergence exist. People with far dark vergence postures tend to position themselves farther away from the display screen than will those with closer postures (Heuer et al., 1989), and they also show more visual fatigue when required to perform close visual work (Jaschinski-Kruza, 1991).

Retina If the focusing system is working properly, the image will be focused on the retina, which lines the back wall of the eye. Objects in the left visual field will be imaged on the right hemi-retina and objects in the right visual field on the left hemi-retina; objects above the point of fixation will be imaged on the lower half of the retina, and vice versa for objects below fixation. The retina contains the photoreceptors that transduce the light energy into a neural signal; their spatial arrangement limits our ability to perceive spatial pattern (see Figure 6). There also are two layers of neurons, and their associated blood vessels, that process the retinal image before information about it is sent along the optic nerve to the brain. These neural layers are in the light path between the lens and the photoreceptors and thus degrade to some extent the clarity of the image at the receptors.

There are two major types of photoreceptors, *rods* and *cones*, with three subtypes of cones. All photoreceptors contain light-sensitive photopigments in their outer segments that operate in basically the same manner. Photons of light are absorbed by the photopigment when they strike it, starting a reaction that leads to the generation of a neural signal. As light is absorbed, the photopigment becomes insensitive and is said to be *bleached*. It must go through a process of regeneration before it is functional again. Because the rod and cone photopigments differ in their absolute sensitivities to light energy, as well as in their differential sensitivities to light across the visual spectrum, the rods and cones have different roles in perception.

Rods are involved primarily in vision under very low levels of illumination, what is called *scotopic vision*. All rods contain the same photopigment, rhodopsin, which is highly sensitive to light. Its spectral sensitivity function shows it to be maximally sensitive to light around 500 nm and to a lesser degree to other wavelengths. One consequence of there being only one rod photopigment is that we cannot perceive color under scotopic conditions. The reason for this is easy to understand. The rods will respond relatively more to stimulation of 500 nm than they will to 560-nm stimulation of equal intensity. However, if the intensity of the 560-nm stimulus is increased, a point would be reached at which the rods responded equally to the two stimuli. In other words, with one photopigment, there is no basis for distinguishing among the wavelength differences associated with color differences from intensity differences.

Cones are responsible for daylight, or *photopic, vision*. Cone photopigments are less sensitive to light than rhodopsin, and hence cones are operative at levels of illumination at which the rod photopigment has been effectively fully bleached. Also, because there are three types of cones, each containing a different photopigment, cones provide color vision. As explained previously, there must be more than one photopigment type if differences in the wavelength of stimulation are to be distinguished from differences in intensity. The spectral sensitivity functions for each of the three cone photopigments span broad ranges of the visual spectrum, but their peak sensitivities are located at different wavelengths. The peak sensitivities are approximately 440 nm for the short-wavelength ("blue") cones, 540 nm for the middle-wavelength ("green") cones, and 565 nm for the long-wavelength ("red") cones. Monochromatic light of a particular wavelength will produce a pattern of activity for the three cone types that is unique from the patterns produced by other wavelengths, allowing each to be distinguished perceptually.

The retina contains two landmarks that are important for visual perception. The first of these is the optic disk, which is located on the nasal side of the retina. This is the region where the optic nerve, composed of the nerve fibers from the neurons in the retina, exits the eye. The significant point is that there are no photoreceptors in this region, which is why it is sometimes called the *blind spot*. We do not normally notice the blind spot because (1) the blind spot for one of the eyes corresponds to part of the normal visual field for the other eye, and (2) with monocular viewing, the perceptual system fills it in with fabricated images based on visual attributes from nearby regions of the visual field. How this filling-in occurs has been the subject of considerable investigation, with current evidence suggesting that it is induced by neurons in the primary visual cortex with large receptive fields that extend outward from the blind spot (Komatsu et al., 2002). If the image of an object falls only partly on the blind spot, the filling-in from the surrounding region will cause the object to appear complete. However, if the image of an object falls entirely within the blind spot, this filling in will cause the object not to be perceived.

The second landmark is the *fovea*, which is a small indentation about the size of a pinhead on which the image of an object at the point of fixation will fall. The fovea is the region of the retina in which visual acuity is highest. Its physical appearance is due primarily to the fact that the neural layers are pulled away, thus allowing the light a straight path to the receptors. Moreover, the fovea contains only cones, which are densely packed in this region.

As shown in Figure 6, the photoreceptors synapse with bipolar cells, which in turn synapse with ganglion cells; the latter cells are the output neurons of the retina, with their axons making up the optic nerve. In addition, horizontal cells and amacrine cells provide interconnections across the retina. The number of ganglion cells is much less than the number of photoreceptors, so considerable convergence of the activity of individual receptors occurs. The neural signals generated by the rods and cones are maintained in distinct pathways until reaching the ganglion cells (Kolb, 1994). In the fovea, each cone has input into

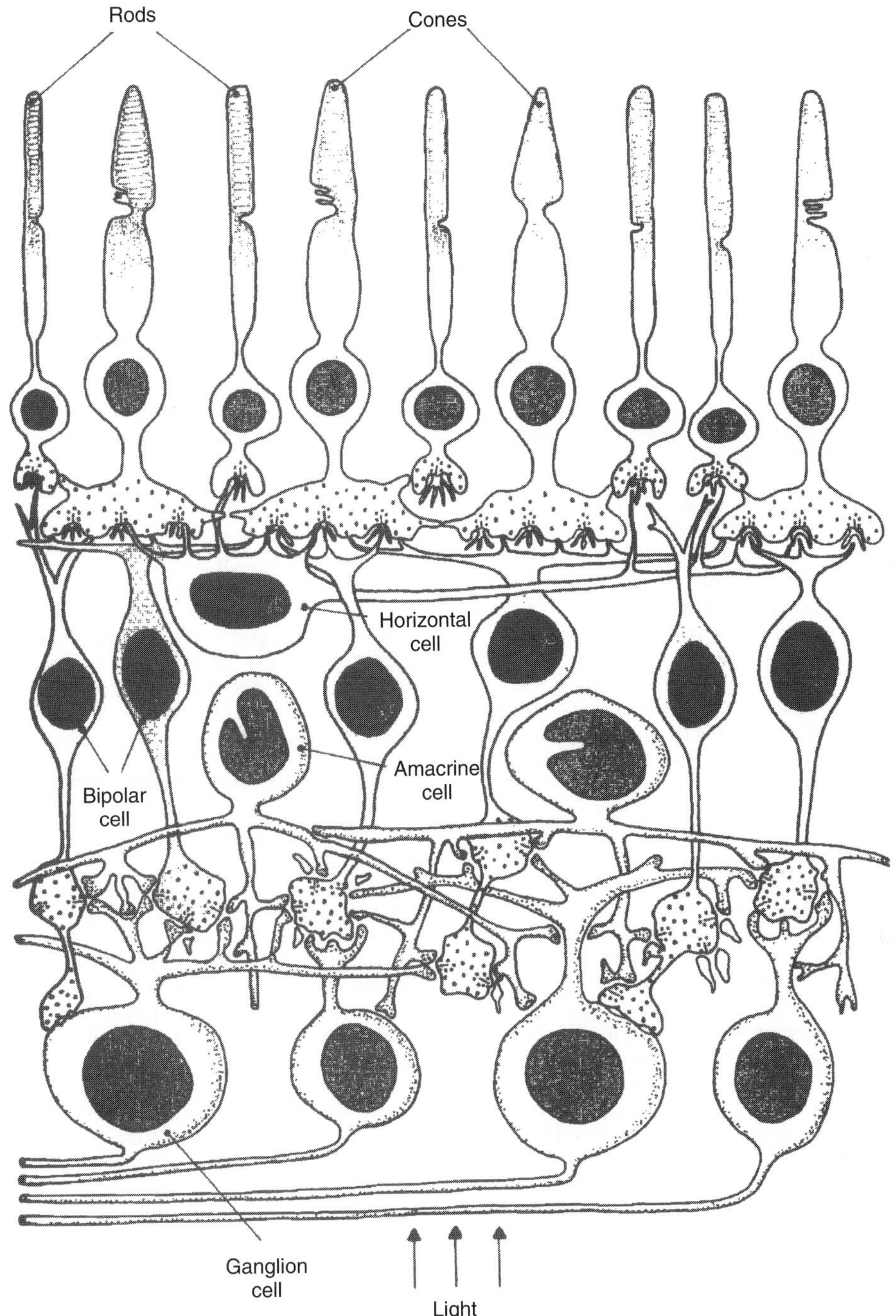

Figure 6 Neural structures and interconnections of the vertebrate retina. (From Dowling and Boycott, 1966.)

more than one ganglion cell. However, convergence is the rule outside the fovea, being an increasing function of distance from the fovea. Overall, the average convergence is 120 : 1 for rods as compared to 6 : 1 for cones. The degree of convergence has two opposing perceptual consequences. Where there is little or no convergence, as in the neurons carrying signals from the fovea, the pattern of stimulation at the retina is maintained effectively complete, thus maximizing spatial detail. When there is considerable convergence, as for the rods, the activity of many photoreceptors in the region is pooled together, optimizing sensitivity to light at the cost of detail. Thus, the wiring of the photoreceptors is consistent with the fact that the rods operate when light energy is at a premium but the cones operate when it is not.

The ganglion cells show several interesting properties pertinent to perception. When single-cell recording techniques are used to measure their receptive fields (i.e., the regions on the retina that when stimulated produce a response in the cell), these fields are found to have a circular, center-surround relation for most cells. If light presented in a circular, center region causes an increase in the firing rate of the neuron, light presented in a surrounding ring region will cause a decrease in the firing rate, or vice versa. What this means is that the ganglion cells are tuned to respond primarily to discontinuities in the light pattern within their receptive fields. If the light energy across the entire receptive field is increased, there will be little if any effect on the firing rate. In short, the information extracted and signaled by these neurons is based principally on contrast, which is important for perceiving objects in the visual scene, and not on absolute intensity, which will vary as a function of the amount of illumination. Not surprisingly, the average receptive field size is larger for ganglion cells receiving their input from rods than for those receiving it solely from cones and increases with increasing distance from the fovea.

Although most ganglion cells have the center-surround receptive field organization, two pathways can be distinguished on the basis of other properties. The ganglion cells in the *parvocellular pathway* have small cell bodies and relatively dense dendritic fields. Many of these ganglion cells, called *midget cells*, receive their input from the fovea. They have relatively small receptive fields, show a sustained response as long as stimulation is present in the receptive field, and have a relatively slow speed of transmission. The ganglion cells in the *magnocellular pathway* have larger cell bodies and sparse dendritic trees. They have their receptive fields at locations across the retina, have relatively large receptive fields, show a transient response to stimulation that dissipates if the stimulus remains on, have a fast speed of transmission, and are sensitive to motion. Because of these unique characteristics and the fact that these channels are kept separated later in the visual pathways, it has been thought that they contribute distinct information to perception. The parvocellular pathway is presumed to be responsible for pattern perception and the magnocellular pathway for high temporal frequency information, such as in motion perception and perception of flicker. The view that different aspects of the sensory stimulus are analyzed in specialized neural pathways has received considerable support in recent years.

Visual Pathways The optic nerve from each eye splits at what is called the *optic chiasma* (see Figure 7). The fibers conveying information from the nasal halves of the retinas cross over and go to the opposite sides of the brain, whereas the fibers conveying information from the temporal halves do not cross over. Functionally, the significance of this is that for both eyes, input from the right visual field is sent to the left half of the brain and input from the left visual field is sent to the right half. A relatively small subset of the fibers (approximately 10%) split off from the main tract and go to structures in the brain stem, the tectum, and then the pulvinar nucleus of the thalamus. This *tectopulvinar pathway* is involved in localization of objects and the control of eye movements.

Approximately 90% of the fibers continue on the primary *geniculostriate pathway*, where the first synapse is at the lateral geniculate nucleus (LGN). The distinction between the parvocellular and magnocellular pathways is maintained here. The LGN is composed of six layers, four parvocellular and two magnocellular, each of which receives input from only a single eye. Hence, at this level the input from the two eyes has yet to be combined. Each layer is laid out in a retinotopic map that provides a spatial representation of the retina. In other respects, the receptive field structure of the LGN neurons is similar to that of the ganglion cells. The LGN also receives input from the cortex—in fact, relatively more synapses in the LGN originate from the cortex than from the retina (Sherman and Koch, 1990)—meaning that this is the first point in the pathway in which the brain can have some effect on the signals arriving from the sensory system.

From the LGN, the fibers go to the primary visual cortex, which is located in the posterior cortex. This region is also called the *striate cortex* (because of its stripes), *area 17*, or *area V1*. The visual cortex consists of six layers. The fibers from the LGN have their synapses in the fourth layer from the outside, with the parvocellular neurons sending their input to the bottom half and the magnocellular neurons to the top half. The neurons in layer 4 then send their output to other layers. In layer 4 the neurons have circular-surround receptive fields, but in other layers, they have more complex patterns of sensitivity. Also, whereas layer 4 neurons receive input from one or the other eye, in other layers most neurons respond to some extent to stimulation at either eye.

A distinction can be made between simple cells and complex cells (e.g., Hubel and Wiesel, 1977). The responses of simple cells to shapes can be determined from their responses to small spots of light (e.g., if the receptive field for the neuron is plotted using spots of light, the neuron will be most sensitive to a stimulus shape that corresponds with that receptive field), whereas those for complex cells cannot be. Simple cells have center-surround receptive fields, but they are more linear than circular; this means that they are orientation selective and will respond optimally to bars in an orientation that corresponds with that of the receptive field. Complex cells have similar linear receptive fields, and so are also orientation selective, but they are movement sensitive as well. These cells respond optimally not only when the bar is at the appropriate orientation, but also when it is moving. Some cells, which receive input from the magnocellular pathway, are also directionally

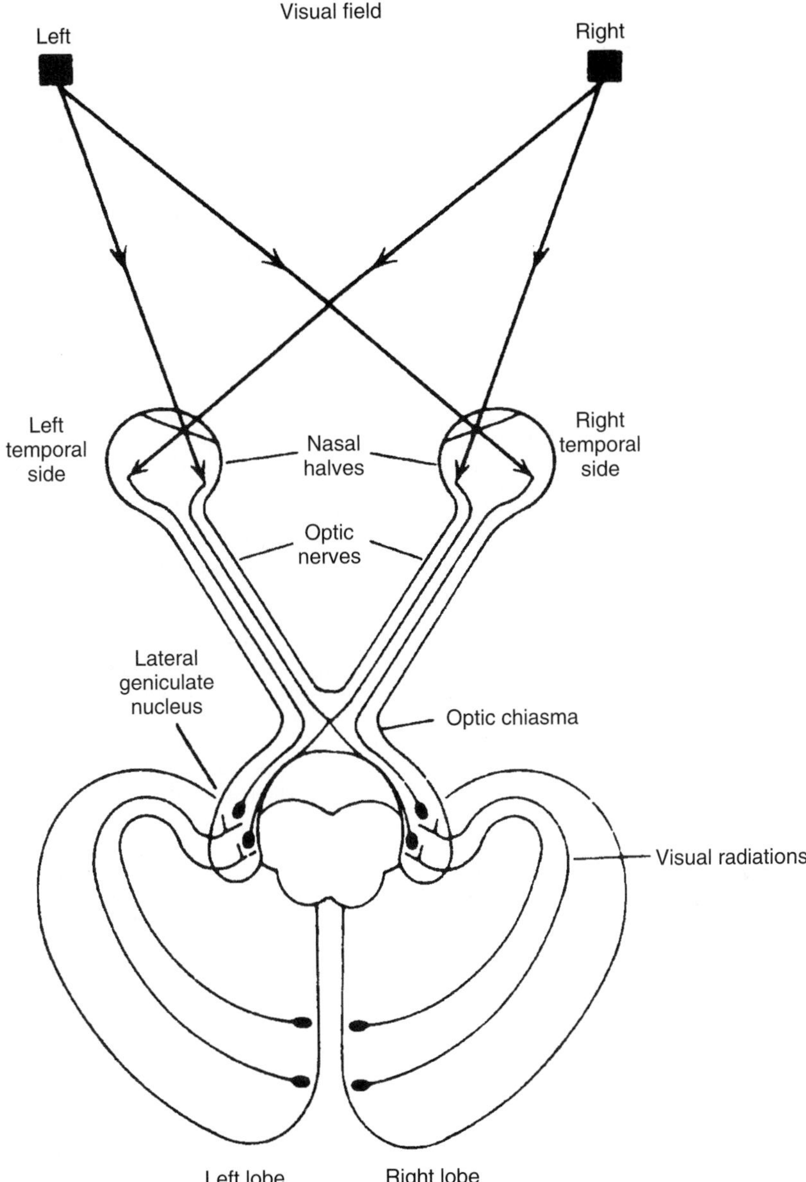

Figure 7 Human visual system showing the projection of the visual fields through the system. (From Schiffman, 1996.)

sensitive: they respond optimally to movement in a particular direction. Certain cells, called *hypercomplex cells*, are sensitive to the length of the bar so that they will not respond if the bar is too long. Some neurons in the visual cortex are also sensitive to disparities in the images at each eye and to motion velocity. In short, the neurons of the visual cortex analyze the sensory input for basic features that provide the information on which higher-level processes operate.

The cortex is composed of columns and hypercolumns arranged in a spatiotopic manner. Within a single column, all of the cells except for those in layer 4 will have the same preferred orientation. The next column will respond to stimulation at the same location on the retina, but will have a preferred orientation that is approximately 10° different than that of the first column. As we proceed through a group of about 20 columns, called a *hypercolumn*, the preferred orientation will rotate 180°. The next hypercolumn will show the same arrangement, but for stimulation at a location on the retina that is adjacent to that of the first. A relatively larger portion of the neural machinery in the visual cortex is devoted to the fovea, which is to be expected because it is the region for which detail is being represented.

More than 30 cortical areas subsequent to area V1 have been found to be involved in the processing of visual information (Frishman, 2001). Some of the neurons in the magnocellular pathway project directly to the medial temporal cortex, suggesting that it may play a role in motion perception. Other neurons in the magnocellular pathway converge with the parvocellular pathway in V1 (Nealey and Maunsell, 1994) and then separate into two pathways. The temporal pathway goes to the inferior portion of the temporal lobe and the parietal pathway goes to the posterior portion of the parietal lobe (Merigan and Maunsell, 1993). As mentioned earlier for the example of visual neglect, it has been suggested that the temporal pathway is involved primarily in determining what the stimulus is, and the parietal pathway is involved in determining where the stimulus is. Barber (1990) found that visual abilities pertaining to detection, identification, acquisition, and tracking of aircraft in air defense simulations clustered into subgroups consistent with the hypothesis that different aspects of visual perception are mediated by distinct subsystems in the brain.

3.1.2 Basic Visual Perception

Brightness Brightness is that aspect of visual perception that corresponds most closely to the intensity of stimulation. To specify the effective intensity of a stimulus, photometric measures, which are calibrated to reflect human spectral sensitivity, should be used. A photometer can be used to measure either the *illuminance*, that is, the amount of light falling on a surface, or *luminance*, that is, the amount of light generated by a surface. To measure illuminance, an illuminance probe is attached to the photometer and placed on the illuminated surface. The resulting measure of illuminance is in lumens per square meter (lm/m^2) or lux (lx). To measure luminance, a lens with a small aperture is attached to the photometer and focused onto the surface from a distance. The resulting measure of luminance is in candelas per square meter (cd/m^2).

Judgments of brightness are related to intensity by the power function

$$B = aI^{0.33}$$

where B is brightness, a is a constant, and I is the physical intensity. However, brightness is not determined by intensity alone but also by several other factors. For example, at brief exposures on the order of 100 ms or less and for small stimuli, temporal and spatial summation occur. That is, a stimulus of the same physical intensity will look brighter if its exposure duration is increased or if its size is increased.

One of the most striking influences on brightness perception and sensitivity to light is the level of *dark adaptation*. When a person first enters a dark room, he or she is relatively insensitive to light energy. However, with time, dark adaptation occurs and sensitivity increases drastically. The time course of dark adaptation is approximately 30 to 45 minutes. Over the first few minutes, the absolute threshold for light decreases and then levels off. However, after approximately 8 minutes in the dark, it begins decreasing again, approaching maximum around the 30-minute point. After 30 minutes in the dark, lights can be seen that were of too low intensity to be visible initially, and stimuli that appeared dim now seem much brighter. Dark adaptation reflects primarily regeneration into a maximally light-sensitive state of the cone photopigments and then the rod photopigment. Jackson et al. (1999) reported a dramatic slowing in the rod-mediated component of dark adaptation that may contribute to increased night vision problems experienced by older adults. After becoming dark-adapted, vision may be impaired momentarily when the person returns to photopic viewing conditions. Providing gradually changing light intensity in regions where light intensity would normally change abruptly, such as at the entrances and exits of tunnels, may help minimize such impairment (e.g., Oyama, 1987).

The brightness of a monochromatic stimulus of constant intensity will vary as a function of its wavelength because the photopigments are differentially sensitive to light of different wavelengths. The scotopic spectral sensitivity function is shifted toward the short-wavelength end of the spectrum, relative to the photopic function. Consequently, if two stimuli, one short-wavelength and one long, appear equally bright at photopic levels, the short-wavelength stimulus will appear brighter at scotopic levels, a phenomenon called the *Purkinje shift*. Little light adaptation will occur when high levels of long-wavelength light are present because the sensitivity of the rods to long-wavelength light is low. Thus, it is customary to use red light sources to provide high illumination for situations in which a person needs to remain dark-adapted.

It has become common practice to distinguish between brightness and lightness as two different aspects of perception: *Judgments of brightness* are of the perceived intensity of a stimulus, whereas *judgments of lightness* are of perceived achromatic color along a black-to-white dimension. Both the brightness and lightness of an object are greatly influenced by the surrounding context. *Lightness contrast* is a phenomenon where the intensity of a surrounding area influences the lightness of a stimulus. The effects can be quite dramatic, with a stimulus of intermediate reflectance ranging in appearance from white to dark gray or black as the reflectance of the surround is increased from low to high. The more common phenomenon of *lightness constancy* occurs when the level of illumination is increased across the entire visual field. In this case, the absolute amount of light reflected to the eye by an object may be quite different, but the percept remains constant. Basically, lightness follows a constant-ratio rule (Wallach, 1972): Lightness will remain the same if the ratio of light energy for a stimulus relative to its surround remains constant. Lightness constancy holds for a broad range of ratios and across a variety of situations, with brightness constancy obtained under a more restricted set of viewing conditions (Jacobsen and Gilchrist, 1988; Arend, 1993).

Although low-level mechanisms early in the sensory system probably contribute at least in part to constancy and contrast, more complex higher-level brain mechanisms do as well. Particularly compelling are demonstrations showing that the lightness and brightness of an object can vary greatly simply as a function of organizational and depth cues. Agostini and Proffitt (1993) demonstrated lightness contrast as a function of whether a target gray circle was organized perceptually with black or white circles, even though the inducing circles were not in close proximity to the target. Gilchrist (1977) used depth cues to cause a piece of white paper to be perceived incorrectly as in a back chamber that was highly illuminated or correctly as in a front chamber that was dimly illuminated. When perceived as in the front chamber, the paper was seen as white; however, when perceived as in the back chamber, the paper appeared to be almost black. Adelson (1993) showed that such effects are not restricted to lightness but also occur for brightness judgments. For example, when instructed to adjust the luminance of square a_1 in Figure 8 to equal that of square a_2, observers set the luminance of a_1 to be 70% higher than that of a_2. Thus, even relatively "sensory" judgments such as brightness are affected by higher-order organizational factors.

Visual Acuity and Sensitivity to Spatial Frequency *Visual acuity* refers to the ability to perceive detail. Acuity is highest in the fovea, and it decreases with increasing eccentricities due in part to the progressively greater convergence of activity from the sensory receptors that occurs in the peripheral retina. Distinctions can be made between different types of acuity. *Identification acuity* is the most commonly measured, using a Snellen eye chart that contains rows of letters that become progressively smaller. The smallest row for which the observer can identify the letters is used as the indicator of acuity. 20/20 vision is regarded as normal. This means that the person being tested is able to identify at a distance of 20 ft letters of a size that a person with normal vision is expected to identify. A person with 20/100 vision can identify letters at 20 ft only as large as those that a person with normal vision could at 100 ft. *Vernier acuity* is a person's ability to discriminate between broken and unbroken lines, and *resolution acuity* is the ability to distinguish gratings from a stimulus that covers the same area but is of the same average intensity throughout. All of these measures are variants of *static acuity*, in that they are based on static displays. *Dynamic acuity* refers to the ability to resolve detail when there is relative motion between the stimulus and the observer. Dynamic acuity is usually poorer than static acuity (Morgan et al., 1983), partly due to an inability to keep a moving image within the fovea (Murphy, 1978). A concern in measuring acuity is that the types are not perfectly correlated, and thus an acuity measure of one type may not be a good predictor of ability to perform a task whose acuity requirements are of a different type. For example, the elderly show typically little loss of identification acuity as measured by a standard test, but they seem to have impaired acuity in dynamic situations and at low levels of illumination (Kosnik et al., 1990; Sturr et al., 1990). Thus, performance on a dynamic acuity test may be a better predictor of driving performance for elderly persons (Wood, 2002).

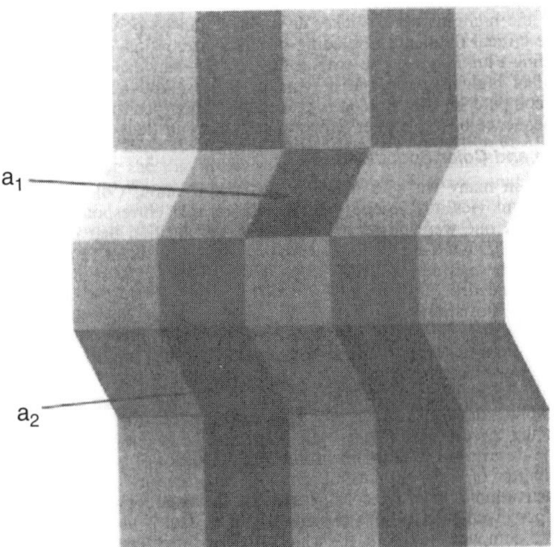

Figure 8 Effect of perceived shading on brightness judgments. The patches a_1 and a_2 are the same shade of gray, but a_1 appears much darker than a_2. (Reprinted with permission from Adelson, 1993. Copyright 1993 AAAS.)

In recent years, spatial contrast sensitivity has been shown to provide an alternative way for characterizing acuity. A *spatial contrast sensitivity function* can be generated by obtaining threshold contrast values for discriminating sine-wave gratings (for which the bars change gradually rather than sharply from light to dark) of different spatial frequencies from a homogeneous field. The contrast sensitivity function for a typical adult shows maximum sensitivity at a spatial frequency of about three to five cycles per degree of visual angle, with relatively sharp drop-offs at high and low spatial frequencies. Basically what this means is that we are not extremely sensitive to very fine or course gratings. Because high spatial frequencies pertain to the ability to perceive detail and low to intermediate frequencies to the more global characteristics of visual stimuli, tests of acuity based on contrast sensitivity may be more analytic concerning aspects of performance that are necessary for performing specific tasks. For example, Evans and Ginsburg (1982) found contrast sensitivity at intermediate and low spatial frequencies to predict the detectability of stop signs at night.

Of particular concern in human factors and ergonomics is temporal acuity. Because many light sources and displays present flickering stimulation, we need to be aware of the rates of stimulation beyond which flicker will not be perceptible. The *critical flicker frequency* is the highest rate of flicker at which it can be perceived. Numerous factors influence the critical frequency, including stimulus size, retinal location, and the level of the surrounding illumination. The critical flicker frequency can be as high as 60 Hz for large stimuli of high intensity, but it typically is less. You may have noticed that the flicker of a computer display screen is perceptible when you are not looking directly at it, as a consequence of the greater temporal sensitivity in the peripheral retina.

Color Vision and Color Specification Color is used in many ways to display visual information. Color can be used, as in television and movies, to provide a representation that corresponds to the colors that would be seen if one were physically present at the location that is depicted. Color is also used to highlight and emphasize, as well as to code different categories of displayed information. In situations such as these, we want to be sure that the colors are perceived as intended.

In a color mixing study, the observer is asked to adjust the amounts of component light sources to match the hue of a comparison stimulus. Human color vision is *trichromatic*, which means that any spectral hue can be matched by a combination of three primary colors, one each from the short-, middle-, and long-wavelength positions of the spectrum. This trichromaticity is a direct consequence of having three types of cones that contain photopigments with distinct spectral sensitivity functions. The pattern of activity generated in the three cone systems will determine what hue is perceived. A specific pattern can be determined by a monochromatic light source of a particular wavelength or by a combination of light sources of different wavelengths. As long as the relative amounts of activation in the three cone systems are the same for different physical stimuli, they will be perceived as being of the same hue. This fact is used in the design of color television sets and computer monitors, for which all colors are generated from combinations of pixels of three different colors.

Another phenomenon of additive color mixing is that blue and yellow, when mixed in approximately equal amounts, yield an achromatic (e.g., white) hue, as do red and green. This stands in contrast to the fact that combinations involving one hue from each of the two complementary pairs are seen as combinations of the two hues. For example, when blue and green are combined additively in similar amounts, the resulting stimulus appears blue-green. The pairs that yield an achromatic additive mixture are called *complementary colors*. That these hues have a special relation is evident in other situations as well. When a background is one of the hues from a pair of complementary colors, it will tend to induce the complementary hue in a stimulus that would otherwise be perceived as a neutral gray or white. Similarly, if a background of one hue is viewed for awhile and then the gaze is shifted to a background of a neutral color, an afterimage of the complementary hue will be seen.

The complementary color relations also appear to have a basis in the visual system, but in the neural pathways rather than in the sensory receptors. That is, considerable evidence indicates that output from the cones is rewired into opponent processes at the level of the ganglion cells and beyond. If a neuron's firing rate increases when a blue stimulus is presented, it decreases when a yellow stimulus is presented. Similarly, if a neuron's firing rate increases to a red stimulus, it decreases to a green stimulus. The pairings in the opponent cells always involve blue with yellow and red with green. Thus, a wide range of color appearance phenomena can be explained by the view that the sensory receptors operate trichromatically but this information is subsequently recoded into an opponent format in the sensory pathways.

The basic color-mixing phenomena are depicted in color appearance systems. A *color circle* can be formed by curving the visual spectrum, as done originally by Isaac Newton. The center of the circle represents white, and its rim represents the spectral colors. A line drawn from a particular location on the rim to the center depicts saturation, the amount of hue that is present. For example, if one picks a monochromatic light source that appears red, points on the line represent progressively decreasing amounts of red as one moves along it to the center. The appearance for a mixture of two spectral colors can be approximated by drawing a chord that connects the two colors. If the two are mixed in equal amounts, the point at the center corresponds to the mixture; if the percentages are unequal, the point is shifted accordingly toward the higher-percentage spectral color. The hue for the mixture point can be determined by drawing a diagonal through it; its hue corresponds to that of the spectral hue at the rim, and its saturation corresponds to the proximity to the rim.

The color circle is too imprecise to be used to specify color stimuli, but a system much like it, the CIE (Commission Internationale de l'Éclairage) system, is the most widely used color specification system. The CIE provided a standardized set of color matching functions, $x(\lambda)$, $y(\lambda)$, and $z(\lambda)$, called the *XYZ tristimulus coordinate system* (see Figure 9). The tristimulus values for a monochromatic stimulus can be used to determine the proportions of three wavelengths (X, Y, and Z, corresponding to red, green, and blue, respectively) needed to match it. For example, the x, y, and z tristimulus values for a 500-nm stimulus are 0.0049, 0.3230, and 0.2720. The proportion of X primary can be determined by dividing the tristimulus value for x by the combined values for x plus y plus z. The proportion of Y primary can be determined in like manner, and the proportion of Z primary is simply 1 minus the X and Y proportions. The spectral stimulus of 500 nm thus has the following proportions: $X = 0.008$, $Y = 0.539$, and $Z = 0.453$.

The *CIE color space*, shown in Figure 9, is triangular rather than circular. Location in the space is specified according to the relative amounts of the three primary colors, X, Y, and Z. Only X and Y are used as the axes for the space because X, Y, and Z sum to 1.0.

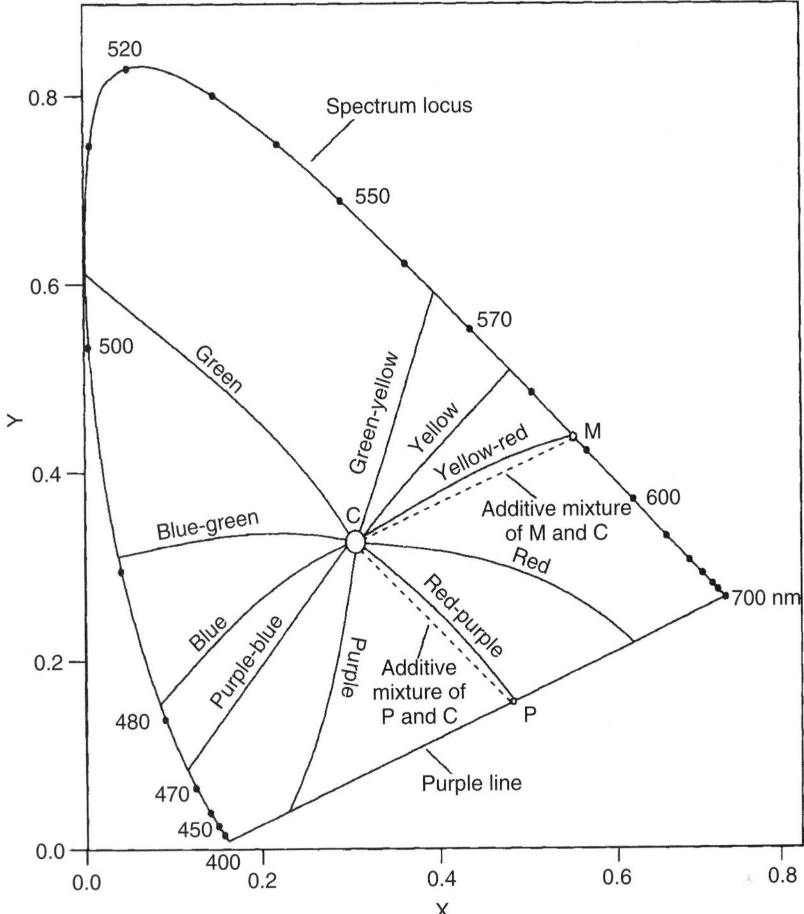

Figure 9 CIE color space. The abscissa and ordinate indicate the proportions of X and Y colors, respectively.

The spectral stimulus of 500 nm would be located on the rim of the triangle, in the upper left of the figure. Saturation decreases as proximity to the rim decreases, to an achromatic point labeled C in Figure 9. The location of mixtures of primaries in the CIE space can either be specified precisely by using the tristimulus values for each component spectral frequency or approximately by determining the coordinates at the location approximating its appearance.

Another widely used color specification system is the *Munsell Book of Colors*. This classification scheme is also a variant of the color circle, but adding in a third dimension that corresponds to lightness. In the Munsell notation, the word *hue* is used as normal, but the words *value* and *chroma* are used to refer to lightness and saturation, respectively. The book contains sheets of color samples organized according to their values on the three dimensions of hue, value, and chroma. Color can be specified by reporting the values of the sample that most closely match those of the stimulus of interest.

When using a colored stimulus, one important consideration is its location in the visual field (Gegenfurtner and Sharpe, 1999). The distribution of cones varies across the retina, resulting in variations in color perception at different retinal locations. For example, because short-wavelength cones are absent in the fovea and only sparsely distributed throughout the periphery, very small blue stimuli imaged in the fovea will be seen as achromatic and the blue component in mixtures will have little impact on the perceived hue. Cones of all three types decrease in density with increasing eccentricity, with the consequence that color perception becomes less sensitive and stimuli must be larger in order for color to be perceived. Red and green discrimination extends only 20 to 30° into the periphery, whereas yellow and blue can be seen up to 40 to 60° peripherally. Color vision is completely absent beyond that point.

Another consideration is that a significant portion of the population has *color blindness*. The most common type of color blindness is dichromatic vision. It is a gender-linked trait, with most dichromats being males. The name arises from the fact that such a person can match any spectral hue with a combination of only two primaries; in most cases this disorder can be attributed to a missing cone photopigment. The names *tritanopia, deuteranopia*, and *protanopia* refer to missing the short-, middle-, or long-wavelength pigment, respectively. The latter two types (commonly known as red–green color blindness) are much more prevalent than the former. The point to keep in mind is that colorblind persons are not able to distinguish all of the colors that a person with trichromatic vision can. Specifically, people with red–green color blindness cannot differentiate middle and long wavelengths (520 to 700 nm), and the resulting perception is composed of short (blue) versus longer (yellow) wavelength hues. O'Brien et al. (2002) found that the inability to discriminate colors in certain ranges of the spectrum reduces their conspicuity (i.e., the ability to attract attention). Deuteranopes performed significantly worse than trichromats at detecting red, orange, and green color-coded traffic control devices in briefly flashed displays, but not at detecting yellow and blue color-coded signs. Tests for color deficiencies include the Ishihara plates, which require differences in color to be perceived if test patterns are to be identified, and the Farnsworth–Munsell 100-hue test, in which colored caps are to be arranged in a continuous series about four anchor-point colors (Wandell, 1995).

3.2 Audition

3.2.1 Auditory System

The sensory receptors for hearing are sensitive to sound waves, which are moment-to-moment fluctuations in air pressure about the atmospheric level. These fluctuations are produced by mechanical disturbances, such as a stereo speaker moving in response to signals that it is receiving from a music source and amplifier. As the speaker moves forward and then back, the disturbances in the air go through phases of compression, in which the density of molecules—and hence the air pressure—is increased, and rarefaction, in which the density and air pressure decrease. With a pure tone, such as that made by a tuning fork, these changes follow a sinusoidal pattern. The frequency of the oscillations (i.e. the number of oscillations per second) is the primary determinant of the sound's pitch, and the amplitude or intensity is the primary determinant of loudness. Intensity is usually specified in *decibels* (dB), which is 20 log (p/p_0), where p is the pressure corresponding to the sound and p_0 is the standard value of 20 micropascal. When two or more pure tones are combined, the resulting sound wave will be an additive combination of the components. In that case, not only frequency and amplitude become important but also the phase relationships between the components, that is, whether the phases of the cycles for each are matched or mismatched. The wave patterns for most sounds encountered in the world are quite complex, but they can be characterized in terms of component sine waves by means of a Fourier analysis. The auditory system must perform something like a Fourier analysis, since we are capable to a large extent of extracting the component frequencies that make up a complex sound signal, so that the pitches of the component tones are heard.

Ear A sound wave propagates outward from its source at the speed of sound (344 m/s), with the amplitude proportional to $1/(\text{distance})^2$. It is the cyclical air pressure changes at the ear as the sound wave propagates past the observer that starts the sensory process. The outer ear (see Figure 10), consisting of the pinna and the auditory canal, serves to funnel the sound into the middle ear; the pinna will amplify or attenuate some sounds as a function of the

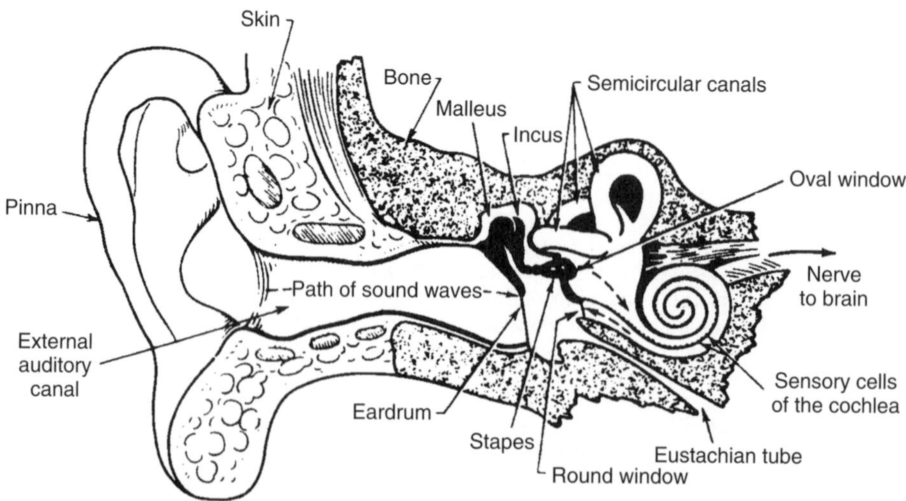

Figure 10 Structure of the human ear. (From Schiffman, 1996.)

direction from which they come and their frequency, and the auditory canal amplifies sounds around 3 kHz because that is its resonant frequency. A flexible membrane, called the *eardrum* or *tympanic membrane*, separates the outer and middle ears. The pressure in the middle ear is maintained at the atmospheric level by means of the Eustachian tube that opens into the throat, so any deviations from this pressure in the outer ear will result in a pressure differential that causes the eardrum to move. Consequently, the eardrum vibrates in a manner that mimics the sound wave that is affecting it. However, changes in altitude, such as those occurring during flight, can produce a pressure differential that impairs hearing until that differential is eliminated, which cannot occur readily if the Eustachian tube is blocked by infection or other causes.

Because the inner ear contains fluid, there is an impedance mismatch between it and the air that would greatly reduce the fluid movement if the eardrum acted on it directly. This impedance mismatch is overcome by a lever system of three bones (the ossicles) in the middle ear: the *malleus, incus,* and *stapes*. The malleus is attached to the eardrum and is connected to the stapes by the incus. The stapes has a footplate that is attached to a much smaller membrane, the oval window, which is at the boundary of the middle ear and the cochlea, the part of the inner ear that is important for hearing. Thus, when the eardrum moves in response to sound, the ossicles move, and the stapes produces movement of the oval window. Muscles attached to the ossicles tighten when sounds exceed 80 dB, thus protecting the inner ear to some extent from loud sounds by lessening their impact. However, because this acoustic reflex takes between 10 and 150 ms to occur, depending on the intensity of the sound, it does not provide protection from percussive sounds such as gunshots.

The *cochlea* is a fluid-filled, coiled structure (see Figure 11). It consists of three chambers, the vestibular and tympanic canals, and the cochlear duct, which separates them except at a small hole at the apex called the *helicotrema*. Part of the wall separating the cochlear duct from the tympanic canal is a flexible membrane called the *basilar membrane*. This membrane is narrower and stiffer nearer the oval window than it is nearer the helicotrema. The organ of Corti, the receptor organ that transduces the pressure changes to neural impulses, sits on the basilar membrane in the cochlear duct. It contains two groups of hair cells whose cilia project into the fluid in the cochlear duct and either touch or approach the tectorial membrane, which is inflexible. When fluid motion occurs in the inner ear, the basilar membrane vibrates, causing the cilia of the hair cells to be bent. It is this bending of the hair cells that initiates a neural signal. One group of hair cells, the inner cells, consists of a single row of approximately 3500; the other group, the outer cells, is composed of approximately 12,000 hair cells arranged in three to five rows. Permanent hearing loss most often is due to hair cell damage that results from excessive exposure to loud sounds or to certain drugs.

Sound causes a wave to move from the base of the basilar membrane, at the end near the oval window, to its apex. Because the width and thickness of the basilar membrane vary along its length, the magnitude of the displacement produced by this traveling wave at different locations will vary. For low-frequency sounds, the greatest movement is produced near the apex; as the frequency increases, the point of maximal displacement shifts toward the base. Thus, not only does the frequency with which the basilar membrane vibrates vary with the frequency of the auditory stimulus, but so does the location.

Auditory Pathways The auditory pathways after sensory transduction show many of the same properties as the visual pathways. The hair cells have synapses with the neurons that make up the auditory nerve.

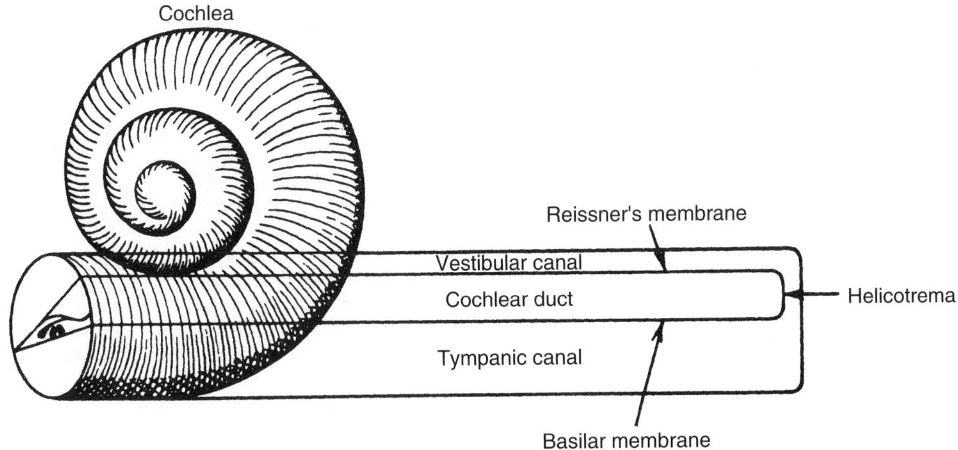

Figure 11 Schematic of the cochlea uncoiled to show the canals. (From Schiffman, 1996.)

The ratio of hair cells to auditory nerve fibers is much greater for the inner hair cells than for the outer hair cells, suggesting that the inner hair cells provide the detailed frequency information for hearing. The neurons in the auditory nerve show frequency tuning. Each has a preferred or characteristic frequency but will fire less strongly to a range of frequencies about the preferred one. Neurons can be found with characteristic frequencies for virtually every frequency in the range of hearing. The contour depicting sensitivity of a neuron to different tone frequencies is called a *tuning curve*. The tuning curves typically are broad, indicating that a neuron is sensitive to a broad range of values, but asymmetric: The sensitivity to frequencies higher than the characteristic frequency is much less than that to frequencies below it. With frequency held constant, there is a dynamic range over which as intensity is increased the neuron's firing rate will increase. This dynamic range is on the order of 25 dB, which is considerably less than the full range of intensities that we can perceive.

The first synapse for the nerve fibers after the ear is the cochlear nucleus, After that point, two separate pathways emerge that seem to have different roles, as in vision. Fibers from the anterior cochlear nucleus go to the superior olive, half to the contralateral side of the brain and half to the ipsilateral side, and then on to the inferior colliculus. This pathway is presumed to be involved in the analysis of spatial information. Fibers from the posterior cochlear nucleus project directly to the contralateral inferior colliculus. This pathway analyzes the frequency of the auditory stimulus. From the inferior colliculus, most of the neurons project to the medial geniculate and then to the auditory cortex. Frequency tuning is evident for neurons in all of these regions, with some neurons responding to relatively complex features of stimulation. The auditory cortex has a tonotopic organization, in which cells responsive to similar frequencies are located in close proximity, and contains neurons tuned to extract complex information.

3.2.2 Basic Auditory Perception

Loudness and Detection of Sounds Loudness for audition is the equivalent of brightness for vision. More intense auditory stimuli produce greater amplitude of movement in the eardrum, which produces higher-amplitude movement of the stapes on the oval window, which leads to bigger waves in the fluid of the inner ear, and hence higher-amplitude movements of the basilar membrane. Thus, loudness is primarily a function of the physical intensity of the stimulus and its effects on the ear, although as with brightness, it is affected by many other factors. The relation between judgments of loudness and intensity follows the power function

$$L = aI^{0.6}$$

where L is loudness, a is a constant, and I is physical intensity.

Just as brightness is affected by the spectral properties of light, loudness is affected by the spectral properties of sound. Figure 12 shows *equal loudness contours* for which a 1000-Hz tone was set at a particular intensity level and tones of other frequencies were adjusted to match its loudness. The contours illustrate that humans are relatively insensitive to low-frequency tones below approximately 200 Hz and, to a lesser extent, to high-frequency tones exceeding approximately 6000 Hz. The curves tend to flatten at high intensity levels, particularly in the low-frequency end, indicating that the insensitivity to low-frequency tones is a factor primarily at low intensity levels. This is why most audio amplifiers include a "loudness" switch for enhancing low-frequency sounds artificially when music is played at low intensities. The curves also show the maximal sensitivity to be in the range 3000 to 4000 Hz, which is critical for speech perception. The two most widely cited sets of equal loudness contours are those of Fletcher and Munson (1933), obtained when listening through earphones, and of Robinson and Dadson (1956), obtained for free-field listening.

Temporal summation can occur over a period of approximately 200 ms, meaning that loudness is a function of the total energy presented for tones of this duration or less. The bandwidth (i.e., the range of the frequencies in a complex tone) is important for determining its loudness. With the intensity held constant, increases in bandwidth have no effect on loudness until a critical bandwidth is reached. Beyond the critical bandwidth, further increases in bandwidth result in increases in loudness.

Extraneous sounds in the environment can mask targeted sounds. This becomes important for situations such as work environments, in which audibility of specific auditory input must be evaluated with respect to the level of background noise. The degree of masking is dependent on the spectral composition of the target and noise stimuli. Masking occurs only from frequencies within the critical bandwidth. Of concern for human factors is that a masking noise will exert a much greater effect on sounds of higher frequency than on sounds of lower frequency. This asymmetry is presumed to arise primarily from the operation of the basilar membrane.

Pitch Perception Pitch is the qualitative aspect of sound that is a function primarily of the frequency of a periodic auditory stimulus. The higher the frequency, the higher the pitch. The pitch of a note played on a musical instrument is determined by what is called its *fundamental frequency*, but the note also contains energy at frequencies that are multiples of the fundamental frequency, called *harmonics* or *overtones*. Observers can resolve perceptually the lower harmonics of a complex tone but have more difficulty resolving the higher harmonics (Plomp, 1964). This is because the perceptual separation of the successive harmonics is progressively less as their frequency increases.

Pitch is also influenced by several factors in addition to frequency. A phenomenon of particular

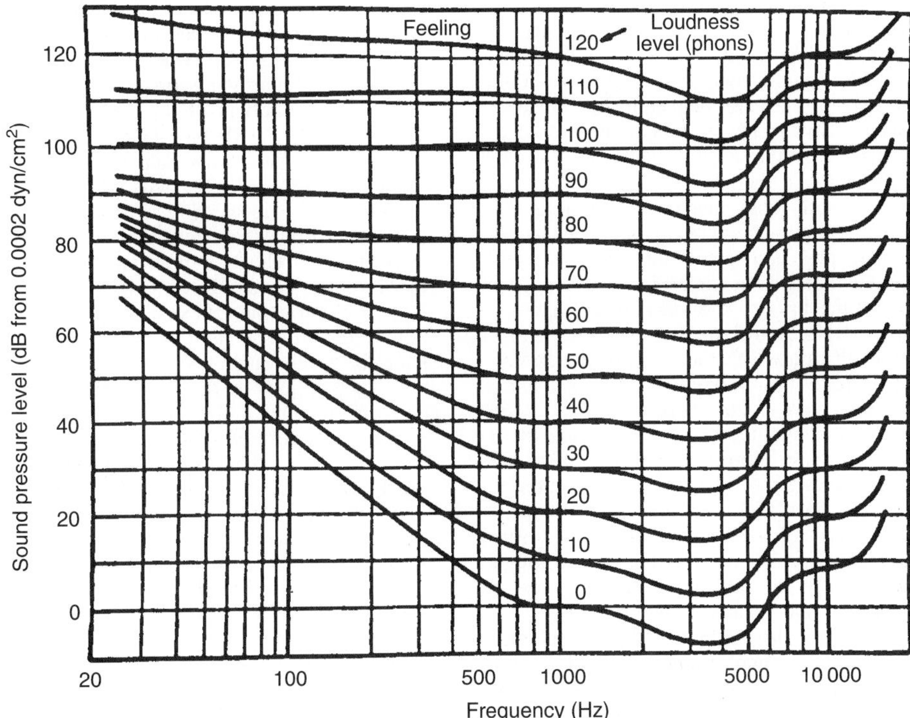

Figure 12 Equal loudness contours. Each contour represents the sound pressure level at which a tone of a given frequency sounds as loud as a 1000-Hz tone of a particular intensity. (From Schiffman, 1996.)

interest in human factors is that of the missing fundamental effect. Here, the fundamental frequency can be removed, yet the pitch of a sound remains unaltered. This suggests that pitch is based on the pattern of harmonics and not just the fundamental frequency. This phenomenon allows a person's voice to be recognizable over the telephone and music to be played over low-fidelity systems without distorting the melody. The pitch of a tone also varies as a function of its loudness. Equal pitch contours can be constructed much like equal loudness contours by holding the stimulus frequency constant and varying its amplitude. Such contours show that as stimulus intensity increases, the pitch of a 3000-Hz tone remains relatively constant. However, tones whose frequencies are lower or higher than 3000 Hz show systematic decreases and increases in pitch, respectively, as intensity increases.

Two different theories were proposed in the nineteenth century to explain pitch perception. According to Ernest Rutherford's (1886) *frequency theory*, the critical factor is that the basilar membrane vibrates at the frequency of an auditory stimulus. This in turn gets transduced into neural signals at the same frequency such that the neurons in the auditory nerve respond at the frequency of the stimulus. Thus, according to this view, it is the frequency of firing that is the neural code for pitch. The primary deficiency of frequency theory is that the maximum firing rate of a neuron is restricted to about 1000 spikes/s. Thus, the firing rate of individual neurons cannot match the frequencies over much of the range of human hearing. Wever and Bray (1937) provided evidence that the range of the auditory spectrum over which frequency coding could occur can be increased by neurons that phase lock and then fire in volleys. The basic idea is that an individual neuron fires at the same phase in the cycle of the stimulus but not on every cycle. Because many neurons are responsive to the stimulus, some neurons will fire on every cycle. Thus, across the group of neurons, distinct volleys of firing will be seen that when taken together match the frequency of the stimulus. Phase locking extends the range for which frequency coding can be effective up to 4000 to 5000 Hz. However, at frequencies beyond this range, phase locking breaks down.

According to Hermann von Helmholtz's (1877) *place theory*, different places on the basilar membrane are affected by different frequencies of auditory stimulation. He based this proposal on his observation that the basilar membrane was tapered from narrow at the base of the cochlea to broad at its apex. This led him to suggest that it was composed of individual fibers, much like piano strings, that would resonate when the frequency of sound to which it was tuned occurred. The neurons that receive their input from a location on the membrane affected by a particular frequency would fire in its presence, whereas the neurons receiving their input from other locations would not. The neural code for frequency thus would correspond to the particular neurons that were being stimulated. However, subsequent physiological evidence showed that the basilar membrane is not composed of individual fibers.

Von Békésy (1960) provided evidence that the basilar membrane operates in a manner consistent with both frequency and place theory. Basically, he demonstrated that waves travel down the basilar membrane from the base to the apex at a frequency corresponding to that of the tone. However, because the width and thickness of the basilar membrane vary along its length, the magnitude of the *traveling wave* is not constant over the entire membrane. The waves increase in magnitude up to a peak and then decrease abruptly. Most important, the location of the peak displacement varies as a function of frequency. Low frequencies have their maximal displacement at the apex; as frequency increases, the peak shifts systematically toward the oval window. Although most frequencies can be differentiated in terms of the place at which the peak of the traveling wave occurs, tones of less than 500 to 1000 Hz cannot be. Frequencies in this range produce a broad pattern of displacement, with the peak of the wave at the apex. Consequently, location coding does not seem to be possible for low-frequency tones. Because of the evidence that frequency and location coding both operate but over somewhat different regions of the auditory spectrum, it is now widely accepted that frequencies less than 4000 Hz are coded in terms of frequency and those above 500 Hz in terms of place, meaning that at frequencies within this range, both mechanisms are involved.

3.3 Vestibular System and the Sense of Balance

The vestibular system provides us with our sense of balance. It contributes to the perception of bodily motion and helps in maintaining an upright posture and the position of the eyes when head movements occur (Highstein et al., 2004). The sense organs for the vestibular system are contained within a part of the inner ear called the *vestibule*, which is a hollow region of bone near the oval window (Goldberg and Hudspeth, 2000). The vestibular system includes the *otolith organs*, one called the *utricle* and the other the *saccule*, and three semicircular ducts (see Figure 10). The otolith organs provide information about the direction of gravity and linear acceleration. The sensory receptors are hair cells lining the organs whose cilia are embedded in a gelatinlike substance that contains otoliths, which are calcium carbonate crystals. Tilting or linear acceleration of the head in any direction causes a shearing action of the otoliths on the cilia in the utricle, and vertical linear acceleration has the same effect in the sacule. The semicircular canals are placed in three perpendicular planes. They also contain hair cells that are stimulated when relative motion between the fluid inside them and the head is created, and thus respond primarily to angular acceleration or deceleration in specific directions.

The vestibular ganglion contains the cell bodies of the afferent fibers of the vestibular system. They project to the vestibular nuclear complex, a part of the medulla that is made up of four distinct nuclei, the lateral vestibular nucleus, the medial vestibular nucleus, the superior vestibular nucleus, and the inferior vestibular nucleus. Each of these nuclei serves a distinct role. The lateral nucleus seems to be involved in the control of posture, the medial and superior nuclei in vestibulo-ocular reflexes, and the inferior nucleus with integration of the vestibular input with inputs from the cerebellum.

Two functions of the vestibular system, one static and one dynamic, can be distinguished. The *static function*, performed primarily by the utricle and saccule, is to monitor the position of the head in space, which is important in the control of posture. The *dynamic function*, performed primarily by the semicircular ducts, is to track the rotation of the head in space. This tracking is necessary for reflexive control of what are called *vestibular eye movements*. If you maintain fixation on an object while rotating your head, the position of the eyes in the sockets will change gradually as the head moves. When your nose is pointing directly toward the object, the eyes will be centered in their sockets, but as you turn your head to the right, the eyes will rotate to the left, and vice versa as the head is turned to the left. These smooth, vestibular eye movements are controlled rapidly and automatically by the brain stem in response to sensing of the head rotation by the vestibular system.

Exposure to motions that have angular and linear accelerations substantially different from those normally encountered, as occurs in aircraft, space vehicles, and ships, can produce erroneous perceptions of attitude and angular motion that result in *spatial disorientation* (Benson, 1990). Spatial disorientation accounts for approximately 35% of all general aviation fatalities, with most occurring at night when visual cues are either absent or degraded and vestibular cues must be relied on heavily. The vestibular sense also is key to producing *motion sickness*, as indicated by the fact that people who do not have a functional vestibular apparatus do not show motion sickness. The dizziness and nausea associated with motion sickness are generally assumed to arise from a mismatch between the motion cues provided by the vestibular system, and possibly vision, with the expectancies of the central nervous system. The vestibular sense also contributes to the related problem of simulator sickness that arises when the visual cues in a simulator or virtual reality environment do not correspond well with the motion cues that are affecting the vestibular system (Draper et al., 2001).

3.4 Somatic Sensory System

The somatic sensory system is composed of four distinct modalities (Gardner et al., 2000). *Touch* is the sensation elicited by mechanical stimulation of the skin; *proprioception* is the sensation elicited by mechanical displacements of the muscles and joints; *pain* is elicited by stimuli of sufficient intensity to damage tissue; and *thermal sensations* are elicited by cool and warm stimuli. The receptors for these senses are the terminals of the peripheral branch of the axons of ganglion cells located in the dorsal root of the spinal cord. The receptors for pain and temperature, called *nociceptors* and *thermoreceptors*, are bare (or free) nerve endings. Three types of nociceptors exist that respond to different types of stimulation. Mechanical nociceptors respond to strong mechanical stimulation; thermal nociceptors respond to extreme heat or cold, and polymodal nociceptors respond to several types of intense stimuli. Distinct thermoreceptors exist for cold and warm stimuli. Those for cold stimuli respond to temperatures between 1 and 20°C below skin temperature, whereas those for warm stimuli respond to temperatures up to 13°C warmer than skin temperature,

The mechanoreceptors for touch have specialized endings that affect the dynamics of the receptor to stimulation. Some mechanoreceptor types are rapidly adapting and respond at the onset and offset of stimulation, whereas others are slow adapting and respond throughout the time that a touch stimulus is present. Hairy skin is innervated primarily by hair follicle receptors. Hairless (glabrous skin) receives innervation from two types: Meissner's corpuscles, which are fast adapting, and Merkel's disks, which are slow adapting, Pacinian corpuscles, which are fast adapting, and Ruffini's corpuscles, which are slow adapting, are located in the dermis, subcutaneous tissue that is below both the hairy and glabrous skin.

The nerve fibers for the skin senses have a center-surround organization of the type found for vision. The receptive fields for the Meissner corpuscles and Ruffini disks are smaller than those for the Pacinian and Ruffini corpuscles, suggesting that the former provide information about fine spatial differences and the latter about coarse spatial differences. The density of mechanoreceptors is greatest for those areas of the skin, such as the fingers and lips, for which two-point thresholds (i.e., the amount of difference needed to tell that two points rather than one are being stimulated) are low. Limb proprioception is mediated by three types of receptors: mechanoreceptors located in the joints, muscle spindle receptors in muscles that respond to stretch, and cutaneous mechanoreceptors. The ability to specify limb positions decreases when the contribution of any of these receptors is removed through experimental manipulation.

The afferent fibers enter the spinal cord at the dorsal roots and follow two major pathways, the dorsal-column medial-lemniscal pathway and the anterolateral (or spinothalamic) pathway. The lemniscal pathway conveys information about touch and proprioception. It receives input primarily from fibers with corpuscles and transmits this information quickly. It ascends along the dorsal part of the spinal column, on the ipsilateral side of the body. At the brainstem, most of its fibers cross over to the contralateral side of the brain and project to the medial lemniscus in the thalamus, and from there to the anterior parietal cortex. The fibers in the anterolateral pathway ascend along the contralateral side of the spinal column and project to the reticular formation, midbrain, or thalamus, and then to the anterior parietal cortex and other cortical regions. This system is primarily responsible for conveying pain and temperature information.

The somatic sensory cortex is organized in a spatiotopic manner, much as is the visual cortex. That is, it is laid out in the form of a homunculus representing the opposite side of the body, with areas of the body for which sensitivity is greater, such as the fingers and lips, having relatively larger areas devoted to them. There are four different, independent spatial maps of this type in the somatic sensory cortex, with each map receiving its inputs primarily from the receptors for one of the four somatic modalities. The modalities are arranged into columns, with any one column receiving input from the same modality. When a specific point on the skin is stimulated, the population of neurons that receive innervation from that location will be activated. Each neuron has a concentric excitatory–inhibitory center-surround receptive field, the size of which varies as a function of the location on the skin. The receptive fields are smaller for regions of the body in which sensitivity to touch is highest. Some of the cells in the somatic cortex respond to complex features of stimulation, such as movement of an object across the skin.

Vibrotaction has proven to be an effective way of transmitting complex information through the tactile sense (Verrillo and Gescheider, 1992). When mechanical vibrations are applied to a region of skin such as the tips of the fingers, the frequency and location of the stimulation can be varied. For frequencies below 40 Hz, the size of the contactor area does not influence the absolute threshold for detecting vibration. For higher frequencies, the threshold decreases with increasing size of the contactor, indicating spatial summation of the energy within the stimulated region. Except for very small contactor areas, sensitivity reaches a maximum for vibrations of 200 to 300 Hz. A similar pattern of less sensitivity for low-frequency vibrations than for high-frequency vibrations is evident in equal sensation magnitude contours (Verrillo et al., 1969), much like the equal-loudness contours for audition. With multicontactor devices, which can present complex spatial patterns of stimulation, masking stimuli presented in close temporal proximity to the target stimulus can degrade identification (e.g., Craig, 1982), as in vision and audition. However, with practice, pattern recognition capabilities with these types of devices can become quite good. As a result, they can be used successfully as reading aids for the blind and to a lesser extent as hearing aids for the hearing impaired (Summers, 1992). Hollins et al. (2002) provide evidence that vibrotaction also plays a necessary and sufficient role in the perception of fine tactile textures.

A distinction is commonly made between active and passive touch (Gibson, 1966; Katz, 1989). *Passive touch* refers to situations in which a person does not move her or his hand, and the touch stimulus is applied passively, as in vibrotaction. *Active touch* refers to situations in which a person moves his or her hand intentionally to manipulate and explore an object. According to Gibson, active touch is the most common mode of acquiring tactile information in the real world and involves a unique perceptual system, which he called *haptics*. Pattern recognition with active touch typically is superior to that with passive touch (Appelle, 1991). However, the success of passive vibrotactile displays for the blind indicates that much information can also be conveyed passively. Passive and active touch can combine in a third type of touch, called *intra-active touch*, in which one body part is used to provide active stimulation to another body part, as when using a finger to roll a ball over the thumb (Bolanowski et al., 2004).

3.5 Gustation and Olfaction

Smell and taste are central to human perceptual experience. The taste of a good meal and the smell of perfume can be quite pleasurable. On the other hand, the taste of rancid potato chips or the smell of manure or of a paper mill can be quite noxious. In fact, odor and taste are quite closely related, in that the taste of a substance is highly dependent on the odor it produces. This is evidenced by the changes in taste that occur when a cold reduces olfactory sensitivity. In human factors, both sensory modalities can be used to convey warnings. For example, ethylmercaptan is added to natural gas to warn of gas leaks because humans are quite sensitive to its odor. Also, as mentioned in Section 2.2.2, there is concern with environmental odors and their influence on people's moods and performance.

The sensory receptors for taste are groups of cells called *taste buds*. They line the walls of bumps on the tongue that are called *papillae*, as well as being located in the throat, the roof of the mouth, and inside the cheeks. Each taste bud is composed of several receptor cells in close arrangement. The receptor mechanism is located in projections from the top end of each cell that lie near an opening called a *taste pore*. Sensory transduction occurs when a taste solution comes in contact with the projections. The fibers from the taste receptors project to several nuclei in the brain and then to the insular cortex, located between the temporal and parietal lobes, and the limbic system.

In 1916, Henning proposed a taste tetrahedron in which all tastes were classified in terms of four primary tastes: sweet, sour, salty, and bitter. This categorization scheme has been accepted since then, although not without opposition. Schiffman and Erickson (1993) summarize research indicating that there are many sensations that fall outside the range of these four tastes. They suggest that there is a broad range of transduction mechanisms, including ion channels and transepithelial currents, in addition to the taste bud receptors.

For smell, molecules in the air that are inhaled affect receptor cells located in the *olfactory epithelium*, a region of the nasal cavity. An *olfactory rod* extends from each receptor and goes to the surface of the epithelium. Near the end of the olfactory rod is a knob from which olfactory cilia project. These cilia are thought to be the receptor elements. Different receptor types apparently have different receptor proteins that bind the odorant molecules to the receptor. The axons from the smell receptors project to the olfactory bulb, located in the front of the brain, via the olfactory nerve. From there, the fibers project to a cluster of neural structures called the *olfactory brain*.

Olfaction shows several functional attributes (Engen, 1991). For one, a novel odor will almost always cause apprehension and anxiety. As a consequence, odors are useful as warnings. However, odors are not very effective at waking someone from sleep, which is illustrated amply by the need for smoke detectors that emit a loud auditory signal, even though the smoke itself has a distinctive odor. There also seems to be a bias to falsely detect the presence of odors and to overestimate the strength when the odor is present. Such a bias ensures that a miss is unlikely to occur when an odor signal is really present. The sense of smell shows considerable plasticity, with associations of odors to events readily learned and habituation occurring to odors of little consequence. Doty (2003) and Rouby et al. (2002) provide detailed treatment of the perceptual and cognitive aspects of smell and taste.

4 HIGHER-LEVEL PROPERTIES OF PERCEPTION

4.1 Perceptual Organization

The stimulus at the retina consists of patches of light energy that affect the photoreceptors. Yet we do not perceive patches of light. Rather, we perceive a structured world of meaningful objects. The organizational processes that affect perception go unnoticed in everyday life, until we encounter a situation in which we initially misperceive the situation in some way. When we realize this and our perception now is more veridical, we become aware that the organizational processes can be misled.

Perceptual organization is particularly important for the design of any visual display. If a symbol on a street sign is organized incorrectly, it may well go unrecognized. Similarly, if a warning signal is grouped perceptually with other displays, its message may be lost. The investigation of perceptual organization was initiated by a group of German psychologists called *Gestalt psychologists*, whose mantra was, "The whole is more than the sum of the parts." The demonstrations they provided to illustrate this point were sufficiently compelling that the concept is now accepted by all perceptual psychologists.

According to the Gestalt psychologists, the overriding principle of perceptual organization, is that of *prägnanz*. The basic idea of this law is that the organizational processes will produce the simplest possible organization allowed by the conditions (Palmer, 2003). The first step in perceiving a figure requires that it be

Figure 13 Ruben's vase, for which two distinct figure–ground organizations are possible. (From Schiffman, 1996.)

separated from the background. Any display that is viewed will be seen as a figure or figures against a background. The importance of *figure–ground* organization is illustrated clearly in figures with ambiguous figure–ground organizations, as in the well-known Ruben's vase (see Figure 13). Such figures can be organized with either the light region or the dark region seen as figure. When a region is seen as figure, the contour appears to be part of it. Also, the region seems to be in front of the background and takes on a distinct form. When the organization changes so that the region is now seen as the ground, its perceived relation with the other region reverses.

Clearly, when designing displays, one wants to construct them such that the figure–ground organization of the observer will correspond with what is intended. Fortunately, research has indicated factors that influence figure–ground organization. Symmetric patterns tend to be seen as figure over asymmetric ones; a region that is surrounded completely by another tends to be seen as figure and the surrounding region as background; convex contours tend to be seen as figure in preference to concave contours; the smaller of two regions tends to be seen as figure and the larger as ground; and a region oriented vertically or horizontally will tend to be seen as figure relative to one that is not so oriented.

In addition to figure–ground segregation being crucial to perception, the way that the figure is organized is important as well (see Figure 14). The most widely recognized *grouping principles* are *proximity*, display elements that are located close together will tend to be grouped together; *similarity*, display elements that are similar in appearance (e.g., orientation or color, will tend to be grouped together); *continuity*, figures will tend to be organized along continuous contours; *closure*, display elements that make up a closed figure will tend to be grouped together; and *common fate*, elements with a common motion will tend to be grouped together. Differences in orientation of stimuli seem to provide a particularly distinctive basis for grouping. As illustrated in Figure 15, when stimuli differ in orientation, those of like orientation are grouped and

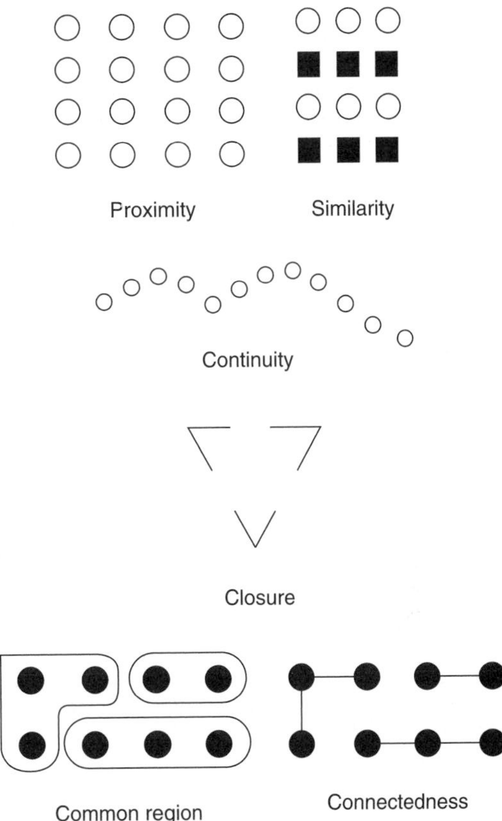

Figure 14 Gestalt organizational principles. (From Proctor and Van Zandt, 1994.)

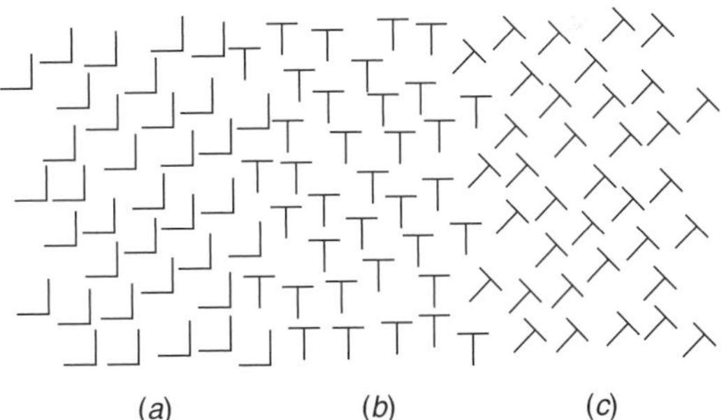

Figure 15 Tilted T's group appears more distinct from upright T's than do backward L characters.

perceived separately from those of a different orientation. This relation lies behind the customary recommendation that displays for check reading be designed so that the pointers on the dials all have the same orientation when working properly. When something is not right, the pointer on the dial will be at an orientation different from that of the rest of the pointers, and it will "jump out" at the operator.

Two additional grouping principles (see Figure 14) have been described by Rock and Palmer (1990). The principle of connectedness is that lines drawn between some elements but not others will cause the connected elements to be grouped perceptually. The principle of common region is that a contour drawn around display elements will cause those elements to be grouped together. Palmer (1992) has demonstrated several important properties of grouping by common region. When multiple, conflicting regions are present, the smaller enclosing region seems to dominate the organization; for nested, consistent regions, the organization appears to be hierarchical. Grouping by common region breaks down when the elements and background region are at different perceived depths, as does grouping by proximity (Rock and Brosgole, 1964), suggesting that such grouping occurs relatively late in processing, after at least some depth perception has occurred.

Although most work on perceptual organization has been conducted with visual stimuli, there are numerous demonstrations that the principles apply as well to auditory stimuli (Julesz and Hirsh, 1972). Grouping by similarity is illustrated in a study by Bregman and Rudnicky (1975) in which listeners had to indicate which of two tones of different frequency occurred first in a sequence. When the two tones were presented in isolation, performance was good. However, when preceded and followed by a single occurrence of a distractor tone of lower frequency, performance was relatively poor. The important finding is that when several occurrences of the distractor tone preceded and followed the critical pair, performance was just as good as when the two tones were presented in isolation. Apparently, the distractor tones were grouped as a distinct stream based on their frequency similarity. Grouping of tones occurs not only with respect to frequency but also on the basis of similarities of their spatial positions, similarities in the fundamental frequencies and harmonics of complex tones, and so on (Bregman, 1990, 1993).

Another distinction that has received considerable interest over the past 30 years is that between *integral* and *separable* stimulus dimensions (Garner, 1974). The basic idea is that stimuli composed from integral dimensions are perceived as unitary wholes, whereas stimuli composed from separable dimensions are perceived in terms of their distinct dimensions. The operations used to distinguish between integral and separable dimensions are that (1) direct similarity scaling should produce a Euclidean metric for integral dimensions (i.e., the psychological distance between two stimuli should be the square root of the sum of the squares of the differences on each dimension) and a city-block metric for separable dimensions (i.e., the psychological distance should be the sum of the differences on the two dimensions); and (2) in free perceptual classification tasks, stimuli from sets with integral dimensions should be classified together if they are close in terms of the Euclidean metric (i.e., in overall similarity), whereas those from sets with separable dimensions should be classified in the same category if they match on one of the dimensions (i.e., the classifications should be in terms of dimensional structure; Garner, 1974). Perhaps most important for human factors, speed of classification with respect to one dimension is unaffected by its relation to the other dimension if the dimensions are separable but shows strong dependencies if they are integral. For integral dimensions, classifications are slowed when the value of the irrelevant dimension is uncorrelated with the value of the relevant dimension but speeded when the two dimensions are correlated.

Based on these criteria, dimensions such as hue, saturation, and lightness, in any combination, or pitch and loudness have been classified as integral; size and lightness or size and angle are classified as separable (e.g., Shepard, 1991). A third classification, called *configural dimensions* (Pomerantz, 1981), has been proposed for dimensions that maintain their separate codes but have a new relational feature that emerges from their specific configuration. For example, as illustrated in Figure 16, a diagonally oriented line can be combined with the context of two other lines to yield an emergent triangle. Configural dimensions behave much like integral dimensions in speeded classification tasks, although the individual dimensions are still relatively accessible. Potts et al. (1998) presented evidence that the distinction between interacting (integral and configural) and noninteracting (separable) dimensions may be oversimplified. They found that with some instructions and spatial arrangements, the dimensions of circle size and tilt of an enclosed line behaved as if they were separable, whereas under others they behaved as if they were integral. Thus, Potts et al. suggest that specific task contexts increase or decrease the salience of dimensional structures and may facilitate or interfere with certain processing strategies.

Wickens and his colleagues have extended the distinction between interactive dimensions (integral and configural) and separable dimensions to display design by advocating what they call the *proximity compatibility principle* (e.g., Wickens and Carswell, 1995). This principle states that if a task requires that information be integrated mentally (i.e., processing proximity

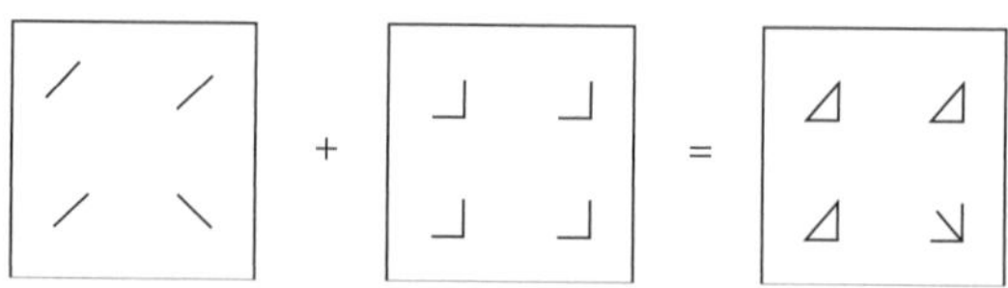

Figure 16 Configural dimensions. The bracket context helps in discriminating the line whose slope is different from the rest.

is high), that information should be presented in an integral or integrated display (i.e., one with high display proximity). High display proximity can be accomplished by, for example, increasing the spatial proximity of the display elements, integrating the elements so that they appear as a distinct object, or combining them in such a way as to yield a new configural feature. The basic idea is to replace the cognitive computations that the operator must perform to combine the separate pieces of information with a much less mentally demanding pattern recognition process. The proximity compatibility principle also implies that if a task requires that the information be kept distinct mentally (i.e., processing proximity is low), the information should be presented in a display with separable dimensions (i.e., one with low display proximity). However, the cost of high display proximity for tasks that do not require integration of displayed information is typically much less than that associated with low display proximity for tasks that do require information integration.

4.2 Spatial Orientation

We live in a three-dimensional world and hence must be able to perceive locations in space relatively accurately if we are to survive. Many sources of information come into play in the perception of distance and spatial relations (Proffitt and Caudek, 2003), and the consensus view is that the perceptual system constructs the three-dimensional representation using this information as cues.

4.2.1 Visual Depth Perception

Vision is a strongly spatial sense and provides us with the most accurate information regarding spatial location. In fact, when visual cues regarding location conflict with those from the other senses, the visual sense typically wins out, a phenomenon called *visual dominance*. There are several areas of human factors in which we need to be concerned about visual depth cues. For example, accurate depth cues are crucial for situations in which navigation in the environment is required; misleading depth cues at a landing strip at an airfield may cause a pilot to land short of the runway. For another, a helmet-mounted display, viewed through a monocle, will eliminate binocular cues and possibly provide information that conflicts with that seen by the other eye. As a final example, it may be desired that a simulator depict three-dimensional relations relatively accurately on a two-dimensional display screen.

One distinction that can be made is between oculomotor cues and visual cues. The *oculomotor cues* are accommodation and vergence angle, both of which we discussed earlier in the chapter. At relatively close distances, vergence and accommodation will vary systematically as a function of the distance of the fixated object from the observer. Therefore, either the signal sent from the brain to control accommodation and vergence angle or feedback from the muscles could provide cues to depth. However, Proffitt and Caudek (2003) conclude that neither oculomotor cue is a particularly effective cue for perceiving absolute depth and that they are easily overridden when other depth cues are available.

Visual cues can be partitioned into binocular and monocular cues. The binocular cue is *retinal disparity*, which arises from the fact that the two eyes view an object from different locations. An object that is fixated falls on corresponding points of the retinas. This object can be regarded as being located on an imaginary curved plane, called the *horopter*; and any other object that is located on this plane will also fall on corresponding points. For objects that are not on the horopter, the images will fall on disparate locations of the retinas. The direction of disparity, uncrossed or crossed (i.e., whether the image from the right eye is located to the right or left of the image from the left eye), is a function of whether the object is in back of or in front of the horopter, respectively, and the magnitude of disparity is a function of how far the object is from the horopter. Thus, retinal disparity provides information with regard to the locations of objects in space with respect to the surface that is being fixated.

Retinal disparity is a strong cue to depth, as witnessed by the effectiveness of three-dimensional (3D) movies and stereoscopic static pictures, which are created by presenting slightly different images to the two eyes to create disparity cues. Anyone who has seen the 3D Muppet Movie at Disney's MGM Studios or California Adventure realizes how compelling these effects can be. In addition to enhancing the perception of depth relations in displays of naturalistic scenes, stereoptic displays may be of value in assisting scientists and others in evaluating multidimensional data sets. Wickens et al. (1994) found that a three-dimensional data set could be processed faster and more accurately to answer questions that required integration of the information if the display was stereoptic than if it was not.

The fundamental problem for theories of stereopsis is that of matching. Disparity can be computed only after corresponding features at the two eyes have been identified. When viewing the natural world, each eye receives the information necessary to perceive contours and identify objects, and stereopsis could occur after monocular form recognition. However, one of the more striking findings of the past 35 years is that there do not have to be contours present in the images seen by the individual eyes in order to perceive objects in three dimensions. This phenomenon was discovered by Julesz (1971), who used random dot stereograms in which a region of dot densities is shifted slightly in one image relative to the other. Although a form cannot be seen if only one of the two images is viewed, when each of the two images is presented to the respective eyes, a three-dimensional form emerges. Random dot stereograms have been popularized recently through figures that utilize the autostereogram variation of this technique, in which the disparity information is incorporated in a single, two-dimensional display. That stereopsis can occur with random dot stereograms suggests that matching of the two images can be based on dot densities.

There are many static, or pictorial, monocular cues to depth. These cues are such that people with only one eye and those who lack the ability to detect disparity differences are still able to interact with the world with relatively little loss in accuracy. The monocular cues include retinal size (i.e., larger images appear to be closer) and familiar size (e.g., a small image of a car provides a cue that the car is far away). The cue of interposition is that an object that appears to block part of the image of another object located in front of it. Although interposition provides information that one object is nearer than another, it does not provide information about how far apart they are. Another cue comes from shading. Because light sources typically project from above, as with the sun, the location of a shadow provides a cue to depth relations. A darker shading at the bottom of a region implies that the region is elevated, whereas one at the top of a region provides a cue that it is depressed. Aerial perspective refers to blue coloration, which appears for objects that are far away, such as is seen when viewing a mountain at a distance. Finally, the cue of linear perspective occurs when parallel lines receding into the distance, such as train tracks, converge to a point in the image.

Gibson (1950) emphasized the importance of texture gradient, which is a combination of linear perspective and relative size, in depth perception. If one looks at a textured surface such as a brick walkway, the parts of the surface (i.e., the bricks) become smaller and more densely packed in the image as they recede into the distance. The rate of this change is a function of the orientation of the surface in depth with respect to the line of sight. This texture change specifies distance on the surface, and an image of a constant size will be perceived to come from a larger object that is farther away if it occludes a larger part of the texture. Certain color gradients, such as a gradual change from red to gray, provide effective cues to depth as well (Truscianko et al., 1991).

For a stationary observer, there are plenty of cues to depth. However, cues become even richer once the observer is allowed to move. When you maintain fixation on an object and change locations, as when looking out a train window, objects in the background will move in the same direction in the image as you are moving, whereas objects in the foreground will move in the opposite direction. This cue is called *motion parallax*. When you move straight ahead, the optical flow pattern conveys information about how fast your position is changing with respect to objects in the environment. There are also numerous ways in which displays with motion can generate depth perception (Braunstein, 1976).

Of particular concern for human factors is how the various depth cues are integrated. Bruno and Cutting (1988) varied the presence or absence of four cues: relative size, height in the projection plane, interposition, and motion parallax. They found that the four cues combined additively in one direct and two indirect scaling tasks. That is, each cue supported depth perception, and the more cues that were present, the more depth was revealed. Bruno and Cutting interpreted these results as suggesting that a separate module processes each source of depth information. Landy et al. (1995) have developed a detailed model of this general nature, according to which interactions among depth cues occur for the purpose of establishing for each cue a map of absolute depth throughout the scene. The estimate of depth at each location is determined by taking a weighted average of the estimates provided by the individual cues.

Because the size of the retinal image of an object varies as a function of the distance of the object from the observer, perception of size is intimately related to perception of distance. When accurate depth cues are present, good *size constancy* results. That is, the perceived size of the object does not vary as a function of the changes in retinal image size that accompany changes in depth. One implication of this view is that size and shape constancy will break down and illusions appear when depth cues are erroneous. There are numerous illusions of size, such as the Ponzo illusion (see Figure 17), in which one of two stimuli of equal physical size appears larger than another, due at least in part to misleading depth cues. Misperceptions of size and distance also can arise when depth cues are minimal, as when flying at night.

4.2.2 Sound Localization

The cues for sound localization involve disparities at the two ears, much as disparities of the images at the two eyes are cues to depth. Two different sources of information, interaural intensity and time differences, have been identified. Both of these cues vary systematically with respect to the position of the sound relative to the listener. At the front and back of the listener, the intensity of the sound and the time at which it reaches the ears will be equal. As the position

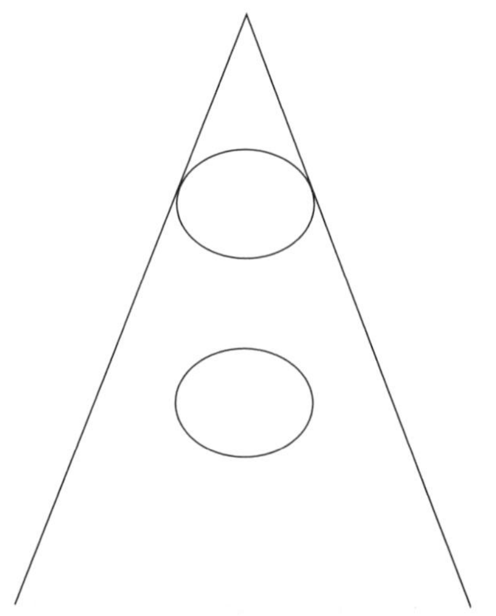

Figure 17 Ponzo illusion. The top circle appears larger than the lower circle, due to the linear perspective cue.

of the sound along the azimuth (i.e., relative to the listener's head) is moved progressively toward one side or the other, the sound will become increasingly louder at the ear closest to it relative to the ear on the opposite side, and it also will reach the ipsilateral ear first. The interaural intensity differences are due primarily to a sound shadow created by the head. Because the head produces no shadow for frequencies less than 1000 Hz, the intensity cue is most effective for relatively high frequency tones. In contrast, interaural time differences are most effective for low-frequency sounds. Localization accuracy is poorest for tones between 2000 and 4000 Hz, and this is thought to be because neither the intensity nor time cue is very effective in this intermediate-frequency range (Stevens and Newman, 1934).

Both the interaural intensity and time difference cues are ambiguous because the same values can be produced by stimuli in more than one location. As an example, both differences are zero for locations directly in front of or behind the listener. Because both cues are ambiguous at the front and back, front–back confusions of the location of brief sounds will often occur. Confusions are relatively rare in the natural world because head movements and reflections of sound make the cues less ambiguous than they are in the typical localization experiment (e.g., Guski, 1990; Makous and Middlebrooks. 1990). As with vision, misleading cues can cause erroneous localization of sounds. Caelli and Porter (1980) illustrated this point by having listeners in a car judge the direction from which a siren occurred. Localization accuracy was particularly poor when all but one window were rolled up, which would alter the normal relation between direction and the cues.

4.3 Eye Movements and Motion Perception

Because details can be perceived well only at the fovea, the location on which the fovea is fixated must be able to be changed regularly and rapidly if we are to maintain an accurate perceptual representation of the environment and to see the details of new stimuli that appear in the peripheral visual field. Such changes in fixation can be brought about by displacement of the body, movements of the head, eye movements, or a combination of the three. Each eye has attached to it a set of extraocular muscle pairs: medial and lateral rectus, superior and inferior rectus, and superior and inferior obliques. Each pair controls a different axis of rotation, with the two members of the pairs acting antagonistically. Fixation is maintained when all of the muscles are active to similar extents. However, even in this case there is a continuous tremor of the eye as well as slow drifts that are corrected with compensatory micromovements, causing small changes in position of the image on the retina. Because the visual system is insensitive to images that are stabilized on the retina, such as the shadows cast by the blood vessels that support the retinal neurons, this tremor prevents images from fading when fixation is maintained on an object for a period of time.

Two broad categories of eye movements are of deepest concern. *Saccadic eye movements* involve a rapid shift in fixation from one point to another. Typically, four or five saccadic movements will be made each second. Saccadic movements can be initiated automatically by the abrupt onset of a stimulus in the peripheral visual field or by conscious intent. The latency of initiation typically is on the order of 200 ms, and the duration of movement less than 100 ms. One of the more interesting phenomena associated with these eye movements is that of saccadic suppression, which is reduced sensitivity to visual stimulation during the time that the eye is moving. Saccadic suppression does not seem to be due to the movement of the retinal image being too rapid to allow perception nor to masking of the image by the stationary images that precede and follow the eye movement. Rather, it seems to have a neurological basis. The loss of sensitivity is much less for high spatial frequency gratings of light and dark lines than for low-spatial-frequency gratings, and is absent for colored edges (Burr et al., 1994). Because lesioning studies suggest that the low spatial frequencies are conveyed primarily by the magnocellular pathway, this pathway is probably the locus of saccadic suppression.

Smooth pursuit movements are those made when a moving stimulus is tracked by the eyes. Such movements require that the direction of motion of the target be decoded by the system in the brain responsible for eye movements. This information must be integrated with cognitive expectancies and then translated into signals that are sent to the appropriate members of the muscle pairs of both eyes, causing them to relax and contract in unison and the eyes to move to maintain fixation on the target. Pursuit is relatively accurate for relatively slow moving targets, with increasingly greater error occurring as movement speed increases.

Eye movement records provide precise information about where a person is looking at any time. Such records have been used to obtain evidence about strategies for determining where successive saccades are directed when scanning a visual scene and about the extraction of information from the display (see Abernethy, 1988, for a review). Because direction of gaze can be recorded online by appropriate eye-tracking systems, eye-gaze computer interface controls have considerable potential applications for persons with physical disabilities and for high workload tasks (e.g., Goldberg and Schryver, 1995). It is tempting to equate direction of fixation with direction of attention, and in many cases that may be appropriate. However, there is considerable evidence that attention can be directed to different locations in space while fixation is held constant (e.g., Sanders and Houtmans, 1985), indicating that direction of fixation and direction of attention are not always one and the same.

Movements of our eyes, head, and body produce changes in position of images on the retina, as does motion of an object in the environment. How we distinguish between motion of objects in the world and our own motion has been an issue of concern for many years. We have already seen that many neurons in the visual cortex are sensitive to motion across the retina. However, detecting changes in position on the retina is not sufficient for motion perception, because those changes could be brought about by our own motion, motion of an object, or a combination of the two. Typically, it has assumed that the position of the eyes is monitored by the brain, and any changes that can be attributed to eye movements are taken into account. According to inflow theory, first suggested by Sherrington (1906), it is the feedback from the muscles controlling the eyes that is monitored. According to outflow theory, first proposed by Helmholtz (1909), it is the command to the eyes to move that is monitored. Evidence, such as that the scene appears to move when an observer who has been paralyzed tries to move her or his eyes (which do not actually move; Stevens et al., 1976; Matin et al., 1982), has tended to support the outflow theory.

Sensitivity to motion is affected by many factors. For one, motion can be detected at a slower speed if a comparison, stationary object is also visible. When a reference object is present, changes of as little as 0.03° per second can be perceived (Palmer, 1986). However, this gain in sensitivity for detecting relative motion is at the potential cost of attributing the motion to the wrong object. For example, it is common for movement of a large region that surrounds a smaller object to be attributed to the object, a phenomenon that is called *induced motion* (Mack, 1986). The possibility for misattribution of motion is a concern for any situation in which one object is moving relative to another.

Induced motion is one example of a phenomenon in which motion of an object is perceived in the absence of motion of its image on the retina. The phenomenon of apparent, or stroboscopic, motion is probably the most important of these. This phenomenon of continuous perceived motion occurs when discrete changes in position of stimulation on the retina take place at appropriate temporal and spatial separations. It appears to be attributable to two processes, a short-range process and a long-range process (Petersik, 1989). The short-range process is presumed to reflect relatively low level directionally sensitive neurons that respond to small spatial changes that occur with short interstimulus intervals. The long-range process is presumed to reflect higher-level processes and to respond to stimuli at relatively large retinal separations presented at interstimulus intervals as long as 500 ms. Apparent motion is not only responsible for the motion produced in flashing signs, but also for motion pictures and television, in which a series of discrete images is presented.

4.4 Pattern Recognition

The organizational principles and depth cues determine form perception, that is, what shapes and objects will be perceived. However, for the information in a display to be conveyed accurately, the objects must be recognized. If there are words, they must be read correctly; if there is a pictograph, the pictograph must be interpreted accurately. In other words, good use of the organizational principles and depth cues by a designer does not ensure that the intended message will be conveyed to the observer.

Concern with the way in which stimuli are recognized and identified is the domain of pattern recognition. Most, but not all research on pattern recognition has been conducted with verbal stimuli. The initial step in pattern recognition is typically presumed to be feature analysis. If visual, alphanumeric characters are presented, they are assumed to be analyzed in terms of features such as a vertical line segment, a horizontal line segment, and so on. Such an assumption is generally consistent with the evidence that neurons in the primary visual cortex respond to specific features of stimulation. Moreover, confusion matrices obtained when letters are misidentified indicate that an incorrect identification is most likely to involve a letter with considerable feature overlap with the one that was actually displayed (e.g., Townsend, 1971).

Letters are composed of features, but they in turn are components of the letter patterns that form syllables and words (see Figure 18). The role played by letter-level information in visual word recognition has been the subject of considerable debate. Numerous findings have suggested that in at least some cases, letter-level information is not available prior to word recognition. For example, Healy and colleagues have found that when people perform a letter detection task while reading a prose passage, the target letter is missed more often when it occurs in a very high frequency

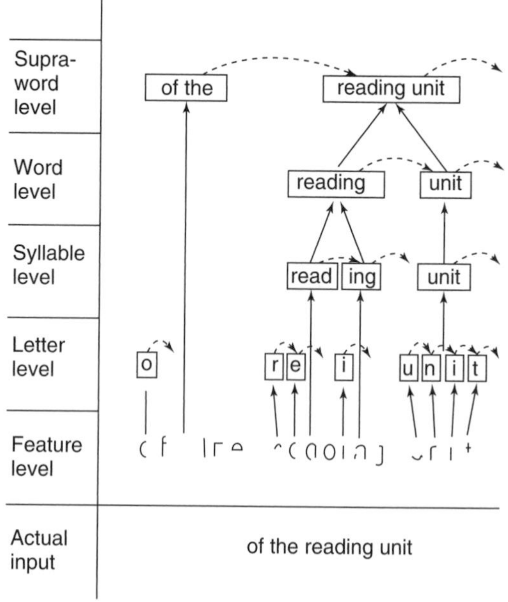

Figure 18 Levels of representation in reading a short passage of text. Operation of the unitization hypothesis is illustrated by the bypassing of levels that occurs for "of the." (From Healy, 1994.)

word such as *the* than when it appears in lower-frequency words (e.g., Healy, 1994; Proctor and Healy, 1995). Their results have shown that this "missing-letter" effect is not just due to skipping over the words while reading. To account for this phenomenon and other data, they have proposed a set of unitization hypotheses, according to which very high frequency words are identified prior to their component letters, with letter processing terminated once the word is identified (see Figure 18). Numerous studies have provided evidence for the need to distinguish the five levels of reading units shown in Figure 18 and for other claims of the unitization hypotheses, such as that the particular units identified by a person depend on the familiarity of those units.

Although there is agreement that letter-level information is often difficult to access within a word context, there is not agreement that this relative inaccessibility is due to word recognition occurring without identification at the letter level. For example, Johnson and Pugh (1994) note that the evidence suggesting that the component letter level is bypassed in word recognition may be an artifact of using tasks that require decisions and responses to be made with regard to the letters that would not be a normal part of reading. They propose a cohort model of visual word recognition in which visual access to lexical entries occurs in a bottom-up sequence that always involves the letter level. It is assumed that there is considerable variation in the rates at which individual letters are encoded, meaning that information about the component letters is made available gradually. The initially encoded feature information activates a cohort of candidate letters for each letter position, and these cohorts get reduced as additional features become available until the letter in a particular position is identified (see Figure 19). The initially identified letter information then activates a cohort of candidate word encodings consistent with this information, and identification of subsequent letters gradually eliminates members of the cohort until the word is identified. Cohort models similar to Johnson and Pugh's have been particularly popular as accounts for speech perception, due to the fact that there are considerable performance data indicating that an initial set of candidates is made available and then reduced (e.g., Marslen-Wilson, 1987), although there is more controversy in speech perception regarding the nature of the component segments and the process that produces them.

The primary emphasis in the models just described is on bottom-up processing from the sensory input to recognition of the pattern, but it is clear that pattern recognition is also influenced by top-down, nonvisual information of several types (Massaro and Cohen, 1994). These include orthographic constraints on the spelling patterns, regularities in the mapping between spelling and spoken sounds, syntactic constraints regarding which parts of speech are permissible, semantic constraints based on coherent meaning, and pragmatic constraints derived from the assumption that the writer is trying to communicate effectively. Interactive activation models, in which lower-level

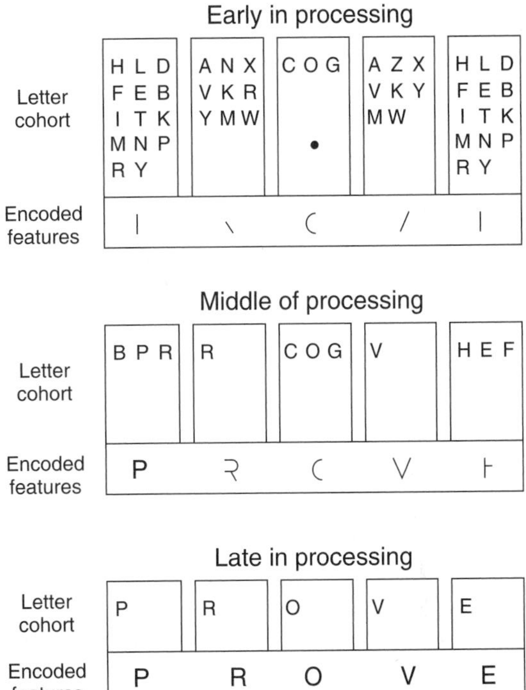

Figure 19 Letter cohorts activated by encoded features from early to late in processing. (Reprinted from Johnson and Pugh, 1994, with permission from Elsevier.)

sources of information are modified by higher levels, have been popular (e.g., McClelland and Rumelhart, 1981). However, Massaro and his colleagues (e.g., Massaro and Cohen, 1994) have been quite successful in accounting for a range of reading phenomena with a model, which they call the *fuzzy logical model* of perception, in which the multiple sources of information are assumed to be processed independently rather than interactively, and then integrated.

Reading can be viewed as a prototypical pattern recognition task. The implications of the analysis of reading are that multiple sources of information, both bottom-up and top-down, are exploited. For accurate pattern recognition, the possible alternatives need to be physically distinct and consistent with expectancies created by the context.

5 SUMMARY

In this chapter we have reviewed much of what is known about sensation and perception. Any such review must necessarily exclude certain topics and be limited in the treatment given to the topics that are covered. Excellent introductory texts that provide more thorough coverage include Schiffman (2001), Goldstein (2002), Sekuler and Blake (2002), and Coren et al. (2003). More advanced treatments of most areas are included in Volume 1 of *Stevens' Handbook of Experimental Psychology* (Pashler and

Yantis, 2002) and Volume 1 of the *Handbook of Perception and Performance* (Boff et al., 1986). The *Engineering Data Compendium: Human Perception and Performance* (Boff and Lincoln, 1988) is an excellent, although now somewhat dated, resource for information pertinent to many human engineering concerns. Also, throughout the text we have provided references to texts and review articles devoted to specific topics. These and related sources should be consulted to get an in-depth understanding of the relevant issues pertaining to any particular application involving perception.

Virtually all concerns in human factors and ergonomics involve perceptual issues to at least some extent. Whether dealing with instructions for a consumer product, control rooms for chemical processing or nuclear power plants, interfaces for computer software, guidance of vehicles, office design, and so on, information of some type must be conveyed to the user or operator. To the extent that the characteristics of the sensory systems and the principles of perception are accommodated in the design of displays and the environments in which the human must work, the transmission of information to the human will be fast and accurate and the possibility for injury low. To the extent that they are not accommodated, the opportunity for error and the potential for damage are increased.

REFERENCES

Abernethy, B. (1988), "Visual Search in Sport and Ergonomics: Its Relationship to Selective Attention and Performer Expertise," *Human Performance*, Vol. 4, pp. 205–235.

Adelson, E. H. (1993), "Perceptual Organization and the Judgment of Brightness," *Science*, Vol. 262, pp. 2042–2044.

Agostini, T., and Proffitt, D. R. (1993), "Perceptual Organization Evokes Simultaneous Lightness Contrast," *Perception*, Vol. 22, pp. 263–272.

Andre, J. (2003), "Controversies Concerning the Resting State of Accommodation," in *Visual Perception: The Influence of H. W. Leibowitz*, J. Andre, D. A. Owens, and L. O. Harvey, Jr., Eds., American Psychological Association, Washington, DC, pp. 69–79.

Appelle, S. (1991), "Haptic Perception of Form: Activity and Stimulus Attributes," in *The Psychology of Touch*, M. A. Heller and W. Schiff, Eds., Lawrence Erlbaum Associates, Mahwah, NJ, pp. 169–188.

Arend, L. E. (1993), "Mesopic Lightness, Brightness, and Brightness Contrast," *Perception and Psychophysics*, Vol. 54, pp. 469–476.

Baird, J. C., and Berglund, B. (1989), "Thesis for Environmental Psychophysics," *Journal of Environmental Psychology*, Vol. 9, pp. 345–356.

Baird, J. C., and Noma, E. (1978), *Fundamentals of Scaling and Psychophysics*, Wiley, New York.

Barber, A. V. (1990), "Visual Mechanisms and Predictors of Far Field Visual Task Performance," *Human Factors*, Vol. 32, pp. 217–233.

Benson, A. J. (1990), "Sensory Functions and Limitations of the Vestibular System," in *Perception and Control of Self-Motion*, R. Warren and A. H. Wertheim, Eds., Lawrence Erlbaum Associates, Mahwah, NJ, pp. 145–170.

Berglund, M. B. (1991), "Quality Assurance in Environmental Psychophysics," in *Ratio Scaling of Psychological Magnitude*, S. J. Bolanowski and G. A. Gescheider, Eds., Wiley, New York, pp. 140–162.

Boff, K. R., and Lincoln, J. E. (1988), *Engineering Data Compendium: Human Perception and Performance*, Harvey G. Armstrong Medical Research Laboratory, Wright-Patterson AFB, Dayton, OH.

Boff, K. R., Kaufman, L., and Thomas, J. P., Eds. (1986), *Handbook of Perception and Performance*, Vol. 1, *Sensory Processes and Perception*, Wiley, New York.

Bolanowski, S. J., and Gescheider, G. A., Eds. (1991), *Ratio Scaling of Psychological Magnitude*, Lawrence Erlbaum Associates, Mahwah, NJ.

Bolanowski, S. J., Verrillo, R. T., and McGlone, F. (2004), "Passive, Active and Intra-active (Self) Touch," *Behavioural Brain Research*, Vol. 148, pp. 41–45.

Braunstein, M. L. (1976), *Depth Perception Through Motion*, Academic Press, New York.

Bregman, A. S. (1990), *Auditory Scene Analysis: The Perceptual Organization of Sound*, MIT Press, Cambridge, MA.

Bregman, A. S. (1993), "Auditory Scene Analysis: Hearing in Complex Environments," in *Thinking in Sound: The Cognitive Psychology of Human Audition*, S. McAdams and E. Bigand, Eds., Oxford University Press, New York, pp. 10–36.

Bregman, A. S., and Rudnicky, A. I. (1975), "Auditory Segregation: Stream or Streams?" *Journal of Experimental Psychology: Human Perception and Performance*, Vol. 1, pp. 263–267.

Bruno, N., and Cutting, J. E. (1988), "Minimodularity and the Perception of Layout," *Journal of Experimental Psychology: General*, Vol. 117, pp. 161–170.

Burr, D. C., Morrone, M. C., and Ross, J. (1994), "Selective Suppression of the Magnocellular Visual Pathway During Saccadic Eye Movement," *Nature*, Vol. 371, pp. 511–513.

Caelli, T., and Porter, D. (1980), "On Difficulties in Localizing Ambulance Sirens," *Human Factors*, Vol. 22, pp. 719–724.

Coren, S., Ward, L. M., and Enns, J. T. (2003), *Sensation and Perception*, 6th ed., Harcourt Brace, Fort Worth, TX.

Cornsweet, T. H. (1962), "The Staircase Method in Psychophysics," *American Journal of Psychology*, Vol. 75, pp. 485–491.

Corwin, J. (1994), "On Measuring Discrimination and Response Bias: Unequal Numbers of Targets and Distractors and Two Classes of Distractors," *Neuropsychology*, Vol. 8, pp. 110–117.

Craig, J. C. (1982), "Vibrotactile Masking: A Comparison of Energy and Pattern Maskers," *Perception and Psychophysics*, Vol. 31, pp. 523–529.

Donders, F. C. (1868/1969), "On the Speed of Mental Processes," W. G. Koster, Trans., *Acta Psychologica*, Vol. 30, pp. 412–431.

Doty, R. L., Ed. (2003), *Handbook of Olfaction and Gustation*, 2nd ed., Marcel Dekker, New York.

Dowling, J. E., and Boycott, B. B. (1966), "Organization of the Primate Retina," *Proceedings of the Royal Society, Series B*, Vol. 166, pp. 80–111.

Draper, M. H., Viirre, E. S., Furness, T. A., and Gawron, V. J. (2001), "Effects of Image Scale and System

Time Delay on Simulator Sickness Within Head-Coupled Virtual Environments," *Human Factors*, Vol. 43, pp. 129–146.

Eadie, T. L., and Doyle, P. C. (2002), "Direct Magnitude Estimation and Interval Scaling of Pleasantness and Severity in Dysphonic and Normal Speakers," *Journal of the Acoustical Society of America*, Vol. 112, pp. 3014–3021.

Ebenholtz, S. M. (1992), "Accommodative Hysteresis as a Function of Target-Dark Focus Separation," *Vision Research*, Vol. 32, pp. 925–929.

Engen, T. (1991), *Odor Sensation and Memory*, Praeger, New York.

Evans, D. W., and Ginsburg, A. P. (1982), "Predicting Age-Related Differences in Discriminating Road Signs Using Contrast Sensitivity," *Journal of the Optical Society of America*, Vol. 72, pp. 1785–1786.

Farah, M. J. (2000), *The Cognitive Neuroscience of Vision*, Blackwell, Oxford.

Fechner, G. T. (1860/1966), *Elements of Psychophysics*, Vol. 1, H. E. Adler, Trans., Holt, Rinehart and Winston, New York.

Fletcher, H., and Munson, W. A. (1933), "Loudness, Its Definition, Measurement, and Calculation," *Journal of the Acoustical Society of America*, Vol. 5, pp. 82–108.

Frishman, L. J. (2001), "Basic Visual Processes," in *Blackwell Handbook of Perception*, E. B. Goldstein, Ed., Blackwell, Malden, MA, pp. 53–91.

Gardner, E. P., and Martin, J. H. (2000), "Coding of Sensory Information," in *Principles of Neural Science*, Vol. 4, E. R. Kandel, J. H. Schwartz, and T. M. Jessell, Eds., Elsevier, Amsterdam, pp. 411–429.

Gardner, E. P., Martin, J. H., and Jessell, T. M. (2000), "The Bodily Senses," in *Principles of Neural Science*, Vol. 4, E. R. Kandel, J. H. Schwartz, and T. M. Jessell, Eds., Elsevier, Amsterdam, pp. 430–450.

Garner, W. (1974), *The Processing of Information and Structure*, Lawrence Erlbaum Associates, Mahwah, NJ.

Gegenfurtner, K. R., and Sharpe, L. T. (1999), *Color Vision: From Genes to Perception*, Cambridge University Press, New York.

Gescheider, G. A. (1988), "Psychophysical Scaling," in *Annual Review of Psychology*, Vol. 39, M. R. Rosenzweig and L. W. Porter, Eds., Annual Reviews, Palo Alto, CA, pp. 169–200.

Gescheider, G. A. (1997), *Psychophysics: The Fundamentals*, 3rd ed., Lawrence Erlbaum Associates, Mahwah, NJ.

Gibson, J. J. (1950), *The Perception of the Visual World*, Houghton Mifflin, Boston.

Gibson, J. J. (1966), *The Senses Considered as Perceptual Systems*, Houghton Mifflin, Boston.

Gilchrist, A. L. (1977), "Perceived Lightness Depends on Perceived Spatial Arrangement," *Science*, Vol. 195, pp. 185–187.

Goldberg, M. E., and Hudspeth, A. J. (2000), "The Vestibular System," in *Principles of Neural Science*, Vol. 4, E. R. Kandel, J. H. Schwartz, and T. M. Jessell, Eds., Elsevier, Amsterdam, pp. 801–831.

Goldberg, J. H., and Schryver, J. C. (1995), "Eye-Gaze-Contingent Control of the Computer Interface: Methodology and Example for Zoom Detection," *Behavior Research Methods, Instruments, and Computers*, Vol. 27, pp. 338–350.

Goldstein, E. B. (2002), *Sensation and Perception*, 6th ed., Wadsworth, Belmont CA.

Green, D. M., and Swets, J. A. (1966), *Signal Detection Theory and Psychophysics*, Wiley, New York; reprinted, 1974, by Krieger, Huntington, NY.

Guski, R. (1990), "Auditory Localization: Effects of Reflecting Surfaces," *Perception*, Vol. 19, pp. 819–830.

Harvey, L. O., Jr. (2003), "Living with Uncertainty in an Uncertain World: Signal Detection Theory in the Real World," in *Visual Perception: The Influence of H. W. Leibowitz*, J. Andre, D. A. Owens, and L. O. Harvey, Jr., Eds., American Psychological Association, Washington, DC, pp. 23–41.

Healy, A. F. (1994), "Letter Detection: A Window to Unitization and Other Cognitive Processes in Reading Text," *Psychonomic Bulletin and Review*, Vol. 1, pp. 333–344.

Helmholtz, H. von (1877/1954), *On the Sensation of Tone as a Psychological Basis for the Theory of Music*, 2nd ed., A. J. Ellis, Trans. and Ed., Dover, New York.

Helmholtz, H. von (1909/1962), *Treatise on Physiological Optics*, J. C. P. Southall, Ed. and Trans., Dover, New York.

Henning, H. (1916), "Die Qualitiitsreibe des Geschmacks," *Zeitschrift für Psychologie*, Vol. 74, pp. 203–219.

Heuer, H., Hollendiek, G., Kroger, H., and Romer, T. (1989), "The Resting Position of the Eyes and the Influence of Observation Distance and Visual Fatigue on VDT Work," *Zeitschrift für Experimentelle und Angewandte Psychologie*, Vol. 36, pp. 538–566.

Highstein, S. M., Fay, R. R., and Popper, A. N., Eds. (2004), *The Vestibular System*, Springer-Verlag, New York.

Hollins, M., Bensmaiea, S. J., and Roy, E. A. (2002), "Vibrotaction and Texture Perception," *Behavioural Brain Research*, Vol. 135, pp. 51–56.

Holmes, T. H., and Rahe, R. H. (1967), "The Social Readjustment Rating Scale," *Journal of Psychosomatic Research*, Vol. 11, pp. 213–218.

Hubel, D. H., and Wiesel, T. N. (1977), "Functional Architecture of Macaque Monkey Visual Cortex," *Proceedings of the Royal Society of London, Series B*, Vol. 198, p. I–59.

Jackson, G. R., Owsley, C., and McGwin, G., Jr. (1999), "Aging and Dark Adaptation," *Vision Research*, Vol. 39, pp. 3975–3982.

Jacobsen, A., and Gilchrist, A. L. (1988), "The Ratio Principle Holds over a Million-to-One Range of Illumination," *Perception and Psychophysics*, Vol. 43, pp. 1–6.

Jaschinski-Kruza, W. (1991), "Eyestrain in VDU Users: Viewing Distance and the Resting Position of Ocular Muscles," *Human Factors*, Vol. 33, pp. 69–83.

Johnson, N. F., and Pugh, K. R. (1994), "A Cohort Model of Visual Word Recognition," *Cognitive Psychology*, Vol. 26, pp. 240–346.

Julesz, B. (1971), *Foundations of Cyclopean Perception*, University of Chicago Press, Chicago.

Julesz, B., and Hirsh, I. J. (1972), "Visual and Auditory Perception: An Essay of Comparison," in *Human Communication: A Unified View*, E. E. David and P. B. Denes, Eds., McGraw-Hill, New York, pp. 283–335.

Kadlec, H. (1999), "MSDA_2: Updated Version of Software for Multidimensional Signal Detection Analyses," *Behavior Research Methods, Instruments, and Computers*, Vol. 31, pp. 384–385.

Kahneman, D. (1973), *Attention and Effort*, Prentice-Hall, Englewood Cliffs, NJ.

Kandel, E. R., and Schwartz, J. H., Eds. (1985), *Principles of Neural Science*, 2nd ed., Elsevier Science, New York.

Kanwisher, N., and Duncan, J., Eds. (2004), *Functional Neuroimaging of Visual Cognition: Attention and Performance XX*, Oxford University Press, New York.

Katz, D. (1989). *The World of Touch* (translated by L. E. Krueger), Lawrence Erlbaum Associates, Mahwah, NJ (original work published 1925).

Kolb, H. (1994), "The Architecture of Functional Neural Circuits in the Vertebrate Retina," *Investigative Opthalmology and Visual Science*, Vol. 35, pp. 2385–2403.

Komatsu, H., Kinoshita, M., and Murakami, I. (2002), "Neural Responses in the Primary Visual Cortex of the Monkey During Perceptual Filling-in at the Blind Spot," *Neuroscience Research*, Vol. 44, pp. 231–236.

Kosnik, W. D., Sekuler, R., and Kline, D. W. (1990), "Self-Reported Visual Problems of Older Drivers," *Human Factors*, Vol. 5, pp. 597–608.

Kramer, A. F., and Weber, T. (2000), "Applications of Psychophysiology to Human Factors," in *Handbook of Psychophysiology*, 2nd ed., J. T. Cacioppo, L. G. Tassinary, and G. Berntson, Eds., Cambridge University Press, New York, pp. 794–814.

Landy, M. S., Maloney, L. T., Johnston, E. B., and Young, M. (1995), "Measurement and Modeling of Depth Cue Combination: In Defense of Weak Fusion," *Vision Research*, Vol. 35, pp. 389–412.

Leibowitz, H. W., and Owens, D. A. (1975), "Anomalous Myopias and the Intermediate Dark Focus of Accommodation," *Science*, Vol. 189, pp. 646–648.

Mack, A. (1986), "Perceptual Aspects of Motion in the Frontal Plane," in *Handbook of Perception and Performance*, Vol. 1, *Sensory Processes and Perception*, K. R. Boff, L. Kaufman, and J. P. Thomas, Eds., Wiley, New York, pp. 17–1 to 17–38.

Macmillan, N. A. (2002), "Signal Detection Theory," in *Stevens' Handbook of Experimental Psychology*, Vol. 4, *Methodology in Experimental Psychology*, H. Pashler and J. Wixted, Eds., Wiley, New York, pp. 43–90.

Macmillan, N. A., and Creelman, C. D. (1990), "Response Bias: Characteristics of Detection Theory, Threshold Theory, and 'Nonparametric' Indexes," *Psychological Bulletin*, Vol. 107, pp. 401–413.

Macmillan, N. A., and Creelman, C. D. (2005), *Detection Theory: A User's Guide*, 2nd ed., Lawrence Erlbaum Associates, Mahwah, NJ.

Makous, J. C., and Middlebrooks, J. C. (1990), "Two-Dimensional Sound Localization by Human Listeners," *Journal of the Acoustical Society of America*, Vol. 87, pp. 2188–2200.

Manning, D. J., and Leach, J. (2002), "Perceptual and Signal Detection Factors in Radiography," *Ergonomics*, Vol. 45, pp. 1103–1116.

Marcos, S., Moreno, E., and Navarro, R. (1999), "The Depth-of-Field of the Human Eye from Objective and Subjective Measurements," *Vision Research*, Vol. 39, pp. 2039–2049.

Marks, L. E., and Gescheider, G. A. (2002), "Psychophysical Scaling," in *Stevens' Handbook of Experimental Psychology*, Vol. 4, *Methodology in Experimental Psychology*, H. Pashler and J. Wixted, Eds., Wiley, New York, pp. 91–138.

Marslen-Wilson, W. D. (1987), "Functional Parallelism in Spoken Word Recognition," *Cognition*, Vol. 25, pp. 75–102.

Massaro, D. W., and Cohen, M. M. (1994), "Visual, Orthographic, Phonological, and Lexical Influences in Reading," *Journal of Experimental Psychology: Human Perception and Performance*, Vol. 20, pp. 1107–1128.

Matin, L., Picoult., E., Stevens, J., Edwards, M., and McArthur, R. (1982), "Oculoparalytic Illusion: Visual-Field Dependent Spatial Mislocations by Humans Partially Paralyzed with Curare," *Science*, Vol. 216, pp. 198–201.

Matlin, M. W. (1988), *Sensation and Perception*, 2nd ed., Allyn and Bacon, New York.

McClelland, J. L., and Rumelhart, D. E. (1981), "An Interactive Activation Model of Context Effects in Letter Perception, I: An Account of Basic Findings," *Psychological Review*, Vol. 88, pp. 375–407.

Merigan, W. H., and Maunsell, J. H. R. (1993), "How Parallel Are the Primate Visual Pathways?" *Annual Review of Neuroscience*, Vol. 16, pp. 369–402.

Milner, A. D., and Goodale, M. A. (1995), *The Visual Brain in Action*, Oxford University Press, New York.

Morgan, M. J., Watt, R. J., and McKee, S. P. (1983), "Exposure Duration Affects the Sensitivity of Vernier Acuity to Target Motion," *Vision Research*, Vol. 23, pp. 541–546.

Mort, D. J., Malhotra, P., Mannan, S. K., Rorden, C., Pambakian, A., Kennard, C., and Husain, M. (2003), "The Anatomy of Visual Neglect," *Brain*, Vol. 126, pp. 1986–1997.

Murphy, B. J. (1978), "Pattern Thresholds for Moving and Stationary Gratings During Smooth Eye Movement," *Vision Research*, Vol. 18, pp. 521–530.

Nealey, T. A., and Maunsell, J. H. R. (1994), "Magnocellular and Parvocellular Contributions to the Responses of Neurons in Macaque Striate Cortex," *Journal of Neuroscience*, Vol. 14, pp. 2069–2079.

O'Brien, K. A., Cole, B. L., Maddocks, J. D., and Forbes, A. B. (2002), "Color and Defective Color Vision as Factors in the Conspicuity of Signs and Signals," *Human Factors*, Vol. 44, pp. 665–675.

Owens, D. A., and Leibowitz, H. W. (1983), "Perceptual and Motor Consequences of Tonic Vergence," in *Vergence Eye Movements: Basic and Clinical Aspects*, C. M. Shor and K. J. Cuiffreda, Eds., Butterworth, Boston, pp. 25–74.

Oyama, T. (1987), "Perception Studies and Their Applications to Environmental Design," *International Journal of Psychology*, Vol. 22, pp. 447–451.

Palmer, J. (1986), "Mechanisms of Displacement Discrimination with and Without Perceived Movement," *Journal of Experimental Psychology: Human Perception and Performance*, Vol. 12, pp. 411–421.

Palmer, S. E. (1992), "Common Region: A New Principle of Perceptual Grouping," *Cognitive Psychology*, Vol. 24, pp. 436–447.

Palmer, S. E. (2003), "Visual Perception of Objects," in *Experimental Psychology*, A. F. Healy and R. W. Proctor, Eds., Vol. 4 in *Handbook of Psychology*, I. B. Weiner, Ed.-in-Chief, Wiley, Hoboken, NJ, pp. 179–211.

Parasuraman, R. (2003), "Neuroergonomics: Research and Practice," *Theoretical Issues in Ergonomics Science*, Vol. 4, pp. 5–20.

Pashler, H., and Yantis, S., Eds. (2002), *Stevens' Handbook of Experimental Psychology*, 3rd ed., Vol. 1, *Sensation and Perception*, Wiley, New York.

Petersik, J. T. (1989), "The Two-Process Distinction in Apparent Motion," *Psychological Bulletin*, Vol. 106, pp. 107–127.

Plateau, J. A. F. (1872), "Sur la me sure des sensations physiques, et sur la loi que lie l'intensité des sensations, II: L'intensité de la cause excitante," *Bulletins de l'Academie Royale des Sciences, des Lettres, et des Beaux-Arts de Belgique*, Vol. 33, pp. 376–388.

Plomp, R. (1964), "The Ear as Frequency Analyzer," *Journal of the Acoustical Society of America*, Vol. 36, pp. 1628–1636,

Pomerantz, J. R. (1981), "Perceptual Organization in Information Processing," in *Perceptual Organization*, M. Kubovy and J. R. Pomerantz, Eds., Lawrence Erlbaum Associates, Mahwah, NJ, pp. 141–180.

Potts, B. C., Melara, R. D., and Marks, L. E. (1998), "Circle Size and Diameter Tilt: A New Look at Integrality and Separability," *Perception and Psychophysics*, Vol. 60, pp. 101–112.

Proctor, J. D., and Healy, A. F. (1995), "Acquisition and Retention of Skilled Letter Detection," in *Learning and Memory of Knowledge and Skills: Durability and Specificity*, A. F. Healy and L. E. Bourne, Jr., Eds., Sage, Thousand Oaks, CA, pp. 282–289.

Proctor, R. W., and Van Zandt, T. (1994). *Human Factors in Simple and Complex Systems*, Allyn and Bacon, New York.

Proffitt, D. R., and Caudek, C. (2003), "Depth Perception and the Perception of Events," in *Experimental Psychology*, A. F. Healy and R. W. Proctor, Eds., Vol. 4 in *Handbook of Psychology*, I. B. Weiner, Ed.-in-Chief, Wiley, Hoboken, NJ, pp. 213–236.

Randle, R. (1988), "Visual Accommodation: Mediated Control and Performance," in *International Reviews of Ergonomics*, Vol. 2, D. J. Oborne, Ed., Taylor & Francis, London.

Robinson, D. W., and Dadson, M. A. (1956), "A Re-determination of the Equal-Loudness Relations for Pure Tones," *British Journal of Applied Physics*, Vol. 7, pp. 166–181,

Rock, I., and Brosgole, L. (1964), "Grouping Based on Phenomenal Proximity," *Journal of Experimental Psychology*, Vol. 67, pp. 531–538.

Rock, I., and Palmer, S. (1990), "The Legacy of Gestalt Psychology," *Scientific American*, Vol. 263, No. 6, pp. 84–90.

Rouby, C., Schaal, B., Dubois, D., Gervais, R., and Holley, A., Eds. (2002), *Olfaction, Taste, and Cognition*, Cambridge University Press, New York.

Rugg, M. D., and Coles, M. G. H., Eds. (1995), *Electrophysiology of Mind: Event-Related Brain Potentials and Cognition*, Oxford University Press, New York.

Rutherford, W. (1886), "A New Theory of Hearing," *Journal of Anatomy and Physiology*, Vol. 21, pp. 166–168.

Sanders, A. F., and Houtmans, M. J. M. (1985), "Perceptual Processing Modes in the Functional Visual Field," *Acta Psychologica*, Vol. 58, pp. 251–261.

Scharff, L. F. V. (2003), "Sensation and Perception Research Methods," in *Handbook of Research Methods in Experimental Psychology*, S. F. Davis, Ed., Blackwell, Malden, MA, pp. 263–284.

Schiffman, H. R. (1996), *Sensation and Perception: An Integrated Approach*, 4th ed., Wiley, New York.

Schiffman, H. R. (2001), *Sensation and Perception: An Integrated Approach*, 5th ed., Wiley, New York.

Schiffman, H. R. (2003), "Psychophysics," in *Handbook of Research Methods in Experimental Psychology*, S. F. Davis, Ed., Blackwell, Malden, MA, pp. 441–469.

Schiffman, S. S., and Erickson, R. P. (1993), "Psychophysics: Insights into Transduction Mechanisms and Neural Coding," in *Mechanisms of Taste Transduction*, S. A. Simon and S. D. Roper, Eds., CRC Press, Boca Raton, FL.

Sekuler, R., and Blake, R. (2002), *Perception*, 4th ed., McGraw-Hill, New York.

Shang, H., and Bishop, I. D. (2000), "Visual Thresholds for Detection, Recognition, and Visual Impact in Landscape Settings," *Journal of Environmental Psychology*, Vol. 20, pp. 125–140.

Shepard, R. N. (1991), "Integrality Versus Separability of Stimulus Dimensions," in *The Perception of Structure*, G. R. Lockhead and J. R. Pomerantz, Eds., American Psychological Association, Washington, DC, pp. 53–71.

Sherman, S. M., and Koch, C. (1990), "Thalamus," in *The Synaptic Organization of the Brain*, 3rd ed., G. M. Shepherd, Ed., Oxford University Press, New York.

Sherrington, C. S. (1906), *Integrative Action of the Nervous System*, Yale University Press, New Haven, CT.

Snodgrass, J. G., and Corwin, J. (1988), "Pragmatics of Measuring Recognition Memory: Applications to Dementia and Amnesia," *Journal of Experimental Psychology: General*, Vol. 117, pp. 34–50.

Sorkin, R. D. (1999), "Spreadsheet Signal Detection," *Behavior Research Methods, Instruments, and Computers*, Vol. 31, pp. 46–54.

Sternberg, S. (1969), "The Discovery of Processing Stages: Extensions of Donders' Method," in *Attention and Performance II*, W. G. Koster, Ed., North-Holland, Amsterdam.

Stevens, S. S. (1956), "The Direct Estimation of Sensory Magnitudes: Loudness," *American Journal of Psychology*, Vol. 69, pp. 1–25.

Stevens, J. K., Emerson, R. C., Gerstein, G. L., Kallos, T., Neufeld, G. R., Nichols, C. W., and Rosenquist, A. C. (1976), "Paralysis of the Awake Human: Visual Perceptions," *Vision Research*, Vol. 16, pp. 93–98.

Stevens, S. S. (1975), *Psychophysics*, Wiley, New York.

Stevens, S. S., and Newman, E. B. (1934), "The Localization of Pure Tone," *Proceedings of the National Academy of Science*, Vol. 20, pp. 593–596.

Sturr, F., Kline, G. E., and Taub, H. A. (1990), "Performance of Young and Older Drivers on a Static Acuity Test Under Photopic and Mesopic Luminance Conditions," *Human Factors*, Vol. 32, pp. 1–8.

Summers, I. R., Ed. (1992), *Tactile Aids for the Hearing Impaired*, Whurr Publishers, London.

Swets, J. A. (1986), "Indices of Discrimination or Diagnostic Accuracy: Their ROCs and Implied Models," *Psychological Bulletin*, Vol. 99, pp. 100–117.

Thurstone, L. L. (1927), "A Law of Comparative Judgment," *Psychological Review*, Vol. 34, pp. 273–286.

Townsend, J. T. (1971), "Theoretical Analysis of an Alphabetic Confusion Matrix," *Perception and Psychophysics*, Vol. 9, pp. 40–50.

Truscianko, T., Montagnon, R., and le Clerc, J. (1991), "The Role of Colour as a Monocular Depth Cue," *Vision Research*, Vol. 31, pp. 1923–1930.

Van Zandt, T. (2002), "Analysis of Response Time Distributions," in *Stevens' Handbook of Experimental Psychology*, Vol. 4, *Methodology in Experimental Psychology*, H. Pashler and J. Wixted, Eds., Wiley, New York, pp. 461–516.

Verrillo, R. T., and Gescheider, G. A. (1992), "Perception via the Sense of Touch," in *Tactile Aids for the Hearing Impaired*, I. R. Summers, Ed., Whurr Publishers, London, pp. 1–36.

Verrillo, R. T., Fraioli, A. J., and Smith, R. L. (1969), "Sensation Magnitude of Vibrotactile Stimuli," *Perception and Psychophysics*, Vol. 6, pp. 366–372.

Von Békésy, G. (1960), *Experiments in Hearing*, McGraw-Hill, New York.

Walker, B. N. (2002), "Magnitude Estimation of Conceptual Data Dimensions for Use in Sonification," *Journal of Experimental Psychology: Applied*, Vol. 8, pp. 211–221.

Wallach, H. (1972), "The Perception of Neutral Colors," in *Perception: Mechanisms and Models: Readings from Scientific American*, T. Held and W. Richards, Eds., W.H. Freeman, San Francisco, pp. 278–285.

Wandell, B. A. (1995), *Foundations of Vision*, Sinauer Associates, Sunderland, MA.

Welford, A. T., Ed. (1980), *Reaction Times*, Academic Press, London.

Wever, E. G., and Bray, C. W. (1937), "The Perception of Low Tones and the Resonance-Volley Theory," *Journal of Psychology*, Vol. 3, pp. 101–114.

Wickens, T. D. (2001), *Elementary Signal Detection Theory*, Oxford University Press, New York.

Wickens, C. D., and Carswell, C. M. (1995), "The Proximity Compatibility Principle: Its Psychological Foundation and Relevance to Display Design," *Human Factors*, Vol. 37, pp. 473–494.

Wickens, C. D., Merwin, D. F., and Lin, E. (1994), "Implications of Graphics Enhancements for the Visualization of Scientific Data: Dimensional Integrality, Stereopsis, Motion, and Mesh," *Human Factors*, Vol. 36, pp. 44–61.

Wilson, G. F. (2002), "Psychophysiological Test Methods and Procedures," in *Handbook of Human Factors Testing and Evaluation*, 2nd ed., S. G. Charlton and T. G. O'Brien, Eds., Lawrence Erlbaum Associates, Mahwah, NJ, pp. 127–156.

Wood, J. M. (2002), "Age and Visual Impairment Decrease Driving Performance as Measured on a Closed-Road Circuit," *Human Factors*, Vol. 44, pp. 482–494.

CHAPTER 4

SELECTION AND CONTROL OF ACTION

Robert W. Proctor
Purdue University
West Lafayette, Indiana

Kim-Phuong L. Vu
California State University–Long Beach
Long Beach, California

Psychology's search for quantitative laws that describe human behavior is long-standing, dating back to the 1850s. A few notable successes have been achieved, including Fitts's law (1954) and the Hick–Hyman law (Hick, 1952; Hyman, 1953).

Delaney et al. (1998)

1 INTRODUCTION

Although research on selection and control of action has a long history, dating to at least the mid nineteenth century, modern-day research in this area has developed contemporaneously with that on human factors and ergonomics. Seminal works in both areas appeared in the period following World War II, and in many instances, people who played important roles in the development of human factors and ergonomics also made significant contributions to our understanding of selection and control of action. Two such contributions are those alluded to in the quote above, the Hick–Hyman law and Fitts's law, involving selection and control of action, respectively, which are among the few well-established quantitative laws of behavior.

Paul M. Fitts, for whom Fitts's law is named, was perhaps the most widely known of those who made significant contributions to the field of human factors and ergonomics and to basic research on human performance. He headed the Psychology Branch of the U.S. Army Airforce Aeromedical Laboratory, now called the Fitts Human Engineering Division, at its founding in 1945, and is honored by a teaching award in his name given annually by the Human Factors and Ergonomics Society. Although Fitts's primary goal was the design of military aircraft and other machines to accommodate the human operator, he fully appreciated that this goal could only be accomplished against a background of knowledge of basic principles of human performance established under controlled laboratory conditions. Consequently, Fitts made many lasting empirical and theoretical contributions to knowledge concerning selection and control of action, including the quantitative law that bears his name and the principle of stimulus–response (S-R) compatibility, both of which are discussed in this chapter.

Since the groundbreaking work of Fitts and others in the 1950s, much research has been conducted on selection and control of action under the headings of human performance, motor learning and control, and motor behavior, among others. Indeed, the relation between perception and action is currently a very active area of research in psychology and associated fields [see, e.g., the special issue of *Psychological Research* devoted to cognitive control of action (Nattkemper and Ziessler, 2004)]. In the present chapter we review some of the major findings, principles, and theories concerning selection and control of action pertinent to designing for human use.

2 SELECTION OF ACTION

2.1 Methods

Selection of action is most often studied in choice–reaction tasks, in which a set of stimulus alternatives is mapped to a set of responses. Simple responses such as keypresses are typically used because the intent is to study the central decisions involved in selecting actions, not the motoric processes involved in carrying out the actions. In a choice–reaction study, response time (RT) is characteristically recorded as the primary dependent measure and error rate as a secondary measure. Among the methods used to interpret the RT data are the additive factors method and related ones that allow examination of the selective influence of variables on various processes (Sternberg, 1998). Analyses based on these methods have suggested that the primary variables affecting the duration of action- or response-selection processes include S-R uncertainty, S-R compatibility, response precuing, and sequential dependencies (Sanders, 1998). Analyses of measures in addition to mean RT and percentage of error, including RT distributions, specific types of errors that are made, and psychophysiological indicators of brain functions can also be used to obtain information about the nature of action selection.

One well-established principle of performance in choice–reaction tasks is that speed can be traded for accuracy (Pachella, 1974). The speed–accuracy trade-off (see Figure 1) can be modeled readily by sequential sampling models of response selection, according to which information accumulates over time after stimulus onset in a later decision stage until a decision is reached. One class of such models, race models, assume that there is a separate decision unit, or counter, for each response, with the response that is ultimately selected being the one for the counter that "wins the race" and reaches threshold first (e.g., Van Zandt et al., 2000). One point of sequential sampling models is that action selection is a function of both the quality of the stimulus information, which affects the rate at which the information accumulates, and the level of the response thresholds, which is affected by instructions and other factors. Speed–accuracy trade-off methods in which subjects are induced to adopt different speed–accuracy criteria in different trial blocks, or in which biases toward one response category or another are introduced, can be used to examine details of the choice process (e.g., Band et al., 2003).

Many situations outside the laboratory require performance of multiple tasks, either in succession or concurrently. Choice RT methods not only can be used to examine action selection under single-task performance, but also under conditions in which two or more task sets must be maintained, and actions to each must be performed concurrently or the person is required to switch between the various tasks periodically. Because considerable research on action selection has been conducted under single- and multiple-task performance, we cover each separately.

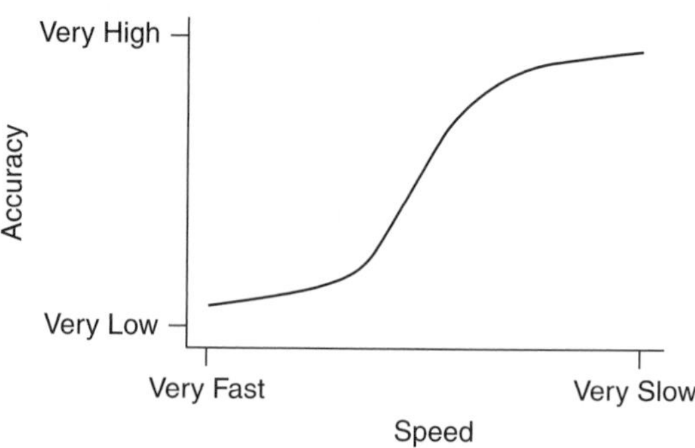

Figure 1 Speed–accuracy trade-off. Depending on instructions, payoffs, and other factors, when a person must choose a response to a stimulus, he or she can vary the combination of response speed and accuracy between the extremes of very fast with low accuracy or very slow with high accuracy.

2.2 Action Selection in Single-Task Performance

2.2.1 Uncertainty and Number of Alternatives: Hick–Hyman Law

Hick (1952) and Hyman (1953), following up on the much earlier work of Merkel (1885; described by Woodworth, 1938), conducted influential studies showing a systematic increase in choice RT as the number of S-R alternatives increased. Both Hick and Hyman were interested in whether effects of S-R uncertainty could be explained in terms of information theory, which Shannon (1948) had recently developed in the field of communication engineering. Information theory provides a metric for information transmission in bits (binary digits), with the number of bits conveyed by an event being a function of uncertainty. The average number of bits for a set of N equally likely stimuli is $\log_2 N$. Because uncertainty also varies as a function of the probabilities with which individual stimuli occur, the average amount of information for stimuli that occur with unequal probability will be less than $\log_2 N$. More generally, the average amount of information (H) conveyed by a stimulus for a set of size N is

$$H = -\sum_{i=1}^{N} p_i \log_2 p_i$$

where p_i is the probability of alternative i. Across trials, all of the information in the stimulus set is transmitted through the responses if no errors are made. However, when errors are made, the amount of transmitted information (H_T) will be less than the average information in the stimulus set.

The stimuli in Hick's (1952) study were 10 lamps arranged in an irregular circle, to which subjects responded by pressing one of the 10 keys, on which the fingers from each hand were placed. Hick served as his own subject in two experiments (and a third control experiment). In experiment 1, Hick performed blocks of trials with set sizes ranging from 2 to 10 in ascending and descending order, maintaining a high level of accuracy. In experiment 2 he used only the set size of 10 but adopted various speed–accuracy criteria in different trial blocks. For both experiments, RT increased as a logarithmic function of the average amount of information transmitted.

In Hyman's (1953) study, the stimuli were eight lights, corresponding to the eight corners of inner and outer squares, to which the subject was to respond with an arbitrarily assigned one-syllable spoken name. As in Hick's (1952) study, RT was a linear function of $\log_2 N$ when the number of equally likely stimulus alternatives ranged from 1 to 8. Hyman also manipulated the probabilities of occurrence of the alternative stimuli and sequential dependencies in other experiments. In both cases, RT increased as a logarithmic function of the average amount of information conveyed by a stimulus, as predicted by information theory.

This relation between RT and the stimulus information that is transmitted in the responses is the Hick–Hyman law (see Figure 2), sometimes called Hick's law, mentioned in the opening quote of the chapter. According to it,

$$\text{RT} = a + bH_T$$

where a is basic processing time and b is the amount that RT increases with increases in the amount of information transmitted (H_T; $\log_2 N$ for equally likely S-R pairs with no errors).

The Hick–Hyman function is obtained in a wide variety of tasks, although the slope of the function is determined by several factors (Teichner and Krebs, 1974). The slope is typically shallower for highly compatible S-R pairings than for less compatible ones (see

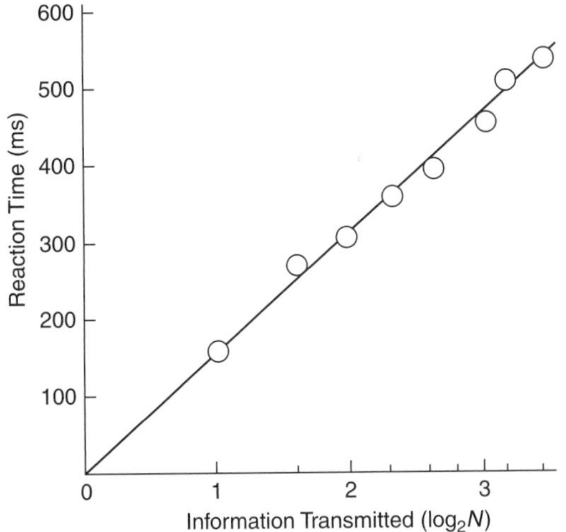

Figure 2 Hick–Hyman law: reaction time increases as a logarithmic function of the amount of information transmitted.

Section 2.2.2), and it decreases as the amount of practice increases. Thus, the cost associated with high event uncertainty can be reduced by using highly compatible display–control arrangements or giving the operators training on the task. In agreement with the view that action selection is a central, cognitive process, the slope of the Hick–Hyman function has been found to correlate highly with standardized measures of intelligence (e.g., Beh et al., 1994). The law is so robust that it even applies to species other than humans, such as pigeons (Vickrey and Neuringer, 2000).

What is the underlying basis for the Hick–Hyman law? Usher et al. (2002; see also Usher and McClelland, 2001) recently provided evidence that the law is a consequence of subjects trying to maintain a constant accuracy for all set sizes. Usher et al. evaluated race models for which, as mentioned earlier, the response-selection process is characterized as involving a separate stochastic accumulator for each S-R alternative. Upon stimulus presentation, activation relevant to each alternative builds up dynamically within the appropriate accumulators, and when the activation in one accumulator reaches a threshold, that response is selected. Response selection is faster with lower than with higher thresholds because a threshold is reached sooner after stimulus presentation. However, this benefit in response speed is obtained at the cost of accuracy because the threshold for an incorrect alternative is more likely to be reached due to the noisy activation process.

With two S-R alternatives there are two accumulators, with four alternatives there are four accumulators, and so on. Each additional accumulator provides an extra chance for an incorrect response to be selected. Consequently, if the error rate is to be held approximately constant as the size of the S-R set increases, the response thresholds must be adjusted upward. Usher et al. (2002) showed that if the increase in the threshold as N increases is logarithmic, the probability of an incorrect response remains approximately constant. This logarithmic increase in criterion results in a logarithmic increase in RT. Based on their model fits, Usher et al. concluded that the major determinant of the Hick–Hyman law is the increase in likelihood of erroneously reaching a response threshold as the number of S-R alternatives increases, coupled with subjects' attempting to hold error rate approximately constant across conditions.

2.2.2 Stimulus–Response Compatibility

Spatial Compatibility S-R compatibility refers to the fact that some arrangements of stimuli and responses, or mappings of individual stimuli to responses, are more natural than others, leading to faster and more accurate responding. S-R compatibility effects were demonstrated by Fitts and colleagues in two classic studies conducted in the 1950s. Specifically, Fitts and Seeger (1953) had subjects perform eight-choice tasks in which subjects moved a stylus (or a combination of two styluses) to a location in response to a stimulus. Subjects performed with each of nine combinations of three display configurations and three control configurations (see Figure 3), using the most compatible mapping of the stimulus and response elements for each combination. The primary finding was that responses were fastest and most accurate when the display and control configurations corresponded spatially than when they did not. Fitts and Deininger (1954) examined different mappings of the stimulus and response elements. In the case of circular display and control arrangements (see Figure 3*a*), performance was much worse with a random mapping of the eight stimulus locations to the eight response locations than with a spatially compatible mapping in which each stimulus was mapped to its spatially corresponding response. This finding demonstrated the basic spatial compatibility effect that has been the subject of many subsequent studies. Almost equally important, performance was also much better with a mirror opposite mapping of stimuli to responses than with the random mapping. This finding implies that action selection benefits from being able to apply the same rule regardless of which stimulus occurs.

Spatial compatibility effects also occur when there are only two alternative stimulus positions, left and right, and two responses, left and right keypresses or movements of a joystick or finger, and regardless of whether the stimuli are lights or tones. Moreover, spatial correspondence not only benefits performance when stimulus location is relevant to the task but also when it is irrelevant. If a person is told to press a right key to the onset of a high pitch tone and a left key to onset of a low pitch tone, the responses are faster when the high pitch tone is in a right location (e.g., the right ear of a headphone) than when it is in a left location, and vice versa for the low pitch tone (Simon, 1990). This effect, which occurs for visual stimuli as well, is known as the *Simon effect* after its discoverer, J. R. Simon. The Simon effect has attracted an immense amount of research interest in the past 10 years because it allows examination of many fundamental issues concerning the relation between perception and action.

Accounts of S-R Compatibility Most accounts of S-R compatibility effects attribute them to two factors. One factor is direct, or automatic, activation of the corresponding response. The other is intentional translation of the stimulus into the desired response according to the instructions that have been provided for the task. The Simon effect is attributed entirely to the automatic activation factor, with intentional translation not considered to be involved because stimulus location is irrelevant to the task. The basic idea is that the corresponding response code is activated automatically by the stimulus at its onset, producing a tendency to select that response regardless of whether it is correct. Evidence suggests that this activation may dissipate across time, either through passive decay or active inhibition, because the Simon effect decreases as overall RT becomes longer (Hommel, 1993b; De Jong et al., 1994). In many situations, stimuli can be coded as left or right with respect to multiple frames of reference, as, for example, when there is a row of eight possible stimulus positions, four in the left hemispace and four in the right, with each of those divided into left and right pairs and left and right elements within the pairs. In such circumstances, stimulus position is coded relative to all frames of reference, with the magnitude of the Simon effect reflecting the combined correspondence effects for each position code (Lamberts et al., 1992; Roswarski and Proctor, 1996).

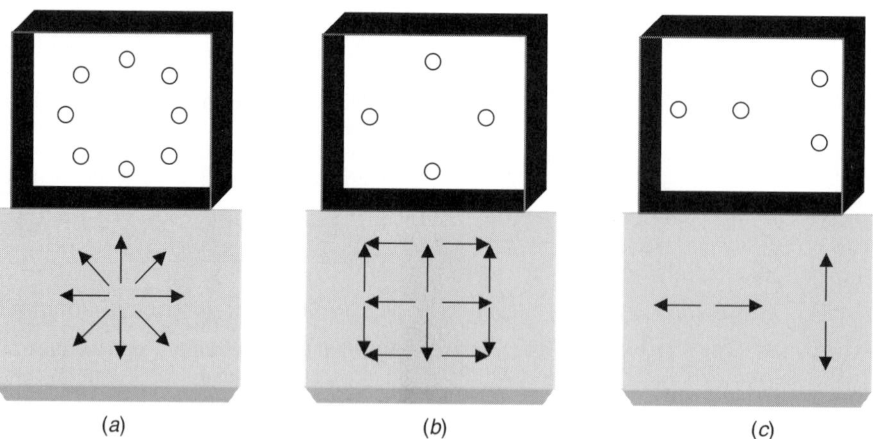

Figure 3 Three configurations of stimulus sets and response sets used by Fitts and Seeger (1953). Displays are shown in black boxes, with the stimulus lights shown as circles. Response panels are in gray, with the directions in which one or two styluses could be moved shown by arrows. As an example, an upper right stimulus location was indicated by the upper right light for stimulus set (*a*), the upper and right lights for stimulus set (*b*), and the right light of the left pair and upper light of the right pair for stimulus set (*c*). The upper right response was a movement of a stylus to the upper right response location for response sets (*a*) and (*b*), directly for response set (*a*) and indirectly through the right or upper position for response set (*b*), or of two styluses for response set (*c*), one to the right and the other up.

S-R compatibility proper is also presumed by many researchers to be determined in part by automatic activation of the corresponding response. As for the Simon effect, the activated response is correct when the mapping is compatible and incorrect when the mapping is incompatible. The most influential dual-route model, that of Kornblum et al. (1990), assumes that this automatic activation occurs regardless of the S-R mapping. However, certain results question this assumption (e.g., Read and Proctor, 2004). The intentional translation route is also presumed to play an important role in S-R compatibility effects, with translation being fastest when a "corresponding" rule can be applied, intermediate when some other rule can be applied (e.g., respond at the opposite position), and slowest when no simple rule is applicable and the specific response assigned to a stimulus must be retrieved.

Dimensional Overlap Although spatial location is an important factor influencing performance, it is by no means the only type of compatibility effect. Kornblum et al. (1990) introduced the term *dimensional overlap* to describe stimulus and response sets that are perceptually or conceptually similar. Left and right stimulus locations overlap with left and right response locations both perceptually and conceptually, and responding is fastest with the S-R mapping that maintains spatial correspondence (left stimulus to left response and right stimulus to right response) than with the mapping that does not. The words "left" and "right" mapped to keypress responses also produce a compatibility effect because of the conceptual correspondence between the words and the response dimension, but the effect is typically smaller than that for physical locations due to the absence of perceptual overlap.

S-R compatibility and Simon effects have been obtained for a number of different stimulus types with location or direction information, for example, direction of stimulus motion (Bosbach et al., 2004), induced direction of stimulus motion (Kerzel et al., 2001), and the direction of gaze of a face stimulus (Ansorge, 2003). They have also been obtained for typing letters on a keyboard, as a function of the positions in which the letters appear on a computer screen relative to the locations of the keys with which they are typed (Logan, 2003), elements in movement sequences (Inhoff et al., 1984), and clockwise versus counterclockwise rotations of a wheel (Wang et al., 2003; Proctor et al., 2004). Properties such as the durations of stimuli and responses (short and long; Kunde and Stöcker, 2002), positive or negative affective valence of a stimulus in relation to that of a response (De Houwer et al., 2001), and pitch of a tone with that of the vowels in syllable sequences (Rosenbaum et al., 1987) also yield compatibility effects. The basic point is that compatibility effects are likely to occur for any situation in which the relevant or irrelevant stimulus dimension has perceptual or conceptual overlap with the response dimension.

Action Goals and Structural Correspondence It is important to understand that S-R compatibility effects are determined largely by action goals and not the actual physical responses. This is illustrated by a study conducted by Hommel (1993a) in which subjects made a left or right keypress to a high or low pitch tone, which could occur in the left or right ear. The closure of the response key produced an action effect of turning on a light on the side opposite that on which the response was made. When instructed to turn on the left light to one tone pitch and the right light to the other, a Simon effect was obtained for which responses were faster when the tone location corresponded with the light location than when it did not, even though this condition was noncorresponding with respect to the key that was pressed. Similarly, when holding a wheel at the bottom, for which the direction of hand movements is incongruent with that of wheel movement, some subjects code the responses as left or right with respect to direction of hand movement and others with respect to direction of wheel movement, and these tendencies can be influenced to some extent by instructions that stress one response coding or the other and by controlled visual events (Guiard, 1983; Wang et al., 2003).

S-R compatibility effects can also occur for situations in which there is no perceptual or conceptual similarity of the stimulus and response sets, solely on the basis of correspondence of structural features. For example, if a stimulus set composed of two letters of two sizes, for which the letter identity feature is salient, and is mapped to a row of four response keys operated by the index and middle fingers of the left and right hands placed adjacent to each other, for which the left–right distinction is salient, performance is best for a mapping in which letter identity distinguishes the two leftmost and two rightmost responses (e.g., a left-to-right mapping of OozZ; Miller, 1982; Proctor and Reeve, 1985). The apparent basis for this S-R compatibility effect is that the salient stimulus feature (letter identity) corresponds to the salient response feature (left versus right side), allowing a simple translation rule to be applied.

Another instance of structural correspondence occurs when stimuli are arrayed along a dimension (e.g., up and down stimulus positions) that is orthogonal to the dimension along which responses are made (e.g., left and right switch movements). In such cases there is no apparent basis for a preference of one S-R mapping over the other. However, S-R compatibility effects occur, with the mapping of up to right and down to left producing better performance than the alternative mapping in many situations (Weeks and Proctor, 1990). Considerable evidence now exists indicating that orthogonal S-R compatibility effects of this type are a consequence of correspondence of bipolar stimulus and response codes (Cho and Proctor, 2003). For the vertical dimension, up tends to be coded as positive polarity and down as negative polarity, whereas for the horizontal dimension, right tends to be coded as positive polarity and left as negative polarity. The benefit for the up–right/down–left mapping is thus due to its maintaining correspondence of the code polarities, whereas the alternative mapping does not. Orthogonal S-R compatibility is affected by factors such as response eccentricity (where the responding limb is located across the frontal plane), whether the left or right hand is used for responding, and whether the hand posture is prone or supine. It is possible to account for all of these effects in a parsimonious manner by assuming that responses are coded relative to multiple frames of reference, the effects of which sum together to determine the overall difference in performance with the alternative mappings.

2.2.3 Sequential Effects

Because many human–machine interactions involve a succession of responses, it is also important to understand how action selection is influenced by immediately preceding events, which can be evaluated by examining the sequential effects that occur in choice–reaction tasks. The most common sequential effect is that the response to a stimulus is faster when the stimulus and response are the same as those on the preceding trial than when they are not (Bertelson, 1961). This repetition benefit increases in size as the number of S-R alternatives becomes larger and is greater for incompatible S-R mappings than for compatible ones (Soetens, 1998). Repetition effects have been attributed to two processes, residual activation from the preceding trial when the current trial is identical to it and intentional preparation for what is expected on the next trial (Soetens, 1998). The former contributes primarily when the interval between a response and onset of the next stimulus is short, whereas the latter contributes primarily when the interval is long. Several results indicate that both of these sequential effects are due primarily to response-selection processes (e.g., Pashler and Baylis, 1991; Campbell and Proctor, 1993).

Although sequential effects with respect to the immediately preceding trial have been most widely studied, higher-order repetition effects, which involve the sequence of the preceding two or three stimuli, also occur (Soetens, 1998). For two-choice tasks, at short response–stimulus intervals, where automatic activation predominates, a string of multiple repetitions is beneficial regardless of whether or not the present trial is a repetition of the immediately preceding one. In contrast, at long response–stimulus intervals, where expectancy is important, a prior string of repetition trials is beneficial if the current trial is also a repetition, and a prior string of alternation trials is beneficial if the current trial is an alternation.

When stimuli contain irrelevant stimulus information, as in the Stroop color-naming task in which the task is to name the ink color in which a conflicting color word is printed, RT is typically longer if the relevant stimulus value on a trial (e.g., the color red) is the same as that of the irrelevant information on the previous trial (e.g., the word *red*). This effect is called *negative priming* (Fox, 1995), with reference to the fact that the "priming" from the preceding trial slows RT compared to a neutral trial, for which there is no repetition of the relevant or irrelevant information from that trial. Negative priming was attributed initially to carryover of inhibition of the response tendency to the irrelevant information from the previous trial. Other suggested explanations include episodic retrieval (Neill and Valdes, 1992) and feature mismatch (Park and Kanwisher, 1994). According to the former, stimulus presentation initiates retrieval of prior episodes involving the stimulus, and, because the most recent episode is most likely to be retrieved, if the relevant stimulus information was irrelevant on the previous trial, the episode retrieved will include an "ignore" tag. According to the latter account, symbol identities are bound to objects and locations, and any change in the bindings from the preceding trial will produce negative priming.

Similar to negative priming, in which the irrelevant information from the previous trial interferes with processing the relevant feature on the current trial, several studies have also shown that in the Simon task, noncorresponding information from the previous trial can alter how the present trial is processed. That is, the Simon effect has been shown to be evident when the preceding trial was one for which the S-R locations corresponded and absent when it was one for which they did not (e.g., Stürmer et al., 1999). A suppression/release hypothesis has been proposed to account for this pattern of results (e.g., Stürmer et al., 1999). According to the suppression/release hypothesis, the Simon effect is absent following a noncorresponding trial because the direct response-selection route is suppressed since automatic activation of the response code corresponding to the stimulus location would lead to the wrong response alternative. This suppression is released, though, following a corresponding trial, which results in the stimulus activating the corresponding response and thus producing a Simon effect.

However, Hommel et al. (2004) noted that the analysis on which the suppression/release hypothesis is based collapses across mapping and location repetitions and nonrepetitions. According to Hommel's (1998b) event-file hypothesis, the stimulus features on a trial and the response made to them are integrated into an *event file*. When both stimulus features are repeated on the next trial, the response with which they were integrated on the previous trial is reactivated, and responding is facilitated. When both stimulus features change, neither feature was associated with the previous response, and the change in stimulus features signals a change in the response. Response selection is more difficult on trials for which one stimulus feature repeats and the other changes because one stimulus feature produces reactivation of the previous response and the other signals a change in response.

Hommel et al. (2004) and Notebaert et al. (2001) provided evidence that the pattern of repetition effects in the Simon task can be attributed to feature integration processes of the type specified by the event-file hypothesis rather than to suppression/release of the automatic route. That is, responses were faster when the relevant stimulus feature and irrelevant stimulus location both repeated or both changed than when only one stimulus feature repeated. The importance of the demonstration that feature integration processes may be the cause of the sequential effects is that those processes are independent from the processes contributing to the mean Simon effect, and it is not necessary to assume that the automatic response-selection route is suppressed and released on a trial-to-trial basis.

2.2.4 Preparation and Advance Information

When a stimulus to which a response is required occurs unexpectedly, the response to it will typically be slower than when it is expected. General preparation is studied in choice-reaction tasks by presenting a neutral warning signal at various intervals prior to the onset of the imperative stimulus. A common finding is that RT first decreases as the warning interval increases and then goes up as the warning interval is increased further, but the error rate first increases and then decreases. Bertelson (1967) demonstrated this relation in a study in which he varied the onset between an auditory warning click and a left or right visual stimulus to which a compatible keypress response was to be made. RT decreased by 20 ms for warning intervals of 0 to 150 ms and increased slightly as the interval increased to 300 ms, but the error rate increased from approximately 7% at the shortest intervals to about 10% at 100- and 150-ms intervals and decreased slightly at the longer intervals. Posner et al. (1973) obtained similar results for a two-choice task in which S-R compatibility was manipulated, and compatibility did not interact with the warning interval. These results suggest that the warning tone alters alertness, or readiness to respond, but does not affect the rate at which the information accumulates in the response-selection system.

People can also use informative cues to prepare for subsets of stimuli and responses. Leonard (1958) performed a task in which six stimulus lights were assigned compatibly to six response keys operated by the index, middle, and ring fingers of each hand. Of most concern was a condition in which either the three left or three right lights came on, precuing that subset as possible on that trial. RT decreased as precuing interval increased, being similar to that of a three-choice task when the precuing interval was 500 ms. Similar results have been obtained using four-choice tasks in which a benefit for precuing the two left or two right locations occurs within the first 500 ms of precue onset (Miller, 1982; Reeve and Proctor, 1984). However, when pairs of alternate locations are precued, a longer period of time is required to attain the maximal benefit of the precue. Reeve and Proctor (1984) showed that the benefit for precuing the two left or two right responses is also obtained when the hands are overlapped such that the index and middle fingers from the two hands are alternated, indicating that it reflects faster translation of the precued stimulus locations into possible response locations. Proctor and Reeve (1986) attributed this pattern of differential precuing benefits to the left–right distinction being salient for both stimulus and response sets, and Adam et al. (2003) have recently proposed a grouping model that expands on this theme.

2.2.5 Acquisition and Transfer of Action-Selection Skill

Response-selection efficiency improves with practice or training on a task. This improvement has been attributed to better pattern recognition or chunking of stimuli and responses (Newell and Rosenbloom, 1981), strengthening of associations between stimuli and responses (e.g., Anderson, 1982), and shifting from an algorithmic mode of processing to one based on retrieval of prior instances (Logan, 1988). The general idea behind all of these accounts is that practice results in performance becoming increasingly automatized. For virtually any task, the absolute benefit of a given amount of additional practice is a decreasing function of the amount of prior practice. Newell and Rosenbloom (1981) showed that the reduction in RT with practice using the mean data from groups of subjects in a variety of tasks is characterized well by a power function:

$$\mathrm{RT} = A + BN^{-\beta}$$

where A is the asymptotic RT, B the performance time on the first trial, N the number of practice trials, and β the learning rate.

Although the power function for practice has been regarded as a law to which any theory or model of skill acquisition must conform (e.g., Logan, 1988), recent evidence indicates that it does not provide the best fit for the practice functions of individual subjects. Heathcote et al. (2000) demonstrated that exponential functions of the following form provided better fits than power functions for individual data sets:

$$\mathrm{RT} = A + Be^{-\alpha N}$$

where α is the rate parameter. In relatively complex cognitive tasks such as mental arithmetic, individual subject data often show one or more abrupt changes (e.g., Haider and Frensch, 2002; Rickard, 2004), suggesting shifts in strategy. Delaney et al. (1998) showed that in such cases the individual improvement in solution time is fit better by separate power functions for each specific strategy than by a single power function for the entire task.

As noted earlier, practice reduces the slope of the Hick–Hyman function (e.g., Hyman, 1953), indicating that the cost associated with increased S-R uncertainty can be offset by allowing more practice. Seibel (1963) showed that after practice with more than 75,000 trials of all combinations of 10 lights mapped directly to 10 keys, RT for a task with 1023 alternatives was only about 25 ms slower than that for a task with 31 alternatives. Practice also benefits performance more for tasks with an incompatible S-R mapping than for ones with a compatible mapping (e.g., Fitts and Seeger, 1953). However, as a general rule, performance with an incompatible mapping does not reach the same level as that with a compatible mapping for the same amount of practice (e.g., Fitts and Seeger, 1953; Dutta and Proctor, 1992). The Simon effect for correspondence of irrelevant location information has also been shown to decrease with practice but not to be eliminated (e.g., Prinz et al., 1995; Proctor and Lu, 1999).

Some evidence suggests that the improvements that occur with practice in spatial choice tasks involve primarily the mappings of the stimuli to spatial response codes and not to the specific motor effectors. Proctor and Dutta (1993) had subjects perform two-choice spatial tasks for 10 blocks of 42 trials each. In alternating trial blocks, subjects performed with their hands uncrossed or crossed such that the right hand operated the left key and the left hand the right key. There was no cost associated with alternating the hand placements for the compatible or incompatible mapping when the mapping of stimulus locations to response locations remained constant across the two hand placements. However, when the mapping of stimulus locations to response locations was switched between blocks so that the hand used to respond to a stimulus remained constant across the two hand placements, there was a substantial cost for participants who alternated hand placements compared to those who did not, indicating the importance of maintaining a constant location mapping.

Although spatial S-R compatibility effects are not eliminated by practice, transfer studies show that changes in processing that occur as one task is practiced can affect performance of a subsequent, unrelated task. Proctor and Lu (1999) had subjects perform with an incompatible spatial mapping of left stimulus to right response and right stimulus to left response for 900 trials. When the subjects subsequently performed a Simon task, for which stimulus location was irrelevant, the Simon effect was reversed: RT was shorter when stimulus location did not correspond with that of the response than when it did. Subsequent studies by Tagliabue et al. (2000) and Vu et al. (2003) showed that as few as 72 practice trials with an incompatible spatial mapping eliminates the Simon effect in a transfer session and that this transfer effect remains present even a week after practice. Thus, a limited amount of practice produces new spatial S-R associations that continue to affect performance at least a week later.

Consistent with the point made earlier that S-R compatibility effects depend on how the stimulus and response sets are coded, Vu (2004) showed that this transfer effect of a spatially incompatible mapping to the Simon task is not an automatic consequence of having executed the spatially incompatible response during practice. Subjects performed 72 trials of a two-choice task for which stimuli occurred in a left or right location and stimulus color was nominally relevant. However, the correct response was always to the side that did not correspond to the stimulus location (i.e., if a left response was to be made to the color red, the red stimulus always occurred in the right location). Thus, the relation between stimulus and response locations was identical to that for a task with an incompatible spatial mapping; if subjects became aware of this spatially noncorresponding relation, they could base response selection on an "opposite" spatial rule instead of on stimulus color. Approximately half of the subjects indicated in a postexperiment interview that they were aware that the noncorresponding response was always correct, whereas half showed no awareness of this relation. The subjects who were aware of the relation showed a reversed Simon effect in the transfer session similar to that obtained from prior practice with an incompatible spatial mapping, whereas those who were unaware showed a normal Simon effect (faster responding when stimulus and response locations corresponded than when they did not) that was comparable to the effect size obtained with a group of subjects who did not perform the prior practice task.

2.3 Action Selection in Multiple-Task Performance

In earlier sections we focused on single-task performance. However, in many activities and jobs, people must engage in multiple tasks concurrently. This is true for an operator of a vehicle, a pilot of an aircraft, a university professor, and a secretary, among others. When more than one task set must be maintained, there is a cost in performance for all tasks even when the person devotes all of his or her attention to only one task at that time. This cost of concurrence that occurs with multiple-task performance has been of considerable interest in human factors and ergonomics. As a result, much research has been devoted to understanding and improving multiple-task performance.

2.3.1 Task Switching and Mixing Costs

Since the mid-1990s, there has been considerable interest in task switching. In task-switching studies, two different tasks are typically performed, one at a time, with the tasks presented in a fixed sequence (e.g., two trials of one task followed by two trials of the other task) or randomly with the current task indicated by a cue or instruction. The interval between successive trials or between the cue and the imperative stimulus can be varied to allow different amounts of time to prepare for the forthcoming task. Four phenomena are commonly obtained (Monsell, 2003):

1. *Mixing cost.* Responses are slower overall compared to when the same task is performed on all trials.
2. *Switch cost.* Responses are slower on trials for which the task switches from the previous trial than for those on which it repeats.
3. *Preparation benefit.* The switch cost is often reduced if the next task is known and adequate time for preparation is allowed.
4. *Residual cost.* Although reduced in magnitude, the switch cost is not eliminated completely, even with adequate time for preparation.

The switch cost is typically attributed to the time needed to change the task set. The fact that the switch cost can be reduced but not eliminated by preparation is often interpreted as evidence for at least two components to the switch cost. One component involves an intentional task-set reconfiguration process, and the other reflects exogenous, stimulus-driven processes, of which several have been suggested as possibilities. Rogers and Monsell (1995) proposed that this second component is a part of task-set reconfiguration that cannot be accomplished until it is initiated by stimulus components related to the task. Allport et al. (1994) attributed this second component to task-set inertia, with the idea that inhibition of the inappropriate task set on the previous trial carries over to the next trial, much as in negative priming. Finally, because the requirement to perform a task a few minutes later can slow performance of the current task, Waszak et al. (2003) proposed that associative retrieval of the task sets associated with the current stimulus is involved in the second component. Monsell (2003) describes the current situation as follows: "Most authors now acknowledge a plurality of causes, while continuing to argue over the exact blend" (p. 137).

One of the more intriguing findings in the task-switching literature is that the costs associated with mixing an easy task with a more difficult one are often larger for the easier task. For example, Allport et al. (1994) found that for Stroop stimuli, in which a color word is printed in an incongruent ink color, the costs were larger for the easy task of naming the word and ignoring the ink color than for the difficult task of naming the ink color and ignoring the word. Similarly, Shaffer (1965) and Vu and Proctor (2004) showed that in a two-choice spatial compatibility task, the advantage for the compatible spatial mapping is eliminated when compatible and incompatible mappings are mixed, with the mapping for the current trial signaled by a separate stimulus or feature. Proctor and Vu (2002) found that the advantage for the spatially compatible mapping is also eliminated when trials for which location is relevant to the task (signaled by white stimuli) are mixed with trials for which location is irrelevant and stimulus color (red or green) is the relevant dimension.

Proctor et al. (2003) used variations of the location relevant/irrelevant mixing task to examine why the spatial compatibility effect is eliminated under mixed conditions. When the colored stimuli were presented in a centered location, distinct from the left and right locations used for the location-relevant stimuli, a normal-sized S-R compatibility effect was obtained. Similarly, a normal-sized Simon effect was obtained when the locations in which the colored stimuli occurred varied along a spatial dimension (top and bottom locations) that was different from the dimension along which the location-relevant stimuli varied (left and right). However, when the stimuli for the two tasks were presented along the horizontal dimension, but in separate rows, the S-R compatibility effect was again absent. Overall, the findings from these experiments indicate that elimination of the spatial compatibility effect is not simply a consequence of having to switch between two tasks but rather to the two tasks sharing the same left–right spatial codes.

It is important to note that the elimination of the spatial compatibility effect under conditions of mixed mappings or tasks does not fully generalize to symbolic and verbal stimulus modes (Proctor and Vu, 2002). For centered arrows that point to the left or right, the compatibility effect is eliminated when compatible and incompatible mappings of arrow direction to left and right keypresses are mixed, but not when trials for which arrow direction is relevant are mixed with trials for which arrow color is relevant. Furthermore, when the stimuli are the centered words "left"–"right," the compatibility effect increases substantially in magnitude both when compatible and incompatible "word"-relevant mappings are mixed and when word-relevant trials are mixed with color-relevant word trials.

Many of the results for the elimination of the S-R compatibility effect with mixing are consistent with a dual-route model of the general type described earlier. According to such an account, response selection can be based on direct activation of the corresponding response when all trials are compatible, but the slower indirect route must be used when compatible trials are mixed with either incompatible trials or trials for which another stimulus dimension is relevant. The increase in the SRC effect that occurs with verbal stimuli under conditions of mixed mappings and tasks seems to reflect increased reliance on verbal mediation, that is, the name of the word. The most important point for application of the compatibility principle is that the benefit for a task with a compatible mapping may not be realized when that task is mixed with other less compatible tasks.

2.3.2 Psychological Refractory Period

Much research on multiple-task performance has focused on what is called the *psychological refractory period* (PRP) *effect* (see Pashler and Johnston, 1998, Lien and Proctor, 2002, for reviews). This phenomenon refers to slowing of RT for the second of two tasks that are performed in rapid succession. Peripheral sensory and motor processes can contribute to decrements in dual-task performance. For example, if you are looking at the display for a compact disk changer in your car, you cannot respond to visual events that occur outside, and if you are holding a cellular phone in one hand, you cannot use that hand to respond to other events. However, research on the PRP effect has indicated that the central processes involved in action selection seem to be the locus of a major limitation in performance.

In the typical PRP study, the subject is required to perform two speeded tasks. Task 1 may be to respond to a high or low pitch tone by saying "high" or "low" out loud, and task 2 may be to respond to the location of a visual stimulus by making a left or right key press. The stimulus-onset asynchrony (SOA, the interval between onsets of the task 1 stimulus, S1, and the task 2 stimulus, S2) is typically varied, either randomly within a block of trials or between blocks. The characteristic PRP effect is that RT is slowed, often considerably, for task 2 when the SOA is short (e.g., 50 ms) compared to when it is long (e.g., 800 ms).

The most widely accepted account of the PRP effect is what has been called the *central bottleneck model* (e.g., Welford, 1952; Pashler and Johnston, 1998). This model assumes that selection of the response for task 2 (R2) cannot begin until response selection for task 1 (R1) is completed (see Figure 4). The central bottleneck model has several testable implications that have tended to be confirmed by the data. First, increasing the duration of response selection for task 2 should not influence the magnitude of the PRP effect because response-selection processes occur after the bottleneck. This result has been obtained in several studies in which manipulations such as S-R compatibility for task 2 have been found to have additive effects with SOA, that is, to affect task 2 RT similarly at all SOAs (e.g., Pashler and Johnston, 1989; McCann and Johnston, 1992). Second, increasing the duration of stimulus-identification processes for task 2 by, for example, degrading S2 should reduce the PRP effect because this increase can be absorbed into the "slack" at short SOAs after which identification of S2 is completed but response selection for the task cannot begin. This predicted underadditive interaction has been obtained in several studies (e.g., Pashler and Johnston, 1989).

Numerous issues concerning the central response-selection bottleneck have been investigated in recent years. One issue is whether all processes associated with action selection are subject to the bottleneck, or only a subset. Consistent with the latter view, several studies have shown crosstalk correspondence effects such that the responses for both tasks 1 and 2 are faster when they correspond than when they do not (e.g., Hommel, 1998a; Lien et al., 2005), which suggests that activation of response codes occurs prior to the response-selection bottleneck. A second issue is whether the bottleneck is better conceived as being of limited capacity than all-or-none. Navon and Miller (2002) and Tombu and Jolicœur (2003) have argued that the evidence is most consistent with a central-capacity sharing model, in which attentional capacity can be allocated in different amounts to response selection for the two tasks. One finding that the capacity sharing account can explain that is difficult for the all-or-none bottleneck model is that RT for task 1, as well as that for task 2, sometimes increases at short SOAs.

Another issue is whether there is a structural bottleneck at all, or whether the bottleneck reflects a strategy adopted to perform the dual tasks as instructed. Meyer and Kieras (1997a,b) developed a computational model, implemented within their EPIC (executive-process interactive control) architecture, which consists of perceptual, cognitive, and motor components, that does not include a limit on central-processing capacity. The specific model developed for the PRP effect, called the strategic response-deferment model, includes an analysis of the processes involved in the performance of each individual task and of the executive control processes that coordinate the joint performance of the two tasks. Attention begins at the perceptual level, orienting focus (i.e., moving the eyes) on sensory input. Limits in the systems are attributed to the sensory and motor effectors, but not to the central processes. Central limitations arise from individuals' strategies for satisfying task demands (e.g., making sure that the responses for the two tasks are made in the instructed order). Specifically, according to the model, the PRP effect occurs when people adopt a conservative strategy of responding with high accuracy at the expense of speed. EPIC computational models can be developed for multitasking in real-world circumstances such as human–computer interaction and military aircraft operation, as well as for the PRP effect.

The view that the bottleneck is strategic implies that it should be possible to bypass its limitations. Greenwald and Shulman (1973) provided evidence suggesting that this is the case when two tasks are "ideomotor" compatible: that is, the feedback from the response is similar to the stimulus. Their ideomotor compatible

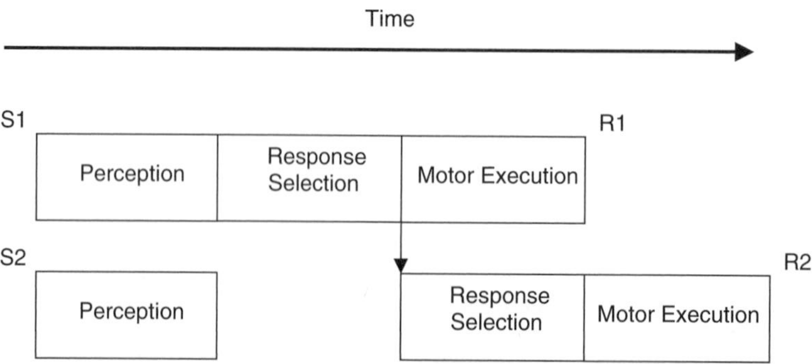

Figure 4 Central bottleneck model. Response selection for task 2 cannot begin until that for task 1 is completed. S1 and S2 are the stimuli for tasks 1 and 2, respectively, and R1 and R2 are the responses.

tasks were moving a joystick to a positioned left- or right-pointing arrow (task 1) and saying the name of an auditorily presented letter (task 2). Greenwald and Shulman's experiment 2 showed no PRP effect when both tasks were ideomotor compatible, although an effect was apparent when only one task was. However, other experiments in which the two tasks were ideomotor compatible, including Greenwald and Shulman's experiment 1, have consistently shown PRP effects (e.g., Lien et al., 2002), although in Greenwald's (2003) study it appeared in the error rates when there was a strong emphasis on response speed at the expense of accuracy. Although the bottleneck does not seem to be eliminated by using two ideomotor tasks, Lien et al. (2005) provided evidence that it may be moved later in the processing stream.

Under certain conditions, the PRP effect can be virtually eliminated with considerable practice, a finding that some authors have interpreted as evidence against a central bottleneck (e.g., Schumacher et al., 2001). However, this elimination is accomplished primarily through the reduction of RT for task 1, which leaves open the possibility that the bottleneck is "latent" and not affecting performance (Ruthruff et al., 2003). That is, because the speed of performing task 1 improves with practice, even at short SOAs, task 1 response selection can be completed prior to the time at which response selection for task 2 is ready to begin. For practical purposes, though, the messages to take from the PRP research is that it is difficult to select different actions concurrently, but many factors can modulate the magnitude of the cost associated with trying to do so.

3 MOTOR CONTROL

3.1 Methods

Whereas action selection focuses primarily on choice between action goals, motor control is concerned mainly with the execution of movements to carry out the desired actions. Tasks used to study motor control typically involve movement of one or more limbs, execution of sequences of events, or control of a cursor following a target that is to be tracked. For example, a person may be asked to make an aimed movement from a start key to a target location under various conditions, and measures such as movement time and accuracy can be recorded. Some issues relevant to human factors examined include the nature of movement representation, the role of sensory feedback in movement execution, and the way in which motor actions are sequenced.

3.2 Control of Action

Motor control is achieved in two different ways, open loop and closed loop. *Open-loop control* is based on an internal model, called a *motor plan* or *motor program*, which provides a set of movement commands. Two pieces of evidence for motor plans include the fact that deafferented monkeys and humans can still make movements (e.g., Taub and Berman, 1968) and the time to initiate a movement increases as the number of elements to be performed increases (e.g., Henry and Rogers, 1960; Monsell, 1986). *Closed-loop control*, in contrast, relies on sensory feedback, comparing the feedback to a desired state, and making the necessary corrections when a difference is detected. The advantages and disadvantages of open-and closed-loop control are the opposite of each other. A movement under open-loop control can be executed quickly, without a delay to process feedback, but at a cost of limited accuracy. In contrast, closed-loop control is slower but more accurate. Not surprisingly, both types of control are often combined, with open-loop control used to approximate a desired action and closed-loop control serving to reduce the deviation of the actual state from the intended state as the action is executed.

3.2.1 Fitts's Law

As indicated in the quote with which the chapter began, Fitts's law, which specifies the time to make aimed movements to a target location (Fitts, 1954), is one of the most widely established quantitative relations in behavioral research. As originally formulated by Fitts, the law is

$$\text{movement time} = a + b\ \log_2(2D/W)$$

where a and b are constants, D is the distance to the target, and W is the target width (see Figure 5). Two important points of Fitts's law are that (1) movement time increases as movement distance increases and (2) movement time decreases as target width increases. It is a speed–accuracy relation in the sense that movement time must be longer when more precise movements are required. Fitts's law provides an accurate description of movement time in many situations, although alternative formulations can provide better fits for certain specific situations. The speed–accuracy

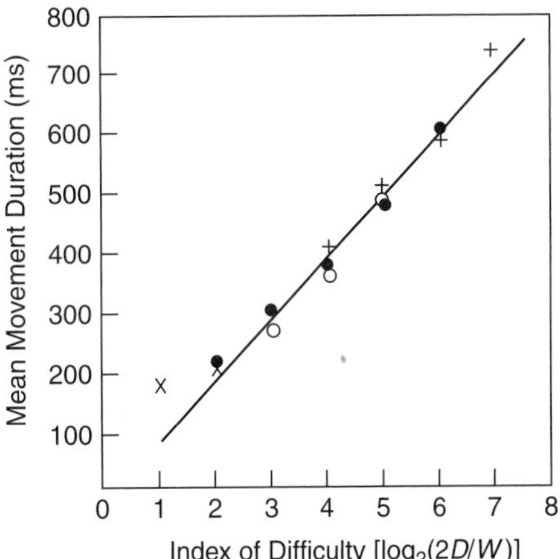

Figure 5 Fitts's law: movement time increases as a function of the index of difficulty [$\log_2(2D/W)$].

relation captured by Fitts's law is a consequence of both open- and closed-loop components. Meyer et al. (1988) provided the most complete account of the relation, a stochastic optimized-submovement model. This model assumes that aimed movements consist of a primary submovement and an optional secondary submovement. Fitts's law arises as a consequence of (1) programming each movement to minimize average movement time while maintaining a high frequency of "hitting" the target, and (2) making the secondary, corrective submovement when the index of difficulty is high.

Fitts's law is of considerable value in human factors because it is quite robust and is applicable to many tasks of concern to human factors professionals. The relation holds not only for tasks that require movement of a finger to a target location (e.g., when using an ATM machine) but also for tasks such as placing washers on pegs and inserting pins into holes (Fitts, 1954), using tweezers under a microscope (Langolf and Hancock, 1975), and making aimed movements underwater (Kerr, 1973). Variants of Fitts's law can also be used to model limb and head movements with extended probes such as screwdrivers and helmet-mounted interfaces (Baird et al., 2002). The slope of the Fitts's law function has been used to evaluate the efficiency of various ways for moving a computer cursor to a target position. For example, Card et al. (1978) showed that a computer mouse produced smaller slopes than text keys, step keys (arrows), and a joystick for the task of positioning a cursor on a desired area of text and pressing a button or key.

Size and distance are only two of many constraints that influence movement time (Heuer, 2003). Movement time will be longer, for example, if the target must be grasped instead of just touched. Moreover, for objects that must be grasped, movement time depends on properties of the object, being longer to one that has to be grasped cautiously (e.g., a knife) than one that does not.

3.2.2 Motor Preparation and Advance Specification of Movement Properties

Movement of a limb is preceded by preparatory processes at various levels of the motor system. For a simple voluntary movement such as a keypress, a negative potential in the electroencephalogram (EEG) begins as much as 1 second before the movement itself, with this potential being stronger over the contralatateral cerebral hemisphere (which controls the finger) 100 to 200 ms before responding. This asymmetry, called the *lateralized readiness potential*, provides an index of being prepared to respond with a limb on one or the other side of the body (Eimer, 1998). In reaction tasks, this preparation may involve what is sometimes called a *response set*, or a readiness to respond, that is, response activation just below the threshold for initiating the response. However, motor preparation depends on the response that is to be performed. As noted, simple RT increases as the number of components of which the to-be-executed movement is composed increases. Also, motor preparation is sensitive to the end state of an action. For example, when executing an action that requires grasping a bar with a pointer and placing it in a specified target position, the bar will be grasped in a manner that minimizes the awkwardness, or maximizes the comfort, of the position in which the arm will end up (Rosenbaum et al., 1990).

Advance specification of movement parameters has been studied using a choice-RT procedure in which subjects must choose between aimed-movement responses that differ in, for example, arm (left or right), direction (toward or away), and extent (near or far). One or more parameters are precued prior to presentation of the stimulus to which the person is to respond, the idea being that RT will decrease if those parameters can be specified in advance (Rosenbaum, 1983). The results of such studies have generally supported the view that movement features can be specified in variable order; that is, there is a benefit of precuing any parameter in isolation or in combination with another. Thus, the results support those described in the section on action selection, which indicated that people can take advantage of virtually any advance information that reduces the possible stimulus and response events. A major drawback of using the movement precuing technique to infer characteristics of parameter specification is that the particular patterns of results may be determined more by S-R compatibility rather than by the motoric preparation process itself (e.g., Goodman and Kelso, 1980; Dornier and Reeve, 1992).

3.2.3 Visual Feedback

Another issue in the control of movements is the role of visual feedback. Woodworth (1899) had people repeatedly draw lines of a specified length on a roll of paper moving through a vertical slot in a tabletop. The rate of movement in drawing the lines was set by a metronome that beat from 20 to 200 times each minute, with one complete movement cycle to be made for each beat. Subjects performed the task with their eyes open or closed. At rates of 180 per minute or greater, movement accuracy was equivalent for the two conditions, indicating that visual feedback had no effect on performance. However, at rates of 140 per minute or less, performance was better with the eyes open. Consequently, Woodworth concluded that the minimum time required to process visual feedback was 450 ms.

Subsequent studies have reduced this estimate substantially. Keele and Posner (1968) had people perform a discrete movement of a stylus to a target that, in separate pacing conditions, was to be approximately 150, 250, 350, or 450 ms in duration. The lights turned off at the initiation of the movement on half of the trials, without foreknowledge of the performer. Movement accuracy was better with the lights turned on than off in all but the fastest pacing condition, leading Keele and Posner to conclude that the minimum duration for processing visual feedback is between 190 and 260 ms. Moreover, when people know in advance whether visual feedback will be present, results indicate that feedback can be used for movements with durations of only slightly longer than 100 ms (Zelaznik et al., 1983).

It might be thought that the role of visual feedback would decrease as a movement task is practiced, but evidence indicates that vision remains important. For example, Proteau and Cournoyer (1992) had people perform 150 trials of a task of moving a stylus to a target with either full vision, vision of both the stylus and the target, or vision of the target only. Performance during these practice trials was best with full vision and worst with vision of the target only. However, when the visual information was eliminated in a subsequent transfer block, performance was worst for those people who had practiced with full vision and worst for those who had practiced with vision of only the target only.

3.3 Coordination of Effectors

To perform many tasks well, it is necessary to coordinate the effectors. For example, when operating a manual transmission, the movements of the foot on the gas pedal must be coordinated with the shifting of gears controlled by the arm and hand. This example illustrates that one factor determining the coordination pattern is the constraints imposed by the task that is to be performed. These coordination patterns are flexible, within the structural constraints imposed by the action system.

For tasks involving bimanual movements, there is a strong tendency toward mirror symmetry; that is, it is generally easy to perform symmetric movements of the arms, as in drawing two circles simultaneously with each hand. Moreover, intended asymmetric movement patterns will tend more toward symmetry in duration and timing than they should. This symmetry tendency has been studied extensively for tasks involving bimanual oscillations of the index fingers: It is easier to maintain the instructed oscillatory pattern if told to make symmetrical movements of the fingers inward and outward together than if told to make parallel movements leftward and rightward together (see Figure 6*a* and *b*). The symmetry tendency in bimanual oscillatory movements and for other bimanual tasks has traditionally been attributed to coactivation of homologous muscles (e.g., Kelso, 1984).

However, Mechsner et al. (2001) recently presented evidence that the bias is toward spatial symmetry and not motor symmetry. To dissociate motor symmetry from spatial symmetry, Mechsner et al. had subjects perform with the palm up for one hand and the palm down for the other (see Figure 6*c* and *d*). A tendency toward coactivation of homologous muscles would predict that, in this case, the bias should be toward parallel oscillation, whereas a tendency toward spatial symmetry should still show the bias toward symmetrical oscillation. The latter result was in fact obtained, with the bias toward symmetrical oscillation being just as strong when one palm faced up and the other down as when both hands were palm down or both palm up. Mechsner et al. and Mechsner and Knoblich (2004) obtained similar results for tasks in which two fingers of each hand are periodically

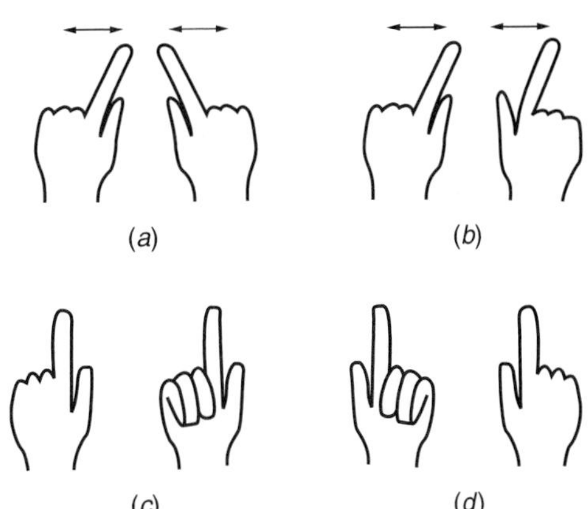

Figure 6 (*a*) Symmetric and (*b*) asymmetric patterns of finger movements, for which symmetric movements are typically easier. Parts (*c*) and (*d*) show conditions examined by Mechsner et al. (2001), in which spatial symmetry and muscle homology are dissociated.

tapped together by comparing congruous conditions for which the fingers from the two hands were the same (e.g., index and middle fingers of each hand, or middle and ring fingers of each hand) and incongruous conditions for which they were different (index and middle finger for one hand, and middle and ring finger for the other). Mechsner and Knoblich concluded: "Homology of active fingers, muscular portions, and thus motor commands plays virtually no role in defining preferred coordination patterns, in particular the symmetry tendency" (p. 502) and that the symmetry advantage "originates at a more abstract level, in connection with planning processes involving perceptual anticipation" (p. 502).

3.4 Sequencing and Timing of Action

How sequences of actions are planned and executed is one of the central problems of concern in the area of motor control (Rosenbaum, 2002). Most discussions of this problem originate with Lashley's (1951) well-known book chapter in which he presented evidence against an associative chaining account of movement sequences, according to which the feedback from each movement in the sequence provides the stimulus for the next movement. Instead, Lashley argued that the sequences are controlled centrally by motor plans.

Considerable evidence is consistent with the idea that these motor plans are structured hierarchically. For example, Povel and Collard (1982) had subjects perform sequences of six taps with the four fingers on a hand (excluding the thumb). A sequence was practiced until it could be performed from memory, and then trials were conducted for which the sequence was to be executed as rapidly as possible. The sequences differed in terms of the nature and extent of their structure. For example, the patterns 1–2–3–2–3–4 and 2–3–4–1–2–3, where the numbers 1, 2, 3, and 4 designate the index, middle, ring, and little fingers, respectively, can each be coded as two separate ordered subsets. Povel and Collard found that the pattern of latencies between each successive tap was predicted well by a model that assumed the memory representation for the sequence was coded in a hierarchical decision tree (see Figure 7), with the movement elements represented at the lowest level, which was then interpreted by a decoding process that traversed the decision tree from left to right. Interresponse latencies were predicted well by the number of links that had to be traversed in the tree between successive responses. For example, for the sequences shown above, the longest latencies were between the start signal and the first tap and between the third and fourth taps, both of which required two levels of the tree to be traversed.

Although many results in tasks requiring execution of sequential actions are in agreement with predictions of hierarchical models, it should be noted that it is not so simple to rule out serial association models. Context-sensitive association models, which allow elements farther back than just the immediately preceding one to affect performance, can generate many of the same result patterns as hierarchical models (e.g., Wickelgren, 1969).

Beginning with a study by Nissen and Bullemer (1987), numerous experiments have been conducted on incidental learning of trial sequences in choice-reaction tasks. Nissen and Bullemer had subjects perform a four-choice RT task in which the stimulus on a trial appeared in one of four horizontal locations, and the response was the corresponding location of one of four buttons also arranged in a row, made with the middle and index fingers of the left and right hands. Subjects received eight blocks of 100 trials for which the stimuli were presented in random order or in a sequence that repeated every 10 trials. There was a slight decrease in RT of about 20 ms across blocks with the random order, but a much larger one of about 150 ms for the repeating sequence. Nissen and Bullemer presented evidence that they interpreted as indicating that such

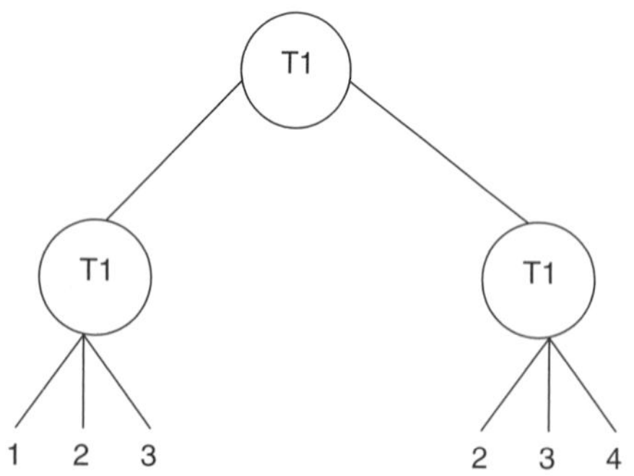

Figure 7 Hierarchical representation of the movement sequence 1–2–3–2–3–4. T1 represents the operation *transpose to an adjacent finger*. The tree traversal model predicts longer latencies for the first and fourth elements in the movement sequence, as Povel and Collard (1982) found.

sequence learning can occur without awareness, but this remains a contentious issue (see, e.g., Destrebecqz and Cleeremans, 2001; Wilkinson and Shanks, 2004).

Of most interest to present concerns is the nature of the representation that is being learned in sequential tasks. In general, the evidence has tended to indicate that the learning is not effector specific, because it can transfer to a different set of effectors (e.g., Cohen et al., 1990), and it does not seem to be perceptual (e.g., Willingham, 1999) either. Willingham et al. (2000) recently concluded that the sequence learning occurs in a part of the motor system involving response locations but not specific effectors or muscle groups. They showed that subjects who practiced the task using a keyboard with one arrangement of response keys during a training phase showed no benefit from the repeating stimulus sequence when subsequently transferred to a keyboard with a different arrangement of response keys. In another experiment, Willingham et al. also showed that subjects who switched from performing the task with the hands crossed in practice to performing with them uncrossed in a transfer session, such that the hand operating each key was switched, showed no cost relative to subjects who used the uncrossed hand placement throughout. Willingham et al. rejected an explanation in terms of S-R associations because Willingham (1999) found excellent transfer as long as the response sequence remained the same in both the practice and transfer sessions even when the stimulus set was changed from digits to spatial locations or the mapping of spatial stimuli to responses was changed. Note that Willingham et al.'s conclusions are similar to those reached by Mechsner and Knoblich (2004) for bimanual coordination in that much of the motor control and learning occurs at a level of spatial response relations rather than the muscles used to execute the actions.

Whereas in some situations, the speed with which a sequence of actions is executed is important, in others the timing of the actions is crucial. One influential model of response timing is that of Wing and Kristofferson (1973), who developed it to explain the timing of successive, discrete tapping responses. According to this model, two processes control the timing of the responses. One is an internal clock that generates trigger pulses that can be used to time the delay (by the number of pulses) and initiate motor responses. The other is a delay process between when a trigger pulse initiates a response and when the movement is actually executed. The interval between successive pulses is assumed to be an independent random variable, as is the interval between a trigger pulse and the response that it initiates. One key prediction of the model is that the variance of the interval between responses should increase as the delay between the responses increases, due to the variability of the internal clock. Another prediction is that adjacent interresponse intervals should be negatively correlated, due to the variability of the delay process. Although these predictions have tended to be confirmed and Wing and Kristofferson's model has generally been successful, there is an ongoing dispute as to whether timing of all motor acts is based on a general-purpose internal clock (see, e.g., Ivry et al., 2002).

3.5 Motor Learning and Acquisition of Skill

Performance of virtually any perceptual-motor skill improves substantially with practice, becoming faster, more accurate, and more fluid. Considerable research has been devoted to understanding the ways in which training and augmented feedback (knowledge of results or performance) should be scheduled to optimize the acquisition of motor skill. Some of this research is described in the following sections.

3.5.1 Practice Schedules

A long-standing issue in the study of motor skill has been whether better learning results from a distributed practice schedule, in which there is a break or rest period between performance attempts, or a massed practice schedule, for which there is not. Although distributed practice often leads to better performance during acquisition of a motor skill, it does not necessarily result in better learning and retention. Lee and Genovese (1988) conducted a meta-analysis of the literature on the distribution of practice and concluded that massed practice depresses learning somewhat in motor tasks that require continuous movements, such as cycling. For discrete tasks, however, massed practice may even be beneficial to learning. Carron (1969) found better retention for massed than distributed practice when the task required the three discrete steps of picking up a dowel out of a hole, flipping the ends of the dowel, and putting the dowel back into the hole. Lee and Genovese (1989) directly compared discrete and continuous versions of a task in which the interval between when a stylus was lifted from one plate and moved to another was to be 500 ms. A single movement was made for the discrete version of the task, whereas 20 movements back and forth were performed in succession for the continuous version. Massed practice produced better retention than distributed practice did for the discrete task, but distributed practice produced better retention for the continuous task.

Lee and Genovese's (1988, 1989) conclusions hold for distribution of practice within a session. However, for performance across practice sessions, evidence suggests that shorter practice sessions spread over more days are more effective than longer sessions spread over fewer days. For example, Baddeley and Longman (1978) gave postal trainees 60 hours of practice learning to operate mail-sorting machines. The trainees who received this practice in 1-hour/day sessions over 12 weeks learned the task much better than those who received the practice in 2-hour sessions, twice daily, over three weeks. One factor contributing to the smaller benefit in learning with the longer, more massed sessions is that the sessions may get tiresome, causing people's attention to wander.

Another issue is whether different tasks or task variations should be practiced individually, in distinct practice blocks, or mixed together within a practice block. Retention and transfer of motor tasks has been shown typically to be better when the tasks are practiced in random order than in distinct blocks, even though performance during the practice session is typically better under blocked conditions. This finding, called the *contextual interference effect*, was first demonstrated by Shea and Morgan (1979). They had subjects knock down three of six barriers in a specified order as quickly as possible when a stimulus light occurred. During the acquisition phase, each subject performed three different versions of the task, which differed with respect to the barriers that were to be knocked down and their order. For half of the subjects, the three barrier conditions were practiced in distinct trial blocks, whereas for the other half, the barrier conditions were practiced in a random order. Although performance during acquisition was consistently faster for the blocked group than for the random group, performance on retention tests conducted 10 min or 10 days later was faster for the random group.

The contextual interference effect has been replicated in numerous studies and tasks (see, e.g., Magill and Hall, 1990; Tsutsui et al., 1998). Shea and Morgan (1979) originally explained the contextual interference effect as follows: Because performance during practice is more difficult for the random group than for the blocked group, the random group is forced to use multiple processing strategies, leading to more elaborate long-term memory representations and better retention. Lee and Magill (1985) proposed instead that the benefit of random practice arises from subjects often forgetting how the task to be performed on the current trial was done previously, requiring that an action plan be reconstructed. This reconstruction process results in a more highly developed memory trace. Although these accounts differ slightly in their details, they make the similar point that random practice schedules lead to better long-term retention because they require deeper or more elaborate processing of the movements required.

Because real-world perceptual-motor skills may be quite complex, another issue that arises is whether it is beneficial for learning to practice parts of a task in isolation before performing the whole, integrated task. Three types of part-task, or part-whole, practice can be distinguished (Wightman and Lintern 1985): *Segmentation* involves decomposing a task into successive subtasks, which are performed in isolation or in groups and then recombined; *fractionation* involves separate performance of subtasks that typically are performed simultaneously and then recombining them; *simplification* involves practicing a reduced version of a task that is easier to perform (such as using training wheels on a bicycle) before performing the complete version. Part-task training is often beneficial, and the results can be striking, as illustrated by Frederiksen and White's (1989) study, involving performance of a video game called Space Fortress that entailed learning and coordinating many perceptual-motor and cognitive components. In that study, subjects who received part-task training on the key task components performed about 25% better over the last five of eight whole-game transfer blocks than did subjects who received whole-game practice (see Figure 8), and this difference showed no sign of diminishing. Although part-task training is beneficial for complex tasks that require learning complex rules and relations and coordinating the components, it is less beneficial for motor skills composed of several elements, such as a tennis serve, where practicing one element in isolation shows at most small transfer to the complete task (e.g., Lersten, 1968).

3.5.2 Provision of Feedback

Intrinsic feedback arises from movement, and this sensory information is a natural consequence of action. For example, as described previously, several types of visual and proprioceptive feedback are typically associated with moving a limb from a beginning location to a target location. Of more concern for motor learning, though, is extrinsic, or augmented, feedback, which is information that is not inherent to performing a task itself. Two types of extrinsic feedback are typically distinguished, *knowledge of results*, which is information about the outcome of the action, and *knowledge of performance*, which is feedback concerning how the action was executed.

Knowledge of results (KR) is particularly important for motor learning when the intrinsic feedback for the task itself does not provide an indication of whether the goal was achieved. For example, in learning to throw darts at a target, the extrinsic KR is not of extreme importance because intrinsic visual feedback provides information about the amount of error in the throws. However, even in this case, KR may provide motivation to the performer and reinforcement of their actions, and knowledge of performance (e.g., whether the throwing motion was appropriate) may also be beneficial. If the task is one of learning to throw darts in the dark, KR increases in importance because there is no longer visual feedback to provide information about the accuracy of the throws. Many issues concerning KR have been investigated, including the precision of the information conveyed and the schedule by which it is conveyed.

Feedback can be given with varying precision. For example, when performing a task that requires contact with a target at a specified time window, say, 490 to 500 ms after movement initiation, the person may be told whether or not the movement was completed within the time window (qualitative KR) or how many milliseconds shorter or longer the movement was than allowed by the window (quantitative KR). Qualitative feedback can be effective, particularly at early stages of practice when the errors are often large, but people tend to learn better when KR is quantitative than when it is just qualitative (Magill and Wood, 1986; Reeve et al., 1990).

Although it may seem that it is best to provide feedback on every trial, research has indicated to the contrary. For example, Winstein and Schmidt (1990) had people learn to produce a lever-movement pattern, consisting of four segments, in 800 ms. Some subjects received KR after every trial during acquisition, whereas others received KR on only half of the trials. The two groups performed similarly during acquisition, but those subjects who received feedback after every trial did substantially worse than the other group on a delayed retention test. Summary KR, for which feedback about a subset of trials is not presented until the subset is completed, has also been

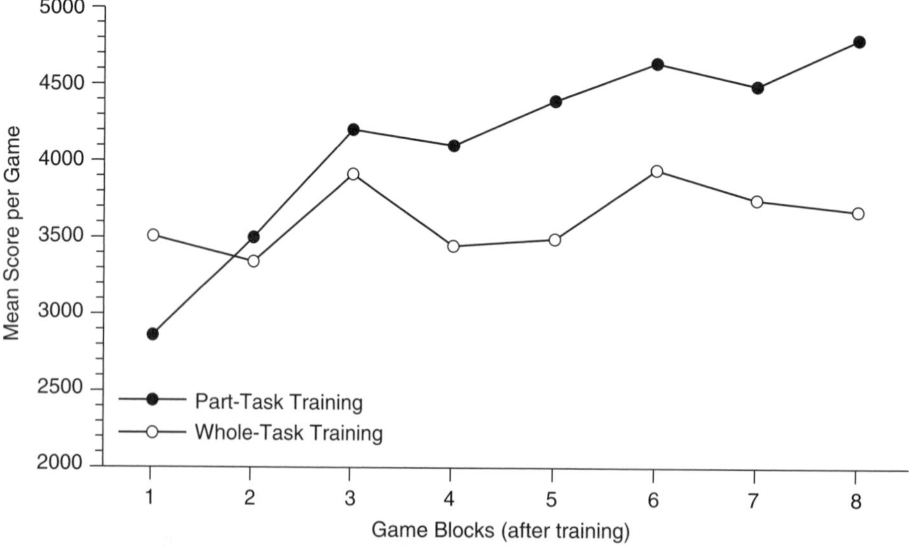

Figure 8 Mean score of Frederiksen and White's (1989) subjects on eight game blocks of Space Fortress in the transfer session after receiving part- or whole-task training in a prior session.

found to be successful (Lavery, 1962; Schmidt et al., 1989). Schmidt et al. had people learn a timed lever-movement task similar to that used by Winstein and Schmidt (1990), providing summary KR after 1, 5, 10, or 15 trials. A delayed retention test showed that learning was best when summary KR was provided every 15 trials and worst when KR was provided every trial. The apparent reason why it is best not to provide feedback on every performance attempt is that the person comes to depend on it. Thus, much like blocked practice of the same task, providing feedback on every trial does not force the person to engage in the more effortful information processing that is necessary to produce enduring memory traces needed for long-term performance.

4 SUMMARY AND CONCLUSIONS

Human–machine interactions involve a succession of reciprocal actions taken by the human and the machine. For performance of the human component to be optimal, it is necessary not only to consider how the machine should display information regarding its states and activities to the human, but also to take into account the processes by which the human selects and executes actions in the sequence of the interaction. Selection and control of action have been studied since the earliest days of research on human performance, and research in these areas continues to produce significant empirical and theoretical advances, several of which have been summarized in this chapter. Because the purpose of the chapter is to provide readers with an overview of the topic of selection and control of action, readers are encouraged to refer to more detailed information on topics of interest in chapters by Rosenbaum (2002), Heuer (2003), and Proctor and Vu (2003), and books by Rosenbaum (1991), Proctor and Dutta (1995), Sanders (1998), and Schmidt and Lee (1999), and other sources.

This chapter showed that the relations between choice uncertainty and response time, captured by the Hick–Hyman law, movement difficulty and movement time, conveyed by Fitts's law, and amount of practice and performance time, depicted by the power law of practice, follow quantitative laws that can be applied to specific research and design issues in human factors and ergonomics. In addition, many qualitative principles are apparent from the research that are directly applicable to human factors. These include:

- The relative speed and accuracy of responding in a situation depends in part on the setting of response thresholds, or how much noisy evidence needs to be sampled before deciding which alternative action to select.
- Although response time increases as the number of alternatives increases, the cost of additional alternatives is reduced when compatibility is high or the performer is highly practiced.
- Compatible spatial relations and mappings typically yield better performance than spatially incompatible ones.
- Compatibility effects are not restricted to spatial relations but occur for stimulus and response sets that have perceptual, conceptual, or structural similarity of any type.
- Compatibility effects occur when an irrelevant dimension of the stimulus set shares similarity with the relevant response dimension.
- For many situations in which compatible mappings are mixed with less compatible ones, the benefit of compatibility is lost.
- When actions are not performed in isolation, the context of preceding events can affect performance significantly.
- Advance information can be used to prepare subsets of responses.
- Improvements in response-selection efficiency with practice that occur in a variety of tasks involve primarily spatial locations of the actions and their relation to the stimuli, not the effectors used to accomplish the actions.
- Small amounts of experience with novel relations may influence performance after a long delay, even when those relations are no longer relevant to the task.
- Costs that are associated with mixing and switching tasks can be only partly overcome by advance preparation.
- It is difficult to select an action for more than one task at a time, although the costs in doing so can be reduced by using highly compatible tasks and with practice.
- Many constraints influence movement time, and the particular way in which an action will be carried out needs to be accommodated when designing for humans.
- Feedback of various types is important for motor control and acquisition of perceptual-motor skills.
- The tendency toward symmetry in preferred bimanual coordination patterns is primarily one of spatial symmetry, not of homologous muscles.
- Practice and feedback schedules that produce the best performance of perceptual-motor skills during the acquisition phase often do not promote learning and retention of the skills.
- Part-task training can be an effective means of teaching someone how to perform complex tasks.

Beyond these general laws and principles, research has yielded many details concerning the factors that are critical to performance in specific situations. Moreover, models of various types, some qualitative and some quantitative, have been developed for various domains of phenomena that provide relatively accurate descriptions and predictions of how performance will be affected by numerous variables. The laws, principles, and model characteristics can be incorporated into cognitive architectures such as EPIC (Meyer and Kieras, 1997 a,b) and ACT-R/PM (Byrne and Anderson, 1998; Byrne, 2001), along with other facts, to develop computational models that enable quantitative predictions to be derived for complex tasks of the type encountered in much of human factors and ergonomics.

REFERENCES

Adam, J. J., Hommel, B., and Umiltà, C. (2003), "Preparing for Perception and Action, I: The Role of Grouping in the Response-Cuing Paradigm," *Cognitive Psychology*, Vol. 46, pp. 302–358.

Allport, A., Styles, E. A., and Hsieh, H. (1994), "Shifting Intentional Set: Exploring the Dynamic Control of Tasks," in *Attention and Performance XV*, C. Umiltà and M. Moscovitch, Eds., MIT Press, Cambridge, MA, pp. 421–452.

Anderson, J. R. (1982), "Acquisition of Cognitive Skill," *Psychological Review*, Vol. 89, pp. 369–406.

Ansorge, U. (2003), "Spatial Simon Effects and Compatibility Effects Induced by Observed Gaze Direction," *Visual Cognition*, Vol. 10, pp. 363–383.

Baddeley, A. D., and Longman, D. J. A. (1978), "The Influence of Length and Frequency of Training Session on the Rate of Learning to Type," *Ergonomics*, Vol. 21, pp. 627–635.

Baird, K. M., Hoffmann, E. R., and Drury, C. G. (2002), "The Effects of Probe Length on Fitts' Law," *Applied Ergonomics*, Vol. 33, pp. 9–14.

Band, G. P. H., Ridderinkhof, K. R., and van der Molen, M. W. (2003), "Speed–Accuracy Modulation in Case of Conflict: The Roles of Activation and Inhibition," *Psychological Research*, Vol. 67, pp. 266–279.

Beh, H. C., Roberts, R. D., and Prichard-Levy, A. (1994), "The Relationship Between Intelligence and Choice Reaction Time Within the Framework of an Extended Model of Hick's Law: A Preliminary Report," *Personality and Individual Differences*, Vol. 16, pp. 891–897.

Bertelson, P. (1961), "Sequential Redundancy and Speed in a Serial Two-Choice Responding Task," *Quarterly Journal of Experimental Psychology*, Vol. 13, pp. 90–102.

Bertelson, P. (1967), "The Time Course of Preparation," *Quarterly Journal of Experimental Psychology*, Vol. 19, pp. 272–279.

Bosbach, S., Prinz, W., and Kerzel, D. (2004), "A Simon Effect with Stationary Moving Stimuli," *Journal of Experimental Psychology: Human Perception and Performance*, Vol. 30, pp. 39–55.

Byrne, M. D. (2001), "ACT-R/PM and Menu Selection: Applying a Cognitive Architecture to HCI," *International Journal of Human–Computer Studies*, Vol. 55, pp. 41–84.

Byrne, M. D., and Anderson, J. R. (1998), "Perception and Action," in *The Atomic Components of Thought*, J. R. Anderson and C. Lebiere, Eds., Lawrence Erlbaum Associates, Mahwah, NJ, pp. 167–200.

Campbell, K. C., and Proctor, R. W. (1993), "Repetition Effects with Categorizable Stimulus and Response Sets," *Journal of Experimental Psychology: Learning, Memory, and Cognition*, Vol. 19, pp. 1345–1362.

Card, S. K., English, W. K., and Burr, B. J. (1978), "Evaluation of the Mouse, Rate-Controlled Isometric Joystick, Step Keys, and Text Keys for Text Selection on a CRT," *Ergonomics*, Vol. 21, pp. 601–613.

Carron, A. (1969), "Performance and Learning in a Discrete Motor Task under Massed and Distributed Practice," *Research Quarterly*, Vol. 40, pp. 481–489.

Cho, Y. S., and Proctor, R. W. (2003), "Stimulus and Response Representations Underlying Orthogonal Stimulus–Response Compatibility Effects," *Psychonomic Bulletin and Review*, Vol. 10, pp. 45–73.

Cohen, A., Ivry, R. I., and Keele, S. W. (1990), "Attention and Structure in Sequence Learning," *Journal of Experimental Psychology: Learning, Memory, and Cognition*, Vol. 16, pp. 17–30.

De Houwer, J., Crombez, G., Baeyens, F., and Hermans, D. (2001), "On the Generality of the Affective Simon Effect," *Cognition and Emotion*, Vol. 15, pp. 189–206.

De Jong, R., Liang, C.-C., and Lauber, E. (1994), "Conditional and Unconditional Automaticity: A Dual-Process Model of Effects of Spatial Stimulus–Response Correspondence," *Journal of Experimental Psychology: Human Perception and Performance*, Vol. 20, pp. 731–750.

Delaney, P. F., Reder, L. M., Staszewski, J. J., and Ritter, F. E. (1998), "The Strategy-Specific Nature of Improvement: The Power Law Applies by Strategy Within Task," *Psychological Science*, Vol. 9, pp. 1–7.

Destrebecqz, A., and Cleeremans, A. (2001), "Can Sequence Learning Be Implicit? New Evidence with the Process Dissociation Procedure," *Psychonomic Bulletin and Review*, Vol. 8, pp. 343–350.

Dornier, L. A., and Reeve, T. G. (1992), "Estimation of Compatibility Effects in Precuing of Arm and Direction Parameters," *Research Quarterly for Exercise and Sport*, Vol. 61, pp. 37–49.

Dutta, A., and Proctor, R. W. (1992), "Persistence of Stimulus–Response Compatibility Effects with Extended Practice," *Journal of Experimental Psychology: Learning, Memory, and Cognition*, Vol. 18, pp. 801–809.

Eimer, M. (1998), "The Lateralized Readiness Potential as an On-Line Measure of Central Response Activation Processes," *Behavior Research Methods, Instruments, and Computers*, Vol. 30, pp. 146–156.

Fitts, P. M. (1954), "The Information Capacity of the Human Motor System in Controlling the Amplitude of Movement," *Journal of Experimental Psychology*, Vol. 47, pp. 381–391.

Fitts, P. M., and Deininger, R. L. (1954), "S–R Compatibility: Correspondence Among Paired Elements Within Stimulus and Response Codes," *Journal of Experimental Psychology*, Vol. 48, pp. 483–492.

Fitts, P. M., and Seeger, C. M. (1953), "S–R Compatibility: Spatial Characteristics of Stimulus and Response Codes," *Journal of Experimental Psychology*, Vol. 46, pp. 199–210.

Fox, E. (1995), "Negative Priming from Ignored Distractors in Visual Selection: A Review," *Psychonomic Bulletin and Review*, Vol. 2, pp. 145–173.

Frederiksen, J. R., and White, B. Y. (1989), "An Approach to Training Based upon Principled Task Decomposition," *Acta Psychologica*, Vol. 71, pp. 89–146.

Goodman, D., and Kelso, J. (1980), "Are Movements Prepared in Parts Not Under Compatible (Naturalized) Conditions?" *Journal of Experimental Psychology: General*, Vol. 109, pp. 475–495.

Greenwald, A. G. (2003), "On Doing Two Things at Once, III: Confirmation of Perfect Timesharing When Simultaneous Tasks Are Ideomotor Compatible," *Journal of*

Experimental Psychology: Human Perception and Performance, Vol. 29, pp. 859–868.

Greenwald, A. G., and Shulman, H. G. (1973), "On Doing Two Things at Once, II: Elimination of the Psychological Refractory Period Effect," *Journal of Experimental Psychology*, Vol. 101, pp. 70–76.

Guiard, Y. (1983), "The Lateral Coding of Rotation: A Study of the Simon Effect with Wheel Rotation Responses," *Journal of Motor Behavior*, Vol. 15, pp. 331–342.

Haider, H., and Frensch, P. A. (2002), "Why Aggregated Learning Follows the Power Law of Practice When Individual Learning Does Not: Comment on Rickard (1997, 1999), Delaney et al. (1998), and Palmeri (1999)," *Journal of Experimental Psychology: Learning, Memory, and Cognition*, Vol. 28, pp. 392–406.

Heathcote, A., Brown, S., and Mewhort, D. J. K. (2000), "The Power Law Repealed: The Case for an Exponential Law of Practice," *Psychonomic Bulletin and Review*, Vol. 7, pp. 185–207.

Henry, F. M., and Rogers, D. E. (1960), "Increased Response Latency for Complicated Movements and a 'Memory Drum' Theory for Neuromotor Reaction," *Research Quarterly*, Vol. 31, pp. 448–458.

Heuer, H. (2003), "Motor Control," in *Experimental Psychology*, A. F. Healy and R. W. Proctor, Eds., Vol. 4 in *Handbook of Psychology*, I. B. Weiner, Ed.-in-Chief, Wiley, Hoboken, NJ, pp. 317–354.

Hick, W. E. (1952), "On the Rate of Gain of Information," *Quarterly Journal of Experimental Psychology*, Vol. 4, pp. 11–26.

Hommel, B. (1993a), "Inverting the Simon Effect by Intention," *Psychological Research*, Vol. 55, pp. 270–279.

Hommel, B. (1993b), "The Relationship Between Stimulus Processing and Response Selection in the Simon Task: Evidence for a Temporal Overlap," *Psychological Research*, Vol. 55, pp. 280–290.

Hommel, B. (1998a), "Automatic Stimulus–Response Translation in Dual-Task Performance," *Journal of Experimental Psychology: Human Perception and Performance*, Vol. 24, pp. 1368–1384.

Hommel, B. (1998b), "Event Files: Evidence for Automatic Integration of Stimulus–Response Episodes," *Visual Cognition*, Vol. 5, pp. 183–216.

Hommel, B., Proctor, R. W., and Vu, K.-P. L. (2004), "A Feature-Integration Account of Sequential Effects in the Simon Task," *Psychological Research*, Vol. 68, pp. 1–17.

Hyman, R. (1953), "Stimulus Information as a Determinant of Reaction Time," *Journal of Experimental Psychology*, Vol. 45, pp. 188–196.

Inhoff, A. W., Rosenbaum, D. A., Gordon, A. M., and Campbell, J. A. (1984), "Stimulus–Response Compatibility and Motor Programming of Manual Response Sequences," *Journal of Experimental Psychology: Human Perception and Performance*, Vol. 10, pp. 724–733.

Ivry, R., Spencer, R. M., Zelaznik, H. N., and Diedrichsen, J. (2002), "The Cerebellum and Event Timing," in *The Cerebellum: Recent Developments in Cerebellar Research*, Vol. 978, S. M. Highstein and W. T. Thach, Eds., New York Academy of Sciences, New York, pp. 302–317.

Keele, S. W., and Posner, M. I. (1968), "Processing of Visual Feedback in Rapid Movements," *Journal of Experimental Psychology*, Vol. 77, pp. 155–158.

Kelso, J. A. S. (1984), "Phase Transitions and Critical Behavior in Human Bimanual Coordination," *American Journal of Physiology: Regulatory, Integrative and Comparative Physiology*, Vol. 15, pp. R1000–R1004.

Kerr, R. (1973), "Movement Time in an Underwater Environment," *Journal of Motor Behavior*, Vol. 5, pp. 175–178.

Kerzel, D., Hommel, B., and Bekkering, H. (2001), "A Simon Effect Induced by Induced Motion and Location: Evidence for a Direct Linkage of Cognitive and Motor Maps," *Perception and Psychophysics*, Vol. 63, pp. 862–874.

Kornblum, S., Hasbroucq, T., and Osman, A. (1990), "Dimensional Overlap: Cognitive Basis for Stimulus–Response Compatibility—A Model and Taxonomy," *Psychological Review*, Vol. 97, pp. 253–270.

Kunde, W., and Stöcker, C. (2002), "A Simon Effect for Stimulus–Response Duration," *Quarterly Journal of Experimental Psychology*, Vol. 55A, pp. 581–592.

Lamberts, K., Tavernier, G., and d'Ydewalle, G. (1992), "Effects of Multiple Reference Points in Spatial Stimulus–Response Compatibility," *Acta Psychologica*, Vol. 79, pp. 115–130.

Langolf, G., and Hancock, W. M. (1975), "Human Performance Times in Microscope Work," *AIEE Transactions*, Vol. 7, pp. 110–117.

Lashley, K. S. (1951), "The Problem of Serial Order Behavior," in *Cerebral Mechanisms in Behavior*, L. A. Jeffress, Ed., Wiley, New York, pp. 112–131.

Lavery, J. J. (1962), "Retention of Simple Motor Skills as a Function of Type of Knowledge of Results," *Canadian Journal of Psychology*, Vol. 16, pp. 300–311.

Lee, T. D., and Genovese, E. D. (1988), "Distribution of Practice in Motor Skill Acquisition: Learning and Performance Effects Reconsidered," *Research Quarterly for Exercise and Sport*, Vol. 59, pp. 277–287.

Lee, T. D., and Genovese, E. D. (1989), "Distribution of Practice in Motor Skill Acquisition: Different Effects for Discrete and Continuous Tasks," *Research Quarterly for Exercise and Sport*, Vol. 70, pp. 59–65.

Lee, T. D., and Magill, R. A. (1985), "Can Forgetting Facilitate Skill Acquisition?" in *Differing Perspectives on Motor Memory, Learning, and Control*, D. Goodman, R. B. Wilberg, and I. M. Franks, Eds., North-Holland, Amsterdam, pp. 3–22.

Leonard, J. A. (1958), "Partial Advance Information in a Choice Reaction Task," *British Journal of Psychology*, Vol. 49, pp. 89–96.

Lersten, K. C. (1968), "Transfer of Movement Components in a Motor Learning Task," *Research Quarterly*, Vol. 39, pp. 575–581.

Lien, M.-C., and Proctor, R. W. (2002), "Stimulus–Response Compatibility and Psychological Refractory Period Effects: Implications for Response Selection," *Psychonomic Bulletin and Review*, Vol. 9, pp. 212–238.

Lien, M.-C., Proctor, R. W., and Allen, P. A. (2002), "Ideomotor Compatibility in the Psychological Refractory Period Effect: 29 Years of Oversimplification," *Journal of Experimental Psychology: Human Perception and Performance*, Vol. 28, pp. 396–409.

Lien, M.-C., McCann, R. E., Ruthruff, E., and Proctor, R. W. (2005), "Dual-Task Performance with Ideomotor Compatible Tasks: Is the Central Processing Bottleneck Intact, Bypassed, or Shifted in Locus?" *Journal of Experimental Psychology: Human Perception and Performance*, Vol. 31, pp. 122–144.

Logan, G. D. (1988), "Toward an Instance Theory of Automatization," *Psychological Review*, Vol. 95, pp. 492–527.

Logan, G. D. (2003), "Simon-Type Effects: Chronometric Evidence for Keypress Schemata in Typewriting," *Journal of Experimental Psychology: Human Perception and Performance*, Vol. 29, pp. 741–757.

Magill, R. A., and Hall, K. G. (1990), "A Review of Contextual Interference in Motor Skill Acquisition," *Human Movement Science*, Vol. 9, pp. 241–289.

Magill, R. A., and Wood, C. A. (1986), "Knowledge of Results Precision as a Learning Variable in Motor Skill Acquisition," *Research Quarterly for Exercise and Sport*, Vol. 57, pp. 170–173.

McCann, R. S., and Johnston, J. C. (1992), "Locus of the Single-Channel Bottleneck in Dual-Task Interference," *Journal of Experimental Psychology: Human Perception and Performance*, Vol. 18, pp. 471–484.

Mechsner, F. and Knoblich, G. (2004), "Do Muscles Matter for Coordinated Action?" *Journal of Experimental Psychology: Human Perception and Performance*, Vol. 30, pp. 490–503.

Mechsner, F., Kerzel, D., Knoblich, G., and Prinz, W. (2001), "Perceptual Basis of Bimanual Coordination," *Nature*, Vol. 414, November 1, pp. 69–73.

Merkel, J. (1885), "Die zeitliche Verhältnisse de Willenstatigkeit" [The Temporal Relations of Activities of the Will], *Philosophische Studien*, Vol. 2, pp. 73–127.

Meyer, D. E., and Kieras, D. E. (1997a), "A Computational Theory of Executive Cognitive Processes and Multiple-Task Performance, 1: Basic Mechanisms," *Psychological Review*, Vol. 104, pp. 3–66.

Meyer, D. E., and Kieras, D. E. (1997b), "A Computational Theory of Executive Cognitive Processes and Multiple-Task Performance, 2: Accounts of Psychological Refractory-Period Phenomena," *Psychological Review*, Vol. 104, pp. 749–791.

Meyer, D. E., Abrams, R. A., Kornblum, S., Wright, C. E., and Smith, J. E. K. (1988), "Optimality in Human Motor Performance: Ideal Control of Rapid Aimed Movements," *Psychological Review*, Vol. 95, pp. 340–370.

Miller, J. (1982), "Discrete Versus Continuous Stage Models of Human Information Processing: In Search of Partial Output," *Journal of Experimental Psychology: Human Perception and Performance*, Vol. 8, pp. 273–296.

Monsell, S. (1986), "Programming of Complex Sequences: Evidence from the Timing of Rapid Speech and Other Productions," in *Generation and Modulation of Action Patterns*, H. Heuer and C. Fromm, Eds., Springer-Verlag, Berlin, pp. 72–86.

Monsell, S. (2003), "Task Switching," *Trends in Cognitive Sciences*, Vol. 7, pp. 134–140.

Nattkemper, D., and Ziessler, M. (2004), "Cognitive Control of Action: The Role of Action Effects," *Psychological Research*, Vol. 68, pp. 71–73.

Navon, D., and Miller, J. (2002), "Queuing or Sharing? A Critical Evaluation of the Single-Bottleneck Notion," *Cognitive Psychology*, Vol. 44, pp. 193–251.

Neill, W. T., and Valdes, L. A. (1992), "Persistence of Negative Priming: Steady State or Decay?" *Journal of Experimental Psychology: Learning, Memory, and Cognition*, Vol. 18, pp. 565–576.

Newell, A., and Rosenbloom, P. S. (1981), "Mechanisms of Skill Acquisition and the Law of Practice," in *Cognitive Skills and Their Acquisition*, J. R. Anderson, Ed., Lawrence Erlbaum Associates, Mahwah, NJ, pp. 1–55.

Nissen, M. J., and Bullemer, P. (1987), "Attentional Requirements of Learning: Evidence from Performance Measures," *Cognitive Psychology*, Vol. 19, pp. 1–32.

Notebaert, W., Soetens, E., and Melis, A. (2001), "Sequential Analysis of a Simon Task: Evidence for an Attention-Shift Account," *Psychological Research*, Vol. 65, pp. 170–184.

Pachella, R. G. (1974), "The Interpretation of Reaction Time in Information-Processing Research," in *Human Information Processing: Tutorials in Performance and Cognition*, B. H. Kantowitz, Ed., Lawrence Erlbaum Associates, Mahwah, NJ, pp. 41–82.

Park, J., and Kanwisher, N. (1994), "Negative Priming for Spatial Locations: Identity Mismatching, Not Distractor Inhibition," *Journal of Experimental Psychology: Human Perception and Performance*, Vol. 20, pp. 613–623.

Pashler, H., and Baylis, G. (1991), "Procedural Learning, 2: Intertrial Repetition Effects in Speeded Choice Task," *Journal of Experimental Psychology: Learning, Memory, and Cognition*, Vol. 17, pp. 33–48.

Pashler, H., and Johnston, J. C. (1989), "Chronometric Evidence for Central Postponement in Temporally Overlapping Tasks," *Quarterly Journal of Experimental Psychology*, Vol. 41A, pp. 19–45.

Pashler, H., and Johnston, J. C. (1998), "Attentional Limitations in Dual-Task Performance," in *Attention*, H. Pashler, Ed., Psychology Press, Hove, East Sussex, England, pp. 155–189.

Posner, M. I., Klein, R., Summers, J., and Buggie, S. (1973), "On the Selection of Signals," *Memory and Cognition*, Vol. 1, pp. 2–12.

Povel, D.-J., and Collard, R. (1982), "Structural Factors in Patterned Finger Tapping," *Acta Psychologica*, Vol. 52, pp. 107–124.

Prinz, W., Aschersleben, G., Hommel, B., and Vogt, S. (1995), "Handlungen als Ereignisse" [Actions as Events], in *Gedächtnis: Trends, Probleme, Perspekitiven*, D. Dörner and E. van der Meer, Eds., Hogrefe, Göttingen, Germany, pp. 129–168.

Proctor, R. W., and Dutta, A. (1993), "Do the Same Stimulus–Response Relations Influence Choice Reactions Initially and After Practice?" *Journal of Experimental Psychology: Learning, Memory, and Cognition*, Vol. 19, pp. 922–930.

Proctor, R. W., and Dutta, A. (1995), *Skill Acquisition and Human Performance*, Sage, Thousand Oaks, CA.

Proctor, R. W., and Lu, C.-H. (1999), "Processing Irrelevant Location Information: Practice and Transfer Effects in Choice-Reaction Tasks," *Memory and Cognition*, Vol. 27, pp. 63–77.

Proctor, R. W. and Reeve, T. G. (1985), "Compatibility Effects in the Assignment of Symbolic Stimuli to Discrete Finger Response," *Journal of Experimental Psychology: Human Perception and Performance*, Vol. 11, pp. 623–639.

Proctor, R. W., and Reeve, T. G. (1986), "Salient-Feature Coding Operations in Spatial Precuing Tasks," *Journal of Experimental Psychology: Human Perception and Performance*, Vol. 12, pp. 277–285.

Proctor, R. W., and Vu, K.-P. L. (2002), "Mixing Location Irrelevant and Relevant Trials: Influence of Stimulus Mode on Spatial Compatibility Effects," *Memory and Cognition*, Vol. 30, pp. 281–294.

Proctor, R. W., and Vu, K.-P. L. (2003), "Action Selection," in *Experimental Psychology*, A. F. Healy and

R. W. Proctor, Eds., Vol. 4 in *Handbook of Psychology*, I. B. Weiner, Ed.-in-Chief, Wiley, Hoboken, NJ, pp. 293–316.

Proctor, R. W., Vu, K.-P. L., and Marble, J. G. (2003), "Spatial Compatibility Effects Are Eliminated When Intermixed Location-Irrelevant Trials Produce the Same Spatial Codes," *Visual Cognition*, Vol. 10, pp. 15–50.

Proctor, R. W., Wang, D.-Y. D., and Pick, D. F. (2004), "Stimulus–Response Compatibility with Wheel-Rotation Responses: Will an Incompatible Response Coding Be Used When a Compatible Coding Is Possible?" *Psychonomic Bulletin and Review*, Vol. 11, pp. 41–48.

Proteau, L., and Cournoyer, J, (1992), "Vision of the Stylus in a Manual Aiming Task," *Quarterly Journal of Experimental Psychology*, Vol. 44A, pp. 811–828.

Read, L. E., and Proctor, R. W. (2004), "Spatial Stimulus–Response Compatibility and Negative Priming," *Psychonomic Bulletin and Review*, Vol. 11, pp. 41–48.

Reeve, T. G., and Proctor, R. W. (1984), "On the Advance Preparation of Discrete Finger Responses," *Journal of Experimental Psychology: Human Perception and Performance*, Vol. 10, pp. 541–553.

Reeve, T. G., Dornier, L., and Weeks, D. J. (1990), "Precision of Knowledge of Results: Consideration of the Accuracy Requirements Imposed by the Task," *Research Quarterly for Exercise and Sport*, Vol. 61, pp. 284–291.

Rickard, T. C. (2004), "Strategy Execution and Cognitive Skill Learning: An Item-Level Test of Candidate Models," *Journal of Experimental Psychology: Learning, Memory, and Cognition*, Vol. 30, pp. 65–82.

Rogers, R. D., and Monsell, S. (1995), "Cost of a Predictable Switch Between Simple Cognitive Tasks," *Journal of Experimental Psychology: General*, Vol. 124, pp. 207–231.

Rosenbaum, D. A. (1983), "The Movement Precuing Technique: Assumptions, Applications, and Extensions," in *Memory and Control of Action*, R. A. Magill, Ed., North-Holland, Amsterdam, pp. 231–274.

Rosenbaum, D. A. (1991), *Human Motor Control*, Academic Press, San Diego, CA.

Rosenbaum, D. A. (2002), "Motor Control," in *Stevens' Handbook of Experimental Psychology*, 3rd ed., H. Pashler and S. Yantis, Eds., Vol. 1, *Sensation and Perception*, Wiley, New York, pp. 315–339.

Rosenbaum, D. A., Gordon, A. M., Stillings, N. A., and Feinstein, M. H. (1987), "Stimulus–Response Compatibility in the Programming of Speech," *Memory and Cognition*, Vol. 15, pp. 217–224.

Rosenbaum, D. A., Marchak, F., Barnes, H. J., Vaughan, J., Slotta, J. D., and Jorgensen, M. J. (1990), "Constraints for Action Selection: Overhand Versus Underhand Grips," in *Attention and Performance XIII: Motor Representation and Control*, M. Jeannerod, Ed., Lawrence Erlbaum Associates, Mahwah, NJ, pp. 321–342.

Roswarski, T. E., and Proctor, R. W. (1996), "Multiple Spatial Codes and Temporal Overlap in Choice-Reaction Tasks," *Psychological Research*, Vol. 63, pp. 148–158.

Ruthruff, E., Johnston, J. C., Van Selst, M., Whitsell, S., and Remington, R. (2003), "Vanishing Dual-Task Interference After Practice: Has the Bottleneck Been Eliminated or Is It Merely Latent?" *Journal of Experimental Psychology: Human Perception and Performance*, Vol. 29, pp. 280–289.

Sanders, A. F. (1998), *Elements of Human Performance*, Lawrence Erlbaum Associates, Mahwah, NJ.

Schmidt, R. A., and Lee, T. D. (1999), *Motor Control and Learning: A Behavioral Emphasis*, 3rd ed., Human Kinetics, Champaign, IL.

Schmidt, R. A., Young, D. E., Swinnen, S., and Shapiro, D. C. (1989), "Summary Knowledge of Results for Skill Acquisition: Support for the Guidance Hypothesis," *Journal of Experimental Psychology: Learning, Memory and Cognition*, Vol. 15, pp. 352–359.

Schumacher, E. H., Seymour, T. L., Glass, J. M., Fencsik, D. E., Lauber, E. J., Kieras, D. E., and Meyer, D. E. (2001), "Virtually Perfect Time Sharing in Dual-Task Performance: Uncorking the Central Cognitive Bottleneck," *Psychological Science*, Vol. 12, pp. 101–108.

Seibel, R. (1963), "Discrimination Reaction Time for a 1,023-Alternative Task," *Journal of Experimental Psychology*, Vol. 66, pp. 215–226.

Shaffer, L. H. (1965), "Choice Reaction with Variable S–R Mapping," *Journal of Experimental Psychology*, Vol. 70, pp. 284–288.

Shannon, C. E. (1948), "A Mathematical Theory of Communication," *Bell System Technical Journal*, Vol. 27, pp. 379–423 and 623–656; http://cm.bell-labs.com/cm/ms/what/shannonday/paper.html.

Shea, J. B., and Morgan, R. L. (1979), "Contextual Interference Effects on Acquisition, Retention, and Transfer of a Motor Skill," *Journal of Experimental Psychology: Human Learning and Memory*, Vol. 5, pp. 179–187.

Simon, J. R. (1990), "The Effects of an Irrelevant Directional Cue on Human Information Processing," in *Stimulus–Response Compatibility: An Integrated Perspective*, R. W. Proctor and T. G. Reeve, Eds., North-Holland, Amsterdam, pp. 31–86.

Soetens, E. (1998), "Localizing Sequential Effects in Serial Choice Reaction Time with the Information Reduction Procedure," *Journal of Experimental Psychology: Human Perception and Performance*, Vol. 24, pp. 547–568.

Sternberg, S. (1998), "Discovering Mental Processing Stages: The Method of Additive Factors," in *An Invitation to Cognitive Science: Methods, Models, and Conceptual Issues*, 2nd ed., Vol. 4, D. Scarborough and S. Sternberg, Eds., MIT Press, Cambridge, MA: pp. 703–863.

Stürmer, B., Leuthold, H., Soetens, E., Schroeter, H., and Sommer, W. (2002), "Control over Location-Based Response Activation in the Simon Task: Behavioral and Electrophysiological Evidence," *Journal of Experimental Psychology: Human Perception and Performance*, Vol. 28, pp. 1345–1363.

Tagliabue, M., Zorzi, M., Umiltà, C., and Bassignani, F. (2000), "The Role of LTM Links and STM Links in the Simon Effect," *Journal of Experimental Psychology: Human Perception and Performance*, Vol. 26, pp. 648–670.

Taub, E., and Berman, A. J. (1968), "Movement and Learning in the Absence of Sensory Feedback," in *The Neuropsychology of Spatially Oriented Behavior*, S. J. Freeman, Ed., Dorsey, Homewood, IL, pp. 173–192.

Teichner, W. H., and Krebs, M. J. (1974), "Laws of Visual-Choice Reaction Time," *Psychological Review*, Vol. 81, pp. 75–98.

Tombu, M., and Jolicœur, P. (2003), "A Central Capacity Sharing Model of Dual-Task Performance," *Journal of Experimental Psychology: Human Perception and Performance*, Vol. 29, pp. 3–18.

Tsutsui, S., Lee, T. D., and Hodges, N. J. (1998), "Contextual Interference in Learning New Patterns of Bimanual Coordination," *Journal of Motor Behavior*, Vol. 30, pp. 151–157.

Usher, M., and McClelland, J. L. (2001), "The Time Course of Perceptual Choice: The Leaky Competing Accumulator Model," *Psychological Review*, Vol. 108, pp. 550–592.

Usher, M., Olami, Z., and McClelland, J. L. (2002), "Hick's Law in a Stochastic Race Model with Speed–Accuracy Tradeoff," *Journal of Mathematical Psychology*, Vol. 46, pp. 704–715.

Van Zandt, T., Colonius, H., and Proctor, R. W. (2000), "A Comparison of Two Response-Time Models Applied to Perceptual Matching," *Psychonomic Bulletin and Review*, Vol. 7, pp. 208–256.

Vickrey, C., and Neuringer, A. (2000), "Pigeon Reaction Time, Hick's Law, and Intelligence," *Psychonomic Bulletin and Review*, Vol. 7, pp. 284–291.

Vu, K.-P. L. (2004), "Influences of a Prior Incompatible Location Mapping on Tasks for Which Stimulus Location is Irrelevant," Unpublished dissertation, Purdue University, West Lafayette, IN.

Vu, K.-P. L., and Proctor, R. W. (2004), "Mixing Compatible and Incompatible Mappings: Elimination, Reduction, and Enhancement of Spatial Compatibility Effects," *Quarterly Journal of Experimental Psychology*, Vol. 57A, pp. 539–556.

Vu, K.-P. L., Proctor, R. W., and Urcuioli, P. (2003), "Transfer Effects of Incompatible Location-Relevant Mappings on Subsequent Visual or Auditory Simon Tasks," *Memory and Cognition*, Vol. 31, pp.1146–1152.

Wang, D.-Y. D., Proctor, R. W., and Pick, D. F. (2003), "The Simon Effect with Wheel-Rotation Responses," *Journal of Motor Behavior*, Vol. 35, pp. 261–273.

Waszak, F., Hommel, B., and Allport, A. (2003), "Task-Switching and Long-Term Priming: Role of Episodic Stimulus-Task Bindings in Task-Shift Costs," *Cognitive Psychology*, Vol. 46, pp. 361–413.

Weeks, D. J. and Proctor, R. W. (1990), "Salient-Features Coding in the Translation Between Orthogonal Stimulus–Response Dimensions," *Journal of Experimental Psychology: General*, Vol. 119, pp. 355–366.

Welford, A. T. (1952), "The 'Psychological Refractory Period' and the Timing of High-Speed Performance—a Review and a Theory," *British Journal of Psychology*, Vol. 43, pp. 2–19.

Wickelgren, W. A. (1969), "Context-Sensitive Coding, Associative Memory, and Serial Order in (Speech) Behavior," *Psychological Review*, Vol. 76, pp. 1–15.

Wightman, D. C., and Lintern, G. (1985), "Part-Task Training for Tracking and Manual Control," *Human Factors*, Vol. 27, pp. 267–283.

Wilkinson, L., and Shanks, D. R. (2004), "Intentional Control and Implicit Sequence Learning," *Journal of Experimental Psychology: Learning, Memory, and Cognition*, Vol. 30, pp. 354–369.

Willingham, D. B. (1999), "Implicit Motor Sequence Learning Is Not Purely Perceptual," *Memory and Cognition*, Vol. 27, pp. 561–572.

Willingham, D. B., Wells, L. A., Farrell, J. M., and Stemwedel, M. E. (2000), "Implicit Motor Sequence Learning Is Represented in Response Locations," *Memory and Cognition*, Vol. 28, pp. 366–375.

Wing, A. M., and Kristofferson, A. B. (1973), "Response Delays and the Timing of Discrete Motor Responses," *Perception and Psychophysics*, Vol. 14, pp. 5–12.

Winstein, C. J., and Schmidt, R. A. (1990), "Reduced Frequency of Knowledge of Results Enhances Motor Skill Learning," *Journal of Experimental Psychology: Learning, Memory, and Cognition*, Vol. 16, pp. 677–691.

Woodworth, R. S. (1899), "The Accuracy of Voluntary Movement," *Psychological Review*, Vol. 3 (Monograph Suppl.), pp. 1–119.

Woodworth, R. S. (1938), *Experimental Psychology*, Holt, New York.

Zelaznik, H. N., Hawkins, B., and Kisselburgh, L. (1983), "Rapid Visual Feedback Processing in Single-Aiming Movements," *Journal of Motor Behavior*, Vol. 15, pp. 217–236.

CHAPTER 5

INFORMATION PROCESSING

Christopher D. Wickens
University of Illinois
Savoy, Illinois

C. Melody Carswell
University of Kentucky
Lexington, Kentucky

1 INTRODUCTION

Information processing lies at the heart of human performance. In a plethora of situations in which humans interact with systems, the operator must perceive information, must transform that information into different forms, must take actions on the basis of the perceived and transformed information, and must process the feedback from that action, assessing its effect on the environment. These characteristics apply no matter whether information processing is defined in terms of the classic open-loop information processing model that derives from much of psychological research (Figure 1*a*), or the closed-loop model of Figure 1*b*, which has its roots both within control engineering (e.g., Pew and Baron, 1978; Baron et al., 1970; McRuer, 1980; Jagacinski and Flach, 2003), and more recent conceptualizing in ecological psychology (Flach et al., 1995; Hancock et al., 1995). In either case, *transformations* must be made on the information as it flows through the human operator. These transformations take time and may be the source of error. Understanding their nature, their time demands, and the kinds of errors that result from their operation is critical to predicting and modeling human–system interaction.

In this chapter we describe characteristics of the different important *stages* of information processing,

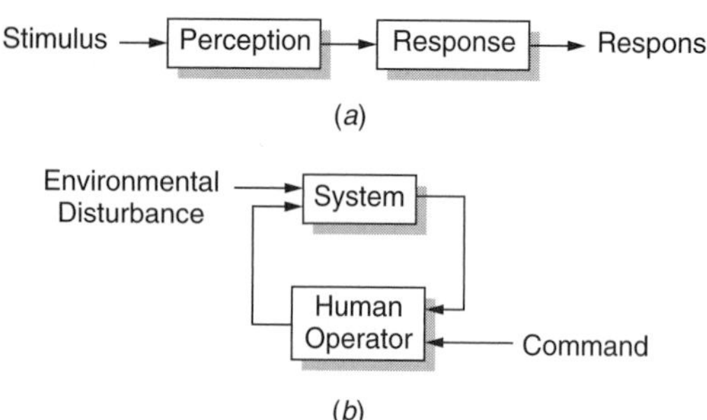

Figure 1 Two representations of information processing: (*a*) traditional open-loop representation from cognitive psychology; (*b*) closed-loop system, following the tradition of engineering feedback models.

from perception of the environment, to acting on that environment. We try to do so in a way that is neither too specific to any particular system nor so generic that the relevance of the information-processing model to system design is not evident. We begin by contrasting three ways in which information processing has been treated in applied psychology, and then we describe processes and transformations related to *attention, perception, memory* and *cognition, action selection*, and *multiple-task performance*.

2 THREE APPROACHES TO INFORMATION PROCESSING

The classic information-processing approach to describing human performance owes much to the seminal work of Broadbent (1958, 1972), Neisser (1967), Sternberg (1969), Posner (1978), and others in the decades of the 1950s, 1960s, and 1970s, who applied the metaphor of the digital computer to human behavior. In particular, as characterized by the representation in Figure 1*a*, information was conceived as passing through a finite number of discrete stages. These stages were identifiable, not only by experimental manipulations, but also by converging evidence from brain physiology. For example, it makes sense to distinguish a perceptual stage from one involving the selection and execution of action, because of the morphological distinctions between perceptual and motor cortex.

There is also a human factors rationale for the stage distinction made by information-processing psychology. This is because different task or environmental factors appear to influence processing differentially at the different stages, a distinction that has certain design implications. For example, the *aging process* appears to affect the selection and execution of actions more than the speed of perceptual encoding (Strayer et al., 1987). Immersed displays may improve perceptual–motor interaction, even as they inhibit the allocation of attention (Olmos et al., 2000), and different sources of workload may have different influences on the different stages (Wickens & Hollands 2000). Decision-making biases can be characterized by whether they influence perception, diagnosis, or action selection (Wallsten, 1980; Wickens and Hollands, 2000), and the different stages may also be responsible for the commission of qualitatively different kinds of errors (Reason, 1990; see Chapter 27).

In contrast to the stage approach, the *ecological approach* to describing human performance provides much greater emphasis on the integrated flow of information through the human rather than on the distinct, analyzable stage sequence (Gibson, 1979; Flach et al., 1995; Hancock et al., 1995). The ecological approach also emphasizes the human's integrated interaction with the *environment* to a greater extent than does the stage approach, which can sometimes characterize information processing in a more context-free manner. Accordingly, the ecological approach focuses very heavily on modeling the perceptual characteristics of the environment to which the user is "tuned" and responds in order to meet the goals of a particular task. Action and perception are closely linked, since to act is to change what is perceived, and to perceive is to change the basis of action in a manner consistent with the closed-loop representation shown in Figure 1*b*.

As a consequence of these properties, the ecological approach is most directly relevant to describing human behavior in interaction with the natural environment (e.g., walking or driving through natural spaces, or manipulating objects directly). However, as a direct outgrowth, this approach is also quite relevant to the design of controls and displays that mimic characteristics of the natural environment—the concept of *direct manipulation interfaces* (Hutchins et al., 1985). As a further outgrowth, the ecological approach is relevant to the design of interfaces that mimic characteristic of how users *think* about a physical process, even if the process itself is not visible in a way that can be represented directly. In this regard, the ecological approach has been used as a basis for designing effective displays of energy conversion processes such as those found in a nuclear reactor (Vicente and Rasmussen, 1992; Moray et al., 1994; Vicente et al., 1995; Burns, 2000; Vicente, 2002; Burns et al., 2004).

Because of its emphasis on interaction with the natural (and thereby familiar) environment, the ecological approach is closely related to other approaches to performance modeling that emphasize people working with domains and systems about which they are experts. This feature characterizes for example, the study of *naturalistic decision making* (Zsambock and Klein, 1997; see Chapter 3), which is often set up in contrast to the representation of decision making within an information-processing framework (Wickens and Hollands, 2000).

Both the stage-based approach and the ecological approach have a great deal to offer to human factors, and the position we take in this chapter is that aspects of each can and should be selected, as they are more appropriate for analysis of the operator in a particular system. For example, the ecological approach is highly appropriate for modeling vehicle control, but less so for describing processes in reading, understanding complex instructions under stress, or dealing with highly symbolic logical systems (e.g., the logic of computers, information retrieval systems, or decision tree analysis; see Chapter 8). Finally, both approaches can be fused harmoniously, as when, for example, the important constraints of the natural environment are analyzed carefully to understand the information available for perception and the control actions allowable for action execution in driving, but the more context-free limits of information processing can be used to understand how performance might break down from a high load that is imposed on memory or dual-task performance requirements in a car.

A final approach, that of *cognitive engineering*, or *cognitive ergonomics* (Rasmussen et al., 1995; Vicente, 1999), is somewhat of a hybrid of the two described above. The emphasis of cognitive engineering is on the one hand, based on a very careful understanding of the environment and task constraints within which an operator works, a characteristic of the ecological approach. On the other hand, as suggested by the prominence of the word *cognitive*, the approach places great emphasis on modeling and understanding the knowledge structures that expert operators have of the domains in which they must work and indeed, the knowledge structures of computer agents in the system. Thus, whereas the ecological approach tends to be more specifically applied to human interaction with physical systems and environments (particularly those that obey the constraints of Newtonian physics), cognitive engineering is relevant to the design of almost any system about which the human operator can acquire knowledge, including the very symbolic computer systems, which have no physical analogy.

Whether human performance is approached from an information-processing, ecological, or cognitive engineering point of view, we assert here that in almost any task, a certain number of mental processes, involved in selecting, interpreting, retaining, or responding to information, may be implemented; and it is understanding the vulnerabilities of these processes and capitalizing, where possible, on their strengths, which can provide an important key to effective human factors of system design.

In this chapter we adopt as a framework the information-processing model depicted in Figure 2 (Wickens and Hollands, 2000). Here stimuli or events are *sensed and attended* (Section 3) and that information received by our sensory system is *perceived*, that is, provided with some meaningful interpretation based on memory of past experience (Section 4). That which is perceived may be responded to directly, through a process of action selection (decision of what act to take) and execution (Section 6). Alternatively, it may be stored temporarily in working memory, a system that may also be involved in thinking about or transforming information that was not sensed and perceived but was generated internally (e.g., mental images, rules, Section 5). Working memory is of limited capacity and heavily demanding of attention in its operation but is closely related to our large-capacity long-term memory, a system that stores vast amounts of information about the world, including both facts and procedures, retains that information without attention; but is not always fully available for retrieval.

As noted in the figure, and highlighted in the ecological approach, actions generally produce feedback, which is then sensed to complete the closed-loop cycle. In addition, human attention, a limited resource, plays two critical roles in the information-processing sequence. As a selective agent, it chooses and constrains this information that will be perceived (Section 3). As a task management agent, it constrains what tasks (or mental operations) can be performed concurrently (Section 7).

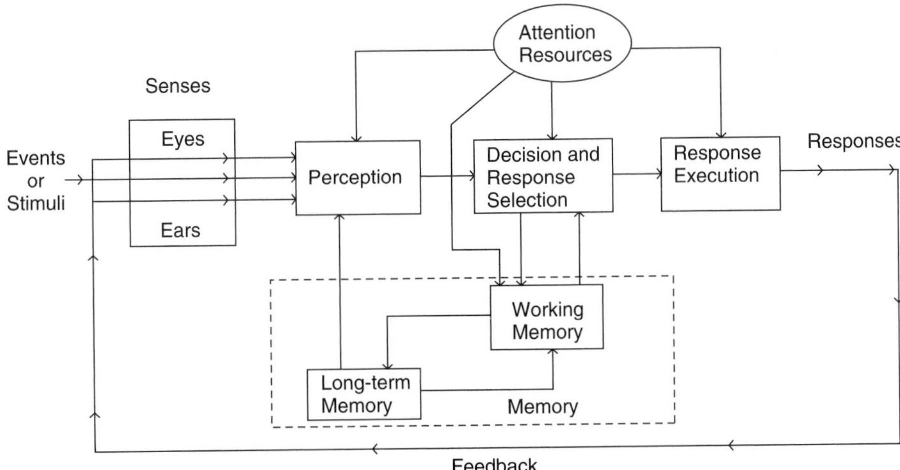

Figure 2 Model of human information processing. (Adapted from Wickens, 1992.)

3 SELECTING INFORMATION

Since Broadbent's (1958) classic book, it has been both conventional and important to model human information processing as, in part, a filtering process. This filtering is assumed to be carried out by the mechanisms of *human attention* (Kahneman, 1973; Parasuraman et al., 1984; Damos, 1991; Pashler, 1998; Johnson and Proctor, 2004). Attention, in turn, may be conceptualized as having three *modes*: Selective attention chooses what to process in the environment, focused attention characterizes the efforts to sustain processing of those elements while avoiding distraction from others, and *divided attention* characterizes the ability to process more than one attribute or element of the environment at a given time.

We discuss below the human factors implications of the selective and focused attention modes and their relevance to visual search and discuss those of divided attention in more detail in Sections 4.6 and Section 7.

3.1 Selective Attention

In complex environments, selective attention may be described in terms of how it is influenced by the combined force of four factors: salience, effort, expectancy, and value (Wickens et al., 2003). These influences can often be revealed by eye movements when visual selective attention is assessed by visual scanning; obviously, however, eye movements cannot reflect the selectivity between vision and the other sensory modalities.

1. *Salient features* of the environment will attract or "capture" attention. Thus, auditory sounds tend to be more attention grabbing than visual events, leading to the choice of sounds to be used in alarms (Stanton, 1994; see also Chapter 24). Within a visual display, the onset of a stimulus (e.g., increase in brightness from zero, or the appearance of an object where one was not present previously) tends to be the most salient or attention-attracting property (Yantis, 1993; Egeth and Yantis, 1997); other features, such as uniqueness, can also attract attention, but these are typically less powerful than onsets. The prominent role of onsets as attention-capture devices can explain the value of repeated onsets ("flashing") as a visual alert. Salient events are sometimes described as those that govern *bottom-up* or stimulus-driven allocation of attention, to be contrasted with knowledge-driven or *top down* features of selective attention, which we describe next, in the context of expectations.

2. *Expectancy* refers to knowledge regarding the probable time and location of information availability. For example, frequently changing areas are scanned more often than slowly changing areas (Senders, 1980). Thus, drivers will keep their eyes on the road more continuously when the car is traveling fast on a curvy road than when traveling slowly on a straight one. The former has a higher "bandwidth." Also, expectancy defines the role of cueing in guiding attention. For example, an auditory warning may direct attention toward the display indicator that the warning is monitoring, because the operator will now expect to see an abnormal reading on that display.

3. High-frequency changes are not, however, sufficient to direct attention. The driver will not look out the side window despite the fact that there is a lot of perceptual "action" there, because information in the side window is generally not relevant to highway safety. It has already passed. Thus, the third factor that drives selective attention is the *value* of information received at different locations. This describes the importance of knowing that information in carrying out useful tasks, or the costs of failing to note important information. It is valuable for the driver to look forward, because of the cost of failing to see a roadway hazard or of changing direction toward the side of the road. Thus, the effect of expectancy (bandwidth or frequency) on the allocation of selective attention is modulated by the value, as if the probability of attending somewhere—$p(A)$ is equal to the *expected value* of information sources to be seen at that location (Moray, 1986; Wickens et al., 2003). Indeed, Moray (1986) and Wickens et al. (2003) find that well-trained, highly skilled operators scan the environment very much as if their attention is driven primarily and nearly exclusively by the multiplicative function of expectancy and value. Thus, we may think of the well-trained operator as developing scanning *habits* that internalize the expectancy and value of sources in the environment, defining an appropriately calibrated *mental model* (Bellenkes et al., 1997).

4. The final factor that may sometimes influence attention allocation is a negative one, and this is the *effort* required to move attention around the environment. Small attention movements, such as scanning from one word to the next in a line of text or a quick glance at the speedometer in a car, require little effort. However, larger movements, such as shifting the eyes *and* head to check the side-view mirrors in a car, or coupling these with rotation of the trunk to check the "blind spot" before changing lanes, requires considerably more information access effort (Wickens, 1993). Indeed, the role of information access effort also generalizes to the effort costs of using the fingers to manipulate a keyboard and access printed material that might otherwise be accessed by a simple scan to a dedicated display (Gray and Fu, 2001). The effort to shift attention to more remote locations may have a minimal effect on the well-rested operator with a well-trained mental model, who knows precisely the expectancy and value of information sources (Wickens et al., 2003). However, the combination of fatigue (depleting effort) and a less well calibrated mental model can seriously inhibit accessing information at effortful locations, even when such information may be particularly valuable.

Collectively, the forces of salience, effort, expectancy, and value on attention can be represented by an attentional model called SEEV, in which $P(A) = \mathrm{sS} - \mathrm{efEF} + \mathrm{exEX} \times \mathrm{vV}$ (Wickens et al., 2003). Good design should try to reduce these four components to two by making valuable information sources salient (correlating salience and value) and by minimizing the effort required to access valuable *and* frequently used (expected) sources. For example, head-up displays (HUDs) in aircraft and automobiles are designed to minimize information access effort of selecting the view outside and the information contained in important instruments (Fadden et al., 2001; Wickens et al., 2004). Reduced information access effort can also be achieved through effective layout of display instruments (Wickens et al., 1997).

3.2 Focused Attention

While selective attention dictates where attention should travel, the goal of focused attention is to maintain processing of the desired source and avoid the distracting influence of potentially competing sources. The primary source of breakdowns in focused attention are certain physical properties of the visual environment (clutter) or the auditory environment (noise), which will nearly guarantee some processing of those environments, whether or not such processing is desired. Thus, any visual information source within about 1° of visual angle of a desired attentional focus will disrupt processing of the latter to some extent (Broadbent, 1982). Any sound within a certain range of frequency and intensity of an attended sound will have a similar disruptive effect on auditory focused attention (Banbury et al., 2001; see Chapter 3). However, even beyond these minimum limits of visual space and auditory frequency, information sources can be disruptive of focused attention if they are salient.

3.3 Discrimination

A key to design that can address issues of both selective and focused attention is concern for *discrimination* between information sources. Making sources discriminable by space, color, intensity, frequency or other physical differences has two benefits. First, it will allow the display viewer to *parse* the world into its meaningful components on the basis of these physical features, thereby allowing selective attention to operate more efficiently (Treisman, 1986; Yeh and Wickens, 2001; Wickens et al., 2004). For example, an air traffic controller who views on her display all of the aircraft within a given altitude range depicted in the same color can easily select all of those aircraft for attention, to ascertain which ones might be on conflicting flight paths. Parsing via a discrimination will be effective as long as all elements that are rendered physically similar (and therefore are parsed together) share in common some characteristic that is relevant for the user's *task* (as in the example above, all aircraft at the same altitude represent potential conflicts).

Second, when elements are made more discriminable by some physical feature, it is considerably easier for the operator to *focus* attention on one and ignore distraction from another, even if the two are close together in space (or are similar in other characteristics). Here again, in our air traffic control example, it will be easier for the controller to focus attention on the converging tracks of two commonly colored aircraft if other aircraft are colored differently than if all are depicted in the same hue.

Naturally, the converse of difference-based discriminability is similarity (or identity)-based confusion between information sources, a property that has many negative implications for design. For example, industrial designers may strive for consistency or uniformity in the style of a particular product interface by making all touchpad controls the same shape or size. Such stylistic uniformity, however, may result in higher rates of errors from users activating the wrong control because it looks so similar to the control intended.

3.4 Visual Search

Discrimination joins with selective and focused attention when the operator is engaged in *visual search*, looking for something in a cluttered environment. The task may characterize looking for a sign by the roadway (Holohan et al., 1978), conflicting aircraft in an air traffic control display (Remington et al., 2000), a weapon in an x-rayed luggage image (McCarley et al., 2004), a feature on a map (Yeh and Wickens, 2001), or an item on a computer menu (Fisher et al., 1989). *Visual search models* are designed to predict the time required to find a target. Such time predictions can be very important for both safety (e.g., if the eyes need to be diverted from vehicle control while searching) and productivity (e.g., if jobs require repeated searches, as in quality control inspection or menu use).

The simplest model of visual search, based on a *serial self-terminating search* (Neisser et al., 1964), assumes that a *search space* is filled with items most of which are *nontargets* or *distracters*. The mean time to find a target is modeled to be $\text{RT} = NT/2$, where N is the number of items in the space, T the time to examine each item and determine that it is *not* a target before moving on to the next, and division by 2 reflects the fact that on average the target will be reached after half of the space is searched, but sometimes earlier and sometimes later. Hence, the variance in search time will also grow with N.

The elegant and simple prediction of the serial self-terminating search model often provides a reasonable accounting for data (Yeh and Wickens 2001; Remington et al., 2000), but is also thwarted (but search performance is improved) by three factors that characterize search in many real-world search tasks: bottom-up parallel processing, top-down processing, and target familiarity. Both of the first two can be accommodated by the concept of a *guided search model* (Wolfe, 2000). Regarding parallel processing, as noted in Section 3.1, certain features (e.g., uniqueness, flashing) will capture attention because they can be preattentively processed or processed in parallel (rather than in series) with all other elements in the search field. Hence, if the target is known to contain such features, it will be found rapidly, and search time will be unaffected by the number of nontarget items in the search field. This is because all nontarget items can be discriminated automatically (as discussed in Section 3.3) and thereby eliminated from imposing any search costs (Yeh and

Wickens, 2001; Wickens et al., 2004). For example, in a police car dispatcher display, all cars currently available for dispatching can be highlighted, and the dispatcher's search for the vehicle closest to a trouble spot can proceed more rapidly. Stated in other terms, search is "guided" to the subset of items containing the single feature which indicates that they are relevant. If there is more than a single such item, the search may be serial between those items that remain. *Highlighting* (Fisher et al., 1989; Wickens et al., 2004; Remington et al., 2000), is a technique that capitalizes on this guided search.

Regarding top-down processing, search may also be guided by the operator's knowledge of *where* the target is most likely to be found. Location expectancy, acquired with practice and expertise, will create search strategies that scan the most likely locations first, to the extent that such predictability exists in the searched environments. For example, tumors may be more likely to appear in some parts of an organ than others, and skilled radiologists capitalize on this in examining an x-ray in a way that novices do not (Kundel and Nodine, 1978). However, such a strategy may not be available to help the scanner of luggage x-rays for weapons, because such weapons may be hidden anywhere in the luggage rather than in a predicable location (McCarley et al., 2004).

A third factor that can speed visual search, target familiarity is like guided search, related to experience and learning, and like parallel search, related to salient features. Here we find that repeated exposures to the same consistent target can speed the search for that target, and in particular, reduce the likelihood that the target may be looked at (fixated) but not actually detected (McCarley et al., 2004). With sufficient repetition looking for the same target (or target possessing the same set of features), the expert tunes his or her sensitivity to discriminate target from nontarget features, and with extensive practice, the target may actually "pop out" of the nontargets, as if its discriminating features are processed preattentively (Schneider and Shiffrin, 1977). Further, even if a target does not become sufficiently salient to pop out when viewed in the visual periphery, repeated exposure can help ensure that it will be detected and recognized once the operator has fixated on it (McCarley et al., 2004).

The information processing involved in visual search culminates in a target detection decision, which sometimes may be every bit as important as the search operations that proceeded it. In the following section we examine this detection process in its own right.

4 PERCEPTION AND DATA INTERPRETATION

4.1 Detection as Decision Making

The first piece of advice offered in most display design checklists is that the designer should be absolutely certain that the display code will be detectable in the environment for which it is intended (e.g., Travis, 1991; Sanders and McCormick, 1993). Assuring the detectability of the relevant information might seem to be simply a matter of knowing enough about the limits of the operator's sensory systems to choose appropriate levels of physical stimulation: for example, appropriate wavelengths of light, frequencies of sound, or concentrations of an odorant. Human sensitivity to the presence and variation of different physical dimensions is reviewed in Chapter 3, and these data must be considered limiting factors in the design of displays. Yet the detectability of any critical signal is also a function of the operator's goals, knowledge, and expectations. We also know that in visual search, there are plenty of opportunities for targets that are clearly above threshold to be missed when the search is hurried and the display is cluttered. New data shows the surprising magnitude with which super threshold changes in the target are not detected (Rensink, 2002).

The interpretive and vulnerable nature of signal detection becomes most apparent when we consider that the signal detector may make two types of detection errors. As shown in Table 1, operators may occasionally miss a signal when it is present, and importantly, they may also respond as if a signal is present when it is not (i.e., a false alarm). *Signal detection theory* (SDT) provides one model of the processing that can lead to false alarms as well as to misses (Tanner and Swets, 1954; Green and Swets, 1988; T. Wickens, 2002). SDT conceptualizes the detection task as one in which the operator is trying to decide whether any momentary amount of sensory stimulation actually reflects the presence of the signal or is simply irrelevant *noise*. The noise may consist of stimulation that is external to the operator, as when a momentary reflection on a windshield is mistaken for an oncoming vehicle, or the noise may be internal to the operator, as when the pilot mistakes a "floater" (debris in the vitreous humor of the eye) for a rapidly moving aircraft. Because noise of both sorts is always present and is always fluctuating in intensity, some detection errors are inevitable.

To deal with the uncertainty that is inherent in detection, SDT proposes that operators choose some level of sensory excitation that will serve as a *response criterion*. If the momentary level of stimulation that they experience is above this criterion, they will respond as if a signal is present, with all the consequences that this action might entail. An operator who sets a low or "risky" criterion will respond as if a signal is present more frequently than one who sets a higher, more "conservative" criterion. Adopting a low criterion will ensure that the operator rarely misses a signal; however, the reduction in misses is at the

Table 1 Joint Contingent Events Used in Signal Detection Theory Analysis

Operator's Decision (Response Criterion)	State of the World: Signal	State of the World: No Signal (Noise)
Signal	Hit	False alarm
No Signal	Miss	Correct rejection

expense of increased false alarms. Setting a higher criterion, in contrast, results in fewer false alarms at the expense of increased misses. SDT provides a way of measuring the criterion set by individual operators, which, in turn, allows the isolation of factors that influence criterion choice.

To the extent that the perfect discrimination of signal from noise is impossible in a system, the designer or supervisor must focus on manipulating those factors that ensure that operators set the optimal response criterion for their given detection task. SDT formally demonstrates that overall detection errors, misses and false alarms combined, are minimized if the response criterion is shifted downward (i.e., is made more risky) when signal likelihood increases. People performing laboratory detection tasks tend to adjust their response criteria to the direction prescribed by SDT; however, they do *not* tend to adjust it far enough (Green and Swets, 1988).

The effects of signal probability observed in the lab have also been observed in operational settings. Lusted (1976) found that physicians' criteria for detecting particular medical conditions were influenced by the base rate of the abnormality (probability of signal). Similarly, sheet metal inspectors adjusted their criteria for fault detection based on estimated defect rates (Drury and Addison, 1973). Yet plant inspectors do not adjust their criteria enough when defect rates fall below 5% (Harris and Chaney, 1969). If an operator's failure to adjust the response criterion results from inadequate knowledge about actual signal probabilities, the presentation of this information may encourage a more optimal criterion choice. Another way to influence the operator's criterion, at least when the criterion appears to be too conservative, is to inflate signal probability by occasionally introducing additional artificial signals into the event stream (Baker, 1961; Wilkinson, 1964): for example, by adding faulty products into the product stream viewed by the quality control inspector (making sure that the added faults are both visible and tagged for later removal in case they are missed by the inspector!).

In addition to probability, a second factor that should influence the setting of the response criterion, according to SDT, is the relative costs associated with misses and false alarms and the relative benefits of correct responses. As an extreme example, if there were dire consequences associated with a miss and absolutely no costs at all for false alarms, the operator should adopt the lowest criterion possible and simply respond as if the signal is there at *every* opportunity. Usually, however, circumstances are not so simple. For example, a missed (or delayed) air space conflict by the air traffic controller or a missed tumor by the radiologist may have enormous costs, possibly in terms of human lives. However, actions taken because of false alarms also have costs. In the previous examples, these might include an unnecessary evasive maneuver by the aircraft or unnecessary surgery. The operator should adjust his or her response criterion downward to the degree that misses are more costly than false alarms, and upward to the extent that avoiding false alarms is more important.

Green and Swets (1988) have found that response criteria are more sensitive to the changes in costs and benefits of the various signal detection outcomes than they are to changes in signal probabilities. Ideally, knowledge of both signal probabilities and the costs and benefits of different action outcomes should be reinforced through explicit training and appropriate incentives in order for the operator to establish the response criterion that maximizes overall detection performance.

It has long been known that humans make fairly poor signal detectors, particularly if vigilance for those signals must be maintained for long periods of time (Warm, 1984). Hence, designers have developed alarms and alerts as a simple form of automation to assist or sometimes replace the human signal detector (Stanton, 1994). Alternatively, target cueing automation can accomplish an analogous function in cluttered visual environments such as those involved in military target detection (Maltz and Shinar, 2003; Yeh et al., 2003). Yet considerable evidence suggests that such automation does not eliminate the detection problem. This is because automated alerts also must confront challenging issues of distinguishing between signals and highly similar noise (e.g., friend and foe on a military image display), and such alert systems can be counted on to create errors. Thus, the alarm designer, rather than the human monitor, is now the agent responsible for adjusting the response criterion of the alarm, to trade off misses versus false alarms; and designers are often tempted to make this adjustment in such a way that signals are never missed by their systems. (Consider the possible litigation if a fire alarm fails to go off.) However, when the human and automation are considered as a total system, the resulting increase in automation false alarms can have serious consequences (Sorkin and Woods, 1985). These consequences arise because a high false-alarm rate can lead to serious issues of *automation mistrust*, in which people may ignore the alarms altogether (Sorkin, 1989) and experience the "cry wolf" phenomenon (Breznitz, 1983).

The analysis of diagnostic systems also reveals that the problems of high false alarm rate will be further amplified to the extent that the signals to be detected themselves occur infrequently, as is often the case with alarm systems (Parasuraman et al., 1997), so that a large majority of the alarms that do sound will be false alarms. Answers to these problems lie in part in making available to human perception the raw data of the signal domain that is the basis of the alarm. There is some evidence that *likelihood alarms* that can signal their own degrees of certainty in graded form (rather than a two- state on–off logic) will assist (Sorkin and Woods, 1985; St. Johns and Manes, 2002). Finally, it is reasonable to assume that training operators as to the nature of the mandatory miss/false alarm trade-off, and the inevitable high false-alarm rate when there are low-base-rate events, should mitigate problems of distrust to some degree.

4.2 Expectancy, Context, and Identification

We have seen that top-down processing influences the operator's response to the question: "Is the target out there?" The efficiency of his or her response is determined in part by (1) beliefs about the likelihood of the target being present, and (2) knowledge about the target's probable location. Similarly, top-down processing will influence the operator's ability to identify an unknown stimulus at a known location: "What's *that*?" In both types of judgments, experience within a given environment, advisories or warnings about likely events, and knowledge about the characteristics of specific targets will generally improve performance by influencing observer expectancy.

Studies of word and letter recognition demonstrate how identification is enhanced by the presence of appropriate contextual cues. Numerous studies have shown that specific words are identified more efficiently when they are embedded in meaningful sentences rather than presented alone (e.g., Tulving et al., 1964; Stanovich and West, 1983). Similarly, researchers have found that individual letters are recognized more efficiently when they are embedded in a word rather than when they stand alone or are embedded in nonwords (Reicher, 1969; Krueger, 1992; Jordan and Bevan, 1994).

The effects of context and expectancy are not limited to verbal displays. Palmer (1975), for example, found that caricature facial features required less physical detail for recognition when they were embedded in a face rather than when they are presented alone. More complex contexts, such as photographs of naturalistic scenes, also seem to enhance the identification of objects typically found there (Biederman et al., 1981).

Explanations for context effects generally assume that added items in the stimulus array will increase the odds that the operator will recognize at least some portion of it. Even if the portion immediately recognized is not the target object, the recognition is still useful because it reduces the likelihood that some stimuli will be encountered while increasing the likelihood of others. In this way, the total set of possible objects, words, or letters from which the observer must choose becomes smaller and more manageable (see Massaro, 1979; McClelland and Rumelhart, 1981; Richman and Simon, 1989; Grainger and Jacobs, 1994, for formal models of context effects).

These findings suggest that the old design maxim that "less is more" may well be wrong when applied to the presentation of task-critical information that must be *identified* with a high degree of accuracy and/or under degraded environmental conditions. Unlike performance in visual search tasks, where additional nontarget stimuli usually cause declines in performance, identification tasks often benefit from the presence of additional stimuli, as long as those stimuli are normally encountered in spatial or temporal proximity to the target. In fact, operator expectancies can be used to offset degraded stimulus conditions such as poor print reproductions, faulty lighting, brief stimulus exposures, presentation to peripheral vision or even the momentary diversion of attention. Therefore, the redundant use of red, an octagonal shape, and the letters S–T–O–P enhances the identification of a stop sign, as does its expected location to the driver's right immediately before an intersection. A "less is more" stop sign that only used the letter S as a distinguishing feature would not be advised!

In addition to the *design* of displays and messages, the critical role of context in identification performance is also important for *evaluating* display designs. For example, for a symbol to be acceptable, the American National Standards Institute (ANSI, 2001) requires that at least 85% of answers from a group of participants correctly identify the symbol and its message. However, Wolff and Wogalter (1998) have found that this criterion is very difficult to meet if the participants are not given contextual cues relevant to the situation in which the symbol is likely to be found.

4.3 Judgments of Two-Dimensional Position and Extent

Both detection and identification are categorical judgments. In the case of detection, the operator must choose from two alternatives: signal or noise. For identification, the operator must choose one from what may be a very large number of alternatives: for example, in identifying which numeral has appeared on the screen or what type of vehicle is on the horizon. In addition to such "what" questions, operators must often answer questions of "how much" and "where." These judgments are critical for manual control and locomotion (see Section 6) as well as for the interpretation of maps, graphs, and dynamic analog indicators. In the next two sections we focus mainly on spatial judgments of static formats before turning to their dynamic counterparts.

It has been known for some time that the spatial judgments required to read even the most familiar graphical formats are prone to systematic distortions. For example, Graham (1937) found that people tended to overestimate the values represented in bar graphs. This finding was especially true with shorter bars and those farthest from the *y*-axis. With line graphs, Poulton (1985) found that point-reading errors seemed to reflect a perceptual flattening of the line as its distance from the *y*-axis increased. Thus, in the typical line graph with the *y*-axis located on the left side, points along an increasing function are underestimated, and this underestimation increases the farther the point is from the *y*-axis. Systematic distortions have also been obtained for pie charts, where the percentage represented by slices subtending 0 to 90° (and 180 to 270°) are overestimated, while other angles are underestimated (Spence and Krizel, 1994).

Movement planning in human–computer interaction may also be influenced by systematic distortions in spatial perception. Phillips et al. (2003), for example, found that cursors shaped like arrowheads were associated with overestimations of the distance to an onscreen target when the arrow pointed in the target's direction. This, in turn, led participants to overshoot the target when positioning the cursor with a mouse.

These distortions in distance and size may be special cases of geometric illusions such as those reviewed by Gillam (1980) and Gregory (1980). Poulton (1985), for example, has ascribed the perceptual flattening of lines in graphs to the Poggendorf illusion (Figure 3*a*). Phillips et al. (2003) refer to the Muller–Lyer illusion (Figure 3*b*) when describing the overestimations of distances between cursor and target. A variety of explanations have been proposed for such illusions, including a perceptual system that exaggerates small differences to enhance discrimination. Whatever the cause of these spatial distortions, design modifications can reduce their severity. For example, Poulton (1985) found that adding a redundant *y*-axis to the right side of his graphs effectively reduced point-reading errors.

It should be noted, however, that the presence of illusions is not always harmful; sometimes they may even increase the probability of desirable behaviors. For example, some designers have used illusions of size to increase traffic safety. Shinar et al. (1980) painted a pattern similar to that used to induce the Wundt illusion (see Figure 3*c*) on a roadway leading to a dangerous obscured curve. After the roadway was painted, drivers tended to reduce their speed before encountering the curve, presumably because the pattern made the road seem more narrow.

The most systematic work on the size and nature of errors in the perception of graphical displays has focused on how precisely we can make comparisons *between* data values. Cleveland and McGill (1984, 1985, 1986; Cleveland, 1985) developed a list of the physical dimensions that are commonly used to code data values in graphs and maps. These dimensions were ordered, as shown in Figure 4, in terms of the accuracy with which they could be used to make relative magnitude judgments (e.g., "What percentage is point A of point B?"). Dimensions at the top of the figure were used more accurately than those at the bottom. Therefore, Cleveland and McGill advise designers to use position on common scales rather than, for example, volumes, or positions on misaligned scales, to represent data whenever possible. A meta-analysis of the comparative graphics

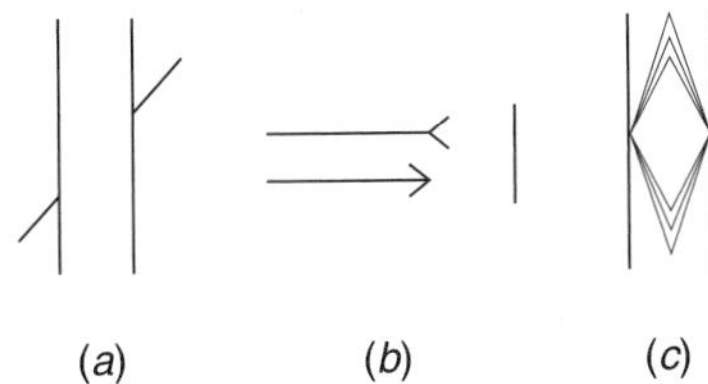

Figure 3 Three perceptual illusions, influencing the perception of location and spatial extent: (*a*) Poggendorf illusion, in which two diagonal-line segments that are actually collinear do not appear so; (*b*) Muller–Lyer illusion, in which the distances between the horizontal line segment and the tips of the two arrowheads appear to be different, even though they are not; (*c*) Wundt illusion, in which two parallel vertical lines appear to curve inward.

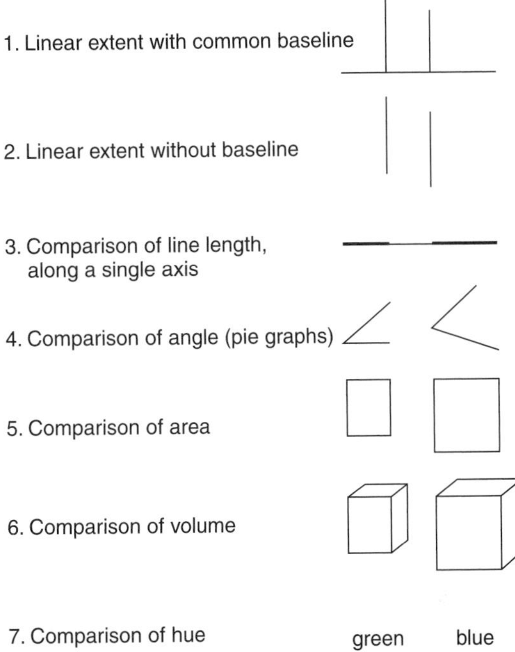

Figure 4 Graphical dimensions for making comparative judgments. (From Wickens, 1992; used by Cleveland and McGill, 1985.)

literature (Carswell, 1992) revealed that the ordering shown in Figure 4 fared well when predicting how well graph users could make simple comparisons or read specific data points. However, this ordering fared less well when used to predict performance in tasks that required the user to identify overall patterns (e.g., "Is there an increasing trend?"). In addition, each step down the ordering did not correspond to an equal decrement in performance. Position, length, and angle judgments were associated with small differences from one another, but were used much more accurately than either area or volume. The misperception of volume, and to a lesser degree the other stimulus dimensions, may reflect ambiguities in the processing of depth information that we discuss next.

4.4 Judgments of Distance and Size in Three-Dimensional Space

Judgments of extent and position, discussed in the preceding section, are also made in three-dimensional space, a space that can either be true space (e.g., judging whether there is adequate spatial separation available to pass a car on a two-lane road), or a display-synthesized three-dimensional space (e.g., evaluating a three-dimensional image, or comparing the volume of cubes in Figure 5). When making judgments in either real or synthesized three-dimensional spaces, human perception depends on a host of depth cues to provide information regarding absolute or relative distance from the viewer (Cutting and Vishton, 1995). Many of

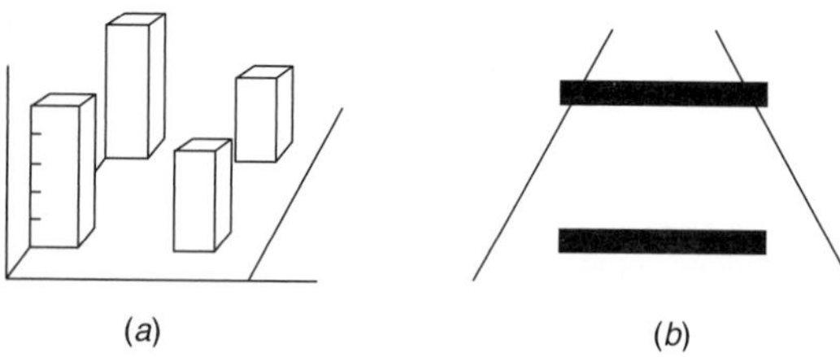

Figure 5 Role of size constancy in depth perception, in creating illusions: (*a*) distorted overestimation of the size of the more distant bar graphs; (*b*) Ponzo illusion, illustrating the greater perceived length of the more distant bar.

these depth cues are called *pictorial cues* because they can be incorporated in static two-dimensional pictures, such as the image shown in Figure 6. Here *linear perspective* suggests that the roadway is receding in depth. The decreasing grain of the texture of the cornfield moving from the bottom to the top of the figure is the cue of *textural gradient*, which also informs the viewer that the surface is receding, as well as revealing the slant angle of the surface and that the viewer is above, not below, the slanting surface. Three depth cues inform the viewer as to which building is closer. The higher building on the *Y* coordinate of the image is signaled to be farther away (*height in the plane*), as is the building that captures the smaller retinal image (*relative size*); and the building that obscures, or interposes in front of, the contours of the other, is seen to be closer (*interposition* or occlusion).

These pictorial depth cues, which can be created by the artist in a static picture, may be combined redundantly with five additional cues. *Motion parallax* is the greater relative motion of closer than of distant objects across the visual field when the head or viewpoint moves. *Binocular disparity* refers to the difference in viewpoint of the two eyes, a difference that diminishes exponentially as objects are farther from the viewer. *Stereopsis*, the use of binocular disparity to perceive depth, can occur in three-dimensional displays when slightly different images are presented to the two eyeballs (Patterson, 1997). Some depth information is also obtained from *accommodation* and *binocular convergence*. These are properties of the eyes themselves that convey some distance information at near viewing distances. Accommodation is the response of the lens required to bring very close objects into focus, and convergence is the "cross-eyed" viewing of the two eyes, also necessary to bring the image of closer objects into focus on the back of both retinas.

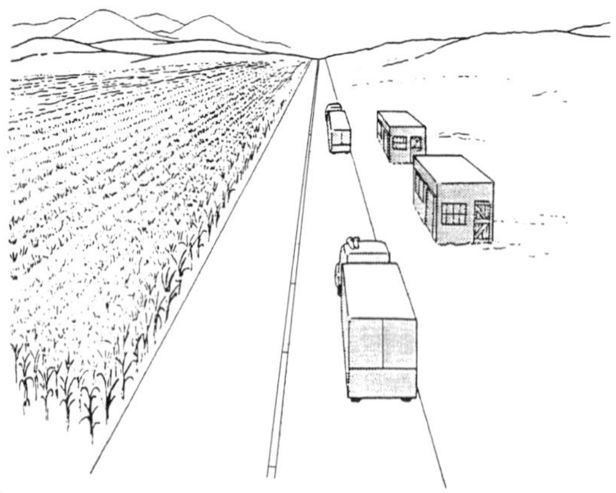

Figure 6 Perceptual cues for depth perception. (From Wickens, 1992.)

In viewing real three-dimensional scenes, most of these depth cues operate redundantly and fairly automatically to give us very precise information about the relative distance of objects in the visual scene and adequate information about the absolute distance (particularly of nearby objects). Such distance judgments are also a necessary component of judgments of three-dimensional shape and form. A host of research studies on depth perception reveal that the depth cues respond in a generally additive fashion to convey a sense of distance (see Wickens et al., 1989). Thus, as the viewer looks at a three-dimensional display, the more depth cues there are available to the viewer, the more perceived separation there is between objects in depth, the *Z-axis* of three-dimensional space. With no cues whatsoever, the image is not viewed as three-dimensional at all, but rather, as one that is fully rotated to be perpendicular to the viewing axis, and parallel with the viewing screen, with all items perceived as equidistant from the observer.

Although all depth cues contribute to the sense of distance, there is evidence that three of those cues (i.e., motion parallax, binocular disparity, and interposition) are the most powerful and will dominate other cues when they are placed in opposition (e.g., Braunstein et al., 1986, Wickens et al., 1989). Hence, in designing a three-dimensional display to synthesize the three-dimensional spatial world, it is a good idea to try to incorporate at least one, if not two, of these dominant cues.

In constructing a three-dimensional perceptual representation from a displayed or real three-dimensional image, people may often be guided by knowledge-driven expectancies and top-down processing when interpreting the bottom-up distance cues. For example, the use of relative size as an effective cue depends on the viewer's knowledge or assumption that the objects to be compared are the same *true size*. But if they are not and a relative size assumption is employed, depth perception can create illusions, sometimes dangerous ones. For example, Eberts and MacMillan (1985) concluded that the high rate of highway accidents suffered by small cars when struck from behind resulted from people's misuse of relative size. The driver behind, seeing the smaller-than-expected retinal image of the small car, perceives the car to be of normal size and farther away. Hence when the small car would brake, the following car would initiate braking later, assuming that the distance was greater than it really was. A similar analysis could be applied to motorcycles, also smaller than the expected vehicle size. Naturally, such illusions, based on inappropriate application of knowledge and expectancies, are more likely to occur when there are fewer depth cues available (i.e., the cues are *impoverished*). This is why pilots are particularly susceptible to illusions of depth (Previc and Ercoline, 2004), since many of the judgments they must make are removed from the natural coordinate framework of the Earth's surface and must sometimes be made in the degraded conditions of haze or night.

Besides the problems of impoverished cues, a second problem with three-dimensional displays is that which can generically be referred to as *line-of-sight ambiguity*. This is illustrated in Figure 7, which depicts a viewer looking into a volume containing three objects, A, B, and C, as we see the views from the side (top view). The view of these objects on the screen is shown below. Here we see that when the viewer makes judgments of position parallel to the viewing axis into the three-dimensional world (or, perpendicular to the display surface on which that world is represented), a given distance in the world is represented by a smaller visual angle than when judgments are made parallel to the display surface, a phenomenon known as *compression*. In Figure 7 the judgment of AB distance along the Z-axis is compressed, whereas the judgment of AC_1 distance along the X-axis is not. As

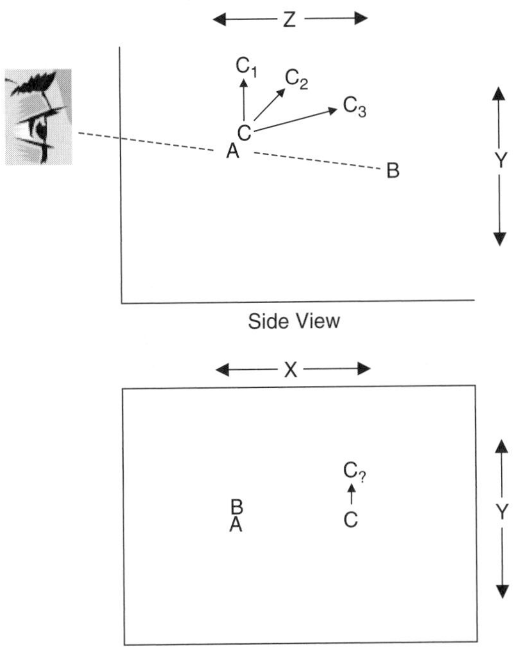

Figure 7 Ambiguity in three-dimensional displays. The actual distance of objects A, B, & C from the observer is shown from the side, in the Top View. The screen viewed by the observer is shown below. From the 3D Screen View it may be ambiguous whether A or B is closer. Movement from C to C_1, C_2 and C_3 may all look identical.

a consequence of this reduced resolution along the viewing axis, it is harder to tell exactly how distant things are: the manifestation of ambiguity.

To make matters worse, it is often difficult in three-dimensional displays to resolve the extent to which an object, displaced to a new location, is moved along the viewing axis (e.g., receding in depth) or is moved to a higher location at the same depth, a further form of ambiguity. For example, the various movements of point C in Figure 7, to points C_1, C_2, and C_3, would all appear nearly equivalent to the viewer of the three-dimensional display with few depth cues, since they all would occupy the same position along the line of sight into the display; the relative contribution of altitude to distance change would be difficult to resolve.

Finally, if the compression makes it difficult to tell exactly how far away an object is, it is difficult to make precise spatial judgments that are orthogonal to the line of sight (e.g.., in the XY-axis, parallel to the display image) because the ratio of true XY distances to displayed XY distances will vary as a function of viewing distance, the perception of which, as we noted, may be ambiguous. Hence, there is some ambiguity in viewing all three axes within a three-dimensional display. This ambiguity is illustrated in the three-dimensional bar graph of Figure 5, in which it is very difficult to tell if the difference in height of the two more distant bar graphs is the same, greater, or smaller than the difference in height of the two closer graphs.

Of course, as we have noted, some of this ambiguity can be resolved in display viewing if the designer incorporates progressively more depth cues. Yet in many circumstances it may be cumbersome or computer intensive to incorporate the most powerful cues of stereo and relative motion realistically; furthermore, there are certain tasks, such as those involved in air traffic control or precise robotic control, in which the requirement for very precise spatial judgments with no ambiguity on any axis is so strong that a set of two-dimensional displays, from orthogonal viewing axes, may provide the best option, even if they present a less natural or realistic view of the world (Wickens, 2000, 2003a).

Concerns for three-dimensional ambiguity notwithstanding, the power of computers to create stereo and motion parallax rapidly and effectively continues to grow, thereby supporting the design of *virtual environments* that capture the natural three-dimensional properties of the world. Such environments have many uses, as discussed in Chapter 40; and their creation must again make effective use of an understanding of the human's natural perception of depth cues. Further discussion of the benefits of different kinds of three-dimensional displays for active navigation are covered in Section 5.4.2.

4.5 Dynamic Displays, Mental Models, and Analog Compatibility

Although many of the displays discussed above are static, many other analog displays are frequently or continuously updated. Dynamic displays are sometimes used for scientific and information visualization (e.g., Cleveland and McGill, 1988), but they have long been a mainstay for air and ground transportation, process control, and manufacturing. Because rapid responses under stressful conditions are often required in these environments, a foremost concern is the compatibility of such displays with the operator's expectations about how the represented system parameters will behave, that is, compatibility with the operator's mental model (Gentner and Stevens, 1983; Norman, 1988; Wickens and Hollands, 2000). There are three ways in which a dynamic analog display can match an operator's mental model successfully. The first two compatibility principles hold for dynamic and static displays alike; the third principle is applicable only to dynamic displays.

Code congruence deals with the fundamental choice of which physical dimension should be used to represent a given variable. For example, how should body temperature readings be presented to an anesthesiologist during surgery? A display designer could choose variation of position along a vertical scale, position along a horizontal scale, angular deflection of a pointer, or any number of other possibilities. We have already seen that the precision with which the dimension can be used to make comparative judgments is one possible factor to consider (Section 4.3). However, we must also consider whether there are any candidate dimensions that are already associated with the concept to be displayed. In the case of temperature, there are associations with both color ("warm" vs. "cool" colors) and position ("high" vs. "low" temperatures). Thus, either color or height might prove to be good candidates.

In the preceding example, temperature is associated with vertical position through training or experience rather than because there is anything inherently spatial about temperature. Many other variables, however, *are* spatial to begin with. In these cases, code congruence is maintained simply by using rescaled versions of the actual spatial dimensions to represent themselves. Thus, the temperature at different locations in our anesthetized patient will be represented most naturally by mapping temperature readings to locations on a three-dimensional model of the human body. When the spatial metaphor is direct, as in this case, the display fulfills what Roscoe (1968) calls the *principle of pictorial realism.*

Choosing the correct physical dimension to represent a variable is a necessary step, but it is not sufficient to ensure that the resulting display supports a user's mental model. It is also important that there be a *congruent mapping* of levels of the code to levels of the variable represented. For example, if color changes are used to represent temperature changes, warmer hues such as orange and red should be chosen to represent warmer temperatures. Using blue to represent warmth would be incongruent with most viewers' expectations. Antes and Chang (1990) provide an example of the detrimental performance effects that result when another color convention—using darker shades to represent higher values of a variable—is violated.

Finally, in dynamic displays, the designer must be concerned with whether motion represented in a display is compatible with the movement expected by the operator. At first it would seem that selecting a congruent code and congruent mapping would automatically take care of *movement compatibility.* However, the static and dynamic aspects of compatibility can sometimes be in opposition. A classic example is that of how an aircraft's bank and pitch are represented in the traditional *moving horizon indicator* (Roscoe, 1968, 1981). This display is compatible with the principle of pictorial realism, in that a horizon line at an orientation opposite that of the plane's orientation is what the pilot actually sees when he or she looks out the windscreen while banking. Yet this same display violates motion compatibility. A leftward bank, in the pilot's mental model of the aircraft, should result in a leftward (or counterclockwise) display rotation. However, the moving horizon moves in the opposite direction. Momentary disorientation may result from this mismatch of mental model and display movement. This disorientation is one possible culprit in crashes involving telltale "graveyard spirals" (Roscoe, 2004).

When display principles come into conflict, as in the case of aviation's moving horizon, hybrid displays may be needed. The frequency-separated display is one example (Fogel, 1959; Johnson and Roscoe, 1972). This display distinguishes dynamic phases of flight, when the pilot is actively changing the bank angle of the aircraft, from relatively static phases, when the pilot is maintaining a particular bank angle when making a turn. The frequency-separated display maintains motion compatibility during the more dynamic phase, whereas static pictorial realism becomes the dominant compatibility concern at other times.

4.6 Perceptual Organization, Display Organization, and Proximity Compatibility

The focus of Section 4.5 was on the compatibility of the operator's mental model with individual indicators representing primarily single values. However, indicators are rarely presented in isolation, so we now consider aspects of perceptual processing with implications for the design and arrangement of multielement displays.

Perceptual organization refers to the relatively rapid processing that parses the multitude of potentially distinct elements in the stimulus array into coherent perceptual groups, configurations, parts, and objects. The laws and processes that underlie such organization were a central concern of the gestalt psychologists in the early part of the twentieth century, and the fundamental questions continue to captivate perceptual researchers (Pomerantz and Kubovy, 1986; Hochberg, 1998). Human factors researchers have also long recognized the implications of perceptual organization for display design (Dashevsky, 1964).

The challenge for the designer of multielement displays, such as those found in cockpits, process control workstations, Web pages, and intensive care units, is to determine how best to organize the various displays so that the natural laws of perceptual organization support rather than hinder the user's acquisition of information. As noted in Section 3, multielement displays have the potential to create a variety of problems for the viewer, including increases in search time, increased information access effort, similarity-based confusion, and challenges to focused attention. Proper display organization seeks to reduce these problems.

The *proximity compatibility principle* provides a framework for applying basic findings about perceptual organization to the problems of display organization (Carswell and Wickens, 1987; Barnett and Wickens, 1988; Wickens and Andre, 1990; see Wickens and Carswell, 1995, for review). In general, the principle holds that those indicators that are related conceptually or that need to be used in combination should belong to the same perceptual group. In short, related information should be bound together perceptually for reasons discussed below (for more detail, see Carswell and Wickens, 1996; Wickens and Carswell, 1995). However, such binding or grouping is not a "free lunch" for the viewer; such perceptual integration has several potential downsides. For example, Kosslyn (1994) describes the difficulty experienced by graph readers when they must isolate an element from one perceptual group in order to compare it with an element from another, a parsing *cost*. Thus, the grouping of information channels must be done with caution and with careful consideration of task demands.

Most applications and evaluations of the proximity compatibility principle compare more integrated arrangements of indicators to control arrangements that are more spatially distributed or involve perceptually dissimilar elements. For example, a panel of indicators arranged so as to combine all information sources into the boundaries of a single perceptual object might be compared to a panel of varied-format indicators such as vertical-pointer displays, semicircular pointers, digital displays, and the like. When such comparisons are made, the more integrated displays are referred to variously as *high-proximity* or *integrated displays*. The more traditional, segregated formats are referred to as *low-proximity* or *separable displays*.

Figure 8 provides a 2 × 2 classification of display arrangements based on the most frequently studied forms of integrated displays: object displays and configural displays. In *object displays*, different variables or information sources are mapped onto the attributes of a single object. Thus, the height and width of a rectangle may be used to represent the average annual return and variability in value, respectively, of a stock mutual fund being considered by an investor. Alternatively, the two variables might be represented by eyebrow slope and hair color of a caricature face. *Configural displays*, by comparison, are any displays in which multiple variables are mapped to spatial dimensions in such a way as to produce additional, higher-level features. For example, several vertical moving pointer displays (top left of Figure 8) if aligned, will produce new *emergent features*, such as the relative smoothness or symmetry of the pattern formed by the pointer symbols. It is important to note that object displays may or may not be configural, and configural

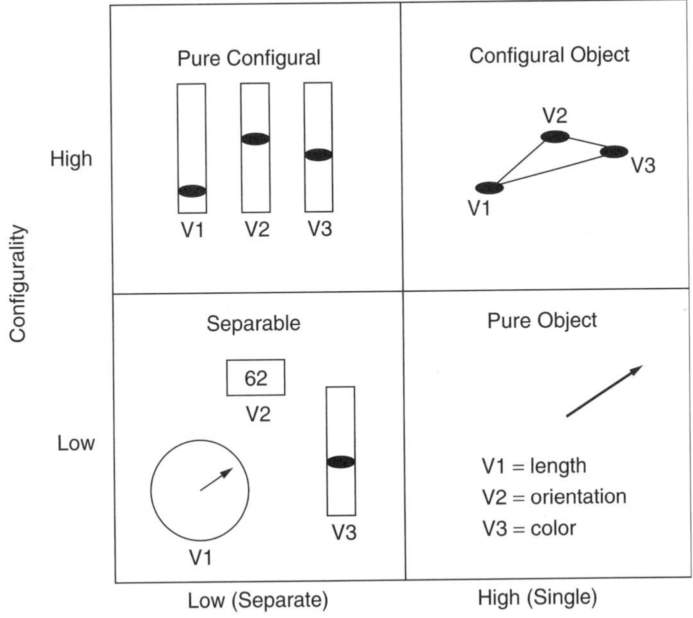

Figure 8 Taxonomy of display configurations based on the concepts of object integration and configurality.

displays may or may not be objects. The exact combination or configurality and object integration, or lack of both, can be used as guidance in determining the appropriate fit between the display arrangement and the task(s) it is designed to support.

Separable displays, exemplified by the three-variable display in the bottom left quadrant of Figure 8, often represent the most traditional arrangement option, in part because of the physical separation imposed by older mechanical displays. As shown, such a display might include nonaligned combinations of digital and analog indicators. Although such an arrangement may be well suited for tasks involving focused attention on individual indicators, integration performance is not well served. Most studies that find difference between integrated and separable displays find advantages favoring the more separable display for focused tasks (e.g., Marino and Mahan, 2005, see Wickens and Carswell, 1995, for a review). The reasons that integrated displays seem to be used less efficiently in these tasks may relate to the additional processing required to parse these forms, to an inability of participants to ignore irrelevant but salient features, or to difficulties of accurately remembering and labeling the component parts of some highly integrated formats.

For tasks involving integration rather than focusing, the to-be-integrated information sources should be combined in one of the ways exemplified by the three remaining quadrants in Figure 8. Turning first to a "pure" object display on the lower right that does not contain configural properties (i.e., emergent features), research has indicated that it does support integration performance but only for one class of integration tasks: integrations that involve identifying different conditions or states based on the co-occurrence of specific values (e.g., Edgell and Morrissey, 1992). Pure object displays can prove to be worse than separable formats for integrations involving quantitative operations such as comparisons, subtractions, or proportion estimates (Carswell and Wickens, 1995).

Nonconfigural objects are formed by spatially integrating dissimilar display codes (features). So, for example, the vertical position and orientation of a single boundary line can be used to represent the pitch and roll of an aircraft, and the color, angular position, and length of a line segment can be used to represent information about an applicant's performance on three different aptitude tests. The advantage for such object displays does not appear to be due solely to the closer physical proximity of features belonging to the same rather than different objects. In addition, it appears that attention can be shifted more efficiently within than between objects, even when controlling for the distance of the shift (Egly et al., 1994a,b). Furthermore, observers can report two features of a single object more rapidly than they can report two features of different objects, even when the two objects overlap spatially (Duncan, 1984; Goldsmith, 1998). Findings of this sort have led some theorists to claim that object displays may encourage (or even force) operators to process information about multiple variables in parallel (Treisman et al., 1983; Kahneman and Treisman, 1984; Treisman, 1993). One cannot, for example, attend to the location of a point in two-dimensional space without also registering its color.

Configural displays, exemplified in the upper half of Figure 8, may be created without forcing codes into unitary objects. What is minimally necessary is that the component indicators be homogeneous and analog (Garner, 1978; Carswell and Wickens, 1990, 1995). For example, they should all be linear moving-pointer displays or all circular moving-pointer displays rather than a combination. The defining feature of such displays is that they include *emergent features*, relational properties of a group of display elements that are not properties of any of the elements in isolation (Garner, 1978; Pomerantz, 1981; Pomerantz and Pristach, 1989). Emergent features are often rapidly detected by the user and in a well-designed configural display may be used as direct response cues for the task at hand (Sanderson et al., 1989; Buttigieg and Sanderson, 1991; Bennett and Flach, 1992). For example, a series of vertical, moving-pointer displays that are placed side by side may produce the emergent feature of pointer alignment. If pointer alignment represents a critical system state (e.g., if alignment indicates that all systems are operating normally), the emergent feature itself becomes a higher-order display code. That is, detecting the emergent feature provides a shortcut to the reading of each individual pointer and the effortful checking of each value against every other. Note that if the indicators were not exactly the same in scale design, had different baselines, or were located at distant parts of the display panel with many intervening displays, alignment would not be available to use as a cue.

Figure 9 illustrates a number of ways that separate indicators can be arranged so as to increase the probability and salience of emergent features and hence be turned into configural displays. These design heuristics are based on the Gestalt laws of perceptual organization which suggest that similarity, spatial proximity, common region, and connectedness, among others factors, will enhance perceptual grouping and make relational patterns more salient.

One way to increase the salience and number of emergent features still further is to make the configural display an object display as well (top right quadrant of Figure 8). This is usually accomplished by adding connecting contours to the spatially varying parts of the display that carry the lower-level ("raw") data. An everyday example involves connecting the tops of bars in a bar graph to form a line graph. Consistent with the proximity compatibility principle, line graphs have been found to facilitate information integration when compared to isolated bars or points (e.g., Schutz, 1961a,b; Hollands and Spence, 1992; Carswell and Wickens, 1996). Multiple indicators can also be arranged so that the addition of line segments creates a closed object. The *polygon*, *star*, or *polar plot display* that has received attention in both process control (e.g., Woods et al., 1981) and aviation (e.g., Beringer and Chrisman, 1991) is an example of

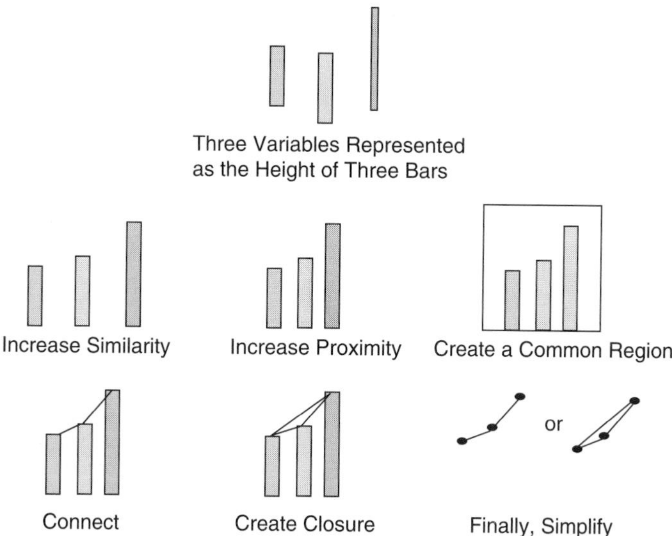

Figure 9 Ways of enhancing display configurality among a triad of indicators. At the top is a triad that uses bar heights to represent data values, but which have a low degree of configurality because they produce no salient emergent features. The remaining triads illustrate ways of combining three bar heights to produce emergent features with varying levels of salience. The most strongly configural displays are those that connect points on the top of the bars to form unitary objects.

this type of object display. Such object displays may contain a wide variety of emergent features, such as global shape, symmetry, and area, that may prove to be useful response cues if mapped directly onto task-relevant variables by the designer.

One potential problem with configural displays, especially configural object displays, is that they are useful only to the extent that the emergent features that they form represent useful characteristics and are not superfluous to the operator's tasks. Configural displays, as defined here, *will* produce emergent features, and if they are irrelevant or, worse, cause patterns that suggest responses inconsistent with those that are appropriate, the user may be better off with separable formats. Using the terms of Section 3, emergent features are salient and capture focused attention, whether wanted or not.

As the foregoing descriptions of display arrangements suggest, application of the proximity compatibility principle to display design requires a detailed understanding of the task or tasks for which the display(s) will be used. For the tasks involved in reading standard data graphics such as scatterplots, bar graphs, and the like, a number of researchers have proposed models that outline the cognitive subtasks involved, giving designers guidance in terms of what types of information integration are necessary (e.g., Simkin and Hastie, 1987; Cleveland, 1990; Pinker, 1990; Lohse, 1991; Hollands, 1992; Gillan and Lewis, 1994; Gillan et al., 1998). For example, reading a single data point from a line graph involves independent processing of the target value in relation to all other values. At the same time, however, it involves information integration (comparison) with an axis. Thus, data points that are far from the axis will be read less accurately than those near it (Poulton, 1985).

Representation aiding is a display design approach in the cognitive engineering tradition (Section 2) that provides guidance on how to optimize the design of configural displays (Vicente and Rasmussen, 1992; Bennett et al., 1993; Kirlik et al., 1993; Bennett and Fritz, 2005). The focus of this approach is directly on understanding the physical constraints of dynamic systems and matching these to the geometric properties of configural formats in a way that is perceptually salient and meaningful to the operator. Thus, knowing that a display needs to contain emergent features is one thing; knowing which emergent features to choose is quite another. Figure 10 provides one such example of a display designed by Moray et al. (1994) to support the perception of constraints in a nuclear power process control plant (see Chapter 3).

5 COMPREHENSION AND COGNITION

In our discussion of perception and display design, we have treated many of the operator's perceptual tasks as decision making, problem solving, or reasoning tasks. Detection involves decisions about criterion setting. Identification involves estimations of stimulus probabilities. Size and distance judgments in three-dimensional space involve the formulation of perceptual hypotheses. For the most part, however, these processes occur rapidly and automatically, and as a result, we are generally not aware of them. In this sense, perceptual reasoning is a far cry from the effortful, deliberate, and often time-consuming process that we are very aware of when trying to troubleshoot a malfunctioning microwave, find our way through an

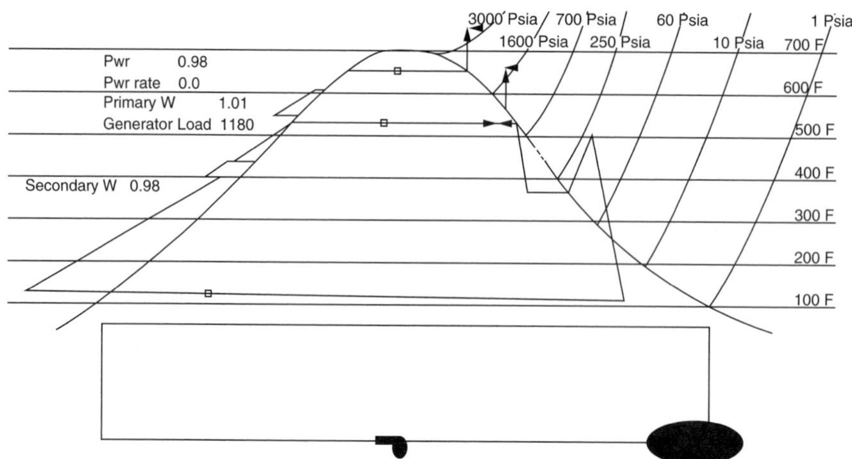

Figure 10 Rankine cycle display for monitoring the health of a nuclear power generating plant. The jagged line indicates the trajectory of the plant parameters (steam pressure and temperature) as they follow the constraints of the thermodynamic laws (proposed but not yet implemented for operational evaluation). (From Moray et al., 1994.)

unfamiliar airport, understand a legal document, or choose among several product designs. Before discussing such higher-order comprehension and cognitive tasks, we first describe the limits of our working memory before discussing its relevance to cognitive tasks of maintaining situation awareness, comprehending text, spatially navigating, planning, and problem solving. As we will see, the parameters of working memory constrain, sometimes severely, the strategies we can deploy to understand and make choices in many dynamic environments.

5.1 Working Memory Limitations

Working memory refers to the limited number of ideas, sounds, or images that we can maintain and manipulate mentally at any point in time. The concept has its roots in William James's (1890) *primary memory* and Atkinson and Shiffrin's (1971) *short-term store*. All three concepts share the distinction between information that is available in the conscious here-and-now (working, short-term, or primary memory) and information that we are not consciously aware of until it is called upon from a more permanent storage system (long-term or secondary memory).

Unlike items in long-term memory, items in working memory are lost rapidly if no effort is made to maintain them (Brown, 1959; Peterson and Peterson, 1959). For example, *decay rates* of less than 20 seconds have been obtained for verbally delivered navigation information (Loftus et al., 1979) as well as for visuospatial radar information (Moray, 1986). However, even when tasks require minimal delays, working memory is still severely limited in terms of its capacity. Miller (1956) suggested that this capacity, the *memory span*, is limited to about five to nine independent items. The qualifier "independent" is critical, however, because physically separate items that are stored together as a unit in long-term memory may be rehearsed and maintained in working memory as a single entity: a *chunk*. Thus, a long-distance telephone number that consists of 11 numbers (e.g., 1-904-638-1803) might seem to be beyond the limits of working memory. However, if the caller is familiar with the area code and the prefix, each of these numeric groups counts as one chunk instead of three, and the entire number is reduced to a more manageable seven chunks.

Taking into account these limitations in duration and capacity as well the nature of errors made in tasks with high working memory demands, Baddeley (1986, 1999) has proposed a three-part model of working memory structure. First, there are two temporary storage systems, the *phonological loop* and *visuospatial sketch pad*. These subsystems are used by a *central executive* that transforms information within each store, transfers information from one store to the other, and integrates information in these stores with information in long-term memory.

Most research on the limits of working memory have focused on the *phonological loop*, sometimes called *verbal working memory*: so named because it is associated with our silent repetition or rehearsal of words, letters, and numbers. The phonological loop stores a limited number of sounds for a short period of time. This description is consistent with the finding that the number of items that can be held in working memory is related to the length of time it takes to pronounce each item (Gathercole and Baddeley, 1993; Gathercole, 1997). Thus, our memory span is slightly lower for words with many syllables.

The *visuospatial sketch pad* holds visual and spatial information as well as visual information that has been transformed from verbal sources (Logie, 1995; Baddeley, 1999). The information held in the sketch pad may be in the form of mental images, and as with the phonological loop, the contents will be lost rapidly if not rehearsed. Research suggests that rehearsal in the visuospatial sketch pad involves repeated switching of selective attention to different positions across these images (Awh et al., 1998).

The central executive is aptly named, because its functions can be compared to those of a business executive (Baddeley, 1999). The central executive's role is not to store information but to coordinate the use of information. This information may come from the phonological loop, the visuospatial sketch pad, or long-term storage. The central executive is presumed to be involved in integrating information from these different stores, and it is involved with selecting information, suppressing irrelevant information, coordinating behavior, and planning (see Section 5.5 for the relevance of these functions for problem solving). It is important to note that there are limits on how many of these operations the central executive can execute at one time.

The concept of working memory and knowledge of its various limitations have a number of implications for design. We describe some of these implications below, and we discuss other implications in the context of higher order tasks such as decision making, problem solving, and creative thinking in subsequent sections.

1. The capacity of either the visuospatial sketch pad or the phonological loop may easily be exceeded, resulting in a loss of information that may be necessary to perform an ongoing task. The design implication is to avoid, whenever possible, codes that infringe on the limits of these systems. However, when longer codes are necessary, there are several ways to reduce memory loss, most involving designs that encourage chunking. For example, parsing material into three- or four-item units may increase chunking and subsequent recall (Wickelgren, 1964). Thus, 3546773 is more difficult to recall than 354–6773. In addition, information for different tasks may be split between the two slave systems so that neither the visuospatial sketch pad nor the phonological loop is overburdened. More will be said about such interventions when we turn to the discussion of multitask performance (see Section 7). In addition, verbal codes composed of more easily pronounced items should be preferred over codes involving longer words. This would suggest that numerical codes that make frequent use of the two-syllable number "seven" will be more prone to loss from the phonological loop than codes that make frequent use of other numbers. It may also suggest that the functional number span for some languages may be larger than those for others.

2. Information from either store may be lost if there are delays longer than a few seconds between receiving the information and using it. Thus, systems should not be designed so that the user must perform several operations before being able to perform a "memory dump." For example, voice mail systems should always allow users to select a menu option as soon as it is presented rather than forcing them to wait until all the options have been read to make their choice. Methods of responding should be simplified as well, so that users do not have to retain their choice for long periods of time while trying to figure out how to execute it.

3. Information may need to be transferred from one subsystem into the other before further transformations or integrations can be made, thus reducing the resources available for the primary processing goal. Wickens et al., (1983, 1984) have provided evidence that the display format should be matched to the working memory subsystem that is used to perform the task. Specifically, visual-analog displays are most compatible with tasks utilizing the visuospatial sketch pad (e.g., air traffic controllers' maintenance of a model of the spatial relations among aircraft) and auditory–verbal displays are most compatible with tasks utilizing the phonological loop (e.g., a nurse keeping track of which medications to administer to a patient).

4. If either working memory subsystem is updated too rapidly, old information may interfere with the new. For alphanumerical information, Loftus et al. (1979) found that a 10-second delay was necessary before information from the last message no longer interfered with the recall of the current material.

5. Any interference in working memory is most likely to occur to the extent that the material is similar in its meaning or sound, thereby creating confusions. Thus, an air traffic controller might have particular difficulties remembering a series of aircraft with similar call signs (UAL 235, UAL 325). Interference will also be greater if there is similarity between material to be remembered and other competing tasks (i.e., listening, speaking) (Banbury et al., 2001).

5.2 Dynamic Working Memory, Keeping Track, and Situation Awareness

Much of the research devoted to working memory has examined tasks in which information is delivered in discrete batches and the goal is to remember as much of the information as possible. However, there are many other tasks in which the operator must deal with continuous information updates with little expectation of perfect retention. Moray (1981) studied several *running memory tasks* that simulated the demands of a more continuous input stream, and he found the typical memory span to be less than five chunks. In some cases it was difficult for subjects to keep track of items more than two places back in the queue. Yntema (1963) demonstrated that the way information is organized has a direct impact on supervisors' abilities to keep track of values of multiple attributes of several objects (e.g., status and descriptions of several aircraft). Supervisors had greater success keeping track of a few objects that varied on many different (and discriminable) attributes than in keeping track of variation in a few attributes for many objects. In the former case there are fewer opportunities for confusion than in the latter case, and confusion is a major source of disruption in working memory (Hess and Detweiler, 1995).

This earlier research on running memory anticipates current interest in *situation awareness* (SA) (Durso and Gronland, 1999; Endsley and Garland, 2000; Banbury and Treselian, 2004 see also Chapter 20). Endsley (1995) defines situation awareness as the perception of the elements of the environment within a volume of time and space, the comprehension of their meaning, and the projection of their status in the near future (Endsley, 1995). Thus, there are three stages to SA: perception or "noticing," understanding or comprehending, and projecting or prediction. Thus, the pilot, chef, or driver must all keep track of a multitude of changing dynamic stimuli and events in their environment, must determine the relevance of those events to their current task and overall goals, and must project the status of the most relevant events in the near future.

The three components of situation awareness can be tied directly to different aspects of information processing. Thus, stage 1, noticing, traces directly to issues of selective attention and attentional capture, discussed in Section 3.4. There we saw that noticing changes in a dynamic environment is not always done well, and indeed, Jones and Endsley (1996) found that a majority of aircraft accidents attributable to loss of SA were related to breakdowns at this first stage. Understanding stage 2 SA depends heavily on working memory, for keeping track of the evolving situation (e.g., the pilot asks: "Where was that traffic aircraft the last time that I looked?"). Such understanding can often be aided by long-term memory, which can generate expectancies that help interpret new data, or *schemas* of typical situations, that can be used to interpret particular events. Furthermore, while effective stage 2 SA depends on working memory, it may also depend on the ability to access information rapidly if needed, from other sources. That is, effective stage 2 SA can be supported by *knowing where to look* to get necessary information (Durso and Gronland, 1999) as well as by appropriate knowledge structures in what Ericsson and Kintsch (1995) have referred to as *long-term working memory*. Finally, we note that stage 2 understanding can also help to drive scanning to acquire updated information at stage 1 (Adams et al., 1995). That is, we look where we expect to find information given our current understanding of the situation.

The stage 3 component, prediction and projection, is perhaps the most complex and may depend to a greater extent than the other two stages on the expertise and training of the operator. Accurate prediction of an evolving situation certainly depends on current perception and understanding (stages 1 and 2), but it also requires a well-calibrated *mental model* of the dynamic process under supervision, a mental model that can be "played" in response to the current data, in order to predict the future state. For example, an air traffic controller with a good mental model of aircraft flight characteristics can examine the display of the current state and turn rate of an aircraft and project when (and whether) that aircraft will intersect a desired approach course, thereby attaining a satisfactory separation. A well-calibrated mental model resides in long-term memory, but to play the model with the current data requires perception of those data as well as the active cognitive operations carried out in working memory. In some cases, prediction can be approximated by using an expert acquired *script* of the way a typical situation unfolds. However, unless active processing (stage 1) of incoming perceptual information is carried out, there is a danger that projection will be based totally on expectancies of typical situations and that unusual or atypical events will be overlooked.

It should be noted, finally, that situation awareness is a construct that is resident within the perceptual–cognitive operations of the brain. It is not itself a part of the action (other than the actions chosen to acquire new information).

5.3 Text Processing and Language Comprehension

Comprehension of language, whether written or spoken, shares many of the processes described for situation awareness. Noticing relevant information, understanding its implications, and to varying degrees, projecting the content of upcoming messages are all part of the active process of language comprehension. The constraints relevant to information processing at each of these stages helps determine why we find some conversations, lectures, journal articles, instructions, and warnings easier to understand than others.

Of course, factors influencing the detectability and discriminability of the individual speech sounds (phonemes) and written symbols (letters) will limit the extent to which language can be processed meaningfully. However, recall that easily comprehended phrases or sentences can also influence the detectability of the individual words. See Section 3 for a dxscxxsion xf thx efxxct xx cxnxext on identificaxxxx. We are typically able to understand sentences such as the last one despite the absence of many of the correct letters in the words, because of the profound effects of context, expectancies, and the redundancies inherent in language. It is worth noting that just as context can help us recognize familiar words, it can also help us understand the meanings of words that we have never encountered before (Sternberg and Powell, 1983).

As our discussion of context suggests, the comprehensibility of text depends on many factors, from the reader's experience, knowledge, and mental models that drive expectations, to the structuring of text so as to make maximum use of these expectancies. It is not surprising, then, that *readability metrics* that attempt to estimate the difficulty of text passages, generally based on average word and sentence length, are not altogether satisfactory. Although it may be true that longer words are generally less familiar and longer sentences place greater demands on our working memory capacities, many other factors influence comprehensibility. Kintsch and Vipond (1979), for example, used traditional readability indices to compare the speeches of candidates in the 1952 presidential campaign. Eisenhower's speeches were generally reputed to be simpler than those of Stevenson, yet formal readability indices indicated that Stevenson's used shorter words and sentences. This contradiction between public opinion and the formal metrics corresponds to our experience that some sentences with a few short words can still be very confusing. We now discuss some additional factors that determine comprehensibility and have implications for message design.

Kintsch and colleagues (e.g., Kintsch and Keenan, 1973; Kintsch and Van Dijk, 1978) argue that the complexity of a sentence is actually determined by the number of underlying ideas, or *propositions*, that it contains rather than by the number of words. Although a few specific words may be carried forward in working memory for brief periods, it is the underlying propositions that are used to relate information in different phrases and sentences. Just as Moray (1981) estimates that running memory carries forward less than five chunks of information, Kintsch and Van Dijk (1978) estimate that only four propositions can be held in working memory at one time. Thus, the reader must be selective in his or her choice of propositions to retain. According to the model, readers tend to favor the most recent propositions and those they believe to be most central to the overall text message.

Problems arise in comprehension when newly encountered propositions cannot easily be related to the propositions active in working memory. Such problems often occur when readers attempt to integrate information across sentence boundaries. Consider, for example, the following sentences:

1. When the battery is weak, a light will appear.
2. You will see it at the top of the display panel.

Readers must make the *bridging inference* that the second sentence is telling them where to look for the light rather than where to find the battery. This inference, in turn, depends on their general knowledge of displays: specifically, the fact that lights rather than batteries tend to appear on display panels. A second type of integration failure occurs when a concept introduced earlier in the text is not actually used again until some sentences, paragraphs, or pages later. For instance, if the battery mentioned above was first introduced a paragraph or two before its current use, with no reference in the intervening text, the reader would have to pause to search long-term memory (or scan the text itself) to determine precisely *what* battery was being discussed. Having to perform such a *reinstatement search* is yet another negative consequence of delaying the use of information in working memory.

One general goal in striving for comprehensibility is to avoid the need to make bridging inferences or perform reinstatement searches. However, it is clearly impossible to remove the need to make some inferences, and it is probably undesirable given that such elaborations may make the information more memorable. One goal of the text designer is simply to assist the reader in making the *appropriate* inferences. One important way that this can be done is by providing adequate context immediately prior to the presentation of target information (McKoon and Ratcliff, 1992). Because inferences draw on the reader's knowledge of particular topics, it is useful to allow the reader to access the relevant knowledge structures in long-term memory at the outset. Bransford and Johnson (1972) provide a powerful demonstration of the importance of providing context in the form of pictures or descriptive titles presented just prior to textual material. A series of instructions on how to wash clothes was presented with and without the prior context of a title, "washing clothes." When the title was removed, the reduction in readers' abilities to understand and recall the instructions was dramatic.

Other factors that increase the processing demands of verbal material include the use of negations and lack of congruence between word orders and logical orders. With regard to negations, research indicates that it takes longer to verify a sentence such as "the circle is not above the star" compared to "the star is above the circle" (Clark and Chase, 1972; Carpenter and Just, 1975). Results suggest further that the delay is due to something other than the time necessary to process an additional word (i.e., "not"). Instead, it appears that listeners or readers first form a representation of the objects in the sentence based on the order of presentation (e.g., circle-before-star in the sentence "the circle is not above the star." However, to make their mental representation congruent with the meaning of the negation, they must perform a transformation of orders (i.e., to end up with a circle that is *not* before/above the star). Similar logic is used to explain why subjects have trouble processing statements in which the logical order represented by the sentence is inconsistent with the physical ordering of the words (DeSoto et al., 1965). Returning to our battery instructions once again, the underlying causal sequence assumed by most people would be that a weak battery would trigger a warning light. To be consistent with this causal order, it would be better to state that "If the battery is weak, the light will come on" rather than "If the light comes on, the battery is weak."

Finally, the physical parsing of sentences on a page, sign, or computer screen can also influence the comprehensibility of verbal messages. Just and Carpenter (1987), have suggested that although meaning is extracted continually as we arrive at each word in a sentence, there is a pause for the overall integration at the end of the constituent phrases. Consistent with this idea, Graf and Torrey (1966) found enhanced comprehension for sentences that were broken into several different lines of text when the end of each line corresponded to the end of a phrase. Thus, instructions or warnings that must appear on several different lines (or as a few words on several successive screens) should be divided by phrases rather than, for example, on the basis of the number of letters. "Watch your step... when exiting... the bus" will be understood more quickly than "Watch your... step when... exiting the bus."

5.4 Spatial Awareness and Navigation

Language comprehension sometimes taxes working memory, particularly the phonological rehearsal loop and central executive. However, as we saw when discussing problems with negation, people may use text to generate representations of spatial relations. This spatial facet of text and language comprehension has been particularly prominent in recent discussion of the "situation models" that we develop when reading or listening to a story (e.g., knowing where in a room all the characters are sitting). We now turn to a task that relies more heavily, for many people, on the capacity limits of the visuospatial sketch pad (Logie, 1995), or more generally, spatial working memory and spatial cognition (Shah and Iyiri, 2005)—navigating through our worlds, both real (finding our way through a maze of looping suburban streets and cul-de-sacs; Whitaker and CuQlock-Knopp, 1995) and virtual (searching a complex computer-displayed multidimensional database).

5.4.1 Geographical Knowledge

Thorndyke (1980) has studied the knowledge that people use when finding their way about. Of particular interest is Thorndyke's claim that increased familiarity with an area causes changes in more than the amount of detail contained in our mental representation of that area stored in long-term memory. In addition, the actual *type* of mental representation (analog versus verbal/symbolic), as well as its frame of reference, may evolve in a predictable way. After an initial encounter with a city, neighborhood, or building, we may develop *landmark knowledge*. If told that his or her destination is beside the "telephone tower," a person with landmark knowledge, will scan the environment visually until spotting something that appears to be the tower and will then strike off in its direction. Thus, the newcomer has the knowledge necessary to recognize the landmark but has no knowledge about its location. For the person with landmark knowledge alone, wayfinding would be impossible if the landmarks were obscured. This problem has become commonplace as once-salient landmarks have become obscured by new and often taller structures. Guidance signs to landmarks have become a familiar antidote to the problem, but these in turn add to the visual clutter and confusion that may greet a first-time visitor to a new area. The problem for urban planners, then, is to ensure that landmarks (both natural and designed) remain easily visible and distinctive in order to serve their navigational function for years to come.

With more experience traveling about an area, we typically develop an ordered series of steps that will get us from one location to another. These sets of directions, called *route knowledge*, tend to be verbal in nature, stated as a series of left–right turns (e.g., "Go left on Woodland until you get to the fire station. Then take a left..."). Navigation along these routes may be rapid and very automatic; however, limited knowledge of the higher-order relations among different routes and landmarks still limits navigational decision making, making it difficult, for example, to figure out short cuts. With still more extensive wayfinding experience, or with specific map study, *survey knowledge* may be acquired. Survey knowledge is an integrated representation of the various routes and landmarks that preserves their spatial relations. This analog representation is usually referred to as a *cognitive map*.

The type of representation—route versus survey—that best supports performance in various wayfinding tasks, like so many other aspects of mental (and display) representation, depends on the nature of the task or problem. Thorndyke and Hayes-Roth (1982) compared *route training* (actual practice navigating between specific points in a large building) to *survey training* (study of the building plan). Route training appeared to facilitate people's estimates of route distance and orientation, while survey training appeared to facilitate judgments of absolute (Euclidean) distance and object localization.

5.4.2 Navigational Aids

Although we can often navigate through environments on the basis of our acquired knowledge stored in long-term memory, whether route, survey, or even landmark, there are many other circumstances in which we require displayed *navigational aids* which are perceived. These aids may take on a wide variety of forms, ranging in the degree to which guidance to a target is supported: from tightly guided flight directors in aircraft, and turn signs on highways, to route lists, to electronic maps that highlight one's current position, to simple paper maps. Furthermore, electronic maps can vary in the extent to which they rotate so that the direction of travel is "up" on the map, and both electronic and paper maps can vary in terms of whether they present the world in planar or three-dimensional perspective view (see Section 4.4).

To understand which forms of maps support the best spatial information processing to accomplish navigation, it is important to consider briefly the stages involved in this process. The navigator must engage in some form of visual search of both the navigational aid (to locate the final destination, intermediate goals, and current location) and of the environment or a displayed representation thereof (to locate features that establish the current location and orientation). The navigator must then establish the extent to which the former and the latter are congruent, determining the extent to which "where I am" (located and oriented) agrees with the intermediate goal of "where I want to be." Establishing this congruence may require any number of different *cognitive transformations* that add both time and effort to the navigational task (Aretz, 1991; Hickox and Wickens, 1999; Gugerty and Brooks, 2001; Wickens et al., 2005).

An example of two of these transformations is represented in Figure 11, which represents the information processing of a pilot flying south through an environment depicted on a north-up contour map. To establish navigational congruence, the pilot must rotate the map mentally to a track-up orientation,

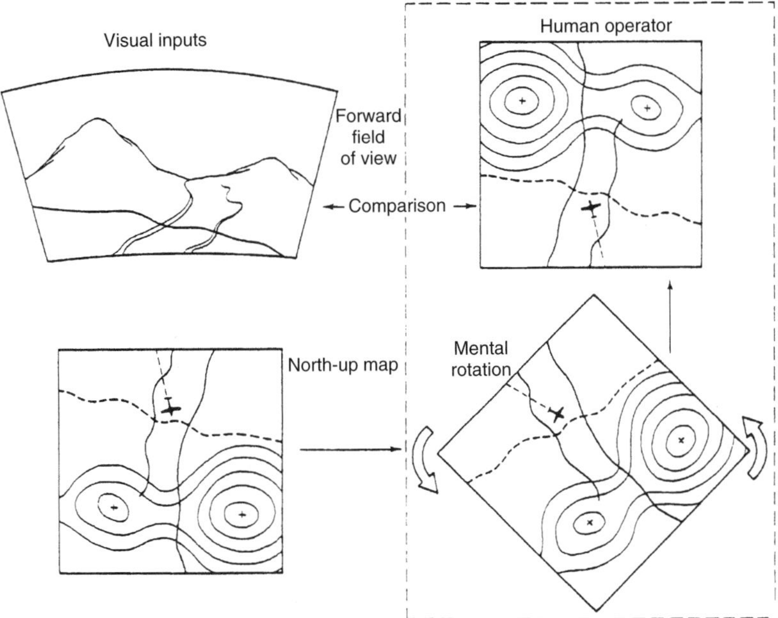

Figure 11 Mental rotation required to compare the image seen in an ego-referenced forward field of view (top) with a world- referenced north-up map (below) when the aircraft is heading south. The map image is mentally rotated (right) to bring it into lateral congruence with the forward field of view. It is then envisioned in three dimensions to compare with the forward field of view.

and then envision the contour representation of the three-dimensional terrain to determine its congruence with the forward view. Both of these information transformations are effortful, time consuming, and provide sources for error. In particular, those sources involved with mental rotation of maps have been well documented (Levine, 1982; Eley, 1988; Warren et al., 1990; Aretz, 1991; Olmos et al., 1997; Gugerty and Brooks, 2001).

Different transformations may be required when other navigational aids are provided. For example, verbal descriptions of landmarks will also require some transformations to evaluate against their visible three-dimensional spatial counterparts. Transformations may also be required to "zoom in" to a large-scale map (Kosslyn, 1987), in order to establish its congruence with a close-in view of a small part of the environment. Modeled in terms of processing operations such as visual search and spatial transformations, one can then determine the form of navigational aids that would be of benefit for certain tasks. For example, electronic maps are beneficial if they highlight the navigator's current location, thus obviating visual search of the map. Highlighting landmarks on the map, which are salient in the visual world, will correspondingly reduce search.

Rotating maps in a track-up orientation will help navigation by eliminating mental rotation (Aretz, 1991; Olmos et al., 1997; Wickens, 1999). Presenting guidance information in a three-dimensional format (Section 4.4), like one would see looking ahead into the environment itself, will also reduce the magnitude of any sort of transformations and considerably improve navigational performance (Wickens and Prevett, 1995; Wickens et al., 2005). The benefits of a three-dimensional view will be enhanced if the viewpoint of the display corresponds to the same zoom-in viewpoint as that occupied by the navigator, looking forward, rather than a viewpoint that is behind and from the outside (Wickens and Prevett, 1995; Olmos et al., 2000). These viewpoint relationships are shown in Figure 12, which depicts the viewpoint location (top) and the view seen by a pilot (bottom) in an immersed or *egocentric view* (*a* and *b*). (These views differ in terms of their geometric field of view.) Panel (*c*) represents an external or exocentric view. Panel (*d*) represents a two-dimensional coplanar view, which was discussed in Section 4.4.

Expressing navigational guidance in terms of command *route lists* (e.g., "turn left at X; go three blocks until Y") will also eliminate the need for many spatial cognitive transformations that may be imposed when spatial maps are used, since the language of command is thereby expressed directly in the language of action. Such congruence can account for the benefits of route lists over spatial maps in certain ground navigation tasks (Wetherell, 1979; Streeter et al., 1985). A second advantage to such route lists is that they can be presented verbally, and represented in working memory mentally in a phonetic or verbal code,

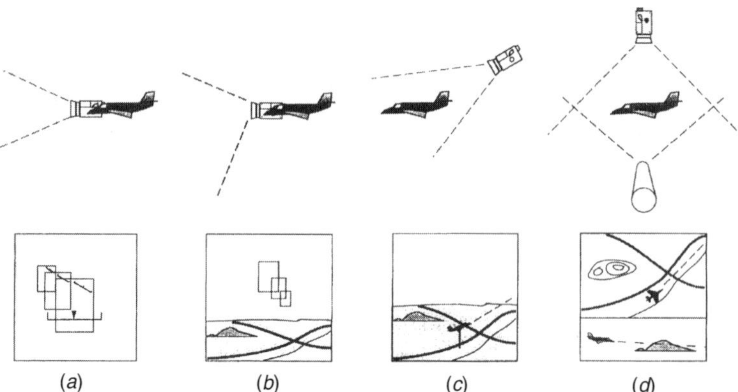

Figure 12 Display viewpoints in an aircraft display that require varying degrees of transformations to compare with a pilot's direct view forward from the cockpit. The lower figures illustrate schematically what would be seen by the pilot with the viewpoint shown above. The transformation in (*a*) is minimal; in (*b*) and (*c*), modest; and in (*d*), large. Views (*a*) and (*b*), however, reduce more global situation awareness.

thus reducing competition for the spatial processing resources involved in many aspects of environmental scanning and vehicle navigation (Section 7). The verbal descriptions inherent in route lists are well suited for some navigational environments, particularly human-designed environments (cities) with objects that are easily labeled and can easily be discriminated by counting (the "fourth house"). In many naturalistic environments, however, where features are defined by continuous, not categorical properties, and may be confusable, route lists are more problematic and should be coupled with redundant spatial or pictorial guidance.

The most direct levels of navigational guidance that eliminate most or all levels of mental transformations (e.g., a flight director, the three-dimensional forward-looking display shown in Figure 12*a*, or a verbal route list) will provide for effective navigation while on route. However, such displays may do a disservice to the navigator who suddenly finds himself lost, disoriented, or required to make a spontaneous departure from the planned route. Very often, those features that make a navigational display best for guidance will harm its effectiveness to support the spatial situation awareness (Section 5.2) that is necessary for a successful recovery from a state of geographical disorientation (Wickens and Prevett, 1995; Wickens, 1999). This is an important trade-off between the immersed three-dimensional view of Figure 12*b*, which is good for guidance, but because of its "keyhole view" of the world, is poor for maintaining global situation awareness, a task better supported by the exocentric view of Figure 12*c*. Finally, we note that the immersed three-dimensional view makes a poor tool for route planning, an activity that we turn to in the following section.

5.5 Planning and Problem Solving

Our previous discussion has focused on cognitive activities that were heavily and directly driven by information in the environment (e.g., text, maps, or material to be retained in working memory). In contrast, the information-processing tasks of planning and problem solving are tied much less directly to perceptual processing and are more critically dependent on the interplay between information available in (and retrieved from) long-term memory, and information-processing transformations carried out in working memory.

5.5.1 Planning

The key to successful operation in many endeavors (Miller et al., 1960) is to develop a good plan of action. When such a plan is formulated, steps toward the goal can be taken smoothly without extensive pauses between subgoals. Furthermore, developing contingency plans will allow selection of alternative courses of actions should primary plans fail. As an example, pilots are habitually reminded to have contingency flight plans available should the planned route to a destination become unavailable because of bad weather.

Planning can typically depend on either of two types of cognitive operations (or a blend of the two). Planners may depend on scripts (Schank and Abelson, 1977), of typical sequences of operations that they have stored in long-term memory on the basis of past experience. In essence, one's plan is either identical to or involves some minor variation on the sequence of operations that one has carried out many times previously. Alternatively, planning may involve a greater degree of guess work, and some level of mental simulation of the intended future activities (Klein and Crandall, 1995; see Chapter 37). For example, in planning how to attack a particular problem, one might play a series of "what if" games, imagining the consequences of action, based again on some degree of past experience. Hence an air traffic controller, in planning how to manage a potential future conflict situation, might mentally simulate the future trajectories of the affected aircraft under different proposed maneuvers to see if the intended commands would resolve the conflict and would stay clear of other aircraft.

Consideration of human performance issues and some amount of experimental data reveals three characteristics of planning activities. First, they place fairly heavy demands on working memory, particularly as plans become less script based, and more simulation based. Hence, planning is a task that is vulnerable to competing demands from other tasks. Under high-workload conditions, planning is often the first task to be dropped, and operators become less proactive and more reactive (Hart and Wickens, 1990). The absence of planning is often a source of poor decision making (Orasanu, 1993 Orasanu and Fischer, 1997). Second, perhaps because of the high-working-memory demands of planning, in many complex settings, people's *planning horizon* tends to be fairly short, working no more than one or two subgoals into the future (Tulga and Sheridan, 1980). To some extent, however, this characteristic may be considered as a reasonably adaptive one in an uncertain world, since many of the contingency plans for a long time horizon in the future would never need to be carried out and hence are probably not worth the workload cost of their formulation. Finally, given the dependency of script-based planning on long-term memory, many aspects of planning may be biased by the *availability heuristic* (Tversky and Kahneman, 1974), discussed in more detail in Chapter 8. That is, one's plans may be biased in favor of trajectories that have been tried with success in the past and therefore easily recalled.

Consideration of such vulnerabilities leads inescapably to the conclusion that human planning is a cognitive information-processing activity that can benefit from automated assistance, and indeed, such planning aids have been well received in the past, for activities such as flight route planning (Layton et al., 1994) and industrial scheduling (Sanderson, 1989). Such automated planners provide assistance that need not necessarily replace the cognitive processes of the human operator but merely provides redundant assistance to those processes in allowing the operator to keep track of plausible courses of future action.

5.5.2 Problem Solving, Diagnosis, and Troubleshooting

The three cognitive activities of problem solving, diagnosis, and troubleshooting all have similar connotations, although there are some distinctions between them. All have in common the characteristic that there is a goal to be obtained by the human operator; that actions, information, or knowledge necessary to achieve that goal is currently missing; and that some physical action or mental operation must be taken to seek these entities (Mayer, 1983; Levine, 1988). To the extent that these actions are not easy or not entirely self-evident, the processes are more demanding.

Like planning, the actual cognitive processes underlying the diagnostic troubleshooting activities can involve some mixture of two extreme approaches. On the one hand, situations can sometimes be diagnosed (or solutions to a problem reached) by a direct match between the features of the problem observed and patterns experienced previously and stored in long-term memory. Such a *pattern-matching technique*, analogous to the role of scripts in planning, can be carried out rapidly, with little cognitive activity, and is often highly accurate (Rasmussen, 1981). This is a pattern of behavior often seen in the study of *naturalistic decision making* (Zsambok and Klein, 1997; see Chapter 8).

At the other extreme, when solving complex and novel problems that one has never experienced before, a series of diagnostic tests must often be performed, their outcomes considered, and based on these outcomes, new tests or actions taken, until the existing state of the world is identified (diagnosis) or the problem is solved. Such an iterative procedure is typical in medical diagnosis (Shalin and Bertram, in press). The updating of belief in the state of the world, on the basis of the test outcomes, may or may not approach prescriptions offered by guidelines for *optimal information integration*, such as Bayes' theorom (Yates, 1990; see Chapter 8).

In between these two extremes are hybrid approaches that depend to varying degrees on information already stored in long-term memory on the basis of experience. For example, the sequence of administering tests (and the procedures for doing so) may be well learned in long-term memory even if the outcome of such tests is unpredictable, and must be retained or aggregated in working memory. Furthermore, the sequence and procedures may be supported by (and therefore directly perceived from) external *checklists*, relieving cognitive demands still further. The tests themselves might be physical tests, such as the blood tests carried out by medical personnel, or they may involve the same mental simulation of "what if" scenarios that was described in the context of planning (Klein et al., 1993).

As with issues of planning, so also with diagnosis and problem solving, there are three characteristics of human cognition that affect the efficiency and accuracy of such processes. First, as these processes become more involved with mental simulation and less with more automatic pattern matching, their cognitive resource demands grow and their vulnerability to interference from other competing tasks increases in a corresponding fashion (see also Chapter 9). Second, as we noted, past experience, reflected in the contents of long-term memory, can often provide a benefit for rapid and accurate diagnosis or problem solutions. But at the same time, such experience can occasionally be hazardous, by trapping the troubleshooter to consider only the most *available* hypotheses: often those that have been experienced recently or frequently, and hence are well represented in long-term memory (Tversky and Kahneman, 1974). In problem solving, this dependence on familiar solutions in long-term memory has sometimes been described as *functional fixedness* (Adamson, 1952; Levine, 1988).

Third, the diagnostic/troubleshooting process is often thwarted by a phenomenon referred to alternatively by such terms as *confirmation bias* and *cognitive tunneling* (Levine, 1988; Woods et al., 1994; Wickens and Hollands, 2000). These terms describe a state in which the troubleshooter tentatively formulates one hypothesis of the true state of affairs (or the best way to solve a problem) and then continues excessively, on that track even when it is no longer warranted. This may be done by actively seeking only evidence to confirm that the hypothesis chosen is correct (the confirmation bias) or simply by ignoring competing and plausible hypotheses (cognitive tunneling).

Collectively, then, the joint cognitive processes of planning and problem solving (or troubleshooting), depending as they do on the interplay between working memory and long-term memory, reflect both the strengths and the weaknesses of human information processing. The output of each process is typically a decision: to undertake a particular course of action, to follow a plan, to choose a treatment based on the diagnosis, or to formulate a solution to the problem. The cognitive processes involved in such decision making are discussed extensively in Chapter 8, as are some of the important biases and heuristics in diagnosis discussed more briefly above.

5.5.3 Creativity

In general, creativity involves human problem solving that is relatively free from the confirmation bias, cognitive tunneling, and functional fixedness, each of which restricts the number of problem solutions we consider. For most theorists, creativity refers to the production of *effective novelty* (Cropley, 1999; Mayer, 1999). This is a process that involves thinking of a variety of previously untried solutions *and* judging their probable effectiveness. Finke et al. (1992) argue that *generating* novel cognitive structures involves retrieving, associating, synthesizing, and transforming information, while *evaluating* novel structures involves inferring, hypothesis testing, and context shifting, among other strategies. It is clear from this analysis that the cognitive load imposed by creative tasks can be immense, and that working memory, including both storage systems and the central executive, will be taxed.

Novelty production may be particularly difficult to maintain for long periods of time, for at least two reasons. First, the cognitive load imposed by creative problem solving, as we have described above, is high from the outset. Second, because novel stimuli often increase arousal levels, it is likely that the *production* of novelty will create a cycle of upward-spiraling arousal in the problem solver. This, in turn, will cause some degree of cognitive tunneling, making continued novelty production and evaluation difficult (Cropley, 1999). This may suggest that unlike some other tasks, where higher levels of arousal may be desirable to maintain performance (e.g., long-duration search tasks for low-probability targets), creativity may be fostered by low initial levels of arousal.

The idea that novelty production may cause spiraling levels of arousal also provides one explanation for the often-discussed benefits of *incubation* for creative problem solving. Smith (1995) describes incubation in terms of the general finding that people are more likely to solve a problem after taking a break rather than working on a solution without interruption. In controlled trials, incubation effects are not invariably found (Nickerson, 1999); however, research continues to focus on the conditions under which incubation works. It is possible that a break from the act of novelty generation may serve to reduce arousal levels to more task-appropriate levels. Another explanation is that the probability of a new *problem representation* being put into action (e.g., the mental image or list of procedural steps being manipulated to generate solutions) is greater when a person disrupts his or her own processing. The person may simply be more likely to have forgotten components of a previous, ineffective representation upon returning to the task.

The importance of the cognitive representation of problems, and the different display formats that support these representations, has been demonstrated for a variety of problem-solving tasks (Davidson and Sternberg, 1998). Flexible scientific and information visualization tools may prove to be particularly valuable for creative problem solving, because changing the orientation, color scheme, format, or level of focus will change the salience of different aspects of the problem and may activate different aspects of the viewer's long-term memory. Again, these changes should combat functional fixedness, but whether they ultimately result in more *creative* solutions is yet to be demonstrated.

In this section we have discussed information-processing tasks that are time consuming, require extensive cognitive resources, and for which the "correct" response is poorly defined and multiple responses are possible, even desirable. We turn now to characteristics of actions that are typically selected rapidly, sometimes without much effort, and often without great uncertainty about their outcome.

6 ACTION SELECTION

6.1 Information and Uncertainty

In earlier sections we discussed different stages at which humans process information about the environment. When we turn to the stage of action selection and execution, a key concern addresses the *speed* with which information is processed, from perception to action. How fast, for example, can we expect the driver to react to the unexpected roadway hazard or pedestrian, or how rapidly we can expect the postal worker to sort letters. Borrowing from terminology in communications, we describe information-processing speed in terms of the *bandwidth*, the amount of information processed per unit time. In this regard, a unit of information is defined as a *bit*. One bit can be thought of as specifying between one of two possible alternatives; two bits as one of four alternatives; three bits as one of eight, or in general, the number of bits (conveyed by an event) $= \log_2 N$, where N is the number of possible environmental events that could occur in the relevant task confronting the operator. In the following pages, after we describe a taxonomy of human actions, we will see how information influences the bandwidth of human processing.

The speed with which people perform a particular action depends jointly on the uncertainty associated with the outcome of that action and on the skill of the operator in the task at hand. Rasmussen (1986; Rasmussen et al., 1995) has defined a behavior-level continuum that characterizes three levels of action selection and execution that is characterized by both uncertainty and skill. *Knowledge-based behavior* describes the action selection of the unskilled operator or of the skilled operator working in a highly complex environment facing a good deal of uncertainty. In the first case, we might consider a vehicle driver trying to figure out how to navigate through an unfamiliar city; in the second case, we consider the nuclear reactor operator trying to diagnose an apparent system failure. This is the sort of behavior discussed in Section 5.5.

Rule-based behavior typically characterizes actions that are selected more rapidly, based on certain well-known rules. These rules map environmental characteristics (and task goals) to actions, and their outcomes are fairly predictable: "If the conditions exist, do *x*, then *y*, then *z*." The operator response in executing rule-based behavior is fairly rapid, but is still "thought through" and may be carried out within the order of a few seconds. Working memory is required. Finally, *skilled-based behavior* is very rapid and nearly automatic in the sense that little working memory is required, performance of concurrent tasks is possible, and the action may be initiated within less than a second of the triggering event. Skill-based behavior, for example, characterizes movement of the fingers to a key to type a letter, the sequence of steering wheel turns used to back out of a familiar driveway or compensate for a wind gust, or the response of the pilot to an emergency ground proximity warning that says "pull up, pull up."

Human factors designers are quite interested in the system variables that affect the speed and accuracy of behavior of all three classes. Typically, those variables affecting knowledge-based behavior are discussed within the realm of problem solving and decision making (see Section 7 and Chapter 8). We discuss below the variables that influence rule- and skill-based behavior (see Wickens and Hollands, 2000, for a more detailed discussion).

6.2 Complexity of Choice

Response times for either rule- or skill-based behavior become longer if there are more possible choices that could be made and therefore more information transmitted per choice (Hick, 1952; Hyman, 1953). The rule-based decision to go left or right at a Y fork in the road is simpler (i.e., 1 bit, and made more rapidly) than at an intersection where there are four alternative paths (i.e., 2 bits). Menu selections take longer on a page where there are more menu options, and each stroke on a typewriter (26 letter options) takes longer to initiate than each depression of a Morse code key (two options). Indeed, the time to select an option is roughly proportional to the number of bits in the choice (Hick, 1952). As a guideline, designers should not give users more choices of action than are essential, particularly if time is critical. Long menus, with lots of rarely chosen options, may not be desirable. The consequences of offering many choices are not only longer response time, but also an increased possibility that the wrong option will be chosen by mistake. More items typically lead to greater similarity between items and hence an invitation for confusion.

The guidance for avoiding very complex choices presented above does not necessarily mean that very simple choices (e.g., 1 bit per choice) are necessarily best. Indeed, generally *an operator can transmit more total information per unit time with a few complex (information-rich) choices than several simple (information-poor) choices*. This conclusion, referred to as the *decision complexity advantage* (Wickens and Hollands, 2000), can be illustrated by two examples: First, an option provided by a single computer menu with eight alternatives (one complex decision) can be selected faster than an option provided by three consecutive selections from three two-item menus (three simple decisions; see Figure 13). Second, voice input, in which each possible word is a choice from a potentially large vocabulary (high complexity) can transmit more information per unit time than typing, with each letter indicating one of only 26 letters (less complex); and typing in turn can transmit more information per unit time than can Morse code. The general conclusion of the decision complexity advantage drawn from these examples and from other studies (Wickens and Hollands, 2000) points to the advantage of incorporating keys or output options that can select from a larger number of possible options, such as special service "macro" keys, keys that represent common words, or "chording" keyboard devices (Baber, 1997), that allow a single action selection (a chord depression using several fingers simultaneously) to select from one of several options.

In conclusion, it may seem that two contradictory messages were offered in the paragraphs above: (1) Keep choices simply, but (2) use a small number of complex choices. In resolving these two guidelines in design, it is best to think that the first guideline pertains to not providing a lot of *rarely* used options, particularly in time-stressed situations and when errors of choice can have high-risk consequences. The second

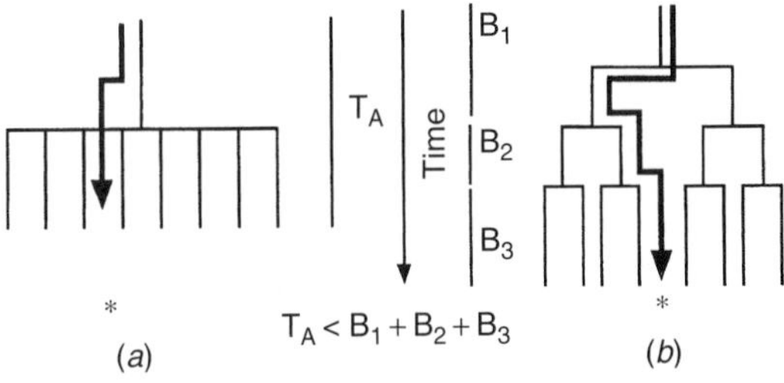

Figure 13 Decision complexity advantage: (*a*) total time required for three "simple" (low-complexity) choices; (*b*) time required for a single high-complexity choice. The total amount of information transmitted is the same in both cases.

guideline pertains to how to structure a choice among a large number of remaining and *plausible* options. A single choice among a larger list is often better than multiple sequential choices among smaller lists.

6.3 Probability and Expectancy

People respond more slowly (and are more likely to respond erroneously) to signals and events that they do not expect. Generally, such events are unexpected (or surprising) because they occur with a low probability in a particular context. This is consistent with our discussion of context effects on object identification in Section 4.2. Low-probability events, such as events with a greater number of alternatives, are also said to *convey more information*. The information in bits, conveyed by a single event that occurs with probability P, is $\log_2 2(1/P)$. As we noted above, greater information content requires more time for processing (Fitts and Peterson, 1964). For example, system failures usually occur rarely, and as such, are often responded to slowly or inappropriately. A similar status may characterize a driver's response to the unexpected appearance of a pedestrian on a freeway or to a traffic light that changes sooner than expected, or a pilot's response to an aircraft that suddenly appears on the runway, which had previously been cleared for landing. The maximum expected response times to truly unexpected events provide important guidance to traffic safety engineers in determining issues related to speed limits and roadway characteristics (Evans, 1991; Summala, 2000). Often more serious than the slower response to the unexpected event is the potential failure to detect that event altogether (see Section 4.1). It is for this reason that designers ensure that annunciators of rare events are made salient and obtrusive or redundant (to the extent that the rare event is also one that is important for the operator's task; see Section 3).

6.4 Practice

Practice has two benefits to action selection. First, practice can move knowledge-based behavior into the domain of rule-based behavior and sometimes move rule-based actions into the domain of skill-based ones. The novice pilot may need to think about what action to take when a stall warning sounds, whereas the expert will respond automatically and instinctively. In this sense, practice increases both speed and accuracy. Second, practice will provide the operator with a sense of expectancy that is more closely calibrated with the actual probabilities and frequencies of events in the real world. Hence, frequent events will be responded to more rapidly by the expert; but ironically, expertise may lead to *less* speedy processing of the rare event than would be the case for the novice, for which the rare event is not perceived as unexpected.

6.5 Spatial Compatibility

The compatibility between a display and its associated control has two components that influence the speed and accuracy of the control response. One relates to the *location* of the control relative to the display; the second to how the display reflects (or commands) control *movement*. In its most general form, the *principle of location compatibility* dictates that the location of a control should correspond to the location of a display. There are several ways of describing this correspondence. Most directly, this correspondence is satisfied by the *principle of collocation*, which dictates that each display should be located adjacent to its appropriate control. But this is not always possible in systems when the displays themselves may be closely grouped (e.g., closely clustered on a display panel) or may not be reached easily by the operator because of other constraints (e.g., common visibility needed by a large group of operators on a group-viewed display, or positioning the control for a display cursor on a head-mounted display).

When collocation cannot be maintained, the spatial compatibility *principle of congruence* takes over, which states that the spatial arrangement of a set of two or more displays should be congruent with the arrangement of their controls. One example of congruence is that left controls should be associated with left displays, and right associated with right. In this regard, the distinction between "left" and "right" in designing for compatibility can be expressed either in relative terms (indicator A is to the left of indicator B) or in absolute terms relative to some prominent axis. This axis may be the body midline (i.e., distinguishing left hand from right hand), or it may be a prominent visual axis of symmetry in the system, such as that bisecting the cockpit on a twin-seat airplane design. When left–right congruence is violated, such that a left display is matched to a right response, the operator may have a tendency to activate the incorrect control, particularly in times of stress (Fitts and Posner, 1967).

Sometimes an array of controls is to be associated with an array of displays (e.g., four engine indicators). Here, congruence can be maintained (or violated) in several ways. Compatibility will best be maintained if the control and display arrays are parallel. It will be reduced if they are orthogonal (Figure 14; i.e., a vertical display array with a horizontal left–right or fore–aft control array). But even where there is orthogonality, compatibility can be improved by adhering to two guidelines: (1) The left end of a horizontal array should map to the near end of a fore–aft array (Figure 14*b*), and (2) the particular display (control) at the end of one array should map to the control (display) at the end of the other array to which it is closest (Andre and Wickens, 1990). It should be noted in closing, however, that the association of the top (or bottom) of a vertical array with the right (or left) level of a horizontal array is not a strong one. Therefore, ordered compatibility effects with orthogonal arrays will not be strong if one of those arrays is vertical. Hence, some augmenting cue should be used to make sure that the association between the appropriate ends of the two arrays is clearly articulated (e.g., a common color code on both, or a painted line between them; Osborne and Ellingstad, 1987).

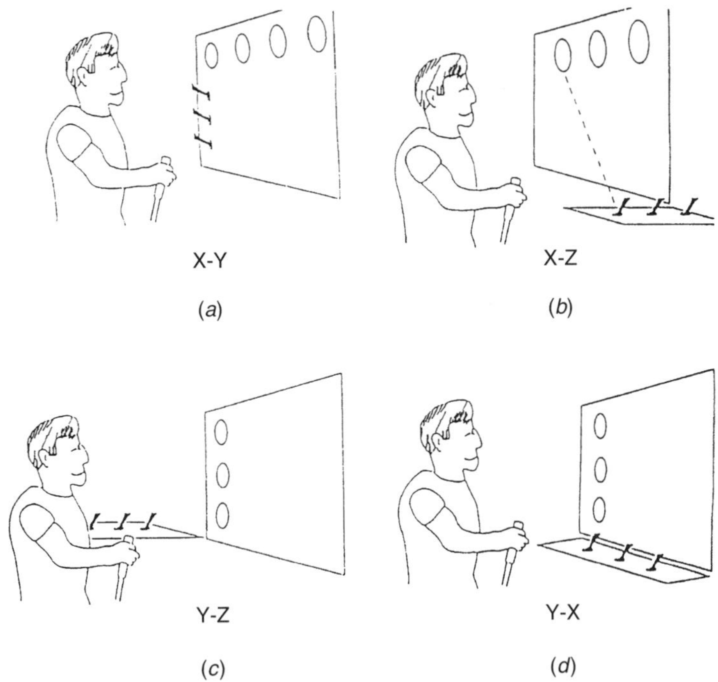

Figure 14 Different possible orthogonal display–control configurations.

The movement aspect of SR compatibility may be defined as *intention–response–stimulus (IRS) compatibility*. This characterizes a situation in which the operator formulates an intention to do something (e.g., increase, activate, set, turn something on, adjust a variable). Given that intention, the operator makes a response or an adjustment. Given that response, some stimulus is (or should be) displayed as feedback from what has been done (Norman, 1988). There is a set of rules for this kind of mapping between an intention to respond, a response, and the display signal. The rules are based on the idea that people generally have a conception of how a quantity is ordered in space. As we noted in Section 4.5, when we think about something increasing, such as temperature, we think about a movement of a display that is upward (or from left to right, or clockwise). Both control and display movement should then be congruent in form and direction with this ordering. These guidelines are shown in Figure 15. Whenever one is dealing, for example, with a rotary control, people have certain expectations (a mental model) about how the movement of that control will be associated with the corresponding movement of a display. These expectancies may be defined as *stereotypes*, and there are three important stereotypes.

The first stereotype is the *clockwise increase stereotype*: A clockwise rotation of a control or display signals an increasing quantity (Figure 15*c* and *d*). The *proximity of movement stereotype* says that with any rotary control, the arc of the rotating element that is closest to the moving display is assumed to move in the same direction as that display. In panel (*c*) of Figure 15, rotating the control clockwise is assumed to move the needle to the right, while rotating it counterclockwise is assumed to move the needle to the left. It is as if the human's mental model is one that assumes that there is a mechanical linkage between the rotating object and the moving element, even though that mechanical linkage may not really be there.

Designers may sometimes develop control display relations that conform to one principle and violate another. Panel (*e*) shows a moving vertical scale display with a rotating indicator. If the operator wants to increase the quantity, he or she rotates the dial clockwise. That will move the needle on the vertical scale up, thus violating the proximity of movement stereotype. The conflict may be resolved by putting the rotary control on the right side rather than the left side of a display. We have now created a display–control relationship that conforms to *both* the proximity of movement stereotype *and* the clockwise to increase stereotype.

The third stereotype of movement compatibility relates to *global congruence*. Just as with location compatibility, movement compatibility is preserved when controls and displays move in a congruent fashion: linear controls parallel to linear displays [(*f*), but not (*g*)], and rotary controls congruent with rotary displays [(*b*) and (*h*)]. Note, however, that (*h*) violates proximity of movement. When displays and controls move in orthogonal directions, as in (*g*), the movement relation between them is ambiguous. Such ambiguity,

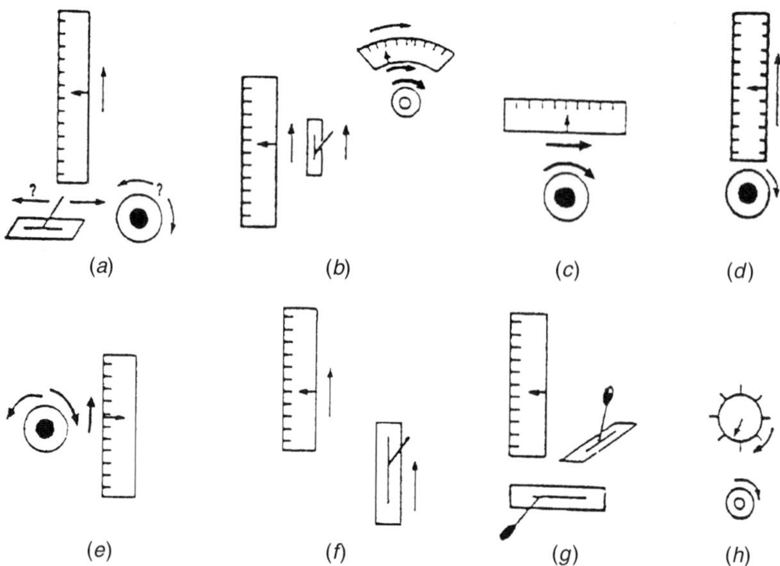

Figure 15 Examples of population stereotypes in control–display relations. (From Wickens, 1984.)

however, can often be reduced by placing a modest "cant" on either the control or display surface, so that some component of the movement axes are parallel, as shown in Figure 16.

6.6 Modality

Skilled responses in most human–machine systems are typically executed by either the hands or the voice. With increasingly sophisticated automated voice recognition systems, the latter option is becoming progressively more feasible. Although the particulars of voice control are addressed in more detail in Chapter 24, at least three characteristics of voice control are relevant here in the context of information processing.

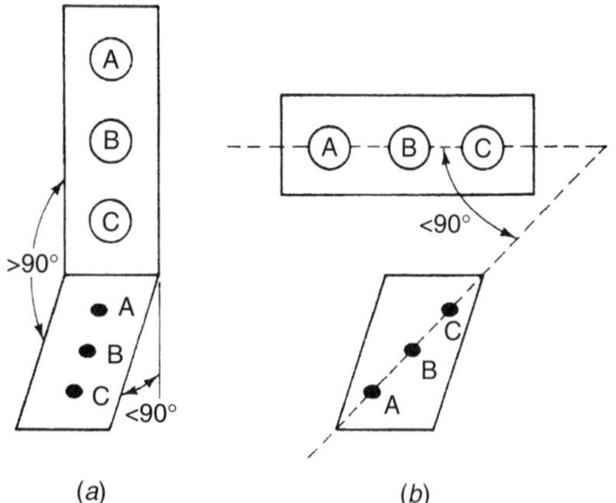

Figure 16 Solutions of location compatibility problems by using cant. (*a*) The control panel slopes downward slightly (an angle greater than 90°), so that control A is clearly above B, and B is above C, just as they are in the display array. (*b*) The controls are slightly angled from left to right across the panel, creating a left–right ordering that is congruent with the display array. (From Wickens and Hollands, 2000.)

1. Voice options allow more possible responses to be given in a shorter period of time without imposing added time-consuming finger movement components (i.e., keys), although this requires more sophisticated software in the voice recognition algorithms. Providing more options, enabling more complex decisions to be selected, is a positive benefit because it exploits the decision complexity advantage, as we saw in Section 6.2.

2. Voice options represent more compatible ways of transmitting symbolic or verbal information than are possible with spatially guided manual options (Wickens et al., 1984), including sequential keypresses. In contrast, voice responses make relatively poor candidates for transmitting continuous analog-spatial information, particularly in dynamic situations (e.g., tracking; Wickens et al., 1985), since spoken vocabulary is better equipped to generate categorical commands (e.g., "left," "right") than continuously modulated closed-loop commands (e.g., "a little more to the left").

3. Voice options are valuable in environments when the eyes, and in particular the hands, are otherwise engaged; but conversely, voice options can be problematic in environments in which a large amount of other verbal activity is required, either by the user or by other people in the nearby workspace. The former

causes competition for processing resources within the operator (see Section 7.2.3), while the latter creates the possibility of confusion on the part of the voice recognizer.

6.7 Response Discriminability

Whenever a set of manual responses are specified, any increases in the similarity between them (decreases in discriminability) will increase the likelihood of confusion. Thus, movement of the control stick to either one of two forward positions is a response choice that has greater opportunity for confusion than movement in either a forward *or* backward direction. Correspondingly, two buttons that look alike are more confusable (and hence, error prone) than are two that are differently colored or shape coded. Although making controls physically distinct from each other may sometimes destroy a sense of aesthetics in design, such distinctions will generally lead to improved human reliability (Norman, 1988). Incidentally, increased similarity between voice control options (Section 6.6) will also produce the same increase in error likelihood, although here the mediating agent is the voice recognition agent (whether human or computer) rather than the human responder. Thus, the vocabulary selected for use in an application should be chosen with a mind to avoiding confusable-sounding articulations, such as "to" and "through."

6.8 Feedback

The quality of feedback provided by control manipulation (or action expression) is often critical to the speed of information transmission (Norman, 1988). Indeed, sometimes the problems of poor response discriminability discussed in Section 6.7 can be addressed, and at least partially remedied, by providing clear, salient, and immediate feedback as to which (of several confusable) response alternatives has been chosen. This feedback may be in the form of a visible light or an auditory or tactile "click" as the control reaches its appropriate destination.

It turns out, however, that salient feedback is not always necessary or even desirable. In particular, expert or highly skilled users rely far less on feedback than do novices (e.g., the skilled typist, when transcribing, rarely looks at the keyboard or the screen). Thus, if the feedback is salient (and hence intrusive), it may be distracting to the expert, even as it is valuable for the novice. This will be particularly true whenever the feedback is *delayed*, a quality that is especially disruptive for relatively continuous tasks such as data transcription or voice translation (Smith, 1962).

6.9 Continuous Control

Our discussion in Section 6.8 focused on the selection of *discrete* actions, such as a keypress or lever movement. Equally important are the *continuous* movements of some controls to reach targets in space. These movements may refer, for example, to the movement of the hand to a point on a touch screen, the movement of a cursor to an icon or word on a computer screen, or the movement of a pointer to a set point on a meter. Generically, then, we can speak of these skills involving movement of a *cursor* to a *target*.

To an even greater extent than the discrete movements discussed in Section 6.8, performance of these continuous movement skills depends greatly on *visual feedback*, depicting the difference between the current cursor location and the desired target. Performance on control tasks in which a cursor is moved a certain distance into a target is well described by *Fitts's law* (Fitts, 1966; Jagacinski and Flach, 2003):

$$\text{movement time} = a + b\ \log_2(2D/W)$$

Where a and b are constants, D is the distance to the target, and W is the target width. This very robust law can accurately predict the movement of all sorts of devices, from microscopic pointers (Langolf et al., 1976), to cursor movements by mice (Card, 1981) to foot movement around a set of pedals (Drury, 1975) to the manipulation of endoscopic surgical instruments (Zheng et al., 2003). The basis of Fitts's law lies in the processing of visual feedback such that movement toward a target is maintained at a rate that is inversely proportional to the momentary distance of the target from the cursor (or other controlled object).

Just as Fitts's law nicely describes continuous movement toward a static target, it can also characterize movement of a cursor toward a *continuously* moving target, a process typically described as *tracking* (McRuer, 1980; Wickens, 1986; Jagacinski and Flach, 2003). When operators engage in continuous tracking, however, whether keeping a car in the center of the highway, flying an airplane down a glide path, or moving one's viewpoint through a virtual environment via some control device, interest is more focused on minimizing the deviation from the target than on the time required to reach the target. Also, concern is less with the amplitude of the required movement than it is with other variables, such as the frequency with which corrections must be made (the *input signal bandwidth*), the complexity and lag of the *system dynamics* mediating between hand movement and cursor or output movement, and the manner in which feedback is displayed. These issues extend well beyond the scope of the current chapter and are covered in more depth in Chapter 5.

6.10 Errors

The previous discussion has focused primarily on the *time* required to process and respond to various items of information. Yet in many systems, the occurrence of *errors* is more critical than the occurrence of delays in processing. That is, the loss of information, rather than its transmission delay, is the factor of greatest concern. Although errors are treated extensively in Chapter 27, we wish here to highlight the manner in which different classes of errors can be categorized in the context of the flow of information as depicted in Figure 1*a* (Norman, 1981; Reason, 1990).

First, *mistakes* represent errors of the earlier stages of information processing, in which incorrect action is carried out as a result of a failure to *understand* the nature of a situation (i.e., a failure of stage 2 or stage 3 situation awareness, as discussed in Section 5.2). This may result from a breakdown in perception, in working memory, or from insufficient knowledge to interpret the available cues (i.e., *knowledge-based errors*). Second, while a situation may be diagnosed and understood correctly, *rule-based errors* may result from a failure to apply the correct rules appropriately for selection of a response (Reason, 1990). Third, errors may result from *slips* of action, when the correct response is intended but an incorrect action is actually released (i.e., an unintended response "slips" out of the hands or mouth) (Norman, 1981). Slips of this sort are typically the result of poor human factors design, such as incompatible control–display relationships (see Section 6.5), confusable displays (Section 3.3) or controls (Section 6.7), coupled with an operator who is well skilled, and performing a task in a highly automated mode, thereby not carefully monitoring his or her own action selections.

Errors can also be attributed directly to a breakdown of *memory*. As noted in Section 2, working memory breakdowns may lead to forgetting or confusion of material, whereas *errors of prospective memory* may lead operators to forget to perform some action that was previously intended. Often described as *errors of omission*, these are typified by leaving the last copied paper on the glass of a photocopier or failing to tighten the bolts after completing a maintenance task (Reason, 1997).

It is usually the case that the conditions that are associated with slower processing are those that also produce more errors, and hence design remediation based on measures of processing speed will be productive in improving overall system accuracy. However, in certain circumstances, a strategic adjustment in how an operator performs a task will lead to an inverse relationship between speed and accuracy: the *speed–accuracy trade-off*. In this case, a "set" to respond rapidly will lead to more rather than fewer errors (Wickelgren, 1977). An example here would be that of the effects of time stress in emergencies, which may lead to hasty but error-prone actions in the processing of information.

7 MULTIPLE-TASK PERFORMANCE

Many task environments require operators to process information from more than one source and to perform more than one task at a time (Damos, 1991). Such environments are as diverse as that confronting the secretary conversing with the supervisor while typing, the vehicle driver placing a cell phone call while searching for a road sign and steering, or the maintenance technician who performs and observes diagnostic tests while keeping active hypotheses about possible faults rehearsed in working memory.

In such multiple-task environments requiring divided attention, we may distinguish between three qualitatively different *modes* of multiple-task behavior: *perfect parallel processing*, in which two (or more) tasks are performed concurrently as well as either is performed alone; *degraded concurrent processing*, in which both tasks are performed concurrently but one or both suffers relative to its single task level; and *strict serial processing*, during which only one task is performed at a time. Each of these modes is observed under different circumstances and has somewhat different implications for design.

7.1 Serial Processing

The concerns of serial processing result when performance of one task or the other is delayed undesirably because of sequential constraints. Such a delay might characterize the behavior of a pilot who fails to check the aircraft altimeter sufficiently often because he is engaged in other visual tasks, leading to dangerous "altitude busts" (Raby and Wickens, 1994; Dismukes, 2001). Typically, the interests of human factors personnel in sequential task performance is in modeling the choice process whereby the operator chooses to perform one task (and, by necessity, neglects another) at any given moment in time. This choice process is often modeled by *queuing theory* (Kleinman and Pattipati, 1991) or variants thereof (Moray, 1986; Wickens et al., 2003), which specify when a task should be sampled (performed) as a function of that task's importance (cost of not performing it) and the frequency with which it should be carried out. When evaluated against these optimal benchmarks, human performance appears to be reasonably optimal, subject to the constraints of working memory. That is, humans may sometimes forget the precise value of a task status at the time that it was last sampled, and hence sample it more frequently than necessary compared to an optimal performer with perfect memory.

However, reasonably optimal is not the same thing as perfectly optimal, and others have focused interest on the occasional breakdowns in optimality that do occur. Thus, a different approach to human multiple-task performance is to focus on the accidents and incidents that have apparently resulted from failures of effective task management (Adams et al., 1991; Raby and Wickens, 1994; Chou et al., 1996; Schutte and Trujillo, 1996; Orasanu and Fischer, 1997; Wickens, 2003b): that is, what causes people to neglect a task.

Here the answers based on empirical research are not entirely clear, although two prominent factors do appear to emerge. First, visible, and in particular audible, reminders to do a task increase the likelihood that that task will be done, compared to circumstances in which task initiation must be based on prospective memory alone (Norman, 1988; Dismukes, 2001). The vulnerability of such memory highlights the value of checklists as visual reminders for people to carry out certain actions at certain times (Degani and Wiener, 1993; Herrmann et al., 1999; Wickens, 2003b). Second, heavy involvement (high workload) with one task may lead an operator to neglect a second task, and perhaps fail to return to an activity at a time when that return should be critical. This deficiency may be addressed through task or workload management training programs (Dismukes, 2001).

Of course, task switching is a two-way street. The desirable properties of having the waiting task call attention to itself may also have the undesirable properties of interrupting an ongoing task, to the detriment of the latter (McFarlane and Latorella, 2002), and some ongoing tasks may be more vulnerable than others to interruptions. For example, certain interrupted tasks (or times of interruptions) may require the interrupted user to "start from scratch" when returning to the task, to his or her considerable annoyance.

7.2 Concurrent Processing

In contrast to sequential processing, an understanding of concurrent processing, whether in degraded mode or perfect parallel mode, depends on somewhat different mechanisms. These mechanisms are as closely related to the structure of the information-processing sequences within the tasks themselves as they are to the operator's knowledge of task importance and priority (although there are interactions between these two influences; Gopher, 1992). Here human factors interest is in the task features that can enable any sort of concurrent processing to emerge from serial processing and that can enable that concurrent processing to be perfect rather than degraded. Three characteristics appear to influence this degree of success: task similarity, task demand, and task structure (Wickens, 1991; Wickens and Hollands, 2000), although how these influences are exerted is somewhat complex.

7.2.1 Task Similarity

A high degree of similarity between two tasks may induce confusion, just as similarity between perceptual signals will cause confusion (Section 3.3), high similarity between items held in working memory will increase the degree of interference between them (Section 5.1), or high similarity between two response devices will increase the possibility of confusion of actions (Section 6.7).

In contrast, however, making certain aspects of two tasks more similar *may* allow the tasks to be better integrated, fostering *more* effective concurrent processing. This may involve combining two axes of motion on a tracking task into the *X* and *Y* deviations of a single cursor (Chernikoff and LeMay, 1963; Fracker and Wickens, 1989), or using similar rules to map stimuli (events) to responses (actions) (Duncan, 1979; Fracker and Wickens, 1989).

The latter aspect of similarity, the similarity of rules, also facilitates the ability of an operator to *switch* attention between two tasks in sequential fashion when in the serial mode of processing. That is, it is easier to keep doing alternative versions of the same task than it is to switch between different tasks, as if there is some "overhead" penalty for switching rules (Rogers and Monsell, 1995).

7.2.2 Task Demand

Easier tasks are more likely to be performed concurrently (and perfectly) than are more difficult or demanding tasks. We argue that easier tasks are generally more automated and consume less mental effort or resources than do more difficult ones (see Chapter 9). Such automaticity can often be achieved by extensive practice on what are called *consistently mapped tasks* (Fisk et al., 1987). These are tasks in which in each encounter by the learner, certain properties of the relation between the displayed elements, cognitive operations, and responses remain constant. These mimic many properties of skill- and rule-based tasks, as described in Section 6.5. Such consistent mapping will not only lead to more rapid performance, but also performance that is relatively attention-free, and hence will allow other tasks to be time shared successfully.

7.2.3 Task Structure

Certain structural differences between two time-shared tasks increase the efficiency of their concurrent processing, as if the two tasks demand entirely (or partially) separated resources within the human processing system, such that it is easier to distribute tasks across *multiple resources* than to focus them within a single resource (Wickens, 1991, 2002; Wickens et al., 2003). These resources appear to be defined by *processing code* (verbal or linguistic versus spatial), *processing stage* (perceptual–cognitive operations versus response operations), *perceptual modality* (auditory versus visual), and *visual subsystems* (focal vision, required for object recognition, versus ambient vision, required for orientation and locomotion; Previc, 1998). It is because of the separate spatial and verbal code resources that spatially guided manual responses may be more effective than vocal responses when operators must also rehearse verbal material, but vocal responses will be more effective than spatially guided manual ones when the operator must concurrently perform another spatial task, such as tracking. As another example, we saw in Section 5.4 that the spatial cognitive processes involved in navigation and vehicle control are better time-shared with the verbal input of a memorized verbal route list than the spatial input of a memorized map. It is because of the separate stage-defined resources that we are often able effectively to time share responding operations (e.g., talking) with perceptual ones (e.g., scanning). It is because of the separate perceptual resources that designers have chosen to offload the heavy visual processing load of pilots and vehicle drivers with some information presented on auditory channels. Finally, it is because of the separate visual channels that we can effectively keep a car centered in the lane while searching for and reading road signs.

8 CONCLUSIONS

In conclusion, the systems with which people must interact vary vastly in their complexity, from the simple graph or tool to things like nuclear reactors or the physiology of a patient under anesthesia. As a consequence, they vary drastically in terms of the type and degree of demands imposed on the varying information-processing components we have discussed in this chapter. In some cases, systems will impose demands on components that are quite vulnerable: working memory, predictive capabilities, and divided attention, for example. At other times they may impose on human capabilities that are a source of great strength, particularly if these sources rely on the vast store of information that we retain in long-term memory; information that assists us in pattern recognition, top-down processing, chunking, and developing plans and scripts on the basis of past experience are examples. The importance of practice and training in the development of this knowledge base cannot be overestimated.

In addition to facilitating the performance of experts in many ways, long-term memory has a second implication for the practice of human factors. This is that predictions of human performance in many systems can be based only partially on an understanding of the generic information-processing components described in this chapter. An equal and sometimes greater partner in this prediction is extensive *domain knowledge* regarding the particular system with which the human is interacting. As several of the chapters in this handbook address, the best prediction of human performance must be based on the intricate interaction between the information-processing components discussed here, the domain knowledge employed by the human operator, and the physical environment within and tools with which the operator works. The reader will find all of these issues covered from multiple perspectives in subsequent chapters of the handbook.

REFERENCES

Adams, M. J., Tenney, Y. J., and Pew, R. W. (1991), "Strategic Workload and the Cognitive Management of Advanced Multi-task Systems," SOAR CSERIAC 91-6, Crew System Ergonomics Information Analysis Center, Wright-Patterson AFB, OH.

Adams, M. J., Tenney, Y. J., and Pew, R. W. (1995), "Situation Awareness and the Cognitive Management of Complex Systems," *Human Factors*, Vol. 37, No. 1, pp. 85–104.

Adamson, R. E. (1952), "Functional Fixedness as Related to Problem Solving: A Repetition of Three Experiments," *Journal of Experimental Psychology*, Vol. 44, pp. 288–291.

Andre, A. D., and Wickens, C. D. (1990), "Display Control Compatibility in the Cockpit: Guidelines for Display Layout Analysis," University of Illinois Institute of Aviation Technical Report ARL-90-12/NASA-A3I-90-1, Aviation Research Laboratory, Savoy, IL.

ANSI (American National Standards Institute) (2001), "Criteria for Safety Symbols," Z535.3-Revised, National Electrical Manufacturers Association, Washington, DC.

Antes, J. R., and Chang, K. (1990), "An Empirical Analysis of the Design Principles for Quantitative and Qualitative Area Symbols," *Cartography and Geographic Information Systems*, Vol. 17, pp. 271–277.

Aretz, A. J. (1991), "The Design of Electronic Map Displays," *Human Factors*, Vol. 33, No. 1, pp. 85–101.

Atkinson, R. C., and Shiffrin, R. M. (1971), "The Control of Short-Term Memory," *Scientific American*, Vol. 225, pp. 82–90.

Awh, E., Jonides, J., and Reuter-Lorenz, P. A. (1998), "Rehearsal in Spatial Working Memory," *Journal of Experimental Psychology: Human Perception and Performance*, Vol. 24, pp. 780–790.

Baber, C. (1997), *Beyond the Desktop*, Academic Press, San Diego, CA.

Baddeley, A. D. (1986), *Working Memory*, Oxford University Press, Oxford.

Baddeley, A. D. (1999), *Essentials of Human Memory*, Psychology Press, Hove, East Sussex, England.

Baker, C. H. (1961), "Maintaining the Level of Vigilance by Means of Knowledge of Results About a Secondary Vigilance Task," *Ergonomics*, Vol. 4, pp. 311–316.

Banbury, S. P., and Treselian, S. (2004), *Cognitive Approaches to Situation Awareness*, Ashgate, Brookfield, VT.

Banbury, S. P., Macken, W. J., Tremblay, S., and Jones, D. M. (2001), "Auditory Distraction and Short-Term Memory: Phenomena and Practical Implications," *Human Factors*, Vol. 43, pp. 12–29.

Barnett, B. J, and Wickens, C. D. (1988), "Display Proximity in Multicue Information Integration: The Benefit of Boxes," *Human Factors*, Vol. 30, pp. 15–24.

Baron, S., Kleinman, D., and Levison, W. (1970), "An Optimal Control Model of Human Response," *Automatica*, Vol. 5, pp. 337–369.

Bellenkes, A. H., Wickens, C. D., and Kramer, A. F. (1997), "Visual Scanning and Pilot Expertise: The Role of Attentional Flexibility and Mental Model Development," *Aviation, Space, and Environmental Medicine*, Vol. 68, No. 7, pp. 569–579.

Bennett, K. B., and Flach, J. M. (1992), "Graphical Displays: Implications for Divided Attention, Focused Attention, and Problem Solving," *Human Factors*, Vol. 34, pp. 513–533.

Bennett, K. B., and Fritz, H. I. (2005), "Objects and Mappings: Incompatible Principles of Display Design" *Human Factors*, Vol. 47, No. 1, pp. 131–137.

Bennett, K. B., Toms, M. L., and Woods, D. D. (1993), "Emergent Features and Graphical Elements: Designing More Effective Configural Displays," *Human Factors*, Vol. 35, No. 1, pp. 71–98.

Beringer, D. B., and Chrisman, S. E. (1991), "Peripheral Polar-Graphic Displays for Signal/Failure Detection," *International Journal of Aviation Psychology*, Vol. 1, pp. 133–148.

Biederman, I., Mezzanotte, R. J., Rabinowitz, J. C., Francolin, C. M., and Plude, D. (1981), "Detecting the Unexpected in Photo Interpretation," *Human Factors*, Vol. 23, pp. 153–163.

Bransford, J. D., and Johnson, M. K. (1972), "Contextual Prerequisites for Understanding: Some Investigations of Comprehension and Recall," *Journal of Verbal Learning and Verbal Behavior*, Vol. 11, pp. 717–726.

Braunstein, M. L., Andersen, G. J., Rouse, M. W., and Tittle, J. S. (1986), "Recovering Viewer Centered Depth from Disparity, Occlusion and Velocity Gradients," *Perception and Psychophysics*, Vol. 40, No. 4, pp. 216–224.

Breznitz, S. (1983), *Cry Wolf: The Psychology of False Alarms*, Lawrence Erlbaum Associates, Mahwah, NJ.

Broadbent, D. (1958), *Perception and Communications*, Permagon Press, New York.

Broadbent, D. (1972), *Decision and Stress*, Academic Press, New York.

Broadbent, D. E. (1982), "Task Combination and Selective Intake of Information," *Acta Psychologica*, Vol. 50, pp. 253–290.

Brown, J. (1959), "Some Tests of the Decay Theory of Immediate Memory," *Quarterly Journal of Experimental Psychology*, Vol. 10, pp. 12–21.

Burns, C. (2000), "Putting It All Together: Improving Integration in Ecologicol Displays," *Human Factors*, Vol. 42, pp. 226–241.

Burns, C. M., Bisantz, A. M., and Roth, E. M. (2004), "Lessons from a Comparison of Work Domain Models," *Human Factors*, Vol. 46, pp. 711–727.

Buttigieg, M. A., and Sanderson, P. M. (1991), "Emergent Features in Visual Display Design for Two Types of Failure Detection Tasks," *Human Factors*, Vol. 33, pp. 631–651.

Card, S. K. (1981), "The Model Human Processor: A Model for Making Engineering Calculations of Human Performance," in *Proceedings of the 25th Annual Meeting of the Human Factors Society*, R. Sugarman, Ed., Human Factors Society, Santa Monica, CA.

Carpenter, P. A., and Just, M. A. (1975), "Sentence Comprehension: A Psycholinguistic Processing Model of Verification," *Psychological Review*, Vol. 82, No. 1, pp. 45–73.

Carswell, C. M. (1992), "Reading Graphs: Interactions of Processing Requirements and Stimulus Structure," in *Precepts, Concepts, and Categories*, B. Burns, Ed., Elsevier Science, Amsterdam, pp. 605–647.

Carswell, C. M., and Wickens, C. D. (1987), "Information Integration and the Object Display: An Interaction of Task Demands and Display Superiority," *Ergonomics*, Vol. 30, pp. 511–527.

Carswell, C. M., and Wickens, C. D. (1990), "The Perceptual Interaction of Graphical Attributes: Configurality, Stimulus Homogeneity, and Object Integration," *Perception and Psychophysics*, Vol. 47, pp. 157–168.

Carswell, C. M., and Wickens, C. D. (1996), "Mixing and Matching Lower-Level Codes for Object Displays: Evidence for Two Sources of Proximity Compatibility," *Human Factors*, Vol. 38, pp. 61–72.

Chernikoff, R., and LeMay, M. (1963), "Effect of Various Display-Control Configurations on Tracking with Identical and Different Coordinate Dynamics," *Journal of Experimental Psychology*, Vol. 6, pp. 95–99.

Chou, C., Madhavan, D., and Funk, K. (1996), "Studies of Cockpit Task Management Errors," *International Journal of Aviation Psychology*, Vol. 6, No. 4, pp. 307–320.

Clark, H. H., and Chase, W. G. (1972), "On the Process of Comparing Sentences Against Pictures," *Cognitive Psychology*, Vol. 3, pp. 472–517.

Cleveland, W. S. (1985), *The Elements of Graphing Data*, Wadsworth, Monterey, CA.

Cleveland, W. S. (1990), "A Model for Graphical Perception," *Proceedings of the Statistical Graphics Section*, American Statistical Association, Alexandria, VA, pp. 30–35.

Cleveland, W. S., and McGill, R. (1984), "Graphical Perception: Theory, Experimentation, and Application to the Development of Graphic Methods," *Journal of the American Statistical Association*, Vol. 70, pp. 531–554.

Cleveland, W. S., and McGill, R. (1985), "Graphical Perception and Graphical Methods for Analyzing Scientific Data," *Science*, Vol. 229, pp. 828–833.

Cleveland, W. S., and McGill, R. (1986), "An Experiment in Graphical Perception," *International Journal of Man–Machine Studies*, Vol. 25, pp. 491–500.

Cleveland, W. S., and McGill, R. (1988), *Dynamic Graphics for Statistics*, Wadsworth, Monterey, CA.

Cropley, A. J. (1999), "Creativity and Cognition: Producing Effective Novelty," *Roeper Review*, Vol. 21, pp. 253–260.

Cutting, J. E., and Vishton, P. M. (1995), "Perceiving Layout and Knowing Distances: The Integration, Relative Potency, and Contextual Use of Different Information About Depth," in *Perception of Space and Motion*, W. Epstein and S. Rogers, Eds., Academic Press, San Diego, CA.

Damos, D., Ed. (1991), *Multiple Task Performance*, Taylor & Francis, London.

Dashevsky, S. G. (1964), "Check-Reading Accuracy as a Function of Pointer Alignment, Patterning, and Viewing Angle," *Journal of Applied Psychology*, Vol. 48, pp. 344–347.

Davidson, J. E., and Sternberg, R. J. (1998), "Smart Problem-Solving: How Metacognition Helps," in *Metacognition in Educational Theory and Practice*, D. J. Hacker, J. Dunlosky, and A. C. Graesser, Eds., Lawrence Erlbaum, Associates, Mahwah, NJ, pp. 47–65.

Degani, A., and Wiener, E. L. (1993), "Cockpit Checklists: Concepts, Design, and Use," *Human Factors*, Vol. 35, No. 4, pp. 345–360.

DeSoto, C. B., London, M., and Handel, S. (1965), "Social Reasoning and Spatial Paralogic," *Journal of Personal and Social Psychology*, Vol. 2, pp. 513–521.

Dismukes, K. (2001), "The Challenge of Managing Interruptions, Distractions, and Deferred Tasks," in *Proceedings of the 11th International Symposium on Aviation Psychology*, Ohio State University, Columbus, OH.

Drury, C. (1975), "Application to Fitts' Law to Foot Pedal Design," *Human Factors*, Vol. 17, pp. 368–373.

Drury, C. G., and Addison, S. L. (1973), "An Industrial Study of the Effects of Feedback and Fault Density on Inspection Performance," *Ergonomics*, Vol. 16, pp. 159–169.

Duncan, J. (1979), "Divided Attention: The Whole Is More Than the Sum of Its Parts," *Journal of Experimental Psychology: Human Perception and Performance*, Vol. 5, pp. 216–228.

Duncan, J. (1984), "Selective Attention and the Organization of Visual Information," *Journal of Experimental Psychology: General*, Vol. 119, pp. 501–517.

Durso, F., and Gronland, S. (1999), "Situation Awareness," in *Handbook of Applied Cognition*, F. T. Durso, Ed., Wiley, New York.

Eberts, R. E., and MacMillan, A. G. (1985), "Misperception of Small Cars," in *Trends in Ergonomics/Human Factors II*, R. E. Eberts and C. G. Eberts, Eds., Elsevier Science, Amsterdam, pp. 33–39.

Edgell, S. E., and Morrissey, J. M. (1992), "Separable and Unitary Stimuli in Nonmetric Multicue Probability Learning," *Organizational Behavior and Human Decision Processes*, Vol. 51, pp. 118–132.

Egeth, H. E., and Yantis, S. (1997), "Visual Attention: Control, Representation, and Time Course," *Annual Review of Psychology*, Vol. 48, pp. 269–297.

Egly, R., Driver, J., and Rafal, R. D. (1994a), "Shifting Visual Attention Between Objects and Locations: Evidence from Normal and Parietal Lesion Subjects," *Journal of Experimental Psychology: General*, Vol. 123, pp. 161–177.

Egly, R., Rafal, R. D., Driver, J., and Starrveveld, Y. (1994b), "Covert Orienting in the Split Brain

Reveals Hemispheric Specialization for Object-Based Attention," *Psychological Science*, Vol. 5, pp. 380–383.

Eley, M. G. (1988), "Determining the Shapes of Landsurfaces from Topographical Maps," *Ergonomics*, Vol. 31, pp. 355–376.

Endsley, M. R. (1995), "Measurement of Situation Awareness in Dynamic Systems," *Human Factors*, Vol. 37, No. 1, pp. 65–84.

Endsley, M. R., and Garland, D. J. (2000), "Situation Awareness Analysis and Measurement," Lawrence Erlbaum Associates, Mahwah, NJ.

Ericsson, K. A., and Kintsch, W. (1995), "Long-Term Working Memory," *Psychological Review*, Vol. 102, pp. 211–245.

Evans, L. (1991), *Traffic Safety and the Driver*, Van Nostrand Reinhold, New York.

Fadden, S., Ververs, P. M., and Wickens, C. D. (2001), "Pathway HUDS: Are They Viable?" *Human Factors*, Vol. 43, No. 2, pp. 173–193.

Finke, R. A., Ward, T. B., and Smith, S. M. (1992), *Creative Cognition*, MIT Press, Cambridge, MA.

Fisher, D. L., Coury, B. G., Tengs, T. O., and Duffy, S. A. (1989), "Minimizing the Time to Search Visual Displays: The Role of Highlighting," *Human Factors*, Vol. 31, No. 2, pp. 167–182.

Fisk, A. D., Ackerman, P. L., and Schneider, W. (1987), "Automatic and Controlled Processing Theory and Its Applications to Human Factors Problems," in *Human Factors Psychology*, P. A. Hancock, Ed., Elsevier Science, New York, pp. 159–197.

Fitts, P. M. (1966), "Cognitive Aspects of Information Processing, III: Set for Speed Versus Accuracy," *Journal of Experimental Psychology*, Vol. 71, pp. 849–857.

Fitts, P. M., and Peterson, J. R. (1964), "Information Capacity of Discrete Motor Responses," *Journal of Experimental Psychology*, Vol. 67, pp. 103–112.

Fitts, P. M., and Posner, M. A. (1967), *Human Performance*, Brooks/Cole, Pacific Palisades, CA.

Flach, J. M., Hancock, P. A., Caird, J., and Vicente, K. J., Eds. (1995), *Global Perspectives on the Ecology of Human–Machine Systems*, Lawrence Erlbaum Associates, Mahwah, NJ.

Fogel, L. J. (1959), "A New Concept: The Kinalog Display System," *Human Factors*, Vol. 1, pp. 30–37.

Fracker, M. L., and Wickens, C. D. (1989), "Resources, Confusions, and Compatibility in Dual Axis Tracking: Displays, Controls, and Dynamics," *Journal of Experimental Psychology: Human Perception and Performance*, Vol. 15, pp. 80–96.

Garner, W. R. (1978), "Selective Attention to Attributes and to Stimuli," *Journal of Experimental Psychology: General*, Vol. 107, pp. 287–308.

Gathercole, S. E. (1997), "Models of Verbal Short-Term Memory," in *Cognitive Models of Memory*, M. A. Conway, Ed., MIT Press, Cambridge, MA, pp. 13–45.

Gathercole, S. E., and Baddeley, A. D. (1993), *Working Memory and Language*, Lawrence Erlbaum Associates Hove, East Sussex, England.

Gentner, D., and Stevens, A. L. (1983), *Mental Models*, Lawrence Erlbaum Associates, Mahwah, NJ.

Gibson, J. J. (1979), *The Ecological Approach to Visual Perception*, Houghton Mifflin, Boston.

Gillam, B. (1980), "Geometrical Illusions," *Scientific American*, Vol. 242, pp. 102–111.

Gillan, D. J., and Lewis, R. (1994), "A Componential Model of Human Interaction with Graphs, I: Linear Regression Modeling," *Human Factors*, Vol. 36, No. 4, pp. 419–440.

Gillan, D. J., Wickens, C. D., Hollands, J. G., and Carswell, C. M. (1998), "Guidelines for Presenting Quantitative Data in HFES Publications," *Human Factors*, Vol. 40, pp. 28–41.

Goldsmith, M. (1998), "What's in a Location? Comparing Object-Based and Space-Based Models of Feature Integration in Visual Search," *Journal of Experimental Psychology: General*, Vol. 127, pp. 189–219.

Gopher, D. (1992), "The Skill of Attention Control: Acquisition and Execution of Attention Strategies," in *Attention and Performance XIV: Synergies in Experimental Psychology, Artificial Intelligence, and Cognitive Neuroscience—A Silver Jubilee*, D. Meyer and S. Kornblum, Eds., MIT Press, Cambridge, MA.

Graf, R., and Torrey, J. W. (1966), "Perception of Phrase Structure in Written Language," *Proceedings of the Annual Convention of the American Psychological Association APA*, Washington, DC, pp. 83–84.

Graham, J. L. (1937), "Illusory Trends in the Observation of Bar Graphs," *Journal of Experimental Psychology*, Vol. 20, pp. 597–608.

Grainger, J., and Jacobs, A. M. (1994), "A Dual Read-Out Model of Word Context Effects in Letter Perception: Further Investigations of the Word Superiority Effect," *Journal of Experimental Psychology: Human Perception and Performance*, Vol. 20, No. 6, pp. 1158–1176.

Gray, W. D., and Fu, W. T. (2001), "Ignoring Perfect Knowledge in-the-World for Imperfect Knowledge in-the-Head: Implications of Rational Analysis for Interface Design," *CHI Letters*, Vol. 3, No. 1.

Green, D. M., and Swets, J. A. (1988), *Signal Detection Theory and Psychophysics*, Wiley, New York.

Gregory, R. L. (1980), "The Confounded Eye," in *Illusion in Nature and Art*, R. E. Gregory and E. H. Gombrich, Eds., Charles Scribner's Sons, New York.

Gugerty, L., and Brooks, J. (2001), "Seeing Where You Are Heading: Integrating Environmental and Egocentric Reference Frames in Cardinal Direction Judgments," *Journal of Experimental Psychology: Applied*, Vol. 7, No. 3, pp. 251–266.

Hancock, P., Flach, J., Caird, J., and Vicente, K., Eds. (1995), *Local Applications of the Ecological Approach to Human–Machine Systems*, Lawrence Erlbaum Associates, Mahwah, NJ.

Harris, D. H., and Chaney, F. D. (1969), *Human Factors in Quality Assurance*, Wiley, New York.

Hart, S. G., and Wickens, C. D. (1990), "Workload Assessment and Prediction," in *MANPRINT: An Approach to Systems Integration*, H. R. Booher, Ed., Van Nostrand Reinhold, New York, pp. 257–296.

Herrmann, D., Brubaker, B., Yoder, C., Sheets, V., and Tio, A. (1999), "Devices That Remind," in *Handbook of Applied Cognition*, F. T. Durso, Ed., Wiley, New York, pp. 377–407.

Hess, S. M., and Detweiler, M. C. (1995), "The Effects of Response Alternatives on Keeping-Track Performance," in *Proceedings of the 39th Annual Meeting of the Human Factors and Ergonomics Society*, Human Factors and Ergonomics Society, Santa Monica, CA, pp. 1390–1394.

Hick, W. E. (1952), "On the Rate of Gain of Information," *Quarterly Journal of Experimental Psychology*, Vol. 4, pp. 11–26.

Hickox, J. C., and Wickens, C. D. (1999), "Effects of Elevation Angle Disparity, Complexity, and Feature Type on Relating Out-of-Cockpit Field of View to an Electronic Cartographic Map," *Journal of Experimental Psychology: Applied*, Vol. 5, No. 3, pp. 284–301.

Hochberg, J. (1998), "Gestalt Theory and Its Legacy," in *Perception and Cognition at Century's End*, J. Hochberg, Ed., Academic Press, New York, pp. 253–306.

Hollands, J. (1992), "Ergonomic Data Display," *Human Factors Bulletin*, Vol. 35, No. 7, pp. 4–5.

Hollands, J. G., and Spence, I. (1992), "Judgments of Change and Proportion in Graphical Perception," *Human Factors*, Vol. 34, No. 3, pp. 313–334.

Holohan, C. J., Culler, R. E., and Wilcox, B. L. (1978), "Effects of Visual Distraction on Reaction Time in a Simulated Traffic Environment," *Human Factors*, Vol. 20, No. 4, pp. 409–413.

Hutchins, E. L., Hollan, J. D., and Norman, D. A. (1985), "Direct Manipulation Interfaces," *Human–Computer Interaction*, Vol. 1, No. 4, pp. 311–338.

Hyman, R. (1953), "Stimulus Information as a Determinant of Reaction Time," *Journal of Experimental Psychology*, Vol. 45, pp. 423–432.

Jagacinski, R. J., and Flach, J. M. (2003), *Control Theory for Humans*, Lawrence Erlbaum Associates, Mahwah, NJ.

James, W. (1890), *The Principles of Psychology*, Vol. 1, Holt, New York.

Johnson, A., and Proctor, R. W. (2004), *Attention Theory and Practice*, Sage, Thousand Oaks, CA.

Johnson, S. L., and Roscoe, S. N. (1972), "What Moves, the Airplane or the World?" *Human Factors*, Vol. 14, pp. 107–129.

Jones, D. G., and Endsley, M. R. (1996), "Sources of Situation Awareness Errors in Aviation," *Aviation, Space, and Environmental Medicine*, Vol. 67, No. 6, pp. 507–512.

Jordan, T. R., and Bevan, K. M. (1994), "Word Superiority over Isolated Letters: The Neglected Case of Forward Masking," *Memory and Cognition*, Vol. 22, pp. 133–144.

Just, M. A., and Carpenter, P. A. (1987), *The Psychology of Reading and Language Comprehension*, Allyn and Bacon, Boston.

Kahneman, D. (1973), *Attention and Effort*, Prentice-Hall, Englewood Cliffs, NJ.

Kahneman, D., and Treisman, D. (1984), "Changing Views of Attention and Automaticity," in *Varieties of Attention*, R. Parasuraman, R. Davies, and J. Beatty, Eds., Academic Press, New York, pp. 29–61.

Kintsch, W., and Keenan, J. (1973), "Reading Rate and Retention as a Function of the Number of Propositions in the Base Structure of Sentences," *Cognitive Psychology*, Vol. 5, pp. 257–274.

Kintsch, W., and Van Dijk, T. A. (1978), "Toward a Model of Text Comprehension and Reproduction," *Psychological Review*, Vol. 85, pp. 363–394.

Kintsch, W., and Vipond, P. (1979), "Reading Comprehension and Readability in Educational Practice and Psychological Theory," in *Perspectives on Memory Research*, L. G. Nilsson, Ed., Lawrence Erlbaum Associates, Mahwah, NJ.

Kirlik, A., Miller, A., and Jagacinski, R. J. (1993), "Supervisory Control in a Dynamic and Uncertain Environment: A Process Model of Skilled Human–Environment Interaction," *IEEE Transactions on Systems, Man, and Cybernetics*, Vol. 23, No. 4, pp. 929–952.

Klein, G., and Crandall, B. W. (1995), "The Role of Simulation in Problem Solving and Decision Making," in *Local Applications of the Ecological Approach to Human–Machine Systems*, P. Hancock, J. Flach, J. Caird, and K. Vicente, Eds., Lawrence Erlbaum Associates, Mahwah, NJ.

Klein, G. A., Orasanu, J., Calderwood, R., and Zsambok, E., Eds. (1993), *Decision Making in Action: Models and Methods*, Ablex, Norwood, NJ.

Kleinman, and Pattipati, (1991), "A Review of Engineering Models of Information Processing and Decision Making in Multitask Supervisory Control," in *Multiple Task Performance*, D. Damos, Ed., Taylor & Francis, London.

Kosslyn, S. M. (1987), "Seeing and Imagining in the Cerebral Hemispheres: A Computational Approach," *Psychological Review*, Vol. 94, pp. 148–175.

Kosslyn, S. M. (1994), *The Elements of Graph Design*, Freeman and Co., New York.

Krueger, L. E. (1992), "The Word-Superiority Effect and Phonological Recoding," *Memory and Cognition*, Vol. 20, pp. 685–694.

Kundel, H. L., and Nodine, C. F. (1978), "Studies of Eye Movements and Visual Search in Radiology," in *Eye Movements and Higher Psychological Functions*. J. Senders, D. Fisher and R. Monty, Eds., Lawrence Erlbaum Associates, Mahwah, NJ, pp. 317–328.

Langolf, C. D., Chaffin, D. B., and Foulke, S. A. (1976), "An Investigation of Fitts' Law Using a Wide Range of Movement Amplitudes," *Journal of Motor Behavior*, Vol. 8, pp. 113–128.

Layton, C., Smith, P. J., and McCoy, C. E. (1994), "Design of a Cooperative Problem-Solving System for En-Route Flight Planning: An Empirical Evaluation," *Human Factors*, Vol. 36, No. 4, pp. 94–119.

Levine, M. (1982), "You-Are-Here Maps: Psychological Considerations," *Environment and Behavior*, Vol. 14, pp. 221–237.

Levine, M. (1988), *Effective Problem Solving*, Prentice-Hall, Englewood Cliffs, NJ.

Loftus, G., Dark, V., and Williams, D. (1979), "Short-Term Memory Factors in Ground Controller/Pilot Communication," *Human Factors*, Vol. 21, pp. 169–181.

Logie, R. H. (1995), *Visuo-spatial Working Memory*, Lawrence Erlbaum Associates, Hove, East Sussex, England.

Lohse, J. (1991), "A Cognitive Model for the Perception and Understanding of Graphs," in *Human Factors in Computing Systems: Reaching Through Technology, CHI >91 Conference Proceedings*, S. P. Robertson, G. M. Olson, and J. S. Olson, Eds., Association for Computing Machinery, New York, pp. 137–144.

Lusted, L. B. (1976), "Clinical Decision Making," in *Decision Making and Medical Care*, D. Dombal and J. Grevy, Eds., North-Holland, Amsterdam.

Maltz, M., and Shinar, D. (2003), "New Alternative Methods of Analyzing Human Behavior in Cued Target Acquisition," *Human Factors*, Vol. 45, No. 2, pp. 281–295.

Marino, C. J., and Mahan, R. P. (2005), "Configural Displays Can Improve Nutrition-Related Decisions: An Application of the Proximity Compatibility Principle," *Human Factors*.

Massaro, D. W. (1979), "Letter Information and Orthographic Context in Word Perception," *Journal of Experimental Psychology: Human Perception and Performance*, Vol. 5, pp. 595–609.

Mayer, R. E. (1983), *Thinking, Problem Solving, Cognition*, W.H. Freeman, San Francisco.

Mayer, R. E. (1999), "Fifty Years of Creativity Research," in *Handbook of Creativity*, R. J. Sternberg, Ed., Cambridge University Press, New York, pp. 449–460.

McCarley, J. M., Kromer, A. F., Wickens, C. D., Vidoni, E. D., and Boot, W. R. (2004), "Visual Skills in Airport Security Screening," *Psychological Science*, Vol. 15, pp. 302–306.

McClelland, J. L., and Rumelhart, D. E. (1981), "An Interactive Activation Model of Context Effects in Letter Perception, I: An Account of Basic Findings," *Psychological Review*, Vol. 88, pp. 375–407.

McFarlane, D. C., and Latorella, K. A. (2002), "The Score and Importance of Human Interruption in Human–Computer Interface Design," *Human–Computer Interaction*, Vol. 17, No. 1, pp. 1–61.

McKoon, G., and Ratcliff, R. (1992), "Inference During Reading," *Psychological Review*, Vol. 99, pp. 440–466.

McRuer, D. (1980), "Human Dynamics in Man–Machine Systems," *Automatica*, Vol. 16, pp. 237–253.

Miller, G. A. (1956), "The Magical Number Seven Plus or Minus Two: Some Limits on Our Capacity for Processing Information," *Psychological Review*, Vol. 63, pp. 81–97.

Miller, G. A., Galanter, E., and Pribram, K. H. (1960), *Plans and the Structure of Behavior*, Holt, Rinehart and Winston, New York.

Moray, N. (1981), "The Role of Attention in the Detection of Errors and the Diagnosis of Errors in Man–Machine Systems," in *Human Detection and Diagnosis of System Failures*, J. Rasmussen and W. Rouse, Eds., Plenum Press, New York.

Moray, N. (1986), "Monitoring Behavior and Supervisory Control," in *Handbook of Perception and Performance*, Vol. 2, K. R. Boff, L. Kaufman, and J. P. Thomas, Eds., Wiley, New York, pp. 40-1–40-51.

Moray, N., Lee, J., Vicente, K., Jones, B. G., and Rasmussen, J. (1994), "A Direct Perception Interface for Nuclear Power Plants," in *Proceedings of the 38th Annual Meeting of the Human Factors and Ergonomics Society*, Human Factors and Ergonomics Society, Santa Monica, CA, pp. 481–485.

Neisser, U. (1967), *Cognitive Psychology*, Prentice-Hall, Englewood Cliffs, NJ.

Neisser, U., Novick, R., and Lazar, R. (1964), "Searching for Novel Targets," *Perceptual and Motor Skills*, Vol. 19, pp. 427–432.

Nickerson, R. S. (1999), "Enhancing Creativity," in *Handbook of Creativity*, R. J. Sternberg, Ed., Cambridge University Press, New York, pp. 392–430.

Norman, D. A. (1981), "Categorization of Action Slips," *Psychological Review*, Vol. 88, pp. 1–15.

Norman, D. (1988), *The Psychology of Everyday Things*, Basic Books, New York.

Olmos, O., Liang, C.-C., and Wickens, C. D. (1997), "Electronic Map Evaluation in Simulated Visual Meteorological Conditions," *International Journal of Aviation Psychology*, Vol. 7, No. 1, pp. 37–66.

Olmos, O., Wickens, C. D., and Chudy, A. (2000), "Tactical Displays for Combat Awareness: An Examination of Dimensionality and Frame of Reference Concepts and the Application of Cognitive Engineering," *International Journal of Aviation Psychology*, Vol. 10, No. 3, pp. 247–271.

Orasanu, J. M. (1993), "Decision-Making in the Cockpit," in *Cockpit Resource Management*, E. L. Wiener, B. G. Kanki, and R. L. Helmreich, Eds., Academic Press, San Diego, CA, pp. 137–173.

Orasanu, J., and Fischer, U. (1997), "Finding Decisions in Natural Environments: The View from the Cockpit," in *Naturalistic Decision Making*, C. Zsambok and G. Klein, Eds., Lawrence Erlbaum Associates, Mahwah, NJ, pp. 343–357.

Osborne, D. W., and Ellingstad, V. S. (1987), "Using Sensor Lines to Show Control–Display Linkages on a Four Burner Stove," in *Proceedings of the 31st Annual Meeting of the Human Factors Society*, Human Factors Society, Santa Monica, CA, pp. 581–584.

Palmer, S. E. (1975), "The Effects of Contextual Scenes on the Identification of Objects," *Memory and Cognition*, Vol. 3, pp. 519–526.

Parasuraman, R., Davies, R., and Beatty, J., Eds. (1984), *Varieties of Attention*, Academic Press, New York.

Parasuraman, R., Hancock, P. A., and Obofinbaba, O. (1997), "Alarm Effectiveness in Driver Centered Collision Warning Systems," *Ergonomics*, Vol. 40, pp. 390–399.

Pashler, H. E. (1998), *The Psychology of Attention*, MIT Press, Cambridge, MA.

Patterson, R. (1997), "Visual Processing of Depth Information in Stereoscopic Displays," *Human Factors*, Vol. 17, pp. 69–74.

Peterson, L. R., and Peterson, M. J. (1959), "Short-Term Retention of Individual Verbal Items," *Journal of Experimental Psychology*, Vol. 58, pp. 193–198.

Pew, R. W., and Baron, S. (1978), "The Components of an Information Processing Theory of Skilled Performance Based on an Optimal Control Perspective," in *Information Processing in Motor Control and Learning*, G. E. Stelmach, Ed., Academic Press, New York.

Phillips, J. G., Triggs, T. J., and Meehan, J. W. (2003), "Conflicting Directional and Locational Cues Afforded by Arrowhead Cursors in Graphical User Interfaces," *Journal of Experimental Psychology: Applied*, Vol. 9, pp. 75–87.

Pinker, S. (1990), "A Theory of Graph Comprehension," in *Artificial Intelligence and the Future of Testing*, R. Freedle, Ed., Lawrence Erlbaum Associates, Mahwah, NJ, pp. 73–126.

Pomerantz, J. R. (1981), "Perceptual Organization in Information Processing," in *Perceptual Organization*, M. Kubovy and J. R. Pomerantz, Eds., Lawrence Erlbaum Associates, Mahwah, NJ, pp. 141–180.

Pomerantz, J. R., and Kubovy, M. (1986), "Theoretical Approaches to Perceptual Organization," in *Handbook of Perception and Human Performance*, Vol. II, K. R. Boff, L. Kaufman, and J. P. Thomas, Eds., Wiley, New York, pp. 36-1–36-46.

Pomerantz, J. R., and Pristach, E. A. (1989), "Emergent Features, Attention, and Perceptual Glue in Visual Form Perception," *Journal of Experimental Psychology: Human Perception and Performance*, Vol. 15, pp. 635–649.

Posner, M. I. (1978), *Chronometric Explorations of the Mind*, Lawrence Erlbaum Associates, Mahwah, NJ.

Poulton, E. C. (1985), "Geometric Illusions in Reading Graphs," *Perception and Psychophysics*, Vol. 37, pp. 543–548.

Previc, F. H. (1998), "The Neuropsychology of 3-D Space," *Psychological Bulletin*, Vol. 124, pp. 123–164.

Previc, F., and Ercoline, W. (2004), *Spatial Orientation in Aviation*, American Institute of Aeronautics and Astronautics, Reston, VA.

Raby, M., and Wickens, C. D. (1994), "Strategic Workload Management and Decision Biases in Aviation," *International Journal of Aviation Psychology*, Vol. 4, No. 3, pp. 211–240.

Rasmussen, J. (1981), "Models of Mental Strategies in Process Control," in *Human Detection and Diagnosis of System Failures*, J. Rasmussen and W. Rouse, Eds., Plenum Press, New York.

Rasmussen, J. (1986), *Information Processing and Human–Machine Interaction: An Approach to Cognitive Engineering*, North-Holland, New York.

Rasmussen, J., Pejtersen, A.-M., and Goodstein, L. (1995), *Cognitive Engineering: Concepts and Applications*, Wiley, New York.

Reason, J. (1990), *Human Error*, Cambridge University Press, New York.

Reason, J. (1997), *Managing the Risks of Organizational Accidents*, Ashgate, Brookfield, VT.

Reicher, G. M. (1969), "Perceptual Recognition as a Function of Meaningfulness of Stimuli Materials," *Journal of Experimental Psychology*, Vol. 81, pp. 275–280.

Remington, R. W., Johnston, J. C., Ruthruff, E., Gold, M., and Romera, M. (2000), "Visual Search in Complex Displays: Factors Affecting Conflict Detection by Air Traffic Controllers," *Human Factors*, Vol. 42, No. 3, pp. 349–366.

Rensink, R. A. (2002), "Change Detection," *Annual Review of Psychology*, Vol. 53, pp. 245–277.

Richman, H. B., and Simon, H. A. (1989), "Context Effects in Letter Perception: Comparison of Two Theories," *Psychological Review*, Vol. 96, No. 3, pp. 417–432.

Rogers, D., and Monsell, S. (1995), "Costs of a Predictable Switch Between Simple Cognitive Tasks," *Journal of Experimental Psychology: General*, Vol. 124, pp. 207–231.

Roscoe, S. N. (1968), "Airborne Displays for Flight and Navigation," *Human Factors*, Vol. 18, pp. 321–332.

Roscoe, S. N. (1981), *Aviation Psychology*, University of Iowa Press, Iowa City, IA.

Roscoe, S. N. (2004), "Moving Horizons, Control Reversals, and Graveyard Spirals," *Ergonomics in Design*, Vol. 12, No. 4, pp. 15–19.

Sanders, M. S., and McCormick, E. J. (1993), *Human Factors in Engineering and Design*, 7th ed., McGraw-Hill, New York.

Sanderson, P. M. (1989), "The Human Planning and Scheduling Role in Advanced Manufacturing Systems: An Emerging Human Factors Domain," *Human Factors*, Vol. 31, pp. 635–666.

Sanderson, P. M., Flach, J. M., Buttigieg, M. A., and Casey, E. J. (1989), "Object Displays Do Not Always Support Better Integrated Task Performance," *Human Factors*, Vol. 31, pp. 183–198.

Schank, R. C., and Abelson, R. (1977), *Scripts, Plans, Goals, and Understanding*, Lawrence Erlbaum Associates, Mahwah, NJ.

Schneider, W., and Shiffrin, R. (1977), "Controlled and Automatic Human Information Processing, I: Detection, Search, and Attention," *Psychological Review*, Vol. 84, pp. 1–66.

Schutte, P. C., and Trujillo, A. C. (1996), "Flight Crew Task Management in Nonnormal Situations," in *Proceedings of the 40th Annual Meeting of the Human Factors and Ergonomics Society*, Human Factors and Ergonomics Society, Santa Monica, CA.

Schutz, H. G. (1961a), "An Evaluation of Formats for Graphic Trend Displays, Experiment II," *Human Factors*, Vol. 3, No. 2, pp. 99–107.

Schutz, H. G. (1961b), "An Evaluation of Methods for Presentation of Graphic Multiple Trends, Experiment III," *Human Factors*, Vol. 3, No. 2, pp. 108–119.

Senders, J. W. (1980), "Visual Scanning Processes," unpublished doctoral thesis, University of Tilburg, Tilburg, The Netherlands.

Shah, P., and Iyiri, A. (2005), *Handbook of Visual Spatial Thinking*, Cambridge University Press, Cambridge.

Shalin, V. L., and Bertram, D. A. (in press), "Functions of Expertise in a Medical Intensive Care Unit," *Journal of Experimental and Theoretical Artificial Intelligence: Special Issue on Expertise*.

Shinar, D., Rockwell, T. H., and Malecki, J. A. (1980), "The Effects of Changes in Driver Perception on Road Curve Negotiation," *Ergonomics*, Vol. 23, pp. 263–275.

Simkin, D., and Hastie, R. (1987), "An Information Processing Analysis of Graph Perception," *Journal of the American Statistical Association*, Vol. 82, pp. 454–465.

Smith, K. U. (1962), *Delayed Sensory Feedback and Balance*, Saunders, Philadelphia.

Smith, S. M. (1995), "Getting into and Out of Mental Ruts: A Theory of Fixation, Incubation, and Insight," in *The Nature of Insight*, R. J. Sternberg and J. F. Davidson, Eds., MIT Press, Cambridge, MA, pp. 229–251.

Sorkin, R. (1989), "Why Are People Turning Off Our Alarms?" *Human Factors Bulletin*, Vol. 32, pp. 3–4.

Sorkin, R. D., and Woods, D. D. (1985), "Systems with Human Monitors, a Signal Detection Analysis," *Human–Computer Interaction*, Vol. 1, pp. 49–75.

Spence, I., and Krizel, P. (1994), "Children's Perceptions of Proportions in Graphs," *Child Development*, Vol. 65, pp. 1189–1209.

St. Johns, M., and Mannes, D. J. (2002), *Proceedings of the 46th Annual Meeting of the Human Factors and Ergonomics Society*, Human Factors and Ergonomics Society, Santa Monica, CA.

Stanovich, K. E., and West, R. F. (1983), "On Priming by a Sentence Context," *Journal of Experimental Psychology: General*, Vol. 112, pp. 1–36.

Stanton, N., Ed. (1994), *Human Factors in Alarm Design*, Taylor & Francis, Bristol, PA.

Sternberg, S. (1969), "The Discovery of Processing Stages: Extension of Donders' Method," *Acta Psychologica*, Vol. 30, pp. 276–315.

Sternberg, R. J., and Powell, J. S. (1983), "Comprehending Verbal Comprehension," *American Psychologist*, Vol. 38, pp. 878–893.

Strayer, D. L., Wickens, C. D., and Braune, R. (1987), "Adult Age Differences in the Speed and Capacity of Information Processing, II: An Electrophysiological Approach," *Psychology and Aging*, Vol. 2, pp. 99–110.

Streeter, L. A., Vitello, D., and Wonsiewicz, S. A. (1985), "How to Tell People Where to Go: Comparing Navigational Aids," *International Journal on Man–Machine Studies*, Vol. 22, pp. 549–562.

Summala, H. (2000), "Brake Reaction Times and Driver Behavior Analysis," *Transportation Human Factors*, Vol. 2, pp. 217–226.

Tanner, W. P., and Swets, J. A. (1954), "A Decision-Making Theory of Visual Detection," *Psychological Review*, Vol. 61, pp. 401–409.

Thorndyke, P. W. (1980), "Performance Models for Spatial and Locational Cognition," Technical Report R-2676-ONR, Rand, Washington, DC, December.

Thorndyke, P. W., and Hayes-Roth, B. (1982), "Differences in Spatial Knowledge Acquired from Maps and Navigation," *Cognitive Psychology*, Vol. 14, pp. 560–589.

Travis, D. (1991), *Effective Color Displays: Theory and Practice*, Academic Press, New York.

Treisman, A. (1986), "Properties, Parts, and Objects," in *Handbook of Perception and Human Performance*, Vol. II, K. R. Boff, L. Kaufman, and J. P. Thomas, Eds., Wiley, New York, pp. 31-1–35-70.

Treisman, A. (1993), "The Perception of Features and Objects," in *Attention, Selection, Awareness, and Control: A Tribute to Donald Broadbent*, A. D. Baddeley and L. Weiskrantz, Eds., Clarendon Press, Oxford.

Treisman, A., Kahneman, D., and Burkell, J. (1983), "Perceptual Objects and the Cost of Filtering," *Perception and Psychophysics*, Vol. 33, pp. 527–532.

Tulga, M. K., and Sheridan, T. B. (1980), "Dynamic Decisions and Workload in Multitask Supervisory Control," *IEEE Transactions on Systems, Man, and Cybernetics*, Vol. 10, pp. 217–232.

Tulving, E., Mandler, G., and Baumal, R. (1964), "Interaction of Two Sources of Information in Tachistoscopic Word Recognition," *Canadian Journal of Psychology*, Vol. 18, pp. 62–71.

Tversky, A., and Kahneman, D. (1974), "Judgment Under Uncertainty: Heuristics and Biases," *Science*, Vol. 185, pp. 1124–1131.

Vicente, K. J. (1999), *Cognitive Work Analysis*, Lawrence Erlbaum Associates, Mahwah, NJ.

Vicente, K. J. (2002), "Ecological Interface Design: Progress and Challenges," *Human Factors*, Vol. 44, pp. 62–78.

Vicente, K., and Rasmussen, J. (1992), "Ecological Interface Design: Theoretical Foundations," *IEEE Transactions on Systems, Man, and Cybernetics*, Vol. 22, pp. 589–606.

Vicente, K. J., Christofferson, K. and Pereklita, A. (1995), "Supporting Operator Problem-Solving through Ecological Interface Designs," *IEEE Transactions on Systems, Man, and Cybernetics*, Vol. 25, No. 4, pp. 529–545.

Wallsten, T. S., Ed. (1980), *Cognitive Processes in Choice and Decision Behavior*, Lawrence Erlbaum Associates, Mahwah, NJ.

Warm, J., Ed., (1984), *Sustained Attention in Human Performance*, Wiley, New York.

Warren, D. H., Rossano, M. J., and Wear, T. D. (1990), "Perception of Map–Environment Correspondence: The Roles of Features and Alignment," *Ecological Psychology*, Vol. 2, pp. 131–150.

Wetherell, A. (1979), "Short-Term Memory for Verbal and Graphic Route Information," in *Proceedings of the 23rd Annual Meeting of the Human Factors Society*, Human Factors Society, Santa Monica, CA.

Whitaker, L. A., and CuQlock-Knopp, V. G. (1995), "Human Exploration and Perception in Off-Road Navigation," in *Local Applications of the Ecological Approach to Human–Machine Systems*, P. A. Hancock, J. M. Flach, J. Caird, and K. J. Vicente, Eds., Lawrence Erlbaum Associates, Mahwah, NJ.

Wickelgren, W. A. (1964), "Size of Rehearsal Group in Short-Term Memory," *Journal of Experimental Psychology*, Vol. 68, pp. 413–419.

Wickelgren, W. A. (1977), "Speed–Accuracy Tradeoff and Information Processing Dynamics," *Acta Psychologica*, Vol. 41, pp. 67–85.

Wickens, C. D. (1984). *Engineering Psychology and Human Performance*, 1st ed., Charles Merrill, Columbus, OHio.

Wickens, C. D. (1986), "The Effects of Control Dynamics on Performance," in *Handbook of Perception and Performance*, Vol. II, K. Boff, L. Kaufman, and J. Thomas, Eds., Wiley, New York, pp. 39-1–39-60.

Wickens, C. D. (1991), "Processing Resources and Attention," in *Multiple Task Performance*, D. Damos, Ed., Taylor & Francis, London.

Wickens, C. D. (1992), *Engineering Psychology and Human Performance*, 2nd ed., HarperCollins, New York.

Wickens, C. D. (1993), "Cognitive Factors in Display Design," *Journal of the Washington Academy of Sciences*, Vol. 83, No. 4, pp. 179–201.

Wickens, C. D. (1999), "Frames of Reference for Navigation," in *Attention and Performance*, Vol. 16, D. Gopher and A. Koriat, Eds., Academic Press, Orlando, FL, pp. 113–144.

Wickens, C. D. (2000), "The When and How of Using 2-D and 3-D Displays for Operational Tasks," in *Proceedings of the IEA2000/HFES2000 Congress*, Human Factors and Ergonomics Society, Santa Monica, CA, pp. 3-403–3-406.

Wickens, C. D. (2002), "Multiple Resources and Performance Prediction," *Theoretical Issues in Ergonomic Science*, Vol. 3, No. 2, pp. 159–177.

Wickens, C. D. (2003a), "Aviation Displays," in *Principles and Practices of Aviation Psychology*, P. Tsang and M. Vidulich, Eds., Lawrence Erlbaum Associates, Mahwah, NJ, pp. 147–199.

Wickens, C. D. (2003b), "Pilot Actions and Tasks: Selections, Execution, and Control," in *Principles and Practices of Aviation Psychology*, P. Tsang and M. Vidulich, Eds., Lawrence Erlbaum Associates, Mahwah, NJ, pp. 239–263.

Wickens, C. D., and Andre, A. D. (1990), "Proximity Compatibility and Information Display: Effects of Color, Space, and Objectness on Information Integration," *Human Factors*, Vol. 32, pp. 61–77.

Wickens, C. D., and Carswell, C. M. (1995), "The Proximity Compatibility Principle: Its Psychological Foundation and Its Relevance to Display Design," *Human Factors*, Vol. 37, No. 3, pp. 473–494.

Wickens, C. D., and Hollands, J. (2000), *Engineering Psychology and Human Performance*, 3rd ed., Prentice-Hall, Upper Saddle River, NJ.

Wickens, C. D., and Prevett, T. (1995), "Exploring the Dimensions of Egocentricity in Aircraft Navigation Displays: Influences on Local Guidance and Global Situation Awareness," *Journal of Experimental Psychology: Applied*, Vol. 1, No. 2, pp. 110–135.

Wickens, C. D., Sandry, D., and Vidulich, M. (1983), "Compatibility and Resource Competition Between Modalities of Input, Output, and Central Processing," *Human Factors*, Vol. 25, pp. 227–248.

Wickens, C. D., Vidulich, M., and Sandry-Garza, D. (1984), "Principles of S–C–R Compatibility with Spatial and Verbal Tasks: The Role of Display-Control Location and Voice-Interactive Display-Control Interfacing," *Human Factors*, Vol. 26, pp. 533–543.

Wickens, C. D., Zenyuh, J., Culp, V., and Marshak, W. (1985), "Voice and Manual Control in Dual Task Situations," in *Proceedings of the 29th Annual Meeting of the*

Human Factors Society, Human Factors Society, Santa Monica, CA.

Wickens, C. D., Todd, S., and Seidler, K. S. (1989), "Three-Dimensional Displays: Perception, Implementation, and Applications," CSERIAC SOAR 89-001, Crew System Ergonomics Information Analysis Center, Wright-Patterson AFB, OH.

Wickens, C. D., Vincow, M. A., Schopper, A. W., and Lincoln, J. E. (1997), "Computational Models of Human Performance in the Design and Layout of Controls and Displays," CSERIAC SOAR 97-22, Crew System Ergonomics Information Analysis Center, Wright Patterson AFB, OH.

Wickens, C. D., Goh, J., Helleberg, J., Horrey, W., and Talleur, D. A. (2003), "Attentional Models of Multitask Pilot Performance Using Advanced Display Technology," *Human Factors*, Vol. 45, No. 3, pp. 360–380.

Wickens, L. D., Alexander, A. L., Martens, M., and Ambinder, M. (2004), "The Role of Highlighting in Visual Search through Maps, *Spatial Vision*, Vol. 37, pp. 373–388.

Wickens, C. D., Vincow, M., and Yeh, M. (2005), "Design Applications of Visual Spatial Thinking: The Importance of Frame of Reference," in *Handbook of Visual Spatial Thinking*, A. Miyaki and P. Shah, Eds., Oxford University Press, Oxford.

Wickens, T. D. (2002) *Signal Detection Theory*, Freeman, San Fransisco.

Wilkinson, R. T. (1964), "Artificial 'Signals' as an Aid to an Inspection Task," *Ergonomics*, Vol. 7, pp. 63–72.

Wolfe, J. M., (1994), "Guided Search 2.0. A revised Model of visual search," *Psychonomics Bulletin and Review*, Vol. 1, No. 2, pp. 202–238.

Wolff, J. S., and Wogalter, M. S. (1998), "Comprehension of Pictorial Symbols: Effects of Context and Test Method," *Human Factors*, Vol. 40, pp. 173–186.

Woods, D., Wise, J., and Hanes, L. (1981), "An Evaluation of Nuclear Power Plant Safety Parameter Display Systems," in *Proceedings of the 25th Annual Meeting of the Human Factors Society*, Human Factors Society, Santa Monica, CA.

Woods, D. D., Johannesen, L. J., Cook, R. I., and Sarter, N. B. (1994), "Behind Human Error: Cognitive Systems, Computers, and Hindsight," CSERIAC #SOAR 94-01, Crew System Engonomics Information Analysis Center, Wright-Patterson AFB, OH.

Yantis, S. (1993), "Stimulus Driven Attentional Capture," *Current Directions in Psychological Science*, Vol. 2, pp. 156–161.

Yates, J. F. (1990), *Judgment and Decision Making*, Prentice-Hall, Englewood Cliffs, NJ.

Yeh, M., and Wickens, C. D. (2001), "Attentional Filtering in the Design of Electronic Map Displays: A Comparison of Color-Coding, Intensity Coding, and Decluttering Techniques," *Human Factors*, Vol. 43, No. 4, pp. 543–562.

Yeh, M., Merlo J. L., Wickens, C. D., and Brandenburg, D. L. (2003), "Head Up Versus Head Down: The Costs of Imprecision, Unreliability, and Visual Clutter on Cue Effectiveness for Display Signaling," *Human Factors*, Vol. 45, No. 3, pp. 390–407.

Yntema, D. (1963), "Keeping Track of Several Things at Once," *Human Factors*, Vol. 6, pp. 7–17.

Zheng, B., Janmohamed, Z., and MacKenzie, C. L. (2003), "Reaction Times and the Decision Making Process in Endoscopic Surgery," *Surgical Endoscopy*, Vol. 17, pp. 1475–1480.

Zsambok, C., and Klein, G. (1997), *Naturalistic Decision Making*, Lawrence Erlbaum Associates, Mahwah, NJ.

CHAPTER 6

COMMUNICATION AND HUMAN FACTORS

Ronald E. Rice and Cynthia Stohl
University of California–Santa Barbara
Santa Barbara, California

1 INTRODUCTION

Human factors is the study of the interaction of humans with systems, products, and the environment. Human factors research is traditionally found at the intersection of engineering, computer science, management, and psychology. In recent years communication scholars have joined in the interdisciplinary study of human factors design, particularly in the aviation and knowledge management industries (e.g., Eiff and Mattson, 1998; Mattson et al., 2001; Armentrout-Brazee and Mattson, 2004). This relatively new and evolving partnership is not surprising. As a science of human performance concerned with the physical, cognitive, and social abilities and limitations of people and a user-focused engineering discipline concerned with the design of interdependent systems for efficiency, safety, and quality (Salvendy, 1997), human factors involves communication within and among participants and technologies as a necessary yet complex condition for effectiveness. Indeed, a Google search in August 2004, using the terms *communication* (or *communications*) and *human factors*, returned over 400,000 Web pages.

In general, communication processes are fundamental to what is designed (hardware and software), how it is designed (the relations among designers and stakeholders, typically conceived of as users), how it is used (the optimization of the interaction between the human and the physical component), and how the designed system influences subsequent communication among users of the system. A communication perspective goes beyond the psychological and physiological, to the social. Making sense of the intentions and suggestions of fellow designers, a system, its features, its uses, and its consequences requires social interaction and a heightened sensitivity to language and the ways in which communicative choices shape the perception and definition of situations.

As a scholarly field, communication has two distinct traditions: one rooted in rhetoric and the humanities, and the other in systems theory and the social sciences. Since the earliest writings of Aristotle, rhetoricians have been concerned with means of persuasion, examining the efficacy of different types of logical, emotional, and ethical appeals, analyzing the credibility of message sources, and identifying the most effective language choices and communication strategies to influence people's attitudes and performance (Griffin, 1997). Since the 1930s, mass media scholarship, taking a behavioral science approach, has addressed the effects of mediated messages across varying communication technologies on a wide range of human activity and attitudes (e.g., Schramm, 1963; McLuhan, 1964). Many interpersonal, group, and organizational communication scholars, using linguistic, observational, survey, network, experimental, and ethnographic methods, focus on the multiple forms of message production, exchange, and interpretation and the pragmatic effects of interaction across a variety of social contexts (Griffin, 1997). Across traditions, many communication researchers focus on messages and their effects on interpretative processes, motivation, cognition, emotion, discrete and collaborative activities, decision making, and overall human performance.

The purpose of this chapter is to identify basic communication issues and processes that arise in pursuit of high-quality performance through the interaction of humans and technology. Communication, defined as the *collective, interactive process of generating and interpreting messages* (Stohl, 1995), focuses our attention on the networks of action and understanding that are created through coordinated activities and relationships. We are thus more concerned with the *relations between*, not the nature of nor the psychology of entities *within*, the interactive system. Although definitions and levels of analysis of the term *communication* can range from conversational interactions to global propaganda, and from dyadic interactions to mass organizing, we limit our discussion to concepts and examples of two or more people interacting (possibly by means of a mediating system) over time to create shared meaning through a common human verbal language. Figure 1 summarizes our area of focus. Note that a system (designed artifact) is considered from a communication perspective only as a possible medium for interaction among and between designers and users rather than from a traditional human factors emphasis on interaction with a system.

Within this constrained domain, the following sections introduce six main communication issues: messages, communication artifacts, communication characteristics of participants, group communication, communication networks, and communication environments. Working from the most basic unit of analysis, a message, to the most abstract, the communication environment, we identify analytic constructs and concepts that are fundamental for developing a communication perspective on human factors. The chapter ends with a brief application of these concepts to general stages of human factors activities in the design process.

2 MESSAGES

From a communication perspective on human factors, a *message* is the fundamental unit of interaction. The interface between humans or between human and machine is rooted in the exchange of messages, embedded in social interactions. Messages are simultaneously the cause and the explanation of many human performance problems. This fundamental dilemma can be seen in the various solutions to one of the most common problems: information overload. *Overload* refers to the transmission of new information at a rate that exceeds the input-processing and output-generating capabilities of individuals. Message overload may result in the intentional or unintentional modification of messages by omitting messages, modifying messages, and/or sampling messages. In each case the messages and patterns of message exchange are changed. The second type of response relates to network specialization. Relevant messages are redefined and delegated to others either in or outside the system, new priority systems are developed in which jobs are redefined in order to apply some selective criteria for message relevance or importance, and/or interpersonal networks are restructured so that there is decentralization and the creation of multiple channels. Each of these solutions makes it less likely that the individual will be saturated. In each of these solutions the interpersonal bonds between persons may be loosened, dissolved, or redefined, and new attachments will develop. In other words, it is impossible to separate relationships from message phenomena; there is no sharp dividing line between sender and receiver. Messages operate at multiple levels, assume multiple forms, and have multiple meanings and functions.

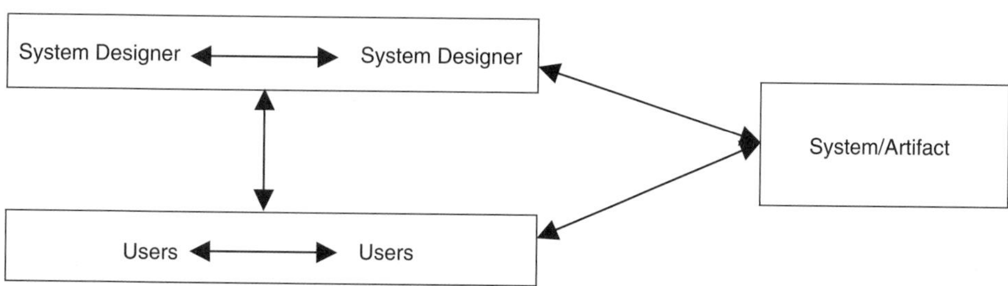

Figure 1 Communication relationships in the human factors emphasized in this chapter.

A message is not just the content or information as typically portrayed in *linear models of communication*. An illustration of these models is Shannon and Weaver's (1949) explication of information theory (designed to compute how much error can be reduced in a transmission given channel and coding constraints). In their model a *source* creates and then *encodes* a message, *transmits* it through a *channel*, and a *receiver decodes* the message, leading to a more informed state. Various forms of *noise* may occur through the system, and depending on its speed and extent, *feedback* may help verify or correct the message. Using this type of model of communication for human factors design focuses on maximizing the clarity of the messages through redundancy, message and channel modifications, and the elimination of environmental distractions. However, as useful as they may seem, transmission models that rely on a conduit metaphor of communication (Reddy, 1993) have several shortcomings that must be considered in the design of human factors.

2.1 What Communication Is Not

First, *communication is not linear*. There are two aspects of messages: ostensive messages, the *actual text* (this can be verbal, nonverbal, paralinguistic cues, or artifacts), and *internally experienced messages*, the interpretations of these ostensive messages (Stohl and Redding, 1987). Human factors analysts need to distinguish among the text, the representative intentions of the sender, and the interpretations of all interactants as they intersubjectively create meaning. For example, what constitutes "support" from a technical help desk is not straightforward. Considerable interaction between technical staff and a user department may be needed before there is shared agreement about the nature of a system problem, how to solve it, and even whether support has been provided.

Second, *communication is not passive*. There are no fixed roles of active sender and passive receiver. Message senders are simultaneously receivers; receivers are simultaneously senders. The message sent is not the message received. Communication requires active engagement on the part of all participants; it is a collaborative and oftentimes dialogic sense-making process in which meanings and interpretations are negotiated through hard work Relational history, social context, and the medium may change the meaning and import of intended messages. Computer error messages such as "abort" or "kill job" may be seen as neutral computer commands, but clearly have additional potential meanings for users that may cause confusion, discomfort, or alienation.

Third, *communication is not unifunctional*. Every communication has a *content* and *relationship* aspect (Watzlawick et al., 1967). Messages not only convey information or content; they simultaneously communicate a relationship between communicants which describes how a message should be taken, thereby providing a context for interpretation. The relationship aspect is often called *metacommunication*, defined as communication about the communication. Metacommunication helps interactants make sense of the message and may be associated with verbal, nonverbal, and/or paralinguistic cues. For example, when human factors engineers label the content of messages as *instructions*, they are providing specific information to participants *and* simultaneously telling the interactants that these messages are of a higher logical type; depending on the wording, the instructions may also convey relative status differentials between the technical expert and the novice user.

Fourth, *communication is not an uninterrupted series of interchanges*. *Punctuation* (not the punctuation of grammar and writing lessons) is the means by which participants organize interchanges and create causal maps (Watzlawick et al., 1967). Participants choose how they punctuate sequences, differentially attributing cause and effect, stimulus and response, and thus create meanings that may or may not be consistent with others in the system and with intended meanings. For example, from a designer's perspective, a series of messages in a training session may be viewed as a continuous sequence of exchanges in which one message is unambiguously the stimulus, the next is the user's response, and the third is the trainer's reinforcement. But from the perspective of the participants in the system, every message in the sequence can potentially and simultaneously be a stimulus, a response, and a reinforcement.

Fifth, *communication is not equivalent to reducing uncertainty*. Traditional approaches to communication view information as the opportunity to reduce uncertainty and lessen ambiguity in a given situation. For example, in information theory, messages contain bits of information that decrease entropy (randomness) and make things more predictable. The semantic aspects of communication are deemed irrelevant to the engineering aspects (Shannon and Weaver, 1949). *Uncertainty reduction theory* (Berger, 1986) also assumes that the main purpose of communication is to make sense of the world and that there is a universal human drive to reduce uncertainty. Behavioral mechanisms of uncertainty reduction are typically addressed through developing and following accepted procedural protocols, while cognitive uncertainty is reduced through information-seeking behaviors designed to acquire the necessary information. However, the reduction of uncertainty, the striving for message clarity, and the design of disambiguating activities cannot be assumed to be the underlying rationale for human activity nor the raison d'être for communication. Significant collaborative work related to identity management, task definition, role negotiation, and other types of interpretive processes are needed to create shared meanings that must be considered in any human factors design.

2.2 Ambiguity and Shared Meaning

Contemporary communication theorists have a complex view of human communication and uncertainty that has important implications for the study of human factors. First, lack of information does not necessarily motivate information seeking. Responses to uncertainty are shaped by appraisals of the situation and emotional reactions to the context in which uncertainty arises (Brashers, 2001). Second, people tend to respond to uncertainty based on their assessments of the valence (positive or negative) and probability of an event happening (Babrow, 2001). In many situations, individuals may choose not to pursue more information, to reframe the issue, to ignore the uncertainty, or even to try to amplify the uncertainty. Thus, *uncertainty management* is a more useful way to consider the potential effects of ambiguity on human performance. Third, uncertainty reduction is not always an appropriate or desirable communication goal. For example, Eisenberg's (1984) work on *strategic ambiguity* reveals many ways in which messages may be designed ambiguously in order to promote unified diversity, preserve privileged positions, enable plausible deniability, and/or facilitate organizational change. Logos, organizational mission statements, and role definitions are often crafted in ways that are open to multiple interpretations and thereby can be defined by individuals within their own conceptual frameworks, enabling them to identify and find their own place within the system, yet simultaneously be considered universally defined across organizations or institutions. From this perspective, explicitly shared meaning is not the primary (or sufficient) goal of interaction; rather, it is organized action (Donnellon et al., 1986; Weick, 1995).

Stohl and Redding (1987) identify two other types of message ambiguity that have important implications for human factors design. Type 1 ambiguity occurs when a receiver is confused or uncertain about the meaning of a message. This may occur because the receiver cannot construct any plausible interpretation of the message (i.e., the message seems meaningless) or the receiver recognizes that there may be multiple possible meanings and cannot select or determine the intended meaning. When we have doubt as to what a message "means" we often seek out others in our network to help us interpret the message. Type 2 ambiguity occurs when there are two or more potential meanings to a message but the creator or receiver is unaware of this possibility. This type of ambiguity is perhaps most problematic for human factors design.

For example, in 1989, an Avianca jet crashed approaching John F. Kennedy Airport, killing 73 of the 158 people aboard. The jet, flying in stormy weather, had aborted a first landing approach and was preparing for another attempt when it crashed into a New York suburb northeast of the airport. The United Press International (UPI) report at the time of the crash highlighted the serious dangers of message type 2 ambiguity and pointed to the importance of shared meanings in safety procedures. "The investigation hinged on exactly what words the crew used to describe the low fuel situation to controllers, Lee Dickson, a member of the National Transportation Safety Board said. Dickson said recorded tapes of cockpit conversations revealed that crew members never used the word *emergency* but did say they needed *priority* when making their second landing attempt after aborting the first." The UPI reported that when pilots declare a fuel emergency, the situation is considered very serious and the plane is given priority to land over all other waiting planes. The air traffic control manual has no listing under terminology for a "fuel priority" (Stohl, 1995, p. 58).

Several strategies exist for the development of shared meanings among diverse interactants. For example, a group of designers brought together for a new product or system development may lack shared understanding about the problem, norms, and behaviors of the members (Clark and Brennan, 1993). Proponents of distributed cognition suggest that shared understanding is created by the team sharing cognitive maps of factors and relationships at various levels of detail and by making tacit knowledge explicit and formalized (Nonaka and Takeuchi, 1995). Another approach is to create *shared artifacts* and *boundary objects* to achieve a collective understanding and the nature and goals of the task (Star, 1993; Suchman, 1995; Majchrzak et al., 2000). In his theory of how media and language are used to develop common ground, Clark (1996) calls these shared artifacts *external representations* or *coordination devices*. Such artifacts must lead to a unique solution and provide a rationale (through an identifiable purpose, individuals' ability to accomplish the purpose, and a mutual belief about the purpose and abilities) for participants to believe that they will converge on the same joint action. Similarly, Krauss and Fussell (1990) describe the communicative process by which descriptions of inanimate objects are transformed into referring expressions that are used in the development of common ground. On the first reference to an objective, people will use a long, rather unwieldy referring expression, but over the course of successive references will shorten the phrase to one or two words if reinforced by "back-channel" (e.g., "uh-uh," "umm," nods) responses that provide indications of what is understood. Olson and Olson (1997) extended this work to suggest that communication systems that foster the development of shared artifacts should have features that support linking conversations to these artifacts (such as allowing message revision and notifying others of relevant messages) and the creation and editing of these artifacts.

2.3 Feedback

Even most conduit models of communication, but certainly more interactive models, emphasize the important role of feedback for developing shared meaning, fostering learning, and repairing misunderstandings. If feedback or control is inaccurate, delayed, repressed, misinterpreted, or if the range through which organizational parameters can safely vary is too narrow, external variations can lead to deviation amplification and chaos (Mees, 1986). The communication processes and content fostering appropriate feedback are central to organizational and project functioning.

Research has shifted from considering feedback as something that a recipient receives passively to two forms of active feedback seeking: eliciting (direct) and monitoring (indirect) (Fedor et al., 1992). These provide different forms of information and may be differentially appropriate. Eliciting gets only what others want to share. Monitoring acquires other information but requires making inferences about that information and its source from nonverbal cues and others' comments. Monitoring would seem more likely when the costs (perceived and actual) of asking directly are high, such as in public situations or low-trust climates. Also, feedback itself may generate uncertainty about its meaning, contingencies, and so on. So monitoring may precede eliciting, to determine costs and appropriateness of eliciting. In Fedor et al.'s (1992) study of 139 student helicopter pilots and 152 instructor pilots, the expected costs of seeking feedback and uncertainty from prior feedback were the most consistent predictors of eliciting and monitoring feedback. There was higher eliciting when the seeker had a high intolerance for ambiguity and the source had a low confrontational style; and higher monitoring under conditions of low self-esteem and low source credibility.

Although user feedback is often proposed as one way of resolving or preventing problems, there are at least two general classes of problems that make feedback much more problematic than presumed: (1) communication problems (such as ignoring information already available, changing usage contexts, and tensions among oppositional competence goals), and (2) organizational problems (such as uses and misuses of organizational memory, information systems designed to support "customer" feedback, the general level of technical ignorance, and overemphasis on rationalistic homeostasis).

The concepts of the two-level communication hierarchy—relational and content—may help explain some problematic feedback routines. "Behavior that may be normatively incompetent may be relationally competent, and vice versa" (Spitzberg, 1993, p. 153). For example, the valued norm of getting along with others (the relational meaning of an episode) may not always be an asset, because it may suppress conflict and resolution (the content meaning of an episode). If one's communication is goal-directed but people are faced with ambiguous, incompatible, or incompetent goals, dialectical motives arise (Argyris, 1986), such as politeness and assertiveness. Repressing conflicts and negatively valenced feedback (say, between a representative of a system design team and a potential user about the appropriate sequence of work processes) avoids impoliteness but potentially leads to relational dissatisfaction (client frustration, job alienation, turnover) or flawed systems.

Culnan (1989) makes the point that all transaction-processing systems are actually organizational message systems. That is, they both generate messages to the user (either a direct user or someone accessing information from the system) and to the organization that may or may not be recognized or intentional and may also foster, collect, and even analyze messages as feedback. Different transaction-processing systems (structured, semistructured, or unstructured) may support or constrain different message processes: routing (selectively distributed, limiting overload); summarizing (reducing amount of message without loss of information); delays (intentional or not, beneficial or not); and modification (distortion, due to motivations or cognitive limitations) (Huber, 1982). For example, semistructured systems such as toll-free telephone lines or Web home pages can be designed to include meta-level tracking of the nature and topic of calls, providing cues that user documentation for certain products or services is inadequate. Because of transaction systems' increased flow and sources of messages, sent, stored, replied to, copied, and forwarded, there arise multiple opportunities for deception and manipulation, which are difficult to detect and even harder to hold any specific person accountable for (Zmud, 1990).

2.4 Sense-Making

Weick (1979, 1995) labels the state of confusion and uncertainty that can stem from complex problems as *equivocality* or *ambiguity*. Uncertainty may be resolved with sufficient information to find the right answer. However, equivocal problems have no single correct answer, although they may be resolved through joint creation of a shared meaning that makes sense to the participants.

Weick's theory of organizing identifies four sets of interlocking processes, constituting organizational sense-making to reduce equivocality. Sense-making begins with *ecological change*, sensed directly or relative to some prior measure, which presents equivocal information. *Enactment* is the process whereby individuals create their environments through attending to and interpreting certain phenomena, while ignoring, avoiding, or not perceiving other phenomena. For example, often the nature of work and organization must be reconceptualized to take advantage of the potential of a new practice or technology [such as word processing (see Johnson and Rice, 1987), or the office copier (see Brown and Duguid, 1996, p. 76; Orr, 1996)]. Then participants *select* a suitable explanation from possible alternative explanations of the enacted environment. Selections that appear to be useful are *retained*, through storage, acting, and evaluating, for future use—thus generating organizational routines. People typically engage in several cycles to reduce equivocality regarding a particular problem or issue. Each of these cycles typically includes an act, an interact, and what Weick (1979, p. 74) refers to as a *double interact*, where the initiator adjusts or responds to the second person's response to the initial action or communication. Thus, within the sense-making framework, Weick conceptualizes organizational structure as a communication activity of interlocked behaviors: an ongoing, cyclical process. The double interact highlights the crucial role of communication in human factors design.

2.5 Messages As Managing Meaning

In summary, the simple transmission models of communication ignore the critical need to understand the *management of meaning*. Messages may complicate rather than simplify, obfuscate rather than clarify: possibly intentionally, and possibly even usefully. They vary in content, interest, and importance, they come through different channels, they create different patterns of response, and they link networks in different ways. A message may have different meanings and effect depending on the source, the channel, the receiver, or the context, including past experience, cultural identification, timing, and relational dynamics. Humans communicate through socially constructed and adapted symbols, and because both those symbols and interpretations have wide flexibility and evolve over time, the extent and nature of the relationship between the sender(s) and the messenger(s) heavily influence what meaning is associated with that process (Axley, 1984). Thus, a "message" really represents a wide variety of possible participants, situations, interpretations, and influences, and their interactions, both within the design process and within the use process. Before we explore the interactive processes associated with joint sense-making, in the next section we address the communication artifacts that produce and are produced through interaction.

3 COMMUNICATION ARTIFACTS

As Figure 1 indicates, we are emphasizing the communication relations among designers, among users, between designers and users, and with the "system" only insofar as the system itself mediates communication among these participants. In this sense, information and communication technologies are particularly interesting human factors artifacts, because they play a role both in communicating about innovations (as media) and are an innovation (a new system) themselves (Rice, 1987). As media, they are channels for communication. Thus, their use influences access to, the timing of, and the meaning socially constructed by, communication. As systems, they are the content (or topic) themselves, such as the "new system" or the innovation an organization is attempting to implementation. Thus, their use can influence the nature, pattern, and extent of social interaction.

3.1 Communication Modes, Media Richness, and Affordances

Potentially significant aspects of new computer-mediated communication systems for human factors include the ability to overcome constraints of time and place, to retrieve and search for associated material, to reprocess and merge different content, and to support many-to-many communication flows (Rice, 1987). However, debate continues as to how suitable or effective new media such as electronic mail, Web pages, mobile phones, and voice mail are for various communication activities, as compared to traditional media such as face-to-face or telephone.

Gibson (1979) introduced the concept of *affordance*, a possibility for action available through characteristics and uses of objects (in particular, technologies and media). Hartson (2003) extended the concept's domain by distinguishing cognitive, physical, sensory, and functional affordances. For example, paper has many different affordances, especially in combination with other technologies (such as pens or thumbtacks), supporting a wide variety of human actions (Sellen and Harper, 2003). Paper documents allow readers to make notes or other marks on them, allow flexible navigation and manipulation, allow users to position or lay out the paper for different purposes, view material in much greater resolution than on-screen, are tangible (involving hands, eyes, and varying position), facilitate the coordination of action among organizational members, provide a medium for information gathering and exchange, support discussion, allow annotation for later discussion, provide a medium for organizing one's thoughts and work processes, and enable storage of information for multiple people, groups, locations, and time periods. Each medium or system thus has a wide range of characteristics and affordances, which may be more or less appropriate and more or less valued, across contexts (Rice, 1987).

Mediated communication faces challenges of supporting both the affordances of familiar media as well as multicue interpersonal interactions and adjustments. Mediation affects first impressions, the extent to which context can be shared, the scalability of participants, the types of interactions, attention, and exposure, formal and informal relationships, access to knowledge and status, accountability, possibilities for reducing misinterpretations, and so on.

Media richness theory (MRT) is frequently applied to this question of choice and effectiveness (Daft and Lengel, 1986; Trevino et al., 1987). MRT proposes that communication media differ in the extent to which they (1) can overcome various communication constraints of time, location, permanence, distribution, and distance; (2) transmit the social, symbolic, and nonverbal cues of human communication; and (3) convey equivocal information. As noted above, uncertainty can potentially be reduced by more information, if only the information can be found, perhaps through a "lean" computer database. However, one "make sense" of an equivocal situation, perhaps through "rich," iterative, negotiated discourse with other actors (Daft and Weick, 1984; Weick, 1995). Use of any communication channel involves both costs and information-processing capabilities, so a medium that is not well matched to the requirements of uncertainty or equivocality will degrade communication performance. Considerable research so far has shown some, although weak, support for this essential proposition. For example, Lind and Zmud's (1991) study of communication between information systems and noninformation systems personnel showed how matches between situation and communication channel affected technological innovation in the firm's business activities.

However, many studies find considerable differences between benefits from, reasons for choosing, applications of, and problems with new media, such as e-mail and voice mail. These studies seriously challenge the simple unidimensionality of "media richness" often used to place e-mail at the lower end of media choice (Rice, 1992). Expanded and alternative models suggest that a communication technology's impact on communication is not simply a function of the task that information processing demands but also of time, critical mass, social influence, group identity and norms, technology adaptation, and experience with both the medium and the communication participants (Hiltz and Turoff, 1981; Rice and Aydin, 1991; Walther, 1992; Markus, 1994; Lea and Spears, 1995; Orlikowski et al., 1995; Carlson and Zmud, 1999).

3.2 Effect of System Design on Communication

Systems, and people's behaviors with and around systems, are deeply embedded in wider social and environmental situations (as argued by activity theory, situated action theory, and distributed cognition theory; Engestrom et al., 1999; O'Hara et al., 2003). Thus, systems, media, space, movement, gesture, people, and work are all interrelated in everyday communication (Chalmers, 2004, p. 74). Artifacts (new systems, as well as familiar furniture, paper, office accessories, and clothes) "communicate information about the organization and the people who work there" (Davis, 1984, p. 277).

Because the work environment affects individuals' attitudes, behavior, perceptions, and communication, people should be able to personalize and change their systems (Barker and Associates, 1978; Zalesny and Farace, 1987). Therefore, system designs should consider how spaces, communication, media, and people can be adapted, coupled, and intercontextualized "to form resources for social interaction and interpretation" (p. 76). For example, a basic attribute such as a display's physical orientation may have considerable implications for human communication. Using a vertical display for group interaction tends to create an authorial atmosphere, with someone controlling the display as a teacher or lecturer; however, horizontal tabletop displays "provide a natural centre for interaction to take place around and encourages collaboration between the users" (Stahl and Wallberg, 2004, p. 53). The location and formatting of office paper (such as binders, or Post-It notes) can signal whether someone should attend to it quickly or whether it can be shared (Sellen and Harper, 2003). Long-term use of video-mediated communication can foster complex patterns of interaction constrained by the features and structures of the new medium (Dourish et al., 1996). Major factors affecting communication interactions through video and audio systems include quality of audio, audio feedback and interference, self-views, awareness of range of shared users' fields of view, "reciprocity of perspective" (Bullock, 2004, p. 147), video refresh rate, conventional eye-gaze direction, proportional size of people in background, and audio or video access by people at periphery. For a more sociological example, the telegraph allowed people to communicate across time and space at a pace and amount never before experienced, but also enabled railroad companies to collect, associate, and analyze information from stations about the dynamics of trains, shipments, and passengers. This transformed how organizations collected and processed information, located headquarters and branches, and how they learned from that information to develop effective schedules, routing algorithms and billing procedures, which changed the domains and design of railroads (Beniger, 1986; Yates and Benjamin, 1991).

Even the physical structures of buildings and offices create considerable constraints on communication, and thus quality of work life, performance, and innovation (Allen, 1977; Johnson, 1993). Physical environments within organizations represent material, although subtle, constraints on behavior, interaction, and possible interpretations. Influential aspects of physical environments include social density, proximity, access, exposure, privacy, mobility, time–space paths, physical structure (architectural and construction choices), physical stimuli (artwork, noise), and symbolic artifacts (office size and windows) (Archea, 1977; Davis, 1984; Johnson, 1993). Physical elements not only facilitate and constrain activities and relations, but often represent particular resources and contexts (consider the familiar concept of the influence of "the water cooler" on emergent relations and communication climate). Physical and temporal distances constrain network relations, increasing the costs of signaling one's interests and of finding people with similar interests (Feldman, 1987). Indeed, some researchers "view space as equivalent to context in providing the medium within which social interaction is embedded" (Johnson, 1993, p. 93).

Developments such as modular offices, shared displays (O'Hara et al., 2003), mobile communication, inhabited information spaces (Snowdon et al., 2004), and personal locator badges (Want et al., 1992) may overcome some of these constraints while generating others. For example, a study of a networked desktop video conference system showed that while it facilitated R&D workers' ability to make contacts and collaborate with others across offices, it still raised issues concerning norms of privacy, interruption, and access (Fish et al., 1993).

Many unanticipated problems and failures with automated systems are related to human–system interactions. Users may have difficulties tracking the system's activities, leading to "automation surprises" for both users and designers (Woods et al., 1997, p. 1927), for a variety of reasons:

1. New systems may help decrease workload in already low-workload conditions, but become distracting or harmful in pressured, critical workload situations, or create new work for different actor roles.
2. Systems may not indicate where or when to look for changes or disconfirming information, especially in rare or crisis situations, so users may not be allocating their attention appropriately, leading to breakdowns in attention to either (or both) the system and the situation.

3. Complex interdependent systems requiring varied inputs may foster mode errors when what might be an appropriate action in one system mode might be inappropriate in another mode and is made worse when the user is not aware of the shift. Indeed, taking action during a second mode while assuming the initial mode may generate internal parameters that activate a third mode, thus generating more unintended and difficult-to-identify consequences.

4. As the increasing demand for coordinating the various automation processes as well as human interactions exceeds one's abilities, supervisors or users may exit the automation system, or the system may reach its autonomy threshold and return full control to the operator, just at the time that the automation system is both generating and monitoring complexities that humans cannot process. The operator is likely not even aware of the underlying crisis, is unprepared or unable to handle the situation, and is thus surprised or shocked to find out the seriousness of the underlying problem.

5. Users come to rely on automation that are highly reliable for expected situations, but may fail in rare situations. Thus, they may overtrust systems, so systems should be designed to communicate its intentions, so that users have realistic and appropriate expectations. A system is a medium through which designers communicate their intention and into which users must responsibly place their trust (p. 1935, referring to Winograd and Flores, 1986).

The greater the levels of system authority and autonomy, complexity and number of components, and coupling among components, the greater the need for communication and coordination among users and between users and system to foster observability or awareness of the primary system as well as of other (social and technical) systems and tasks.

4 COMMUNICATION CHARACTERISTICS OF PARTICIPANTS

The role of individual differences in human performance has been well documented in the physiology, psychology, and human factors literature (Salvendy, 1997). Communication traits, attributes, skills, and knowledge that differentiate communicative performance thereby have significance for user and design profiles are plentiful. These individual differences can range from communicative apprehension (Richmond and McCroskey, 1998) to nonverbal sensitivity (Hall et al., 1997), bargaining styles (Putnam, 1994; Putnam and Kolb, 2000) to framing (Fairhurst and Sarr, 1996), conflict styles and argumentativeness (Infante et al., 1984) to general communicative competence (Westmyer et al., 1998). Indeed, there is strong empirical evidence that personal communicative profiles influence perceptions of and performance with systems, and among designers and users. However, within this vast theoretical, research, and applied domain, the relationship between particular communicative characteristics and human factors is relatively undeveloped. Using illustrative exemplars, we suggest behavioral impacts of several communication characteristics.

4.1 Demographic Differences

Differences in communication styles, decoding and encoding nonverbal behavior, eye contact, proxemics, haptics, verbal aggressiveness, conversational style, and so on, have been well documented in demographic groups according to gender, race, ethnicity, nationality, and age (Tannen, 1990; Hofstede, 1991; Allen, 2004). Although the emergence and range of these demographic differences in communication is hotly debated (e.g., Buzzanell, 2000), and the artificial construction of these social identities has been critiqued (Allen, 2004), the practical importance of understanding the relationship between communication and demography is undisputed. We consider only gender and culture here.

4.1.1 Gender

The recent growth in studies of gender, communication, and technology have important implications for human factors work. In general, these studies suggest that the gendered differences in communication found in face-to-face communication, such as nonverbal sensitivity preferences for collaboration and power sharing, are typically replicated in the interface between humans and technology (Ebben, et al., 1993). Dennis et al. (1999) found that matching richness of media to task equivocality results in better performance only for all-female teams, because, they suggest, females tend to be more sensitive to nonverbal communication and more affected by its absence in computer-mediated communication. Further, the gendered patterns of interaction associated with both formal and informal group meetings such as men gaining the speaking "floor" more often, and keeping the floor for longer periods of time regardless of their status in the organization, are also found in computer-mediated communication (CMC). Sutcliff (1998) found that messages on female-oriented Web sites emphasized communality, stressed personal experience, resisted authoritative language, and encouraged emotional interaction. In contrast, male-oriented sites emphasized privacy, stressed professionalism, relied on authoritative language, and minimized personal interaction. Herring (1994) also found that contrary to the claim that CMC neutralizes distinctions of gender, the patterns of interaction in postings to the Internet were recognizably different for women and men. Electronic messages posted to discussion groups by men were more likely to be written in an aggressive, competitive style, while women's messages tended to be more supportive. Furthermore, she found that women and men have different "communicative ethics"; that is, they value different kinds of online interactions as appropriate and desirable.

It is important to note, however, that there are many studies in which the communication context may be more important in predicting and modifying performance than demographic variables such as gender. For example, in a series of studies, Buzzanell and Burrell (1997) found that metaphorical schema and linguistic cues used by conflict participants was associated with the type and context of conflict rather than demographic characteristics.

From a human factors perspective, the demographic stereotypes that users and designers use when entering a communication situation are also critical to consider. For example, although research has not consistently confirmed that women and men differ in the frequency of their use of tag questions, disclaimers, and other communicative forms which decrease assertiveness and effectiveness, women are perceived to communicate in this more deferential style, so violations of those expectations have serious implications for women's power and efficacy in a system (Tannen, 1995). Stereotypes of Asian Americans as passive and docile have also been found to impede career advancement (Woo, 2000).

4.1.2 Culture

Globalization and the increasing diversity of today's workplace make the study of cultural differences and communication processes ever more critical for human factors work (Stohl, 2001). Hofstede (1980, 1991), the primary proponent and developer of measures of *national cultures* and organizational values, defines culture as "the collective programming of the mind which distinguishes the members of one human group from another" (Hofstede, 1984, p. 210). He originally identified four dimensions of cultural variability (power distance, uncertainty avoidance, masculinity, and individualism) and in 1988, added a fifth dimension, Confucian dynamism (Hofstede and Bond, 1988), rooted in principles of stability, status, thrift, and shame.

All the dimensions of cultural variability reflect the differing values given to issues of equality, ambiguity, instrumentalism, and community and are strongly associated with the ways in which people across the world perform roles, communicate, relate to one another, and utilize communication technologies (Teboul et al., 1994). In high-power-distance countries such as Singapore, the Philippines, France, India, Venezuela, and Portugal, employers and employees are more likely to consider that violating the chain of command constitutes serious insubordination. Low-power-distance countries such as Denmark, New Zealand, and Israel expect people to work around hierarchical chains and do not see hierarchy as an essential part of organizational life. When working in or with high-power-distance countries, Hofstede suggests, it is important to respect the authority structure and show deference to the formal hierarchy. In low-power-distance countries, organizations tend to be less formal and have more open communication across the social system (Hofstede, 1984).

Driskill (1995) substantiated these conclusions in a study of Euro-American and Asian-Indian engineers. American and Indian co-workers identified situations involving authority, role duties, and supervision as the most salient contexts for the emergence of strong cultural differences. Asian Indians felt that competent supervisors should provide daily and direct surveillance, were very comfortable with an authoritarian decision-making style, and accustomed to strict adherence to job descriptions and titles.

It is important to note that despite the valenced stereotypes people may hold about particular national cultures, no particular level of cultural values is better or worse than others, although they may be associated with norms or behaviors inappropriate in other cultures. For example, Hofstede notes the negatives of both ends of the individualism/collectivism dimension: the selfishness of individualism and the tyranny of collectivity.

In more individualist societies, "the ties between individuals are loose; everyone is expected to look after himself or herself and his or her immediate family. ... Collectivism ... pertains to societies in which people from birth onwards are integrated into strong, cohesive ingroups, which throughout people's lifetime continue to protect them in exchange for unquestioning loyalty" (Hofstede, 1991, p. 51). Earley (1993) found that collectivists' performance was lower than that of individualists' under conditions of working alone, or working in an out-group, compared to working with an in-group. Collectivism's emphasis on tradition may also limit the extent of new technology transfer (Hofstede, 1980, p. 218), such as implementing and using new computer-based communication media.

Individualism and collectivism also are related to concepts of low-context and high-context cultures (Gudykunst and Kim, 1992). In *high-context cultures*, most of the content of messages is embedded in the communicative context rather than the denotative content, in nonverbal cues, in relational cues and group membership, or in indirect messages. Meaning is often internalized, creating a greater responsibility for the receiver to intuit the appropriate interpretation. These differences would imply that people from collectivist cultures might prefer richer media and possibly interpret situations as being more equivocal, unless the situations place the responsibility for interpretation on the receiver (such as when the receiver is a subordinate). Rice et al. (1998) found that approximately 400 respondents from two "collectivist" countries rated the telephone as less rich, and the business memo as richer, than did respondents from two "individualist" countries, but there were no significant differences in evaluations of e-mail, meetings, and face-to-face communication, and no differences in relationships between task equivocality and media preferences across the two cultural types.

4.1.3 Intercultural Communication

The communicative adaptation and accommodation necessary in cross-cultural encounters has long been studied in psychology, communication, and business (Hall 1981; Black and Mendenhall 1990; Gudykunst 1991; Triandis, 1994; Giles and Noels 2001). Participants of multinational groups—groups that transcend any single culture and produce new, effective, and efficient systems of interaction—must recognize, empathize with, understand, and address cultural differences, develop a shared vision or superordinate goal, develop mutual respect, and provide feedback in culturally sensitive ways (Adler, 1980; Harris and Moran, 1999).

With the increasing globalization of organizational and technological processes, intercultural communication has become an even greater focus of employee training and design (Galarneault, 2003; Landis et al., 2003) and for participants in global e-learning (Marinetti and Dunn, 2002). Most training in intercultural communication has cognitive, affective, and behavioral components designed to sensitize individuals to their own cultural blinders and to increase awareness of cultural differences. Training modules tend to include a combination of lectures, cultural assimilators, case studies, role-playing, self-awareness and attributional questionnaires, experiential simulations, analyses of popular culture, and live demonstrations of intercultural encounters (Brislin and Yoshida, 1994; Galarneault, 2003). Generally, intercultural communicative competence consists of knowledge about and sensitivity toward cultural differences, awareness of one's own and others cultural responses to interactive sequences, attitudes that promote openness, and basic behavioral and relational skills (Porter and Samovar, 1991; Chen and Starosta, 1997). A Google search of "measures of intercultural competence" yielded 16,100 entries alone!

It is also important in the design and implementation of technologies and quality and safety processes to take account of the native language of participants, not just the language in which interaction is taking place. For example, Choong and Salvendy (1999) have explored the design of icons and computer interfaces for Chinese language users. Bantz (1993) found that although cross-national research teams usually agree on a working language, differences in language competence, comfort in working in a nonnative tongue, and nontransferability of some abstract concepts, sometimes minimized the contributions members could make to the team and strongly affected conflict management and the emergence of norms. Analyzing the discourse of both English and Spanish versions of a meeting between a hotel general manager and 75 workers whose dominant language is Spanish, Banks and Banks (1991) illustrate the many ways in which the translation process not only provides a conduit for information transfer but creates a set of interactive and relational issues based on power and context which influences how messages are interpreted.

Clearly, intercultural communication is not only about the accurate transfer of messages. It entails sense-making and meaning construction as well as active production of interpretive frames.

4.2 Communication Styles, Cognitive Complexity, and Message Design Logics

Communicator style reflects recurring patterns of communication behavior by which one verbally, nonverbally, and paraverbally interacts to signal how literal meaning should be taken, interpreted, filtered, or understood (Norton, 1983). Within the communication literature, 11 communicator styles have been grouped according to two underlying dimensions: nondirective (attentive/friendly) vs. directive (dominant/argumentative) and low energy (relaxed) vs. high energy (dramatic/animated). Communication styles are related to leadership performance, teaching effectiveness, conflict management, decision making, relational development, and gender equity. To the extent that interactants prefer or exhibit specific styles, they may have more or less effective and satisfying communication with each other or interactions with and through systems. Communication styles are also associated with individuals' attitudes toward technologies. For example, a study of two organizations found that people with more relaxed and friendly styles were more likely to adopt and use e-mail. "Individuals with communicator styles that involve taking the time to think about and respond to others ... may find email systems supportive ... because they can consider the others' communication without the pressure to respond immediately or lose conversational turn. Aspects of the friendly style (such as 'habitually acknowledging others' and 'encouraging others') and the attentive style ('repeating' and 'deliberately react') represent aspects of email message responses that could motivate other users to return more messages" (Rice et al., 1992, p. 23).

A highly developed and focused research program related to individual differences and communicative performance is associated with constructivism and *cognitive complexity* (Delia et al., 1982). Early constructivist research demonstrated that people differ in the degree to which they develop complexity in social cognition, which in turn affects their ability to take the perspective of the other, manage interactions and be effective at accomplishing both instrumental and relationship goals. Over the years, cognitive complexity has been associated with the ability to acquire, organize, and integrate social information (Sypher, 1991); the ability to produce sophisticated, behaviorally complex message forms that are both interpersonally sensitive and pragmatically effective; the ability to interpret and comprehend the messages of others (Hynds, 1985); and the ability to manage conversational interactions in a coherent and effective fashion. Several studies (e.g., Sypher et al., 1989; Penley et al., 1991; Zorn, 1991; Zimmermann, 1994) suggest that cognitive complexity is an asset in the work environment. Cognitive complexity has been linked to a person's ability to design effective and persuasive messages that are adapted to different audiences and address multiple goals (O'Keefe et al., 1993).

Humans use three *message design logics* (O'Keefe, 1988). People who employ an *expressive design logic* believe that communication is a little more than a process to express what they feel. They create verbal messages with few goals in mind other than the declaration of fact, emotion or attitude. People who operative within this logic often seem to lack an "edit" function, their communication tends to be socially inappropriate, overly blunt, and may express shockingly personal remarks. O'Keefe (1988) finds that expressive message producers often do not see messages as open to multiple interpretations or able to accomplish multiple goals. In a *conventional design logic*, communication is considered to be a cooperative enterprise based on socially conventional rules and procedures, intended to achieve desired social effects. Within interactions, people employing this logic are quite responsive to the social and normative "demands" and expectations created within the sociotechnical system. In a *rhetorical design logic*, communication is the construction and negotiation of social selves and situations. Instead of altering their communicative actions to fit the situation as a conventional message producer might, rhetorical logic users alter the situation to fit the action they want to perform. Thus, people employing conventional logic react to context, but rhetorical logic users understand that they help create the context to which they then respond. Interactants may use different message design logics. Thus, two expressives with different opinions or goals may find it difficult to connect, while a rhetorical receiver may overestimate the systematicity of an expressive's message, possibly perceiving inconsiderateness and uncooperativeness (O'Keefe et al., 1997, p. 42). Making sense and coordinating action can become even more difficult when design team conflicts lead some expressive members to use only e-mail as a way to avoid interacting with specific others and to attempt to send clear messages, but thereby allowing others with conventional or rhetorical design logics to interpret the messages in multiple ways.

4.3 Communication and Work Roles

Roles comprise the actions and activities assigned to or required or expected of a person or group. People occupy different work and social roles. These may be the familiar *organizational position* or division of labor, such as manager, secretary, shopworker, or systems analyst, whereby duties, goals, and obligations differ because of organizational hierarchy, job tasks, and experience. Roles may involve belonging to different professions—such as engineer, accountant, consultant—or organizations—such as members of a cross-organizational project. Roles may involve different stages of problem solving, such as designer, coder, marketer, implementer, trainer, or user. Or as discussed below, roles may be locations in a communication network such as a gatekeeper or isolate.

People in these different work roles are likely to have different communication needs and media use patterns. There are at least three general reasons for this. First, different positions have different tasks and roles. So, for example, people higher up the organizational hierarchical, especially higher-level managers, have more equivocal activities involving multiple roles, so they engage in much more oral (vs. written) communication—the position affects their communication (Mintzberg, 1980; Rice and Shook, 1990). Mintzberg (1980) identified three categories of 10 interrelated managerial roles: interpersonal, informational, and decisional. *Interpersonal roles* (figurehead, liaison, leader) use authority and status in interpersonal relationships. Managers also perform crucial *informational roles* (monitors, disseminators, spokespeople) pertaining to the flow of non-routine information crucial to decision making. Being an "information nerve center" enables and requires the manager to carry out *decisional roles* (entrepreneur, disturbance handler, resource allocator, negotiator). The organization's environment may foster particular roles (Gibbs, 1994).

Second, people may have different communication skills, abilities, and preferences that are more or less appropriate for types of positions—the person is drawn to or selected by the position because of this fit. Early work on computer displays found that people from different professions or positions tended to prefer different graphical data formats (such as tables, charts, and figures; Ives, 1982). Or consider the issues of communicator style, cognitive complexity, or message design logics describe above.

Third, one's organization and profession may socialize or train the person into appropriate communication behaviors and preferences; the context helps the person adapt to the position (and vice versa). Organizations may well have both underlying common (such as meetings) as well as different (such as e-mail vs. voice mail) relationships between positions and media use patterns (Rice and Shook, 1990).

4.4 Organizational Culture

As suggested above, all communication is embedded in the larger social system that must be considered in any human factors design. *Organizational culture* is often used to mean the shared meanings, patterns of belief, symbols, rituals, and myths that both support and limit organizational participants in defining and responding to situations, fostering a communal identity and shared expectations for generally approved behavior. Generally, these aspects of culture are latent and implicit rather than overt and explicit, although socialization processes, regulations, codes, policies, and manuals may attempt to represent, or maintain, some aspects of culture. Schein (1994) says that the strength of any particular culture is a function of (1) initial convictions of organizational founders, (2) the stability of the group or organization, (3) the variety and intensity of learning experience, and (4) the extent to which the learning process avoids anxiety vs. positively reinforced change. And, of course, there may be multiple subgroup cultures, inside and outside the organization, and individuals may belong to multiple subgroup cultures.

Although most interpretive or critical researchers would dispute that culture can, or should, be measured through surveys, or even managed in any way, there are many reliable multidimensional culture assessments, usually grounded in a specific conceptualization of organizational culture (Glaser et al., 1987; O'Reilly et al., 1991; Cameron and Quinn, 1999). For example, Cooke and Szumal (1993) propose a dozen normative beliefs that are associated with different behavioral expectations, leading to a dozen different cultural styles: affiliative, approval, conventional, dependent, avoidance, oppositional, power, competitive, perfectionist, achievement.

Organizational culture may fundamentally influence how systems are used, or even whether and how they are designed and adopted. Schein (1994) first makes the familiar claim that an innovative culture is necessary for achieving the potential of new information systems (whether new products or services, or new ways of doing things and defining roles) in organizations. But he then argues that the underlying vision of system designers—their culture—significantly affects how systems are conceptualized, designed, and used. For example, systems can be used simply to automate, or to informate (provide information from system processes to users and managers to learn about the processes; Zuboff, 1985), or to transform the organization (work, communication, authority relations). As we have suggested, information and communication technologies can be conceptualized as both media and content of innovation. Thus, they can also be used to help unfreeze organizational cultures to promote more innovative systems, through the characteristics of accessibility, rapidity, simultaneity, presentational flexibility, complexity, system awareness/informating, system and network accountability, teamwork capacity, task-based authority, and self-designing capacity, thus supporting collaborative teamwork and enabling alternative coordination methods (Schein, 1994).

Organizations may have more or less supportive information-sharing cultures, affecting what kinds of systems can even be discussed, how they are designed, what features are available to different users, and how learning about a new system is facilitated or suppressed (Dewhirst, 1970–1971; Pettigrew, 1972; Johnson and Rice, 1987; Rice et al., 1999). These norms affect knowledge sharing, whether the knowledge is personally or organizationally owned and whether the sharing is internal or external (Jarvenpaa and Staples, 2001). Indeed, there is no specific isolated knowledge, but rather, a *situated knowledge web*, as knowledge is embedded in individuals, connections between individuals, and artifacts (Nidumlolu et al., 2001). A "quality management" corporate culture would influence the nature, form, and frequency of communication within an organization and with its customers (Fairhurst, 1993), and both locates and legitimizes measures such as customer satisfaction or error rates.

Perhaps a more fundamentally communication-oriented approach to organizational culture is Lakoff and Johnson's (1980) argument that how we perceive the world, and thus how we behave, is influenced fundamentally by the metaphors we have access to and use (see Stohl, 1995, for a discussion of organizational metaphors). Three main groups of metaphors shape these perceptions and behaviors: structural, orientational, and ontological. For example, orientational metaphors organize concepts relative to each other through spatial or geographical forms (the system is "up"; "downloading" data).

5 GROUP COMMUNICATION

Given their importance and the advantages of groups/teams for organizational effectiveness, it is not surprising that human factors is concerned with how to design and implement effective teams. Rather than focusing on member selection and group composition, we emphasize the communication processes that comprise the group experience. For example, many practitioners utilize specific communication techniques that are designed to control the process and lead groups to more effective interaction. These processes can be supported and improved through *hardware* and *software technologies* (group decision support and communication systems, e-mail, collaborative authoring, etc.). But more important is understanding the range of *social technologies* also available to support and improve group communication (such as among designers, or between designers and users).

5.1 Group Process Losses and Gains

Group communication has both advantages (gains) and disadvantages (losses) compared to dyadic communication (Steiner, 1972; Nunamaker et al., 1991). *Group process gains* include more information (greater access and diversity of knowledge), synergy (fosters different uses of information), more objective evaluations (as others can catch errors), stimulation (greater motivation and encouragement), and learning (via observation of and practice with skilled members).

Group process losses include air time fragmentation (dividing up speaking time); attention blocking (group members are prevented from contributing comments as they are occur, forget or suppress them later in the meeting); cognitive inertia (bias toward one train of thought); conformance pressure (due to politeness norms or reprisal fear); coordination problems (difficult to integrate members' contributions, leading to cycling or premature decisions); domination by some member(s); evaluation apprehension (members criticize and censor their own ideas before stating them); failure to remember (others' contributions, or topic); free-riding/social loafing (as not everyone can or must participate, whereas all can benefit, others conclude that their contributions are dispensable, and reduce their effort); incomplete task analysis; incomplete use of information; information overload (speed, amount, and diversity); production blocking (only one member can speak at a time, blocking others' ideas, either because the others forget their ideas, are spending time rehearsing their own idea, which reduces new ideas, or most commonly, simply prevents other ideas from being discussed because of time limitations); production matching (performance norms develop, so exceptional ideas are suppressed); social facilitation/inhibition (some may facilitate performance if the task is simple but inhibit it if it is complex); and socializing (although some is necessary for effective group maintenance).

Groupthink can be thought of as a group process loss whereby members of highly cohesive groups that place greater value in mutual attraction, or on acceptance by the leader, than on high-quality decisions. This drive for internal group maintenance suppresses alternative positions or information, ignores flaws in solutions, and labels challengers as betraying group norms (Janis, 1982; Whyte, 1989). Groupthink generates an overestimation of the invulnerability and morality of the group, negative stereotypes of out-groups, joint rationalizations, and pressures toward uniformity (via self-censorship, false sense of unanimity, and pressures against challengers). It is easy to imagine how a system design group with considerable history and professional allegiance can fall prey to such a process.

5.2 Communication Skills

Group members should be trained in a wide range of skills, such as initiating, questioning, interpreting, suggesting, facilitating, evaluating, receiving and giving feedback, clarifying, summarizing, closing, active listening, confronting, blocking counterproductive behaviors, modeling, reflecting feelings and supporting, and empathizing (Shockley-Zalabak, 1998). Gouran (1982) develops an agenda for *vigilance* that is designed to prevent symptoms of groupthink from arising. Stevens and Campion (1994) identified five areas (involving 14 different sets of knowledge, skills, and abilities) affecting group member performance: conflict resolution, collaborative problem solving, communication, goal setting and performance management, and planning and task coordination. Group leaders may use, exhibit, or be unaware of a variety of communication tactics, such as authoritarian (block ideas, control process, announce goals, punish others), participative (seek ideas, facilitate, encourage disagreement, seek idea evaluation, verbalize consensus), avoidance (ignore conflict, change subjects, agree with others, refuse responsibility), vision setting (visualize abstract ideas, state desired outcomes, articulate reasons), meaning management (solicit feedback, generate symbolism, manage meaning), trust (communicate constancy, identify values, encourage access), and positive regard (provide support, offer praise, avoid blame, identify challenges) (Schockley-Zalabak, 1998). These skills are needed, in various combinations, in both *group task roles* (problem analysis, idea generation, idea evaluation, vision identification, solution generation, solution implementation, goal setting, agenda making, discussion clarification, disagreement identification, and consensus identification) and *group maintenance roles* (group participation, group climate, conflict management). Several group social technologies are discussed below.

5.3 Brainstorming

Both users and designers have a wide range of potential contributions, ideas, and contingencies that are relevant to developing and implementing a new system, but group process losses and poor group communication skills may suppress most of these. *Brainstorming* is designed to stimulate many, and diverse, ideas for consideration. The goal here is simply to generate as many ideas as possible as quickly as possible, without any evaluation or regard for feasibility. Group members are encouraged to offer ideas during a short period of time—with no editing, evaluation, or comment by any member—which the group facilitator records. In subsequent group processes, this rich resource can be grouped, evaluated, ranked/prioritized, and allocated to committees for elaboration (Moore, 1994).

5.4 Nominal Group Technique

Extending the brainstorming technique, but avoiding public attributions or hesitancy, here participants are asked to generate silently and write down as many ideas as possible on the particular topic or problem. Then each person is asked, in turn, to contribute one (new) idea, which a facilitator records and may combine into related ideas. After all the ideas are recorded, the group then clarifies and discusses each idea, one at a time. Then the ideas are prioritized through ranking or voting (such as each member ranking the five most important ideas), possibly initially for indicating acceptance of the ideas. The final prioritized ideas are then the basis for further action (Wilson, 2002).

5.5 Delphi Technique

The *Delphi technique* elicits experts' views (about relevant criteria for a problem, or forecasting predictions, threats, or trends), through anonymous, asynchronous responses (now commonly done via Web pages or e-mail surveys) in several rounds until general consensus is reached (Linstone and Turoff, 1975). Communication among the experts is intentionally anonymous in order to encourage openness and equality, avoid group dynamics and individual domination, and reduce performance anxiety and the importance of oral communication skills. A summary of the results (means and standard deviations) for each question from the prior round, showing the distribution of opinions, is provided to identify areas of agreement and disagreement, which helps participants to reassess their own opinions, fostering a convergence to consensus. The *nominal group technique* is related, but the face-to-face group meeting usually involves just one round of anonymous voting.

5.6 Functional Theory of Group Performance

Besides particular techniques designed to improve group effectiveness, there are several theoretical frameworks that are useful for understanding and improving group performance. *Functional theory* (Gouran and Hirokawa, 1983) focuses on the systematic procedures that groups use to accomplish their tasks. The basic premise is that communication serves task functions, and the accomplishment of those functions should be associated with effective group problem solving and decisions. The general model is based on four decision-making functions: (1) the problem must be understood, (2) the requirements for the decision must be assessed, (3) the positive alternatives to the choice must also be assessed, and (4) the negative consequences must be analyzed. Communication agendas or problem management sequences (Gouran and Hirokawa, 1983) are designed to make sure that group members' communication fulfills the needed functions so that they can choose an alternative based on the group's evaluation. Promotive communication calls the group's attention to the performance of a functional requisite. Counteractive communication refocuses a group's attention on a functional requisite after it has deviated away from it. A functional analysis can also be used to determine if the group's decision is "faulty" or inappropriate. There are five communication factors that can potentially lead a group to a low-quality decision: (1) the improper assessment of a choice-making solution, (2) the establishment of inappropriate goals and objectives, (3) the improper assessment of positive and negative qualities associated with the various alternatives, (4) the establishment of a flawed information base, and (5) faulty reasoning based on the group's information base.

Other communication theorists suggest that the assumptions on which functional theory rests limit their usefulness in understanding and improving group processes. For example, Stohl and Holmes (1993) argue that the theory needs to be reframed in ways that more fully articulate important features of bona fide groups (Putnam and Stohl, 1996) and thereby reflect the intricate and interlaced texture of collaborative work. For example, functional approaches take for granted that decision quality is an objective characteristic or attribute that is apparent at the time of the production of the decision, yet standards of effectiveness may depend on who does the evaluation and when the evaluation is done. Furthermore, the requisite task functions are only those that accomplish logical, rational group decisions. Yet communication functions to situate and embed a group and its definition of its task within its context as well as to create an interactive climate in which group cohesion may or may not develop. Hence, there are institutional, historical, and maintenance functions that also must be considered.

5.7 Adaptive Structuration

Poole and his colleagues (Seibold and Meyers, 1986; Poole and DeSanctis, 1990; DeSanctis and Poole, 1994) have applied *structuration theory* (Giddens, 1984) to group decision making in computer-mediated contexts. This work provides a significantly different perspective on group interaction. In structuration theory, a system is the social entity that gives rise to observable patterns of relations. Structures are the rules and resources that actors use to generate and sustain the system and are institutionalized by human action: "They are both the medium and the outcome of action. Interaction patterns—human action—become structural properties through repeated, habitual action, which are then referred to or applied through subsequent action. Structuration is the process by which systems are produced and reproduced through members' use of rules and resources" (p. 117). Structures include procedural structures (such as computerized group support systems), argument structures, and decision rules. This research program has been able to identify the ways in which structures emerge, are appropriated and used by groups, and explore the ways in which structures help guide, constrain, and enable effective group action. Several schemes for identifying structuring moves, modalities of structuration, and general types of appropriations have been developed (DeSanctis and Poole, 1994).

Three sources of structures as preexisting conditions form the context in which the technology is implemented, and as such, affect appropriations, which in turn affect decision processes and outcomes. *Technology structures* include the restrictiveness, sophistication, and comprehensiveness of its features as well as the technology's spirit, the general intent of the technology with regard to values and goals. *Task* and *organizational environment* refers to the nature of the task (such as complexity and interdependence) and the organizational setting (such as hierarchy, corporate information, and cultural beliefs). The *group's structure* includes the interaction patterns and decision-making processes of its members. *Appropriations*, which may be subtle and difficult to observe, are the immediate, visible actions that evidence deeper structuration processes. Appropriations can be more or less faithful (the extent to which appropriations are in line with the technology's spirit), enable more or fewer instrumental uses, follow more or less the "spirit" of the designed features, and reflect more or less the users' attitudes. DeSanctis and Poole (1994) proposed that the more faithful the appropriation, the more likely the team's decision processes will lead to successful outcomes.

Majchrzak et al.'s (2000) analysis of a year-long virtual team designing a new rocket fuel thruster lead to a revised or extended model of structural adaptation. Structures suggested include technology, group, and organizational environment. Appropriation moves lead to decision processes, which in turn lead to (ideally, but not always, positive) outcomes. In addition, however, the effect of preexisting structures on appropriation moves is not direct, but is instead mediated by three factors: the degree of misalignment, the malleability of the structures, and the occurrence of discrepant events. However, these discrepant events are not necessarily discontinuous but, rather, occur potentially continuously over the life of an adaptation process (depending on the size, cost, and time frame). Moreover, the discrepant events do not necessarily result from preexisting structures, but may instead, arise from emerging events. The discrepant events can lead to increased misalignments instead of a necessarily gradual reduction in misalignments. Emergent structures are likely to occur, but these emergent structures may themselves create new discrepant events. Any of these structures—technology, organizational environment, or group—are inherently able to change in this structuration process; one should not be seen as necessarily any more constraining than another, although in a particular context any particular structure's malleability may be restricted.

Considering human factors, then, attitudes toward and uses of current organizational systems become structured by acceptable norms, evaluations, and resources; attitudes and users are mediated by the systems themselves; and these structures in turn constrain or facilitate interpretations and uses of new systems (Orlikowski, 1992). The interpretations of new systems are constrained by earlier interpretations, perhaps by exaggeration or misunderstanding of its potential characteristics, comparisons to media artifacts, even by rationales for design choices that are now unknown by new users (such as reduced labor costs, a visionary supervisor, or strategic initiatives; see Johnson and Rice, 1987; Rice and Gattiker, 2000).

6 COMMUNICATION NETWORKS AND COLLABORATION

Networks are patterns of relations among entities within a system, considered by many to be the quintessential organizational form of the twenty-first century and the embodiment of flexibility, responsiveness, and efficiency. Network analysis is an important tool for enhancing human performance. A "new science of networks" has been heralded within the biological, physical, and communication sciences (Watts, 1999; Barabasi, 2002; Buchanan, 2002; Monge and Contractor, 2003) insofar as it has uncovered an underlying mathematical dynamic of interconnectedness, a common architecture of shared deep structural properties that strongly influence how we think and how we organize. The overall structure of a network, the emergence of new linkages, the relationships among the network members, and the location of a member within the network are critical factors in understanding and enhancing access to key resources, the distribution of social and organizational power, the spread of new ideas, the development and utilization of expertise and knowledge, and the identifications and motivation of workers (Miles and Snow, 1986; Monge and Contractor, 2003). The emergent structures of egocentric and organizational networks also strongly influence sense-making and interpretive processes (Stohl, 1995). For example, when people are enmeshed in a highly interconnected network, they tend to receive the same information and reading of a social situation over and over again, reinforcing the view that there is only one correct way to interpret messages and events. In contrast, when a person has a number of *weak ties* (i.e., people who they may not often communicate with and are not linked to many others in their own network), they often receive new and unique information (Granovetter, 1973). The diverse set of opinions, rationales, and positions suggest that there is more than one view of the situation. Weak ties can produce a more complex and complete view of information and messages.

6.1 Semantic Information Distance

Researchers in organizational communication have long been aware of the gap in understanding and/or information among specified homogeneous groups in organizations such as between management and labor, supervisory and hourly workers, rank-and-file union members, and union headquarters staff. Tompkins (1962) termed this disparity of interpretation *semantic information distance*. Many scholars believe that homogeneity of attitudes and interpretations which produces serious differences in understanding is a logical by-product of cohesive groups because as members interact frequently with each other, they socialize one another, creating strong group norms, reinforcing like-minded opinions and conforming interpretations while challenging deviant opinions. Thus, designers see the world differently from hourly workers because each group is enmeshed in a very different set of network links and share their experiences. Other researchers suggest that people who are in the same types of groups (e.g., upper management, union) develop similar interpretations, not necessarily because they talk to one another but because they are *structurally equivalent* (Burt, 1983). That is, they develop similar behaviors and attitudes because they tend to interact with the same types of people in the same manner in the same types of context, so even though their networks may not include the same people, they include links in the same position with similar configurations and hence are similarly socialized by others.

6.2 Communication Network Nodes, Links, and Properties

Network nodes may be people, organizations, words, systems, events, and so on; the relations may be communication, trade, task interdependencies, system features, hierarchies, or others. The strength of such relations among the entities might be measured by frequency, attraction, length, dependency, or other entities. Typical organizational communication network roles include being a member of a group (or clique), the liaison (who connects groups but is not a member of any group), the bridge (who belongs to one group but provides a direct link to another group; this may include the *gatekeeper*), the isolate (who does not belong to any particular group), the opinion leader (to whom others turn for leadership and legitimization of group norms), and the boundary spanner, environmental scanner, or cosmopolite (who provides a link between the organization and the environment). Other roles include the broker (who passes information or resources along), a follower (who provides links to but not from others), a leader (who receives links from but may not provide links to other), and *structurally equivalent actors* (people who have relations similar to others in the network).

Network analysis can also characterize network properties of dyads (such as reciprocity and similarity in the network), triads (such as transitiveness, the extent to which communication between project leader A and software engineer B, and a relation between A and software engineer C, also involves a relation between B and C), the network as a whole (overall density, centrality, integrativeness, power/prestige, reciprocity, transitiveness), and other measures of structure, as well as cliques or positions within the network. Each of these variables provides important information regarding the capacity of the system to develop, process, and utilize information, perform tasks efficiently, safely, and effectively, and respond to disruption and nonroutine events. For example, Krikorian et al. (1997) studied a satellite manufacturer as they attempted to improve performance by "reengineering" five processes generating specific problems: "redo" processes, non-value-added processes, splintered processes, lack of standardized metrics, and treating customers as outsiders. An analysis of the emergent technical, social, and organizational networks showed that these apparently separate problems were in fact quite interrelated; so any one isolated improvement might create further difficulties for one or more other processes.

6.3 Collaboration, Transactive Memory, and Knowledge Networks

Successful and effective work, especially research and development, requires obtaining, being exposed to, and interpreting information from and about others and about orientation contexts, whether known or unknown, intended or unintended, through multiple channels (Churchill et al., 2004). Software development especially requires coordination (Kraut and Streeter, 1995). Software development involves challenges of the scale of large projects (leading to a multiplicity of actors and their associated expertise, physical obstacles, problems of ownership, and unwillingness to trade information), the uncertainty in designing new products or software (identifying and solving errors, nonexistent or unavailable information, changing requirements, competing schedules), interdependence (because of the tight integration of project components), and the need for informal communication (to support spontaneous information exchange and context development, often provided inadequately via technical tools or formal procedures).

Thus, spaces and media should foster both awareness of, and spontaneous interaction among, both co-located and distributed users. Awareness (both direct and peripheral), notification, and spontaneous interaction media have received considerable, although not well-diffused research (Krauss and Fussell, 1990; Dourish and Bellotti, 1992; Jang et al., 2002; Kraut et al., 2003; O'Hara et al., 2003). Systems and interfaces must support awareness without being too explicit about it, provide summaries compiled relative to each user's "observation rhythm," indicate when something unusual happens that requires the user's attention, or allow overviews of past and current activities and participants (Prinz et al., 2004). These activities could be folders, Web sites, workspaces, procedures, sequences, documents, tasks, and so on.

In designing high-performing systems, it is important to build knowledge assets and to make sure that this knowledge is effectively identified, distributed, shared, and used. Contractor et al. (2002) have found that communication among group members not only provides the starting point for learning about others' expertise, but network dynamics are the basis for coordinating who will learn what. Communication increases efficiency and reduces the redundancy of knowledge in the *transactive memory system* (Hollingshead, 1998).

Combining the tools of *network analysis* with their theory of transactive memory, they are able to predict how actors in a knowledge network acquire and select knowledge, based on what they think others know. For example, they identified knowledge brokers as those actors with high "betweenness" scores. In addition, they suggest that an actor whose cognitive knowledge network maps accurately onto the observable knowledge network is more likely to be identified as the one "who knows who knows what." One knowledge network system is IKNOW (Inquiring Knowledge Networks on the Web), a Web-based e-solution created to help systems manage their knowledge assets (Contractor et al., 1998). IKNOW "helps an organization by putting in place a mapping, visualization, and measurement system that can help organizations to study the patterns of knowledge and information flow though an organization's informal network. IKNOW will answer the following: Who knows who? Who knows what? Who knows who knows who? Who knows who knows?" (http://www.spcomm.uiuc.edu/Projects/TECLAB/IKNOW).

7 COMMUNICATION ENVIRONMENTS AND INNOVATION

7.1 Assessing Communication Environments

As organizations are systems of interdependent subsystems—in Weick's terms, organizing processes constituted through interaction—research shows clear relationships between appropriate organizational communication and performance, innovation, and working conditions. These relations operate through information sharing, participation and commitment, seeking suggestions, delegation, connectedness and social support, fewer worker disputes, more positive organizational climate, communication satisfaction, job satisfaction, better understanding of work, and increased certainty (Hargie and Tourish, 2000). Thus, organizational designers (i.e., designers of systems, the social groups designing systems, and managers of relationships with users) must understand, assess, and improve organizational communication.

There are well-established and widely used surveys to assess these organizational concepts. Components of communication satisfaction may include climate, relation to superiors, organizational integration, media quality, horizontal relations, organizational perspective, relation with subordinates, personal feedback, job satisfaction, productivity, managerial communication competence and style, conflict communication, group and team communication, communication load, mentoring, organizational commitment, organizational identity, subordinate trust for supervisor, gatekeeping, job characteristics, leadership style and development, and so on (Roberts and O'Reilly, 1979; Mills et al., 1988; Downs 1994b).

They may be integrated as part of what is usually called an *organizational communication audit* (Downs, 1994a; Hargie and Tourish, 2000). Organizational communication concepts, especially audits, are part of a systematic approach to constructing both internal and external communications strategies, and require rigorous evaluation of all steps (Hargie and Tourish, 2000). Goldhaber and Rogers's (1979; Goldhaber, 2002) approach includes amount of information received, amount of information sent, amount of follow-up, amount received from a range of sources, amount received through channels, timeliness of information, organizational relations, and satisfaction with organizational outcomes. Other audits include different sets of such concepts (Wiio, in Downs, 1994a; Roberts and O'Reilly, 1979). Such surveys may be administered regularly, with results reported back in aggregated form (such as by department), with open discussions of the implications of the results and how to improve the domains.

There are surveys designed specifically to assess internal user satisfaction with information systems (Ives et al., 1983) (including dimensions of satisfaction with IT staff and services, information product, vendor support, information product, and knowledge or involvement) and of task-oriented uses of Web interfaces (D'Ambra and Rice, 2001) (including dimensions of training, resources for interests, finding information, avoiding shopping costs, finding hard-to-locate information, entertainment, social influence, identity control, and use control).

External communication with users should also be conceptualized as part of the system design and implementation process. System designers need to identify and assess attitudes, needs, and expectations both as part of initial *user needs assessments* (Whyte et al., 1997) or even marketing, but also as ongoing formative evaluation, because the interpretation of any particular system, as well as the users' attitudes and needs, will change over time. A central principle of the customer relationship management literature is that "every contact point is a communication opportunity," so a broader conceptualization of a system would include interfaces and processes for fostering and maintaining ongoing interaction and feedback with users.

7.2 Communication of Innovations and Critical Mass

The subject of human factors generally involves the design or creation of some new technical artifact, or at least new features of a preexisting artifact. In this sense, human factors involves developing and diffusing innovations. However, typical conceptualization of system implementation ends once the innovation has entered the marketplace: that is, has left the design stage and has entered the production, marketing, and distribution phase. Occasionally, fitting the new system into its organizational setting (configuring, networking, adjusting) may also be considered part of this process. However, the diffusion of an innovation is a significantly more extensive process.

The diffusion of an innovation is the spread of a product, process, or idea perceived as new, through interpersonal and mediated communication channels, among the members of a social system, over time (Rogers, 1996). Innovations can be a new product or output, a new process or way of doing something, or a new idea or concept.

7.2.1 Diffusion Processes

Generally, the diffusion, or cumulative adoption, of an innovation over time follows an S curve: that is, growing slowing initially, then accumulating quickly, then flattening out as the maximum level of adoption is reached. Portions of this diffusion curve (i.e., standard deviations of the normal curve) can be characterized as types of adopters: innovators, early adopters, early majority, late majority, and laggards. Innovators and early adopters are usually distinguished by high levels of innovativeness and greater communication, education, and income, among other factors.

Diffusion and adoption can be measured in a variety of ways, such as number or percent of adopters at a certain time, number or percent of organizational units adopting, average duration of usage, number of innovation components adopted, number of units sold or implemented, level of system usage (such as number of log-ons, message sent, files stored, records processed), level of satisfaction, acceptance, diversity of planned uses, and number of new uses.

Crucial to diffusion of interactive communication innovations, such as an organizational intranet, is the achievement of a "critical mass," the number of adopters sufficient to foster sustained adoption beyond that point (Markus, 1987). This is because the value of the overall system grows exponentially as each additional user adopts, uses, and contributes to the system [the number of directed potential relations for N users is $N(N-1)$, and the more contributions, the greater likelihood that one will be useful to another user], so that later adopters perceive and obtain much greater value than do early adopters. Further, with communication innovations, there are typically competing channels already in place, so that before critical mass is achieved, early adopters have to use multiple channels, while nonadopters, or late adopters, have to choose only one of the competing channels. Thus, it is important to provide early adopters with extra incentives, or to target clusters of early adopters who have special needs for, or who can gain particular benefits from, the new innovation. Unless critical mass is achieved early, the new communication channel will probably falter, and similar new systems may have to compete for users (Kraut et al., 1998).

Positive feedbacks, positive network externalities, and complementarities are benefits associated with innovations that accrue to later adopters rather than early adopters, benefits that increase the value of early versus later innovations, or services and other innovations that arise due to the success and features of an earlier innovation (Arthur, 1990). For example, the Microsoft Windows operating system has extensive positive externalities because, since it is the dominant personal computer operating system, most other companies design their software applications for use under Windows. This, in turn, raises the value and market centrality of Windows. Thus, initial adoption patterns can heavily constrain or influence later diffusion, often institutionalizing initial innovations that are in fact less technologically or socially innovative or effective.

An intriguing extension of critical mass is the concept of adoption thresholds (Valente, 1995). The idea here is that each person has a (possibly variable) threshold for adopting a particular innovation. From a social and critical mass perspective, innovators have low thresholds: They may have sufficient resources, high innovativeness, unique relative advantages, and a low need for social influence. Later adopters have higher thresholds, but as more and more innovators adopt an innovation, the innovation becomes more commodified and there are greater social pressures to adopt, so these higher thresholds are more likely to be met. The implication here is that innovation implementers must be able to identify those with low initial thresholds and enable those to communicate soon after with those having slightly higher thresholds; also, different features need to be designed for earlier vs. later adopters, so as to lower thresholds more quickly.

Another time-based factor in the diffusion process is the "chasm" between early and later innovation design and adoption (Norman, 1998; Moore, 2002). Initial development of an innovation tends to be technology-driven, as widespread uses and critical mass have not yet been established. Here, developers attempt to design sufficient performance, features, and quality to satisfy early adopters, who are often willing to pay more (and become initial subscribers), and to tolerate poorer design, in return for new technological features and the status of innovators and early adopters. However, early and late majority adopters are not typically interested in the technological aspects but are more concerned about relative advantage, compatibility, and low complexity. Thus, the technology itself is not perceived as important; rather, usable devices, commodities, services, and content become more valued. The challenge for the developer and implementer, then, is to cross this chasm, knowing when to emphasize technology and when to emphasize the general marketplace.

There are several interim stages in a person's adoption decision process. These include knowledge or awareness of the innovation, persuasion (reactions to and evaluations of the innovation), decision (to obtain, purchase, try out), implementation (acquiring, adjusting, applying, including a "fair trial" period), and confirmation (including public display of the adoption and recommending the innovation to others). Within organizations, there are five major stages as well. These include agenda setting (a general definition of the initial rationale or problem statement, which may be more or less "rational" or well-informed), matching (alternative solutions are identified, evaluated, and compared to the agenda), redefining (the innovation's attributes are defined relative to the organization's needs, but the alternative solutions may also lead to recasting the initial agenda), structuring and interconnecting (where elements of the current social system and/or the innovation are redesigned to integrate the innovation within appropriate procedures and processes, through both formal and informal negotiations and peer pressure), and routinization (where the innovation becomes a part of normal organizational operations). Initial understandings of a new system and its social applications may set an agenda that is then used as the criterion for design and implementation but which may, in fact, hinder ongoing design innovation and innovative uses (Johnson and Rice, 1987).

There are, of course, many other factors influencing the success, failure, or rate of diffusion of an organizational innovation. These include the justification for the initial agenda rationale; the geographic location and closeness of potential adopters; the complexity, size, and culture of the organization (decentralized, small organizations may be much better at initiating innovations, whereas centralized, large organizations may be more successful at implementing them); the personalities and power bases of the organizational actors; changes in political agendas, resources, and goals that affect the nature and evaluation of the innovation; different stakeholders becoming activated by different stages in the life cycle of the innovation; external organizational environments, including changing competitors, regulatory environments, and economic resources; and technological changes, rendering a current innovation incompatible or inappropriate (O'Callaghan, 1998). Because of these several stages in the individual and organizational adoption process, and the wide and complex range of factors affecting diffusion, an innovation may not be rejected initially but still may be discontinued at any stage of the diffusion process.

7.2.2 Innovation Attributes

Generally, potential adopters assess five main attributes of an innovation. Relative advantage is the extent to which the innovation provides greater benefits, and/or fewer costs, than the current product or process. *Compatibility* is the extent to which the innovation fits in with existing habits, norms, procedures, and technical standards. *Trialability* is the extent to which potential adopters can try out components of an innovation instead of the entire innovation, or can try out the innovation through pilot demonstrations or trial periods but decide to return to their prior conditions without great cost. *Complexity* is the extent to which potential adopters perceive the innovation as difficult to understand or use. Finally, *observability* or *communicability* is the extent to which potential adopters can observe or find out about the properties and benefits of the innovation. Every innovation has positive and negative aspects of each of these attributes.

Consider, for example, electronic mail. Clearly, a general critical mass of users has been achieved, especially within communities of certain online information services, but certain subgroups have low overall levels of adoption so would not experience critical mass. E-mail has relative advantages over a face-to-face interaction because one can send the message at any time, regardless of where the other person is or how difficult it would be to meet with them. To many people, e-mail is still somewhat incompatible with traditional social norms such as sending holiday greetings but is highly compatible with other work and computer applications. With trial subscriptions or even free e-mail now offered, it is relatively inexpensive to try out electronic messaging, but one still has to have purchased a modem and communication software. Regardless of how simple advertisements make using e-mail appear, the various functions and interconnections with other applications still make e-mail fairly complex to understand and use. It is fairly easy to communicate the basic features, uses, and benefits of e-mail to others, but it might be hard to actually observe some of those benefits, or even one's own e-mail without taking the time and effort to check the e-mail system.

An innovation is not a fixed, static, objective entity. It may be adapted and reinvented. *Reinvention* is the degree to which an innovation is changed by the adopter(s) in the process of adoption and implementation, after its original development. A reinvention may involve a new use or application of an already adopted innovation or an alteration in the innovation to fit a current use. Reinventions may be categorized as to intentionality—planned (intentional) or vicarious (learning by other's mistakes)—and source—reactive (solving a problem generated by the innovation itself) or secondary (solving unintended consequences elsewhere in the organization or innovation due to the reinvention). The four levels of reinvention include unsuccessful adoption, successful adoption, local adaptation, and systemwide adaptation (Johnson and Rice, 1987).

7.2.3 Communication Channels

Communication channels also play an important role in diffusion. As the innovation is a new product, process, or idea, it must be discussed among developers and implementers and communicated to potential adopters in order for them to assess its attributes and decide whether to try out and eventually adopt it, and what the initial rationales might be. Very broadly speaking, mediated communication and interpersonal communication play complementary, but different roles. Electronic mass media channels such as television and radio are useful for raising awareness about the innovation but cannot provide much detail (except for specialty radio programs). They can provide images and brand name identification, helping the attributes of compatibility and observability. Print mass-media channels such as newspapers and magazines (and, to some extent, the Internet) are useful for explaining conceptual and technical details, helping out with the attributes of relative advantage and complexity. New media such as the Web can provide interesting mixtures of image, explanation, and demonstrations, thus also fostering trialability.

Interpersonal communication is especially important in changing opinions and reducing uncertainty about the innovation, as potential adopters turn to credible and important sources to provide firsthand experiences and legitimization of the new idea. Much innovation research shows the significant role that social influence, peer pressure, and social learning—all operating through potential adopters' communication networks and network roles—play in affecting the evaluation of the innovation's attributes, the adoption decision, and applications and reinvention. The *cosmopolite* is a member of a network who travels more, communicates and uses the media more, attends more conferences, and is generally more aware of the external environment than are other members. Thus, the cosmopolite is a valuable source to the social system for innovations. Within particular groups or organizational units, this role may be filled by a *technical gatekeeper*, who seeks out and brings into the group relevant facts and practices, freeing the rest of the group to focus on the group's task but also keeping it informed of innovative ideas. Within a social system, the *opinion leader* plays the valuable role of evaluating and legitimizing new ideas, especially normative ideas that fit in with the general social context of the group. The opinion leader must be fairly similar to the rest of the group in order to represent the central norms and values of the group, but tends to be just slightly more educated and experienced, and receives more communication, than the other members (Rogers, 1996). Different types of innovations or social norms may be regulated by different opinion leaders. For example, organizationally distributed task software, online games, and downloaded music would probably be discussed and evaluated by different (if somewhat overlapping) social groupings and opinion leaders. Thus, an important diffusion strategy is to identify the appropriate opinion leader for the type of innovation, communicate the relative advantage and compatibility of the innovation for that social system, and then provide incentives and communication channels for the opinion leader to diffuse the idea to other members.

An example application of combining the diffusion and network approaches to human factors is identifying the extent to which members' use of a new communication medium, such as the Internet, is due to (1) task/individual factors, (2) social/structural influences such as managers' and co-workers' opinions and use of the new medium, (3) the extent to which other resources and systems are connected through the medium (critical mass), and (4) the extent to which other valuable members (both inside and outside the organization) are connected. These kinds of studies have shown that adoption of a new system is influenced by one's co-workers (although also from managers when that person is a role model or opinion leader), is more likely if the medium is perceived as highly innovative or uncertain, if there is a potential critical mass of adopters, involves stronger network influences early on in the process while giving way to more individual-level influences such as task demands, perceived benefits, access, and ease of use (Rice et al.,

1990; Kraut et al., 1998), and is stimulated by greater network density and size (Papa and Papa, 1992).

8 IMPLICATIONS

In this chapter we have discussed six communication issues associated with human factors practice and research: messages, communication artifacts, communication characteristics of participants, group communication, communication networks and collaboration, and communication environments and innovation. The main theme is that communication is an interactive sense-making process constituted by messages and relationships among designers, users, and systems. Human factors must take into account the interpretive processes that produce shared meanings and collective action.

Engleke and Oliver's (2002) discussion of human factors activities in the design process provides a simple framework for suggesting how the communication issues discussed in this chapter pervade human factors issues. They identify five phases of activity, each of which is dependent on effective communication and intersubjective sense-making among designers, users, and machines. During *definitional activities* they suggest that designers must "solicit input from user focus groups to obtain preferences." *Designing device and human interface activities* need to "use metaphors that are familiar to the user." During the *implementation and testing phase* designers should "prepare to use interfaces sketches for early review and input by user groups." During the *refining stage* they must obtain "concrete feedback from users." Test activities include ensuring that "the full range of user expertise is integrated into test procedures." During the *user documentation phase*, the provision of "descriptive text and error messages" is critical.

Suggestions follow for how one or two example communication issues are embedded in each of these five design activity phases, highlighting some of the chapter's central concepts. Other communicative issues will clearly arise as human factors designers and researchers strive to improve the efficiency, effectiveness, quality, and safety of products and processes.

Definitional Activities (Needs Assessment, Preferences)

- *Messages and sense-making.* Support sequences of interactions, and *metacommunication*, between designers and users, and among users, to make sense of what may be an unfamiliar feature or system concept.
- *Artifact as channel and content.* Find out how users may wish to use the system to communicate spontaneously with designers and implementers to maintain ongoing *interpretation* of the system and its possible applications; assess needs for possible changes in work and social interactions.
- *Communication differences.* Understand *cultural differences* in meanings and symbolism of terms and practices, including methods of communicating about these (such as focus groups vs. individual interviews), and of selection, enactment, and retention sequences.
- *Group communication.* For potentially transformative systems, consider the *nominal* or *Delphi technique* to stimulate ideas without risk; identify group process gains and losses in project meetings.
- *Networks, collaboration.* Involve current and probable future stakeholders; identify current *expertise and knowledge networks*.
- *Communication environments and processes.* Identify what innovation attributes are especially salient for this problem, such as compatibility, and the extent of preexisting *critical mass*.

Designing Device and Human Interface Activities

- *Messages and sense-making.* Identify *boundary objects* to support shared knowledge among different design professions.
- *Artifact as channel and content.* Build in support and *feedback* for designer communication, such as commentary, referent links, version archives, mobile participation, and *affordances*.
- *Communication differences.* Understand whether interface (such as searching vs. browsing) supports or biases different *message design logics*.
- *Group communication.* Project groups need to apply *social technologies* for both *task accomplishment* and *group maintenance*; *brainstorming* among designers and users.
- *Networks, collaboration.* Different media use, professional identities, *PERT* processes, and design preferences may create latent *cliques* and pockets of conflict; new systems are social as well as technical artifacts, so users must be involved in design and *assessment*; assess forms, *thresholds*, and simultaneity of *feedback* from the system to the operator.
- *Communication environments and processes.* Designer groups may wish to evaluate their own group communication processes during the project life cycle, such as *groupthink*.

Implementation and Testing Activities

- *Messages and sense-making.* Some *strategic ambiguity* in marketed features and in outcome expectations may be helpful for early users to create their own sense of what is feasible.
- *Artifact as channel and content.* Provide *multiple channels* for users to discuss the new system among themselves and with implementers; seek out problems in task flow and work relations occasioned by the new system.

- *Communication differences.* Different *content or relational communication*, *work roles*, and *organizational and national cultures* may interpret or support implementation differently.
- *Group communication.* Offer *group training* and *mentoring*; identify *group norms* and attitudes that influence interpretation and use of the new system; obtain feedback in culturally sensitive ways.
- *Networks, collaboration.* Seed some systems in *supportive networks* with critical mass; promote *knowledge sharing* among *early and late adopters*.
- *Communication environments and processes.* Prevent early closure by initial innovation agenda; consider costs, benefits, and thresholds of groups of potential adopters.

Refining Activities (Feedback)

- *Messages and sense-making.* Identify where *retained understandings* become rigid and out-of-date; continually assess how both *relational and content communication* from designers and implementers are interpreted by users and clients.
- *Artifact as channel and content.* Provide opportunities for interaction and commentary within modules and components, using different formats; assess whether a new system has negatively affected *social and work networks*, such as through problems of new access or expertise requirements.
- *Communication differences.* Some work roles or cultural members, or some organizational cultures, may suppress explicit criticism, so silence is not an indicator of no problems.
- *Group communication.* Reward emergent *network roles* about ways to improve the system continually; manage negotiations over system revisions.
- *Networks, collaboration.* Seek out and reward negative feedback in order to avoid closed social systems and *dysfunctional systems*; encourage user networks to develop.
- *Communication environments and processes.* Evaluate *structuring processes* beyond initial implementation; identify, evaluate, and diffuse *reinventions* and adaptations; foster *knowledge sharing* among different user groups.

User and Error Documentation Activities

- *Messages and sense-making.* Technical and designer language will probably be highly ambiguous (both types 1 and 2) to users; consider any documentation or error messages as initial stages in *sense-making interactions*, so subject to ongoing editing, revision, reformatting, and updating.
- *Artifact as channel and content.* Avoid relying on a single channel or feature of a system to report or seek help about errors, as that may be the site or source of the error; build in detection of *automation surprises*.
- *Communication differences.* Technical and staff support should be aware of differences in their own *message design logics* and *communication styles*, and those of users, to avoid misinterpreting interactions.
- *Group communication.* New errors and explanations arise through system use over time, because of new users, interactions among system components, and *changed work processes or roles*, so maintain group interactions among designers and users.
- *Networks, collaboration.* Understand what *network positions* are the most helpful and knowledgeable, and what network positions others turn to for system help and guidance; they are often not the formally identified help positions.
- *Communication environments and processes.* Understand the social rationale for some kinds of *errors*, such as *ironic or unfaithful adaptations* of a system, or inadequate *information load* and *low communication satisfaction*.

REFERENCES

Adler, N. (1980), *Cultural Synergy: The Management of Cross-Cultural Organizations*, University Associates, San Diego, CA.

Allen, B. (2004), *Difference Matters: Communicating Social Identity*, Waveland Press, Long Grove, IL.

Allen, T. (1977), *Managing the Flow of Technology*, MIT Press, Cambridge, MA.

Archea, J. (1977), "The Place of Architectural Factors in Behavioral Theories of Privacy," *Journal of Social Issues*, Vol. 33, pp. 116–137.

Argyris, C. (1986), "Skilled Incompetence," *Harvard Business Review*, Vol. 64, pp. 74–79.

Armentrout-Brazee, C., and Mattson, M. (2004), "Clash of Subcultures in On-Gate Communication," in *Tapping Diverse Talent in Aviation: Culture, Gender, and Diversity*, M. A. Turney, Ed., Ashgate, Aldershot, Hampshire, England, pp. 182–198.

Arthur, W. B. (1990), "Positive Feedbacks in the Economy," *Scientific American*, February, pp. 92–99.

Axley, S. (1984), "Managerial and Organizational Communication in Terms of the Conduit Metaphor," *Academy of Management Review*, Vol. 9, pp. 428–437.

Babrow, A. (2001), "Uncertainty, Value, Communication, and Problematic Integration," *Journal of Communication*, Vol. 51, pp. 553–573.

Banks, S., and Banks, A. (1991), "Translation as Problematic Discourse in Organizations," *Journal of Applied Communication Research*, Vol. 3, pp. 223–241.

Bantz, C. (1993), "Cultural Diversity and Group Cross-Cultural Team Research," *Journal of Applied Communication Research*, Vol. 20, pp. 1–19.

Barabasi, A.-L. (2002), *Linked: The New Science of Networks*, Perseus Publishing, Cambridge, MA.

Barker, R. G., and Associates (1978), *Habitats, Environments, and Human Behavior*, Jossey-Bass, San Francisco.

Beniger, J. (1986), *The Control Revolution: Technological and Economic Origins of the Information Society*, Harvard University Press, Cambridge, MA.

Berger, C. R. (1986), "Uncertain Outcome Values in Predicted Relationships: Uncertainty Reduction Theory Then and Now," *Human Communication Research*, Vol. 13, pp. 34–38.

Black, J. S., and Mendenhall, M. E. (1990), "Cross-Cultural Training Effectiveness: A Review and a Theoretical Framework for Future Research," *Academy of Management Review*, Vol. 15, No. 1, pp. 113–136.

Brashers, D. (2001), "Communication and Uncertainty Management," *Journal of Communication*, Vol. 51, pp. 477–497.

Brislin, R. W., and Yoshida, T. (1994), *Intercultural Communication Training: An Introduction*, Sage, Thousand Oaks, CA.

Brown, J., and Duguid, P. (1996), "Organizational Learning and Communities-of-Practice: Toward a Unified View of Working, Learning, and Innovation," in *Organizational Learning*, M. Cohen and L. Sproull, Eds., Sage, Thousand Oaks, CA, pp. 58–82.

Buchanan, M. (2002), *Nexus: Small Worlds and the Groundbreaking Science of Networks*, W.W. Norton, New York.

Bullock, A. (2004), "Communicating in an IIS: Virtual Conferencing," in *Inhabited Information Spaces: Living with Your Data*, D. Snowdon, E. Churchill, and E. Frecon, Eds., Springer-Verlag, London, pp. 115–131.

Burt, R. (1983), "Cohesion Versus Structural Equivalence as a Basis for Network Subgroups," in *Applied Network Analysis*, R. Burt and M. Minor, Eds., Sage, Beverly Hills, CA, pp. 262–282.

Buzzanell, P. M., and Burrell, N. A. (1997), "Family and Workplace Conflict: Examining Metaphorical Conflict Schemas and Expressions Across Context and Sex," *Human Communication Research*, Vol. 24, pp.109–146.

Cameron, K. S., and Quinn, R. E. (1999), *Diagnosing and Changing Organizational Culture*, Prentice-Hall, Upper Saddle River, NJ.

Carlson, J., and Zmud, R. (1999), "Channel Expansion Theory and the Experiential Nature of Media Richness Perceptions," *Academy of Management Journal*, Vol. 42, No. 2, pp. 153–170.

Chalmers, M. (2004), "City: A Mixture of Old and New Media," in *Inhabited Information Spaces: Living with Your Data*, D. Snowdon, E. Churchill, and E. Frecon, Eds., Springer-Verlag, London, pp. 71–88.

Chen, G. M., and Starosta, W. J. (1997), "A Review of the Concept of Intercultural Sensitivity," *Human Communication*, Vol. 1, pp. 1–16.

Choong, Y. Y., and Salvendy, G. (1999), "Implications for Design of Computer Interfaces for Chinese Users in Mainland China," *International Journal of Human–Computer Interaction*, Vol. 11, No. 1, pp. 29–46.

Churchill, E., Snowdon, D., and Frecon, E. (2004), "Inhabited Information Spaces: An Introduction," in *Inhabited Information Spaces: Living with Your Data*, D. Snowdon, E. Churchill, and E. Frecon, Eds., Springer-Verlag, London, pp. 3–8.

Clark, H. (1996), *Using Language*, Cambridge University Press, Cambridge.

Clark, H., and Brennan, S. (1993), "Grounding in Communication," in *Groupware and Computer-Supported Cooperative Work*, R. M. Baecker, Ed., Morgan Kaufmann, San Francisco, pp. 222–233.

Contractor, N., Zink, D., and Chan, M. (1998), "IKNOW: A Tool to Assist and Study the Creation, Maintenance, and Dissolution of Knowledge Networks," in *Community Computing and Support Systems*, T. Ishida, Ed., *Lecture Notes in Computer Science* 1519, Springer-Verlag, Berlin, pp. 201–217.

Contractor, N., Carley, K., Levitt, R., Monge, P., Wasserman, S., Bar, F., Fulk, J., Hollingshead, A., and Kunz, J. (2002), *Co-evolution of Knowledge Networks and 21st Century Organizational Forms: Computational Modeling and Empirical Testing*, NSF Grant: IIS-9980109.

Cooke, R. A., and Szumal, J. L. (1993), "Measuring Normative Beliefs and Shared Behavioral Expectations in Organizations: The Reliability and Validity of the Organizational Culture Inventory," *Psychological Reports*, Vol. 72, No. 3, pp. 1299–1330.

Culnan, M. (1989), "Designing Information Systems to Support Customer Feedback: An Organizational Message System Perspective," in *Proceedings of the 10th International Conference on Information Systems*, J. DeGross, J. Henderson, and B. Konsynski, Eds., Boston, December, pp. 305–313.

Daft, R., and Lengel, R. (1986), "Organizational Information Requirements, Media Richness and Structural Design," *Management Science*, Vol. 32, No. 5, pp. 554–571.

Daft, R., and Weick, K. (1984), "Toward a Model of Organizations as Interpretation Systems," *Academy of Management Review*, Vol. 9, pp. 284–295.

D'Ambra, J., and Rice, R. E. (2001), "Emerging Factors in User Evaluation of the World Wide Web," *Information and Management*, Vol. 38, No. 6, pp. 373–384.

Davis, T. R. V. (1984), "The Influence of the Physical Environment in Offices," *Academy of Management Review*, Vol. 9, No. 2, pp. 271–283.

Delia, J. G., O'Keefe, B. J., and O'Keefe, D. J. (1982), "The Constructivist Approach to Communication," in *Human Communication Theory: Comparative Essays*, F. Dance, Ed., Harper & Row, New York, pp. 89–101.

Dennis, A., Kinney, S., and Hung Y. (1999), "Gender Differences in the Effects of Media Richness," *Small Group Research*, Vol. 30, No. 4, pp. 405–437.

DeSanctis, G., and Poole, M. S. (1994), "Capturing the Complexity in Advanced Technology Use: Adaptive Structuration Theory," *Organization Science*, Vol. 5, No. 2, pp. 121–147.

Dewhirst, H. D. (1970–1971), "Influence of Perceived Information-Sharing Norms on Communication Channel Utilization," *Academy of Management Journal*, Vol. 13–14, pp. 305–315.

Donnellon, A., Gray, B., and Bougon, M. (1986), "Communication, Meaning, and Organized Action," *Administrative Science Quarterly*, Vol. 31, pp. 43–55.

Dourish, P., and Bellotti, V. (1992), "Awareness and Coordination in Shared Workspace," in *CSCW 92: Sharing Perspectives*, J. Turner and R. Kraut, Eds., ACM Press, Toronto, CA, pp. 107–114.

Dourish, P., Adler, A., Bellotti, V., and Henderson, A. (1996), "Your Place or Mine? Learning from Long-Term Use of Audio-Video Communication," *Journal of Computer Supported Co-operative Work*, Vol. 5, No. 1, pp. 33–62.

Downs, C. W. (1994a), "Organizational Communication Audit Questionnaire," in *Communication Research Measures: A Sourcebook*, R. B. Rubin, P. Palmgreen, and H. E. Sypher, Eds., Guilford, New York, pp. 247–253.

Downs, C. W. (1994b), "Communication Satisfaction Questionnaire," in *Communication Research Measures: A Sourcebook*, R. B. Ruben, P. Palmgreen, and H. E. Sypher, Eds., Guilford, New York, pp. 114–119.

Driskill, G. W. (1995), "Managing Cultural Differences: A Rules Analysis in a Bicultural Organization," *The Howard Journal of Communications*, Vol. 5, No. 4, pp. 353–372.

Earley, P. (1993), "East Meets West Meets MidEast: Further Explorations of Collectivistic and Individualistic Work Groups," *Academy of Management Journal*, Vol. 36, No. 2, pp. 319–348.

Ebben, M., Kramarae, C. and Taylor, J., Eds. (1993), *Women, Information Technology, and Scholarship*, University of Illinois Press, Urbana, IL.

Eiff, G., and Mattson, M. (1998), "Moving Toward an Organizational Safety Culture," *Society of Automotive Engineers Transactions: Journal of Aerospace*, Vol. 107, pp. 1310–1327.

Eisenberg, E. (1984), "Ambiguity as Strategy in Organizational Communication," *Communication Monographs*, Vol. 51, pp. 227–242.

Engestrom, Y., Miettinen, R., and Punamaki, R. (1999), *Perspectives on Activity Theory*, Cambridge University Press, Cambridge.

Engleke, C., and Oliver, D. (2002), "Putting Human Factors Engineering into Practice," *Medical Device and Diagnostic Industry*, July; http://www.devicelink.com/mddi/archive/02/07/003.html.

Fairhurst, G. T. (1993), "Echoes of the Vision: When the Rest of the Organization Talks Total Quality," *Management Communication Quarterly*, Vol. 6, No. 4, pp. 331–371.

Fairhurst, G., and Sarr, R. (1996), *The Art of Framing: Managing the Language of Leadership*, Jossey-Bass, San Francisco.

Fedor, D., Rensvold, R., and Adams, S. (1992), "An Investigation of Factors Expected to Affect Feedback Seeking: A Longitudinal Field Study," *Personnel Psychology*, Vol. 45, No. 4, pp. 779–806.

Feldman, M. S. (1987), "Electronic Mail and Weak Ties in Organizations," *Office: Technology and People*, Vol. 3, pp. 83–101.

Fish, R., Kraut, R., Root, R., and Rice, R. E. (1993), "Video as a Technology for Informal Communication," *Communications of the ACM*, Vol. 36, No. 1, pp. 48–61.

Galarneault, S. (2003), "Communicating Complex Connectivity: Global Training for Managers," unpublished dissertation, Purdue University, West Lafayette, IN.

Gibbs, B. (1994), "The Effects of Environment and Technology on Managerial Roles," *Journal of Management*, Vol. 20, No. 3, pp. 581–605.

Gibson, J. (1979), *The Ecological Approach to Visual Perception*, Houghton Mifflin, Boston.

Giddens, A. (1984), *The Constitution of Society: Outline of the Theory of Structuration*, University of California Press, Berkeley, CA.

Giles, H., and Noels, K. (2001), "Communication Accommodation in Intercultural Encounters," in *Readings in Cultural Contexts*, 2nd ed., J. Martin, T. Nakayama, and L. Flores, Eds., Mayfield, Mountain View, CA, pp. 139–149.

Glaser, S. R., Zamanou, S., and Hacker, K. (1987), "Measuring and Interpreting Organizational Culture," *Management Communication Quarterly*, Vol. 1, No. 2, pp. 173–198.

Goldhaber, G. M. (2002), "Communication Audits in the Age of the Internet," *Management Communication Quarterly*, Vol. 15, pp. 451–457.

Goldhaber, G. M., and Rogers, D. (1979), *Auditing Organizational Communication Systems: The ICA Communication Audit*, Kendall-Hunt, Dubuque, IA.

Gouran, D. (1982) *Making Decisions in Groups: Choice and Consequences*, Scott Foresman, Glenview, IL.

Gouran, D., and Hirokawa, R. (1983), "The Role of Communication in Decision-Making Groups: A Functional Perspective," in *Communications in Transition* M. S. Mander, Ed., Praeger, New York, pp. 168–185.

Granovetter, M. (1973), "The Strength of Weak Ties," *American Journal of Sociology*, Vol. 78, pp. 1360–1380.

Griffin, E. (1997), *A First Look at Communication Theory*, McGraw-Hill, New York.

Gudykunst, W. (1991), *Bridging Differences: Effective Intergroup Communication*, Sage, Newbury Park, CA.

Gudykunst, W., and Kim, Y. (1992), *Communicating with Strangers: An Approach to Intercultural Communication*, 2nd ed., McGraw-Hill, New York.

Hall, E. T. (1981), *Beyond Culture*, Anchor Press, Garden City, NY.

Hall, J., Halberstadt, A., and O'Brien, C. (1997), " 'Subordination' and Nonverbal Sensitivity: A Study and Synthesis of Findings Based on Trait Measures," *Sex Roles: A Journal of Research*, Vol. 21, pp. 295–317.

Hargie, O., and Tourish, D., Eds. (2000), *Handbook of Communication Audits for Organisations*, Routledge, London.

Harris, P., and Moran, R. (1999), *Managing Cultural Differences*, 5th ed., Butterworth-Heinemann, Woburn, MA.

Hartson, R. (2003), "Cognitive, Physical, Sensory, and Functional Affordances in Interaction Design," *Behaviour, and Information Technology*, Vol.22, No. 5, pp.315–338.

Herring, S. (1994), "Gender Differences in Computer-Mediated Communication: Bringing Familiar Baggage to the New Frontier," keynote address at Making the Net*Work, American Library Association Annual Convention, Miami, FL, June 27.

Hiltz, S. R., and Turoff, M. (1981), "The Evolution of User Behavior in a Computerized Conferencing System," *Communications of the ACM*, Vol. 24, No. 11, pp. 739–751.

Hofstede, G. (1980), *Culture's Consequences*, Sage, Beverly Hills, CA.

Hofstede, G. (1984). *Culture's Consequences: International Differences in Work-Related Values*. Sage, Beverly Hills, CA.

Hofstede, G. (1991), *Cultures and Organizations: Software of the Mind*, McGraw-Hill, London.

Hofstede, G., and Bond, M. H. (1988). "The Confucius Connection: From Cultural Roots to Economic Growth." *Organizational Dynamics*, Vol. 16, No. 4, pp. 5–21.

Hollingshead, A. B. (1998), "Communication, Learning and Retrieval in Transactive Memory Systems," *Journal of Experimental Social Psychology*, Vol. 34, pp. 423–442.

Huber, G. P. (1982), "Organizational Information Systems: Determinants of Their Performance and Behavior," *Management Science*, Vol. 28, February, pp. 138–155.

Infante, D., Trebring, J., Sheperd, P., and Seeds, D. (1984), "The Relationship of Argumentativeness and Verbal Aggression," *Southern Speech Communication Journal*, Vol. 50, pp. 67–77.

Ives, B. (1982), "Graphical User Interfaces for Business Information Systems," *MIS Quarterly*, special issue, pp. 15–47.

Ives, B., Olson, M. H., and Baroudi, J. J. (1983), "The Measurement of User Information Satisfaction," *Communications of the ACM*, Vol. 26, No. 10, pp. 785–793.

Jang, C., Steinfield, C., and Pfaff, B. (2002), "Virtual Team Awareness and Groupware Support: An Evaluation of the TeamSCOPE System," *International Journal of Human Computer Studies*, Vol. 56, No. 1, pp. 109–126.

Janis, I. L. (1982), *Groupthink*, Houghton Mifflin, Boston.

Jarvenpaa, S., and Staples, D. S. (2001), "Exploring Perceptions of Organizational Ownership of Information and Expertise," *Journal of Management Information Systems*, Vol. 18, No. 1, pp. 151–183.

Johnson, B., and Rice, R. (1987), *Managing Organizational Innovation: The Evolution from Word Processing to Office Information Systems*, Columbia University Press, New York.

Johnson, J. D. (1993), *Organizational Communication Structure*, Ablex, Norwood, NJ.

Krauss, R., and Fussell, S. (1990), "Mutual Knowledge and Communicative Effectiveness," in *Intellectual Teamwork: The Social and Technological Bases of Cooperative Work*, J. Galegher, R. E. Kraut, and C. Egido, Eds., Laurence Erlbaum Associates, Mahwah, NJ, pp. 111–144.

Kraut, R., and Streeter, L. (1995), "Coordination in Software Development," *Communications of the ACM*, Vol. 38, No. 3, pp. 69–81.

Kraut, R., Rice, R. E., Cool, C., and Fish, R. (1998), "Varieties of Social Influence: The Role of Utility and Norms in the Success of a New Communication Medium," *Organization Science*, Vol. 9, No. 4, pp. 437–453.

Kraut, R., Fussell, S. R., and Siegel, J. (2003), "Visual Information as a Conversational Resource in Collaborative Physical Tasks," *Human–Computer Interaction*, Vol. 18, pp. 13–49.

Krikorian, D., Seibold, D., and Goode, P. (1997), "Reengineering at LAC: A Case Study of Emergent Network Processes," in *Case Studies in Organizational Communition*, Vol. 2, *Perspectives on Contemporary Work Life*, B. D. Sypher, Ed., Guilford Press, New York, pp. 129–144.

Lakoff, G., and Johnson M. (1980), "The Metaphorical Structure of the Human Conceptual System," *Cognitive Science*, Vol. 4, No. 2, pp. 195–208.

Landis, D., Bennett, J., and Bennett, M. (2003), *Handbook of Intercultural Training*, 3rd ed., Sage, Thousand Oaks, CA.

Lea, M., and Spears, R. (1995), "Love at First Byte? Building Personal Relationships over Computer Networks," in *Understudied Relationships: Off the Beaten Track*, J. Wood and S. Duck, Eds., Sage, Beverly Hills, CA, pp. 197–233.

Lind, M., and Zmud, R. (1991), "The Influence of a Convergence in Understanding Between Technology Providers and Users on Information Technology Innovativeness," *Organization Science*, Vol. 2, No. 2, pp. 195–209.

Linstone, H., and Turoff, M., Eds. (1975), *The Delphi Method: Techniques and Applications*, Addison-Wesley, Reading, MA.

Majchrzak, A., Rice, R. E., Malhotra, A., King, N., and Ba, S. (2000), "Technology Adaptation: The Case of a Computer-Supported Inter-organizational Virtual Team," *MIS Quarterly*, Vol. 24, No. 4, pp. 569–600.

Marinetti, A., and Dunn, P. (2002), "Cultural Adaptation: A Necessity for Global E-Learning," retrieved March 27 2004, from http://www.learningbites.net/learningbites/main_feature1a.htm.

Markus, M. L. (1987), "Toward a 'Critical Mass' Theory of Interactive Media: Universal Access, Interdependence and Diffusion," *Communication Research*, Vol. 14, pp. 491–511.

Markus, M. L. (1994), "Electronic Mail as the Medium of Managerial Choice," *Organization Science*, Vol. 5, No. 4, pp. 502–527.

Mattson, M., Petrin, D. A., and Young, J. P. (2001), "Integrating Safety in the Aviation System: Interdepartmental Training for Pilots and Maintenance Technicians," *Journal of Air Transportation World Wide*, Vol. 6, pp. 37–64.

McLuhan, M. (1964), *Understanding Media: The Extensions of Man*, McGraw-Hill, New York.

Mees, A. (1986), "Chaos in Feedback Systems," in *Chaos*, A. V. Holden, Ed., Manchester University Press, Manchester, Lancashire, England, pp. 99–110.

Miles, R. E., and Snow C. C. (1986), "Organizations: New Concepts for New Forms," *California Management Review*, Vol. 28, pp. 62–73.

Mills, G. E., Pace, R. W., and Peterson, B. D. (1988), *ANALYSIS in Human Resource Training, and Organization Development*, Addison-Wesley, Reading MA.

Mintzberg, H. (1980), *The Nature of Managerial Work*, Prentice-Hall, Englewood Cliffs, NJ.

Monge, P., and Contractor, N. (2003), *Theories of Communication Networks*, Oxford University Press, Oxford.

Moore, G. (2002), *Crossing the Chasm: Marketing and Selling Hi-Tech Products to Mainstream Customers*, Harper-Collins, New York.

Moore, C. (1994), *Group Techniques for Idea Building*, 2nd ed., Sage, Thousand Oaks, CA.

Nidumlolu, S., Subramani, M., and Aldrich, A. (2001), "Situated Learning and the Situated Knowledge Web: Exploring the Ground Beneath Knowledge Management," *Journal of Management Information Systems*, Vol. 18, No. 1, pp. 115–150.

Nonaka, I., and Takeuchi, H. (1995), *The Knowledge Creating Company*, Oxford University Press, New York.

Norman, D. (1998), *The Invisible Computer: Why Good Products Can Fail, the Personal Computer Is So Complex, and Information Appliances Are The Solution*, MIT Press, Cambridge, MA.

Norton, R. (1983), *Communicator Style: Theory, Applications, and Measures*. Sage, Newbury Park, CA.

Nunamaker, J. F., Dennis, A. R., Valacich, J. S., Vogel, D. R., and George, J. F. (1991), "Electronic Meeting Systems to Support Group Work," *Communications of the ACM*, Vol. 34, No. 7, pp. 40–61.

O'Callaghan, R. (1998), "Technology Diffusion and Organizational Transformation: An Integrative Framework," in *Information Systems Innovation and Diffusion: Issues and Directions*, T. Larsen and E. McGuire, Eds., Idea Group Publishing, Hershey, PA, pp. 390–410.

O'Hara, K., Perry, M., Churchill, E., and Russell, D. (2003), *Public and Situated Displays: Social and Interactional Aspects of Shared Display Technologies*, Kluwer, Dordrecht, The Netherlands.

O'Keefe, B. (1988), "The Logic of Message Design: Individual Differences in Reasoning About Communication," *Communication Monographs*, Vol. 55, pp. 80–103.

O'Keefe, B., Lambert, B., and Lambert, C. (1997), "Conflict and Communication in a Research and Development Unit," in *Case Studies in Organizational Communition*, Vol. 2, *Perspectives on Contemporary Work Life*, B. D. Sypher, Ed., Guilford Press, New York, pp. 31–52.

Olson, G. M., and Olson, J. R. (1997), "Research on Computer Supported Cooperative Work," in *Handbook of Human–Computer Interaction*, M. Helander, T. K. Landauer, and P. Prabhu, Eds., Elsevier, New York, pp. 1433–1457.

O'Reilly, C. A., Chatman, J. A., and Caldwell, D. F. (1991), "People and Organizational Culture: A Profile Comparison Approach to Person–Organization Fit," *Academy of Management Journal*, Vol. 34, No. 3, pp. 487–516.

Orlikowski, W. J. (1992), "The Duality of Technology: Rethinking the Concept of Technology in Organizations," *Organization Science*, Vol. 3, No. 3, pp. 398–427.

Orlikowski, W. J., Yates, J., Okamura, K., and Fujimoto, M. (1995), "Shaping Electronic Communication: The Metastructuring of Technology in the Context of Use," *Organization Science*, Vol. 6, No. 4, pp. 423–443.

Orr, J. (1996), *Talking About Machines: An Ethnography of a Modern Job*, Cornell University Press, Ithaca, NY.

Papa, W., and Papa, M. (1992), "Communication Network Patterns and the Re-invention of New Technology," *Journal of Business Communication*, Vol. 29, No. 1, pp. 41–61.

Penley, L. E., Alexander, E. R., Jernigan, T. E., and Henwood, C. I. (1991), "Communication Abilities of Managers: The Relationship to Performance," *Journal of Management*, Vol. 17, pp. 57–76.

Pettigrew, A. (1972), "Information Control as a Power Resource," *Sociology*, Vol. 6, No. 2, pp. 187–204.

Poole, M. S., and DeSanctis, G. (1990), "Understanding the Use of Group Decision Support Systems: The Theory of Adaptive Structuration," in *Organizations and Communication Technology*, J. Fulk and C. Steinfield, Eds., Sage, Newbury Park, CA, pp. 173–193.

Porter, R., and Samovar, L. (1991), "Basic Principles of Intercultural Communication," in *Intercultural Communication: A Reader*, L. Samovar and R. E. Porter, Eds., Wadsworth, Belmont, CA, pp. 5–22.

Prinz, W., Pankoke-Babatz, U., Grather, W., Gross, T., Kolvenbach, S., and Schafer, L. (2004), "Presenting Activity Information in an Inhabited Information Space," in *Inhabited Information Spaces: Living with Your Data*, D. Snowdon, E. Churchill, and E. Frecon, Eds., Springer-Verlag, London, pp. 181–208.

Putnam, L. (1994), "Productive Conflict: Negotiation as Implicit Coordination," *International Journal of Conflict Management*, Vol. 5, pp. 285–298.

Putnam, L., and Kolb, D. (2000), "Rethinking Negotiation: Feminist Views of Communication and Exchange," in *Rethinking Organizational Communication from Feminist Perspectives*, P. Buzzanell, Ed., Sage, Newbury Park, CA, pp. 76–104.

Putnam, L., and Stohl, C. (1996), "Bona Fide Groups: An Alternative Perspective for Communication and Small Group Decision Making," in *Communication and Group Decision Making*, R. Hirokawa and M. Poole, Eds., Sage, Thousand Oaks, CA, pp. 147–178.

Reddy, M. (1993), "The Conduit Metaphor: A Case of Frame Conflict in Our Language About Language," in *Metaphor and Thought*, 2nd ed., A. Ortony, Ed., Cambridge University Press, Cambridge, pp. 164–201.

Rice, R. E. (1987), "Computer-Mediated Communication and Organizational Innovation," *Journal of Communication*, Vol. 37, No. 4, pp. 65–94.

Rice, R. E. (with Hart, P., Torobin, J., Shook, D., Tyler, J., Svenning, L., and Ruchinskas, J.) (1992), "Task Analyzability, Use of New Media, and Effectiveness: A Multi-site Exploration of Media Richness," *Organization Science*, Vol. 3, No. 4, pp. 475–500.

Rice, R. E., and Aydin, C. (1991), "Attitudes Towards New Organizational Technology: Network Proximity as a Mechanism for Social Information Processing," *Administrative Science Quarterly*, Vol. 36, pp. 219–244.

Rice, R. E., and Gattiker, U. (2000), "New Media and Organizational Structuring," in *New Handbook of Organizational Communication*, F. Jablin and L. Putnam, Eds., Sage, Newbury Park, CA, pp. 544–581.

Rice, R. E., and Shook, D. (1990), "Relationships of Job Categories and Organizational Levels to Use of Communication Channels, Including Electronic Mail: A Meta-analysis and Extension," *Journal of Management Studies*, Vol. 27, No. 2, pp. 195–229.

Rice, R. E., Grant, A., Schmitz, J., and Torobin, J. (1990), "Individual and Network Influences on the Adoption and Perceived Outcomes of Electronic Messaging," *Social Networks*, Vol. 12, No. 1, pp. 27–55.

Rice, R. E., Chang, S.-J., and Torobin, J. (1992), "Communicator Style, Media Use, Organizational Level, and Use and Evaluation of Electronic Messaging," *Management Communication Quarterly*, Vol. 6, No. 1, pp. 3–33.

Rice, R. E., D'Ambra, J., and More, E. (1998), "Cross-Cultural Comparison of Organizational Media Evaluation and Choice," *Journal of Communication*, Vol. 48, No. 3, pp. 3–26.

Rice, R. E., Collins-Jarvis, L., and Zydney-Walker, S. (1999), "Individual and Structural Influences on Information Technology Helping Relationships," *Journal of Applied Communication Research*, Vol. 27, No. 4, pp. 285–303.

Richmond, V., and McCroskey, J. (1998), *Communication: Apprehension, Avoidance, and Effectiveness*, 5th ed., Allyn and Bacon, New York.

Roberts, K. H., and O'Reilly, C. A. (1979), "Some Correlations of Communication Roles in Organizations," *Academy of Management Journal*, Vol. 22, pp. 42–57.

Rogers, E. M. (1996), *Diffusion of Innovations*, 4th ed., Free Press, New York.

Salvendy, G., Ed. (1997), *Handbook of Human Factors and Ergonomics*, 2nd ed., Wiley, New York.

Schein, E. (1994), "Innovative Cultures and Organizations," in *Information Technology and the Corporation of the 1990s: Research Studies*, T. Allen and M. Scott-Morton, Eds., Oxford University Press, New York, pp. 125–146.

Schramm, W. (1963), *The Science of Human Communication*, Basic Books, New York.

Seibold, D., and Meyers, R. A. (1986), "Communication and Influence in Group Decision-Making," in *Communication and Group Decision-Making*, R. Y. Hirokawa, and M. S. Pool, Eds., Sage, Beverly Hills, Ca, pp. 133–155.

Sellen, A., and Harper, R. (2003), *The Myth of the Paperless Office*, MIT Press, Cambridge, MA.

Shannon, C., and Weaver, W. (1949), *The Mathematical Theory of Communication*, University of Illinois Press, Urbana, IL.

Shockley-Zalabak, P. (1998), *Fundamentals of Organizational Communication: Knowledge, Sensitivity, Skills, Values*, 4th ed., Longman, New York.

Snowdon, D., Churchill, E. F., and Frecon, E. (2004), *Inhabited Information Spaces: Living with Your Data*. Springer-Verlag, London.

Spitzberg, B. (1993), "The Dialectics of (In)competence," *Journal of Social and Personal Relationships*, Vol. 19, pp. 137–158.

Stahl, O., and Wallberg, A. (2004), "Using a Pond Metaphor for Information Visualization and Exploration," in *Inhabited Information Spaces: Living with Your Data*, D. Snowdon, E. Churchill, and E. Frecon, Eds., Springer-Verlag, London, pp. 51–68.

Star, S. L. (1993), "Cooperation Without Consensus in Scientific Problem Solving: Dynamics of Closure in Open Systems," in *CSCW: Cooperation or Conflict*, S. Easterbrook, Ed., Springer-Verlag, London, pp. 93–106.

Steiner, I. D. (1972), *Group Process and Productivity*, Academic Press, New York.

Stevens, M. J., and Campion, M. A. (1994), "The Knowledge, Skill, and Ability Requirements for Teamwork: Implications for Human Resource Management," *Journal of Management*, Vol. 20, pp. 503–530.

Stohl, C. (1995), *Organizational Communication: Connectedness in Action*, Sage, Newbury Park, CA.

Stohl, C. (2001), "Globalizing Organizational Communication," in *The New Handbook of Organizational Communication*, F. Jablin and L. Putnam, Eds., Sage, Thousand Oaks, CA, pp. 323–375.

Stohl, C., and Holmes, M. (1993), "A Functional Perspective for Bona Fide Groups," in *Communication Yearbook 16*, S. Deetz, Ed., Sage, Newbury Park, CA, pp. 601–614.

Stohl, C., and Redding, W. (1987), "Messages and Message Exchange Processes," in *Handbook of Organizational Communication: An Interdisciplinary Approach*, F. Jablin, L. Putnam, K. Roberts, and L. Porter, Eds., Sage, Newbury Park, CA, pp. 451–502.

Suchman, L. (1995), "Making Work Visible," *Communications of the ACM*, Vol. 38, No. 9, pp. 56–65.

Sutcliff, T. (1998), "Gender and Communication Styles on the World Wide Web," http://homepages.waymark.net/~bikechic/abstract.html.

Sypher, B. D. (1991), "A Message-Centered Approach to Leadership," in *Communication Yearbook 14*, J. A. Anderson, Ed., Sage, Newbury Park, CA, pp. 547–557.

Sypher, B. D., Bostrom, R. N., and Seibert, J. H. (1989), "Listening, Communication Abilities, and Success at Work," *Journal of Business Communication*, Vol. 26, pp. 293–303.

Tannen, D. (1990), *You Just Don't Understand*, William Morrow, New York.

Tannen, D. (1995), "The Power of Talk: Who Gets Heard and Why," *Harvard Business Review*, Vol. 73, September, pp. 12–17.

Teboul, J., Chen, L., and Fritz, L. (1994). "Communication in multinational organizations in the United States and Western Europe." In *Communicating in Multinational Organizations*, R. Wiseman and R. Shuter, Eds., Sage, Thousand Oaks, CA, Vol. 28, pp. 12–29.

Tompkins, P. (1962), "An Analysis of Communication Between Headquarters and Selected Units of a National Labor Union," unpublished dissertation, Purdue University, Lafayette, IN.

Trevino, L., Lengel, R., and Daft, R. (1987), "Media Symbolism, Media Richness and Media Choice in Organizations: A Symbolic Interactionist Perspective," *Communication Research*, Vol. 14, No. 5, pp. 553–575.

Triandis, H. (1994), *Culture and Social Behavior*, McGraw-Hill, New York.

Walther, J. (1992), "Interpersonal Effects in Computer-Mediated Interaction: A Relational Perspective," *Communication Research*, Vol. 19, No. 1, pp. 52–90.

Want, R., Hopper, A., Falcao, V., and Gibbons, J. (1992), "The Active Badge Location System," *ACM Transactions on Information Systems*, Vol. 10, No. 1, pp. 91–102.

Watts, D. (1999), *Small Worlds: The Dynamics of Networks Between Order and Randomness*, Princeton University Press, Princeton, NJ.

Watzlawick, P., Beavin, J., and Jackson, D. (1967), *Pragmatics of Human Communication*, W.W. Norton, New York.

Weick, K. (1979), *The Social Psychology of Organizing*, 2nd ed., Addison-Wesley, Reading, MA.

Weick, K. (1995), *Sensemaking in Organizations*, Sage, Thousand Oaks, CA.

Westmyer, S., DiCioccio, R., and Rubin, R. (1998), "Appropriateness and Effectiveness of Communication Channels in Competent Interpersonal Communication," *Journal of Communication*, Vol. 48, pp. 27–48.

Whyte, G. (1989), "Groupthink Reconsidered," *Academy of Management Review*, Vol. 14, pp. 40–56.

Whyte, G., Bytheway, A., and Edwards, C. (1997), "Understanding User Perceptions of Information Systems Success," *Journal of Strategic Information Systems*, Vol. 6, pp. 35–68.

Wilson, G. (2002), *Groups in Context: Leadership and Participation in Small Groups*, 6th ed., McGraw-Hill, New York.

Winograd, T., and Flores, F. (1986), *Understanding Computers and Cognition*, Addison-Wesley, Reading, MA.

Woo, D. (2000), *Glass Ceilings and Asian Americans: The New Face of Workplace Barriers*, Altamira Press, Walnut Creek, CA.

Woods, D., Sarter, N., and Billings, C. (1997), "Automation Surprises," in *Handbook of Human Factors and Ergonomics*, 2nd ed., G. Salvendy, Ed., Wiley, New York, pp. 1926–1943.

Yates, J., and Benjamin, R. (1991), "The Past and Present as a Window on the Future," in *The Corporation of the 1990s: Information Technology and Organizational Transformation*, M. S. Scott-Morton, Ed., Oxford University Press, New York, pp. 61–92.

Zalesny, M. D., and Farace, R. V. (1987), "Traditional Versus Open Offices: A Comparison of Sociotechnical, Social Relations, and Symbolic Meaning Perspectives," *Academy of Management Journal*, Vol. 30, No. 2, pp. 240–259.

Zimmermann, S. (1994), "Social Cognition and Evaluations of Health Care Team Communication Effectiveness," *Western Journal of Communication*, Vol. 58, pp. 116–141.

Zmud, R. (1990), "Opportunities for Strategic Information Manipulation Through New Information Technology," in *Organizations and Communication Technology*, J. Fulk and C. Steinfield, Eds., Sage, Newbury Park, CA, pp. 95–116.

Zorn, T. E. (1991), "Construct System Development, Transformational Leadership, and Leadership Messages," *Southern Communication Journal*, Vol. 56, pp.178–193.

Zuboff, S. (1985), "Automate/Informate: The Two Faces of Intelligent Technology," *Organizational Dynamics*, Autumn, pp. 5–18.

CHAPTER 7
CULTURAL ERGONOMICS

Nuray Aykin
The New School
New York, New York

Pia Honold Quaet-Faslem
Siemens AG
Munich, Germany

Allen E. Milewski
Monmouth University
West Long Branch, New Jersey

1 INTERNATIONALIZATION AND LOCALIZATION IN HUMAN FACTORS DESIGN

In his speech at the 2004 International Geographical Union Congress, Microsoft's Tom Edwards explains the recall of 200,000 copies of Windows 95 after India banned the product because it represented the disputed Kashmiri territory in a different shade of green—suggesting visually that it was non-Indian (Best, 2004). Similarly, Microsoft used chanting of the Koran as a soundtrack for a computer game, which greatly offended the Saudi Arabia government.

In our age of instant communication and global markets, every facet of a product is exposed to a wide range of users whose cultural beliefs, history, and practices can differ in striking ways. In addition, these users care about how products are designed. DePalma et al. (1998) reports that visitors linger twice as long at Web sites in their local language as they do at English-only sites. Furthermore, business users are three times more likely to buy when addressed in their language. Sales can suffer if a Web site does not allow the purchaser to buy using local currency. There has been a groundswell in recent years to make technology products usable across the globe, but results are mixed. Eighty percent of European-based corporate Web sites may be multilingual (Dunlop, 1999). But more than 50% of U.S. companies do nothing to localize; less than 25% allow a choice of language on their Web sites, and less than 10% allow purchases to be made with a foreign currency.

For nearly any technology-driven market, the fulcrum for ensuring globally usable products is the human factors designer. It is the designer who must have the expertise, the commitment, and the enthusiasm to carry out a successful globalization process. All three—expertise, commitment, and enthusiasm—are required because globalization is not an easy task. In this chapter we outline the basic issues of globalization in order to introduce them to human factors designers who might be faced with an opportunity to take a product international either currently or in the future.

We describe the terminology and the three main aspects of product globalization: (1) management of internationalization and localization, (2) guidelines to facilitate successful global design, and (3) processes required to efficiently implement global products.

As can be seen by this outline, the human factors designer cannot achieve globalization alone; many other organization members are required, as are sound processes to organize their activities. But it is the designer who has the expertise and often drives the processes. This chapter can provide an initial base of this expertise. The commitment and enthusiasm required can only come from a deep desire to create products that can be used worldwide and an understanding of the cultures for which the product is targeted.

2 TERMINOLOGY

As defined by the Localization Industry Standards Association (LISA),

Globalization refers to the general process of worldwide economic, political, technological, and social integration. More specifically, it is the process of making all the necessary technical, managerial, personnel, marketing, and other enterprise decisions necessary to facilitate localization.

Internationalization is the process of ensuring, at a technical and design level, that a product can be easily localized. It helps define the core content and processes so that they can be modifiable for localization.

Localization is a follow-up process after internationalization and is the process of modifying products or services so that they are usable in, and acceptable by, target cultures.

3 MANAGEMENT OF INTERNATIONALIZATION AND LOCALIZATION

The management of product internationalization and localization is, in many respects, similar to general project management with a modified set of processes to follow. What is really different, however, is the necessity to cooperate in a multinational and multicultural team. Setting up such team requires an investment of substantial time, money, and personal enthusiasm, all of these which are absolutely necessary.

Beu et al. (2000) describe their first experience during the setup of a usability lab in Beijing. "Whereas the German colleagues were focusing on the technical details of the new laboratory, their Chinese counterparts were extremely keen to make contact with as many people as possible at the new site and to forge new contacts both inside and outside the company. From the German point of view, this behavior was initially baffling; the right sequence of events would have been to set up the laboratory first and then establish contacts. From the Chinese viewpoint, building up a relationship and a positive image are enormously important because these relationships form the basis for all future dealings" (p. 355).

It takes time to create a dynamic team to manage the internationalization process when product internationalization is introduced into an organization. This literally requires a culture change in the organization: starting from telling everyone that internationalization is much more than simply a translation to other languages and a few format changes.

3.1 Internationalization and Localization Teams

Aykin (2005) describes the roles of interdisciplinary and intercultural teams in an organization to manage successful product internationalization.

1. *Management team.* The management team consists of all top managers in an organization who are related to any globalization efforts, including in-country managers. Although the management team does not engage in day-to-day operations of globalization efforts, it is very essential that they understand what globalization means and what it takes to move a product into global markets. Without their support, the globalization effort will face many roadblocks, especially if the effort affects the time and cost of deployment. One of the primary reasons that globalization efforts fail is because of management's lack of understanding and therefore failure to promote full globalization processes.
2. *Marketing team.* A truly successful marketing team can move an organization's globalization efforts. Their market analysis can provide crucial data regarding global market trends, customer needs and preferences, and return on investment (ROI). In many cases, the management team is more apt to listen to the marketing team's reports on the importance of designing products for the global markets than to listen to a product designer or human factors engineer. This is the team that any designer should work very closely with, and they should use the power that the marketing team provides.
3. *Implementation team.* Depending on the size of the organization, the implementation team can consist of from a handful of individuals to large teams. The team includes:
   a. *Project managers*: manage the process of internationalization and localization and have strong contacts with the marketing and management teams.
   b. *Human factors engineers/user interface designers*: know how to study user needs in different cultures, design international user interfaces, and localize them according to cultural needs. Engineers and designers need to understand cultural needs, preferences, daily life, learning styles, religion, purchasing habits, customs, ethics, and relationships.
   c. *Software engineers and developers*: define the internationalized software architecture, develop software that is ready to be localized for target cultures, and provide localization for target markets.

d. *Technical writers and translators*: create documents that are ready to be translated at minimal cost, and customize and translate them for the target cultures.

3.2 Factors Affecting Internationalization and Localization Levels

As in any decision making, time to market and cost of development have a significant impact on company's decisions regarding (1) whether or not to go global; (2) if globalization is a "go," selecting target countries, and choosing the methodologies for internationalization and localization; (3) the level of effort to spend on understanding the target market and user needs; and, (4) the degree of internationalization and localization to incorporate into product development. When making the decision to go global, one of the biggest factors that defines the level of effort is the cost. The companies face two types of costs: (1) cost of internationalization, and (2) cost of localization.

3.2.1 Cost of Internationalization

If a company makes an early decision and takes the initial steps well before the product is designed and developed, internationalization cost can be virtually zero. This is achieved by designing the user interface flexibly enough to accommodate cultural differences, including considerations of layout for text expansion, flexibility in writing directions, differences in labeling and font usage, global data formatting, and use of global icons. A single core code base should be provided for all versions of the application, with all user interface elements and other localizable elements in separate resource files. The operating system should be allowed to handle all localizable elements and character encoding to handle target languages. However, if the initial application software plan and design are not internationalized, the first step in the process of internationalization and localization involves a large investment to internationalize the software, which can cost from $60,000 to $800,000, depending on the effort required (Simultrans, 2004).

3.2.2 Cost of Localization

The cost of localization depends on how well the core product is internationalized. Spending a good deal of effort at the beginning to ensure that the core product is internationalized can dramatically reduce the time and effort spent in getting a product localized for different target cultures. The localization cost involves project management, requirements gathering from target locales, design and software modifications to core product, usability evaluation during and after design, software testing (streamlining testing processes for target locales reduces cost and time for each target locale), and translation cost (reducing documentation and using translation memory can save up to 50% for subsequent releases).

4 GLOBAL DESIGN GUIDELINES AND PRACTICES

The second major issue involved in internationalization and localization has to do with the actual design of products. One goal of design is to produce a product that is useful and at the same time satisfies all the expectations and norms of the target cultures. This is not an easy task since it is unlikely that one designer would have extensive experience with all of the cultures for which a product is targeted. Generally, the way this has been handled by global design practitioners is to have guidelines available that list problems and issues that are likely to arise in the course of global design, along with sample solutions that have worked in the past. There are advantages and disadvantages to the use of guidelines in global design. On the negative side, there are simply so many countries and critical differences that the number of guidelines can be huge. Given this, it is sometimes difficult to keep accurate. There are examples of "urban legends" in the internationalization community (i.e., stories and guidelines purported and believed to be true for lengthy periods of time, only later to be exposed as inaccurate) (Aykin, 2005). Perhaps most important, it is true that every design issue for every new product has its own unique characteristics. What has worked for an early international design may not work for a new one. This may be because the new product has a different use or because the users differ or because the environment of use is different, all of which are problems for design in general but are magnified in the global context. As we describe in Section 4, the only true remedy is carefully done but often-costly international usability studies. This points to the overwhelming advantage of guidelines: They are inexpensive and efficient to use. Therefore, the use of global design guidelines is a good choice, especially (1) when used as an initial basis for design, followed by usability studies; and (2) when cost and time factors do not permit further study.

4.1 Internationalization Guidelines for Information Displays

One of the most prominent aspects of many products is its display. Aykin and Milewski (2005) have categorized guidelines according to the mnemonic GLOCL. The mnemonic itself points to the continual tension between global and local requirements in all cross-cultural design projects. The elements of this mnemonic stand for key internationalization problem areas:

- Graphics
- Language
- data Object formatting
- Color
- Layout

Table 1 summarizes some of these most notable guidelines. For a more complete listing, see Aykin and Milewski (2005).

Table 1 GLOCL Guidelines

	Guidelines	Examples	References
Graphics and icons	Use abstract human figures. Consider modest clothing (no bare arms, legs, feet). Avoid hand and body gestures. Leave text out of graphics, or layer the text so that it can be translated without altering the underlying graphics. Use universally recognized objects. Limit use of human, animals, maps, flags, and the directional reading (in case, the language becomes bidirectional).	In Arabic and Hebrew, the images are also read from right to left, which can cause confusion while telling a story in subsequent frames.	Horton (2005)
Language Character sets	There are several character sets to deal with the processing and rendering of languages. The most commonly used ones are ASCII, ISO-8859 Series, Unicode (ISO 10646), UTF-8.	UTF-8 is an addendum to Unicode where characters are represented with 8 bits; provides compatibility with the ASCII set.	Aykin and Milewski (2005)
Fonts	Choose fonts to accommodate target languages, provide enough space for changes in line heights, accommodate underlining. Provide enough space between lines to accommodate changes in line heights, to ensure clear separation between lines, and to accommodate underlining.	Most non-Latin languages require proportional spacing.	Aykin and Milewski (2005)
Text direction	Accommodate bidirectional (left-to-right and right-to-left), and top-to-bottom text in design.	Arabic and Hebrew are bidirectional languages, where Arabic and Hebrew characters are read from right to left, non-Arabic/Hebrew and numbers are read from left to right.	Aykin and Milewski (2005)
Paper size	Select margins to accommodate different paper and envelope sizes (e.g., A4 vs. standard letter). Ensure that printers can accommodate different paper and envelope sizes. ISO 216 provides the international paper sizes.	Not only do U.S. paper and envelope sizes differ from those in Europe, business card sizes also differ.	Kuhn (2003)
Translation	Identify the content in the code to be translated. Provide instructions for the translators, and translation glossaries. Use translation memory to eliminate replicates of the same text translation.	If not marked properly, the code itself or file names may get translated, causing a large problem with perfectly working software.	Microsoft (2002), Hoft (1995)
Abbreviations and acronyms	Provide full text for abbreviations and acronyms Instruct translators on what and what not to translate (some organization names are known worldwide).	Although the United Nations is a well-known organization, many countries still use the translated version of the organization [e.g., BM (Birleşmiş Milletler) in Turkish].	Aykin and Milewski (2005)
Spelling	Same language can have different spelling rules in different countries.	*Internationalisation* in UK English is spelled *internationalization* in U.S. English.	Aykin and Milewski (2005)

Table 1 ***(continued)***

	Guidelines	Examples	References
Text expansion	Layout should be designed to accommodate text expansion in labels and the body text. Can use larger font types for the source language and smaller fonts for the target language (given the high legibility), can increase the margins, or can provide dynamic layouts to accommodate text expansions. Placing labels above the fields, using automatically expanding text objects, use of graphics, and changing the space between lines and paragraphs could help to accommodate text expansions.	Text on short labels can expand up to 400% when translated from English to other languages. Paragraphs over 70 characters can expand an average 40% when translated.	Hoft (1995)
Sorting	Sorting rules should be able to handle accents, character combinations, case differences, non-Latin scripts, and Far-Eastern languages. ISO/IEC 14651:2000 provides default collation orders.	In German, the character ß is sorted as "ss."	Microsoft (2002)
Writing practices	Use short sentences, active prose, formal vocabulary, and present tense, per organizers such as graphics/icons or lists. Avoid joined sentences, indefinite pronouns, personal pronouns, semicolons, use of *may*.	Instead of "You may turn on the computer," use "You can turn on the computer."	GNOME (2003), Hoft (1995)
Terminology	Avoid culture-specific metaphors, acronyms, abbreviations, jokes, humor, and idioms, gender-specific references, colloquial language. Use standard terminology within the target language, while considering terminology differences between countries. Ensure that the translated version retains the technical accuracy.	*Bathroom* in U.S. English means *toilet* in UK English.	Aykin and Milewski (2005)
Data object formatting Date	Although there is an international date notation (ISO 8601), countries still prefer their own accustomed way of representing dates. Ensure that the application software and layout accommodate different date representations. The date-formatting differences include the order of day, month, and year, and the delimiters separating these.	In Saudi Arabia, the date is shown as 1425/04/14, in the United States as 04/14/2004, and in France as 14/04/2004.	Aykin and Milewski (2005)
Time	The international standard notation for time is hh:mm:ss (ISO 8601). However, there are differences between countries on how to represent the hours: AM/PM or 24-hour notation.	In the United States, the time is represented as 2 : 34 : 60 PM, and in most of Europe it is represented as 14 : 34 : 60.	Kuhn (2001)
Calendar/ holidays	Although most of the world uses Gregorian calendar, there exceptions could have an impact on the localization. Other calendars include the Arabic, Jewish, Iranian, and Japanese Imperial calendars.	In the Arabic calendar, the date changes with the sunset, making calculations and conversions complicated.	Aykin and Milewski (2005)

(*continued overleaf*)

Table 1 ***(continued)***

	Guidelines	Examples	References
	It is also important to know the holidays of a target culture, especially if you are conducting studies there and interacting with the local offices. On schedules, the start of the week may change from country to country.	Islamic countries do not celebrate Christmas, so sending Happy Holidays card during Christmas time may not be appropriate. Start of the week is Monday in most European countries, and it is Sunday in the United States.	
Numeric formatting	The elements of numeric formatting include thousands and decimal separator, number of digits between separators, and negativity placement.	123,456, 789.00 in the United States is shown as 12,34,56,789.00 in Hindi. Negative numbers may be indicated by using a minus (–) sign before or after a number, or by enclosing the number in parentheses or brackets.	Aykin and Milewski (2005)
Names and addresses	The name and address formatting can change the layout of forms drastically. The name format can include title, gender, first/given, middle, last/family/surname (even two last names). The address format can include (not necessarily in the same order) street number, building number, street name, city or town, state/province/region, country, zip/postal code. Zip/postal codes can have alphanumeric characters and more than five-digit codes; the fields should be flexible to accommodate these differences.		Aykin and Milewski (2005)
Telephone numbers	Format for telephone numbers varies from country to country: including the total number of digits, separators, grouping of numbers, long-distance access codes, and extensions. Provide flexible labeling and the capability to handle at least 15 digits for entering phone numbers. Allow free format entry, with no separate fields as "area codes." Follow ITU-T recommendation E.164.	Separators used in phone numbering can include hyphens (-), period (.), parentheses [(·)], and spaces. Long-distance access codes could be 001, 011, or 00. Many countries do not have area codes, but have city codes.	Aykin and Milewski (2005)
Currency	Symbols representing currency differ from country to country. Use ISO 4127 three-letter abbreviations of world currencies. Well-known symbols such as the dollar ($) could replace the ISO symbols. The format for placing the monetary symbol differs also ($123.45, €123.45).	Monetary symbol placement: $123.45 in the United States, and 123.45 TL in Turkey. Some countries use parentheses to indicate negative value [United States: ($1,234.56), France: €–1,234.56]	Microsoft (2002)
Measurements	Measurements include distance and weight units (Imperial or metric systems), typographic units (point sizes), temperature (Fahrenheit, Celsius), clothing sizes (European, U.S.).	1 point (Didot) equals 0.3759 mm, 1 point (ATA) equals 0.3514 mm, 1 point (Postcript) equals 0.3527 mm. Women's size 6 in the United States is equivalent size 36 in Europe.	Kuhn (1999)

Table 1 *(continued)*

	Guidelines	Examples	References
Color	Color preferences still exist, not due to their meanings, but due to aesthetic preferences. Although the meanings of colors are becoming less apparent across cultures, avoid prime colors for design purposes. As a rule of thumb, use muted, pastel colors for Asian countries and bright, bold colors for Central America.	Green is the national color for Ireland and a religious color in Islamic cultures, while it means family, harmony, peace, life in Asia. It also means "safe, go" in traffic lights all around the world. Red means good luck, prosperity in China; it means death in Egypt.	Morton (2003), Spartan (1999)
Layout	Ensure flexibility and expansion space, since layout elements vary, including text direction, placement, and alignment on the screen. Use dynamic layout managers instead of fixed-position layout.	East Asian vertical writing requires different layout arrangement and text placement.	Microsoft (2002)

4.1.1 Graphics

Graphics include everything from icons to full-screen illustrations and maps (Horton, 2005). It may seem surprising that graphics can cause problems in global design given the widespread notion that icons and pictures are the road to solving the problem of multiple languages. But although it is true that some graphics have universal meanings and can be used across all interfaces, this is not the case for many. Even when a graphic enjoys worldwide and cross-cultural generality, it is useful only for solving the internationalization part of the global design process. Many cultures have their own preferred way of visually representing things, and an astute designer will also be sensitive to optimizing localization by using the culturally preferred version. As always, decisions about the balance of these two forces have to carefully consider cost and effort. These factors can be reduced if fewer, more generic product versions are marketed.

It is usually not feasible to craft prescriptive graphics design guidelines of the form "express this concept this way in these cultures" or "do not express this concept this way in these cultures" because there are so many cultures and potential concepts. Instead, most guidelines pose more general methods of avoiding problems. Horton (2005), for example, suggests that when representing something for which all cultures have specific versions (e.g., paper currency), it is best to depict a range of examples together in one graphic rather than assuming that all cultures will recognize one example (e.g., the U.S. dollar).

Nonetheless, there is one form of prescriptive guideline that is common. These are the guidelines that say never to use a certain representation under *any* conditions. They usually occur when a particular graphic has so many culturally specific meanings, sometimes both positive and negative, that using it is always dangerous. A well-known example of this is the rule that hand gestures should always be avoided in graphic displays because nearly every hand gesture has a negative meaning someplace. For example, the gesture for "OK" in the United States, can mean "great!" in Lebanon and Germany, and "zero" in China, but can indicate a threat of murder in Tunisia (Aykin, 2003).

4.1.2 Language

Language is pervasive in user interfaces and poses a variety of issues for global design. One of the complexities is that although languages and countries are related, the relationship is not one-to-one. Some countries share the same language. Other countries may have a mix of languages, some being shared with other countries and others not. For example, Belgium and Canada are both multilingual and share French, but not Dutch and English. Finally, even when counties share a language, dialect differences can make a phrase understandable in one country and not in another. Designers have to consider both language *and* country, and there are notable problems when this gets neglected. An example is the use of country flags (e.g., Great Britain flag) as icons to switch a Web site's language (e.g., English). Although not quite so common anymore, this design feature caused confusion for several years.

Language issues in design actually fall into at least two categories: rendering and translating. Guidelines on rendering try to help the designer around the fact that languages differ in such things as font characters, number of characters, and text direction as well as printing standards and keyboard differences. Guidelines focused on language translation highlight issues ranging from how to write in a way that simplifies translation all the way down to how to handle differing standards for spelling and sorting text—which often differ even in countries that use the same language. What is clear is that language has a remarkable effect on the success of international products. One anecdote, for example, describes the difficulties some years ago when a well-known running shoe company named their new product line Incubus, which means "an evil spirit that lies on persons in their sleep" (*Merriam-Webster's Collegiate Dictionary*, 11th ed.). When discovered, the company had to redo all its packaging and advertising.

4.1.3 Data Object Formatting

Data objects such as form fields and cells in a report constitute a language in themselves, and it is not surprising that cultures have developed their own standards and practices for them. Objects such as date, time, numbers, sizes, temperatures, telephone numbers, and currency all differ widely in format across the globe. Names and addresses differ not only in format but also in the content required, for example, for successful postal delivery. In some countries, names include only last names. In others, both paternal and maternal names are standard. Some countries have states or regions, whereas others do not. The designer of a product has a lot to consider to prepare for globalization of these data objects.

The critical impact of these format differences can be felt by considering the scenario of being invited to a global design team meeting on "05/03/04." U.S. participants might attend in May; those from Germany might come in March. It is unlikely, but possible, that Swedish participants would show up a year late!

Data object formatting is one area in which international standards have often been established. The International Standards Organization (ISO) maintains standards for a variety of common metrics, and many global design guidelines are driven by them. Unfortunately, both the people and governments in many countries still prefer their own versions, so additional guidelines are still needed to provide the flexibility to handle these differences and still make their meaning clear.

4.1.4 Color

Color affects the user interface in many ways. It creates an aesthetic quality, which can determine whether or not a product appears professional. Colors also have very specific meanings, and these meanings differ across cultures (Morton, 2003). There are many anecdotes speculating on how the careless use of colors has reduced the effectiveness of products designed without cultural differences in mind. A major computer vendor, for example, had its Japanese Web site's content bordered heavily in black, which is a sign of negativity. A well-known amusement park, designed in the United States but built in Europe, featured the color purple as its signature even though purple is often interpreted as a sign of death in some parts of Europe. It is clear that emotional content has a significant effect on usability (Norman, 2003), so that guidelines about the use of colors in global products can help prevent a designer from inadvertently degrading a customer's experience.

4.1.5 Layout

Layout issues are usually a secondary kind of problem, often resulting from cultural differences in the other areas already described. Any change in graphics, language, or data formatting that is required to globalize a product can have large effects on the layout of the user interface. For example, when translating from English to other languages, text can expand anywhere from 30 to 200%, even reaching 400% for short labels (Hoft, 1995). Without some planning during the design phase, text expansion can ruin an interface. Even color can interact with layout when one considers the aesthetics and legibility of adjacent colors. The goal of layout guidelines is to avoid situations where the layout of the entire user interface must be redone every time the target country changes. With careful design, such redesigns can, in fact, become rare.

5 INTEGRATING INTERNATIONALIZATION IN THE USER-CENTERED DESIGN PROCESS

5.1 Internationalization — Beyond the Surface

Guidelines on internationalization are a good starting point when thinking about the most obvious and visible differences that have to be considered for making products usable globally. But they do not cover differences in what we can conceptualize as "culture": work practices and organizational structures, interests and values, education and preexperience (e.g., with competitor products), infrastructure and tools, all of which have considerable influence on the acceptance of products in a foreign market.

Honold (2000a) conducted in-depth interviews with seven product managers in a global-acting German company. When asked about "critical incidents" with their products on foreign markets, the interviewees reported 27 incidents. Eleven of these incidents covered difficulties in localization or internationalization of user interface elements. Although unpleasant, these incidents were assessed as foreseeable and relatively easy to improve; in the project described they had a minor effect on the overall success. Interestingly, 16 critical incidents were reported on culture-specific user requirements. Here, local work practice and user expectations did not fit the product planned. Interviewees emphasized that the requirements came completely unexpectedly and that quick fixes on the user interface alone often did not help recovery. Products that don't fulfill culture-specific requirements are often failures and are hard to fix.

The success of a project starts in the very early phases, during elicitation of and setting of priorities on user requirements. Here, a good enough understanding of the relationship between work practice, values, roles, organizational structures, and any other aspects that are relevant for the use of systems and products has to be developed.

Beu et al. (2000) give an example of a missing fit of user requirements between German and American users. In a project on automation software, American engineers called for features like "beginner mode" and "easy start function." The German engineers rated these requirements as being of low priority. None of the operators they had talked to in Germany had ever shown interest in such features. They interpreted the

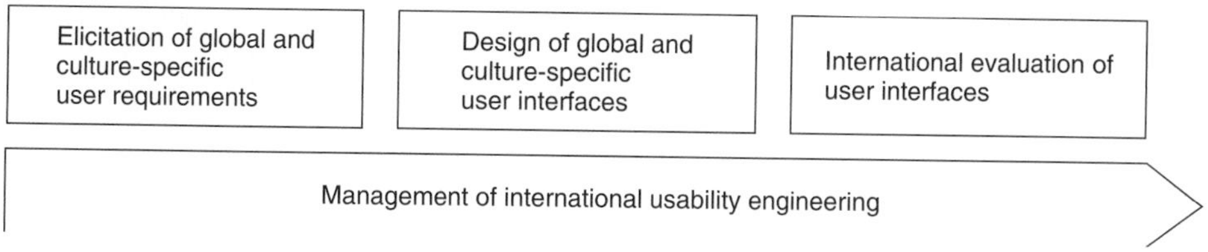

Figure 1 Human factors tasks in the process of internationalization.

requirement as a "display highlights" function, which they expected to be used for marketing reasons only. Nevertheless, after discussions with local user interface designers, basic differences in work practice became apparent. German engineers in the automation industry very often have a formal training background and work in one company for a comparatively long time. They tend to prefer systems that offer a maximum of flexibility, even if it leads to greater complexity and a shallower learning curve. On the other hand, the staff in the American automation industry generally has a higher turnover rate; learning on the job is the preferred method of training. American users therefore often more readily accept a restriction in flexibility if for standard procedures they gain an easy start and instant success. With this background information, the requirement for a beginner mode had to be interpreted and assessed in a completely new light.

The fundamental question to answer therefore is: Which areas of the product or system need to be internationalized and localized so that the product is appropriate for the target users and the target context, under the given schedule and within budget? This question will lead to different answers in different contexts, depending on target markets, industrial domains, product visions, and business strategies. Cookbook solutions will not help: Unfortunately (or luckily), cultures as well as technologies and individuals are not static entities but instead, develop and change. So instead of answering how to design correctly for a foreign market, we can only describe the processes and methods that have proved to be helpful in internationalization.

5.2 Role of Human Factors Experts in the Process of Internationalization

Human factors experts are involved throughout the product development process to assure that the final products and systems will be usable in the target context of use. In an international setting, a huge diversity of users must be satisfied not only on the basis of functional needs, but also on emotional and aesthetic aspects. The tasks involved can be grouped into the following three main fields: (1) getting the right requirements (and interpreting them correctly), (2) designing based on the requirements, and (3) Ensuring that the requirements fit the target cultures.

Of course, the success of all three of these task categories depends on solid management of the process of international usability engineering and the ability to work effectively in an international team. Figure 1 shows the human factors tasks in the internationalization process.

In the following sections, requirements engineering, designing, and evaluation in an international context are looked at in more detail.

5.3 International Requirements Engineering

Getting the "right" requirements (i.e., requirements that are relevant for a given user group) is the main success factor for any kind of product development. Information about culture-specific user requirements provides a basis for decisions about the proper degree of internationalization and an appropriate balance between user needs and business needs. Misunderstandings at the beginning will be perpetuated throughout the project. So how can culture-specific user requirements be elicited?

5.3.1 Qualitative versus Quantitative Methods

In the early definition phase of product development, it seems to be most important to understand the users, their intentions, and the context of use. Therefore, a hermeneutic approach and the use of qualitative methods seem to be most fruitful for the elicitation of culture-specific user requirements. Thus, the generation of meaningful questions is already one of the most important results. Methods such as observation, interviews, or focus groups help us to experience why certain features or designs are relevant for a given user group and open our eyes for needs that we had not thought of before.

For example, Konkka (2003) describes how functional and emotional needs for mobile devices were collected in India following the basic rules of contextual inquiry methods (Beyer and Holtzblatt, 1998). This work was perceived as a key for market acceptance. As Konkka states: "If we don't understand our markets well enough from the perspective of end-user needs, the new features we create will not be accepted and—even worse—we won't necessarily know why" (p. 98). Results from her study showed that often only seemingly little adaptations of the user interface could have an enormous influence on the perceived usefulness and usability. As Konkka says: "For example, the user's ability to add Hindi greetings or religious symbols to a message adds a great deal of emotional value to short message service (SMS) in India among its Hindu people. Hindi music in ringing tones and other alerts would do the same" (p. 98).

Harel and Prabhu (1999) conducted studies on "global user experience" for Japan, China, and India for imaging products and software. Their research methods included "in-depth ... ethnological studies, intensive interviews with indigenous persons in their local environment, ... prototype development and validation focus groups in major cultural centers in all three countries" (p. 205). They could identify a set of cultural characteristics and appearance qualities for each of the cultures investigated. These not only served as guidelines for the design of new products but also helped in discovering similarities between cultures and gave a basis for informed decisions for global design of specific products. Similar studies (Dray and Mrazek, 1996; Ruuska, 1999; Jokinen et al., 2003) all describe the value of a qualitative approach in getting the requirements that were relevant for the target users, and on which design decisions could be based. After a research team has a rich and detailed understanding about the context of use and users need, getting more data about relevant questions (e.g., via questionnaires) is sometimes a useful next step. Nevertheless, the validity of standard research tools in a different cultural context needs to be reconsidered.

5.3.2 Methodological Considerations in Requirements Engineering

Social sciences have considered the question of how a functional equivalence can be achieved when conducting empirical research in different cultural settings (e.g., Helfrich, 1993). The following two criteria seem to be especially important for assessing the quality of a method.

1. *Conceptual equivalence.* This criterion asks whether a concept does have the same meaning in different countries. For example, when speaking about cars, does the concept "convenient" have the same meaning for customers in the United States, Germany, and China? Differences in constructs can be detected and minimized by translating a concept into the language of the target culture, translating is back into the original language, and comparing the outcome. In addition to looking at the similarity of concepts, their relative importance needs to be considered as well. Although the concept of "convenience" may be rather similar for Americans and Germans, its relative importance compared to other concepts (e.g., "sportive", "safe") might differ significantly.
2. *Operational equivalence.* It is not enough to check underlying concepts for equivalence; their operationalization for validity in different cultural contexts must be checked as well. To give an example: The number of negative comments reported by a subject in a focus group or usability test does not provide a stable or valid indicator for the quality of a user interface design. Stated the other way around: The absence of negative comments or criticism is probably a better indicator for an indirect and face-saving communication style than for the quality of a user interface. The criteria for conceptual and operational equivalence draw our attention to the need to reevaluate standard tools and procedures such as questionnaires, interview guidelines, or testing procedures with local experts.

5.3.3 Recruitment of Users: Asking the Right People

The importance of getting feedback from a representative group of users is common knowledge in user interface design. When it comes to internationalization, real users are, unfortunately, often hard to get. As a start, the following workarounds are frequently observed in industry:

1. *Studying members of other cultures in one's own culture.* It is relatively easy to invite people of other cultures who currently live in one's own country. Although a practical first step, this approach is not unproblematic. First, the population e.g., of Indian or Chinese people living in Germany, for example, is not very typical. Often, they are better educated than the average person in their home country. Second, depending on the length of their stay outside their home country, these people have already acculturated themselves to the country they live in and have lost contact with the day-to-day context of their country of origin. Third, they might not be aware of recent cultural and social changes in their home country and stick to viewpoints that are more traditional and conservative than those prevailing at present in the target country.
2. *Getting feedback from local colleagues.* Another source for local requirements is local colleagues within the same organization. These people have knowledge of the domain, they are relatively easily accessible, and often can provide quick feedback. Despite these advantages, we have to be aware that local colleagues are not end users. So although feedback from them can be helpful for minor questions, it can never substitute for information from real end users. When you are lucky enough to have the opportunity to conduct user requirement studies in local markets, it can also be a challenge to recruit local participants.
3. *Recruiting local participants.* Although we do expect a certain dropout rate in Europe or the United States when scheduling participants in usability studies, the dropout can increase dramatically when going, for example, to Latin America or Asia. On the one hand, one might have misunderstood the answer of a potential participant. In Asian cultures such as Japan or China, "yes" might be translated as a polite

form of "maybe" or "no." On the other hand, especially in collectivistic cultures, participants might not feel a real commitment to an appointment, as there are no personal bonds and past and future commitments between you and them. The support of a local company and/or local colleagues is therefore absolutely necessary. They will know the right strategy to find and motivate participants. In China, for example, it makes sense that local colleagues recruit target users from their wide network of relationships (outside the company), as the exploitation of relationships (guanxi) is a key characteristic of Chinese society (Bond and Hwang, 1986). Locals can also help in deciding whether paying money is the best way to compensate people for their time and effort, or whether this can heavily bias the feedback of the participants. Other means of compensation can include giving incentives to an entire community rather than to the individual participant, or giving presents instead of money.

5.3.4 Interpretation of Local Requirements

While conducting user studies abroad, it is important to cooperate with local experts not only during data gathering but also during the phase of interpreting the data. Otherwise, there is the danger that we either overestimate the relevance of observations because they are exotic and new, or that we just miss the point. Another risk lies in misinterpretation of data, because we integrate them in our own culturally biased cognitive schemes and scripts and look for meaning according to our own everyday experience.

Honold (2000a,b) dealt with the use of washing machines in urban middle-class Indian households. One of the first things that the German engineers on the team noticed was that many of the machines they saw were standing on trolleys. They also noticed that the flats they visited were comparatively small. They concluded that there was a need for "mobile" washing machines based on these observations: To save space, a washing machine might be stored (e.g., on the balcony) when not in use. For laundry, it might be rolled into the bathroom for water connection. However, when users were asked about this mobility, they stared blankly: Nobody ever wanted to move their washing machine through the house. In the end, it was clear: The trolleys were needed to be able to clean the floor under the washing machine—a requirement that made a lot of sense given the hot and humid climate in Mumbai.

An international team can help to broaden the base of experiences during the interpretation phase. Furthermore, design decisions have to be made more explicit than would be necessary in a team sharing one common culture. Though at the beginning this can be strenuous, the quality of decisions will be improved in the long run.

5.4 Designing Global and Local User Interfaces

Everybody involved in design decisions should have sufficient knowledge of international requirements and their rationales to come up with sensible conclusions for design. Such a conclusion could even be: One standard design for all users is acceptable at the moment. We believe that the most important quality of internationalization is whether or not a product or system is built on informed and aware decisions.

5.4.1 Danger of Clichés

Many articles on internationalization give examples of do's and don'ts for specific countries, such as the meaning of colors (white is the color of mourning in India, black in Europe), traditional use of symbols (the pig is a symbol of luck in Europe, a dirty animal in Arabic countries), and so on. Such examples help to raise the general awareness for diversity in user needs. But the danger is that they can sometimes lead designers to concentrate on stereotypes and clichés without looking at the actual context. This problem of lack of empirical data has been discussed from the very beginning in the human factors community (Marcus, 1993; Stiff, 1995, Leventhal et al., 1996). With little or superficial information available, well-meant attempts of guessing what is appropriate for a user group can lead to design that is similarly inappropriate to that of ignoring culture-specific user needs from the start.

5.4.2 Involving Decision Makers

To transfer the richness of international user studies, a good solution is to involve important decision makers in the core team. Ketola (2003) describes an early prototype study with smart phones in Rome. By setting up a team of designer, usability expert, Italian marketing expert, translator, and internationalization expert, all relevant decision makers were heavily involved in the user studies and got firsthand experience on user feedback. "In other words, having first-hand access to the development process from test results to product design optimized the effectiveness and efficiency of the test.. . . In a way, we ended by testing the user interface twice: first, by walking through it with local staff and again in administering the formal tests" (p. 177).

5.4.3 Documentation and Communication

Not every member of a design team will have the opportunity to observe target users abroad. If information from international usability studies is available, it is crucial that they not be stripped from their context. *Grounded scenarios*, that is, prototypical stories that are grounded in empirical data, together with audiovisual material from the field are a good way to share experiences.

5.5 Usability Evaluation

Usability evaluation can be conducted very early in the development process (e.g., using paper prototypes) as well as very late (e.g., with the finished product). Again, it is important to know what information is needed at the given time of a project and what the objective of the testing is. In the international context the following have to be considered:

- Is the method chosen (e.g., usability testing) the right instrument for the usability evaluation?
- In which aspects do country-specific user groups differ in the tests, and what is the reason for it?
- In the following, we concentrate on the technique of usability testing.

5.5.1 Adaptation of Usability Testing Methodology

High organizational demands are often highlighted in the literature on international usability testing (Fernandes, 1995; del Galdo and Nielsen, 1996). New strategies have to be developed if the usability experts do not speak the language of the test persons, if there is no software or documentation in the target language, or when local usability experts are not available. There is still the common experience that simulating the use of a product delivers valuable information in virtually every country in the world.

Influence of Cultural Communication Patterns

Nevertheless, the implications of what tested subjects say in a usability test can differ between cultures. Having the chance to observe actual use of a product is therefore extremely important, as the verbal feedback of users is influenced by communicative patterns and cultural values.

German test persons showed more signs of stress than did Indian test persons and were less confident in their own abilities to master the product. This was especially true for German women. At the same time, the German sample was much more negative and critical about the usability of the product than was the Indian sample. Indian subjects stayed relaxed and were positive in their verbal feedback and in the questionnaires, even if they experienced severe difficulties with the task presented (Honold, 2000a).

Positive verbal feedback alone does not say much about the real usability of a product. A creative approach to usability evaluation is therefore needed when testing abroad. Chavan (2002) gave the following example for a "Bollywood style" evaluation during a CHI 2002 workshop on "Usability Innovations for Global Marketplaces." Chavan (2002) reported that "under normal circumstances an [Indian] individual would find it difficult to critique any product.... Therefore she made use of a widely known method of critique in the Indian culture, the film review.... Since the film review format is perhaps the only popular and accepted format for critique and comparing 'products,' we experimented with the use of this form for critiquing websites." Use of a Web site for booking flights was interwoven in the story line of a typical Indian movie. The user had to help the hero of the story in booking a flight from Bangalore to Delhi online in order to prevent an arranged marriage, a plot that is extremely common in Indian movies. This procedure was compared to a "standard" usability test with a comparable Web site. According to Chavan, the differences in feedback from standard usability tests were striking when a dramatic storyline was presented. "All subjects were unexpectedly forthcoming with their criticism ...; many even offered design solutions. All subjects also accomplished their task in a much shorter time frame."

5.5.2 Planning International User Studies

International user studies, used either to gather requirements or to evaluate usability, follow the same basic principles of any user study, with one difference. You are no longer within the familiar territory of your own culture. Instead, you are in a totally different setting and your own cultural interpretations and expectations could alter your interpretation of your study.

Dray and Siegel (2005) document their experiences in detail regarding international user studies. These include:

1. *Selection of cities/towns/user groups in target cultures.* This involves how to balance the cost of a study with the depth of coverage planned. Of course, several factors should be taken into consideration while selecting where to go and who the users would be, including market share, new markets, technology penetration (city vs. towns), sales/field support, window of opportunity, and other special challenges.
2. *Arranging a study in a target country.* It is crucial to find contacts (company offices, sales representatives, vendors, usability testing facilities) who could assist you before your arrival and during your stay. The work involves renting equipment, facilities, recruitment of users, schedule, help conducting studies, and translation. At this step, researchers should prepare or adapt the study plan and protocol, including the translation of materials that will be used in the study.
3. *Preparations in a target country before the study.* This involves meeting and training the local facilitators and/or interpreters for the study, and checking equipment and facilities. Extra time should be allotted to accommodate local work practices, potential equipment failures, and schedule changes.
4. *Conducting a study in a target culture.* Working together with the local support involves running studies and interpreting results with cultural differences in mind. Test facilitators may change the protocol due to differences in how people express their experience and thoughts and react to the study. Similarly, interpretation of results may differ from culture to culture, reflecting the in-depth study of findings based on cultural differences.

5.5.3 Interpretation of Usability Data

The feedback of test persons is not only influenced, by cultural patterns of communication. The perceived relevance of a function can also be confounded with the usability of a product or service. Although the observed use of a short phone message was very similar for Indian and German test users, in the feedback questionnaire the Indians rated the usability of this task significantly more positively than did the Germans. They also evaluated this functionality as more useful and relevant for their everyday life than did the German users (Honold, 2000a). When relying only on observations, differences between user groups were generally fewer. In areas where German users had problems, users in other countries generally had problems as well. Usability problems could often be tracked to inaccurate translations. Early evaluation of a translation within the context of the product (e.g., via a paper prototype) can improve the quality of the localized version.

6 SUMMARY

Cultural ergonomics and global human factors design are areas that have been growing rapidly and will keep on growing as international markets continue to mature. It is impossible to describe these topics with any completeness in a single review. We have described briefly the essential role of management for internationalization and localization. We have illustrated the part that guidelines play in global design activities. Finally, we have shown how real contact with global users is critical in gathering requirements, designing products, and testing their usability. Hopefully, we have described enough to capture the interest of designers and others required for successful product globalization such that this review can serve as a springboard for continued improvement of global usability.

REFERENCES

Aykin, N. (2003), *Internationalization and Localization of Web Sites*, tutorial presented at the Nielsen-Norman Group User Experience 2003 Conference.

Aykin, N. (2005), "Overview: Where to Start and What to Consider," in *Usability and Internationalization for Information Technology*, N. Aykin, Ed., Lawrence Earlbaum Associates, Mahwah, NJ.

Aykin, N., and Milewski, A. E. (2005), "Practical Issues and Guidelines for International Information Display," in *Usability and Internationalization for Information Technology*, N. Aykin, Ed., Lawrence Earlbaum Associates, Mahwah, NJ.

Best, J. (2004), "How Eight Pixels Cost Microsoft Millions," published August 19, 2004; retrieved August 29, 2004, from http://news.com.com/How%20eight%20pixels%20cost%20Microsoft%20millions/2100-1014_3-5316664.html?tag=nefd.top.

Beu, A., Honold, P., and Yuan, X. (2000), "How to Build Up an Infrastructure for Intercultural Usability Engineering," *International Journal of Human–Computer Interaction*, Vol. 12, No. 3–4, pp. 347–358.

Beyer, H., and Holtzblatt, K. (1998), *Contextual Design: Defining Customer-Centered Systems*, Morgan Kaufmann, San Francisco.

Bond, M. H., and Hwang, K. (1986), "The Social Psychology of the Chinese People," in *The Psychology of the Chinese People*, M. H. Bond, Hrsg., Oxford University Press, New York, pp. 213–295.

Chavan, A. L. (2002), "Usability Evaluation—Indian 'Ishtyle,'" CHI 2002 Workshop 15, "It's a Global Economy Out There: Usability Innovations for Global Marketplaces," unpublished paper.

del Galdo, E. M., and J. Nielsen (Eds.) (1996), *International User Interfaces*, Wiley, New York.

DePalma, D., McCarthy, J. C., and Armstrong, A. (1998), "Strategies for Global Sites," Forrester Research Report No. 3, Cambridge, MA.

Dray, S., and Mrazek, D. (1996), "A Day in the Life: Studying Context Across Culture," in *International User Interfaces*, E. M. del Galdo and J. Nielsen, Eds., Wiley, New York, pp. 242–256.

Dray, S. M., and Siegel, D. A. (2005), "Sunday in Shangai, Monday in Madrid?! Key Issues and Decisions in Planning International User Studies," in *Usability and Internationalization of Information Technology*, N. Aykin, Ed., Lawrence Earlbaum Associates, Mahwah, NJ.

Dunlop, B. (1999), "Success in Any Language," retrieved March 25, 2005, from http://www.clickz.com/experts/archives/int_mkt/int_comm/article.php/812791.

Fernandes, T. (1995), *Global Interface Design: A Guide to Designing International User Interfaces*, AP Professional, London.

GNOME Documentation Style Guide (2003), *Writing for Localization*, Sun Microsystems; retrieved January 22, 2004, from http://developer.gnome.org/documents/styleguide.

Harel, D., and Prabhu, G. V. (1999), "Global User Experience (GLUE): Design for Cultural Diversity: Japan, China and India," in *Designing for Global Markets, First International Workshop on Internationalization of Products and Systems*, G. V. Prabhu and E. DelGaldo, Eds., Backhouse Press, Rochester, NY, pp. 205–216.

Helfrich, H. (1993), "Methodologie Kulturvergleichender Psychologischer Forschung," in *Kulturvergleichende Psychologie: Eine Einführung*, A. Thomas, Ed., Hogrefe, Göttingen, Germany, pp. 81–102.

Hoft, N. (1995), *International Technical Communication: How to Export Information About High Technology*, Wiley, New York.

Honold, P. (2000a), "Interkulturelles Usability Engineering," in *Eine Untersuchung zu kulturellen Einflüsse auf die Gestaltung und Nutzung technischer Produkte*, VDI-Verlag, Düsseldorf, Germany, 2000.

Honold, P. (2000b), "Culture and Context: An Empirical Study for the Development of a Framework for the Elicitation of Cultural Influence in Product Usage," *International Journal of Human–Computer Interaction*, Vol. 12, No. 3–4, pp. 327–345.

Horton, W. (2005), "Graphics: The Not Quite Universal Language," in *Usability and Internationalization of Information Technology*, N. Aykin, Ed., Lawrence Erlbaum Associates, Mahwah, NJ.

Jokinen, P., Karimäki, K., and Kangas, A. (2003), "Demanding Needs for Mobile Phones: A Qualitative User Study on the Young Urban Lower Middle Class in China," in *Designing for Global Markets 5, Proceedings of the 5th International Workshop on Internationalisation of*

Products and Systems, V. Evers, K. Röse, P. Honold, J. Coronado, and D. Day, eds., University of Kaiserslautern, Kaiserslautern, Germany.

Ketola, P. (2003), "Series 60 Voice Mailbox in Rome: A Case Study of a Paper Prototyping Tour," in *Mobile Usability: How Nokia Changed the Face of the Mobile Phone*, C. Lindholm, T. Keinonen, and H. Kiljander, Eds., McGraw-Hill, New York.

Konkka, K. (2003), "Indian Needs–Cultural End-User Research in Mombai," in *Mobile Usability: How Nokia Changed the Face of the Mobile Phone*, C. Lindholm, T. Keinonen, and H. Kiljander, eds., McGraw-Hill, New York.

Kuhn, M. (1999), "Metric Typographic Units," retrieved January 22, 2004, from http://www.cl.cam.ac.uk/~mgk25/metric-typo/.

Kuhn, M. (2001), "A Summary of the International Date and Time Notation," retrieved January 22, 2004, from http://www.cl.cam.ac.uk/~mgk25/iso-time.html.

Kuhn, M. (2003), "International Standard Paper Size," retrieved January 22, 2004, from http://www.cl.cam.ac.uk/~mgk25/iso-paper.html.

Leventhal, L., Teasley, B., and Blumenthal, B. (1996), "Assessing User Interfaces for Divers User Groups: Evaluation Strategies and Defining Characteristics," *Behaviour and Information Technology*, Vol. 15, No. 3, pp. 127–137.

Marcus, A. (1993), "Human Communication Issues in Advanced UIs," *Communications of the ACM*, Vol. 36, No. 4, pp. 101–109.

Microsoft (Dr. International Team) (2002), *Developing International Software*, Microsoft Press, Redmond, WA.

Morton, J. I. (2003), "Color Matters," retrieved June 10, 2003, from http://www.colormatters.com.

Norman, D. A. (2003), *Emotional Design: Why We Love (or Hate) Everyday Things*, Basic Books, New York.

Ruuska, S. (1999), "Mobile Communication Devices for International Users: Exploring Cultural Diversity Through Contextual Inquiry," in *Designing for Global Markets, First International Workshop on Internationalization of Products and Systems*, G. V. Prabhu and E. DelGaldo, Eds., Backhouse Press, Rochester, NY, pp. 217–226.

Simultrans (2004), "Localization Return on Investment," retrieved August 30, 2004, from http://www.simultrans.com/Articledetail.cfm?PostingID=27.

Spartan (1999), "Multimedia Presentation on the Colors of Culture," retrieved January 22, 2004, from http://www.mastep.sjsu.edu/Alquist/workshop2/color_and_culture_files/frame.htm.

Stiff, P. (1995), "Design Methods, Cultural Diversity and the Limits of Design," *Information Design Journal*, Vol. 8, No. 1, pp 35–47.

CHAPTER 8

DECISION-MAKING MODELS AND DECISION SUPPORT

Mark R. Lehto
Purdue University
West Lafayette, Indiana

Fiona Nah
University of Nebraska–Lincoln
Lincoln, Nebraska

1 INTRODUCTION

This chapter focuses on the broad topic of human decision making. Decision making is often viewed as a stage of human information processing because people must gather, organize, and combine information from various sources to make decisions. However, as decisions grow more complex, information processing actually becomes part of decision making, and methods of decision support that help decision makers process information become of growing importance. Decision making also overlaps with problem solving. The point where decision making becomes problem solving is fuzzy, but many decisions require problem solving, and the opposite is true as well. Cognitive models of problem solving are consequently relevant for describing many aspects of human decision making. They become especially relevant for describing steps taken in the early stages of a decision where choices are formulated and alternatives are identified.

A complete treatment of human decision making is well beyond the scope of a single chapter.* The topic has its roots in economics and is currently a focus of operations research and management science, psychology, sociology, and cognitive engineering.

*No single book covers all the topics addressed here. More detailed sources of information are referenced throughout the chapter. Sources such as von Neumann and Morgenstern (1947), Savage (1954), Luce and Raiffa (1957), Shafer (1976), and Friedman (1990), are useful texts for people desiring an introduction to normative decision theory. Raiffa (1968), Keeney and Raiffa (1976), Saaty (1988), Buck (1989), and Clemen (1996) are applied texts on decision analysis. Kahneman et al. (1982), Winterfeldt and Edwards (1986), Payne et al. (1993), Svenson and Maule (1993), Heath et al. (1994), and Yates (1992), among numerous others, are texts addressing elements of behavioral decision theory. Klein et al. (1993) and Klein (1998) provide introductions to naturalistic decision making.

These fields have produced numerous models and a substantial body of research on human decision making. At least three objectives have motivated this work: to develop normative prescriptions that can guide decision makers, to describe how people make decisions and compare the results to normative prescriptions, and to determine how to help people apply their "natural" decision-making methods more successfully. The goals of this chapter are to synthesize the elements of this work into a single picture and to provide some depth of coverage in particularly important areas. The integrative model presented in Section 1.3 focuses on the first goal. The remaining sections address the second goal.

1.1 Role and Utility of the Chapter

This chapter is intended to provide an overall perspective on human decision making to human factors practitioners, developers of decision tools (such as expert systems), product designers, researchers in related areas, and others who are interested both in how people make decisions and in how decision making might be improved. Consequently, we present a broad set of prescriptive and descriptive approaches. Numerous applications are presented and strengths and weaknesses of particular approaches are noted. Emphasis is also placed on providing useful references containing additional information on topics that the reader may find of special interest.

In Section 2 we address topics grouped under the somewhat arbitrary heading of classical decision theory. The material presented provides a normative and prescriptive framework for making decisions. In Section 3 we summarize decision analysis, which refers to the application of normative decision theory to improve decisions. The discussion considers the advantages of the various approaches, how they can be applied, and what problems might arise during their application. Section 4 addresses topics grouped under the heading of behavioral decision theory. The material compares human decision making to the normative models discussed earlier. Several descriptive models of human judgment, preference, and choice are also discussed. In Section 5 we explore topics falling under the heading of dynamic and naturalistic decision theory. This material should be of interest to practitioners interested in the process followed when many real-world decisions are made, the quality of these decisions, and why people use particular methods to make decisions. The discussion provides insight into how people perform diagnostic tasks, make decisions to take risks when using products, and develop expertise. In Section 6 we introduce the topic of group decision making. The discussion addresses conflict resolution both within and between groups, group performance and biases, and methods of group decision making. Section 7 covers methods of assisting or supporting the decision making of individuals, groups, and organizations.

1.2 Elements of Decision Making

Decision making requires that the decision maker make a choice between two or more alternatives (note that doing nothing can be viewed as making a choice). The alternative selected results in real or imaginary consequences to the decision maker. Judgment is a closely related process in which a person rates or assigns values to attributes of the alternatives considered. For example, a person might judge both the safety and attractiveness of a car being considered for purchase. Obtaining an attractive car is a desirable consequence of the decision, while obtaining an unsafe car is an undesirable consequence. A rational decision maker seeks desirable consequences and attempts to avoid undesirable consequences.

The nature of decision making can vary greatly, depending on the decision context. Certain decisions, such as deciding where and what to eat for lunch, are routine and repeated often. Other choices, such as purchasing a house, choosing a spouse, or selecting a form of medical treatment for a serious disease, occur seldom, may involve much deliberation, and take place over a longer period. Decisions may also be required under severe time pressure and involve potentially catastrophic consequences, such as when a fire chief decides whether to send firefighters into a burning building. Previous choices may constrain or otherwise influence subsequent choices (e.g., a decision to enter graduate school might constrain a future employment-related decision to particular job types and locations). The outcomes of choices may be uncertain and in certain instances are determined by the actions of potentially adverse parties, such as competing manufacturers of a similar product. Decisions may be made by a single person or by a group. Within a group, there may be conflicting opinions and differing degrees of power between individuals or factions. Decision makers may also vary greatly in their knowledge and degree of aversion to risk.

Conflict occurs when a single decision maker is not sure which choice should be selected or when there is lack of consensus within a group regarding the choice. Both for groups and single decision makers, conflict occurs, at the most fundamental level, because of uncertainty or conflicting objectives. Uncertainty can take many forms and is one of the primary reasons that decisions can be difficult. In ill-structured decisions, decision makers may not have identified the current condition, alternatives to choose between, or their consequences. Decision makers also may be unsure what their aspirations or objectives are, or how to choose between alternatives. After a decision has been structured, at least four reasons for conflict may exist. First, when alternatives have both undesirable and desirable consequences, decision makers may experience conflict due to conflicting objectives. For example, a decision maker considering the purchase of an air bag–equipped car may experience conflict because an air bag increases cost as well as safety. Second, decision makers may be unsure of their reaction to a consequence. For example, people considering whether to enter a raffle where the prize is a sailboat may be unsure how much they want a sailboat. Third, decision makers may not know whether a consequence will be sure to happen. Even worse, they may be unsure what the probability of the consequences is, or may not have enough time to evaluate the situation carefully. They may also be uncertain about the reliability of their information. For example, it may be difficult to determine the truth of a sales person's claim regarding the probability that a product will break down immediately after the warranty expires.

To resolve conflicts, decision makers must deal appropriately with uncertainty, conflicting objectives, or a lack of consensus. Conflict resolution therefore becomes a primary focus of decision theory. In the following section we present an integrative model of decision making that relates conflict resolution to the elements of decision making discussed above. This model considers specifically how decision making changes when different sources of conflict are present. It also matches methods of conflict resolution to particular sources of conflict and decision rules.

1.3 Integrative Model of Decision Making

Human decision making can be viewed as a stage of information processing that falls between perception and response execution (Welford, 1976). The integrative model of human decision making, presented in Figure 1, shows how the elements of decision making discussed above fit into this perspective. From this view, decision making is the process followed when a response to a perceived stimulus is chosen. The process followed depends on what decision strategy is applied and can vary greatly between decision contexts.* Decision strategies in Figure 1 correspond to different paths between situation assessment and executing an action. The particular decision strategy followed depends on both the decision context and on whether or not the decision maker experiences conflict.†

At least four, sometimes overlapping categories of decision making can be distinguished. *Group decision making* occurs when multiple decision makers interact and is represented at the highest level of the model as a source of conflict that might be resolved through debate, bargaining, or voting. For example, members of a university faculty committee might debate and bargain before voting between candidates for a job opening.

Dynamic decision making occurs in a changing environment, in which the results of earlier decisions affect future decisions. The decisions made in such settings often make use of feedback and are multistage in nature. For example, a decision to take a medical test almost always requires a subsequent decision regarding what to do after receiving the test results. Dynamic decision making is represented at the lowest level of the model by the presence of two feedback loops, which show how the action taken and its effects can feed forward to the assessment of a new decision or feed back to the reassessment of the current decision.

Routine decision making occurs when decision makers use knowledge and past experience to decide quickly what to do and is especially prevalent in dynamic decision-making contexts. Routine decision making is represented in Figure 1 as a single pattern-matching step or associative leap between situation assessment and executing an action. For example, a driver after perceiving a stop sign decides to stop, or the user of a word-processing system after perceiving a misspelled word decides to activate the spell checker. Since routine decisions are often made in dynamic task environments, routine decision making is discussed in this chapter as a subtopic of dynamic decision making.

Conflict-driven decision making occurs when various forms of conflict must be resolved before an alternative action can be chosen and often involves a complicated path between situation assessment and executing an action.‡ Before executing an action, the decision maker experiences conflict, somehow resolves it, and then either recognizes the best action (conflict resolution might transform the decision to a routine one) or applies a decision rule. Applying the decision rule leads ideally to a choice which is then executed. Attempting to apply the decision rule may, however, cause additional conflicts, leading to more conflict resolution. For example, decision makers may realize that they need more information to apply a particular decision rule. In response, they might decide to use a different decision rule that requires less information. Along these lines, when choosing a home, a decision maker might decide to use a satisficing decision rule after seeing that hundreds of homes are listed in the classified ads of the local newspaper.

Potential sources of conflict, methods of conflict resolution, and the results of conflict resolution are listed at the top of Figure 1. Each source of conflict maps to a particular method of conflict resolution, which then provides a result necessary to apply a decision rule, as illustrated schematically in the figure.§ Table 1 presents a set of decision rules, briefly describes their procedural nature and required

*The notion that the best decision strategy varies between decision contexts is a fundamental assumption of the theory of contingent decision making (Payne et al., 1993), cognitive continuum theory (Hammond, 1980), and other approaches discussed later in the chapter.

†Conflict has been recognized as an important determinant of what people will do in risky decision-making contexts (Janis and Mann, 1977). Janis and Mann focus on the stressful nature of conflict and on how affective reactions in stressful situations can affect the decision strategies followed.

‡The distinction between routine and conflict-driven decision making made here is similar to Rasmussen's (1983) distinction between (1) routine skill or rule-based levels of control and (2) nonroutine knowledge-based levels of control in information-processing tasks.

§Note that multiple sources of conflict are possible for a given decision context. An attempt to resolve one source of conflict

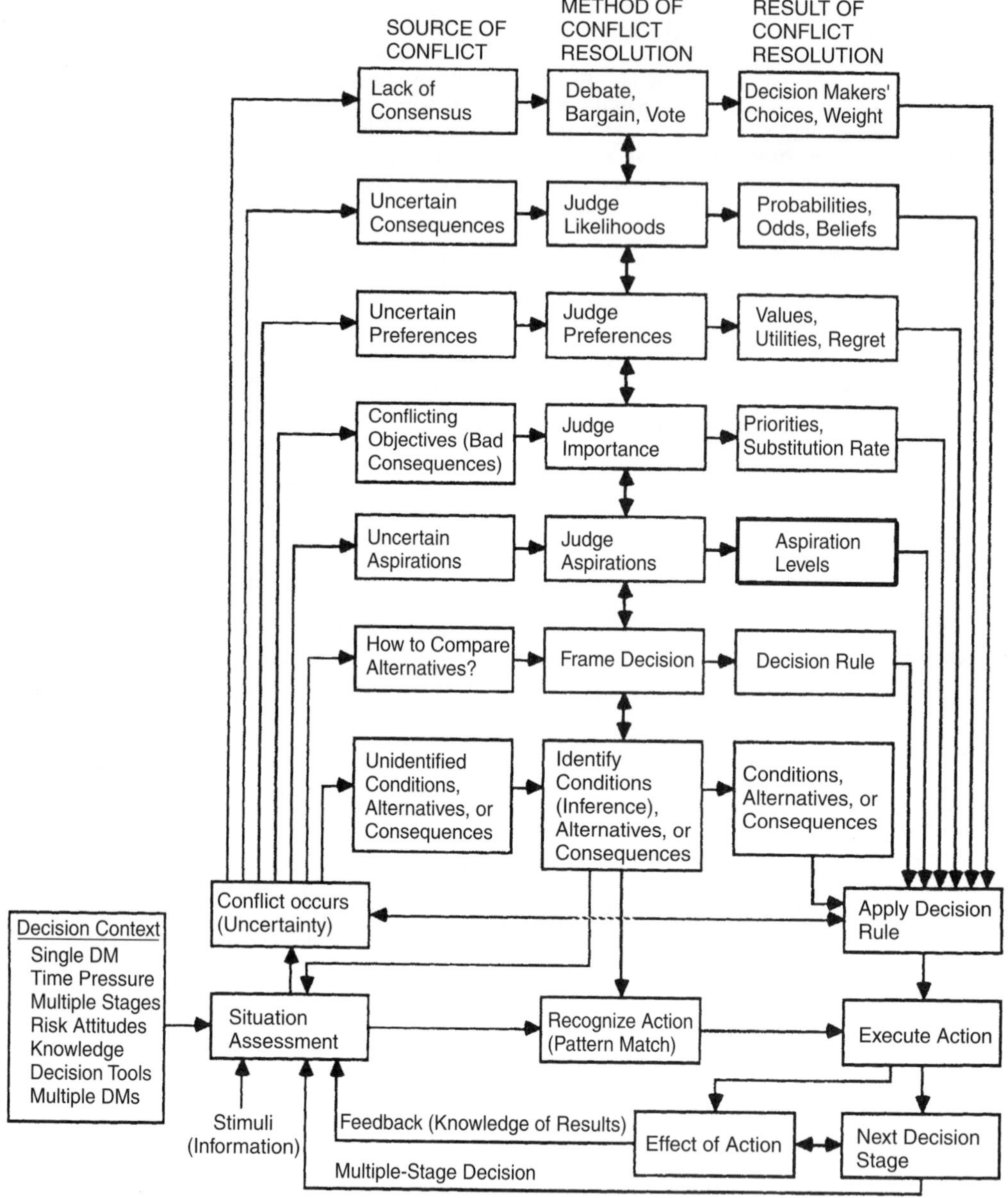

Figure 1 Integrative model of human decision making (DM, decision maker).

inputs, and lists the sections of this chapter where they are covered. The required inputs of particular decision rules can easily be mapped to sources of conflict. As shown in the table, each decision rule requires that alternatives and their consequences be identified. Other decision rules require measures of aspiration, importance, preference, and uncertainty for each consequence or consequence dimension. For example, to compare alternatives using expected value, the probability and value of each consequence must also be known. Certain decision rules also accept inputs describing the degree of consensus between decision makers.

may also make the decision maker aware of other conflicts that must first be resolved. For example, decision makers may realize they need to know what the alternatives are before they can determine their aspiration levels.

Table 1 Decision Rules, Required Inputs, and Procedure Applied by the Rule

Decision Rule	Inputs Required	Procedure Applied	Section
Dominance	All alternatives, value of each consequence	Select alternative best on all consequences	2.1.2
EBA	All alternatives, value of each consequence	Select first alternative found to be best on a consequence dimension, random order of consequences.	2.1.3
Lexicographic	All alternatives, value of each consequence, priority	Order consequences by priority, select first alternative found to be best on a consequence dimension	2.1.3
Satisficing	At least one and up to all alternatives, aspiration level and value of each consequence	Sequentially evaluate each alternative, stop if each consequence of an alternative equals or exceeds the aspiration level	2.1.4
Minimax cost	All alternatives, value of each consequence	Compare the worst consequence values of each alternative	2.1.5
Minimax regret	All alternatives, regret for each consequence	Compare largest regrets of each alternative	2.1.5
EV	All alternatives, probability and value of each consequence	Weight value of each consequence by its probability, for each alternative	2.1.6
Laplace	All alternatives, value or utility of each consequence	Weight value or utility of each consequence equally, for each alternative	2.1.7
SEU	All alternatives, probability and utility of each consequence	Weight utility of each consequence by its probability, for each alternative	2.1.7
MAUT	All alternatives, value or utility of each consequence, priority	Weight value or utility of each consequence by priority, for each alternative	2.1.8
Holistic	All alternatives and consequences	Wholistically compare the consequences of each alternative	2.1.9

Accordingly, conflict occurs at the most fundamental level when the current condition, alternative actions, or their consequences have not been identified. At the next most fundamental level, conflict occurs when the decision maker is unsure how to compare the alternatives. In other words, the decision maker has not yet selected a decision rule. Given that the decision maker has a decision rule, conflict can still occur if the needed inputs are not available. These sources of conflict and associated methods of conflict resolution are addressed briefly below in relation to the remainder of this chapter.

Identifying the current condition, alternative actions, and their consequences is an important part of decision making. This topic is emphasized in both naturalistic decision theory (Klein et al., 1993) and decision analysis* (Raiffa, 1968; Clemen, 1996). Decision trees, influence diagrams, and other tools for structuring decisions are covered in Section 3. Normative methods of identifying the current condition falling under the topic of inference (or diagnosis) are presented in Section 2.2. In Section 4.1 we describe several descriptive models of human inference and discuss their limitations. Section 6 covers discussion group decision-making methods that may be useful at this decision-making stage.

*Clemen (1996) includes a chapter on creativity and decision structuring. Some practitioners claim that structuring the decision is the greatest contribution of the decision analysis process.

When decision makers are unsure how to compare alternatives, they must consider what information is available and then frame a decision appropriately. The way the decision is framed then determines (1) which decision rules are appropriate, (2) what information is needed to make the decision using the rules given (as discussed earlier in reference to Table 1), and (3) the choices selected. As discussed in Section 5.1, there is reason to believe that people apply different decision-making strategies in different decision contexts. In Section 2.1 we discuss the appropriateness of decision rules and how the particular rule used can affect choices. When the specific inputs needed by a decision rule are not available, the resulting conflict might be resolved by judging aspirations, importance, preference, or likelihood. It might also be resolved by choosing a different decision rule or strategy. As noted in Section 5.1, there is a prevalent tendency among decision makers in naturalistic settings to minimize analysis and the cognitive effort required. In group situations, conflict due to a lack of consensus among decision makers might be resolved through debate, bargaining, or voting (Section 6).

2 CLASSICAL DECISION THEORY

Classical decision theory began with the development of normative models in economics and statistics that specified optimal decisions (von Neumann and Morgenstern, 1947; Savage, 1954). Classical decision theory focuses heavily on the notion of rationality (Savage, 1954; Winterfeldt and Edwards 1986).

Emphasis is placed on the quality of the process followed when making a decision rather than on the ultimate outcome. Accordingly, a rational decision maker must think logically about the decision. To do this, the decision maker must first formally describe what is known about the decision. The decision is then made by applying principles of logic and Bayesian probability theory (Savage, 1954). This approach is therefore quantitative, and also normative or prescriptive if the numerical inputs needed are available.

The classical approach has been applied to two related problems: (1) preference and choice, and (2) statistical inference.

2.1 Choice Procedures

Classical decision theory represents preference and choice problems in terms of four basic elements: (1) a set of potential actions (A_i) to choose between, (2) a set of events or world states (E_j), (3) a set of consequences (C_{ij}) obtained for each combination of action and event, and (4) a set of probabilities (P_{ij}) for each combination of action and event. For example, a decision maker might be deciding whether to wear a seat belt when traveling in an automobile. Wearing or not wearing a seat belt corresponds to two actions, A_1 and A_2. The expected consequence of either action depends on whether an accident occurs. Having or not having an accident corresponds to two events, E_1 and E_2. Wearing a seat belt reduces the expected consequences of having an accident (E_1). As the probability of having an accident increases, use of a belt should therefore become more attractive.

Once a decision has been represented in terms of these basic elements, the choice is then made by applying decision rules. Numerous decision rules have been developed. Decision rules are based on basic axioms (or what are felt to be self-evident assumptions) of rational choice. Not all rules, however, make use of the same axioms. Different rules make different assumptions and can provide different preference orderings for the same basic decision. In the following discussion we first present some of the most basic axioms. Then several well-known decision rules are covered briefly.

2.1.1 Axioms of Rational Choice

Numerous axioms have been proposed that are essential either for a particular model of choice or for the method of eliciting numbers used for a particular model (Winterfeldt and Edwards, 1986). The best known set of axioms (Table 2) establishes the normative principle of *subjective expected utility* (SEU) as a basis for making decisions (see Savage, 1954, and Luce and Raiffa, 1957, for a more rigorous description of the axioms). On an individual basis, these axioms are intuitively appealing (Stukey and Zeckhauser, 1978), but as discussed in Section 4, people's preferences can deviate significantly from the SEU model in ways that conflict with certain axioms. Consequently, there has been a movement toward developing other, less restrictive standards of normative decision making (Zey, 1992; Frisch and Clemen, 1994).

Frisch and Clemen propose that "a good decision should (a) be based on the relevant consequences of the different options (*consequentialism*), (b) be based on an accurate assessment of the world and a consideration of all relevant consequences (*thorough structuring*), and (c) make tradeoffs of some form (*compensatory decision rule*)." Consequentialism and the need for thorough structuring are both assumed by all normative decision rules. Most normative rules are also compensatory. However, when people make routine habitual decisions, they often don't consider the consequences of their choices, as discussed in Section 5. Also, because of cognitive limitations and the difficulty of obtaining information, it becomes unrealistic in many settings for the decision maker to consider all the options and possible consequences.

Table 2 Basic Axioms of Subjective Expected Utility Theory

A. *Ordering/quantification of preference.* Preferences of decision makers between alternatives can be quantified and ordered using the relations:

$>$, where $A > B$ means that A is preferred to B

$=$, where A = B means that A and B are equivalent

$\geq$, where $A \geq B$ means that B is not preferred to A

B. *Transitivity of preference.* If $A_1 \geq A_2$ and $A_2 \geq A_3$, then $A_1 \geq A_3$.

C. *Quantification of judgment.* The relative likelihood of each possible consequence that might result from an alternative action can be specified.

D. *Comparison of alternatives.* If two alternatives yield the same consequences, the alternative yielding the greater chance of the preferred consequence is preferred.

E. *Substitution.* If $A_1 > A_2 > A_3$, the decision maker will be willing to accept a gamble $[p(A_1)$ and $(1-p)(A_3)]$ as a substitute for A_2 for some value of $p \geq 0$.

F. *Sure thing principle.* If $A_1 \geq A_2$, then for all p, the gamble $[p(A_1)$ and $(1-p)(A_3)] \geq [p(A_2)$ and $(1-p)(A_3)]$.

To make a decision under such conditions, decision makers may limit the scope of the analysis by applying principles such as satisficing and other noncompensatory decision rules discussed below. They may also apply heuristics, based on their knowledge or experience, leading to performance that can approximate the results of applying compensatory decision rules (Section 4).

2.1.2 Dominance

Dominance is perhaps the most fundamental normative decision rule. *Dominance* is said to occur between two alternative actions, A_i and A_j, when A_i is at least as good as A_j for all events E, and for at least one event E_k, A_i is preferred to A_j. For example, one investment might yield a better return than another regardless of whether the stock market goes up or down. Dominance can also be described for the case where the consequences are multidimensional. This occurs when for all events E, the kth consequence associated with action i (C_{ik}) and action j (C_{jk}) satisfies the relation $C_{ik} \geq C_{jk}$ for all k and for at least one consequence $C_{ik} > C_{jk}$. For example, a physician choosing between alternative treatments has an easy decision if one treatment is *both* cheaper and more effective for all patients.

Dominance is obviously a normative decision rule, since a dominated alternative can never be better than the alternative that dominates it. Dominance is also conceptually simple, but it can be difficult to detect when there are many alternatives to consider or many possible consequences. The use of tests for dominance by decision makers in naturalistic settings in discussed further in Section 5.1.5.

2.1.3 Lexicographic Ordering and EBA

The *lexicographic ordering principle* (see Fishburn, 1974) considers the case where alternatives have multiple consequences. For example, a purchasing decision might be based on both the cost and performance of the product considered. The various consequences are first ordered in terms of their importance. Returning to the example above, performance might be considered more important than cost. The decision maker then compares each alternative sequentially, beginning with the most important consequence. If an alternative is found that is better than the others on the first consequence, it is selected immediately. If no alternative is best on the first dimension, the alternatives are compared for the next-most-important consequence. This process continues until an alternative is selected or all the consequences have been considered without making a choice. The latter situation can happen only if the alternatives have the same consequences.

The *elimination by aspects* (EBA) *rule* (Tversky, 1972) is similar to the lexicographic decision rule. It differs in that the consequences used to compare the alternatives are selected in random order, where the probability of selecting a consequence dimension is proportional to its importance. Both EBA and lexicographic ordering are noncompensatory decision rules, since the decision is made using a single consequence dimension. Returning to the example above, the lexicographic principle would result in selecting a product with slightly better performance, even if it costs much more. EBA would select either product, depending on which of the consequences was selected first.

2.1.4 Minimum Aspiration Level and Satisficing

The *minimum aspiration level* or *satisficing decision rule* assumes that the decision maker screens alternative actions sequentially until an action is found that is good enough. For example, a person considering the purchase of a car might stop looking once he or she has found an attractive deal, instead of comparing every model on the market. More formally, the comparison of alternatives stops once a choice is found that exceeds a minimum aspiration level S_{ik} for each of its consequences C_{ik} over the possible events E_k.

Satisficing can be a normative decision rule when (1) the expected benefit of exceeding the aspiration level is small, (2) the cost of evaluating alternatives is high, or (3) the cost of finding new alternatives is high. More often, however, it is viewed as an alternative to maximizing decision rules. From this view, people cope with incomplete or uncertain information and their limited rationality by satisficing in many settings instead of optimizing (Simon, 1955, 1983).

2.1.5 Minimax (Cost and Regret) and the Value of Information

Minimax cost selects the best alternative (A_i) by first identifying the worst possible outcome for each alternative. The worst outcomes are then compared between alternatives. The alternative with the minimum worst-case cost is selected. Formally, the preferred action A_i is the action for which over the events k, $\text{Max}_k(C_{ik}) = \text{Min}_i[\text{Max}_k(C_{ik})]$. For example, in Table 3, the maximum cost is 5 for alternative A_1, 7 for A_2, and 8 for A_3. A_1 would be chosen since it has the smallest maximum cost. Minimax cost corresponds to assuming the worst and therefore makes sense as a strategy where an adverse opponent is able to control the events (von Neumann and Morgenstern, 1947). Along these lines, an airline executive considering whether to reduce fares might assume that a competitor will also cut prices, leading to a no-win situation.

Minimax regret involves a similar process, but the calculations are performed using regret instead

Table 3 Example Comparison of Minimax Cost and Minimax Regret[a]

	E_1	E_2	E_3	Max. Cost	Max. Regret
A_1	5	5	5	**5**	3
A_2	2	7	2	7	**2**
A_3	6	8	4	8	4

[a]Minimax cost selects A_1 and minimax regret selects A_2

of cost (Savage, 1954). Regret is calculated by first identifying which alternative is best for each possible event. The regret R_{ik} associated with each consequence C_{ik} for the combination of event E_k and alternative A_i then becomes $R_{ik} = \text{Max}_i(C_{ik}) - C_{ik}$. Returning to an earlier example, if E_1 occurs, alternative A_2 with a cost of 2 is best, resulting in a regret of 0 (2 minus 2). A_1 has a cost of 5, resulting in a regret of 3 (5 minus 2). A_3 has a cost of 6, resulting in a regret of 4 (6 minus 2). These calculations are repeated for events E_2 and E_3, resulting in regret values for each combination of events and alternative actions. The preferred action A_i is the action for which over the events k, $\text{Max}_k(R_{ik}) = \text{Min}_i[\text{Max}_k(R_{ik})]$. Returning to the example, the maximum regret for A_1 (a value of 3) and A_3 (a value of 4) are both found when event E_1 occurs. The maximum regret for A_2 (a value of 2) is found when event E_2 occurs. Alternative A_2 is then selected because it has the minimum maximum regret.

Note that the minimax cost and minimax regret principles do not always suggest the same choice (Table 3). Minimax cost is easily interpreted as a conservative strategy. Minimax regret is more difficult to judge from an objective or normative perspective (Savage, 1954). As shown by the example, minimax regret can be less conservative than minimax cost. Alternatives that were not chosen can also affect choices made using minimax regret. For example, if alternative A_3 is removed from consideration, minimax regret and minimax cost will both select A_1. The interesting conclusion is that comparative and absolute measures of preference can result in different choices.

Bell (1982) argues persuasively that regret plays a very prominent role in decision making under uncertainty. For example, the purchaser of a new car might be happy until finding out that a neighbor got the same car for $200 less from a different dealer. It is interesting to observe that regret is closely related to the value of information. This follows, since with hindsight, decision makers may regret their choice if they did not select the alternative giving the best result for the event E_k that actually took place. With perfect information, the decision maker would have chosen E_k. Consequently, the regret R_{ik} associated with having chosen alternative A_i is a measure of the value of having perfect information, or of knowing ahead of time that event E_k would occur. When each of the events E_k occurs with probability P_k, it becomes possible to calculate the expected value of perfect information [EVPI(A_i)] given that the decision maker would chose action A_i before receiving this information with the following expression:

$$\text{EVPI}(A_i) = \sum_k P_k R_{ik} \tag{1}$$

The approach above can be extended to the case of imperfect information (Raiffa, 1968) by replacing P_k in equation (1) with the probability of event k (E_k) given the imperfect sample information (I). This results in an expression for the expected value of sample information [EVSI(A_i, I)] given that the decision maker would chose action A_i before receiving this information:

$$\text{EVSI}(A_i, I) = \sum_k (P_k|I) R_{ik} \tag{2}$$

The value of imperfect (or sample) information provides a normative rule for deciding whether to collect additional information. For example, a decision to perform a survey before introducing a product can be made by comparing the cost of the survey to the expected value of the information obtained. It is often assumed that decision makers are biased when they fail to seek out additional information. The discussion above shows that *not* obtaining information is justified when the information costs too much. From a practical perspective, the value of information can guide decisions to provide information to product users (Lehto and Papastavrou, 1991).

2.1.6 Maximizing Expected Value

From elementary probability theory, return is maximized by selecting the alternative with the greatest expected value. The expected value of an action A_i is calculated by weighting its consequences C_{ik} over all events k, by the probability P_{ik} that the event will occur. The expected value of a given action A_i is therefore

$$\text{EV}[A_i] = \sum_k P_{ik} C_{ik} \tag{3}$$

More generally, the decision maker's preference for a given consequence C_{ik} might be defined by a value function $\text{V}(C_{ik})$, which transforms consequences into preference values. The preference values are then weighted using the same equation. The expected value of a given action A_i becomes

$$\text{EV}[A_i] = \sum_k P_{ik} \text{V}(C_{ik}) \tag{4}$$

Monetary value is a common value function. For example, lives lost, units sold, or air quality might all be converted into monetary values. More generally, however, value reflects preference, as illustrated by ordinary concepts such as the value of money or the attractiveness of a work setting. Given that the decision maker has large resources and is given repeated opportunities to make the choice, choices made on the basis of expected monetary value are intuitively justifiable. A large company might make nearly all of its decisions on the basis of expected monetary value. Insurance buying and many other rational forms of behavior cannot, however, be justified on the basis of expected monetary value. It has long been recognized that rational decision makers made choices not easily explained by expected monetary value (Bernoulli, 1738). Bernoulli cited the St. Petersburg paradox, in which the prize received in a lottery was 2^n, n being the number of times a flipped coin turned up heads before a tails was observed. The probability of n flips before the first tail is observed is 0.5^n. The expected value of this lottery becomes

$$\mathrm{EV}[L] = \sum_k P_{ik}\mathrm{V}(C_{ik}) = \sum_{n=0}^{\infty} 0.5^n 2^n = \sum_{n=1}^{\infty} 1 \rightarrow \infty \tag{5}$$

The interesting twist is that the expected value of the lottery above is infinite. Bernoulli's conclusion was that preference cannot be a linear function of monetary value since a rational decision maker would never pay more than a finite amount to play the lottery. Furthermore, the value of the lottery can vary between decision makers. According to utility theory, this variability reflects rational differences in preference between decision makers for uncertain consequences.

2.1.7 Subjective Expected Utility (SEU) Theory

Expected utility theory extended expected value theory to describe better how people make uncertain economic choices (von Neumann and Morgenstern, 1947). In their approach, monetary values are first transformed into utilities, using a utility function $u(x)$. The utilities of each outcome are then weighted by their probability of occurrence to obtain an expected utility. Subjective utility theory (SEU) added the notion that uncertainty about outcomes could be represented with subjective probabilities (Savage, 1954). It was postulated that these subjective estimates could be combined with evidence using Bayes' rule to infer the probabilities of outcomes* (see Section 2.2). This group of assumptions corresponds to the Bayesian approach to statistics. Following this approach, the SEU of an alternative A_i, given subjective probabilities S_{ik} and consequences C_{ik} over events E_k, becomes

$$\mathrm{SEU}[A_i] = \sum_k S_{ik} U(C_{ik}) \tag{6}$$

Note the similarity between formulation (6) for SEU and equation (3) for expected value. EV and SEU are equivalent if the value function equals the utility function. Methods for eliciting value and utility functions differ in nature (Section 3). Preferences elicited for uncertain outcomes measure utility.† Preferences elicited for certain outcomes measure value. It has, accordingly, often been assumed that value functions differ from utility functions, but there are reasons to treat value and utility functions as equivalent (Winterfeldt and Edwards, 1986). The latter authors claim that the differences between elicited value and utility functions are small and that "severe limitations constrain those relationships, and only a few possibilities exist, one of which is that they are the same."

When people are presented with choices that have uncertain outcomes, they react in different ways. In some situations, people find gambling to be pleasurable. In others, people will pay money to reduce uncertainty: for example, when people buy insurance. SEU theory distinguishes between risk neutral, risk averse, risk seeking, and mixed forms of behavior. These different types of behavior are described by the shape of the utility function (Figure 2).

A risk-neutral decision maker will find the expected utility of a gamble to be the same as the utility of the gamble's expected value. That is, expected $u(\text{gamble}) = u(\text{gamble's expected value})$. For a risk-averse decision maker, expected $u(\text{gamble}) < u$ (gamble's expected value); for a risk-seeking decision maker, expected $u(\text{gamble}) > u(\text{gamble's expected value})$. On any given point of a utility function, attitudes toward risk are described formally by the coefficient of risk aversion:

$$C_{\mathrm{RA}} = \frac{u''(x)}{u'(x)} \tag{7}$$

where $u'(x)$ and $u''(x)$ are, respectively, the first and second derivatives of $u(x)$ taken with respect to x. Note that when $u(x)$ is a linear function of x [i.e., $u(x) = ax + b$], then $C_{\mathrm{RA}} = 0$. For any point of the utility function, if $C_{\mathrm{RA}} < 0$, the utility function depicts risk-averse behavior, and if $C_{\mathrm{RA}} > 0$, the utility function depicts risk-seeking behavior. The coefficient of risk aversion therefore describes attitudes toward risk at each point of the utility function, given that the utility function is continuous. SEU theory consequently

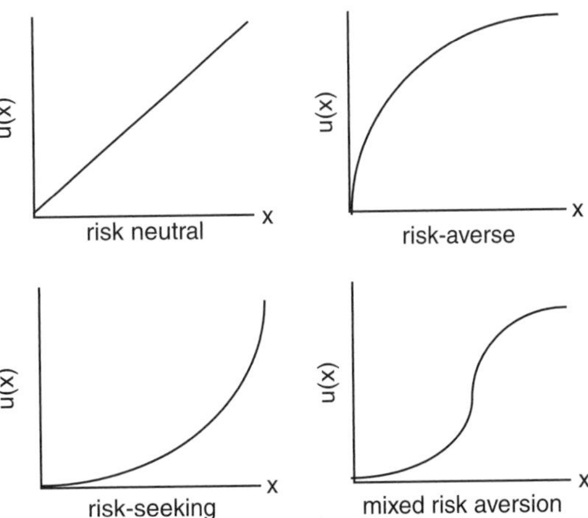

Figure 2 Utility functions for differing risk attitudes.

*When no evidence is available concerning the likelihood of different events, it was postulated that each consequence should be assumed to be equally likely. The Laplace decision rule makes this assumption and then compares alternatives on the basis of expected value or utility.

†Note that classical utility theory assumes that utilities are constant. Utilities may, of course, fluctuate. The random utility model (Bock and Jones, 1968) allows such fluctuation.

provides a powerful tool for describing how people might react to uncertain or risky outcomes. However, some commonly observed preferences between risky alternatives cannot be explained by SEU. Section 4.2 focuses on experimental findings showing deviations from the predictions of SEU.

A major contribution of SEU is that it represents differing attitudes toward risk and provides a normative model of decision making under uncertainty. The prescriptions of SEU are also clear and testable. Consequently, SEU has played a major role in fields other than economics, both as a tool for improving human decision making and as a stepping stone for developing models that describe how people make decisions when outcomes are uncertain. As discussed further in Section 4, much of this work has been done in psychology.

2.1.8 Multiattribute Utility Theory

Multiattribute utility theory (Keeney and Raiffa, 1976) extends SEU to the case where the decision maker has multiple objectives. The approach is equally applicable for describing utility and value functions. Following this approach, the utility (or value) of an alternative A, with multiple attributes x, is described with the multiattribute utility (or value) function $u(x_1 \cdots x_n)$, where $u(x_1 \cdots x_n)$ is some function $f(x_1 \cdots x_n)$ of the attributes x. In the simplest case, multiattribute utility theory (MAUT) describes the utility of an alternative as an additive function of the single-attribute utility functions $u_n(x_n)$. That is,

$$u(x_1 \cdots x_n) = \sum_n k_n u_n(x_n) \tag{8}$$

where the constants k_n are used to weight each single-attribute utility function (u_n) in terms of its importance. Assuming that an alternative has three attributes, x, y, and z, an additive utility function is $u(x, y, z) = k_x u_x(x) + k_y u_y(y) + k_z u_z(z)$. Along these lines, a community considering building a bridge across a river vs. building a tunnel or continuing to use the existing ferry system might consider the attractiveness of each option in terms of the attributes of economic benefits, social benefits, and environmental benefits.*

More complex multiattribute utility functions include multiplicative forms and functions that combine utility functions for subsets of two or more attributes (Keeney and Raiffa, 1976). An example of a simple multiplicative function would be $u(x, y) = u_x(x)u_y(y)$. A function that combines utility functions for subsets, would be $u(x, y, z) = k_{xy}u_{xy}(x, y) + k_z u_z(z)$. The latter type of function becomes useful when utility independence is violated. Utility independence is violated when the utility function for one attribute depends on the value of another attribute. Along these lines, when assessing $u_{xy}(x, y)$ it might be found that $u_x(x)$ depends on the value of y. For example, people's reaction to the level of crime in their own neighborhood might depend on the level of crime in a nearby suburb. In the latter case, it is probably better to measure u_{xy} (x is crime in one's own neighborhood and y is crime in a nearby suburb) directly than to estimate it from the single-attribute functions. Assessment of utility and value functions is discussed in Section 3.

MAUT has been applied to a wide variety of problems (Keeney and Raiffa, 1976; Winterfeldt and Edwards, 1986; Saaty 1988; Clemen, 1996). An advantage of MAUT is that it helps structure complex decisions in a meaningful way. Alternative choices and their attributes often naturally divide into hierarchies. The MAUT approach encourages such divide-and-conquer strategies and, especially in its additive form, provides a straightforward means of recombining weights into a final ranking of alternatives. The MAUT approach is also a compensatory strategy that allows normative trade-offs between attributes in terms of their importance.

2.1.9 Holistic Comparison

Holistic comparison is a nonanalytical method of comparing alternatives. This process involves a holistic comparison of the consequences for each alternative instead of measuring separately and then recombining measures of probability, value, or utility (Janis and Mann, 1977; Sage, 1981; Stanoulov, 1994). A preference ordering between alternatives is thus obtained. For example, the decision maker might rank in order of preference a set of automobiles that vary on objectively measurable attributes such as color, size, and price. Mathematical tools can then be used to derive the relationship between the ordering and attribute values observed and ultimately predict preferences for unevaluated alternatives, as discussed in Section 3.3.4.

One advantage of holistic comparison is that it requires no formal consideration of probability or utility. Consequently, decision makers unfamiliar with these concepts may find holistic comparison to be more intuitive, and potential violations of the axioms underlying SEU and MAUT, due to their lack of understanding, become of lesser concern. People seem to find the holistic approach helpful when they compare complex alternatives (Janis and Mann, 1977). In fact, people may feel that there is little additional benefit to be obtained from analyzing separately the probability and value attached to each attribute. This tendency becomes prevalent in naturalistic decision making, as addressed further in Section 5.

2.2 Statistical Inference

Inference is the procedure followed when a decision maker uses information to determine whether a hypothesis about the world is true. Hypotheses can specify past, present, or future states of the world, or causal relationships between variables. Diagnosis

*To develop the multiattribute utility function, the single-attribute utility functions (u_n) and the importance weights (k_n) are determined by assessing preferences between alternatives. Methods of doing so are discussed in Section 3.4.

is concerned with determining past and present states of the world. Prediction is concerned with determining future states. Inference or diagnosis is required in many decision contexts. For example, before deciding on a treatment, a physician must diagnose the illness.

From the classical perspective, the decision maker is concerned with determining the likelihood that a hypothesis H_i is true. Bayesian inference is the best-known technique, but signal detection theory and fundamentally different approaches such as the Dempster–Schafer method have seen application. Each of these approaches is discussed below.

2.2.1 Bayesian Inference

Bayesian inference is a well-defined procedure for inferring the probability P_i that a hypothesis H_i is true, from evidence E_j linking the hypothesis to other observed states of the world. The approach makes use of Bayes' rule to combine the various sources of evidence (Savage, 1954). *Bayes' rule* states that the posterior probability of hypothesis H_i given that evidence E_j is present, $P(H_i|E_j)$, is given by the equation

$$P(H_i|E_j) = \frac{P(E_j|H_i)P(H_i)}{P(E_j)} \qquad (9)$$

where $P(H_i)$ is the probability of the hypothesis being true prior to obtaining the evidence E_j and $P(E_j|H_i)$ is the probability of obtaining the evidence E_j given that the hypothesis H_i is true. For example, consider the case where a physician is attempting to determine whether a patient has a disease that is present in 10% of the general population. The physician has a test available that gives a positive result 90% of the time when administered to patients who actually have the disease. The test also gives a positive result 20% of the time when administered to patients who don't have the disease. If the test were to be administered to a member of the general population, equation (9) predicts that the probability of having the disease given a positive test result is

$$P(\text{disease}|\text{positive test}) = \frac{\begin{array}{c}P(\text{positive test}|\text{disease})\\ P(\text{disease in general population})\end{array}}{P(\text{positive test})}$$

Also,

$$\begin{aligned}P(\text{positive test}) &= P(\text{positive test}|\text{disease}) \\ &\quad \times P(\text{disease in general population}) \\ &\quad + P(\text{positive test}|\text{no disease}) \\ &\quad \times P(\text{no disease in general population})\end{aligned}$$

$$P(\text{disease}|\text{positive test}) = \frac{(0.9)(0.1)}{(0.9)(0.1) + (0.2)(0.9)} = 0.33$$

As discussed further in Section 4.1, people often fail to combine evidence consistently with the predictions of Bayes' rule above. A common finding is that people fail to consider adequately the base rate of the hypothesis. In the example above, this would correspond to focusing on P(positive test|disease) = 0.9 and not considering P(disease in general population) = 0.1. As a consequence, many people might be surprised that P(disease|positive test) = 0.33 rather than a number close to 0.9.

When the evidence E_j consists of multiple states $E_1, \ldots, E_n$, each of which is conditionally independent, Bayes' rule can be expanded into the expression

$$P(H_i|E_j) = \frac{\prod_{j=1}^{n} P(E_j|H_i)P(H_i)}{P(E_j)} \qquad (10)$$

Calculating $P(E_j)$ can be somewhat difficult, due to the fact that each piece of evidence must be dependent* or else it would not be related to the hypothesis. The odds forms of Bayes' rule provides a convenient way of looking at the evidence for and against a hypothesis that does not require $P(E_j)$ to be calculated. This results in the expression

$$\theta(H_i|E_j) = \frac{P(H_i|E_j)}{P(\sim H_i|E_j)} = \frac{\prod_{j=1}^{n} P(E_j|H_i)P(H_i)}{\prod_{j=1}^{n} P(E_j|\sim H_i)P(\sim H_i)} \qquad (11)$$

where $\theta(H_i|E_j)$ refers to the posterior odds for hypothesis H_i, $P(\sim H_i)$ is the prior probability that hypothesis H_i is not true, and $P(\sim H_i|E_j)$ is the posterior probability that hypothesis H_i is not true.

The two latter forms of Bayes' rule provide an analytically simple way of combining multiple sources of evidence. Bayesian inference becomes much more difficult when the evidence is not certain or when the conditional independence assumption is not met. When evidence is not certain, complex multistage forms of Bayesian analysis are required that consider the probability of the evidence being true (Winterfeldt and Edwards, 1986). When conditional independence is not true, the expanded form of Bayes' rule must be modified. For example, consider the case where the evidence consists of three events (E_1, E_2, E_3), where E_1 and E_2 are conditionally dependent and E_3 is conditionally independent of the other two events. The posterior probability, $P(H_i|E_1, E_2, E_3)$, then becomes

$$P(H_i|E_1, E_2, E_3) = \frac{P(E_1, E_2|H_i)P(E_3|H_i)P(H_i)}{P(E_1, E_2)P(E_3|E_1, E_2)} \qquad (12)$$

*Note that conditional independence between E_1 and E_2 implies that $P(E_1|H_i, E_2) = P(E_1|H_i)$ and that $P(E_2|H_i, E_1) = P(E_2|H_i)$. This is very different from simple independence, which implies that $P(E_1) = P(E_1|E_2)$ and that $P(E_2) = P(E_2|E_1)$.

where $P(E_1, E_2|H_i)$ is the conditional probability of obtaining E_1 and E_2 given the hypothesis H_i, $P(E_3|H_i)$ is the conditional probability of obtaining E_3 given H_i, and $P(E_1, E_2)P(E_3|E_1, E_2)$ is the probability of obtaining the evidence (E_1, E_2, E_3).

2.2.2 Signal-Detection Theory

Bayesian inference combined with SEU leads to signal-detection theory (Tanner and Swets 1954), which has been applied in a large variety of contexts to model human performance (Wickens, 1992). In signal-detection theory, the human operator is assumed to use Bayes' rule to estimate the probability that a signal actually is present from a noisy observation of the system. For example, an operator might estimate the probability that a machine is going out of tolerance from a warning signal. The responses of the operator and the true state of the system together determine a set of four outcomes (Table 4).

The signal-detection model assumes that an operator receives evidence from the environment regarding the true state of the world. The relationship between the signal (S) and the evidence (E) is measured by the conditional probability $[P(E|S)]$ of obtaining the evidence observed, given that the signal is there. The decision maker is assumed to select a criterion value (x_c) that the evidence must exceed before saying yes. It is assumed that the value chosen will maximize utility. If the evidence is represented with a variable x, the expected utility of the operator can be described in terms of x, x_c, and the four outcomes in Table 4. The expected utility for a given probability cutoff x_c is given by the expression

$$\begin{aligned}\text{SEU}[x_c] &= P(x \geq x_c|S)P(S)u(h) \\ &\quad + P(x \geq x_c|N)P(N)u(\text{fa}) \\ &\quad + P(x < x_c|S)P(S)u(\text{m}) \\ &\quad + P(x < x_c|N)P(N)u(\text{cr}) \qquad (13)\end{aligned}$$

Expression (13) can be maximized by first substituting $1 - P(x \geq x_c|N)$ for $P(x < x_c|N)$ and substituting $1 - P(x \geq x_c|S)$ for $P(x < x_c|S)$ into the equation for SEU$[x_c]$, and then setting the derivative of SEU$[x_c]$ with respect to x_c to zero. The result at the cutoff x_c is

$$\frac{P(x = x_c|S)}{P(x = x_c|N)} = \beta^* = \frac{P(N)[u(\text{cr}) - u(\text{fa})]}{P(S)[u(\text{h}) - u(\text{m})]} \qquad (14)$$

Table 4 Potential Outcomes Considered by Signal Detection Theory

	State of the World	
Response	Noise (*N*)	Signal (*S*)
Yes	False alarm (fa)	Hit (h)
No	Miss (m)	Correct rejection (cr)

Substituting back the relation $P(E|S) = x$, the optimal decision rule is to say yes if

$$\frac{P(E|S)}{P(E|N)} \geq \beta^* \qquad (15)$$

Equation (15) can be extended to multiple operators or multiple sources of evidence (Lehto and Papastavrou, 1991). The resulting expression takes into account the probability of a false alarm and the probability of detection for the other source of information. Lehto and Papastavrou use this approach to analyze situations where the other source of information is a warning signal. The extent to which human judgments correspond to the predictions of Bayes' rule is discussed further in Section 4.1.

2.2.3 Dempster–Schafer Method

The Dempster–Schafer method (Schafer, 1976; Fedrizzi et al., 1994) is an alternative to Bayesian inference for accumulating evidence for or against a hypothesis that has been proposed for use in decision analysis (Strat, 1994). In this approach, the relation of hypotheses (H) to evidence (e) is described by a basic probability assignment (bpa) function p. Given evidence e, the function $p_e(n)$ assigns a value between 0 and 1 to each subset of H such that the sum of the values assigned is 1. For example, consider the case where there are three hypotheses (A,B,C). When no evidence is available, the vacuous bpa assigns a value of 1 to the set of hypotheses $H = (A, B, C)$ and a 0 to all subsets. That is, subsets (A), (B), (C), (A, B), and (A, C) are each assigned a value of 0. The Bayesian approach would instead assign a probability of 0.33 to A, B, and C, respectively.

Also, given that evidence $p_e(A) = x$ supporting a specific hypothesis A is found, the Dempster–Schafer approach assigns $1 - p_e(A)$ to H. The Bayesian approach, of course, assigns $1 - p_e(A)$ to the complement of A. Returning to the example above, suppose that the evidence supports hypothesis A to the degree $p_e(A) = 0.6$. Using the Dempster–Schafer approach, $p_e(A, B, C) = 0.4$. This, of course, is very different from the Bayes' interpretation, where $P(A) = 0.6$ and $P(\text{not } A) = 0.4$. The Dempster–Schafer method uses a belief function B(n) to assign a total belief to n, where n is a subset of the set of possible hypotheses H, as the sum of the beliefs assigned to m, where m is the set of possible subsets of n. In the example above, the belief in (A, B, C) after receiving evidence e is

$$\begin{aligned}\text{B}(A, B, C) &= p_e(A, B, C) + p_e(A, B) + p_e(A, C) \\ &\quad + p_e(B, C) + p_e(A) + p_e(B) + p_e(C) \\ &= 0.4 + 0 + 0 + 0 + 0.6 + 0 + 0 \\ &= 1.0 \qquad (16)\end{aligned}$$

Similarly, the belief in A,B after receiving the evidence e is

$$\text{B}(A, B) = p_e(A, B) + p_e(A) + p_e(B)$$

$$= 0 + 0.6 + 0$$
$$= 0.6$$
$$= \mathrm{B}(A) \tag{17}$$

To combine evidence from multiple sources e and f, Dempster–Schafer theory uses the combining function $c(p_e(X), p_f(Y))$, where X and Y are both sets of subsets of H. For example, we might have $X = [(A), (A, B, C)]$ and $Y = [(A, B), (A, B, C)]$. The combining function then assigns a value to each subset n of H. The value assigned is determined first by describing the set of subsets n' within n defined by the intersection of subsets within X and subsets within Y. A value of 0 is assigned to all subsets of n not within n'. The products $p_e(X)p_f(Y)$ are then summed and assigned to each subset within n'. Returning to the example above, we can calculate $c(n')$ using the values given in Table 5. First note that the set of subsets n' for the example is defined by the inner elements of the table. Specifically, $n' = [(A), (A, B), (A, B, C)]$. The values used by the combining function $c(n')$ are also shown. Using these numbers, the values of $c(n')$ become $c(A) = 0.24 + 0.36 = 0.6$, $c(A, B) = 0.16$, $c(A, B, C) = 0.24$. All remaining subsets, for this evidence are assigned a value of 0.

It has been argued that the Dempster–Schafer method of assigning evidence is better suited than the Bayesian method for diagnosing medical problems (Gordon and Shortliffe, 1984). These researchers criticize particularly the Bayesian assumption that evidence partially supporting a hypothesis should also support its negation. Gordon and Shortliffe note that the Dempster–Schafer method shows promise as a means of accumulating belief in expert diagnostic systems used in medicine.

3 DECISION ANALYSIS

The application of classical decision theory to improve human decision making is the goal of decision analysis (Howard, 1968, 1988; Raiffa, 1968; Keeney and Raiffa, 1976). Decision analysis requires inputs from decision makers, such as goals, preference and importance measures, and subjective probabilities. Elicitation techniques have consequently been developed that help decision makers provide these inputs. Particular focus has been placed on methods of quantifying preferences, trade-offs between conflicting objectives, and uncertainty (Raiffa, 1968; Keeney and Raiffa, 1976). As a first step in decision analysis, it is necessary to do some preliminary structuring of the decision, which then guides the elicitation process. The following discussion first presents methods of structuring decisions and then covers techniques for assessing subjective probabilities, utility functions, and preferences.

Table 5 Tableau for Dempster–Shafer Method of Combining Evidence

	$Y = [(A, B), (A, B, C)]$	
$X = [(A), (A, B, C)]$	$p_f(A, B) = 0.4$	$p_f(A, B, C) = 0.6$
$p_e(A) = 0.6$	$A; p_e p_f = 0.24$	$A; p_e p_f = 0.36$
$p_e(A, B, C) = 0.4$	$A, B; p_e p_f = 0.16$	$A, B, C; p_e p_f = 0.24$

3.1 Structuring Decisions

The field of decision analysis has developed many useful frameworks for representing what is known about a decision (Howard, 1968; Winterfelt and Edwards, 1986; Clemen, 1996). In fact, these authors and others have stated that the process of structuring decisions is often the greatest contribution of going through the process of decision analysis. Among the many tools used, decision matrices and trees provide a convenient framework for comparing decisions on the basis of expected value or utility. Value trees provide a helpful method of structuring the sometimes complex relationships among objectives, attributes, goals, and values and are used extensively in multiattribute decision-making problems. Event trees, fault trees, inference trees, and influence diagrams are useful for describing probabilistic relationships between events and decisions. Each of these approaches is discussed briefly below.

3.1.1 Decision Matrices and Trees

Decision matrices are often used to represent single-stage decisions (Figure 3). The simplicity of decision matrices is their primary advantage. They also provide a very convenient format for applying the decision rules discussed in Section 2. Decision trees are also commonly used to represent single-stage decisions (Figure 4) and are particularly useful for describing multistage decisions (Raiffa, 1968). Note that in a multistage decision tree, the probabilities of later events are conditioned on the result of earlier events. This leads to the important insight that the results of earlier events provide information regarding future events.* Following this approach, decisions may be stated in conditional form. An optimal decision, for example, might be to do a market survey first, then market the product only if the survey is positive.

Analysis of a single- or multistage decision tree involves two basic steps, averaging out and folding

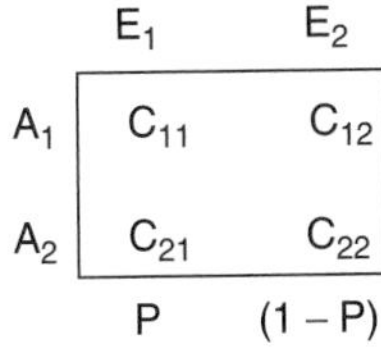

Figure 3 Decision matrix representation of a single-stage decision.

*For example, the first event in a decision tree might be the result of a test. The test result then provides information useful in making the final decision.

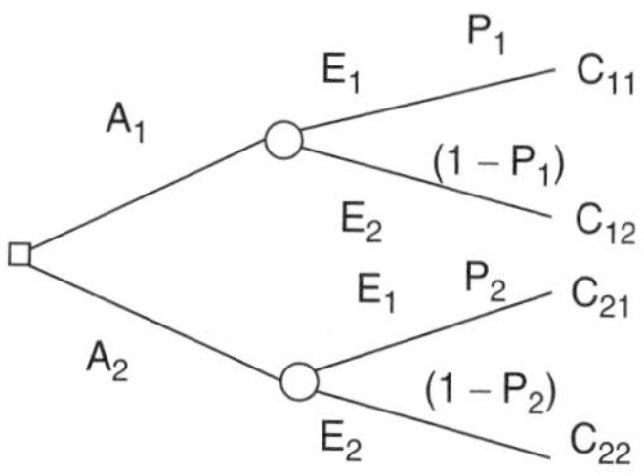

Figure 4 Decision tree representation of a single-stage decision.

back (Raiffa, 1968). These steps occur at chance and decision nodes respectively.* *Averaging out* occurs when the expected value (or utility) at each chance node is calculated. In Figure 4 this corresponds to calculating the expected value of A_1 and A_2, respectively. *Folding back* refers to choosing the action with the greatest value expected at each decision node.

Decision trees thus provide a straightforward way of comparing alternatives in terms of expected value or SEU. However, their development requires significant simplification of most decisions and the provision of numbers, such as measures of preference and subjective probabilities, that decision makers may have difficulty determining. In certain contexts, decision makers struggling with this issue may find it helpful to develop value trees, event trees, or influence diagrams, as expanded on below.

3.1.2 Value Trees

Value trees hierarchically organize objectives, attributes, goals, and values (Figure 5). From this perspective, an objective corresponds to satisficing or maximizing a goal or set of goals. When there is more than one goal, the decision maker will have multiple objectives, which may differ in importance. Objectives

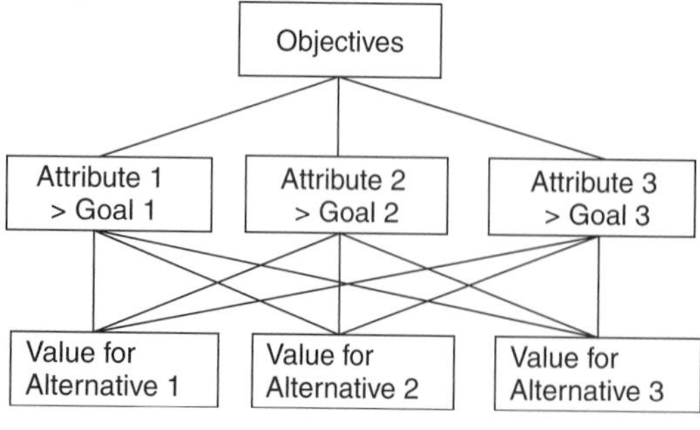

Figure 5 Generic value tree.

*Note that the standard convention uses circles to denote chance nodes and squares to denote decision nodes (Raiffa, 1968).

and goals are both measured on a set of attributes. Attributes may provide (1) objective measures of an goal, such as when fatalities and injuries are used as a measure of highway safety; (2) subjective measures of an goal, such as when people are asked to rate the quality of life in the suburbs vs. the city; or (3) proxy or indirect measures of a goal, such as when the quality of ambulance service is measured in terms of response time.

In generating objectives and attributes, it becomes important to consider their relevance, completeness, and independence. Desirable properties of attributes (Keeney and Raiffa, 1976) include:

1. *Completeness*: the extent to which the attributes measure whether an objective is met.
2. *Operationality*: the degree to which the attributes are meaningful and feasible to measure.
3. *Decomposability*: whether the whole is described by its parts.
4. *Nonredundancy*: the fact that correlated attributes give misleading results.
5. *Minimum size*: the fact that considering irrelevant attributes is expensive and may be misleading.

Once a value tree has been generated, various methods can be used to assess preferences directly between the alternatives.

3.1.3 Event Trees or Networks

Event trees or networks show how a sequence of events can lead from primary events to one or more outcomes. Human reliability analysis (HRA) event trees are a classic example of this approach (Figure 6). If probabilities are attached to the primary events, it becomes possible to calculate the probability of outcomes, as illustrated in Section 3.2.4. This approach has been used in the field of risk assessment to estimate the reliability of human operators and other elements of complex systems (Gertman and Blackman, 1994). Chapter 27 provides additional information on human reliability analysis and other methods of risk assessment.

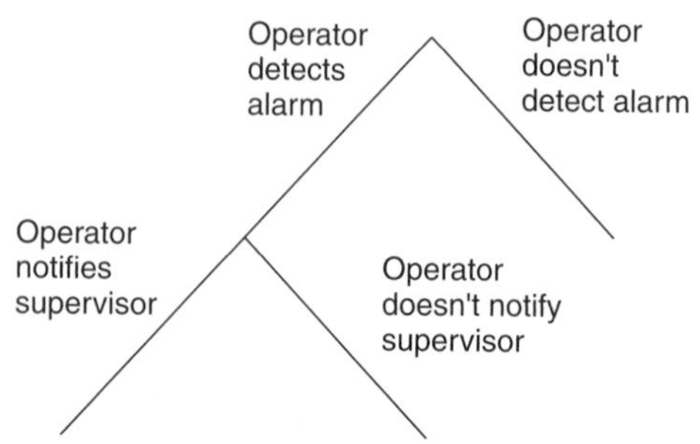

Figure 6 HRA event tree. (Adapted from Gertman and Blackman, 1994.)

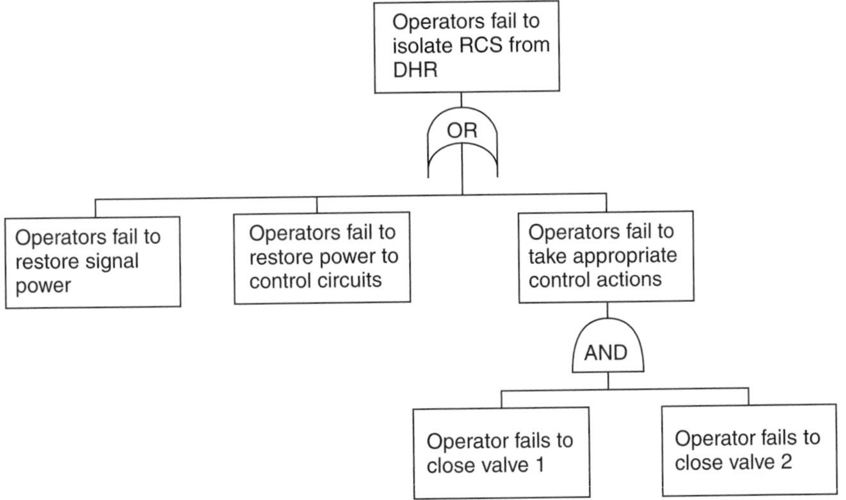

Figure 7 Fault tree for operators. (Adapted from Gertman and Blackman, 1994.)

Fault trees work backward from a single undesired event to its causes (Figure 7). Fault trees are commonly used in risk assessment to help infer the chance of an accident occurring (Hammer, 1993; Gertman and Blackman, 1994). Inference trees relate a set of hypotheses at the top level of the tree to evidence depicted at the lower levels. The latter approach has been used by expert systems such as Prospector (Duda et al., 1979). Prospector applies a Bayesian approach to infer the presence of a mineral deposit from uncertain evidence.

3.1.4 Influence Diagrams and Cognitive Mapping

Influence diagrams are often used in the early stages of a decision to show how events and actions are related. Their use in the early stages of a decision is referred to as *knowledge* (or *cognitive*) *mapping* (Howard, 1988). Links in an inference diagram depict causal and temporal relations between events and decision stages.* A link leading from event *A* to event *B* implies that the probability of obtaining event *B* depends on whether event *A* has occurred. A link leading from a decision to an event implies that the probability of the event depends on the choice made at that decision stage. A link leading from an event to a decision implies that the decision maker knows the outcome of the event at the time the decision is made.

One advantage of influence diagrams in comparison to decision trees is that influence diagrams show the relationships between events more explicitly. Consequently, influence diagrams are often used to represent complicated decisions where events interactively influence the outcomes. For example, the influence diagram

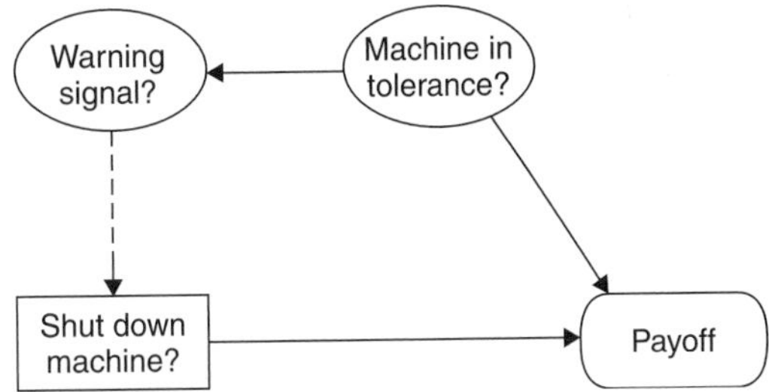

Figure 8 Influence diagram representation of a single-stage decision.

in Figure 8 shows that the true state of the machine affects both the probability of the warning signal and the consequence of the operator's decision. This linkage would be hidden within a decision tree.† Influence diagrams have been used to structure medical decision-making problems (Holtzman, 1989) and are emphasized in modern texts on decision analysis (Clemen, 1996). Howard (1988) states that influence diagrams are the greatest advance he has seen in the communication, elicitation, and detailed representation of human knowledge. Part of the issue is that influence diagrams allow people who do not have deep knowledge of

*As for decision trees, the convention for influence diagrams is to depict events with circles and decisions with squares.

†The conditional probabilities in a decision tree would reflect this linkage, but the structure of the tree itself does not show the linkage directly. Also, the decision tree would use the flipped probability tree using P(warning) at the first stage and P(machine down|warning) at the second stage. It seems more natural for operators to think about the problem in terms of P(machine down) and P(warning|machine down), which is the way the influence diagram in Figure 7 depicts the relationship.

probability to describe complex conditional relationships with simple linkages between events. Once these linkages are defined, the decision becomes well defined and can be formally analyzed.

3.2 Probability Assessment

Several approaches have been used in decision analysis to assess subjective probabilities. In this section several of the better known techniques are summarized. These techniques include (1) direct numerical assessment, (2) fitting subjective belief forms, (3) the bisection method, (4) conditioning arguments, (5) preferences between reference gambles, and (6) scaling methods. Techniques proposed for improving the accuracy of assessed probabilities, including scoring rules, calibration, and group assessment, are then presented.

3.2.1 Direct Numerical Assessment

In direct numerical estimation, decision makers are asked to give a numerical estimate of how likely they think an event is to happen. These estimates can be probabilities, odds, log odds, or words (Winterfeldt and Edwards, 1986). Winterfeldt and Edwards argue that log odds have certain advantages over the other measures. Gertman and Blackman (1994) note that log odds are normally used in risk assessment for nuclear power applications because human error probabilities (HEPs) vary greatly in value. HEPs between 1 and 0.00001 are typical.

3.2.2 Fitting a Subjective Belief Form

Fitting a subjective belief form requires that the questions be posed in terms of statistical parameters. That is, decision makers could be asked first to consider their uncertainty regarding the true value of a given probability and then estimate their mean, mode, or median belief. This approach can be extended by asking decision makers to describe how certain they are of their estimate. For example, a worker might subjectively estimate the mean and variance of the proportion of defective circuit boards before inspecting a small sample of circuit boards. If the best estimate corresponds to a mean, mode, or median, and the estimate of certainty to a confidence interval or standard deviation, a functional form such as the beta $-$ 1 probability density function (pdf) can then be used to fit a subjective probability distribution (Buck, 1989; Clemen, 1996).

In other words, a distribution is specified that describes a subject's belief that the true probability equals particular values. This type of distribution can be said to express uncertainty about uncertainty (Raiffa, 1968). Given that the subject's belief can be described with a beta $-$ 1 pdf, Bayesian methods can be used to combine binomially distributed evidence easily with the subject's prior belief (Buck, 1989; Clemen, 1996). Returning to the example above, the worker's prior subjective belief can be combined with the results of inspecting the small sample of circuit boards, using Bayes' rule. As more evidence is collected, the weight given to the subject's initial belief becomes smaller compared to the evidence collected. The use of prior belief forms also reduces the amount of sample information that must be collected to show that a proportion, such as the percentage of defective items, has changed (Buck, 1989).

3.2.3 Bisection Method

The bisection method (Raiffa, 1968) is another direct technique for attempting to estimate a subjective pdf. This technique is somewhat more general than fitting the subject's belief with a functional form, such as the beta $-$ 1, since it makes no parametric assumptions. The bisection method involves two steps that are repeated until the subject's belief is adequately described. Following this approach, the first step is to determine the median ($p_{0.5}$) of the subjective pdf. This question is posed to the decision maker in a form such as: "For what value of p do you feel it is equally likely that the true value p^* is greater than or less than p?" This step is then repeated for subintervals to obtain the desired level of detail.

3.2.4 Conditioning Arguments

Statistical conditioning arguments are based on the idea that the probability of a complicated event, such as the chance of having an accident, can be determined by estimating the probability of simpler events (or subsets). From a more formal perspective, a conditioning argument determines the probability of event A by considering the possible conditions C_i under which A might happen, the associated conditional probabilities $[P(A|C_i)]$, and the probability of each condition, $P(C_i)$. The probability of A can then be represented as

$$P(A) = \sum_i P(A|C_i)P(C_i) \tag{18}$$

This approach is illustrated by the development of event trees and fault tree analysis. In fault tree analysis, the probability of an accident is estimated by considering the probability of human errors, component failures, and other events. This approach has been applied extensively in the field of risk analysis* (Gertman and Blackman, 1994). Therp (Swain and Guttman, 1983) extends the conditioning approach to the evaluation of human reliability in complex systems.

Slim-Maud (Embrey, 1984) implements a related approach in which expert ratings are used to estimate human error probabilities (HEPs) in various environments. The experts first rate a set of tasks in terms of performance-shaping factors (PSFs) that are present. Tasks with known HEPs are used as upper and lower anchor values. The experts also judge the importance of individual PSFs. A subjective likelihood index (SLI) is then calculated for each task in terms of the PSFs. A logarithmic relationship is assumed between the HEP

*Note that it has been shown that people viewing fault trees can be insensitive to missing information (Fischhoff et al, 1978).

and SLI, allowing calculation of the human error probability for task j (HEP_j) from the subjective likelihood index assigned to task j (SLI_j). More specifically,

$$\log(1 - \text{HEP}_j) = a\text{SLI}_j + b \qquad (19)$$

where

$$\text{SLI}_j = \sum_i \text{PSF}_{ij}\text{I}(\text{PSF}_i) \qquad (20)$$

where $\text{I}(\text{PSF}_i)$ is the importance of PSF_i, and PSF_{ij} is the rating given to PSF_i for task j. Gertman and Blackman (1994) provide guidelines regarding the use of this method and have generally positive conclusions. Slim-Maud is interesting in that it uses multiattribute utility theory as a basis for generating probability estimates.

3.2.5 Reference Lotteries

Reference lottery methods take a less direct approach to obtaining point estimates of the decision maker's subjective probabilities. When the objective is to measure how likely event A is to occur, the approach asks decision makers to consider a lottery where they will receive a prize x if event A occurs, and a prize y if it does not. They are then asked how much they would be willing to pay for the lottery. The amount they are willing to pay, z, is then equated to the lottery, using the relation $z = P(A)x + [1 - P(A)]y$. From this expression it becomes possible to estimate the decision maker's subjective estimate of $P(A)$. Specifically, $P(A) = (z - y)/(x - y)$. A variant of this approach that asks decision makers to compare two lotteries over the same range of preferences might be preferable because it removes the potential effect of risk aversion (Winterfelt and Edwards, 1986).

3.2.6 Scaling Methods

Scaling methods ask subjects to rate or rank the probabilities to be assessed. Likert scales with verbal anchors have been used to obtain estimates of how likely people feel certain risks are (Kraus and Slovic, 1988). Another approach has been to ask subjects to do pairwise comparisons of the likelihoods of alternative events (Saaty, 1988). Pairwise comparisons of probabilities on a ratio scale correspond to relative odds and consequently have high construct validity. In fact, much of risk assessment focuses on determining order-of-magnitude differences in probability. Saaty (1988), however, argues that the psychometric literature indicates that people's ability to distinguish items on the same scale is limited to 7 ± 2 categories. Consequently, he proposes use of a relative scale to measure differences in importance, preference, and probability that uses verbal anchors corresponding to equal, weak, strong, very strong, and absolute differences between rated items. In perhaps the most controversial aspect of his approach, these five verbal anchors are assigned the numbers 1, 3, 5, 7, and 9. Using these numbers, subjective probabilities can then be calculated from pairwise ratings on his verbal scale.

3.2.7 Scoring Rules, Calibration, and Group Assessment

A number of approaches have been developed for improving the accuracy of assessed probabilities (Lichtenstein et al., 1982; Winterfelt and Edwards, 1986). Two desirable properties of elicited probabilities include extremeness and calibration. More extreme probabilities [e.g., $P(\text{good sales}) = 0.9$ vs. $P(\text{good sales}) = 0.5$] make decisions easier since the decision maker can be more sure of what is really going to happen. Well-calibrated probability estimates match the actual frequencies of observed events. Scoring rules provide a means of evaluating assessed probabilities in terms of both extremeness and calibration. If decision makers assess probabilities on a routine basis, feedback can be provided using scoring rules. Such feedback seems to be associated with the highly calibrated subjective probabilities provided by weather forecasters (Murphy and Winkler, 1974).

Group assessment of subjective probabilities is another often-followed approach, as alluded to earlier in reference to Slim-Maud. There is evidence that group judgments are usually more accurate than individual judgments and that groups tend to be more confident in their estimates (Sniezek and Henry, 1989; Sniezek, 1992; see also Section 6.2). Assuming that individuals within a group provide independent estimates, which are then averaged, the benefit of group judgment is easily shown to have a mathematical basis. Simply put, a mean should be more reliable than an individual observation. Group dynamics, however, can lead to a tendency toward conformity (Janis, 1972). Winterfeldt and Edwards (1986) therefore recommend that members of a group be polled independently.

3.3 Utility Function Assessment

Standard methods for assessing utility functions (Raiffa, 1968) include (1) the variable probability method, and (2) the certainty equivalent method. In the variable probability method, the decision maker is asked to give the value for the probability of winning at which they are indifferent between a gamble and a certain outcome (Figure 9). A utility function is then mapped out when the value of the certainty equivalent (CE) is changed over the range of outcomes. Returning to Figure 9, the value of P at which the decision maker is indifferent between the gamble and the certain loss of \$50 gives the value for $u(-\$50)$. In the

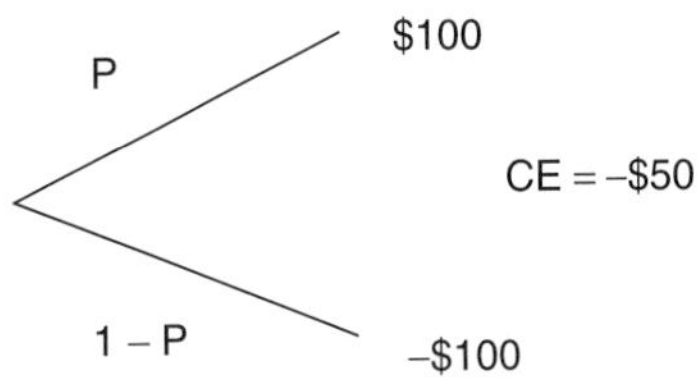

Figure 9 Standard gamble used in the variable probability method of eliciting utility functions.

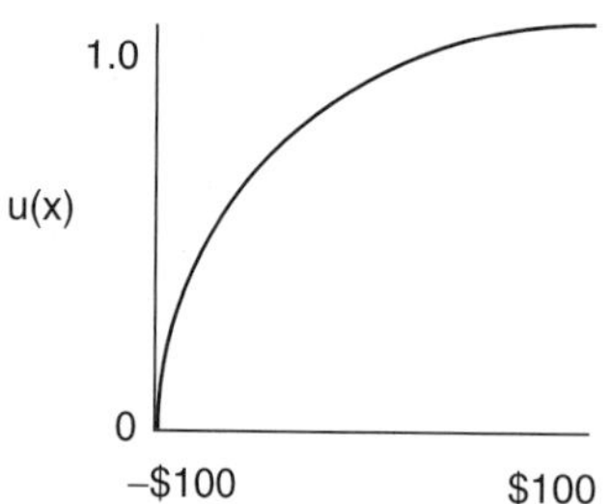

Figure 10 Typical utility function.

utility function in Figure 10, the decision maker gave a value of about 0.5 in response to this question.

The certainty equivalent method uses lotteries in a similar way. The major change is that the probability of winning or losing the lottery is held constant while the amount won or lost is changed. In most cases the lottery provides an equal chance of winning and losing. The method begins by asking the decision maker to give a certainty equivalent for the original lottery (CE_1). The value chosen has a utility of 0.5. This follows since the utility of the best outcome is assigned a value of 1 and the worst is given a utility of 0. The utility of the original gamble is therefore

$$\begin{aligned} u(\mathrm{CE}_1) &= pu(\text{best}) + (1-p)u(\text{worst}) \\ &= p(1) + (1-p)(0) = p = 0.5 \quad (21) \end{aligned}$$

The decision maker is then asked to give certainty equivalents for two new lotteries. Each uses the CE from the previous lottery as one of the potential prizes. The other prizes used in the two lotteries are the best and worst outcomes from the original lottery, respectively. The utility of the certainty equivalent (CE_2) for the lottery using the best outcome and CE_1 is given by

$$\begin{aligned} u(\mathrm{CE}_2) &= pu(\text{best}) + (1-p)u(\mathrm{CE}_1) \\ &= p(1) + (1-p)(0.5) = 0.75 \quad (22) \end{aligned}$$

The utility of the certainty equivalent (CE_3) given for the lottery using the worst outcome and CE_1 is given by

$$\begin{aligned} u(\mathrm{CE}_3) &= pu(\mathrm{CE}_1) + (1-p)u(\text{worst}) \\ &= p(0.5) + (1-p)(0) = 0.25 \quad (23) \end{aligned}$$

This process is continued until the utility function is specified in sufficient detail. A problem with the certainty equivalent method is that errors are compounded as the analysis proceeds. This follows since the utility assigned in the first preference assessment [i.e., $u(\mathrm{CE}_1)$] is used throughout the subsequent preference assessments. A second issue is that the CE method uses different ranges in the indifference lotteries, meaning that the CEs are compared against different reference values. This might create inconsistencies since, as discussed in Section 4, attitudes toward risk usually change depending on whether outcomes are viewed as gains or losses. The use of different reference points may, of course, cause the same outcome to be viewed as either a loss or a gain. Utilities may also vary over time. In Section 4.2 we discuss some of these issues further.

3.4 Preference Assessment

Methods for measuring strength of preference include indifference methods, direct assessment, and indirect measurement (Keeney and Raiffa, 1976; Winterfeldt and Edwards, 1986). Indifference methods modify one of two sets of stimuli until subjects feel that they are indifferent between the two. Direct-assessment methods ask subjects to rate or otherwise assign numerical values to attributes, which are then used to obtain preferences for alternatives. Indirect-measurement techniques avoid decomposition and simply ask for preference orderings between alternatives. There has been some movement toward evaluating the effectiveness of particular methods for measuring preferences (Huber et al., 1993; Birnbaum et al., 1992).

3.4.1 Indifference Methods

Indifference methods are illustrated by the variable probability and certainty equivalent methods of eliciting utility functions presented in Section 2. There, indifference points were obtained by varying either probabilities or values of outcomes. Similar approaches have been applied to develop multiattribute utility or value functions. This approach involves four steps: (1) develop the single attribute utility or value functions, (2) assume a functional form for the multiattribute function, (3) assess the indifference point between various multiattribute alternatives, and (4) calculate the substitution rate or relative importance of one attribute compared to the other. The single-attribute functions might be developed by indifference methods (i.e., the variable probability or certainty equivalent methods) or direct-assessment methods, as discussed later. Indifference points between multiattribute outcomes are obtained through an interactive process in which the values of attributes are increased or decreased systematically. Substitution rates are then obtained from the indifference points.

For example, consider the case for two alternative traffic safety policies, A_1 and A_2. Each policy has two attributes, x = lives lost and y = money spent. Assume that the decision maker is indifferent between A_1 and A_2, meaning the decision maker feels that $v(x_1, y_1) = v$(20,000 deaths; \$1 Trillion) is equivalent to $v(x_2, y_2) = v$(10,000 deaths; \$1.5 T). For the sake of simplicity, assume an additive value function, where $v(x, y) = kv_x(x) + (1-k)v_y(y)$. Given this functional form, the indifference point $A_1 = A_2$ is used to derive the relation

$$\begin{aligned} &(1-k)kv_x(20{,}000 \text{ deaths}) + kv_y(\$1 \text{ T}) \\ &\quad = (1-k)v_x(10{,}000 \text{ deaths}) + kv_y(\$1.5 \text{ T}) \quad (24) \end{aligned}$$

This results in the substitution rate

$$\frac{k}{1-k} = \frac{v_x(20{,}000 \text{ deaths}) - v_x(10{,}000 \text{ deaths})}{v_y(\$1.5 \text{ T}) - v_y(\$1 \text{ T})} \quad (25)$$

If $v_x = -x$ and $v_y = -y$, a value of approximately 2^{-5} is obtained for k. The procedure becomes somewhat more complex when nonadditive forms are assumed for the multiattribute function (Keeney and Raiffa, 1976).

3.4.2 Direct-Assessment Methods

Direct-assessment methods include curve fitting and various numerical rating methods (Winterfeldt and Edwards, 1986). Curve fitting is perhaps the simplest approach. Here, the decision maker first orders the various attributes and then simply draws a curve assigning values to them. For example, an expert might draw a curve relating levels of traffic noise (measured in decibels) to their level of annoyance (on a scale of 0 to 1). Rating methods, as discussed earlier in reference to subjective probability assessment, include direct numerical measures on rating scales and relative ratings.

The analytic hierarchy process (AHP) provides one of the more implementable methods of this type (Saaty, 1988). In this approach, the decision is first structured as a value tree (Figure 5). Then each of the attributes is compared in terms of importance in a pairwise rating process. When entering the ratings, decision makers can enter numerical ratios (e.g., an attribute might be twice as important as another) or use the subjective verbal anchors mentioned earlier in reference to subjective probability assessment. The AHP program uses the ratings to calculate a normalized eigenvector assigning importance or preference weights to each attribute. Each alternative is then compared on the separate attributes. For example, two houses might first be compared in terms of cost and then be compared in terms of attractiveness. This results in another eigenvector describing how well each alternative satisfies each attribute. These two sets of eigenvectors are then combined into a single vector that orders alternatives in terms of preference. The subjective multiattribute rating technique (Smart) developed by Edwards (see Winterfeldt and Edwards, 1986) provides a similar, easily implemented approach. Both techniques are computerized, making the assessment process relatively painless.

3.4.3 Indirect Measurement

Indirect-measurement techniques avoid asking people to rate or rank directly the importance of factors that affect their preferences. Instead, subjects simply state or order their preferences for different alternatives. A variety of approaches can then be used to determine how individual factors influence preference. *Conjoint measurement theory* provides one such approach for separating the effects of multiple factors when only their joint effects are known. Application of the approach entails asking subjects to develop an ordered set of preferences for a set of alternatives that systematically vary attributes felt to be related to preference. The relationship between preferences and values of the attributes is then assumed to follow some functional form. The most common functional form assumed is a simple *additive-weighting model*. Preference orderings obtained using the model are then compared to the original rankings. Example applications of conjoint measurement theory to describe preferences between multiattribute alternatives are discussed in Winterfeldt and Edwards (1986). Related applications include the dichotomy-cut method, used to obtain decision rules for individuals and groups from ordinal rankings of multiattribute alternatives (Stanoulov, 1994).

The *policy-capturing approach* used in social judgment theory (Hammond et al., 1975; Hammond, 1993) is another indirect approach for describing human judgments of both preferences and probability. The policy-capturing approach uses multivariate regression or similar techniques to relate preferences to attributes for one or more decision makers. The equations obtained correspond to policies followed by particular decision makers. An example equation might relate medical symptoms to a physician's diagnosis. It has been argued that the policy-capturing approach measures the influence of factors on human judgments more accurately than do decomposition methods. Captured weights might be more accurate because decision makers may have little insight into the factors that affect their judgments (Valenzi and Andrews, 1973). People may also weigh certain factors in ways that reflect social desirability rather than influence on their judgments (Brookhouse et al., 1986). For example, people comparing jobs might rate pay as being lower in importance than intellectual challenge, whereas their preferences between jobs might be predicted entirely by pay. Caution must also be taken when interpreting regression weights as indicating importance, since regression coefficients are influenced by correlations between factors, their variability, and their validity (Stevenson et al., 1993).

4 BEHAVIORAL DECISION THEORY

As a normative ideal, classical decision theory has influenced the study of decision making in a major way. Much of the earlier work in behavioral decision theory compared human behavior to the prescriptions of classical decision theory (Edwards, 1954; Slovic et al., 1977; Einhorn and Hogarth, 1981). Numerous departures were found, including the influential finding that people use heuristics during judgment tasks (Tversky and Kahneman, 1974). On the basis of such research, pyschologists have concluded that other approaches are needed to describe the process of human decision making. Descriptive models that relax assumptions of the normative models but retain much of their essence are now being evaluated in the field of judgment and decision theory (Stevenson et al., 1993). The following discussion summarizes findings from this broad body of literature. The discussion begins by considering research on statistical estimation and inference. Attention then shifts to the topic of decision making under uncertainty and risk.

4.1 Statistical Estimation and Inference

The ability of people to perceive, learn, and draw inferences accurately from uncertain sources of information has been a topic of much research. In the following discussion we first consider briefly human abilities and limitations on such tasks. Attention then shifts to several heuristics that people may use to cope with their limitations and how their use can cause certain biases. In the next section we then consider briefly the role of memory and selective processing of information from a similar perspective. Attention then shifts to mathematical models of human judgment that provide insight into how people judge probabilities, the biases that might occur, and how people learn to perform probability judgment tasks. In the final section we summarize briefly findings on debiasing human judgments.

4.1.1 Human Abilities and Limitations

Research conducted in the early 1960s tested the notion that people behave as "intuitive statisticians" who gather evidence and apply it in accordance with the Bayesian model of inference (Peterson and Beach, 1967). Much of the earlier work focused on how good people are at estimating statistical parameters such as means, variances, and proportions. Other studies have compared human inferences obtained from probabilistic evidence to the prescriptions of Bayes' rule.

A number of interesting results were obtained (Table 6). The research first shows that people can be fairly good at estimating means, variances, or proportions from sample data. Hertwig et al. (1999) point out that "there seems to be broad agreement with the conclusion" of Jonides and Jones (1992) that people can give answers that reflect the actual relative frequencies of many kinds of events with great fidelity. However, as discussed by Winterfelt and Edwards (1986), like other psychophysical measures, subjective

Table 6 Sample Findings on the Ability of People to Estimate and Infer Statistical Quantities

	Reference
Statistical estimation	
Accurate estimation of sample means	Peterson and Beach (1967)
Variance estimates correlated with mean	Lathrop (1967)
Variance biases not found	Levin (1975)
Variance estimates based on range	Pitz (1980)
Accurate estimation of event frequency	Estes (1976), Hasher and Zacks (1984); Jonides and Jones (1992)
Accurate estimates of sample proportions between 0.75 and 0.25	Edwards (1954)
Severe overestimates of high probabilities; severe underestimates of low proportions	Fischhoff et al. (1977), Lichtenstein et al. (1982)
Reluctance to report extreme events	Du Charme (1970)
Weather forecasters provided accurate probabilities	Winkler and Murphy (1973)
Poor estimates of expected severity	Dorris and Tabrizi (1978)
Correlation of 0.72 between subjective and objective measures of injury frequency	Rethans (1980)
Risk estimates lower for self than for others	Weinstein (1980, 1987)
Risk estimates related to catastrophic potential, degree of control, familiarity	Lichtenstein et al. (1978)
Evaluations of outcomes and probabilities are dependent	Weber (1994)
Statistical inference	
Conservative aggregation of evidence	Edwards (1968)
Nearly optimal aggregation of evidence in naturalistic setting	Lehto et al. (2000)
Failure to consider base rates	Tversky and Kahneman (1974)
Base rates considered	Birnbaum and Mellers (1983)
Overestimation of conjunctive events	Bar-Hillel (1973)
Underestimation of disjunctive events	
Tendency to seek confirming evidence, tendency to discount disconfirming evidence, tendency to ignore reliability of the evidence	Einhorn and Hogarth (1978), Baron (1985)
Subjects considered variability of data when judging probabilities	Kahneman and Tversky (1973)
People insensitive to information missing from fault trees	Evans and Pollard (1985)
Overconfidence in estimates	Fischhoff et al. (1978)
Hindsight bias	Fischhoff et al. (1977)
Illusionary correlations	Fischhoff (1982), Christensen-Szalanski and Willham (1991)
Gambler's fallacy	Tversky and Kahneman (1974)
Misestimation of covariance between items	Arkes (1981)
Misinterpretation of regression to the mean	Tversky and Kahneman (1974)

probability estimates are noisy. Their accuracy will depend on how carefully they are elicited and on many other factors. Studies have shown that people are especially likely to have trouble estimating accurately the probability of unlikely events, such as nuclear plant explosions. For example, when people were asked to estimate the risk associated with the use of consumer products (Dorris and Tabrizi, 1978; Rethans, 1980) or various technologies (Lichtenstein et al., 1978), the estimates obtained were often weakly related to accident data. Weather forecasters are one of the few groups of people that have been documented as being able to estimate high and low probabilities accurately (Winkler and Murphy, 1973).

Part of the issue is that when events occur rarely, people will not be able to base their judgments on a representative sample of their own observations. Most of the information they receive about unlikely events will come from secondary sources, such as media reports, rather than from their own experience. This tendency might explain why risk estimates are often related more strongly to factors other than likelihood, such as catastrophic potential or familiarity (Lichtenstein et al., 1978; Slovic 1978, 1987; Lehto et al., 1994). Media reporting focuses on "newsworthy" events, which tend to be more catastrophic and unfamiliar. Consequently, judgments based on media reports might reflect the latter factors instead of likelihood. Weber (1994) provides additional evidence that subjective probabilities are related to factors other than likelihood and argues that people will overestimate the chance of a highly positive outcome because of their desire to obtain it. Weber also argues that people will overestimate the chance of a highly undesirable outcome because of their fear of receiving it. Traditional methods of decision analysis separately elicit and then recombine subjective probabilities with utilities, as discussed earlier, and assume that subjective probabilities are independent of consequences. A finding of dependency therefore casts serious doubt on the normative validity of this commonly accepted approach.

When studies of human inference are considered, several other trends become apparent (Table 6). In particular, several significant deviations from the Bayesian model have been found.

1. Decision makers tend to be conservative in that they don't give as much weight to probabilistic evidence as does Bayes' rule (Edwards, 1968).
2. Decision makers don't consider base rates or prior probabilities adequately (Tversky and Kahneman, 1974).
3. Decision makers tend to ignore the reliability of the evidence (Tversky and Kahneman, 1974).
4. Decision makers tend to overestimate the probability of conjunctive events and underestimate the probability of disjunctive events (Bar-Hillel, 1973).
5. Decision makers tend to seek out confirming evidence rather than disconfirming evidence and place more emphasis on confirming evidence when it is available (Einhorn and Hogarth, 1978; Baron, 1985). The order in which the evidence is presented has an influence on human judgments (Hogarth and Einhorn, 1992).
6. Decision makers are overconfident in their predictions (Fischhoff et al. 1977), especially in hindsight (Fischhoff, 1982; Christensen-Szalanski and Willham, 1991).
7. Decision makers show a tendency to infer illusionary causal relations (Tversky and Kahneman, 1973).

A lively literature has developed regarding these deviations and their significance* (Evans, 1989; Caverni et al., 1990; Wickens, 1992; Klein et al., 1993; Doherty, 2003). From one perspective, these deviations demonstrate inadequacies of human reason and are a source of societal problems (Baron, 1998; and many others). From the opposite perspective, it has been held that the foregoing findings are more or less experimental artifacts that do not reflect the true complexity of the world (Cohen, 1993). A compelling argument for the latter point of view is given by Simon (1955, 1983). From this perspective, people do not use Bayes' rule to compute probabilities in their natural environments because it makes unrealistic assumptions about what is known or knowable. Simply put, the limitations of the human mind and time constraints make it nearly impossible for people to use principles such as Bayes' rule to make inferences in their natural environments. To compensate for their limitations, people use simple heuristics or decision rules that are adapted to particular environments. The use of such strategies does not mean that people will not be able to make accurate inferences, as emphasized by both Simon and researchers embracing the ecological† (i.e., Hammond, 1996; Gigerenzer et al., 1999) and naturalistic (i.e., Klein et al., 1993) models of decision making. In fact, as discussed further in Section 4.1.4, the use of simple heuristics in rich environments can lead to inferences that are in many cases more accurate than those made using naïve Bayes, or linear regression (Gigerenzer et al., 1999).

There is an emerging body of literature that, on the one hand, shows that deviations from Bayes' rule can in fact be justified in certain cases from a normative

*Doherty (2003) groups researchers on human judgment and decision making into two camps. The optimists focus on the success of imperfect human beings in a complex world. The pessimists focus on the deficiencies of human reasoning compared to normative models.

†As noted by Gigerenzer et al. (1999, p. 18), because of environmental challenges, "organisms must be able to make inferences that are fast, frugal, and accurate." Similarly, Hammond (1996) notes that a close correspondence between subjective beliefs and environmental states will provide an adaptive advantage.

view and, on the other hand, shows that these deviations may disappear when people are provided with richer information or problems in more natural contexts. For example, drivers performing a simulated passing task combined their own observations of the driving environment with imperfect information provided by a collision-warning system, as predicted by a distributed signal detection theoretic model of optimal team decision making (Lehto et al., 2000). Other researchers have pointed out that:

1. A tendency toward conservatism can be justified when evidence is not conditionally independence (Navon, 1979).
2. Subjects do use base rate information and consider the reliability of evidence in slightly modified experimental settings (Birnbaum and Mellers, 1983; Koehler, 1996). In particular, providing natural frequencies instead of probabilities to subjects can improve performance greatly (Gigerenzer and Hoffrage, 1995; Krauss et al., 1999).
3. A tendency to seek out confirming evidence can offer practical advantages (Cohen, 1993) and may reflect cognitive failures, due to a lack of understanding of how to falsify hypotheses, rather than an entirely motivational basis (Klayman and Ha, 1987; Evans, 1989).
4. Subjects prefer stating subjective probabilities with vague verbal expressions rather than precise numerical values (Wallsten et al., 1993), demonstrating that they are not necessarily overconfident in their predictions.
5. There is evidence that the hindsight bias can be moderated by familiarity with both the task and the type of outcome information provided (Christensen-Szalanski and Willham, 1991).

Based on such results, numerous researchers have questioned the practical relevance of the large literature showing different types of biases. One reason that this literature may be misleading us is that researchers overreport findings of bias (Evans, 1989; Cohen, 1993). A more significant concern is that studies showing bias are almost always conducted in artificial settings where people are provided information about an unfamiliar topic. Furthermore, the information is often given in a form that forces use of Bayes' rule or other form of abstract reasoning to get the correct answer. For example, consider the simple case where a person is asked to predict how likely it is that a woman has breast cancer given a positive mammogram (Martignon and Krauss, 2003). In the typical study looking for bias, the subject might be told to assume (1) the probability that a 40-year-old woman has breast cancer is 1%, (2) the probability of a positive mammogram given that a women has cancer is 0.9, and (3) the probability of a positive mammogram given that a women does not have cancer is 0.1. Although the correct answer can be easily calculated using Bayes' rule, it is not at all surprising that people unfamiliar with probability theory will have difficulty determining it. In the real world, it seems much more likely that a person would simply keep track of how many women receiving a mammogram actually had breast cancer. The probability that a woman has breast cancer, given a positive mammogram, is then determined by dividing the number of women receiving a mammogram who actually had breast cancer by the number of women receiving a mammogram. The latter calculation gives exactly the same answer as using Bayes' rule, and is much easier to do.

The implications of the example above are obvious: First, people can duplicate the predictions of the Bayes rule by keeping track of the right relative frequencies. Second, if the right relative frequencies are known, accurate inferences can be made using very simple decision rules. Third, people will have trouble making accurate inferences if they don't know the right relative frequencies. Recent studies and reevaluations of older studies provide additional perspective. The finding that subjects are much better at integrating information when they are provided data in the form of natural frequencies instead of probabilities (Gigerenzer and Hoffrage, 1995; Krauss et al., 1999) is particularly interesting.

One conclusion that might be drawn from the latter work is that people are Bayesians after all if they are provided adequate information in appropriate representations (Martignon and Krauss, 2003). Other support for the proposition that people aren't as bad at inference as it once seemed includes Dawes and Mulford's (1996) review of the literature supporting the overconfidence effect or bias, in which they conclude that the methods used to measure this effect are logically flawed and that the empirical support is inadequate to conclude that it really exists. Part of the issue is that much of the psychological research on the overconfidence effect "overrepresents those situations where cue-based inferences fail" (Juslin and Olsson, 1999). When people rate objects that are selected randomly from a natural environment, overconfidence is reduced. Koehler (1996) provides a similarly compelling reexamination of the base rate fallacy. He concludes that the literature does not support the conventional wisdom that people routinely ignore base rates. To the contrary, he states that base rates are almost always used and that their degree of use depends on task structure and representation as well as their reliability compared to other sources of information.

Because such conflicting results can be obtained, depending on the setting in which human decision making is observed, researchers embracing the ecological (i.e., Hammond, 1996; Gigerenzer et al., 1999), and naturalistic (Klein et al., 1993; Klein, 1998) models of decision making strongly emphasize the need to conduct ecologically valid research in rich realistic decision environments.

4.1.2 Heuristics and Biases

Tversky and Kahneman (1973, 1974) made a key contribution to the field when they showed that many of the above-mentioned discrepancies between human estimates of probability and Bayes' rule could be explained by the use of three heuristics. The three heuristics they proposed were those of representativeness, availability, and anchoring and adjustment.

The *representativeness* heuristic holds that the probability of an item A belonging to some category B is judged by considering how representative A is of B. For example, a person is typically judged more likely to be a librarian than a farmer when described as "a meek and tidy soul, who has a desire for order and structure and a passion for detail." Application of this heuristic will often lead to good probability estimates but can lead to systematic biases. Tversky and Kahneman (1974) give several examples of such biases. In each case, representativeness influenced estimates more than other, more statistically oriented information. In the first study, subjects ignored base rate information (given by the experimenter) about how likely a person was to be either a lawyer or an engineer. Their judgments seemed to be based entirely on how representative the description seemed to be of either occupation. Tversky and Kahneman (1983) found people overestimated conjunctive probabilities in a similar experiment. Here, after being told that "Linda is 31 years old, single, outspoken, and very bright," most subjects said it was more likely she was both a bank teller and active as a feminist than simply a bank teller. In a third study, most subjects felt that the probability of more than 60% male births on a given day was about the same for both large and small hospitals (Tversky and Kahneman, 1974). Apparently, the subjects felt that large and small hospitals were equally representative of the population.

Other behaviors explained in terms of representativeness by Tversky and Kahneman included gambler's fallacy, insensitivity to predictability, illusions of validity, and misconceptions of statistical regression to the mean. With regard to gambler's fallacy, they note that people may feel that long sequences of heads or tails when flipping coins are unrepresentative of normal behavior. After a sequence of heads, a tail therefore seems more representative. Insensitivity to predictability refers to a tendency for people to predict future performance without considering the reliability of the information on which they base the prediction. For example, a person might expect an investment to be profitable solely on the basis of a favorable description without considering whether the description has any predictive value. In other words, a good description is believed to be representative of high profits, even if it states nothing about profitability. The illusion of validity occurs when people use highly correlated evidence to make a conclusion. Despite the fact that the evidence is redundant, the presence of many representative pieces of evidence increases confidence greatly. Misconception of regression to the mean occurs when people react to unusual events and then infer a causal linkage when the process returns to normality on its own. For example, a manager might incorrectly conclude that punishment works after seeing that unusually poor performance improves to normal levels following punishment. The same manager might also conclude that rewards don't work after seeing that unusually good performance drops after receiving a reward.

The *availability* heuristic holds that the probability of an event is determined by how easy it is to remember the event happening. Tversky and Kahneman state that perceived probabilities will therefore depend on familiarity, salience, effectiveness of memory search, and imaginability. The implication is that people will judge events as more likely when the events are familiar, highly salient (such as an airplane crash), or easily imaginable. Events will also be judged more likely if there is a simple way to search memory. For example, it is much easier to search for words in memory by the first letter rather than the third letter. It is easy to see how each item above affecting the availability of information can influence judgments. Biases should increase when people lack experience or when their experiences are too focused.

The *anchoring and adjustment* heuristic holds that people start from an initial estimate and then adjust it to reach a final value. The point chosen initially has a major impact on the final value selected when adjustments are insufficient. Tversky and Kahneman refer to this source of bias as an anchoring effect. They show how this effect can explain under- and overestimates of disjunctive and conjunctive events. This happens if the subject starts with a probability estimate of a single event. The probability of a single event is, of course, less than that for the disjunctive event and greater than that for the conjunctive event. If adjustment is too small, under- and overestimates occur, respectively, for the disjunctive and conjunctive events. Tversky and Kahneman also discuss how anchoring and adjustment may cause biases in subjective probability distributions.

The notion of heuristics and biases has had a particularly formative influence on decision theory. A substantial recent body of work has emerged that focuses on applying research on heuristics and biases (Kahneman et al., 1982; Heath et al., 1994). Applications include medical judgment and decision making, affirmative action, education, personality assessment, legal decision making, mediation, and policy making. It seems clear that this approach is excellent for describing many general aspects of decision making in the real world. However, research on heuristics and biases has been criticized as being pretheoretical (Slovic et al., 1977), and as pointed out earlier, has contributed to overselling of the view that people are biased. The latter point is interesting, as Tversky and Kahneman have claimed all along that using these heuristics can lead to good results. However, nearly all the research conducted in this framework has focused on when they might go wrong.

4.1.3 Memory Effects and Selective Processing of Information

The heuristics and biases framework has been criticized by many researchers for its failure to adequately address more fundamental cognitive processes that might explain biases (Dougherty et al., 2003). This follows because the availability and representativeness heuristics can both be described in terms of more fundamental memory processes. For example, the availability heuristic proposes that the probability of an event is determined by how easy it is to remember the event happening. Ease of recall, however, depends on many things, such as what is stored in memory, how it is represented, how well it is encoded, and how well a cue item matches the memory representation.

Dougherty et al. (2003) note that three aspects of memory can explain many of the findings on human judgment: (1) how information is stored or represented, (2) how information is retrieved, and (3) experience and domain knowledge. The first aspect pertains to what is actually stored when people experience events. The simplest models assume that people store a record of each instance of an experienced event, and in some cases, additional information, such as the frequency of the event (Hasher and Zacks, 1984) or ecological cue validities (Brehmer and Joyce, 1988). More complex models assume that people store an abstract representation or summary of the event (Pennington and Hastie, 1988), in some cases at multiple levels of abstraction (Reyna and Brainerd, 1995). The way information is stored or represented can explain several of the observed findings on human judgment.

First, there is strong evidence that people are often excellent at storing frequency information* and that the process by which this is done is fairly automatic (Hasher and Zacks, 1984; Gigerenzer et al., 1991). Gigerenzer et al. conclude that with repeated experience people should also be able to store ecological cue validities. The accuracy of these stored representations would, of course, depend on how large and representative the sample of encoded observations is. Such effects can be modeled with simple adding models that might include the effects of forgetting (or memory trace degradation), or other factors, such as selective sampling or the amount of attention devoted to the information at the time it is received. As pointed out by Dougherty et al. and many others, many of the biases in human judgment follow directly from considering how well the events are encoded in memory. In particular, except for certain sensory qualities which are encoded automatically, encoding quality is assumed to depend on attention. Consequently, some biases should reflect the tendency of highly salient stimuli to capture attention. Another completely different type of bias might reflect the fact that the person was exposed to an unrepresentative sample of events. Lumping these two very different biases together, as is done by the availability heuristic, is obviously debatable.

*This point directly confirms Tversky and Kahneman's original assumption that the availability heuristic should often result in good predictions (see Tversky and Kahneman, 1973).

Other aspects of human memory mentioned by Dougherty et al. (2003) that can explain certain findings on human judgment include the level of abstraction of the stored representation and retrieval methods. One interesting observation is that people often find it preferable to reason with gist-based representations rather than verbatim descriptions of events (Reyna and Brainerd, 1995). When the gist does not contain adequate detail, the reasoning may lead to flawed conclusions. Some of the differences observed between highly skilled experts and novices might correspond to situations where experts have stored a large number of relevant instances and their solutions in memory, whereas novices have only gist-based representations. In such situations, novices will be forced to reason using the information provided. Experts, on the other hand, might be able to solve the problem with little or no reasoning, simply by retrieving the solution from memory. The latter situation would correspond to Klein's recognition-primed decision making (Klein, 1989, 1998). However, there is also reason to believe that people are more likely to develop abstract gist-type representations of events with experience (Reyna and Brainerd, 1995). This might explain the findings in some studies that people with less knowledge and experience sometimes outperform experts. A particularly interesting demonstration is given by Gigerenzer et al. (1999), who discuss a study where a simple recognition heuristic based on the collective recognition of the names of companies by 180 German laypeople resulted in a phenomenally high yield of 47%, and outperformed the Dax 30 market index by 10%. It outperformed several mutual funds managed by professionals by an even greater margin.

Memory models and processes can also be used to explain primacy and recency effects in human judgment (Hogarth and Einhorn, 1992). Such effects seem similar to the well-known serial position effect.† Given that human judgment involves retrieval of information from memory, it seems reasonable that judgments would also show primacy and recency effects. Several mathematical models have been developed that show how the order in which evidence is presented to people might affect their judgments. For example, Hogarth and Einhorn (1992) present an anchoring and adjustment model of how people update beliefs that predicts both primacy and recency effects. The latter model holds that the degree of belief in a hypothesis after collecting k pieces of evidence can be described as

$$S_k = S_{k-1} + w_k[s(x_k) - R] \qquad (26)$$

where S_k is the degree of belief after collecting k pieces of evidence, S_{k-1} is the anchor or prior belief, w_k is

†When people are asked to memorize lists of information, they almost always are able to remember items at the beginning or end of a list better than items in the middle of the list (Ebbinghaus, 1913). The improved ability for words at the start of the list is called the *primacy effect*. The improved ability for words at the end of the list is called the *recency effect*.

the adjustment weight for the kth piece of evidence, $s(x_k)$ is the subjective evaluation of the kth piece of evidence, and R is the reference point against which the kth piece of evidence is compared. In evaluation tasks, $R = 0$. This corresponds to the case where evidence is either for or against a hypothesis.[‡] For estimation tasks, $R \neq 0$. The different values of R result in an additive model for evaluation tasks and an averaging model for estimation tasks. Also, if the quantity $s(x_k) - R$ is evaluated for several pieces of evidence at a time, the model predicts primacy effects. If single pieces of evidence are evaluated individually in a step-by-step sequence, recency effects become more likely.

Biases in human judgment, which in some but not all cases are memory related, can also be explained by models of how information is processed during task performance. Along these lines, Evans (1989) argues that factors which cause people to process information in a selective manner or attend to irrelevant information are the major cause of biases in human judgment. Factors assumed to influence selective processing include the availability, vividness, and relevance of information, and working memory limitations. The notion of availability refers to the information actually attended to by a person while performing a task. Evans's model assumes that information elements attended to are determined during a heuristic, preattentional stage. This stage is assumed to involve unconscious processes and is influenced by stimulus salience (or vividness) and the effects of prior knowledge.

In the next stage of his model, inferences are drawn from the information selected. This is done using rules for reasoning and action developed for particular types of problems. Working memory influences performance at this stage by limiting the amount of information that can be attended to consciously while performing a task. Evans assumes that the knowledge used during the inference process might be organized in schemas that are retrieved from memory and fit to specific problems (Cheng and Holyoak, 1985). Support for the latter conclusion is provided by studies showing that people are able to develop skills in inference tasks but may fail to transfer these skills (inference related) from one setting to another. Evans also provides evidence that prior knowledge can cause biases when it is inconsistent with information provided and that improving knowledge can reduce or eliminate biases.

Evans's model of selective processing of information is consistent with other explanations of biases. Among such explanations, information overload has been cited as a reason for impaired decision making by consumers (Jacoby, 1977). The tendency of highly salient stimuli to capture attention during inference tasks has also been noted by several researchers (Nisbett and Ross, 1980; Payne, 1980). Nisbett and Ross suggest that vividness of information is determined by its emotional content, concreteness and imagability, and temporal and spatial proximity. As noted by Evans and many others, these factors have also been shown to affect the memorability of information. The conclusion is that biases due to salience can occur in at least two different ways: (1) People might focus on salient but irrelevant items while performing the task, and (2) people might draw incorrect inferences when the contents of memory are biased due to salience effects during earlier task performance.

4.1.4 Mathematical Models of Human Judgment

A number of approaches have been developed for describing human judgments mathematically. These approaches include the use of policy-capturing models in social judgment theory, probabilistic mental models, multiple-cue probability learning models, and information integration theory. Much of the work in this area builds on the Brunswik lens model (Brunswik, 1952), developed originally to describe how people perceive their environment. The lens model is used in these approaches to describe human judgment on some criterion in terms of two symmetric concepts: (1) the ecological validity of probabilistic cues in the task environment, and (2) cue utilization. The ecological validity of a cue is defined in terms of the correlation or probabilistic relation between a cue and the criterion; cue utilization is defined in terms of the correlation or probabilistic relation between the cue and the judgment. The emphasis on ecological validity in this approach is one of its key contributions. The focus on ecological validity results in a clear, measurable definition of domain-specific expertise or knowledge and can also be used to specify limits as to how well people can perform. More specifically, a perfectly calibrated person will know the ecological validities of each environmental cue. A second issue is that a good decision maker will utilize cues in a way that reflects ecological validity appropriately. That is, cues with higher ecological validity should be emphasized more heavily. The ultimate limit to performance is described in terms of the maximum performance possible given a set of cues and their ecological validities. The latter quantity is often estimated in terms of the variance explained by a linear regression model that predicts the criterion using the cues (Hammond et al., 1975).

In social judgment theory (SJT) (Hammond et al., 1975; Brehmer and Joyce 1988; Hammond, 1993), the approach described above is followed to develop policy-capturing models that describe how people use probabilistic environmental cues to make judgments. As mentioned earlier with regard to preference assessment, linear or nonlinear forms of regression are used in this approach to relate judgments to environment cues. This approach has been applied to a wide number of real-world applications to describe expert judgments (Brehmer and Joyce, 1988). For example, policy-capturing models have been applied to describe software selection by management information system managers (Martocchio

[‡]It is easy to see that equation (26) approximates the log-odds form of Bayes' rule, where evidence for or against the hypothesis is combined additively.

et al., 1993), medical decisions (Brehmer and Joyce, 1988), and highway safety (Hammond, 1993). Policy-capturing models provide surprisingly good fits to expert judgments. In fact, there is evidence, and consequently much debate, over whether the models can actually do better than experts on many judgment tasks (Slovic et al., 1977; Brehmer, 1981; Kleinmuntz, 1984).

Cognitive continuum theory (Hammond, 1980) builds on social judgment theory by distinguishing judgments on a cognitive continuum varying from highly intuitive decisions to highly analytical decisions. Hammond (1993) summarizes earlier research showing that task characteristics cause decision makers to vary on this continuum. A tendency toward analysis increases, and reliance on intuition decreases, when (1) the number of cues increases, (2) cues are measured objectively instead of subjectively, (3) cues are of low redundancy, (4) decomposition of the task is high, (5) certainty is high, (6) cues are weighted unequally in the environmental model, (7) relations are nonlinear, (8) an organizing principle is available, (9) cues are displayed sequentially instead of simultaneously, and (10) the time period for evaluation is long. One of the conclusions developed from this work is that intuitive methods can be better than analytical methods in some situations (Hammond et al., 1987).

The theory of probabilistic mental models (Gigerenzer et al., 1991) is another Brunswikian model that has attracted a lot of attention. As in SJT, this approach holds that human knowledge can be described as a set of cues, their values, and their ecological validities. However, the ecological validity of a cue is defined as the relative frequency with which the cue correctly predicts how well an object does on some criterion measure, instead of as a correlation. Inference is assumed to be a cue-based process involving one or more pairwise comparisons of the cue values associated with particular objects. Several different heuristics have been proposed within the framework of probabilistic mental models that describe how the inference process might be performed. One of the better-performing heuristics is called the *Take the Best* heuristic. This simple heuristic begins by comparing a pair of objects using the most valid cue. If one of the objects has a positive cue value and the other does not, the object with the higher positive cue value is given a higher value on the criterion, and the inference process stops. Otherwise, the heuristic moves on to the next most valid cue. This process continues until a choice is made, or all cues have been evaluated.

Take the Best is an example of what Gigerenzer et al. (1999) call a fast and frugal heuristic. This follows because the heuristic will generally make a choice without considering all of the cues. As such, Take the Best differs in a major way from the multiple linear equations normally (but not always, as emphasized by Hammond) used in the policy-capturing approach and elsewhere. One of the most interesting results reported by Gigerenzer et al. (1999) was the finding that Take the Best was always as good and normally outperformed multiple linear regression, Dawes' rule (which sums the number of positively related cues and subtracts the number of negatively related cues), and naïve Bayes when making predictions for several different data sets drawn from very different domains. The latter procedures used all the cues and involve more complex statistical operations, which makes the performance of Take the Best quite impressive. Gigerenzer et al. (1999) conclude that this performance demonstrates that fast and frugal heuristics can be highly effective. At this point it should be mentioned that this conclusion provides support for Klein's model of recognition-primed decision making (Section 5.1.2). It also provides a possible explanation of Hammond's observation mentioned above that intuitive methods can lead to better results than those of more analytical methods, and contradicts the assumption often made that people will perform better when they are use more information to make a judgment.

Multiple-cue probability learning models extend the lens model to the psychology of learning (Brehmer and Joyce, 1988). Research on multiple-cue probability learning has provided valuable insight into factors affecting learning of inference tasks. One major finding is that providing cognitive feedback about cues and their relationship to the effects inferred leads to quicker learning than with feedback about outcomes (Balzer et al., 1989). Stevenson et al. (1993) summarize a number of other findings, including that (1) subjects can learn to use valid cues, even when they are unreliable; (2) subjects are better able to learn linear relationships than nonlinear or inverse relationships; (3) subjects do not consider redundancy when using multiple cues; (4) source credibility and cue validity are considered; and (5) the relative effectiveness of cognitive and outcome feedback depends on the formal, substantive, and contextual characteristics of the task.

Information-integration theory (Anderson, 1981) takes a somewhat different approach than SJT or the lens model to develop similar models of how cue information is used when making judgments. The main difference is that information-integration theory emphasizes the use of factorial experimental designs where cues are systematically manipulated rather than trying to duplicate the cue distribution of the naturalistic environment. The goal of this approach is to determine (1) how people scale cues when determining their subjective values, and (2) how these scaled values are combined to form overall judgments. Various functional forms of how information is integrated are considered, including additive and averaging functions. A substantial body of research follows this approach to test various ways that people might combine probabilistic information. A primary conclusion is that people tend to integrate information using simple averaging, adding, subtracting, and multiplying models. Conjoint measurement approaches (Wallsten, 1972, 1976), in particular, provide a convenient way of both scaling subjective values assigned to cues and testing different functional forms describing how these values are combined to develop global judgments.

4.1.5 Debiasing or Aiding Human Judgments

The notion that many biases (or deviations from normative models) in statistical estimation and inference can be explained has led researchers to consider the possibility of *debiasing* (a better term might be *improving*) human judgments (Keren, 1990). Part of the issue is that the heuristics people use often work very well. The nature of the heuristics also suggests some obvious generic strategies for improving decision making. One conclusion that follows directly from the earlier discussion is that biases related to the availability and representativeness heuristics might be reduced if people were provided better, more representative samples of information. Other strategies that follow directly from the earlier discussion include making ecologically valid cues more salient, providing both outcome and cognitive feedback, and helping people do analysis. These strategies can be implemented in training programs or guide the development of decision aids.*

Emerging results from the field of naturalistic decision making support the conclusion that decision-making skills can be improved through training (Fallesen and Pounds, 2001; Pliske et al., 2001; Pliske and Klein, 2003). The use of computer-based training to develop task-specific decision-making skills is one very interesting development (Sniezek et al., 2002). Decision-making games (Pliske et al., 2001) and cognitive simulation (Satish and Streufert, 2002) are other approaches that have been applied successfully to improve decision-making skills. Other research shows that training in statistics reduces biases in judgment (Fong et al., 1986). In the latter study, people were significantly more likely to consider sample size after training.

These results supplement some of the findings discussed in Section 4.1.1, indicating that judgment biases can be moderated by familiarity with the task and the type of outcome information provided. Some of these results discussed earlier included evidence that providing feedback on the accuracy of weather forecasts may help weather forecasters (Winkler and Murphy, 1973), and research showing that cognitive feedback about cues and their relationship to the effects inferred leads to quicker learning than does feedback about outcomes (Balzer et al., 1989). Other studies have shown that simply asking people to write down reasons for and against their estimates of probabilities can improve calibration and reduce overconfidence (Koriat et al., 1980). This, of course, supports the conclusion that judgments will be less likely to be biased if people think carefully about their answers. Other research showed that subjects were less likely to be overconfident if they expressed subjective probabilities verbally instead of numerically (Zimmer, 1983; Wallsten et al., 1993). Conservatism, or the failure to modify probabilities adequately after obtaining evidence, was also reduced in Zimmer's study.

The results above support the conclusion that it might be possible to improve or aid human judgment. On the other hand, many biases, such as optimistic beliefs regarding health risks, have been difficult to modify (Weinstein and Klein, 1995). People show a tendency to seek out information that supports their personal views (Weinstein, 1979) and are quite resistant to information that contradicts strongly held beliefs (McGuire, 1966; Nisbett and Ross, 1980). Evans (1989) concludes that "pre-conceived notions are likely to prejudice the construction and evaluation of arguments." Other evidence shows that experts may have difficulty providing accurate estimates of subjective probabilities even when they receive feedback. For example, many efforts to reduce both overconfidence in probability estimates and the hindsight bias have been unsuccessful (Fischhoff, 1982). One problem is that people may not pay attention to feedback (Fischhoff and MacGregor, 1982). They also may attend only to feedback that supports their hypothesis, leading to poorer performance and at the same time, greater confidence (Einhorn and Hogarth, 1978). Several efforts to reduce confirmation biases, the tendency to search for confirming rather than disconfirming evidence, through training have also been unsuccessful (Evans, 1989).†

The conclusion is that debiasing human judgments is difficult but not impossible. Some perspective can be obtained by considering that most studies showing biases have focused on statistical inference and generally involved people not particularly knowledgeable about statistics, who are not using decision aids such as computers or calculators. It naturally may be expected that people will perform poorly on such tasks, given their lack of training and forced reliance on mental calculations (Winterfeldt and Edwards, 1986). The finding that people can improve their abilities on such tasks after training in statistics is particularly telling, and also encouraging. Another encouraging finding is that biases are occasionally reduced when people process information verbally instead of numerically. This result might be expected given that most people are more comfortable with words than with numbers.

4.2 Preference and Choice

Much of the research on human preference and choice has focused on comparing observed preferences to the predictions of subjective utility theory (SEU) (Goldstein and Hogarth, 1997). Early work

*These strategies seem to be especially applicable to the design of information displays and decision support systems. Chapter 45 addresses the issue of display design. Computer-based decision support is addressed in Section 7. These strategies also overlap with decision analysis. As discussed in Section 3, decision analysis focuses on the use of analytic methods to improve decision quality.

†Engineers, designers, and other real-world decision makers will find it very debatable whether the confirmation bias is really a bias. Searching for disconfirming evidence obviously makes sense in hypothesis testing. That is, a single negative instance is enough to disprove a logical conjecture. In real-world settings, however, checking for evidence that supports a hypothesis can be very efficient.

examining SEU as a descriptive theory drew generally positive conclusions. However, it soon became apparent that people's preferences for risky or uncertain alternatives often violated basic axioms of SEU theory. The finding that people's preferences change when the outcomes are framed in terms of costs, as opposed to benefits, has been particularly influential. Several other common deviations from SEU have been observed. One potentially serious deviation is that preferences can be influenced by sunk costs or prior commitment to a particular alternative. Preferences change over time and may depend on which alternatives are being compared, or even the order in which they are compared. The regret associated with making the "wrong" choice seems to play a major role when people compare alternatives. Accordingly, the satisfaction people derive from obtaining particular outcomes after making a decision is influenced by positive and negative expectations prior to making the decision. Other research on human preference and choice has shown that people choose between and apply different decision strategies depending on the cognitive effort required to apply a decision strategy successfully, the needed level of accuracy, and time pressure. Certain strategies are more likely than others to lead to choices consistent with those prescribed by SEU theory.

Alternative models, such as prospect theory and random utility theory, were consequently developed to explain human preferences under risk or uncertainty.* The following discussion will first summarize some common violations of the axioms underlying SEU theory before moving on to framing effects and preference reversals. Attention will then shift to models of choice and preference. The latter discussion will begin with prospect theory before addressing other models of labile or conditional preferences. Decision-making strategies, and how people choose between them, are covered in Section 6.

4.2.1 Violation of the Rationality Axioms

Several studies have shown that people's preferences between uncertain alternatives can be inconsistent with the axioms underlying subjective expected utility (SEU) theory. One fundamental violation of the assumptions is that preferences can be intransitive (Tversky, 1969; Budescu and Weiss, 1987). Also, as mentioned earlier, subjective probabilities may depend on the values of consequences (violating the independence axiom), and as discussed in the next section, the framing of a choice can affect preference. Another violation is given by the *Myers effect* (Myers et al., 1965), where preference reversals between high (H) and low (L) variance gambles can occur when the gambles are compared to a certain outcome, depending on whether the certain outcome is positive (H preferred to L) or negative (L preferred to H). The latter effect violates the assumption of independence because the ordering of the two gambles depends on the certain outcome.

Another commonly cited violation of SEU theory is that people show a tendency toward uncertainty avoidance, which can lead to behavior inconsistent with the "sure-thing" axiom. The Ellsberg and Allais paradoxes (Allais, 1953; Ellsberg, 1961) both involve violations of the sure-thing axiom (see Table 2) and seem to be caused by people's desire to avoid uncertainty. The *Allais paradox* is illustrated by the following set of gambles. In the first gamble, a person is asked to choose between gambles $A1$ and $B1$, where:

> Gamble $A1$ results in \$1 million for sure. Gamble $B1$ results in \$2.5 million with a probability of 0.1, \$1 million with a probability of 0.89, and \$0 with a probability of 0.01.

In the second gamble, the person is asked to choose between gambles $A2$ and $B2$, where:

> $A2$ results in \$1 million with a probability of 0.11 and \$0 with a probability of 0.89. Gamble $B2$ results in \$2.5 million with a probability of 0.1 and \$0 with a probability of 0.9.

Most people prefer gamble $A1$ to $B1$ and gamble $B2$ to $A2$. It is easy to see that this set of preferences violates expected utility theory. First, if $A1 > B1$, then $u(A1) > u(B1)$, meaning that $u(\$1 \text{ million}) > 0.1u(\$2.5 \text{ million}) + 0.89u(\$1 \text{ million}) + 0.01u(\$0)$. If a utility of 0 is assigned to receiving \$0 and a utility of 1 to receiving \$2.5 million, then $u(\$1 \text{ million}) > 1/11$. However, from the preference $A2 > B2$, it follows that $u(\$1 \text{ million}) < 1/11$. Obviously, no utility function can satisfy this requirement of assigning a value both greater than and less than 1/11 to \$1 million.

Savage (1954) mentioned that the set of gambles above can be reframed in a way that shows that these preferences violate the sure-thing principle. After doing so, Savage found that his initial tendency toward choosing $A1$ over $B1$ and $A2$ over $B2$ disappeared. As noted by Stevenson et al. (1993), this example is one of the first cases cited of a preference reversal caused by reframing a decision, the topic discussed below.

4.2.2 Framing of Decisions and Preference Reversals

A substantial body of research has shown that people's preferences can shift dramatically depending on the way a decision is represented. The best known work on this topic was conducted by Tversky and Kahneman (1981), who showed that preferences between medical intervention strategies changed dramatically depending on whether the outcomes were posed as losses or gains. The following question, worded in terms of benefits, was presented to one set of subjects:

> Imagine that the U.S. is preparing for the outbreak of an unusual Asian disease, which is expected to

*Singleton and Hovden (1987) and Yates (1992) are useful sources for the reader interested in additional details on risk perception, risk acceptability, and risk taking behavior. Section 5.1.1 is also relevant to this topic.

kill 600 people. Two alternative programs to combat the disease have been proposed. Assume that the exact scientific estimate of the consequences of the programs are as follows:

If program A is adopted, 200 people will be saved.

If program B is adopted, there is a 1/3 probability that 600 people will be saved, and a 2/3 probability that no people will be saved.

Which of the two programs would you favor?

The results showed that 72% of subjects preferred program A. The second set of subjects, was given the same cover story, but worded in terms of costs:

If program C is adopted, 400 people will die.

If program D is adopted, there is a 1/3 probability that nobody will die, and a 2/3 probability that 600 people will die.

Which of the two programs would you favor?

The results now showed that 78% of subjects preferred program D. Since program D is equivalent to B and program A is equivalent to C, the preferences for the two groups of subjects were strongly reversed. Tversky and Kahneman concluded that this reversal illustrated a common pattern in which choices involving gains are risk averse and choices involving losses are risk seeking. The interesting result was that the way the outcomes were worded caused a shift in preference for identical alternatives. Tversky and Kahneman called this tendency the *reflection effect*. A body of literature has since developed showing that the framing of decisions can have practical effects for both individual decision makers (Kahneman et al., 1982; Heath et al., 1994) and group decisions (Paese et al., 1993). On the other hand, recent research shows that the reflection effects can be reversed by certain outcome wordings (Kuhberger, 1995); more important, Kuhberger provides evidence that the reflection effect observed in the classic experiments can be eliminated by fully describing the outcomes (i.e., referring to the paragraph above, a more complete description would state: "If program C is adopted, 400 people will die *and* 200 *will live*").

Other recent research has explored the theory that perceived risk and perceived attractiveness of risky outcomes are psychologically distinct constructs (Weber et al., 1992). In the latter study, it was concluded that perceived risk and attractiveness are "closely related but distinct phenomena." Related research has shown weak negative correlations between the perceived risk and value of indulging in alcohol-related behavior for adolescent subjects (Lehto et al., 1994). The latter study also showed that the rated propensity to indulge in alcohol-related behavior was strongly correlated with perceived value ($R = 0.8$) but weakly correlated with perceived risk ($R = -0.15$). Both findings are consistent with the theory that perceived risk and attractiveness are distinct constructs, but the latter finding indicates that perceived attractiveness may be the better predictor of behavior. Lehto et al. conclude that intervention methods attempting to lower preferences for alcohol-related behavior should focus on lowering perceived value rather than on increasing perceived risk.

4.2.3 Prospect Theory

Prospect theory (Kahneman and Tversky, 1979) attempts to account for behavior not consistent with the SEU model by including the framing of decisions as a step in the judgment of preference between risky alternatives. Prospect theory assumes that decision makers tend to be risk averse with regard to gains and risk seeking with regard to losses. This leads to a value function that weights losses disproportionately. As such, the model is still equivalent to SEU, assuming a utility function expressing mixed risk aversion and risk seeking. Prospect theory, however, assumes that the decision maker's reference point can change. With shifts in the reference point, the same returns can be viewed as either gains or losses.* The latter feature of prospect theory, of course, is an attempt to account for the framing effect discussed above. Prospect theory also deviates significantly from SEU theory in the way in which probabilities are addressed. To describe human preferences more closely, perceived values are weighted by a function $\pi(p)$ instead of the true probability, p. Compared to the untransformed form of p, $\pi(p)$ overweights very low probabilities and underweights moderate and high probabilities. The function $\pi(p)$ is also generally assumed to be discontinuous and poorly defined for probability values close to 0 or 1.

Prospect theory assumes that the choice process involves an editing phase and an evaluation phase. The editing phase involves reformulation of the options to simplify subsequent evaluation and choice. Much of this editing process is concerned with determining an appropriate reference point in a step called *coding*. Other steps that may occur include the segregation of riskless components of the decision, combining probabilities for events with identical outcomes, simplification by rounding off probabilities and outcome measures, and search for dominance. In the evaluation phase, the perceived values are then weighed by the function $\pi(p)$. The alternative with the greatest weighed value is then selected. Several other modeling approaches that differentially weigh utilities in risky decision making have been proposed (Goldstein and Hogarth, 1997). As in prospect theory, such models often assume that the subjective probabilities, or decision weights, are a function of outcome sign (i.e.

*The notion of a reference point against which outcomes are compared has similarities to the notion of making decisions on the basis of regret (Bell, 1982). Regret, however, assumes comparison to the best outcome. The notion of different reference points is also related to the well-known trend that the buying and selling price of assets often differ for a decision maker (Raiffa, 1968).

positive, neutral, or negative), rank (i.e., first, second, etc.), or magnitude. Other models focus on display effects (i.e., single-stage vs. multistage arrangements) and distribution effects (i.e., two outcome lotteries vs. multiple-outcome lotteries). Prospect theory and other approaches also address how the value or utility of particular outcomes can change between decision contexts, as discussed below.

4.2.4 Labile Preferences

There is no doubt that human preferences often change after receiving some outcome. After losing money, an investor may become risk averse. In other cases, an investor may escalate her commitment to an alternative after an initial loss, even if better alternatives are available. From the most general perspective, any biological organism becomes satiated after satisfying a basic need, such as hunger. Preferences also change over time or between decision contexts. For example, a 30-year-old decision maker considering whether to put money into a retirement fund may currently have a very different utility function than at retirement. The latter case is consistent with SEU theory but obviously complicates analysis.

Economists and behavioral researchers have both focused on mathematically modeling choice processes to explain intransitive or inconsistent preference orderings of alternatives (Goldstein and Hogarth, 1997). Game theory provides interesting insight into this issue. From this perspective, preferences of the human decision maker are modeled as the collective decisions obtained by a group of internal agents, or selves, each of which is assumed to have distinct preferences (see Elster, 1986). Intransitive preferences and other violations of rationality on the part of the human decision maker then arise from interactions between competing selves.* Along these lines, Ainslie (1975) proposed that impulsive preference switches (often resulting in risky or unhealthy choices) arise as the outcome of a struggle between selves representing conflicting short- and long-term interests, respectively.

Another area of active research has focused on how experiencing outcomes can cause shifts in preference. One robust finding is that people tend to be more satisfied if an outcome exceeds their expectations and less satisfied if it does not (i.e., Feather, 1966; Connolly et al., 1997). Expectations therefore provide a reference point against which outcomes are compared. Numerous studies have also shown that people in a wide variety of settings often consider sunk costs when deciding whether to escalate their commitment to an alternative by investing additional resources (Arkes and Blumer, 1985; Arkes and Hutzel, 2000). From the perspective of prospect theory, sunk costs cause people to frame their choice in terms of losses instead of gains, resulting in risk-taking behavior and consequently, escalating commitment. Other plausible explanations for escalating commitment include a desire to avoid waste or to avoid blame for an initially bad decision to invest in the first place. Interestingly, some recent evidence suggests that people may deescalate commitment in response to sunk costs (Heath, 1995). The latter effect is also contrary to classical economic theory, which holds that decisions should be based solely on marginal costs and benefits. Heath explains such effects in terms of mental accounting. Escalation is held to occur when a mental budget is not set or expenses are difficult to track. Deescalation is held to occur when people exceed their mental budget, even if the marginal benefits exceed the marginal costs.

Other approaches include value or utility as random variables within models of choice to explain intransitive or inconsistent preference orderings of alternatives. Random utility models (Iverson and Luce, 1998) describe the probability $P_{a,A}$ of choosing a given alternative a from a set of options A as

$$P_{a,A} = \text{Prob}(U_{a,} \geq U_b \text{ for all } b \text{ in } A) \qquad (27)$$

where $U_{a,}$ is the uncertain utility of alternative a and U_b is the uncertain utility of alternative b. The most basic random utility models assign a utility to each alternative by sampling a single value from a known distribution. The sampled utility of each alternative then remains constant throughout the choice process. Basic random utility models can predict a variety of preference reversals and intransitive preferences for single- and multiple-attribute comparisons of alternatives (i.e., Tversky, 1972).

Sequential sampling models extend this approach by assuming that preferences can be based on more than one observation. Preferences for particular alternatives are accumulated over time, by integrating or otherwise summing the sampled utilities. The utility of an alternative at a particular time is proportional to the latter sum. A choice is made when the summed preferences for a particular alternative exceed some threshold, which itself may vary over time or depend on situational factors (Busemeyer and Townsend, 1993; Wallsten, 1995). It is interesting to observe that sequential sampling models can explain speed–accuracy tradeoffs in signal-detection tasks (Stone, 1960), as well as shifts in preferences due to time pressure (Busemeyer and Townsend, 1993; Wallsten, 1995) if it is assumed that people adjust their threshold downward under time pressure. That is, under time pressure, people sample less information before making a choice. In the following section we explore further how and why decision strategies might change over time and between decision contexts.

*As discussed further in Section 6, group decisions, even though they are made by rational members, are subject to numerous violations of rationality. For example, consider the case where the decision maker has three selves that are, respectively, risk averse, risk neutral, and risk seeking. Assume that the decision maker is choosing between alternatives A, B, and C. Suppose that the risk-averse self rates the alternatives in the order A, B, C; the risk-neutral self rates them in the order B, C, A; and the risk-seeking self rates them in the order C, A, B. Also assume that the selves are equally powerful. Then two of the three agents always agree that $A > B$, $B > C$, and $C > A$. This ordering is, of course, nontransitive.

5 DYNAMIC AND NATURALISTIC DECISION MAKING

In dynamic decision making, actions taken by a decision maker are made sequentially in time. Taking actions can change the environment, resulting in a new set of decisions. The decisions might be made under time pressure and stress, by groups or by single decision makers. This process might be performed on a routine basis or might involve severe conflict. For example, either a group of soldiers or an individual officer might routinely identify marked vehicles as friends or foes. When a vehicle has unknown or ambiguous marking, the decision changes to a conflict-driven process. Naturalistic decision theory has emerged as a new field that focuses on such decisions in real-world environments (Klein, 1998; Klein et al., 1993). The notion that most decisions are made in a routine, nonanalytical way is the driving force of this approach.* Areas where such behavior seems prominent include juror decision making, troubleshooting of complex systems, medical diagnosis, management decisions, and numerous other examples.

In the following discussion we first address models of dynamic and naturalistic decision making. These models both illustrate naturalistic decision-making strategies and explain their relation to experience and task familiarity. A brief discussion is also provided on teams and team leadership, in naturalistic settings. Attention will then shift to the issue of time pressure and stress and how this factor influences performance in naturalistic decision making.

5.1 Naturalistic Decision Making

In recent years it has been recognized that decision making in natural environments often differs greatly between decision contexts (Beach, 1993; Hammond, 1993). In addressing this topic, the researchers involved often question the relevance and validity of both classical decision theory and behavioral research not conducted in real-world settings (Cohen, 1993). Numerous naturalistic models have been proposed (Klein et al., 1993). These models assume that people rarely weigh alternatives and compare them in terms of expected value or utility. Each model is also descriptive rather than prescriptive. Perhaps the most general conclusion that can be drawn from this work is that people use different decision strategies, depending on their experience, the task, and the decision context. Several of the models also postulate that people choose between decision strategies by trading off effectiveness against the effort required.

In the following discussion we briefly review seven modeling perspectives that fit into this framework: (1) levels of task performance (Rasmussen, 1983), (2) recognition-primed decisions (Klein, 1989), (3) image theory (Beach, 1990), (4) contingent decision making (Payne et al., 1993), (5) dominance structuring (Montgomery, 1989), (6) explanation-based decision making (Pennington and Hastie, 1988), and (7) shared mental models and awareness. Attention then shifts to leadership and its impact on team performance in naturalistic settings.

5.1.1 Levels of Task Performance

There is growing recognition that most decisions are made on a routine basis in which people simply follow past behavior patterns (Rasmussen, 1983; Svenson, 1990; Beach, 1993). Rasmussen (1983) follows this approach to distinguish among skill-based, rule-based, and knowledge-based levels of task performance. Lehto (1991) further considers judgment-based behavior as a fourth level of performance. Performance is said to be at either a skill-based or a rule-based level when tasks are routine in nature. *Skill-based performance* involves the smooth, automatic flow of actions without conscious decision points. As such, skill-based performance describes the decisions made by highly trained operators performing familiar tasks. *Rule-based performance* involves the conscious perception of environmental cues, which trigger the application of rules learned on the basis of experience. As such, rule-based performance corresponds closely to recognition-primed decisions (Klein, 1989). *Knowledge-based performance* is said to occur during learning or problem-solving activity during which people cognitively simulate the influence of various actions and develop plans for what to do. The judgment-based level of performance occurs when affective reactions of a decision maker cause a change in goals or priorities between goals (Janis and Mann, 1977; Etzioni, 1988; Lehto, 1991). Distinctive types of errors in decision making occur at each of the four levels (Reason, 1990; Lehto, 1991).

At the skill-based level, errors occur due to perceptual variability and when people fail to shift up to rule-based or higher levels of performance. At the rule-based level, errors occur when people apply faulty rules or fail to shift up to a knowledge-based level in unusual situations where the rules they normally use are no longer appropriate. The use of faulty rules leads to an important distinction between running and taking risks. Along these lines, Wagenaar (1992) discusses several case studies in which people following risky forms of behavior do not seem to be consciously evaluating the risk. Drivers, in particular, seem habitually to take risks. Wagenaar explains such behavior in terms of faulty rules derived on the basis of benign experience. In other words, drivers get away with providing small safety margins most of the time and consequently learn to run risks on a routine basis. Drucker (1985) points out several cases where organizational decision makers have failed to recognize that the generic principles they used to apply were no longer appropriate, resulting in catastrophic consequences.

At the knowledge-based level, errors occur because of cognitive limitations or faulty mental models or

*In discussing ways of improving the effectiveness of executive decision makers, emphasizes the importance of establishing a generic principle or policy that can be applied to specific cases in a routine way. This recommendation is interesting, as it prescribes a naturalistic form of behavior.

when the testing of hypotheses cause unforeseen changes to systems. At judgment-based levels, errors (or violations) occur because of inappropriate affective reactions, such as anger or fear (Lehto, 1991). As noted by Isen (1993), there also is growing recognition that positive affect can influence decision making. For example, positive affect can promote the efficiency and thoroughness of decision making, but may cause people to avoid negative materials. Positive affect also seems to encourage risk-averse preferences. Decision making itself can be anxiety provoking, resulting in violations of rationality (Janis and Mann, 1977).

A recent study involving drivers arrested for drinking and driving (McKnight et al., 1995) provides an interesting perspective on how the sequential nature of naturalistic decisions can lead people into traps. The study also shows how errors can occur at multiple levels of performance. In this example, decisions made well in advance of the final decision to drive while impaired played a major role in creating situations where drivers were almost certain to drive impaired. For example, the driver may have chosen to bring along friends and therefore have felt pressured to drive home because the friends were dependent on him or her. This initial failure by drivers to predict the future situation could be described as a failure to shift up from a rule-based level to a knowledge-based level of performance. In other words, the driver never stopped to think about what might happen if he or she drank too much. The final decision to drive, however, would correspond to an error (or violation) at the judgment-based level if the driver's choice was influenced by an affective reaction (perceived pressure) to the presence of friends wanting a ride.

5.1.2 Recognition-Primed Decision Making

Klein (1989, 1998) developed the theory of recognition-primed decision making on the basis of observations of firefighters and other professionals in their naturalistic environments. He found that up to 80% of the decisions made by firefighters involved some sort of situation recognition, where the decision makers simply followed a past behavior pattern once they recognized the situation.

The model he developed distinguishes between three basic conditions. In the simplest case, the decision maker recognizes the situation and takes the obvious action. A second case occurs when the decision maker consciously simulates the action to check whether it should work before taking it. In the third and most complex case, the action is found to be deficient during the mental simulation and is consequently rejected. An important point of the model is that decision makers do not begin by comparing all the options. Instead, they begin with options that seem feasible based on their experience. This tendency, of course, differs from the SEU approach but is comparable to applying the satisficing decision rule (Simon, 1955) discussed earlier.

Situation assessment is well recognized as an important element of decision making in naturalistic environments (Klein et al., 1993). Recent research by Klein and his colleagues has examined the possibility of enhancing situation awareness through training (Klein and Wolf, 1995). Klein and his colleagues have also applied methods of cognitive task analysis to naturalistic decision-making problems. In these efforts they have focused on identifying (1) critical decisions, (2) the elements of situation awareness, (3) critical cues indicating changes in situations, and (4) alternative courses of action (Klein, 1995). Accordingly, practitioners of naturalistic decision making tend to focus on process-tracing methods and behavioral protocols (Ericsson and Simon, 1984) to document the processes people follow when they make decisions.*

5.1.3 Image Theory

Image theory (Beach, 1990) is a descriptive theory of decision making. Beach theorizes that knowledge used to make decisions falls into three categories: value images, trajectory images, and strategic images. The value image describes the decision maker's values and principles; the trajectory image describes goals; the strategic image describes plans to attain the goals. He also theorizes that there are two types of decisions: adoption decisions and progress decisions. Adoption decisions first involve a screening process where alternatives are eliminated from consideration. The most promising alternative is then selected from the screened set. Progress decisions involve a comparison between goals and the expected result of choosing the alternative.

Two means of evaluating decisions are applied. One test compares the compatibility of the generated alternatives to value images, trajectory images, and strategic images. The profitability test is used to evaluate screened options further in adoption decisions when more than one option survives the initial screening. Beach (1993) argues strongly for the primacy of screening as a characteristic of most real-world decision-making activity.

5.1.4 Contingent Decision Making

The theory of contingent decision making (Beach and Mitchell, 1978; Payne et al., 1993) is similar to image theory and cognitive continuum theory (see Section 4.1.4) in that it holds that people use different decision strategies depending on the characteristics of the task and the decision context. Payne et al. limit their modeling approach to tasks that require choices to be made (simple memory tasks are excluded from consideration). They also add the assumption that people make choices about how to make choices.†

*Goldstein and Hogarth (1997) describe a similar trend in judgment and decision-making research.

†As such, the theory of contingent decision making directly addresses a potential source of conflict shown in the integrative model of decision making presented earlier (Figure 1). That is, it states that decision makers must choose between decision strategies when they are uncertain how to compare alternatives.

Choices between decision strategies are assumed to be made rationally by comparing their cost (in terms of cognitive effort) against their benefits (in terms of accuracy). Cognitive effort and accuracy (of a decision strategy) are both assumed to depend on task characteristics, such as task complexity, response mode, and method of information display. Cognitive effort and accuracy also are assumed to depend on contextual characteristics, such as the similarity of the compared alternatives, attribute ranges and correlations, the quality of the options considered, reference points, and decision frames. Payne et al. place much emphasis on measuring the cognitive effort of different decision strategies in terms of the number of elemental information elements that must be processed for different tasks and contexts. They relate the accuracy of different decision strategies to task characteristics and contexts and also present research showing that people will shift decision strategies to reduce cognitive effort, increase accuracy, or in response to time pressure.

5.1.5 Dominance Structuring

Dominance structuring (Montgomery, 1989; Montgomery and Willen, 1999) holds that decision making in real contexts involves a sequence of four steps. The process begins with a preediting stage in which alternatives are screened from further analysis. The next step involves selecting a promising alternative from the set of alternatives that survive the initial screening. A test is then made to check whether the promising alternative dominates the other surviving alternations. If dominance is not found, the information regarding the alternatives is restructured in an attempt to force dominance. This process involves both the bolstering and deemphasizing of information in a way that eliminates disadvantages of the promising alternative.

Empirical support can be found for each of the four stages of the bolstering process (Montgomery and Willen, 1999). Consequently, this theory may have value as a description of how people make nonroutine decisions.

5.1.6 Explanation-Based Decision Making

Explanation-based decision making (Pennington and Hastie, 1986, 1988) assumes that people begin their decision-making process by constructing a mental model that explains the facts they have received. While constructing this explanatory model, people are also assumed to be generating potential alternatives to choose between. The alternatives are then compared to the explanatory model rather than to the facts from which it was constructed.

Pennington and Hastie have applied this model to juror decision making and obtained experimental evidence that many of its assumptions seem to hold. They note that juror decision making requires consideration of a massive amount of data that is often presented in haphazard order over a long time period. Jurors seem to organize this information in terms of stories describing causation and intent. As part of this process, jurors are assumed to evaluate stories in terms of their uniqueness, plausibility, completeness, or consistency. To determine a verdict, jurors then judge the fit between choices provided by the trial judge and the various stories they use to organize the information. Jurors' certainty about their verdict is assumed to be influenced both by evaluation of stories and by the perceived goodness of fit between the stories and the verdict.

5.1.7 Shared Mental Models and Awareness

Orasanu and Salas (1993) discuss two closely related frameworks for describing the knowledge used by teams in naturalistic settings. These are referred to as *shared mental models* and the *team mind*. The common element of these two frameworks is that the members of teams hold knowledge in common and organize it in the same way. Orasanu and Salas claim that this improves and minimizes the need for communication between team members, enables team members to carry out their functions in a coordinated way, and minimizes negotiation over who should do what at what time. Under emergency conditions, Orasanu and Salas claim that there is a critical need for members to develop a shared situation model. As evidence for the notion of shared mental models and the team mind, the authors cite research in which firefighting teams and individual firefighters developed the same solution strategies for situations typical of their jobs.

This notion of shared mental models and the team mind can be related to the notion discussed earlier of schemas containing problem-specific rules and facts (Cheng and Holyoak, 1985). It might also be reasonable to consider other team members as a form of external memory (Newell and Simon, 1972). This approach would have similarities to Wegner's (1987) concept of transactive memory, where people in a group know who has specialized information of one kind or another. Klein (1998) provides an interesting discussion of how this metaphor of the team mind corresponds to thinking by individuals. Teams, like people, have a working memory that contains information for a limited time, a long-term or permanent memory, and limited attention. Like people, they also filter out and process information and learn in many ways.

5.1.8 Team Leadership

Torrance (1953) describes retrospective accounts of military survivors lost behind enemy lines, indicating that survival depended on the leader's leadership skills. Important elements of leadership skills included keeping the members of the group focused on a common goal, making sure that they knew what needed to be done, and keeping them informed of the current status. Related conclusions concerning the value of keeping people informed have been obtained in retrospective accounts of survivors of mining accidents (Mallet et al., 1993). Orasanu and Salas (1993) cite research in which captains of high-performing air crews explicitly stated more plans, strategies, and intentions to the other members of the crew. They also gave more warnings and predictions to the crew members. Orasanu and Salas cite other work showing that crews performed better with captains who were task oriented and had good personal skills. Performance dropped when captains had negative expressive styles and low task orientation.

A complementary literature has been developed on leadership theory (Chemers and Ayman, 1993). Most of this research is based on leaders in organizational contexts. A sampling of factors that have been shown to be related to the effectiveness of leadership includes legitimacy, charisma, individualized attention to group members, and clear definitions of goals. These results seem quite compatible with the foregoing findings for leadership in naturalistic, dynamic contexts.

5.2 Time Pressure and Stress

Time pressure and stress are defining characteristics of naturalistic decision making. Jobs requiring high levels of skill or expertise, such as firefighting, nursing, emergency care, and flying an airplane, are especially likely to involve high stakes, extreme time pressure, uncertainty, or risk to life. The effect of stressors such as those mentioned above on performance has traditionally been defined in terms of physiological arousal.* The *Yerkes–Dodson law* (Yerkes and Dodson, 1908) states that the relation between performance and arousal is an inverted U. Either too much or too little arousal causes performance to drop. Too little arousal makes it difficult for people to maintain focused attention. Too much arousal results in errors, more focused attention (and filtering of low-priority information), reduced working memory capacity, and shifts in decision strategies.† One explanation of why performance drops when arousal levels are too high is that arousal consumes cognitive resources that could be allocated to task performance (Mandler, 1979).

Time pressure is a commonly studied stressor assumed to affect decision making. Maule and Hockey (1993) note that people tend to filter out low-priority types of information, omit processing information, and accelerate mental activity when they are under time pressure. Variable state activation theory (VSAT) provides a potential explanation of the foregoing effects in terms of a control model of stress regulation (Maule and Hockey, 1993). Sequential sampling models provide a compatible perspective on how time pressure can cause changes in performance, such as speed–accuracy trade-offs (see Section 4.2.4). The two approaches are compatible, because VSAT provides a means of modeling how the decision thresholds used within a sequential sampling model might change as a function of time pressure. VSAT also proposes that disequilibriums between control processes and the demands of particular situations can lead to strong affective reactions or feelings of time pressure. Such reactions could, of course, lead to attentional narrowing or reduced working memory capacity and therefore result in poorer task performance. Alternatively, performance might change when decision thresholds are adjusted.

Time pressure can also cause shifts between the cognitive strategies used in judgment and decision-making situations (Edland and Svenson, 1993; Maule and Hockey, 1993; Payne et al., 1993). People show a strong tendency to shift to noncompensatory decision rules when they are under time pressure. This finding is consistent with contingency theories of strategy selection (Section 5.1.4). In other words, this shift may be justified when little time is available, because a noncompensatory rule can be applied more quickly. Compensatory decision rules also require more analysis and cognitive effort. Intuitive decision strategies require much less effort because people can rely on their experience or knowledge and can lead to better decisions in some situations (Hammond et al., 1987). As Klein (1998) points out, stress should affect performance if people use analytical choice procedures.

Novices and experts in novel, unexpected situations will lack domain experience and knowledge and therefore will have to rely on analytical choice procedures. Consequently, it is not surprising that time pressure and stress have a major negative impact on novice decision makers performing unfamiliar tasks. Interestingly, there is little evidence that stress or time pressure causes experienced personnel to make decision errors in real-world tasks (Klein, 1996; Orasanu, 1997). The latter finding is consistent with research indicating that experts rely on their experience and intuition when they are under stress and time pressure (Klein, 1998). The obvious implication is that training and experience are essential if people are to make good decisions under time pressure and stress.

6 GROUP DECISION MAKING

Much research has been done over the past 25 years or so on decision making by groups and teams. Most of this work has focused on groups as opposed to teams. In a team it is assumed that the members are working toward a common goal and have some degree of interdependence, defined roles and responsibilities, and task-specific knowledge (Orasanu and Salas, 1993). Team performance is a major area of interest in the field of naturalistic decision theory (Klein et al., 1993; Klein, 1998), as discussed earlier. Group performance has traditionally been an area of study in the fields of organizational behavior and industrial psychology. Traditional decision theory has also devoted some attention to group decision making (Raiffa, 1968; Keeney and Raiffa, 1976). In the following discussion we first discuss briefly some of the ways that group decisions differ from those made by isolated decision makers who need to consider only their own preferences. That is, ethics and social norms play a much more prominent role when decisions are made

*The general adaptation syndrome (Selye, 1936, 1979) describes three stages of the human response to stressors. In simplified form, this sequence corresponds to (1) arousal, (2) resistance, and (3) exhaustion.

†The literature on stress and its effects on decision making is not surveyed here. Books edited by Hamilton and Warburton (1979), Svenson and Maule (1993), Driskell and Salas (1996), and Flin et al. (1997) provide a good introduction to the area.

by or within groups. Attention will then shift to group processes and how they affect group decisions. In the last section we address methods of supporting or improving group decision making.

6.1 Ethics and Social Norms

When decisions are made by or within groups, a number of issues arise that have not been touched on in the earlier portions of this chapter. To start, there is the complication that preferences may vary between members of a group. It often is impossible to maximize the preferences of all members of the group, meaning that trade-offs must be made and issues such as fairness must be addressed to obtain acceptable group decisions. Another complication is that the return to individual decision makers can depend on the actions of others. Game theory* distinguishes two common variations of this situation. In competitive games, individuals are likely to take "self-centered" actions that maximize their own return but reduce returns to other members of the group. Behavior of group members in this situation may be well described by the minimax decision rule discussed in Section 2.1.5. In cooperative games, the members of the group take actions that maximize returns to the group as a whole.

Members of groups may choose cooperative solutions that are better for the group as a whole, for many different reasons (Dawes et al., 1988). Groups may apply numerous forms of coercion to punish members who deviate from the cooperative solutions. Group members may apply decision strategies such as reciprocal altruism. They also might conform because of their social conscience, a need for self-esteem, or feelings of group identity. Fairness considerations can in some case explain preferences and choices that seem to be in conflict with economic self-interest (Bazerman, 1998). Changes in the status quo, such as increasing the price of bottled water immediately after a hurricane, may be viewed as unfair even if they are economically justifiable based on supply and demand. People are often willing to incur substantial costs to punish "unfair" opponents and reward their friends or allies. The notion that costs and benefits should be shared equally is one fairness-related heuristic that people use (Messick, 1991). Consistent results were found by Guth et al. (1982) in a simple bargaining game where player 1 proposes a split of a fixed amount of cash and player 2 either accepts the offer or rejects it. If player 2 rejects the offer, both players receive nothing. Classical economics predicts that player 2 will accept any positive amount (i.e., player 2 should always prefer something to nothing). Consequently, player 1 should offer player 2 a very small amount greater than zero. The results showed that contrary to predictions of classical economics, subjects tended to offer a substantial proportion of the cash (the average offer was 30%). Some of the subjects rejected positive offers. Others accepted offers of zero. Further research, summarized by Bolton and Chatterjee (1996), confirms these findings that people seem to care about whether they receive their fair share.

Ethics clearly plays an important role in decision making. Some choices are viewed by nearly everyone as being immoral or wrong (i.e., violations of the law, dishonesty, and numerous other behaviors that conflict with basic societal values or behavioral norms). Many corporations and other institutions formally specify codes of ethics prescribing values such as honesty, fairness, compliance with the law, reliability, considerance or sensitivity to cultural differences, courtesy, loyalty, respect for the environment, and avoiding waste. It is easy to visualize scenarios where it is in the best interest of a decision maker to choose economically undesirable options (at least in the short term) to comply with ethical codes. According to Kidder (1995), the "really tough choices ... don't center on right versus wrong. They involve right versus right." Kidder refers to four dilemmas of right vs. right that he feels qualify as paradigms: (1) truth vs. loyalty (i.e., whether to divulge information provided in confidence), (2) individual vs. community, (3) short term vs. long term, and (4) justice vs. mercy. At least three principles, which in some cases provide conflicting solutions, have been proposed for resolving ethical dilemmas. These include (1) utilitarianism, selecting the option with the best overall consequences; (2) rule-based, following a rule regardless of its current consequences (i.e., waiting for a stop light to turn green even if no cars are coming); and (3) fairness, doing what you would want others to do for you.

Numerous social dilemmas also occur in which the payoffs to each participant result in individual decision strategies harmful to the group as a whole. The tragedy of the commons (Hardin, 1968) is illustrative of social dilemmas in general. For a recent example, discussed in detail by Baron (1998), consider the recent crash of the east coast commercial fishing industry, brought about by overfishing. Here, the fishing industry as a whole is damaged by overfishing, but individual fishers gain a short-term advantage by catching as many fish as possible. Individual fishers may reason that if they don't catch the fish, someone else will. Each fisher attempts to catch as many fish as possible, even if this will cause the fish stocks to crash. Despite the fact that cooperative solutions, such as regulating the catch, are obviously better than the current situation, individual fishers continue to resist such solutions. Regulations are claimed to infringe on personal autonomy, to be unfair, or to based on inadequate knowledge.

Similar examples include littering, wasteful use of natural resources, pollution, or social free riding. These behaviors can all be explained in terms of the choices faced by the offending individual decision maker (Schelling, 1978). Simply put, the individual decision maker enjoys the benefits of the offensive behavior, as small as they may be, but the costs are incurred by the entire group.

6.2 Group Processes

A large amount of research has focused on groups and their behavior. Accordingly, many models have been

*Friedman (1990) provides an excellent introduction to game theory.

developed that describe how groups make decisions. A common observation is that groups tend to move through several phases as they go through the decision-making process (Ellis and Fisher, 1994). One of the more classic models (Tuckman, 1965) describes this process with four words: forming, storming, norming, and performing. *Forming* corresponds to initial orientation, *storming* to conflict, *norming* to developing group cohesion and expressing opinions, and *performing* to obtaining solutions. As implied by Tuckman's choice of terms, there is a continual interplay between socioemotive factors and rational, task-oriented behavior throughout the group decision-making process. Conflict, despite its negative connotations, is a normal, expected aspect of the group decision process and can in fact serve a positive role (Ellis and Fisher, 1994). In the following discussion we first address causes and effects of group conflict, then shift to conflict resolution.

6.2.1 Conflict

Whenever people or groups have different preferences, conflict can occur. As pointed out by Zander (1994), conflict between groups becomes more likely when groups have fuzzy or potentially antagonistic roles or when one group is disadvantaged (or perceives that it is not being treated fairly). A lack of conflict-settling procedures and separation or lack of contact between groups can also contribute to conflict. Conflict becomes especially likely during a crisis and often escalates when the issues are perceived to be important, or after resistance or retaliation occurs. Polarization, loyalty to one's own group, lack of trust, and cultural and socioeconomic factors are often contributing factors to conflict and conflict escalation.

Ellis and Fisher (1994) distinguish between affective and substantive forms of conflict. *Affective conflict* corresponds to emotional clashes between individuals or groups; *substantive conflict* involves opposition at the intellectual level. Substantive conflict is especially likely to have positive effects on group decisions by promoting better understanding of the issues involved. Affective conflict can also improve group decisions by increasing interest, involvement, and motivation among group members and, in some cases, cohesiveness. On the other hand, affective conflict may cause significant ill-will, reduced cohesiveness, and withdrawal by some members from the group process. Baron (1998) provides an interesting discussion of violent conflict and how it is related to polarized beliefs, group loyalty, and other biases.

Defection and the formation of coalitions is a commonly observed effect of conflict, or power struggles, within groups. Coalitions often form when certain members of the group can gain by following a common course of action at the expense of the long-run objectives of the group as a whole. Rapidly changing coalitions between politicians and political parties are obviously a fact of life. Another typical example is when a subgroup of technical employees leave a corporation to form their own small company, producing a product similar to one they had been working on. Coalitions, and their formation, have been examined from decision-analytic and game theory perspectives (Raiffa, 1982; Bolton and Chatterjee, 1996). These approaches make predictions regarding what coalitions will form, depending on whether the parties are cooperating or competing, which have been tested in a variety of experiments (Bolton and Chatterjee, 1996). These experiments have revealed that the formation of coalitions is influenced by expected payoffs, equity issues, and the ease of communication. However, Bazerman (1998) notes that the availability heuristic, overconfidence, and sunk cost effects are likely to explain how coalitions actually form in the real world.

6.2.2 Conflict Resolution

Groups resolve conflict in many different ways. Discussion and argument, voting, negotiation, arbitration, and other forms of third-party intervention are all methods of resolving disputes. Discussion and argument are clearly the most common methods followed within groups to resolve conflict. Other methods of conflict resolution normally play a complementary rather than a primary role in the decision process. That is, the latter methods are relied on when groups fail to reach consensus after discussion and argument, or they simply serve as the final step in the process.

Group discussion and argument are often viewed as constituting a less than rational process. Along these lines, Brashers et al. (1994) state that the literature suggests "that argument in groups is a social activity, constructed and maintained in interaction, and guided perhaps by different rules and norms than those that govern the practice of ideal or rational argument. Subgroups speaking with a single voice appear to be a significant force.... Displays of support, repetitive agreement, and persistence all appear to function as influence mechanisms in consort with, or perhaps in place of, the quality or rationality of the arguments offered." Brashers et al. also suggest that members of groups appear uncritical because their arguments tend to be consistent with social norms rather than the rules of logic: "[S]ocial rules such as: (a) submission to higher status individuals, (b) experts' opinions are accepted as facts on all matters, (c) the majority should be allowed to rule, (d) conflict and confrontation are to be avoided whenever possible."

A number of approaches for conflict management have been suggested that attempt to address many of the issues raised by Brashers et al. These approaches include seeking consensus rather than allowing decisions to be posed as win–lose propositions, encouraging and training group members to be supportive listeners, deemphasizing status, depersonalizing decision making, and using facilitators (Likert and Likert, 1976). Other approaches that have been proposed include directing discussion toward clarifying the issues, promoting an open and positive climate for discussion, facilitating face-saving communications, and promoting the development of common goals (Ellis and Fisher, 1994).

Conflicts can also be resolved through voting and negotiation, as discussed further in Section 6.3.

Negotiation becomes especially appropriate when the people involved have competing goals and some form of compromise is required. A typical example would be a dispute over pay between a labor union and management. Strategic concerns play a major role in negotiation and bargaining (Schelling, 1960). Self-interest on the part of the involved parties is the driving force throughout a process involving threats and promises, proposals and counterproposals, and attempts to discern how the opposing party will respond. Threats and promises are a means of signaling what the response will be to actions taken by an opponent and consequently become rational elements of a decision strategy (Raiffa 1982). Establishing the credibility of signals sent to an opponent becomes important because if they are not believed, they will not have any influence.

Methods of attaining credibility include establishing a reputation, the use of contracts, cutting off communication, burning bridges, leaving an outcome beyond control, moving in small steps, and using negotiating agents (Dixit and Nalebuff, 1991). Given the fundamentally adversarial nature of negotiation, conflict may move from a substantive basis to an affective, highly emotional state. At this stage, arbitration and other forms of third-party intervention may become appropriate, due to a corresponding tendency for the negotiating parties to take extreme, inflexible positions.

6.3 Group Performance and Biases

The quality of the decisions made by groups in a variety of different settings has been seriously questioned. Part of the issue here is the phenomenon of *groupthink*, which has been blamed for several disastrous public policy decisions (Janis, 1972; Hart et al., 1997). Eight symptoms of groupthink cited by Janis and Mann (1977) are the illusion of invulnerability, rationalization (discounting of warnings and negative feedback), belief in the inherent morality of the group, stereotyping of outsiders, pressure on dissenters within the group, self-censorship, illusion of unanimity, and the presence of mindguards who shield the group from negative information. Janis and Mann proposed that the results of groupthink include failure to consider all the objectives and alternatives, failure to reexamine choices and rejected alternatives, incomplete or poor search for information, failure to adequately consider negative information, and failure to develop contingency plans. Groupthink is one of the most cited characteristics of how group decision processes can go wrong. Given the prominence of groupthink as an explanation of group behavior, it is somewhat surprising that only a few studies have evaluated this theory empirically. Empirical evaluation of the groupthink effect and the development of alternative modeling approaches continue to be an active area of research (Hart et al., 1997).

Other research has attempted to measure the quality of group decisions in the real world against rational, or normative, standards. Viscusi (1991) cites several examples of apparent regulatory complacency and regulatory excess in government safety standards in the United States. He also discusses a variety of inconsistencies in the amounts awarded in product liability cases. Baron (1998) provides a long list of what he views as errors in public decision making and their very serious effects on society. These examples include collective decisions resulting in the destruction of natural resources and overpopulation, strong opposition to useful products such as vaccines, violent conflict between groups, and overzealous regulations, such as the Delaney clause. He attributes these problems to commonly held, and at first glance innocuous, intuitions such as Do no harm, Nature knows best, and Be loyal to your own group, the need for retribution (an eye for an eye), and a desire for fairness.

A significant amount of laboratory research is available that compares the performance of groups to that of individual decision makers (Davis, 1992; Kerr et al., 1996). Much of the early work showed that groups were better than individuals on some tasks. Later research indicated that group performance is less than the sum of its parts. Groups tend to be better than individuals on tasks where the solution is obvious once it is advocated by a single member of the group (Davis, 1992; Kerr et al., 1996). Another commonly cited finding is that groups tend to be more willing than individuals, to select risky alternatives, but in some cases the opposite is true. One explanation is that group interactions cause people within the group to adopt more polarized opinions (Moscovici, 1976). Large groups seem especially likely to reach polarized, or extreme, conclusions (Isenberg, 1986). Groups also tend to overemphasize the common knowledge of members, at the expense of underemphasizing the unique knowledge certain members have (Stasser and Titus, 1985; Gruenfeld et al., 1996). A more recent finding indicates that groups were more rational than individuals when playing the ultimatum game (Bornstein and Yaniv, 1998).

Duffy (1993) notes that teams can be viewed as information processes and cites team biases and errors that can be related to information-processing limitations and the use of heuristics, such as framing. Topics such as mediation and negotiation, jury decision making, and public policy are now being evaluated from the latter perspective (Heath et al., 1994). Much of this research has focused on whether groups use the same types of heuristics and are subject to the same biases of individuals. This research has shown (1) framing effects and preference reversals (Paese et al., 1993), (2) overconfidence (Sniezek, 1992), (3) use of heuristics in negotiation (Bazerman and Neale, 1983), and (4) increased performance with cognitive feedback (Harmon and Rohrbaugh, 1990). One study indicated that biasing effects of the representativeness heuristic were greater for groups than for individuals (Argote et al., 1986). The conclusion is that group decisions may be better than those of individuals in some situations but are subject to many of the same problems.

6.4 Prescriptive Approaches

A wide variety of prescriptive approaches have been proposed for improving group decision making. The approaches address some of the foregoing issues, including the use of agendas and rules of order, idea-generating techniques such as brainstorming, nominal group and Delphi techniques, decision structuring, and methods of computer-mediated decision making. As noted by Ellis and Fisher (1994), there is conflicting evidence regarding the effectiveness of such approaches. On the negative side, prescriptive approaches might stifle creativity in some situations and can be sabotaged by dissenting members of groups. On the positive side, prescriptive approaches make the decision process more orderly and efficient, promote rational analysis and participation by all members of the group, and help ensure implementation of group decisions. In the following discussion we review briefly some of these tools for improving group decision making.

6.4.1 Agendas and Rules of Order

Agendas and rules of order are often essential to the orderly functioning of groups. As noted by Welch (1994), an agenda "conveys information about the structure of a meeting: time, place, persons involved, topics to be addressed, perhaps suggestions about background material or preparatory work." Agendas are especially important when the members of a group are loosely coupled or do not have common expectations. Without an agenda, group meetings are likely to dissolve into chaos (Welch, 1994). Rules of order, such as *Robert's Rules of Order* (Robert, 1990), play a similarly important role, by regulating the conduct of groups to ensure fair participation by all group members, including absentees. Rules of order also specify voting rules and means of determining consensus. Decision rules may require unanimity, plurality, or majority vote for an alternative.

Attaining consensus poses an advantage over voting, because voting encourages the development of coalitions, by posing the decision as a win–lose proposition (Ellis and Fisher, 1994). Members of the group who voted against an alternative are often unlikely to support it. Voting procedures can also play an important role (Davis, 1992).

6.4.2 Idea-Generation Techniques

A variety of approaches have been developed for improving the creativity of groups in the early stages of decision making. Brainstorming is a popular technique for quickly generating ideas (Osborn, 1937). In this approach, a small group (of no more than 10 people) is given a problem to solve. The members are asked to generate as many ideas as possible. Members are told that no idea is too wild and are encouraged to build on the ideas submitted by others. No evaluation or criticism of the ideas is allowed until after the brainstorming session is finished. Buzz group analysis is a similar approach, more appropriate for large groups (Ellis and Fisher, 1994). Here, a large group is first divided into small groups of four to six members. Each small group goes through a brainstorming-like process to generate ideas. They then present their best ideas to the entire group for discussion. Other commonly applied idea-generating techniques include focus group analysis and group exercises intended to inspire creative thinking through role playing (Ellis and Fisher, 1994; Clemen, 1996).

The use of brainstorming and the other idea-generating methods mentioned above will normally provide a substantial amount of, in some cases, creative suggestions, especially when participants build on each other's ideas. However, personality factors and group dynamics can also lead to undesirable results. Simply put, some people are much more willing than others to participate in such exercises. Group discussions consequently tend to center around the ideas put forth by certain more forceful individuals. Group norms, such as deferring to participants with higher status and power, may also lead to undue emphasis on the opinions of certain members.

6.4.3 Nominal Group and Delphi Technique

Nominal group technique (NGT) and the Delphi technique attempt to alleviate some of the disadvantages of working in groups (Delbecq et al., 1975). The nominal group technique consists of asking each member of a group to write down and think about his or her ideas independently. A group moderator then asks each member to present one or more of his or her ideas. Once all of the ideas have been posted, the moderator allows discussion to begin. After the discussion is finished, each participant rates or ranks the ideas presented. The subject ratings are then used to develop a score for each idea. Nominal group technique is intended to increase participation by group members and is based on the idea that people will be more comfortable presenting their ideas if they have a chance to think about them first (Delbecq et al., 1975).

The Delphi technique allows participants to comment anonymously, at their leisure, on proposals made by other group members. Normally, the participants do not know who proposed the ideas they are commenting on. The first step is to send an open-ended questionnaire to members of the group. The results are then used to generate a series of follow-up questionnaires in which more specific questions are asked. The anonymous nature of the Delphi process theoretically reduces the effect of participant status and power. Separating the participants also increases the chance that members will provide opinions "uncontaminated" by the opinions of others.

6.4.4 Structuring Group Decisions

As discussed earlier in this chapter, the field of decision analysis has devised several methods for organizing or structuring the decision-making process. The rational reflection model (Siebold, 1992) is a less formal, six-step procedure that serves a similar function. Group members are asked first to define and limit the problem by identifying goals, available resources, and procedural constraints. After defining and limiting the problem, the group is asked to analyze the problem, collect relevant information, and establish the criteria that a solution must meet. Potential solutions are then discussed in terms of the agreed-upon decision criteria. After further discussion, the group selects a solution and determines how it should be implemented. The focus of this approach is on forcing the group to confine its discussion to the issues that arise at each step in the decision-making process. As such, this method is similar to specifying an agenda.

Raiffa (1982) provides a somewhat more formal decision-analytic approach for structuring negotiations. The approach begins by assessing (1) the alternatives to a negotiated settlement, (2) the interests of the involved parties, and (3) the relative importance of each issue. This assessment allows the negotiators to think analytically about mutually acceptable solutions. In certain cases, a bargaining zone is available. For example, an employer may be willing to pay more than the minimum salary acceptable to a potential employee. In this case, the bargaining zone is the difference between the maximum salary the employer is willing to pay and the minimum salary a potential employee is willing to accept. The negotiator may also think about means of expanding the available resources to be divided, potential trading issues, or new options that satisfy the interests of the concerned parties.

Other methods for structuring group preferences are discussed in Keeney and Raiffa (1976). The development of group utility functions is one such approach. A variety of computer-mediated methods for structuring group decisions are also available.

7 DECISION SUPPORT

The preceding sections of this chapter have much to say about how to help decision makers make better decisions. To summarize that discussion briefly: (1) classical decision theory provides optimal prescriptions for how decisions should be made, (2) decision analysis provides a set of tools for structuring decisions and evaluating alternatives, and (3) studies of human judgment and decision making, in both laboratory settings and naturalistic environments, help identify the strengths and weakness of human decision makers. These topics directly mirror important elements of decision support. That is, decision support should have an objective (i.e., optimal or satisfactory choices, easier choices, more justifiable choices, etc.). Also, it must have a means (i.e., decision analysis or other method of decision support) and it must have a current state (i.e., decision quality, effort expended, knowledge, etc., of the supported decision makers). The effectiveness of decision support can then be defined in terms of how well the means move the current state toward the objective.

The focus of this section is on providing an overview of commonly used methods of computer-based decision support* for individuals, groups, and organizations. Throughout this discussion, an effort is made to address the objectives of each method of support and its effectiveness. Somewhat surprisingly, less information is available on the effectiveness of these approaches than might be expected given their prevalence (see also Yates et al., 2003), so the latter topic is not addressed in a lot of detail.

The discussion begins with decision support systems (DSSs), expert systems, and neural networks. These systems can be designed to support the intelligence, design, or choice phases of decision making (Simon, 1977). The intelligence phase involves scanning and searching the environment to identify problems or opportunities. The design phase entails formulating models for generating possible courses of action. The choice phase refers to finding an appropriate course of action for the problem or opportunity. Hence, the boundary between the design and choice phases is often unclear. Decision support systems and expert systems can be used to support all three phases of decision making, whereas neural networks tend to be better suited for design and choice phases. For example, decision support systems can be designed to help with interpreting economic conditions, while expert systems can diagnose problems. Neural networks can learn a problem domain, after which they can serve as a powerful aid for decision making.

Attention then shifts to methods of supporting decisions by groups and organizations. The latter discussion first addresses the use by groups of DSSs and other tools designed to support decisions made by individuals. In the sections that follow we address approaches specifically designed for use by groups.

7.1 Decision Support Systems

The concept of decision support systems (DSSs) dates back to the early 1970s. It was first articulated by Little (1970) under the term *decision calculus* and by Scott-Morton (1977) under the term *management decision systems*. DSSs are interactive computer-based systems that help decision makers utilize *data* and *models* to solve unstructured or semistructured problems (Scott-Morton, 1977; Keen and Scott-Morton, 1978). Given the unstructured nature of these problems, the goal of such systems is to *support*, rather than replace, human decision making.

The three key components of a DSS are (1) a model base, (2) a database, and (3) a user interface.

*Over the years, many different approaches have been developed for aiding or supporting decision makers (see Winterfelt and Edwards, 1986; Yates et al., 2003). Some of these approaches have already been covered earlier in this chapter and consequently are not addressed further in this section. In particular, decision analysis provides both tools and perspectives on how to structure a decision and evaluate alternatives. Decision analysis software is also available and commonly used. In fact, textbooks on decision analysis normally discuss the use of spreadsheets and other software; software may even be made available along with the textbook (for an example, see Clemen, 1996). Debiasing, discussed earlier in this chapter, is another technique for aiding or supporting decision makers.

The model base comprises quantitative models (e.g., financial or statistical models) that provide the analysis capabilities of DSSs. The database manages and organizes the data in meaningful formats that can be extracted or queried. The user interface component manages the dialogue or interface between DSS and the users. For example, visualization tools can be used to facilitate communication between the DSS and the users.

DSSs are generally classified into two types: model-driven and data-driven. *Model-driven DSSs* utilize a collection of mathematical and analytical models for the decision analysis. Examples include forecasting and planning models, optimization models, and sensitivity analysis models (i.e., for asking "what-if" questions). The analytical capabilities of such systems are powerful because they are based on strong theories or models. On the other hand, *data-driven DSSs* are capable of analyzing large quantities of data to extract useful information. The data may be derived from transaction processing systems, enterprise systems, data warehouses, or web warehouses. Online analytical processing and datamining can be used to analyze the data. Multidimensional data analysis enables users to view the same data in different ways using multiple dimensions. The dimensions could be product, salesperson, price, region, and time period. *Data mining* refers to a variety of techniques that can be used to find hidden patterns and relationships in large databases and to infer rules from them to guide decision making and predict future behavior. Data mining can yield information on associations, sequences, classifications, clusters, and forecasts (Laudon and Laudon, 2003). *Associations* are occurrences linked to a single event (e.g., beer is purchased along with diapers); *sequences* are events linked over time (e.g., the purchase of a new oven after the purchase of a house). *Classifications* refer to recognizing patterns and rules to categorize an item or object into its predefined group (e.g., customers who are likely to default on loans); *clustering* refers to categorizing items or objects into groups that have yet been defined (e.g., identifying customers with similar preferences). Data mining can also be used for forecasting (e.g., projecting sales demand).

Despite the popularity of DSSs not a lot of data are available documenting that they improve decision making (Yates et al., 2003). It does seem logical that DSSs should play a useful role in reducing biases (see Section 4.1.5) and otherwise improving decision quality. This follows because a well-designed DSS will increase both the amount and quality of information available to the decision maker. A well-designed DSS will also make it easier to analyze the information with sophisticated modeling techniques. Ease of use is another important consideration. As discussed earlier, Payne et al. (1993) identify two factors influencing the selection of a decision strategy: (1) cognitive effort required of a strategy in making the decision, and (2) the accuracy of the strategy in yielding a "good" decision. Todd and Benbasat (1991, 1992) found that DSS users adapted their strategy selection to the type of decision aids available in such a way as to reduce effort. In other words, effort minimization is a primary or more important consideration to DSS users than is the quality of decisions. More specifically, the role of effort may have a direct impact on DSS effectiveness and must be taken into account in the design of DSSs.

In a follow-up study, Todd and Benbasat (1999) studied the moderating effect of incentives and cognitive effort required to utilize a more effortful decision strategy that would lead to a better decision outcome (i.e., additive compensatory vs. elimination by aspects; the former strategy requires more effort but leads to a better outcome). Although the results show that the level of incentives has no effect on decision strategy, the additive compensatory (i.e., 'better') strategy was used more frequently when its level of support was increased from no or little support to moderate or high support. The increased support decreased the amount of effort needed to utilize the additive compensatory strategy, thus inducing a strategy change. When designing DSSs, effort minimization should be given considerable attention, as it can drive the choice of decision strategy, which in turn influences the decision accuracy.

7.1.1 Expert Systems

Expert systems are developed to capture knowledge for a very specific and limited domain of human expertise. Expert systems can provide the following benefits: cost reduction, increased output, improved quality, consistency of employee output, reduced downtime, captured scarce expertise, flexibility in providing services, easier operation of equipment, increased reliability, faster response, ability to work with incomplete and uncertain information, improved training, increased ability to solve complex problems, and better use of expert time.

Organizations routinely use expert systems to enhance the productivity and skill of human knowledge workers across a spectrum of business and professional domains. They are computer programs capable of performing specialized tasks based on an understanding of how human experts perform the same tasks. They typically operate in narrowly defined task domains. Despite the name *expert systems*, few of these systems are targeted at replacing their human counterparts; *most of them are designed to function as assistants or advisers to human decision makers*. Indeed, the most successful expert systems—those that actually address mission-critical business problems—are not "experts" as much as "advisors" (LaPlante, 1990).

An expert system is organized in such a way that the knowledge about the problem domain is separated from general problem-solving knowledge. The collection of domain knowledge is called the *knowledge base*, whereas the general problem-solving knowledge is called the *inference engine*. The knowledge base stores domain-specific knowledge in the form of facts and rules. The inference engine operates on the knowledge base by performing logical inferences and deducing new knowledge when it applies rules to facts. Expert systems are also capable of providing explanations to users.

Examples of expert systems include the PlanPower system used by the Financial Collaborative for financial planning (Sviokla, 1989), Digital's XCON for computer configurations (Sviokla, 1990), and Baroid's MUDMAN for drilling decisions (Sviokla, 1986). As pointed out by Yates et al. (2003), the large number of expert systems that are now in actual use suggests that expert systems are by far the most popular form of computer-based decision support. However, as for DSS, not a lot of data are available showing that expert systems improve decision quality. Ease of use is probably one of the main reasons for their popularity. This follows, because the user of an expert system can take a relatively passive role in problem-solving process. That is, the expert system asks a series of questions which the user simply answers if he or she can. The ability of most expert systems to answer questions and explain their reasoning can also help users understand what the system is doing and confirm the validity of the system's recommendations. Such give and take may make users more comfortable with an expert system than they are with models that make sophistical mathematical calculations that are difficult to verify.

7.1.2 Neural Networks

Neural networks consist of hardware or software that is designed to emulate the processing patterns of the biological brain. There are eight components in a neural network (Rumelhart et al., 1986):

1. A set of *processing units*
2. A *state of activation*
3. An *output function* for each unit
4. A *pattern of connectivity* among units
5. A *propagation rule* for propagating patterns of activities through the network
6. An *activation rule* for combining inputs impinging on a unit with its current state
7. A *learning* (or training) *rule* to modify patterns of connectivity
8. An *environment* within which the system must operate

A neural network comprises many interconnected processing elements that operate in parallel. One key characteristic of neural networks is their ability to learn. There are two types of learning algorithms: supervised learning and unsupervised learning. In supervised learning, the desired outputs for each set of inputs are known. Hence, the neural network learns by adjusting its weights in such a way that it minimizes the difference between the desired and actual outputs. Examples of supervised learning algorithms are backpropagation and the Hopfield network. Unsupervised learning is similar to cluster analysis in that only input stimuli are available. The neural network self-organizes itself to produce clusters or categories. Examples of unsupervised learning algorithms are adaptive resonance theory and Kohenen self-organizing feature maps.

By applying a training set such as historical cases, learning algorithms can be used to teach a neural network to solve or analyze problems. The outputs or recommendations from the system can be used to support human decision making. For example, neural networks have been developed to predict customer responses to direct marketing (Cui and Wong, 2004), to forecast stock returns (Olson and Mossman, 2003; Sapena et al., 2003; Jasic and Wood, 2004). to assess product quality in the metallurgical industry (Zhou and Xu, 1999), and to support decision making on sales forecasting (Kuo and Xue, 1998).

7.1.3 Other Forms of Individual Decision Support

Other forms of individual decision support can be developed using fuzzy logic, intelligent agents, case-based reasoning, and genetic algorithms. *Fuzzy logic* refers to the use of membership functions to express imprecision and an approach to approximate reasoning in which the rules of inference are approximate rather than exact. Garavelli and Gorgoglione (1999) used fuzzy logic to design a DSS to improve its robustness under uncertainty, and Coma et al. (2004) developed a fuzzy DSS to support a design for assembly methodology. Collan and Liu (2003) combined fuzzy logic with agent technologies to develop a fuzzy agent–based DSS for capital budgeting. *Intelligent agents* use built-in or learned rules to make decisions. In a multiagent marketing DSS, the final solution is obtained through cooperative and competitive interactions among intelligent agents acting in a distributed mode (Aliev et al., 2000). Intelligent agents can also be used to provide real-time decision support on airport gate assignment (Lam et al., 2003). *Case-based reasoning*, which replies on past cases to derive at a decision, has been used by Lari (2003) to assist in making corrective and preventive actions for solving quality problems and by Belecheanu et al. (2003) to support decision making on new product development. *Genetic algorithms* are robust algorithms that can search through large spaces quickly by mimicking the Darwinian "survival of the fittest" law. They can be used to increase the effectiveness of simulation-based DSSs (Fazlollahi and Vahidov, 2001).

7.2 Group and Organizational Decision Support

Computer tools have been developed to assist in group and organizational decision making. Some of them implement the approaches discussed in Section 6. The spectrum of such tools ranges from traditional tools used in decision analysis, such as the analytic hierarchy process (Saaty, 1988; Basak and Saaty, 1993), to electronic meetingplaces or group decision support systems (DeSanctis and Gallupe, 1987; Nunamaker et al., 1991), to negotiation support systems (Bui et al., 1990; Lim and Benbasat, 1993). We will discuss the use of individual decision support tools for group support, group decision support systems, negotiation support systems, enterprise system support, and other forms of group and organizational support.

7.2.1 Using Individual Decision Support Tools for Group Support

Traditional single-user tools can be used to support groups in decision making. A survey by Satzinger and Olfman (1995) found that traditional single-user tools were perceived by groups to be more useful than group support tools. Sharda et al. (1988) assessed the effectiveness of a DSS for supporting business simulation game and found that groups with access to the DSS made significantly more effective decisions than their non-DSS counterparts. The DSS groups took more time to make their decisions than the non-DSS groups at the beginning of the experiment, but decision times converged in a later period. The DSS teams also exhibited a higher confidence level in their decisions than the non-DSS groups. Knowledge-based systems (or expert support systems) are effective in supporting group decision making, particularly so with novices than experts (Nah and Benbasat, 2004). Groups using the system also make better decisions than individuals provided with the same system (Nah et al., 1999). Hence, empirical findings have shown that traditional single-user tools can be effective in supporting group decision making.

7.2.2 Group Decision Support Systems

Group decision support systems (GDSSs) combine communication, computing, and decision support technologies to facilitate formulation and solution of unstructured problems by a group of people (DeSanctis and Gallupe, 1987). DeSanctis and Gallupe defined three levels of GDSS. Level 1 GDSSs provide technical features aimed at removing common communication barriers, such as large screens for instantaneous display of ideas, voting solicitation and compilation, anonymous input of ideas and preferences, and electronic message exchange among members. In other words, a level 1 GDSS is a communication medium only. Level 2 GDSSs provide decision modeling or group decision techniques aimed at reducing uncertainty and "noise" that occur in the group's decision process. These techniques include automated planning tools (e.g., PERT, CPM, Gantt), structured decision aids for the group process (e.g., automation of Delphi, nominal, or other idea-gathering and compilation techniques), and decision analytic aids for the task (e.g., statistical methods, social judgment models). Level 3 GDSSs are characterized by machine-induced group communication patterns and can include expert advice in the selecting and arranging of *rules* to be applied during a meeting. To date, there has been little research in level 3 GDSSs because of the difficulty and challenges in automating the process of group decision making.

GDSSs facilitate computer-mediated group decision making and provide several potential benefits (Brashers et al., 1994), including (1) enabling all participants to work simultaneously (e.g., they don't have to wait for their turn to speak, thus eliminating the need to compete for air time), (2) enabling participants to stay focused and be very productive in idea generation (i.e., eliminating production blocking caused by attending to others), (3) providing a more equal and potentially anonymous opportunity to be heard (i.e., reducing the negative effects caused by power distance), (4) providing a more systematic and structured decision-making environment (i.e., facilitating a more linear process and better control of the agenda). GDSSs also make it easier to control and manage conflict, through the use of facilitators and convenient voting procedures.

The meta-analysis by Dennis et al. (1996) suggest that in general, GDSSs improve decision quality, increases time to make decisions, and has no effect on participant satisfaction. They also found that larger groups provided with a GDSS had higher satisfaction and experienced greater improvement in performance than smaller groups with GSSs. The findings from McLeod's (1992) and Benbasat and Lim's (1993) meta-analyses show that GDSSs increase decision quality, time to reach decisions, and equality of participation, but decrease consensus and satisfaction. To resolve inconsistencies in the GDSS literature (such as those relating to satisfaction), Dennis and his colleagues (Dennis et al., 2001; Dennis and Wixom, 2002) carried out further meta-analyses to test a fit-appropriation model and identify further moderators for these effects. The result shows that both fit (between GSS structures and task) and appropriation support (i.e., training, facilitation, and software restrictiveness) are necessary for GDSSs to yield an increased number of ideas generated, reduce the time taken for the task, and increase satisfaction of users (Dennis et al., 2001). The fit-appropriation profile is adapted from Zigurs and Buckland (1998).

Computer-supported collaborative systems provide features beyond GDSSs, such as project and calendar management, group authoring, audio and video conferencing, and group and organizational memory management. They facilitate collaborative work beyond simply decision making and are typically referred to as computer-supported collaborative work. These systems are particularly helpful for supporting group decision making in a distributed and asynchronous manner.

7.2.3 Negotiation Support Systems

Negotiation support systems (NSSs) are used to assist people in activities that are competitive or involve conflicts of interest. The need for negotiation can arise from differences in interest or in objectives, or even from cognitive limitations. To understand and analyze a negotiation activity, eight elements must be taken into account (Holsapple et al., 1998): (1) the issue or matter of contention, (2) the set of participants involved, (3) participants' regions of acceptance, (4) participants' location (preference) within region of acceptance, (5) strategies for negotiation (e.g., coalition), (6) participants' movements from one location to another, (7) rules of negotiation, and (8) assistance from an intervenor (e.g., mediator, arbitrator, or facilitator). NSSs should be designed with these eight components in mind by supporting these components.

The configuration of basic NSSs comprises two main components (Lim and Benbasat, 1993): (1) a

DSS for each negotiating party, and (2) an electronic linkage between these systems to enable electronic communication between the negotiators. Full-feature session-oriented NSSs should also offer group process structuring techniques, support for an intervenor, and documentation of the negotiation (Foroughi, 1998). Nego-Plan is an expert system shell that can be used to represent negotiation issues and decompose negotiation goals to help analyze consequences of negotiation scenarios (Matwin et al., 1989; Holsapple and Whinston, 1996). A Web-based NSS called Inspire is used in teaching and training (Kersten and Noronha, 1999). Espinasse et al. (1997) developed a multiagent NSS architecture that can support a mediator in managing the negotiation process. To provide comprehensive negotiation support, NSSs should provide features of level 3 GDSSs, such as the ability to (1) perform analysis of conflict contingencies, (2) suggest appropriate process structuring formats or analytical models, (3) monitor the semantic content of electronic communications, (4) suggest settlements with high joint benefits, and (5) provide automatic mediation (Foroughi, 1998). In general, NSSs can support negotiation either by assisting participants or by serving as a participant (intervenor).

7.2.4 Enterprise Systems for Decision Support

Enterprise-wide support can be provided by enterprise systems (ESs) and executive support systems (ESSs) (Turban and Aronson, 2001). ESSs are designed to support top executives, whereas ESs can be designed to support top executives or to serve a wider community of users. ESSs are comprehensive support systems that go beyond flexible DSSs by providing communication capabilities, office automation tools, decision analysis support, advanced graphics and visualization capabilities, and access to external databases and information in order to facilitate business intelligence and environmental scanning. For example, intelligent agents can be used to assist in environmental scanning.

The ability to use ESs, also known as enterprise resource planning (ERP) systems, for decision support is made possible by data warehousing and online analytical processing. ESs integrate all the functions as well as the transaction processing and information needs of an organization. These systems can bring significant competitive advantage to organizations if they are integrated with supply chain management and customer relationship management systems, thus providing comprehensive information along the entire value chain to key decision makers and facilitating their planning and forecasting. Advanced planning and scheduling packages can be incorporated to help optimize production and ensure that the right materials are in the right warehouse at the right time to meet customers' demands (Turban and Aronson, 2001).

7.2.5 Other Forms of Group and Organizational Decision Support

We have discussed how individual decision support tools, GDSSs, NSSs, ESSs, and ESs can facilitate and support group and organizational decision making. Other techniques drawn from the field of artificial intelligence, such as neural networks, expert systems, fuzzy logic, genetic algorithms, case-based reasoning, and intelligent agents, can also be used to enhance the decision support capabilities of these systems. It should also be noted that knowledge management practices can benefit groups and organizations by capitalizing on existing knowledge to create new knowledge, codifying existing knowledge in ways that are readily accessible to others, and facilitating knowledge sharing and distribution throughout an enterprise (Davenport and Prusak, 2000). Since knowledge is a key asset of organizations and is regarded as the only source of sustainable competitive strength (Drucker, 1995), the use of technologies for knowledge management purposes is a high priority in most organizations. For example, knowledge repositories (e.g., intranets) can be created to facilitate knowledge sharing and distribution, focused knowledge environments (e.g., expert systems) can be developed to codify expert knowledge to support decision making, and knowledge work systems (e.g., computer-aided design, virtual reality simulation systems, and powerful investment workstations) can be used to facilitate knowledge creation. By making existing knowledge more available, these systems can help groups and organizations make more informed and better decisions.

8 SUMMARY AND CONCLUSIONS

Beach (1993) discusses four revolutions in behavioral decision theory. The first took place when it was recognized that the evaluation of alternatives is seldom extensive. It is illustrated by use of the satisficing rule (Simon, 1955) and heuristics (Tversky and Kahneman, 1974; Gigerenzer et al., 1999) rather than optimizing. The second occurred when it was recognized that people choose between strategies to make decisions. It is marked by the development of contingency theory (Beach, 1990) and cognitive continuum theory (Hammond, 1980). The third is currently occurring. It involves the realization that people rarely make choices and instead rely on prelearned procedures. This perspective is illustrated by the levels-of-processing approach (Rasmussen, 1983) and recognition-primed decisions (Klein, 1989). The fourth is just beginning. It involves recognization that decision-making research must abandon a singleminded focus on the economic view of decision making and include approaches drawn from relevant developments and research in cognitive psychology, organizational behavior, and systems theory.

The discussion within this chapter parallels this view of decision making. The integrative model presented at the beginning of the chapter shows how the various approaches fit together as a whole. Each path through the model is distinguished by specific sources of conflict, the methods of conflict resolution followed, and the types of decision rules used to analyze the results of conflict-resolution processes. The different paths through the model correspond to fundamentally different ways of making decisions, ranging from routine situation assessment-driven decisions to satisficing, analysis of single- and multiattribute expected utility, and even obtaining consensus of multiple decision makers in group contexts. Numerous other strategies and potential methods of decision support discussed in this chapter are also described by particular paths through the model.

This chapter goes beyond simply describing methods of decision making by pointing out reasons that people and groups may have difficulty making good decisions. These include cognitive limitations, inadequacies of various heuristics used, biases and inadequate knowledge of decision makers, and task-related factors such as risk, time pressure, and stress. The discussion also provides insight into the effectiveness of approaches for improving human decision making. The models of selective attention point to the value of providing only truly relevant information to decision makers. Irrelevant information might be considered simply because it is there, especially if it is highly salient. Methods of highlighting or emphasizing relevant information are therefore warranted. The models of selective information also indicate that methods of helping decision makers cope with working memory limitations will be of value. There also is reason to believe that providing feedback to decision makers in dynamic decision-making situations will be useful. Cognitive rather than outcome feedback is indicated as being particularly helpful when decision makers are learning. Training decision makers also seems to offer potentially large benefits. One reason for this conclusion is that the studies of naturalistic decision making revealed that most decisions are made on a routine, nonanalytical basis.

Studies of debiasing also partially support the potential benefits of training and feedback. On the other hand, the many failures to debias expert decision makers imply that decision aids, methods of persuasion, and other approaches intended to improve decision making are no panacea. Part of the problem is that people tend to start with preconceived notions about what they should do and show a tendency to seek out and bolster confirming evidence. Consequently, people may become overconfident with experience and develop strongly held beliefs that are difficult to modify, even if they are hard to defend rationally.

REFERENCES

Ainslie, G. (1975), "Specious Reward: A Behavioral Theory of Impulsiveness and Impulse Control," *Psychological Bulletin*, Vol. 82, pp. 463–509.

Aliev, R. A., Fazlollahi, B. and Vahidov, R. M. (2000), "Soft Computing Based Multi-agent Marketing Decision Support System," *Journal of Intelligent and Fuzzy Systems*, Vol. 9, No. 1–2, pp. 1–9.

Allais, M. (1953), "Le comportement de l'homme rationel devant le risque: critique des postulateset axioms de l'école américaine," *Econometrica*, Vol. 21, pp. 503–546.

Anderson, N. H. (1981), *Foundations of Information Integration Theory*, Academic Press, New York.

Argote, L., Seabright, M. A., and Dyer, L. (1986), "Individual Versus Group: Use of Base-Rate and Individuating Information," *Organizational Behavior and Human Decision Making Processes*, Vol. 38, pp. 65–75.

Arkes, H. R. (1981), "Impediments to Accurate Clinical Judgment and Possible Ways to Minimize Their Impact," *Journal of Consulting and Clinical Psychology*, Vol. 49, pp. 323–330.

Arkes, H. R., and Blumer, C. (1985), "The Psychology of Sunk Cost," *Organizational Behavior and Human Decision Processes*, Vol. 35, pp. 124–140.

Arkes, H. R., and Hutzel, L. (2000), "The Role of Probability of Success Estimates in the Sunk Cost Effect," *Journal of Behavioral Decision Making*, Vol. 13, No. 3, pp. 295–306.

Balzer, W. K., Doherty, M. E., and O'Connor, R. O., Jr. (1989), "Effects of Cognitive Feedback on Performance," *Psychological Bulletin*, Vol. 106, pp. 41–43.

Bar-Hillel, M. (1973), "On the Subjective Probability of Compound Events," *Organizational Behavior and Human Performance*, Vol. 9, pp. 396–406.

Baron, J. (1985), *Rationality and Intelligence*, Cambridge University Press, Cambridge.

Baron, J. (1998), *Judgment Misguided: Intuition and Error in Public Decision Making*, Oxford University Press, New York.

Basak, I., and Saaty, T. (1993), "Group Decision Making Using the Analytic Hierarchy Process," *Mathematical and Computer Modeling*, Vol. 17, pp. 101–109.

Bazermen, M. (1998), *Judgment in Managerial Decision Making*, 4th ed., Wiley, New York.

Bazerman, M. H., and Neale, M. A. (1983), "Heuristics in Negotiation: Limitations to Effective Dispute Resolution," in *Negotiating in Organizations*, M. H. Bazerman and R. Lewicki, Eds., Sage, Beverly Hills, CA.

Beach, L. R. (1990), *Image Theory: Decision Making in Personal and Organizational Contexts*, Wiley, Chichester, West Sussex, England.

Beach, L. R. (1993), "Four Revolutions in Behavioral Decision Theory," in *Leadership Theory and Research*, M. M. Chemers and R. Ayman, Eds., Academic Press, San Diego, CA.

Beach, L. R., and Mitchell, T. R. (1978), "A Contingency Model for the Selection of Decision Strategies," *Academy of Management Journal*, Vol. 3, pp. 439–449.

Belecheanu, R., Pawar, K. S., Barson, R. J., and Bredehorst, B. (2003), "The Application of Case-Based Reasoning to Decision Support in New Product Development," *Integrated Manufacturing Systems*, Vol. 14, No. 1, pp. 36–45.

Bell, D. (1982), "Regret in Decision Making Under Uncertainty," *Operations Research*, Vol. 30, pp. 961–981.

Benbasat, I., and Lim, L. H. (1993), "The Effects of Group, Task, Context, Technology Variables on the Usefulness of Group Support Systems: A Meta-analysis of Experimental Studies," *Small Group Research*, Vol. 24, No. 4, pp. 430–462.

Bergus, G. R., Levin, I. P., and Santee, A. S. (2002), "Presenting Risks and Benefits to Patients: The Effects of Information Order on Decision Making," *Journal of General Internal Medicine*, Vol. 17, No. 8, pp. 612–617.

Bernoulli, D. (1738), *Exposition of a New Theory of the Measurement of Risk*, Imperial Academy of Science, St. Petersburg, Russia.

Birnbaum, M. H., and Mellers, B. A. (1983), "Bayesian Inference: Combining Base Rates with Opinions of Sources Who Vary in Credibility," *Journal of Personality and Social Psychology*, Vol. 37, pp. 792–804.

Birnbaum, M. H., Coffey, G., Mellers, B. A., and Weiss, R. (1992), "Utility Measurement: Configural-Weight Theory and the Judge's Point of View," *Journal of Experimental Psychology: Human Perception and Performance*, Vol. 18, pp. 331–346.

Bock, R. D., and Jones, L. V. (1968), *The Measurement and Prediction of Judgment and Choice*, Holden-Day, San Francisco.

Bolton, G. E., and Chatterjee, K. (1996), "Coalition Formation, Communication, and Coordination: An Exploratory Experiment," in *Wise Choices: Decisions, Games, and Negotiations*, R. J. Zeckhauser, Keeney, R. C. and Sebanies, J. K. (Eds.), Harvard University Press, Boston, MA.

Bornstein, G., and Yaniv, I. (1998), "Individual and Group Behavior in the Ultimatum Game: Are Groups More Rational Players?" *Experimental Economics*, Vol. 1, pp. 101–108.

Brashers, D. E., Adkins, M., and Meyers, R. A. (1994), "Argumentation and Computer-Mediated Group Decision Making," in *Group Communication in Context*, L. R. Frey, Ed., Lawrence Erlbaum Associates, Mahwah, NJ.

Brehmer, B. (1981), "Models of Diagnostic Judgment," in *Human Detection and Diagnosis of System Failures*, J. Rasmussen and W. Rouse, Eds., Plenum Press, New York.

Brehmer, B., and Joyce, C. R. B. (1988), *Human Judgment: The SJT View*, North-Holland, Amsterdam.

Brookhouse, J. K., Guion, R. M., and Doherty, M. E. (1986), "Social Desirability Response Bias as One Source of the Discrepancy Between Subjective Weights and Regression Weights," *Organizational Behavior and Human Decision Processes*, Vol. 37, pp. 316–328.

Brunswik, E. (1952), *The Conceptual Framework of Psychology*, University of Chicago Press, Chicago.

Buck, J. R. (1989), *Economic Risk Decisions in Engineering and Management*, Iowa State University Press, Ames, IA.

Budescu, D., and Weiss, W. (1987), "Reflection of Transitive and Intransitive Preferences: A Test of Prospect Theory," *Organizational Behavior and Human Performance*, Vol. 39, pp. 184–202.

Bui, T. X., Jelassi, T. M. and Shakun, M. F. (1990), "Group Decision and Negotiation Support Systems," *European Journal of Operational Research*, Vol. 46, No. 2, pp. 141–142.

Busemeyer, J. R., and Townsend, J. T. (1993), "Decision Field Theory: A Dynamic–Cognitive Approach to Decision Making in an Uncertain Environment," *Psychological Review*, Vol. 100, pp. 432–459.

Caverni, J. P., Fabre, J. M., and Gonzalez, M. (1990), *Cognitive Biases*, North-Holland, Amsterdam.

Chemers, M. M., and Ayman, R., Eds. (1993), *Leadership Theory and Research*, Academic Press, San Diego, CA.

Cheng, P. E., and Holyoak, K. J. (1985), "Pragmatic Reasoning Schemas," *Cognitive Psychology*, Vol. 17, pp. 391–416.

Christensen-Szalanski, J. J., and Willham, C. F. (1991), "The Hindsight Bias: A Meta-analysis," *Organizational Behavior and Human Decision Processes*, Vol. 48, pp. 147–168.

Clemen, R. T. (1996), *Making Hard Decisions: An Introduction to Decision Analysis*, 2nd ed., Duxbury Press, Belmont, CA.

Cohen, M. S. (1993), "The Naturalistic Basis of Decision Biases," in *Decision Making in Action: Models and Methods*, G. A. Klein, J. Orasanu, R. Calderwood, and E. Zsambok, Eds., Ablex, Norwood, NJ, pp. 51–99.

Collan, M., and Liu, S. (2003), "Fuzzy Logic and Intelligent Agents: Towards the Next Step of Capital Budgeting Decision Support," *Industrial Management and Data Systems*, Vol. 103, No. 6, pp. 410–422.

Coma, O., Mascle, O., and Balazinski, M. (2004), "Application of a Fuzzy Decision Support System in a Design for Assembly Methodology," *International Journal of Computer Integrated Manufacturing*, Vol. 17, No. 1, pp. 83–94.

Connolly, T., Ordonez, L. D., and Coughlan, R. (1997), "Regret and Responsibility in the Evaluation of Decision Outcomes," *Organizational Behavior and Human Decision Processes*, Vol. 70, pp. 73–85.

Cosmides, L., and Tooby, J. (1996), "Are Humans Good Intuitive Statisticians After All? Rethinking Some Conclusions from the Literature on Judgment Under Uncertainty", *Cognition*, Vol. 58, pp. 1–73.

Cui, G., and Wong, M. L. (2004), "Implementing Neural Networks for Decision Support in Direct Marketing," *International Journal of Market Research*, Vol. 46, No. 2, pp. 235–254.

Davenport, T. H., and Prusak, L. (2000), *Working Knowledge: How Organizations Manage What They Know*, Harvard Business School Press, Boston.

Davis, J. H. (1992), "Some Compelling Intuitions About Group Consensus Decisions, Theoretical and Empirical Research, and Interperson Aggregation Phenomena: Selected Examples, 1950–1990," *Organizational Behavior and Human Decision Processes*, Vol. 52, pp. 3–38.

Dawes, R. M., and Mulford, M. (1996), "The False Consensus Effect and Overconfidence: Flaws in Judgement or Flaws in How We Study Judgement?" *Organizational Behavior and Human Decision Processes*, Vol. 65, pp. 201–211.

Dawes, R. M., van de Kragt, A. J. C., and Orbell, J. M. (1988), "Not Me or Thee but We: The Importance of Group Identity in Eliciting Cooperation in Dilemma Situations: Experimental Manipulations," *Acta Psychologia*, Vol. 68, pp. 83–97.

Delbecq, A. L., Van de Ven, A. H., and Gustafson, D. H. (1975), *Group Techniques for Program Planning*, Scott, Foresman, Glenview, IL.

Dennis, A. R., and Wixom, B. H. (2002), "Investigators the Moderators of the Group Support Systems Use with Meta-analysis," *Journal of Management Information Systems*, Vol. 18, No. 3, pp. 235–258.

Dennis, A. R., Haley, B. J., and Vandenberg, R. J. (1996), "A Meta-analysis of Effectiveness, Efficiency, and Participant Satisfaction in Group Support Systems Research," *Proceedings of the International Conference on Information Systems*, pp. 278–289.

Dennis, A. R., Wixom, B. H., and Vandenberg, R. J. (2001), "Understanding Fit and Appropriation Effects in Group Support Systems via Meta-analysis," *MIS Quarterly*, Vol. 25, No. 2, pp. 167–193.

DeSanctis, G., and Gallupe, R. B. (1987), "A Foundation for the Study of Group Decision Support Systems," *Management Science*, Vol. 33, No. 5, pp. 589–609.

Dixit, A., and Nalebuff, B. (1991), "Making Strategies Credible," in *Strategy and Choice*, R. J. Zechhauser, Ed., MIT Press, Cambridge, MA, pp. 161–184.

Doherty, M. E. (2003), "Optimists, Pessimists, and Realists," in *Emerging Perspectives on Judgment and Decision Research*, S. Schnieder and J. Shanteau, Eds., Cambridge University Press, New York, pp. 643–679.

Dorris, A. L., and Tabrizi, J. L. (1978), "An Empirical Investigation of Consumer Perception of Product Safety," *Journal of Products Liability*, Vol. 2, pp. 155–163.

Dougherty, M. R. P., Gronlund, S. D., and Gettys, C. F. (2003), "Memory as a Fundamental Heuristic for Decision Making," in *Emerging Perspectives on Judgment and Decision Research*, S. Schnieder and J. Shanteau, Eds., Cambridge University Press, New York, pp. 125–164.

Driskell, J. E., and Salas, E., Eds. (1996), *Stress and Human Performance*, Lawrence Erlbaum Associates, Mahwah, NJ.

Drucker, P. F. (1985), *The Effective Executive*, Harper & Row, New York.

Drucker, P. (1995), "The Information Executives Truly Need," *Harvard Business Review*, Vol. 73, No. 1, pp. 54–62.

Du Charme, W. (1970), "Response Bias Explanation of Conservative Human Inference," *Journal of Experimental Psychology*, Vol. 85, pp. 66–74.

Duda, R. O., Hart, K., Konolige, K., and Reboh, R. (1979), "A Computer-Based Consultant for Mineral Exploration," Technical Report, SRI International, Stanford, CA.

Duffy, L. (1993), "Team Decision Making Biases: An Information Processing Perspective," in *Decision Making in Action: Models and Methods*, G. A. Klein, J. Orasanu, R. Calderwood, and E. Zsambok, Eds., Ablex, Norwood, NJ.

Ebbinghaus, H. (1913), *Memory: A Contribution to Experimental Psychology*, Teachers College, Columbia University, New York.

Edland, E., and Svenson, O. (1993), "Judgment and Decision Making Under Time Pressure," in *Time Pressure and Stress in Human Judgment and Decision Making*, O. Svenson and A. J. Maule, Eds., Plenum Press, New York.

Edwards, W. (1954), "The Theory of Decision Making," *Psychological Bulletin*, Vol. 41, pp. 380–417.

Edwards, W. (1968), "Conservatism in Human Information Processing," in *Formal Representation of Human Judgment*, B. Kleinmuntz, Ed., Wiley, New York, pp. 17–52.

Einhorn, H. J., and Hogarth, R. M. (1978), "Confidence in Judgment: Persistence of the Illusion of Validity," *Psychological Review*, Vol. 70, pp. 193–242.

Einhorn, H. J., and Hogarth, R. M. (1981), "Behavioral Decision Theory: Processes of Judgment and Choice," *Annual Review of Psychology*, Vol. 32, pp. 53–88.

Ellis, D. G., and B. A. Fisher (1994), *Small Group Decision Making: Communication and the Group Process*, 4th ed., McGraw-Hill, New York.

Ellsberg, D. (1961), "Risk, Ambiguity, and the Savage Axioms," *Quarterly Journal of Economics*, Vol. 75, pp. 643–699.

Elster, J., Ed. (1986), *The Multiple Self*, Cambridge University Press, Cambridge.

Embrey, D. E. (1984), *SLIM-MAUD: An Approach to Assessing Human Error Probabilities Using Structured Expert Judgment*, NUREG/CR-3518, Vols. 1 and 2, U.S. Nuclear Regulatory Commission, Washington, DC.

Ericsson, K. A., and Simon, H. A. (1984), *Protocol Analysis: Verbal Reports as Data*, MIT Press, Cambridge, MA.

Espinasse, B., Picolet, G., and Chouraqui, E. (1997), "Negotiation Support Systems: A Multi-criteria and Multi-agent Approach," *European Journal of Operational Research*, Vol. 103, No. 2, pp. 389–409.

Estes, W. (1976), "The Cognitive Side of Probability Learning," *Psychological Review*, Vol. 83, pp. 37–64.

Etzioni, A. (1988), "Normative-Affective Factors: Toward a New Decision-Making Model," *Journal of Economic Psychology*, Vol. 9, pp. 125–150.

Evans, J. B. T. (1989), *Bias in Human Reasoning: Causes and Consequences*, Lawrence Erlbaum Associates, London.

Evans, J. B. T., and Pollard, P. (1985), "Intuitive Statistical Inferences About Normally Distributed Data," *Acta Psychologica*, Vol. 60, pp. 57–71.

Fallesen, J. J., and Pounds, J. (2001), "Identifying and Testing a Naturalistic Approach for Cognitive Skill Training," in *Linking Expertise and Naturalistic Decision Making*, E. Salas and G. Klien, Eds., Lawrence Erlbaum Associates, Mahwah, NJ, pp. 55–70.

Fazlollahi, B., and Vahidov, R. (2001), "Extending the Effectiveness of Simulation-Based DSS Through Genetic Algorithms," *Information and Management*, Vol. 39, No. 1, pp. 53–64.

Feather, N. T. (1966), "Effects of Prior Success and Failure on Expectations of Success and Failure," *Journal of Personality and Social Psychology*, Vol. 3, pp. 287–298.

Fedrizzi, M., Kacprzyk, J., and Yager, R. R., Eds. (1994), *Decision Making Under Dempster–Shafer Uncertainties*, Wiley, New York.

Fischhoff, B. (1982), "For Those Condemned to Study the Past: Heuristics and Biases in Hindsight," in *Judgment Under Uncertainty: Heuristics and Biases*, D. Kahneman, P. Slovic, and A. Tversky, Eds., Cambridge University Press, Cambridge.

Fischhoff, B., and MacGregor, D. (1982), "Subjective Confidence in Forecasts," *Journal of Forecasting*, Vol. 1, pp. 155–172.

Fischhoff, B., Slovic, P., and Lichtenstein, S. (1977), "Knowing with Certainty: The Appropriateness of Extreme Confidence," *Journal of Experimental Psychology: Human Perception and Performance*, Vol. 3, pp. 552–564.

Fischhoff, B., Slovic, P., and Lichtenstein, S. (1978), "Fault Trees: Sensitivity of Estimated Failure Probabilities to Problem Representation," *Journal of Experimental Psychology: Human Perception and Performance*, Vol. 4, pp. 330–344.

Fishburn, P. C. (1974), "Lexicographic Orders, Utilities, and Decision Rules: A Survey," *Management Science*, Vol. 20, pp. 1442–1471.

Flin, R., Salas, E., Strub, M., and Martin, L., Eds. (1997), *Decision Making Under Stress: Emerging Themes and Applications*, Ashgate, Aldershot, Hampshire, England.

Fong, G. T., Krantz, D. H., and Nisbett, R. E. (1986), "The Effects of Statistical Training on Thinking About Everyday Problems," *Cognitive Psychology*, Vol. 18, pp. 253–292.

Foroughi, A. (1998), "Minimizing Negotiation Process Losses with Computerized Negotiation Support Systems," *Journal of Applied Business Research*, Vol. 14, No. 4, pp. 15–26.

Friedman, J. W. (1990), *Game Theory with Applications to Economics*, Oxford University Press, New York.

Frisch, D., and Clemen, R. T. (1994), "Beyond Expected Utility: Rethinking Behavioral Decision Research," *Psychological Bulletin*, Vol. 116, No. 1, pp. 46–54.

Garavelli, A. C., and Gorgoglione, M. (1999), "Fuzzy Logic to Improve the Robustness of Decision Support Systems Under Uncertainty," *Computers and Industrial Engineering*, Vol. 27, No. 1–2, pp. 477–480.

Gertman, D. I., and Blackman, H. S. (1994), *Human Reliability and Safety Analysis Data Handbook*, Wiley, New York.

Gigerenzer, G., and Hoffrage, U. (1995), "How to Improve Bayesian Reasoning Without Instruction: Frequency Formats," *Psychological Review*, Vol. 102, pp. 684–704.

Gigerenzer, G., Hoffrage, U., and Kleinbolting, H. (1991), "Probabilistic Mental Models: A Brunswikian Theory of Confidence," *Psychological Review*, Vol. 98, pp. 506–528.

Gigerenzer, G., Todd, P., and the ABC Research Group (1999), *Simple Heuristics That Make Us Smart*, Oxford University Press, New York.

Goldstein, W. M., and Hogarth, R. M. (1997), "Judgment and Decision Research: Some Historical Context," in *Research on Judgment and Decision Making: Currents, Connections, and Controversies*, W. M. Goldstein and R. M. Hogarth, Eds., Cambridge University Press, Cambridge, pp. 3–65.

Gordon, J., and Shortliffe, E. H. (1984), "The Dempster–Schafer Theory of Evidence," in *Rule-Based Expert Systems: The MYCIN Experiments of the Stanford Heuristic Programming Project*, Addison-Wesley, Reading, MA.

Gruenfeld, D. H., Mannix, E. A., Williams, K. Y., and Neale, M. A. (1996), "Group Composition and Decision Making: How Member Familiarity and Information Distribution Affect Process and Performance," *Organizational Behavior and Human Decision Making Processes*, Vol. 67; No. 1, pp. 1–15.

Guth, W., Schmittberger, R., and Schwarze, B. (1982), "An Experimental Analysis of Ultimatum Bargaining," *Journal of Economic Behavior and Organization*, Vol. 3, pp. 367–388.

Hamilton, V., and Warburton, D. M., Eds. (1979), *Human Stress and Cognition*, Wiley, New York.

Hammer, W. (1993), *Product Safety Management and Engineering*, 2nd ed., American Society of Safety Engineers, Chicago.

Hammond, K. R. (1980), "Introduction to Brunswikian Theory and Methods," in *Realizations of Brunswick's Experimental Design*, K. R. Hammond and N. E. Wascoe, Eds., Jossey-Bass, San Francisco.

Hammond, K. R. (1993), "Naturalistic Decision Making from a Brunswikian Viewpoint: Its Past, Present, Future," in *Decision Making in Action: Models and Methods*, G. A. Klein, J. Orasanu, R. Calderwood, and E. Zsambok, Eds., Ablex, Norwood, NJ, pp. 205–227.

Hammond, K. R. (1996), *Human Judgment and Social Policy: Irreducible Uncertainty, Inevitable Error, Unavoidable Injustice*, Oxford University Press, New York.

Hammond, K. R, Stewart, T. R., Brehmer, B., and Steinmann, D. O. (1975), "Social Judgment Theory," in *Human Judgment and Decision Processes*, M. F. Kaplan and S. Schwartz, Eds., Academic Press, New York, pp. 271–312.

Hammond, K. R., Hamm, R. M., Grassia, J., and Pearson, T. (1987), "Direct Comparison of the Efficacy of Intuitive and Analytical Cognition in Expert Judgment," *IEEE Transactions on Systems, Man, and Cybernetics*, Vol. 17, pp. 753–770.

Hardin, G. (1968), "The Tragedy of the Commons," *Science*, Vol. 162, pp. 1243–1248.

Harmon, J., and Rohrbaugh, J. (1990), "Social Judgement Analysis and Small Group Decision Making: Cognitive Feedback Effects on Individual and Collective Performance," *Organizational Behavior and Human Decision Processes*, Vol. 46, pp. 34–54.

Hart, P., Stern, E. K., and Sundelius, B. (1997), *Beyond Groupthink: Political Group Dynamics and Foreign Policy-Making*, University of Michigan Press, Ann Arbor, MI.

Hasher, L., and Zacks, R. T. (1984), "Automatic Processing of Fundamental Information: The Case of Frequency of Occurrence," *American Psychologist*, Vol. 39, pp. 1372–1388.

Heath, C. (1995), "Escalation and De-escalation of Commitment in Response to Sunk Costs: The Role of Budgeting in Mental Accounting," *Organizational Behavior and Human Decision Processes*, Vol. 62, pp. 38–54.

Heath, L., Tindale, R. S., Edwards, J., Posavac, E. J., Bryant, F. B., Henderson-King, E., Suarez-Balcazar, Y., and Myers, J. (1994), *Applications of Heuristics and Biases to Social Issues*, Plenum Press, New York.

Hertwig, R., Hoffrage, U., and Martignon, L. (1999). "Quick Estimation: Letting the Environment Do the Work," in *Simple Heuristics That Make Us Smart*, G. Gigerenzer, P. Todd, and the ABC Research Group, Eds., Oxford University Press, New York; pp. 209–234.

Hogarth, R. M., and Einhorn, H. J. (1992), "Order Effects in Belief Updating: The Belief-Adjustment Model," *Cognitive Psychology*, Vol. 24, pp. 1–55.

Holsapple, C. W., and Whinston, A. B. (1996), *Decision Support Systems: A Knowledge-Based Approach*, West Publishing, St. Paul, MN.

Holsapple, C. W., Lai, H., and Whinston, A. B. (1998), "A Formal Basis for Negotiation Support System Research," *Group Decision and Negotiation*, Vol. 7, No. 3, pp. 203–227.

Holtzman, S. (1989), *Intelligent Decision Systems*, Addison-Wesley, Reading, MA.

Howard, R. A. (1968), "The Foundations of Decision Analysis," *IEEE Transactions on Systems, Science, and Cybernetics*, Vol. 4, pp. 211–219.

Howard, R. A. (1988), "Decision Analysis: Practice and Promise," *Management Science*, Vol. 34, pp. 679–695.

Huber, J., Wittink, D. R., Fiedler, J. A., and Miller, R. (1993), "The Effectiveness of Alternative Preference Elicitation Procedures in Predicting Choice," *Journal of Marketing Research*, Vol. 30, pp. 105–114.

Isen, A. M. (1993), "Positive Affect and Decision Making," in *Handbook of Emotions*, M. Lewis, and J. M. Haviland, Eds., Guilford Press, New York, pp. 261–277.

Isenberg, D. J. (1986), "Group Polarization: A Critical Review and Meta Analysis," *Journal of Personality and Social Psychology*, Vol. 50, pp. 1141–1151.

Iverson, G., and Luce, R. D. (1998), "The Representational Measurement Approach to Psychophysical and Judgmental Problems," in *Measurement, Judgement, and Decision Making*, M. H. Birnbaum, Ed., Academic Press, San Diego, CA, pp. 1–79.

Jacoby, J. (1977), "Information Load and Decision Quality: Some Contested Issues," *Journal of Marketing Research*, Vol. 14, pp. 569–573.

Janis, I. L. (1972), *Victims of Groupthink*, Houghton Mifflin, Boston.

Janis, I. L., and Mann, L. (1977), *Decision Making: A Psychological Analysis of Conflict, Choice, and Commitment*, Free Press, New York.

Jasic, T., and Wood, D. (2004), "The Profitability of Daily Stock Market Indices Trades Based on Neural Network Predictions: Case Study for the S&P 5000, the DAX, the TOPIX and the FTSE in the Period 1965–1999," *Applied Financial Economics*, Vol. 14, No. 4, pp. 285–297.

Johansen, R. (1988), *Groupware: Computer Support for Business Teams*, Free Press, New York.

Johnson, E. J., and Payne, J. W. (1985), "Effort and Accuracy in Choice," *Management Science*, Vol. 31, No. 4, pp. 395–414.

Jonides, J., and Jones, C. M. (1992), "Direct Coding for Frequency of Occurrence," *Journal of Experimental Psychology: Learning, Memory, and Cognition*, Vol. 18, pp. 368–378.

Juslin, P., and Olsson, H. (1999), "Computational Models of Subjective Probability Calibration," in *Judgment and Decision Making: Neo-Brunswikian and Process-Tracing Approaches*, Lawrence Erlbaum Associates, Mahwah, NJ, pp. 67–95.

Kahneman, D., and Tversky, A. (1973), "On the Psychology of Prediction," *Psychological Review*, Vol. 80, pp. 251–273.

Kahneman, D., and Tversky, A. (1979), "Prospect Theory: An Analysis of Decision Under Risk," *Econometrica*, Vol. 47, pp. 263–291.

Kahneman, D., Slovic, P., and Tversky, A., Eds. (1982), *Judgment Under Uncertainty: Heuristics and Biases*, Cambridge University Press, Cambridge.

Keen, P. G. W., and Scott-Morton, M. S. (1978), *Decision Support Systems: An Organizational Perspective*, Addison-Wesley, Reading, MA.

Keeney, R. L., and Raiffa, H. (1976), *Decisions with Multiple Objectives: Preferences and Value Tradeoffs*, Wiley, New York.

Keren, G. (1990), "Cognitive Aids and Debiasing Methods: Can Cognitive Pills Cure Cognitive Ills?" in *Cognitive Biases*, J. P. Caverni, J. M. Fabre, and M. Gonzalez, Eds., North-Holland, Amsterdam.

Kerr, L. N., MacCoun, R. J., and Kramer, G. P. (1996), "Bias in Judgment: Comparing Individuals and Groups," *Psychological Review*, Vol. 103, pp. 687–719.

Kersten, G. E., and Noronha, S. J. (1999), "WWW-Based Negotiation Support Systems: Design, Implementation, and Use," *Decision Support Systems*, Vol. 25, No. 2, pp. 135–154.

Kidder, R. M. (1995), *How Good People Make Tough Choices*, Harper Collins, New York.

Kieras, D. E. (1985), "The Role of Prior Knowledge in Operating Equipment from Written Instructions," Report 19 (FR-85/ONR-19), Department of Industrial and Operations Engineering, University of Michigan, Ann Arbor, MI.

Klayman, J., and Ha, Y. W. (1987), "Confirmation, Disconfirmation, and Information in Hypothesis Testing," *Journal of Experimental Psychology: Human Learning and Memory*, pp. 211–228.

Klein, G. A. (1989), "Recognition-Primed Decisions," in *Advances in Man–Machine System Research*, W. Rouse, Ed., JAI Press, Greenwich, CT.

Klein, G. A. (1995), "The Value Added by Cognitive Analysis," in *Proceedings of the Human Factors and Ergonomics Society 39th Annual Meeting*, pp. 530–533.

Klein, G. A. (1996), "The Effect of Acute Stressors on Decision Making," in *Stress and Human Performance*, J. E. Driskell, and E. Salas, Eds., Lawrence Erlbaum Associates, Mahwah, NJ.

Klein, G. A. (1998), *Sources of Power: How People Make Decisions*, MIT Press, Cambridge, MA.

Klein, G. A., and Wolf, S. (1995), "Decision-Centered Training," *Proceedings of the Human Factors and Ergonomics Society 39th Annual Meeting*, pp. 1249–1252.

Klein, G. A., Orasanu, J., Calderwood, R., and Zsambok, E., Eds. (1993), *Decision Making in Action: Models and Methods*, Ablex, Norwood, NJ.

Kleinmuntz, B. (1984), "The Scientific Study of Clinical Judgment in Psychology and Medicine," *Clinical Psychology Review*, Vol. 4, pp. 111–126.

Koehler, J. J. (1996), "The Base Rate Fallacy Reconsidered: Descriptive, Normative, and Methodological Challenges," *Behavioral and Brain Sciences*, Vol. 19, pp. 1–53.

Koriat, A., Lichtenstein, S., and Fischhoff, B. (1980), "Reasons for Confidence," *Journal of Experimental Psychology: Human Learning and Memory*, Vol. 6, pp. 107–118.

Kraus, N. N., and Slovic, P. (1988), "Taxonomic Analysis of Perceived Risk: Modeling Individual and Group Perceptions Within Homogenous Hazards Domains," *Risk Analysis*, Vol. 8, pp. 435–455.

Krauss, S., Martignon, L., and Hoffrage, U. (1999), "Simplifying Bayesian Inference: The General Case," in *Model-Based Reasoning in Scientific Discovery*, L. Magnani, N. Nersessian, and P. Thagard, Eds., Plenum Press, New York, pp. 165–179.

Kuhberger, A. (1995), "The Framing of Decisions: A New Look at Old Problems," *Organizational Behavior and Human Decision Processes*, Vol. 62, pp. 230–240.

Kuo, R. J., and Xue, K. C. (1998), "A Decision Support System for Sales Forecasting Through Fuzzy Neural Networks with Asymmetric Fuzzy Weights," *Decision Support Systems*, Vol. 24, No. 2, pp. 105–126.

Lam, S. -H., Cao, J. -M. and Fan, H. (2003), "Development of an Intelligent Agent for Airport Gate Assignment," *Journal of Air Transportation*, Vol. 8, No. 1, pp. 103–114.

LaPlante, A. (1990), "Bring in the Expert: Expert Systems Can't Solve All Problems, But They're Learning," *InfoWorld*, Vol. 12, No. 40, pp. 55–64.

Lari, A. (2003), "A Decision Support System for Solving Quality Problems Using Case-Based Reasoning," *Total Quality Management and Business Excellence*, Vol. 14, No. 6, pp. 733–745.

Lathrop, R. G. (1967), "Perceived Variability," *Journal of Experimental Psychology*, Vol. 23, pp. 498–502.

Laudon, K. C., and Laudon, J. P. (2003), *Management Information Systems: Managing the Digital Firm*, 8th ed., Prentice Hall, Upper Saddle River, NJ.

Lehto, M. R. (1991), "A Proposed Conceptual Model of Human Behavior and Its Implications for Design of Warnings," *Perceptual and Motor Skills*, Vol. 73, pp. 595–611.

Lehto, M. R., and Papastavrou, J. (1991), "A Distributed Signal Detection Theory Model: Implications to the Design of Warnings," in *Proceedings of the 1991 Automatic Control Conference*, Boston, pp. 2586–2590.

Lehto, M. R., James, D. S., and Foley, J. P. (1994), "Exploratory Factor Analysis of Adolescent Attitudes Toward Alcohol and Risk," *Journal of Safety Research*, Vol. 25, pp. 197–213.

Lehto, M. R., Papastavrou, J. P., Ranney, T. A., and Simmons, L. (2000), "An Experimental Comparison of Conservative vs. Optimal Collision Avoidance System Thresholds," *Safety Science*, Vol. 36, No. 3, pp. 185–209.

Levin, L. P. (1975), "Information Integration in Numerical Judgements and Decision Processes," *Journal of Experimental Psychology: General*, Vol. 104, pp. 39–53.

Lichtenstein, S., Slovic, P., Fischhoff, B., Layman, M., and Coombs, B. (1978), "Judged Frequency of Lethal Events," *Journal of Experimental Psychology: Human Learning and Memory*, Vol. 4, pp. 551–578.

Lichtenstein, S., Fischhoff, B., and Phillips, L. D. (1982), "Calibration of Probabilities: The State of the Art to 1980," in *Judgment Under Uncertainty: Heuristics and Biases*, D. Kahneman, P. Slovic, and A. Tversky, Eds., Cambridge University Press, Cambridge, pp. 306–334.

Likert, R., and Likert, J. G. (1976), *New Ways of Managing Conflict*, McGraw-Hill, New York.

Lim, L. -H., and Benbasat, I. (1993), "A Theoretical Perspective of Negotiation Support Systems," *Journal of Management Information Systems*, Vol. 9, No. 3, pp. 27–44.

Little, J. D. C. (1970), "Models and Managers: The Concept of a Decision Calculus," *Management Science*, Vol. 16, No. 8, pp. 466–485.

Luce, R. D., and Raiffa, H. (1957), *Games and Decisions*, Wiley, New York.

Mallet, L., Vaught, C., and Brnich, M. J., Jr. (1993), "Sociotechnical Communication in an Underground Mine Fire: A Study of Warning Messages During an Emergency Evacuation," *Safety Science*, Vol. 16, pp. 709–728.

Mandler, D. D. (1979), "Thought Processes, Consciousness and Stress," in *Human Stress and Cognition: An Information Processing Approach*, V. Hamilton and D. M. Warburton, Eds., Wiley, West Sussex, Chichester, England.

Martignon, L., and Krauss, S. (2003), "Can L'Homme Eclaire be Fast and Frugal? Reconciling Bayesianism and Bounded Rationality," in *Emerging Perspectives on Judgment and Decision Research*, S. Schnieder and J. Shanteau, Eds., Cambridge University Press, New York, pp. 108–122.

Martocchio, J. J., Webster, J., and Baker, C. R. (1993), "Decision-Making in Management Information Systems Research: The Utility of Policy Capturing Methodology," *Behaviour and Information Technology*, Vol. 12, pp. 238–248.

Matwin, S., Szpakowicz, S., Koperczak, Z., Kersten, G. E., and Michalowski, W. (1989), "Negoplan: An Expert System Shell for Negotiation Support," *IEEE Expert*, Vol. 4, No. 4, pp. 50–62.

Maule, A. J., and Hockey, G. R. J. (1993), "State, Stress, and Time Pressure," in *Time Pressure and Stress in Human Judgment and Decision Making*, O. Svenson and A. J. Maule, Eds., Plenum Press, New York, pp. 27–40.

McGuire, W. J. (1966), "Attitudes and Opinions," *Annual Review of Psychology*, Vol. 17, pp. 475–514.

McKnight, A. J., Langston, E. A., McKnight, A. S., and Lange, J. E. (1995), "The Bases of Decisions Leading to Alcohol Impaired Driving," in *Proceedings of the 13th International Conference on Alcohol, Drugs, and Traffic Safety*, C. N. Kloeden and A. J. McLean, Eds., Adelaide, Australia, August 13–18, pp. 143–147.

McLeod, P. L. (1992), "An Assessment of the Experimental Literature on Electronic Support of Group Work: Results of a Meta-analysis," *Human–Computer Interaction*, Vol. 7, pp. 257–280.

Messick, D. M. (1991), "Equality as a Decision Heuristic," in *Psychological Issues in Distributive Justice*, B. Mellers, Ed., Cambridge University Press, New York.

Mockler, R. L., and Dologite, D. G. (1991), "Using Computer Software to Improve Group Decision-Making," *Long Range Planning*, Vol. 24, pp. 44–57.

Montgomery, H. (1989), "From Cognition to Action: The Search for Dominance in Decision Making," in *Process and Structure in Human Decision Making*, H. Montgomery and O. Svenson, Eds., Wiley, Chichester, West Sussex, England.

Montgomery, H. and Willen, H. (1999), "Decision Making and Action: The Search for a Good Structure," in *Judgment and Decision Making: Neo-Brunswikian and Process-Tracing Approaches*, Lawrence Erlbaum Associates, Mahwah, NJ, pp. 147–173.

Moscovici, S. (1976), *Social Influence and Social Change*, Academic Press, London.

Murphy, A. H., and Winkler, R. L. (1974), "Probability Forecasts: A Survey of National Weather Service Forecasters," *Bulletin of the American Meteorological Society*, Vol. 55, pp. 1449–1453.

Myers, J. L., Suydam, M. M., and Gambino, B. (1965), "Contingent Gains and Losses in a Risky Decision Situation," *Journal of Mathematical Psychology*, Vol. 2, pp. 363–370.

Nah, F., and Benbasat, I. (2004), "Knowledge-Based Support in a Group Decision Making Context: An Expert–Novice Comparison," *Journal of Association for Information Systems*, Vol. 5, No. 3, pp. 125–150.

Nah, F. H., Mao, J., and Benbasat, I. (1999), "The Effectiveness of Expert Support Technology for Decision Making: Individuals Versus Small Groups," *Journal of Information Technology*, Vol. 14, No. 2, pp. 137–147.

Naruo, N., Lehto, M., and Salvendy, G. (1990), "Development of a Knowledge Based Decision Support System for Diagnosing Malfunctions of Advanced Production Equipment," *International Journal of Production Research*, Vol. 28, pp. 2259–2276.

Navon, D. (1979), "The Importance of Being Conservative," *British Journal of Mathematical and Statistical Psychology*, Vol. 31, pp. 33–48.

Newell, A., and Simon, H. A. (1972), *Human Problem Solving*, Prentice-Hall, Englewood Cliffs, NJ.

Nisbett, R., and Ross, L. (1980), *Human Inference: Strategies and Shortcomings of Social Judgment*, Prentice-Hall, Englewood Cliffs, NJ.

Nunamaker, J. F., Dennis, A. R., Valacich, J. S., Vogel, D. R., and George, J. F. (1991), "Electronic Meeting Systems to Support Group Work: Theory and Practice at Arizona," *Communications of the ACM*, Vol. 34, No. 7, pp. 40–61.

Olson, D., and Mossman, C. (2003), "Neural Network Forecasts of Canadian Stock Returns Using Accounting Ratios," *International Journal of Forecasting*, Vol. 19, No. 3, pp. 453–465.

Orasanu, J. (1997), "Stress and Naturalistic Decision Making: Strengthening the Weak Links," in *Decision Making Under Stress: Emerging Themes and Applications*, R. Flin, E. Salas, M. Strub, and L. Martin, Eds., Ashgate, Aldershot, Hampshire, England.

Orasanu, J., and Salas, E. (1993), "Team Decision Making in Complex Environments," in *Decision Making in Action: Models and Methods*, G. A. Klein, J. Orasanu, R. Calderwood, and E. Zsambok, Eds., Ablex, Norwood, NJ.

Osborn, F. (1937), *Applied Imagination*, Charles Scribner & Sons, New York.

Paese, P. W., Bieser, M., and Tubbs, M. E. (1993), "Framing Effects and Choice Shifts in Group Decision Making," *Organizational Behavior and Human Decision Processes*, Vol. 56, pp. 149–165.

Payne, J. W. (1980), "Information Processing Theory: Some Concepts and Methods Applied to Decision Research," in *Cognitive Processes in Choice and Decision Research*, T. S. Wallsten, Ed., Lawrence Erlbaum Associates, Hillsdale, NJ.

Payne, J. W., Bettman, J. R., and Johnson, E. J. (1993), *The Adaptive Decision Maker*, Cambridge University Press, Cambridge.

Pennington, N., and Hastie, R. (1986), "Evidence Evaluation in Complex Decision Making," *Journal of Personality and Social Psychology*, Vol. 51, pp. 242–258.

Pennington, N., and Hastie, R. (1988), "Explanation-Based Decision Making: Effects of Memory Structure on Judgment," *Journal of Experimental Psychology: Learning, Memory, and Cognition*, Vol. 14, pp. 521–533.

Peterson, C. R., and Beach L. R. (1967), "Man as an Infinitive Statistician," *Psychological Balletin*, Vol. 68, pp. 29–46.

Pitz, G. F. (1980), "The Very Guide of Life: The Use of Probabilistic Information for Making Decisions," in *Cognitive Processes in Choice and Decision Behavior*, T. S. Wallsten, Ed., Lawrence Erlbaum Associates, Mahwah, NJ.

Pliske, R., and Klein, G. (2003), "The Naturalistic Decision-Making Perspective," in *Emerging Perspectives on Judgment and Decision Research*, S. Schnieder and J. Shanteau, Eds., Cambridge University Press, New York, pp. 108–122.

Pliske, R. M., McCloskey, M. J., and Klein, G. (2001), "Decision Skills Training: Facilitating Learning from Experience," in *Linking Expertise and Naturalistic Decision Making*, E. Salas and G. Klien, Eds., Lawrence Erlbaum Associates, Mahwah, NJ, pp. 37–53.

Raiffa, H. (1968), *Decision Analysis*, Addison-Wesley, Reading, MA.

Raiffa, H. (1982), *The Art and Science of Negotiation*, Harvard University Press, Cambridge, MA.

Rasmussen, J. (1983), "Skills, Rules, Knowledge: Signals, Signs, and Symbols and Other Distinctions in Human Performance Models," *IEEE Transactions on Systems, Man, and Cybernetics*, Vol. 13, No. 3, pp. 257–267.

Reason, J. (1990), *Human Error*, Cambridge University Press, Cambridge.

Rethans, A. J. (1980), "Consumer Perceptions of Hazards," in *PLP-80 Proceedings*, pp. 25–29.

Reyna, V. F., and Brainerd, C. F. (1995), "Fuzzy-Trace Theory: An Interim Synthesis," *Learning and Individual Differences*, Vol. 7, pp. 1–75.

Robert, H. M. (1990), *Robert's Rules of Order Newly Revised*, 9th ed., Scott, Foresman, Glenview, IL.

Rumelhart, D. E., and McClelland, J. L. (1986), *Parallel Distributed Processing: Explorations in the Microstructure of Cognition*, Vol. 1, MIT Press, Cambridge, MA.

Saaty, T. L. (1988), *Multicriteria Decision Making: The Analytic Hierarchy Process*, T. Saaty, Pittsburgh, PA.

Sage, A. (1981), "Behavioral and Organizational Considerations in the Design of Information Systems and Processes for Planning and Decision Support," *IEEE Transactions on Systems, Man, and Cybernetics*, Vol. 11, pp. 61–70.

Sapena, O., Botti, V., and Argente, E. (2003), "Application of Neural Networks to Stock Prediction in 'Pool' Companies," *Applied Artificial Intelligence*, Vol. 17, No. 7, pp. 661–673.

Satish, U., and Streufert, S. (2002), "Value of a Cognitive Simulation in Medicine: Towards Optimizing Decision Making Performance of Healthcare Personnel," *Quality and Safety in Health Care*, Vol. 11, No. 2, pp. 163–167.

Satzinger, J., and Olfmann, L. (1995), "Computer Support for Group Work: Perceptions of the Usefulness of Support Scenarios and Specific Tools," *Journal of Management Systems*, Vol. 11, No. 4, pp. 115–148.

Savage, L. J. (1954), *The Foundations of Statistics*, Wiley, New York.

Schafer, G. (1976), *A Mathematical Theory of Evidence*, Princeton University Press, Princeton, NJ.

Schelling, T. (1960), *The Strategy of Conflict*, Harvard University Press, Cambridge, MA.

Schelling, T. (1978), *Micromotives and Macrobehavior*, W.W. Norton, New York.

Scott-Morton, M. S. (1977), *Management Decision Systems: Computer-Based Support for Decision Making*, Harvard University Press, Cambridge, MA.

Selye, H. (1936), "A Syndrome Produced by Diverse Noxious Agents," *Nature*, Vol. 138, p. 32.

Selye, H. (1979), "The Stress Concept and Some of Its Implications," in *Human Stress and Cognition*, V. Hamilton and D. M. Warburton, Eds., Wiley, New York.

Sharda, R., Barr, S. H., and McDonnell, J. C. (1988), "Decision Support System Effectiveness: A Review and an Empirical Test," *Management Science*, Vol. 34, No. 2, pp. 139–159.

Siebold, D. R. (1992), "Making Meetings More Successful: Plans, Formats, and Procedures for Group Problem-Solving," in *Small Group Communication*, 6th ed., R. Cathcart and L. Samovar, Eds., Wm. C. Brown, Dubuque, IA, pp. 178–191.

Simon, H. A. (1955), "A Behavioral Model of Rational Choice," *Quarterly Journal of Economics*, Vol. 69, pp. 99–118.

Simon, H. (1977), *The New Science of Management Decisions*, Prentice-Hall, Englewood Cliffs, NJ.

Simon, H. A. (1983), "Alternative Visions of Rationality," in *Reason in Human Affairs*, Stanford University Press, Stanford, CA.

Singleton, W. T., and Hovden, J. (1987), *Risk and Decisions*, Wiley, New York.

Slovic, P. (1978), "The Psychology of Protective Behavior," *Journal of Safety Research*, Vol. 10, pp. 58–68.

Slovic, P. (1987), "Perception of Risk," *Science*, Vol. 236, pp. 280–285.

Slovic, P., Fischhoff, B., and Lichtenstein, S. (1977), "Behavioral Decision Theory," *Annual Review of Psychology*, Vol. 28, pp. 1–39.

Sniezek, J. A. (1992), "Groups Under Uncertainty: An Examination of Confidence in Group Decision Making,"

Organizational Behavior and Human Decision Processes, Vol. 52, pp. 124–155.

Sniezek, J. A., and Henry, R. A. (1989), "Accuracy and Confidence in Group Judgment," *Organizational Behavior and Human Decision Processes*, Vol. 43, pp. 1–28.

Sniezek, J. A.,, Wilkins, D. C., Wadlington, P. L., and Baumann, M. R. (2002), "Training for Crisis Decision-Making: Psychological Issues and Computer-Based Solutions," *Journal of Management Information Systems*, Vol. 18, No. 4, pp. 147–168.

Stanoulov, N. (1994), "Expert Knowledge and Computer-Aided Group Decision Making: Some Pragmatic Reflections," *Annals of Operations Research*, Vol. 51, pp. 141–162.

Stasser, G., and Titus, W. (1985), "Pooling of Unshared Information in Group Decision Making: Biased Information Sampling During Discussion," *Journal of Personality and Social Psychology*, Vol. 48, pp. 1467–1478.

Stevenson, M. K., Busemeyer, J. R., and Naylor, J. C. (1993), "Judgment and Decision-Making Theory," in *Handbook of Industrial and Organizational Psychology*, 2nd ed., Vol. 1, M. D. Dunnette and L. M. Hough, Eds., Consulting Press, Palo Alto, CA.

Stone, M. (1960), "Models for Reaction Time," *Psychometrika*, Vol. 25, pp. 251–260.

Strat, T. M. (1994), "Decision Analysis Using Belief Functions," in *Decision Making Under Dempster–Shafer Uncertainties*, M. Fedrizzi, J. Kacprzyk, and R. R. Yager, Eds., Wiley, New York.

Stukey, E., and Zeckhauser, R. (1978), "Decision Analysis," in *A Primer for Policy Analysis*, W.W. Norton, New York, pp. 201–254.

Svenson, O. (1990), "Some Propositions for the Classification of Decision Situations," in *Contemporary Issues in Decision Making*, K. Borcherding, O. Larichev, and D. Messick, Eds., North-Holland, Amsterdam, pp. 17–31.

Svenson, O., and Maule, A. J. (1993), *Time Pressure and Stress in Human Judgment and Decision Making*, Plenum Press, New York.

Sviokla, J. J. (1986), "PlanPower, Xcon and Mudman: An In-Depth Analysis into Three Commercial Expert Systems in Use," unpublished Ph.D. dissertation, Harvard University, Cambridge, MA.

Sviokla, J. J. (1989), "Expert Systems and Their Impact on the Firm: The Effects of PlanPower Use on the Information Processing Capacity of the Financial Collaborative," *Journal of Management Information Systems*, Vol. 6, No. 3, pp. 65–84.

Sviokla, J. J. (1990), "An Examination of the Impact of Expert Systems on the Firm: The Case of XCON," *MIS Quarterly*, Vol. 14, No. 2, pp. 127–140.

Swain, A. D., and Guttman, H. (1983), *Handbook for Human Reliability Analysis with Emphasis on Nuclear Power Plant Applications*, NUREG/CR-1278, U.S. Nuclear Regulatory Commission, Washington, DC.

Tanner, W. P., and Swets, J. A. (1954), "A Decision Making Theory of Visual Detection," *Psychological Review*, Vol. 61, pp. 401–409.

Todd, P., and Benbasat, I. (1991), "An Experimental Investigation of the Impact of Computer Based Decision Aids on Decision Making Strategies," *Information Systems Research*, Vol. 2, No. 2, pp. 87–115.

Todd, P., and Benbasat, I. (1992), "The Use of Information in Decision Making: An Experimental Investigation of the Impact of Computer Based DSS on Processing Effort," *MIS Quarterly*, Vol. 16, No. 3, pp. 373–393.

Todd, P., and Benbasat, I. (1999), "Evaluating the Impact of DSS, Cognitive Effort, and Incentives on Strategy Selection," *Information Systems Research*, Vol. 10, No. 4, pp. 356–374.

Torrance, E. P. (1953), "The Behavior of Small Groups Under the Stress Conditions of 'Survival,' " *American Sociological Review*, Vol. 19, pp. 751–755.

Tuckman, B. W. (1965), "Development Sequence in Small Groups," *Psychological Bulletin*, Vol. 63, pp. 289–399.

Turban, E., and Aronson, J. E. (2001), *Decision Support Systems and Intelligent Systems*, 6th ed., Prentice Hall, Upper Saddle River, NJ.

Tversky, A. (1969), "Intransitivity of Preferences," *Psychological Review*, Vol. 76, pp. 31–48.

Tversky, A. (1972), "Elimination by Aspects: A Theory of Choice," *Psychological Review*, Vol. 79, pp. 281–289.

Tversky, A., and Kahneman, D. (1973), "Availability: A Heuristic for Judging Frequency and Probability," *Cognitive Psychology*, Vol. 5, pp. 207–232.

Tversky, A., and Kahneman, D. (1974), "Judgment under Uncertainty: Heuristics and Biases," *Science*, Vol. 185, pp. 1124–1131.

Tversky, A., and Kahneman, D. (1981), "The Framing of Decisions and the Psychology of Choice," *Science*, Vol. 211, pp. 453–458.

Valenzi, E., and Andrews, I. R. (1973), "Individual Differences in the Decision Processes of Employment Interviews," *Journal of Applied Psychology*, Vol. 58, pp. 49–53.

Viscusi, W. K. (1991), *Reforming Products Liability*, Harvard University Press, Cambridge, MA.

von Neumann, J., and Morgenstern, O. (1947), *Theory of Games and Economic Behavior*, Princeton University Press, Princeton, NJ.

Wagenaar, W. A. (1992), "Risk Taking and Accident Causation," in *Risk-Taking Behavior*, J. F. Yates, Ed., Wiley, New York, pp. 257–281.

Wallsten, T. S. (1972), "Conjoint-Measurement Framework for the Study of Probabilistic Information Processing," *Psychological Review*, Vol. 79, pp. 245–260.

Wallsten, T. S. (1976), "Using Conjoint-Measurement Models to Investigate a Theory About Probabilistic Information Processing," *Journal of Mathematical Psychology*, Vol. 14, pp. 144–185.

Wallsten, T. S. (1995), "Time Pressure and Payoff Effects on Multidimensional Probabilistic Inference," in *Time Pressure and Stress in Human Judgment*, O. Svenson and J. Maule, Eds., Plenum Press, New York.

Wallsten, T. S., Zwick, R., Kemp, S., and Budescu, D. V. (1993), "Preferences and Reasons for Communicating Probabilistic Information in Verbal and Numerical Terms," *Bulletin of the Psychonomic Society*, Vol. 31, pp. 135–138.

Weber, E. (1994), "From Subjective Probabilities to Decision Weights: The Effect of Asymmetric Loss Functions on the Evaluation of Uncertain Outcomes and Events," *Psychological Bulletin*, Vol. 115, pp. 228–242.

Weber, E., Anderson, C. J., and Birnbaum, M. H. (1992), "A Theory of Perceived Risk and Attractiveness," *Organizational Behavior and Human Decision Processes*, Vol. 52, pp. 492–523.

Wegner, D. (1987), "Transactive Memory: A Contemporary Analysis of Group Mind," in *Theories of Group Behavior*, B. Mullen and G. R. Goethals, Eds., Springer-Verlag, New York, pp. 185–208.

Weinstein, N. D. (1979), "Seeking Reassuring or Threatening Information About Environmental Cancer," *Journal of Behavioral Medicine*, Vol. 2, pp. 125–139.

Weinstein, N. D. (1980), "Unrealistic Optimism About Future Life Events," *Journal of Personality and Social Psychology*, Vol. 39, pp. 806–820.

Weinstein, N. D. (1987), "Unrealistic Optimism About Illness Susceptibility: Conclusions from a Community-Wide Sample," *Journal of Behavioral Medicine*, Vol. 10, pp. 481–500.

Weinstein, N. D., and Klein, W. M. (1995), "Resistance of Personal Risk Perceptions to Debiasing Interventions," *Health Psychology*, Vol. 14, pp. 132–140.

Welch, D. D. (1994), *Conflicting Agendas: Personal Morality in Institutional Settings*, Pilgrim Press, Cleveland, OH.

Welford, A. T. (1976), *Skilled Performance*, Scott, Foresman, Glenview, IL.

Wickens, C. D. (1992), *Engineering Psychology and Human Performance*, HarperCollins, New York.

Winkler, R. L., and Murphy, A. H. (1973), "Experiments in the Laboratory and the Real World," *Organizational Behavior and Human Performance*, Vol. 10, pp. 252–270.

Winterfeldt, D. V., and Edwards, W. (1986), *Decision Analysis and Behavioral Research*, Cambridge University Press, Cambridge.

Yates, J. F., Ed. (1992), *Risk-Taking Behavior*, Wiley, New York.

Yates, J. F., Veinott, E. S., and Patalano, A. L. (2003). "Hard Decisions, Bad Decisions: On Decision Quality and Decision Aiding," in *Emerging Perspectives on Judgment and Decision Research*, S. Schnieder and J. Shanteau, Eds., Cambridge University Press, New York, pp. 13–63.

Yakes, R. M. and Dodson J. O. (1908), "The Relative Strength of Stimulus to Rapidity of Habit Formation," *Journal of Comparative and Neurological Psychology*, Vol. 18, pp. 459–482.

Zander, A. (1994), *Making Groups Effective*, 2nd ed., Jossey-Bass, San Francisco.

Zey, M., Ed. (1992), *Decision Making: Alternatives to Rational Choice Models*, Sage, London.

Zhou, M., and Xu, S. (1999), "Dynamic Recurrent Neural Networks for a Hybrid Intelligent Decision Support System for the Metallurgical Industry," *Expert Systems*, Vol. 16, No. 4, pp. 240–247.

Zigurs, I., and Buckland, B. K. (1998), "A Theory of Task/Technology Fit and Group Support Systems Effectiveness," *MIS Quarterly*, Vol. 22, No. 3, pp. 313–334.

Zimmer, A. (1983), "Verbal Versus Numerical Processing of Subjective Probabilities," in *Decision Making Under Uncertainty*, R. W. Scholtz, Ed., North-Holland, Amsterdam.

CHAPTER 9

MENTAL WORKLOAD AND SITUATION AWARENESS

Pamela S. Tsang
Wright State University
Dayton, Ohio

Michael A. Vidulich
Air Force Research Laboratory
Wright-Patterson Air Force Base, Ohio

1 INTRODUCTION

Human engineering seeks to understand and improve human interactions with machines to perform tasks. This goal can be especially difficult to achieve in dynamic complex systems that characterize much of modern work. A good example of the problem can be seen in considering the human operator's reaction to automation. Many modern tasks, such as controlling the complex reactions in a process control plant (Wickens and Hollands, 2000), would not be possible without the assistance of automation. However, many researchers (e.g., Kessel and Wickens, 1982; Moray, 1986; Wiener and Curry, 1986; Bainbridge, 1987; Tsang and Vidulich, 1989; Adams et al., 1991) have pointed out that automated assistance can come at a high price. Moray (1986) expressed this issue well when he noted that as the computer automation did more, the operator would do less and therefore experience less mental workload, but: "Is there a price for the advantages? It could be said that the information processing demands become so alien to the operator that if called upon to reenter the control loop such reentry is no longer possible. . . . The system will be poorly understood and the operator will lack practice in exercising control, so the possibility of human error in emergencies will increase" (Moray, 1986, p. 40–5).

Naturally, there is an issue of how well the requirements of using any machine matches the capabilities of the human operators. Some mismatches should be easy for an observer to see, especially physical mismatches. For example, Fitts and Jones (1961a) found that 3% of 460 "pilot errors" were due to the pilot being physically unable to reach the required control. But other aspects of how well the human operator can match the requirements of using a machine to accomplish a task may not be so obvious, especially mental mismatches. For example, Fitts and Jones (1961b) found that 18% of errors in reading and interpreting aircraft instruments were due to problems associated with integrating the information from multirevolution indicators (e.g., altitude displays with different pointers for one, tens, and hundreds of feet). An outside observer would not necessarily know by watching the pilot that such a misinterpretation had occurred.

Given the impossibility of seeing the mental processes of an operator performing a task, it is not surprising that human engineering specialists have developed concepts to relate the impact of various task demands on the human operator and on system performance. Vidulich (2003) identified two such concepts: mental workload and situation awareness (SA). Vidulich echoed the argument of other researchers that mental workload and SA had taken on the quality of meta-measures for system evaluation (Hardiman et al., 1996; Selcon et al., 1996). Selcon et al. (1996) pointed out that it is seldom possible to evaluate all of the mission-relevant uses of a display. Thus, any evaluation must generalize from the evaluation environment to the real world. It has been argued that rather than focusing on task-specific performance, it might be preferable to examine meta-measures that encapsulate the cognitive reaction to performance with a given interface. So, for example, an interface that allows a task to be performed with a more comfortable level of mental workload and better SA would be preferable to one that did not.

To appreciate the potential roles that workload and SA might play in supporting system development, several visions of future systems will be considered. Fallows (2001 a,b) presented an intriguing vision of the future of aviation. Examining the increasing bottlenecks and delays inherent in the existing airline industry, Fallows proposed that a simple scaling-up of the existing system with more planes and more runways at existing airports would not be a practical nor economically feasible approach to keep pace with the projected increases in airline travel. Fallows made a compelling argument that the increased reliability of aircraft mechanical systems combined with innovative research on cockpit interfaces will not only revitalize general aviation, but also lead to the emergence of a much more extensive small aircraft "taxi" service. Such an expansion would naturally lead to more pilots flying that lack the extensive training of current professional pilots. To maintain an acceptable level of safety, Fallows assumes that interface changes will occur to make the demands of piloting an aircraft tolerable for a wide range of pilots.

In the current context it is important to emphasize the role that human engineering is expected to assume in the development of the expanded air system. Fallows (2001 a,b) points out that current research, such as that performed by the National Aeronautics and Space Administration's (NASA) Advanced General Aviation Transport Experiments (AGATE) alliance, will decrease the difficulty of piloting and increase flight safety for a new generation of more user-friendly aircraft. NASA is aggressively pursuing the goal of making advanced cockpit technologies effective and affordable even to general aviation pilots. These technologies include highway-in-the-sky displays, head-mounted displays (Fiorino, 2001), and synthetic vision systems that use advanced sensors to present a view of the world to the pilot during degraded visual conditions (Wynbrandt, 2001). In other words, Fallows expects that future general aircraft cockpits will take advantage of advanced interface technologies to reduce the mental workload and increase the SA of the general aviation pilot. To this end, understanding the human pilot and building systems that best accommodate the human's cognitive strengths while supporting human frailty will remain a vital component of making those systems effective and safe.

In addition to changes to the cockpit, air traffic control (ATC) is also expected to face changing demands. Jha et al. (2003) discuss the proposed future architecture of the National Airspace System (NAS). Most proposed changes are expected to increase the pilot's autonomy in controlling the route of their aircraft, including the requirement to maintain required separation between aircraft. Flight time, delays, and fuel expenditures are all expected to profit from such changes. Despite the increase in pilot autonomy, controllers are still responsible for running the ATC system safely. In other words, the controller's role would be shifted from that of an active controller to one more like that of a monitor. Assuming that the equipment and the pilots all perform correctly, the controller's workload would be expected to decrease since they are interacting with fewer aircraft. It also seems plausible that the controller's SA of the airspace would decrease as well, since they would not be focusing as much attention as they previously did on many of the aircraft. However, as Metzger et al. (2003) point out, this may not be the case. For an unknown period of time, certain aircraft in the system would be properly equipped to participate according to the new rules, allowing more autonomy, but others would not be and would require the controller to manage their flight path. Metzger et al. (2003) explored the effect of the proportions of equipped and unequipped aircraft in the controller's airspace and the availability of a decision support system. The decision support was largely effective. Controllers were more able to control mixed types of aircraft, and rated their workload lower when decision support was available. Although Metzger et al. (2003) did not discuss their results in terms of SA, there were interesting findings of relevance. When the decision support tool was available, the controllers' attention (measured by eye fixations) to the radar display was partially drawn away to the decision support tools. This behavior could presumably reduce the controller's SA of information on the radar screen. The impact of such automated decision support tools will require careful analysis to ensure that important information from other sources is not lost due to their use. It is important to consider the impact of any technological support in the larger context of overall system performance and workload imposed on the human operator.

So technological changes constantly bring about a practical need to know about the cognitive processing of the operator. There seems to be a consensus that the concepts of SA and mental workload are useful for assessing the impact of such changes on the human operator. But although mental workload and SA are both concepts for understanding the human reaction to interacting with a system to achieve a goal, as we shall see, they are separate concepts. The utility of assessing SA therefore does not diminish that of mental workload. In fact, parallel studies of both are not only good for applied evaluations but could also help sharpen their respective definitions and stimulate new understanding (Endsley, 1993; Wickens, 2001; Vidulich, 2003).

In the next section we review a framework for understanding both mental workload and SA as parts of human information processing. Then select approaches for assessing mental workload and SA are reviewed. Finally, the roles of mental workload and SA in meeting the demands of future systems are considered.

2 THEORETICAL UNDERPINNINGS OF MENTAL WORKLOAD AND SITUATION AWARENESS

In 1994, Pew stated that *situation awareness* had replaced *workload* as the buzzword of the 1990s. But can the concept of situation awareness replace that of mental workload? Hendy (1995) and Wickens (1995) argued that the concepts of situation awareness and mental workload are clearly distinct but are also intricately related to each other. A decade later, there now seems to be a consensus that one concept does not replace, or encompass, the other, even though the two concepts are affected by many of the same human variables (such as limited processing capacity and the severe limit of working memory) and system variables (such as task demands and technological support).

Figure 1 provides a conceptual sketch of the relationship between mental workload and situation awareness and is not intended to be a complete representation of all the processes that are involved. There are two main components in this figure: the attention and mental workload loop and the memory and situation awareness loop. The ensuing portrayal will make clear that mental workload and situation awareness are intricately intertwined, as one affects and is affected by the other. Although convention would bias us in thinking that elements on top or on the left in the figure might have temporal precedence over those at the bottom or on the right, this is not necessarily the case with the dynamic interplay between workload and SA. For example, task demands could be initiated by an external task (such as the need to respond to an air traffic controller's request) as well as by an internal decision to engage in solving a nagging problem. Despite the seemingly discrete and linear depiction of the relations among the elements in the figure, the elements are actually thought of to be mutually interacting adaptively in response to both exogenous demands and endogenous states (e.g., Hockey, 1997).

2.1 Attention and Workload

Since the 1970s, much has been debated and written about the concept of mental workload (e.g., Welford, 1978; Moray, 1979; O'Donnell and Eggemeier, 1986; Adams et al., 1991; Huey and Wickens, 1993; Gopher, 1994; Kramer et al., 1996; Tsang and Wilson, 1997). Gopher and Donchin (1986) offered the following description of the role of mental workload:

> The term workload is used to describe aspects of the interaction between an operator and an assigned task. Tasks are specified in terms of their structural properties; a set of stimuli and responses are specified with a set of rules that map responses to stimuli. There are, in addition, expectations regarding the quality of the performance, which derive from knowledge of the relation between the structure of the task and the nature of human capacities and skills. . . . [W]orkload is invoked to account for those aspects of the interaction between a person and a task that cause task demands to exceed the person's capacity to deliver. . . . [M]ental workload is clearly an attribute of the information processing and control systems that mediate between stimuli, rules, and responses. (p. 41–3)

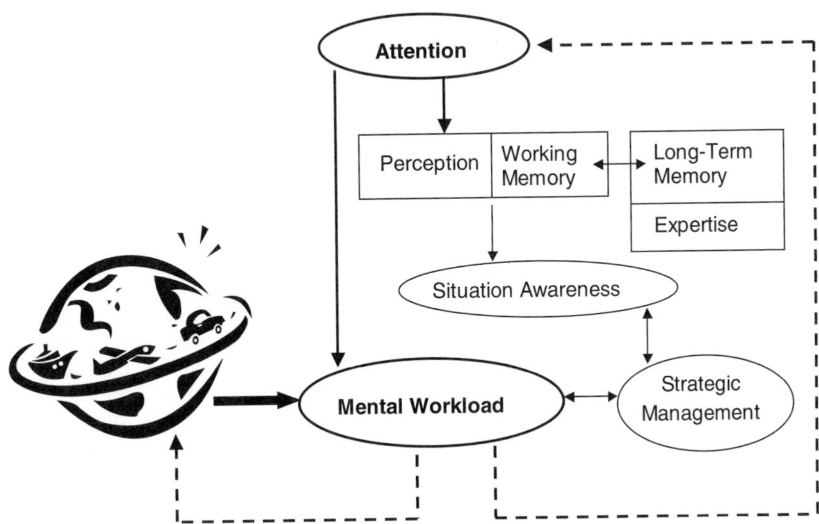

Figure 1 Theoretical framework illustrating the relationship between mental workload and situation awareness.

A commonly accepted notion is that mental workload is very much a function of the *supply and demand* of attentional or processing resources. The attention–workload loop in Figure 1 is minimally sketched but is described in more detail here. There are two main determinants of workload: the exogenous task demands as specified by factors such as task difficulty, task priority, and situational contingencies (represented by the globe in Figure 1); and the endogenous supply of attentional or processing resources to support information processing such as perceiving, updating memory, planning, decision making, and response processing. Further, this supply is modulated by individual differences such as one's skill level or expertise. The ultimate interest in measuring workload, of course, is how it might affect system performance represented via the feedback loop to the globe in Figure 1. Mental workload can be expressed in subjective experience, performance, and physiological manifestations. A host of assessment techniques have now been developed and are used in both laboratories and applied settings. They are reviewed in Section 3.

Although there are numerous theoretical accounts of attention, the one readily embraced and adopted in the workload literature is the energetics account (e.g., Hockey et al., 1986). Central to the present discussion is the notion that attentional resources are demanded for task processing, but they are of limited supply. Performance improves monotonically with increased investment of resources up to the limit of resource availability (Norman and Bobrow, 1975). An important implication of this relationship is that performance could be the basis of inference for the amount of resources used and remained. The latter, referred to as *spare capacity*, could serve as reserve fuel for emergencies and unexpected added demands. Further, attentional resources are subject to voluntary and strategic allocation. According to Kahneman (1973), attention is allocated via a closed feedback loop with continuous monitoring of the efficacy of the allocation policy that is governed by enduring dispositions (of lasting importance, such as one's own name and well-learned rules), momentary intentions (pertinent to the task at hand), and evaluation of the performance (involving self-monitoring of the adequacy of performance in relation to task demands). Because attention can be deployed flexibly, researchers advocate the need to examine the allocation policy in conjunction with the joint performance in a multitask situation in order to assess the workload and spare capacity involved (e.g., Gopher, 1994).

Among the most convincing support for the limited, energetic, and allocatable property of attentional resources are the reciprocity effects in performance and certain neuroindices observed between time-shared tasks. As the demand or priority of one task changes, the increase in performance, P300 amplitude, or PET-measured activity in one task has been observed to be accompanied by a decrease in the corresponding measures in the other task (Gopher et al., 1982; Wickens et al., 1983; Kramer et al., 1987; Sirevaag et al., 1989; Fowler, 1994; Tsang et al., 1996; Parasuraman and Caggiano, 2002; Just et al., 2003).

By the late 1970s, the notion of Kahneman's undifferentiated or all-purpose attentional resource was challenged by a body of data that suggested multiple specialized resources for different types of processing (see Allport et al., 1972; Kinsbourne and Hicks, 1978; Navon and Gopher, 1979; Friedman and Polson, 1981; Wickens, 1984). Based on an expansive systematic review of the interference pattern in the extant dual-task data, Wickens (1980, 1987) proposed a multiple resource model. According to this model, attentional resources are defined along three dichotomous dimensions: (1) stages of processing with perceptual/central processing requiring resources different from those used for response processing, (2) processing codes with spatial processing requiring resources different from those used for verbal processing, and (3) input/output modalities with visual and auditory processing requiring different processing resources and manual and speech responses also requiring different processing resources. An important application of this model is its prediction of multiple-task performance that is common in many modern complex work environment. The higher the similarity in the resource demands among the task components, the more severe the competition for similar resources, the less spare capacity, and the higher the level of workload that would result. The other side of the coin is that it would be less feasible to exchange and reallocate resources among task components that utilize highly dissimilar resources (see Wickens, 2002). That is, it would be more difficult to manage the workload levels dynamically between tasks that rely on dissimilar resources. According to the multiple-resource model, the intensity aspect and the structural aspect are intrinsically intertwined when characterizing the processing demand of a task. Similarity in the resource demand among the time-shared tasks effectively determines resource availability as well as resource exchangeability. An increased degree of resource similarity promotes resource sharability but effectively reduces resource availability that leads to increased resource competition.

The energetic and specificity aspects of the attentional resources are receiving converging support from subjective (e.g., Tsang and Velazquez, 1996; Rubio et al., 2004), performance (e.g., Tsang et al., 1996; Wickens, 2002), and neurophysiological (e.g., Just et al., 2003; Parasuraman, 2003) measures. First, parametric manipulation of task demands have been found to produce systematic and graded changes in the level of subjective workload ratings, performances, and amount of neuronal activation. Second, all of these measures have been found to be sensitive to the competition for specific resource demands. Further, increased neuronal activation associated with different types of processing (e.g., spatial processing and verbal processing) are found to be localized in different cortical regions.

As mentioned above, the supply or availability of processing resources is subject to individual differences such as one's ability and skill level. Recently, Just et al. (2003) pointed out a set of neurophysiological data that appear to support the notion that a higher level of skill or ability effectively constitutes a larger resource supply. Parks et al. (1988) used a verbal fluency task that required subjects to generate as many words as possible that began with a given stimulus letter. Those who were more proficient at this task, and presumably had higher verbal ability, exhibited a lower level of positron emission tomography (PET) measures of brain activity. Just et al. (2003) proposed that the difference between the more and less proficient subjects lay in the proportion of resources they needed to perform the task. Since the same task was performed, the task demand objectively should be the same for all the subjects. The lower level of brain activity therefore would suggest that the more proficient subjects had a larger supply of resources. In another study, Haier et al. (1992) found that weeks of practice in the spatial computer game Tetris led to improved performance and a reduced amount of PET-measured activity. Just et al. proposed that practice improved the subjects' procedural knowledge, and the newly acquired, more efficient procedures entailed a lower level of resource use. In practice, a reduced level of resource requirement by one task would translate to increased spare resources for processing other tasks.

2.2 Memory and Situation Awareness

Like the concept of mental workload (and many other psychological concepts, such as intelligence), situation awareness (SA) is difficult to define precisely (e.g., Durso and Gironlund, 1999). Pew (1994) defines a situation as "a set of environmental conditions and system states with which the participant is interacting that can be characterized uniquely by a set of information, knowledge and resource options" (p. 18). A commonly referenced working definition for SA came from Endsley (1990, p. 1-3): situation awareness is "the perception of the elements in the environment within a volume of time and space, the comprehension of their meaning, and the projection of their status in the near future." An on-line dictionary states that aware "implies knowledge gained through one's own perceptions or by means of information." These definitions connote both perception of the now and present and connection with knowledge gained in the past. As Figure 1 denotes, SA is most closely linked to the perceptual and the working memory processes. Certainly, it is not sufficient just for the information relevant to the situation to be available; the information needs to be perceived by the operator, and perception entails far more than the detection of signals or changes. For pattern recognition, object categorization, and comprehension of meaning to occur, contact with knowledge is necessary. But knowledge stored in long-term memory is accessible only through short-term or working memory. Baddeley (1990) introduced the term working memory to emphasize that short-term memory is far more than a temporary depository for information: It is an active process involved in maintaining information available in short-term memory. Working memory is an effortful process subject to capacity as well as attentional limits.

Adams et al. (1995) make a distinction between the process and product of SA: "*product* refers to the state of awareness with respect to information and knowledge, whereas *process* refers to the various perceptual and cognitive activities involved in constructing, updating, and revising the state of awareness" (p. 88). To elaborate, although only the perceptual and working memory processes are explicitly linked to SA in Figure 1, SA is supported by other processes that are subject to attentional limits. The product of SA is a distillation of the ongoing processing of the interchange between information perceived from the now and present (working memory) and knowledge and experience gained from the past (long-term memory). Both the process and product are influenced by one's experience. As will be made clear below, this distinction between the process and product of SA has profound implications on the interaction of mental workload and SA and on the appropriate assessment techniques.

Just as given the same objective task demand, mental workload could vary due to individual differences in resource supply as a result of skill and ability differences, given the same situation, SA could vary due to individual differences. Although long-term memory is not linked directly to SA in Figure 1, it plays a critical role since it stores the knowledge and experience associated with skill development. Ericsson and Simon (1993) pointed out: "Recognition and retrieval processes are determined in part by information in LTM [long-term memory], because the information in STM [short-term memory] is not sufficiently specific to determine a unique product of recognition and retrieval" (p. 197). The extant view of the nature of expertise further expounds on the role of memory in determining the content of SA.

Expertise is mostly learned, acquired through many hours of deliberate practice (e.g., Glaser, 1987; Chi et al., 1988; Druckman and Bjork, 1991; Adams and Ericsson, 1992; Ericsson, 1996). A fundamental difference between novices and experts is the amount of acquired domain-specific knowledge. In addition to having acquired declarative knowledge (facts), experts have a large body of procedural (how-to) knowledge. With practice, many procedural rules (productions) become concatenated into larger rules that can produce action sequences efficiently (Druckman and Bjork, 1991). However, the expertise advantage goes beyond a quantitative difference. The organization of knowledge is fundamentally different between experts and novices. An expert's knowledge is highly organized and well structured, so that retrieving information is much facilitated. The large body of organized knowledge enables experts readily, to see meaningful patterns, to make inferences from partial information, to constrain search, to frame the problem, to apprehend the situation, to update perception of the current situation continuously, and to anticipate future events, including eventual retrieval conditions (Glaser, 1987; Charness, 1995; Vidulich, 2003). An accurate account of the current situation allows an experienced operator to retrieve rapidly the appropriate course of action directly from memory, enabling swift suitable responses.

In addition, Ericsson and Kintsch (1995) proposed that a long-term working memory (LTWM) emerges as expertise develops and is a defining feature of an advanced level of skill (Ericsson and Delaney, 1998). Whereas working memory has severe capacity and temporal limits, LTWM is hypothesized to have a larger capacity that persists for a period of minutes (or even hours). Experts do not have a larger memory capacity, the critical aspect is *how* information is stored and indexed in long-term memory. With a meaningful system for organizing information that already would have been built in LTM, even very briefly seen, seemingly random, incoming information might be organized similarly. Retrieval cues can then be devised and used to access information in LTM quickly. One caveat is that skilled memory is highly domain specific (Ericsson and Charness, 1997). For example, Chase and Ericsson (1982) and Staszewski (1988) reported three people who, after extensive practice, developed a digit span in the neighborhood of 100 numbers. Being avid runners, the subjects associated the random digits to facts related to running that already existed in their LTM (e.g., date of the Boston Marathon). These subjects had a normal short-term memory span when the studies began, and after practice demonstrated the normal span when materials other than digits were tested. Charness (1995) pointed out that such escapes from normal limits also have been observed in perceptual processing. Reingold and Charness (1995) found that when chess symbols (as opposed to letters designating chess pieces) were used, highly skilled players could make their decision in some cases without moving their eye from the initial fixation point at the center of the display. In contrast, weaker players had to make direct fixations. When letter symbols instead of chess piece symbols were used, even the experts were forced to fixate on the pieces directly much more often. Charness (1995) also pointed out that these observations show that experts can both accurately encode a situation and prepare an appropriate response much more quickly than their less skilled counterparts, but only in the domain of their expertise.

To further illustrate the workings of LTWM, Ericsson and Kintsch (1995) described the medical diagnosis process that requires one to store numerous individual facts in working memory. Having developed a retrieval structure in LTM that would foster accurate encoding of patient information and effective reasoning, medical experts were found to be better able to recall important information at a higher conceptual level that subsumed specific facts and to produce more effective diagnosis. A very important function of LTWM appeared then to be providing working memory support for reasoning about and evaluating diagnostic alternatives (Norman et al., 1989; Patel and Groen, 1991). That is, expert problem solving is more than just quick retrieval of stored solutions to old problems. Expertise is also associated with effective application of a large amount of knowledge in reasoning to cope with novel problems (Charness, 1989; Horn and Masunaga, 2000).

Finally, experts show metacognitive capabilities that are not present in novices (Druckman and Bjork, 1991). These capabilities include knowing what one knows and does not know, planning ahead, efficiently apportioning one's time and attentional resources, and monitoring and editing one's efforts to solve a problem (Glaser, 1987).

2.3 Mental Workload and Situation Awareness

As several researchers have emphasized, the two concepts are intricately intertwined (e.g., Wickens, 2002; Vidulich, 2003). In this section we attempt to sharpen their distinction and to examine their interactions more closely. Wickens (2001, p. 446) contrasts the two concepts in the following way: "Mental workload is fundamentally an *energetic* construct, in which the quantitative properties ("how much") are dominant over the qualitative properties ("what kind"), as the most important element. In contrast, situation awareness is fundamentally a *cognitive* concept, in which the critical issue is the operator's accuracy of ongoing understanding of the situation (i.e., a qualitative property)." In practice, one assesses the amount and type of workload and the quality (scope, depth, and accuracy) of the content of SA (Vidulich, 2003).

Both the level of workload and the quality of SA are shaped by exogenous and endogenous factors. Exogenous factors are inherent in the situation (e.g., task demands and situation complexity and uncertainty). Endogenous factors are inherent in a person's ability and skill. The same level of task demands could impose different levels of workload on the operator, depending on her ability or skill level. As discussed above, a high skill level is functionally equivalent to having a larger processing resource supply. A moderate crosswind could be a challenge for a student pilot trying to land a plane but a rather routine task for a seasoned pilot. An overly sensitive warning alarm could be exceedingly disruptive to assessment of the situation by a new operator but could safely be ignored by an experienced operator with intimate knowledge of the workings of the system. Although calibrating the exogenous demands is not always straightforward, their influences on workload is obvious. Less apparent is the endogenous influences on the interplay between the level of workload and the quality of SA.

To the extent that workload is caused and SA supported by many of the same cognitive processes, they are enabled by, and subject to the limits of, many of the same processes. The more demanding the task, the more complex the situation and the more "work" is required to get the job done and the situation assessed. By our definition, the higher the level of workload, the more attention is needed for task performance and the less is left for keeping abreast of the situation. The SA process could actually compete with task performance for the limited resource supply, and therefore a high level of workload could lead to poor SA. On the other hand, SA could be improved by working harder (e.g., more frequent sampling and updating of information). That is, a high-level workload is sometimes necessary to maintain a good SA. Thus, a high level of workload could be associated with either a low or high degree of SA (Endsley, 1993). But poor SA may or may not impose more workload. One could simply not be doing the work necessary to attain and maintain SA, and if one is not aware of the dire situation that one is in and takes no action to correct the situation, no additional work would be initiated. Although a low degree of SA is never desirable, an awareness of one's lack of SA could start a course of action that could increase the level of workload in the process of attaining or restoring SA. The ideal scenario is one where a high degree of SA would support more efficient use of resources and thereby producing a low level of workload. In short, mental workload and SA could support each other as well as compete with each other.

Strategic management is proposed to be needed for the balancing act of maintaining adequate SA without incurring excessive workload. Strategic management is also referred to as executive control and is a much discussed topic in the literature. One point of contention is what exactly constitutes executive control, since a host of higher-level cognitive functions have been included under the rubric of executive control. The coordinating of multiple tasks (including the allocation of limited processing resources), planning, chunking or the reorganizing of information to increase the amount of materials that can be remembered, and the inhibiting of irrelevant information have all been labeled as part of the executive control. As Figure 1 indicates, strategic management is skills-based and is highly dependent on one's apprehension of the situation. For example, a beginner tennis player would be content to have made contact with the tennis ball and would not have the spare resources or the knowledge to ponder game strategies. After having mastered the basic strokes (which have become more automatic), however, the strategic component would take on more central importance. But strategic management is not attention free. Even though declarative and procedural knowledge develops as expertise develops and are used to support performance, there are components in many complex performances that are never automatic. High-performing athletes, chess players, musicians, and command and control officers expend considerable effort to perform at the level that they display.

Recent neurophysiological evidence provides some support for the notion that executive control is a distinct construct and consumes processing resources. Just et al. (2003) point out that the executive system is identified primarily with the prefrontal cortex which does not receive direct sensory input but has widespread connections with a number of cortical areas associated with various types of processing (e.g., spatial and verbal processing). Further, neuropsychological patients with lesions in the frontal lobe show impairments in planning and other higher-level cognitive functions (Shallice, 1988). Importantly, a number of functional magnetic resonance imaging studies show a higher level of activation in the prefrontal cortex in (1) a problem-solving task that requires more planning than one that requires less (Baker et al., 1996), (2) a working memory task that requires more updating of a larger amount of information (Braver et al., 1997), and (3) a dual-task (a semantic category judgment and a mental rotation task) than the single-task performance (D'Exposito et al., 1999). These results show that the activation in the prefrontal cortex vary systematically with the task demand.

Returning to Figure 1, strategic management competes directly with all the processes that generate mental workload for processing resources. But strategic management could optimize performance by planning and by smartly allocating the limited resources to the processes that need resources the most to meet system requirements. An efficacious strategic management would, of course, require a high-quality situation assessment. In the last section we discuss potential human factors support (such as display support, automation aids, training) that would improve the potential of attaining this ideal scenario of a high level of SA without an exceedingly high level of workload.

3 METRICS OF MENTAL WORKLOAD AND SITUATION AWARENESS

Measures of mental workload and SA have often been divided into three categories, based on the nature of the data collected: subjective ratings, operator performance, and psychophysiological measures. There are several properties that should be considered when selecting measures of cognitive activity: sensitivity, diagnosticity, intrusiveness, validity, reliability, ease of use, and operator acceptance. In addition to the foregoing concerns, Tenny et al. (1992) caution that there is an additional matter to consider in the case of SA measures. They distinguish between the *process* of building SA and the actual awareness that is the *product* of that process. SA measures should be designed and selected based on which aspect (i.e., process or product) the evaluator wishes to assess.

As outlined below, each group of measures has its strengths and weaknesses and a thoughtful combination of measures can lead to a more complete picture. Since the various workload and SA measures have different properties, one should have a good understanding of the properties of each measure so that the most appropriate choice(s) can be made. Readers are encouraged to consult more in-depth coverage of the metrics presented here (Gopher and Donchin, 1986; O'Donnell and Eggemeier, 1986; Lysaght et al., 1989; Vidulich et al., 1994a; Bryne, 1995; Tsang and Wilson, 1997; Gawron, 2000; Vidulich, 2003).

3.1 Performance Measures

System designers are typically most concerned with system performance. Some might say that the workload or SA experienced by an operator can be important only if it affects system performance. Consequently, performance-based measures might be the most valuable to system designers. There are two main categories of performance-based workload measures: primary task performance and secondary task performance. SA assessment also has made use of primary task performance. But instead of assessing secondary task performance, SA researchers have often employed recall-based memory probe performance or real-time performance. Although primary task performance is obviously the measure that is most strongly linked to the system designer's goal of optimizing system performance, Vidulich (2003) suggested that the secondary task method of workload assessment and the memory probe method of SA assessment are prototypical measures of the theoretical concepts behind workload and SA.

3.1.1 Primary Task Performance

Primary Task Workload Assessment The primary task method of workload assessment consists of monitoring the operator's performance and noting what changes occur as the task demands are varied. This methodology is grounded in the framework presented above. Since human operators have a finite capacity to deal with the demands of a task, as that task's demands continue to increase, task performance would be expected to deteriorate, and at some point the operator will no longer able to perform the task adequately. For example, an automobile driver might have more difficulty maintaining a proper course as the weather becomes more windy, and if the wind increases even more when the road is slippery, the driver may fail completely to keep the car in the proper lane.

It should be noted that mental workload is not the only thing that can influence operator performance. The operator's level of motivation might change, for example. Kahneman (1973) suggested that the human's capacity to perform mental work is related to the person's arousal level. In addition, Kahneman also pointed out that humans can monitor their own performance, and if the performance is found wanting and there are resources available (perhaps by increasing arousal), more resources can be allocated to the task to maintain or augment performance. Recently, Salvendy and his colleagues found that including a factor that reflected a person's skill, attitude, and personality contributed significantly to the predictive value of their projective modeling technique (Bi and Salvendy, 1994; Xie and Salvendy, 2000a,b).

In considering all of these issues, Gopher and Donchin (1986, p. 41–25) concluded: "In summary, direct measures of performance on the task of interest are usually a poor indicator of mental workload because they often do not reflect variation in resource investment due to difficulty changes, they do not diagnose the source of load, and they do not make possible a systematic conversion of performance units into measures of relative demands or load on the processing system." Thus, although the primary task performance is, clearly, very important to system evaluators as a test of whether design goals have been achieved, primary task performance by itself typically does not provide an adequate test of an operator's mental workload.

Primary Task SA Assessment Despite the problems in using primary task performance as a workload measure, it has become a common tool for assessing the impact of human–machine interface modifications intended to improve SA. For example, Vidulich (2000) found that it was common for researchers to propose an interface alteration that would improve SA and test it by determining if performance improved when the alteration was in place. The logic behind primary task performance based measures of SA is well illustrated by Andre et al. (1991). In a study of aircraft cockpit design, Andre et al. (1991) postulated that the pilot's ability to recover from disorienting events was a direct measure of how well the attitude information provided by the cockpit supported the pilots' SA of current and future attitudes of the aircraft. Following their stated logic, Andre et al. (1991) concluded that the display incorporating inside-out reference frames produced better performance and supported superior SA than the alternatives studied.

3.1.2 Secondary Task Measures of Workload

The secondary task measure of workload has been considered the prototypical measure of mental workload (e.g., Ogden et al., 1970; Gopher, 1994; Vidulich, 2003). A system evaluator would usually desire to assess primary task performance even if it was not being interpreted as a mental workload indicator. In contrast, a secondary task is usually only incorporated in a system assessment for assessing mental workload. More important, the secondary task technique is a procedure that is optimally suited to reflect the commonly accepted concept of mental workload (see the theoretical framework above). Workload is often assessed to determine whether the human operator is working within a tolerable information-processing capacity while performing the required task. It follows logically that if there is unused capacity, the operator could perform another task. For example, it is expected that spare capacity would be very valuable in emergencies or when under stress (Wickens, 2001; Hockey et al., 2003).

The secondary task measure of mental workload also offers some practical advantages for workload assessment in comparison to primary task performance assessment. Many secondary tasks have been developed and calibrated for use in different evaluations (Gawron, 2000). Many of these tasks vary in the resources demanded as characterized by multiple-resource theories. Presumably, this allows an evaluator to select secondary tasks to compete for specific resources of the primary task. Thus, a well selected secondary can be diagnostic of the primary task's resource demands.

The secondary task measure can be assessed in environments where primary task performance is difficult to obtain or is not available. This is often the case in many real-world systems, such as automobiles, ships, and airplanes that do not have performance-recording capability. Also, with highly automated systems in which the primary role of the operator is that of monitoring and supervising, little observable performance would be available for analysis. Finally, as noted above, primary task measures may not be sensitive to very low workload levels because operators could increase their efforts to maintain a stable level of performance (e.g., O'Donnell and Eggemeier, 1986). Adding a secondary task will increase the overall task demand to a level that performance measures may be more sensitive.

With the secondary task method, the operator is required to perform a second task concurrently with the primary task of interest. It is explained to the operators that the primary task is more important and the primary task performance must be performed to the best of their ability whether or not it is performed with the secondary task. Operators are to use only their spare capacity to perform the secondary task. Since the primary and secondary tasks would compete for the limited processing resources, changes in the primary task demand should result in changes in the secondary task performance as more or less resources become available for the secondary task.

These changes can be interpreted within a performance operating characteristic (POC) representation. As illustrated in Figure 2*a*, the performance trade-off between tasks can be examined by plotting them within a space representing their joint performance. In other words, the POC reflects the subject's allocation strategy for distributing attention between the time-shared tasks. Figure 2*a* shows possible trade-off between Tasks A and B performed as a dual-task combination. If the two tasks did not compete for any resources, perfect time-sharing could be observed. This is represented by the "X" on the figure, which would indicate that both tasks were performed at their respective single-task levels when performed together. Such perfect time-sharing is rare, but has been observed (e.g., Allport et al., 1972). The data lines in Figure 2*a*, being much closer to the origin of the graph than the perfect time-sharing point, indicate substantial interference in dual-task conditions. Such interference would typically show up in dual-task studies that manipulated the relative priorities of the two tasks. The dashed line shows a perfect trade-off pattern between the tasks. As one task's performance improves by a certain amount, a comparable degradation is observed in the other task. The dotted line shows the two task's joint performance being somewhat better than the perfect trade-off case. This would be expected to occur if the two tasks required at least some different types of information-processing resources.

The secondary task procedure differs from the standard laboratory dual-task study in that the priorities are not usually manipulated in the latter. The primary task's performance must be defended. Figure 2*b* illustrates what can be expected in this situation. In this example, the primary task performance (x-axis) of two possible interfaces is being evaluated. In this hypothetical example, the subjects have done a good job of following the secondary task (y-axis) instructions and are performing the primary task at very near the single-task level. Notice that both primary task interfaces (Y and Z) are maintaining the same level of primary task performance. However, interface Y's secondary task performance is substantially better than interface Z's secondary task performance. This result would be interpreted as interface Y inflicting less workload on the operator while performing the primary task than would interface Z.

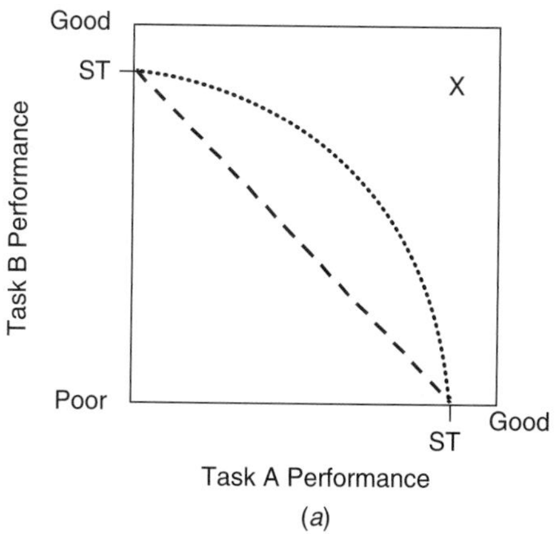

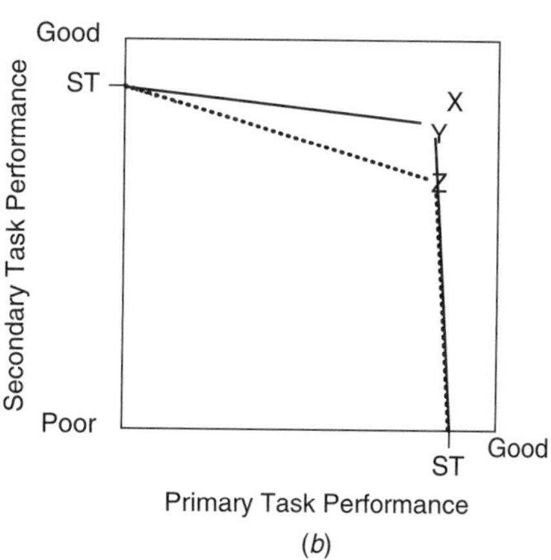

Figure 2 (*a*) Hypothetical performance operating characteristic (POC). Tasks A and B are two tasks that have been performed both independently and together. ST = level of single-task performance. The dashed and dotted lines illustrate possible joint performance when the two tasks are performed together. X, perfect time-sharing. (*b*) Hypothetical secondary task POC. The primary task was performed with two different interfaces. Y and Z, joint performances observed when the secondary task is performed with the two versions of the primary task.

An important consideration in the selection of a secondary task is the type of task demand of both the primary and secondary tasks, according to the logic of multiple-resource theory. Secondary task performance will be a sensitive workload measure of the primary task demand only if the two tasks compete for the same processing resources. The greater the dissimilarity of the resource demands of the time-shared tasks, the lower the degree of the interference there would be between the two tasks. Although a low degree of interference usually translates to a higher level of performance (which of course is desirable), this is not compatible with the goal of workload assessment. A fundamental assumption of the secondary task method is that the secondary task will compete with the primary task for limited processing resources. It is the degree of interference that is used for inferring the level of workload. Care must therefore be taken to assure that the secondary task selected demands resources similar to those of the primary task.

One drawback of the secondary task method is that the addition of an extraneous task to the operational environment may not only add to the workload, but may fundamentally change the processing of the primary task. The resulting workload metric would then be nothing more than an experimental artifact. The *embedded secondary task technique* was proposed to circumvent this difficulty (Shingledecker, 1984; Vidulich and Bortolussi, 1988). With this method, a normally occurring part of the overall task is used as the secondary task. In some situations, such as piloting a jet fighter aircraft, task shedding is an accepted and taught strategy that is used when primary task workload becomes excessive. Tasks that can be shed can perhaps serve as naturally lower-priority embedded secondary tasks in a less intense workload evaluation situation. However, a naturally lower-priority operational task may not always be available. Another drawback is that using the secondary task method requires considerable background knowledge and experience to properly conduct a secondary task evaluation and to interpret the results. For example, care must be taken to control for the operator's attention allocation strategy, so as to assure that the operator is treating the primary task as a high-priority task. The use of secondary tasks may also entail additional software and hardware development.

Despite the drawbacks and challenges of using the secondary task procedure, it is still used profitably for system assessment. For example, Ververs and Wickens (2000) used a set of secondary tasks to assess a simulated flight path following and taxiing performance with different sets of HUD symbology. One set of symbology presented a "tunnel in the sky" for the subjects to follow during landing approaches. The other display was a more traditional presentation of flight director information. The tunnel display reduced the subject's flight path error during landing. Subjects also responded more quickly and accurately to secondary task airspeed changes and were more accurate at detecting intruders on the runway. However, the other display was associated with faster detections of the runway intruder and quicker identification of the runway. The authors concluded that although the tunnel display produced a lower workload during the landing task, it also caused cognitive tunneling that reduced sensitivity to unexpected outside events. More recently, Leyman et al. (2004) used a secondary task to assess the workload associated with various simulated office tasks of varying complexity. The subjects in the experiment typed a practiced paragraph as the secondary task which was time-shared with a random word memory task of varying list lengths, a geographical reasoning task, or a scheduling task. The secondary task typing performance showed significant degradation. The secondary task results validated the workload assessments of a new electromyographic (EMG) office workload measure.

3.1.3 Memory Probe Measures of Situation Awareness

The first popular and standardized procedure for assessing SA was the memory probe technique. It can be considered the prototypical SA measurement tool (Vidulich, 2003). Any memory probe technique attempts to assess at least part of the contents of memory at a specific time during task performance, so it assesses the product of SA processes. As represented by the Situation Awareness Global Assessment Technique (SAGAT) (Endsley, 1988, 1990), the typical memory probe procedure consists of unexpectedly stopping the subject's task, blanking the displays, and asking the subject to answer questions to assess his or her knowledge of the current situation. The questions asked are typically drawn from a large set of questions that correspond to experimenter's assessment of the SA requirements for task performance. The subject's answers are compared to the true situation to determine the SAGAT score. Vidulich (2000) found that this SAGAT-style approach with unpredictable measurement times and random selection of queries from large sets of possible questions was generally sensitive to interface manipulations designed to affect SA. In contrast, as memory probes were made more specific or predictable, the sensitivity to interface manipulations appeared to be diminished. For example, Vidulich et al. (1994b) used a memory probe procedure in which the memory probe, if it appeared, was at a predictable time and the same question was always used. This procedure failed to detect a beneficial effect of display augmentation to highlight targets in a simulated air-to-ground attack, even though there was a significant benefit in task performance (i.e., more targets destroyed) and a significant increase in SA ratings. In contrast, Vidulich et al. (1995) used a SAGAT-like approach with many different questions that were asked during unpredictable trial stoppages. In this case, the memory probe data showed a significant SA benefit of the presence of a tactical situation display.

Although the memory probe procedure is attractive due to its assessing the information possessed by the subject at specific moment in time, it does have practical constraints that limit its applicability. First, it is clearly intrusive to stop task performance unexpectedly to ask questions. Endsley (1988) demonstrated that the performance of a simulated air-to-air combat task in trials that included SAGAT stoppages did not significantly differ from trials that did not. But even if the SAGAT stoppages are always unobtrusive to simulator performance, there are assessment environments where such stoppages are impossible (e.g., actual airplane flight tests). Also, the number of questions required that must be selected randomly and presented unpredictably can result in a large number of trials being needed to assess all questions.

The research of Strater et al. (2001) can be considered a typical SAGAT evaluation. They assessed U.S. Army platoon leaders in simulated Military Operations on Urbanized Terrain (MOUT) exercises. The platoon leaders varied from relatively inexperienced lieutenants to relatively experienced captains. SAGAT data were collected in a scenario that had the soldiers assaulting an enemy position and a scenario that involved defending a position. SAGAT probe questions were developed that could be used in either scenario. Results showed that the soldiers were more sensitive to different information depending on the scenario type. For example, in the assault scenarios the soldiers were more sensitive to the location of adjacent friendly forces than they were in the defend scenario. In the defend scenario, soldiers were more sensitive to the location of exposed friendly elements. Significant effects of soldier experience level were also detected. For example, experienced soldiers were more sensitive to enemy locations and strength than were inexperienced soldiers. The authors suggested that the data collected from such experimentation could be used to improve training programs, by helping to identify better information-seeking behaviors for the novices.

3.1.4 Situation Awareness Real-Time Performance Assessment

Real-time performance has been used as a potential indicator of SA (Durso et al., 1995; Pritchett and Hansman, 2000; Vidulich and McMillan, 2000). The logic of assessing real-time performance is based on the assumption that if an operator is aware of task demands and opportunities, she will react appropriately to them in a timely manner. This approach, if successful, would be unintrusive to task performance, diagnostic of operator success or failure, and potentially useful for guiding automated aiding. Since the continuous stream of operator performance is assessed, real-time performance should illuminate the SA processes of the operator.

The Global Implicit Measure (GIM) (Vidulich and McMillan, 2000) is an example of this approach. The GIM is based on the assumption that the operator of a human–machine system is attempting to accomplish known goals at various priority levels. Therefore, it is possible to consider the momentary progress toward accomplishing these goals as a performance-based measure of SA. Development of the GIM was an attempt to develop a real-time SA measurement that could effectively guide automated pilot aiding (Brickman et al., 1995, 1999; Vidulich, 1995; Shaw et al., 2004). In this approach, a detailed task analysis was used to link measurable behaviors to the accomplishment of mission goals. The goals will be varied depending on the mission phase. For example, during a combat air patrol, a pilot might be instructed to maintain a specific altitude and to use a specific mode of the on-board radar, but during an intercept the optimal altitude might be defined in relation to the aircraft being intercepted, and a different radar mode might be appropriate. For each phase, these measurable behaviors that logically affect goal accomplishment are identified and scored. The scoring was based on the contribution to goal accomplishment. The proportion of mission-specific goals being accomplished successfully according to the GIM algorithms indicated how well the pilot was accomplishing the goals of that mission phase. More important, the behavioral components scored as failing should identify the portions of the task that the pilot was either unaware of or unable to perform at the moment. Thus, GIM scores could potentially provide a real-time indication of the person's SA as reflected by of the quality of task performance and a diagnosis of the problem if task performance deviates from the ideal, as specified by the GIM task analysis and scoring algorithms.

Vidulich and McMillan (2000) tested the GIM metric in a simulated air-to-air combat task using two cockpit designs that were known from previous evaluations to produce different levels of mission performance, mental workload, and rated SA. The subjects were seven U.S. military pilots or weapons systems officers. The real-time GIM scores distinguished successfully between the two cockpits and the different phases of the mission. No attempt was made to guide adaptation on the basis of the GIM scores, but the results suggested that such an approach has promise.

3.2 Subjective Measures

Subjective measures consist primarily of using techniques that usually require subjects to quantify their experience of workload or SA. Many researchers are suspicious of subjective data, perhaps as a holdover from the behaviorists' rejection of introspection as an unscientific research method (Watson, 1913). However, Annett (2002a,b) argued that subjective ratings are maligned unfairly. In an in-depth discussion of the issues, he contended that the lack of precision associated with subjective measures was expected to prohibit their use in setting design standards. However, he also concluded that subjective ratings could be useful for evaluating the mechanism underlying performance or for the comparative evaluation of competing interface designs. Such a comparative process is how subjective ratings of workload and SA are typically used.

Vidulich and Tsang (1987) and Tsang and Vidulich (1994) found three variables that were useful for categorizing subjective rating techniques: dimensionality, evaluation style, and immediacy. *Dimensionality* refers to whether the metric required the subjects to rate their experiences along a single dimension or multiple dimensions. *Evaluation style* refers to whether the subjects were asked to provide an absolute rating of an experience or a relative rating comparing one experience to another. *Immediacy* distinguishes between subjective metrics that were designed to be used as soon as possible after the to-be-rated experience and those that were used at the end of a session or even at the end of an experiment.

Although it is theoretically possible to create a subjective technique that combines any level of the three variables, in practice two basic combinations have dominated. The most common techniques combine multidimensionality, the absolute evaluation style, and immediacy. The typical alternative to the multidimensional-absolute-immediate approach, are techniques that are usually unidimensional, use a relative comparison evaluation style, and are collected retrospectively rather than immediately.

3.2.1 Multidimensional Absolute Immediate Ratings

The subject's immediate assessment after trial completion should minimize the potentially damaging effects of any bias the subject may have regarding the task conditions and the likelihood that the ratings are based on between-condition comparisons (Tsang and Vidulich, 1994). Also, ratings after the trial should benefit from the freshest memory for the experience of performing the trial. The absolute scale design should also encourage the subjects to consider each trial condition individually. The multidimensional aspect supports diagnosticity, because the subjects can be more precise in describing how experimental conditions influence their experience.

Workload Ratings Although numerous scales have been developed, two popular multidimensional, absolute, and immediate ratings scales are the National Aeronautics and Space Administration's Task Load Index (NASA-TLX) (Hart and Staveland, 1988) and the Subjective Workload Assessment Technique (SWAT) (Reid and Nygren, 1988). NASA-TLX is based on six scales (i.e., mental demand, physical demand, temporal demand, performance, effort, and frustration level), and the ratings on the six scales are weighted according to the subject's evaluation of their relative importance. SWAT is based on three rating scales (i.e., time load, mental effort load, and psychological stress load). The relative roles of the three scales is determined by the subjects' rankings of the workload inflicted by each combination of the various levels of workload (1 to 3) in each of the three workload scales. A conjoint analysis is then conducted to produce a look-up table that translates the ordinal rankings to ratings with interval-scale properties. Both NASA-TLX and SWAT ultimately produce a workload rating from 0 to 100 for each trial rated.

NASA-TLX and SWAT have been compared to each other and to a number of other rating scales a number of times (e.g., Battiste and Bortolussi, 1988; Hill et al., 1992; Rubio et al., 2004). In reviewing the comparisons, Rubio et al. (2004) noted that SWAT and NASA-TLX both offer diagnosticity, due to their multiple scales, and have generally demonstrated good concurrent validity with performance. Rubio et al. (2004) also pointed out that both NASA-TLX and SWAT have demonstrated sensitivity to difficulty manipulations, although some researchers have found NASA-TLX to be slightly more sensitive, especially for low levels of workload (e.g., Battiste and Bortolussi, 1988; Hill et al., 1992).

NASA-TLX and SWAT have also been compared in terms of their ease of use. Each technique is composed of two major parts: the scales that the subjects fill out after each trial and a procedure for converting the raw scale ratings into the final workload scale. The actual scales used by SWAT are fewer than NASA-TLX (three vs. six) and only require the subject to choose one of three possible levels instead of rating on a 0-to-100 scale as NASA-TLX incorporates. That means that SWAT would be easier to collect in a prolonged task, such as flying, while the task performance actually continues. On the other hand, NASA-TLX's paired comparison technique to generate scale weights is much easier than SWAT's card-sorting procedure for both the subject to complete and the experimenter to process. The NASA-TLX procedure only requires the subject to make 15 forced choices of importance between the individual scales. The raw count of the number of times that each scale was considered more important than another is then used to weigh the individual scale ratings provided by the subject. In contrast, the SWAT card sort requires each subject to consider and sort 27 cards (each representing a possible combination of rating scale selections), and then the experimenter must use specialized software to convert the card sort data into an overall workload scale.

Some researchers have investigated simpler methods of generating weights for SWAT. The simple sum of the three SWAT dimensions has been shown to exhibit the same pattern of significant findings as SWAT ratings using the conjoint analysis of card-sort data (Biers and Maseline, 1987; Biers and McInerney, 1988; Luximon and Goonetilleke, 2001). Additionally, Luximon and Goonetilleke (2001) found that SWAT sensitivity could be improved by using a continuous scale rather than a three-level discrete scale. As with SWAT, the weighting procedure of NASA-TLX has undergone testing. Both Nygren (1991) and Hendy et al. (1993) have argued that the NASA-TLX weighting procedure does not add to NASA-TLX's effectiveness.

Lee and Liu (2003) provide an example of the use of NASA-TLX to assessing the workload of 10 China Airline pilots flying a Boeing 747 aircraft in a high-fidelity 747 simulator. Lee and Liu found that the overall NASA-TLX ratings discriminated successfully among four flight segments: takeoff, cruise, approach, and landing. As expected, takeoff, landing, and approach were all rated higher than cruise in mental workload. Lee and Liu also used the multidimensional scales of NASA-TLX to diagnose the causes of the higher workload. For example, they found that temporal demand was an important contributor to the takeoff and approach segments, but effort was a more important contributor to landing. The authors concluded that training programs should be designed to help the pilot cope with the specific expected stresses of different flight segments.

SA Ratings Multidimensional, absolute, and immediate ratings have also been a popular approach for assessing SA. Probably the most commonly used subjective rating tool for SA has been the Situation Awareness Rating Technique (SART), developed by Taylor (1990). The SART technique characterizes SA as having three main dimensions: attentional demands (D), attentional supply (S), and understanding (U). The ratings on each of the three dimensions are combined into a single SART value according to a formula (Selcon et al., 1992): $SA = U - (D - S)$.

Inasmuch as SART contains ratings of attentional supply and demand, it can be seen to incorporate elements of mental workload in its evaluation. However, in a direct comparison of NASA-TLX and SART, it was found that although both were sensitive to task demand level, SART was also sensitive to the experience level of the 12 Royal Air Force pilot subjects (Selcon et al., 1991).

3.2.2 Unidimensional Relative Retrospective Judgments

The unidimensional, relative, retrospective judgment approach is based on the assumption that the subject who has experienced all of the task conditions is considered a subject matter expert with knowledge about the subjective experience of performing the various task conditions under consideration. This approach attempts to extract and quantify subjects' opinions about the experiences associated with task performance.

Workload Judgments The use of unidimensional, relative, retrospective judgments was strongly supported by the work of Gopher and Braune (1984). Inspired by Stevens's (1957, 1966) psychophysical measurement theory, Gopher and Braune adapted it to the measurement of subjective workload. The procedure used one task as a reference task with an arbitrarily assigned workload value. All of the other tasks' subjective workload values were evaluated relative to that of the reference task. The resulting ratings were found to be highly sensitive in a number of studies (e.g., Tsang and Vidulich, 1994; Tsang and Shaner, 1998). In addition, high reliability of these ratings was revealed by split-half correlations of repeated ratings of the task conditions.

Another approach to collecting unidimensional, relative, retrospective judgments was developed by a mathematician, Thomas Saaty (1980). Saaty's technique was named the *Analytic Hierarchy Process* (AHP) and was developed to aid decision making. When applied to workload assessment, the AHP requires subjects to perform all pairwise comparisons of all task conditions. These comparisons fill a dominance matrix, which is then solved to provide the ratings for each task condition. Saaty's AHP was originally designed to evaluate all dimensions relevant to a decision and then combine the multiple dimensions to support selection of one option in a decision-making task. However, Lidderdale (1987) demonstrated that a unidimensional version of the AHP could be an effective workload assessment tool and inspired further investigations using the tool. Vidulich and Tsang (1987) compared the AHP to NASA-TLX and a unidimensional, absolute, immediate rating of overall workload in assessing the workload of selected laboratory tasks. The AHP was found to be both more sensitive and more reliable than the other techniques. Vidulich (1989) compared several methods for converting dominance matrices to the final ratings and used the results to create the Subjective Workload Dominance (SWORD) technique. In one application, Toms et al. (1997) used SWORD to evaluate a prototype decision aid for landing an aircraft. The participating pilots performed landings with and without the decision aid and in both low- and high-task-load conditions. Task load was varied by changing the information available to the pilot. Overall, the results showed that the decision aid improved landing performance while lowering mental workload.

Vidulich and Tsang performed a series of studies to examine the various approaches to subjective assessment. Although specific instruments were compared in these studies, the goal was not to determine which instrument was superior. Rather, the objective was to determine which assessment approach can elicit the most and accurate workload information. Tsang and Vidulich (1994) found that the unidimensional, relative, retrospective SWORD technique with the highly redundant pairwise comparisons superior to a procedure using relative comparisons to a single reference task. Tsang and Velazquez (1996) found that compared to an immediate absolute instrument, a relative, retrospective psychophysical scaling was more sensitive to task demand manipulation and had higher concurrent validity with performance. Tsang and Velazquez (1996) also found that a subjective multidimensional retrospective technique, the Workload Profile, provided diagnostic workload information that could be subjected to quantitative analysis. Rubio et al. (2004) confirmed the diagnostic power of the Workload Profile technique. They found the Workload Profile more diagnostic than either NASA-TLX or SWAT. Collectively, these studies suggested a relative-retrospective approach advantage.

SA Judgments Unidimensional, relative, retrospective judgments have also been applied to SA assessment. For example, the SWORD workload technique was adapted to measure SA (SA-SWORD; Vidulich and Hughes, 1991). Vidulich and Hughes used the SA-SWORD to evaluate the effect of data-linked information in an advanced fighter aircraft simulation. The technique demonstrated good sensitivity to the experimental manipulation and good reliability. Toms et al. (1997) used the SA-SWORD to assess SA along with SWORD to measure workload. Their results showed that a decision aid's benefits to landing performance and mental workload were also associated with improved SA. In this case, workload assessment and SA assessment both showed that the decision aiding was valuable.

3.3 Physiological Measures

A host of physiological measures have been used to assess mental workload with the assumption that there are physiological correlates to mental work. The most common measures include cardiovascular (e.g., heart rate and heart rate variability), ocular (e.g., pupil dilation, eye movement measures), and measures of brain activity. The present review focuses on the brain measures because (1) it would seem that brain activity could most directly reflect mental work; (2) in line with our framework that hypothesizes both an intensity aspect and a structural aspect to mental work, many of the brain measures have been demonstrated to be sensitive to parametric manipulation of task demands and to be diagnostic with regard to the types of cognitive demands involved in certain task performance; and (3) there already exist reviews of the nonbrain measures (e.g., Beatty, 1982; Stern et al., 1984; Wilson and Eggemeier, 1991; Jorna, 1992; Mulder, 1992; Backs and Boucsein, 2000; Kramer and Weber, 2000; Kramer and McCarley, 2003), but the brain-imaging workload studies are relatively new.

3.3.1 Electroencephalographic Measures

Electroencephalographic (EEG) measures are recorded from surface electrodes placed directly on the scalp and have been shown to be sensitive to momentary changes in task demands in laboratory studies (e.g., Glass, 1966), simulated environments (e.g., Fournier et al., 1999; Gevins and Smith, 2003), and real-world settings (e.g., Wilson, 2002b). Spectral power in two major frequency bands of the EEG have been identified as being sensitive to workload manipulations: the alpha (7 to 14 Hz) and theta (4 to 7 Hz) bands. Spectral power in the alpha band that arises in widespread cortical areas is inversely related to the attentional resources allocated to the task, whereas theta power recorded over the frontal cortex increases with increased task difficulty and higher memory load (Parasuraman and Caggiano, 2002). Sterman and Mann (1995) reported a series of EEG studies conducted in simulated and operational military flights. A systematic decrease in power in the alpha band of the EEG activity was observed with a degraded control responsiveness of a T4 aircraft. A graded decrease in the alpha band power was also observed as U.S. Air Force pilots flew more difficult in-flight refueling missions in a B2 aircraft simulator. Brookings et al. (1996) had Air Force air traffic controllers perform computer-based air traffic control simulation (TRACON). Task difficulty was manipulated by varying the traffic volume (number of aircraft to be handled), traffic complexity (arriving to departing flight ratios, pilot skill, and aircraft types), and time pressure. Brookings et al. found the alpha power to decrease with increases in traffic complexity and the theta power to increase with traffic volume.

Kramer and Weber (2000) point out that a distinct advantage of EEG measures is their sensitivity to variations of mental workload and their potential to track momentary fluctuations in mental workload associated with rapid changes in task demands. However, the sensitivity of EEG measures to numerous artifacts such as head and body movements pose special difficulties for extralaboratory applications. Kramer and Weber also point out that it is not yet entirely clear whether the EEG measures reflect changes in the level of general arousal or changes in more specific cognitive operations. That is, EEG measures may not be particularly diagnostic with regard to the specific types of demand incurred. More in-depth discussion can be found in Gevins et al. (1995) and Davidson et al. (2000).

3.3.2 Event-Related Potentials

Evoked potentials are embedded in the background of EEG and are responsive to discrete environmental events. There are several different positive and negative voltage peaks and troughs that occur 100 to 600 ms following stimulus presentation. The P300 component has been extensively studied as a mental workload measure (e.g., Gopher and Donchin, 1986; Parasuraman, 1990; Wickens, 1990; Kramer and Weber, 2000). The P300 is typically examined in a dual-task condition with either the oddball paradigm or the irrelevant probe paradigm. In the *oddball paradigm*, the P300 is elicited by the subject keeping track of an infrequent signal (e.g., counting infrequent tones among frequent tones). One drawback of the oddball paradigm is that the additional processing of the oddball and having to respond to it could inflate the true workload of interest artifactually. As an alternative, additional stimuli (e.g., tones) are presented, but subjects are not required to keep track of them in the *irrelevant probe paradigm*. Needless to say, creating artifactual workload in the assessment process is of particular concern in applied settings.

With the oddball paradigm, the amplitude of the P300 has been found to decrease with increased task difficulty manipulated in a variety of laboratory tasks (e.g., Hoffman et al., 1985; Strayer and Kramer, 1990; Backs, 1997). Importantly, P300 is found to be selectively sensitive to perceptual and central processing demands. For example, Isreal et al. (1980) found that the amplitude of P300 elicited by a series of counted tones was not sensitive to response-related manipulations of tracking difficulty but was affected by manipulations of display perceptual load. Many of the laboratory-based findings have been replicated in simulator studies. For example, Kramer et al., (1987) had student pilots flew an instrument flight plan in a single-engine aircraft simulator. The P300s elicited by the secondary tone-counting task decreased in amplitude with increasing turbulence and subsystem failures (see also Fowler, 1994). Using the irrelevant probe paradigm, Sirevaag et al. (1993) had senior helicopter pilots fly low-level high speed flight in a high-fidelity helicopter simulator. The P300 amplitude was found to increase with increased difficulty in the primary tracking task. In addition, the P300 amplitude elicited by the secondary irrelevant probes decreased with increases in the communication load. Kramer and Weber (2000) point out further that the irrelevant probe paradigm would work only if the irrelevant probes are presented in a channel that would be monitored anyway.

In short, event-related potential (ERP) measures have been found to be sensitive to, and diagnostic of, changes in the perceptual and central processing task demands. One potential drawback is the possibility of artifactually augmenting the real workload of interest if the ERP measures are elicited from a secondary task. As with the performance, it would be ideal if a secondary task naturally embedded in the test environment could be used. But because ERP signals are relatively small and ensemble averaging across many stimuli is necessary for meaningful interpretation, there are not always sufficient stimuli available from the embedded secondary task or the primary task.

3.3.3 Brain Imaging Measures

Two measures of the brain's metabolic responses are considered here: positron emission tomography (PET) and functional magnetic resonance imaging (fMRI) measures. These measures most notably have been used to localize cortical regions associated with various cognitive processing (e.g., D'Exposito et al., 1999; Posner and DiGirolamo, 2000). Equally important, recent research has shown that these measures exhibit systematic variations with parametric manipulation of task difficulty (Parasuraman and Caggiano, 2002). That is, they have been shown to be both diagnostic and sensitive measures.

As an example, Corbetta et al. (1990) had subjects determine whether two stimuli presented in two frames separated by a blank display were the same or different. Between the two frames, the stimuli could vary in one of three dimensions: shape, color, and velocity. In the selective-attention condition, one of the dimensions would be designated as the relevant dimension, and zero, one, or two irrelevant dimensions could covary with the relevant dimension. In the divided-attention condition, any one of the dimensions could vary. The behavioral data (d') indicated that the divided-attention condition was more difficult. In the selective-attention condition, increased blood flow was observed in the region of the visual cortex known to be related to the processing of the relevant dimension designated. Corbetta et al. proposed that the increased neural activity in the specialized regions for the different dimensions was a result of a top-down attentional control since the sensory information should be the same across selective- and divided-attention conditions.

Just et al. (2003) reviewed a series of PET studies and found lower brain metabolic rate to be associated with higher language proficiency (Parks et al., 1988) and increased practice with a spatial computer game (Haier et al. 1992). Just et al. interpreted these results to mean that high-ability, high-skill persons could process more efficiently, thereby requiring a smaller amount of their total amount of processing resources available. They effectively would have a larger supply of processing resources (for other processing). Just and Carpenter (1992) proposed that the computational work underlying thinking must be accompanied by resource utilization. In their 3CAPS model, a brain region is considered a resource pool. Computational activities are resource consuming in the sense that they all operate by consuming an entity called activation. The intensity and volume of brain activation in a given cortical area is expected to increase in a graded fashion with increased computational load. Indeed, Just et al. (1996) found that with increasing sentence complexity, the level of neuronal activation and the volume of neural tissue activated increased in four cortical areas associated with language processing (Wernicke's, Broca's, and their right hemisphere homologues). With a spatial mental rotation task, Carpenter et al. (1999) found a monotonic increase in signal intensity and volume activation in the parietal region as a function of increased angular disparity between the two stimuli whose similarity was to be judged.

Highlighting the results from a number of studies that use an array of behavioral and neurophysiological measures (ERPs, PET, and fMRI), Just et al. (2003) propose cognitive workload to be a function of resource consumption and availability. Several similarities between the 3CAPS model and Wickens' multiple resource model are apparent. According to both models, (1) mental workload is a function of supply and demand of processing resources, (2) resources can be modulated in a graded fashion, (3) specific resources are used for different types of cognitive processing (e.g., verbal and spatial task demands bring about activations in different cortical regions), and (4) supply or availability of resources can be modulated by individual differences in ability and skill or expertise.

The PET and fMRI studies described so far were all conducted in the laboratory. Although more applied studies would certainly be desirable, the methodologies involved with these newly available technologies are still being refined. Practical concerns aside, results of these laboratory studies are encouraging and at the same time point to the need for continual learning about interpreting these brain images and validating the interpretations. Notwithstanding, one simulated study on pilot performance can be presented. Pérès et al. (2000) had expert (with at least 3000 flight hours and flight instructor qualifications) and novice (with less than 50 flight hours) French Air Force pilots perform a continuous simulated flight control task at two speeds (100 and 200 knots) while fMRI measures were collected. The fMRI measures showed that neuronal activation was dominant in the right hemisphere, as would be expected for a visual spatial task. Further, novice pilots exhibited more intense and more extensive activation than expert pilots. At the high-speed condition, the expert pilots exhibited increased activation in the frontal and prefrontal cortical areas and reduced activity in visual and motor regions. This suggested to researchers that the expert pilots were better able to use their knowledge to focus their resources for the higher-level functions in working memory, planning, attention, and decision making. In contrast, novice pilots' increased activation in the high-speed condition was more widespread and extended across the frontal, parietal, and occipital areas, suggesting that they were engaged in nonspecific perceptual processing. Interestingly, when the expert pilots were asked to track at an even higher speed (400 knots), their pattern of activation resembled that of the novice pilots tracking at 200 knots.

Notably, there is paucity of physiological studies on SA included in this chapter. This is partly because compared to the concept of mental workload, the concept of SA is relatively new (Pew, 1994; Wickens, 2001) and both its theoretical and methodological development have not reached the level of maturity that the concept of mental workload has. It is also the case that the concept of SA does not refer to a specific process. Although complex performance generally entails multiple processes, it is often possible to identify many of the processes and hence the type of workload involved. However, whereas SA is supported by the many of the same processes, SA is an emergent property that has not been hypothesized to be associated with specific cortical regions or other physiological responses.

As a class of measures, many of the physiological measures have the distinct ability to serve as a continuous measure for on-line assessment. Some of them are diagnostic with regard to the type of cognitive demands entailed (e.g., ERPs, PET, fMRI). Most of them are not cognitively intrusive, as in having to perform additional work in order to provide a measure. The main drawback with most physiological measures is that they are equipment intensive, which makes real-world assessment impractical. But it is not impossible. Some physiological measures are or will become more feasible with the present rapid technological advances (e.g., see Gevins et al., 1995; Wilson, 2000; Wilson, 2002a; Kramer and McCarley, 2003; Parasuraman, 2003). Even for the costly fMRI studies, attempts have been made to assess the mental workload of simulated flight performance (Pèrés et al., 2000).

3.4 Multiple Measures of Workload and Situation Awareness

There are several facets to the undertaking of assessing workload and SA of a complex, dynamic human–machine system. First, there are a number of candidate measures to choose from, each with strengths and weaknesses. Measures that provide global information about the mental workload of these tasks may fail to provide more specific information about the nature of the demand. Measures that could provide more diagnostic information may be intrusive or insensitive to other aspects of interest, and certain sensitive measures may be collected only under restrictive conditions. Still, many workload measures often associate. For example, many subjective measures have been found to correlate with performance (e.g., Tsang and Vidulich, 1994; Hockey et al., 2003; Rubio et al., 2004). Just et al. (2003) present a convincing account of how the associations among a number of behavioral and neurophysiological measures support the extant understanding of many cognitive concepts relevant to mental workload (see also Wickens, 1990; Fournier et al., 1999; Lee and Liu, 2003). Importantly, when the measures do not associate, they do not do so in haphazard ways. The association and dissociation patterns among measures should therefore be evaluated carefully rather than treated as unreliable randomness. Below we discuss in greater detail the dissociation between the subjective and performance workload measures and the relation between the workload and SA measures.

3.4.1 Dissociations among Workload Measures

When different types of workload measures suggest different trends for the same workload situation, the workload measures are said to *dissociate*. Given that mental workload is a multidimensional concept, and the various workload measures may be differentially sensitive to the different workload dimensions, dissociations among workload measures are to be expected. Measures having qualities of general sensitivity (such as certain unidimensional subjective estimates) respond to a wide range of task manipulations but may not provide diagnostic information about the individual contributors to workload. Measures having selective sensitivity (such as secondary task measures) respond only to specific manipulations. In fact, the nature of the dissociation should be particularly revealing with regard to the characteristics of the workload in the task under evaluation.

Several conditions for the dissociation of performance and subjective measures have been identified (Vidulich and Wickens, 1986; Vidulich, 1988;Yeh and Wickens, 1988):

1. Dissociation tends to occur under low-workload conditions (e.g., Eggemeier et al., 1982). Performance could already be optimal when the workload is low and thus would not change further with additional effort that would be reflected in the subjective measures.
2. Dissociation would occur when subjects are performing data-limited tasks (when performance is governed by the quality of the data rather than by the availability of resources). If subjects are already expending their maximum resources, increasing task demand would further degrade performance but would not affect the subjective ratings.
3. Greater effort would generally result in higher subjective ratings, however, greater effort could also improve performance (e.g., Vidulich and Wickens, 1986).
4. Subjective ratings are particularly sensitive to the number of tasks that subjects have to time-share. For example, performing an easy dual task (that results in good performance) tends to produce higher ratings than does performing a difficult single task (that results in poor performance) (e.g., Yeh and Wickens, 1988).
5. Performance measures are sensitive to the severity of the resource competition (or similarity

of resource demand) between the time-shared tasks, but subjective measures are less so (Yeh and Wickens, 1988).

6. Given that subjects only have access to information available in their consciousness (Ericsson and Simon, 1993), subjective ratings are more sensitive to central processing demand (such as working memory demand) than to demands that are not represented well consciously, such as response execution processing demand. Dissociation would therefore tend to occur when the main task demands lie in response execution processing (Vidulich, 1988). McCoy et al. (1983) provided an excellent list of realistic examples of how performance and subjective ratings may dissociate in system evaluations and discussed how the dissociations can be interpreted in meaningful ways.

Hockey (1997) offers a more general conceptual account for the relations among performance, subjective, and physiological measures. Hockey proposes a compensatory control mechanism that allocates resources dynamically through an internal monitor very much like the one proposed by Kahneman (1973). Performance may be protected (as in primary task performance) by recruiting further resources, but only at the expense of increased subjective effort and physiological costs and degraded secondary task performance or strategic management of the overall system performance. Alternatively, performance goals may be lowered. Although performance will then degrade, no additional effort or physiological cost will be incurred. Hockey emphasizes that the efficacy of the control mechanism hinges on the accuracy of the perception of the situation. For example, Sperandio (1978) found air traffic controllers to switch strategy when the traffic load increased. Beyond a certain number of aircraft that the controllers handled, controllers would switch to a uniform strategy across aircraft as opposed to paying more individual attention to the various aircraft. Although this strategy should reduce the cognitive resources needed for dynamic planning, it would also probably produce less optimal scheduling. That is, the primary task performance might have been preserved with the strategy switch, but some secondary goals would have suffered (Hockey, 1997).

3.4.2 Relations of Workload and Situation Awareness Measures

Wickens (2001) point out that due to the energetic properties of workload, many physiological and subjective rating measures are suited for capturing the quantitative aspects of workload. In contrast, physiological measures are likely to be poor candidates for assessing the quality or content of SA. Self-ratings of one's awareness are unlikely to be informative since one cannot be aware of what one is not aware. However, subjective SA ratings could still be useful if they are used for system evaluative purposes. As illustrated earlier, subjects often could indicate reliably which system design affords greater SA. Last, Wickens pointed out that explicit performance measures designed to examine what one is aware of (content of SA) have no parallel use for workload assessment. However, implicit performance measures such as those used to check for reaction to unexpected events can be used to assess both workload and SA.

As discussed earlier, there is not one fixed relationship between workload and SA. Although high SA and an acceptable level of workload is always desirable, workload and SA can correlate positively or negatively with each other, depending on a host of exogenous and endogenous factors. Two sample studies will be described to illustrate their potential relationships. Vidulich (2000) reviewed a set of studies that examined SA sensitivity to interface manipulations. Of the nine studies that manipulated the interface by providing additional information on the display, seven showed an increase in SA, four showed a concomitant reduction in workload, and three showed a concomitant increase in workload. In contrast, of another nine studies that manipulated the interface by reformatting the display, all nine showed an increase in SA, six showed a concomitant reduction in workload, and none showed an increase in workload. In short, although different patterns in the relationship between the workload and SA measures were observed, the various patterns were reasonably interpretable given the experimental manipulations. In another study, Alexander et al. (2000) examined the relationship between mental workload and situation awareness in a simulated air-to-air combat task. Seven pilots flew simulated air intercepts against four bombers supported by two fighters. The main manipulations were two cockpit designs (the conventional cockpit with independent gauges and a virtually augmented cockpit designed by a subject-matter expert) and four mission phases of various degrees of difficulty and complexity. A negative correlation between the workload and SA measures were observed for both the cockpit design and the mission complexity manipulations. The augmented cockpit improved SA while reduced workload, whereas increased mission complexity decreased SA and increased workload. These results underscore the value of assessing both the mental workload and SA involved in any test and evaluation.

3.4.3 Need for Multiple Measures

There are however several broad guiding principles that would be helpful in measures selection. Muckler and Seven (1992) hold that "the distinction between 'objective' and 'subjective' measurement is neither meaningful nor useful in human performance studies" (p. 441). They contend that all measurements contain a subjective element as long as the human is part of the assessment. Not only is there subjectivity in the data obtained from the human subject, the human experimenter also imparts his or her subjectivity in the data collection, analysis, and interpretation. Thus, performance measures are not all objective, nor are subjective measures entirely subjective (see also, Annett, 2002a,b; Salvendy, 2002). Muckler and Seven advocate that the selecting of a measure (or a set of measures) be guided by the information needs. Candidate measures can be evaluated by considering their relative strengths (such as diagnosticity) and weaknesses (such as intrusiveness). In addition, Kantowitz (1992) advocates using theory to select the measures. Kantowitz made an analogy between theory and the blueprint of a building. Trying to interpret data without the guidance of a theory is like assembling bricks randomly when constructing a building. To elaborate, Kantowitz points out that an understanding of both the substantive theory of human information processing and the psychometric theory of the measurements is helpful. The former dictates what one should measure, and the latter suggests ways of measuring them. Another useful (if not required) strategy is to use multiple measures as much as feasible. As discussed above, even seemingly dissociate measures are informative (and sometimes especially so) if one is cognizant of the idiosyncratic properties of the different measures. In fact, Wickens (2001) points out that converging evidence from multiple measures is needed to ensure an accurate assessment of the level of workload incurred and the quality of SA attained.

To emphasize the value of assessing multiple measures, Parasuraman (1990) reported a study that examined the effectiveness of safety monitoring devices in high-speed electric trains in Europe (Fruhstorfer et al., 1977). Drivers were required to perform a secondary task by responding to the occurrence of a target light in a cab within 2.5 seconds. If no response was made, a loud buzzer would be activated. If the buzzer was not responded to within an additional 2.5 seconds, the train's braking system was activated automatically. Over a number of train journeys, onset of the warning buzzer was rare, and the automatic brake was activated only once. However, the EEG spectra showed that the secondary task performance could remain normal even when the drivers were transiently in stage 1 sleep.

4 DESIGN FOR MENTAL WORKLOAD AND SITUATION AWARENESS: INTEGRATED APPROACH TO OPTIMIZING SYSTEM PERFORMANCE

It is probably fair to say that after a decade of debate, there is now a general agreement that mental workload and SA are distinct concepts and yet are intricately intertwined. Both can be affected by very many of the same exogenous and endogenous factors and have a significant impact on each other and on system performance. One implication is that fairly well understood psychological principles can be applied to both concepts. For example, in the framework presented above, both workload and SA are subject to attentional and memory limits, and both can be supported by expertise. There exists an established body of knowledge about the effects of these limits and the enabling power of expertise to allow fairly reliable performance predictions. But the fact that the two concepts are distinct also means that they each contribute uniquely to the functioning of a human–machine system. Below we review three research areas that could be exploited for developing support that would manage workload and SA cooperatively to optimize system performance.

4.1 Adaptive Automation

Automation is often introduced to alleviate the heavy demand on an operator or to augment system performance and reduce error. Many modern complex systems simply cannot be operated by humans alone without some form of automation aids. However, it is now recognized that automation often redistributes, rather than reduces, the workload within a system (e.g., Wiener, 1988; Lee and Moray, 1992). Further, an increasing level of automation could distance the operator from the control system (e.g., Adams et al., 1991; Billings, 1997). The upshot of this is that even if automation reduces mental workload successfully, it could reduce SA and diminish an operator's ability to recover from unusual events. The idea of adaptive automation was introduced as a means of achieving the delicate balance of a manageable workload level and an adequate SA level. This idea has been around for some time (e.g., Rouse, 1977, 1988) and is receiving much attention in recent research (e.g., Rothrock et al., 2002; Parasuraman and Bryne, 2003). Proponents of adaptive automation argue that static automation that entails predetermined fixed task allocation will not serve complex dynamic systems well. Workloads can change dynamically due to environmental and individual factors (e.g., skill level and effectiveness of strategies used). It has been proposed that a major environmental determinant of workload is rapid (Huey and Wickens, 1993) and unexpected (Hockey et al., 2003) changes in task load. So ideally, more or fewer tasks should be delegated to automation dynamically. More automation would be introduced during moments of high workload, but as the level of workload eases more tasks would be returned to the operator, thereby keeping the operator in the loop without overloading the person. The key issue is the development of an implementation algorithm that could efficaciously adapt the level of automation to the operator's state of workload and situation awareness.

Parasuraman and Bryne (2003) describe several adaptation techniques that rely on different inputs to trigger an increase or decrease in the extent of automation in the system. One technique relies on physiological measures and another relies on performance measures. One obvious advantage of physiological measures is their continuous availability and noninvasive nature. A number of physiological measures have been evaluated for their potential to provide real-time assessment of workload. They include heart rate variability (e.g., Jorna, 1999), EEG (e.g., Gevins et al., 1998; Prinzel et al., 2000; Gevins and Smith, 2003), ERPs (Parasuraman, 1990; Kramer et al., 1996) and eye movement measures (Hilburn et al., 1997; Kramer and McCarley, 2003).

Although performance-based measures are not used as much as physiological-based measures, recent studies have demonstrated their potential promise as well. In one study, Kaber and Riley (1999) used a secondary monitoring task along with a target acquisition task. Adaptive computer aiding based on secondary-task performance was found to enhance primary-task performance. Notice that the tasks used here afford fairly continual performance measures, a property that not many performance-based measures possess. Also, as discussed above, for the secondary-task methodology to provide useful workload information, the time-shared tasks would need to be competing for some common resources, which of course could add to the workload. In another set of studies, Vidulich and colleagues propose that the Global Implicit Measure (GIM, described above) could be developed as a real-time situation awareness measurement that could guide effective automated pilot aiding based on real-time scoring of both continuous and discrete tasks.

4.2 Display Design

To the extent that excessive workload could reduce SA, any display that supports performance without incurring excessive workload would at least indirectly support SA as well (see, e.g., Previc, 2000). Wickens (1995) propose that displays that do not overtax working memory and selective attention are particularly attractive because SA depends heavily on these processes. Wickens (1995, 2002, 2003) discusses various display principles (e.g., proximity compatibility principle, visual momentum) that have been shown to support various types of performance (e.g., flight control as opposed to navigation) and display features (e.g., frame of reference) that would lend support to SA.

While display formats that facilitate information-processing support performance and thereby free up resources for SA maintenance, Wickens (2002) shows that display formats could also affect the product (type) of the SA. For example, a display with an egocentric frame of reference (an inside-out view with a fixed aircraft and a moving environment) provides better support for flight control, whereas an egocentric frame of reference (an outside-in view with a moving aircraft and fixed environment) provides better support for noticing hazards and general awareness of one's location. Wickens points out further that there are often trade-offs between alternative display formats. For example, whereas an integrated, ecological display generally provides better information about three-dimensional motion flow, a three-dimensional representation on a two-dimensional viewing surface tends to create ambiguity in locating objects in the environment. Such ambiguity is less of a problem in a two-dimensional display format. But it would take more than one two-dimensional display to present the same information in a three-dimensional display. It has been shown that it can be more cognitively demanding in trying to integrate information from two separate two-dimensional displays. The trade-off between promoting SA for objects in the environment and accomplishing other tasks at a lower workload level could only be resolved with regard to the specific goals or the priorities of competing goals of the system.

4.3 Training

Given the role that expertise plays in one's workload and situation awareness, there is great potential in training to support SA and to permit tasks to be accomplished with less resources at a lower level of workload. The issue is: What does one train for? That expertise is based largely on a large body of domain-specific knowledge suggests that a thorough understanding of the workings of the system would be helpful, particularly in nonroutine situations. Although expertise speeds up performance and experts generally perform at a high level under normal situations, their expertise is particularly useful in unexpected circumstances because of their ability to use their acquired knowledge to recognize and solve problems. The concern that operators trained on automated systems (which would be especially helpful for novices because of the presumed lower level of workload involved) might never acquire the needed knowledge and experience to build up their expertise is certainly a valid one until there exist automated systems with total reliability that would never operate outside a perfectly orchestrated environment. One possibility might be to provide some initial and refresher training in a nonautomated or less automated simulated system.

Although there exists in the literature a large body of training research that aims at accelerating the learning process and there is much evidence to support the advantages of not subjecting a trainee to an excessive level of workload, there are additional considerations when the goal is to build SA as well. One, it would be most useful to have some ideas about the knowledge structure that experts have so that the training program can build upon reinforcing this structure. After all, it is the structure and organization of information that support fast and accurate pattern recognition and information retrieval. Two, experts do not merely possess more knowledge, they are better at using it. This would suggest that training should extend to strategic training. Given the growing body of evidence to support that strategic task management (or executive control) is a higher-level generalizable skill, much of the strategic training could be accomplished with low-cost low-physical-fidelity simulated systems such as a complex computer game (see Haier et al., 1992; Gopher, 1993). The strategic training can be at odds with the goal of keeping the level of workload down while the operators are in training. However, research has shown that the eventual benefits outweigh the initial cost in mental workload. As desirable as it is to train to develop automatic processing that is characterized as fast, accurate, and attention free, this training strategy may have only limited utility in training operators who have to function within a dynamic complex system. This is because there would be relatively few task components in these systems that would have an invariant stimulus–response mapping (a requirement for automatic processing to be developed and applied).

All three research areas underscore the interdependence of the concepts of workload and SA. The design of any efficacious technical support or training program would need to take into account the interplay of the two. Any evaluation of the effectiveness of these supports would need to assess both the operator's workload and situation awareness in order to have a clear picture of their impact on system performance.

5 CONCLUSIONS

So the years of research into mental workload and situation awareness have been profitable. The research has developed a multitude of metric techniques, and although the results of different mental workload or SA assessment techniques sometimes show dissociations, they seem to fit within the theoretical constructs behind the measures. Workload is primarily a result of the limited attentional resources of humans, whereas SA is a cognitive phenomenon emerging from perception, memory, and expertise. The concepts of workload and SA have been studied extensively in the laboratory and have been transitioned successfully to real-world system evaluation. Indeed, workload and SA have been useful tools of system evaluators for years, and now they are providing vital guidance for shaping future automation, display, and training programs. In short, these concepts have been, and should continue to be, essential tools for human factors researchers and practitioners.

REFERENCES

Adams, R. J., and Ericsson, K. A. (1992), "Introduction to Cognitive Processes of Expert Pilots," DOT/FAA/RD-92/12, U.S. Department of Transportation, Federal Aviation Administration, Washington, DC.

Adams, M. J., Tenny, Y. J., and Pew, R. W. (1991), "Strategic Workload and the Cognitive Management of Advanced Multi-task System," State of the Art Report CSERIAC 91-6, Crew System Ergonomics Information Evaluation Center, Wright-Patterson Air Force Base, OH.

Adams, M. J., Tenny, Y. J., and Pew, R. W. (1995), "Situation Awareness and the Cognitive Management of Complex Systems," *Human Factors*, Vol. 37, pp. 85–104.

Alexander, A. L., Nygren, T. E., and Vidulich, M. A. (2000), "Examining the Relationship Between Mental Workload and Situation Awareness in a Simulated Air Combat Task," Technical Report AFRL-HE-WP-TR-2000-0094, Air Force Research Laboratory, Wright-Patterson Air Force Base, OH.

Allport, D. A., Antonis, B., and Reynolds, P. (1972), "On the Division of Attention: A Disproof of the Single Channel Hypothesis," *Quarterly Journal of Experimental Psychology*, Vol. 24, pp. 255–265.

Andre, A. D., Wickens, C. D., Moorman, L., and Boschelli, M. M. (1991), "Display Formatting Techniques for Improving Situation Awareness in the Aircraft Cockpit," *International Journal of Aviation Psychology*, Vol. 1, pp. 205–218.

Annett, J. (2002a), "Subjective Rating Scales: Science or Art?" *Ergonomics*, Vol. 45, pp. 966–987.

Annett, J. (2002b), "Subjective Rating Scales in Ergonomics: A Reply," *Ergonomics*, Vol. 45, pp. 1042–1046.

Backs, R. W. (1997), "Psychophysiological Aspects of Selective and Divided Attention during Continuous Manual Tracking," *Acta Psychologica*, Vol. 96, pp. 167–191.

Backs, R. W., and Boucsein, W. (2000), *Engineering Psychophysiology: Issues and Applications*, Lawrence Erlbaum Associates, Mahwah, NJ.

Baddeley, A. (1990), *Human Memory: Theory and Practice*, Allyn, & Bacon, Boston.

Bainbridge, L. (1987), "Ironies of Automation," in *New Technology and Human Error*, J. Rassmusen, K. Duncan, and J. Leplat, Eds., Wiley, New York, pp. 271–283.

Baker, S. C., Rogers, R. D., Owen, A. M., Frith, C. D., Dolan, R. J., Frackowiak, R. S., et al. (1996), "Neural Systems Engaged by Planning: A PET Study of the Tower of London Task," *Neuropsychologia*, Vol. 34, pp. 515–526.

Battiste, V., and Bortolussi, M. (1988), "Transport Pilot Workload: A Comparison of Two Subjective Techniques," in *Proceedings of the Human Factors Society 32nd Annual Meeting*, Human Factors Society, Santa Monica, CA, pp. 150–154.

Beatty, J. (1982), "Task-Evoked Pupillary Responses, Processing Load, and the Structure of Processing Resources," *Psychological Bulletin*, Vol. 91, pp. 276–292.

Bi, S., and Salvendy, G. (1994), "Analytical Modeling and Experiment Study of Human Workload in Scheduling of Advanced Manufacturing Systems," *International Journal of Human Factors in Manufacturing*, Vol. 4, pp. 205–235.

Biers, D. W., and Maseline, P. J. (1987), "Alternative Approaches to Analyzing SWAT Data," in *Proceedings of the Human Factors Society 31st Annual Meeting*, Human Factors Society, Santa Monica, CA, pp. 63–66.

Biers, D. W., and McInerney, P. (1988), "An Alternative to Measuring Subjective Workload: Use of SWAT Without the Card Sort," in *Proceedings of the Human Factors Society 32nd Annual Meeting*, Human Factors Society, Santa Monica, CA, pp. 1136–1139.

Billings, C. E. (1997), *Aviation Automation: The Search for a Human-Centered Approach*, Lawrence Erlbaum Associates, Mahwah, NJ.

Braver, T. S., Cohen, J. D., Nystrom, L. E., Jonides, J., Smith, E. E., and Noll, D. C. (1997), "A Parametric Study of Prefrontal Cortex Involvement in Human Working Memory," *NeuroImage*, Vol. 5, pp. 49–62.

Brickman, B. J., Hettinger, L. J., Roe, M. M., Stautberg, D., Vidulich, M. A., Haas, M. W., and Shaw, R. L. (1995), "An Assessment of Situation Awareness in an Air Combat Task: The Global Implicit Measure Approach," in *Experimental Analysis and Measurement of Situation Awareness*, D. J. Garland and M. R. Endsley, Eds., Embry-Riddle Aeronautical University Press, Daytona Beach, FL, pp. 339–344.

Brickman, B. J., Hettinger, L. J., Stautberg, D., Haas, M. W., Vidulich, M. A., and Shaw, R. L. (1999), "The Global Implicit Measurement of Situation Awareness: Implications for Design and Adaptive Interface Technologies," in *Automation Technology and Human Performance: Current Research and Trends*, M. W. Scerbo and M. Mouloua, Eds., Lawrence Erlbaum Associates, Mahwah, NJ, pp. 160–164.

Brookings, J., Wilson, G. F., and Swain, C. (1996), "Psychophysiological Responses to Changes in Workload During Simulated Air Traffic Control," *Biological Psychology*, Vol. 42, pp. 361–378.

Bryne, E. A. (1995), "Role of Volitional Effort in the Application of Psychophysiological Measures to Situation Awareness," in *Experimental Analysis and Measurement of Situation Awareness*, D. J. Garland and M. R. Endsley, Eds., Embry-Riddle Aeronautical University Press, Daytona Beach, FL, pp. 147–153.

Carpenter, P. A., Just, M. A., Keller, T. A., Eddy, W. F., and Thulborn, K. R. (1999), "Graded Functional Activation in the Visuospatial System with the Amount of Task Demand," *Journal of Cognitive Neuroscience*, Vol. 11, pp. 9–24.

Charness, N. (1989), "Age and Expertise: Responding to Talland's Challenge," in *Everyday Cognition in Adulthood and Late Life*, L. W. Poon, D. C. Rubin, and B. A. Wilson, Eds., Cambridge University Press, Cambridge. pp. 437–456.

Charness, N. (1995), "Expert Performance and Situation Awareness," in *Experimental Analysis and Measurement of Situation Awareness*, D. J. Garland and M. R. Endsley, Eds., Embry-Riddle Aeronautical University Press, Daytona Beach, FL, pp. 35–42.

Chase, W. G., and Ericsson, K. A. (1982), "Skill and Working Memory," in *The Psychology of Learning and Motivation*, Vol. 16, G. H. Bower, Ed., Academic Press, New York.

Chi, M., Glaser, R., and Farr, M., Eds. (1988), *On the Nature of Expertise*, Lawrence Erlbaum Associates, Mahwah, NJ.

Corbetta, M., Miezin, F. M., Dobmeyer, S., Shulman, G. L., and Petersen, S. E. (1990), "Attentional Modulation of Neural Processing of Shape, Color, and Velocity in Humans," *Science*, Vol. 248, pp. 1556–1559.

Davidson, R. J., Jackson, D. C., and Larson, C. L. (2000), "Human Electroencephalography," in *Handbook of Psychophysiology*, J. T. Cacioppo, L. G., Tassinary, and G. G., Berntson, Eds., Cambridge University Press, Cambridge, pp. 27–52.

D'Exposito, M., Zarahn, E., and Aguire, G. K. (1999), "Event-Related Functional MRI: Implications for Cognitive Psychology," *Psychological Bulletin*, Vol. 125, pp. 155–164.

Druckman, D., and Bjork, R. A., Eds. (1991), *In the Mind's Eye: Enhancing Human Performance*, National Academy Press, Washington, DC, Chap. 4.

Durso, F. T., and Gironlund, S. D. (1999). "Situation Awareness," in F. T. Durso, Ed., *Handbook of Applied Cognition*, Wiley, New York, pp. 283–314.

Durso, F. T., Truitt, T. R., Hackworth, C. A., Ohrt, D., Hamic, J. M., Crutchfield, J. M., and Manning, C. A. (1995), "Expertise and Chess: A Pilot Study Comparing Situation Awareness Methodologies," in *Experimental Analysis and Measurement of Situation Awareness*, D. J. Garland and M. R. Endsley, Eds., Embry-Riddle Aeronautical University Press, Daytona Beach, FL, pp. 189–195.

Eggemeier, F. T., Crabtree, M. S., Zingg, J. J., Reid, G. B., and Shingledecker, C. A. (1982), "Subjective Workload Assessment in a Memory Update Task," in *Proceedings of the 26th Human Factors Society Annual Meeting*, Human Factors Society, Santa Monica, CA, pp. 643–647.

Endsley, M. R. (1988), "Situation Awareness Global Assessment Technique (SAGAT)," in *Proceedings of the IEEE 1988 National Aerospace and Electronics Conference, NAECON 1988*, Vol. 3, Institute of Electrical and Electronics Engineers, New York, pp. 789–795.

Endsley, M. (1990), "A Methodology for the Objective Measurement of Pilot Situation Awareness," in *Situation Awareness in Aerospace Operations*, AGARD-CP-478, AGARD, Neuilly-sur-Seine, France, pp. 1–1 to 1–9.

Endsley, M. (1993), "Situation Awareness and Workload: Flip Sides of the Same Coin," in *Proceedings of the 7th International Symposium on Aviation Psychology*, Ohio State University, Columbus, OH, pp. 906–911.

Ericsson, K. A. (1996), "The Acquisition of Expert Performance: An Introduction to Some of the Issues," in *The Road to Excellence*, K. A. Ericsson, Ed., Lawrence Erlbaum Associates, Mahwah, NJ, pp. 1–50.

Ericsson, K. A., and Charness, N. (1997), "Cognitive and Developmental Factors in Expert Performance," in *Expertise in Context*, P. J. Feltovich, K. M. Ford, and R. R. Hoffman, Eds., MIT Press, Cambridge, MA, pp. 3–41.

Ericsson, K. A., and Delaney, P. F. (1998), "Working Memory and Expert Performance," in *Working Memory and Thinking*, R. H. Logie and K. J. Gilhooly, Eds., Lawrence Erlbaum Associates, Mahwah, NJ, pp. 93–114.

Ericsson, K. A., and Kintsch, W. (1995), "Long-Term Working Memory," *Psychological Review*, Vol. 105, pp. 211–245.

Ericsson, K. A., and Simon, H. A. (1993), *Protocol Analysis: Verbal Reports As Data*, rev. ed., MIT Press, Cambridge, MA.

Fallows, J. (2001a), *Free Flight: From Airline Hell to a New Age of Travel*, Public Affairs, New York.

Fallows, J. (2001b), "Freedom of the Skies," *The Atlantic*, Vol. 287, No. 6, June, pp. 37–49.

Fiorino, F. (2001), "Tech Trickle-Down Enhances GA Safety," *Aviation Week and Space Technology*, Vol. 155, No. 7, August 13, p. 52.

Fitts, P. M., and Jones, R. E. (1961a), "Analysis of Factors Contributing to 460 'Pilot-Error' Experiences in Operating Aircraft Controls," in *Selected Papers on Human Factors in the Design and Use of Control Systems*, H. W. Sinaiko, Ed., Dover, New York, pp. 332–358; reprinted from "Analysis of Factors Contributing to 460 'Pilot-Error' Experiences in Operating Aircraft Controls," Memorandum Report TSEAA-694-12, Aero Medical Laboratory, Air Materiel Command, Wright-Patterson Air Force Base, OH.

Fitts, P. M., and Jones, R. E. (1961b), "Psychological Aspects of Instrument Display," in *Selected Papers on Human Factors in the Design and Use of Control Systems*, H. W. Sinaiko, Ed., Dover, New York, pp. 359–396; reprinted from "Psychological Aspects of Instrument Display," Memorandum Report TSEAA-694-12A, Aero Medical Laboratory, Air Materiel Command, Wright-Patterson Air Force Base, OH.

Fournier, L. R., Wilson, G. F., and Swain, C. R. (1999), "Electrophysiological, Behavioral, and Subjective Indexes of Workload When Performing Multiple Tasks: Manipulations of Task Difficulty and Training," *International Journal of Psychophysiology*, Vol. 31, pp. 129–145.

Fowler, B. (1994), "P300 as a Measure of Workload During a Simulated Aircraft Landing Task," *Human Factors*, Vol. 36, pp. 670–683.

Friedman, A., and Polson, M. C. (1981), "Hemispheres as Independent Resources Systems: Limited-Capacity Processing and Cerebral Specialization," *Journal of Experimental Psychology*, Vol. 7, pp. 1031–1058.

Fruhstorfer, H., Langanke, P., Meinzer, K., Peter, J., and Plaff, U. (1977), "Neurophysiological Vigilance Indicators and Operational Analysis of a Train Vigilance Monitoring Device: A Laboratory and Field Study," *in Vigilance: Theory, Physiological Correlates, and Operational Performance*, R. R. Mackie, Ed., Plenum Press, New York, pp. 147–162.

Gawron, V. J. (2000), *Human Performance Measures Handbook*, Lawrence Erlbaum Associates, Mahwah, NJ.

Gevins, A., and Smith, M. E. (2003), "Neurophysiological Measures of Cognitive Workload During Human–Computer Interaction," *Theoretical Issues in Ergonomics Science*, Vol. 4, pp. 113–131.

Gevins, A., Leong, H., Du R., Smith, M. E., Le, J., DuRousseau, D., et al. (1995), "Towards Measurement of Brain Function in Operational Environments," *Biological Psychology*, Vol. 40, pp. 169–186.

Gevins, A., Smith, M. E., Leong, H., McEvoy, L., Whitfield, S., Du, R., et al. (1998), "Monitoring Working Memory During Computer-Based Tasks with EEG Pattern Recognition," *Human Factors*, Vol. 40, pp. 79–91.

Glaser, R. (1987), "Thoughts on Expertise," in *Cognitive Functioning and Social Structure over the Life Course*, C. Schooler and K. W. Schaie, Eds., Ablex, Norwood, NJ, pp. 81–94.

Glass, A. (1966), "Comparison of the Effect of Hard and Easy Mental Arithmetic upon Blocking of the Occipital Alpha Rhythm," *Quarterly Journal of Experimental Psychology*, Vol. 8, pp. 142–152.

Gopher, D. (1993), "The Skill of Attention Control: Acquisition and Execution of Attention Strategies," in *Attention and Performance XIV*, D. Meyer and S. Kornblum, Eds., Lawrence Erlbaum Associates, Mahwah, NJ, pp. 299–322.

Gopher, D. (1994), "Analysis and Measurement of Mental Load," in *International Perspectives on Psychological Science*, Vol. 2, *The State of the Art*, G. d'Ydewalle, P. Eelen, and P. Bertelson, Eds., Lawrence Erlbaum Associates, Hove, East Sussex, England, pp. 265–291.

Gopher, D., and Braune, R. (1984), "On the Psychophysics of Workload: Why Bother with Subjective Ratings?" *Human Factors*, Vol. 26, pp. 519–532.

Gopher, D., and Donchin, E. (1986), "Workload: An Examination of the Concept," in *Handbook of Perception and Human Performance*, Vol. II, *Cognitive Processes and Performance*, K. R. Boff, L. Kaufman, and J. P. Thomas, Eds., Wiley, New York, pp. 41–1 to 41–49.

Gopher, D., Brickner, M., and Navon, D. (1982), "Different Difficulty Manipulation Interacts Differently with Task Emphasis: Evidence for Multiple Resources," *Journal of Experimental Psychology: Human Perception and Performance*, Vol. 8, pp. 146–157.

Haier, R. J., Siegel, B., Tang, C., Abel, L., and Buchsbaum, M. (1992), "Intelligence and Changes in Regional Cerebral Glucose Metabolic Rate Following Learning," *Intelligence*, Vol. 16, pp. 415–426.

Hardiman, T. D., Dudfield, H. J., Selcon, S. J., and Smith, F. J. (1996), "Designing Novel Head-Up Displays to Promote Situational Awareness," in *Situation Awareness: Limitations and Enhancement in the Aviation Environment*, AGARD-CP-575, Advisory Group for Aerospace Research and Development, Neuilly-sur-Seine, France, pp. 15–1 to 15–7.

Hart, S. G., and Staveland, L. E. (1988), "Development of NASA-TLX (Task Load Index): Results of Empirical and Theoretical Research," in *Human Mental Workload*, P. A. Hancock and N. Meshkati, Eds., Elsevier, Amsterdam, The Netherlands, pp. 139–183.

Hendy, K. D. (1995), "Situation Awareness and Workload: Birds of a Feather?" in *AGARD Conference Proceedings 575: Situation Awareness: Limitations and Enhancements in the Aviation Environment, Advisory Group for Aerospace Research and Development*, Neuilly-sur-Seine, France, pp. 21–1 to 21-7.

Hendy, K. C., Hamilton, K. M., and Landry, L. N. (1993), "Measuring Subjective Workload: When Is One Scale Better Than Many?" *Human Factors*, Vol. 35, pp. 579–601.

Hilburn, B., Joma, P. G., Byrne, E. A., and Parasuraman, R. (1997). "The Effect of Adaptive Air Traffic Control (ATC) Decision Aiding on Controller Mental Workload," in M. Mouloua and J. Koonce, Eds., *Human–Automation Interaction*, Lawrence Erlbaum Associates, Mahwah, NJ, pp. 84–91.

Hill, S. G., Iavecchia, H. P., Byers, J. C., Bittner, A. C., Zaklad, A. L., and Christ, R. E. (1992), "Comparison of Four Subjective Workload Rating Scales," *Human Factors*, Vol. 34, pp. 429–439.

Hockey, G. R. J. (1997), "Compensatory Control in the Regulation of Human Performance Under Stress and High Workload: A Cognitive–Energetic Framework," *Biological Psychology*, Vol. 45, pp. 73–93.

Hockey, R., Gaillard, A., and Coles, M. (1986), *Energetics and Human Information Processing*, Matinus Nijhoff, Dordrecht, The Netherlands.

Hockey, R. J., Healey, A., Crawshaw, M., Wastell, D. G., and Sauer, J. (2003), "Cognitive Demands of Collision Avoidance in Simulated Ship Control," *Human Factors*, Vol. 45, pp. 252–265.

Hoffman, J., Houck, M., MacMilliam, F., Simons, R., and Oatman, L. (1985), "Event Related Potentials Elicited by Automatic Targets: A Dual-Task Analysis," *Journal of Experimental Psychology: Human Perception and Performance*, Vol. 11, pp. 50–61.

Horn, J. L., and Masunaga, H. (2000), "New Directions for Research into Aging and Intelligence: The Development of Expertise," in *Models of Cognitive Aging*, T. J. Perfect and E. A. Maylor, Eds., Oxford University Press, New York, pp. 125–159.

Huey, B. M., and Wickens, C. D., Eds. (1993), *Workload Transition*, National Academy Press, Washington, DC.

Isreal, J. B., Chesney, G. L., Wickens, C. D., and Donchin, E. (1980), "P300 and Tracking Difficulty: Evidence for Multiple Resources in Dual Task Performance," *Psychobiology*, Vol. 17, pp. 57–70.

Jha, P. D., Bisantz, A. M., and Parasuraman, R. (2003), "A Lens Model Analysis of Pilot and Controller Decision Making Under Free Flight," in *Proceedings of the 12th International Symposium on Aviation Psychology*, Wright State University, Dayton, OH, pp. 619–624.

Jorna, P. G. A. M. (1992), "Spectral Analysis of Heart Rate and Psychological State: A Review of Its Validity as a Workload Index," *Biological Psychology*, Vol. 34, pp. 237–258.

Jorna, P. G. A. M. (1999), "Automation and Free Flight: Exploring the Unexpected, in *Automation Technology and Human Performance*, M. W. Scerbo and M. Mouloua, Eds., Lawrence Erlbaum Associates, Mahway, NJ, pp. 107–111.

Just, M. A., and Carpenter, P. A. (1992), "A Capacity Theory of Comprehension: Individual Differences in

Working Memory," *Psychological Review*, Vol. 99, pp. 122–149.

Just, M. A., Carpenter, P. A., Keller, T. A., Eddy, W. F., and Thulborn, K. R. (1996), "Brain Activation Modulated by Sentence Comprehension," *Science*, Vol. 274, pp. 114–116.

Just, M. A., Carpenter, P. A., and Miyake, A. (2003), "Neuroindices of Cognitive Workload: Neuroimaging, Pupillometric and Event-Related Potential Studies of Brain Work," *Theoretical Issues in Ergonomics Science*, Vol. 4, pp. 56–88.

Kaber, D. B., and Riley, J. M. (1999), "Adaptive Automation of a Dynamic Control Task Based on Secondary Task Workload Measurement," *International Journal of Cognitive Ergonomics*, Vol. 3, pp. 169–187.

Kahneman, D. (1973), *Attention and Effort*, Prentice-Hall, Englewood Cliffs, NJ.

Kantowitz, B. H. (1992), "Selecting Measures for Human Factors Research," *Human Factors*, Vol. 34, pp. 387–398.

Kessel, C. J. and Wickens, C. D. (1982), "The Transfer of Failure Detection Skills Between Monitoring and Controlling Dynamic Systems," *Human Factors*, Vol. 24, pp. 49–60.

Kinsbourne, M., and Hicks, R. (1978), "Functional Cerebral Space," in *Attention and Performance VII*, J. Requin, Ed., Lawrence Erlbaum Associates, Mahwah, NJ, pp. 345–362.

Kramer, A. F., and McCarley, J. S. (2003), "Oculomotor Behaviour as a Reflection of Attention and Memory Processes: Neural Mechanisms and Applications to Human Factors," *Theoretical Issues in Ergonomics Science*, Vol. 4, pp. 21–55.

Kramer, A. F., and Weber, T. (2000), "Applications of Psychophysiology to Human Factors," in *Handbook of Psychophysiology*, 2nd ed., J. T. Cacioppo, L. G. Tassinary, and G. G. Berntson, Eds., Cambridge University Press, New York, pp. 794–814.

Kramer, A. F., Sirevaag, E., and Braune, R. (1987), "A Psychophysiological Assessment of Operator Workload During Simulated Flight Missions," *Human Factors*, Vol. 29, pp. 145–160.

Kramer, A. F., Trejo, L. J., and Humphrey, D. G. (1996), "Psychophysiological Measures of Workload: Potential Applications to Adaptively Automated Systems," in *Automation and Human Performance: Theory and Application*, R. Parasuraman and M. Mouloua, Eds., Lawrence Erlbaum Associates, Mahwah, NJ, pp. 137–161.

Lee, Y., and Liu, B. (2003), "Inflight Workload Assessment: Comparison of Subjective and Physiological Measurements," *Aviation, Space, and Environmental Medicine*, Vol. 74, pp. 1078–1084.

Lee, J. D., and Moray, N. (1994), "Trust, Self-Confidence, and Operators' Adaptation to Automation," *International Journal of Human–Computer Studies*, Vol. 40, pp. 153–184.

Leyman, E. L. C., Mirka, G. A., Kaber, D. B., and Sommerich, C. M. (2004), "Cervicobrachial Muscle Response to Cognitive Load in a Dual-Task Scenario," *Ergonomics*, Vol. 47, pp. 625–645.

Lidderdale, I. G. (1987), "Measurement of Aircrew Workload in Low-Level Flight," in *The Practical Assessment of Pilot Workload, AGARDograph* 282, A. Roscoe, Ed., *Advisory Group for Aerospace Research and Development*, Neuilly-sur-Seine, France, pp. 67–77.

Luximon, A., and Goonetilleke, R. S. (2001), "Simplified Subjective Workload Assessment Technique," *Ergonomics*, Vol. 44, pp. 229–243.

Lysaght, R. J., Hill, S. G., Dick, A. O., Plamondon, B. D., Linton, P. M., Wierwille, W. W., Zaklad, A. L., Bittner, A. C., Jr., and Wherry, R. J. Jr. (1989), "Operator Workload: Comprehensive Review and Evaluation of Workload Methodologies," ARI Technical Report 851, U.S. Army Research Institute for the Behavioral and Social Sciences, Alexandria, VA.

McCoy, T. M., Derrick, W. L., and Wickens, C. D. (1983). "Workload Assessment Metrics: What Happens When They Dissociate?" in *Proceedings of the 2nd Aerospace Behavioral Engineering Technology Conference, P-132*, Society of Automotive Engineers, Warrendale, PA, pp. 37–42.

Metzger, U., Rovira, E., and Parasuraman, R. (2003), "Controller Performance, Workload and Attention Allocation in Distributed Air–Ground Traffic Management: Effects of Mixed Equipage and Decision Support," in *Proceedings of the 12th International Symposium on Aviation Psychology*, Wright State University, Dayton, OH, pp. 803–809.

Moray, N., Ed. (1979), *Mental Workload: Theory and Measurement*, Plenum Press, New York.

Moray, N. (1986), "Monitoring Behavior and Supervisory Control," in *Handbook of Perception and Human Performance, Vol. II, Cognitive Processes and Performance*, K. R. Boff, L. Kaufman, and J. P. Thomas, Eds., Wiley, New York, pp. 40–1 to 40-51.

Muckler, F. A., and Seven, S. A. (1992), "Selecting Performance Measures: 'Objective' Versus 'Subjective' Measurement," *Human Factors*, Vol. 34, pp. 441–456.

Mulder, L. J. M. (1992), "Measurement and Analysis Methods of Heart Rate and Respiration for Use in Applied Environments," *Biological Psychology*, Vol. 34, pp. 205–236.

Navon, D., and Gopher, D. (1979), "On the Economy of the Human-Processing System," *Psychology Review*, Vol. 86, pp. 214–255.

Norman, D. A., and Bobrow, D. G. (1975), "On Data-Limited and Resource-Limited Processes," *Cognitive Psychology*, Vol. 7, pp. 44–64.

Norman, G. R., Brooks, L. R., and Allen, S. W. (1989), "Recall by Expert Medical Practitioners and Novices as a Record of Processing Attention," *Journal of Experimental Psychology: Learning, Memory, and Cognition*, Vol. 15, pp. 1166–1174.

Nygren, T. E. (1991), "Psychometric Properties of Subjective Workload Techniques: Implications for Their Use in the Assessment of Perceived Mental Workload," *Human Factors*, Vol. 33, pp. 17–33.

O'Donnell, R., and Eggemeier, F. T. (1986), "Workload Assessment Methodology," in *Handbook of Perception and Human Performance*, Vol. II, *Cognitive Processes and Performance*, K. R. Boff, L. Kaufman, and J. P. Thomas, Eds., Wiley, New York, pp. 42–1 to 42-49.

Ogden, G. D., Levine, J. M., and Eisner, E. J. (1970), "Measurement of Workload by Secondary Tasks," *Human Factors*, Vol. 21, pp. 529–548.

Parasuraman, R. (1990), "Event-Related Brain Potentials and Human Factors Research," in *Event-Related Brain Potentials: Basic Issues and Applications*, J. W. Rohrbaugh, R. Parasuraman, and R. Johnson, Jr., Eds., Oxford University Press, New York, pp. 279–299.

Parasuraman, R. (2003), "Neuroergonomics: Research and Practice," *Theoretical Issues in Ergonomics Science*, Vol. 4, pp. 5–20.

Parasuraman, R., and Bryne, E. A. (2003), "Automation and Human Performance in Aviation," in *Principles and Practice of Aviation Psychology*, P. S. Tsang and M. A. Vidulich, Eds., Lawrence Erlbaum Associates, Mahwah, NJ, pp. 311–356.

Parasuraman, R., and Caggiano, D. (2002), "Mental Workload," in *Encyclopedia of the Human Brain*, V. S. Ramachandran, Ed., Academic Press, San Diego, CA, Vol. 3, pp. 17–27.

Parks, R. W., Lowenstein, D. A., Dodrill, K. L., Barker, W. W., Yoshi, F., Chang, J. Y., et al. (1988), "Cerebral Metabolic Effects of a Verbal Fluency Test: A PET Scan Study," *Journal of Clinical and Experimental Neurophysiology*, Vol. 10, pp. 565–575.

Patel, V. L., and Groen, G. J. (1991), "The General and Specific Nature of Medical Expertise: A Critical Look," in K. A. Ericsson and J. Smith, Eds., *Toward a General Theory of Expertise* Cambridge University Press, Cambridge, MA, pp. 93–125.

Pérès, M., Van De Moortele, P. F., Pierard, C., Lehericy, S., Satabin, P., Le Bihan, D. et al. (2000), "Functional Magnetic Resonance Imaging of Mental Strategy in a Simulated Aviation Performance Task," *Aviation, Space, and Environmental Medicine*, Vol. 71, pp. 1218–1231.

Pew, R. A. (1994), "An Introduction to the Concept of Situation Awareness," in *Situational Awareness in Complex Systems*, R. D. Gilson, D. J. Garland, and J. M. Koonce, Eds., Embry-Riddle Aeronautical University Press, Daytona Beach, FL, pp. 17–23.

Posner, M. I., and DiGirolamo, G. J. (2000), "Cognitive Neuroscience: Origin and Promise," *Psychological Bulletin*, Vol. 126, pp. 873–889.

Previc, F. H. (2000), "Neuropsychological Guidelines for Aircraft Control Stations," *IEEE Engineering in Medicine and Biology*, March–April, pp. 81–88.

Prinzel, L. J., Freeman, F. G., Scerbo, M. W., Mikulka, P. J., and Pope, A. T. (2000), "A Closed-Loop System for Examining Psychophysiological Measures for Adaptive Automation, *International Journal of Aviation Psychology*, Vol. 10, pp. 393–410.

Pritchett, A. R., and Hansman, R. J. (2000), "Use of Testable Responses for Performance-Based Measurement of Situation Awareness," in *Situation Awareness Analysis and Measurement*, M. R. Endsley and D. J. Garland, Eds., Lawrence Erlbaum Associates, Mahwah, NJ, pp. 189–209.

Reid, G. B., and Nygren, T. E. (1988), "The Subjective Workload Assessment Technique: A Scaling Procedure for Measuring Mental Workload," in *Human Mental Workload*, P. A. Hancock and N. Meshkati, Eds., Elsevier, Amsterdam, The Netherlands, pp. 185–218.

Reingold, E., and Charness, N. (1995), "Perceptual Automaticity in Chess Skill: Evidence from Eye Movements," poster presented at the 36th Annual Meeting of the Psychonomic Society, Los Angeles.

Rothrock, L., Koubek, R., Fuchs, F., Haas, M., and Salvendy, G. (2002), "Review and Reappraisal of Adaptive Interfaces: Toward Biologically Inspired Paradigms," *Theoretical Issues in Ergonomics Science*, Vol. 3, pp. 47–84.

Rouse, W. B. (1977), "Human–Computer Interaction in Multitask Situations," *IEEE Transactions on Systems, Man and Cybernetics*, Vol. 7, pp. 293–300.

Rouse, W. B. (1988), "Adaptive Aiding for Human/Computer Control," *Human Factors*, Vol. 30, pp. 431–443.

Rubio, S., Díaz, E., Martín, J., and Puente, J. M. (2004), "Evaluation of Subjective Mental Workload: A Comparison of SWAT, NASA-TLX, and Workload Profile Methods," *Applied Psychology: An International Review*, Vol. 53, pp. 61–86.

Saaty, T. L. (1980), *The Analytic Hierarchy Process: Planning, Priority Setting, Resource Allocation*, McGraw-Hill, New York.

Salvendy, G. (2002), "Use of Subjective Rating Scores in Ergonomics Research and Practice," *Ergonomics*, Vol. 45, pp. 1005–1007.

Selcon, S. J., Taylor, R. M., and Koritsas, E. (1991), "Workload or Situational Awareness? TLX vs. SART for Aerospace System Design Evaluation," in *Proceedings of the Human Factors Society 35th Annual Meeting*, Human Factors Society, Santa Monica, CA, pp. 62–66.

Selcon, S. J., Taylor, R. M., and Shadrake, R. M. (1992), "Multi-modal Cockpit Warnings: Pictures, Words, or Both?" in *Proceedings of the Human Factors Society 36th Annual Meeting*, Human Factors Society, Santa Monica, CA, pp. 57–61.

Selcon, S. J., Hardiman, T. D., Croft, D. G., and Endsley, M. R. (1996), "A Test-Battery Approach to Cognitive Engineering: To Meta-measure or Not to Meta-measure, That Is the Question!" in *Proceedings of the Human Factors and Ergonomics Society 40th Annual Meeting*, Vol. 1, Human Factors and Ergonomics Society, Santa Monica, CA, pp. 228–232.

Shallice, T. (1988), *From Neuropsychology to Mental Structure*, Cambridge University Press, New York.

Shaw, R. L., Brickman, B. J., and Hettinger, L. J. (2004), "The Global Implicit Measure (GIM): Concept and Experience," in *Human Performance, Situation Awareness and Automation*, D. A. Vincenzi, M. Mouloua, and P. A. Hancock, Eds., Lawrence Erlbaum Associates, Mahwah, NJ, pp. 97–102.

Shingledecker, C. A. (1984), "A Task Battery for Applied Human Performance Assessment Research," Technical Report AFAMRL-TR-84-071, Air Force Aerospace Medical Research Laboratory, Wright-Patterson Air Force Base, OH.

Sirevaag, E. J., Kramer, A. F., Coles, M. G. H., and Donchin, E. (1989), "Resource Reciprocity: An Event-Related Brain Potentials Analysis," *Acta Psychologica*, Vol. 70, pp. 77–97.

Sirevaag, E. J., Kramer, A. F., Wickens, C. D., Reisweber, M., Strayer, D. L., and Grenell, J. (1993), "Assessment of Pilot Performance and Mental Workload in Rotary Wing Aircraft," *Ergonomics*, Vol. 36, pp. 1121–1140.

Sperandio, A. (1978), "The Regulation of Working Methods as a Function of Workload Among Air Traffic Controllers," *Ergonomics*, Vol. 21, pp. 367–390.

Staszewski, J. J. (1988), "Skilled Memory and Expert Mental Calculation," in *The Nature of Expertise*, M. T. H. Chi and R. Glaser, Eds., Lawrence Erlbaum Associates, Hillsdale, NJ, pp. 71–128.

Sterman, B., and Mann, C. (1995), "Concepts and Applications of EEG Analysis in Aviation Performance Evaluation," *Biological Psychology*, Vol. 40, pp. 115–130.

Stern, J. A., Walrath, L. C., and Goldstein, R. (1984), "The Endogenous Eyeblink," *Psychophysiology*, Vol. 21, pp. 22–33.

Stevens, S. S. (1957), "On the Psychophysical Law," *Psychological Review*, Vol. 64, pp. 153–181.

Stevens, S. S. (1966), "On the Operation Known as Judgment," *American Scientist*, Vol. 54, pp. 385–401.

Strater, L. D., Endsley, M. R., Pleban, R. J., and Matthews, M. D. (2001), "Measures of Platoon Leader Situation Awareness in Virtual Decision-Making Exercises," Technical Report 1770, U.S. Army Research Institute for the Behavioral and Social Sciences, Alexandria, VA.

Strayer, D., and Kramer, A. F. (1990), "Attentional Requirments of Automatic and Controlled Processes," *Journal of Experimental Psychology: Learning, Memory and Cognition*, Vol. 16, pp. 67–82.

Taylor, R. M. (1990), "Situational Awareness Rating Technique (SART): The Development of a Tool for Aircrew Systems Design," in *Situational Awareness in Aerospace Operations, AGARD-CP-478*, Advisory Group for Aerospace Research and Development, Neuilly-sur-Seine, France, pp. 3–1 to 3–17.

Tenny, Y. J., Adams, M. J., Pew, R. W. Huggins, A. W. F., and Rodgers, W. H. (1992), "A Principled Approach to the Measurement of Situation Awareness in Commercial Aviation," NASA Contractor Report 4451, Langley Research Center, National Aeronautics and Space Administration, Hampton, VA.

Toms, M. L., Cavallo, J. J., Cone, S. M., Moore, F. W., and Gonzalez-Garcia, A. (1997), "Pilot-in-the-Loop Evaluation of the Approach Procedures Expert System (APES)," Technical Report WL-TR-97-3102, Wright Laboratory, Wright-Patterson Air Force Base, OH.

Tsang, P. S., and Shaner, T. L. (1998), "Age, Attention, Expertise, and Time-Sharing Performance," *Psychology and Aging*, Vol. 13, pp. 323–347.

Tsang, P. S., and Velazquez, V. L. (1996), "Diagnosticity and Multidimensional Workload Ratings," *Ergonomics*, Vol. 39, pp. 358–381.

Tsang, P. S., and Vidulich, M. A. (1989), "Cognitive Demands of Automation in Aviation," in *Aviation Psychology*, R. S. Jensen, Ed., Gower Publishing, Aldershot, Hampshire, England, pp. 66–95.

Tsang, P. S., and Vidulich, M. A. (1994), "The Roles of Immediacy and Redundancy in Relative Subjective Workload Assessment," *Human Factors*, Vol. 36, pp. 503–513.

Tsang, P. S., and Wilson, G. (1997), "Mental Workload," in *Handbook of Human Factors and Ergonomics*, 2nd ed., G. Salvendy, Ed., Wiley, New York, pp. 417–449.

Tsang, P. S., Velazquez, V. L., and Vidulich, M. A. (1996), "The Viability of Resource Theories in Explaining Time-Sharing Performance," *Acta Psychologica*, Vol. 91, pp. 175–206.

Ververs, P. M., and Wickens, C. D. (2000), "Designing Head-Up Displays (HUDs) to Support Flight Path Guidance While Minimizing Effects of Cognitive Tunneling," in *Proceedings of the IEA 2000/HFES 2000 Congress*, Vol. 3, Human Factors and Ergonomics Society, Santa Monica, CA, pp. 3–45 to 3–48.

Vidulich, M. A. (1988), "The Cognitive Psychology of Subjective Mental Workload," in *Human Mental Workload*, P. Hancock and N. Meshkati, Eds., North-Holland, Amsterdam, The Netherlands, pp. 219–229.

Vidulich, M. (1989), "The Use of Judgment Matrices in Subjective Workload Assessment: The Subjective WORkload Dominance (SWORD) Technique," in *Proceedings of the Human Factors Society 33rd Annual Meeting*, Human Factors Society, Santa Monica, CA, pp. 1406–1410.

Vidulich, M. A. (1995), "The Role of Scope as a Feature of Situation Awareness Metrics," in *Experimental Analysis and Measurement of Situation Awareness*, D. J. Garland and M. R. Endsley, Eds., Embry-Riddle Aeronautical University Press, Daytona Beach, FL, pp. 69–74.

Vidulich, M. A. (2000), "Testing the Sensitivity of Situation Awareness Metrics in Interface Evaluations," in *Situation Awareness Analysis and Measurement*, M. R. Endsley and D. J. Garland, Eds., Lawrence Erlbaum Associates, Mahwah, NJ, pp. 227–246.

Vidulich, M. A. (2003), "Mental Workload and Situation Awareness: Essential Concepts for Aviation Psychology Practice," in *Principles and Practice of Aviation Psychology*, P. S. Tsang and M. A. Vidulich, Eds., Lawrence Erlbaum Associates, Mahwah, NJ, pp. 115–146.

Vidulich, M. A., and Bortolussi, M. R., (1988), "Speech Recognition in Advanced Rotorcraft: Using Speech Controls to Reduce Manual Control Overload," in *Proceedings of the American Helicopter Society National Specialist's Meeting: Automation Applications for Rotorcraft*, American Helicopter Society, Atlanta Southeast Region, Atlanta, GA, pp. 1–10.

Vidulich, M. A., and Hughes, E. R. (1991), "Testing a Subjective Metric of Situation Awareness," in *Proceedings of the Human Factors Society 35th Annual Meeting*, Human Factors Society, Santa Monica, CA, pp. 1307–1311.

Vidulich, M. A., and McMillan, G. (2000), "The Global Implicit Measure: Evaluation of Metrics for Cockpit Adaptation," in *Contemporary Ergonomics 2000*, P. T. McCabe, M. A. Hanson, and S. A. Robertson, Eds., Taylor & Francis, London, pp. 75–80.

Vidulich, M. A., and Tsang, P. S. (1987), "Absolute Magnitude Estimation and Relative Judgment Approaches to Subjective Workload Assessment," in *Proceedings of the Human Factors Society 31st Annual Meeting*, Human Factors Society, Santa Monica, CA, pp. 1057–1061.

Vidulich, M. A., and Wickens, C. D. (1986), "Causes of Dissociation Between Subjective Workload Measures and Performance: Caveats for the Use of Subjective Assessments," *Applied Ergonomics*, Vol. 17, pp. 291–296.

Vidulich, M., Dominguez, C., Vogel, E., and McMillan, G. (1994a), "Situation Awareness: Papers and Annotated Bibliography," Technical Report AL/CF-TR-1994-0085, Armstrong Laboratory, Wright-Patterson Air Force Base, OH.

Vidulich, M. A., Stratton, M., Crabtree, M., and Wilson, G. (1994b), "Performanced-Based and Physiological Measures of Situational Awareness," *Aviation, Space, and Environmental Medicine*, Vol. 65 (Suppl.), pp. A7–A12.

Vidulich, M. A., McCoy, A. L., and Crabtree, M. S. (1995), "The Effect of a Situation Display on Memory Probe and Subjective Situational Awareness Metrics," in *Proceedings of the 8th International Symposium on Aviation Psychology*, Ohio State University, Columbus, OH, pp. 765–768.

Watson, J. B. (1913), "Psychology as the Behaviorist Views It," *Psychological Review*, Vol. 20, pp. 158–177.

Welford, A. T. (1978), "Mental Workload as a Function of Demand, Capacity, Strategy and Skill," *Ergonomics*, Vol. 21, pp. 151–167.

Wickens, C. D. (1980), "The Structure of Attentional Resources," in *Attention and Performance VIII*, R. S. Nickerson, Ed., Lawrence Erlbaum Associates, Mahwah, NJ, pp. 239–257.

Wickens, C. D. (1984), "Processing Resources in Attention," in *Varieties of Attention*, R. Parasuraman and D. R. Davies, Eds., Academic Press, San Diego, CA, pp. 63–102.

Wickens, C. D. (1987), "Attention," in *Human Mental Workload*, P. A. Hancock and N. Meshkati, Eds., North-Holland, New York, pp. 29–80.

Wickens, C. D. (1990), "Applications of Event-Related Potential Research to Problems in Human Factors," in *Event-Related Brain Potentials: Basic Issues and Applications*, J. W. Rohrbaugh, R. Parasuraman, and R. Johnson, Jr., Eds., Oxford University Press, New York, pp. 301–309.

Wickens, C. D. (1995), "Situation Awareness: Impact of Automation and Display Technology," in *AGARD Conference Proceedings 575: Situation Awareness: Limitations and Enhancements in the Aviation Environment, Advisory Group for Aerospace Research and Development*, Neuilly-sur-Seine, France, pp. K2-1 to K2-13.

Wickens, C. (2001), "Workload and Situation Awareness," in *Stress, Workload, and Fatigue*, P. A. Hancock and P. A. Desmond, Eds., Lawrence Erlbaum Associates, Mahwah, NJ, pp. 443–450.

Wickens, C. D. (2002), "Multiple Resources and Performance Prediction," *Theoretical Issues in Ergonomics Science*, Vol. 3, pp. 159–177.

Wickens, C. D. (2003), "Aviation Displays," in *Principles and Practice of Aviation Psychology*, P. S. Tsang and M. A. Vidulich, Eds., Lawrence Erlbaum Associates, Mahwah, NJ, pp. 147–200.

Wickens, C. D. and Hollands, J. G. (2000), *Engineering Psychology and Human Performance*, 3rd ed., Charles E. Merrill, Columbus, OH.

Wickens, C. D., Kramer, A. F., Vanasse, L., and Donchin, E. (1983), "The Performance of Concurrent Tasks: A Psychophysiological Analysis of the Reciprocity of Information Processing Resources," *Science*, Vol. 221, pp. 1080–1082.

Wiener, E. L. (1988), "Cockpit Automation," in *Human Factors in Aviation*, E. L. Wiener and D. C. Nagel, Eds., Academic Press, San Diego, CA, pp. 433–461.

Wiener, E. L., and Curry, R. E. (1980), "Flight-Deck Automation: Promises and Problems," *Ergonomics*, Vol. 23, pp. 955–1011.

Wilson, G. F. (2000), "Strategies for Psychophysiological Assessment of Situation Awareness," in *Situation Awareness: Analysis and Measurement*, M. R. Endsley and D. J. Garland, Eds., Lawrence Erlbaum Associates, Mahwah, NJ, pp. 175–188.

Wilson, G. F. (2002a), "A Comparison of Three Cardiac Ambulatory Recorders Using Flight Data," *International Journal of Aviation Psychology*, Vol. 12, pp. 111–119.

Wilson, G. F. (2002b), "An Analysis of Mental Workload in Pilots During Flight Using Multiple Psychophysiological Measures," *International Journal of Aviation Psychology*, Vol. 12, pp. 3–18.

Wilson, G. F., and Eggemeier, F. T. (1991), "Physiological Measures of Workload in Multi-task Environments," in *Multiple-Task Performance*, D. Damos, Ed., Taylor & Francis, London, pp. 329–360.

Wynbrandt, J. (2001), "Seeing the Flightpath to the Future: Synthetic Vision and Heads-Up Displays Are on the Way to a Cockpit near You," *Pilot Journal*, Vol. 2, No. 2, Spring, pp. 30–34, 94.

Xie, B., and Salvendy, G. (2000a), "Prediction of Mental Workload in Single and Multiple Tasks Environments," *International Journal of Cognitive Ergonomics*, Vol. 4, pp. 213–242.

Xie, B., and Salvendy, G. (2000b), "Review and Reappraisal of Modelling and Predicting Mental Workload in Single- and Multi-task Environments," *Work and Stress*, Vol. 14, pp. 74–99.

Yeh, Y. Y., and Wickens, C. D. (1988), "Dissociation of Performance and Subjective Measures of Workload," *Human Factors*, Vol. 30, pp. 111–120.

CHAPTER 10

SOCIAL AND ORGANIZATIONAL FOUNDATIONS OF ERGONOMICS

Alvaro D. Taveira
University of Wisconsin–Whitewater
Whitewater, Wisconsin

Michael J. Smith
University of Wisconsin–Madison
Madison, Wisconsin

1 INTRODUCTION

The importance and influence of social context has been debated recurrently in attempts to define the ergonomics discipline. Wilson (2000) defines *ergonomics* "as the theoretical and fundamental understanding of human behavior and performance in purposeful interacting socio-technical systems, and the application of that understanding to the design of interactions in the context of real settings." Ergonomics focuses on interactive behavior, on the central role of human behavior in complex interacting systems. In this chapter we argue that these behaviors are deeply immersed in and cannot be separated from their social context.

When one examines the assumptions held by researchers and practitioners about the fundamentals of ergonomics, it is no surprise that the role of social context remains a focus of discussion. Daniellou (2001) describes some of these diverging tacit conceptualizations on *human nature, health*, and *work*, which reflect on distinct research models and practices and on different consequences to the public. He says that when referring to the *human nature*, ergonomists, may have in mind a biomechanical entity, an information-processing system, a subjective person with unique psychological traits, or a social creature member of groups that influence his or her behaviors and values. Similarly, the concept of *health* may be thought of as the absence of recognized pathologies, which would exclude any notion of discomfort, fatigue, or poverty, or could be defined in more comprehensive fashion as a general state of well-being (i.e., physical, mental, and social), or even interpreted as a process in a homeostatic state. Daniellou (2001) observes that although many ergonomists today embrace a definition of *work* that includes both its physical and cognitive aspects, other important distinctions remain. Work is often considered as the task and work environment requirements, a specific quantifiable definition of what is demanded from all workers to accomplish a given target. Another perspective on work is given by examination of the workload from the perspective of each worker, a vantage point that emphasizes the individual and collective strategies to manage the work situation in a dynamic fashion. Finally, some ergonomists focus on the ethical aspects of work, on its role on the shaping of individual and societal character. Daniellou reminds us of the challenge of producing work system models, which are necessarily reductions of the actual world, but ensuring that the reduction process does not eliminate the essential nature of the larger context.

Discussion of the role of social context in ergonomics can be described from two opposite perspectives. On one side there is a view of ergonomics as an applied science, a branch of engineering, a practice, even an art, which is deeply embedded in and influenced by a larger social context (Perrow, 1983; Oborne et al., 1993; Moray, 1994; Shackel, 1996; Badham, 2001). On the other side, we have authors who see ergonomics as applying the same standards of the natural and physical sciences, and as a *science* is able to produce overarching laws and principles that are able to predict human performance. It is a laboratory-based discipline in which contextual factors have minimal influence (e.g., Meister, 2001). We maintain that the latter position poses significant limitations for the relevance of ergonomics as a practice because it severely limits its usefulness and impact in the real world.

Ergonomics is a discipline with a set of fundamentals derived from the scientific method application, but these cannot be isolated from a vital applied (situational) component. We understand that the development of ergonomics is strongly affected by the social environment, and its permanence as a workable modern discipline depends on its sensitivity to economic, legal, technological, and organizational contexts.

Ergonomics is an integral part of a social enterprise to optimize work systems by improving productivity and quality. It contributes to the complete and uniform utilization of human and technical resources and to the reliability of production systems. It is central to the quest for better work conditions, by reducing exposure to physical hazards and the occurrence of fatigue. Ergonomic efforts play a fundamental role in the maintenance of workforce health by reducing the prevalence of musculoskeletal disorders and controlling occupational stress. It answers to market needs of increased product value, useful product features, and increased sales appeal. Ergonomics is ultimately instrumental to the advance and welfare of individuals, organizations, and broader society.

One could draw an analogy between the role of social forces in technological development and of these social forces in the evolution of the ergonomics discipline. Technological development has been seen as an important source of social change. In the traditional view, technology was considered to be the progressive application of science to solve immediate technical problems, and their release in the marketplace in turn had social impacts. In summary, "technological determinism" saw technology as a natural consequence of basic science progress, and in many circumstances, a cause of changes in society. Conversely, a more recent view of technological development sees it as part of a social system with specific needs and structure and not driven exclusively by basic science (Hughes, 1991). The latter perspective is also germane to the development of ergonomics (technology), in which social demands and structure marked its evolution. For example, the introduction of an (ergonomic) technological improvement such as stirrups allowed horse riders to hold themselves steady and freed both hands for work or warfare (Pacey, 1991). From a deterministic view to this innovation can be ascribed some of the success of nomadic groups in India and China in the thirteenth century and for the permanence of feudalism in Western Europe. From a systems perspective, on the other hand, this innovation can be conceived as part of a social system and one of the by-products of the critical reliance on horses by these nomadic groups, which later fit the needs of feudalism in Western Europe (White, 1962).

In the next section we examine the path of ergonomics to become an established discipline and how this course was shaped by social context.

2 HISTORICAL PERSPECTIVE

The development over time of the ergonomics discipline can be clearly traced to evolving societal needs. Ergonomics as a practice started at the moment the first human groups selected or shaped pieces of rock, wood, or bone to perform specific tasks necessary to their survival (K. Smith, 1965). Tools as extensions of their hands allowed these early human beings to act on their environment. Group survival relied on this ability, and the fit between hands and tools played an important role in the success of the human enterprise. The dawning of the discipline is therefore marked by our early ancestors' attempts to improve the fit between their hands and their rudimentary tools, to find shapes that increase efficiency, dexterity, and the capacity to perform their immediate tasks effectively.

The practice of ergonomics has been apparent as different civilizations across history used ergonomics methods in their projects. Historically, body motions have been a primary source of mechanical power, and its optimal utilization was paramount to the feasibility of many individual and collective projects. Handles, harnesses, and other fixtures that match human anatomy and tasks requirements allowed the use of human power for innumerous endeavors. Tools and later simple machines were created and gradually improved by enhancing their fit to users' and tasks' characteristics (K. Smith, 1965).

Although human work was early recognized as the source of all economic growth, workplaces have always been plagued with risks for workers. The groundbreaking work of Ramazzinni (1700) pointed out the connections between work and the development of illnesses and injuries. In contrast to the medical practitioners of his time who focused exclusively on the patients' symptoms, Ramazzini argued for the analysis of work and its environment as well as the workers themselves and their clinical symptoms. In his observations he noted that some common (*musculoskeletal*) disorders seem to follow from prolonged, strenuous, and unnatural physical motions and protracted stationary postures of the worker's body. Ramazzini's contribution to the development of modern ergonomics is enormous, in particular his argument for a systematic approach to work analysis.

Concomitantly with the increasing effectiveness of human work obtained through improved tools and machines, a growing specialization and intensification of work was also realized. The advent of the factory, a landmark in the advancement of production systems, was rather a social rather than a purely technical phenomenon. By housing a number of previously autonomous artisans under the same roof it was possible to obtain gains in productivity by dividing the work process into smaller, simpler tasks (A. Smith, 1776; Babbage, 1832). This process of narrowing down the types of tasks (and motions) performed by each worker while increasing the pace of the work activities had tremendous economic impact but was gradually associated with health issues for the workforce. This became more noticeable with the beginning of the Industrial Revolution, when significant changes in technology and work organization occurred and larger portions of the population engaged in industrial work.

Industrialization brought along an accelerated urbanization process and an increased access to manufactured goods. In the United States, in particular, this endeavor allowed for the tremendous growth in the supply and demand for capital and consumer goods. Some concern with the incidence of injuries and illnesses related to the industrialization process and its influence on the national economy (i.e., national human resources) led to some early attempts to improve safety in factories in the nineteenth century (Owen, 1816).

During the Industrial Revolution almost exclusive attention was paid to production output increase. This *productivity enhancement* thrust crystallized in the work of Taylor (1912) and his *scientific management* approach. Taylor compiled and put in practice one of the first systematic methods for work analysis and design, focusing on improving efficiency. His approach emphasized the analysis of the work situation and the identification of the "one best way" to perform a task. This technique called for the fragmentation of the work into small, simple tasks that could be performed by most people. It required analysis and design of tasks, development of specialized tools, determination of lead times, establishment of work pace, specification of breaks, and work schedules.

This effort to better utilize technology and personnel had profound effects on the interactions between humans and their work. It could also be argued that expansion of output through increased efficiency was socially preferable to long work hours (Stanney et al., 1997, 2001), but one should not lose sight of the fact that the higher economic gains can be derived from the former, which provides a more compelling justification for work intensification.

Taylor's work represented a crucial moment for the ergonomics discipline, although this development had a lopsided emphasis on the shop-floor efficiency aspects. Frank and Lilian Gilbreth (1917) further elaborate the study of human motions (e.g., micro-motions or Therbligs) and provided the basis for time-and-motion methods. Concurrently with the focus on the worker and workstation efficiency, the need to consider the larger organizational structure to optimize organizational output was also recognized. Published at almost the same time as Taylor's *The Principles of Scientific Management*, Fayol's *General and Industrial Management* (1916) defined the basics of work organization. Fayol's principles were in fact very complementary to Taylor's, as the latter focused on production aspects whereas the former addressed organizational issues.

Although preceded by Ransom Olds by almost 10 years, it is ultimately Henry Ford's assembly line that best epitomized this early twentieth-century work rationalization drive. In Ford's factories, tasks were narrowly defined, work cycles drastically shortened, and the pace of work accelerated. Time required per unit produced dropped from 13 hours to 1 hour, and costs of production were sharply reduced, making the Ford Model T an affordable and highly popular product (Konz and Johnson, 2004). Despite the high hourly rates offered, about double the prevalent rate of the time, turnover at Ford's facilities was very high in its early years. Concern with the workers' health and well-being at this juncture was limited at preventing acute injury and some initial concern with fatigue (Gilbreth and Gilbreth, 1920).

The institution of ergonomics as a modern discipline has often been associated with the World War II period, when human–complex systems interactions revealed themselves as a serious vulnerability (Christensen, 1987; Sanders and McCormick, 1993; Chapanis, 1999). The operation of increasingly sophisticated military equipment was being compromised recurrently by lack of consideration to the human–machine interface (Chapanis, 1999). Endeavors to address these issues in the early postwar period led to the development and utilization of applied anthropometrics, biomechanics, and the study of human perception in the context of display and control designs, among others. The ensuing Cold War and Space Race period provided a powerful impetus for rapid expansion of ergonomics in the defense and aerospace arenas, with a gradual transfer of that knowledge to civil applications.

Increasingly during the twentieth century, the scientific management approach became the dominant paradigm, with several nations reaping large economic benefits from its widespread adoption. In this period, ergonomics practice experienced a tremendous growth of its knowledge base at the task and human–machine interface levels. Over time, however, this *work rationalization* led to a prevalence of narrowly defined jobs, with typical cycle times reduced to a few seconds. Repetition rates soared, and work paces became very intense. This increased specialization also resulted in the need for large support staffs and in a general underutilization of abilities of the workforce. In fact, many jobs became mostly devoid of any meaningful content, characterized by fast and endless repetition of a small number of motions and by overwhelming monotony. This "dehumanization" of work was associated with low worker morale, increasing labor–management conflicts, as well as by the growing problem of musculoskeletal disorders (although not recognized until later).

It has been argued that much of the success of the scientific management approach was due to its suitability to social conditions in the early twentieth century, when large masses of uneducated and economically deprived workers were available (Lawler, 1986). Decades of industrialization and economic development changed this reality significantly, creating a much more educated and sophisticated workforce dealing with a much more complex work environment. These workers had higher expectations for their work content and environment than Taylor's approach could provide. In addition, there was growing evidence that this rigid work organization was preventing the adoption and effective utilization of emerging technologies (Trist, 1981). Some researchers and practitioners started to realize that this mechanicist work organization provided little opportunity for workers to learn and to contribute to the improvement of organizational performance (McGregor, 1960; Argyris, 1964; K. Smith, 1965). One could infer that during that period a micro-optimization of work (i.e., at the task and workstation level) was actually going against broader systems performance since it underutilized technological and human resources, exposed the workforce to physical and psychological stressors, burdened national health systems, and ultimately jeopardized national economic development (Hendrick, 1991).

3 BRINGING SOCIAL CONTEXT TO THE FOREFRONT OF WORK SYSTEMS

The social consequences of managerial decisions in the workplace were addressed systematically for the first time by Lewin starting in the 1920s (Marrow, 1969). In his research he emphasized the human(e) aspects of management, the need to reconcile scientific thinking and democratic values, and the possibility of actual labor–management cooperation. He revolutionized management (consulting) practice by advocating an *intervention orientation* to work analysis. Lewin believed that work situations should be studied in ways that make participants ready and committed to act (Weisbord, 2004). He was concerned with how workers find meaning in their work. This search for meaning led him to argue for the use of ethnographic methods in the study of work. He was a pioneer in proposing worker involvement in the analysis and (re)design of work and in defining job satisfaction as a central outcome for work systems. Lewin laid the seeds for the concept of interactive systems and highlighted the need for strategies to reduce resistance to change in organizations. Lewin could rightly be named as a precursor of the study of psychosocial factors in the workplace.

Lewin's research set the ground for the development of macroergonomics decades later. In particular, Lewin's work was a forerunner to participatory ergonomics, as he defined workers as legitimate knowledge producers and agents of change. Lewin understood that workers could "learn how to learn" and had a genuine interest in improving working conditions and by consequence, human–systems interactions. He saw work improvement not simply as a matter of shortening the workday but as one of increasing the human value of work (Weisbord, 2004).

Introduction of general systems theory by Bertalanffy (1950, 1968) and its subsequent application to several branches of science had profound implications for ergonomics theory and practice. The seminal work by researchers at the Tavistock Institute (Emery and Trist, 1960; Trist, 1981) and establishment of sociotechnical systems theory, especially, made clear the importance of the social context in work systems optimization.

The proponents of sociotechnical systems considered the consequences of organizational choices on technical, social, and environmental aspects. Their studies indicated that the prevailing work organization (i.e., scientific management) failed to fully utilized workers' skills, and was associated with high absenteeism and turnover, low productivity, and poor worker morale (Emery and Trist, 1960). They saw Taylorism as creating an imbalance between the social and technical components of the work system and leading to diminishing returns on technical investments. Sociotechnical researchers advocated the reversal of what they saw as extreme job specialization, which often led to underutilized work crews and equipment and to high economic and social costs. They argued for organizational structures based on flexible, multiskilled workers (i.e., knowledgeable in multiple aspects of the work system), operating within self-regulated or semiautonomous groups.

According to sociotechnical principles, high performance of the technical component at the expense of the social component would lead to dehumanization of work, possibly to a situation where some segments of society would enjoy the economic benefits of work, and another (larger) portion would bear its costs (i.e., work itself!). The opposite situation, where the social component becomes preponderant, would be equally troubling because it would lead to system output reduction, with negative effects on organizational and national economies. In summary, a total system output decline could be expected in both cases of suboptimization.

Sociotechnical systems (STS) emphasized the match between social needs and technology, or more specifically, improvement in the interaction between work systems' technical and social components. Sociotechnical systems focused on the choice of technologies suitable to the social and psychological needs of humans. The path to improved integration was pursued, with the maximization of worker well-being as its primary system optimization criteria. An alternative path to the same end was proposed decades later in the macroergonomic approach, which posited the quick and effective adoption of new technology as the crux of organizational survival and the primary optimization criteria (Hendrick, 1991, 1997). The macroergonomic approach urged organizations to implement job and organization designs that increased the chances of successful technology implementation by utilizing its human resources fully. Although providing different rationales, the two propositions have similar visions (i.e., effective and healthy work systems), achieved through analogous work organization designs.

A major contribution of sociotechnical systems to an understanding of work systems (organizations) is that they are intrinsically open systems which to survive need to exchange information, energy, or materials with their environment (Bertalanffy, 1950). A second major contribution was the recognition of fast and unpredictable changes in the environmental contexts themselves. Emery and Trist (1965) describe four different types of causal texture, which refer to different possible dispositions of interconnected events producing conditions that call for very diverse organizational responses. The *placid, randomized environment* is characterized by relatively stable and randomly distributed *rewards* and *penalties*. This implies a strategy that is indistinguishable from tactics, where the entity or organization attempts to do its best in a purely local basis. Organizations survive adaptively and tend to remain small and independent in this environment. The *placid, clustered environment* implies some type of knowable (i.e., nonrandom) distribution of rewards and penalties. This environment justifies the adoption of a strategy distinct from tactics, and rewards organizations that anticipate these clusters. Successful organizations in this context tend to expand and to become hierarchical. The *disturbed-reactive environment* indicates the existence of multiple organizations competing for the same resources and trying to move, in the long term, to the same place in the environment. Competitive advantage is the dominant strategy in this situation. Finally, *turbulent fields* are marked by accelerating changes, increasing uncertainty, and unpredictable connections of environmental components (Weisbord, 2004). In Emery and Trist's (1965) words: "The ground is in motion," which creates a situation of *relevant uncertainty*. This circumstance demands from organizations the development of new forms of data collection, problem solving, and planning.

A significant offshoot of the sociotechnical theory was the *industrial democracy* movement, which had considerable influence in organizational and governmental policies and labor–management relations in several Scandinavian and Western European countries in the 1970s and 1980s. Industrial democracy proponents argue that workers should be entitled to a significant voice in decisions affecting the companies in which they work. The term has several connotations, but essentially it advocated union rights to representation in the boards of directors of large companies. It has also been used to describe various forms of consultation, employee involvement, and participation (Broedling, 1977).

A related concept, *codetermination*, achieved prominent status in Germany in the 1970s, when legislation expanded the right of workers to participate in management decisions. Codetermination typically implied having an organization's board of directors with its composition made up to 50% of employee representatives. In the German experience, codetermination applies to both the plant and industry levels, and the proportion of worker representation in executive boards varies depending on the business size and type. Whereas industrial democracy urged increased workforce participation across the entire organization, codetermination concerned primarily worker influence at the top of the organization.

Sociotechnical theory deeply influenced another important societal and organizational initiative, the *quality of work life* (QWL) movement. The QWL concept has its roots in the somewhat widespread dissatisfaction with the organization of work in the late 1960s and early 1970s. Taylorism inflexibility, extreme task specialization, and the resulting low morale were identified as the primary sources of these issues. A number of models of job and organization design attempting to improve utilization of worker initiative and to reduce job dissatisfaction were developed and put in practice, particularly in the Scandinavian countries. QWL initiatives called for organizational changes involving increased task variety and responsibility, making the case for increasing worker participation as a potent intrinsic motivator.

Of particular relevance was the experimental assembly plant established in 1989 by Volvo in Uddevalla, Sweden. The project was an attempt to apply sociotechnical principles to a mass production facility on a wide scale. In the context of a very tight job market, this initiative tried to make jobs and the work environment more attractive to workers by increasing autonomy and group cohesiveness and by recovering some of the meaning lost by extreme task fragmentation. Despite tremendous expectations from company, labor, and sociotechnical researchers, the plant was closed in 1993 because of inferior performance. Although some debate remains about the causes of the plant closure, one could infer that increased pressure for short-term financial returns, larger labor availability, globalization, and fierce international competition are primary contributing factors. Most of all, the end of the Uddevalla plant highlighted the difficulties of achieving acceptable compromises or, preferably, converging strategies between enhanced quality of work life and organizational competitiveness (Huzzard, 2003). The challenge of competitiveness remains an important one for the viability of QWL. In Huzzard's words: "Despite the evidence that firms can reap considerable performance advantages through attempts at increasing the quality of working life through greater job enlargement, job enrichment, competence development and participation, there is also considerable evidence that some firms are actually eschewing such approaches in deference to short-run pressure for immediate results on the 'bottom-line' of the profit and loss account and rapid increases in stock market valuation."

The sociotechnical theory helped to consolidate the objectives of ergonomics by virtue of its *joint-optimization principle*, which establishes the possibility and the need for work systems to achieve concurrently high social and technical performance. In other words, it advocated that through a work organization focused on human physical and social needs (i.e., free of hazards, egalitarian, team based, semiautonomous) it would be possible to attain high productivity, quality, and reliability while reaching high levels of job satisfaction, organizational cohesiveness, and mental and physical well-being. The affinities between ergonomics and the sociotechnical theory are quite evident, as the former is a keystone to the feasibility of joint optimization. In fact, contemporary ergonomics interventions are deeply influenced by sociotechnical theory calling for worker involvement, delegation of operational decisions, and a decentralized management structure (Cohen et al., 1997; Chengalur et al., 2004; Konz and Johnson, 2004).

The sociotechnical approach inspired a number of successful work systems performance improvement initiatives over recent decades. These initiatives often accentuated the need for decision-making decentralization and the enlargement of workers' understanding of the entire system and of the technical and economic effects of their actions on it. These approaches advocated jobs with increased amounts of variety, control, feedback, and opportunity for growth (Emery and Trist, 1960; Hackman and Oldham, 1980). Workers needed to recover their perception of the whole and to be encouraged to regulate their own affairs in collaboration with their peers and supervisors. These STS-inspired initiatives aimed at eliminating the inefficiencies and bottlenecks created by decades of scientific management practice, which shaped an inflexible and highly compartmentalized workforce.

Not all STS-inspired programs were equally successful, and many of them were withdrawn after some years. Proposals that relied heavily on worker participation but did not provide adequate support and guidance were found to be ineffective and stressful to workers. Fittingly, Kanter (1983) observed that many of the participatory management failures are caused by too much emphasis on *participation* and too little on *management*. Similarly, in some cases workers found themselves overwhelmed by excessive job variety. Studies have also shown that job control is not always sufficient to attenuate workload effects (Jackson, 1989). Finally, experiments with the concept of industrial democracy in Sweden, Norway, and Germany, where labor representatives participated in executive boards, produced little in the way of increasing rank-and-file workers' involvement and improving the social meaning of their jobs, and did not result in competitive advantage in productivity or quality, at least not in the short term.

4 ERGONOMICS AND THE ORGANIZATION

In the 1980s a number of practitioners and researchers started to realize that in some circumstances, what were considered superior ergonomic solutions at the workstation level failed to produce relevant outcomes at the organizational level (Dray, 1985; Carlopio, 1986; Brown, 1990; Hendrick, 1991, 1997). These failures were attributed to a narrow scope of analysis typical of ergonomics interventions, which neglected to consider the overall organizational structure (DeGreene, 1986). A new area within the ergonomics discipline was conceived, under the term *macroergonomics* (Hendrick, 1986). According to Hendrick (1991, 1997), macroergonomics emphasizes the interface between organizational design and technology with the purpose of optimizing work systems performance. Hendrick sees macroergonomics as a top-down sociotechnical approach to the design of work systems. Imada (1991) states that this approach recognizes that organizational, political, social, and psychological factors of work have the same influence on the adoption of new concepts as that of the merits of the concepts themselves. Finally, Brown (1991) conceives macroergonomics addressing the interaction between the organizational and psychosocial contexts of a work system with, and emphasis on, the fit between organizational design and technology. Effective absorption of technology is at the forefront of macroergonomic endeavors.

Although having most of its concepts derived from sociotechnical theory, macroergonomics diverges from the former in some significant aspects. Whereas macroergonomics is a top-down approach, sociotechnical systems embraces a bottom-up approach, where the workstation is the building block for organizational design. Macroergonomics sees macro-level decisions as a prerequisite to micro-level decisions (Brown, 1991), in sharp contrast with STS, which affirms that joint optimization must first be constructed into the primary work system, otherwise it would not become a property of the organization as whole (Trist, 1981).

Hendrick (1991) defines organizational design around three concepts: complexity, formalization, and centralization. *Complexity* refers to an organization's degree of internal differentiation and extent of use of integration and coordination mechanisms. Internal differential is further elaborated into horizontal differentiation, which refers to job specialization and departmentalization; vertical differentiation, which relates the number of hierarchical levels in the organization; and spatial dispersion. *Formalization* conveys the reliance on written rules and procedures. *Centralization* refers to the degree of dispersion of decision-making authority. Although some given combinations of these structural elements seem to be suitable to some types of organizations pursuing specific goals, a number of possible interactions may have results difficult to predict.

Hendrick (1997) points out three work system design practices that characteristically undermine ergonomic efforts. The first problematic practice relates to a situation where technology (hardware or software) is taken as a given and user consideration comes as an afterthought. These are efforts that typically overlook motivational and psychosocial aspects of users. The second practice is also related to the overriding attention to the technical component, where technical feasibility is the only criterion for function allocation. In this situation, optimization of the technical component forces on the social component the left-over functions (DeGreene, 1986). In other words, users are forced to accommodate the remaining tasks, frequently resulting in work situations void of any human value. Finally, there is failure to consider adequately the four elements (subsystems) of sociotechnical systems: personnel, technology, organizational structure, and external environmental.

Methods employed in macroergonomics are numerous and are typically embedded in a four-step process: analysis/assessment, design, implementation, and evaluation (Brown, 1991). For a review of methods in macroergonomics, see Hendrick and Kleiner (2001) and Hendrick (2002).

5 PARTICIPATORY ERGONOMICS

As discussed previously, sociotechnical theory provides several arguments and examples supporting worker participation, particularly through teamwork, as an effective work system design strategy. The germinal study, which gave the essential evidence for development of the sociotechnical field, was a group experience observed in the English mining industry during the 1950s. In this new form of work organization a set of relatively autonomous work groups performed a complete collection of tasks, interchanging roles and shifts and regulating their affairs with a minimum of supervision. This experience was considered a way of recovering group cohesion and self-regulation concomitantly with a higher level of mechanization (Trist, 1981). The group had the power to participate in decisions concerning work arrangements, and these changes resulted in increased cooperation between task groups, personal commitment from participants, reduction in absenteeism, and fewer accidents. From the sociotechnical perspective, participation may permit ordering and utilization of worker-accumulated experience; it validates and legitimizes this experiential knowledge.

Another strong defense of employee participation is made by Sashkin (1984), arguing that "participatory management has positive effects on performance, productivity, and employee satisfaction because it fulfills three basic human needs: increased autonomy, increased meaningfulness, and decreased isolation." Perhaps the most original and polemic of the Sashkin's contributions to the subject is the statement that participation is an "ethical imperative." His reasoning is that since basic human needs are met by participation, denial of the process will produce psychological and physical harm to the workers. Research on worker participation is extensive and started as early as the 1940s (Coch and French, 1948) and is punctuated by controversial and often by ideological debate (Locke and Schweiger, 1979; Sashkin, 1984; Cotton et al., 1988; Leana et al., 1990). For an extensive review on participation and teamwork, refer to Medsker and Campion (2001) and Sainfort et al. (2001).

Participatory ergonomics can be understood, with some caution, as a spin-off of another STS-inspired approach, total quality management (TQM). TQM relies strongly on teamwork for problem solving and change implementation related to quality and production issues (Dean and Bowen, 1994). Over time, some of those teams also started to focus on working conditions. The term *participatory ergonomics* (PE) originated from discussions among Noro, Kogi, and Imada in the 1980s (Noro, 1999). It assumes that ergonomics is bounded by the degree to which people are involved in conducting its practice. According to Imada (1991), PE requires users (the real beneficiaries of ergonomics) to be involved directly in developing and implementing ergonomics. Wilson (1995) defines PE as "the involvement of people in planning and controlling a significant amount of their own work activities, with sufficient knowledge and power to influence both processes and outcomes in order to achieve desirable goals." PE is identified with the use of participative techniques in the analysis and implementation of ergonomics solutions (Wilson and Haines, 1997) and is recognized as an approach to disseminate ergonomics knowledge throughout the organization (Noro, 1991).

Imada (1991) points out three major arguments in support of worker involvement in ergonomics. First, because ergonomics is an intuitive science, which in many cases simply organizes knowledge that workers are already using, it can validate workers' accumulated experience. Second, people are more likely to support and adopt solutions for which they feel responsible. Involving users and workers in the ergonomic process has the potential to transform them into makers and supporters of the process rather than passive recipients. Finally, developing and implementing technology enables workers to modify and correct problems continuously.

Participatory ergonomics sees end users' contributions as indispensable elements of its scientific methodology. It stresses the validity of simple tools and workers' experience in problem solution and denies that these characteristics result in nonscientific outcomes (Imada, 1991; Taveira and Hajnal, 1997). In most situations, employees or end users are in the best position to identify the strengths and weaknesses of work situations. Their involvement in the analysis and redesign of their workplace can lead to better designs as well as to increase their and the company's knowledge of the process.

Participatory ergonomics was also conceived as an approach to enhance the human–work systems fit. Work environments have become highly complex, often beyond the capacity of individual workers. This mismatch between workers' capabilities and work systems requirements has been pointed out as an important factor in organizational failures (Weick, 1987; Reason, 1990, 1997; Reason and Hobbs, 2003). A possible strategy to address this imbalance is to pool workers' abilities through group teamwork, making them collectively more sophisticated (Imada, 1991). Other beneficial outcomes of PE include increased commitment to changes (Lawler, 1986; Imada and Robertson,

1987), increased learning experiences (i.e., reduced training costs) and improved performance (Wilson and Grey Taylor, 1995), and increased job control and skills (Karasek and Theorell, 1990).

6 ERGONOMICS AND QUALITY IMPROVEMENT EFFORTS

A growing amount of attention among ergonomics scholars and practitioners concerns whether and how ergonomics is affected by organizational transformations. In particular, the integration of ergonomics and quality management programs has been discussed extensively (Drury, 1997, 1999; Eklund, 1997, 1999; Axelsson et al., 1999; Taveira et al., 2003). It seems clear, at least in principle, that the two are related and interact in a variety of applications, such as inspection, process control, safety, and environmental design (e.g., Drury, 1978; Eklund, 1995; Rahimi, 1995; Stuebbe and Houshmand, 1995; Warrack and Sinha, 1999; Axelsson, 2000). There is some consensus that the form of this relationship is one in which "good ergonomics" (e.g., appropriate workstation, job, and organization design) leads to improved human performance and reduced risk of injury, which in turn leads to improved product and process quality. Eklund (1995), for example, found that the odds of having quality deficiencies among ergonomically demanding tasks at a Swedish car assembly plant were 2.95 times more likely than for other tasks.

TQM practice seems to enlarge the employees' role by increasing their control over their activities and by providing them with information and skills and the opportunity to apply them. On the other hand, this positive evaluation of TQM is not shared unanimously. Some believe that TQM is essentially a new package for the old Taylorism, with the difference that *work rationalization* is being made by the employees themselves (Parker and Slaughter, 1994). Lawler et al. (1992), commenting on work simplification in the context of TQM, noticed that even though this emphasis seems to work at cross purposes with job enrichment, most TQM concepts have not referred methodically to the issue of how jobs should be designed. Lawler and his colleagues propose that work process simplification can otherwise occur concurrently with the creation of jobs that have motivating characteristics (i.e., significance, autonomy providing, feedback, etc.).

Implementation by the manufacturing industry of flexible production systems (e.g., just-in-time, lean manufacturing) has created a renewed demand for ergonomics. Flexible production systems are based on the assumption that regarding workers as repetitive mechanical devices will not provide any competitive advantage and that only skilled and motivated workers are able to add value to production. They embrace the concept of continuous improvement, although without totally rejecting Taylor's *one best way* (Reeves and Bednar, 1994). These systems see line workers as capable of performing most functions better than specialists, allowing for a lean organization stripped from most personnel redundancies. These systems require that each step of the fabrication process be conducted perfectly every time, thus reducing the need for buffer stocks and producing a higher-quality end product.

Information technology has been connected to changes in internal organization; it facilitates the work of multidisciplinary teams whose members work together from the start of a job to its completion. Information systems make it efficient to push decision making down in the organization—to the teams that perform an organization's work. Efficient operations in the modern workplace call for a more equal distribution of knowledge, authority, and responsibility.

Social context, along with characteristics of the production process (i.e., type of industry), have a central role in the production philosophy embraced by organizations. The choice of a particular production philosophy has in turn a substantial impact on the work organization, technology, human resources practices and policies, organizational efficiency, and in the general quality of work life. Eklund and Berggren (2001) identify four generic production philosophies: Taylorism, sociotechnical systems, flexible production systems, and modern craft. Most organizations use some mix of these approaches and rarely does one find an organization strictly employing a specific philosophy. Taylorism, Fordism, or scientific management is characterized by high levels of standardization, job and equipment specialization, close control over worker activities, machine-paced work, and short cycle times. This production philosophy allowed for tremendous efficiency gains but produced poor work conditions, marked by fatigue, monotony, and poor utilization of workers capabilities. The second approach, sociotechnical systems (STS), addressed some of the Taylorism shortcomings by expanding job content in terms of its variety and meaningfulness, by reducing the rigidity of the production flow by adopting alternative production layouts and using buffers, and by transferring to workers some decision-making power. Although lessening some of the ergonomics issues associated with Taylorism, STS did not effectively addressed some of workload-related issues, had difficulty in maintaining high levels of output, and the use of buffers came at the cost of additional work in process. The third philosophy of production is associated with the quality movement discussed above and is characterized by flexible equipment with quick setup times, a multiskilled workforce able to perform a variety of tasks, teamwork, and continuous improvement. Finally, modern craft is typically limited to large, specialized capital equipment production involving highly skilled workers and high-precision tools in very long work cycles. In many respects it is very representative of pre–Industrial Revolution work organization.

7 SOCIALLY CENTERED DESIGN

As defined by Stanney et al. (1997), socially centered design is an approach concerned with system design variables that "reflect socially constructed and maintained world views which both drive and constrain how people can and will react to and interact with a system or its elements." It is a strategy aimed at filling the gap between system (outcome)- centered approaches and user-centered approaches to job and organization design. In other words, it is conceived of as a bridge between a macro- and a micro-orientation to ergonomics.

Although conceding that the macroergonomic approach attempts to alleviate some of the issues associated with widespread adoption of scientific management, these authors argue that it does not sufficiently address group and intergroup interactions. In fact, the authors tend to group Taylorism and macroergonomics at one end of the technical versus social spectrum, as both consider the worker or user as a resource to be optimized (Stanney et al., 1997). User-centered design is positioned at the other end of the spectrum, as it regards the users' abilities, limitations, and preferences as the key design objectives (i.e., enabling user control as the primary goal). User-centered methodologies emphasize human error avoidance but generally ignore the social context of work, focusing instead on the individual and his or her workstation.

Socially centered design focus on real-time interactions among people in the context of their work practices. In the Lewin tradition, the use of naturalistic methods are advocated to identify informal skills employed in the work process. It is assumed that the identification of these "unofficial" organizational and social factors adds value to system design (Grudin, 1990). It sees workers' roles and responsibilities as characterized dynamically by ever-changing local interactions. Similarly, the concept of optimization is local and contingent, being defined by the system context. Socially centered design looks at artifacts (i.e., objects and systems) as solutions to problems. Considering that these problems are situated in a social context, one could expect that different groups in different situations may use the same artifacts in dissimilar ways. It leads to the conclusion that artifacts must be examined from both a physical and a social perspective.

The methodologies for data acquisition advocated by this approach focus on group processes related to artifact use. These methods take into consideration the influence of the social work environment on system design effectiveness. Socially centered design sees everyday work interactions as relevant to effective system design and that these factors must be extracted through contextual inquiries (Stanney et al., 1997). The most common techniques include group studies and ethnographic studies (Jirotka and Goguen, 1994). These methods can potentially be time and resource consuming, and transference of acquired knowledge to other situations may require further refinement, since these are locally generated explanations.

So far we have examined some of the critical aspects of the social foundations of ergonomics. The views described were rooted primarily in the seminal work of Kurt Lewin and of researchers at the Tavistock Institute, particularly Eric Trist and Fred Emery. Next we review approaches that although related to the same tradition, put a stronger emphasis on how the social environment affects workers' mental and physical health.

8 PSYCHOSOCIAL FOUNDATIONS

For more than a century, medical practitioners have known that social, psychological, and stress factors can influence the course of recovery from disease. During World Wars I and II, much attention was placed on how social, physical, and psychological stress affected soldier, sailor, and pilot performance, motivation, and health. Then, beginning in the 1950s and continuing until today, much attention has been paid to the relationship between job stress and employee ill health (Caplan et al., 1975; Cooper and Marshall, 1976; M. Smith, 1987a; NIOSH, 1992; M. Smith et al., 1992; Kalimo et al., 1997). What has emerged is an understanding that "psychosocial" attributes of the environment can influence human behavior, motivation, performance, and health, and that psychosocial factors have implications for human factors design considerations. Several conceptualizations have been proposed over the years to explain the human factors aspects of psychosocial factors and ways to deal with them. Next we discuss some foundational considerations for social influences on human factors.

9 JOB STRESS AND THE PSYCHOSOCIAL ENVIRONMENT

Selye (1956) defined stress as a biological process created by social influences. The environment (physical and psychosocial) produces stressors that lead to adaptive bodily reactions by mobilizing energy, disease fighting, and survival responses. The individual's reactions to the environment are automatic survival responses (autonomic nervous system) and can be mediated by cognitive processes that are built on social learning. In Selye's concept, an organism undergoes three stages leading to illness. In the first, the state of *alarm*, the body mobilizes biological defenses to resist the assault of an environmental demand. This stage is characterized by high levels of hormone production, energy release, muscle tension, and increased heart rate. In the second stage, *adaptation*, the body's biological processes return to normal, as it seems that the environmental threat has been defeated successfully. In this second stage, the body is taking compensatory actions to maintain its homeostatic balance. These compensatory actions often carry a heavy physiological cost, which ultimately leads to the third stage. In the third and final phase, *exhaustion*, the physiological integrity of the organism is in danger. In this stage several biological systems begin to fail from the overwork of trying to adapt. These biological system failures can result in serious illness or death.

Much research has demonstrated that when stress occurs, there are changes in body chemistry that may increase the risk of illness. Changes in body chemistry include higher blood pressure, increases in corticosteroids and peripheral neurotransmitters in the blood, increased muscle tension, and increased immune system responses (Selye, 1956; Levi, 1972; Frankenhaeuser and Gardell, 1976; Frankenhaeuser, 1986; Karasek et al., 1988). Selye's pioneering research defined the importance of environmental stressors and the medical consequences of stress on the immune system, the gastrointestinal system, and the adrenal glands. However, Selye emphasized the physiological consequences of stress and paid little attention to the psychological aspects of the process or the psychological outcomes of stress.

Lazarus (1974, 1977, 1993) proposed that physiological changes caused by stressors came from a need for action resulting from emotions in response to the stressors (environment). The quality and intensity of the emotional reactions that lead to physiological changes depend on cognitive appraisal of the "threat" posed by the environment to personal security and safety. From Lazarus's perspective the threat appraisal process determined the quality and intensity of the emotional reaction and defined coping activities that affected the emotional reactions. The extent of the emotion reaction influenced the number of physiological reactions.

Levi (1972) tied together the psychological and physiological aspects of stress. He recognized the importance of psychological factors as primary determinants of the stress sources (perception of stressors). He proposed a model that linked psychosocial stimuli with disease. In this model any psychosocial stimulus, that is, any event that takes place in the social environment, can act as a stressor. In accordance with a *psychobiological program*, psychosocial stimuli may evoke physiological responses similar to those described by Selye (1956). In turn, when these physiological responses occur chronically, they can lead to disease. Several intervening variables (individual susceptibility, personality, coping strategies, or social support) can moderate the link between psychosocial stimuli and disease. Levi emphasized that the sequence of events described by the model is not a one-way process but a *cybernetic system with continuous feedback*, a significant application of a human factors concept (K. Smith, 1966). The physiological responses to stress and disease can influence the psychosocial stimuli as well as the individual's psychobiological program. These feedback loops are important to an understanding of how stress reactions and disease states themselves can, in turn, act as additional stressors or mediate a person's response to environmental stressors.

Frankenhaeuser and her colleagues emphasized a psychobiological model of stress that defined the specific environmental factors most likely to induce increased levels of cortisol and catecholamines (Lundberg and Frankenhaeuser, 1980; Frankenhaeuser and Johansson, 1986). There are two different neuroendocrine reactions in response to a psychosocial environment: (1) secretion of catecholamines via the sympathetic–adrenal medullary system, and (2) secretion of corticosteroids via the pituitary–adrenal–cortical system. Frankenhaeuser and her colleagues observed that different patterns of neuroendocrine stress responses occurred depending on the particular characteristics of the environment. They considered the most important environmental factors to be effort and the individual factors to be distress. The effort factor "involves elements of interest, engagements, and determination"; the distress factor "involves elements of dissatisfaction, boredom, uncertainty, and anxiety" (Frankenhaeuser and Johansson, 1986). Effort with distress is accompanied by increases in both catecholamine and cortisol secretion.

What Frankenhaeuser and her associates found was that effort without distress was characterized by increased catecholamine secretion but no change in cortisol secretion. Distress without effort was generally accompanied by increased cortisol secretion, with a slight elevation of catecholamines. Their approach emphasized the role of personal control in mediating the biological responses to stress. A lack of personal control over the stressors was almost always related to distress, whereas having personal control tended to stimulate greater effort. Studies performed by Frankenhaeuser and her colleagues showed that work overload led to increased catecholamine secretion but not to increased cortisol secretion when the employee had a high degree of control over the environment (Frankenhaeuser and Johansson, 1986).

Several studies have shown a link between elevated blood pressure and job stressors, especially workload, work pressure, and lack of job control. Rose et al. (1978) found that workload was associated with increased systolic and diastolic blood pressure. Van Ameringen et al. (1988) found that intrinsic pressures related to job content were related to increased standing diastolic blood pressure. The index of intrinsic pressures included a measure of quantitative workload (demands) and a measure of job participation (job control). Matthews et al. (1987) found that having few opportunities for participating in decisions at work was related to increased diastolic blood pressure. Longitudinal studies of job stress and blood pressure show that blood pressure increases were related to the introduction of new technologies at work and to work complexity (Kawakami et al., 1989). Schnall et al. (1990) demonstrated the link between hypertension and job stress, and between emotions (e.g., anger and anxiety). James et al. (1986) demonstrated a relationship between emotions and increased blood pressure.

French (1963) and Caplan et al. (1975) have proposed a stress model that defined the interaction between the environment and the individual, including coping processes for controlling the external stressors. In their approach the development of stress is an outcome of an imbalanced interaction between the environmental resources that are available and the person's needs for resources. If the environmental demands are greater than a person's capacities and/or if the person's expectations are greater than the environmental supplies, stress will occur (Caplan et al., 1975; Cooper and Payne, 1988; Kalimo, 1990; Ganster and Schaubroeck, 1991; Johnson and Johansson, 1991;

Cox and Ferguson, 1994). They proposed that social support from family and colleagues is a coping process that can mitigate the stress effects on health.

10 WORK ORGANIZATION AND PSYCHOSOCIAL INFLUENCES

The characteristics of a work organization are often sources of occupational stress that can lead to health consequences. Cooper and Marshall (1976) categorized these job stress factors into groups as those intrinsic to the job, the role in the organization, career development, the relationships at work, and the organizational structure and climate. Factors intrinsic to the job were similar to those studied and by Frankenhaeuser's group and French and Caplan. They included (1) physical working conditions; (2) workload, both quantitative and qualitative, and time pressure; (3) responsibilities (for lives, economic values, safety of other persons); (4) job content; (5) decision making; and (6) perceived control over the job. M. Smith and Carayon-Sainfort (1989) believed that the work organization level of psychosocial factors dictated the extent of environmental exposure in terms of workload, work pace, work schedule, work–rest cycle, design of equipment and workstations, product and materials design, and environmental design. In addition, the psychosocial work environment affected a person's motivation to work safely, the attitude toward personal health and safety, and the willingness to seek health care. They postulated that the work organization defined the job stressors through the task demands, personal skill requirements, extent and nature of personal training, and supervisory methods that influenced the work methods used by employees (Carayon and Smith, 2000).

Many work organization factors have been linked to short- and long-term stress reactions. Short-term stress reactions include increased blood pressure, adverse mood states, and job dissatisfaction. Studies have shown a link between overload, lack of control and work pressure, and increased blood pressure (Matthews et al., 1987; Van Ameringen et al., 1988; Schnall et al., 1990). Other studies have found a link between job future uncertainty, lack of social support and lack of job control, and adverse mood states and job dissatisfaction (Karasek, 1979; Sainfort, 1989; M. Smith et al., 1992). Long-term stress reactions include cardiovascular disease and depression. Studies have shown that job stressors are related to increased risk for cardiovascular disease (Karasek, 1979; Karasek et al., 1988; Johnson, 1989). Carayon et al. (2000) have proposed that work organization factors can define or influence ergonomic risk factors for musculoskeletal: for example, the extent of repetition, force, and posture. Work organization policies and practices define the nature, strength, and exposure time to ergonomic risk factors.

According to M. Smith and Carayon-Sainfort (1989), stress results from an imbalance between various elements of the work system. This imbalance produces a load on the human response mechanisms that can produce adverse reactions, both psychological and physiological. The human response mechanisms, which include behavior, physiological reactions, and perception/cognition, act to exert control over the environmental factors that are creating the imbalance. These efforts to bring about balance, coupled with an inability to achieve a proper balance, produce overloading of the response mechanisms that leads to mental and physical fatigue. Chronic exposure to these "fatigues" leads to stress, strain, and disease. This model emphasizes the effects of the environment (stressors), which can be manipulated to produce proper balance in the work system. These stressors can be categorized into one of the following elements of the work system: (1) the task, (2) the organizational context, (3) technology, (4) physical and social environment, and (5) the individual (see Figure 1).

The organizational context in which work is done often can influence worker stress and health (Landy, 1992). Career considerations such as over- and underpromotion, status incongruence, job future ambiguity, and lack of job security have been linked to worker stress (Cooper and Marshall, 1976; Cobb and Kasl, 1977; Jackson and Schuler, 1985; Sainfort, 1989; Heaney et al., 1992). In particular, companies that have the potential for reductions in the labor force (layoff or job loss) may be more susceptible to employees

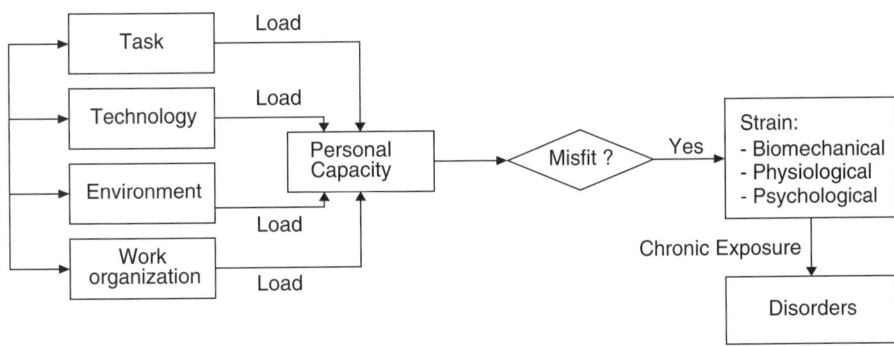

Figure 1 Model of work–system misfit.

reporting more problems and more serious problems as an economic defense. These conditions may create a working climate of distrust, fear, and confusion that could lead employees to perceive a higher level of aches and pains.

Other organizational considerations that act as environmental stressors with social elements are work schedule and overtime. Shift work has been shown to have negative social, mental, and physical health consequences (Tasto et al., 1978; Monk and Tepas, 1985). In particular, night and rotating shift regimens affect worker sleeping and eating patterns, family and social life satisfaction, and injury incidence (Rutenfranz et al., 1977; M. Smith et al., 1982). Caplan et al. (1975) found that unwanted overtime was a far greater problem than simply the amount of overtime. Overtime may also have an indirect effect on worker stress and health because it reduces the amount of rest and recovery time, takes time away from relaxation with family and friends, and reduces time with these sources of social support that can buffer stress.

Technology is an environmental influence that can produce stress: for example, physical and mental requirements that do not match employee competencies, poorly designed software, and poor system performance, such as crashes and breakdowns (Turner and Karasek, 1984; Carayon-Sainfort, 1992). There is some evidence showing that computers can be a source of physical and mental stress (M. Smith et al., 1981, 1992; NAS, 1983). New technology may exacerbate worker fears of job loss due to increased efficiency of technology (Ostberg and Nilsson, 1985; M. Smith 1987b). The way the technology is introduced may also influence worker stress and health (M. Smith and Carayon, 1995). For instance, when workers are not given enough time to get accustomed to the technology, they may develop unhealthy work methods.

11 COMMUNITY ERGONOMICS

Recently, human factors has looked beyond work systems to examine more complex environments where multiple systems interact with each other. Community ergonomics (CE) is an approach to applying human factors principles to the interaction of multiple systems in a community setting for the improvement of its quality of life (M. Smith et al., 1994; Cohen and J. Smith, 1994; J. Smith et al., 1996; J. Smith et al., 2002). CE evolved from two parallel directions, one being theories and principles in human factors and ergonomics, the other the evaluation of specific improvements in communities that led to theories and principles at the societal level of analysis. The CE approach focuses on distressed community settings characterized by poverty, social isolation, dependency, and low levels of self-regulation (and control). Deteriorating inner-city areas in the United States are examples of such communities, as are underdeveloped countries and countries devastated by war and poverty. The practice of CE seeks to identify and implement interventions that provide the disadvantaged residents of a community with the resources within their social environment that resolve their problems systematically in a holistic way. In macroergonomic terms this means achieving a good fit among people, the community, and the social environment. McCormick (1970) explained the difficulty in transforming the human factors discipline to deal with communities. He stated that the human factors aspects of this environment required a significant jump from the conventional human factors context of airplanes, industrial machines, and automobiles. In the first place, the systems of concern (the community) are amorphous and less well defined than are pieces of hardware. Lodge and Glass (1982) recognized the need for a systems approach to dealing with issues at a societal level. For improvement to occur, they advocated a cooperative, holistic approach with multiple reinforcing links from several directions. They pointed out correctly that providing jobs is an ineffective way to introduce change if training, day care, and other support systems are unavailable to the disadvantaged residents who are employed.

Several concepts of CE are extrapolated from behavioral cybernetic principles proposed by K. Smith (1966) regarding how a person interacts with her or his environment. CE proposes that community residents be able to track competently their environment and other community members within it. In addition, it is important that people be able to exert control over their lives within the environment and in social interactions. Residents need to be taught how to develop an awareness of the impact of their own actions on the environment and on others in their community. This understanding and control over their interactions with the environment and people enables residents to build self-regulating mechanisms for learning, social tracking, and feedback control of their lives. An effective self-regulating process is one that helps community residents identify situations of misfit between community residents and their environment, as well as allowing for the generation and implementation of solutions to improve fit and to deal with emerging challenges in a continuously changing and turbulent environment.

To put the situation in distressed communities in context, it can be likened to the cumulative trauma injuries and stress observed at the workplace. Many residents in a distressed inner city suffer from what can be termed *cumulative social trauma* (CST). Like work-related musculoskeletal disorders, CST results from long-term chronic exposure to detrimental circumstances in this case, societal conditions that create a cycle of dependency, social isolation, and learned helplessness. CST results from repeated exposure to poorly designed environments and/or long-term social isolation that leads to ineffective individual performance or coping abilities. Harrington (1962) observed a personality of poverty, a type of human being produced by the grinding, wearing life of the slums. Cumulative social trauma is the repetition of an activity or combinations of activities, environmental interactions, and daily life routines that develop gradually over a period of time and produce wear and tear on motivation, skill, and emotions, leading to social and behavioral disruption, psychosomatic disorders, and mental distress.

The obstacles encountered on the path to progress in the declining areas of inner cities can be defined in terms of human errors by community residents and by the public institution employees who serve them that lead to a lack of systems effectiveness and reliability. Human errors are decisions, behaviors, and actions that do not result in desired and expected outcomes: for example, failing to keep a job, relying on public assistance, failing to pay bills, being involved in illegal activities for the residents, and a lack of understanding, compassion, and effective services for the public employees. Social (and economic) systems effectiveness and reliability are measured by the ability of public (and private) institutions to achieve positive community performance outcomes (J. Smith and M. Smith, 1994).

Social and economic achievements are enhanced by the level of the goodness of the fit between the characteristics and needs of community residents and those of the community environment. In the case of substantial misfit between people and environment, poor urban residents are likely to make repeated "errors" in life which will result in poor economic and societal outcomes. Institutions with low systems effectiveness and reliability do not provide adequate feedback (performance information or direction) and/or services to enable residents to correct their errors. One of the fundamental purposes of community ergonomics is to improve the goodness of fit between environmental conditions and residents' behaviors to reduce residents' errors. To be effective, interventions have to deal with the total system, including the multifaceted elements of the environment and community residents. The theory and practice of community ergonomics is based on the assumptions that individuals or community groups must attain and maintain some level of self-regulation, and that individuals or groups need to have control over their lives and their environments in order to succeed. This can be achieved by having residents participate actively in their own self-improvement and in improvement of their environment.

The resident–community–environment system has multilateral and continuous interactions among residents, groups, living conditions, public institutions, stores, and workplaces. These are linked through interactions and feedback from the interactions. Other communities and external public and private institutions that may have very different beliefs, values, and modes of behavior surround the community. Similarly, the community is surrounded by communication systems, architecture, transportation, energy systems, and other technology that influence and act on the community. These affect the life quality of individuals and groups in the community. This resident–community–environment system includes institutions for education, financial transactions, government and politics, commerce and business, law enforcement, transportation, and housing. The organizational complexity of this system affects the ways by which individuals and groups try to control the environment through their behavior.

According to K. Smith and Kao (1971), a *social habit* is defined as self-governed learning in the context of the control of the social environment. The self-governance of learning becomes patterned through sustained and persistent performance and reward, and these are critically dependent on time schedules. As these timed patterns become habituated, the individual can predict and anticipate social events, a critical characteristic in the management of social environments. Self-control and self-guidance in social situations further enhances the ability of the individual to adjust to various environments (old and new) by following or tracking the activities of other persons or groups. Social tracking patterns during habit cycles determine what is significant behavior for success or failure. Residents, groups, and communities develop and maintain their identity through the establishment and organization of social habits (i.e., accepted behaviors). In addition, the maintenance of group patterns and adherence to this process are achieved through social yoking (or mutual tracking), so that people can sense each other's social patterns and respond appropriately.

A resident–community–environment management process should build from the social habits of the community, coupled with a human-centered concept of community design. This seeks to achieve better community fit by making public and private institutions more responsive to resident capabilities, needs, and desires, and residents more responsive to community norms. K. Smith and Kao (1971) indicated that cultural design could aid and promote development of individuals and communities if it is compliant with people's built-in or learned behavior. It can adversely affect behavior, learning, and development of individuals and communities if it is noncompliant with their needs and built-in makeup. Management of the integration of residents, the community, and the environment must be approached as a total system enterprise. The aim is to build proper compliancies among the residents, the community, and the environment using social habits and individual control as key elements in compliance. This leads to the design of a resident–community–environment system that improves residents' perceptions, feedback, level of control, adherence to social habits, and performance through improved services and opportunities provided by public and private institutions.

The management process seeks to establish positive social tracking between the residents and public and private institutions. Residents can be viewed as being nested within the community. The guidance of a resident's behavior is determined by her or his ability to develop reciprocal control over the economic, social, and cultural institutions in the community using feedback from interactions with these institutions. The feedback concept is a significant aspect because it shows the resident the effectiveness of self-generated, self-controlled activity when interacting with the community and the environment. The quality of the feedback determines the course, rate, and degree of individual learning and behavior improvement in relation to interaction with the community and the environment.

The performance of the resident–community–environment system is dependent on critical timing considerations, such as work, school, and, public services schedules. Successful participation in these activities can act as a developmental aspect of residents acquiring social habits for daily living. Some difficulties faced by the community are due to the misfit of critical timing factors between residents and institutions. There are many reasons for disruption: for example, residents not being motivated to adhere to a fixed schedule, or the poor scheduling of the public services. An important element of a CE management process is to develop a tracking system to aid residents and public agencies in the sensing of critical aspects of timing and in the development of a "memory" for determining future events. The tracking system is aimed at building good habits of timing for the residents and the agencies. The absence of good timing habits results in delays in functioning, social disruption, organizational adjustment problems, emotional difficulty, and poor performance.

J. Smith et al. (2002) described seven principles that constituted the philosophy of the CE approach:

1. *Action orientation.* Rather than trying to change community residents in order to cure them of unproductive behavior, CE believes in getting them actively involved in changing the environmental factors that lead to misfit. The CE approach strives to reach collective aims and perspectives on issues of concern, and to meet specific goals and aspirations through specific actions developed by all involved parties. This process requires an organized and structured evaluation process and the formulation of plans for solutions and their implementation. The approach is based on new purposes, goals, and aspirations developed through community–environment reciprocal exchange rather than on existing community skills, needs assessments, or external resources. Other approaches have tended to become bogged down trying to fulfill institutional rules and requirements or institutional directives that pursue a good solution but to the wrong problems (Nadler and Hibino, 1994).
2. *Participation by everyone.* Community improvements often fail because residents are not substantially involved in the process of selecting the aims, objectives, and goals. It is essential to have resident participation from start to finish. Such participation is a source of ideas, a means of motivation for the residents, and a way to educate residents to new ideas and modes of behavior. There are many mechanisms for participation, including individual involvement, action groups, and committees. Early involvement of strategic persons and institutions in the process brings the necessary concepts, technical expertise, and capital into the process and the solutions. Although participation by every resident of the community may not be possible at first, the goal is to get everyone involved at some point in some way. It is expected that reluctant and passive involvement will be minimized and will decrease continually throughout a project or activity. Effective information transfer among individuals, organizations, and institutions is essential for success.
3. *Diversity and conflict management.* Communities are made up residents who may have differing perspectives, values, cultures, habits, and interests. Problems in a community are never neat and compartmentalized, and because the community system is complex, it is difficult to comprehend all of the perspectives. However, it is necessary to recognize the diversity in perspectives and opinions and find ways to work toward consensus. It is essential to formulate a process for managing diversity, conflict, and confusion that will occur in a community with many cultures and perspectives. Distressed communities typically have a low level of self-regulatory capability and high levels of diversity and dissention. It is important to spend the necessary time designing a process for handling diversity and conflict. Nadler and Hibino (1994) and Cohen (1997) have developed methodologies for working with diversity and conflict in designing solutions to problems.
4. *Encouraging learning.* A well-designed process will allow residents and institutions to interact positively and effectively even in the conditions of a highly turbulent environment. It is expected that community residents and planners will learn from each other and from participating in the process. In addition, these learning effects will be transferred to subsequent related endeavors, with or without the presence of a formal community ergonomics process to facilitate the interaction. Thus, there is a transfer of "technology" (control, knowledge, skills) to the community in the form of the process and the learning experiences. Furthermore, participants will enhance their abilities in leadership, management, group activities, evaluation, and design learned while being involved in the CE process. Formal documentation of the system management process within the group setting as it occurs provides for better understanding and management of the community ergonomics process, and provides historical documentation for future endeavors of a similar nature.
5. *Building self-regulation.* One aspiration of the community ergonomics process is to provide participants with an increased ability and capacity for self-regulation. *Self-regulation*, defined as the ability of a person or group to exert influence over the environmental context,

is enhanced by creating specific tasks, actions, and learning opportunities that lead directly to the successful development of skill. When community members participate in a project that achieves specific goals as part of its evaluation, design, and implementation processes, they develop new abilities and skills to self-regulate themselves. This serves as motivation toward more community improvement activity.

6. *Feedback triad.* Feedback is a critical aspect of the CE improvement process. Different levels of feedback provide opportunities for learning. K. Smith (1966) and K. Smith and Kao (1971) defined three levels of feedback for individual performances: reactive, instrumental, and operational feedback. *Reactive feedback* is an understanding of the response of the muscles in taking action, *instrumental feedback* is the feel of the tool being used to take the action, and *operational feedback* is the resulting change in the environment when action of the tool occurs. K. Smith (1966) stated that these are integrated into a feedback triad allowing for closed-loop control of the activity, social tracking, and self-regulation. Feedback that provides these levels of information at the community level is important in designing and implementing community environment improvements. Thus, mutatis mutandis, feedback on resident perspectives and actions provides reactive feedback, feedback on institutional perspectives and action provides instrumental feedback, and feedback on the success of community improvements provides operational feedback. Reactive feedback is the personal sense that one's actions (or the group's actions) result in a perceived outcome on the environment. Instrumental feedback is sensed from subsequent movement of an institution or group in the form of milestones achieved and output produced. Operational feedback comes from the results of such activities as planning, designing, installing, and managing group intentions, as well as the new policies, laws, buildings, and institutions resulting from such activities. These are examples of persisting results that can be sensed directly by community environment social tracking systems. Without the feedback triad, self-regulation by group participants would not be very effective and the system could quickly degenerate. Participants must sense that their personal actions, words, and participation have effects on themselves, on others, and on the environment.
7. *Continuous improvement and innovation.* The community ergonomics approach recognizes the need for continuous improvement, which can be achieved by continuous planning and monitoring of results of projects implemented. Private organizations can be encouraged to provide guidance and feedback on the purpose, goals, and management of improvement initiatives. Inputs from private organizations and governmental programs can be utilized to promote an entrepreneurial spirit that encourages effective community habits. These can be benchmarked against other communities and other programs. Valuable information can be elicited by studying the effects of a solution over a period of many years, to prevent the redevelopment of a dysfunctional system and to give members of troubled communities opportunities for better lives. This implies the need for ongoing monitoring to evaluate the operational requirements for implementation, measuring effectiveness, and use of feedback to alter programs already in existence. Consistent monitoring of citizen needs, desires, and values must be established to verify that programs and products are accessible, usable, useful, and helpful to community residents.

Community ergonomics is a long-overdue answer to the application of human factors engineering principles to address complex societal problems. Community ergonomics is a philosophy, a theory, a practice, a solution-finding approach, and a process, all in one. CE is a way to improve complex societal systems that are showing signs of CST (cumulative social trauma). CST is not to be taken lightly, as the costs are immense in every respect: financial, human, social, and developmental. When any group of people, a community, or a region is isolated, alienated, and blocked from access to resources needed to prosper, the consequences are long lasting and deeply detrimental.

Recently, concepts of CE have been developed specifically for corporations engaged in international development and trade (Derjani-Bayeh and Smith, 2000). During the last half of the twentieth century, a struggle began for fairness, equality, freedom, and justice for people in many developing nations. This has occurred during a time of expansion of the global economy. Companies now operate in a complex world economy characterized by continuous change; a heterogeneous (often international) workforce at all job levels; increased spatial dispersion of their financial, physical, and human assets; increased diversification of products and markets; uneven distribution of resources; variable performance within and between locales; increased operational and safety standards; and differences in economic, social, political, and legal conditions. In this climate of increased international trade, important issues of social and cultural values need to be examined carefully. There is need for an understanding of how specific cultures and cultural values can affect corporate operations.

Companies have developed management practices in response to numerous obstacles encountered when they expand abroad or when they transfer processes abroad. Difficulties that they have encountered in their growth, expansion, internationalization, and globalization are due primarily to the following factors: (1) the lack of a process for effective transnational transfers; (2) the lack of knowledge of operational requirements and specifications in newly entered markets; (3) ignorance of cultural norms and values in different countries; (4) the lack of adaptability mechanisms; (5) a low tolerance for uncertainty, ambiguity, change and diversity; and (6) a lack of acceptance by segments in the populations in which they are starting operations. One of the most important issues in acceptance by the local people is a company's commitment to social responsibility has the same intensity as those striving for large profits and the use of cheap local natural resources and labor. In fact, some managers believe that social responsibility may conflict with corporate financial goals and tactics.

However, CE postulates that long-term corporate stability and success (profits) will occur only if there is local community support for the enterprise and its products. Social responsibility nurtures the mutual benefits for the enterprise and the community that lead to stability and success. CE proposes that multinational corporations must accept a corporate social responsibility that recognizes the universal rights of respect and fairness for all employees, neighbors, purchasers, and communities. Whether these rights are profitable or not in the short term, or difficult to attain, such corporate social responsibility is a requirement if global ventures are going to be successful in the long term. The survivability, acceptability, and long-term success of a corporation will not depend only on quick profits but also on social responsibility that builds long-term community acceptance and support of the enterprise. This will lead to greater accessibility to worldwide markets. Organizations must address multicultural design, a comfortable corporate culture, and principles of respect and fairness for employees, customers, and neighbors. This will improve the fit with the international cultures in which the corporation operates.

CE has developed principles for successful multinational organizational design (Derjani-Bayeh and Smith, 2000). These principles focus on a goal of social responsibility, fairness, and social justice, and do not threaten the prosperity of a company or organization. They are built on the premise that the society and communities in which a company operates should benefit from the presence of the company in the society. Thus, the corporation should contribute to the development, growth, and progress of local communities. These principles propose a reciprocal relationship between the hosting community or society and the outside corporation.

11.1 Fit Principle

Accommodations need to be made by companies for a diverse workforce. That is, companies need to design for cultural diversity. Corporations must understand and incorporate the norms, customs, beliefs, traditions, and modes of behavior of the local community into their everyday operations. In some communities multiple cultures will need to be included in this process. Often, there is a need to strike a balance among the various cultures, as they are not always compatible. The corporation's organizational structure and operational style need to be flexible to bridge the gap between the corporate and local cultures.

11.2 Balance Principle

The balance principle defines the need to find the proper relationship among components of a larger system. Based on this concept we believe that there is a need for balancing corporate financial goals and objectives with societal rights and corporate social responsibility. Companies are a part of a larger community, and through their employment, purchasing, civic, and charity activities they can influence community development and prosperity. As an integral part of this larger system, companies have a responsibility to promote positive balance for the benefit of the community and the corporation.

11.3 Sharing Principle

Traditionally, a corporation's success has been measured in terms of its financial growth, but there are other factors that will become more critical as social awareness becomes more prominent. For instance, customer loyalty, community support, and acceptability of products will be critically related to the corporation's long-term financial success. If a corporation chooses to invest some of its profits back into a community in ways that are significant to that community, the corporation may be viewed not only as a business but also as a community partner. In giving something back to the community the corporation is developing loyalty to its products and protecting its long-term profitability.

11.4 Reciprocity Principle

The reciprocity principle deals with the mutual commitment, loyalty, respect, and gain between producers and consumers. A bond results from the corporation giving something back to the community, which builds loyalty from the consumers to the company, and eventually leads to a genuine sense of loyalty from the organization back to the community. In this respect, what might have started as responsibility, will over time become mutual loyalty and commitment. Within the corporate organization, the same phenomenon takes place when the organization shows responsibility toward its employees (producers), who in turn become loyal and committed partners with the corporation.

11.5 Self-Regulation Principle

Corporations should be viewed as catalysts of self-regulation and socioeconomic development in host communities. Communities and countries in disadvantaged economic conditions typically show symptoms of learned helplessness, dependency, isolation, and cumulative social trauma. Instead of perpetuating conditions that weaken people and institutions, an effort should be made to help people to self-regulate, grow, flourish, and become productive. In this effort, corporations are very important because they provide employment, training, and professional development opportunities that give people the tools to help themselves. Corporations can also invest in the community infrastructure, such as schools, clinics, and hospitals, which leads to stronger, healthier, and more independent communities in the future.

11.6 Social Tracking Principle

Awareness of the environment, institutional processes, and social interaction is necessary for people and corporations to navigate through their daily lives and for communities to fit into the broader world. Clear awareness helps to control the external world and leads to more robust, flexible, open system design. It is important for community members, employees, and corporations to be aware of their surroundings to be able to predict potential outcomes of actions taken. Similarly, it is important for corporations to develop a certain level of awareness regarding the workforce, the community, and the social, economic, and political environment within which they operate. This includes the cultural values of the people affected by a corporation's presence in a particular community.

11.7 Human Rights Principle

The human rights principle underscores the belief that every person has the right to respect, a reasonable quality of life, fair treatment, a safe environment, cultural identity, respect, and dignity. There is no reason for anyone not to be able to breathe fresh air, preserve their natural resources, achieve a comfortable standard of living, feel safe and dignified while working, and be productive. People should not be assigned a difference in worth based on class, gender, race, nationality, or age. The workplace is a good starting point to bring about fairness and justice in societies where these do not exist as a norm.

11.8 Partnership Principle

This principle proposes a partnership among the key players in a system in order to achieve the best possible solution: corporation, community, government, employees, and international links. By doing this, balance may be achieved between the interests of all parties involved, and everyone is treated fairly. In addition, partnership assures commitment to common objectives and goals.

The essence of internationalization, globalization, and multiculturalism is in the culture and social climate that the corporation develops to be sensitive to the community and the diversity of the workforce. This includes respect, partnership, reciprocity, and social corporate responsibility toward employees, the community, and society as a whole. It requires seeking a balance between the corporate culture and that of the community where the business operates and the cultures brought into the company by the diverse employees. In the past, corporations have entered new markets all over the world, profiting from cheap labor and operating freely with little or no safety or environmental liability. However, the level of social awareness has increased all over the world, exposing sweat shops, inhumane working conditions, labor exploitation, and environmental violations across the board. The focus in the future will be in doing business with a social conscience. By doing this, corporations will become welcome in any part of the world they wish to enter.

12 CONCLUDING REMARKS

As ergonomics has matured as a science, the emphasis has broadened from looking primarily at the individual worker (user, consumer) and her or his interaction with tools and technology to encompass larger systems. A natural progression has led to an examination of how the social environment and processes affect an individual and groups using technology, as well as how the behavior and uses of technology affect society. We have described aspects of these reciprocal effects by emphasizing select theories and perspectives where social considerations have made a contribution to system design and operation.

We see ergonomics as an essential aspect of the continuous human effort to survive and prosper. It is central to our collective endeavor to improve work systems. Ergonomics answers to the social needs of effective utilization of human talent and skills and of respect and support for their different abilities. By reducing exposure to physical hazards, particularly those associated with the onset of musculoskeletal disorders, by controlling occupational stress and fatigue, ergonomics is essential for the improvement of work conditions and the overall health of the population. It is instrumental to the safety and functionality of consumer products and key to the reliability of systems we depend on. In summary, ergonomics is a critical aspect for the well-being of individuals, for the effectiveness of organizations, and for the prosperity of national economies.

Ergonomics can be seen as a technology and part of a broader social context. As such, it responds to the needs, conditions, and structure of that society. Ergonomics technology evolution is shaped by ever-changing societal motivations. The relationships between ergonomics and societal demands can be understood as reciprocal: where social needs determine the direction of ergonomics development, and ergonomic innovations once introduced in the environment allow the fruition and reinforcement of some aspects or drives of the social process.

Social needs for increased productivity, higher quality, better working conditions, and reliability changed over time as some of these drives become more prominent. As work systems become more complex and the workforce more educated and sophisticated, the consideration by the ergonomics discipline of broader social, political, and financial aspects has been heightened. These changes have been answered by ergonomics in different but ultimately interrelated approaches.

Earlier we highlighted the seminal contribution of Kurt Lewin to an understanding of the social aspects of work with emphasis on the humane side of the organization. Lewin focused on reconciling scientific thinking and democratic values in the workplace and on recovering the meaning of work. He was an early advocate of worker involvement and saw job satisfaction as an essential goal to be met by work systems. Lewin's ideas and later the application of general systems theory had deep implications for the ergonomics discipline. Work by researchers at the Tavistock Institute, also inspired by Lewin, led to establishment of the sociotechnical systems theory, which confirmed the importance of the social context in work systems optimization.

While Kurt Lewin was one of the leaders in defining the critical need for work to provide psychological and social benefits to the employees, others who followed him carried through with these ideas by turning them into reality at the workplace. At the heart of most of these ideas was the concept of participation by employees in the design and control of their own work and workplaces. French, Kahn, Katz, McGregor, K. U. Smith, Emory, Trist, Davis, Drucker, Deming, Juron, Lawler, M. J. Smith, Hendrick, Noro, Imada, Karasek, Wilson, Carayon, and Sainfort have all promoted use of employee participation as a process or vehicle to engage employees more fully in their work as a source of motivation, as a means for providing more employee control over the work process and decision making, or as a mechanism to enlist employee expertise in improving work processes and products. The nature of employee participation has varied from something as simple as asking for employee suggestions for product improvements, to something as complex as semiautonomous work groups, where the employees make critical decisions about production and resource issues. The critical feature of participation is the active engagement of employees in providing input into decisions about how things are done at the workplace. The social benefits of participation are in the development of company, group, and team cohesiveness and a cooperative spirit in fulfilling company and individual goals and needs. Individuals learn that they can contribute to something bigger than themselves and their jobs, how to interact with other employees and managers positively, and receive social recognition for their contributions. This "socialization" process leads to higher ego fulfillment, greater motivation, higher job satisfaction, higher performance, less absenteeism, and fewer labor grievances. Participatory ergonomics has become an essential aspect of work improvement.

The social side of work is more than just the positive motivation and ego enhancement that can occur with good workplace design; it is also the stress that can develop when there is poor workplace design. Work design theorists and practitioners have put substantial emphasis on the psychosocial aspects of work and how these aspects influence employee productivity and health (M. Smith, 1987; Kalimo et al., 1997; Carayon and Smith, 2000). This tradition grew out of work democracy approaches in Scandinavia and Germany (Levi, 1972; Gardell, 1982) that emphasized the role of employee participation and codetermination in providing "satisfying" work. The absence of satisfying work and/or combinations of chronic exposure to high physical and psychological work demands were shown to lead to ill health and reduced motivation at work. Strong ties between poor working conditions, employee dissatisfaction, poor quality of working life, and negative outcomes for motivation, production, and health were documented (M. Smith, 1987). Thus, social and psychological aspects of working conditions were shown to influence not only personal and group satisfaction, ego, motivation, and social and productive outcomes, but also the health, safety, and welfare of the workforce. This tradition defined critical social and organizational design features that could be improved through organizational and job design strategies. Many of the strategies included considerations of employee involvement in the work process, social mechanisms such as laws and rules defining workplace democracy and codetermination, and health care approaches that encompass psychosocial aspects of work as a consideration.

Macroergonomics grew out of several traditions in organizational design, systems theory, employee participation, and psychosocial considerations in work design. Championed by Hendrick (1984, 1986, 2002), and Brown (1986, 2002), this tradition expanded the focus of work design from the individual employee and work group to a higher systems level that examined the interrelationship among various subsystems of an enterprise. By definition there are structures (organizational, operational, social) that define the nature of the interaction among the subsystems, and macroergonomics aims for the joint optimization of these subsystems. Hendrick (2002) describes how macroergonomics was a response to bad organizational design and management practices that minimized the importance and contributions of the human component of the work system. This is in line with and an elaboration of the prior traditions described above but with a primary focus on the systems nature of the work process and integration of the subsystems. At the heart of macroergonomics is the use of employee participation to achieve system integration and balance (Carayon and Smith, 2000; Brown, 2002).

Community ergonomics is a natural extension of macroergonomics to a higher level above the enterprise, in this case to the community. Like many of the preceding theories and concepts, an important aspect of community ergonomics is improvement in the quality of life for the people in the system. Such improvement concerns economic benefits, but there is also a strong emphasis on developing the social and psychosocial aspects of individual and community life that enhance overall well-being. Like macroergonomics, community ergonomics examines the subsystems in an enterprise and how to optimize them jointly to achieve benefits for the people. At a higher level, community ergonomics has provided advice to multinational enterprises in how to provide reciprocal benefits to the enterprise and the several communities in which the enterprise operates. Among the considerations is the critically important concept of the enterprise's sensitivity to, and accommodation of, social and cultural differences (Derjani-Bayeh and Smith, 2000; J. Smith et al., 2002).

REFERENCES

Argyris, C. (1964), *Integrating the Individual and the Organization*, Wiley, New York.

Axelsson, J. R. C. (2000), "Quality and Ergonomics: Towards a Successful Integration," Dissertation 616, Linköping Studies in Science and Technology, Linköping University, Linköping, Sweden.

Axelsson, J. R. C., Bergman, B., and Eklund, J., Eds. (1999), *Proceedings of the International Conference on TQM and Human Factors*, Linköping, Sweden.

Babbage, C. (1832), *On the Economy of Machinery and Manufactures*, 4th ed.; A. M. Kelley, Publishers, New York, reprinted 1980.

Badham, R. J. (2001), "Human Factor, Politics and Power," in *International Encyclopedia of Ergonomics and Human Factors*, W. Karwowski, Ed., Taylor & Francis, New York, pp. 94–96.

Bertalanffy, L. von (1950), "The Theory of Open Systems in Physics and Biology," *Science*, Vol. 111, pp. 23–29.

Bertalanffy, L. von (1968), *General Systems Theory: Foundations, Development, Applications*, George Brazilier, New York; rev. ed., 1976.

Broedling, L. A. (1977), "Industrial Democracy and the Future Management of the United States Armed Forces," *Air University Review*, Vol. 28, No. 6; retrieved August 15, 2004, from http://www.airpower.maxwell.af.mil/airchronicles/aureview/1977/sep-oct/broedling.html.

Brown, O., Jr. (1986), "Participatory Ergonomics: Historical Perspectives, Trends, and Effectiveness of QWL Programs," in *Human Factors in Organizational Design and Management II*, O. Brown, Jr., and H. W. Hendrick, Eds., North-Holland, Amsterdam, pp. 433–437.

Brown, O., Jr. (1990), "Macroergonomics: A Review," in *Human Factors in Organizational Design and Management III*, K. Noro and O. Brown, Eds., North-Holland, Amsterdam, pp. 15–20.

Brown, O., Jr. (1991), "The Evolution and Development of Macroergonomics," in *Designing for Everyone: Proceeding of the 11th Congress of the International Ergonomics Association*, Y. Queinnec and F. Daniellou, Eds., Taylor & Francis, Paris, pp. 1175–1177.

Brown, O., Jr. (2002), "Macroergonomic Methods: Participation," in *Macroergonomics: Theory, Methods, and Applications*, H. W. Hendrick and B. M. Kleiner, Eds., Lawrence Erlbaum Associates, Mahwah, NJ, pp. 25–44.

Caplan, R. D., Cobb, S., French, J. R., Jr., van Harrison, R., and Pinneau, S. R. (1975), *Job Demands and Worker Health*, U.S. Government Printing Office, Washington, DC.

Carayon-Sainfort, P. (1992), "The Use of Computer in Offices: Impact on Task Characteristics and Worker Stress," *International Journal of Human–Computer Interaction*, Vol. 4, No. 3, pp. 245–261.

Carayon, P., and Smith, M. J. (2000), "Work Organization and Ergonomics," *Applied Ergonomics*, Vol. 31, pp. 649–662.

Carlopio, J. (1986), "Macroergonomics: A New Approach to the Implementation of Advanced Technology," in *Human Factors in Organizational Design and Management II*, O. Brown and H. Hendrick, Eds., North-Holland, Amsterdam, pp. 581–591.

Chapanis, A. (1999), *The Chapanis Chronicles: 50 Years of Human Factors Research, Education, and Design*, Aegean, Santa Barbara, CA.

Chengalur, S. N., Rodgers, S. H., and Bernard, T. E. (2004), *Kodak's Ergonomic Design for People at Work*, 2nd ed., Wiley, New York.

Christensen, J. M. (1987), "The Human Factors Function," in *Handbook of Human Factors*, G. Salvendy, Ed., Wiley, New York, pp. 1–16.

Cobb, S., and Kasl, S. (1977), *Termination: The Consequences of Job Loss*, U.S. Government Printing Office, Washington, DC.

Coch, L., and French, J. R. (1948), "Overcoming Resistance to Change," *Human Relations*, Vol. 1, pp. 512–532.

Cohen, W. J. (1997), *Community Ergonomics: Design Practice and Operational Requirements*, University of Wisconsin Library, Madison, WI.

Cohen, W. J., and Smith, J. H. (1994), "Community Ergonomics: Past Attempts and Future Prospects Toward America's Urban Crisis," in *Proceedings of the Human Factors and Ergonomics Society 38th Annual Meeting*, Human Factors and Ergonomics Society, Santa Monica, CA, pp. 734–738.

Cohen, A. L., Gjessing, C. C., Fine, L. J., Bernard, B. P., and McGlothlin, J. D. (1997), *Elements of Ergonomics Programs: A Primer Based on Workplace Evaluations of Musculoskeletal Disorders*, National Institute for Occupational Safety and Health, Cincinnati, OH.

Cooper, C. L., and Marshall, J. (1976), "Occupational Sources of Stress: A Review of the Literature Relating to Coronary Heart Disease and Mental Ill Health," *Journal of Occupational Psychology*, Vol. 49, pp. 11–28.

Cooper, C. L., and Payne, R., Eds. (1988), *Causes, Coping and Consequences of Stress at Work*, Wiley, Chichester, West Sussey, England.

Cotton, J. L., Vollrath, D. A., Frogatt, K. L., Lengnick-Hall, M. L., and Jennings, K. R. (1988), "Employee Participation: Diverse Forms and Different Outcomes," *Academy of Management Review*, Vol. 13, pp. 8–22.

Cox, T., and Ferguson, E. (1994), "Measurement of the Subjective Work Environment," *Work and Stress*, Vol. 8, pp. 98–109.

Daniellou, F. (2001), "Epistemological Issues About Ergonomics and Human Factors," in *International Encyclopedia of Ergonomics and Human Factors*, W. Karwowski, Ed., Taylor & Francis, New York, pp. 43–46.

Dean, J. W., and Bowen, D. E. (1994), "Management Theory and Total Quality: Improving Research and Practice Through Theory Development," *Academy of Management Review*, Vol. 19, pp. 392–418.

DeGreene, K. B. (1986), "Systems Theory, Macroergonomics, and the Design of Adaptive Organizations," in *Human Factors in Organizational Design and Management II*, O. Brown and H. Hendrick, Eds., North-Holland, Amsterdam, pp. 479–491.

Derjani-Bayeh, A., and Smith, M. J. (2000), "Application of Community Ergonomics Theory to International Corporations," in *Proceedings of the IEA 2000/HFES 2000 Congress*, Vol. 2, Human Factors and Ergonomics Society, Santa Monica, CA, pp. 788–791.

Dray, S. M. (1985), "Macroergonomics in Organizations: An Introduction," *Ergonomics International*, Vol. 85, pp. 520–522.

Drury, C. G. (1978), "Integrating Human Factors Models into Statistical Quality Control," *Human Factors*, Vol. 20, No. 5, pp. 561–572.

Drury, C. G. (1997), "Ergonomics and the Quality Movement," *Ergonomics*, Vol. 40, No. 3, pp. 249–264.

Drury, C. G. (1999), "Human Factors and TQM," in *The Occupational Ergonomics Handbook*, W. Karwowski and W. Marras, Eds., CRC Press, Boca Raton, FL, pp. 1411–1419.

Eklund, J. (1995), "Relationships Between Ergonomics and Quality in Assembly Work," *Applied Ergonomics*, Vol. 26, No. 1, pp. 15–20.

Eklund, J. (1997), "Ergonomics, Quality, and Continuous Improvement: Conceptual and Empirical Relationships in an Industrial Context," *Ergonomics*, Vol. 40, No. 10, pp. 982–1001.

Eklund, J. (1999), "Ergonomics and Quality Management: Humans in Interaction with Technology, Work Environment, and Organization," *International Journal of Occupational Safety and Ergonomics*, Vol. 5, No. 2, pp. 143–160.

Eklund, J., and Berggren, C. (2001), "Ergonomics and Production Philosophies," in *International Encyclopedia of Ergonomics and Human Factors*, W. Karwowski, Ed., Taylor & Francis, New York, pp. 1227–1229.

Emery, F. E., and Trist, E. L. (1960), "Sociotechnical Systems," in *Management Sciences: Models and Techniques*, C. W. Churchman et al., Eds., Pergamon Press, London.

Emery, F. E., and Trist, E. L. (1965), "The Causal Texture of Organizational Environments," *Human Relations*, Vol. 18, No. 1, pp. 21–32.

Fayol, H. (1916), *General and Industrial Management*, revised by Irwin Gray, 1987, David S. Lake Publications, Belmont, CA.

Frankenhaeuser, M., and Gardell, B. (1976), "Underload and Overload in Working Life: Outline of a Multidisciplinary Approach," *Journal of Human Stress*, Vol. 2, No. 3, pp. 35–46.

Frankenhaeuser, M. (1986), "A Psychobiological Framework for Research on Human Stress and Coping," in M. H. Appley and R. Trumbull, Eds., *Dynamics of Stress—Physiological, Psychological and Social Perspectives*. Plenum, New York, pp. 101–116.

Frankenhaeuser, M., and Johansson, G. (1986), "Stress at Work: Psychobiological and Psychosocial Aspects," *International Review of Applied Psychology*, Vol. 35, pp. 287–299.

French, J. R. P. (1963), "The Social Environment and Mental Health," *Journal of Social Issues*, Vol. 19, pp. 39–56.

French, J. R. P., and Caplan, R. D. (1973), "Organizational Stress and Individual Strain," in *The Failure of Success*, A. J. Marrow, Ed., AMACOM, New York, pp. 30–66.

Ganster, D. C., and Schaubroeck, J. (1991), "Work Stress and Employee Health," *Journal of Management*, Vol. 17, No. 2, pp. 235–271.

Gardell, B. (1982). "Scandinavian Research on Stress in Working Life," *International Journal of Health Services*, Vol 12, pp. 31–41.

Gilbreth, F., and Gilbreth, L. (1917), *Applied Motion Study*, Sturgis & Walton, New York; reprinted, Hive Publishing, Easton, PA, 1973.

Gilbreth, F., and Gilbreth, L. (1920), *Fatigue Study*, Sturgis & Walton, New York; revised edition, Macmillan, New York; reprinted, Hive Publishing, Easton, PA, 1973.

Grudin, J. (1990), "The Computer Reaches Out: The Historical Continuity of Interface Design," in *Proceedings of CHI '90: Empowering People*, J. C. Chew and J. Whiteside, Eds., ACM, New York, pp. 261–268.

Hackman, R. J., and Oldham, G. R. (1980), *Work Redesign*, Addison-Wesley, Reading, MA.

Harrington, M. (1962), *The Other America: Poverty in the United States*, Macmillan, New York.

Hendrick, H. W. (1984), "Wagging the Tail with the Dog: Organizational Design Considerations in Ergonomics," in *Proceedings of the Human Factors Society 28th Annual Meeting*, Human Factors Society, Santa Monica, CA, pp. 899–903.

Hendrick, H. W. (1986), "Macroergonomics: A Concept Whose Time Has Come," in *Human Factors in Organizational Design and Management II*, O. Brown, Jr., and H. W. Hendrick, Eds., North-Holland, Amsterdam, pp. 467–478.

Hendrick, H. W. (1991), "Human Factors in Organizational Design and Management," *Ergonomics*, Vol. 34, pp. 743–756.

Hendrick, H. W. (1997), "Organizational Design and Macroergonomics," in *Handbook of Human Factors and Ergonomics*, 2nd ed., G. Salvendy, Ed., Wiley, New York.

Hendrick, H. W. (2002), "An Overview of Macroergonomics," in *Macroergonomics: Theory, Methods, and Applications*, Lawrence Erlbaum Associates, Mahwah, NJ, pp. 1–23.

Hendrick, H. W., and Kleiner, B. M. (2001). *Macroergonomics: An Introduction to Work System Design*, Human Factors and Ergonomics Society, Santa Monica, CA.

Hughes, T. P. (1991), "From Deterministic Dynamos to Seamless-Web Systems," in *Engineering as a Social Enterprise*, H. E. Sladovich, Ed., National Academy Press, Washington, DC, pp. 7–25.

Huzzard, T. (2003), "The Convergence of Quality of Working Life and Competitiveness: A Current Swedish Literature Review," National Institute for Working Life; retrieved August 10/2004, from http://ebib.arbetslivsinstitutet.se/aio/2003/aio2003_09.pdf.

Imada, A. S. (1991), "The Rationale for Participatory Ergonomics," in *Participatory Ergonomics*, K. Noro and A. Imada, Eds., Taylor & Francis, London.

Imada, A. S., and Robertson, M. M. (1987), "Cultural Perspectives in Participatory Ergonomics," in *Proceedings of the Human Factors Society 31st Annual Meeting*, Human Factors Society, Santa Monica, CA, pp. 1018–1022.

Jackson, S. E. (1989), "Does Job Control Control Job Stress?" in *Job Control and Worker Health*, S. L. Sauter, J. J. Hurrel, and C. L. Cooper, Eds., Wiley, Chichester, West Sussex, England, pp. 25–53.

Jackson, S. E., and Schuler, R. S. (1985), "A Meta-analysis and Conceptual Critique of Research on Role Ambiguity and Role Conflict in Work Settings," *Organizational Behavior and Human Decision Processes*, Vol. 36, pp. 16–78.

James, G. D., Yee, L. S., Harshfield, G. A., Blank, S. G., and Pickering, T. G. (1986), "The Influence of Happiness, Anger, and Anxiety on the Blood Pressure of Borderline Hypertensives," *Psychosomatic Medicine*, Vol. 48, No. 7, pp. 502–508.

Jirotka, M., and Goguen, J. (1994), *Requirements Engineering: Social and Technical Issues*, Academic Press, London.

Johnson, J. V., and Johansson, G. (1991), "Work Organisation, Occupational Health, and Social Change: The Legacy of Bertil Gardell," in *The Psychosocial Work Environment: Work Organization, Democratization and Health*, J. V. Johnson and G. Johansson, Eds., Baywood, Amityville, NY.

Kalimo, R. (1990), "Stress in Work," *Scandinavian Journal of Work, Environment and Health*, Vol. 6, Suppl. 3.

Kalimo, R., Lindstrom, K., and Smith, M. J. (1997), "Psychosocial Approach in Occupational Health," in *Handbook of Human Factors and Ergonomics*, 2nd ed., G. Salvendy, Ed., Wiley, New York, pp. 1059–1084.

Kanter, R. M. (1983), *The Change Masters*, Simon and Schuster, New York.

Karasek, R. A. (1979), "Job Demands, Job Decision Latitude, and Mental Strain: Implications for Job Redesign," *Administrative Science Quarterly*, Vol. 4, pp. 285–308.

Karasek, R., and Theorell, T. (1990), *Healthy Work*, Basic Books, New York.

Karasek, R. A., Theorell, T., Schwartz, J. E., Schnall, P. L., Pieper, C. F., and Michela, J. L. (1988), "Job Characteristics in Relation to the Prevalence of Myocardial Infarction in the U.S. Health Examination Survey (HES) and the Health and Nutrition Examination Survey (HANES)," *American Journal of Public Health*, Vol. 78, pp. 910–918.

Kawakami, N., Haratani, T., Kaneko, T., and Araki, S. (1989), "Perceived Job-Stress and Blood Pressure Increase Among Japanese Blue Collar Workers: One-Year Follow-up Study," *Industrial Health*, Vol. 27, No. 2, pp. 71–81.

Konz, S., and Johnson, S. (2004). *Work Design: Occupational Ergonomics*, Holcomb Hethaway, Scottsdale, AZ.

Landy, F. J. (1992), "Work Design and Stress," in *Work and Well-Being*, G. P. Keita and S. L. Sauter, Eds., American Psychological Association, Washington, DC, pp. 119–158.

Lawler, E. E., III (1986), *High-Involvement Management*, Jossey-Bass, San Francisco.

Lawler, E. E., III., Morhman, S. A., and Ledford, G. E., Jr. (1992), *Employee Participation and Total Quality Management*, Jossey-Bass, San Francisco.

Lazarus, R. S. (1974), "Psychological Stress and Coping in Adaptation and Illness," *International Journal of Psychiatry in Medicine*, Vol. 5, pp. 321–333.

Lazarus, R. S. (1977), "Cognitive and Coping Processes in Emotion," in *Stress and Coping: An Anthology*, A. Monat and R. S. Lazarus, Eds., Columbia University Press, New York, pp. 145–158.

Lazarus, R. S. (1993), "From Psychological Stress to the Emotions: A History of Changing Outlooks," *Annual Reviews in Psychology*, Vol. 44, pp. 1–21.

Leana, C. R., Locke, E. A., and Schweiger, D. M. (1990), "Fact and Fiction in Analyzing Research on Participative Decision Making: A Critique of Cotton, Vollrath, Frogatt, Lengnick-Hall, and Jennings," *Academy of Management Review*, Vol. 15, pp. 137–146.

Levi, L. (1972), "Stress and Distress in Response to Psychosocial Stimuli," *Acta Medica Scandinavia*, Vol. 191, Suppl. 528.

Locke, E. A., and Schweiger, D. M. (1979), "Participation in Decision Making: One More Look," *Research in Organizational Behavior*, Vol. 1, pp. 265–339.

Lodge, G. C., and Glass, W. R. (1982), "The Desperate Plight of the Underclass: What a Business–Government Partnership Can Do About Our Disintegrated Urban Communities," *Harvard Business Review*, Vol. 60, pp. 60–71.

Lundberg, U., and Frankenhaeuser, M. (1980), "Pituitary–Adrenal and Sympathetic–Adrenal Correlates of Distress and Effort," *Journal of Psychosomatic Research*, Vol. 24, pp. 125–130.

Marrow, A. F. (1969), *The Practical Theorist: The Life and Work of Kurt Lewin*, Basic Books, New York.

Matthews, K. A., Cottington, E. M., Talbott, E., Kuller, L. H., and Siegel, J. M. (1987), "Stressful Work Conditions and Diastolic Blood Pressure Among Blue Collar Factory Workers," *American Journal of Epidemiology*, Vol. 126, No. 2, pp. 280–291.

McCormick, E. J. (1970), *Human Factors Engineering*, 3rd ed., McGraw-Hill, New York.

McGregor, D. (1960), *The Human Side of Enterprise*, McGraw-Hill, New York.

Medsker, G. J., and Campion, A. (2001), "Job and Team Design," in *Handbook of Industrial Engineering*, G. Salvendy, Ed., Wiley, New York.

Meister, D. (2001), "Fundamental Concepts of Human Factors," in *International Encyclopedia of Ergonomics and Human Factors*, W. Karwowski, Ed., Taylor & Francis, New York, pp. 68–70.

Monk, T. H., and Tepas, D. I. (1985), "Shift Work," in *Job Stress and Blue-Collar Work*, C. L. Cooper and M. J. Smith, Eds., Wiley, New York, pp. 65–84.

Moray, N. (1994), "De Maximus non Curat Lex" or "How Context Reduces Science to Art in the Practice of Human Factors," in *Proceedings of the Human Factors and Ergonomics Society 38th Annual Meeting*, Human Factors and Ergonomics Society, Santa Monica, CA, pp. 526–530.

Nadler, G., and Hibino, S. (1994), *Breakthrough Thinking: The Seven Principles of Creative Problem Solving*, 2nd ed., Prima Communications, Rocklin, CA.

NAS (1983), *Video Displays, Work and Vision*, National Academy Press, National Academy of Sciences, Washington, DC.

NIOSH (1992), *Health Hazard Evaluation Report: HETA 89-299-2230-US West Communications*, U.S. Department of Health and Human Services, Washington, DC.

Noro, K. (1991), "Concepts, Methods, and People," in *Participatory Ergonomics*, K. Noro and A. Imada, Eds., Taylor & Francis, London.

Noro, K. (1999), "Participatory Ergonomics," in *Occupational Ergonomics Handbook*, W. Karwowski and W. S. Marras, Eds., CRC Press, Boca Raton, FL, pp. 1421–1429.

Oborne, D. J., Branton, R., Leal, F., Shipley, P., and Stewart, T. (1993), *Person-Centered Ergonomics: A Brantonian View of Human Factors*, Taylor & Francis, London.

Ostberg, O., and Nilsson, C. (1985), "Emerging Technology and Stress," in *Job Stress and Blue-Collar Work*, C. L. Cooper and M. J. Smith, Eds., Wiley, New York, pp. 149–169.

Owen, R. (1816), *A New View of Society*; 2004 edition, Kessinger Publishing, Kila, MT.

Pacey, A. (1991), *Technology in World Civilization: A Thousand-Year History*, MIT Press, Cambridge, MA.

Parker, M., and Slaughter, J. (1994), *Working Smart: A Union Guide to Participation Programs and Reengineering*, Labor Notes, Detroit, MI.

Perrow, C. (1983), "The Organizational Context of Human Factors Engineering," *Administrative Science Quarterly*, Vol. 28, pp. 521–541.

Rahimi, M. (1995), "Merging Strategic Safety, Health, and Environment into Total Quality Management," *International Journal of Industrial Ergonomics*, Vol. 16, pp. 83–94.

Ramazzinni, B. (1700), *De Morbis Artificum Diatriba. Modena: Antonii Capponi* [Diseases of Workers], translation by W. C. Wright, 1940, University of Chicago Press, Chicago.

Reason, J. (1990), *Human Error*, Cambridge University Press, New York.

Reason, J. (1997), *Managing the Risks of Organizational Accidents*, Ashgate, Brookfield, VT.

Reason, J., and Hobbs, A. (2003), *Managing Maintenance Error: A Practical Guide*, Ashgate, Brookfield, VT.

Reeves, C. A., and Bednar, D. A. (1994), "Defining Quality: Alternatives and Implications," *Academy of Management Review*, Vol. 19, pp. 419–445.

Rose, R. M., Jenkins, C. D., and Hurst, M. W. (1978), *Air Traffic Controller Health Change Study*, U.S. Department of Transportation, Federal Aviation Administration, Office of Aviation Medicine, Washington, DC.

Rutenfranz, J., Colquhoun, W. P., Knauth, P., and Ghata, J. N. (1977), "Biomedical and Psychosocial Aspects of Shift Work," *Scandinavian Journal of Work Environment and Health*, Vol. 3, pp. 165–182.

Sainfort, P. C. (1989), "Job Design Predictors of Stress in Automated Offices," *Behaviour and Information Technology*, Vol. 9, No. 1, pp. 3–16.

Sainfort, F., Taveira, A. D., Arora, N. K., and Smith, M. J. (2001), "Teams and Team Management and Leadership," in *Handbook of Industrial Engineering*, G. Salvendy, Ed., Wiley, New York.

Sanders, M. S., and McCormick, E. J. (1993), *Human Factors in Engineering and Design*, McGraw-Hill, New York.

Sashkin, M. (1984), "Participative Management Is an Ethical Imperative," *Organizational Dynamics*, Spring, pp. 5–22.

Schnall, P. L., Pieper, C., Schwartz, J. E., Karasek, R. A., Schlussel, Y., Devereux, R. B., Ganau, A., Alderman, M., Warren, K., and Pickering, T. G. (1990), "The Relationship Between 'Job Strain,' Workplace Diastolic Blood Pressure, and Left Ventricular Mass Index," *Journal of the American Medical Association*, Vol. 263, pp. 1929–1935.

Selye, H. (1956), *The Stress of Life*, McGraw-Hill, New York.

Shackel, B. (1996), "Ergonomics: Scope, Contribution and Future Possibilities," *The Psychologist*, Vol. 9, No. 7, pp. 304–308.

Smith, A. (1776), *An Inquiry into the Nature and Causes of the Wealth of Nations*; 2003 edition, Bantam Classics, New York.

Smith, J. H., and Smith, M. J. (1994), "Community Ergonomics: An Emerging Theory and Engineering Practice," in *Proceedings of the Human Factors and Ergonomics Society, 38th Annual Meeting*, Human Factors and Ergonomics Society, Santa Monica, CA, pp. 729–733.

Smith, J. H., Cohen, W., Conway, F., and Smith, M. J. (1996), "Human Centered Community Ergonomic Design," in *Human Factors in Organizational Design and Management V*, J. O. Brown and H. W. Hendrick, Eds., Elsevier Science, Amsterdam, pp. 529–534.

Smith, J. H., Cohen, W. J., Conway, F. T., Carayon, P., Derjani-Bayeh, A., and Smith, M. J. (2002), "Community Ergonomics," in *Macroergonomics: Theory, Methods, and Applications*, H. W. Hendrick and B. M. Kleiner, Eds., Lawrence Erlbaum Associates, Mahwah, NJ, pp. 289–309.

Smith, K. U. (1965), *Behavior Organization and Work*, rev. ed., College Printing and Typing Company, Madison, WI.

Smith, K. U. (1966), "Cybernetic Theory and Analysis of Learning," in *Acquisition of Skill*, Academic Press, New York.

Smith, K. U., and Kao, H. (1971), "Social Feedback: Determination of Social Learning," *Journal of Nervous and Mental Disease*, Vol. 152, No. 4, pp. 289–297.

Smith, M. J., Cohen, B. G. F., Stammenjohn, L. W., Jr., and Happ, A. (1981), "An Investigation of Health Complaints and Job Stress in Video Display Operations," *Human Factors*, Vol. 23, pp. 389–400.

Smith, M. J. (1987a), "Occupational Stress," in *Handbook of Human Factors*, G. Salvendy, Ed., Wiley, New York, pp. 844–860.

Smith, M. J. (1987b). "Mental and Physical Strain at Computer Workstations," *Behaviour and Information Technology*, Vol. 6, pp. 243–255.

Smith, M. J., and Carayon, P. C. (1995), "New Technology, Automation and Work Organization: Stress Problems and Improved Technology Implementation Strategies," *International Journal of Human Factors in Manufacturing*, Vol. 5, pp. 99–116.

Smith, M. J., and Carayon-Sainfort, P. (1989), "A Balance Theory of Job Design for Stress Reduction," *International Journal of Industrial Ergonomics*, Vol. 4, pp. 67–79.

Smith, M. J., Colligan, M. J., and Tasto, D. L. (1982), "Health and Safety Consequences of Shift Work in the Food Processing Industry," *Ergonomics*, Vol. 25, No. 2, pp. 133–144.

Smith, M. J., Carayon, P., Sanders, K. J., Lim, S. -Y., and LeGrande, D. (1992), "Electronic Performance Monitoring, Job Design and Worker Stress," *Applied Ergonomics*, Vol. 23, No. 1, pp. 17–27.

Smith, M. J., Carayon, P., Smith, J. H., Cohen, W., and Upton, J. (1994), "Community Ergonomics: A Theoretical Model for Rebuilding the Inner City," in *Proceedings of the Human Factors and Ergonomics Society 38th Annual Meeting*, Human Factors and Ergonomics Society, Santa Monica, CA, pp. 724–728.

Stanney, K. M., Maxey, J., and Salvendy, G. (1997), "Socially Centered Design," in *Handbook of Human Factors and Ergonomics*, 2nd ed., G. Salvendy, Ed., Wiley, New York, pp. 637–656.

Stanney, K. M., Maxey, J., and Salvendy, G. (2001), "Socially Centered Design," in *International Encyclopedia of Ergonomics and Human Factors*, W. Karwowski, Ed., Taylor & Francis, New York, pp. 1712–1714.

Stuebbe, P. A., and Houshmand, A. A. (1995), "Quality and Ergonomics," *Quality Management Journal*, Winter, pp. 52–64.

Tasto, D. L., Colligan, M. J., Skjei, E. W., and Polly, S. J. (1978), *Health Consequences of Shift Work*, U.S. Department of Health, Education and Welfare, Publication NIOSH-78-154, U.S. Government Printing Office, Washington, DC.

Taveira, A. D., and Hajnal, C. A. (1997), "The Bondage and Heritage of Common Sense for the Field of

Ergonomics," in *Proceedings of the 13th Congress of the International Ergonomics Association*, Tampere, Finland.

Taveira, A. D., James, C. A., Karsh, B., and Sainfort, F. (2003), "Quality Management and the Work Environment: An Empirical Investigation in a Public Sector Organization," *Applied Ergonomics*, Vol. 34, pp. 281–291.

Taylor, F. W. (1912), *The Principles of Scientific Management*; 1998 reprint, Dover Publications, New York.

Trist, E. (1981), *The Evolution of Socio-technical Systems*, Ontario Quality of Working Life Centre, Toronto, Ontario, Canada.

Turner, J., and Karasek, R. A. (1984), "Software Ergonomics: Effects of Computer Application Design Parameters on Operator Task Performance and Health," *Ergonomics*, Vol. 27, No. 6, pp. 663–690.

Van Ameringen, M. R., Arsenault, A., and Dolan, S. L. (1988), "Intrinsic Job Stress and Diastolic Blood Pressure Among Female Hospital Workers," *Journal of Occupational Medicine*, Vol. 30, No. 2, pp. 93–97.

Warrack, B. J., and Sinha, M. N. (1999), "Integrating Safety and Quality: Building to Achieve Excellence in the Workplace," *Total Quality Management*, Vol. 10, No. 4–5, pp. S779–S785.

Weick, K. E. (1987), "Organizational Culture as a Source of High Reliability," *California Management Review*, Vol. 24, pp. 112–127.

Weisbord, M. R. (2004), *Productive Workplaces Revisited: Dignity, Meaning, and Community in the 21st Century*, Jossey-Bass, San Francisco.

White, L. (1962), *Medieval Technology and Social Change*, Clarendon Press, Oxford.

Wilson, J. R. (1995), "Ergonomics and Participation," in *Evaluation of Human Work: A Practical Ergonomics Methodology*, Taylor & Francis, London, pp. 1071–1096.

Wilson, J. R. (2000), "Fundamentals of Ergonomics in Theory and Practice," *Applied Ergonomics*, Vol. 31, pp. 557–567.

Wilson, J. R., and Grey Taylor, S. M. (1995), "Simultaneous Engineering for Self Directed Teams Implementation: A Case Study in the Electronics Industry," *International Journal of Industrial Ergonomics*, Vol. 16, November–December, pp. 353–366.

Wilson, J. R., and Haines, H. M. (1997), "Participatory Ergonomics," in *Handbook of Human Factors and Ergonomics*, G. Salvendy, Ed., Wiley-Interscience, New York, pp. 353–366.

CHAPTER 11
HUMAN FACTORS AND ERGONOMIC METHODS

V. Kathlene Leonard, Julie A. Jacko, Ji Soo Yi, and François Sainfort
Georgia Institute of Technology
Atlanta, Georgia

1 INTRODUCTION

Methods are a core component in the successful practice of human factors and ergonomics (HF/E). Methods are necessary to (1) collect data about people, (2) develop new and improved systems, (3) evaluate system performance, (4) evaluate the demands and effects of work on people, (5) understand why things fail, and (6) develop programs to manage HF/E. A primary concern for the disciplines of HF/E resides in the ability to make generalizations and predictions about human interactions for improved productivity, safety, and overall user satisfaction. Accordingly, HF/E methods play a critical role in the corroboration of these generalizations and predictions. Without validated methods, predictions and generalizations would be approximate at best, and HF/E principles would present unfounded theories informed by common sense and by anecdotal observations and conclusions. HF/E methods are the investigative toolkits used to assess user and system characteristics as well as the resulting requirements imposed on the abilities, limitations, and requirements of each. HF/E methods are implemented via scientifically grounded empirical investigative techniques that are categorized into experimental research, descriptive studies, and evaluation research. Further discrimination is made based on psychometric properties, practical issues, and descriptive, empirical, and evaluation research methodological processes.

By the very nature of its origin, HF/E is an interdisciplinary field of study comprising aspects of psychology, physiology, engineering, statistics, computer science, and other physical and social sciences. In fact, the back cover of the journal *Human Factors* in 2004 listed 23 topics and 59 subtopics addressed in the papers published therein (although admittedly not an exhaustive list). Why, then, bother with a chapter devoted to HF/E methods when it is an obvious union of well-documented methods offered by this wide array of subject matters? The answer is that the discipline of HF/E is concerned with understanding human-*integrated* systems. That is, HF/E researchers and practitioners strive to understand how the body, mind, machine, software, systems, rules, environment, and so on, work in harmony or dissonance and how to improve those relationships and outcomes. Although HF/E is a hybrid of other disciplines, it demonstrates unique characteristics that make it a distinct field of its own (Meister, 2004).

Fitts's law provides a classic example of the application of basic psychological principles to HF/E problems. Fitts's law is a model of psychomotor behavior, predicting human reaction time. The model relates movement time, the size of two targets, and the distance between them (Fitts, 1954). Fitts's law is a principle that has been adapted and embedded within several HF/E studies of systems that employ similar types of psychomotor responses of the user. Examples include interface design selection, performance prediction, allocation of operators, and even movement time prediction for assembly line work.

However, the original Fitts's law was developed under certain assumptions in its application. This includes error-free performance, the assumption of one-dimensional movement along a single axis, and specifies no guidelines for the input devices that control the movement (e.g., mouse, lever, etc.). This is due to the fact that the original Fitts's law was developed *independent* of any specific system, human variability, and other contextual factors. HF/E specialists therefore have to take these factors into account for the application of Fitts's law to various systems. An example of this is the work of MacKenzie, which extends Fitts's law to human–computer interaction (HCI) research and design. MacKenzie has, in basic research, manipulated Fitts's law to account for the use of a mouse and pointer, two-dimensional tasks, and more (MacKenzie, 1992a,b). Clearly, concessions were made in order to apply Fitts's law within the context of the HCI system.

Perhaps the most demanding challenge encountered by HF/E investigators is deciding the most appropriate methodology to address their goals and research questions (Wickens et al., 2003a). Multitudes of basic scientific methodologies, principles, and metrics are available to HF/E researchers and practitioners. An accompanying challenge of HF/E methods emerges as how to use these methods to assess the human factor within the system context, a feature that is often absent in the traditionally basic research methods. The transformation of physical and social sciences into methodologies applicable to HF/E often generates conflicting goals and assumptions. The HF/E investigator is faced with trade-offs in how they select and actuate the methods. The distinctive nature of the HF/E disciplines is certainly reflected in the methods used by both researchers and practitioners.

The appendix to this chapter is a set of tables that summarize the methods employed in a variety of HF/E investigations. This collection provides a snapshot of the current application of HF/E methods in the published scholarly journals *Ergonomics* and *Human Factors* for the years 2000–2004. Although the tables are not comprehensive, it is representative of the types of methods selected and how they are applied in the various subdisciplines of HF/E. The discipline(s), goal, and method(s) used by authors are underscored and labeled according to different methodological characteristics that are defined in this chapter. One can appreciate the variety of basic methods that HF/E investigators encounter in the generalization and prediction of human interaction with systems and physical machines.

The studies featured apply a range of methods, from highly controlled laboratory settings to loosely structured observational field studies. The variety of goals presented by these authors fit into two categories: (1) HF/E methods that develop and test scientific principles and theories as they apply to human-integrated systems, and (2) HF/E methods that focus on applied problems, incorporating specific features of the target population, task environment, and/or the system. In both cases the investigators are concerned with the human performance or behavior embedded in some system. HF/E researchers and practitioners should look regularly to the scientific literature to glean the types of methods that investigators select to achieve their goals.

No absolute right or wrong exists in method selection and application. However, some methods are more appropriate than others, influenced greatly by circumstantial factors. This is what makes this discipline of HF/E both exciting and frustrating at times for researchers and practitioners alike. The most common answer to the appropriateness of a method for a given HF/E objective is: *It depends*. The specific combination of methods used and control exerted depend heavily on task factors and study goals, among other key contextual factors. The studies listed in the HF/E literature sample, and a handful of others, are referred to provide readers with real examples of the application of HF/E methods employed by investigators in an effort to realize a variety of goals under an assortment of assumptions. Four of the studies from the review have been selected as specific case studies of field studies, survey methodologies, empirical methods, and evaluation studies. Experience is one of the best tools to apply in the selection of methods. Much can be learned from the experiences of others, as their goals, methodologies, and procedures are published in the HF/E literature.

In this chapter we present the reader with critical issues in the selection and execution of HF/E methods such as the ethical issues of working with people, psychometrics issues, and practical constraints. We do not aim to provide readers with all the answers to problems in applying HF/E methods. Instead, readers should gain an improved sense of which questions they should ask prior to implementation of HF/E methods in research or practice. Moreover, we do not provide instructions for the implementation of specific methodologies. Instead, we introduce the most relevant facets of methods for addressing critical issues. Classifications of methods are provided in ways that can direct HF/E investigators in their selection and planning. The data and information gathered through HF/E methods are analyzed and applied in ways unlike traditional psychology or engineering. Chapter 44 is devoted to HF/E outcomes.

2 HF/E RESEARCH PROCESS

Although a significant amount of improvisation is required of HF/E researchers to account for contextual factors while preserving experimental control, a general framework for HF/E investigations can be constructed. This framework is supported by psychometric attributes, previous research methods, principles, outcomes, and ethics of investigations. The foundation of this framework is the specific goals established for a given investigation. Figure 1 presents a schematic of the framework. Each decision point and its relative attributes are addressed throughout the chapter, beginning with assertion of HF/E goals.

2.1 Problem Definition

What motivates HF/E investigations in addition to the underlying desire to improve human-integrated system safety and efficiency is usually the recognition of a problem by the HF/E researcher or specialist, management, or a funding agency. For example, the management team at a software call center may ask the usability group to evaluate the problematic issues of the installation of their software package. Alternatively, a government funding agency may issue a call for proposals to discover the source of errors in hospital staffs' distribution of medication to patients. HF/E investigators also come across ideas from reading relevant scientific literature, networking with colleagues and peers at work and conferences, observing some novel problem, or even through attempting to reveal

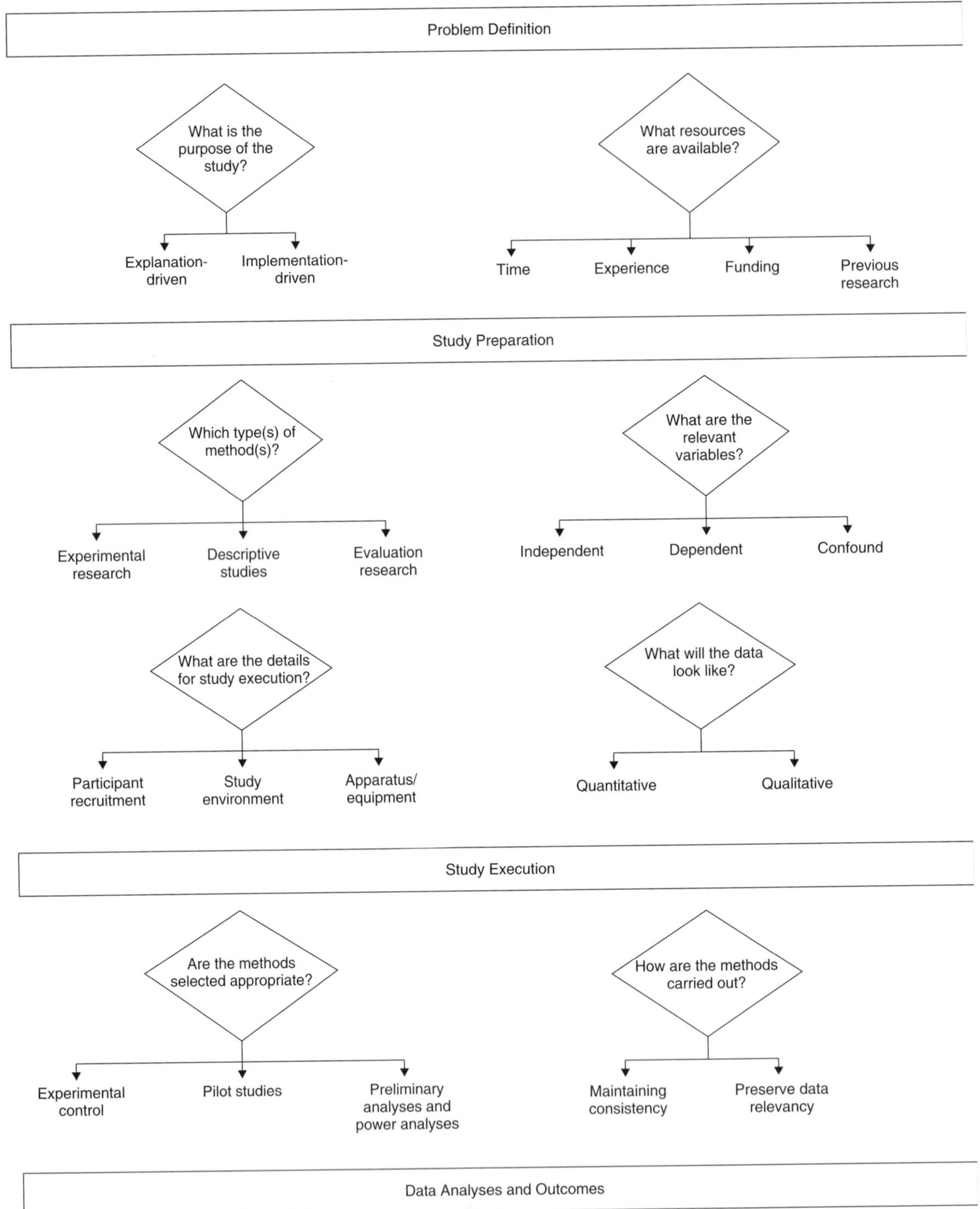

Figure 1 Framework of steps in the selection and applications of HF/E methods.

the source of unexplained variance in their or someone else's research. Problems usually stem from a gap in the research, contradictory sets of results, or the occurrence of unexplained facts (Weimer, 1995). These problems are influential in defining the purpose of the investigation at hand as well as subsequent decisions throughout the application of HF/E methodologies. The first important criterion is to determine the purpose or scope of the investigation. The goal of the investigation is critical. Methods selected have to be relevantly linked to the goal for the investigation to succeed.

Investigations may be classified as *basic* or *applied* (Weimer, 1995). Of course, as with most HF/E theories and principles, these are not completely dichotomous. Studies that are basic are explanation driven, with the purpose of contributing to the advancement of scientific knowledge. Basic investigations may seem out of place because HF/E is so applied in nature. However, explanation-driven basic methods serve a critical role presenting solutions to real-world H/FE problems. Basic research is aimed at advancing the understanding of factors that influence human performance in human-integrated systems. The desired outcome associated with this line of research is some generalizable theory of performance (Weimer, 1995). An example of basic research is a study of the impact of multimodal feedback on user performance in a drag-and drop task with a desktop computer (Vitense et al., 2003). This publication reported the efficacy of eight different supplemental multimodal feedback conditions on user reaction time. Basic research may employ a variety of experimental methodologies. Basic investigations in HF/E are comparable to the nomothetic approaches used in psychology. The nomothetic approach uses investigations of large groups of people in order to find general laws that apply to everyone (Cohen and Swerdlik, 2002).

Pollatsek and Rayner (1998) present classifications and explanations for several basic methodologies of tracking human behavior. These include psychophysical methods (subjective, discrimination, and tachiscopic methods), reaction-time methods, processing-time methods, eye movement methods, physiological methods, memory methods, question–answering methods, and observational methods. A characteristic of basic investigations is that the majority of these methods are operationalized in highly controlled settings, usually in an academic setting (Weimer, 1995). Most commonly, basic studies incorporate theories and principles stemming from behavioral research, especially experimental psychology. HF/E basic research goes beyond basic experimental psychology, conceiving the basic theories that explain the human–system interaction, not just the human in isolation (Meister, 1971). The development of human models of performance, such as ACT-R/PM (Byrne, 2001), demonstrate the integration of basic theories of human information and physiological processes to create an improved understanding of human–system interactions relevant to a variety of applications.

Applied research directs the knowledge from basic research into real-world problems. Work in applied research is focused on system definition, design, development, and evaluation. Applied investigations are, in a sense, supplements to basic research. They would lack merit in the absence of basic research. A characteristic of applied investigations is that the problems identified are typically too specific for their solutions to be generalizable. These investigations are implementation driven, with very specific goals to apply the outcomes relevant to specific applications, populations, environments, and conditions. Applied studies focus on system definition, system design, development, and evaluation. Investigations of the applied nature are used to assess problems, develop requirements for the human and/or machine, and evaluate performance (Committee on Human Factors, 1983).

Applied or implementation-driven work is typically associated with work in the field. Participants, tasks, environments, and other extraneous variables usually need to closely match the actual real-world situation to truly answer the problems at hand. That said, applied methods are abundant and diverse and very much reflect the vast number of HF/E disciplines (mentioned above). Investigators often modify their techniques due to external demands and constraints of the operational system and environment (losing control over what is available for assessment and allowing confounding interactions to occur) (Wixon and Ramey, 1996). An example of this class of applied investigations from the literature summary is found in the analysis of interorganizational coordination after a railway accident (Smith and Dowell, 2000). Stanton and Young (1999) present a comparison of two car stereos to compare several applied methodologies. Some examples of applied methodologies include keystroke-level models (KLM) (Card et al., 1983), checklists, predictive human error analyses, observations, questionnaires, task analyses, error analyses, interviews, heuristic evaluations, and contextual design.

Based on the framework presented in Figure 1, the development of a hypothesis is an important first step in both basic and applied research. Between the two types of research, the difference in the hypotheses is granularity. Hypotheses formulated for applied research are much more specific in terms of applied context. In either case, a hypothesis should be in the form of a proposition: *If* A, *then* B. A problem must be testable if HF/E methods are to be applied. Generally, a problem is testable if it can be translated into the hypothesis format, and the likeliness of truth or falsity of that statement is attainable (Weimer, 1995). However, just the fact that a problem is testable does not ensure that results will be widely applicable or useful. Factors that can affect the applicability and acceptability of the results are discussed in subsequent sections.

2.2 Choosing the Best Method

The choice of HF/E method is influenced by several factors, as the decision to employ a specific methodology elicits several consequences relevant to the efficacy of that method in meeting the established goals. The ability to generalize the results of investigations is shaped by both the design/selection of methods and statistical analysis (see Chapter 44). The "study preparation" section of the framework presented in Figure 1 illustrates the various factors influencing the selection of methods. The judicious selection and implementation of HF/E methods entails a clear understanding of what information will be collected or what will provide the information, how it will be collected, how it is analyzed, and how the method is presented as relevant to the predetermined objectives and hypotheses.

Several authors offer opinions on what the most important considerations should be in the selection of methods. Stanton and Young (1999) provide one of a handful of comparative examinations looking into the utility of various different descriptive methodologies for HF/E. They present case studies evaluating 12 different methodologies based on their use in the investigation of automobile radio controls. The authors evaluated the methods on the criteria of reliability, validity, resources required, ease of use, and efficacy. The accuracy required, criteria to be evaluated, acceptability of the method (to both participants and investigators), abilities of those involved in the process, and a cost–benefits analysis of the method are additional deciding factors for implementation. At the very least, an investigator needs to be conscious of the attributes that are present in their chosen methods and the possible impact, to avoid misrepresenting result and forming ill-conceived conclusions.

Kantowitz (1992) focuses on reliability and validity by looking to problem representation, problem uniqueness, participant representativeness, variable representativeness, and setting representativeness (ecological validity). The specific selection of methodology is rarely covered in HF/E texts. Instead, most authors jump directly past into method selection (e.g., descriptive, experimental, and evaluative) and look at variable and metric definition. However, variable selection, definition, and the resulting validity are intertwined decisively with the methods selected and must link back in a relevant way to the investigation's objectives. In this section, methodological constraints are broken down into two categories:

1. *Practical concerns*
    - Intrusiveness
    - Acceptability
    - Resources
    - Utility
2. *Psychometric concerns*
    - Validity (uniqueness)
        - Construct validity
        - Content validity
        - Face validity
    - Reliability (representation)
        - Accuracy and precision
    - Theoretical foundation
    - Objectivity

Humans are by nature complex, unreliable systems. Kantowitz (1992) asserts that considered as a stand-alone system, the human complexity supersedes that of a nuclear power plant. This creates an abundance of convolutions when considering the human system embedded within another system (social, manufacturing, technological). Humans can be inconsistent in their external and internal behaviors, which are often sensitive to extraneous factors (overt and covert). Undeniably, this creates a conundrum for the HF/E investigator. The impact of this variability can, to a certain extent, be mitigated through various strategies in method selection and implementation. This usually entails close examination and careful attention by the investigator to the relevant variables, including their definition, collection, and analysis. In addition to human fallibility, the selection of any given method and its execution is critically influenced by the objective of the study (basic/applied), objective clarification (i.e., hypothesis), experience of the investigator(s), resources (money, time, staff, equipment, etc.), and previously validated (relevant) research. Returning to the human side of method selection, the selection of method is informed by the ethical and legal requirements of working with human beings as participants.

In practice, it is not feasible to comply with all of the psychometric and practical issues that occur in conjunction with HF/E methodologies. More often than not, the investigator must weigh the implications of their method choice on the desired outcome of the study. Investigators also prioritize the requirements placed on their work with respect to potential impact on the study. For example, the ethical treatment of human participants is of high priority, because an investigator's institutional review board (IRB) or funding agency may choose to terminate the study if risks are posed by participation. In this section we illustrate these potential issues further. It is important for the reader to have an awareness of these issues before our discussion of various methods. In this way, novices can examine the HFE methods more critically with respect to practical constraints most relevant to their research and work.

2.2.1 Practical Concerns

Practical concerns for the application of HF/E methods should be fairly obvious to the HF/E investigator. However, they must be taken into account early in the planning process and revisited continually. Brief definitions for the practical concerns follow.

Intrusiveness This is an appraisal of the extent to which the methodology used interferes with the system being measured. A measure that distracts the participant, or interferes with their performance in other ways, is intrusive. The extent to which

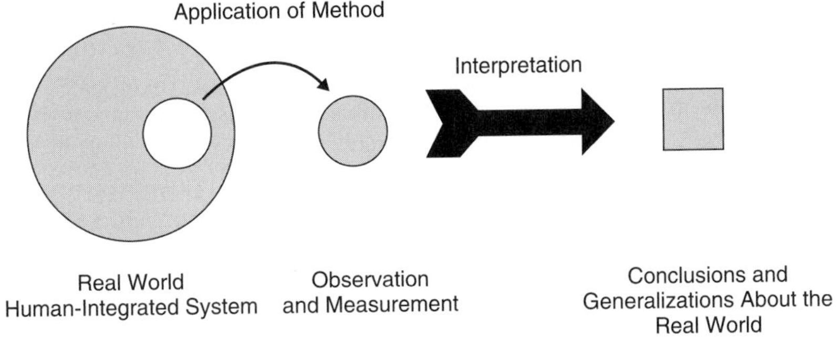

Figure 2 Application of HF/E methods.

an intrusive method causes covariance in recorded observations differs when applied to different scenarios (Rehmann, 1995).

Acceptability This includes the appropriateness and relevance of the method as perceived by investigators, participants, and the HF/E community. For this, the investigator needs to perform an extensive literature review and also network with peers to understand their opinions of the method. Those who fund the research must also be accepting of the method (Meister, 2004).

Resources This refers to the fact that many methods place prerequisites on investigator's resources. These include the time it takes for investigators to train and practice using the method, the number of people needed to apply the method, and any preliminary work that is needed before application of the method. In addition, the method may require the purchase of hardware and software, or other special measurement instruments (Stanton and Young, 1999). Investigations typically have limited financial assistance, which must be considered prior to the adoption of any methodology.

Utility There are two types of utility relevant to HF/E methods: conceptual and physical utility (Meister, 2004). Research with conceptual utility yields results that are applicable in future research on human-integrated systems. Research that holds physical utility proves useful in the design and use of human-integrated systems. In general, investigators need to ensure usefulness and applicability of their proposed methods for the responsiveness of others to their results, and easier dissemination of their finding in conference proceedings, journals, and texts.

2.2.2 Psychometric Concerns

In investigations of human-integrated systems, the methods used should possess certain psychometric attributes, including reliability, validity, and objectivity. Methods are typically used to apply some criteria or metric to a sample to derive a representation of the real world, and subsequently, link conclusions back to the established goals. As depicted in Figure 2, inferences are applied to the measurements to make generalizable conclusions about the real world. Interpretation is inclusive of statistical analyses, generalizations, and explanation of results (Weimer, 1995). The assignment of these inferences should be made to a unique set of attributes in the real world. How well these conclusions match the real world depends on a give and take between controlling for extraneous factors, without disrupting the important representative factors. For example, Figure 2 exemplifies a set of inferences in the shape of a square, which will not easily be matched up to the initial population sampled.

Several of the issues emergent in the selection and application of methods are attributable to representation and uniqueness (Kantowitz, 1992). Representation tends to inform issues of reliability, or the "consistency or stability of the measures of a variable over time or across representative samples" (Sanders and McCormick, 1993, p. 37). A highly reliable method will capture metrics with relatively low errors repeatedly over time. Attributes of reliability include accuracy, precision, detail, and resolution. Human and system reliability, which is a reference to failures in performance, is a topic apart from methodological issues of reliability.

Validity is typically informed by the issues of uniqueness. Validity is the index of truth of a measure, or in other words, if it actually captured what it set out to, and not observing the extraneous (Kantowitz, 1992; Sanders and McCormick, 1993; Kanis, 2000). Both concepts are alluded to by many, defined by few, and measured by an even more select group of HF/E researchers and practitioners, with many different interpretations for this basic concept (Kanis, 2000). Despite the disparity in definitions and interpretations of the terms, it is generally agreed that these concepts are multifaceted.

Reliability and validity in HF/E are not dichotomous, but lucid concepts, as they may appear in the social sciences. The evolutionary nature of the HF/E discipline does not support such a "neatly organized" practice (Kanis, 2000). A method must be reliable to be valid, but the reverse is not always true (i.e., reliable methods are not necessarily valid) (Gawron,

2000). Stanton and Young (1999) found this to be the case in their evaluation of hierarchical task analyses. The predictive validity of this method was found to be robust, but the reliability was less so. The authors concluded that the validity of the technique could not be accepted because of the underlying shortcomings in terms of reliability. Types of validity include face, content, and construct. In this section we discuss different psychometric properties of both reliability and validity, how to control for them in the selection of methods, limitations imposed by the practical issues, and the resulting trade-offs. One disclaimer before the discussion of validity and reliability that ensues: Although validity and reliability of methods enhance acceptance of the conclusion, they do not guarantee widespread utility of the conclusion (Kantowitz, 1992).

Key Characteristics of Reliability Characteristics of reliability include accuracy and precision, which influence the consistency of the methodology over representative samples and the degree to which the methods and results are free from error. Accuracy is a description of how near a measure is to a standard or true value. Precision details the degree to which several methods provide closely related results, observable through distribution of the results. Test–retest reliability is a way to assess the precision of a given method. This is simply an assessment of correlations between separate applications of the methods. Sanders and McCormick (1993) report that for HF/E, test–retest reliability scores of 0.80 and above are usually satisfactory. This score should be taken in context, however, because what determines an acceptable test–retest reliability score is intertwined with the specific contextual factors of the investigation.

The level of precision and/or accuracy sought in HF/E method selection and implementation is heavily contextually dependent. The investigator needs to select the method with reliability that is consistent with the requirements alluded to in the goals and problems. The keystroke-level model (KLM) introduced by Card et al. (1983) was one of the first predictive methods for the field of HCI. This method predicts the time to execute a task given error-free performance using four motor operators, one mental operator, and one system response operator. KLM predicts error-free behavior, so the functions to calculate the time for the operators would probably be consistent between the applications of the method. The accuracy of the KLM method is purportedly high for certain tasks (Stanton and Young, 1999). However, the accuracy of the method could deviate drastically from what Stanton and Young observed in their evaluation of car stereo designs, when KLM is applied to a different scenario, with the overall precision of the method constant.

Face Validity Face validity is defined as the extent to which the results look as though the method captured what is intended (Sanders and McCormick, 1993). It is a gauge of the perceived relevance of the methods to the identified goals of the investigation, without any explanation by the investigator. Not only is face validity important with respect to acceptance of the results reported by the scientific and practicing communities, but it is also important from the perspective of the participants in the study. If a measure seems irrelevant or inappropriate to the participants, it may affect their motivation in a negative way. People may not take their participation seriously if the methods seem disconnected from the purported goals. This can be mitigated first by briefing the participants on the purpose of the methods used or by collecting measures of the performance in the background, so the participant is not exposed to the specifics of the study.

Content Validity The content validity of a method is essentially the scope of the assessment relevant to the domain of the established goals of the investigation. The analysis of Web logs provides an example of content validity. For example, consider an investigation with the goal to report employee use of a corporate intranet portal, which provides information on insurance benefits. Simply reporting the number of hits the portal receives does not possess high content validity. This is because this method provides no indication if the employees are actually pulling content from the intranet site. The Web logging methodology could instead look at various facets of employee activity on the intranet site to illustrate a more complete representation of use, as well as actually to talk to some employees to get verbalizations and perceptions about the intranet site.

Construct Validity Construct validity is best defined as the degree to which a method can be attributed to the underlying paradigm of interest. Figure 3 exemplifies the concept of construct validity and other relevant features. The gray-shaded circle on the left represents the model or theory under scrutiny, and the white circle on the right represents the space that is assessed by the selected measures. The star marks the intersection of these two spaces, which represents the construct validity of the measure. It represents the aspects of the target construct that are actually captured by the methods used. This measure leaves out elements of the construct (because it cannot account for the entirety of the concept), called the *deficiency* of a measurement. Equally, there are areas unrelated to the construct that the measure captures. This undesired or unintended area measured is termed the *contaminant*.

Physiological methods such as heart rate are especially prone to construct validity issues. For example, heart rate may be affected by caffeine, medications, increased mental workload, age, or physical stressors such as exercise. Consider a study that aims to measure the mental workload experienced by drivers while driving under different weather conditions, by recording each driver's heart rate. The investigation will not necessarily detect the changes in mental workload as affected by the conditions. Instead, the investigator will also have captured a measure of the effects of coffee consumption, age, recent physical activity,

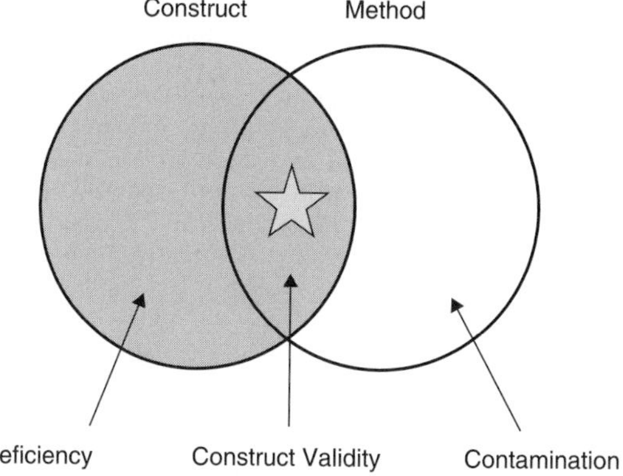

Figure 3 Construct validity. (Adapted from Sanders and McCormick, 1993.)

medications, and mental workload, so that the effects of mental workload are virtually inseparable from the other *extraneous* contamination variants. The investigator can mitigate these contaminants through exercising control in the applications of their methods. In this case, specifying inclusion criteria for subjects' selection and participation in the study would be advantageous. Methods of control are discussed further in this section.

Controlling for Reliability and Validity If not controlled during the selection and application of HF/E methods, problems with validity and reliability can prove detrimental to the generalizability and predictive value of conclusion. The following issues in the results are probably ascribable to matters of validity and reliability:

- A lack of correlation between reality and the criteria used
- A correlation of the criteria with unknown bias(es), so even if changes are detected the absolute value of the factor(s) cannot be determined
- Multivariate correlations, because the construct of interest is actually affected by several factors
- Interference from extraneous factors may inappropriately suggest causal relationships when it is in fact just a correlation

Psychometric issues may be mitigated through the control of extraneous factors that can affect the construct, and collecting data/observations in representative environments. That being said, fundamental conflicts often arise in trying to ascertain control without sacrificing critically representative aspects of the system, task, or population. Furthermore, time, financial, and practical constraints can make it impossible to ascertain desired levels of validity of H/FE methods. There are methods and approaches for the analysis of HF/E outcomes, discussed in Chapter 44, which can potentially account for some of the validity and reliability issues. Yet much like HF/E in the design process, the earlier changes are made in the selection and applications of a methodology to correct for validity and reliability issues, the more easily the changes are implemented and the greater positive impact they will have on the methodological outcome.

Control Control in the selection and application of methods strives to challenge the sources of variance to which HF/E is highly prone. Sources of variance can include noise from the measurements, unexpected variance of the construct, and unexpected participant behavior (Meister, 2004). It is the regulation of standard conditions to reduce the sources of variance. Variance is detrimental to HF/E methods because it restricts the certainty of inferences made. Control removes known confounding variables by making sure that the extraneous factors do not vary freely during the investigation (Wickens et al., 2003a). Control is not absolute. An investigator can exercise various levels of control to ensure minimal effects of confounding variables. Methods that lack control can lead to data that are virtually uninterpretable (Meister, 2004).

Ways to reduce variance include choosing appropriate participants, tasks, contexts, and measures; eliminating confounding variables to reduce covariance effects; implementing methods consistently; and increasing the structure with which the methodology is utilized. For example, control was exercised in research conducted to predict performance on a computer-based task for people with age-related macular degeneration (AMD). Great control was exercised in the selection of participants who had AMD and age-matched controls (Jacko et al., 2005). The selection of participants controlled for the exclusion of any person who had any ocular dysfunction other than AMD. Great care was also taken in ensuring that the variation of age between the experimental and control groups was consistent. If age had not been controlled for in recruitment, the differences between the two groups could be a result of interactions between age and ocular disease. Methods of control will be introduced in relation to experimental studies, descriptive studies, and evaluations.

Participant Representativeness As stated earlier, human behavior is sensitive to a variety of factors, and interactions often surface between specific characteristics of the participant and the environment. The extent to which the results of investigations are generalizable depends on those connections between the characteristics of those observed and the actual population. Although it is not always necessary only to sample participants from the actual population (Kantowitz, 1992), consistency checks should be made. In a study of age-related differences in training on home medical devices, presented in the literature summary table (Mykityshyn et al., 2002), the investigators needed to recruit persons from the aging population so that their age-related capabilities, mental and physical, would be consistent with that in the general population. The relevant aspects of the population investigated should be in the same proportion as what is found in the real population (Sanders and McCormick, 1993).

Variable Representativeness The selection of methods mandates the selection of necessary measures and variables. For HF/E, measurement is the assignment of value to attributes of human-integrated systems. Assigning value can be accomplished through various methods, such as nominal, ordinal, interval, and ratio scales. In Chapter 44 we discuss HF/E outcome measurements and their analysis in detail. A given measure affords a specific set of statistical summary techniques and inferences that have implications on validity and reliability.

Measurement selection and its assignment of value to events in human-integrated systems should be guided by theory and previous studies. In HF/E it is most germane to include more than one measure. Three classes of variables have been identified as necessary to capture human-integrated systems: (1) *system descriptive criteria*, which evaluate the engineering aspects of a system, (2) *task performance criteria*, which indicate the global measure of the interaction such as performance time, output quantities, and output qualities, and (3) *human criteria*, which capture the human's behavior and reactions throughout task performance through performance measures (e.g., intensity measures, latency measures, duration measures), physiological measures, and subjective responses (Sanders and McCormick, 1993).

Table 1 presents examples of task performance, human criteria, and system criteria. System descriptive criteria tend to possess the highest reliability and validity, followed by task performance criteria. Human criteria are the noisiest, with the most validity and reliability issues. Note that human criteria demonstrate the broadest classification of measurements. This is due to the inherent variability (and noise) in human data. These metrics—performance, physiological, and subjective responses—portray a more complete characterization of human experience when observed in combination. Performance optimization should not be pursued, say, at the cost of high levels of workload observed through heart rate and subjective measures using the NASA-TLX subjective assessment of mental workload. A useful guide in the selection of specific human measures is Gawron's *Human Performance Measures Handbook* (2000), where over 100 performance, workload, and situational awareness measures are defined operationally for application in different methodologies.

"The utility of human factors research is linked intimately to the selection of measures" (Kantowitz, 1992, p. 387). Validity and reliability of results are best substantiated when the three classes of criterion are addressed in the methodology. In fact, the selection of measurements or metrics is second in importance to method selection. Measurement and methodology are not independent concepts, and the selection of both are closely related and often iterative (Drury, 1995). The interpretation of results is greatly influenced by the combination of measures chosen in a particular test plan. Robust measurement techniques can make the interpretation process a much more streamlined process.

Table 1 Classification of Criteria Addressed in HF/E Methods

System Descriptive Criteria	Task Performance Criteria	Human Criteria
Reliability	Quantity of output	Performance
Quality measures	Output rate	Frequency
Operation cost	Event frequency	Latency
Capacity	Quality of output	Duration
Weight	Errors	Reliability
Bandwidth	Accidents	Physiological
	Variation	Cardiovascular
	Completion time	Nervous system
	Entire task	Sensory
	Subtask time	Subjective opinion
		Situation awareness
		Mental workload
		Comfort
		Ease of use
		Design preference

Objectivity A second issue of variable representativeness is the level of objectivity in its definition and measurement. Objectivity is a function of the specific techniques employed in collecting and recording data and observations. Data and observations recorded automatically are the most objective approach. Objective variables can be captured without probing the participant directly. In contrast, in the collection of highly subjective variables, the participant is the *medium of expression* for the variable (Meister, 2004). The investigator may also interject subjectivity. The investigator can impose subjectivity and bias in how they conduct the investigation, which participants they choose to collect data from, and what they attend to, observe, and report.

For instance, three levels of objectivity can be demonstrated in capturing task time for a person to complete the assembly of widgets on a manufacturing line. A highly objective method may involve the use of the computer to register and store task time automatically based on certain events in the process (e.g., the product passes by a sensor on the manufacturing line). A less objective method would be to have the investigator capture assembly with a stopwatch, where the investigator determines the perceived start and completion of the assembly. Finally, the least objective, most subjective method would be to ask the participant, without using a clock, to estimate the assembly time. Clearly, the level of objectivity in the methods influences the accuracy and precision of the outcomes. Subjective measures are not unwarranted and are in fact quite important. In the assembly example, if the worker perceives the assembly time for a certain component as very long, even if a more objective assessment method does not detect it as long, the workers perceptions still affect the quality of the work they produce and the amount of workload they perceive, ultimately affecting the quality of the system and its resulting widgets. In eliciting covert mental processes from people for measurement and assessment, the investigator does, however, need to ensure that the results obtained will adequately answer the questions defined in the formulation of a project's goals.

Finally, how researchers choose to interpret the results of methods may also induce subjectivity into the outcome. By decreasing the level of involvement of either participant or investigator in the expression of performance, an increase in the overall objectivity of the method and outcomes can be realized. Tables 2 and 3 provide a taxonomy of methods and measures that summarize Meister's (2004) work. Subjective and objective methods are outlined with a specific example of each class of methods. In conclusion, both subjective and objective measurements have their place in HF/E investigations (Wickens et al., 2003a).

Setting Representativeness Setting representativeness is the coherence between the environment where methods are performed and the real-world environment of the target situation where the results are to be applied. This is not necessarily a judgment of realism, but rather, the level of comparability between how the participants' physical and psychological processes are affected by the context of the study (Kantowitz, 1992). This informs the investigator's consideration for collecting data in the laboratory versus in the field. Another term for setting representativeness is the *ecological validity* of the study. Ecological validity influences the generalizability of the results but can also influence the behavior of those who participate in the investigation. Research has shown that participants can exhibit different behavior when they know they are being observed, a phenomenon known as the *Hawthorne effect* [for a complete overview of the Hawthorne effect, see Gillespie (1991)]. The more representative the task and environment, and the less intrusive the investigations to the participants' behaviors, the better this effect can be mitigated. The investigator may retain a more complete picture of human behavior with a complex system

Table 2 Taxonomy of HF/E Objective Method

Objective Methods			
Outcome Measure	Method	Description	Example(s)
Performance measures	Unmanaged performance	Measures of human-integrated system performance through unnoticed observations *Assumptions*: (1) System of measurement is completely functional; (2) mission, procedures, and goals of the system are fully documented and available; (3) expected system performance is available in quantitative criteria to link human performance with system criteria.	Evaluation of an advance brake warning system, in government fleet vehicles (Shinar, 2000)
	Empirical assessment	Comparison of conditions of different system and human characteristic in terms of treatment conditions *Assumption*: Experimental control of variables and representativeness will enable conclusions about the correlation between manipulated conditions for valid, generalizable results.	Investigation of multimodal feedback conditions on performance in a computer-based task (Vitense et al., 2003)
	Predictive models of human performance	Applying theories of cognition and physiological processes and statistics to predict human performance (involves no participation by human participants) *Assumptions*: (1) The model explains the cognitive processes to a reasonable degree; (2) the model incorporates contextual factors relevant to the operational environment.	ACT-R/PM, a cognitive architecture to predict human performance, using drop-down menus (Byrne, 2001)
	Analysis of archival data	Aggregated of data sets aimed at the representation of a particular facet of human-integrated systems; HF/E archival data from journal articles, subdivided into subtopics such as computers, health systems, safety, and aging *Assumptions*: (1) Differences between individual study situations are small in nature; (2) error rate predicts performance with validity; (3) the models will be informed continually by new data studies.	Anthropometric differences among occupational groups (Hsiao et al., 2002)

Table 3 Taxonomy of HF/E Subjective Methods

Subjective Methods			
Outcome Measure	Method	Description	Example(s)
Observational measures	Observations	Information about what happened and is happening in the human-integrated system; status of the person(s) and system components and their characteristics of the outcomes *Assumptions*: (1) What the observation recorded is the essence of what actually happened; (2) observers record the situation veridically; (3) interobserver differences are minimal with respect to reliability; (4) no questions are probed during observation	Task analyses of automated systems with which humans interact (Sheridan, 2002)
	Inspection	Similar to observation, but objects have a role in what is considered; comparisons between the object at hand and a predetermined guideline for the required characteristics of both the object and the target user *Assumption*: The object of inspection has some deficiency, and the standards provided are accurate in representing what is truly required	Ergonomic redesign of a poultry plant facility; data collected on tracking employee musculoskeletal disorders are compared against OSHA guidelines (Ramcharan, 2001)
Self-reported measures	Interviews and questionnaires	Direct questioning for the participant(s) to express convert mental processes, including reasoning of perceptions of their interaction *Assumptions*: (1) People can validly describe their response to different stimuli; (2) the words and phrasing used in the questions accurately capture what is intended; (3) credibility of the respondent in their ability to answer the questions; (4) formality of structure required in responses	Case study of disaster management (Smith and Dowell, 2000)
	Concurrent verbal protocol	Verbal protocols to elicit covert participant information processing while executing a task; participant explains and justifies actions while performing the task *Assumption*: People can better explain processes when there are aspects of the ecological validity.	Study of mental fatigue on a complex computer task (van der Linden et al., 2003)
Judgmental measures	Psychophysical methods	Questions that determine thresholds for discrimination perceptual qualities and quantities; size weight, distance, loudness, and so on. *Assumptions*: (1) The participant has a conceptual frame of reference for evaluations; (2) the judgment is a result of analysis of internal stimuli.	Investigation of time estimation during sleep deprivation (Miro et al., 2003)

in the actual, operational environment (Meister, 2004), which supports fewer objective measures.

HF/E investigators must ultimately decide where the best location is to collect data: in the field or in the laboratory. The collection of data in a field study versus in a highly controlled laboratory setting is a trade-off that HF/E practitioners and researchers continually debate in the execution of methods. Field research typically provides an investigation of the means to look at the system in order to shape their assumptions about the construct, in a way that informs the selection and implementation of other methods (Wixon and Ramey, 1996). Wixon and Ramey (1996) claim further that most field studies are best suited for situations about which little is known, saving time in the laboratory studying the wrong problem. Conversely, fieldwork serves the purpose as an executable setting to validate theories and principles developed in more controlled, laboratory environment environments. Case Study 1 provides a summary of HF/E work from the literature in which field studies have been used.

CASE STUDY 1: Effects of Task Complexity and Experience on Learning and Forgetting

Goal The goal of this study by Nembhard (2000) was to investigate how task complexity and experience affect individual learning and forgetting in manual sewing tasks using worker-paced machinery that placed high demands on manual dexterity and hand–eye coordination.

Methods A notable amount of related research had been conducted in this domain through laboratory studies. Although the previous research was strong in finding causal inferences, it could not validate the findings for real-world situations. This gap in the knowledge base motivated the author to study the effects of task complexity and experience on performance in the factory. In the design of the study, the author ascribed task complexity and worker experience (e.g., training for their task) as two prominent factors determining the trends of learning and forgetting during task performance. Task complexity was measured by three variables: complexity of the method, machine, and material. Over the course of one year, the study captured 2853 episodes of learning/forgetting from all the workers. User performance was sampled 10 times per week, and averaged to derive the learning/forgetting. The complexity variables and the worker experience variable were recorded in combination with each learning and forgetting episode.

Analyses Based on the data collected, the parameters, and the variables derived (e.g., prior expertise, steady-state productivity, rate of learning, and degree of forgetting), a mathematical model of learning and forgetting was developed. Then, using statistical methods such as Kolmogorov–Smirnov, ANOVA, pairwise comparisons, and regression, the effects of learning and the effects of task complexity and experience on learning/forgetting were extrapolated.

Methodological Implications

1. *Expensive to conduct.* As this study shows, a field study requires a larger number of samples (i.e., 2853 episodes) or observations to mitigate extraneous confounds. Thus, it can take more time and be costly.

2. *Strongly valid.* Because the research hypotheses and questions are tested under real situations, the validity of the argument is usually strong. In fact, this served as the major motivation for this study. This enables improved implementation of the results of this study back into the field more easily than those laboratory-based conjectures of potential causal relationships.

3. *Complex analyses.* The data from a field study are naturally large and complex because they were captured under real situations. That is, they are subject to a lot of extraneous noise in the system observed. Therefore, strategic analytical methods are quite useful. For example, this study simplified the presentation of data by introducing a mathematical model of learning/forgetting.

Typically, the data from field methods are more subjective (coming mostly from surveys and observations), which affects the analysis of the outcomes. Three widely recognized field methods include (1) ethnography (Ford and Wood, 1996; Woods, 1996), (2) participatory design (Wixon and Ramey, 1996), and (3) contextual design (Holtzblatt and Beyer, 1996). These types of methods typically illustrate the big picture and do not provide the investigator with a simple yes or no answer (Wixon and Ramey, 1996). Instead, the data gathered tend to be information rich, somewhat subjective, and highly qualitative.

2.2.3 Trade-offs

Control versus Representation The need for experimental control and representative environments, tasks, and participants creates a fundamental conflict. It is impossible to have both full control and completely representative environments (Kantowitz, 1992). This is because in representative environments participants control their environment, as they want to, barring any artificial constraints. Both highly controlled and representative methods serve important roles in HF/E. The answer of which to sacrifice, when faced with this predicament, is entirely dependent on the objectives of the study and the availability of resources. Ideally, an investigation should be able to incorporate aspects of both. The selection of H/FE methodologies is largely directed by the investigator's needs and abilities in terms of objectivity and control and how the results are to be applied (Meister, 2004). In fact, investigators often include specific aspects of the operational system while controlling aspects of the testing environment. The applied nature of HF/E (even in basic research) leads investigators to simulate as much as they can in a study while maintaining control on extraneous factors (Meister, 2004).

Ideally, HF/E researchers and practitioners strive to generalize the results of investigations to a range of tasks, people, and contexts with confidence. Intuitively, it becomes necessary to apply methods to a range of tasks, people, and contexts to achieve this. Conflict often arises in terms of the available resources for the investigations (time and money). That said, the level of representation achieved through a method should be selected consciously, addressing the set goals, what is known about the method, and the practical limitations. Of course, the larger the sample size, the more confidence in results, but the larger sample usually entails higher financial and time investments. Furthermore, human limitations such as fatigue, attention span, and endurance may impose constraints on the amount of information to be gathered. It is also critical to consider the implications of method choice in terms of the analysis used. For example, investigations, which collect and manipulate a large numbers of variables, can take months to mine and analyze the data. Qualitative data or videos can take a significant amount of time to code for analysis. For this reason, the reader is encouraged to review Chapter 44 prior to using HF/E methods.

There is a point of diminishing returns when it comes to increasing the size of the sample or number of observations. In other words, the amount of certainty or knowledge gained from the additional observation may or may not be worth the time and effort spent in its collection and analysis. Sanders and McCormick (1993) introduced three factors that can influence sample size:

1. *Degree of accuracy required.* The greater the required accuracy, the larger the sample size required.
2. *Variance in the sample population.* The greater the variance, the larger the sample size required.
3. *Statistic to be estimated.* A greater number of samples are required to estimate the median than the mean with the same degree of accuracy and certainty.

2.2.4 Incorporating Theory and Previous Work

The number of factors to consider in the selection of HF/E methodologies may seem an impossible task. However, method selection is greatly informed by theory, as well as previous applications of the methodologies, as documented in the literature. The investigator needs to examine the existing knowledge base critically, as well as talk with other HF/E investors to gain practical insight. This serves as one of the best ways to justify the selection of methods for the problem at hand. In the selection of methods, only those that offer evidence of practicality and validity should be selected.

When conducting a critical examination of literature, readers should be cognizant of the following factors, adapted from Weimer (1995):

- What are the authors' goals, both explicit and inferred from the text?
- What prior research do they reference, and how do they interpret it?
- What are their hypotheses?
- How are the methods linked to the hypothesis?
- What are the variables (independent, dependent, and control), and operationally, how are they defined?
- Are extraneous variables controlled, and how?
- What are the relevant characteristics of the participant population?
- How did the authors recruit the participant population, and how many people did they use?
- Was the research done in a laboratory or in the field?
- Did they use any special measurement equipment, technologies, surveys, or questionnaires?
- What statistical tests were run?
- What was the resulting statistical power?
- How do the authors interpret the results?
- How well do the authors' results fit with the existing knowledge base?
- Are there any conflicts in the interpretation of data between authors of different studies?

Investigators who are able to find literature relevant to their objective problem(s) can apply methods based on what others have applied successfully. However, because the subject matter and context of HF/E are so varied, care should be taken in this extrapolation. In the application of historically successful methods, the investigator must justify any deviation he or she made from the accepted status quo of the method. Basic research typically has the most to gain from such literature reviews.

Applied research is somewhat more problematic, as the methods are more diversified according to the conditions specified in the target application. Additionally, there are issues in the documentation of applied methods (Committee on Human Factors, 1983). The historical memory of human factors methods resides largely in the heads and thick report files of the practitioners. Probing colleagues and other HF/E practitioners for their practical experience in using applied methods is therefore useful.

The survey of relevant literature is such a critical step in investigation that it can, by itself, serve as a complete method. A thorough literature search for theory and practice combined with discussions of methodological issues prevent investigators from reinventing the methodological approach. It may also save time through the avoidance of the common pitfalls in the execution of certain methods and analyses. In fact, literature reviews can substitute the need for the application of methods if the experimental questions have already been addressed. A *meta-analysis* is a specific method for the combination of statistical results from several studies (Wickens et al., 2003a).

Basic HF/E investigations, theories, and principles from psychology, physiology, and engineering all merit review. Theory is especially important because it can direct attention in complex systems as to where to focus resources. Knowledge of existing theories provides blueprints for the selection and application of methods as well as the explanation of results (Kantowitz, 1992). Theory is essential in the planning process because of the need to link the methodological processes and hypotheses strongly to the problem and goals of the investigation. However, when applying theories and principles, the investigator must ensure that the end results are in line with the theories employed.

The guidelines for critical examinations of literature serve another role for the selection and application of HF/E methods. Investigators should realize that others will, one day, examine their work critically in a similar fashion. That said, consideration of these systematic evaluations in the design of methods can save undue hardship later during analysis and especially in the course of result interpretation. Similar to the design of systems, it is easier to make changes in the beginning steps of method formulation than to make changes to, or draw logical and meaningful conclusions from, ill-conceived outcomes of the method.

The decision of what methods are generally accepted by the community as standard and valid is difficult. The field of human factors, relative to more basic research, is much less grounded in terms of methodology. With the continued introduction of new technologies and systems, investigators are continually deriving new methods, metrics, and inferences. Despite this ongoing development, it is the investigator's responsibility to consider the issues of validity, reliability, and practical issues of a method before implementing it and reporting the ensuing results. Furthermore, HF/E investigators, in the report of results need to clearly mark what is informed by theory and what, in actual fact, is their own speculation. Speculations are easily mistaken for fact when not labeled explicitly as such (Meister, 2004). This is true even when examples of a method exist in the literature, as the new method of application must be validated. What should be considered is *how* the authors in the scientific literature justified a method (and those who do not justify should be looked at with some skepticism).

2.3 Working with Humans as Research Participants

Many HF/E methodologies require humans to serve as participants, providing data needed in the analysis of the system. As researchers (and humans beings ourselves), we are bound to the ethical handling of participants and their data. The foundation of ethical concerns is to ensure that the investigators do not sacrifice participants' general health, welfare, or well-being in lieu of achieving results for their research goals. Professional and federal agencies have assembled specific guidelines aimed at the appropriate treatment of people and their data in research and analysis funded by U.S. federal monies. The federal code of regulations for the protection of human subject (U.S. Department of Health and Human Services, 2001) (for investigators in the United States) and the American Psychological Association ethical guidelines for research with human participants (American Psychology Association, 2002) should be familiar to anyone conducting research with people as participants. Basically, these principles entail (1) guarding participants from mental or physical harm, (2) guarding participants' privacy with respect to their actions and behavior during the study, (3) ensuring that participation in the research is voluntary, and (4) allowing the participant the right to be informed about the nature of the experimental procedure and any potential risks (Wickens et al., 2003a).

Although the associated risk of HF/E investigations may seem minor, the rights of participants should not be taken lightly. Several historical events have informed the development of codes of conduct under which participants experienced undue mental and/or physical harm. Perhaps the most widely known is the Nuremberg Code, written by American judges in response to scientific experiments (mental and physical) in which prisoners were exposed to extreme medical and psychological tests in Nazi concentration camps. The Nuremberg Code was the first of its kind and mandates that the duties of those conducting research have the responsibility to protect the welfare of the participant ["Nuremberg Code (1947)," 1996].

Even after the Nuremberg Code, several instances of unethical treatment of human participants were documented. In 1964 the Declaration of Helsinki was developed to provide guidance to those conducting medical trials (World Medical Association, 2002). Finally, in 1979 the Belmont Report (Office for Human Research Protections, 1979) was released partly in response to inappropriately conducted U.S. human radiation experiments (U.S. Department of Energy, 2004). The three principles emergent from the Belmont Report included (1) *respect* in recognition of the personal dignity and autonomy of individuals and special protection for those with diminished autonomy, (2) *beneficence* by maximizing the anticipate benefits of participation and minimizing the risks of harm, and (3) *justice* in the fair distribution of research benefits and burdens to participants (Office for Human Research Protections, 1979).

To aid researchers in ethical conduct, many institutions have what is called an institutional review board (IRB) to provide guidance and approval for the use of human participants in research. Each protocol must be approved by the IRB before experimentation can begin. The IRB may review, approve, disapprove, or require changes in the research activities proposed. Approval is based on the disclosure of experimental details by the investigator(s). For example, many IRBs request the following information:

- Completion of educational training for research involving human participants by all persons (investigators and support staff) involved in the study
- A description of the research in lay language, including the scientific significance and goals:
    - Description of participant recruitment procedures (even copies of advertisements)
    - Inclusion/exclusion criteria for participant entry and justification for the specific exclusion of minorities, women, or minors
    - Highlights of the potential benefits and risks
    - Copies of all surveys and questionnaires
    - Vulnerable groups such as minors
- Funding of the research
- Location of the research
- How the data will be archived and secured to ensure participant privacy

In addition, researchers are instructed to create an informed consent form for the participants to sign. This document, approved by the IRB, explains the nature and risks of the study, noting voluntary participation, and stating that withdrawal from the study is possible at any time without penalty.

Although the documentation and certification to ensure the welfare of participants may impose a lot of paperwork, these factors do have implications as to the quality of results in HF/E. The more comfortable the participant is, the more likely they are to cooperate with the investigator during human–system investigations (and return for subsequent sessions). This contributes to the acceptability of a method by participants, one of the practical criteria to be used in method selection described in Section 2.2.

2.4 Next Steps in Method Selection

Operational methods are most commonly classified into three categories: (1) experimental studies, (2) descriptive studies, and (3) evaluative studies. The selection of methodology from one of these categories will lead the investigator through a series of directed choices, as depicted in Figure 1. These decisions include:

- What are the relevant variables?
  - How are they defined?
  - How are they captured?
- What will the actual measurements look like?
  - Is it qualitative?
  - Is it quantitative?
  - Is it a combination of both?
- What levels of experimental control and representativeness will be exercised?
  - Who are the participants?
  - Where will the study be conducted?
  - What equipment and measurement tools are needed?

The selection of method type, variables, measurements, and experimental control factors are very intertwined. There is no specific order to be followed in answering these questions except what is directed by the priorities established in the problem definition phase. The choices available in response to each question are limited, according to the method that is applied. Furthermore, the psychometric and practical issues introduced in this section must be verified routinely during the selection of specific plans, for improved robustness of predictability and generalizability of the investigation outcomes.

In the remainder of this chapter we introduce the three different operational approaches. Specific examples of methodologies in each category are provided, along with answers to the questions outlined above and in Figure 1. The execution of each method will also serve as a point of discussion. Although the number of issues to consider in HF/E methods is sizable, the implications of improper use of methods can be far reaching. The careless application of HF/E methods may result in lost time, lost money, health detriments, discomfort, dissatisfaction, injury, stress, and loss of competitiveness (Wilson and Corlett, 1995).

3 TYPES OF METHODS AND APPROACHES

The taxonomy of HF/E methodologies is not straightforward, as there are areas that overlap within these defining characterizations (Meister, 2004). However, a classification enables guidance in methodology selection. There are several different classifications of methods in the literature, each author presenting the field in different scope, point of view, and even terminology. One of more detailed and comprehensive taxonomies is that of Wilson and Corlett (1995). The authors classify methods as (1) general methods, (2) collection of information about people, (3) analysis and design, (4) evaluation of human–machine system performance, (5) evaluation of demands on people, and (6) management and implementation of ergonomics into group and subgroup. The authors then detail 35 groups of methods each with subgroup classifiers. Finally, the authors present techniques that are used in each method, and common measures and outcomes.

Other authors, this handbook included, present a more simplified classification of methodological processes. Although the taxonomy presented by Wilson and Corlett (1995) has utility, that level of detail is beyond the scope of this chapter. Instead, methods are broken down in a manner similar to those of Meister (1971), Sanders and McCormick (1993), and Wickens et al. (2003a).

Methodologies will be classified as *descriptive, experimental*, and *evaluation-based*. Thus far, classifications of basic and applied research goals and attributes that methods can possess in terms of validity, reliability, and objectiveness have been covered. Each of the three classes of research best serves a different goal while directing the selection of research setting, variables, and participants (to meet the demands of validity, reliability, and objectiveness). Although some overlap exists between descriptive, experimental, and evaluation-based methods, HF/E research can usefully be classified into one of the three (Sanders and McCormick, 1993).

3.1 Descriptive Methods

Descriptive methods assign certain attributes to features, events, and conditions in an attempt to characterize a specific population (Sanders and McCormick, 1993). The investigator is typically interested in describing a population in terms of attributes, identifying any possible parallels between attributes (or variables). The variables of interest encompass who, what, when, where, and how. The objective of descriptive research is to obtain a "snapshot" of the status of an attribute or phenomenon. The results of descriptive methods do not provide causal explanation of attributes. Correlation is the only relationship between variables that can be determined unless the specific attributes of relationships are captured.

The utility of descriptive research is that it provides a basis for conducting additional, more specific investigations. Descriptive methods are identified by the characterization of system states, populations, or interactions in its most natural form, without manipulation of conditions (as in the case of empirical methods). The results of descriptive research methods often serve as motivation for experimental or evaluative research. Furthermore, assumptions of populations, environments, and systems underlie just about any research. These assumptions may be implicit or explicit, well founded, and in some cases unfounded.

Descriptive methodologies clear up assumptions by providing investigators with an improved characterization of the target population, environment, or system. Descriptive studies can increase the probability of being well informed (Wixon and Ramey, 1996).

Descriptive studies may be cross-sectional or longitudinal. *Cross-sectional descriptive studies* take a one-time snapshot of the attributes of interest. The collection of anthropometric data from schoolchildren and dimensions of their school furniture was a cross-sectional assessment of these two attributes (Milanese and Grimmer, 2004). The majority of available anthropometric data in the scientific knowledge base is, in fact, cross-sectional, representative of a single population (usually military) at one point in time (the 1950s).

Longitudinal studies follow a sample population over time and track changes in the attributes of that population. A longitudinal study asks the same question or involves observations at two or more times. There are four different types of longitudinal studies, dictated by the type of sampling used in the repeated methodology (Menard, 2002):

1. *Trend studies.* The same inquiries are made to different samples of the target population over time.
2. *Cohort studies.* Tracks changes in individuals with membership in an identified group that experiences similar life events (e.g., organizational, geographical groups) over time.
3. *Panel.* The same inquiries are made to the same people over time.
4. *Follow-up.* Inquiries are made to the participants after a significant amount of time has passed.

3.1.1 Variables

Descriptive studies ascribe values to characteristics, behaviors, or events of interest in a human-integrated system. The variables captured can be qualitative (such as a person's perceived comfort) and/or quantitative (such as the number of female employees). These variables sort out into two classes: (1) criterion variables and (2) stratification variables. Criterion variables summarize characteristic behaviors and events of interest for a given group (such as the number of lost-time accidents for a given shift). Stratification variables are predictive variables that are aimed at the segmentation of the population into subgroups (e.g., age, gender, and experience).

3.1.2 Key Concern: Sampling

As noted by the classification of longitudinal descriptive studies, the approach to selecting participants is a critical factor in descriptive studies. The plan used in sampling or acquiring data points directs the overall validity of the method. To establish a highly representative sample, the investigator can try to ensure equal probability for the inclusion of each member of a population in a study through random sampling of the target population. However, this is not always feasible to do, as monetary and time constraints sometimes compel investigators to "take what they can get" in terms of participants. Still, if sampling biased has occurred, it can skew data analysis and suggest inferences that lack validity and reliability.

The solution is to review prior research, theories, and their experience to estimate the potential impact of bias factors on the variables of interest. A classic example of sampling bias occurs in telephone-administered surveys. This method neglects the proportion of the population whose socioeconomic status does not afford a home telephone. This can translate into bias in the variables gathered. Investigators need to weigh the potential impact on this "participant misrepresentation" in the potential confounding of their data.

Common types of bias issues include the following (Arleck and Settle, 1995):

- *Visibility bias.* Bias results when some units of a population are more visible than others (e.g., the telephone example provided above).
- *Order bias.* This occurs when the log of potential participants is in a specific order, such as birth dates or alphabetical order.
- *Accessibility bias.* When measures are collected in the field, certain persons in the population are more accessible than others (e.g., teachers vs. the administrative staff).
- *Cluster bias.* When a method targets clusters of participants from the sample frame, some clusters may be interrelated such that they share similar opinions, experiences, and values (e.g., workers from the third shift of a manufacturing operation).
- *Affinity bias.* Usually a problem in fieldwork, the investigator may be more likely to select people based on extraneous physical and personality traits (e.g., approaching only those who seem to be friendly and cooperative).
- *Self-selection bias.* Persons in the population can, by choice, elect to participate in the descriptive methodology (e.g., people who respond to customer feedback surveys are only those who have a complaint).
- *Nonresponse bias.* Typically associated with mail or e-mail surveys, those who elect not to respond could do so at random or due to some feature of the survey (e.g., the amount of personal information requested was too intrusive for some participants).

3.1.3 Techniques Employed

Observational techniques, surveys, and questionnaires are techniques most commonly associated with descriptive research methods. Descriptive methods may collect data in the field, laboratory, or through survey methods. Participants must be recruited from the real world for representation sake, but the actual methods may be carried out in the laboratory (Sanders and McCormick, 1993). Typically, methods are conducted in a laboratory when the measurement equipment is too difficult to transport to a participant. This is often the case for anthropometric studies.

Surveys, questionnaires, and *interviews* embody the second class of methods used most often in descriptive studies. They are information-gathering tools to characterize user and system features. The data collected with the surveys can be qualitative, from open-ended response questions, or employ quantitative scales. Surveys and questionnaires are very challenging to design with the assurance of valid, reliable results (Wickens et al., 2003a). They are susceptible to bias attributable to the investigator's wording and administration of questions as well as to the subjective opinions of those being questioned.

Survey and questionnaire design and administration are topics complicated enough for an entire handbook of its own. In fact, for more detailed explanations of questionnaire and survey design, readers are encouraged to review texts such as *The Survey Research Handbook* (Arleck and Settle, 1995). In the scope of this chapter, the advantages and common pitfalls of surveys and questionnaires will be introduced. Interviews can be considered similar to questionnaires and surveys because they share the element of question and response (Meister, 2004), with the exception of aural administration (in most cases). Interviews are characterized by their ability to be conducted with more than one investigator or respondent and the range of formality they may take on. Interviews typically take more time, and for this reason, questionnaires are often used in lieu of interviews.

Interviews and questionnaires are useful for their ability to extract the respondent's perceptions of the system and their performance and behaviors for descriptive studies. That said, the construct validity of participant responses, and both the inter- and intrarater reliability of responses, are difficult to validate and confirm before analysis of the data collected. Case Study 2 provides examples of surveys employed in descriptive research.

CASE STUDY 2: Survey of World Wide Web Use in Middle-Aged and Older Adults

Goal The goal of this study by Morrell et al. (2000) was to capture World Wide Web (WWW) usage patterns of various age groups, with particular interest in older adults' perceptions and use of the Web.

Methods Surveys were distributed to a diverse sample population to meet the study goal to capture behavioral patterns from a wide range of individuals. To sample users and nonusers of WWW effectively, two types of questionnaires were distributed via mail (and not via e-mail or other electronic media*). The survey was distributed to 550 adults aged 40 years or older in southeastern Michigan. The response rate was approximately 71% (392 responses were collected). Demographic data were collected from the sample population and compared with those of the general population in the area for consistency. This helped investigators to verify the absence of sampling biases.

Analyses Various analytical techniques were used, such as paired t-test, hierarchical regression analysis, and analysis of variance (ANOVA) to extract meaningful patterns from data. Particularly, hierarchical regression analysis was used to identify chronological age–related predictors in terms of patterns of WWW use.

Methodological Implications

1. *Less expensive to conduct.* Because the survey can be distributed and filled out by each participant simultaneously, the cost per participant is relatively low in terms of time and money. This makes it possible to sample a large population.

2. *Complications imposed by survey mode.* The means by which a survey is distributed and administered are also critically important. Although electronic mail and telephone calls can be efficient means of survey administration, they can inadvertently introduce sampling bias. The assumption that everybody in your target population has e-mail accessibility should be contemplated carefully. In this study, the authors used mail instead of electronic media in order to capture people who did and did not use the World Wide Web on a regular basis.

3. *Judicious design of survey content.* Once a survey has been distributed, it is almost impossible to modify the questionnaire. Thus, the questions should be designed carefully to meet the goal of the research and be easily interpretable to provide high interrespondent reliability. For example, in this study two different types of questionnaires were designed to cover two groups of participants, WWW users and nonusers, because the nonusers would not consistently understand Web terminologies and concepts.

The selection of which questions to ask; how, procedurally, the questions are presented; and the responses collected are therefore critical in a situation so prone to bias. The responses given are dependent on the participants' psychological abilities, especially their memory (Meister, 2004). Respondents may interpret words and phrases in ways that may produce invalid responses. They may also have trouble in the affirmation and expression of their internal states. Finally, the level of control exercised over responses is important. Structured responses, such as multiple-response questions, can pigeonhole respondents into an ill-fit self-categorization. An alternative is the use of free-response answers and less formal interviews, which create data that are highly subjective and qualitative, making analysis and comparisons difficult. Specific scaling techniques for self-reported measures are detailed in Chapter 44.

*For an example of surveys through electronic media, see Rau and Salvendy (2001).

Observation techniques consist of an investigator sensing, recording, and interpreting behavior. They capture covert behavior with clarity but demonstrate difficulty in estimates of tacit behavior. Meister (2004) asserts: "Observation is the most 'natural' of the measurement methods, because it taps a human function that is instinctive" (p. 131). Observations may be casual and undirected, or direct observation with a definite measurement goal and highly structured. In planning observational methods, Wickens et al. (2003a) suggest that the investigator predefine the variables of interest, the observational methods, how each variable will be recorded, the specific conditions that afford a specific observation, and the observational time frame. These observational categories form a taxonomy of specific pieces of information to be collected during the course of the observation. Furthermore, defining observational scenarios can enable the investigator to sample only at times when events are occurring that are relevant to the study's goals. This prevents the investigator from filtering through extraneous events and data related with the situation. Observational methods often use some data recording equipment to better enable the investigator to return to specific events, code data postobservation, and archive raw data for future investigations. Commercial software programs are available that enable the investigator to flag certain events in the video stream for frequency counts and to return for a closer observation.

Observation-based methods are appropriate when working under constraints that limit contact with the participant or interference with the task, such as observing a team of surgeons in an operating room. There are also times when observations are useful because the population cannot express their experience accurately in alternative terms. This is especially the case when working with children, who are sometimes unable to use written surveys or respond to questionnaires. As with most descriptive methods, observation is useful in conceptual research, as the precursor to empirical or evaluative research.

Some important factors in the implementation of observation include the amount of training required of observers, inter- and intraobserver reliability of the recorded data, the intrusiveness of the observation on the situation of interest, and how directly observable the variable of interest is (e.g., caller frustration is more difficult to measure than the frequency of calls someone makes to a technical support center). Table A1 in the appendix to this chapter provides the reader with real-world examples of descriptive studies that have been published in the past five years.

3.2 Empirical Methods

Empirical research methods, also known as *experimental methods*, assess whether relationships between system, performance, and human measures are due to random error or if there is a causal relationship. The question in empirical research is: "If x is changed, what will happen to y?" at different levels of complexity. In empirical research the investigator typically manipulates one or more variables to appraise the effects on human, performance, or system criteria. The investigator manipulates the system directly, to invoke an observable change (Drury, 1995).

Empirical methods are beneficial because the manipulations of variables enable the observation of circumstances that may occur infrequently in the operational (i.e., real) world. What's more, this manipulated situation allows for the application of more robust measurement approaches by removing the negative consequences of employing invasive implications to safety- or time-critical situations. The ability to exercise control in the situation to reduce variability may also provide the advantage of smaller sample sizes.

However, these benefits are not without cost. Drury (1995) asserts that face validity is sacrificed in the use of empirical methods; much more persuasion is necessary for acceptance of the studies. Sanders and McCormick (1993) state that increases in precision, control, and replication are coupled with a loss of generalizability. It is then difficult to make an argument for applicability when dealing with theoretical questions. For this reason, many HF/E practitioners take their inferences and theories developed in highly controlled empirical research and confirm them via field-based descriptive and evaluation-based studies (Meister, 2004). Representation and validity of results are often the most problematic concerns with conducting empirical methods.

3.2.1 Variables

For investigators to hypothesize potential relationships between human and system components, they must select variables. *Independent variables* are those factors that are manipulated or controlled by the investigator and are expected to illicit some change in system and/or human behavior in an observable way. Independent variables can be classified as task related, environmental, or participant related, and occur at more than one level. *Dependent variables* are measures of the change imposed by the independent variable(s). Extraneous variables are those factors that are not relevant to the hypotheses but that may influence the dependent variable. If extraneous variables are not controlled, their effect on the dependent variable could confound the observed changes triggered by the dependent variable.

Dependent variables are much like the criterion variables used in descriptive studies, with the exception that physical traits such as height, weight, and age are uncommon. Of course, the independent and dependent variables should be linked back to the hypotheses and goals. The best approach, when possible, is the assessment of human behavior in terms of performance, physiological, and subjective dependent measurements, to tap accurately into the construct of interest. The goal of empirical research is to detect variance in the dependent variables triggered by different levels of the independent variable(s). Variable selection and definition play a key part in the structuring of an experimental plan.

3.2.2 Selecting Participants

While descriptive methods typically require sampling from an actual population, empirical research directs the investigator to select participants who are representative of those in the target population. Certain traits of the population are more important than others, depending on the task and the physical and mental traits exhibited by the target population. The HF/E investigator needs to seriously contemplate if the participant population will be influenced by the independent variable in the same ways as the target population, and which factors are extraneous. To determine this, a review of previous theory and literature is once again valuable. Of additional value in narrowing the scope of participant characteristics are descriptive studies: observations, interviews, and questionnaires. These studies can characterize the target population and help an investigator to incorporate the necessary subjective features.

In some circumstances, members of the target population are so highly skilled and trained in their behavior and activities that it is difficult to match participants of similar skill levels. The investigator may find that they can circumvent this by training. Learning is typically estimated by an exponential model; there is an asymptotic point where there is little improvement in the knowledge or skill acquired (Gawron, 2000). The investigator may train participants to a certain point so that their interactions can match more closely those of the target population. Training can be provided to the participants through a specific regimen (e.g., subjects are exposed to three practice trials), they may be trained until they ascertain a specific performance level (e.g., accuracy, time to complete), or the training can be self-directed (i.e., the participant trains until he or she has attained a self-perceived comfort level). Circumstances exist where the amount of time and money to train is prohibitive or training for the construct of interest is simply unrealistic. For example, in studies that employ flight simulators it is not feasible to train a group of undergraduates to the same level as that of rated pilots (e.g., Pritchett, 2002).

Another issue that surfaces in conjunction with the selection of participants is interparticipant variability. This could be age, experience, formal training, or skill. When variability is attributable to differences in knowledge and skill level, the same approach can be taken as mentioned above to train participants to a certain skill level. Variability among participants can create confounding variables for the analysis and interpretation of the data. If this variability is indicative of the actual population (and is desired), the investigator can take specific measures in assigning participants to the experimental conditions.

3.2.3 Key Concern: Experimental Plan

The experimental plan is the blueprint for empirical research. It outlines in detail how the experiment will be implemented (Wickens et al., 2003a). The key components of the experimental plan include:

- Defining variables in quantifiable terms in order to determine:
  - The experimental task
  - The levels of manipulation for the independent variable (e.g., the experimental conditions)
  - Which aspects of the behavior to measure: the dependent variable
  - The strategy for controlling confounding variables
  - The type of equipment used for data collection (e.g., pencil and paper, video, computer, eye-tracker)
  - The types of analytical methods that can be applied (e.g., parametric vs. nonparametric statistical)
- Specification of the experimental design in order to determine:
  - Which participants will be exposed to different experimental conditions
  - Order of exposure to treatments
  - How many replications will be used
  - The number of participants required and recruitment methods (e.g., statistical power)

Experimental designs represent (1) different methods for describing variation in treatment conditions, (2) the assignment of participants to those conditions, and (3) the order in which participants are exposed to treatments (Williges, 1995; Meister, 2004). The basic concept of experimental design is discussed here, but for a more thorough discussion the reader is encouraged to review Williges (1995). A more statistically based account may be found in Box et al. (1978). The assignment of participants to treatment conditions is accomplished by means of two-group designs, multiple-group designs, factorial designs, within-subject designs, and between-subject designs. Two-group, multiple-group, and factorial designs describe ways in which the independent variable(s) of interest are broken down into quantifiable, determinant units. Between- and within-subject designs detail how the levels are assigned to the participants. Following is a brief description for each type of design and the conditions that are best supported by each design [Wickens et al. (2003a), compiled from Williges (1995)].

1. *Two-group design.* An evaluation is conducted using one independent variable with two conditions or treatment levels. The dependent variable is compared between the two conditions. Sometimes there is a *control condition*, in which no treatment is given. Thus the two levels are the presence or absence of treatment.
2. *Multiple-group design.* One independent variable is specified at more than two levels to gain more information (often, more diagnostic) on the impact of the independent variable.

3. *Factorial design.* An evaluation of two or more independent variables is conducted so that all-possible combinations of the variables are evaluated to assess the effect of each variable in isolation and in interaction.

4. *Between-subject design.* Each experimental condition is given to a unique group of participants, and participants experience only one condition. This is used widely when it is problematic to expose participants to more than one condition and if time is an issue (e.g., fatigue, learning, order effects).

5. *Within-subject design.* Each participant is exposed to every experimental condition. This is called *repeated measure design* because each participant is observed more than once. It typically reduces the number of participants required.

6. *Mixed-subject design.* Variables are explored within and between subjects.

Factorial designs are the most comprehensive type of experimental design. Furthermore, this design enables the variation of more than one system attribute during a single experiment; it is more representative of the real-world complexity of keeping track of the interactions between factors. Factorial designs are common in HF/E empirical work. Terminology is used to explain factorial designs quickly in the reporting of results. For example, if an empirical result has employed four independent variables, the design would be described as a four-way factorial design. If each of three of the independent variables has two levels, and the fourth has three levels, these levels are disclosed by describing the experiment as using a $2 \times 2 \times 2 \times 3$ factorial design.

Between-subject designs are highly susceptible to variation between groups on extraneous factors. This variation can impose constraints on the interpretation and generalizability of the results and can impose the risk of concluding a difference between the two groups based on the independent variable, when in fact it is the other intragroup variation. The randomized allocation of participant to group does not ensure the absence of intergroup variation on factors such as education, gender, age, and experience. These factors are identified through experience, preliminary research, or literature reviews. If extraneous factors have a potential influence on the dependent variable, the investigators should do their best to distribute the variation among the experimental groups. Randomized blocking is a two-step process of separating participants based on the intervening factors; an *equal* number of participants from each block is randomly assigned a condition.

Within-subject designs are prone to order effects. That is, participants might exhibit different behaviors depending on the sequence in which they are exposed to the conditions. Participants may exhibit improved performance over consecutive trials due to learning effects, or degraded performance over the consecutive trials due to fatigue or boredom. Unfortunately, fatigue and learning effects do not tend to balance each other out. In terms of fatigue, the investigator may offer the participants rest breaks between sessions or schedule several individual sessions with each participant over time (Gawron, 2000). Learning effects can be mitigated if the participants are trained to a specified point using the techniques mentioned previously.

Investigators may also use specific strategies for assigning the order of conditions to participants. If each condition is run at a different place in the sequence among participants, the potential learning effects may be averaged out; this is called *counterbalancing* (Wickens et al., 2003a). This can be accomplished through randomization, which requires a large number of participants to be effective (imbalance of assignment is likely with a smaller sample). Alternatively, there are structured randomization techniques, which ensure that each "sequence" is experienced. However, to mitigate the learning impact effectively, the number of participants needs to be a multiplier of the number of sequences (which is difficult in studies with many variables), and this may be implausible, depending on the constraints of the study.

Carryover effects are also a possible effect in with-subject designs if conditions are consecutively run repeatedly. Say that in an experimental design, there are four conditions: a, b, c, and d, and the following orders have been determined for the experimental runs of four participants:

- Order 1: $a–b–c–d$
- Order 2: $b–c–d–a$
- Order 3: $c–d–a–b$
- Order 4: $d–a–b–c$

Note that in these four orders, each condition is in a sequentially unique position each time. However, condition a always precedes b, b always precedes c, and c always precedes d. Features of one condition could potentially influence changes in the participants' behaviors under subsequent conditions. As an example, consider an empirical study that strives to understand the visual search strategies employed by quality control inspectors in the detection of errors under a variety of environmental conditions. If one condition is more challenging and it takes an inspector longer to find an error, the inspectors might well change their visual search strategies in reaction, based on the prior difficulties. An investigator may therefore have to employ a combination of random assignment and structured assignment of conditions.

Empirical research methods are typically conducted in the lab, but can be gathered in the field as well. The field offers investigators higher representation, but their control of independent and extraneous variables diminishes significantly. The advantages of working in a laboratory setting include the high level of control an investigator can exercise in the specification of independent variable levels and the blocking of potential confounding variables.

The importance in empirical research of running a pilot study cannot be understated. This provides the investigator with a preview of potential issues with equipment, participants, and even the analysis of data. Even with a thorough experimental plan, investigators can encounter unplanned sources of variability in data, or unknown confounding variables. This "practice run" can help an investigator to circumvent such problems when collecting actual data. The potential sunk cost of experimental trials that yield contaminated data drives the need for pilot studies.

Empirical investigations possess many advantages in terms of isolating the construct of interest, but the amount of control applied to the empirical setting can drastically limit the generalizability of the results. Empirical research is typically more *basic* in nature, for it drives the understanding of principles and theories which can then be applied to (and validated by) real-world systems. Case Study 3 provides readers with a review of one empirical investigation using a mixed factorial design. In addition, Table A2 provides several more examples of contemporary empirical work in HF/E.

CASE STUDY 3: Multimodal Feedback Assessment of Performance and Mental Workload

Goal The goal of this study by Vitense et al. (2003) was to establish recommendations for multimodal interfaces using auditory, haptic, and visual feedback.

Methods To extract and assess the complexity of HCI with multimodal feedback in a quantifiable way, the authors conducted a highly controlled empirical study. Thirty-two participants were selected carefully in order to control extraneous factors and to meet hardware requirements. These inclusion criteria were right-handedness, normal visual acuity, and near-normal hearing capability. Appropriate software and hardware were developed and purchased to generate the multimodal feedback to match both the real world and research published previously.

To investigate three different modalities and all possible combinations of the modalities, this study used a $2 \times 2 \times 2$ factorial, within-subject design. Participants used a computer to perform drag-and-drop tasks while being exposed to various combinations of multimodal feedback. Training sessions were conducted to familiarize participants with the experimental tasks, equipment, and each feedback condition. NASA-TLX and time measurement were employed to capture the workload and task performance of participants quantitatively.

Analyses A general linear model repeated-measures analysis was run to analyze the various performance measures and the workload. Interaction plots were also used to present and explain some significant interaction among visual, auditory, and haptic feedback.

Methodological Implications

1. *A small number of observations is required.* By controlling uninteresting factors from an experiment, unnecessary variability can be decreased. Thus, as you can see in this case study, empirical studies generally employ smaller numbers of participants than do other types of studies (e.g., descriptive).

2. *Factors are difficult to control.* Controlling extraneous factors is not an easy task. As this case study shows, careful selection of participants and training were necessary to reduce contaminant variability.

3. *A covert, dynamic HF/E phenomenon is easier to capture.* Human subjects are easily affected by various extraneous factors, making isolated appraisal of the construct difficult. In this example, the authors conducted a highly controlled experiment in an attempt to extract subtle differences in the interactions among feedback conditions.

3.3 Evaluation Methods

Evaluation methods are probably the most difficult to classify because they embody features of both descriptive and empirical studies. Many of the techniques and tools used in evaluation methods overlap with descriptive and empirical methods. Evaluation methods are chosen specifically because the objectives mandate the evaluation of a design or product, the evaluation of competing designs or products, or even the evaluation of methodologies or measurement tools. The goals of evaluation methods also match closely both descriptive and field methods, but with more of an applied flavor. Evaluation methods are a critical part of system designs, and the specific evaluation methodology used depends on the stage of the design. These methods are highly *applied* in nature, as they typically reference real-world systems.

The purpose of evaluation research embodies (1) understanding the effect of interactions for system or product use (akin to empirical research), (2) descriptions of people using the system (akin to descriptive research), and (3) assessment of the outcomes of system or product use compared to the system or product goal (akin to descriptive research), to confirm intended and unintended outcomes of use (unique to evaluation methods).

Evaluation research is part of the design process. Evaluations assess the integrity of a design and make recommendations for iterative improvements. Therefore, they can be used at a number of points during the design process. The stage of the product or system, including concept, design, prototype, and operational products, is the authority in mandating which techniques to use. Stanton and Young (1999) usefully categorized 12 evaluation methods according to applicability to the various product stages. Table 4 presents a summary of their classification. It is interesting to note that the further along the design process is, the greater the number of applicable techniques. The ease

Table 4 Assessment Techniques in the Product Design Process

Product Phase	Assessment Techniques
Concept	*5/12 methods applicable:* checklists, hierarchical task analysis (HTA), repertory grids, interviews, heuristics
Design	*10/12 methods applicable:* KLM, link analysis, checklists, protective human error analysis (PHEA), HTA, repertory grids, task analysis for error identification (TAFEI), layout analysis, interviews, heuristics
Prototype	*12/12 techniques applicable:* KLM, link analysis, checklists, PHEA, observation, questionnaires, HTA, repertory grids, TAFEI, layout analysis, interviews, heuristics
Operational	*12/12 techniques applicable:* KLM, link analysis, checklists, PHEA, observation, questionnaires, HTA, repertory grids, TAFEI, layout analysis, interviews, heuristics

with which methods are applied is therefore a function of the abstraction in the design process. Those products with a physical presence or systems that are tangible are compatible with a wider variety of methodological techniques. This does not imply greater importance in using evaluation methods at later design stages. In fact, evaluations can have the greatest impact in the conceptual stage of product design, when designers express the most flexibility and acceptance of change.

Evaluation methods typically have significant constraints placed on their resources in terms of time, money, and staff. Therefore, these factors, combined with the goal of the evaluation, direct the selection of methods. The relevant questions to consider when selecting a method for evaluation include:

- Resource-specific criteria
  - What is the cost–benefit ratio for using this method?
  - How much time is available for the study?
  - How much money is available for the study?
  - How many staff members are available for implementation and analysis of the study?
  - How can designers be involved in the evaluation?
- Method-specific criteria
  - What is the purpose of the evaluation?
  - What is the state of the product or system?
  - What will the outcome of the evaluation be? (e.g., a report, presentation, design selection)

Evaluation methods typically serve three roles: (1) functional analysis, (2) scenario analysis, and (3) structural analysis (Stanton and Young, 1999). *Functional analyses* seek to understand the scope of functions that a product or system supports. *Scenario analyses* seek to evaluate the actual sequence of activities that users of the system must step through to achieve the desired outcome. *Structural analysis* is the deconstruction of the design from a user's perspective. The selection of variables for evaluation research methods is influenced largely by the same factors that influence variable selection in both descriptive and empirical studies. Quantifiable, objective criteria of system and human performance are most useful in making comparisons of competing designs, systems, or products.

3.3.1 Key Concern: Representation

The research setting, tasks, and participants need to be as close to the real world as possible. A lack of generalizability of evaluation research to the actual design, users, tasks, and environment would mean significant gaps in the inferences and recommendations to be made. Sampling of participants should follow those guidelines outlined previously for descriptive studies.

The research setting should be selected based on the constraints listed above. The research needs to ask: "Do you gain more from watching the interactions in context than what you lose from lack of control (Woods, 1996)?" In evaluation studies, field research can provide an in-depth understanding of the goals, needs, and activities of users. But pure field methods such as ethnographic interviews create extensive challenges in terms of budgets, scheduling, and logistics (Woods, 1996). Evaluation methods often succumb to constraints of time, financial support, and expectations as to outcomes. That said, investigators must leverage their resources to best meet those expectations. Practitioners must adopt creative techniques to deal with low-fidelity prototypes (and sometimes no prototype) and limited population samples. The prioritization of methodological decisions must be clearly aligned with the goals and expectations of the study.

Case Study 4 provides readers with an example of one evaluation study. Additionally, Table A3 provides several more examples of evaluation studies.

CASE STUDY 4: Fleet Study Evaluation of an Advance Brake Warning System

Goal The goal of this study by Shinar (2000) is to evaluate the effectiveness of an advance brake warning system (ABWS) under true driving conditions.

Methods This case study is one of several evaluation studies of the ABWS. Prior to study execution, a simulation study proved the ABWS effective in decreasing the possibility of rear-end crashes (from 73% to 18%). However, the assumptions made in developing the simulation caused limitations in the applicability of its results and conclusions to real-world situations. The inadequacies in the previous study motivated this longitudinal field study investigating 764 government vehicles. Half of the vehicles were equipped with an ABWS, and the other vehicles were without an ABWS. Over four years, the 764 vehicles were used as government fleet vehicles and all crashes involving the vehicles were tracked. This tracking process was carried out unbeknownst to the vehicle drivers. Because the accidents happened for a variety of reasons, it was difficult to distinguish whether a collision was relevant or not. Although the assessment of causality could not be objective, judgments by the investigator were made conservatively in order to improve the validity and integrity of study results.

Analyses A paired t-test was used to detect a statistically significant difference in the number of accidents between the two automobile systems. Because the data were gathered under real circumstances, some uncontrolled factors potentially confounded the results. For example, the average distance driven of the control group was different from that of the treatment group. Those factors were accounted for by introducing more diagnostic evaluative measures, such as the number of rear-end collisions per kilometer for a specific region.

Methodological Implications

1. *Specific to a certain design or system.* One salient characteristic of evaluation studies is that they target a specific design or system. In this case study the target system is an ABWS.

2. *Emphasizing representation.* The major reason that the author was not satisfied with the simulation method was that it lacked representation of the real world, due to some assumptions. Thus, the second study was conducted using actual operating vehicles. Often, investigators must do both for a more complete understanding of the HF/E related to the systems of interest.

3. *Lacking control.* Representation is not obtained without cost. Under real situations, experimenters cannot control extraneous, confounding factors. As a result, this author had difficulty distinguishing relevant crashes from irrelevant ones. To compensate, the experiment took place over the course of four years, which increased the sample size to a more acceptable level for analysis.

4 CONCLUSIONS

The selection and application of HF/E methods are part art, part science. There is a certain creative skill for the effective application of HF/E methods. Furthermore, that creative skill is acquired through practice and experience. HF/E investigators must be knowledgeable in several areas, be able to interpret theories and principles of other sciences, and integrate them with their own knowledge and creativity in valid, reliable ways to meet the investigation's goals. Of course, all this is to be accomplished within the constraints of time and resources encountered by researchers and practitioners. An awareness of HF/E methods—their limitations, strengths, and prior uses—provides an investigator with a valuable toolkit of knowledge. This and practical experience lend the investigator the ability to delve into the complex phenomena associated with HF/E.

APPENDIX: EXEMPLARY STUDIES OF HF/E METHODOLOGIES

Table A1 Examples of Descriptive Studies

Study	HF/E Subdiscipline	Goal	Methodology	Analysis Methods
School furniture and the user population: an anthropometric perspective (Milanese and Grimmer, 2004)	Anthropometry	To determine the relationship between reported spinal symptoms in students and the match between their individual anthropometric dimensions and their school furniture.	*Field study*. measured anthropometric data from randomly selected 1269 schoolchildren was used to assess ratio measures of stature.	Stature categorized in quartiles; and derivative ratios calculated to obtain a range of anthropometric proportions
Differences in safety climate between hospital personnel and naval aviators (Gaba et al., 2003)	Safety	To compare results of safety climate survey questions from health care respondents with those from naval aviation, a high-reliability organization.	*Survey*. Survey questions were derived from the Command Safety Assessment Survey (CSAS) and the Patient Safety Cultures in Healthcare Organizations Survey, two previously validated surveys on safety and health care. A preliminary form of the survey was tested in a pilot survey. In total, 6312 survey packages were collected across all 15 hospitals.	Chi-square analysis recognizing $p < 0.001$ as significant in order to correct for multiple comparisons

Table A1 ***(continued)***

Study	HF/E Subdiscipline	Goal	Methodology	Analysis Methods
Truck driver fatigue risk assessment and management: a multinational survey (Adams–Guppy and Guppy, 2003)	Fatigue	To determine the relationship between fatigue and near-miss and accident experiences of drivers	*Field study*. In-depth interviews were conducted to develop appropriate questions for the survey. Questions were asked of drivers concerning their feelings of fatigue, accidents and close calls, work environments, and life-styles. Surveys were administered to 640 truck drivers across 17 countries.	ANOVA; ANCOVA
Anthropometric differences among occupational groups (Hsiao et al., 2002)	Anthropometry	To identify differences in various body measurements between occupational groups in the United States	Archival data collected in the third National Health and Nutrition Examination Survey (NHANES III 1988–94) were analyzed.	Two-tailed *t*-test
An evaluation of warning habits and beliefs across the adult life span (Hancock et al., 2001)	Aging, warning perception	To show age-effect warning habits and beliefs	*Survey*. Participants were shown various illustrative warning signs and asked questions about their comprehension and familiarity with respect to each sign. Additionally, participants were asked their attitudes toward warning on a variety of consumer products. Questionnaires were administered by mail. Of 4250 surveys mailed, 863 were returned and valid.	Regression analysis; ANOVA
Ergonomics of electronic mail address systems: related literature review and survey of users (Rau and Salvendy, 2001)	E-mail, user-centered design	To obtain information on preferences, dislikes, and difficulties associated with the e-mail address system	*Survey*. A survey was conducted through e-mail and a newsgroup. Seventy questions were administered regarding respondents' use of and attitude toward their electronic mail systems. 160 electronic questionnaires were returned.	Analysis of correlations
The effects of task complexity and experience on learning and forgetting: a field study (Nembhard, 2000)	Cognitive processes (learning, memory)	To examine the effects of task complexity and experience on parameters of individual learning and forgetting	*Longitudinal and field study*. 2853 learning and forgetting episodes were captured over the course of a year. The episodes captured and averaged performance data, task complexity, and task experience for each employee on a weekly basis.	Mathematical model development; Kolmogorov–Smirnov; ANOVA; pairwise comparison; regression analysis
A survey of World Wide Web use in middle-aged and older adults (Morrell et al., 2000)	HCI, aging	To reveal Web usage patterns for middle-aged and older adults	*Survey*. Demographic data were collected using the Older American Resources and Services Assessment Questionnaire (OARS), a multidimensional functional assessment questionnaire. Questions were also directed toward respondent familiarity with the Web environment.	*t*-test; regression analysis; ANOVA

(*continued overleaf*)

Table A1 (*continued*)

Study	HF/E Subdiscipline	Goal	Methodology	Analysis Methods
A case study of co-ordinative decision-making in disaster management (Smith and Dowell, 2000)	Organizational behavior	To report a case study of interagency coordination during the response to a railway accident in the UK	Interviews were conducted to capture workers' accounts of a railway incident. Interviews were audio-recorded. Investigators also reviewed documentation taken in relation to the accident.	Critical decision method
Traffic sign symbol comprehension: a cross-cultural study (Shinar et al., 2003)	Surface transportation system (highway design), warning perception	To understand the cultural difference in comprehending sign symbol among five countries	1000 unpaid participants were recruited. Participants were presented with 31 multinational traffic signs and asked their comprehension of each.	ANOVA; arcsin $\sqrt{p}$ transformation

Table A2 Examples of Empirical Studies

Study	HF/E Subdiscipline	Goal	Methodology	Analysis Methods
Detection of temporal delays in visual–haptic interfaces (Vogels, 2004)	Displays and controls, multimodality	To address the question of how large the temporal delay between a visual and a haptic stimulus can be for the stimuli to be perceived as synchronous	Three different experiments were conducted to remove unintended methodological factors and to investigate deeply. Learning effect was controlled through training.	Mathematical model developed and applied to quantify the data; ANOVA used to analyze the quantified mathematical model
Adjustable typography: an approach to enhancing low vision text accessibility (Arditi, 2004)	Accessibility, typography	To show that adjustable typography enhances text accessibility	Participants who had low vision were allowed to adjust key font parameters (e.g., size and spacing) of text on a computer display monitor. After adjustment, the participants' accuracy on a reading task was collected. Participants completed the experiment within a single experimental session due to the fatigue experienced in this predominantly older population (mean age = 68.6 years).	Box-and-whisker plot; regression analysis
No evidence for prolonged latency of saccadic eye movements due to intermittent light of a CRT computer screen (Jainta et al., 2004)	Psychomotor processes (eye movement)	To show that there is no clear relationship between latency of saccadic eye movements and the intermittency of light of cathode-ray tubes	A special fluorescent lamp display was used to control the refresh rate. An eye tracker captured saccadic eye movements.	ANOVA with repeated measures; Greenhouse–Geisser adjusted error probabilities
Physical workload during use of speech recognition and traditional computer input devices (Juul-Kristensen et al., 2004)	Work physiology (physical workload), HCI	To investigate musculoskeletal workload during computer work using speech recognition and traditional computer input devices	The workload of 10 participants while performing text entry, editing, and reading aloud with and without the speech recognition program was studied. Workload was measured using muscle activity (EMG).	Nonparametric statistics (e.g., Wilcoxon's ranked-sign test Mann–Whitney test)

Table A2 ***(continued)***

Study	HF/E Subdiscipline	Goal	Methodology	Analysis Methods
Attentional models of multitask pilot performance using advanced display technology (Wickens et al., 2003b)	Aerospace systems, attention	To compare air traffic control presentation of auditory (voice) information regarding traffic and flight parameters with advanced display technology presentation of equivalent information	Pilots were exposed to both auditory and advanced display technology conditions. Performance with the information presented in each condition was assessed. A Latin Square design was used to counterbalance order effects.	Within-subjects ANOVA and regression analysis
Effects of age, speech rate, and environmental support in using telephone voice menu systems (Sharit et al., 2003)	Aging, displays, and controls	To investigate age difference in the use of telephone menu systems in two experiments	Over the course of two days, prescreened participants were given cognitive battery tests and ability tests concerning their use of telephone menu systems; menu type, and voice recognition. Three different measures were collected: score, navigation pattern, and subjective rating.	ANOVA
Measuring the fit between human judgments and automated alerting algorithms: a study of collision detection (Bisantz and Pritchett, 2003)	Aviation, automation	To evaluate the impact of displays on human judgment using the n-system lens model; to explicitly assess the similarity between human judgments and a set of potential judgment algorithms for use in automated systems	Using a flight simulator, the approach of an oncoming aircraft was manipulated. Data were collected on the performance of the automation system and its effect on pilot judgments. A time-sliced approach was used to capture wider environmental conditions.	Within-subjects ANOVA used on a transformation of the data
Bimodal displays improve speech comprehension in environments with multiple speakers (Rudmann et al., 2003)	Displays and controls	To prove that showing additional visual cues from a speaker can improve speech comprehension	Twenty-four participants were exposed to voice recordings with and without visual cues. In some trials, noise distracters were introduced. The level of participant comprehension was assessed while listening to the recording, and eye movement data were collected.	Within-subjects ANOVA
Performance in a complex task and breathing under odor exposure (Danuser et al., 2003)	Displays and controls (olfactory displays)	To investigate the influence of odor exposure on performance and breathing	15 healthy individuals were each exposed to different odors. To capture the emotional status, a self-assessment manikin (SAM) was used.	ANOVA with repeated measures; Wilcoxon's ranked-sign test
Time estimation during prolonged sleep deprivation and its relationship to activation measures (Miro et al., 2003)	Fatigue	To investigate the effect of prolonged sleep deprivation for 60 hours on time estimation	*Longitudinal*. For 60 hours of sleep deprivation, time estimations were measured every 2 hours. Skin resistance level, body temperature, and Stanford sleepiness scale scores were collected.	ANOVA with repeated measures; regression analysis (linear, quadratic, quintic, and sextic)

(*continued overleaf*)

Table A2 (*continued*)

Study	HF/E Subdiscipline	Goal	Methodology	Analysis Methods
The impact of mental fatigue on exploration in a complex computer task: rigidity and loss of systematic strategies (Van Der Linden et al., 2003)	Fatigue; HCI	To investigate the impact of mental fatigue on how people explore in a complex computer task	68 participants (psychology students) performed a complex computer test using the think-aloud protocol. Data were collected on mental fatigue and performance through observation, videotaping, the activation–deactivation checklist, and the rating scale mental effort.	Multivariate test ($\alpha = 0.10$); univariate/post hoc tests ($\alpha = 0.05$)
What to expect from immersive virtual environment exposure: influences of gender, body mass index, and past experience (Stanney et al., 2003)	Virtual reality	To investigate potential adverse effects, including sickness, associated with exposure to virtual reality and extreme responses	Of the 1102 subjects recruited for participation, 142 (12.9%) dropped out because of sickness. Qualitative measurement tools were used to assess motion sickness with the motion history questionnaire and simulator sickness. Sessions were videotaped for archival purposes.	Spearman's correlation test; Kruskal–Wallis nonparametric test; chi-squared test
Control and perception of balance at elevated and sloped surfaces (Simeonov et al., 2003)	Work physiology	To investigate the effects of the environment characteristics of roof work (e.g., surface slope, height, and visual reference) on standing balance in construction workers	24 participants were recruited. The slope of a platform, on which they stood, was varied. At each slope the participant performed a manual task and were asked afterward to rate their perceived balance. Instrumentation measured the central pressure movement. Each subjects received the same 16 treatments ($4 \times 2 \times 2$). Balanced to control order effects.	ANOVA with repeated measures and the Student–Newman–Keuls multiple-range test used when ANOVA indicated significance
Multimodal feedback: an assessment of performance and mental workload (Vitense et al., 2003)	Displays and controls (multimodality), HCI	To establish recommendations for the incorporation of multimodal feedback in a drag-and-drop task	The NASA-TLX was used to assess workload. Time measures, such as trial completion time and target highlight time, were used to capture performance as it was affected by multimodal feedback.	Interaction plots
The contribution of apparent and inherent usability to a user's satisfaction in a searching and browsing task on the Web (Fu and Salvendy, 2002)	Usability, WWW	To investigate the impact of inherent and apparent usability on user's satisfaction of Web page designs	The questionnaire for user interaction satisfaction was used to measure the levels of users' satisfaction with a browsing task completed on one of four interfaces.	ANOVA; stepwise regression analysis

Table A3 Examples of Evaluation Studies

Study	HF/E subdiscipline	Goal	Methodology	Analysis Methods
Handle dynamics predictions for selected power hand tool applications (Lin et al., 2003)	Biomechanics	To test a previously developed model of handle dynamics by collecting muscle activity (EMG) data	Muscle activity (EMG) was collected to calculate the magnitude of torque that participants experienced.	Regression analysis
Learning to use a home medical device: mediating age-related differences with training (Mykityshyn et al., 2002)	Aging, training	To examine the differential benefits of instructional materials, such as a user manual or an instructional video, for younger and older adults learning to use a home medical device	*Longitudinal*. The NASA-task load index (TLX) was used to assess the workload associated with instructional methods. A longitudinal study, there was a two-week retention session used between training and measurement.	ANOVA
Effects of an in-vehicle collision avoidance warning system on short- and long-term driving performance (Ben-Yaacov et al., 2002)	Surface transportation system (driver behavior)	To determine how drivers adapt to an in-vehicle collision avoidance warning system over time	*Field study*. Experiments were conducted on the real highway. In this longitudinal study, a six-month retention session was placed between the third and fourth trials.	ANOVA; interaction plots
Fleet study evaluation of an advance brake warning system (Shinar, 2000)	Accident, surface transportation system	To prove the effectiveness of an advanced brake warning system (ABWS)	*Longitudinal and field study*. This study traced accidents of vehicles with ABWS-installed government vehicles (382) and non-ABWS vehicles (382) for 4 years.	*t*-test; chi-square test
Continuous assessment of back stress (CABS): a new method to quantify low-back stress in jobs with variable biomechanical demands (Mirka et al., 2000)	Work physiology	To compare three different back-stress modeling techniques for the continuous assessment of lower-back stress in various situations and to incorporate them into a hybrid model	*Field study and observation*. 28 construction workers were observed and videotaped while on the job. Based on this videotaped observation, three models of each task that induced biomechanical stress were created. Each model was evaluated based on the extent to which it modeled the actual biomechanical stress experienced.	Time-weighted histograms to aggregate and compare the output from three different models

REFERENCES

Adams-Guppy, J., and Guppy, A. (2003),"Truck Driver Fatigue Risk Assessment and Management: A Multinational Survey,"*Ergonomics*,Vol. 46, No.8, pp. 763–779.

American Psychological Association (2002), "Ethical Principles of Psychologists and Code of Conduct," retrieved June 18, 2004, from http://www.apa.org/ethics/code2002.html.

Arditi, A. (2004), "Adjustable Typography: An Approach to Enhancing Low Vision Text Accessibility," *Ergonomics*, Vol. 47, No. 5, pp. 469–482.

Arleck, P. L., and Settle, R. B. (1995), *The Survey Research Handbook*, Irwin, Chicago.

Ben-Yaacov, A., Maltz, M., and Shinar, D. (2002), "Effects of an In-Vehicle Collision Avoidance Warning System on Short- and Long-Term Driving Performance," *Human Factors*, Vol. 44, No. 2, pp. 335–342.

Bisantz, A. M., and Pritchett, A. R. (2003), "Measuring the Fit Between Human Judgments and Automated Alerting Algorithms: A Study of Collision Detection," *Human Factors*, Vol. 45, No. 2, pp. 266–280.

Box, G. E., Hunter, W. G., and Hunter, J. S. (1978), *Statistics for Experimenters: An Introduction to Design, Data Analysis, and Model Building*, Wiley, New York.

Byrne, M. D. (2001), "ACT-R/PM and Menu Selection: Applying a Cognitive Architecture to HCI," *International Journal of Human–Computer Studies*, Vol. 55, pp. 41–84.

Card, S., Moran, T., and Newell, A. (1983), *The Psychology of Human–Computer Interaction*, Lawrence Erlbaum Associates, Mahwah, NJ.

Cohen, R. J., and Swerdlik, M. (2002), *Psychological Testing and Assessment: An Introduction to Tests and Measurement*, 5th ed., McGraw-Hill, Burr Ridge, IL.

Committee on Human Factors (1983), *Research Needs for Human Factors*, Commission on Behavioral and Social Sciences and Education, National Academy of Sciences, Washington DC.

Danuser, B., Moser, D., Vitale-Sethre, T., Hirsig, R., and Krueger, H. (2003), "Performance in a Complex Task and Breathing Under Odor Exposure," *Human Factors*, Vol. 45, No. 4, pp. 549–562.

Drury, C. G. (1995), "Designing Ergonomic Studies and Experiments," in *Evaluation of Human Work*, J. R. Wilson and E. N. Corlett, Eds., Taylor & Francis, Bristol, PA, pp. 113–140.

Fitts, P. M. (1954), "The Information Capacity of the Human Motor System in Controlling the Amplitude of Movement," *Journal of Experimental Psychology*, Vol. 47, No. 6 pp. 381–391.

Ford, J. M., and Wood, L. E. (1996), "An Overview of Ethnography and System Design," in *Field Methods Casebook for Software Design*, D. Wixon and J. Ramey, Eds., Wiley Computer Publishing, New York, pp. 269–282.

Fu, L., and Salvendy, G. (2002), "The Contribution of Apparent and Inherent Usability to a User's Satisfaction in a Searching and Browsing Task on the Web," *Ergonomics*, Vol. 45, No. 6, pp. 415–424.

Gaba, D. M., Singer, S. J., Sinaiko, A. D., Bowen, J. D., and Ciavarelli, A. P. (2003), "Differences in Safety Climate Between Hospital Personnel and Naval Aviators," *Human Factors*, Vol. 45, No. 2, pp. 173–185.

Gawron, V. (2000), *Human Performance Measures Handbook*, Lawrence Erlbaum Associates, Mahwah, NJ.

Gillespie, R. (1991), *Manufacturing Knowledge: A History of the Hawthorne Experiments*, Cambridge University Press, New York.

Hancock, H. E., Rogers, W. A., and Fisk, A. D. (2001), "An Evaluation of Warning Habits and Beliefs Across the Adult Life Span," *Human Factors*, Vol. 43, No. 3, pp. 343–354.

Holtzblatt, K., and Beyer, H. (1996), "Contextual Design: Principles and Practice," in *Field Methods Casebook for Software Design*, D. Wixon and J. Ramey, Eds., Wiley Computer Publishing, New York, pp. 301–354.

Hsiao, H., Long, D., and Snyder, K. (2002), "Anthropometric Differences among Occupational Groups," *Ergonomics*, Vol. 45, No. 2, pp. 136–152.

Jacko, J. A., Moloney, K. P., Kongnakorn, T., Barnard, L., Edwards, P., Lenard, V. K., Sainfort, F., and Scott, I. U. (2005), "Multimodal Feedback as a Solution to Ocular Disease-Based User Performance Decrements in the Absence of Functional Visual Loss," *International Journal of Human–Computer Interaction.*, Vol. 18, No. 2, pp. 183–218.

Jainta, S., Jaschinski, W., and Baccino, T. (2004), "No Evidence for Prolonged Latency of Saccadic Eye Movements Due to Intermittent Light of a CRT Computer Screen," *Ergonomics*, Vol. 47, No. 1, pp. 105–114.

Juul-Kristensen, B., Laursen, B., Pilegaard, M., and Jensen, B. R. (2004), "Physical Workload During Use of Speech Recognition and Traditional Computer Input Devices," *Ergonomics*, Vol. 47, No. 2, pp. 119–133.

Kanis, H. (2000), "Questioning Validity in the Area of Ergonomics/Human Factors," *Ergonomics*, Vol. 43, No. 12, pp. 1947–1965.

Kantowitz, B. H. (1992), "Selecting Measures for Human Factors Research," *Human Factors*, Vol. 34, No. 4, pp. 387–398.

Lin, J.-H., Radwin, R. G., and Richard, T. G. (2003), "Handle Dynamics Predictions for Selected Power Hand Tool Applications," *Human Factors*, Vol. 45, No. 4, pp. 645–656.

MacKenzie, I. S. (1992a), "Fitts' Law as a Research Design Tool in Human–Computer Interaction," *Human–Computer Interaction*, Vol. 7, No. 1992, pp. 91–139.

MacKenzie, I. S. (1992b), "Extending Fitts' Law to Two Dimensional Tasks," in *Human Factors in Computing Systems*, ACM Press, New York, pp. 219–226.

Meister, D. (1971), *Human Factors: Theory and Practice*, Wiley-Interscience, New York.

Meister, D. (2004), *Conceptual Foundations of Human Factors Measurement*, Lawrence Erlbaum Associates, Mahwah, NJ.

Menard, S. (2002), *Longitudinal Research*, 2nd ed., Sage Publications, Newbury Park, CA.

Milanese, S., and Grimmer, K. (2004), "School Furniture and the User Population: An Anthropometric Perspective," *Ergonomics*, Vol. 47, No. 4, pp. 416–426.

Mirka, G. A., Kelaher, D. P., Nay, D. T., and Lawrence, B. M. (2000), "Continuous Assessment of Back Stress (CABS): A New Method to Quantify Low-Back Stress in Jobs with Variable Biomechanical Demands," *Human Factors,* Vol. 42, No. 2, pp. 209–225.

Miro, E., Cano, M. C., Espinosa-Fernandez, L., and Buela-Casal, G. (2003), "Time Estimation During Prolonged Sleep Deprivation and Its Relation to Activation Measures," *Human Factors,* Vol. 45, No. 1, pp. 148–159.

Morrell, R. W., Mayhorn, C. B., and Bennett, J. (2000), "A Survey of World Wide Web Use in Middle-Aged and Older Adults," *Human Factors,* Vol. 42, No. 2, pp. 175–182.

Mykityshyn, A. L., Fisk, A. D., and Rogers, W. A. (2002), "Learning to Use a Home Medical Device: Mediating Age-Related Differences with Training," *Human Factors,* Vol. 44, No. 3, pp. 354–364.

Nembhard, D. A. (2000), "The Effects of Task Complexity and Experience on Learning and Forgetting: A Field Study," *Human Factors,* Vol. 42, No. 2, pp. 272–286.

"Nuremberg Code (1947)," (1996), *British Medical Journal,* Vol. 313, No. 7070, p. 1448.

Office for Human Research Protections (1979), "The Belmont Report," retrieved May 5, 2005, from http://ohsr.od.nih.gov/guidelines/belmont.html.

Pollatsek, A., and Rayner, K. (1998), "Behavioral Experimentation," in *A Companion to Cognitive Science,* W. Bechtel and G. Graham, Eds., Blackwell, Malden, MA, pp. 352–370.

Pritchett, A. R. (2002), "Human–Computer Interaction in Aerospace," in *Handbook of Human–Computer Interaction,* J. A. Jacko and A. Sears, Eds., Lawrence Erlbaum Associates, Mahwah, NJ, pp. 861–882.

Ramcharan, D. (2001), "Ergonomic Redesign of a Hanging Table for Poultry Processing," in *Applied Ergonomics,* D. Alexander and R. Rabourn, Eds., Taylor & Francis, New York, pp. 126–134.

Rau, P.-L. P., and Salvendy, G. (2001), "Ergonomics of Electronic Mail Address Systems: Related Literature Review and Survey of Users," *Ergonomics,* Vol. 44, No. 4, pp. 382–401.

Rehmann, A. J. (1995), *Handbook of Human Performance Measures and Crew Requirements for Flightdeck Research*, Federal Aviation Administration, Washington, DC.

Rudmann, D. S., McCarley, J. S., and Kramer, A. F. (2003), "Bimodal Displays Improve Speech Comprehension in Environments with Multiple Speakers," *Human Factors,* Vol. 45, No. 2, pp. 329–336.

Sanders, M. S., and McCormick, E. J. (1993), *Human Factors in Engineering and Design,* McGraw-Hill, New York.

Sharit, J., Czaja, S. J., Nair, S., and Lee, C. C. (2003), "Effects of Age, Speech Rate, and Environmental Support in Using Telephone Voice Menu Systems," *Human Factors,* Vol. 45, No. 2, pp. 234–251.

Sheridan, T. B. (2002), *Humans and Automation: System Design and Research Issues*, Wiley, New York.

Shinar, D. (2000), "Fleet Study Evaluation of an Advance Brake Warning System," *Human Factors,* Vol. 42, No. 3, pp. 482–489.

Shinar, D., Dewar, R. E., Summala, H., and Zakowska, L. (2003), "Traffic Sign Symbol Comprehension: A Cross-Cultural Study," *Ergonomics,* Vol. 46, No. 15, pp. 1549–1565.

Simeonov, P. I., Hsiao, H., Dotson, B. W., and Ammons, D. E. (2003), "Control and Perception of Balance at Elevated and Sloped Surfaces," *Human Factors,* Vol. 45, No. 1, pp. 136–147.

Smith, W., and Dowell, J. (2000), "A Case-Study of Coordinative Decision-Making in Disaster Management," *Ergonomics,* Vol. 43, No. 8, pp. 1153–1166.

Stanney, K. M., Hale, K. S., Nahmens, I., and Kennedy, R. S. (2003), "What to Expect from Immersive Virtual Environment Exposure: Influences of Gender, Body Mass Index, and Past Experience," *Human Factors,* Vol. 45, No. 3, pp. 504–520.

Stanton, N. A., and Young, M. A. (1999), *A Guide to Methodology in Ergonomics,* Taylor & Francis, New York.

U.S. Department of Energy (2004), "DOE Openness: Human Radiation Experiments," retrieved May 5, 2005, from http://tis.eh.doe.gov/ohre/.

U.S. Department of Health and Human Services (2001), "Code of Federal Regulations, Part 46: Protection of Human Subjects," retrieved June 20, 2004, from http://www.hhs.gov/ohrp/humansubjects/guidance/45cfr46.htm.

van der Linden, D., Frese, M., and Sonnentag, S. (2003), "The Impact of Mental Fatigue on Exploration in a Complex Computer Task: Rigidity and Loss of Systematic Strategies," *Human Factors,* Vol. 45, No. 3, pp. 483–494.

Vitense, H. S., Jacko, J. A., and Emery, V. K. (2003), "Multimodal Feedback: An Assessment of Performance and Mental Workload," *Ergonomics,* Vol. 46, No. 1–3, pp. 68–87.

Vogels, I. M. (2004), "Detection of Temporal Delays in Visual–Haptic Interfaces," *Human Factors,* Vol. 46, No. 1, pp. 118–134.

Weimer, J. (1995), "Developing a Research Project," in *Research Techniques in Human Engineering,* J. Weimer, Ed., Prentice Hall, Englewood Cliffs, NJ, pp. 35–54.

Wickens, C. D., Lee, J. D., Liu, Y., and Gordon-Becker, S. (2003a), *An Introduction to Human Factors Engineering,* Prentice Hall, Englewood Cliffs, NJ.

Wickens, C. D., Goh, J., Helleberg, J., Horrey, W. J., and Talleur, D. A. (2003b), "Attentional Models of Multitask Pilot Performance Using Advanced Display Technology," *Human Factors,* Vol. 45, No. 3, pp. 360–380.

Williges, R. C. (1995), "Review of Experimental Design," in *Research Techniques in Human Engineering,* J. Weimer, Ed., Prentice Hall, Englewood Cliffs, NJ, pp. 49–71.

Wilson, J. R., and Corlett, E. N. (1995), *Evaluation of Human Work,* Taylor & Francis, Bristol, PA.

Wixon, D., and Ramey, J. (1996), *Field Methods Casebook for Software Design,* Wiley Computer Publishing, New York.

Woods, L. (1996), "The Ethnographic Interviews in User Centered Work/Task Analysis," in *Field Methods Casebook for Software Design,* D. Wixon and J. Ramey, Eds., Wiley Computer Publishing, New York, pp. 35–56.

World Medical Association (2002), "Police: World Medical Association Declaration of Helsinki Ethical Principles for Medical Research Involving Human Subjects," retrieved June 23, 2004, from http://www.wma.net/e/policy/b3.htm.

CHAPTER 12
ANTHROPOMETRY

Kathleen M. Robinette
Air Force Research Laboratory
Wright-Patterson Air Force Base, Ohio

Jeffrey A. Hudson
General Dynamics AIS
Dayton, Ohio

1 INTRODUCTION

Did you know that designing for the 5th percentile female to the 95th percentile male can lead to poor and unsafe designs? If not, you are not alone. These and similar percentile cases, such as the 99th percentile male, are the only cases presented as anthropometry solutions to many engineers and ergonomics professionals. In this chapter we review this and other anthropometry issues and present an overview of practical effective methods for incorporating the human body in design.

Anthropometry, the study of human body measurement, is used in engineering to ensure the maximum benefit and capability of products that people use. The use of anthropometric data in the early concept stage can minimize the size and shape changes needed later, when modifications can be very expensive. To use anthropometry knowledge effectively, it is also important to have knowledge of the relationships between the body and the items worn or used. The study of these relationships is called *fit mapping*. Databases containing both anthropometry and fit-mapping data can be used as a lessons-learned source for development of new products. Therefore, anthropometry and fit mapping can be thought of as an information core around which products are designed, as illustrated in Figure 1.

A *case* is a combination of body measurements, such as a set of measurements on one subject, the average measurements from a sample, a three-dimensional scan of a person, or a two- or three-dimensional human model. If the relationship between the anthropometry and the fit of a product is simple or known, cases may be all that is needed to arrive at an effective design. However, if the relationship between the anthropometry and the fit is complex or unknown, cases alone may not suffice. In these situations, fit mapping with a prototype, mock-up, or similar product is needed to determine how to accommodate the cases and to predict accommodation for any case.

The chapter is divided into three sections. Section 2 deals with the selection of cases for characterizing anthropometric variability. Section 3 covers fit-mapping methods. Section 4 is devoted to some of the benefits of the newest method in anthropometric data collection, three-dimensional anthropometry.

2 ANTHROPOMETRY ALTERNATIVES, PITFALLS, AND RISKS

When a designer or engineer asks the question "What anthropometry should I use in the design?", he or she is essentially asking, "What cases should I design around?" Of course, the person would always like to be given one case or one list of measurements, and be told that nothing else is needed. That would make it simple. However, if the question has to be asked, the design is probably more complicated than that, and the answer is correspondingly more complicated as well. In this section we discuss alternative ways to determine which cases to use.

2.1 Averages and Percentiles

Since as early as 1952, when Daniels (1952) presented the argument that no one is average, we have known that anthropometric averages are not acceptable for many applications. For example, an average human head is not appropriate to use for helmet sizing, and an average female shape is not appropriate for sizing apparel. In addition, we have known since Searle and Haslegrave (1969) presented their debate with Ed Hertzberg that the 5th and 95th percentile people are no better. Robinette and McConville (1982) demonstrated that it is not even possible to construct a 5th or 95th

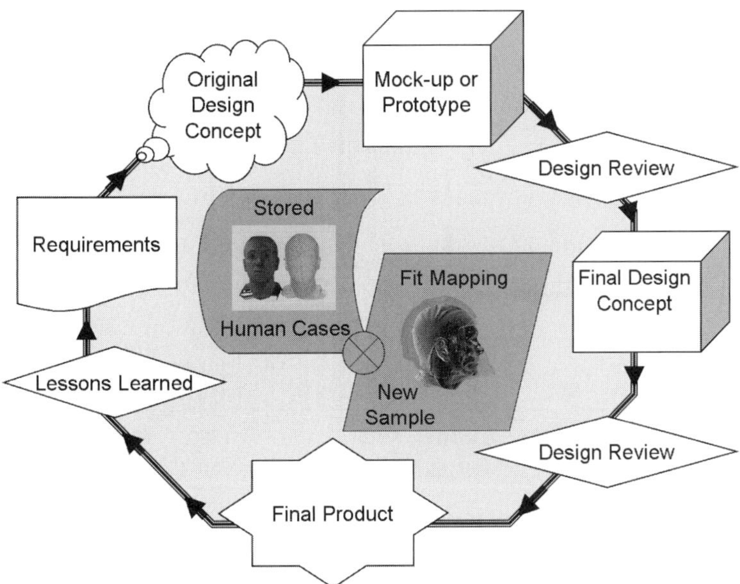

Figure 1 Anthropometric information as a design core.

percentile human figure: The values do not add up. This means that 5th or 95th percentile values can produce very unhumanlike figures that do not have the desired 5th or 95th percentile size for some of their dimensions.

The impact of using percentiles can be huge. For example, for one candidate aircraft for the T-1 program, the use of the 1st percentile female and 99th percentile male resulted in an aircraft that 90% of females, 80% of African-American males, and 30% of white males could not fly. The problem is illustrated in Figure 2. The pilots needed to be able, simultaneously, to see over the nose of the plane and operate the yoke, a control that is similar to a steering wheel in a car. For the 99th percentile seated eye height, the seat would be adjusted all the way down to enable the pilot to see over the nose. For the 1st percentile seated eye height, the seat would be adjusted all the way up. Since the design used all 1st percentile values for the full-up seat position, it accounted for only a 1st percentile or smaller female thigh size when the seat was all the way up. As a result, it did not accommodate most female pilots' thigh size without having the yoke interference as pictured in Figure 2. For designs such as this, where there are conflicting or interacting measurements or requirements, percentiles will not be effective. Cases that have combinations of small and large dimensions are needed.

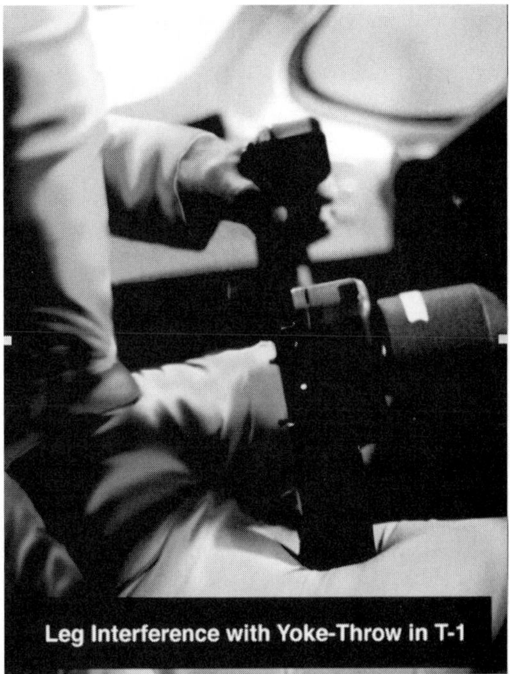

Figure 2 A problem that occurred when using 1st percentile female and 99th percentile male.

To understand when to use and when not to use averages and percentiles, it is important to understand what they are and what they are not. Figure 3 illustrates average and percentile values for stature and weight. Sample frequency distributions for these two measurements are shown for the female North American data from the CAESAR survey (Harrison and Robinette, 2002) in the form of histograms. The averages for 5th and 95th percentiles are indicated. The *frequency* is the count of the number of times that a value or range of values occurs, and the vertical bars

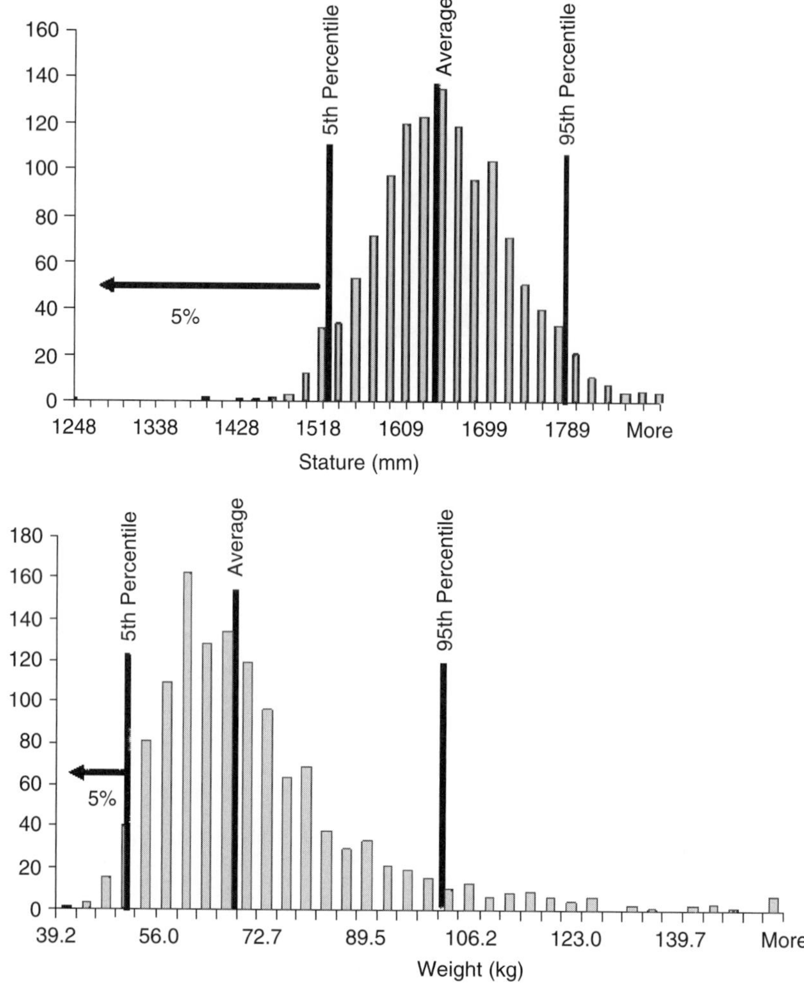

Figure 3 Stature and weight univariate frequencies, CAESAR, U.S. females.

in Figure 3 indicate the number of people who had a stature or weight of the size indicated. For example, the one vertical bar to the right of the 95th percentile weight indicates that approximately 10 people have a weight between 103 and 105 kg. Percentiles indicate the location of a particular cumulative frequency. For example, the 50th percentile is the point at which 50% have a smaller value, and the 95th percentile is the point at which 95% have a smaller value.

The *average* is a value for one measurement that falls near the middle of the distribution for that measurement. In this case the arithmetic average is shown. Another kind of central value is the 50th percentile, which will be the same as the arithmetic average when the frequency distribution is symmetric. The stature distribution shown in Figure 3 is approximately symmetrical, so the average and the 50th percentile differ by just 0.5%, whereas the weight distribution is not symmetric, so the average and the 50th percentile differ by 5.4%.

2.1.1 Percentile Issues

Percentile values refer only to the location of the cumulative frequency of one measurement. This means that the 95th percentile weight has no relationship to the 95th percentile stature. This is illustrated in Figure 4, which shows the two-dimensional frequency distribution for stature and weight along with the one-dimensional frequency distributions that appeared in Figure 3. Stature values are represented by the vertical axis and weight by the horizontal. The histogram from Figure 3 for stature is shown to the right of the plot, and the histogram from Figure 3 for weight is shown at the top of the plot, each with its respective 5th and 95th percentile values. Each of the circular dots in the center of the plot indicates the location of one subject from the sample of CAESAR U.S. females. The ellipse

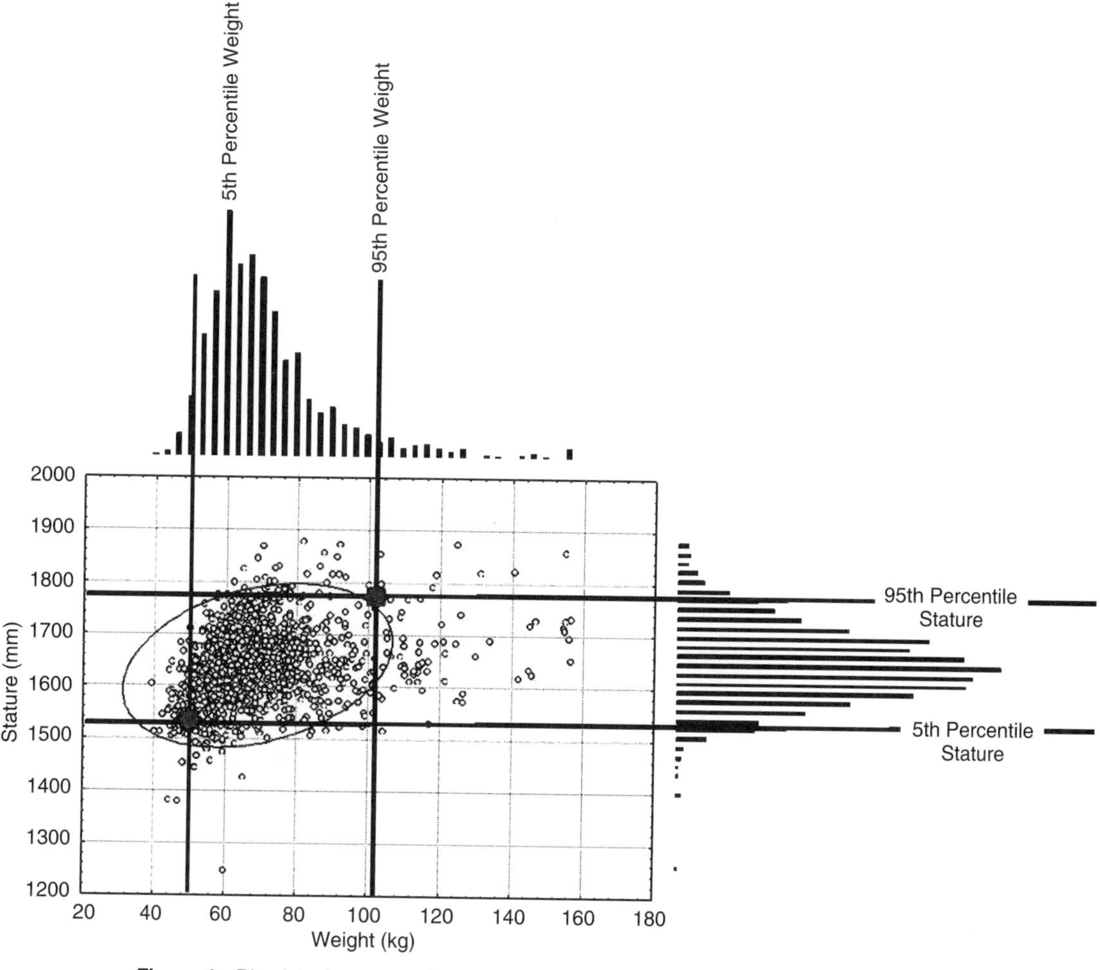

Figure 4 Bivariate frequency distribution of stature and weight, CAESAR, U.S. females.

toward the center that surrounds many of the dots is the 90% ellipse; in other words, it encircles 90% of the subjects.

If a designer uses the "5th percentile female to the 95th percentile female" approach, only two cases are being used. These two cases are indicated as black squares, one at the lower left and the other at the upper right of the two-dimensional plot in Figure 4. The one at the lower left is the intersection of the 5th percentile stature and the 5th percentile weight. The one at the upper right is the intersection of the 95th percentile stature and the 95th percentile weight. The stature range from 5th to 95th percentile falls between the two horizontal 5th and 95th percentile lines and contains approximately 90% of the population. The weight range from 5th to 95th percentile falls between the two vertical 5th and 95th percentile lines and contains approximately 90% of the population. The people who fall between the 5th and 95th percentiles for both stature and weight are only those people who fall within the intersection of the vertical and horizontal bands. The intersection contains only approximately 82%. If a third measurement is added, it makes the frequency distribution three-dimensional, and the percentage accommodated between the 5th and 95th percentiles for all three measurements will be fewer still.

The bar in the weight histogram in Figure 3 that is just to the right of the 95th percentile is the same as the bar that is just to the right of the 95th percentile weight in Figure 4. As stated above, approximately 10 subjects fall at this point. If you look at the two-dimensional plot, you can see that these 10 subjects who are at approximately the 95th percentile weight have stature values from as small as 1500 mm to as large as 1850 mm. In other words, women in this sample who have a 95th percentile weight have a range of statures that extends from below the stature 5th percentile to above the stature 95th percentile. This means that if the product has conflicting requirements,

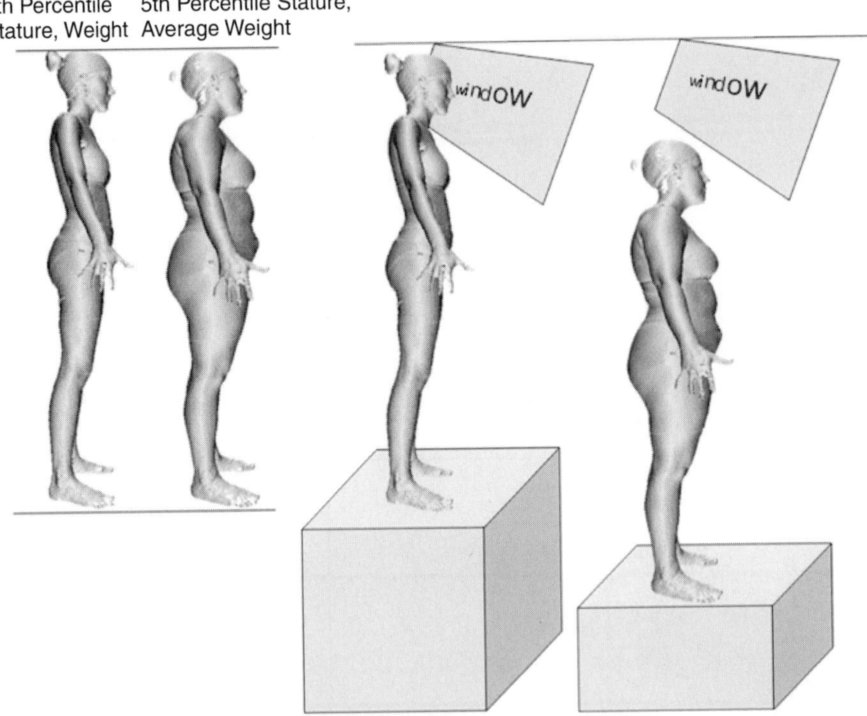

Figure 5 Woman with 5th percentile stature and average weight is not accommodated.

the 5th to 95th percentile cases would not work effectively. For example, suppose that a zoo has an automatically adjusting platform for an exhibit that adjusts its height based on the person's weight, and the exhibit designers want to make a window or display large enough so that the population can see the exhibit. If this is designed for the 5th percentile person to the 95th percentile person, they would design for (1) the 5th percentile stature with the 5th percentile weight as one case, and (2) the 95th percentile weight with the 95th percentile stature as the other case. Let us use the same female population data to see what the 5th and 95th percentiles would accommodate. At the 5th percentile case the platform would be full-up and the stature accommodated would be 1525 mm. At the 95th percentile case the platform would be full-down and the stature accommodated would be 1767 mm. This range of stature is 1767 − 1525 = 242 mm. This will accommodate the 5th percentile female to the 95th percentile female, but not the 5th percentile stature with an average weight or the 5th percentile weight with the average stature. The female who has a 5th percentile weight of 49.2 kg but an average stature of 1639 mm would need 114 mm more headroom. The female who has an average weight but a 5th percentile stature may not be able to see the display because the weight-adjusted platform is half way down. This is illustrated in Figure 5.

2.1.2 When to Use Averages and Percentiles

Averages, percentiles, and other one-dimensional summary statistics such as the standard deviation, minimum, and maximum are very useful for comparing samples from different populations to determine if there are size and variability differences. For example, in Table 1, one-dimensional summary statistics from the U.S. CAESAR sample (Harrison and Robinette, 2002) are compared with summary statistics from the U.S. ANSUR survey (Gordon et al., 1989). The CAESAR sample was taken from a civilian population, whereas the ANSUR sample was taken from a military population, the U.S. Army. The U.S. Army has fitness and weight limitations for its personnel. As a result, the ANSUR sample has a more limited range of variability for weight-related measurements. The effect of this can be seen by examining the differences in the weight and Buttock–Knee length ranges (minimum to maximum) vs. the ranges for the other measurements that are less affected by weight.

You might also notice that the difference between CAESAR and ANSUR females in buttock–knee length is greater than the difference in CAESAR and ANSUR males in buttock–knee length. This highlights a key difference between men and women. Women tend to gain weight in their hips, buttocks, and thighs, whereas men tend to gain weight or bulk in their waists and shoulders.

Table 1 Comparison of U.S. Civilian Summary Statistics (CAESAR Survey) with U.S. Army Statistics (ANSUR Survey)

	N	Mean	Minimum	Maximum	Std. Dev.
Acromion height, sitting (mm)					
Females					
CAESAR	1264	567.42	467.00	672.00	29.76
ANSUR	2208	555.54	464.06	663.96	28.65
Males					
CAESAR	1127	607.21	489.00	727.00	34.19
ANSUR	1774	597.76	500.89	694.94	29.59
Buttock–knee length (mm)					
Females					
CAESAR	1263	586.97	489.00	805.00	37.43
ANSUR	2208	588.93	490.98	690.88	29.63
Males					
CAESAR	1127	618.93	433.00	761.00	35.93
ANSUR	1774	616.41	505.97	722.88	29.87
Sitting eye height (mm)					
Females					
CAESAR	1263	755.34	625.00	878.00	34.29
ANSUR	2208	738.71	640.08	864.11	33.24
Males					
CAESAR	1127	808.07	681.00	995.00	39.24
ANSUR	1774	791.97	673.10	902.97	34.21
Sitting knee height (mm)					
Females					
CAESAR	1264	509.06	401.00	649.00	28.28
ANSUR	2208	515.41	405.89	632.97	26.33
Males					
CAESAR	1127	562.35	464.00	671.00	31.27
ANSUR	1774	558.79	453.90	674.88	27.91
Sitting height (mm)					
Females					
CAESAR	1263	865.02	720.00	994.00	36.25
ANSUR	2208	851.96	748.03	971.04	34.90
Males					
CAESAR	1127	925.77	791.00	1093.00	40.37
ANSUR	1774	913.93	807.97	1032.00	35.58
Stature (mm)					
Females					
CAESAR	1264	1639.66	1248.00	1879.00	73.23
ANSUR	2208	1629.38	1427.99	1869.95	63.61
Males					
CAESAR	1127	1777.53	1497.00	2084.00	79.19
ANSUR	1774	1755.81	1497.08	2041.91	66.81
Thumb tip reach (mm)					
Females					
CAESAR	1264	738.65	603.30	888.00	39.55
ANSUR	2208	734.61	605.03	897.89	36.45
Males					
CAESAR	1127	813.99	694.60	1027.00	44.11
ANSUR	1774	800.84	661.92	979.93	39.17
Weight (kg)					
Females					
CAESAR	1264	68.84	39.23	156.46	17.60
ANSUR	2208	62.00	41.29	96.68	8.35
Males					
CAESAR	1127	86.24	45.80	181.41	18.00
ANSUR	1774	78.47	47.59	127.78	11.10

Table 2 Comparison of the U.S. (US) and Dutch (TN) Statistics from the CAESAR Survey

	N	Mean	Minimum	Maximum	Std. Dev.
Acromion height, sitting (mm)					
Females					
US	1264	567.42	467.00	672.00	29.76
TN	687	589.53	490.98	709.93	33.20
Males					
US	1127	607.21	489.00	727.00	34.19
TN	559	629.89	544.07	739.90	35.96
Buttock–knee length (mm)					
Females					
US	1263	586.97	489.00	805.00	37.43
TN	688	608.07	515.87	728.98	31.15
Males					
US	1127	618.93	433.00	761.00	35.93
TN	558	636.22	393.95	766.06	37.44
Sitting eye height (mm)					
Females					
US	1263	755.34	625.00	878.00	34.29
TN	676	774.46	664.97	942.09	35.82
Males					
US	1127	808.07	681.00	995.00	39.24
TN	593	825.35	736.09	957.07	39.71
Sitting knee height (mm)					
Females					
US	1264	509.06	401.00	649.00	28.28
TN	676	510.67	407.92	600.96	28.78
Males					
US	1127	562.35	464.00	671.00	31.27
TN	549	557.09	369.06	680.97	35.97
Sitting height (mm)					
Females					
US	1263	865.02	720.00	994.00	36.25
TN	687	884.64	766.06	1049.02	38.06
Males					
US	1127	925.77	791.00	1093.00	40.37
TN	559	941.85	823.98	1105.92	42.54
Stature (mm)					
Females					
US	1264	1639.66	1248.00	1879.00	73.23
TN	679	1672.29	1436.12	1947.93	79.02
Males					
US	1127	1777.53	1497.00	2084.00	79.19
TN	593	1808.08	1314.96	2182.88	92.81
Thumb tip reach (mm)					
Females					
US	1264	738.65	603.30	888.00	39.55
TN	690	751.22	632.97	889.25	37.71
Males					
US	1127	813.99	694.60	1027.00	44.11
TN	564	826.7	488.70	1055.62	53.55
Weight (kg)					
Females					
US	1264	68.84	39.23	156.46	17.60
TN	690	73.91	37.31	143.23	15.81
Males					
U.S.	1127	86.24	45.80	181.41	18.00
TN	564	85.57	50.01	149.73	17.28

The ANSUR/U.S. CAESAR differences in Table 1 are contrasted with another comparison of anthropometric data in Table 2. This compares the U.S. CAESAR data with those collected in The Netherlands on the Dutch population (TN). The Dutch claim to be the tallest people in Europe, and this is reflected in all the heights and limb lengths. Both the male and female Dutch subjects are more than 30 mm taller on average than their U.S. counterparts.

Averages and percentiles and other one-dimensional statistics can also be very useful for products that do not have conflicting requirements. In these instances the loss in accommodation with each additional dimension can be compensated for by increasing the percentile range for each dimension. For example, if you want to ensure 90% accommodation for a simple design problem (one that has no interactive measurements) with five key measurements, you can use the 1st and 99th percentile values instead of the 5th and 95th. Each of the five measurements restricts 2% of the population, so at most you would have $5 \times 2\% = 10\%$ disaccommodated. This approach is illustrated in Figure 6. The bars represented by the 5th percentile values would be moved to where the stars are in the figure.

To summarize, percentiles represent the proportion accommodated for one dimension only. When used for more than one dimension, the combination of measurements will accommodate less than the proportion indicated by the percentiles. If the design has no conflicting requirements, you can sometimes compensate by moving out the percentiles. However, if the design has conflicting requirements, using percentiles may accommodate very few people, and an alternative set of cases is required.

2.2 Alternative Methods

There are two categories of alternatives to percentiles: (1) select a set of cases with relevant combinations of dimensions, or (2) use a random or Monte Carlo sample of 30 or more subjects. Generally, using a random sample with lots of subjects is not practical, although new three-dimensional modeling and CAD technologies may soon change this. Therefore, the selection of a small number of cases that effectively represent the population is preferable.

2.2.1 Case Selection

The purpose of using a small number of cases is to simplify the problem by reducing to a minimum the amount of information needed. Generally, the first thing reduced is the number of dimensions. This is done using knowledge of the product and by examining the correlation of the dimensions that are related to the product. The goal is to keep just those that are critical

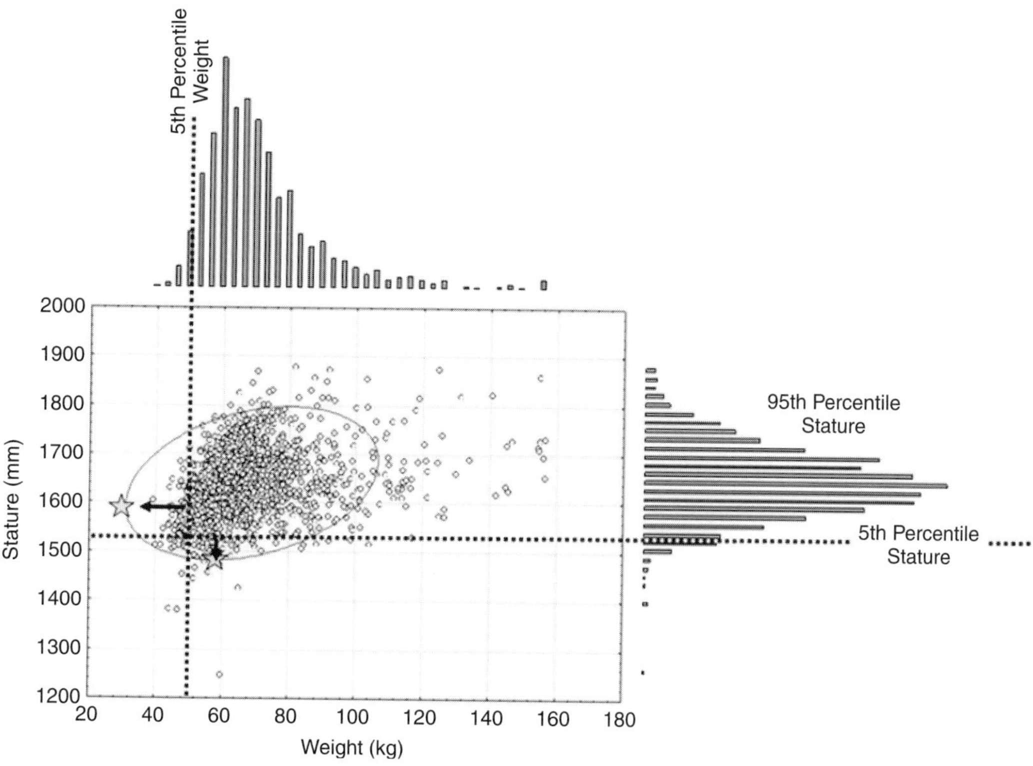

Figure 6 Moving the percentile values out farther to get a desired joint accommodation of 90%.

and have as little redundant information as possible. For example, eye height, sitting, and sitting height are highly correlated; therefore, accommodating one could accommodate the other, and only one would be needed in case selection.

It is easiest if the number of critical dimensions can be reduced to four or fewer, because all combinations of small and large proportions need to be considered. If there are two critical dimensions, the minimum number of small and large combinations is four: small–small, small–large, large–small, and large–large. If there are three critical dimensions, the minimum number of small and large combinations is eight: small–large–large, small–small–large, small–large–small, small–small–small, large–small–small, large–small–large, large–large–small, and large–large–large. With more than four the problem gets quite complex, and a random sample may be easier to use.

The next simplification is a reduction in the combinations used. Often in a design, only the small or large size of a dimension is needed. For example, for a chair hip breadth, sitting might be one of the critical dimensions but only the large size is needed to define the minimum width of the seat. If it does not have any interactive or conflicting effect with the other critical dimensions, it can be used as a stand-alone single value. Also, if two or more groups have overlapping cases, such as males and females in some instances, it is possible to drop some of the cases.

This process is best explained using an example of a seated workstation with three critical dimensions: eye height, sitting; buttock–knee length; and hip breadth, sitting. The minimum seat width for this design should be the largest hip breadth, sitting, but this is the only seat element that is affected by hip breadth, sitting, so it interferes with no other dimension. The desired accommodation overall is 90% of the male and female population. First the designer selected the hip breadth, sitting value by examining its summary statistics for the large end of its distribution. These are shown in Table 3. As can be seen from the table, the women have a larger hip breadth, sitting than the men. Therefore, the women's maximum value will be used. If the 99th percentile is used, approximately 1% of U.S. civilian women would be estimated to be larger. If 90% are accommodated with the remaining two dimensions, only 89% would be expected to be accommodated for all three. It would be simplest to use the maximum and then accommodate 90% in the other two. This was the approach used by Zehner (1996) for the JPATS aircraft. An alternative is to assume some risk in the design and to select a smaller number than the maximum. This is a judgment to be made by the manufacturer or customer.

Next we examine the two-dimensional (also called *bivariate*) frequency distribution for eye height, sitting and buttock–knee length. The distribution for female subjects from the CAESAR database is shown in Figure 7, and the distribution for male subjects is

Table 3 Hip Breadth, Sitting Statistics from the U.S. CAESAR Survey (mm)

	N	Mean	95th Percentile	99th Percentile	Maximum
Females	1264	410	501	556	663
Males	1127	376	435	483	635

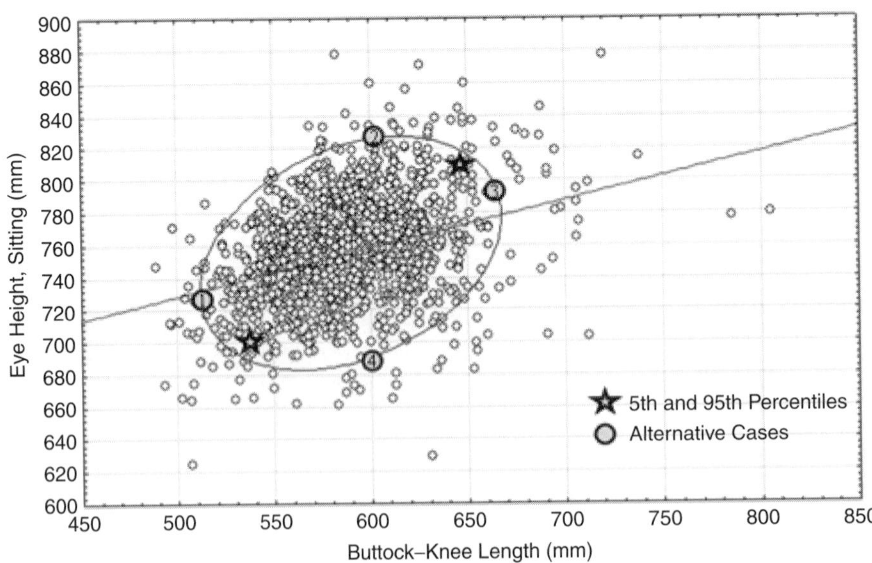

Figure 7 Bivariate frequency distribution for eye height, sitting, and buttock–knee length for CAESAR U.S. female sample ($N = 1263$) with four female cases shown. 90% Probability ellipse, NHANES weighted.

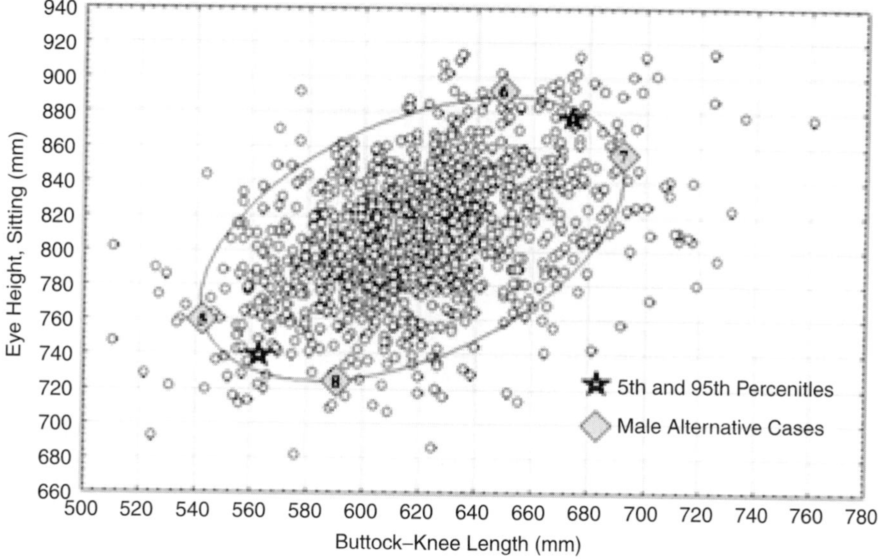

Figure 8 Bivariate frequency distribution for eye height, sitting, and buttock–knee length for U.S. male sample ($N = 1125$). 90% Probability ellipse, NHANES weighted.

shown in Figure 8. The stars in Figures 7 and 8 represent the location of the 5th and 95th percentiles, and the probability ellipses enclose 90% of each sample. To achieve the target 90% accommodation, cases that lie on the elliptical boundary are selected. Boundary cases chosen in this way represent extreme combinations of the two measurements. For example, in Figure 7, cases 1 and 3 represent the two extremes for buttock–knee length, and cases 2 and 4 represent the two extremes for eye height, sitting. Note that the cases are moderate in size for one dimension but extreme for the other. The boundary ellipse provides combinations that are not captured in the range between small–small (5th/5th) or large–large (95th/95th) percentiles. Case selection of this type makes the assumption that if the boundary cases are accommodated by the design, so are all those within the probability ellipse. Although this assumption is valid for workspace design, where vision, reach, and clearance from obstruction are key issues, it is not necessarily true for design of clothing or other gear worn on the body. In the latter application, an adequate number of cases must be selected to represent the inner distribution of anthropometric combinations, which should be given much more emphasis than the boundary cases.

The dimensions for the eight cases represented in Figures 7 and 8 are shown in Table 4. Note that this table includes the same hip breadth, sitting for all cases. This is the hip breadth, sitting taken from Table 3 and represents the smallest breadth that should be used in the design. The dimensions for each case must be applied to the design as a set. For example, the seat must be adjustable to accommodate a buttock–knee length of 510 mm at the same time that it is adjusted to accommodate an eye height, sitting of 725 mm and a hip breadth, sitting of 663 mm to accommodate case 1.

An option for reducing the number of cases is to drop those that are overlapping or redundant. If the risk is small that differences in men and women will affect the design significantly, it is possible to drop some of the overlapping cases and still accommodate the desired proportion of the population. For example, male cases 5 and 8 are not as extreme as female cases 1 and 4, and the accommodation risk due to dropping them is small. The bivariate distribution in Figure 9 illustrates buttock–knee length and eye height, sitting for both men and women. The final set of anthropometric cases is shown, as well as the location of the dropped cases, 5 and 8.

2.2.2 Distributing Cases

As introduced in Section 2.2.1, all of the prior examples make the assumption that if the outer boundaries of the distribution are accommodated, all of the people within the boundaries will also be

Table 4 Case Dimensions for Seated Workstation Example (mm)

Females	Case 1	Case 2	Case 3	Case 4
Buttock–knee length	510	600	660	600
Eye height, sitting	725	820	795	690
Hip breadth, sitting	663	663	663	663
Males	**Case 5**	**Case 6**	**Case 7**	**Case 8**
Buttock–knee length	541	655	690	595
Eye height, sitting	760	890	855	725
Hip breadth, sitting	663	663	663	663

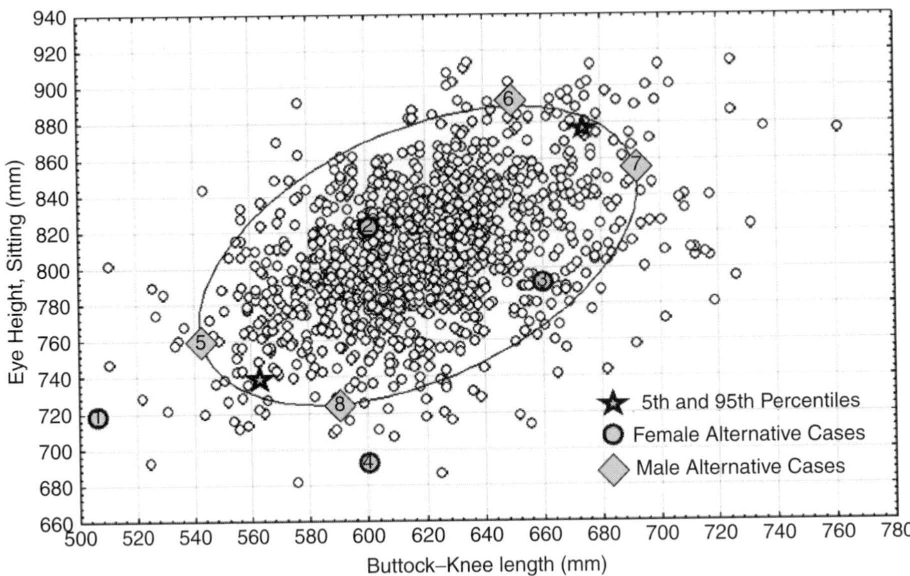

Figure 9 Bivariate frequency distribution for eye height, sitting, and buttock–knee length for U.S. male ($N = 1127$) and female ($N = 1263$) sample. Cases 5 and 8 were not included in the final set due to proximity to cases 1 and 4.

accommodated. This is true for both the univariate case approach (upper and lower percentile values) and the multivariate case approach (e.g., bivariate ellipse cases, as above). For products that come in sizes or with adjustments that are stepped rather than continuous, this may not be a valid assumption. Imagine a T-shirt that comes in only X-small and XX-large sizes. Few people would be accommodated. For these kinds of products, it is necessary to select, or distribute, cases both *at* and *within* the boundaries.

For distributing cases it is important that there be more cases than expected sizes or adjustment steps to ensure that people are not missed between sizes or steps. A good example of distributed cases is shown by Harrison et al. (2000) in their selection of cases for laser eye protection (LEP) spectacles. They used three key dimensions: face breadth for the spectacle width, nose depth for the distance of the spectacle forward from the eye, and eye orbit height for the spectacle height. They used bivariate plots for each of these dimensions with the other two, and selected 30 cases to characterize the variability for all three. They also took into account the different ethnicities of subjects when selecting cases, to ensure adequate accommodation of all groups. One of their bivariate plots with the cases selected is shown in Figure 10.

For the LEP effort, the critical dimensions were used to select individual subjects, and their three-dimensional scans were used to characterize them as a case for implementation in the spectacle design. Figure 11 illustrates the side view of the three-dimensional scan for one of the cases. By using distributed cases throughout the critical dimension distribution, a broader range is covered than using the equivalent number of subjects in a random sample, and no assumption about the range of accommodation within one size is made. This permits evaluation of the range of fit within a size and the degree of size overlap during the design process.

2.2.3 Principal Components Analysis

In the examples above, the set of dimensions was reduced using judgment based on knowledge of the problem and the relationship between measurements. Principal components analysis (PCA) can be helpful in both understanding the relationship between relevant measurements and in reducing the set of dimensions to a small, manageable number. This technique has been used effectively for aircraft cockpit crew station design (Bittner et al., 1987; Zehner et al., 1993; Zehner, 1996).

Generally, for any design problem the important human dimensions have some relationship with each other. For example, sitting height and eye height, sitting are highly correlated. The paired relationships between a set of dimensions can be expressed as either a correlation or a covariance matrix. PCA uses a correlation or covariance matrix and creates a new set of variables called *components*. The total number of components is equal to the number of original variables, and the first component will always represent the greatest amount of variation in the distribution. The second component describes the second greatest, and so on. An examination of the relative contributions, or correlations, of each original dimension and a particular component can be used to interpret and "name" the component. For example, the first component usually describes overall body size

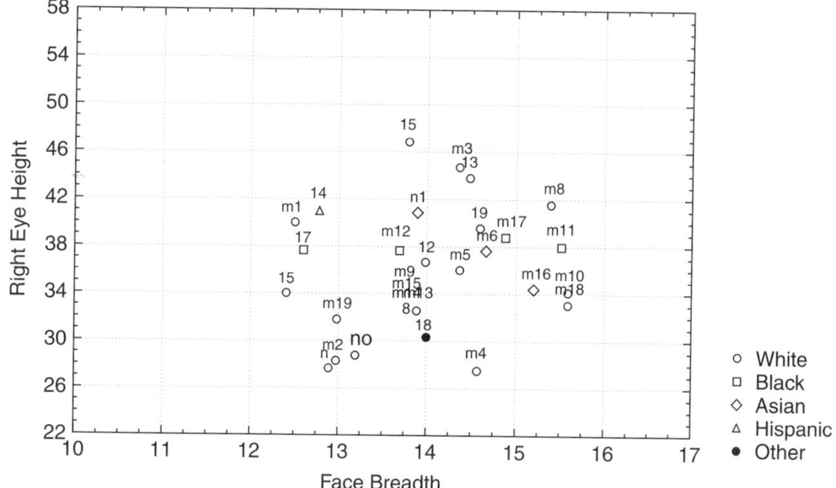

Figure 10 Plot 2 of the three critical dimensions and the cases for the LEP.

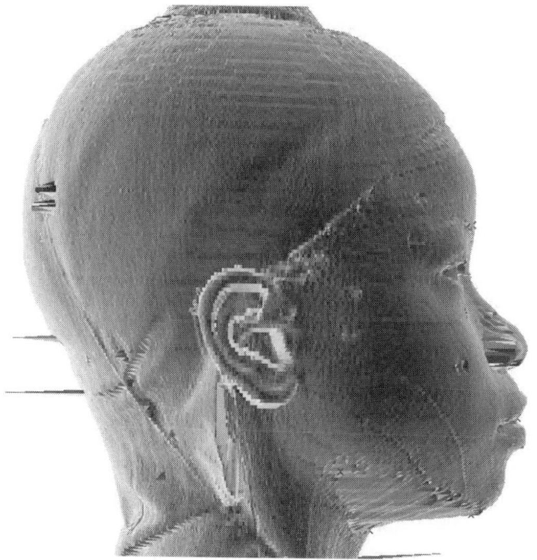

Figure 11 Side view of one LEP case.

and is defined by observing a general increase in the values for the original anthropometric dimensions as the value, or score, of the first component increases.

The premise in using PCA for accommodation case selection is that if most of the total variability in the distribution can be represented in the first two or three components, the PCA approach can be used to select your cases to achieve a designated level of accommodation. (If just two or three components explain most of the variability, they are called the *principal components*, hence the name of the method.) For example, to write the anthropometric specifications for cockpit design in Joint Primary Air Training System (JPATS) aircraft, Zehner (1996) used the first and second components from a PCA on six cockpit-relevant anthropometric dimensions. The first two components explained 90% of the total variability for all six combined measurements. This was approximately the same for each gender (conducted in separate analyses). Zehner then used a 99.5% probability ellipse on the first two principal components to select the initial boundary cases. One of the genders is shown in Figure 12. Combining the initial set of cases from both genders (with some modification) resulted in a final set of JPATS cases that offered an accommodation of 95% for the women and 99.9% for the men. The first principal component was defined as size; the second was a contrast between limb length and torso height (short limbs/tall torso vs. long limbs/short torso).

Unlike compiled percentile methods (or compiled bivariate approaches when there are more than two variables), multivariate PCA takes into account the simultaneous relationship of three or more variables. However, with PCA the interpretation of the components may not always be clear, and it can be more difficult to understand what aspect of size is being accommodated. An alternative way to use PCA is to use it only to understand which dimensions are correlated with others, and then select the most important single dimensions to represent the set as a key dimension. In this way, the key dimension is easier to understand.

The chief limitation of PCA is that all of the dimensions are accepted into the analysis as if they have equal design value and PCA has no way to know the design value. As a result, accommodating the components will accommodate some of the variability of the less important dimensions at the expense of the more important ones. Also, PCA is affected by the number of correlated dimensions used of each type.

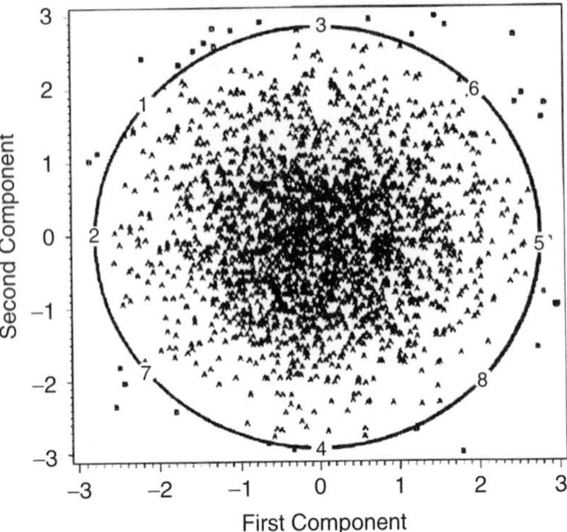

Figure 12 PCA bivariate and 99.5% boundary for the first and second principal components for one gender of the JPATS population. Initial cases 1 to 8 for this gender are regularly distributed around the boundary shape.

For example, if 10 dimensions are used and nine of them are strongly correlated with one another, the one dimension that is not correlated with any other may end up being component 4. Since it is one of 10 dimensions, it represents 10% of the total variability. So it is possible to accommodate 90% of the variability in the first three components and not accommodate the most important dimension. Therefore, when using PCA it is important to (1) include only dimensions that are both relevant and important, and (2) check the range of accommodation achieved in the cases for each individual dimension.

3 FIT MAPPING

Fit mapping is a type of design guidance study that provides information about who a product fits well and who it does not. When anthropometry is used in product design without the knowledge of fit, many assumptions must be made about how to place the anthropometry in the design space and the range of accommodation. As a result, even with digital human models and computer-aided design, it is often the case that the first prototypes do not accommodate the full range of the population and may accommodate body-size regions that do not exist in the population.

Fit mapping involves using prototypes or mock-ups and performance testing in conjunction with anthropometric measurement to determine the fit effectiveness of a product for different body sizes and shapes. Fit effectiveness means that the desired population is accommodated without wasted sizes or wasted accommodation regions. Because most performance-based fit tests cannot be done on digital models, fit mapping involves using human subjects to do the assessments. The following is a list of things needed for a fit-mapping study:

1. Human subjects drawn to represent the broadest variability (even broader than the target population is okay)
2. A prototype or sample of the product and, if it is a sized product, at least one of each size of garment, and preferably two so that more than one subject can be run at a time and you have a spare
3. A concept-of-fit definition
4. An expert fit evaluator or one who is trained to be consistent using the concept-of-fit definition
5. Anthropometry measuring equipment
   a. Traditional tools
   b. Scanner capable of recognizing landmarks
6. Statistical software
7. Statistical analysis expert, preferably one experienced with fit analysis
8. Survey data from the target population with relevant measurements

The study process consists of:

1. Scoring the fit for each size that the subject can don against the concept of fit
2. Measuring the subjects
3. Analyzing the data to determine:
   a. The key size-determining dimensions
   b. The range of accommodation for each size with respect to the key dimensions
   c. General design or shaping issues
   d. Size or shape gaps in target population coverage
   e. Size or shape overlaps in target population coverage

The end result of the study is that wasted sizes or adjustment ranges are dropped, sizes or adjustment ranges are added where there are gaps, and design and reshaping recommendations are provided to make the product fit better overall. One example of the magnitude of the improvement that can be achieved with the use of fit mapping was demonstrated in the Navy women's uniform study (Mellian et al., 1990; Robinette et al., 1990). The Navy women's uniform consisted of two jackets, two skirts, and two pairs of slacks. The fit mapping consisted of measuring body size and assessing the fit of each of the garments on more than 1000 Navy women. Prior to the study, the Navy had added odd-numbered sizes in an attempt to improve fit, because 75% of all Navy recruits had to have major alterations. The sizes included sizes 6, 7, 8, 9, 10, 11, 12, 13, 14, 15, 16, 18, 20, and 22, with three lengths for each, for a total of 42 sizes.

The results indicated three important facts. First, there was 100% overlap in some of the sizes. For each of the items, sizes 7 and 8, 9 and 10, 11 and 12, 13 and 14, and 15 and 16 fit the same subjects equally well. Second, the size of best fit was different

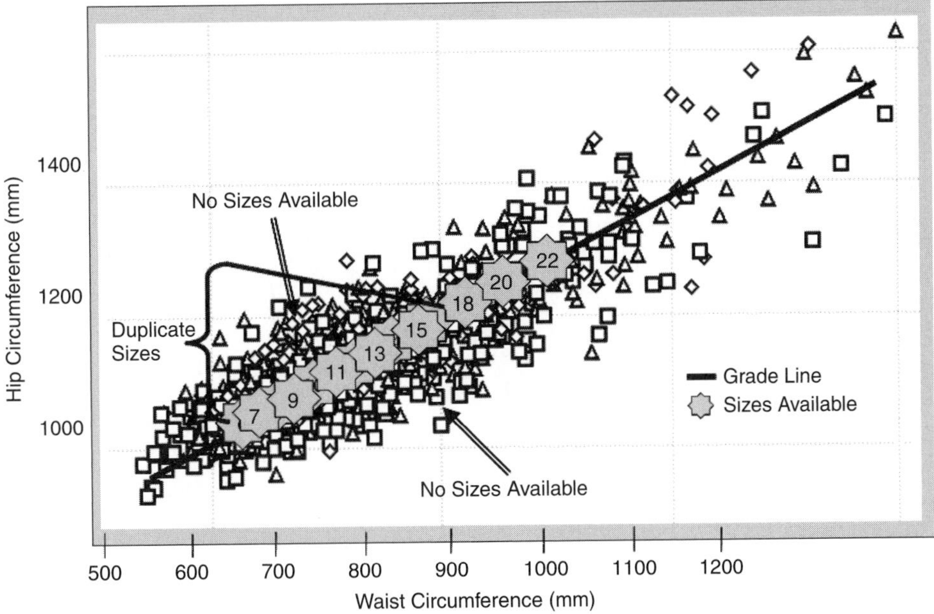

Figure 13 Before fit mapping apparel sizes.

for nearly every garment, with some women wearing up to four different sizes. For example, one woman had the best fit in a size 8 for the blue skirt, size 10 for the white skirt, size 12 for the blue slacks, and size 14 for the white slacks. Third, most women did not get an acceptable fit in any size.

The size overlap was examined and it was determined that the difference between the sizes was less than the manufacturing tolerance for a size, which was $\frac{1}{2}$ inch. Therefore, the manufacturers had actually used exactly the same pattern for sizes 7 and 8, 9 and 10, 11 and 12, 13 and 14, and 15 and 16. Therefore, sizes 7, 9, 11, 13, and 15 in all three of their lengths could be removed with no effect on accommodation.

The difference in which size fits a given body was resolved by renaming the sizes for some of the garments, to make them consistent. This highlights the fact that the size something is designed to be is not necessarily the size it actually is. Fabric, style, concept of fit, function, and many other factors affect fit. Many of these cannot be known without fit testing on human subjects.

Finally, the women who did not get an acceptable fit in any size were proportioned differently than the size range. They had either a larger hip for the same waist or a smaller hip for the same waist as the Navy size range. This is an example of an interaction or conflict in the dimensions. All of the sizes were in a line consisting of the same shape scaled up and down. This is consistent with common apparel sizing practice. Most apparel companies start with a base size, such as a 10 or a 12, and scale it up and down along a line. The scaling is called *grading*. This is illustrated in Figure 13. The grading line is shown in bold in Figure 13. The sizes that fall along this line are similar to those used in the Navy women's uniform. Note the overlapping of the odd-numbered sizes with the even-numbered sizes in one area. This is the area where there were more sizes than necessary. Also, note that above and below the grading line, no sizes are available.

Figure 14 illustrates the types of changes made to the sizing to make it more effective. (Note that the sample of women shown is that of the civilian CAESAR survey, not Navy women, who do not have the larger waist sizes.) The overlapping sizes have been dropped. The sizes shown above the grade line have a larger hip for the same waist and are called plus hip (+), and the sizes below the line have a smaller hip for the same waist and are called minus hip (−). Before these sizes were added, women who fell in the plus hip region had to wear a size with a very large waist in order to fit their hip. Then they had to have the entire waist-to-hip region altered. Women who fell in the minus hip region previously had to get a garment that was way too large in the hips in order to get a fit for their waist. Adding sizes with the modified hip-to-waist proportion resulted in accommodating 99% of the women without needing alternations. The end result of adjusting the sizing based on fit mapping was to improve accommodation from 25% to 99%, with the same number of sizes (Figure 14).

4 THREE-DIMENSIONAL ANTHROPOMETRY

Three-dimensional anthropometry has been around since the advent of stereo photography. Originally stereo pairs had to be viewed through a stereo viewer and digitized manually, and it was very time

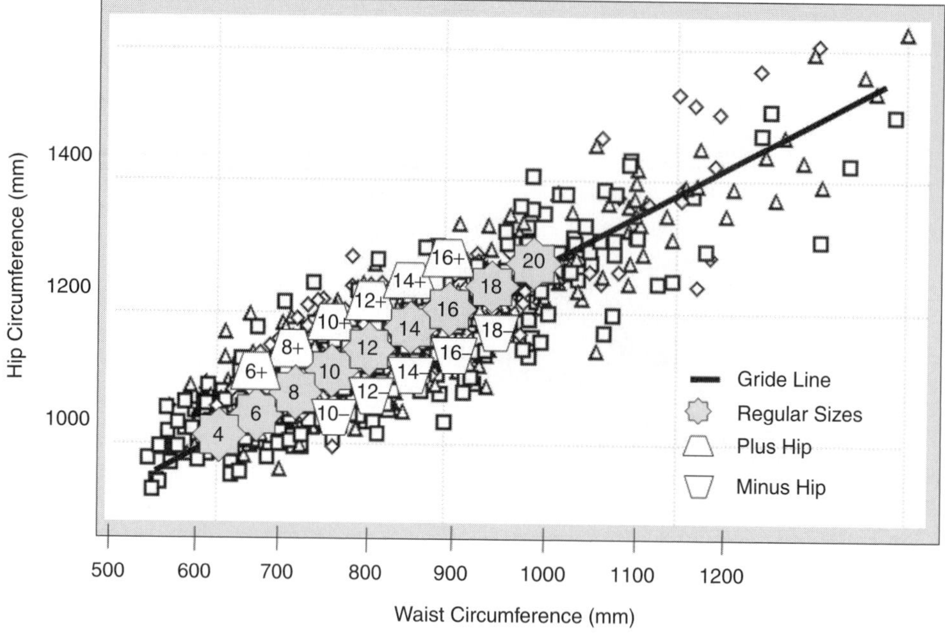

Figure 14 After fit mapping sizes.

consuming. This process is described by Herron (1972). However, digital photography allowed us to automate the process, and this has dramatically affected our ability to design effectively. Automated three-dimensional scanning began to take off in the 1980s (Robinette, 1986). Now there are many tools available to use and analyze three-dimensional scan data, and the first civilian survey to provide whole-body scans of all subjects, CAESAR, was completed in 2002 (Blackwell et al., 2002; Harrison and Robinette, 2002; Robinette et al., 2002). We describe briefly here some of the benefits of the new technology.

4.1 Why Three-Dimensional Scans?

By far the biggest advantage of three-dimensional surface anthropometry is visualization of cases, particularly the ability to visualize them with respect to the equipment or apparel they wear or use. When cases are selected, some assumptions are made about the measurements that are critical for the design. Three-dimensional scans of the subjects often reveal other important information that might otherwise have been overlooked. An example of this is illustrated in Figure 15. When designing airplane, stadium, or theater seats, two common assumptions are made: (1) that the minimum width of the seat should be based on hip breadth, sitting, and (2) that the minimum width of the seat should be based on the large male. In Figure 15 we see the scans of two figures overlaid, a male with a 99th percentile hip breadth, sitting and a female with a 99th percentile hip breadth, sitting. The male figure is in green and the female in red. It is immediately apparent that the female figure has broader hips than the male. Although she is shorter and has smaller shoulders, her hips are wider by more than 75 mm (almost 3 inches). Second, it is also clear that the shoulders and arms of the male figure extend out beyond the female hips. The breadth across the arms when seated comfortably is clearly a more appropriate measure for the spacing of seats.

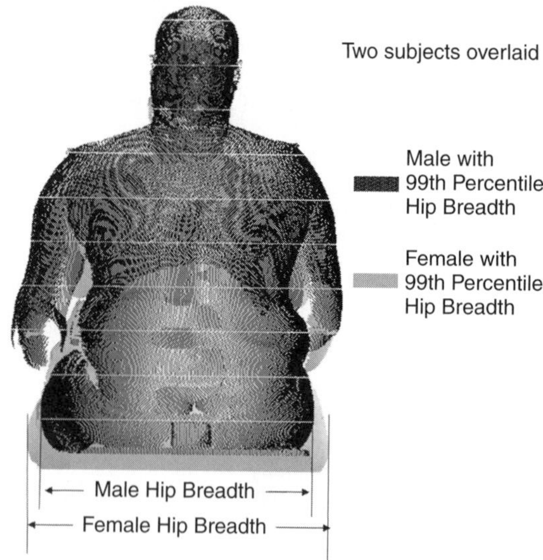

Figure 15 View of male and female with 99th percentile hip breadth, sitting.

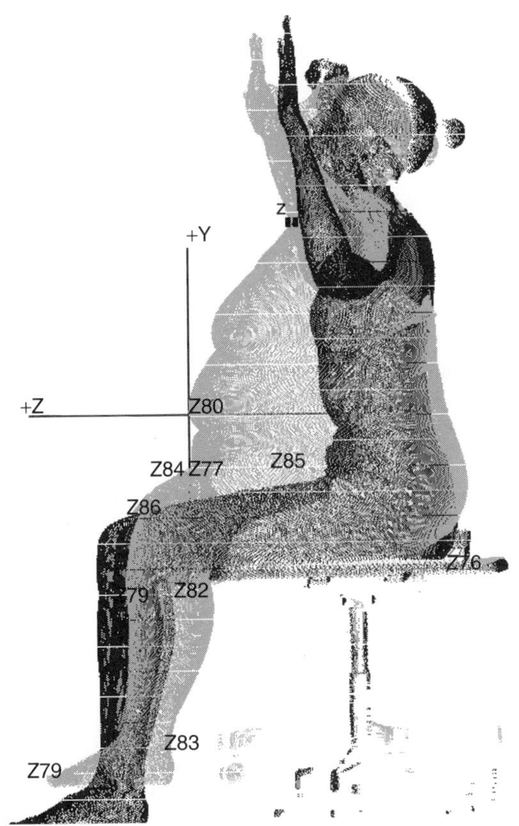

Figure 16 Two women with the same buttock–knee length and eye height, sitting.

For the design of a vehicle interior, measures such as buttock–knee length and eye height, sitting are often considered to be key. Figure 16 shows two women who have the same buttock–knee length and eye height, sitting. However, it is immediately clear from the image that because of the difference in their soft tissue distribution, they would have very different needs in terms of steering wheel placement. These things are much more difficult to comprehend by looking at tables of numbers. Three-dimensional anthropometry captures some measurements, such as contour change, three-dimensional landmark locations, or soft tissue distribution that cannot be captured adequately with traditional anthropometry. Finally, three-dimensional anthropometry offers the opportunity to measure the location of a person with respect to a product for use in identifying fit problems during fit mapping, and even for creating custom fit apparel or equipment. For example, by scanning subjects with and without a flight helmet and examining the range of ear locations within the helmet, fit problems due to ear misplacement can be identified. This is illustrated in Figure 17.

Figure 17 shows four examples of using three-dimensional anthropometric measurement to visualize and quantify fit. This figure was created using scans of the subjects with and without the helmet and superimposing the two images in three dimensions using software called Integrate (Burnsides et al., 1996). The image at the upper left of Figure 17 shows the location of the ears of eight subjects in the helmet being tested. The red curved lines show the point at which the subjects complained of ear pain. The image at the lower left shows the locations of two different subjects in the same helmet as they actually wore it to fly, demonstrating different head orientations. The image at the upper right shows the 90% and 95% accommodation ellipses for the point on the ear called *tragion*, for those subjects who did not complain of ear pain. The image at the lower right shows the spread of the *tragion* points for those subjects who did not complain of ear pain, along with one of the subject's ears (subject 4). It can be seen that the points are not elliptical but seem to have a concave shape, indicating a rotational difference between ear locations. These four images together with the fit and comfort evaluations completed by the subjects enable an understanding of the geometry of ear fit in that helmet. Without the three-dimensional images, the fit and comfort scores are difficult to interpret.

5 SUMMARY

Whether a product is personal gear (such as clothing or safety equipment), the crew station of a vehicle, or the layout of an office workspace, accommodating the variation in shape and size of the future user population will have an impact on a product's ultimate success. In this chapter we demonstrate the use of cases, fit mapping, and three-dimensional anthropometry to design effectively, simultaneously minimizing cost and maximizing accommodation.

The commonly used averages and percentiles are examples of cases, but they can be insufficient or incorrect for many design problems. Percentiles represent the proportion accommodated for one dimension only. When used for more than one dimension, the combination of measurements will accommodate less than the proportion indicated by the percentiles. If the design has no conflicting requirements, you can sometimes compensate by moving out the percentiles. However, if the design has conflicting requirements, using percentiles may accommodate very few people, and an alternative set of cases is required. The chapter covers alternative cases and how to select them, including the use of principal components analysis.

The best anthropometric data in the world are not sufficient to create a good design if the relationship between the anthropometry and the product proportions that accommodate it is not known. Fit mapping is the study of this relationship. The fit-mapping process is described with examples to demonstrate its benefits.

Finally, for complex multidimensional design problems, three-dimensional imaging technology provides an opportunity to visualize and contrast the variation in a sample and to quantify the differences between locations of a product on subjects who are accommodated vs. those who are not. The technology can also

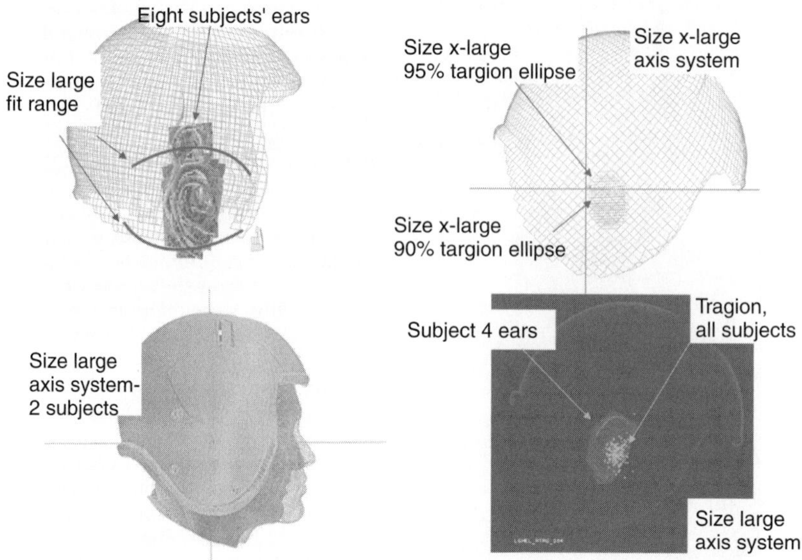

Figure 17 Three-dimensional scan visualizations to relate to fit mapping data for ear fit within a helmet.

be used to capture shape or morphometric data, such as contour change, three-dimensional landmark locations, or soft tissue distribution that cannot be captured adequately with traditional anthropometry. Therefore, three-dimensional anthropometry offers comprehension of accommodation issues to a degree not possible previously.

The study of anthropometry is not just about body measurements, it is about how to employ body measurement information effectively. In this chapter we have summarized the latest methods for doing so.

REFERENCES

Bittner, A. C., Glenn, F. A., Harris, R. M., Iavecchia, H. P., and Wherry, R. J. (1987), "CADRE: A Family of Mannikins for Workstation Design," in *Trends in Ergonomics/Human Factors IV*, S. S. Asfour, Ed., North-Holland/Elsevier, Amsterdam, pp. 733–740.

Blackwell, S., Robinette, K., Daanen, H., Boehmer, M., Fleming, S., Kelly, S., Brill, T., Hoeferlin, D., and Burnsides, D. (2002), *Civilian American and European Surface Anthropometry Resource (CAESAR), Final Report*, Vol. II, *Descriptions*, AFRL-HE-WP-TR-2002-0173, U.S. Air Force Research Laboratory, Human Effectiveness Directorate, Crew System Interface Division, Wright-Patterson Air Force Base, OH.

Burnsides, D. B., Files, P., and Whitestone, J. J. (1996), "Integrate 1.25: A Prototype for Evaluating Three-Dimensional Visualization, Analysis and Manipulation Functionality (U)," Technical Report AL/CF-TR-1996-0095, Armstrong Laboratory, Wright-Patterson Air Force Base, OH.

Daniels, G. S. (1952), "The Average Man," TN-WCRD 53–7 (AD 10 203), Wright Air Development Center, Wright-Patterson Air Force Base, OH.

Gordon, C. C., Churchill, T., Clauser, C. C., Bradtmiller, B., McConville, J. T., Tebbets, I., and Walker, R. (1989), "1988 Anthropometric Survey of U.S. Army Personnel: Methods and Summary Statistics," Technical Report NATICK/TR-89/044 (AD A225 094), U.S. Army Natick Research, Development and Engineering Center, Natick, MA.

Harrison, C. R., and Robinette, K. M. (2002), "CAESAR: Summary Statistics for the Adult Population (Ages 18–65) of the United States of America," AFRL-HE-WP-TR-2002-0170, U.S. Air Force Research Laboratory, Human Effectiveness Directorate, Crew System Interface Division, Wright-Patterson Air Force Base, OH.

Harrison, C., Robinette, K., and DeVilbiss, C. (2000), "Anthropometric Variable Selection for Laser Eye Protection Spectacles," in *SAFE Symposium Proceedings 2000*.

Herron, R. E. (1972), "Biostereometric Measurement of Body Forms," *Yearbook of Physical Anthropology*, Vol. 16, pp. 80–121.

Mellian, S. A., Ervin, C., and Robinette, K. M. (1990), "Sizing Evaluation of Navy Women's Uniforms," Technical Report 182, Navy Clothing and Textile Research Facility, Natick MA, and AL-TR-1991-0116 (AD A249 782), Armstrong Laboratory, Air Force Systems Command, Wright-Patterson Air Force Base, OH.

Robinette, K. M. (1986), "Three-Dimensional Anthropometry: Shaping the Future," in *Proceedings of the Human Factors Society 30th Annual Meeting*, Vol. 1, Human Factors Society, Santa Monica, CA, p. 205.

Robinette, K. M., and McConville, J. T. (1982), "An Alternative to Percentile Models," SAE Technical Paper 810217, in *1981 SAE Transactions*, Society of Automotive Engineers, Warrendale, PA, pp. 938–946.

Robinette, K. M., Mellian, S., and Ervin, C. (1990), "Development of Sizing Systems for Navy Women's Uniforms," Technical Report 183, Navy Clothing and Textile Research Facility, Natick MA, and AL-TR-1991-

0117, Armstrong Laboratory, Air Force Systems Command, Wright-Patterson Air Force Base, OH.

Robinette, K., Blackwell, S., Daanen, H., Fleming, S., Boehmer, M., Brill, T., Hoeferlin, D., and Burnsides, D. (2002), *Civilian American and European Surface Anthropometry Resource (CAESAR), Final Report, Vol. I, Summary*, AFRL-HE-WP-TR-2002-0169, U.S. Air Force Research Laboratory, Human Effectiveness Directorate, Crew System Interface Division, Wright-Patterson Air Force Base, OH.

Searle, J. A., and Haslegrave, C. M. (1969), "Anthropometric Dummies for Crash Research," MIRA Bulletin 5, Monthly Summary of Automobile Engineering Literature, Motor Industry Research Association, Lindley, near Neuneaton, Warwickshire, England, pp. 25–30.

Zehner, G. F. (1996), "Cockpit Anthropometric Accommodation and the JPATS Program," *SAFE Journal*, Vol. 26, No. 3, pp. 19–24.

Zehner, G. F., Meindl, R. S., and Hudson, J. A. (1993), "A Multivariate Anthropometric Method for Crew Stations Design: Abridged," AL-TR-1992-0164, Armstrong Laboratory, Air Force Systems Command, Wright-Patterson Air Force Base, OH.

CHAPTER 13

BASIC BIOMECHANICS AND WORKSTATION DESIGN

W. S. Marras
The Ohio State University
Columbus, Ohio

1 DEFINITIONS

Occupational biomechanics is an interdisciplinary field in which information from both the biological sciences and engineering mechanics is used to quantify the forces present on the human body during work. *Biomechanics* assumes that *the body behaves according to the laws of Newtonian mechanics.* Kroemer has defined *mechanics* as "the study of forces and their effects on masses" (Kroemer, 1987). The object of interest in occupational ergonomics is a quantitative assessment of mechanical loading occurring within the musculoskeletal system. The goal of such an assessment is to describe quantitatively the musculoskeletal loading that occurs during work so that one can derive an appreciation for the degree of risk associated with work-related task. This high degree of precision and quantification is the characteristic that distinguishes occupational biomechanics analyses from other types of ergonomic analyses. Thus, with biomechanical techniques the ergonomics can address the issue of how much exposure to the occupational risk factors is too much exposure.

The workplace biomechanical approach is often called *industrial or occupational biomechanics.* Chaffin et al. (1999) defined occupational biomechanics as "the study of the physical interaction of workers with their tools, machines, and materials so as to enhance the worker's performance while minimizing the risk of musculoskeletal disorders." We address occupational biomechanical issues concepts as they apply to work design.

2 ROLE OF BIOMECHANICS IN ERGONOMICS

The approach to a biomechanical assessment is to characterize the human–work system situation through a mathematical representation or model. The model is intended to represent the various underlying biomechanical concepts through a series of rules or equations in a system that helps us understand how the human body is affected by the various main effects and interactions associated with risk factor exposure. One can think of a biomechanical systems model as the glue that holds our logic together when considering the various factors that would affect risk in a specific work situation.

The advantage of representing the worker in a biomechanical model is that the model permits one to consider quantitatively the *trade-offs* associated with workplace risk factors to various parts of the body in the design of a workplace. It is difficult to accommodate all parts of the body in an ideal biomechanical environment since improving the conditions for one body segment often make things worse for another part of the body. Therefore, the key to the proper application of biomechanical principles is to consider the appropriate biomechanical trade-offs associated with various parts of the body as a function of the work requirements and the various workplace design options and constraints. Ultimately, biomechanical analyses would be most effective in predicting workplace risk during the design stage before physical construction of the workplace has begun.

This chapter focuses on the information required to develop proper biomechanical reasoning when assessing physical demands of a workplace. We first present and explain a series of key biomechanical concepts that constitute the underpinning of biomechanical reasoning. Second, these concepts are applied to the various parts of the body that are often affected during work. Once this reasoning is established, we examine how the various biomechanical concepts must be considered collectively in terms of trade-off when designing a workplace from an ergonomic perspective under realistic conditions. The logic in this chapter demonstrates that one *cannot* practice ergonomics successfully simply by memorizing a set of ergonomic rules (e.g., keep the wrist straight, don't bend from the waist when lifting) or applying a generic checklist to a workplace situation. These types of rule-based design strategies often result in suboptimizing the workplace ergonomic conditions or changing workplaces with no payoff.

3 BIOMECHANICAL CONCEPTS

A fundamental concept in the application of occupational biomechanics to ergonomics is that one should design workplaces so that the load imposed on a structure does not exceed the tolerance of the structure. Figure 1 illustrates the traditional concept of biomechanical risk in occupational biomechanics. This figure illustrates how a loading pattern is developed on a body structure that is repeated as the work cycles recur during a job. Structure tolerance is also shown in this figure. When the magnitude of the load imposed on a structure is less than the tissue tolerance, the task is considered safe and the magnitude of the difference between the load and the tolerance is considered the safety margin. Implicit in this figure is the idea that risk occurs when the load imposed exceeds the tissue tolerance. Although *tissue tolerance* is defined as the ability of a tissue to withstand a load without damage, ergonomists are beginning to expand the concept of tolerance to include not only mechanical tolerance of the tissue but also the point at which the tissue exhibits an inflammatory reaction.

A trend in industrial tasks is that tasks are becoming increasingly repetitive, yet involve lighter loads. The conceptual load-tolerance model can also be adjusted to account for this type of risk exposure. Figure 2 shows that occupational biomechanics logic can account for this trend by decreasing tissue tolerance over time. Hence, occupational biomechanics models and logic are moving toward systems that consider manufacturing and work trends in the workplace and attempt to represent these observations (such as cumulative trauma disorders) in the model logic.

3.1 Acute versus Cumulative Trauma

In occupational settings two types of trauma can affect the human body and lead to musculoskeletal disorders

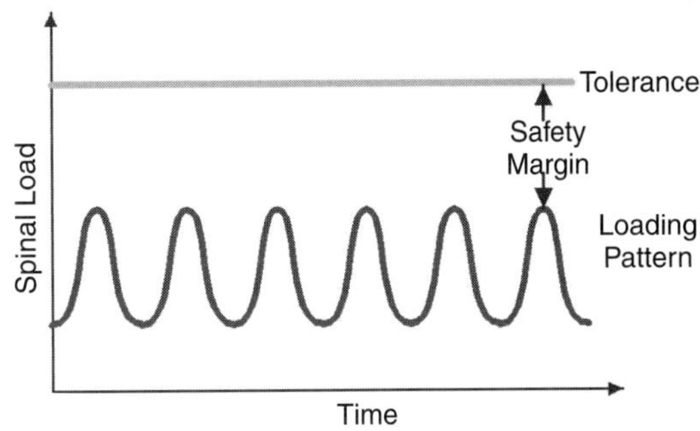

Figure 1 Traditional concept of biomechanical risk. (From McGill, 1997.)

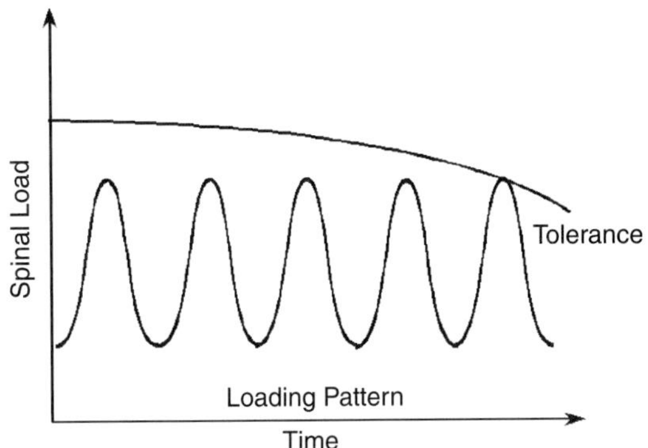

Figure 2 Realistic scenario of biomechanical risk.

in occupational settings. First, *acute trauma* can occur when a single application of force is so large that it exceeds the tolerance of the body structure during an occupational task. Acute trauma is associated with large exertions of force that would be expected to occur infrequently, such as when a worker lifts an extremely heavy object. This situation would result in a peak load that exceeds the load tolerance.

Cumulative trauma, on the other hand, refers to repeated application of force to a structure that tends to wear down the structure, thus, lowering its tolerance to the point where it is exceeded through a reduction of this tolerance limit (Figure 2). Cumulative trauma is more representative of wear and tear on the structure. This type of trauma is becoming more common in occupational settings as more repetitive jobs requiring lower force exertions become more prevalent in industry.

The cumulative trauma process can initiate a response resulting in a cycle that is extremely difficult to break. As shown in Figure 3, the cumulative trauma process begins by exposing a worker to manual exertions that are either frequent (repetitive) or prolonged. The repetitive application of force can affect either the tendons or the muscles of the body. If the tendons are affected, the tendons are subject to mechanical irritation as they are repeatedly exposed

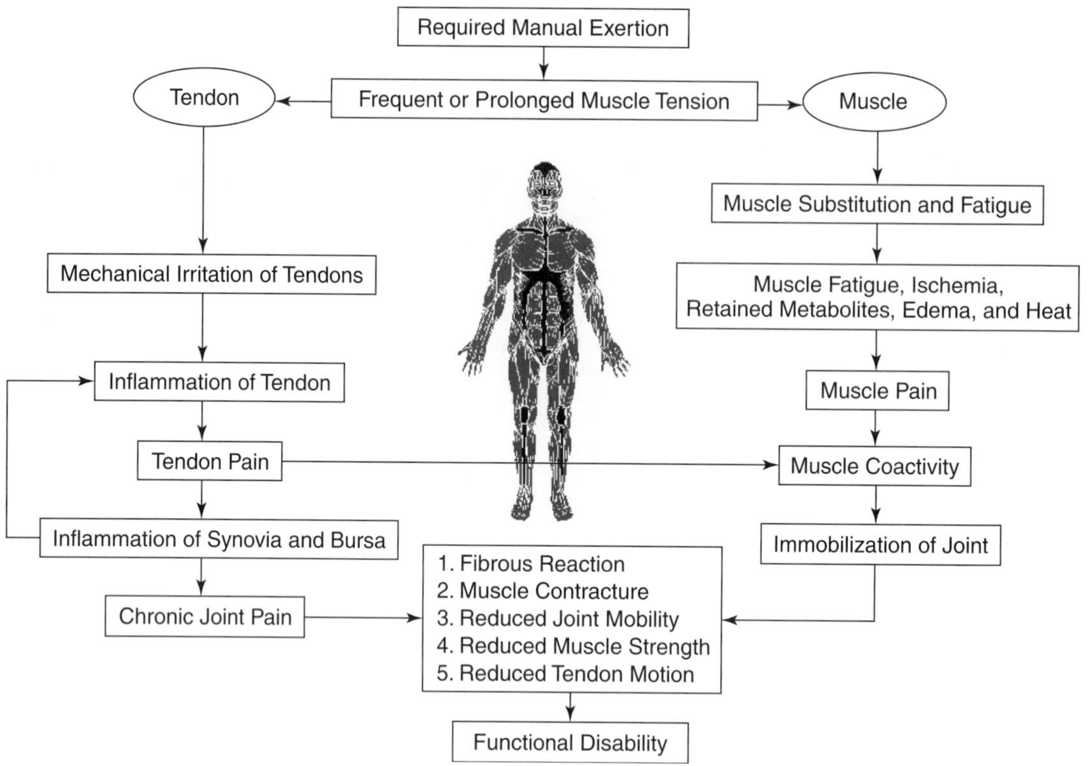

Figure 3 Sequence of events in cumulative trauma disorders.

to high levels of tension. Groups of tendons may rub against each other. The physiological response to this mechanical irritation can result in inflammation and swelling of the tendon(s). The swelling will stimulate the nociceptors surrounding the structure and signal the central control mechanism (brain), via pain perception, that a problem exists. In response to this pain, the body attempts to control the problem via two mechanisms. First, the muscles surrounding the irritated area will coactivate in an attempt to stabilize the joint and prevent motion of the tendons. Since motion will further stimulate the nociceptors and result in further pain, motion avoidance is indicative of the start of a cumulative trauma disorder and is often indicated when workers shorten their motion cycle and move more slowly. Second, in an attempt to reduce the friction occurring within the tendon, the body can increase its production of lubricants (synovial fluid) within the tendon sheath. However, given the limited space available between the tendon and the tendon sheath, increased production of synovial fluid often exacerbates the problem by further expanding the tendon sheath. This action further stimulates the surrounding nociceptors. This initiates a viscous cycle where the response of the tendon to the increased friction results in a reaction (inflammation and the increased production of synovial fluid) that exacerbates the problem (see Figure 3). Once this cycle is initiated, it is very difficult to stop, and often, anti-inflammatory agents are prescribed in order to break the cycle. The process results in chronic joint pain and a series of musculoskeletal reactions such as reduced strength, reduced tendon motion, and reduced mobility. Together, these reactions result in a functional disability.

Cumulative trauma can also affect the muscles. Muscles are overloaded when they become fatigued. Fatigue lowers the tolerance to stress and can result in microtrauma to the muscle fibers. This typically means that the muscle is partially torn, which causes capillaries to rupture and results in swelling, edema, or inflammation near the site of the tear. The inflammation can stimulate nociceptors and result in pain. Once again, the body reacts by cocontracting the surrounding musculature and minimizing the joint motion. However, since muscles do not rely on synovial fluid for their motion, there is no increased production of synovial fluid. However, the end result of this process is the same as that for tendons (i.e., reduced strength, reduced tendon motion, and reduced mobility). The ultimate consequence of this process is, once again, a functional disability.

Although the stimulus associated with the cumulative trauma process is somewhat similar between tendons and muscles, there is a significant difference in the time required to heal from damage to a tendon compared with damage to a muscle. The mechanism of repair for both tendons and muscles depends on blood flow. Blood flow provides nutrients for repair as well as dissipating waste materials. However, the blood supply to a tendon is a fraction (typically, about 5% in an adult) of that supplied to a muscle. Thus, given an equivalent strain to a muscle and a tendon, the muscle will heal rapidly (in about 10 days if not reinjured), whereas the tendon could take months (20 times longer) to reach the same level of repair. For this reason, ergonomists must be particularly vigilant in the assessment of workplaces that could pose a danger to the tendons of the body. This lengthy repair process also explains why many ergonomic processes place a high value on identifying potentially risky jobs, through mechanisms such as discomfort surveys, before a lost-time incident occurs.

3.2 Moments and Levers

Biomechanical loads are only partially defined by the magnitude of weight supported by the body. The position of the weight (or mass of the body segment) relative to the axis of rotation of the joint of interest defines the imposed load on the body and is referred to as a *moment*. A moment is defined as the product of force and distance. As an example, a 50-newton (N) mass held at a horizontal distance of 75 cm (0.75 m) from the shoulder joint imposes a moment of 37.5 N · m (50 N × 0.75 m) on the shoulder joint, whereas the same weight held at a horizontal distance of 25 cm from the shoulder joint imposes a moment or load of only 12.5 N · m (50 N × 0.25 m) on the shoulder. Thus, the joint load is a function of where the load is held relative to the joint axis and the mass of the weight held. Hence, load is not simply a function of weight.

As implied in the example above, moments are a function of the mechanical lever systems of the body. In biomechanics, the musculoskeletal system is represented by a system of levers, and it is the lever systems that are used to describe the tissue loads with a biomechanical model. Three types of lever systems are common in the human body. *First-class levers* are those that have a fulcrum placed between the imposed load (on one end of the system) and an opposing force (*internal to the body*) imposed on the opposite end of the system. The back or trunk is an example of a first-class lever. In this case, the spine serves as the fulcrum. As the human lifts, a moment (*load imposed external to the body*) is imposed anterior to the spine, due to the object weight times the distance of the object from the spine. This moment is counterbalanced by the activity of the back muscles; however, they are located in such a way that they are at a mechanical disadvantage since the distance between the back muscles and the spine is much less than the distance between the object lifted and the spine.

A *second-class lever* system can be seen in the lower extremity. In this situation the fulcrum is located at one end of the lever, the opposing force (internal to the body) is located at the other end of the system, and the applied load is in between the two. The foot is a good example of this lever system. The ball of the foot acts as the fulcrum, the load is applied through the tibia or bone of the lower leg, and the restorative force is applied through gastrocnemius or calf muscle. The muscle activates and causes the body to rotate about the fulcrum or ball of the foot and move the body forward.

Finally, a *third-class lever* is one where the fulcrum is located at one end of the system, the applied load acts at the other end of the system, and the opposing (internal) force acts in between the two. An example of such a lever system in the human body, the elbow joint, is shown in Figure 4.

3.3 External versus Internal Loading

Based on these lever systems it is evident that two types of forces can impose loads on a tissue during work. *External loads* refer to those forces that are imposed on the body as a direct result of gravity acting on an external object being manipulated by the worker. For example, Figure 4*a* shows a tool held in a worker's hand that is subject to the forces of gravity. This situation imposes a 44.5-N (10-lb) external load at a distance from the joint of 30.5 cm (12 in.) on the elbow joint. However, to maintain equilibrium, this external force must be counteracted by an *internal force* that is generated by the muscles of the body. Figure 4*a* also shows that the internal load (muscle) acts at a distance relative to the elbow joint that is at much closer to the fulcrum than the external load (tool). Thus, the internal force must be supplied at a biomechanical disadvantage (because of the smaller lever arm) and must be much larger (534 N or 120 lb) than the external load (44.5 N or 10 lb) to keep the musculoskeletal system in equilibrium. It is not unusual for the magnitude of the internal load to be much greater (often, 10 times greater) than the external load. Thus, it is the internal loading that contributes most to cumulative trauma of the musculoskeletal system during work. The net sum of the external load and the internal load defines the total loading experienced at the joint. Therefore, when evaluating a workstation the ergonomist must not only consider the externally applied load but must be particularly sensitive to the magnitude of the internal forces that can load the musculoskeletal system.

3.4 Modifying Internal Loads

In Section 3.3 we emphasized the importance of understanding the relationship between the external loads imposed on the body and the internal loads generated by the force-generating mechanisms within the body. The key to proper ergonomic design is based on the principle of designing workplaces so that the internal loads are minimized. Internal forces can be thought of as both the component that loads the tissue as well as a structure that can be subject to overexertion. Thus, muscle strength or capacity can be considered as a tolerance measure. If the forces imposed on the muscles and tendons as a result of a task exceed the strength (tolerance) of the muscle or tendon, a potential injury is possible. Generally, three components of the physical

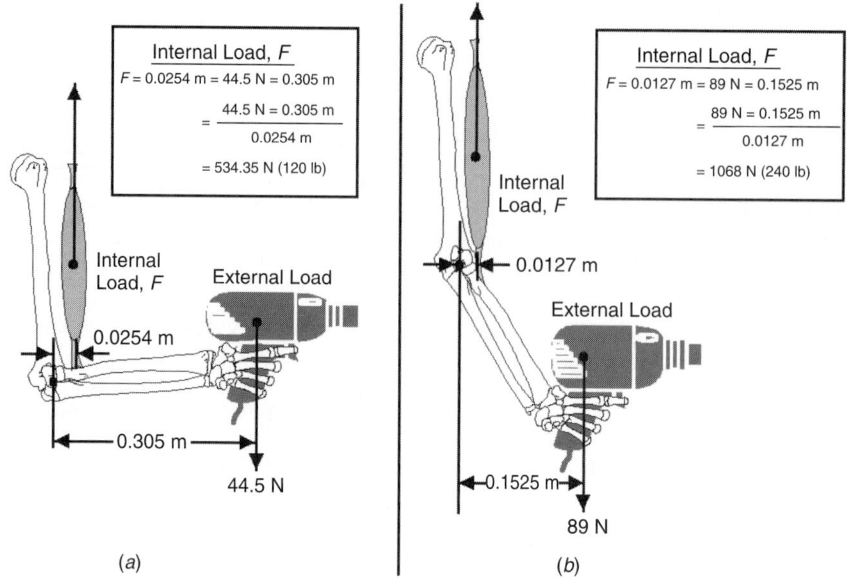

Figure 4 Anatomical third-class lever (*a*) demonstrating how the mechanical advantage changes as the elbow position changes (*b*).

work environment (biomechanical arrangement of the musculoskeletal lever system, length–strength relationships, and temporal relationships) can be manipulated to facilitate this goal and serve as the basis for many ergonomic recommendations.

3.4.1 Biomechanical Arrangement of the Musculoskeletal Lever System

The posture imposed via the design of a workplace can affect the arrangement of a body's lever system and thus can affect the magnitude of the internal load required to support an external load. The arrangement of the lever system could influence the magnitude of the external moment imposed on the body as well as dictate the magnitude of the internal forces and the subsequent risk of either acute or cumulative trauma. If one considers the biomechanical arrangement of the elbow joint (shown in Figure 4*a*), it is evident that the mechanical advantage of the internal force generated by the biceps muscle and tendon is defined by a posture that keeping one's arm bent at a 90° angle. If one palpates the tendon and inserts the index finger between the elbow joint center and the tendon, one can gain an appreciation for the internal moment arm distance. It is also possible to appreciate how this internal mechanical advantage can change with posture. With the index finger still inserted between the elbow joint and the tendon, if the elbow joint is extended, one can appreciate how the distance between the tendon and the joint center of rotation is significantly reduced. If the moment imposed about the elbow joint is held constant (shown in Figure 4*b* by a heavier tool), the mechanical advantage of the internal force generator is reduced significantly. Thus, the bicep muscle must generate greater force to support the external load. This greater force is transmitted through the tendon and can increase the risk of cumulative trauma. Hence, the positioning of the mechanical lever system (which can be accomplished though work design) can greatly affect internal load transmission within the body. A task can be performed in a variety of ways, but some of these positions are much more costly than others in terms of loading of the musculoskeletal system.

3.4.2 Optimizing the Length–Strength Relationship

Another important relationship that influences a load on the musculoskeletal system is the length–strength relationship of the muscles. This relationship is shown in Figure 5. The active portion of this figure refers to active force-generating structures such as muscles. When muscles are at their resting length (generally, seen in the fetal position), they have the greatest capacity to generate force. However, when the muscle length deviates from this resting position, the muscle's capacity to generate force is greatly reduced because the cross-bridges between the components of the muscle proteins become inefficient. When a muscle stretches or when a muscle attempts to generate force while at a short length, the ability to generate force is greatly diminished. As indicated in Figure 5, passive tissues in the muscle (and ligaments) can also generate tension when muscles are stretched. Thus, the length of a muscle during task performance can greatly influence the force available to perform work and can influence risk by altering the internal force available within the system.

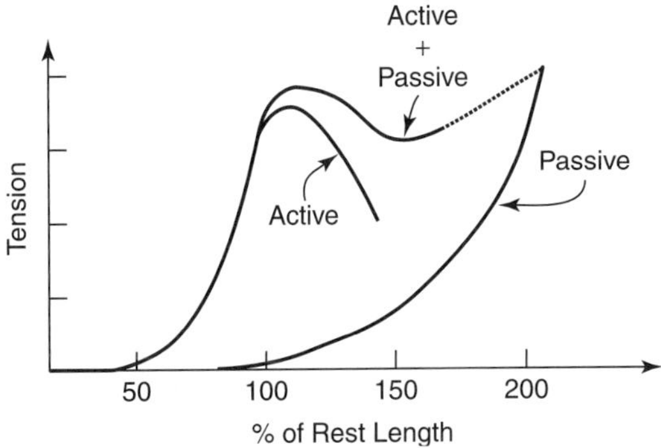

Figure 5 Length–tension relationship for a human muscle. (Adapted from Basmajian and De Luca, 1985.)

Therefore, what might be considered a moderate force for a muscle at resting length can become the maximum force a muscle can produce when it is in a stretched or contracted position, thus increasing the risk of muscle strain. When this relationship is considered in combination with the mechanical load placed on the muscle and tendon (by means of the arrangement of the lever system), the position of the joint arrangement becomes a major factor in the design of the work environment. Typically, the length–strength relationship interacts synergistically with the lever system. Figure 6 indicates the effect of elbow position on the force generation capability of the elbow. The joint position can have a dramatic effect on force generation and can greatly affect the internal loading of the joint and the subsequent risk of cumulative trauma.

3.4.3 Impact of Velocity on Muscle Force

Motion can also influence the ability of a muscle to generate force and therefore load the biomechanical system. Motion can be a benefit to the biomechanical system if momentum is employed properly, or it can

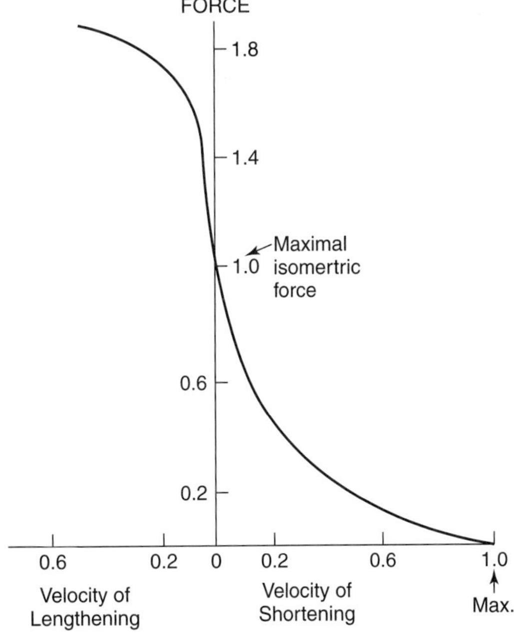

Figure 7 Influence of velocity on muscle force. (Adapted from Astrand and Rodahl, 1977.)

increase the load on the system if the worker is not taking advantage of momentum. This relationship between muscle velocity and force generation is shown in Figure 7. The figure indicates that in general, the faster the muscle is moving, the greater the reduction in force capability of the muscle. This reduction in muscle capacity can result in the muscle strain that may occur at a lower level of external loading, with a subsequent increase in the risk of cumulative trauma. In addition, this effect is considered in dynamic ergonomic biomechanical models.

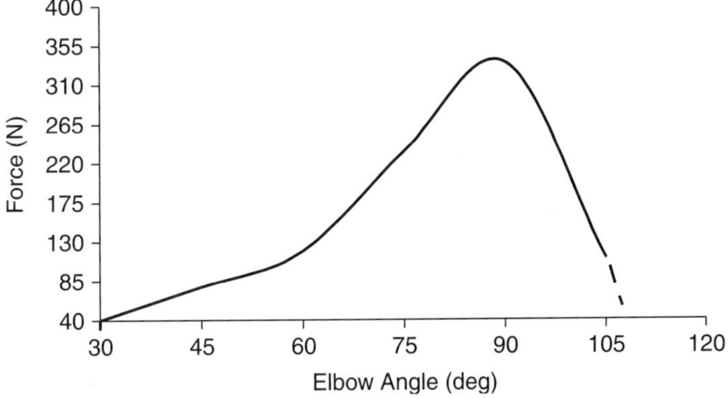

Figure 6 Position–force diagram produced by flexion of the forearm in pronation. "Angle" refers to the included angle between the longitudinal axes of the forearm and upper arm. The highest parts of the curve indicate the configurations where the biomechanical lever system is most effective. (Adapted from Chaffin and Andersson, 1991.)

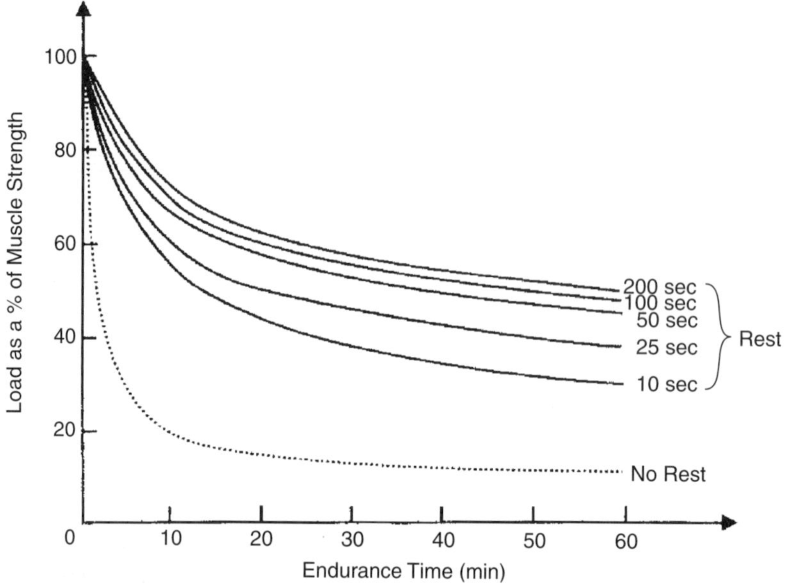

Figure 8 Forearm flexor muscle endurance times in consecutive static contractions of 2.5-second duration with varied rest periods. (Adapted from Chaffin and Andersson, 1991.)

3.4.4 Temporal Relationships

Strength–Endurance Strength must be considered as both an internal force and a tolerance. However, it is important to realize that strength is transient. A worker may generate a great amount of strength during a one-time exertion; however, if the worker is required to exert such strength either repeatedly or for a prolonged period of time, the amount of force that the worker can generate can be reduced dramatically. Figure 8 demonstrates this relationship. The dashed line in this figure indicates the maximum force generation capacity of a static exertion over time. Maximum force is generated only for a very brief period of time. As time advances, strength output decreases exponentially and levels off at about 20% of maximum after about 7 minutes. Similar trends occur during repeated dynamic conditions. If a task requires a large portion of a worker's strength, one must consider how long that portion of the strength must be exerted to ensure that the work does not strain the musculoskeletal system.

Rest Time As discussed earlier, the risk of cumulative trauma increases when the capacity to exert force is exceeded by the force requirements of a job. Another factor that may influence strength capacity (and tolerance to muscle strain) is rest time, which has a profound effect on a worker's ability to exert force. Figure 9 summarizes how energy for a muscular contraction is regenerated during work. Adenosine triphosphate (ATP) is required to produce a power-producing muscular contraction. ATP changes into adenosine diphosphate (ADP) once a muscular contraction has occurred; however, ADP is not capable of producing a significant muscular contraction. The ADP must be converted to ATP to enable another muscular contraction. This conversion to ATP can occur with the addition of oxygen to the system. If oxygen is not available, the system goes into oxygen debt and insufficient ATP is available for a muscular contraction. Figure 9 indicates that oxygen is a key ingredient in maintaining a high level of muscular exertion. Oxygen is delivered to the target muscles via the blood. Under static exertions the blood flow is reduced and there is a subsequent reduction in the blood available to the muscle. This restriction of blood flow and subsequent oxygen deficit are responsible for the rapid decrease in force generation over time, as shown in Figure 8. The solid lines in Figure 8 indicate how the force generation capacity of the muscles increase when different amounts of rest are permitted during prolonged exertion. As more rest time is permitted, increases in force generation are achieved when more oxygen is delivered to the muscle, and more ADP can be converted to ATP. This relationship indicates that any more than about 50 seconds of rest, under these conditions, does not result in a significant increase in the force generation capacity of the muscle. Practically, this relationship indicates that to optimize the strength capacity of a worker and minimize the risk of muscle strain, a schedule of frequent and brief rest periods would be more beneficial than lengthy infrequent rest periods.

3.5 Load Tolerance

Biomechanical analyses must consider not only the loads imposed on a structure but also the ability of the structure to withstand or tolerate a load during work. In this section we briefly review the knowledge base associated with human structure tolerances.

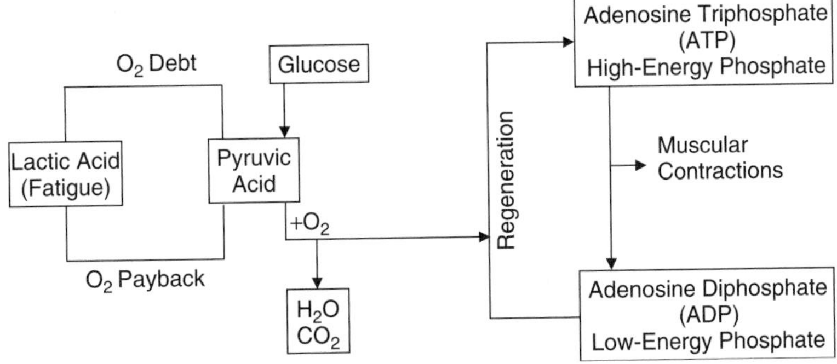

Figure 9 The body's energy system during work. (Adapted from Grandjean, 1982.)

3.5.1 Muscle, Ligament, Tendon, and Bone Capacity

The precise tolerance characteristics of human tissues such as muscles, ligaments, tendons, and bones loaded under various working conditions is difficult to estimate. Tolerances of these structures vary greatly under similar loading conditions. In addition, tolerance depends on many other factors, such as strain rate, age of the structure, frequency of loading, physiologic influences, heredity, conditioning, as well as other unknown factors. Furthermore, it is not possible to measure these tolerances under in vivo conditions. Therefore, most estimates of tissue tolerance have been derived from various animal and/or theoretical sources.

3.5.2 Muscle and Tendon Strain

The muscle is the structure within the musculoskeletal system that has the lowest tolerance. The ultimate strength of a muscle has been estimated to be 32 MPa (Hoy et al., 1990). In general, it is believed that a muscle will rupture prior to a (healthy) tendon (Nordin and Frankel, 1989), since tendon stress has been estimated at between 60 and 100 MPa (Nordin and Frankel, 1989; Hoy et al., 1990). As indicated in Table 1, there is a safety margin between the muscle failure point and the failure point of the tendon of about twofold (Nordin and Frankel, 1989) to threefold (Hoy et al., 1990).

3.5.3 Ligament and Bone Tolerance

Ligament and bone tolerances have also been estimated. Ultimate ligament stress has been estimated to be approximately 20 MPa. The ultimate stress of bone depends on the direction of loading. Bone tolerance can range from 51 MPa in transverse tension to over 190 MPa in longitudinal compression. Table 1 also indicates the ultimate stress of bone loaded under both longitudinal and transverse loading conditions.

A strong temporal component to ligament recovery has also been identified. Solomonow found that ligaments require long periods of time to regain structural integrity, during which compensatory muscle activities are observed (Solomonow et al., 1998, 1999, 2000, 2002; Stubbs et al., 1998; Gedalia et al., 1999; Wang et al., 2000; Solomonow, 2004). Recovery time has been observed to be several times the loading duration and can easily exceed the typical work–rest cycles observed in industry.

Table 1 Tissue Tolerance of the Musculoskeletal System

Structure	Estimated Ultimate Stress, σ_u (MPa)
Muscle	32–60
Ligament	20
Tendon	60–100
Bone longitudinal loading	
Tension	133
Compression	193
Shear	68
Bone transverse loading	
Tension	51
Compression	133

Source: Adapted from Ozkaya and Nordin (1991).

3.5.4 Disc/End Plate and Vertebrae Tolerance

The mechanism of cumulative trauma to the vertebral disc is believed to be related to repeated trauma to the vertebral end plate. The end plate is a very thin (about 1 mm thick) structure that facilitates nutrient flow from the vertebrae to the disc fibers (annulus fibrosis). The disc has no direct blood supply, so it relies heavily on this nutrient flow for disc viability. Repeated microfracture of this vertebral end plate is thought to lead to the development of scar tissue, which can impair nutrient flow to the disc fibers. This, in turn, leads to atrophy of the fiber and fiber degeneration. Since the disc contains few nociceptors, the development of microfractures is typically unnoticed by a person. Given this process, if one can determine the level at which the end plate experiences a microfracture, one can minimize the effects of cumulative trauma and disc degeneration within the spine.

Several studies of in vitro disc/end plate tolerance have been reported in the literature. Figure 10 indicates

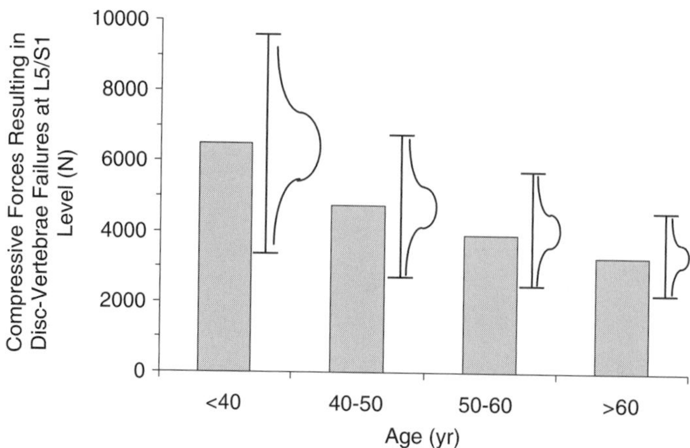

Figure 10 Mean and range of disc compression failures by age. (Adapted from NIOSH, 1981.)

the levels of end plate compressive loading tolerance that have been used to establish safe lifting situations at the work site. This figure shows the compressive force mean (column value) as well as the compression force distribution (thin line and normal distribution curve) that would result in vertebral end plate microfracture. The figure indicates that for those under 40 years of age, end plate microfracture damage begins to occur at about 3432 N of compressive load on the spine. If the compressive load is increased to 6375 N, approximately 50% of those exposed will experience vertebral end plate microfracture. Finally, when the compressive load on the spine reaches a value of 9317 N, almost all of those exposed to the loading will experience a vertebral end plate microfracture. It is also obvious from this figure that the tolerance distribution shifts to lower levels with increasing age (Adams et al., 2000). In addition, it should be recognized that this tolerance is based on compression of the vertebral end plate alone. Shear and torsional forces in combination with compressive loading would further lower the tolerance of the end plate.

The vertebral end plate tolerance distribution has been used widely to set limits for spine loading and to define risk. It should also be noted that others have identified different limits of vertebral end plate tolerance. Jager et al. (1991) have reviewed the spine tolerance literature and suggested different compression value limits. Their spine tolerance summary is shown in Table 2. They have also been able to describe vertebral compressive strength based on an analysis of 262 values collected from 120 samples. According to their data, the compressive strength of the lumbar spine can be described according to a regression equation:

$$\text{compressive strength (kN)} = (7.26 + 1.88G) - (0.494 + 0.468G)A + (0.042 + 0.106G)C - 0.145L - 0.749S$$

Table 2 Lumbar Spine Compressive Strength

		Strength (kN)	
Population	*n*	Mean	Std. Dev.
Females	132	3.97	1.50
Males	174	5.81	2.58
Total	507	4.96	2.20

Source: Jager et al. (1991).

where A is age (decades); G represents gender (coded as 0 for female or 1 for male); C is the cross-sectional area of the vertebra (cm^2); L the lumbar-level unit, where 0 is the L5/S1 disc, 1 represents the L5 vertebra, etc. through 10, which represents the T10/L1 disc; and S the structure of interest (where 0 is a disc and 1 is a vertebra). This equation suggests that the decrease in strength within a lumbar level is about 0.15 kN of that of the adjacent vertebra and that the strength of the vertebrae is about 0.8 kN lower than the strength of the discs (Jager et al., 1991). This equation can account for 62% of the variability among the samples.

It has also been suggested that spine tolerance limits vary as a function of frequency of loading (Brinckmann et al., 1988). Figure 11 indicates how spine tolerance varies as a function of spine load level and frequency of loading.

3.5.5 Pain Tolerance

Over the past decade we have learned that there are numerous pathways to pain perception associated with musculoskeletal disorders (Cavanaugh, 1995; Cavanaugh et al., 1997; Khalsa, 2004). It is important to understand these pathways since they may be able to be used as tissue tolerance limits as opposed to tissue damage limits. Hence, one might be able to consider the quantitative limits above which a pain pathway is initiated as a tolerance limit for ergonomic purposes. Although none of these pathways have been defined

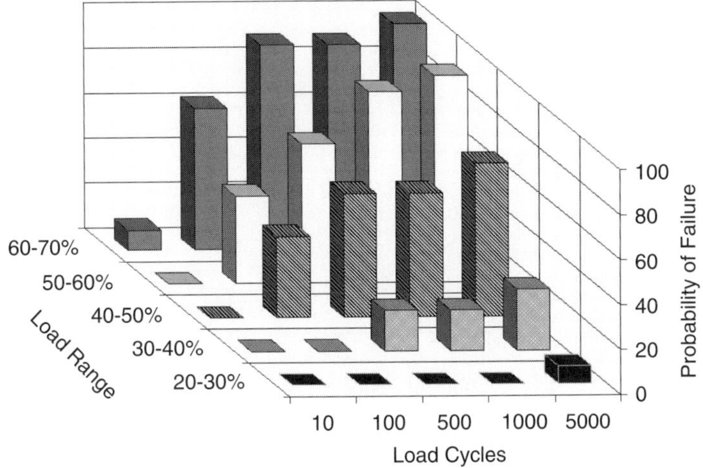

Figure 11 Probability of a motion segment to be fractured in dependence on the load range and the number of load cycles. (Adapted from Brinckmann et al., 1988.)

quantitatively, they represent an appealing approach since they represent biologically plausible mechanisms that complement the view of injury association derived from the epidemiologic literature.

Several categories of pain pathways are believed to exist that might be used as tolerance limits in workplace design: (1) structural disruption, (2) tissue stimulation and pro-inflammatory response, (3) physiologic limits, and (4) psychophysical acceptance. Each of these pathways is expected to respond differently to mechanical loading of the tissue and thus serve as tolerance limits. Although many of these limits have yet to be defined quantitatively, current biomechanical research is attempting to define these tolerances, and it is expected that one will someday be able to use these limits to identify the characteristics of a dose–response relationship.

4 APPLICATION OF BIOMECHANICAL PRINCIPLES TO REDUCING STRESS IN THE WORKPLACE

The basic concepts and principles of biomechanics can now be applied to workplace design situations. Due to differences in structure, different body parts are affected by work design in different ways. In this section we discuss, in general, how established biomechanical principles relate to biomechanical loading of the parts of the body often affected by work.

4.1 Shoulder

Shoulder pain is believed to be one of the most underrecognized occupationally related musculoskeletal disorders. Shoulder disorders are increasingly being recognized as a major workplace problem by those organizations that have reporting systems sensitive enough to detect such trends. The shoulder is one of the more complex structures of the body, with numerous muscles and ligaments crossing the shoulder joint girdle complex. Because of this biomechanical complexity, surgical repair can be problematic. During shoulder surgeries it is often necessary to damage much of the surrounding tissue in an attempt to reach the structure in need of repair. The target structure is often small (e.g., a joint capsule) and difficult to reach. Thus, damage is often done to surrounding tissues that may offset the benefits of surgery. Hence, the best course of action is to design workstations ergonomically so that the risk of initial injury is minimized.

Since the shoulder joint is biomechanically complex, much of our biomechanical knowledge is derived from empirical evidence. The shoulder represents a statically indeterminate system in that we can typically measure six external moments and forces acting about the point of rotation, yet there are far more internal forces (over 30 muscles and ligaments that are capable of counteracting the external moments). Thus, quantitative estimates of shoulder joint loading are not common for ergonomic purposes.

When shoulder-intensive work is considered, optimal workplace design is typically defined in terms of preferred posture during work. Shoulder *abduction*, defined as the elevation of the shoulder in the lateral direction, is often a problematic posture when work is performed overhead. Figure 12 indicates shoulder performance measures in terms of both available strength and perceived fatigue when the shoulder is held at varying degrees of abduction. The figure indicates that the shoulder can produce a considerable amount of force throughout shoulder abduction angles between 30 and 90°. However, when comparing reported fatigue at these same abduction angles, it is apparent that fatigue increases rapidly as the shoulder is abducted above 30°. Thus, even though strength is not an issue at shoulder abduction angles up to 90°, fatigue becomes a limiting factor. Therefore, the only position of the shoulder that is acceptable from both a strength and

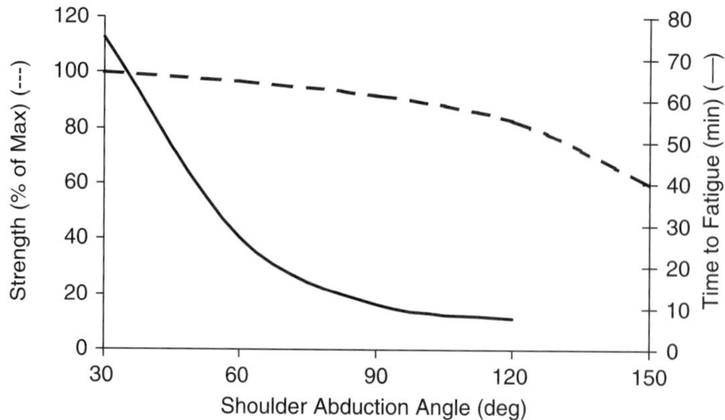

Figure 12 Shoulder abduction strength and fatigue time as a function of shoulder abducted from the torso. (Adapted from Chaffin and Andersson, 1991.)

a fatigue standpoint is a shoulder abduction of at most 30°.

Shoulder *flexion* has been examined almost exclusively as a function of reported fatigue. Chaffin (1973) has shown that even slight shoulder flexion can influence fatigue of the shoulder musculature. Figures 13 and 14 indicate the effects of vertical and horizontal positioning of the work, respectively, during shoulder flexion while seated, upon fatigability of the shoulder musculature. Fatigue occurs more rapidly as a worker's arm becomes more elevated (Figure 13). This trend is probably due to the fact that the muscles are deviated from the neutral position as the shoulder becomes more elevated, thus affecting the length–strength relationship (Figure 5) of the shoulder muscles. Figure 14 indicates that as the horizontal distance between the work and the body is increased, the time to reach significant fatigue is decreased. This is due to the fact that as a load is held farther from the body, more of the external moment (force × distance) must be supported by the shoulder. Thus, the shoulder muscles must produce greater internal force when the load is held farther from the body. With this increased force, the muscles fatigue more quickly. Elbow supports can significantly increase endurance time in these postures. In addition, an elbow support changes the biomechanical situation by providing a fulcrum at the elbow. Thus, the axis of rotation becomes the elbow instead of the shoulder, and this reduces the external moment. This not only increases the time that one can maintain a posture, but also significantly increases the external load that one can hold in the hand (Figure 15).

4.2 Neck

Neck disorders may also be associated with sustained work postures. Generally, the more upright the posture of the head, the less muscle activity and neck strength are required to maintain the posture. Upright neck positions also have the advantage of reducing the extent of fatigue experienced in the neck (Figure 16). This figure indicates that when the head is tilted forward 30° or more from the vertical position, the time to experience significant neck fatigue decreases rapidly. From a biomechanical standpoint, as the head is flexed, the center of mass of the head moves forward relative to the base of support of the head (spine). Therefore, as the head is moved forward, more of a moment is imposed about the spine, which necessitates increased activation of the neck musculature and greater probability of fatigue since a static posture is maintained by the neck muscles. On the other hand, when the head is not flexed forward and is relatively upright, the neck can be positioned such that minimal muscle activity is required of the neck muscles, and thus fatigue is minimized.

4.2.1 Trade-offs in Work Design

The key to proper ergonomic design of a workplace from a biomechanical standpoint is to consider the biomechanical trade-offs associated with a particular work situation. These trade-offs are necessary because it is often the case that a situation that is advantageous for one part of the body is disadvantageous for another part. Thus, many biomechanical considerations in the ergonomic design of the workplace require one to consider the differing trade-offs and rationales for various design options.

One of the most common trade-off situations encountered in ergonomic design is that between accommodating the shoulders and accommodating the neck. This trade-off is often resolved by considering the hierarchy of needs required by the task. Figure 17 illustrates this logic. It shows the recommended height of the work as a function of the type of work that is to be performed. Precision work requires a high level of visual acuity, which is of utmost importance in accomplishing a work task. If the work is performed at too low a level, the head must be flexed to accommodate the visual requirements of the job. This situation could result in significant neck discomfort. Therefore, in this situation, visual accommodation is at the top of the hierarchy of task needs and the work is

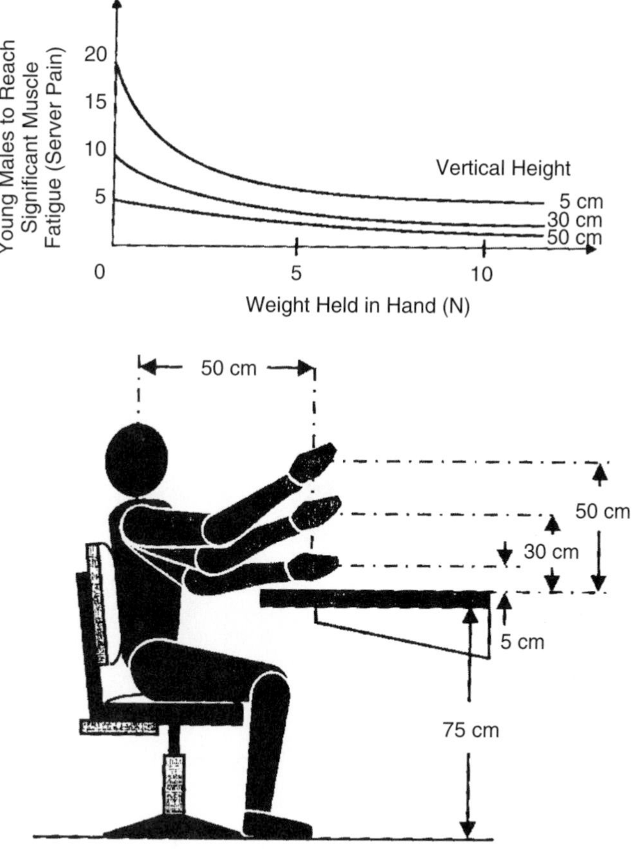

Figure 13 Expected time to reach significant shoulder muscle fatigue for varied arm flexion postures. (Adapted from Chaffin and Andersson, 1991.)

typically raised to a relatively high level (95 to 110 cm above the floor). This position accommodates the neck but creates a problem for the shoulders, since they must be abducted when the work level is high. Thus, a tradeoff must be considered. In this instance, ideal shoulder posture is sacrificed to accommodate the neck since the visual requirements of the job are great whereas the shoulder strength required for precision work is low. Thus, visual accommodation is given a higher priority in the hierarchy of task needs. In addition, shoulder problems can be minimized by providing wrist or elbow supports at the workplace.

The other extreme of the working-height situation involves heavy work. The greatest demand on a worker in heavy work is for a high degree of arm strength, whereas visual requirements in this type of work are typically minimal. Thus, the shoulder position is higher on the hierarchy of task needs in this situation. Therefore, in this situation ideal neck posture is typically sacrificed in favor of more favorable shoulder and arm postures. Hence, heavy work is performed at a height of 70 to 90 cm above floor level. With the work set at this height, the elbow angles are close to 90°, which maximizes strength (Figure 6), and the shoulders are close to 30° of abduction, which minimizes fatigue. In this situation, the neck is not in an optimal position, but logic dictates that the visual demands of a heavy task would not be substantial and thus the neck should not be flexed for prolonged periods of time.

A third work-height situation involves light work. Light work is a mixture of moderate visual demands with moderate strength requirements. In such a situation, work is a compromise between shoulder position and visual accommodation, and neither the visual demands of the job nor the strength requirements dominate the hierarchy of job demands; both are important considerations. The solution is to minimize the negative aspects of both the strength and neck posture situations by splitting the difference between extreme situations. Thus, the height of the work is set between the precision work height level and the heavy work height level. This situation leads to a situation where the work is performed at a level between 85 and 95 cm off the floor under light work conditions.

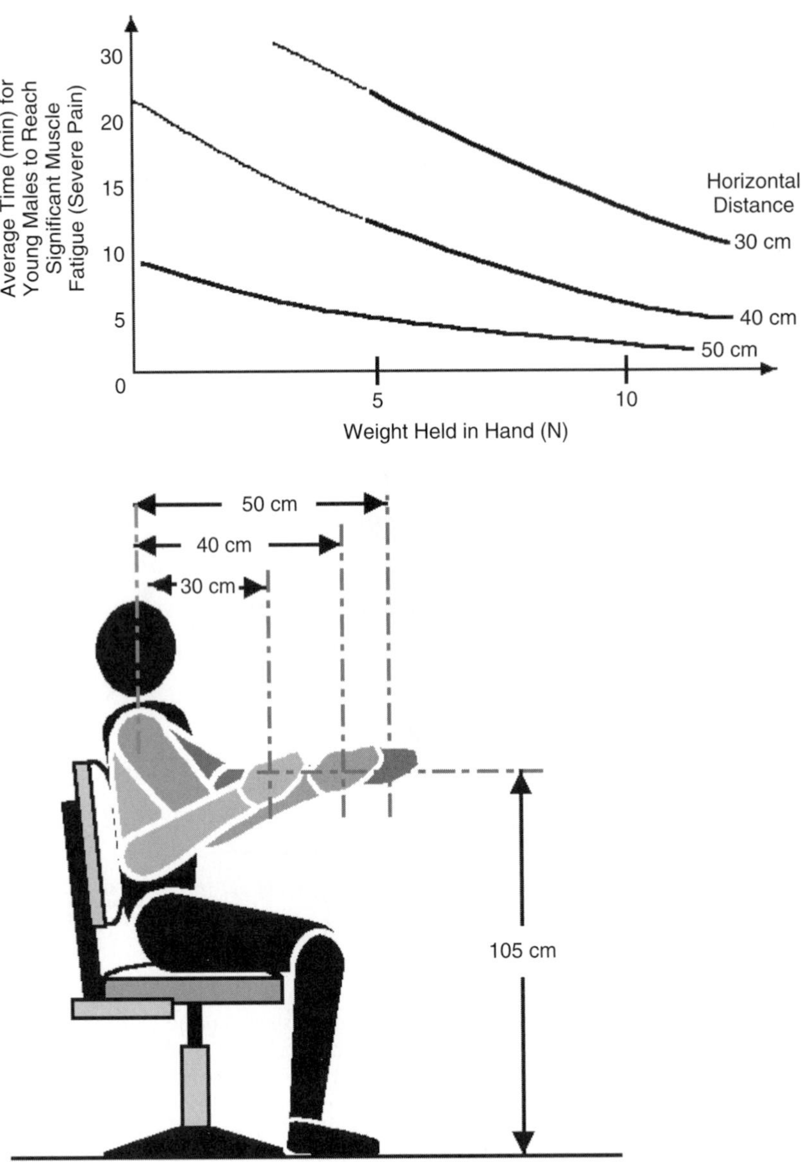

Figure 14 Expected time to reach significant shoulder muscle fatigue for different forward arm-reach postures. (Adapted from Chaffin and Andersson, 1991.)

4.3 Back

Low back disorders (LBDs), one of the most common and significant musculoskeletal problems in the United States, result in substantial amounts of morbidity, disability, and economic loss (Hollbrook et al., 1984; Praemer et al., 1992). Low back disorders are one of the most common reasons for missing work. Back disorders were responsible for more than 100 million lost workdays in 1988, with 22 million cases reported that year (Guo, 1993; Guo et al., 1999). Among those under 45 years of age, LBD is the leading cause of activity limitations, and it can affect up to 47% of workers with physically demanding jobs (Andersson, 1997). The prevalence of LBDs is also on the rise. They are reported to have increased by 2700% since 1980 (Pope, 1993). Costs associated with LBD are also significant; health care expenditures in the United States incurred by those with back pain exceeded $90 billion in 1998 (Luo et al., 2004).

It is clear that the risk of LBD is associated with occupational tasks (NRC, 1999, 2001). Thirty percent of occupation injuries in the United States are related

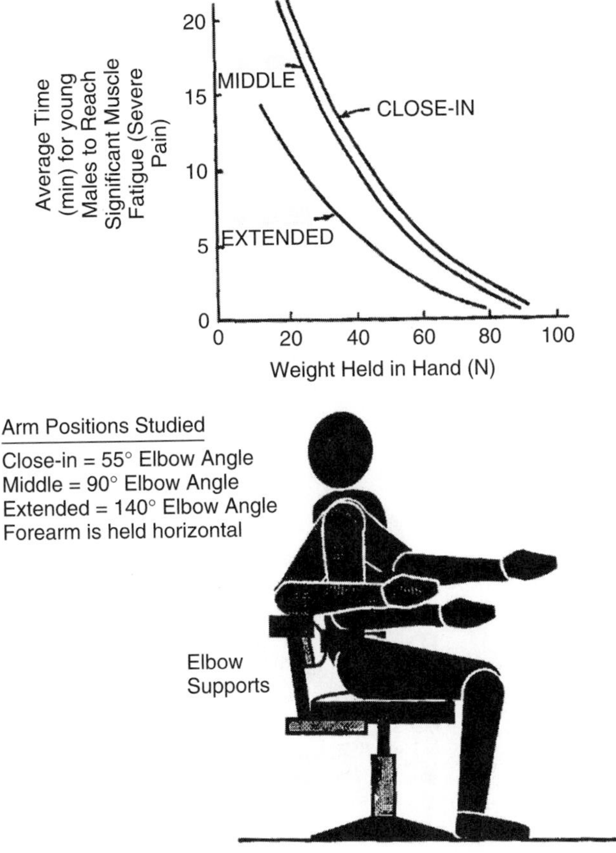

Figure 15 Expected time to reach significant shoulder and arm muscle fatigue for different arm postures and hand loads with the elbow supported. The greater the reach, the shorter the endurance time. (Adapted from Chaffin and Andersson, 1991.)

to overexertion, lifting, throwing, holding, carrying, pushing, and or pulling objects that weigh 50 lb or less. Around 20% of all workplace injuries and illnesses are back injuries, which account for up to 40% of compensation costs. Estimates of occupational annual LBD prevalence vary from 1 to 15%, depending on occupation, and over a career can seriously affect 56% of workers.

Manual materials handling (MMH) activities, specifically lifting, are most often associated with occupationally related LBD risk. It is estimated that lifting and MMH account for up to two-thirds of work-related back injuries (NRC, 2001). Biomechanical assessments target disc-related problems, since disc problems are the most serious and costly type of back pain and have a mechanical origin (Nachemson, 1975). The literature reports increased degeneration in the spines of cadavers of persons who had been exposed to physically heavy work (Videman et al., 1990). These findings suggest that occupationally related low back disorders are closely associated with spine loading.

4.3.1 Significance of Moments

The most important component of occupationally related LBD risk is that of the external moments imposed about the spine (Marras et al., 1993, 1995). As with most biomechanical systems, loading is influenced greatly by the external moment imposed on the system. However, because of the biomechanical disadvantage at which the torso muscles operate relative to the trunk fulcrum during lifting, very large loads can be generated by the muscles and imposed on the spine. Figure 18 shows an idealized biomechanical lever system arrangement. The back musculature is at a severe biomechanical disadvantage in many manual materials handling situations. Supporting an external load of 222 N (about 50 lb) at a distance of 1 m from the spine imposes a 222-N·m external moment load about the spine. However, since the spine's supporting musculature is in fairly close proximity to the external load, the trunk musculature must exert extremely large forces [4440 N (998 lb)] simply to hold the external load in equilibrium. The internal loads can increase greatly if dynamic motion of the body is

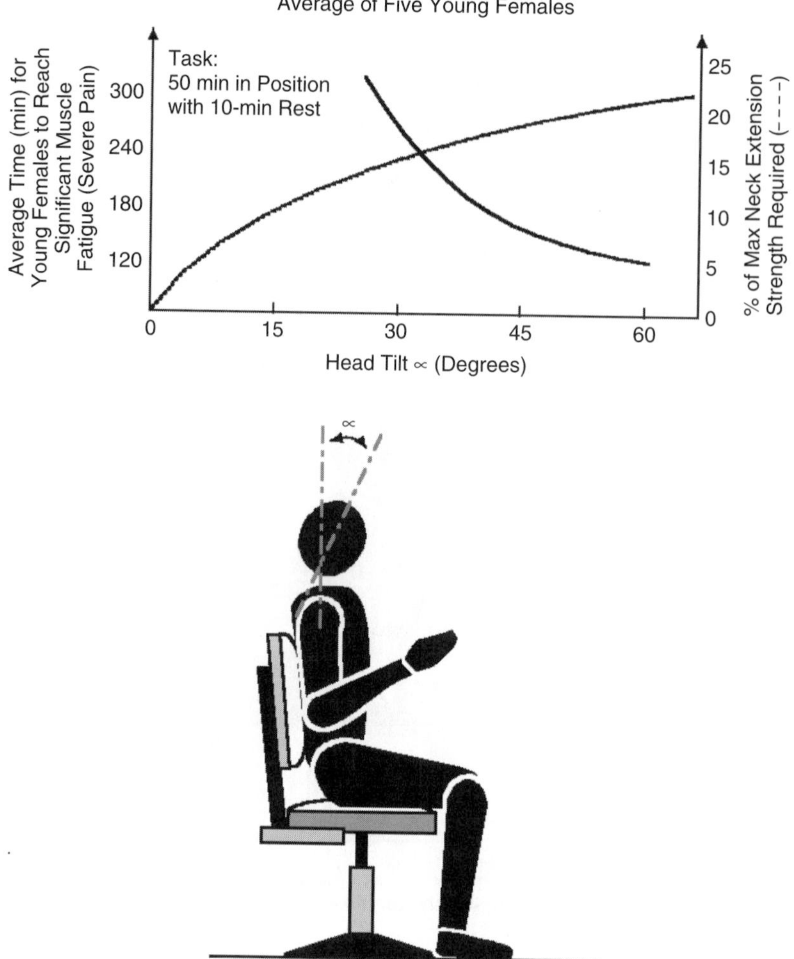

Figure 16 Neck extensor fatigue and muscle strength required vs. head tilt angle. (Adapted from Chaffin and Andersson, 1991.)

considered (since force is a product of mass and acceleration). Thus, this moment concept dominates risk interpretation in workplace design from a back protection standpoint. Thus, a fundamental issue is to keep the external load's moment arm at a minimum.

The concept of minimizing the external moment during lifting has major implications for lifting styles (i.e., the best way to lift). Since the externally applied moment during a lift influences the internal loading significantly, the lifting style is of far less concern then is the magnitude of the applied moment. Some have suggested that proper lifting involves lifting "using the legs" or using the "stoop" lift method (bending from the waist). In addition, research has shown that spine load is a function of anthropometry as well as lifting style (Chaffin et al., 1999). Hence, biomechanical analyses (Park and Chaffin, 1974; van Dieen et al., 1999) have demonstrated that no single lift style is correct for all body types. For this reason, the National Institute of Occupational Safety and Health (NIOSH, 1981) has concluded that lift style need not be a consideration when assessing risk due to materials handling. Some have suggested that the internal moment within the torso is optimized when lumbar lordosis is preserved during the lift (NIOSH, 1981; Anderson et al., 1985; McGill et al., 2000; McGill, 2002). However, from a practical, biomechanical standpoint, the primary indicator of spine loading, and thus the correct lifting style, is whatever style permits the worker to bring the center of mass of the load as close to the spine as possible.

4.3.2 Seated versus Standing Workplaces

Seated workplaces have become more prominent in modern work, especially with the aging of the workforce and the introduction of service-oriented

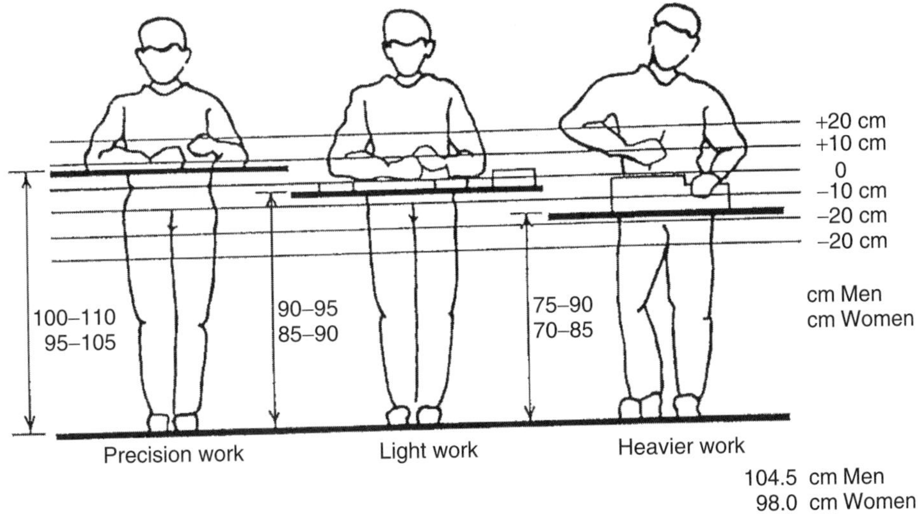

Figure 17 Recommended heights of bench for standing work. The reference line (+0) is the height of the elbows above the floor. (From Grandjean, 1982.)

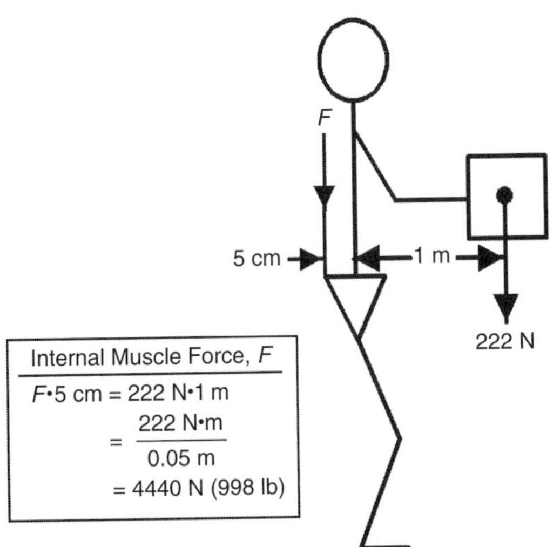

Figure 18 Internal muscle force required to counterbalance an external load during lifting.

and data-processing jobs. It has been documented that loads on the lumbar spine are greater when a worker is seated than when standing (Andersson et al., 1975). This is true since the posterior (bony) elements of the spine form an active load path when one is standing. However, when seated these elements are disengaged, and more of the load passes through the intervertebral disc. Thus, work performed in a seated position puts the worker at greater risk of spine loading and therefore at greater risk of damaging a disc. Given this mechanism of spine loading, it is important to consider the design features of a chair since it may be possible to influence disc loading through chair design. Figure 19 shows the results of a study involving pressure measurements taken within the intervertebral disc of persons as the back angle of the chair and magnitude of lumbar support were varied (Andersson et al., 1975). It is not feasible to measure the forces in the spine directly in vivo. Therefore, disc pressure measures have traditionally been used as a rough approximation of loads imposed on the spine. Figure 19 indicates that both the seat back angle and lumbar support features have a significant impact on disc pressure. Disc pressure decreases as the backrest angle is increased. However, increasing the backrest angle in the workplace is often not practical, since it can also move the worker farther away from the work, thereby increasing external moment. Figure 19 also indicates that increasing lumbar support can reduce disc pressure significantly, due to the fact that as lumbar curvature (lordosis) is reestablished (with lumbar support), the posterior elements play more of a role in providing an alternative load path, as is the case when standing in the upright position.

Less is known about risk to the lower back relative to prolonged standing. The trunk muscles may experience low-level static exertion conditions and may be subject to static overload through the muscle static fatigue process, shown in Figure 9. Muscle fatigue can result in lowered muscle force generation capacity and can thus initiate the cumulative trauma sequence of events (Figure 3). The fatigue and cumulative trauma sequence can be minimized through two actions. First, foot rails can provide a mechanism to allow relaxation of the large back muscles and thus increased blood flow to the muscle. This reduces the static load and subsequent fatigue in the muscle by the process illustrated in Figure 9. When a leg is rested on the footrest, the large back muscles are relaxed on

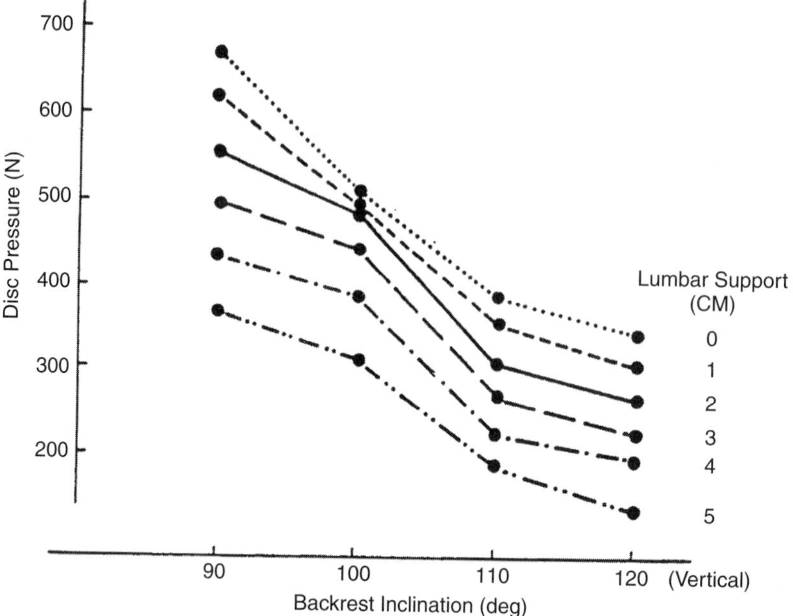

Figure 19 Disc pressures measured with different backrest inclinations and lumbar supports of different sizes. (From Chaffin and Andersson, 1991.)

one side of the body and the muscle can be supplied with oxygen. Alternating legs placed on the footrest provides a mechanism to minimize back muscle fatigue throughout the day. Second, floor mats can decrease the fatigue in the back muscles provided that the mats have proper compression characteristics (Kim et al., 1994). Floor mats are believed to facilitate body sway, which enhances the pumping of blood through back muscles, thereby minimizing fatigue.

Knowledge of when standing workplaces are preferable to seated workplaces is dictated mainly by work performance criteria. In general, standing workplaces are preferred when (1) the task requires a high degree of mobility (reaching and monitoring in positions that exceed the employee's reach envelope or when performing tasks at different heights or different locations), (2) precise manual control actions are not required, (3) legroom is not available (when legroom is not available, the moment arm distance between the external load and the back is increased, resulting in greater internal back muscle force and spinal load), and (4) heavy weights are handled or large forces are applied. When jobs must accommodate both sitting and standing postures, it is important to ensure that the positions and orientations of the body, especially the upper extremity, are in the same location under both standing and sitting conditions.

4.4 Wrists

The Bureau of Labor Statistics reports that repetitive trauma increased in prevalence from 18% of occupational illnesses in 1981 to 63% in 1993. Based on these figures, repetitive trauma has been described as the fastest-growing occupationally related problem. Although these numbers and statements appear alarming, one must realize that occupational illnesses represent only 6% of all occupational injuries and illnesses. Furthermore, the statistics for illness include illnesses unrelated to musculoskeletal disorders, such as noise-induced hearing loss. Thus, the magnitude of the cumulative trauma problem should not be overstated. Nonetheless, there are specific industries (e.g., meatpacking, poultry processing) where cumulative trauma to the wrist is a major problem, reaching epidemic proportions within these industries.

4.4.1 Wrist Anatomy and Loading

To understand the biomechanics of the wrist and how cumulative trauma occurs, one must appreciate the anatomy of the upper extremity. Figure 20 is a simplified anatomical drawing of the wrist joint complex. The hand has few power-producing muscles in the hand itself. The thenar muscle that activates the thumb is one of the few power-producing muscles located in the hand. The vast majority of the power-producing muscles are located in the forearm. Force is transmitted from these forearm muscles to the fingers through a series of tendons (tendons attach muscles to bone). The tendons originate at the muscles in the forearm, transverse the wrist (with many of them passing through the carpal canal), pass through the hand, and culminate at the fingers. These tendons are secured or "strapped down" at various points along this path with ligaments that keep the tendons in close proximity to the bones, forming a pulley system around the joints. This results in a system (the

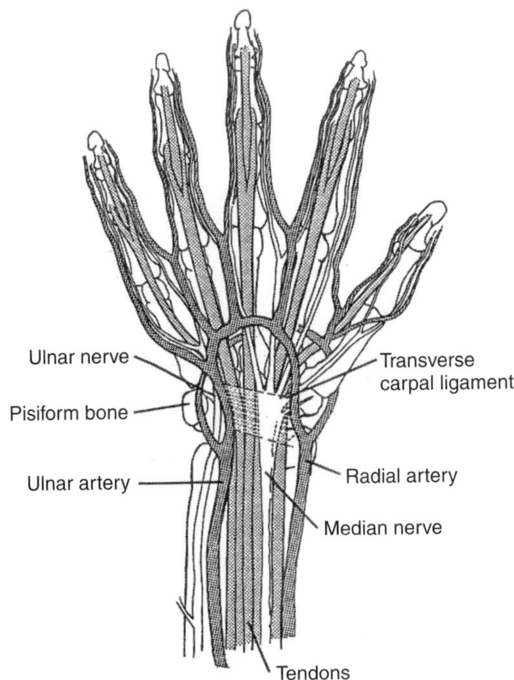

Figure 20 Important anatomical structures in the wrist.

hand) that is very small and compact, yet capable of generating large amounts of force. However, the price the musculoskeletal system pays for this design is friction. The forearm muscles must transmit force over a long distance to supply internal forces to the fingers. Thus, a great deal of tendon travel must occur, and this can result in tendon friction under repetitive motion conditions, thereby initiating the events outlined in Figure 3. The key to controlling wrist cumulative trauma is embedded in an understanding of those workplace factors that adversely affect the internal force generating (muscles) and transmitting (tendons) structures.

4.4.2 Biomechanical Risk Factors

A number of risk factors for upper extremity cumulative trauma disorders have been documented in the literature. Most of these risk factors have a biomechanical basis for their risk. First, deviated wrist postures reduce the volume of the carpal tunnel and thus increase tendon friction. In addition, grip strength is reduced dramatically once wrist posture deviates from its neutral position. Figure 21 demonstrates the magnitude of grip strength decrement due to any deviation from the wrist's neutral position. The reduction in strength is caused by a change in the length–strength relationship (Figure 5) of the forearm muscles when the wrist deviates from the neutral posture. Hence, the muscles must work at level lengths that are nonoptimal when the wrist is bent. This reduced strength associated with deviated wrist positions can therefore more easily initiate the sequence of events associated with cumulative trauma (Figure 3). Therefore, deviated wrist postures not only increase tendon travel and friction but also increase the amount of muscle strength necessary to perform a gripping task.

Second, increasing the frequency or repetition of the work cycle has also been identified as a risk factor for cumulative trauma disorders (CTDs) (Silverstein et al., 1996, 1997). Studies have shown that increased frequency of wrist motions increase the risk of CTD reporting. Repeated motions that require a cycle time of less than 30 seconds are considered candidates for cumulative trauma. Increased frequency is believed to increase the friction within the tendons, thereby accelerating the cumulative trauma progression described in Figure 3.

Third, the force applied by the hands and fingers during a work cycle has been identified as a cumulative trauma risk factor. In general, the greater the force required by the work, the greater the risk of CTD. Greater hand forces result in greater tension within the tendons and greater tendon friction and tendon travel. Another factor related to force is that of wrist acceleration. Industrial surveillance studies report that repetitive jobs resulting in greater wrist acceleration are associated with greater CTD incident rates (Marras and Schoenmarklin, 1993; Schoenmarklin et al., 1994). Force is a product of mass and acceleration. Thus, jobs that increase the angular acceleration of the wrist joint result in greater tension and force being transmitted through the tendons. Therefore, wrist acceleration can be another mechanism to impose force on wrist structures.

Finally, as shown in Figure 20, the anatomy of the hand is such that the median nerve becomes very superficial at the palm. Direct impacts to the palm through pounding or striking an object (with the palm) can directly assault the median nerve and initiate symptoms of cumulative trauma even though the work may not be repetitive.

4.4.3 Grip Design

The design of a tool's gripping surface can affect the activity of the internal force transmission system (tendon travel and tension). Grip opening and shape have a major influence on the grip strength available. Figure 22 indicate how grip strength capacity changes as a function of the separation distance of the grip opening. This figure indicates that maximum grip strength occurs within a very narrow range of grip span. If the grip opening deviates from this ideal range by as little as an inch (a couple of centimeters), grip strength is reduced markedly. This reduction in strength is, once again, due to the length–strength relationship of the forearm muscles. Also indicted in Figure 22 are the effects of hand size. The worker's hand anthropometry as well as hand preference can influence grip strength and risk. Therefore, proper design of tool handles is crucial in optimizing ergonomic workplace design.

Handle shape can also influence the strength of the wrist. Figure 23 shows how changes in the design of

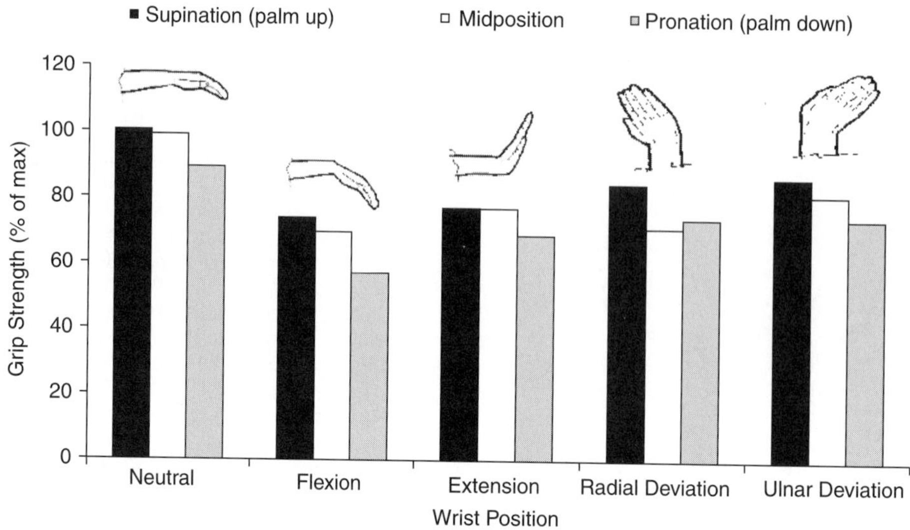

Figure 21 Grip strength as a function of wrist and forearm position. (Adapted from Sanders and McCormick, 1993.)

screwdriver handles can affect the maximum force that can be exerted on the tool. The biomechanical origin of these differences in strength capacity is believed to be related to the length–strength relationship of the forearm muscles as well as the area in contact with the tool. A handle design's resulting diminished strength permits the wrist to twist or permits the grip to slip, resulting in a deviation from the ideal length–strength position in the forearm muscles.

4.4.4 Gloves

The use of gloves can also significantly influence the generation of grip strength and may play a role in the development of cumulative trauma disorders. When gloves are worn during work, three factors must be considered. First, the grip strength that is generated is often reduced. Typically, a 10 to 20% reduction in grip strength is noted when gloves are worn. Gloves reduce the coefficient of friction between the hand and the tool, which in turn permits some slippage of the hand on the tool surface. This slippage may result in a deviation from the ideal muscle length and thus a reduction in available strength. The degree of slippage and the subsequent degree of strength loss depends on how well the gloves fit the hand as well as the type of material used in the glove. Poorly fitting gloves probably result in greater strength loss. Figure 24 indicates how the glove material and glove fit can influence grip force potential.

Second, while wearing gloves, even though the force applied externally (grip strength) is often reduced, the internal forces are often very large relative to a bare-handed condition. For a given grip force application, the muscle activity is significantly greater when using gloves compared to a bare-handed condition (Kovacs et al., 2002). Thus, the musculoskeletal system is less efficient when wearing a glove, due to the fact that the hand typically slips within the glove, thereby altering the length–strength relationship of the muscle.

Third, the ability to perform a task is significantly affected when wearing gloves. Figure 25 shows the increase in time required to perform work tasks when wearing gloves composed of various materials compared to performing the same task bare-handed. The figure indicates that task performance can increase up to 70% when wearing certain types of gloves.

These effects have indicated that there are biomechanical costs associated with the use of gloves. Less strength capacity is available to the worker, more internal force is generated, less force output is available, and worker productivity is reduced when wearing gloves. These negative effects of glove use do not mean that gloves should never be worn at work. When hand protection is required, gloves should be considered as a potential solution. However, protection should only be provided to the parts of the hand that are at risk. For example, if the palm of the hand requires protection but not the fingers, fingerless gloves might provide an acceptable solution. If the fingers require protection but there is little risk to the palm of the hand, grip tape wrapped around the fingers might be considered as a potential solution. Additionally, different styles, materials, and sizes of gloves will fit workers differently. Thus, gloves produced by different manufacturers and of different sizes should be available to workers to minimize the negative effects mentioned above.

4.4.5 Design Guidelines

This discussion has indicated that many factors can affect the biomechanics of the wrist and the subsequent risk of cumulative trauma disorders. Proper ergonomic

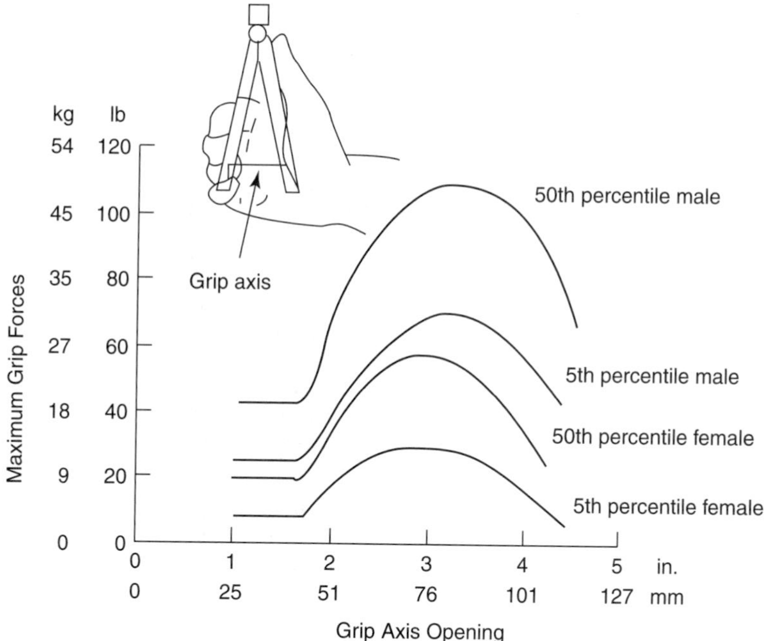

Figure 22 Grip strength as a function of grip opening and hand anthropometry. (Adapted from Sanders and McCormick, 1993.)

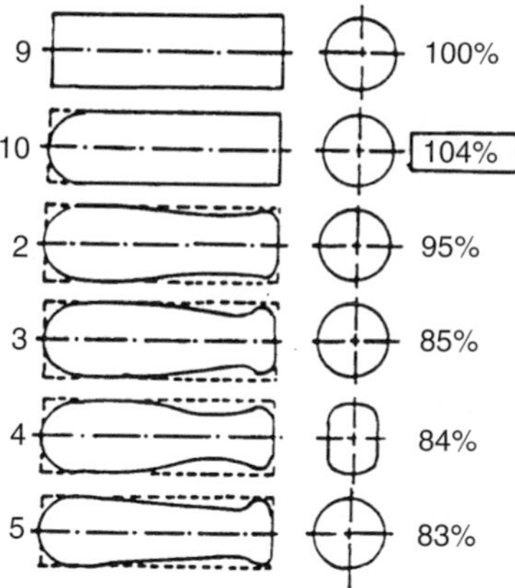

Figure 23 Maximum force that could be exerted on a screwdriver as a function of handle shape. (From Konz, 1983.)

design of a work task cannot be accomplished simply by providing a worker with an "ergonomically designed" tool. Ergonomics is associated with matching the workplace design to the workers' capabilities, and it is not possible to design an ergonomic tool without considering workplace design and task requirements simultaneously. What might be an ergonomic tool for one work condition may be improper for use when a worker is assuming another work posture. For example, an *in-line tool* may keep the wrist straight when inserting a bolt into a horizontal surface. However, if the bolts are to be inserted into a vertical surface, a *pistol grip tool* may be more appropriate. Using an in-line tool in this situation (inserting a bolt into a vertical surface) may cause the wrist to be deviated. This illustrates that there are no ergonomic tools; there are just ergonomic *situations*. A tool that is considered ergonomically correct in one situation may be totally incorrect in another work situation. Thus, a workplace should be designed with care, and one should be alert to the trade-offs between different parts of the body that must be considered, by taking into consideration the various biomechanical trade-offs.

Given these considerations, the following components of the workplace should be considered when designing a workplace to minimize cumulative trauma risk. First, keep the wrist in a neutral posture. A neutral posture is a relaxed wrist posture with a slight extension (which optimizes the forearm's length–strength relationship), not a rigid linear posture. Second, minimize tissue compression on the hand. Third, avoid tasks and actions that repeatedly impose force on the internal structures. Fourth, minimize required wrist accelerations and motions through the design of the work. Fifth, be sensitive to the impact of glove use, hand size, and left-handed workers.

5 ANALYSIS AND CONTROL MEASURES

Several analysis and control measures have been developed to evaluate and control biomechanical loading of the body during work tasks. Since low

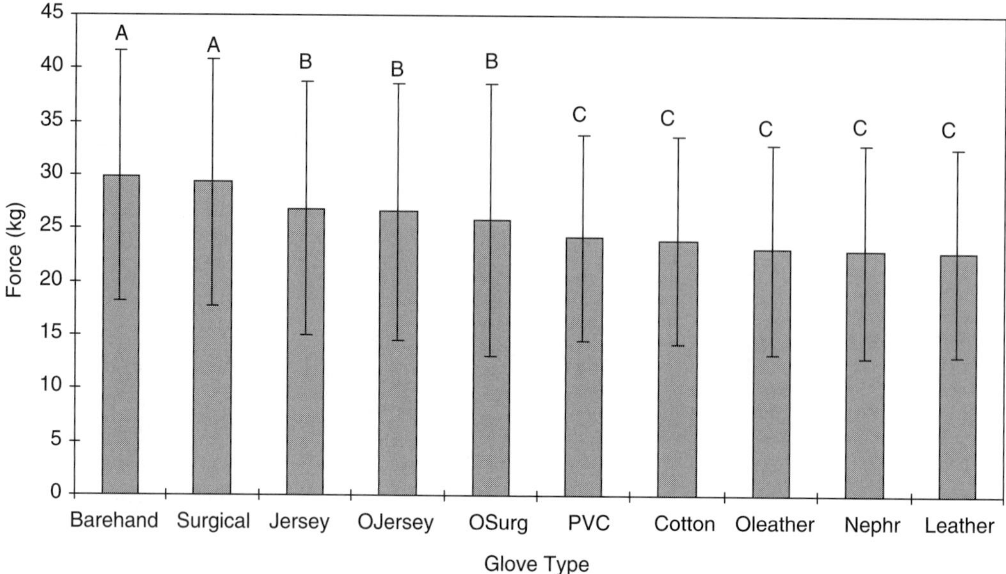

Figure 24 Peak grip force shown as a function of type of glove. Different letters above the columns indicate statistically significant differences.

back disorders are often associated with spine loading magnitude, most analysis methods have focused on risk to the back. However, several of the measures also include analyses of risk to other body parts.

5.1 NIOSH Lifting Guide and Revised Equation

The National Institute for Occupational Safety and Health (NIOSH) has developed two assessment tools or guides to help determine the risk associated with manual materials handling tasks. The lifting guide was originally developed in 1981 (NIOSH, 1981) and applied to lifting situations where the lifts were performed in the sagittal plane and to motions that are slow and smooth. Two benchmarks or limits were defined by this guide. The first limit, called the *action limit* (AL), represents a magnitude of weight in a given lifting situation which would impose a spine load corresponding to the beginning of low back disorder risk along a risk continuum. The AL was associated with the point at which people under 40 years of age just begin to experience a risk of vertebral end plate microfracture (3400 N of compressive load) (see Figure 10). The guide estimates the force imposed on the spine of a worker as a result of lifting a weight, and compares the spine load to the AL. If the weight of the object results in a spine load that is below the AL, the job is considered safe. If the weight lifted by the worker is larger than the AL, there is some level of risk associated with the task. The general form of the AL formula is

$$\text{AL} = k(\text{HF})(\text{VF})(\text{DF})(\text{FF}) \qquad (1)$$

where AL is the action limit (kilograms or pounds); k the load constant (40 kg or 90 lb), which is the greatest weight a subject could lift under optimal lifting conditions; HF the horizontal factor, defined as the horizontal distance from a point bisecting the ankles to the center of gravity of the load at the lift origin, defined algebraically as $15/H$ (metric) or $6/H$ (U.S. units); VF the vertical factor or height of the load at the lift origin, defined algebraically as $(0.004)\ |V - 75|$ (metric) or $1 - (0.01)|V - 30|$ (U.S. units); DF the distance factor or vertical travel distance of the load, defined algebraically as $0.7 + 7.5/D$ (metric) or $0.7 + 3/D$ (U.S. units); and FF the frequency factor or lifting rate, defined algebraically as $1 - F/F_{\max}$ (F is the average frequency of lift; $F_{\max}$ is shown in Table 3). This equation assumes that if the lifting conditions are ideal, a worker could safely hold (and implies "lift") the load constant, k (40 kg or 90 lb). However, if the lifting conditions are not ideal, the allowable weight is discounted according to the four factors HF, VF, DF, and FF. These four discounting factors in (Figures 26 to 29) relate to many of the biomechanical principles discussed earlier. According to the relationships indicated in these figures, the HF associated with the external moment has the most dramatic effect on acceptable lifting conditions. Both VF and DF are associated with the back muscle's length–strength relationship. Finally, FF attempts to account for the cumulative effects of repetitive lifting.

The second benchmark associated with the 1981 lifting guide is the *maximum permissible limit* (MPL). MPL represents the point at which significant risk, defined in part as a significant risk of vertebral

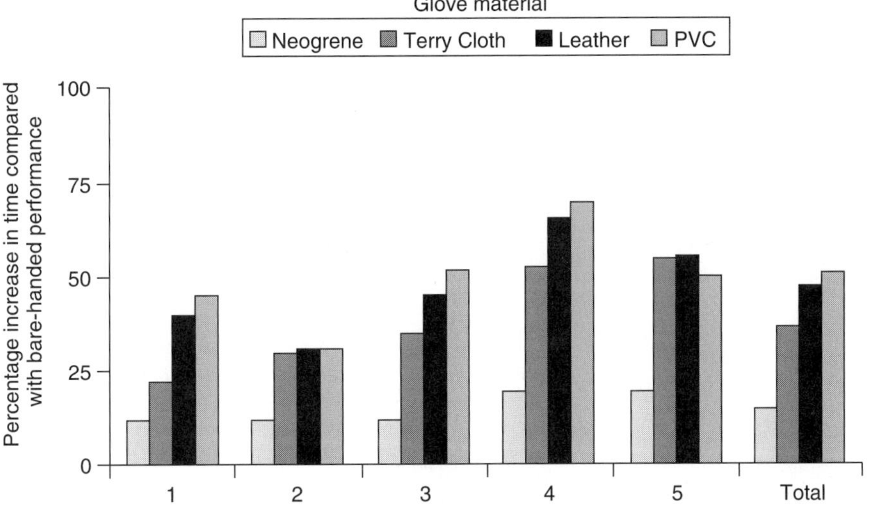

Figure 25 Performance (time to complete) on a maintenance-type task while wearing gloves constructed of five different materials. (From Sanders and McCormick, 1993.)

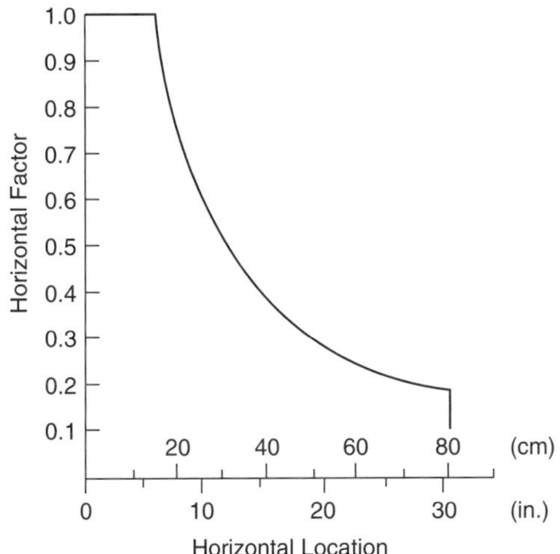

Figure 26 Horizontal factor varies between the body interference limit and the limit of functional reach. (Adapted from NIOSH, 1981.)

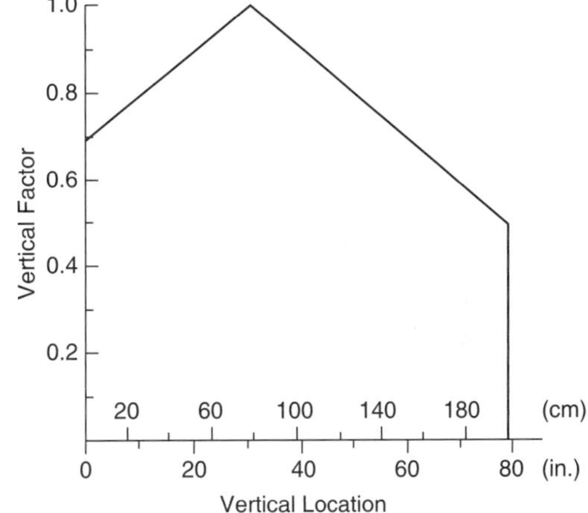

Figure 27 Vertical factor varies both ways from knuckle height. (Adapted from NIOSH, 1981.)

end plate microfracture (Figure 10), occurs. MPL is associated with a compressive load on the spine of 6400 N, which corresponds to the point at which 50% of the people would be expected to suffer a vertebral end plate microfracture. MPL is a function of AL and is defined as

$$\text{MPL} = 3(\text{AL}) \tag{2}$$

The weight that the worker is expected to lift in a work situation is compared to AL and MPL. If the magnitude of weight falls below AL, the work is considered safe and no work adjustments are necessary. If the magnitude of the weight falls above

Table 3 F_{max} **Values**

	Average Vertical Location	
Period (h)	Standing $V > 75$ cm (3 in.)	Stooped $V \leq 75$ cm (3 in.)
1	18	15
8	15	12

Source: NIOSH (1981).

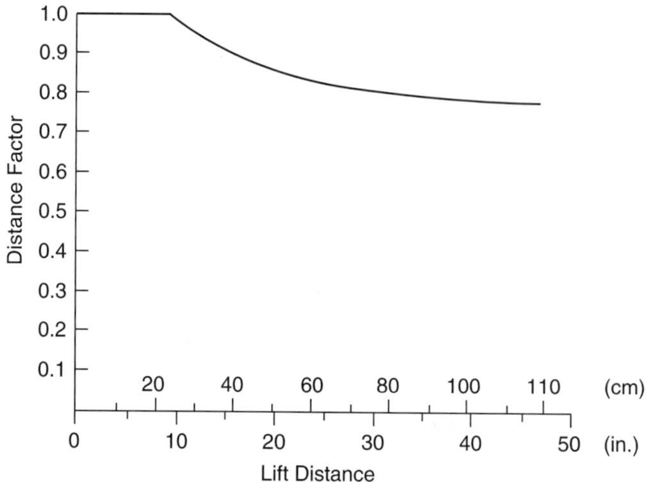

Figure 28 Distance factor varies between a minimum vertical distance moved of 25 cm (10 in.) to a maximum distance of 200 cm (80 in.). (Adapted from NIOSH, 1981.)

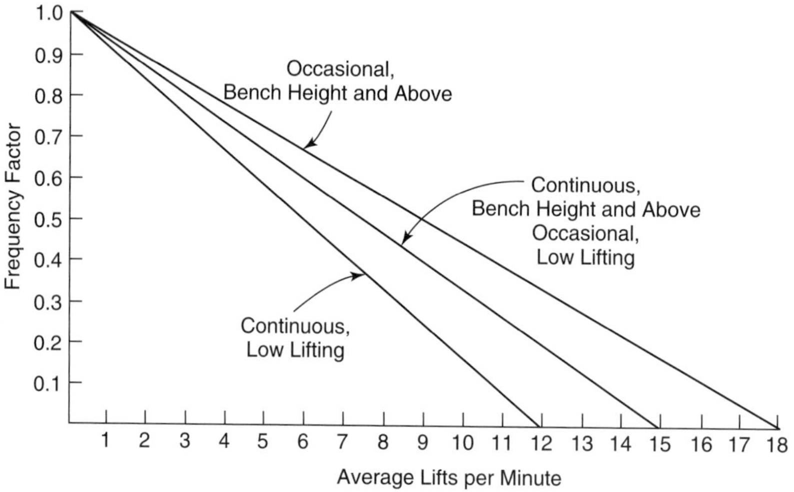

Figure 29 Frequency factor varies with lifts/min and the F_{max} curve. F_{max} depends on lifting posture and lifting time. (From NIOSH, 1981.)

MPL, the work is considered to represent a significant risk and engineering changes involving the adjustment of HF, VF, and/or DF are required to reduce AL and MPL. If the weight falls between AL and MPL, either engineering changes or administrative changes, defined as selecting workers who are less likely to be injured, or rotating workers, would be appropriate. AL and MPL were also indexed relative to nonbiomechanical benchmarks. NIOSH (1981) states that these limits also correspond to strength, energy expenditure, and psychophysical acceptance points.

The 1993 NIOSH revised lifting equation was introduced to address those lifting jobs that violate the sagittally symmetric lifting assumption of the original 1981 lifting guide (Waters et al., 1993). The concepts of AL and MPL were replaced by the concept of a *lifting index* (LI). LI is defined as

$$\text{LI} = \frac{L}{\text{RWL}} \tag{3}$$

where LI is the lifting index, used to estimate the relative magnitude of physical stress for a particular job, L the load weight or weight of the object to be lifted, and RWL the recommended weight limit for the particular lifting situation. If LI is greater than

1.0, an increased risk of suffering a lifting-related low back disorder exists. RWL is similar in concept to the 1981 Lifting Guide AL equation [equation (1)] in that it contains factors that discount the allowable load according to the horizontal distance, vertical location of the load, vertical travel distance, and frequency of lift. However, the form of these discounting factors was adjusted. In addition, two discounting factors have been included. These additional factors include a lift asymmetry factor, which accounts for asymmetric lifting conditions, and a coupling factor, which accounts for whether or not the load lifted has handles. RWL is given by

$$\text{RWL(kg)} = 23(25/H)\ [1 - (0.003 \mid V - 75 \mid)] \times [(0.82 + 4.5/D)]\ (\text{FM}) \times [1 - (0.0032A)]\ (\text{CM}) \quad (4)$$

$$\text{RWL(lb)} = 51(10/H)\ [1 - (0.0075 \mid V - 30 \mid)] \times [(0.82 + 1.8/D)]\ (\text{FM}) \times [1 - (0.0032A)]\ (\text{CM}) \quad (5)$$

where H is the horizontal location forward of the midpoint between the ankles at the origin of the lift (if significant control is required at the destination, H should be measured at both the origin and destination of the lift), V the vertical location at the origin of the lift, D the vertical travel distance between the origin and destination of the lift, FM the frequency multiplier shown in Table 4, A the angle between the midpoint of the ankles and the midpoint between the hands at the origin of the lift, and CM the coupling multiplier, ranked as good, fair, or poor, and described in Table 5.

In this revised equation the load constant has been reduced significantly relative to the 1981 equation. The discounting adjustments for load moment, muscle length–strength relationships, and cumulative loading are still integral parts of this equation. However, these adjustments relationships have been changed (compared to the 1981 Lifting Guide) to reflect the most conservative value of the biomechanical, physiological, psychophysical, or strength data on which they are based. Effectiveness studies report that the 1993 revised equation yields a more conservative (protective) prediction of work-related low back disorder risk (Marras et al., 1999).

5.2 Static Biomechanical Models

Biomechanically based spine models have been developed to assess occupationally related manual materials handling tasks. These models assess the task based

Table 5 Coupling Multiplier Values

Coupling Type	$V < 75$ cm (30 in.)	$V \geq 75$ cm (30 in.)
Good	1.00	1.00
Fair	0.95	1.00
Poor	0.90	0.90

Source: NIOSH (1994).

Table 4 Frequency Multiplier Values

	Work Duration (h)					
	≤ 1		> 1 but ≤ 2		>2 but ≤ 8	
Frequency, F (lifts/min)[a]	$V < 75$ cm (30 in.)	$V \geq 75$ cm (30 in.)	$V < 75$ cm (30 in.)	$V \geq 75$ cm (30 in.)	$V < 75$ cm (30 in.)	$V \geq 75$ cm (30 in.)
≥0.2	1.00	1.00	0.95	0.95	0.85	0.85
0.5	0.97	0.97	0.92	0.92	0.81	0.81
1	0.94	0.94	0.88	0.88	0.75	0.75
2	0.91	0.91	0.84	0.84	0.65	0.65
3	0.88	0.88	0.79	0.79	0.55	0.55
4	0.84	0.84	0.72	0.72	0.45	0.45
5	0.80	0.80	0.60	0.60	0.35	0.35
6	0.75	0.75	0.50	0.50	0.27	0.27
7	0.70	0.70	0.42	0.42	0.22	0.22
8	0.60	0.60	0.35	0.35	0.18	0.18
9	0.52	0.52	0.30	0.30	0.00	0.15
10	0.45	0.45	0.26	0.26	0.00	0.13
11	0.41	0.41	0.00	0.23	0.00	0.00
12	0.37	0.37	0.00	0.21	0.00	0.00
13	0.00	0.34	0.00	0.00	0.00	0.00
14	0.00	0.31	0.00	0.00	0.00	0.00
15	0.00	0.28	0.00	0.00	0.00	0.00
>15	0.00	0.00	0.00	0.00	0.00	0.00

Source: NIOSH (1994).

[a]For lifting less frequently than once per 5 minutes, set $F = 0.2$ lifts/min.

on both spine loading criteria and through a strength assessment of task requirements. One of the early static assessment models was developed by Don Chaffin at the University of Michigan (Chaffin, 1969). The original two-dimensional model has been expanded to a three-dimensional static model (Chaffin and Muzaffer, 1991; Chaffin et al., 1999). In this model, the moments imposed on the various joints of the body due to the object lifted are evaluated assuming that a static posture is representative of the instantaneous loading of the body. These models compare the imposed moments about each joint with the static strength capacity derived from a working population. The static strength capacity required of the major joint articulations used in this model have been documented in a database of over 3000 workers. In this manner, the proportion of the population capable of performing a particular static exertion is estimated. The joint that limits the capacity to perform the task can be identified using this method. The model assumes that a single equivalent muscle (internal force) supports the external moment about each joint. By considering the contribution of the externally applied load and the internally generated single muscle equivalent, spine compression at the lumbar discs is predicted. The compression predicted can then be compared to the tolerance limits for the vertebral end plate (Figure 10). An important assumption of this model is that no significant motion occurs during exertion since it is a static model. Figure 30 shows the output screen for this computer model, where the lifting posture, lifting distances, strength predictions, and spine compression are shown.

5.3 Multiple-Muscle-System Models

One significant simplifying assumption from a biomechanical standpoint in most static models is that one internal force counteracts the external moment. In reality, a great deal of coactivity (simultaneous recruitment of multiple muscles) occurs in trunk muscles during exertion, and the more complex the exertion, the greater the coactivity. Hence, the trunk is truly a multiple-muscle system, with many major muscle groups supporting and loading the spine (Schultz and Andersson, 1981). This arrangement can be seen in the cross section of the trunk shown in Figure 31. Significant coactivation also occurs in many of the major muscle groups in the trunk during realistic *dynamic* lifting (Marras and Mirka, 1993). Accounting for coactivation in these models is important because all the trunk muscles have the ability to load the spine since antagonist muscles can oppose each other during occupational tasks, thereby increasing the total load on the spine. Ignoring the coactivation of the trunk muscles during dynamic lifting can misrepresent spine loading by 45 to 70% (Granata and Marras, 1995b; Thelen et al., 1995). To more accurately estimate the loads on the lumbar spine, especially under complex, changing (dynamic) postures, multiple-muscle-system models of

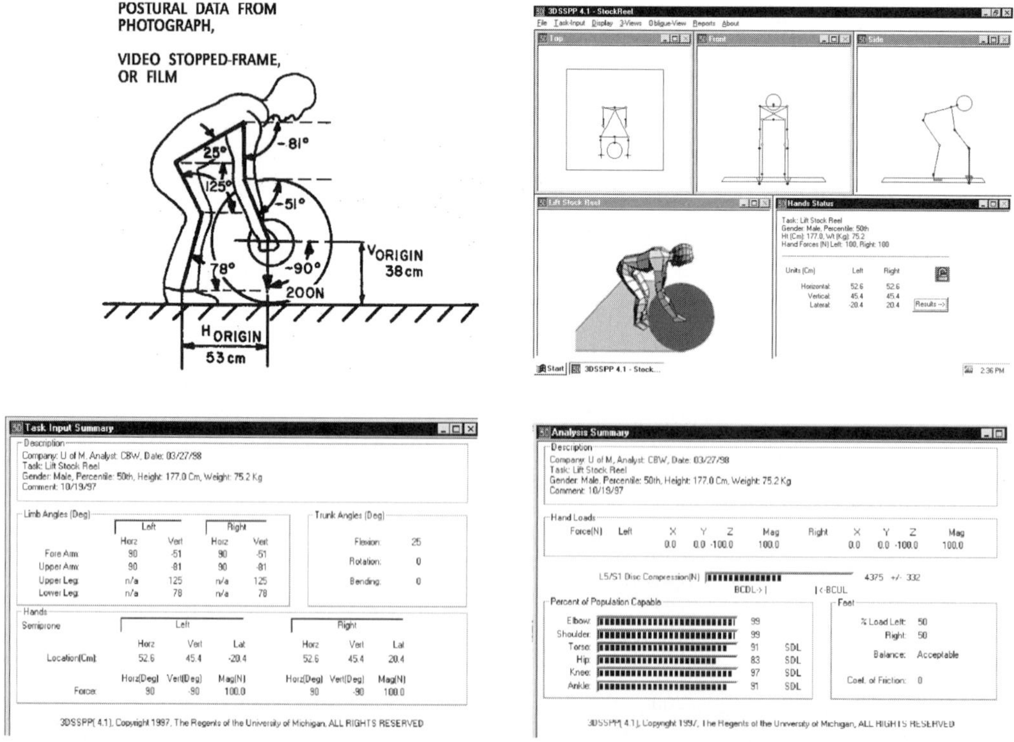

Figure 30 Two-dimensional static strength prediction model. (From Chaffin and Andersson, 1991.)

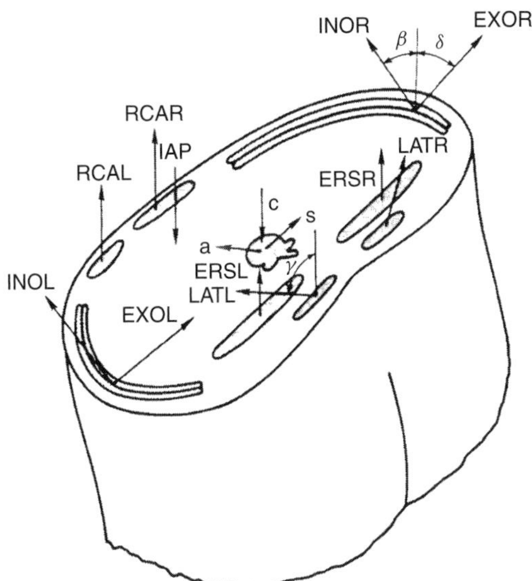

Figure 31 Cross-sectional view of the human trunk at the lumbrosacral junction. (Adapted from Schultz and Andersson, 1981.)

the trunk have been developed. However, predicting the activity of the muscles is the key to accurate low-back-loading assessment.

One way of assessing the degree of activation of the trunk muscles during a task is to monitor the muscle through electromyography or EMG and use this information as input to a biomechanical model. EMG-assisted models take into account the individual recruitment patterns of the muscles during a specific lift for a specific person. By monitoring muscle activity directly, the EMG-assisted model is capable of determining individual muscle force and subsequent spine loading. These models have been developed and tested under bending and twisting dynamic motion conditions and have been validated (McGill and Norman, 1985, 1986; Marras and Reilly, 1988; Reilly and Marras, 1989; Marras and Sommerich, 1991a,b; Granata and Marras, 1993, 1995a; Marras and Granata, 1995, 1997a,b; Marras et al., 2001). These models are the only biomechanical models that can predict *multidimensional loads* on the lumbar spine under many three-dimensional complex dynamic lifting conditions. However, the limitation of such models is that they require significant instrumentation of the worker.

5.4 Dynamic Motion Assessment at the Workplace

It is clear that dynamic activity may significantly increase the risk of low back disorder, yet there are few assessment tools available to assess the biomechanical demands associated with workplace dynamics and the risk of LBD. To assess this biomechanical situation at the work site, one must know the type of motion that increases biomechanical load and determine "how much motion exposure is too much motion exposure" from a biomechanical standpoint. These issues were the focus of several industrial studies performed over a six-year period in 68 industrial environments. Trunk motion and workplace conditions were assessed in workers exposed to a high risk of low-back-disorder jobs and compared to trunk motions and workplace conditions associated with low-risk jobs (Marras et al., 1993, 1995) (Table 6). A trunk goniometer [lumbar motion monitor (LMM)] has been used to document the trunk motion patterns of workers at the workplace and is shown in Figure 32. Based on this study, a five-factor multiple logistic regression model was developed that is capable of discriminating between task exposures that indicate probability of high-risk group membership. These risk factors include (1) frequency of lifting, (2) load moment (load weight multiplied by the distance of the load from the spine), (3) average twisting velocity (measured by the LMM), (4) maximum sagittal flexion angle through the job cycle (measured by the LMM), and (5) maximum lateral velocity (measured by the LMM). This LMM risk assessment model is the only model capable of

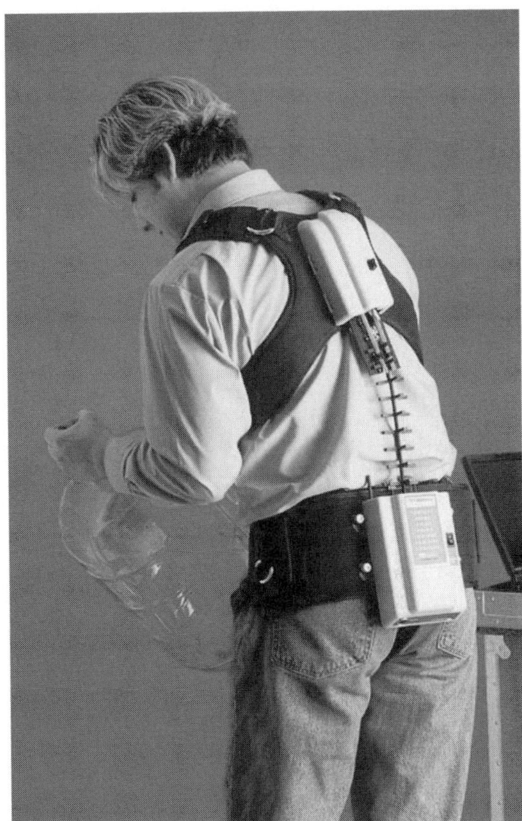

Figure 32 Lumbar motion monitor.

Table 6 Descriptive Statistics of the Workplace and Trunk Motion Factors in Each of the Risk Groups

	High Risk ($N = 111$)				Low Risk ($N = 124$)				
Factor	Mean	Std. Dev.	Minimum	Maximum	Mean	Std. Dev.	Minimum	Maximum	Statistics t
Workplace Factors									
Lift rate (lifts/h)	175.89	8.65	15.30	900.00	118.83	169.09	5.40	1500.00	2.1[a]
Vertical load location at origin (m)	1.00	0.21	0.38	1.80	1.05	0.27	0.18	2.18	1.4
Vertical load location at destination (m)	1.04	0.22	0.55	1.79	1.15	0.26	0.25	1.88	3.2[b]
Vertical distance traveled by load (m)	0.23	0.17	0.00	0.76	0.25	0.22	0.00	1.04	0.8
Average weight handled (N)	84.74	79.39	0.45	423.61	29.30	48.87	0.45	280.92	6.4[b]
Maximum weight handled (N)	104.36	88.81	0.45	423.61	37.15	60.83	0.45	325.51	6.7[b]
Average horizontal distance between load and L_5-S_1 (N)	0.66	0.12	0.30	0.99	0.61	0.14	0.33	1.12	2.5[a]
Maximum horizontal distance between load and L_5-S_1 (N)	0.76	0.17	0.38	1.24	0.67	0.19	0.33	1.17	3.7[b]
Average moment (N · m)	55.26	51.41	0.16	258.23	17.70	29.18	0.17	150.72	6.8[b]
Maximum moment (N · m)	73.65	60.65	0.19	275.90	23.64	38.62	0.17	198.21	7.4[b]
Job satisfaction	5.96	2.26	1.00	10.00	7.28	1.95	1.00	10.00	4.7[b]
Trunk Motion Factors									
Sagittal plane									
Maximum extension position (deg)	−8.30	9.10	−30.82	18.96	−10.19	10.58	−30.00	33.12	3.5[b]
Maximum flexion position (deg)	17.85	16.63	−13.96	45.00	10.37	16.02	−25.23	45.00	1.5
Range of motion (deg)	31.50	15.67	7.50	75.00	23.82	14.22	399.00	67.74	3.8[b]
Average velocity (deg/s)	11.74	8.14	3.27	48.88	6.55	4.28	1.40	35.73	6.0[b]
Maximum velocity (deg/s)	55.00	38.23	14.20	207.55	38.69	26.52	9.02	193.29	3.7[b]
Maximum acceleration (deg/s^2)	316.73	224.57	80.61	1341.92	226.04	173.88	59.10	1120.10	4.2[b]
Maximum deceleration (deg/s^2)	−92.45	63.55	−514.08	−18.45	−83.32	47.71	−227.12	−4.57	1.2
Lateral plane									
Maximum left bend (deg)	−1.47	6.02	−16.80	24.49	−2.54	5.46	−23.80	13.96	1.4
Maximum right bend (deg)	15.60	7.61	3.65	43.11	13.24	6.32	0.34	34.14	2.6[a]
Range of motion (deg)	24.44	9.77	7.10	47.54	21.59	10.34	5.42	62.41	2.2[a]
Average velocity (deg/s)	10.28	4.54	3.12	33.11	7.15	3.16	2.13	18.86	6.1[b]

Table 6 ***(continued)***

Factor	High Risk ($N = 111$)				Low Risk ($N = 124$)				
	Mean	Std. Dev.	Minimum	Maximum	Mean	Std. Dev.	Minimum	Maximum	Statistics t
Maximum velocity (t/s)	46.36	19.12	13.51	119.94	35.45	12.88	11.97	76.25	4.9[b]
Maximum acceleration (deg/s^2)	301.41	166.69	82.64	1030.29	229.29	90.90	66.72	495.88	4.1[b]
Maximum deceleration (deg/s^2)	−103.65	60.31	−376.75	0.00	−106.20	58.27	−294.83	0.00	0.3
Twisting plane									
Maximum left twist (deg)	1.21	9.08	−27.56	29.54	−1.92	5.36	−30.00	11.44	3.2[b]
Maximum right twist (deg)	13.95	8.69	−13.45	30.00	10.83	6.08	−11.20	30.00	2.2[a]
Range of motion (deg)	20.71	10.61	3.28	53.30	17.08	8.13	1.74	38.59	2.9[b]
Average velocity (deg/s)	8.71	6.61	1.02	34.77	5.44	3.19	0.66	17.44	3.8[b]
Maximum velocity (deg/s)	46.36	25.61	8.06	136.72	38.04	17.51	5.93	91.97	4.7[a]
Maximum acceleration (deg/s^2)	304.55	175.31	54.48	853.93	269.49	146.65	44.17	940.27	2.9[b]
Maximum deceleration (deg/s^2)	−88.52	70.30	−428.94	−5.84	−100.32	72.40	−325.93	−2.74	1.6[a]

[a]Significant at $\alpha \leq 0.05$ (two-sided).
[b]Significant at $\alpha \leq 0.01$ (two-sided).

assessing the *risk* associated with three-dimensional trunk motion on the job. This model has a high degree of predictability (odds ratio = 10.7) compared to previous attempts to assess work-related LBD risk. The advantage of such an assessment is that the evaluation provides information about risk that would take years to derive from historical accounts of incidence rates. The model has also been validated prospectively (Marras et al., 2000).

5.5 Threshold Limit Values

Threshold limit values (TLVs) have recently been introduced as a means for controlling biomechanical risk to the back in the workplace. TLVs have been introduced through the American Conference of Governmental Industrial Hygienists (ACGIH) and provide lifting weight limits as a function of *lift origin zones* and repetitions associated with occupational tasks. Lift origin zones are defined by the lift height off the ground and lift distance from the spine associated with the lift origin. Twelve zones are defined that relate to lifts within ± 30° of asymmetry from the sagittal plane. These zones are represented in a series of figures, with each figure corresponding to different lift frequency and time exposures. Within each zone, weight lifting limits are specified based on the best information available from several sources: (1) EMG-assisted biomechanical models, (2) the 1993 revised lifting equation, and (3) the historical risk data associated with the LMM database. The weight lifted by the worker is compared to these limits. Weights exceeding the zone limit are considered hazards.

6 SUMMARY

In this chapter we have shown that biomechanics provides a means to consider the implications of workplace design *quantitatively*. Biomechanical design considerations are important when a particular job is suspected of imposing large or repetitive forces on the structures of the body. It is particularly important to recognize that the internal structures of the body, such as muscles, are the primary generators of force within the joint and tendon structures. To evaluate the risk of injury due to a particular task, one must consider the contribution of both external and internal loads on a structure and how they relate to the tolerance of the structure. Armed with an understanding of some general biomechanical concepts (presented in this chapter) and how they apply to different parts of the body (affected by work), one can logically reason through the design considerations and trade-offs so that musculoskeletal disorders are minimized due to the design of the work.

REFERENCES

Adams, M. A., Freeman, B. J., Morrison, H. P., Nelson, I. W., and Dolan, P. (2000), "Mechanical Initiation of Intervertebral Disc Degeneration," *Spine*, Vol. 25, pp. 1625–1636.

Anderson, C. K., Chaffin, D. B., Herrin, G. D., and Matthews, L. S. (1985), "A Biomechanical Model of the Lumbosacral Joint During Lifting Activities," *Journal of Biomechanics*, Vol. 18, pp. 571–584.

Andersson, G. B. (1997), "The Epidemiology of Spinal Disorders," in *The Adult Spine: Principles and Practice*, Vol. 1, 2nd ed., J. W. Frymoyer, Ed., Lippincott-Raven, Philadelphia, pp. 93–141.

Andersson, B. J., Ortengren, R., Nachemson, A. L., Elfstrom, G., and Broman, H. (1975), "The Sitting Posture: An Electromyographic and Discometric Study," *Orthopedic Clinics of North America*, Vol. 6, pp. 105–120.

Astrand, P. O., and Rodahl, K. (1977), *Textbook of Work Physiology: Physiological Basis of Exercises*, 2nd ed., McGraw-Hill, New York.

Basmajian, J. V., and De Luca, C. J. (1985), *Muscles Alive: Their Functions Revealed by Electromyography*, 5th ed., Williams & Wilkins, Baltimore, MD.

Brinckmann, P., Biggermann, M., and Hilweg, D. (1988), "Fatigue Fracture of Human Lumbar Vertebrae," *Clinical Biomechanics (Bristol, Avon)*, Vol. 3, pp. S1–S23.

Cavanaugh, J. M. (1995), "Neural Mechanisms of Lumbar Pain," *Spine*, Vol. 20, pp. 1804–1809.

Cavanaugh, J. M., Ozaktay, A. C., Yamashita, T., Avramov, A., Getchell, T. V., and King, A. I. (1997), "Mechanisms of Low Back Pain: A Neurophysiologic and Neuroanatomic Study," *Clinical Orthopaedics*, Vol. pp. 166–180.

Chaffin, D. B. (1969), "A Computerized Biomechanical Model: Development of and Use in Studying Gross Body Actions," *Journal of Biomechanics*, Vol. 2, pp. 429–441.

Chaffin, D. B. (1973), "Localized Muscle Fatigue: Definition and Measurement," *Journal of Occupational Medicine*, Vol. 15, pp. 346–354.

Chaffin, D. B., and Andersson, G. B. (1991), *Occupational Biomechanics*, Wiley, New York.

Chaffin, D. B., and Muzaffer, E. (1991), "Three-Dimensional Biomechanical Static Strength Prediction Model Sensitivity to Postural and Anthropometric Inaccuracies," *IIE Transactions*, Vol. 23, pp. 215–227.

Chaffin, D. B., Andersson, G. B. J., and Martin, B. J. (1999), *Occupational Biomechanics*, 3rd ed., Wiley, New York.

Gedalia, U., Solomonow, M., Zhou, B. H., Baratta, R. V., Lu, Y., and Harris, M. (1999), "Biomechanics of Increased Exposure to Lumbar Injury Caused by Cyclic Loading, 2: Recovery of Reflexive Muscular Stability with Rest," *Spine*, Vol. 24, pp. 2461–2467.

Grandjean, E. (1982), *Fitting the Task to the Man: An Ergonomic Approach*, Taylor & Francis, London.

Granata, K. P., and Marras, W. S. (1993), "An EMG-Assisted Model of Loads on the Lumbar Spine During Asymmetric Trunk Extensions," *Journal of Biomechanics*, Vol. 26, pp. 1429–1438.

Granata, K. P., and Marras, W. S. (1995a), "An EMG-Assisted Model of Trunk Loading During Free-Dynamic Lifting," *Journal of Biomechanics*, Vol. 28, pp. 1309–1317.

Granata, K. P., and Marras, W. S. (1995b), "The Influence of Trunk Muscle Coactivity on Dynamic Spinal Loads," *Spine*, Vol. 20, pp. 913–919.

Guo, H. R. (1993), "Back Pain and U.S. Workers," paper presented at the American Occupational Health Conference, April 29.

Guo, H. R., Tanaka, S., Halperin, W. E., and Cameron, L. L. (1999), "Back Pain Prevalence in U.S. Industry and Estimates of Lost Workdays," *American Journal of Public Health*, Vol. 89, pp. 1029–1035.

Hollbrook, T. L., Grazier, K., Kelsey, J. L., and Stauffer, R. N. (1984), *The Frequency of Occurrence, Impact and Cost of Selected Musculoskeletal Conditions in the United States*, American Academy of Orthopaedic Surgeons, Chicago.

Hoy, M. G., Zajac, F. E., and Gordon, M. E. (1990), "A Musculoskeletal Model of the Human Lower Extremity: The Effect of Muscle, Tendon, and Moment Arm on the Moment–Angle Relationship of Musculotendon Actuators at the Hip, Knee, and Ankle," *Journal of Biomechanics*, Vol. 23, pp. 157–169.

Jager, M., Luttmann, A., and Laurig, W. (1991), "Lumbar Load During One-Hand Bricklaying," *International Journal of Industrial Ergonomics*, Vol. 8, pp. 261–277.

Khalsa, P. S. (2004), "Biomechanics of Musculoskeletal Pain: Dynamics of the Neuromatrix," *Journal of Electromyographic Kinesiology*, Vol. 14, pp. 109–120.

Kim, J., Stuart-Buttle, C., and Marras, W. S. (1994), "The Effects of Mats on Back and Leg Fatigue," *Applied Ergonomics*, Vol. 25, pp. 29–34.

Konz, S. A. (1983), *Work Design: Industrial Ergonomics*, 2nd ed., Grid Publishing, Columbus, OH.

Kovacs, K., Splittstoesser, R., Maronitis, A., and Marras, W. S. (2002), "Grip Force and Muscle Activity Differences Due to Glove Type," *AIHA Journal (Fairfax, Va.)*, Vol. 63, pp. 269–274.

Kroemer, K. H. E. (1987), "Biomechanics of the Human Body," in *Handbook of Human Factors*, 2nd ed., G. Salvendy, Ed., Wiley, New York.

Luo, X., Pietrobon, R. S. X. S., Liu, G. G., and Hey, L. (2004), "Estimates and Patterns of Direct Health Care Expenditures Among Individuals with Back Pain in the United States," *Spine*, Vol. 29, pp. 79–86.

Marras, W. S., and Granata, K. P. (1995), "A Biomechanical Assessment and Model of Axial Twisting in the Thoracolumbar Spine," *Spine*, Vol. 20, pp. 1440–1451.

Marras, W. S., and Granata, K. P. (1997a), "The Development of an EMG-Assisted Model to Assess Spine Loading During Whole-Body Free-Dynamic Lifting," *Journal of Electromyography and Kinesiology*, Vol. 7, pp. 259–268.

Marras, W. S., and Granata, K. P. (1997b), "Spine Loading During Trunk Lateral Bending Motions," *Journal of Biomechanics*, Vol. 30, pp. 697–703.

Marras, W. S., and Mirka, G. A. (1993), "Electromyographic Studies of the Lumbar Trunk Musculature During the Generation of Low-Level Trunk Acceleration," *Journal of Orthopaedic Research*, Vol. 11, pp. 811–817.

Marras, W. S., and Reilly, C. H. (1988), "Networks of Internal Trunk-Loading Activities Under Controlled Trunk-Motion Conditions," *Spine*, Vol. 13, pp. 661–667.

Marras, W. S., and Schoenmarklin, R. W. (1993), "Wrist Motions in Industry," *Ergonomics*, Vol. 36, pp. 341–351.

Marras, W. S., and Sommerich, C. M. (1991a), "A Three-Dimensional Motion Model of Loads on the Lumbar Spine, I: Model Structure," *Human Factors*, Vol. 33, pp. 123–137.

Marras, W. S., and Sommerich, C. M. (1991b), "A Three-Dimensional Motion Model of Loads on the Lumbar Spine, II: Model Validation," *Human Factors*, Vol. 33, pp. 139–149.

Marras, W. S., Lavender, S. A., Leurgans, S. E., Rajulu, S. L., Allread, W. G., Fathallah, F. A., and Ferguson, S. A. (1993), "The Role of Dynamic Three-Dimensional Trunk Motion in Occupationally-Related Low Back Disorders: The Effects of Workplace Factors, Trunk Position, and Trunk Motion Characteristics on Risk of Injury," *Spine*, Vol. 18, pp. 617–628.

Marras, W. S., Lavender, S. A., Leurgans, S. E., Fathallah, F. A., Ferguson, S. A., Allread, W. G., and Rajulu, S. L. (1995). "Biomechanical Risk Factors for Occupationally Related Low Back Disorders," *Ergonomics*, Vol. 38, pp. 377–410.

Marras, W. S., Fine, L. J., Ferguson, S. A., and Waters, T. R. (1999), "The Effectiveness of Commonly Used Lifting Assessment Methods to Identify Industrial Jobs Associated with Elevated Risk of Low-Back Disorders," *Ergonomics*, Vol. 42, pp. 229–245.

Marras, W. S., Allread, W. G., Burr, D. L., and Fathallah, F. A. (2000), "Prospective Validation of a Low-Back Disorder Risk Model and Assessment of Ergonomic Interventions Associated with Manual Materials Handling Tasks," *Ergonomics*, Vol. 43, pp. 1866–1886.

Marras, W. S., Davis, K. G., and Splittstoesser, R. E. (2001), *Spine Loading During Whole Body Free Dynamic Lifting*, Ohio State University, Columbus, OH.

McGill, S. (2002), *Low Back Disorders: Evidence-Based Prevention and Rehabilitation*, Human Kinetics, Champaign, IL.

McGill, S. M. (1997), "The Biomechanics of Low Back Injury: Implications on Current Practice in Industry and the Clinic," *Journal of Biomechanics*, Vol. 30, pp. 465–475.

McGill, S. M., and Norman, R. W. (1985), "Dynamically and Statically Determined Low Back Moments During Lifting," *Journal of Biomechanics*, Vol. 18, pp. 877–885.

McGill, S. M., and Norman, R. W. (1986), "Partitioning of the L4–L5 Dynamic Moment into Disc, Ligamentous, and Muscular Components During Lifting," *Spine*, Vol. 11, pp. 666–678.

McGill, S. M., Hughson, R. L., and Parks, K. (2000), "Changes in Lumbar Lordosis Modify the Role of the Extensor Muscles," *Clinical Biomechanics (Bristol, Avon)*, Vol. 15, pp. 777–780.

Nachemson, A. (1975), "Towards a Better Understanding of Low-Back Pain: A Review of the Mechanics of the Lumbar Disc," *Rheumatology and Rehabilitation*, Vol. 14, pp. 129–143.

NIOSH (1981), *Work Practices Guide for Manual Lifting*, Publication 81-122, National Institute for Occupational Safety and Health, U.S. Department of Health and Human Services, Cincinnati, OH.

NIOSH (1994), *Applications Manual for the Revised NIOSH Lifting Equation*, Publication 94-110, National Institute for Occupational Safety and Health, U.S. Department of Health and Human Services, Cincinnati, OH.

Nordin, M., and Frankel, V. (1989), *Basic Biomechanics of the Musculoskeletal System*, 2nd ed., Lea & Febiger, Philadelphia.

NRC (1999), *Work-Related Musculoskeletal Disorders: Report, Workshop Summary, and Workshop Papers*, National Academy Press, Washington DC.

NRC (2001), *Musculoskeletal Disorders and the Workplace: Low Back and Upper Extremity*, National Academy Press, Washington DC.

Ozkaya, N., and Nordin, M. (1991), *Fundamentals of Biomechanics: Equilibrium, Motion, and Defermation*, Van Nostrand Reinhold, New York.

Park, K., and Chaffin, D. (1974), "A Biomechanical Evaluation of Two Methods of Manual Load Lifting," *AIIE Transactions*, Vol. 6, pp. 105–113.

Pope, M. H. (1993), "*Muybridge Lecture*," paper presented at the International Society of Biomechanics XIVth Congress, Paris, July 5.

Praemer, A., Furner, S., and Rice, D. P. (1992), "Musculoskeletal Conditions in the United States," in American Academy of Orthopaedic Surgeons, Park Ridge, IL.

Reilly, C. H., and Marras, W. S. (1989), "Simulift: A Simulation Model of Human Trunk Motion," *Spine*, Vol. 14, pp. 5–11.

Sanders, M. S., and McCormick, E. J. (1993), *Human Factors in Engineering and Design*, McGraw-Hill, New York.

Schoenmarklin, R. W., Marras, W. S., and Leurgans, S. E. (1994), "Industrial Wrist Motions and Incidence of Hand/Wrist Cumulative Trauma Disorders," *Ergonomics*, Vol. 37, pp. 1449–1459.

Schultz, A. B., and Andersson, G. B. (1981), "Analysis of Loads on the Lumbar Spine," *Spine*, Vol. 6, pp. 76–82.

Silverstein, M. A., Silverstein, B. A., and Franklin, G. M. (1996), "Evidence for Work-Related Musculoskeletal Disorders: A Scientific Counterargument," *Journal of Occupational and Environmental Medicine*, Vol. 38, pp. 477–484.

Silverstein, B. A., Stetson, D. S., Keyserling, W. M., and Fine, L. J. (1997), "Work-Related Musculoskeletal Disorders: Comparison of Data Sources for Surveillance," *American Journal of Industrial Medicine*, Vol. 31, pp. 600–608.

Solomonow, M. (2004), "Ligaments: A Source of Work-Related Musculoskeletal Disorders," *Journal of Electromyographic Kinesiology*, Vol. 14, pp. 49–60.

Solomonow, M., Zhou, B. H., Harris, M., Lu, Y., and Baratta, R. V. (1998). "The Ligamento-Muscular Stabilizing System of the Spine," *Spine*, Vol. 23, pp. 2552–2562.

Solomonow, M., Zhou, B. H., Baratta, R. V., Lu, Y., and Harris, M. (1999), "Biomechanics of Increased Exposure to Lumbar Injury Caused by Cyclic Loading, 1: Loss of Reflexive Muscular Stabilization," *Spine*, Vol. 24, pp. 2426–2434.

Solomonow, M., He Zhou, B., Baratta, R. V., Lu, Y., Zhu, M., and Harris, M. (2000), "Biexponential Recovery Model of Lumbar Viscoelastic Laxity and Reflexive Muscular Activity After Prolonged Cyclic Loading," *Clinical Biomechanics (Bristol, Avon)*, Vol. 15, pp. 167–175.

Solomonow, M., Zhou, B., Baratta, R. V., Zhu, M., and Lu, Y. (2002), "Neuromuscular Disorders Associated with Static Lumbar Flexion: A Feline Model," *Journal of Electromyographic Kinesiology*, Vol. 12, pp. 81–90.

Stubbs, M., Harris, M., Solomonow, M., Zhou, B., Lu, Y., and Baratta, R. V. (1998). "Ligamento-Muscular Protective Reflex in the Lumbar Spine of the Feline," *Journal of Electromyographic Kinesiology*, Vol. 8, pp. 197–204.

Thelen, D. G., Schultz, A. B., and Ashton-Miller, J. A. (1995), "Co-contraction of Lumbar Muscles During the Development of Time-Varying Triaxial Moments," *Journal of Orthopaedic Research*, Vol. 13, pp. 390–398.

van Dieen, J. H., Hoozemans, M. J., and Toussaint, H. M. (1999), "Stoop or Squat: A Review of Biomechanical Studies on Lifting Technique," *Clinical Biomechanics (Bristol, Avon)*, Vol. 14, pp. 685–696.

Videman, T., Nurminen, M., and Troup, J. D. (1990), "1990 Volvo Award in Clinical Sciences: Lumbar Spinal Pathology in Cadaveric Material in Relation to History of Back Pain, Occupation, and Physical Loading," *Spine*, Vol. 15, pp. 728–740.

Wang, J. L., Parnianpour, M., Shirazi-Adl, A., and Engin, A. E. (2000), "Viscoelastic Finite-Element Analysis of a Lumbar Motion Segment in Combined Compression and Sagittal Flexion: Effect of Loading Rate," *Spine*, Vol. 25, pp. 310–318.

Waters, T. R., Putz-Anderson, V., Garg, A., and Fine, L. J. (1993), "Revised NIOSH Equation for the Design and Evaluation of Manual Lifting Tasks," *Ergonomics*, Vol. 36, pp. 749–776.

PART 3
DESIGN OF TASKS AND JOBS

CHAPTER 14

TASK ANALYSIS: WHY, WHAT, AND HOW

Erik Hollnagel
University of Linköping
Linköping, Sweden

1 THE NEED TO KNOW

The purpose of task analysis is to describe tasks, and more particularly, to identify and characterize the fundamental characteristics of a specific activity or set of activities. According to the *Shorter Oxford Dictionary*, a *task* is "any piece of work that has to be done," which is generally taken to mean one or more functions or activities that must be carried out to achieve a specific goal. Task analysis can therefore be defined as the study of what an operator (or a team of operators) is required to do, in terms of actions and/or cognitive processes, to achieve a given goal.

Since a task is always a directed activity, in the sense that it has a purpose or an objective, there is little or no methodological merit in speaking simply of activities and tasks without taking into account both their goals and the context in which they occur. Task analysis should therefore be defined as the study of what a person or a team is required to do to achieve a specific goal—or, simply put, as *who* does *what* and *why*.

Who *Who* refers to the people who carry out a task. This is often a single person, and task analysis then describes individual work. It may, however, also be two or more people—a team or a group—in which case task analysis describes what the team does. Moving from the realm of human work to artifacts or agents (such as robots), task analysis becomes the analysis of functions (movements, etc.) that an artifact must carry out to achieve a goal. In industrialized societies, tasks are in most cases accomplished by one or more humans using some kind of technological artifact: in other words, a human–machine system. Task analysis is therefore often focused on the performance of the human–machine system: for instance, as task analysis for human–computer interaction (e.g., Diaper and Stanton, 2003). More generally, humans and machines working together can be described as *cognitive systems* or *joint cognitive systems* (Hollnagel and Woods, 2005). The built-in assumptions about the nature of who carries out the task have important consequences for task analysis, as will be clear from the following. The use of the pronoun *who* should not be understood to mean that task analysis is only about what humans do, although that was the original objective. In contemporary terms it would probably be more appropriate to refer to the *system* that carries out the task.

What *What* refers to the contents of the task and is usually described in terms of the activities that constitute the task. Task analysis started by focusing on physical tasks (i.e., manifest work), but has since the 1970s enlarged its scope to include cognitive or mental tasks. The contents of the task thus comprises a systematic description of the activities or functions that make up the task, either in terms of observable actions (e.g., grasping, holding, moving, assembling) or in terms of the usually unobservable functions that may lie behind these actions, commonly referred to as *cognition* or *cognitive functions*.

Why Finally, *why* refers to the purpose or goal of the task: for instance, as a specific system state or condition that is to be achieved. A goal may be something that is objective and physically measurable (a product) but also something that is subjective: for instance, a psychological state or objective, such as "having done a good job." The task analysis literature has usually eschewed the subjective and affective aspects of tasks and goals, although they clearly are essential for understanding human performance as well as for designing artifacts and work environments.

Task analysis is necessary because of a *practical* need to know in detail how things should be done or are done. When dealing with work, and more generally with how people use sociotechnical artifacts such as in human–machine interaction, it is necessary to know both what activities (functions) are required to accomplish a specified objective and how people habitually go about doing them, particularly since the latter is sometimes significantly different from the former. Such knowledge is necessary to design, implement, and operate human–machine systems, and task analysis looks specifically at how the interaction takes place and how it can be facilitated. Task analysis, however, has applications that go well beyond interface and interaction design and may be used to address issues such as training, performance assessment, event reporting and analysis, function allocation and automation, procedure writing, maintenance planning, risk assessment, staffing and job organization, personnel selection, and work management.

The term *task analysis* is commonly used as a generic label. A survey of task analysis methods shows that they represent many different meanings of the term (Kirwan and Ainsworth, 1992). A little closer inspection, however, reveals that they fall into a few main categories:

- The analysis and description of tasks or working situations that do not yet exist or are based on hypothetical events
- The description and analysis of observations of how work is carried out or of event reports (e.g., accident investigations)
- The representation of either of the above, in the sense of the notation used to capture the results (of interest due to the increasing use of computers to support task analysis)
- The various ways of further analysis or refinement of data about tasks (from either of the foregoing sources)
- The modes of presentation of results and the various ways of documenting the outcomes

Methods of task analysis should in principle be distinguished from methods of task description. A task description produces a generalized account or summary of activities as they have been carried out. It is based on empirical data or observations rather than on design data and specifications. A typical example is link analysis or even hierarchical task analysis (Annett et al., 1971). Properly speaking, task description or performance analysis deals with actions rather than with tasks. This distinction, by the way, is comparable to the French ergonomic tradition, where the task described is seen as different from the effective task. The latter usually is called the *activity*. The task described is thus what the organization assigns to the person or what the person should do, while the effective task or activity is the person's response to the prescribed task or what the person actually does. Understanding the task accordingly requires an answer to the question of *what* the person does, while understanding the activity requires an answer regarding *how* the person performs the task. An important difference between tasks and activities is that the latter are dynamic and may change depending on the circumstances, such as fluctuations in demands and resources, varying physical working conditions, the occurrence of unexpected events, and so on. The distinction between the task described and the effective task can be applied to both individual and collective tasks (Leplat, 1991).

1.1 Role of Task Analysis in Human Factors

Task analysis has over the years developed into a stable set of methods that constitute an essential part of human factors and ergonomics as applied disciplines. The focus of human factors engineering or ergonomics is humans at work, and more particularly the human use of technology in work, although it sometimes may look more like technology's use of humans. The aim of *human factors* (which in what follows is used as a common denominator for human factors engineering and ergonomics) is to apply knowledge about human behavior, abilities, and limitations to design tools, machines, tasks, and work environments to be as productive, safe, healthy, and effective as possible. From the beginning, ergonomics was defined broadly as the science of work (Jastrzebowski, 1857). At that time work was predominantly manual work, and tools were relatively few and simple. Human factors, which originally was called human factors engineering, came into existence around the mid-1940s as a way of solving problems brought on by emerging technologies such as computerization and automation. Whereas ergonomics and human factors thus may have had different perspectives in their original versions, they are now practically synonymous. At the present time, practically every type of work involves the use of technology, and the difference between ergonomics and human factors engineering is at present rather nominal.

Throughout most of history, technology in the form of artifacts has constituted a genuine tool for users, such as the painter's brush or the blacksmith's hammer. As long as users were artisans rather than workers, and as long as they worked individually rather than collectively, the need of prior planning, or of anything resembling task analysis, was limited or even nonexistent. The demand for task analysis arose when the use of technology became more widespread, especially when tools changed from being simple to being complex. More generally, the need of a formal task analysis arises when one or more of the following three conditions are met:

1. The accomplishment of a goal requires more effort than one person can provide or a combination of skills that go beyond what one person can be expected to master. In such cases task analysis serves to break down a complex and collective activity to descriptions

of a number of simpler and more elementary activities. For example, building a ship, in contrast to building a dinghy, requires the collaboration of many individuals and the coordination of many different types of work. In such cases people have to collaborate and must therefore adjust their own work to match the progress and demands of others. Task analysis is needed to identify the task components that correspond to what a person can achieve or provide over a reasonable period of time, as well as to provide a way of organizing or integrating the components to an overall whole.

2. Tasks become so complex that one person can no longer control or comprehend them. This may happen when the task becomes so large or takes so long that a single person is unable to complete it, (i.e., the transition from individual to collective tasks). It may happen when the task requires the use of technological artifacts where the use of the artifact becomes a task in its own right. This is often the case when the artifacts themselves become so complex that they involve independent or semiautonomous functions (i.e., that they begin to regulate themselves partially rather than passively carrying out an explicit function under the user's control).

3. A similar argument goes when technology itself—machines—become so complex that the situation changes from simply being one of *using* the technology, to learning how to *master* or *control* the technology. In other words, being in control of the technology becomes a goal in itself, as a means to achieve the original goal. Examples are driving a car in contrast to riding an ordinary bicycle, using a food processor instead of a knife, using a computer (as in writing this chapter) rather than paper and pencil, and so on. In these cases, and of course also in cases of far more complex work, use of the technology is no longer straightforward but requires preparation and prior thought either by the person who does the task or work, or by those who prepare tasks or tools for others. Task analysis can in these cases be used to describe situations where the task itself is very complex because it involves interaction and dependencies with other people. It can similarly be used to describe situations where use of the technology is no longer straightforward, but requires mastery of the system to such a degree that not everyone can apply it directly as intended and designed.

In summary, task analysis became necessary when work changed from something that could be done by an unaided individual to becoming something requiring the collective efforts either of people or of people and machines. Although collective work has existed since the beginning of history, its presence became more conspicuous after the Industrial Revolution. In this new situation the work of the individual became a mere part of the work of the collective, and individual control of work was consequently lost. The worker became part of a larger context, a cog in complex social machinery that defined the demands and constraints to work. One important effect of that was that people no longer could work at a pace suitable for them and pause whenever needed, but instead, had to comply with the pace set by others—and increasingly, by the pace set by machines.

1.2 Artifacts and Tools

It is common to talk about humans and machines, or humans and technology, and to use expressions such as *human–machine systems*. In the context of tasks and task analysis, the term *technological artifact*, or simply *artifact*, will be used to denote that which is being applied to achieve a goal. Although it is common to see computers and information technology as primary constituents of the work environment, it should be remembered that not all tools are or include computers, and that task analysis is therefore more than human–computer interaction. That something is an artifact means that it has been constructed or designed by someone, hence that it expresses or embodies a specific intention or purpose. In contrast to that, a natural object does not have an intended use, but is the outcome of evolution—or happenstance—rather than design. Examples of natural objects are stones used to hammer or break something, sticks used to poke for something, and so on.

A natural object may be seen as being instrumental to doing something, hence used for that purpose. In the terminology of Gibson (1979), the natural object is perceived as having an actionable property, which means that it is seen as being useful for a specific purpose. An artifact is designed with a specific purpose (or set of uses) in mind, and should ideally offer a similar perceived affordance. To the extent that this is the case, the artifact has been designed successfully. Task analysis (i.e., describing and understanding in advance the uses of artifacts) is obviously one of the ways in which that can be achieved. Although such success may not always be the case for a designed piece of technology or an artifact, it is of course in the best interest of the designer—and the producer—to ensure that this is so.

When a person designs or constructs something, an artifact or a complex task, for himself or herself there is no need to ask either what the person is capable of or what the artifact should be used for or how it should be used. This is the case for a single but complex artifact, where the use requires a series of coordinated or ordered actions. It is also the case for more complex, organized work processes, where the activities or tasks of an individual must fit into a larger whole. Indeed, just as the designer of a complex artifact considers its components and how they must work together for the artifact to be able to provide its function, so must the work process designer consider the human components and how they must work together to deliver the desired end product or result. It is, indeed, no coincidence that the first task analyses were for organized work processes rather than for the use of artifacts or machines.

The person's knowledge of what he or she can do corresponds to the designer's assumptions about the user, while the person's knowledge of how the artifact should be used corresponds to the user's assumptions about the artifact and the designer's intentions. Yet as the artifact or the work processes are built around the person, there is little need to make any of these assumptions explicit or indeed to be very formal about them. The user and the designer are effectively the same person. There is also little need of prior thought or prior analysis since the development is an integral part of work rather than a separate activity. But when the artifact is designed by one person to be used by someone else, he or she needs to be very careful and explicit in making the assumptions and to consider what the future user may be able to do and will do. In other words, it is necessary in these cases to analyze how the artifact shall be used or to perform a task analysis.

2 TASK TYPES AND COMPONENTS

Task analysis is in the main a collection of methods, which describe (or prescribe) how the analysis shall be performed, preferably by describing each step of the analysis as well as how they are organized. Each method should also describe the *stop rule* or criterion (i.e., to define the principles to determine when the analysis has come to an end when the level of elementary tasks has been reached).

An important part of the method is to name and identify the main constituents of a task and how they are organized. As described later in this chapter, task analysis has through its development embraced several different principles of task organization, of which the main ones are the sequential principle, the hierarchical principle, and the functional dependency principle. To do so, the method must obviously refer to a classification scheme or set of categories that can be used to describe and represent the essential aspects of a task. The hallmark of a good method is that the classification scheme is applied consistently and uniformly, thereby limiting the opportunities for subjective interpretations and variations. Task analysis should depend not on personal experience and skills but on generalized public knowledge and common sense. A method is also important as a way of documenting how the analysis has been done and of describing the knowledge that was used to achieve the results. It helps to ensure that the analysis is done in a systematic fashion so that, if needed, it can be repeated with the same results. This reduces the variability between analysts, hence improves the reliability.

The outcome of the task analysis accounts for the organization or structure of constituent tasks. A critical issue is the identification or determination of elementary activities or task components. The task analysis serves among other things to explain how something should be done for a user who does not know or remember (in the situation). There is therefore no need to describe things or tasks that the user does know. The problem, is, however, where that level is.

Task analysis has from the very start tried to demarcate the basic components. It would clearly be useful if it was possible to find a set of basic tasks—or activity atoms—that could be applied in all contexts. This is akin to finding a set of elementary processes or functions from which complex behavior can be built. Such endeavors are widespread in the behavioral and cognitive sciences, although the success rate usually is limited. The main reason for that is that the level of an elementary task depends on the person as well as on the domain. Even if a common denominator could be found, it would probably be at a level of detail such that it had little practical value (e.g., for training or scheduling).

2.1 Changing Views of Elementary Tasks

Although the search for all-purpose task components is bound to fail, it is nevertheless instructive to take a brief look at the attempts of doing so throughout the short history of task analysis. Probably the first—and probably also the most ambitious—attempt was made by Frank Bunker Gilbreth, who was one of the pioneers of task analysis. The categorization, first reported about 1919, evolved from the observation by trained motion-and-time specialists of human movement, specifically of the fundamental motions of the hands of a worker. Gilbreth found that it was possible to distinguish among the following 17 types of motion: search, select, grasp, reach, move, hold, release, position, preposition, inspect, assemble, disassemble, use, unavoidable delay, wait (avoidable delay), plan, and rest (to overcome fatigue). (The basic motions are know as *therbligs*, using an anagram of the developer's name.)

A more contemporary version is a list of typical process control tasks suggested by Rouse (1981). This comprises 11 functions, which are in alphabetical order: communicating, coordinating tasks, executing procedures, maintaining, planning, problem solving, recognizing, recording, regulating, scanning, and steering. In contrast to the therbligs, it is possible to organize these functions in several ways: for instance, in relation to an input–output model of information processing, in relation to a control model, in relation to a decision-making model, and so on. The functions proposed by Rouse are characteristically on a higher level of abstraction than the therbligs and refer to cognitive functions, or cognitive tasks, rather than to physical movements.

A final example is the GOMS model proposed by Card et al. (1983). The purpose of GOMS, which is an acronym that stands for "goals, operators, methods, and selection rules," was to provide a system for modeling and describing human task performance. Of the four components of GOMS, operators denote the set of atomic-level operations with which a user can compose a solution to a goal, while methods represent sequences of operators grouped together to accomplish a single goal. For example, the manual operators of GOMS are: Keystroke key_name, Type_in string_of_characters, Click mouse_button, Double_click mouse_button, Hold_down mouse button, Release mouse_button, Point_to target_object, and Home_to destination. These operators all refer to simple motions but differ from the *therbligs* in the sense that they are only meaningful relative to a mental rather than a physical task.

The definition of elementary tasks in scientific management could comfortably refer to what people did, hence to what could be reported by independent observers. The problem with defining elementary cognitive or mental tasks is that no such independent verification is possible. Although GOMS was successful in defining elementary tasks on the keystroke level, it was more difficult to do the same for the cognitive or mental aspects (i.e., the methods and selection rules). The physical reality of elementary tasks such as *grasp, reach, move*, and *hold* has no parallel when it comes to cognitive functions. The problems in identifying elementary mental tasks are not due to a lack of trying. This has indeed been a favorite topic of psychology from Donders (1969; orig. 1868) to Simon (1972). The problems come about because the "smallest" unit is defined by the theory being used rather than by intersubjective reality. In practice, this means that elementary tasks must be defined relative to the domain at the time of the analysis (i.e., in terms of the context rather than as absolutes or context-free components).

3 BRIEF HISTORY OF TASK ANALYSIS

Task analysis has a relatively short history, starting around the beginning of the twentieth century. The first major publications were Gilbreth (1911) and Taylor (1911), which introduced the principles of scientific management. The developments that followed reflected both the changing view of human nature, for instance in McGregor's Theory X and Theory Y (McGregor, 1960), and the changes in psychological schools, specifically the models of the human mind. Of the three examples of a classification system mentioned above, Gilbreth (1911) represents the scientific management view, Rouse (1981) represents the supervisory control view (human–machine interaction), and Card et al. (1983) represent the information-processing view (human–computer interaction). These views can be seen as alternative ways of describing the same reality: namely, human work and human activities. One standpoint is that human nature has not changed significantly for thousands of years and that different descriptions of the human mind and of work therefore only represent changes in the available models and concepts. Although this undoubtedly is true, it is also a fact that the nature of work has changed due to developments in technology. Gilbreth's description in terms of physical movements would therefore be as inapplicable to today's work as a description of cognitive functions would have been in 1911.

3.1 Sequential Task Analysis

The dawn of task analysis is usually linked to the proposal of a system of scientific management (Taylor, 1911). This approach was based on the notion that tasks should be specified and designed in minute detail and that workers should receive precise instructions about how their tasks should be carried out. To do so, it was necessary that tasks could be analyzed unequivocally or "scientifically," if possible in quantitative terms, so that it could be determined how each task step should be done in the most efficient way and how the task steps should be distributed among the people involved.

One of the classical studies is Taylor's (1911) analysis of the handling of pig iron, where the work was done by men with no "tools" other than their hands. A pig-iron handler would stoop down, pick up a pig weighing about 92 pounds, walk up an inclined plank, and dropped it on the end of a railroad car. Taylor and his associates found that a gang of pig-iron handlers was loading on the average about $12\frac{1}{2}$ long tons per man per day. The aim of the study was to find ways in which to raise this output to 47 tons a day, not by making the men work harder but by reducing the number of unnecessary movements. This was achieved both by careful motion-and-time studies, and by a system of incentives that would benefit workers as well as management.

Scientific management was based on four elements or principles, which were used in studies of work.

1. The development of the science of work with *rigid* rules for each motion of every person, and the perfection and *standardization* of all implements and working conditions.
2. The careful *selection* and subsequent *training* of workers into first-class people, and the elimination of all people who refuse to or are unable to adopt the best methods.
3. Bringing the first-class workers and the science of working together, through the constant help and watchfulness of the management, and through paying each person a large daily bonus for working fast and doing what he is told to do.
4. An almost equal division of the work and responsibility between workers and management.

Of these four elements, the first (the development of the science of work) is the most interesting and spectacular. It was essentially an analysis of a task into its components, using, for example, the list of *therbligs* mentioned above. In the case of manual work this was entirely feasible, since the task could be described as a single sequence of more detailed actions or motions. The motion-and-time study method was, however, unable to cope with the growing complexity of tasks that followed developments in electronics, control theory, and computing during the 1940s and 1950s. Due to the increasing capabilities of machines, people were asked—and tasked—to engage in multiple activities at the same time, either because individual tasks became more complex or because simpler tasks were combined into larger units. An important consequence of this was that tasks changed from being a sequence of activities referring to a single goal to become an organized set of activities referring to a hierarchy of goals. The use of machines and technology also became more prevalent, so that simple manual work such as pig-iron handling was taken over by machines, which in turn were operated or controlled by workers.

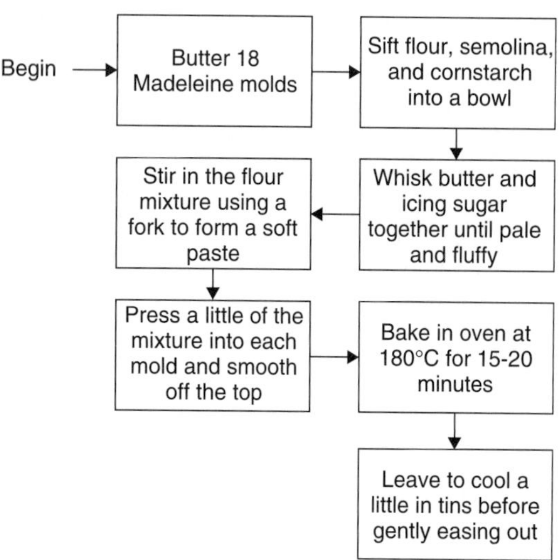

Figure 1 Sequential task description.

Since the use of technology has made work environments more complex, relatively few tasks today are sequential tasks. Examples of sequential tasks are therefore most easily found in the world of cooking. Recipes are typically short and describe the steps as a simple sequence of actions, although novice cooks sometimes find that recipes are underspecified. As an example of a sequential task analysis, Figure 1 shows the process for baking Madeleines.

3.2 From Sequential to Hierarchical Task Organization

The technological development meant that the nature of work changed from being predominantly manual to becoming more dependent on mental capabilities (comprehension, monitoring, planning). After awhile, human factors engineering, or classical ergonomics, recognized that traditional methods of breaking the task down into small pieces, which each could be performed by a person, were no longer adequate. Since the nature of work had changed, the human capacity for processing information became decisive for the capacity of the human–machine system. This capacity could not be extended beyond its natural upper limit, and it soon became clear that the human capacity for learning and adaptation was insufficient to meet technological demands.

To capture the more complex task organization, Miller (1953) developed a method for human–machine task analysis in which main task functions could be decomposed into subtasks. Each subtask could then be described in detail, for instance by focusing on information display requirements and control actions. This led to the following relatively simple and informal procedure for task analysis:

1. Specify the human–machine system criterion output.
2. Determine the system functions.
3. Trace each system function to the machine input or control established for the operator to activate.
4. For each respective function, determine what information is displayed by the machine to the operator whereby he or she is directed to appropriate control activation (or monitoring) for that function.
5. Determine what indications of response adequacy in the control of each function will be fed back to the operator.
6. Determine what information will be available and necessary to the operator from the human–machine "environment."
7. Determine what functions of the system must be modulated by the operator at or about the same time, or in close sequence, or in cycles.
8. In reviewing the analysis, be sure that each stimulus is linked to a response and that each response is linked to a stimulus.

The behavior groups associated with the combinations of functions that the operator should carry out were the tasks. These were labeled according to the subpurpose they fulfilled within the system. Point 8 reflects the then-current psychological thinking, which was that of stimulus and response couplings. The operator was, in other words, seen as a transducer or a machine that was coupled to the "real" machine. For the human–machine system to work, it was necessary that the operator interpreted the machine's output in the proper way and that he or she responded with the correct input. The purpose of task analysis was to determine what the operator had to do to enable the machine to function as efficiently as possible.

3.2.1 Task–Subtask Relation

The task–subtask decomposition was a significant change from sequential task analysis and was necessitated by the growing complexity of work. The development was undoubtedly influenced by the emerging practice—and later science—of computer programming, where one of the major innovations was the subroutine. Arguably the most famous example of a task–subtask relation is the TOTE (test–operate–test–exit), which was proposed as a building block of human behavior (Miller et al., 1960). This introduced into the psychological vocabulary the concept of a plan, which is logically necessary to organize combinations of tasks and subtasks. Whereas a subroutine can be composed of motions and physical actions, hence in principle can be found even in scientific management, a plan is obviously a cognitive or mental component. The very introduction of the task–subtask relation, and of plans, therefore changed task analysis from describing only what happened in the physical world to describing what happened in the minds of the people who carried out the work.

Miller's task–subtask analysis method clearly implied the existence of a hierarchy of tasks and subtasks, although this was never a prominent feature of the method. As the technological environments developed further, the organization of tasks and subtasks became increasingly important for task analysis, culminating by the development of hierarchical task analysis (HTA) (Annett and Duncan, 1967; Annett et al., 1971). HTA has since its introduction in practice become the standard method for task analysis and task description and is widely used in a variety of contexts, including interface design.

The process of HTA is to decompose tasks into subtasks and to repeat this process until a level of elementary tasks has been reached. Each subtask or operation is specified by its goal, the conditions under which the goal becomes relevant or "active," the actions required to attain the goal, and the criteria that mark the attainment of the goal. The relationship between a set of subtasks and the superordinate task is governed by plans expressed as, for instance, procedures, selection rules, or time-sharing principles. A simple example of a HTA is a description of how to get money from a bank account using an ATM (see Figure 2). In this description, there is an upper level of tasks (marked, 1, 2, 3), which describe the order of the main segments, and a lower level of subtasks (marked 1.1, 1.2, etc.), which provide the details. It is clearly possible to break down each of the subtasks into further detail, for instance by describing the steps comprised by *1.2 Enter PIN code*. This raises the question of when the HTA should stop (i.e., what the elementary subtasks or task components are; cf. below).

The overall aim of HTA is to describe a task in sufficient detail, where the level of resolution required depends on the specific purposes (e.g., interaction design, training requirements, interface design, risk analysis, etc.). HTA can be seen as a systematic search strategy, which is adaptable for use in a variety of different contexts and purposes within the field of human factors (Shepherd, 1998). In practice, performing a HTA comprises the following steps. (Note, by the way, that this is a sequential description of hierarchical task analysis!)

1. Decide the purpose of the analysis.
2. Get agreement between stakeholders on the definition of task goals and criterion measures.
3. Identify sources of task information and select means of data acquisition.
4. Acquire data and draft a decomposition table or diagram.
5. Recheck the validity of the decomposition with the stakeholders.
6. Identify significant operations in light of the purpose of the analysis.
7. Generate and, if possible, test hypotheses concerning factors affecting learning and performance.

Whereas Miller's description of human–machine task analysis concentrated on how to analyze the required interactions between human and machine, the HTA extends the scope to consider the context of the analysis, in particular which purpose it serves. Although this is a welcome and weighty development, it leaves the actual HTA somewhat underspecified. Indeed, in the description above it is only the fourth step that is the actual task analysis.

3.2.2 Tasks and Cognitive Tasks

In addition to the change that led from sequential motion-and-time descriptions to hierarchical task organization, a further change occurred in the late 1980s to emphasize the cognitive nature of tasks. The need to consider the organization of tasks was partly a consequence of changing from a sequential to a hierarchical description, as argued above. The changes in the nature of work also meant that "thinking" tasks became more important than "doing." The need to understand the cognitive activities of the human–machine system, first identified by Hollnagel and Woods (1983), soon developed a widespread interest in cognitive task analysis, defined as the extension of traditional task analysis techniques to yield information about the knowledge, thought processes, and goal structures that underlie observable task performance (e.g., Schraagen et al., 2000). As such, it represents a change in

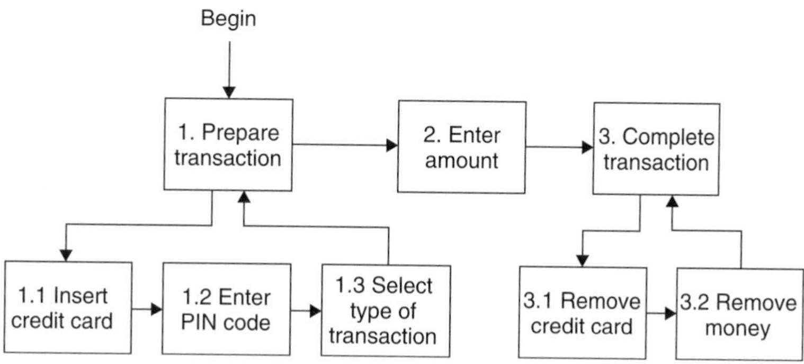

Figure 2 Hierarchical task description.

emphasis from overt to covert activities. An example is that of the task analysis principles described by Miller et al., (1960), which refer to mental actions as much as to motor behavior. Since many tasks require a considerable amount of mental functions and effort, in particular in retrieving and understanding the information available and in planning and preparing what to do (including monitoring of what happens), much of what is essential for successful performance is covert. Whereas classical task analysis relies very much on observable actions or activities, the need to find out what goes on in other peoples' minds requires other approaches.

3.2.3 Elementary Task

All task analysis methods require an answer to what the elementary task is. As long as task analysis was occupied mainly with physical work, the question could be resolved in a pragmatic manner. But when task analysis changed to include the cognitive aspects of work, the answer became more contentious. This is obvious from the simple example of a HTA shown in Figure 2. For a person living in a developed or industrialized society, entering a PIN code can be assumed to be an elementary task. Yet either a motion-and-time study or a GOMS-type interaction analysis could break this down into more detailed elements. The determination of what an elementary task is clearly cannot be done separately from assumptions about who the users are, what the conditions of use (or work) are, and what the purpose of the task analysis is. If the purpose is to develop a procedure or a set of instructions such as the instructions that appear on the screen of an ATM, there may be no need to go further than "enter PIN code" or possibly "enter PIN code and press ACCEPT." Given the population of users, it is reasonable for the system designer to take for granted that they will know how to do this. If, however, the purpose is to design the physical interface itself or to perform a risk analysis, it will be necessary to continue the analysis at least one more step. GOMS is a good example of this, as would be the development of instructions for a robot to use an ATM.

In the contexts of work, assumptions about elementary task can be satisfied by ensuring that users have the requisite skills (e.g., through training and instruction). A task analysis may indeed be performed with the explicit purpose of defining training requirements. Designers can therefore, in a sense, afford the luxury of dictating what an elementary task is, as long as the requirements can be fulfilled by training. In the context of artifacts with a more widespread use, typically in the public service domain, greater care must be taken in making assumptions about an elementary task, since users in these situations often are "accidental" (Marsden and Hollnagel, 1996).

3.3 Functional Dependency and Goals–Means Task Analysis

Both sequential and hierarchical task analysis are structural in the sense that they describe the order in which the prescribed activities are to be carried out. A hierarchy is by definition the description of how something is ordered, and the very representation of a hierarchy (as in Figure 2) emphasizes the structure. As an alternative, it is possible to analyze and describe tasks from a functional point of view (i.e., in terms of how tasks relate to or depend on each other). This changes the emphasis from how tasks and activities are ordered, to what the tasks and activities are supposed to achieve.

Whereas task analysis in practice stems from the beginning of the twentieth century, the principle of functional decomposition can be traced back at least to Aristotle (Book III of the *Nicomachean Ethics*). This is not really surprising, since the focus of a functional task analysis is the reasoning about tasks rather than the way in which they are carried out (i.e., the physical performance). Whereas the physical nature of tasks has changed throughout history, and especially after the start of the Industrial Revolution, thinking about how to do things is largely independent on how things are actually done.

In relation to task analysis, functional dependency means thinking about tasks in terms of goals and means. The strength of a goals–means, or means–ends, decomposition principle is that it is ubiquitous, important, and powerful (Miller et al., 1960, p. 189). It has therefore been used widely most famously as the basis for the general problem solver (Newell and Simon, 1961).

The starting point of a functional task analysis is a goal or an end, defined as a specified condition or state of the system. A description of the goal usually includes or implies the criteria of achievement (i.e., the conditions that determine when the goal has been reached). To achieve the goal, certain means are required. These are typically one or more activities that need to be carried out (i.e., a task). Yet most tasks are possible only if specific conditions are fulfilled. For instance, you can work on your laptop only if you have access to an external power source or if the batteries are charged sufficiently. When these conditions are met, the task can be carried out. If not, bringing about these preconditions becomes a new goal, denoted a subgoal. In this way goals are decomposed recursively, thereby defining a set of goal–subgoal dependencies that also serves to structure or organize the associated tasks.

An illustration of that are the functions or tasks needed to start up an industrial boiler shown in Figure 3 (see Lind and Larsen, 1995). The diagram illustrates how the top goal, "St1 established," requires that a number of conditions have been established, where each of these in turn can be described as subgoals. Although the overall structure is a hierarchical ordering of goals and means, it differs from a HTA because the components of the diagram are goals rather than tasks. The goals–means decomposition can be used as a basis for identifying the tasks that are necessary to start the boiler, but this may not necessarily fit into the same representation.

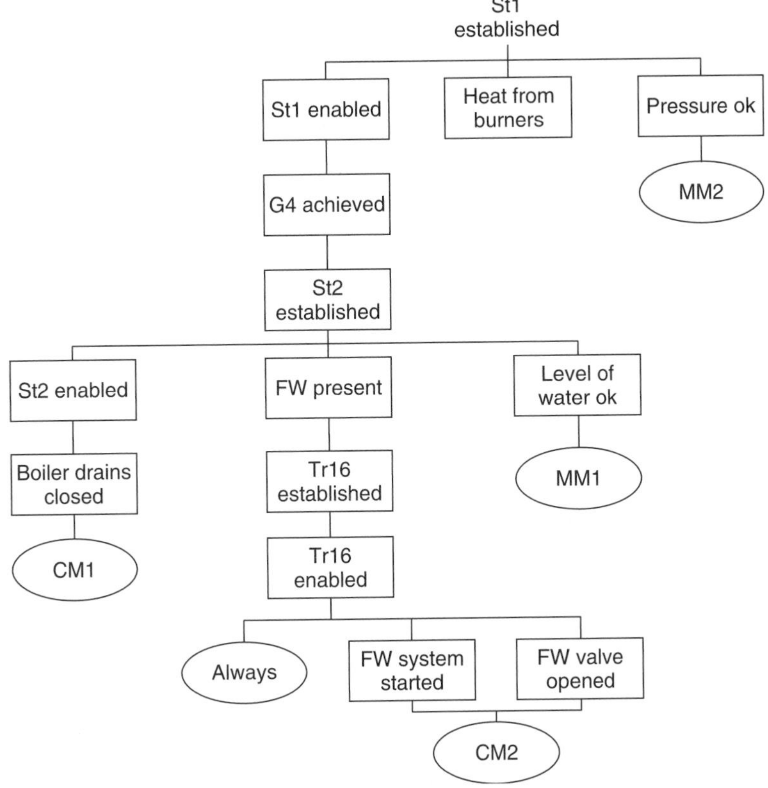

Figure 3 Goals–means task description.

4 PRACTICE OF TASK ANALYSIS

As already mentioned, task analysis can be used for a variety of purposes. Although the direct interaction between humans and computers got the lion's share of attention in the 1990s, task analysis is necessary for practically any aspect of a human–machine system's functioning. Task analysis textbooks, such as Kirwan and Ainsworth (1992), provide detailed information and excellent descriptions of the many varieties of task analysis. More recent works, such as Hollnagel (2003), extend the scope from task analysis to task design, emphasizing the constructive use of task knowledge. Regardless of which method an investigator decides to use, there are a number of general aspects that deserve consideration.

4.1 Task Data Collection Techniques

The first challenge in task analysis is to know where relevant data can be found and to collect them. The behavioral sciences have developed many ways of doing this, such as activity sampling, critical incident technique, field observations, questionnaire, structured interview, and verbal protocols. In many cases, data collection can be supported by various technologies, such as audio and video recording, measurements of movements, and so on, although the ease of mechanical data collection often is offset by the efforts needed to analyze the data.

As task analysis extended its scope from physical work to include cognitive functions, methods were needed to get data about the unobservable parts of a task. The main techniques used to overcome this were "think-aloud" protocols and introspection (i.e., extrapolating from one's own experience to what others may do). The issue of thinking aloud has been hotly debated, as has the issue of introspection (Nisbett and Wilson, 1977). Other structured techniques rely on controlled tasks, questionnaires, and so on. Yet in the end the problem is that of making inferences from some set of observable data to what goes on behind. This raises interesting issues of methods for data collection to support task analysis and leads to an increasing reliance on models of the tasks. As long as task analysis is based on observation of actions or performance, it is possible to establish some kind of objectivity or intersubjective agreement or verification. As more and more of the data refer to the unobservable, the dependence on interpretations, hence on models, increase.

4.2 Task Description Techniques

When the data have been collected, the next challenge is to represent them in a suitable fashion. It is important for a task analysis that it can represent the information about the task in a manner that can easily be comprehended. For some purposes, the outcome of a task analysis may simply be rendered as a written description of the tasks and how they are organized. In most cases this is supplemented by some kind of graphical representation or diagram, since this makes it considerably easier to grasp the overall relations. Examples of that are the diagrams shown in Figures 1 to 3. Other staple solutions are charting and networking techniques, decomposition methods, HTA, link analysis, operational sequence diagrams (OSD), and timeline analyses.

4.3 Task Simulation Methods

For a number of other purposes, specifically those that have to do with design, it is useful if the task can be represented in other ways, specifically as some kind of description or model that can be manipulated. The benefit is clearly that putative changes to the task can be implemented in the model and the consequences explored. This has led to the development of a range of methods that rely on some kind of symbolic model of the task or activity, going from the production rule systems to task networks (e.g., Petri nets). This development often goes hand in hand with user models (i.e., symbolic representation of users that can be used to simulate responses to what happens in the work environment). In principle, such models can carry out the task as specified by the task description, but the strength of the results depends critically on the validity of the model assumptions. Other solutions, which do not require the use of computers, are mock-ups, walk-throughs, and talk-throughs.

4.4 Task Behavior Assessment Methods

Task analyses are in many cases used as a starting point to look at a specific aspect of the task execution, usually risk or consequences for system safety. One specific type of assessment looks at the possibility for humans to carry out a task incorrectly (i.e., the issue of human reliability). Approaches to human reliability analysis that are based on structural task descriptions are generally oversimplified not only because humans are not machines but also because there is an essential difference between described and effective tasks, or between tasks and activities. Task descriptions, in the form of event trees or as procedural prototype models, represent an idealized sequence or hierarchy of steps. Tasks as they are carried out or as they are perceived by the person are more often series of activities whose scope and sequence are adjusted to meet the demands—perceived or real—of the current situation. It can be argued that task descriptions used for risk and reliability analyses on the whole are inadequate and unable to capture the real nature of human work. The decomposition principle has encouraged—or even enforced—a specific form of task description (the event tree), and this formalism has been self-sustaining. It has, however, led human reliability analysis into a cul-de-sac.

4.5 Future of Task Analysis

We started this chapter by pointing out that task analysis is the study of *who* does *what* and *why*, where the *who* should be broadened to include individual work, collective work, and joint cognitive systems. The future of task analysis is bright in the sense that there will always be a practical need to know how things should be done. The question is whether task analysis as it is currently practiced is capable of meeting this need in the long run. There are several reasons why the reply need not be unequivocally positive.

1. Task analysis has from the beginning been concerned mostly with individuals, whether as single workers or single users, despite the fact that most work involves multiple users (collaboration, distributed work) in complex systems. Although the importance of distributed cognition and collective work is generally acknowledged, only few methods are capable of analyzing that, over and above representing explicit interactions such as in link analysis and operational sequence diagrams (OSDs).
2. Many task analysis methods are adequate for describing single lines of activity. Unfortunately, most work involves multiple threads and timelines. Although HTA represents a hierarchy of tasks, each subtask or activity is carried out on its own. There is little possibility of describing two or more simultaneous tasks, even though that is often what people have to cope with in reality. Another shortcoming is the difficulty of representing temporal relations other than simple durations of activities.
3. There is a significant difference between described and effective tasks. Work in practice is characterized by ongoing adaptations and improvizations rather than the straightforward carrying out of a procedure or an instruction. The reasons for this are that demands and resources rarely correspond to what was anticipated when the task was developed, and that the actual situation may differ considerably from that which is assumed by the task description, thereby rendering the latter unworkable.

The problem in a nutshell is that task analysis was developed to deal with linear work environments, where effects were proportional to causes and where orderliness and regularity on the whole could be assured. For reasons that this chapter cannot go into, work environments have changed to become nonlinear, in the sense that it is practically impossible to make reliable predictions of what the conditions will be and how events will develop. In consequence of that, work has become a coping with complexity that defies the capability of traditional task analysis. Effective tasks are often so different from described tasks that the analyses have little practical impact. One solution would be to find ways of constraining the complexity of work environments to match the analytical power of existing methods. Since this is neither very probable or desirable, the alternative is to reconsider the meaning of task analysis and to develop methods that are powerful enough to match the sociotechnical reality in which we find ourselves today.

REFERENCES

Annett, J., and Duncan, K. D. (1967), "Task Analysis and Training Design," *Occupational Psychology*, Vol. 41, pp. 211–221.

Annett, J., Duncan, K. D., Stammers, R. B., and Gray, M. J. (1971), "Task Analysis," Training Information Paper 6, HMSO, London.

Card, S., Moran, T., and Newell, A. (1983), *The Psychology of Human–Computer Interaction*, Lawrence Erlbaum Associates, Mahwah, NJ.

Diaper, D., and Stanton, N., Eds. (2003), *The Handbook of Task Analysis for Human–Computer Interaction*, Lawrence Erlbaum Associates, Mahwah, NJ.

Donders, F. C. (1969), "On the Speed of Mental Processes," *Acta Psychologica*, Vol. 30, pp. 412–431; translated from "Over de snelheid van psychische processen: Onderzoekingen gedaan in het Physiologisch Laboratorium der Utrechtsche Hoogeschool, 1868–1869," *Tweede Reeks*, Vol. II, pp. 92–120.

Gibson, J. J. (1979), *The Ecological Approach to Visual Perception*, Lawrence Erlbaum Associates, Mahwah, NJ.

Gilbreth, F. B. (1911), *Motion Study*, Van Nostrand, Princeton, NJ.

Hollnagel, E., Ed. (2003), *Handbook of Cognitive Task Design*, Lawrence Erlbaum Associates, Mahwah, NJ.

Hollnagel, E., and Woods, D. D. (1983), "Cognitive Systems Engineering: New Wine in New Bottles," *International Journal of Man–Machine Studies*, Vol. 18, pp. 583–600.

Hollnagel, E., and Woods, D. D. (2005), *Joint Cognitive Systems: Foundations of Cognitive Systems Engineering*, CRC Press, Boca Raton, FL.

Jastrzebowski, W. (1857), "Rys ergonomiji czyli Nauki o Pracy, opartej naprawdach poczerpnietych z Nauki Przyrody" [An Outline of Ergonomics or the Science of Work Based on the Truths Drawn from the Science of Nature], *Przyoda i Przemysl*, Vol. 29, pp. 227–231.

Kirwan, B., and Ainsworth, L. K., Eds. (1992), *A Guide to Task Analysis*, Taylor & Francis, London.

Leplat, J. (1991), "Organization of Activity in Collective Tasks," in *Distributed Decision Making: Cognitive Models for Cooperative Work*, J. Rasmussen, B. Brehmer, and J. Leplat, Eds., Wiley, Chichester, West Sussex, England.

Lind,M., and Larsen, M. N. (1995), "Planning Support and the Intentionality of Dynamic Environments," in *Expertise and Technology: Cognition and Human–Computer Interaction*, J. M. Hoc, P. C. Cacciabue, and E. Hollnagel, Eds.,Lawrence Erlbaum Associates, Mahwah, NJ.

Marsden, P., and Hollnagel, E. (1996), "Human Interaction with Technology: The Accidental User," *Acta Psychologica*, Vol. 91, pp. 345–358.

McGregor, D. (1960), *The Human Side of Enterprise*, McGraw-Hill, New York.

Miller, G. A., Galanter, E., and Pribram, K. H. (1960), *Plans and the Structure of Behavior*, Holt, Rinehart & Winston, New York.

Miller, R. B. (1953), "A Method for Man–Machine Task Analysis," Technical Report 53–137, Wright AF Development Center, Dayton, OH.

Newell, A., and Simon, H. A. (1961), "GPS: A Program That Simulates Human Problem-Solving," in *Proceedings of a Conference on Learning Automata*, Technische Hochschule, Karlsruhe, Germany, April 11–14.

Nisbett, R. E., and Wilson, T. D. (1977), "Telling More Than We Can Know: Verbal Reports on Processes," *Psychological Review*, Vol. 74, pp. 231–259.

Rouse, W. B. (1981), "Human–Computer Interaction in the Control of Dynamic Systems," *ACM Computing Survey*, Vol. 13, No. 1, pp. 71–99.

Schraagen, J. M., Chipman, S. F., and Shalin, V. L., Eds. (2000), *Cognitive Task Analysis*, Lawrence Erlbaum Associates, Mahwah, NJ.

Shepherd, A. (1998), "HTA as a Framework for Task Analysis," *Ergonomics*, Vol. 41, No. 11, pp. 1537–1552.

Simon, H. A. (1972), *The Sciences of the Artificial*, MIT Press, Cambridge, MA.

Taylor, F. W. (1911), *The Principles of Scientific Management*, Harper, New York.

CHAPTER 15

TASK DESIGN AND MOTIVATION

Holger Luczak
Research Institute for Rationalization and Operations Management
Aachen, Germany

Tanja Kabel and Torsten Licht
RWTH Aachen University
Aachen, Germany

1 MEANING AND IMPACT OF WORK

Work has been and still is an object of study in many scientific disciplines. Work science, pedagogy, jurisprudence, industrial engineering, and psychology—to name just a few—have all made significant contributions to this field. Therefore, today there is a wide variety of perspectives on work and the corresponding perceptions of human beings.

As an introduction, perhaps a look into another philosophical tradition is necessary—a tradition that is not likely to be accused of offering merely a one-sided and reduced view (Luczak and Rohmert, 1985). The authors of the *Encyclica Laborem Exercens* (Pope John Paul II, 1981) regard work as a positive human good, because through work, people not only reshape nature to fit their needs but also fulfill themselves in a spiritual sense. They become more human and thus realize creation's divine mandate. According to Luczak and Rohmert, the uniqueness of this definition lies in its ability to stand for itself in every work-related scientific discipline.

The term *work* has always been associated with aspects of burden as well as with those of pride. In history, priority was once given to the first aspect, at other times to the latter (Schmale, 1983). In ancient times work was avoided by people who could afford it; in contrast, Christianity looked upon work as a task intended by God and elevated successful working within the scope of the *Protestant work ethic* to the standard of salvation, a perception that has often been made responsible for the development of the great advances made during the Industrial Revolution (Weber, 1904/1905).

In connection with the population's attitude toward work, a shift from material to postmaterial values (Inglehart, 1977, 1989) appeared. This silent revolution consists of a slow change from industrial society's appreciation of safety and security to postmaterial society's emphasis on personal liberty. According to Inglehart, this trend is explained largely by development of the welfare state and improvements in education. Inglehart's theory of the silent revolution played an important role in shaping the perception of changing values. Not only the value that has been assigned to work during the various historical eras but also the concept of work itself is different, depending on the ideas of society and humankind as a whole (Hoyos, 1974; Schmale, 1983; Frei and Udris, 1990).

Waged work fulfills, in addition to the assurance of income, a series of psychosocial functions. Research, especially the work on effects of unemployment, indicates the high mental and social benefits of work. The most important functions are as follows (Jahoda, 1983; Warr, 1984):

1. *Activity and competence.* The activity resulting from work is an important precondition for the development of skills. While accomplishing work tasks one acquires skills and knowledge and at the same time, the cognition of skills and knowledge (i.e., a sense of competency).

2. *Structure.* Work structures the daily, weekly, and yearly cycle as well as whole life planning. This is

reflected by the fact that many terms referring to time, such as *leisure, vacation*, and *pension*, are definable only in relation to work.

3. *Cooperation and contact.* Most professional tasks can be executed only in collaboration with others. This forms an important basis for the development of cooperative skills and creates an essential social field of contact.

4. *Social appreciation.* One's own efforts, as well as cooperation with others, lead to social appreciation that in turn produces the feeling of making a useful contribution to society.

5. *Identity.* The professional role and the task, as well as the experience of possessing the necessary skills and knowledge to master a certain job, serve as a fundamental basis for the development of identity and self-esteem.

How important these functions are is often observed when people lose their jobs or have not yet had the chance to gain working experience. But also in the definition of work by employees themselves, these functions become evident. Despite some contrary claims, work still takes a central position in the lives of many people (Ruiz Quintanilla, 1984). At the same time, distinctions can be observed. Ranking work first does not remain unquestioned any more: Values have become more pluralistic, life concepts more flexible. The number of persons is increasing who are neither clearly work oriented or not work oriented; they show a flexible attitude, and this is especially the case for younger people (Udris, 1979a,b). This should not be misjudged as a devaluation of work as a sphere of life and least of all as a disappearance of the Protestant work ethic. Young people today differ in several respects from the generation of their parents, which is clearly motivated by the values of industrial society. Young people have been shaped by the global communication society in which they now live. There is a shift away from values such as obedience or pure fulfillment of duty, to values that favor the assertion of needs for self-development and fulfillment (Klages, 1983).

The aspects of work that people consider as most important can be summarized with the help of five keywords. Such listings (e.g., Kaufmann et al., 1982; Hacker, 1986) vary in detail and their degree of differentiation. Essentially, however, they are very much alike.

1. *Content of work*: completeness of tasks, diversity, interesting tasks, possibility to employ one's knowledge and skills, possibility to learn something new, possibility to take decisions

2. *Working conditions*: time (duration and position), stress factors (noise, heat, etc.); and adequacy of furniture, tools, and spatial circumstances, demanding working speed

3. *Organizational environment*: job security, promotion prospects, possibilities of further education, information management of the organization

4. *Social conditions*: opportunities for contact, relations to co-workers and superiors, working atmosphere

5. *Financial conditions*: wage, social benefits

However, opinions about the weighting of these aspects vary more fundamentally. Schools of work and organizational psychology often differ in the significance they attribute to the various characteristics (Neuberger, 1989). Taylor (1911) gives priority to the economic motive, whereas the human relations movement emphasizes the social aspect (Greif, 1983). Hackman and Oldham (1980) particularly stress the content of work; for representatives of the sociotechnical system approach (Emery, 1972; Ulich, 1991) the integration of social and technical aspects plays an essential role.

The notion that only the financial aspect is important to workers is widespread. Asking workers themselves, a more differentiated pattern results (Ruiz Quintanilla, 1984; MOW, 1987). When asked about the meaning of various aspects of work in general (keeps me busy, facilitates contacts, is interesting, gives me an income, gives me prestige and status, allows me to serve society), income ranks first, followed by possibilities for contact. Few think that the work itself is interesting and satisfactory. An evaluation of the roles that the tasks, the company, the product, the people, the career, and the money play in one's working life shows that money ranks first again, this time coequally followed by tasks and social contacts. Finally, when asked which of 11 aspects (possibilities to learn, working time, variety, interesting task, job security, remuneration, physical circumstances, and others) is most important to employees, the interesting task has top priority. How should these—at first sight, contrary results—be interpreted?

1. The great importance that is attached to payment in the first two questions shows that ensuring one's living is seen as a fundamental function of work. It is not surprising that answering the first question, very few people judge work itself as being interesting if one considers that it is about work in general. Work has to fulfill certain conditions to be interesting; it does not obtain this quality automatically.

2. The second question refers directly to the meaning that the various aspects have in one's own working life. Here, the tasks that one fulfills (i.e., the aspect of content) gain considerable significance.

3. The third question is directed toward expectations. Here, payment plays an important role, too, but it can now be found on the same level with other aspects; all are exceeded by one attribute: the desire for an interesting task.

Therefore, remuneration as a fundamental function in terms of income maintenance is of overriding importance for waged work. When talking about one's own working life, it remains the most important aspect, followed by the content of work and the social conditions. Beyond the fundamental function of income maintenance, an exceptionally high remuneration does not have priority; in this context, the aspect of an interesting task ranks first.

2 MOTIVATION TO WORK

Science and practice are equally interested in finding out which forces are motivating people to invest energy in a task or a job, in taking up a job at all, in being at the workplace every day, or in working with initiative and interest on the completion of a task. The understanding of work motivation makes it possible to explain why people direct their forces and energy in a certain direction, pursue a set goal, and show certain patterns of behavior and reactions in the job environment of an organization (Heckhausen, 1980; Phillips and Lord, 1980; Wiswede, 1980; McClelland, 1985; Weiner, 1985, 1992; Staw et al., 1986; Katzell and Thompson, 1990; Nicholson et al., 2004).

These considerations about motivation processes in the job environment are predominated by the assumption that behavior and work performance within an organization are influenced and determined by these motivational processes. Although one cannot doubt this observation, it must be noted that motivation cannot be the only determinant of working performance and behavior. Other variables affect the working process of the individual organizational member, too, and a motivation theory also has to take into account such variables as efforts, abilities, expectations, values, and former experiences, to name only a few, if it wants to explain working behavior.

While considering how working behavior is initiated, sustained, stopped, and directed, on which energies it is based and which subjective reactions it can trigger in the organism, most motivation researchers fall back on the two psychological concepts of needs and goals. In doing so, needs are seen as suppliers of energy and trigger the mechanism of the behavioral pattern of a working person. Therefore, a lack of necessities felt by a person at a particular time activates a search process with the intention of eliminating this deficit. Moreover, many theorists assume that the motivation process is goal oriented. Thus, the goal or final result that an employee is striving for in the working process possesses a certain attraction for the person. As soon as the goal is achieved, the lack of necessities is reduced. Thereby, the size of the deficiency is influenced by the individual characteristics of the organizational member (perceptions, personality, attitudes, experiences, values) as well as by a variety of organizational variables (structure, level, span of control, workplace, technology, leadership, team). The individual characteristics and organizational variables influence the searching behavior shown by an employee trying to attain a goal and determine the amount of energy invested by a person to achieve a goal. The employee will deliver a certain performance that in turn will lead to the expected financial and nonfinancial rewards, and finally, to job satisfaction. It is true that motivation and satisfaction are closely linked, but they are not synonymous. Whereas motivation is to be understood as a predisposition to a specific, goal-oriented way of acting in the working process, job satisfaction is to be seen as a consequence of performance-based rewards. Therefore, the employee can be unsatisfied with the present relation between working behavior, performance, and rewards but might still be highly motivated to fulfill a task, showing initiative, good performance, and working extra hours. In other words, a motivated employee does not have to be satisfied with the various aspects of his or her work.

To a large extent, productivity of organizations depends on the willingness of its employees to use their qualifications in a goal- and task-oriented way. This motivation characterizes an important factor of human productivity in organizations and becomes apparent as quantitative or qualitative work performance as well as in low rates of absenteeism and personnel turnover. As soon as the goals of the organization correspond to those of the employees, the individual motivational states fit into the organizational frame and thus facilitate employee satisfaction. Job and organizational design strive for the creation of conditions under which people can work productively and be satisfied at the same time.

3 THEORIES OF WORK MOTIVATION

To explain motivated behavior in the work situation as well as the relationship between behavior and outcome or performance, a series of alternative motivation theories have been developed; some of them are described below. These theories are subdivided into two groups: content theories and process models (Campbell and Pritchard, 1976). The motivation theories of the first category concentrate on the description of the factors motivating people to work. They analyze, among other things, the needs and rewards that drive behavior. In contrast, the process models of work motivation deal primarily with the processes determining execution or omission as well as with the type of execution of an action.

3.1 Content Theories

3.1.1 Maslow's Hierarchy of Needs

As a result of psychological experiments and of his own observations, Maslow (1954) formulated a theory that was meant to explain the structure and dynamics of motivation of healthy human beings. In doing so, he distinguished five different levels of needs (Figure 1):

1. *Physiological needs.* These serve to maintain bodily functions (e.g., thirst, hunger, sexuality, tranquility); they manifest themselves as physical deficiencies and are therefore easy to detect.

2. *Safety needs.* These appear as the desire for safety and constancy, stability, shelter, law, and order; in industrial nations they emerge in their original form only in disaster situations; however, in a culture-specific form they are omnipresent here as well: as a need for a secure job, a savings account, or several kinds of insurances; as resistance to change; and as a tendency to take on a philosophy of life that allows orientation.

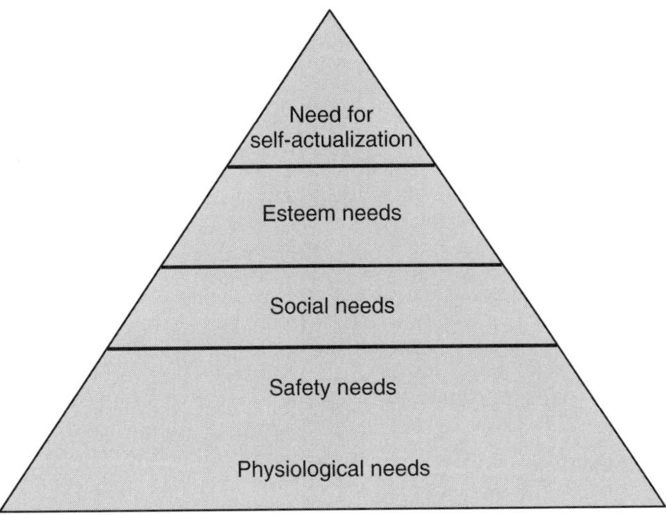

Figure 1 Maslow's levels of needs.

3. *Social needs.* These needs, such as the desire for affection and belonging, aim at the give and take of sympathy and admission to society.
4. *Esteem needs.* The satisfaction of these needs results in self-confidence and recognition; their frustration leads to feelings of inferiority and helplessness.
5. *Need for self-actualization.* This is the desire of human beings to realize their potential abilities and skills.

According to Maslow, these needs are integrated into a hierarchical structure, with physiological needs as the lowest level and the need for self-actualization as the highest. Maslow combines this hierarchy with the thesis that the elementary needs will take effect first; the contents of the higher levels will become important only when the needs of lower levels are satisfied to a certain extent. Only when need levels 1 to 4 are satisfied does the need for self-actualization get into focus. The single stages show repletion points (i.e., with adequate satisfaction of needs, motivation by means of this need is no longer possible). Striving for self-actualization is the only motive without satisfaction limits and thus remains effective indefinitely; in contrast to deficiency needs 1 to 4, it is a need for growth serving the perfection of human personality.

The importance of Maslow's concept has to be seen primarily in his verbalization of self-actualization as a human objective; he thus provoked an ongoing discussion (also in industrial companies). The weak spots of this theory lie especially in the difficulties of operationalization and verification and in the fact that the central concept of self-actualization is kept surprisingly vague. Nevertheless, this model is still very popular among practitioners; in the field of research, a critical reception dominates. The practical effectiveness of Maslow's approach is linked more to its plausibility than to its stringency. Still, it has initiated a number of other concepts or has at least influenced them (e.g., Barnes, 1960; McGregor, 1960; Alderfer, 1969).

The extensive criticism of this model (Neuberger, 1974) shows that especially the fascinating claim for universality cannot be confirmed. Apart from the fact that the categories of needs often cannot be distinguished sufficiently, the criticism alludes primarily to the hierarchical order of the needs and the dynamics of their satisfaction. Summarizing some research projects evaluating the validity of Maslow's model (e.g., Salancik and Pfeffer, 1977; Staw et al., 1986), there is little support for the existence of a need hierarchy:

- Most people differ very clearly regarding the degree to which they want a lower need to be satisfied before they concentrate on the satisfaction of a higher need.
- Several categories of needs overlap, and thus an individual need can fall into various categories at the same time.
- Within certain limits, working people are able to find substitutes for the satisfaction of some needs.
- The opportunities and chances given to an employee in the world of work are of great importance for the striving and intention to satisfy certain needs.

Finally, a number of researchers substantiated the fact that the types of needs that working persons are trying to satisfy within their organizations depends on the occupational group they belong to and their values, goals, and standards, as well as on the options for need satisfaction offered within their occupational group. These studies have shown that unskilled workers who are offered few possibilities for autonomous work and promotion within an organization stress job security and physical working conditions a lot more than do members of other professional categories. In comparison, skilled workers emphasize the type of work that satisfies or dissatisfies them. Employees in service companies usually focus on the satisfaction of social needs and, as a consequence, on the job satisfaction derived from their social interactions with colleagues and customers.

Engineers concentrate more on the performance needs at the workplace, whereas accountants in the same companies were more concerned about their promotion, even if the promotion did not result in a financial or other material benefit (e.g., Herzberg et al., 1959). Differences of this type in the pursuit of need satisfaction can be explained partially by the fact that the possibilities for the accountant to be creative at work and to develop self-initiative are a lot more restricted than those of an engineer. A study by Porter (1964) shows that the degree of job satisfaction among executives compared to other professions in the same organization was above average, but that executives of lower ranks within the organizational hierarchy were a lot less satisfied with their opportunities to work independently, autonomously, and creatively than were employees of higher ranks in the same organizational hierarchy.

3.1.2 Alderfer's ERG Theory

Since numerous analyses demonstrate that an excessive differentiation of needs is difficult to operationalize and that their hierarchical structure can be falsified easily, Alderfer (1972) is of the opinion that Maslow's theory is not fully applicable to employees in organizations. In his ERG theory, he reduced the number of need categories—according to Alderfer, Maslow's model shows overlaps among the safety, social, and appreciation needs—and developed a well-elaborated system of relations between these needs. This contains only three levels of basic needs (Figure 2):

1. *Existence needs* (E needs). These needs comprise the desire for physiological and material well-being and include Maslow's physiological needs, financial and nonfinancial rewards and remuneration, and working conditions.

2. *Relatedness needs* (R needs). These needs can be subsumed as desire for satisfying interpersonal relationships and contain Maslow's social needs as well as the esteem needs.

3. *Growth needs* (G needs). These needs can be described as a desire for continued personal growth and development and as the pursuit of self-realization and productivity; therefore, this category forms an overlap of Maslow's esteem needs and self-actualization.

Maslow defines the motivation process of humans, who are always aspiring for the next-higher need level, as a type of progression by satisfaction and fulfillment of the particular needs: The person has to satisfy a specific need first, before the next higher one is activated. Alderfer includes a component of frustration and regression in his model. Thereby, Alderfer's theoretical assumptions are opposed to Maslow's model in several fundamental aspects: ERG theory does not claim that lower-level needs have to be satisfied as a precondition for the effectiveness of higher-level needs. Moreover, the need hierarchy now works in the opposite direction as well (i.e., if the satisfaction of higher-level needs is blocked, the underlying need is reactivated). ERG theory acknowledges that if a higher-level need remains unfulfilled, the person might regress to lower-level needs that appear easier to satisfy. According to Alderfer's hypothesis of frustration, the power of a need is increased by its frustration, but there is no mandatory association between the needs of the various categories. Therefore, needs already satisfied also serve as motivators, as long as they are a substitute for still unsatisfied needs. Another difference is that unlike Maslow's theory, ERG theory contends that more than one need may be activated at the same time. Finally, ERG theory allows the order of the needs to be different for different people.

The two concepts of fulfillment progression and frustration regression are both constituents of the dynamics of ERG theory. As a consequence of the diverging assumptions of Maslow and Alderfer, different explanations and prognoses about the behavior

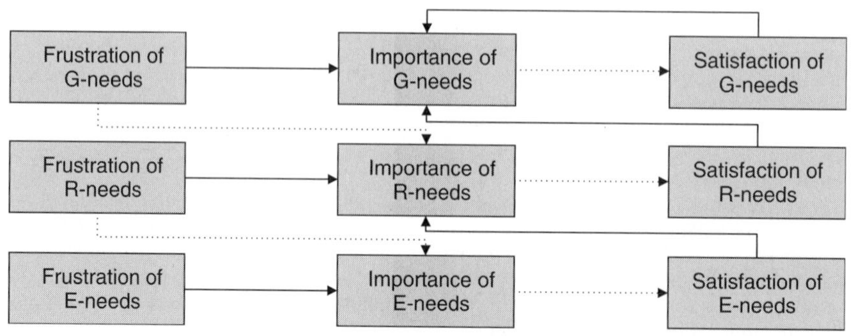

Figure 2 Alderfer's ERG theory.

of employees at their workplace are possible (e.g., Schneider and Alderfer, 1973; Guest, 1984). So far, a completely satisfying proof, especially of the psychological progression and regression processes, has not been successful. On the other hand, the relatively primitive classification of the three needs shows it to be surprisingly acceptable and able to discriminate in several international studies (Elizur et al., 1991; Borg et al., 1993). According to the authors' results, international comparison demonstrates that in case of a lack of financial means and poverty of employees in a certain region, the phenomena that are attributed to the complex "existence" come to the fore, whereas in saturated societies especially, growth needs are dominant.

3.1.3 McGregor's X- and Y-Theory

McGregor (1960) supported a direct transfer of Maslow's theory to job motivation. He objected to a Theory X that was derived from managerial practice. It starts from the following assumptions: The tasks of management concerning personnel consist in the steering of its performance and motivation as well as in the control and enforcement of company goals. Since without these activities employees face a company's goals in a passive way or resist them, it is necessary to reward, punish, and control. Therefore, a principally negative view of employees prevails among managers. In detail, this view is determined by the wrong ideas that the average person is lazy and inactive, lacks ambition, dislikes responsibility, is egocentric by nature and indifferent to the organization, is greedy and money oriented, objects to changes from the outset, and is credulous and not very clever.

According to McGregor, this concept of management is destructive for employees' motivation. Hence, McGregor drafts an antithesis building on Maslow's need hierarchy, naming it Theory Y. Essentially, it contains the following assumptions: (1) observable idleness, unreliability, dislike of responsibility, and material orientation are consequences of the traditional treatment of the working person by management; and (2) motivation in terms of potential for development, the willingness to adapt to organizational goals, and the option to assume responsibility exists in every person, and it is the fundamental task of management to create organizational conditions and to point out ways that allow employees to reach their own goals best when bringing them into agreement with the company's goals.

McGregor proposes the following measures that can ease the restrictions of the possibilities of satisfaction for the employees and facilitate responsible employment for the purpose of Maslow's ideal conception: (1) decentralization of responsibility at the workplace, (2) participation and a consulting management, and (3) involvement of employees in control and evaluation of their own work.

3.1.4 Herzberg's Two-Factor Theory

The motivation theory developed by Herzberg et al. (1959) is probably the most popular theory of work motivation (Figure 3). Its central topic is job satisfaction. Results of empirical studies led Herzberg and his colleagues to the opinion that satisfaction and dissatisfaction at the workplace are influenced by various groups of factors. Dissatisfaction does not occur simply because of the lack or insufficient value of parameters that otherwise cause satisfaction. Herzberg called the factors that lead to satisfaction *satisfiers*.

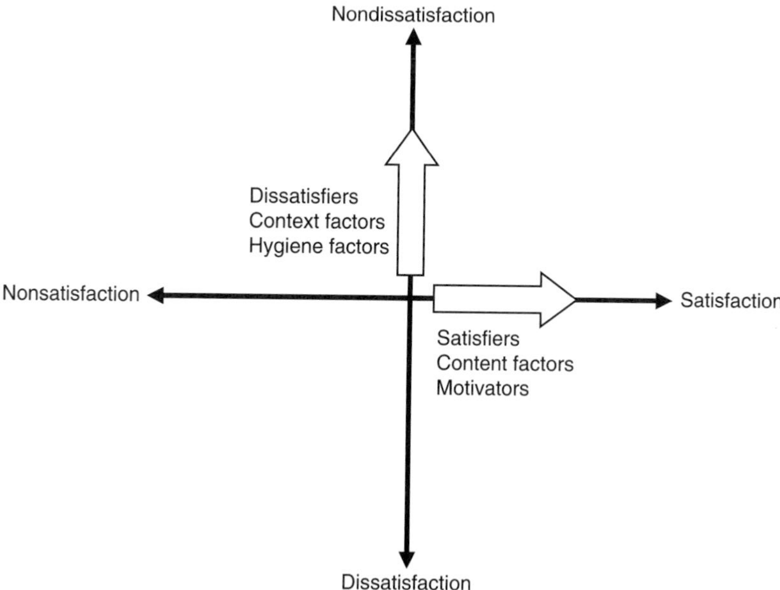

Figure 3 Herzberg's two-factor theory.

These are primarily (1) the task itself, (2) the possibility to achieve something, (3) the opportunity to develop oneself, (4) responsibility at work, (5) promotion possibilities, and (6) recognition.

Since these factors are linked directly to the content of work, Herzberg also referred to them as *content factors*. Due to the fact that their positive value leads to satisfaction and consequently motivates for performance, they were seen as the actual motivators. According to Herzberg, for the majority of employees, motivators serve the purpose of developing their professional occupation as a source of personal growth.

In contrast, *dissatisfiers* have to be assigned to the working environment and are therefore also called *context factors*. According to Herzberg, they include especially (1) the design of the surrounding working conditions, (2) the relationship with colleagues, (3) the relationship with superiors, (4) company policy and administration, (5) remuneration (including social benefits), and (6) job security. Since the positive values of these parameters accommodate the employees' need to avoid unpleasant situations in a preventive way, they were also referred to as *hygiene factors*. In many cases content factors allude to intrinsic motivation and context factors to extrinsic motivation.

Since Herzberg's theory also suggests a number of practical solutions to organizational problems, allows predictions about behavior at the workplace to a certain extent, and because of the impressive simplicity of the model and its orientation toward the terminology of organizational processes, a large variety of empirical studies have been undertaken to examine the underlying postulates and assumptions (e.g., Lawler, 1973; Kerr et al., 1974; Caston and Braito, 1985). However, many of these studies raised additional questions. The essential objections that can be put forward against Herzberg's model are:

- The restricted validity of data, being based on a small number of occupational groups (only engineers and accountants)
- The oversimplification of Herzberg's construct of motivation or job satisfaction (e.g., satisfaction and dissatisfaction could be based on the working context as well as on the task itself, or on both equally)
- The division of satisfaction and dissatisfaction into two separate dimensions
- Lack of consideration of unconscious factors that can have an effect on motivation and dissatisfaction
- Lack of an explanation of why different extrinsic and intrinsic work factors are to influence the performance in a negative or positive way, and why various work factors are important
- Lack of consideration of situational variables
- No measurement of job satisfaction as a whole (it is very possible that somebody dislikes parts of his or her job but still thinks that the work is acceptable as a whole)

Existing studies about Herzberg's theory have left many problems unsolved, and it is doubtful whether a relatively simple theory such as Herzberg's can ever shed light on all the questions that are raised here. Despite this criticism, the theory, although being an explanation for job satisfaction rather than a motivation theory, is still very popular. Despite restrictions regarding the methodology and content of this theory, it can be stated that motivators offer a greater motivational potential to intrinsically motivated employees than extrinsic incentives do (e.g., Zink, 1979). Especially when described in a shortened way, Herzberg's concept obviously has such a high plausibility that in industry it still has an astonishing repercussion. The importance of Herzberg's approach is to be seen primarily in the fact that he set the content of work as the focus of attention. This gave numerous companies food for thought and induced manifold change processes. Last but not least, the emphasis on work content had an effect on the dissemination of the so-called *new forms of work design* (Miner, 1980; Ulich, 1991).

3.1.5 Hackman and Oldham's Job Characteristics Model

Behavior and attitudes of employees and managers can be influenced to a great extent by a multitude of context variables. Moreover, during recent years the connection and adequate fit of task characteristics or work environment, on the one hand, and the psychological characteristics of the person, on the other (person–job fit), have been studied very intensively and from many different perspectives. Task and work design, especially, formed the focus of interest. These studies were initiated particularly by a motivation model of job and task characteristics (the job characteristics model of work motivation) (Figure 4) developed by Hackman and Oldham (1976, 1980). The model postulates that certain core dimensions of work lead to certain psychological states of the working person which result in specific organizational or personal outcomes. Hackman and Oldham list five job characteristics that cause enhancement of motivation and a higher degree of performance and job satisfaction. Furthermore, they propose that persons with a strong psychological growth need react in a more positive way to tasks containing many core dimensions than do people with a weak growth need. Here, direct reference to Maslow's need hierarchy becomes obvious. A person who currently prioritizes the need for self-actualization is categorized as strong with regard to his or her growth need, whereas somebody operating on the level of safety needs would be seen as somebody with a weak growth need.

The core dimensions of work are the following:

- *Skill variety*: degree to which tasks require different skills or abilities
- *Task identity*: degree to which a person completes a connected piece of work or a task instead of parts or facets of it

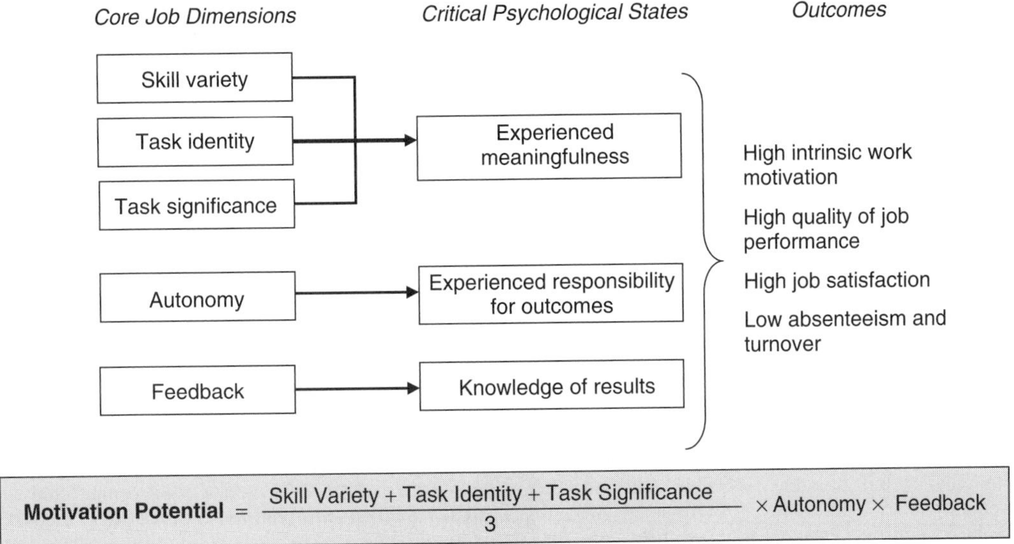

$$\textbf{Motivation Potential} = \frac{\text{Skill Variety} + \text{Task Identity} + \text{Task Significance}}{3} \times \text{Autonomy} \times \text{Feedback}$$

Figure 4 Job characteristics model.

- *Task significance*: degree to which work has an effect on the lives and jobs of others
- *Autonomy*: freedom and independence in the accomplishment of work
- *Feedback*: degree to which work provides clear and direct information about success and effectiveness of the performing person

Summarizing these core dimensions, they result in an index of motivational potential of a job by which jobs can be assessed regarding their possibilities to motivate. But this also makes it possible to compare and classify entire work processes depending on whether they are able to help a person to achieve professional and intellectual growth as well as further development. The score can be seen as a measure of adequacy, quality, and success of the present work design. The formula postulates an additive relation for skill variety, task identity, and task significance whereby the single dimensions can compensate for each other. The multiplicative connection between autonomy and feedback does not allow this and can in an extreme case reduce the motivational potential score to zero. Although other authors affirm the strong emphasis on feedback only with reservations, the formula sharply covers important effects on motivation. In many cases an additive combination of the variables shows it to be on a par with the postulated multiplicative formula.

In the center of the model are the *critical psychological states*. These are defined as follows:

- *Meaningfulness of work*: degree to which a person experiences work in general as meaningful, valuable, and worthwhile
- *Responsibility for outcomes*: degree to which a person feels personally responsible for the work that he or she is doing
- *Knowledge of results*: degree to which a person is continually informed about how successful and effective the job done is

When these states are on an acceptable level, it can be presumed that the person feels good and reacts in a positive way toward the job. It is expected that the dimensions skill variety, task identity, and task significance influence the meaningfulness of work experienced by a person. The dimension autonomy presumably affects the experienced responsibility for outcomes. Feedback contributes to knowledge of the actual results. The critical psychological states for their part determine a variety of personal and work outcomes. Several studies show correlations with intrinsic motivation, job satisfaction, absenteeism, and turnover, but only low correlations with quality of job performance. Finally, it is assumed that the growth need strength of an employee mediates relations among the other elements of the theory.

To test the theory empirically, Hackman and Oldham (1975) developed the Job Diagnostic Survey (JDS). This instrument allows objective measurement of job dimensions and measures the various psychological states caused by these characteristics, the affective responses to these job characteristics, and the strength of the personal need to grow and to develop oneself. The JDS can be used to identify workplaces or work structures with a high or low motivational potential. Since the motivational potential score provides a comprising index for the overall motivational potential of a job, a low score points to jobs that deserve redesign. For implementation of the theory in practice, a set of action principles has been developed that give instructions about how the core dimensions of work can be improved (Hackman and Suttle, 1977). In practice, the model has reached a comparably high degree of popularity and is used for the improvement of tasks and job design as often as Herzberg's model. Therefore, scientific literature concerning this model is very extensive (e.g., Fried and Ferris, 1987; Fried, 1991).

Although research has so far supported the model on the whole, the model contains a number of unsolved problems and methodological weaknesses on which further research will have to focus (Roberts and Glick, 1981). There are questions about the tools that measure the various components of the model, about the significance and modality of computation of the motivational potential score, and about the theoretical foundation on which the model is based. It has to be asked whether the five core dimensions measured with the JDS are really independent (Idaszak and Drasgow, 1987; Williams and Bunker, 1993). In addition, the assumed impact of employees' growth need strength has to be examined, especially its significance and direction. Empirical evidence has to be provided as to which way the critical psychological states are caused or influenced by the five core dimensions of work. Although the formula for the computation of motivational potential suggests considerable interactions among the different job characteristics, it is just as possible that employees tend to ascribe more weight to job characteristics that are most beneficial to them (e.g., autonomy, variety). Furthermore, the mediating function of the psychological states is still quite indistinct (Hackman and Oldham, 1980; Miner, 1980; Udris, 1981).

3.1.6 McClelland's Theory of Acquired Needs

McClelland's theory of acquired needs (McClelland, 1984, 1985; McClelland et al., 1989) is closely linked to the psychological concepts of learning and is based to a great extent on the works of Murray (1938). McClelland holds the view that many needs are learned by dealing with and mastering the cultural environment in which a person is living (McClelland, 1961). Since these needs are learned from early childhood on, working behavior that is rewarded will occur more often. Applied to an organization, this means that employees can be motivated, by financial and nonfinancial rewards, to be at their workplace on time and regularly as long as these rewards are linked directly to the favored working behavior. As a result of this learning process, people develop certain need configurations that influence their working behavior as well as their job performance.

Together with other researchers (e.g., Atkinson and Feather, 1966), McClelland filtered those needs out of Murray's list of human needs that in his opinion represent the three key needs in human life: (1) the need for achievement (n-ach), (2) the need for affiliation (n-affil), and (3) the need for power (n-pow). These three unconscious motives have a considerable effect on both the short- long-term behavior of a person (McClelland et al., 1989). The need for achievement is relevant for change behavior and contains continuous improvement of performance. The need for affiliation is important for group cohesion, cooperation, support, and attractiveness in groups. The need for power is of importance for persuasiveness, orientation toward contest and competition, and for readiness to combat. There is an important link to the role of a manager that consists essentially of activating these motivations and thus to energize and guide the behavior of subordinates.

The main interest focuses on the need for achievement, and as a consequence, a theory of achievement motivation was formulated (Atkinson and Feather, 1966) that was applied widely in organizational psychology (Stahl, 1986). Achievement motivation corresponds to a relatively stable disposition of behavior or a potential tendency of behavior of an employee in an organization to strive for achievement and success. However, this motivation becomes effective only when being stimulated by certain situational constellations or incentives that lead a person to assume that a certain working behavior will produce the feeling of achievement. The final result is an inner feeling of satisfaction and pride of achievement. The model developed for this purpose is in effect an expectancy-valence model of motivation processes. Working behavior is understood as the resultant of (1) motivation strength, (2) the valence or attractiveness of the incentive that activates motivation, and (3) a person's expectancy that a certain behavior will lead to gaining the incentive. Hence, the corresponding motivation model can be designed as follows:

$$T_S = M_s \times P_s \times I_s$$

A person's tendency (T_S) to approach a task is a multiplicative function of the person's strength of the achievement motivation (M_s), the subjective probability of success (P_s), and the valence or degree of attractiveness of this success or reward (I_s). From this assumption, a set of conclusions can be deduced for the processes of job design not only for the selection and promotion of organizational members, but also for the preference of certain leadership styles and for the motivation of risk behavior among managers in decision situations (McClelland et al., 1989).

Empirical research on McClelland's model is extensive and shows a set of consistent results. Here are some examples: People who are highly achievement motivated (have high scores on n-ach) prefer job situations in which they bear responsibility, get feedback on a regular basis, and that ask for a moderate attitude toward risks. Such a constellation has a very motivating effect on high achievers. These kinds of people are active mainly in self-dependent fields of activities. Restrictively, it has to be added, though, that high achievers often show less interest in influencing the achievement of others than in personally seeking high achievements. This is why they often are not good as managers. Reciprocally, managers of big companies rarely are n-ach people. The constellation looks a lot different when considering the need for affiliation or for power. Research shows that successful managers frequently have a high need for power and a rather low need for affiliation, a constellation that probably is necessary for efficiency of leadership and that can possibly be deduced from the function or role within the organizational context (Parker and Chusmir, 1992). Finally, Miron and McClelland (1979) point out that for the filling of positions that require a high need for achievement, a combination of selection and training is advised. People with high n-ach scores are selected and developed by means of achievement trainings with the goal of imparting to these persons a pattern of thought in terms of achievement, success, and profit and to act upon this pattern. The three secondary or learned motivations investigated proved to be very informative and stable with regard to the explanation of working behavior and leadership. However, it is possible that the explication of organizational behavior can be improved by other secondary motives. This applies especially in the area of managerial functions. Therefore, Yukl (1990) adds two more secondary motives to his description of skills, characteristics, and goals of successful managers: the need for security and the need for status. For a better illustration, the five key motivations and their descriptors are listed below.

1. Need for achievement
   - To excel others
   - To attain challenging goals
   - To solve complex problems
   - To develop a better method to do a job
2. Need for power
   - To influence others to change their attitudes and behavior
   - To control people and things
   - To have a position of authority
   - To control information and resources
3. Need for affiliation
   - To need to be liked by others
   - To be accepted as part of a group
   - To relate to others in a harmonious way and to avoid conflicts
   - To take part in enjoyable social activities
4. Need for security
   - To have a secure job
   - To be protected against loss of income
   - To avoid tasks and decisions that include risks or failures
   - To be protected against illness and incapacity for work
5. Need for status
   - To have the right car and to wear the right clothes
   - To work for the right company in the right position
   - To have the priviledges of leaders
   - To live in the right neighborhood and to belong to the right club

3.1.7 Argyris's Concept

An approach that joins different concepts together is that of Argyris (1964). According to Argyris, work motivation, the competency to solve problems, and emotional well-being are facilitated primarily by a feeling of self-esteem based on *psychological success*. The possibility of defining one's own goals according to one's own needs and values and to control goals in a self-dependent way operates as an important precondition for psychological success. From this emanates a contradiction to the structures of the formal organization that work such that a single employee can control his or her own working conditions to a minimal extent only, that he or she can bring in only a few or very limited skills in his or her work, and that he or she can behave only in a very dependent way.

Only if organizations believe that employees want to apply their skills in the framework of the company's goals and that they want to get involved in relevant decisions will employees be able to behave like grown-ups. On the other hand, if companies are structured differently, employees will behave accordingly: dependent, with little interest, with a short-term perspective; independence of thinking and acting might find their expression only in the development of defense.

Argyris's contribution contains a variety of unclarified points (Greif, 1983). As for Maslow and Herzberg, it is also true for Argyris that interindividual differences are largely disregarded in their concrete meaning for the development of job and organizational structures.

3.1.8 Deci and Ryan's Self-Determination Theory

Another meta-theory of motivation and personality, the self-determination theory (SDT) of Deci and Ryan (1985), is based on the assumption that humans naturally strive for psychological growth and development. While mastering continuous challenges, the social context plays a vital role; its interaction with the active organism allows predictions about behavior, experience, and development. Deci and Ryan postulate three motivating factors that influence human development: (1) the need for autonomy, (2) the need for competence, and (3) the need for relatedness. They are referred to as basic psychological needs that are innate and universal (i.e., they apply to all people, regardless of gender, group, or culture).

The need for autonomy refers to people's striving for self-determination of goals and actions. Only when perceiving of oneself as the origin of one's own actions and not being at the mercy of one's environment can one feel motivated. The successful handling of a task will be perceived as the confirmation of one's own competence only if it was solved mainly autonomously. The need for competence expresses the ambition to perceive oneself as capable of acting effectively in interaction with the environment. The need for relatedness has a direct evolutionary basis and comprises the close emotional bond with another person.

When these three needs are supported by social contexts and are able to be fulfilled by individuals, well-being is reinforced. Conversely, when cultural, contextual, or intrapsychic forces inhibit the fulfillment of the three basic needs, well-being is reduced. In this theory, motivation is seen as a continuum from amotivation to extrinsic motivation to intrinsic motivation. According to SDT, autonomy, competence, and relatedness are three psychological nutriments that facilitate the progression from amotivation to intrinsic motivation (Ryan and Deci, 2000; Deci, 2002).

SDT contains four subtheories that have been developed to explain a set of motivational phenomena that has emerged from laboratory and field research:

- *Cognitive evaluation theory*: deals with the effects of social contexts on intrinsic motivation
- *Organismic integration theory*: helps to specify the various forms of extrinsic motivation and the contextual factors that either promote or prevent internalization
- *Causality orientations theory*: pictures individual varieties in people's tendencies toward self-determined behavior and toward directing to the environment in a mode that supports their self-determination
- *Basic needs theory*: develops the idea of basic needs and their connection to psychological health and well-being

The cognitive evaluation theory (CET), for example, aims at specifying factors that explain variability in intrinsic motivation and focuses on the need for competence and autonomy. *Intrinsic motivation* means to do an activity for the inherent satisfaction of the activity itself and thus differs from *extrinsic motivation*, referring to the performance of an activity to attain some separable outcome (Ryan and Deci, 2000). According to the CET, social-contextual events (e.g., feedback, communication, rewards) conducive to feelings of competence and autonomy during action can enhance intrinsic motivation. Studies showed that intrinsic motivation is facilitated by optimal challenges, effectance-promoting feedback, and freedom from demeaning evaluations (e.g., Deci, 1975; Vallerand and Reid, 1984). However, the principles of the CET do not apply to those activities that do not hold intrinsic interest and that do not have the appeal of novelty, challenge, or aesthetic value (Ryan and Deci, 2000).

Contrary to some studies, Deci does not assume an additive connection between intrinsic and extrinsic motivation. Rather, he postulates an interaction between both types of motivation, which means that extrinsic incentives can replace intrinsic motivation (Vansteenkiste and Deci, 2003). In his studies he tested the following hypotheses:

1. If intrinsic motivation makes a person perform an action and this action is recompensed with an extrinsic reward (e.g., money), his or her intrinsic motivation for the particular action decreases.
2. If intrinsic motivation makes a person perform an action and this action is recompensed with verbal encouragement and positive feedback, his or her intrinsic motivation for this particular action increases.

The design of Deci's studies has always been the same: An extrinsic incentive is added to an interesting activity. Then the variance of the intrinsic motivation has been measured on the basis of the dependent variable. He concludes that his experimental results support the hypotheses mentioned above. Money seems to have a negative impact on intrinsic motivation. Verbal encouragement and positive feedback, on the other hand, have a positive impact.

The self-determination theory has been applied within very diverse domains, such as health care, education, sports, religion, and psychotherapy, as well as in industrial work situations. For example, Deci et al. (1989) found that managers' interpersonal orientations toward supporting subordinates' self-determination vs. controlling their behavior, correlated with their subordinates' perceptions, affects, and satisfactions. Moreover, the evaluation of an organizational development program focusing on the concept of supporting subordinates' self-determination showed a clearly positive impact on managers' orientations but a less conclusive impaction subordinates. Later studies on the topic of supervisory style support these findings: Participants experienced higher levels of intrinsic motivation under conditions of an autonomy-supportive style than of nonpunitive controlling and punitive controlling supervisory styles (Richer and Vallerand, 1995). Researchers were also able to show that the constructs of self-determination theory were equivalent across countries as well. Deci et al. (2001) found that a model derived from the SDT in which autonomy-supportive work climates predict satisfaction of the intrinsic needs for competence, autonomy, and relatedness, which in turn predict task motivation and psychological adjustment on the job, was true in work organizations in the United States as well as in state-owned companies in Bulgaria.

Among the three needs, autonomy is the most controversial. Iyengar and Lepper (1999) presume that cultural values for autonomy are opposed to those of relatedness and group cohesion. They provided experimental evidence showing that the imposition of choices by an experimenter relative to personal choice undermined intrinsic motivation in both Asian

Americans and Anglo Americans. However, they also showed that adopting choices made by trusted others uniquely enhanced intrinsic motivation for the Asian group. Their interpretation focused on the latter findings, which they portrayed as challenging the notion that autonomy is important across cultures. Oishi (2000) measured autonomy by assessing people's individualistic values, apparently assuming them to represent autonomy as defined within SDT. On the basis of this measure, Oishi reported that outside of a very few highly individualistic Western nations, autonomous persons were not more satisfied with their lives. Finally, Miller (1997) suggested that in some cultures, adherence to controlling pressures yields more satisfaction than does autonomy. Her characterizations of autonomy, like those of Iyengar and Lepper and Oishi, do not concur with SDT's definition.

3.1.9 Summary of the Content Theories

So far it has not been possible to provide evidence that certain motives are universal. More promising seems the identification of dominant motives or constellations for certain groups of persons (Six and Kleinbeck, 1989). French (1958) shows, for example, that praise is more effective when its content addresses the dominant motive: Praise for efficiency leads to a better performance in achievement-motivated people, praise for good cooperation, in affiliation-motivated people. Other studies support the importance of the congruence of the person's motivation structure and the organization's structure of incentives.

It is neither possible nor reasonable to fix the number of motivating factors at work once and for all; the differentiation has to differ depending on the purpose of analysis and design. It is striking, though, that from many analyses two factors of higher order result that correspond to a large extent to Herzberg's factors (content and context) (Campbell and Pritchard, 1976; Ruiz Quintanilla, 1984). All concepts mentioned have pointed out the importance of intrinsic motivation by means of holistic and stimulating work contents (Hacker, 1986; Volpert, 1987; Ulich, 1991). The complexity of the problem makes it impossible to choose a theory that is able to serve as an exclusive basis for a satisfying explanation of working behavior in organizations. All theories have in common that they try to explain the "what" of energized behavior, that they recognize that all people possess either congenital or learned and acquired needs, and finally, that they reveal nothing about "how" behavior is energized and directed.

3.2 Process Models

These motivation theories try to answer the question of how human behavior is energized, directed, and stopped and why humans choose certain ways of behavior to reach their goals. They differ from the content theories especially by stressing cognitive aspects of human behavior and postulating that people have cognitive expectations concerning the goal or final result that is to be reached. According to these instrumentality theories, humans only decide to act if they can achieve something that is valuable for them, and thus an action becomes instrumental for the achievement of a result to which a certain value is attached.

3.2.1 Vroom's VIE Theory

Vroom (1964) refers to his instrumentality theory as VIE theory. The central part of this theory is constituted by the three concepts of valence (V), instrumentality (I), and expectancy (E). *Valence* describes the component comprising the attracting and repellent properties of the psychological object in the working environment—payment has a positive, danger a negative valence. Thus, before a job action is initiated, the person is interested in the value of the final result. This valence reflects the strength of the individual desire or the attractiveness of the goal or final result for the person that can be reached by different means. To be able to explain the processes in this situation of selection of alternative actions, Vroom establishes the idea of a result on a first level and a result on a second level. Thereby, the functions of the two other components, instrumentality and expectancy, are defined. An employee may assume that if he or she does a good job, he or she will be promoted. The degree to which the employee believes in this is an estimation of subjective probability, referred to as *expectancy*. Finally, the expectancy of a person relates to his or her assumed probability that a certain effort will lead to a certain outcome. Thus, in attaining the goal of the first level, the person sees a way to reach the goal of the second level. According to this, the considerations of Vroom's motivation model can be phrased as follows: The valence of the result of the first level (e.g., effort) is determined by the person's estimation of the probability that this first-level result will lead to a row of second-level results (e.g., pay rise or promotion) and the valences linked to it.

Thus, Vroom's model (Figure 5) states that motivation or effort put in by a person to reach his or her goals are a function of his or her expectancy that as a result of his or her behavior, a certain outcome will be achieved and of the valence that the result has. If one of the two factors is zero, there is no motivational force and an action for the achievement of a certain result will not take place. Hence, this motivation model provides very concrete explanations for the working behavior of employees in organizational practice. In this context it is important to emphasize that the direct explanations of the VIE model do not refer to working results or job performance but to motivated behavior—to decision making and effort in particular. As to this, the model can be seen as confirmed to a rather wide extent; expected connections to performance are not that strong because there are several mediators between effort and result that have to be taken into account. To make some good explanations with the help of this model, there has to be an appropriate organizational environment; outcomes of actions and their consequences for the persons concerned have to be transparent and calculable consistently. This is where to find important practical

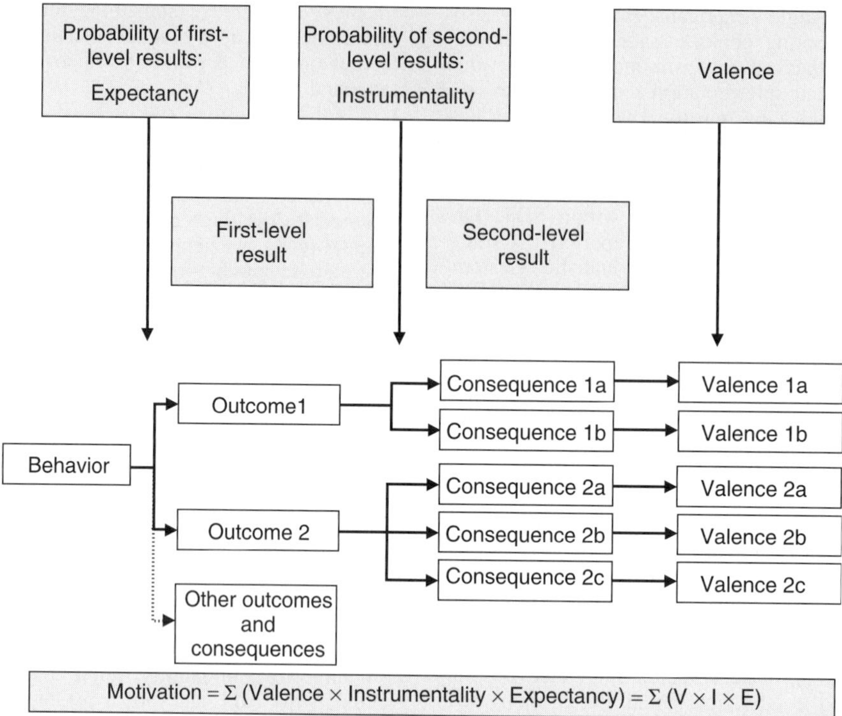

Figure 5 Vroom's VIE theory.

implications of the model for leadership and organizational design.

However, despite the advantages offered by Vroom's motivation model, there are some unsolved problems inherent in the model as well that restrict its explanatory power. For example, Vroom provides no information about the effect of those factors that influence the expectancy of an organizational member (e.g., self-esteem, former experiences in similar situations, abilities, leadership). It is also possible that employees misjudge a work situation, possibly because of their needs, emotions, values, or assumptions. This situation may result in employees choosing a nonadequate behavior and in not considering all factors that are relevant. In addition, it could be put forth against Vroom's ideas that to date there has been no research effort to determine how expectations and instrumentalities develop and by which factors they are influenced. Furthermore, the specific operation mode of the model is too complex and too rational to represent human calculations in a realistic way. Besides, an additive model works just as well as a multiplicative one. Altogether, the particular value of the model is that it points out the importance of multiple results or consequences with their probabilities and appraisal being different for each person. In this respect, it turns against perceptions that are too simple and according to which few motives virtually lead directly to actions and suggests a more complex strategy of analysis.

3.2.2 Porter and Lawler's Motivation Model

Among the process models, Porter and Lawler's (1968) model and in succession those of Zink (1979) and Wiswede (1980) have to be pointed out. They consider satisfaction, among other things, as a consequence of external rewards or, with intrinsic motivation, of self-reward for results of actions. The value of these models, which try to integrate various social-psychological principles (e.g., social matching processes, aspiration level, self-esteem, attribution, role perceptions, achievement motivation) is to offer a heuristically effective framework in which relevant psychological theories are related in a systematic way to explain performance and satisfaction.

Porter and Lawler's motivation model (Figure 6) is closely related to Vroom's ideas but focuses more on the special circumstances in industrial organizations. It is a *circulation model* of the relationship between job performance and job satisfaction. With this model, Porter and Lawler describe working behavior in organizations by emphasizing the rational and cognitive elements of human behavior that have been ignored, especially by the content theories. This is particularly true with regard to planning and decision making regarding anticipated future events at the work place. The two crucial points in this model are:

- *The subjective probability* $E \rightarrow P$: the expectation to achieve a goal with greater effort

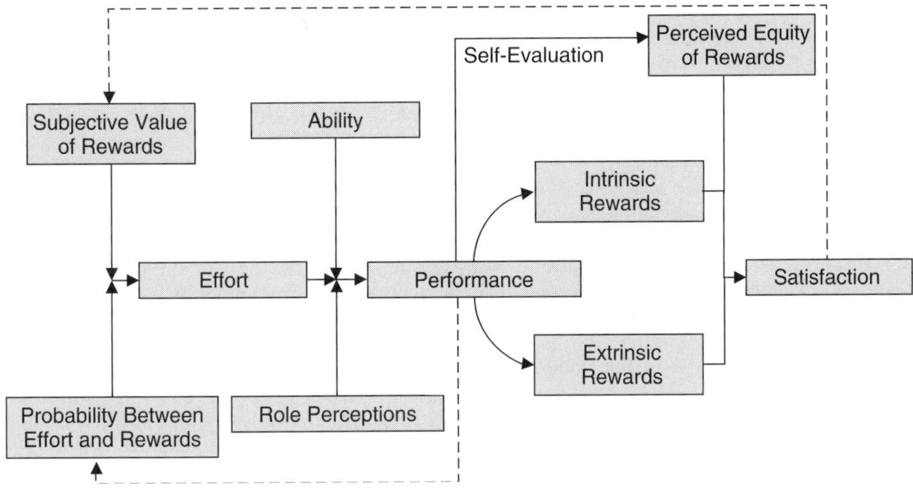

Figure 6 Porter and Lawler's motivation model.

- *The probability* $P \rightarrow O$: a good performance will lead to the desired output, considering the valences of these goals

Thus, Porter and Lawler postulate that the motivation of an organizational member to do a good job is determined essentially by two probability factors: by the subjective estimated values $E \rightarrow P$ and $P \rightarrow O$. In other words the individual motivation at the workplace is determined by the probabilities that increased effort leads to better performance and that better performance leads to goals and results that have a positive valence for the person. Moreover, Porter and Lawler state that the two probabilities $E \rightarrow P$ and $P \rightarrow O$ are linked to each other in a multiplicative way. But this multiplicative relationship says as well that as soon as one of the two factors is zero, the probability relation between effort and final result also decreases to zero. Explications for observable behavior at the workplace that can be derived from this are evident.

In this model the first component, subjective value of rewards, describes the valence or attractiveness that different outcomes and results of the work done have for the person—different employees have different values for different goals or results. The second component, probability between effort and rewards, refers to the subjective probability with which a person assumes that an increase in effort leads to the receipt of certain results of rewards and remuneration considered as valuable and useful by the person. This estimated probability contains in the broader sense the two subjective probabilities specified above: $E \rightarrow P$ and $P \rightarrow O$. The third component of the model consists of the effort of an organizational member to perform on a certain level. In this point, Porter and Lawler's motivation model differs from former theories because it distinguishes between effort (applied energy) and work actually performed (efficiency of work performance). The fourth component, the area of individual abilities and characteristics (e.g., intelligence or psychomotoric skills), has to be mentioned. It sets limits on an employee's accomplishment on a task. These individual characteristics, which are relatively stable, constitute a separate source of interindividual variation in job performance in this model. Role perceptions are based primarily on how an employee interprets success or successful accomplishment of a task at the workplace. They depend on what and in which direction a person will focus his or her efforts. In other words, role perceptions directly influence the relationship between effort and quality of job performance. This is why inadequate role perceptions lead to a situation where an employee obtains wrong or useless work results while showing great effort. Finally, the accomplishment of a job constitutes a sixth component that refers to the level of work performance an employee achieves. Although task execution and work performance play such an important role in organizations, the components themselves, as well as their interactions, are often misunderstood, oversimplified, and wrongly interpreted. Successful accomplishment of a task is, as Porter and Lawler's model shows, influenced by multiple variables and their interactions. It is the resultant of a variety of components and a combination of various parameters and their effects. The component reward consists of two parts: The intrinsic reward, given by oneself, and the extrinsic reward, essentially (but not exclusively) given by a superior. An intrinsic reward is perceived only if the person believes that he or she has mastered a difficult task. On the other hand, an extrinsic reward can be perceived only, if the successful execution of a task is noticed and valued accordingly by a superior, which is often not the case. The last two components of the model are the reward seen as appropriate by the employee and the "satisfaction" of the employee. This component, perceived equity of rewards, refers to the amount of the reward that the employee, based on performance, expects as appropriate and fair from the organization. The degree of satisfaction can be understood as the result of the employee's comparison of the reward actually obtained to the reward considered as appropriate and fair as compensation for the job done. The greater the difference between these two values, the higher will be the degree of satisfaction or dissatisfaction.

The model developed by Porter and Lawler shows in a striking way that a happy employee is not necessarily a productive one. Numerous empirical studies prove the correctness of the assumptions of the model in its essential points (Podsakoff and Williams, 1986; Locke and Latham, 1990; Thompson et al., 1993; Blau, 1993).

Finally, it has to be stated that in the underlying formula of performance, $P = f(M \times A)$ (i.e., performance is a function of the interaction of motivation and ability), another important parameter has been ignored. If we want to explain and predict work performance, it seems more realistic to consider the possibility of achieving a certain goal and showing a certain performance as well. Even if a person is willing and able to do a good job, there might be obstacles reducing or even thwarting success. This is why the formula has to be broadened to $P = f(M \times A \times O)$. A lack of possibilities or options to achieving maximum performance can be found in any work environment. They range from defective material, tools, devices, and machines to a lack of support by superiors and colleagues and inhibiting rules and processes or incomplete information while making decisions.

The model of Porter and Lawler points out precise fields of application in practice: Each component is practicable and the processes are just as easy to understand and to see through. Organizational management can influence the relationship between effort and reward by linking reward directly to work performance. Moreover, it is possible to influence virtually every component in a systematic way because according to Porter and Lawler, the effects are predictable within certain limits.

3.2.3 Adams's Equity Theory

In contrast to the instrumentality theories, which deal essentially with the expectancies of a person at the workplace and with how these expectations influence behavior, the balance theories of motivation focus on interindividual comparisons and on states of tension and their reduction. These are already the general assumptions of this type of motivation model: behavior is initiated, directed, and sustained by people's attempts to find a kind of internal balance (i.e., to keep their psychological budget balanced). Festinger's (1957) theory of cognitive dissonance serves as a basis for the various versions of balance theories especially designed for work organizations. Simplified, Festinger postulates that discrepant cognitions cause psychological tensions that are perceived as unpleasant, and that humans act to reduce these tensions. Hence, if a person has two inconsistent cognitions, the person falls into an aversive state of motivation called *cognitive dissonance*.

The central idea of Adams' equity theory (Adams, 1963, 1965), which has been applied primarily in work organizations and has often been examined (Mikula, 1980), is that employees of an organization make comparisons on the one hand, between their contributions and the rewards received for it, and on the other, between the contributions and the rewards of relevant other persons in a similar work situation. The choice of the person or group of comparison increases the complexity of the theory. It is an important variable because it can be a person or group within or outside the present organization. As a result, the employee may also compare himself or herself with friends or colleagues in other organizations or from former employments. The choice of the referent is influenced predominantly by information that the employee has about the respective person as well as by the attractiveness of this person. The pertinent research has therefore been interested particularly in the following moderating variables:

- *Gender.* Usually, a person of the same gender is preferred.
- *Duration of membership in a company.* The longer the duration of membership, the more frequently a colleague in the same company is preferred.
- *Organizational level and training.* The higher the level and the more training a person has, the more likely he or she is to make comparisons with persons outside the organization.

For employees the principle of equity at the workplace is kept when they perceive the ratio of their own contributions (input $= I$) and the rewards obtained (outcome $= O$) as being equivalent to the respective ratio of other persons in the same work situation. Inequity and therefore tension exist for a person when these two ratios are not equivalent. Hence, if this ratio of contribution and reward is smaller or greater for a person than for the referent, the person is motivated to reduce the internal tension produced; one will count as one's contributions everything that one adds personally to a given work situation: psychomotoric or intellectual skills, expertise, traits, or experience. Accordingly, everything that a person obtains and considers as valuable is counted as a reward: remuneration, commendation, appreciation, or promotion. The inner tension perceived by a person pushes for the reestablishment of equity and therefore justice. According to Adams, the strength of the motivated behavior is directly proportional to the amount or strength of tension produced by inequity. Depending on the causes and the strength of the perceived inequity, the person can now choose different alternatives of action that can be predicted by means of the referent. For example, a person can try to get a raise in reward if this is lower than that of the referent chosen. On the other hand, one could increase or decrease one's input by intensifying or reducing one's contributions. If in the case of a perceived inequity, neither of these reactions is possible, the person's reaction might be frequent absenteeism from the workplace or even quitting the job. Besides, there are other options as to how to respond: (1) distortion of self-perception, (2) distortion of the perception of others, or (3) variation of the chosen referent.

It is important to state that the contributions as well as the estimation of the rewards and the ratio of the two variables are subject to the perception and judgment of the employee (i.e., they do not necessarily correspond to reality). The majority of research studies on the evaluation of Adams's motivation theory have dealt with the choice of the referent and with payment as a category of reward in work organizations (Husemann et al., 1987; Greenberg, 1988; Summers and DeNisi, 1990; Kulik and Ambrose, 1992). Usually, in these studies for the assessment of the effect of a state of unequal payment in laboratory experiments as well as in field studies, four different conditions have been created: (1) overpayment of the hourly wage, (2) underpayment of the hourly wage, (3) overpayment of the piece wage, and (4) underpayment of the piece wage.

Subjects were assigned in a randomized way to these conditions of inequity, dissonance, and tension. Thus, an employee working under the condition of overpayment/piece wage (i.e., perception of inequity) will improve the quality and reduce the quantity in order to reduce the state of tension because another increase in quantity would augment the state of inequity. In many surveys it turned out to be problematic that the results generally verify the model only under conditions of overpayment and of hourly wage. However, the model could not be supported convincingly concerning overpayment and piece wage (Steers et al., 1996). Moreover, the question of how a person chooses the referent is largely unsolved: whether a person is chosen within or outside the company and whether the referent is exchanged over the years. Furthermore, there is little knowledge about the strategy of the reduction of tension chosen. Because of a lack of research in this area, it is not yet possible to generalize the applicability of the equity theory with regard to the effect of nonfinancial rewards. Finally, it seems as if the model's predictable and expectable possibilities to react to solve the problem of inequity that a person can choose from are too limited. Despite these limitations, Adams's motivation model offers the option of explaining and predicting attitudes and reactions of employees (job satisfaction, quitting the job, or absenteeism) on the basis of their rewards and contributions.

3.2.4 Locke's Goal-Setting Theory

Locke (1968) holds the view that the conscious goals and aims of persons are the essential cognitive determinants of their behavior. Thereby, values and value judgments play an important role. Humans strive for achievement of their goals to satisfy their emotions and desires. Goals give direction to human behavior, and they guide thoughts and actions. The effect of such goal-setting processes is seen primarily in the fact that they (1) guide attention and action, (2) mobilize effort, (3) increase perseverance, and (4) facilitate the search for adequate strategies of action.

Meanwhile, the goal-setting theory (Figure 7) has attracted wide interest among theorists and practitioners and has received convincing and sustained support by recent research (Tubbs, 1986; Mento et al., 1987). Locke and Latham (1990) point to almost 400 studies dealing solely with the aspect of the difficulty of goals. Through increasing insights into the effect of goal setting on work performance, the original model could be widened.

In organizational psychology, goals serve two different purposes under a motivational perspective: (1) They are set jointly by employees and superiors to serve as a motivational general agreement and mark of orientation that can be aimed at; and (2) goals can serve as an instrument of control and mechanism of leadership to reach the overall goal of the organization with the help of employees' individual goals.

According to Locke, from a motivational perspective, a goal is something desirable that has to be obtained. As a result, the original theory of 1968 postulates that work performance is determined by two specific factors: the difficulty and specificity of the

Figure 7 Locke's goal-setting theory.

goal. *Difficulty of the goal* relates to the degree to which a goal represents a challenge and requires an effort. Thereby, to work as an incentive, the goal has to be realistic and achievable. The correctness of this assumption has already been proven by a large number of early studies (Latham and Baldes, 1975; Latham and Yukl, 1975). Locke postulates that the performance of a person can be increased in proportion to the increase in the goal's difficulty, until performance reaches a maximum level. *Specificity of goal determination* refers to the degree of distinctness and accuracy with which a goal is set. Correspondingly, the goals to give one's best or to increase productivity are not very specific, whereas the goal to increase turnover by 4% during the next six months is very specific. Goals referring to a certain output or profit or to a reduction in costs are easy to specify. In contrast, goals concerning ethical or social problems and their improvement such as job satisfaction, organizational culture, image, or working atmosphere, are difficult to grasp in exact terms (Latham and Yukl, 1975).

Set goals not always result directly in actions but can exist within the person over a longer period of time without a perceivable effect. To become effective, a commitment of the person toward such goals is necessary. The greater the commitment (i.e., the greater the wish to achieve the goal), the more intensive and persistent the person's performance will be influenced by it. It facilitates concentration on action processes and at the same time insulates the person from distractions by potential disturbing variables (e.g., alternative goals). Today, there is abundant research pointing to an extension of the model that would make it possible to meet the complexity of the motivational process concerning setting goals in organizations (Locke and Latham, 1990). A newer version of the theory states that goal-oriented effort is not only a function of the difficulty and specificity of the goal, but also of two other goal properties: acceptance and commitment. *Acceptance* refers to the degree to which one views a goal as one's own; *commitment* says something about the degree to which an employee or a superior is personally interested in achieving a goal. Tubbs (1993) pointed out that factors such as active participation while setting goals, challenging and at the same time realistic goals, or the certainty that the achievement of the goal will lead to the personally appreciated rewards, particularly facilitate goal acceptance and commitment (Kristof, 1996).

The model is being enhanced and improved continuously (Mento et al., 1992; Wofford et al., 1992; Tubbs et al., 1993; Austin and Klein, 1996). The latest developments analyze particularly the role of expectancies, including the differentiation pointed out by Bandura between the *outcome expectation* (the expectation that an action will lead to a certain outcome) and the *efficacy expectation* (the expectation that a person is able to conduct the necessary action successfully) (Bandura, 1982, 1989).

Four central insights are relevant for the practice of organizational psychology in particular:

1. Specific goals (e.g., quotas, marks, or exact numbers) are more effective than vague and general goals (e.g., do your best).

2. Difficult, challenging goals are more effective than relatively easy and common goals. Such goals have to be reachable, though; otherwise, they have a frustrating effect.

3. Accepted goals set in participation are to be preferred over assigned goals.

4. Objective feedback about the advances attained in respect to the goal is absolutely necessary but is not a sufficient condition for the successful implementation of goal setting.

Goal-setting offers a useful and important method to motivate employees and junior managers to achieve their goals. These persons will work toward their set goals in a motivated way as long as these are defined exactly and are of a medium difficulty, if they are accepted by employees, and if they show themselves committed to the set goals. The correctness of the assumptions of the theory has been tested in a variety of situations. It turns out that the variables "difficulty" and "specificity" of the goal stand in close relation to performance. Other elements of the theory, such as goal acceptance or commitment, have not been examined that often. Besides, there is little knowledge about how humans accept their goals and how they develop a commitment to certain goals. The question of whether this is a real theory or simply represents an effective technique of motivation has been discussed many times. It has been argued that the process of setting goals constitutes too narrow and rigid a perspective on the employee's behavior. Moreover, it is essential to state that important aspects cannot be quantified that easily. Additionally, goal setting may focus attention on short-term goals, leading to a detriment of long-term considerations. Furthermore, there are other critical appraisals of the theory of goal setting as a motivational instrument of organizational psychology. Setting difficult goals may lead to a higher probability that managers and employees will develop a higher tendency toward risk, which could possibly be counterproductive. In addition, difficult goals can cause stress. Other working areas for which no goals were set might be neglected, and in extreme cases, goal setting may lead to dishonesty and deception.

A practical application of the goal-setting theory is benchmarking. *Benchmarking* refers to the process of comparing the working and/or service processes of one's own company to the best types of processing and production and the results that are detectable in this branch of business. The objective is to identify necessary changes to improve the quality of one's own products and services. The technique of goal setting is used to initiate the necessary activities, to identify the goals that have to be pursued, and to use these as a basis of future actions. To improve one's own work, production, and fabrication, not only the processes within but also those outside the organization are inspected. Through this procedure, several advantages emerge for the company: (1) Benchmarking enables the company to learn from others, (2) the technique places the company in a position to compare itself with a successful competitor, with the objective of identifying strategies of improvement, and (3) benchmarking helps to make a need for change visible in the company by showing how one's own procedures and the assignment of tasks have to be changed and how resources have to be reallocated.

A broader perspective of goal setting in terms of a motivational function is the process of *management by objectives* (MBO). The term refers to a joint process of setting goals with the participation of subordinates and superiors in a company; by doing this, the company's objectives circulate and are communicated top down (Rodgers et al., 1993). Today, MBO is a widespread technique of leadership and motivation featuring substantial advantages: MBO (1) possesses a good potential for motivation, helping to implement the theory of goal setting systematically into the organizational process of a company, (2) stimulates communication, (3) clarifies the system of rewards, (4) simplifies performance review, and (5) can serve managers as a controlling instrument.

Although the technique has to be adjusted to the specific needs and circumstances of the company, there is a general way of proceeding. Top management has to draw up the global objectives of the company and has to stand personally for implementation of the MBO program. After top management has set these goals and has communicated them to the members of the company, superiors and the respective assigned or subordinate employees have to decide jointly on appropriate objectives. Thereby, each superior meets with each employee and communicates the corresponding goals of the division or department. Both have to determine how the employee can contribute to the achievement of these goals in the most effective way. Here, the superior works as a consultant to ensure that the employee sets realistic and challenging goals that are at the same time exactly measurable and verifiable. Finally, it has to be ensured that all resources the employee needs for the achievement of the goals are available. Usually, there are four basic components of a MBO model: (1) exact description of the objective, (2) participation in the decision making, (3) an explicit period of time, and (4) feedback on the work performed.

Generally, the time frame set for achievement of the objectives is one year. During this period the superior meets on a regular basis with each employee to check progress. It can turn out that because of new information, goals have to be modified or that additional resources are necessary. At the end of the set time frame, each superior meets with each employee for a final appraisal conversation to assess to what degree the goals have been reached and the reasons for this conclusion. Such a meeting also serves to revise the proposed figures and performance levels, to determine changes in payment, and as a starting session for a new MBO cycle in the following year.

Overall, goal-setting theory can be considered as being confirmed rather well for individual behavior (Kleinbeck et al., 1990), while its effect on groups or even entire organizations seems to depend on additional conditions that are not yet clarified sufficiently (Miner, 1980). Results are showing that goal setting works better if linked to information about reasonable strategies of action and that both goal setting and strategic information facilitate effort as well as planning behavior (Earley et al., 1987). Moreover, it has to be stressed that the achievement of goals can itself be motivating (Bandura, 1989). Here the thesis of Hacker (1986) is confirmed, stating that tasks are not only directed by motives but can modify motives and needs.

3.2.5 Kelley's Attribution Theory

Research about behavior in organizations has shown that attributions made by managers and employees provide very useful explanations for work motivation. Thus, the theory of attribution (Myers, 1990; Stroebe et al., 1997) offers a better understanding of human behavior in organizations. It is important to point out that in contrast to other motivation theories, the theory of attribution is, rather, a theory of the relation between personal perception and interpersonal behavior.

As one of the main representatives of this direction of research, Kelley (1967) emphasizes that attribution theory deals with the cognitive processes with which a person interprets behavior as caused by the environment or by characteristics of the actor. Attribution theory mainly asks questions about the "why" of motivation and behavior. Apparently, most causes for human behavior are not observable directly. For this reason, one has to rely on cognitions, especially on perception. Kelley postulates that humans are rational and motivated to identify and understand structures of reasons in their relevant environment. Hence, the main characteristic of attribution theory consists in the search for attributions. At present, one of the most frequently used attributions in organizational psychology is the locus of control. By means of the dimension internal/external, it can be explained whether an employee views his or her work outcome as dependent on either internal or external control (i.e., whether the employee considers himself or herself as able to influence the result personally, e.g., through his or her skills and effort, or thinks that the result lies beyond his or her own possibilities of influence and control). It is therefore important that this perceived locus of control has an effect on the performance and satisfaction of the employee. Research studies by Spector (1982) and Kren (1992) point to a correlation of the locus of control and work performance and job satisfaction that is not only statistical in nature. Concerning the relation between motivation and work incentives, it also seems to take a moderating position. Despite the fact that until now the locus of the control dimensions internality and externality for the explanation of work motivation has been the only link on the part of organizational psychology to the attributional approach, it has been suggested repeatedly that other dimensions be examined as well. Weiner (1985) proposes a dimension of stability (fixed vs. variable, e.g., in terms of the stability of internal attribution concerning one's own abilities).

Kelley suggests the dimensions consensus (refers to persons: do others act similarly in this situation?), distinctiveness (refers to tasks: does the person act on this task as on other tasks?), and consistency (refers to time: does the person always act in the same way over time?). These dimensions will influence the type of attribution that is made, for example, by a superior (Kelley, 1973). If consensus, consistency, and distinctiveness are high, the superior will attribute the working behavior of an employee to external (i.e., situational or environmental) causes (e.g., the task cannot be fulfilled better because of external circumstances). If consensus is low, consistency high, and distinctiveness low, causes are attributed to internal or personal factors (e.g., the employee lacks skills, effort, or motivation) (Mitchell and Wood, 1980).

The attributional approach offers promising possibilities for organizational psychology to explain work behavior. However, it has to be mentioned that various sources of error have to be taken into account. For example, there is an obvious tendency of managers to attribute situational difficulties to personal factors (skills, motivation, attitudes). But the reverse also occurs quite frequently: In too many cases, failure is attributed to external factors although personal factors are actually responsible.

3.2.6 Summary of the Process Models

Process theories refer more closely to actual behavior; they take into account connections between appraisals and results and thereby fill a gap left open by the content theories. Moreover, they do not assume that all people are led by the same motives but emphasize, instead, that each person can have his or her own configuration of desired and undesired facts. They thus open a perspective toward conflicting motives; they can explain why a highly valued behavior might not be executed (e.g., because another one is valued even more or because the expectation to be successful is too small); to the tempting tendency to presume human motives, they oppose the demand not to assume but to study these; and they point out processes that support the connection of highly valued facts and concrete actions (e.g., goal setting, feedback, clear presentation of consequences).

Nevertheless, the question remains unanswered as to what typically is valued as high by which persons—hence, the issue of the contents cannot be evaded. And these are, despite big differences between persons, not so arbitrary that there is nothing that could be said about them. Therefore, the great strength of not assuming any contents becomes a weakness as well. Nevertheless, process models have received a lot of support and they have practical implications. But concerning the contents, not everything should be set aside unseen.

Ultimately, the two approaches deal with completely different aspects. The link between basic motives and specific actions is manifold and indirect, influenced by lots of aspects (e.g., expectations, abilities, situational restrictions); that is why a close direct relationship cannot be expected. To consider the contents (which do not differ much in the various approaches) as usually being effective without presuming them stiffly for each person and to take into account at the same time, the characteristics specified by the process models can provide guidance for practice and theoretical integration possibilities that seem already to have emerged (Locke and Henne, 1986; Six and Kleinbeck, 1989).

4 POSSIBLE APPLICATIONS OF MOTIVATION MODELS

A question arises as to ways in which the theories of motivation can be applied to the world of work. All of these models can contribute in different ways to a better understanding and predetermination of human behavior and reactions in organizations. This has already been proven for each model. In the following sections we examine the possible applications more closely: involvement and empowerment, remuneration, work and task design, working time, and motivation of various target groups.

4.1 Involvement and Empowerment

The role of involvement and empowerment as motivational processes can be seen with regard to need-oriented actions as well as with regard to the postulates of the expectancy theory of motivation. Employee involvement is creating an environment in which people have an impact on decisions and actions that affect their jobs (Argyris, 2001). Employee involvement is not the goal nor is it a tool, as practiced in many organizations. Rather, it is a management and leadership philosophy about how people are most enabled to contribute to continuous improvement and the ongoing success of their work organization. One form of involvement is *empowerment*, the process of enabling or authorizing a person to think, behave, take action, and control work and decision making in autonomous ways. It is the state of feeling empowered to take control of one's own destiny. Thus, empowerment is a broad concept that aims at reaching involvement in a whole array of different areas of occupation.

Empowerment does not primarily use involvement to reach higher satisfaction and to increase personal performance. The vital point within this process is, rather, the overall contribution of human resources to the efficiency of a company. Employees who can participate in the decision-making process show higher commitment when carrying out these decisions. This addresses simultaneously the two factors in the need for achievement: The person feels appreciated and accepted, and responsibility and self-esteem are increasing.

Furthermore, involvement in decision making helps to clear up expectations and to make the connection between performance and compensation more transparent. Involvement and empowerment can be implemented in various fields, which may concern work itself (e.g., execution, tools, material) or administrative processes (e.g., work planning). Moreover, involvement means participation in relevant decisions concerning the entire company. Naturally, the underlying idea is that employees who are involved in decisions have a considerable impact on their work life, and that workers who have more control and autonomy over their work life are motivated to a higher extent, are more productive, and are more satisfied with their work and will thus show higher commitment.

Organizations have experimented for years with lots of different techniques to stimulate involvement and empowerment among their employees and executives.

Quality circles as an integrative approach have attracted particular attention. Their task is to solve work problems with the help of regular discussions among a number of employees. In connection with quality management and qualification programs, quality circles play a crucial role in learning organizations. Not only do they make possible a continuous improvement process and an ideal integration of the experiences of every employee, but they also allow participants to acquire various technical skills and social competences.

To ensure these positive effects, it is crucial that quality circles not pursue solely economic goals. Specific needs and requests of employees regarding improvements in job quality or humanization of work must be considered equally. However, the evidence is somehow restricted. Although a lot of organizations say that quality circles lead to positive results, there is little research about long-term efficiency (Griffin, 1988).

Another commonly used method of involvement and empowerment of employees is that of work teams. These can be, for example, committees, cross-functional work teams, or interfunctional management teams. They comprise representatives of different departments (e.g., finance, marketing, production) that work together on various projects and activities (e.g., launch of a new product) (Saavedra et al., 1993).

4.2 Remuneration

The issue of financial compensation for both executives and employees plays a crucial role in practice (Ash et al., 1985; Judge and Welbourne, 1994; Schettgen, 1996). According to Maslow, payment satisfies the lower needs (physiological and safety needs), and Herzberg considers it as one of the hygienes. Vroom regards payment as a result of the second level that is of particular valence for the employee. According to his VIE theory, the resulting commitment will be high when work leads to a result and a reward highly appreciated by the employee. Adams's equity theory puts emphasis on the relation between commitment and return of work performance. This input/output relation is correlated with a comparison person called a *referent*, a comparison position, or a comparison profession. Perceived inequality leads to attempts at its reduction by changing work commitment or return. Perceived underpayment has, therefore, a negative impact on performance in terms of quality as well as quantity.

The same applies to fringe benefits such as promotions or job security. Three different factors are to be considered which determine the efficiency of payment schemes with regard to the theories of motivation mentioned above: (1) the type of relation between the payment scheme and a person's work performance, (2) the subjective perception of these connections, and (3) different assessments of payment schemes by employees in the same work situation. In addition to that, there are internal performance-oriented benefits, staff shares, and employee suggestion schemes either within or outside the department. Eventually, various additional premiums, such as an increased Christmas bonus or a company pension, have to be mentioned; the latter, especially, is a vital factor in times of diminishing governmental provision.

4.3 Work Design

The way in which tasks are combined, the degree of flexibility that executives and employees have, and the existence or lack of important support systems in a company have a direct impact on motivation and on work performance and satisfaction. Motives can be generated within a job by extending work contents and demands.

Organizational psychological research since the 1960s has shown repeatedly that a certain task complexity positively affects work behavior and that lots of employees prefer jobs that contain complexity and challenge. But taking a closer look at the actual development of work tasks, one has to conclude that the results noted above have, due largely to economic considerations, generally been ignored, and that most work flows have been organized on the basis of economic and technical efficiency. They were primarily attempts to create very specialized work roles and to control employees. However, the situation today has changed significantly.

Executives have acknowledged that the most valuable resource available is employee commitment, motivation, and creativity. Only this lets a company stay healthy and competitive in global markets. Technology and downsizing alone can achieve neither flexibility nor new, customer-oriented products and services. It is motivation that must be improved. This can be realized partially by restructuring work tasks and working processes.

4.3.1 Strategies of Work Design

Corrective Work Design It is a widespread experience that working systems and work flows have to be changed after their introduction into a company in order to adapt them to specific human needs. Often, these corrections are necessary because of insufficient consideration of anthropometric or ergonomic demands. Such procedures, called *corrective work design*, are always necessary when ergonomic, physiological, psychological, safety–technical, or judicial requirements are not (or not sufficiently) met by planners, design engineers, machine producers, software engineers, organizers, and other responsible authorities. Corrective work design that is at least somewhat effective often causes considerable economic costs. However, its omission can potentially cause physical, psychophysical, or psychosocial harm. Expenditures on corrective work design have to be borne by the companies, whereas the latter costs are carried by the employees affected and thus indirectly by the economy. Both types of costs can be avoided or at least reduced considerably if corrective work design is replaced as far as possible by preventive work design.

Preventive Work Design Preventive work design means that concepts and rules of work science are considered when working systems and work flows are being developed. Hence, possible damage to health and well-being are taken into account when job division between humans and machines is being determined.

Prospective Work Design The strategy of prospective work design arises due to the demand for personality-developing jobs. The criterion of personality development puts an emphasis on the fact that the adult personality develops mainly by dealing with its job. Jobs and working conditions that are personality developing ensure that a person's characteristic strengths can be kept and further developed. Prospective work design means that possibilities of personal development are created intentionally at the stage of planning or reengineering of work systems. This is done by creating a scope of action that can be used and, if possible, extended by employees in different ways. It is crucial for the strategy of prospective work design not to regard it as equivalent to future-oriented work design. Instead, the creation of work that provides possibilities of development for employees should be seen as a vital feature.

Differential Work Design The principle of differential work design (Ulich, 1990, 1991) takes into account differences between employees: that is, it considers interindividual differences (e.g., different working styles) in dealing with jobs. Employees can choose between different work structures according to their individual preferences and abilities. Since human beings develop through dealing with their jobs, changes among work structures and altering of the structures should be made possible.

The possibility of choosing among alternatives and of correcting a choice, if necessary, means that there will be no need to look for the one best way of organizing jobs and work flows. However, this implies a considerable increase in autonomy and control over one's working conditions. Furthermore, such possibilities of job change lead to a reduction in unbalanced strains.

Dynamic Work Design Dynamic work design does not mean a choice between different existing structures but deals with the possibility of continuously changing and extending existing work structures and creating new ones. Dynamic work design takes into account intraindividual differences (e.g., different learning experiences) in dealing with jobs.

Empirical Evidence Concerning Differential and Dynamic Work Design Without considering interindividual differences, neither optimal personal development nor optimal efficiency can be guaranteed. Differences in cognitive complexity and memory organization may play a role that is just as important as differences in the degree of anticipation, the motivational orientation, the style of learning, or the style of information processing. Empirical data support the assumption that the concept of the one best way, which only needs to be found, constitutes a fundamental and far-reaching error of traditional work design. Moreover, it becomes clear that a standard job structure that is optimal for every employee cannot exist (Zink, 1978). This is consistent with Triebe's (1980, 1981) investigations: In the absence of detailed work schedules, there are interindividually different possibilities as to how to assemble car engines. He observed that workers developed a whole array of different strategies and that these do by no means necessarily lead to differing efficiency or effectiveness. Conversely, such results mean that strict work schedules for an operating sequence, which are supposed to be optimal, may sometimes even lead to inefficient work. Differential work design stands out deliberately against the classic search for the one best way in designing work flows. Considering interindividual differences, it is especially appropriate to offer alternative work structures to guarantee optimal personal development in the job.

To take into account processes of personal development (i.e., intraindividual differences in time), the principle of differential work design must be complemented by the principle of dynamic work design. Steinmann and Schreyögg (1980) have pointed out that when facing choices, some employees might choose the conventional working conditions they are used to. These employees have developed a resigned general attitude and a state of more or less apathetic helplessness because of unchallenging tasks and missing prospects. Hence, it is necessary to develop procedures that make it possible to emphasize a worker's subject position, to reduce barriers to qualification, and to promote readiness for qualification (Alioth, 1980; Ulich, 1981; Baitsch, 1985; Duell and Frei, 1986).

More generally, differential work design can form a link between work design measures, different conditions, and needs of individuals.

It is an important principle of modern work design to offer different work structures. Zülch and Starringer (1984) describe its realization in business by examining the production of electronic flat modules. A macro-work system was created in which differently skilled and motivated employees were simultaneously offered different forms of work organization with different work items. The authors conclude that these new work structures were seen as interesting and motivating.

According to Grob (1985), who provides data and hints for possible extensions, this structure can be applied not only to the production of flat modules but to all jobs in the company that (1) require several (normally, 4 to 10) employees, (2) have to be carried out frequently in different types and variants, (3) have to be managed with few workshop supplies, and (4) may have a crucial impact on reducing the duration of the cycle time. In this context it is especially important that Zülch and Starringer were able to prove theoretically that the concept of differential work design can even be realized when facing progressing automation. The production of electronic flat modules for communication devices can serve as an example. It turned out that a useful division into automatic and human operations can be facilitated by not automating all possible operations.

In the beginning, interindividual differences concerning the interaction between individuals and computers were examined almost entirely with regard to the user's role as beginner, advanced user, or expert when dealing with technical systems. It became more and more obvious, however, that the impact of differential concepts goes far beyond that. Hence, Paetau and Pieper (1985) report, with reference to the concept of differential work design, laboratory experiments that examined whether test subjects with approximately the same skills and experiences and given the same work items develop the same preferences for certain systems. Given various office applications, individuals at first preferred a high degree of menu prompting. But with increasing experience, accordance in preferences declined significantly. Due to their results and the experiences and concepts of other authors, Paetau and Pieper (1985) do not see any point in looking for an optimal dialogue design. Demands for programmable software systems, flexible information systems, adaptability of groupware, adaptability of user interfaces, or choices between alternative forms of dialogue put emphasis on the necessary consideration of inter- and intraindividual differences by means of differential and dynamic work design. This has been reflected in the EC guideline 90/270/EWG and in ISO 9241 (Haaks, 1992; Oberquelle, 1993; Rauterberg et al., 1994). Triebe et al. (1987) conclude that creating possibilities for individualization with the help of individually adaptable user interfaces will probably be one of the most important means to optimize strain, prevent stress, and develop personality.

Possible achievements and results of the creation of these scopes of development have mainly been examined experimentally. Results of these studies (e.g., Ackermann and Ulich, 1987; Greif and Gediga, 1987; Morrison and Noble, 1987) support the postulate of abolishing generalizing one-best-way concepts in favor of differential work design. Both the participative development of scopes for action and the possible choice between different job structures will objectively increase control in the sense of being able to influence relevant working conditions. This shows at the same time that possibilities of individualization and differential work design are determined at the stage of software development. This is similar for production, where the scope for action is determined mainly by design engineers and planners. Interindividually differing proceedings also matter in design engineers' work. This is revealed in a study by von der Weth (1988) examining the application of concepts and methods of psychological problem solving in design engineering. Design engineers, who acted as test subjects, had to solve a construction task while thinking aloud. Their behavior was registered by video cameras. One of the results that applies to this context says that test persons with adequate problem-solving skills do not show homogeneous procedures. Both the strategy of putting a draft gradually into concrete terms as well as joining solutions of single detailed problems together into a total solution led to success. Thus, the author concludes that an optimal way of reaching a solution that is equally efficient for everybody does not exist. He assumes that this is caused by different styles of behavior, which are linked to motivational components such as control needs. As a conclusion, design engineers must be offered a whole array of visual and linguistic possibilities for presenting and linking information. If the system forces a special procedure on users, a lot of creative potential is lost.

Participative Work Design Another strategy, participative work design, has to be considered (Ulich, 1981). In participative work design, all persons concerned with work design measures are included. This participation must focus on all stages of a measure (e.g., including preliminary activities such as evaluation of the actual situation). Participative work design must not result in participation of persons concerned only when one is stuck (e.g., due to technical problems). It must not be a pure measure to get decisions accepted. The various principles overlap and thus cannot often be identified unequivocally. Different strategies of work design pursue different goals that differ not only qualitatively but also in terms of range and time horizon (Ulich, 1980).

The design of work structures implies change in technical, organizational, and social working conditions to adapt structures to workers' qualifications. In this way, they aim at promoting personality development and well-being of workers within the scope of efficient and productive work flows. Criteria are needed to assess jobs and the relating design measures concerning this aim. Work psychology and work science can provide a large number of findings and offer support for this (e.g., Oppolzer, 1989; Greif et al., 1991; Leitner, 1993).

4.3.2 Characteristics of Task Design

From the point of view of industrial psychology, task design can be seen as an interface between technical or organizational demands and human capabilities (Volpert, 1987). As a consequence, work content and work routine will be determined fundamentally by the design of tasks. Therefore, task design plays a key role in effectiveness, work load, and personality development. Hence, task design takes precedence over the design of work materials and technology, since their use is determined fundamentally by work content and work routine.

Task analysis methods can be used to improve task design. Luczak (1997) sees the fundamental idea of task analysis "in a science-based and purpose-oriented method or procedure to determine, what kind of elements the respective task is composed of, how these elements are arranged and structured in a logical or/and timely order, how the existence of a task can be explained or justified ... and how the task or its elements can be aggregated to another entity, composition or compound." Their aim is to transform the task into complete activities or actions.

Design Criteria Tasks should be workable, harmless, free of impairment, and sustain the development of the working person's personality (Luczak et al., 2003). The fundamental aim of task design is to abolish the Tayloristic separation of preparing, planning, performing, and controlling activities (Locke, 2003) so as to make complete activities or actions possible (Hacker, 1986; Volpert, 1987). The results are tasks that offer goal-setting possibilities, the choice between different work modes, and the control of work results. Essentially, it is about granting people a certain scope in decision making.

The attempt in sociotechnical system design (e.g., Alioth, 1980; Trist, 1990) to name other design criteria can be seen as important for the development of abilities and motivation (Ulich, 1993). In addition to the aforementioned autonomy, these are (1) completeness of a task, (2) skill variety, (3) possibilities of social interaction, (4) room for decision making, and (5) possibilities of learning and development. Tasks designed according to these guidelines promote employee motivation, qualifications, and flexibility and are therefore an excellent way to provide and promote the personnel resources of a company in a sensible and economical manner.

Similar dimensions can also be found in the *job characteristics model* of Hackman and Oldham (1976). In the concept of *human strong points* (Dunckel et al., 1993), these criteria are taken up and widened. Human criteria for the assessment and design of tasks and work systems are formulated: (1) Work tasks should have a wide scope concerning actions, decisions, and time; (2) the working conditions and especially the technology should be easily comprehensible and changeable in accordance with one's own aims; (3) task fulfillment should not be hindered by organizational or technical conditions; (4) the work tasks should require sufficient physical activity; (5) the work tasks should enable dealing with real objects and/or direct access to social situations; (6) human work tasks should offer possibilities for variation; and (7) human work tasks should enable and promote social cooperation as well as direct interpersonal contacts.

According to the definition of work science (Luczak et al., 1989), human criteria are only one theoretically justified way for the assessment and design of work under human aspects (Dunckel, 1996). Neuberger (1985), for example, names other aims of humanization, such as dignity, meaning, security, and beauty. It has to be considered that personality-supporting task design also has a positive effect on the use of technology and on customer orientation; in addition to that, it promises clear economic benefits (Landau et al., 2003). Furthermore, it is interesting to know how work tasks should be designed to create task orientation. Task orientation promotes the development of personality in the process of work and motivates employees to perform tasks without requiring permanent compensation and stimulation from the outside.

Task Orientation Task orientation describes a state of interest and commitment that is created by certain characteristics of the task. Emery (1959) names two conditions for the creation of task orientation: (1) The working person must have the control over the work process and the equipment needed for it; and (2) the structural characteristics of a task need to be of a type that sets off in the working person the strength for completing or continuing the work. The extent of control over the work process depends not only on the characteristics of the task or the delegated authority, but above all, on the knowledge and competence that are brought into dealing with a task.

For those motivational powers that have a stimulating effect on completing or continuing the work, the task itself has to appear to be a challenge with realistic demands (Alioth, 1980). Apart from that, it should be neither too simple nor too complex. Summing up the statements of Emery and Emery (1974), Cherns (1976), and Emery and Thorsrud (1976), the following characteristics of work tasks encourage the process of a task orientation: completeness, skill variety, possibilities for social interaction, autonomy, possibilities for learning and development. Furthermore, Emery and Thorsrud mention the aspect of meaning. Therefore, work should make a visible contribution to the usefulness of a product for the consumer.

These characteristics correspond so well with the characteristics of tasks derived theoretically by Hackman and Lawler (1971) and Hackman and Oldham (1976) that Trist (1981) points out that this degree of agreement is exceptional in such a new field and has placed work redesign on a firmer foundation than is commonly realized.

Task Completeness An early description of what is now called a *complete task* can be found in Hellpach's article about group fabrication (Hellpach, 1922). He came to the conclusion that it should be a main objective to overcome fragmentation in favor of complete tasks, in the sense of the at least partial restoration of the unity of planning, performing, and controlling. Incomplete tasks show a lack of possibilities for individual goal setting and decision making, for the development of individual working methods, or for sufficiently exact feedback (Hacker, 1987). Research could show, furthermore, that the fragmentation that goes together with classic rationalization strategy can have negative effects on a person in many areas. Restrictions on the scope of action can lead to indisposition and to continuous mental and physical problems. It can also possibly result in the reduction of individual efficiency, especially of mental activity, and passive leisure behavior, as well as in a lower commitment in the areas of politics and trade unions.

Specific consequences for production design resulting from the principle of the complete task may be outlined at this point with the help of some examples:

1. The independent setting of aims requires a turning away from central control to decentralized workshop control; this creates the possibility of individual decision making within defined periods of time.

2. Individual preparations for actions require the integration of planning tasks into the workshop.

3. Choice of equipment can mean, for example, leaving to the constructor the decision of using a drawing board (or forming models by hand) instead of using computer-aided design for the execution of certain construction tasks.

4. Isolated working processes require feedback as to progress, to minimize the distance and to make corrections possible.

5. Control with feedback as to the results means to transfer the functions of quality control to the workshop itself.

First, a complete task is complete in a sequential sense. Besides mere execution functions, it contains preparation functions (goal setting, development of the way of processing, choosing useful variations in the work mode), coordination functions (divide the tasks among a variety of people), and control functions (get feedback about the achievement of the goals set). Second, complete tasks are complete in a hierarchical regard. They make demands on different alternating levels of work regulation. It should be noted that complete tasks, because of their complexity, can often be designed only as group tasks.

Job Enrichment *Job enrichment* is commonly described as changes regarding the content of an employee's work process. Herzberg (1968) points out that with the help of this method, motivation is being integrated into an employee's work process to improve his or her satisfaction and performance. Job enrichment refers to the *vertical* widening of a work role. The aim is to give employees more control concerning planning, performing, and appraisal of their work. Tasks will be organized such that they appear to be a complete module to heighten identity with the task and to create greater variety. Work is to be experienced as meaningful, interesting, and important. Employees will receive more independence, freedom, and heightened responsibility. On top of that, employees will get regular feedback, enabling them to assess their own performance and, if necessary, to adjust it. In this context, customer orientation is of special importance. It will result nearly automatically in direct and regular feedback. The customer can be either internal or external, but the relationship must be direct.

Reports from large companies such as Imperial Chemical Industries (Paul et al., 1969) and Texas Instruments (Myers, 1970) tell of the success of job enrichment programs. Ford (1969) mentioned about 19 job enrichment projects from the American Telephone and Telegraph Company; nine were called extraordinary successful, nine successful, and one a failure. The success was often rated by means of productivity and quality reference numbers, the rate of times absent, and of turnover, as well as by examinations of the attitude of employees.

Job Enlargement *Job enlargement* refers to the *horizontal* expansion of a work role. In this process the number and variety of tasks may be increased to diversify and achieve a motivational effect. Employees will perform several operations on the same product or service. Job enlargement therefore intends to string together several equally structured or simple task elements and, by doing that, to enlarge the work cycle. It becomes obvious that job enlargement touches primarily on the work process, whereas an attempt at job enrichment also concerns the organizational structure. However, it is only the realization of concepts of vertical work expansion that can contribute to overcoming the Tayloristic principle of separating planning and performing and therefore to a work arrangement that develops the personality of an employee. On the other hand, research results have shown that outcomes regarding a heightened challenge or motivation have been rather disappointing (Campion and McClelland, 1993).

Job Rotation *Job rotation* deals with lateral exchange of a work role. If a strong routine in the work becomes a problem for employees, if the tasks are no longer challenging, and the employee is no longer motivated, many companies make use of the principle of rotation to avoid boredom. An employee will usually be transferred from one task to an other periodically. This principle is also favorable for the company: Employees with a wider span of experience and abilities allow for more flexibility regarding adaptation to change and the filling of vacancies. On the other hand, this process has disadvantages: Job rotation increases the cost of training, employees will always be transferred to a new position when they are on the highest productive level and have thus reached the highest efficiency, the process can have negative consequences on a well-operating team, and the process can have negative effects on ambitious employees, who would like to take on particular responsibilities in a chosen position.

Group Tasks Currently, companies employ project teams, quality circles, and working groups as typical forms of teamwork. The main task of *project teams* is to solve interdisciplinary problems. Unlike quality circles and working groups, however, they work together for only a limited period of time and will be dissolved after having found the solution to a certain problem (Rosenstiel et al., 1994). The group will therefore be put together on the basis of professional criteria and will consist mostly of employees in middle and upper management. Well-founded evaluations about the concept of project teams are still missing. That is why Bungard et al. (1993) discover a clear deficiency of research, although project teams gain more and more importance in companies.

It is typical of *quality circles* that groups do not work together continuously; instead, they meet only at regular intervals. Employees get the chance to think about improvements systematically. Attendance is explicitly optional (i.e., employees need to wish to deal with these questions). Other requirements for the success of quality circles are a usable infrastructure in the company (e.g., a conference room and moderation equipment); company support, especially from middle management; and a business culture that is characterized by participation and comprehensive quality thoughts. Behind this concept is the idea that the people affected are better able than anyone else to recognize and solve their own problems. As a side effect, communication among employees will also improve (Wiendieck, 1986a,b).

Working groups are organizational units that can regulate themselves within defined boundaries. It therefore is a group that is supposed to solve essential problems with sole responsibility. This work form is, among other things, meant to create motivating work contents and working conditions. The concepts of job enlargement, job enrichment, and job rotation are transferred to the group situation (Rosenstiel et al., 1994; Hackman, 2002).

Psychologically, work in a group has two principal intertwined reasons: (1) The experience of a complete task is possible in modern work processes only where interdependent parts are combined to complete group tasks; and (2) the combination of interdependent parts to a common group task makes a higher degree of self-regulation and social support possible. Concerning the first point, Wilson and Trist (1951) as well as Rice (1958) found out that in cases in which the individual task does not allow this, satisfaction can result from cooperation in completing a group task. Concerning the second point, Wilson and Trist are of the opinion that the possible degree of group autonomy can be characterized by how far the group task shows an independent and complete unit. Incidentally, Emery (1959) found out that a common work orientation in a group develops only if the group has a common task for which it can take over responsibility as a group, and if it is able to control the work process inside the group.

The common and complete task is what practically all supporters of the sociotechnical approach call a central characteristic of group work (Weber and Ulich, 1993). The existence of a common task and task orientation also have an essential influence on the intensity and length of group cohesion. Work groups whose cohesion is based mainly on socioemotional relations therefore show less stability than do work groups that have a common task orientation (Alioth et al., 1976).

Hackman's (1987) considerations about group work make clear that the organization of work in groups contributes not only to the support of work motivation but also to an increase in work efficiency and therefore in productivity. However, work motivation and efficiency will not develop without organizational efforts and will not remain without any kind of endeavor. A study conducted by a German university together with six well-known companies and more than 200 employees revealed that insufficient adaptation of the organization to the requirements of teamwork is the biggest problem area (Windel and Zimolong, 1998). Another problem is that companies at first sight believe that teamwork is a concept of better value compared to the acquisition of expensive technical systems. Windel and Zimolong stress that even with teamwork, investments (e.g., into the qualification of employees) are necessary before the concept pays off in the medium and long terms. For the management of a company, this means that in addition to endurance there needs to be trust in the concept.

Teamwork is associated with a variety of dangers for which a company needs to be prepared (e.g., group targets do not orient themselves toward the overall goals of the company). Therefore, systems are needed that are able to develop complete tasks and that can also be used to orient the motivation potentials toward organizational goals. Arbitrarily used scopes of action and motivation potentials can endanger an organization fundamentally. Such instruments need to include the various areas of responsibility and to turn the work of the group into something measurable in order to compare it to the goals set. In addition, it should offer the possibility of assessing the working results of the group regarding their importance for the entire organization and to give feedback to employees. In this way, the productivity of the organization can be increased because at the same time, a high work motivation arises. As a consequence, individual preconditions for performance are used optimally, absenteeism decreases, and work satisfaction increases.

4.3.3 Working Time

Motivation, satisfaction, absenteeism, and work performance can be improved within certain limits by means of the implementation of alternative models of working time. It is also possible to explain this improvement in terms of the reduction in need deficiencies, the achievement of a second-level result and its instrumentality, the principle of affiliation, or the motivators developed. The model that is being used most frequently is *flextime*. Here the employee has to stick to certain mandatory working hours, beyond which it is up to him or her how the rest of the workday is arranged. Such a model can be expanded such that extra hours can be saved to create a free day each month. The advantages of this popular model are considerable for both sides and range from the reduction of absenteeism to cost saving and higher productivity and satisfaction to increased autonomy and responsibility at the workplace (Ralston and Flanagan, 1985;

Ralston et al., 1985). However, there are a lot of professions for which the model is not applicable. *Job sharing* refers to the division of a job or workweek between two or more persons. This approach offers a maximum of flexibility for the employees and for the company.

In the search for new work structures, *teleworking* has been a main point of issue for many years. The term refers to employees who do their job at home at a computer that is connected to their office or their company. The tasks range from programming to processing and analysis of data to the acceptance of orders, reservations, and bookings by telephone. The advantages for the employees and the organization are obvious: For the former it means no traveling to and from work and flexible working hours, and for the latter there is an immense reduction in costs. The disadvantages include a lack of important social contacts and sources of information, and because the employee is no longer integrated into important processes, they might suffer from disadvantages concerning promotions and salary increases.

Other models of working time are also being discussed, such as the *compressed workweek*, with the same number of working hours completed in four days of nine hours each. Supporting this approach, it has been argued that there would be extended leisure time and that employees would not have to travel to and from work during rush hour. It has also been stated that with the help of this model, commitment, job satisfaction, and productivity would be increased and costs would be reduced. Extra hours would no longer be necessary, and rates of absenteeism would be lower. Undoubtedly, the acceptance of this model among employees is rather high, but there are also opponents of the approach. They consider the workday as being too long and believe that problems will arise in trying to structure the demands of private life with those of the job.

A working time model that seems to be especially beneficial for older employees or to fight high unemployment is a reduction in the weekly working time without pay compensation. For older employees this eases the transition to retirement. In the scope of measures against unemployment, this model would stand for a fairer allocation of existing work to a greater number of people without increasing total costs. However, for employees it is most important how they are affected personally rather than the positive effects that the model has on a country's unemployment problem.

Another solution that is being discussed, especially for large projects where considerable overtime can be necessary, is to give the employee a time account. As soon as the project is finished, he or she can take up to two months off while receiving his or her regular pay.

4.3.4 Motivation of Various Target Groups

In an increasingly diversifying working world (Jackson, 1992; Jackson and Ruderman, 1995) individual differences with respect to needs and expectations are wide. On the part of companies, they are hardly taken into account or taken seriously. For this reason, some motivating measures have little effect. This is why organizations of the future will have to strive for more flexibility regarding the structuring of work and work processes. If they want to maintain the commitment, motivation, and creativity of their managers and employees, they will have to consider family-oriented employees as well as dual-career couples. Organizations of the future will have to show just as much interest in fast-trackers pushing early for leadership responsibility as in employees entering a field that is different from their educational background or mid-career persons wanting to take on a completely new profession.

Different needs and values in various professional groups are crucial for whether or not a motivational measure is effective. In academic professions, especially, employees and managers obtain a considerable part of their intrinsic satisfaction out of their job. They have a very strong and long-lasting commitment toward their field of work, and their loyalty to their special subject is often stronger than that toward their employer. For them, remaining up to date with their knowledge is more important than financial aspects, and they will not insist on a working day with only seven or eight hours and free weekends. What they do possesses a central value in their lives. That is why it is important for these people to focus on their work as their central interest in life. What is motivating them are challenging projects rather than money or leisure time. They wish for autonomy to pursuit their own interests and to go their own way. Such persons can be motivated very well by means of further education, training, participation in workshops and conferences—and far less by money.

In future organizations, many employees will be working only temporarily, by project, or part time. People may experience the same working conditions in very different ways. Part-time, project, or temporary work is seen by one group as lacking security or stability, and such employees will not identify themselves with an organization and will show little commitment. But there are also a lot of people for whom this status is convenient. They need a lot of personal flexibility and freedom and are often mothers, older employees, or persons who dislike restrictions by organizational structures. For this group of persons, the long-term prospect of permanent status is more important, and therefore more motivating, than momentary financial incentives. Just as motivating, probably, is the offer of continuous education and training that helps to augment one's market value.

5 PRACTICAL EXPERIENCES FROM EUROPEAN STUDIES

Extreme working conditions, which could have been met in the early days of industrialization, with excessive daily hours, children's work, high risks of accidents, and the nonexistence of social security programs, are features of the past, at least in most industrialized countries. In these countries, however, a successive change in attitude with respect to working conditions took place in the early 1960s. Conditions of the working environment such as noise, toxic substances, heat, cold, high physical loads, high levels of concentration, monotonous short-cycle repetitive work, or impaired communication met decreasingly with the pretensions of working people (Kreikebaum and Herbert, 1988; Staehle, 1992). As a result, working people increasingly depreciated shortcomings in work design and answered with "work to rule" (Schmidtchen, 1984) or hidden withdrawal. In Germany, an increased number of complaints about working conditions, increasing dissatisfaction, and decreasing working moral were observed, for instance, by Noelle-Neumann and Strümpel (1984).

Since the early 1970s, the term *quality of working life* (QWL) has developed into a popular issue both in research and in practice. In its early phase it has often been defined as the degree to which employees are able to satisfy important personal needs through their work and experience with the organization. According to the ideas of organizational psychology, projects of organizational development should particularly create a working environment in which the needs of the employees are satisfied. Management saw this movement as a good possibility of increasing productivity and therefore supported it. What was needed were satisfied, dedicated, motivated, and competent employees who were connected emotionally to the organization. These early perceptions of quality of working life have been concretized in the meantime, but also diversified. Like any other movement with broadly and diffusely defined goals, this one produced programs, claims, and procedures differing in respect to content and a variety of at least overlapping terms often being used synonymously. Besides *quality of working life*, there are terms such as *humanization of work* (Schmidt, 1982; Hettlage, 1983), *sociotechnical Systems* (Cummings, 1978), *industrial democracy* (Andriessen and Coetsier, 1984; Wilpert and Sorge, 1984), *structuring of work*, and others.

The typical characteristics of a high-quality work environment can be summarized as follows: (1) adequate and fair payment; (2) a secure and healthy work environment; (3) guaranteed basic rights, including the principle of equality; (4) possibilities for advancement and promotion; (5) social integration; (6) integration of the entire life-time or life span; (7) an environment that fosters human relations; (8) an organization with social relevance; and (9) an environment that allows employees to have a say or control of decisions concerning them.

Pursuit of these goals for the creation of a work environment with a high-quality work structure reaches back to Herzberg et al. (1959), who assigned them an extremely important role. In research on extrinsic or hygiene factors, Herzberg came to the conclusion that motivation of employees is not increased through higher payment, different leadership styles, or social relations at the workplace but only through considerable change in the type and nature of the work itself (i.e., task design). Certainly, today, there are still a lot of workplaces (e.g., in mass production and in the service sector) that do not meet these requirements, although they have been altered or even replaced to an increasing degree by means of new technologies.

In the 1970s and 1980s, in the light of the quality of working life discussion, many European countries initiated programs, of considerable breadth to support research for the humanization of working life. From a motivational point of view, the focus of interest in these action research projects was on (1) avoiding demotivation caused by inadequate workplace and task designs, and (2) supporting intrinsic motivation by identifying and designing factors that provide such potential. Researchers basically built on earlier empirical studies that investigated informal groups of workers. Studies were carried out in the United States, where groups in the Hawthorne plant of Western Electric were surveyed in the late 1920s (Roethlisberger and Dickson, 1939), and in Great Britain at the Durham coal mines, where the influence of technology-driven changes on autonomous groups were surveyed in the 1940s (Trist and Bamforth, 1951). In fact, these studies had a considerable influence on such concepts and movements as sociotechnical systems, industrial democracy, quality of working life, and humanization of work, which all are more or less centered around task design and motivation. In the following sections several programs and projects are outlined. This selection is meant neither as a rating of projects nor as a devaluation of programs or projects not discussed.

5.1 German Humanization of Working Life Approach

As early as 1922, Hellpach conducted a survey of group work experiments in a German car manufacturing company (Böhrs, 1978). Shifting task design from timed assembly lines toward small groups which were able to organize a certain scope of work autonomously, the company enhanced the content and extent of tasks, and therefore fostered intrinsic motivation of the workers. In 1966, the German company Klöckner-Moeller abolished automated assembly lines and changed over to assembling their electrical products at single workplaces and within work groups. Other companies partially followed Klöckner-Moeller, including Bosch, Siemens, BMW, Daimler-Benz, Audi, and Volkswagen (Kreikebaum and Herbert, 1988). However, changes in these companies were by no means as far-reaching as those of some Norwegian or Swedish companies (see Section 5.2.2).

Organizational players in German companies can be divided into (1) representatives of the owners (management), (2) representatives of the employees (works councils), and (3) employees. A works council has to be informed by management of several issues of "information rights." The works council can enter objections to other issues of "participation possibilities," and regarding issues of "participation rights" the works council has to be asked for permission by management: for instance, when it comes to employing new persons. In the German Occupational Constitution

Act (Betriebsverfassungsgesetz) of 1972, management and works councils were obligated to regard results of work science in order to design work humanely. Compared with other countries, this was novel.

In 1974, the German Federal Secretary of Research and Technology [Bundesministerium für Forschung und Technologie (BMFT)] initiated the program "Research for the Humanization of Working Life" [Humanisierung des Arbeitslebens (HdA)] and funded about 1600 studies from 1974 to 1988. In 1989, this program was followed by the strategic approach "Work and Technology" [Arbeit und Technik (AuT)], which besides humanization focused particularly on technology and rationalization aspects. In 1999, this program was succeeded by the program "Innovative Work Design—Work of the Future" [Zukunft der Arbeit (ZdA)], where (among others) humanization aspects with respect to the service sector have been of interest. So far, about 3400 single projects have been funded with the help of these three German programs. These projects were accomplished with the assistance of one or more research institutes, which carried out accompanying engineering, psychological, medical, or sociological research. They usually received between 10 and 20% of the funds cited in Figure 8. Objects of research were the survey of actual conditions as well as the implementation and evaluation of solutions with respect to workplace design, task design, or environmental influences in one or more companies or organizations out of all branches and fields that one could imagine.

Therefore, these case studies were carried out primarily with industrial partners from mechanical engineering, the mining and steel-producing industry, the electrical industry, the automotive industry, the clothing industry, the chemical industry, the food-processing industry, and the building industry. However, workplaces were also surveyed in service branches such as the hotel and catering industry, the transport industry (including trucking, harbor, and airports), the retail industry, railway transportation, postal services, merchant shipping, and health care. In recent years, aspects of humanization have become less important in these projects, which is very unfortunate in light of the fact that employees in the automotive industry, for example, and in the "new economy" faced increasingly inhumane working conditions.

The HdA program, including its successors, AuT and ZdA, is not the only German program that funded research on humanization and quality of working life projects. Regarding the qualification of employees, the program of the Federal Secretary of Economy [Bundesministerium für Wirtschaft (BMFW)] funded projects with 350 million euros from 1994 until 1999, with about 150 million euros being provided by the European Social Fund (ESF) (BMWA, 2003). Additionally, several other programs could be mentioned, such as those initiated by the federal states in Germany.

5.1.1 Goals of the HdA Program

In terms of task design and motivation, the HdA program focused initially on avoiding demotivation

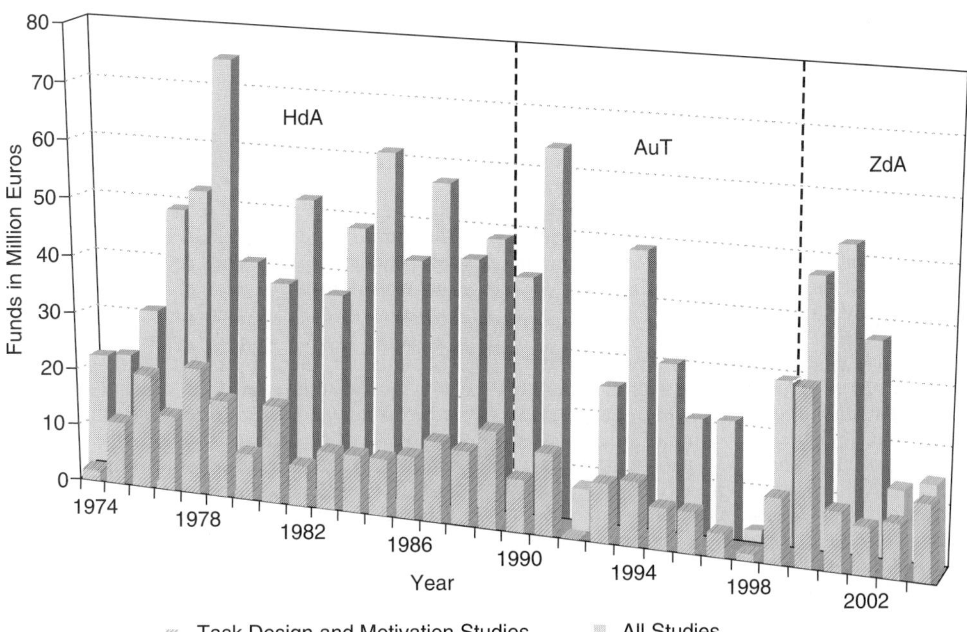

Figure 8 Funds provided in German "Humanization of Working Life" (HdA) studies and the succeeding programs "Work and Technology" (AuT) and "Innovative Work Design – Future of the Work" (ZdA), million euros.

caused by shortcomings in fulfillment of basic needs of workers. Later, issues such as increasing motivation of workers by introducing new forms of work organization became more and more important. Basic goals as propagated by the Federal Secretary were as follows (Keil and Oster, 1976): (1) development of standards of hazard prevention, reference values, and minimum requirements regarding machines, installations, and workplaces; (2) development of humane working techniques; (3) development of exemplary recommendations and models for work organization and task design; (4) distribution and application of scientific results and insights; and (5) supporting economy in implementing these insights practically. With respect to particular areas of work design, the focus of interest was on assessing and reducing risks of accidents; environmental influences such as noise, vibration and concussion, and hazardous substances; as well as physical and psychical stress and strain at work.

In the scope of this chapter, the following HdA goal (Keil and Oster, 1976) is of particular interest: Influences of task design should be surveyed with respect to the organization of work processes, structures of decision making and participation, planning of labor utilization, remuneration, and occupational careers, as well as satisfaction and motivation. Around 1975, the focus of interest in industrial research projects shifted. Before the shift, projects with a strong focus on reducing risks of accidents or physical stress and strain of workers were funded. After an increase in funding volume in 1975, studies with a focus on introducing new organizational structures in the fields of production or administration began to play a more important role (Kreikebaum and Herbert, 1988).

5.1.2 Quantitative Analysis of HdA Studies

Aspects of motivation were of special interest in many of the HdA projects. Even though these aspects played an only partial starring role, they can be considered as a common ground for all studies. Unfortunately, in the beginning of the HdA program, the central funding organization, the German Aerospace Center [Deutsches Zentrum für Luft- und Raumfahrt (DLR)], neglected a systematic documentation of all funded projects by not forcing the receiving organizations to standardize the documenting of results. Therefore, today a complete investigation of these funded projects is almost impossible.

The quantitative analysis of funded action research studies as depicted in Figures 8 to 10 draws on the work of Brüggmann et al. (2001), who gathered a database of about 35,000 sets of journal articles from various fields of occupational safety and health (OSH), aiming to identify tendencies in OSH particularly between research work done in different countries. Additionally, project descriptions of more than 4000 projects of HdA and AuT and of other German project-funding institutions, such as the Federal Agency of Occupational Safety and Health [Bundesanstalt für Arbeitsschutz und Arbeitsmedizin (BauA)] or the legally obligated Mutual Indemnity Association [Berufsgenossenschaft (BG)] have been gathered. Since only the HdA, AuT, and ZdA studies are of interest in this chapter, the remaining studies as well as the OSH literature have been ignored. For simplicity, the term "HdA studies" may be used for both AuT and ZdA studies.

In the database there were fields available such as title, accomplishing institutes, funding period, and keywords for studies. Titles of studies combined with keywords provide a high information density which can be compared to articles with available title and abstract. To be able to classify the data sets correctly, an elaborate, hierarchical system of criteria with manifold combinations of logical AND, OR, and NOT matches of thousands of buzzwords was developed by Brüggmann. The classification hierarchy was then validated empirically by several experts.

In this chapter the set of criteria has been expanded and adapted to the scope of task design and motivation. Since the number of available data sets of studies (carried out between 1974 and 2004) as well as the amount of available relevant information varied over time, a representation in relative percentages has been chosen for depicting time-based characteristics; 100% in the diagrams indicates 100% of all data sets that have at least one hit for each criterion.

The results of the quantitative analysis can be used to balance HdA studies with respect to several aspects of task design and motivation. To do so, three boundaries for the respective scope have been chosen. These boundaries, in turn, contain several criteria, which might be of special interest and therefore build up a hierarchical system. In terms of *system theory*, a criterion in one scope can again build up its own scope, such as "task design" in Figure 9. The first scope of interest covered all funded studies vs. studies that are relevant for task design and motivation, as depicted in Figure 8. The second scope covered all studies that are relevant for task design and motivation, as depicted in Figure 9.

The first thing that stands out is the continually decreasing proportion of studies that cover aspects of task design, even though the hits for "leadership/autonomy" could also be counted as "task design" (interpreting them as autonomy according to Hackman and Oldham's job characteristics model). The second remarkable characteristic is the increased proportion of both "leadership/autonomy" and "incentives." In recent years, especially incentives have become a big issue in funded studies. But aspects of leadership increasingly gained influence. Two representative projects that address questions of leadership are "Modern Services by Innovative Processes of Organization" (MoveOn) and "Flexible Cooperation with Information and Communication Technologies" (SPICE). Breaking "task design" down into "feedback," "task significance," "task identity," or "skill variety," the proportions depicted in Figure 10 can be observed.

Again, light trends can be deduced from the characteristics pictured. The courses of "skill variety" and of "task identity" seem to follow a steady downward tendency. This could be explained by the trend of tasks becoming more and more demanding,

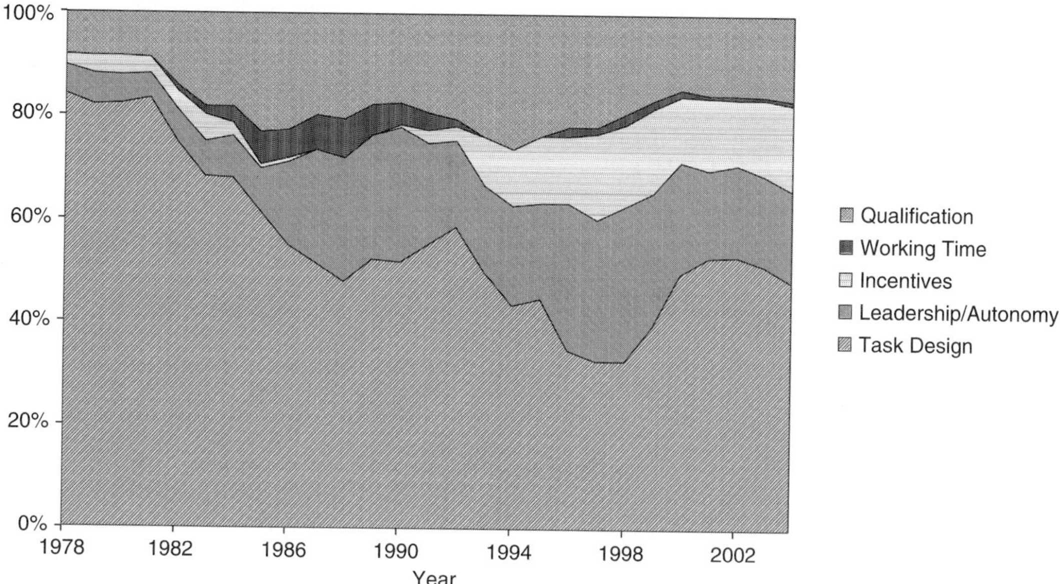

Figure 9 Proportions (five-year moving averages) of several criteria of HdA studies relevant for task design and motivation.

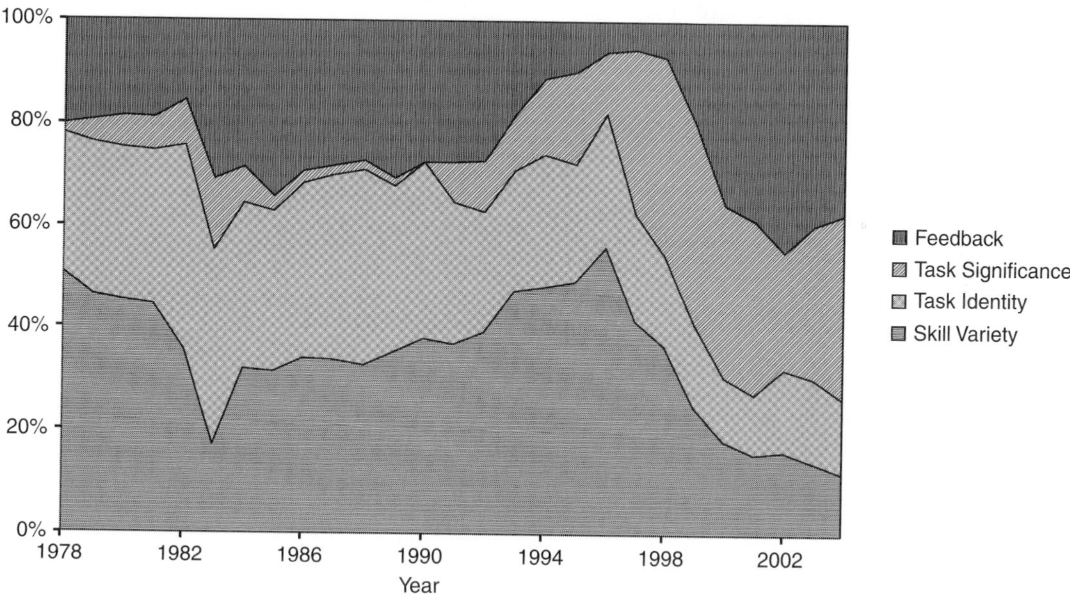

Figure 10 Proportions (five-year moving averages) of several criteria of task design studies.

making research in this field redundant. An aspect that has become more and more important in recent years is feedback from the task carried out. Aspects of "task significance" gained influence similarly. After all, quantitative analysis gives hints about which fields have been in researchers' and funding organizations' focuses of interest and how these have been objects of shifts over the years.

5.1.3 Selected Case Studies

The following is a small selection of typical HdA-funded projects. Aim of this selection is to give an idea of how diversified—in terms of surveyed types of workplaces—the scope of this program was. Furthermore, only case studies from the first years of this program have been selected, since these studies reached a certain degree of recognition within the German scientific community, and since all of these studies had the character of a role model with respect to succeeding HdA projects. However, many other studies within the HdA program revealed countless scientifically and practically valuable results and insights as well. For more details regarding these studies, the reader is referred to BMFT (1981).

Electrical Components Industry: The Case of Bosch The original title of the Bosch 1 study was "Personalentwicklungsorientierte Arbeitsstrukturierung" (Structuring of Work with Focus on the Development of Employees). It was accomplished by the Institute of Work Science of the Technical University of Darmstadt (IAD), the Institute of Production Technology and Automation of the Fraunhofer-Gesellschaft in Stuttgart (IPA), the Institute of Sociology of the University of Karlsruhe (IfS), and the Working Group of Empirical Research in Education in Heidelberg (AfEB), in cooperation with the Robert Bosch GmbH, a global player in the electric and automotive industry. The project was funded from 1974 to 1980 with about 10 million euros.

The primary goal was the development of new forms of work organization in the field of assembling products of varying complexity, such as car radios, cassette decks, TV sets, speakers, electrical tools, and dishwashers. With respect to task design and motivation in many of the plants considered, the contents of work were expanded in order to realize the concept of job enrichment. Technically, this was done by decreasing the degree of automation. In the plant at Herne, tasks for the assembly of car speakers were considered. The time for a task for which an employee was responsible increased from 0.3 minute to 1.5 minutes on average. In the plant at Hildesheim, the assembly of car cassette decks was considered. Here, the increase in time was from about 1.0 minute to about 5.5 minutes. Considering a flow assembly with an average working time of up to 1 minute for each worker, the time spend for one assembly could be extended to about 1 hour by combining functionally different tasks. Additionally, logistical tasks were done by the group. Other projects in the plants of Hildesheim, Leinfelden, and Dillingen were concluded with similar results. The project team established a "learning on the job" concept in these projects, which was close to the job rotation approach. Additionally, a qualification approach for implementing advanced social structures was applied and evaluated. The scope of actions and decision making was increased by a decoupling of conveyor belts, and the cycle of tasks by means of buffers. Therefore, employees could, to a certain degree, dispose of their own work. The researchers predicted a great potential in the self-disposing of work systems. However, strong participation by employees was considered essential. In some of the groups surveyed, they even found that workers took over disposing activities without any authority.

Automotive Industry: The Case of Volkswagen The German title of this case study was "Untersuchung von Arbeitsstrukturen im Bereich der Aggregatefertigung der Volkswagen AG Salzgitter" (Survey of Structures of Work in the Field of Aggregate-Assembly of Volkswagen). It was accomplished by the Institute of Work Science of the Technical University of Darmstadt (IAD), the Institute of Work Psychology of the ETH Zurich (IfAP), and the Institute of Production Technology and Automation of the Fraunhofer-Gesellschaft in Stuttgart (IPA) in cooperation with Volkswagen AG, one of the biggest car manufacturers in the world. The project was funded from 1975 to 1978 with about 5.5 million euros.

The primary goal of this project was the analysis of conventional and new forms of work structures in the field of manufacturing aggregates in the automotive industry. Therefore, a qualitative as well as a quantitative evaluation of person-specific as well as monetary criteria was aimed at, both cross-sectioned and longitudinal-sectioned, with a time frame of 3 years. The object of research were the concepts of (1) conventional assembly line with pallets, (2) intermittent transfer assembly, and (3) assembly groups. In all three alternatives, the variables of feasibility, tolerability, reasonability, and satisfaction were investigated. Remarkable at this point were the open-minded employees of Volkswagen: 268 of 450 potential participants volunteered to participate in the project.

The alternative of assembly groups embodied the concept of job enrichment, where tasks formerly automated were now handled additionally by the group. Even though this implied partially increased stress for the persons involved, the overall distribution of stress was perceived as being more favorable. After all, psychologists found that satisfaction of workers in assembly groups was significantly higher than in the other alternatives. Furthermore, this work was perceived as being more demanding.

Finally, Volkswagen evaluated the results and came to the following conclusions:

1. Many improvements in work and task design will be accounted for in future corporate planning.
2. Stress resulting from work in all three alternatives was on a tolerable level; only partially significant differences could be verified.
3. The proposed new form of work organization competed with established rules, agreements, and legal regulations for distribution of tasks in companies.
4. Mandatory preconditions for implementing a comprehensive process of qualification were individual skills and a methodical proceeding incorporating adequate tools and techniques.
5. Decisions regarding the introduction of new forms of work organization depended on expectations about long-term improvements resulting from that change; the evaluation of economical issues was crucial in that respect.
6. The research project made clear that, in opposition to common positions, a change in working conditions does not necessarily lead to substantial

improvement. But it appeared that increased consideration of employees' desires and capabilities regarding assignment to tasks is leading to motivation of these employees.

From an economical point of view, assembly groups were considered to be cost-effective only for small lot sizes of up to 500 motors a day.

Clothing Industry: A Case of a Total Branch The German title of this branch project was "Neue Arbeitsstrukturen in der Bekleidungsindustrie—Branchenvorhaben" (New Structures of Work in the Clothing Industry—Branch Projects). It was accomplished by the Institute of Operations Research in Berlin (AWF), the Institute of Economical Research in Munich (IFO), the Country's Institute of Social Research in Dortmund, the Institute of Stress Research of the University of Heidelberg, and the Research Institute of Hochenstein, in cooperation with the German clothing union, the German Association of Clothing Industry as well as the companies Weber, Bierbaum & Proenen, Bogner, Patzer, and Windsor. The project was funded from 1977 to 1993 with about 24.5 million euros. In fact, this was the first German project that investigated an entire branch of industry.

The primary goal of this project was the identification of possible and convertible improvements in clothing production processes. With respect to task design and motivation, changes in organizational structures and participation of employees were considered as a necessary condition to realize these improvements rather than as the actual object of research. However, it was found that shortcomings regarding task design were evident before the project started: unchallenging work; short cycle times; one-sided physical stress; piece-rate-, quality-, and time-based remuneration; low scope of disposing one's own tasks; social isolation; and demotivating leadership by directing and controlling. Only a few months after improvements in workplace and task design had been implemented (e.g., setting up groups with an enhanced scope of action), the first positive results regarding job satisfaction, communicativeness, increased qualification, and degree of performance were observed. Hence, it was possible to prove that the changes did indeed lead to increased motivation of employees. These new structures also proved to be useful in economic terms. The involved companies soon began to implement these structures in other fields as well.

Services in Public Administration: The Case of the Legal Authority of the City of Hamburg The German title of this project was "Verbesserung der Arbeitbedingungen bei gleichzeitiger Steigerung der Effizienz des Gerichts durch Einführung von Gruppenarbeit in den Geschäftstellen" (Improvement of Working Conditions and Increase of Efficiency of the Legal Authority by Introduction of Group Work in Offices). It was accomplished by the Consortium for Organizational Development in Hamburg and the Research Group for Legal-Sociology in Hannover in cooperation with the Legal Authority of the City of Hamburg. The project was funded from 1977 to 1981 with about 0.8 million euros.

The primary goals of the project were improving working conditions, improving employee job satisfaction, guaranteeing efficient operation, and decreasing the duration of processes. Therefore, group offices were implemented which could react more flexibly in the case of a varying workload. Additionally, a cutback of hierarchical structures combined with improved participation and qualification of employees should be reached. Together with a better design of communication structures and flows of information, improved service should be reached. Therefore, the researchers established four model groups: two in the field of civil law and two in the field of criminal law. The results implied that introduction of group work in large courts would be an adequate way of encountering dysfunctions by means of division of labor. Regarding the introduction of novel office technology, the results implied that an appropriate accompaniment by organizational changes can be seen as necessary. Otherwise, partial overload of employees would be possible.

Metal Work Industry: The Case of Peiner AG In terms of employee participation, the *Peiner model* attracted considerable attention within the German scientific community. Therefore, this project is presented here in more depth. The German Research Institute of the Friedrich-Ebert-Stiftung accomplished this action research study from 1975 to 1979 in cooperation with Peiner AG, an incorporated company in the metal work industry with about 2000 employees at that time. In the period of 1973–1974, one year before the project started, Peiner AG closed its balance sheet with 10 million euros of losses. Up to 1977–1978, these losses were reduced by 80%. The following description of the project is based on the final report of the Research Institute of the Friedrich-Ebert-Stiftung (Fricke et al., 1981).

At the beginning of the study, machinery and installation at plant I, which was mainly producing screws, were in bad shape. Production in plant I was characterized by small lot sizes and short delivery times. Due to an unsteady supply of incoming orders, utilization of both workers and machines was changing with some degree of uncertainty. Therefore, employees had an average employment guarantee of only a few days. Furthermore, wages were coupled with particular activities in the production process. Since workers had to be extremely flexible concerning the tasks they had to perform in one day (which was not meant as sanitizing work in the sense of job rotation but was born from the necessity of the production situation), wages could differ from day to day. Altogether the situation represented considerable uncertainty for the workers. The focus of interest of the team of researchers from the Friedrich-Ebert-Stiftung was division ZII of plant I. In terms of the production flow, chipping, which was carried out in ZII, succeeded the warm-forming and cold-forming divisions. At the start of the project, 47 employees worked at ZII, where high rates of fluctuation to other divisions were observed.

With respect to workplace design, tremendous shortcomings could be found: Due to the nonergonomical shapes of machines, machine workplaces did not leave workers a choice of whether they prefer to work standing or sitting. The working heights of machine workplaces were not personally adjustable. For small persons it was not even possible to place a small platform in front of the machines. Reaching spaces for machines in ZII were designed without any consideration of percentiles of human arms. Therefore, while working in a standing position, joints, muscles, ligaments, and the vertebral column could easily be overstrained. Some control pedals on a special machine forced workers to stay solely on one foot during an entire shift of eight hours. Each raw part had to be placed into the machines manually. Especially with heavy parts, this was painful for workers. Due to boxes, bins, hand gears, actuators, raw parts, parts of machines, or tools around the machine and behind the workers, freedom of movement was cut down dramatically. These types of enforced bearing caused tremendous impairments of the human body. To actuate machines, workers had to expend enormous force. Changing from machine to machine made these workers suffer from adjustment pain, which could last a few days. Furthermore, these workers often suffered from inflammation of the synovial sheath of tendon. To keep track for piece-rate purposes, frequent clearance of chipping boxes was neglected until the boxes were overly full. Emptying 35-kg heavy boxes at a height of 1.50 m often lead to overstraining by female workers and sometimes by male workers, too. To refill the cutting oil emulsion in machines, workers had to tow heavy buckets of this fluid. Various other findings indicated tremendous shortcomings in terms of today's standards of occupational safety and health.

With respect to task design, several factors that promoted physical and psychical overstrain could be observed at ZII in 1975. In addition to socially unfavorable conditions, this led to systematic demotivation and therefore to considerable withdrawal of workers at ZII, manifesting in a high rate of fluctuation. The situation in 1975 could be characterized by a high degree of division of work. A foreman was responsible for assigning jobs to machines or workers. He received all necessary papers a few weeks before the start of production from the division of job preparation. However, when the jobs would arrive at ZII, nobody could tell. Usually, if the job arrived in the preliminary division, the respective craftsman sent a status message. If jobs were urgent, deadlines were short, or utilization of ZII was low, the craftsman headed in person for the next job. Division of labor was then realized by providing several services, which were carried out by setters, mechanics, weighing machine workers, pallet jack drivers, inspectors, and shop floor typists.

Machinery workers did not have fixed workplaces and therefore changed on the basis of demand between different machines. Maintenance work on machines was carried out only if it was really necessary. The pallet jack drivers got direct orders from the foremen as to where to provide which raw materials and where to put which final materials. The setters got direct orders from the foremen as to which tools to prepare and in which machines tools had to be changed. Machinery workers got to know on which machines to work next only during the actual shift. Beginning work on a particular machine, the piece-rate ticket had to be stamped. The actual work consisted of depositing up to five parts in parallel into the machines in which they were processed. Each of these processes had to be controlled particularly. If processing of a particular piece was finished, the carriages had to be reset in starting position by the worker. Work cycles were up to 1.8 seconds per piece. If the target was 100 pieces within 5 minutes, the performance rate was about 140%. Short cycles carried out over an entire shift of 8 hours caused tremendous stress, resulting from monotony. Additionally, the workers had to spent a certain degree of attention to avoid injury to their hands, to control the quality of processed parts and to track the even operation of the machines, to refill cooling emulsion if necessary, and to request a setter if tools were about to lose sharpness. Even though workers spent this degree of attention, they soon lapsed into an automated working mode in where they basically reacted habitually. Another source of demotivation arose out of the enforced cooperation between machinery workers and setters. Since machinery workers' wages depended on piece rates, the workers usually reacted angrily if the setters did not appear instantly or did not work quickly enough (from their point of view). Setters, in turn, felt provoked by machinery workers.

This unchallenging work, short cycle times, permanent attention in combination with piece-rate based remuneration, authoritarian leadership, and the other environmental conditions mentioned constituted an enormous source of stress for these employees. In consequence, this led to a cliff-hanging atmosphere, with separation of single workers, as well as competition and conflicts between co-workers.

In carrying out the project, researchers aimed basically at the following goals: First, the source of imagination would come from committed employees' ideas about how to improve work and task design, which Fricke (1975) called *qualification to innovate*. One of the goals was determining the social conditions and requirements that are necessary to mediate, apply, and unfold qualifications to innovate. At the same time, these processes of mediating and applying should indicate how these qualifications to innovate could look and what impact they could have. This goal was based on empirical findings that employees may, in fact, have qualifications enabling them to formulate innovative changes, yet they were hindered by a variety of environmental factors, including firmly established organizational structures or resistance from colleagues or superiors. Since from a scientific point of view, qualification to innovate is a potential for action, it would have to be observed to enable drawing conclusions about influencing variables. In a normal work environment, however, this observation would not have been possible, for the reasons mentioned above. Therefore, the project was planned as an action research project in which participants were enabled and encouraged to formulate and express such potential.

Furthermore, the researchers aimed at developing and testing approaches for organizing systematic processes of employees' participation in changing work design and task design. This procedure was meant to provide a frame of action for any employee to contribute to the design of working conditions. Therefore, again an action research approach had to be chosen. Finally, together with workers at ZII, actual improvements in workplace and task design were developed, which could be implemented upon approval by management. Hence, the three goals could be pursued in an integrative, simultaneous fashion.

The project could be sectioned into seven phases. The core of the project's phases were workshop weeks, where employees discussed solutions together with researchers. Additionally, project groups were built consisting of employees, ombudsmen, members of the works council, and experts. Actual solutions and suggestions for improvements were proposed to the plant's management and when approved, were implemented. In each of these phases, countless discussions and meetings took place with superiors and management as well as with the works council. Additionally, several economic, ergonomic, and medical surveys were conducted.

The systematic procedure for employees' participation made clear that even unskilled workers are both willing and capable of participating in the design of workplaces and tasks, and therefore employees do have qualifications to innovate. With respect to the second goal, the following results could be presented. The systematic procedure of employees' participation turned out to be one possible way toward decentralization of decision-making structures, not only in industrial organizations. The research revealed preventable problems occurring when employees are not involved in workplace and task design processes. Participatory workplace and task design can lead to an increase in productivity. Fricke et al. (1981) note that such gains in productivity must not be misused. Therefore, agreements regarding distribution of resulting time in the form of rest periods, reduction in working time, and looking ahead to technical-organizational changes have to be met. From their point of view, these new approaches could be useful completions to existing legal forms of organizational participation. With respect to the third goal, the results of the six workshops should be mentioned. Altogether, 150 pages of suggestions by employees at ZII are evidence of the innovative potential and motivation of these workers.

A major result of this project was an official agreement between Peiner AG and employees: "Participation of Employees in Designing Work Places, Tasks and Environment" in 1979 (Fricke et al., 1981). The results of the project with respect to task design and motivation of employees at ZII were not overwhelming but could be seen as a starting point for further research projects. Employees confirmed that, in addition to improvements in physical working conditions, they gained considerably in self-confidence, everyone felt "free," and everyone was more capable of discussing needs and ideas, and everybody talked increasingly about working conditions and task design. Employees learned how to express their needs and who to contact in certain situations. Employee motivation to innovate in their direct work environment and task design increased significantly.

5.2 European Approaches to Humanization of Work

5.2.1 Employee Participation in Europe

Especially during the Industrial Revolution, voices that criticized inhumane working conditions in industry gained increasing influence, leading to the development of unions and labor parties. In many European countries this political influence resulted over time in legal regulations that at least assured a minimum of human rights and human dignity for industrial employees.

Scandinavian countries particularly, but also France and the Netherlands, legally codified these rights, including several forms of employee participation, in specific labor acts. With the renewed European Union guideline RL 89/391, a sort of constitution for occupational safety and health was enacted in 1989, which for the first time contained the concept of employee participation (Kohte, 1999). In this guideline as well as in country-specific legislations, participation is always implemented in the form of democratic institutions within companies.

Regarding distribution of power, two types of participation can be distinguished: (1) unilateral participation, where rules of communication and decision making are implemented either by management or by employee representatives (usually, by unionlike institutions), or (2) multilateral participation, where rules of communication and decision making may have been negotiated between management and employee representatives (Kißler, 1992). In neither of these approaches do actual workers have much influence on, or are in charge of, their actual work environment, including task design.

In opposition to legally implemented forms of participation, a management-driven form of participation, called *quality cycles* or *quality circles* (Kahan and Goodstadt, 1999), have become widespread in recent years. In these quality circles, employees gather on a regular basis to discuss actual work-related problems such as work organization, task design, or qualification matters. In many cases these cycles are well accepted by employees since they are considered as opportunities for advancement. However, programs and institutions that are implemented parallel to the actual work processes are likely not to benefit sufficiently from the inherent potential and intrinsic motivation of employees (Kißler, 1992; Sprenger, 2002). However, organizing work within autonomous groups can be seen

Table 1 Some Necessary Conditions for Empowering Participation

Norwegian Model	Other Models	Significant Common Features
Institutional and political support at "higher" levels High levels of cooperation and conflict A vision of how work should be organized "Do-it-yourself" participative research Researchers act as "colearners", not as experts in charge of change	Some parity of power prior to participation Systematic development of bases of power Overcoming resistance to empowerment by the powerless	A rejection of conventional organizational design and sociotechnical systems as a source of empowerment Recognition that participation can be either cooperative or empowering Recognition of significant differences between organizational and political democracy Empowerment as learning legitimates new realities and possibilities for action from the bottom up

Source: Based on Elden (2002).

as one possible way to overcome these shortcomings, although both really humanize work in terms of needs and dignity as well as meeting economic demands.

5.2.2 European QWL Programs

In 1975, the European Foundation for the Improvement of Living and Working Conditions was established by the European Union. This organization comprises members of a respective country's governments, economies, and unions. In the years 1993–1998, this foundation carried out "Employee Direct Participation in Organisational Change" (EPOC), a major program of research dealing with the nature and extent of direct participation and new forms of work organization (Sisson, 2000). Major results were (1) the insight that a significant number of managers consider new forms of work organization as beneficial for reaching conventional business performance goals, such as output, quality, and reduction in throughput time, as well as reducing sickness and absenteeism; (2) that companies adapting new forms of work organization will probably stabilize themselves in long-term perspective, and therefore employment may increase in these companies; and (3) that there are surprisingly only a handful of organizations that actually practice integrated approaches. Uncountable programs and projects aiming at the improvement of QWL have been carried out in most European countries in recent decades.

Norway In the 1960s, Norway's social partners, under the guidance of the psychologist Einar Thorsrud, initiated the "Norwegian Industrial Democracy Project" (NIDP) (Gustavsen, 1983; Kreikebaum and Herbert, 1988; Elden, 2002). Researchers around Thorsrud further developed basic ideas that had originated at the Tavistock Institute in London in the 1950s and invented new conceptual tools (Elden, 2002). The dominating form of research was the action research approach (Gustavsen, 1983). In the focus of interest was the survey of new forms of work organization. In fact, Thorsrud and his colleagues introduced the concept of autonomous groups in several industrial companies, such as Christiana Spigerverk, Hunsfos, and Norsk Hydro (Kreikebaum and Herbert, 1988). In the course of a project with Norsk Hydro, groups were in charge of an entire process beginning with the actual production of a up to shipping of the product. Since the new tasks were less physically straining but technically more demanding, an enhancement in employee qualifications became necessary.

The common feature in the Norwegian projects was the systematical empowerment of employees to design their own work environment. Some necessary conditions of empowerment are depicted in Table 1.

In 1977, as a "spin-off product" of the effort of all groups and organizations involved, the government enacted the Norwegian Worker Protection and Work Environment Act. This law contained regulations regarding participation of employees. In contrast to conventional legislation, participation in designing their own work environment was mentioned explicitly. In the following years, several agreements between players in the Norwegian economy and endorsements in legislation were aimed at improving QWL further in Norway.

Sweden From the late 1960s to the present, Sweden has made considerable efforts to improve conditions of work, therefore humanizing work environments and contents. However, these reforms were in fact born in a debate between ideological voices that focused on humanization aspects and practical voices that focused on rationalization aspects in order to countervail increasing employment of foreign workers. In the 1970s, several legislative acts regarding humanization of work were enacted: Act on Employee Representation on Boards, Security of Employment Act, Promotion of Employment Act, Act on the Status of Shop Stewards, Worker Protection or Safety Act, Act on Employee Participation in Decision Making, and the Work Environment Act (Albrecht and Deutsch, 2002).

The majority of Swedish research programs draw on the Swedish Fund of Work Environment [Arbetsmiljöfonden, (AMFO)]. The AMFO is a state authority that is financed by the Swedish employers. Regarding task design and motivation, two large projects can be mentioned that gained considerable recognition within the international scientific community. The first is the case of Saab-Scania, where in the late 1960s and early 1970s, 130 production groups and 60 development groups were established. As a result of these steps in the plant at Södertälje, the rate of fluctuation decreased from 100% in 1968–1969 to 20% in 1972 (Kreikebaum and Herbert, 1988). Furthermore, in the case of motor assembly, the work cycles were decoupled from automated assembly lines. The company soon began to introduce these new forms of work organization in other plants as well.

The second project we mention is the case of Volvo. No other company in the world was as radical at that time in terms of abolishing automated assembly lines. Research experiments focusing on aspects of autonomous groups were conducted in seven plants, of which the plant of Torslanda was the largest. Uncountable experiments with the 8500 employees at Torslanda provided valuable insights into how best to introduce such groups. The new plants at Skövde and Kalmar were later built with the knowledge gathered in Torslanda.

In recent years, AMFO has carried out several succeeding programs for improving QWL. One of these programs was the program for "Leadership, Organization and Participation" (LOM), which from 1985 to 1991 accomplished 72 change projects in 148 public and private organizations. The program was funded with 5 million euros (Gustavsen, 1990; Naschold, 1992). In the period 1990–1995, the "Working Life Fund" [Arbetslivsfonden (ALF)] funded 24,000 workplace programs with about 1500 million euros (Hofmaier and Riegler, 1995). To recognize the weight of this program, one should note that Sweden's overall population was only about 8 million people at that time. The Swedish government established this fund last (but not least) in fear of an overheating economy. Therefore, several regenerating funds were established by AMFO in which employers had to spend up to 10% of their profits. Out of these funds they were able to finance, for example, development programs for their employees for a period of five years.

Actually, a national program for "Sustainable Work Systems and Health" is being carried out by the Swedish Agency of Innovative Systems (VINNOVA) from 1999 to 2006. The main goals of this program regarding QWL are keeping sustainability of organizational structures as well as integrating job design with organizational design. The program is accomplished as an action research approach as well as an action learning approach. The program will be funded with about 23 million euros (Brödner and Latniak, 2002).

France Traditionally, employees' participation in France was based on interaction between the two organizational players: (1) committees of employees (delegués du personnel), and (2) representatives of the employer (comité d'entreprise). Role of the delegués du personnel was to formulate complaints of employees about working conditions, mainly regarding issues of occupational safety and health (Kißler, 1988). In 1982, the French Secretary of Labour enacted "Auroux's Act" (Lois Auroux). With that a third player gained influence: the employee itself. Based on co-determination-groups (groupes d'expression) employees got the right to directly influence their own working conditions.

In 1973, the French government founded the National Agency for Improvement of Work Conditions [Agence Nationale pour l'Amélioration des Conditions de Travail (ANACT)], which consists of representatives of the government, the economy, and the unions. ANACT, often in association with other French organizations, such as the Improvement of Work Condition Fund [Le Fonds pour l'Amélioration des Conditions de Travail (FACT)], funded several research activities with a focus on occupational health and safety and issues of QWL. Additionally, ANACT provides offices all over France where companies can be consulted in questions of workplace design, work organization, and so on.

In 1983, the French Ministry of Research launched the "Mobilize Technology, Employment, Work Program" [Mobilisateur Technology, Emploi, Travail (TET)] (Tanguy, 1986). This program aimed at establishing research potential and an academic community for investigating, among others, forms of work organization. In 1989, the "Mobilize Man, Work and Technology" [Homme, Travail et Technologie (HTT)] program succeeded. The aim of this program was to investigate all dimensions of work, such as physical, physiological, psychological, social, and organizational. The second goal of this program was to conduct increasing action research.

In 1984, the National Center for Scientific Research [Centre National de la Recherche Scientifique (CNRS)] launched the "Interdisciplinary Research Program on Technology, Work, and Lifestyles" [Programme Interdisciplinaire de Recherche sur les Technologies, le Travail et les Modes de Vie (PIRTTEM)]. This program focused mainly on projects on the development of technology and respective influences on work organization, especially on employees accepting or not accepting new technologies.

From 1983 to 1985, for example, a consortium of ANACT, TET, and the "Action for Improving Work Conditions in the Alsace" [Action pour l'Amérilation des Conditions de Travail en Alsace (ACTAL)], as well as the CNRS research group of Group Lyonnais de Sociologie Industrielle (GLYSI) and a management consultant accomplished an action research project in the Moulhouse plant of the car manufacturer Peugeot. The project was called "Social and Organizational Impact of Automation and Robotics" [Impact Social et Organisationnel des Automatismes et de la Robotique (ISOAR)] (Coffineau and Sarraz, 1992). The project basically aimed at preparing the company for future investment in automation technology. New organizational and social equilibriums resulting from that change were to be identified. Employees in the affected parts of the plant were to be developed and qualified accordingly. Therefore, a new concept of participation consisting of three hierarchical levels was established. The first level dealt with shop floor working groups comprising superiors, foremen, workers, and union members. The second level consisted of superiors, a production engineering group, union members, and public representatives. Finally, the third level consisted of representatives from management, the unions, and public organizations.

Great Britain In the advent of European humanization approaches, Great Britain's companies could not provide legally codified forms of employee participation (i.e., there were neither any works councils nor any forms of written agreements on the management level) (Heller et al., 1980). However, some of the roots of humanization of working life can be found in Great Britain: namely, in research at the Tavistock Institute of Human Relations in London, where the concepts of sociotechnical systems and the quality of working life (QWL) emerged. Most notably, the Tavistock coal-mining study of Trist and Bamforth (1951) contributed to the insight that technical innovations in the workplace and task design cannot be applied without regard to the social impact these changes can have on employees. Trist discovered that workers in the Durham coal mines who for decades had worked in autonomous groups barely accepted new, technology-driven forms of work which forced them to abandon the social relations that had grown up among the old group. Furthermore, the impact of concepts such as content of work, extent of work, order of tasks, and degrees of control and feedback on the work's output and on the motivation of workers was investigated. The results can be seen as the foundations of well-known concepts: job enrichment, job enlargement, job rotation, and work organization in autonomous groups.

Even though we could mention other projects that investigated humanization and participation aspects in Great Britain (e.g., several British car manufacturers, the aerospace industry, Indian textile mills), no governmental programs comparable to HdA, AuT, AMFO, or ANACT were accomplished at that time. In 1999, the British Prime Minister carried out The "Partnership at Work Fund," which until 2004 has funded projects with about 14 million pounds sterling. The program focuses on improving relations among organizational players where issues of employee participation are of particular interest (Brödner and Latniak 2002).

5.3 Summary of the European Studies

Changing actual working conditions as well as scientific insights and implications for work design naturally resulted in conflicts among the involved organizational and societal players. Employer federations argued that humanization must not lead to a shift in responsibilities and power. Beyond that, however, voices from this side admitted that humanization goals do not necessarily compete with economical goals. Employee federations criticized many of the studies as leading to an implementation of measures of rationalization, with the consequence of increased unemployment. Apart from that, employee federations such as unions or works committees widely supported the humanization efforts.

Although these humanization studies have been carried out considering different forms of work and workplaces in different types of companies and branches, one common aspect can be identified in all the projects: task design and motivation. Whether surveying the Durham coal mines, the assembly lines of Bosch, Saab, Volvo, or Volkswagen, the clothing industry, or authorities of cities and countries in retrospect, if one talks about enlarging work contents and extents, the scope of responsibilities, the possibilities for organizing work in groups, and employee qualifications, one also talks about factors that may influence the intrinsic motivation of employees. Speaking with Maslow: social or esteem needs; speaking with Alderfer: growth needs; speaking with Herzberg: satisfiers and dissatisfiers—these were all addressed substantially in these studies. Herzberg's two-factor theory and Hackman and Oldham's job characteristics model can be seen especially as a basic source of inspiration for most of the practical task design solutions that have been surveyed in the action research studies mentioned.

But also in those studies that surveyed primarily the influences of work environment on employee safety and health (which have not been taken into account in Figure 8), issues of motivation actually provide the common ground. That is, whether these studies focused on development of standards of hazard prevention or on reducing physical stress and strain for workers, these issues can be connected to fulfillment of basic needs of employees according to the content theories of motivation.

Recapitulating, one could argue that from an employer's point of view, the humanization of work (as well as the quality of working life, work life balance, etc.) debates led to a better understanding of the needs and motives of employees. Facing the tremendous change in attitude with respect to working conditions that occurred in the 1960s, these insights provided room for reducing demotivation of employees substantially, therefore improving productivity and quality of work results. However, there are voices (Sprenger, 2002) that postulate a new shift of employee attitudes with respect to motivational techniques and incentive systems. They give warning of focusing on manipulating employees' extrinsic motivation. In their opinion, such forms of leadership can easily lead to incentive-dependent employees in the best case—or to demotivated employees who feel that they are being treated as immature and are not taken seriously in the worst case. From an employee's point of view, these debates and the resulting changes can be assessed as a noteworthy contribution to improvement in the quality of working life. For single employees, however, these new forms of work organization often come together with increased stress and strain, which can be partially balanced by increased efforts at qualification.

Considering all the new challenges that the information age has evoked for working conditions, one can recognize that there is still a lot of research to be conducted. The development of new forms of work organization leading to increasingly demanding, highly complex tasks has generated a new source of stress for employees which cannot simply be countervailed by measures of qualification. Along with inhumane pressure of time and a competitive culture, this results more and more in psychosocial diseases such as the well-known burnout syndrome. Furthermore, there is little knowledge of the long-term effects of manipulative motivational techniques on employee motivation and achievement potential. From a practical point of view, commonsense task design that accounts for both motives and the dignity of human beings appears to be one of the keys in facing these challenges.

REFERENCES

Ackermann, D., and Ulich, E. (1987), "On the Question of Possibilities and Consequences of Individualisation of Human–Computer Interaction," in *Psychological Issues of Human–Computer Interaction in the Work Place*, M. Frese, E. Ulich, and W. Dzida, Eds., North-Holland, Amsterdam, pp. 131–145.

Adams, J. S. (1963), "Toward an Understanding of Inequity," *Journal of Abnormal and Social Psychology*, Vol. 67, pp. 422–436.

Adams, J. S. (1965), "Inequity in Social Change," in *Advances in Experimental Social Psychology*, L. Berkowitz, Ed., Academic Press, New York.

Albrecht, S., and Deutsch, S. (2002), "The Challenge of Economic Democracy: The Case of Sweden," in *Economic Democracy: Essays and Research on Workers' Empowerment*, W. P. Woodworth, Ed., Sledgehammer Press, Pittsburgh, PA.

Alderfer, C. P. (1969), "An Empirical Test of a New Theory of Human Needs," *Organizational Behavior and Human Performance,* Vol. 4, pp. 142–175.

Alderfer, C. P. (1972), *Existence, Relatedness, and Growth: Human Needs in Organizational Settings*, Free Press, New York.

Alioth, A. (1980), *Entwicklung und Einführung alternativer Arbeitsformen*, Hans Huber, Bern, Switzerland.

Alioth, A., Martin, E., and Ulich, E. (1976), "Semi-autonomous Work Groups in Warehousing," in *Proceedings of the 6th Congress of the International Ergonomics Association*, Washington, DC, pp. 187–191.

Andriessen, E. J. H., and Coetsier, P. L. (1984), "Industrial Democratization," in *Handbook of Work and Organizational Psychology*, P. J. D. Drenth, H. Thierry, P. J. Willems, and C. J. de Wolff, Eds., Wiley, Chichester, West Sussex, England.

Argyris, C. (1964), *Integrating the Individual and the Organization*, Wiley, New York.

Argyris, C. (2001), "Empowerment: The Emperor's New Clothes," in *Creative Management*, J. Henry, Ed., Sage Publications, London, pp. 195–201.

Ash, R. A., Lee, Y.-L., and Dreher, G. F. (1985), "Exploring Determinants of Pay Satisfaction," *Proceedings of the APA Meeting*, Los Angeles, pp. 22–27.

Atkinson, J. W., and Feather, N. T., Eds. (1966), *A Theory of Achievement Motivation*, Wiley, New York.

Austin, J. T., and Klein, H. J. (1996), "Work Motivation and Goal Striving," in *Individual Differences and Behavior in Organizations*, K. R. Murphy, Ed., Jossey-Bass, San Francisco.

Baitsch, C. (1985), *Kompetenzentwicklung und partizipative Arbeitsgestaltung*, Lang, Bern, Switzerland.

Bandura, A. (1982), "Self-Efficacy Mechanism in Human Agency," *American Psychologist,* Vol. 37, pp. 122–147.

Bandura, A. (1989), "Self-Regulation of Motivation and Action Through Internal Standards and Goal Systems," in *Goal Concepts in Personality and Social Psychology*, L. A. Pervin, Ed., Lawrence Erlbaum Associates, Mahwah, NJ, pp. 19–85.

Barnes, L. B. (1960), *Organizational Systems and Engineering Groups: A Comparative Study of Two Technical Groups in Industry*, Division of Research, Harvard Business School, Boston.

Blau, G. (1993), "Operationalizing Direction and Level of Effort and Testing Their Relationships to Individual Job Performance," *Organizational Behavior and Human Decision Processes*, Vol. 55, No. 1, pp. 152–170.

BMFT (1981), "Das Programm 'Forschung zur Humanisierung des Arbeitslebens': Ergebnisse und Erfahrungen arbeitsorientierter Forschung, 1974–1980," in *Schriftenreihe "Humanisierung des Arbeitslebens*," Vol. 1, *Bundesminister für Forschung und Technologie*, Ed., Campus Verlag, Frankfurt on Main, Germany.

BMWA (2003), "Abschlussbericht zum operationellen Programm in der Bundesrepublik im Rahmen der Gemeinschaftsinitiative ADAPT im Förderzeitraum, 1994–1999," ARINCO-Nr. 94.DE.05.077, ESF-Nr. 946001 D8, Bonn, Germany; available at http://www.equal-de.de/download/Abschlussbericht-GI%20BESCHAEFTIGUNG. doc [07-28-2004].

Böhrs, H. (1978), "Gruppenfabrikation 1922 in einer deutschen Automobilfabrik," in *Veröffentlichungen des REFA zur Humanisierung der Arbeit 2: Artikel aus REFA-Nachrichten und fortgeschrittene Betriebsführung*, REFA, Darmstadt, Germany, *1976–1978*.

Borg, I., Braun, M., and Haeder, M. (1993), "Arbeitswerte in Ost- und Westdeutschland, unterschiedliche Gewichte, aber gleiche Struktur," *ZUMA Nachrichten*, Vol. 33, pp. 64–82.

Brödner P., and Latniak, E. (2002), "Moderne Arbeitsformen für Innovation und Wettbewerbsfähigkeit: Nationale Förderprogramme zur Entwicklung neuer Formen der Arbeitsorganisation," European Union Technical Report EMPL-2002-12380-00-00-DE-TRA-00 (EN); available at http://iat-info.iatge.de/aktuell/veroeff/ps/broedner02c.pdf [07-28-2004].

Brüggmann, M., Rötting, M., and Luczak, H. (2001), "International Comparison of Occupational Safety and Health Research: A Review Based on Published Articles," *International Journal of Occupational Safety and Ergonomics*, Vol. 7, No. 4, pp. 387–401.

Bungard, W., Antoni, C. H., and Lehnert, E. (1993), *Gruppenarbeitskonzepte mittlerer Industriebetriebe: Forschungsbericht*, Ehrenhof-Verlag, Ludwigshafen, Germany.

Campbell, J. P., and Pritchard, R. D. (1976), "Motivation Theory in Industrial and Organizational Psychology," In *Handbook of Industrial and Organizational Psychology*, M. D. Dunette, Ed., Rand McNally, Chicago, pp. 63–130.

Campion, M. A., and McClelland, C. L. (1993), "Follow-up and Extension of the Interdisciplinary Costs and Benefits of Enlarged Jobs," *Journal of Applied Psychology*, Vol. 78, No. 3, pp. 339–351.

Caston, R. J., and Braito, R. (1985), "A Specification Issue in Job Satisfaction Research," *Sociological Perspectives*, April, pp. 175–197.

Cherns, A. (1976), "The Principles of Organizational Design," *Human Relations*, Vol. 29, pp. 783–792.

Coffineau, A., and Sarraz, J. P. (1992), "Partizipatives Management und Unternehmensberatung: Das Projekt ISOAR bei Peugeot-Mulhouse," in *Management und Partizipation in der Automobilindustrie: Zum Wandel der Arbeitsbeziehungen in Deutschland und Frankreich*, L. Kißler, Ed., Centre d'Information et de Recherche sur l'Allemagne Contemporaine, Paris.

Cummings, T. G. (1978), "Self-Regulating Work Groups: A Socio-technical Synthesis," *Academy of Management Review*, Vol. 3, pp. 625–634.

Deci, E. L. (1975), *Intrinsic Motivation*, Plenum Publishing, New York.

Deci, E. L. (2002), *Handbook of Self-Determination Research*, University of Rochester Press, Rochester, NY.

Deci, E. L., and Ryan, R. M. (1985), *Intrinsic Motivation and Self-Determination in Human Behavior*, Plenum Publishing, New York.

Deci, E. L., Connell, J. P., and Ryan, R. M. (1989), "Self-Determination in a Work Organization," *Journal of Applied Psychology*, Vol. 74, pp. 580–590.

Deci, E. L., Ryan, R. M., Gagné, M., Leone, D. R., Usunov, J., and Kornazheva, B. P. (2001), "Need Satisfaction, Motivation, and Well-Being in the Work Organizations of a Former Eastern Bloc Country," *Personality and Social Psychology Bulletin*, Vol. 27, pp. 930–942.

Duell, W., and Frei, F. (1986), *Leitfaden für qualifizierende Arbeitsgestaltung*, TÜV Rheinland, Köln, Germany.

Dunckel, H. (1996), *Psychologisch orientierte Systemanalyse im Büro*, Hans Huber, Bern, Switzerland.

Dunckel, H., Volpert, W., Zölch, M., Kreutner, U., Pleiss, C., and Hennes, K. (1993), *Kontrastive Aufgabenanalyse im Büro: Der KABA-Leitfaden*, Teubner, Stuttgart, Germany.

Earley, P. Ch., Wojnaroski, P., and Prest, W. (1987), "Task Planning and Energy Expended: Exploration of How Goals Influence Performance," *Journal of Personality and Social Psychology*, Vol. 72, pp. 107–114.

Elden, M. (2002), "Socio-Technical Systems Ideas as Public Policy in Norway: Empowering Participation Through Worker-Managed Change," in *Economic Democracy: Essays and Research on Workers' Empowerment*, W. P. Woodworth, Ed., Sledgehammer Press, Pittsburgh, PA.

Elizur, D., Borg, I., Hunt, R., and Magyari-Beck, I. (1991), "The Structure of Work Values: A Cross Cultural Comparison," *Journal of Organizational Behavior*, Vol. 12, pp. 21–38.

Emery, F. E. (1959), *Characteristics of Socio-technical Systems*, Document 527, Tavistock Institute of Human Relations, London.

Emery, F. E. (1972), "Characteristics of Socio-technical Systems," in *Job Design*, L. E. Davis and J. C. Taylor, Eds., (pp. 177–198). Penguin Books, Harmondsworth, Middlesex, England.

Emery, F. E., and Emery, M. (1974), *Participative Design*, Australian National University, Canberra, Australia.

Emery, F. E., and Thorsrud, E. (1976), *Democracy at Work*, Martinus Nijhoff, Leiden, The Netherlands.

Festinger, L. (1957), *A Theory of Cognitive Dissonance*, Row, Peterson, Evanston, IL.

Ford, N. (1969), *Motivation Through the Work Itself*, American Management Association, New York.

Frei, F., and Udris, I., Eds. (1990), *Das Bild der Arbeit*, Hans Huber, Bern, Switzerland.

French, E. G. (1958), "Effects of the Interaction of Motivation and Feedback on Task Performance," in *Motives in Fantasy, Action, and Society*, J. W. Atkinson, Ed., Van Nostrand, Princeton, NJ, pp. 400–408.

Fricke, W. (1975), *Arbeitsorganisation und Qualifikation: Ein industriesoziologischer Beitrag zur Humanisierung der Arbeit, Neue Schriftenreihe des Forschungsinstituts der Friedrich-Ebert-Stiftung*, Vol. 119, Gesellschaft, Bonn Bad-Godesberg, Germany.

Fricke, E., Fricke, W., Schönwälder, M., and Stiegler, B. (1981), "Qualifikation und Beteiligung, 'Das Peiner Modell,'" in *Schriftenreihe Humanisierung des Arbeitslebens*, Vol. 12, *Bundesminister für Forschung und Technologie*, Ed., Campus Verlag, Frankfurt, on Main, Germany.

Fried, Y. (1991), "Meta-analytic Comparison of the Job Diagnostic Survey and Job Characteristics Inventory as Correlates of Work Satisfaction and Performance," *Journal of Applied Psychology*, Vol. 76, pp. 690–697.

Fried, Y., and Ferris, G. R. (1987), "The Validity of the Job Characteristics Model: A Review and Meta-analysis," *Personnel Psychology*, Vol. 40, pp. 287–322.

Greenberg, J. (1988), "Equity and Workplace Status," *Journal of Applied Psychology*, Vol. 73, No. 4, pp. 606–613.

Greif, S. (1983), *Konzepte der Organisationspsychologie: Eine Einführung in grundlegende theoretische Ansätze*, Hans Huber, Bern, Switzerland.

Greif, S., and Gediga, G. (1987), "A Critique and Empirical Investigation of the 'One-Best-Way-Models' in Human–Computer Interaction," in *Psychological Issues of Human–Computer Interaction in the Work Place*, M. Frese, E. Ulich, and W. Dzida, Eds., North-Holland, Amsterdam, pp. 357–377.

Greif, S., Bamberg, E., and Semmer, N., Eds. (1991), *Psychischer Stress am Arbeitsplatz*, Hogrefe Verlag, Göttingen, Germany.

Griffin, R. W. (1988), "A Longitudinal Assessment of the Consequences of Quality Circles in an Industrial Setting," *Academy and Management Journal*, Vol. 31, pp. 338–358.

Grob, R. (1985), *Flexibilität in der Fertigung*, Springer-Verlag, Berlin.

Guest, D. (1984), "What's New in Motivation?" *Personnel Management*, Vol. 16, pp. 20–23.

Gustavsen, B. (1983), *Some Aspects of the Development of Social Science Work Research in Scandinavia*, Internationales Institut für Vergleichende Gesellschaftsforschung und Arbeitspolitik, Berlin.

Gustavsen, B. (1990), "Demokratische Arbeitspolitik mit LOM in Schweden," in *Jahrbuch Arbeit + Technik*, W. Fricke, Ed., Verlag J. H. W. Dietz, Bonn, Germany, pp. 213–228.

Haaks, D. (1992), *Anpaßbare Informationssysteme*, Verlag für Angewandte Psychologie, Göttingen, Germany.

Hacker, W. (1986), *Arbeitspsychologie: Psychische Regulation von Arbeitstätigkeiten*, Hans Huber, Bern, Switzerland.

Hacker, W. (1987), "Software-Ergonomie: Gestalten rechnergestützter Arbeit?" in *Software-Ergonomie '87: Nützen Informationssysteme dem Benutzer?* W. Schönpflug and M. Wittstock, Eds., Teubner, Stuttgart, Germany, pp. 31–54.

Hackman, J. R. (1987), "The Design of Work Teams," in *Handbook of Organizational Behavior*, J. W. Lorsch, Ed., Prentice Hall, Englewood Cliffs, NJ, pp. 315–342.

Hackman, J. R. (2002), *Leading Teams: Setting the Stage for Great Performances*, Harvard Business School Press, Boston.

Hackman, J. R., and Lawler, E. E. (1971), "Employee Reactions to Job Characteristics," *Journal of Applied Psychology*, Vol. 55, pp. 259–286.

Hackman, J. R., and Oldham, G. R. (1975), "Development of the Job Diagnostic Survey," *Journal of Applied Psychology*, Vol. 60, pp. 159–170.

Hackman, J. R., and Oldham, G. R. (1976), "Motivation Through the Design of Work: Test of a Theory," *Organizational Behavior and Human Performance*, Vol. 16, pp. 250–279.

Hackman, J. R., and Oldham, G. R. (1980), *Work Redesign*, Addison-Wesley, Reading, MA.

Hackman, J. R., and Suttle, J. L., Eds. (1977), *Improving Life at Work: Behavioral Science Approaches to Organizational Change*, Goodyear, Santa Monica, CA.

Heckhausen, H. (1980), *Motivation und Handeln: Lehrbuch der Motivationspsychologie*, Springer-Verlag, Berlin.

Heller, F. A., Tynan, O., and Hitchon, B. (1980), Der Beitrag der Arbeitsbestimmung zur Arbeitsgestaltung: am Beispiel von British Leyland, in *Arbeitsstrukturierung durch Verhandlung*, H. W. Hetzler, Ed., Forschungsstelle für Betriebswirtschaft und Sozialpraxis, Munich–Mannheim.

Hellpach, W. (1922), "Sozialpsychologische Analyse des betriebstechnischen Tatbestandes 'Gruppenfabrikation,'" in *Gruppenfabrikation*, R. Lang and W. Hellpach, Eds., Springer-Verlag, Berlin, pp. 5–186.

Herzberg, F. (1968), "One More Time: How Do You Motivate Employees?" *Harvard Business Review*, Vol. 46, pp. 53–62.

Herzberg, F., Mausner, B., and Snyderman, B. (1959), *The Motivation to Work*, Wiley, New York.

Hettlage, R. (1983), "Humanisierung der Arbeit: Über einige Zusammenhänge zwischen Wirklichkeitsbildern und Wirklichkeit," *Die Betriebswirtschaft*, Vol. 43, No. 3, pp. 395–406.

Hofmaier, B., and Riegler, C. H. (1995), "Staatlich geförderte Unternehmensentwicklung in Schweden," *Arbeit*, Vol. 3, No. 4, pp. 249–270.

Hoyos, C. Graf (1974), *Arbeitspsychologie*, Kohlhammer, Stuttgart, Germany.

Husemann, R. C., Hatfield, J. D., and Miles, E. W. (1987), "A New Perspective on Equity Theory," *Academy of Management Review*, Vol. 12, pp. 222–234.

Idaszak, J. R., and Drasgow, F. (1987), "A Review of the Job Diagnostic Survey: Elimination of a Measurement Artifact," *Journal of Applied Psychology*, Vol. 72, pp. 69–74.

Inglehart, R. (1977), *The Silent Revolution: Changing Values and Political Styles Among Western Publics*, Princeton University Press, Princeton, NJ.

Inglehart, R. (1989), *Kultureller Umbruch: Wertwandel in der westlichen Welt*, Campus Verlag, Frankfurt on Main, Germany.

Iyengar, S. S., and Lepper, M. R. (1999), "Rethinking the Value of Choice: A Cultural Perspective on Intrinsic Motivation," *Journal of Personality and Social Psychology*, Vol. 76, pp. 349–366.

Jackson, S. E. (1992), *Diversity in the Workplace: Human Resources Initiatives*, Guilford Press, New York.

Jackson, S. E., and Ruderman, M. N. (1995), *Diversity on Work Teams*, American Psychological Association, Washington, DC.

Jahoda, M. (1983), *Wieviel Arbeit braucht der Mensch? Arbeit und Arbeitslosigkeit im 20. Jahrhundert*, Julius Beltz, Weinheim, Germany.

Judge, T. A., and Welbourne, T. M. (1994), "A Confirmatory Investigation of the Dimensionality of the Pay Satisfaction Questionnaire," *Journal of Applied Psychology*, Vol. 79, pp. 461–466.

Kahan, B., and Goodstadt, M. (1999), "Continuous Quality Improvement and Health Promotion: Can CQI Lead to Better Outcomes?" *Health Promotion International*, Vol. 14, No. 1, pp. 83–91.

Katzell, R. A., and Thompson, D. E. (1990), "Work Motivation: Theory and Practice," *American Psychologist*, Vol. 45, pp. 149–150.

Kaufmann, I., Pornschlegel, H., and Udris, I. (1982), "Arbeitsbelastung und Beanspruchung," in *Belastungen und Streß bei der Arbeit*, L. Zimmermann, Ed., Rowohlt Verlag, Reinbek, Germany, pp. 13–48.

Keil, G., and Oster, A. (1976), *Humanisierung des Arbeitslebens*, Honnefer Buchverlag, Bad Honnef, Germany.

Kelley, H. H. (1967), "Attribution Theory in Social Psychology," in *Nebraska Symposium on Motivation*, D. Levine, Ed., University of Nebraska Press, Lincoln, NE, pp. 192–238.

Kelley, H. H. (1973), "The Process of Causal Attribution," *American Psychologist*, Vol. 28, pp. 107–128.

Kerr, S., Harlan, A., and Stogdill, R. M. (1974), "Preference for Motivator and Hygiene Factors in a Hypothetical Interview Situation," *Personnel Psychology*, Vol. 25, pp. 109–124.

Kißler, L. (1988), "Die 'Software' der Modernisierung: Partizipationsexperimente in deutschen und französischen Betrieben," in *Modernisierung der Arbeitsbeziehungen: Direkte Arbeitnehmerbeteiligung in deutschen und französischen Betrieben*, L. Kißler, Ed., Centre d'Information et de Recherche sur l'Allemagne Contemporaine, Paris.

Kißler, L. (1992), "Direkte Arbeitnehmerbeteiligung und Wandel der betrieblichen Arbeitsbeziehungen als Managementaufgabe: Erste Ergebnisse einer empirischen Untersuchung in der deutschen und französischen Automobilindustrie, Das Beispiel Peugeot-Mulhouse," in *Modernisierung der Arbeitsbeziehungen: Direkte Arbeitnehmerbeteiligung in deutschen und französischen Betrieben*, L. Kißler, Ed., Centre d'Information et de Recherche sur l'Allemagne Contemporaine, Paris.

Klages, H. (1983), "Wertwandel und Gesellschaftskrise in der sozialstaatlichen Demokratie," in *Krise der Arbeitsgesellschaft?*, H. Matthes, Ed., Campus Verlag, Frankfurt on Main, Germany, pp. 341–352.

Kleinbeck, U., Quast, H.-H., Thierry, H., and Häcker, H., Eds. (1990), *Work Motivation*, Lawrence Erlbaum Associates, Mahwah, NJ.

Kohte, W. (1999), *Die Stärkung der Partizipation der Beschäftigten im betrieblichen Arbeitsschutz*, Hans-Böckler-Stiftung, Ed., Vol. 9, Hans-Böckler-Stiftung, Düsseldorf, Germany.

Kreikebaum, H., and Herbert, K. J. (1988), *Humanisierung der Arbeit: Arbeitsgestaltung im Spannungsfeld ökonomischer, technologischer und humanitärer Ziele*, Gabler Verlag, Wiesbaden, Germany.

Kren, L. (1992), "The Moderating Effects of Locus of Control on Performance Incentives and Participation," *Human Relations*, September, pp. 991–999.

Kristof, A. L. (1996), "Person–Organization Fit: An Integrative Review of Its Conceptualizations, Measurement, and Implications," *Personnel Psychology*, Vol. 49, pp. 1–49.

Kulik, C. T., and Ambrose, M. L. (1992), "Personal and Situational Determinants of Referent Choice," *Academy of Management Review*, Vol. 17, pp. 212–237.

Landau, K., Luczak, H., Keith, H., Rösler, D., Schaub, K., and Winter, G. (2003), *Innovative Konzepte: Bilanz erfolgreicher Veränderungen in der Arbeitsgestaltung und Unternehmensorganisation*, Ergonomia-Verlag, Stuttgart, Germany, pp. 1–8.

Latham, G. P., and Baldes, J. J. (1975), "The Practical Significance of Locke's Theory of Goal Setting," *Journal of Applied Psychology*, Vol. 60, pp. 187–191.

Latham, G. P., and Yukl, G. A. (1975), "A Review of Research on the Application of Goal Setting in Organizations," *Academy of Management Journal*, Vol. 18, pp. 824–845.

Lawler, E. E., III (1973), *Motivation in Work Situations*, Brooks/Cole, Monterey, CA.

Leitner, K. (1993), "Auswirkungen von Arbeitsbedingungen auf die psychosoziale Gesundheit," *Zeitschrift für Arbeitswissenschaft*, Vol. 47, No. 2, pp. 98–107.

Locke, E. A. (1968), "Toward a Theory of Task Motivation and Incentives," *Organizational Behavior and Human Performance*, 3, 157–189.

Locke, E. A. (2003), "The Ideas of Frederick W. Taylor: An Evaluation," in *Operations Management: A Supply Chain Approach*, D. L. Waller, Ed., Thomson, London, pp. 73–91.

Locke, E., and Henne, D. (1986), "Work Motivation Theories," in *International Review of Industrial and Organizational Psychology*, C. L. Cooper and I. T. Robertson, Eds., Wiley, Chichester, West Sussex, England, pp. 1–35.

Locke, E. A., and Latham, G. P. (1990), *A Theory of Goal Setting and Task Performance*, Prentice Hall, Englewood Cliffs, NJ.

Luczak, H. (1997), "Task Analysis," in *Handbook of Human Factors and Ergonomics*, 2nd ed., G. Salvendy, Ed., Wiley, New York, pp. 340–416.

Luczak, H., and Rohmert, W. (1985), "Ansätze zu einer anthropologischen Systematik arbeitswissenschaftlicher Erkenntnisse," *Zeitschrift für Arbeitswissenschaft*, Vol. 39, No. 3, pp. 129–142.

Luczak, H., Volpert, W., Raeithel, A., and Schwier, W. (1989), *Arbeitswissenschaft. Kerndefinition–Gegenstandkatalog–Forschungsgebiete*, TÜV Rheinland, Köln, Germany.

Luczak, H., Schmidt, L., and Springer, J. (2003), "Gestaltung von Arbeitssystemen nach ergonomischen und gesundheitsförderlichen Prinzipien," in *Neue Organisationsformen im Unternehmen: Ein Handbuch für das moderne Management*, H.-J. Bullinger, H.-J. Warnecke, and E. Westkämper, Eds., Springer-Verlag, Berlin, pp. 421–458.

Maslow, A. (1954), *Motivation and Personality*, Harper & Row, New York.

McClelland, D. C. (1961), *The Achieving Society*, Van Nostrand, Princeton, NJ.

McClelland, D. C. (1984), *Motives, Personality, and Society*, Praeger, New York.

McClelland, D. C. (1985), *Human Motivation*, Scott, Foresman, Glenview, IL.

McClelland, D. C., Koestner, R., and Weinberger, J. (1989), "How Do Self-Attributed and Implicit Motives Differ?" *Psychological Review*, Vol. 96, No. 4, pp. 690–702.

McGregor, D. M. (1960), *The Human Side of Enterprise*, McGraw-Hill, New York.

Mento, A. J., Steel, R. P., and Karren, R. J. (1987), "A Meta-analytic Study of the Effects of Goal Setting on Task Performance," *Organizational Behavior and Human Decision Processes*, Vol. 39, pp. 52–83.

Mento, A. J., Klein, H. J., and Locke, E. A. (1992), "Relationship of Goal Level to Valence and Instrumentality," *Journal of Applied Psychology*, Vol. 77, No. 4, pp. 395–405.

Mikula, G., Ed. (1980), *Gerechtigkeit und soziale Interaktion*, Hans Huber, Bern, Switzerland.

Miller, J. G. (1997), "Cultural Conceptions of Duty: Implications for Motivation and Morality," in *Motivation and Culture*, D. Munro, J. F. Schuhmaker, and A. C. Carr, Eds., Routledge, New York, pp. 178–192.

Miner, J. B. (1980), *Theories of Organizational Behavior*, Dryden Press, Hinsdale, IL.

Miron, D., and McClelland, D. C. (1979), "The Impact of Achievement Motivation Training on Small Business," *California Management Review*, Vol. 21, pp. 13–28.

Mitchell, T. R., and Wood, R. E. (1980), "Supervisor's Responses to Subordinate Poor Performance: A Test of an Attribution Model," *Organizational Behavior and Human Performance*, Vol. 25, pp. 123–138.

Morrison, P. R., and Noble, G. (1987), "Individual Differences and Ergonomic Factors in Performance on a Videotex-Type Task," *Behavior and Information Technology*, Vol. 6, pp. 69–88.

MOW International Research Team (1987), *The Meaning of Work*, Academic Press, London.

Murray, H. A. (1938), *Exploration in Personality*, Oxford University Press, New York.

Myers, M. S. (1970), *Every Employee a Manager: More Meaningful Work Through Job Enrichment*, McGraw-Hill, New York.

Myers, D. G. (1990), *Social Psychology*, McGraw-Hill, New York.

Naschold, F. (1992), *Den Wandel organisieren: Erfahrungen des schwedischen Entwicklungsprogramms 'Leitung, Organisation, Mitbestimmung' (LOM) im internationalen Wettbewerb*, Rainer Bohn Verlag, Berlin.

Neuberger, O. (1974), *Theorien der Arbeitszufriedenheit*, Kohlhammer, Stuttgart, Germany.

Neuberger, O. (1985), *Arbeit*, Enke Verlag, Stuttgart, Germany.

Neuberger, O. (1989), Organisationstheorien, in *Organisationspsychologie: Enzyklopädie der Psychologie*, Vol. D/III/3, E. Roth, Ed., Hogrefe Verlag, Göttingen, Germany, pp. 205–250.

Nicholson, N., Herzberg, F., Simon, W., Sprenger, R., and Kühl, S. (2004), *Motivation: Was Manager und Mitarbeiter antreibt*, Wirtschaftsverlag Carl Ueberreuter, Frankfurt, Germany.

Noelle-Neumann, E., and Strümpel, B. (1984), *Macht Arbeit krank? Macht Arbeit glücklich?* Piper Verlag, Munich.

Oberquelle, H. (1993), "Anpaßbarkeit von Groupware als Basis für die dynamische Gestaltung von computergestützter Gruppenarbeit," in *Software-Ergonomie in der Gruppenarbeit*, U. Konradt and L. Drisis, Eds., Westdeutscher Verlag, Opladen, Germany, pp. 37–54.

Oishi, S. (2000), "Goals as Cornerstones of Subjective Well-Being: Linking Individuals and Cultures," in *Culture and Subjective Well-Being*, E. Diener and E. Suh, Eds., MIT Press, Cambridge, MA, pp. 87–112.

Oppolzer, A. (1989), *Handbuch Arbeitsgestaltung*, VSA-Verlag, Hamburg, Germany.

Paetau, M., and Pieper, M. (1985), "Differentiell-dynamische Gestaltung der Mensch–Maschine–Kommunikation," in *Software-Ergonomie '85: Mensch–Computer-Interaktion*, H. J. Bullinger, Ed., *Berichte des German Chapter of the ACM*, Vol. 24, Teubner, Stuttgart, Germany, pp. 316–324.

Parker, B., and Chusmir, L. H. (1992), "Development and Validation of a Life-Success Measures Scale," *Psychological Reports*, Vol. 70, No. 1, pp. 627–637.

Paul, P. J., Robertson, K. B., and Herzberg, F. (1969), "Job Enrichment Pays Off," *Harvard Business Review*, Vol. 47, pp. 61–78.

Phillips, J. S., and Lord, R. G. (1980), "Determinants of Intrinsic Motivation: Locus of Control and Competence Information as Components of Deci's Cognitive Evaluation Theory," *Journal of Applied Psychology*, Vol. 65, pp. 211–218.

Podsakoff, P. M., and Williams, L. J. (1986), "The Relationship Between Job Performance and Job Satisfaction," in *Generalizing from Laboratory to Field Settings: Research Findings from Industrial-Organizational Psychology, Organizational Behavior, and Human Resource Management*, E. Locke, Ed., Lexington Books, Lexington, MA, pp. 207–254.

Pope John Paul II (1981), *Enzyklika Laborem Exercens: Über die menschliche Arbeit*, Georg Bitter, Recklinghausen, Germany.

Porter, L. W. (1964), *Organizational Patterns of Managerial Job Attitudes*, American Foundation for Management Research, New York.

Porter, L. W., and Lawler, E. E. (1968), *Managerial Attitudes and Performance*, Dorsey Press, Homewood, IL.

Ralston, D. A., and Flanagan, M. F. (1985), "The Effect of Flextime on Absenteeism and Turnover for Male and Female Employees," *Journal of Vocational Behavior*, April, pp. 206–217.

Ralston, D. A., Anthony, W. P., and Gustafson, D. J. (1985), "Employees May Love Flextime, but What Does It Do to the Organization's Productivity?" *Journal of Applied Psychology*, Vol. 70, pp. 272–279.

Rauterberg, M., Spinas, P., Strohm, O., Ulich, E., and Waeber, D. (1994), *Benutzerorientierte Software-Entwicklung*, Teubner, Stuttgart, Germany.

Rice, A. K. (1958), *Productivity and Social Organization: The Ahmedabad Experiment*, Tavistock Institute of Human Relations, London.

Richer, S. F., and Vallerand, R. J. (1995), "Supervisors' Interactional Styles and Subordinates' Intrinsic and Extrinsic Motivation," *Journal of Social Psychology*, Vol. 135, pp. 707–722.

Roberts, K. H., and Glick, W. (1981), "The Job Characteristics Approach to Task Design: A Critical Review," *Journal of Applied Psychology*, Vol. 66, pp. 193–217.

Rodgers, R., Hunter, J. E., and Rodgers, D. I. (1993), "Influence of Top Management Commitment on Management Program Success," *Journal of Applied Psychology*, Vol. 78, pp. 151–155.

Roethlisberger, F., and Dickson, W. (1939), *Management and the Worker: An Account of a Research Program Conducted by the Western Electric Company, Chicago*, Harvard University Press, Cambridge, MA.

Rosenstiel, L. V., Hockel, C., and Molt, W., Eds. (1994), *Handbuch der angewandten Psychologie: Grundlagen, Methoden, Praxis*, Ecomed, Landsberg, Germany.

Ruiz Quintanilla, S. A. (1984), "Bedeutung des Arbeitens: Entwicklung und empirische Erprobung eines sozialwissenschaftlichen Modells zur Erfassung arbeitsrelevanter Werthaltungen und Kognitionen," dissertation, Technische Universität Berlin.

Ryan, R. M., and Deci, E. L. (2000), "Self-Determination Theory and the Facilitation of Intrinsic Motivation, Social Development, and Well-Being," *American Psychologist*, Vol. 55, No. 1, pp. 68–79.

Saavedra, R., Earley, P., and Van Dyne, L. (1993), "Complex Interdependence in Task-Performing Groups," *Journal of Applied Psychology*, Vol. 78, pp. 61–72.

Salancik, G. R., and Pfeffer, J. (1977), "An Examination of Need-Satisfaction Models of Job Attitudes," *Administrative Science Quarterly*, Vol. 22, pp. 427–456.

Schettgen, P. (1996), *Arbeit, Leistung, Lohn: Analyse- und Bewertungsmethoden aus sozioökonomischer Perspektive*, Enke Verlag, Stuttgart, Germany.

Schmale, H. (1983), *Psychologie der Arbeit*, Klett-Cotta, Stuttgart, Germany.

Schmidt, G. (1982), "Humanisierung der Arbeit," in *Einführung in die Arbeits- und Industriesoziologie*, W. Litek, W. Rammert, and G. Wachtler, Eds., Campus Verlag, Frankfurt on Main, Germany.

Schmidtchen, G. (1984), *Neue Technik–Neue Arbeitsmoral: Eine sozialpsychologische Untersuchung über Motivation in der Metallindustrie*, Deutscher Institutsverlag, Köln, Germany.

Schneider, B., and Alderfer, C. P. (1973), "Three Studies of Measures of Need Satisfaction in Organizations," *Administrative Science Quarterly*, Vol. 18, pp. 489–505.

Sisson, K. (2000), "Direct Participation and the Modernisation of Work Organisation," European Foundation for the Improvement of Living and Working Conditions; available at http://www.eurofound.eu.int/publications/files/EF0029EN.pdf [28.05.2004].

Six, B., and Kleinbeck, U. (1989), "Motivation und Zufriedenheit in Organisationen," in *Organisationspsychologie: Enzyklopädie der Psychologie*, Vol. D/III/3, E. Roth, Ed., Hogrefe Verlag, Göttingen, Germany, pp. 348–398.

Spector, P. E. (1982), "Behavior in Organizations as a Function of Employee's Locus of Control," *Psychological Bulletin*, Vol. 91, pp. 482–497.

Sprenger R. K. (2002), *Mythos Motivation: Wege aus einer Sackgasse*, Campus Verlag, Frankfurt on Main, Germany.

Staehle W. H. (1992), "Das Bild vom Arbeitnehmer im Wandel der Arbeitgeber-Arbeitnehmer-Beziehungen," in *Modernisierung der Arbeitsbeziehungen: Direkte Arbeitnehmerbeteiligung in deutschen und französischen Betrieben*, L. Kißler, Ed., Centre d'Information et de Recherche sur l'Allemagne Contemporaine, Paris.

Stahl, M. J. (1986), *Managerial and Technical Motivation: Assessing Needs for Achievement, Power, and Affiliation*, Praeger, New York.

Staw, B. M., Bell, N. E., and Clausen, J. A. (1986), "The Dispositional Approach to Job Attitudes," *Administrative Science Quarterly*, Vol. 31, pp. 56–77.

Steers, R. M., Porter, L. W., and Bigley, G. A. (1996), *Motivation and Leadership at Work*, McGraw-Hill, New York.

Steinmann, H., and Schreyögg, G. (1980), "Arbeitsstrukturierung am Scheideweg," *Zeitschrift für Arbeitswissenschaft*, Vol. 34, pp. 75–78.

Stroebe, W., Hewstone, M., and Stephenson, G. M., Eds. (1997), *Sozialpsychologie*, Springer-Verlag, Heidelberg, Germany.

Summers, T. P., and DeNisi, A. S. (1990), "In Search of Adam's Other: Reexamination of Referents Used in the Evaluation of Pay," *Human Relations*, Vol. 43, No. 6, pp. 120–133.

Tanguy, L. (1986), *L'introuvable relation formation/emploi: un état des recherches en France*, Ministère de la Recherche et de l'enseignement superieur, programme mobilisateur technologie, emploi, travail, La Documentation Française, Paris.

Taylor, F. W. (1911), *The Principles of Scientific Management*, Harper, New York.

Thompson, E. P., Chaiken, S., and Hazlewood, J. D. (1993), "Need for Cognition and Desire for Control as Moderators of Extrinsic Reward Effects," *Journal of Personality and Social Psychology*, Vol. 64, No. 6, pp. 987–999.

Triebe, J. K. (1980), "Untersuchungen zum Lernprozeß während des Erwerbs der Grundqualifikationen (Montage eines kompletten Motors): Arbeits- und sozialpsychologische Untersuchungen von Arbeitsstrukturen im Bereich der Aggregatefertigung der Volkswagenwerk AG, Bonn," BMFT, HA 80-019.

Triebe, J. K. (1981), "Aspekte beruflichen Handelns und Lernens," dissertation. Universität Bern.

Triebe, J. K., Wittstock, M., and Schiele, F. (1987), *Arbeitswissenschaftliche Grundlagen der Software-Ergonomie*, Wirtschaftsverlag NW, Bremerhaven, Germany.

Trist, E. L. (1981), *The Evaluation of Sociotechnical Systems: Issues in the Quality of Working Life*, Ontario Quality of Working Life Centre, Toronto, Ontario, Canada.

Trist, E. (1990), "Sozio-technische Systeme: Ursprünge und Konzepte," *Organisationsentwicklung*, Vol. 4, pp. 10–26.

Trist, E., and Bamforth, K. (1951), "Some Social and Psychological Consequences of the Longwall Method of Coal-Getting," *Human Relations*, Vol. 4, pp. 3–38.

Tubbs, M. E. (1986), "Goal Setting: A Meta-analytic Examination of the Empirical Evidence," *Journal of Applied Psychology*, Vol. 71, pp. 474–483.

Tubbs, M. E. (1993), "Commitment as a Moderator of the Goal–Performance Relation: A Case for Clearer Construct Definition," *Journal of Applied Psychology*, Vol. 78, pp. 86–97.

Tubbs, M. E., Boehme, D. M., and Dahl, J. G. (1993), "Expectancy, Valence, and Motivational Force Functions in Goal-Setting Research," *Journal of Applied Psychology*, Vol. 78, No. 3, pp. 361–373.

Udris, I. (1979a), "Ist Arbeit noch länger zentrales Lebensinteresse?" *Psychosozial*, Vol. 2, No. 1, pp. 100–120.

Udris, I. (1979b), "Central Life Interests: Werte, Wertwandel, Arbeitspsychologie," *Schweizerische Zeitschrift für Psychologie*, Vol. 38, pp. 252–259.

Udris, I. (1981), "Streß in arbeitspsychologischer Sicht," in *Stress*, J. R. Nitsch, Ed., Hans Huber, Bern, Switzerland, pp. 391–440.

Ulich, E. (1980), "Psychologische Aspekte der Arbeit mit elektronischen Datenverarbeitungssystemen," *Schweizerische Technische Zeitschrift*, Vol. 75, pp. 66–68.

Ulich, E. (1981), "Subjektive Tätigkeitsanalyse als Voraussetzung autonomieorientierter Arbeitsgestaltung," in *Beiträge zur psychologischen Arbeitsanalyse*, F. Frei and E. Ulich, Eds., Hans Huber, Bern, Switzerland, pp. 327–347.

Ulich, E. (1990), Individualisierung und differentielle Arbeitsgestaltung. *Ingenieurpsychologie: Enzyklopädie der Psychologie*, Vol. D/III/2, In C. Graf Hoyos and B. Zimolong, Eds., Hogrefe Verlag, Göttingen, Germany, pp. 511–535.

Ulich, E. (1991), *Arbeitspsychologie*, Poeschel, Stuttgart, Germany.

Ulich, E. (1993), "Gestaltung von Arbeitstätigkeiten," in *Lehrbuch Organisationspsychologie*, H. Schuler, Ed., Hans Huber, Bern, Switzerland, pp. 189–208.

Vallerand, R. J., and Reid, G. (1984), "On the Causal Effects of Perceived Competence on Intrinsic Motivation: A Test of Cognitive Evaluation Theory," *Journal of Sport Psychology*, Vol. 6, pp. 94–102.

Vansteenkiste, M., and Deci, E. L. (2003), "Competitively Contingent Rewards and Intrinsic Motivation: Can Losers Remain Motivated?" *Motivation and Emotion*, Vol. 27, No. 4, pp. 273–299.

Volpert, W. (1987), "Psychische Regulation von Arbeitstätigkeiten," in *Arbeitspsychologie: Enzyklopädie der Psychologie*, Vol. D/III/1, U. Kleinbeck and J. Rutenfranz, Eds., Hogrefe Verlag, Göttingen, Germany, pp. 1–42.

von der Weth, R. (1988), "Konstruktionstätigkeit und Problemlösen," in *Rechnerunterstützte Konstruktion*, E. Frieling and H. Klein, Eds., Hans Huber, Bern, Switzerland, pp. 32–39.

Vroom, V. H. (1964), *Work and Motivation*, Wiley, New York.

Warr, P. (1984), "Work and Unemployment," in *Handbook of Work and Organizational Psychology*, P. J. D. Drenth, H. Thierry, P. J. Williams, and C. J. de Wolff, Eds., Wiley, Chichester, West Sussex, England, pp. 413–443.

Weber, M. (1904/1905), "Die protestantische Ethik und der Geist des Kapitalismus," *Archiv für Sozialwissenschaft und Sozialpolitik*, Vol. 20, pp. 1–54 and Vol. 21, pp. 1–110.

Weber, W., and Ulich, E. (1993), "Psychological Criteria for the Evaluation of Different Forms of Group-Work in Advanced Manufacturing Systems," in *Human–Computer Interaction: Application and Case Studies*, M. J. Smith and G. Salvendy, Eds., Elsevier, Amsterdam, pp. 26–31.

Weiner, B. (1985), *Human Motivation*, Springer-Verlag, New York.

Weiner, B. (1992), *Human Motivation: Metaphors, Theories, and Research*. Sage Publications, Newbury Park, CA.

Wiendieck, G. (1986a), "Warum Qualitätszirkel? Zum organisationspsychologischen Hintergrund eines neuen Management-Konzeptes," in *Qualitätszirkel als Instrument zeitgemäßer Betriebsführung*, G. Wiendieck and W. Bungard, Eds., Verlag Moderne Industrie, Landsberg on Lech, Germany, pp. 61–75.

Wiendieck, G. (1986b), "Widerstand gegen Qualitätszirkel: Eine Idee und Ihre Feinde," in *Qualitätszirkel als Instrument zeitgemäßer Betriebsführung*, G. Wiendieck and W. Bungard, Eds., Verlag Moderne Industrie, Landsberg on Lech, Germany, pp. 207–223.

Williams, E. S., and Bunker, D. R. (1993), "Sorting Outcomes: A Revision of the JCM," in *The Job Characteristics Model: Recent Work and New Directions*, D. R. Bunker and E. D. Williams, Eds., Academy of Management, Atlanta, GA.

Wilpert, B., and Sorge, A. (1984), *International Perspectives in Organizational Democracy*, Wiley, Chichester, West Sussex, England.

Wilson, A. T. M., and Trist, E. L. (1951), *The Bolsover System of Continuous Mining*, Document 290, Tavistock Institute of Human Relations, London.

Windel, A., and Zimolong, B. (1998), "Nicht zum Nulltarif: Erfolgskonzept Gruppenarbeit," *Rubin*, Vol. 2, pp. 46–51.

Wiswede, G. (1980), *Motivation und Arbeitsverhalten*, E. Reinhardt, Munich.

Wofford, J. C., Goodwin, V. L., and Premack, S. (1992), "Meta-analysis of the Antecedents of Personal Goal Level and of the Antecedents and Consequences of Goal Commitment," *Journal of Management*, Vol. 18, No. 3, pp. 595–615.

Yukl, G. A. (1990), *Skills for Managers and Leaders*, Prentice Hall, Englewood Cliffs, NJ.

Zink, K. (1978), "Zur Begründung einer zielgruppenspezifischen Organisationsentwicklung," *Zeitschrift für Arbeitswissenschaft*, Vol. 32, pp. 42–48.

Zink, K. J. (1979), *Begründung einer zielgruppenspezifischen Organisationsentwicklung auf der Basis von Untersuchungen von Arbeitszufriedenheit und Arbeitsmotivation*, O. Schmidt, Köln, Germany.

Zülch, G., and Starringer, M. (1984), "Differentielle Arbeitsgestaltung in Fertigungen für elektronische Flachbaugruppen," *Zeitschrift für Arbeitswissenschaft*, Vol. 38, pp. 211–216.

CHAPTER 16

JOB AND TEAM DESIGN

Frederick P. Morgeson
Michigan State University
East Lansing, Michigan

Gina J. Medsker
Human Resources Research Organization
Alexandria, Virginia

Michael A. Campion
Purdue University
West Lafayette, Indiana

1 INTRODUCTION

1.1 Job Design

Job design is an aspect of managing organizations that is so commonplace it often goes unnoticed. Most people realize the importance of job design when an organization or new plant is starting up, and some recognize the importance of job design when organizations are restructuring or changing processes. But fewer people realize that job design may be affected as organizations change markets or strategies, managers use their discretion in the assignment of tasks on a daily basis, people in the jobs or their managers change, the workforce or labor markets change, or there are performance, safety, or satisfaction problems. Fewer yet realize that job design change can be used as an intervention to enhance organizational goals (Campion and Medsker, 1992).

It is clear that many different aspects of an organization influence job design, especially an organization's structure, technology, processes, and environment. These influences are beyond the scope of this chapter, but they are dealt with in other references (e.g.,Davis, 1982; Davis and Wacker, 1982). These influences impose constraints on how jobs are designed and will play a major role in any practical application. However, it is the assumption of this chapter that considerable discretion exists in the design of jobs in most situations, and the job (defined as a set of tasks performed by a worker) is a convenient unit of analysis in both developing new organizations and changing existing ones (Campion and Medsker, 1992).

The importance of job design lies in its strong influence on a broad range of important efficiency and human resource outcomes. Job design has predictable consequences for outcomes, including the following (Campion and Medsker, 1992): productivity, quality, job satisfaction, training times, intrinsic work motivation, staffing, error rates, accident rates, mental fatigue, physical fatigue, stress, mental ability requirements, physical ability requirements, job involvement, absenteeism, medical incidents, turnover, and compensation rates.

According to Louis Davis, one of the most prolific writers on job design in the engineering literature over the last 35 years, many of the personnel and productivity problems in industry may be the direct result of the design of jobs (Davis et al., 1955; Davis, 1957; Davis and Valfer, 1965; Davis and Taylor, 1979; Davis and Wacker, 1982, 1987). Unfortunately, people mistakenly view the design of jobs as technologically determined and inalterable. However, job designs are actually social inventions. They reflect the values of the era in which they were constructed. These values include the economic goal of minimizing immediate costs (Davis et al., 1955; Taylor, 1979) and theories of human motivation (Warr and Wall, 1975; Steers and Mowday, 1977). These values, and the designs they influence, are not immutable givens, but are subject to modification (Campion and Thayer, 1985; Campion and Medsker, 1992).

The question then becomes: What is the best way to design a job? In fact, there is no single best way. There are several major approaches to job design, each derived from a different discipline and reflecting different theoretical orientations and values. This chapter covers these approaches, their costs and benefits, and tools and procedures for developing and assessing jobs in all types of organizations. We highlights trade-offs that must be made when choosing among different approaches to job design. We also compare the design of jobs for people working independently to the design of work for teams, which is an alternative to designing jobs at the level of individual workers. We present the advantages and disadvantages of designing work around individuals compared to designing work for teams and provide advice on implementing and evaluating the various work design approaches.

1.2 Team Design

The major approaches to job design typically focus on designing jobs for individual workers. However, the approach to work design at the level of the group or team, rather than at the level of individual workers, is gaining substantially in popularity, and many U.S. organizations are now using teams (Campion et al., 1996; Parker, 2003; Ilgen et al., 2005). New manufacturing systems (e.g., flexible, cellular) and advancements in our understanding of team processes not only allow designers to consider the use of work teams, but often seem to encourage the use of team approaches (Gallagher and Knight, 1986; Majchrzak, 1988).

In designing jobs for teams, one assigns a task or set of tasks to a team of workers rather than to an individual, and considers the team to be the primary unit of performance. Objectives and rewards focus on team, not individual, behavior. Depending on the nature of its tasks, a team's workers may be performing the same tasks simultaneously or they may break tasks into subtasks to be performed by individuals within the team. Subtasks can be assigned on the basis of expertise or interest, or team members might rotate from one subtask to another to provide variety and increase breadth of skills and flexibility in the workforce (Campion and Medsker, 1992; Campion et al., 1994b).

Some tasks are of a size, complexity, or, otherwise seem to naturally fit into a team job design, whereas others may seem to be appropriate only at the individual job level. In many cases, though, there may be a considerable degree of choice regarding whether one organizes work around teams or individuals. In such situations, the designer should consider advantages and disadvantages of the use of the job and team design approaches with respect to an organization's goals, policies, technologies, and constraints (Campion et al., 1993).

2 JOB DESIGN APPROACHES

In this chapter we adopt an interdisciplinary perspective on job design. Interdisciplinary research on job design has shown that different approaches to job design exist. Each is oriented toward a particular subset of outcomes, each has disadvantages as well as advantages, and trade-offs among approaches are required in most job design situations (Campion and Thayer, 1985; Campion, 1988, 1989; Campion and Berger, 1990; Campion and McClelland, 1991, 1993; Edwards et al., 1999, 2000; Morgeson and Campion, 2002, 2003). The four major approaches to job design are reviewed below. Table 1 summarizes the job design approaches, and Table 2 provides specific recommendations. The team design approach is reviewed in Section 3.

2.1 Mechanistic Job Design Approach

2.1.1 Historical Development

The historical roots of job design can be traced back to the idea of the division of labor, which was very important to early thinking on the economies of manufacturing (Babbage, 1835; Smith, 1776). Division of labor led to job designs characterized by specialization and simplification. Jobs designed in this fashion had many advantages, including reduced learning time, saved time from not having to change tasks or tools, increased proficiency from repeating tasks, and development of specialized tools and equipment.

A very influential person for this perspective was Frederick Taylor (Taylor, 1911; Hammond, 1971). He explicated the principles of scientific management, which encouraged the study of jobs to determine the "one best way" to perform each task. Movements of skilled workers were studied using a stopwatch and simple analysis. The best and quickest methods and tools were selected, and all workers were trained to perform the job the same way. Standard performance levels were set, and incentive pay was tied to the standards. Gilbreth also contributed to this design approach (Gilbreth, 1911). With time and motion study, he tried to eliminate wasted movements by the appropriate design of equipment and placement of tools and materials.

Table 1 Advantages and Disadvantages of Various Job Design Approaches[a]

Approach/Discipline Base References)	Recommendations	Benefits	Costs
Mechanistic/classic industrial engineering (Gilbreth, 1911; Taylor, 1911; Niebel, 1988)	Increase in: • Specialization • Simplification • Repetition • Automation Decrease in: • Spare time	Decrease in: • Training • Staffing difficulty • Making errors • Mental overload and fatigue • Mental skills and abilities • Compensation	Increase in: • Absenteeism • Boredom Decrease in: • Satisfaction • Motivation
Motivational/organizational psychology (Hackman and Oldham, 1980; Herzberg, 1966)	Increase in: • Variety • Autonomy • Significance • Skill usage • Participation • Feedback • Recognition • Growth • Achievement	Increase in: • Satisfaction • Motivation • Involvement • Performance • Customer Service • Catching errors Decrease in: • Absenteeism • Turnover	Increase in: • Training time/cost • Staffing difficulty • Making errors • Mental overload • Stress • Mental skills and abilities • Compensation
Perceptual-Motor/experimental psychology, human factors (Salvendy, 1987; Sanders and McCormick, 1987)	Increase in: • Lighting quality • Display and control quality • User-friendly equipment Decrease in: • Information processing requirements	Decrease in: • Making errors • Accidents • Mental overload • Stress • Training time/cost • Staffing difficulty • Compensation • Mental skills and abilities	Increase in: • Boredom Decrease in: • Satisfaction
Biological/physiology, biomechanics, ergonomics (Astrand and Rodahl, 1977; Tichauer, 1978; Grandjean, 1980)	Increase in: • Seating comfort • Postural comfort Decrease in: • Strength requirements • Endurance requirements • Environmental stressors	Decrease in: • Physical abilities • Physical fatigue • Aches and pains • Medical incidents	Increase in: • Financial cost • Inactivity

Source: Adapted from Campion and Medsker (1992).
[a]Advantages and disadvantages are based on findings in previous interdisciplinary research (Campion and Thayer, 1985; Campion, 1988, 1989; Campion and Berger, 1990; Campion and McClelland, 1991, 1993).

Surveys of industrial job designers indicate that this "mechanistic" approach to job design has been the prevailing practice throughout the twentieth century (Davis et al., 1955; Taylor, 1979). These characteristics are also the primary focus of many modern-day writers on job design (e.g., Mundel, 1985; Niebel, 1988) and are present in such newer techniques as lean production (Parker, 2003). The discipline base for this approach is early or "classic" industrial engineering.

2.1.2 Design Recommendations

Table 2 provides a brief list of statements that describe the essential recommendations of the mechanistic approach. In essence, jobs should be studied to determine the most efficient work methods and techniques. The total work in an area (e.g., department) should be broken down into highly specialized jobs assigned to different employees. The tasks should be simplified so that skill requirements are minimized. There should also be repetition to gain improvement

Table 2 Multimethod Job Design Questionnaire[a]

Instructions: Indicate the extent to which each statement is descriptive of the job using the scale below.
Circle answers to the right of each statement

Please Use the Following Scale:
(5) Strongly agree
(4) Agree
(3) Neither agree nor disagree
(2) Disagree
(1) Strongly disagree
() Leave blank if do not know or not applicable

Mechanistic Approach					
1. *Job specialization:* The job is highly specialized in terms of purpose, tasks, or activities.	1	2	3	4	5
2. *Specialization of tools and procedures:* The tools, procedures, materials, etc., used on this job are highly specialized in terms of purpose.	1	2	3	4	5
3. *Task simplification:* The tasks are simple and uncomplicated.	1	2	3	4	5
4. *Single activities:* The job requires you to do only one task or activity at a time.	1	2	3	4	5
5. *Skill simplification:* The job requires relatively little skill and training time.	1	2	3	4	5
6. *Repetition:* The job requires performing the same activity(s) repeatedly.	1	2	3	4	5
7. *Spare time:* There is very little spare time between activities on this job.	1	2	3	4	5
8. *Automation:* Many of the activities of this job are automated or assisted by automation.	1	2	3	4	5
Motivational Approach					
9. *Autonomy:* The job allows freedom, independence, or discretion in work scheduling, sequence, methods, procedures, quality control, or other decision making.	1	2	3	4	5
10. *Intrinsic job feedback:* The work activities themselves provide direct and clear information as to the effectiveness (e.g., quality and quantity) of job performance.	1	2	3	4	5
11. *Extrinsic job feedback:* Other people in the organization, such as managers and co-workers, provide information as to the effectiveness (e.g., quality and quantity) of job performance.	1	2	3	4	5
12. *Social interaction:* The job provides for positive social interaction such as team work or co-worker assistance.	1	2	3	4	5
13. *Task/goal clarity:* The job duties, requirements, and goals are clear and specific.	1	2	3	4	5
14. *Task variety:* The job has a variety of duties, tasks, and activities.	1	2	3	4	5
15. *Task identity:* The job requires completion of a whole and identifiable piece of work. It gives you a chance to do an entire piece of work from beginning to end.	1	2	3	4	5
16. *Ability/skill level requirements:* The job requires a high level of knowledge, skills, and abilities.	1	2	3	4	5
17. *Ability/skill variety:* The job requires a variety of knowledge, skills, and abilities.	1	2	3	4	5
18. *Task significance:* The job is significant and important compared with other jobs in the organization.	1	2	3	4	5
19. *Growth/learning:* The job allows opportunities for learning and growth in competence and proficiency.	1	2	3	4	5
20. *Promotion:* There are opportunities for advancement to higher level jobs.	1	2	3	4	5
21. *Achievement:* The job provides for feelings of achievement and task accomplishment.	1	2	3	4	5
22. *Participation:* The job allows participation in work-related decision making.	1	2	3	4	5
23. *Communication:* The job has access to relevant communication channels and information flows.	1	2	3	4	5
24. *Pay adequacy:* The pay on this job is adequate compared with the job requirements and with the pay in similar jobs.	1	2	3	4	5
25. *Recognition:* The job provides acknowledgment and recognition from others.	1	2	3	4	5
26. *Job security:* People on this job have high job security.	1	2	3	4	5
Perceptual/Motor Approach					
27. *Lighting:* The lighting in the workplace is adequate and free from glare.	1	2	3	4	5
28. *Displays:* The displays, gauges, meters, and computerized equipment on this job are easy to read and understand.	1	2	3	4	5

(continued overleaf)

Table 2 (*continued*)

Biological Approach					
29. *Programs:* The programs in the computerized equipment on this job are easy to learn and use.	1	2	3	4	5
30. *Other equipment:* The other equipment (all types) used on this job is easy to learn and use.	1	2	3	4	5
31. *Printed job materials:* The printed materials used on this job are easy to read and interpret.	1	2	3	4	5
32. *Workplace layout:* The workplace is laid out such that you can see and hear well to perform the job.	1	2	3	4	5
33. *Information input requirements:* The amount of information you must attend to in order to perform this job is fairly minimal.	1	2	3	4	5
34. *Information output requirements:* The amount of information you must output on this job, in terms of both action and communication, is fairly minimal.	1	2	3	4	5
35. *Information processing requirements:* The amount of information you must process, in terms of thinking and problem solving, is fairly minimal.	1	2	3	4	5
36. *Memory requirements:* The amount of information you must remember on this job is fairly minimal.	1	2	3	4	5
37. *Stress:* There is relatively little stress on this job.	1	2	3	4	5
38. *Strength:* The job requires fairly little muscular strength.	1	2	3	4	5
39. *Lifting:* The job requires fairly little lifting, and/or the lifting is of very light weights.	1	2	3	4	5
40. *Endurance:* The job requires fairly little muscular endurance.	1	2	3	4	5
41. *Seating:* The seating arrangements on the job are adequate (e.g., ample opportunities to sit, comfortable chairs, good postural support).	1	2	3	4	5
42. *Size differences:* The workplace allows for all size differences between people in terms of clearance, reach, eye height, leg room, etc.	1	2	3	4	5
43. *Wrist movement:* The job allows the wrists to remain straight without excessive movement.	1	2	3	4	5
44. *Noise:* The workplace is free from excessive noise.	1	2	3	4	5
45. *Climate:* The climate at the workplace is comfortable in terms of temperature and humidity, and it is free of excessive dust and fumes.	1	2	3	4	5
46. *Work breaks:* There is adequate time for work breaks given the demands of the job.	1	2	3	4	5
47. *Shift work:* The job does not require shift work or excessive overtime.	1	2	3	4	5
For jobs with little physical activity due to single workstation, add:					
48. *Exercise opportunities:* During the day, there are enough opportunities to get up from the workstation and walk around.	1	2	3	4	5
49. *Constraint:* While at the workstation, the worker is not constrained to a single position.	1	2	3	4	5
50. *Furniture:* At the workstation, the worker can adjust or arrange the furniture to be comfortable (e.g., adequate legroom, foot rests if needed, proper keyboard or work surface height).	1	2	3	4	5

Source: Adapted from Campion (1988).

[a]Specific recommendations from each job design approach. See source and related research (e.g., Campion and McClelland, 1991, 1993; Campion and Thayer, 1985) for reliability and validity information. Scores for each approach are calculated by averaging applicable items.

from practice. Idle time should be minimized. Finally, activities should be automated or assisted by automation to the extent possible and economically feasible.

2.1.3 Advantages and Disadvantages

The goal of this approach is to maximize efficiency, in terms of both productivity and utilization of human resources. Table 1 summarizes some human resource advantages and disadvantages that have been observed in research. Jobs designed according to the mechanistic approach are easier and less expensive to staff. Training times are reduced. Compensation requirements may be less because skill and responsibility are reduced. And because mental demands are less, errors may be less common. Disadvantages include the fact that extreme use of the mechanistic approach may result in jobs so simple and routine that employees experience low job satisfaction and motivation. Overly mechanistic, repetitive work can lead to health problems such as repetitive motion disorders.

2.2 Motivational Job Design Approach

2.2.1 Historical Development

Encouraged by the human relations movement of the 1930s (Mayo, 1933; Hoppock, 1935), people began to point out the negative effects on worker attitudes and health of the overuse of mechanistic design (Argyris, 1964; Blauner, 1964). Overly specialized, simplified jobs were found to lead to dissatisfaction (Caplan et al., 1975) and adverse physiological consequences for workers (Johansson et al., 1978; Weber et al., 1980). Jobs on assembly lines and other machine-paced work were especially troublesome in this regard (Walker and Guest, 1952; Salvendy and Smith, 1981). These trends led to an increasing awareness of employees' psychological needs.

The first efforts to enhance the meaningfulness of jobs involved the opposite of specialization. It was recommended that tasks be added to jobs, either at the same level of responsibility (i.e., job enlargement) or at a higher level (i.e., job enrichment) (Herzberg, 1966; Ford, 1969); This trend expanded into a pursuit of identifying and validating characteristics of jobs that make them motivating and satisfying (Turner and Lawrence, 1965; Hackman and Oldham, 1980; Griffin, 1982). This approach considers the psychological theories of work motivation (e.g., Vroom, 1964; Steers and Mowday, 1977). Thus, this "motivational" approach draws primarily from organizational psychology as a discipline base.

A related trend following later but somewhat comparable in content is the sociotechnical approach (Emory and Trist, 1960; Rousseau, 1977; Pasmore, 1988). It focuses not only on the work, but also on the technology itself and the relationship of the environment to work and organizational design. Interest is less on the job and more on roles and systems. Keys to this approach are work system and job designs that fit their external environment and the joint optimization of both social and technical systems in the organization's internal environment. Although this approach differs somewhat in that consideration is also given to the technical system and external environment, it is similar in that it draws on the same psychological job characteristics that affect satisfaction and motivation. It suggests that as organizations' environments are becoming increasingly turbulent and complex, organizational and job design should involve greater flexibility, employee involvement, employee training, and decentralization of decision making and control, and a reduction in hierarchical structures and the formalization of procedures and relationships (Pasmore, 1988).

Surveys of industrial job designers have consistently indicated that the mechanistic approach represents the dominant theme of job design (Davis et al., 1955; Taylor, 1979). Other approaches to job design, such as the motivational approach, have not been given as much explicit consideration. This is not surprising because the surveys included only job designers trained in engineering-related disciplines, such as industrial engineering and systems analysis. It is not necessarily certain that other specialists or line managers would adopt the same philosophies, especially in recent times. Nevertheless, there is evidence that even fairly naive job designers (i.e., college students in management classes) also adopt the mechanistic approach in job design simulations. That is, their strategies for grouping tasks were primarily the similarity of such factors as activities, skills, equipment, procedures, or location. Even though the mechanistic approach may be the most natural and intuitive, this research has also revealed that people can be trained to apply all four approaches to job design (Campion and Stevens, 1991).

2.2.2 Design Recommendations

Table 2 provides a list of statements that describe recommendations for the motivational approach. It suggests that a job should allow a worker autonomy to make decisions about how and when tasks are to be done. A worker should believe that his or her work is important to the overall mission of an organization or department. This is often done by allowing a worker to perform a larger unit of work or to perform an entire piece of work from beginning to end. Feedback on job performance should be given to workers from the task itself, as well as from the supervisor and others. Workers should be able to use a variety of skills and to grow personally on the job. This approach also considers the social, or people-interaction, aspects of the job: Jobs should have opportunities for participation, communication, and recognition. Finally, other human resource systems should contribute to the motivating atmosphere, such as adequate pay, promotion, and job security systems.

2.2.3 Advantages and Disadvantages

The goal of this approach is to enhance psychological meaningfulness of jobs, thus influencing a variety of attitudinal and behavioral outcomes. Table 1 summarizes some of the advantages and disadvantages found in research. Jobs designed according to the motivational approach have more satisfied, motivated, and involved employees who tend to have higher performance and lower absenteeism. Customer service may be improved, because employees take more pride in work and can catch their own errors by performing a larger part of the work. In terms of disadvantages, jobs too high on the motivational approach require more training, have greater skill and ability requirements for staffing, and may require higher compensation. Overly motivating jobs may also be so stimulating that workers become predisposed to mental overload, fatigue, errors, and occupational stress.

2.3 Perceptual/Motor Job Design Approach

2.3.1 Historical Development

The perceptual/motor design approach draws on a scientific discipline that goes by many names, including human factors, human factors engineering, human engineering, human–machine systems engineering, and engineering psychology. It developed from a number of other disciplines, primarily experimental psychology, but also industrial engineering (Meister, 1971). Within experimental psychology, job design recommendations draw heavily from knowledge of human skilled performance (Welford, 1976) and the analysis of humans as information processors (see Chapters 3 to 6). The main concern of this approach is efficient and safe utilization of humans in human–machine systems, with emphasis on selection, design, and arrangement of system components to take account of both human abilities and limitations (Pearson, 1971). It is more concerned with equipment than with psychology, and more concerned with human abilities than with engineering.

This approach received public attention with the Three Mile Island incident, where it was concluded that the control room operator job in the nuclear power plant may have placed too many demands on the operator in an emergency situation, thus predisposing errors of judgment (Campion and Thayer, 1987). Government regulations issued since then require nuclear plants to consider "human factors" in their design (U.S. Nuclear Regulatory Commission, 1981). The primary emphasis of this approach is on perceptual and motor abilities of people. (See Chapters 22 to 25 for more information on equipment design.)

2.3.2 Design Recommendations

Table 2 provides a list of statements describing important recommendations of the perceptual/motor approach. They refer to either equipment and environment or to information-processing requirements. Their thrust is to consider mental abilities and limitations of humans, such that the attention and concentration requirements of the job do not exceed the abilities of the least capable potential worker. Focus is on the limits of the least capable worker because this approach is concerned with the effectiveness of the total system, which is no better than its "weakest link." Jobs should be designed to limit the amount of information workers to which must pay attention and remember. Lighting levels should be appropriate, displays and controls should be logical and clear, workplaces should be well laid out and safe, and equipment should be easy to use. (See Chapters 58 to 61 for more information on human factors applications.)

2.3.3 Advantages and Disadvantages

The goals of this approach are to enhance reliability, safety, and positive user reactions. Table 1 summarizes advantages and disadvantages found in research. Jobs designed according to the perceptual/motor approach have lower errors and accidents. Like the mechanistic approach, it reduces the mental ability requirements of the job; thus, employees may be less stressed and mentally fatigued. It may also create some efficiencies, such as reduced training time and staffing requirements. On the other hand, costs from excessive use of the perceptual/motor approach can include low satisfaction, low motivation, and boredom due to inadequate mental stimulation. This problem is exacerbated by the fact that designs based on the least capable worker essentially lower a job's mental requirements.

2.4 Biological Job Design Approach

2.4.1 Historical Development

The biological job design approach and the perceptual/motor approach share a joint concern for proper person–machine fit. The major difference is that this approach is more oriented toward biological considerations and stems from such disciplines as work physiology (see Chapter 10), biomechanics (i.e., study of body movements, see Chapter 9) and anthropometry (i.e., study of body sizes; see Chapters 8 and 23). Although many specialists probably practice both approaches together, as reflected in many texts in the area (Konz, 1983), a split does exist between Americans, who are more psychologically oriented and use the title "human factors engineer," and Europeans, who are more physiologically oriented and use the title "ergonomist" (Chapanis, 1970). Like the perceptual-motor approach, the biological approach is concerned with the design of equipment and workplaces as well as the design of tasks (Grandjean, 1980).

2.4.2 Design Recommendations

Table 2 lists important recommendations from the biological approach. This approach tries to design jobs to reduce physical demands to avoid exceeding people's physical capabilities and limitations. Jobs should not require excessive strength and lifting, and again, abilities of the least physically able potential worker set the maximum level. Chairs should be designed for good postural support. Excessive wrist movement should be reduced by redesigning tasks and equipment. Noise, temperature, and atmosphere should be controlled within reasonable limits. Proper work/rest schedules should be provided so that employees can recuperate from the physical demands.

2.4.3 Advantages and Disadvantages

The goals of this approach are to maintain employees' comfort and physical well-being. Table 1 summarizes some advantages and disadvantages observed in research. Jobs designed according to this approach require less physical effort, result in less fatigue, and create fewer injuries and aches and pains than jobs low on this approach. Occupational illnesses, such as lower back pain and carpal tunnel syndrome, are fewer on jobs designed with this approach. There may be lower absenteeism and higher job satisfaction on jobs that are not physically arduous. However, a direct cost of this approach may be the expense of changes in equipment or job environments needed to implement the recommendations. At the extreme, costs may include jobs with so few physical demands that workers become drowsy or lethargic, thus reducing performance. Clearly, extremes of physical activity and inactivity should be avoided, and an optimal level of physical activity should be developed.

3 TEAM DESIGN APPROACH

3.1 Historical Development

An alternative to designing work around individual jobs is to design work for teams of workers. Teams can vary a great deal in how they are designed and can conceivably incorporate elements from any of the job design approaches discussed. However, the focus here is on the self-managing, autonomous type of team design approach, which has gained considerable popularity in organizations and substantial research attention today (Hoerr, 1989; Sundstrom et al., 1990; Guzzo and Shea, 1992; Swezey and Salas, 1992; Campion et al., 1996; Parker, 2003; Ilgen et al., 2005). Autonomous work teams derive their conceptual basis from motivational job design and from sociotechnical systems theory, which in turn reflect social and organizational psychology and organizational behavior (Davis and Valfer, 1965; Davis, 1971; Cummings, 1978; Morgeson and Campion, 2003). The Hawthorne studies (Homans, 1950) and European experiments with autonomous work groups (Kelly, 1982; Pasmore et al., 1982) called attention to the benefits of applying work teams in other than sports and military settings. Although enthusiasm for the use of teams had waned in the 1960s and 1970s due to research discovering some disadvantages of teams (Buys, 1978; Zander, 1979), the 1980s brought a resurgence of interest in the use of work teams and it has become an extremely popular work design in organizations today (Hoerr, 1989; Sundstrom et al., 1990; Hackman, 2002; Ilgen et al., 2005) This renewed interest may be due to the cost advantages of having fewer supervisors with self-managed teams or the apparent logic of the benefits of teamwork.

3.2 Design Recommendations

Teams can vary in the degree of authority and autonomy they have (Banker et al., 1996). For example, manager-led teams have responsibility only for the execution of their work. Management designs the work, designs the teams, and provides an organizational context for the teams. However, in autonomous work teams, or self-managing teams, team members design and monitor their own work and performance. They may also design their own team structure (e.g., delineating interrelationships among members) and composition (e.g., selecting members). In such self-designing teams, management is only responsible for the teams' organizational context (Hackman, 1987). Although team design could incorporate elements of either mechanistic or motivational approaches to design, narrow and simplistic mechanistically designed jobs would be less consistent with other suggested aspects of the team approach to design than motivationally designed jobs. Mechanistically designed jobs would not allow an organization to gain as much of the advantages from placing workers in teams.

Figure 1 and Table 3 provide important recommendations from the self-managing team design approach. Many of the advantages of work teams depend on how teams are designed and supported by their organization. According to the theory behind self-managing team design, decision making and responsibility should be pushed down to the team members (Hackman, 1987). If management is willing to follow this philosophy, teams can provide several additional advantages. By pushing decision making down to the team and requiring consensus, the organization will find greater acceptance, understanding, and ownership of decisions (Porter et al., 1987). The perceived autonomy resulting from making work decisions should be both satisfying and motivating. Thus, this approach tries to design teams so they have a high degree of self-management and all team members participate in decision making.

The team design approach also suggests that the set of tasks assigned to a team should provide a whole and meaningful piece of work (i.e., have task identity as in the motivational approach to job design). This allows team members to see how their work contributes to a whole product or process, which might not be possible with individuals working alone. This can give workers a better idea of the significance of their work and create greater identification with the finished product or service. If team workers rotate among a variety of subtasks and cross-train on different operations, workers should also perceive greater variety in the work (Campion et al., 1994b).

Interdependent tasks, goals, feedback, and rewards should be provided to create feelings of team interdependence among members and focus on the team as the unit of performance rather than on the individual. It is suggested that team members be heterogeneous in terms of areas of expertise and background so that their varied knowledge, skills, and abilities (KSAs) complement one another. Teams also need adequate training, managerial support, and organizational resources to carry out their tasks. Managers should encourage positive group processes, including open communication and cooperation within and between work groups, supportiveness and sharing of the workload among team members, and development of positive team spirit and confidence in the team's ability to perform effectively.

3.3 Advantages and Disadvantages

Table 4 summarizes advantages and disadvantages of team design relative to individual job design. To begin with, teams designed so that members have heterogeneity of KSAs can help team members learn by working with others who have different KSAs. Cross-training on different tasks can occur, and the workforce can become more flexible (Goodman et al., 1986). Teams with heterogeneous KSAs also allow for synergistic combinations of ideas and abilities not possible with individuals working alone, and such teams have generally shown higher performance, especially when task requirements are diverse (Shaw, 1983; Goodman et al., 1986).

Social support can be especially important when teams face difficult decisions and deal with difficult psychological aspects of tasks, such as in military squads, medical teams, or police units (Campion and Medsker, 1992). In addition, the simple presence of others can be psychologically arousing. Research has shown that such arousal can have a positive effect on performance when the task is well learned (Zajonc, 1965) and when other team members are perceived as evaluating the performer (Harkins, 1987; Porter et al., 1987). With routine jobs, this arousal effect

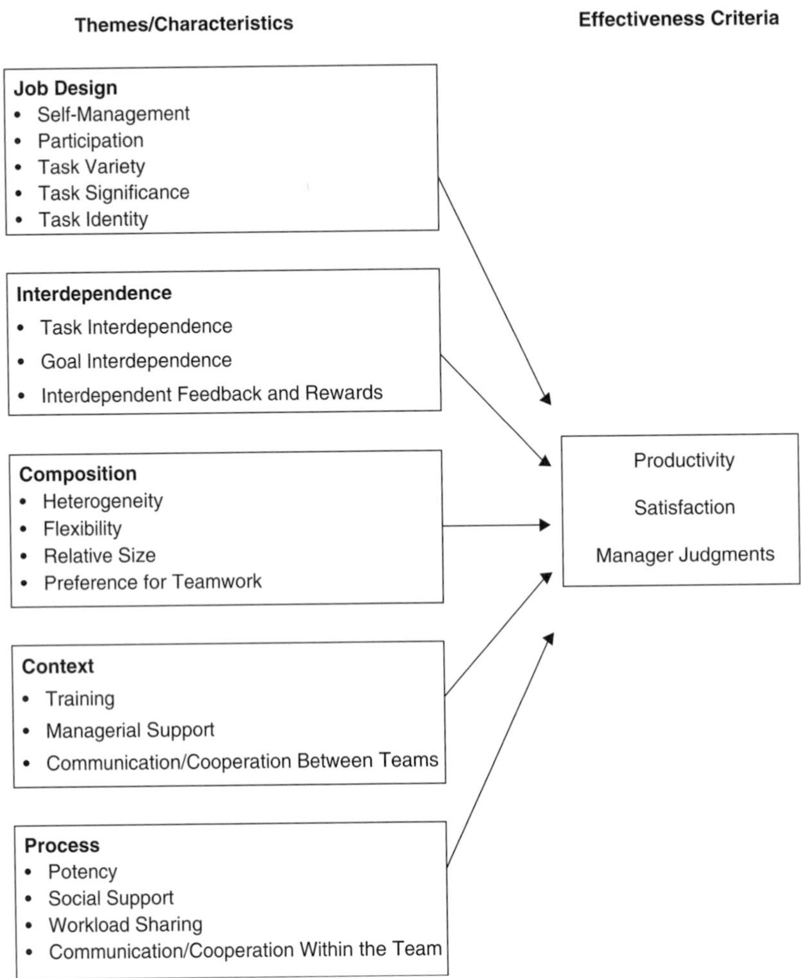

Figure 1 Characteristics related to team effectiveness.

may counteract boredom and performance decrements (Cartwright, 1968).

Another advantage of teams is that they can increase information exchanged between members through proximity and shared tasks (McGrath, 1984). Increased cooperation and communication within teams can be particularly useful when workers' jobs are highly interrelated, such as when workers whose tasks come later in the process must depend on the performance of workers whose tasks come earlier or when workers exchange work back and forth among themselves (Thompson, 1967; Mintzberg, 1979).

In addition, if teams are rewarded for team effort rather than individual effort, members will have an incentive to cooperate with one another (Leventhal, 1976). The desire to maintain power by controlling information may be reduced. More experienced workers may be more willing to train the less experienced when they are not in competition with them. Team design and rewards can also be helpful in situations where it is difficult to measure individual performance or where workers mistrust supervisors' assessments of performance (Milkovich and Newman, 1993).

Finally, teams can be beneficial if team members develop a feeling of commitment and loyalty to their team (Cartwright, 1968). For workers who do not develop high commitment to their organization or management and who do not become highly involved in their job, work teams can provide a source of commitment. That is, members may feel responsible to attend work, cooperate with others, and perform well because of commitment to their work team, even though they are not strongly committed to the organization or the work itself.

Thus, designing work around teams can provide several advantages to organizations and their workers. Unfortunately, there are also disadvantages to using

Table 3 Team Design Measure[a]

Instructions: This questionnaire consists of statements about your team and how your team functions as a group. Please indicate the extent to which each statement describes your team by circling a number to the right of each statement.

Please Use the Following Scale:
(5) Strongly agree
(4) Agree
(3) Neither agree nor disagree
(2) Disagree
(1) Strongly disagree
() Leave blank if do not know or not applicable

Self-Management					
1. The members of my team are responsible for determining the methods, procedures, and schedules with which the work gets done.	1	2	3	4	5
2. My team rather than my manager decides who does what tasks within the team.	1	2	3	4	5
3. Most work-related decisions are made by the members of my team rather than by my manager.	1	2	3	4	5
Participation					
4. As a member of a team, I have a real say in how the team carries out its work.	1	2	3	4	5
5. Most members of my team get a chance to participate in decision making.	1	2	3	4	5
6. My team is designed to let everyone participate in decision making.	1	2	3	4	5
Task Variety					
7. Most members of my team get a chance to learn the different tasks the team performs.	1	2	3	4	5
8. Most everyone on my team gets a chance to do the more interesting tasks.	1	2	3	4	5
9. Task assignments often change from day to day to meet the workload needs of the team.	1	2	3	4	5
Task Significance (Importance)					
10. The work performed by my team is important to the customers in my area.	1	2	3	4	5
11. My team makes an important contribution to serving the company's customers.	1	2	3	4	5
12. My team helps me feel that my work is important to the company.	1	2	3	4	5
Task Identity (Mission)					
13. The team concept allows all the work on a given product to be completed by the same set of people.	1	2	3	4	5
14. My team is responsible for all aspects of a product for its area.	1	2	3	4	5
15. My team is responsible for its own unique area or segment of the business.	1	2	3	4	5
Task Interdependence (Interdependence)					
16. I cannot accomplish my tasks without information or materials from other members of my team.	1	2	3	4	5
17. Other members of my team depend on me for information or materials needed to perform their tasks.	1	2	3	4	5
18. Within my team, jobs performed by team members are related to one another.	1	2	3	4	5
Goal Interdependence (Goals)					
19. My work goals come directly from the goals of my team.	1	2	3	4	5
20. My work activities on any given day are determined by my team's goals for that day.	1	2	3	4	5
21. I do very few activities on my job that are not related to the goals of my team.	1	2	3	4	5
Interdependent Feedback and Rewards (Feedback and Rewards)					
22. Feedback about how well I am doing my job comes primarily from information about how well the entire team is doing.	1	2	3	4	5
23. My performance evaluation is strongly influenced by how well my team performs.	1	2	3	4	5
24. Many rewards from my job (pay, promotion, etc.) are determined in large part by my contributions as a team member.	1	2	3	4	5

(continued overleaf)

Table 3 (*continued*)

Heterogeneity (Membership)					
25. The members of my team vary widely in their areas of expertise.	1	2	3	4	5
26. The members of my team have a variety of different backgrounds and experiences.	1	2	3	4	5
27. The members of my team have skills and abilities that complement each other.	1	2	3	4	5
Flexibility (Member Flexibility)					
28. Most members of my team know each other's jobs.	1	2	3	4	5
29. It is easy for the members of my team to fill in for one another.	1	2	3	4	5
30. My team is very flexible in terms of membership.	1	2	3	4	5
Relative Size (Size)					
31. The number of people in my team is too small for the work to be accomplished. (Reverse scored)	1	2	3	4	5
Preference for Team Work (Team Work Preferences)					
32. If given the choice, I would prefer to work as part of a team rather than work alone.	1	2	3	4	5
33. I find that working as a member of a team increases my ability to perform effectively.	1	2	3	4	5
34. I generally prefer to work as part of a team.	1	2	3	4	5
Training					
35. The company provides adequate technical training for my team.	1	2	3	4	5
36. The company provides adequate quality and customer service training for my team.	1	2	3	4	5
37. The company provides adequate team skills training for my team (communication, organization, interpersonal, etc.).	1	2	3	4	5
Managerial Support					
38. Higher management in the company supports the concept of teams.	1	2	3	4	5
39. My manager supports the concept of teams.	1	2	3	4	5
Communication/Cooperation Between Work Groups					
40. I frequently talk to other people in the company besides the people on my team.	1	2	3	4	5
41. There is little competition between my team and other teams in the company.	1	2	3	4	5
42. Teams in the company cooperate to get the work done.	1	2	3	4	5
Potency (Spirit)					
43. Members of my team have great confidence that the team can perform effectively.	1	2	3	4	5
44. My team can take on nearly any task and complete it.	1	2	3	4	5
45. My team has a lot of team spirit.	1	2	3	4	5
Social Support					
46. Being in my team gives me the opportunity to work in a team and provide support to other team members.	1	2	3	4	5
47. My team increases my opportunities for positive social interaction.	1	2	3	4	5
48. Members of my team help each other out at work when needed.	1	2	3	4	5
Workload Sharing (Sharing the Work)					
49. Everyone on my team does his or her fair share of the work.	1	2	3	4	5
50. No one in my team depends on other team members to do the work for them.	1	2	3	4	5
51. Nearly all the members of my team contribute equally to the work.	1	2	3	4	5
Communication/Cooperation within the Work Group					
52. Members of my team are very willing to share information with other team members about our work.	1	2	3	4	5
53. Teams enhance the communications among people working on the same product.	1	2	3	4	5
54. Members of my team cooperate to get the work done.	1	2	3	4	5

Source: Adapted from Campion et al. (1993).

[a]See source and related research (Campion et al., 1995) for reliability and validity information. Scores for each team characteristic are calculated by averaging applicable items.

Table 4 Advantages and Disadvantages of Work Teams

Advantages	Disadvantages
• Team members learn from one another • Possibility of greater work force flexibility with cross-training • Opportunity for synergistic combinations of ideas and abilities • New approaches to tasks may be discovered • Social facilitation and arousal • Social support for difficult tasks and situations • Increased communication and information exchange between team members • Greater cooperation among team members • Beneficial for interdependent work flows • Greater acceptance and understanding of decisions when team makes decisions • Greater autonomy, variety, identity, significance, and feedback possible for workers • Commitment to the team may stimulate performance and attendance	• Lack of compatibility of some individuals with team work • Additional need to select workers to fit team as well as job • Possibility some members will experience less motivating jobs • Possible incompatibility with cultural, organizational, or labor–management norms • Increased competition and conflict between teams • More time consuming due to socializing, coordination losses, and need for consensus • Inhibition of creativity and decision-making processes; possibility of groupthink • Less powerful evaluation and rewards; social loafing or free-riding may occur • Less flexibility in cases of replacement, turnover, or transfer

Source: Adapted from Campion and Medsker (1992).

work teams and situations in which individual-level design is preferable to team design. For example, some individuals may dislike team work and may not have necessary interpersonal skills or desire to work in a team. When selecting team members, one has the additional requirement of selecting workers to fit the team as well as the job. (Section 4.3 provides more information on the selection of team members; see also Chapter 17 for general information on personnel selection.)

Individuals can experience less autonomy and less personal identification when working on a team. Designing work around teams does not guarantee workers greater variety, significance, and identity. If members within the team do not rotate among tasks or if some members are assigned exclusively to less desirable tasks, not all members will benefit from team design. Members can still have fractionated, demotivating jobs.

Teamwork can also be incompatible with cultural norms. The United States has a very individualistic culture (Hofstede, 1980). Applying team methods that have been successful in collectivistic societies such as Japan may be problematic in the United States. In addition, organizational norms and labor–management relations may be incompatible with team design, making its use more difficult.

Some advantages of team design can create disadvantages as well. First, though team rewards can increase communication and cooperation and reduce competition within a team, they may cause greater competition and reduced communication between teams. If members identify too strongly with a team, they may not realize when behaviors that benefit the team detract from organizational goals and create conflicts detrimental to productivity. Increased communication within teams may not always be task-relevant either. Teams may spend work time socializing. Team decision making can take longer than individual decision making, and the need for coordination within teams can be time consuming.

Decision making and creativity can also be inhibited by team processes. When teams become highly cohesive, they may become so alike in their views that they develop *groupthink* (Janis, 1972). When groupthink occurs, teams tend to underestimate their competition, fail to adequately critique fellow team members' suggestions, not appraise alternatives adequately, and fail to work out contingency plans. In addition, team pressures distort judgments. Decisions may be based more on persuasiveness of dominant individuals or the power of majorities rather than on the quality of decisions. Research has found a tendency for team judgments to be more extreme than the average of individual members' predecision judgments (Janis, 1972; McGrath, 1984; Morgeson and Campion, 1997). Although evidence shows that highly cohesive teams are more satisfied with their teams, cohesiveness is not necessarily related to high productivity. Whether cohesiveness is related to performance depends on a team's norms and goals. If a team's norm is to be productive, cohesiveness will enhance productivity; however, if the norm is not one of commitment to productivity, cohesiveness can have a negative influence (Zajonc, 1965).

The use of teams and team-level rewards can also decrease the motivating power of evaluation and reward systems. If team members are not evaluated for individual performance, do not believe their output can be distinguished from the team's, or do not perceive a link between their personal performance and outcomes, social loafing (Harkins, 1987) can occur. In such situations, teams do not perform up to the potential expected from combining individual efforts.

Finally, teams may be less flexible in some respects because they are more difficult to move or transfer as a unit than individuals are to transport (Sundstrom et al., 1990). Turnover, replacements, and employee transfers may disrupt teams; and members may not readily accept new members.

Thus, whether work teams are advantageous depends to a great extent on the composition, structure, reward systems, environment, and task of the team. Table 5 presents questions that can help determine whether work should be designed around teams rather than individuals. The more questions that are answered in the affirmative, the more likely teams are to be beneficial. If one chooses to design work around teams, suggestions for designing effective teams are presented in Section 4.3.

Table 5 When to Design Jobs around Work Teams[a]

1. Are workers' tasks highly interdependent, or could they be made to be so? Would this interdependence enhance efficiency or quality?
2. Do the tasks require a variety of knowledge, skills, and abilities such that combining individuals with different backgrounds would make a difference in performance?
3. Is cross-training desired? Would breadth of skills and workforce flexibility be essential to the organization?
4. Could increased arousal, motivation, and effort to perform make a difference in effectiveness?
5. Can social support help workers deal with job stresses?
6. Could increased communication and information exchange improve performance rather than interfere?
7. Could increased cooperation aid performance?
8. Are individual evaluation and rewards difficult or impossible to make, or are they mistrusted by workers?
9. Could common measures of performance be developed and used?
10. Is it technically possible to group tasks in a meaningful, efficient way?
11. Would individuals be willing to work in teams?
12. Does the labor force have the interpersonal skills needed to work in teams?
13. Would team members have the capacity and willingness to be trained in interpersonal and technical skills required for teamwork?
14. Would teamwork be compatible with cultural norms, organizational policies, and leadership styles?
15. Would labor–management relations be favorable to team job design?
16. Would the amount of time taken to reach decisions, consensus, and coordination not be detrimental to performance?
17. Can turnover be kept to a minimum?
18. Can teams be defined as a meaningful unit of the organization with identifiable inputs, outputs, and buffer areas which give them a separate identity from other teams?
19. Would members share common resources, facilities, or equipment?
20. Would top management support team job design?

Source: Adapted from Campion and Medsker (1992).
[a]Affirmative answers support the use of team job design.

4 IMPLEMENTATION ADVICE FOR JOB AND TEAM DESIGN

4.1 General Implementation Advice

4.1.1 Procedures

Several general philosophies are helpful when designing or redesigning jobs or teams:

1. As noted previously, designs are not inalterable or dictated by technology. There is some discretion in the design of all work situations, and considerable discretion in most.
2. There is no single best design, there are simply better and worse designs depending on one's design perspective.
3. Design is iterative and evolutionary and should continue to change and improve over time.
4. Participation of workers affected generally improves the quality of the resulting design and acceptance of suggested changes.
5. The process of the project, or how it is conducted, is important in terms of involvement of all interested parties, consideration of alternative motivations, and awareness of territorial boundaries.

Procedures for the Initial Design of Jobs or Teams In consideration of process aspects of design, Davis and Wacker (1982) suggest four steps:

1. *Form a steering committee.* This committee usually consists of a team of high-level executives who have a direct stake in the new jobs or teams. The purposes of the committee are to (a) bring into focus the project's objective, (b) provide resources and support for the project, (c) help gain the cooperation of all parties affected, and (d) oversee and guide the project.
2. *Form a design task force.* The task force may include engineers, managers, job or team design experts, architects, specialists, and others with relevant knowledge or responsibility. The task force is to gather data, generate and evaluate design alternatives, and help implement recommended designs.
3. *Develop a philosophy statement.* The first goal of the task force is to develop a philosophy statement to guide decisions involved in the project. The philosophy statement is developed with input from the steering committee and may include the project's purposes, organization's strategic goals, assumptions about workers and the nature of work, and process considerations.
4. *Proceed in an evolutionary manner.* Jobs should not be overspecified. With considerable input from eventual jobholders or team members, the work design will continue to change and improve over time.

According to Davis and Wacker (1982), the process of redesigning existing jobs is much the same as designing original jobs with two additions. First, existing job incumbents must be involved. Second, more attention needs to be given to implementation issues. Those involved in the implementation must feel ownership of and commitment to the change and believe the redesign represents their own interests.

Potential Steps to Follow Along with the steps discussed above, a redesign project should include the following five steps:

1. *Measuring the design of the existing job or teams.* The questionnaire methodology and other analysis tools described in Section 5 may be used to measure current jobs or teams.

2. *Diagnosing potential design problems.* Based on data collected in step 1, the current design is analyzed for potential problems. The task force and employee involvement are important. Focused team meetings are a useful vehicle for identifying and evaluating problems.

3. *Determining job or team design changes.* Changes will be guided by project goals, problems identified in step 2, and one or more of the approaches to work design. Often, several potential changes are generated and evaluated. Evaluation of alternative changes may involve consideration of advantages and disadvantages identified in previous research (see Table 1) and opinions of engineers, managers, and employees.

4. *Making design changes.* Implementation plans should be developed in detail along with backup plans in case there are difficulties with the new design. Communication and training are keys to implementation. Changes might also be pilot tested before widespread implementation.

5. *Conducting a follow-up evaluation.* Evaluating the new design after implementation is probably the most neglected part of the process in most applications. The evaluation might include the collection of design measurements on the redesigned jobs/teams using the same instruments as in step 1. Evaluation may also be conducted on outcomes, such as employee satisfaction, error rates, and training time (Table 1). Scientifically valid evaluations require experimental research strategies with control groups. Such studies may not always be possible in organizations, but quasiexperimental and other field research designs are often possible (Cook and Campbell, 1979). Finally, the need for adjustments are identified through the follow-up evaluation. (For examples of evaluations, see Section 5.8 and Campion and McClelland, 1991, 1993.)

4.1.2 Individual Differences among Workers

It is a common observation that not all employees respond the same to the same job. Some people on a job have high satisfaction, whereas others on the same job have low satisfaction. Clearly, there are individual differences in how people respond to work. Considerable research has looked at individual differences in reaction to the motivational design approach. It has been found that some people respond more positively than others to highly motivational work. These differences are generally viewed as differences in needs for personal growth and development (Hackman and Oldham, 1980).

Using the broader notion of preferences or tolerances for types of work, the consideration of individual differences has been expanded to all four approaches to job design (Campion, 1988; Campion and McClelland, 1991) and to the team design approach (Campion et al., 1993, 1995). Table 6 provides scales that can be used to determine job incumbents' preferences or tolerances. These scales can be administered in the same manner as the questionnaire measures of job and team design discussed in Section 5.

Although consideration of individual differences is encouraged, there are often limits to which such differences can be accommodated. Jobs or teams may have to be designed for people who are not yet known or who differ in their preferences. Fortunately, although evidence indicates individual differences moderate reactions to the motivational approach (Fried and Ferris, 1987), the differences are of degree but not direction. That is, some people respond more positively than others to motivational work, but few respond negatively. It is likely that this also applies to the other design approaches.

4.1.3 Some Basic Choices

Hackman and Oldham (1980) have provided five strategic choices that relate to implementing job redesign. They note that little research exists indicating the exact consequences of each choice, and correct choices may differ by organization. The basic choices are:

1. *Individual versus team designs for work.* An initial decision is either to enrich individual jobs or create teams. This also includes consideration of whether any redesign should be undertaken and its likelihood of success.

2. *Theory based versus intuitive changes.* This choice was basically defined as the motivational (theory) approach vs. no particular (atheoretical) approach. In the present chapter, this choice may be better framed as choosing among the four approaches to job design. However, as argued earlier, consideration of only one approach may lead to some costs or additional benefits being ignored.

3. *Tailored versus broadside installation.* This choice is between tailoring changes to individuals or making the changes for all in a given job.

4. *Participative versus top-down change processes.* The most common orientation is that participative is best. However, costs of participation include the time involved and incumbents' possible lack of a broad knowledge of the business.

5. *Consultation versus collaboration with stakeholders.* The effects of job design changes often extend

Table 6 Preferences/Tolerances for the Design Approaches[a]

Instructions: Indicate the extent to which each statement is descriptive of your preferences and tolerances for types of work on the scale below. Circle answers to the right of each statement.

Please Use the Following Scale:
(5) Strongly agree
(4) Agree
(3) Neither agree nor disagree
(2) Disagree
(1) Strongly disagree
() Leave blank if do not know or not applicable

Preferences/Tolerances for Mechanistic Design						
1.	I have a high tolerance for routine work.	1	2	3	4	5
2.	I prefer to work on one task at a time.	1	2	3	4	5
3.	I have a high tolerance for repetitive work.	1	2	3	4	5
4.	I prefer work that is easy to learn.	1	2	3	4	5
Preferences/Tolerances for Motivational Design						
5.	I prefer highly challenging work that taxes my skills and abilities.	1	2	3	4	5
6.	I have a high tolerance for mentally demanding work.	1	2	3	4	5
7.	I prefer work that gives a great amount of feedback as to how I am doing.	1	2	3	4	5
8.	I prefer work that regularly requires the learning of new skills.	1	2	3	4	5
9.	I prefer work that requires me to develop my own methods, procedures, goals, and schedules.	1	2	3	4	5
10.	I prefer work that has a great amount of variety in duties and responsibilities.	1	2	3	4	5
Preferences/Tolerances for Perceptual/Motor Design						
11.	I prefer work that is very fast paced and stimulating.	1	2	3	4	5
12.	I have a high tolerance for stressful work.	1	2	3	4	5
13.	I have a high tolerance for complicated work.	1	2	3	4	5
14.	I have a high tolerance for work where there are frequently too many things to do at one time.	1	2	3	4	5
Preferences/Tolerances for Biological Design						
15.	I have a high tolerance for physically demanding work.	1	2	3	4	5
16.	I have a fairly high tolerance for hot, noisy, or dirty work.	1	2	3	4	5
17.	I prefer work that gives me some physical exercise.	1	2	3	4	5
18.	I prefer work that gives me some opportunities to use my muscles.	1	2	3	4	5
Preferences/Tolerances for Team Work						
19.	If given the choice, I would prefer to work as part of a team rather than work alone.	1	2	3	4	5
20.	I find that working as a member of a team increases my ability to perform effectively.	1	2	3	4	5
21.	I generally prefer to work as part of a team.	1	2	3	4	5

Source: Adapted from Campion (1988) and Campion et al. (1993).
[a]See source for reliability and validity information. Scores for each preference/tolerance are calculated by averaging applicable items. Interpretations differ slightly across the scales. For the mechanistic and motivational designs, higher scores suggest more favorable reactions from incumbents to well-designed jobs. For the perceptual/motor and biological approaches, higher scores suggest less unfavorable reactions from incumbents to poorly designed jobs.

far beyond the individual incumbent and department. For example, a job's output may be an input to a job elsewhere in the organization. The presence of a union also requires additional collaboration. Depending on considerations, participation of stakeholders may range from no involvement, through consultation, to full collaboration.

4.1.4 Overcoming Resistance to Change in Redesign Projects

Resistance to change can be a problem in any project involving major changes (Morgeson et al., 1997). Failure rates of new technology implementations demonstrate a need to give more attention to the human aspects of change projects. This concern has also been reflected in the area of participatory ergonomics, which encourages the use of participatory techniques when undertaking an ergonomics intervention (Wilson and Haines, 1997). It has been estimated that between 50 and 75% of newly implemented manufacturing technologies in the United States have failed, with a disregard for human and organizational issues considered to be a bigger cause for the failures than technical problems (Majchrzak, 1988; Turnage, 1990). The number one obstacle to implementation was considered to be human resistance to change (Hyer, 1984).

Based on the work of Gallagher and Knight (1986), Majchrzak (1988), and Turnage (1990), guidelines for reducing resistance to change include the following:

1. *Involve workers in planning the change.* Workers should be informed of changes in advance and involved in the process of diagnosing current problems and developing solutions. Resistance is decreased if participants feel that the project is their own and not imposed from outside and if the project is adopted by consensus.

2. *Top management should strongly support the change.* If workers feel that management is not strongly committed, they are less likely to take the project seriously.

3. *Create change consistent with worker needs and existing values.* Resistance is less if change is seen to reduce present burdens, offer interesting experience, not threaten worker autonomy or security, or be inconsistent with other goals and values in the organization. Workers need to see the advantages to them of the change. Resistance is less if proponents of change can empathize with opponents (recognize valid objections and relieve unnecessary fears).

4. *Create an environment of open, supportive communication.* Resistance will be lessened if participants experience support and have trust in each other. Resistance can be reduced if misunderstandings and conflicts are expected as natural to the innovation process. Provision should be made for clarification.

5. *Allow for flexibility.* Resistance is reduced if the project is kept open to revision and reconsideration with experience.

4.2 Implementation Advice for Job Design and Redesign

4.2.1 Methods for Combining Tasks

In many cases, designing jobs is largely a function of combining tasks. Some guidance can be gained by extrapolating from specific design recommendations in Table 2. For example, variety in the motivational approach can be increased simply by combining different tasks in the same job. Conversely, specialization from the mechanistic approach can be increased by including only very similar tasks in the same job. It is also possible when designing jobs first to generate alternative task combinations, then evaluate them using the design approaches in Table 2.

A small amount of research within the motivational approach has focused explicitly on predicting relationships between combinations of tasks and the design of resulting jobs (Wong, 1989; Wong and Campion, 1991). This research suggests that a job's motivational quality is a function of three task-level variables, as illustrated in Figure 2.

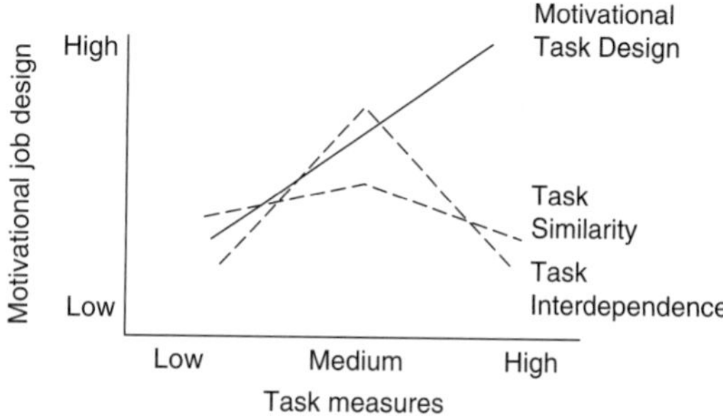

Figure 2 Effects of task design, interdependence, and similarity on motivational job design.

1. *Task design.* The higher the motivational quality of individual tasks, the higher the motivational quality of a job. Table 2 can be used to evaluate individual tasks, then motivational scores for individual tasks can be summed together. Summing is recommended rather than averaging because both the motivational quality of the tasks and the number of tasks are important in determining a job's motivational quality (Globerson and Crossman, 1976).

2. *Task interdependence.* Interdependence among tasks has been shown to be positively related to motivational value up to some moderate point; beyond that point, increasing interdependence has been shown to lead to lower motivational value. Thus, for motivational jobs, the total amount of interdependence among tasks should be kept at a moderate level. Both complete independence and excessively high interdependence should be avoided. Table 7 contains the dimension of task interdependence and provides a questionnaire to measure it. Table 7 can be used to judge the interdependence of each pair of tasks that are being evaluated for inclusion in a job.

3. *Task similarity.* Similarity among tasks may be the oldest rule of job design, but beyond a moderate level, it tends to decrease a job's motivational value. Thus, to design motivational jobs, high levels of similarity should be avoided. Similarity at the task pair level can be judged in much the same manner as interdependence by using dimensions in Table 7 (see the footnote to Table 7).

4.2.2 Trade-offs among Job Design Approaches

Although one should strive to construct jobs that are well designed on all the approaches, it is clear that design approaches conflict. As Table 1 illustrates, the benefits of some approaches are the costs of others. No single approach satisfies all outcomes. The greatest potential conflicts are between the motivational and the mechanistic and perceptual/motor approaches. They produce nearly opposite outcomes. The mechanistic and perceptual/motor approaches recommend jobs that are simple, safe, and reliable, with minimal mental demands on workers. The motivational approach encourages more complicated and stimulating jobs, with greater mental demands. The team approach is consistent with the motivational approach, and therefore may also conflict with the mechanistic and perceptual/motor approaches.

Table 7 Dimensions of Task Interdependence[a]

Instructions: Indicate the extent to which each statement is descriptive of the pair of tasks using the scale below. Circle answers to the right of each statement. Scores are calculated by averaging applicable items.

Please Use the Following Scale:
(5) Strongly agree
(4) Agree
(3) Neither agree nor disagree
(2) Disagree
(1) Strongly disagree
() Leave blank if do not know or not applicable

Inputs of the Tasks					
1. *Materials/supplies:* One task obtains, stores, or prepares the materials or supplies necessary to perform the other task.	1	2	3	4	5
2. *Information:* One task obtains or generates information for the other task.	1	2	3	4	5
3. *Product/service:* One task stores, implements, or handles the products or services produced by the other task.	1	2	3	4	5
Processes of the Tasks					
4. *Input–output relationship:* The products (or outputs) of one task are the supplies (or inputs) necessary to perform the other task.	1	2	3	4	5
5. *Method and procedure:* One task plans the procedures or work methods for the other task.	1	2	3	4	5
6. *Scheduling:* One task schedules the activities of the other task.	1	2	3	4	5
7. Supervision: One task reviews or checks the quality of products or services produced by the other task.	1	2	3	4	5
8. *Sequencing:* One task needs to be performed before the other task.	1	2	3	4	5
9. *Time sharing:* Some of the work activities of the two tasks must be performed at the same time.	1	2	3	4	5
10. *Support service:* The purpose of one task is to support or otherwise help the other task get performed.	1	2	3	4	5
11. *Tools/equipment:* One task produces or maintains the tools or equipment used by the other task.	1	2	3	4	5
Outputs of the Tasks					
12. *Goal:* One task can only be accomplished when the other task is properly performed.	1	2	3	4	5
13. *Performance:* How well one task is performed has a great impact on how well the other task can be performed.	1	2	3	4	5
14. *Quality:* The quality of the product or service produced by one task depends on how well the other task is performed.	1	2	3	4	5

Source: Adapted from Wong and Campion (1991).
[a]See source and Wong (1989) for reliability and validity information. The task similarity measure contains 10 comparable items (excluding items 4, 6, 8, 9, and 14, and including an item on customer/client). Scores for each dimension are calculated by averaging applicable items.

Because of these conflicts, trade-offs may be necessary. Major trade-offs will be in the mental demands created by the alternative design strategies. Making jobs more mentally demanding increases the likelihood of achieving workers' goals of satisfaction and motivation, but decreases the chances of reaching the organization's goals of reduced training, staffing costs, and errors. Which trade-offs will be made depends on outcomes one prefers to maximize. Generally, a compromise may be optimal.

Trade-offs may not always be needed, however. Jobs can often be improved on one approach while maintaining their quality on other approaches. For example, in one redesign study, the motivational approach was applied to clerical jobs to improve employee satisfaction and customer service (Campion and McClelland, 1991). Expected benefits occurred along with some expected costs (e.g., increased training and compensation requirements), but not all potential costs occurred (e.g., quality and efficiency did not decrease).

In another redesign study, Morgeson and Campion (2002) sought to increase both satisfaction and efficiency in jobs at a pharmaceutical company. They found that when jobs were designed to increase only satisfaction or only efficiency, the common trade-offs were present (e.g., increased or decreased satisfaction, training requirements). When jobs were designed to increase both satisfaction and efficiency, however, these trade-offs were reduced. They suggested that a work design process that explicitly considers both motivational and mechanistic aspects of work is key to avoiding the trade-offs.

Another strategy for minimizing trade-offs is to avoid design decisions that influence the mental demands of jobs. An example of this is to enhance motivational design by focusing on social aspects (e.g., communication, participation, recognition, feedback). These design features can be raised without incurring costs of increased mental demands. Moreover, many of these features are under the direct control of managers.

The independence of the biological approach provides another opportunity to improve design without incurring trade-offs with other approaches. One can reduce physical demands without affecting mental demands of a job. Of course, the cost of equipment may need to be considered.

Adverse effects of trade-offs can often be reduced by avoiding designs that are extremely high or low on any approach. Or, alternatively, one might require minimum acceptable levels on each approach. Knowing all approaches and their corresponding outcomes will help one make more informed decisions and avoid unanticipated consequences.

4.2.3 Other Implementation Advice for Job Design

Davis and Wacker (1982, 1987) have provided a list of criteria for grouping tasks, part of which is reproduced below. The list represents a collection of criteria from both the motivational (e.g., 1, 5, 9) and mechanistic (e.g., 2, 8) approaches. Many of the recommendations could also be applied to designing work for teams.

1. Each task group is a meaningful unit of the organization.
2. Task groups are separated by stable buffer areas.
3. Each task group has definite, identifiable inputs and outputs.
4. Each task group has associated with it definite criteria for performance evaluation.
5. Timely feedback about output states and feedforward about input states are available.
6. Each task group has resources to measure and control variances that occur within its area of responsibility.
7. Tasks are grouped around mutual cause–effect relationships.
8. Tasks are grouped around common skills, knowledge, or data.
9. Task groups incorporate opportunities for skill acquisition relevant to career advancement.

Based on experience redesigning jobs in AT&T, Ford (1969) advocated *work-itself workshops*. These are basically workshops of managers and employees trained in motivational job design who then attempt to come up with ways to improve jobs. Ford provides the following advice for these workshops:

1. Start with a meeting with senior management.
2. Work within a single department at first.
3. Gain commitment.
4. Pick a job to focus on.
5. Conduct workshop meetings.
6. Facilitate creative thinking.
7. Deal with visitors to the job site.
8. Search for a natural module of work.
9. Deal with resistance due to expense.
10. Individualize feedback.

Griffin's (1982) advice is geared toward the manager considering a job redesign intervention in his or her area. He notes that the manager may also rely on consultants, task forces, or informal discussion groups. Griffin suggests nine steps:

1. Recognition of a need for change
2. Selection of job redesign as a potential intervention
3. Diagnosis of the work system and content on the following factors:
   a. Existing jobs
   b. Existing workforce
   c. Technology
   d. Organization design
   e. Leader behaviors
   f. Team and social processes
4. Cost–benefit analysis of proposed changes
5. Go/no-go decision
6. Establishment of a strategy for redesign
7. Implementation of the job changes
8. Implementation of any needed supplemental changes
9. Evaluation of the redesigned jobs

4.3 Implementation Advice for Team Design

4.3.1 Deciding on Team Composition

Research encourages heterogeneous teams in terms of skills, personality, and attitudes because it increases the range of competencies in teams (Gladstein, 1984) and is related to effectiveness (Campion et al., 1995). However, homogeneity is preferred if team morale is the main criterion, and heterogeneous attributes must be complementary if they are to contribute to effectiveness. Heterogeneity for its own sake is unlikely to enhance effectiveness (Campion et al., 1993). Another composition characteristic of effective teams is whether members have flexible job assignments (Sundstrom et al., 1990; Campion et al., 1993). If members can perform different jobs, effectiveness is enhanced because they can fill in as needed.

A third important aspect of composition is team size. Evidence suggests the importance of optimally matching team size to team tasks to achieve high performance and satisfaction. Teams need to be large enough to accomplish work assigned to them, but may be dysfunctional when too large due to heightened coordination needs (Steiner, 1972; O'Reilly and Roberts, 1977) or increased social loafing (Wicker et al., 1976; McGrath, 1984). Thus, groups should be staffed to the smallest number needed to do the work (Goodman et al., 1986; Hackman, 1987; Sundstrom et al., 1990).

4.3.2 Selecting Team Members

With team design, interpersonal demands appear to be much greater than with traditional individual-based job design (Lawler, 1986). A team-based setting highlights the importance of employees being capable of interacting in an effective manner with peers, because the amount of interpersonal interactions required is higher in teams (Stevens and Campion, 1994a, b, 1999). Team effectiveness can depend heavily on members' *interpersonal competence*, their ability to maintain healthy working relationships and react to others with respect for their viewpoints (Perkins and Abramis, 1990). There is a greater need for team members to be capable of effective interpersonal communication, collaborative problem solving, and conflict management (Stevens and Campion, 1994a, b, 1999).

The process of employment selection for team members places greater stress on adequately evaluating interpersonal competence than is normally required in the selection of workers for individual jobs. To create a selection instrument for evaluating potential team members' ability to work successfully in teams, Stevens and Campion (1994a, b) reviewed literature in areas of sociotechnical systems theory (e.g., Cummings, 1978; Wall et al., 1986), organizational behavior (e.g., Hackman, 1987; Shea and Guzzo, 1987; Sundstrom et al., 1990), industrial engineering (e.g., Davis and Wacker, 1987; Majchrzak, 1988), and social psychology (e.g., Steiner, 1972; McGrath, 1984) to identify relevant knowledge, skills, and abilities (KSAs). Table 8 shows the 14 KSAs identified as important for teamwork.

These KSAs have been used to develop a 35-item, multiple-choice employment test, which was validated in two studies to determine how highly related it was to team members' job performance (Stevens and Campion, 1999). The job performance of team members in two different companies was rated by both supervisors and co-workers. Correlations between the test and job performance ratings were significantly high, with some correlations exceeding 0.50. The test was also able to add to the ability to predict job performance beyond that provided by a large battery of traditional employment aptitude tests. Thus, these findings provide support for the value of the teamwork KSAs and a selection test based on them (Stevens and Campion, 1994a). Table 9 shows some example items from the test.

Aside from written tests, there may be other ways that teamwork KSAs could be measured for purposes of selection. For example, interviews may be especially suited to measuring interpersonal attributes (e.g., Posthuma et al., 2002). There is evidence that a structured interview specifically designed to measure social (i.e., nontechnical) KSAs can have validity with job performance and predict incrementally beyond traditional employment tests (Campion et al., 1994a).

Table 8 Knowledge, Skill, and Ability (KSA) Requirements for Teamwork

I. Interpersonal KSAs
   A. Conflict Resolution KSAs
      1. The KSA to recognize and encourage desirable, but discourage undesirable, team conflict
      2. The KSA to recognize the type and source of conflict confronting the team and to implement an appropriate conflict resolution strategy
      3. The KSA to employ an integrative (win–win) negotiation strategy rather than the traditional distributive (win–lose) strategy
   B. Collaborative Problem-Solving KSAs
      4. The KSA to identify situations requiring participative group problem solving and to utilize the proper degree and type of participation
      5. The KSA to recognize the obstacles to collaborative group problem solving and implement appropriate corrective actions
   C. Communication KSAs
      6. The KSA to understand communication networks, and to utilize decentralized networks to enhance communication where possible
      7. The KSA to communicate openly and supportively, that is, to send messages that are (a) behavior- or event-oriented, (b) congruent, (c) validating, (d) conjunctive, and (e) owned
      8. The KSA to listen nonevaluatively and to use active listing techniques appropriately
      9. The KSA to maximize consonance between nonverbal and verbal messages, and to recognize and interpret the nonverbal messages of others
      10. The KSA to engage in ritual greetings and small talk, and a recognition of their importance
II. Self-Management KSAs
   D. Goal-Setting and Performance Management KSAs
      11. The KSA to help establish specific, challenging, and accepted team goals
      12. The KSA to monitor, evaluate, and provide feedback on both overall team performance and individual team member performance
   E. Planning and Task Coordination KSAs
      13. The KSA to coordinate and synchronize activities, information, and task interdependencies between team members
      14. The KSA to help establish task and role expectations of individual team members, and to ensure proper balancing of workload in the team

Table 9 Example Items from the Teamwork KSA Test[a]

1. Suppose that you find yourself in an argument with several co-workers who should do a very disagreeable but routine task. Which of the following would probably be the most effective way to resolve this situation?
   A. Have your supervisor decide, because this would avoid any personal bias.
   *B. Arrange for a rotating schedule so that everyone shares the chore.
   C. Let the workers who show up earliest choose on a first-come, first-served basis.
   D. Randomly assign a person to do the task, and don't change it.
2. Your team wants to improve the quality and flow of the conversations among its members. Your team should:
   *A. use comments that build upon and connect to what others have said.
   B. set up a specific order for everyone to speak and then follow it.
   C. let team members with more to say determine the direction and topic of conversation.
   D. do all of the above.
3. Suppose that you are presented with the following types of goals. You are asked to pick one for your team to work on. Which would you choose?
   A. An easy goal to ensure that the team reaches it, thus creating a feeling of success.
   B. A goal of average difficulty so that the team will be somewhat challenged but successful, without too much effort.
   *C. A difficult and challenging goal that will stretch the team to perform at a high level, but attainable so that effort will not be seen as futile.
   D. A very difficult, or even impossible goal so that even if the team falls short, it will at least have a very high target to aim for.

[a]Asterisks denote correct answers.

Assessment center techniques might also lend themselves to measuring teamwork KSAs. Group exercises have been used to measure leadership and other social skills with good success (Gaugler et al., 1987). It is likely that existing team exercises, such as group problem-solving tasks, could also be modified to score teamwork KSAs.

Selection techniques using biodata may be another way to measure teamwork KSAs. Many items in biodata instruments reflect previous life experiences of a social nature, and recruiters interpret biodata information on applications and resumes as reflecting attributes such as interpersonal skills (Brown and Campion, 1994). A biodata measure developed to focus on teamwork KSAs might include items on teamwork in previous jobs, team experiences in school (e.g., college clubs, class projects), and recreational activities of a team nature (e.g., sports teams, social groups).

4.3.3 Designing a Team's Jobs

This aspect of team design involves team characteristics derived from the motivational job design approach. The main distinction is in level of application rather than content (Wall et al., 1986; Shea and Guzzo, 1987; Campion and Medsker, 1992). All the job characteristics of the motivational approach to job design can be applied to team design.

One such characteristic is self-management, which is the team-level analogy to autonomy at the individual job level. It is central to many definitions of effective work teams (e.g., Cummings, 1978, 1981; Hackman, 1987). A related characteristic is participation. Regardless of management involvement in decision making, teams can still be distinguished in terms of the degree to which all members are allowed to participate in decisions (McGrath, 1984; Porter et al., 1987). Self-management and participation are presumed to enhance effectiveness by increasing members' sense of responsibility and ownership of the work. These characteristics may also enhance decision quality by increasing relevant information and by putting decisions as near as possible to the point of operational problems and uncertainties.

Other important characteristics are task variety, task significance, and task identity. Variety motivates by allowing members to use different skills (Hackman, 1987) and by allowing both interesting and dull tasks to be shared among members (Davis and Wacker, 1987). Task significance refers to the perceived significance of the consequences of the team's work, either for others inside the organization or its customers. Task identity (Hackman, 1987), or task differentiation (Cummings, 1978), refers to the degree to which the team completes a whole and meaningful piece of work. These suggested characteristics of team design have been found to be positively related to team productivity, team member satisfaction, and managers' and employees' judgments of their teams' performance (Campion et al., 1993, 1995).

4.3.4 Developing Interdependent Relations

Interdependence is often the reason that teams are formed (Mintzberg, 1979) and is a defining characteristic of teams (Wall et al., 1986; Salas et al., 1992). Interdependence has been found to be related to team members' satisfaction and team productivity and effectiveness (Campion et al., 1993, 1995).

One form of interdependence is task interdependence. Team members interact and depend on one another to accomplish their work. Interdependence varies across teams, depending on whether the work flow in a team is pooled, sequential, or reciprocal (Thompson, 1967). Interdependence among tasks in the same job (Wong and Campion, 1991) or between jobs (Kiggundu, 1983) has been related to increased motivation. It can also increase team effectiveness because it enhances the sense of responsibility for others' work (Kiggundu, 1983) or because it enhances the reward value of a team's accomplishments (Shea and Guzzo, 1987).

Another form of interdependence is goal interdependence. Goal setting is a well-documented, individual-level performance improvement technique (Locke and Latham, 1990). A clearly defined mission or purpose is considered to be critical to team effectiveness (Davis and Wacker, 1987; Hackman, 1987; Sundstrom et al., 1990; Campion et al., 1993, 1995). Its importance has also been shown in empirical studies on teams (e.g., Woodman and Sherwood, 1980; Buller and Bell, 1986). Not only should goals exist for teams, but individual members' goals must be linked to team goals to be maximally effective.

Finally, interdependent feedback and rewards have also been found to be important for team effectiveness and team member satisfaction (Campion et al., 1993, 1995). Individual feedback and rewards should be linked to a team's performance to motivate team-oriented behavior. This characteristic is recognized in many theoretical treatments (e.g., Steiner, 1972; Leventhal, 1976; Hackman, 1987; Sundstrom et al., 1990) and research studies (e.g., Pasmore et al., 1982; Wall et al., 1986).

4.3.5 Creating the Organizational Context

Organizational context and resources are considered in all recent models of work team effectiveness (e.g., Hackman, 1987; Guzzo and Shea, 1992). One important aspect of context and resources for teams is adequate training. Training is an extensively researched determinant of team performance (for reviews, see Dyer, 1984; Salas et al., 1992), and training is included in most interventions (e.g., Pasmore et al., 1982; Wall et al., 1986). Training is related to team members' satisfaction, and managers' and employees' judgments of their teams' effectiveness (Campion et al., 1993, 1995).

Training content often includes team philosophy, group decision making, and interpersonal skills, as well as technical knowledge. Many team-building interventions focus on aspects of team functioning that are related to the teamwork KSAs shown in Table 8. A recent review of this literature divided such interventions into four approaches (Tannenbaum et al., 1992)—goal setting, interpersonal, role, and problem solving—which are similar to the teamwork KSA categories. Thus, these interventions could be viewed as training programs on teamwork KSAs. Reviews indicate that the evidence for the effectiveness of this training appears positive despite the methodological limitations that plague this research (Woodman and Sherwood, 1980; Buller and Bell, 1986; Tannenbaum et al., 1992). It appears that workers can be trained in teamwork KSAs. (See Chapter 16 for more information on team training.)

Regarding how such training should be conducted, there is substantial guidance on training teams in the human factors and military literatures (Dyer, 1984; Salas et al., 1992; Swezey and Salas, 1992). Because these topics are addressed thoroughly in the sources cited, they are not reviewed here.

Managers of teams also need to be trained in teamwork KSAs, regardless of whether the teams are manager-led or self-managed. The KSAs are needed for interacting with employee teams and for participating on management teams. It has been noted that managers of teams, especially autonomous work teams, need to develop their employees (Cummings, 1978; Hackman and Oldham, 1980; Manz and Sims, 1987). Thus, training must not only ensure that managers possess teamwork KSAs, but that they know how to train employees on these KSAs.

Managerial support is another contextual characteristic (Morgeson, 2005). Management controls resources (e.g., material and information) required to make team functioning possible (Shea and Guzzo, 1987), and an organization's culture and top management must support the use of teams (Sundstrom et al., 1990). Teaching facilitative leadership to managers is often a feature of team interventions (Pasmore et al., 1982). Finally, communication and cooperation between teams is a contextual characteristic because it is often the responsibility of managers. Supervising team boundaries (Cummings, 1978) and externally integrating teams with the rest of the organization (Sundstrom et al., 1990) enhance effectiveness. Research indicates that managerial support and communication and cooperation between work teams are related to team productivity and effectiveness and to team members' satisfaction with their work (Campion et al., 1993, 1995).

4.3.6 Developing Effective Team Processes

Process describes those things that go on in the group that influence effectiveness. One process characteristic is potency, the belief of a team that it can be effective (Shea and Guzzo, 1987; Guzzo and Shea, 1992). It is similar to the lay term *team spirit*. Hackman (1987) argues that groups with high potency are more committed and willing to work hard for the group, and evidence indicates that potency is highly related to team members' satisfaction with work, team productivity, and members' and managers' judgments of their teams' effectiveness (Campion et al., 1993, 1995).

Another process characteristic found to be related to team satisfaction, productivity, and effectiveness is social support (Campion et al., 1993, 1995). Effectiveness can be enhanced when members help each other and have positive social interactions. Like social facilitation (Zajonc, 1965; Harkins, 1987), social support can be arousing and may enhance effectiveness by sustaining effort on mundane tasks.

Another process characteristic related to satisfaction, productivity, and effectiveness is workload sharing (Campion et al., 1993, 1995). Workload sharing enhances effectiveness by preventing social loafing or free-riding (Harkins, 1987). To enhance sharing, group members should believe that their individual performance can be distinguished from the group's, and that there is a link between their performance and outcomes.

Finally, communication and cooperation within the work group are also important to team effectiveness, productivity, and satisfaction (Campion et al.,

1993, 1995). Management should help teams foster open communication, supportiveness, and discussions of strategy. Informal, rather than formal communication channels and mechanisms of control should be promoted to ease coordination (Bass and Klubeck, 1952; Majchrzak, 1988). Managers should encourage self-evaluation, self-observation, self-reinforcement, self-management, and self-goal setting by teams. Self-criticism for purposes of recrimination should be discouraged (Manz and Sims, 1987).

5 MEASUREMENT AND EVALUATION OF JOB AND TEAM DESIGN

The purpose of an evaluation study for either a job or team design is to provide an objective evaluation of success and to create a tracking and feedback system to make adjustments during the course of the design project. An evaluation study can provide objective data to make informed decisions, help tailor the process to the organization, and give those affected by the design or redesign an opportunity to provide input (see Morgeson and Campion, 2002). An evaluation study should include measures that describe the characteristics of the jobs or teams so that it can be determined whether or not jobs or teams ended up having the characteristics they were intended to have. An evaluation study should also include measures of effectiveness outcomes an organization hoped to achieve with a design project. Measures of effectiveness could include such *subjective* outcomes as employee job satisfaction or employee, manager, or customer perceptions of effectiveness. Measures of effectiveness should include *objective* outcomes such as cost, productivity, rework/scrap, turnover, accident rates, or absenteeism. Additional information on measurement and evaluation of such outcomes may be found in Part VII of this handbook.

5.1 Using Questionnaires to Measure Job and Team Design

One way to measure job or team design is by using questionnaires or checklists. This method of measuring job or team design is highlighted because it has been used widely in research on job design, especially on the motivational approach. More important, questionnaires are a very inexpensive, easy, and flexible way to measure work design characteristics. Moreover, they gather information from job experts, such as incumbents, supervisors, and engineers and other analysts.

Several questionnaires exist for measuring the motivational approach to job design (Sims et al., 1976; Hackman and Oldham, 1980), but only one questionnaire, the *Multimethod Job Design Questionnaire*, measures characteristics for all four approaches to job design. This questionnaire (presented in Table 2) evaluates the quality of a job's characteristics based on each of the four approaches. The *Team Design Measure* (presented in Table 3) evaluates the quality of work design based on the team approach.

Questionnaires can be administered in a variety of ways. Employees can complete them individually at their convenience at their workstation or some other designated area, or they can complete them in a group setting. Group administration allows greater standardization of instructions and provides the opportunity to answer questions and clarify ambiguities. Managers and engineers can also complete the questionnaires either individually or in a group session. Engineers and analysts usually find that observation of the work site, examination of the equipment and procedures, and discussions with any incumbents or managers are important methods of gaining information on the work before completing the questionnaires.

Scoring for each job design approach or for each team characteristic on the questionnaires is usually accomplished simply by averaging the applicable items. Then scores from different incumbents, managers, or engineers describing the same job or team are combined by averaging. Multiple items and multiple respondents are used to improve the reliability and accuracy of the results. The implicit assumption is that slight differences among respondents are to be expected because of legitimate differences in viewpoint. However, absolute differences in scores should be examined on an item-by-item basis, and large discrepancies (e.g., more than one point) should be discussed to clarify possible differences in interpretation. It may be useful to discuss each item until a consensus rating is reached.

The higher the score on a particular job design scale or work team characteristic scale, the better the quality of the design in terms of that approach or characteristic. Similarly, the higher the score on a particular item, the better the design is on that dimension. How high a score is needed or necessary cannot be stated in isolation. Some jobs or teams are naturally higher or lower on the various approaches, and there may be limits to the potential of some jobs. The scores have most value in comparing different jobs, teams, or design approaches rather than evaluating the absolute level of the quality of a job or team design. However, a simple rule of thumb is that if the score for an approach is smaller than three, the job or team is poorly designed on that approach and it should be reconsidered. Even if the average score on an approach is greater than three, examine any individual dimension scores that are at two or one.

Uses of Questionnaires in Different Contexts

1. *Designing new jobs or teams.* When jobs or teams do not yet exist, the questionnaire is used to evaluate proposed job or team descriptions, workstations, equipment, and so on. In this role, it often serves as a simple design checklist. Additional administrations of the questionnaire in later months or years can be used to assess the longer-term effects of the job or team design.

2. *Redesigning existing jobs or teams or switching from job to team design.* When jobs or teams already exist, there is a much greater wealth of information. Questionnaires can be completed by incumbents, managers, and engineers. Questionnaires can be used to measure design both before and after changes

are made to compare the redesign with the previous design approach. A premeasure before the redesign can be used as a baseline measurement against which to compare a postmeasure conducted right after the redesign implementation. A follow-up measure can be used in later months or years to assess the long-term difference between the previous design approach and the new approach.

If other sites or plants with the same types of jobs or teams are not immediately included in the redesign but are maintained with the older design approach, they can be used as a comparison or control group to enable analysts to draw even stronger conclusions about the effectiveness of the redesign. Such a control group allows one to control for the possibilities that changes in effectiveness were not due to the redesign but were in fact due to some other causes, such as increases in workers' knowledge and skills with the passage of time, changes in workers' economic environment (i.e., job security, wages, etc.), or workers trying to give socially desirable responses to questionnaire items.

3. *Diagnosing problem job or team designs.* When problems occur, regardless of the apparent source of the problem, the job or team design questionnaires can be used as a diagnostic device to determine if any problems exist with the design of the jobs or teams.

5.2 Choosing Sources of Data

1. *Incumbents.* Incumbents are probably the best source of information for existing jobs or teams. Having input can enhance the likelihood that changes will be accepted, and involvement in such decisions can enhance feelings of participation, thus increasing motivational job design in itself (see item 22 of the motivational scale in Table 2). One should include a large number of incumbents for each job or team because there can be slight differences in perceptions of the same job or team due to individual differences (discussed in Section 4.1). Evidence suggests that one should include at least five incumbents for each job or team, but more are preferable (Campion, 1988; Campion and McClelland, 1991; Campion et al., 1993, 1995).

2. *Managers or supervisors.* First-level managers or supervisors may be the next most knowledgeable persons about an existing work design. They may also provide information on jobs or teams under development. Some differences in perceptions of the same job or team will exist among managers, so multiple managers should be used.

3. *Engineers or analysts.* Engineers may be the only source of information if the jobs or teams are not yet developed. But also for existing jobs or teams, an outside perspective of an engineer, analyst, or consultant may provide a more objective viewpoint. Again, there can be differences among engineers, so several should evaluate each job or team.

It is desirable to get multiple inputs and perspectives from different sources in order to get the most reliable and accurate picture of the results of the job or team design.

5.3 Long-Term Effects and Potential Biases

It is important to recognize that some effects of job or team design may not be immediate, others may not be long lasting, and still others may not be obvious. Initially, when jobs or teams are designed, or right after they are redesigned, there may be a short-term period of positive attitudes (often called a *honeymoon effect*). As the legendary Hawthorne studies indicated, changes in jobs or increased attention paid to workers tends to create novel stimulation and positive attitudes (Mayo, 1933). Such transitory elevations in affect should not be mistaken for long-term improvements in satisfaction, as they may wear off over time. In fact, with time, employees may realize that their work is now more complex and believe that they should be paid higher compensation (Campion and Berger, 1990).

Costs that are likely to lag in time also include stress and fatigue, which may take awhile to build up if mental demands have been increased excessively. Boredom may take awhile to set in if mental demands have been overly decreased. In terms of lagged benefits, productivity and quality are likely to improve with practice and learning on the new job or team. And some benefits, such as reduced turnover, simply take time to estimate accurately.

Benefits that may potentially dissipate with time include satisfaction, especially if the elevated satisfaction is a function of novelty rather than basic changes to the motivating value of the work. Short-term increases in productivity due to heightened effort rather than better design may not last. Costs that may dissipate include training requirements and staffing difficulties. Once jobs are staffed and everyone is trained, these costs disappear until turnover occurs. So these costs will not go away completely, but they may be less after initial start-up. Dissipating heightened satisfaction but long-term increases in productivity were observed in a recent motivational job redesign study (Griffin, 1989). These are only examples to illustrate how dissipating and lagged effects might occur. A more detailed example of long-term effects is given in Section 5.8.

A potential bias that may confuse the proper evaluation of benefits and costs is spillover. Laboratory research has shown that the job satisfaction of employees can bias perceptions of the motivational value of their jobs (O'Reilly et al., 1980). Similarly, the level of morale in the organization can have a spillover effect onto employees' perceptions of job or team design. If morale is particularly high, it may have an elevating effect on how employees or analysts view the jobs or teams; conversely, low morale may have a depressing effect on views. The term *morale* refers to the general level of job satisfaction across employees, and it may be a function of many factors, including management, working conditions, wages, and so on. Another factor that has an especially strong effect on employee reactions to work design changes is *employment security*. Obviously, employee enthusiasm for work design changes will be negative if employees view them as potentially decreasing their job security. Every effort should be made to eliminate these fears. The best method of addressing these effects is to be attentive to their potential existence and to conduct longitudinal evaluations of job and team design.

In addition to questionnaires, many other analytical tools are useful for work design. The disciplines that contributed the different approaches to work design have also contributed different techniques for analyzing tasks, jobs, and processes for design and redesign purposes. These techniques include job analysis methods created by specialists in industrial psychology, variance analysis methods created by specialists in sociotechnical design, time and motion analysis methods created by specialists in industrial engineering, and linkage analysis methods created by specialists in human factors. In this section we describe briefly a few of these techniques to illustrate the range of options. The reader is referred to the citations for detail on how to use the techniques.

5.4 Job Analysis

Job analysis can be defined broadly as a number of systematic techniques for collecting and making judgments about job information (Morgeson and Campion, 1997, 2000). Information derived from job analysis can be used to aid in recruitment and selection decisions, determine training and development needs, develop performance appraisal systems, and evaluate jobs for compensation, as well as to analyze tasks and jobs for job design. Job analysis may also focus on tasks, worker characteristics, worker functions, work fields, working conditions, tools and methods, products and services, and so on. Job analysis data can come from job incumbents, supervisors, and analysts who specialize in the analysis of jobs. Data may also be provided in some cases by higher management levels or subordinates.

Considerable literature has been published on the topic of job analysis (U.S. Department of Labor, 1972; Ash et al., 1983; Gael, 1983; Harvey, 1991; Morgeson and Campion, 1997; Peterson et al., 2001; Dierdorff and Wilson, 2003; Morgeson et al., 2004). Some of the more typical methods of analysis are described briefly below.

1. *Conferences and interviews.* Conferences or interviews with job experts, such as incumbents and supervisors, are often the first step. During such meetings, information collected typically includes job duties and tasks, and knowledge, skill, ability (KSA), and other worker characteristics.

2. *Questionnaires.* Questionnaires are used to collect information efficiently from a large number of people. Questionnaires require considerable prior knowledge of the job to form the basis of the items (e.g., primary tasks). Often, this information is first collected through conferences and interviews, and then the questionnaire is constructed and used to collect judgments about the job (e.g., importance and time spent on each task). Some standardized questionnaires have been developed that can be applied to all jobs to collect basic information on tasks and requirements. Examples of standardized questionnaires are the Position Analysis Questionnaire (McCormick et al., 1972) and the Occupational Information Network (O*NET) (Peterson et al., 2001).

3. *Inventories.* Inventories are much like questionnaires, except that they are simpler in format. They are usually simple checklists where the job expert checks whether a task is performed or an attribute is required.

4. *Critical incidents.* This form of job analysis focuses on aspects of worker behavior that are especially effective or ineffective.

5. *Work observation and activity sampling.* Quite often, job analysis includes the actual observation of work performed. More sophisticated technologies involve statistical sampling of work activities.

6. *Diaries.* Sometimes it is useful or necessary to collect data by having the employee keep a diary of activities on his or her job.

7. *Functional job analysis.* Task statements can be written in a standardized fashion. Functional job analysis suggests how to write task statements (e.g., start with a verb, be as simple and discrete as possible). It also involves rating jobs on the degree of data, people, and things requirements. This form of job analysis was developed by the U.S. Department of Labor and has been used to describe over 12,000 jobs as documented in the Dictionary of Occupational Titles (Fine and Wiley, 1971; U.S. Department of Labor, 1977).

Very limited research has been done to evaluate the practicality and quality of various job analysis methods for different purposes. But analysts seem to agree that combinations of methods are preferable to single methods (Levine et al., 1983; Morgeson and Campion, 1997).

Current approaches to job analysis do not give much attention to analyzing teams. For example, the *Dictionary of Occupational Titles* (U.S. Department of Labor, 1972) considers "people" requirements of jobs, but does not address specific teamwork KSAs. Similarly, recent reviews of the literature mention some components of teamwork, such as communication and coordination (e.g., Harvey, 1991), but give little attention to other teamwork KSAs. Thus, job analysis systems may need to be revised. The recent O*NET reflects a major new job analysis system designed to replace the DOT (Peterson et al., 2001). Although not explicitly addressing the issue of teamwork KSAs, it does contain a large number of worker attribute domains that may prove useful. Teamwork KSAs are more likely to emerge with conventional approaches to job analysis because of their unstructured nature (e.g., interviews), but structured approaches (e.g., questionnaires) will have to be modified to query teamwork KSAs.

5.5 Variance Analysis

Variance analysis is a tool of sociotechnical design used to identify areas of technological uncertainty in a production process. Variance analysis aids the organization in designing jobs so that jobholders can control variability in their work. A *variance* is defined as an unwanted discrepancy between a desired state and an actual state and is a deviation that falls outside a specified range of tolerance. The variance concept is applied to the technical system and involves five steps (Davis and Wacker, 1982):

1. List variances that could impede the production or service process.
2. Identify causal relationships among variables. Job designers can use information about dependencies and points of interrelatedness to cluster tasks and link jobs.
3. Identify and focus on key variances whose control is most critical to successful outcomes.
4. Construct a table of key variance control that contains brief descriptions of variances.
5. Construct a table of skills, knowledge, information, and authority needed so that workers can control key variances.

Chapters 9 and 14 provide more information about task and workload analysis.

5.6 Time and Motion Analysis

Industrial engineers have created many techniques for use in the study of job design that help job designers visualize operations in order to improve efficiencies. A considerable literature exists on the topic (e.g., Mundel, 1985; Niebel, 1988). Some of the methods are described briefly below.

Process charts graphically represent separate steps or events that occur during performance of a task or series of actions. Charts usually begin with inputs of raw materials and follow the inputs through transportation, storage, inspection, production, and finishing. Charts use symbols for different types of operations. Examples of different types of process charts include operation process charts, which show a chronological sequence of operations, inspections, time allowances, and materials used in a process from arrival of raw material to packaging of the finished product. Another type of process chart is a worker and machine process chart, which combines operations of both the worker and equipment and shows idle time and active time for both. These charts are used to analyze only one workstation at a time.

Flow diagrams differ from process charts because they utilize drawings of an area or building in which an activity takes place. Flow diagrams help designers visualize the physical layout of the work. Lines are drawn to show the path of travel. Process chart symbols and notations can be included to describe the process.

Possibility guides are tools for listing systematically all possible changes suggested for a particular activity or output. They assist in examining consequences of suggestions to aid in selecting the most feasible changes. Suggestions are recorded and are coded as to what classes of change they affect: job, equipment, process, product design, or raw materials.

Network diagrams are better for use in describing complex relationships than the techniques described above. They are useful for situations where (1) dependencies are tangled and do not progress uniformly, (2) the output has many components, (3) many of the components are service-type outputs, (4) the relationships among the steps of the process with respect to time are of vital importance, or (5) the process is too complex or large in scope for the usual process chart analysis. In network diagrams, a circle or square represents a *status*, which is a partial or complete service or substantive output. Heavy lines are *critical paths*, which determine the minimum time in which a project can be expected to be completed.

5.7 Linkage Analysis

Linkage analysis is a technique used by human factors specialists to represent relationships between components in a work system (Sanders and McCormick, 1987). Components can be either people or things and the relationships between them are called *links*. Links fall into three classes:

1. Communication links
   a. Visual (person to person or equipment to person)
   b. Auditory, voice (person to person, person to equipment, or equipment to person)
   c. Auditory, nonvoice (equipment to person)
   d. Touch (person to equipment)
2. Control links
   a. Control (person to equipment)
3. Movement links (movements from one location to another)
   a. Eye movements
   b. Manual movements, foot movements, or both
   c. Body movements

Information collected about links generally includes how often components are linked, in what sequence links occur, and the importance of links. Once obtained, linkage data can be summarized in link tables, adjacency layout diagrams, and spatial operational sequences (SOS) diagrams. Designers of physical work arrangements use these tools to represent relationships between components so that they can better understand how to place these components in advantageous locations to minimize lengths between frequent or important links. With complex systems involving many components, quantitative analysis techniques, such as linear programming, can be used.

5.8 Example of Evaluation of a Job Design

Studies conducted by Campion and McClelland (1991, 1993) are described as an illustration of an evaluation of a job redesign project. They illustrate the value of considering an interdisciplinary perspective. The setting was a large financial services company. The units under study processed the paperwork in support of other units that sold the company's products. Jobs had been designed in a mechanistic manner such that individual employees prepared, sorted, coded, and computer-input the paper flow.

The organization viewed the jobs as designed too mechanistically. Guided by the motivational approach, the project intended to enlarge jobs by combining existing jobs to attain three objectives: (1) enhance motivation and satisfaction of employees; (2) increase incumbent feelings of ownership of the work, thus increasing customer service; and (3) maintain productivity despite potential lost efficiencies from the motivational approach. The consequences of all approaches to job design were considered. It was anticipated that the project would increase motivational consequences, decrease mechanistic and perceptual/motor consequences, and have no effect on biological consequences (Table 1).

The evaluation consisted of collecting detailed data on job design and a broad spectrum of potential benefits and costs of enlarged jobs. The research strategy involved comparing several varieties of enlarged jobs with each other and with unenlarged jobs. Questionnaire data were collected and focused team meetings were conducted with incumbents, managers, and analysts. The study was repeated at five geographic sites.

Results indicated that enlarged jobs had the benefits of more employee satisfaction, less boredom, better quality, and better customer service; but they also had the costs of slightly higher training, skill, and compensation requirements. Another finding was that all potential costs of enlarging jobs were not observed, suggesting that redesign can lead to benefits without incurring every cost in a one-to-one fashion.

In a two-year follow-up evaluation study, it was found that the costs and benefits of job enlargement changed substantially over time, depending on the type of enlargement. Task enlargement, which was the focus of the original study, had mostly long-term costs (e.g., lower satisfaction, efficiency, and customer service, and more mental overload and errors). Conversely, knowledge enlargement, which emerged as a form of job design since the original study, had mostly benefits (e.g., higher satisfaction and customer service, lower overload and errors).

There are several important implications of the latter study. First, it illustrates that the long-term effects of job design changes can be different than the short-term effects. Second, it shows the classic distinction between enlargement and enrichment (Herzberg, 1966) in that simply adding more tasks did not improve the job, but adding more knowledge opportunities did. Third, it illustrates how the job design process is iterative. In this setting, the more favorable knowledge enlargement was discovered only after gaining experience with task enlargement. Fourth, as in the previous study, it shows that it is possible in some situations to gain the benefits of job design without incurring all the potential costs, thus minimizing the trade-offs between the motivational and mechanistic approaches to job design.

5.9 Example of Evaluation of a Team Design

Studies conducted by the authors and their colleagues are described here as an illustration of an evaluation of a team design project (Campion et al., 1993, 1995). They illustrate the use of multiple sources of data and multiple types of team effectiveness outcomes. The setting was the same financial services company as in the example job design evaluation above. Questionnaires based on Table 3 were administered to 391 clerical employees in 80 teams and 70 team managers in the first study (Campion et al., 1993) and to 357 professional workers in 60 teams (e.g., systems analysts, claims specialists, underwriters) and 93 managers in the second study (Campion et al., 1995) to measure teams' design characteristics. Thus, two sources of data were used, team members and team managers, to measure the team design characteristics.

In both studies, effectiveness outcomes included the organization's employee satisfaction survey, which had been administered at a different time than the team design characteristics questionnaire, and managers' judgments of teams' effectiveness, measured at the same time as the team design characteristics. In the first study, several months of records of team productivity were also used to measure effectiveness. Additional effectiveness measures in the second study were employees' judgments of their team's effectiveness, measured at the same time as the team design characteristics, managers' judgments of teams' effectiveness, measured a second time three months after the team design characteristics, and the average of team members' most recent performance ratings.

Results indicated that all of the team design characteristics had positive relationships with at least some of the outcomes. Relationships were strongest for process characteristics, followed by job design, context, interdependence, and composition characteristics (see Figure 1). Results also indicated that when teams were well designed according to the team design approach, they were higher on both employee satisfaction and team effectiveness ratings than were less well-designed teams.

Results were stronger when the team design characteristics data were from team members rather than from the team managers. This illustrates the importance of collecting data from different sources to gain different perspectives on the results of a team design project. Collecting data from only a single source may lead one to draw different conclusions about a design project than if one obtains a broader picture of the team design results from multiple sources.

Results were also stronger when outcome measures came from employees (employee satisfaction, team member judgments of their teams), managers rating their own teams, or productivity records than when they came from other managers or from performance appraisal ratings. This illustrates the use of different types of outcome measures to avoid drawing conclusions from overly limited data. This example also illustrates the use of separate data collection methods and times for collecting team design characteristics data vs. team outcomes data. A single data collection method and time in which team design characteristics and outcomes are collected from the same source (e.g., team members only) on the same day can create an illusion of higher relationships between design characteristics and outcomes than really exist. Although it is more costly to use multiple sources, methods, and administration times, the ability to draw conclusions from the results is far stronger if one does.

REFERENCES

Argyris, C. (1964), *Integrating the Individual and the Organization*, Wiley, New York.

Ash, R. A., Levine, E. L., and Sistrunk, F. (1983), "The Role of Jobs and Job-Based Methods in Personnel and Human Resources Management," in *Research in Personnel and Human Resources Management*, Vol. 1, K. M. Rowland and G. R. Ferris, Eds., JAI Press, Greenwich, CT.

Astrand, P. O., and Rodahl, K. (1977), *Textbook of Work Physiology: Physiological Bases of Exercise*, 2nd ed., McGraw-Hill, New York.

Babbage, C. (1835), "On the Economy of Machinery and Manufacturers," reprinted in *Design of Jobs*, 2nd ed., L. E. Davis and J. C. Taylor, Eds., Goodyear, Santa Monica, CA.

Banker, R. D., Field, J. M., Schroeder, R. G., and Sinha, K. K. (1996), "Impact of Work Teams on Manufacturing Performance: A Longitudinal Study," *Academy of Management Journal*, Vol. 39, pp. 867–890.

Bass, B. M., and Klubeck, S. (1952), "Effects of Seating Arrangements and Leaderless Team Discussions," *Journal of Abnormal and Social Psychology*, Vol. 47, pp. 724–727.

Blauner, R. (1964), *Alienation and Freedom*, University of Chicago Press, Chicago.

Brown, B. K., and Campion, M. A. (1994), "Biodata Phenomenology: Recruiters' Perceptions and Use of Biographical Information in Personnel Selection," *Journal of Applied Psychology*, Vol. 79, pp. 897–908.

Buller, P. F., and Bell, C. H. (1986), "Effects of Team Building and Goal Setting on Productivity: A Field Experiment," *Academy of Management Journal*, Vol. 29, pp. 305–328.

Buys, C. J. (1978), "Humans Would Do Better Without Groups," *Personality and Social Psychology Bulletin*, Vol. 4, pp. 123–125.

Campion, M. A. (1988), "Interdisciplinary Approaches to Job Design: A Constructive Replication with Extensions," *Journal of Applied Psychology*, Vol. 73, pp. 467–481.

Campion, M. A. (1989), "Ability Requirement Implications of Job Design: An Interdisciplinary Perspective," *Personnel Psychology*, Vol. 42, pp. 1–24.

Campion, M. A., and Berger, C. J. (1990), "Conceptual Integration and Empirical Test of Job Design and Compensation Relationships," *Personnel Psychology*, Vol. 43, pp. 525–554.

Campion, M. A., and McClelland, C. L. (1991), "Interdisciplinary Examination of the Costs and Benefits of Enlarged Jobs: A Job Design Quasi-experiment," *Journal of Applied Psychology*, Vol. 76, pp. 186–198.

Campion, M. A., and McClelland, C. L. (1993), "Follow-up and Extension of the Interdisciplinary Costs and Benefits of Enlarged Jobs," *Journal of Applied Psychology*, Vol. 78, pp. 339–351.

Campion, M. A., and Medsker, G. J. (1992), "Job Design," in *Handbook of Industrial Engineering*, G. Salvendy, Ed., Wiley, New York.

Campion, M. A., and Stevens, M. J. (1991), "Neglected Questions in Job Design: How People Design Jobs, Influence of Training, and Task–Job Predictability," *Journal of Business and Psychology*, Vol. 6, pp. 169–191.

Campion, M. A., and Thayer, P. W. (1985), "Development and Field Evaluation of an Interdisciplinary Measure of Job Design," *Journal of Applied Psychology*, Vol. 70, pp. 29–43.

Campion, M. A., and Thayer, P. W. (1987), "Job Design: Approaches, Outcomes, and Trade-offs," *Organizational Dynamics*, Vol. 15, No. 3, pp. 66–79.

Campion, M. A., Medsker, G. J., and Higgs, A. C. (1993), "Relations Between Work Group Characteristics and Effectiveness: Implications for Designing Effective Work Groups," *Personnel Psychology*, Vol. 46, pp. 823–850.

Campion, M. A., Campion, J. E., and Hudson, J. P. (1994a), "Structured Interviewing: A Note on Incremental Validity and Alternative Question Types," *Journal of Applied Psychology*, Vol. 79, pp. 998–1002.

Campion, M. A., Cheraskin, L., and Stevens, M. J. (1994b), "Job Rotation and Career Development: Career-Related Antecedents and Outcomes of Job Rotation," *Academy of Management Journal*, Vol. 37, pp. 1518–1542.

Campion, M. A., Papper, E. M., and Medsker, G. J. (1996), "Relations Between Work Team Characteristics and Effectiveness: A Replication and Extension," *Personnel Psychology*, Vol. 49, pp. 429–452.

Caplan, R. D., Cobb, S., French, J. R. P., Van Harrison, R., and Pinneau, S. R. (1975), *Job Demands and Worker Health: Main Effects and Occupational Differences, HEW Publication (NIOSH)* 75–160, U.S. Government Printing Office, Washington, DC.

Cartwright, D. (1968), "The Nature of Team Cohesiveness," in *Team Dynamics: Research and Theory*, 3rd ed., D. Cartwright and A. Zander, Eds., Harper & Row, New York.

Chapanis, A. (1970), "Relevance of Physiological and Psychological Criteria to Man–Machine Systems: The Present State of the Art," *Ergonomics*, Vol. 13, pp. 337–346.

Cook, T. D., and Campbell, D. T. (1979), *Quasi-experimentation: Design and Analysis Issues for Field Settings*, Rand McNally, Chicago.

Cummings, T. G. (1978), "Self-Regulating Work Teams: A Sociotechnical Synthesis," *Academy of Management Review*, Vol. 3, pp. 625–634.

Cummings, T. G. (1981), "Designing Effective Work Groups," in *Handbook of Organization Design*, Vol. 2, P. C. Nystrom and W. H. Starbuck, Eds., Oxford University Press, New York.

Davis, L. E. (1957), "Toward a Theory of Job Design," *Journal of Industrial Engineering*, Vol. 8, pp. 305–309.

Davis, L. E. (1971), "The Coming Crisis for Production Management: Technology and Organization," *International Journal of Production Research*, Vol. 9, pp. 65–82.

Davis, L. E. (1982), "Organization Design," in *Handbook of Industrial Engineering*, G. Salvendy, Ed., Wiley, New York.

Davis, L. E., and Taylor, J. C. (1979), *Design of Jobs*, 2nd ed., Goodyear, Santa Monica, CA.

Davis, L. E., and Valfer, E. S. (1965), "Intervening Responses to Changes in Supervisor Job Design," *Occupational Psychology*, Vol. 39, pp. 171–189.

Davis, L. E., and Wacker, G. L. (1982), "Job Design," in *Handbook of Industrial Engineering*, G. Salvendy, Ed., Wiley, New York.

Davis, L. E., and Wacker, G. L. (1987), "Job Design," in *Handbook of Human Factors*, G. Salvendy, Ed., Wiley, New York.

Davis, L. E., Canter, R. R., and Hoffman, J. (1955), "Current Job Design Criteria," *Journal of Industrial Engineering*, Vol. 6, No. 2, pp. 5–8, 21–23.

Dierdorff, E. C., and Wilson, M. A. (2003), "A Meta-analysis of Job Analysis Reliability," *Journal of Applied Psychology*, Vol. 88, pp. 635–646.

Dyer, J. (1984), "Team Research and Team Training: A State-of-the-Art Review," in *Human Factors Review*, F. A. Muckler, Ed., Human Factors Society, Santa Monica, CA.

Edwards, J. R., Scully, J. A., and Brtek, M. D. (1999), "The Measurement of Work: Hierarchical Representation of the Multimethod Job Design Questionnaire," *Personnel Psychology*, Vol. 52, pp. 305–334.

Edwards, J. R., Scully, J. A., and Brtek, M. D. (2000), "The Nature and Outcomes of Work: A Replication and Extension of Interdisciplinary Work-Design Research," *Journal of Applied Psychology*, Vol. 85, pp. 860–868.

Emory, F. E., and Trist, E. L. (1960), "Sociotechnical Systems," in *Management Sciences, Models, and Techniques*, Vol. 2, C. W. Churchman and M. Verhulst, Eds., Pergamon Press, London.

Fine, S. A., and Wiley, W. W. (1971), *An Introduction to Functional Job Analysis*, W.E. Upjohn Institute for Employment Research, Kalamazoo, MI.

Ford, R. N. (1969), *Motivation Through the Work Itself*, American Management Association, New York.

Fried, Y., and Ferris, G. R. (1987), "The Validity of the Job Characteristics Model: A Review and Metaanalysis," *Personnel Psychology*, Vol. 40, pp. 287–322.

Gael, S. (1983), *Job Analysis: A Guide to Assessing Work Activities*, Jossey-Bass, San Francisco.

Gallagher, C. C., and Knight, W. A. (1986), *Team Technology Production Methods in Manufacture*, Ellis Horwood, Chichester, West Sussex, England.

Gaugler, B. B., Rosenthal, D. B., Thornton, G. C., and Benston, C. (1987), "Metaanalysis of Assessment Center Validity" (monograph), *Journal of Applied Psychology*, Vol. 72, pp. 493–511.

Gilbreth, F. B. (1911), *Motion Study: A Method for Increasing the Efficiency of the Workman*, Van Nostrand, New York.

Gladstein, D. L. (1984), "Groups in Context: A Model of Task Group Effectiveness," *Administrative Science Quarterly*, Vol. 29, pp. 499–517.

Globerson, S., and Crossman, E. R. (1976), "Nonrepetitive Time: An Objective Index of Job Variety," *Organizational Behavior and Human Performance*, Vol. 17, pp. 231–240.

Goodman, P. S., Ravlin, E. C., and Argote, L. (1986), "Current Thinking About Teams: Setting the Stage for New Ideas," in *Designing Effective Work Teams*, P. S. Goodman and Associates, Eds., Jossey-Bass, San Francisco.

Grandjean, E. (1980), *Fitting the Tasks to the Man: An Ergonomic Approach*, Taylor & Francis, London.

Griffin, R. W. (1982), *Task Design: An Integrative Approach*, Scott, Foresman, Glenview, IL.

Griffin, R. W. (1989), "Work Redesign Effects on Employee Attitudes and Behavior: A Long-Term Experiment," *Academy of Management Best Papers Proceedings*, Washington, DC, pp. 214–219.

Guzzo, R. A., and Shea, G. P. (1992), "Group Performance and Intergroup Relations in Organizations," in *Handbook of Industrial and Organizational Psychology*, Vol. 3, M. D. Dunnette and L. M. Hough, Eds., Consulting Psychologists Press, Palo Alto, CA.

Hackman, J. R. (1987), "The Design of Work Teams," in *Handbook of Organizational Behavior*, J. Lorsch, Ed., Prentice-Hall, Englewood Cliffs, NJ.

Hackman, J. R. (2002), *Leading Teams: Setting the Stage for Great Performances*, Harvard Business School Press, Boston.

Hackman, J. R., and Oldham, G. R. (1980), *Work Redesign*, Addison-Wesley, Reading, MA.

Hammond, R. W. (1971), "The History and Development of Industrial Engineering," in *Industrial Engineering Handbook*, 3rd ed., H. B. Maynard, Ed., McGraw-Hill, New York.

Harkins, S. G. (1987), "Social Loafing and Social Facilitation," *Journal of Experimental Social Psychology*, Vol. 23, pp. 1–18.

Harvey, R. J. (1991), "Job Analysis," in *Handbook of Industrial and Organizational Psychology*, Vol. 2, 2nd ed., M. D. Dunnette and L. M. Hough, Eds., Consulting Psychologists Press, Palo Alto, CA.

Herzberg, F. (1966), *Work and the Nature of Man*, World Publishing, Cleveland, OH.

Hoerr, J. (1989), "The Payoff from Teamwork," *Business Week*, July 10, pp. 56–62.

Hofstede, G. (1980), *Culture's Consequences*, Sage Publications, Beverly Hills, CA.

Homans, G. C. (1950), *The Human Group*, Harcourt, Brace & World, New York.

Hoppock, R. (1935), *Job Satisfaction*, Harper & Row, New York.

Hyer, N. L. (1984), "Management's Guide to Team Technology," in *Team Technology at Work*, N. L. Hyer, Ed., Society of Manufacturing Engineers, Dearborn, MI.

Ilgen, D. R., Hollenbeck, J. R., Johnson, M. D., and Jundt, D. (2005), "Teams in Organizations: From I-P-O Models to IMOI Models, *Annual Review of Psychology*, Vol. 56, pp. 517–543.

Janis, I. L. (1972), *Victims of Groupthink*, Houghton Mifflin, Boston.

Johansson, G., Aronsson, G., and Lindstrom, B. O. (1978). "Social Psychological and Neuroendocrine Stress Reactions in Highly Mechanised Work," *Ergonomics*, Vol. 21, pp. 583–599.

Kelly, J. (1982), *Scientific Management, Job Redesign, and Work Performance*, Academic Press, London.

Kiggundu, M. N. (1983), "Task Interdependence and Job Design: Test of a Theory," *Organizational Behavior and Human Performance*, Vol. 31, pp. 145–172.

Konz, S. (1983), *Work Design: Industrial Ergonomics*, 2nd ed., Grid Publishing, Columbus, OH.

Lawler, E. E. (1986), *High-Involvement Management: Participative Strategies for Improving Organizational Performance*, Jossey-Bass, San Francisco.

Leventhal, G. S. (1976), "The Distribution of Rewards and Resources in Teams and Organizations," in *Advances in Experimental Social Psychology*, Vol. 9, L. Berkowitz and E. Walster, Eds., Academic Press, New York.

Levine, E. L., Ash, R. A., Hall, H., and Sistrunk, F. (1983), "Evaluation of Job Analysis Methods by Experienced Job Analysts," *Academy of Management Journal*, Vol. 26, pp. 339–348.

Locke, E. A., and Latham, G. P. (1990), *A Theory of Goal Setting and Task Performance*, Prentice-Hall, Englewood Cliffs, NJ.

Majchrzak, A. (1988), *The Human Side of Factory Automation*, Jossey-Bass, San Francisco.

Manz, C. C., and Sims, H. P. (1987), "Leading Workers to Lead Themselves: The External Leadership of Self-Managing Work Teams," *Administrative Science Quarterly*, Vol. 32, pp. 106–129.

Mayo, E. (1933), *The Human Problems of an Industrial Civilization*, Macmillan, New York.

McCormick, E. J., Jeanneret, P. R., and Mecham, R. C. (1972), "A Study of Job Characteristics and Job Dimensions as Based on the Position Analysis Questionnaire (PAQ)," *Journal of Applied Psychology*, Vol. 56, pp. 347–368.

McGrath, J. E. (1984), *Teams: Interaction and Performance*, Prentice-Hall, Englewood Cliffs, NJ.

Meister, D. (1971), *Human Factors: Theory and Practice*, Wiley, New York.

Milkovich, G. T., and Newman, J. M. (1993), *Compensation*, 4th ed., Business Publications, Homewood, IL.

Mintzberg, H. (1979), *The Structuring of Organizations: A Synthesis of the Research*, Prentice-Hall, Englewood Cliffs, NJ.

Morgeson, F. P. (2005), "The External Leadership of Self-Managing Teams: Intervening in the Context of Novel and Disruptive Events," *Journal of Applied Psychology*, Vol. 90, pp. 497–508.

Morgeson, F. P., and Campion, M. A. (1997), "Social and Cognitive Sources of Potential Inaccuracy in Job Analysis," *Journal of Applied Psychology*, Vol. 82, pp. 627–655.

Morgeson, F. P., and Campion, M. A. (2000), "Accuracy in Job Analysis: Toward an Inference-Based Model," *Journal of Organizational Behavior*, Vol. 21, pp. 819–827.

Morgeson, F. P., and Campion, M. A. (2002), "Minimizing Tradeoffs When Redesigning Work: Evidence from a Longitudinal Quasi-experiment," *Personnel Psychology*, Vol. 55, pp. 589–612.

Morgeson, F. P., and Campion, M. A. (2003), "Work Design," in *Handbook of Psychology: Industrial and Organizational Psychology*, Vol. 12, W. C. Borman, D. R. Ilgen, and R. J. Klimoski, Eds., Wiley, Hoboken, NJ, pp. 423–452.

Morgeson, F. P., Aiman-Smith, L. D., and Campion, M. A. (1997), "Implementing Work Teams: Recommendations from Organizational Behavior and Development Theories," In *Advances in Interdisciplinary Studies of Work Teams: Issues in the Implementation of Work Teams*, Vol. 4, M. Beyerlein, D. Johnson, and S. Beyerlein, Eds., JAI Press, Greenwich, CT, pp. 1–44.

Morgeson, F. P., Delaney-Klinger, K. A., Mayfield, M. S., Ferrara, P., and Campion, M. A. (2004), Self-Presentation Processes in Job Analysis: A Field Experiment Investigating Inflation in Abilities, Tasks, and Competencies," *Journal of Applied Psychology*, Vol. 89, pp. 674–686.

Mundel, M. E. (1985), *Motion and Time Study: Improving Productivity*, 6th ed., Prentice-Hall, Englewood Cliffs, NJ.

Niebel, B. W. (1988), *Motion and Time Study*, 8th ed., Richard D. Irwin, Homewood, IL.

O'Reilly, C. A., and Roberts, K. H. (1977), "Task Group Structure, Communication, and Effectiveness," *Journal of Applied Psychology*, Vol. 62, pp. 674–681.

O'Reilly, C., Parlette, G., and Bloom, J. (1980), "Perceptual Measures of Task Characteristics: The Biasing Effects of Differing Frames of Reference and Job Attitudes," *Academy of Management Journal*, Vol. 23, pp. 118–131.

Parker, S. K. (2003), "Longitudinal Effects of Lean Production on Employee Outcomes and the Mediating Role of Work Characteristics," *Journal of Applied Psychology*, Vol. 88, pp. 620–634.

Pasmore, W. A. (1988), *Designing Effective Organizations: The Sociotechnical Systems Perspective*, Wiley, New York.

Pasmore, W., Francis, C., and Haldeman, J. (1982), "Sociotechnical Systems: A North American Reflection on Empirical Studies of the Seventies," *Human Relations*, Vol. 35, pp. 1179–1204.

Pearson, R. G. (1971), "Human Factors Engineering," in *Industrial Engineering Handbook*, 3rd ed., H. B. Maynard, Ed., McGraw-Hill, New York.

Perkins, A. L., and Abramis, D. J. (1990), "Midwest Federal Correctional Institution," in *Groups That Work (and Those That Don't)*, J. R. Hackman, Ed., Jossey-Bass, San Francisco.

Peterson, N. G., Mumford, M. D., Borman, W. C., Jeanneret, P. R., Fleishman, E. A., Campion, M. A., Levin, K. Y., Mayfield, M. S., Morgeson, F. P., Pearlman, K., Gowing, M. K., Lancaster, A., and Dye, D. (2001), "Understanding Work Using the Occupational Information Network (O*NET): Implications for Practice and Research," *Personnel Psychology*, Vol. 54, pp. 451–492.

Porter, L. W., Lawler, E. E., and Hackman, J. R. (1987), "Ways Teams Influence Individual Work Effectiveness," in *Motivation and Work Behavior*, 4th ed., R. M. Steers and L. W. Porter, Eds., McGraw-Hill, New York.

Posthuma, R. A., Morgeson, F. P., and Campion, M. A. (2002), "Beyond Employment Interview Validity: A Comprehensive Narrative Review of Recent Research and Trends over Time," *Personnel Psychology*, Vol. 55, pp. 1–81.

Rousseau, D. M. (1977), "Technological Differences in Job Characteristics, Employee Satisfaction, and Motivation: A Synthesis of Job Design Research and Sociotechnical Systems Theory," *Organizational Behavior and Human Performance*, Vol. 19, pp. 18–42.

Salas, E., Dickinson, T. L., Converse, S. A., and Tannenbaum, S. I. (1992), "Toward an Understanding of Team Performance and Training," in *Teams: Their Training and Performance*, R. W. Swezey and E. Salas, Eds., Ablex, Norwood, NJ.

Salvendy, G., Ed. (1987), *Handbook of Human Factors*, Wiley, New York.

Salvendy, G., and Smith, M. J., Eds. (1981), *Machine Pacing and Occupational Stress*, Taylor and Francis, London.

Sanders, M. S., and McCormick, E. J. (1987), *Human Factors in Engineering and Design*, 6th ed., McGraw-Hill, New York.

Shaw, M. E. (1983), "Team Composition," in *Small Teams and Social Interaction*, Vol. 1, H. H. Blumberg, A. P. Hare, V. Kent, and M. Davies, Eds., Wiley, New York.

Shea, G. P., and Guzzo, R. A. (1987), "Teams as Human Resources," in *Research in Personnel and Human Resources*, Vol. 5, K. M. Rowland and G. R. Ferris, Eds., JAI Press, Greenwich, CT.

Sims, H. P., Szilagyi, A. D., and Keller, R. T. (1976), "The Measurement of Job Characteristics," *Academy of Management Journal*, Vol. 19, pp. 195–212.

Smith, A. (1776), *An Inquiry into the Nature and Causes of the Wealth of Nations*, reprinted by R. H. Campbell and A. S. Skinner, Eds., Liberty Classics, Indianapolis, IN.

Steers, R. M., and Mowday, R. T. (1977), "The Motivational Properties of Tasks," *Academy of Management Review*, Vol. 2, pp. 645–658.

Steiner, I. D. (1972), *Group Process and Productivity*, Academic Press, New York.

Stevens, M. J., and Campion, M. A. (1994a), "Staffing Teams: Development and Validation of the Teamwork-KSA Test," paper presented at the annual meeting of the Society of Industrial and Organizational Psychology, Nashville, TN.

Stevens, M. J., and Campion, M. A. (1994b), "The Knowledge, Skill, and Ability Requirements for Teamwork: Implications for Human Resource Management," *Journal of Management*, Vol. 20, pp. 503–530.

Stevens, M. J., and Campion, M. A. (1999), "Staffing Work Teams: Development and Validation of a Selection Test for Teamwork Settings," *Journal of Management*, Vol. 25, pp. 207–228.

Sundstrom, E., DeMeuse, K. P., and Futrell, D. (1990), "Work Teams: Applications and Effectiveness," *American Psychologist*, Vol. 45, pp. 120–133.

Swezey, R. W., and Salas, E. (1992), *Teams: Their Training and Performance*, Ablex, Norwood, NJ.

Tannenbaum, S. I., Beard, R. L., and Salas, E. (1992), "Team Building and Its Influence on Team Effectiveness: An Examination of Conceptual and Empirical Developments," in *Issues, Theory, and Research in Industrial and Organizational Psychology*, K. Kelley, Ed., Elsevier, Amsterdam.

Taylor, F. W. (1911), *The Principles of Scientific Management*, W.W. Norton, New York.

Taylor, J. C. (1979), "Job Design Criteria Twenty Years Later," in *Design of Jobs*, 2nd ed., L. E. Davis and J. C. Taylor, Eds., Wiley, New York.

Thompson, J. D. (1967), *Organizations in Action*, McGraw-Hill, New York.

Tichauer, E. R. (1978), *The Biomechanical Basis of Ergonomics: Anatomy Applied to the Design of Work Situations*, Wiley, New York.

Turnage, J. J. (1990), "The Challenge of New Workplace Technology for Psychology," *American Psychologist*, Vol. 45, pp. 171–178.

Turner, A. N., and Lawrence, P. R. (1965), *Industrial Jobs and the Worker: An Investigation of Response to Task Attributes*, Harvard Graduate School of Business Administration, Boston.

U.S. Department of Labor (1972), *Handbook for Analyzing Jobs*, U.S. Government Printing Office, Washington, DC.

U.S. Department of Labor (1977), *Dictionary of Occupational Titles*, 4th ed., U.S. Government Printing Office, Washington, DC.

U.S. Nuclear Regulatory Commission (1981), *Guidelines for Control Room Design Reviews*, NUREG 0700, Nuclear Regulatory Commission, Washington, DC.

Vroom, V. H. (1964), *Work and Motivation*, Wiley, New York.

Walker, C. R., and Guest, R. H. (1952), *The Man on the Assembly Line*, Harvard University Press, Cambridge, MA.

Wall, T. B., Kemp, N. J., Jackson, P. R., and Clegg, C. W. (1986), "Outcomes of Autonomous Workgroups: A Long-Term Field Experiment," *Academy of Management Journal*, Vol. 29, pp. 280–304.

Warr, P., and Wall, T. (1975), *Work and Well-Being*, Penguin Books, Baltimore.

Weber, A., Fussler, C., O'Hanlon, J. F., Gierer, R., and Grandjean, E. (1980), "Psychophysiological Effects of Repetitive Tasks," *Ergonomics*, Vol. 23, pp. 1033–1046.

Welford, A. T. (1976), *Skilled Performance: Perceptual and Motor Skills*, Scott, Foresman, Glenview, IL.

Wicker, A., Kirmeyer, S. L., Hanson, L., and Alexander, D. (1976), "Effects of Manning Levels on Subjective Experiences, Performance, and Verbal Interaction in Groups," *Organizational Behavior and Human Performance*, Vol. 17, pp. 251–274.

Wilson, J. R., and Haines, H. M. (1997), "Participatory Ergonomics," in *Handbook of Human Factors and Ergonomics*, 2nd ed., G. Salvendy, Ed., Wiley, New York.

Wong, C. S. (1989), "Task Interdependence: The Link Between Task Design and Job Design," Ph.D. dissertation, Purdue University, West Lafayette, IN.

Wong, C. S., and Campion, M. A. (1991), "Development and Test of a Task Level Model of Job Design," *Journal of Applied Psychology*, Vol. 76, pp. 825–837.

Woodman, R. W., and Sherwood, J. J. (1980), "The Role of Team Development in Organizational Effectiveness: A Critical Review," *Psychological Bulletin*, Vol. 88, pp. 166–186.

Zajonc, R. B. (1965), "Social Facilitation," *Science*, Vol. 149, pp. 269–274.

Zander, A. (1979), "The Study of Group Behavior over Four Decades," *Journal of Applied Behavioral Science*, Vol. 15, pp. 272–282.

CHAPTER 17

PERSONNEL SELECTION

Jerry W. Hedge
Organizational Solutions Group
Holly Hill, South Carolina

Walter C. Borman
University of South Florida and
Personnel Decisions Research Institutes, Inc.
Tampa, Florida

1 INTRODUCTION

The world of work is in the midst of profound changes that will require new thinking about personnel selection. Workers will probably need to be more versatile, handle a wider variety of diverse and complex tasks, and have more sophisticated technological knowledge and skills. The aim of this chapter is to examine the state of the science concerning personnel selection, with an emphasis on recent developments that seem especially relevant. Accordingly, we discuss work performance predictors that tap such domains as ability, personality, vocational interests, and biodata. We also expand this individual-level perspective to encompass broader organizational issues related to person–job match and person–organization fit. Before we look more closely at the predictor domain, some discussion of workforce and occupational trends should provide a useful context.

2 WORKFORCE TRENDS

Major changes have taken place in the workplace over the last several decades and continue today. The globalization of numerous companies and industries; organizational downsizing, right-sizing, and restructuring; expansion of information technology use at work; changes in work contracts; and increased use of alternative work strategies and schedules have transformed the nature of work in many organizations. The workforce itself is also changing, with a growing number of older workers, females, and dual-career couples.

Some researchers and workplace futurists have speculated that the world of work is in the midst of such fundamental change that there may be no stable jobs in the future. Instead, work activities may be organized around projects and initiatives, with flexible task forces to deal with organizational requirements (e.g., Bridges, 1994). In such an environment, adequate job performance will require increased flexibility and adaptability in order to deal effectively with these less well-defined roles. Regardless of whether one supports this perspective, the trend toward more flexible organizational structures and more adaptable organization members seems inevitable.

At the very least, these workforce changes are likely to result in different occupational and organizational structures in the future. Technological modernization will require workers to adjust to new equipment and procedures, adapt to ever-changing environments, and continue to enhance their job-related skills. Technology changes the nature of work, and as Czaja (2001) has suggested, it will have a major impact on the future structure of the labor force, transforming the jobs that are available, and how they are performed. Indeed, according to the U.S. Department of Labor, Bureau of Labor Statistics (2003), almost two-thirds of the projected job openings in the next 10 years will require some on-the-job training.

Changing technology is almost certain to change the structure of the labor force in the future. For example, Czaja and Moen (2004) noted recently that in 2001 more than half of the labor force used a computer at work. This number is expected to increase as developments in technology continue. In fact, computer occupations such as computer software engineers, computer support specialists, and network and computer systems administrators will account for eight of the 20 fastest-growing jobs; and the use of computers and other forms of technology is becoming more prevalent in other occupations. Not surprisingly, most workers will need to interact with some type of technology to perform their jobs.

Given the spreading use of technology in most occupations, it will also be important to understand how this will affect employment opportunities for workers. Technology will create new jobs and opportunities for employment for some, and eliminate jobs and create conditions of unemployment for other workers. It will also change the ways in which jobs are performed, and alter job content and job demands.

With such changes becoming a more routine part of the work environment, workers will need to upgrade their knowledge, skills, and abilities to avoid obsolescence, probably learning new systems and new activities at multiple points during their working life. Because organizations increasingly operate in a wide and varied set of situations, cultures, and environments, not only will workers probably need to be more versatile and able to handle a wider variety of diverse and complex tasks, but employers will need to deal with an increasingly diverse workplace.

3 OCCUPATIONAL TRENDS

Because the Census Bureau and the Bureau of Labor Statistics routinely collect data on population and workforce trends, including occupational areas of growth and decline, it is relatively easy to project, with some semblance of accuracy, future employment patterns. For example, Hecker (2001) examined data collected for calendar year 2000 across 10 primary occupational groups: management/business/financial; professional and related occupations; services; sales and related occupations; office/administrative support; farming/fishing/forestry; construction/extraction; installation/maintenance/repair; production; transportation/material moving occupations. He noted that among these occupational groups, two in particular, professional occupations and services occupations, are expected to grow the fastest and add the most jobs between 2000 and 2010. Together, these two occupational groups should provide more than half of the total job growth in the economy for the first decade of the twentieth-first century. On the other end of the continuum, the three slowest-growing groups are expected to be office and administrative support occupations; production occupations; and farming, fishing, and forestry occupations.

Within the professional and related occupations group, which expects to add almost 7 million workers, nearly three-fourths of the projected growth is expected to occur within three subgroups: computer and mathematical occupations; health care practitioners and technical occupations; and education, training, and library occupations. Computer and mathematical occupations (e.g., computer programmer, systems analyst, database administrator) are projected to add roughly 2 million employees and to grow most rapidly among the eight professional and related occupations subgroups. This demand for computer-related occupations should continue for the foreseeable future. Health care practitioners and technical occupations are projected to add another 1.6 million jobs, as the demand for health care services continues to grow rapidly as well.

During the 2000–2010 time frame, employment in services occupations is expected to grow by over 5 million jobs. Of the subgroups making up the services occupations, food preparation and serving subgroup occupations was largest in 2000, and is also projected to add the most jobs by 2010. Health care support occupations (e.g., medical assistant, nursing aid) are expected to add over 1 million jobs; protective services (e.g., security guard, law enforcement worker) are also projected to grow rapidly.

3.1 Changes in the Structure of Work

The relative increases and decreases associated with the broad occupational categories used by the BLS provide a useful picture of the changing occupational landscape. BLS data clearly show the rise of professional, technical, and what has traditionally been called white-collar (managerial/administrative, marketing/sales) workers, as well as the decline of farm and blue-collar workers. As has been suggested by a recent National Academy of Sciences report, *The Changing Nature of Work* (Committee on Techniques for the Enhancement of Human Performance, 2000), the nature of work is changing not only within these categories but also in ways that tend to blur the traditional distinctions among them.

For example, this committee suggested that blue-collar *production work* in many organizations is expanding to include more decision-making tasks than traditionally would have been part of a supervisory/managerial job. In addition, for some production workers, relatively narrow parameters of the job are giving way to broader involvement in work teams as well as interactions with external customers, clients, and patients. As team-based work structures have been used more widely, a number of studies have suggested that both cognitive and interactive skills are becoming more important in blue-collar jobs.

Technology is also having a significant impact on blue-collar jobs: for example, in some situations, replacing physical activity with mental and more abstract forms of responding. Generally, then, the implication of these changes is that information technology changes the mix of skills that are required, often creating jobs that require less sensory and physical skill and more "intellective" skills, such as abstract reasoning, inference, and cause–effect analysis (Committee on Techniques for the Enhancement of Human Performance, 2000).

To summarize the trends in blue-collar employment, then, four developments seem relatively widespread. Compared with the past, an increasing number of blue-collar jobs appear to (1) offer workers more autonomy and control over their work activities, (2) cover a wide range of tasks, (3) require more interpersonal skill, and (4) have become more analytic and cognitively complex. The adoption of lean production techniques, the acceptance of team-based work systems, and the growth of computer-integrated manufacturing technologies appear to be primary drivers of these changes in the content of blue-collar jobs.

The nature of most professional and technical work does not appear to be evolving in dramatic ways, even though technical advances can change completely what professional and technical workers need to know in a relatively short span of time. Professional and technical jobs continue to allow considerable autonomy and control over work processes, to demand high levels of interpersonal skills, and to offer cognitively complex challenges. Certainly, cognitive complexity is a central feature of professional and technical work, and increasingly, the use of cross-functional teams requires professional and technical workers to have the cognitive and interactive skills needed to communicate, negotiate, and solve problems across horizontal boundaries.

For managerial jobs, the report by the Committee (2000) suggested a serious dearth of research on the changing nature of managerial work. Nonetheless, two interesting developments seem evident. First, at least lower-level managers appear to be experiencing some loss in authority and control. Second, the need to communicate horizontally both within and across organizations may be becoming even more important than the supervision of an employee's work. There is also considerable discussion about the substantive content of managers' jobs, shifting toward the procurement and coordination of resources, toward coaching as opposed to commanding employees, and toward project management skills.

Within the service industry, the content of work is also evolving. First, a significant percentage of service jobs are probably becoming more routinized, in large measure because new information technologies enable greater centralization and control over work activities. Second, there is a tendency toward the blurring of sales and clerical jobs. Although the heterogeneity of work within specific service occupations appears to be increasing, this heterogeneity reflects, at least in part, the tendency to structure work differently according to market segments.

Interestingly, studies suggest a trend toward overall increase in technical skill requirements and cognitive complexity of service jobs. Although the initial impact of information technology involved a shift from manual to computer-mediated information processing, more recent applications involve the manipulation of a variety of software programs and databases. In addition, the rapid diffusion of access to the Internet has increased the potential for greater information-processing and cognitively complex activities. Finally, interpersonal interactions remain critical to service work, requiring skills in communications, problem solving, and negotiations.

4 NECESSARY COMPONENTS OF PERSONNEL SELECTION

Recognizing that fundamentally new ways of thinking and acting will be necessary to meet the changing nature of work and worker requirements, one critically important intervention—and the focus of this chapter—is personnel selection. Historically, entry-level selection has centered on identifying skills important for performance early in a career. However, because finding and training workers in the future will be much more complex and costly than it is today, success on the job during and beyond the first few years will be increasingly important. The prediction of such long-term success indicators as retention and long-term performance will require the use of more complex sets of predictor variables that include such measures as personality, motivation, and vocational interest. To develop effective measures to predict long-term performance, it will be crucial to better understand the context of the workplace of the future, including the environmental, social, and group structural characteristics. Ultimately, combining the personal and organizational characteristics should lead to improved personnel selection models that go beyond the usual person–job relations, encouraging a closer look at theories of person–organization (P-O) fit.

Consequently, what follows is a detailed examination of the state of the science relevant to the changing nature of jobs, and what this means for the future of workplace selection, with sections on recent research on job analysis and the job performance domain; predictor measurement, in particular, ability, personality, vocational interests, and biodata; and an alternative model to the traditional person–job fit selection strategy: namely, person–organization fit.

5 JOB ANALYSIS AND THE JOB PERFORMANCE DOMAIN

The role of personnel selection is to identify persons most likely to succeed in particular jobs, with the overall purpose of enhancing organizational effectiveness. There are several basic steps involved in matching people and jobs. The proper first step is to identify the critical performance requirements of the job and the knowledge, skills, abilities, and other characteristics (KSAOs) that might be important to effective performance on the job. In turn, these KSAOs suggest the kinds of selection measures to be targeted for use. The final step is to validate these predictor measures against job performance.

5.1 Job Analysis

Job analysis identifies the critical performance requirements of the job and the KSAOs important to effective performance on the job; thus, it tells us what we should be looking for in a job candidate. Job analysis is a common activity, with a well-defined methodology for conducting such analyses. As noted by Cascio (1995), terms such as *job element, task, job description*, and *job family* are well understood. Although job analysis continues to be an important first step in selection research and practice, the changing nature of work may suggest some movement in the future from a focus on discrete job components to a more process-oriented approach.

A recent major development in job analysis is a Department of Labor initiative to analyze virtually all jobs in the U.S. economy in order to build a database of occupational information (O*NET). This database may be used by organizations and individuals to help match people with jobs. The person–job fit feature of the O*NET enable comparisons between personal attributes and targeted occupational requirements. There is also an organizational-characteristics component that facilitates P-O matches. The hope is that O*NET will help unemployed workers and students entering the workforce to find more appropriate jobs and careers, and employers to identify more highly qualified employees. These matches should be realized more systematically and with more precision than has been possible heretofore. An additional hope is that this initiative will encourage research that further advances the effectiveness of person–job matching, person–organization fit, and the science of personnel selection (Borman et al., 1997).

5.2 Performance Criteria

A central construct of concern in work psychology is job performance, because performance criteria are often what we attempt to predict from our major interventions, including personnel selection, training, and job design. Traditionally, while most attention has focused on models related to predictors (e.g., models of cognitive ability, personality, and vocational interests), job performance models and research associated with them are beginning to foster more scientific understanding of criteria.

For example, Hunter (1983), using a path analytic approach, found that cognitive ability has primarily a direct effect on individuals' acquisition of job knowledge. Job knowledge, in turn, influenced technical proficiency. Supervisory performance ratings were a function of both job knowledge and technical proficiency, with the job knowledge-ratings path coefficient three times as large as the technical proficiency-ratings coefficient. This line of research continued, with additional variables being added to the models. Schmidt et al. (1986) added job experience to the mix; they found that job experience had a direct effect on the acquisition of job knowledge and an indirect effect on task proficiency through job knowledge.

Later, Borman et al. (1991) included two personality variables, achievement and dependability, and behavioral indicators of achievement and dependability. The path model results showed that the personality variables had indirect effects on the supervisory performance ratings through their respective behavioral indicators. The best-fitting model also had paths from ability to acquisition of job knowledge, job knowledge to technical proficiency, and technical proficiency to the supervisory job performance ratings, arguably the most comprehensive measure of overall performance. Perhaps the most important result of this study was that the variance accounted for in the performance rating exogenous (dependent) variable increased substantially with the addition of personality and the behavioral indicators of personality beyond that found with previous models, including ability along with job knowledge and technical proficiency.

5.2.1 Task and Contextual Performance

Another useful way to divide the job performance domain has been according to task and contextual performance. Borman and Motowidlo (1993) argued that organization members may engage in activities that are not directly related to their main task functions but nonetheless are important for organizational effectiveness because they support the "organizational, social, and psychological context that serves as the critical catalyst for task activities and processes" (p. 71). Borman and colleagues have settled on a three-dimensional system: (1) personal support, (2) organizational support, and (3) conscientious initiative (Coleman and Borman, 2000; Borman et al., 2001). The notion is to characterize the citizenship performance construct according to the recipient or target of the behavior: other persons, the organization, and oneself, respectively.

Additional research directions have emerged that use as a starting point these distinctions between criterion constructs. First, building on the task–citizenship performance distinction, research has examined especially the weights that supervisor raters place on rate task and citizenship performance when making overall performance or similar types of global worth-to-the-organization judgments. This research, using various types of methodologies, has consistently found that task and citizenship performance are weighted roughly the same when raters make these overall effectiveness judgments. Podsakoff et al. (2000) have summarized these studies. Across eight studies they found that an average of 9.3% of the variance in overall performance ratings was accounted for by task performance; for citizenship performance, the average was 12.0%. Further, Conway (1999) conducted a meta-analysis targeted toward the same issue and found that collapsing across supervisor and peer raters, the respective percentage of variance accounted for in overall performance ratings by task and citizenship performance was 11.5% and 12.0%.

5.2.2 Summary

Thus, criteria have always been important in personnel selection research, but a recent trend has been to study job performance in its own right in an attempt to develop substantive models of performance. The vision has been to learn more about the nature of job performance, including its components and dimensions, so that performance itself, as well as predictor–performance links, will be better understood. Again, if we can make more progress in this direction, the cumulative evidence for individual predictor construct/performance construct relationships will continue to progress and the science of personnel selection will be enhanced substantially.

6 PREDICTOR MEASUREMENT

6.1 Ability

Abilities are relatively stable individual differences that are related to performance on some set of tasks, problems, or other goal-oriented activities (Murphy, 1996). Another definition, offered by Carroll (1993), conceptualizes abilities as relatively enduring attributes of a person's capability for performing a particular range of tasks. Although the term *ability* is widely used in both the academic and applied literature, several other terms have been related loosely to abilities. For example, the term *competency* has been used to describe individual attributes associated with the quality of work performance. In practice, lists of competencies often include a mixture of knowledges, skills, abilities, motivation, beliefs, values, and interests (Fleishman et al., 1999).

Another term that is often confused in the literature with abilities is *skills*. Whereas abilities are general traits inferred from relationships among performances of individuals observed across a range of tasks, skills are more dependent on learning and represent the product of training on particular tasks. In general, skills are more situational, but the development of a skill is, to a large extent, predicted by a person's possession of relevant underlying abilities, usually mediated by the acquisition of the requisite knowledge. That is, these underlying abilities are related to the rate of acquisition and final levels of performance that a person can achieve in particular skill areas (Fleishman et al., 1999).

Ability tests usually measure mental or cognitive ability but may also measure other constructs, such as physical abilities. Ability tests have almost always been paper-and-pencil tests administered to applicants in a standardized manner, although recent advances in computerized testing have led to more ability tests being administered by computer (e.g., Drasgow and Olson-Buchanon, 1999). In the following sections we discuss recent developments and the most current topics in the areas of cognitive ability, tacit knowledge or practical intelligence, and physical ability testing.

6.1.1 Cognitive Ability

In the early twentieth century, Cattell, Scott, Bingham, Viteles, and other applied psychologists used cognitive ability tests to lay the foundation for the current practice of personnel selection (Landy, 1993). Recent attention has focused on the usefulness of general cognitive ability *g* vs. more specific cognitive abilities for predicting training and job performance, and the contributions of information-processing models of cognitive abilities for learning more about ability constructs. First, there has been a debate concerning the "ubiquitousness" of the role of general cognitive ability, or *g*, in the prediction of training and job performance. Several recent studies have demonstrated that psychometric *g*, generally operationalized as the common variance in a battery of cognitive ability tests (e.g., the first principal component), accounts for the majority of the predictive power in the test battery, and that the remaining variance (often referred to in this research as "specific abilities") accounts for little or no additional variance in the criterion (e.g., Ree et al., 1994; Larson and Wolfe, 1995).

Other researchers have expressed concern related to the statistical model often used to define *g*. A general factor or *g* represents the correlations between specific ability tests, so specific abilities will, by definition, be correlated with the general factor. Thus, it could be argued that it is just as valid to enter specific abilities *first* and then say that *g* does not contribute beyond the prediction found with specific abilities alone (e.g., Murphy, 1996). In fact, Muchinsky (1993) found this to be the case for a sample of manufacturing jobs, where mechanical ability was the single best predictor of performance, and an intelligence test had no incremental validity beyond the mechanical test alone.

We know very little about specific abilities when they are defined as the variance remaining once a general factor is extracted statistically. Interestingly, what little information is available suggests that these specific-ability components tend to be most strongly related to cognitive ability tests that have a large knowledge component (e.g., aviation information) (Olea and Ree, 1994). This is consistent with previous research showing that job knowledge tests tend to be slightly more valid than ability tests (Hunter and Hunter, 1984), and also with research demonstrating that job knowledge appears to mediate the relationship between abilities and job performance (e.g., Borman et al., 1993). Meta-analysis has demonstrated the generality of job knowledge tests as predictors of job performance (Dye et al., 1993). In addition, these authors found that the validity of job knowledge tests was moderated by job complexity and by job-test similarity, with validities significantly higher for studies involving high-complexity jobs and those with high job–test similarity.

6.1.2 Tacit or Practical Intelligence

Sternberg and colleagues have attempted to broaden the discussion of general intelligence (Sternberg and Wagner, 1992). Based on a triarchic theory of intelligence, Sternberg (1985) suggested that practical intelligence and tacit knowledge play a role in job success. Practical intelligence is often described as the ability to respond effectively to practical problems or demands in situations that people commonly encounter in their jobs (Wagner and Sternberg, 1985; Sternberg et al., 1993). Conceptually, practical intelligence is distinct from cognitive ability. In fact, some research (Chan, 2001) has found measures of practical intelligence to be uncorrelated with traditional measures of cognitive ability.

Sternberg and his colleagues have repeatedly found significant correlations and some incremental validity (over general intelligence) for measures of tacit knowledge in predicting job performance or success (Sternberg et al., 1995). Tacit knowledge has been shown to be trainable and to differ in level according to relevant expertise. Certainly, tacit knowledge measures deal with content that is quite different from that found in traditional job knowledge tests (e.g., knowledge related to managing oneself and others).

6.1.3 Physical Abilities

Another important area for selection into many jobs that require manual labor or other physical demands is the use of physical ability tests. Most physical ability tests are performance tests (i.e., not paper and pencil) that involve demonstration of attributes such as strength, cardiovascular fitness, or coordination. Although physical ability tests are reported to be used widely for selection (J. Hogan and Quigley, 1994), not much new information has been published in this area in the past few years. In one study, Blakley et al. (1994) provided evidence that isometric strength tests are valid predictors across a variety of different physically demanding jobs, and that females scored substantially lower than males on these isometric strength tests. In light of these findings, there is a recent and growing interest in reducing adverse impact through pretest preparation. Hogan and Quigley (1994) demonstrated that participation in a physical training program can improve females' upper body strength and muscular endurance, and that participation in a pretest physical training program was significantly related to the likelihood of passing a firefighter physical ability test.

6.1.4 Summary

A cumulation of 85 years of research demonstrates that if we want to hire people without previous experience in a job, the most valid predictor of future performance is general cognitive ability (Schmidt and Hunter, 1998). General cognitive ability measures have many advantages in personnel selection: (1) they show the highest validity for predicting training and job performance, (2) they may be used for all jobs from entry level to advanced, and (3) they are relatively inexpensive to administer. In addition, there is some evidence that measures of "practical intelligence" or tacit knowledge may under certain conditions provide incremental validity beyond general cognitive ability for predicting job performance (Sternberg et al., 1995). Finally, physical ability tests may be useful in predicting performance for jobs that are physically demanding.

6.2 Personality

Interest in personality stems from the desire to predict the motivational aspects of work behavior. Nevertheless, until recently, the prevalent view was that personality variables were a dead end for predicting job performance. Some of the factors fueling this belief were (1) the view that a person's behavior is not consistent across situations and thus that traits do not exist, (2) literature reviews concluding that personality variables lack predictive validity in selection contexts, and (3) concern about dishonest responding on personality inventories.

However, by the late 1980s, favorable opinions about personality regarding personnel selection began to grow (R. Hogan, 1991). Evidence accumulated to refute the notion that traits are not real (Kenrick and Funder, 1988) or stable (Conley, 1984). Research showed at least modest validity for some personality traits in predicting job performance (e.g., Barrick and Mount, 1991; McHenry et al., 1990; Ones et al., 1993). Further, evidence mounts that personality measures produce small, if any, differences between majority and protected classes of people (Ones and Viswesvaran, 1998b) and that response distortion does not necessarily destroy criterion-related validity (e.g., Ones et al., 1996b; Ones and Viswesvaran, 1998a).

Today's well-known, hierarchical, five-factor model (FFM) of personality (alternatively, "the Big Five") was first documented in 1961 by Tupes and Christal (see Tupes and Christal, 1992). The five factors were labeled *surgency*, *agreeableness*, *dependability*, *emotional stability*, and *culture*. Following Tupes and Christal, McCrae and Costa (1987) replicated a similar model of the FFM. In their version, *extraversion* is comprised of traits such as talkative, assertive, and active; *conscientiousness* includes traits such as organized, thorough, and reliable; *agreeableness* includes the traits kind, trusting, and warm; *neuroticism* includes traits such as nervous, moody, and temperamental; and *openness* incorporates such traits as imaginative, curious, and creative (Goldberg, 1992).

A large amount of evidence supports the generalizability and robustness of the Big Five. Others argue that the theoretical value and the practical usefulness of the Big Five factors are severely limited by their breadth, *and* that important variables are missing from the model. From an applied perspective, important questions pertain to the criterion-related validity of personality variables and the extent to which it matters whether we focus on broad factors or narrower facets.

Although the Big Five model of personality is not universally accepted, considerable research has been conducted using this framework. For example, Barrick et al. (2001) conducted a second-order meta-analysis including 11 meta-analyses of the relationship between Big Five personality dimensions and job performance. They included six performance criteria and five occupational groups. Conscientiousness was a valid predictor across all criteria and occupations

($\rho = 0.19$ to 0.26),* with the highest overall validity of the Big Five dimensions. Emotional stability was predictive for four criteria and two occupational groups. Extraversion, agreeableness, and openness did not predict overall job performance, but each was predictive for some criteria and some occupations. Barrick et al.'s results echoed conclusions of previous research on this topic (e.g., Barrick and Mount, 1991; Hurtz and Donovan, 2000).

Despite the low to moderate magnitude of the Big Five's predictive validity, optimism regarding personality's usefulness in selection contexts remains high. One reason is that even a modestly predictive variable, if uncorrelated with other predictors, offers incremental validity. Personality variables tend to be unrelated to tests of general cognitive ability (g) and thus they have incremental validity over g alone (Day and Silverman, 1989; J. Hogan and Hogan, 1989; McHenry et al., 1990; Schmitt et al., 1997). For example, conscientiousness produces gains in validity of 11 to 18% compared to using g alone (Salgado, 1998; Schmidt and Hunter, 1998). Salgado (1998) reported that emotional stability measures produced a 10% increment in validity over g for European civilian and military samples combined. For military samples alone, the incremental validity was 38%.

6.2.1 Compound Traits: Integrity, Adaptability, and Core Self-Evaluation

In contrast to considering links between narrow personality traits and individual job performance criteria, a very different approach for personality applications in personnel selection is the development of *compound traits*. Compound traits have the potential to show even stronger relationships with criteria because they are often constructed by identifying the criterion first and then selecting a heterogeneous group of variables expected to predict it (J. Hogan and Ones, 1997). Integrity, adaptability, and core self-evaluation are three compound traits that may be especially useful in a selection context.

Integrity consists of facets from all Big Five factors: mainly conscientiousness, agreeableness, and emotional stability (Ones et al., 1994). Integrity tests have several advantages as part of selection systems, including considerable appeal to employers. Because of the enormous costs of employee theft and other counterproductive behaviors (U.S. Department of Health and Human Services, 1997; Durhart, 2001), it is understandable that employers want to avoid hiring dishonest applicants. Also, paper-and-pencil integrity measures have grown in popularity since the 1988 Federal Polygraph Protection Act banned most preemployment uses of the polygraph test. Empirical evidence shows that integrity tests are better than any of the Big Five at predicting job performance ratings ($\rho = 0.41$) (Ones et al., 1993). Under some conditions, integrity tests also predict various counterproductive behaviors at work. Integrity tests are uncorrelated with tests of g (J. Hogan and Hogan, 1989; Ones et al., 1993) and produce a 27% increase in predictive validity over g alone (Schmidt and Hunter, 1998).

Adaptive job performance has become increasingly important in today's workplace. Existing personality scales carrying the adaptability label tend to be narrowly focused on a particular aspect of adaptability (e.g., International Personality Item Pool, 2001) and would probably not be sufficient to predict the multiple dimensions of adaptive job performance. One would expect that someone who readily adapts on the job is likely to be patient, even-tempered, and confident (facets of emotional stability); open to new ideas, values, and experiences (facets of openness); and determined to do what it takes to achieve goals (a facet of conscientiousness). It is possible that high levels of other aspects of conscientiousness, such as dutifulness and orderliness, are detrimental to adaptive performance because they involve overcommitment to established ways of functioning. Again, although speculative, these relationships are in line with the results of several different studies that linked personality variables to adaptive performance (e.g., Mumford et al., 1993; Le Pine et al., 2000).

Core self-evaluation (CSE) is a fundamental, global appraisal of oneself (Judge et al., 1998). We should note that whereas some have called CSE a compound trait, Judge and his colleagues might not agree with this characterization. Erez and Judge (2001) described CSE as a single higher-order factor explaining the association among four more specific traits: self-esteem, generalized self-efficacy, internal locus of control, and emotional stability. A meta-analysis showed that CSE's constituent traits are impressively predictive of overall job performance (Judge and Bono, 2001). Estimated ρ values, corrected for sampling error and criterion unreliability, ranged from 0.19 (emotional stability) to 0.26 (self-esteem).

CSE is even more strongly related to job satisfaction than it is to overall job performance. Predictive validity estimates ranged from 0.24 (emotional stability) to 0.45 (self-efficacy) (Judge and Bono, 2001). People with high CSE seem to seek out challenging jobs and apply a positive mindset to the perception of their jobs (Judge et al., 1998, 2000). The result is desirable job characteristics, both real and perceived, which contribute to high job satisfaction. The link between CSE and job satisfaction has important implications, because satisfied employees are more likely to stay on the job than are dissatisfied employees (Tett and Meyer, 1993; Harter et al., 2002). Low turnover rates are especially important in organizations that invest heavily in the training of new employees.

6.2.2 Summary

Over the past two decades, personality has enjoyed a well-deserved resurgence in research and applied use, aided in part by the wide acceptance of the Big

*Unless otherwise noted, ρ represents true operational validity, an estimate corrected for sampling error, range restriction, and criterion unreliability.

Five model of personality. These broad traits offer incremental predictive validity over cognitive ability alone (Salgado, 1998; Schmidt and Hunter, 1998). Using personality predictors narrower than the Big Five shows promise for revealing stronger criterion-related validities, especially when the criteria are relatively specific. Compound traits may be especially important in organizational environments that are more dynamic and team oriented (Edwards and Morrison, 1994). Complex, heterogeneous predictors are needed to predict the complex, heterogeneous performance criteria that are likely in such an environment.

Recent research has shown not only the criterion-related validity of personality, it has also addressed certain persistent objections to personality testing. Adverse impact appears not to be a problem. Response distortion, although certainly problematic in selection settings, may be alleviated with targeted interventions, although it still may be problematic with small selection ratios (i.e., a relatively low ratio of selectees) (Rosse et al., 1998). Also, considerable evidence exists to suggest that personality is impressively stable over the entire life span (Roberts and Del Vecchio, 2000). Thus, we can be confident that personality variables used in selection will predict performance of selectees throughout their tenure with an organization.

6.3 Vocational Interests

Needs, drives, values, and interests are closely related motivational concepts that refer to the intentions or goals of a person's actions. Interests are generally thought of as the most specific and least abstract of these concepts (R. Hogan and Blake, 1996). Hogan and Blake pointed out that vocational interests have not often been studied in relation to other motivational constructs. Holland and Hough (1976) suggested that a likely reason for this lack of attention to theoretical links with other constructs is the early empirical successes of vocational interest inventories, predicting outcomes such as vocational choice. In a sense, there was little reason to relate to the rest of psychology because of these successes. Accordingly, many psychologists have regarded the area of vocational interest measurement as theoretically and conceptually barren (Strong, 1943).

The most obvious link between vocational interests and relevant criteria appears to be between vocational interest responses and occupational tenure, and by extension, job satisfaction. For the most part, the tenure relation has been confirmed. For example, regarding occupational tenure, Strong (1943) demonstrated that occupational membership could be predicted by vocational interest scores on the Strong Vocational Interest Blank administered between 5 and 18 years previously. This finding at least implies that persons suited for an occupation on the basis of their interests tend to gravitate to and stay in that occupation.

For job satisfaction, the relationships are more mixed, but at best, the vocational interest/job satisfaction correlations are moderate (0.31). Vocational interests are not usually thought of in a personnel selection context, and in fact, there are not many studies linking vocational interests and job performance. Those that do exist find a median validity for the interest predictors against job performance around 0.20. Although this level of validity is not very high and the number of studies represented is not large, the correlation observed compares favorably to the validities of personality constructs (e.g., Barrick and Mount, 1991).

Analogous to personality measures, vocational interest inventories used for selection have serious potential problems with slanting of responses or faking. It has long been evident that people *can* fake interest inventories (e.g., Campbell, 1971). However, some research has shown that in an actual selection setting, applicants may not slant their responses very much (e.g., Abrahams et al., 1971).

Although the personality and vocational interest may be closely linked conceptually, it is evident that inventories measuring the two constructs are quite different. Personality inventories present items that presumably reflect the respondent's tendency to act in a certain way in a particular situation. Vocational interest items elicit like–dislike responses to objects or activities. There has been a fair amount of empirical research correlating personality and vocational interest responses. Hogan and Blake (1996) summarized the findings of several studies linking personality and vocational interests at the level of the Big 5 personality factors and the six Holland types. Although many of these correlations are significant, the magnitude of the relationships is not very large. Thus, it appears that personality constructs and vocational interest constructs have theoretically and conceptually reasonable and coherent relationships but that these linkages are relatively small. What does this mean for selection research? Overall, examining vocational interests separately as a predictor of job performance (*and* job satisfaction and attrition) seems warranted.

6.3.1 Summary

Vocational interest measures, similar to personality measures, tap motivation-related constructs. Interests have substantial conceptual similarity to personality, but empirical links between the two sets of constructs are modest. Vocational interest measures are most often used in counseling settings, and have been linked primarily to job satisfaction criteria. However, although not often used in a selection context, limited data suggest reasonable levels of validity. Accordingly, vocational interests may show some promise for predicting job performance.

6.4 Biodata

The primary principle behind the use of biodata is that the best predictor of future behavior is past behavior. In fact, biodata offer a number of advantages when used in personnel selection. Among the most significant is their power as a predictor across a number of work-related criteria. For example, in a recent meta-analytic review of over 85 years of personnel psychology research, Schmidt and Hunter (1998) reported mean biodata validity coefficients of 0.35 and 0.30 against job and training success, respectively. These findings support previous research reporting validities ranging from 0.30 to 0.40 between biodata and a range of criteria, such as turnover, absenteeism, job proficiency, and performance appraisal ratings (e.g., Asher, 1972; Reilly and Chao, 1982; Hunter and Hunter, 1984). Based on these meta-analytic results, researchers have concluded that biographical inventories have almost as high validities as cognitive ability tests (Reilly and Chao, 1982). In addition, research indicates that biodata show less adverse impact than that of cognitive ability tests (Wigdor and Garner, 1982). Importantly, the high predictability associated with biodata, the ease of administration of biodata instruments, the low cost, and the lack of adverse impact have led to the widespread use of biodata in both the public and private sectors (Farmer, 2001).

Mael (1991) reviewed certain ethical and legal concerns that have been raised about biodata. The first of these deals with the controllability of events. That is, there are actions that respondents choose to engage in (controllable events), whereas other events are either imposed upon them or happened to them (noncontrollable events). Despite the belief held by numerous biodata researchers that all events, whether or not controllable, have the potential to influence later behavior, some researchers (e.g., Stricker, 1987, 1988) argue that it is unethical to evaluate individuals based on events that are out of their control (e.g., parental behavior, socioeconomic status). As a result, some have either deleted all noncontrollable items from their biodata scales, or have even created new measures with the exclusion of these items. A frequent consequence, however, is that using only controllable items reduces the validity of the biodata instrument (Mael, 1991).

Two other ethical and legal concerns that have been raised include equal accessibility and invasion of privacy. That is, some researchers (e.g., Stricker 1987, 1988) argue that items dealing with events not equally accessible to all individuals (e.g., playing varsity football) are inherently unfair and should not be included. Similarly, the current legal climate does not encourage the use of items perceived as personally invasive. Overall, minimizing such issues as invasiveness might be encouraged. What should be especially avoided, however, is a reliance on subjective and less verifiable items that compromises the primary goal of retrieving relatively objective, historical data from applicants.

6.4.1 Summary

Biodata predictors are a powerful noncognitive alternative to cognitive ability tests that have shown significant promise as a predictor in selection. The principle relative to biodata is that past behaviors matter and should be taken into account when criteria such as performance, absenteeism, and other work-related outcomes are being predicted. In addition, efforts are currently under way to develop a more theoretical understanding of the constructs involved with biodata. Finally, ethical and legal concerns are being addressed in hopes of creating an acceptable compromise between high predictability and overall fairness. Thus, although there is still much to be done in understanding how past behaviors can be used in personnel selection, evidence suggests that enhancing biodata techniques seems like a step in the right direction. Some examples of published tests that measure many of the predictor constructs discussed in the preceding section are displayed in Table 1.

7 PERSON–ORGANIZATION FIT

Conventional selection practices are geared toward hiring employees whose KSAs provide the greatest fit with clearly defined requirements of specific jobs. The characteristics of the organization in which the jobs reside, those characteristics of the person relative to the organization as a whole, are rarely considered. The basic notion with person–organization fit (P-O fit) is that a fit between personal attributes and characteristics of the target organization contributes to important positive individual and organizational outcomes.

7.1 Schneider's Attraction–Selection–Attrition Framework

Much of the recent interest in the concept of P-O fit can be traced to the attraction–selection–attrition (ASA)

Table 1 Sampling of Tests That Measure Predictor Constructs Discussed in This Chapter

Test	Publisher	Construct/Focus
Achievement and Success Index	LIMRA	Biodata
Adaptability Test	Pearson Reid London House	Adaptability
Basic Skills Test	Psychological Services, Inc.	Cognitive abilities
California Psychological Inventory	CPP	Personality
Differential Aptitude Tests	Harcourt Assessment	Cognitive abilities
Employee Aptitude Survey	Harcourt Assessment	Cognitive abilities
Hogan Personality Inventory	Hogan Assessment Systems	Personality
Management Interest Inventory	SHL	Managerial interests
NEO Five-Factor Inventory	PAR	Personality
Myers–Briggs Type Indicator	CPP	Personality
MPTQ	Proctor & Gamble	Biodata
Physical Ability Test	FSI	Physical abilities
Self-Directed Search	PAR	Vocational interest
Strong Interest Inventory	CPP	Vocational interest
Watson–Glaser Critical Thinking	Harcourt Assessment	Cognitive abilities

framework proposed by Schneider (e.g., Schneider, 1987, 1989; Schneider et al., 2000). Schneider (1987) outlined a theoretical framework of organizational behavior based on the mechanism of person–environment fit that integrates both individual and organizational theories. It suggests that certain types of people are attracted to, and prefer, particular types of organizations; organizations formally and informally seek certain types of employees to join the organization; and attrition occurs when employees who do not fit a particular organization leave. Those people who stay with the organization, in turn, define the structure, processes, and culture of the organization.

Van Vianen (2000) argued that although many aspects of organizational life may be influenced by the attitudes and personality of the employees in the organization, this does not necessarily require that the culture of a work setting originate in the characteristics of people. He suggested, instead, that cultural dimensions reflecting the human side of organizational life are more adaptable to characteristics of people, whereas cultural dimensions that reflect the production side of organizational life are more determined by organizational goals and the external environment.

Similarly, Schaubroeck et al., (1998) proposed that a more complex conceptualization of the ASA process that incorporates the distinction between occupational and organizational influences should be examined more closely. These researchers investigated the role of personality facets and P-O fit and found that personality homogenization occurs differently and more strongly within particular occupational subgroups within an organization. Similarly, Haptonstahl and Buckley (2002) suggested that as work teams become more widely used in the corporate world, person–group fit becomes an increasingly relevant construct.

7.2 Toward an Expanded Model of Fit and a Broader Perspective of Selection

The research and theorizing reported in this section has suggested that selection theory should consider making fit assessments based on person–job fit, person–organization fit, and person–group fit. Traditional selection theory considers person–job fit as the basis for selecting job applicants, with the primary predictor measures being KSAs and the criterion targets being job proficiency and technical understanding of the job.

To include P-O fit as a component of the selection process, one would evaluate applicants' needs, goals, and values. The assumption here is that the greater the match between the needs of the applicant and organizational reward systems, the greater the willingness to perform for the organization; and the greater the match between a person's goals and values and an organization's expectations and culture, the greater the satisfaction and commitment.

Finally, at a more detailed level of fit, there is the expectation that suborganizational units such as groups may have different norms and values than the organization in which they are embedded. Thus, the degree of fit between an individual and a group may differ significantly from the fit between the person and the organization. P-G fit has not received as much research attention as either P-J fit or P-O fit, but it is clearly different from these other types of fit (Kristof, 1996; Borman et al., 1997).

Werbel and Gilliland (1999) suggested the following tenets about the three types of fit and employee selection: (1) the greater the technical job requirements, the greater the importance of P-J fit; (2) the more distinctive the organizational culture, the greater the need for P-O fit; (3) the lengthier the career ladder associated with an entry-level job, the greater the importance of P-O fit; (4) the more frequent the use of team-based systems within a work unit, the greater the importance of P-G fit; and (5) the greater the work flexibility within the organization, the greater the importance of P-O fit and P-G fit.

7.3 Summary

When jobs and tasks are changing constantly, the process of matching a person to some fixed job requirements becomes less relevant. Whereas traditional selection models focused primarily on P-J fit, several have argued that with new organizational structures and ways of functioning, individual–organizational fit and individual–group fit become more relevant concepts. In our judgment, selection models in the future should incorporate all three types of fit as appropriate for the target job and organization. We can conceive of a hybrid selection model where two or even all three types of fit are considered simultaneously in making selection decisions. For example, consider the possibility of a special multiple-hurdle application in which in order to be hired, applicants must have above a level of fit for the initial job, the team to which they will first be assigned, *and* the target organization. *Or*, depending on the particular selection context for an organization, the three types of fit might be weighted differentially in selecting applicants. Obviously, the details of hybrid selection models such as these have yet to be worked out. However, the notion of using more than one fit concept seems to hold promise for a more flexible and sophisticated approach to making selection decisions.

8 CONCLUSIONS

The world of work is in the midst of profound changes that will require new thinking about personnel selection. Workers will probably need to be more versatile, handle a wider variety of diverse and complex tasks, and have more sophisticated technological knowledge and skills. The aim of this chapter was to provide a review of the state of the science on predictor research and thinking. Accordingly, we reviewed research on performance criteria and provided a detailed review of predictor space issues, including predictors that tap such domains as ability, personality, vocational interests, and biodata. Finally, we felt that it was important to expand this individual-level perspective to encompass broader organizational issues related to person–job match and person–organization fit.

In terms of criterion measurement, research has demonstrated that short-term, technical performance criteria are best predicted by general cognitive ability, whereas longer-term criteria such as nontechnical job performance, retention, and promotion rates are better predicted by other measures, including personality, vocational interest, and motivation constructs. To select and retain the best possible applicants, it would seem critical to understand, develop, and evaluate multiple measures of short- and long-term performance as well as other indicators of organizational effectiveness, such as turnover/retention.

On the predictor side, advances in the last decade have shown that we can reliably measure personality, motivational, and interest facets of human behavior, and that under certain conditions these can add substantially to our ability to predict turnover, retention, and job performance. The reemergence of personality and related volitional constructs as predictors is a positive sign, in that this trend should result in a more complete mapping of the KSAO requirements for jobs and organizations, beyond general cognitive ability.

Finally, we would recommend that organizations consider ways of expanding the predictor and criterion domains that result in selecting applicants with a greater chance of long-term career success, and when doing so, it will be important to extend the perspective to broader implementation issues that involve classification of personnel and person–organization (P-O) fit. As organizational flexibility in effectively utilizing employees increasingly becomes an issue, the P-O fit model may be more relevant compared to the traditional person–job match approach.

REFERENCES

Abrahams, N. M., Neumann, I., and Githens, W. H. (1971), "Faking Vocational Interests: Simulated Versus Real-Life Motivation," *Personnel Psychology*, Vol. 24, pp. 5–12.

Asher, J. J. (1972), "The Biographical Item: Can It Be Improved?" *Personnel Psychology*, Vol. 25, pp. 251–269.

Barrick, M. R., and Mount, M. K. (1991), "The Big Five Personality Dimensions and Job Performance: A Meta-analysis," *Personnel Psychology*, Vol. 44, pp. 1–26.

Barrick, M. R., Mount, M. K., and Judge, T. A. (2001), "Personality and Performance at the Beginning of the New Millennium: What Do We Know and Where Do We Go Next?" *International Journal of Selection and Assessment*, Vol. 9, pp. 52–69.

Blakley, B. R., Quinones, M. A., Crawford, M. S., and Jago, I. A. (1994), "The Validity of Isometric Strength Tests," *Personnel Psychology*, Vol. 47, pp. 247–274.

Borman, W. C., and Motowidlo, S. M. (1993), "Expanding the Criterion Domain to Include Elements of Contextual Performance," in *Personnel Selection,* N. Schmitt and W. C. Borman Eds., Jossey-Bass, San Francisco, pp. 71–98.

Borman, W. C., White, L. A., Pulakos, E. D., and Oppler, S. H. (1991), "Models of Supervisory Job Performance Ratings," *Journal of Applied Psychology*, Vol. 76, pp. 863–872.

Borman, W. C., Hanson, M. A., Oppler, S. H., Pulakos, E. D., and White, L. A. (1993), "The Role of Early Supervisory Experience in Supervisor Performance," *Journal of Applied Psychology*, Vol. 78, pp. 443–449.

Borman, W. C., Hanson, M. A., and Hedge, J. W. (1997), "Personnel Selection," in *Annual Review of Psychology*, Vol. 48, J. T. Spence, J. M. Darley, and D. J. Foss, Eds., Annual Reviews, Palo Alto, CA, pp. 299–337.

Borman, W. C., Buck, D. E., Hanson, M. A., Motowidlo, S. J., Stark, S., and Drasgow, F. (2001), "An Examination of the Comparative Reliability, Validity, and Accuracy of Performance Ratings Made Using Computerized Adaptive Rating Scales," *Journal of Applied Psychology*, Vol. 86, pp. 965–973.

Bridges, W. (1994), "The End of the Job," *Fortune*, Vol. 130, pp. 62–74.

Campbell, D. P. (1971), *Handbook for the Strong Vocational Interest Blank*, Stanford University Press, Stanford, CA.

Carroll, J. B. (1993), *Human Cognitive Abilities: A Survey of Factor-Analytic Studies*, Cambridge University Press, New York.

Cascio, W. F. (1995), "Whither Industrial and Organizational Psychology in a Changing World of Work?" *American Psychologist*, Vol. 50, pp. 928–939.

Chan, D. (2001), "Practical Intelligence and Job Performance," paper presented at the 16th Annual Conference of the Society for Industrial and Organizational Psychology, San Diego, CA.

Coleman, V. I, and Borman, W. C. (2000), "Investigating the Underlying Structure of the Citizenship Performance Domain," *Human Resource Management Review*, Vol. 10, pp. 25–44.

Committee on Techniques for the Enhancement of Human Performance: Occupational Analysis, Commission on Behavioral and Social Sciences and Education, National Research Council (2000), *The Changing Nature of Work: Implications for Occupational Analysis,* National Academy Press, Washington, DC.

Conley, J. J. (1984), "Longitudinal Consistency of Adult Personality: Self-Reported Psychological Characteristics Across 45 Years," *Journal of Personality and Social Psychology*, Vol. 47, pp. 1325–1333.

Conway, J. M. (1999), "Distinguishing Contextual Performance from Task Performance for Managerial Jobs," *Journal of Applied Psychology*, Vol. 84, pp. 3–13.

Czaja, S. J. (2001), "Technological Change and the Older Worker," in *Handbook of the Psychology of Aging*, J. E. Birren and K. W. Schaie, Eds., Academic Press, San Diego, CA, pp. 547–568.

Czaja, S. J., and Moen, P. (2004), "Technology and Employment," in *Technology for Adaptive Aging: Report and Papers*, R. Pew and S. Van Hemel, Eds., National Academies Press, Washington, DC.

Day, D. V. and Silverman, S. B., (1989), "Personality and Job Performance: Evidence of Incremental Validity," *Personnel Psychology*, Vol. 42, pp. 25–36.

Drasgow, F. and Olson-Buchanon, J. B., Eds. (1999), *Innovations in Computerized Assessment*, Lawrence Erlbaum Associates, Hillsdale, NJ.

Durhart, D. T. (2001), *Violence in the Workplace, 1993–1999*, Publication 190076, U.S. Department of Justice, Washington, DC.

Dye, D. A., Reck, M., and Murphy, M. A. (1993), "The Validity of Job Knowledge Measures," *International Journal of Selection and Assessment*, Vol. 1, pp. 153–517.

Edwards, J. E., and Morrison, R. F. (1994), "Selecting and Classifying Future Naval Officers: The Paradox of Greater Specialization in Broader Areas," in *Personnel Selection and Classification,* M. G. Rumsey, C. B. Walker, and J. H. Harris, Eds., Lawrence Erlbaum Associates, Mahwah, NJ, pp. 69–84.

Erez, A., and Judge, T. A. (2001), "Relationship of Core Self-Evaluations to Goal Setting, Motivation, and Performance," *Journal of Applied Psychology*, Vol. 86, pp. 1270–1279.

Farmer, W. L. (2001), *A Brief Review of Biodata History, Research, and Applications*, unpublished manuscript, Navy Personnel Research, Studies, and Technology.

Fleishman, E. A., Constanza, D. P., and Marshall-Mies, J. (1999), "Abilities," in *An Occupational Information System for the 21st Century: The Development of O*NET*, N. G. Peterson, M. D. Mumford, W. C. Borman, R. P. Jeanneret, and E. A. Fleishman, Eds., American Psychological Association, Washington D.C.

Goldberg, L. R. (1992), "The Structure of Phenotypic Personality Traits (or the Magical Number Five, Plus or Minus Zero)," keynote address to the Sixth European Conference on Personality. Groningen, The Netherlands.

Haptonstahl, D. E., and Buckley, M. R. (2002), "Applicant Fit: A Three-Dimensional Investigation of Recruiter Perceptions," paper presented at the Annual Meeting of the Society for Industrial and Organizational Psychology, Toronto, Ontario, Canada.

Harter, J. K., Schmidt, F. L., and Hayes, T. L. (2002), "Business-Unit-Level Relationship Between Employee Satisfaction, Employee Engagement, and Business Outcomes: A Meta-analysis," *Journal of Applied Psychology*, Vol. 87, pp. 268–279.

Hecker, D. E. (2001), "Occupational Employment Projections to 2010," *Monthly Labor Review*, Vol. 124, pp. 57–84.

Hogan, R., and Blake, R. J. (1996), "Vocational Interests: Matching Self-Concept with the Work Environment," in *Individual Differences and Behavior in Organizations*, K. R. Murphy (Ed.), Jossey-Bass, San Francisco, pp. 89–144.

Hogan, J., and Hogan, R. (1989), "How to Measure Employee Reliability," *Journal of Applied Psychology*, Vol. 74, pp. 273–279.

Hogan, J., and Ones, D. S. (1997), "Conscientiousness and Integrity at Work," in *Handbook of Personality Psychology*, R. Hogan, J. A. Johnson, and S. R. Briggs, Eds., Academic Press, San Diego, CA, pp. 849–870.

Hogan, J., and Quigley, A. (1994), "Effects of Preparing for Physical Ability Tests," *Public Personnel Management*, Vol. 23, pp. 85–104.

Hogan, R. (1991), "Personality and Personality Measurement," in *Handbook of Industrial and Organizational Psychology*, 2nd ed., Vol. 2, M. D. Dunnette and L. M. Hough, Eds., Consulting Psychologists Press, Palo Alto, CA, pp. 873–919.

Hogan, R., and Blake, R. J. (1996), "Vocational Interests: Matching Self-Concept with the Work Environment," in *Individual Differences and Behavior in Organizations*, K. R. Murphy, Ed., Jossey-Bass, San Francisco, pp. 89–144.

Holland, J. L., and Hough, L. M. (1976), "Vocational Preferences," in *Handbook of Industrial and Organizational Psychology*, M. D. Dunnette, Ed., Rand McNally, Skokie, IL, pp. 521–570.

Hunter, J. E. (1983), "A Causal Analysis of Cognitive Ability, Job Knowledge, Job Performance and Supervisory Ratings," in *Performance Measurement and Theory*, F. Landy, S. Zedeck, and J. Cleveland, Eds., Lawrence Erlbaum Associates, Mahwah, NJ, pp. 257–266.

Hunter, J. E., and Hunter, R. F. (1984), "Validity and Utility of Alternative Predictors of Job Performance," *Psychological Bulletin*, Vol. 96, pp. 72–98.

Hurtz, G. M., and Donovan, J. J. (2000), "Personality and Job Performance: The Big Five Revisited," *Journal of Applied Psychology*, Vol. 85, pp. 869–879.

International Personality Item Pool (2001), "A Scientific Collaboratory for the Development of Advanced Measures of Personality Traits and Other Individual Differences," http://www.ipip. ori. org/.

Judge, T. A., and Bono, J. E. (2001), "Relationship of Core Self-Evaluations Traits—Self-Esteem, Generalized Self-Efficacy, Locus of Control, and Emotional Stability—with Job Satisfaction and Job Performance: A Meta-analysis," *Journal of Applied Psychology*, Vol. 86, pp. 80–92.

Judge, T. A., Locke, E. A., Durham, C. C., and Kluger, A. N. (1998), "Dispositional Effects on Job and Life Satisfaction: The Role of Core Evaluations," *Journal of Applied Psychology*, Vol. 83, pp. 17–34.

Judge, T. A., Bono, J. E., and Locke, E. A. (2000), "Personality and Job Satisfaction: The Mediating Role of Job Characteristics," *Journal of Applied Psychology*, Vol. 85, pp. 237–249.

Kenrick, D. T., and Funder, D. C. (1988), "Profiting from Controversy: Lessons from the Personsituation Debate," *American Psychologist*, Vol. 43, pp. 23–34.

Kristof, A. L. (1996), "Person–Organization Fit: An Integrative Review of Its Conceptualizations, Measurement, and Implications," *Personnel Psychology*, Vol. 49, pp. 1–49.

Landy, F. J. (1993), "Early Influences on the Development of Industrial/Organizational Psychology," in *Explaining Applied Psychology: Origins and Critical Analyses*, T. K. Fagan and G. R. Vardenbos, Eds., American Psychological Association, Washington, DC, pp. 79–118.

Larson, G. E., and Wolfe, J. H. (1995), "Validity Results for *g* from an Expanded Test Base," *Intelligence*, Vol. 20, pp. 15–25.

Le Pine, J. A., Colquitt, J. A., and Erez, A. (2000), "Adaptability to Changing Task Contexts: Effects of General Cognitive Ability, Conscientiousness, and Openness to Experience," *Personnel Psychology*, Vol. 53, pp. 563–593.

Mael, F. A. (1991), "A Conceptual Rationale for the Domain and Attributes of Biodata Items," *Personnel Psychology*, Vol. 44, pp. 763–792.

McCrae, R. R., and Costa, P. T., Jr. (1987), "Validation of the Five-Factor Model of Personality Across Instruments and Observers," *Journal of Personality and Social Psychology*, Vol. 52, pp. 81–90.

McHenry, J. J., Hough, L. M., Toquam, J. L., Hanson, M., and Ashworth, S. (1990), "Project A Validity Results: The Relationship Between Predictor and Criterion Domains," *Personnel Psychology*, Vol. 43, pp. 335–354.

Muchinsky, P. M. (1993), "Validation of Intelligence and Mechanical Aptitude Tests in Selecting Employees for Manufacturing Jobs," *Journal of Business Psychology*, Vol. 7, pp. 373–82.

Mumford, M. D., Baughman, W. A., Threlfall, K. V., Uhlman, C. E., and Constanza, D. P. (1993), "Personality, Adaptability, and Performance: Performance on Well-Defined and Ill-Defined Problem-Solving Tasks," *Human Performance*, Vol. 6, pp. 241–285.

Murphy, K. R. (1996), "Individual Differences and Behavior in Organizations: Much More Than *g*," in *Individual Differences and Behavior in Organizations*, K. Murphy, Ed., Jossey-Bass, San Francisco.

Olea, M. M., and Ree, M. J. (1994), "Predicting Pilot and Navigator Criteria: Not Much More Than *g*," *Journal of Applied Psychology*, Vol. 79, pp. 845–851.

Ones, D. S., and Viswesvaran, C. (1998a), "The Effects of Social Desirability and Faking on Personality and Integrity Assessment for Personnel Selection," *Human Performance*, Vol. 11, pp. 245–269.

Ones, D. S., and Viswesvaran, C. (1998b), "Gender, Age, and Race Differences on Overt Integrity Tests: Results Across Four Large-Scale Job Applicant Data Sets," *Journal of Applied Psychology*, Vol. 83, pp. 35–42.

Ones, D. S., Viswesvaran, C., and Schmidt, F. L. (1993), "Comprehensive Meta-Analysis of Integrity Test Validities: Findings and Implications for Personnel Selection and Theories of Job Performance, *Journal of Applied Psychology (Monograph)*, Vol. 78, pp. 679–703.

Ones, D. S., Schmidt, F. L., and Viswesvaran, C. (1994), "Do Broader Personality Variables Predict Job Performance with Higher Validity?" in *Personality and Job Performance: Big Five Versus Specific Traits*, R. Page, Chair, symposium conducted at the 9th Annual Conference of the Society for Industrial and Organizational Psychology, Nashville, TN.

Ones, D. S., Viswesvaran, C., and Reiss, A. D. (1996b), "Role of Social Desirability in Personality Testing for Personnel Selection: The Red Herring," *Journal of Applied Psychology*, Vol. 81, pp. 660–679.

Podsakoff, P. M., MacKenzie, S. B., Paine, J. B., and Bachrach, D. G. (2000), "Organizational Citizenship Behaviors: A Critical Review of the Theoretical and Empirical Literature and Suggestions for Future Research," *Journal of Management*, Vol. 26, pp. 513–563.

Ree, M. J., Earles, J. A., and Teachout, M. S. (1994), "Predicting Job Performance: Not Much More than *g*," *Journal of Applied Psychology*, Vol. 79, pp. 518–524.

Reilly, R. R., and Chao, G. T. (1982), "Validity and Fairness of Some Alternative Employee Selection Procedures," *Personnel Psychology*, Vol. 35, pp. 1–62.

Roberts, B. W., and Del Vecchio, W. F. (2000), "The Rank Order Consistency of Personality Traits from Childhood to Old Age: A Quantitative Review of Longitudinal Studies," *Psychological Bulletin*, Vol. 126, pp. 3–25.

Rosse, J. G., Stecher, M. D., Miller, J. L., and Levin, R. A. (1998), "The Impact of Response Distortion on Preemployment Personality Testing and Hiring Decisions," *Journal of Applied Psychology*, Vol. 83, pp. 634–644.

Salgado, J. F. (1998), "Big Five Personality Dimensions and Job Performance in Army and Civil Occupations: A European Perspective," *Human Performance*, Vol. 11, pp. 271–288.

Schaubroeck, J., Ganster, D. C., and Jones, J. R. (1998), "Organization and Occupation Influences in the Attraction–Selection–Attrition Process," *Journal of Applied Psychology*, Vol. 83, pp. 869–891.

Schmidt, F. L., and Hunter, J. E. (1998), "The Validity and Utility of Selection Methods in Personnel Psychology: Practical and Theoretical Implications of 85 Years of Research Findings," *Psychological Bulletin*, Vol. 124, pp. 262–274.

Schmidt, F. L., Hunter, J. G., and Outerbridge, A. N. (1986), "Impact of Job Experience and Ability on Job Knowledge, Work Sample Performance, and Supervisory Ratings of Job Performance," *Journal of Applied Psychology*, Vol. 71, pp. 432–439.

Schmitt, N., Rogers, W., Chan, D., Sheppard, L., and Jennings, D. (1997), "Adverse Impact and Predictive Efficiency of Various Predictor Combinations," *Journal of Applied Psychology*, Vol. 82, pp. 719–730.

Schneider, B. (1987), "The People Make the Place," *Personnel Psychology*, Vol. 40, pp. 437–453.

Schneider, B. (1989), "$E = f(P, B)$: The Road to a Radical Approach to $P-E$ Fit," *Journal of Vocational Behavior*, Vol. 31, pp. 353–361.

Schneider, B., Smith, D. B., and Goldstein, H. W. (2000), "Toward a Person–Environment Psychology of Organizations," in *Person–Environment Psychology*, 2nd ed., W. B. Walsh, K. H. Craik, and R. H. Price, Eds., Lawrence Erlbaum Associates, Mahwah, NJ.

Sternberg, R. J. (1985), *Beyond IQ: A Triarchic Theory of Human Intelligence*, Cambridge University Press, New York.

Sternberg, R. J., and Wagner, R. K. (1992), "Tacit Knowledge: An Unspoken Key to Managerial Success," *Tacit Knowledge*, Vol. 1, pp. 5–13.

Sternberg, R. J., Wagner, R. K., and Okagaki, L. (1993), "Practical Intelligence: The Nature and Role of Tacit Knowledge in Work and at School," in *Advances in Lifespan Development*, H. Reese and J. Puckett, Eds., Lawrence Erlbaum Associates, Mahwah, NJ, pp. 205–227.

Sternberg. R. J., Wagner, R. K., Williams, W. M., and Horvath, J. A. (1995), "Testing Common Sense," *American Psychologist*, Vol. 50, pp. 912–927.

Stricker, L. J. (1987), "Developing a Biographical Measure to Assess Leadership Potential," paper presented at the Annual Meeting of the Military Testing Association, Ottawa, Ontario, Canada.

Stricker, L. J. (1988). "Assessing Leadership Potential at the Naval Academy with a Biographical Measure," paper presented at the Annual Meeting of the Military Testing Association, San Antonio, TX.

Strong, E. K. (1943), *Vocational Interests of Men and Women*, Stanford University Press, Stanford, CA.

Tett, R., and Meyer J. (1993), "Job Satisfaction, Organizational Commitment, Turnover Intention, and Turnover: Path Analysis Based on Meta-analytic Findings," *Personnel Psychology*, Vol. 46, pp. 259–293.

Tupes, E. C., and Christal, R. E. (1992), "Recurrent Personality Factors Based on Trait Ratings," *Journal of Personality*, Vol. 60, pp. 225–251.

U.S. Department of Health and Human Services (1997), *An Analysis of Worker Drug Use and Workplace Policies and Programs*, U.S. Government Printing Office, Washington, DC.

U.S. Department of Labor, Bureau of Labor Statistics (2003), *Occupational Outlook Handbook*, U.S Government Printing Office, Washington, DC.

Van Vianen, A. E. M. (2000), "Person–Organization Fit: The Match Between Newcomers' and Recruiters' Preferences for Organizational Cultures," *Personnel Psychology*, Vol. 53, pp. 1–32.

Wagner, R. K., and Sternberg, R. J. (1985), "Practical Intelligence in Real-World Pursuits: The Role of Tacit Knowledge," *Journal of Personality and Social Psychology*, Vol. 49, pp. 436–458.

Werbel, J., and Gilliland, S. W. (1999), "Person–Environment Fit in the Selection Process," in *Research in Personnel and Human Resources Management*, Vol. 17, G. R. Ferris, Ed., JAI Press, Stanford, CT, pp. 209–243.

Wigdor, A. K., and Garner, W. R., Eds. (1982), *Ability Testing: Uses, Consequences, and Controversies*, (Parts I and II), National Academy Press, Washington, DC.

CHAPTER 18

DESIGN, DELIVERY, AND EVALUATION OF TRAINING SYSTEMS

Eduardo Salas, Katherine A. Wilson, Heather A. Priest, and Joseph W. Guthrie
University of Central Florida
Orlando, Florida

1 INTRODUCTION

Training has become a way of life in many organizations. Organizations depend on learning methodologies, training technology, and learning and development efforts to prepare its workforce. Without a knowledgeable and skillful workforce, organizations will be less likely to succeed, so training in organizations has become big business. The American Society for Training and Development's "State of the Industry Report" (ASTD, 2003) states that on average, organizations spend $826 per employee for training and employees spend an estimated 28 hours in training each year. These costs result in an estimated $54.2 to $200 billion being spent each year in training and developing employees (Salas and Cannon-Bowers, 2001; Galvin, 2002). With this in mind, designing and delivering effective training systems should be of utmost importance to any organization. Why? Because the benefits of properly designed training systems can be a competitive advantage. A skilled and prepared workforce can yield higher productivity and better service, improved quality, higher motivation and commitment, fewer errors, increased safety, and higher morale and teamwork. Furthermore, a poorly developed workforce can cost organizations billions of dollars in legal fees (e.g., Goldman, 2000). In addition, the National Safety Council (2001) reports that $131.7 billion is spent in private-sector firms as a result of worker-related injuries and deaths, costs that can be mitigated with training efforts.

Although the purpose of training in organizations varies (e.g., improve safety, increase competencies, better product quality, error reduction), the overarching theme is to create a high-quality workforce and better products for consumers. To accomplish this, organizations, we suggest, must rely on the science of training (Tannenbaum and Yukl, 1992; Salas and Cannon-Bowers, 2001), a science that now offers tools, techniques, strategies, and methodologies that if applied systematically can yield desired outcomes. This science has produced an enormous quantity of information, all of which must be applied during the design, delivery, and evaluation (as well as transfer) of training systems. Therefore, the purpose of this chapter is to highlight the scientifically derived information necessary for designing and delivering an effective training system. To do this, we reviewed the available training literature (e.g., Goldstein, 1980, 1993; Tannenbaum and Yukl, 1992; Ford et al., 1994; Swezey and Llaneras, 1997; Salas and Cannon-Bowers, 2001) and extracted from it principles, concepts, and suggestions that scientists and practitioners can use, apply, and explore.

1.1 Training Defined

Training can be defined as the systematic acquisition of knowledge (i.e., what we *need to know*), skills (i.e., what we *need to do*), and attitudes* (i.e., what we *need to feel*) (KSAs) that together lead to improved performance in a particular environment. Training is about cognitive and behavioral change. Training is about a permanent change in people's behaviors and actions. Training is about people getting the right competencies to do a job. This change occurs when training events are carefully crafted. We submit that the design and delivery of a training system should be done systematically (see Salas and Cannon-Bowers, 2000a). Effective training should create an environment where trainees can (1) learn the requisite KSAs, (2) practice applying the learned KSAs, and (3) receive constructive and timely feedback to improve performance in the future. This is accomplished by creating instructional strategies that are focused on the specific needs of the organization (Salas and Cannon-Bowers, 1997, 2001).

We note that training should not be equated by simulations or technology. Simulations and technology are just tools available to enhance the learning environment, but these are not training systems in and of themselves. These need to be augmented with instructional features (e.g., diagnosis, performance measurement) that facilitate the learning process.

Campbell (1971) reviewed the training and development literature and concluded that "by and large, the training and development literature is voluminous, nonempirical, nontheoretical, poorly written, and dull" (p. 565). This indictment of the training literature has led to an explosion of research—specifically, on theory development. Three decades later, the science of training brought a number of theoretical frameworks to guide the science of the design, delivery, and evaluation of training systems as well as what leads to the transfer of the newly acquired skills. This progress in training research can be attributed to more profound, more comprehensive, and more focused thought. Specifically, research has produced more and better theories, models, and frameworks (Salas and Cannon-Bowers, 2001). We illustrate some next.

2 SCIENCE OF TRAINING: THEORETICAL DEVELOPMENTS

Transfer of training has become an increasingly important aspect of training research. Thayer and Teachout (1995) proposed a model outlining the proper climate necessary for transfer of training to the job (see Figure 1). This model outlines several variables that can influence learning, which, in turn, has a direct effect on transfer. Specifically, Thayer and Teachout (1995) describe seven pretraining variables, supported by the literature: (1) reactions to previous training (Baldwin and Ford, 1988; Mathieu et al., 1992), (2) previous education (Mathieu et al., 1992), (3) self-efficacy (pretraining) (Ford et al., 1992), (4) ability (Ghiselli, 1966), (5) locus of control (Williams et al., 1991), (6) job involvement (Noe and Schmitt, 1986), and (7) career/job attitudes (Williams et al., 1991). All of these create a pretraining environment that can promote or inhibit learning. In addition, learning can be affected by trainees' reactions to the training intervention (Kirkpatrick, 1976). As shown in the model, self-efficacy is also a posttraining factor. Research supports the role of self-efficacy as both an antecedent and outcome of training (Latham, 1989; Tannenbaum et al., 1991). The final set of variables and a major focus of this model are the transfer-enhancing training activities and climate for transfer. Transfer-enhancing strategies (e.g., goal setting, relapse prevention) have been identified as factors that can enhance transfer through such phenomena as overlearning (e.g., McGhee and Thayer, 1961). The climate for transfer (i.e., environmental favorability/unfavorability), both cues and consequences, can contribute significantly to transfer (e.g., Rouiller, 1989; Williams et al., 1991; Ford et al., 1992). These variables include positive transfer cues, as well as consequence variables, as listed in the model. The final factor in the model is the results. The relationship between transfer and results refers to the appearance of learned knowledge of skills and behaviors in the workplace (i.e., transfer) and the outputs of that transfer (i.e., results).

In addition to Thayer and Teachout's model, Kozlowski and Salas (1997) proposed an integrative multilevel framework for the implementation and transfer of training within an organization (see Figure 2). Based on organizational theory, this framework characterizes the factors and processes that comprise the training and transfer environment. First, this framework distinguishes between different levels within the organization (i.e., organization, team, and individual levels) and discusses the process linkages between these levels. Next, this framework identifies relevant features that comprise the contexts and distinguishes between technostructural and enabling process content. Finally, a critical mechanism of the framework

*Although much of the literature refers to the "A" of KSAs as abilities (e.g., Goldstein, 1993), in this chapter we refer to the "A" as attitudes.

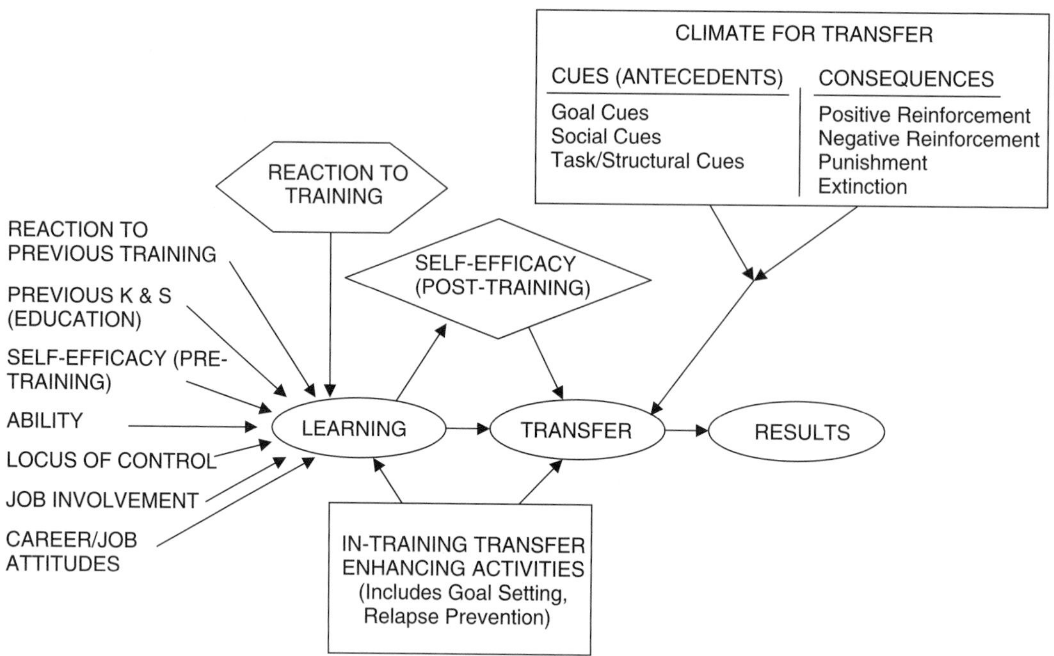

Figure 1 Model of training transfer. (From Thayer and Teachout, 1995.)

is the congruence that conceptualizes configurations among key variables comprising the organizational system. Taken together, this framework addresses how variables within and between content domains, and between levels, are connected.

More comprehensive frameworks of training effectiveness have also been proposed and investigated by Tannenbaum and colleagues (1993; Cannon-Bowers et al., 1995). The model presented in Figure 3 takes a longitudinal, process-oriented approach that integrates the variables affecting the acquisition of competencies and its transfer prior to, during, and after training. In addition, this model recognizes the influence of factors outside the training that affect its effectiveness (e.g., individual, organizational, and situational factors). Finally, this model specifies the critical steps that should be taken for training to be a success (e.g., training needs analysis, training evaluation).

Other researchers have examined specific issues within training, such as training motivation (Colquitt et al., 2000), individual characteristics and work environment (Tracey et al., 2001), training evaluation (Kraiger et al., 1993), and transfer of training (Quinones, 1997). The shift toward using teams in organizations has also resulted in a number of theoretical developments. Kozlowski and colleagues (2000) investigated the organizational factors and training issues that affect the vertical transfer from individual-level processes to organizational and team-level products. Tannenbaum and colleagues developed the team effectiveness model, which provides a framework for all the variables and factors that may inspire team performance. This framework helped organize the literature into input process/output factors that influence team functioning. More recently, Salas and colleagues (in press) have undertaken an effort to integrate the more than 50 models available in the team literature. They proposed a much more comprehensive model of the factors, variables, processes, and mediators that affect individual and team outcomes.

2.1 Summary

The steps that researchers have taken to overcome the criticism in Campbell's (1971) review have resulted in great progress toward creating a better science of training and at putting to rest the fact that the training field is atheoretical. Furthermore, the models and frameworks that have been advanced have provided a solid foundation on which empirical research can now be conducted. And much has.

3 INSTRUCTIONAL SYSTEMS DEVELOPMENT MODEL

For the purpose of this chapter, we take a macrolevel, systems approach to discussing the design, delivery, and evaluation of training. We argue for the systems approach suggested by Goldstein (1993), which includes four components: (1) training program design is iterative and thus feedback is used to update and modify the program continuously; (2) complex interactions are formed between training components (e.g., trainees, tools, instructional strategies); (3) a framework for reference to planning is provided; and (4) recognition that training systems are just a

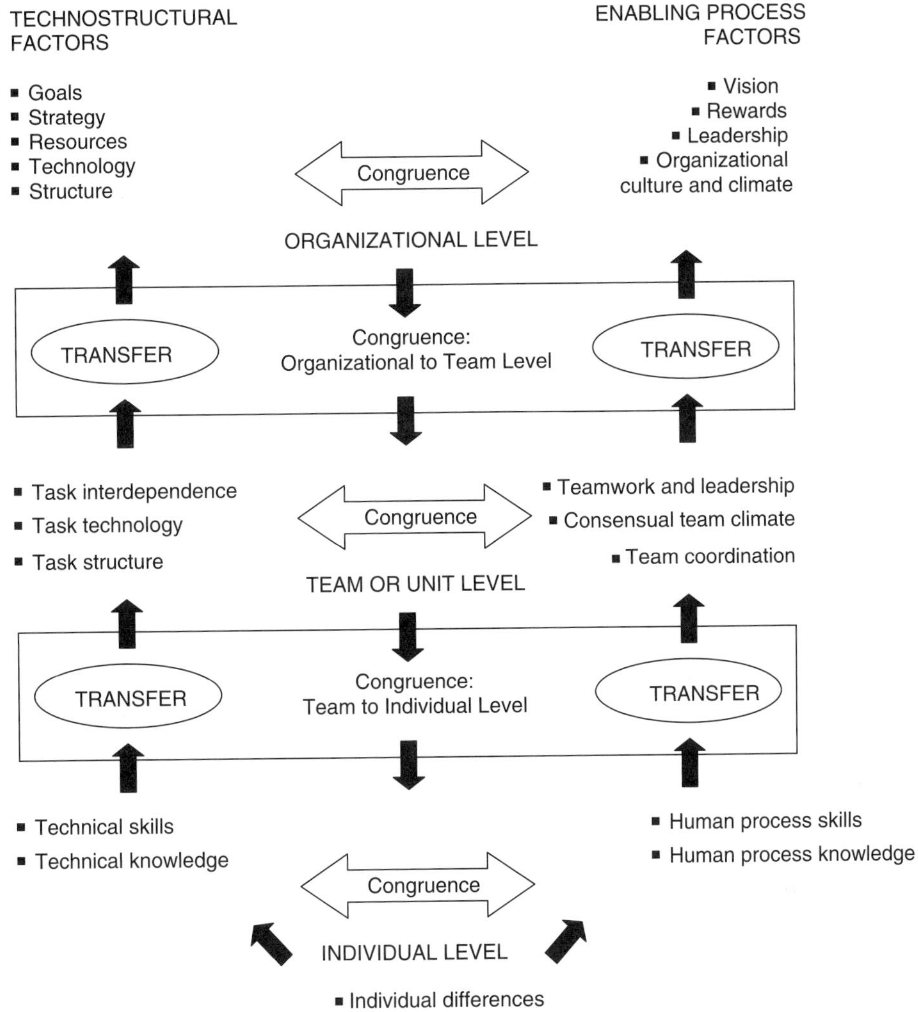

Figure 2 Training implementation and transfer model. (From Koslowski and Salas, 1997.)

small component of the overall organizational system, and as such, components of the organization, task, and person need to be considered in the design. We use the instructional systems development (ISD) model (Branson et al., 1975) to guide much of this chapter. The ISD model was developed over 30 years ago (commonly used in the development of U.S. military training programs) and continues to be widely used today, as some have argued that it is the most comprehensive training design and implementation model available in the literature (Swezey and Llaneras, 1997). Because this model has its critics as well, we use the model to organize the literature that we have reviewed rather than as an endorsement of its superiority over other models in the literature.

There are five basic steps to the ISD model: analysis, design, development, implementation, and evaluation. We added a sixth step to the process: transfer of training because we feel this, too, is an important aspect of any training system. We will also utilize what we know about the science of training to delineate each phase and provide a comprehensive discussion of how training should be designed, delivered, and evaluated (see Salas and Cannon-Bowers, 2000a, 2001). We discuss briefly the six phases before delving more deeply into each of them. The purpose of the first phase, *training analysis*, is to identify the needs of the organization, task, and person, develop training goals, and develop a plan for the design, implementation, and evaluation of training. Phase two, *training design*, encompasses the development of learning objectives and performance measures, and storyboarding the progression of training. It is in this phase when the principle of learning and training are used to guide the design of the learning events. *Training development*,

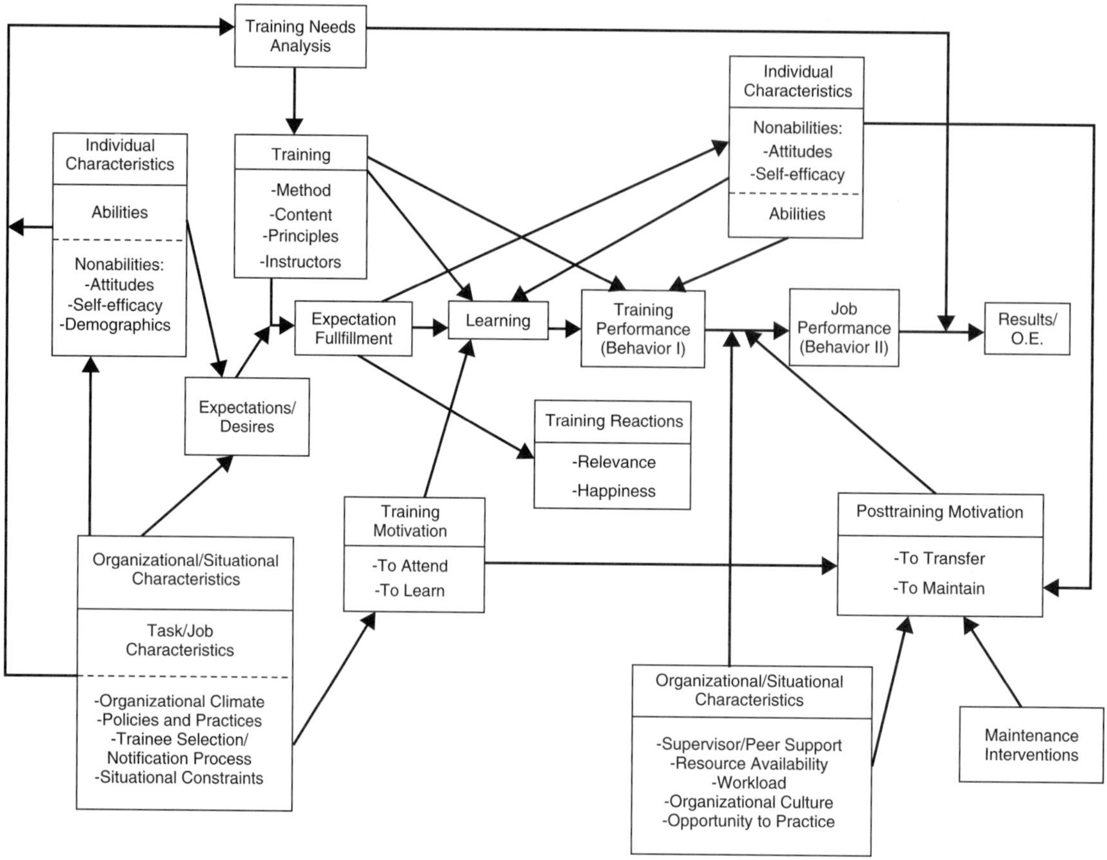

Figure 3 Training effectiveness model. (From Tannenbaum et al., 1993; Cannon-Bowers et al., 1995.)

phase three, is the actual development of the training program. This phase is important in the identification of any weaknesses in the training program to correct them before the training program is implemented. The next phase, *training implementation*, is the actual execution and delivery of the training. In this phase, the mechanics (e.g., location) of the delivery are prepared. Once this phase is completed, the training program should (must) be evaluated. We argue that the *training evaluation* phase must take a multilevel approach to determine training's effectiveness. Finally, once training has been evaluated, the *training transfer* phase ensures that the trained competencies are transferred (or applied) to the actual task environment. The reader will see as we progress through each of these phases that they are not mutually exclusive and thus there will be some overlap between them.

3.1 Summary

There are numerous models in the literature that can be used to guide the design, delivery, and evaluation of training. The reader must take these models at face value and recognize that no single model is perfect. Therefore, the effective design, delivery, and evaluation of training may require a combination of several models. We argue that a systems approach needs to be taken and use the instructional systems design model as a framework by which to organize this chapter.

4 TRAINING ANALYSIS

One of the most important and first steps when designing a training program is conducting a training needs analysis (Goldstein and Ford, 2002). A training needs analysis allows training designers to understand several critical aspects of the training program: (1) where training is needed, (2) what needs to be trained, and (3) who needs to be trained (Goldstein, 1993). There are several outcomes of a training needs analysis: (1) specification of learning objectives, (2) shaping of the design and delivery of training, and (3) criterion development. As noted by Tannenbaum and Yukl (1992), limited empirical evidence can be found on training needs analysis. Three types of analyses must be conducted: organizational, job/task, and person analyses. We discuss each of these next.

4.1 Organizational Analysis

The first step of a training needs analysis is to conduct an organizational analysis. The purpose of an organizational analysis is to ascertain the system-wide components (e.g., climate, norms, goals), available resources, constraints, and support for transfer of the organization that may affect how the training program is delivered (Goldstein, 1993). Furthermore, the organizational analysis focuses on determining how well the training objectives (see Section 5.1) fit with these organizational factors. In addition, an organization's environment must support the training program. For example, if the purpose of a training program is to improve safety on the job, the organizational environment must be designed such that it supports safe behaviors (e.g., error management). Later we discuss the need for a positive transfer climate to support the application of newly acquired skills on the job (Rouiller and Goldstein, 1993; Tracey et al., 1995). As the organizational factors, especially constraints and conflicts, may significantly affect the effectiveness of training, it is important that organizations pay close attention to them prior to implementing a training program.

It was not until the early 1990s that organizational analyses were recognized as important by researchers. Several studies have been well cited in the literature as demonstrating the importance of the organization. For example, research by Rouiller and Goldstein (1993) suggested that the climate of the organization (e.g., situational cues, consequences) was a powerful predictor of whether trainees at fast-food chain restaurants would transfer the learned skills. Similar findings were demonstrated by Tracey and colleagues (1995), whose research suggested that the culture and climate of the organization were directly related to behaviors following training (i.e., transfer of training). These studies, and others, provide a clearer picture of the influential effects of organizational factors on the transfer of trained competencies (i.e., knowledge, skills, and attitudes) to the job.

4.2 Job/Task Analysis

In addition to the organization analysis, the job/task analysis is another necessary component of the training needs analysis. The purpose of a job/task analysis is to uncover characteristics of the tasks being trained, which in turn create the learning objectives (Goldstein, 1993). First, the job/task description needs to be clarified. This description generally includes information pertaining to the essential work functions of the job and the resources needed to complete the job effectively (e.g., materials, equipment). Once the job description is complete, the next step of the job/task analysis is to determine the specifications of the task. Task specifications include the specific tasks to be performed by trainees and the conditions under which the tasks are completed. The final step of the job task analysis is the determination of the task requirements or competencies (i.e., knowledge, skills, and attitudes) needed by trainees to complete the job. This could be considered one of the more difficult of the steps to complete because knowledge and attitudes are more difficult than skills to observe and thus are more likely to be ignored by training designers.

Furthermore, as for more complex tasks that require high cognitive demands (e.g., decision making, problem solving), it is important that these less observable competencies be focused on. A cognitive task analysis has been developed to help uncover these competencies.

4.2.1 Cognitive Task Analysis

Recent research has been aimed at understanding how trainees acquire and develop knowledge and how they mentally organize rules, concepts, and associations (see Dubois et al., 1997–1998; Zsambok and Klein, 1997; Schraagen et al., 2000). Furthermore, research has examined the nature of expertise and how experts make decisions in natural, complex environments (Gordon and Gill, 1997). This research has led to the development of tools such as cognitive task analysis (CTA) (see Salas and Klein 2000; Schraagen et al., 2000). CTA allows training designers to gain insight into the cognitive processes and requirements for job performance of subject matter experts through the use of knowledge elicitation techniques (e.g., verbal protocols, observation, interviews, conceptual methods) (Cooke, 1999; Klein and Militello, 2001). The information generated from the CTA provides templates for expert mental model development, cues for promoting complex decision-making skills, cues for developing simulation- and scenario-based training, and information for the design of performance measurement and feedback protocols.

Some argue that three criteria need to be met if CTA is to be successful (Klein and Militello, 2001). First, the CTA needs to uncover new information relating to trainee judgments (e.g., patterns), decisions (e.g., strategies), and/or other cognitive demands (e.g., cue patterns). Next, there must be effective communication between the cognitive scientists conducting the CTA and training designers. In other words, the information obtained from the CTA must be translated and provided to training designers. Last, the findings from the CTA must be put into action successfully (i.e., it must have an impact). See Table 1 for steps to take to conduct a CTA.

4.3 Person Analysis

The third and final stage of a training needs analysis is the person analysis. The purpose of this analysis is to determine who needs to be trained and what training is needed by each person (Tannenbaum and Yukl, 1992; Goldstein, 1993). In other words, the person analysis makes certain that the right people get the right training. Not all people need the same training. Taking the aviation community as an example, flight crews need different training from that of cabin crews, who need different training from that of maintenance crews. Additionally, pilots new to an airline require different training from that experienced pilots receiving recurrent training (e.g., Feldman, 1988). Research examining the training of managers suggests that managers at

Table 1 Steps in Conducting Cognitive Task Analysis

Step	Guidelines
1. Select experts.	• Consider how many subject matter experts (SMEs) would be appropriate. • Select SMEs who have experience with domain. • Select SMEs who have experience with task.
2. Develop scenarios based on task analysis.	• Develop scenarios that are task relevant with problem statements. • Pretest scenarios to determine if they are complete.
3. Choose a knowledge elicitation method.	• Determine the information you are trying to get at.
• Interviews	• Ask SMEs about cognitive processes needed to complete task scenarios.
• Verbal protocols	• Require SMEs to think out loud as they complete task scenarios.
• Observations	• Observe SMEs performing tasks.
• Conceptual methods	• Develop inferences based on SME input and relatedness judgments.

4. Implement chosen knowledge elicitation method.
• Interviews
	• Decide how many sessions to be recorded.
	• Obtain consent and provide task scenario to SMEs.
	• Provide each scenario to each SME.
	• Ask SMEs relevant questions:
	  • What rules/strategies would they use to complete scenarios?
	  • What knowledge and cognitive skills would they use to complete scenarios?
	  • If A or B happens, what would you do?
	• Keep a record of SME responses: use video, audio, pen and paper, etc.

• Verbal protocols	• Provide each scenario to each SME. • Require SMEs to think out loud as they complete task scenarios. • Keep a record of SME responses: use video, audio, pen and paper, etc.
• Observations	• Provide each scenario to each SME. • Be as unobtrusive as possible, but don't be afraid to ask for clarification when necessary. • Keep a record of SME responses: use video, audio, pen and paper, etc.
• Conceptual methods	• Provide each scenario to each SME. • Present pairs of tasks to experts. • Ask SMEs for relatedness judgments regarding task pairs.
5. Organize and analyze data.	• Keep a record of similarities between SMEs. • Interview additional experts if questions arise from data. • Identify the rules and strategies applied to the scenario. • Identify knowledge and cognitive skills required. • Generate a list of task requirements. • Verify list with additional SMEs.

different levels within an organization require different skills and thus require different training. Specifically, it was suggested that lower-level managers require more administrative skills than do high-level managers (Ford and Noe, 1987). Whereas the job/task analysis identified the competencies required to complete the tasks effectively, the person analysis identifies whether or not trainees have the requisite competencies. In addition, trainees' motivation to learn and participate in training is determined from this analysis.

4.4 Summary

The training needs analysis is especially critical if training is to be effective: that is, if trainees do not meet the requirements for training (e.g., have the right competencies), if training is not specific to the job, or if the organization is not prepared for training, training will fail to be a success. Our advice: Conduct a thorough training needs analysis in the design of any training system.

5 TRAINING DESIGN

The next step in the ISD framework is training design. Driven by the outcomes of the training analysis phase, training design ensures that training is developed systematically and produces a blueprint or model of what the training program will look like. During training design, training developers need to focus on several things: development of training objectives, factors external to the training program (i.e., individual and organizational characteristics and resources), selection of instructional strategies and methods, and specification of program content.

5.1 Training Objectives

Information obtained from the training needs analysis drives the development of the training or learning objectives. It is important when developing these objectives that they are specific, measurable, and task relevant. This will ensure that they can be evaluated when training is completed. The training objectives have three general characteristics that help guide training (Goldstein, 1993). First, training objectives provide both trainers and trainees with expectations as to what trainees should be able to do as a result of the training (i.e., performance). Specifically, training objectives should state the competencies that trainees are expected to acquire and demonstrate once the training is complete. Second, objectives describe the conditions under which the performance, stated by the first guideline, should occur. Finally, objectives provide a description of acceptable performance criteria. In short, training objectives describe at what level the trainee should perform and when to be judged acceptable. Once clearly defined, training objectives are used to guide what instructional strategies should be implemented. These strategies, discussed later, should be selected based on their capability to promote the task-relevant behaviors and competencies determined in the objectives. See Table 2 for guidance on developing training objectives.

Table 2 Steps in Developing Training Objectives

Step	Guidelines
1. Review existing documents to determine job tasks and competencies required.	Examine your sources: • Performance standards for the organization • Essential task lists • Past training objectives • SMEs and instructors to mine their previous experiences

2. Translate identified competencies into training objectives.	Include objectives that:
	• Specify targeted behaviors.
	  • Use "action" verbs (e.g., "provide," "prepare," "locate," and "decide").
	  • Outline specific behavior(s) that demonstrate the appropriate skill or knowledge.
	  • Say it in a way that can be easily understood.
	• Clearly outline the conditions under which skills and behaviors should be seen.
	• Standards to which they will be held when behaviors or skills are performed or demonstrated.
	  • Make sure that standards are realistic.
	  • Make sure that standards are clear.
	  • Make sure that standards are complete, accurate, timely, and performance-rated.

3. Organize training objectives.	Make sure to categorize: • General objectives that specify the end state that trainees should attain/strive for. • Specific objectives that identify the tasks that trainees must perform to meet the general objectives.
4. Implement training objectives.	Use training objectives to: • Design exercise training events (e.g., scenarios). • Use events as opportunities to evaluate how well trainees exhibit training objectives. • Develop performance measurement tools (e.g., checklists). • Brief trainees on training event.

5.2 Individual Characteristics

When designing any training program, training designers must focus on characteristics that each trainee brings to the training program; these have been shown recently to influence learning outcomes. Individual characteristics suggested to influence training's outcomes are cognitive abilities, self-efficacy, goal orientation, and motivation. We discuss each of these next.

5.2.1 Cognitive Ability

Cognitive ability (i.e., *g* or general ability) is one trainee characteristic that has been shown to influence the outcomes of training. Research investigating how cognitive ability influences training outcomes indicates that it influences the attainment of knowledge about the job (see Ree et al., 1995; Colquitt et al., 2000), is a strong determinant of training success (e.g., Ree and Earles, 1991; Randel et al., 1992; Colquitt et al., 2000), and promotes self-efficacy and skill acquisition (e.g., Hunter, 1986). Taken together, these findings suggest that trainees high in cognitive ability are likely to learn more and be more successful in training when all other things are equal.

5.2.2 Self-Efficacy

Another individual characteristic influencing the outcomes of training is self-efficacy or one's belief in his or her own ability. Research examining self-efficacy and learning has been extensive throughout the past decade. These studies indicate that self-efficacy, whether acquired during training or held prior to it, is influenced by cognitive ability (see Hunter, 1986), influences reactions to training (Mathieu et al., 1992), has motivational effects (Quinones, 1995), leads to better performance (e.g., Martocchio and Webster, 1992; Ford et al., 1997; Stevens and Gist, 1997), and dictates whether trainees will or will not use training technology (Christoph et al., 1998). Furthermore, self-efficacy has been shown to mediate numerous individual variables, including the relationship between conscientiousness and learning (Martocchio and Judge, 1997), job satisfaction, intention to quit a job, commitment to the organization, and the relationship between training and the adjustment in newcomers (Saks, 1995).

5.2.3 Goal Orientation

In recent years, goal orientation has received considerable attention regarding its ability to influence trainees' learning (e.g., Phillips and Gully, 1997; Ford et al., 1998; Brett and Vande Valle, 1999). Defined as the mental framework used to interpret and shape how to behave in achievement- or learning-oriented environments, goal orientation takes two forms: mastery or performance (Dweck, 1986; Dweck and Leggett, 1988). Mastery-oriented (or learning-oriented) persons seek to acquire new skills and master novel situations. Research examining mastery-level goal orientation suggests that it is a strong predictor of knowledge-based learning outcomes (Fisher and Ford, 1998) and is related positively to a person's metacognition (Ford et al., 1997) and self-efficacy (Phillips and Gully, 1997). Performance-oriented persons seek to achieve high performance ratings, as this assures them of their own competence, and to circumvent low ones. Despite what we know about goal orientation, the debate continues as to whether goal orientation is a disposition and/or a state (e.g., Stevens and Gist, 1997), whether the construct is multidimensional (e.g., Elliot and Church, 1997; VandeValle, 1997) or whether mastery and performance goal orientations are mutually exclusive (Buttom et al., 1996). Continued research in this area will provide the necessary clarity.

5.2.4 Motivation

Trainee motivation can be conceptualized as the direction, effort, intensity, and persistence that people put forth toward learning-oriented activities pre-, during, and posttraining (Naylor et al., 1980, as cited in Goldstein, 1993; Kanfer 1991; Tannenbaum and Yukl 1992). Trainee motivation is affected by characteristics of the person (e.g., self-efficacy) and the organization (e.g., notification of participation). Research suggests that trainees' motivation to learn and participate in training has an effect on their acquisition, retention, and willingness to apply trained competencies (i.e., KSAs) on the job (e.g., Martocchio and Webster, 1992; Mathieu et al., 1992; Tannenbaum and Yukl, 1992; Quinones 1995). Furthermore, the greater the motivation of trainees before training, the greater the learning and positive reactions to training that will result (see Baldwin et al., 1991; Tannenbaum et al., 1991; Williams et al., 1991; Mathieu et al., 1992). Finally, research examining trainee motivation suggests that trainees who believe that training outcomes are relevant to their job performance will be more likely to apply trained KSAs on the job (Noe, 1986).

The literature looking at trainee motivation is fairly clear, however, lacking is conceptual precision and specificity—the literature is somewhat piecemeal (Salas and Cannon-Bowers, 2001). Recently, however, Colquitt and colleagues (2000) conducted a meta-analysis to understand the underlying processes and variables influencing trainee motivation throughout the training process. This effort indicated that trainees' motivation to learn is influenced by individual (e.g., self-efficacy, valence, anxiety, cognitive ability, and age) and situational (e.g., positive climate, supervisor and peer support, and organizational support) characteristics. One of the most important findings from this research was that trainee motivation is multifaceted, suggesting the need to expand the training needs analysis phase of training design to consider a wider array of individual characteristics and how they might affect learning outcomes.

5.3 Organizational Characteristics

In addition to individual characteristics of trainees, organizational characteristics (i.e., those present within the organization to which the newly acquired KSAs are to be applied) may also influence the outcomes of training and need to be considered during its development. Examples of these characteristics are organizational culture, policies and procedures, situational influences (e.g., improper equipment), and prepractice conditions (see Salas et al., 1995, for more detail).

5.3.1 Organizational Culture

The term *organizational culture* was not discussed in the literature until the 1980s (Guldenmund, 2000). Since then, research examining this concept has suggested that organizational culture is critical to an organization's success (Glendon and Stanton, 2000). Organizational culture can be defined as "a pattern of shared basic assumptions that the group learned as it solved its problems of external adaptation and internal integration ..." (Schein, 1992, p. 12). Furthermore, culture has been argued to consist of norms, values, behavior patterns, rituals, and traditions that influence all aspects of the organization, including training outcomes. These assumptions must "therefore, be taught to new members as the correct way" to perceive, think, and feel about a number of issues, problems, and so on, in the organization (Schein, 1992, p. 12). Similarly, Burke (1997) asserts that once developed, organizational culture is transmitted by organizational leaders to others through the socialization process. Part of the socialization process is training. In addition to leadership and management support of training, policies and procedures of the organization will influence the outcomes of training. We discuss these next.

5.3.2 Policies and Procedures

Underlying an organizational culture are the organization's policies and procedures. Policies can be described as broad requirements that management have established to provide employees with a set of expectations regarding various things (e.g., performance on the job) (Degani and Wiener, 1997). Related to set policies are procedures that provide employees with guidance on how to meet these expectations. For example, if management has an established policy that all employees perform a series of quality control checks, they may provide employees with a checklist to follow to make sure that all quality control checks are completed correctly. A well-designed training program can be used to make trainees aware of these policies and procedures and allow them to practice applying them so that they will adhere to them on the job. Concerns arise, however, when there are social pressures within the organization that are more influential on employees than the formalized policies and procedures (e.g., Hofmann and Stetzer, 1996). Continuing the quality control example discussed previously, if there are unwritten policies that encourage (or rather, don't discourage) employees to make shortcuts and deviate from established checklists, the written policies and procedures will be less effective. This phenomenon has been observed in the oil production industry, for example, which led to a significant number of accidents and incidents (Wright, as cited in Hofmann et al., 1995). As such, if an organization wants to ensure that policies and procedures are followed, it is important that the desired attitudes and behaviors be developed during training and promoted on the job.

5.3.3 Situational Influences

There are two situational influences that have been argued to influence trainees' motivation to learn during training and transfer of training to the job: framing of training participation (i.e., voluntary vs. mandatory), the work environment (e.g., improper equipment), and previous training experience. First, the framing of training has been shown to influence the outcomes of training. For example, attendance policies have been shown to affect trainees' motivation to attend training. Specifically, research suggests that when training attendance is voluntary, trainees are more willing to attend training than when training is mandatory (Baldwin and Magjuka, 1997). In addition to attendance policies, training can be framed such that it is remedial or advanced. Research by Quinones (1995, 1997) indicates that this type of framing can influence the motivation and learning of trainees. Similar findings were found by Martocchio (1992) when training assignment was labeled as an "opportunity" for trainees. Second, perceptions of the work environment can also influence trainees' motivation to learn and whether or not trainees will transfer the learned competencies to the job (e.g., Goldstein, 1993). Also within the work environment, situational constraints such as lack of proper materials and information can lead to employee frustration and poor performance on the job (e.g., Peters et al., 1985). Finally, trainees' previous training experiences have been shown to influence their learning and retention. Previous training that is viewed by trainees as a negative experience will hinder their ability to learn and retain information in future training programs (Smith-Jentsch et al., 1996a).

5.3.4 Prepractice Conditions

Prepractice conditions can be described as elements in the pretraining environment that serve to prepare trainees for practice during training. The research suggesting the benefits of practice for skill acquisition is well documented. However, not all practice is the same. For example, task exposure or repetition alone is not enough, as practice for skill acquisition is a complex process (Schmidt and Bjork 1992; Shute and Gawlick, 1995; Ehrenstein et al., 1997). Cannon-Bowers and colleagues (1998) delineate which conditions might lead to enhanced utility and efficacy of practice. Their review of the literature suggests that interventions applied prior to practice, such as preparatory information, advanced organizers, or metacognitive strategies, can help prepare trainees for training, thus leading to more learning. Although these interventions are just beginning to be investigated empirically, their benefits appear promising.

5.4 Practice Opportunities

Important to any successful training program are the practice opportunities provided to trainees during training. It has been suggested in the literature that experience with different situations (either simulated or real) will improve performance on the job by generating knowledge structures within a meaningful context (i.e., mental models) (Satish and Streufert, 2002). The use of practice scenarios that are scripted a priori will ensure that trainees are practicing the correct competencies and will allow for better performance assessment as well. Although it is widely accepted among laypersons that practice contributes to better performance, training studies have also focused on how practice relates to learning. For example, Goettl et al. (1996) found support for and advantage in learning when alternating task modules (i.e., video game practice and algebra word problems) with a massed protocol, which blocked sessions on the tasks. The findings showed improvement in learning and retention when practice was provided. Further evidence is provided by Bjork and colleagues (Schmidt and Bjork, 1992; Ghodsian et al., 1997). These researchers looked at practice schedules and argued that introducing difficulties during practice will enhance transfer of training for the trainee, although there may not be an improvement in performance immediately following training. These studies reconceptualized interpretation of data from several studies and proposed a new protocol for practice scheduling. Specifically, Schmidt and Bjork (1992) recommended introducing variations in the ordering of tasks during practice, in the nature and scheduling of feedback, and in the versions of the task to be practiced, providing less frequent feedback. In all the versions above, researchers found that this method enhances retention and generalization, even though an initial decrease in acquisition (i.e., immediate) performance was observed, suggesting deeper, more meaningful information processing. These findings were supported by Shute and Gawlick (1995) in an investigation of computer-based training for flight engineering knowledge and skill.

5.5 Feedback

Providing feedback to trainees in a constructive and timely manner is important to the success of any training program because it allows trainees to know how they did during training and where improvements are needed (Cannon-Bowers and Salas, 1997). This requires that several criteria be met. First, feedback should be based on the person's or team's performance during practice and on the training outcomes. Next, feedback provided to trainees should be specific to the skill performance of trainees but not critical of the person. Third, feedback should provide trainees with the necessary knowledge that allows them to adjust their learning strategies to meet the expected performance levels. Finally, feedback must be meaningful to trainees and focus on both individual and team performance (if applicable). Without feedback, breakdowns in performance may go unnoticed by trainees, corrective strategies will not be developed, and errors will probably occur on the job.

5.6 Instructional Strategies

The design of training includes the selection of instructional strategies that are appropriate for the trainees and the organization. A number of instructional strategies are discussed in the literature that can be used to train both individuals and teams (see Table 3). Depending on the needs and resources of the organization, the desired outcomes of training can range from classroom based to simulation driven.

To ensure effective training, strategies should be follow four basic guidelines: (1) Strategies should present information or concepts to be learned, (2) strategies should demonstrate the KSAs to be learned/ targeted, (3) strategies should create opportunities for practice, and (4) strategies should provide opportunities for feedback to trainees during and postpractice. Although much progress has been made in the investigation of effective training strategies, it is also true that there is no single method to deliver all training. Therefore, researchers continue to address how best to present targeted information to trainees, based on a number of factors (e.g., organizational resources, who is being trained, what needs to be trained). The goal of current research is to develop and test cost-effective, content-valid, easy-to-use, engaging, and technology-based methods of training (e.g., Bretz and Thompsett 1992; Baker et al. 1993; Steele-Johnson and Hyde, 1997).

As organizations progress into the twenty-first century, they will be faced with several issues influencing the training strategies that will be chosen: increasing use of teams, improvements in technology, and internationalization. As such, we organize the instructional strategies that we discuss here around those three issues. Taking these issues into consideration but regardless of the instructional strategy chosen by the organization, for training to be a success we argue that the strategy must address three main issues. First, the instructional strategy should encourage trainees to be adaptable to changing situations and to recognize when things go wrong. For example, as technology is introduced into the workplace, employees will need to maintain vigilance and be able to adapt quickly if the technology system is to fail. The training of flexible knowledge structures (i.e., mental models) will allow employees to adjust their behavior to compensate for any changes in their environment. Second, all training strategies must include constructive feedback for trainees. When trainees are provided feedback, they can readjust or correct their strategies and compensate for incorrect behaviors, which will result in better performance. The third and final issue is that training must be dynamic or interactive. Unfortunately, recent reports have suggested that only 10% of delivery methods are interactive, digital technologies (Bassi and Van Buren, 1998). In contrast, the most commonly used delivery methods (approximately 90% of the time) were videotapes and workbooks, and 84% of all companies surveyed used classroom-based and instructor-led training. Computer-based or other technology-based training was used only about 35% of the time.

5.6.1 Teams

The past 40 years has witnessed much change in the areas of organizational theory, structure, and business practice. As noted in the literature, teams are used heavily in industry, government, and the military (e.g., Tannenbaum and Yukl, 1992; Guzzo and Dickson, 1996). Leading up to this are technological advances, geopolitical stability, and free-trade agreements that have increased organizational competition within a global economy. To remain adaptive and to prosper under these circumstances, many organizations are witnessing a flattening of traditional hierarchical structures in favor of teams (Kozlowski and Bell, 2002; Zaccaro et al., 2002). It is estimated that at least 50% of all organizations and 80% of organizations with 100 or more employees use teams in some form (Banker et al., 1996). Similarly, 80% of surveyed workers report that they are currently members of at least one team, and this estimate will continue to increase in step with evolving environmental complexities (Fiore et al., 2001). Ultimately, organizations believe that teams are the answer to many of their problems and are implementing them more readily into their daily business practices.

As a result, researchers have invested a large amount of resources in the study of teams and team training (Salas and Cannon-Bowers, 2001). The focus on team training evolved from problems in the real world, especially in aviation. For example, during the 1970s there were an increasing number of accidents that could be attributed to failures of teamwork in flight crews. To address this, a form of team coordination training known as *cockpit* (and now *crew*) *resource management* (CRM) was introduced as a way to mitigate these failures (Weiner et al., 1993). Since its inception in the early 1980s, CRM training has evolved and spread to the military and a host of other organizations (e.g., health care, nuclear power, offshore oil production), making it one of the most successful team training strategies in use today.

The commercial and military aviation communities have been a driving force in team and team training research (e.g., Salas et al., 1995), investing significant resources to explore these areas further (Tannenbaum, 1997). As a result, a number of theoretically driven team training strategies beyond that of team coordination training (e.g., CRM) have emerged and been validated that are useful for all organizations (e.g., cross-training and team self-correction training). These team training strategies are nothing new and have been discussed widely in the literature. We feel that a lengthy discussion of these strategies here would be redundant, and we encourage the reader to look to others for this review (see Salas and Cannon-Bowers, 2001; Wilson et al., 2005). We do, however, provide a brief description of these team training strategies and additional resources in Table 3. In addition, throughout the next two sections we refer at times to team training strategies that are less common in the literature but are still important to the future success of organizations.

5.6.2 Technology

Not surprisingly, training is slowly turning toward the use of learning technologies as a delivery method vs. traditional classroom training; however, the percentage of training using the classroom method still exceeds that of learning technologies (72% and 15.4%, respectively). There has been an explosion of technological advancements in recent decades. As such, organizations will face two challenges: training employees to interact with technology or utilizing technology to train employees. In today's technology-driven environment, it is difficult to remember a time when certain key tools were not the norms. E-mail, tele- and videoconferencing, text messages, and chat rooms are now the norm. Everyone has become accustomed to the countless "help" icons in various software programs. We not only accept them, we expect them. Thus, training has readily accepted both the benefits and challenges inherent to the utilization of technology.

Advances in technology have enabled training to incorporate technological tools into interventions, and in recent years we have experienced an upsurge in the use of technology (e.g., computer- and Web-based) to train employees (Goldstein and Ford, 2002; ASTD, 2003). The development of intelligent tutoring systems has the potential to reduce or eliminate the need for human instructors for certain types of learning tasks. For example, Anderson and colleagues (Anderson et al., 1995) has proposed that intelligent software can be programmed to monitor, assess, diagnose, and remediate performance successfully in computer programming and to solve algebra problems, and may be extended to other tasks. Additionally, technology-based training offers trainees added control over their learning by allowing them to choose what method to use, the amount of time taken to learn, how to practice, and/or when to receive feedback during training (e.g., Milheim and Martin, 1991). Thus, learner control has been shown to improve trainees' attitudes and motivation toward learning (e.g., Morrison et al., 1992), leading to better learning and performance (Schmidt and Ford, 2003). In this section we discuss some of the technology-based instructional strategies as well as some strategies that organizations can use to help employees better adapt to the use of technology in the workplace. Although some of these strategies still require significant investigation, they have begun to change the face of training. Most important, as this technology becomes more widely available and less costly to develop, it may provide organizations with better alternatives to the traditional classroom training that we may be used to.

Simulation-Based Training and Games One technological advance that has revolutionized aviation and military training for decades is simulation and game-based training. When manuscripts and theories were first being developed in the late 1960s for publications on simulations and learning, the idea was novel (Ruben, 1999). Up until then, a majority of training theory assumed that the learning process consisted of articles, lectures, and books being related to trainees by an instructor. But times have changed. Technology-assisted learning is now emerging as the leader in training innovations. Computer-based instruction and simulations are now readily accepted (and even preferred by younger, tech-savvy participants). Practice and application of KSAs, often difficult in classrooms settings, are now enabled by technology. Computer simulations and games are now being used for instruction, practice, and feedback in training at a rapidly expanding pace.

Simulators continue to be a popular training method in business, education, and the military (Jacobs and Dempsey, 1993), especially within the military and in commercial aviation, which are probably the biggest investors in simulation-based training. Although evidence supports the position that simulation is an effective training tool, the reasons why simulators are so effective remain vague. A few studies have provided preliminary data (e.g., Ortiz 1994; Bell and Waag 1998; Jentsch and Bowers 1998). However, it may be a misnomer to say confidently that simulation (in and of itself) leads to learning. One reason for this is that a majority of the studies addressing these issues rely on trainee reactions and not on actual performance data (see Salas et al., 1998). Although more rigorous and systematic evaluation of simulators is needed to make definitive connections between technology and performance, the use of simulation continues unchecked at a rapid pace in medicine, maintenance, law enforcement, and emergency management settings.

The most popular application of simulators is flight simulation in commercial and military aviation. Flight simulation has been a front-runner in training and evaluation technology, covering such topics as adaptive decision making (e.g., Gillan, 2003), discrimination, performance (Aiba et al., 2002), response time (Harris and Khan, 2003), performance under workload (Wickens et al., 2002), and team issues (Prince and Jentsch, 2001). Driving simulators, originally used primarily for evaluation and assessment for factors such as fatigue and age effects, have begun to explore driving training. Driving simulators have now become popular methods of driver training, safety training, and driver assessment (e.g., Fisher et al., 2002; Roenker et al., 2003). In addition, the medical community has recently begun to utilize simulations for training skills (e.g., the METI doll) as well as team training.

Table 3 Instructional Strategies

Strategy	Definition	References
Technology-based strategies		
Simulation-based training and games	Uses technology to provide opportunities for practice and instruction in realistic settings: lifelike terrain, interaction, and dynamic situations. Used in business, the military, and research. Simulation and games vary according to their fidelity, immersion, and cost.	Tannenbaum and Yukl, (1992),Marks (2000)
Distance learning	Uses technology to facilitate training between instructors and students separated by time and/or space. Uses synchronous or asynchronous and technology, including the World Wide Web, CD-ROM, e- or online learning, videoconferencing, and interactive TV. Broader than e-learning and encompasses nonelectronic learning.	Moe and Blodget (2000)
E-learning	Uses similar methodology to distance learning but is more specific and uses electronic and mostly computer-based methods. Uses Web-based learning, computer-based learning, virtual classrooms, and digital collaboration. Ideal for employees who lack the time for formal training.	Kaplan-Leiserson (2002)
Learner control	Provides learners with the opportunity to make one or more key instructional decisions in the learning process. The structure is ideal for e-learning and other distance learning strategies.	Wydra (1980),Brown and Ford (2002)
Scenario-based training	Provides embedded scenarios with dynamic, complex, and realistic environments. These scenarios trigger targeted behaviors in complex environments. Provides guidelines and steps for training objectives, trigger events, measures of performance, scenario generation, exercise conduct and control, data collection, and feedback. Incorporates technology and uses a meaningful framework to embed opportunities for practice and feedback within meaningful learning events.	Fowlkes et al. (1998),Oser et al. (1999a, b)
Collaborative learning	Incorporates technology tools to facilitate training in groups. Utilizes group interaction to facilitate training, while not necessarily focusing on training for tasks. Focus is on group interaction.	Arthur et al. (1996, 1997)
Error training	Promotes learning through trainees experiencing errors, seeing the consequences of such actions, and receiving feedback. May be especially useful with new technology, where errors may be more prevalent.	Dormann and Frese (1994),Ivancic and Hesketh (1995)
Stress exposure training	Provides information-based instruction that links stressors, trainee affect, and performance. Can be useful in overcoming stress caused by technology or promote the use of technology to mediate stressor effects. Provides coping strategies for trainees in dealing with stressors.	Johnston and Cannon-Bowers (1996),Driskell and Johnston, (1998)
On-the-job training	Provides an opportunity to practice actual required behaviors needed to do a task, including interacting with new technology performing actual tasks. Targets team members procedurally based cognitive skills and psychomotor development. Training is provided in the same environment in which they will be working.	Goldstein (1993),Ford et al. (1997)
Team-based strategies		
Team coordination training	Provides training and practice opportunities for team coordination, communication (both explicit and implicit), backup behavior, and other KSAs that lead to effective coordination. This may be especially important with the introduction of new technology lessening traditional team, face-to-face (FTF) interaction.	Bowers et al. (1998),Entin and Serfaty (1999),Serfaty et al. (1998)
Cross-training	Provides team members with training and practice for performing other team members' roles and tasks. Leads to a better understanding of other team members' responsibilities and task work. Leads to enhanced shared mental models and interpositional knowledge. Also helps understanding the technology used by team members and how it may relate to their own roles and tasks.	Volpe et al. (1996),Salas et al. (1997)

Table 3 ***(continued)***

Strategy	Definition	References
Team self-correction training	Helps team members assess themselves by training team members to correct and evaluate their own behavior to assess the effectiveness of the behavior. Team members also learn to assess the other team members. Encourages constructive feedback and correction of discrepancies. Can help compensate for miscommunications and errors due to new technology.	Blickensderfer et al. (1997),Smith-Jentsch et al. (1998)
Distributed team training	Provides team training to teams distributed by time and/or space that must rely on some type of technology to communicate and coordinate. Uses strategies to promote team competencies and performance by utilizing technology when teams are not colocated, but can encounter problems.	Townsend et al. (1996),Carroll (1999)
Internationalization-based strategies		
Individual-level strategies	Traditionally, multicultural training was focused on two broad categories of learning focused on the individual: didactic or information-giving and experiential learning activities.	Deshpande and Viswesvaran (1992),Kealey and Protheroe (1996)
Attribution training	Provides training for expatriates and others that allows them to make attributions for behaviors more similar to the point of view of members from other cultures. Provides understanding and knowledge of other cultures.	Befus (1988),Bhawuk (2001)
Cultural awareness training	Employs many strategies to teach trainees about their own feelings, concerns, emotions, and unconscious responses. Also, gives information about these areas in other cultures. Goes on the assumption that by knowing their own cultures, people can better understand other cultures.	Bennett (1986),Befus (1988)
Didactic training	Involves information-giving activities, including factual information regarding working conditions, living conditions of the other cultures, and cultural differences, travel, shopping, and appropriate attire. Also includes information about the political and economic structure of other countries.	Kealey and Protheroe (1996),Morris and Robie (2001)
Experiential training	Involves learning by doing where trainees are put through scenarios, often in the form of simulations, so that they can practice their responses to realistic situations. Provides trainees with knowledge (i.e., cognitive tools) and attribution skills for working and interacting with people from different cultures.	Kealey and Protheroe (1996),Morris and Robie (2001)
Team-level strategies	Multicultural team strategies have been developed or adapted to target the complexity of team interactions. Promotes effective performance in multicultural teams involving both team leaders and members. Five common components: (1) a goal to enhance teamwork, (2) reliance on a traditional team performance framework, (3) application of appropriate tools and feedback, (4) the combination of more than one delivery methods, and (5) a short duration to deliver.	Salas and Cannon-Bowers (2001)
Team leader training	Provides leaders with strategies (e.g., coaching) that help promote effective team performance in heterogeneous teams. Organizations can apply a number of guidelines to identify what role team leaders should play within multicultural teams.	Kozlowski et al. (1996),Thomas (1999)
Team building	Allows team members to be a part of the planning and implementation of change rather than having it forced upon them. Relies on one of four team building models: goal setting, interpersonal relations, problem solving, and role clarification.	Dyer (1977),Beer (1980),Buller (1986),Salas et al. (1999)
Role-playing	Provides team members with scripted scenarios that they must act out together. Creates awareness and can be adapted to fit any cultural interactions. Two types of exercises: (1) learning about one's own culture and biases (i.e., enculturation), and (2) learning about other cultures (i.e., acculturation).	Bennett (1986),Roosa et al. (2002)

Another example of simulation training that has received significant attention in recent years is behavior role modeling. For example, Skarlicki and Latham (1997) found that a behavioral role-modeling training approach (e.g., role-playing,) was successful in training organizational citizenship behavior in a labor union setting. In a similar study, Smith-Jentsch et al. (1996b) found that emphasizing practice through role-playing and performance feedback through a behavior modeling approach was more effective in training assertiveness skills than a lecture-only or lecture-with-demonstration format. In another assertiveness training study, Baldwin (1992) found that the best way to achieve behavioral reproduction (i.e., demonstrating assertiveness in a situation that was similar to the training environment) was to expose trainees to positive model displays alone. However, exposing trainees to both positive and negative model displays was most effective in achieving behavioral generalization (i.e., applying the skill outside the training simulation) four weeks later.

Simulations contain a great deal of variability with regard to cost, fidelity, and functionality. Although early simulators were sometimes archaic by today's standards, these environments have come a long way. With the amazing technological advances in recent years, many simulation systems (e.g., simulators, virtual environments) have the ability to replicate detailed terrain, equipment failures, motion, vibration, and visual cues about a situation. However, some widely used simulators are less sophisticated and have less physical fidelity, but represent well the KSAs to be trained (e.g., Jentsch and Bowers, 1998). Although the first instinct of trainers would be to go to the highest fidelity, most realistic-looking simulation, with all the bells and whistles, so to speak, low-fidelity, low-cost simulations can be just as effective.

In fact, recent trends have been to use more of these low-fidelity (e.g., computer-based) devices to train complex skills. Some research even proposes that these low-fidelity simulators result in more skills transfer after training (e.g., Gopher et al., 1994). Therefore, many researchers are examining the feasibility of using low-fidelity off-the-shelf computer games for training KSAs. For example, Gopher et al. (1994) proposed that context-relevant games can be effective in training complex skills. These researchers tested the transfer of skills from a complex computer game to the flight performance of cadets in the Israeli Air Force flight school. The game was based on a skill-oriented task analysis, which used information provided by contemporary models of the human processing system as the framework. Results showed that flight performance scores of two groups of cadets who received 10 hours of training in the computer game performed much better when they were compared with a matched group with no game experience. These results support the use of lower-fidelity computer games for improving performance after the training of complex skills. Similar results may be found in Goettl et al. (1996) and Jentsch and Bowers (1998).

Similarly, Ricci et al. (1995) investigated the use of a computer-based game to train chemical, biological, and radiological defense procedures. Although not a full-scale simulation, the computer-based slot machine that presented trainees with questions about the material enabled trainees to earn points for correct answers and receive corrective feedback for incorrect ones. As hypothesized by the authors, motivation to engage in this type of presentation over text-based material (e.g., books) resulted in higher learning, indicating that reactions and retention (but not immediate training performance) were higher for the game condition.

Although simulations have become more and more prevalent in the training community, some have noted (e.g., Salas et al., 1998) that simulation and simulators are being used without much consideration of what has been learned about cognition, training design, or effectiveness. Therefore, it is up to researchers to integrate the science of training with simulator application, design, and practice. One possible solution to this dilemma is to incorporate the event-based training approach, a scientifically based training strategy, with simulations (Cannon-Bowers et al., 1998; Fowlkes et al., 1998; Oser et al., 1999b). This means that simulation training should incorporate training objectives, diagnostic measures of processes and outcomes, feedback, and guided practice. Ultimately, we can ascertain that simulators and computer games work as a training tool only when the training is theoretically driven, focused on required competencies, and designed to provide trainees with realistic opportunities to practice and receive feedback. More research must be done to verify that simulators being used today will follow the science of training.

Distance Learning It would be hard to argue that developing technology and our expanded abilities resulting from that technology is not a major driver of training methodologies and strategies in recent years. Moe and Blodget (2000) estimate that $8.2 million will be spent by organizations on technology-driven training in 2001. Although organizations have not done away with traditional or "old-fashioned" training (e.g., classroom lectures), organizations are integrating training technologies such as videoconferencing, electronic performance support systems, videodisks, and online Internet/intranet courses at an accelerated pace. The evidence of Web-based training alone can be found by doing a simple Internet search. People can get certified in programming languages, learn cultural sensitivity, and even earn a Ph.D. online. However, trainees must approach this technology-driven explosion cautiously, since it soon becomes obvious that this implementation is happening without much reliance on the science of training. With organizations expanding globally, technology evolving rapidly, and the workloads placed on employees increasingly rapidly, training has begun to go in the direction of distance learning. *Distance learning* is the result of distance training, although the two terms are often used interchangeably. *Distance training* refers to a training situation where instructors and students are separated by time and/or space. Training can be synchronous or asynchronous and utilizes such technology as the World Wide Web, CD-ROM, e- or online learning, videoconferencing, and interactive TV.

There are many issues concerning the design of distance learning that remain to be addressed. Like all training, the application of distance learning, must be steeped in theory in order to develop principles and guidelines that will guide the instructional design of such interventions. Some research has already begun to focus on the issues surrounding this topic (e.g., Schreiber and Berge, 1998), but a science of distance learning and training must still be secured. Specifically, research must explore what level of interaction is needed when applying collaborative learning tools or between trainers and trainees, along with the nature of their interaction. Some important questions to be addressed include (1) do instructors need to have visual contact with trainees to conduct effective instruction? (2) do trainees need to see instructors, or is it better for them to view other material? (3) how do you best address trainee questions (e.g., through chat rooms or e-mail) or provide feedback to trainees? (4) should learner control be used? [Some evidence from studies of computer-based training support the use of learner control (see Shute et al., 1998), but the extent of its benefits for distance learning is not known.] Future research must address these and other questions.

Distributed Team Training With the increases in technology, organizations are quickly moving toward a more global workplace. Because of this, team training often cannot be conducted with team members physically co-located. Instead, team members are distributed, or "mediated by time, space or technology" (Driskell et al., 2003, p. 3) and rely on some technological medium to communicate (i.e., e-mail, videoconferencing, telephone, or fax machine) (Townsend et al., 1996). Because team members are dispersed, traditional training programs are not effective. The military has been most affected with the problem of training distributed teams and has made the greatest strides in designing training programs specifically for distributed teams.

Military-based distributed mission training (DMT) involves real, virtual, and constructive components to create an interactive virtual reality system that allows trainees to engage in real-time scenario-based training that requires coordination and communication with teammates both real and virtual (Carroll, 1999). After training scenarios are completed, users are given feedback on their performance, and the information from that training scenario is saved in the user's file so the next training scenario can be completed.

The development of DMT programs in the military has resulted in two training platforms that utilize simulated training scenarios and virtual teammates, or cognitive agents, and actual military personnel: Synthetic Cognition for Operational Team Training (SCOTT) (Zachary et al., 2001) and Synthetic Teammates for Realtime Anywhere Training and Assessment (STRATA) (Bell, 2003). Users are able to log in to a secure network using a user name and password from anywhere in the world and are placed into a training scenario that will develop the knowledge and skills that are needed according to their training profile. They will then complete the training scenario in conjunction with either real teammates or cognitive agents acting as teammates. Although the majority of work involving DMT is military related, the skills trained in DMT are critical for all teams regardless of type (Bell, 1999).

E-Learning Although e-learning is a related concept, distance learning is much broader and encompasses nonelectronic learning; e-learning is electronic and mostly computer-based (Kaplan-Leiserson, 2002). With the advancements in technology and a busier workforce, many organizations have turned to e-learning as a tool to train distributed and/or "time-crunched" employees. E-learning can be defined as "a wide set of applications and processes, such as Web-based learning, computer-based learning, virtual classrooms, and digital collaboration" (Kaplan-Leiserson, 2002, p. 85). E-learning offers organizations a training tool that is affordable and can be used any time anywhere. The biggest advantage of this tool appears to be the fact that not only is it cost-efficient, but it offers training developers a great deal of freedom. E-learning can be structured to be used by distributed trainees, to be collaborative from remote environments, can be used synchronously or asynchronously, can be structured or learner controlled, and can be used continuously by different employees in different locations (DeRouin et al., in press). However, research (Brown and Ford, 2002) has shown that there are a few limitations on e-learning. Workers who are ideal for e-learning as a training tool typically lack the time for formal training. Therefore, training structured in e-learning environments must be able to be offered quickly, on-demand, and should be built in a way that is accessible from multiple locations (e.g., over the Web).

Learner Control One aspect of e-learning is learner control, explored mostly in the literature in general (Steinberg, 1977; Chung and Reigeluth, 1992; Goforth, 1994; Hamel and Ryan-Jones, 1997; Brown and Ford, 2002). Learner control refers to "a mode of instruction in which one or more key instructional decisions are delegated to the learner" (Wydra, 1980, p. 3). The structure of e-learning makes it ideal for the opportunity of learner control. Research has found both benefits (e.g., improved learning outcomes, increased satisfaction with training, and an increased amount of time trainees choose to spend training) (Ellermann and Free, 1990; Shyu and Brown, 1992; Freitag and Sullivan, 1995). However, conflicting evidence has also been found supporting negative results of learner control (Tennyson, 1980; Murphy and Davidson, 1991; Lai, 2001). DeRouin and colleagues (2004) suggest that this is due to the training offered and the relevance of the instructional material to trainees.

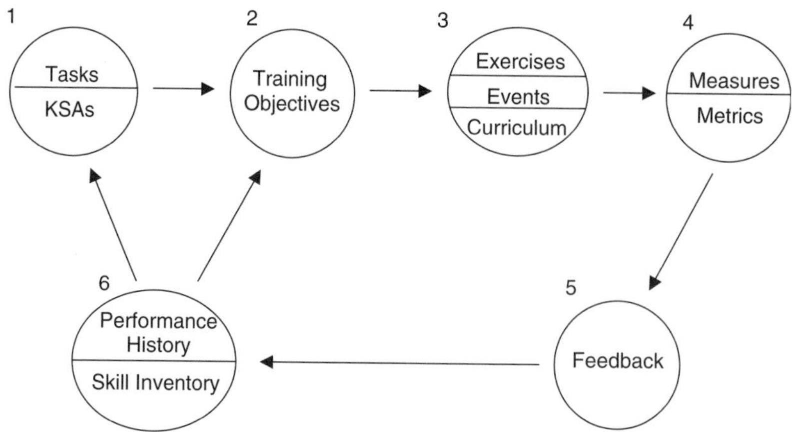

Figure 4 Components of scenario-based training. (From Cannon-Bowers et al., 1998.)

Scenario-Based Training *Scenario-based training* (SBT; Figure 4), also known as *event-based training*, is one instructional strategy that can be used to train both individuals and teams (Fowlkes et al., 1998). Unlike most traditional instructional strategies, there is no formal curriculum to scenario-based training. Rather, the scenario acts as the curriculum. This strategy uses embedded "trigger" events (i.e., learning events) to elicit targeted knowledge and skills through structured and guided practice scenarios (Karl et al., 1993; Dormann and Frese, 1994; Ivancic and Hesketh, 1995). Scenario-based training, when applied appropriately, is an effective training strategy because it is practice-based and provides trainees with a meaningful framework by which to learn (Fowlkes et al., 1998; Salas and Cannon-Bowers, 2000b). These events are defined a priori, determined from critical incidents data, and provide trainers with up-front knowledge of what competencies should be demonstrated and when. This allows for more effective and accurate performance measurement. Feedback is then provided immediately following the completion of practice to improve trainees' performance in future practice and on the job.

In addition to being practice-based, there are several additional benefits to using SBT for training individuals and teams. First, scenario-based training has a flexible architecture that can be adapted to a wide variety of situations that require different responses from trainees. By providing trainees with different opportunities to learn, they can build templates of what to expect and how to react in these situations (Richman et al., 1995). These templates or mental models will allow trainees to recall more rapidly the correct actions and decisions to make when similar situations are faced in the future (Klein, 1997). This is critical for teams and individuals operating in complex decisions where timeliness is of the utmost importance. Finally, scenario-based training is a broad instructional strategy that can be combined with other instructional strategies. For example, scenario-based training can be used in conjunction with assertiveness training to train junior team members to assert themselves to more senior members. Scenarios requiring that junior members assert themselves could be devised allowing these team members to practice speaking up to their superiors (e.g., a junior flight crew member letting the captain know that he or she skipped a step on the before-takeoff checklist). This combination of training not only benefits junior team members, but also creates an awareness to higher-ranking team members of the consequences of not considering the concerns of others, regardless of rank.

Collaborative Learning Additional research has focused on the development of collaborative tools to facilitate training. *Collaborative learning* is defined as situations where trainees are trained in groups. It should be noted that this should be distinguished from team training, in that the latter refers to training focused on training competencies necessary for performance of team tasks, and the former is not necessarily focusing on training for tasks performed as a group. Support has been found for the idea that certain features of group interaction facilitate the learning process (e.g., interaction with peers can facilitate learning. Specifically, Arthur et al. (1996, 1997) showed support for the ongoing use of innovative dyadic tools in training for both military and nonmilitary pilots and navigators. However, they found that the effectiveness of these collaborative tools with computer-based training is moderated by trainees' level of interaction anxiety, with only low-interaction-anxiety trainees benefiting from dyadic protocols. In addition, Shebilske et al. (1992) reported evidence that collaborative protocols have also been shown to reduce required instructor time and resources by half and found supportive evidence for the social learning theory that observational learning opportunities compensate for hands-on practice efficiently and effectively (Shebilske et al., 1998).

Error Training Error training, although relatively neglected in the literature, has in recent years gained more attention. Error training is especially critical as we move further into the technological age, in which workers switch their roles from active members to passive observers. The premise behind error training is that trainees will experience errors, see the consequences of such actions, and receive feedback to improve their learning (Frese and Altman, 1989; Karl et al., 1993; Heimbeck et al., 2003; Lorenzet et al., 2003). Once feedback has been received, trainees are taught strategies for avoiding or minimizing the consequences of errors and are provided with opportunities to practice them. Findings in the literature indicate that providing error training improves performance on the job more than when error-free training is provided (e.g., Dormann and Frese, 1994; Ivancic and Hesketh, 1995).

There are two key components of errors used as a part of training that need to be considered: error occurrence and error correction (Lorenzet et al., 2003). *Error occurrence*, how or whether errors occur, can be approached in four different ways: (1) avoid, (2) allow, (3) induce, or (4) guide. First, errors can be avoided by designing training to prevent errors from occurring. For example, trainees are provided only with the information that they need to know; no other information (e.g., errors) is included in training. Although this improves trainee learning and motivation, it does little to help trainees who may be faced with errors on the job. Second, errors can be allowed to occur by chance. This type of approach allows trainees to make errors if such a situation should occur but can be challenging for trainers in that it is unclear what errors may occur and when (Gully et al., 2002). Next, errors can be induced or evoked by making changes in the training program (e.g., increasing complexity) with the hopes of errors occurring (Dormann and Frese, 1994). Although there may be benefits to trainees experiencing errors, approaches such as allowed and induced can have negative consequences if trainees make internal attributions about why errors are occurring rather than assuming that it is a part of the training program itself. Finally, guided error occurrence takes an approach in which trainees are intentionally guided to errors to encourage learning. Guided error occurrence has benefits over the other approaches because training becomes more standardized (i.e., all trainees experience the same errors).

The second component of error training is *error correction*, how trainees correct errors that occur during training. There are two subcomponents to error correction. First, trainees can work through the errors alone without the aid of an instructor or other mechanism (i.e., self-correction) (Frese et al., 1991). This approach is beneficial to trainees in that they can learn their own strategies for dealing with errors (i.e., self-discovery). On the other hand, trainees can be offered support and feedback to help them correct the errors (i.e., supported correction) (Carlson et al., 1992). Support can be provided in the form of a computer-based training aid or instructor intervention (Lorenzet et al., 2003).

Taking this together, Lorenzet et al. (2003) suggest that guided error training in conjunction with supported correction may be best at improving the skill development of trainees. Results of their research indicate that trainees given error training performed more accurately and in a quicker manner than those provided with error-free training. In addition, these trainees also reported higher levels of self-efficacy following error training, due to their ability to practice what they learned. These findings lend further support to the importance of providing practice and feedback to trainees.

Stress Exposure Training Another instructional strategy used to train people to deal with the uncertainties often associated with technology is stress exposure training (SET). The purpose of SET is to provide trainees with the abilities and tools necessary to maintain effective performance when operating in high-stress environments (Driskell and Johnston, 1998). This training is especially important in environments where the consequences for errors are high, as stress increases the likelihood of errors. There are three phases of SET training. The first phase involves providing information regarding different types of stressors that may be encountered on the job and the effects of those stressors on performance. Once a basic awareness and knowledge has been created, phase two follows with trainees acquiring skills (behavioral and cognitive) that are necessary to effectively manage and adapt to the potential stressors. Phase three concludes SET with trainees applying and practicing the knowledge and skills acquired during the previous phases in an environment that gradually approximates a high-stress situation followed by feedback on their performance. In addition to preparing trainees to perform effectively under a particular stressor on a specific task (e.g., time pressure while landing at an airport), SET has proven to generalize from stressor to stressor (e.g., time pressure to high workload) and task to task (e.g., landing an aircraft to handling an emergency situation) (Driskell et al., 2001).

On-the-Job Training On-the-job training (OJT), one of the most widely used instructional strategies in organizations, can be implemented in isolation or in conjunction with an off-the-job training strategy (such as those discussed throughout this chapter) (Goldstein, 1993). OJT, as the name suggests, takes place in the actual physical and social environment where the tasks being trained are performed and consists of experts or supervisors training specific job-related tasks (e.g., what to do, how to do it) (Wehrenberg, 1987; Sacks, 1994; De Jong and Versloot, 1999). Unlike most other training programs, OJT offers several benefits to organizations. First, trained skills will more likely transfer to the job following training because they were learned and practiced in the work environment. Additionally, the fact that trainees practice the trained skills while under supervision will result in incorrect behaviors being trapped and corrected before the trainee is allowed to conduct tasks without supervision.

There are two types of OJT: apprenticeship training and mentoring. *Apprenticeship training* has typically been associated with training trade skills (e.g., blacksmithing); however, more nontraditional trade organizations have begun to show an interest in this type of OJT (Goldstein, 1993). In general, apprenticeship training involves both classroom-based instruction and supervision on the job by an experienced employee. Once a predetermined amount of time has past and the necessary skills have been acquired, an apprentice is advanced to a "journeyman," where he or she is able to perform the learned tasks without supervision (Goldstein, 1993; Lewis, 1998; Hendricks, 2001).

The second type of OJT is *mentoring*. Mentoring, similar to apprenticeship training, involves creating a relationship between a less experienced person and an experienced person (Wilson and Johnson, 2001). In this relationship, the experienced person trains and develops the trainee such that he or she can perform the job accurately and effectively. Studies indicate that building a mentoring relationship fosters communication, job satisfaction, and success on the job (Mobley et al., 1994; Forret et al., 1996). Like apprenticeship training, mentoring allows trainees to practice the learned skills under supervision, allowing appropriate behaviors to be encouraged while inappropriate behaviors are discouraged (Scandura et al., 1996). A final benefit of mentoring is that it provides the mentor (in addition to the trainee) the opportunity to improve his or her skills that may have deteriorated with time (Forret et al., 1996).

5.6.3 Internationalization

With global markets expanding to over 10,000 companies across the world, the interaction between cultures in the business world and in daily life is increasing (Adler, 1997). In addition, teams are becoming a more integral component of organizational goals within all companies, increasing the presence of interdependent, multicultural teams. At the same time, new technologies are emerging that enable team members to engage in teamwork while being temporally and geographically distributed (Bell and Kozlowski, 2002). As a result, more individuals and teams are interacting with others from different cultures. However, despite the application of multicultural interactions, an understanding of how those interactions affect employees and the organization is still in the early stages. Although this multicultural training is still in its infancy, a great deal of training research has focused on the strategies of multicultural training, mostly with a focus on preparing expatriates for foreign jobs. However, with the global nature of the modern organization, cross-cultural training is a necessary intervention with employees of varied backgrounds in team settings, not just for employees going to foreign destinations. Therefore, the following section outlines some cultural implications important to individuals and teams, and describes a number of individual and team strategies being applied as a result.

Training for culture has been called many things in the literature: intercultural training, diversity training, multicultural training, and cross-cultural training. Although these titles are often used interchangeably, it may be argued that there are differences. However, this is not the forum to explore this differentiation (see Gudykunst et al., 1996). Therefore, for the current purpose, we refer to this type of training as *multicultural training*, to reflect the nature of training across different cultures. Multicultural training has been defined by researchers (Landis and Brislin, 1996; Morris and Robie, 2001) as the process of educating individuals or teams on behavioral, cognitive, and affective patterns that promote successful interaction across cultures. The goal of multicultural training is not just the acquisition of information, but the changing of trainee attitudes toward different cultures, which ultimately affect their behaviors (Bhagat and Prien, 1996). Research has also provided additional support for this type of training as a major technique in improving managers in multicultural environments (e.g., Deshpande and Viswesvaran, 1992; Bhagat and Prien, 1996; Bhawuk and Brislin, 2000).

Individual Cultural Training Strategies Traditionally, multicultural training was focused on two broad categories of learning focused on the individual: didactic or information-giving and experiential learning activities (Deshpande and Viswesvaran, 1992; Kealey and Protheroe, 1996). However, recent methodology has expanded to include seven approaches to multicultural training: attribution, culture awareness, didactic, experiential, and cognitive behavior modification, interaction, and language training (Bennett, 1986; Befus, 1988). The most popular and often used methods are as follows:

1. *Attribution training*. Attribution training is an intervention that aims to teach expatriates and others to make attributions for behaviors more similar to the point of view of members of other cultures (Befus, 1988). Ultimately, this training will help people from one culture learn about other cultures so that they can interpret behaviors they observe in a similar manner. This not only provides trainees with knowledge, but serves to deepen their understanding of culture perspectives (Bhawuk, 2001).
2. *Cultural awareness training*. Cultural awareness training operates on the assumption that by knowing their own culture, trainees can better understand the differences between their own culture and other cultures they may encounter (Befus, 1988). This general approach employs many strategies, including T-groups, which involves trainees learning about their own feelings, concerns, emotions, and unconscious responses (Bennett, 1986). The ultimate goal of awareness training is to learn about your own culture, as well as others, to better recognize the differences and improve the successful interactions you have with those from other cultures (Bennett, 1986).
3. *Didactic training*. Didactic training involves information-giving activities, including factual information regarding working conditions, living conditions

of the other cultures, and cultural differences, travel, shopping, and appropriate attire. This approach focuses on cognitive goals, culture-specific content, and traditional education (Bennett, 1986). In addition to the foregoing topics, trainees receive information regarding the political and economic structure of other countries (Kealey and Protheroe, 1996; Morris and Robie, 2001). The information provided is designed to enable trainees to identify differences between their own culture and others, the ultimate goal being to enhance cognitive skills so that trainees can better understand and evaluate cultures other than their own (Morris and Robie, 2001). When done properly, didactic training provides trainees with a frame of reference when they encounter new situations (Kealey and Protheroe, 1996; Morris and Robie, 2001).

Didactic training can be applied in a number of ways (e.g., informal briefings, cultural assimilators) (Brewster, 1995), which can take several forms. For example, informal briefings can be used which take the form of casual conversations with experts or past trainees who have interacted successfully in other cultural environments (Brewster, 1995; Kealey and Protheroe, 1996). Informal briefings can also include lectures, videotapes, workbooks, and Q and A sessions (Grove and Torbiörn, 1993; Kealey and Protheroe, 1996). Another example of didactic training involves cultural assimilators. According to Bhawuk (1998, 2001), cultural assimilator training requires trainees to read scenarios or critical incidents and choose one of four options of how they would react. After they make their choice, an expert's view is provided as the most appropriate response (Kealey and Protheroe, 1996; Bhawuk, 1998, 2001; Morris and Robie, 2001). All of these can be classified as didactic training because they involve information giving.

4. *Experiential training*. Experiential learning involves learning by doing where trainees are put through scenarios, often in the form of simulations, so that they can practice their responses to realistic situations. This method provides trainees with knowledge (i.e., cognitive tools) and attribution skills for working and interacting with people from different cultures (Kealey and Protheroe, 1996; Morris and Robie, 2001). When applied appropriately, experiential training will result in an improvement in multicultural communication skills and the application of knowledge about what to do in certain situations (Kealey and Protheroe, 1996; Morris and Robie, 2001). The development of cognitive skills also enables trainees to see the perspective of people from other cultures (Morris and Robie, 2001). In addition to simulations, role-playing (discussed later as a multicultural team training strategy) and workshops are often used to promote experiential training (Grove and Torbiörn, 1993; Kealey and Protheroe, 1996; Morris and Robie, 2001).

An example can be found in simulations. A simulation game often used for multicultural experiential training is BAFA BAFA (Bhawuk and Brislin, 2000). This simulation revolves around two hypothetical countries: alpha (i.e., masculine and collectivistic culture) and beta (i.e., feminine and individualistic culture). After being assigned to one of the two cultures, each participant is required to "go to" the other country and come back and explain their experiences to their group. Following the exercise, trainees are debriefed and told that the purpose of the simulation was to show them the differences that they may encounter in different cultures (Gudykunst et al., 1996).

Although these strategies are useful in training multicultural issues, there is some concern in the literature (Black and Mendenhall, 1990; Deshpande and Viswesvaran, 1992; Morris and Robie, 2001). Specifically, empirical results are lacking to support claims that this type of training is effective (Selmer, 2001). In addition, it must be noted that a number of moderating factors could explain the positive results found following multicultural training. As a result, the field of multicultural training is in need of more empirical, quantitative support for the implementation of training, which appears to work at the anecdotal level. Ultimately, it is not an exaggeration to say that empirical studies are the most drastically needed aspect of research in this domain (Black and Mendenhall, 1990; Selmer et al., 1998; Morris and Robie, 2001; Selmer, 2001). It is necessary to decrease the theoretical, anecdotal arguments and begin performing empirical investigations to provide stronger support for multicultural training techniques. In addition, variables that may influence the relationship between multicultural training effectiveness and performance must be included in these analyses (Black and Mendenhall, 1990; Morris and Robie, 2001). These moderators include organization-level attributes, job-level attributes, and individual-level attributes (Bird and Dunbar, 1991; Bhagat and Prien, 1996).

Multicultural Team Training Strategies With the increase in teams, multicultural team strategies have been developed or adapted to target the complexity of team interactions. Building on traditional team training strategies, these three strategies can be used to promote effective performance in multicultural teams involving both team leaders and members. There are generally five common components present in multicultural team training strategies: (1) a goal to enhance teamwork within specific settings with a focus on general team objectives, (2) a reliance on a traditional team performance framework, (3) the application of appropriate tools and feedback, (4) the combination of more than one delivery method (i.e., information, demonstration, and practice), and (5) a short duration to deliver their message (e.g., 2 to 5 days) (Salas and Cannon-Bowers, 2001). Although a number of the team training strategies can be adapted to fit multicultural team training, the following are three examples of multicultural team training strategies that have been applied within organizations:

1. *Team leader training*. Poor management has been identified as a reason that multicultural teams fail (Moore, 1999). Leaders of multicultural teams

must confront a number of challenges not encountered within homogeneous teams (Salas et al., 2004). Although team leader training does not consist of a formal training technique, organizations can apply a number of guidelines to identify what role team leaders should play within multicultural teams. One role often associated with multicultural training is that of leader as coach. For example, process losses, such as communication, have been identified as a major contributor to poor multicultural team performance (Thomas, 1999). If the leader is able to act as a coach to multicultural team members, they can help in the development of strategies to overcome this process loss (Kozlowski et al., 1996). Therefore, it can be argued that coaching is an effective team leader role within multicultural teams. There are several strategies that can be used to promote effective coaching by leaders (see Martin and Lumsden, 1987). One strategy involves the leader offering praise for appropriate or desirable effort or processes exhibited by team members (e.g., openly communicating). A second strategy entails rewards for team members who exhibit the desired behaviors. A third strategy would be to have the leader encourage positive interactions among the team members (e.g., avoiding stereotypes). Team leaders can helps motivate team members, inspire them to work together, and help them overcome differences by following these coaching guidelines.

2. *Team building*. While team building is a general team training strategy, it may also be used to target multicultural team issues. At the core of team building, teams are allowed to be a part of the planning and implementation of change rather than having it forced upon them (Salas et al., 1999). The literature provides a number of sources to guide team building. Specifically, organizations typically rely on one of four team building models: goal setting, interpersonal relations, problem solving, and role clarification (Dyer, 1977; Beer, 1980; Buller, 1986). No single model is more appropriate than another across the board. The model or combination of models that an organization selects should be determined by organizational goals and team tasks (Salas et al., 1999). Goal setting provides team members with opportunities to set their own individual and team goals, develop team objectives, and develop strategies for achieving set goals and objectives. This allows them to show a greater range of perspectives (Watson et al., 1993) and promotes more comprehensive goals. Interpersonal relations models can be used to promote team skills, develop confidence, and increase trust among team members. This can be helpful because multicultural teams have been shown to possess less trust, which can lead to decreased performance (Distefano and Maznevski, 2000; Triandis, 2000). The problem-solving model can be applied to identify problems that may exist within the team. This enables team members to develop strategies for solving these problems as a cohesive unit and then to evaluate these strategies. In this way, teams can better define the problems they may encounter (Adler, 1997) and generate more solutions to problems (Daily et al., 1996), which leads to better solutions are better than are available from homogeneous teams (Hoffman and Maier, 1961). Finally, role clarification models focus on an increase in team member understanding of members' roles and increases their ability to communicate. This can be particularly helpful in multicultural teams since they have been shown to have communication difficulties (Steiner, 1972; Thomas, 1999) and differences in role classifications (Hofstede, 1980).

3. *Role playing*. Role playing is another strategy that can increase effectiveness in multicultural teams. In role playing, members are given scripted scenarios that they must act out together. The goal of role playing is to create awareness and can be adapted to fit any cultural interactions. To apply this strategy appropriately, two types of exercises should be implemented: (1) learning about one's own culture and biases (i.e., enculturation) (Roosa et al., 2002), and (2) learning about other cultures (i.e., acculturation) (Bennett, 1986). *Enculturation* is a vital part of the multicultural training process. Many of us are not aware of how our own biases, beliefs, and customs affect our behaviors and influence how we decide to treat others. Without an understanding of our own tendencies, we cannot truly understand other cultural tendencies, which is a very important step in the functioning of multicultural teams. The second component of role playing is *acculturation*, which focuses on providing team members with awareness of other cultures. When applying this two-faceted approach, role-playing can serve to promote awareness of their own and other cultural biases, customs, traditions, and tendencies by allowing team members to act out both dimensions. Furthermore, to promote an understanding of the different perspectives in multicultural settings, team members should be required to play multiple roles within the scripted scenarios. An additional benefit of role-playing is that it provides team members with practice for skills that can help them overcome the differences between cultural biases and tendencies. To ensure that this awareness and practice are effective, team members should be provided with feedback regarding their performance and additional training when necessary following these scenarios.

5.7 Program Content

The final step in the training design phase is to lay out the program content. This step entails determining the sequence and structure of the training program (Clark, 2000). The sequence should be logical and easy to follow by trainees and should be structured such that all training objectives are met. Furthermore, each learning activity (or practice) should have a definite purpose and be provided in a meaningful context. This process will also provide standardization of the training program from one implementation to the next as the material to be covered and when will be clearly specified.

5.8 Summary

The research exploring the how training systems need to be designed is abundant. It is important to remember that there is more to training than just the training system itself and its content. Rather, designers must also consider factors external to the training system (e.g., organizational and individual characteristics) that may influence its effectiveness. This is a requirement if the training system is to be a success.

6 TRAINING DEVELOPMENT

The third phase of the ISD model involves the actual development of the training program. As a part of this phase, course materials, including lesson plans, should be developed, learning activities (i.e., practice) should be specified, and tests and performance measures should be developed.

6.1 Practice Scenario Development

Critical to the success of training is the availability of practice opportunities during training. These opportunities will not only help to identify deficiencies in individual and team performance, but will help trainees with the transfer and retention of learned competencies. Practice scenarios, however, must be laid out carefully and storyboarded prior to training. This provides trainers and researchers added control over the practice portion of training by standardizing what competencies are being training, how the competencies are being presented to trainees, and when. It is often believe that practice scenarios should have all the "bells and whistles" of the real-world environment. Although realism is important and the small details should not be ignored, low-fidelity simulations (such as role-playing) offer trainees benefits similar to those of high-fidelity practice (e.g., full motion simulators). The level of realism will undoubtedly be related to what is being trained and what trainees are to get out of training. In addition to being realistic, practice scenarios should be developed so that they challenge trainees by varying in their levels of difficulty. Next, scenarios should be scripted such that they allow trainees to respond during practice in different ways (i.e., there should not be one right or wrong answer). Finally, multiple practice opportunities should be made available to allow trainees to practice the trained knowledge and skills on multiple occasions (Prince et al., 1993). By developing scenarios that engage trainees, they will build confidence (Richman et al., 1995) and be more likely to transfer and apply what they have learned on the job—and this can only benefit organizations.

6.2 Performance Measures

Performance measurement is a must; without measurement and feedback, there are no opportunities for learning. It is well known that diagnosis is critical for learning. It can be argued that training will be effective only to the extent that trainee competence can be assessed. For this to be true, three criteria must be met. First, measurement opportunities must be provided that ease the burden on those responsible for performance measurement. In other words, the use of prescripted, learner-focused scenarios ensures that the significant competencies are being prompted. In this way instructors know a priori when these "trigger" events will occur and can observe and record performance.

Second, a basis for diagnosing performance trends and providing feedback must be established, and this is more challenging than one would expect. For example, automated technology as a part of a simulation is a great way to capture performance outcomes (e.g., time, errors). This technology is limited, however, in that it cannot easily capture data related to the real-time processes that trainees progress through to attain these outcomes (e.g., communication, decision making). This is especially true when assessing team performance, as teams are very dynamic in nature and teamwork processes are difficult to capture. For example, during periods of high workload (such as those experienced by trauma teams), teams will often communicate and coordinate implicitly, which is impossible for a simulation-based system to detect. A human observer, on the other hand, will be more able to make inferences from observing the behaviors to diagnose teamwork issues using checklists or observation forms (e.g., TARGETS) (see Fowlkes et al., 1994). It is recommended that training programs utilize several (at least two) observers or evaluators, who can more readily diagnose performance and provide strategies for improving future performance (Brannick et al., 1995). The use of evaluators to provide ratings, unfortunately, is not free of error and bias. Thus, training designers must focus on improving the reliability (i.e., are evaluators' ratings consistent with each other, and are each evaluator's ratings consistent over time?) and validity (i.e., are evaluators rating the right things?) of evaluators through training to ensure consistency and accuracy (for a description, see Brannick et al., 2002; Holt et al., 2002; O'Connor et al., 2002). A final challenge faced by training designers is ensuring that multiple measurements are taken throughout the simulation to gather a truly representative picture. Again, this is especially important when one is interested in obtaining process-related data.

6.3 Summary

The development of practice scenarios and feedback are critical to the success of a training system. It is with these opportunities to practice applying the knowledge and skills learned and receiving feedback on one's performance that long-lasting learning can occur.

7 TRAINING IMPLEMENTATION

At this point, the training program has been developed and the organization should be ready to implement the training program. As a part of this stage, an adequate training location needs to be identified that has the necessary resources for training (e.g., Internet access for Web-based training). In addition, instructors need to be trained, training should be pilot-tested, feedback should be received from trainees, and necessary revisions should be made (Clark, 2000). Once this is completed, delivery of the full-scale training system is ready to go.

8 TRAINING EVALUATION

Just as important as events that occur before and during training, posttraining conditions can significantly influence the effectiveness of training. The evaluation of training, the environment following training, and the effective application of KSAs acquired in training to work environments (i.e., transfer of training) continue to be of interest to training researchers. The most significant progress has been made in the areas of training evaluation and transfer of training. Furthermore, within these areas of study there are theoretical, methodological, empirical, and practical advances. Although not all issues have yet been addressed, meaningful advances have been made in the last decade. In the following sections we document the advances made in the last few decades and the suggestions for steps that may still need to be taken. We first look at training evaluation and the prevailing theories that have driven it.

8.1 Evaluation Design Concerns

Once the instructional strategy has been chosen and the training program has been implemented, it is imperative that the training be evaluated. Few organizations conduct systematic evaluations of their training programs. Although we acknowledge that evaluation can be resource intensive, it is the only way to truly assess training's effectiveness. *Training evaluation* refers to a system for measuring the intended outcomes of training. It is concerned with issues of measurement, design, learning objectives, and the attainment of desired knowledge, skills, and abilities. Ultimately, training evaluation asks, "Did the training work?" and effective evaluation models are necessary to inform trainers and researchers of the added value of their training program.

Everyone can agree that training evaluation is a good idea, but it also becomes apparent as one begins to tackle it that the evaluation process is labor intensive, costly, political, and often gives the organization bad news. Furthermore, some training evaluation procedures must be conducted in the field or on the job, which is always a difficult undertaking. The process, however, has benefited in recent decades based on some empirically tested thoughtful, innovative, and practical approaches to aid the evaluation process. General evaluation research includes Sackett and Mullen (1993), who proposed other alternatives (e.g., posttesting only, no control group) to formal experimental designs when answering evaluation questions. They suggested that each evaluation mechanism needed, which is driven by the evaluation questions, requires a different design. Furthermore, Haccoun and Hamtiaux (1994) proposed the *internal referencing strategy*, which tests the implicit training evaluation notion that training-relevant content should show more change (pre–post) than training-irrelevant content. This is a simple procedure for estimating effectiveness of training in improving trainee knowledge. This method was based on an experiment that tested an empirical evaluation using internal referencing strategy vs. a more traditional experimental evaluation Findings suggested that the internal referencing strategy approach might permit inferences that mirror those obtained by more complex designs.

It is hopeful that more evaluations are being reported in the literature. This is a much needed and valuable addition. This literature allows us to learn from past evaluations and ensures that the design and delivery of training will continue to progress. Training evaluation research has also spanned numerous training areas, including team training settings (e.g., Leedom and Simon, 1995; Salas et al., 1999), sales training (e.g., Morrow et al., 1997), stress training (e.g., Friedland and Keinan 1992), cross-cultural management training (e.g., Harrison, 1992), transformational leadership training (e.g., Barling et al., 1996), career self-management training (e.g., Kossek et al., 1998), workforce diversity training (e.g., Hanover and Cellar 1998), and approaches to computer training (e.g., Simon and Werner, 1996). All research reinforces what we already know: Training works. What is not decided is the best way to go about evaluating training in general. Unfortunately, several surveys of public and private organizations (e.g., Catalanello and Kirkpatrick, 1968; Ralphs and Stephan, 1986) indicate that high-quality multifaceted evaluations of training programs are rarely conducted. To date, the best accepted evaluation "model" was that proposed by Kirkpatrick (1959, 1976, 1987) (see Section 8.3).

8.2 Costs of Training Evaluations

There are also practical considerations in training (e.g., organizational resources, costs). Recent research has begun to address the costs of training evaluation. Yang et al. (1996) examined two ways to reduce costs: (1) assigning different numbers of subjects to training and control groups, and (2) substituting a less expensive proxy criterion measure in place of the target criterion when evaluating the training effectiveness. First, an unequal group size design with a larger total sample size may achieve the same level of statistical power at lower cost. In addition, using a proxy increases the sample size needed to achieve a given level of statistical power. Furthermore, the authors described procedures that examined the trade-off between the savings from using the less expensive proxy criterion and costs incurred by the larger sample size. See Arvey et al. (1992) for similar suggestions.

8.3 Kirkpatrick's Typology and Beyond

While there are numerous methods of assessment that can be applied to training, Kirkpatrick (1976) discussed evaluation in terms of four steps or criterion types: (1) trainee reactions (i.e., what trainees think of the training), (2) learning (i.e., what trainees learned), (3) behavior (i.e., how trainees' behavior changes), and (4) organizational results (i.e., impact on organization) (see Table 4). Reactions are assessed by asking trainees how well they liked the program. Learning is measured by examining the extent to which trainees have acquired the principles, facts, and/or skills trained. Changes in behavior are assessed by evaluating trainees' performance back on the job. Finally, organizational results that are assessed include

Table 4 Kirkpatrick's (1976) Multilevel Training Evaluation Typology

Level	What Is Being Measured/Evaluated	Measurement	Sample Questions
1. Reactions	• Learner and/or instructor reactions after training • Satisfaction with training • Ratings of course materials • Effectiveness of content delivery	• Self-report survey • Evaluation or critique	• Did you like the training? • Did you think the trainer was helpful? • How helpful were the training objectives?
2. Learning	• Attainment of trained competencies (i.e., knowledge, skills, and attitudes) • Mastery of learning objectives	• Final examination • Performance exercise • Knowledge pre- and posttests	• True or false: Large training departments are essential for effective training. • Supervisors are closer to employees than is upper management.
3. Behaviour	• Application of learned competencies on the job • Transfer of training • Improvement in individual and/or team performance	• Observation of job performance	• Do the trainees perform learned behaviors? • Are the trainees paying attention and being observant? • Have the trainees shown patience?
4. Results	• Operational outcomes • Return on training investment • Benefits to organization	• Longitudinal data • Cost–benefits analysis • Organizational outcomes	• Have there been observable changes in employee turnover, employee attitudes, and safety since the training?

Source: Adapted from Childs and Bell (2002) and Wilson et al. (2005).

reduced turnover, reduced costs, improved efficiency, and improved quality.

Kirkpatrick's (1976) multilevel approach has been applied numerous times in recent years and has been shown to be effective (Cohen and Ledford, 1994; Field, 1995) as a framework for evaluation efforts in both individual studies (e.g., Noe and Schmitt, 1986; Wexley and Baldwin, 1986) and meta-analytic reviews of training literature (Burke and Day, 1986). However, most studies have elicited measures at only a single level, typically trainee reactions (e.g., Catalanello and Kirkpatrick, 1968; Bunker and Cohen, 1977) or trainee learning (e.g., Alliger and Horowitz, 1989), exposing some problems with not only the original typology, but in how it is applied: namely, a lack of multilevel diagnostic measures.

One example comes from Salas and colleagues (2001), who examined the success of crew resource management (CRM) training. Researchers examined the available literature and found that whereas 41% of the studies collected information at multiple levels of Kirkpatrick's framework, a majority of those only collected information pertaining to two of the levels, usually reaction/learning or reaction/behavior. Findings support the suggestion that CRM has been successful in improving safety in the aviation community, but the results are ambiguous and incomplete, further emphasizing the importance of organizations to evaluate their training programs at all levels.

These problems have led to a reevaluation of the original typology and its application. It should be recognized that Kirkpatrick's typology was written for a relatively unsophisticated audience, with respect to measurement. Consequently, there are conceptual flaws and ambiguities in the model (Snyder et al., 1980; Clement, 1982; Alliger and Janak, 1989). In addition, this model ignores other potentially relevant trainee outcome measures such as trainee motivation and self-efficacy (Gist et al., 1988; Gist, 1989; Tannenbaum et al., 1991), as well as other indications of the value of training, such as the program's content validity (Ford and Wroten, 1984) and cost-effectiveness or utility (Schmidt et al., 1982; Cascio, 1989). More critically, Kirkpatrick's typology did not anticipate later developments in learning theory. The typology discusses learning primarily as a function of increased declarative knowledge and thus ignores modern theories of cognitive skill acquisition (Anderson, 1982; Ackerman, 1987). Accordingly, the model has little value as a conceptual heuristic.

In an effort to begin to address these issues while building on Kirkpatrick's framework, several researchers have reviewed and revised Kirkpatrick's original typology. Kraiger and colleagues (1993) outlined three similar outcomes: (1) affective (i.e., reactions), (2) cognitive (i.e., learning), and (3) skill-based (i.e., behavior) outcome. These three levels were similar to what Kirkpatrick proposed, but addressed the cognitive dimension as well. The first level, reactions, measures trainee affect or how well they like the program, as well as program utility or how useful the trainees thought the program was. This last addition is based on the findings that a significant problem with training evaluation was its overreliance on self-report measures of trainees' reactions to the training. Although self-report measures are popular because of their ease of use, self-report alone do not give evaluators a full picture of training effectiveness. How well the training was conducted and the competencies that it targeted are often not reflected in whether or not trainees "liked" the training. In fact, researchers found that "liking does not equate to learning or to performing" (Alliger et al., 1997, p. 344). Therefore, the utility aspect was added to at least in part combat this fallacy. The trainees' opinions of whether or not what they learned will help them in their jobs are referred to as their *utility judgments*. However, the researchers still included the trainees' affective reactions to measure how much the trainees liked the training. This was done because while liking, as noted earlier, does not necessarily translate into desired outcomes, evaluators can gain valuable information about organizational factors (e.g., organizational support for training) based on the trainees' feelings toward the training. Therefore, within this model, affective outcomes were referred to as attitudinal outcomes and motivational states (i.e., motivational disposition, self-efficacy, and goal setting) that were produced by training.

Learning evaluates the principles, skills, and knowledge gained from training. The learning level is used to determine whether or not the trainees actually learned the targeted KSAs presented in the program (i.e., training validity), while not evaluating whether or not their behaviors changed due to training. The behavioral evaluation level assesses the changes in behavior exhibited by trainees following the training program in their actual work environment (i.e., transfer validity). Finally, organizational outcomes of the training (e.g., reduced costs, improved quality), the highest level of evaluation, are assessed. This final level of evaluation determines two types of training validity: (1) intraorganizational validity (i.e., are the performances of multiple groups of trainees consistent?), and (2) interorganizational validity (i.e., will the training program in one organization or department be effective in another?). Therefore, it is disappointing that this last level is rarely evaluated, out of necessity based on the difficulty of collecting such data.

Another incarnation of Kirkpatrick's original typology was proposed by Alliger and colleagues (1997) (see Table 18.5). The researchers reviewed Kirkpatrick's approach through meta-analysis of 34 articles used in an earlier study (Alliger and Janak, 1989). Their findings led to an augmented framework, shown in Figure 5 compared to the Kirkpatrick model, which further expanded and clarified Kirkpatrick's original method of evaluation. Alliger and Janak classified different types of reactions and learning and focused more on transfer of training to the work environment rather than the more general behavior category.

They did this by dividing the learning phase into three categories: (1) immediate knowledge, (2) knowledge retention, and (3) behavior/skill demonstration. Immediate knowledge involves trainees indicating

Table 5 Alliger et al.'s (1997) Augmented Kirkpatrick Training Taxonomy

Step	Definition
1. Reactions	
a. Affective reactions	Measures emotional self-report of trainees given immediately, with little if any thought; impressions.
b. Utility judgments	Evaluates trainee opinions or judgments about the transferability and utility of the training; behaviorally based opinions.
2. Learning	
a. Immediate knowledge	Assesses how much trainees learned from training (i.e., how much they know about what they were trained). Uses multiple choice, open-ended questions, lists, etc.
b. Knowledge retention	Used to assess what trainees know about training, much like immediate knowledge tests, but are administered after some time has passed, to test retention. Used in combination with or instead of immediate knowledge tests.
c. Behavior/skill demonstration	Measures behaviors/skills indicators of performance exhibited during training as opposed to on the job. Uses simulations, behavioral reproduction, ratings of training performance, and performance-centered scorings in classes.
3. Transfer	Measures output, outcomes, and work samples to assess on-the-job performance. Measured some time after training to assess some measurable aspect of job performance. Assess transfer of training to job setting.
4. Results	Assesses what organizational impact training had after the fact. Uses measurement of productivity gains, customer satisfaction, any change in cost, an improvement in employee morale, and profit margin, among others. Measurement is often difficult, due to organizational limitations and because results are the most distal from training. Caution should be used, however, because results are often regarded as the basis for judging training success, but judgments are often based on false expectations.

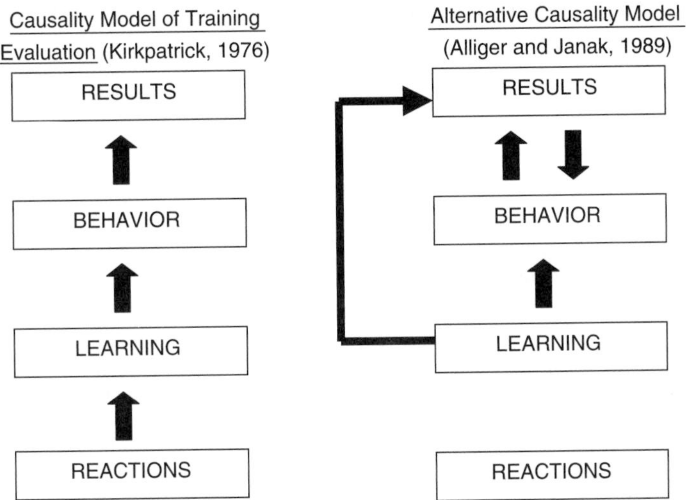

Figure 5 Comparison of two training evaluation models.

how much they know about the training content (i.e., what they were taught during training). Furthermore, Alliger and Janak (1997) applied multiple methods of evaluation (i.e., multiple-choice tests, open-ended questions, listing of facts). Knowledge retention, the second category, involves the same types of tests that would be given to test immediate knowledge gain, but trainees are given these assessments after more time has elapsed. Evaluators can use both immediate knowledge assessment and knowledge retention tests to provide a more in-depth evaluation.

Finally, behavior/skill evaluation involves the trainees' performances within training, as opposed to their performances on the job. Methods such as simulations, SME ratings of team performance, and behavioral role-plays are used to measure behaviors and skills obtained in training. These three components are used to assess trainee learning. When combined with trainee reactions, these assessments of trainee learning provide a stronger evaluation of training and give reactions a more predictive power of performance (Alliger et al., 1997). The last two ingredients for the revised typology proposed by Alliger and Janak (1997) are transfer and results. Regarding transfer, measurements are given some time after training when trainees are back on the job. Transfer evaluations differ from knowledge retention measures since they are more behaviorally based and focus on on-the-job performance, outputs, outcomes, and work samples. The last component of this model focuses on the bottom-line variables, typically referred to as *training outcomes* (e.g., measurement of productivity gains, customer satisfaction, changes in cost of production, employee morale, and profits). However, the final assessment of results should be viewed cautiously. Evaluators must keep in mind that these variables can be difficult to capture accurately and can be a result of factors other than training. In addition, management can have inflated expectations of the impact of training on bottom-line results, leading to disappointment and harsh judgment of training. Therefore, caution should be used when dealing with result variables.

Although the foregoing models represent major strides in fixing the problems inherent in Kirkpatrick's typology, other recent work has either expanded it or pointed out weaknesses, such as the need to develop more diagnostic measures. For example, Kraiger and Jung (1997) suggested several processes by which learning outcomes can be derived from instructional objectives of training. Goldsmith and Kraiger (1997) proposed a method for structural assessment of an individual learner's knowledge and skill in a specific domain. This model has been used with some success in several domains (e.g., Kraiger et al., 1995; Stout et al., 1997).

Despite all the controversy, Kirkpatrick's (1976) typology continues to be the most popular framework for guiding evaluations. Clearly, the original typology has served as a good foundation for training evaluation for the last several decades and it will probably continue to do so (Kirkpatrick, 1976). Although it has been both used and misused extensively, serving as a jumping-off point for effective evaluation, research must continue to find better, more rigorous diagnostic assessments for training evaluation. Especially as technology evolves and becomes more and more a part of the training environment, researchers must strive to improve the ease and utility of evaluation both during and after training (e.g., Ghodsian et al., 1997).

8.4 Summary

At the conclusion of any training program, it is critical an organization evaluate its effectiveness. Furthermore, organizations must look beyond the reactions of trainees, as research suggests that positive attitudes do not necessarily indicate that learning has taken place or that behaviors will be exhibited on the job. As training evaluations take time and resources, the benefits of understanding the impact of training will be crucial to its future success.

9 TRANSFER OF TRAINING

Transfer of training can be defined as the extent to which competencies targeted in training are applied, generalized, and maintained over some time in the work environment or on the job (Baldwin and Ford, 1988). Many studies and a great deal of theoretical musings have focused on the transfer of training (see Ford and Weissbein, 1997). Based on research emerging in the recent past, several conclusions and propositions have been laid out. For example, researchers have concluded that context matters (Quinones, 1997). Context helps determine trainee motivations, expectations, perceptions, and attitudes toward transfer. Furthermore, it is possible to measure the organizational learning environment, which can vary in meaningful ways across organizations (Tannenbaum, 1997). In addition, several studies have proposed that the transfer "climate" can have a powerful impact on the extent to which newly acquired KSAs are used back on the job (e.g., Thayer and Teachout, 1995; Tracey et al., 1995). Ultimately, factors found to affect the transfer of training include opportunities provided to trainees to perform KSAs (Ford et al., 1992; Quinones et al., 1995); delays between training and actual use on the job, which can create significant skill decay (Arthur et al., 1998); situational cues and consequences, which predict the extent to which transfer occurs (Rouiller and Goldstein, 1993); and social, peer, subordinate, and supervisor support (e.g., Facteau et al., 1995; Tracey et al., 1995). Conclusions also suggest that training can be generalized from one context to another (e.g., Tesluk et al., 1995) and intervention strategies can be designed such as to improve the probability of transfer (e.g., Brinkerhoff and Montesino, 1995; Kraiger et al., 1995). Regarding teams and their leaders, informal reinforcement (or punishment) of transfer activities by team leaders can shape the degree of transfer (Smith-Jentsch et al., 2000).

To design training effectively and ensure the transfer of competencies, training transfer needs to be conceptualized as a multidimensional construct, meaning that it differs depending on the type of training and closeness of supervision on the job (Yelon and Ford, 1999). Ford and Weissbein (1997) pointed out that there are more studies using complex tasks with diverse samples that actually measure transfer over time. Although the study of transfer of training has come a long way, much more needs to be done. For example, more research should be done that actually manipulates the transfer climate as opposed to observing one climate and making assumptions that findings will generalize across organizations (e.g., Smith-Jentsch et al., 2000). Since a majority of studies still utilize surveys as the preferred method of measurement, many problems still exist. Additional methods, which tap other measurement points, should be developed. For example, vertical transfer of training (i.e., upward transfer across different levels of the organizational system) should be examined further. Individual outcomes can have implications at the higher level (i.e., the organization). Therefore, the construct of vertical transfer may be used as a leverage point for strengthening the links between learning outcomes and organizational effectiveness (see Kozlowski et al., 2000).Although a great deal still needs to be done, these studies, taken together, emphasize the importance of the organizational environment in training. Therefore, a focus of transfer of training research needs to include environmental factors in posttraining. In the next section we address some of the issues inherent in posttraining environments.

9.1 Posttraining Environment

Another important aspect following training, which is essential in encouraging the transfer of training, is the posttraining environment. Whether or not a training program is effective depends heavily on the trainee's ability to use the competencies in their real-world job. Environmental factors helps determine whether trainees will exhibit learned behaviors when transferred back to their work settings. Even when a program is designed well and implemented effectively, without an environment that encourages transfer of the targeted competencies, their will be no positive outcomes. This variable helps determine whether or not competencies learned during training will transfer to the actual job. Although supporting empirical evidence is limited, there are several characteristics of posttraining environments that seem to contribute to training effectiveness: (1) supervisor support, (2) organizational transfer climate, and (3) continuous-learning culture (see Baldwin and Ford, 1988; Rouiller and Goldstein, 1993; Tracey et al., 1995; Ford and Weissbein, 1997). Furthermore, the transfer of training can be enhanced or hindered by some elements of the transfer climate(e.g., rewards; positive transfer climate; lack of peer or supervisor support; lack of resources) (Tannenbaum and Yukl, 1992; Rouiller and Goldstein, 1993). The effects on transfer of supervisor support organizational transfer climate, and continuous learning environments will be explored next.

The transfer of training has been linked to supervisor support. Several studies have provided evidence that the transfer of trained skills is a result of discussions with supervisors prior to and following training, as well as supervisor sponsorship (Huczynski and Lewis, 1980; Brinkerhoff and Montesino, 1995). Further research has suggested that opportunities to perform learned skills provided by supervisors also encourage the transfer of training (Ford et al., 1991). However, Baldwin and Ford (1988) found that there may be some misunderstandings regarding what constitutes supervisor support (i.e., what behaviors are perceived as supportive by workers). However, some supportive behaviors suggested by researchers include goal-setting activities (e.g., minimize number of accidents), reinforcement (e.g., error reporting), and modeling of trained behaviors (e.g., safe behaviors) (Tannenbaum and Yukl, 1992). While this list is helpful, further research is required to adequately define supervisor behaviors that are universally perceived as supportive.

Even more than supervisor support, organizational transfer climate has been shown to affect the transfer of training significantly. Organizational climate is defined by the interaction between elements within the organizational setting that are observable as well as those that are perceived by trainees (Hellreigel and Slocum, 1974; James and Jones, 1974). Specifically, when trainees perceive a positive organizational climate (e.g., organizational support, rewards, safety policies, nonpunitive error-reporting systems), they appear to apply learned competencies more readily on the job (Baumgartel et al., 1984; Rouiller and Goldstein, 1993; Tracey et al., 1995).

Finally, when organizations promote a continuous learning culture, an aspect of the work environment, competencies are transferred more readily to the workplace. Researchers define a continuous-learning work environment as an environment that encourages the acquisition of knowledge, skills, and attitudes with opportunities to practice, achievement that is reinforced, and the encouragement of innovation and competition (Rosow and Zager, 1988; Dubin, 1990; Tracey et al., 1995). This climate is entrenched in an organization that recognizes that learning is part of their daily work environment. Support of this is found by Tracey et al., who observed more posttraining behaviors in trainees who perceived a continuous learning environment. Therefore, it can be concluded that to reduce errors and encourage improved performance from training, employees must perceive that a continuous learning climate is intrinsic to their organization, thus encouraging the transfer of learned competencies and safe behaviors to the actual workplace.

9.2 Job Aids

Another tool that organizations can utilize to foster transfer of training is job aids. Job aids are developed to assist the user in the actual performance of a job or task (Swezey, 1987). Benefits to the use of job aids include a reduction in the amount of time that employees have to spend away from their job in training and improving performance by minimizing the cognitive load required to memorize various aspects of the job. For example, checklists are a type of job aid that can be provided to employees which will walk them through the steps necessary to complete a task without requiring that the employees have these steps memorized. Job aids can be especially critical in stressful environments where critical items might be omitted from a task. There are several types of job aids, including informational, procedural, and decision making and coaching (Rossett and Gautier-Downes, 1991) (see Table 6). After describing the types of job aids and the process of developing job aids, we discuss how job aids can be used in training and provide examples of both job and training aids that organizations can utilize.

Table 6 Types of Job Aids

Type	Description	When to Use
Informational	Provides access to large amounts of information, such as telephone directory or online database.	During task
Procedural	Provides step-by-step instructions for completing a task, such as directions for installing a faucet.	During task
Decision-making and coaching	Provides a heuristic to guide the user through a thought process to choose the best solution.	Before, during, and after task

9.2.1 Informational Aids

Informational job aids are similar to on-the-job reference manuals. These manuals are provided to employees to reduce mental workload (e.g., recall of memorized information) or in situations where it would be impossible for an employee to remember this information (e.g., an aircraft maintenance manual). Informational job aids may include facts about names, places, dates, and times that would be relevant to the job (Rossett and Gautier-Downes, 1991). Traditionally, these informational job aids have been in paper form, but more recently, informational job aids have been created as computer databases. As technology advances, laptop computers and personal data assistants (PDAs) can augment one's performance on the job by making this type of information readily available.

9.2.2 Procedural Aids

Procedural job aids provide step-by-step instructions for completing a task (Swezey, 1987). Procedural job aids tell the user which actions to take in sequential order and often provide feedback for what the result of that step should look like. The aviation community, for example, uses procedural job aids frequently (in the form of checklists) to assist aircrews with their required tasks. Like informational job aids, procedural job aids have traditionally been produced on paper; however, some companies now provide procedural job aids online to assist employees in completing their tasks. Procedural job aids can also be provided to consumers. For example, the Home Depot provides step-by-step directions online for installing various household items, such as faucets, doorknobs, and locks.

9.2.3 Decision-Making and Coaching Aids

The third type of job aid is decision making and coaching job aids. Also referred to as *heuristics*, decision-making and coaching job aids provide a reference for employees to consult that will help an employee think along the right lines to determine the best decision or solution to a problem (Rossett and Gautier-Downes, 1991). Decision-making aids differ from procedural aids in that they do not provide steps in a sequential order. Instead, they provide ideas or questions that will simply keep the user along the path that will lead to the best solution while allowing the order of steps to vary.

Traditionally, job aids have only been used during the time that employees are unsure about a piece of information or the next step to take on a task. With the advent and development of decision-making and coaching aids, employees are also able to use job aids prior to and after the specific time they are needed. Decision-making and coaching aids are tied much more closely to training aids than are the other types of job aids because employees can learn different decision-making processes that can be used if similar problems arise on a future task. More specifically, Rossett and Gautier-Downes (1991) provide guidelines for situations in which job aids should be used. These situations are shown in Table 7.

9.2.4 Development of Job Aids

Once you have determined that a specific task requires a job aid, an appropriate and effective job aid must be developed. The first step in developing a job aid is to perform a task analysis, which results in a set of knowledge and skills necessary for the task, equipment necessary to perform the task, technical data required to perform the task, and discrete and critical steps required to perform the task as well as the sequence of those steps (Swezey, 1987). The information garnered from the task analysis will allow development of the type(s) of job aids necessary for the task. Once the job aid is developed, it should be tested and modified to produce the best results. After modifications and revisions are made, job aids should be updated as information, procedures, or decision-making processes change (Rossett and Gautier-Downes, 1991).

Table 7 Situations When Job Aids Should Be Used

- The performance of a task is infrequent and the information is not expected to be remembered.
- The task is complex or has several steps.
- The costs of errors are high.
- Task performance is dependant on knowing a large amount of information.
- Performance depends on dynamic information or procedures.
- Performance can be improved through self-correction.
- The task is simple and there is high turnover volume.
- There is not enough time for training or training resources are not available.

Source: Adapted from Rossett and Gautier-Downes (1991).

Over the last decade, some researchers have suggested that job aids can be used in training and knowledge acquisition (Tillman, 1985). Spaulding and Dwyer (2001) found that the use of job aids is useful for instruction and knowledge acquisition although not all types of job aids are equally useful. These findings have prompted a great deal of research on the use of training aids.

9.2.5 Training Aids

Some job aids can also be used or modified to serve as training aids. Training aids are different from job aids in that they aid in skill and knowledge acquisition and are not used specifically to complete a task while on the job. Training aids are considered to be papers, documents, manuals, or devices that are designed to assist the user in learning the appropriate skills and/or knowledge that is associated with a task or job (Swezey, 1987). Training aids come in many different forms but have generally been associated with manuals given to trainees to supplement the normal training program that companies offer new employees and aid in knowledge acquisition. However, the technological advances of the last decade have provided organizations with the means and opportunity to produce computer-based training aids.

9.2.6 Examples of Job Aids

A number of job aids are available and are used in organizations. We discuss two commonly used job aids: manuals and decision support systems.

Manuals Both informational and procedural job aids can be organized into a manual. Manuals can provide information that is too long or too technical to be memorized, or can provide information on subjects that are not often encountered by an employee. For example, most organizations provide a directory listing employees' phone extensions. This allows people to look for contact information without having to bombard their memories with perhaps hundreds of four-digit numbers. In addition, manuals can be provided that offer procedures for tasks that are not completed often, such as steps in creating posters for presentations. Many organizations are now creating databases or other computer documents with the same information, as it is generally more convenient to use than are large paper manuals.

Manuals have also often been used as training aids. As newly hired employees sit down for their first day of training, they were inevitably handed a large bound notebook entitled "Training Manual." Inside, employees could find all the information they needed to know about the company and information on how to perform their job. Although the training program was designed to teach the major components of the job, the manual was provided as a supplement to classroom training so that the employee could learn the nuances and finer details of the job on his or her own. As technological advances have been made and classroom training programs replaced by other learning strategies discussed earlier (i.e., simulation,

Table 8 Steps in Designing, Delivering, and Evaluating Training Systems

Step	Outcome
	Training Analysis
1. Organizational analysis	Identifies: • Where training is needed • When training is needed • Resources and constraints • Support for transfer
2. Job/task analysis	Identifies: • Task specifications (e.g., what tasks, under what conditions) • Task characteristics (e.g., equipment needed for task) • Competencies (KSAs) needed to perform task
2a. Cognitive task analysis	Identifies: • Cognitive processes and requirements for the task
3. Person analysis	Identifies: • Who needs training • What they need to be trained on
	Training Design
4. Develop training objectives.	• Desired outcomes/goals are identified. • Assumptions about training are identified. • Objectives are documented. • Competencies are established.
5. Consider individual characteristics.	Identifies trainee characteristics that may affect training: • Cognitive ability • Self-efficacy • Goal orientation • Motivation
6. Consider organizational characteristics.	Identifies organizational characteristics that may affect training: • Organizational culture • Policies and procedures • Situational influences • Prepractice conditions
7. Establish practice opportunities.	• Practice opportunities are specified (e.g., when they will occur during training, number of opportunities provided, levels of difficulty).
8. Establish feedback opportunities.	• When feedback will be provided (e.g., immediately after training) and at what level (e.g., individual, team, both) are specified. • Trainees know how they did. • Trainees know where improvements are necessary.
9. Select an instructional strategy.	• The best instructional strategy or combination of strategies will be selected to train competencies of interest based on the needs of the organization (e.g., teams, technology, internationalization).
10. Outline the program content.	• Sequence and structure of the training program is laid out.
	Training Development
11. Develop practices scenarios.	• Realistic practice scenarios are scripted that engage trainees. • Scenarios of varying difficulty are scripted.
12. Develop performance measures.	• The measurement plan is identified. • Criteria for success are developed. • Performance measures are established. • Tools for assisting performance measurement are developed (e.g., observation checklists).
	Training Implementation
13. Select the instructional setting/location.	• Available training site is identified. • Training environment is prepared.
14. Train instructors.	• Instructors are adequately trained to conduct the instruction. • Instructors are knowledgeable in terms of the program content to handle questions and/or problems that may arise.

(*continued overleaf*)

Table 8 ***(continued)***

Step	Outcome
15. Conduct a pilot test.	• Issues or concerns with training are identified. • Feedback is received from trainees. • Necessary adjustments are made to the training program.
16. Conduct the instruction.	• Developed instructional materials are put into practice. • Training program is live and functional. • Training program is completed.
	Training Evaluation
17. Consider evaluation design issues.	• Experimental plan is laid out (e.g., posttest only, control group vs. no control group). • Where evaluations will be conducted is specified (i.e., in the field, on the job, both).
18. Consider costs of training evaluations.	• Low-cost alternatives are explored (e.g., unequal sample sizes between trained and untrained groups; low-cost proxy criterion measure selected).
19. Evaluate training system at multiple levels.	• Data on training's effectiveness are collected at multiple levels and analyzed. • Data on job performance are collected and analyzed.
	Transfer of Training
20. Establish a positive posttraining environment.	• Organization and supervisors support competencies on the job. • Continuous learning climate is established. • Trainees are rewarded. • Behaviors that contradict those that are trained are discouraged.
21. Use job aids.	• Performance on the job is enhanced.

scenario-based training, e-learning, distance learning, etc.), the use of manuals as training aids has become less common.

Decision Support Systems Because computers can be found at nearly every workstation in many companies and thus a switch from classroom to computer-based training, companies have developed computer-based job and training aids such as decision support systems (DSSs) and intelligent tutoring systems (ITSs) to replace paper-based job aids and to complement computer-based training programs.

DSSs are designed to improve and support human decision making (Brody et al., 2003) and can be used as both job aids and training aids. As job aids, DSSs are provided to aid in making better decisions during an actual task. For example, the design and implementation of the Navy's DSS associated with TADMUS provided much needed guidance for decision making during training and resulted in increased situation awareness, lower workload, more confidence in the decisions made, and more effective performance (Zachary et al., 1998). Zaklad and Zachary (1992) provided a set of general standards and principles around which a DSS should be designed. A company's specific needs can then be included with the general principles they outlined.

As training aids, DSSs are provided in conjunction with simulated scenarios to aid in teaching better critical thinking and decision-making skills and are just one element in the overall training program. Organizations that use scenario-based training could be particularly benefited by a complementary DSS. During each of the simulated exercises, the DSS could aid the trainee in the decision-making process and provide feedback for each decision that is made. This strategy has been implemented most recently in the military with the development of Synthetic Cognition for Operational Team Training (SCOTT) (Zachary et al., 2001) and Synthetic Teammates for Realtime Anywhere Training and Assessment (STRATA) (Bell, 2003). These distributed-based training programs offer real-time strategy training and feedback via DSS in a secure networked training environment.

Another more specific type of DSS are intelligent tutoring systems (ITSs) (Ong and Ramachandran, 2003). Woolf and colleagues (2001) discuss some of the abilities of various ITSs and how ITSs can be improved to increase the number of people who can use them. ITSs can teach a variety of knowledge domains; however, they require an extensive knowledge of the subject as well as strategies for error diagnosis and decision making and examples and analogies of relevant topics. This requires a great deal of time and work to design and program, but the benefits for organizations of running training programs without the need for a facilitator to be there physically are worth many times the cost. It allows trainees to be trained at their own pace, which should facilitate better transfer of training. The development of ITSs as job aids does not require the same amount of time because the user has already acquired the basic knowledge and normally needs only small amounts of specific information to complete the task or to make a decision.

The job and training aids discussed above are being upgraded and improved constantly by further advances in technology: specifically, cognitive modeling, simulation, and the ongoing development of more realistic and humanlike cognitive agents. Developments in the area of job and training aids give employees the opportunity to acquire knowledge and skills and put that information to practice while being directed by a DSS or an ITS. As the technology develops and further research is conducted, the costs associated with developing and implementing technologically based job and training aids will be driven down.

9.3 Summary

Once training has been evaluated, the training system is not complete. Rather, a climate for transfer must be established and tools (e.g., job aids) provided to help trainees apply what they have learned on the job. Without transfer of the learned competencies on the job, training will not be a success, even if trainees liked the training, learned from the training, and applied what they learned in practice.

10 CONCLUSIONS

Without a knowledgeable and skillful workforce, organizations are likely to suffer. With that in mind, training in organizations should be of the utmost importance. The purpose of this chapter was to provide the human factors community with guidance on designing, delivering, and evaluating a training system (see Table 8). Throughout the chapter we argue that training designers must take a systematic approach to training by considering carefully all aspects of the training program (e.g., individual and organizational characteristics). Organizations must remember that training is more than just a program. Rather, there is a science of training that has been developed by human factors, industrial/organizational, and educational scientists that needs to be exploited. By doing so, organizations can reap the benefits of what the science has to offer.

REFERENCES

Ackerman, P. (1987), "Individual Differences in Skill Learning: An Integration of Psychometric and Information Processing Perspectives," *Psychological Bulletin*, Vol. 102, No. 1, pp. 3–27.

Adler, N. J. (1997), *International Dimensions of Organizational Behavior*, 3rd ed., International Thomson Publishing, Cincinnati, OH.

Aiba, K., Kadoo, A., and Nomiyama, T. (2002), "Pilot Performance of the Night Approach and Landing Under the Limited Runway Light Conditions," *Reports of Aeromedical Laboratory*, Vol. 42, No. 2–3, pp. 19–29.

Alliger, G. M., and Horowitz, H. M. (1989), "IBM Takes the Guessing Out of Testing," *Training and Development Journal*, April, pp. 69–73.

Alliger, G. M., and Janak, E. A. (1989), "Kirkpatrick's Levels of Training Criteria: Thirty Years Later," *Personnel Psychology*, Vol. 42, pp. 331–342.

Alliger, G. M., Tannenbaum, S. I., and Bennett, W., Jr. (1997), "A Meta-analysis of the Relations Among Training Criteria," *Personnel Psychology*, Vol. 50, pp. 341–358.

Anderson, J. R. (1982), "Acquisition of Cognitive Skill," *Psychological Review*, Vol. 89, No. 4, pp. 369–406.

Anderson, J. R., Corbett A. T., Koedinger K. R., and Pelletier, R. (1995), "Cognitive Tutors: Lessons Learned," *Journal of Learning Science*, Vol. 4, pp. 167–207.

Arthur, W., Young, B., Jordan, J. A., and Shebilske, W. L. (1996), "Effectiveness of Individual and Dyadic Training Protocols: The Influence of Trainee Interaction Anxiety," *Human Factors*, Vol. 38, pp. 79–86.

Arthur, W., Day, E. A., Bennett, W., McNelly, T. L., and Jordan, J. A. (1997), "Dyadic Versus Individual Training Protocols: Loss and Reacquisition of a Complex Skill," *Journal of Applied Psychology*, Vol. 82, pp. 783–791.

Arthur, W., Bennett, W., Stanush, P. L., and McNelly, T. L. (1998), "Factors That Influence Skill Decay and Retention: A Quantitative Review and Analysis," *Human Performance*, Vol. 11, pp. 79–86.

Arvey, R. D., Salas, E., and Gialluca, K. A. (1992), "Using Task Inventories to Forecast Skills and Abilities," *Human Performance*, Vol. 5, pp. 171–190.

ASTD (2003), "ASTD 2003 State of the Industry Report Executive Summary," Retrieved June 3, 2004, from http://www.astd.org/NR/rdonlyres/6EBE2E82-1D29-48A7-8A3A-357649BB6DB6/0/SOIR_2003_Executive_Summary.pdf.

Baker, D., Prince, C., Shrestha, L., Oser, R., and Salas, E. (1993), "Aviation Computer Games for Crew Resource Management Training," *International Journal of Aviation Psychology*, Vol. 3, pp. 143–156.

Baldwin, T. T. (1992), "Effects of Alternative Modeling Strategies on Outcomes of Interpersonal Skills Training," *Journal of Applied Psychology*, Vol. 77, pp. 147–154.

Baldwin, T. T., and Ford, J. K. (1988), "Transfer of Training: A Review and Directions for Future Research," *Personnel Psychology*, Vol. 41, pp. 63–105.

Baldwin, T. T., and Magjuka, R. J. (1997), "Training as an Organizational Episode: Pretraining Influences on Trainee Motivation," in *Improving Training Effectiveness in Work Organizations*, J. K. Ford, S. Kozlowski, K. Kraiger, E. Salas, and M. Teachout, Eds., Lawrence Erlbaum Associates, Mahwah, NJ, pp. 99–127.

Baldwin, T. T., Magjuka, R. J., and Loher, B. T. (1991), "The Perils of Participation: Effects of Choice of Training on Trainee Motivation and Learning," *Personnel Psychology*, Vol. 44, pp. 51–65.

Banker, R. D., Field, J. M., Schroeder, R. G., and Sinha, K. K. (1996), "Impact of Work Teams on Manufacturing Performance: A Longitudinal Study," *Academy of Management Journal*, Vol. 39, No. 4, pp. 867–890.

Barling, J., Weber, T., and Kelloway, E. K. (1996), "Effects of Transformational Leadership Training on Attitudinal and Financial Outcomes: A Field Experiment," *Journal of Applied Psychology*, Vol. 81, pp. 827–832.

Bassi, L. J., and Van Buren, M. E. (1998), "The 1998 ASTD State of the Industry Report," *Training and Development*, Vol. 52, No. 1, pp. 21–43.

Baumgartel, H., Reynolds, M., and Pathan, R. (1984), "How Personality and Organizational Climate Variables Moderate the Effectiveness of Management Development Programmes: A Review and Some Recent Research Findings," *Management and Labour Studies*, Vol. 9, pp. 1–16.

Beer, M. (1980), *Organization Change and Development: A Systems View*, Scott, Foresman, Glenview, IL.

Befus, C. P. (1988), "A Multilevel Treatment Approach for Culture Shock Experience by Sojourners," *International Journal of Intercultural Relations*, Vol. 12, pp. 381–400.

Bell, H. H. (1999), "The Effectiveness of Distributed Mission Training," *Communications of the ACM*, Vol. 42, No. 9, pp. 72–78.

Bell, B. (2003), "On-Demand Team Training with Simulated Teammates: Some Preliminary Thoughts on Cognitive Model Reuse," *TTG News*, pp. 6–8.

Bell, B. S. and Kozlowski, S. W. J. (2002), "A Typology of Virtual Teams: Implications for Effective Leadership," *Group and Organization Management*, Vol. 27, No. 1, pp. 14–49.

Bell, H., and Waag, W. (1998), "Evaluating the Effectiveness of Flight Simulators for Training Combat Skills: A Review," *International Journal of Aviation Psychology*, Vol. 8, pp. 223–242.

Bennett, J. M. (1986), "Modes of Cross-Cultural Training: Conceptualizing Cross-Cultural Training as Education," *International Journal of Intercultural Relations*, Vol. 10, pp. 117–134.

Bhagat, R., and Prien, K. O. (1996), "Cross-Cultural Training in Organizational Contexts," in *Handbook of Intercultural Training*, 2nd ed., D. Landis and R. S. Bhagat, Eds., Sage Publications, Thousand Oaks, CA, pp. 216–230.

Bhawuk, D. P. S. (1998), "The Role of Culture Theory in Cross-Cultural Training: A Multimethod Study of Culture-Specific, Culture-General, and Culture Theory-Based Assimilators," *Journal of Cross-Cultural Psychology*, Vol. 29, No. 5, pp. 630–655.

Bhawuk, D. P. S. (2001), "Evolution of Culture Assimilators: Toward Theory-Based Assimilators," *International Journal of Intercultural Relations*, Vol. 25, pp. 141–163.

Bhawuk, D. P. S., and Brislin, R. W. (2000), "Cross-Cultural Training: A Review," *Applied Psychology: An International Review*, Vol. 49, pp. 162–191.

Bird, A., and Dunbar, R. (1991), "Getting the Job Done over There: Improving Expatriate Productivity," *National Productivity Review*, Vol. 10, pp. 145–156.

Black, J. S., and Mendenhall, M. E. (1990), "Cross-Cultural Training Effectiveness: A Review and a Theoretical Framework for Future Research," *Academy of Management Review*, Vol. 15, pp. 113–136.

Blickensderfer, E. L., Cannon-Bowers, J. A., and Salas, E. (1997), "Theoretical Bases for Team Self-Correction: Fostering Shared Mental Models," in *Advances in Interdisciplinary Studies in Work Teams Series*, Vol. 4, M. Beyerlein, D. Johnson, and S. Beyerlein, Eds., JAI Press, Greenwich, CT, pp. 249–279.

Bowers, C. A., Blickensderfer, E. L., and Morgan, B. B. (1998), "Air Traffic Control Specialist Team Coordination," in *Human Factors in Air Traffic Control*, M. W. Smolensky and E. S. Stein, Eds., Academic Press, San Diego, CA, pp. 215–236.

Brannick, M. T., Prince, A., Prince, C., and Salas, E. (1995), "The Measurement of Team Process," *Human Factors*, Vol. 37, pp. 641–651.

Brannick, M. T., Prince, C., and Salas, E. (2002), "The Reliability of Instructor Evaluations of Crew Performance: Good News and Not So Good News," *International Journal of Aviation Psychology*, Vol. 12, pp. 241–261.

Branson, R. K., Rayner, G. T., Cox, J. L., Furman, J. P., King, F. J., and Hannum, W. H. (1975), *Interservice Procedures for Instructional Systems Development*, 5 vols., *TRADOC Pam 350–30 NAVEDTRA 106A*, U.S. Army Training and Doctrine Command, Ft. Monroe, VA.

Brett, J. F., and VandeValle, D. (1999), "Goal Orientation and Goal Content as Predictors of Performance in a Training Program," *Journal of Applied Psychology*, Vol. 84, pp. 863–873.

Bretz, R. D., Jr., and Thompsett, R. E. (1992), "Comparing Traditional and Integrative Learning Methods in Organizational Training Programs," *Journal of Applied Psychology*, Vol. 77, pp. 941–951.

Brewster, C. (1995), "Effective Expatriate Training," in *Expatriate Management: New Ideas for International Business*, J. Selmer, Ed., Quorem Books, Westport, CT, pp. 57–71.

Brinkerhoff, R. O., and Montesino, M. U. (1995), "Partnership for Training Transfer: Lessons from a Corporate Study," *Human Resource Development Quarterly*, Vol. 6, pp. 263–274.

Brody, R. G., Kowalczyk, T. K., and Coulter, J. M. (2003), "The Effect of a Computerized Decision Aid on the Development of Knowledge," *Journal of Business and Psychology*, Vol. 18, No. 2, pp. 157–174.

Brown, K. G., and Ford, J. K. (2002), "Using Computer Technology in Training: Building an Infrastructure for Active Learning," in *Creating, Implementing, and Managing Effective Training and Development*, K. Kraiger, Ed., Jossey-Bass, San Francisco, pp. 192–233.

Buller, P. F. (1986), "The Team Building–Task Performance Relation: Some Conceptual and Methodological Refinements," *Group and Organizational Studies,* Vol. 11, pp. 147–168.

Bunker, K. A., and Cohen, S. L. (1977), "The Rigors of Training Evaluation: A Discussion and Field Demonstration," *Personnel Psychology*, Vol. 30, No. 4, pp. 525–541.

Burke, R. J. (1997), "Organizational Hierarchy and Cultural Values," *Psychological Reports*, Vol. 81, pp. 832–834.

Burke, M. J., and Day, R. R. (1986), "A Cumulative Study of the Effectiveness of Managerial Training," *Journal of Applied Psychology*, Vol. 71, No. 2, pp. 232–245.

Buttom, S. B., Mathieu, J. E., and Zajac, D. M. (1996), "Goal Orientation in Organizational Research: A Conceptual and Empirical Foundation," *Organizational Behavior and Human Decision Processes*, Vol. 67, pp. 26–48.

Campbell, J. P. (1971), "Personnel Training and Development," *Annual Review of Psychology*, Vol. 22, pp. 565–602.

Cannon-Bowers, J. A., and Salas, E. (1997), "A Framework for Measuring Team Performance Measures in Training," in *Team Performance Assessment and Measurement: Theory, Methods, and Applications*, M. T. Brannick, E. Salas, and C. Prince, Eds., Lawrence Erlbaum Associates, Mahwah, NJ, pp. 45–62.

Cannon-Bowers, J. A., Salas, E., Tannenbaum, S. I., and Mathieu, J. E. (1995), "Toward Theoretically-Based Principles of Trainee Effectiveness: A Model and Initial Empirical Investigation," *Military Psychology*, Vol. 7, pp. 141–164.

Cannon-Bowers, J. A., Burns, J. J., Salas, E., and Pruitt, J. S. (1998), "Advanced Technology in Scenario-Based Training," in *Making Decisions Under Stress: Implications for Individual and Team Training*, J. A. Cannon-Bowers and E. Salas, Eds., American Psychological Association, Washington, DC, pp. 365–374.

Carlson, R. A., Lundy, D. H., and Schneider, W. (1992), "Strategy Guidance and Memory Aiding in Learning Problem-Solving Skill," *Human Factors*, Vol. 34, pp. 129–145.

Carroll, L. A. (1999), "Multimodal Integrated Team Training," *Communications of the ACM*, Vol. 42, No. 9, pp. 68–71.

Cascio, W. F. (1989), "Using Utility Analysis to Assess Training Outcomes," in *Training and Development in Organizations*, I. L. Goldstein, Ed., Jossey-Bass, San Francisco, pp. 63–88.

Catalanello, F., and Kirkpatrick, D. L. (1968), "Evaluating Training Programs: The State of Art," *Training and Development Journal*, Vol. 22, No. 5, pp. 2–9.

Childs, J. M., and Bell, H. H. (2002), "Training Systems Evaluation," in *Handbook of Human Factors Testing and Evaluation*, S. G. Charlton, and T. G. O'Brien, Eds., Lawrence Erlbaum Associates, Mahwah, NJ, pp. 473–509.

Christoph, R. T., Schoenfeld, G. A., Jr., and Tansky, J. W. (1998), "Overcoming Barriers to Training Utilizing Technology: The Influence of Self-Efficacy Factors on Multimedia-Based Training Receptiveness," *Human Resources Development Quarterly*, Vol. 9, pp. 25–38.

Chung, J., and Reigeluth, C. M. (1992), "Instructional Prescriptions for Learner Control," *Educational Technology*, Vol. 32, pp. 14–20.

Clark, D. (2000), "Introduction to Instructional System design," retrieved from http://www.nwlink.com/~donclark/hrd/sat1.html#model.

Clement, R. W. (1982), "Testing the Hierarchy Theory of Training Evaluation: An Expanded Role for Trainee Reactions," *Public Personnel Management*, Vol. 11, No. 2, pp. 176–184.

Cohen, S. G., and Ledford, G. E. (1994), "The Effectiveness of Self-Directed Teams," *Human Relations*, Vol. 47, pp. 13–43.

Colquitt, J. A., LePine, J. A., and Noe, R. A. (2000), "Toward an Integrative Theory of Training Motivation: A Meta-analytic Path Analysis of 20 Years of Research," *Journal of Applied Psychology*, Vol. 85, No. 5, pp. 678–707.

Cooke, N. J. (1999), "Knowledge Elicitation," in *Handbook of Applied Cognition*, F. T. Durso, R. S. Nickerson, R. W. Schvaneveldt, S. T. Dumais, D. S. Lindsay, and M. T. H. Chi, Eds., Wiley, New York, pp. 479–509.

Daily, B., Whatley, A., Ash, S. R., and Steiner, R. L. (1996), "The Effects of a Group Decision Support System on Culturally Diverse and Culturally Homogeneous Group Decision Making," *Information and Management*, Vol. 30, pp. 281–289.

Degani, A., and Wiener, E. L. (1997), "Philosophy, Policies, Procedures and Practices: The Four 'P's of Flight Deck Operations," in *Aviation Psychology in Practice*, N. Johnston, N. McDonald, and R. Fuller, Eds., Avebury, Aldershot, Hampshire, England, pp. 44–67.

De Jong, J. A., and Versloot, B. (1999), "Structuring On-the-Job Training: Report of a Multiple Case Study," *International Journal of Training and Development*, Vol. 3, No. 3, pp. 186–199.

DeRouin, R. E., Fritzsche, B. A., and Salas E. (2004), "Optimizing E-Learning: Research Based Guidelines for Learner-Controlled Training," *Human Resource Management Journal*, Vol. 43, No. 2/3, pp. 147–162.

Deshpande, S. P., and Viswesvaran, C. (1992), "Is Cross-Cultural Training of Expatriate Managers Effective: A Meta Analysis," *International Journal of Intercultural Relations*, Vol. 16, pp. 295–310.

Distefano, J. J., and Maznevski, M. L. (2000), "Creating Value with Diverse Teams in Global Management," *Organizational Dynamics*, Vol. 29, No. 1, pp. 45–63.

Dormann, T., and Frese, M. (1994), "Error Training: Replication and the Function of Exploratory Behavior," *International Journal of Human–Computer Interaction*, Vol. 6, No. 4, pp. 365–372.

Driskell, J. E., and Johnston, J. H. (1998), "Stress Exposure Training," in *Making Decisions Under Stress: Implications for Individual and Team Training*, J. A. Cannon-Bowers and E. Salas, Eds., American Psychological Association, Washington, DC, pp. 191–217.

Driskell, J. E., Johnston, J. H., and Salas, E. (2001), "Does Stress Training Generalize to Novel Settings?" *Human Factors*, Vol. 43, No. 1, pp. 99–110.

Driskell, J. E., Radtke, P. H., and Salas, E. (2003), "Virtual Teams: Effects of Technological Mediation on Team Processes," *Group Dynamics*, Vol. 7, No. 4, pp. 297–323.

Dubin, S. S. (1990), "Maintaining Competence Through Updating," in *Maintaining Professional Competence*, S. L. Willis and S. S. Dubin, Eds., Jossey-Bass, San Francisco, pp. 9–43.

Dubois, D. A, Shalin, V. L., Levi, K. R., and Borman, W. C. (1997–1998), "A Cognitively-Oriented Approach to Task Analysis," *Training Research Journal,* Vol. 3, pp. 103–141.

Dweck, C. S. (1986), "Motivational Processes Affecting Learning," *American Psychology*, Vol. 41, pp. 1040–1048.

Dweck, C. S., and Leggett, E. L. (1988), "A Social-Cognitive Approach to Motivation and Personality," *Psychological Review*, Vol. 95, pp. 256–273.

Dyer, W. G. (1977). *Team Building: Issues and Alternatives*, Addison-Wesley, Reading, MA.

Ehrenstein, A., Walker, B., Czerwinski, M., and Feldman, E. (1997), "Some Fundamentals of Training and Transfer: Practice Benefits Are Not Automatic," in *Training for a Rapidly Changing Workplace: Applications of Psychological Research*, American Psychological Association, Washington, DC, pp. 31–60.

Ellermann, H. H., and Free, E. L. (1990), "A Subject-Controlled Environment for Paired Associate Learning," *Journal of Computer-Based Instruction*, Vol. 17, pp. 97–102.

Elliot, A. J., and Church, M. A. (1997), "A Hierarchical Model of Approach and Avoidance Achievement Motivation," *Journal of Personality and Social Psychology*, Vol. 72, pp. 218–232.

Entin, E. E., and Serfaty, D. (1999), "Adaptive Team Coordination," *Human Factors*, Vol. 41, No. 2, pp. 312–325.

Facteau, J. D., Dobbins, G. H., Russel, J. E. A., Ladd, R. T., and Kudisch, J. D. (1995), "The Influence of General Perceptions for the Training Environment on Pretraining Motivation and Perceived Training Transfer," *Journal of Management*, Vol. 21, pp. 1–25.

Feldman, D. (1988), *Managing Careers in Organizations*, Scott, Foresman, Glenview, IL.

Field, L. (1995), "Organizational Learning: Basic Concepts," in *Understanding Adult Education and Training*, G. Foley, Ed., Allen & Unwin, Sydney, Australia, pp. 159–173.

Fiore, S. M., Salas, E., and Cannon-Bowers, J. A. (2001), "Group Dynamics and Shared Mental Model Development," in *How People Evaluate Others in Organizations*, M. London, Ed., Lawrence Erlbaum Associates, Mahwah, NJ, pp. 309–336.

Fisher, S. L., and Ford, J. K. (1998), "Differential Effects of Learner Efforts and Goal Orientation on Two Learning Outcomes," *Personnel Psychology*, Vol. 51, pp. 397–420.

Fisher, D. L., Laurie, N. E., and Glaser, R. (2002), "Use of a Fixed-Based Driving Simulator to Evaluate the Effects of Experience and PC-Based Risk Assessment Training on Drivers' Decision," *Human Factors*, Vol. 44, No. 2, pp. 287–302.

Ford, J. K., and Noe, R. A. (1987), "Self-Assessed Training Needs: The Effects of Attitudes Toward Training, Managerial Level, and Function," *Personnel Psychology*, Vol. 40, pp. 39–53.

Ford, J. K., and Weissbein, D. A. (1997), "Transfer of Training: An Updated Review and Analysis," *Performance Improvement Quarterly*, Vol. 10, pp. 22–41.

Ford, J. K., and Wroten, S. P. (1984), "Introducing New Methods for Conducting Training Evaluation and for Linking Training Evaluation to Program Redesign," *Personnel Psychology*, Vol. 37, pp. 651–665.

Ford, J. K., Quinones, M., Sego, D., and Speer, J. (1991), "Factors Affecting the Opportunity to Use Trained Skills on the Job," paper presented at the 6th Annual Conference for the Society of Industrial and Organizational Psychology, St. Louis, MO.

Ford, J. K., Quinones, M. A., Sego, D. J., and Sorra, J. S. (1992), "Factors Affecting the Opportunity to Perform Trained Tasks on the Job," *Personnel Psychology*, Vol. 45, pp. 511–527.

Ford, J. K., Kozlowski, S., Kraiger, K., Salas, E., and Teachout, M., Eds. (1997), *Improving Training Effectiveness in Work Organizations*, Lawrence Erlbaum Associates, Mahwah, NJ.

Ford, J. K., Smith, E. M., Weissbein, D. A., Gully, S. M., and Salas, E. (1998), "Relationships of Goal Orientation, Metacognitive Activity, and Practice Strategies with Learning Outcomes and Transfer," *Journal of Applied Psychology*, Vol. 83, pp. 218–233.

Forret, M. L., Turban, D. B., and Dougherty, T. W. (1996), "Issues Facing Organizations When Implementing Formal Mentoring Programmes," *Leadership and Organization Development Journal*, Vol. 17, No. 3, pp. 27–30.

Fowlkes, J. E., Lane, N. E., Salas, E., Franz, T., and Oser, R. (1994), "Improving the Measurement of Team Performance: The Targets Methodology," *Military Psychology*, Vol. 6, No. 1, pp. 47–61.

Fowlkes, J., Dwyer, D. J., Oser, R. L., and Salas, E. (1998), "Event-Based Approach to Training (EBAT)," *International Journal of Aviation Psychology*, Vol. 8, No. 3, pp. 209–221.

Freitag, E. T., and Sullivan, H. J. (1995), "Matching Learner Preference to Amount of Instruction: An Alternative Form of Learner Control," *Educational Technology, Research and Development*, Vol. 43, pp. 5–14.

Frese, M., and Altman, A. (1989), "The Treatment of Errors in Learning and Training," in *Developing Skills with Information Technology*, L. Bainbridge and S. A. Quintanilla, Eds., Wiley, Chichester, West Sussex, England, pp. 65–86.

Frese, M., Brodbeck, F., Heinnbokel, T., Mooser, C., Schleiffenbaum, E., and Thiemann, P. (1991), "Errors in Training Computer Skills: On the Positive Functions of Errors," *Human–Computer Interaction*, Vol. 6, pp. 77–93.

Friedland, N., and Keinan, G. (1992), "Training Effective Performance in Stressful Situations: Three Approaches and Implications for Combat Training," *Military Psychology*, Vol. 4. pp. 157–174.

Galvin, T. (2002), "2002 Industry Report," *Training*, October, pp. 24–33.

Ghiselli, E. E. (1966), *The Validity of Occupational Aptitude Tests*, Wiley, New York.

Ghodsian, D., Bjork, R., and Benjamin, A. (1997), "Evaluating Training During Training: Obstacles and Opportunities," in *Training for a Rapidly Changing Workplace: Applications of Psychological Research*, M. A. Quinones and A. Ehrenstein, Eds., American Psychological Association, Washington, DC, pp. 63–88.

Gillan, C. A. (2003), "Aircrew Adaptive Decision Making: A Cross-Case Analysis," *Dissertation Abstracts International, Section B: The Sciences and Engineering*, Vol. 64, No. 1-B, p. 438.

Gist, M. E. (1989), "Effects of Alternative Training Methods on Self-Efficacy and Performance in Computer Software Training," *Journal of Applied Psychology*, Vol. 74, No. 6, pp. 884–891.

Gist, M., Rosen, B., and Schwoerer, C. (1988), "The Influence of Training Method and Trainee Age on the Acquisition of Computer Skills," *Personnel Psychology*, Vol. 41, No. 2, pp. 255–265.

Glendon, A. I., and Stanton, N. A. (2000), "Perspectives on Safety Culture," *Safety Science*, Vol. 34, pp. 193–214.

Goettl, B. P., Yadrick, R. M., Connolly-Gomez, C., Regian, W. J., and Shebilske, W. L. (1996), "Alternating Task Modules in Isochronal Distributed Training of Complex Tasks," *Human Factors*, Vol. 38, pp. 330–346.

Goforth, D. (1994), "Learner Control = Decision Making + Information: A Model and Meta Analysis," *Journal of Educational Computing Research*, Vol. 11, pp. 1–26.

Goldman, D. (2000), "Legal Landmines to Avoid in Employment Training," retrieved June 14, 2002, from http://startribune.hr.com.

Goldsmith, T., and Kraiger, K. (1997), "Structural Knowledge Assessment and Training Evaluation," in *Improving Training Effectiveness in Work Organizations*, J. K. Ford, S. Kozlowski, K. Kraiger, E. Salas, and M. Teachout, Eds., Lawrence Erlbaum Associates, Mahwah, NJ, pp. 19–46.

Goldstein, I. L. (1980), "Training in Work Organizations," *Annual Review of Psychology*, Vol. 31, pp. 229–272.

Goldstein, I. L. (1993), *Training in Organizations*, 3rd ed., Brooks/Cole, Pacific Grove, CA.

Goldstein, I. L., and Ford, J. K. (2002), *Training in Organizations: Needs Assessment, Development, and Evaluation*, 4th ed., Wadsworth, Belmont, CA.

Gopher, D., Weil, M., and Bareket, T. (1994), "Transfer of Skill from a Computer Game Trainer to Flight," *Human Factors*, Vol. 36, pp. 387–405.

Gordon, S. E., and Gill, R. T. (1997), "Cognitive Task Analysis," in *Naturalistic Decision Making*, C. E. Zsambok and G. Klein, Eds., Lawrence Erlbaum Associates, Mahwah, NJ, pp. 131–140.

Grove, C., and Torbiörn, I. (1993), "A New Conceptualization of Intercultural Adjustment and the Goals of Training," in *Education for the Intercultural Experience*, R. M. Paige, Ed., Intercultural Press, Yarmouth, ME, pp. 205–233.

Gudykunst, W. B., Guzley, R. M., and Hammer, M. R. (1996), "Designing Intercultural Training," in *Handbook of Intercultural Training*, 2nd ed., D. Landis and R. W. Brislin, Eds., Sage Publications, Thousand Oaks, CA, pp. 61–80.

Guldenmund, F. W. (2000), "The Nature of Safety Culture: A Review of Theory and Research," *Safety Science*, Vol. 34, pp. 215–257.

Gully, S. M., Payne, S. C., Kiechel Koles, K. L., and Whiteman, J. K. (2002), "The Impact of Error Training and Individual Differences on Training Outcomes: An Attribute–Treatment Interaction Perspective," *Journal of Applied Psychology*, Vol. 87, No. 1, pp. 143–155.

Guzzo, R. A., and Dickson, M. W. (1996), "Teams in Organizations: Recent Research on Performance and Effectiveness," *Annual Review of Psychology*, Vol. 47, pp. 307–338.

Haccoun, R. R., and Hamtiaux, T. (1994), "Optimizing Knowledge Tests for Inferring Learning Acquisition Levels in Single Group Training Evaluation Designs: The Internal Referencing Strategy," *Personnel Psychology*, Vol. 47, pp. 593–604.

Hamel, C. J., and Ryan-Jones, D. L. (1997), "Using Three-Dimensional Interactive Graphics to Teach Equipment Procedures," *Educational Technology, Research and Development*, Vol. 45, pp. 77–87.

Hanover, J. M. B., and Cellar, D. F. (1998), "Environmental Factors and the Effectiveness of Workforce Diversity Training," *Human Resource Development Quarterly*, Vol. 9, pp. 105–124.

Harris, D., and Khan, H. (2003), "Response Time to Reject a Takeoff," *Human Factors and Aerospace Safety*, Vol. 3, No. 2, pp 165–175.

Harrison, J. K. (1992), "Individual and Combined Effects of Behavior Modeling and the Cultural Assimilator in Cross-Cultural Management Training," *Journal of Applied Psychology*, Vol. 77, pp. 952–962.

Heimbeck, D., Frese, M., Sonnentag, S., and Keith, N. (2003), "Integrating Errors into the Training Process: The Function of Error Management Instructions and the Role of Goal Orientation," *Personnel Psychology*, Vol. 56, pp. 333–361.

Hellreigel, D., and Slocum, J. W. (1974), "Organizational Climate: Measures, Research, and Contingencies," *Academy of Management Journal*, Vol. 17, pp. 255–280.

Hendricks, C. C. (2001), "Teaching Causal Reasoning Through Cognitive Apprenticeship: What Are the Results from Situated Learning?" *Journal of Educational Research*, Vol. 94, No. 5, pp. 302–311.

Hoffman, L. R., and Maier, N. R. F. (1961), "Quality and Acceptance of Problem Solutions by Members of Homogeneous and Heterogeneous Groups," *Journal of Abnormal and Social Psychology*, Vol. 62, pp. 401–407.

Hofmann, D. A., and Stetzer, A. (1996), "A Cross-Level Investigation of Factors Influencing Unsafe Behaviors and Accidents," *Personnel Psychology*, Vol. 49, pp. 307–339.

Hofmann, D. A., Jacobs, R., and Landy, F. (1995), "High Reliability Process Industries: Individual, Micro, and Macro Organizational Influences on Safety Performance," *Journal of Safety Research*, Vol. 26, pp. 131–149.

Hofstede, G. (1980), *Culture's Consequences: International Differences in Work Related Values*, Sage Publications, Beverly Hills, CA.

Holt, R. W., Hansberger, J. T., and Boehm-Davis, D. A. (2002), "Improving Rater Calibration in Aviation: A Case Study," *International Journal of Aviation Psychology*, Vol. 12, pp. 305–330.

Huczynski, A. A., and Lewis, J. W. (1980), "An Empirical Study into the Learning Transfer Process Management Training," *Journal of Management Studies*, Vol. 17, pp. 227–240.

Hunter, J. E. (1986), "Cognitive Ability, Cognitive Aptitudes, Job Knowledge, and Job Performance," *Journal of Vocational Behavior*, Vol. 29, pp. 340–362.

Ivancic, K., and Hesketh, B. (1995), "Making the Best of Errors During Training," *Training Research Journal*, Vol. 1, pp. 103–125.

Jacobs, J. W., and Dempsey, J. V. (1993), "Simulation and Gaming: Fidelity, Feedback, and Motivation," in *Interactive Instruction and Feedback*, J. V. Dempsey, and G. C. Sales, Eds., Educational Technology, Englewood Cliffs, NJ, pp. 197–229.

James, L. R., and Jones, A. P. (1974), "Organizational Climate: A Review of Theory and Research," *Psychological Bulletin*, Vol. 81, pp. 1096–1112.

Jentsch, F., and Bowers, C. (1998), "Evidence for the Validity of PC-Based Simulations in Studying Aircrew Coordination," *International Journal of Aviation Psychology*, Vol. 8, pp. 243–260.

Johnston, J. H., and Cannon-Bowers, J. A. (1996), "Training for Stress Exposure," in *Stress and Human Performance*, J. E. Driskell and E. Salas, Eds., Lawrence Erlbaum Associates, Mahwah, NJ, pp. 223–256.

Kanfer, R. (1991), "Motivational Theory and Industrial and Organizational Psychology," in *Handbook of Industrial and Organizational Psychology*, 2nd ed., M. D. Dunnette, L. M. Hough, and M. Leaetta, Eds., Consulting Psychologists Press, Palo Alto, CA, pp. 75–170.

Kaplan-Leiserson, E. (2002), "E-Learning Glossary," retrieved July 1, 2002, from http://www.learningcircuits.org/glossary.html.

Karl, K. A., O'Leary-Kelly, A. M., and Martocchio, J. J. (1993), "The Impact of Feedback and Self-Efficacy on Performance in Training," *Journal of Organizational Behavior*, Vol. 14, No. 4, pp. 379–394.

Kealey, D. J., and Protheroe, D. R. (1996), "The Effectiveness of Cross-Cultural Training for Expatriates: An Assessment of the Literature on the Issue," *International Journal of Intercultural Relations*, Vol. 20, pp. 141–165.

Kirkpatrick, D. L. (1959), "Techniques for Evaluating Training Programs," *Journal of the American Society of Training Directors*, Vol. 13, No. 3–9, pp. 21–26.

Kirkpatrick, D. L. (1976), "Evaluation of Training," in *Training and Development Handbook: A Guide to Human Resource Development*, 2nd ed., R. L. Craig, Ed., McGraw-Hill, New York, pp. 1–26.

Kirkpatrick, D. L. (1987), "Evaluation," in *Training and Development Handbook: A Guide to Human Resource Development*, 3rd ed., R. L. Craig, Ed., McGraw-Hill, New York.

Klein, G. A. (1997), "Developing Expertise in Decision Making," unpublished draft.

Klein, G., and Militello, L. (2001), "Some Guidelines for Conducting a Cognitive Task Analysis," in *Advances in Human Performance and Cognitive Engineering Research*, Vol. 1, E. Salas, Ed., Elsevier Science, Amsterdam, pp. 161–199.

Kossek, E. E., Roberts, K., Fisher, S., and Demarr, B. (1998), "Career Self-Management: A Quasi-experimental Assessment of the Effects of a Training Intervention," *Personnel Psychology*, Vol. 51, pp. 935–962.

Kozlowski, S. J., and Bell, B. S. (2002), "A Typology of Virtual Teams: Implications for Effective Leadership," *Group and Organization Management*, Vol. 27, No. 1, pp. 14–49.

Kozlowski, S. W. J., and Salas, E. (1997), "A Multilevel Organizational Systems Approach for the Implementation and Transfer of Training," in *Improving Training Effectiveness in Work Organizations*, J. K. Ford, S. Kozlowski, K. Kraiger, E. Salas, and M. Teachout, Eds., Lawrence Erlbaum Associates, Mahwah, NJ, pp. 247–287.

Kozlowski, S. W. J., Gully, S. M., Salas, E., and Cannon-Bowers, J. A. (1996), "Team Leadership and Development: Theory, Principles, and Guidelines for Training Leaders and Teams," in *Advances in Interdisciplinary Studies of Work Teams: Team Leadership*, M. Beyerlein, S. Beyerlein, and D. Johnson, Eds., JAI Press, Greenwich, CT, pp. 3:253–292.

Kozlowski, S. W. J., Brown, K., Weissbein D., Cannon-Bowers J., and Salas, E. (2000), "A Multilevel Approach to Training Effectiveness: Enhancing Horizontal and Vertical Transfer," in *Multilevel Theory, Research and Methods in Organization*, K. Klein and S. W. J. Kozlowski, Eds., Jossey-Bass, San Francisco, pp. 157–210.

Kraiger, K., and Jung, K. (1997), "Linking Training Objectives to Evaluation Criteria," in *Training for a Rapidly Changing Workplace: Applications of Psychological Research*, M. A. Quinones and A. Ehrenstein, Eds., American Psychological Association, Washington, DC, pp. 151–176.

Kraiger, K., Ford, J. K., and Salas, E. (1993), "Application of Cognitive, Skill-Based, and Affective Theories of Learning Outcomes to New Methods of Training Evaluation," *Journal of Applied Psychology*, Vol. 78, No. 2, pp. 311–328.

Kraiger, K., Salas, E., and Cannon-Bowers, J. A. (1995), "Measuring Knowledge Organization as a Method for Assessing Learning During Training," *Human Factors*, Vol. 37, No. 4, pp. 804–816.

Lai, Shu-Ling (2001), "Controlling the Display of Animation for Better Understanding," *Journal of Research on Computing in Education*, Vol. 33; retrieved June 13, 2001 from http://www.iste.org/jrte/33/5/lai.html.

Landis, D., and Brislin, R. W., Eds. (1996), *Handbook of Intercultural Training*, 2nd ed., Sage Publications, Thousand Oaks, CA.

Latham, G. P. (1989), "Behavioral Approaches to the Training and Learning Process," in *Training and Development in Organizations*, L. L. Goldstein, Ed., Jossey-Bass, San Francisco, pp. 256–295.

Leedom, D. K., and Simon, R. (1995), "Improving Team Coordination: A Case for Behavior-Based Training," *Military Psychology*, Vol. 7, pp. 109–122.

Lewis, M. (1998), "What Workers Need to Succeed," *American Machinist*, Vol. 142, No. 3, pp. 88–98.

Lorenzet, S. J., Salas, E., and Tannenbaum, S. I. (2003), "The Impact of Guided Errors on Skill Development and Self-Efficacy," manuscript submitted for publication.

Marks, M. A. (2000), "A Critical Analysis of Computer Simulations for Conducting Team Research," *Small Group Research*, Vol. 31, No. 6, pp. 653–675.

Martin, G. L. and Lumsden, J. A. (1987), *Coaching: An Effective Behavioral Approach*, C.V. Mosby, St. Louis, MO.

Martocchio, J. J. (1992), "Microcomputer Usage as an Opportunity: The Influence of Context in Employee Training," *Personnel Psychology*, Vol. 45, pp. 529–551.

Martocchio, J. J., and Judge, T. A. (1997), "Relationship Between Conscientiousness and Learning in Employee Training: Mediating Influences of Self-Deception and Self-Efficacy," *Journal of Applied Psychology*, Vol. 82, pp. 764–773.

Martocchio, J. J., and Webster, J. (1992), "Effects of Feedback and Cognitive Playfulness on Performance in Microcomputer Software Training," *Personality Psychology*, Vol. 45, pp. 553–578.

Mathieu, J. E., Tannenbaum, S. I., and Salas, E. (1992), "Influences of Individual and Situational Characteristics on Measures of Training Effectiveness," *Academy of Management Review*, Vol. 35, pp. 828–847.

McGhee, W., and Thayer, P. W. (1961), *Training in Business and Industry*, Wiley, New York.

Milheim, W. D., and Martin, B. L. (1991), "Theoretical Bases for the Use of Learner Control: Three Different Perspectives," *Journal of Computer-Based Instruction*, Vol. 18, pp. 99–105.

Mobley, G. M., Jaret, C., Marsh, K., and Lim, Y. Y. (1994), "Mentoring, Job Satisfaction, Gender, and the Legal Profession," *Sex Roles: A Journal of Research*, Vol. 31, pp. 79–98.

Moe, M. T., and Blodget, H. (2000), *The Knowledge Web*, Part 3, *Higher Web Universities Online*, Merrill Lynch & Co., New York.

Moore, S. (1999), "Understanding and Managing Diversity Among Groups at Work: Key Issues for Organisational Training and Development," *Journal of European Industrial Training*, Vol. 23, No. 4–5, pp. 208–217.

Morris, M. A., and Robie, C. (2001), "A Meta-analysis of the Effects of Cross-Cultural Training on Expatriate Performance and Adjustment," *International Journal of Training and Development*, Vol. 5, pp. 112–125.

Morrison, G. R., Ross, S. M., and Baldwin, W. (1992), "Learner Control of Context and Instructional Support in Learning Elementary School Mathematics," *Educational Technology Research and Development*, Vol. 40, pp. 5–13.

Morrow, C. C., Jarrett, M. Q., and Rupinski, M. T. (1997), "An Investigation of the Effect and Economic Utility of Corporate-Wide Training," *Personnel Psychology*, Vol. 50, pp. 91–119.

Murphy, M. A. and Devidson, G. V. (1991), "Computer-Based Adaptive Instruction: Effects of Learner Control on Concept Learning," *Journal of Computer-Based Instruction*, Vol. 18, No. 2, pp. 51–56.

National Safety Council (2001), *Accident Facts*. NSC, Chicago.

Noe, R. A. (1986), "Trainee Attributes and Attitudes: Neglected Influences on Training Effectiveness," *Academy of Management Review*, Vol. 4, pp. 736–749.

Noe, R. A., and Schmitt, N. (1986), "The Influence of Trainee Attitudes on Training Effectiveness: Test of a Model," *Personnel Psychology*, Vol. 39, pp. 497–523.

O'Connor, P., Hörmann, H., and Flin, R. (2002), "Developing a Method for Evaluating Crew Resource Management Skills: A European Perspective," *International Journal of Aviation Psychology*, Vol. 12, pp. 263–285.

Ong, J., and Ramachandran, S. (2003), "Intelligent Tutoring Systems: Using AI to Improve Training Performance and ROI," retrieved July 18, from http://www.shai.com/papers.

Ortiz, G. A. (1994), "Effectiveness of PC-Based Flight Simulation," *International Journal of Aviation Psychology*, Vol. 4, pp. 285–291.

Oser, R. L., Cannon-Bowers, J. A., Salas, E., and Dwyer, D. J. (1999a), "Enhancing Human Performance in Technology-Rich Environments: Guidelines for Scenario-Based Training," in *Human/Technology Interaction in Complex Systems*, E. Salas, Ed., JAI Press, Greenwich, CT, pp. 175–202.

Oser, R. L., Gualtieri, J. W., Cannon-Bowers, J. A., and Salas, E. (1999b), "Training Team Problem Solving Skills: An Event-Based Approach," *Computers in Human Behavior*, Vol. 15, No. 3–4, pp. 441–462.

Peters, L. H., O'Connor, E. J. and Eulberg, J. R. (1985), "Situational Constraints: Sources, Consequences, and Future Considerations," in *Research in Personnel and Human Resources Management*, Vol. 3, K. Rowland and G. Ferris, Eds., JAI Press, Greenwich, CT, pp. 79–113.

Phillips, J. M., and Gully, S. M. (1997), "Role of Goal Orientation, Ability, Need for Achievement, and Locus of Control in the Self-Efficacy and Goal Setting Process," *Journal of Applied Psychology*, Vol. 82, pp. 792–802.

Prince, C., and Jentsch, F. (2001), "Aviation Crew Resource Management Training with Low-Fidelity Devices," in *Improving Teamwork in Organizations*, E. Salas, C. A. Bowers, and E. Edens, Eds., Lawrence Erlbaum Associates, Mahwah, NJ, pp. 147–164.

Prince, C., Oser, R, Salas, E., and Woodruff, W. (1993), "Increasing Hits and Reducing Misses in CRM/LOS Scenarios: Guidelines for Simulator Scenario Development," *International Journal of Aviation Psychology*, Vol. 3, No. 1, pp. 69–82.

Quinones, M. A. (1995), "Pretraining Context Effects: Training Assignment as Feedback," *Journal of Applied Psychology*, Vol. 80, pp. 226–238.

Quinones, M. A. (1997), "Contextual Influencing on Training Effectiveness," in *Training for a Rapidly Changing Workplace: Applications of Psychological Research*, M. A. Quinones and A. Ehrenstein, Eds., American Psychological Association, Washington, DC, pp. 177–200.

Quinones, M. A., Ford, J. K., Sego, D. J., and Smith, E. M. (1995), "The Effects of Individual and Transfer Environment Characteristics on the Opportunity to Perform Trained Tasks," *Training Research Journal*, Vol. 1, pp. 29–48.

Ralphs, L. T., and Stephan, E. (1986), "HRD in the Fortune 500," *Training and Development Journal*, Vol. 40, pp. 69–76.

Randel, J. M., Main, R. E., Seymour, G. E., and Morris, B. A. (1992), "Relation of Study Factors to Performance in Navy Technical Schools," *Military Psychology*, Vol. 4, pp. 75–86.

Ree, M. J., and Earles, J. A. (1991), "Predicting Training Success: Not Much More Than *G*," *Personality Psychology*, Vol. 44, pp. 321–332.

Ree, M. J., Carretta, T. R., and Teachout, M. S. (1995), "Role of Ability and Prior Job Knowledge in Complex Training Performance," *Journal of Applied Psychology*, Vol. 80, pp. 721–730.

Ricci, K. E., Salas, E., and Cannon-Bowers, J. A. (1995), "Do Computer Based Games Facilitate Knowledge Acquisition and Retention?" *Military Psychology*, Vol. 8, pp. 295–307.

Richman, H. B., Staszewski, J. J., and Simon, H. A. (1995), "Simulation of Expert Memory Using EPAM IV," *Psychological Review*, Vol. 102, pp. 305–330.

Roenker, D. L., Cissell, G. M., Ball, K. K., Wadley, V. G., and Edwards, J. E. (2003), "Speed-of Processing and Driving Simulator in Training Result in Improved Driving Performance," *Human Factors*, Vol. 45, No. 2, pp. 218–233.

Roosa, M. W., Dumka, L. E., Gonzales, N. A., and Knight, G. P. (2002), "Cultural/Ethnic Issues and the Prevention Scientist in the 21st Century," *Prevention and Treatment*, Vol. 5; retrieved from http://journals.apa.org/prevention/volume5/pre0050005a.html.

Rosow, J. M., and Zager, R. (1988), *Training the Competitive Edge*, Jossey-Bass, San Francisco.

Rossett, A., and Gautier-Downes, J. (1991), *A Handbook of Job Aids*, Jossey-Bass/Pfeiffer, San Francisco.

Rouiller, J. Z. (1989), "Detriments to the Climate of Transfer of Training," unpublished doctoral dissertation, University of Maryland.

Rouiller, J. Z., and Goldstein, I. L. (1993), "The Relationship Between Organizational Transfer Climate and Positive Transfer of Training," *Human Resource Development Quarterly*, Vol. 4, pp. 377–390.

Ruben, B. D. (1999), "Simulations, Games, and Experience-Based Learning: The Quest for a New Paradigm for Teaching and Learning," *Simulation and Gaming*, Vol. 30, No. 4, pp. 498–505.

Sackett, P. R., and Mullen, E. J. (1993), "Beyond Formal Experimental Design: Towards an Expanded View of the Training Evaluation Process," *Personnel Psychology*, Vol. 46, pp. 613–627.

Sacks, M. (1994), *On-the-Job Learning in the Software Industry*, Quorum Books, Westport, CT.

Saks, A. M. (1995), "Longitudinal Field Investigation of the Moderating and Mediating Effects of Self-Efficacy on the Relationship Between Training and Newcomer Adjustment," *Journal of Applied Psychology*, Vol. 80, pp. 221–225.

Salas, E., and Cannon-Bowers, J. A. (1997), "Methods, Tools, and Strategies for Team Training," in *Training for a Rapidly Changing Workplace: Applications of Psychological Research*, M. A. Quinones and A. Ehrenstein, Eds., American Psychological Association, Washington, DC, pp. 249–279.

Salas, E., and Cannon-Bowers, J. A. (2000a), "Designing Training Systematically," in *The Blackwell Handbook of Principles of Organizational Behavior*, E. A. Locke, Ed., Blackwell Publishers, Malden, MA, pp. 43–59.

Salas, E., and Cannon-Bowers, J. A. (2000b), "The Anatomy of Team Training," in *Training and Retraining: A Handbook for Business, Industry, Government, and the Military*, S. Tobias and J. D. Fletcher, Eds., Macmillan, New York, pp. 312–335.

Salas, E., and Cannon-Bowers, J. A. (2001), "The Science of Training: A Decade of Progress," *Annual Review of Psychology*, Vol. 52, pp. 471–499.

Salas, E., and Klein, G., Eds. (2000), *Linking Expertise and Naturalistic Decision Making*, Lawrence Erlbaum Associates, Mahwah, NJ.

Salas, E., Burgess, K. A., and Cannon-Bowers, J. A. (1995), "Training Effectiveness Techniques," in *Research Techniques in Human Engineering*, J. Weiner, Ed., Prentice-Hall, Englewood Cliffs, NJ, pp. 439–471.

Salas, E., Bowers, C. A., and Blickensderfer, E. (1997a), "Enhancing Reciprocity Between Training Theory and Training Practice: Principles, Guidelines, and Specifications," in *Improving Training Effectiveness in Work Organizations*, J. K. Ford, S. Kozlowski, K. Kraiger, E. Salas, and M. Teachout, Eds., Lawrence Erlbaum Associates, Mahwah, NJ, pp. 19–46.

Salas, E., Cannon-Bowers, J. A., and Johnston, J. H. (1997b), "How Can You Turn a Team of Experts into an Expert Team?: Emerging Training Strategies," in *Naturalistic Decision Making*, C. E. Zsambok and G. Klein, Eds., Lawrence Erlbaum Associates, Mahwah, NJ, pp. 359–370.

Salas, E., Bowers, C. A., and Rhodenizer, L. (1998), "It Is Not How Much You Have but How You Use It: Toward a Rational Use of Simulation to Support Aviation Training," *International Journal of Aviation Psychology*, Vol. 8, pp. 197–208.

Salas, E., Rozell, D., Mullen, B., and Driskell, J. E. (1999), "The Effect of Team Building on Performance: An Integration," *Small Group Research*, Vol. 30, No. 3, pp. 309–329.

Salas, E., Burke, C. S. Bowers, C. A., and Wilson, K. A. (2001), "Team Training in the Skies: Does Crew Resource Management (CRM) Training Work?" *Human Factors*, Vol. 43, No. 4, pp. 641–674.

Salas, E., Burke, C. S., Fowlkes, J. E., and Wilson, K. A. (2004), "Challenges and Approaches to Understanding Leadership Efficacy in Multi-cultural Teams," in *Advances in Human Performance and Cognitive Engineering Research*, Vol. 4, M. Kaplan, Ed., Elsevier, Oxford, pp. 341–384.

Salas, E., Stagl, K. C., Burke, C. S., and Goodwin, G. F. (in press). "Fostering Team Effectiveness in Organizations: Toward an Integrative Theoretical Framework of Team Performance," in *Modeling Complex Systems: Motivation, Cognition and Social Processes*, J. W. Stuart, W. Spaulding, and J. Poland, Eds., *Nebraska Symposium on Motivation*, 51, University of Nebraska Press, Lincoln, NE.

Satish, U., and Streufert, S. (2002), "Value of a Cognitive Simulation in Medicine: Towards Optimizing Decision Making Performance of Healthcare Personnel," *Quality and Safety in Health Care*, Vol. 11, pp. 163–167.

Scandura, T. A., Tejeda, M. J., Werther, W. B., and Lankau, M. J. (1996), "Perspectives on Mentoring," *Leadership and Organizational Development Journal*, Vol. 17, pp. 50–56.

Schein, E. H. (1992), *Organizational Culture and Leadership*, 2nd ed., Jossey-Bass, San Francisco.

Schmidt, R. A., and Bjork, R. A. (1992), "New Conceptualizations of Practice: Common Principles in Three Paradigms Suggest New Concepts for Training," *Psychological Science*, Vol. 3, pp. 207–217.

Schmidt, A. M., and Ford, J. K. (2003), "Learning Within a Learner Control Training Environment: The Interactive Effects of Goal Orientation and Metacognitive Instruction on Learning Outcomes," *Personnel Psychology*, Vol. 56, pp. 405–429.

Schmidt, F. L., Hunter, J. E., and Pearlman, K. (1982), "Assessing the Economic Impact of Personnel Programs on Workforce Productivity," *Personnel Psychology*, Vol. 35, pp. 333–347.

Schraagen, J. M., Chipman, S. F., and Shalin, V. L., Eds. (2000), *Cognitive Task Analysis*, Lawrence Erlbaum Associates, Mahwah, NJ.

Schreiber, D., and Berge, Z., Eds. (1998), *Distance Training: How Innovative Organizations Are Using Technology to Maximize Learning and Meet Business Objectives*, Jossey-Bass, San Francisco.

Selmer, J. (2001), "The Preference for Predeparture or Postarrival Cross-Cultural Training: An Exploratory Approach," *Journal of Managerial Psychology*, Vol. 16, pp. 50–58.

Selmer, J., Torbiörn, I., and de Leon, C. T. (1998), "Sequential Cross-Cultural Training for Expatriate Business Managers: Pre-departure and Post-arrival," *International Journal of Human Resource Management*, Vol. 9, pp. 831–840.

Serfaty, D., Entin, E. E., Johnston, J. H., and Cannon-Bowers,J. A (1998), "Team Coordination Training," in *Making Decisions Under Stress: Implications for Individual and Team Training*, J. A. Cannon-Bowers and E. Salas, Eds., American Psychological Association, Washington, DC, pp. 221–245.

Shebilske, W. L., Regian, J. W., Arthur, W., Jr., and Jordan, J. A. (1992), "A Dyadic Protocol for Training Complex Skills," *Human Factors* Vol. 34, pp. 369–374.

Shebilske, W. L., Jordan, J. A., Goettl, B. P., and Paulus, L. E. (1998), "Observation Versus Hands-on Practice of Complex Skills in Dyadic, Triadic, and Tetradic Training-Teams," *Human Factors* Vol. 40, pp. 525–540.

Shute, V. J., and Gawlick, L. A. (1995), "Practice Effects on Skill Acquisition, Learning Outcome, Retention, and Sensitivity to Relearning," *Human Factors*, Vol. 37, pp. 781–803.

Shute, V. J., Gawlick, L. A., and Gluck, K. A. (1998), "Effects of Practice and Learner Control on Short- and Long-Term Gain," *Human Factors*, Vol. 40, pp. 296–310.

Shyu, H., and Brown, S. W. (1992), "Learner Control Versus Program Control in Interactive Videodisc Instruction: What Are the Effects in Procedural Learning?" (electronic version), *International Journal of Instructional Media*, Vol. 19, pp. 85–96.

Simon, S. J., and Werner, J. M. (1996), "Computer Training Through Behavior Modeling, Self-Paced, and Instructional Approaches: A Field Experiment," *Journal of Applied Psychology*, Vol. 81, pp. 648–659.

Skarlicki, D. P., and Latham, G. P. (1997), "Leadership Training in Organizational Justice to Increase Citizenship Behavior Within a Labor Union: A Replication," *Personnel Psychology*, Vol. 50, pp. 617–633.

Smith-Jentsch, K. A., Jentsch, F. G., Payne, S. C., and Salas, E. (1996a), "Can Pretraining Experiences Explain Individual Differences in Learning?" *Journal of Applied Psychology*, Vol. 81, pp. 909–936.

Smith-Jentsch, K., Salas, E., and Baker, D. P. (1996b), "Training Team Performance-Related Assertiveness," *Personnel Psychology*, Vol. 49, pp. 909–936.

Smith-Jentsch, K. A., Blickensderfer, E., Salas, E. and Cannon-Bowers, J. A. (2000), "Helping Team Members Help Themselves: Propositions for Facilitating Guided Them Self-Correction," in *Advances in Interdisciplinary Studies of Work Teams*, Vol. 6, M. M. Beyerlein, D. A. Johnson, & S. T. Beyerlein, (Eds.), JAI Press, Greenwich, CT, pp. 55–72.

Smith-Jentsch, K. A., Zeisig, R. L., Acton, B., and McPherson, J. A. (1998), "Team Dimensional Training: A Strategy for Guided Team Self-Correction," in *Making Decisions Under Stress: Implications for Individual and Team Training*, J. A. Cannon-Bowers and E. Salas, Eds., American Psychological Association, Washington, DC, pp. 271–297.

Snyder, R., Raben, C., and Farr, J. (1980), "A Model for the Systematic Evaluation of Human Resource Development Programs," *Academy of Management Review*, Vol. 5, No. 3, pp. 431–444.

Spaulding, K., and Dwyer, F. (2001), "The Effect of Time-on-Task When Using Job Aids as an Instructional Strategy," *International Journal of Instructional Media*, Vol. 28, No. 4, pp. 437–447.

Steele-Johnson, D., and Hyde, B. G. (1997), "Advanced Technologies in Training: Intelligent Tutoring Systems and Virtual Reality," in *Training for a Rapidly Changing Workplace: Applications of Psychological Research*, M. A. Quinones and A. Ehrenstein, Eds., American Psychological Association, Washington, DC, pp. 225–248.

Steinberg, E. R. (1977), "Review of Student Control in Computer-Assisted Instruction," *Journal of Computer-Based Instruction*, Vol. 3, pp. 84–90.

Steiner, I. D. (1972), *Group Process and Productivity*, Academic Press, New York.

Stevens, C. K., and Gist, M. E. (1997), "Effects of Self-Efficacy and Goal-Orientation Training on Negotiation Skill Maintenance: What Are the Mechanisms?" *Personnel Psychology*, Vol. 50, pp. 955–978.

Stout, R. J., Salas, E., and Fowlkes, J. (1997), "Enhancing Teamwork in Complex Environments Through Team Training," *Group Dynamics: Theory Research and Practice*, Vol. 1, pp. 169–182.

Swezey, R. W. (1987), "Design of Job Aids and Procedure Writing," in *Handbook of Human Factors and Ergonomics*, G. Salvendy, Ed., Wiley, New York, pp. 1039–1057.

Swezey, R. W., and Llaneras, R. E. (1997), "Models in Training and Instruction," in *Handbook of Human Factors and Ergonomics*, 2nd ed., G. Salvendy, Ed., Wiley, New York, pp. 514–577.

Tannenbaum, S. I. (1997), "Enhancing Continuous Learning: Diagnostic Findings from Multiple Companies," *Human Resource Management*, Vol. 36, pp. 437–452.

Tannenbaum, S. I., and Yukl, G. (1992), "Training and Development in Work Organizations," *Annual Review of Psychology*, Vol. 43, pp. 399–441.

Tannenbaum, S. I., Mathieu, J. E., Salas, E., and Cannon-Bowers, J. A. (1991), "Meeting Trainees' Expectations: The Influence of Training Fulfillment on the Development of Commitment," *Journal of Applied Psychology*, Vol. 76, pp. 759–769.

Tannenbaum, S. I., Beard, R. L., and Salas, E. (1992), "Team Building and Its Influence on Team Effectiveness: An Examination of Conceptual and Empirical Developments," in *Issue, Theory, and Research in Industrial/Organizational Psychology*, K. Kelley, Ed., Elsevier, Amsterdam, pp. 117–153.

Tannenbaum, S. I., Cannon-Bowers, J. A., and Mathieu, J. E. (1993), "Factors That Influence Training Effectiveness: A Conceptual Model and Longitudinal Analysis," Report 93-011, Naval Training Systems Center, Orlando, FL.

Tennyson, R. D. (1980), "Instructional Control Strategies and Content Structure as Design Variables in Concept Acquisition Using Computer-Based Instruction," *Journal of Educational Psychology*, Vol. 72, No. 4, pp. 525–532.

Tesluk, P. E., Farr, J. L., Mathieu, J. E., and Vance, R. J. (1995), "Generalization of Employee Involvement Training to the Job Setting: Individual and Situational Effects," *Personnel Psychology*, Vol. 8, pp. 607–632.

Thayer, P. W., and Teachout, M. S. (1995), "A Climate for Transfer Model," Report AL/HR-TP-1995-0035, Air Force Material Command, Brooks Air Force Base, TX.

Thomas, D. C. (1999), "Cultural Diversity and Work Group Effectiveness," *Journal of Cross-Cultural Psychology*, Vol. 30, No. 2, 242–263.

Tillman, M. (1985), "How Technology Is Changing Training," *Data Training*, Vol. 9, pp. 18–23.

Townsend, A. M., DeMarie, S. M., and Hendrickson, A. R. (1996), "A Multi-disciplinary Approach to the Study of Virtual Workgroups," in *Proceedings of the 1996 Decision Sciences Institute*.

Tracey, B. J., Tannenbaum, S. I., and Kavanagh, M. J. (1995), "Applying Trained Skills on the Job: The Importance of the Work Environment," *Journal of Applied Psychology*, Vol. 80, pp. 239–252.

Tracey, J. B., Hinkin, T. R., Tannenbaum, S. I., and Mathieu, J. E. (2001), "The Influence of Individual Characteristics and the Work Environment on Varying Levels of Training Outcomes," *Human Resource Development Quarterly*, Vol. 12, pp. 5–24.

Triandis, H. C. (2000), "Culture and Conflict," *International Journal of Psychology*, Vol. 35, No. 2, pp. 145–152.

VandeValle, D. (1997), "Development and Validation of a Work Domain Goal Orientation Instrument," *Educational Psychology Measurement*, Vol. 57, pp. 995–1015.

Volpe, C. E., Cannon-Bowers, J. A., Salas, E., and Spector, P. E. (1996), "The Impact of Cross-Training on Team Functioning: An Empirical Investigation," *Human Factors*, Vol. 38, pp. 87–100.

Watson, W. E., Kumar, K., and Michaelson, K. K. (1993), "Cultural Diversity's Impact on Interaction Process and Performance: Comparing Homogeneous and Diverse Task Groups," *Academy of Management Journal*, Vol. 36, pp. 590–602.

Wehrenberg, S. B. (1987), "Supervisors as Trainers: The Long-Term Gains of OJT," *Personnel Journal*, Vol. 66, No. 4, pp. 48–51.

Wexley, K. N., and Baldwin, T. T. (1986), "Posttraining Strategies for Facilitating Positive Transfer: An Empirical Exploration," *Academy of Management Journal*, Vol. 29, No. 3, pp. 503–520.

Wickens, C. D., Helleberg, J., and Xu, X. (2002), "Pilot Maneuver Choice and Workload in Free Flight," *Human Factors*, Vol. 44, No. 2, pp. 171–188.

Wiener, E. L., Kanki, B. G., and Helmreich, R. L., Eds. (1993), *Cockpit Resource Management*, Academic Press, New York.

Williams, T. C., Thayer, P. W., and Pond, S. B. (1991), "Test of a Model of Motivational Influences on Reactions to Training and Learning," paper presented at the 6th Annual Conference of the Society for Industrial and Organizational Psychology, St. Louis, MO.

Wilson, P. F., and Johnson, W. B. (2001), "Core Virtues for the Practice of Mentoring," *Journal of Psychology and Theology*, Vol. 29, No. 2, pp. 121–130.

Wilson, K. A., Priest, H. A., Salas, E., and Burke, C. S. (2005), "Can Training for Safe Practices Reduce the Risk of Organizational Liability?" in *Handbook of Human Factors in Litigation*, I. Noy and W. Karwowski, Eds., Taylor & Francis, London, pp. 6-1–6-32.

Woolf, B. P., Beck, J., Eliot, C., and Stern, M. (2001), "Growth and Maturity of Intelligent Tutoring Systems: A Status Report," in *Smart Machines in Education: The Coming Evolution in Educational Technology*, K. D. Forbus and P. J. Feltovich, Eds., MIT Press, Cambridge, MA, pp. 99–144.

Wydra, F. T. (1980), *Learner Controlled Instruction*, Educational Technology Publications, Englewood Cliffs, NJ.

Yang, H., Sackett, P. R., and Arvey, R. D. (1996), "Statistical Power and Cost in Training Evaluation: Some New Considerations," *Personnel Psychology*, Vol. 49, pp. 651–668.

Yelon, S. L., and Ford, J. K. (1999), "Pursuing a Multidimensional Model of Transfer," *Performance Improvement Quarterly*, Vol. 12, pp. 58–78.

Zaccaro, S. J. and Bader, P. (2002), "E-Leadership and the Challenges of Leading E-Teams: Minimizing the Bad and Maximizing the Good," *Organizational Dynamics*, Vol 31, No. 4, pp. 377–387.

Zachary, W. W., Ryder, J. M., and Hicinbothom, J. H. (1998), "Cognitive Task Analysis and Modeling of Decision Making in Complex Environments," in *Making Decisions Under Stress: Implications for Individual and Team Training*, J. A. Cannon-Bowers and E. Salas, Eds., American Psychological Association, Washington, DC, pp. 313–344.

Zachary, W., Santarelli, T., Lyons, D., Bergondy, M., and Johnston, J. (2001), "Using a Community of Intelligent Synthetic Entities to Support Operational Team Training," in *Proceedings of the 10th Conference on Computer Generated Forces and Behavioral Representation*, pp. 215–224.

Zaklad, A., and Zachary, W. W. (1992), "Decision Support Design Principles for Tactical Decision-Making in Ship-Based Anti-air Warfare," Technical Report 20930.9000, CHI Systems, Springhouse, PA.

Zsambok, C., and Klein, G., Eds. (1997), *Naturalistic Decision Making*, Lawrence Erlbaum Associates, Mahwah, NJ.

CHAPTER 19

HUMAN FACTORS IN ORGANIZATIONAL DESIGN AND MANAGEMENT

Karen M. Dettinger
University of Wisconsin Hospital and Clinics
Madison, Wisconsin

Michael J. Smith
University of Wisconsin–Madison
Madison, Wisconsin

1 INTRODUCTION

Human factors engineering focuses on designing systems and products to meet the needs, capabilities, and motivations of people. Organizations are one example of systems that benefit from a human factors perspective. Definitions of organizations typically focus on their business purpose; generally, this purpose is to provide the structure and processes for operations that make and deliver goods and services to people. In this chapter we look at organizations from a human factors perspective and put emphasis on the human element and the notion that organizations are communities of people; organizations are communities of collective effort that yield more than the sum of each person's contributions (Galbraith et al., 2002).

Most business and academic discussions about organizational design focus on aligning a series of organizational elements that include leadership and decision making, structure, strategy, management processes, work processes, human resources policies, and interaction among people within and outside the organization, such as unions, customers, and the government. There is substantial technical and philosophical information about these elements, and there are many good references available that provide specific guidance on their content and implementation strategies. However, in this chapter we do not focus on these elements. Rather, we focus on what it takes to create an organizational foundation and management philosophy that fosters a healthy work experience for employees while creating an organization that is sustainable for the long term. Our assumption is that meeting the needs of the people in an organization is necessary for both the short- and long-term financial success of the organization.

Consistent with many organizational design and management authors (McGregor, 1960; K. Smith, 1965; Deming, 1986; Hendrick, 1986, 1987; Lawler, 1986, 1996, 2003; M. Smith and Sainfort, 1989; M. Smith and Carayon, 1995; Carayon and Smith, 2000; Hackman, 2002), we view employees as the foundation on which to build a successful organization. As we proceed, we present our personal perspectives on organizational design, operations, and management. Our beliefs are grounded in theory, research, and practice, and that gives us confidence that they can be helpful to many organizations.

This chapter begins by providing some background context about the history of organizations and organizational design and management. We then make a significant shift in tone and content to describe a societal reawakening we see toward a greater focus on the importance of finding meaning in work and life. We address what we believe are four key attributes of healthy and sustainable organizations, organizations that are able to respond to and even lead this reawakening. Following the attributes, we present five principles that provide more specific guidance on actions that can be taken to create a healthy and sustainable organization. The chapter concludes with two examples and two case studies that highlight successful organizational responses to challenges.

2 BACKGROUND

The background in organizational design and management discussed here comes from the traditions in North American conceptualizations of organizations and how they operate. We recognize that there are many other traditions and concepts with a rich history. However, we believe that our focus is appropriate for this chapter since the societal shift we discuss in later sections is based on the behaviors of organizations and individuals in the United States.

Organizational structure has always been a primary consideration in organizational design and management. Since the dawn of human history there have been various organizational structures for groups of people who have come together around common interests. Over time, the structures of these groups have increased in their complexity as the size of the organizations grew and as varied interests had to be accommodated. As organizations became more complex, they formed multiple sectors or departments of specialization. This mirrored the growing complexity of society with multiple sectors, such as government, military, judicial, religious, marketplace, trading, financial, construction, artisan/professional, and labor. Work organizations developed a similar makeup, with management, finance, security, manufacturing, marketing, sales, customer services, production, transportation, and human resources. This complexity led to a need for structure, rules, and procedures to provide effective and efficient operations.

As K. Smith (1965) reported, very early in human history a dominant organizational structure, the hierarchy, was developed. One leader at the top had absolute power, and then delegated authority and responsibility downward in a pyramid-type progression. The greatest power and authority was held by those few near the top, while little was present near the base. This is the well-known top-down power structure of an organization. Independent of the nature of the leadership, the structure of the organization typically followed a military structure with a top leader (CEO), generals (vice presidents), colonels (division managers), lieutenants (department managers), sergeants (supervisors), corporals (lead workers), and privates (employees), and orders flowing downward. To this day, this type of hierarchy and military command structure remains a dominant structure of small and large organizations. One of the few changes that has commonly been implemented is the board of directors representing stockholders at the top of the pyramid.

Characteristic of this structure is the chain of command, with the orders for action flowing from the top down through the organization to the bottom. Functions, knowledge, and skills tend to be specialized in specific units within the organization, sometimes referred to as divisions or departments. This structure requires the coordination and integration of expertise within and across the various units into a unified operation, and this is accomplished through the management process. At the higher levels in the hierarchy, the management functions are more similar than they are different, but specialized knowledge may still differentiate one unit from another. At the lowest level in the hierarchy, the activities of the people are quite different across departments, and front-line supervisors are well versed in the day-to-day details of how specific activities are carried out.

In the early twentieth century much emphasis was put on the specialization of function and knowledge at the department level, the supervisor level, and the individual employee level. The purpose was to develop greater expertise by focusing the attention and skills of the workforce. To build competence and skill, researchers and practitioners used scientific measurement methods and motivational theory to improve employee performance and productivity substantially. This led to highly structured work activities that required focused knowledge and intelligence and highly developed perceptual-motor skills. The workforce responded positively to the specialization of function. People learned new skills, took pride in the quality of what they produced, and their wages and standard of living increased substantially.

Over the next several decades the workforce became more educated and less satisfied with the focused and specialized nature of their work. Routine, boring work and the realization that opportunities for growth were limited set in. This led to problems in productivity and serious concerns for employee physical and mental health. New organizational structures and approaches to job design started to develop in the middle of the twentieth century and progressed through the rest of the century (Lawler, 1986, 1996, 2003; Black and Porter, 2000; Porter et al., 2002). However, even with the growth of new management approaches and job designs, the most dominant organizational approach in the United States remained a hierarchical militarylike structure with top-down power, authority, and decision making. This structure was difficult to discard because it was effective in getting the orders followed, departments coordinated, and products and services produced and delivered in a consistent manner.

Some of the important lessons that we believe emerged from studying the organization and management of work over the past 100 years are that (1) the hierarchy structure of management and control produces predictable results; (2) an effective hierarchy structure does best when there is a top-down power structure with a strong leader; (3) other forms of organizational structure can be effective, but primarily in small organizations or in large organizations that operate as an integrated network of small businesses; (4) people at all levels of the hierarchy are the most critical resources to the success of the organization, even more important to success than capital, technology, materials, or products and services offered, and (5) the best organizational designs incorporate the needs and the knowledge of the workforce in managing the organization.

Starting around the beginning of the twentieth century and up to today, much has been learned about how organizational structure and management affects employee satisfaction and performance. Early on we learned that employees responded well to taking orders from supervisors if they trusted the orders (Taylor, 1911; McGregor, 1960; K. Smith, 1965). They followed explicit instructions in what tasks to do and the directions and specifications in how to do the tasks. This created clear roles and responsibilities for managers and employees, removed role ambiguity, and let employees know explicitly what their performance requirements were. Employees responded positively to efforts to improve their skills through training. They appreciated new tools that reduced the effort needed to do their tasks, and increased the rewards that came with high achievement. The application of detailed and careful evaluation of work management, operations, and tasks led to a "scientific" basis for establishing guidelines for employee selection, supervision, the design of work tasks, and performance requirements (Taylor, 1911; McGregor, 1960; K. Smith, 1965; Drucker, 2001).

The consistent use of scientific work evaluation and design methods was well received by employees when the methods were perceived as unbiased. Employees felt that they were treated fairly. This then established some important human factors considerations in organizational management that led to increased employee satisfaction with their work and less stress. Human factors considerations included developing reasonable work standards, and creating an environment in which employees trust the organization's decisions and feel treated fairly.

For the fair treatment of employees and their perceptions of fairness and trust, "scientific" analytical methods and design criteria were used as the basis for work design and management. These scientific methods and criteria were based on sound evidence of validity and reliability, and are simple and clear enough to be understood and accepted by employees. When such "fair," "trustworthy," and "scientific" requirements were applied to managing and designing work, employees were more satisfied, less stressed, and performed better than when arbitrary requirements were applied (Lawler, 1986, 2003; K. Smith, 1965; M. Smith, 1987; Carayon and Smith, 2000).

Although the formal structure of the organization and the management process are important for organizational success, we have learned that the informal and social aspects also need to be considered since they can influence the effectiveness of the formal elements (Roethlisberger and Dickson, 1939; Lawler, 1986). The informal hierarchy of leadership and management that exists in work organizations can substantially influence the attitudes and behavior of employees. Social processes at work can influence cohesiveness within the work unit and how close employees feel toward the organization. Both of these factors can influence employee satisfaction, stress, and productivity. Informal leaders can facilitate management of the organization by their conformity with formal directives or can inhibit organizational management by presenting contrary perspectives and directives to employees. The informal influence can be so subtle that it does not appear to confront management directives directly. Informal influence can also provide assistance to employees in obstructing the goals of management. In organizations that have labor unions, the informal processes are well organized and are easier to identify, and it is easier to understand their perspectives. The work group can buffer stressful aspects of work by providing social support and technical assistance to co-workers (French, 1963; Caplan et al., 1975;House, 1981).

To take advantage of the social aspects of work, many companies use special programs or processes to get employees more involved in the improvement of products and the company. For example, they use quality circles and other techniques of total quality improvement to get employees involved in the design of their own work (Deming, 1986; Lawler, 1986; M. Smith et al., 1989). Some companies use climate questionnaires to assess the status of employee satisfaction and stress and to define specific areas of employee concern (Lawler, 1986). These approaches provide data to help management align formal and informal structures. Successful organizations recognize the importance of aligning the formal management process with informal social processes to get employees positively involved in organizational success. An important human factors consideration is for companies to recognize the importance of the informal social aspects of work groups, and to provide formal structures and processes that harness the informal group process to benefit the management of the organization and the satisfaction and success of the employees.

The power of the social process in organizations has been recognized and is one of the drivers in the shift from individually based work to using groups, or teams, of employees working together to achieve a goal (Sainfort et al., 2001; Hackman, 2002). Although it is commonly recognized that many complex issues require a cross-functional team-based approach, teams have also been found to be beneficial for other reasons in jobs that were previously done in isolation or on assembly lines. By creating work teams, an environment that provides social support can be fostered. Social support has been shown to reduce stress for employees (House, 1981). Team-based work processes are observed in a wide variety of organizations, from manufacturing production and assembly processes to service activities and in new product or service design processes. Companies teach employees about the importance of teamwork, how to interact in a team, and how to coordinate with other teams.

Teams consisting of managers, marketing, sales, engineering, production, and labor are used to solve critical product design problems and to develop new products. When team operations can be achieved for making products, assembling products, providing services, and selling, they provide social and motivational benefits that lead to greater employee satisfaction and performance (Sainfort et al., 2001). However, social pressures from the group can sometimes increase employee stress, and this requires careful monitoring by the organization. Many management approaches call for the inclusion of front-line employees as part of a team formed to resolve product and service quality problems. Teams capitalize on the mixing of perspectives and knowledge to define problems and to develop preferred solutions. Teams capitalize on multiple expertise, and they also capitalize on the social aspects of the process that provide positive feelings of recognition and respect to the individual participants.

An important consideration in managing work is that employees like opportunities to participate in the decision-making process (McGregor, 1960; Lawler et al., 1986; M. Smith and Sainfort, 1989; Sainfort et al., 2001). Employees like to feel important, respected, and appreciated. Positive participation experiences address an employee's social and ego needs. Organizational management processes that incorporate employee participation into production, problem-solving, designing, and/or opinion-sharing activities provide benefits to both the organization and the individual employee.

3 CHANGING SOCIAL VALUES

In Section 2 we described some of the theories and research that has been developed over the past century about how organizations manage their workforce to provide benefits to the organization and the employees. In general, we have moved from command and control, to a focus on educating employees, and to recognizing the value of engaging employees' hearts and minds in addressing the challenges of the organization (McGregor, 1960; Hackman, 2002; M. Smith and Carayon, 1995; Lawler, 1986, 1996, 2003; Carayon and Smith, 2000). To take it a step further, we believe the United States is now in the midst of a major societal shift that has implications for how effective organizations will be in attracting the most talented and creative people in the workforce. These individuals have high expectations for themselves; enough talent to find alternatives to the traditional career in a large organization; and a desire to find meaning, balance, authenticity, and spiritual fulfillment in their lives. Organizations that are viewed as hampering these human desires will find it difficult to attract and keep the best employees, to the long-term detriment of the organization.

In seeking to gain insight into the next major shift in workforce beliefs and needs, it quickly becomes apparent that many different realities exist today. For example, in reviewing current books and articles on organizations and the workforce, a wide variety of experiences and trends are described, each of which is reality to many people: *Management: Meeting New Challenges* (Black and Porter, 2000); *Motivation and Work Behavior* (Porter et al., 2002); *The Betrayal of Work: How Low-Wage Jobs Fail 30 Million Americans* (Shulman, 2003); *The Rise of the Creative Class . . . and How It's Transforming Work, Leisure, Community, and Everyday Life* (Florida, 2002); *White-Collar Sweatshop: The Deterioration of Work and Its Rewards in Corporate America* (Fraser, 2001); *Leading the Learning Organization: Communication and Competencies for Managing Change* (Belasen, 2000); *Nickel and Dimed: On (Not) Getting by in America* (Ehrenreich, 2001); "What Do Men Want?" (Kimmel, 2000) and *The Cultural Creatives: How 50 Million People Are Changing the World* (Ray and Anderson, 2000).

These books are based on research and experience. A number of them refer to a "new truth" about what is happening in some organizations and some sectors of society. They, along with other academic, social science, and business writings, support our belief that we are in the midst of a major societal shift with changing and diverse perspectives about the nature and meaning of work. After decades of downsizing, "sweatshop management," and corporate fraud in traditional manufacturing operations, service companies, and some of our most prestigious white-collar corporations, both corporate and employee loyalty are at an all-time low. For educated and creative people, start-ups and entrepreneurial ventures appear no more risky than working for a Fortune 100 firm. Fears about terrorism, the environment, and the economy have added to an already growing movement toward finding balance, meaning, and spiritual fulfillment in life. If organizations are to continue to attract top-level talent, they must provide leadership in creating workplaces that foster human development and higher-level human needs that have been well documented in past decades (McGregor, 1960; Lawler, 1986, 2003; M. Smith and Sainfort, 1989; M. Smith and Carayon, 1995; Carayon and Smith, 2000).

Florida (2002) believes that human creativity is the current driving force of economic growth, and as such, the direction of society is being strongly influenced by a culture of creativity. This is not an isolated perspective, as others have noted the critical role of employee creativity for the success and survival of companies (Lawler, 1996, 2003; Porter et al., 2002; Hackman, 2002). Florida talks about the rise of a creative class in the United States that will be at the heart of future new business opportunities. For an organization, access to these talented and creative people has much the same meaning as access to coal and iron had to businesses 100 years ago. The creative class in the United States is estimated to be 38 million people who are working as architects, artists, engineers, designers, entrepreneurs, computer designers, software developers, and entertainers, to name a few. These people do not currently view themselves as a single group, and as such, do not have a collective voice. They make choices that are driven more by quality-of-life issues and life-style desires, than by the location or mandates of a particular employer. Florida makes a compelling argument that creativity is the driving force behind competitiveness and success for individuals and organizations.

Similarly, *cultural creatives* is a term coined by Ray and Anderson (2000) to describe roughly 50 million people in the United States. The cultural creatives, another population that does not define itself as a group, are choosing to be authentic and to act in ways they believe are consistent with who they are and what they believe. They are actively engaged in learning and want to understand the big picture, whether this be world events or their own workplaces. They value diversity and the environment, and they are interested in self-actualization, spirituality, and in creating balanced lives.

Florida, and Ray and Anderson describe trends that we believe are just starting to affect organizations in the United States. Over a decade ago some organizations recognized the power of the *learning organization*, a type of organization in which employees are an integral aspect of the management process (Senge, 1990). It was found that sharing power with employees and encouraging them to fully participate in solving workplace challenges led to fuller employee engagement with their work organizations and their jobs (Deming, 1986; Lawler, 1986, 1996; Senge, 1990). In this approach employees were considered more than their physical strength or task knowledge. They were encouraged to use their heads as well as their bodies.

We believe that the next evolution of organization management will require organizations to engage employees' hearts as well as their heads and bodies. This will be essential for those organizations that want to attract and keep the talented and creative employees that will be crucial to the success of the organization. If organizations do not create an environment that encourages respect, loyalty, authenticity, self-reflection, and learning, these most talented persons are creative enough to find alternatives to the organization to take care of themselves and their families economically. The need for this improved environment is not limited to the creative elite who are pursued by companies, but is important to the rank-and-file employees as well. Without the efforts and support of all levels, the organization will not reach its full potential. There is a compelling argument that organizations need people with the attributes described here if they are to be competitive in the marketplace (Lawler, 1996, 2003), and these people can be found at all levels of the organization (McGregor, 1960; Lawler, 1986, 2003; Drucker, 2001). Yet many companies continue to engage in organizational practices that discourage and disenfranchise many of their employees. Such practices will have long-term consequences for the success and survival of these companies.

The remainder of this chapter focuses on what it takes to create an organizational foundation and management philosophy that fosters a healthy work experience for employees while creating an organization that is sustainable for the long term. Although we have laid out a business case for creating healthy and sustainable organizations, we do want to emphasize that it is our belief that organizations should be designed to provide a positive experience for employees not only because it is a necessity for financial success, but because it is the right thing to do.

4 ATTRIBUTES OF HEALTHY AND SUSTAINABLE ORGANIZATIONS

In this section we present a description of the attributes of a healthy and sustainable organization. By a *healthy organization* we mean that the organization supports people's need to find meaning, balance, authenticity, and spiritual fulfillment in work and life. Employee burnout is not good for an employee or an organization (M. Smith, 1987; Kalimo et al., 1997), and people need to lead lives they can sustain in the long run if they are to be mentally and physically healthy. A healthy environment engages people's minds and hearts in meeting the needs of the organization. An organization cannot be sustainable unless it fully engages employees. It follows that a healthy environment is a prerequisite to a sustainable organization.

A sustainable organization has the flexibility and adaptability to respond readily to both internal and external influences. This might include shifts in the business climate, market opportunities, or the labor market as well as new ideas or challenges that may develop internally. Without full employee engagement, the organization will lack the flexibility and adaptability to respond in a timely manner to changing circumstances. A sustainable organization has employee stability and engagement as well as effective formal and informal communication channels and work processes that enable it to adjust to changing circumstances in an ongoing manner.

The following are four attributes that a healthy and sustainable organization embodies: (1) has a clearly articulated vision that it strives consistently to meet through a well-defined strategy; (2) has a culture of respect, including respect for employees, clients, competitors, partners, the community, the environment, and so on; (3) is flexible and able to respond to changing circumstances in an effective and timely manner while minimizing disruption; and (4) makes difficult decisions in a timely and balanced manner. Whether an organization manufactures small appliances, operates amusement parks, provides support services to single parents, develops the next generation of consumer electronics, provides housekeeping services, manages utility distribution, or is an arm of a government agency, these four attributes provide a foundation for the healthy and sustainable organization. Yet we do not suggest that these attributes are simple to achieve or easy to live by. For-profit organizations must make good profits to succeed, high-tech firms face fierce competition and great unknowns, nonprofit organizations continually seek gifts and grants to survive, and government agencies have little to no control over their mission and budgets. These are real constraints that lead organizations to give up even on thinking about organizational management changes. However, the key is to strive consistently to be an organization that values and lives by these attributes. The four attributes of a healthy and sustainable organization are expanded on in the following sections.

4.1 Clearly Articulated Vision and Well-Defined Strategy

A clearly articulated vision and well-defined strategy foster trust and faith in the organization's leadership. They provide a high-level road map, an essential part of leadership that ensures that those responsible for implementation have "big picture" clarity. Once this context is clear, individual executives, managers, and staff can think for themselves to make decisions within their own spans of control, which might include hiring, training, inventory management, capital purchases, work processes, marketing tactics, partnering, and others.

Related to this is the necessity for an organization to ensure that all decisions that are made maintain the organization's focus. The vision and strategy provide a philosophical background as well as guidelines that decision makers use to keep the organization focused. Although organizations need to be flexible in responding to new opportunities, decisions to move in directions that are not consistent with the current vision and strategy should be made carefully, thoughtfully, and with sufficient knowledge that the opportunity will succeed. For example, if a nonprofit provider of legal services wishes to begin providing financial services to the poor, the organization must consider the multitude of implications, such as sources of funding, the ability to reach their constituents, the probability that constituents want and will use the service, the current staff's skills, the availability of new people with the necessary skills, new knowledge requirements, and the impact on current structure and processes. Without careful evaluation, planning, and implementation, the new service could jeopardize the current services that are valuable and effective for the constituents. Failure of the new enterprise would mean that the organization, employees, and constituents would all lose.

A clear vision and strategy also lays the foundation for people to learn how they can personally connect and find meaning for themselves within the organization. Whether the person working in a hotel is a housekeeper or the concierge, knowing that the vision of the organization is to provide a pleasant night's stay that takes away all the burdens of home for the guest helps both employees better understand their role and performance requirements, and to find meaning and satisfaction in their work.

One area that bears mentioning given the actions taken by many organizations over the past few decades is that of downsizing. A healthy and sustainable organization does not view layoffs or downsizing as a business strategy. Rather, it views them as a failure of leadership and management, and something to be avoided. A layoff can be a strategy to correct past errors, but by itself it is not an effective organizational strategy. Although there are circumstances in which a layoff cannot be avoided, no one in a leadership role should be viewed as a hero or savior for cutting jobs of those who may not have contributed to the problems at hand. Layoffs decimate the informal social structure and communication channels, lower the morale of the remaining employees, and destroy trust. If layoffs must occur, the first step toward rebuilding employee trust and motivation is for the leader to take responsibility for the errors in judgment that led to the problems. The leader must let remaining employees know that such errors will not occur again, and the company is committed to doing all it can to maintain jobs. Even such a commitment has a low probability of gaining employees' trust, loyalty, or hard work in the short term. The rebuilding and refocusing of the organization requires significant organizational resources to develop new processes and structure and to rebuild employee trust and loyalty.

4.2 Culture of Respect

The healthy and sustainable organization views developing a culture of respect as both the right thing to do and the smart thing to do. The culture respects people, the community, and the environment and does not readily sacrifice one or all for quick profits. This attribute is not about everyone being nice to each other, sponsoring excessive rewards, or paying higher-than-market salaries. Rather, in term of employees, it is about respecting them for what they know, empowering them, and finding ways for them to bring in new ideas (Herzberg, 1970). It is also about respecting their humanity and reaching a joint agreement between the organization and individual employees that does not hamper employees' ability to have healthy and fulfilling lives. Instead, the organization helps employees lead fuller lives by fully engaging them in addressing workplace challenges.

By respecting the specific local knowledge possessed by employees doing the work, you encourage those with the most information to participate in improving the work process. This knowledge cannot be replaced by that of supervisors or consultants. When people are respected, it is much more likely that they will share ideas and routinely make improvements in how they do their work. In contrast, when the culture does not respect employees, a culture of fear and distrust, where people do not take chances with new ideas, is more likely to develop.

Even in this type of culture, difficult decisions must still be made, and sometimes there will be negative consequences that will be felt by both organizations and individuals. For example, a loss of a large account, a merger of two organizations to save one of them, or the firing of an ineffective employee can all still occur for very good reasons, although each has consequences that must be worked through. However, none is taken lightly, and all are done in a mindful and respectful manner.

In *White-Collar Sweatshop: The Deterioration of Work and Its Rewards in Corporate America*, Fraser (2001) describes working conditions in some of our most prestigious and profitable financial and high-technology corporations: employing "perma-temps" to avoid paying benefits, requiring salaried employees to work a minimum of 60 hours per week, drastic and repeated downsizing, heavy workload, and unrelenting work pressure. People still seek these positions because of the potential, which has been a reality for many, for significant financial gains. However, the work environments are becoming increasingly negative, morale is low, and people are getting burned out routinely. In these organizations, profits and competitive advantage are respected, at the expense of employees. Given the societal shift that we see occurring, over the longer term these firms will find it difficult to attract and keep the most talented employees, and at the very least, they will not be reaching their full potential if they maintain a cycle of stress and burnout.

The firms that Fraser (2001) describes are not fostering a culture of respect. In a culture of respect, each person and what he or she brings to the organization are valued. While this includes each person's skills, ideas, and specific job knowledge, it also refers to their idiosyncrasies, emotions, and worldview. Diversity is valued and people are encouraged to bring their whole selves to the organization and to find personal meaning in the work they do. Employees' hearts as well as their heads are engaged in their work, leading to a sense of community and enhanced creativity. Collaboration and innovation are fostered, which in turn can lead to competitive advantage.

This culture creates opportunities for every employee, including those in low-skilled, traditionally unrewarding work, to feel respected and find ways to enjoy his or her work. By encouraging interpersonal interactions and social support, and treating people with respect, the camaraderie co-workers develop, and the support a supervisor can provide, can make even the most mundane or difficult jobs manageable or even enjoyable.

4.3 Flexibility and Responsiveness to Changing Circumstances

Through flexibility and responsiveness, an organization can ensure that it is responsive to minor as well as major challenges along the way. We have observed from personal experience, as well as the literature on case studies, that the cycles in a company's profitability often occur because the "little things" were not addressed as they occurred. Examples from a human relations perspective include lack of feedback to employees (both positive and negative), confusing roles, or insufficient training. By not tending to these things, problems that require significant effort may arise: employee lawsuits, high accident rates, stressed workforce, and so on. All can lead to significant reductions in productivity and increased indirect costs, such as health insurance increases, lost work time, and increased workers' compensation claims.

The clearly articulated vision and well-defined strategy set the stage for a flexible and responsive organization. For example, take the example of a firm that provides diagnostic products to the health care industry. Their vision is to be the leading provider of high-end biomedical products. If marketing detects a broad shift in the market in terms of where in the health care process procurement decisions will be made in the future, this should not affect overall corporate vision or strategy. However, a mechanism should be in place internally to assess the impact effectively, make decisions, and implement a shift in whatever areas are affected, such as in marketing tactics and possibly, broader marketing strategy. This mechanism is dependent on the specific knowledge of a great many people, those with broad corporate knowledge as well as those with local knowledge about what will be happening with specific clients. To be able to adapt quickly and respond to this change, this variety of ideas and input is critical. Flexibility and adaptability are dependent on the local knowledge of employees on the front line of the issue, whether the issue is related to marketing or to a manufacturing line.

A flexible and responsive organization encourages communication without boundaries. Employees are encouraged to interact with whomever they need to on whatever issues need to be addressed. There are no repercussions for moving outside the chain of command, and there are no sacred cows. As challenges are identified, they are addressed. The organization also ensures that effective formal communication channels are in place among those with relevant information about the climate the organization works in: client feedback, competitors' actions, changes in laws or regulations, trends in the marketplace, quality of products and services, and so on.

While many employees welcome the opportunity to participate and make decisions, many others lack the knowledge, confidence, or skills to participate effectively. Organizations need to be creative in helping employees develop. Participatory programs that build in opportunities for employees to read and discuss new ideas, self-reflect, learn to trust others, and experiment with what they are learning have been shown to be successful (Sanders et al., 1997) (see Section 6.4) in helping raise employee awareness and set the stage for them to learn how to take action effectively.

A flexible and responsive organization encourages employee input and creative thinking. When rote thinking is encouraged, when people are afraid to ask questions, and when new knowledge and ideas are not incorporated, groupthink can become institutionalized. The General Electric (GE) case study (Section 6.3) describes a culture in which employees did not feel empowered to contribute ideas. This culture ultimately contributed, at least in part, to massive layoffs and a major restructuring. It was only after the Work-Out process was implemented and the culture made a dramatic shift to empower employees that GE management became more responsive to new ideas, and less bureaucratic.

4.4 Timely and Effective Decision Making

Creating and maintaining a healthy and sustainable organization is not an ideal that puts all the emphasis on everyone being happy all of the time. We are not proposing a feel-good panacea. Organizations could not sustain themselves with this vision. Rather, this attribute highlights the need to make and implement difficult decisions effectively and in a timely manner. Decisions may be difficult for a number of reasons: Stakeholders have widely varying opinions and needs; the amount of information available may not feel sufficient, even though the decision needs to be made in a time frame that precludes more information; cost may be a problem; or the decision may affect certain people negatively.

For example, we do not believe that people should be guaranteed a job for life, although they should have some sense of security in their jobs as long as they continue to learn and contribute. Yet the organization does itself and its employees a disservice by not addressing challenges posed by employees whose skills, abilities, or behaviors do not complement the organization's needs. Keeping people employed who do not effectively contribute, or who create additional work or disruptions for others, can lower the morale of employees who are committed to doing a good job.

Employees who are not meeting expectations should be told so in an honest, supportive manner. Development plans should be established jointly with the employee, supervisor, and someone from human resources or employee development and supported by the organization. If the employee still does not meet expectations following a reasonable period of focused effort, terminating his or her employment may be the only solution available. However, outplacement services, severance packages, and notification periods can all assist in easing this transition. Terminating employees should never be taken lightly, but sometimes it is unavoidable. Organizations must do the difficult work of mentoring, developing, and providing feedback to employees. Avoiding the difficult task of terminating an employee can result in an organization with too many people who do not perform effectively, and the organization may find itself in a position of laying off larger groups of employees because the problems were not addressed effectively along the way.

Difficult decisions can show up in any area, including whether to continue to pursue a long-term market, finding an appropriate response to stockholder pressures, acquiring another firm, dealing with internal conflicts, or changing elements of the organizational structure. Not addressing issues as they arise can lead to the need to take more drastic measures, as each has associated consequences, including downsizing, restructuring, and bankruptcy.

5 ORGANIZATIONAL DESIGN PRINCIPLES

In Section 4 we described attributes of a healthy and sustainable organization and what it would feel like to be in such an organization. The following principles provide more specific guidance into the types of actions that can lead to creating such a healthy and sustainable organization: (1) Individual managers exhibit a commitment to assist individual employees in creating a balance between their work and personal lives, (2) formal and informal structures and processes are in place to provide a framework for employees to succeed, (3) a range of opportunities for participation and connection to the organization are available, (4) professional development activities designed to foster self-reflection and engagement are in place, and (5) the need for balance between work and personal life is openly supported across the organization.

The principles are written with the assumptions that (1) individual managers reading this chapter who have a desire to create cultural change within their organization may not yet feel they have broad organizational support but still may be looking for ways they can make an impact on their own; and (2) there are numerous ways to bring about change in an organization. Although there is significant literature on the benefits of organizational change being championed at the top, there are also meaningful examples of how it can occur through grass roots activities and pockets of action. Some of the principles identified here can be implemented on their own, even under the radar, and still affect positively the lives of a number of people in the organization, while building momentum for change. In contrast, the fifth principle describes a broader cultural understanding and acceptance across the organization of what it means to create a healthy and sustainable organization. The five principles are described in more detail below.

5.1 Commitment of Managers to Assist Individual Employees

According to House (1981), a positive relationship between an employee and his or her direct supervisor is one of the most effective means of alleviating stress. However, this principle takes it a few steps further and says that individual supervisors, or managers, should be actively involved in working with employees to find specific ways to balance work and personal priorities, thus creating a more healthy work environment. In research drawn from several dozen U.S. companies Friedman et al. (2000) identified three principles that progressive managers are guided by with respect to working with employees to create balance across their work and personal lives. They describe a one-on-one partnership between a manager and an employee in which work priorities and personal priorities are openly shared. The intent is to get all priorities on the table and then jointly create a plan for meeting all of them. The manager must share clear information about the organization's priorities with a focus on expected outcomes, as opposed to time commitment expected. For the partnership to work, the employee must also share specifics about his or her personal priorities. Second, the manager must view the employee as a whole person, acknowledging and showing appreciation for who the person is outside the job. Third, the manager must be willing to experiment continually with ways of getting the work done that help both the organization and the employee meet their goals. The three elements reinforce each other. For example, by learning about the employee's personal priorities, the manager sees the employee as a whole person, and by experimenting continually with ways of working, the manager and employee are both open to addressing shifts in priorities that occur over time.

In many organizations, this principle is something that an individual manager can implement on his or her own without a mandate or blessing from the organization. In fact, the researchers found that many managers do just that and are able to affect the lives of many employees positively.

5.2 Framework for Employees to Succeed

Every organization has formal as well as informal structures and processes. For example, the structure, communication channels, human resources policies, reward systems, decision-making methods, work processes, and means of using technology all affect employees' ability to succeed in their jobs. There are a number of excellent resources on this, and we do not attempt to cover that ground in this chapter. However, by listing this as the second principle, our intent is to highlight the idea that individual managers can take some action within an established organization to make changes that positively affect an employee's ability to create a healthy work environment. Actions such as modifying reporting structures, taking advantage of informal communication channels and using communications technologies to allow more flexibility in when and where work occurs may all be within the realm of a particular manager's responsibilities and can all contribute to creating a healthier work environment.

5.3 Opportunities for Participation and Connection to the Organization

Organizations that have the attributes of a healthy and sustainable organization actively engage employees' hearts and minds in the work of the organization. By engaging those who do the work, the organization not only gains the benefit of the local knowledge but also empowers employees to take ownership and make their jobs personally meaningful. Examples include opportunities to participate on an ad hoc committee developed to address a business challenge, including an entire team in a planning session (see Section 6.4), activities such as GE's Work-Out (see Section 6.3), and having decision-making authority about how a particular process is carried out. The opportunities referred to here require employees to use more of their entire selves. The focus of this principle is to present opportunities that will challenge and fully engage the individual employee.

5.4 Professional Development Activities

This principle refers to professional development activities that go beyond skills training for a specific job and encourage employees to test their beliefs, self-reflect, and find ways to bring themselves more fully to work. Most organizations provide training in specific job skills, either on the job or in a formal classroom, or some combination of the two. Depending on how well the training is designed and implemented, there may be added benefits in that the employee may gain additional insights and benefits beyond specific skills training. However, this is a happy outcome, and not an intentional plan. Adhering to this principle means that the organization provides opportunities intentionally designed to focus on encouraging people to look inward, and to self-reflect, and that this greater self understanding allows the employee to find greater personal satisfaction, and hopefully, to contribute ultimately at a higher level. However, the focus on contribution to the organization is secondary to the focus on personal growth. Examples of activities include leadership development activities that create space for journaling and reflection, programs that focus on reading and discussing writings that stimulate or inspire, incorporating quiet self-reflection or writing into a routine meeting, or integrating yoga or meditation into daily activities. The essence of this principle is that people are encouraged to think in ways that they might not traditionally have viewed as appropriate for the work setting.

5.5 Support for Balance between Work and Personal Life

This principle is consistent with the first principle, but the difference is that the actions described here occur more broadly and openly across the organization. A healthy and sustainable organization recognizes that employees need to have a balance across their work and personal lives. Jobs that create balance at work provide working conditions that control the level of stress (M. Smith and Sainfort, 1989; M. Smith and Carayon, 1995; Carayon and Smith, 2000). This balance can be achieved by balancing job stressors with positive aspects of work to limit or neutralize the effects of the stressors. Long-term imbalance leads to employee stress, reduced motivation, reduced capability to perform over the long term, and ill health. We recommend that organizations put employee assistance programs in place and have policies and procedures that encourage employees to spend time with their families and friends and that put limits on overtime work and extended work assignments away from home. SSI, a small high-tech firm, is a good example of a firm that openly acknowledges the need to balance work and personal issues. They hold weekly planning meetings in which activities and assignments for the week, as well as upcoming personal commitments, are discussed. Team members readily share that they have a child's soccer game that they would like to attend, are refinancing a mortgage this week, or are trying to help an elderly parent adjust to a new living situation. The team environment has advanced to the point that people pitch in to help each other and trust that no one will take advantage of the situation. There is a positive give and take in which each person's personal life, as well as his or her business contributions, are valued. Although things do not always go smoothly, each person on the team works to best integrate their personal life with their work life, and it works because SSI encourages it. Although SSI is a small firm, with essentially one team, larger organizations can implement similar activities within work groups or departments.

6 EXAMPLES AND CASE STUDIES

Following are two examples and two case studies that illustrate organizational design and management actions implemented in response to a need for change. The two examples are brief and make one or two points; the two case studies are more extensive and provide greater opportunity for discussion and interpretation. Each is followed by our comments on the measures the organization took as well how those measures relate to healthy and sustainable organizations. When reviewing the examples and case studies, it is helpful to consider the attributes of a healthy and sustainable organization and the guiding principles provided above and consider how these elements are, or are not, exhibited both prior to and following the organization taking some action. The name of the organization is used in the previously published examples and case studies. In those that have not been previously published, a pseudonym is used.

6.1 Example 1: Six Flags Entertainment*

When Bob Pittman was the CEO of Six Flags Entertainment, he once worked incognito as a janitor in an effort to gain some insights into problems that were occurring at the parks. Janitors were being rude to park guests. He realized that the guests were getting in the way of the janitors meeting the goals of their job. Managers had instructed the janitors to keep the park clean, and the guests kept messing it up. According to Pittman: "So we had to go back and redefine their jobs. We said, 'Your main job isn't to keep the park clean. Your main job is to make sure that people have the greatest day of their lives when they come to Six Flags.' Oh, and by the way, what would prevent customers from enjoying themselves? A dirty park" (Hayashi, 2001).

Organizational Design Principles Highlighted in This Example This example illustrates the importance of clearly articulating the organization's vision and strategy in a way that employees relate to and can translate to their day-to-day activities. By doing so, all employees can feel connected to the goals of the larger organization and can find some greater meaning in their work. The janitors' understanding of what "doing a good job is" will vary significantly based on whether their performance is measured by how clean the park is or by how much fun customers are having.

*This example was adapted from Hayashi (2001).

6.2 Example 2: Plymouth†

A 1956 Plymouth was developed by a team of designers to have "a sculpted panel with many vertical ribs indented between the two taillights." The door for the gas cap was located between two of the vertical ribs so as not to interfere with the lines of the automobile's design. The door was not easily noticed since the cracks at the edge of the gas cap lined up where the eye expected a vertical line. The design worked well from an aesthetic perspective. However, when the design was reviewed by a manager outside the design team, the manager decided the customer would have too much difficulty finding the door to the gas cap, and ordered the door chromed, thus "ruining the whole sweep of the design with an anomalous chrome square" (Pinchot and Pinchot, 1996).

Organizational Design Principles Highlighted in This Example When a team goes through a design process, the members become very familiar with their customers' needs and desires as well as the constraints and trade-offs associated with the many decisions they make. They draw on their specific, local knowledge to develop solutions for the challenge they are working on (Pinchot and Pinchot, 1996;Jensen and Meckling, 1996). Although it is important to ensure that outside reviews take place, the reviewers must respect the work completed to date and recognize that there is no way an outsider can be knowledgeable on every detail or compromise decided to date. In this situation, it would have been appropriate for the manager from outside the team to raise the potential problem he saw, and for the design team to develop a modified solution if appropriate. For teams to work at a high level, they must have local decision-making authority and feel confident that their decisions will not be overruled.

6.3 Case Study 1: General Electric‡

When Jack Welch took over as CEO of General Electric (GE) in 1981, GE was considered one of the most successful companies in the world and was routinely held up as a model for others to emulate. With over 400,000 employees, a value of $25 billion, earnings of $1.5 billion annually, and a triple-A rating, "some employees in the firm proudly described the company as a 'supertanker'—strong and steady in the water" (Welch and Byrne, 2001). Although Welch respected that view, he saw decision making that was slow and frequently ineffective; too much energy being spent on internal processes, to the point that marketplace shifts were not receiving the necessary attention; and too many management layers, resulting in a company mired in bureaucracy and cumbersome internal processes. The culture of the company was one of fear, in which senior executives and managers were afraid to stick their necks out. In his early days Welch

†This example was adapted from Pinchot and Pinchot (1996).

‡This case was adapted from Welch and Byrne (2001) and Ulrich et al. (2002).

found it nearly impossible to overcome GE's formal management culture that focused on audit and control, to engage his team in a free-form discussion. This reluctance to speak out was systemic throughout the company. Welch felt that the company's organizational bureaucracy posed a threat to its own survival, so he set out to remake GE "to be more like a speedboat, fast and agile, able to turn on a dime" (Welch and Byrne, 2001). He believed GE was facing an impending crisis that could not be averted without core changes to the way that business was carried out.

Welch's stated philosophy is that developing people must be a core competency of GE and that the key to this is finding ways to foster a sense of empowerment and entrepreneurialism. On the surface, this philosophy would seem at odds with many of the actions that Welch took. Welch's vision for GE was that each of the 150 business units would be number one or two in their markets. Any unit manager whose business was not in that position would be required to "fix, sell, or close" the unit. Following significant consolidations, sell-offs, and internal reconfigurations, 13 business units remained. Concurrently, Welch eliminated organizational layers and functions, reduced staff size, expanded spans of control, and located decision making within the business units. If a position or employee did not fit within the strategy of making the business unit first or second in the marketplace, it was eliminated. Over a five-year period, 118,000 jobs at GE disappeared, 37,000 through units that were sold and the remaining through layoffs. Jack Welch became known as "Neutron Jack," the guy who got rid of people but left the buildings standing.

As the company downsized and reorganized, a new organizational infrastructure was developed. New processes, including mechanisms for strategic planning, budgeting, employee performance reviews, and internal audits, were established; and a corporate executive council for running the company was put into place. The company's management development center was revitalized through the hiring of a new leader and a refocus on strategic business needs. By the late 1980s, financial performance was improving. However, even with the new structures in place, GE continued to experience problems. The management culture still focused on command and control, informal communication networks had been decimated with the downsizing, many employees were overloaded with what they saw as non-value-added work, morale was low, decision making was still slow, and in general, too much bureaucracy remained. Welch realized that "the wrenching transformation of the past several years had changed the strategy, structure, and cost-base of GE, but not the way day-to-day work got done" (Ulrich et al., 2002).

Throughout 1988, with input from the head of the leadership and management development function, a team of hand-picked consultants and others from within the firm, a new concept for reducing bureaucracy, eliminating unnecessary work, and transforming the management culture was developed. The process was called Work-Out and was patterned after the concept of a town meeting. Welch described Work-Out as a means to create a "liberated" organization that would free up employees to put their energies into innovation and serving customers. This was an ambitious proposition. GE employed over 275,000 people across 13 business units that were each the size of a Fortune 50 firm, most with operations scattered around the world. Each unit had a strong ambitious leader who had learned how to run a business using traditional management methods.

As Welch and his team of lead consultants (each assigned to a business or functional unit) met with leaders and introduced the concept, there was a great deal of skepticism and resistance. Welch, this champion of creating a liberated organization, was the same "Neutron Jack" who had laid off tens of thousands. Within days, Work-Out was dubbed Heads-Out within one of the business units, because it was viewed as merely a disguise for another round of downsizing. In addition, many of the business unit leaders had their own visions of how to improve their businesses and had already begun rolling out initiatives consistent with their own visions. They did not take kindly to a top-down mandate. However, Welch was relentless in pushing the initiative. He removed barriers, providing funding for the first year, and required that every GE employee "have a taste of Work-Out by the end of 1989." Although the unit leaders were skeptical, the employees within the units were enthusiastic about the opportunities they saw with Work-Out, and were eager to get started. In April 1989 the first Work-Out sessions were held.

Trained facilitators, many of them from academia, were brought in from the outside to lead the Work-Out workshops. Workshops were held over two to three days. They started with a presentation by a senior manager, who would describe a business problem, or possibly issue a broad challenge to the group of 40 to 100 participants. He or she would leave the workshop, and the participants would go through a process of brainstorming solutions, developing recommendations and presenting ideas to each other. The workshop concluded with a town meeting in which the teams presented proposals for improvements to the senior manager, who was expected to make a yes–no decision on the spot on at least 75% of the proposals. For each "yes" decision, an implementer was assigned on the spot and given the authority and resources to implement the recommendation. If the senior manager could not make a decision immediately, a date would be set for the decision. No proposal could be buried or ignored.

Many of the issues addressed in the first Work-Out sessions focused on the mundane: copy machines, timesheets, and approval processes, but as the sessions progressed, mission-critical processes such as product development and customer service were also addressed. According to Welch, "as people saw their ideas getting instantly implemented, it [Work-Out] became a true bureaucracy buster" (Welch and Byrne, 2001). Across GE, Work-Out's implementation was idiosyncratic to the individual business units and the people. Interest grew through word of mouth and as people who trusted each other talked. Curiosity grew, more people became open to trying it, and the resulting successes became apparent. Between April 1989 and January 1990, 160 Work-Out workshops, with approximately 8000 participants, had taken place, and another 5000 employees had been exposed to the Work-Out philosophy. By mid-1992, more than 200,000 employees had participated in Work-Outs. Work-Out served to foster a dramatic culture shift from one of command and control, to one where managers led more and controlled less, and everyone's ideas could be heard. Welch believes that every good thing that has happened in the company since can be traced to the liberation of a business unit, a team, or an individual.

GE is a very large company with many stories that can be instructive in issues related to organizational design: creating a boundaryless organization, implementing Six Sigma, entering e-business, acquisitions, globalization, and enhancing services, to name a few. However, each of these other initiatives that are not included in this case study also has at its core the idea of empowering every person to generate ideas. Through implementing Work-Out, GE was able to revitalize itself successfully following a reorganization that resulted in 30% fewer employees.

Organizational Design Principles Incorporated by GE

- When drastic measures were deemed necessary, GE linked downsizing to a business vision and strategy that few would argue with: Every business unit needed to be first or second in its market.
- Executive leadership provided a consistent message through downsizing, reorganization, and the development of Work-Out.
- The culture was used to change the culture. The top-down mandates achieved the desired results in a hierarchical, bureaucratic organization, and paradoxically, were the first steps toward empowering employees and respecting their ideas.
- Decision making was moved to the local level, closest to where the knowledge resides. For many employees, the Work-Out sessions were the first time they felt their knowledge was respected within GE. Empowering people through the Work-Out process was the key factor in transforming the command and control management culture.
- Cross-functional teams were involved in the Work-Out sessions.
- GE has a process (Work-Out) in place that is now intrinsic to how work is done. If it continues, it should ensure the company stays as dynamic and flexible as possible for a firm its size.

6.4 Case Study 2: Carson Engineering

Carson Engineering is a facilities consulting firm of 400 people that has been in business for approximately 80 years. Consistent with many professional services firms, Carson has a strong focus on individual entrepreneurialism, and its corporate strategy provides a lot of room for employee innovation and creativity. New markets and services are typically developed because of one person's passion to develop the area. Beyond the overall corporate strategy, little is dictated from above. In the 1980s, following some significant changes in the business, the CEO researched organizational models. He ultimately decided to implement a matrix structure that continues to this day (Figure 1). The x-axis of the matrix is comprised of market groups, or facility types, each led by a senior project manager. Each of these project managers has ultimate responsibility for marketing efforts within their areas, maintaining client relationships, ensuring that the design solution meets the specific needs and desires of the client, and for leading the planning and overall design on specific projects. The y-axis is made up of the engineering disciplines, each led by a discipline leader, who fulfills the role of project engineer on specific projects. Within their engineering disciplines, the discipline leaders each have ultimate responsibility for ensuring that every project is delivered effectively: documents are consistently high in quality, projects are staffed with appropriate expertise, and employee development is occurring.

In design firms, there is an inherent tension between the project management and project delivery functions. Transforming this tension into a positive for the company is what made the matrix structure particularly appealing to the CEO. He believed that neither project management nor project delivery should be optimized at the expense of the other, and that providing high-quality products and services can only be achieved if the matrix is in balance.

Over the next three decades, Carson became known for excellent engineering and for high-quality documents and service. Employees were proud to be part of Carson, and a strong culture of teamwork and caring for colleagues developed. At the same time, a gradual, though unintended shift away from the original intent of the company's structure began to occur. Employees who were interested in understanding the big picture and who enjoyed the interpersonal communication aspects of consulting gravitated toward project management, while people who were more interested in doing the high-level detailed engineering work gravitated toward project engineering and discipline leadership. Project managers tended to better see big-picture strategy and were not only better skilled at presenting their views but also more assertive in doing so.

The executive leadership of Carson evolved to be made up almost entirely of project managers. Consequently, talented employees gravitated there because they saw it as the path toward leadership and success within the firm. Over time, discipline leadership weakened to the point that department leaders were passive implementers, taking orders from project managers. The discipline leaders frequently worked by exception in response to project managers' statements that "every project is different, so every project requires a unique solution." Market groups

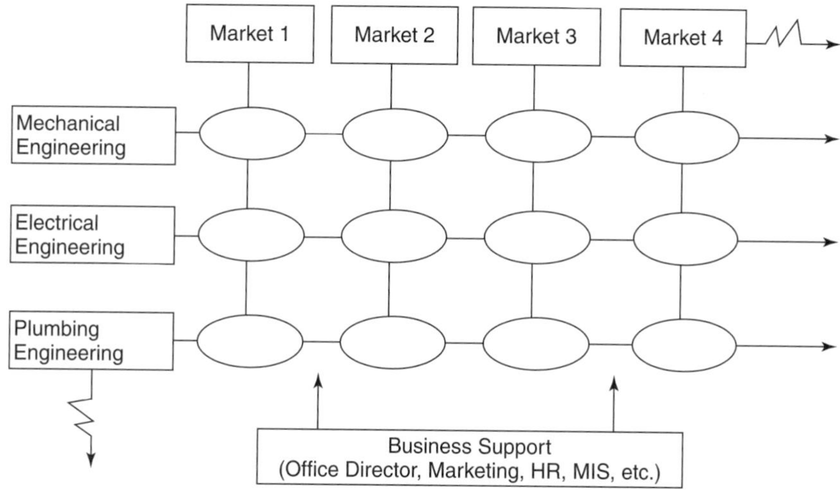

Figure 1 Organizational structure for Carson Engineering.

evolved into silos and project managers became more aggressive in "fighting" for the staff they believed they needed for their projects. The challenges were exacerbated by the entrepreneurial culture as well as the lack of management experience of many employees. Engineers in general have little formal education in management and human relations. Add to that the underlying assumption within Carson that those people with ability would simply rise to the top. Consequently, little effort was put into training in management and leadership skills for employees.

Although the original CEO's vision of an entrepreneurial culture was based on a desire for innovation in the market, customization was the norm in all aspects of the business, even those in which creativity did not add value. Every project had its own method of logging requests for information from contractors; quality reviews were inconsistent in terms of criteria, timing, and knowledge required to conduct them; project assignments were made based on the experience of project managers with specific individuals, meaning that good experienced engineers were overwhelmed with work while less known but talented people were not being developed. At the same time, the department leaders, who did not believe they were valued within the organization, felt bruised and were reluctant to speak up when something did not seem right to them. They had internalized the idea that the project managers led everything and that was the only way it could be.

Following the late 1990s, a time in which workload was high and the architectural engineering industry was experiencing a severe shortage of skilled employees, Carson, a firm that had built its reputation on excellent engineering and service began experiencing some difficulties. Clients complained about document quality and inconsistent service. Lack of management oversight on projects led to significant rework, which resulted in projects with financial difficulties. It became apparent that the entrepreneurial approach to developing people who could lead was not working.

Project managers and project engineers had each developed their own methods of managing their projects, staff, and time. These methods were idiosyncratic and discussed infrequently, meaning that during this intense period of work, the design process itself was not being taught effectively to newer employees. Over a two-month period, the office director and business analyst assessed the quality issues within Carson. This was accomplished through interviewing employees, reviewing project logs, and engaging senior engineers in assessing documents on projects with significant problems. Following this assessment, it became clear that there was a lack of understanding of the design process; that the lack of understanding was creating the alienation of some younger employees and a lack of ownership in work; that some discipline leaders were ineffective decision makers; and that Carson was not drawing adequately on the knowledge of its discipline leaders.

Carson implemented a number of solutions over the next several months. Some examples are:

- An organizational development specialist was hired to work closely with the office director and to create employee development programs that connected employee development needs and career desires with the needs of the company.
- Over a several-month period, two discipline leaders were effectively transitioned out of formal leadership roles and into engineering roles within the firm.
- The office director began holding biweekly meetings with discipline leaders. The purpose was to develop the leaders' management and

leadership abilities and to develop them as a team. The agendas included a combination of addressing current management issues in the office, and group discussions focused around readings regarding self-reflection and learning on broader strategic, cultural, leadership, and human relations issues.

- The office director put more emphasis on one-on-one coaching with several of the leaders, making it clear the discipline leaders' voices needed to be heard and that project managers and discipline leaders are expected to treat each other with respect.
- A cross-disciplinary and cross-functional team documented the design process. Subsequently, this documentation was used in new employee orientations and led to a process for developing project teamwork plans as a group. The entire design team for a particular project would meet for up to half a day to develop a plan for their project. This served as both a useful planning tool for the project and as a training tool for employees unfamiliar with the design process.
- Over a three-month period, multiple workshops were held across the office with the goal of every employee participating in two to three (95% were reached). The agendas included brainstorming and developing solutions for the current quality challenges.
- A mentoring program was developed and will be piloted within the next year.
- A series of self-reflective leadership development activities have begun and will be continued. The organizational development specialist facilitates teams of eight to 10 employees over a period of a few months in readings and discussions that focus on leadership, addressing day-to-day challenges, and finding meaning in work.

The solutions above are all still in the early stages of implementation but have established a foundation for Carson to transform and develop. Improvements in employee motivation and engagement, management and leadership skills, and document quality have already begun.

Organizational Design Principles Incorporated by Carson Engineering

- The matrix structure supports the company's vision and mission, strategy, entrepreneurial culture, and client focus.
- Difficult decisions that had not been made with respect to moving out ineffective leaders were made and implemented successfully. The people who had been ineffective as discipline leaders were treated with respect, and new roles consistent with their strengths were found for them.
- Flexible training and development opportunities for employees are being implemented. The programs focus on encouraging employees to think, to connect what is meaningful to them to their work life, and to participate in improving how the organization operates.
- By increasing the quality and frequency of lateral communications (such as between discipline leaders), many day-to-day decisions are made in a more timely manner and with more information.

REFERENCES

Belasen, A. T. (2000), *Leading the Learning Organization: Communication and Competencies for Managing Change*, State University of New York Press, Albany, NY.

Black, J. S., and Porter, L. W. (2000), *Management: Meeting New Challenges*, Prentice-Hall, Upper Saddle River, NJ.

Caplan, R. D., Cobb, S., French, J. R. P., Harrison, R. V., and Pinneau, S. R. (1975), *Job Demands and Worker Health*, U.S. Government Printing Office, Washington, DC.

Carayon, P., and Smith, M. J. (2000), "Work Organization and Ergonomics," *Applied Ergonomics*, Vol. 31, pp. 649–662.

Deming, W. E. (1986), *Out of the Crisis*, MIT Business School, Cambridge, MA.

Drucker, P. F. (2001), *The Essential Drucker: Selections from the Management Works of Peter F. Drucker*, HarperCollins, New York.

Ehrenreich, B. (2001), *Nickel and Dimed: On (Not) Getting by in America*, Henry Holt, New York.

Florida, R. (2002), *The Rise of the Creative Class ... and How It's Transforming Work, Leisure, Community, and Everyday Life*, Basic Books, New York.

Fraser, J. A. (2001), *White Collar Sweatshop: The Deterioration of Work and Its Rewards in Corporate America*, W.W. Norton, New York.

French, J. R. P., Jr. (1963), "The Social Environment and Mental Health," *Journal of Social Issues*, Vol. 19, No. 4, pp. 39–56.

Friedman, S. D., Christensen, P., and Degroot, J. (2000), "Work and Life: The End of the Zero-Sum Game," in *Harvard Business Review on Work and Life Balance,* Harvard Business School Press, Boston.

Galbraith, J., Downey, D., and Kates, A. (2002), *Designing Dynamic Organizations: A Hands-on Guide for Leaders at All Levels,* AMACOM, New York.

Hackman, J. R. (2002), *Leading Teams: Setting the Stage for Great Performances*, Harvard Business School Press, Cambridge, MA.

Hayashi, A. M. (2001), "When to Trust Your Gut," *Harvard Business Review*, Vol. 79, No. 2, pp. 59–65.

Hendrick, H. W. (1986), "Macroergonomics: A Conceptual Model for Integrating Human Factors with Organizational Design," in *Human Factors in Organizational Design and Management II*, O. Brown, Jr., and H. W. Hendricks, Eds., North Holland, Amsterdam, pp. 467–478.

Hendrick, H. W. (1987), "Organizational Design," In *Handbook of Human Factors*, G. Salvendy, Ed., Wiley, New York, pp. 470–494.

Herzberg, F. (1970), "One More Time: How Do You Motivate Employees?" *Harvard Business Review,* September–October, pp. 109–120.

House, J. S. (1981), *Work Stress and Social Support*, Addison-Wesley, Reading, MA.

Jensen, M. C., and Meckling, W. H. (1996), "Specific and General Knowledge, and Organizational Structure," in *Knowledge Management and Organizational Design*, Butterworth-Heinemann, Boston.

Kalimo, R., Lindstrom, K., and Smith, M. J. (1997), "Psychosocial Approach in Occupational Health," in *Handbook of Human Factors and Ergonomics*, 2nd ed., G. Salvendy, Ed., Wiley, New York, pp. 1059–1084.

Kimmel, M. S. (2000), "What Do Men Want?" in *Harvard Business Review on Work and Life Balance*, Harvard Business School Press, Boston.

Lawler, E. E., III (1986), *High Involvement Management*, Jossey-Bass, San Francisco.

Lawler, E. E., III (1996), *From the Ground Up: Six Principles for Building the New Logic Corporation*, Wiley (Jossey-Bass), New York.

Lawler, E. E., III (2003), *Treat People Right: How Organizations and Individuals Can Propel Each Other into a Virtuous Spiral of Success*, Wiley, New York.

McGregor, D. (1960), *The Human Side of Enterprise*, McGraw-Hill, New York.

Pinchot, G., and Pinchot, E. (1996), "The Rise and Fall of Bureaucracy," in *Knowledge Management and Organizational Design*, Butterworth-Heinemann, Boston.

Porter, L. W., Steers, R., and Bigley, G. (2002), *Motivation and Work Behavior*, 7th ed., McGraw-Hill, New York.

Ray, P. H., and Anderson, S. R. (2000), *The Cultural Creatives: How 50 Million People Are Changing the World*, Three Rivers Press, New York.

Roethlisberger, F. J., and Dickson, W. J. (1939), *Management and the Worker*, Harvard University Press, Cambridge, MA.

Sainfort, F., Taveira, A. D., and Smith, M. J. (2001), "Teams and Team Management and Leadership," in *Handbook of Industrial Engineering: Technology and Operations Management*, G. Salvendy, Ed., Wiley, New York, pp. 975–994.

Sanders, K. J., Carlson-Dakes, C., Dettinger, K., Hajnal, C., Laedtke, M., and Squire, L. (1997), "A New Starting Point for Faculty Development in Higher Education: Creating a Collaborative Learning Environment," in *To Improve the Academy: Resources for Faculty, Instructional, and Organizational Development*, Vol. 16, New Forums Press, Inc., Stillwater, OK, pp. 117–150.

Senge, P. M. (1990), *The Fifth Discipline*, Doubleday Currency, New York.

Shulman, B. (2003), *The Betrayal of Work: How Low-Wage Jobs Fail 30 Million Americans*, New Press, New York.

Smith, K. U. (1965), *Behavior Organization and Work*, rev. Ed., College Printing & Typing Co., Madison, WI.

Smith, M. J. (1987), "Occupational Stress," in *Handbook of Human Factors*, G. Salvendy, Ed., Wiley, New York, pp. 844–860.

Smith, M. J., and Carayon, P. C. (1995), "New Technology, Automation and Work Organization: Stress Problems and Improved Technology Implementation Strategies," *International Journal of Human Factors in Manufacturing*, Vol. 5, pp. 99–116.

Smith, M. J., and Sainfort, P. C. (1989), "A Balance Theory of Job Design for Stress Reduction," *International Journal of Industrial Ergonomics*, Vol. 4, pp. 67–79.

Smith, M. J., Sainfort, F. C., Sainfort, P. C., and Fung, C. (1989), "Efforts to Solve Quality Problems," in *Investing in People: A Strategy to Address America's Workforce Crisis*, Vol. II, Superintendent of Documents, Washington, DC, pp. 1949–2002.

Taylor, F. W. (1911), *The Principles of Scientific Management*, Harper & Brothers, New York.

Ulrich, D., Kerr, S., and Ashkenas, R. (2002), *The GE Work-out: How to Implement GE's Revolutionary Method for Busting Bureaucracy and Attacking Organizational Problems—Fast!* McGraw-Hill, New York.

Welch, J. F., and Byrne, J. A. (2001), *Jack: Straight from the Gut*, Warner Business Books, New York.

CHAPTER 20

SITUATION AWARENESS

Mica R. Endsley
SA Technologies, Inc.
Marietta, Georgia

1 INTRODUCTION

As we move into the twenty-first century, the biggest challenge within most industries and the most likely cause of an accident receives the label of human error. This is a most misleading term, however, that has done much to sweep the real problems under the rug. It implies that people are merely careless or poorly trained or somehow not very reliable in general. In fact, in the vast majority of these accidents the human operator was striving against significant challenges. On a day-to-day basis, they cope with hugely demanding complex systems. They face both data overload and the challenge of working with a complex system. They are drilled with long lists of procedures and checklists designed to cope with some of these difficulties, but from time to time they are apt to fail. Industry's typical response to such failures has been more procedures and more systems, but unfortunately, this only adds to the complexity of the system. In reality, the person is not the cause of these errors but is the final dumping ground for the inherent problems and difficulties in the technologies we have created. The operator is usually the one who must bring it all together and overcome whatever failures and inefficiencies exist in the system.

So why are people having trouble coping with the present technology and data explosion? The answer lies in understanding how people process the vast amount of data around them to arrive at effective performance. If these accidents are examined in detail, one finds that the operators generally have no difficulty in performing their tasks physically, and no difficulty in knowing what is the correct thing to do, but they continue to be stressed by the task of *understanding what is going on in the situation*. Developing and maintaining a high level of *situation awareness* is the most difficult part of many jobs and is one of the most critical and challenging tasks in many domains today.

Situation awareness (SA) can be thought of as an internalized mental model of the current state of the operator's environment. All of the incoming data from the many systems, the outside environment, fellow team members, and others (e.g., other aircraft and ATC) must all be brought together into an integrated whole. This integrated picture forms the central organizing feature from which all decision making and action takes place (Figure 1).

A vast portion of the operator's job is involved in developing SA and keeping it up to date in a rapidly changing environment. This is a task that is not simple in light of the complexity and sheer number of factors that must be taken into account to make effective decisions. The key to coping in the information age is developing systems that support this process, yet where current technologies have left human operators the most vulnerable to error. Problems with SA were found to be the leading causal factor in a review of military aviation mishaps (Hartel et al., 1991) and in a study of accidents among major air carriers; 88% of those involving human error could be attributed to problems with SA (Endsley, 1995c). A similar review of errors in other domains, such as air traffic control (Rodgers et al., 2000) or nuclear power (Hogg et al., 1993; Mumaw et al., 1993), showed that this is not a

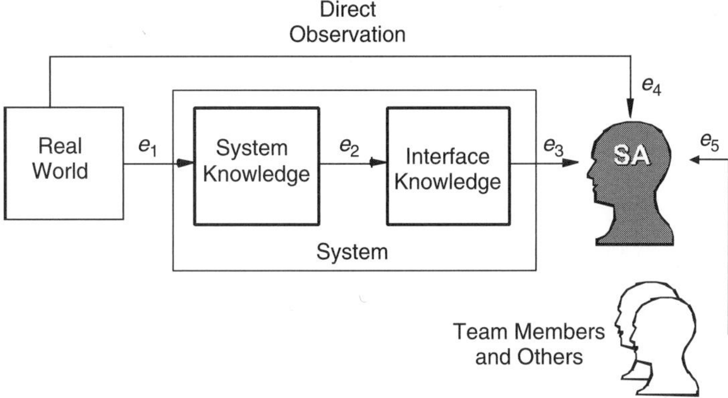

Figure 1 Sources of SA. (From Endsley, 1995d, 1997.)

problem limited to aviation, but one faced by many complex systems.

Successful system designs must deal with the challenge of combining and presenting vast amounts of data now available from many technological systems in order to provide true SA (whether it is to a pilot, a physician, a business manager, or an automobile driver). An important key to the development of complex technologies is understanding that true SA exists only in the mind of the human operator. Therefore, presenting a ton of data will do no good unless the data are transmitted, absorbed, and assimilated successfully and in a timely manner by the human in order to form SA. Unfortunately, most systems fail in this regard, leaving significant SA problems in their wake (Figure 2).

1.1 Definition of Situation Awareness

Although much SA research (and the term) originated within the aviation domain, SA as a construct is widely studied and exists as a basis of performance across many different domains, including air traffic control, military operations, education, driving, train dispatching, maintenance, and weather forecasting. One of the earliest and most widely applicable SA definitions describes it as "the perception of the elements in the environment within a volume of time and space, the comprehension of their meaning and the projection of their status in the near future" (Endsley, 1988). SA therefore involves perceiving critical factors in the environment (level 1 SA), understanding what those factors mean, particularly when integrated together in relation to the operator's goals (level 2), and at the highest level, an understanding of what will happen with the system in the near future (level 3). These higher levels of SA allow people to function in a timely and effective manner, even with very complex and challenging tasks. Each of these levels will be discussed in more detail.

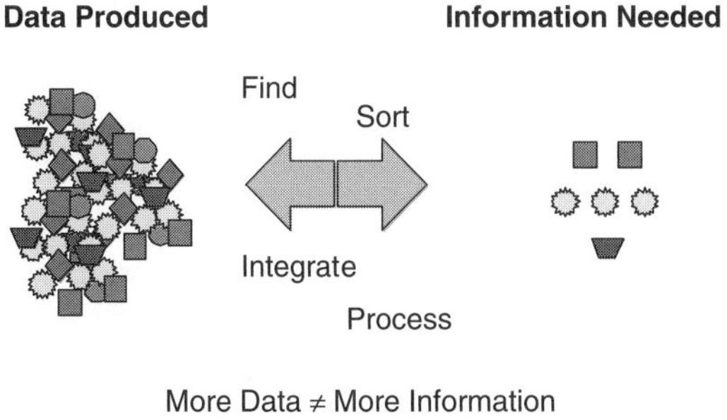

1.1.1 Level 1: Perception of the Elements in the Environment

The first step in achieving SA is to perceive the status, attributes, and dynamics of relevant elements in the environment. A pilot needs to perceive important elements such as other aircraft, terrain, system status, and warning lights along with their relevant characteristics. In the cockpit, just keeping up with all of the relevant system and flight data, other aircraft, and navigational data can be quite taxing. An army officer needs to detect enemy, civilian, and friendly positions and actions, terrain features, obstacles, and weather. An air traffic controller or automobile driver has a different set of information that is needed for SA.

Figure 2 Information gap. (From Endsley, 2000b.)

1.1.2 Level 2: Comprehension of the Current Situation

Comprehension of the situation is based on a synthesis of disjointed level 1 elements. Level 2 SA goes beyond simply being aware of the elements that are present, to include an understanding of the significance of those elements in light of one's goals. The operators put together level 1 data to form a holistic picture of the environment, including a comprehension of the significance of objects and events. For example, upon seeing warning lights indicating a problem during takeoff, the pilot must quickly determine the seriousness of the problem in terms of the immediate air worthiness of the aircraft and combine this with knowledge on the amount of runway remaining in order to know whether or not it is an abort situation. A novice operator may be capable of achieving the same level 1 SA as more experienced ones, but may fall far short of being able to integrate various data elements along with pertinent goals in order to comprehend the situation.

1.1.3 Level 3: Projection of the Future Status

It is the ability to project the future actions of the elements in the environment, at least in the very near term, that forms the third and highest level of SA. This is achieved through knowledge of the status and dynamics of the elements and a comprehension of the situation (both levels 1 and 2 SA). Amalberti and Deblon (1992) found that a significant portion of experienced pilots' time was spent in anticipating possible future occurrences. This gives them the knowledge (and time) necessary to decide on the most favorable course of action to meet their objectives. This ability to project can similarly be critical in many other domains, including driving, plant control, and sports.

1.2 Elements of Situation Awareness

The "elements" of SA in the definition are very domain specific. Examples for air traffic control are shown in Table 1. These elements are clearly observable, meaningful pieces of information for an air traffic controller. Things such as aircraft type, altitude, heading, and flight plan, and restrictions in effect at an airport or conformance to a clearance each comprise meaningful elements of the situation for an air traffic controller. The elements that are relevant for SA in other domains can be delineated similarly. Cognitive task analyses have been conducted to determine SA requirements in commercial aviation (Farley et al., 2000), fighter aircraft (Endsley, 1993), bomber aircraft (Endsley, 1989), and infantry operations (Matthews et al., 2004), among others.

2 DEVELOPING SITUATION AWARENESS

Several researchers have developed theoretical formulations for depicting the role of numerous cognitive processes and constructs on SA (Endsley, 1988, 1995d; Fracker, 1988; Taylor, 1990; Tenney et al., 1992; Taylor and Selcon, 1994; Adams et al., 1995; Smith and Hancock, 1995). There are many commonalties in these efforts, pointing to essential mechanisms that are important for SA. The key points are discussed here; however, more details on each model may be found in these readings. Reviews of these theoretical models of SA are also provided in Pew (1995), Durso and Gronlund (1999), and Endsley (2000b).

Endsley (1988, 1990b, 1995d) describes a theoretical framework model of SA which is summarized in Figure 3. In combination, the mechanisms of short-term sensory memory, perception, working memory, and long-term memory form the basic structures on which SA is based. According to this model, which is formulated in terms of information-processing theory, elements in the environment may initially be processed in parallel through preattentive sensory stores, where certain properties are detected, such as spatial proximity, color, simple properties of shapes, and movement, providing cues for further focalized attention. Those objects that are most salient are processed further using focalized attention to achieve perception. Limited attention creates a major constraint on an operator's ability to perceive multiple items accurately in parallel, and as such, is a major limiting factor on a person's ability to maintain SA in complex environments.

The description thus far accurately depicts only simple data-driven processing, however, the model also shows a number of other factors that affect this process. First, attention and the perception process can be directed by the contents of both working and long-term memory. For instance, advance knowledge regarding the location of information, the form of the information, the spatial frequency, the color, or the overall familiarity and appropriateness of the information can all significantly facilitate perception. Long-term memory also serves to shape the perception of objects in terms of known categories or mental representations. Categorization tends to occur almost instantly.

For operators who have not developed other cognitive mechanisms (novices and those in novel situations), the perception of the elements in the environment (the first level of SA) is significantly limited by attention and working memory. In the absence of other mechanisms, most of the operator's active processing of information must occur in working memory. New information must be combined with existing knowledge and a composite picture of the situation developed. Projections of future status and subsequent decisions as to appropriate courses of action will also occur in working memory. Working memory will be significantly taxed while simultaneously achieving the higher levels of SA, formulating and selecting responses, and carrying out subsequent actions.

In actual practice, however, goal-directed processing and long-term memory (often in the form of mental models and schema) can be used to circumvent the limitations of working memory and direct attention more effectively. First, much relevant knowledge about a system is hypothesized to be stored in mental models. Rouse and Morris (1985) define mental models as "mechanisms whereby humans are able to generate descriptions of system purpose and form, explanations of system functioning and observed system states, and predictions of future states."

Mental models are cognitive mechanisms that embody information about system form and function; often, they are relevant to a physical system (e.g., a car, computer, or power plant) or an organizational system (e.g., how a university, company, or military unit works). They typically contain information not only about the components of a particular system, but also how those components interact to produce various system states and events. Mental models can significantly aid SA as people recognize key features

Table 1 Elements of Situation Awareness for Air Traffic Control

Level 1

- Aircraft
  - Aircraft ID, CID, beacon code
  - Current route (position, heading, aircraft turn rate, altitude, climb/descent rate, groundspeed)
  - Current flight plan (destination, filed plan)
  - Aircraft capabilities (turn rate, climb/descent rate, cruising speed, max/min speed)
  - Equipment on board
  - Aircraft type
  - Fuel/loading
  - Aircraft status
  - Activity (enroute, arriving, departing, handed off, pointed out)
    - Level of control (IFR, VFR, flight following, VFR-on top, uncontrolled object)
    - Aircraft contact established
    - Aircraft descent established
    - Communications (present/frequency)
    - Responsible controller
    - Aircraft priority
      - Special conditions
      - Equipment malfunctions
      - Emergencies
    - Pilot capability/state/intentions
    - Altimeter setting
- Emergencies
  - Type of emergency
  - Time on fuel remaining
  - Souls on board
- Requests
  - Pilot/controller requests
  - Reason for request
- Clearances
  - Assignment given
  - Received by correct aircraft
  - Readback correct/complete
  - Pilot acceptance of clearance
  - Flight progress strip current
- Sector
  - Special airspace status
  - Equipment functioning
  - Restrictions in effect
  - Changes to standard procedures
- Special Operations
  - Type of special operation
  - Time begin/terminate operations
  - Projected duration
  - Area and altitude effected
- ATC equipment malfunctions
  - Equipment affected
  - Alternate equipment available
  - Equipment position/range
  - Aircraft in outage area
- Airports
  - Operational status
  - Restrictions in effect
  - Direction of departures
  - Current aircraft arrival rate
  - Arrival requirements
  - Active runways/approach
  - Sector saturation
  - Aircraft in holding (time, number, direction, leg length)
- Weather
  - Area affected
  - Altitudes affected
  - Conditions (snow, icing, fog, hail, rain, turbulence, overhangs)
  - Temperatures
  - Intensity
  - Visibility
  - Turbulence
  - Winds
  - IFR/VFR conditions
  - Airport conditions

Level 2

- Conformance
  - Amount of deviation (altitude, airspeed, route)
  - Time until aircraft reaches assigned altitude, speed, route/heading
- Current separation
  - Amount of separation between aircraft/objects/airspace/ground along route
  - Deviation between separation and prescribed
  - Limits
  - Number/timing aircraft on routes
  - Altitudes available
- Timing
  - Projected time in airspace
  - Projected time until clear of airspace
  - Time until aircraft landing expected
  - Time/distance aircraft to airport
  - Time/distance until visual contact
  - Order/sequencing of aircraft
- Deviations
  - Deviation aircraft/landing request
  - Deviation aircraft/flight plan
  - Deviation aircraft/pilot requests
- Other sector/airspace
  - Radio frequency
  - Aircraft duration/reason for use
- Significance
  - Impact of requests/clearances on:
    - Aircraft separation/safety
    - Own/other sector workload
  - Impact of weather on:
    - Aircraft safety/flight comfort
    - Own/other sector workload
    - Aircraft flow/routing (airport arrival rates, Flow rates, holding requirements aircraft routes, separation procedures)
    - Altitudes available
    - Traffic advisories
  - Impact of special operations on sector
  - Operations/procedures
  - Location of nearest capable airport for aircraft type/emergency
  - Impact of malfunction on: routing, communications, flow control, aircraft, coordination procedures, other sectors, own workload
  - Impact on workload of number of aircraft sector demand vs. own capabilities
- Confidence level/accuracy of information
  - Aircraft ID, position, altitude, airspeed, heading
  - Weather
  - Altimeter setting

Level 3

- Projected aircraft route (current)
  - Position, fight plan, destination, heading, route, altitude, climb/descent rate, airspeed, winds, groundspeed, intentions, assignments
- Projected aircraft route (potential)
  - Projected position *x* at time *t*
  - Potential assignments
- Projected separation
  - Amount of separation along route (aircraft/objects/airspace/ground)
  - Deviation between separation and prescribed limits
  - Relative projected aircraft routes
  - Relative timing along route
- Predicted changes in weather
  - Direction/speed of movement
  - Increasing/decreasing in intensity
- Impact of potential route changes
  - Type of change required
  - Time and distance until turn aircraft amount of turn/new heading, altitude, route change required
  - Aircraft ability to make change
  - Projected number of changes necessary
  - Increase/decrease length of route
  - Cost/benefit of new clearance
  - Impact of proposed change on aircraft separation

Source: Endsley and Rogers (1994a).

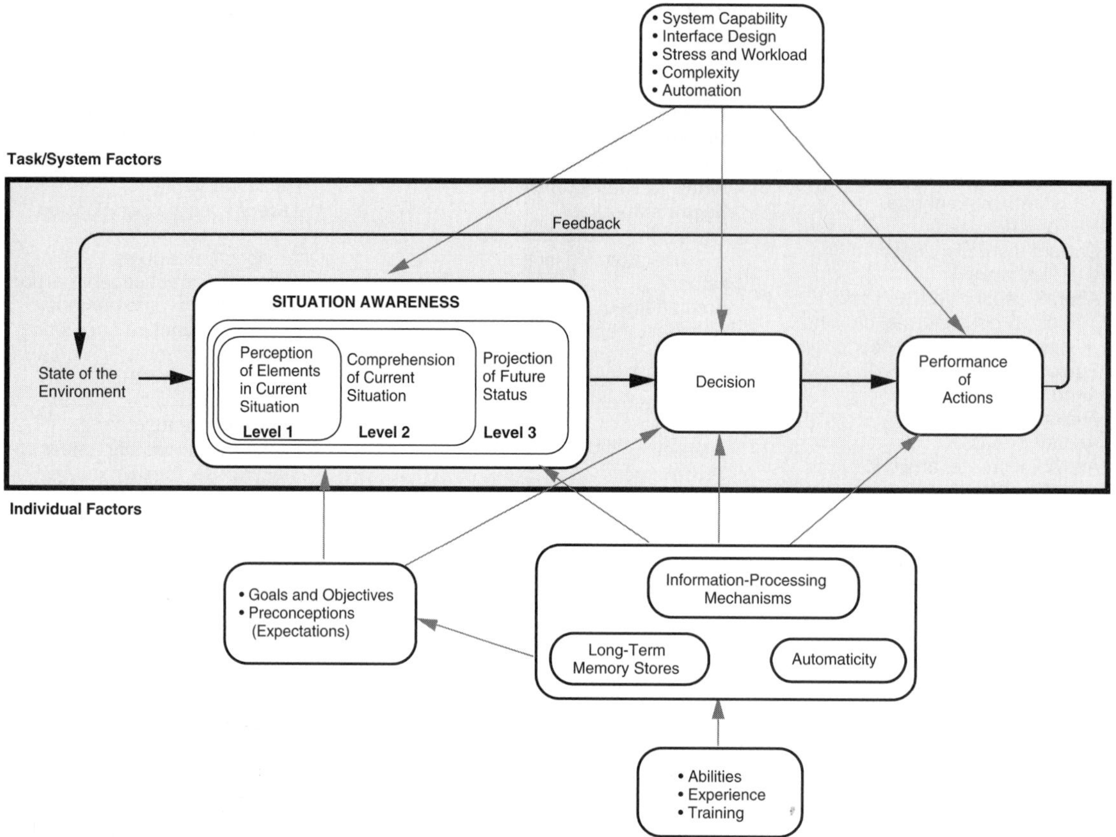

Figure 3 Model of situation awareness in dynamic decision making. (From Endsley, 1995d, 1997.)

in the environment that map to key features in the model. The model then creates a mechanism for determining associations between observed states of components (comprehension) and predictions of the behavior and status of these elements over time. Thus, mental models can provide much of the higher levels of SA (comprehension and projection) without loading working memory.

Also associated with mental models are schema: prototypical classes of states of the system (e.g., an engine failure, an enemy attack formation, or a dangerous weather formation). These schema are even more useful to the formation of SA since these recognized classes of situations provide an immediate one-step retrieval of the higher levels of SA, based on pattern matching between situation cues and known schema in memory. Very often scripts, set sequences of actions, have also been developed for schema, so that much of the load on working memory for generating alternative behaviors and selecting among them is also diminished. These mechanisms allow the operator simply to execute a predetermined action for a given recognized class of situations (based on their SA). The current situation does not need to be exactly like the one encountered previously, due to the use of categorization mapping; as long as a close-enough mapping can be made into relevant categories, a situation can be recognized, comprehended in terms of the model, predictions made, and appropriate actions selected. Since people have very good pattern-matching abilities, this process can be almost instantaneous and produce a much lower load on working memory, which makes high levels of SA possible, even in very demanding situations.

Expertise therefore plays a major role in the SA process. For novices or those dealing with novel situations, decision making in complex and dynamic systems can be very demanding or impossible to accomplish successfully in that it requires detailed mental calculations based on rules or heuristics, placing a heavy burden on working memory. Where experience has allowed the development of mental models and schema, pattern matching between the perceived elements in the environment and existing schema/mental models can occur on the basis of pertinent cues that have been learned. Thus, the comprehension and future projection required for the higher levels of SA can be developed with far less effort and within the constraints of working memory. When scripts have been developed, tied to these schema, the entire decision-making process will be greatly simplified.

The operator's goals also play an important part in the process. These goals can be thought of as ideal states of the system model that the operator wishes to achieve. In what Casson (1983) has termed a *top-down* decision-making process, the operator's goals and plans will direct which environmental aspects are attended to in the development of SA. Goal-driven or top-down processing is very important in the effective information process and development of SA. Conversely, in a *bottom-up* or data-driven process, patterns in the environment may be recognized which will indicate to the operator that different plans will be necessary to meet goals or that different goals should be activated.

Alternating between "goal driven" and "data driven" is characteristic of much human information processing and underpins much of the SA development in complex worlds. People who are purely data driven are very inefficient at processing complex information sets; there is too much information, so they are simply reactive to the cues that are most salient. People who have clearly developed goals, however, will search for information that is relevant to those goals (on the basis of the associated mental model, which contains information on which aspects of the system are relevant to goal attainment), allowing the information search to be more efficient and providing a mechanism for determining the relevance of the information that is perceived. If people are only goal driven, however, they are likely to miss key information that would indicate that a change in goals is needed (e.g., no longer the goal "land the airplane" but the goal "execute a go-around"). Thus, effective information processing is characterized by alternating between these modes: using goal-driven processing to find and process efficiently the information needed for achieving goals, using data-driven processing to regulate the selection of which goals should be most important at any given time.

The development of SA is a dynamic and ongoing process that is effected by these key cognitive mechanisms. Although it can be very challenging in many environments, with mechanisms that can be developed through experience (schema and mental models), we find that people are able to circumvent certain limitations (working memory and attention) to develop sufficient levels of SA to function very effectively. Nevertheless, developing accurate SA remains a very challenging feature in many complex settings and demands a significant portion of an operator's time and resources. Thus, developing selection batteries, training program, and system designs to enhance SA is a major goal in many domains.

3 SITUATION AWARENESS CHALLENGES

Building and maintaining SA can be a difficult process for people in many different jobs and environments. Pilots report that the majority of their time is generally spent trying to ensure that their mental picture of what is happening is current and correct. The same can be said for people in many other domains, where systems are complex and there is a great deal of information to understand, where information changes rapidly, and where information is difficult to obtain. The reasons for this have been captured in terms of eight SA demons, factors that work to undermine SA in many systems and environments. (Endsley et al., 2003).

3.1 Attentional Tunneling

Successful SA is highly dependent on constantly juggling one's attention between different aspects of the environment. Unfortunately, there are significant limits on people's ability to divide their attention across multiple aspects of the environment, particularly within a single modality, such as vision or sound, and thus attention sharing can occur only to a limited extent (Wickens, 1992). They can often get trapped in a phenomenon called *attentional narrowing* or *tunneling* (Bartlett, 1943; Broadbent, 1954; Baddeley, 1972). When succumbing to attentional tunneling, they lock in on certain aspects or features of the environment they are trying to process, and will intentionally or inadvertently drop their scanning behavior. In this case, their SA may be very good on the part of the environment of their concentration but will quickly become outdated on other aspects they are not watching. Attentional narrowing has been found to undermine SA in tasks such as flying and driving and poses one of the most significant challenges to SA in many domains.

3.2 Requisite Memory Trap

The limitations of working memory also create a significant SA demon. Many features of the situation may need to be held in memory. As a person scans different information from the environment, information accessed previously must be remembered and combined with new information. Auditory information must also be remembered, as it cannot be revisited in the way that visual displays can. Given the complexity and sheer volume of information required for SA in many systems, these memory limits create a significant problem for SA. System designs that necessitate that people remember information, even short term, increase the likelihood of SA error.

3.3 Workload, Anxiety, Fatigue, and Other Stressors

Stressors such as anxiety, time pressure, mental workload, uncertainty, noise or vibration, excessive heat or cold, poor lighting, physical fatigue, and working against one's circadian rhythms are unfortunately an unavoidable part of many work environments. These stressors can act to reduce SA significantly by further reducing an already limited working memory and reducing the efficiency of information gathering. It has been found that people may pay less attention to peripheral information, become more disorganized in scanning information, and are more likely to succumb to attentional tunneling when affected by these stressors. People are also more likely to arrive at a decision without taking into account all available information (premature closure).

3.4 Data Overload

Data overload is a significant problem in many systems. The volume of data and the rapid rate of change of that data create a need for information intake that quickly outpaces one's ability to gather and assimilate the data. As people can take in and process only a limited amount of information at a time, significant lapses in SA can occur. While it is easy to think of this problem as simply a human limitation, in reality it often occurs because data are processed, stored, and presented ineffectively in many systems. This problem is not just one of volume but also one of bandwidth, the bandwidth provided by a person's sensory and information-processing mechanisms. The rate that data can flow through the pipeline can be increased significantly based on the form of information presentation employed in the interface.

3.5 Misplaced Salience

The human perceptual system is more sensitive to certain features than others, including the color red, movement, and flashing lights. Similarly, loud noises, larger shapes, and things that are physically nearer have the advantage of catching a person's attention. These natural salient properties can be used to promote SA or to hinder it. When used carefully, properties such as movement or color can be used to draw attention to critical and highly important information and are thus important tools for designing to enhance SA. Unfortunately, these features are often overused or used inappropriately. The unnecessary distractions of misplaced salience can act to degrade SA of the other information the person is attempting to assimilate. Unfortunately, in many systems there is a proliferation of lights, buzzers, alarms, and other signals that work actively to draw people's attention, frequently either misleading or overwhelming them.

3.6 Complexity Creep

Over time, systems have become more and more complex, often through a misguided attempt to add more features or capabilities. Unfortunately, this complexity makes it difficult for people to form sufficient internal representations of how these systems work. The more features, and the more complicated and branching the rules that govern a system's behavior, the greater the complexity. Although system complexity can slow down a person's ability to take in information, it works primarily to undermine the person's ability to correctly interpret the information presented and to project what is likely to happen (levels 2 and 3 SA). A cue that should indicate one thing can be completely misinterpreted, as the internal mental model will be developed inadequately to encompass the full characteristics of the system.

3.7 Errant Mental Models

Mental models are important mechanisms for building and maintaining SA, providing key interpretation mechanisms for information collected. They tell a person how to combine disparate pieces of information, how to interpret the significance of that information, and how to develop reasonable projections of what will happen in the future. If an incomplete mental model is used, however, or if the wrong mental model is relied on for the situation, poor comprehension and projection (levels 2 and 3 SA) can result. Also called a *representational error*, it can be very difficult for people to realize that they are working on the basis of an errant mental model and break out of it. Mode errors, in which people misunderstand information because they believe that the system is in one mode when it is really in another, are a special case of this problem.

3.8 Out-of-the-Loop Syndrome

Automation creates a final SA demon. While in some cases, automation can help SA by eliminating excessive workload, it can also act to lower SA by putting people *out of the loop*. In this state, they develop poor SA as to both how the automation is performing and the state of the elements the automation is supposed to be controlling. When the automation is performing well, being out-of-the-loop may not be a problem, but when the automation fails or, more frequently, reaches situational conditions that it is not equipped to handle, the person is out of the loop and often unable to detect the problem, properly interpret the information presented, and intervene in a timely manner.

4 LEVELS OF SITUATION AWARENESS

There is some evidence that some people are significantly better than others at developing SA. In one study of experienced military fighter pilots, Endsley and Bolstad (1994) found a tenfold difference in SA between the pilot with the lowest SA and the one with the highest SA. They also found this to be highly stable, with test–retest reliability rates exceeding 0.94 for those evaluated. Others (Secrist and Hartman, 1993; Bell and Waag, 1995) have similarly noted consistent individual differences, with some pilots routinely having better SA than their compatriots. These individual differences appear even when people operate with the same system capabilities and displays and in the same environment, subject to the same demands.

A number of studies have sought to find the focus of these individual differences in SA abilities. Are they due simply to the effects of expertise and experience, or are they indicative of the better cognitive mechanisms or capabilities that some people have? Endsley and Bolstad (1994) found that military pilots with better SA were better at attention sharing, pattern matching, spatial abilities, and perceptual speed. O'Hare (1997) also found evidence that elite pilots (defined as consistently superior in gliding competitions) performed better on a divided-attention task purported to measure SA. Gugerty and Tirre (1997) found evidence that people with better SA performed better on measures of working memory, visual processing, temporal processing, and time-sharing ability.

Although many of these studies have examined individual differences in only a few domains (e.g. piloting and driving), some of these attributes may also be relevant to SA differences in other arenas. If reliable markers can be found that differentiate those who will eventually be most successful at SA, more valid selection batteries can be developed for critical jobs such as air traffic controller, pilot, or military commander.

More important, there has also been research to examine what skills differentiate those with high SA from those with low SA that might be trainable, thus significantly improving SA in the existing population of operators in a domain. For instance, SA differences between those at different levels of expertise have been examined in groups of pilots (Prince and Salas, 1998; Endsley et al., 2000), military officers (Strater et al., 2003), aircraft mechanics (Endsley and Robertson, 2000a), power plant operators (Collier and Folleso, 1995), and drivers (Horswill and McKenna, 2004). These studies have found many systematic differences, some of which may relate to underlying abilities, but many of which also point to learned skills or behaviors that may be trainable.

Training programs directed at improving SA are in their infancy, but a number are being developed. They seek to train skills related to developing SA (at the individual or team level) in classroom settings, simulated scenarios or case studies, or through computer-based training. These include training programs for commercial aviation pilots (Robinson, 2000; Hormann et al., 2004), general aviation pilots (Prince, 1998; Endsley and Garland, 2000a; Bolstad et al., 2002; Endsley et al., 2002), drivers (Sexton, 1988; McKenna and Crick, 1994), aircraft mechanics (Endsley and Robertson, 2000a,b), and army officers (Strater et al., 2003, 2004). The preliminary findings reported by the majority of these efforts show some successes in improving SA and performance in their respective settings. In general, more longitudinal studies are needed to ascertain the degree to which such efforts can be successful in improving the SA of persons in the wide variety of challenging situations that are common in these domains.

5 SYSTEM DESIGN TO SUPPORT SITUATION AWARENESS

In addition to training to improve SA at the individual level, efforts to improve SA through better sensors, information processing, and display approaches have characterized much of the past 20 years. Unfortunately, a significant portion of these efforts have stopped short of really addressing SA; instead, they simply add a new sensor or black box that is purported to improve SA. While insuring that operators have the data needed to meet their level 1 SA requirements is undoubtedly important, a rampant increase in such data may inadvertently hurt SA as much as it helps. Simply increasing the amount of data available to an operator instead adds to the information gap, overloading the operator without necessarily improving the level of SA the person can develop and maintain.

As a construct, however, SA provides a key mechanism for overcoming this data overload. SA specifies how all the data in an environment need to be combined and understood. Therefore, instead of loading the operator down with 100 pieces of miscellaneous data provided in a haphazard fashion, SA requirements provide guidance as to what the real comprehension and projection needs are. Therefore, it provides the system designer with key guidance on how to bring the various pieces of data together to form meaningful integrations and groupings of data that can be absorbed and assimilated easily in time-critical situations. This type of systems integration usually requires very unique combinations of information and portrayals of information that go far beyond the black-box technology-oriented approaches of the past. In the past it was up to the operator to do it all. This task left him or her overloaded and susceptible to missing critical factors. If system designers work to develop systems that support the SA process, however, they can alleviate this bottleneck significantly.

So how should systems be designed to meet the challenge of providing high levels of SA? Over the past decade a significant amount of research has been focused on this topic, developing an initial understanding of the basic mechanisms that are important for SA and the design features that will support those mechanisms. Based on this research, the SA-oriented design process has been established (Endsley et al., 2003) to guide the development of systems that support SA (Figure 4). This structured approach incorporates SA considerations into the design process, including a determination of SA requirements, design principles for SA enhancement, and measurement of SA in design evaluation.

5.1 Requirements Analysis

The problem of determining what aspects of the situation are important for a particular operator's SA has frequently been approached using a form of cognitive task analysis called *goal-directed task analysis*, illustrated in Figure 5. In such analysis, the major goals of a particular job class are identified, along with the major subgoals necessary for meeting each goal. Associated with each subgoal, the major decisions that need to be made are then identified. The SA needed for making these decisions and carrying out each subgoal are identified. These SA requirements focus not only on what data the operator needs, but also on how that information is integrated or combined to address each decision. In this analysis process, SA requirements are defined as those dynamic information needs associated with the major goals or subgoals of the operator in performing his or her job (as opposed to more static knowledge, such as rules, procedures, and general system knowledge). This type of analysis is based on goals or objectives, not tasks (as a traditional task analysis might). This is because goals form the basis for decision making in many complex environments. Conducting such an analysis is usually carried out using a combination of cognitive engineering procedures. Expert elicitation, observation

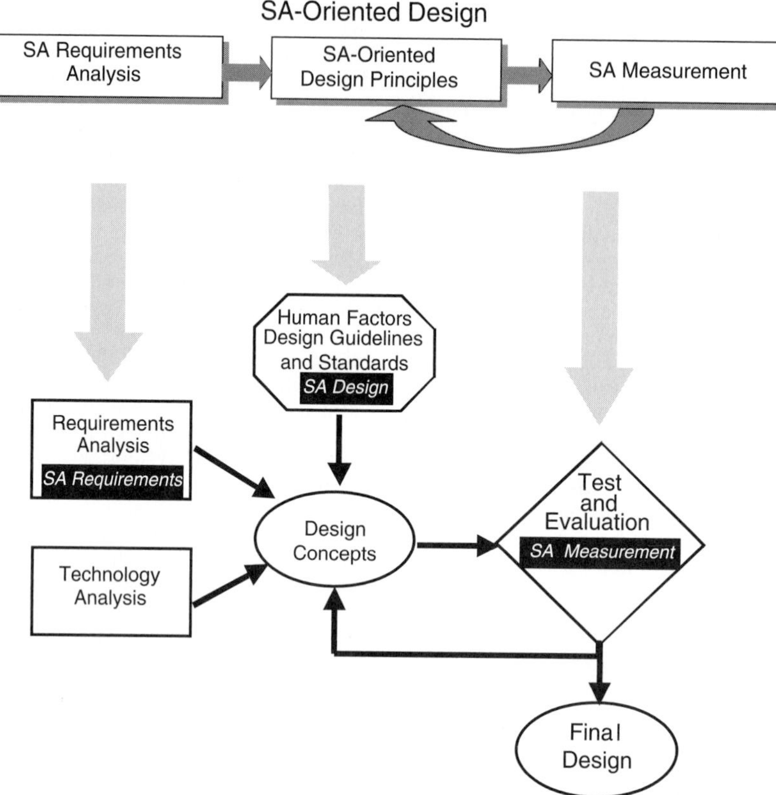

Figure 4 SA-oriented design process. (From Endsley et al., 2003.)

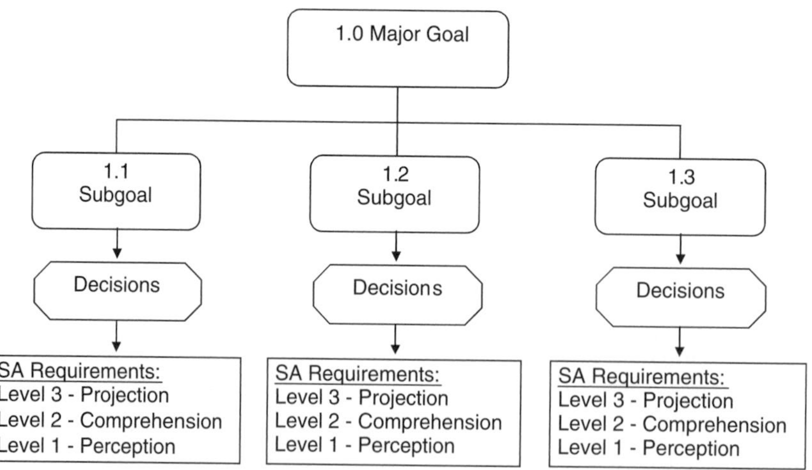

Figure 5 Goal-directed task analysis for determining SA requirements.

of operator performance of tasks, verbal protocols, analysis of written materials and documentation, and formal questionnaires have formed the basis for the analyses. In general, the analysis has been conducted with a number of operators, who are interviewed, observed, and recorded individually, with the resulting analyses pooled and then validated overall by a larger number of operators.

Table 2 Example of Goal-Directed Task Analysis for En Route Air Traffic Control

1.3 Maintain aircraft conformance
- 1.3.1 Assess aircraft conformance to assigned parameters
  - Aircraft at/proceeding to assigned altitude?
  - Aircraft proceeding to assigned altitude fast enough?
    - Time until aircraft reaches assigned altitude
    - Amount of altitude deviation
    - Climb/descent
      - Altitude (current)
      - Altitude (assigned)
      - Altitude rate of change (ascending/descending)
  - Aircraft at/proceeding to assigned airspeed?
  - Aircraft proceeding to assigned airspeed fast enough?
    - Time until aircraft reaches assigned airspeed
    - Amount of airspeed deviation
      - Airspeed (indicated)
      - Airspeed (assigned)
      - Groundspeed
  - Aircraft on/proceeding to assigned route?
  - Aircraft proceeding to assigned route fast enough?
  - Aircraft turning?
    - Time until aircraft reaches assigned route/heading
    - Amount of route deviation
      - Aircraft position (current)
      - Aircraft heading (current)
      - Route/heading (assigned)
    - Aircraft turn rate (current)
      - Aircraft heading (current)
      - Aircraft heading (past)
      - Aircraft turn capabilities
        - Aircraft type
        - Altitude
      - Aircraft groundspeed
      - Weather
      - Winds (direction, magnitude)

Source: Endsley and Rodgers (1994b).

An example of the output of this process is shown in Table 2. This example shows the SA requirements analysis for the subgoal "maintain aircraft conformance" for the major goal "avoid conflictions" for an air traffic controller. In this example, the subgoal is divided even further into lower-level subgoals prior to the decisions and SA requirements being listed. In some cases, addressing a particular subgoal occurs through reference to another subgoal in other parts of the analysis, such as the need to readdress aircraft separation in this example. This shows the degree to which a particular operator's goals and resultant SA needs may be very interrelated. The example in Table 2 shows just one major subgoal out of four that are relevant for the major goal "avoid conflictions," which is just one of three major goals for an air traffic controller.

This analysis defines systematically the SA requirements (at all three levels of SA) that are needed to effectively make the decisions required by the operator's goals. Many of the same SA requirements appear throughout the analysis. In this manner, the way in which pieces of data are used together and combined to form what the operator really wants to know is determined. Although the analysis will typically include many goals and subgoals, they may all be active at once. In practice, at any given time more than one goal or subgoal may be operational, although they will not always have the same prioritization. The analysis does not indicate any prioritization among the goals (which can vary over time), or that each subgoal within a goal will always be active. Unless particular events are triggered (e.g., the subgoal of assuring aircraft conformance in this example), a subgoal may not be active for a given controller.

The analysis strives to be as technology-free as possible. How the information is acquired is not addressed, as this can vary considerably from person to person, from system to system, and from time to time. In some cases it may be through system displays, verbal communications, other operators, or internally generated from within the operator. Many of the higher-level SA requirements fall into this category. The way in which information is acquired can vary widely between persons, over time, and between system designs.

The analysis seeks to determine what operators would ideally like to know to meet each goal. It is recognized that they often must operate on the basis of incomplete information and that some desired information may not be available at all with today's system. However, for purposes of design and evaluation of systems, we need to set the yardstick to measure against what they ideally need to know, so that artificial ceiling effects, based on today's technology, are not induced in the process. Finally, it should be noted that static knowledge, such as procedures or rules for performing tasks, is outside the bounds of an SA requirements analysis. The analysis focuses on the dynamic situational information that affects what the operators do.

To date, these analyses have been completed for many domains of common concern, including en route air traffic control (Endsley and Rodgers, 1994b), TRACON air traffic control (Endsley and Jones, 1995), fighter pilots (Endsley, 1993), bomber pilots (Endsley, 1989), commercial transport pilots (Endsley et al., 1998b), aircraft mechanics (Endsley and Robertson, 1996), and airway facilities maintenance (Endsley and Kiris, 1994). A similar process was employed by Hogg et al. (1993) to determine appropriate queries for a nuclear reactor domain.

5.2 SA-Oriented Design Principles

The development of a system design for successfully providing the multitude of SA requirements that exist in complex systems is a significant challenge. A set of design principles have been developed based on the theoretical model of the mechanisms and processes involved in acquiring and maintaining SA in dynamic complex systems (Endsley, 1988, 1990b, 1995d; Endsley et al., 2003). The 50 design principles include (1) general guidelines for supporting SA, (2) guidelines for coping with automation and complexity, (3) guidelines for the design of alarm systems, (4) guidelines for the presentation of information uncertainty, and (5) guidelines for supporting SA in team operations. Some of the general principles include the following: (1) Direct presentation of higher-level SA needs (comprehension and projection) is recommended, rather than supplying only low-level data that operators must integrate and interpret manually; (2) goal-oriented information displays should be provided, organized so that the information needed for a particular goal is colocated and answers directly the major decisions associated with the goal; (3) support for global SA is critical, providing an overview of the situation across the operator's goals at all times (with detailed information for goals of current interest) and enabling efficient and timely goal switching and projection; (4) critical cues related to key features of schemata need to be determined and made salient in the interface design (in particular, those cues that will indicate the presence of prototypical situations will be of prime importance and will facilitate goal switching in critical conditions); (5) extraneous information not related to SA needs should be removed (while carefully ensuring that such information is not needed for broader SA needs); and (6) support for parallel processing, such as multimodal displays, should be provided in data-rich environments.

SA-oriented design is applicable to a wide variety of system designs. It has been used successfully as a design philosophy for systems involving remote maintenance operations, medical systems, flexible manufacturing cells, and command and control for distributed teams.

5.3 Design Evaluation

Many concepts and technologies are currently being developed and touted as enhancing SA. Prototyping and simulation of new technologies, new displays, and new automation concepts is extremely important for evaluating the actual effects of proposed concepts within the context of the task domain and using domain-knowledgeable subjects. If SA is to be a design objective, it is critical that it be evaluated specifically during the design process. Without this, it will be impossible to tell if a proposed concept actually helps SA, does not affect it, or inadvertently compromises it in some way. A primary benefit of examining system design from the perspective of operator SA is that the impact of design decisions on SA can be assessed objectively as a measure of the quality of the integrated system design when used within the actual challenges of the operational environment.

SA measurement has been approached in a number of ways. See Endsley and Garland (2000b) for details on these methods. A review of the advantages and disadvantages of these methods may be found in Endsley (1996) and Endsley and Smolensky (1998). In general, direct measurement of SA can be very advantageous in providing more sensitivity and diagnosticity in the test and evaluation process. This provides a significant addition to performance measurement and workload measurement in determining the utility of new design concepts. Whereas workload measures provide insight into how hard an operator must work to perform tasks with a new design, SA measurement provides insight into the level of understanding gained from that work.

Direct measurement of SA has generally been approached either through subjective ratings or by objective techniques. Although subjective ratings are simple and easy to administer, research has shown that they correlate poorly with objective SA measures, indicating they more closely capture a person's confidence in his or her SA rather than the actual level or accuracy of that SA (Endsley et al., 1998a).

One of the most widely used objective measures of SA is the situation awareness global assessment technique (SAGAT) (Endsley, 1988, 1995b, 2000a). SAGAT has been used successfully to measure operator SA directly and objectively when evaluating avionics concepts, display designs, and interface technologies (Endsley, 1995b). Using SAGAT, a simulated test scenario employing the design of interest is frozen at randomly selected times, the system displays are blanked, and the simulation is suspended while operators quickly answer questions about their current perceptions of the situation. The questions correspond to their SA requirements as determined from an SA requirements analysis for that domain. Operator perceptions are then compared to the real situation based on simulation computer databases to provide an objective measure of SA.

Multiple "snapshots" of operators' SA can be acquired in this way, providing an index of the quality of SA provided by a particular design. The collection of SA data in this manner provides an objective, unbiased assessment of SA that overcomes the problems incurred when collecting such data after the fact and minimizes biasing of controller SA due to secondary task loading or cuing the controller's attention artificially, which real-time probes may do. By including queries across the full spectrum of an operator's SA requirements, this approach minimizes possible biasing of attention, as subjects cannot prepare for the queries in advance since they could be queried over almost every aspect of the situation to which they would normally attend. The primary disadvantage of this technique involves the temporary halt in the simulation.

The method is not without some costs, however, as a detailed analysis of SA requirements is required in order to develop the battery of queries to be administered. SAGAT is a global tool developed to assess SA across all of its elements based on a comprehensive assessment of operator SA. As a global measure, SAGAT includes queries about all operator

SA requirements, including level 1 (perception of data), level 2 (comprehension of meaning), and level 3 (projection of the near future) components. This includes a consideration of system functioning and status as well as relevant features of the external environment.

SAGAT has also been shown to have predictive validity, with SAGAT scores indicative of pilot performance in a combat simulation (Endsley, 1990a). It is also sensitive to changes in task load and to factors that affect operator attention (Endsley, 2000a), demonstrating construct validity. It has been found to produce high levels of reliability (Endsley and Bolstad, 1994; Collier and Folleso, 1995; Gugerty, 1997). Studies examining the intrusiveness of the freezes to collect SAGAT data have generally found there to be no effect on operator performance (Endsley, 1995a, 2000a)

An example of the use of SAGAT for evaluating the impact of new system concepts may be found in (Endsley et al., 1997a). A totally new form of distributing roles and responsibilities between pilots and air traffic controllers was examined. Termed *free flight*, this concept was originally developed as a major change in the operation of the national airspace. It may include pilots filing direct routes to destinations rather than along predefined fixed airways, and authority for the pilot to deviate from that route either with air traffic controllers' permission or perhaps even fully autonomously (RTCA, 1995). As it was felt that such changes could have a marked effect on the ability of the controller to keep up as monitor in such a new system, a study was conducted to examine this possibility (Endsley et al., 1997b).

Results showed a trend toward poorer controller performance in detecting and intervening in aircraft separation errors with these changes in the operational concept and poorer subjective ratings of performance. Finding statistically significant changes in separation errors during ATC simulation testing is quite rare, however. More detailed analysis of the SAGAT results provided more diagnostic detail as well as backing up this finding. As shown in Figure 6, controllers were aware of significantly fewer aircraft in the simulation under free-flight conditions. Attending to fewer aircraft under a higher workload has also been found in other studies (Endsley and Rodgers, 1998).

In addition to reduced level 1 SA, however, controllers had a significantly reduced understanding (level 2 SA) of what was happening in the traffic situation, as evidenced by lower SA regarding which aircraft weather would affect the situation and a reduced awareness of those aircraft that were in a transitionary state. They were less aware of which aircraft had not yet completed a clearance, and for those aircraft, whether the instruction was received correctly and whether they were conforming. Controllers also demonstrated lower level 3 SA with free flight. Their knowledge of where the aircraft was going (to the next sector) was significantly lower under free-flight conditions.

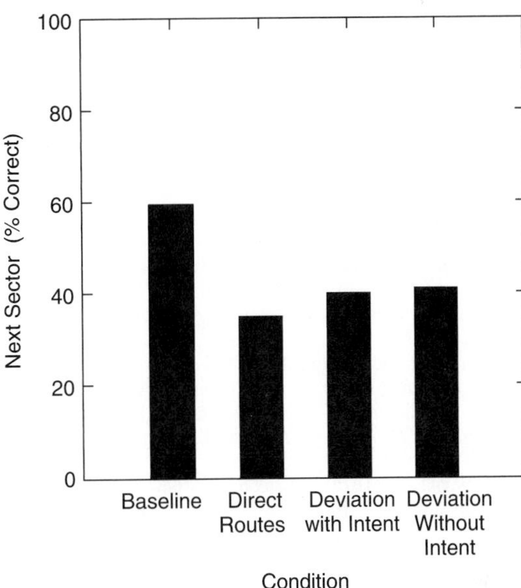

Figure 6 SAGAT results. (From Endsley et al., 1997b.)

These findings were useful in pinpointing whether concerns over this new and very different concept were justified or whether they merely represented resistance to change. The SAGAT results showed not only that the new concept did indeed induce problems for controller SA that would prevent them from performing effectively as monitors to back up pilots with separation assistance; it also showed in what ways these problems were manifested. This information is very useful diagnostically in that it allows one to determine what sort of aid might be needed for operators to assist them in overcoming these deficiencies.

For instance, in this example, a display that provides enhanced information on flight paths for aircraft in transitionary states may be recommended as a way of compensating for the lower SA observed. Far from just providing a thumbs-up or thumbs-down input on a concept under evaluation, this rich source of data is very useful in developing iterative design modifications and making trade-off decisions.

6 CONCLUSIONS

A firm theoretical foundation has been laid for understanding the factors that affect SA in complex environments. This foundation can be used to guide the development of training programs and the development of system designs that go beyond data presentation to provide higher levels of SA. In either case, validation of the effectiveness of the proposed solutions through detailed, objective testing is paramount to ensuring that the approach is actually successful in improving SA.

The need to process and understand large volumes of data is critical for many endeavors, from the cockpit to military missions, from power plants to automobiles, and from space stations to day-to-day business operations. It is likely that the potential benefits of the information age will not be realized until we come to grips with the challenges of managing this dynamic information base to provide people with the SA they need on a real-time basis. Doing so is the primary challenge of the next decade of technology.

REFERENCES

Adams, M. J., Tenney, Y. J., and Pew, R. W. (1995), "Situation Awareness and the Cognitive Management of Complex Systems," *Human Factors*, Vol. 37, No. 1, pp. 85–104.

Amalberti, R., and Deblon, F. (1992), "Cognitive Modeling of Fighter Aircraft Process Control: A Step Towards an Intelligent On-board Assistance System," *International Journal of Man–Machine Systems*, Vol. 36, pp. 639–671.

Baddeley, A. D. (1972), "Selective Attention and Performance in Dangerous Environments," *British Journal of Psychology*, Vol. 63, pp. 537–546.

Bartlett, F. C. (1943), "Fatigue Following Highly Skilled Work," *Proceedings of the Royal Society (B)*, Vol. 131, pp. 147–257.

Bell, H. H., and Waag, W. L. (1995), "Using Observer Ratings to Assess Situational Awareness in Tactical Air Environments," in *Experimental Analysis and Measurement of Situation Awareness*, D. J. Garland and M. R. Endsley, Eds., Embry-Riddle Aeronautical University Press, Daytona Beach, FL, pp. 93–99.

Bolstad, C. A., Endsley, M. R., Howell, C., and Costello, A. (2002), "General Aviation Pilot Training for Situation Awareness: An Evaluation," in *Proceedings of the 46th Annual Meeting of the Human Factors and Ergonomics Society*, Human Factors and Ergonomics Society, Santa Monica, CA, pp. 21–25.

Broadbent, D. E. (1954), "Some Effects of Noise on Visual Performance," *Quarterly Journal of Experimental Psychology*, Vol. 6, pp. 1–5.

Casson, R. W. (1983), "Schema in Cognitive Anthropology," *Annual Review of Anthropology*, Vol. 12, pp. 429–462.

Collier, S. G., and Folleso, K. (1995), "SACRI: A Measure of Situation Awareness for Nuclear Power Plant Control Rooms," in *Experimental Analysis and Measurement of Situation Awareness*, D. J. Garland and M. R. Endsley, Eds., Embry-Riddle Aeronautical University Press, Daytona Beach, FL, pp. 115–122.

Durso, F. T., and Gronlund, S. D. (1999), "Situation Awareness," in *Handbook of Applied Cognition*, F. T. Durso, R. Nickerson, R. Schvaneveldt, S. Dumais, S. Lindsay, and M. Chi, Eds., Wiley, New York, pp. 284–314.

Endsley, M. R. (1988), "Design and Evaluation for Situation Awareness Enhancement," in *Proceedings of the Human Factors Society 32nd Annual Meeting*, Human Factors Society, Santa Monica, CA, pp. 97–101.

Endsley, M. R. (1989), "Final Report: Situation Awareness in an Advanced Strategic Mission," NOR DOC 89-32, Northrop Corporation, Hawthorne, CA.

Endsley, M. R. (1990a), "Predictive Utility of an Objective Measure of Situation Awareness," in *Proceedings of the Human Factors Society 34th Annual Meeting*, Human Factors Society, Santa Monica, CA, pp. 41–45.

Endsley, M. R. (1990b), "Situation Awareness in Dynamic Human Decision Making: Theory and Measurement," Ph.D. dissertation, University of Southern California, Los Angeles.

Endsley, M. R. (1993), "A Survey of Situation Awareness Requirements in Air-to-Air Combat Fighters," *International Journal of Aviation Psychology*, Vol. 3, No. 2, pp. 157–168.

Endsley, M. R. (1995a), "Direct Measurement of Situation Awareness in Simulations of Dynamic Systems: Validity and Use of SAGAT," in *Experimental Analysis and Measurement of Situation Awareness*, D. J. Garland and M. R. Endsley, Eds., Embry-Riddle Aeronautical University Press, Daytona Beach, FL, pp. 107–113.

Endsley, M. R. (1995b), "Measurement of Situation Awareness in Dynamic Systems," *Human Factors*, Vol. 37, No. 1, pp. 65–84.

Endsley, M. R. (1995c), "A Taxonomy of Situation Awareness Errors," in *Human Factors in Aviation Operations*, R. Fuller, N. Johnston, and N. McDonald, Eds., Avebury Aviation, Ashgate Publishing, Aldershot, Hampshire, England, pp. 287–292.

Endsley, M. R. (1995d), "Toward a Theory of Situation Awareness in Dynamic Systems," *Human Factors*, Vol. 37, No. 1, pp. 32–64.

Endsley, M. R. (1996), "Situation Awareness Measurement in Test and Evaluation," in *Handbook of Human Factors Testing and Evaluation*, T. G. O'Brien and S. G. Charlton, Eds., Lawrence Erlbaum Associates, Mahwah, NJ, pp. 159–180.

Endsley, M. R. (1997), "Communication and Situation Awareness in the Aviation System," *A presentation to the Aviation Communication: A Multi-cultural Forum*, Embry-Riddle University, Prescott, AZ.

Endsley, M. R. (2000a), "Direct Measurement of Situation Awareness: Validity and Use of SAGAT," in *Situation Awareness Analysis and Measurement*, M. R. Endsley and D. J. Garland, Eds., Lawrence Erlbaum Associates, Mahwah, NJ, pp. 147–174.

Endsley, M. R. (2000b), "Theoretical Underpinnings of Situation Awareness: A Critical Review," in *Situation Awareness Analysis and Measurement*, M. R. Endsley and D. J. Garland, Eds., Lawrence Erlbaum Associates, Mahwah, NJ, pp. 3–32.

Endsley, M. R., and Bolstad, C. A. (1994), "Individual Differences in Pilot Situation Awareness," *International Journal of Aviation Psychology*, Vol. 4, No. 3, pp. 241–264.

Endsley, M. R., and Garland, D. G. (2000a), "Pilot Situation Awareness Training in General Aviation," in *Proceedings of the 14th Triennial Congress of the International Ergonomics Association and the 44th Annual Meeting of the Human Factors and Ergonomics Society*, Human Factors and Ergonomics Society, Santa Monica, CA, pp. 357–360.

Endsley, M. R., and Garland, D. J., Eds. (2000b), *Situation Awareness Analysis and Measurement*, Lawrence Erlbaum Associates, Mahwah, NJ.

Endsley, M. R., and Jones, D. G. (1995), "Situation Awareness Requirements Analysis for TRACON Air Traffic Control," TTU-IE-95-01, Texas Tech University, Lubbock, TX.

Endsley, M. R., and Kiris, E. O. (1994), "Situation Awareness in FAA Airway Facilities Maintenance Control Centers (MCC): Final Report," Texas Tech University, Lubbock, TX.

Endsley, M. R., and Robertson, M. M. (1996), *Team Situation Awareness in Aircraft Maintenance*, Texas Tech University, Lubbock, TX.

Endsley, M. R., and Robertson, M. M. (2000a), "Situation Awareness in Aircraft Maintenance Teams," *International Journal of Industrial Ergonomics*, Vol. 26, pp. 301–325.

Endsley, M. R., and Robertson, M. M. (2000b), "Training for Situation Awareness in Individuals and Teams," in *Situation Awareness Analysis and Measurement*, M. R. Endsley and D. J. Garland, Eds., Lawrence Erlbaum Associates, Mahwah, NJ.

Endsley, M. R., and Rodgers, M. D. (1994a), "Situation Awareness Information Requirements for En Route Air Traffic Control," DOT/FAA/AM-94/27, Federal Aviation Administration Office of Aviation Medicine, Washington, DC.

Endsley, M. R., and Rodgers, M. D. (1994b), "Situation Awareness Information Requirements for En Route Air Traffic Control," in *Proceedings of the Human Factors and Ergonomics Society 38th Annual Meeting*, Human Factors and Ergonomics Society, Santa Monica, CA, pp. 71–75.

Endsley, M. R., and Rodgers, M. D. (1998), "Distribution of Attention, Situation Awareness, and Workload in a Passive Air Traffic Control Task: Implications for Operational Errors and Automation," *Air Traffic Control Quarterly*, Vol. 6, No. 1, pp. 21–44.

Endsley, M. R., and Smolensky, M. (1998), "Situation Awareness in Air Traffic Control: The Picture," in *Human Factors in Air Traffic Control*, M. Smolensky and E. Stein, Eds., Academic Press, New York, pp. 115–154.

Endsley, M. R., Mogford, R., Allendoerfer, K., Snyder, M. D., and Stein, E. S. (1997a), Effect of Free Flight Conditions on Controller Performance, Workload and Situation Awareness: A Preliminary Investigation of Changes in Locus of Control Using Existing Technology," DOT/FAA/CT-TN 97/12, Federal Aviation Administration William J. Hughes Technical Center, Atlantic City, NJ.

Endsley, M. R., Mogford, R. H., and Stein, E. S. (1997b), "Controller Situation Awareness in Free Flight," in *Proceedings of the Human Factors and Ergonomics Society 41st Annual Meeting*, Human Factors and Ergonomics Society, Santa Monica, CA, pp. 4–8.

Endsley, M. R., Selcon, S. J., Hardiman, T. D., and Croft, D. G. (1998a), "A Comparative Evaluation of SAGAT and SART for Evaluations of Situation Awareness," in *Proceedings of the Human Factors and Ergonomics Society Annual Meeting*, Human Factors and Ergonomics Society, Santa Monica, CA, pp. 82–86.

Endsley, M. R., Farley, T. C., Jones, W. M., Midkiff, A. H., and Hansman, R. J. (1998b), "Situation Awareness Information Requirements for Commercial Airline Pilots," ICAT-98-1, MIT International Center for Air Transportation, Cambridge, MA.

Endsley, M. R., Garland, D. J., Shook, R. W. C., Coello, J., and Bandiero, M. (2000), "Situation Awareness Problems in General Aviation," SATech 00–01, SA Technologies, Marietta, GA.

Endsley, M. R., Bolstad, C. A., Garland, D., Howell, C., Shook, R. W. C., Costello, A., et al. (2002), "Situation Awareness Training for General Aviation Pilots: Final Report," SATECH 02–04, SA Technologies, Marietta, GA.

Endsley, M. R., Bolte, B., and Jones, D. G. (2003), *Designing for Situation Awareness: An Approach to Human-Centered Design*, Taylor & Francis, London.

Farley, T. C., Hansman, R. J., Amonlirdviman, K., and Endsley, M. R. (2000), "Shared Information Between Pilots and Controllers in Tactical Air Traffic Control," *Journal of Guidance, Control and Dynamics*, Vol. 23, No. 5, pp. 826–836.

Fracker, M. L. (1988), "A Theory of Situation Assessment: Implications for Measuring Situation Awareness," in *Proceedings of the Human Factors Society 32nd Annual Meeting*, Human Factors Society, Santa Monica, CA, pp. 102–106.

Gugerty, L. J. (1997), "Situation Awareness During Driving: Explicit and Implicit Knowledge in Dynamic Spatial Memory," *Journal of Experimental Psychology: Applied*, Vol. 3, pp. 42–66.

Gugerty, L., and Tirre, W. (1997), "Situation Awareness: A Validation Study and Investigation of Individual Differences," in *Proceedings of the Human Factors and Ergonomics Society 40th Annual Meeting*, Human Factors and Ergonomics Society, Santa Monica, CA, pp. 564–568.

Hartel, C. E., Smith, K., and Prince, C. (1991), "Defining Aircrew Coordination: Searching Mishaps for Meaning," presented at the 6th International Symposium on Aviation Psychology, Columbus, OH.

Hogg, D. N., Torralba, B., and Volden, F. S. (1993), "A Situation Awareness Methodology for the Evaluation of Process Control Systems: Studies of Feasibility and the Implication of Use," 1993-03-05, OECD Halden Reactor Project, Storefjell, Norway.

Hormann, H. J., Blokzijl, C., and Polo, L. (2004), "ESSAI: A European Training Solution for Enhancing Situation Awareness and Threat Management on Modern Aircraft Flight Decks, in *Proceedings of the 16th Annual European Aviation Safety Seminar of the Flight Safety Foundation and European Regions Airline Association*, Flight Safety Foundation, Barcelona, Spain.

Horswill, M. S., and McKenna, F. P. (2004), "Drivers Hazard Perception Ability: Situation Awareness on the Road," in *A Cognitive Approach to Situation Awareness: Theory, Measurement and Application*, S. Banbury and S. Tremblay, Eds., Ashgate Publishing, Aldershot, Hampshire, England, pp. 155–175.

Matthews, M. D., Strater, L. D., and Endsley, M. R. (2004), "Situation Awareness Requirements for Infantry Platoon Leaders," *Military Psychology*, Vol. 16, No. 3, pp. 149–161.

McKenna, F., and Crick, J. L. (1994), *Developments in Hazard Perception*, Department of Transport, London.

Mumaw, R. J., Roth, E. M., and Schoenfeld, I. (1993), "Analysis of Complexity in Nuclear Power Severe Accidents Management," in *Proceedings of the Human Factors and Ergonomics Society 37th Annual Meeting*, Human Factors and Ergonomics Society, Santa Monica, CA, pp. 377–381.

O'Hare, D. (1997), "Cognitive Ability Determinants of Elite Pilot Performance," *Human Factors*, Vol. 39, No. 4, pp. 540–552.

Pew, R. W. (1995), "The State of Situation Awareness Measurement: Circa 1995," in *Experimental Analysis and Measurement of Situation Awareness*, M. R. Endsley and D. J. Garland, Eds., Embry-Riddle Aeronautical University Press, Daytona Beach, FL, pp. 7–16.

Prince, C. (1998), *Guidelines for Situation Awareness Training*, Naval Air Warfare Center Training Systems Division, Orlando, FL.

Prince, C., and Salas, E. (1998), "Situation Assessment for Routine Flight and Decision Making," *International Journal of Cognitive Ergonomics*, Vol. 1, No. 4, pp. 315–324.

Robinson, D. (2000), "The Development of Flight Crew Situation Awareness in Commercial Transport Aircraft," in *Proceedings of the Human Performance, Situation Awareness and Automation: User-Centered Design for a New Millennium Conference*, SA Technologies, Marietta, GA, pp. 88–93.

Rodgers, M. D., Mogford, R. H., and Strauch, B. (2000), "Post-hoc Assessment of Situation Awareness in Air Traffic Control Incidents and Major Aircraft Accidents," in *Situation Awareness Analysis and Measurement*, M. R. Endsley and D. J. Garland, Eds., Lawrence Erlbaum Associates, Mahwah, NJ.

Rouse, W. B., and Morris, N. M. (1985), "On Looking into the Black Box: Prospects and Limits in the Search for Mental Models," DTIC AD-A159080, Center for Man–Machine Systems Research, Georgia Institute of Technology, Atlanta, GA.

RTCA (1995), *Report of the RTCA Board of Directors Select Committee on Free Flight*, RTCA, Washington, DC.

Secrist, G. E., and Hartman, B. O. (1993), "Situational Awareness: The Trainability of Near-Threshold Information Acquisition Dimension," *Aviation, Space and Environmental Medicine*, Vol. 64, pp. 885–892.

Sexton, G. A. (1988), "Cockpit-Crew Systems Design and Integration," in *Human Factors in Aviation*, E. L. Wiener and D. C. Nagel, Eds., Academic Press, San Diego, CA, pp. 495–526.

Smith, K., and Hancock, P. A. (1995), "Situation Awareness Is Adaptive, Externally Directed Consciousness," *Human Factors*, Vol. 37, No. 1, pp. 137–148.

Strater, L. D., Jones, D., and Endsley, M. R. (2003), "Improving SA: Training Challenges for Infantry Platoon Leaders," in *Proceedings of the 47th Annual Meeting of the Human Factors and Ergonomics Society*, Human Factors and Ergonomics Society, Santa Monica, CA, pp. 2045–2049.

Strater, L. G., Reynolds, J. P., Faulkner, L. A., Birch, K., Hyatt, J., Swetnam, S., et al. (2004), "PC-Based Tools to Improve Infantry Situation Awareness," in *Proceedings of the Human Factors and Ergonomics Society Annual Meeting*, Human Factors and Ergonomics Society, Santa Monica, CA, pp. 668–672.

Taylor, R. M. (1990), "Situational Awareness Rating Technique (SART): The Development of a Tool for Aircrew Systems Design," in *Situational Awareness in Aerospace Operations*, AGARD-CP-478, NATO-AGARD, Neuilly sur Seine, France, pp. 3/1–3/17.

Taylor, R. M., and Selcon, S. J. (1994), "Situation in Mind: Theory, Application and Measurement of Situational Awareness," in *Situational Awareness in Complex Settings*, R. D. Gilson, D. J. Garland, and J. M. Koonce, Eds., Embry-Riddle Aeronautical University Press, Daytona Beach, FL, pp. 69–78.

Tenney, Y. T., Adams, M. J., Pew, R. W., Huggins, A. W. F., and Rogers, W. H. (1992), "A Principled Approach to the Measurement of Situation Awareness in Commercial Aviation," NASA Contractor Report 4451, NASA Langely Research Center, Langely, VA.

Wickens, C. D. (1992), *Engineering Psychology and Human Performance*, 2nd ed., HarperCollins, New York.

CHAPTER 21

AFFECTIVE AND PLEASURABLE DESIGN

Martin G. Helander
Nanyang Technological University
Singapore

Halimahtun M. Khalid
Damai Sciences
Kuala Lumpur, Malaysia

1 INTRODUCTION

During the last 10 years there has been a rapid growth in research concerning affect and pleasure. Considering the lack of interest from the psychological community during much of the twentieth century, this comes as a surprise; behaviorism and cognitivism dealt with other issues. One exception in the early part of the century was Titchener (1910), who considered pleasure an irreducible fundamental component of human emotion.

Advances in psychological research were elegantly summarized by Kahneman et al. (1999) in their edited volume: *Well-Being: Foundations of Hedonic Psychology*. In human factors and industrial design there are publications by Helander et al. (2001), Nagamachi (2001), Green and Jordan (2002), and Norman (2004). In human–computer interaction (HCI) there is the classic book *Affective Computing* by Picard (1997) and a recent review by Brave and Nass (2003). New trends include *funology* in HCI design (Carroll, 2004) and *hedonomics* in human factors (Helander and Tham, 2003; Khalid, 2004; Hancock et al., 2005). As Nielsen (1996) observed, one important challenge in theory as well as application is the design of seductive and fun interfaces. Research in this area is just beginning.

Emotions have since becoming increasingly important in product semantics. The question of which emotions are invoked while using artifacts naturally follows the question of what artifacts could mean to the users (Krippendorff, 2005). In emotional design, pleasure and usability should go hand in hand, as well as aesthetics, attractiveness, and beauty (Norman, 2004). The interplay between user-perceived usability (i.e., pragmatic attributes), hedonic attributes (e.g., stimulation, identification), goodness (i.e., satisfaction), and beauty was considered in the design of MP3-player skins (Hassenzahl, 2004). He found that goodness depended on both perceived usability and hedonic attributes. The findings are not surprising; the use and user experience of a product are important in product evaluation (Khalid and Helander, 2004).

Jordan (2002) noted that a product or service offering should engage the people for whom it is designed at three abstraction levels: First, it has to be able to perform the task for which it was designed. For example, a car has to be able to take the user from point A to B. The product's functionality should work well and it should be easy to use (i.e., usability function). The second level relates to the emotions associated with the product or service, in the context of the associated tasks. These emotions are part of the "user experience." For example, when using an automated teller machine, feelings of trust and security might be appropriate. Driving a sports car should be exciting, but there should also be a sense of safety. The third level reflects the aspirational qualities associated with the product or service (i.e., persona or social factors). What does owning the product or using the service say about the user? For example, owning the latest, smallest mobile phone may suggest a pretty cool person. Meeting these requirements makes a case not only for the ergonomics of the product or service, but for emotional design and achievement of social status as well.

Emotion affects how we feel, how we behave and think; and it has gained significant attention in interaction design. For example, the iPod is the runaway best seller of MP3 players, although it was marketed late and is more expensive than competing models. To consumers, the iPod is easy to use and aesthetically appealing—it is *cool*, it *feels* good. *Affect* is said to be the customer's psychological response to the design details of the product, while *pleasure* is the emotion that accompanies the acquisition or possession of something good or desirable (Demirbilek and Sener, 2003).

Beyond pleasure, a new debate focused on "fun" is emerging in the human–computer interaction literature. Things are fun when they attract, capture, and hold our attention by provoking new or unusual perceptions and arousing emotions. They are fun when they surprise us, and present challenges or puzzles as we try to make sense of them.

A review of the psychological literature on emotions by Fredrickson (1998) showed that positive emotions such as joy, interest, contentment, and love that share a pleasant subjective feeling have inadvertently been marginalized in research compared to negative emotions. Two reasons for this are (1) that positive emotions are few in number and rather diffuse; and (2) that negative emotions pose problems that demand attention. For example, anger and its management have been implicated in the etiology of heart disease, and so on. Positive emotions should therefore, be tapped to promote individual and collective well-being and health (Fredrickson, 1998).

Affective appreciation is, of course, not new—just the research. People have affective reactions toward tasks, artifacts, and interfaces. These are caused by design features that operate either through the perceptual system (looking at) or from a sense of controlling (touching and activating) or from reflection and experience. These reactions are difficult if not impossible to control; the limbic system in the brain is in operation whether we want it or not. The reactions are in operation whenever we look at beautiful objects, and they are particularly obvious when we try "emotional matching," such as buying clothes or selecting a birthday card for someone else.

Affective evaluations provide a new and different perspective in human factors engineering. It is not how to evaluate users—it is how the user evaluates. The research on hedonic values and seductive interfaces is, in fact, a welcome contrast to safety and productivity, which have dominated human factors and ergonomics. Consequently, emotions and affect have received increasing attention in recent years (Velásquez, 1998a). Approaches to emotions and affect have been studied at many different levels, and several models have been proposed for a variety of domains and environments.

This raises many research issues: (1) how we can measure and analyze human reactions to affective and pleasurable design, and (2) how we can assess the corresponding affective design features of products. In the end, we need to develop theories and predictive models for affective- and pleasure-based design.

The purpose of this chapter is to summarize various perspectives that have evolved in psychology, human factors, and neuroscience. We provide an overview of the basic neurological functions; define terms such as affect, emotions (the terms *affect* and *emotion* are used interchangeably), and sentiments; review couplings between the cognitive and affective systems in processing information and evaluating decision alternatives; summarize theories dealing with affect and design; and provide an overview of some of the most common measurement methodologies to measure affect and pleasure in design. The main focus is on design: design activates and design evaluation from the user's perspective as well as the designer's.

1.1 Neurological Basis of Emotions

The neurological mechanisms are illustrated in Figure 1. In the brain there are three main areas: the thalamus, the limbic system, and the cortex. The thalamus receives sensory input from the environment, which is then sent to the cortex for fine analysis. It is also sent to the limbic system, the main location for emotions, where the relevance of the information is determined (LeDoux, 1995). The limbic system coordinates the physiological response and directs the attention (in cortex) and various cognitive functions. Primitive emotions (e.g., the startle effect) are handled directly through the thalamus–limbic pathway. In this case the physiological responses are mobilized, such as for fight and flight. Reflective emotions, such as pondering over a beautiful painting, are handled by the cortex. In this case there are not necessarily any physiological responses—they are not required to deal with the situation. According to Kubovy

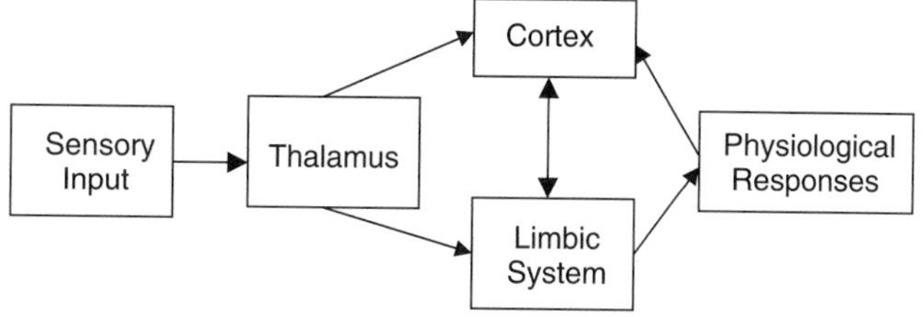

Figure 1 Neurological mechanisms in affect and emotions. (From LeDoux 1995.)

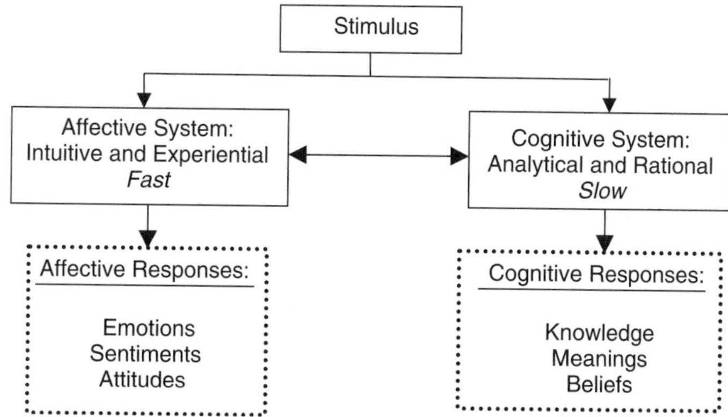

Figure 2 Cross-coupling of affect and cognition.

(1999), pleasures of the mind do not give rise to a physiological response nor to facial expressions.

1.2 Cross-Coupling of Affective and Cognitive Systems

The correlation between cognition and affect is an old philosophical problem, but it has not been dealt with to any great extent in cognitive psychology. In our view, cognition must consider affect or emotion, and human behavior is guided by cognition as well as emotions.

Figure 2 denotes the relationship between affect and cognition. Whereas affect refers to feeling responses, cognition is used to interpret, make sense of, and understand user experience. To do so, symbolic, subjective concepts are created that represent the personal interpretations of the stimuli. Cognitive interpretations may include a deeper, symbolic understanding of products and behaviors.

One of the most important accounts of affect in decision making comes from Damasio (1994). In his book *Descartes' Error*, he described observations of patients with damage to the ventromedial frontal cortex of the brain. This left their intelligence and memory intact but impaired their emotional assessments. The patients were socially incompetent, although their intellect and ability to analyze and reason about solutions worked well.

Damasio argued that thought is largely made up from a mix of images, sounds, smells, words, and visual impressions. During a lifetime of learning, these become "marked" with affective information: positive or negative feelings. These *somatic markers* are helpful in predicting decision making and behavior.

Damasio tested the somatic marker theory in a game of cards where normal subjects and patients drew a card from one of four piles. Each card resulted in a gain or loss of a sum of money, as revealed on the back of the cards. Normal subjects learned to avoid cards with attractive large payoffs but occasional disastrous losses, but the patients did not learn to anticipate future events, and they lost much money on this game.

Clearly, the patients were unable to make decisions effectively: They could not determine where to live, what to buy, and what to eat. Emotions are necessary to enable selection among alternatives, particularly when there is no rational basis.

One of the authors asked a person who suffers from similar problems: "Would you like to go for a walk with me?" The answer was "Yes." "Where would you like to go? We can go to the city park, or to the old town, or to the botanical garden." The answer was "I don't know."

Affect plays a central role in dual-process theories of thinking, knowing, and information processing (Epstein, 1994). There is much evidence that people perceive reality in at least two ways: affective (intuitive and experiential) and cognitive (analytical and rational). Formal decision making relies on the analytical and cognitive abilities; unfortunately, this mode is slow. The experiential and affective system is much quicker. When a person seeks to respond to an emotional event, he or she will search the experiential system automatically. This is like searching a memory bank for related events, including their emotional contents (Epstein, 1994) (see Figure 2).

Emotions do not cause thinking to be nonrational; they can motivate a passionate concern for objectivity, such as anger at injustice. Rational thinking entails feelings, and affective thinking entails cognition. Rational thinking is more precise, comprehensive, and insightful than is nonrational thinking. However, it is just as emotional.

Separating emotion from cognition is a major weakness of psychology and cognitive science (Vygotsky, 1962). New breakthroughs in neuroscience using functional magnetic resonance imaging (fMRI) validated the assertions that cognition and emotions are unified and contribute to the control of thought and behavior conjointly and equally (LeDoux, 1995). Additionally, cognition contributes to the regulation of emotion. Contemporary views in artificial intelligence are also embracing an integrated view of emotion and cognition. In *Emotion Machine*, Minsky (2004) claimed: "Our traditional idea is that there is something called 'thinking' and that it is contaminated, modulated or affected by emotions. What I am saying is that emotions aren't separate."

Combining the description from contemporary psychology and neuroscience, Camerer et al. (2003) illustrated the two distinctions between controlled and automatic processes (Schneider and Shiffrin, 1977), and between cognition and affect as in Table 1. As described in Table 1, controlled processes have several characteristics. They tend (1) to be serial (employing a step-by-step logic or computations), (2) to be invoked deliberately by the agent when encountering a challenge or surprise, (3) to be associated with a subjective feeling of effort, and (4) typically, to occur consciously. As such, people often have reasonably good introspective access to controlled processes. If people are asked how they solved a math problem or choose a new car, they can usually provide a good account of the decision-making process.

Table 1 Two-Dimensional Characterization of Neural Functioning

Type of Process	Cognitive	Affective
Controlled processes Serial Evoked deliberately Effortful Occurs consciously	I	II
Automatic processes Parallel Effortless Reflexive No introspective access	III	IV

Source: Camerer et al. (2003).

Automatic processes are the opposite of controlled processes. Automatic processes (1) tend to operate in parallel, (2) are not associated with any subjective feeling of effort, and (3) operate outside conscious awareness. As a result, people often have little introspective access as to why the automatic choices or judgments were made. For example, a face is perceived as "attractive" or a verbal remark as "sarcastic" automatically and effortlessly. It is only in retrospect that the controlled system may reflect on the judgments and try to substantiate it logically.

The second distinction, represented by the two columns of Table 1, is between cognitive and affective processes. This distinction is pervasive in contemporary psychology (e.g., Zajonc, 1998), and neuroscience (Damasio, 1994; LeDoux, 1995). Zajonc (1998) defined *cognitive processes* as those that answer true–false questions and *affective processes* as those that motivate approach–avoidance behavior. Affective processes include emotions such as anger, sadness, and shame, as well as "biological affects" such as hunger, pain, and the sex drive (Buck, 1999).

Elaborating this further, quadrant I, for example, is in charge when one considers entering a business deal. Quadrant II can be used by "method actors," such as stand-up comedians, who replay previous emotional experiences to fool an audience into thinking that they are experiencing these emotions. Quadrant III deals with motor control and governs the movements of the limbs, such as a tennis player when he returns a serve. Quadrant IV applies when a person jumps because someone says "boo." The four categories are often not so easy to distinguish; most behavior results from a combination of several quadrants.

1.3 Positive Effect of Positive Emotions

Research has shown that even moderate fluctuations in positive feelings (emotions) can systematically affect cognitive processing. Isen (1999) found that mild positive affect improves creative problem solving, facilitates recall of neutral and positive material, and systematically changes strategies used in decision-making tasks. Despite these well-documented effects, there are few theories of how positive affect influences cognition.

For a complete theory of positive affect, it is necessary to understand why certain things make people happy. For example, the dopaminergic theory of positive affect postulated by Ashby et al. (1999) assumes that during periods of mild positive affect, there is a concomitant increased dopamine release in the mesocorticlimbic system. The theory assumes further that the resulting elevated dopamine levels influence performance on a variety of cognitive functions and tasks (e.g., olfactory, episodic memory, working memory, creative problem solving).

Zajonc (1980) found that when a stimulus is presented to subjects repeatedly, the exposure leads to positive affect, and the more frequent the exposure of a stimulus, the greater the affect. Winkelman et al. (1977) primed participants in an experiment with a very brief 1/250-second exposure to affective

Table 2 Types of Affect

Type of Affective Response	Examples of Positive and Negative Affect	Level of Physiological Arousal	Intensity or Strength of Feeling
Emotions	Joy, love Fear, guilt, anger	Higher arousal and activation	Stronger
Specific feelings	Warmth, appreciation Disgust, sadness	↕	↕
Moods	Alert, relaxed, calm Blue, listless, bored		
Evaluations	Like, good, favorable Dislike, bad, unfavorable	Weaker	Lower arousal and activation

Source: Peter and Olson (1996).

stimuli. The time was so short that there could be no recognition or recall of the stimuli. Following this they were exposed for 2 seconds to an ideograph. The mean liking for the ideograph was greater when it was preceded by a smiling face.

1.4 Understanding Affect and Pleasure in Different Disciplines

Several definitions and classifications of affect and pleasure exist in the literature, stemming from different traditions: marketing, product design, and psychology. We mention a few that have relevance to human factors design.

1.4.1 Marketing

Peter and Olson (1996), with a background in marketing, defined four different types of affective responses: emotions, feelings, moods, and evaluations, and offered a classification (Table 2). These responses are associated with different levels of physiological arousal as well as different intensities of feeling. There are both positive and negative responses. Some examples are given below.

Types of Affect	Positive Response	Negative Response
Emotion	Love	Fear
Feeling	Warmth	Disgust
Mood	Alert	Bored
Evaluation	Like	Dislike

Emotions are said to be associated with the physiological arousal, while evaluations (e.g., reflections) of products typically encompass weak affective responses, which are accompanied by a low level of arousal.

1.4.2 Product Design

Tiger (1992) identified four conceptually distinct types of pleasure from a product, which were further elaborated by Jordan (see Blythe, 2004). We extended the taxonomy to five. Whether they are used as a source for pleasure depends on the person's needs.

1. *Physical pleasure* has to do with the body and the senses. It includes such things as feeling good physically (e.g., eating, drinking), pleasure from relief (e.g., sneezing, sex), as well as sensual pleasures (e.g., touching a pleasant surface).
2. *Sociopleasures* include social interaction with family, friends and co-workers. This includes the way we are perceived by others, our persona, and status.
3. *Psychological pleasure* has to do with pleasures of the mind, reflective as well as emotional. It may come from doing things that interest and engage us (e.g., playing in an orchestra or listening to a concert), including being creative (e.g., painting) or enjoying the creativity of other people.
4. *Reflective pleasure* has to do with reflection on our knowledge and experiences. The value of many products comes from this and includes aesthetics and quality.
5. *Normative pleasure* has to do with societal values such as moral judgment, caring for the environment, and religious beliefs. These can make us feel better about ourselves when we act in line with the expectation of others as well as our beliefs.

Jordan (1998) defined pleasure with products as the emotional and hedonic *benefits* associated with product use. Coelho and Dahlman (2000) defined displeasure as the emotional and hedonic *penalties* associated with product use. This argument makes an interesting point. Could it be that to understand pleasure, we also need to understand displeasure? However, we think it is not necessary since they may not be related. Take, for example, chair comfort, which has to do with feeling relaxed, whereas chair discomfort has to do with poor biomechanics. The two entities should be measured on different scales. Helander and Zhang (1997) observed that in understanding comfort there is little we can learn from discomfort; they are two different dimensions. Like discomfort, displeasure operates like a design constraint—we know what to avoid—but that does not mean that we understand how to design a pleasurable product. Take the example of chair design. Chair comfort has to do with feeling relaxed, while chair discomfort has to do with poor biomechanics. Fixing the poor biomechanics and getting rid of displeasure does not automatically generate a sense of relaxation and pleasure. The two entities should be measured on different scales.

With increasing experience the repertoire for emotions becomes larger. In fact, many researchers think that only the startle reflex is innate, whereas most emotions and definitely sentiments are learned over time. Many persons are particularly attracted to the complexity in music and in art. One can listen to a piece of music many times; each time one discovers something new. Similarly with a painting; many modern paintings are difficult to comprehend—each time you look, the interpretation changes. Some pleasures are difficult to appreciate, hence induce an interesting challenge for many people.

1.4.3 Psychology

The term *affect* has several meanings in psychology and human factors (Leontjev, 1978; Norman, 2004). *Affect* is the general term for the judgmental system; *emotion* is the conscious experience of affect. Much of human behavior is subconscious, beneath conscious awareness. Consciousness came late in human evolution and also in the way the brain processes information (Norman, 2004). The affective system makes quick and efficient judgments which help in determining if an environment is dangerous—shall I fight or flight? For instance, I may have an uneasy feeling (affect) about a colleague at work, but I don't understand why, since I am not conscious about what I am reacting to. However, I am certainly aware of my strong emotions about finishing this chapter within the deadline given.

Pleasure, on the other hand, is a good feeling coming from satisfaction of homeostatic needs such as hunger, sex, and bodily comfort (Seligman and Csikszentmihalyi, 2000). This is differentiated from *enjoyment*, which is a good feeling coming from breaking through the limits of homeostasis of people's experiences: for example, performing in an athletic event or playing in a string quartet. Enjoyment could lead to more personal growth and long-term happiness than pleasure, but people usually prefer pleasure over enjoyment, perhaps because it is less effortful.

Although each discipline has a unique definition, their goals are quite similar. We elaborate further later when we discuss relevant theories of affect and pleasure.

2 FRAMEWORK FOR EVALUATING AFFECTIVE DESIGN

Design is a problem-solving discipline. It considers not only the appearance of the designed product, but also the underlying structure of the solution and its anticipated reception by users. Besides specifying characteristics of the solution, a design theory helps designers in identifying the problem and in developing their instincts in choosing the "right" solutions (Cross, 2000).

Affect, as discussed above, is the basis of beliefs, human values, and human judgment. For this reason it might be argued that models of design process that do not include affect are essentially weakened. Until recently, the affective aspects of designing and design cognition have been substantially absent from formal theories of design process. Affective design, then, is the inclusion or representation of affect (emotions, feelings, etc.) in design processes.

2.1 Affective User–Designer Model

The systems model in Figure 3 provides a framework for issues that must be addressed in affective design. There are two parts of the model: the designer's environment and the affective user. The purpose of the model is to illustrate how a designer may achieve affective design and how the user of the design will perceive and react to the design.

In the *designer's environment* there are three main subsystems: artifact, context of use, and society trends. In the *artifact* subsystem we consider that the designed object can incorporate several characteristics, each of which leads to emotional responses: visceral design, behavioral design, and reflective design (Norman, 2004). The designer needs to consider and if possible, predict the user's needs and reaction to all three aspects.

Visceral design (also called *reactive* design), appeals to the perceptual senses. It deals with appearance. Although there are no firm guidelines for visceral design, much is known from arts and graphics about what constitutes good design: the golden ratio, symmetry, appropriate use of colors, and visual balance (the use of white space). A beautiful face, a sunset, and rolling hills are examples of this. Everybody seems to agree on this aspect of affective design (Norman, 2004).

Behavioral design focuses on what a person can do with an object. If the object affords manipulation, we can develop good design rules. This is where most of the activity in HCI and the usability community is directed. Behavioral design also incorporates Csikszentmihalyi's (1975, 1990) concept of *flow*. An example is when a person manipulates the controls in a computer game. While turning a knob or touching a control, the user feels fully in control and the device always responds as expected. Artists rarely talk about this because their focus is on visual appearance. To them the visual appearance of an interface is more important than the smoothness of operating controls.

Reflective design considers a designer's (or user's) thoughts and evaluations of the current design. This is intellectually driven and is influenced greatly by the knowledge and experience of the designer (user), including the person's culture and idiosyncrasies. For judgment of taste and fashion, people of different cultures think differently; it all depends on upbringing, traditions, needs, and expectations.

Some of the best reflective designs are loved by some and hated by others. Such contrasts may be desirable, since controversial designs have often proven to be very successful. This is where the skills and intuition of designers play a large part. One example is Volkswagen Bora; user evaluations

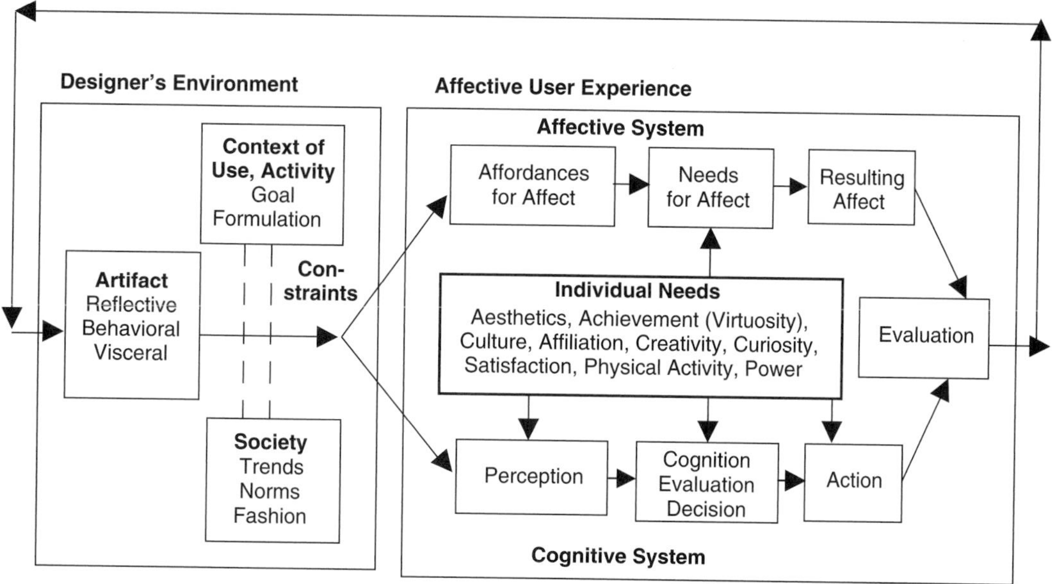

Figure 3 Framework for evaluation of affective design.

include great looks, stupid looks, looks like a baby, not masculine enough, no prestige, environmentally friendly, beautiful symmetry.

Constraints and Filters The next issue in the designer's environment deals with design constraints and filters. These are marked in Figure 3 with dashed lines connecting the "context of use" and "society." In the evaluation of an artifact, a designer will consider the context of use and the context of activity of the artifact. For example, will the product be used at work or at home? At work, the aesthetics of an office chair may be less important than pleasant interactions with colleagues. At home, however, a chair with an inspired design may be a means to express one's personality through aesthetic preferences. In addition, the designer must consider society trends, norms, and fashions. These aspects operate like constraints and modify the design of the artifact.

The context of use can be understood through a task analysis. This is a tool often applied in human factors. For example, the design of the new Duet washer and dryer manufactured by Whirlpool (see http://duet.whirlpool.com/) applied cognitive task analysis to analyze the context of use and needs of housewives. Although the product is twice as expensive as other washers/dryers, it sells extremely well. Washing clothes has become entertaining.

The context of use is not always easy to consider in design (Mäkelä and Fulton-Suri, 2001). This is because people's experiences result from motivated action in a context; as such, the designer can neither know nor control the user's experience. Similarly, it is not always possible to predict needs, motivation, context, and action which are relevant for the creation of user experience, leading to design features of an artifact.

Moreover, people have different motivations and needs for using a product. Take, for example, the mobile phone: to keep in touch with loved ones, be efficient at work, and avoid boredom. There are value-added activities of mobile phones, such as games, short messaging system (SMS), alarm clock, and Internet connections. The phone is also used in many contexts: while commuting to work, at home, in recreation, and so on. Use of new digital products is like a situated cognition—the context of use will determine the user experience. It is also impossible to predict the use of the artifact in the individual case (Mäkelä and Fulton-Suri, 2001). The designer's goal then should be to design products that support user creativity in using the product. Equip the mobile phone with many features. It is up to the user to test them out and decide what is important.

Returning to the affective user in Figure 3, we note that perception, cognition, and action are influenced by differences in needs and other idiosyncratic characteristics (knowledge, education, gender, etc.). An experienced person will see things differently from an inexperienced person, resulting in different decisions. In Figure 3 the assumption is that there are two simultaneously operating systems for evaluating design: an affective system and a cognitive system. These are both influenced by individual needs.

Below we examine the consumer process and its relationship to customer satisfaction and pleasure. Studies in industrial design relating to affective user needs are discussed. The aim is to highlight considerations in the design of products that address both customer needs and affect.

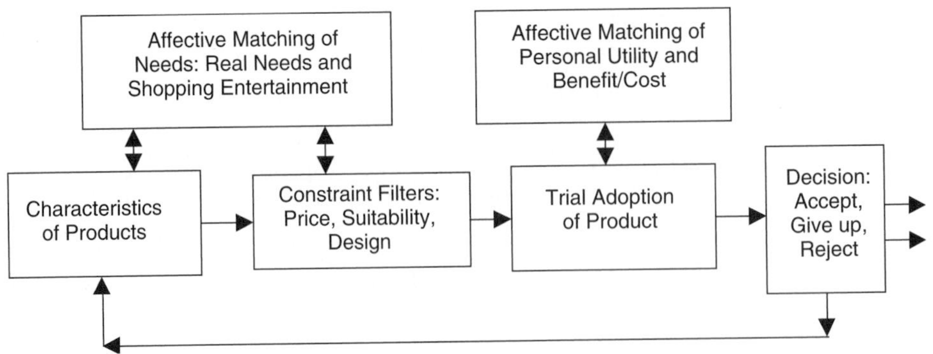

Figure 4 Consumer process.

2.2 Consumer Process

The process of buying a product is influenced by two affective processes: (1) affective matching of needs, and (2) affective matching of personal utility (see Figure 4). In the first instance a consumer matches the features of several alternative products to his or her perceived needs. At the same time, the customer has constraints that eliminate many products, due to price, suitability, and aesthetics design. Assume that you are buying a shirt for a friend. You will consider the price, size, style, and color. You will try to imagine how well it fits his personal needs and if he will appreciate the shirt. This "emotional matching" also occurs when you buy a shirt or a blouse for yourself, except that the process is more automatic and you may not consciously reflect on all the details because you understand your own needs much better than you understand your friend's needs. The evaluation process is therefore quicker and sometimes subconscious. Although you are aware of why you like something, you may not have reflected on exactly what made you reject an item. Consider, for example, going through a rack of blouses in a store. The rejection of an item may take only a second. The affective matching of a blouse is a pattern-matching process with well-developed criteria for aesthetics and suitability.

The constraint filter helps in decision making by eliminating alternatives. It operates in a fashion similar to the "elimination by aspect" decision heuristic (Tversky, 1972). Some products are rejected at an early stage. This can happen for many reasons, such as that: the price is too high, the color is ugly, or the quality is poor. A quick decision is made to reject the product and consider the next product.

If the product is accepted, there will be a trial adoption. A customer may try a blouse or a shirt. A second affective matching takes place, where the personal utility and the benefit–cost trade-off of the purchase are judged. There can be three decision outcomes: reject (search for another product), accept (pay and leave), or give up (walk out of the store).

Customer emotional needs drive designers, but the needs are difficult to measure and analyze. Based on the results of sales patterns and customer surveys, an existing product may be refined. However, for a new product, it is often not possible to predict customer emotions and the ensuing sales. Below we focus on methods for measuring emotion response to artifacts, an important factor in determining the success of a product.

2.3 Satisfying the Customer

Understanding customer needs is the first step in any product development (Chapanis, 1995). We present here models that deal with functional and affective needs.

Kano (1984) was among the first to address the discrepancy between functional and pleasurable design features. He distinguished between two principal types of product features: "must-haves" and "delighters" (Helander and Du, 1999). His approach was actually inspired by Hertzberg's (1966) two-factor theory, which was formulated to predict job satisfaction. According to Hertzberg, two types of factors affect job satisfaction: motivation factors and hygiene factors. Good salary and a good ergonomic chair are hygiene factors, and if they are present, they may prevent job dissatisfaction but will not increase job satisfaction. Codetermination and good relationships at work are examples of motivation factors, and they create job satisfaction.

Kano developed an analogy with product design features. Some design features are expected; these are called *must-haves* (Figure 5). Their presence in the design does not create satisfaction; they merely avoid dissatisfaction. In Figure 5 these are represented by the transition from A to B. Some features are not expected—they create a surprising effect and are called *delighters*. Although they are not necessarily crucial to product functionality, they create customer satisfaction. This corresponds to the transition from X to Y.

Kano exemplified his model by evaluating product features of a television set using a questionnaire. Several response categories were given, including: like it, must-be, no feeling, do not like, and give-up. To evaluate image quality there were two questions:

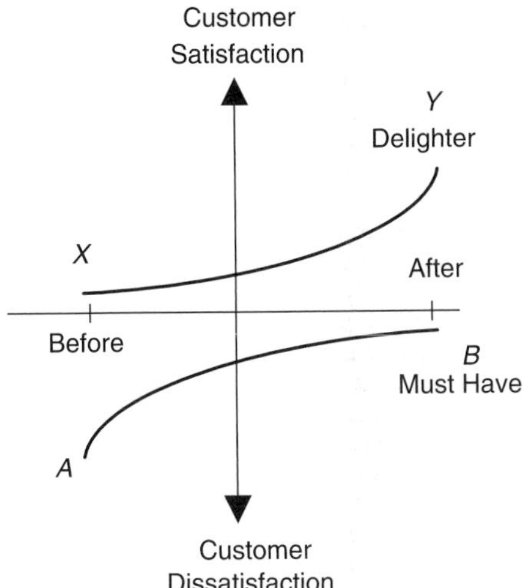

Figure 5 Kano's model for product satisfaction.

(1) How would you feel if the television picture was poor? As expected, most users (774 of 899) responded "do not like." These users then answered a second question: (2) How would you feel if the television picture was good? The results are as follows (Helander and Du, 1999) (Table 3). Of the total respondents, 497 or 64% answered "must-be." These people go from point A to point B in Figure 5, from dissatisfaction to indifference. There were no delighters. The typical response for a delighter would be "no feeling" to question 1, combined with "like" to question 2. In the evaluation of the TV set, there were two product features that were deemed delighters: "remote control" and "feather touch." Both were unexpected (in 1980) and created much satisfaction. These users went from point X to point Y.

Table 3 Responses to Question 2

Of the 774 who responded "do not like" to question 1, the following distribution was obtained for question 2:

Question 2 Response	Number of Responses	Category
Like	277	Satisfied when fulfilled, dissatisfied when not fulfilled
Must-be	497	No feeling when fulfilled, dissatisfied when not fulfilled
Total	774	

Researchers on job satisfaction would argue that the parallel between job satisfaction and product satisfaction is farfetched, since job satisfaction has a very different set of motivational factors than that of product satisfaction. There is nothing cursory about motivation factors at work, whereas users may forego delighters without thinking twice about it. The parallel between the two models may therefore be coincidental. What seems important, however, is the consideration in product design of satisfying and dissatisfying product features in two (orthogonal) dimensions. Product designers should strive to reduce customer dissatisfaction as well as to increase satisfaction. The means for achieving these goals are quite different.

Similarly, Faulkner and Caplan (1985) at Eastman Kodak categorized product design features using two variables: importance and satisfaction. Test persons rated 14 design features of a clock radio using 5-point scales (Figure 6). They concluded that features that are unimportant but received high customer satisfaction should be promoted through advertising to enhance customers' perception of product quality. Ultimately, they increase in importance, and the items can be transferred from quadrant C to quadrant B. Chances are that such features are the only ones that are different from the competition, and they can therefore

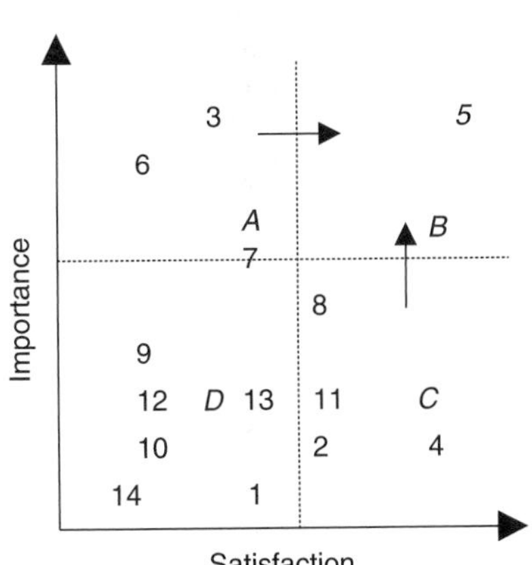

1. Sound quality AM
2. Sound quality FM
3. Ease of setting clock
4. Ease of recovery from power failure
5. Snooze alarm features
6. Regular alarm features
7. Overall size
8. Method for selecting station
9. Sleep timer feature
10. Integral electrical outlet
11. Color and finish
12. Area of footprint
13. Styling lines
14. Accommodates headset

Figure 6 Evaluation of importance and satisfaction of design features of an alarm clock.

be used to promote sales. For example, the item "recovery from power failure" would be a "delighter" in Kano's terms. One should try to move the items from quadrant A to quadrant B. High-importance items can also become high-satisfaction items. Consider dropping the items in quadrant D, especially the more expensive ones.

2.4 Need Structures of an Affective User

The need for affect varies greatly among people. Some persons have a well-developed sense of aesthetics and will seek opportunities to satisfy their needs for beautiful things. Other people do not care about aesthetics. Some people have a need to prove their virtuosity in games and will take challenges as opportunities arise. They will seek to develop great skills so that the game will "flow" effortlessly. Thus, a person's need structure is essential for purposeful activity. People's needs therefore drive design. Needs, however, are very different among different persons.

The *need for virtuosity* is a proven basic human need (Kubovy, 1999). An act is performed with virtuosity when it is difficult for most people to do but is carried out with ease and economy. This is what drives many of us to play computer games. However, virtuosity as a source of pleasure does not require extraordinary performance. Kubovy (1999) made reference to a situation where a person learned to improvise jazz on the piano over a period of six years; despite the slow development, it was a very satisfactory experience.

Take another scenario, a skiing resort. While waiting in the lift lines, the latest fashion in skiing outfit and equipment will be much appreciated, but the activity of skiing cannot be displayed, so it is not relevant. While skiing downhill, a great skiing performance with superb control and flow is admired, but in this case, a trendy outfit is beside the point.

On the basis of customer needs, designers select, organize, and size product design variables to satisfy these needs. Understanding customer needs can help companies to develop products that will sell. Information about the needs may be gauged from different consumer groups. Nevertheless, customer needs are difficult to capture. Based on the results of sales patterns and customer surveys, an existing product may be refined. But there are differences between seasoned products and new products. For seasoned products such as cars, radios, and mobile phones, customer needs are well understood because of the past sales record. Companies will often improve such products in incremental steps, and customers follow along. However, for a new product it may not be possible to predict sales with any accuracy. One common example is software design. A new software version 1.0 is put on the market without many expectations, but there will be future upgrades once the sales patterns and customer needs are understood.

A common problem is that although prospective customers may respond in a survey that they like to buy a product, they may change their minds at the point of purchase. There is a long mental step between intention and behavior (Fishbein and Ajzen, 1972). Hence, the information on customer needs may be sketchy, and designers will proceed by ignoring customer needs and estimate functional requirements as well as they can. The mapping from the designer's environment to the affective user (Figure 3) will then be based on incomplete information.

In sum, pleasure with products is viewed from three theory-based perspectives: (1) the context of use and activity; (2) categories of pleasure with products, including visceral, behavioral, and reflective; and (3) the centrality of human needs structure in driving both the cognitive and affective evaluation systems. With reference to Figure 3, pleasure with products should be considered in the context of product use—the activity context. The same product can bring forth different levels of pleasure, depending on the goals and expectations of the user and the activity that is being performed. The system has a feedback loop. This implies that if the evaluation of an artifact does not lead to satisfactory experience or to a purchase, the designer may modify the design and the user will again evaluate the design.

2.5 Shift from Usability to User Experience

In evaluating affective design, the relationship to functionality and usability must be considered. In Childs's (2001) framework there are three parts to design: functionality, usability, and affect (Figure 7). Design for performance and usability is no longer sufficient. Affective design can give a competitive edge and will enhance the design of products as well as user interfaces. So usability is not the ultimate paradigm. Currently, the shift is to user experience models that combine cognition and emotion. In other words, there is a need to understand the gestalt impression—the sum total of the user's experience with the product, rather than just simplicity and ease of use.

A strong claim to consider the joint effect of usability and affect comes from the research of Kurosu and Kishimura (1995) and Tractinsky et al. (2000). Kurosu and Kishimura experimented with different layouts for automatic teller machines (ATMs). Several versions of the ATM, with controls and buttons in different locations, were compared. Some were arranged in an attractive manner and some were unattractive. They found that the attractive ATMs were easier to use. Tractinsky et al. (2000) replicated the experiment in Israel. They, too, found that usability and aesthetics correlated.

The findings are in agreement with Carroll's (2004) proposed redefinition of usability to incorporate fun and other significant aspects of user experience. The new concept should rely on an integrated analysis of the *user's experience*. This is likely to lead to greater technological progress than is merely itemizing a variety of complementary aspects of usability. By so doing, the user experience is unified—it is a gestalt impression.

The importance of user experience and customer needs has been well documented in the design

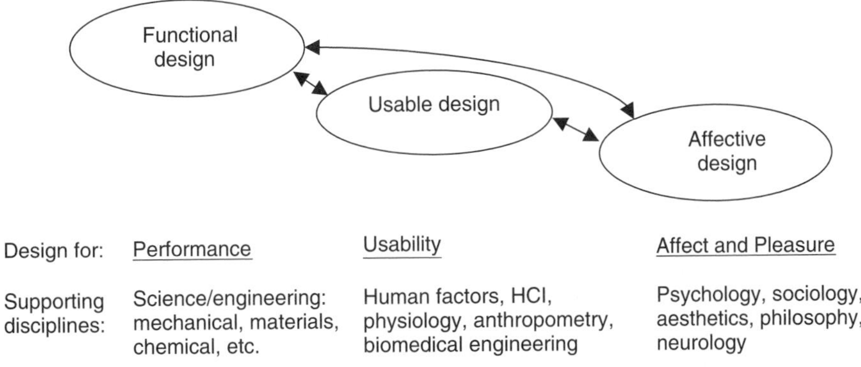

Figure 7 Contributing disciplines to evaluating affective design. (From Childs, 2001.)

literature. The next section highlights some past studies that focused on aesthetics, among other needs.

3 RESEARCH IN INDUSTRIAL DESIGN AND AESTHETICS

The research on pleasurable design in human factors is fairly recent, although there is a long research tradition in industrial design and aesthetics. We summarize here some issues that have been well investigated: (1) aesthetics and symbolic association, (2) context of use and product semantics, (3) holistic and gestalt design, and (4) familiarity and information value.

3.1 Aesthetics and Symbolic Association

A general concern for aesthetics, that is, for an attractive look, touch, feel, and attention to detail, is common in many cultures. In Asia, for example, three principles guide aesthetic appreciation: (1) complexity and decoration, (2) harmony, and (3) naturalism (Schütte and Ciarlante, 1998). The display of multiple forms, shapes, and colors is found to be highly pleasing to the Chinese, Malays, Thais, and Indonesians, as they value complexity and decoration. But harmony among the elements is regarded as one of the highest goals of aesthetic expression. Japanese and Koreans value naturalism, and images of nature are frequently depicted in the packaging of consumer goods.

There are symbolic associations in all cultures for colors, shapes, numbers, and so on (McManus, 1981). Red means happiness and good luck to the Chinese and is therefore the most appealing, whereas Indians identify red with power and energy. Yellow is considered pleasant and signifies authority, whereas white is linked to death.

To achieve a "good" design, the attributes must be relatively stable across time and cultures. A good design is a visual statement that draws on shared symbolic expression of a certain subset of people in a given culture and that maximizes life goals (Csikszentmihalyi, 1995). In each culture, public taste develops, as visual qualities are eventually linked with values. Visual values can be unanimous or contested, elite or popular, strong or vulnerable, depending on the integration of the culture.

3.2 Context of Use and Product Semantics

Products are always seen in a context, and the context is construed cognitively. Product semantics connects the attributes of a product and its context of use into a coherent unity (Krippendorff, 1995). A starting point for developing product semantics is to observe what objects people surround themselves with, what the objects are used for, and how they are referred to (Krippendorff, 1995). Objects can have several meanings to customers, as identified by Bih (1992): as functional and utilitarian (e.g., radio), with religious cultural value (e.g., statues), to mark personal achievements (e.g., degree), to extend memory (e.g., photographs), for social exchange (e.g., gifts), to illustrate shared experience (e.g., travel), and to extend self and personal values (e.g., antiques).

Khalid (1999) uncovered product semantics for watches. In addition to measure time, watches are used to enhance prestige, to portray aesthetics values of the user, as fashion statements, and costume accessory. Preference for a watch depends on a holistic assessment of attributes, such as the type of casing, precision in time, and strength of material.

3.3 Holistic and Gestalt Features in Design

A *gestalt* is a whole, an organization of parts, whether these are graphic elements, tones, or colors. There is an inherent tendency in perception toward achieving a parsimonious gestalt, that is, to render a structure as simple rather than complex. Simple structures have characteristics such as unity, symmetry, regularity, and harmony (Crozier, 1994). Kreitler and Kreitler (1972) claimed that there is widespread preference for good gestalt, particularly in children's products. Objects that are symmetrical and have an even weight distribution are regarded as more balanced and are preferred over other objects (Margolin and Buchanan, 1995). This is termed here *holistic design*, a global organization of the form. It is clearly important to explore people's judgments of attributes, such as simplicity, balance, and symmetry, as these attributes have long been considered to be basic to design. However, people are also drawn to complexity: They look more at complex figures than at simple ones (Berlyne, 1974).

3.4 Familiarity and Information Value

Crozier (1994) observed that preference is correlated with exposure; the more experience we have with an object, the better we like it; people tend to prefer what is familiar. But Martindale and Uemura (1983) claimed that preference diminishes with increased familiarity because of habituation, and that aesthetic preference is linked to the physiological arousal potential of an artifact. Purcell (1986) also argued that emotional responses to objects are greater if the object is different from what was expected. To some, informativeness is more important than either novelty or familiarity. Teigen (1987) proposed that intrinsic interest in an object is related to its information value, which is a function of the joint presence of novel and familiar elements.

4 THEORIES OF AFFECT AND PLEASURE

Several theories in psychology support the notions that we have raised, and some provide directions for future research and methods development. These theories are summarized below.

4.1 Activity Theory

Activity theory employs a set of basic principles and tools—object-orientedness, dual concepts of internalization/externalization, tool mediation, hierarchical structure of activity, and continuous development—which together constitute a general conceptual system (Bannon, 1993). In human activity theory, the basic unit of analysis is human (work) activity. Human activities are driven by certain needs, where people wish to achieve a certain purpose (Bannon and Bødker, 1991). The activities are usually mediated by one or more instruments or tools, such as a photographer using a camera. Thus, the concept of mediation is central to activity theory.

Leontjev (1978) distinguished between three different types of cognitive activities: (1) simple activity, which corresponds to automated stimulus–response; (2) operational activity, which entails perception and an adaptation to the existing conditions; and (3) intellectual activity, which makes it possible to evaluate and consider alternative activities. Note that these activities are in agreement with Rasmussen's model of skill-based, rule-based, and knowledge-based behavior (Rasmussen et al., 1994). For each of the cognitive stages there are corresponding emotional expressions: affect, emotion, and sentiments.

Affect is an intensive and relatively short-lasting emotional state. For instance, as I walk down colorful Orchard Road in Singapore and look at items displayed in the shop windows, there are instantaneous reactions to the displayed items; most of these reactions are unconscious, and I have no recollection of them afterward. Through affect, we can monitor routine events. Many events are purely perceptual and do not require decision making but there is an affective matching of events that are stored in memory. This helps in understanding and interpreting their significance.

Emotions are conscious. When I stop to look at some item in one of the shop windows, I am aware of why I stopped. Emotions go beyond the single situation and typically remain in memory for one or several days.

Sentiments or *attitudes*, according to Leontjev (1978), are longer lasting and include intellectual and aesthetic sentiments, which also affect my excursion along Orchard Road. I know from experience that some stores are impossible; on the other hand, there are a few that are clearly very interesting. Sentiments and attitudes are learned responses.

Feelings are an integral aspect of human activity and must be investigated as psychological processes that emerge in a person's interaction with his or her objective world. Their processes and states guide people toward achieving their goals (Aboulafia and Bannon, 2004). Feelings should not be viewed merely as perturbances of underlying cognitive processes. Predicting affect is likely to be easier than predicting emotions or sentiments. To evoke affective reactions in a user, the artifact could be designed to provide people with a variety of sudden and unexpected changes (visual or auditory) that cause excitement and joy or alarm. Designing toys for children has given us ideas about such design space.

Predicting emotional responses that extend over several situations can be more difficult. Emotions are not dependent on the immediate perceptual situation. The emotional state of a computer user is not usually oriented toward the mediating device itself but to the overall activity in general (either work activity or pleasure). The artifact is merely a mediating tool between the motive and the goal of the user (Aboulafia and Bannon, 2004).

Leontjev (1978) emphasized that emotions are relevant to activity, not to the actions or operations that realize it. In other words, several work or pleasure situations influence the emotion of the user. Even a successful accomplishment of one action or another does not always lead to positive emotions. For example, the act of sneezing in itself usually evokes satisfaction. However, it may also evoke fear of infecting another person. Thus, the affective and emotional aspects of objects are capable of changing, depending on the nature of the human activity (the overall motive and goal). As such, stressed Aboulafia and Bannon (2004, p. 12), "objects or artifacts—in and of themselves—should not be seen as affective, just as objects in and of themselves should not be defined as 'cognitive' artifacts, in Norman's (1991) sense. The relation between the object (the artifact) and the human is influenced by the motive and the goal of the user, and hereby the meaning or personal sense of the action and operation that realize the activity."

We note that Norman (2004) would object to these notions. In fact, he proposed that domestic robots need affect in order to make complex decisions, and Velásquez (1998b) talked about robots that weep. Equipped with only pure logical functions, a robot would not be able to make decisions—just like Damasio's (1994) patients.

4.2 Emotions versus Pleasures of the Mind

Ekman (1992, 1994) stated that there are a number of fundamental emotions that differ from one another in important ways: anger, fear, sadness, disgust, happiness. Evolution played an important role in shaping the features among these emotions as well as their current function.

The pleasures of the mind have been neglected by contemporary psychology (Cabanac, 1992). Kubovy (1999) argued that pleasures of the mind are different from basic emotions. Pleasures of the mind are not accompanied by any distinctive facial expression. Take for example, a person viewing a painting. She may feel elated, but nothing is revealed on her face, and there is no distinctive physiological response pattern. This is very different from social interaction, such as a conversation with a colleague at work, where half of the message is in the person's face. Since one may not be able to use either physiological measures or facial expressions, one is left with subjective measures. There is nothing wrong with asking people; subjective methods, interviews, questionnaires, and verbal protocols provide valuable information. The problem is: What questions should be asked in order to differentiate between products?

The notion of the pleasures of the mind dates back to Epicurus (341–270 b.c.), who regarded pleasures of the mind as superior to pleasures of the body, because they were more varied and durable. Kubovy (1999) also noted that pleasures of the mind are quite different from pleasures of the body—tonic pleasures and relief pleasures. Ekman's eight features of emotion are summarized in the left-hand column of Table 4 the right-hand column shows pleasures of the mind.

Table 4 Features of Emotions and Pleasures of the Mind

Emotions ...	Pleasures of the mind ...
• have a distinctive universal signal (such as a facial expression).	• do not have a distinctive universal (facial) signal.
• are almost all present in other primates.	• may be present at least some of them in other primates.
• are accompanied by a distinctive physiological response.	• are not accompanied by a distinctive physiological response.
• give rise to coherent responses in the autonomic and expressive systems.	• do not give rise to coherent responses.
• can develop rapidly and may happen before one is aware of them.	• are relatively extended in time.
• are of brief duration (on the order of seconds).	• are usually not of brief duration.
• are quick and brief; they imply the existence of an automatic appraisal mechanism.	• even though neither quick nor brief, may be generated by an automatic appraisal mechanism.

Source: Kubovy (1999, p. 137).

4.3 Reversal Theory: Relationship between Arousal and Hedonic Tone

Arousal is a general drive rooted in the central nervous system. According to common arousal theories, organisms fluctuate slightly about a single preferred point. Reversal theory, on the other hand, focuses on the subjective experiences of humans. The central concept of reversal theory is that the preferred arousal level fluctuates (Apter, 1989). Reversal theory claims that people have two preferred points, and they frequently switch or reverse between them. The theory therefore posits bistability rather than homeostasis. People can be in one of two states. In the first state, which is called *telic*, low arousal is preferred, whereas high arousal is experienced as unpleasant. In the telic state, calmness (low arousal, pleasant) is contrasted with anxiety (high arousal, unpleasant). The opposite is true when the person is in the *paratelic* state. In the paratelic state, low arousal is experienced as boredom (unpleasant) and high arousal as excitement (pleasant).

A given level of arousal may therefore be experienced as either positive or negative. One may experience a quiet Sunday afternoon as serene or dull. One may also experience a crowded and noisy party as exciting or anxiety provoking. The perceived level of pleasantness, called *hedonic tone*, is different for the two states. The paratelic state is characterized as an arousal-seeking state and the telic state as arousal avoiding. When in the telic state, people are goal oriented; they are serious-minded and try to finish their current activity to attain their goal. On the other hand, to have a good time, the paratelic state is appropriate. Goals and achievements are not of interest; rather, this is the time to play, have fun, and be spontaneous.

4.4 Theory of Flow

Flow is a state of optimal experience, concentration, deep enjoyment, and total absorption in an activity (Csikszentmihalyi, 1992). Csikzentmihalyi (1975) described the flow state accordingly: "Players shift into a common mode of experience when they become absorbed by their activity. This mode is characterized by a narrowing of the focus of awareness so that irrelevant perceptions are filtered out; by loss of self consciousness, by responsiveness to clear goals and unambiguous feedback, and by a sense of control over the environment—it is this common flow experience that people adduce as the main reason for performing an activity." The experience of flow is associated with positive affect; people remember these situations as pleasurable. It may be participation as a violin player in an orchestra, solving math problems, or playing chess. All of these cases may involve a sense of total attention and accomplishment which the person thinks of as a pleasurable experience.

Flow has been studied in a broad range of contexts, including sports, work, shopping, games, hobbies, and computer use. It has been found useful by psychologists, who study life satisfaction, happiness, and intrinsic motivations; by sociologists, who see in it the opposite of anomie and alienation; by anthropologists, who are interested in the phenomenon of rituals.

Webster et al. (1993) suggested that flow is a useful construct for describing human–computer interactions. They claimed that "flow represents the extent to which (1) the individual perceives a sense of control over the interactions with technology; (2) the individual perceives that his or her attention is focused on the interaction; (3) the individual's curiosity is aroused during the interaction; and (4) the individual finds the interaction interesting."

In e-commerce, a compelling design of a Web site should facilitate a state of flow for its customers. Hoffman and Novak (1996) defined flow as "the state occurring during network navigation which is: (1) characterized by a seamless sequence of responses facilitated by machine interactivity, (2) intrinsically enjoyable, (3) accompanied by a loss of self-consciousness, and (4) self-reinforcing." To experience flow while engaged in any activity, people must perceive a balance between their skills and the challenges of the interaction, and both their skills and challenges must be above a critical threshold.

Games promote flow and positive affect (Johnson and Wiles, 2003). The study of games can inform the design of nonleisure software for positive affect. Bergman (2000) noted that "the pleasure of mastery only occurs by overcoming obstacles whose level of frustration has been carefully placed and tuned to not be excessive or annoying yet sufficient to give a sense of accomplishment." Thus, an interface may be designed that can improve a user's attention and make the user feel in total control as well as free of distractions from nonrelated tasks, including poor usability.

4.5 Affect Heuristic

Research in decision making was for many years dominated by normative models, where probabilities of outcome and the associated values in terms of losses and gains could be optimized. Lately, there has been much research on how people make decisions in real life. It turns out that people usually do not try to maximize the outcome but are often driven by their intuition, which may or may not be optimal. The reason is that people cannot hold all the facts and figures in short-term memory, which is easily crowded by all the detailed information. People could, of course, calculate the optimal solution using a computer, but in real life this is not done. Decisions are made on the spin. Since the capacity of the short-term memory is easily exceeded, there are other ways of coping with decisions: namely, to use heuristics or "rules of thumb." In other words, people try to "wing" decisions.

In most cases the quality of decision making produced by heuristics is good enough; there is rarely a need for exacting decisions, but there are also exceptions (Gilovich et al., 2002). Several common heuristics have been identified, including the availability heuristic, the anchoring heuristic, the confirmation bias, the framing effect, and the as-if heuristic. The most famous is prospect theory, which was the basis for Kahneman's Nobel Prize in Economics in 2002 (Kahneman and Tversky, 1984) (see Figure 8).

According to this heuristic, people will make unexpected choices. The positive utility of increasing your wealth is fairly small. But the negative utility of losing money is much greater. Note the asymptotic curve for gain—once you have $1 million, the potential gain of another million has less utility, and the opposite is true for losses. To illustrate these notions, assume that you are given the choice of gambling. There are two options:

A. You will obtain $10 with 100% certainty.

B. You will obtain $20 with 50% certainty and nothing with 50% certainty.

Would you choose option *A* or *B*? Although both options carry equal value according to the normative school, option *A* is chosen by 75%, and option *B* with 25% probability. People will rather take a certain win than gamble.

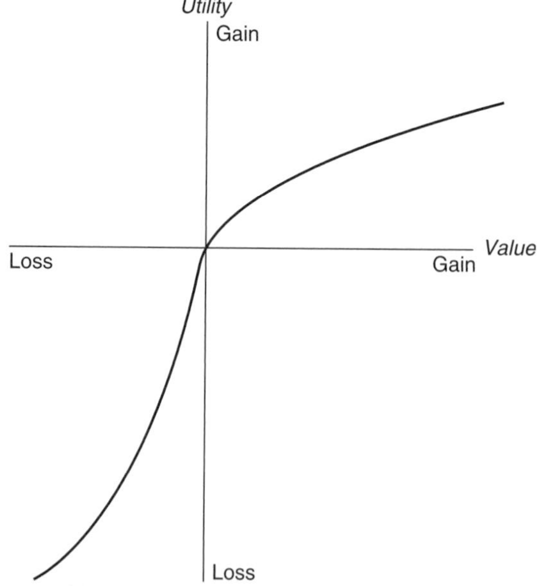

Figure 8 Prospect Theory. A marginal gain in wealth has fairly small positive utility. Losses are however much more consequential and carry heavy negative utility.

Assume that you are given another choice of gambling. There are two options:

A. You will lose \$10 with 100% certainty.

B. You will either lose \$20 with 50% certainty or nothing with 50% certainty.

About 30% of a population selects option *A* and 70% option *B*. The prospect of losing 10% for sure makes alternative *A* less attractive.

These findings have great implications for financial decision making; people are not logical decision makers and they need advice. In proposing the *affect heuristic*, Slovic et al. (2002) claimed that many decisions are based on emotional criteria. Kahneman (2002), in agreement, said that had he understood about the emotional basis of decision making, he would have reformulated the prospect theory.

Slovic et al. (2002) explained on how decisions are made. The *analytic system* uses algorithms and normative rules, such as probability calculus, formal logic, and risk assessment. It is relatively slow, effortful, and requires conscious control. The *experiential system* is not very accessible to conscious awareness but is intuitive, fast, and mostly automatic. The challenge now is to understand how we can design information systems so that they appeal to the emotional and experiential system with fast and intuitive processing of information as a result.

4.6 Endowment Effect

Research on the *endowment effect* has shown that people tend to become attached to objects they are endowed with, even if they did not have any desire to own the object before they got possession of it (Thaler, 1980). Once a person comes to possess a good, he or she values it more than before possessing it. This psychology works well for companies that sell a product and offer a two-week return policy. Very few return the product. Put simply, this means that people place an extra value on the product once they own it.

Lerner et al. (2004) extended the endowment effect by examining the impact of negative emotions on the assessment of goods. As predicted by appraisal-tendency theory, disgust induced by a prior, irrelevant situation carried over to unrelated economic decisions; thereby reducing selling and choice prices and eliminating the endowment effect. Sadness also carried over, reducing selling prices but increasing choice prices. In other words, the feeling of sadness produces a *reverse endowment effect* in which choice prices exceeded selling prices. Their study demonstrates that incidental emotions can influence decisions even when real money is at stake, and that emotions of the same valence can have opposing effects on such decisions.

4.7 Hierarchy of Needs

According to Maslow (1968), people have hierarchies of needs that are ordered from physiological needs through safety, love/belonging, and esteem, to self-actualization. They are usually depicted using a pyramid or a staircase, such as in Figure 9. The hierarchy affects how needs are prioritized. Once a person has fulfilled a need at a lower level, he or she can progress to the next level. To satisfy the need for self-actualization, a person would have to fulfill the lower four needs, which Maslow (1968) referred to as *deficiency needs*. These needs are different than self-actualization in nature. Many authors have pointed out that the hierarchy is not a strict progression. For example, some people may deemphasize safety but emphasize the needs for love/belonging.

Hancock et al. (2005) presented a hierarchy of needs for ergonomics and hedonomics (Figure 9). The ergonomic needs address safety, functionality, and usability; in Maslow's reasoning they would be referred to as deficiency needs. The two upper levels, pleasure and individuation, deal with self-actualization. Individuation, at the top of the pyramid, is concerned with ways in which a person customizes his or her engagement and priorities, thereby optimizing pleasure as well as efficiency.

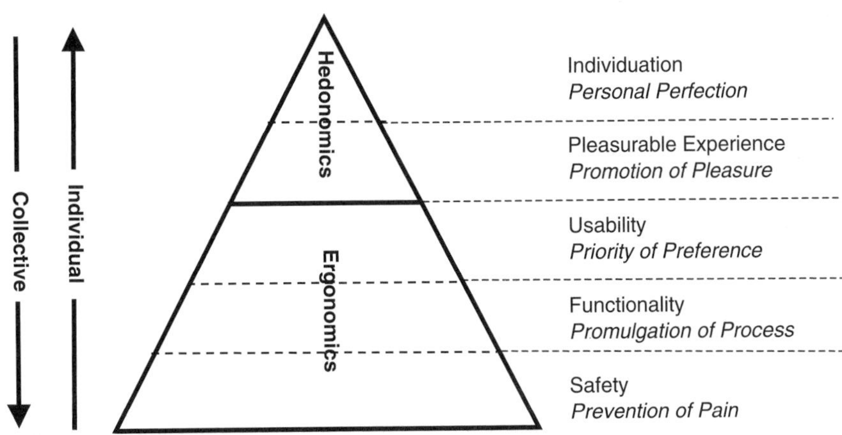

Figure 9 Hierarchy of ergonomics and hedonomics needs. (From Hancock et al., 2005.)

One may question if there is really a hierarchy or if the elements of Figure 9 are independent of each other. If so, there would not be a progression from bottom to top, but rather, in parallel. Helander and Zhang (1997) found that comfort and discomfort are orthogonal concepts, and it is necessary to use two different scales to measure them. Similarly, it may be necessary to use several scales in Figure 9 to measure each of the five concepts. Essentially, a combination of subjective and objective measures is needed to capture the various dimensions of emotion.

5 MEASUREMENT OF AFFECT AND PLEASURE

Several studies have proposed methods for measuring affect. Nagamachi (1989) is in the forefront with the development of questionnaire scales for assessment of what he refers to as *Kansei* (feeling) engineering. Picard (1997) has published extensively on the use of various methodologies for measurement of affect in computing. More recently, Jordan (2000) and Desmet (2003) introduced questionnaires and measurement instruments to assess emotional responses to consumer products. Whatever the method, emotions are difficult to assess, especially when several emotions blend (Scherer, 1998). At this stage the measurement of emotions poses one of the most challenging aspects to human factors.

5.1 Measurement Issues

Larsen and Fredrickson (1999) proposed four pertinent aspects of emotion assessment: dynamics, context, reliability, and validity. We add a fifth issue, measurement error.

5.1.1 Dynamics

Emotions are generated by different systems in the brain with different timing mechanisms, and they evolve over time. Therefore, they are difficult to capture. This raises three critical measurement issues: (1) how to identify the onset of a particular emotion (when does it start and when does it end?), (2) how to ensure that a measure of emotion can capture the dynamic aspects, and (3) how to relate in time the subjective emotion experience to the experience measured.

5.1.2 Context

Emotions occur in a context. Activity theory, for example, emphasizes the ongoing work activity. Therefore, it is important to capture the context and the peculiarities of the scenario in which the emotions were generated. Emotions also vary from person to person and are related to personality, experience, mood, and physiological arousal. In addition, the 24-hour circadian rhythms influence the emotion experience.

5.1.3 Reliability

The purpose here is to find measurements that are stable from time to time. For some situations, a test–retest correlation is a good estimate of reliability. However, in estimating reliability, we must consider that a person's mood changes frequently and it may be difficult to reproduce the emotive experience a second time for a retest. Emotion can also be measured for members in a group. The interest here may be differences between people in their reactions to emotion-provoking events.

5.1.4 Validity

The question here is whether a measure that we use to evaluate emotion(s) measures what we intend to measure. One complicating factor is that emotions are complex responses. Larsen and Fredrickson (1999) were of the opinion that measurement of an emotion cannot be reduced to one single measure.

Construct validity is important to consider; is there a theory that drives our research interest? If so, we need to define measure(s) that can be linked to the theory. This simplifies measurement since we have an "excuse" to focus on only a few types of measures. For example, let us assume that we would like to measure pleasures of the mind. From what we understand, these do not necessarily generate a facial expression or physiological response. Therefore, we would neither consider physiological variables nor facial measures. In this case, a theory of emotional expression restricts the selection of dependent variables.

5.1.5 Measurement Error

There are two types of measurement error: random error and systematic error. To overcome random error, one can take many measures instead of a single measure and estimate a mean value. Therefore, multiple items or mathematical measurement models can be used to control or eliminate random measurement error. However, this approach is not suitable for methods, which require assessments at certain times, such as experience sampling.

Another problem is that some types of assessments are intrusive (Schimmack, 2003). By asking a person to respond to a question, the contextual scenario of the emotional experience is disrupted, which may reduce the validity of the data. To minimize disruption, one can reduce the number of questions. Another way is to seek measures that are less intrusive: for example, physiological responses and facial expressions.

For heterogeneous scales that sample a broad range of affects (e.g., PANAS scales), many items are needed. Watson et al. (1988) used 10 items and obtained item-factor correlations ranging from 0.75 to 0.52. Systematic measurement error does not pose a problem for within-subject analysis, because the error is constant across repeated measurements. However, it can be misleading to use average values for calculation of correlation coefficients.

5.2 Measurement Methods

Despite much development in human factors research, the methods for measuring affect are entrenched in psychology. Various research in consumer behavior, marketing, and advertising have developed instruments

Table 5 Overview of Human Factors Methods to Measure Affect

Methods	Techniques	Research Examples
Subjective measures		
Subjective rating of emotional product attributes	Kansei engineering	Nagamachi (2001), Helander and Tay (2003)
	Semantic scales	Küller (1975), Chen and Liang (2001), Karlsson et al. (2003), Khalid and Helander (2004)
Subjective rating of emotions — general	Self-report	Rosenberg and Ekman (1994)
	Emotional well-being report	Brown and Schwartz (1980), Sandvik et al. (1993)
	Experience sampling method	Larson and Csikszentmihalyi (1983), Singer and Salovey (1988), Feldman-Barrett (1998), Eid and Diener (1999), Schimmack (2003)
	Affect grid	Russell et al. (1989), Warr (1999)
	Checklist (MACL)	Nowlis and Green (1957)
	Multiple affect adjective check list	Zuckerman and Lubin (1965)
	Activation–deactivation adjective check list	Thayer (1967)
	Differential emotional scale	Izard (1977)
	Interview	Jordan (2000)
	Aesthetic development interview	Housen (1992)
Subjective rating of emotions induced by artifact	PANAS scale	Watson et al. (1988)
	Philips questionnaire	Jordan (2000)
	Product emotion measurement instrument	Desmet (2003)
Objective measures		
	Facial action coding system	Ekman and Friesen (1976), Ekman (1982)
	Maximally discriminative affect coding system	Izard (1979)
	Facial electromyography	Davis et al. (1995)
	Emotion judgment in speech	Scherer (1986), Maffiolo and Chateau (2003)
	Psychoacoustics and psychophonetics	Larsen and Fredrickson (1999)
Psychophysiological measures		
	Galvanic skin response and other ANS measures	Larsen and Fredrickson (1999)
	Wearable sensors	Picard (2000)
Performance Measures		
	Judgment task involving probability estimates	Mayer and Bremer (1985), Ketelaar (1989)
	Lexical decision task	Challis and Krane (1988), Niedenthal and Setterlund (1994)

for measuring emotional responses to advertisement and consumer experiences of products. Here we focus on methods that may be applied to affective design of products. We classify the methods into four broad categories: (1) subjective, (2) objective, (3) physiological, and (4) performance. The subjective methods are further categorized into three classes of measures: (1) user ratings of product characteristics, (2) user ratings of emotions and/or reporting of user experience without specific reference to an artifact, and (3) user ratings of emotions as induced by artifacts. The methods are summarized in Table 5.

5.2.1 Subjective Measures

Ratings of Product Characteristics These subjective methods involve user evaluations of products. There are two established techniques: Kansei engineering and semantic scales.

Kansei Engineering Developed by Mitsuo Nagamachi 20 years ago, Kansei engineering centers on the notion of *Kansei*, customer's feelings for a product (Nagamachi, 1989, 2001). The word *Kansei* encompasses various concepts, including sensitivity, sense, sensibility, feeling, aesthetics, emotion, affection, and intuition—all of which are conceived in Japanese as mental responses to external stimuli, often summarized as psychological feelings (Krippendorff, 2005). Nagamachi validated several scales for assessment of different products. To build a scale, the following procedure was used:(1) collect Kansei words; (2) correlate design characteristics with Kansei words (e.g., using Osgood's semantic differential technique); and (3) perform factor analysis on Kansei words to determine similarity; (4)analyze product features to predict emotions.

A Kansei database of descriptors has been developed for various products: beautiful, cheerful, citylike,

Table 6 Results from Factor Analyses of Four Kitchen Appliances[a]

Descriptor	Curvy Toaster		Square Toaster		Round Coffeemaker		Tubelike Coffeemaker	
Stylish	1	0.587	1	0.839	2	0.706	1	0.839
Modern	1	0.544	1	0.801	1	0.736	1	0.756
Fashionable	1	0.771	1	0.702	2	0.603	1	0.631
Cool	1	0.540	1	0.805	1	0.697	1	0.606
Attractive	1	0.798	1	0.821	1	0.708	1	0.844
Beautiful	1	0.798	1	0.702	1	0.697	1	0.814
Elegant	1	0.533	1	0.645	1	0.608	1	0.717
Likeable	1	0.598	1	0.723	1	0.737	1	0.675
Luxurious	4	0.724	6	0.602	2	0.707	1	0.564
Lively	1	0.779	1	0.642	1	0.628	1	0.691
Interesting	1	0.578	1	0.546	2	0.743	1	0.718
Unique	3	0.699	1	0.548	2	0.800	1	0.569
Cheerful	1	0.734	1	0.656	1	0.593	1	0.689
Urban	5	0.516	1	0.762	1	0.681	3	0.713
Unusual	3	0.666	2	0.531	2	0.714	2	0.365
Cute	2	0.767	5	0.558	2	0.522	1	0.580
Curvy	2	0.789	3	0.430	1	0.710	3	0.590
Homely	2	0.541	3	0.729	3	0.568	3	0.630
Natural	6	0.722	3	0.776	3	0.684	4	0.670
Friendly	2	0.568	4	0.658	3	0.565	1	0.592
Compact	5	0.604	3	0.639	3	0.576	4	0.509
Ordinary	6	0.689	2	0.744	3	0.699	3	0.592
Loud	1	0.491	5	0.800	5	0.821	2	0.492
Complex	3	0.735	2	0.697	2	0.495	5	0.804
Masculine	4	0.708	4	0.699	4	0.859	4	0.636
Feminine	2	0.656	4	0.684	4	−0.525	2	0.747
Explained variance(%)		68.0		68.7		66.2		65.0

[a]Varimax rotation was used. In each column the number of the factor and the factor loadings are given.

delightful, enjoyable, fashionable, friendly, happy, natural, pretty, rich, and stimulating. One questions if the descriptors are generic enough to be used across products. Also, can Kansei engineering predict feelings of cultures other than Japanese? The qualities of feelings are not universal or literally caused by particular forms (Krippendorff, 2005).

Helander and Tay (2003) investigated if the same Kansei words could be used to describe four different types of kitchen appliances. A list of 26 descriptors was generated from an original list of 200 descriptors. Table 6 shows the set of Kansei words that was used. Each product was then rated by 100 test persons using a 7-point Likert scale ranging from 1 (absolutely not) to 3 (not really) to 5 (much) to 7 (very much) (see Table 6). From the table we can see that the factor analyses for the four kitchen appliances generated similar factors, and it explained about 68% of the variance. From this analysis we can conclude that for kitchen appliances it is possible to use the same set of Kansei words.

Semantic Scales Semantic scales are similar to Kansei scales. The main difference is that the scales rely on the methodology proposed in Osgood's semantic differential technique (Osgood et al., 1967). This technique makes it possible to assess semantic differences between objects. Adjective pairs of opposite meanings are created, such as light—heavy, open—closed, and fun—boring. Subjects then rate objects using, for example, a 5-point scale, such as 1 (very fun), 2 (fun), 3 (neutral), 4 (boring), and 5 (very boring). A main problem is to validate the word pairs. In the first place, it is not trivial to assess if the two words constitute semantic opposites. One would also need to demonstrate that the word pair chosen is appropriate to evaluate the artifact in question.

Küller (1975) was one of the first to develop semantic scales for design, in his case for architectural appreciation. His interest was driven by the observation that a pleasant environment increases calmness and security and reduces aggressiveness. He validated 36 adjectives (in Swedish) which constituted seven factors. The factors and adjectives were *pleasantness* (stimulating, secure, idyllic, good, pleasant, ugly, boring, brutal); *complexity* (motley, lively, composite, subdued); *unity* (functional, of pure style, consistent, whole); *enclosedness* (closed, demarcated, open, airy); *potency* (masculine, potent, fragile, feminine); social status (expensive, well-kept, lavish, simple); *affection* (modern, new, timeless, aged); and *originality* (curious, surprising, special, and ordinary).

Karlsson et al. (2003) used Küller's method for evaluation of automobiles. They obtained significant results that discriminated among the designs of four passenger cars: BMW 318 (more complex and potent), Volvo S80 (more original and higher social status),

Audi A6 (less enclosed), and VW Bora (greater affect). Considering the significant results, one may debate whether formal validation is necessary; the significant results carried much face validity! Obviously, this methodology works well with cars as well as architecture.

Chen and Liang (2001) evaluated 19 cars using six adjectives: streamlined, futuristic, cute, dazzling, comfortable, dignified, sturdy, powerful, and mature. Each car was rated by 48 subjects using a scale. The adjectives used by subjects were factor analyzed to extract the underlying dimensions of the attributions. This technique was used to obtain measures of semantic differences between existing cars, but it could be used to measure how a proposed design deviates from those on the market. The data were evaluated using multidimensional scaling, and the results as represented in Figure 10. Chen and Liang then used image morphing to evaluate the various dimensions. For example, visualization of the vector "cute" can be calculated as one goes from a source image (Volkswagen Bora) to the destination image (Mercedes-Benz). This affective evaluation of the attractiveness of a car, combined with an interpolation of shapes, has proven to be a valuable design tool.

Still on cars, Khalid and Helander (2004) developed a rating tool to measure user responses to four future electronic devices [cell phone, personal digital

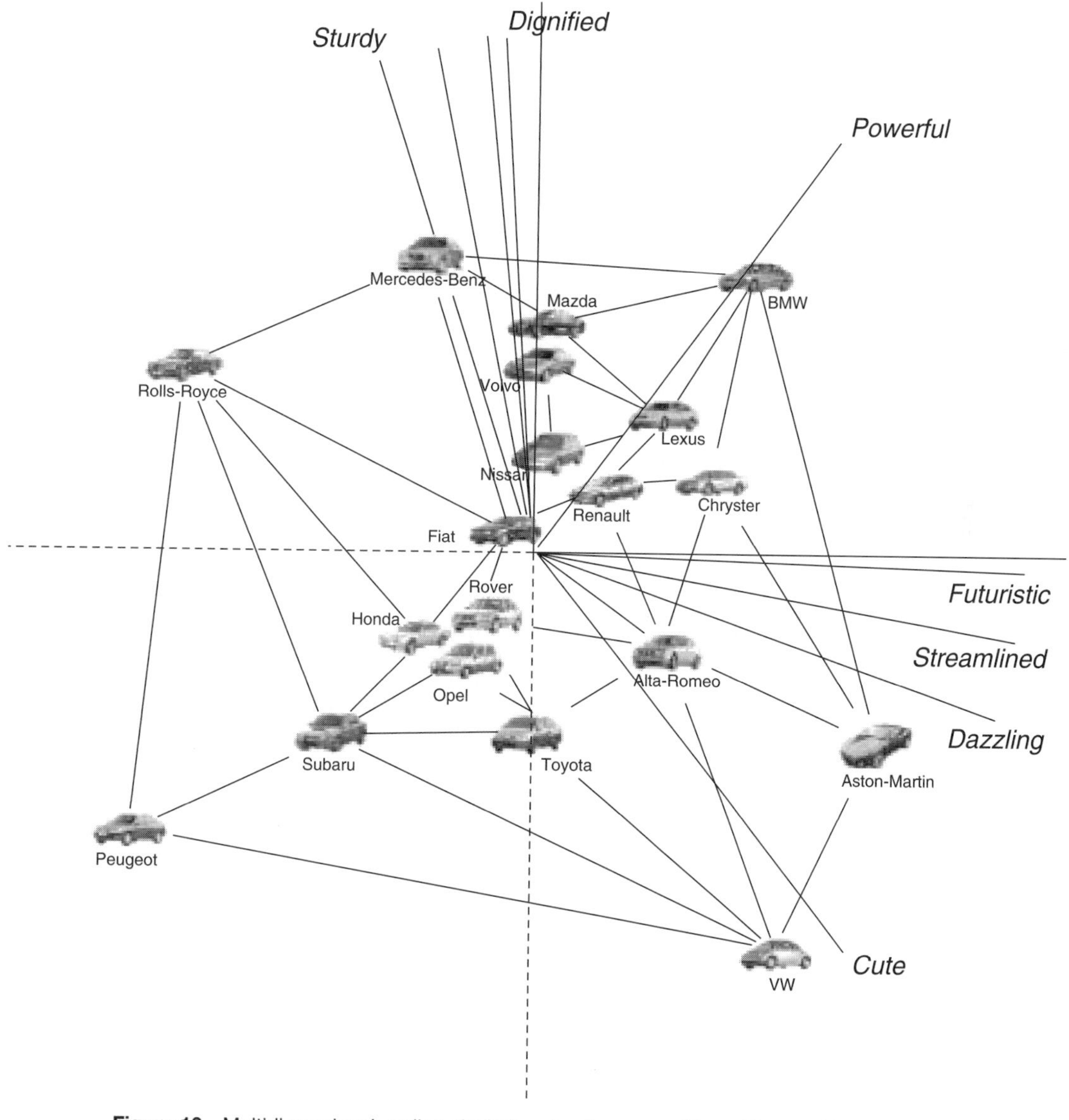

Figure 10 Multidimensional scaling similarity of various cars. (From Chen and Liang, 2001.)

assistant, radio, and geographical positioning system (GPS)] for an instrument panel of a car. Users had to imagine the products and rated their affective preferences for 15 product attributes on 10-point semantic differential scales. These attributes comprised functional and affective customer needs derived from a customer survey. Using factor analysis, three generic factors were extracted: *holistic attributes, styling*, and *functional design*. Depending on the familiarity of the device, there were clear differences among users. Devices that were unfamiliar to the test persons, such as GPS, were assessed using holistic attributes. Familiar designs, such as car radio and cell phone, were assessed using styling and functionality attributes.

Subjective Ratings of Emotions These include techniques that report a person's subjective experience, such as self-reports and experience sampling method, or rating of one's own emotions in the form of affect grid, checklist, and interview. These methods have more general applicability to products as well as tasks and scenarios.

Self-Reports This technique requires participants to document their subjective experiences of the current situation. A self-report can reflect on one's present state and compare it to the past state. As such, the self-reporting technique relies on the participant's ability to report experiences and to reflect accurately on their experiences. The measures may be instantaneous or retroactive. Instantaneous reports refer to the emotion as first experienced, whereas retrospective reports refer a situation after-the-fact. Such assessments can be accomplished using a video as a reminder. For example, Rosenberg and Ekman (1994) asked participants to stop a video when they wanted to describe their emotions. In addition to a verbal report, they also responded to a questionnaire.

Because self-report procedures ask people to remember and summarize their experiences over longer or shorter intervals of time, a major limitation of self-report is that it relies exclusively on the person's cognitive labels of his or her emotions. But emotion, as argued above, is a multichannel phenomenon and is not limited to the cognition of emotion. In addition, there are physiological, facial, nonverbal, behavioral, and experiential elements (Diener, 1994). Self-reports of emotional well-being, such as happiness, tend to reveal greater consistency than do many other types of emotion (Brown and Schwartz, 1980). There is also agreement between self-reports of emotional well-being and interview ratings, peer reports, and memory for pleasant events (Sandvik et al., 1993).

Experience Sampling Method (ESM) Coined by Larson and Csikszentmihalyi (1983), ESM measures people's self-reported experiences close in time to the occurrence of the scenario that evoked the emotion. Typically, ESM uses a combination of online and short-term retrospective question formats in which people report what is presently or recently occurring (e.g., "How do you feel *right now*? How did you feel *this past hour*?"). As such, these procedures measure

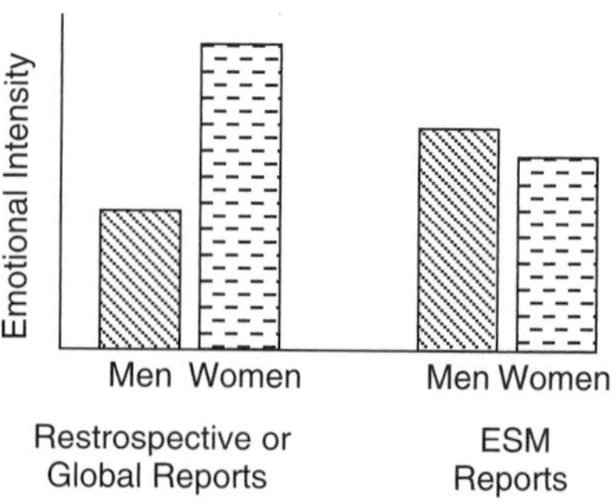

Figure 11 Comparisons between other self-reports and ESM reports. (From Feldman-Barrett, 1998.)

subjective experience that is episodic in nature. A comparison of retrospective emotion reports with ESM reports for men and women is shown in Figure 11. In retrospective reports of emotional intensity, women give higher ratings than a concurrent ESM rating, while men give a lower rating.

Advantages of ESM over other self-report methods are:

- *Immediacy:* reduces retrospective memory bias of mood (Singer and Salovey, 1988) and reduces beliefs and theories about experiences (Ross, 1989).
- *Multiple assessments over time:* makes it possible to study ongoing processes in a person. Thereby, one can (1) understand the patterns and relations among variables for a given person, and (2) understand how a person reacts to the situation.
- *Natural reporting context:* improves the validity of the reports and makes it possible to model experiences that would not show up in a controlled laboratory setting.

The ESM method has been applied to studies on flow (Csikszentmihalyi, 1990), mood variability (Eid and Diener, 1999), and hedonic balance (Schimmack, 2003). In particular, Schimmack (2003) found that pleasant affects and unpleasant affects had high discriminant validity. Extraversion is highly related to aspects of pleasant affects, and neuroticism to unpleasant affects.

Affect Grid Developed by Russell et al. (1989) the technique measures single-item affect in the form of a grid. On the basis of subjective feelings, a subject places an X along two dimensions: pleasantness and arousal. Both aspects will be rated; if both are rated highly, the subject feels great excitement. Similarly, there are feelings of depression, stress, and relaxation. The affect grid displays strong evidence of discriminant validity between the dimensions of pleasure and arousal. Studies that used the affect grid to assess mood provided further evidence of construct validity. However, the scale is not an all-purpose scale and is slightly less reliable than a multiple-item questionnaire for self-reported mood.

Similarly, Warr (1999) used the same scale to measure well-being along a two-dimensional framework of well-being, as shown in Figure 12. A person's well-being may be described in terms of the location in this two-dimensional space of arousal and pleasure. A particular degree of pleasure or displeasure may be accompanied by high levels of mental arousal or a low level of arousal (sleepiness), and a particular level of mental arousal may be either pleasurable (pleasant) or unpleasurable (unpleasant) (Warr, 1999).

Checklists Mood checklists comprise lists of adjectives that describe emotional states. Subjects are required to check their emotions. The mood adjective check list (MACL) developed by Nowlis and Green (1957) contains 130 adjectives with a 4-point scale: "definitely like it," "slightly," "cannot decide," and "definitely not."

Zuckerman and Lubin (1965) developed the multiple-affect adjective checklist (MAACL) comprising 132 items, which they revised in 1985 (MAACL-R). The revised version allowed scoring of several pleasant emotions, taking into account global positive and negative affect as well as sensation seeking.

Thayer (1967) then developed the activation–deactivation adjective checklist (A-DACL), which contained adjectives relating to valenced arousal states (i.e., energetic, lively, active, sleepy, tired-tense, clutched-up, fearful jittery, calm, quiet, and at rest). They used a 4-point scale from "definitely do not feel" to "definitely feel." Izard (1977) developed the multi-item differential emotional scale (DES) with the purpose of assessing multiple discrete emotions.

Interviews Interviews may be performed to assess product pleasure or pleasure from activities or tasks. It is a versatile method and can be performed face to face or through phone conversation. Subjects are asked questions that can be structured, unstructured, and semistructured (Jordan, 2000). A structured interview has a predetermined set of questions, whereas an unstructured interview uses a series of open-ended questions. A semistructured interview allows the investigator to improvise by making unplanned diversions into topics of interest. This method requires that the investigator master the topic

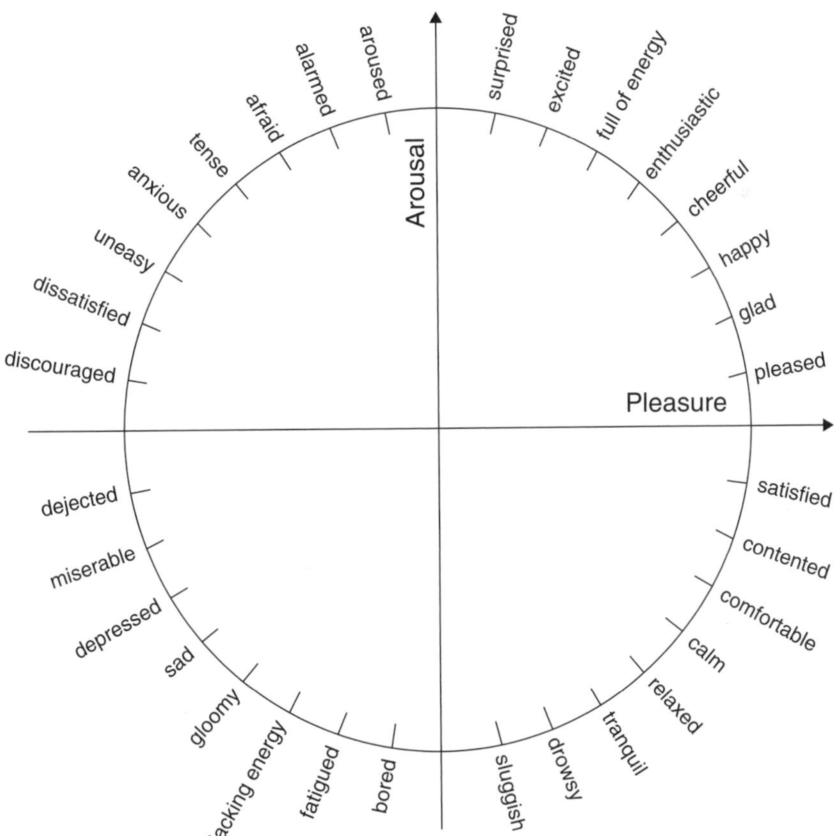

Figure 12 Two-dimensional view of well-being. (From Warr, 1999.)

so that he or she can improvise and understands what to ask for.

Housen (1992) used a nondirective, stream-of-consciousness interview. Participants are asked simply to talk about anything they see as they look at a work of art, to say whatever comes into their minds. There are no directed questions or other prompts to influence the viewer's process. Called the *aesthetic development interview*, it provides a window into a person's thinking processes, and in addition to being empirical, minimizes researcher biases or assumptions. The interviews are often examined by two independent coders to ensure reliability and consistency, and the coding is then charted graphically by computer to enable a comprehensive representation of all thoughts, which also depicts the subject's pattern of thinking.

Subjective Ratings of Emotions Induced by Artifacts These rating scales have been used to document how artifacts make a person feel. By asking a question such as "What does the look of this car makes you feel?" the user is expected to evaluate his or her emotions in relation to the artifact. This approach clearly differs from the general rating methods as used by Helander and Tay (2003) and Khalid and Helander (2004).

PANAS Scales Watson et al. (1988) developed the positive affect negative affect schedule (PANAS). The purpose of PANAS is to measure positive and negative mood states of a person during different times or contexts: today, in a week, a year, and so on. PANAS uses mood adjectives on a 5-point scale: "not at all" or "slight," "a little," "moderately," "quite a bit," and "very much." *Positive affect* (PA) refers to feelings of enthusiasm, alertness, and activeness. A high PA score reflects a state of "high energy, full concentration and pleasurable engagement" (Watson et al., 1988). *Negative affect* (NA), on the other hand, refers to feelings of distress and unpleasurable engagement.

To describe PA, 10 descriptors were used: attentive, interested, alert, excited, enthusiastic, inspired, proud, determined, strong, and active. NA is measured on the following 10 descriptors: distressed, upset, hostile, irritable, scared, afraid, ashamed, guilty, nervous, and jittery. The 10-item scales are shown to be highly internally consistent, largely uncorrelated, and stable at appropriate levels over a period of two months. When used with short-term instructions (e.g., *right now* or *today*) they are sensitive to fluctuations in mood, but when longer-term instructions are used (e.g. *past year* or *general*) they exhibit traitlike stability.

Philip's Questionnaire Jordan (2000) developed a questionnaire for measuring pleasure from products. It has been used by Philips Corporate Design in evaluating their products. The questionnaire has 14 questions, focusing on user's feelings: stimulated, entertained, attached, sense of freedom, excited, satisfaction, rely, miss, confidence, proud, enjoy, relax, enthusiastic, and looking after the product. Using a 5-point scale, ranging from disagree (0) to neutral (2) and strongly agree (4), the close-ended items in the questionnaire covered most of a user's possible responses. To measure pleasure, open-ended items were used as an option. This was particularly useful when the investigator does not know how a product affects the user's evaluation of pleasure.

Product Emotion Measurement Instrument Desmet (2003) developed the product emotion measurement instrument (PrEmo) to assess emotional responses to consumer products. PrEmo is a nonverbal, self-report instrument that measures 14 emotions that are elicited by product design. Participants report their emotions by selecting animations that correspond to their felt emotions. Each emotion is portrayed using an animated cartoon character, with a dynamic facial and bodily expression. It is presented on a computer interface, as illustrated in Figure 13. There are seven faces with positive expressions: inspiration, desire, satisfaction, pleasant surprise, fascination, amusement, admiration, and seven negative faces: disgust, indignant, contempt, disappointment, dissatisfaction, boredom, and unpleasant surprise. These animations were developed with the aim of making them unambiguous and recognizable across cultures. As such, PrEmo was validated in the Netherlands, Japan, Finland, and the United States.

From the results of PrEmo, a tool, called the *emotion navigator*, was developed to assist designers in grasping the emotional potency of their designs (Desmet and Hekkert, 2002). The [product & emotion] navigator is an anecdotal database of some 250 photos of products that elicit emotions. The tool is structured in accordance with a model of product emotion and visualized in an open-ended manner that aims to be inviting and alluring. The model distinguishes important variables in the eliciting conditions of product emotions that can be used to explain how products elicit emotions, and why particular products elicit particular emotions.

5.2.2 Objective Measures

Objective measurements can be obtained either directly or indirectly using measurement techniques. We present two popular methods to record emotions: analysis of facial expressions, and vocal content of speech or voice expressions.

Facial Expressions Numerous methods exist for measuring facial expressions (Ekman, 1982). Facial expressions provide information about (1) *affective state*, including emotions such as fear, anger, enjoyment, surprise, sadness, and disgust, and more enduring moods, such as euphoria, dysphoria, and irritableness; (2) *cognitive state*, such as perplexity, concentration, and boredom; and (3) *temperament and personality*, including such traits as hostility, sociability, and shyness.

Ekman and Friesen (1976) identified five types of messages conveyed by rapid facial signals:

1. *Emotions:* including happiness, sadness, anger, disgust, surprise, and fear

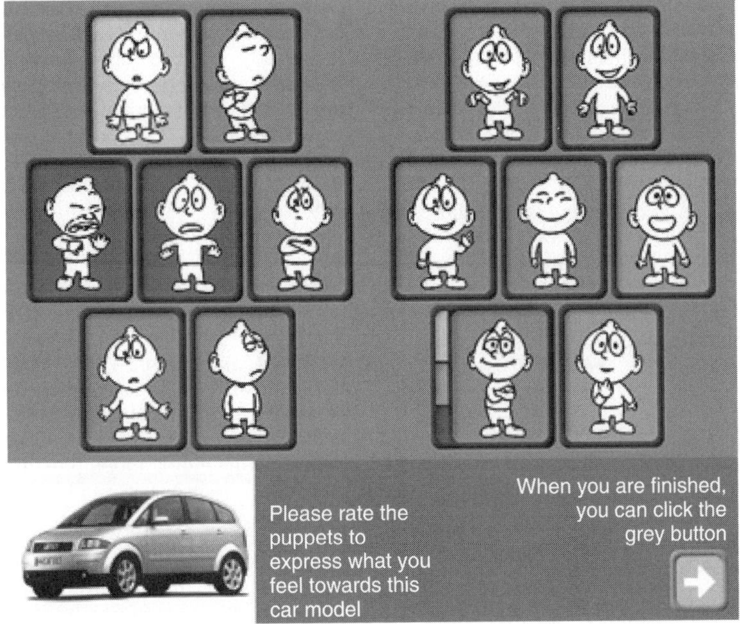
Please rate the puppets to express what you feel towards this car model
When you are finished, you can click the grey button

Figure 13 Product emotion measurement tool. (From Desmet, 2003.)

2. *Emblems:* culture-specific symbolic communicators such as the wink
3. *Manipulators:* self-manipulative associated movements such as lip biting
4. *Illustrators:* actions accompanying and highlighting speech such as a raised brow
5. *Regulators:* nonverbal conversational mediators such as nods or smiles

Measurement of facial expressions may be accomplished by using the facial action coding system (FACS). The method, developed by Ekman and Friesen (1975, 1976), captures the facial changes that accompany an emotional response to an event. FACS was developed by determining how the contraction of each facial muscle (singly and in combination with other muscles) changes the appearance of the face. Videotapes of more than 5000 different combinations of muscular actions were examined to determine the specific changes in appearance and how best to differentiate one appearance from another.

Measurement with FACS is done in terms of action units (AUs) rather than muscular units, for two reasons. First, for a few changes in appearance, more than one muscle is used to produce a single AU. Second, FACS distinguishes between two AUs for the activity of the frontalis muscle that produces wrinkles on the forehead. This is because the inner and outer portion of this muscle can act independently, producing different changes in appearance. There are 46 AUs that account for changes in facial expression, and 12 AUs that describe gross changes in gaze direction and head orientation. To use FACS, the investigator must learn about the appearance and the muscles of the face for each AU. This demands much time and effort.

The maximally discriminative affect coding system (MAX) developed by Izard (1979) measures visible appearance changes in the face. The MAX units are formulated in terms of facial expressions that are relevant to eight specific emotions rather than in terms of individual muscles. Unlike FACS, MAX does not measure all facial actions, but scores only facial movements that relate to the eight emotions.

Facial changes can also be registered using electromyography (EMG). EMG measures nerve impulses to muscles, which produce facial changes or expressions. This measure assumes that emotions are visible through facial expressions, which is the case when people interact with each other.

Davis et al. (1995) compared facial electromyography with standard self-report of affect. He obtained a good correlation between activity of facial muscles and self-report of affect. The pattern of muscular activation could be used to indicate categories of affect, such as happy and sad, and the amplitude of electromyographic signals gave information on degree of emotions. In other words, Davis et al. (1995) was able to categorize as well as quantify affective states using facial electromyography.

Vocal Measures of Emotion Most of the emotions conveyed in speech are from the verbal content. Additionally, the style of the voice, such as pitch, loudness, tone, and timing, can convey information about the speaker's emotional state. This is to be expected because vocalization is "a bodily process sensitive to emotion-related changes" (Larsen and Fredrickson, 1999). A simple and perhaps also the best way to analyze the emotional content would be to listen to recordings of voice messages. Scherer (1986) noted that judges seem to be rather accurate in decoding emotional meaning from vocal cues. Some emotions are easier to recognize than others. Sadness and anger are easiest to recognize, whereas joy, disgust, and contempt are difficult to recognize and distinguish from one another.

Maffiolo and Chateau (2003) investigated the emotional quality of speech messages used by the France Telecom Orange. Each year, vocal servers were used to respond to hundreds of millions of phone calls. The audio messages can be help messages, navigation messages, and information messages. The purpose of the study was to create a set of voice messages that were perceived as friendly, sincere, and helpful.

In their experiment, listening tests were conducted using messages with 20 female speakers, who pronounced two sentences in five elocution styles. Twenty criteria were used to characterize the speech: welcoming, pleasant, aggressive, authoritative, ordinary, warm, clear, shrill, dynamic, exaggerated, expressive, happy, young, natural, professional, speedy, reassuring, sensual, smiling, and stressful. The speech was classified in terms of global impression as well as the hedonic impressions experienced by listeners.

For example, one of the 20 speakers was classified as cheerful, pleasant, clear, and lively, lots of changes that make the message friendly, something is missing, a bit too sharp (but not too much), sympathetic (in a good way, not exaggerated), simple, and convincing. In short, this person's voice was viewed as "pleasant and lively, with expression."

A higher-tech method is to digitize voice recordings and analyze the voice by decomposing the speech sound waves into a set of acoustic parameters and then analyzing the psychoacoustics and psychophonetics content (Larsen and Fredrickson, 1999). This includes analysis of pitch, small deviations in pitch, speaking rate, use of pauses, and intensity. *Emotive Alert*, a voicemail system designed by Inanoglu and Caneel of the Media Lab at the Massachusetts Institute of Technology (Biever, 2005), labels messages according to the caller's tone of voice. It can be installed in a telephone exchange or in an intelligent answering machine. It will analyze incoming messages and send the recipient a text message along with an emoticon indicating whether the message is urgent, happy, excited, or formal. In tests on real-life messages, the software was able to tell the difference between excited and calm and between happy and sad, but found it harder to distinguish between formal and informal, and urgent and nonurgent. This is because excitement and happiness are often conveyed through speech rate and volume, which are easy to measure, whereas formality and urgency are normally expressed through the choice of words and are not easy to measure (Biever, 2005). At the present time the first method, listening to speech, is probably the more reliable.

Regardless of the method used, vocal measures of emotion are sometimes difficult to use since (1) voice is not a continuous variable (people do not speak continuously, thus, vocal indicators of emotion are not always present), (2) positive and negative emotions are sometimes difficult to distinguish and (3) the voice can reflect both emotional/physiological and sociocultural habits, which are difficult to distinguish (Scherer, 1998).

5.2.3 Psychophysiological Measures

Emotions often affect the activity of the autonomic nervous system (ANS) and thereby the activation level and arousal. At the same time, there are increases and decreases in bodily functions, such as in heart functions, electrodermal activity, and respiratory activity (Picard, 1997). Thus, there is a variety of physiological responses that can be measured, including blood pressure, skin conductivity, pupil size, brain waves, and heart rate frequency and variability. For example, in situations of surprise and startle, the electrical conductivity of certain sweat glands is increased momentarily. This is referred to as a galvanic skin response (GSR). These sweat glands are primarily found on the inside of the hands and on the soles of the feet. Electrodes are then attached to measure the electrical conductivity (Helander, 1978). The nerve signals take about 1.5 seconds to travel from the brain to the hand; therefore, the response is a bit delayed.

Researchers in the field of affective computing are actively developing "ANS instruments," such as IBM's emotion mouse (Ark et al., 1999) and a variety of wearable sensors (e.g., Picard, 2000). With these instruments, computers can gather a multitude of psychophysiological information while a person is experiencing an emotion, and learn which pattern is most indicative of which emotion.

ANS responses can be investigated in experiments, for example, by using film clips to induce the type of emotions investigated (e.g., amusement, anger, contentment, disgust, fear, sadness) while electrodermal activity, blood pressure, and electrocardiogram (ECG) are recorded. Therefore, it is possible to associate a variety of emotions with specific physiological reactions.

Although autonomic measures are fruitful, it is important to note the following consideration.

1. Autonomic measures vary widely in how invasive they are. The less invasive measures include pulse rate and skin conductance, whereas measures of blood pressure are often invasive since they use pressure cuffs which are deflated. This may distract a person, so that the emotion is lost.

2. The temporal resolution of various autonomic measures varies widely. Some measures are instantaneous, such as GSR, whereas impedance cardiography, for example, requires longer duration for reliable measurement (Larsen and Fredrickson, 1999).

3. Different measures have different sensitivity. Depending on the emotion that is recorded, it is best

first to validate the particular physiological measures so as to understand if it is sensitive enough to record differences in the intensity of the emotion.

5.2.4 Performance Measures

Performance measures typically indicate the effect of emotions on decision making. Emotion-sensitive performance measures may be obtained through judgment tasks. One popular task is to have participants make probability estimates of the likelihood of various good and bad events. It has been shown that persons in unpleasant emotional states tend to overestimate the probability of bad events (Johnson and Tversky, 1983). Ketelaar (1989) showed that people in a good mood also overestimated the probability of pleasant events. Another useful performance task is to ask participants to generate associations to positive, neutral, and negative stimuli. Mayer and Bremer (1985) showed that a change in a person's mood correlated with changes in performance in affect-sensitive tasks involving cognitive and psychomotor skills.

A second category of performance measures involves information-processing parameters. Reaction times in lexical decision tasks have been shown to be sensitive to affective states (Challis and Krane, 1988). The task involves judging if a string of letters presented on the computer screen represents a word or nonword. Participants in positive affective states are quicker and sometimes more accurate at judging positive words as words compared to participants in neutral states, and vice versa for unpleasant moods (Niedenthal and Setterlund, 1994).

5.3 Conclusions

The use of subjective methods such as self-report, single-item, and multiple-item measures has their drawbacks. Such methods rely heavily on the use of words and adjectives. The subject's vocabulary should be taken into account, because some people may have little comprehension of some of the words used in the methods mentioned above. The subject should be allowed to use his or her principal language, or else some feelings might be misinterpreted. The words or adjectives must be concise and easy to understand and take into account cultural as well as contextual factors (Larsen and Fredrickson, 1999).

A major advantage of physiological methods (or nonverbal instruments) is that they are language independent and can be used in different cultures. A second advantage is that they are unobtrusive and do not disturb participants during the measurement (Desmet, 2003). There are however limitations to the physiological measures. They can assess reliably only a limited set of "basic" emotions and cannot assess mixed emotions. For pleasures of the mind, it is doubtful if any of the psychophysiological methods will be sensitive enough to capture the subtleness of emotions.

Objective methods such as vocal content and facial expressions can be used to measure mixed emotions, but they are difficult to apply between cultures. It would be important to make cultural comparisons between vocal and facial expressions. For this purpose a multimedia database can be developed and shared by the research community. The database could contain images of faces (still and motion), vocalizations and speech, psychophysiological correlates of specific facial actions, and interpretations of facial scores in terms of emotional state, cognitive process, and other internal processes. This would facilitate an integration of research efforts by highlighting contradictions and consistencies, and suggest fruitful avenues for new research.

6 DISCUSSION

Emotions are often elicited by products, such as art, clothing, and consumer goods; therefore, designers must consider affect and emotion in design. Today, many corporations challenge designers to manipulate the emotional impact of designs. Nokia design is an example. Emotional responses induce customers to pick a particular model among many; emotions thus influence purchase decisions. In practice, user emotions toward products are well established and sometimes difficult to manipulate.

Emotion may be the strongest differentiator in user experience. It triggers both conscious and unconscious responses to a product or an interface. There are many important reasons to consider emotion in product design, such as to increase sales and keep customers happy. This is done by maximizing positive emotions while minimizing negative emotions. Understanding and reducing users' anxiety and fears (negative emotions) can help to increase satisfaction with products. Poor usability will also induce negative responses such as frustration, annoyance, anger, and confusion.

On the other hand, even moderate fluctuations in positive emotions can systematically improve cognitive processing. A happy person has an open mind, whereas a negative person is restrained in processing of information. When products result in positive user experiences, the emotional effects are often more important to the customer than gains in productivity, efficiency, and effectiveness. Negative emotions such as frustration, anxiety, and so on, should definitely be avoided in conceptualizing product design.

However, emotion is not an exclusive factor in defining a successful user experience. Every single product feature affects the "experience," which can be complex and multifaceted. Furthermore, emotions are culturally specific and variable. Because there is no such thing as a neutral interface, any design will elicit emotions from the user and the designer (Gaver, 1996). The designer should aim to "control" the user experience through a deliberate design effort, thus bridging the gap between the affective user and the designer's environment, as outlined in our framework. However, measuring affective responses to designed objects can be problematic. So is designing affect into a product. Desmet's (2003) PrEmo tool is a good start at supporting designers.

Separating emotion from cognitive functions does not seem helpful from a research perspective or from a design perspective. Instead, an integrated view of emotion and cognition is taking hold, not

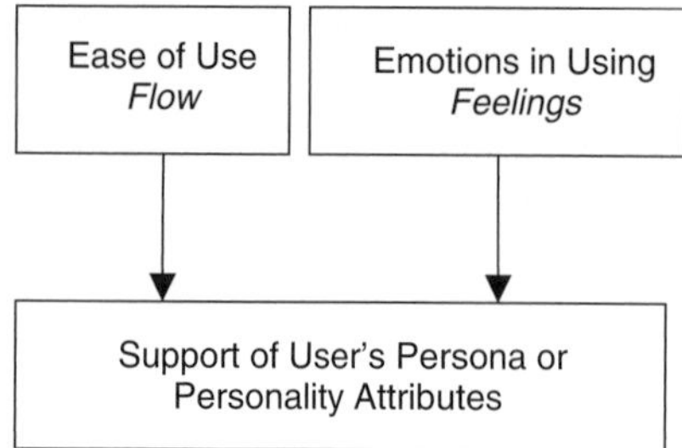

Figure 14 Ease of use and emotions in using support the user's persona.

only in neuroscience but also in product design. A product should be designed to support customer needs, including the customer's persona or personality attributes. This can be done by providing flow—or ease of use—and inducing feelings—or emotions—in interacting with the product (Figure 14).

We can illustrate the model in Figure 14 by using a computer game as an example. An application software that is designed with a good flow will be easy to use and has novel and adaptive controls enabling quick moves and perceived situational control, thus inducing positive emotions of fun, enjoyment, and satisfaction. The pleasures derived from the optimal experience of seamless interaction and usable functions are considered *hedonic benefits* that can enhance the "skillful" and "confident" persona of the user. They also enhance the sense of virtuosity in interaction, a common need among users.

On the other hand, a poorly designed application with complex user interface and controls impedes natural interaction, leading to negative emotions such as quitting the game, becoming moody, swearing at the machine, or even kicking it (Luczak et al., 2003). Such displeasures constitute *hedonic penalties* and will lead to sentiments of angriness and frustration. In other words, hedonic affordances invite an emotional reaction from the user through the product's physical attributes that result in the user's appraisal or frustration. Pleasurable interaction may be derived by integrating adaptability into designs and providing design features that enhance user control.

In sum, customers tend to make decisions based on their feelings, perceptions, values, and reflections that usually come from gut feelings rather than logical or rational thinking. As such, designers and manufacturers should consider emotional design a bottom line in product design.

Measuring affective responses to designed objects can be problematic. Further research is needed to develop expressions of emotions that are quantifiable so that they can easily be verified. Various issues warrant further investigation, and we raise a few below.

1. *Lack of facial and physiological signals.* Pleasures of the mind are not accompanied by any distinctive facial expression (Kubovy, 1999). For example, a person viewing the *Mona Lisa* may feel elated, but nothing is revealed on his or her face, and there is no distinctive physiological response pattern. This is very different from social interaction, such as a conversation with a colleague at work, where half of the message is in the person's face. Since one may not be able to use either physiological measures or facial expressions, one is left with subjective measures. There is nothing wrong with subjective methods; the data come directly from the user. In many instances, verbal or written reports provide the most valuable information. Asking people in interviews and questionnaires can generate very informative data. The forever-pressing problem is what questions we should ask that can differentiate between products.

2. *Design for context of use and activity.* In affective design, our primary goal may be to design a product. However, products are used in a context—an envelope of product semantics surrounds the product itself. Products communicate with users and can never be contextually neutral. This makes a designer's task an even greater challenge. To increase human enjoyment or engagement, perhaps the entire setting (activity, context of use) needs to be designed, not just the artifact (Aboulafia and Bannon, 2004). Is it realistic to ask designers to live up to this challenging perspective?

3. *Tapping positive emotions.* The expectations of users in terms of customer needs are changing: Functionality, attractiveness, ease of use, affordability, and safety are taken for granted. The new trends are for objects or artifacts that inspire users, enhance their lives, and evoke emotions and dreams in their minds (Demirbilek and Sener, 2003). This requires research into the thoughts and dreams that are related to positive emotions and pleasurable experiences. The "method" as described by Jordan (Blythe, 2004) may provide an introspective approach. The product concept prospecting approach, based on realistic scenarios, may be another (Cayol and Bonhoure, 2004).

4. *Designing affect into a product.* The notion that human feel or touch can be designed into a product has led to Kreifeldt's (2001) "physics" of emotion. He claimed that it is necessary to build sensors that can transduce the weight and moment of inertia for the person to feel an object, thereby making it pleasurable. But researchers (e.g., Loewenstein and Schkade, 1999) are quick to lament that the mechanics of hedonics (what makes people happy) are not fully understood. For example, the effects of satiation and ownership of objects in relation to pleasure have been reasonably well researched. Yet there are conflicting results. Some studies found that subjects' feelings did change substantially over time, but they had little idea, at the outset, about how they would change.

In conclusion, then, this chapter is just a beginning of a very promising and challenging research area. It opens a minefield of conceptual and methodological issues for research and development. Much needs to be done to develop predictive models of affect and pleasure for design of products and interfaces. The current methodology is still immature. In the future, when we understand the mechanics of hedonomics, there will be significant rewards in terms of monetary benefits for developers of pleasurable products, as well as many happy customers and users.

The desire for informed knowledge drew human factors professionals, industrial designers, engineers, and behavioral and social scientists to the 2001 International Conference on Affective Human Factors Design in Singapore (Helander et al., 2001). An outcome of a meeting of minds is a change of hearts—a welcome paradigm shift in human factors and ergonomics: from pain and performance to pleasure.

REFERENCES

Aboulafia, A., and Bannon, L. J. (2004), "Understanding Affect in Design: An Outline Conceptual Framework," *Theoretical Issues on Ergonomics Science*, Vol. 5, No. 1, pp. 4–15.

Apter, M. J. (1989), *Reversal Theory: Motivation, Emotion and Personality*, Routledge, London.

Ark, W., Dryer, D. C., and Lu, D. J. (1999), "The Emotion Mouse," *in Proceedings of HCI International '99*, Munich, Germany, August.

Ashby, F. G., Isen, A. M., and Turken, U. (1999), "A Neuropsychological Theory of Positive Affect and Its Influence on Cognition," *Psychological Review*, Vol. 106, No. 3, pp. 529–550.

Bannon, L. (1993), "CSCW: An Initial Exploration," *Scandinavian Journal of Information Systems*, Vol. 5, pp. 3–24.

Bannon, L., and Bødker, S. (1991), "Beyond the Interface: Encountering Artifacts in Use, in *Designing Interaction: Psychology at the Human-Computer Interface*, J. M. Carroll, Ed., Cambridge University Press, New York, pp. 227–253.

Bergman, E. (2000), *Information Appliances and Beyond*, Academic Press, San Diego, CA.

Berlyne, D. E. (1974), *Studies in the New Experimental Aesthetics*, Hemisphere Publishing, Washington, DC.

Biever, C. (2005), "Voicemail Software Recognises Callers' Emotions," *New Scientist*, Vol. 2481, p. 21.

Bih, H. D. (1992), "The Meaning of Objects in Environmental Transitions: Experiences of Chinese Students in the United States," *Journal of Environmental Psychology*, Vol. 12, pp. 135–147.

Blythe, M. (2004), "Interview with Patrick Jordan," *Interactions*, Vol. 11, No. 4, pp. 40–41.

Brave, S., and Nass, C. (2003), "Emotion in Human–Computer Interaction," in *The Human–Computer Interaction Handbook: Fundamentals, Evolving Technologies and Emerging Applications*, J. Jacko and A. Sears, Eds., Lawrence Erlbaum Associates, Mahwah, NJ, pp. 81–96.

Brown, S. L., and Schwartz, G. E. (1980), "Relationships Between Facial Electromyography and Subjective Experience During Affective Imagery," *Biological Psychology.*, Vol. 11, No. 1, pp. 49–62.

Buck, R. (1999), "The Biological Affects: A Typology," *Psychological Review*, Vol. 106, No. 2, pp. 301–336.

Cabanac, M. (1992), "Pleasure: The Common Currency," *Journal of Theoretical Biology*, Vol. 155, pp. 173–200.

Camerer, C., Loewenstein, G., and Prelec, D. (2003), "Neuroeconomics: How Neuroscience Can Inform Economics," working paper.

Carroll, J. M. (2004), "Beyond Fun," *Interactions*, Vol. 11, No. 4, pp. 38–40.

Cayol, A., and Bonhoure, P. (2004), "User Pleasure in Product Concept Prospecting," *Theoretical Issues in Ergonomics Science*, Vol. 5, No. 1, pp. 16–26.

Challis, B. H., and Krane, R. V. (1988), "Mood Induction and the Priming of Semantic Memory in a Lexical Decision Task: Assymetric Effects of Elation and Depression," *Bulletin of the Psychonomic Society*, Vol. 26, pp. 309–312.

Chapanis, A. (1995), "Ergonomics in Product Development: A Personal View," *Ergonomics*, Vol. 38, No. 8, pp. 1625–1638.

Chen, L. L., and Liang, J. (2001), "Image Interpolation for Synthesizing Affective Product Shapes," in *Proceedings of the International Conference on Affective Human Factors Design*, M. G. Helander, H. M. Khalid, and M. P. Tham, Eds., ASEAN Academic Press, London, pp. 531–537.

Childs, T. (2001), "An Essay on Affective Design: Applied Science or Art? And Opportunities for Enhanced Packaging Development," unpublished paper, University of Leeds, Leeds, Yorkshire, England.

Coelho, D. A., and Dahlman, S. (2000), "Comfort and Pleasure," in *Pleasure in Product Use*, P. W. Jordan and B. Green, Eds., Taylor & Francis, London, pp. 321–331.

Cross, N. (2000), *Engineering Design Methods: Strategies for Product Design*, 3rd ed., Wiley, Chichester, West Sussex, England.

Crozier, R. (1994), *Manufactured Pleasures: Psychological Responses to Design*, Manchester University Press, Manchester, Lancashire, England.

Csikszentmihalyi, M. (1975), *Beyond Boredom and Anxiety*, Jossey-Bass, San Francisco.

Csikszentmihalyi, M. (1990), *Flow: The Psychology of Optimal Experience*, HarperCollins, New York.

Csikszentmihalyi, M. (1992), "Imagining the Self: An Evolutionary Excursion," *Poetics*, Vol. 21, No. 3, pp. 153–167.

Csikszentmihalyi, M. (1995), "Design and Order in Everyday Life," in *The Idea of Design*, V. Margolin and R. Buchanan, Eds., MIT Press, Cambridge, MA, pp. 118–126.

Damasio, A. R. (1994), *Descartes' Error: Emotion, Reason, and the Human Brain*, Grosset/Putnam, New York.

Davis, W. J., Rahman, M. A., Smith, L. J., Burns, A., Senecal, L., McArthur, D., Halpern, J. A., Perlmutter, A., Sickels, W., and Wagner, W. (1995), "Properties of Human Affect Induced by Static Color Slides (IAPS): Dimensional, Categorical, and Electromyographic Analysis," *Biological Psychology*, Vol. 41, No. 3, pp. 229–253.

Demirbilek, O., and Sener, B. (2003), "Product Design, Semantics, and Emotional Response," *Ergonomics*, Vol. 46, No. 13–14, pp. 1346–1360.

Desmet, P. M. A. (2003), "Measuring Emotion: Development and Application of an Instrument to Measure Emotional Responses to Products," in *Funology: From Usability to Enjoyment*, M. A. Blythe, A. F. Monk, K. Overbeeke, and P. C. Wright, Eds., Kluwer Academic, Dordrecht, The Netherlands,, pp. 111–123.

Desmet, P. M. A., and Hekkert, P. (2002), "The Basis of Product Emotions," in *Pleasure with Products, Beyond Usability*, W. Green and P. Jordan, Eds., Taylor & Francis, London, pp. 60–68.

Diener, E. (1994), "Assessing Subjective Well-Being: Progress and Opportunities," *Social Indicators Research*, Vol. 31, pp. 103–157.

Eid, M., and Diener, E. (1999), "Intraindividual Variability in Affect: Reliability, Validity, and Personality Correlates," *Journal of Personality and Social Psychology*, Vol. 76, No. 4, pp. 662–676.

Ekman, P. (1982), *Emotion in the Human Face*, 2nd ed., Cambridge University Press, New York.

Ekman, P. (1992), "An Argument for Basic Emotions," *Cognition and Emotion*, Vol. 6, pp. 169–200.

Ekman, P. (1994), "Strong Evidence for Universals in Facial Expressions: A Reply to Russell's Mistaken Critique," *Psychological Bulletin*, Vol. 115, No. 2, pp. 268–287.

Ekman, P., and Friesen, W. V. (1975), *Unmasking the Face: A Guide to Recognizing Emotions from Facial Clues*, Prentice-Hall, Englewood Cliffs, NJ.

Ekman, P., and Friesen, W. V. (1976), "Measuring Facial Movement," *Environmental Psychology and Nonverbal Behavior*, Vol. 1, No. 1, pp. 56–75.

Epstein, S. (1994), "Integration of the Cognitive and Psychodynamic Unconscious," *American Psychologist*, Vol. 49, No. 8, pp. 709–724.

Faulkner, T., and Caplan, S. (1985), "The Role of Human Factors Specialists in the Development of Consumer/Commercial Products," workshop presented at the 4th Symposium on Human Factors and Industrial Design in Consumer Products, St. Paul, MN.

Feldman-Barrett, L. (1998), "Discrete Emotions or Dimensions: The Role of Valence Focus and Arousal Focus," *Cognition and Emotion*, Vol. 12, pp. 579–599.

Fishbein, M., and Ajzen, L. (1972), "Attitudes and Opinions," *Annual Review of Psychology*, Vol. 23, pp. 487–554.

Fredrickson, B. L. (1998), "What Good Are Positive Emotions?" *Review of General Psychology*, Vol. 2, No. 3, pp. 300–319.

Gaver, W. (1996), "Affordances for Interaction: The Social Is Material for Design," *Ecological Psychology*, Vol. 8, No. 2, pp. 111–129.

Green, W. S., and Jordan, P. W. (2002), *Pleasure with Products: Beyond Usability*, Taylor & Francis, London.

Gilovich, T., Griffin, D., and Kahneman, D. (2002), *Heuristics and Biases: The Psychology of Intuitive Decision Making*, Cambridge University Press, Cambridge.

Hancock, P. A., Pepe, A. A., and Murphy, L. L. (2005), "Hedonomics: The Power of Positive and Pleasurable Ergonomics," *Ergonomics in Design*, Vol. 13, No. 1, pp. 8–14.

Hassenzahl, M. (2004), "The Interplay of Beauty, Goodness, and Usability in Interactive Products," *Human–Computer Interaction*, Vol. 19, No. 4, pp. 319–349.

Helander, M. G. (1978), "Applicability of Drivers' Electrodermal Response to the Design of the Traffic Environment," *Journal of Applied Psychology*, Vol. 9, pp. 481–488.

Helander, M. G., and Du, X. (1999), "From Kahneman to Kano: A Comparison of Models to Predict Customer Needs," in *Proceedings of the International Conference on TQM and Human Factors*, Centre for Human Technology Organization, Linköping University, Linköping, Sweden, pp. 315–321.

Helander, M. G., and Tay, D. W. L. (2003), "What Is in a Word? Describing Affect in Product Design, in *Proceedings of the 15th Triennial Congress of the International Ergonomics Association*, Ergonomics Society of Korea, Seoul, Korea.

Helander, M. G., and Tham, M. P. (2003), "Hedonomics: Affective Human Factors Design," *Ergonomics*, Vol. 46, No. 13–14, pp. 1269–1272.

Helander, M. G., and Zhang L. (1997), "Field Studies of Comfort and Discomfort in Sitting," *Ergonomics*, Vol. 40, No. 9, pp. 895–915.

Helander, M. G., Khalid, H. M., and Tham, M. P. (2001), *Proceedings of the International Conference on Affective Human Factors Design*, ASEAN Academic Press, London.

Hertzberg, F. (1966), *Work and Nature of Man*, Thomas Y. Crowell, New York.

Hoffman, D. L., and Novak, T. P. (1996), "Marketing in Hypermedia Computer-Mediated Environments: Conceptual Foundations," *Journal of Marketing*, Vol. 60, No. 3, pp. 50–68.

Housen, A. (1992), "Validating a Measure of Aesthetic Development for Museums and Schools," *ILVS Review*, p. 2.

Isen, A. M. (1999), "On the Relationship Between Affect and Creative Problem Solving," in *Affect, Creative Experience, and Psychological Adjustment*, S. Russ, Ed., Taylor & Francis, Philadelphia, pp. 3–17.

Izard, C. E. (1977), *Human Emotions*, Plenum Press, New York.

Izard, C. E. (1979), *The Maximally Discriminative Facial Movement Coding System (MAX)*, Instructional Resources Centre, University of Delaware, Newark, DE.

Johnson, D., and Wiles J. (2003), "Effective Affective User Interface Design in Games," *Ergonomics*, Vol. 46, No. 13–14, pp. 1332–1345.

Johnson, E. J., and Tversky, A. (1983), "Affect, Generalization, and the Perception of Risk," *Journal of Personality and Social Psychology*, Vol. 45, pp. 21–31.

Jordan, P. W. (1998), "Human Factors for Pleasure in Product Use," *Applied Ergonomics*, Vol. 29, No. 1, pp. 25–33.

Jordan, P. W. (2000), "The Four Pleasures: A Framework for Pleasures in Design," in *Proceedings of the Conference on Pleasure Based Human Factors Design*, P. W. Jordan, Ed., Philips Design, Groningen, The Netherlands.

Jordan, P. W. (2002), *How to Make Brilliant Stuff That People Love and Make Big Money Out of It*, Wiley, Chichester, West Sussex, England.

Kahneman, D. (2002), "Maps of Bounded Rationality: A Perspective on Intuitive Judgment and Choice," Nobel Prize Lecture, Princeton University, Princeton, NJ, December 8.

Kahneman, D., and Tversky, A. (1984), "Choices, Values and Frames," *American Psychologist*, Vol. 39, pp. 341–350.

Kahneman, D., Diener, E., and Schwarz, N., Eds. (1999), *Well-Being: Foundations of Hedonic Psychology*, Russell Sage Foundation, New York.

Kano, N. (1984), "Attractive Quality and Must-Be Quality" (in Japanese), *Hinshitsu*, Vol. 14, No. 2, pp. 39–48.

Karlsson, B. S. A., Aronsson, N., and Svensson, K. A. (2003), "Using Semantic Environment Description as a Tool to Evaluate Car Interiors," *Ergonomics*, Vol. 46, No. 13–14, pp. 1408–1422.

Ketelaar, T. (1989), "Examining the Circumplex Model of Affect in the Domain of Mood-Sensitive Tasks," master's thesis, Purdue University, West Lafayette, IN.

Khalid, H. M. (1999), "Uncovering Customer Needs for Web-Based DIY Product Design," *Proceedings of the International Conference on TQM and Human Factors*, Centre for Human Technology Organization, Linköping University, Linköping, Sweden, pp. 343–348.

Khalid, H. M. (2004), "Conceptualizing Affective Human Factors Design," *Theoretical Issues in Ergonomics Science*, Vol. 5, No. 1, pp. 1–3.

Khalid, H. M., and Helander, M. G. (2004), "A Framework for Affective Customer Needs in Product Design," *Theoretical Issues in Ergonomics Science*, Vol. 5, No. 1, pp. 27–42.

Kreifeldt, J. (2001), "Designing 'Feel' into a Product," in *Proceedings of the International Conference on Affective Human Factors Design*, in M. G. Helander, H. M. Khalid, and M. P. Tham, Eds., ASEAN Academic Press, London, pp. 25–30.

Kreitler, H., and Kreitler, S. (1972), *Psychology and the Arts*, Duke University Press, Durham, NC.

Krippendorff, K. (1995), "On the Essential Contexts of Artifacts or on the Proposition That 'Design Is Making Sense (of Things),' " in *The Idea of Design*, V. Margolin and R. Buchanan, Eds., MIT Press, Cambridge, MA, pp. 156–184.

Krippendorff, K. (2005), *The Semantic Turn: A New Foundation for Design*, Taylor & Francis/CRC Press, Boca Raton, FL.

Kubovy, M. (1999), "On the Pleasures of the Mind," in *Well-Being: The Foundations of Hedonic Psychology*, D. Kahneman, E. Diener, and N. Schwarz, Eds., Russell Sage Foundation, New York, pp. 134–154.

Küller, R. (1975), *Semantisk Milöbeskriving (SMB)* [*Semantic Descriptions of Environments* (Impression of Colors and Colored Environments)], ByggforskningsrÅdet, Stockholm, Sweden.

Kurosu, M., and Kishimura, K. (1995), "Apparent Usability Versus Inherent Usability: Experimental Analysis on the Determinants of the Apparent Usability," in *Conference Companion on Human Factors in Computing Systems*, ACM Press, New York, pp. 292–293.

Larsen, R. J., and Fredrickson B. (1999), "Measurement Issues in Emotion Research," in *Well-Being: The Foundations of Hedonic Psychology*, D. Kahneman, E. Diener, and N. Schwarz, Eds., Russell Sage Foundation, New York, pp. 40–60.

Larson, R., and Csikszentmihalyi, M. (1983), "The Experience Sampling Method," *New Directions for Methodology of Social and Behavioral Science*, Vol. 15, pp. 41–56.

LeDoux, J. E. (1995), "Emotion: Clues from the Brain," *Annual Review of Psychology*, Vol. 46, pp. 209–235.

Leontjev, A. N. (1978), *Activity, Consciousness and Personality*, Prentice-Hall, London.

Lerner, J. S., Small, D. A., and Loewenstein, G. (2004), "Heart Strings and Purse Strings: Carryover Effects of Emotions on Economic Decisions," *Psychological Science*, Vol. 15, No. 5, pp. 337–341.

Loewenstein, G., and Schkade, D. (1999), "Wouldn't It Be Nice? Predicting Future Feelings," in *Well-Being: Foundations of Hedonic Psychology*, D. Kahneman, E. Diener, and N. Schwarz, Eds., Russell Sage Foundation, New York, pp. 85–105.

Luczak, H., Roetting, M., and Schmidt, L. (2003), "Let's Talk: Anthropomorphization as Means to Cope with Stress of Interacting with Technical Devices," *Ergonomics*, Vol. 46, No. 13–14, pp. 1361–1374.

Maffiolo, V., and Chateau, N. (2003), "The Emotional Quality of Speech in Voice Services," *Ergonomics*, Vol. 46, No. 13–14, pp. 1375–1385.

Mäkelä, A., and Fulton-Suri, J. (2001), "Supporting Users' Creativity: Design to Induce Pleasurable Experiences," in *Proceedings of the International Conference on Affective Human Factors Design*, M. G. Helander, H. M. Khalid, and M. P. Tham, Eds., ASEAN Academic Press, London, pp. 387–394.

Margolin, V., and Buchanan, R. (1995), *The Idea of Design*, MIT Press, Cambridge, MA.

Martindale, C., and Uemura, A. (1983), "Stylistic Change in European Music," *Leonardo*, Vol. 16, pp. 225–228.

Maslow, A. (1968), *Towards a Psychology of Being*, Van Nostrand, Princeton, NJ.

Mayer, J. D., and Bremer, D. (1985), "Assessing Mood with Affect-Sensitive Tasks," *Journal of Personality Assessment*, Vol. 49, pp. 95–99.

McManus, I. C. (1981), "The Aesthetics of Color," *Perception*, Vol. 10, pp. 651–666.

Minsky, M. (2004), "Emotion Machine," retrieved April 23, 2005 from http://web.media.mit.edu/~ minsky/E1/eb1.html.

Nagamachi, M. (1989), *Kansei Engineering*, Kaibundo Publisher, Tokyo.

Nagamachi, M. (2001), *Research on Kansei Engineering: Selected Papers on Kansei Engineering*, Nakamoto Printing, Hiroshima, Japan.

Niedenthal, P. M., and Setterlund, M. B. (1994), "Emotion Congruence in Perception," *Personality and Social Psychology Bulletin*, Vol. 20, pp. 401–411.

Nielsen, J. (1996), "Seductive Interfaces," retrieved August 5, 2004, from http://www.useit.com/papers/.

Norman, D. (1991), "Cognitive Artifacts," in *Designing Interaction: Psychology at the Human–Computer Interface*, J. Carroll, Ed., Cambridge University Press, Cambridge, pp. 17–38.

Norman, D. A. (2004), *Emotional Design: Why Do We Love (or Hate) Everyday Things*, Basic Books, New York.

Nowlis, V., and Green, R. (1957), "The Experimental Analysis of Mood," Technical Report Nonr-668(12), Office of Naval Research, Washington, DC.

Osgood, C., Suci, G., and Tannenbaum, P. (1967), *The Measurement of Meaning*, University of Illinois Press, Urbana–Champaign, IL.

Peter, J. P., and Olson, J. C. (1996), *Consumer Behavior and Marketing Strategy*, Richard D. Irwin, Chicago.

Picard, R. W. (1997), *Affective Computing*, MIT Press, Cambridge, MA.

Picard, R. W. (2000), "Towards Computers That Recognize and Respond to User Emotion," *IBM Systems Journal*, Vol. 39, pp. 3–4.

Purcell, A. T. (1986), "Environmental Perception and Affect: A Schema Discrepancy Model," *Environment and Behavior*, Vol. 18, pp. 3–30.

Rasmussen, J., Pejtersen, M. A., and Goodstein, L. P. (1994), *Cognitive Systems Engineering*, Wiley, New York.

Rosenberg, E. L., and Ekman, P. (1994), "Coherence Between Expressive and Experiential Systems in Emotion," *Cognition and Emotion*, Vol. 8, pp. 201–229.

Ross, M. (1989), "Relation of Implicit Theories to the Construction of Personal Histories," *Psychological Review*, Vol. 96, pp. 341–357.

Russell, J. A., Weiss, A., and Mendelsohn, G. A. (1989), "The Affect Grid: A Single-Item Scale of Pleasure and Arousal," *Journal of Personality and Social Psychology*, Vol. 57, pp. 493–502.

Sandvik, E., Diener, E., and Seidlitz, L. (1993), "Subjective Well-Being: The Convergence and Stability of Self-Report and Non Self-Report Measures," *Journal of Personality*, Vol. 61, pp. 317–342.

Scherer, K. R. (1986), "Vocal Affect Expression: A Review and a Model for Future Research," *Psychological Bulletin*, Vol. 99, pp. 143–165.

Scherer, K. R. (1998), "Analyzing Emotion Blends," in *Proceedings of the 10th Conference of the International Society for Research on Emotions*, A. Fischer, Ed., Würzburg, Germany, pp. 142–148.

Schimmack, U. (2003), "Affect Measurement in Experience Sampling Research," *Journal of Happiness Studies*, Vol. 4, pp. 79–106.

Schneider, W., and Shiffrin, R. M. (1997), "Controlled and Automatic Human Information Processing, I: Detection, Search and Attention," *Psychological Review*, Vol. 84, No. 1, pp. 1–66.

Schütte, H., and Ciarlante, D. (1998), *Consumer Behavior in Asia*, Macmillan, London.

Seligman, M. E. P., and Csikszentmihalyi, M. (2000), "Positive Psychology: An Introduction," *American Psychologist*, Vol. 55, pp. 5–14.

Singer, J. A., and Salovey, P. (1988), "Mood and Memory: Evaluating the Network Theory of Affect," *Clinical Psychology Review*, Vol. 8, No. 2, pp. 211–251.

Slovic, P., Finucane, M., Peters, E., and MacGregor, D. G. (2002), "The Affect Heuristic," in *Intuitive Judgment: Heuristics and Biases*, T. Gilovic, D. Griffin, and D. Kahneman, Eds., Cambridge University Press, Cambridge, pp. 397–420.

Teigen, K. H. (1987), "Intrinsic Interest and the Novelty-Familiarity Interaction," *Scandinavian Journal of Psychology*, Vol. 28, pp. 199–210.

Thaler, R. H. (1980), "Toward a Positive Theory of Consumer Choice," *Journal of Economic Behavior and Organizations*, Vol. 1, pp. 39–60.

Thayer, R. E. (1967), "Measurement of Activation Through Self-Report," *Psychological Reports*, Vol. 20, pp. 663–678.

Tiger, L. (1992), *The Pursuit of Pleasure*, Little, Brown, Boston.

Titchener, E. B. (1910), "The Past Decade in Experimental Psychology," *American Journal of Psychology*, Vol. 21, pp. 404–421.

Tractinsky, N., Katz, A. S., and Ikar, D. (2000), "What Is Beautiful Is Usable," *Interacting with Computers*, Vol. 13, pp. 127–145.

Tversky, A. (1972), "Elimination by Aspect: A Theory of Choice," *Psychological Review*, Vol. 61, pp. 281–289.

Velásquez, J. D. (1998a), "From Affect Programs to Higher Cognitive Emotions: An Emotion-Based Control Approach," retrieved January 1, 2002, from http://www.ai.mit.edu/people/jvelas/ebaa99/.

Velásquez, J. D. (1998b), "When Robots Weep: Emotional Memories and Decision-Making," in *Proceedings of the 15th National Conference on Artificial Intelligence*, AAAI Press, Menlo Park, CA, pp. 70–75.

Vygotsky, L. (1962), *Thought and Language*, MIT Press, Cambridge, MA.

Warr, P. (1999), in "Well-Being and the Workplace," *Well-Being: Foundations of Hedonic Psychology*, D. Kahneman, E. Diener, and N. Schwarz, Eds., Russell Sage Foundation, New York, pp. 392–412.

Watson, D., Lee, A. C., and Tellegen, A. (1988), "Development and Validation of Brief Measures of Positive and Negative Affect: The PANAS Scales," *Journal of Personality and Social Psychology*, Vol. 54, No. 6, pp. 1063–1070.

Webster, J., Trevino, L. K., and Ryan, L. (1993), "The Dimensionality and Correlates of Flow in Human–Computer Interactions," *Computers in Human Behavior*, Vol. 9, pp. 411–426.

Winkelman, P., Zajonc, R. B., and Schwarz, N. (1977), "Subliminal Affective Priming Resists Attributional Interventions," *Cognition and Emotion*, Vol. 11, No. 4, pp. 433–465.

Zajonc, R. B. (1980), "Feeling and Thinking: Preferences Need No Inferences," *American Psychologist*, Vol. 35, No. 2, pp. 151–175.

Zajonc, R. B. (1998), "Emotions," in *The Handbook of Social Psychology*, D. Gilbert, S. Fiske, and G. Lindzey, Eds., Oxford University Press, New York, pp. 591–632.

Zuckerman, M., and Lubin B. (1965), *The Multiple Affect Adjective Check List*, Educational and Industrial Testing Service, San Diego, CA.

PART 4

EQUIPMENT, WORKPLACE, AND ENVIRONMENTAL DESIGN

CHAPTER 22

WORKPLACE DESIGN

Nicolas Marmaras and Dimitris Nathanael
National Technical University of Athens
Athens, Greece

1 INTRODUCTION

Workplace design deals with the shape, dimensions, and layout (i.e., the placement and orientation) of the various material elements that surround one or more working persons. Examples of such elements are the seat, the working surfaces, the desk, the equipment, the tools, and the controls and displays used during work, but also the passages, the windows, the heating/cooling equipment, and so on.

Ergonomic workplace design aims at improving work performance (in both quantity and quality), through (1) minimizing the physical strain and workload of the working person, (2) facilitating task execution (i.e., ensuring effortless information exchange with the environment, minimization of physical constraints, etc.), (3) ensuring occupational health and safety, and (4) achieving ease of use of the various workplace elements.

Designing a workplace that meets ergonomics principles is a difficult problem, as one should consider an important number of interacting and variable elements and try to meet an important number of requirements, some of which may be contradictory. In fact, there is interdependence among the workplace components, the working person, the task requirements, the environment, and the habitual body movements and postures that working persons adopt (Figure 1).

Consider, for example, a person working in a computerized office (task requirement: work with a computer). If the desk (workplace component 1) is too low and the seat (workplace component 2) is too high for the anthropometric characteristics of the worker (characteristics of the working person), the worker will lean forward (awkward posture), with negative effects on his or her physical workload, health (particularly if working for a long period in this workplace), and finally, on overall performance. Furthermore, if behind the worker there is a window causing glare on the computer screen (characteristic of the environment), he or she will probably bend sideways (awkward posture) to see what is presented on the screen (task requirement), causing similar effects. Consequently, when designing a workplace, one has to adopt a systemic view, considering the characteristics of the working person, the task requirements, and the environment in which the task will be performed.

Furthermore, the elements of the work system are variable. In fact, the working person may be short or tall, massively built or slim; young or elderly; with specific needs; and so on. Also, he or she may be refreshed or tired, depending on the time of day. The task requirements may also be multiple and variable. For example, at a secretarial workstation, the task may require exclusive use of the computer for a period of time, then the secretary may enter data from paper forms to the computer, and then he or she may serve a customer. At the same time, the workstation should be oriented such that the secretary is able to watch both the entry and the executive's doors. Finally, the workplace environment may be noisy or quiet; warm or cool, with annoying airstreams; illuminated by natural or artificial light; with all of these factors changing during the working day.

If to the complexity of the work system and the multiplicity of ergonomic criteria one adds financial

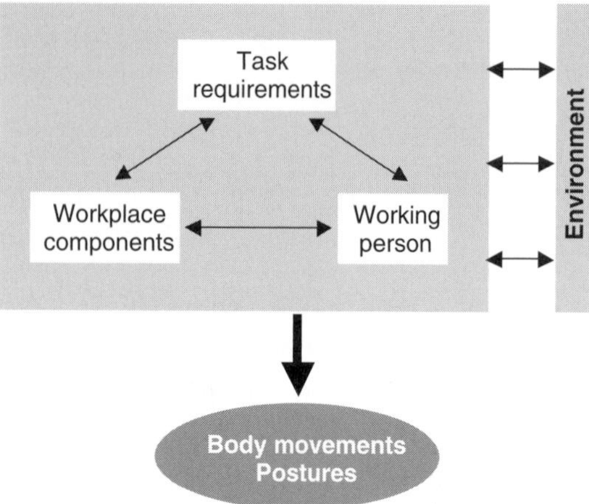

Figure 1 There is interdependence between a working person and the task requirements, workplace components, environment, and body movements and postures.

and aesthetic issues, successful design of a workplace becomes extremely difficult. Hence, some people maintain that designing a good workplace is rather an "art" than a "discipline," as there is no standard theory or method that ensures a successful result, the output depending heavily on the designer's "inspiration." Although this is true to a certain extent, good knowledge of the characteristics of the working persons who will occupy the workplace, of their tasks, and of the broader environment, combined with an effort for rigueur during the design process, contribute decisively to a successful design.

The present chapter is mainly methodological; we present and discuss a number of methods, techniques, guidelines, and design solutions that aim to support the decisions to be taken during the workplace design process. In the next section we discuss the problem of working postures and stress the fact that there is no *one best posture* that can be assumed for long periods of time. Consequently, the effort should be put on designing the components of the workplace in such a way as to form a "malleable envelope" that permits workers to adopt various healthy postures. The two other sections deal with the design of individual workstations and with the layout of groups of workstations in a given space.

2 PROBLEM OF WORKING POSTURES

A central issue of ergonomic workplace design relates to the postures the working person will adopt. In fact, the decisions that will be taken during workplace design will affect to a great extent the postures that the working person will or will not be able to adopt. The two most common working postures are sitting and standing. Between the two, sitting posture is, of course, more comfortable. However, there is research evidence that sitting adopted for prolonged periods of time results in discomfort, aching, or even irreversible injuries. Figure 2 shows the most common musculoskeletal disorders encountered at office workstations.

Studying the effects of *postural fixity* while sitting, Grieco (1986) found that it causes, among other things, (1) reduction of nutritional exchanges at the spinal disks and in the long term may promote their degeneration; (2) static loading of the back and shoulder muscles, which can result in aching and cramping; and (3) restriction in blood flow to the legs, which can cause swelling (edema) and discomfort. Consequently, the following conclusion can be drawn: The workplace should permit alteration between various postures, because there is no "ideal" posture that can be adopted for a long period of time.

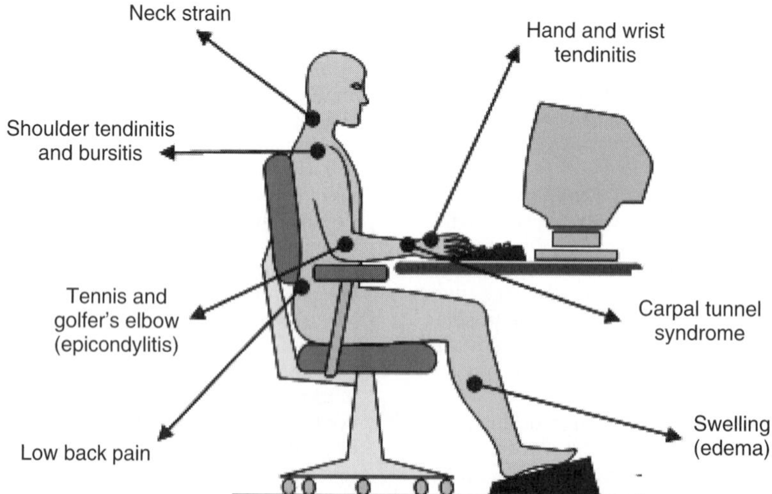

Figure 2 Common musculoskeletal disorders encountered at office workstations.

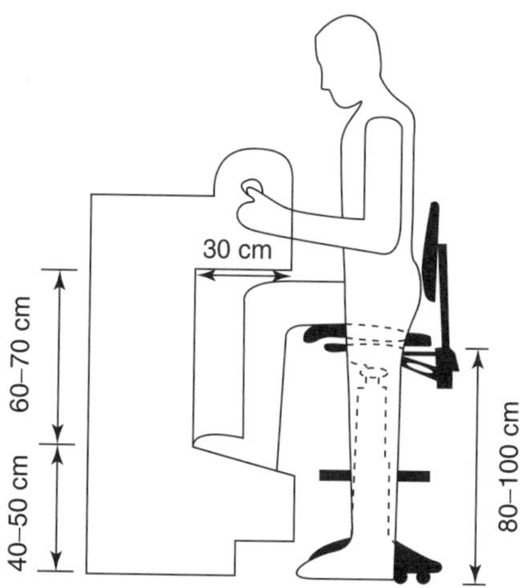

Figure 3 Standing–sitting workstation.

Based on this conclusion, a standing–sitting workstation has been proposed, especially for cases where tasks require long periods of continuous work (e.g., bank tellers or assembly workstations). Such a workstation (Figure 3) permits a worker to perform a job alternating standing with sitting.

Despite the absence of an ideal posture, however, there are postures that are more comfortable and healthy than others. Ergonomic research aims at identifying these postures and formulating requirements and principles that should be considered during the design of the components of a workplace. In this way, the resulting design will promote healthy work postures and constrain prolonged adoption of unhealthy postures.

2.1 Sitting Posture and Seats

The problem of designing seats that are appropriate for work is far from solved. In recent decades the sitting posture and design of seats have attracted the interest of researchers, designers, and manufacturers, due to the ever-increasing number of clerical workers and the importance of musculoskeletal problems encountered by them. This has resulted in the emergence of a science, the *science of seating*, and subsequently to a plethora of publications and design solutions (see, e.g., Mandal, 1985; Lueder and Noro, 1994).

Sitting posture poses a number of problems at a musculoskeletal level. The most important of these is *lumbar kyphosis*. When one is sitting, the lumbar region of the back flattens out and may even assume an outward bend. This shape of the spine, called *kyphotic*, is somewhat opposite to the *lordotic* shape of the spine when someone is standing erect (Figure 4). The less the angle between the thighs and the body, the greater the kyphosis. This occurs because of the restrained rotation of the hip joint, which forces the pelvis to rotate backward. Kyphosis provokes increased pressure on the spinal disks. Nachemson and Elfstrom (1970), for example, found that unsupported

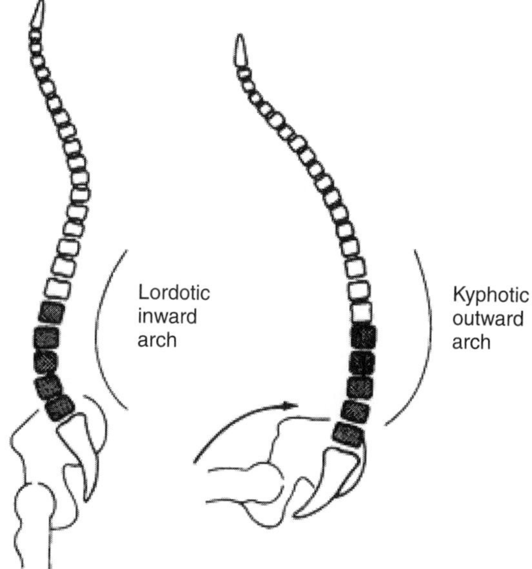

Figure 4 Lordotic and kyphotic postures of the spine. (From Grandjean, 1987.)

sitting upright resulted in a 40% increase in the disks' pressure compared to the pressure when standing. There are three complementary ways to minimize lumbar kyphosis: (1) by using a thick lumbar support, (2) by reclining a backrest, and (3) by providing a forward-tilting seat. Andersson et al. (1979) found that use of a 4-cm-thick lumbar support, combined with a backrest recline of 110° resulted in a lumbar curve closely resembling the lumbar curve of a standing person. Another finding of Andersson et al. (1979) was that the exact location of the support within the lumbar region did not influence significantly any of the angles measured in the lumbar region. The studies of Bendix (1986) and Bridger (1988) support Mandal's (1985) proposition for the forward-tilting seat.

Considering the above, the following ergonomic requirements should be met: (1) the seats should include a backrest that can recline, (2) the backrest should provide lumbar support, and (3) the seat should provide a forward-tilting option. However, as Dainoff (1994) observes, when tasks require close attention to the objects on the working surface or the computer screen, people usually bend forward and the backrest support becomes useless.

A design solution that aims to minimize lumbar kyphosis is the *kneeling* or *balance chair* (Figure 5), where the seat is inclined more than 20° from the horizontal plane. Besides the somewhat unusual way of sitting, this chair also has the drawbacks of loading the knee area, as the knees receive a great part of the body's load, and of constraining leg movement. On the other hand, it enforces lumbar lordosis very close to that adopted while standing and does not constrain the torso to move freely forward, backward, or sideways.

Figure 5 Kneeling chair. (From www.comcare.gov.au/officewise.html.)

There are quite a lot of detailed ergonomic requirements concerning the design of seats used at work. For example; (1) the seat should be adjustable so as to fit the various anthropometric characteristics of their users as well as to different working heights; (2) the seat should offer stability to the user; (3) the seat should offer freedom of movement to the user; (4) the seat should be equipped with armrests; and (5) the seat lining material should be water absorbent, to absorb body perspiration. Detailed requirements will not be presented extensively here, as the interested reader can find them easily in any specialized handbook. Furthermore, these requirements became "classical" and have been transformed into regulatory documents such as health and safety or design standards and legislation (see, e.g., EN 1335, ISO 9241, ANSI/HFS 100-1988, and DIN 4543 standards for office work, or EN 1729 for chairs and tables for educational institutions, and ISO/DIS 16121 for the driver's workplace in line-service buses).

Although most modern seats for office work meet the basic ergonomic requirements, design of their controls does not meet usability principles. This fact, combined with users' knowledge of healthy sitting, often results in nonuse of the adjustment possibilities offered by seats (Vitalis et al., 2000). Lueder (1986) provides the following guidelines for increasing the usability of controls: (1) controls should be easy to find and interpret, (2) controls should be easily reached and adjusted from the standard seated work position, (3) controls should provide immediate feedback (e.g., seats that adjust in height by use of a rotating pan delay feedback because a user must get up and down repeatedly to determine the correct position), (4) the direction of operation of controls should be logical and consistent with their effect, (5) few motions should be required to use the controls, (6) adjustments should require the use of only one hand, (7) special tools should not be necessary for the adjustment, and (8) labels and instructions on furniture should be easy to understand.

2.2 Sitting Posture and Work Surface Height

In addition to the problem of lumbar kyphosis, sitting working posture may provoke excessive muscle strain at the level of the back and shoulders. For example, if the working surface is too low, a person will bend forward too far; if it is too high, he or she will be forced to raise the shoulders.

To minimize these problems, appropriate design of the workplace is required. More specifically, the working surface should be at a height that permits a person to work with the shoulders at the relaxed posture. It should be noted here that the working height is not always the work surface height. The former depends on what one is working on (e.g., the keyboard of a computer), whereas the later is the height of the upper surface of the table, desk, or bench. Furthermore, to define the appropriate work surface height, one should consider the angles between the upper arms and elbows and the angle between the elbows and wrists. To increase comfort and to minimize the occupational risks, the first of the two angles should be about 90° if no force is required, and a little bit broader if application of force is required. The wrists should be as straight as possible, to avoid carpal tunnel syndrome.

Two other common problems encountered by people working in a sitting posture are neck aches and dry eye syndrome. These problems are related to prolonged gazing at objects placed too high: for example, when the visual display terminal or visual display unit (VDU) of a computer workstation is placed too high (Ankrum, 1997). Research that aims at determining the optimal placement of such objects, considering the mechanisms of both the visual and musculoskeletal systems, is still active (for a review, see Ankrum and Nemeth, 2000). However, most research findings agree that (1) neck flexion is more comfortable than extension, with the zero point (dividing flexion from extension) described as the posture of the head/neck when standing erect and looking at a visual target 15° below eye level; and (2) the visual system prefers downward gaze angles. Furthermore, there is evidence that when assuming an erect posture, people prefer to tilt their head, with the ear–eye Line (i.e., the line that crosses by the cartilaginous protrusion in front of the ear hole and the outer slit in the eyelid) about 15° below the horizontal plane (Grey et al., 1966; Jampel and Shi, 1992). Based on these findings, many authors propose the following rule of thumb for the placement of a VDU: The center of the monitor should be placed a minimum of 15° below eye level, with the top and bottom an equal distance from the eyes (i.e., the screen plane should be facing slightly upward).

Sanders and McCormick (1992) propose, in addition, the following general ergonomic recommendations for work surfaces: (1) if at all possible, the work surface height should be adjustable to fit individual physical dimensions and preferences; (2) the work surface should be at a level that places the working height at elbow height, with shoulders at relaxed posture; and (3) the work surface should provide adequate clearance for a person's thighs.

2.3 Spatial Arrangement of Work Artifacts

While working one uses a number of artifacts: for example, the controls and displays on a control panel, the various parts of an assembled object at an assembly workstation, or the keyboard, the mouse, the visual display terminal, the hard-copy documents, and the telephone at an office workstation. Application of the following ergonomic recommendations for the arrangement of these artifacts helps to decrease workload, facilitate work flow, and improve overall performance:

1. *Frequency of use and criticality.* Artifacts that are frequently used, or are of special importance, should be placed in prominent positions: for example, in the center of the work surface or near the right hand for right-handed people, and vice versa for left-handed people.
2. *Sequential consistency.* When a particular procedure is always executed in sequential order, the artifacts involved should be arranged according to this order.
3. *Topological consistency.* Where the physical location of controlled elements is important for the work, the layout of the controlling artifacts should reflect the geographical arrangement of the former.
4. *Functional grouping.* Artifacts (e.g., dials, controls, visual displays) that are related to a particular function should be grouped together.

Application of the recommendations above requires knowledge of the work activities to be performed at the workplace designed. Task analysis provides enough data to apply these recommendations appropriately, as well as to solve eventual contradictions between them, by deciding which arrangement best fits the situation at hand.

3 DESIGNING INDIVIDUAL WORKSTATIONS

Figure 6 presents a generic process for the ergonomic design of individual workstations, with the various phases, the data or sources of data that have to be considered at each phase, and methods that could be applied. It should be noted that certain phases of the process may be carried out concurrently, or in a different order, depending on the particularities of the workstation to design, or the preferences and experience of the designers.

3.1 Phase 1: Decisions Regarding Resources and High-Level Requirements

The first phase of the design process is to decide about the time to spend and the people who will participate in it (the design team). These decisions depend on the high-level requirements of stakeholders (e.g., improvement of working conditions, increase of productivity, innovation, occupational safety and health protection, as well as the money they are ready to spend), and the importance of the project (e.g., number of identical workstations, significance of the tasks carried out, particularities of the working persons). An additional issue that has to be dealt with at this phase is to ensure the participation in the design process of the people who will occupy the future workstations. Access to workstations where similar jobs are being performed is also advisable. The remainder of the design process will be influenced significantly by the decisions taken at this phase.

3.2 Phase 2: Identification of Work System Constraints and Requirements

The aim of phase 2 is to identify the various constraints and requirements posed by the work system that have to be considered during workstation design. More specifically, during this phase one has to collect data about the types of tasks to be carried out at the workstation designed; the work organization (i.e., the interdependency between the tasks to be carried out in the workstation and others in the proximal environment); the various technological equipment and tools that will be used, their functions, user interfaces, shape, and dimensions; the environmental conditions of the broader area in which the workstation will be placed (e.g., illumination and sources of light, level of noise and noise sources, thermal conditions and sources of warm or cold drafts); normal as well as exceptional situations in which working persons could be found (e.g., electricity breakdowns, fire); and any other element of or situation related to the work system that may interfere directly or indirectly with the workstation designed. These data can be collected by questioning the appropriate people as well as by observation and analysis of similar work situations. Specific design standards (e.g., ANSI, EC, DIN, or ISO), as well as legislation related to the type of the workstation designed, should be collected and studied during this phase.

3.3 Phase 3: Identification of User Needs

The needs of future workstation occupants are identified during phase 3, considering the tasks to be performed at the workstation designed as well as the characteristics of persons who will occupy it. Consequently, task analysis (see Chapter 14) and analysis of user characteristics should be carried out at this phase.

Of particular importance are the characteristics of the user population, which depend on their gender, age, nationality, or particular disabilities, and concern the size of body parts (anthropometry) (see Chapter 12), the ability and limitations of their movements (biomechanics) (see Chapter 13), visual and auditory perception abilities and limitations, previous experiences and work practices, and cultural or religious obligations (e.g., women in certain countries are obliged to wear particular costumes).

Task analysis aims at identifying mainly the work processes that will take place and the workstation elements implicated in them; the physical actions that will be carried out (e.g., fine manipulations, whole-body movement, force exertion), the information exchange required (visual, auditory, kinesthetic, etc.) and information sources; the privacy required; and the necessary

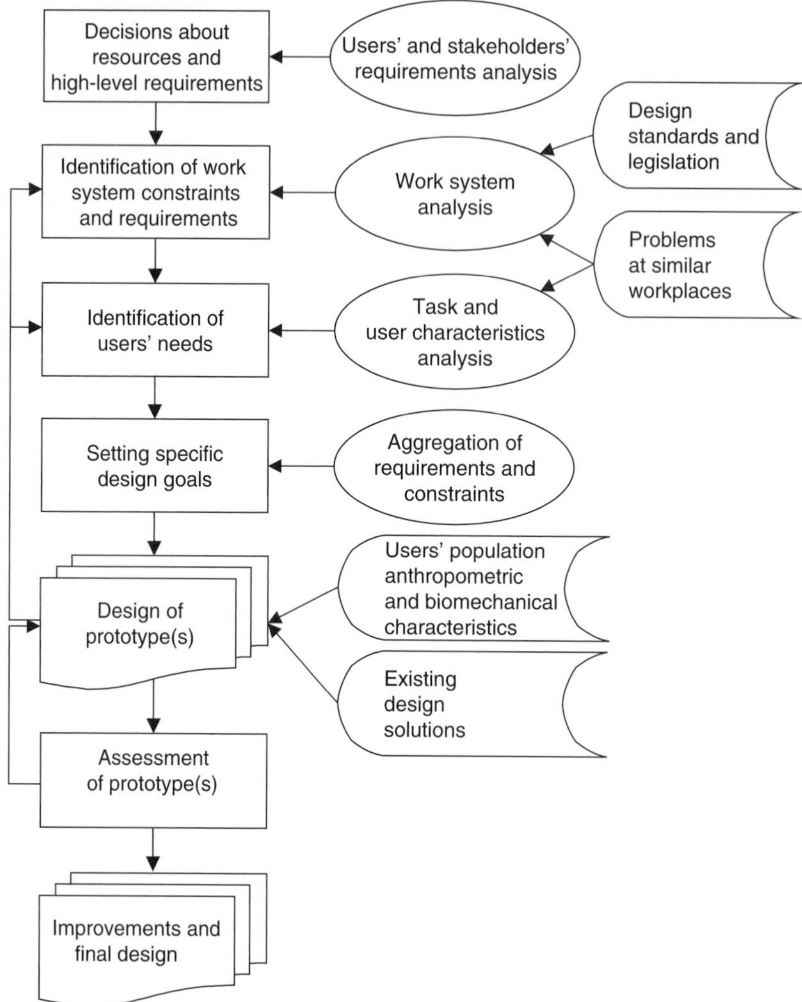

Figure 6 Generic process for the ergonomic design of individual workstations.

proximity with other workstations, equipment, or elements of the broader working environment. The more the design team can analyze work situations similar to the workstation designed, the more valuable the results of the task analysis will be.

At this phase, data about performance and health problems of persons working in similar work situations should be collected. Ergonomic and occupational safety and health literature may be used as the main source for the collection of such data. Finally, as in the preceding phase, user needs should be identified not only for normal but also for exceptional situations in which workstation occupants may be found (e.g., electricity blackout, fire).

3.4 Phase 4: Setting Specific Design Goals

Considering the outputs of preceding phases, the design team is now able to transform the generic ergonomic requirements of workstation design into a set of specific goals. These will guide the choices and decisions to be made in the next phase. Furthermore, they will be used as criteria for assessment of the prototype designed and will guide its improvement. The specific goals are an aggregation of *shoulds* and consist of the requirements of the stakeholders (e.g., the workstation should be convenient for 95% of the user population, should cost a maximum of $\$X$, should increase productivity at least 10%); the constraints and requirements posed by the work system in which the designed workstation(s) will be installed [e.g., the workstation(s) should not exceed X centimeters of length and Y centimeters of width, should offer working conditions not exceeding X decibels of noise and $Y°$ of wet bulb globe temperature], user needs (e.g., the workstation should accommodate elderly people, should be appropriate for prolonged computer work, should facilitate cooperation with neighboring workstations), the requirement to avoid common health problems associated with similar situations (e.g., the workstation should minimize upper limb musculoskeletal problems), and design standards and related legislation (e.g., the workstation should ensure absence of glare and of cold drafts). The systematic record of all specific design goals is very helpful for the next phases. It is important to note that agreement on these specific goals among the design team, management, and users' representatives is indispensable.

3.5 Phase 5: Design of Prototypes

Phase 5 is the most demanding in the design process. In fact, the design team has to generate design solutions that meet all the specific design goals identified in phase 4. Given the large number of design goals as well the fact that some of them may be conflicting, the design team has to make appropriate compromises, considering some goals as more important than others, and eventually passing by some of them. Good knowledge of the particularities of the task that will be performed at the workstation designed, as well as the specific user characteristics, is the only way to set the right priorities and avoid serious mistakes.

A first decision to make is the working posture(s) that will be assumed by users of the workstation designed. Table 1 provides some recommendations for this. Once the working posture has been decided, the design may continue to define the shape, dimensions, and arrangement of the various elements of the workstation. To do so, one has to consider the anthropometric and biomechanical characteristics of the user population as well as the working actions to be performed. In addition to the ergonomic recommendations presented previously, some additional recommendations for the design of the workstation are the following:

Table 1 Recommendations for Choosing the Working Posture

Working Posture	Task Requirements
Working person's choice	It is preferable to arrange for both sitting and standing (see Figure 3).
Sitting	Where a stable body is needed: • For accurate control, fine manipulation • For light manipulation work (continuous) • For close visual work with prolonged attention • For limited headroom, low work heights Where foot controls are necessary (unless of infrequent or short duration) Where a large proportion of the working day requires standing
Standing	For heavy, bulky loads Where there are frequent moves from the workplace Where there is no knee room under the equipment Where there is limited front–rear space Where there are a large number of controls and displays Where a large proportion of the working day requires sitting
Support seat (see Figure 7)	Where there is no room for a normal seat but support is desirable

Source: Corlett and Clark (1995).

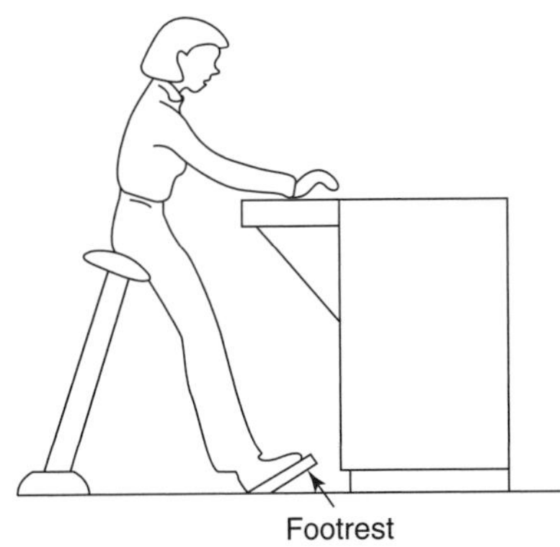

Figure 7 Where there is no room for a normal seat, a support is desirable. (From Helander, 1995.)

1. To define the *clearance*, the minimum required free space for placement of the body, one has to consider the largest user (usually, the anthropometric dimensions corresponding to the 97.5th percentile). In fact, providing free space for these users, all shorter users will also have enough space to place their bodies. For example, if the vertical, lateral, and forward clearance below the working desk are designed considering the height of the upper surface of the thigh of a sitting person, the hip width, and the thigh length corresponding to the 97.5th percentile of the user population (plus an allowance of 1 or 2 cm), 97.5% of the users of this desk will be able to approach the desk easily while sitting.

2. To position the various elements of the workplace that have to be *reached* by users, consider the smaller users. In fact, if smaller users reach the various workstation elements easily, without leaning forward or bending sideways, all larger users will also reach them easily.

3. Draw the common kinetospheres or comfort zones for larger and smaller users, and add the various elements of the workstation that have to be manipulated (e.g., controls) (Figure 8).

4. When necessary, provide the various elements of the workstation with appropriate adjustability to fit the anthropometric characteristics of the user

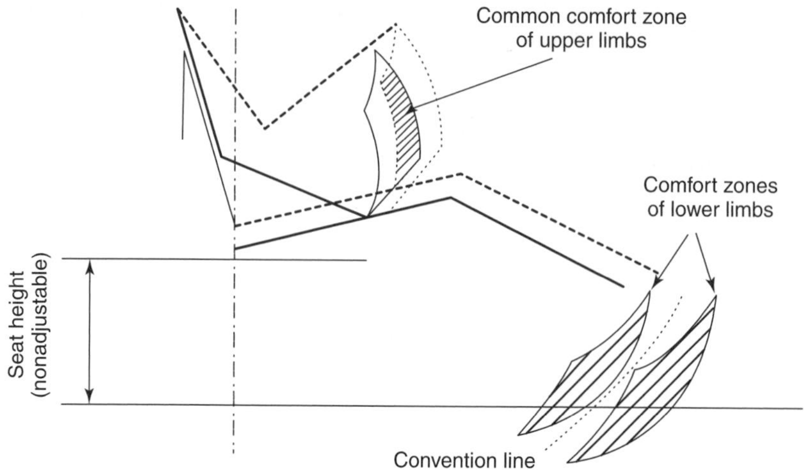

Figure 8 Common comfort zones of the hands and legs for large and small users of a driving workplace with a nonadjustable chair.

population. (Care should be given to the usability of adjustability controls.)

5. While envisioning design solutions, check continuously to ensure that the workstation elements do not obstruct users' courses of action (e.g., perception of necessary visual information, manipulation of controls).

It should be stressed that at least some iterations among phases 2, 3, and 5 of the design process are unavoidable. In fact, it is almost impossible to identify from the start all the constraints and requirements of the work system, user characteristics, or task requirements that intertwine with elements of the workstation anticipated. Another issue to deal with during this phase is designing to protect the working person from possible annoying or hazardous environmental factors. If the workstation has to be installed in a harsh environment (noisy, cold or warm, hazardous atmosphere, etc.), one has to provide it with appropriate protection. Again, attention must be paid to the design of such protective elements. These should take into consideration the anthropometric characteristics of the user population and the special requirements of the task, in order not to obstruct the processes involved in both normal and degraded operation (e.g., maintenance, breakdowns). Other important issues that have to be resolved at this phase are workstation maintainability, its unrestricted evacuation, its stability and robustness, and such safety issues as rough corners.

A search for already existing design ideas and solutions is quite useful. However, they should be examined carefully before adoption. In fact, although valuable for anticipation, such design ideas may not be readily applicable for the specific user population, the particularities of their tasks, or the environment in which the workplace designed will be installed. Furthermore, many existing design solutions may disregard important ergonomic issues. Finally, although the adoption of already existing design solutions exploits the design community's experience and saves time, it deprives the design team of generating their own innovative solutions.

The use of computer-aided design (CAD) applications with human models is very helpful at this phase (see Chapter 39). If such software is not available, appropriate drawings and mock-ups should be developed for the generation of design solutions as well as for their assessment (see phase 6).

Given the complexity of generating good design solutions, the search for alternatives is valuable. Members of the design team should not be anchored at the first design solution that comes to their minds. They should try to generate as many alternative ideas as possible, gradually converging to the one or ones that best satisfy the design goals.

3.6 Phase 6: Assessment of Prototypes

Assessment of the designed prototype(s) is required to check how well the specific design goals set at the fourth phase have been met, as well as to uncover possible omissions during the identification of work system constraints and requirements and the user needs analysis (second and third design phases). The assessment can be performed analytically or/and experimentally, depending on the importance of the project. At the analytical assessment, the design team assesses the designed workplace, considering exhaustively the specific design goals using the drawings and mock-ups as support. Applying a multicriteria method, the design team may rank the degree to which the design goals have been met. This ranking may be used as a basis for the next phase of the design process (improvement of the prototype) as well as a means to chose among alternative design solutions.

The experimental assessment (or user testing) is performed with the participation of a sample of future users, simulating the work with a full-scale mock-up of the designed workstation prototype(s). The assessment should be made in conditions as close as possible to those of the real work. Development of use scenarios of both normal and exceptional work situations is useful for this reason. Experimental assessment is indispensable for the identification of problematic aspects that are difficult, if not impossible, to realize before having a real workplace with real users. Furthermore, this type of assessment provides valuable insights for eventual needs during implementation (e.g., the training needed, the eventual need for a user's manual).

3.7 Phase 7: Improvements and Final Design

In phase 7, considering the outputs of the assessment, the design team proceeds with the necessary modifications of the design prototype. The opinions of other specialists, such as architects and decorators, which have more to do with the aesthetics, or production engineers and industrial designers, which have more to do with production, materials, and robustness matters, should be considered at this phase, in case such specialists are not members of the design team. The final design should be complemented by drawings for production and appropriate documentation, including the rationale behind the solutions adopted; cost estimation for the production of the workstation(s) designed; and implementation requirements, such as the training needed and the user's manual, if required.

3.8 Final Remarks

The reason for conducting a user needs and requirements analysis is to anticipate the future work situation in order to design a workstation that fits its users, their tasks, and the surrounding environment. However, it is impossible to anticipate a future work situation completely, in all its specificity, as work situations are complex, dynamic, and evolving. Furthermore, if the workstation designed is destined to form part of an already existing work system, it might affect the overall work ecology, something that is also very difficult to anticipate. Therefore, a number of modifications will eventually be needed some time after workstation installation and use. Thus, it is strongly suggested to conduct a new assessment of the designed workstation once users have been familiarized with the new work situation.

4 ERGONOMIC LAYOUT OF WORKSTATIONS

Ergonomic layout deals with the placement and orientation of individual workstations in a given space (building). The main ergonomic requirements to meet concern the tasks performed, the work organization, and environmental factors. More specifically, such requirements are as follows:

- The layout of the workstations should facilitate the work flow.
- The layout of the workstations should facilitate the cooperation of both personnel and external persons.
- The layout of the workstations should conform to the organizational structure.
- The layout should ensure the necessary privacy.
- There should be appropriate lighting, conforming to task needs.
- The lighting should be uniform throughout the working person's visual field.
- There should be no annoying reflections or glare in the working area.
- There should be no annoying hot or cold drafts in the workplace.
- Access to the workstations should be unobstructed and safe.

In this section we focus on the ergonomic layout of workplaces for office work. The choice to focus on the ergonomic layout of workstations in offices has been made for the following reasons: First, office layout is an exemplary case for the arrangement of a number of individual workstations in a given space, encompassing all major ergonomic requirements found in most types of workplaces (with the exception of workplaces where the technology involved determines to a large extent the layout, such as workstations in front of machinery). Second, office workplaces concern a growing percentage of the working population worldwide. For example, during the twentieth century the percentage of office workers increased from 17% to over 50% of the U.S. workforce, with the rest working in agriculture, sales, industrial production, and transportation (Czaja, 1987). With the spread of information technologies, the proportion of office workers is expected to increase further. Third, health problems encountered by today's office workers are to a great extent related to inappropriate layout of their workplaces (Marmaras and Papadopoulos, 2002).

4.1 Generic Office Layouts

There are a number of generic types of office layouts (Shoshkes, 1976; Zelinsky, 1998). The two extremes are the *private office*, where each worker has his or her own personal closed space or room, and the *open-plan office*, where all the workstations are placed in a common open space. In between are a multitude of combinations of private offices and open plans. Workstation arrangements in open plans can be either orthogonal, with single, double, or fourfold desks forming parallel rows, or with workstations arranged in groups, matching the organizational or functional structure of the work. A recent layout philosophy is the *flexible office*, where the furniture and equipment are designed to be easily movable in order to be able to modify the workstation arrangement depending on the number of the people present in the office as well as the projects or work schemes themselves (Brunnberg, 2000). Finally, to respond to the current need for flexibility in organizations and the structuring of enterprises, as well as to reduce costs, a new trend in office management is the *free address office* or *nonterritorial office*, where workers do not have a proper workstation, but whenever at the office, use the workstation they find free.

Each type of layout has strengths and weaknesses. Private offices offer increased privacy and better control of environmental conditions, being easily fitted to the particular preferences and needs of their users. On the other hand, they are more expensive to both construct and maintain, not easily modifiable to match changing organizational needs, and render cooperation and supervision difficult. Open-plan offices offer flexibility in changing organizational needs and facilitate cooperation between co-workers but tend to suffer from environmental annoyances such as noise and suboptimal climatic conditions as well as lack of privacy. To minimize the noise level as well as to create some sense of privacy in open plans, movable barriers may be used. To be effective, the barriers have to be at least 1.5 m high and 2.5 m wide. Furthermore, Wichman (1984) proposes the following specific design recommendations to enhance the working conditions in an open-plan office:

- Use sound-absorbing materials on all major surfaces wherever possible. Noise is often more of a problem than expected.
- Equip workstations with low-noise technological devices (e.g., printers, photocopy machines, telephones). For example, provide telephones that flash a light for the first two "rings" before emitting an auditory signal.
- Leave some elements of design for the workstation user. People need to have control over their environments, so leave some opportunities for changing or rearranging things.
- Provide both vertical and horizontal surfaces for the display of personal belongings. People like to personalize their workstations.
- Provide several easily accessible islands of privacy. This would include small rooms with full walls and doors that can be used for conferences and private or long-distance telephone calls.
- Provide all private work areas with a way to signal willingness of the occupant to be disturbed.
- Have clearly marked flow paths for visitors. For example, hang signs from the ceiling showing where secretaries and department boundaries are located.
- Design workstations so that it is easy for drop-in visitors to sit down while speaking. This will tend to reduce disturbances to other workers.
- Plan for ventilation airflow. Most traditional offices have ventilation ducting. This is usually not the case with open-plan cubicles, so they become dead-air cul-de-sacs that are extremely resistant to post hoc resolution.
- Overplan for storage space. Designers of open-plan systems, which emphasize tidiness, seem to chronically underestimate people's storage needs.

The decision as to the generic type of layout should be made by the stakeholders. The role of the ergonomist here is to indicate the strengths and weaknesses of each alternative to facilitate adoption of the most appropriate type of layout for the specific situation. After this decision has been made, the design team should proceed to a detailed layout of the workstations. In the next section we describe a systematic method for this purpose.

4.2 Ergonomic Method for Office Layout

The ergonomic method proposes a systematic way to design workplaces for office work. The method aims at alleviating the design process for arranging the workstations by reducing the entire problem to a number of stages during which only a limited number of ergonomic requirements are considered. Another characteristic of the method is that the ergonomic requirements to be considered have been converted to design guidelines (Margaritis and Marmaras, 2003). Figure 9 presents the main stages of the method.

Before starting the layout design, the design team should collect data concerning the activities that will be performed in the workplace and the needs of the workers. More specifically, the following information should be gathered:

- The number of people who will work permanently or occasionally.
- The organizational structure and the organizational units that it comprises.
- The activities carried out by each organizational unit. Of particular interest are the needs for cooperation among the various units (and consequently, the desired relative proximity between them), the need for reception of external visitors (and consequently, the need to provide easy access to them), as well as any other need related to the particularities of the unit (e.g., security requirements).
- The activities carried out by each worker. Of particular interest are the need for cooperation with other workers, privacy needs, the reception of external visitors, and the specific needs such as for lighting.
- The equipment required for each work activity (e.g., computer, printer, storage).

At this stage the design team should also request detailed ground plan drawings of the space concerned, including all elements that should be considered as fixed (e.g., structural walls, heating systems).

4.2.1 Stage 1: Determination of Space Available

The aim of stage 1 is to determine the space where no furniture should be placed, to ensure free passage by the doors and to allow the necessary room for elements such as windows and radiators, for manipulation and maintenance purposes. Following are suggestions for

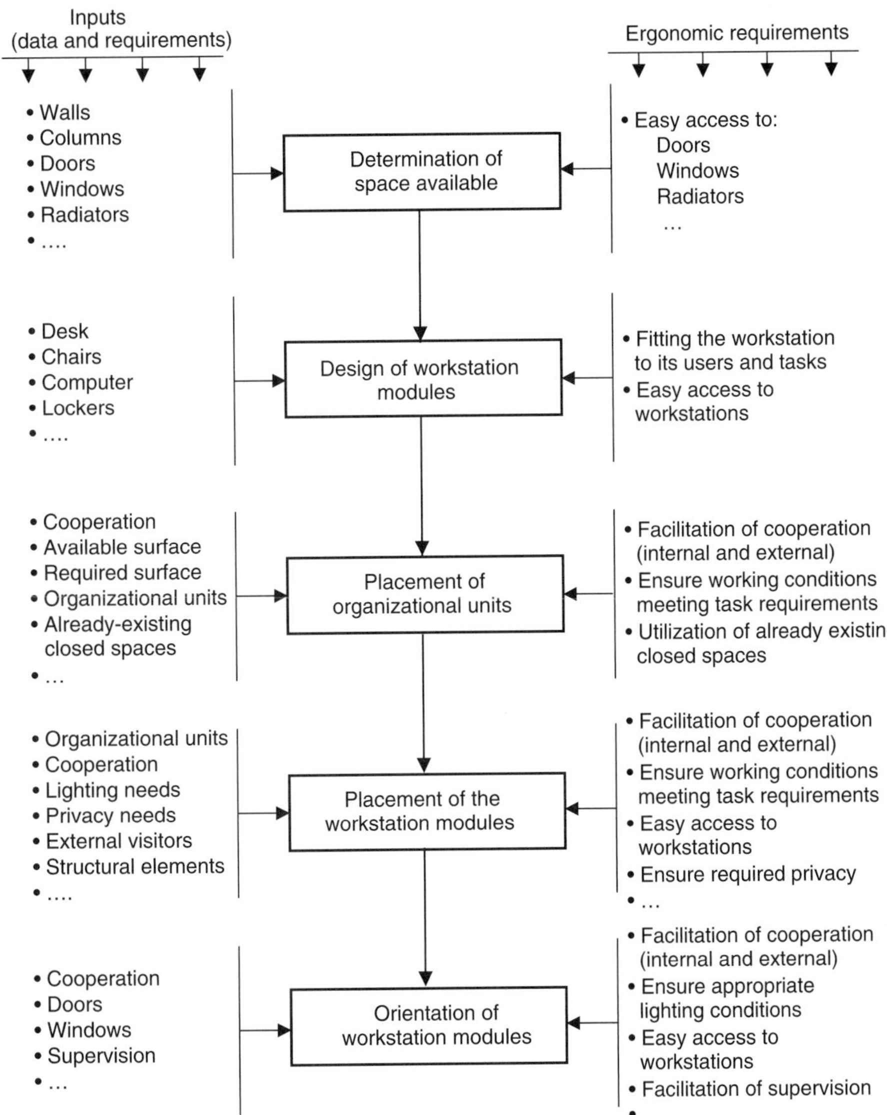

Figure 9 Main stages of a method for office layout meeting ergonomic requirements.

determining spaces that to remain free of furniture (Figure 10). Allow for an area of 50 cm in front of any window, an area of 3 m in front and 1 m at both sides of the main entrance door, an area of 1.50 m in front and 50 cm at both sides of any other door, and an area of 50 cm around any radiator.

4.2.2 Stage 2: Design of Workstation Modules

The aim of stage 2 is to design workstation modules that meet the needs of workers. Each module is composed of the appropriate elements for the working activities: desk, seat, storage cabinets, visitors' seats, and any other equipment required for the work. A free space should be provided around the furniture for passages between workstations as well as for unobstructed sitting and for getting up from the seat. This free space may be delimited in the following way (minimum areas). Allow for an area of 55 cm along the front side of the desk or the outer edge of the visitor's seat; an area of 50 cm along the entry side of the workstation; an area of 75 cm along the back side of the desk (seat side); and an area of 100 cm along the back side of the desk if there are storage cabinets behind the desk.

A number of different modules will result from this stage, depending on the particular work requirements

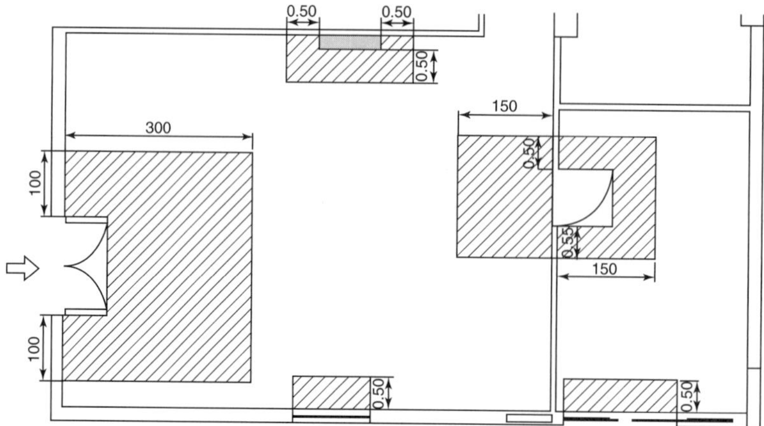

Figure 10 Determining the available space.

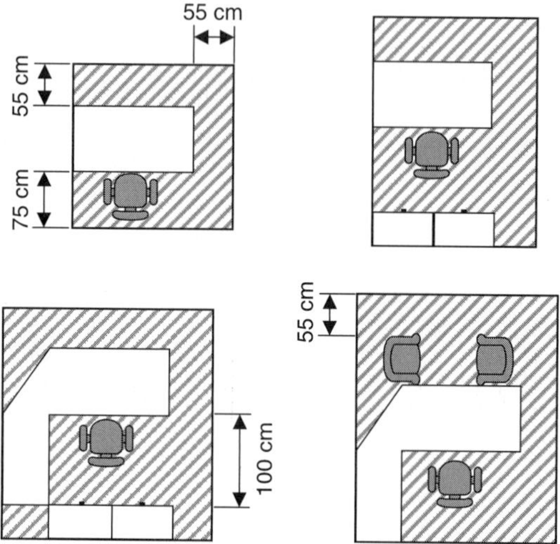

Figure 11 Workstation modules.

(e.g., secretarial module, head of unit module, client service module) (Figure 11). Layout by workstation modules instead of by individual elements such as desks and seats permits the designer to focus on the requirements related to the overall layout of the workplace, ensuring at the same time compliance with requirements related to individual workstations.

4.2.3 Stage 3: Placement of Organizational Units

The aim of stage 3 is to decide about the placement of the various organizational units (i.e., departments, working teams, etc.) within the various free spaces of the building. There are five primary issues to be considered here: (1) the shape of each space, (2) the exploitable area of each space (i.e., the area where workstations can be placed), (3) the area required for each unit, (4) the desired proximity between units, and (5) eventual particular requirements of each unit, which may determine their absolute placement within the building (e.g., the reception area should be placed right next to the main entrance).

The exploitable area of each space is an approximation of the areas free of furniture defined in stage 1, also considering narrow shapes where modules cannot fit. Specifically, this area can be calculated as follows:

$$A_{\text{exploitable}} = A_{\text{total}} - A_{\text{where no modules can be placed}}$$

where A_{total} is the total area of each space and $A_{\text{where no modules can be placed}}$ is the nonexploitable area, where workstation modules should not or cannot be placed.

The area required for each organizational unit can be estimated considering the number of workstation modules needed and the area required for each module. Specifically, to estimate the area required for each organizational unit, A_{required}, one has to calculate the sum of the areas of the various workstation modules of the unit. Comparing the exploitable area of the various spaces with the area required for each unit, the candidate spaces for placing the units can be defined. Specifically, the candidate spaces for the placement of a particular unit are the spaces where

$$A_{\text{exploitable}} \geq A_{\text{required}}$$

Once the candidate spaces for each unit have been defined, final decisions about the placement of organizational units can be made. This is done in two steps. In the first step the designer designates spaces for eventual units which present particular placement requirements (e.g., reception). In the second step the designer positions the remaining units, considering their desirable relative proximity plus additional criteria, such as the need for natural lighting or the reception of external visitors. To facilitate placement of the organizational units according to their proximity requirements, a proximity table and proximity diagrams may be drawn.

A *proximity table* represents the desired proximity of each unit with any other unit, rated using the following scale:

9 The two units cooperate fully and should be placed close together.

3 The two units cooperate from time to time, and it would be desirable to place them in proximity.

1 The two units do not cooperate frequently, and it does not matter if they are placed in proximity.

Figure 12 is the proximity table of a hypothetical firm consisting of nine organizational units. At the right bottom, the total proximity rate (TPR) has been calculated for each unit as a sum of its individual proximity rates. The TPR value is an indication of the cooperation needs of each unit with all the others. The designer should try to place the units with high TPRs at a central position.

Proximity diagrams are a graphical method for the relative placement of organizational units. They facilitate the heuristic search for configurations that minimize the distance between units that should demonstrate close cooperation. Proximity diagrams are drawn on a sheet of paper with equidistant points, such as the one shown in Figure 13. The various units alternated at different points, in an attempt to locate arrangements in which the units that require close cooperation will be as close to each other as possible. The following rules may be used to obtain a first configuration:

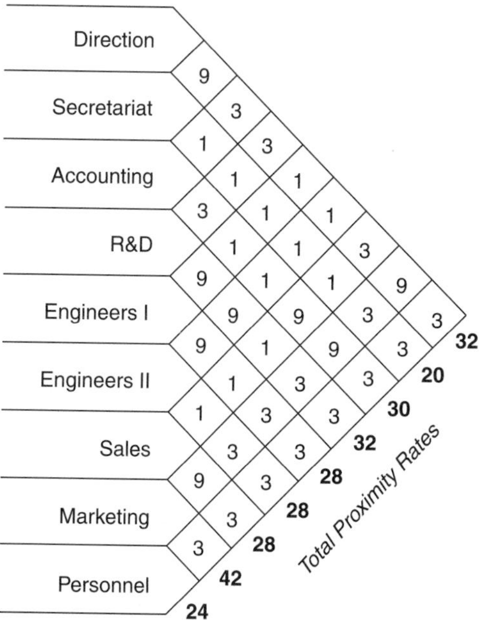

	Direction	Secretariat	Accounting	R&D	Engineers I	Engineers II	Sales	Marketing	Personnel	Total Proximity Rates
Direction		9	3	3	1	1	3	9	3	32
Secretariat	9		1	1	1	1	1	3	3	20
Accounting	3	1		3	1	1	9	9	3	30
R&D	3	1	3		9	9	1	3	3	32
Engineers I	1	1	1	9		9	1	3	3	28
Engineers II	1	1	1	9	9		1	3	3	28
Sales	3	1	9	1	1	1		9	3	28
Marketing	9	3	9	3	3	3	9		3	42
Personnel	3	3	3	3	3	3	3	3		24

Figure 12 Proximity table of a hypothetical firm.

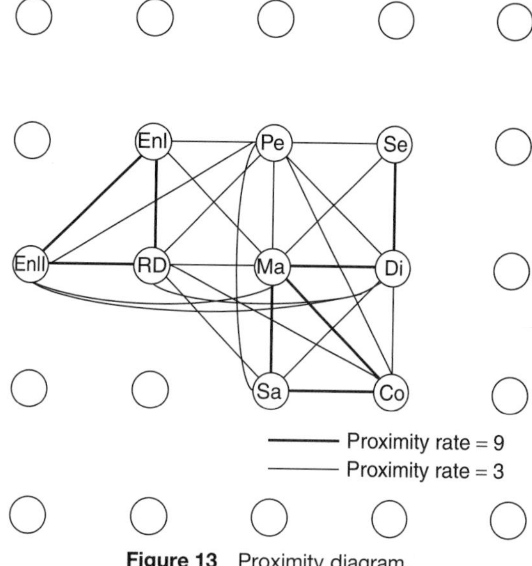

Figure 13 Proximity diagram.

- Place the unit with the highest TPR value at the central point.
- If there is more that one unit with the same TPR value, place first the unit that has the closest proximity rates (9's).
- Continue placing units that have the higher proximity rates with those that are already positioned.
- In cases where more than one unit has a proximity rate equal to that of a unit already positioned, place the unit with the higher TPR value first.
- Continue in this manner until all the units have been positioned.

More than one alternative arrangement may be obtained in this way. It should be stressed that proximity diagrams are drawn without taking into account the area required for each unit and the exploitable area of the spaces in which the units may be placed. Consequently, the arrangements drawn cannot be transposed to the ground plan of the building without modifications. Drawing proximity diagrams is a means to facilitate decisions concerning the relative positioning between organizational units. As a method it becomes useful in cases where the number of units is important.

4.2.4 Stage 4: Placement of Workstation Modules

Once the areas where the various organizational units will be placed have been determined, placement of workstation modules for each unit can begin. The following guidelines provide help in meeting the ergonomic requirements:

1. Place the workstations in a way that facilitates cooperation between co-workers. In other words,

workers who cooperate closely should be placed near each other.

2. Place workstations at which external visitors will be received near the entrance doors.

3. Place as many workstations as possible near the windows. Windows may provide benefits in addition to variety in lighting and a view (Hall, 1966). They permit fine adjustment of light through curtains or venetian blinds and provide distant points of visual focus, which can relieve eye fatigue. Furthermore, related research has found that people strongly prefer workstations that are placed near windows (Manning, 1965; Sanders and McCormick, 1992).

4. Avoid placing workers in airstreams created by air conditioners and by open windows and doors.

5. Place workstation modules to form straight corridors leading to doors. The corridor width allowing for one-person passage should be at least 60 cm, and for two-person passage, at least 120 cm (Alder, 1999).

6. Leave the necessary space in front and to the sides of electric switches and wall plugs.

7. Leave the necessary space for waiting visitors. In cases where waiting queues are expected, provide at least a free space of 120 cm width and $n \times 45$ cm length, where n is the maximum expected number of waiting people. Add to this length another 50 cm in front of the queue.

4.2.5 Stage 5: Orientation of Workstation Modules

The aim of stage 5 is to define the direction of the workstation modules for each unit so as to meet the ergonomic requirements. This stage can be carried out either concurrent with or following the preceding phase. The following guidelines support this phase and should be applied judiciously, as it may not always be possible to satisfy all of them.

1. Orient workstations such that there are no windows directly in front or behind workers when they are looking toward a visual display unit (VDU). In offices, windows play a role similar to that of lights: a window right in front of a worker disturbs through direct glare; one directly behind the worker produces reflected glare. For this reason, ideally, VDU workstations should be placed at right angles to windows (Grandjean, 1987) (Figure 14).

2. To avoid direct glare, orient workstations such that there are no direct lighting sources within $\pm 40^\circ$ vertically and horizontally from the line of sight (Kroemer et al., 1994).

3. Orient workstations to allow workers to observe entrance doors.

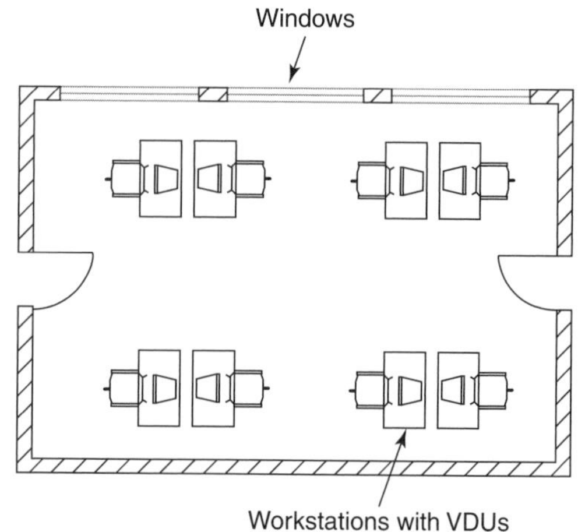

Figure 14 Workstations with VDT ideally should be placed at right angles to windows.

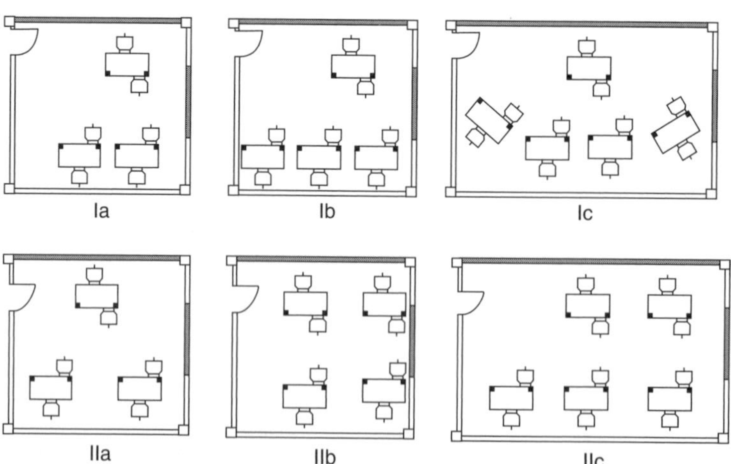

Figure 15 Alternative orientations of workstations, depending on the number of team members and the presence or absence of a leader. Ia, Ib, and Ic: arrangements with a leader; IIa, IIb, and IIc: arrangements without a leader.

4. Orient workstations to facilitate cooperation between members of work teams. Figure 15 shows alternative orientations of workstations, depending on the number of team members and the presence or absence of a leader (Cummings et al., 1974).

4.3 Concluding Comments

Given the complexity of workplace layout design, a designer trying to apply the various ergonomic guidelines in the various phases will almost definitely encounter contradictions. To resolve them, he or she should be able to focus on the guidelines that seem most important for the case at hand and pay less attention (eventually, even ignore) others. Good knowledge of generic human abilities and limitations, the specific characteristics of the people who will work in the designed workplace, and the specificities of the work that will be carried out are prerequisites for successful decisions. Furthermore, the designer should demonstrate an open and innovative mind and try as many solutions as possible. A systematic assessment of these alternative solutions is advisable to decide on the most satisfactory solution. The participation of the various stakeholders in this process is strongly recommended. The use of specialized computer-aided design (CAD) tools may prove very helpful when using the method presented, greatly facilitating the generation and assessment of alternative design solutions.

REFERENCES

Alder, D. (1999), *Metric Handbook: Planning and Design Data*, 2nd ed., Architectural Press, New York.

Andersson, G., Murphy, R., Ortengren, R., and Nachemson, A. (1979), "The Influence of Backrest Inclination and Lumbar Support on the Lumbar Lordosis in Sitting," *Spine*, Vol. 4, pp. 52–58.

Ankrum, D. R. (1997), "Integrating Neck Posture and Vision at VDT Workstations," in *Proceedings of the 5th International Scientific Conference on Work with Display Units*, pp. 63–64.

Ankrum, D. R., and Nemeth, K. J. (2000), "Head and Neck Posture at Computer Workstations: What's Neutral?" in *Proceedings of the 14th Triennial Congress of the International Ergonomics Association*, Vol. 5, pp. 565–568.

Bendix, T. (1986), "Seated Trunk Posture at Various Seat Inclinations, Seat Heights, and Table Heights," *Human Factors*, Vol. 26, pp. 695–703.

Bridger, R. (1988), "Postural Adaptations to a Sloping Chair and Work-Surface," *Human Factors*, Vol. 30, pp. 237–247.

Brunnberg, H. (2000), "Evaluation of Flexible Offices," in *Proceedings of the IEA 2000/HFES 2000 Congress*, Vol. 1, Human Factors and Ergonomics Society, San Diego, CA, pp. 667–670.

Corlett, E. N., and Clark, T. S. (1995), *The Ergonomics of Workspaces and Machines*, Taylor & Francis, London.

Cummings, L., Huber, G. P., and Arendt, E. (1974), "Effects of Size and Spatial Arrangements on Group Decision-Making," *Academy of Management Journal*, Vol. 17, No. 3, pp. 460–475.

Czaja, S. J. (1987), "Human Factors in Office Automation," in *Handbook of Human Factors*, G. Salvendy, Ed., Wiley, New York.

Dainoff, M. (1994), "Three Myths of Ergonomic Seating," in *Hard Facts About Soft Machines: The Ergonomics of Seating*, R. Lueder and K. Noro, Eds., Taylor & Francis, London.

Grandjean, E. (1987), *Ergonomics in Computerized Offices*, Taylor & Francis, London.

Grey, F. E., Hanson, J. A., and Jones, F. P. (1966), "Postural Aspects of Neck Muscle Tension," *Ergonomics*, Vol. 9, No. 3, pp. 245–256.

Grieco, A. (1986), "Sitting Posture: An Old Problem and a New One," *Ergonomics*, Vol. 29, No. 3, pp. 345–362.

Hall, E. T. (1966), *The Hidden Dimension*, Doubleday, New York.

Helander, M. (1995), *A Guide to the Ergonomics of Manufacturing*, Taylor & Francis, London.

Jampel, R. S., and Shi, D. X. (1992), "The Primary Position of the Eyes, the Resetting Saccade, and the Transverse Visual Head Plane," *Investigative Ophthalmology and Visual Science*, Vol. 33, pp. 2501–2510.

Kroemer, K., and Grandjean, E. (1997), *Fitting the Task to the Human: A Textbook of Occupational Ergonomics*, 5th ed., Taylor & Francis, London.

Kroemer, K., Kroemer, H., and Kroemer-Elbert, K. (1994), *How to Design for Ease and Efficiency*, Prentice-Hall, Englewood Cliffs, NJ.

Lueder, R. (1986), "Work Station Design," in *The Ergonomics Payoff: Designing the Electronic Office*, R. Lueder, Ed., Holt, Rinehart and Winston, Toronto, Ontario, Canada.

Lueder, R., and Noro, K. (1994), *Hard Facts About Soft Machines: The Ergonomics of Seating*, Taylor & Francis, London.

Mandal, A. (1985), *The Seated Man*, Dafnia Publications, Klampenborg, Denmark.

Manning, P. (1965), *Office Design: A Study of Environment by the Pilkington Research Unit*, University of Liverpool Press, Liverpool, Lancashire, England.

Margaritis, S., and Marmaras, N. (2003), "Making the Ergonomic Requirements Functional: The Case of Computerized Office Layout," in *Proceedings of the 15th Triennial Congress of the International Ergonomics Association and the 7th Conference of the Ergonomics Society of Korea/Japan Ergonomics Society*.

Marmaras, N. and Papadopoulos, S. (2002), "A study of Computerized Offices in Greece: Are Ergonomic Design Requirements Met?" *International Journal of Human-Computer Interaction*. Vol. 16, No. 2, pp. 261–281.

Nachemson, A., and Elfstrom, G. (1970), "Intravital Dynamic Pressure Measurements in Lumbar Disks," *Scandinavian Journal of Rehabilitation Medicine*, Suppl., p. 1.

Sanders, S. M., and McCormick, J. E. (1992), *Human Factors in Engineering and Design*, 7th ed., McGraw-Hill, New York.

Shoshkes, L. (1976), *Space Planning: Designing the Office Environment*, Architectural Record Books, New York.

Vitalis, A., Marmaras, N., Legg, S., and Poulakakis, G. (2000), "Please Be Seated," in *Proceedings of the 14th Triennial Congress of the International Ergonomics Association*, Vol. 6, San Diego, CA, pp. 43–45.

Wichman, H. (1984), "Shifting from Traditional to Open Offices: Problems and Suggested Design Principles," in *Human Factors in Organizational Design and Management*, H. Hendrick and O. Brown, Jr., Eds., Elsevier, Amsterdam.

Zelinsky, M. (1998), *New Workplaces for New Work Styles*, McGraw-Hill, New York.

CHAPTER 23

VIBRATION AND MOTION

Michael J. Griffin
University of Southampton
Southampton, England

1 INTRODUCTION

In work and leisure activities the human body experiences movement. The motion may be voluntary (as in some sports) or involuntary (as for passengers in vehicles). Movements may occur simultaneously in six different directions: three translational directions (fore and aft, lateral and vertical) and three rotational directions (roll, pitch, and yaw). Translational movements at constant velocity (i.e., with no change of speed or direction) are mostly imperceptible, except where exteroceptors (e.g., the eyes or ears) detect a change of position relative to other objects. Translational motion is detected primarily when the velocity changes, causing acceleration or deceleration of the body which may be detected by interoceptors (e.g., the vestibular, cutaneous, kinesthetic, or visceral sensory systems). Rotation of the body at constant velocity may be detected because it gives rise to translational acceleration in the body, because it reorients the body relative to the gravitational force of Earth or because the changing orientation relative to other objects is perceptible through exteroceptors. Vibration is oscillatory motion: the velocity is changing constantly so the movement is detectable by interoceptors and exteroceptors.

Vibration of the body may be desirable or undesirable. It can be described as pleasant or unpleasant; it can interfere with the performance of various tasks and cause injury and disease. Low-frequency oscillations of the body and movements of visual displays can cause motion sickness. It is convenient to consider human exposure to oscillatory motion in three categories.

1. *Whole-body vibration* occurs when the body is supported on a surface that is vibrating (e.g., sitting on a seat that vibrates, standing on a vibrating floor, lying on a vibrating surface). Whole-body vibration occurs in transport (e.g., road, off-road, rail, air, and marine transport) and when near some machinery.
2. *Motion sickness* can occur when real or illusory movements of the body or the environment lead to ambiguous inferences as to the movement or orientation of the human body. The movements associated with motion sickness are always of very low frequency, usually below 1 Hz.
3. *Hand-transmitted vibration* is caused by various processes in industry, agriculture, mining, construction, and transport where vibrating tools or workpieces are grasped or pushed by the hands or fingers.

There are many different effects of oscillatory motion on the body and many variables influencing each effect. The variables may be categorized as extrinsic variables (those occurring outside the human body) and intrinsic variables (the variability that occurs between and within people), as in Table 1. Some variables, especially intersubject variability, have large effects but are not easily measured. Consequently, it is often not practicable to make highly accurate predictions of the discomfort, interference with activities,

Table 1 Variables Influencing Human Responses to Oscillatory Motion

Extrinsic Variables	Intrinsic Variables
Vibration variables	Intrasubject variability
Vibration magnitude	Body posture
Vibration frequency	Body position
Vibration direction	Body orientation (sitting, standing, recumbent)
Vibration input positions	
Vibration duration	Intersubject variability
Other variables	Body size and weight
Other stressors (noise, temperature, etc.)	Body dynamic response
	Age
Seat dynamics	Gender
	Experience, expectation, attitude, and personality
	Fitness

or health effects for an individual. However, methods exist for predicting the average effect, or the probability of an effect, for groups of people. In this chapter we introduce human responses to oscillatory motion, summarize current methods of evaluating exposures to oscillatory motion, and identify some methods of minimizing unwanted effects of vibration.

2 MEASUREMENT OF VIBRATION AND MOTION

2.1 Vibration Magnitude

When vibrating, an object has, alternately, a velocity in one direction and then a velocity in the opposite direction. This change in velocity means that the object is constantly accelerating, first in one direction and then in the opposite direction. Figure 1 shows the displacement, velocity, and acceleration waveforms for a movement occurring at a single frequency (i.e., a sinusoidal oscillation). The magnitude of a vibration can be quantified by its displacement, its velocity, or its acceleration. For practical convenience, the magnitude of vibration is now usually expressed in terms of acceleration and is measured using accelerometers. The units of acceleration are meters per second per second (i.e., m/s^2 or m s^{-2}). The acceleration due to gravity on Earth is approximately 9.81 m/s^2.

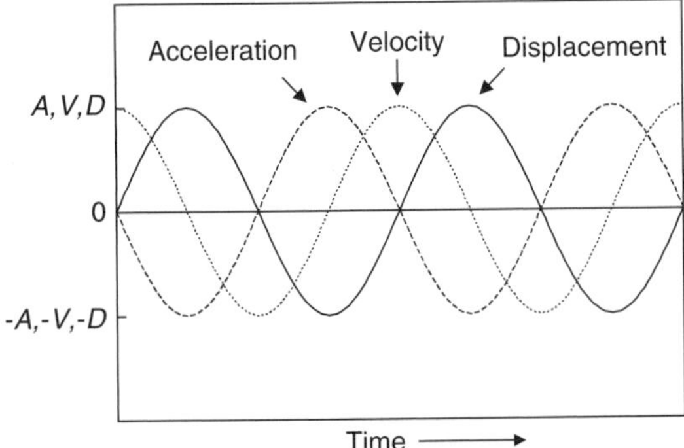

Figure 1 Displacement, velocity, and acceleration waveforms for a sinusoidal vibration. (If the vibration has frequency, f (in hertz), and peak displacement, D (in meters), the peak velocity is $V = 2\pi fD$ (in m/s), and the peak acceleration is $A = (2\pi f)^2 D$ (in m/s^2).

The magnitude of an oscillation can be expressed as the distance between the extremities reached by the motion (i.e., the peak-to-peak acceleration) or the maximum deviation from a central point (i.e., the peak acceleration). Magnitudes of vibration are very often expressed in terms of an average measure of the acceleration of the oscillatory motion, usually the root-mean-square (rms) value (i.e., m/s^2 rms for translational acceleration, rad/s^2 rms for rotational acceleration). (For a sinusoidal motion, the rms value is the peak value divided by $\sqrt{2}$, i.e., approximately 1.4.)

When observing vibration it is sometimes possible to estimate the displacement caused by motion. For sinusoidal motion, the acceleration, a, can be calculated from the frequency, f, in hertz, and the displacement, d:

$$a = (2\pi f)^2 d$$

For example, a sinusoidal motion with a frequency of 1 Hz and a peak-to-peak displacement of 0.1 m will have an acceleration of 3.95 m/s^2 peak to peak, 1.97 m/s^2 peak, and 1.40 m/s^2 rms. Although this expression can be used to convert acceleration measurements to corresponding displacements, it is accurate only when the motion occurs at a single frequency (i.e., it has a sinusoidal waveform as shown in Figure 1).

Logarithmic scales for quantifying vibration magnitudes in decibels are sometimes used. When using the reference level in International Standard 1683 (ISO, 1983), the acceleration level, L_a is expressed by $L_a = 20\log_{10}(a/a_0)$, where a is the measured acceleration (in m/s^2) and a_0 is the reference level of 10^{-6} m/s^2. With this reference, an acceleration of 1 m/s^2 corresponds to 120 dB an acceleration of 10 m/s^2 corresponds to 140 dB. Other reference levels are also in use.

2.2 Vibration Frequency

The frequency of vibration is expressed in cycles per second using the SI unit, hertz (Hz). The frequency of vibration influences the extent to which vibration is transmitted to the surface of the body (e.g., through seating), the extent to which it is transmitted through the body (e.g., from seat to head), and the responses to vibration within the body. From Section 2.1 it will be seen that the relation between the displacement and the acceleration of a motion also depends on the frequency of oscillation: A displacement of 1 millimeter corresponds to a low acceleration at low frequencies (e.g., 0.039 m/s^2 at 1 Hz) but a very high acceleration at high frequencies (e.g., 394 m/s^2 at 100 Hz).

2.3 Vibration Direction

The responses of the body differ according to the direction of the motion. Vibration is often measured at the interfaces between the body and the vibrating surfaces in three orthogonal directions. Figure 2 shows a coordinate system used when measuring the vibration of a hand holding a tool. The three principal directions of whole-body vibration for seated and standing persons are fore-and-aft (x-axis), lateral (y-axis), and vertical (z-axis). The vibration is measured at the interface between the body and the surface supporting the body (e.g., on the seat beneath the ischial tuberosities for a seated person, beneath the feet for a standing person). Figure 3 illustrates the translational and rotational axes for an origin at the ischial tuberosities on a seat, and the translational axes at a backrest and the feet of a seated person.

2.4 Vibration Duration

Some human responses to vibration depend on the duration of exposure. Additionally, the duration of measurement may affect the measured magnitude of the vibration. The rms acceleration may not provide a good indication of vibration severity if the vibration is intermittent, contains shocks, or otherwise varies in magnitude from time to time (see Section 3.3.1).

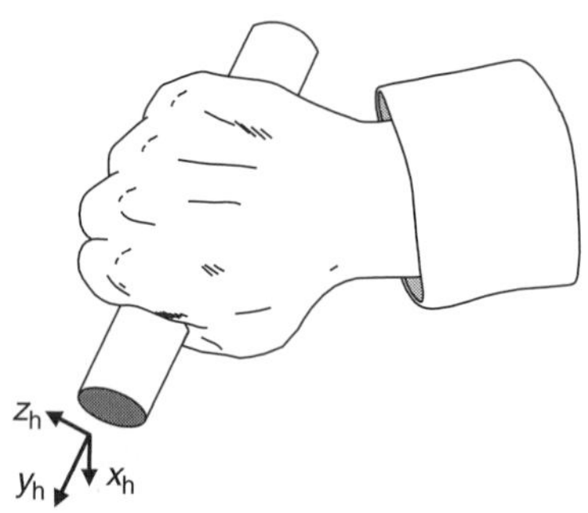

Figure 2 Axes of vibration used to measure exposures to hand-transmitted vibration.

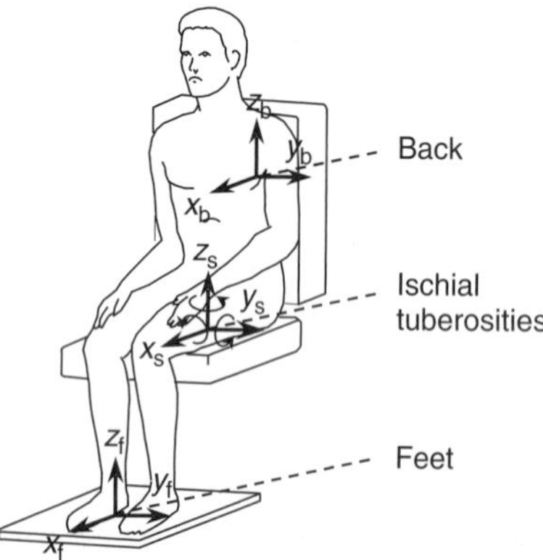

Figure 3 Axes of vibration used to measure exposures to whole-body vibration.

3 WHOLE-BODY VIBRATION

Whole-body vibration may affect health, comfort, and the performance of activities. The comments of persons exposed to vibration derive primarily from the sensations produced by vibration rather than knowledge that the vibration is causing harm or interfering with their activities. Vibration of the whole body is produced by various types of industrial machinery and by all forms of transport (including road, off-road, rail, sea, and air transport).

3.1 Vibration Discomfort

The relative discomfort caused by different oscillatory motions can be predicted from measurements of the vibration. For very low magnitude motions, it is possible to estimate the percentage of persons who will be able to feel vibration and the percentage who will not be able to feel the vibration. For higher vibration magnitudes, an approximate indication of the extent of subjective reactions is available in a semantic scale of discomfort.

Limits appropriate to the prevention of vibration discomfort vary between different environments (e.g., between buildings and transport) and between different types of transport (e.g., between cars and trucks) and within types of vehicle (e.g., between sports cars and limousines). The design limit depends on external factors (e.g., cost and speed) and the comfort in alternative environments (e.g., competitive vehicles).

3.1.1 Effects of Vibration Magnitude

The absolute threshold for the perception of vertical whole-body vibration in the frequency range 1 to 100 Hz is very approximately 0.01 m/s^2 rms; a magnitude of 0.1 m/s^2 will be easily noticeable; magnitudes around 1 m/s^2 rms are usually considered uncomfortable; magnitudes of 10 m/s^2 rms are usually dangerous. The precise values depend on vibration frequency and the exposure duration, and they are different for other axes of vibration (Griffin, 1990). A doubling of vibration magnitude (expressed in m/s^2) produces an approximate doubling of the sensation of discomfort; the precise increase depends on the frequency and direction of vibration. For many motions, a halving of vibration magnitude greatly reduces discomfort.

3.1.2 Effects of Vibration Frequency and Direction

The dynamic responses of the body and the relevant physiological and psychological processes dictate that subjective reactions to vibration depend on vibration frequency and vibration direction. The extent to which a given acceleration will cause a larger or smaller effect on the body at different frequencies is reflected in *frequency weightings*: frequencies capable of causing the greatest effect are given the greatest

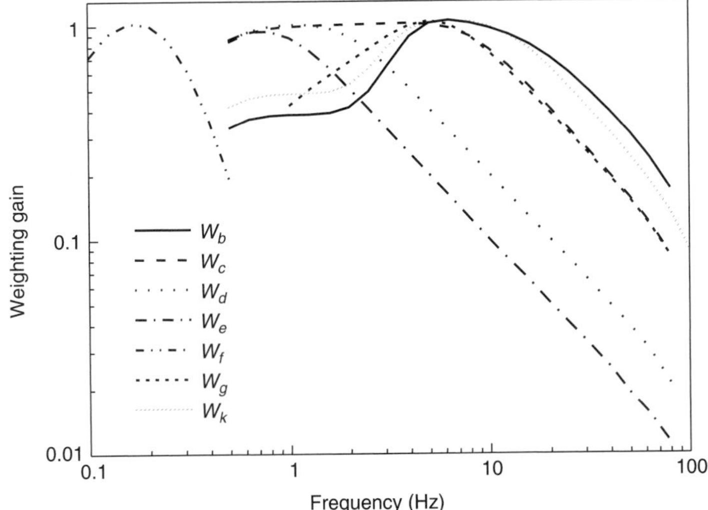

Figure 4 Acceleration frequency weightings for whole-body vibration and motion sickness as defined by BSI (1987a) and ISO (1997).

weight, and others are attenuated in accord with their relative importance.

Frequency weightings for human response to vibration have been derived from laboratory experiments in which volunteer subjects have been exposed to a set of motions having different frequencies. The subjects' responses are used to determine *equivalent comfort contours*. The reciprocal of such a curve forms the shape of the frequency weighting. Figure 4 shows frequency weightings W_b to W_f as defined in British Standard 6841 (BSI, 1987a). International Standard 2631 (ISO, 1997) allows the same weightings for evaluating vibration with respect to comfort but suggests the use of W_k in place of the almost identical weighting W_b when evaluating vertical vibration at the seat with respect to health effects. Table 2 defines simple asymptotic (i.e., straight line) approximations to these weightings, and Table 3 shows how the weightings should be applied to the 12 axes of vibration illustrated in Figure 3. [The weightings W_g and W_f are not required to predict vibration discomfort: W_g has been used for assessing interference with activities and is similar to the weighting for vertical vibration in an outdated International Standard (ISO, 1974, 1985); W_f is used to predict motion sickness caused by vertical oscillation; see Section 4.]

To minimize the number of frequency weightings, some are used for more than one axis of vibration, with different *multiplying factors* allowing for overall differences in sensitivity between axes (see Table 3). The frequency-weighted acceleration should be multiplied by the multiplying factor before the component is compared with components in other axes, or included in any summation over axes. The rms value of this acceleration (i.e., after frequency weighting and after being multiplied by the multiplying factor) is sometimes called a *component ride value* (Griffin, 1990).

Table 2 Asymptotic Approximations to Frequency Weightings, $W(f)$, for Comfort, Health, Activities, and Motion Sickness[a]

Weighting Name	Weighting Definition	
W_b	$0.5 < f < 2.0$	$W(f) = 0.4$
	$2.0 < f < 5.0$	$W(f) = f/5.0$
	$5.0 < f < 16.0$	$W(f) = 1.00$
	$16.0 < f < 80.0$	$W(f) = 16.0/f$
W_c	$0.5 < f < 8.0$	$W(f) = 1.0$
	$8.0 < f < 80.0$	$W(f) = 8.0/f$
W_d	$0.5 < f < 2.0$	$W(f) = 1.00$
	$2.0 < f < 80.0$	$W(f) = 2.0/f$
W_e	$0.5 < f < 1.0$	$W(f) = 1.00$
	$1.0 < f < 20.0$	$W(f) = 1.00/f$
W_f	$0.100 < f < 0.125$	$W(f) = f/0.125$
	$0.125 < f < 0.250$	$W(f) = 1.0$
	$0.250 < f < 0.500$	$W(f) = (0.25/f)^2$
W_g	$1.0 < f < 4.0$	$W(f) = (f/4)^{1/2}$
	$4.0 < f < 8.0$	$W(f) = 1.00$
	$8.0 < f < 80.0$	$W(f) = 8.0/f$

Source: BSI (1987a).
[a] f = frequency (Hz); $W(f) = 0$ where not defined.

Vibration occurring in several axes is more uncomfortable than vibration occurring in a single axis. To obtain an *overall ride value*, the root sums of squares of the component ride values is calculated:

$$\text{overall ride value} = \left[\sum(\text{component ride values})^2\right]^{1/2}$$

Overall ride values from different environments can be compared: a vehicle having the highest overall ride

Table 3 Application of Frequency Weightings for the Evaluation of Vibration with Respect to Discomfort

Input Position	Axis	Frequency Weighting	Axis Multiplying Factor
Seat	x	W_d	1.0
	y	W_d	1.0
	z	W_b	1.0
	r_x (roll)	W_e	0.63
	r_y (pitch)	W_e	0.40
	r_z (yaw)	W_e	0.20
Seat back	x	W_c	0.80
	y	W_d	0.50
	z	W_d	0.40
Feet	x	W_b	0.25
	y	W_b	0.25
	z	W_b	0.40

value would be expected to be the most uncomfortable with respect to vibration. The overall ride values can also be compared with the discomfort scale shown in Table 4. This scale indicates the approximate range of vibration magnitudes which are significant in relation to the range of vibration discomfort that might be experienced in vehicles.

3.1.3 Effects of Vibration Duration

Vibration discomfort tends to increase with increasing duration of exposure to vibration. The rate of increase may depend on many factors, but a simple fourth-power time dependency is used to approximate how discomfort varies with duration of exposure from the shortest possible shock to a full day of vibration exposure [i.e., $(\text{acceleration})^4 \times \text{duration} = \text{constant}$] (see Section 3.3.1).

3.2 Interference with Activities

Vibration and motion can interfere with the acquisition of information (e.g., by the eyes), the output of information (e.g., by hand or foot movements), or the complex central processes that relate input to output (e.g., learning, memory, decision making). Effects of oscillatory motion on human performance may impair safety.

There is most evidence of whole-body vibration affecting performance for input processes (mainly vision) and output processes (mainly continuous hand control). In both cases there may be a disturbance occurring entirely outside the body (e.g., vibration of a viewed display or vibration of a handheld control), a disturbance at the input or output (e.g., movement of the eye or hand), and a disturbance within the body affecting the peripheral nervous system (i.e., afferent or efferent system). Central processes may also be affected by vibration, but understanding is currently too limited to make confident generalized statements (see Figure 5).

The effects of vibration on vision and manual control are usually caused by the movement of the part of the body affected (i.e., eye or hand). The effects may be decreased by reducing the transmission of vibration to the eye or to the hand, or by making the task less susceptible to disturbance (e.g., increasing the size of a display or reducing the sensitivity of a control). Often, the effects of vibration on vision and manual control can be much reduced by redesign of the task.

3.2.1 Vision

Reading a newspaper in a moving vehicle may be difficult because the paper is moving, the eye is moving, or both the paper and the eye are moving. There are many variables that affect visual performance in these conditions: it is not possible to represent adequately the effects of vibration on vision without considering the effects of these variables.

Stationary Observer When a stationary observer views a moving display, the eye may be able to track the position of the display using *pursuit eye movements*. This closed-loop reflex will give smooth pursuit movements of the eye and clear vision if the display is moving at frequencies below about 1 Hz and with low velocity. At slightly higher frequencies of oscillation, the precise value depending on the predictability of the motion waveform, the eye will make saccadic eye movements to redirect the eye with small jumps. At frequencies above about 3 Hz, the eye will best be directed to one extreme of the oscillation and attempt to view the image as it is temporarily stationary while reversing the direction of movement (i.e., at the nodes of the motion).

In some conditions, the absolute threshold for the visual detection of the vibration of an object occurs when the peak-to-peak oscillatory motion gives an angular displacement at the eye of approximately 1 minute of arc. The acceleration required to achieve this threshold is very low at low frequencies but increases in proportion to the square of the frequency to become very high at high frequencies. When the vibration displacement is above the visual detection threshold, there will be perceptible blur if the vibration

Table 4 Scale of Vibration Discomfort

	rms Weighted Acceleration (m/s^2)	
Extremely uncomfortable (3.15–2.0)	3.15	
	2.5	Very uncomfortable (2.5–1.25)
	2.0	
Uncomfortable (1.6–0.8)	1.6	
	1.25	
	1.0	Fairly uncomfortable (1.0–0.5)
	0.8	
A little uncomfortable (0.63–0.315)	0.63	
	0.5	
	0.4	
	0.315	Not uncomfortable (0.315–0.25)
	0.25	

Source: BSI (1987a) and ISO (1997).

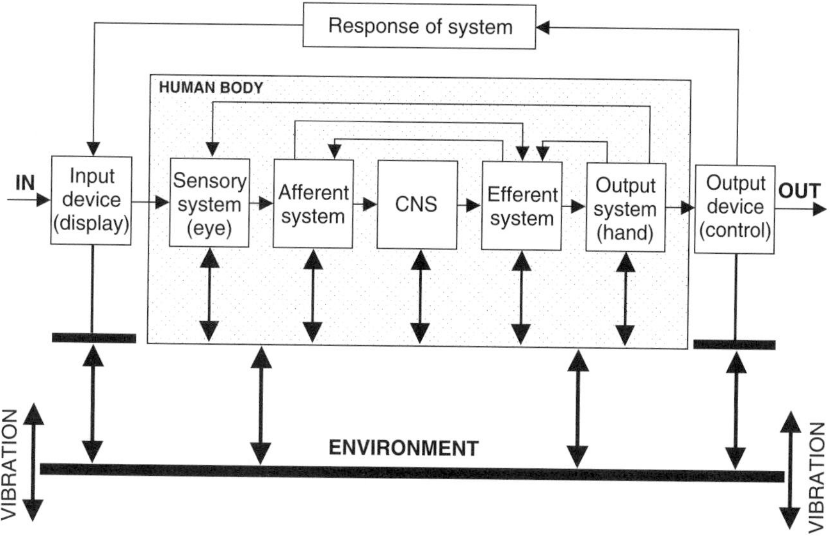

Figure 5 Information flow in a simple system and areas where vibration may affect human activities.

frequency is above about 3 Hz. The effects of vibration on visual performance (e.g., effects on reading speed and reading accuracy) may then be estimated from the maximum time that the image spends over some small area of the retina (e.g., the period of time spent near the nodes of the motion with sinusoidal vibration). For sinusoidal vibration this time decreases (so reading errors increase) in linear proportion to the frequency of vibration and in proportion to the square root of the displacement of vibration (O'Hanlon and Griffin, 1971). With dual-axis vibration (e.g., combined vertical and lateral vibration of a display), this time is greatly reduced and reading performance drops greatly (Meddick and Griffin, 1976). With narrowband random vibration there is a greater probability of low image velocity than with sinusoidal vibration of the same magnitude and predominant frequency, so reading performance tends to be less affected by random vibration than by sinusoidal vibration (Moseley et al., 1982). Display vibration reduces the ability to see fine detail in displays while having little effect on the clarity of larger forms.

Vibrating Observer If an observer is sitting or standing on a vibrating surface, the effects of vibration depend on the extent to which the vibration is transmitted to the eye. The motion of the head is highly dependent on body posture but is likely to occur in both translational axes (i.e., in the *x*-, *y*-, and *z*-axes) and rotational axes (i.e., in the roll, pitch, and yaw axes). Often, the predominant head motions affecting vision are in the vertical and pitch axes of the head. The dynamic response of the body may result in greatest head acceleration in these axes at frequencies around 5 Hz, but vibration at higher and lower frequencies can also have large effects on vision.

The pitch motion of the head is well compensated by the *vestibulo-ocular reflex*, which serves to help stabilize the line of sight of the eyes at frequencies below about 10 Hz (e.g., Benson and Barnes, 1978). Although there is often pitch motion of the head at 5 Hz, there is less pitch motion of the eyes at this frequency. Pitch motion of the head therefore has a less than expected effect on vision, unless the display is attached to the head, as with a helmet-mounted display (see Wells and Griffin, 1984).

The effects on vision of translational motion of the head depend on viewing distance: the effects are greatest when close to a display. As the viewing distance increases, the retinal image motions produced by translational displacements of the head decrease until when viewing an object at infinite distance, there is no retinal image motion produced by translational head displacement (Griffin, 1976).

For a vibrating observer there may be little difficulty with low-frequency pitch head motions when viewing a fixed display and no difficulty with translational head motions when viewing a distant display. The greatest problems occur with pitch head motion when the display is attached to the head and with translational head motion when viewing near displays. Additionally, there may be resonances of the eye within the head, but these are highly variable between individuals and often occur at high frequencies (e.g., 30 Hz and greater), where it is often possible to attenuate the vibration entering the body.

Observer and Display Vibrating When an observer and a display oscillate together, in phase, at low frequencies, the retinal image motions (and decrements in visual performance) are less than when either the observer or the display oscillate separately (Moseley and Griffin, 1986b). However, the advantage is lost as the vibration frequency is increased

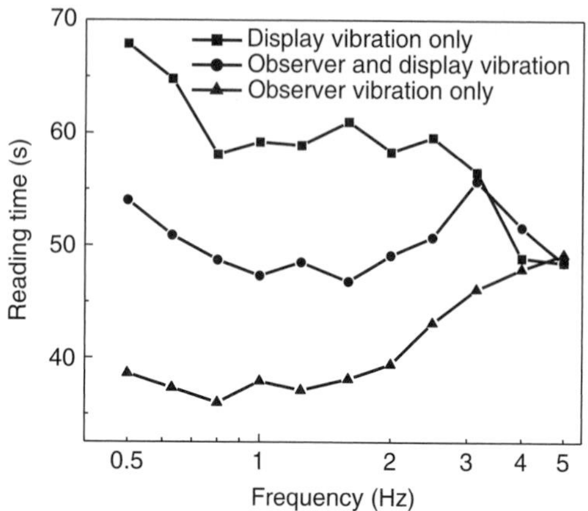

Figure 6 Average percentage reading times for display vibration only, observer vibration only, and simultaneous observer and display vibration. Data obtained with sinusoidal vertical vibration at 2.0 m/s^2 rms.

since there is then an increasing phase difference between the motion of the head and the motion of the display. At frequencies around 5 Hz, the phase lags between seat motion and head motion may be 90° or more (depending on seating conditions) and sufficient to eliminate any advantage of moving the seat and the display together. Figure 6 shows an example of how reading times were affected for the three viewing conditions with sinusoidal vibration in the frequency range 0.5 to 5 Hz.

Other Variables Some common situations in which vibration affects vision do not fall into one of the three categories in Figure 6. For example, when reading a newspaper on a train, the motion of the arms may result in the motion of the paper being different in magnitude and phase from the motions of both the seat and the head of the observer. The dominant axis of motion of the newspaper may be different from the dominant axis of motion of the person (Griffin and Hayward, 1994).

Increasing the size of detail in a display will often greatly reduce adverse effects of vibration on vision (Lewis and Griffin, 1979). In one experiment a 75% reduction in reading errors was achieved with only a 25% increase in the size of Landolt C targets (O'Hanlon and Griffin, 1971). Increasing the spacing between rows of letters and choosing appropriate character fonts can also be beneficial. The contrast of the display, or other reading material, also has an effect, but maximum performance may not occur with maximum contrast. The known influence of all such factors are summarized in a design guide for visual displays to be used in vibration environments (Moseley and Griffin, 1986a).

Optical devices may increase or decrease the effects of vibration on vision. Simple optical magnification of a vibrating object will increase both the apparent size of the object and the apparent magnitude of the vibration. Sometimes this will be beneficial, since the benefits of increasing the size of the detail may more than offset the effects of increased magnitude of vibration. The effect is similar to reducing the viewing distance, which can be beneficial for stationary observers viewing vibrating displays. If the observer is vibrating, the use of binoculars (and other magnifying devices) can be detrimental if the vibration (e.g., rotation in the hand holding the binoculars) causes such an increase in the image movement that it is not sufficiently compensated by the increase in image size. The use of binoculars and telescopes in moving vehicles becomes difficult for these reasons.

3.2.2 Manual Control

Reading a newspaper in a moving vehicle can be impeded by the action of vibration on vision; writing and other complex control tasks can also be impeded by vibration. Studies of the effects of whole-body vibration on the performance of hand-tracking tasks have been reviewed elsewhere (e.g., McLeod and Griffin, 1989). The characteristics of the task and the characteristics of the vibration combine to determine effects of vibration on performance: A given vibration may greatly effect one type of tracking task but have little effect on another.

Effects Produced by Vibration The most obvious consequence of vibration on a continuous manual control task is direct mechanical jostling of the hand, causing unwanted movement of the control. This is sometimes called *breakthrough, feedthrough*, or *vibration-correlated error*. The inadvertent movement of the pencil caused by jostling while writing in a vehicle is a form of *vibration-correlated error*. In a simple tracking task where the operator is required to follow movements of a target, some of the error will also be correlated with the target movements. This is called *input-correlated error* and often reflects primarily the inability of an operator to follow the target without the delays inherent in visual, cognitive, and motor activity. The part of the tracking error that is not correlated with either the vibration or the tracking task is called the *remnant*. This includes operator-generated noise and any source of nonlinearity: drawing a freehand straight line does not result in a perfect straight line even in the absence of environmental vibration. The effects of vibration on vision can result in increased remnant with some tracking tasks, and some studies show that vibration, usually at frequencies above about 20 Hz, interferes with neuromuscular processes, which may be expected to result in increased remnant. The cause of the three components of the tracking error are shown in the model presented as Figure 7.

Effects of Task Variables The gain (i.e., sensitivity) of a control determines the control output corresponding to a given force, or displacement, applied to the control by the operator. The optimum gain in static conditions (high enough not to cause fatigue but low enough to prevent inadvertent movement) is likely to

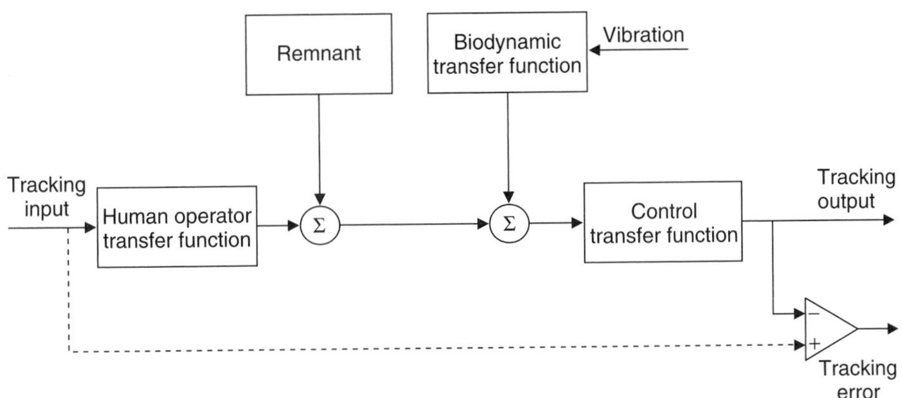

Figure 7 Linear model of a pursuit manual control system showing how tracking errors may be caused by the vibration (vibration-correlated error), the task (input correlated error), or some other cause (remnant).

be too great during exposure to vibration where inadvertent movement is more likely (Lewis and Griffin, 1977). First- and second-order control tasks (i.e., rate and acceleration control tasks) are more difficult than zero-order tasks (i.e., displacement control tasks), so tend to give more errors. However, there may sometimes be advantages with such controls which are less affected by vibration breakthrough at higher vibration frequencies.

In static conditions, isometric controls (which respond to force without movement) tend to result in better tracking performance than isotonic controls (which respond to movement but require the application of no force). However, several studies show that isometric controls may suffer more from the effects of vibration (e.g., Allen et al., 1973; Levison and Harrah, 1977). The relative merits of the two types of control and the optimum characteristics of a spring-centered control will depend on control gain and control order.

The results of studies investigating the influence of the position of a control appear consistent, with differences being dependent on the transmission of vibration to the hand in different positions (e.g., Shoenberger and Wilburn, 1973). Torle (1965) showed that the provision of an armrest could substantially reduce the effects of vibration on the performance of a task with a sidearm controller. The shape and orientation of controls may also be expected to affect performance, either by modifying the amount of vibration breakthrough or by altering the proprioceptive feedback to the operator.

Vibration may affect the performance of tracking tasks by reducing the visual performance of the operator. Wilson (1974) and McLeod and Griffin (1990) have shown that collimating a display by means of a lens so that it appears to be at infinity can reduce, or even eliminate, errors with some tasks. It is possible that visual disruption has played a significant part in the performance decrements reported in other experimental studies of the effects of vibration on manual control.

With some simple tasks, performance may be so easy as to be immune to disruption by vibration. At the other extreme, a task may be so difficult that any additional difficulty caused by vibration may be insignificant. Some studies suggest that with tasks having moderate ranges of difficulty, the effects of vibration may increase as the task difficulty increases (see McLeod and Griffin, 1989).

Effects of Vibration Variables The vibration transmissibility of the body is approximately linear (i.e., doubling the magnitude of vibration at the seat may be expected to approximately double the magnitude of vibration at the head or at the hand). Vibration-correlated error may therefore increase in approximately linear proportion to vibration magnitude.

There is no simple relation between the frequency of vibration and its effects on control performance. The effects of frequency depend on the control order (which varies between tasks) and the biodynamic response of the body (which varies with posture and between operators). With zero-order tasks and the same magnitude of acceleration at each frequency, the effects of vertical seat vibration may be greatest in the range 3 to 8 Hz since transmissibility to the shoulders is greatest in this range (see McLeod and Griffin, 1989). In the horizontal axes (i.e. the x- and y-axes of the seated body) the greatest effects appear to occur at lower frequencies: around 2 Hz or below. Again, this corresponds to the frequencies at which there is greatest transmission of vibration to the shoulders. The axis of the control task most affected by vibration may not be the same axis as that in which most vibration occurs at the seat. Often, fore-and-aft movements of the control (which generally correspond to vertical movements on the display) are most affected by vertical whole-body vibration. Few controls are sensitive to vertical hand movements, and these have rarely been studied. Multiple frequency vibration causes more disruption to performance than the presentation of any one of the constituent single frequencies alone. Similarly, the effects of multiple axis vibration are greater than the effects of any one of the single axes alone.

The impression that prolonged exposure to vibration causes fatigue gave rise to the fatigue-decreased proficiency boundary in International Standard 2631, first published in 1974 (ISO, 1974, 1985). This standard proposed a complex time-dependent magnitude of vibration which is said to be "a limit beyond which exposure to vibration can be regarded as carrying a significant risk of impaired working efficiency in many kinds of tasks, particularly those in which time-dependent effects ('fatigue') are known to worsen performance as, for example, in vehicle driving." Reviews of experimental studies show time-dependent effects of performance in only a few cases, with performance sometimes improving with time. It may be concluded that experimental evidence supporting the ISO fatigue-decreased proficiency boundary is very weak. There are certainly no substantial data justifying a time-dependent limit for the effects of vibration on performance with the complexity included in International Standard 2631 (ISO, 1974, 1985). Any duration-dependent effects of vibration may be influenced by complex central factors, including motivation, arousal, and similar concepts that depend on the form of the task: they may not lend themselves to satisfactory representation by a single time-dependent limit in an International Standard. The most common and most easily understood "direct" effects of vibration on vision and manual control are probably not intrinsically dependent on the duration of vibration exposure.

Other Variables Repeated exposure to vibration may allow subjects to develop techniques for minimizing vibration effects by, for example, adjusting body posture to reduce the transmission of vibration to the head or the hand or by learning how to recognize images blurred by vibration. Results of experiments performed in one experimental session of vibration exposure may not necessarily apply to situations where operators have an opportunity to learn techniques to ameliorate the effects of vibration.

There have been few investigations of the effects of vibration on common everyday tasks. Corbridge and Griffin (1991) found that the effects of vertical whole-body vibration on spilling liquid from a handheld cup were greatest close to 4 Hz. They also found that the effects of vibration on writing speed and subjective estimates of writing difficulty were most affected by vertical vibration in the range 4 to 8 Hz. Although 4 Hz was a sensitive frequency for both the "drinking" and the writing task, the dependence on frequency of the effects of vibration were different for the two activities.

Whole-body vibration can cause a warbling of speech due to fluctuations in the airflow through the larynx. Greatest effects may occur with vertical vibration in the range 5 to 20 Hz, but they are not usually sufficient to appreciably reduce the intelligibility of speech (e.g., Nixon and Sommer, 1963). Some studies suggest that exposure to vibration may contribute to noise-induced hearing loss, but further study is required to allow a full interpretation of these data.

3.2.3 Cognitive Tasks

To be useful, studies of cognitive effects of vibration must be able to show that any effects were not caused by vibration affecting input processes (e.g., vision) or output processes (e.g., hand control). Only a few investigators have addressed possible cognitive effects of vibration with care and considered such problems. For example, Shoenberger (1974) found that with the Sternberg memory-reaction-time task, the time taken for subjects to recall letters presented on a display depended on the angular size of the letters. He was able to conclude that performance was degraded by visual effects of vibration and not by cognitive effects of vibration. In most other studies there has been little attempt to develop hypotheses to explain any significant effects of vibration in terms of the component processes involved in cognitive processing.

Simple cognitive tasks (e.g., simple reaction time) appear to be unaffected by vibration other than by changes in arousal or motivation or by direct effects on input and output processes. This may also be true for some complex cognitive tasks. However, the scarcity and diversity of experimental studies allow the possibility of real and significant cognitive effects of vibration (see Sherwood and Griffin, 1990, 1992). Vibration may influence fatigue, but there is little relevant scientific evidence to provide a foundation for the complex form of the *fatigue-decreased proficiency limit* that was offered in International Standard 2631 (ISO, 1974, 1985).

3.3 Health Effects

Epidemiological studies have reported disorders among persons exposed to vibration from occupational, sport, and leisure activities (see Dupuis and Zerlett, 1986; Hulshof and van Zanten, 1987; Bongers and Boshuizen, 1990; Griffin, 1990; Bovenzi and Zadini, 1992; Bovenzi and Hulshof, 1998). The studies do not all agree on either the type or the extent of disorders, and rarely have the findings been related to measurements of the vibration exposures. However, the incidence of some disorders of the back (back pain, displacement of intervertebral disks, degeneration of spinal vertebrae, osteoarthritis, etc.) appear to be greater in some groups of vehicle operators, and it is thought that this is sometimes associated with their vibration exposure. There may be several alternative causes of an increase in disorders of the back among persons exposed to vibration (e.g., poor sitting postures, heavy lifting). It is not always possible to conclude confidently that a back disorder is solely, or primarily, caused by vibration.

Other disorders that have been claimed to be due to occupational exposures to whole-body vibration include abdominal pain, digestive disorders, urinary frequency, prostatitis, hemorrhoids, balance and visual disorders, headaches, and sleeplessness. Further research is required to confirm whether these signs and symptoms are causally related to exposure to vibration.

3.3.1 Vibration Evaluation

Epidemiological data alone are not sufficient to define how to evaluate whole-body vibration so as to predict the relative risks to health from the different types of vibration exposure. A consideration of such data in combination with an understanding of biodynamic responses and subjective responses is used to provide current guidance. The manner in which the health effects of oscillatory motions depend on the frequency, direction, and duration of motion is currently assumed to be similar to that for vibration discomfort (see Section 3.1). However, it is assumed that the "total" exposure, rather than the "average" exposure, is important, so a "dose" measure is used. British Standard 6841 (BSI, 1987a) and International Standard 2631 (ISO, 1997) can be interpreted as providing similar guidance, but there is more than one method within ISO 2631 and they are not consistent (Griffin, 1998b).

3.3.2 British Standard 6841

British Standard 6841 (BSI, 1987a) defines an *action level* for vertical vibration using *vibration dose values*. The vibration dose value employs a fourth-power time dependency to accumulate vibration severity over the exposure period from the shortest possible shock to a full day of vibration:

$$\text{vibration dose value} = \left[\int_{t=0}^{t=T} a^4(t)\,dt\right]^{1/4} \quad (1)$$

where $a(t)$ is the frequency-weighted acceleration. If the exposure duration (t, seconds) and the frequency-weighted rms acceleration (a_{rms}, m/s^2 rms) are known for conditions in which the vibration characteristics are statistically stationary, it can be useful to calculate the *estimated vibration dose value* (eVDV):

$$\text{estimated vibration dose value} = 1.4 a_{\text{rms}} t^{1/4} \quad (2)$$

The eVDV is not applicable to transients, shocks, and repeated shock motions in which the crest factor (peak value divided by the rms value) is high.

No precise limit can be offered to prevent disorders caused by whole-body vibration, but British Standard 6841 (BSI, 1987a) offers the following guidance: "High vibration dose values will cause severe discomfort, pain and injury. Vibration dose values also indicate, in a general way, the severity of the vibration exposures which caused them. However there is currently no consensus of opinion on the precise relation between vibration dose values and the risk of injury. It is known that vibration magnitudes and durations which produce vibration dose values in the region of 15 m/s$^{1.75}$ will usually cause severe discomfort. It is reasonable to assume that increased exposure to vibration will be accompanied by increased risk of injury." An action level might be set higher or lower than 15 m/s$^{1.75}$. Figure 8 shows this action level for exposure durations from 1 second to 1 day.

3.3.3 International Standard 2631

International Standard 2631 (ISO, 1997) offers two different methods of evaluating vibration severity

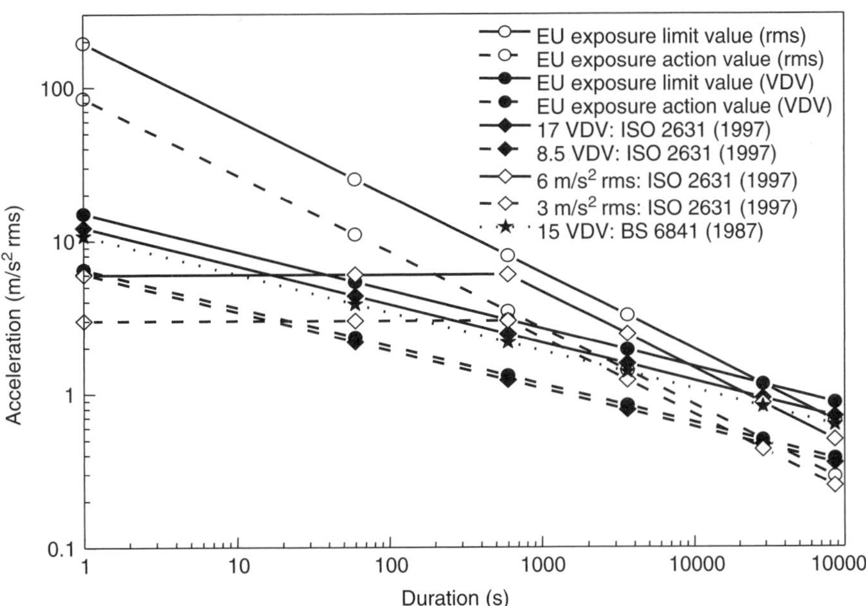

Figure 8 Comparison among the health guidance caution zones for whole-body vibration in ISO (1997) (3 to 6 m/s^2 rms; 8.5 to 17 m/s$^{1.75}$), 15 m/s$^{1.75}$ action level implied in BSI (1987a), and the rms and VDV exposure limit values and exposure action values for whole-body vibration in the EU Physical Agents (Vibration) Directive.

with respect to health effects, and for both methods there are two boundaries. When evaluating vibration using the vibration dose value, it is suggested that below a boundary corresponding to a vibration dose value of 8.5 $m/s^{1.75}$, "health risks have not been objectively observed," between 8.5 and 17 $m/s^{1.75}$, "caution with respect to health risks is indicated," and above 17 $m/s^{1.75}$, "health risks are likely." The two boundaries define a *VDV health guidance caution zone*. The alternative method of evaluation in ISO 2631 (1997) uses a time dependency in which the acceptable vibration does not vary with duration between 1 and 10 minutes and then decreases in inverse proportion to the square root of duration from 10 minutes to 24 hours. This method suggests an rms health guidance caution zone, but the method is not fully defined in the text, it allows very high accelerations at short durations, it conflicts with the vibration dose value method, and it cannot be extended to exposure durations below 1 minute (Figure 8). With severe vibration exposures, prior consideration of the fitness of the exposed persons and the design of adequate safety precautions may be required. The need for regular health surveillance of routinely exposed persons may also be considered.

3.3.4 EU Machinery Safety Directive

The Machinery Safety Directive of the European Community (EEC, 1989) states that machinery must be designed and constructed so that hazards resulting from vibration produced by the machinery are reduced to the lowest practicable level, taking into account technical progress and the availability of means of reducing vibration. The instruction handbooks for handheld and hand-guided machinery must specify the equivalent acceleration to which the hands or arms are subjected where this exceeds some stated value (for whole-body vibration this is currently a frequency-weighted acceleration of 0.5 m/s^2 rms). The relevance of any such value will depend on the test conditions to be specified in other standards. Many work vehicles exceed this value at some stage during an operation or journey. Standardized procedures for testing work vehicles are being prepared; the values currently quoted by manufacturers may not always be representative of the operating conditions in the work for which the machinery is used.

3.3.5 EU Physical Agents Directive

In 2002, the Parliament and Commission of the European Community agreed on minimum health and safety requirements for the exposure of workers to the risks arising from vibration. For whole-body vibration, the directive defines an 8-h equivalent *exposure action value* of 0.5 m/s^2 rms (or a vibration dose value of 9.1 $m/s^{1.75}$) and an 8-h equivalent *exposure limit value* of 1.15 m/s^2 rms (or a vibration dose value of 21 $m/s^{1.75}$). Member states of the European Union must bring into force by July 6, 2005 laws to comply with the directive.

The directive says that workers must not be exposed above the exposure limit value. If the exposure action values are exceeded, the employer must establish and implement a program of technical and/or organizational measures intended to reduce to a minimum exposure to mechanical vibration and the attendant risks. The directive says that workers exposed to vibration in excess of the exposure action values are entitled to appropriate health surveillance. Health surveillance is also required if there is any reason to suspect that workers may be injured by the vibration even if the exposure action value is not exceeded.

The probability of injury arising from occupational exposures to whole-body vibration at the exposure action value and the exposure limit value cannot be estimated because epidemiological studies have not yet produced dose–response relationships. However, it seems clear that the directive does not define safe exposures to whole-body vibration since the rms values are associated with extraordinarily high magnitudes of vibration (and shock) when the exposures are short: these exposures may be assumed to be hazardous (see Figure 8) (Griffin, 2004). The vibration dose value procedure suggests more reasonable vibration magnitudes for short-duration exposures.

3.4 Disturbance in Buildings

Acceptable magnitudes of vibration in buildings are close to vibration perception thresholds. The effects of vibration in buildings are assumed to depend on the use of the building in addition to the vibration frequency, direction, and duration. Guidance is given in various standards [e.g., International Standard 2631, part 2 (ISO, 1989); American National Standard S3.29 (ANSI, 1983); British Standard 6472 (BSI, 1992)]. Using the guidance contained in ISO 2631, part 2, it is possible to summarize the acceptability of vibration in different types of building in a single table of vibration dose values (see Table 5 and British Standard 6472). The vibration dose values in Table 5 are applicable irrespective of whether the vibration occurs as a continuous vibration, an intermittent vibration, or repeated shocks.

3.5 Biodynamics

The human body is a complex mechanical system that does not, in general, respond to vibration in the same

Table 5 Vibration Dose Values at Which Various Degrees of Adverse Comment May Be Expected in Buildings

Place	Low Probability of Adverse Comment	Adverse Comment Possible	Adverse Comment Probable
Critical working areas	0.1	0.2	0.4
Residential	0.2–0.4	0.4–0.8	0.8–1.6
Office	0.4	0.8	1.6
Workshops	0.8	1.6	3.2

Source: Based on ISO (1989) and BSI (1992); see Griffin (1990).

manner as a rigid mass: there are relative motions between the body parts that vary with the frequency and the direction of the vibration applied. Although there are resonances in the body, it is oversimplistic to summarize the dynamic response of the body merely by mentioning one or two resonance frequencies. The biodynamics of the body affect human responses to vibration, but the discomfort, the interference with activities, and the health effects of vibration cannot be well predicted solely by considering the body as a mechanical system.

3.5.1 Transmissibility of the Human Body

The extent to which the vibration at the input to the body (e.g., the vertical vibration at a seat) is transmitted to a part of the body (e.g., vertical vibration at the head or the hand) is described by the transmissibility. At low frequencies of oscillation (e.g., below about 1 Hz), the oscillations of the seat and the body are very similar, so the transmissibility is approximately 1.0. With increasing frequency of oscillation, the motions of the body increase above that measured at the seat; the ratio of the motion of the body to the motion of the seat will reach a peak at one or more frequencies (i.e., resonance frequencies). At high frequencies the body motion will be less than that at the seat.

The resonance frequencies and the transmissibilities at resonance vary according to where the vibration is measured on the body and the posture of the body. For seated persons there may be resonances to the head and the hand at frequencies in the range 4 to 12 Hz for vertical vibration, below 4 Hz with x-axis vibration, and below 2 Hz with lateral vibration (see Paddan and Griffin, 1988a, b). A seat back can greatly increase the transmission of x-axis vibration to the heads and upper bodies of seated people, and bending of the legs can greatly affect the transmission of vertical vibration to the heads of standing persons.

3.5.2 Mechanical Impedance of the Human Body

Mechanical impedance reflects the relation between the driving force at the input to the body and the resulting movement of the body. If the human body were rigid, the ratio of force to acceleration applied to the body would be constant and indicate the mass of the subject. Because the body is not rigid, the ratio of force to acceleration is close to the body mass only at very low frequencies (below about 2 Hz with vertical vibration; below about 1 Hz with horizontal vibration).

Measures of mechanical impedance usually show a principal resonance for vertical vibration of seated subjects at about 5 Hz, and sometimes a second resonance in the range 7 to 12 Hz (Fairley and Griffin, 1989; Matsumoto and Griffin, 2000; Nawayseh and Griffin, 2003). Unlike some of the resonances affecting the transmissibility of the body, these resonance are influenced only by movement of large masses close to the input of vibration to the body. The large difference in impedance between that of a rigid mass and that of the human body means that the body cannot usually be represented by a rigid mass when measuring the vibration transmitted through seats.

3.5.3 Biodynamic Models

Various mathematical models of the responses of the body to vibration have been developed. A simple model with one or two degrees of freedom can represent the impedance of the body, and a dummy might be constructed to represent this impedance for seat testing. Compared with impedance, the transmissibility of the body is affected by many more variables and so requires a more complex model reflecting the posture of the body and the translation and rotation associated with the various modes of vibration.

3.6 Protection from Whole-Body Vibration

Wherever possible, vibration should be reduced at the source. This may involve reducing the undulations of the terrain, reducing the speed of travel of vehicles, or improving the balance of rotating parts. Methods of reducing the transmission of vibration to operators require an understanding of the characteristics of the vibration environment and the route for the transmission of vibration to the body. For example, the magnitude of vibration often varies with location: lower magnitudes are experienced in some areas adjacent to machinery or in different parts of vehicles.

3.6.1 Seating Dynamics

Most seats exhibit a resonance at low frequencies that results in higher magnitudes of vertical vibration occurring on the seat than on the floor! At high frequencies there is usually attenuation of vibration. The resonance frequencies of common seats are usually in the region of 4 Hz (see Figure 9). The amplification at resonance is partially determined by the *damping* in the seat. Increases in the damping of a seat cushion tend to reduce the amplification at resonance but increase the transmission of vibration at high frequencies. The variations in transmissibility between seats are sufficient to result in significant differences in the vibration experienced by people supported by different seats.

A simple numerical indication of the isolation efficiency of a seat for a specific application is provided by the *seat effective amplitude transmissibility* (SEAT) (Griffin, 1990). A SEAT value greater than 100% indicates that overall, the vibration on the seat is worse than the vibration on the floor beneath the seat:

$$\text{SEAT } (\%) = \frac{\text{ride comfort on seat}}{\text{ride comfort on floor}} \times 100$$

Values below 100% indicate that the seat has provided some useful attenuation. Seats should be designed to have the lowest SEAT value compatible with other constraints.

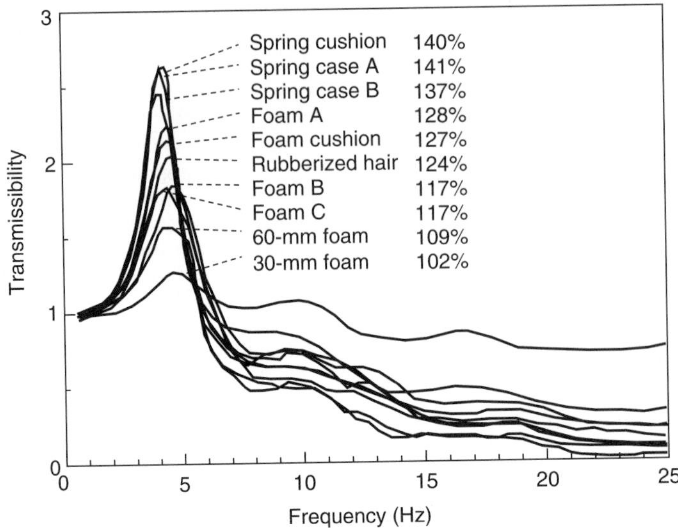

Figure 9 Comparison of the vertical transmissibilities and SEAT values for 10 alternative cushions of passenger railway seats. (Data from Corbridge et al., 1989.)

In practice, the SEAT value is a mathematical procedure for predicting the effect of a seat on ride comfort. The ride comfort that would result from sitting on the seat or on the floor can be predicted using the frequency weightings in the appropriate standard. The SEAT value may be calculated from the rms values or the vibration dose values of the frequency-weighted acceleration on the seat and the floor:

$$\text{SEAT (\%)} = \frac{\text{vibration dose value on seat}}{\text{vibration dose value on floor}} \times 100$$

The SEAT value is a characteristic of the vibration input and not merely a description of the dynamics of the seat: different values are obtained with the same seat in different vehicles. The SEAT value indicates the suitability of a seat for a particular type of vibration.

A separate suspension mechanism is provided beneath the seat pan in *suspension seats*. These seats, used in some off-road vehicles, trucks, and coaches, have low resonance frequencies (often below about 2 Hz) and so can attenuate vibration at frequencies above about 2 Hz. The transmissibilities of these seats are usually determined by the seat manufacturer, but their isolation efficiencies vary with operating conditions.

4 MOTION SICKNESS

Motion sickness is not an illness but a normal response to motion that is experienced by many fit and healthy people. A variety of different motions can cause sickness and reduce the comfort, impede the activities, and degrade the well-being of both those affected directly and those associated with the motion sick. Although vomiting can be the most inconvenient consequence, other effects (e.g., yawning, cold sweating, nausea, stomach awareness, dry mouth, increased salivation, headaches, bodily warmth, dizziness, drowsiness) can also be unpleasant. In some cases the symptoms can be so severe as to result in reduced motivation to survive difficult situations.

4.1 Causes of Motion Sickness

Motion sickness can be caused by many different movements of the body (e.g., translational and rotational oscillation, constant speed rotation about an off-vertical axis, Coriolis stimulation), movements of the visual scene, and various other stimuli-producing sensations associated with movement of the body (see Table 6 and Griffin, 1991). Motion sickness is neither

Table 6 Examples of Environments, Activities, and Devices That Can Cause Symptoms of Motion Sickness

Boats	Camel rides
Ships	Elephant rides
Submarines	
Hydrofoils	Vehicle simulators
Hovercraft	
Swimming	Fairground devices
Fixed-wing aircraft	Cinerama
Helicopters	Inverting or distorting spectacles
Spacecraft	Microfiche readers
	Head-coupled visual displays
Cars	
Coaches	Rotation about off-vertical axis
Buses	Coriolis stimulation
Trains	Low-frequency translational oscillation
Tanks	

explained nor predicted solely by the physical characteristics of the motion, although some motions can be predicted reliably as being more nauseogenic than others.

Motions of the body may be detected by three basic sensory systems: the vestibular system, the visual system, and the somatosensory systems. The *vestibular system* is located in the inner ear and comprises the semicircular canals, which respond to the rotation of the head, and the otoliths, which respond to translational forces (either translational acceleration or rotation of the head relative to an acceleration field, such as the force of gravity). The eyes may detect relative motion between the head and the environment, caused by either head movements (in translation or rotation), movements of the environment, or a combination of the movements of the head and the environment. The *somatosensory systems* respond to force and displacement of parts of the body and give rise to sensations of body movement, or force.

It is assumed that in normal environments, the movements of the body are detected by all three sensory systems and that this leads to an unambiguous indication of the movements of the body in space. In some other environments the three sensory systems may give signals corresponding to different motions (or motions that are not realistic) and lead to some form of conflict. This leads to the idea of a *sensory conflict theory* of motion sickness, in which sickness occurs when the sensory systems disagree on the motions that are occurring. However, this implies some absolute significance to sensory information, whereas the meaning of the information is probably learned. This led to the *sensory rearrangement theory* of motion sickness, which states that "all situations which provoke motion sickness are characterized by a condition of sensory rearrangement in which the motion signals transmitted by the eyes, the vestibular system and the non-vestibular proprioceptors are at variance either with one another or with what is expected from previous experience" (Reason, 1970, 1978). Reason and Brand (1975) suggest that the conflict may be considered sufficiently well in two categories: *intermodality* (between vision and the vestibular receptors) and *intramodality* (between the semicircular canals and the otoliths within the vestibular system). For both categories it is possible to identify three types of situation in which conflict can occur (see Table 7). The theory implies that all situations that provoke motion sickness can be fitted into one of the six conditions shown in Table 7 (see Griffin,

Table 7 Type of Motion Cue Mismatch Produced by Various Provocative Stimuli

	Category of Motion Cue Mismatch	
	Visual (A)/Vestibular (B)	Canal (A)/Otolith (B)
Type 1: A and B give contradictory or uncorrelated information simultaneously	Watching waves from a ship	Making head movements while rotating (Coriolis or cross-coupled stimulation
	Use of binoculars in a moving vehicle	Making head movements in an abnormal environment that may be constant (e.g. hyper- or hypogravity) or fluctuating (e.g., linear oscillation)
	Making head movements when vision is distorted by an optical device	Space sickness
	Pseudo-Coriolis stimulation	Vestibular disorders (e.g., Ménière's disease, acute labyrinthitis, trauma labyrinthectomy)
Type IIa: A signals in absence of expected B signals	Cinerama sickness	Positional alcohol nystagmus
	Simulator sickness	Caloric stimulation of semicircular canals
	"Haunted swing"	Vestibular disorders (e.g., pressure vertigo, cupulolithiasis)
	Circular vection	
Type IIb: B signals in absence of expected A signals	Looking inside a moving vehicle without external visual reference (e.g., below deck in a boat)	Low-frequency (<0.5 Hz) translational oscillation
	Reading in a moving vehicle	Rotating linear acceleration vector (e.g., "barbecue spit" rotation, rotation about an off-vertical axis)

Source: Adapted from Benson (1984).

1990). There is evidence that the average susceptibility to sickness among males is less than that among females, and susceptibility decreases with increased age among both males and females (Lawther and Griffin, 1988a). However, there are larger individual differences within any group of either gender at any age: some people are easily made ill by motions that can be endured indefinitely by others. The reasons for these differences are not properly understood.

4.2 Sickness Caused by Oscillatory Motion

Motion sickness is not caused by oscillation (however violent) at frequencies much above about 1 Hz: the phenomenon arises from motions at the low frequencies associated with normal postural control of the body. Various experimental investigations have explored the extent to which vertical oscillation causes sickness at different frequencies. These studies have allowed the formulation of a frequency weighting, W_f (see Figure 4), and the definition of a *motion sickness dose value*. The frequency weighting W_f reflects greatest sensitivity to acceleration in the range 0.125 to 0.25 Hz, with a rapid reduction in sensitivity at higher frequencies. The motion sickness dose value predicts the probability of sickness from knowledge of the frequency and magnitude of vertical oscillation (see Lawther and Griffin, 1987; BSI, 1987; ISO, 1997):

$$\text{motion sickness dose value} = a_{\text{rms}} t^{1/2}$$

where a_{rms} is the root-mean-square value of the frequency-weighted acceleration (m/s^2) and t is the exposure period (seconds). The percentage of unadapted adults who are expected to vomit is given by $\frac{1}{3}$MSDV. (These relationships have been derived from exposures in which up to 70% of persons vomited during exposures lasting between 20 minutes and 6 hours.)

The motion sickness dose value has been used for the prediction of sickness on various marine craft (ships, hovercraft, and hydrofoil) in which vertical oscillation has been shown to be a prime cause of sickness (Lawther and Griffin, 1988b). Vertical oscillation is not the principal cause of sickness in many road vehicles (Turner and Griffin, 1999; Griffin and Newman, 2004) and some other environments; the expression above should not be assumed to be applicable to the prediction of sickness in all environments.

5 HAND-TRANSMITTED VIBRATION

Prolonged and regular exposure of the fingers or the hands to vibration or repeated shock can give rise to various signs and symptoms of disorder. The precise extent and interrelation between the signs and symptoms are not fully understood, but five types of disorder may be identified (see Table 8). The various disorders may be interconnected: more than one disorder can affect a person at the same time and it is possible that the presence of one disorder facilitates the appearance of another. The onset of each disorder is dependent on several variables, such as the vibration characteristics, the dynamic response of the fingers or hand, individual susceptibility to damage, and other aspects of the environment. The terms *vibration syndrome* or *hand–arm vibration syndrome* (HAVS) are sometimes used to refer to one or more of the effects listed in Table 8.

Table 8 Types of Disorders Associated with Hand-Transmitted Vibration Exposures[a]

Type	Disorder
A	Circulatory disorders
B	Bone and joint disorders
C	Neurological disorders
D	Muscle disorders
E	Other general disorders (e.g., central nervous system)

Source: Griffin (1990).
[a]Some combinations of these disorders are sometimes referred to as the hand–arm vibration syndrome (HAVS).

5.1 Sources of Hand-Transmitted Vibration

The vibration on tools varies greatly depending on tool design and method of use, so it is not possible to categorize individual tool types as safe or dangerous. However, Table 9 lists tools and processes that are sometimes a cause for concern.

5.2 Effects of Hand-Transmitted Vibration

5.2.1 Vascular Disorders

The first published cases of the condition now most commonly known as *vibration-induced white finger* (VWF) are acknowledged to be those reported in Italy by Loriga (1911). A few years later, cases were documented at limestone quarries in Indiana. Vibration-induced white finger has subsequently been reported to occur in many other widely varied occupations in which there is exposure of the fingers to vibration (see Taylor and Pelmear, 1975; Wasserman et al., 1982; Griffin, 1990).

Signs and Symptoms Vibration-induced white finger (VWF) is characterized by intermittent whitening (i.e., blanching) of the fingers (Griffin and Bovenzi, 2002). The fingertips are usually the first to blanch, but the affected area may extend to all of one or more fingers with continued vibration exposure. Attacks of blanching are precipitated by cold and therefore usually occur in cold conditions or when handling cold objects. The blanching lasts until the fingers are rewarmed and vasodilation allows the return of the blood circulation. Many years of vibration exposure often occur before the first attack of blanching is noticed. Affected persons often have other signs and symptoms, such as numbness and tingling. Cyanosis

Table 9 Examples of Tools and Processes Potentially Associated with Vibration Injuries

Tool or Process	Examples
Percussive metal-working tools	Riveting tools, caulking tools, chipping tools, chipping hammers, fettling tools, hammer drills, clinching and flanging tools, impact wrenches, swaging, needle guns
Grinders and other rotary tools	Pedestal grinders, handheld grinders, handheld sanders, handheld polishers, flex-driven grinders/polishers, rotary burring tools
Percussive hammers and drills used in mining, demolition, and road construction	Hammers, rock drills, road drills
Forest and garden machinery	Chain saws, antivibration chain saws, brush saws, mowers and shears, barking machines
Other processes and tools	Nut runners, shoe-pounding-up machines, concrete vibro-thickeners, concrete leveling vibrotables, motorcycle handlebars

and, rarely, gangrene, have also been reported. It is not yet clear to what extent these other signs and symptoms are causes of, caused by, or unrelated to attacks of white finger.

Diagnosis There are other conditions that can cause similar signs and symptoms to those associated with VWF. Vibration-induced white finger cannot be assumed to be present merely because there are attacks of blanching. It will be necessary to exclude other known causes of similar symptoms (by medical examination) and also to exclude *primary Raynaud's disease* (also called *constitutional white finger*). This exclusion cannot yet be achieved with complete confidence, but if there is no family history of the symptoms, if the symptoms did not occur before the first significant exposure to vibration, and if the symptoms and signs are confined to areas in contact with the vibration (e.g., the fingers, not the feet), they will often be assumed to indicate vibration-induced white finger.

Diagnostic tests for vibration-induced white finger can be useful, but at present they are not infallible indicators of the disease. The measurement of finger systolic blood pressure following finger cooling and the measurement of finger rewarming times following cooling can be useful, but many others tests are in use (see Griffin and Bovenzi, 2002). The severity of the effects of vibration are sometimes recorded by reference to the *stage* of the disorder. The staging of vibration-induced white finger is based on verbal statements made by the affected person. In the Stockholm Workshop staging system, the staging is influenced by both the frequency of attacks of blanching and the areas of the digits affected by blanching (see Table 10).

A *scoring system* is used to record the areas of the digits affected by blanching (see Figure 10). The scores correspond to areas of blanching on the digits beginning with the thumb. On the fingers a score of 1 is given for blanching on the distal phalanx, a score of 2 for blanching on the middle phalanx, and a score of 3 for blanching on the proximal phalanx. On the thumbs the scores are 4 for the distal phalanx and 5 for the proximal phalanx. The blanching score may be based on statements from the affected person or on the visual observations of a designated observer (e.g., a nurse).

Table 10 Stockholm Workshop Scale for the Classification of Vibration-Induced White Finger[a]

Stage	Grade	Description
0	—	No attacks
1	Mild	Occasional attacks affecting only the tips of one or more fingers
2	Moderate	Occasional attacks affecting distal and middle (rarely also proximal) phalanges of one or more fingers
3	Severe	Frequent attacks affecting all phalanges of most fingers
4	Very severe	As in stage 3, with trophic skin changes in the fingertips

Source: Gemne et al. (1987).

[a]If a person has stage 2 in two fingers of the left hand and stage 1 in a finger on the right hand, the condition may be reported as 2L(2)/1R(1). There is no defined means of reporting the condition of digits when this varies between digits on the same hand. The scoring system is more helpful when the extent of blanching is to be recorded.

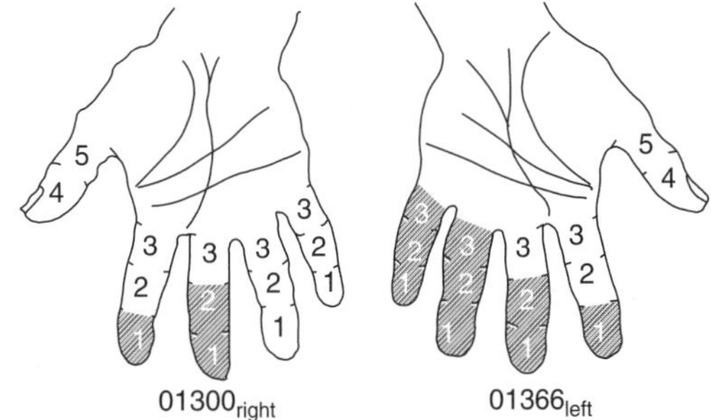

Figure 10 Method of scoring the areas of digits affected by blanching. The blanching scores for the hands shown are 01300_{right} and 01366_{left}. (From Griffin, 1990.)

Table 11 Proposed Sensorineural Stages of the Effects of Hand-Transmitted Vibration

Stage	Symptoms
0_{SN}	Exposed to vibration but no symptoms
1_{SN}	Intermittent numbness with or without tingling
2_{SN}	Intermittent or persistent numbness, reduced sensory perception
3_{SN}	Intermittent or persistent numbness, reduced tactile discrimination and/or manipulative dexterity

Source: Brammer et al. (1987).

5.2.2 Neurological Disorders

Neurological effects of hand-transmitted vibration (e.g., numbness, tingling, elevated sensory thresholds for touch, vibration, temperature, and pain, and reduced nerve conduction velocity) are considered to be separate effects of vibration and not merely signs of vibration-induced white finger (Griffin and Bovenzi, 2002). A method of reporting the extent of vibration-induced neurological effects of vibration has been proposed (see Table 11). This staging is not currently related to the results of any specific objective test: the *sensorineural stage* is a subjective impression of a physician based on the statements of the person affected or the results of any available clinical or scientific testing. Neurological disorders are sometimes identified by screening tests using measures of sensory function, such as the thresholds for feeling vibration, heat, or cold on the fingers.

5.2.3 Muscular Effects

The research literature includes reports of muscle atrophy among users of vibrating tools. Workers exposed to hand-transmitted vibration sometimes report difficulty with their grip, including reduced dexterity, reduced grip strength, and locked grip. Many of the reports are derived from symptoms reported by exposed persons rather than signs detected by physicians and could be a reflection of neurological problems (Griffin and Bovenzi, 2002).

Muscle activity may be of great importance to tool users since a secure grip can be essential to the performance of the job and the safe control of the tool. The presence of vibration on a handle may encourage the adoption of a tighter grip than would otherwise occur, and a tight grip may increase the transmission of vibration to the hand. If the chronic effects of vibration result in reduced grip, this may sometimes help to protect operators from further effects of vibration, but interfere with both work and leisure activities.

5.2.4 Articular Disorders

Many surveys of the users of handheld tools have found evidence of bone and joint problems, most often among men operating percussive tools such as those used in metal-working jobs and mining and quarrying. It is speculated that some characteristic of such tools, possibly the low-frequency shocks, is responsible. Some of the injuries reported relate to specific bones and suggest the existence of cysts, vacuoles, decalcification, or other osteolysis, or degeneration or deformity of the carpal, metacarpal, or phalangeal bones. Osteoarthrosis and olecranon spurs at the elbow, and other problems at the wrist and shoulder are also documented.

Notwithstanding the evidence of many research publications, there is not universal acceptance that vibration is a common cause of articular problems, and there is currently no dose–effect relation which predicts their occurrence. In the absence of specific information, it seems that adherence to current guidance for the prevention of vibration-induced white finger may provide reasonable protection.

5.2.5 Other Effects

Effects of hand-transmitted vibration may not be confined to the fingers, hands, and arms; many studies have found a high incidence of problems such as headaches and sleeplessness among tool users and have concluded that these symptoms are caused by hand-transmitted vibration. Although these are real problems to those affected, they are "subjective" effects that are not accepted as real by all researchers. Some current research is seeking a physiological basis for such symptoms. It would appear that caution is appropriate, but it is reasonable to assume that the adoption of modern guidance to prevent vibration-induced white finger will also provide some protection from any other effects of hand-transmitted vibration within, or distant from, the hand.

5.3 Preventive Measures

Protection from the effects of hand-transmitted vibration requires actions from management, tool manufacturers, technicians, and physicians at the workplace and from tool users. Table 12 summarizes some of the actions that may be appropriate. When there is reason to suspect that hand-transmitted vibration may cause injury, the vibration at tool–hand interfaces should be measured. This will help to predict whether the tool or process is likely to cause injury and whether any other tool or process could give a lower vibration severity. The duration of exposure to vibration should also be quantified. Reduction of exposure time may include the provision of exposure breaks during the day and, if possible, prolonged periods away from vibration exposure. For any tool or process having a vibration magnitude sufficient to cause injury, or otherwise known to be associated with injury, there should be a system to quantify and control the maximum daily duration of exposure of any person.

Gloves are sometimes recommended as a means of reducing the adverse effects of vibration on the hands. When using the frequency weightings in current standards, most gloves commonly available do *not* normally provide effective attenuation of the vibration on most tools (Griffin, 1998a). Gloves and "cushioned" handles may reduce the transmission of high frequencies of vibration, but current standards

Table 12 Some Preventive Measures to Consider When Persons Are Exposed to Hand-Transmitted Vibration

Group	Actions
Management	Seek technical advice, seek medical advice, warn exposed persons, train exposed persons, review exposure times, policy on removal from work
Tool manufacturers	Measure tool vibration, design tools to minimize vibration, ergonomic design to reduce grip force, design to keep hands warm, provide guidance on tool maintenance, provide warning of dangerous vibration
Technical at workplace	Measure vibration exposure, provide appropriate tools, maintain tools, inform management
Medical	Preemployment screening, routine medical checks, record all signs and symptoms reported, warn workers with predisposition, advise on consequences of exposure, inform management
Tool user	Use tool properly, avoid unnecessary vibration exposure, minimize grip and push forces, check condition of tool, inform supervisor of tool problems, keep warm, wear gloves when safe to do so, minimize smoking, seek medical advice if symptoms appear, inform employer of relevant disorders

Source: Adapted from Griffin (1990).

imply that these frequencies are not usually the primary cause of disorders. Gloves may protect the hand from other forms of mechanical injury (e.g., cuts and scratches) and protect the fingers from temperature extremes. Warm hands are less likely to suffer an attack of finger blanching, and some believe that maintaining warm hands while exposed to vibration may also lessen the damage caused by vibration.

Workers who are exposed to vibration magnitudes sufficient to cause injury should be warned of the possibility of vibration injuries and educated on the ways of reducing the severity of their vibration exposures. They should be advised of the symptoms to look out for and told to seek medical attention if the symptoms appear. There should be preemployment medical screening wherever a subsequent exposure to hand-transmitted vibration may reasonably be expected to cause vibration injury. Medical supervision of each exposed person should continue throughout employment at suitable intervals, possibly annually.

5.4 Standards for Evaluation of Hand-Transmitted Vibration

There are standard methods for measuring, evaluating, and assessing hand-transmitted vibration.

5.4.1 Vibration Measurement

International Standards 5349-1 (ISO, 2001) and 5349-2 (ISO, 2002) give general methods of measuring hand-transmitted vibration on tools and processes. Care is required to obtain representative measurements of tool vibration with appropriate operating conditions. There can be difficulties in obtaining valid measurements using some commercial instrumentation (especially when there are high shock levels). It is wise to determine acceleration spectra and inspect the acceleration time histories before accepting the validity of any measurements.

5.4.2 Vibration Evaluation

All current national and international standards use the same frequency weighting (called W_h) to evaluate

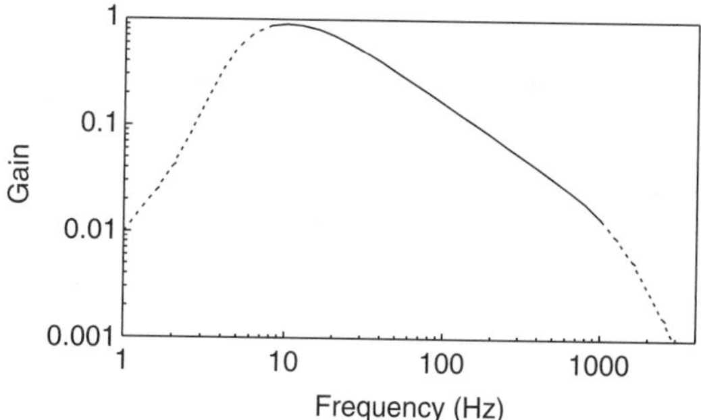

Figure 11 Frequency weighting W_h for the evaluation of hand-transmitted vibration.

hand-transmitted vibration over the approximate frequency range 8 to 1000 Hz (Figure 11) (Griffin, 1997). This weighting is applied to measurements of vibration acceleration in each of the three axes of vibration at the point of entry of vibration to the hand. More recent standards suggest that the overall severity of hand-transmitted vibration should be calculated from the root sums of squares of the frequency-weighted acceleration in the three axes. The standards imply that if two tools expose the hand to vibration for the same period of time, the tool having the lowest frequency-weighted acceleration will be least likely to cause injury or disease.

Occupational exposures to hand-transmitted vibration can have widely varying daily exposure durations, from a few seconds to many hours. Often, exposures are intermittent. To enable a daily exposure to be reported simply, the standards refer to an equivalent 8-h exposure:

$$a_{\mathrm{hw(eq,8h)}} = A(8) = a_{\mathrm{hw}} \left[\frac{t}{T(8)} \right]^{1/2}$$

where t is the exposure duration to an rms frequency-weighted acceleration, a_{hw}, and $T(8)$ is 8 hours (in the same units as t).

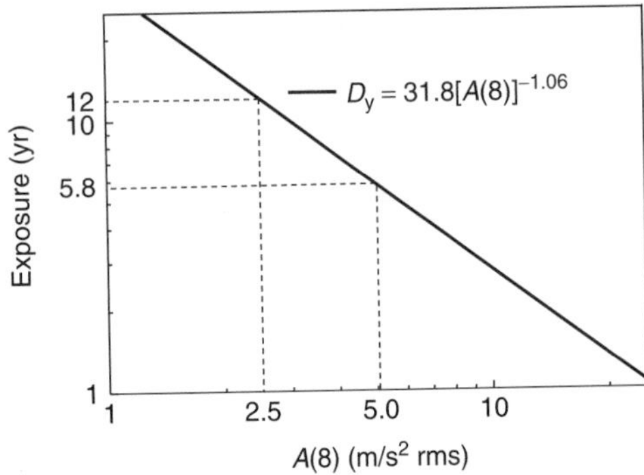

Figure 12 Relation between daily $A(8)$ and years of exposure expected to result in 10% incidence of finger blanching according to ISO (2001). A 10% probability of finger blanching is predicted after 12 years at the EU exposure action value and after 5.8 years at the EU exposure limit value.

5.4.3 Vibration Assessment According to ISO 5349

In an informative annex of ISO 5349-1 (ISO, 2001) there is a suggested relation between the lifetime exposure to hand-transmitted vibration, D_y (in years), and the 8-h energy-equivalent daily exposure $A(8)$ for the conditions expected to cause 10% prevalence of finger blanching (Figure 12):

$$D_y = 31.8[A(8)]^{-1.06}$$

The percentage of persons affected in any group of persons exposed will not always correspond to the values shown in Figure 12: the frequency weighting, the time dependency, and the dose–effect information are based on less than complete information and they have been simplified for practical convenience. Additionally, the number of persons affected by vibration will depend on the rate at which persons enter and leave the exposed group. The complexity of the equation above implies far greater precision than is possible; a more convenient estimate of the years of exposure (in the range 1 to 25 years) required for 10% incidence of finger blanching is

$$D_y = \frac{30.0}{A(8)}$$

This equation gives the same result as the equation in the standard (to within 14%), and there is no information suggesting that it is less accurate.

The informative annex to ISO 5349 (ISO, 2001) states: "Studies suggest that symptoms of the hand–arm vibration syndrome are rare in persons exposed with an 8-h energy-equivalent vibration total value, A(8), at a surface in contact with the hand, of less than 2 m/s^2 and unreported for A(8) values less than 1 m/s^2." However, this sentence should be interpreted with caution in view of the very considerable doubts over the frequency weighting and time dependency in the standard (Griffin et al., 2003).

5.4.4 EU Machinery Safety Directive

The Machinery Safety Directive of the European Community (EEC, 1989) requires that instruction handbooks for handheld and hand-guided machinery specify the equivalent acceleration to which the hands or arms are subjected where this exceeds a stated value (currently, a frequency-weighted acceleration of 2.5 m/s^2 rms). Many handheld vibrating tools can exceed this value. Standard test conditions for the measurement of vibration on many tools (e.g., chipping and riveting hammers, rotary hammers and rock drills, grinding machines, pavement breakers, chain saws) have been defined (e.g., ISO, 1988).

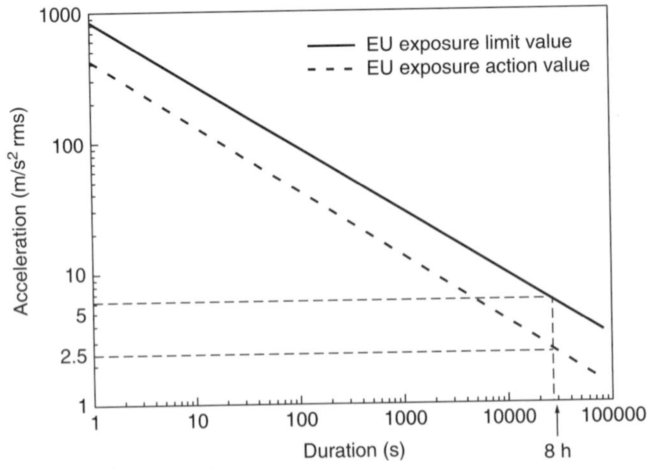

Figure 13 Hand-transmitted vibration exposure limit value [$A(8) = 5.0$ m/s^2 rms] and exposure action value [$A(8) = 2.5$ m/s^2 rms] in the EU Physical Agents (Vibration) Directive.

5.4.5 EU Physical Agents Directive

For hand-transmitted vibration, the EU Physical Agents Directive (2002) defines an 8-h equivalent exposure action value of 2.5 m/s^2 rms and an 8-h equivalent exposure limit value of 5.0 m/s^2 rms (Figure 13). The directive says that workers must not be exposed above the exposure limit value. If the exposure action values are exceeded, the employer must establish and implement a program of technical and/or organizational measures intended to reduce to a minimum exposure to mechanical vibration and the attendant risks. The directive mandates that workers exposed to mechanical vibration in excess of the exposure action values are entitled to appropriate health surveillance. However, health surveillance is not restricted to situations where the exposure action value is exceeded; it is required if there is any reason to suspect that workers may be injured by the vibration even if the action value is not exceeded.

According to ISO 5349-1 (2001), the onset of finger blanching would be expected in 10% of persons after 12 years at the EU exposure action value and after 5.8 years at the exposure limit value. The exposure action value and the exposure limit value in the directive do not define "safe exposures" to hand-transmitted vibration (Griffin, 2004).

REFERENCES

Allen, R. W., Jex, H. R., and Magdaleno, R. E. (1973), "Manual Control Performance and Dynamic Response During Sinusoidal Vibration," AMRL-TR-73-78, Aerospace Medical Research Laboratory, Wright-Patterson Air Force Base, OH.

ANSI (1983), "Guide to the Evaluation of Human Exposure to Vibration in Buildings," ANSI S3.29-1983 (ASA 48-1983), American National Standards Institute, New York.

Benson, A. J. (1984), "Motion Sickness," in *Vertigo*, M. R. Dix and J. S. Hood, Eds., Wiley, New York.

Benson, A. J., and Barnes, G. R. (1978), "Vision During Angular Oscillation: The Dynamic Interaction of Visual and Vestibular Mechanisms," *Aviation, Space and Environmental Medicine*, Vol. 49, No. 1, Sec. II, pp. 340–345.

Bongers, P. M., and Boshuizen, H. C. (1990), "Back Disorders and Whole-Body Vibration at Work," Thesis, University of Amsterdam.

Bovenzi, M., and Hulshof, C. T. J. (1998), "An Updated Review of Epidemiologic Studies on the Relationship Between Exposure to Whole-Body Vibration and Low Back Pain," *Journal of Sound and Vibration*, Vol. 215, No. 4, pp. 595–611.

Bovenzi, M., and Zadini, A. (1992), "Self-Reported Back Symptoms in Urban Bus Drivers Exposed to Whole-Body Vibration," *Spine*, Vol. 17, No. 9, pp. 1048–1059.

Brammer, A. J., Taylor, W., and Lundborg, G. (1987), "Sensorineural Stages of the Hand–Arm Vibration Syndrome," *Scandinavian Journal of Work, Environment and Health*, Vol. 13, No. 4, pp. 279–283.

BSI (1987a), "Measurement and Evaluation of Human Exposure to Whole-Body Mechanical Vibration and Repeated Shock," BS 6841, British Standards Institution, London.

BSI (1987b), "Measurement and Evaluation of Human Exposure to Vibration Transmitted to the Hand," BS 6842, British Standards Institution, London.

BSI (1992), "Evaluation of Human Exposure to Vibration in Buildings (1 Hz to 80 Hz)," BS 6472, British Standards Institution, London.

Corbridge, C., and Griffin, M. J. (1991), "Effects of Vertical Vibration on Passenger Activities: Writing and Drinking," *Ergonomics*, Vol. 34, No. 10, pp. 1313–1332.

Corbridge, C., Griffin, M. J., and Harborough, P. (1989), "Seat Dynamics and Passenger Comfort," *Proceedings of the Institution of Mechanical Engineers*, Vol. 203, pp. 57–64.

Dupuis, H., and Zerlett, G. (1986), *The Effects of Whole-Body Vibration*, Springer-Verlag, New York.

EEC (1989), "On the Approximation of the Laws of the Member States Relating to Machinery," Council Directive (89/392/EEC), *Official Journal of the European Communities*, June, 9–32, Council of the European Communities, Brussels.

Fairley, T. E., and Griffin, M. J. (1989), "The Apparent Mass of the Seated Human Body: Vertical Vibration," *Journal of Biomechanics*, Vol. 22, No. 2, pp. 81–94.

Gemne, G., Pyykko, I., Taylor, W., and Pelmear, P. (1987), "The Stockholm Workshop Scale for the Classification of Cold-Induced Raynaud's Phenomenon in the Hand–Arm Vibration Syndrome (Revision of the Taylor–Pelmear Scale)," *Scandinavian Journal of Work*, Environment and Health, Vol. 13, No. 4, pp. 275–278.

Griffin, M. J. (1976), "Eye Motion During Whole-Body Vertical Vibration," *Human Factors*, Vol. 18, No. 6, pp. 601–606.

Griffin, M. J. (1990), *Handbook of Human Vibration*, Academic Press, London.

Griffin, M. J. (1991), "Physical Characteristics of Stimuli Provoking Motion Sickness," Paper 3, in *Motion Sickness: Significance in Aerospace Operations and Prophylaxis*, AGARD Lecture Series LS-175, *Advisory Group for Aerospace Research and Development*, Neuilly-sur-Seine, France.

Griffin, M. J. (1997), "Measurement, Evaluation, and Assessment of Occupational Exposures to Hand-Transmitted Vibration," *Occupational and Environmental Medicine*, Vol. 54, No. 2, pp. 73–89.

Griffin, M. J. (1998a), "Evaluating the Effectiveness of Gloves in Reducing the Hazards of Hand-Transmitted Vibration," *Occupational and Environmental Medicine*, Vol. 55, No. 5, pp. 340–348.

Griffin, M. J. (1998b), "A Comparison of Standardized Methods for Predicting the Hazards of Whole-Body Vibration and Repeated Shocks," *Journal of Sound and Vibration*, Vol. 215, No. 4, pp. 883–914.

Griffin, M. J. (2004), "Minimum Health and Safety Requirements for Workers Exposed to Hand-Transmitted Vibration and Whole-Body Vibration in the European Union: A Review," *Occupational and Environmental Medicine*, Vol. 61, pp. 387–397.

Griffin, M. J., and Bovenzi, M. (2002), "The Diagnosis of Disorders Caused by Hand-Transmitted Vibration: Southampton Workshop 2000," *International Archives of Occupational and Environmental Health*, Vol. 75, No. 1–2, pp. 1–5.

Griffin, M. J., and Hayward, R. A. (1994), "Effects of Horizontal Whole-Body Vibration on Reading," *Applied Ergonomics*, Vol. 25, No. 3, pp. 165–169.

Griffin, M. J., and Newman, M. M. (2004), "Effects of the Visual Field on Motion Sickness in Cars," *Aviation, Space and Environmental Medicine*, Vol. 75, pp. 739–748.

Griffin, M. J., Bovenzi, M., and Nelson, C. M. (2003), "Dose Response Patterns for Vibration-Induced White Finger," *Occupational and Environmental Medicine*, Vol. 60, pp. 16–26.

Hulshof, C., and van Zanten, B. V. (1987), "Whole-Body Vibration and Low-Back Pain," *International Archives of Occupational and Environmental Health*, Vol. 59, pp. 205–220.

ISO (1974), "Guide for the Evaluation of Human Exposure to Whole-Body Vibration," ISO 2631(E), International Organization for Standardization, Geneva.

ISO (1983), "Acoustics: Preferred Reference Quantities for Acoustic Levels," ISO 1683, International Organization for Standardization, Geneva.

ISO (1985), "Evaluation of Human Exposure to Whole-Body Vibration, Part 1: General Requirements," 2631/1, International Organization for Standardization, Geneva.

ISO (1988), "Hand-Held Portable Tools: Measurement of Vibration at the Handle, Part 1: General." ISO 8662-1, International Organization for Standardization, Geneva.

ISO (1989), "Evaluation of Human Exposure to Whole-Body Vibration, Part 2: Continuous and Shock-Induced Vibration in Buildings," ISO 2631-2, International Organization for Standardization, Geneva.

ISO (1997), "Mechanical Vibration and Shock: Evaluation of Human Exposure to Whole-Body Vibration. Part 1: General requirements." ISO 2631-1, International Organization for Standardization, Geneva.

ISO (2001), "Mechanical Vibration: Measurement and Evaluation of Human Exposure to Hand-Transmitted Vibration, Part 1: General Requirements," ISO 5349-1 : 2001(E), International Organization for Standardization, Geneva.

ISO (2002), "Mechanical Vibration: Measurement and Evaluation of Human Exposure to Hand-Transmitted Vibration, Part 2: Practical Guidance for Measurement at the Workplace," ISO 5349-2 : 2001(E), International Organization for Standardization, Geneva.

Lawther, A., and Griffin, M. J. (1987), "Prediction of the Incidence of Motion Sickness from the Magnitude, Frequency, and Duration of Vertical Oscillation," *Journal of the Acoustical Society of America*, Vol. 82, No. 3, pp. 957–966.

Lawther, A., and Griffin, M. J. (1988a), "A Survey of the Occurrence of Motion Sickness Amongst Passengers at Sea," *Aviation, Space and Environmental Medicine*, Vol. 59, No. 5, pp. 399–406.

Lawther, A., and Griffin, M. J. (1988b), "Motion Sickness and Motion Characteristics of Vessels at Sea," *Ergonomics*, Vol. 31, No. 10, pp. 1373–1394.

Levison, W. H., and Harrah, C. B. (1977), "Biomechanical and Performance Response of Man in Six Different Directional Axis Vibration Environments," AMRL-TR-77-71, Aerospace Medical Research Laboratory, Wright-Patterson Air Force Base, OH.

Lewis, C. H., and Griffin, M. J. (1977), "The Interaction of Control Gain and Vibration with Continuous Manual Control Performance," *Journal of Sound and Vibration*, Vol. 55, No. 4, pp. 553–562.

Lewis, C. H., and Griffin, M. J. (1979), "The Effect of Character Size on the Legibility of Numeric Displays During Vertical Whole-Body Vibration," *Journal of Sound and Vibration*, Vol. 67, No. 4, pp. 562–565.

Loriga, G. (1911), "Il Lavoro con i Martelli Pneumatici" [The Use of Pneumatic Hammers], *Bolletino dell' lspettorato del Lavoro*, Vol. 2, pp. 35–60.

Matsumoto, Y., and Griffin, M. J. (2000), "Comparison of Biodynamic Responses in Standing and Seated Human Bodies," *Journal of Sound and Vibration*, Vol. 238, No. 4, pp. 691–704.

McLeod, R. W., and Griffin, M. J. (1989), "A Review of the Effects of Translational Whole-Body Vibration on Continuous Manual Control Performance," *Journal of Sound and Vibration*, Vol. 133, No. 1, pp. 55–115.

McLeod, R. W., and Griffin, M. J. (1990), "Effects of Whole-Body Vibration Waveform and Display Collimation on the Performance of a Complex Manual Control Task," *Aviation, Space and Environmental Medicine*, Vol. 61, No. 3, pp. 211–219.

Meddick, R. D. L., and Griffin, M. J. (1976), "The Effect of Two-Axis Vibration on the Legibility of Reading Material," *Ergonomics*, Vol. 19, No. 1, pp. 21–33.

Moseley, M. J., and Griffin, M. J. (1986a), "A Design Guide for Visual Displays and Manual Tasks in Vibration Environments, Part I: Visual Displays," Technical Report 133, Institute of Sound and Vibration Research, University of Southampton, Southampton, Hampshire, England.

Moseley, M. J., and Griffin, M. J. (1986b), "Effects of Display Vibration and Whole-Body Vibration on Visual Performance," *Ergonomics*, Vol. 29, No. 8, pp. 977–983.

Moseley, M. J., Lewis, C. H., and Griffin, M. J. (1982), "Sinusoidal and Random Whole-Body Vibration: Comparative Effects on Visual Performance," *Aviation, Space and Environmental Medicine*, Vol. 53, No. 10, pp. 1000–1005.

Nawayseh, N., and Griffin, M. J. (2003), "Non-linear Dual-Axis Biodynamic Response to Vertical Whole-Body Vibration," *Journal of Sound and Vibration*, Vol. 268, pp. 503–523.

Nixon, C. W., and Sommer, H. C. (1963), "Influence of Selected Vibrations upon Speech (Range of 2 cps–20 cps and Random)," AMRL-TDR-63-49, Aerospace Medical Research Laboratories, Wright-Patterson Air Force Base, OH.

O'Hanlon, J. G., and Griffin, M. J. (1971), "Some Effects of the Vibration of Reading Material upon Visual Performance," Technical Report 49, Institute of Sound and Vibration Research, Southampton, Hampshire, England.

Paddan, G. S., and Griffin, M. J. (1988a), "The Transmission of Translational Seat Vibration to the Head, I: Vertical Seat Vibration," *Journal of Biomechanics*, Vol. 21, No. 3, pp. 191–197.

Paddan, G. S., and Griffin, M. J. (1988b), "The Transmission of Translational Seat Vibration to the Head, II: Horizontal Seat Vibration," *Journal of Biomechanics*, Vol. 21, No. 3, pp. 199–206.

Reason, J. T. (1970), "Motion Sickness: A Special Case of Sensory Rearrangement," *Advancement of Science*, Vol. 26, June, pp. 386–393.

Reason, J. T. (1978), "Motion Sickness Adaptation: A Neural Mismatch Model," *Journal of the Royal Society of Medicine*, Vol. 71, pp. 819–829.

Reason, J. T., and Brand, J. J. (1975), *Motion Sickness*, Academic Press, London.

Sherwood, N., and Griffin, M. J. (1990), "Effects of Whole-Body Vibration on Short-Term Memory," *Aviation,*

Space and Environmental Medicine, Vol. 61, No. 12, pp. 1092–1097.

Sherwood, N., and Griffin, M. J. (1992), "Evidence of Impaired Learning During Whole-Body Vibration," *Journal of Sound and Vibration*, Vol. 152, No. 2, pp. 219–225.

Shoenberger, R. W. (1974), "An Investigation of Human Information Processing During Whole-Body Vibration," *Aerospace Medicine*, Vol. 45, No. 2, pp. 143–153.

Shoenberger, R. W., and Wilburn, D. L. (1973), "Tracking Performance During Whole-Body Vibration with Side-Mounted and Centre-Mounted Control Sticks," AMRL-TR-72-120, Aerospace Medical Research Laboratory, Wright-Patterson Air Force Base, OH.

Taylor, W., and Pelmear, P. L., Eds., (1975), *Vibration White Finger in Industry*, Academic Press, New York.

Torle, G. (1965), "Tracking Performance Under Random Acceleration: Effects of Control Dynamics," *Ergonomics*, Vol. 8, No. 4, pp. 481–486.

Turner, M., and Griffin, M. J. (1999), "Motion Sickness in Public Road Transport: The Relative Importance of Motion, Vision and Individual Differences," *British Journal of Psychology*, Vol. 90, pp. 519–530.

Wasserman, D., Taylor, W., Behrens, V., Samueloff, S., and Reynolds, D. (1982), "Vibration White Finger Disease in U.S. Workers Using Pneumatic Chipping and Grinding Handtools, I: Epidemiology," DHSS (NIOSH) Publication 82-118, U.S. Department of Health and Human Services, National Institute for Occupational Safety and Health, Washington, DC.

Wells, M. J., and Griffin, M. J. (1984), "Benefits of Helmet-Mounted Display Image Stabilisation Under Whole-Body Vibration," *Aviation, Space and Environmental Medicine*, Vol. 55, No. 1, pp. 13–18.

Wilson, R. V. (1974), "Display Collimation Under Whole-Body Vibration," *Human Factors*, Vol. 16, No. 2, pp. 186–195.

CHAPTER 24
SOUND AND NOISE*

John G. Casali
Virginia Polytechnic Institute and State University
Blacksburg, Virginia

1 INTRODUCTION

Sound, and its subset, noise, which is often defined as unwanted sound, is a phenomenon that confronts human factors professionals in many settings and applications. A few examples of this are in order: (1) an auditory warning signal, for which the proper sound parameters must be selected for maximizing detection, identification, and localization; (2) a situation wherein the speech communication that is critical between operators is compromised in its intelligibility by environmental noise, and therefore redesign of the communications system and/or acoustic environment is needed; (3) a residential community is intruded upon by the noise from vehicular traffic or a nearby industrial plant, causing annoyance and sleep arousal, and necessitating abatement; (4) an in-vehicle auditory display that warns of dangerous conditions must convey urgency and localization cues; or (5) a worker is exposed to hazardous noise on the job, and to prevent hearing loss, an appropriate hearing protection device (hereafter, HPD) must be selected. To deal effectively with examples of these types, the human factors engineer must understand the basics of sound, instrumentation, and techniques for its measurement and quantification, analyses of acoustical measurements for ascertaining the audibility of signals and speech as well as the risks to hearing, and countermeasures to combat the deleterious effects of noise. In this chapter we address these and related matters from a human factors engineering perspective while also delving into a few noise-related standards and regulations.

At the outset it should be noted that the science of acoustics, and sound and noise within it, is very broad and comprises a vast body of research and standards literature. Thus, as the subject of a single chapter, this topic cannot be covered in great depth herein. It is therefore an intent of this chapter to introduce several major topics concerning sound/noise, particularly as it impacts humans, and to point the reader to other publications for detail on specific topics. As for the area of

*Sections of this chapter are based in part on Robinson and Casali (2003) and Casali and Robinson (1999).

sound/noise as a whole, three excellent, broad coverage texts are Kryter (1994), Crocker (1998), and Berger et al. (2003).

2 SOUND AND NOISE

Most aspects of acoustics rely on accurate quantification and evaluation of the sound itself; therefore, a basic understanding of sound parameters and sound measurement is needed before delving into application-oriented issues.

2.1 Basic Parameters

Sound is a disturbance in a medium (in industry, home, or recreational settings, most commonly air or a conductive structure such as a floor or wall) that has mass and elasticity. For example, an exhaust fan on the roof of an industrial plant has blades that rotate in the air, creating noise which may propagate into the surrounding community. Because the blades are coupled to the air medium, they produce pressure waves that consists of alternating compressions (above ambient air pressure) and rarefactions (below ambient pressure) of air molecules, the *frequency (f)* of which is the number of above/below ambient pressure cycles per second, or *hertz (Hz)*. The reciprocal of frequency, $1/f$, is the *period* of the waveform. The waveform propagates outward from the fan as long as it continues to rotate, and the disturbance in air pressure that occurs in relation to ambient air pressure is heard as sound, in this case "fan roar." The linear distance traversed by the sound wave in one complete cycle of vibration is the *wavelength*:

$$\lambda = c/f \tag{1}$$

Wavelength (λ in meters or feet) depends on the sound frequency (f in Hz) and velocity (c in m/s or ft/sec; in air at 68°F and pressure of 1 atmosphere (atm), 344 m/s or 1127 ft/sec) in the medium. The speed of sound is influenced by the temperature of the medium and in air, increases about 1.1 ft/sec for each increase of 1°F.

Vibrations are oscillations in solid media, and are often associated with the production of sound waves. *Noise* can be loosely defined as a subset of sound; that is, noise is sound that is undesirable or offensive in some aspect. However, the distinction is largely situation- and listener-specific, as perhaps best stated in the old adage "one person's music is another's noise."

Unlike some common ergonomics-related stressors such as repetitive motions or awkward lifting maneuvers, noise is a physical stimulus that is readily measurable and quantifiable using transducers (microphones) and instrumentation (sound level meters and their variants) that are commonly available. Aural exposure to noise, and the damage potential therefrom, is a function of the *total energy* transmitted to the ear. In other words, the energy is equivalent to the product of the noise intensity and duration of the exposure. Several metrics that relate to the energy of the noise exposure have been developed, most with an eye toward accurately reflecting the exposures that occur in industrial or community settings. These metrics are covered in Section 3.2, but first, the most basic unit of measurement must be understood, the *decibel*.

2.2 Physical Quantification: Sound Levels and the Decibel Scale

The unit of *decibel*, $\frac{1}{10}$ of a *bel*, is the most common metric applied to the quantification of noise amplitude. The decibel (dB) is a measure of *level*, defined as the logarithm of the ratio of a quantity to a reference quantity of the same type. In acoustics, it is applied to sound level, of which there are three types.

Sound power level, the most basic quantity, is typically expressed in decibels and is defined as

$$\text{sound power level (dB)} = 10\ \log_{10} \text{Pw}_1/\text{Pw}_r \tag{2}$$

where Pw_1 is the acoustic power of the sound in watts or other power unit, and Pw_r is the acoustic power of a reference sound in watts, usually taken to be the acoustic power at hearing threshold for a young, healthy ear at the frequency of maximum sensitivity, the quantity 10^{-12} W.

Sound intensity level, following from power level, is typically expressed in decibels and is defined as

$$\text{sound intensity level (dB)} = 10\ \log_{10} I_1/I_r \tag{3}$$

where I_1 is the acoustic intensity of the sound in W/m^2 or other intensity unit, and I_r is the acoustic intensity of a reference sound in W/m^2, usually taken to be the acoustic intensity at hearing threshold, or the quantity 10^{-12} W/m^2.

Within the last decade, sound measurement instruments to measure sound *intensity* level have become commonplace, albeit expensive and relatively complex. Sound *power* level, by contrast, is not directly measurable but can be computed from empirical measures of sound intensity level or sound pressure level. On the other hand, sound *pressure* level is directly measurable by using relatively straightforward instruments and is by far the most common metric used in practice.

Sound pressure level, (SPL), abbreviated in formulas as L_P, is also typically expressed in decibels. Since power is directly proportional to the square of the pressure, SPL is defined as

$$\begin{aligned} &\text{sound pressure level (SPL or } L_P\text{; dB)} \\ &\quad = 10\ \log_{10} P_1^2/P_r^2 = 20\ \log_{10} P_1/P_r \end{aligned} \tag{4}$$

where P_1 is the pressure level of the sound in micropascals (μPa) or other pressure unit, and P_r is the pressure level of a reference sound in μPa, usually taken to be the pressure at hearing threshold, or the quantity 20 μPa, or 0.00002 Pa. Other equivalent reference quantities are 0.0002 dyn/cm^2 and 20 μbar.

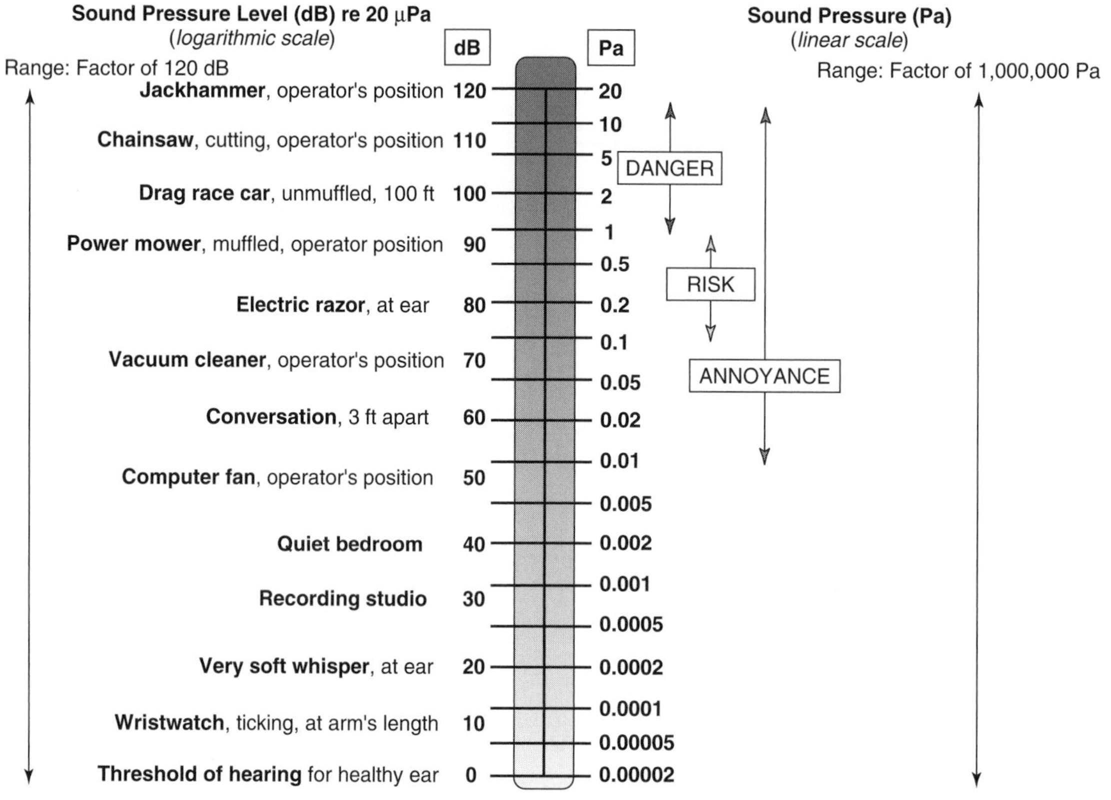

Figure 1 Sound pressure level in decibels and sound pressure in pascals for typical sounds.

The application of the decibel scale to acoustical measurements yields a convenient means of collapsing the vast range of sound pressures which would be required to accommodate sounds that can be encountered into a more manageable, compact range. As shown in Figure 1, using the logarithmic compression produced by the decibel scale, the range of typical sounds is 120 dB, while the linear pressure scale applied to the same sounds produces a range of 1,000,000 Pa. Of course, sounds do occur that are higher than 120 dB (e.g., artillery fire) or lower than 0 dB (below normal threshold on an audiometer). A comparison of decibel values of example sounds to their pressure values (in pascals) is also depicted in Figure 1.

In considering changes in sound level measured in decibels, a few numerical relationships emanating from the decibel formulas above are often helpful in practice. An increase (decrease) in SPL by 6 dB is equivalent to a doubling (halving) of the sound pressure. Similarly, on the power or intensity scales, an increase (decrease) of 3 dB is equivalent to a doubling (halving) of the sound power or intensity. The latter relationship gives rise to what is known as the *equal energy rule* or *trading relationship*. Because sound represents energy which is itself a product of intensity and duration, an original sound that increases (decreases) by 3 dB is equivalent in total energy to the same original sound that does not change in decibel value but decreases (increases) in its duration by half (twice).

2.3 Computations with Decibels

There are many practical instances in which it is helpful to predict the combined result of several individual sound sources that have been measured separately in decibels. This can be performed for random, uncorrelated sound sources using the equation

$$L_{P\text{combo}}(\text{dB}) = 10 \log_{10}(10^{L_{P1}/10} + 10^{L_{P2}/10} + \cdots + 10^{L_{Pn}/10}) \qquad (5)$$

and it applies for any decibel weighting (dBA, dBC, etc., as explained later), or for any bandwidth (such as $\frac{1}{3}$ octave, full octave, etc.). For example, suppose that an industrial plant currently exposes workers in a work area to a time-weighted average (TWA) of 83.0 dBA, which is below the OSHA Action Level (85.0 dBA) at which a hearing conservation program would be necessary (OSHA, 1983). Two new pieces of equipment are proposed for purchase and installation in this area: a new single-speed conveyor that has a constant noise output of 78.0 dBA, and a new compressor that has a constant output of 82.5 dBA. The combined sound level will be approximately

$$L_{P\text{combo}}(\text{dB}) = 10\ \log_{10}(10^{83.0/10} + 10^{78.0/10} + 10^{82.5/10}) = 86.4\ \text{dBA}$$

Thus, by purchasing this conveyor, the plant would move from a noise exposure level (83.0 dBA) that is in compliance with OSHA to one that is not (86.4 dBA). This is one illustration why industries should adhere to a "buy quiet" policy, so that noise exposure problems are not created unknowingly by equipment purchases.

Subtraction of decibels works in the same manner as addition:

$$L_{P\text{difference}}(\text{dB}) = 10\ \log_{10}(10^{L_{P1}/10} - 10^{L_{P2}/10}) \quad (6)$$

Using the example above, if the compressor were eliminated from the situation, the overall combined noise level would be the combination of the three sources as computed to be 86.4 dBA, reduced by the absence of the compressor at 82.5 dBA:

$$L_{P\text{difference}}(\text{dB}) = 10\ \log_{10}(10^{86.4/10} - 10^{82.5/10}) = 84.1\ \text{dBA}$$

With this result, the plant area noise level moves back into OSHA compliance under the Action Level of 85.0 dBA, but just by about 1.0 dBA. To err on the safe side, especially to accommodate the potential of any upward fluctuations in noise level, this plant's management should still look to reduce the noise further, or install a hearing conservation program.

There are a few rules of thumb that arise from the computations shown above. One is that when two sound sources are approximately equivalent in SPL, the combination of the two will be about 3 dB larger than the decibel level of the *higher* source. Another is that as the difference between two sounds exceeds about 13 dB, the contribution of the lower-level sound to the combined sound level is negligible (i.e., about 0.2 dB). Relatedly, when it is desirable to measure a sound of interest in isolation but it cannot be physically separated from a background noise, the question becomes: To what extent is the background noise influencing the accuracy of the measurement? In many cases, such as in some manufacturing plants, the background noise cannot be turned off but the sound of interest can. If this is the case, then the sound of interest is measured in the background noise, and then the background noise is measured alone. If the background noise measurement differs from the combined measurement by more than 13 dB, then it has not influenced the measurement of the sound of interest in a significant manner. If the difference is smaller, then equation (6) can be applied to correct the measurement, effectively by removing the background noise's contribution.

Finally, it is important to recognize that due to the limits in precision and reliability of decibel measurements, for the applications discussed in this chapter (and most others in acoustics as well), it is unnecessary to record decibel calculations that result from the formulas herein to greater than one decimal point, and it is usually sufficient to round final results to the nearest 0.5 dB, or even to integer values. However, to avoid interim rounding error, it is important to carry the significant figures through each step of the formulas until the end result is obtained (Ostergaard, 2003).

3 MEASUREMENT AND QUANTIFICATION OF SOUND AND NOISE EXPOSURES

3.1 Basic Instrumentation

Measurement and quantification of sound levels and noise exposure levels provide the fundamental data for assessing hearing exposure risk, speech and signal-masking effects, hearing conservation program needs, and engineering noise control strategies. A vast array of instrumentation is available; however, for most of the aforementioned applications, a basic understanding of three primary instruments (sound level meters, dosimeters, and real-time spectrum analyzers) and their data output will suffice. In instances where noise is highly impulsive in nature, such as gunfire, and/or development of situation-specific engineering noise control solutions is anticipated, more specialized instruments may be necessary.

Because sound is propagated as pressure waves that vary over space and in time, a complete acoustic record of a noise exposure or a sound event that has a prolonged duration requires simultaneous measurements at all points of interest in the sound field over a representative, continuous time period to document the noise level exhaustively in the space. Obviously, this is typically cost- and time-prohibitive, so one must resort to sampling strategies for establishing the observation points and intervals. The analyst must also decide whether detailed, discrete-time histories with averaging over time and space are needed (such as with a noise-logging *dosimeter*), if discrete samples taken with a short-duration moving time average (with a basic *sound level meter*) will suffice, or if frequency-band-specific SPLs are needed for selecting noise abatement materials (with a *spectrum analyzer*). A discussion of these three primary types of sound measurement instruments and the noise descriptors that can be obtained therefrom follows.

3.1.1 Sound Level Meter

Most sound measurement instruments derive from the basic sound level meter (SLM), a device for which four grades and associated performance tolerances that become more stringent as the grade number increases, and are described by ANSI S1.4-1983 (ANSI, 2001). Type 0 instruments have the most stringent tolerances and are for laboratory use only. Other grades include

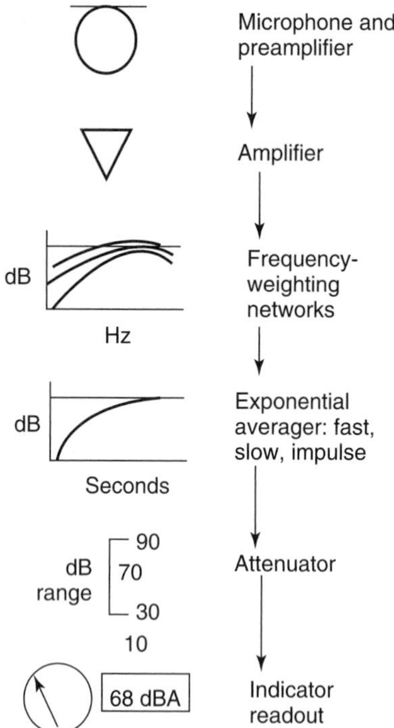

Figure 2 Functional components of a sound level meter.

Type 1, intended for precision measurement in the field or laboratory; Type 2, intended for general field use, especially where frequencies above 10,000 Hz are not prevalent; and Type S, a special-purpose meter that may perform at grade 1, 2, or 3, but may not include all of the operational functions of the particular grade. A grade of Type 2 or better is needed for measuring occupational exposures and community noise and to obtain data for most court proceedings.

A block diagram of the functional components of a generic SLM appears in Figure 2. At the top, a microphone/preamplifier senses the pressure changes caused by an airborne sound wave and converts the pressure signal into a voltage signal. Because the pressure fluctuations of a sound wave are small in magnitude, the corresponding voltage signal must be preamplified and then input to an amplifier which boosts the signal before it is processed further. The passband, the range of frequencies that are passed through and processed, of a high-quality SLM contains frequencies from about 10 to 20,000 Hz, but depending on the frequency weighting used, not all frequencies are treated in the same way. A selectable frequency-weighting network, or filter, is then applied to the signal. These networks most commonly include the A-, B-, and C-weighting functions shown in Figure 3*b*. For OSHA noise monitoring measurements and for many community noise applications, the A-scale, which deemphasizes the low frequencies and to a smaller extent the high frequencies, is used. In addition to the common A-scale (which approximates the 40-phon level of hearing) and C-scale (100-phon level), other selections may be available. If no weighting function is selected on the meter, the notation *dB* or *dB(linear)* is used, and all frequencies are processed without weighting factors. The actual weighting functions for the three suffix notations A, B, and C are superimposed on the phon contours of Figure 3*a*, and are also depicted in Figure 3*b* as actual frequency-weighting functions.

Next (not shown), the signal is squared to reflect the fact that sound pressure level in decibels is a function of the square of the sound pressure. The signal is then applied to an exponential averaging network, which defines the meter's dynamic response characteristics. In effect, this response creates a moving-window, short-time average display of the sound waveform. The two most common settings are FAST, which has a time constant of 0.125 s, and SLOW, which has a time constant of 1.0 s. These time constants were established decades ago to give analog needle indicators a rather sluggish response (particularly on the SLOW setting) so that they could be read by the human eye even when highly fluctuating sound pressures were measured. Under the FAST or SLOW dynamics, the meter indicator rises exponentially toward the decibel value of an applied constant SPL. For OSHA measurements, the SLOW setting is used, and this setting is also best when the average value (as it is changing over time) is desired. The FAST setting is more appropriate when the variability or range of fluctuations of a time-varying sound is desired. On certain SLMs, a third time constant, IMPULSE, may also be included for measurement of sounds that have sharp transient characteristics over time and are generally less than 1 s in duration, exemplified by gunshots or impact machinery such as drop forges. The IMPULSE setting has an exponential rise-time constant of 35 ms and a decay time of 1.5 s. It is useful to afford the observer the time to view the maximum value of a burst of sound before it decays and is more commonly applied in community and business machine noise measurements than in industrial settings.

Because sound often consists of symmetrical pressure fluctuations above and below ambient air pressure for which the arithmetic average is zero, a *root mean square* (rms) averaging procedure is applied when FAST, SLOW, or IMPULSE measurements are taken, and the result is displayed in decibels. In effect, each pressure (or converted voltage) value is squared, the arithmetic mean of all squared values is then obtained, and finally, the square root of the mean is computed to provide the rms value.

Some SLMs include an unweighted TRUE PEAK setting that does not utilize the rms computation, but instead, provides an indication of the actual peak SPL reached during a pressure impulse. This measurement mode is necessary for certain applications: for instance, to determine if the OSHA limit of 140 dB for impulsive exposure is exceeded. A Type 1 or 2 meter must be capable of measuring a 50-μs pulse. It is important to note that the aforementioned rms-based IMPULSE

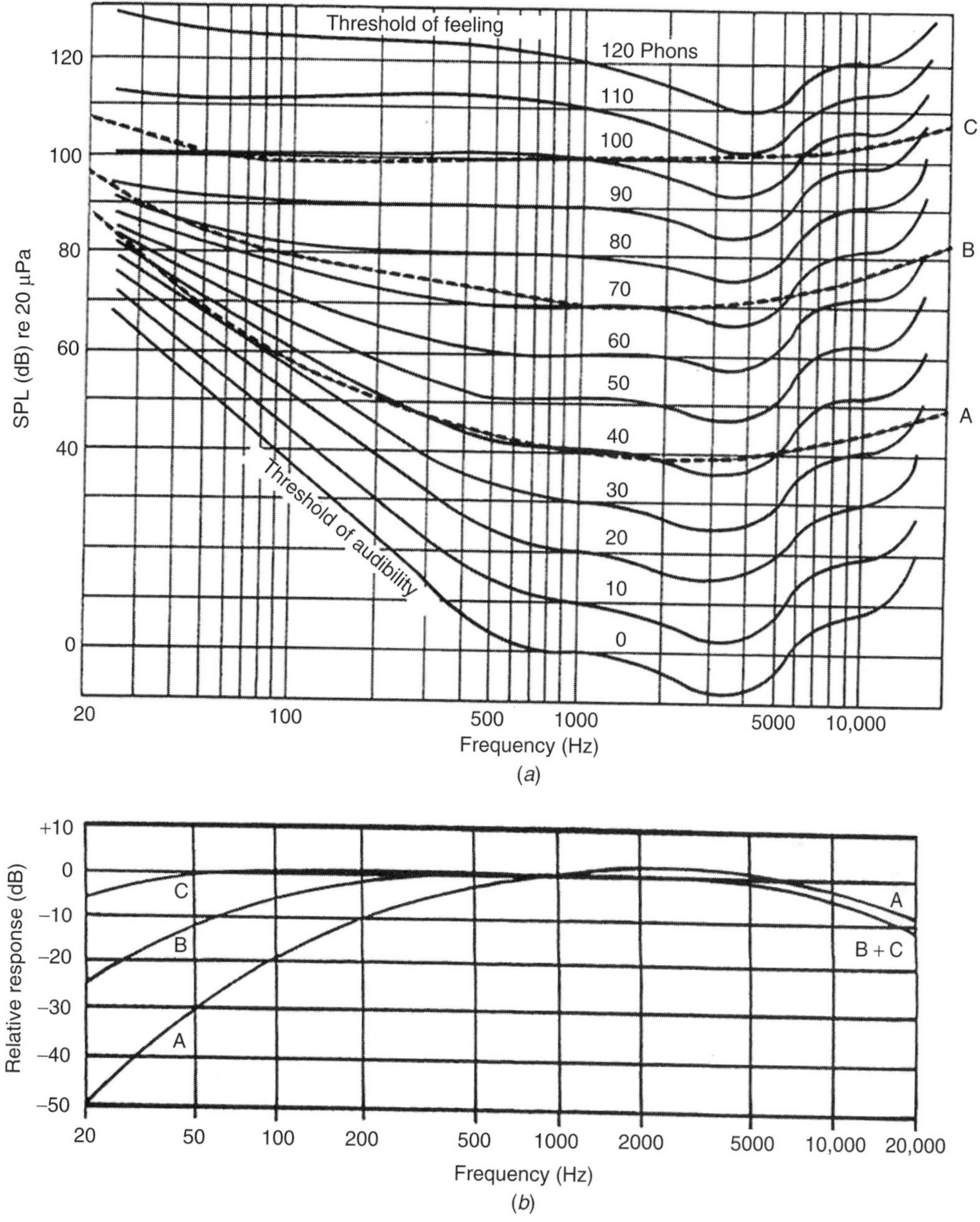

Figure 3 (*a*) Equal-loudness contours based on the psychophysical phon scale, with sound level meter frequency-weighting curves superimposed; (*b*) decibels vs. frequency values of A, B, and C sound level meter weighting curves. (Adapted with permission from Earshen, 1986.)

dynamics setting is unsuitable for measurement of TRUE PEAK SPLs.

With regard to the final component of a SLM shown in Figure 2, the indicator display or readout, much debate has existed over whether an analog (needle pointer or bar "thermometer-type" linear display) or digital (numeric) display is best. Ergonomics research indicates that although the digital readout affords higher precision of information to be presented in a smaller space, its disadvantage is that the least significant digit becomes impossible to read when the sound level is fluctuating rapidly. Also, it is more difficult with a digital readout for the observer to capture the maximum and minimum values of a sound, as is often desirable using the FAST or IMPULSE response. On the other hand, if very precise measurements down to a fraction of a decibel are needed, the digital indicator is preferable as long as the meter incorporates an appropriate time integrating/averaging feature or "hold" setting so that the data values can be captured. Because of the advantages and disadvantages of each type of display, some contemporary SLMs include both analog and digital readouts.

Microphone Considerations Most SLMs have interchangeable microphones which offer varying frequency response, sensitivity, and directivity characteristics (Peterson, 1979). The *response* of the microphone is the ratio of electrical output (in volts) to the sound pressure at the diaphragm of the microphone. Sound pressure is commonly expressed in pascals for free-field conditions (where there are no sound reflections resulting in reverberation), and the free-field voltage response of the microphone is given as mV/Pa. When specifications for *sensitivity* or *output level* are given, the response is usually based on a pure-tone sound-wave input. Typically, the output level is provided in decibels re 1 V at the microphone electrical terminals, and the reference sensitivity is 1 V/Pa.

Most microphones that are intended for general sound measurements are essentially *omnidirectional* (i.e., nondirectional) in their response for frequencies below about 1000 Hz. The 360° response pattern of a microphone is called its *polar response*, and the pattern is generally symmetrical about the axis perpendicular to the diaphragm. Some microphones are designed to be highly directional, of which one example is the cardiode design, which has a heart-shaped polar response wherein the maximum sensitivity is for sounds whose direction of travel causes them to enter the microphone at 0° (or the *perpendicular incidence response*), and minimum sensitivity is for sounds entering at 180° behind the microphone. The response at 90°, where sound waves travel and enter parallel to the diaphragm, is known as the *grazing incidence response*. Another response pattern, the *random incidence response*, represents the mean response of the microphone for sound waves that strike the diaphragm from all angles with equal probability. This response characteristic is the most versatile, and thus it is the response pattern used most often in the United States. Frequency responses for various microphone incidence patterns are depicted in Figure 4.

Because most U.S. SLM microphones are omnidirectional and utilize the random-incidence response, it is best for an observer to point the microphone at the primary noise source and hold it at an angle of incidence from the source at approximately 70°. This will produce a measurement most closely corresponding to the random-incidence response. Care must be taken to avoid shielding the microphone with the body or other structures. The response of microphones can also vary with temperature, atmospheric pressure,

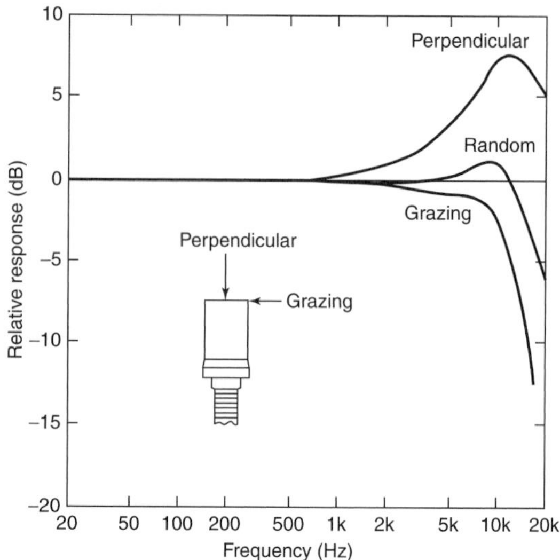

Figure 4 Frequency response of a hypothetical microphone for three angles of incidence. (Adapted with permission from Peterson, 1979.)

and humidity, with temperature usually being the most critical factor. Correction factors for variations in decibel readout due to temperature effects are supplied by most microphone manufacturers. Atmospheric effects are generally significant only when measurements are made in aircraft or at very high altitudes, and humidity has a negligible effect except at very high levels. In any case, microphones must not be exposed to moisture or large magnetic fields, such as those produced by transformers. When used in windy conditions, a foam windscreen should be placed over the microphone. This will reduce the contaminating effects of wind noise while influencing the frequency response of the microphone only slightly at high frequencies. The windscreen offers the additional benefit of protection of the microphone from damage due to being struck and/or from airborne foreign matter.

Sound Level Meter Applications It is important to note that the basic SLM is intended to measure sound levels at a given moment in time, although certain specialized devices can perform integration or averaging of sound levels over an extended period of time. When the nonintegrating/averaging SLM is used for long-term noise measurements, such as over a workday, it is necessary to sample and make multiple manual data entries on a record to characterize the exposure. This technique is usually best limited to area measurements, not an individual's exposure sampling. Furthermore, the sampling process becomes more difficult as the fluctuations in a noise become more rapid and/or random in nature. SLMs are useful for determining the levels of human speech in rms or peak values, calibration of laboratory experiments, calibration of audiometers (with special attachments), and community noise event-related measurements.

3.1.2 Dosimeter

The *audio-dosimeter* is a portable battery-powered device that is derived directly from a SLM but also features the ability to obtain special measures of noise exposure (discussed later) that relate to regulatory compliance and hearing hazard risk. Some versions are weather resistant and can be used outdoors to log a record of noise in a community setting, including both event-related, short-term measures and long-term averages and other statistical data.

Dosimeters for industrial use are very compact and are generally worn on the belt or in the pocket of an employee, with the microphone generally clipped to the lapel or shoulder of a shirt or blouse. The intent is to obtain a noise exposure history over the course of a full or partial workshift, and to obtain, at a minimum, a readout of the TWA exposure and noise dose for the period measured. Depending on the features, the dosimeter may provide a running histogram of noise levels on a short-time-interval (such as 1-minute) basis, compute statistical distributions of the noise exposures for the period, flag and record exposures that exceed OSHA maxima of 115 dBA continuous or 140 dB TRUE PEAK, and compute average metrics using 3 dB, 5 dB, or even other time-versus-level exchange rates. The dosimeter eliminates the need for the observer to set up a discrete sampling scheme, follow a noise-exposed worker, or monitor continuously an instrument that is staged outdoors, all of which are necessary with a conventional SLM.

3.1.3 Spectrum Analyzer

A *spectrum analyzer* is an advanced SLM which incorporates selective frequency-filtering capabilities to provide an analysis of the noise level as a function of frequency. In other words, the noise is broken down into its frequency components and a distribution of the noise energy in all measured frequency bands is available. Bands are delineated by upper and lower edge or cutoff frequencies and a center frequency. Different widths and types of filters are available, with the most common width being the *octave filter*, wherein the center frequencies of the filters are related by multiples of 2 (i.e., 31.5, 63, 125, 250, …, 4000, 8000, and 16,000 Hz), with the most common type being the center-frequency proportional, wherein the width of the filter depends on the center frequency (as in an octave filter set, in which the passband width equals the center frequency divided by $2^{1/2}$). The *octave band*, commonly called the 1/1-octave filter, has a center frequency (CF) that is equal to the geometric mean of the upper (f_u) and lower (f_l) cutoff frequencies. The formulas to compute the center frequency for the octave filter, as well as the band-edge frequencies, are

$$\begin{aligned} &\text{center frequency (CF)} = (f_u f_l)^{1/2} \\ &\text{upper cutoff, } f_u = \text{CF} \cdot 2^{1/2} \\ &\text{lower cutoff, } f_l = \text{CF}/2^{1/2} \end{aligned} \tag{7}$$

More precise spectral resolution can be obtained with other center-frequency proportional filter sets with narrower bandwidths, the most common being the 1/3-octave, and with constant-percentage bandwidth filter sets, such as 1 or 2% filters. Note that in both types, the filter bandwidth increases as the center frequency increases. Still other analyzers have constant-bandwidth filters, such as 20-Hz-wide bandwidths which are of constant width regardless of center frequency. Whereas in the past, most spectrum analyzer filters have been analog devices with "skirts" or overshoots extending slightly beyond the cutoff frequencies, digital computer-based analyzers are now very common. These "computational" filters use fast Fourier transform (FFT) or other algorithms to compute sound level in a prespecified band of fixed resolution. FFT devices can be used to obtain very high resolutions of noise spectral characteristics using bandwidths as low as 1 Hz. However, in most measurement applications, a 1/1- or 1/3-octave analyzer will suffice unless the noise has considerable power in near-tonal components that must be isolated. One caution is in order: If a noise fluctuates in time and/or frequency, an integrating/averaging analyzer should be used to achieve good accuracy of measurements. It is important that the averaging period be long in comparison to the variability of the noise being sampled.

Real-time analyzers incorporate parallel banks of filters (not FFT-driven) that can process all frequency bands simultaneously, and the signal output may be controlled by a SLOW, FAST, or other time constant setting, or it may be integrated or averaged over a fixed time period to provide L_{OSHA}, L_{eq}, or other average-type measurements to be discussed later.

Spectrum Analyzer Applications While occupational noise is monitored with a dosimeter or SLM for the purpose of noise exposure compliance (using A-weighted broadband measurement) or the assessment of hearing protection adequacy (using C-weighted broadband measurement), both of these applications can also be addressed (in some cases more accurately) with the use of spectral measurements of the noise level. For instance, the OSHA occupational noise exposure standard (OSHA, 1983) allows the use of octave-band measurements reduced to broadband dBA values to determine if noise exposures exceed dBA limits defined in Table G-9 of the standard. Furthermore, Appendix B of the standard concerns hearing protector adequacy and allows the use of an octave-band method for determining, on a spectral rather than a broadband basis, whether a hearing protector is adequate for a particular noise spectrum. It is also noteworthy that spectral analysis can help the hearing conservationist discriminate noises as to their hazard potential even though they may have similar A-weighted SPLs. This is illustrated in Figure 5, where both noises would be considered to be of equal hazard by the OSHA-required dBA measurements (since they both are 90 dBA), but the 1/3-octave analysis demonstrates that the lowermost noise is more hazardous, as evidenced by the heavy concentration of energy in the midrange and high frequencies.

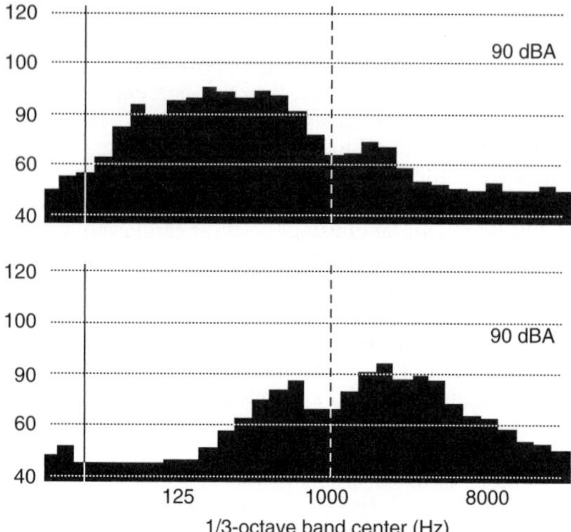

Figure 5 Spectral differences for two different noises that have the same dBA value.

One of the most important applications of the spectrum analyzer is to obtain data that will provide the basis for engineering noise control solutions. For instance, to select an absorption material for lining interior surfaces of a workplace, the spectral content of the noise must be known so that the appropriate density, porosity, and thickness of material may be selected. Spectrum analyzers are also necessary for performing the frequency-specific measurements needed to predict either signal audibility or speech intelligibility in noisy situations, according to the techniques discussed in Section 7. Furthermore, they can be applied for calibration of signals for laboratory experiments and audiometers, for determining the frequency response and other quality-related metrics for systems designed for music and speech rendition, and for determining certain acoustical parameters of indoor spaces, such as reverberation time.

dBC–dBA Lacking a spectrum analyzer, one can obtain a very rough indication of the dominant spectral content of a noise by using a SLM and taking measurements in both dBA and dBC for the same noise. If the (dBC − dBA) value is large, that is, about 5 dB or more, then it can be concluded that the noise has considerable low-frequency content. If, on the other hand the (dBC − dBA) value is negative, the noise clearly has strong midrange components, since the A-weighting curve exhibits slight amplification in the range 2000 to 4000 Hz. Such rules of thumb rely on the differences in the C- and A- weighting curves shown in Figure 3*b*. However, they should not be relied upon in lieu of a spectrum analysis if the noise is believed to have high-frequency or narrowband components that need noise control attention.

3.1.4 Acoustical Calibrators

Each of the instruments described above contains a microphone that transduces the changes in pressure and inputs this signal into the electronics. Although modern sound measurement equipment is generally stable and reliable, calibration is necessary to match the microphone to the instrument so that the accuracy of the measurement is assured. Because of its susceptibility to varying environmental conditions and damage due to rough handling, moisture, and magnetic fields, the microphone is generally the weakest link in the measurement equipment chain. Therefore, an acoustical calibrator should be applied before and after each measurement with a SLM. The pretest calibration ensures that the instrument is indicating the correct SPL for a standard reference calibrator output at a specified SPL and frequency (e.g., 94 dB at 1000 Hz). The posttest calibration is done to determine if the instrumentation, including the microphone, has drifted during the measurement, and if so, if the drift is large enough to invalidate the data obtained. Calibrators may be electronic transducer-type devices with loudspeaker outputs from an internal oscillator, or "pistonphones," which use a reciprocating piston in a closed cavity to produce sinusoidal pressure variations as the cylinder volume changes. Both types include adapters that allow the device to be mated to microphones of different diameters. Calibrators should be sent to the manufacturer at least annually for bench calibration and certification.

There are many other issues that bear on the proper application of sound level measurement equipment, such as microphone selection and placement, averaging-time and sampling schemes, and statistical data reduction techniques, all of which are beyond the scope of this chapter. For further coverage of these topics, the reader is referred to Harris (1991) and Berger et al. (2003).

3.2 Sound and Noise Metrics

3.2.1 Exchange or Trading Rates

Because both sound amplitude and sound duration determine the energy of an *exposure*, average-type measures are based on simple algorithms or *exchange rates*, which trade amplitude for time, and vice versa. For example, most noise regulations, OSHA (1983) or otherwise, stipulate that a worker's exposure may not exceed a maximum daily accumulation of noise energy. In other words, in OSHA terms the product of duration and intensity must remain under the regulatory cap or *permissible exposure limit* (PEL) of 90 dBA time-weighted average (TWA) for an 8-h work period, which is equivalent to a 100% noise dose. Much debate has occurred over the past several decades about which exchange rate is most appropriate for prediction of hearing damage risk, and most countries currently use either a 3- or 5-dB relationship. The OSHA exchange rate is 5 dB, which means that an increase (decrease) in decibel level by 5 dB is equivalent (in exposure) to a doubling (halving) of time. For instance, using the OSHA PEL of 90 dBA for 8 h, if a noise is at 95 dBA, the allowable exposure per workday is half of 8 h, or 4 h. If a noise is at 85 dBA, the allowable exposure time is twice 8 h, or

16 h. These allowable reference exposure durations (T values) are provided in Table A-1 of the OSHA (1983) regulation, or they may be computed using the formula for T which appears below as equation (14). The 5-dB exchange rate is predicated in part on the theory that intermittent noise is less damaging than continuous noise because some recovery from temporary hearing loss occurs during quiet periods. Arguments against it include the fact that an exchange of 5 dB for a factor of 2 in time duration has no real physical basis in terms of energy equivalence. Furthermore, there is some evidence that the quiet periods of intermittent noise exposures are insufficient in length to allow for recovery to occur. The 5-dB exchange rate is used for all measures associated with OSHA regulations, including the most general average measure of L_{OSHA}, the TWA referenced to an 8-h duration, and noise dose in percent.

Most European countries use a 3-dB exchange rate, also known as the aforementioned *equal energy rule*. In this instance, a doubling (halving) of sound intensity, which corresponds to a 3-dB increase (decrease), equates (in energy) to a doubling (halving) of exposure duration. The equal energy concept stems from the fact that if sound intensity is doubled or halved, the equivalent sound intensity level change is 3 dB. An exposure to 90 dBA for 8 h using a 3-dB exchange rate is equivalent to a 120-dBA exposure of only 0.48 min. Because each increase in decibels by 10 corresponds to a 10-fold increase in intensity, the 30-dB increase from 90 to 120 dBA represents a 1000-fold (10^3) increase in sound intensity, from 0.001 W/m^2 to 1 W/m^2. The 90-dBA exposure period is 8 h or 480 min, and this must be reduced by the same factor as the SPL increase, so 480/1000 equals 0.48 min or 29 s. The 3-dB exchange rate is used for all measures associated with the equivalent continuous sound level, L_{eq}.

3.2.2 Average and Integrated SPLs

As discussed earlier, conventional SLMs provide "momentary" decibel measurements that are based on a very short moving-window exponential averages using FAST, SLOW, or IMPULSE time constants. However, since the majority of noises fluctuate over time, one of several types of average measurements, discussed below, is usually most appropriate as a descriptor of the central tendency of the noise. Averages may be obtained in one of two ways: (1) by observing and recording conventional SLM readouts using a short-time-interval sampling scheme, and then manually computing the average value from the discrete values, or (2) by using a SLM or dosimeter which automatically calculates a running-average value using microprocessor circuitry which provides either a true continuous integration of the area under the sound pressure curve or which obtains discrete samples of the sound at a very fast rate and computes the average. Generally, average measures obtained by method 2 yield more representative values because they are based on continuous or near-continuous sampling of the waveform, which the human observer cannot perform well even with continuous vigilance.

The average metrics discussed below are generally considered as the most useful for evaluating noise hazards in industry, annoyance potential in the community, and other sounds in the laboratory or in the field which fluctuate over time. In most cases for industrial hearing conservation as well as community noise annoyance purposes, the metrics utilize the A-weighting scale. For precise spectral measurements with no frequency weighting, the decibel unweighted (linear) scale may be applied in the measurements. The equations are all in a form where the data values are considered to be discrete sound levels. Thus, they can be applied to data from conventional SLMs or dosimeters. For continuous sound levels (or when the equations are used to describe true integrating meter functioning), the $\sum$ sign in the equations would be replaced by the integral sign, $\int_0^T$, and the t_i replaced by dt. Variables used in the equations are as follows:

L_i = decibel level in measurement interval i

N = number of intervals

T = total measurement time period

t_i = length of measurement interval i

Q = exchange rate (decibels)

$$\text{q} = Q/\log_{10}(2)\begin{cases}\text{for 3-dB exchange, } q = 10.0\\ \text{for 4-dB exchange, } q = 13.3\\ \text{for 5-dB exchange, } q = 16.6\end{cases}$$

The general form equation for *average SPL*, or L_{average}, L_{av}, is

$$L_{\text{av}}(Q) = q\log_{10}\left[\frac{1}{T}\sum_{i=1}^{N}(10^{L_i/q}t_i)\right] \tag{8}$$

The *equivalent continuous sound level*, L_{eq}, equals the continuous sound level which when integrated or averaged over a specific time would result in the same energy as a variable sound level over the same time period. The equation for L_{eq}, which uses a 3-dB exchange rate, is

$$L_{\text{eq}} = L_{\text{av}}(3) = 10\log_{10}\left[\frac{1}{T}\sum_{i=1}^{N}(10^{L_i/10}t_i)\right] \tag{9}$$

In applying the L_{eq}, the individual L_i values are usually in dBA. Equation (9) may also be used to compute the overall equivalent continuous sound level (for a single site or worker) from individual L_{eq} values that are obtained over contiguous time intervals by substituting the L_{eq} values in the L_i variable. L_{eq} values are often expressed with the time period over which the average is obtained; for instance, L_{eq} (24) is an equivalent continuous level measured over a 24-h period. Another average measure that is derived from L_{eq} and often used for community noise quantification is L_{dn}, which is simply a 24-h L_{eq} measurement with a 10-dB penalty added to all nighttime noise levels from 10 P.M. to 7 A.M. The rationale for the penalty is that humans are more disturbed by noise, especially due to sleep arousal, during nighttime periods.

The equation for the *OSHA average noise level*, L_{OSHA}, which uses a 5-dB exchange rate, is

$$L_{OSHA}(5) = 16.61 \log_{10} \left[\frac{1}{T} \sum_{i=1}^{N} (10^{L_{iA}/16.61} t_i) \right] \quad (10)$$

where L_{iA} is in dBA, slow response.

OSHA's *time-weighted average* (TWA) is a special case of L_{OSHA} which requires that the total time period always be 8 h, that time is expressed in hours, and that sound levels below 80 dBA, termed the *threshold level*, are not included in the measurement:

$$\text{TWA} = 16.61 \log_{10} \left[\frac{1}{8} \sum_{i=1}^{N} (10^{L_{iA}/16.61} t_i) \right] \quad (11)$$

where L_{iA} is in dBA, slow response, and T is always 8 h. Only $L_{iA} \geq 80$ dBA is included.

OSHA's *noise dose* is a percentage representation of the noise exposure, where 100% is the maximum allowable dose, corresponding to a 90-dBA TWA referenced to 8 h. Dose utilizes a *criterion sound level*, which is presently 90 dBA, and a *criterion exposure period*, which is presently 8 h. A noise dose of 50% corresponds to a TWA of 85 dBA, and this is known as the OSHA *action level*. Calculation of dose, D, is as follows:

$$D = \frac{100}{T_c} \sum_{i=1}^{N} (10^{(L_{iA} - L_c)/q} t_i) \quad (12)$$

where L_{iA} is in dBA, slow response, L_c is the criterion sound level, and T_c is the criterion exposure duration. Only $L_{iA} \geq 80$ dBA is included.

Noise dose, D, can also be expressed as follows, for a constant sound level over the workday:

$$D = 100 \left(\frac{C_1}{T_1} + \frac{C_2}{T_2} + \cdots + \frac{C_n}{T_n} \right) \quad (13)$$

where C_i is the total time (h) of actual exposure at L_i, T_i is total time (h) of reference allowed exposure at L_i, from Table G-16a of OSHA, (1983), and C_i/T_i represents a partial dose at sound level i.

T, the *reference allowable* exposure for a given sound level, can also, in lieu of consulting Table G-16a in OSHA (1983), be computed as

$$T = 8/2^{(L-90)/5} \quad (14)$$

where L is the measured dBA level.

Two other useful equations to compute dose, D, from TWA, and vice versa, are

$$D = 100(10^{(TWA-90)/16.61}) \quad (15)$$

$$\text{TWA} = [16.61 \log_{10}(D/100)] + 90 \quad (16)$$

where D is the dose in %. TWA can also be found for each value of dose, D, in Table A-1 of OSHA (1983).

A final measure that is particularly useful for quantifying the exposure due to single or multiple occurrences of an acoustical event (such as a complete operating cycle of a machine, a vehicle driveby, or an aircraft flyover), is the *sound exposure level* (SEL). The SEL represents a sound of 1-s length that imparts the same acoustical energy as a varying or constant sound that is integrated over a specified time interval, t_i, in seconds. Over t_i, an L_{eq} is obtained which indicates that SEL is used only with a 3-dB exchange rate. A reference duration of 1 s is applied for t_0 in the following equation for SEL:

$$\text{SEL} = L_{eq} + 10 \log_{10}(t_i/t_0) \quad (17)$$

where L_{eq} is the equivalent sound pressure level measured over time period t_i.

Detailed example problems and solutions using the formulas above may be found in Casali and Robinson (1999).

4 INDUSTRIAL NOISE REGULATION AND ABATEMENT

4.1 The Need for Attention to Noise

In this section the discussion will concentrate on the management of noise in industry, because that is the major source of noise exposure for most people, and as such, it constitutes a very common threat toward noise-induced hearing loss (NIHL). Many of the techniques for measurement, engineering control, and hearing protection also apply to other exposures, such as those encountered in recreational or military settings. The need for attention to industrial noise is indicated when (1) noise creates sufficient intrusion and operator distraction such that job performance (and even job satisfaction) are compromised; (2) noise creates interference with important communications and signals, such as interoperator communications, machine- or process-related aural cues, and/or alerting/emergency signals; and/or (3) noise exposures constitute a hazard for noise-induced hearing loss in workers.

4.2 OSHA Noise Exposure Limits

In regard to combating the hearing loss problem, in OSHA terms if the noise dose exceeds the OSHA action level of 50%, which corresponds to an 85-dBA TWA, the employer must institute a *hearing conservation program* (HCP) which consists of several facets (OSHA, 1983). If the criterion level of 100% dose is exceeded [which corresponds to the permissible exposure level (PEL) of 90-dBA TWA for an 8-h day], the regulations specifically state that steps must be taken to reduce the employee's exposure to the PEL or below via administrative work scheduling and/or the use of engineering controls. It is stated specifically that hearing protection devices (HPDs) must be provided if administrative and/or engineering controls fail to reduce the noise to the PEL. Therefore, in applying the letter of the law, HPDs are only intended to be relied on when administrative or engineering controls are infeasible or ineffective. The final OSHA noise-level requirement pertains to impulsive or impact noise, which is not to exceed a TRUE PEAK SPL limit of 140 dB.

4.3 Hearing Conservation Programs

4.3.1 Shared Responsibility: Management, Workers, and Government

A successful hearing conservation program (HCP) depends on the shared commitment of management and labor as well as the quality of services and products provided by external noise control consultants, audiology or medical personnel who conduct the hearing measurement program, and vendors (e.g., hearing protection suppliers). Involvement and interaction of corporate positions such as the plant safety engineer, ergonomist, occupational nurse, noise control engineer, purchasing director, and manufacturing supervisor are important. Furthermore, government regulatory agencies, such as OSHA and the Mine Safety and Health Administration (MSHA), have a responsibility to maintain and disseminate up-to-date noise exposure regulations and HCP guidance, to conduct regular in-plant compliance checks of noise exposure and quality of HCPs, and to provide enforcement where noise control and/or hearing protection is inadequate. Finally, the "end user" of the HCP, that is, the worker, must be an informed and motivated participant. For instance, if a fundamental component of the HCP is the personal use of HPDs, the effectiveness of the program in preventing NIHL will depend most heavily on the worker's commitment to wear the HPD properly and consistently. Failure by any of these groups to carry out their responsibilities can result in HCP failure and worker hearing loss. Side benefits of a successful HCP may include a marked reduction in noise-induced distractions and interference on the job, and an improvement in worker comfort and morale.

4.3.2 Hearing Conservation Program Components

Hearing conservation in industry should be thought of as a strategic, programmatic effort that is initiated, organized, implemented, and maintained by the employer, with cooperation from other parties as indicated above. A well-accepted approach is to address the noise exposure problem from a *systems* perspective, wherein empirical noise measurements provide data input which drives the implementation of countermeasures against the noise (including engineering controls, administrative strategies, and personal hearing protection). Subsequently, noise and audiometric data, which reflect the effectiveness of those countermeasures, serves as feedback for program adjustments and improvements. A brief discussion of the major elements of a HCP, as dictated by OSHA (1983), follows.

Monitoring Noise exposure monitoring is intended to identify employees for inclusion in the HCP and to provide data for the selection of HPDs. The data are also useful for identifying areas where engineering noise control solutions and/or administrative work scheduling may be necessary. All OSHA-related measurements, with the exception of the TRUE PEAK SPL limit, are to be made using a SLM or dosimeter (of at least ANSI Type 2) set on the dBA scale, SLOW response, using a 5-dB exchange rate, and incorporating all sounds whose levels are from 80 to 130 dBA. It is unspecified, but it must be assumed that sounds above 130 dBA should also be monitored. (Of course, such noise levels represent OSHA noncompliance since the maximum allowable continuous sound level is 115 dBA.) Appendix G of the OSHA regulation suggests that monitoring be conducted at least once every one or two years. Relating to the noise monitoring requirement is that of notification. Employees must be given the opportunity to observe the noise monitoring process, and they must be notified when their exposures exceed the 50% dose (85-dBA TWA) level.

Audiometric Testing Program All employees whose noise exposures are at the 50% dose level or above must be included in a pure-tone audiometric testing program wherein a baseline audiogram is completed within six months of the first exposure, and subsequent tests are done on an annual basis. Annual audiograms are compared against the baseline to determine if the worker has experienced a *standard threshold shift* (STS), which is defined by OSHA (1983). The annual audiogram may be adjusted for age-induced hearing loss (presbycusis) using the gender-specific correction data in Appendix F of the regulation. All OSHA-related audiograms must include 500, 1000, 2000, 3000, 4000, and 6000 Hz, in comparison to most clinical audiograms, which extend from 125 to 8000 Hz. If an STS is revealed, a licensed physician or audiologist must review the audiogram and determine the need for further audiological or otological evaluation, the employee must be notified of the STS, and the selection and proper use of HPDs must be revisited.

Training Program and Record Keeping An essential component of an HCP is a training program for all noise-exposed workers. Training elements to be covered include the effects of noise on hearing; purpose, selection, and use of HPDs; and purpose and procedures of audiometric testing. Also, accurate records must be kept of all noise exposure measurements, at least from the last two years, as well as audiometric test results for the duration of the worker's employment. It is important, but not required by OSHA, that noise and audiometric data be used as feedback for improving the program. For example, noise exposure records may be used to identify machines that need maintenance attention, to assist in the relocation of noisy equipment during plant layout efforts, to provide information for future equipment procurement decisions, and to target plant areas that are in need of noise control intervention. Some employers plot noise levels on a "contour map," delineating floor areas by their decibel levels. When monitoring indicates that the noise level in a particular contour has changed, it is taken as a sign that the machinery and/or work process has changed in the area and that further evaluation may be needed.

Hearing Protection Devices OSHA (1983) requires that a selection of HPDs that are suitable for the noise and work situation must be made available to all employees whose TWA exposures meet or exceed 85 dBA. Such HPDs are also useful outside the workplace, for the protection of hearing against noises produced by power tools, lawn care equipment, recreational vehicles, target shooting and hunting, spectator events, and other exposures.

Earplugs consist of vinyl, silicone, spun fiberglass, cotton/wax combinations, and closed-cell foam products that are inserted into the ear canal to form a noise-blocking seal. Proper fit to the user's ears and training in insertion procedures are critical to the success of earplugs. A related device is the *semi-insert* or *ear canal cap*, which consists of earplug-like pods that are positioned at the rim of the ear canal and held in place by a lightweight headband. The headband is useful for storing the device around the neck when the user moves out of the noise. *Earmuffs* consist of earcups, usually of a rigid plastic material with an absorptive liner, that enclose the outer ear completely and seal around it with foam- or fluid-filled cushions. A headband connects the earcups, and on some models this band is adjustable so that it can be worn over the head, behind the neck, or under the chin, depending on the presence of other headgear, such as a welder's mask. In general terms, as a group, earplugs provide better attenuation than earmuffs below about 500 Hz and equivalent or greater protection above 2000 Hz. At intermediate frequencies, earmuffs sometimes have the advantage in attenuation. Earmuffs are generally more easily fit by the user than either earplugs or canal caps, and depending on the temperature and humidity of the environment, the earmuff can be uncomfortable (in hot or high-humidity environs) or a welcome ear insulator (in a cold environ). Semi-inserts generally offer less attenuation and comfort than earplugs or earmuffs, but because they are readily storable around the neck, they are convenient for those workers who frequently move in and out of noise. A thorough review of HPDs and their applications may be found in Berger and Casali (1997). Recent new technologies in hearing protection have emerged, including electronic devices offering active noise cancellation, communications capabilities, and noise-level-dependent attenuation, as well as passive, mechanical HPDs which offer level-dependent attenuation and near flat or uniform attenuation spectra; these devices are reviewed in Casali and Berger (1996).

Regardless of type, HPD effectiveness depends heavily on the proper fitting and use of the devices (Park and Casali, 1991). Therefore, the employer is required to provide training in the fitting, care, and use of HPDs to all employees affected (OSHA, 1983). Hearing protector use becomes mandatory when the worker has not undergone the baseline audiogram, has experienced an STS, or has a TWA exposure that meets or exceeds 90 dBA. In the case of the worker with an STS, the HPD must attenuate the noise to 85 dBA TWA or below. Otherwise, the HPD must reduce the noise to at least 90 dBA TWA.

The protective effectiveness or adequacy of an HPD for a given noise exposure must be determined by applying the attenuation data required by the EPA (1979) to be included on protector packaging. These data are obtained from psychophysical threshold tests at nine 1/3-octave bands with centers from 125 to 8000 Hz that are performed on human subjects, and the difference between the thresholds with the HPD on and without it constitutes the attenuation at a given frequency. Spectral attenuation statistics (means and standard deviations) and the single-number noise reduction rating (NRR), which is computed therefrom, are provided. The ratings are the primary means by which end users compare different HPDs on a common basis and make determinations of whether adequate protection and OSHA compliance will be attained for a given noise environment.

The most accurate method of determining HPD adequacy is to use octave-band measurements of the noise and the spectral mean and standard deviation attenuation data to determine the protected exposure level under the HPD. This is called the *NIOSH long method* or *octave-band method*. Computational procedures appear in NIOSH (1975). Because this method requires octave-band measurements of the noise, preferably with each noise band's data in TWA form, the data requirements are large and the method is not widely applied in industry. However, because the noise spectrum is compared against the attenuation spectrum of the HPD, a "matching" of exposure to protector can be obtained; therefore, the method is considered to be the most accurate available.

The NRR represents a means of collapsing the spectral attenuation data into one broadband attenuation estimate that can easily be applied against broadband dBC or dBA TWA noise exposure measurements. In calculation of the NRR, the mean attenuation is reduced by two standard deviations; this translates into an estimate of protection theoretically achievable by 98% of the population (EPA, 1979). The NRR is intended primarily to be subtracted from the dBC exposure TWA to estimate the protected exposure level in dBA:

$$\text{workplace TWA (dBC)} - \text{NRR} = \text{protected TWA (dBA)} \tag{18}$$

Unfortunately, because OSHA regulations require that noise exposure monitoring be performed in dBA, the dBC values may not be readily available to the hearing conservationist. In the case where the TWA values

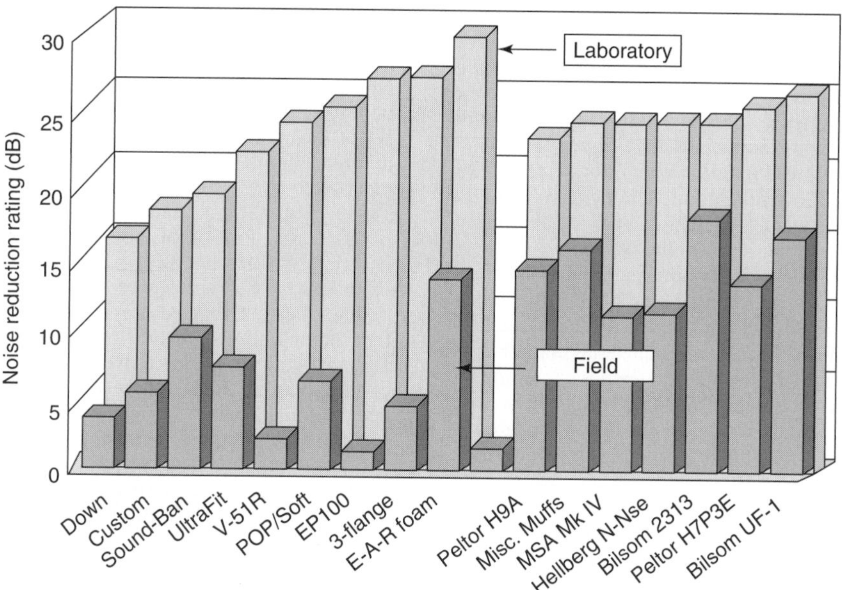

Figure 6 Comparison of hearing protection device NRRs by device type: manufacturers' laboratory data vs. real-world "field" data. (Adapted with permission from Berger, 2003.)

are in dBA, the NRR can still be applied, albeit with some loss of accuracy. With dBA data, a 7-dB "safety" correction is applied to the NRR to account for the largest typical differences between C- and A-weighted measurements of industrial noise, and the equation is

$$\text{Workplace TWA (dBA)} - (\text{NRR} - 7) = \text{protected TWA (dBA)} \tag{19}$$

Although the methods above are promulgated by OSHA (1983) for determining HPD adequacy for a given noise situation, a word of caution is needed. The data appearing on HPD packaging are obtained under optimal laboratory conditions with properly fitted protectors and trained human subjects. In no way does the "experimenter-fit" protocol and other aspects of the currently required (by the EPA) test procedure, ANSI S3.19-1974 (ANSI, 1974) represent the conditions under which HPDs are selected, fit, and used in the workplace (Park and Casali, 1991). *Therefore, the attenuation data used in the octave-band or NRR formulas shown above are, in general, inflated and cannot be assumed as representative of the protection that will be achieved in the field.* The results of a review of research studies in which manufacturers' on-package NRRs were compared against NRRs computed from actual subjects taken with their HPDs from field settings are shown in Figure 6. Clearly, the differences between laboratory and field estimates of HPD attenuation are large and the hearing conservationist must take this into account when selecting protectors. Efforts by ANSI Working Group S12/WG11 has focused on the development of a new testing standard, ANSI, S12.6-1997(R2002) (ANSI, 2002b) which has a provision for subject (not experimenter) fitting of the HPD and relatively naive (not trained) subjects. This testing protocol has been demonstrated to yield attenuation data that are more representative of those achievable under workplace conditions wherein a high-quality HCP is operated. However, at press time this new standard had not been adopted into law promulgating its use in producing the data to be utilized in labeling HPD performance. More information on HPD testing, labeling, and selection for particular noise exposures may be found in Berger (2003).

4.4 Engineering Noise Control

As discussed above, hearing protection and/or administrative controls are not a panacea for combating the risks posed by noise. They should not supplant noise control engineering; in fact, the best solution, in part because it does not rely on employee behavior, is to reduce the noise itself, preferably at the emission source. The physical reduction of the noise energy, either at its source, in its path, or at the worker, should be a major focus of noise management programs. However, in many cases where noise control is ineffective, infeasible (as on an airport taxi area), or prohibitively expensive, HPDs become the primary countermeasure.

There are many techniques used in noise control, and the specific approach must be tailored to the noise problem at hand. Spectrum analyzer measurements are typically used by noise control engineers in the selection of control strategies. Example noise control strategies include (1) *isolation of the source* via relocation, enclosure, or vibration damping using metal or air springs (below about 30 Hz) or elastomer (above 30 Hz) supports; (2) *reduction at the source or in the path* using mufflers or silencers on exhausts, reducing cutting, fan, or impact speeds, dynamically balancing rotating components, reducing fluid flow speeds and turbulence, absorptive foam or fiberglass on reflective surfaces to reduce reverberation, shields to reflect and redirect noise (especially high frequencies), and lining or wrapping of pipes and ducts; (3) *replacement or alteration of machinery*, examples including belt drives as opposed to noisier gears, electrical rather than pneumatic tools, and shifting frequency outputs such as by using centrifugal fans (low frequencies) rather than propeller or axial fans (high frequencies), keeping in mind that low frequencies propagate further than high frequencies, but high frequencies are more hazardous to hearing; and (4) *application of quieter materials*, such as rubber liners in parts bins, conveyors, and vibrators, resilient hammer faces and bumpers on materials handling equipment, nylon slides or rubber tires rather than metal rollers, and fiber rather than metal gears. Further discussion of these techniques may be found in Driscoll and Royster (2003), and an illustration of implementation possibilities in an industrial plant appears in Figure 7. A final approach that has recently become available to industry is *active noise reduction* (ANR), in which an electronic system is used to transduce an offensive noise in a sound field and then process and reintroduce the noise into the same sound field such that it is exactly 180° out of phase with, but of equal amplitude to, the original noise. The superposition of the out-of-phase *antinoise* with the original noise causes physical cancellation of the noise in a target zone of the workplace. For highly repetitive, predictable noises, synthesis of the antinoise, as opposed to transduction and reintroduction, may also be used. At frequencies below about 1000 Hz, the ANR technique is most effective, which is fortuitous since the passive noise control materials to combat low-frequency noise, such as absorptive liners and barriers, are typically heavy, bulky, and expensive. At higher frequencies and their corresponding shorter wavelengths, the processing and phase relationships become more complex and cancellation is less successful, although the technology is improving rapidly (Casali et al., 2004).

In designing and implementing noise control hardware, it is important that ergonomics be taken into account. For instance, in a sound-treated booth to house an operator, the ventilation system, lighting, visibility outward to the surrounding work area, and other considerations relating to operator comfort and performance must be considered. With regard to noise-isolating machine enclosures, access provisions should be designed so as not to compromise the operator–machine interface. In this regard, it is important that production and maintenance needs be met. If noise control hardware creates difficulties for the operators in carrying out their jobs, they may tend to modify or remove it, rendering it ineffective.

5 AUDITORY EFFECTS OF NOISE

5.1 Hearing Loss in the United States

Noise-induced hearing loss (NIHL) is one of the most widespread occupational maladies in the United States, if not the world. In the early 1980s, it was estimated that over 9 million workers are exposed to noise levels averaging over 85 dBA for an 8-h workday (EPA, 1981). Today, this number is likely to be higher because the control of noise sources, both in type and number, has not kept pace with the proliferation of industrial and service sector development. Due in part to the fact that before 1971 there were no U.S. federal regulations governing noise exposure in general industry, many workers over 50 years of age now exhibit hearing loss that results from the effects of occupational noise.

Of course, the total noise exposure from both occupational and nonoccupational sources determines the NIHL that a victim experiences. Of the estimated 28 million Americans who exhibit significant hearing loss due to a variety of etiologies, such as pathology of the ear and hereditary tendencies, over 10 million have losses that are directly attributable to noise exposure (National Institutes of Health, 1990). Therefore, the noise-related losses are *preventable* in nearly all cases. The majority of losses are due to on-the-job exposures, but leisure noise sources do contribute a significant amount of energy to the total noise exposure of some people. Although the effects of noise exposure are serious and must be reckoned with by the safety professional, one fact is encouraging: Process/machine-produced noise, as well as most sources of leisure noise, are physical stimuli that can be avoided, reduced, or eliminated; therefore, NIHL is preventable with effective abatement and protection strategies. Total elimination of NIHL should thus be the only acceptable goal.

5.2 Types and Etiologies of Noise-Induced Hearing Loss

Although the major concern of the industrial hearing conservationist is to prevent employee hearing loss that stems from occupational noise exposure, it is important to recognize that hearing loss may also emanate from a number of sources other than noise, including infections and diseases specific to the ear, most frequently originating in the middle or conductive portion; other bodily diseases, such as multiple sclerosis, which injures the neural part of the ear; ototoxic drugs, of which the mycin family is a prominent member; exposure to certain chemicals and industrial solvents; hereditary factors; head trauma; sudden hyperbaric- or altitude-induced pressure changes; and, aging of the ear (presbycusis). Furthermore, not all noise exposure occurs on the job. Many workers are exposed to hazardous levels during leisure activities, from such sources as automobile/motorcycle racing, personal stereo headsets and car stereos, firearms, and power tools. The effects of noise on hearing are generally subdivided into acoustic trauma and temporary or permanent threshold shifts (Melnick, 1991).

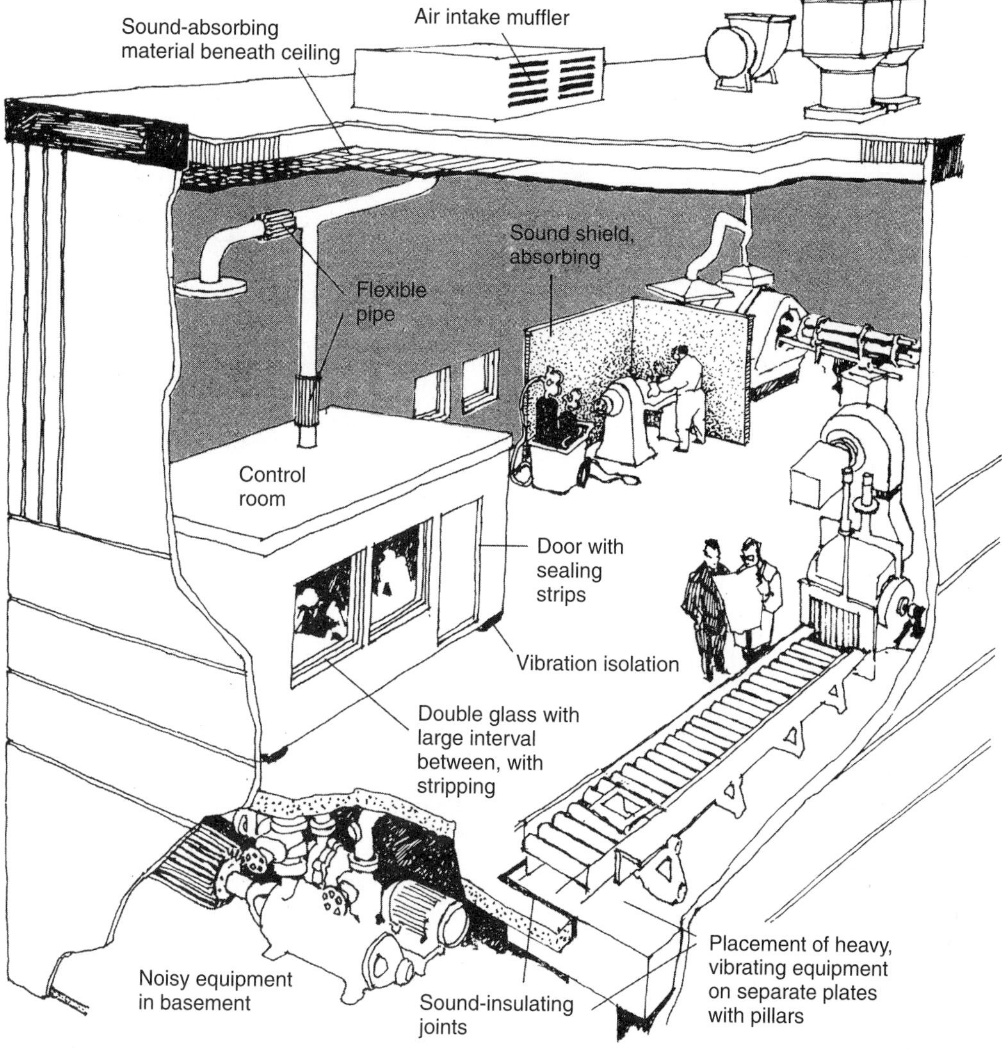

Figure 7 Noise control implementation in an industrial plant. (Adapted with permission from OSHA, 1980.)

5.2.1 Acoustic Trauma

Immediate organic damage to the ear from an extremely intense acoustic event such as an explosion is known as *acoustic trauma*. The victim will notice the loss immediately, and it often constitutes a permanent injury. The damage may be to the conductive chain of the ear, including rupture of the eardrum or dislodging of the ossicles (small bones) of the middle ear. Conductive losses can, in many cases, be compensated for with a hearing aid and/or surgically corrected. Neural damage may also occur, involving a dislodging of the hair cells and/or breakdown of the neural organ (Organ of Corti) itself. Unfortunately, neural loss is irrecoverable and not typically compensable with a hearing aid. Acoustic trauma represents a severe injury, but fortunately, its occurrence is relatively uncommon, even in the industrial setting.

5.2.2 Noise-Induced Threshold Shift

A *threshold shift* is defined as an elevation of hearing level from a person's baseline hearing level and it constitutes a loss of hearing sensitivity. *Noise-induced temporary threshold shift* (NITTS), sometimes referred to as "auditory fatigue," is by definition recoverable with time away from the noise. Thus, elevation of threshold is temporary and usually can be traced to an overstimulation of the neural hair cells (actually, the stereocilia) in the Organ of Corti. Although the person may not notice the temporary loss of sensitivity,

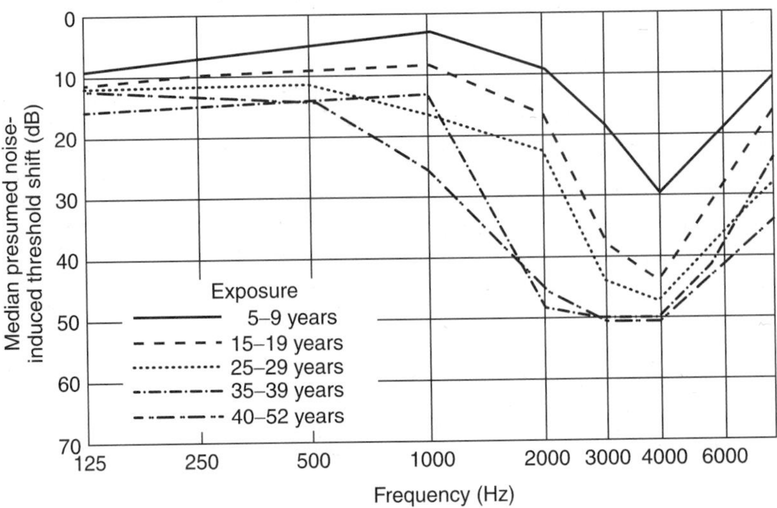

Figure 8 Cumulative auditory effects of years of noise exposure in a jute-weaving industry. (Adapted with permission from Taylor et al., 1964.)

NITTS is a cardinal sign of overexposure to noise. It may occur over the course of a full workday in noise or even after a few minutes of exposure to very intense noise. Although the relationships are somewhat complex and individual differences are rather large, NITTS does depend on the level, duration, and spectrum of the noise, as well as on the audiometric test frequency in question (Melnick, 1991).

With *noise-induced permanent threshold shift* (NIPTS), there is no possibility of recovery. NIPTS can manifest suddenly as a result of acoustic trauma; however, noises that cause NIPTS most typically constitute exposures that are repeated over a long period of time and have a cumulative effect on hearing sensitivity. In fact, the losses are often quite insidious in that they occur in small steps over a number of years of overexposure and the person may not be aware until it is too late. This type of exposure produces permanent neural damage, and although there are some individual differences as to magnitude of loss and audiometric frequencies affected, the typical pattern for NIPTS is a prominent elevation of threshold at the 4000-Hz audiometric frequency (sometimes called the 4-kHz notch), followed by a spreading of loss to adjacent frequencies of 3000 and 6000 Hz. From a classic study on workers in the jute weaver industry, Figure 8 depicts the temporal profile of NIPTS as the family of audiometric threshold shift curves, with each curve representing a different number of years of exposure. As noise exposure continues over time, the hearing loss spreads over a wider-frequency bandwidth inclusive of midrange and high frequencies and encompassing the range of most auditory warning signals. In some cases, the hearing loss renders it unsafe or unproductive for the victim to work in certain occupational settings where the hearing of certain signals are requisite to the job. Unfortunately, the power of the consonants of speech sounds, which heavily influence the intelligibility of human speech, also lie in the frequency range that is typically affected by NIPTS, compromising the victim's ability to understand speech. This is the tragedy of NIPTS in that the worker's ability to communicate is hampered, often severely and always irrecoverably. Hearing loss is a particularly troubling disability because its presence is not overt; therefore, the victim is often unintentionally excluded from conversations and may miss important auditory signals because others either are unaware of the loss or simply forget about the need to compensate for it.

5.3 Concomitant Auditory Injuries

Following exposure to high-intensity noise, some people will notice that ordinary sounds are perceived as "muffled," and in some cases, they may experience a ringing or whistling sound in the ears, known as *tinnitus*. These manifestations should be taken as serious indications that overexposure has occurred and that protective action should be taken if similar exposures are encountered in the future. Tinnitus may also occur by itself or in conjunction with NIPTS, but in any case it is thought to be the result of *otoacoustic emissions*, which are essentially acoustic outputs from the inner ear that are audible to the victim, apparently resulting from mechanical activity of the neural cells. Some people report that tinnitus is always present, pervading their lives. It thus has the potential to be quite disruptive and in severe cases, debilitating.

More rare than tinnitus, but typically quite debilitating, is the malady known as *hyperacusis*, which refers to hearing that is extremely sensitive to sound. Hyperacusis can manifest in many ways, but a number of victims report that their hearing became painfully sensitive to sounds of even normal levels after exposure to a particular noise event. Therefore, at least for some, hyperacusis can be traced directly to noise exposure. Sufferers often must use HPDs when performing normal activities, such as walking on city streets, visiting movie theaters, or washing dishes in a sink, because such activities produce sounds that are painfully loud to them. It should be noted that hyperacusis sufferers often exhibit normal audiograms, even though their reaction to sound is one of hypersensitivity.

6 PERFORMANCE, NONAUDITORY, AND PERCEPTUAL EFFECTS OF NOISE

6.1 Performance and Nonauditory Health Effects of Noise

6.1.1 Task Performance Effects

It is important to recognize that among other deleterious effects, noise can degrade operator task performance. Research studies concerning the effects of noise on performance are primarily laboratory-based and task/noise specific; therefore, extrapolation of the results to actual industrial settings is somewhat risky (Sanders and McCormick, 1993). Nonetheless, on the negative side, noise is known to mask task-related acoustic cues as well as to cause distraction and disruption of "inner speech"; on the positive side, noise may at least initially heighten operator arousal and thereby improve performance on tasks that do not require substantial cognitive processing (Poulton, 1978). To obtain reliable effects of noise on performance, except on tasks that rely heavily on short-term memory, the level of noise must be fairly high, usually 95 dBA or greater. Tasks that are simple and repetitive often show no deleterious performance effects (and sometimes improvements) in the presence of noise, whereas difficult tasks that rely on perception and information processing on the part of the operator will often exhibit performance degradation (Sanders and McCormick, 1993). It is generally accepted that unexpected or aperiodic noise causes greater degradation than predictable, periodic, or continuous noise, and that the startle response created by sudden noise can be disruptive.

6.1.2 Nonauditory Health Effects

Noise has been linked to physiological problems other than those of the hearing sense, including hypertension, heart irregularities, extreme fatigue, and digestive disorders. Most physiological responses of this nature are symptomatic of stress-related disorders. Because the presence of high noise levels often induces other stressful feelings (such as sleep disturbance and interference with conversing in the home, and fear of missing oncoming vehicles or warning signals on the job), there are second-order effects of noise on physiological functioning that are difficult to predict. The reader is referred to Kryter (1994) for a detailed discussion of nonauditory health effects of noise.

6.2 Annoyance Effects of Noise

Noise has frequently given rise to vigorous complaints in many settings, ranging from office environments to aircraft cabins to homes. Such complaints are manifestations of what is known as noise-induced *annoyance*, which has given rise to a host of products, such as white/pink noise generators for masking undesirable noise sources, noise-canceling headsets, and noise barriers for reducing sound propagation over distances and through walls. In the populated community, noise is a common source of disturbance, and for this reason many communities, both urban and rural, have noise ordinances and/or zoning restrictions which regulate the maximum noise levels that can result from certain sources and/or in certain land areas. In communities that have no such regulations, residents who are disturbed by noise sources such as industrial plants or spectator events often have no other recourse than to bring civil lawsuits for remedy (Casali, 1999). The principal rationale for limiting noise in communities is to reduce sleep and speech interference, and to avoid annoyance (Driscoll et al., 2003). Some of the measurement units and instrumentation discussed in this chapter are useful for community and other noise annoyance applications, while more detailed information on the subject may be found in Fidell and Pearsons (1997), Casali (1999), and Driscoll et al. (2003).

6.3 Loudness and Related Scales of Measurement

One of the most readily identified aspects of a sound or noise, and one that relates to a majority of complaints, be it a theater actor's voice which is too quiet or a background noise which is too intense, is that of *loudness*. As discussed above, the decibel is useful for quantifying the amplitude of a sound on a physical scale; however, it does not yield an absolute or relative basis for quantifying the human *perception* of sound amplitude, commonly called *loudness*. However, there are several psychophysical scales that are useful for measuring loudness, the two most prominent being *phons* and *sones*.

6.3.1 Phons

The decibel level of a 1000-Hz tone which is judged by human listeners to be equally loud to a sound in question is the phon level of the sound. The phon levels of sounds of different intensities are shown in Figure 3*a*; this family of curves is referred to as the *equal loudness contours*. On any given curve, the combinations of sound level and frequency along the curve produce sound experiences of equal loudness to the normal-hearing listener. Note that at 1000 Hz on each curve the phon level is equal to the decibel level. The threshold of hearing for a young, healthy ear is represented by the 0-phon-level curve. The young, healthy ear is sensitive to sounds between about 20 and 20,000 Hz, although, as shown by the curve, it is not equally sensitive to all frequencies. At low- and midlevel sound intensities, low frequency and to a lesser extent, high-frequency sounds are perceived as less intense than sounds in the range 1000 to 4000 Hz, where the undamaged ear is most sensitive. But as phon levels move to higher values, the ear becomes more linear in its loudness perception for sounds of different frequencies. It is because the ear exhibits this nonlinear behavior that the frequency-weighting responses for dBA, dBC, and so on, were developed, as discussed in Section 3.1.1.

6.3.2 Sones

Although the phon scale provides the ability to equate the loudness of sounds of various frequencies, it does not afford an ability to describe how much louder one sound is than another. For this, the *sone* scale is needed (Stevens, 1936). One sone is defined as the loudness of a 1000-Hz tone of 40-dB SPL. In relation to 1 sone, 2 sones are twice as loud, 3 sones are three times as loud, $\frac{1}{2}$ sone is half as loud, and so on. Phon level (L_P) and sones are related by the following formula for sounds at or above a 40-phon level:

$$\text{loudness (sones)} = 2^{(L_P-40)/10} \tag{20}$$

According to equation (20), 1 sone equals 40 phons and the number of sones doubles with each 10-phon increase above 40; therefore, it is straightforward to conduct a comparative estimate of loudness levels of sounds with different decibel levels. The rule of thumb is that each 10-dB increase in a sound (i.e., one that is above 40 dB to begin with) will result in a doubling of its loudness. For instance, a home theater room that is currently at 50 dBA may be comfortable for listening to movies and classical music. However, if a new air-conditioning system increases the noise level in the room by 10 dBA, the occupants will experience a perceptual doubling of loudness and will probably complain about the interference with speech and music in the room. Once again, the compression effect of the decibel scale yields a measure that does not reflect the much larger influence that an increase in sound level will have on the human perception of loudness.

Precise Calculation of Sone Levels by the Stevens Method It should be evident that sone levels can be calculated directly from psychological measurements in phons [per equation (20)] but not from physical measurements of SPL in decibels without special conversions. This is because the phon-based loudness and SPL relationship changes as a function of the sound frequency, and the magnitude of this change depends on the intensity of the sound. The *Stevens method*, also known as the *ISO spectral method*, is fully described in Rossing (1990). Briefly, this method requires measurement of the dB(linear) level in 10 standard octave or 1/3-octave bands, with centers at 31, 63, 125, 250, 500, 1000, 2000, 4000, 8000, and 16,000 Hz. Then, for each band measurement, the loudness index, S_i, is computed from Figure 9 as follows:

$$\text{loudness level (sones)} = S_{\max} + 0.3 \sum S_i \tag{21}$$

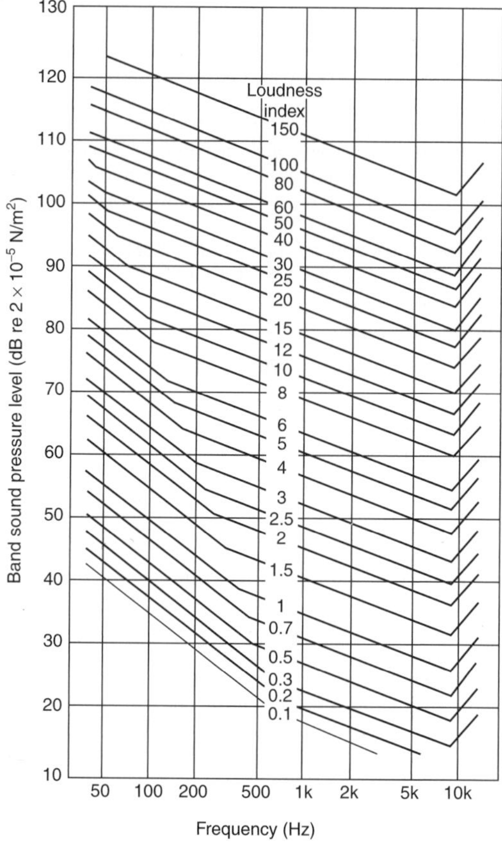

Figure 9 Chart for calculating loudness indices from decibel levels in various frequency bands, for use in computing sone levels. (Adapted with permission from Rossing, 1990.)

where S_i is the loudness index from Figure 9, $S_{\max}$ is the largest of the loudness indices, and $\sum S_i$ is the sum of the loudness indices for all bands except $S_{\max}$. Using this "precise" method, the effect is to include the loudest band of noise at 100%, while the totality of the other bands is included at 30%. Obviously, because the noise must be measured in octave or 1/3-octave bands, the method is measurement intensive and requires special instrumentation (i.e., a real-time spectrum analyzer).

Approximation of Sone Levels from dBA In contrast to the Stevens method, the loudness of a sound in sones can be computed from dBA values, albeit with less spectral precision. In this method, only a sound level meter (as compared to a spectrum analyzer) is needed, and measurements are captured in dBA. Then 1.5 sones is equated to 30 dBA, and the number of sones is doubled for each 10-dBA increase over 30 dBA. For example, 40 dBA = 3 sones, 50 dBA = 6 sones, 55 dBA = 8 sones, 60 dBA = 12 sones, 65 dBA = 16 sones, 70 dBA = 24 sones, 75 dBA =

32 sones, 80 dBA = 48 sones, 85 dBA = 64 sones, and 90 dBA = 96 sones (Rossing, 1990). This method is particularly accurate at low to moderate sound levels since the ear responds in similar sensitivity to the A-weighting curve at these levels.

Practical Applications of the Sone Despite its practicality, the sone scale is not widely used (an exception is that household ventilation fans typically have voluntary sone ratings). However, it is the most useful scale for comparing different sounds as to their loudnesses as perceived by humans. Given its interval qualities, the sone is more useful than decibel measurements when attempting to compare the loudness of different products' emissions; for example, a vacuum cleaner that emits 60 sones is twice as loud as one of 30 sones. The sone also is useful in conveying sound loudness experiences to lay groups. An example of such use for illustrating the perceptual impacts of a community noise disturbance (automobile racetrack) to a civil court jury may be found in Casali (1999).

6.3.3 Modifications of the Sone

A modification of the sone scale (Mark VI and subsequently, Mark VII sones) was proposed by Stevens (1972) to account for the fact that most real sounds are more complex than pure tones. Utilizing the general form equation (22) below, this method incorporates octave-band, $\frac{1}{2}$-octave-band, or $\frac{1}{3}$-octave-band noise measurements and adds to the sone value of the most intense frequency band a fractional portion of the sum of the sone values of the other bands ($\sum$S).

$$\text{loudness (sones)} = S_m + k\left(\sum S - S_m\right) \qquad (22)$$

where S_m is the maximum sone value in any band, k is a fractional multiplier that varies with bandwidth (octave, $k = 0.3$; 1/2 octave, $k = 0.2$; 1/3 octave, $k = 0.15$), and $\sum S$ is the sum of the sone values of the other bands.

6.3.4 Zwicker's Method of Loudness

The concept of the critical band for loudness formed the basis for Zwicker's method of loudness quantification (Zwicker, 1960). The critical band is the frequency band within which the loudness of a band of continuously distributed sound of equal SPL is independent of the width of the band. The critical bands widen as frequency increases. A graphical method is used for computing the loudness of a complex sound based on critical band results obtained and graphed by Zwicker. The noise spectrum is plotted and lines are drawn to depict the spread of a masking effect. The result is a bounded area on the graph which is proportional to total loudness. The method is relatively complex, and Zwicker (1960) should be consulted for computational detail.

6.3.5 Noisiness Units

As descriptive terms, *noisiness* and *loudness* are related but not synonymous. Noisiness can be defined as the "subjective unwantedness" of a sound. Perceived noisiness may be influenced by a sound's loudness, tonality, duration, impulsiveness, and variability (Kryter, 1994). Whereas a low level of loudness might be perceived as enjoyable or pleasing, a low level of unwantedness (i.e., noisiness) is by definition undesirable. Equal noisiness contours, analogous to equal loudness contours, have been developed based on a unit (analogous to the phon) called the *perceived noise level* (PN_{dB}), which is the SPL in decibels of a 1/3-octave band of random noise centered at 1000 Hz which sounds equally noisy to the sound in question. Also, an N- (later D-) sound level meter weighting curve was developed for measuring the perceived noise level of a sound. A subjective noisiness unit analogous to the sone, the *noy*, is used for comparing sounds as to their relative noisiness. One noy is equal to 40 PN_{dB}, and 2 noys are twice as noisy as one, 5 noys are 5 times as noisy, and so on. Similar to the behavior of sones as discussed above for loudness, an increase of about 10 PN_{dB} is equivalent to a doubling of the perceived noisiness of a sound.

7 SIGNAL DETECTION AND SPEECH COMMUNICATIONS IN NOISE

7.1 General Concepts in Signal and Speech Audibility

7.1.1 Signal-to-Noise Ratio Influence

One of the most noticeable effects of noise is its interference with speech communications and the hearing of nonverbal signals. Operators often complain that they must shout to be heard and that they cannot hear others trying to communicate with them. Similarly, noise interferes with the detection of signals such as alarms for general area evacuation and warnings in buildings, annunciators, on-equipment alarms, and machine-related sounds which are relied upon for feedback to industrial workers. In a car or truck, the hearing of external signals, such as emergency vehicle sirens or train horns, or in-vehicle warning alarms or messages, may be compromised by the ambient noise levels. The ratio (actually the algebraic difference) of the speech or signal level to the noise level, termed the *signal* (or *speech)-to-noise ratio* (S/N) is a critical parameter in determining whether speech or signals will be heard in noise. A S/N value of 5 dB means that the signal is 5 dB greater than the noise; a S/N value of −5 dB means that the signal is 5 dB lower than the noise.

7.1.2 Masking and Masked Threshold

Technically, *masking* is defined as the increase (in decibels) of the threshold of a desired signal or speech (the *masked sound)* to be raised in the presence of an interfering sound (the *masking sound* or *masker)*. For example, in the presence of noisy traffic alongside a busy street, an auditory pedestrian crossing signal's volume must be sufficiently louder than the traffic to enable a pedestrian to hear it, whereas a lower volume will be audible (and possibly more comfortable) when no traffic is present. It is also possible for one signal to mask another signal if both are active at the same time. The *masked threshold* is often defined in psychophysical terms as the SPL required for 75% correct detection of a signal when that signal is presented in a two-interval task wherein, on a random basis, one of the two intervals of each task trial contains the signal and the noise and the other contains only noise. In a controlled laboratory test scenario, a signal that is about 6 dB above the masked threshold will result in nearly perfect detection performance (Sorkin, 1987). In the remainder of this chapter, various aspects of the masking phenomenon are discussed and methods for calculating a masked signal threshold or, in the case of speech, an estimate of intelligibility are presented. Throughout, it is important to remember that the masked threshold is, in fact, a *threshold*; it is *not* the level at which the signal is clearly audible. For the ensuing discussion, a functional definition of an auditory threshold is the SPL at which the stimulus is *just audible* to a person listening intently for it in the specified conditions. If the threshold is determined in "silence," as is the case during an audiometric examination, it is referred to as an *absolute threshold*. If, on the other hand, the threshold is determined in the presence of noise, it is referred to as a *masked threshold*.

7.2 Analysis of Signal Detectability in Noise

Fundamentally, *detection* of an auditory signal is prerequisite to any other function performed on or about that signal, such as *discrimination* of it from other signals, *identification* of its source, *recognition* of its intended meaning or urgency, *localization* of its placement in space, and/or *judgment* of its speed. Although the S/N ratio is one of the most critical parameters that determine a signal's detectability in a noise, there are many other factors as well. These include the spectral content of the signal and noise (affecting *critical bandwidth*), temporal characteristics of signal and noise, duration of the signal's presentation, listener's hearing ability, demands on the listener's attention, criticality of the situation at hand, and the attenuation of hearing protectors, if used. These factors are discussed in detail in Robinson and Casali (2003). The ensuing discussion concentrates primarily on the most important issue of spectral content of the signal and noise and how that content impacts masking effects.

7.2.1 Spectral Considerations and Masking

Generally speaking, the greater the decibel level of the background noise relative to the signal (inclusive of speech), the more difficult it will be to hear the signal. Conversely, if the level of the background noise is reduced and/or the level of the signal is increased, the masked signal will be more readily audible. In some cases, ambient noise can be reduced through engineering controls, and in the same or other cases, it may be possible to increase the intensity of the signals. Although most off-the-shelf auditory warning devices have a preset output level, it is possible to increase the effective level of the devices by placing multiple alarms or warning devices throughout a coverage area instead of relying on one centrally located device. This approach can also be used for variable-output systems such as public-address loudspeakers since simply increasing the output of such systems often results in distortion of the amplified speech signal, thereby reducing intelligibility. Simply increasing the signal level without adding more sound sources can have the undesirable side effect of increasing the noise exposures of people in the area of the signal if it is used too often. If the signal levels are extremely high (e.g., over 105 dB), exposed persons could experience temporary threshold shifts or tinnitus if they are in the vicinity of the device when it is sounding.

One problem directly related to the level of the background noise is distortion within the inner ear. At very high noise levels, the cochlea becomes overloaded and cannot accurately transduce/discriminate different forms of acoustic energy (e.g., signal and noise) reaching it, resulting in the phenomenon known as *cochlear distortion*. In order for a signal, including speech, to be audible at very high noise levels, it must be presented at a higher level, relative to the background noise, than would be necessary at lower noise levels. This is one reason why it is best to make reduction of the background noise a high priority in occupational or other environments.

In addition to manipulating the levels of the auditory displays, alarms, warnings, and background noise, it is also possible to increase the likelihood of detection of an auditory display or alarm by manipulating its spectrum so that it contrasts with the background noise and other common workplace sounds. In a series of experiments, Wilkins and Martin (1982, 1985) found that the contrast of a signal with both the background noise and irrelevant signals was an important parameter in determining the detectability of a signal. For example, in an environment characterized by high-frequency noise such as sawing and/or planing operations in a wood mill, it might be best to select a warning device with strong low-frequency components, perhaps in the range of 700 to 800 Hz. On the other hand, for low-frequency noise such as might be encountered in the vicinity of large-capacity ventilation fans, an alarm with strong midfrequency components in the range of 1000 to 1500 Hz might be a better choice.

Upward Spread of Masking When considering masking of a tonal signal by a tonal noise or a narrow band of noise, masking is greatest in the immediate vicinity of the masking tone or, in the case of a bandlimited noise, the center frequency of the band. (This is one reason why increasing the contrast in frequency between the signal and noise can increase the audibility of a signal.) However, the masking effect does spread out above and below this frequency, being greater at the frequencies above the frequency of the masking noise than at frequencies below the frequency of the masking noise (Wegel and Lane, 1924; Egan and Hake, 1950). This phenomenon, referred to as the *upward spread of masking*, becomes more pronounced as the level of the masking noise increases, probably

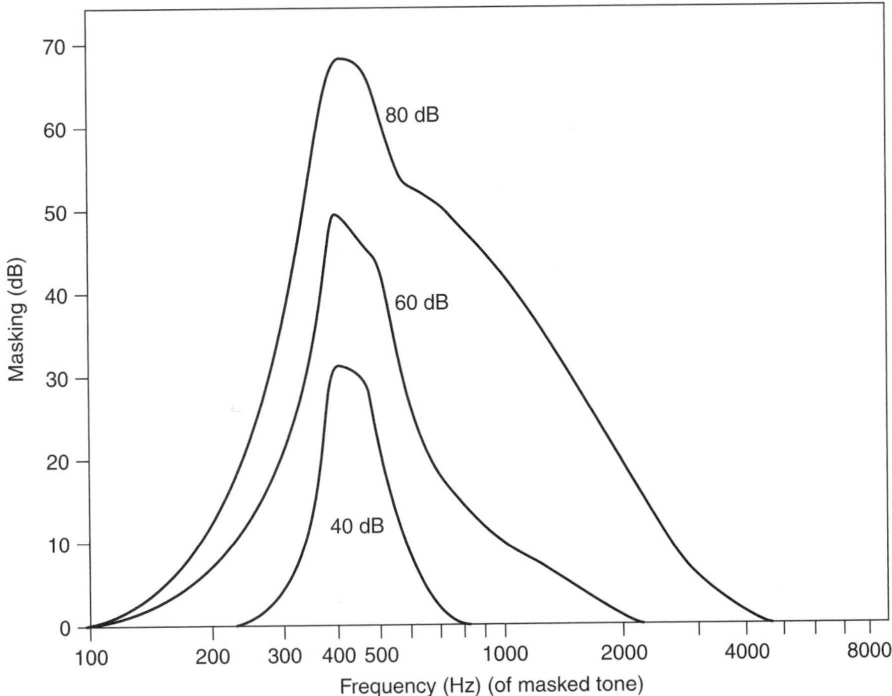

Figure 10 Upward spread of masking of a pure tone by three levels (40, 60, and 80 dB) of a 90-Hz-wide band of noise centered at 410 Hz. The ordinate (*y*-axis) is the amount (in decibels) by which the absolute threshold of the masked tone is raised by the masking noise, and the abscissa is the frequency of the masked tone. (Adapted with permission from Egan and Hake, 1950.)

due to cochlear distortion. In practical situations, masking by pure tones would seldom be a problem, except in instances where the noise contains strong tonal components, or if two warnings with similar frequencies were activated simultaneously. Although less pronounced, upward spread of masking does occur when both broadband and bandlimited noise are used as maskers. This phenomenon is illustrated in Figure 10.

Masking with Broadband Noise A very common form of masking characteristic of typical industrial workplaces or building spaces such as conference rooms or auditoria occurs when a signal or speech is masked by a broadband noise. In examining the masking of pure-tone stimuli by white noise (white noise sounds very much like static on a radio or TV tuned to a frequency or channel with no signal), Hawkins and Stevens (1950) found that masking was directly proportional to the level of the masking noise, irrespective of the frequency of the masked tone. In other words, if a given background white noise level increased the threshold of a 2500-Hz tone by 35 dB, the threshold of a 1000-Hz tone would also be increased by 35 dB. Furthermore, they found that for the noise levels investigated, masking increased linearly with the level of the white noise, meaning that if the level of the masking noise were increased 10 dB, the masked thresholds of the tones also increased by 10 dB. The bottom line is that broadband noise such as white, due to its inclusion of all frequencies, serves as a very effective masker of tonal signals and speech. Thus, its abatement often needs to be of high priority.

7.2.2 Signal Audibility Analysis Method Based on Critical Band Masking

Fletcher (1940) developed what would become *critical band theory*, which has formed the fundamental basis for explaining how signals are masked by narrowband noise. According to this theory, the ear behaves as if it contains a series of overlapping auditory filters, with the bandwidth of each filter being proportional to its center frequency. When masking of pure tones by broadband noise is considered, only a narrow "critical band" of the noise centered at the frequency of the tone is effective as a masker and the width of the band is dependent only on the frequency of the tone being masked. In other words, the masked threshold of a pure tone could be predicted simply by knowing the frequency of the tone and the spectrum level (decibels per hertz) of the masking noise, assuming that the noise spectrum is reasonably flat in the region around the tone. Thus, the masked threshold of a tone in white noise would simply be

$$L_{\text{mt}} = L_{\text{ps}} + 10\log(\text{BW}) \qquad (23)$$

where L_{mt} is the masked threshold, L_{ps} is the spectrum level of the masking noise, and BW is the width of the auditory filter centered around the tone. Strictly speaking, this relationship applies only when the masking noise is flat (equal energy per hertz) and when the masked signal has a duration greater than 0.1 s. However, an acceptable approximation may be obtained for other noise conditions as long as the spectrum level in the critical band does not vary by more than 6 dB (Sorkin, 1987). In many environments, the background noise is likely to be sufficiently constant and can often be presumed to be flat in the critical band for a given signal. The exception to this assumption is a situation where the noise has prominent tonal components and/or fluctuates a great deal.

The spectrum level of the noise in each of the 1/3-octave bands containing the signal components *is not* the same as the band level measured using an octave-band or 1/3-octave-band analyzer. Spectrum level refers to the level per hertz, or the level that would be measured if the noise were measured using a filter 1 Hz wide. If it is assumed that the noise is flat within the bandwidth of the 1/3-octave band filter, the spectrum level can be estimated using the equation

$$L_{\text{ps}} = 10\log(10^{L_{\text{pb}}/10}/\text{BW}_{1/3}) \qquad (24)$$

where L_{ps} is the spectrum level of the noise within the 1/3-octave band, L_{pb} is the sound pressure level measured in the 1/3-octave band in question, and $\text{BW}_{1/3}$ is the bandwidth of the 1/3-octave band, calculated by multiplying the center frequency (f_c) of the band by 0.232.

Finally, the bandwidth of the auditory filter can be approximated by multiplying the frequency of the masked signal/tone by 0.15 (Patterson, 1982; Sorkin, 1987). If the signal levels measured in one or more of the 1/3-octave bands considered exceed these masked threshold levels, the signal should be audible. A lengthy computational example using the critical band method appears in Robinson and Casali (2003).

7.2.3 Signal Audibility Analysis Method Based on ISO 7731-1986(E)

The Department of Defense, National Fire Protection Association, Society of Automotive Engineers, Underwriters' Laboratories, ANSI, and ISO are examples of organizations that have promulgated standards to guide the design of auditory warning signals for specific applications, such as on-vehicle alarms, sirens, evacuation alarms, and fire alarms. However, for performing an *analysis* of most any acoustic alarm as to its predicted audibility in a specific noise, perhaps the most comprehensive standard is ISO 7731-1986(E), *Danger Signals for Work Places—Auditory Danger Signals* (ISO, 1986). (At publication date, this standard was in draft revision as ISO/DIS 7731.) This standard provides guidelines for calculation of the masked threshold of audibility but also specifies the spectral content and minimum signal-to-noise ratios (S/N) of the signals, and requires special considerations for people suffering from hearing loss or those wearing HPDs.

Application of ISO 7731 is best illustrated by an example. A warning signal that is quite common is a standard backup alarm typically found on commercial trucks and construction/industrial equipment. It has strong tonal components at 1000 and 1250 Hz and strong harmonic components at 2000 and 2500 Hz. The alarm has a 1-s period and a 50% duty cycle (i.e., it is "on" for 50% of its period). The levels in all other 1/3-octave bands are sufficiently below those in the bands mentioned as to be inconsequential. The levels needed for audibility of this signal will be determined for application in a hypothetical noise spectrum represented by its 1/3-octave and octave band levels, shown in columns (2) and (4) respectively, in Table 1.

1. Starting at the lowest octave-band or 1/3-octave-band level available, the masked threshold (L_{mt1}) for a signal in that band is

$$L_{\text{mt1}} = L_{\text{pb1}} \qquad (25)$$

where L_{pb1} is the sound pressure level measured in the octave band or 1/3-octave band in question.

2. For each successive octave-band or 1/3-octave-band filter n, the masked threshold (L_{mtn}) is the noise level in that band or the masked threshold in the preceding band, less a constant; whichever is *greater*:

$$L_{\text{mtn}} = \max(L_{\text{pbn}}; L_{\text{mtn}-1} - C) \qquad (26)$$

where $C = 7.5$ dB for octave-band data or 2.5 dB for 1/3-octave-band data.

This procedure (unlike the aforementioned critical band procedure) presumes that the auditory filter width is equal to the 1/3-octave band or to the octave band, and also takes upward spread of masking into account by comparing the level in the band in question to the level in the preceding band. The masked thresholds for each 1/3-octave band and octave band of noise for the example are shown in columns (3) and (5) respectively, in Table 1. For the purposes of the example signal (backup alarm), only the thresholds for the 1/3-octave bands centered at 1000, 1250, 2000, and 2500 Hz and the threshold for the octave bands centered 1000 and 2000 Hz are relevant, because these are the signal component bands. The conclusion is that if the signal levels measured in one or more of these bands exceed the calculated masked threshold levels (indicated by

Table 1 Masked Threshold Calculations according to ISO 7731-1986(E) for 1/3-Octave-Band and Octave-Band Methods

(1) Center Frequency (Hz)[a]	(2) $\frac{1}{3}$-Octave-Band Level (dB)	(3) Masked Threshold (dB)[b]	(4) Octave-Band Level (dB)	(5) Masked Threshold (dB)[b]
25	52.0	52.0		
31.5	50.7	50.7	54.7	54.7
40	42.9	48.2		
50	56.4	56.4		
63	86.8	86.8	88.5	88.5
80	83.7	84.3		
100	79.7	81.8		
125	83.7	83.7	87.1	87.1
160	82.8	82.8		
200	76.5	80.3		
250	81.4	81.4	85.1	85.1
315	81.6	81.6		
400	76.3	79.1		
500	77.3	77.3	80.7	80.7
630	73.1	74.8		
800	74.4	74.4		
1,000	79.6	**79.6**	81.5	**81.5**
1,250	73.4	**77.1**		
1,600	82.6	82.6		
2,000	80.1	**80.1**	87.9	**87.9**
2,500	85.3	**85.3**		
3,150	83.7	83.7		
4,000	85.7	85.7	90.9	90.9
5,000	88.0	88.0		
6,300	74.2	85.5		
8,000	77.3	83.0	79.1	83.4
10,000	58.7	80.5		
12,500	67.4	78.0		
16,000	48.7	75.5	67.6	75.9
20,000	53.3	73.0		

Source: ISO (1986).
[a]Frequencies in boldface type are octave-band center frequencies.
[b]Thresholds in boldface type are the masked thresholds for the signal components of the backup alarm described in the text.

boldface type), the backup alarm is predicted to be audible.

Both the critical band and ISO methods are based on critical band theory and may be used to calculate masked thresholds with and without hearing protection devices (HPDs). Calculating a protected masked threshold for a particular signal requires (1) subtracting the attenuation of the HPD from the noise spectrum to obtain the noise spectrum effective when the HPD is worn; (2) calculation of a masked threshold for each signal component using the procedures outlined in the preceding discussion, which results in the signal-component levels that would be *just audible* to the listener when the HPD is worn; and (3) adding the attenuation of the HPD to the signal-component thresholds to provide an estimate of the environmental (exterior to the HPD) signal-component levels that would be required to produce the under-HPD threshold levels calculated in step 2. Although not difficult, this procedure does require a reasonably reliable estimate of the actual attenuation provided by the HPD. The manufacturer's data supplied with the HPD are unsuitable for this purpose because they overestimate the real-world performance of the HPD, as explained in Section 4.3.2. Furthermore, if a 1/3-octave band masking computation is desired, the manufacturer's attenuation data, which are available for only nine selected 1/3-octave bands, are insufficient for the computation. Finally, both methods fail to take the listener's hearing level into account. It is simply assumed that if the calculated masked thresholds are above the listeners' absolute thresholds, the signals should be audible.

Use of the ISO 7731 Standard for prediction of masked threshold for auditory signals is not limited to the octave and 1/3-octave calculations discussed herein, although the latter is the most precise method. As a much less accurate method (which is advocated by this author only as a last resort), ISO 7731 also offers a broadband analysis that can be performed by obtaining the dBA level of the ambient noise, and if the signal exceeds this level by 15 dB, it is said to be audible in most circumstances. However, this does not take into account upward masking or other spectrally specific effects, and it may result in (unnecessarily) higher signal levels than computed by either spectral technique. ISO 7731 also includes recommendations for signal temporal characteristics, unambiguous meaning, discriminability, inclusion of signal energy below 1500 Hz to accommodate high-frequency hearing loss, and addition of redundant visual signals if the ambient noise exceeds 100 dBA.

The broadband S/N recommendation of 15 dB of ISO 7731 is generally in keeping with those of auditory researchers. For example, Sorkin (1987) suggests that signal levels 6 to 10 dB above masked threshold are adequate to ensure 100% detectability, whereas signals which are approximately 15 dB above their masked threshold will elicit rapid operator response. He also suggests that signals more than 30 dB above the masked threshold could result in an unwanted startle response and that no signal should exceed 115 dB. [This suggested upper limit on signal level is consistent with OSHA hearing conservation requirements (OSHA, 1983), which prohibits exposure to continuous noise levels greater than 115 dBA.] These recommendations are in line with those of other authors (Deatherage, 1972; Wilkins and Martin, 1982).

Masked thresholds estimated via ISO 7731 are not necessarily exact, nor are they intended to be. They do provide conservative masked threshold estimates for a large segment of the population representing a wide range of hearing levels for nonspecific noise environments and signals. Further information on the design of auditory warnings and alarms, including relevant technical standards and guidelines, appears in Robinson and Casali (2003). Before embarking on the design of any auditory signal that is associated with a safety issue, the designer should first determine if there are any standards or regulations that have bearing. In this area of acoustics, the coverage of consensus standards is fairly broad and in depth.

7.3 Analysis of Speech Intelligibility in Noise

Many of the concepts presented above that relate to the masking of nonspeech signals by noise apply equally well to the masking of speech, so they will not be repeated in this section. However, for the spoken message, the concern is not simply audibility or detection, but rather, intelligibility. The listener must understand *what* was said, not simply know that something was said. Furthermore, speech is a very complex broadband signal whose components are not only differentially susceptible to noise, but are also highly dependent on vocal effort, the gender of the speaker, and the content and context of the message. In addition, other factors must be considered, such as the effects of HPD use by the speaker and/or listener, hearing loss of the listener, or speech signal degradation occurring in a communications system.

7.3.1 Speech-to-Noise Ratio Influence

Similar to the case with nonverbal signals, the signed difference between the speech level and the background noise level is referred to as the *speech-to-noise* (S/N) *ratio*. The speech level referred to is usually the long-term rms level measured in decibels. When background noise levels are between 35 and 110 dB, an S/N ratio of 12 dB is usually adequate to reach a normal-hearing person's threshold of intelligibility (Sanders and McCormick, 1993); however, it is quite impossible for anyone to sustain the vocal efforts required in the higher noise levels without electronic amplification (i.e., a public address system). The *threshold of intelligibility* is defined as the level at which the listener is just able to obtain without perceptible effort the meaning of almost every sentence and phrase of continuous speech (Hawkins and Stevens, 1950, p. 11); essentially, this is 100% intelligibility. Intelligibility decreases as S/N decreases, reaching 70 to 75% (as measured using phonetically balanced words) at an S/N of 5 dB, 45 to 50% at an S/N of 0 dB, and 25 to 30% at an S/N of −5 dB (Acton, 1970).

At least in low to moderate noise levels, people seem to modulate their vocal effort automatically, using the *Lombard reflex*, to maintain S/N ratios in increasing background noise so that they can communicate with other people. However, there is an upper limit to this ability, and speech levels cannot be maintained at more than 90 dB for long periods (Kryter, 1994). Since a relatively high S/N ratio (12 dB or so) is necessary for reliable speech communications in noise, it should be obvious that in high noise levels (greater than about 75 to 80 dB), unaided speech cannot be relied upon except for short durations over short distances. Furthermore, since speech levels for females tend to be about 2 to 7 dB less than for males, depending on vocal effort, the female voice is at a disadvantage in high levels of background noise.

7.3.2 Speech Bandwidth Influence

The speech bandwidth extends from 200 to 8000 Hz, with male voices generally having more energy than female voices at the low frequencies (Kryter, 1974); however, the region between 600 and 4000 Hz is most critical to intelligibility (Sanders and McCormick, 1993). This also happens to be the frequency range at which most auditory alarms are presented, providing an opportunity for the direct masking of speech by an alarm or warning. Therefore, speech communications in the vicinity of an activated alarm can be difficult.

Consonant sounds, which are generally higher than vowel sounds in the 600 to 4000 Hz bandwidth are also more critical than vowels to intelligibility. This fact renders speech differentially susceptible to masking by bandlimited noise, depending on the level of the noise. At low levels, bands of noise in the mid- to high-frequency ranges mask consonant sounds directly, thus impairing speech intelligibility more than would low-frequency sounds presented at similar levels. However, at high levels, low-frequency bands of noise can also adversely affect intelligibility due to upward spread of masking into the critical speech bandwidth.

When electronic transmission/amplification systems are used to overcome problems associated with speech intelligibility, it is important to understand that the systems themselves may exacerbate the problem if they are not designed properly. Most industrial telecommunications systems [i.e., intercoms, telephones, personal assistant (PA) systems] do not transmit the full speech bandwidth, nor do they reproduce the entire dynamic range of the human voice. To reduce costs and simplify the electronics, such systems often filter the signal and pass (transmit) only a portion of the speech bandwidth (e.g., the telephone passband is generally 300 to 3600 Hz). If the frequencies above 4000 Hz or the frequencies below 600 Hz are filtered out (not transmitted), there is little negative impact on speech intelligibility. However, if the frequencies between 1000 and 3000 Hz are filtered out of the signal, intelligibility is severely impaired (Sanders and McCormick, 1993).

In addition to filtering the speech signal, it is possible to clip the speech peaks so that the full dynamic range of a speaker's voice is not transmitted to a listener. This clipping may be intentional on the part of the designer to reduce the cost of the system, or it may be an artifact of the amplitude distortion caused by an overloaded amplifier. Either way, the effects on intelligibility are the same. Since the speech peaks contain primarily vowel sounds and intelligibility relies predominantly on the recognition of consonants, there is little loss in intelligibility due strictly to peak clipping. However, if the clipping is caused by distortion within the amplifier, there may be ancillary distortion of the speech signal in other ways that could affect intelligibility adversely.

7.3.3 Acoustic Environment Influence

The acoustic environment (room volume, distances, barriers, reverberation, etc.) can also have a dramatic effect on speech intelligibility. This is a complex subject in acoustics and a detailed treatment is beyond the scope of this chapter, but more information may be found in Kryter (1974, 1994). One fairly obvious point is that as the distance between the listener and the speech source (person or loudspeaker) increases, the ability to understand the speech can be affected adversely if the S/N ratio decreases sufficiently. In the same vein, barriers in the source–receiver path can create shadow zones in which the S/N ratio is insufficient for reliable intelligibility. Finally, speech intelligibility decreases linearly as reverberation time increases. Reverberation time (RT_{60}) is defined for a given space as the time (in seconds) required for a steady sound to decay by 60 dB from its original value after being shut off. Each 1-s increase in reverberation time will result in a loss of approximately 5% in intelligibility (Sanders and McCormick, 1993). Thus, rooms with long reverberation times, producing an echo effect, will not provide good conditions for speech reception.

7.3.4 Speech Intelligibility Analysis Method Based on the Preferred Speech Interference Level

There are a number of techniques to analyze, or in some cases, to predict accurately the intelligibility of a speech communications system based on empirical measurements of an incident noise and, in some cases, additional measurements of the system's speech output, be it amplified or live unamplified voice. A variety of techniques are covered in Sanders and McCormick (1993) and Kryter (1994). However, one of the better known techniques, the Preferred Speech Interference Level (PSIL), which involves only measurements of the noise and is straightforward to administer (although limited in its predictive ability), warrants discussion here.

The PSIL is the arithmetic average of the noise levels measured in three octave bands centered at 500, 1000, and 2000 Hz. It is most useful when the spectrum of the background noise is relatively flat, and intended only as an indication of whether or not there is likely to be a communications problem, not as a predictor of intelligibility. If the background noise is not flat, is predominated by or contains strong tonal components, or fluctuates a great deal, the utility of the PSIL is lessened. As an example of PSIL application, the hypothetical octave-band noise spectrum presented earlier in column (4) of Table 1 can be used. The PSIL for this spectrum is $(80.7 + 81.5 + 87.9)/3 = 83$. With this information, Figure 11 can be consulted to determine how difficult verbal communication is likely to be in this noise. At a PSIL of 83, verbal communications will be "difficult" at any speaker–listener distance greater than about 18 inches. Even at closer distances, a "raised" or "very loud" voice must be used. If octave-band levels are not available, the A-weighted sound level may also provide rough guidance concerning the speech-interfering effects of background noise, also shown in Figure 11. In summary, the PSIL is a useful, simple tool for estimating the degree of difficulty that can be expected when verbal communications are attempted in a steady, flat background noise.

7.3.5 Speech Intelligibility Analysis Method Based on the Speech Intelligibility Index

In contrast to the PSIL, a more precise analytical prediction of the interfering effects of noise on speech communications may be conducted using the Speech Intelligibility Index (SII) technique defined in ANSI S3.5-1997 (R2002) (ANSI, 2002a). Essentially, this well-known standardized technique utilizes a weighted sum of the S/N ratios in specified frequency bands to compute an SII score ranging between 0.0 and

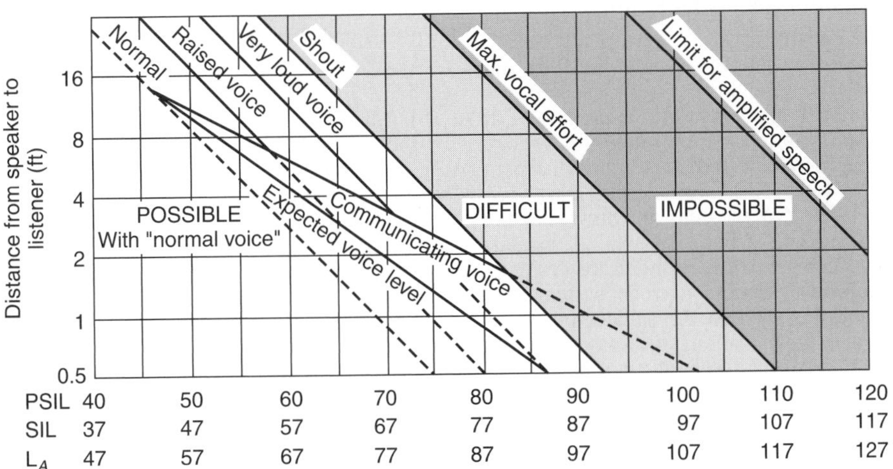

Figure 11 Relationship among PSIL, speech difficulty, vocal effort, and speaker–listener separation. (Adapted with permission from Sanders and McCormick, 1993.)

1.0, with higher scores indicative of greater predicted speech intelligibility. While the end result is an SII score on a simple scale or 0.0 to 1.0, the process of measurement and calculation is complex by comparison to the PSIL. However, the SII is much more accurate, broader in its coverage, and can account for many additional factors, such as speaker vocal effort, room reverberation, monaural versus binaural listening, hearing loss, varying message content, hearing protector effects, communications system gain, and the existence of external masking noise.

Four calculation methods are available with the SII: the critical band method (most accurate), the 1/3-octave-band method, the equally contributing critical band method, and the octave-band method (least accurate). At a minimum, the calculations require knowledge of the spectrum level of the speech and noise as well as the listeners' hearing thresholds. Where speech spectrum level(s) are unavailable or unknown, the standard offers guidance in their estimation. Although quite flexible in the number and types of conditions to which it can be applied, application of the standard is limited to natural speech, otologically normal listeners with no linguistic or cognitive deficiencies, and situations that do not include sharply filtered bands of speech or noise.

The SII "score" actually represents the proportion of the speech cues that would be available to the listener for "average speech" under the noise/speech conditions for which the calculations were performed. Hence, intelligibility is predicted to be greatest when the SII = 1.0, indicating that all of the speech cues are reaching the listener, and poorest when the SII = 0.0, indicating that none of the speech cues are reaching the listener. The general steps used in calculating the SII and estimating intelligibility are beyond the scope of this chapter, but they may be found in the standard itself, ANSI S3.5-1997 (R2002) (ANSI, 2002a), or in paraphrased terms with examples in Robinson and Casali (2003).

7.3.6 Speech Intelligibility Experimental Test Methods

In lieu of analytical techniques such as the PSIL and the SII, both of which require spectral measurements, an alternative (or complementary) approach is to conduct an experiment to measure intelligibility for a given set of conditions with a group of human listeners. For this purpose, there exists a standard, (ANSI S3.2-1989) (ANSI, 1989) that provides not only guidance for conducting such tests but also sets of standard speech stimuli. The standard is intended for designers and manufacturers of communications systems and provides valuable insight into the subject of speech intelligibility and how various factors associated with the speaker, transmission path/environment, and listener can affect it. Although space does not permit a detailed description of the procedures, the strategy involves presenting speech stimuli to a listener in an environment that replicates the conditions of concern and measuring how much of the speech message is understood. The speech stimuli may be produced by a trained talker speaking directly to the listener while in the same environment or via an intercom system. Alternatively, the materials may be recorded and presented electronically. Use of recorded stimuli and/or electronic presentation of the stimuli offers the greatest control over the speech levels presented to the listener.

7.4 Other Considerations for Signal Detectability and Speech Intelligibility

7.4.1 Distance Effects

It cannot be overemphasized that the noise and signal levels referred to in the analysis techniques above refer to the levels measured *at the listener's location*. Measurement made at some central location or the specified output levels of the alarm or warning devices are not representative of the levels present at a given workstation and cannot be used for masked threshold calculations. In a free-field, isotropic environment, the sound level of an alarm or warning will decrease in inverse relationship to the distance from the source, in accord with the formula

$$p_1/p_2 = d_2/d_1 \tag{27}$$

where p_1 and p_2 are the sound pressure of signal at distances d_2 and d_1, respectively, in μPa or dyn/cm^2, and d_1 and d_2 are, respectively, distance 1 (near point) and distance 2 (far point) at which signal is measured, in linear distance units; or, alternatively, where the drop between distance 1 and 2 in SPL of the signal in dB is given by

$$\text{SPL}_{\text{drop}} = 20\ \log_{10}(d_2/d_1) \tag{28}$$

These formulas provide accurate results in outdoor environments where there are no barriers, such as trees, or highly reflecting planes, such as paved parking lots. Indoors, the formulas will typically overestimate the drop in signal level, where reflective surfaces reinforce the signal propagation.

7.4.2 Barrier Effects

Furthermore, buildings or other large structures in the source–receiver path can create "shadow zones" in which little or no sound is audible. It is for these reasons that the U.S. Department of Defense (1981) recommends that frequencies below 1000 Hz be used for outdoor alarms since low frequencies are less susceptible to atmospheric absorption and diffract more readily around barriers. Similar problems can be encountered indoors as well. Problems associated with the general decrease in SPL with increasing distance as well as shadow zones created by walls, partitions, screens, and machinery/vehicles must be considered. Since different materials reflect and absorb sound depending on its frequency, not only do the sound levels change from position to position, but the spectra of both the noise and signals/speech can change as well. Finally, since most interior spaces reverberate to some degree, the designer should also be concerned with phase differences between reflected sounds, which can result in superposition effects of enhancement or cancellation of the signals and speech from location to location. It is for all these reasons that it is necessary to know the SPL at the *listener's location* when considering masked thresholds.

7.4.3 Hearing Protection Device Effects

HPDs are often blamed for exacerbating the effects of noise on the audibility of speech and signals; although, at least for people with normal hearing, protectors may actually facilitate hearing in some noisy situations. Overall, the research evidence on normal hearers generally suggests that conventional passive HPDs have little or no degrading effect on the wearer's understanding of external speech and signals in ambient noise levels above about 80 dBA, and may even yield some improvements, with a crossover between disadvantage and advantage between 80 and 90 dBA. However, HPDs do often cause increased misunderstanding and poorer detection (compared to unprotected conditions) in lower sound levels, where HPDs are not typically needed for hearing defense anyway, but may be applied for reduction of annoyance (Casali and Berger, 1996). In intermittent noise, HPDs may be worn during quiet periods so that when a loud noise occurs, the wearer will be protected. However, during those quiet periods, the conventional passive HPDs typically reduce hearing acuity. In certain of these cases, the family of *level-dependent augmented HPDs*, those that provide minimal or moderate attenuation (or alternatively, more amplification of external sounds) during quiet but increased attenuation (or less amplification) as noise increases can be beneficial (Casali and Berger, 1996).

Noise- and age-induced hearing losses generally occur in the high-frequency regions first, and for those so impaired, the effects of HPDs on speech perception and signal detection are not clear-cut. Due to their already elevated thresholds for mid- to high-frequency speech sounds being raised further by the protector, hearing-impaired persons are usually disadvantaged in their hearing by conventional HPDs. Although there is no consensus across studies, certain reviews have concluded that sufficiently hearing-impaired persons will usually experience additional reductions in communications abilities with conventional HPDs worn in noise. In some instances, HPDs with electronic hearing-assistive circuits, sometimes called *electronic sound-transmission or sound restoration HPDs*, can be offered to hearing-impaired persons to determine if their hearing, especially in quiet to moderate noise levels below about 85 dBA, may be improved with such devices while still receiving a measure of protection (Casali and Berger, 1996).

Conventional passive HPDs cannot differentiate or selectively pass speech or nonverbal signal (or speech) energy vs. noise energy at a given frequency. Therefore, conventional HPDs do not improve the S/N ratio in a given frequency band, which is the most important factor for achieving reliable signal detection or intelligibility. Conventional HPDs attenuate high-frequency sound more than low-frequency sound, thereby attenuating the power of consonant sounds that are important for word discrimination, as well as most warning signals, both of which lie in the higher frequency range, while also allowing low-frequency noise through. Thus, the HPD can enable an associated upward spread of masking to occur if the penetrating noise levels are high enough. Certain augmented HPD technologies help to overcome the weaknesses of conventional HPDs as to low-frequency attenuation in particular; these include a variety of *active noise reduction* (ANR) devices, which through electronic phase-derived cancellation of noises below about 1000 Hz improve the low-frequency attenuation of passive HPDs. Concomitant benefits of ANR-based HPDs can include the reduction of upward spread of masking of low-frequency noise into the speech and warning signal bandwidths, as well as reduction of noise annoyance in certain environments that are dominated by low frequencies, such as jet aircraft cockpits and passenger cabins (Casali et al., 2004).

7.4.4 Hearing-Aided Users

People with a hearing loss sufficient to require the use of hearing aids are already at a disadvantage when attempting to hear auditory alarms, warnings, or speech, and this disadvantage is exacerbated when noise levels are high. Activation of hearing aids in high levels of noise so as to improve hearing of speech or signals can increase the risk of additional damage to hearing due to amplification of the ambient noise (Humes and Bess, 1981). But shutting off the hearing aids increases the chance that the signals will be missed, and since it has been shown that vented hearing aid inserts do not function well as hearing protectors (Berger, 2003), there is still a risk of further hearing damage by doing so. Recommendations for accommodating hearing-aided users in the workplace appear in Robinson and Casali (2003).

7.5 Summary of Guidance for Reducing Effects of Noise on Signals and Speech

The following principles regarding masking effects on nonverbal signals and speech are offered as a summary for general guidance.

1. Due to *direct masking*, the greatest increase in masked threshold occurs for nonverbal signal frequencies that are equal or near the predominant frequencies of the masking noise. Therefore, warning signals should not utilize tonal frequencies equivalent to those of the masker. Preferably, the signal should contain energy in the most sensitive range of human hearing, approximately 1000 to 4000 Hz, unless the noise energy is intense at these frequencies.

2. If the signal and masker are tonal in nature, the primary masking effect is at the fundamental frequency of the masker and at its harmonics. For instance, if a masking noise has primary frequency content at 1000 Hz, this frequency and its harmonics

(2000, 3000, 4000, etc.) should be avoided as signal frequencies.

3. The greater the SPL of the masker, the more the increase in masked threshold of the signal. A general rule of thumb is that the S/N ratio at the listener's ear should at a minimum be about 15 dB above masked threshold for reliable signal detection. However, in noise levels above about 80 dBA, the signal levels required to maintain a S/N ratio of 15 dB above masked threshold may increase the hearing exposure risk, especially if signal presentation occurs frequently. Therefore, if lower S/N values become necessary, it is best to design contrasting signals which are unlike the masker in frequency and have modulated or alternating frequencies to grab attention.

4. Warning signals should not exceed the masked threshold by more than 30 dB, to avoid verbal communications interference and operator annoyance (Sorkin, 1987).

5. As the SPL of the masker increases, the primary change in the masking effect is that it spreads upward in frequency, often causing signal frequencies which are higher than the masker to be missed (i.e., *upward masking*). Since most warning signal guidelines recommend that midrange and high-frequency signals (about 1000 to 4000 Hz) be used for detectability, it is important to consider that the masking effects of noise of lower frequencies can spread upward and cause interference in this range. Therefore, if the noise has its most significant energy in this range, a lower-frequency signal, say 500 Hz, may be necessary. However, as shown in Figure 3*a*, it must be kept in mind that the ear is not as sensitive to low frequencies, so the signal level must be set carefully to ensure reliable audibility.

6. Masking effects can also spread downward in frequency, causing signal frequencies below those of the masker to be raised in threshold (i.e., *remote masking*). The effect is most prominent at signal frequencies that are subharmonics of the masker. With typical industrial noise sources, remote masking is generally less of a problem than direct or upward masking.

7. When a signal must be localized, it is advantageous to include signal energy content below 1000 Hz and above 3000 Hz to maximize one's ability to locate the signal.

8. In extremely loud environments of about 110 dB and above, nonauditory signal channels such as visual and vibro-tactile should be considered as alternatives to auditory displays. They should also be used for redundancy in some lower-level noises where the auditory signal may be overlooked or it blends in as the background noise varies, and also where people who have hearing loss must attend to the signal.

9. Speech intelligibility in noise depends on a combination of complex factors and, as such, predictions based on simple S/N ratios should not be relied on. However, in very general terms, S/N ratios of 15 dB or higher should result in intelligibility performance above about 80% words correct for normal-hearing persons in broadband noise (Acton, 1970). Above speech levels of about 85 dBA, there is some decline in intelligibility even if the S/N ratio is held constant (Pollack, 1958). In very high noise levels, it is impractical and may pose additional hearing hazard risk to amplify the voice to maintain the high S/N ratios necessary for good intelligibility performance. The S/N ratio required for reliable intelligibility may be reduced via the use of certain techniques, such as reduction of speaker-to-listener distances, use of smaller vocabularies, provision of contextual cues in the message, use of the phonetic alphabet, and use of noise-attenuating headphones and noise-canceling microphones in electronic systems.

10. Electronic speech communications systems should reproduce speech frequencies accurately in the range 500 to 5000 Hz, which encompasses the most sensitive range of hearing and includes the speech sounds important for message understandability. More specifically, because much of the information required for word discrimination lies in the consonants, which are in the higher end of the frequency range and of low power (while the power of the vowels is in the peaks of the speech waveform), the use of electronic peak clipping and reamplification of the waveform may improve intelligibility because the power of the consonants is thereby boosted relative to the vowels. Furthermore, to maintain intelligibility it is critical that frequencies in the region 1000 to 4000 Hz be faithfully reproduced in electronic communication systems. Filtering out of frequencies outside this range will not appreciably affect word intelligibility but will influence the quality of the speech.

11. Actual human speech typically results in higher intelligibility in noise than that of computer-generated speech, and there are also differences among synthesizers as to their intelligibility. Especially for critical message displays and annunciators, live, recorded, or digitized human speech may be preferable to synthesized speech (Morrison and Casali, 1994), and if synthesized speech is used, the selection of synthesizer must be made carefully (Lancaster, et al., 2004). Furthermore, there are differences in intelligibility between modern speech synthesizers, so the selection of the synthesizer is important (Lancaster et al., 2004).

REFERENCES

Acton, W. I. (1970), "Speech Intelligibility in a Background Noise and Noise-Induced Hearing Loss," *Ergonomics*, Vol. 13, No. 5, pp. 546–554.

ANSI (1974), "Method for the Measurement of Real-Ear Protection of Hearing Protectors and Physical Attenuation of Earmuffs," ANSI S3.19-1974, American National Standards Institute, New York.

ANSI (1989), "American National Standard Methods for Measuring the Intelligibility of Speech over Communications Systems," ANSI S3.2-1989, American National Standards Institute, New York.

ANSI (2001), "Specification for Sound Level Meters," ANSI S1.4-1983 (R2001), American National Standards Institute, New York.

ANSI (2002a), "Methods for the Calculation of the Speech Intelligibility Index," ANSI S3.5-1997 (R2002), American National Standards Institute, New York.

ANSI (2002b), "Methods for Measuring the Real-Ear Attenuation of Hearing Protectors," ANSI S12.6-1997 (R2002), American National Standards Institute, New York.

Berger, E. H. (2003), "Hearing Protection Devices," in *The Noise Manual*, rev. 5th ed., E. H. Berger, L. H. Royster, J. D. Royster, D. P. Driscoll, and M. Layne, Eds., American Industrial Hygiene Association, Fairfax, VA, pp. 379–454.

Berger, E. H., and Casali, J. G. (1997), "Hearing Protection Devices," in *Encyclopedia of Acoustics*, M. J. Crocker, Ed., Wiley, New York, pp. 967–981.

Berger, E. H., Royster, L. H., Royster, J. D., Driscoll, D. P., and Layne. M., Eds. (2003), *The Noise Manual*, rev. 5th ed., American Industrial Hygiene Association, Fairfax, VA.

Casali, J. G. (1999), "Litigating Community Noise Annoyance: A Human Factors Perspective," in *Proceedings of the 1999 Human Factors and Ergonomics Society 42nd Annual Conference*, Houston, TX, September 27–October 1, pp. 612–616.

Casali, J. G., and Berger, E. H. (1996), "Technology Advancements in Hearing Protection: Active Noise Reduction, Frequency/Amplitude-Sensitivity, and Uniform Attenuation," *American Industrial Hygiene Association Journal*, Vol. 57, pp. 175–185.

Casali, J. G., and Robinson, G. S. (1999), "Noise in Industry: Auditory Effects, Measurement, Regulations, and Management," in *Handbook of Occupational Ergonomics*, W. Karwowski and W. Marras, Eds., CRC Press, Boca Raton, FL, pp. 1661–1692.

Casali, J. G., Robinson, G. S., Dabney, E. C., and Gauger, D. (2004), "Effect of Electronic ANR and Conventional Hearing Protectors on Vehicle Backup Alarm Detection in Noise," *Human Factors*, Vol. 46, No. 1, pp. 1–10.

Crocker, M., Ed. (1998), *Handbook of Acoustics*, Wiley, New York.

Deatherage, B. H. (1972), "Auditory and Other Sensory Forms of Information Presentation," in *Human Engineering Guide to Equipment Design*, H. P. Van Cott and R. G. Kincade, Eds., Wiley, New York, pp. 123–160.

Driscoll, D. P., and Royster, L. H. (2003), "Noise Control Engineering," in *The Noise Manual*, rev. 5th ed., E. H. Berger, L. H. Royster, J. D. Royster, D. P. Driscoll, and M. Layne, Eds., American Industrial Hygiene Association, Fairfax, VA, pp. 279–378.

Driscoll, D. P., Stewart, N. D., and Anderson, R. R. (2003), "Community Noise," in *The Noise Manual*, rev. 5th ed., E. H. Berger, L. H. Royster, J. D. Royster, D. P. Driscoll, and M. Layne, Eds., American Industrial Hygiene Association, Fairfax, VA, pp. 601–637.

Earshen, J. J. (1986), "Sound Measurement: Instrumentation and Noise Descriptors," in *Noise and Hearing Conservation Manual*, E. H. Berger, W. D. Ward, J. C. Morrill, and L. H. Royster, Eds., American Industrial Hygiene Association, Akron, OH, pp. 38–95.

Egan, J. P., and Hake, H. W. (1950), "On the Masking Pattern of a Simple Auditory Stimulus," *Journal of the Acoustical Society of America*, Vol. 22, No. 5, pp. 622–630.

EPA (1979), 40CFR211, "Noise Labeling Requirements for Hearing Protectors," *Federal Register*, Vol. 44, No. 190, pp. 56130–56147.

EPA (1981), "Noise in America: The Extent of the Noise Problem," Report 550/9-81-101, U.S. Environmental Protection Agency, Washington, DC.

Fidell, S. M., and Pearsons, K. S. (1997), "Community Response to Environmental Noise," in *Encyclopedia of Acoustics*, M. Crocker, Ed., Wiley, New York, pp. 1083–1091.

Fletcher, H. (1940), "Auditory Patterns," *Reviews of Modern Physics*, Vol. 12, pp. 47–65.

Harris, C. M., Ed. (1991), *Handbook of Acoustical Measurements and Noise Control*, McGraw-Hill, New York.

Hawkins, J. E., and Stevens, S. S. (1950), "The Masking of Pure Tones and of Speech by White Noise," *Journal of the Acoustical Society of America*, Vol. 22, No. 1, pp. 6–13.

Humes, L. E., and Bess, F. H. (1981), "Tutorial on the Potential Deterioration in Hearing Due to Hearing Aid Usage," *Journal of Speech and Hearing Research*, Vol. 24, No. 1, pp. 3–15.

ISO (1986), "Danger Signals for Work Places—Auditory Danger Signals," ISO 7731-1986(E), International Organization for Standardization, Geneva.

Kryter, K. D. (1974), "Speech Communication," in *Human Engineering Guide to Equipment Design*, H. P. Van Cott and R. G. Kincade, Eds., Wiley, New York, pp. 161–226.

Kryter, K. D. (1994), *The Handbook of Hearing and the Effects of Noise*, Academic Press, New York.

Lancaster, J. A., Robinson, G. A., and Casali, J. G. (2004), "Comparison of Two Voice Synthesis Systems as to Speech Intelligibility in Aircraft Cockpit Noise," in *Proceedings of the Human Factors and Ergonomics Society 48th Annual Meeting*, New Orleans, LA, September 20–24, pp. 127–131.

Melnick, W. (1991), "Hearing Loss from Noise Exposure," in *Handbook of Acoustical Measurements and Noise Control*, C. M. Harris, Ed., McGraw-Hill, New York, pp. 18.1–18.19.

Morrison, H. B., and Casali, J. G. (1994), "Intelligibility of Synthesized Voice Messages in Commercial Truck Cab Noise for Normal-Hearing and Hearing-Impaired Listeners," in *Proceedings of the 1994 Human Factors and Ergonomics Society 38th Annual Conference*, Nashville, TN, October 24–28, pp. 801–805.

NIH (National Institutes of Health) Consensus Development Panel (1990), "Noise and Hearing Loss," *Journal of the American Medical Association*, Vol. 263, No. 23, pp. 3185–3190.

NIOSH (1975), "List of Personal Hearing Protectors and Attenuation Data," Publication 76-120, National Institute for Occupational Safety and Health–HEW, Washington, DC, pp. 21–37.

OSHA (1980), "Noise Control: A Guide for Workers and Employers," OSHA 3048, Occupational Safety and Health Administration, U.S. Department of Labor, Washington, DC.

OSHA (1983), "Occupational Noise Exposure; Hearing Conservation Amendment; Final Rule," 29CFR1910.95, Occupational Safety and Health Administration, *Code of Federal Regulations*, Title 29, Chapter XVII, Part 1910, Subpart G, 48 FR 9776-9785, Federal Register, Washington, DC.

Ostergaard, P. (2003), "Physics of Sound and Vibration," in *The Noise Manual*, rev 5th ed., E. H. Berger, L. H. Royster, J. D. Royster, D. P. Driscoll, and M.

Layne, Eds., American Industrial Hygiene Association, Fairfax, VA, pp. 19–39.

Park, M. Y., and Casali, J. G. (1991), "A Controlled Investigation of In-Field Attenuation Performance of Selected Insert, Earmuff, and Canal Cap Hearing Protectors," *Human Factors*, Vol. 33, No. 6, pp. 693–714.

Patterson, R. D. (1982), "Guidelines for Auditory Warning Systems on Civil Aircraft," Paper 82017, Civil Aviation Authority, Airworthiness Division, Cheltenham, Gloucestershire, England.

Peterson, A. P. G. (1979), "Noise Measurements: Instruments," in *Handbook of Noise Control*, C. M. Harris, Ed., McGraw-Hill, New York, pp. 5-1 to 5-19.

Pollack, I. (1958), "Speech Intelligibility at High Noise Levels: Effects of Short-Term Exposure," *Journal of the Acoustical Society of America*, Vol. 30, pp. 282–285; ANSI S3.19-1974.

Poulton, E. (1978), "A New Look at the Effects of Noise: A Rejoinder," *Psychological Bulletin*, Vol. 85, pp. 1068–1079.

Robinson, G. S., and Casali, J. G. (2003), "Speech Communications and Signal Detection in Noise," in *The Noise Manual*, rev. 5th ed., E. H. Berger, L. H. Royster, J. D. Royster, D. P. Driscoll, and M. Layne, Eds., American Industrial Hygiene Association, Fairfax, VA, pp. 567–600.

Rossing, T. (1990), *The Science of Sound*, Addison-Wesley, Reading, MA.

Sanders, M. S., and McCormick, E. J. (1993), *Human Factors in Engineering and Design*, 7th ed., McGraw-Hill, New York.

Sorkin, R. D. (1987), "Design of Auditory and Tactile Displays," in *Handbook of Human Factors*, G. Salvendy, Ed., McGraw-Hill, New York, pp. 549–576.

Stevens, S. S. (1936), "A Scale for the Measurement of a Psychological Magnitude: Loudness," *Psychological Review*, Vol. 43, pp. 405–416.

Stevens, S. S. (1972), "Perceived Level of Noise by Mark VII and Decibels (E)," *Journal of the Acoustical Society of America*, Vol. 51, No. 2, Pt. 2, pp. 575–601.

Taylor, W., Pearson, J., Mair, A., and Burns, W. (1964), "Study of Noise and Hearing in Jute Weavers," *Journal of the Acoustical Society of America*, Vol. 38, pp. 113–120.

U.S. Department of Defense, (1981), *Human Engineering Design Criteria for Military Systems, Equipment and Facilities*, MIL-STD-1472C, DoD, Washington, DC.

Wegel, R. L., and Lane, C. E. (1924), "The Auditory Masking of One Pure Tone by Another and Its Probable Relation to the Dynamics of the Inner Ear," *Physiological Review*, Vol. 23, p. 266.

Wilkins, P. A., and Martin, A. M. (1982), "The Effects of Hearing Protection on the Perception of Warning Sounds," in *Personal Hearing Protection in Industry*, P. W. Alberti, Ed., Raven Press, New York, pp. 339–369.

Wilkins, P. A., and Martin, A. M. (1985), "The Role of Acoustical Characteristics in the Perception of Warning Sounds and the Effects of Wearing Hearing Protection," *Journal of Sound and Vibration*, Vol. 100, No. 2, pp. 181–190.

Zwicker, E. (1960), "En Verfahren zur Berechnung der Lautstärke," *Acustica*, Vol. 10, pp. 304–308.

CHAPTER 25

ILLUMINATION

Peter R. Boyce
Consultant
Canterbury, Great Britain

1 INTRODUCTION

Illumination is the act of placing light on an object. By providing illumination, stimuli for the human visual system are produced and the sense of sight is allowed to function. With light, we can see; without light, we cannot see. This chapter is devoted to describing how to measure and produce illumination, the effects of different lighting conditions on visual performance and visual comfort, the photobiological and psychological effects of illumination and the risks inherent in exposure to light.

2 MEASUREMENT OF ILLUMINATION

2.1 Photometric Quantities

Light is a part of the electromagnetic spectrum lying between the wavelength limits 380 to 780 nm. What separates this wavelength region from the rest is that radiation in this region is absorbed by the photoreceptors of the human visual system, which initiates the process of seeing. The most fundamental measure of the electromagnetic radiation emitted by a source is its *radiant flux*. This is a measure of the rate of flow of energy emitted and is measured in watts. The most fundamental quantity used to measure light is *luminous flux*. Luminous flux is radiant flux multiplied by the relative spectral sensitivity of the human visual system over the wavelength range 380 to 780 nm.

The relative spectral sensitivity of the human visual system is based on the perception of brightness associated with each wavelength. In fact, there are two different relative spectral sensitivities, sanctified by international agreement arranged through the Commission Internationale de l'Eclairage (CIE, 1983, 1990). There are two relative spectral sensitivities because the human visual system has two classes of photoreceptor: *cones*, which operate primarily when light is plentiful, and *rods*, which operate when light is very limited. These two photoreceptor types have different spectral sensitivities: the day photoreceptor, the cones, being characterized by the CIE standard photopic observer, and the night photoreceptor, the rods, being characterized by the CIE standard scotopic observer (Figure 1).

Luminous flux is used to quantify the total light output of a light source in all directions. Although this is important, for lighting practice it is also important to be able to quantify the luminous flux emitted in a given direction. The measure that quantifies this concept is *luminous intensity*. Luminous intensity is the luminous flux emitted per unit solid angle, in a specified direction. The unit of measurement is the candela, which is equivalent to a lumen per steradian. Luminous intensity is used to quantify the distribution of light from a luminaire.

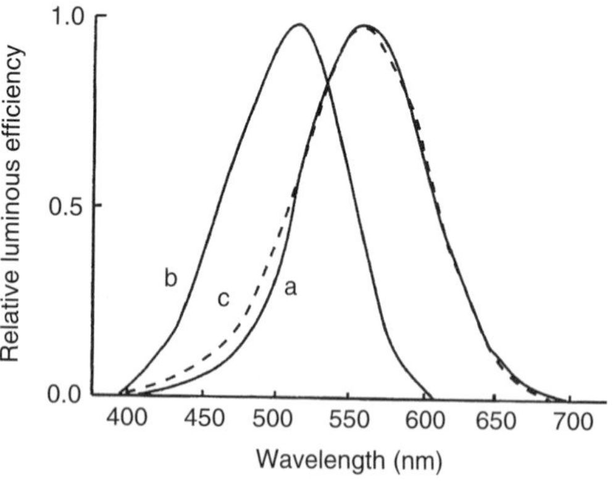

Figure 1 Relative luminous efficiency functions for (curve a) the CIE standard photopic observer and (curve b) the CIE standard scotopic observer. The CIE standard photopic observer is based on a 2° field of view. Also shown (curve c) is the relative luminous efficiency function for a 10° field of view in photopic conditions.

Both luminous flux and luminous intensity have associated area measures. The luminous flux falling on a unit area of surface is called the *illuminance*. The unit of measurement of illuminance is the lumen per square meter or lux. The luminous intensity emitted per unit projected area in a given direction is the *luminance*. The unit of measurement of luminance is the candela per square meter. The illuminance incident on a surface is the most widely used electric lighting design criterion. The luminance of a surface is a correlate of its brightness. Table 1 summarizes these photometric quantities and the relationship between illuminance and luminance.

Unfortunately for consistency, photometry has a long history that has generated a number of different units of measurement for illuminance and luminance. Table 2 lists some of the alternative units, together with the multiplying factors necessary to convert from the alternative unit to the Systém International (SI) units of lumens per square meter for illuminance and candela per square meter for luminance. SI units will be used throughout this chapter. Table 3 shows some illuminances and luminances typical of commonly occurring situations.

2.2 Colorimetric Quantities

The photometric quantities described above do not take into account the wavelength combination (i.e., the color) of the light being measured. There are two approaches to characterizing color, the color atlas and the CIE colorimetry system.

2.2.1 Color Atlases

The color atlas, as its name implies, is a physical, three-dimensional representation of color space. It is three-dimensional because colors have three separate subjective attributes: hue, brightness, and strength. Hue tells us whether the color is primarily red or yellow or green or blue. Brightness tells us to what extent the color transmits or reflects light. Strength tells us whether the color is strong or weak. Several different color atlas systems are used in different parts of the world (Wyszecki and Stiles, 1982). Probably the most widely used atlas is the *Munsell Book of Color* available from the Munsell Color Company. Figure 2 shows the three-dimensional color space of the Munsell atlas. The position of any color is identified by an alphanumeric code made up of three terms: hue, value, and chroma (e.g., a strong red is given the alphanumeric 7.5R/4/12). Hue, value, and chroma are related to the three attributes of color: hue, brightness, and strength, respectively. Building materials, such as paints, plastic, and ceramics, are commonly classified in terms of a color atlas.

2.2.2 CIE Colorimetric System

Sometimes, it is necessary to quantify the color of a light or a surface before either exists. To meet this

Table 1 Photometric Quantities

Quantity	Definition	Units
Luminous flux	That quantity of radiant flux which expresses its capacity to produce visual sensation	lumen (lm)
Luminous intensity	The luminous flux emitted in a very narrow cone containing the given direction divided by the solid angle of the cone (i.e., luminous flux/unit solid angle)	candela (cd)
Illuminance	The luminous flux/unit area at a point on a surface	lumen meter^{-2} (lm m^{-2})
Luminance	The luminous flux emitted in a given direction divided by the product of the projected area of the source element perpendicular to the direction and the solid angle containing that direction, i.e. luminous flux/unit solid angle/unit area	candela meter^{-2} (cd m^{-2})
Reflectance	The ratio of the luminous flux reflected from a surface to the luminous flux incident on it:	
For a matte surface	$\text{luminance} = \dfrac{\text{illuminance} \times \text{reflectance}}{\pi}$	
Luminance factor	The ratio of the luminance of a reflecting surface, viewed in a given direction to that of a perfect white uniform diffusing surface identically illuminated:	
For a nonmatte surface for a specific viewing direction and lighting geometry	$\text{luminance} = \dfrac{\text{illuminance} \times \text{luminance factor}}{\pi}$	

Table 2 Common Photometric Units of Measurement for Illuminance and Luminance and the Factors Necessary to Change Them to SI Units

Quantity	Unit	Dimensions	Multiplying Factor to Convert to SI Unit
Illuminance (SI unit = lumen meter^{-2})	lux	lumen meter^{-2}	1.00
	meter candle	lumen meter^{-2}	1.00
	phot	lumen centimeter^{-2}	10,000.00
	foot candle	lumen foot^{-2}	10.76
Luminance (SI unit = candela meter^{-2})	nit	candela meter^{-2}	1.00
	stilb	candela centimeter^{-2}	10,000.00
	—	candela inch^{-2}	1,550.00
	—	candela foot^{-2}	10.76
	apostilb[a]	lumen meter^{-2}	0.32
	blondel[a]	lumen meter^{-2}	0.32
	lambert[a]	lumen centimeter^{-2}	3,183.00
	foot-lambert[a]	lumen foot^{-2}	3.43

[a]These four items are based on an alternative definition of luminance. This definition is that if the surface can be considered as perfectly matte, its luminance in any direction is the product of the illuminance on the surface and its reflectance. Thus, the luminance is described in lumens per unit area. This definition is deprecated in the SI system.

Table 3 Typical Illuminance and Luminance Values

Situation	Illuminance on Horizontal Surface (lm m^{-2})	Typical Surface	Luminance (cd m^{-2})
Clear sky in summer in northern temperate zones	150000	Grass	2900
Overcast sky in summer in northern temperate zones	16000	Grass	300
Textile inspection	1500	Light gray cloth	140
Office work	500	White paper	120
Heavy engineering	300	Steel	20
Good street lighting	10	Concrete road surface	1.0
Moonlight	0.5	Asphalt road surface	0.01

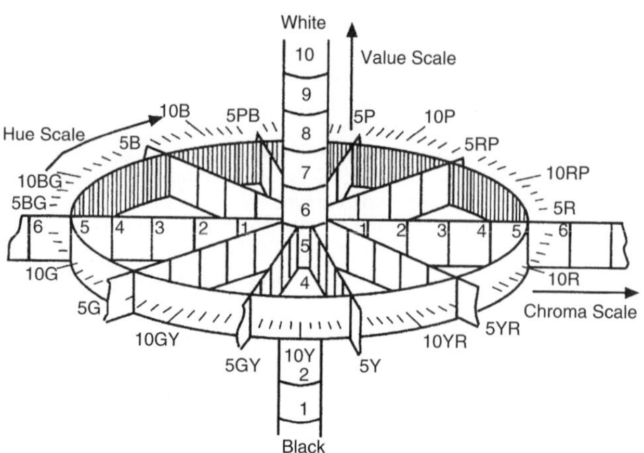

Figure 2 Organization of the Munsell color system. The hue letters are B, blue; PB, purple/blue; P, purple; RP, red/purple; R, red; YR, yellow/red; Y, yellow; GY, green/yellow; G, green; BG, blue/green.

need and to provide a more accurate characterization of color, the CIE has developed a system of colorimetry ranging from the complex to the relatively simple (CIE, 1971, 1972, 1978, 1986, 1995, 1998a). The most fundamental characteristic of light is its spectral power distribution reaching the eye. It is this spectral power distribution that largely determines the color seen. Unfortunately, comparisons between spectral power distributions are difficult to comprehend. The CIE has developed two three-dimensional color spaces, both based on mathematical manipulations applied to spectral power distributions (Robertson, 1977; CIE, 1978). These two three-dimensional color spaces, L_{ab} and L_{uv}, are the most comprehensive means of quantifying color, the L_{ab} space being used mainly for object colors and the L_{uv} space being used mainly for self-luminous colors. If two colors have the same coordinates in one of these color spaces, under the same observing conditions they will appear the same. The distance two colors are apart in color space is related to how easily they can be distinguished.

An earlier CIE color space, the 1964 Uniform Color Space, is used in the calculation of the CIE General Color Rendering Index, a single-number index which is applied to light sources to indicate how accurately they render colors relative to some standard (CIE, 1995). Specifically, the positions in color space of eight test colors, under a reference light source and under the light source of interest, are calculated. The separation between the two positions of each test color are calculated, and the separations for all the test colors are summed and scaled to give a value of 100 when there is no separation for any of the test colors (i.e., for perfect color rendering). It should be noted that this is a very crude system. Different light sources have different reference light sources, and the summation means that light sources that render the test colors differently can have the same Color Rendering Index. Nonetheless, the Color Rendering Index is widely used as a means of classifying the color-rendering capabilities of light sources.

Three two-dimensional color surfaces are still widely used to characterize the color appearance of light sources and to define the acceptable color characteristics of light signals (CIE, 1994). The most commonly used color surface is the CIE 1931 chromaticity diagram (CIE, 1971) (Figure 3). Essentially, it is a slice through color space at a fixed luminance. The curved boundary of the chromaticity diagram consists of the colors produced by single wavelengths. The equal energy point in the center of the diagram corresponds to a colorless surface. The farther the coordinates of a color are from the equal energy point and the closer they are to the boundary, the greater the strength of the color. Figure 3 also shows several areas in which a signal light needs to fall if it is to be perceived as the color specified. The color appearance of light sources is conventionally described by their correlated color temperature. This is the temperature of the full radiator that is closest to the coordinates of the light source on the CIE 1931 chromaticity diagram (Wyszecki and Stiles, 1982).

The two other two-dimensional chromaticity diagrams are the CIE 1960 and the CIE 1976 Uniform Chromaticity Scale diagrams. These are linear transformations of the CIE 1931 chromaticity diagram intended to make the surface more perceptually uniform. Whenever chromaticity coordinates are quoted, care should be taken to state the chromaticity diagram being used. A useful summary of these colorimetry systems is given in the 9th edition of the *Lighting Handbook* of the Illuminating Engineering Society of North America (IESNA, 2000).

2.3 Instrumentation

The instrumentation for measuring photometric and colorimetric quantities can be divided into laboratory and field equipment. Laboratory equipment tends to be

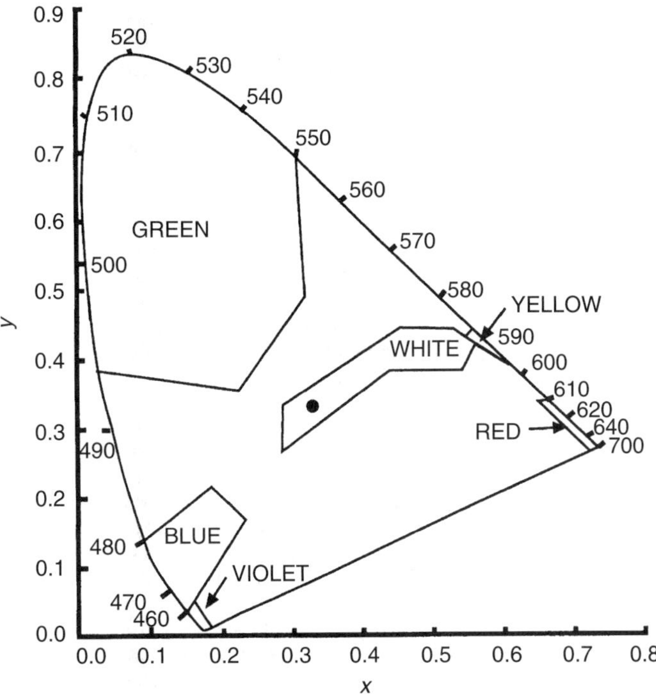

Figure 3 CIE 1931 chromaticity diagram. The boundary curve is the spectrum locus with the wavelengths (nm) marked. The filled circle is the equal energy point. The enclosed areas indicate the chromaticity coordinates of light signals that will be identified as the specified colors.

large and/or sophisticated and hence expensive. Field equipment is small and portable. The luminous flux from a light source, the luminous intensity distribution of a luminaire, and light source color properties are measured conventionally in the laboratory.

The two most commonly occurring field instruments are the illuminance meter and the luminance meter. Illuminance meters have three important characteristics: sensitivity, color correction, and cosine correction. *Sensitivity* refers to the range of illuminances covered, the range desired being dependent on whether the instrument is to be used to measure daylight, interior lighting, or nighttime exterior lighting. *Color correction* means that the illuminance meter has a spectral sensitivity matching the CIE standard photopic observer. *Cosine correction* means that the illuminance meter's response to light striking it from directions other than the normal follows a cosine law.

The luminance meter is designed to measure the average luminance over a specified area. The luminance meter has an optical system that focuses an image on a detector. Looking through the optical system allows the operator to identify the area being measured and usually displays the luminance of the area. The important characteristics of a luminance meter are its spectral response, its sensitivity, and the quality of its optical system. Again a good luminance meter has a spectral response matching the CIE standard photopic observer. The sensitivity needed depends on the conditions under which it will be used. The quality of its optical system can be measured by its sensitivity to light from outside the measurement area (CIE, 1987).

Recently, imaging photometers have become more widely available (Rea and Jeffrey, 1990; Ashdown and Franck, 1995). These instruments are based around a digitized image captured from a video camera. Such instruments are expensive but do provide a means for measuring the luminance of detailed or rapidly changing scenes. Procedures for using illuminance or luminance meters in the field and for light measurements in the laboratory are described and referenced in the guidance published by national bodies (IESNA, 2000; CIBSE, 2002). It should be noted that virtually all commercial instrumentation used to measure illuminance and luminance uses the CIE standard photopic observer as the basis of the instrument's spectral sensitivity, even when the instrument is designed to be used in mesopic and scotopic conditions.

3 PRODUCTION OF ILLUMINATION

Illumination is produced naturally, by the sun and artificially, by electric light sources. The development and growth in use of electric light sources over the last century has fundamentally changed the pattern of life for everyone.

3.1 Daylight, Sunlight, and Skylight

Natural light is light received on Earth from the sun, either directly or after reflection from the moon. The prime characteristic of natural light is its variability. Natural light varies in magnitude, spectral content, and distribution with different meteorological conditions, at different times of day and year, at different latitudes. Moonlight is of little interest as a source of illumination but daylight is used, and strongly desired, for the lighting of buildings. Daylight can be divided into two components, sunlight and skylight. Sunlight is light received at Earth's surface, directly from the sun. Sunlight produces strong, sharp-edged shadows. Skylight is light from the sun received at Earth's surface after scattering in the atmosphere. Skylight produces only weak, diffuse shadows. The balance between sunlight and skylight is determined by the nature of the atmosphere and the distance that the light passes through it. The greater the amount of water vapor and the longer the distance, the higher is the proportion of skylight.

The illuminances on Earth's surface produced by daylight can cover a large range, from 150,000 lx on a sunny summer's day to 1000 lx on a heavily overcast day in winter. Several models exist for predicting the daylight incident on a plane, at different locations, for different atmospheric conditions (Robbins, 1986). These models can be used to predict the contribution of daylight to the lighting of interiors. The spectral composition of daylight also varies with the nature of the atmosphere and the path length through it. The correlated color temperature of daylight can vary from 4000 K for an overcast day to 40,000 K for a clear blue sky. For calculating the appearance of objects under natural light, the CIE recommends the use of one of three different spectral distributions corresponding to correlated color temperatures of 5503, 6504, and 7504 K (Wyszecki and Stiles, 1982).

3.2 Electric Light Sources

The lighting industry makes several thousand different types of electric lamps. Those used for providing illumination can be divided into two classes: incandescent lamps and discharge lamps. Incandescent lamps produce light by heating a filament. Discharge lamps produce light by an electric discharge in a gas. Incandescent lamps operate directly from mains electricity. Discharge lamps all require control gear between the lamp and the electricity supply, because different electrical conditions are required to initiate the discharge and to sustain it.

Electric light sources can be characterized on several different dimensions. They are:

- *Luminous efficacy*: the ratio of luminous flux produced to power supplied (lumens watt^{-1}). If the lamp needs control gear, the watts supplied should include the power demand of the control gear.
- *Correlated color temperature*: a measure of the color appearance of the light produced, measured in degrees Kelvin (see Section 2.2.2).
- *CIE general color rendering index*: a measure of the ability to render colors accurately (see Section 2.2.2).
- *Lamp life*: the number of burning hours until either lamp failure or a stated percentage reduction in light output occurs. Lamp life can vary widely with switching cycle.
- *Run-up time*: the time from switch-on to full light output.
- *Restrike time*: the time delay between the lamp being switched off before it will reignite.

Table 4 summarizes these characteristics for two incandescent lamp types and six discharge lamp types that are widely used and gives the most common applications for each lamp type. The values in Table 4 should be treated as indicative only. Details about the characteristics of any specific lamp should always be obtained from the manufacturer. Many of the lamp types described in the table are also used for internally and externally illuminated signs and signals, but there are other lamp types, operating on different principles, that are used for this purpose (Boyce, 2003). Of these, the one of most interest is the light-emitting diode (LED). The LED has become the lamp type of choice for traffic signals and exit signs. The LED is a semiconductor that emits light when a current is passed through it. The spectral emission of the LED depends on the materials used to form the semiconductor, and typically is narrowband, giving a highly saturated color. LEDs have been developed to produce white light, either by combining red, green, and blue LEDs or by attaching a phosphor to a blue LED. White LEDs currently have a luminous efficacy comparable with an incandescent lamp, but a much longer life and greater durability. They are being used as sources of local illumination (e.g., aircraft reading lights). Considerable time and money is being devoted to enhancing the luminous efficacy and reducing the cost of white LEDs so they are competitive with other lamp types for general illumination.

3.3 Control of Light Distribution

Being able to produce light is only part of what is necessary to produce illumination. The other part is to control the distribution of light from the light source. For daylight, this is done by means of window shape, placement, and glass transmittance (Robbins, 1986). For electric light sources, it is done by placing the light source in a luminaire. The luminaire provides electrical and mechanical support for the light source and controls the light distribution. The light distribution is controlled by using reflection, refraction, or diffusion, individually or in combination (Simons and Bean, 2000). One factor in the choice of which method of light control to adopt in a luminaire is the balance desired between the reduction in the luminance of the

Table 4 Properties of Some Widely Used Electric Light Sources

Source	Luminous Efficacy (lm/W)	Correlated Color Temperature (K)	CIE General Color Rendering Index	Lamp Life (hr)	Run-up Time (min)	Restrike Time (min)	Applications
Incandescent							
Tungsten	8–19	2700	100	750–2000	Instant	Instant	Residential, retail
Tungsten-halogen	8–20	2900	100	2000–6000	Instant	Instant	Display
Discharge							
Low-pressure mercury (fluorescent lamp)	60–110	3000–5000	50–95	9000–20,000	10	Instant	Commercial
Compact fluorescent lamp	50–70	2700–4100	80–85	9000–20,000	10	Instant	Commercial, retail
High-pressure mercury (vapor)	30–60	3200–7000	15–50	16,000–24,000	4	3–10	Older industrial agricultural
High-pressure mercury (metal halide)	50–110	3000–6500	65–95	3000–20,000	6	5–20	Industrial, commercial, retail
Low-pressure sodium	100–180	1800	n/a	16,000–18,000	4–6	1	Security, road
High-pressure sodium	60–140	2100–2500	20–70	10,000–24,000	10–12	0–1	Industrial, road

light source and the precision required in light distribution. Highly specular reflectors can provide precise control of light distribution, but do little to reduce source luminance. Conversely, diffusers make precise control of light distribution impossible but do reduce the luminance of the luminaire. Refractors are an intermediate case. The light distribution provided by a specific luminaire is quantified by the luminous intensity distribution. All reputable luminaire manufacturers provide luminous intensity distributions for their luminaires. With luminaires, you tend to get what you pay for. Luminaires, well constructed from quality materials, cost more.

3.4 Control of Light Output

The control of daylight admitted through a window is achieved by mechanical structures, such as light shelves, or by adjustable blinds (Littlefair, 1990). Whenever the sun, or a very bright sky, is likely to be directly visible through a widow, some form of blind will be required. Blinds can take various forms; horizontal, venetian, vertical, and roller being the most common. Blinds can also be manually operated or motorized, either under manual control or under photocell control. Probably the most important feature to consider when selecting a blind is the extent to which it preserves a view of the outside. Roller blinds that can be drawn down to a position where the sun and/or sky is hidden but the lower part of the widow is still open are an attractive option. Roller blinds made of a mesh material can preserve a view through the whole window while reducing the luminance of the view out. Such blinds are an attractive option where the problem is an overbright sky but will be of limited value when a direct view of the sun is the problem. The same applies to low-transmission glass.

For electric light sources, control of light output is provided by switching or dimming systems. Switching systems can vary from the conventional manual switch to sophisticated daylight control systems that dim lamps near windows when there is sufficient daylight. Time switches are used to switch off all or parts of a lighting installation at the end of the working day. Occupancy sensors are used to switch off lighting when there is nobody in the space. Such switching systems can reduce electricity waste but will be irritating if they switch lighting off when it is required, and they may shorten lamp life if switching occurs frequently. The factors to be considered when selecting a switching system are whether to rely on a manual or an automatic system, and if it is automatic, how to match the switching to the activities in the space. If your interest is primarily in reducing electricity consumption, a good principle is to use automatic switch off and manual switch on. This principle uses human inertia for the benefit of reducing energy consumption. If you wish to rely on voluntary manual switching of lighting, care should be taken to make the lighting being switched visible from the control panel and to label the switches so that the operator knows which lamps are being switched. Labels asking people to switch off the lighting when it is not needed can be effective.

As for dimming systems, these all reduce light output and energy consumption, but a different system is required for each lamp type. The factors to consider when evaluating a dimming system are the range over

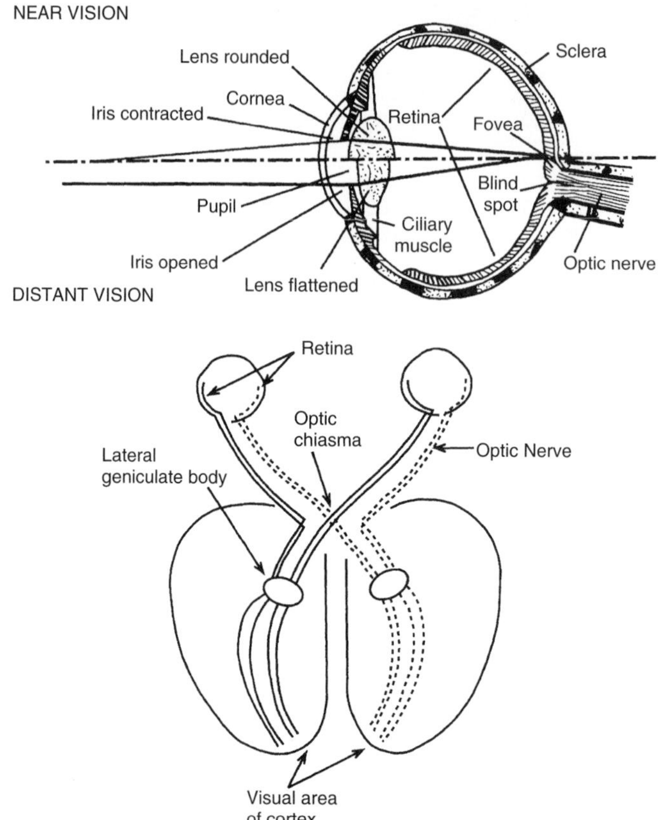

Figure 4 Section through the eye adjusted for near and distant vision, and the binocular nerve pathways of the visual system.

which dimming can be achieved without flicker or the lamp extinguishing, the extent to which the color properties of the lamp change as the light output is reduced, and any effect that dimming has on lamp life and energy consumption. Sophisticated lighting control systems are available for some light sources which allow the user to have a number of preset scenes. These systems use dimming and switching to alter the lighting of a space. They are commonly used in rooms with multiple functions, such as conference rooms.

4 FUNCTIONAL CHARACTERISTICS OF THE HUMAN VISUAL SYSTEM

4.1 Visual System Structure

Illumination is important to humans because it alters the stimuli to the visual system and the operating state of the visual system itself. Therefore, an understanding of the capabilities of the visual system and how they vary with illumination is important to an understanding of the effects of illumination. The visual system is composed of the eye and brain working together. Light entering the eye is brought to focus on the retina by the combined optical power of the air/cornea surface and the lens of the eye. The retina is really an extension of the brain, consisting of two different types of photoreceptors and numerous nerve interconnections. At the photoreceptors, the incident photons of light are absorbed and converted to electrical signals. The nerve interconnections take these signals and carry out some basic image processing. The processed image is transmitted up the optic nerve of each eye to the optic chiasma, where nerve fibers from the two eyes are combined and transmitted to the left and right parts of the visual cortex. It is in the visual cortex that the signals from the eye are interpreted in terms of past experience (Figure 4).

Many of the capabilities of the visual system can be understood from the organization of the retina. The two types of visual photoreceptors, rods and cones from their anatomical appearance, have different wavelength sensitivities and different absolute sensitivities to light and are distributed differently across the retina. Rods are the more sensitive of the two and effectively provide a night retina. Cones are less sensitive to light and operate during daytime. In fact, there are three types of cones, each with a different spectral sensitivity. These cones are commonly called

UNIVERSITY OF STRATHCLYDE

ISBN	Qty	Sales Order
9780471449171	1	F 8363288 1

Customer P/O No	Cust P/O List
79692/VLI85161RY	155.00 GBP

Fund: ALR404

Title: Handbook of human factors and ergonomics / edited by

Format: Cloth/HB

Author:

Publisher: John Wiley,

Volume:

Edition: 3rd ed. Year: c2006.

Order Specific Instructions

Ship To: 28995001 F
UNIVERSITY OF STRATHCLYDE
CURRAN BUILDING
101 ST. JAMES' ROAD
GLASGOW
STRATHCLYDE
G4 0NS

Bill To: 28995000
UNIVERSITY OF STRATHCLYDE
CURRAN BUILDING
101 ST JAMES ROAD
GLASGOW
STRATHCLYDE
G4 0NS

Routing

Sorting
Y04G06X
Covering – BXAXX
Despatch

RUSH

COUTTS LIBRARY SERVICES: UK 138281691 LGUK001R001 UKRWLG14

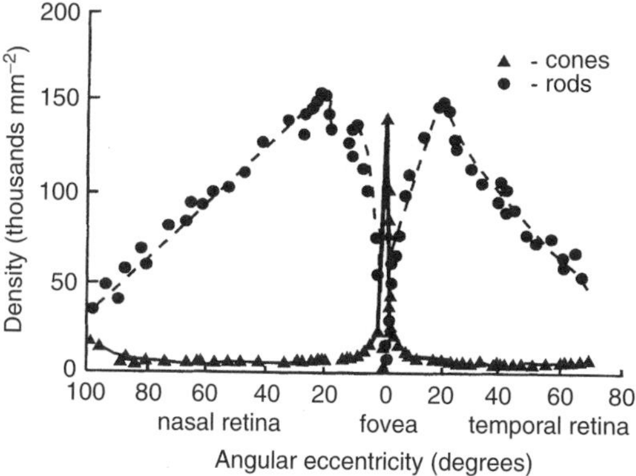

Figure 5 Density of rod and cone photoreceptors across the retina on a horizontal meridian. (After Osterberg, 1935.)

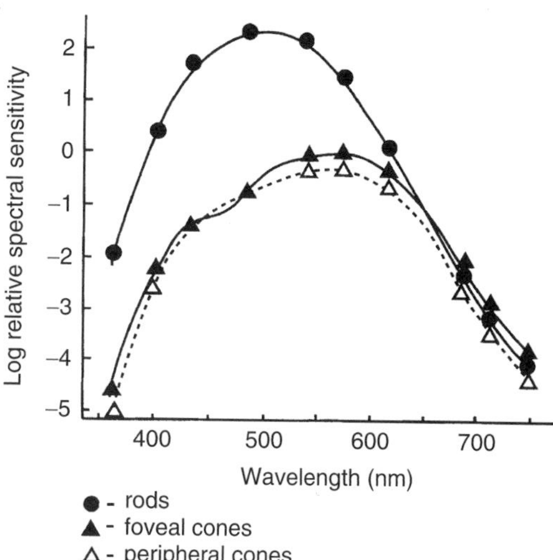

Figure 6 Log relative spectral sensitivity of rod and cone photoreceptors plotted vs. wavelength. (After Wald, 1945.)

long-, middle-, and short-wavelength cones, from their regions of maximum spectral sensitivity. These three cone types combine to give the perception of color. Figure 5 shows the distribution of rods and cones across the retina. Cones are concentrated in a small central area of the retina called the *fovea* that lies where the visual axis of the eye meets the retina, although there are cones distributed evenly across the rest of the retina. Rods are absent from the fovea, reaching their maximum concentration about 20° from the fovea. This variation in concentration of rods and cones with deviation for the fovea is amplified by the number of photoreceptors connected to each optic nerve fiber. In the fovea, the ratio of photoreceptors to optic nerve fibers is close to 1 but increases rapidly as the deviation from the fovea increases. The net effect of this structure is to provide different functions for the fovea and the periphery. The fovea is the part of the retina that provides fine discrimination of detail. The rest of the retina is devoted primarily to detecting changes in the visual environment which require the attention of the fovea.

4.2 Wavelength Sensitivity

The rod and cone photoreceptors have different absolute spectral sensitivities (Figure 6). The spectral response of the cones lies between 380 and 780 nm, with the peak sensitivity occurring at 555 nm. The spectral response of the rods lies between 380 and 780 nm, with the peak at 507 nm. The peak sensitivity of the rods is much greater than that of the cones. These spectral sensitivities form the basis of the CIE standard observers and hence the photometric quantities discussed in Section 2.1. By adjusting the spectral emission of a light source to lie within the most sensitive part of the spectral response of the visual system, lamp manufacturers are able to vary the luminous efficacy of their light sources (i.e., to change the number of lumens emitted for each watt of power applied).

4.3 Adaptation

The visual system can operate over a range of about 12 log units of luminance, from a luminance of 10^{-6} to 10^{6} cd/m^2, from starlight to bright sunlight. But it cannot cover this range simultaneously. At any instant in time, the visual system can cover a range of two or three log units of luminance. Luminances above this limited range are seen as glaringly bright, those below as undifferentiated black. The capabilities of the visual system depend on where in the complete range of luminances it is adapted. Three different functional ranges of luminance are conventionally identified: the photopic, mesopic, and scotopic. Table 5 summarizes the visual system capabilities in each of these functional ranges.

The visual system adjusts its state of adaptation continuously through three mechanisms: neural, mechanical, and photochemical. These three mechanisms differ in their speed and range of adjustment. The neural mechanism, which is based in the retina, operates in milliseconds and covers a range of two to three log units in luminance. The mechanical mechanism involves the expansion and contraction of the iris. The consequent changes in pupil size take about a second but cover less than one log unit in luminance. The photochemical mechanism covers the entire range of luminance but is slow, the changes taking minutes. The exact time will depend on the starting and finishing luminances for the adaptation. If both are greater than 3 cd/m^2, only cones are involved. As the time constant for cones is of the order of 2 to 3 minutes, adaptation takes only a few minutes. When the starting luminance is in the operating range of the cones and the finishing luminance is within the operating range of the rods, a two-stage adaptation process occurs, involving both cones and rods. As rods have a time constant around

Table 5 Functional Ranges of Visual System Capabilities

Name	Luminance Range (cd m^{-2})	Photoreceptor Active	Wavelength Range (nm)	Capabilities
Photopic	>3	Cones	380–780	Color vision, good detail discrimination
Scotopic	<0.001	Rods	380–780	No color vision, poor detail discrimination
Mesopic	>0.001 and <3	Cones and rods	380–780	Diminished color vision, reduced detail discrimination, and a shift in spectral sensitivity as adaptation luminance moves from photopic to scotopic

7 to 8 minutes, the adaptation time is much longer. Complete adaptation from a high photopic luminance to darkness can take up to an hour.

Interior lighting is almost always sufficient for the visual system to be operating in the photopic region. Exterior lighting on roads and in urban areas is usually sufficient to keep the visual system operating in the low photopic or mesopic regions. It is in very rural areas, at sea, or underground, where there is neither exterior lighting nor moonlight that the visual system reaches scotopic adaptation. The speed of adaptation is important where a large and sudden change in the luminance occurs. Examples of situations where this happens are the entrance to road tunnels during daytime (Bourdy et al., 1987) and the onset of emergency lighting during a power failure (Boyce, 1985). These problems are overcome either by installing a gradual reduction in luminance which allows more time for adaptation to occur or by setting a minimum luminance within the neural adaptation range.

4.4 Color Vision

When photopically adapted, the visual system can discriminate many thousands of colors. This ability to discriminate colors reduces as the adaptation luminance decreases through the mesopic region and vanishes in the scotopic vision. This is because color vision is mediated by the cone photoreceptors. Different light sources have different spectral emissions and hence render colors differently. To ensure good color discrimination, it is necessary to use a light source that has a high CIE General Color Rendering Index and produces sufficient light to ensure that the visual system is operating in the photopic region. However, it is important to note that light sources with the same CIE Color Rendering Index do not necessarily render all colors in the same way. For example, an incandescent lamp and a fluorescent lamp, both of which can have CIE Color Rendering Index values in the 90s, make blue and green colors appear very different. If you are concerned about color appearance as well as color discrimination, you will have to chose a light source that gives both good color discrimination and the desired color appearance.

4.5 Receptive Field Size and Eccentricity

The retina is organized such that increasing numbers of photoreceptors are connected to each optic nerve fiber as the deviation from the fovea increases. This feature of the visual system is important when detection of a stimulus is necessary and it can occur anywhere in the visual field. The visual system will normally operate by first detecting the stimulus off-axis (i.e., in the peripheral visual field) and then turning the eye so that the stimulus is brought onto the fovea for detailed examination. To identify a stimulus off-axis, the stimulus should be clearly different from its background, in luminance or color, and should change in space or time (i.e., it should either move or flicker). A flickering light is commonly used to draw drivers' attention to important signs placed beside or above the road.

4.6 Meaningful Stimulus Parameters

Any stimulus to the visual system can be described by five parameters: visual size, luminance contrast, chromatic contrast, retinal image quality, and retinal illumination. These parameters are important in determining the extent to which the visual system can detect and identify the stimulus.

4.6.1 Visual Size

The visual size of a stimulus describes how big the stimulus is. The larger a stimulus is, the easier it is to detect. There are several different ways to express the size of a stimulus presented to the visual system, but all of them are angular measures. The visual size of a stimulus for detection is best given by the solid angle the stimulus subtends at the eye. The solid angle is given by the quotient of the areal extent of the object and the square of the distance from which it is viewed. The larger the solid angle, the easier the stimulus is to detect.

The visual size for resolution is usually given as the angle the critical dimension of the stimulus subtends at the eye. What the critical dimension is depends on the stimulus. For two points, the critical dimension is the distance between the two points. For two lines it is the separation between the two lines. For a Landolt ring, it is the gap size. The larger the visual size of detail in a stimulus, the easier it is to resolve the detail.

For complex stimuli, the measure used to express their dimensions is the spatial frequency distribution. Spatial frequency is the reciprocal of the angular subtense of a critical detail, in cycles per degree. Complex stimuli have many spatial frequencies and hence a spatial frequency distribution. The match between the spatial frequency distribution of the stimulus and the contrast sensitivity function of the visual system (see Section 5.2) determines if the stimulus will be seen and what detail will be resolved. Lighting can change the visual size of three-dimensional stimuli by casting shadows that extend or diminish the apparent visual size of the stimulus.

4.6.2 Luminance Contrast

The luminance contrast of a stimulus quantifies its luminance relative to its background. The higher the luminance contrast, the easier it is to detect the stimulus. There are two different forms of luminance contrast. For stimuli that are seen against a uniform background, luminance contrast is defined as

$$C = |L_t - L_b|/L_b$$

where C is the luminance contrast, L_t the luminance of the detail, and L_b the luminance of the background. This formula gives luminance contrasts that range from 0 to 1 for stimuli that have details darker than the background and from 0 to infinity for stimuli that have details brighter than the background. It is widely used for the former (e.g., printed text).

For stimuli that have a periodic pattern (e.g., a grating), the luminance contrast or modulation is given by

$$C = (L_{\max} - L_{\min})/(L_{\max} + L_{\min})$$

where C is the luminance contrast, $L_{\max}$ the maximum luminance, and $L_{\min}$ the minimum luminance. This formula gives luminance contrast that ranges from 0 to 1. Lighting can change the luminance contrast of a stimulus by producing disability glare in the eye or veiling reflections from the stimulus or by changing the incident spectral radiation when colored stimuli are involved.

4.6.3 Chromatic Contrast

Luminance contrast uses the total amount of light emitted from a stimulus and ignores the wavelengths of the light emitted. It is the wavelengths emitted from the stimulus that largely determine its color. It is possible to have a stimulus with zero luminance contrast that can still be detected because it differs from its background in color (i.e., it has chromatic contrast). There is no widely accepted measure of chromatic contrast, although various suggestions have been made (Tansley and Boynton, 1978). Fortunately, chromatic contrast becomes important for detection only when luminance contrast has reached a low level. Lighting can alter chromatic contrast by using light sources with different spectral emission characteristics.

4.6.4 Retinal Image Quality

As with all image-processing systems, the visual system works best when it is presented with a clear, sharp image. The sharpness of the stimulus can be quantified by the spatial frequency distribution of the stimulus: A sharp image will have high spatial frequency components present; a blurred image will not. The sharpness of the retinal image is determined by the stimulus itself, the extent to which medium through which it is transmitted scatters light, and the ability of the visual system to focus the image on the retina. Lighting can do little to alter any of these factors, although it has been shown that light sources that are rich in the short wavelengths produce smaller pupil sizes, and these tend to improve visual acuity for briefly presented low-contrast targets. The explanation suggested is that the smaller pupil sizes produce greater depth of field and hence better retinal image quality (Berman et al., 1993).

4.6.5 Retinal Illumination

The retinal illumination determines the state of adaptation of the visual system and therefore alters its capabilities. The retinal illumination is determined by the luminance in the visual field, modified by pupil size. Retinal illumination is measured in trolands, a quantity formed from the product of the luminance of the visual field and the pupil size (Wyszecki and Stiles, 1982). Illuminances and surface reflectances determine the luminances of the visual field. Luminances and light spectrum determine pupil size.

5 EFFECTS ON THRESHOLD VISUAL PERFORMANCE

Qualitatively, threshold visual performance is the performance of a visual task close to the limits of what is possible. Quantitatively, it is the performance of a task at a level such that it can be carried out correctly on 50% of the occasions it is undertaken. Threshold visual performance is affected by many different variables. For example, visual acuity is affected by the form of the target used, the luminance contrast of the target, the duration for which it is presented, where in the visual field it appears, and the luminance of the surround relative to the luminance of the immediate background. In this discussion of threshold visual performance, attention will be limited to the effects of variables that are controlled by the lighting system (i.e., the adaptation luminance and the spectral content of the light). Information on the influence of other variables can be obtained from Boff and Lincoln (1988). In the data presented it will be assumed that the observer is fully adapted to the prevailing luminance, that the image of the target is on the fovea, that the target is presented for an unlimited time, and that the observer is correctly refracted. Again, the influence of departures from these assumptions can be estimated from the data given by Boff and Lincoln (1988).

5.1 Visual Acuity

Visual acuity is the limit in the ability to resolve detail. Visual acuity has frequently been measured using gratings or Landolt C's. Visual acuity can be quantified as the angle subtended at the eye by the size of detail that can be detected correctly on 50% of the occasions it is presented. No matter what target is used, visual acuity improves (i.e., the size of detail that can be resolved decreases) as adaptation luminance

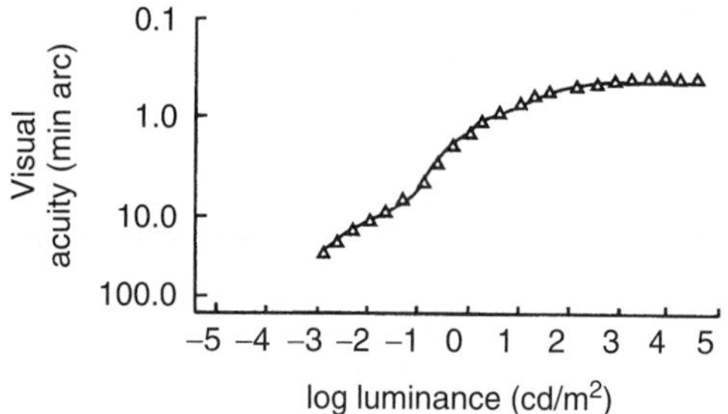

Figure 7 Effect of adaptation luminance on the gap size of a Landolt C target which can just be resolved. (After Shlaer, 1937.)

increases. Figure 7 shows that as adaptation luminance increases from scotopic to photopic conditions, the visual acuity increases, asymptotically approaching a maximum at high luminances. The adaptation luminance produced by a lighting installation will depend on the illuminances produced on different surfaces and the reflectance of those surfaces. Table 3 gives some luminances typically found in interior and exterior lighting installations. Given a value for the adaptation luminance, Figure 7 can be used to determine if detail of a given size can be resolved. A useful rule of thumb is that the detail needs to be four times bigger than the visual acuity limit if it is to be resolved sufficiently quickly to avoid affecting visual performance (Bailey et al., 1993). As for the light spectrum, provided that the lamp produces white light rather than an emission in a narrow spectral region, the effect on visual acuity is very small, certainly much less than the effect of adaptation luminance (Shlaer et al., 1941).

5.2 Contrast Sensitivity Function

Contrast sensitivity is the reciprocal of the luminance contrast that can be detected on 50% of the occasions it is presented. Contrast sensitivity is usually measured using a sinusoidal grating target. The contrast sensitivity function is contrast sensitivity plotted against the spatial frequency of the sinusoidal target. Figure 8 shows the effect of adaptation luminance on the contrast sensitivity function. It shows that as the adaptation luminance increases from scotopic to photopic conditions, the contrast sensitivity increases for all spatial frequencies; the spatial frequency at which the peak contrast sensitivity occurs increases and the highest spatial frequency that can be detected also increases. Figure 8 can be used to determine if a given target will be visible by breaking the target into its spatial frequency components and determining if any of the components are within the limit set by the contrast sensitivity function (Sekular and Blake, 1994). The target will only be visible if at least one of its components falls within this limit, although it should be noted that the appearance of the target will be different depending on which component or components are visible. As a rule of thumb, for a target to be easily seen, it is necessary for the luminance contrast to be at least twice the contrast threshold. As for the light spectrum, there is no evidence that the contrast sensitivity function is influenced by different white light spectra, provided that the luminances are the same.

5.3 Temporal Sensitivity Function

The temporal sensitivity function shows percentage modulation amplitude plotted against the frequency of the modulation. Figure 9 shows the effect of adaptation luminance on the temporal sensitivity function. It shows that as the adaptation luminance increases

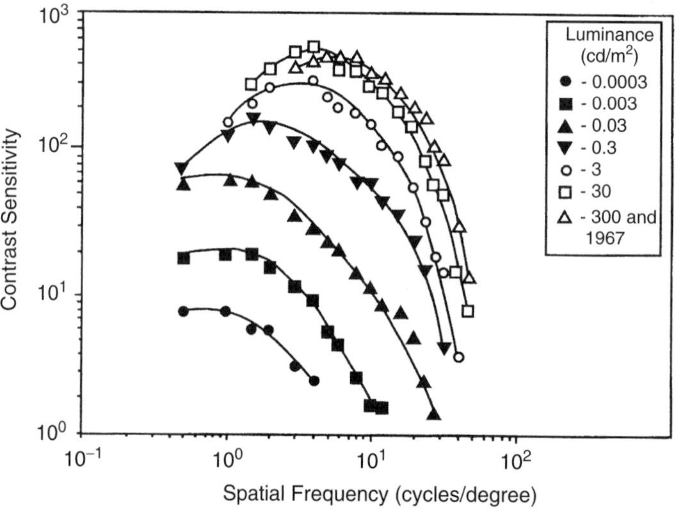

Figure 8 Effect of adaptation luminance on the contrast sensitivity function. Contrast sensitivity is plotted vs. spatial frequency (cycles/degree) for various adaptation luminances. (After Boff and Lincoln, 1988.)

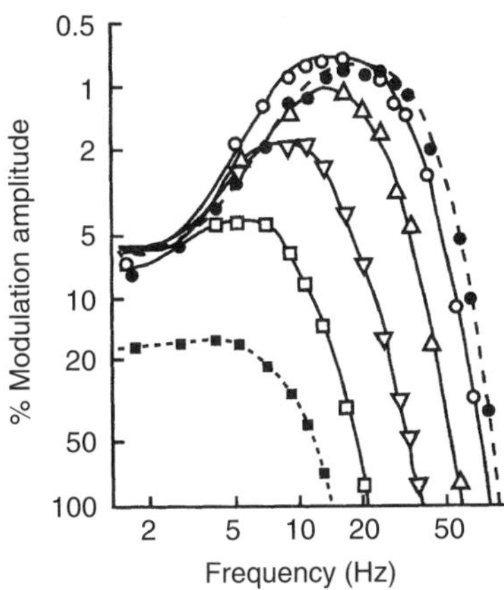

Figure 9 Effect of adaptation luminance on the temporal sensitivity function. Percentage modulation amplitude is plotted vs. frequency (in hertz) for various levels of retinal illumination. The retinal illuminations are: filled square, 0.06 troland; open square, 0.65 troland; open, inverted triangle, 7.1 trolands; open, upright triangle, 77 trolands; open circle, 850 trolands; filled circle, 9300 trolands (After Kelly, 1961.)

from mesopic to photopic conditions, the temporal sensitivity increases for all frequencies; the frequency at which the peak temporal sensitivity occurs increases, and the highest frequency that can be detected also increases. Figure 9 can be used to determine if a given temporal variation will be visible by breaking the waveform representing the light fluctuation into its frequency components and determining if any of the components are within the limit set by the temporal sensitivity function. The fluctuation will be visible only if at least one of its frequency components falls within this limit.

Temporal fluctuation in luminous flux (i.e., flicker) is undesirable in lighting installations. To eliminate flicker, it is necessary to increase the frequency and/or decrease the percentage modulation sufficiently to take their combination outside the limits set by the temporal sensitivity function. In practice, this is easily done. Incandescent lamps have sufficient thermal inertia to ensure that even though the frequency of the fluctuation is only twice the supply frequency (120 Hz for a 60-Hz electrical supply), the percentage modulation is small, so there is little chance of seeing flicker from such a lamp. Discharge lamps, such as the fluorescent lamp, do not have thermal inertia, so their percentage modulation can be high. To ensure that fluorescent lamps do not produce visible flicker, it is best to use an electronic ballast to control the lamp. Electronic ballasts typically operate at frequencies in the tens of kilohertz, with small percentage modulations, and consequently, are very unlikely to produce visible flicker.

5.4 Color Discrimination

The ability to discriminate between two colors of the same luminance depends on the difference in spectral power distribution of the light received at the eye. Figure 10 shows the MacAdam ellipses, the area around a number of chromaticities, each magnified 10 times, within which no discrimination of color can be made, even under side-by-side comparison conditions (Wyszecki and Stiles, 1982). The effect of illuminance on the ability to discriminate between colors is limited in the photopic region, an illuminance of 300 lx being sufficient for good color judgment work (Cornu and Harlay, 1969). As the visual system enters the mesopic region, the ability to discriminate colors deteriorates and ultimately fails as the scotopic region is reached. The effect of light spectrum is much more important. The position of a color on the CIE 1931 Chromaticity Diagram is determined by the spectrum of the light, and if it is reflected from or transmitted through a surface, the spectral reflectance or transmittance of that surface. Therefore, by changing the light spectrum emitted by the lamp, it is possible to make colors easily discriminable or difficult to discriminate. The careful choice of light source is important wherever good color discrimination is important.

5.5 Interactions

The fact that there are many other variables besides adaptation luminance and light spectrum that influence threshold visual performance has been mentioned earlier. It is now necessary to introduce another complication: interaction between the various components of visual system performance. As an example, consider the effect of luminance contrast on visual acuity. Visual acuity is conventionally measured using targets with a high luminance contrast. However, as the luminance contrast of the target is decreased, visual acuity also worsens. Similarly, the temporal sensitivity function as presented applies to a uniform luminance field. If the field has a pattern and hence a distribution of spatial frequencies, the temporal sensitivity function may be changed (Koenderink and Van Doorn, 1979). Put crudely, what this means is that as visual performance gets closer to threshold, almost everything about the stimulus presented to the visual system becomes important. Further details on some of the interactions that occur are given in Boff and Lincoln (1988).

5.6 Approaches to Improving Threshold Visual Performance

Working close to threshold is not easy. In fact, it can be argued that the main function of anyone designing lighting is to provide conditions that avoid the need to use the visual system close to threshold. However, if this is the situation, the following steps can be taken to improve threshold visual performance. Not all of the following steps will be possible in every situation, and not all are appropriate for every problem. The discussion above should indicate which approach is likely to be most effective.

Changing the Task

- Increase the size of the detail in the task.

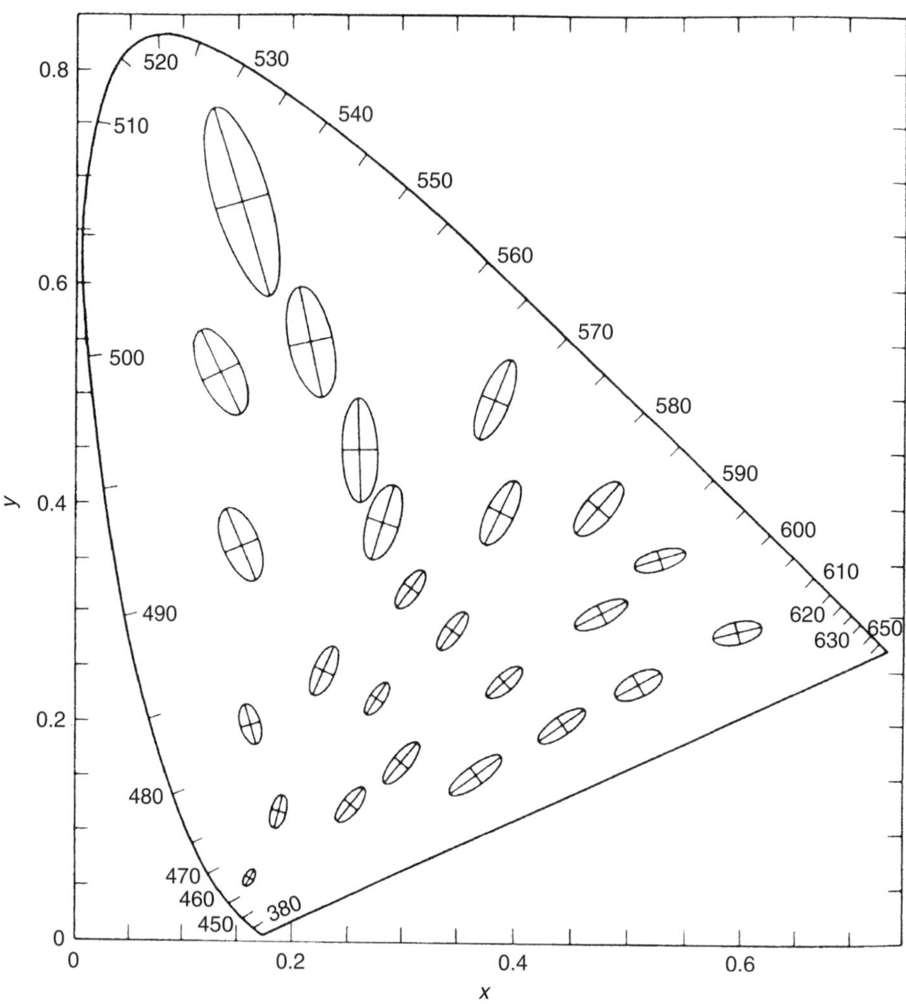

Figure 10 MacAdam ellipses plotted on the CIE 1931 chromaticity diagram. The boundary of each ellipse represents 10 times the standard deviation of color matches made for the chromaticity indicated. (After MacAdam, 1942.)

- Increase the luminance contrast of the detail in the task.
- Present the task so that it can be looked at directly (i.e., with the fovea).
- Change the color of the target to make it more conspicuous.
- Reduce the velocity of the task.
- Present the task for a longer time.

Changing the Environment

- Increase the adaptation luminance.
- Select a lamp with better color properties.
- Design the lighting so that it is free from disability glare and veiling reflections (see Section 7).

6 EFFECTS ON SUPRATHRESHOLD VISUAL PERFORMANCE

Suprathreshold visual performance is the performance of tasks that are easily visible because the stimuli they present to the visual system are well above those associated with threshold conditions. This raises the question as to why lighting conditions make a difference to task performance once what has to be seen is clearly visible. The answer is that although the stimuli are clearly visible, lighting influences the speed with which the visual information extracted from the stimuli can be processed. The aspect of lighting that determines this effect is the *retinal illumination*, which is determined by the luminance of the visual field that is viewed and hence by the illuminance on the surfaces which form that field.

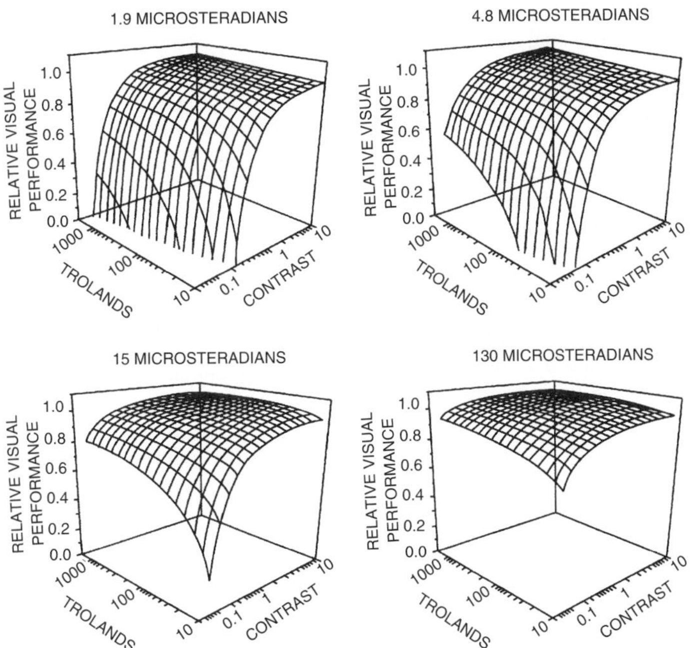

Figure 11 Relative visual performance surfaces plotted vs. retinal illumination, in trolands, and luminance contrast, for four stimuli subtending four different solid angles, measured in microsteradians. (After Rea and Ouellette, 1991.)

6.1 Relative Visual Performance Model for On-Axis Detection

The relative visual performance (RVP) model of visual performance is an empirical model of the reaction time for the detection of different visual stimuli seen on the fovea, for a range of adaptation luminances, luminance contrasts, and visual sizes (Rea and Ouellette, 1988, 1991). Figure 11 shows the form of the relative visual performance (RVP) model for four different visual size tasks, each surface being for a range of contrasts and retinal illuminances. The overall shape of the relative visual performance surface has been described as a plateau and an escarpment (Boyce and Rea, 1987). In essence, what it shows is that the visual system is capable of a high level of visual performance over a wide range of visual sizes, luminance contrasts, and retinal illuminations (the plateau), but at some point either visual size or luminance contrast or retinal illumination will become insufficient and visual performance will rapidly collapse (the escarpment) toward threshold. The existence of a plateau of visual performance, or rather, a near plateau because there is really a slight improvement in visual performance across the plateau, implies that for a wide range of visual conditions, visual performance changes very little with changes in the lighting conditions. To put it bluntly, what this means is that for many visual tasks, visual performance is insensitive to lighting conditions, the visual system being flexible enough to cope equally well with a wide variety of visual stimuli.

The RVP model of suprathreshold visual performance provides a quantitative means of predicting the effects of changing either task size or contrast or the adaptation luminance on visual performance. It has been developed using rigorous methodology and has been validated against data collected independently (Eklund et al., 2001; Boyce, 2003). However, it is important to note that it should only be applied to a limited range of tasks. Specifically, it is most appropriate for tasks where task performance is dominated by the visual component (see Section 6.3); that do not require the use of off-axis vision to any extent; that present stimuli to the visual system that can be completely characterized by their visual size, luminance contrast, and background luminance, and that have values for these variables that fall within the ranges used to develop the model. Where the task involves chromatic contrast as well as luminance contrast, the RVP model is likely to be misleading and the light spectrum used for illumination will be important. Where the task is achromatic, the light spectrum is not likely to be important for suprathreshold visual performance (Boyce et al., 2003a).

6.2 Visual Search

One class of tasks for which the RVP model is not applicable are those in which the object to be detected can appear anywhere in the visual field. These tasks involve visual search. Visual search is typically undertaken through a series of eye fixations, the fixation pattern being guided either by expectations about where the object to be seen is most likely to appear or by what part of the visual scene is most important. Typically, the object to be detected is first detected off-axis and then confirmed or resolved by an on-axis fixation. The speed with which a visual search task is completed depends on the visibility of the object to be found, the presence of other objects in the search area, and the extent to which the object to be found is different from the other objects. The simplest visual search task is one in which the object to be found appears somewhere in an otherwise empty field (e.g., paint defects on a car body). The most difficult visual search task is one where the object to be found is situated in a cluttered field, and the clutter is very similar to the object to be found (e.g., searching for a face in a crowd).

The lighting conditions necessary to achieve fast visual search are similar to those used to improve foveal threshold visual performance. By improving foveal threshold visual performance, the peripheral threshold visual performance is also improved, so the object to be found is made more visible. The lighting required for fast visual search will have to be matched to the physical characteristics of the object to be found. For example, if the object is two-dimensional and of matte reflectance located on a matte background, increasing the adaptation luminance is about the only option. However, if the object is three-dimensional and has a specular reflectance component, light distribution can be used to increase the apparent size by casting shadows, and the luminance contrast of the object by producing highlights on or around the object, changes that will be much more effective than simply increasing the adaptation luminance. Similarly, if the object is distinguished from its background primarily by color, the light spectrum used is an important consideration. It is this need to match the lighting conditions to the nature of the objects to be found that makes the design of lighting installations for visual inspection tasks so difficult and diverse (IESNA, 2000; Boyce, 2003).

The extent to which a lighting installation is effective in revealing an object can be estimated from the object's *visibility lobe* (Inditsky et al., 1982), the distribution of the probability of detecting the object within one fixation pause. This probability is a maximum when the object is viewed on-axis and decreases with increasing deviation from the fovea. The probability distribution is assumed to be radially symmetrical about the visual axis, resulting in circles around the fixation points, each circle having a given probability of detection within one fixation pause. For objects that appear on a uniform field, the visibility lobe is based on the detection of the object. For objects that appear among other similar objects, the visibility lobe is based on the discriminability of the object from the others surrounding it. Visual search will be fastest for objects that have the largest visibility lobe.

6.3 Visual Performance, Task Performance, and Productivity

Figure 12 shows the relationships between the stimuli to the visual system and their impact on visual performance, task performance, and productivity. The stimuli to the visual system, including the retinal illumination, determine the operation of the visual system and hence the level of visual performance achieved. This visual performance then contributes to task performance. It is important to point out that visual performance and task performance are not necessarily the same. Task performance is the performance of the complete task. Visual performance is the performance of the visual component of the task. Task performance is what is needed to measure productivity and to establish cost/benefit ratios comparing the costs of providing a lighting installation with the resulting benefits in terms of better task performance. Visual performance is the only thing that changing the lighting conditions can affect directly.

Most apparently, visual tasks have three components: visual, cognitive, and motor. The *visual component* is the process of extracting information relevant to the performance of the task using the sense of sight. The *cognitive component* is the process by which sensory stimuli are interpreted and the appropriate action determined. The *motor component* is the process by which the stimuli are manipulated to extract information and/or the actions decided upon are carried out. Every task is unique in its balance among visual, cognitive, and motor components, and hence in the effect that lighting conditions have on task performance. It is this uniqueness that makes it impossible to generalize from the effect of lighting on the performance of one task to the effect of lighting on the performance of another. The RVP model for on-axis tasks and the visual search models discussed above can be used to quantify the effects of lighting conditions on visual performance, but there is no general model to translate those results to task performance.

6.4 Approaches to Improving Suprathreshold Visual Performance

The main purpose of lighting installations is to ensure that people can perform the work they need to do quickly, easily, comfortably, and safely. To achieve this desirable aim, it is necessary to provide lighting which ensures that people are working on the plateau of visual performance and not on the escarpment. The RVP model of visual performance provides a simple means of checking whether lighting is adequate for the visual performance of many on-axis tasks. The visibility lobe provides an approach to quantifying the effect of lighting conditions on visual search tasks. Alternatively, most countries have well-established recommendations for the illuminances to be provided for working interiors (IESNA, 2000; CIBSE, 2002). Most of these recommendations easily exceed what would be deduced as necessary from a consideration of visual performance alone.

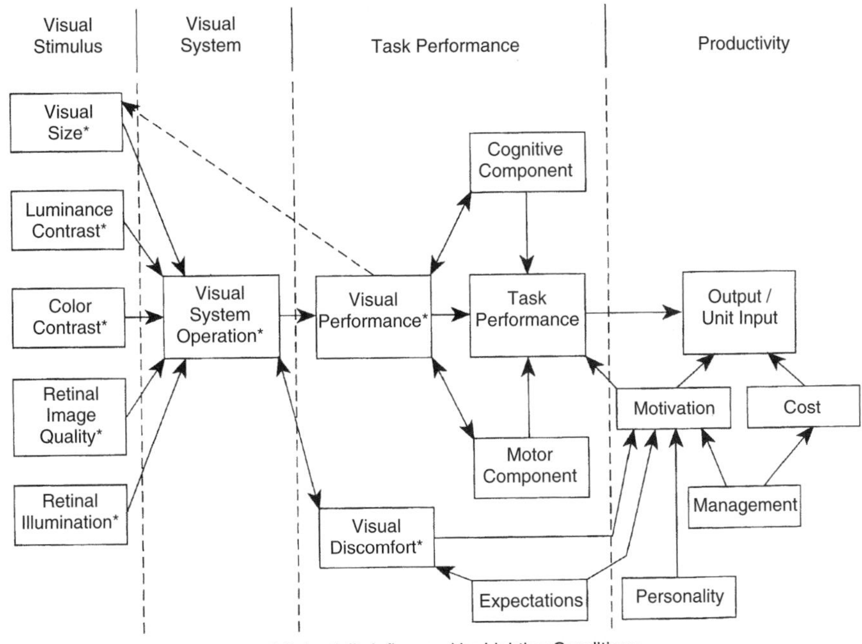

Figure 12 Schematic relationships between the stimuli to the visual system and their impact on visual performance, task performance, and productivity. The arrows indicate the direction of the effects. The dashed arrow between visual performance and visual size indicates that if visual performance is poor, a common response is to move closer to the stimulus to increase its visual size.

Although the discussion above has been focused on lighting conditions, it is important to recognize that improving suprathreshold visual performance can be achieved by changing the characteristics of the task as well as the lighting. The following list is divided into two parts, task changes and lighting changes. Not all of the following suggestions will be possible in every situation, and not all are appropriate for every problem.

Changing the Task

- Increase the size of the detail in the task.
- Increase the luminance contrast of the detail in the task.
- For off-axis tasks in a cluttered field, make the object to be detected clearly differ from the surrounding objects on as many different dimensions as possible (e.g., size, contrast, color, shape).
- Ensure that the object presents a clear sharp image on the retina.

Changing the Environment

- Increase the adaptation luminance.
- Where the task involves color, select a lamp with better color properties.
- Design the lighting so that it is free of disability glare and veiling reflections (see Section 7).
- Design the lighting to increase the apparent size or luminance contrast of the object.

7 EFFECTS ON COMFORT

Lighting installations are rarely designed for visual performance alone. Visual comfort is almost always a consideration. The aspects of lighting that cause visual discomfort include those relevant to visual performance and extend beyond them. This is because the factors relevant to visual performance are generally restricted to the task and its immediate area, whereas the factors affecting visual discomfort can occur anywhere within the lit space.

7.1 Symptoms and Causes of Visual Discomfort

Visual discomfort can give rise to an extensive list of symptoms. Among the more common are red, sore, itchy, and watering eyes; headaches and migraine attacks; gastrointestinal problems; and aches and pains associated with poor posture. Visual discomfort is not the only possible source of these symptoms. All can have other causes. It is this vagueness that makes it essential to consider the nature of the visual environment before ascribing any of these symptoms to the lighting conditions.

Features of the visual environment that can cause visual discomfort are as follows:

1. *Visual task difficulty*. The visual system is designed to extract information from the visual environment. Any visual task that is close to threshold contains information that is difficult to extract. The usual reaction to a high level of visual difficulty is to bring the task closer to increase its visual size. As the task is brought closer, the accommodation mechanism of the eye has to adjust to keep the retinal image sharp. This adjustment can lead to muscle fatigue and hence symptoms of visual discomfort.

2. *Under- and Over-stimulation*. The visual system is designed to extract information from the visual environment. Discomfort occurs either when there is no information to be extracted or when there is an excessive amount of repetitive information. Examples of no information occur when driving in fog or in a "white-out" snowstorm. In both cases, the visual system is searching for information that is hidden but which may appear suddenly and require a rapid response. The stress experienced while driving in these conditions is a common experience. As for overstimulation, the important point is not the total amount of visual information, but rather, the presence of large areas of the same spatial frequency. Wilkins (1993) has associated the presence of large areas of specific spatial frequencies in printed text with the occurrence of headaches, migraines, and reading difficulties.

3. *Distraction*. The visual system is designed to extract information from the visual environment. To do this, it has a large peripheral field that detects the presence of objects which are then examined using the small, high-resolution fovea. For this system to work, objects in the peripheral field that are bright, moving, or flickering have to be detected easily. If, upon examination, these bright, moving, or flickering objects prove to be of little interest, they become sources of distraction because their attention-gathering power is not diminished after one examination. Ignoring objects that attract attention automatically is stressful and can lead to symptoms of visual discomfort.

4. *Perceptual confusion*. The visual system is designed to extract information from the visual environment. The visual environment consists of a pattern of luminances, developed from the differences in reflectance of the surfaces in the field of view and the distribution of illuminance on those surfaces. Perceptual confusion occurs when a pattern of luminances is present that is related solely to the illuminance distribution and conflicts with the pattern of luminance associated with the reflectances of surfaces.

7.2 Lighting Conditions That Can Cause Discomfort

Many different aspects of lighting can cause discomfort. Insufficient light for the performance of a task has been discussed earlier and will not be discussed again. Rather, attention will be devoted to flicker, glare, shadows, and veiling reflections. It should be noted that whether these aspects of lighting cause discomfort will depend on the context. All can be used to positive effect in some contexts.

1. *Flicker*. A lighting installation that produces visible flicker will be almost universally disliked unless it is being used for entertainment. Individual differences, and the fact that electrical signals associated with flicker can be detected in the retina, even when there is no visible flicker (Berman et al., 1991), imply that a clear safety margin is necessary. This can be achieved by the use of high-frequency control gear for discharge lamps and/or the mixing of light from lamps powered from different phases of the electricity supply. The same approaches, which will result in a changed frequency and/or a reduced percentage modulation, can be used to diminish any stroboscopic illusions. The use of high-frequency control gear has been associated with a reduction in the prevalence of headaches (Wilkins et al., 1989).

2. *Glare*. Glare occurs in two ways. First, it is possible to have too much light. Too much light produces a simple photophobic response in which the observer screws up his eyes, blinks, or looks away. Too much light is rare indoors but is common in full sunlight. Second, glare occurs when the range of luminance in a visual environment is too large. Glare of this sort can have two effects, a reduction in threshold visual performance and a feeling of discomfort. Glare that reduces threshold visual performance, called *disability glare*, is due to light scattered in the eye, reducing the luminance contrast of the retinal image on the fovea. The magnitude of disability glare can be estimated by calculating the equivalent veiling luminance (IESNA, 2000).

The effect of disability glare on the luminance contrast of the object being looked at can be determined by adding the equivalent veiling luminance to all elements in the formulas for luminance contrast (see Section 4.6.2). Disability glare is rare in interior lighting but is common on roads at night from oncoming headlights and during the day from the sun. Usually, disability glare also causes discomfort, but it is possible to have disability glare without discomfort when the glare source is large in area. This can be seen by looking at a picture hung on a wall adjacent to a window. The picture will usually be much easier to see when the eye is shielded from the window.

As for discomfort glare, this, by definition, does not cause any shift in threshold visual performance but does cause discomfort. There are many different national systems for predicting the magnitude of discomfort glare produced by interior lighting installations (IESNA, 2000; CIBSE, 2002; CIE, 2002). All these systems are based on a formula that implies that discomfort glare increases as the luminance and solid angle of the glare source increase and decreases as the luminance of the background and the deviation from the glare source increase. Lighting equipment manufacturers use these formulas to produce tabular estimates of the level of discomfort glare produced by a regular array of their luminaires for a range of standard interiors. These tables provide all the precision necessary for estimating the average level of discomfort glare likely to occur in an interior, although the precision with which they predict a person's sense of discomfort is low (Stone and Harker, 1973).

3. *Shadows.* Shadows are cast when light coming from a particular direction is intercepted by an opaque object. If the object is big enough, the effect is to reduce the illuminance over a large area. This is typically the problem in industrial lighting, where large pieces of machinery cast shadows in adjacent areas. The effect of these shadows can be overcome either by increasing the proportion of interreflected light by using high reflectance surfaces or by providing local lighting in the shadowed area. If the object is smaller, the shadow can be cast over a meaningful area, which in turn can cause perceptual confusion, particularly if the shadow moves. An example of this is the shadow of a hand cast on a blueprint. This problem can be reduced by increasing the interreflected light in the space or by providing local lighting which can be adjusted in position.

Although shadows can cause visual discomfort, it should be noted that they are also an essential element in revealing the form of three-dimensional objects. Techniques of display lighting are based around the idea of creating highlights and shadows to change the perceived form of the object being displayed. The number and nature of shadows produced by a lighting installation depends on the size and number of light sources and the extent to which light is interreflected around the space. The strongest shadow is produced from a single point source in a black room. Weak shadows are produced when the light sources are large in area and the degree of interreflection is high.

4. *Veiling Reflections.* Veiling reflections occur when a source of high luminance, usually a luminaire or a window, is reflected from a specularly reflecting surface such as a glossy printed page or a computer screen. The luminance of the reflected image changes the luminance contrast of the printed text or the display. The extent to which this changes visual performance can be estimated using the RVP model, but the extent to which it causes discomfort is different. Bjorset and Frederiksen (1979) have shown that a 20% reduction in luminance contrast is the limit of what is acceptable, regardless of the luminance contrast without veiling reflections (Figure 13).

The two factors that determine the magnitude of veiling reflections are the specularity of the material being viewed and the geometry between the observer, the object, and any sources of high luminance. If the object is completely diffusely reflecting, no veiling reflections occur, but if it has a specular reflection component, veiling reflections can occur. The positions where they occur are those where the incident ray corresponding to the reflected ray that reaches the observer's eye from the object comes from a source of high luminance. This means that the strength and magnitude of veiling reflections can vary dramatically within a single lighting installation (Boyce and Slater, 1981). Like shadows, veiling reflections can also be used positively, but when they are, they are conventionally called *highlights*. Display lighting of specularly reflecting objects is all about producing highlights to reveal the specular nature of the surface.

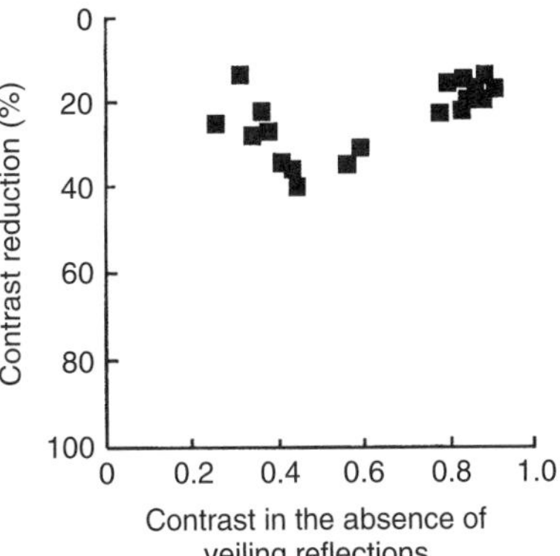

Figure 13 Luminance contrast reduction considered acceptable by 90% of observers plotted against the luminance contrast of the materials when no veiling reflections occurred. (After Bjorset and Frederiksen, 1979.)

7.3 Comfort, Performance, and Expectations

While lighting conditions that make it difficult to achieve good visual performance will almost always be considered uncomfortable, lighting conditions that allow a high level of visual performance may also be considered uncomfortable. Figure 14 shows the mean detection speed for finding a number from many laid out at random on a table and the percentage of people considering the lighting good. As might be expected, increasing the illuminance on the table increases mean detection speed and the percentage considering the lighting good. However, as the illuminance exceeds 2000 lx, the percentage considering the lighting good declines even though the mean detection speed continues to increase. This result indicates that if you wish to achieve a satisfactory lighting installation, it is necessary to provide lighting that allows easy visual performance and avoids discomfort and that visual discomfort is more sensitive to lighting conditions than visual performance.

There is another aspect of visual comfort that distinguishes it from visual performance. Visual performance is determined solely by the capabilities of the visual system. Visual comfort is linked to people's expectations. Any lighting installation that does not meet expectations may be considered uncomfortable even though visual performance is adequate, and expectations can change over time. Figure 12 also demonstrates another potential impact of visual comfort. Lighting conditions that are considered uncomfortable may influence task performance by changing motivation even when they have no effect on the stimuli presented to the visual system and hence on visual performance.

7.4 Approaches to Improving Visual Comfort

To ensure visual comfort it is necessary to ensure that the lighting allows a good level of visual performance,

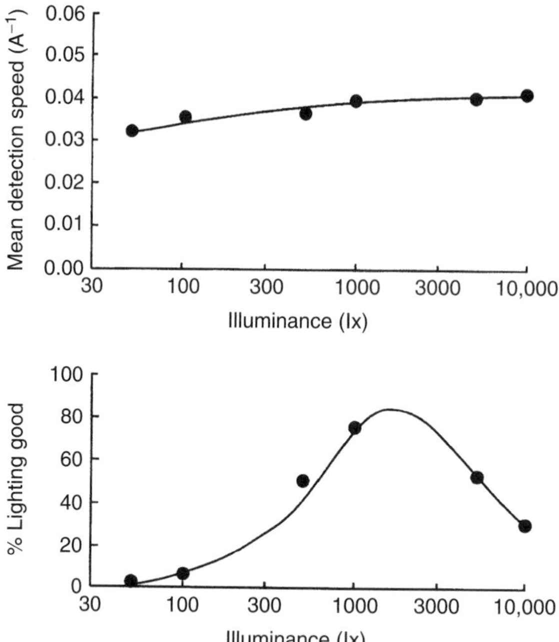

Figure 14 Mean detection speed for locating a specified number from among others at different illuminances, and the percentage of observers who consider the lighting good at each illuminance. (After Muck and Bodmann, 1961.)

does not cause distraction, and allows sufficient stimulation without perceptual confusion. This can be done by:

- Identifying the visual tasks to be performed and then determining the characteristics of the lighting needed to allow a high level of visual performance of the tasks (see Sections 5 and 6).
- Eliminating flicker from the lighting by using appropriate control gear for discharge lamps. If this is not possible, reduce the percentage modulation of the flicker by mixing light from sources operating on different phases of the electricity supply.
- Reducing disability glare by careful selection, placing, and aiming of luminaires so as to reduce the luminous intensity of the luminaires close to the common lines of sight.
- Reducing discomfort glare by careful selection and layout of luminaires. Use the appropriate national discomfort glare system to estimate the magnitude of discomfort glare. Using high reflectance surfaces in the space will help reduce discomfort glare by increasing the background luminance against which the luminaires are seen.
- Considering the density and areal extent of any shadows that are likely to occur. If shadows are undesirable and large area shadows are likely to occur, use high reflectance surfaces in the space to increase the amount of interreflected light and use more lower-wattage lamps to supply the desired illuminance. If shadows cannot be avoided because of the extent of obstruction in the space, be prepared to provide supplementary task lighting in the shadowed areas. If dense, small area shadows occur in the immediate work area, use adjustable task lighting to moderate their impact.
- Considering the extent to which veiling reflections (or highlights) are desirable. If they are undesirable, veiling reflections can be reduced by:
  - Reducing the specular reflectance of the surface being viewed
  - Changing the geometry between the viewer, the surface being viewed, and any zones of high luminance
  - Reducing the maximum luminances in the space
  - Increasing the amount of interreflected light in the space

If the veiling reflections are occurring on a self-luminous surface such as a computer screen, all the foregoing approaches apply, but it is also possible to increase the luminance of the display by using dark letters on a bright background. This will reduce the impact of any veiling reflections seen on the screen (Boyce, 1991).

8 INDIVIDUAL DIFFERENCES

Differences between persons in visual capabilities are common and are usually dealt with by providing lighting that is more than adequate for visual performance and visual comfort. However, there are three sources of individual differences that are both common and consistent enough in direction to deserve special consideration: the effects of age, partial sight, and defective color vision.

8.1 Changes with Age

As the visual system ages, a number of changes in its structure and capabilities occur. Usually, the first to occur is an increase in the near point (i.e., the shortest distance at which a clear, sharp retinal image can be achieved). This increase occurs due to an increase in the rigidity of the lens with age. This change, called *presbyopia*, is why the majority of people over 50 have to wear glasses or contact lenses to read. Whereas the increasing rigidity of the lens, and other forms of focusing difficulty, can be compensated by adjusting the optical power of an eye's optical system with spectacle lenses, the other changes that occur in the eye cannot. As the visual system ages, the amount of light reaching the retina is reduced, more of the light entering the eye is scattered, and the color of the light is altered by preferential absorption of the short, visible wavelengths. The rate at which these changes occur accelerates after about 60. The consequences of these changes with age are reduced visual acuity, reduced contrast sensitivity, reduced color discrimination, increased time taken to adapt to large and sudden changes in adaptation luminance, and increased sensitivity to glare (Boyce, 2003).

Lighting can be used to compensate for these changes to some extent. Older people benefit from higher illuminances than are needed by young people (Smith and Rea, 1979), but simply providing more light may not be enough. The light has to be provided in such a way that both disability and discomfort glare are carefully controlled and veiling reflections are avoided. Where elderly people are likely to be moving from a well-lit area to a dark area, such as from a supermarket to a parking lot, a transition zone with a gradually reducing illuminance is desirable. Such a transition zone allows their visual system more time to make the necessary changes in adaptation.

8.2 Helping People with Partial Sight

Partial sight is a state of vision that falls between normal vision and total blindness. Different countries use different capabilities to define the state of partial sight and blindness. The factors considered are visual acuity and the extent of the visual field. The World Health Organization accepts that a distance visual acuity of 6/18 implies that people are visually disabled and then grades the extent of disability in five steps. A visual acuity of 6/18 means that a person can just resolve details at 6 m which people with normal vision can resolve at 18 m.

Although some people are born with partial sight, the majority of people with partial sight are elderly. Kahn and Moorhead (1973) found that among the partially sighted, 20% became partially sighted between birth and 40 years, 21% between 41 and 60 years, and 59% after 60 years of age. Surveys in the United States and the United Kingdom suggest that the percentage of the total population who are classified as partially sighted are in the range 0.5 to 1%. This percentage increases markedly in less developed countries (Tielsch, 2000).

The three most common causes of partial sight are cataract, macular degeneration, and glaucoma. These causes involve different parts of the eye and have different implications for how lighting might be used to help people with partial sight. Cataract is an opacity developing in the lens. The effect of cataract is to absorb and scatter more light as the light passes through the lens. This increased absorption results in reduced visual acuity and reduced contrast sensitivity over the entire visual field, as well as greater sensitivity to glare. The extent to which more light can help a person with cataract depends on the balance between absorption and scattering. More light will help overcome the increased absorption, but if scattering is high, the consequent deterioration in the luminance contrast of the retinal image will reduce visual capabilities. There is really little alternative to testing the effectiveness of additional light on an individual basis. What is true for everyone with cataract is that they will be very sensitive to glare from luminaires and windows. Careful selection of luminaires and window treatments to limit glare is desirable. The use of dark backgrounds against which objects are to be seen will also help.

Macular degeneration occurs when the macular of the retina, which is just above the fovea, becomes opaque due to bleeding or atrophy. An opacity immediately in front of the fovea implies a serious reduction in visual acuity and in contrast sensitivity at high spatial frequencies. It also implies that the ability to discriminate colors will be reduced. Typically, these changes make reading difficult, if not impossible. However, peripheral vision is unaffected, so the ability to orient oneself in space and to find one's way around is unchanged. Providing more light, usually by way of a task light, will help people in the early stage of macular deregulation to read, although as the deterioration progresses, additional light will be less effective. Increasing the size of the retinal image by magnification or by getting closer is helpful at all stages.

Glaucoma is shown by a progressive narrowing of the visual field. Glaucoma is due to an increase in intraocular pressure that damages the retina and the anterior optic nerve. Glaucoma will continue until complete blindness occurs unless the intraocular pressure is reduced. As glaucoma develops, it leads to a reduction in visual field size, reduced contrast sensitivity, poor night vision, and slowed transient adaptation, but the resolution of detail seen on-axis is unaffected until the final stage. Lighting has limited value in helping people in the early stages of glaucoma, because where damage has occurred, the retina has been destroyed. However, consideration should be given to providing enough light for exterior lighting at night to enable the fovea to operate.

Although the extent to which providing more light is helpful will depend on the specific cause of partial sight, there is one approach that is generally useful. This approach is to simplify the visual environment and to make its salient details more visible. As an example, consider the problem of how to set a table so that a person with partial sight can eat with confidence. The plate containing the food and the associated cutlery can be made more visible by using a contrasting tablecloth (e.g., a dark tablecloth with a white plate and cutlery). The food on the plate can be made easier to identify by using an overlarge plate so that individual food items can be separated from each other. The entire scene can be simplified by using solid colors rather than patterns. This same approach of simplification and enhanced visibility can be applied to entire rooms, for example, by painting a door frame in a contrasting color to the door so that the door is easily identified.

8.3 Consequences of Defective Color Vision

About 8% of males and 0.5% of females have some form of defective color vision (McIntyre, 2002). For most activities this causes few problems, either because the exact identification of color is unnecessary or because there are other cues by which the necessary information can be obtained. Where defective color vision does become a problem is where color is the sole means used to identify significant information, as, for example, in some forms of electrical wiring. People with defective color vision will have difficulty with such activities (Steward and Cole, 1989).

Where self-luminous colors are used as signals, care should be taken to restrict the range of colored lights used to those that can be distinguished by people with the most common forms of color defect. For example, the CIE has recommended areas on the CIE 1931 Chromaticity Diagram within which red, green, yellow, blue, and white signal lights should lie. These areas are designed so that the red signal will be named as red and the green as green by people with the most common forms of defective color vision (CIE, 1994).

9 OTHER EFFECTS OF LIGHT ENTERING THE EYE

Although making things visible is the most obvious effect of light entering the eye, there are two other ways in which light can affect us. The first is through the circadian system. The second is through the psychological impact of what is visible.

9.1 Circadian System

Light entering the eye does more than stimulate the visual system. It also influences the human circadian system and hence our biological rhythms. These rhythms occur primarily over one of the three geophysical periods found in nature: the day–night cycle, the lunar cycle, and the seasonal cycle. These cycles are important because human physiology, mood, and capabilities vary over each cycle.

The human circadian system has three components: an internal oscillator, a number of external oscillators that entrain the internal oscillator, and a messenger hormone, melatonin, that carries the internal "time" information to all parts of the body through the bloodstream (Dijk et al., 1995). The light–dark cycle is one of the most potent of the external stimuli for entrainment. Also, by varying the amount of light exposure and when it is presented, it is possible to shift the phase of the circadian system clock, either forward or backward, as required. In addition, it is possible to have an immediate alerting effect by exposure to light during circadian night (Boyce, 2003). The amount of light required to cause phase shifts or an immediate alerting effect is within the range of current lighting practice. However, the spectral sensitivity of the retinal photoreceptor(s) that provide signals to the internal oscillator are not the same as the CIE Standard Photopic Observer, the peak sensitivity being about 465 nm (Brainard et al., 2001; Thapan et al., 2001). This means that the effectiveness of light sources for stimulating the circadian system cannot be evaluated using the CIE photometry system (see Section 2).

The growing understanding of the importance of the light–dark cycle for circadian rhythms has significance for human health and well-being. One application where exposure to light has been of interest is shift work. The short-term problems of shift work are fatigue, produced by poor-quality sleep, and maintaining alertness during work. Long term, there is evidence that shift workers have a higher risk of cardiovascular disease, gastrointestinal ailments, and emotional and social problems. The short-term problems are believed to occur because of a mismatch between the demands of the work and the state of the worker's circadian rhythm. Put plainly, the workers are expected to work when their physiology is telling them to sleep and sleep when their physiology is telling them to be awake. Light is useful in alleviating this problem because it can shift the human circadian rhythm so that it better matches the functional requirements, but to do this requires control of light exposure for a complete 24-h period (Eastman et al., 1994). As for the immediate alerting affect, improvements in alertness and cognitive performance have been found following exposure to high light levels during night shift work, together with physiological changes indicative of the state of the circadian rhythm (French et al., 1990; Boyce et al., 1997).

Another human health problem that has been shown to be sensitive to light exposure is seasonal affective disorder (SAD). People with this condition typically experience decreased energy and stamina, depression, feelings of despair, and a greater need for sleep during the winter months. Light therapy, in which the patient is exposed to a high illuminance for a set period each day, has been shown to alleviate these symptoms in many patients (Lam and Levitt, 1999).

The use of light to alleviate the problems of shift work and to treat seasonal depression are just the most advanced examples of the influence of light on human well-being. Other potential applications of light therapy include treatment of sleep disorders; more general, nonseasonal depression, and jet lag.

Also of interest is what side effects exposure to light during circadian night might have (Brainard et al., 1999). Until a clearer understanding of the positive and negative impacts of exposure to light during circadian night is achieved, it would be wise to treat with caution attempts to use light exposure to manipulate such a fundamental part of our physiology as the circadian system.

9.2 Positive and Negative Affect

Psychology is a vast field and the psychology of lighting is only a small part of it. The area relevant to lighting practice that has been studied most consistently is that of perception. Studies have been undertaken in abstract situations and have lead to quantitative relations being proposed between simple sensations such as brightness and photometric measurements such as luminance (Boyce, 2003). Other studies have been undertaken in rooms with complete lighting installations and have lead to an understanding of the link between the perception of gloom and such photometric characteristics of the room as reflectance and illuminance distributions (Shepherd et al., 1989). Yet others have tried to establish if lighting generates cues by which people interpret a room: for example, whether lighting the walls enhances the perception of spaciousness (Flynn et al., 1973).

Although such studies have certainly influenced lighting design, they cannot be said to constitute a coherent body of knowledge. Further, they cannot form a basis for lighting practice until the impacts of specific perceptions are understood. To understand the consequences of perception of lighting, it is necessary to take a broader view. This view centers around positive affect. *Positive affect*, defined as pleasant feelings induced by commonplace events or circumstances, has been found to influence cognition and social behavior (Isen and Baron, 1991). Specifically, positive affect has been shown to increase efficiency in making some type of decisions and to promote innovation and creative problem solving. It also changes the choices people make and the judgments they deliver. For example, it has been shown to alter people's preference for resolving conflict by collaboration rather than avoidance and also to change their opinions of the tasks they perform.

Given these usually desirable outcomes of positive affect, it is necessary to ask what can generate positive affect. The answer is both small and wide. Small, because the stimuli that have been shown to generate positive affect are low-level stimuli, ranging from receiving a small but unexpected gift from a manufacturer's representative to being given positive feedback about task performance. Wide, because positive affect can be influenced by the physical environment, the organizational structure, and the organizational culture. Lighting is clearly a part of the physical environment and has been shown to influence positive affect (Baron et al., 1992; McCloughan et al., 1999), but it is only one of many factors that can do that.

As would be expected, it is also possible to generate negative affect. There is considerable information on the influence of frustration or anger on aggression and on the relationship between anxiety and performance (Baron, 1977). It seems reasonable to propose that lighting conditions that cause visual discomfort could generate negative affect.

Positive and negative affect provide plausible routes whereby the perception of the visual environment might influence the efficiency and effectiveness of organizations. As such, they represent a very different approach to identifying what is the most appropriate form of lighting for organizations to the visibility-based recommendations used in lighting practice today. The possibility that improving the quality of lighting beyond that required for good visibility without discomfort would lead to enhanced organizational performance is a topic of current interest (CIE, 1998b; Boyce et al., 2003b).

10 TISSUE DAMAGE

The part of the electromagnetic spectrum from 100 nm to 1 mm is called *optical radiation*. This part of the electromagnetic spectrum covers ultraviolet (100 to 400 nm), visible (400 to 760 nm), and infrared radiation (760 nm to 1 mm). Sunlight and electric light sources all emit optical radiation. In sufficient quantities, optical radiation can cause damage to the eye and the skin. Details are given in CIE (1998c).

10.1 Mechanisms for Damage to the Eye and Skin

There are two mechanisms for tissue damage to occur: photochemical and thermal. They are not mutually exclusive; both can occur for the same incident optical radiation, but one will have a lower damage threshold than the other. *Photochemical damage* is related to the energy absorbed by the tissue within the repair or replacement time of the cells of the tissue. *Thermal damage* is determined by the magnitude and duration of the temperature rise. The factors that determine the likelihood of tissue damage are the spectral irradiance incident on the tissue, the spectral sensitivity of the tissue, the time for which the radiation is incident and for thermal damage, the area over which the irradiance occurs. Spectral irradiance will be determined by the spectral radiant intensity of the source of optical radiation; the spectral reflectance and/or the spectral transmittance of materials from which the optical radiation is reflected or through which it is transmitted; and the distance from the source of optical radiation. Area is important for thermal tissue damage because the potential for dissipating heat gain is greater for a small area than a large area.

The visual system provides automatic protection from tissue damage in the eye for all but the highest levels of visible radiation. This is the involuntary aversion response produced when viewing bright light. The response is to blink and look away, thereby reducing the duration of exposure. Of course, this involuntary response works only for sources that have a high visible radiation component, such as the sun. Sources that produce large amounts of ultraviolet and infrared radiation with little visible radiation are particularly dangerous because they do not trigger the aversion response.

10.2 Acute and Chronic Damage to the Eye and Skin

Tissue damage can be classified according to the duration of exposure that it takes to produce the damage. Acute forms of damage are detectable immediately or at least within a few hours of exposure. Chronic forms of damage only become apparent after many years. Ultraviolet radiation incident on the skin produces immediate pigment darkening, followed a few hours latter by erythema (reddening of the skin), and ultimately, by a tan, produced by an increase in the number, size, and pigmentation of melanin granules. Excessive ultraviolet radiation incident on the eye can produce, a few hours later, an inflammation of the cornea called *photokeratitis*. This typically lasts a few days, followed by recovery. As for chronic damage, prolonged exposure to ultraviolet radiation has been shown to be associated with various forms of skin cancer and cataract.

Visible radiation incident on the skin will produce erythema but not tanning, and, in sufficient quantity, skin burns. Visible radiation incident in the eye reaches the retina. This irradiance represents both an acute photochemical and an acute thermal hazard to the eye. Photochemical damage to the retina is associated with short-wavelength light (blue light). The thermal damage covers retinal burns. As for chronic damage, it may be that prolonged and repeated exposure to light is involved in the retinal aging process.

Infrared radiation incident on the skin again initially produces skin reddening and, at a high enough irradiance, burns. Infrared radiation incident on the eye will cause heating of various elements of the eye, depending on the spectral content of the irradiance and the spectral transmittance of the various components of the eye. Infrared radiation from 760 to 1400 nm will reach the retina and can cause retinal burns. Longer wavelengths will be absorbed by other components of the eye. Prolonged heating of the lens is believed to be involved in the incidence of cataract.

10.3 Damage Potential of Various Light Sources

The light source with the greatest potential for tissue damage is the sun. The sun produces copious amounts of ultraviolet, visible, and infrared radiation. Voluntary staring at the sun is a common cause of retinal burns. Voluntary exposure of the skin to the sun commonly produces sunburn. However, there exist some electric light sources that can be hazardous, some being intended for lighting and others being used as a source of optical radiation for industrial processes.

The extent to which a light source represents a hazard can be evaluated by applying the recommendations of the American Conference of Governmental Industrial Hygienists (ACGIH, 2001), using the procedures described by the Illuminating Engineering Society of North America (IESNA, 1996). These recommendations take several different forms, ranging from maximum permissible exposure times to irradiance limits. Application of these standards to various electric light sources indicate that such sources, as conventionally used for interior lighting, rarely represent a hazard (McKinlay et al., 1988; Bergman et al., 1995; Kohmoto, 1999).

10.4 Approaches to Limiting Damage

The approach to minimizing the damage caused by optical radiation is to limit the irradiance and/or the time of exposure. Whether any such action is necessary can be determined by applying the ACGIH recommendations to the situation. For sources of optical radiation used for lighting, if the threshold limiting values are exceeded, it will often be possible to use a different light source that is less hazardous. If this is not possible, it is necessary to filter the source to eliminate some of the hazardous wavelengths or to use some form of eye or skin protection to attenuate the optical radiation or to limit the exposure time. For sources of optical radiation used in industrial processes, the source should be installed in an enclosure, with an interlock so that opening the enclosure extinguishes the source. If this is not possible, appropriate forms of eye and skin protection are required.

11 EPILOGUE

Illumination has been a subject of study for more than 90 years. The result has been a growing understanding of how lighting conditions and the visual system interact to facilitate visual performance and diminish visual discomfort. This knowledge has formed the framework around which many national illuminating engineering organizations have built recommendations for lighting practice (IESNA, 2000; CIBSE, 2002). These recommendations provide a firm basis for designing everyday lighting installations, provided always that the recommendations are applied with thought and not by rote.

There are three current areas of study with considerable potential: (1) the value of better lighting quality for the efficiency of organizations, (2) the effect of light spectrum on visual performance in mesopic conditions, and (3) the nonvisual effects of light, particularly the photobiological. Knowledge gained in these areas has the potential to change how lighting is designed and the technology that is used to provide it.

REFERENCES

ACGIH (2001), *TLVs and BEIs Threshold Limit Values for Chemical Substances and Physical Agents, Biological Exposure Indices*, American Conference of Governmental Industrial Hygienists, Cincinnati, OH.

Ashdown, I., and Franck, P. J. (1995), "Luminance Gradient: Photometric Analysis and Perceptual Reproduction," in *Proceedings of the IESNA Annual Conference*, IESNA, New York, pp. 128–150.

Bailey, I., Clear, R., and Berman, S. (1993), "Size as a Determinant of Reading Speed," *Journal of the Illuminating Engineering Society*, Vol. 22, pp. 102–117.

Baron, R. A. (1977), *Human Aggression*, Plenum Press, New York.

Baron, R. A., Rea, M. S., and Daniels S. G. (1992), "Effects of Indoor Lighting (Illuminance and Spectral Distribution) on the Performance of Cognitive Tasks and Interpersonal Behaviors: The Potential Mediating Role of Positive Affect," *Motivation and Emotion*, Vol. 16, pp. 1–33.

Bergman, R. S., Parham, T. G., and McGowan, T. K. (1995), "UV Emission from General Lighting Lamps," *Journal of the Illuminating Engineering Society*, Vol. 24, pp. 13–24.

Berman, S. M., Greenhouse, D. S., Bailey, I. L., Clear, R. D., and Raasch, T. W. (1991), "Human Electroretinogram Responses to Video Displays, Fluorescent Lighting and Other High Frequency Sources," *Optometry and Vision Science*, Vol. 68, pp. 645–662.

Berman, S. M., Fein, G., Jewett, D. L., and Ashford, F. (1993), "Luminance-Controlled Pupil Size Affects Landolt C Task Performance," *Journal of the Illuminating Engineering Society*, Vol. 22, pp. 150–165.

Bjorset, H. H., and Frederiksen, E. (1979), "A Proposal for Recommendations for the Limitation of the Contrast Reduction in Office Lighting," in *Proceedings of the 19th Session of the CIE*, Kyoto, Japan, Commission Internationale de l'Éclairage, Vienna, pp. 310–314.

Boff, K. R., and Lincoln, J. E. (1988), *Engineering Data Compendium: Human Perception and Performance*, Harry G. Armstrong Aerospace Medical Research Laboratory, Wright-Patterson Air Force Base, OH.

Bourdy, C., Chiron, A., Cottin, C., and Monor, A. (1987), "Visibility at a Tunnel Entrance: Effect of Temporal Adaptation," *Lighting Research and Technology*, Vol. 19, pp. 35–44.

Boyce, P. R. (1985), "Movement Under Emergency Lighting: The Effect of Illuminance," *Lighting Research and Technology*, Vol. 17, pp. 51–71.

Boyce, P. R. (1991), "Lighting and Lighting Conditions," in *The Man–Machine Interface*, Vol. 15 of *Vision Visual and Dysfunction*, J. A. J. Roufs, Ed., Macmillan, London.

Boyce, P. R. (2003), *Human Factors in Lighting*, Taylor & Francis, London.

Boyce, P. R., and Rea, M. S. (1987), "Plateau and Escarpment: The Shape of Visual Performance," in *Proceedings of the 21st Session of the CIE*, Venice, Italy, Commission Internationale de l'Éclairage, Vienna, pp. 82–85.

Boyce, P. R., and Slater, A. I. (1981), "The Application of Contrast Rendering Factor to Office Lighting Design," *Lighting Research and Technology*, Vol. 13, pp. 65–79.

Boyce, P. R., Beckstead, J. W., Eklund, N. H., Strobel, R. W., and Rea, M. S. (1997), "Lighting the Graveyard Shift: The Influence of a Daylight-Simulating Skylight on the Task Performance and Mood of Night Shift Workers," *Lighting Research and Technology*, Vol. 29, pp. 105–142.

Boyce, P. R., Akashi, Y., Hunter, C. M., and Bullough, J. D. (2003a), "The Impact of Spectral Power Distribution on the Performance of an Achromatic Visual Task," *Lighting Research and Technology*, Vol. 35, pp. 141–161.

Boyce, P. R., Veitch, J. A., Newsham, G. R., Myer, M. A., and Hunter, C. M. (2003b), *Lighting Quality and Office Work: A Field Simulation Study*, Light Right Consortium, Battelle, Richland, WA.

Brainard, G. C., Kavet, R., and Kheifets, L. I. (1999), "The Relationship Between Electromagnetic Field and Light Exposures to Melatonin and Breast Cancer: A Review of the Relevant Literature," *Journal of Pineal Research*, Vol. 26, pp. 65–100.

Brainard. G. C., Hanifin, J. P., Greeson, J. M., Byrne, B., Glickman, G., Gerner, E., and Rollag, M. D. (2001), "Action Spectrum for Melatonin Regulation in Humans: Evidence for a Novel Circadian Photoreceptor," *Journal of Neuroscience*, Vol. 21, pp. 6405–6412.

CIBSE (2002), *CIBSE Code for Lighting*, Chartered Institution of Building Services Engineers, London.

CIE (1971), *Colorimetry*, Publication 15, Commission Internationale de l'Éclairage, Vienna.

CIE (1972), *Special Metamerism Index: Change in Illuminant*, Supplement 1 to Publication 15, Commission Internationale de l'Éclairage, Vienna.

CIE (1978), *Recommendations on Uniform Color Spaces, Color-Difference Equations, Psychometric Color Terms*, Supplement 2 to Publication 15, Commission Internationale de l'Éclairage, Vienna.

CIE (1983), *The Basis of Physical Photometry,* Publication 18.2, Commission Internationale de l'Éclairage, Vienna.

CIE (1986), *Colorimetry*, Publication 15.2, Commission Internationale de l'Éclairage, Vienna.

CIE (1987), *Methods of Characterizing Illuminance Meters and Luminance Meters: Performance, Characteristics and Specification*, Standard 69, Commission Internationale de l'Éclairage, Vienna.

CIE (1990), *CIE 1988 2° Spectral Luminous Efficiency Function for Photopic Vision*, Publication 86, Commission Internationale de l'Éclairage, Vienna.

CIE (1994), *Review of the Official Recommendations of the CIE for the Colours of Signal Lights*, Technical Report 107, Commission Internationale de l'Éclairage, Vienna.

CIE (1995), *Method of Measuring and Specifying Color Rendering Properties of Light Sources*, Publication 13.3, Commission Internationale de l'Éclairage, Vienna.

CIE (1998a), *The CIE 1997 Interim Colour Appearance Model CIECAM97*, Publication 131-1998, Commission Internationale de l'Éclairage, Vienna.

CIE (1998b), *Proceedings of the First CIE Symposium on Lighting Quality*, Publication x015-1998, Commission Internationale de l'Éclairage, Vienna.

CIE (1998c), *Measurements of Optical Radiation Hazards*, Publication x016-1998, Commission Internationale de l'Éclairage, Vienna.

CIE (2002), *CIE Collection on Glare*, Publication 146-2002, Commission Internationale de l'Éclairage, Vienna.

Cornu, L., and Harlay, F. (1969), "Modifications de la discrimination chromatique en fonction de l'éclairement," *Vision Research*, Vol. 9, pp. 1273–1280.

Dijk, D.-J., Boulos, Z., Eastman, C. I., Lewy, A. J., Campbell, S. S., and Terman, M. (1995), Light Treatment for Sleep Disorders, Consensus Report II: Basic Properties of Circadian Physiology and Sleep Regulation," *Journal of Biological Rhythms*, Vol. 10, pp. 113–125.

Eastman, C. I., Stewart, K. T., Mahoney, M. P., Liu, L., and Fogg, L. F. (1994), "Dark Goggles and Bright Light Improve Circadian Rhythm Adaptation to Night Shift Work," *Sleep*, Vol. 17, pp. 535–543.

Eklund, N. H., Boyce, P. R., and Simpson, S. N. (2001), "Lighting and Sustained Performance: Modeling Data-Entry Task Performance," *Journal of the Illuminating Engineering Society*, Vol. 30, pp. 126–141.

Flynn, J. E., Spencer, T. J., Martyniuck, O., and Hendrick, C. (1973), "Interim Study of Procedures for Investigating the Effect of Light on Impression and Behavior," *Journal of the Illuminating Engineering Society*, Vol. 3, pp. 87–94.

French, J., Hannon, P., and Brainard, G. C. (1990), "Effects of Bright Illuminance on Body Temperature and Human Performance," *Annual Review of Chronopharmacology*, Vol. 7, pp. 37–40.

IESNA (1996), *Recommended Practice for Photobiological Safety for Lamps and Lamp Systems*, ANSI/IESNA RP-27-96, Illuminating Engineering Society of North America, New York.

IESNA (2000), *Lighting Handbook*, 9th ed., Illuminating Engineering Society of North America, New York.

Inditsky, B., Bodmann, H. W. and Fleck, H. J. (1982), "Elements of Visual Performance, Contrast Metric–Visibility Lobe–Eye Movements," *Lighting Research and Technology*, Vol. 14, pp. 218–231.

Isen, A. M., and Baron, R. A. (1991), "Positive Affect as a Factor in Organizational Behavior," *Research in Organizational Behavior*, Vol. 13, pp. 1–53.

Kahn, H. A., and Moorhead, H. B. (1973), *Statistics on Blindness in the Model Reporting Area, 1969–70*, Publication (NIH) 73-427, U.S. Department of Health, Education and Welfare, Washington, DC.

Kelly, D. H. (1961), "Visual Response to Time-Dependent Stimuli: 1. Amplitude Sensitivity Measurements," *Journal of the Optical Society of America*, Vol. 51, pp. 422–429.

Koenderink, J. J., and van Doorn, A. J. (1979), "Spatiotemporal Contrast Detection Threshold Surface Is Bimodal," *Optics Letters*, Vol. 4, pp. 32–34.

Kohmoto, K. (1999), "Evaluation of Actual Light Sources with Proposed Photobiological Lamp Safety Standard and Its Applicability to Guide on Lighted Environment," in *Proceedings of the CIE, 24th Session*, Warsaw, Commission Internationale de l'Éclairage, Vienna.

Lam, R. W., and Levitt, A. J. (1999), *Canadian Consensus Guidelines for the Treatment of Seasonal Affective Disorder*, Clinical and Academic Publishing, Vancouver, British Columbia, Canada.

Littlefair, P. J. (1990), "Innovative Daylighting: Review of Systems and Evaluation Methods," *Lighting Research and Technology,* Vol. 22, pp. 1–17.

McCloughan, C. L. B., Aspinall, P. A., and Webb, R. S. (1999), "The Impact of Lighting on Mood," *Lighting Research and Technology*, Vol. 31, pp. 81–88.

McIntyre, D. A. (2002), *Colour Blindness: Causes and Effects*, Dalton Publishing, Chester, UK.

MacAdam, D. L. (1942), "Visual Sensitivity to Color Differences in Daylight," *Journal of the Optical Society of America*, Vol. 32, pp. 247–274.

McKinlay, A. F., Harlen, F., and Whillock, M. J. (1988), *Hazards of Optical Radiation*, Adam Hilger, Bristol, England.

Muck, E., and Bodmann, H. W. (1961), "Die Bedeutung des Beleuchtungsniveaus bei Praktische Sehtatigkeit," *Lichttechnik*, Vol. 13, pp. 502–507.

Osterberg, G. (1935), "Topography of the Layer of Rods and Cones in the Human Retina," *Acta Ophthalmologica*, Supplement 6, pp. 1–103.

Rea, M. S., and Jeffrey, I. G. (1990), "A New Luminance and Image Analysis System for Lighting and Vision, 1: Equipment and Calibration," *Journal of the Illuminating Engineering Society,* Vol. 19, pp. 64–72.

Rea, M. S., and Ouellette, M. J. (1988), "Visual Performance Using Reaction Times," *Lighting Research and Technology*, Vol. 20, pp. 139–153.

Rea, M. S., and Ouellette, M. J. (1991), "Relative Visual Performance: A Basis for Application," *Lighting Research and Technology*, Vol. 23, pp. 135–144.

Robbins, C. L. (1986), *Daylighting: Design and Analysis*, Van Nostrand Reinhold, New York.

Robertson, A. R. (1977), "The CIE 1976 Color-Difference Formulae," *Color Research and Application,* Vol. 2, pp. 7–11.

Sekular, R., and Blake, R. (1994), *Perception*, McGraw-Hill, New York.

Shepherd, A. J., Julian, W. G., and Purcell, A. T. (1989), "Gloom as a Psychophysical Phenomenon," *Lighting Research and Technology*, Vol. 21, pp. 89–97.

Shlaer, S. (1937), "The Relation Between Visual Acuity and Illumination," *Journal of General Physiology*, Vol. 21, pp. 165–168.

Shlaer, S., Smith, E. L., and Chase, A. M. (1941), "Visual Acuity and Illumination in Different Spectral Regions," *Journal of General Physiology*, Vol. 25, pp. 553–569.

Simons, R. H., and Bean, A. R. (2000), *Lighting Engineering*, Butterworth-Heinemann, London.

Smith, S. W., and Rea, M. S. (1979), "Relationships Between Office Task Performance and Ratings of Feelings and Task Evaluations Under Different Light Sources and Levels," in *Proceedings of the 19th Session of the CIE*, Kyoto, Japan, Commission Internationale the l'Éclairage, Vienna, pp. 207–211.

Steward, J. M., and Cole, B. L. (1989), "What Do Colour Defectives Say About Everyday Tasks," *Optometry and Vision Science*, Vol. 66, pp. 288–295.

Stone, P. T., and Harker, S. P. D. (1973), "Individual and Group Differences in Discomfort Glare Responses," *Lighting Research and Technology*, Vol. 5, pp. 41–49.

Tansley, B. W., and Boynton, R. M. (1978), "Chromatic Border Perception: The Role of Red- and Green-Sensitive Cones," *Vision Research,* Vol. 18, pp. 683–697.

Thapan, K., Arendt, J., and Skene, D. J. (2001), "An Action Spectrum for Melatonin Suppression: Evidence for a Novel Non-rod, Non-cone Photoreceptor System in Humans," *Journal of Physiology*, Vol. 535, pp. 261–267.

Tielsch, J. M. (2000), "The Epidemiology of Vision Impairment," in *The Lighthouse Handbook on Vision Impairment and Vision Rehabilitation*, B. Silverstone, M. A. Lang, B. P. Rosenthal, and E. E. Faye, Eds., Oxford University Press, New York.

Wald, G. (1945), "Human Vision and the Spectrum," *Science*, Vol. 101, pp. 653–658.

Wilkins, A. (1993), "Reading and Visual Discomfort," in *Visual Process in Reading and Reading Disabilities*, D. M. Willows, R. S. Kruk, and E. Corcos, Eds., Lawrence Erlbaum Associates, Hillsdale, NJ.

Wilkins, A. J., Nimmo-Smith, I., Slater, A. I., and Bedocs, L. (1989), "Fluorescent Lighting, Headaches and Eye- strain," *Lighting Research and Technology*, Vol. 21, pp. 11–18.

Wyszecki, G., and Stiles, W. S. (1982), *Color Science: Concepts and Methods, Quantitative Data and Formulae*, Wiley, New York.

PART 5

DESIGN FOR HEALTH, SAFETY, AND COMFORT

CHAPTER 26

OCCUPATIONAL HEALTH AND SAFETY MANAGEMENT

Bernhard M. Zimolong and Gabriele Elke
Ruhr University–Bochum
Bochum, Germany

1 RISK MANAGEMENT PRINCIPLES

Risk management may be defined as the reduction and control of the adverse effects of the risks to which an organization is exposed. Risks include all aspects of accidental losses that may lead to any wastage of the organization's, society's and environmental assets. These assets cover personnel, materials, machinery, procedures, products, money, and natural resources: soil, water, energy, natural areas. Losses may result from the presence of potential harm to one or more elements of the system, either because of the interactions with other elements inside the system or with the environment outside the system. *Risk* is the measuring stick for this potential, which may be defined as the probability that harm will occur within a certain period.

Management as a function comprises all processes and functions resulting from the division of labor in an organization, such as planning, organizing, leading, and controlling. In most organizations more or less formalized management systems serve to structure, develop, and direct business processes. Systems differ with respect to branches, nature of business, company size, and human factors such as culture and policy. As firms grow in size, management systems gain complexity and become difficult to use, thus resulting in domain-specific systems such as management of health, safety, environmental resources, quality, or personnel. Since health, safety, and environmental (HSE) management overlaps in several ways and is often practiced by the same people in an integrated manner, companies are now moving toward integrated HSE management systems as a subsystem of business and operations management.

Risk management can be understood as an approach to reintegrate domain-specific management systems into an integrative management concept. Many of the features of risk management are indistinguishable from the sound management practices advocated by proponents of quality and business excellence. This is reflected in standards usually based on the ISO 9000 (e.g., BS 7750) and ISO 14000 series, and in legislative developments in many countries: for example, in the EU, the Environmental Management and Audit Scheme (EMAS) (European Union, 2001) regulation and the Control of Major Hazards (COMAH) directive (European Community, 1999), and in the United States, the Risk Management Program (RMP) of the Environmental Protection Agency. The British environmental standard BS 7750 and EMAS contributed to the development of the ISO 14000 series Environmental Management Systems standards. Initiatives to launch an international standard on occupational health and safety (OHS) management systems have been delayed. In many countries, national guidelines give guidance on OHS management systems.

The essence of risk management is to prepare, protect, and preserve the resources of the enterprise. This approach demands analyzing the current and past operating hazard, risk, and loss-producing patterns and forecasting expected hazard, risk, and loss-operating patterns. According to Bamber (2003), risk control strategies may be classified into four main areas: risk avoidance, risk retention, risk transfer, and risk reduction. *Risk avoidance* means involves a deliberate decision on an part of an organization to avoid a particular risk. *Risk retention* relates to the decision of an organization to meet any resulting loss from within the organization's financial resources. *Risk transfer* refers to the legal assignment of the costs of potential losses from one party to another. The most common approach is by insurance. The principles of *risk reduction* or *risk control* rely on the implementation of a health, safety, and environment (HSE) program, whose basic aim is to protect a company's assets from waste caused by accidental loss.

The system elements to be managed include, among others, (1) the health and safety of employees, suppliers, contractors, customers, and residents of the community (e.g., improvement of public health and safety); (2) the reliability and safety of products and services, of materials, equipment, work systems, and plants, and of the transport of hazardous goods; (3) integrated pollution control, radiation protection, waste minimization, recycling, and waste disposal; and (4) sustainable management of natural resources (soil, water, natural areas, and coastal zones) and reduction in the consumption of nonrenewable energy.

Sustainable development is an improvement in the quality of life that does not impair the ability of the ecosystem to maintain life. Managing for sustainability is based predominantly on the principles of intergenerational and intrageneral equity as well as social and ecological balance (Hutchinson and Hutchinson, 1997). The management approach of the ISO standards is based on generic management principles which are derived from different theoretical and organizational perspectives. The elements of the systems are considered to present *best practices* of successful enterprises. They are designed to be used by organizations of all sizes and regardless of the nature of their activities. The key elements of a generic management system are integrated in the management control cycle outlined in Figure 1 (see Section 3). The cycle is based on ISO 14000 (environmental management) and on standard BS 8800 designed for an OHS management system. The key elements of such a management system are set out by the Health and Safety Executive (1997).

1. *Effective OHS management* involves developing, coordinating, and controlling a continuous improvement process by setting and adjusting OHS standards. The corporate policy is summarized in a vision, containing perspectives for the future and providing an idea of identification for all members of the organization. Policy and strategy are translated into planning processes. Both internal and external assessment methods should be used for the evaluation of a strategy's effectivity and efficiency. Regular reviews of performance based on data from monitoring activities and from audits of the OHS management system may serve as instruments.

2. *Formulation of an OHS policy* addresses the preservation and development of physical and human resources and reductions in financial losses and liabilities. The policy provides guidance on the allocation of responsibilities and the organization of people, of resources, communication, and documentation. It influences design and operation of working systems, the design and delivery of products and services, and the control and disposal of waste. OHS policy should be aligned with people management policy to secure the commitment, involvement, and well-being of employees, suppliers, contractors, and customers.

3. *OHS planning* is an organizational approach that emphasizes prevention and involves risk identification, evaluation, and control. Proactive planning means that hazards are identified and risks assessed and controlled according to a systematic plan, before anyone or anything could be affected adversely. Reactive planning means that measures are considered only after the occurrence of incidents such as loss damages, accidents, or deficient safety performance.

4. *Implementation and operation* Designing OHS structures concerns the divisions of responsibility and distribution of formal authority, the creation of hierarchical or lean structures, the degree of self-regulation of work groups and units, and the formal relations between groups and leaders. Establishing and maintaining control is central to all management functions including OHS. The allocation of OHS responsibilities to line managers, team leaders, and self-managed work groups serves as an important tool to foster the integration of OHS into the daily work activities with specialists acting as advisers.

5. *Checking and corrective actions* are the final steps in the OHS management control cycle and part of the feedback loop needed to enable the organization to maintain and develop its ability to control risks successfully. Both qualitative and quantitative measures provide information on the effectiveness of the OHS system. Learning from experience is supported through performance reviews and independent audits. This needs to be done systematically through regular reviews of performance based on data from monitoring activities and from audits of the HSE management system.

6. *Management review* is a periodic status review of the OHS management system and considers the overall performance of the OHS management system, of individual elements, and of the findings of audits. The review identifies the actions that need to be taken to adjust any deviations.

Safety* management is more than a "paper system" of policies and procedures. An audit of the official safety management system (SMS) may start and end with an analysis of what is contained in the paperwork,

*The term *safety*, used in short for "health, safety, and environment," refers to damage to hardware and environment as well as to people.

but it therefore says little about how the system is being transferred into practice. Such an analysis identifies what an organization should be doing to protect its workers, the public, and the environment from harm, but it does not reveal what is actually happening at the work site and whether or not people and the environment are being protected and adverse events are not occurring.

2 HEALTH AND SAFETY MANAGEMENT

2.1 Scope and Dimensions

Recent promotion of scientific studies on safety management can be related to different events and trends (Hale and Hovden, 1998):

1. *Major disasters.* Several major disasters in the nuclear, petrochemical, and transport industries have caused strong public concerns over the management of hazardous activities (e.g., Seveso, Bhopal, Chernobyl, Piper Alpha, Challenger, Herald of Free Enterprise) (see Reason, 1990). People had trusted until then that these high-technology industries had been managed appropriately by well-developed safety management systems. Official reports found that the root causes involved more than technical or human failures and pointed to faults in management and organization. Turner (1978) provided an analysis of human-made disasters and pointed beyond the technical and human factors to the organizational and cultural factors.
2. *Probabilistic risk assessment (PRA).* As a consequence of the disasters, mandatory quantified risk assessments in the nuclear industry, followed in some countries by the petrochemical industry, were required by regulators. Early PRAs were almost exclusively technical. The Three Mile Island accident emphasized the human factor. Human reliability analysis (HRA) was added to PRAs. Norman Rasmussen (NRC, 1975) developed its use for process safety management at nuclear facilities.
3. *Self-regulation and certification.* Under the philosophy of self-regulation, that those who create risks and pollution should be responsible for their control, regulatory emphasis changed in the 1970s. The central responsibility was placed on each company's management for devising, installing, and monitoring its own safety management system. This led to a withdrawal of government from detailed technical regulation and close shop floor inspection of health and safety. Regulators adopted indicators for assessing how the potentialities were being used by companies. Thus, certification, audits, and other periodic assessments are used to assess company performance.

The safety management principles of the ISO standards and of standard textbooks on safety management seem to suggest that science and industry have reasonable models of how safe and reliable organizations work. However, this is not the case. As Roberts (1990) points out, the organizational literature fails to deal specifically with either hazardous organizations or high levels of performance reliability. The standard texts on safety management (e.g., Heinrich et al. (1980); Bird and Germain, 1987) present neither specific models of the safety management system nor provide empirical evidence of how particular aspects of the suggested frameworks contribute to the overall level of HSE. Hale and Baram (1998) conducted a thorough literature review on HSE management and revealed a number of lines of research and isolated studies which seem to have few links with each other. They concluded that the literature on SMS can be characterized, at least until the 1980s, as the accumulated experience of common sense and as general management principles applied to the specific field of safety.

One of the earliest studies was that of Cohen (1977). He reviewed seven studies that dealt with critical determinants in different industrial settings. Some of the factors associated with high safety performance were strong management commitment to safety; close contact and interaction between workers, supervisors, and management, enabling open communications on safety as on other job matters; workforce subject to less turnover, including a large core of married, older workers with significant length of service in their jobs; high level of housekeeping, orderly workplace conditions, and effective environmental quality control; well-developed selection, job placement, and advancement procedures and other employee support services; training practices emphasizing early indoctrination and follow-up instruction in job safety procedures; and evidence of added features or variations in conventional safety practices serving to enhance their effectiveness.

Shafai-Sahrai (1971) examined 11 matched pairs of companies conducting on-site interviews and site inspections at each. Factors prevalent in low-injury-rate companies were senior management involvement in safety; prioritization of safety in meetings and in decisions concerning work practice; better injury record-keeping systems; use of accident cost analysis; reduced span of supervisor responsibility; spacious and clean workplace environment; and improved safety devices on machinery. Additionally, Cohen and Cleveland (1983) reported findings from a linked series of studies examining health and safety management in organizations with good safety performance across different industries. Methods included a questionnaire survey of 42 matched pairs of plants with low and high accident rates, with seven pairs of these subject to detailed site surveys. Those with lower accident rates were characterized by a strong management commitment to safety; a humanistic approach to dealing with employees, with frequent positive contact and interaction; encouragement of hazard identification by workers; better housekeeping and general plant cleanliness; presence of both informal and formal workplace inspections; greater availability and use of personal protection equipment; improved employee selection procedures; low turnover and absenteeism; and better plant environment.

Referring to these studies, Chew (1988) compared safety activities in 18 pairs of low- and high-injury-rate companies drawn from three Asian countries. Prevalent factors were supervisory involvement in safety activities; safety inspection; safety training; use of accident record analysis for prevention purposes; carefully applied safety rules; machine guarding; supply of personal protection equipment; and standard of housekeeping. Shannon et al. (1996) conducted a postal survey of over 400 manufacturing companies, each having at least 50 employees. The defining features of organizations with lower rates of lost-time injuries included managers who perceived more participation in decision making by the workforce and more harmonious management–worker relations; encouragement of long-term career commitment; provision of short- and long-term disability plans; definition of health and safety responsibilities in every manager's job description; performance appraisals with topics related to health and safety; and more frequent attendance of senior managers at health and safety meetings.

Finally, Shannon et al. (1997) reviewed 10 studies each including at least 20 separate workplaces or organizational units and using injury rates as an outcome variable. Forty-eight variables representing areas of management practices were examined. The study only listed the practices consistently associated with performance (i.e., the association was significant in one direction in at least two-thirds of studies in which it appeared, and the direction of relationship was consistent for all studies). Those practices are as follows:

1. *Joint health and safety committee*: health and safety professionals on the committee, longer duration of training of committee members
2. *Managerial style and culture*: direct channels of communication and information, empowerment of the workforce, good relations between management and workers
3. *Organizational philosophy on health and safety*: delegation of safety activities, active role of top management in safety, more thorough safety audits, lengthier duration of safety training for employees, safety training on regular basis, employee health screening

Although there may be methodological weaknesses with the empirical studies (Dufort and Infante-Rivard, 1998), these studies, together with accounts of successful safety initiatives (Griffiths, 1985; Harper et al., 1997; DePasquale and Geller, 1999; Hine et al., 1999) are in some level of agreement about the ideal safety management practices. According to Mearns et al. (2003, p. 644), the general themes that emerge are (1) genuine and consistent management commitment to safety (including prioritization of safety over production, maintaining a high profile for safety in meetings, personal attendance of managers at safety meetings and in walk-abouts, face-to-face meetings with employees that feature safety as a topic, and job descriptions that include safety contracts), (2) communication about safety issues (including pervasive channels of formal and informal communication and regular communications between management, supervisors, and the workforce), and (3) involvement of employees (including empowerment, delegation of responsibility for safety, and encouraging commitment to the organization).

An extensive review on the literature dealing with internal management system of organizations was provided by Hale and Hovden (1998). Literature on risk management at the national or industry level dealing with regulation, standard setting, risk policies, enforcement, and the management of individual workplaces and work groups was excluded. These concern notably participative management studies and studies of high-reliability organizations, which concern themselves with online management of risk, as opposed to the off-line concern with management structure found in much of the literature. Not included are studies of safety analysis and information systems in companies or of feedback of safety performance as a method of safety improvement (Kjellén and Larsson, 1981; Saari and Näsänen, 1989; see the review in Komaki, 1998).

The review summarized its findings using a management classification scheme proposed by Bolman and Deal (1984). The four frames represent four different perspectives of an organization: (1) structural topics emphasizing organization, information systems and procedures, hierarchy, rules, and how goals are achieved in analogy to quality assurance; (2) human resource factors focusing on human needs, motivation, commitment, competence, participation, and motivation; (3) political factors relating to or dealing with control authorities, responsibility, power, allocation of resources, negotiation, and conflict; and (4) symbolic topics emphasizing values, culture, role-playing, metaphors, and heroes.

2.1.1 Structural Topics

Associated with success in developing internal control systems were measurable goals and standards, competence in organizational development, and access to external expertise and available resources. Positive relation to low accident rate or good environmental performance was found for availability of financial resources, problem-solving approach, stable workforce, safety training systems, good communication channels, good coordination and centralization of safety control, specialist safety service in the company, good records, a small span of control for and time to plan by supervisors, presence of accident reporting and analysis system, and evaluation and review systems. Many studies advocate systems for deviation control, modeled on quality management systems. On the subject of rules, procedures, and formalization, the studies are most contradictory. Some of the NIOSH studies (Smith et al., 1978) have shown the positive value of well-defined and detailed rules in stable, predictable situations; others in the same series showed no discrimination effect of good rules on safety performance; still others (see, e.g., Rasmussen and Batstone, 1991) have warned of their negative effect, especially in online management of complex, dynamic technologies. The difference would appear to relate to the type of organization studied or the state of development of its system of rules.

2.1.2 Human Resource Factors

The human resource factors have been reasonably well researched. Positive relations with safety performance were found for the following topics: participation, empowerment, encouraging innovation, group norms, leadership style, feeling of control, efficacy, autonomy, social policy, quality of work life, and career progression. In the area of communication these emphasize the content and quality of the communication as opposed to the structural aspects of communication channels.

2.1.3 Political Factors

Company profitability and availability of resources were found to be related positively to high safety in two studies but unrelated in a third. Inconsistent results were found for absence of incentive payment schemes for production and presence of safety incentive schemes. Sanctioning of violations was related to high accidents. Unclear results were obtained for the role of discipline as opposed to counseling. Openness to criticism, good labor relations, low stress, and low grievance rates were all related to low accident rate. External pressure from regulators and the presence of an order-seeking management (as opposed to crisis management) were related to good performance. The importance of good industrial relations, openness to admit mistakes and criticism, and a constructive rather than disciplinary approach to violations has been reasonable supported.

2.1.4 Symbolic Factors

Top management commitment and real visibility in that commitment are found several times positively related to safety. Supervisor's and individual commitment, importance of safety as a value, safety attitudes of coworkers, and work as a source of pride were positively related to safety performance. The evidence of safety promotion is inconsistent.

In their summary of the literature review, Hale and Hovden (1998) concluded (p. 9ff.):

1. The field of safety management is young and not yet adequately based on empirical studies. A considerable amount of the state of the art rests on expert opinion and the analysis of case studies. An exception is the safety climate research.

2. There has been an overemphasis on the structural aspects of management; the rules, responsibilities, formal hierarchies, plans, and social policies. In contrast, too little attention has been paid to the internal issues of human resources, of coping with potential conflicts and power-plays within the organization, of different perceptions and objectives among the various parties in the organization, and of integrating the values and commitment of all parties concerned.

3. Literature is dominated by studies carried out in large organizations, often with a very considerable investment in sophisticated defense-in-depth management systems in safety and environment. It has led to the idea that safety management systems must be rule bound and rule dominated. Audit systems, certification regimes, and regulatory checks reinforce this view with their search for indicators that can easily be detected and proven.

Hospitals, research organizations, small innovative companies such as designer and information technology companies, or simple conventional small companies have different structures and different means of coordinating their activities and of setting and monitoring their standards, which sometimes hardly overlap with those of large organization. Many do not rely for their normal management control on extensive explicit rules, but much more on competence and communication. Existing assessment techniques are not very well equipped to measure these.

The survey reveals the lack of research on the dimension of organizational learning and change. Studies of the Berkeley School (Roberts, 1990; Rochlin, 1996) on high-reliability organizations is work considering online management of dynamic processes enclosed within military command and control cultures, not the robustness of safety management systems in companies subject to the sorts of reorganization available in a wide-open business environment.

2.2 Leadership

Occupational health and safety has not been recognized by academics as a managerial and organizational research domain (Fahlbruch and Wilpert, 1999). Less than 1% of organizational research published in top journals has focused on occupational safety, a situation that has not changed for more than two decades (Barling et al., 2002). Contrary to academic neglect, safety management has been practiced successfully worldwide by a great number of enterprises for decades. Policies, strategies, procedures, and practices of excellent enterprises have been reviewed by business consultants, safety practitioners, and academics (Johnson, 1975; Heinrich et al., 1980; Bird and Germain, 1987; Hale and Glendon, 1987; Hoyos and Zimolong, 1988; Zimolong, 1997; Zimolong and Elke, 2001b).

2.2.1 Role of Managers and Supervisors

For the last decades, businesses have reorganized continuously to cut costs, improve productivity, and remain competitive. As a result, there has been a proliferation of management delayering to move responsibilities to those people carrying out the operations and to focus on teamwork. Teams can be managed in different ways: by supervisors, team leaders, or self-managed (Vassie and Lucas, 2001). A supervisor who is considered to be the accountable manager of the team is responsible for planning, organizing, and controlling the members of the group but will often not undertake any work within the group. In contrast, a team leader is normally not accountable for the work but relies on leadership skills to motivate and coordinate the work of others and facilitate their self-development. A team leader is often a working member of the group. Where a work group has no team leader, the team becomes self-managed. Surveys conducted in Britain and Germany (Windel and Zimolong, 1998; Cully et al., 1999) found that 60 to 80% of the workforce worked in teams. However, in both countries, only 3% worked in self-managed work groups.

In strongly hierarchically structured organizations, senior management defines organizational policies, from which in turn strategic goals and means of goal attainment are derived. Procedures provide tactical guidelines for action related to these goals and means. There is a continuous process of adapting policies to internal and external requirements and subsequently adapting internally strategic goals, procedures, and practices. In delayered organizations, the policy development process may be characterized as a ping-pong process; (i.e., as an active exchange process across the levels of the organization). On the operational level, lower management (i.e., supervisors and team leaders) execute procedures by turning them into practices. From this perspective, senior managers are concerned with policy making and the establishment of procedures to support policy implementation. On the other hand, managers do execute actions, which should be in alliance with policies and procedures but often fail to comply with procedures.

Upper management generally indicates their safety support indirectly. They establish priorities for policies, procedures, and goals, set production schedules that may accommodate safe operations, and they control the incentives for complying with those priorities (e.g., compensation, rewards, discipline). Group leaders indicate management support for safety more directly than do higher-level managers. They monitor compliance with higher management's policies, and they provide feedback to employees regarding the adequacy of their behaviors. Management policy and practice often sent inconsistent messages. Inconsistencies result from comparison of officially declared safety aims and goals with actual practices. For instance, if management is perceived as willing to set aside safe practices to meet production goals, employees are likely to attribute management's support for safety as paying lip service. This could lead some employees to conclude that cutting corners will be rewarded, or at least not be punished.

Thompson et al. (1998) explored three factors that might be strongly influencing perception of policy–practice incongruence: goal incongruence (i.e., confusion over safety goals and the significance of competing organizational goals such as timeliness and customer orientation); organizational politics of "not sending disagreeable messages to management"; and manager fairness (i.e., perception that elevated safety concerns might not be given a fair hearing). Authors tested a model that linked management support, safety perceptions, and self-reported safety outcomes. Confirmatory factor analysis results indicated that organizational politics was related to perceptions of upper management support, which in turn influenced perception of workplace safety conditions. Supervisor fairness influenced perceived supervisor support for workplace safety, which in turn influenced perceptions of workplace safety compliance. Finally, upper manager support for workplace safety was also found to influence perceptions of supervisor support.

There has been done little empirical research to understand how managers on different levels of hierarchy, including supervisors, team leaders, and self-managed groups, promote workplace safety. The role of management and leadership and their influence on safety performance have been analyzed by comparing organizations with high and low rates of injuries and/or absenteeism (Zohar, 1980; Hofmann et al., 1995; Shannon et al., 1997; Zimolong, 2001a; Mearns et al., 2003), by analyzing accident and breakdown rates (Reason, 1990; Feyer et al., 1997), by case studies from the process industry (Hofmann et al., 1995), or by the examination of correlations between organizational characteristics and safety-oriented and health-promoting behavior (Butler et al., 1990; Hofmann and Stetzer, 1996; Zohar, 2002a, b).

Safety management practices of upper management were studied by Mearns et al. (2003). Authors conducted safety surveys on 13 offshore oil and gas installations in separate years and compared safety management practices with safety performance rates. Data on safety management practices were collected by questionnaire from health and safety manager of each participating company or business unit, or the asset manager for each installation. Appropriate indicators have been identified in different industries (Hurst et al., 1996; Miller and Cox, 1997; Lee, 1998; Fuller, 1999). The items were organized thematically in line with the Health and Safety Executive (1997) classification of management practices. Safety performance data represented among other indicators the official accident and incident rates and the proportion of individually reported accidents. In both years, the following practices were consistently negatively associated with the rate of lost time injuries: (1) policies for health and safety, (2) organizing for health and safety, (3) management commitment and involvement, (4) health promotion and surveillance, and (5) health and safety auditing.

Overall scores of safety management practices were associated with lower accident rates in both years. Probably due to small numbers, only health promotion was related significantly to the injury rate in both years. In total, proficiency in safety management practices was associated with lower official accident rates and fewer respondents reporting accidents. Some results showed inconsistent outcomes for the two years, including significant positive coefficients that contraindicate certain practices presumed favorable. For example, management commitment was positively associated with the rate of dangerous occurrences in the first year.

The influence of individual senior managers on the safety behavior of supervisors was investigated in a study by Zohar (2002b). Intervention takes place at the level above target behavior, that is, on the level of direct superiors of the subordinate. This is a cross-level approach whereby processes introduced at one hierarchical level influence a lower subordinate level. Zohar (2002a) conducted a leadership-based intervention study in a maintenance center of heavy-duty equipment. Line supervisors received weekly personal feedback concerning the frequency of safety-related interactions with subordinates. Feedback was based on repeated episodic interviews with subordinates concerning the cumulative frequency of their safety-oriented interactions. Immediate superiors (section managers) received the same information and used it to communicate high safety priority. Results indicated that supervisory safety practices changed over a short period from a baseline rate of 9% to a new plateau averaging 58%. This change, in turn, resulted in significant decrease of minor injury rate.

The moderating effect of assigned safety priority by the direct superior indicates the importance of managerial goal setting and feedback. A leader whose immediate superior emphasizes safety will be more concerned with safety issues than he or she would have been otherwise. Even though leader roles entail considerable discretion, the expectations communicated by an immediate superior will influence a leader's practice. As a consequence, a leader's emphasis on safety issues is an interactive function of personally assigned safety priorities derived from the interaction with group members and externally assigned safety priorities by upper management.

Research has focused primarily on supervisors as role models for promoting safety awareness and supporting safe behavior (Mattila et al., 1994; see the review in Komaki, 1998). Effective group leaders (e.g., line supervisors, team leaders) continually provide training and goal setting (antecedents) and feedback and incentives (consequences). A series of field studies conducted by Komaki and colleagues (Komaki et al., 1978) revealed two primary attributes of effective supervision: performance-based monitoring and timely communication of consequences. Effective supervisors monitor work in progress, particularly through work sampling (i.e., direct observation), and act accordingly. This practice clarifies both supervisory directives and expectations (i.e., antecedents) and behavior–outcome contingencies. For example, Bentley and Haslam (2001) compared safety practices of supervisors of high- and low-accident-rate postal delivery offices, particularly with respect to slip, trip, and fall accidents. Supervisors from low-accident-rate offices appeared to have improved performance with respect to quality of safety communication, dealing with hazards reported on delivery walks, and accident investigation and remedial action.

Behavior safety is often at odds with other performance aspects, particularly speed and productivity. As a consequence, a leader's effectiveness will be strongly influenced by his or her skill and willingness to deal simultaneously with competing goals and practices and to communicate safety commitment and practices to his or her followers. Results of a study by Hofmann and Morgeson (1999) indicated that quality of relationships between group leaders and their superiors (i.e., leader–member–exchange–LMX level) predicted injury records in work groups through the mediating effects of safety communication (i.e., frequency of raising safety concerns with a superior and the leader' s declared commitment to safety).

2.2.2 Transactional and Transformational Leadership

Several recent studies have suggested that transformational leadership is associated with better safety records (O'Dea and Flin, 2001; Zohar, 2002a). Hofmann and Morgeson (1999) showed that the relations in 49 dyads between leader–member exchanges and occupational accidents were mediated by safety communication and safety commitment. This is consistent with results of other studies that show that the effects of transformational leadership on performance are mediated by different aspects of employee morale, such as organizational commitment, trust in management, and fairness (Jung and Avolio, 2000). Hence, transformational leadership is characterized by value-based and individualized interaction, resulting in better exchange quality and greater concern for welfare (Bass and Avolio, 1997). It affects critical subordinate attitudes and work-related outcomes such as satisfaction with leadership, work performance, and consolidated business unit performance. Barling et al. (2002) tested a model linking safety-specific transformational leadership to occupational injuries. Results of the study provided strong support for the mediation model linking safety specific transformational leadership to occupational injuries through the effects of perceived safety climate, safety consciousness, and safety-related events. A second study replicated and extended this model by role overload.

Transactional leadership, the other global dimension, concerns planning and organization of tasks and getting people trained and motivated to do actions more reliable and efficiently. Transactional supervision influences safety due to effective monitoring and rewarding practices, which are needed to maintain reliable performance during routine job operations. On the other hand, transformational leadership role relates to development and motivating people to commit themselves to more challenging goals. As noted by several authors (e.g., Bass, 1990; Bass and Avolio, 1997), the relationship between the two is that of augmentation (i.e., transactional supervision provides reliability and predictability, and transformational leadership provides better motivation and development orientation). That means that effective managers must excel in both.

The transactional role of managers was further subdivided into constructive, corrective, and laissez-faire dimensions (Bass and Avolio, 1997). Constructive leadership implies an intermediate level of concern for members' welfare. Leaders must identify needs, desires, and individual capabilities to offer motivationally relevant rewards. Transformational and constructive dimensions often merge into a single factor due to such individualized considerations (Avolio et al., 1999). Corrective leadership (i.e., management by exception) mainly includes error detection and correction based on monitoring of subordinates' performance in relation to required standards. This results in poorer, nonindividualized interactions. Finally, laissez-faire leadership implies the lowest level of concern for members' welfare (i.e., nonleadership or disowning supervisory responsibilities despite rank in the hierarchy).

Zohar (2002a) studied the influence of transformational and transactional leadership on safety behavior of group members. The setting was the same as described in Zohar (2002b). Leadership was measured with the Multifactor Leadership Questionnaire (MLQ-5X-Revised) (Bass and Avolio, 1997). Results indicated that transformational and constructive leadership were highly intercorrelated, as was the passive role of corrective leadership and laissez-faire. Transformational and constructive leadership predicted injury rate, whereas corrective leadership provided indirect, conditional prediction. Although corrective leadership failed to predict injury rate directly, it could be predicted with a two-part linkage: corrective leadership predicted climate, which then predicted injury rate. Leadership effects were moderated by assigned safety priorities of superiors and mediated by commensurate safety-climate variables. The type of interaction depended on leadership dimensions. Climate perceptions, transformational and constructive leadership, and assigned priority of superior were significantly correlated with injury rate in the direction expected.

The fact that the effects of leadership dimensions on injury records varied depending on assigned managerial priorities is important for an efficient safety management system. Depending on departmental policies and the relative priority of competing goals, supervisors monitor some performance aspects closely while paying less attention to others. In the case of unfavorable safety priorities, highly corrective supervisors seem to actively monitor speed, quality, or productivity but to neglect safety deviations. Thus, assigned safety priorities of superiors are particularly important to reinforce safety priorities of supervisors with corrective or laisser-faire leadership styles. Zohar (2002a) argued that improved transactional supervision enhances performance reliability of shop floor employees and that transformational qualities should result in incremental effects, particularly under high production pressures. This augmentation implies that leadership-based intervention should be expanded to include both leadership factors.

Leadership styles must be placed into the context of policies, structure, and underlying culture of organizations. Simard and Marchand (1994) illustrated the influence of what they call *micro organizational factors* on safety initiatives. Their results showed that a participatory leadership style shaped the propensity of work groups to take such initiatives. Participatory involvement of supervisors was associated significantly with accident prevention activities and lost-time accident rates. In other types of organizations with a different workforce, such a type of leadership and its effect on safety performance may be less effective. Leadership style and safety performance is probably a product of the underlying culture.

2.3 Safety Culture and Climate

2.3.1 Definitions and Relations

Research on safety climate emerged from studies on organizational culture and climate. The construct of organizational climate refers to shared perceptions among members of an organization with regard to organizational policies, procedures, and practices (Reichers and Schneider, 1990; Rentsch, 1990). Climate is a multidimensional construct that covers a wide range of individual assessments of the work environment. These assessments may be directed toward general dimensions of the environment, such as leadership, roles, and communication, or to specific dimensions such as the climate for safety or the climate for customer service (James et al., 1990).

The concept of safety culture has developed largely since the OECD Nuclear Agency (1987) observed that the errors and violations of operating procedures occurring prior to the Chernobyl disaster were evidence of a poor safety culture at the plant and within the former Soviet nuclear industry in general (Pidgeon and O'Leary, 2000). Safety culture has been defined as "that assembly of characteristics and attitudes in organizations and individuals, which establishes that, as an overriding priority, plant safety issues receive the attention warranted by their significance" (IAEA, 1986). Safety culture is important because it forms the context within which individual safety attitudes develop and persist and safety behaviors are promoted. Pidgeon (1991) considers culture as a system of meanings and defines *culture* as the collection of beliefs, norms, attitudes, roles, and practices shared within a given social grouping or organization. Turner 1991 characterizes safety culture as the set of beliefs, norms, attitudes, roles, and social and technical practices that are concerned with minimizing the exposure of employees, managers, customers, and members of the public to conditions considered dangerous or injurious. According to Pidgeon (1991), a "good safety culture can be characterized by three attributes: norms and rules for handling hazards, attitudes towards safety, and reflexivity on safety practice" (p. 135). When industrial and national cultures are also embraced, organizations may be considered as subcultures within societies. Safety climate may reflect employees' perceptions of the organization's policies, procedures, and practices concerning safety and helps employees to make sense of the priority accorded to safety within the organization (Barling et al., 2002).

The relationship between safety culture and climate is unclear, and considerable confusion exists in the literature about the cause, the content, and the consequence of safety culture and climate (Guldenmund, 2000). Referring to the culture concept of Schein (1992), who conceives climate as a reflection and manifestation of cultural assumptions, Guldenmund (2000) suggested the following definition: "Safety culture is defined as those aspects of the organizational culture which will impact on attitudes and behavior related to increasing or decreasing risk" (p. 251). Hence, safety climate is considered as a manifestation of safety culture in the behavior and expressed attitude of employees.

Most definitions of safety culture invoke shared norms or attitudes so that the level of aggregations is considered to be the group. Group-level climate perceptions reflect shared patterns of practices rather than isolated supervisory actions (Zohar, 2000). Some authors assert that only safety culture is being assessed when the attitude object is the organization (Cabrera and Isla, 1998).

Zohar (2000) suggests regarding climate perceptions as a relatively stable pattern of actions, in contrast to individual actions. For example, if managers and workers emphasize speed over safety, this may result in safety violation actions or in dangerous work activities. The focal issue is the overriding priority of production versus safety, derived as a conclusion from frequent individual observations and/or conversations. Managers, particularly supervisors, act as a role model, reflecting in individual role episodes overall emphasis or deemphasis on safety issues.

The link between organizational climate and specific safety climate was studied by Neal et al. (2000). Results from 32 work groups in a large Australian hospital indicated that general organizational climate exerted a significant impact on safety climate, and safety climate in turn was related to self-report of compliance with safety regulations and procedures as well as participation in self-related activities within the workplace. Safety climate was found to operate as a mediating variable between organizational climate and safety performance, while the effect of safety climate on safety performance was partially mediated by safety knowledge and motivation.

Over 30 studies using safety climate questionnaires have been published so far (Glendon and Litherland, 2001). Flin et al. (2000) and Guldenmund (2000) have summarized most of them in an approach to identify key factors of safety climate. Flin et al. (2000). found that the number of factors varied between 2 and 19 in the studies they reviewed, while Lee and Harrison (2000) extracted 28 factors from their analysis. However, if the same questionnaire was applied in different settings or even in similar organizations, different factor structures have been found (Brown and Holmes, 1986; Dedobbeleer and Béland, 1991; Coyle et al., 1995). One explanation for the difference observed is the variety of questionnaires, samples, and methodologies used in different studies. Glendon and Litherland (2001) argue that labeling of factors play an important role. In most cases, factor analysis is applied to derive one-dimensional factors. This methodology relies extensively on researcher discretion, in particular in factor labeling. It seems possible that more similarities exist between factor structures than is apparent from the comparisons conducted to date.

The number of dimensions of safety climate remains disputed, although recurring themes across safety climate surveys include management commitment, commitment of leaders, supervisory competence, priority of safety over production, and time pressure (Elke, 2001). Flin et al. (2000) reported three main factors: management/supervision, safety system, and risk; also frequently emerging were work pressure and competence. It might be well assumed that factors of safety climate and culture not only differ between subunits and organizations, but also between industrial and national culture (Ludborzs, 1995).

Conceptual confusion may arise in differentiating the concept of safety management from safety culture and climate. Kennedy and Kirwan (1998) assert that "safety climate and safety management are at lower levels of abstraction and are considered to be a manifestation of the overall safety culture" (p. 251). In this sense the safety culture is reflected in the strength of the SMS and the safety climate. Mearns et al. (2003) take the view that safety management practice is an indicator of the safety culture of upper management. More favorable safety management practices are expected to result in an improved safety climate of the general workforce, and vice versa. Examination of safety management practices should be considered an adjunct to the assessment of safety climate within an organization. From a functional point of view, safety culture represents an implicit control system in the form of a socialization programme. It is a value-based system based on the organization's mission, values, business details, customers, and the expectations of the employees.

2.3.2 Safety Climate and Performance

Various studies have revealed that safety climate factors can predict safety-related outcomes, such as accidents or injuries (Zohar, 1980, 2003; Brown and Holmes, 1986; Dedobbeleer and Béland, 1991; DeJoy, 1994, Niskanen 1994; Hofmann and Stetzer, 1996; Diaz and Cabrera, 1997; for an overview, see Glendon and Litherland, 2001). Factors of safety climate emerge as predictors of unsafe behavior or accidents in numerous structural models (Thompson et al., 1998; Cheyne et al., 1999; Tomas et al., 1999; Brown et al., 2000) and nonlinear models (Guastello, 1989; Guastello et al., 1999). Neal et al. (2000) found that safety climate influenced self-reported components of safety performance. Cheyne et al. (1998), using structural equation modeling (SEM), revealed that safety activity was influenced by five safety climate factors as antecedents: safety management, communication, individual responsibility, safety standards and goals, and personal involvement. Workplace hazards and the physical work environment were also components of the model. Results of SEM approaches support the notion of Hofmann and Stetzer (1996) that the influence of safety climate upon safety performance is mediated through the work conditions and environment.

An expected, negative correlation between a positive degree of safety culture and accident and injury rates as well as a positive correlation with safety-oriented actions can also be confirmed (Zimolong and Stapp, 2001; Barling et al., 2002). Inconsistent results revealed the aforementioned study of Mearns et al. (2003). Safety climate surveys were conducted parallel to management practice surveys on offshore oil and gas installations in separate years. Installations were assessed on their safety climate, safety management practice, and safety performance. Results indicated that (1) differences between installations in their accident rates were reflected in differences in safety climate scores, (2) favorable safety climate at installation level was not consistently associated with lower proportion of employees experiencing an accident (not consistently with lower official accident reports either), and (3) favorable safety climate at the individual level was associated with a lower likelihood of accident involvement.

There were inconsistent results for the expectation that installation safety climate predicts the proportion of respondents reporting an accident on each installation. This expectation was confined to year 1, not to year 2. Similarly, involvement in health and safety decision making was significantly associated with injury rate in year 1, not in year 2. The safety climate scores were also expected at the individual level as predictors of self-reported accidents. The discriminant function exceeded 68% correct classifications. However, as Mearns et al. (2003) admitted, this value is not high and there was little consistency in the set of best predictors.

What are the antecedent factors that promote a favorable climate? Hofmann et al. (1995) denote the individual attitudes and behaviors noticeable in safety climate as the microelements of an organization, which themselves are determined by macroelements of the SMS and practices. Safety philosophy of upper management, attitudes and behavior permeate down through the levels of organization to the workforce. As a consequence, research has focused on supervisors as role models for promoting safety awareness and supporting safe behavior (Mattila et al., 1994). Involvement of the workforce in safety decision making has also received attention (Simard and Marchand, 1994). Several studies identified various procedure-based criteria through which employees could assess the relative priority of safety, resulting in organizational-level perceptions (Brown and Holmes, 1986; Coyle et al., 1995). These criteria include nonproductive investment in safety technology (e.g., protective devices, safety inspections) or practices of personnel management (safety training, safety-related incentive system).

A safety climate may possibly have a mediator function between leadership and safety-oriented behavior (Simard and Marchand, 1997; Hofmann and Morgeson, 1999; Zohar, 2002b). The mediation role of safety climate was supported in a study of restaurant employees (Barling et al., 2002). Safety-specific transformational leadership predicted injuries through the effects of perceived safety climate, safety consciousness, and safety-related events. A second study replicated and extended this model by role overload.

Results of the study by Hofmann and Morgeson (1999) indicated that quality of relationships between group leaders and their superiors (i.e., leader–member–exchange–LMX level) predicted injury records in work groups through the mediating effects of safety communication (i.e., frequency of raising safety concerns with a superior and the leader's declared commitment to safety). Both of these mediators are related to safety climate (i.e., they provide behavioral and declarative evidence for workers to assess supervisory concern for their safety and the importance of safe behavior). Outcomes on the effects of leadership dimensions, safety climate, and assigned priorities on minor injuries in work groups (Zohar, 2002a) indicated that leadership effects were moderated by assigned safety priorities and mediated by commensurate safety-climate variables. Climate perceptions, transformational and constructive leadership, and assigned priority of superior were correlated significantly with injury rate in the expected direction. Transformational and corrective leadership were correlated with one aspect of climate perception: preventive action. Similarly, corrective and laissez-faire leadership and climate correlated negatively with this type of climate perception.

3 ORGANIZATIONAL CONTROL OF HAZARD AND RISK

3.1 Planning and Controlling for Safety

Safety management efforts have traditionally been directed at the prevention of repetitions of accidents that have already occurred. Strategy has concentrated on reactive prevention rather than proactive prevention. Measures to control hazards have been based on information derived from detailed accident investigations. Assessing risks and devising preventive measures without the help of accident data is difficult: It involves assessing the probabilities of a wide range of unwanted outcomes. A hazard-effect control plan must be developed and maintained to cope with all the hazards and adverse effects detected. Range of measures has focused almost exclusively on the errors made by people who were involved in accidents, not the managers and engineers whose errors may have created a climate, and a physical environment where failures occur more likely or more serious. Unsafe acts create conditions where further unsafe acts may lead to incidents or accidents. The remote errors by managers and designers have been described by Reason (1987a) as latent or decision failures, and the errors performed by people directly at risk as active failures.

Proactive safety management must address the following distinctive elements of the accident causation process:

1. *Multicausality of accidents.* Accidents happen as a result of a chance concatenation of many distinctive causative factors, each one necessary but not sufficient to cause a final breakdown (Reason, 1987a).

2. *Active and latent failures.* Active failures are events that have an immediate adverse effect. In

contrast, latent failures lie dormant in an organization for some time, becoming evident only when they are triggered by active failures: unsafe acts and unsafe conditions. The dichotomy stresses the importance of management responsibilities for safety and draws attention to the scope for detecting latent failures in the system well before they are revealed by active failures.

3. *Different modes of human errors.* According to the framework of Rasmussen (1983), human errors may be triggered on different levels of awareness. Skill-based errors involve slips and lapses in highly practiced and routine behavior; rule-based errors are mismatches between a situation and the required actions to be taken; and knowledge-based errors occur due simply to missing information or wrong inference. Violations, sometimes referred to as *risk taking*, may occur when people deliberately carry out actions in a way contrary to a known or unknown rule. The success of training programs depends on the nature of the errors likely to be made.

Risk management and safety management are both allied to the system safety approach. Similar disciplines are the total quality management and the environmental management systems, which are based on the same considerations but work with different operational processes and objectives. Quality management systems are designed to detect and correct deviations from quality standards. This principle is known as the *deviation concept* in safety sciences (Kjellén, 1984). Hazards are defined as deviations from a standard or ideal situation. They will lead to damage of the system elements (human, equipment, material) and/or to the system's environment (water, soil, air, people), if they are not prevented, discovered, and corrected.

The overall objective of system safety is the reliability of functions: The system is to work the way it was planned and nobody should be harmed by an accident, toxic substance, or malfunction. The term *system* is used to describe clearly defined activities, processes, and equipment during the lifetime of the system (Hoyos, 1992). Events such as exploding spray cans, magnesium burning on dumps for weeks, or chlorofluorocarbon (CFC) that goes up into the atmosphere are examples of the need of a total system management of safety. First, systems were regarded as systems with their subcomponents; soon the systems of humans–machines and the extension of the term to organizational systems with the subcomponents human, group, organization, methods, and processes were included. According to Bird and Loftus (1976), the stages associated with system safety are as follows: (1) the preaccident (proactive) identification of potential hazards; (2) the timely incorporation of effective safety-related design and operational specification, provisions, and criteria; (3) the early evaluation of design and procedures for compliance with applicable safety requirements and criteria; and (4) the continued surveillance over all safety aspects throughout the total life span, including disposal of the system.

Depending on the objective of the system, each system will have a series of phases, which follows a chronological pattern. The overall life span of a manufactured product may comprise the following phases: conception and planning; design and engineering; use/operation; modification and maintenance; demolition and disposal. Each of these phases can be further subdivided. The risk management of HSE for an organization must consider all phases of the life cycle of the plant or facility that it exploits and the goods and services that it delivers. Each phase is an activity that must be managed safely in its own right and which must feed forward into subsequent phases and have a feedback loop to facilitate organizational learning processes.

This concept is illustrated in Figure 1, taking into consideration the life span of a system (Zimolong and Hale, 1989). The model emphasizes that hazards built into a technology or activity, and the preventive measures to eliminate and control them are largely conditioned by the decisions made in the planning and design phases of the activity. The life-cycle phases are shown horizontally, the control and rescue measures are placed vertically: the elimination, reduction, and control of hazards as well as the limitations of the consequences of damage.

During the operational phase, the system may move from time to time outside the defined parameters or standards. The flexibility and built-in controls of the sociotechnical system (i.e., of personnel, organization, and technical protective measures) normally allow the system to return to normal operation, and no near miss, breakdown, or accident is experienced. In the overwhelming number of industrial settings, however, only a few of the deviations are discovered and controlled effectively. The system remains in an unstable situation. This state may last anywhere from seconds to months, years, and lifetime. During this time, deviations might be detected by the operator, maintenance and repair personnel, safety inspector, or a safety audit team. If no recovery actions are taken and the defenses are inappropriate for this case, the latent failures eventually turn into active failures, which suddenly become potent with some conditions that are harmless in themselves (Reason, 1993). Now only secondary protection measures to contain or divert the energy that was released accidentally are appropriate measures to limiting damage to less important system elements. Examples are safety goggles to guard against foreign objects, fall-arrest harnesses to stop fall, and hard hats to protect against falling objects. Further damage-reducing measures are rescue actions that limit the time of exposure to harmful energy, for example, disconnecting the current in electrocution cases or washing off a caustic chemical, administering timely first aid and proper treatment by the rescue system.

System safety essentially is planning for health and safety. The process includes establishing performance standards by which to measure and assess the HSE policy, the organizational arrangements for developing and maintaining it, the physical controls needed to

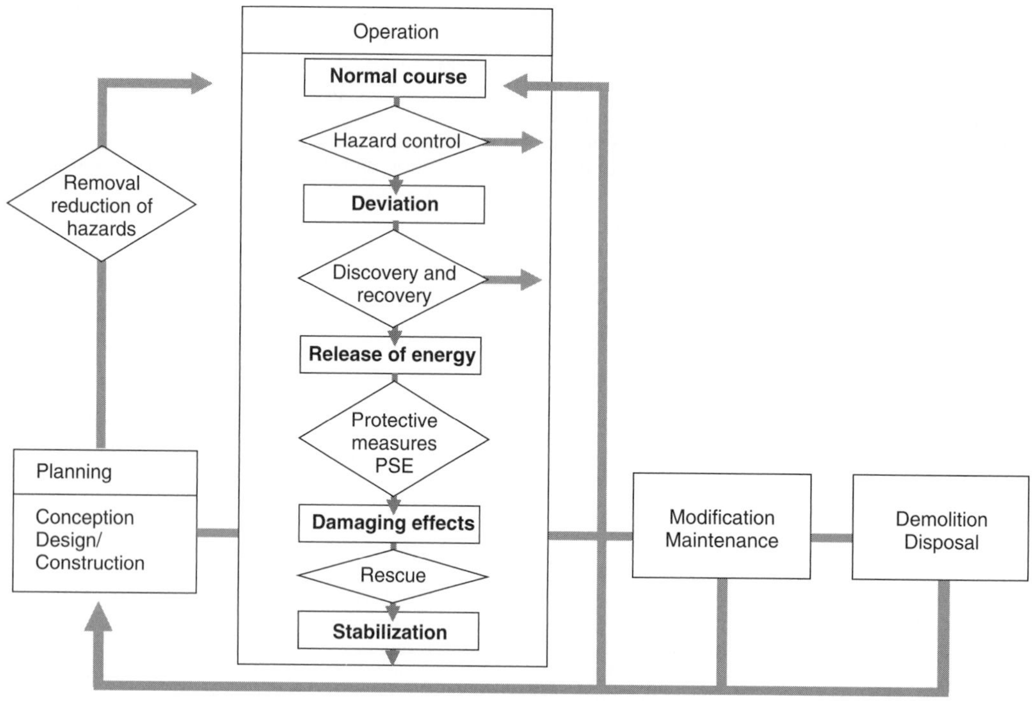

Figure 1 Risk assessment and effect management through the life cycle of business activities. The process consists of four steps: identify, assess, control, and recover. Hazards are interpreted as deviations from standards. Four generic life-cycle phases (design/planning, operation, modification/maintenance, and demolition) are shown from left to right. For the operation phase, the normal, deviation, and rescue phase is depicted as a model.

meet the requirements of the performance standards (hardware control), and the systems and procedures required by the performance standards for managers, supervisors, and other employees (software control).

The Health and Safety Executive (1998) provides a framework for identifying key areas in the operational phase of a system for which performance standards are important. First-stage control includes the elimination and reduction of hazards and risks entering the organization. Performance standards cover the physical resources (i.e., workplaces, materials and substances, plants), the human resources (recruitment, selection of personnel and contracting organizations), and the information related to health and safety, risk control, and positive health and safety culture. Second-stage control eliminates or minimizes risks arising inside the organization. Performance standards cover hazards and risks arising from premises, plant and substances, procedures, and people. Third-stage controls minimize risks outside the organization arising from work activities, products, and services. Performance standards are required to control these risks, including both organizational procedures and the control of specific risks.

Table 1 Forms of Possible Health Hazards

Category	Examples
Physical hazard	Noise, temperature and humidity extremes, illumination, vibration, infrared and ultraviolet radiation
Biological agent	Microorganisms, germs, viruses, toxins
Chemical hazard	Mists, vapors, gases, fumes, dusts, liquids, pastes whose chemical composition can create health problems
Physical workload	Working postures, physiological load, movements and exertion of forces, manual material handling and lifting
Mental workload	Perceptive and cognitive workload
Stress	Resulting from control demands and individual control capabilities, from role conflicts and ambiguity; emotional strain resulting from aggression, loss of feedback, loss of control, helplessness

3.2 Origins of Deviations

From a safety point of view, hazard represents a source of energy with the potential of causing immediate injury to personnel and damage to equipment or structure (Table 1). Employees are further exposed to diverse toxic substances, such as chemicals, gases, or radioactivity, some of which cause health problems.

Unlike hazardous energies, which have an immediate effect on the body, toxic substances have quite different temporal characteristics, ranging from immediate effects to delays over months and years. Ozone is a good example. It causes inflammation along the entire respiratory tract. This is like sunburn deep in the lung. For people with asthma, it increases sensitivity to allergies, frequently requiring hospitalization. Often, there is an accumulating effect of small doses of toxic substances which are imperceptible to humans.

The harmful effects of health hazards, such as hearing loss, cancer, liver damage, and silicoses, are regarded as illnesses. However, back pains may result from improperly designed chairs, headaches from poor ergonomic layout of VDT workplace. This exceeds the traditional view on accidents. Consequently, controls are not always identical: The prevention of contact or its reduction to a level where no harm is done is valid only for hazardous or toxic materials, whereas illnesses resulting from poor design requires ergonomic standards, planning, and sometimes complete reinstallation of the working system. Under the total loss approach, accidents are taken as undesired events that result in harm to people, damage to property, or loss to process (Bird and Germain, 1987). They include not only those circumstances that actually cause health problems or injury, but also every event involving damage to property, plant, products or the environment, production losses, or increased liabilities. The severity of an injury that results from an accident is often a matter of chance. It depends on many factors, such as dexterity, reflexes, physical condition, the portion of the body injured; as well as the amount of energy exchanged, what barriers were in place, whether or not protective equipment was worn. The "no injury" incident or "near miss" often has the potential to become events with more serious consequences. Analysis of the more frequently occurring property damage incidents and the near misses provides more information for guidance in the work of prevention and clearer understanding of the causes of accident problems.

Several studies have been undertaken to establish the relationship between serious and minor accidents and other dangerous events (Bird and Germain, 1987; Health and Safety Executive, 1997). In industry, the pyramid of accident ratios is used by many companies, which is a statistical ratio between the number of fatalities, injuries, no-injury accidents, and incidents. It is by no means a causal relationship (i.e., preventing all minor injuries would not result in the prevention of all serious or fatal accidents). The actual ratios in different pyramids differ significantly, indicating the problem of reliable measurement and different ratios for different locations.

Traffic studies conducted by means of the traffic conflicts technique have clearly proved different ratios of fatalities, injuries, and conflicts (incidents) at various types of intersections (Zimolong, 1981). Not surprising, at least to the expert, is the result that apparently dangerous-looking traffic-light-regulated junctions have a vast amount of conflicts. The ratio of conflicts to accidents equals 1170 : 1. "Safe"-looking nonsignalized urban junctions have a smaller ratio of 470 : 1. At safe junctions, conflicts turn into accidents three times more often than at signalized junctions (Erke and Zimolong, 1978). Most accidents happen because people commit active failures, which are called *unsafe acts*. Not wearing safety glasses is one example. In terms of system safety, unsafe acts and unsafe conditions are substandard practices and substandard conditions (i.e., deviation from an accepted standard or practice). A vast number of substandard conditions involve poor ergonomic design of machine, equipment, and the work environment.

It is essential to consider these practices and conditions only as symptoms, which point to the latent failures (Wagenaar et al., 1994) or basic causes behind the symptoms (Bird and Germain, 1987). Incidents usually start with relatively insignificant and common failures of design, operating and maintaining of equipment, with human errors or degraded performance. In combination with circumstances and the reactions of equipment and people, hazards can be released and escalate to cause injuries or damage to environment and assets. The equipment failures and exacerbating circumstances themselves are also generally the result of human failures long before the incident (e.g., during design, construction, and planning). Human failures may also have a long history of bad habits and wrong work methods that were not corrected and of procedures that were not enforced. The interactions are complex between equipment, people, and surrounding environment.

The prevention of unsafe acts and conditions to minimize incidents will be quite troublesome if their systemic nature is overlooked. They are not random events, but logical and systematic consequences of psychological states. Examples are lack of attention, haste, inexperience, reasoning errors, and misperceived risk. Psychological states are again not random events. They are caused by latent errors related to managerial and organizational failures and omissions; errors that were made long before the accident, and which have been present all the time. "Haste may be caused by any one of the following: too rigorous planning, a reward system that stresses speed, lack of personnel, frequent breakdown of equipment, a motivation to complete more than the normal portion of work, exceptional emergencies that had never been foreseen" (Wagenaar et al., 1993, p. 159). In all these examples the cause of haste is a latent failure that has been present for a long time. Telling people not to be hasty is pointless. Haste can only be prevented by removal of latent failures that cause haste.

Various substandard practices relate to the deficiencies in communication and information between functional units of the company. Equipment and materials that are inadequate or hazardous will be purchased if there are no adequate standards and if compliance with standards is not managed. Poor work process layouts and interfaces will be designed and built if there are no adequate standards and compliance for design and construction. Equipment will wear out and produce products with quality deficiencies, create waste, or break down and cause property damage if that equipment is not properly selected, used, and maintained (Timpe, 1993).

The origins of the deviations from standard are deficiencies in management and organization. Wagenaar et al. (1994) have suggested 11 types of latent failures, which have emerged on the basis of studies of hundreds of accidents and incidents. They are related to the work environment, to the people doing their jobs, and to the management. Detailed lists that cover different factors are provided in Petersen (1978) and Bird and Germain (1987).

System safety engineering involves the application of scientific and engineering principles for the timely identification of hazards and initiation of those actions necessary to prevent or control hazards within the system (Leveson, 2002). It draws upon professional knowledge and specialized skills in the mathematical, physical, and related scientific disciplines, together with the principles and methods of engineering design and analysis to specify, predict, and evaluate the safety of the system. Although much of occupational safety is now being recognized as being behavioral, safety engineering still has a major role in occupational safety. General topics and methods are concerned with guarding energy sources, design and redesign of machinery, equipment and processes, application of environmental standards, and establishment of inspection systems, such as statutory engineering inspections of pressure vessels, cranes and lifting machines, or electrical installations. Comprehensive coverage of those topics provides Bird and Germain (1987) and Ridley and Channing (2003).

Workplace designs that do not take ergonomic principles into account are likely to lead to an increase in errors and accidents and a decrease in safety and efficiency. Error-prone designs place demands on performance that exceed the capabilities of the user, violate the user's expectancies based on his or her past experience, and make the task unnecessarily difficult, unpleasant, or dangerous. The systems approach applied to ergonomics treats humans and machines/computers as components interacting together to bring about some desired objective. The role of the individual is characterized by his or her capabilities and limitations: mainly, the human sensory capabilities, the perceptual/cognitive processes, and the human performance abilities. For the workplace designer or safety practitioner, reliable anthropometric data (i.e., data concerning the measurement of physical features and functions of the body are found in human factors handbooks) (Van Cott and Kinkade, 1972; Woodson, 1981). Ergonomic principles for enhancing the design and safe operation of work facilities and equipment may be organized at the component, the workstation, and the work-space level, including the work environment. At the component level, visual displays, various types of controls, and visually or auditory warnings are considered. Workstation designs are based on anthropometric data, which are available for designing cabinets, consoles, desks, and other workstations (Woodson, 1981; Corlett and Clark, 1995). Particularly, special considerations have been given to computer workplace design and to software user interface design (Hix and Hartson, 1993).

Overall workspace design is concerned with the integration of several work areas and how to ensure that ambient environmental conditions fall within acceptable ranges. Thermal comfort, noise, and lighting are among the most important environmental factors to assess in occupational settings. The essentially multidisciplinary nature of the subject is covered in a number of useful chapters and books, among others in Grandjean (1980), Shackel (1984), Corlett and Clark (1995), and Meister and Enderwick (2002).

3.3 Hazards and Effects Management

A vital part of safety management is the hazards and effects management (HEM) process (e.g., identifying and managing hazards and adverse effects of activities). It consists of four steps: identify, assess, control, and recover. The process should be applied to current and new activities, operations, products, and services, and involves the assessment of HSE impacts or potential impacts on people, environment, and assets. It should include the full life cycle of the business from inception to termination (Visser, 1998). The types of risk assessment and measures to be taken are directed by the potential consequences of risks.

Most injuries result from minor hazards at the workplace. Slip, trip, and fall incidents are the most common incidents when performing normal activities. Qualitative risk assessment, housekeeping rules, procedures and behavior rules, awareness programs, and training are convenient measures to manage these hazards. Workplace hazard management is part of the overall HEM process. A smaller proportion of injuries results from hazardous activities; the performer of the activity or people in the immediate vicinity are the potential targets for these activities. These hazards are generic for the type of work performed; working at height, with electricity, with gases under pressure. Principles on identification of hazards and barriers to prevent progression to consequences can also be applied to these hazardous activities. Precautions to be taken are not specific for a particular location or enterprise. They can be dealt with by using procedures and checklists, adopted by the company, which may originate from the regulator, from industry, or be generated in house.

HSE-critical activities in the business process are defined as the activities required establishing and maintaining the barriers along the paths that lead to major consequences. Safety cases deal with critical operating procedures, design, and engineering studies. They are communicated for guidance and reference for supervisory staff and engineers. The hazards and effects management process requires that hazards with the potential of intolerable consequences be identified and fully assessed, necessary controls be provided, and recovery preparedness measures be in place to take care of any potential loss of control. Two types of barriers are used: measures taken to prevent the circumstance or initiating an event from occurring. Recovery measures and equipment to prevent the event causing harm and damage are barriers to limit the consequences and reinstate the process. For each initiating event, one or more barriers can be deployed, depending on their effectiveness and the potential severity of the consequences. Barriers can be hardware, instrumentation, procedures, competencies, and others.

Various types of activities are necessary to recognize the need for barriers, to deploy them, and to keep them effective. Design activities specify the necessary hardware and equipment and inspection and maintenance activities to ensure that this hardware and equipment retains its integrity and reliability, operational activities ensure that equipment is used within the defined limits of the controls provided, and administrative activities provide the necessary training, awareness, and attitudes to ensure that people perform predictably in all normal and abnormal situations.

In the simplest case, workplace hazards can be identified by observation, comparing the circumstances with the legal standards and guidance. In more complex cases, measurements such as air sampling or examining the methods of machine operation may be necessary to identify the presence of hazards presented by chemicals or machinery. The complexity of many health risks means that the identification of health hazards and risks will generally require the measurement of exposure, calling for specific monitoring and assessment techniques and the competence to use them. While health risks arising from the use of substances can be controlled by physical control measures, systems of work and personal protective equipment, confirmation of the adequacy of control, will often require measurements of the working environment to check that exposures are within preset limits. In special cases, surveillance of those at risk to detect excessive uptake of a substance (i.e. biological monitoring) or early signs of harm (i.e., health surveillance) may also be necessary.

Assessing risks is necessary to identify their relative importance and to obtain information about their extent and nature. This will help to identify where to place the major effort in prevention and control and to make decisions on the adequacy of control measures. The relative importance of risks may be determined by taking into consideration both the severity of the hazard and the likelihood of occurrence. There is no general formula for rating risks in relative importance, but a number of simple risk estimation techniques have been developed to assist in decision making (Bird and Germain, 1987; Steel, 1990). They involve only some means of estimating the likelihood of occurrence and the severity of a hazard. An example is given in Figure 2.

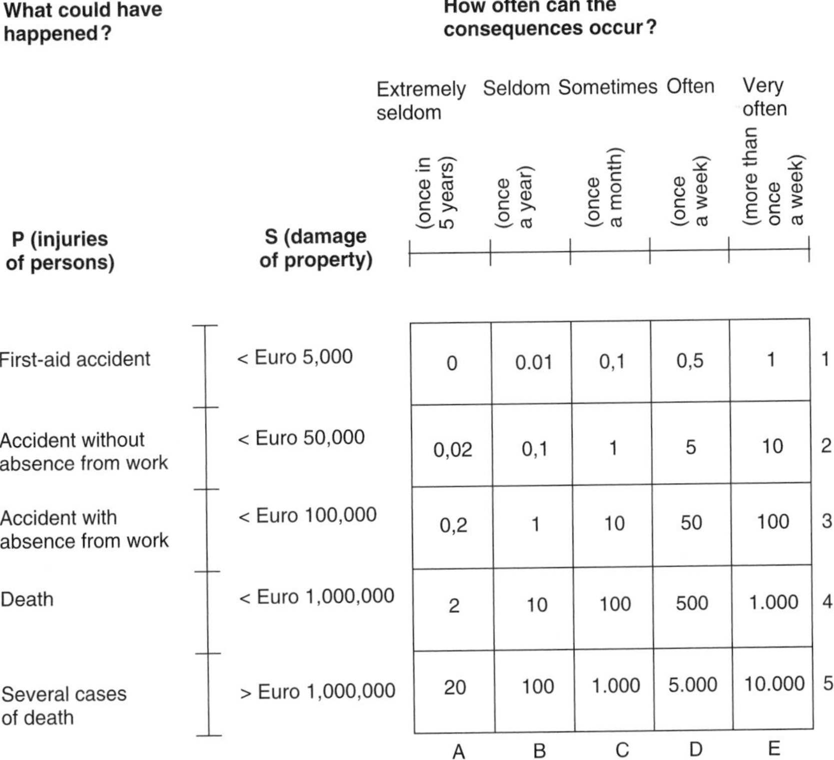

What could have happened?

How often can the consequences occur?

P (injuries of persons)	S (damage of property)	Extremely seldom (once in 5 years)	Seldom (once a year)	Sometimes (once a month)	Often (once a week)	Very often (more than once a week)	
First-aid accident	< Euro 5,000	0	0.01	0,1	0,5	1	1
Accident without absence from work	< Euro 50,000	0,02	0,1	1	5	10	2
Accident with absence from work	< Euro 100,000	0,2	1	10	50	100	3
Death	< Euro 1,000,000	2	10	100	500	1.000	4
Several cases of death	> Euro 1,000,000	20	100	1.000	5.000	10.000	5
		A	B	C	D	E	

Figure 2 Determination of a risk index.

Usually, the amount of harm resulting from a critical event or accident is a matter of chance. The hazard potential can be assessed in a risk matrix. Two questions have to be answered: What could have happened? and How often can the consequences occur? Hazards, the potential to cause harm, will vary in severity. The effect is rated in terms of both injuries and property damage. In Figure 2 the likelihood of harm is rated on a five-level scale ranging from once in five years to more than once a week. The cells of the matrix may contain some arbitrary risk numbers. The organization decides what kind of measure has to be taken by which kind of risk number. Health hazards, malfunctions, and near accidents and even accidents can be assessed in the same way. Systems of assessing relative risks can contribute not only to establishing risk control priorities but also assist in prioritizing other activities and ranking departments and units according to their safety levels. Hoyos and Ruppert (1993) developed a hazard potential analysis for the identification and assessment of hazards at the workplace. The safety diagnosis questionnaire (SDQ) relates the hazards identified to a broad repertoire of abilities and skills that are considered the basis of hazard control by the worker.

The identification of deficiencies and deviations from standards related to managerial and organizational arrangements is usually done in an audit and review process. Some topics are, for example, organizational arrangements to secure the involvement of all employees in the HSE efforts, the acceptance of HSE responsibilities by line managers, the communication of policy and relevant information, and the timely consideration of HSE-related standards into the planning and design process. Audits are normally performed by competent people, independent of the area or activities being audited. This can be achieved either by using external consultants or by using staff from different sections, departments, or sites to audit their colleagues. On the basis of structured work group discussions, Wagenaar et al. (1994) integrated the employees into the review process. The authors developed a method for predicting the causal structure of future accidents on the basis of symptoms already visible. In more complex situations or in high-risk systems, qualitative or quantitative risk assessments may be required by legal standards and guidance. For example, in the chemical or nuclear industry, special techniques and systems may be applied when planning a new system or major changes of an existing system.

Hazard and operability studies (HAZOPS) and hazard analysis systems such as event or fault tree analysis have repeatedly been summed up and presented, among others, by Hoyos and Zimolong (1988) and Ridley and Channing (2003). In Table 2 an overview is given on the techniques suitable

Table 2 Hazard Analysis and Risk Assessment Techniques

Instrument	Description	Use
Formal methods (hazard and operability studies, failure mode and effects analysis, fault tree analysis)	HAZOP and FMEA belong to the inductive techniques: What happens if a coolant pump fails? System and its components are analyzed with regard to operability and process to identify hazards, malfunctions, weak points. Mostly complex and time consuming, multidisciplinary brainstorming sessions. Requires high levels of expertise. FTA belongs to the deductive techniques: How can the failure of the coolant pump happen? Graphical presentation and analysis of logical contributions of errors and failures to one type of malfunction of a system.	Suited for plants and situations with high risk or severe consequences. The objective is to design out risk at early stages of planning or when changes are to be made.
Standards, thresholds, limit values	Simple way of evaluation whether situation is acceptable. Powerful design technique. Some standards or threshold limits are based on epidemiological data. Gives little information about severity or probabilities of hazards.	Evaluation of health hazards: for example, noise and vibration levels, exposure to chemical substances; evaluation of several aspects of workplace, especially ergonomic standards for workplace layout, space, environment, tools, equipment, machines.
Risk classification and ranking	Simple classification scheme for various kinds of hazards. A rank ordering of the identified hazards is drawn up, severity and frequency of occurrences are assessed in terms of qualitative effect classes. Outcome gives information of the risk and the priority on precautions to be taken.	Evaluation of safety and health hazards. Simple technique to be used at all levels of organization and at workplaces. Relative ranking of risks. An example is shown in Figure 2.

for the planning and design phase of a system. Probabilistic risk assessment (PRA) is a systematic quantitative assessment of the likelihood of the levels of damage from operating industrial settings. The assessments are derived from combining the likelihood of occurrence of hardware and human-related errors. One well-documented application of a procedure to obtain human error probability estimates is THERP (Swain and Guttmann, 1983). Other methods, such as the human cognitive reliability model and the maintenance personnel performance simulation model, have been discussed by Svenson (1989). Recently, new techniques have been developed, among others the dynamic reliability technique, which identifies the origin of HEPs in the dynamic interaction of the operator and plant control system (Cacciabue et al., 1993), A technique for human error analysis (ATHEANA) (Cooper et al., 1996), and the cognitive reliability and error analysis method (Hollnagel, 1998). Reviews on current trends in reliability analysis, systems, and cognitive engineering are given in Giesa and Timpe (2000), Wickens and Hollands (2000), Leveson (2002), and Strauch (2004).

Swain and Guttmann (1983) recommend the following procedure to conduct a PRA: (1) dividing up the tasks in subtasks and elements, (2) analysis of possible failure with the help of a fault or event tree, (3) determining the probabilities of faults for the corresponding task element from a database or through subjective estimation, and (4) computing of the reliability of tasks or frequency of failures with the rules of probability calculus.

The most difficult problem still remains the availability of data. A number of attempts have been made to collect human reliability data (Miller and Swain, 1987). Sometimes expert judgment is calibrated in some way, for example, through psychophysical methods such as rating, ranking, or magnitude estimation, or through analytical methods such as SMART (Edwards, 1977) or MAUD (Humphreys and Wisudha, 1983). Research has shown that experts are liable to errors of judgment in just the same way as laypersons, especially when they are forced to go beyond their experience (Kahneman et al., 1982). Zimolong (1992) conducted an experiment that compared three different estimation techniques, including an expert ranking and a decision-aiding technique (SLIM) (Embrey, 1987) and THERP. A satisfactory match between estimated and empirically derived human error data was yielded only for THERP for routine tasks. The application of SLIM and ranking led to a mismatch between estimated and actual HEPs. At present, no reliable technique for the quantitative assessment of error probabilities is available.

The challenges for safety management are to devise appropriate indicators of operational and economic impact and to be able to attribute types and amounts of operational and economic impact on formulated strategies and activities. Active monitoring systems provide essential feedback on performance before an accident, ill health, or an incident happens. The methods involved in monitoring ongoing safety strategies and activities vary widely, ranging from periodic checks of compliance with performance standards, to relatively straightforward tracking of training and education services delivered, to checking on how hazardous exposures are controlled. Monitoring can include serious reexamination of whether the needs of the safety strategy and related projects as originally intended still exist, or it may suggest modification, updating, or revitalization. Reactive systems monitor accidents, ill health, and incidents. Securing the reporting of serious injuries and ill health generally presents few problems for most organizations. However, the reporting of minor injuries, other loss events, incidents, and hazards requires special efforts and attention and creates difficulties in most organizations.

Most companies are set up to measure accountability through analysis of results. Monthly accident reports at most plants suggest that the supervisor and manager should be judged by the number and cost of accidents that occur in his or her department. Petersen (1978) strongly emphasized that they should be judged by what they do to control losses. Numerous techniques have been introduced for guiding and measuring accident prevention efforts of supervisors and line managers (Bird and Germain, 1987).

An alternative approach to measuring and evaluating the efficacy of safety management is the use of safety-related financial/economic performance ratios. These ratios are performance indicators that are used to show the strategic value and operating leverage (i.e., the ratio of the percentage change in operating income to the percentage change in operating costs associated with safety management efforts) (Imada, 1990). The direct costs of accidents can be determined by using one of the suggested models: for example, by Veltri (1990) or the Health and Safety Executive (1993). Models disclose the direct costs of accidents and the economic impact on cost–volume–profit performance standards and profitability potential of the company. The indirect costs of accidents are difficult to calculate, chiefly because no suitable model for verifying any results in reliable and valid ways has been developed. However, Grimaldi and Simonds (1989) presented an uninsured cost model that provides a reasonable measure for determining uninsured costs of accidents.

There are a growing number of well-known auditing methods (i.e., an inspection of the organizations and administrative procedures on a scheduled basis to determine their safety relative to a standard criterion). Among these, the International Safety Rating System (ISRS) (Bird and Germain, 1987) is widely used. ISRS is a safety auditing program for assessing the various parts of a company's health and safety activities. The safety management factors in ISRS are ranked and assigned a numeric value, based on a qualitative judgment of the relative importance of the elements. The MORT (management oversight and risk tree) (Johnson, 1980) concept has formed a basis for further developments of safety analyses and safety assurance methodology in industry. MORT is a logic tree that provides a disciplined method to analyzing an accident and provides a format for safety program evaluation. The SMORT (safety management and organization review technique) (Kjellén, 2000) method provides a systematic and stepwise means of unfolding relevant causal factors by starting with identification of risk influencing factors at the workplace level and proceeding through the various managerial levels of the organization.

Common to the majority of audit methods used is the characteristic that they have only to a minor extent been validated scientifically (Eisner and Leger, 1988; Guastello, 1991; Rouhiainen, 1992). Methods with the exception of SMORT (Tinmannsvik and Hovden, 2003) have been developed and tested through practical approaches rather than a scientific approach. Such audit systems are based largely on the collected experience of long years of consultancy or management. As Hale et al. (1997) claim, they can give the impression of arbitrary lists of topics clustered under convenient headings which vary from one instrument to another. They do not have an explicit model of management system, nor do they provide an empirically based weighting system of the items or topics covered. It is not clear whether they are too detailed or not complete enough. Even the merits are unclear. As an example, Eisner and Leger (1988) very much questioned the assertion that use of the International Loss Control Institute's audit instrument for the mining industry was the primary cause of accident reductions in South African mines (ILCI, 1990).

4 INDIVIDUAL CONTROL OF HAZARD AND RISK

4.1 Hazard Perception

Most safety- and health-related hazards in industry cannot be eliminated, reduced, or minimized. People have to perceive, detect, and control hazards and dangers if confronted with them at work. For individual control of hazards, careful analysis of hazards is required as well as what it takes to control hazards. The perception of hazards is essential to subsequent phases of action in hazardous situations: personal risk assessment, decision making and selection of the appropriate protective, and/or preventive action.

Saari (1976) defines the information processed during the accomplishment of a task in terms of the following two components: (1) the information required to execute a task, and (2) the information required to keep existing risks under control.

For example, a construction worker perched on the top of a ladder, who is required to drill a hole in the wall, has both to keep the balance and coordinate the body–hand movements automatically while drilling a hole; a driver searching for route information ahead simultaneously adjusts distance and speed of the car relative to the vehicles in front of the car. In both cases, hazard perception is crucial to coordinate body movement to keep hazards under control, whereas conscious risk assessment plays only a minor role, if at all.

Not all hazards are directly perceptible to human senses. Examples are electricity; colorless, odorless gases such as methane and carbon monoxide; x-rays and other forms of radioactivity; and oxygen-deficient atmospheres. Their very presence must be signaled by devices that translate the presence of the hazard into something recognizable. Most of the toxic substances are not visible at all. Ruppert (1987) found in an investigation in an iron and steel factory, in municipal garbage collecting, and in medical laboratories that from 2230 hazard indicators only 42% were perceptible to the human senses; 22% of the indicators have to be inferred from comparisons with standards. For example, an increase in the noise level coming from the press of the garbage truck indicates the risk of being hit by particles bursting from the container. Hazard perception is based in 23% of cases on clearly perceptible events that have to be interpreted with respect to knowledge about the type of hazardousness (e.g., a glossy surface of a wet floor indicates slippery conditions). In 13% of reports, hazard indicators can only be retrieved from memory; for example, current in a wall socket can be made perceivable only by the proper checking device. There are also situations where hazards exist that are not perceivable at all and cannot be made perceivable at a given time. One example is the risk of infection when opening blood samples for medical tests; another is the risk of material falling from scaffoldings at construction sites. The knowledge that hazards exist must be deduced from one's knowledge of general principles of causality or acquired by experience.

The results demonstrate that the requirements of hazard perception range from pure detection and perception to elaborate cognitive inference processes of anticipation and assessment. Delayed or accumulating effects of health hazards (e.g., toxic substances are likely to impose additional burdens on people). Cause-and-effect relationships are sometimes unclear, scarcely detectable, or misinterpreted. In Table 3 a list of requirements on perceptual processes of hazard detection and perception is presented based on studies of 391 work sites in industry and public services. Experienced raters had to assess all hazards at a particular site which resulted in 2373 hazards identified (Hoyos,

Table 3 Detection and Perception of Hazard Indicators

Requirement	Rated Total[a] (%)
1. Visual recognition	77.3
2. Selective attention	63.0
3. Division of attention	57.5
4. Rapid identification and responsiveness	56.3
5. Perception of incessant hazards	51.5
6. Observation and maintenance of distance	44.2
7. Detection of potentially dangerous objects	28.0
8. Vigilance	25.0
9. Auditory detection	21.2
10. Recognition of changing danger zones	19.9
11. Directed attention (distance)	17.9
12. Auditory recognition (e.g., of warnings)	15.9
13. Visual recognition (e.g., of warnings, labels)	7.3

Source: After Hoyos and Ruppert (1993).
[a]Frequency of demands per hazard.

1995). On the average, people have to cope with six perceptual and cognitive demands per hazard in order to control the risk at work. They include visual recognition, selective attention, auditory recognition, and vigilance. As expected, visual recognition dominates auditory recognition. About three-fourths of the hazards have to be detected visually; in only 21.2% of the cases it is by auditory detection. In more than half of all hazards observed, people had to divide attention between task and hazard control, for example, observing crane movements while working at construction sites. This is mentally strenuous and likely to be error prone. Even more alarming is the findings that in 56% of all hazards, employees have to cope with rapid activities and responsiveness to avoid being hit and injured (e.g., by sudden sidestepping to avoid an oncoming vehicle).

Table 4 shows the cognitive demands of anticipation and assessment which are required to control hazards at the work site. The core characteristic of all activities summarized in this table is the requirement of knowledge and experience to cope with hazards. It emphasizes the need of establishing signs, warnings, introducing personal counseling, training, and qualification efforts. As Hoyos et al., (1991) have demonstrated, employees have little knowledge of hazards, safety rules, and proper personal protective behavior. In some cases (16.1%), perception of hazards is supported by signs and warnings, usually, however, people rely on knowledge, training, and work experience. It is without doubt mandatory to improve the indication of hazards and risks by warning signs and labels. The use of labels and warnings to combat potential hazards, however, is a controversial procedure for managing risks. Too often, they are seen as a way for manufacturers to avoid responsibility for unreasonably risky products. Obviously, labels and warning signs will be successful only if the information they contain is read and understood by members of the intended group of people. Frantz and Rhoades (1993) found that 40% of clerical personnel filling a filing cabinet noticed a warning label placed on the top drawer of the cabinet, 33% read part of it, and no one read the entire label. Contrary to expectation, 20% complied completely by not placing any material in the top drawer first. Lehto and Papastavrou (1993) provided a thorough analysis of findings pertaining to warning signs and labels by examining receiver-, task-, product-, and message-related factors. The editorial work of Wogalter et al. (2001) covers a selection of articles on warnings and hazard communication since 1973.

Table 4 Prediction and Evaluation of Hazard Indicators

Requirements	Rated Total[a] (%)
1. Estimates of physical units (e.g., weight, force, and energy)	32.7
2. Identification (screening) of defects and inadequacies	29.6
3. Prediction of structural weaknesses	25.1
4. Expectancy of warning stimuli (e.g., railway signal lights)	19.8
5. Perception of visual cues (e.g., flags, traffic or warning signs)	19.6
6. Subjectively perceived somatic symptoms (e.g., dizziness, breathlessness, nausea)	19.5
7. Predictions of nonobvious dangers (e.g., radioactive contamination, bacterial infection)	17.7
8. Recognition of instable storage (e.g., open paint containers, excessively high piles of bricks)	16.3
9. Comprehension of warning signals (e.g., symbols, colors)	16.1
10. Evaluation of material stress (e.g., material pressures, efficacy of heat-resisting clothing)	16.0
11. Interpretation of displays and data (e.g., gauges, switches, monitors)	5.4

Source: After Hoyos and Ruppert (1993).

[a]Frequency of demands per hazard.

4.2 Personal Risk Assessment

Personal risk assessment refers to the decision process as to whether and to what extent the person will be exposed to hazard: for instance, working on a high scaffolding or driving a car at high speed. It seems people must decide in the face of danger. However, people doing their jobs on a routine basis rarely consider these hazards or accidents in advance: They run risks, but they do not take them. Much of the time there will be no conscious perception or consideration of hazards as such. "The lack of safety consciousness is both a normal and a healthy state of affairs, despite what has been said in countless books, articles and speeches. Being constantly conscious of danger is a reasonable definition of paranoia" (Hale and Glendon, 1987, p. 41).

There is a wide variety of interpretations of the term *risk*. On the one hand, risk is interpreted to mean "probability of an undesired event." It is an expression of the likelihood that something unpleasant will happen. A more neutral definition of risk is used by Yates (1992), who argued that risk should be perceived as a multidimensional concept that as a whole refers to the prospect of loss. Technical risk assessment usually focuses on the potential for loss, which includes the probability of the loss occurring and the magnitude of the loss in terms of death, injury, or monetary costs. Laypeople's sense of risk depends on more than the probability and magnitude of loss. It may depend on such factors as potential degree of damage as well as on dimensions such as the unfamiliarity of the consequences, the involuntary nature of exposure to risk, the uncontrollability of damage, and the biased media coverage. The feeling of control in a situation may be a particularly important factor (Slovic, 1987). A different research direction has addressed emotional reactions to risky situations. The potential for serious loss generates a variety of emotional reactions, not all of which are necessarily unpleasant. There is a fine line between fear and

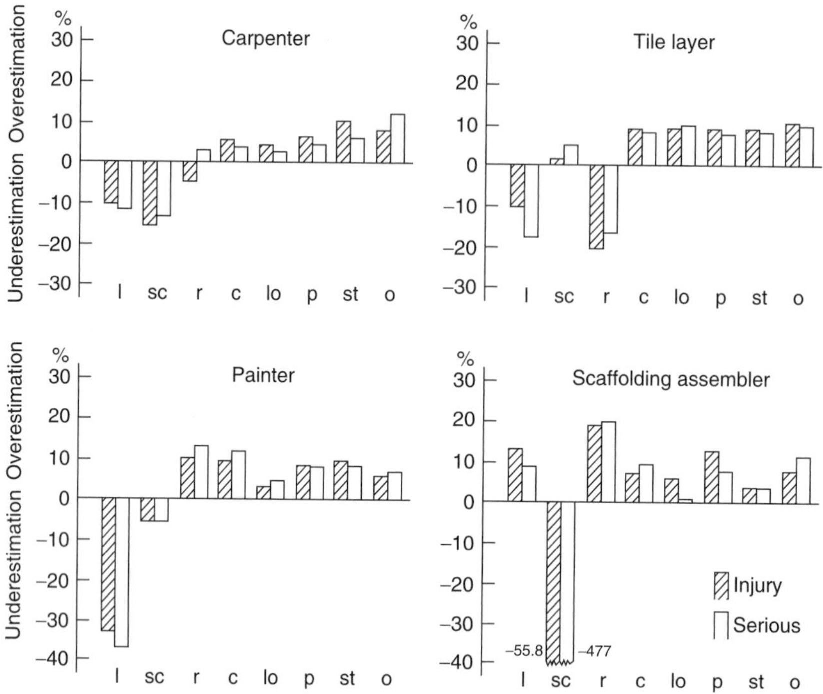

Figure 3 Subjective risk estimation of accidental falls from various work sites. Columns depict positive percent figures (overestimation) and negative figures (underestimation) of specific workplace risks. Assessments were made separately for eight different work sites, including ladder (l), scaffolding (sc), roof (r), and stairs (st). (From Hoyos and Zimolong, 1988.)

excitement. Again, a major determinant of perceived risk and of affective reactions to risky situations seems to be a person's feeling of control or lack thereof. As a consequence, for many people, risk may be nothing more than a feeling (Trimpop, 1994).

The overwhelming evidence that people often make poor choices in risky situations also seems related to inappropriate risk assessment. In particular, research on judgment and choice has shown that people have methodological deficiencies, such as in understanding probabilities; negligence of the effect of sample sizes; reliance on misleading personal experience; holding judgments of fact with unwarranted confidence; and misjudging risks. People are more likely to underestimate risks if they have been exposed to risks voluntarily over a longer period, such as people living in the neighborhood of reservoirs facing the risk of flooding; or in areas where earthquakes are not uncommon. Similar results have been reported from industry. In a field study (Zimolong, 1985) 153 members of six occupational groups mainly from the building industry and auxiliary building trade, among them carpenters, tile layers, and construction workers, rated the frequencies of falls resulting in minor injuries and in serious/fatal injuries, respectively. Ratings were carried out for eight different work sites: ladder, scaffolding, roof, building under construction, and four others. Estimates were compared with frequencies based on fall accident statistics. The results clearly indicated a job-specific underestimation or overestimation of the frequencies. As can be seen from Figure 3, there is a mismatch between the risk as assessed subjectively and as measured objectively.

Generally speaking, employees underestimate high-risk activities and overestimate low-risk activities, however, underestimation of risks is ruled by the exposure time to the specific risk. For instance, carpenters and painters frequently use ladders, and they considerably underestimate the risk of accidental fall. Tile layers typically underestimate the risk of fall from roofs, whereas scaffolding assemblers obviously seem to believe that falls from scaffoldings will never happen to them, although actually, it is the most frequent cause of minor and severe injuries. Shunters, miners, and forest and construction workers all dramatically underestimate the riskiness of their most common work activities compared to objective accident statistics. However, they tend to overestimate obvious dangerous activities of fellow workers when required to rate them.

Unfortunately, experts' judgments appear to be prone to many of the same biases as those of laypersons, particularly when experts are forced to go beyond the limits of available data and rely on their intuitions. Research indicates further that disagreements about risk will not disappear completely when sufficient evidence is available. Strong initial views are resistant to change because they influence the way that subsequent information is interpreted. New evidence appears reliable and informative if it is consistent with one's initial beliefs; contrary evidence tends to be dismissed as unreliable, erroneous, or unrepresentative (Nisbett and Ross, 1980). When people lack strong prior opinions, the opposite situation occurs; they are at the mercy of the problem formulation. Presenting the same information about risk in different ways—for example, mortality rates as opposed to survival rates—alters their perspectives and their actions (Tversky and Kahneman, 1981).

Most of the personal risk decisions in everyday life are not conscious decisions at all. People are not even aware of risk. Most of our daily behavior is automated and runs smoothly without continuous attentional control and conscious risk taking. Reason's (1987b) GEMS model describes how the transition from automatic control to conscious problem solving takes place when exceptional circumstances arise or novel situations are encountered. In normal work routines, however, conscious risk assessment and decision making is just not present. Therefore, it cannot be argued that people's way of evaluating risk is inaccurate and need to be improved. The notion that the acceptance of risks, identified after the occurrence of accidents, is the primary cause of the incident does not take into account that in most cases no conscious risk assessment was undertaken. In research as in practice, less attention has been paid to conditions in which people will act automatically, follow their good feeling, or accept the first choice that is offered (Wagenaar, 1992). Contrary to these findings, there is a widespread acceptance in society and among safety and health professionals that risk taking is a prime factor for causing mishaps and errors. In a representative sample of Swedes aged between 18 and 70 years, 90% agreed that risk taking is the major source of accidents (Hovden and Larsson, 1987).

Preventive activities to cope with hazards comprise among others: planning work procedures and steps ahead; regular checking of equipment and material for defective parts; provision of adequate storage; selection of safe work procedures by means of selecting proper material and tools; setting an appropriate work pace; and inspection of facilities, equipment, machinery, and tools. The most frequent protective measure required is the usage of personal protective equipment(PPE) (Hoyos and Ruppert, 1993). Together with correct handling and maintenance, it is by far the most important requirement in industry. There are major differences in the use of PPE among companies. In some of the best companies, mainly in the chemical and mineral oil industry, use of PPE approaches 100%. In contrast, in the construction industry, safety representatives have problems even in attempts to introduce particular PPE on a regular basis. Is risk perception and assessment the major factor that makes the difference? Again, this is doubtful. Some of the companies have successfully enforced the use of PPE, which then becomes habituated (e.g., the wearing of safety helmets). They have established the "right safety culture" and thus have subsequently altered personal risk assessment. In his short discussion on the use of seat belts, Slovic (1987) shows that about 20% of road users wear seat belts voluntarily, 50% would use them if it was made mandatory by law, and beyond this number, only control, incentives, and punishment will serve to improve automatic use. Thus, it is important to understand what factors govern risk perception. However, it is equally important to know what the company can do to change behavior and, subsequently, how to alter risk perception.

5 BEHAVIORAL RISK MANAGEMENT

5.1 Scientific Foundations

Most successful behavioral programs have tried to modify the value function for safe behavior by introducing short-term rewards that outweigh immediate costs. Literature reviews reveal that most documented interventions have used the operant perspective of role behavior and the attendant ABC framework (i.e., antecedents–behavior–consequences) (see Luthans and Kreitner, 1985; Stajkovic and Luthans, 1997, 2003). Mainly two types of antecedents were used—goal setting and training—and three types of consequences: feedback, incentive, and social recognition. Antecedents have mostly been used in combination with positive consequences of some type (O'Hara et al., 1985; McAfee and Winn, 1989; Geller, 1996).

Behavioral management is based on the behavioral model of learning, with its roots in Skinner's operant conditioning and Thorndike's law of effect. Learning is understood as a relatively enduring change in behavior that results from the behavior and its consequences and antecedents. Contextual and antecedent events, such as the physical and social factors of the work environment, key features of job requirement, prevailing hazards, work rules, and cues and prompts of superiors set the stage for a given performance. Consequences exert the strongest influence on the rate and durability of behaviors. The main premise of behavioral management is that employee behavior is a function of contingent consequences (Bandura, 1969; Komaki, 1998).

Goal setting is widely accepted as to be one of the most powerful behavioral motivation techniques. Goal-setting theory was formulated inductively largely on the basis of empirical research conducted over nearly four decades (Locke and Latham, 2002). The essential elements of goal-setting theory—specificity and difficulty of goals; goal effects on the individual, group, and organization levels; the proper use of learning; and the moderators and mediators of goals—are summarized in the high-performance cycle (Locke and Latham, 1990). Goal setting affects performance by directing the attention and actions of individuals and/or groups, mobilizing efforts, increasing persistence, and by motivating the search for appropriate performance strategies. Goals have a directive function. They direct attention and effort toward goal-relevant activities and away from goal-irrelevant activities. They also have an energizing function. High goals lead to greater effort than low goals. They also affect persistence. Difficult goals prolong effort. Finally, goals affect action indirectly by leading to the search for and use of task-relevant knowledge: setting difficult, yet achievable goals, and providing performance feedback in relation to them. The goal-performance relation is mediated through the concept of self-efficacy (Bandura, 1997). People with high self-efficacy set higher goals, and vice versa. They also find and use better task strategies to attain the goals. The goal–performance relationship is strongest when people are committed to their goals. This is particularly important when goals are difficult (Klein et al., 1999). Goal commitment can be enhanced by leaders communicating an inspiring vision and behaving supportively. An alternative to assigning goals is to allow subordinates to participate in decision making. Meta-analyses of the effects of participation in decision making on performance yielded an effect size of only $d = 0.11$ (Wagner and Gooding, 1987). Subsequently, Locke et al. (1997) found that the primary benefit of participation in decision making is cognitive rather than motivational. Participation stimulates information exchange and leads to a better informational basis (e.g., on efficient task strategies).

Training is one of the most pervasive methods for enhancing the skills and performance of individuals. Over the past decades, there have been several cumulative reviews of the training and development literature. The most recent meta-analytic review was conducted by Arthur et al. (2003). Authors examined the relationship between specified training design and evaluation features and the effectiveness of training in organizations. Depending on the criterion type, the sample-weighted effect size for organizational training was $d = 0.60 - 0.63$, which is a medium to large effect. Results indicated a substantial decrease in effect sizes from individual learning outcomes to manifestations in subsequent job behaviors and organizational performance outcomes. This decrease in effect size points to the critical issues of favorability of the posttraining environment for the performance of learned skills. Trained and learned skills will not be demonstrated if incumbents do not have the opportunity to perform them (Noe, 1986; Ford et al., 1992).

In the field of safety, the development of employee's competencies through training programs may increase the knowledge of hazards and how to control them, improve skill levels, and help develop better hazard control strategies. However, safety-related training is often flawed because it is conducted in an environment where the members of the group are unknown to each other. Hence different group norms, standards, and behavior emerge as compared to the original work setting. Therefore, knowledge acquired and safety attitudes developed in the training environment will not be used or reinforced by superiors and colleagues in the original work setting. Additionally, objectives and aims of the trainings are mostly unclear to the trainee and the organization, respectively. Hence, superiors and colleagues do not know how to capitalize on the new skills. Personal development programs that identify health and safety needs of the employees and link those needs to strategic requirements of the organization may overcome some of these constraints.

Performance feedback is usually defined as information about the effectiveness of particular work behaviors and is thought to fulfill several functions. For example, it is directive, by clarifying specific behaviors that ought to be performed, it is motivational, as it stimulates greater effort, and it is error correcting, as it provides information about the deviations from a prescribed standard or level. The relationship between goals and performance is complex, but goals have been demonstrated to mediate the effectiveness of feedback, while feedback has been shown to moderate the effectiveness of goals. The relationship between goals and feedback can be construed as the joint effects of motivation and cognition that control action (Guzzo et al., 1985). Goals, for example, inform people to achieve particular levels of performance in order to direct and evaluate their actions and effort, while performance feedback allows a person to track how well he or she is doing in relation to the goal, so that if necessary, adjustments in effort, direction, or possibly task strategies can be made.

Feedback has the potential to function in a variety of ways: as a reinforcer when it conveys success, as a punisher when it conveys failure, and as an antecedent when it prompts or cues the conditions under which responses will be reinforced and/or punished. Perhaps it is because it can operate in all these ways at once that feedback has been found to be an especially powerful modification technique. In a meta-analysis of psychologically based interventions, Guzzo et al. (1985) found goal setting to increase performance with an average effect size of $d = 0.12$ for management by objective, and $d = 0.75$ for goal setting on productivity. They reported a mean effect of $d = 0.35$ for appraisal and feedback. Kluger and DeNisi (1996) found a mean effect size of $d = 0.41$ for feedback and performance. That does not mean, however, that feedback is less powerful than goal setting. They complement each other, and, in combination, should be far more effective than either one alone, thereby providing a powerful management tool for effecting change.

Behaviors that positively affect performance must be contingently reinforced. Contingently administered money, feedback, and social recognition are the most recognized reinforcers in behavioral management at work (Bandura, 1986). Stajkovic and Luthans (2003) conducted a meta-analysis on behavioral management studies in organizational settings. They examined whether combined reinforcement effects of money, feedback, and social recognition are additive or synergistic (i.e., the combined effects are greater than the sum of the individual effects). In particular, authors found a significant main effect of behavioral management on performance ($d = 0.47$). This average effect size represents a 16% improvement in performance and

63% probability of success. The effects of the individual reinforcers on task performance was 23% ($d = 0.68$) for money, 17% ($d = 0.51$) for social recognition, and 10% ($d = 0.29$) for feedback. When these three reinforcers were used in combination, they produced the strongest synergistic effect on performance of 45% ($d = 1.88$). The result that financial incentives strongly affect motivation and performance was found in numerous studies (for a review, see Cameron et al., 2001). In a meta-analytic review carried out by Jenkins et al. (1998), financial incentives were related to performance quantity (effect size of $d = 0.34$) but not related to performance quality. Tasks had been coded into intrinsic and extrinsic tasks. Contrary to expectations on intrinsic motivation (Deci, 1971), the task type did not moderate the strength of the relationship between financial incentive and performance quantity.

It is quite clear that financial and nonfinancial incentives can indeed increase performance when the incentive system is designed properly. Guzzo et al. (1985) reported an average effect size for financial incentives of $d = 0.57$ with a broad confidence interval that included zero. Authors concluded that the strength of incentive effects depends heavily on the circumstances and methods of applying them. The major findings of 24 studies that have examined the effectiveness of the use of positive reinforcement and feedback was that all studies found that incentives or feedback were successful in improving safety conditions or reducing accidents, at least on a short term (McAfee and Winn, 1989). However, there are also some constraints to be noticed. Several studies reported situations in which safety indices did not improve. For example, Hopkins et al. (1986) found that training and praise improved the use of respirator use of only one of four sprayers (gelcoaters). Apparently, respirator usage was disagreeable to the other three workers because of the discomfort and inconvenience involved. A managerial challenge is to discriminate between reinforcing qualified people versus inflating the competence perceptions of unqualified employees.

Also, the question remains of whether some incentives are more effective than others. Many consequences have been found to function effectively as reinforcers for many people. Among others are recognition, praise, privileges, and material or monetary rewards, which can be embedded within compensation or performance appraisal systems, and preferred activities or assignments. Unless used with care, prizes and awards soon begin to lose their appeal, while having one's efforts specifically and sincerely acknowledged or praised by a respected peer or supervisor will tend to keep serving as a reinforcing function. Once high performance has been demonstrated, rewards can become important as inducements to continue, but not all rewards are external. Internal self-administered rewards that can occur following high performance include a sense of achievement based on attaining a certain level of excellence, pride in accomplishment, and feelings of success and efficacy. The experience of success will depend on reaching one's goal or level of aspiration.

Another issue to consider with respect to the effects of internal incentives is the nature of the task. Hackman and Oldham's (1980) job characteristics theory states that the degree to which the work is seen as rewarding is dependent on the degree to which the task possesses four core attributes: personal significance, feedback, responsibility/autonomy, and identity as a whole piece of work. These core attributes are growth producing, and they fulfill important needs. A review of the literature on the relation of the core attributes, critical psychological states, and the outcomes supports the approach despite the fact that some issues have been risen concerning the method of asking the same people for the core attributes, psychological states, and outcomes (Algera, 1990).

5.2 Behavioral Safety Programs

Behavior sampling has been used successfully by several researchers, implementing behavior modification safety programs (e.g., Komaki et al., 1978; Reber and Wallin, 1984; Reber et al., 1993; Sulzer-Azaroff et al., 1994; Shannon et al., 1999; Vassie and Lucas, 2001). This method is based on randomly sampled observations of a person's behavior, and evaluating whether observed behaviors are safe or unsafe. Lingard and Rowlinson (1997) introduced their intervention on construction sites with a series of joint goal-setting meetings, leading to specific performance safety goals concerning housekeeping activities, access to heights, and scaffolding construction. These meetings were followed by publicly displayed feedback charts for eight weeks, based on observations conducted by trained observers. During that time, the gap between the baseline level and the designated goals had to be minimized.

Goal setting in a paper mill improved performance if goals were assigned by supervisors but not if they were set by the workers (Fellner and Sulzer-Azaroff, 1985). McCarthy (1978) used a time series design to reduce the number of "high bobbins" in a textile mill. Bobbins are spindles of thread which if not pushed down far enough, cause tangles. Introducing goals of increasing difficulty plus feedback led to a steady decrease in the number of high bobbins. When feedback was removed, the number of high bobbins increased and then decreased again when feedback was reintroduced. Saari (1987) improved housekeeping in a shipyard significantly through feedback and implicit goal setting. Employees were given a written list of correct work practices, shown slides illustrating correct and incorrect practices, and given a one-hour training seminar. Correct practices observed on the shop floor were presented with posted boards showing the performance on a quantitative index. Chhokar and Wallin (1984) used a six-stage time-series design for machine shop workers: baseline, training plus goal setting, weekly feedback added, monthly feedback added, training and goal setting only, and bimonthly feedback added. Safe behavior of the employees was defined as the percentage of employees performing their jobs in a completely safe manner. They found an increase in performance that reached 95% of maximum possible safe performance if goal setting and feedback was provided. This was an increase of 30% compared to the baseline of 65%. Alavosius and Sulzer-Azaroff (1985) introduced feedback on safe performance in patient transfer by staff in a hospital, thereby improving proper lifting techniques significantly.

The effects of goal setting are applicable not only to the individual but to groups (O'Leary-Kelly et al., 1994) and organizational units (Rodgers and Hunter, 1991). A growing amount of research has been done showing positive effects of goal setting and feedback for groups. Among others, Latham and Kinne (1974) and Nadler (1979) found that specific goals led to better group performance than unspecified, vague goals. Other researchers reported (Latham and Yukl, 1975) that groups performed better if their goals were difficult than if they were easy. The effectiveness of task feedback in group goal setting will be maximized if feedback involves both individual and group performance information. In addition, to maximize the productivity of such teams, information and control systems should contain information on team progress and individual progress.

An instructive example of the interaction of various elements of a behavioral safety management program is the study of Komaki et al. (1978). Authors introduced goals and feedback to improve worker safety in two departments in a food manufacturing plant. The intervention consisted of identifying safe and unsafe practices, the safe practice was introduced as the desired behavior goal, a board was posted with the baseline data of safe behavior, and the departmental goal of 90% was suggested and agreed to by the employees. Thereafter, whenever observers collected data, they posted on the graph the percentage of incidents performed safely by the group as a whole. In addition to feedback, another planned component of the intervention program was for supervisors to recognize workers when they performed selected incidents safely. To ensure the participation of the supervisors, the president and the plant manager were asked to talk to each supervisor about the safety program at least once a week.

Following the intervention in both departments, the percentage of safe practices increased significantly. By the first week, the wrapping department of the bakery had obtained their first 100% score. During the entire intervention, no score fell below the baseline, and over half of the time, the department obtained 100% scores. Scores in the makeup department immediately rose to 100% and, with one exception, continued at this level. During the reversal phase, however, performance dropped to baseline levels (see Figure 4). Difficulties were encountered in implementing the recognition component of the intervention. Only 15% and 54% of the recognition checklists of the supervisors were turned in, respectively. Although management continued to give their verbal support, there were few indications that management was communicating their support to supervisory personnel.

Overall, some important issues of implementing a behavioral safety program can be derived from this study. A requirement analysis was performed

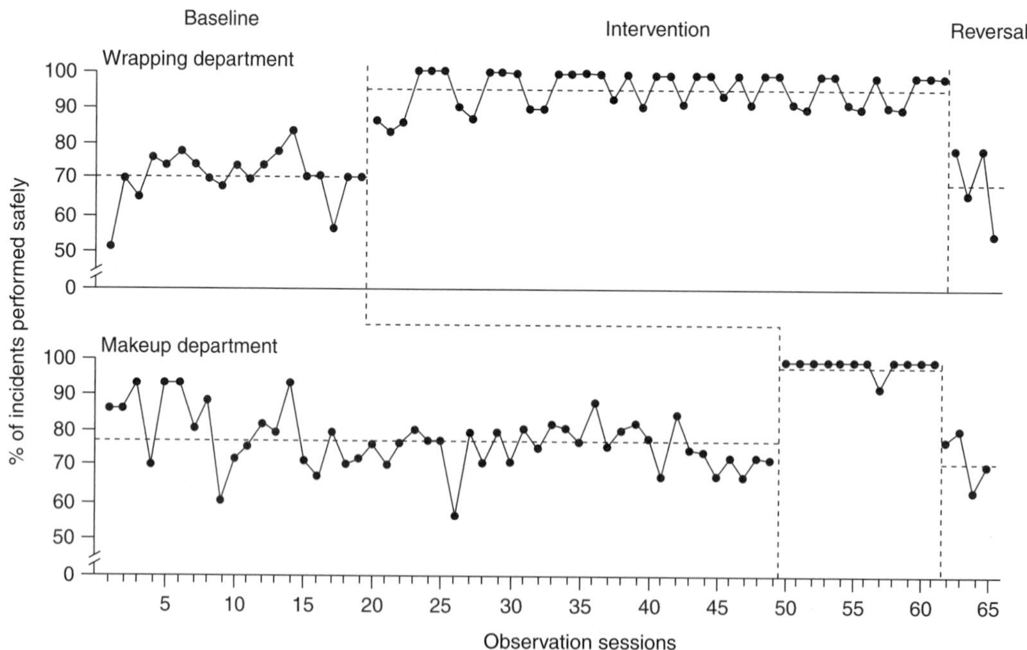

Figure 4 Effects of goal setting and feedback on safe performance in two departments of a food manufacturing plant. (From Komaki et al., 1978.)

jointly with supervisors, a difficult goal of safe performance score was suggested and agreed upon by workers, observations of critical actions prior to the intervention served as baseline level, safe conduct rules (i.e., behavioral standards) were stated explicitly, and specific feedback consisting of the weekly percentage of safe practices was displayed publicly. The fact that performance returned to baseline levels during the reversal phase indicates that the program was not supported sufficiently by employees, supervisors, and management.

Participative approaches look more promising to ensure commitment of workers and of superiors. They include participative goal setting, establishment of problem-solving groups, quality circles, and use of group discussion techniques (e.g., Marks et al., 1986). Participative goal setting was applied to set safety improvement goals for critical behavior in a three-shift production plant (Cooper et al., 1994). All factory personnel, including senior management, attended their respective department's goal-setting meetings. Performance feedback was presented graphically in each department on a weekly basis. The results indicate significant improvements in safety performance, with a corresponding reduction in the plant's accident rate. There was a lack of support, however, for an inverse relationship between actual safety performance and accident rates. It was found that the nature of the tasks mediates the relationship between safety performance and accident rates, including such factors as time pressure, manning level, and sickness absenteeism, which created greater time pressure for workers. Nature of the task and sickness absenteeism explained 70% of the overall variance on accidents. These findings reinforce further the importance of ascertaining the impact that managerial, organizational, and job-related variables have on workplace safety.

An intervention program using safety circles to reduce accidents with respect to housekeeping was introduced in a Finnish shipyard (Saarela, 1990). Two main tasks were given to the small groups: identifying and eliminating obstacles of good order and safety and establishing new housekeeping practices in the department. Members of top management and the safety organization coordinated the program and disseminated general information concerning the program. Results indicated a 20% decrease of occupational accidents related to housekeeping during the intervention period and one year later after the implementation. Gregersen et al. (1996) reported a study using the group decision technique to improve the commitment in safety of the drivers of a Swedish telecommunication company. Drivers met in groups of 10 to 15 persons to discuss general and specific problems related to safety. After the third meeting, each driver decided about his or her own future activities to solve the safety problem in question. This was only one of the conditions in the safety program, which included driver training, safety campaigns, and a bonus system for safe driving. There was a significant decrease in accident rates in the experimental group which performed under the condition of group discussion, but none in the control group.

As Sulzer-Azaroff (1978, 1982) points out, safe and unsafe practices probably persist because they are in some way naturally reinforced. In addition, although there are natural punishers for committing risky acts (e.g., injuries, pains, sicknesses), these are often delayed, weak, or infrequent. Most health hazards are not perceivable at all, and negative consequences have to be inferred from knowledge, memory, or experience with a considerable risk of drawing wrong conclusions. Safety programs often have turned to heavily emphasizing behavioral antecedents directly at the workplace. Signs, guidance, threats, incentives, goals, training, work design, and ergonomic design of workplaces are among the more frequently utilized antecedents. The antecedents of risky performance frequently are difficult or impossible to control. No one can alter a person's history of reinforcement, and modifying such conditions as hazard perception, risk assessment, or turning risky performance into preventive behavior is at best difficult.

The key practical implication from the results on behavioral safety programs is that behavioral management strongly affects performance in organizations. However, managers have to take into account that organizational process and design problems may limit the effectiveness of such programs. Aside from the difficulties of proper installation of ABC principles into procedures, techniques, and actions, they are contingent upon the characteristics of the individual organization. They depend on function, policies and strategic goals of the organization, type of technology used, requirements of the market in terms of stability and flexibility, and sociocultural background of employees. High-reliability organizations such as in the nuclear industry require different procedures and techniques then those of companies in the construction industry. Another managerial concern refers to commitment and durability. One possible way to ensure long-term commitment of employees and superiors is to address both behavior and attitudes simultaneously by fostering active employee involvement and utilizing feedback in long-term programs. This requires employees to become involved and participate actively in the identification of hazards, substandard practices, and conditions, in the setting of goals, and in monitoring the safety performance of their colleagues. Management commitment and support are essential prerequisites to facilitate this type of safety intervention.

5.3 Human Resource Management

The last two decades have witnessed a remarkable shift in human resource management (HRM) research from a microanalytical approach to a macrostrategic perspective. People are recognized as the strategic resource, shifting the focus of attention on linking HRM practices with business strategy and organizational performance (Pfeffer, 1994; Huselid, 1995; Paul and Anantharaman, 2003). Although the HRM–performance relationship has been proved, the linkage process remains unclear (Becker and Gerhart, 1996). Some researchers have proposed competence, commitment, congruence, and cost effectiveness as intermediary variables (Beer et al., 1984). Becker and Huselid (1998) have suggested that intervening variables such as employee skills, motivation, job design, leadership, and work structure link operating performance.

Management of an organization's human resources includes such activities as personnel planning, recruitment, placement, development, performance appraisal, training and competency assessment, counseling and guiding of individuals and groups (leadership), payment concepts, back-to-work programs, and rehabilitation at the workplace. Such personnel systems include valid selection procedures for hiring and promotion purposes, performance appraisal and review systems to ensure that the person is measured on the right goals and standards and receives accurate feedback, effective training procedures for the development process, and labor relations that are conducive to employee motivation. Personnel development of people's capabilities means in-company training focused on company-related targets, closing the gap between the actual state of the art and the target state through qualitative personnel planning, and supporting and realizing career planning with personnel-related targets. Contractors, suppliers, and temporary workers have to be included in the training program according to the level of risk to which they may be exposed or could cause.

With respect to safety, a major purpose of the performance appraisal process is to modify behavior: to feed back information to the employee for counseling and development purposes so that the person will start doing or continue doing the activities critical to performing effectively on the job. The feedback must lead to the setting of and commitment to specific HSE goals. External and internal rewards play a significant role in getting people to accept goals and motivating them to maintain goal-relevant behavior in the long term. The key reward in organizational settings is probably not feedback but rather, the consequences to which feedback leads, such as recognition, praise, raises, financial rewards and promotions, privileges, material or monetary rewards (which can be embedded within compensation or performance appraisal systems), and preferred activities or assignments. Internal, self-administered rewards that can occur following high performance include a sense of achievement based on attaining a certain level of excellence, pride in accomplishment, and feelings of success and efficacy.

Recently, new concepts and frameworks of HRM systems with respect to health, safety, and environment have been proposed by Hutchinson and Hutchinson (1997), Zimolong (2001b), and Zimolong and Elke (2001a). Key elements of the systems are people management, management of information and communication, (re)design of work and technology, and management of an HSE supporting culture. Generic management activities include those of the management control loops of the ISO standards. To date, the relationship between HRM practices, procedures, and organizational health and safety performance has not been thoroughly investigated. Reviews on HRM literature with respect to health, safety, and environment (Zimolong, 1997; Hale and Baram, 1998; Zimolong and Elke, 2001a) revealed a number of lines of research and isolated studies on HRM topics. However, the linkage between HRM and organization's HS performance remains a black box.

Zimolong and Elke conducted a cross-organizational study in the chemical industry to identify key practices and systems of HRM that are connected to an organization's HSE performance (Elke, 2000). Practices, processes, and structures in HSE-related planning and design of work systems, in human resource management, in information and communication management, and in cultural aspects were sampled through interviews and questionnaires. Other topics addressed the control strategies of the human resource subsystems, such as lead, training, and incentive systems, and the types of substitutes that companies have developed to maintain an efficient control loop. In total, 18 plants participated, their size ranging from 200 to 1500 employees. Managers from top to lower level ($n = 292$), HSE experts (144), and union representatives (55) were personally interviewed. A sample of 10 to 15% of the workforce responded to an HS management questionnaire. In sum, 686 interviews were conducted and 1.536 questionnaires returned. The study covered excellent and poor companies as well as firms getting better or becoming worse within a period of four years. The HS performance level of companies was measured by the frequency of lost time injuries (LTI > 3 days) and ill-health-related lost workdays. LTI figures were converted into relative accident rates, taking into account levels of risks associated with the production of chemicals (products). The excellent group included all companies that stayed under the median of their comparable risk group for four years (range between 1.5 and 13.1 accidents per 1000 employees). In the progress group, plants improved during the last two years of the study period. They showed LTIs similar to those of the excellent group. In the falloff group, companies got worse in their HS performance; their median was equal to the median of the overall risk group. Finally, the poor group included all plants that always reported LTIs above the median of the risk group (range from 13.7 to 125.4).

In each company the actual status of the HRM procedures and systems with respect to control strategies, recruitment, selection and placement, performance appraisal, training, leadership style, and incentive programs was raised. Members of the human resource departments documented official use of the human resource (HR) systems. This was called the *documented status* of the systems. Additionally, 100 interviews with the managers of different departments (production, planning, maintenance, and personnel) were carried out to compare the documented status with the systems actually in use (i.e., applied by the managers). A rather strictly defined criterion for the *practiced application* was chosen: More than 50% of the plant managers interviewed had personally to confirm regular use of a system; otherwise, it was coded as not practiced.

The most frequently documented HR systems were appraisal systems (53/31%) (the second figure indicates the practiced systems), career development and promotion systems (53/26%), training systems (47/24%), and reward systems (40/34%). As expected, there is a considerable difference between the documented state and the actual application of the systems. Most differences between documented systems and systems in use exceed 20 percentage points. Organizations differ fundamentally in how they control the use of a HR system. A complete control loop (a Deming loop) has three phases: monitoring, measuring, and reviewing and taking action, if necessary. Most companies monitor the application of systems (65%); only 16% measure it; however, 31% review the outcomes. As compared to the poor group, excellent companies show a higher percentage of reviews (57/20%), but have similar figures in monitoring (57/60%) and measuring (14/0%).

Leadership related to HS, performance appraisal, reward system, career development and promotion, and training systems are the most frequently documented as well as practiced applications of HR systems (see Figure 5). In general, poor companies use only one or two systems: mainly leadership, occasional appraisal, or development systems. The most remarkable difference between the excellent and the poor groups is the adoption of reward systems related to HS performance for managers and workers. Companies striving to improve their HS performance invest many activities into HRM systems. The application rate is higher than those of the excellent firms.

The combination of systems creates a further distinction between the groups. Excellent companies generally combine strong leadership accountability in HS with appraisal and reward systems. The adoption of the triad seems to be a key characteristic of successful companies. No particular pattern of combined systems emerges from the progressing or poor group. Companies striving to get better do not yield typical patterns of HRM systems; they invest into all systems. To summarize, companies with excellent records in HS performance rely mainly on strong leadership responsibility in HS, on appraisal, and on reward systems that are combined in a *holistic HRM system*. These systems not only include indicators of business performance, they also address HS indicators and performance. No stand-alone or parallel management system based on HS criteria has been found. Companies of the progressing group have adopted a variety of systems. Eventually, they will reduce the multitude of systems to a few combined systems, which will form their typical HRM profile.

Systems have to be monitored, reviewed, and changed, if necessary. Typical control strategies of the excellent companies include monitoring, reviewing, and taking actions. However, it seems that some companies have developed substitution measures for superior-based control loops. One approach is the introduction of self-managed teams, whose members have managerial responsibilities for HS.

5.4 Design of Work and Technology

The outcomes of safety, health, and well-being depend to a significant degree on the installation and maintenance of safe and reliable technology as well as on

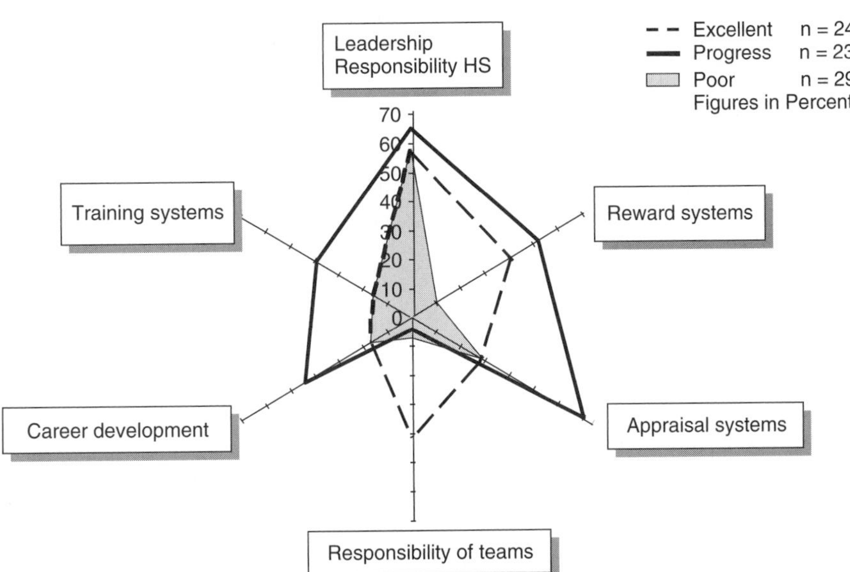

Figure 5 Application rate of human resource systems for the management of health and safety: excellent, progressing and poor. Excellent management uses a combination of systems: strong leadership accountability, appraisal, and reward systems.

ergonomic and psychosocial aspects of work design. General topics and methods of systems safety engineering are concerned with guarding energy sources; design and redesign of machinery, equipment, and processes; application of environmental standards; and establishment of inspection systems, such as statutory engineering inspections of pressure vessels, cranes and lifting machines, or electrical installations (see Ridley and Channing, 2003).

Ergonomic and psychosocial aspects of work are potential contributors to the health and well-being of employees and organizations. Their health effects can also be regarded as contributors to work motivation, or work performance. From an ergonomic perspective, design of the environment, workstations, tasks, work organizations, and the tools or technology should reasonably accommodate employee capacities, dimensions, strengths, and skills. Particularly, it should accommodate the human sensory capabilities, support the decision-making processes, and adapt to the human performance abilities. Error-prone designs place demands on performance that exceed the capabilities of the user, violate the user's expectancies based on his or her past experience, and make the task unnecessarily difficult, unpleasant, and error-prone. One example is back injuries: High physical energy expenditure, no variation in work movements and postures, prolonged standing, sitting, or stooping, or repetitive work can contribute to the development of musculoskeletal problems, low-back pain, and back disorders. A critical aspect of the success of ergonomic interventions for providing health and psychosocial benefits is strong management support for the ergonomic program. Further aspects are the involvement of managers, supervisors, and employees; the willingness to participate in defining problems, proposing solutions, and improving work practices; and the application of engineering improvement to reduce biomechanical and physiological loads.

There are various characteristics of working that have generally been shown to have negative physical and/or psychological consequences: for example, machine-paced work, a lack of task control, high job demands, shift work, time pressure, and poor supervisory relations. A person normally copes with transactional periods of stress by either altering the situation or controlling his or her reactions. Problems arise when work conditions are in conflict with human capacities and expectations over a long period of time, and when coping fails. The extent of the negative consequences varies from person to person depending on the perceived threat, individual constitution, and coping mechanisms. Tension, boredom, worry, anxiety, and irritability are inevitably some of the first indicators of strain. Emotional stress reactions are quite normal responses. Short-term stress reactions may include increased blood pressure, adverse mood states, and job dissatisfaction. Depression and apathy are later symptoms. Long-term stress reactions may even develop into cardiovascular diseases and upper extremity disorders (Kalimo et al., 1997).

Job control and social support are beneficial factors for well-being and job satisfaction. A person can have control over various job demands, such as the task itself, pacing of the work, work scheduling, the physical environment, decision making, other people, or mobility. When job control is high, the other job demands tend to have less potential adverse effects on health. Social support is thought to exert a protective function during conditions of stress (i.e., to buffer a person against the harmful effects of the social environment). Evidence for the beneficial effect of job control and social support is conclusive enough to promote better job designs and interpersonal interaction at the workplace.

The criteria in job design deal with the physical work environment, compensation systems, institutional rights and decisions, job content, internal and external social relations, and career development. According to quality of working life principles, the following characteristics of job content are of primary interest: variety of tasks and task identity; feedback from the job; perceived contribution to product or service; challenges and opportunities to use one's own skills; and individual autonomy. Depending on the degree of autonomy, the team design approach allows members to regulate their work activities by themselves. People get a better idea of the significance of their work and create greater identification with the finished product or service. If team members rotate among a variety of subtasks and cross-train on different operations, the team can become more flexible. Teams with heterogeneous backgrounds also allow for synergistic combinations of ideas and abilities not possible with people working alone, and such teams have generally shown higher satisfaction, better involvement, and superior performance, especially when task requirements are diverse.

In any redesign process there are trade-offs among specific improvements and achieving the best overall job design solution. There is no perfect job design that provides complete psychological satisfaction and health for all employees and maximizes the outcomes of the organization. Making jobs more mentally demanding increases the likelihood of achieving people's goal of satisfaction and motivation, but may decrease the chances of reaching the organization's goals of reduced training, staffing costs, decent wages, and error-free products and services. Which trade-offs will be made depends on the outcomes the organization prefers to maximize.

6 CONCLUDING REMARKS

The past two decades have witnessed a significant transformation in how firms are structured. "Tall" organizations with many management levels have become flatter; competitors that have adopted a modular organizational structure have gained market share. Organizational delayering and the rise of smaller, often entrepreneur-based firms gives self-management new meaning, covering personal self-management, self-leading teams, and semiautonomous units. Companies and public services adopt cooperative forms of work at a very fast pace. Teleworks provides flexibility in both working hours and the location of work and allow employees to cultivate tailored life-styles while working a full-time job. These "boundaryless" organizations (e.g., organizations whose membership, departmental identity, and job responsibility are flexible) create new challenges for safety management, particularly for people management.

The traditional approach to managing people focuses on selection, training, performance appraisal, and compensation for persons in specific jobs. It also presumes a hierarchy of control loops rather than horizontal work-flow sequences. When tall organizations become flatter and/or are restructured around teamwork, different forms of team autonomy and HS responsibilities emerge. Selection, performance appraisal, and reward policies are the most likely candidates for change. Contingent pay and peer pressure generated by teams are emerging as substitutes for both managerial influence and internalized member commitment. HS criteria, rules, procedures, and achievements have to be reformulated and integrated into the systems. Delayered organizations also need to develop combined systems, particularly a recruitment system that includes participatory concepts of job analysis and assessment, and a team management system focusing on performance appraisal, compensation, rewards and benefits, and personal career development. Systems are based on peer reviews instead of managerial appraisal and on team-based output measures instead of individual performance. This requires an intensive qualification process for employees and new cognitive and social skills to run the systems.

REFERENCES

Alavosius, M. P., and Sulzer-Azaroff, B. (1985), "An On-the-Job-Method to Evaluate Patient Lifting Technique," *Applied Ergonomics*, Vol. 16, pp. 307–311.

Algera, J. A. (1990), "The Job Characteristics Model of Work Motivation Revisited," in *Work Motivation*, U. Kleinbeck, H.-H. Quast, H. Thierry, and H. Häcker, Eds., Lawrence Erlbaum Associates, Mahwah, NJ, pp. 85–104.

Arthur, W., Bennett, W., Edens, P. S., and Bell, S. T. (2003), "Effectiveness of Training in Organizations: A Meta-analysis of Design and Evaluation Features," *Journal of Applied Psychology*, Vol. 88, pp. 234–245.

Avolio, B. J., Bass, B. M., and Jung, D. I. (1999), "Re-examining the Components of Transformational and Transactional Leadership Using the Multifactor Leadership Questionnaire," *Journal of Occupational and Organizational Psychology*, Vol. 72, pp. 441–462.

Bamber, L. (2003), "Principles of the Management of Risk," in *Safety at Work*, J. Ridley and J. Channing, Eds., Butterworth-Heinemann, Oxford, pp. 187–204.

Bandura, A. (1969), *Principles of Behavior Modification*, Holt, Rinehart & Winston, New York.

Bandura, A. (1986), *Social Foundation of Thought and Action*, Prentice-Hall, Englewood Cliffs, NJ.

Bandura, A. (1997), *Self-Efficacy: The Exercise of Control*, W.H. Freemann, New York.

Barling, J., Loughlin, C., and Kelloway, E. K. (2002), "Development and Test of a Model Linking Safety-Specific Transformational Leadership and Occupational Safety," *Journal of Applied Psychology*, Vol. 87, pp. 488–496.

Bass, B. M. (1990), *Bass & Stogdill's Handbook of Leadership*, Free Press, New York.

Bass, B. M., and Avolio, B. J. (1997), *Full Range Leadership Development: Manual for the MLQ*, Mind Garden, Palo Alto, CA.

Becker, B., and Gerhart, B. (1996), "The Impact of Human Resource Management on Organizational Performance: Progress and Prospects," *Academy of Management Journal*, Vol. 39, pp. 779–801.

Becker, B. E., and Huselid, M. A. (1998), "High Performance Work Systems and Firm Performance: A Synthesis of Research and Managerial Implications," *Research in Personnel and Human Resources Management*, Vol. 16, pp. 53–101.

Beer, K., Spector, B., Lawrence, P., Mills, D., and Walton, R. (1984), *Managing Human Assets*, Macmillan, New York.

Bentley, T. A., and Haslam, R. A. (2001), "A Comparison of Safety Practices Used by Managers of High and Low Accident Rate Postal Delivery Offices," *Safety Science*, Vol. 37, pp. 19–37.

Bird, F. E., and Germain, L. E. (1987), *Practical Loss Control Leadership*, Institute Publishing, Loganville, PA.

Bird, F. E., and Loftus, R. G. (1976), *Loss Control Management*, Institute Press, Loganville, PA.

Bolman, L. G., and Deal, T. E. (1984), *Modern Approaches to Understanding and Managing Organizations*, Jossey-Bass, San Francisco.

Brown, R. L., and Holmes, H. (1986), "The Use of Factor Analytic Procedure for Assessing the Validity of an Employee Safety Climate Model," *Accident Analysis and Prevention*, Vol. 18, pp. 455–470.

Brown, K. A., Willis, P. G., and Prussia, G. E. (2000), "Predicting Safe Employee Behaviour in the Steel Industry: Development and Test of a Sociotechnical Model," *Journal of Operations Management*, Vol. 18, pp. 445–465.

Butler, J. E., Ferris, G. R., and Napier, N. K. (1990), *Strategy and Human Resource Management*, Southwestern Publishing, Cincinnati, OH.

Cabrera, D. D., and Isla, R. (1998), "The Role of Safety Climate in a Safety Management System, in *Safety Management: The Challenge of Change*, A. R. Hale and M. Baram, Eds., Elsevier, Oxford.

Cacciabue, P. C., Caprignano, A., and Vivalda, C. (1993), "A Dynamic Reliability Technique for Error Assessment in Man–Machine Systems," *International Journal of Man–Machine Studies*, Vol. 38, pp. 403–428.

Cameron, J., Banko, K. M., and Pierce, W. D. (2001), "Pervasive Negative Effects of Rewards on Intrinsic Motivation: The Myth Continues," *Behavior Analyst*, Vol. 24, pp. 1–44.

Chew, D. C. E. (1988), "Effective Occupational Safety Activities: Findings in Three Asian Developing Countries," *International Labour Review*, Vol. 127, pp. 111–124.

Cheyne, A., Cox, S., Oliver, A., and Tomas, J. M. (1998), "Modelling Safety Climate in the Prediction of Levels of Safety Activity," *Work and Stress*, Vol. 12, pp. 255–271.

Cheyne, A., Tomas, J. M., Cox, S., and Oliver, A. (1999), "Modelling Employee Attitudes to Safety: A Comparison Across Sectors," *European Psychologist*, Vol. 1, pp. 4–10.

Chhokar, J. S., and Wallin, J. A. (1984), "Improving Safety Through Applied Behavior Analysis," *Journal of Safety Research*, Vol. 15, pp. 141–151.

Cohen, A. (1977), "Factors of Successful Occupational Safety," *Journal of Safety Research*, Vol. 9, pp. 168–178.

Cohen, A., and Cleveland, R. J. (1983), "Safety Practices in Record-Holding Plant," *Professional Safety*, Vol. 9, pp. 26–33.

Cooper, M. D., Phillips, R. A., Sutherland, V. J., and Makin, P. J. (1994), "Reducing Accidents Using Goal Setting and Feedback: A Field Study," *Journal of Occupational and Organizational Psychology*, Vol. 67, pp. 219–240.

Cooper, S., Ramey-Smith, A., Wreathall, J., Parry, G., Bley, D., Luckas, W., Taylor, J., and Barriere, M. (1996),"A Technique for Human Error Analysis (ATHEANA) : Technical Basis and Methodology Description," NUREG/CR-6350, NRC, Washington, DC.

Corlett, E. N., and Clark, T. S. (1995), *The Ergonomics of Workspaces and Machines*, Taylor & Francis, London.

Coyle, I. R., Sleeman, S. D., and Adams, N. (1995), "Safety Climate," *Journal of Safety Research*, Vol. 26, pp. 247–254.

Cully, M, Woodland, S., O'Reilly, A., and Dix, G. (1999), *Britain at Work: As Depicted by the 1998 Workplace Employee Relations Survey*, Routledge, London.

Deci, E. L. (1971), "Effects of Externally Mediated Rewards on Intrinsic Motivation," *Journal of Personality and Social Psychology*, Vol. 18, pp. 105–115.

Dedobbeleer, N., and Béland, F. (1991), "A Safety Climate Measure for Construction Sites," *Journal of Safety Research*, Vol. 22, pp. 97–103.

DeJoy, D. M. (1994), "Managing Safety in the Workplace: An Attribution Theory Analysis and Model," *Journal of Safety Research*, Vol. 25, pp. 3–17.

DePasquale, J. P., and Geller, E. (1999), "Critical Success Factors for Behavior-Based Safety: A Study of Twenty Industry-wide Applications," *Journal of Safety Research*, Vol. 30, pp. 237–249.

Diaz, R. I., and Cabrera, D. D. (1997), "Safety Climate and Attitude as Evaluation Measures of Organizational Safety," *Accident Analysis and Prevention*, Vol. 29, pp. 643–650.

Dufort, V. M., and Infante-Rivard, C. (1998), "Housekeeping and Safety: An Epidemiological Review," *Safety Science*, Vol. 28, pp. 127–138.

Edwards, W. (1977), "How to Use Multiattribute Utility Measurement for Social Decision Making," *IEEE Transactions on Systems, Man and Cybernetics*, Vol. 7, pp. 326–340.

Eisner, H. S., and Leger, J. P. (1988), "The International Safety Rating System in South African Mining," *Journal of Occupational Accidents*, Vol. 10, pp. 141–160.

Elke, G. (2000), *Management des Arbeitsschutzes [Management of Health and Safety]*, Deutscher Universitaets-Verlag, Wiesbaden, Germany.

Elke, G. (2001), Sicherheits- und Gesundheitskultur, I: Handlungs- und Wertorientierung im betrieblichen Alltag [Safety and Health Culture, I: Behavioral and Value Orientation in Daily Work Routine], in *Management des Arbeits- und Gesundheitsschutzes: Die erfolgreichen Strategien der Unternehmen [Health and Safety Management: The Successful Strategies of Enterprises]*, B. Zimolong, Ed., Gabler Verlag, Wiesbaden, Germany, pp. 171–200.

Embrey, D. E. (1987), "SLIM-MAUD: The Assessment of Human Error Probabilities Using an Interactive Computer Based Approach," in *Effective Decision Support Systems*, J. Hawgood and P. Humphreys, Eds., Technical Press, Aldershot, Hampshire, England. pp. 20–32.

Erke, H., and Zimolong, B. (1978), "Verkehrskonflikte im Innerortsbereich" [Traffic Conflicts in Urban Areas], in *Unfall und Sicherheitsforschung Straßenverkehr*, Vol. 15, Bundesanstalt für Straßenwesen, Cologne, Germany.

European Community (1999), *The Control of Major Accident Hazard Regulations, 1999 (COMAH)*, Her Majesty's Stationery Office, London.

European Union (2001), "Regulation No. 761/2001 of the European Parliament and of the Council of 19 March 2001 Allowing Voluntary Participation by Organizations in a Community Eco-Management and Audit Scheme (EMAS)," European Union, Luxembourg.

Fahlbruch, B., and Wilpert, B. (1999), "System Safety: An Emerging Field for I/O Psychology," in *International Review of Industrial and Organizational Psychology*, C. L. Cooper and I. v. T. Robertson, Eds., Wiley, Chichester, West Sussex, England, pp. 55–93.

Fellner, D. J., and Sulzer-Azaroff, B. (1985), "Occupational Safety: Assessing the Impact of Adding Assigned or Participative Goal-Setting," *Journal of Organizational Behavior Management*, Vol. 7, pp. 3–24.

Feyer, A.-M., Williamson, A. M., and Cairns, D. R. (1997), "The Involvement of Human Behavior in Occupational Accident: Errors in Context," *Safety Science*, Vol. 25, pp. 55–65.

Flin, R., Mearns, K., O'Connor, P., and Bryden, R. (2000), "Measuring Safety Climate: Identifying the Common Features," *Safety Science*, Vol. 34, pp. 177–192.

Ford, J. K., Quinones, M., Sego, D. J., and Speer Sorra, J. S. (1992), "Factors Affecting the Opportunity to Perform Trained Tasks on the Job," *Personnel Psychology*, Vol. 45, pp. 511–527.

Frantz, J. P., and Rhoades, T. P. (1993), "A Task Analytic Approach to the Temporal and Spacial Placement of Product Warnings," *Human Factors*, Vol. 35, pp. 719–730.

Fuller, C. W. (1999), "An Employee–Management Consensus Approach to Continuous Improvement in Safety Management," *Employee Relations*, Vol. 21, pp. 405–417.

Geller, E. S. (1996), *The Psychology of Safety*, Chilton Enterprises, Radnor, PA.

Giesa, H.-G., and Timpe, K.-P. (2000), "Technisches Versagen und menschliche Zuverlässigkeit: Bewertung der Verlässlichkeit in Mensch–Maschine-Systemen" [Design Failures and Human Reliability: Assessment of the Reliability of Human–Machine Systems], In, *Mensch–Maschine–Systemtechnik*, S. 63–106, K.-P. Timpe, T. Jürgensohn, and H. Kolrep, Eds., Symposion Publishing, Düsseldorf, Germany.

Glendon, A. I., and Litherland, D. K. (2001), "Safety Climate Factors, Group Differences and Safety Behaviour in Road Construction," *Safety Science*, Vol. 39, pp. 157–188.

Grandjean, E. (1980), *Fitting the Task to the Man: An Ergonomic Approach*, Taylor & Francis, London.

Gregersen, N. P., Brehmer, B., and Moren, B. (1996), "Road Safety Improvement in Large Companies: An Experimental Comparison of Different Measures," *Accident Analysis and Prevention*, Vol. 28, pp. 297–306.

Griffiths, D. K. (1985), "Safety Attitudes of Management," *Ergonomics*, Vol. 28, pp. 61–67.

Grimaldi, J. V., and Simonds, R. H. (1989), *Safety Management*, Richard D. Irwin, Homewood, IL.

Guastello, S. J. (1989), "Catastrophe Modelling of the Accident Process: Evaluation of an Accident Reduction Program Using the Occupational Hazards Survey," *Accident Analysis and Prevention*, Vol. 21, pp. 61–77.

Guastello, S. J. (1991), "Psychosocial Variables Related to Transit Safety: The Application of Catastrophe Theory," *Work and Stress*, Vol. 5, pp. 17–28.

Guastello, S. J., Gershon, R. R. M., and Murphy, L. R. (1999), "Catastrophe Model for the Exposure to Blood-borne Pathogens and Other Accidents in Health Care Settings," *Accident Analysis and Prevention*, Vol. 31, pp. 739–749.

Guldenmund, F. W. (2000), "The Nature of Safety Culture: A Review of Theory and Research," *Safety Science*, Vol. 34, pp. 215–257.

Guzzo, R. A., Jette, R. D., and Katzell, R. A. (1985), "The Effects of Psychologically Based Intervention Programs on Worker Productivity: A Meta-analysis," *Personnel Psychology*, Vol. 38, pp. 275–291.

Hackman, J. R., and Oldham, G. R. (1980), *Work Redesign*, Addison-Wesley, London.

Hale, A. R., and Baram, M., Eds. (1998), *Safety Management and the Challenge of Organizational Change*, Elsevier, Oxford.

Hale, A. R., and Glendon, A. I. (1987), *Individual Behaviour in the Control of Danger*, Elsevier, Amsterdam.

Hale, A. R., and Hovden, J. (1998), "Management and Culture: The Third Age of Safety. A Review of Approaches to Organizational Aspects of Safety, Health and Environment," in *Occupational Injury: Risk, Prevention and Intervention*, A.-M. Feyer and A. Williamson, Eds., Taylor & Francis, London, pp. 129–165.

Hale, A. R., Heming, B. H., Carthey, J., and Kirwan, B. (1997), "Modelling of Safety Management Systems," *Safety Science*, Vol. 26, pp. 121–140.

Harper, A. C., Cordery, J. L., de Klerk, N. H., Sevastos, P., Geelhoed, E., Gunson, C., Robinson, L., Sutherland,M., Osborn, D., and Colquhoun, J. (1997), "Curtin Industrial Safety Trial: Managerial Behaviour and Program Effectiveness," *Safety Science*, Vol. 24, pp. 173–179.

Health and Safety Executive (1993), *The Costs of Accidents at Work*, Her Majesty's Stationery Office, London.

Health and Safety Executive (1997), *Successful Health and Safety Management*, HS(G) 65, London.

Heinrich, H. W., Petersen, D., and Roos, N. (1980), *Industrial Accident Prevention: A Safety Management Approach*, McGraw-Hill, New York.

Hine, D. W., Lewko, J., and Blanko, J. (1999), "Alignment to Workplace Safety Principles: An Application to Mining," *Journal of Safety Research*, Vol. 30, pp. 173–185.

Hix, D., and Hartson, H. R. (1993), *Developing User Interfaces*, Wiley, New York.

Hofmann, D. A., and Morgeson, F. P. (1999), "Safety-Related Behavior as a Social Exchange: The Role of Perceived Organizational Support and Leader–Member Exchange," *Journal of Applied Psychology*, Vol. 84, pp. 286–296.

Hofmann, D. A., and Stetzer, A. (1996), "A Cross-Level Investigation of Factors Influencing Unsafe Behaviors and Accidents," *Personnel Psychology*, Vol. 49, pp. 307–339.

Hofmann, D. A., Jacobs, R., and Landy, F. (1995), "High Reliability Process Industries: Individual, Micro, and Macro Organizational Influences on Safety Performance," *Journal of Safety Research*, Vol. 26, pp. 131–149.

Hollnagel, E. (1998), *Cognitive Reliability and Error Analysis Method (CREAM)*, Elsevier, Oxford.

Hopkins, B. L., Conrad, R. J., Dangel, R. F., Fitch, H. G., Smith, M. J., and Anger, W. K. (1986), "Behavioral Technology for Reducing Occupational Exposures to Styrene," *Journal of Applied Behavior Analysis*, Vol. 19, pp. 3–11.

Hovden, J., and Larsson, T. J. (1987), "Risk: Culture and Concepts," in *Risk and Decisions*, W. T. Singleton and J. Hovden Eds., Wiley, New York, pp. 47–66.

Hoyos, C. (1992), "A Change in Perspective: Safety Psychology Replaces the Traditional Field of Accident Research," *German Journal of Psychology*, Vol. 16, pp. 1–23.

Hoyos, C. (1995), "Occupational Safety: Progress in Understanding the Basic Aspects of Safe and Unsafe Behaviour," *Applied Psychology: An International Review*, Vol. 44, pp. 233–250.

Hoyos, C., Bernhardt, U., Hirsch, G., and Arnhold, T. (1991), "Vorhandenes und erwünschtes Wissen in Industriebetrieben" [Actual and Required Knowledge at Industrial Workplaces], *Zeitschrift für Arbeits- und Organisationspsychologie*, Vol. 35, pp. 68–76.

Hoyos, C., and Ruppert, F. (1993), *Der Fragebogen zur Sicherheitsdiagnose [Safety Diagnosis Questionnaire] (FSD)*, Hans Huber, Bern, Switzerland.

Hoyos, C., and Zimolong, B. (1988), *Occupational Safety and Accident Prevention: Behavioral Strategies and Methods*, Elsevier, Amsterdam.

Humphreys, P. C., and Wisudha, A. (1983), *MAUD: An Interaction Computer Program for the Structuring Decomposition and Recomposition of Preferences Between Multiattributed Alternatives*, Decision Analysis Unit, London.

Hurst, N. W., Young, S., Donald, I., Gibson, H., and Muyselaar, A. (1996), "Measures of Safety Management Performance and Attitudes to Safety at Major Hazard Sites," *Journal of Loss Prevention in the Process Industries*, Vol. 9, pp. 161–172.

Huselid, M. A. (1995), "The Impact of Human Resource Management Practices on Turnover, Productivity, and Corporate Financial Performance," *Academy of Management Journal*, Vol. 38, pp. 635–672.

Hutchinson, A., and Hutchinson, F., Eds. (1997), *Environmental Business Management*, McGraw-Hill, London.

IAEA (1986), "Summary Report on the Post Accident Review Meeting on the Chernobyl Accident," 75-INSAG-1, International Atomic Energy Authority, Vienna.

ILCI (1990), *International Safety Rating System (ISRS)*, International Loss Control Institute, Loganville, PA.

Imada, A. S. (1990), "Ergonomics: Influencing Management Behavior," *Ergonomics*, Vol. 33, pp. 621–628.

James, L. R., James, L. A., and Ashe, D. K. (1990), "The Meaning of Organizations: The Role of Cognition and Values," in *Organizational Climate and Culture*, B. Schneider, Ed., Jossey-Bass, San Francisco, pp. 40–84.

Jenkins, G. D., Mitra, A., Gupta, N., and Shaw, J. D. (1998), "Are Financial Incentives Related to Performance? A Meta-analytic Review of Empirical Research," *Journal of Applied Psychology*, Vol. 83, pp. 777–787.

Johnson, W. G. (1975), "MORT: The Management Oversight and Risk Tree," *Journal of Safety Research*, Vol. 7, pp. 4–15.

Johnson, W. G. (1980), *MORT Safety Assurance Systems*, National Safety Council, Chicago.

Jung, D. I., and Avolio, B. J. (2000), "Opening the Black Box: An Experimental Investigation of the Mediating Effect of Trust and Value Congruence on Transformational and Transactional Leadership," *Journal of Organizational Behaviour*, Vol. 21, pp. 949–964.

Kahneman, D., Slovic, P., and Tversky, A., Eds. (1982), *Judgment Under Uncertainty*, Cambridge University Press, New York.

Kalimo, R., Lindström, K., and Smith, M. J. (1997), "Psychosocial Approach in Occupational Health," in *Handbook of Human Factors and Ergonomics*, 2nd ed., G. Salvendy, Ed., Wiley, New York, pp. 1059–1084.

Kennedy, R., and Kirwan, B. (1998), "Development of a Hazard and Operability-Based Method for Identifying Safety Management Vulnerabilities in High Risk Systems," *Safety Science*, Vol. 30, pp. 249–274.

Kjellén, U. (1984), "The Deviation Concept in Occupational Accident Control, II: Data Collection and Assessment of Significance," *Accident Analysis and Prevention*, Vol. 16, pp. 307–323.

Kjellén, U. (2000), *Prevention of Accidents Through Experience Feedback*, Taylor & Francis, London.

Kjellén, U., and Larsson, T. (1981), "Investigating Accidents and Reducing Risks: A Dynamic Approach," *Journal of Occupational Accidents*, Vol. 3, pp. 129–140.

Klein, H. J., Wesson, M. J., Hollenbeck, J. R., and Alge, B. J. (1999), "Goal Commitment and the Goal-Setting Process: Conceptual Clarification and Empirical Synthesis," *Journal of Applied Psychology*, Vol. 84, pp. 885–896.

Kluger, A. N., and DeNisi, A. (1996), "The Effects of Feedback Interventions on Performance: A Historical Review, a Meta-analysis, and a Preliminary Feedback Intervention Theory," *Psychological Bulletin*, Vol. 119, pp. 254–284.

Komaki, J. L. (1998), *Leadership from an Operant Perspective*, Routledge, New York.

Komaki, J., Barwick, K. D., and Scott, L. R. (1978), "A Behavioral Approach to Occupational Safety: Pinpointing and Reinforcing Safe Performance in a Food Manufacturing Plant," *Journal of Applied Psychology*, Vol. 63, pp. 434–445.

Latham, G. P., and Kinne, S. B. (1974), "Improving Job Performance Through Training in Goal Setting," *Journal of Applied Psychology*, Vol. 59, pp. 187–191.

Latham, G. P., and Yukl, G. A. (1975), "Assigned Versus Participative Goal Setting with Educated and Uneducated Wood Workers," *Journal of Applied Psychology*, Vol. 60, pp. 299–302.

Lee, T. (1998), "Assessment of Safety Culture at a Nuclear Reprocessing Plant," *Work and Stress*, Vol. 12, pp. 217–237.

Lee, T., and Harrsion, K. (2000), "Assessing Safety Culture in Nuclear Power Stations," *Safety Science*, Vol. 34, pp. 61–97.

Lehto, M. R., and Papastavrou, J. D. (1993), "Models of the Warning Process: Important Implications Toward Effectiveness," *Safety Science*, Vol. 16, pp. 569–595.

Leveson, N. (2002), *A New Approach to System Safety Engineering*, MIT Press, Cambridge, MA.

Lingard, H., and Rowlinson, S. (1997), "Behavior-Based Safety Management in Hong Kong," *Journal of Safety Research*, Vol. 24, pp. 243–256.

Locke, E. A., and Latham, G. P. (1990), *A Theory of Goal Setting and Task Performance*, Prentice-Hall, Englewood Cliffs, NJ.

Locke, E. A., and Latham, G. P. (2002), "Building a Practically Useful Theory of Goal Setting and Task Motivation," *American Psychologist*, Vol. 57, pp. 705–717.

Locke, E. A., Alavi, M., and Wagner, J. (1997), "Participation in Decision-Making: An Information Exchange Perspective," in *Research in Personnel and Human Resources Management*, Vol. 15, G. Ferris, Ed., JAI Press, Greenwich, CT, pp. 293–331.

Ludborzs, B. (1995), "Surveying and Assessing 'Safety Culture' Within the Framework of Safety Audits," in *Loss Prevention and Safety Promotion in the Process Industries*, Vol. 1, J. J. Lewis, H. J. Pasman, and E. E. De Rademaecker, Eds., Elsevier, Amsterdam, pp. 83–92.

Luthans, F., and Kreitner, R. (1985), *Organizational Behavior Modification and Beyond*, Scott, Foresman, Glenview, IL.

Marks, M. L., Mirvis, P. H., Hackett, E. J., and Grady, J. F. (1986), "Employee Participation in a Quality Circle Program: Impact on Quality of Work Life, Productivity, and Absenteeism," *Journal of Applied Psychology*, Vol. 71, pp. 61–69.

Mattila, M., Hyttinen, M., and Rantanen, E. (1994), "Effective Supervisory Behavior and Safety at the Building Site," *International Journal of Industrial Ergonomics*, Vol. 13, pp. 85–93.

McAfee, R. B., and Winn, A. R. (1989), "The Use of Incentives/Feedback to Enhance Work Place Safety: A Critique of the Literature," *Journal of Safety Research*, Vol. 20, pp. 7–19.

McCarthy, M. (1978), "Decreasing the Incidence of 'High Bobbins' in a Textile Spinning Department Through a Group Feedback Procedure," *Journal of Organizational Behavior Management*, Vol. 1, pp. 150–54.

Mearns, K., Whitaker, S. M., and Flin, R. (2003), "Safety Climate, Safety Management Practice and Safety Performance in Offshore Environments," *Safety Science*, Vol. 41, pp. 641–680.

Meister, D., and Enderwick, T. (2002), *Human Factors in System Design, Development, and Testing*, Lawrence Erlbaum Associates, Mahwah, NJ.

Miller, I., and Cox, S. (1997), "Benchmarking for Loss Control," *Journal of the Institute of Occupational Safety and Health*, Vol. 1, pp. 39–47.

Miller, D. P., and Swain, A. D. (1987), "Human Error and Human Reliability," in *Handbook of Human Factors*, G. Salvendy, Ed., Wiley, New York, pp. 219–250.

Nadler, D. A. (1979), "The Effects of Feedback on Task Group Behavior: A Review of the Experimental Research," *Organizational Behavior and Human Performance*, Vol. 23, pp. 309–338.

Neal, A., Griffin, M. A., and Hart, P. M. (2000), "The Impact of Organizational Climate on Safety Climate and Individual Behavior," *Safety Science*, Vol. 34, pp. 99–109.

Nisbett, R., and Ross, L. (1980), *Human Inference: Strategies and Shortcomings of Social Judgment*, Prentice-Hall, Englewood Cliffs, NJ.

Niskanen, T. (1994), "Assessing the Safety Environment in Work Organization of Road Maintenance Jobs," *Accident Analysis and Prevention*, Vol. 26, pp. 27–39.

Noe, R. A. (1986), "Trainee's Attributes and Attitudes: Neglected Influences on Training Effectiveness," *Academy of Management Review*, Vol. 11, pp. 736–749.

NRC (1975), "Reactor Safety Study: An Assessment of Accident Risks in U.S. Commercial Nuclear Power Plants (WASH 1400)," NUREG 75/014, U.S. Nuclear Regulatory Commission, Washington, DC.

O'Dea, A., and Flin, R. (2001), "Site Managers, Supervisors, and Safety in the Offshore Oil and Gas Industry," *Safety Science*, Vol. 37, pp. 39–57.

OECD Nuclear Agency (1987), *Chernobyl and the Safety of Nuclear Reactors on OECD Countries*, Organization for Economic Co-operation and Development, Paris.

O'Hara, K., Johnson, C. M., and Beehr, T. A. (1985), "Organizational Behavior Management in the Private Sector: A Review of Empirical Research," *Academy of Management Review*, Vol. 10, pp. 848–864.

O'Leary-Kelly, A., Martocchio, J., and Frink, D. (1994), "A Review of the Influence of Group Goals on Group Performance," *Academy of Management Journal*, Vol. 37, pp. 1285–1301.

Paul, A. K., and Anantharaman, R. N. (2003), "Impact of People Management Practices on Organizational Performance: Analysis of a Causal Model," *International Journal of Human Resource Management*, Vol. 14, pp. 1246–1266.

Petersen, D. C. (1978), *Techniques of Safety Management*, McGraw-Hill Kogakusha, Kyobashi, Japan.

Pfeffer, J. (1994), *Competitive Advantage Through People*, Harvard Business School Press, Boston.

Pidgeon, N. F. (1991). "Safety Culture and Risk Management in Organizations," *Journal of Cross-Cultural Psychology*, Vol. 22, pp. 129–140.

Pidgeon, N. F., and O'Leary, M. (2000), "Man-Made Disasters: Why Technology and Organizations (Sometimes) Fail," *Safety Sciences*, Vol. 34, pp. 15–30.

Rasmussen, J. (1983), "Skills, Rules, Knowledge, Signals, Signs and Symbols and Other Distinctions in Human Performance Models," *IEEE Transactions on Systems, Man and Cybernetics*, Vol. 3, pp. 266–275.

Rasmussen, J., and Batstone, R. (1991), *Safety Control and Risk Management: Toward Improved Low Risk Operation of High Hazard Systems*, World Bank, Washington, DC.

Reason, J. T. (1987a), "The Chernobyl Errors," *Bulletin of the British Psychology Society*, Vol. 40, pp. 201–206.

Reason, J. T. (1987b), "Generic Error-Modelling System (GEMS): A Cognitive Framework for Locating Common Human Error Forms," in *New Technology and Human Error*, K. D. Rasmussen and J. Leplat, Eds., Wiley, New York, pp. 63–83.

Reason, J. T. (1990), *Human Error*, Cambridge University Press, Cambridge.

Reason, J. T. (1993), "Managing the Management Risk: New Approaches to Organizational Safety," in *Reliability and Safety in Hazardous Work Systems: Approaches to Analysis and Design*, B. Wilpert and T. U. Qvale, Eds., Lawrence Erlbaum Associates, Hove, England, East Sussex, pp. 7–22.

Reber, R. A., and Wallin, J. A. (1984), "The Effects of Training, Goal Setting, and Knowledge of Results on Safety Behaviour: A Component Analysis," *Academy of Management Journal*, Vol. 27, pp. 544–560.

Reber, R. A., Wallin, J. A., and Duhon, D. L. (1993), "Preventing Occupational Injuries Through Performance Management," *Public Personnel Management*, Vol. 22, pp. 301–311.

Reichers, A. E., and Schneider, B. (1990), "Climate and Culture: An Evolution of Constructs," in *Organizational Climate and Culture*, B. Schneider, Ed., Jossey-Bass, San Francisco.

Rentsch, J. R. (1990), "Climate and Culture: Interaction and Qualitative Differences in Organizational Meanings," *Journal of Applied Psychology*, Vol. 75, pp. 668–681.

Ridley, J., and Channing, J., Eds. (2003), "*Safety at Work*, 6th ed., Butterworth-Heinemann, Oxford.

Roberts, K. H. (1990), "Some Characteristics of One Type of High Reliability Organization," *Organization Science*, Vol. 1, pp. 160–176.

Rochlin, G. I. (1996), "Reliable Organizations: Present Research and Future Directions," *Journal of Contingencies and Crisis Management*, Vol. 4, pp. 5–59.

Rodgers, R., and Hunter, J. E. (1991), "Impact of Management by Objectives on Organizational Productivity," *Journal of Applied Psychology*, Vol. 76, pp. 322–336.

Rouhiainen, V. (1992), "QUASA: A Method for Assessing the Quality of Safety Analysis," *Safety Science*, Vol. 15, pp. 155–172.

Ruppert, F. (1987), "Gefahrenwahrnehmung: ein Modell zur Anforderungsanalyse für die verhaltensabhängige Kontrolle von Arbeitsplatzgefahren" [Hazard Perception: A Model for the Requirement Analysis of Workplace Hazards], *Zeitschrift für Arbeitswissenschaft*, Vol. 41, pp. 84–87.

Saarela, K. L. (1990), "An Intervention Program Utilizing Small Groups: A Comparative Study," *Journal of Safety Research*, Vol. 21, pp. 149–156.

Saari, J. (1976), "Characteristics of Tasks Associated with the Occurrence of Accidents," *Journal of Occupational Accidents*, Vol. 1, pp. 273–279.

Saari, J. (1987), "Management of Housekeeping by Feedback," *Ergonomics*, Vol. 30, pp. 313–317.

Saari, J., and Näsänen, M. (1989), "The Effect of Positive Feedback on Industrial Housekeeping and Accidents: A Long Term Study at a Shipyard," *International Journal of Industrial Ergonomics*, Vol. 4, pp. 201–211.

Schein, E. H. (1992), *Organizational Culture and Leadership*, 2nd ed., Jossey-Bass, San Francisco.

Shackel, B., Ed. (1984), *Applied Ergonomics Handbook*, Butterworth Scientific, London.

Shafai-Sahrai, Y. (1971), "An Inquiry into Factors That Might Explain Differences in Occupational Accident Experience of Similar Size Firms in the Same Industry," Division of Research, Graduate School of Business Administration, Michigan State University, East Lansing, MI; cited in Cohen (1977).

Shannon, H. S., Walters, V., Lewchuk, W., Richardson, J., Moran, L. A., Haines, T. A., and Verma, D. (1996), "Workplace Organizational Correlates of Lost Time Accident Rates in Manufacturing," *American Journal of Industrial Medicine*, Vol. 29, pp. 258–268.

Shannon, H. S., Mayr, J., and Haines, T. (1997), "Overview of the Relationship Between Organizational and Workplace Factors and Injury Rates," *Safety Science*, Vol. 26, pp. 210–127.

Shannon, H. S., Robson, L. S., and Guastello, S. J. (1999), "Methodological Criteria for Evaluating Occupational Safety Intervention Research," *Safety Science*, Vol. 31, pp. 161–179.

Simard, M., and Marchand, A. (1994), "The Behavior of First-Line Supervisors in Accident Prevention and

Effectiveness in Occupational Safety," *Safety Science*, Vol. 17, pp. 169–185.

Simard, M., and Marchand, A. (1997), "Workgroups Propensity to Comply with Safety Rules: The Influence of Micro–Macro Organisation Factors," *Ergonomics*, Vol. 40, pp. 172–188.

Slovic, P. (1987), "Perception of Risk," *Science*, Vol. 236, pp. 280–285.

Smith, M. J., Cohen, H., Cohen, A., and Cleveland, R. (1978), "Characteristics of Successful Safety Programs," *Journal of Safety Research*, Vol. 10, pp. 5–15.

Stajkovic, A. D., and Luthans, F. (1997), "A Meta-analysis of the Effects of Organization Behavior Modification on Task Performance," *Academy of Management Journal*, Vol. 40, pp. 1122–1149.

Stajkovic, A. D., and Luthans, F. (2003), "Behavioral Management and Task Performance in Organizations: Conceptual Background, Meta-analysis, and Test of Alternative Models," *Personnel Psychology*, Vol. 56, pp. 155–194.

Steel, C. (1990), "Risk Estimation," *Safety Practitioner*, Vol. 8, pp. 20–21.

Strauch, B. (2004), *Investigating Human Error: Incidents, Accidents, and Complex Systems*, Ashgate Publishing, Burlington, VT.

Sulzer-Azaroff, B. (1978), "The Modification of Occupational Behavior," *Journal of Occupational Accidents*, Vol. 9, pp. 177–197.

Sulzer-Azaroff, B. (1982), "Behavioral Approaches to Occupational Health and Safety," in *Handbook of Organizational Behavior Management*, D. R. Frederiksen, Ed., Wiley, New York, pp. 505–537.

Sulzer-Azaroff, B., Harris, T. C., and McCann, K. B. (1994), "Beyond Training: Organizational Performance Management Techniques," *Occupational Medicine*, Vol. 9, pp. 321–339.

Svenson, O. (1989), "On Expert Judgment in Safety Analyses in the Process Industries," *Reliability Engineering and System Safety*, Vol. 25, pp. 219–256.

Swain, A. D., and Guttmann, H. E. (1983), *Handbook of Human Reliability Analysis with Emphasis on Nuclear Power Plant Applications*," U.S. Nuclear Regulatory Commission, Washington, DC.

Thompson, R. C., Hilton, T. F., and Witt, L. A. (1998), "Where the Safety Rubber Meets the Shop Floor: A Confirmatory Model of Management Influence on Workplace Safety," *Journal of Safety Research*, Vol. 29, pp. 15–24.

Timpe, K.-P. (1993), "Psychology's Contributions to the Improvement of Safety and Reliability in the Man–Machine System," in *Reliability and Safety in Hazardous Work Systems*, B. Wilpert and T. Qvale, Eds., Lawrence Erlbaum Associates, Mahwah, NJ, pp. 119–132.

Tinmannsvik, R. K., and Hovden, J. (2003), "Safety Diagnosis Criteria: Development and Testing," *Safety Science*, Vol. 41, pp. 575–590.

Tomas, J. M., Melia, J. L., and Oliver, A. (1999), "A Cross-Validation of a Structural Equation Model of Accidents: Organizational and Psychological Variables as Predictors of Work Safety," *Work and Stress*, Vol. 13, pp. 49–58.

Trimpop, R. M. (1994), *The Psychology of Risk Taking Behavior*, Elsevier, Amsterdam.

Turner, B. A. (1978), *Man-Made Disasters*, Wykeham, London.

Turner, B. A. (1991), "The Development of a Safety Culture," *Chemistry and Industry*, Vol. 7, pp. 241–243.

Tversky, A., and Kahneman, D. (1981), "The Framing of Decisions and the Psychology of Choice," *Science*, Vol. 211, pp. 453–458.

Van Cott, H. P., and Kinkade, R. G., Eds. (1972), *Human Engineering Guide to Equipment Design*, American Institutes for Research, Washington, DC.

Vassie, L. H., and Lucas, W. R. (2001), "An Assessment of Health and Safety Management Within Groups in the UK Manufacturing Sector," *Journal of Safety Research*, Vol. 32, pp. 479–490.

Veltri, A. (1990), "An Accident Cost Impact Model: The Direct Cost Component," *Journal of Safety Research*, Vol. 21, pp. 67–73.

Visser, J. P. (1998), "Development in HSE Management in Oil and Gas Exploration and Production," in *Safety Management*, A. Hale and M. Baram, Eds., Elsevier, Oxford, pp. 43–66.

Wagenaar, W. A. (1992), "Risk Taking and Accident Causation," in *Risk-Taking Behavior*, J. F. Yates, Ed., Wiley, Chichester, West Sussex, England, pp. 257–281.

Wagenaar, W. A., Souverijn, A. M., and Hudson, P. T. (1993), "Safety Management in Intensive Care Wards," in *Reliability and Safety in Hazardous Work Systems*, B. Wilpert and T. Qvale, Eds., Lawrence Erlbaum Associates, Mahwah, NJ, pp. 157–169.

Wagenaar, W. A., Groeneweg, J., Hudson, P. T., and Reason, J. T. (1994), "Promoting Safety in the Oil Industry," *Ergonomics*, Vol. 37, pp. 1999–2013.

Wagner, J., and Gooding, R. (1987), "Effects of Societal Trends on Participation Research," *Administrative Science Quarterly*, Vol. 32, pp. 241–262.

Wickens, C. D., and Hollands, J. G. (2000), *Engineering Psychology and Human Performance*, 3rd ed., Prentice-Hall, Upper Saddle River, NJ.

Windel, A., and Zimolong, B. (1998), "Nicht zum Nulltarif: Erfolgskonzept Gruppenarbeit" [Worth the Effort: Group Work Is a Successful Concept], *RUBIN*, Vol. 2, pp. 46–51.

Wogalter, M. S., Young, S. L., and Laughery, K. R. (2001), *Human Factors Perspectives on Warnings*, Vol. 2, Human Factors and Ergonomics Society, Santa Monica, CA.

Woodson, W. E. (1981), *Human Factors Design Handbook*, McGraw-Hill, New York.

Yates, J. F., Ed. (1992), "The Risk Construct," in *Risk-Taking Behavior*, J. F. Yates, Ed., Wiley, Chichester, West Sussex, England, pp. 1–25.

Zimolong, B. (1981), "Traffic Conflicts: A Measure of Road Safety," in *Road Safety: Research and Practice*, H. C. Foot, A. J. Chapmann, and F. M. Wade, Eds., Praeger, imprint of Holt-Saunders, Eastbourne, East Sussex, England, pp. 35–41.

Zimolong, B. (1985), "Hazard Perception and Risk Estimation in Accident Causation," in *Trends in Ergonomics/ Human Factors II*, R. E. Eberts and C. G. Eberts, Eds., Elsevier, Amsterdam, pp. 463–470.

Zimolong, B. (1992), "Empirical Evaluation of THERP, SLIM and Ranking to Estimate HEPs," *Reliability Engineering and System Safety*, Vol. 35, pp. 1–11.

Zimolong, B. (1997), "Occupational Risk Management," in *Handbook of Human Factors and Ergonomics*, 2nd ed., G. Salvendy, Ed., Wiley, New York, pp. 989–1020.

Zimolong, B., Ed. (2001a), *Management des Arbeits- und Gesundheitsschutzes: Die erfolgreichen Strategien der*

Unternehmen [Health and Safety Management: The Successful Strategies of Enterprises], Gabler Verlag, Wiesbaden, Germany.

Zimolong, B. (2001b), "Arbeitsschutz-Management-Systeme" [Safety Management Systems], in *Management des Arbeits- und Gesundheitsschutzes: Die erfolgreichen Strategien der Unternehmen* [*Health and Safety Management: The Successful Strategies of Enterprises*], B. Zimolong (Ed.), Gabler Verlag, Wiesbaden, Germany, pp. 13–30.

Zimolong, B., and Elke, G. (2001a), "Risk Management," in *International Encyclopedia of Ergonomics and Human Factors*, W. Karwowski, Ed., Taylor & Francis, London, pp. 1327–1333.

Zimolong, B., and Elke, G. (2001b), "Die erfolgreichen Strategien und Praktiken der Unternehmen" (The Successful Strategies and Practices of Enterprises)," in *Management des Arbeits- und Gesundheitsschutzes: Die erfolgreichen Strategien der Unternehmen* [*Health and Safety Management: The Successful Strategies of Enterprises*] B. Zimolong, Ed., Gabler Verlag, Wiesbaden, Germany, pp. 235–268.

Zimolong, B., and Hale, A. R. (1989), "Arbeitssicherheit" [Workplace Safety], in *Europäisches Handbuch der Arbeits- und Organisationspsychologie*, S. Greif, W. Holling, and N. Nicholson, Eds., Psychologie Verlagsunion, Munich, pp. 126–131.

Zimolong, B., and Stapp, M. (2001), "Psychosoziale Gesundheitsförderung" [Psychosocial Health Promotion], in *Management des Arbeits- und Gesundheitsschutzes: Die erfolgreichen Strategien der Unternehmen* [*Health and Safety Management: The Successful Strategies of Enterprises*] B. Zimolong, Ed., Gabler Verlag, Wiesbaden, Germany, pp. 141–169.

Zohar, D. (1980), "Safety Climate in Industrial Organizations: Theoretical and Applied Implications," *Journal of Applied Psychology*, Vol. 65, pp. 96–102.

Zohar, D. (2000), "A Group-Level Model of Safety Climate: Testing the Effect of Group Climate on Microaccidents in Manufacturing Jobs," *Journal of Applied Psychology*, Vol. 85, pp. 587–596.

Zohar, D. (2002a), "Modifying Supervisory Practices to Improve Subunit Safety: A Leadership-Based Intervention Model," *Journal of Applied Psychology*, Vol. 87, pp. 156–163.

Zohar, D. (2002b), "The Effects of Leadership Dimensions, Safety Climate, and Assigned Priorities on Minor Injuries in Work Groups," *Journal of Organizational Behavior*, Vol. 23, pp. 75–92.

Zohar, D. (2003), "Safety Climate: Conceptual and Measurement Issues," in *Handbook of Occupational Health Psychology*, J. C. Quick and L. E. Tetrick, Eds., American Psychological Association, Washington, DC, pp. 123–142.

CHAPTER 27

HUMAN ERROR

Joseph Sharit
University of Miami
Coral Gables, Florida

1 INTRODUCTION

1.1 Does Human Error Exist?

Human error in human–system interaction was a major influence in establishing the area of human factors (Helander, 1997). Recognition of human error as an area in its own right came some years later, when an appreciation for its implications in complex high-risk systems became more widespread. Human error has since become inextricably linked to safety science, and together these areas have strongly influenced the design, reliability, risk assessment, and risk management programs that have become pivotal to the success of many organizations.

For a good part of the twentieth century the dominant perspective on human error by many U.S. industries was to attribute adverse outcomes to the persons whose actions were most closely associated to these events. We are now witnessing an almost complete turnaround in this perspective, to the extent that it has even become fashionable to reject the notion of human error. In this view the human is deemed to be a reasonable entity at the mercy of an array of design, organizational, and situational factors that can lead to behaviors external observers come to regard, although often unfairly, as human errors.

The appeal of this view should be readily apparent in each of the following two cases. The first case involves a worker who is subjected to performing a task in a restricted space. While attempting to reach for a tool, the worker's forearm inadvertently activates a switch, resulting in the emission of heat. Visual feedback concerning the activation is not possible, due to the awkward posture the worker must assume; tactile cues are not detectable due to requirements for wearing protective clothing; and although present, auditory feedback from the switch's activation is not audible, due to high noise levels. Residual vapors originating from a rarely performed procedure during the previous shift ignite, resulting in an explosion. In the second case, a worker adapts the relatively rigid and unrealistic procedural requirements dictated in a written work procedure to demands that continually materialize in the form of shifting objectives, constraints on resources, and changes in production schedules. Management tacitly condones these procedural adaptations, in effect relying on the resourcefulness of the worker for ensuring that its goals are met (Section 8). However, when an unanticipated scenario causes the worker's adaptations to result in an accident, management is swift to renounce any support of actions in violation of work procedures.

In the first case the worker's action that led to the accident was unintentional; in the second case the worker's actions were intentional. In both cases the issue of whether the "actor" committed an error is debatable. One variant on the position that rejects the notion of human error would shift the blame for the adverse consequences from the actor to management or the designers. Latent management or latent designer errors (Section 7.3.1) would thus absolve the actor from human error in each of these cases. The worker, after all, was in the heat of the battle, performing "normal work," responding to the contextual features of the situation in reasonable, even skillful ways.

A second variant on this position would cast doubt on the process by which the attribution of error is made (Dekker, 2005). By virtue of having knowledge of events, especially bad events such as accidents, outside observers are able—perhaps even motivated—to invoke a backward series of rationalizations and logical connections that has neatly filtered out the subtle and complex situational details that are likely to be the basis for the perpetrating actions. Whether this process of establishing causality (Section 10) is due to convenience, or derives from the inability to determine or comprehend the perceptions and assessments made by the actor that interlace the more prominently observable events, the end result is a considerable underestimation of the influence of context. Even the workers themselves, if given the opportunity in each of these cases to examine or reflect upon their performance, may acknowledge their actions as errors, easily spotting all the poor decisions and improperly executed actions, when in reality, within the frames of references at the time the behaviors occurred, their actions were in fact reasonable, and constituted "mostly normal work." The challenge, according to Dekker (2005), is "to understand how assessments and actions that from the outside look like errors become neutralized or normalized so that from the inside they appear unremarkable, routine, normal" (p. 75).

These views, which essentially deny the existence of human error (at least on the part of the actors) are appealing and to some extent justified. The issue, however, is not so much whether these views should be dismissed, but whether they should be embraced. The position taken here is that human error is a real phenomenon that has at its roots many of the same attentional processes and architectural features of memory that enable the human to adapt, abstract, infer, and create, but that also subject the human to various kinds of information-processing constraints that can provoke unintended or mistaken actions. Thus, although it may be convenient to explain unintended *action slips* (Section 3.2.5) such as the activation of an incorrect control or the selection of the wrong medication as rational responses in contexts characterized by pressures, conflicts, ambiguities, and fatigue, a closer inspection of the work context can, in theory, reveal the increased possibility for certain types of errors as compared to others. It is human fallibility, in all its guises, that infiltrates these contexts, and by failing to acknowledge the interplay between human fallibility and context—for instance, the tendency for a context to induce "capture" by the wrong control or the wrong medication—we are left with a shoddier picture of the context. Granted, the contextual details comprising dynamic work activities are difficult enough to establish, let alone their interplay with human fallibility. However, this fact attests only to the difficulty of predicting human error (Section 4), especially complex errors, not to the dismissal of its existence. Whereas rejecting the notion of human error may represent a gracious gesture toward the human's underlying disposition, it can also dangerously downplay aspects of human fallibility that need to be understood for implementing error reduction and error management strategies.

1.2 New Directions

Much of the practical knowledge that has been accumulated on human error in the last half century has derived primarily from industries requiring hazardous operations that are capable of producing catastrophic events. Not surprisingly, the textbook scenarios typically used for studying human error came from domains such as nuclear power, chemical processing, and aviation. With the publication of *To Err Is Human* (Kohn et al., 1999) came the revelation of shocking data that formally announced the new scourge in human error—medical error. According to this report, between 44,000 and 98,000 hospitalized patients die annually as a result of human error. These figures were extrapolated from studies that included the relatively well known Harvard Medical Practice Study in New York (Leape et al., 1991). Although these figures have been contested on the grounds that many of the patients whose deaths were attributed to medical error were predisposed to die due to the severity of their illnesses, there is also an opposing belief that these errors were underreported by as much as a factor of 10 (Cullen et al., 1995). If true, the number of preventable hospital deaths attributable to human error is staggering, even if adjustments are made for deaths that were likely to occur due to illness alone.

It also signals the need for heightened concern for the many mistakes in health care that are probably occurring outside hospital environments.

Fear of blame and retribution through litigation accounts for much of the underreporting in health care and reflects an industry that is still mired in the blame culture of traditional mid-twentieth-century American industry. What truly dissociates medical error from human error in other high-risk work domains is the belief by many people that they can assume the role of expert based on the experiences they, a family member, or a close friend have had, and that they have the right to hold an industry that is extracting high premiums for their services accountable for its actions. It remains to be seen if this attitude will carry over to other industries. In any case, medical error presents unique challenges, and although we do not intend to diminish the significance of human error in industries with relatively long-standing traditions for addressing the role of human error in safety, due emphasis will also be given to medical error.

2 DEFINING HUMAN ERROR

The presumption of human error generally occurs when various types of committed or omitted human actions appear, when viewed in retrospect, to be linked to undesirable consequences, although unwanted consequences do not necessarily imply the occurrence of human error. Following the distinctions proposed by Norman (1981) and Reason (1990), the term *error* usually applies only to those situations where there was an intention to perform some type of action, and would include cases where there was no prior intention to act. Thus, a very well practiced routine that is performed without any prior intention, such as swiping dirt from a tool, may constitute an error depending on the effect of that action. More typically, errors are associated with prior intentions to act, in which case two situations can be differentiated. If for whatever reason the actions did not proceed as planned, any unwanted consequences resulting from these actions would be attributed to an error arising from an unintentional action. In the case where the actions did proceed as intended but did not achieve their intended result, any unwanted outcomes stemming from these actions would be associated with an error resulting from intended but mistaken actions.

In each of these situations the common element is the occurrence of unwanted or adverse outcomes. Whether intended or not, negative outcomes need not be directly associated with these actions. Human error thus also subsumes actions whose unwanted outcomes may occur at much later points in time or following the interjection of many other actions by other people. It can also be argued that even if these actions did not result in adverse outcomes but had the potential to, they should be viewed as errors, in line with the current emphasis on near misses and the recognition that what separates many accidents from events with no visibly apparent negative consequences is chance alone. Acts of sabotage, although capable of bringing about adverse consequences, are not actions that deviate from expectations and thus do not constitute human error. Similarly, intentional violations of procedures, although also of great concern, are typically excluded from definitions of human error when the actions have gone as planned. For example, violations in rigid "ultrasafe and ultraregulated systems" are often required for effectively managing work constraints (Amalberti, 2001). However, when violations result in unforeseen and potentially hazardous conditions, these actions would constitute human error. Exploratory behavior under presumably protective or kind conditions as encountered in formal training programs or trial-and-error self-learning situations, which leads either to unintentional actions or mistaken actions should also be dissociated from human error. This distinction highlights the need to acknowledge the role of error—indeed, even the need for encouraging errors—in adaptation and creativity and in the acquisition of knowledge and skill associated with learning.

The situation becomes more blurred when humans knowingly implement strategies in performance that will result in some degree of error, as when a supervisor encourages workers to adopt shortcuts that trade off accuracy for speed, or when the human reasons that the effort needed to eliminate the possibility of some types of errors may increase the likelihood of more harmful errors. As with procedural violations, if these strategies come off as intended, the actor would not consider the attendant negative outcomes as having resulted from human error. However, depending on the boundaries of acceptable outcomes established or perceived by external observers such as managers or the public, the human's actions may in fact be considered to be in error. Accordingly, a person's ability to provide a reasonable argument for behaviors that resulted in unwanted consequences does not necessarily exonerate the person from having committed an error. What of actions the person intends to commit that are normally associated with acceptable outcomes but which result in adverse outcomes? These would generally not be considered to be human error except perhaps by unforgiving stakeholders who are compelled to exact blame.

The lack of consensus in arriving at a satisfying definition of human error is troubling in that it can undermine efforts to identify, control, and mitigate errors across different work domains and organizations. In fact, some authors have abandoned the term *human error* altogether. Hollnagel (1993) prefers the term *erroneous action* to human error, which he defines as "an action which fails to produce the expected result and which therefore leads to an unwanted consequence" (p. 67). Dekker's (2005) view of errors as "*ex post facto* constructs rather than as objective, observed facts" (p. 67) is based on the accumulated evidence on hindsight bias (Section 10.1). Specifically, the predisposition for this bias has repeatedly demonstrated how observers, including people who may have been recent participants of the experiences being investigated, impose their knowledge (in the form of assumptions and facts), past experiences, and future intentions

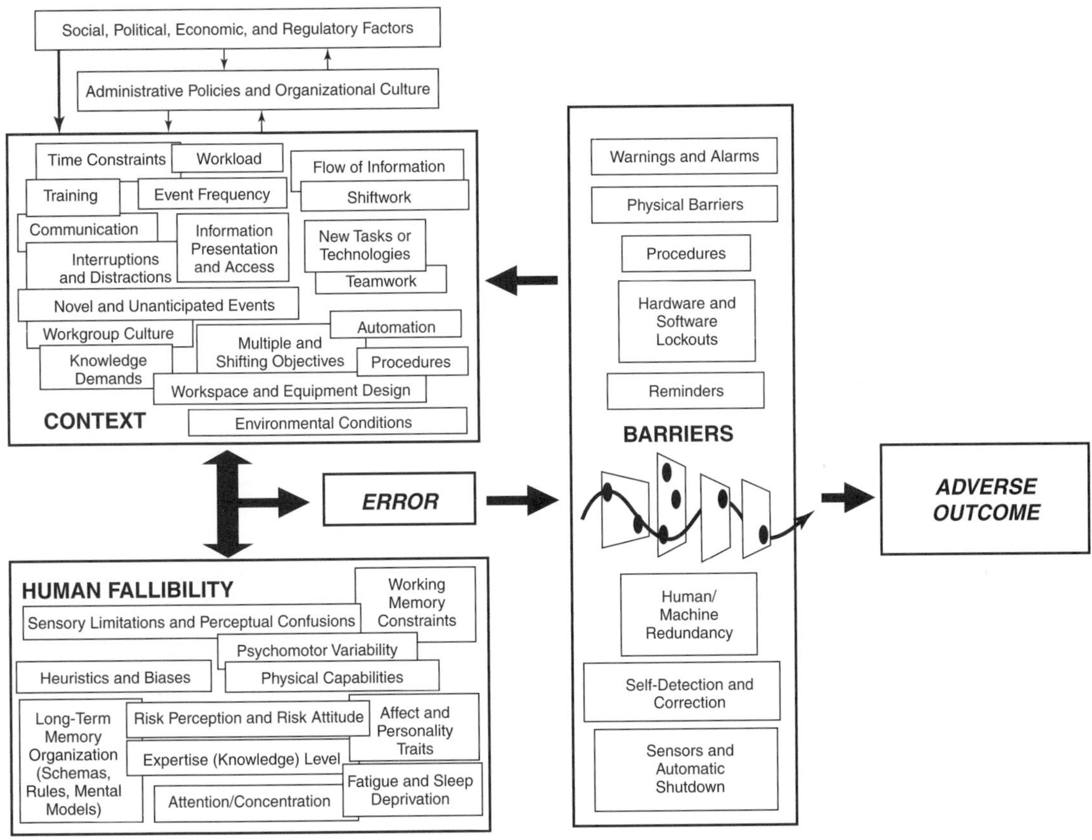

Figure 1 Framework for understanding human error.

to transform what was in fact inaccessible information at the time into neatly unfolding sequences of events and deterministic schemes that are capable of explaining any adverse consequence. These observer and hindsight biases presumably do not bring us any closer to understanding the experiences of the actor in the actual situation for whom there is no error—"the error only exists by virtue of the observer and his or her position on the outside of the stream of experience" (p. 66).

Although this view is enlightening in its ability to draw attention to the limitations of empiricist-based paradigms that underlie many human factors methods, it is also subject to some of the same criticisms that were raised in Section 1.1 in response to the current trend toward perspectives that negate the existence of human error. Understanding both human fallibility and the contexts in which humans must act keeps us on a pragmatic path capable of shaping design and safety-related interventions, even as we strive to find methods that can close the gaps between objective and reconstructed experiences. As we shall see in Section 4, the problems associated with defining human error can be partly overcome by shifting the emphasis to classification schemes that are capable of establishing links between human psychological processes and the manifestation of adverse outcomes across different work domains.

3 UNDERSTANDING HUMAN ERROR

3.1 A Modeling Framework: Human Fallibility, Context, and Barriers

Figure 1 presents a simple modeling framework for demonstrating how human error arises and can result in adverse outcomes. There are three major components in this model. The first component, *human fallibility*, addresses the fundamental sensory, cognitive, and motor limitations of humans that predispose them to error. The second component, *context*, refers to situational variables that can affect the way in which human fallibility becomes manifest. The third component, *barriers*, concerns the various ways in which human errors can be contained.

A number of general observations concerning this modeling framework are worth noting. First, human error is viewed as arising from an interplay between human fallibility and context. This is probably the most intuitive way for practitioners to understand the causality of human error. Interventions that minimize human dispositions to fallibility, for example by placing fewer memory demands on the human, are helpful only to the extent that they do not create new contexts that can, in turn, create new opportunities for human fallibility to become manifest. Similarly, interventions intended to reduce the error-producing potential of work contexts, for instance, by introducing new protocols for communication, could unsuspectingly produce new ways in which human fallibility can exert itself. Second, the depiction of overlapping elements in the human fallibility and context components of the model (Figure 1) is intended to convey the interactive complexity that may exist among these factors. For example, memory constraints may result in the use of heuristics that, in certain contexts, may predispose the human to error; these same memory constraints may also produce misguided perceptions of risk likelihood. Similarly, training programs that dictate how work procedures should be implemented could lead to antagonistic work group cultures whose doctrines afford increased opportunities for operational errors.

Third, barriers capable of preventing the propagation of errors to adverse outcomes could also affect the context. This potential interplay between barriers and context is often ignored or misunderstood in evaluating a system's risk potential. Fourth, system states or conditions that result from errors can propagate into adverse outcomes such as accidents, but only if the gaps in existing barriers are aligned to expose such windows of opportunity (Reason, 1990). The likelihood that errors will penetrate these juxtaposed barriers, especially in high-risk work activities, is generally low and is the basis for the much larger number of near misses that are observed compared to events with serious consequences. Finally, this modeling framework is intended to encompass various perspectives on human error that have been proposed (CCPS, 1994)—in particular, the human factors and ergonomics, cognitive engineering, and sociotechnical perspectives.

In the human factors perspective, error is the result of a mismatch between task demands and human mental and physical capabilities. Presumably this perspective allows only general predictions of human error to be made—primarily predictions of errors that are based on their external characteristics. For example, cluttered displays or interfaces that impose heavy demands on working memory are likely to overload perceptual and memory processes (Section 3.2) and thus possibly lead to the omission of actions or the confusion of one control with another. Guidelines that have been proposed for designing displays (Wickens et al., 2004) are offered as a means for diminishing mismatches between demands and capabilities and thus the potential for error. In contrast, the cognitive engineering perspective emphasizes detailed analysis of work contexts (Section 4) coupled with analysis of the human's intentions and goals. Although both the human factors and cognitive engineering perspectives on human error are very concerned with human information processing, cognitive engineering approaches attempt to derive more detailed information about how humans acquire and represent information and how they use it to guide actions. This emphasis provides a stronger basis for linking underlying cognitive processes with the external form of the error, and thus should lead to more effective classifications of human performance and human errors. As a simple illustration of the cognitive engineering perspective, Table 1 demonstrates how the same external expression of an error could derive from various underlying causes.

Sociotechnical perspectives on human error focus on the potential impact of management policies and organizational culture on shaping the contexts within which people act. These "higher-order" contextual factors are capable of exacting considerable influence

Table 1 Examples of Different Underlying Causes of the Same External Error Mode

Situation: A worker in a chemical processing plant closes valve B instead of nearby valve A, which is the required action as set out in the procedures. Although there are many possible causes of this error, consider the following five possible explanations.

1. The valves were close together and badly labeled. The worker was not familiar with the valves and therefore chose the wrong one.
   Possible cause: wrong identification compounded by lack of familiarity leading to wrong intention (once the wrong identification had occurred the worker intended to close the wrong valve).
2. The worker may have misheard instructions issued by the supervisor and thought that valve B was the required valve.
   Possible cause: communications failure giving rise to a mistaken intention.
3. Because of the close proximity of the valves, even though he intended to close valve A, he inadvertently operated valve B when he reached for the valves.
   Possible cause: correct intention but wrong execution of action.
4. The worker closed valve B very frequently as part of his everyday job. The operation of A was embedded within a long sequence of other operations that were similar to those normally associated with valve B. The worker knew that he had to close A in this case, but he was distracted by a colleague and reverted back to the strong habit of operating B.
   Possible cause: intrusion of a strong habit due to external distraction (correct intention but wrong execution).
5. The worker believed that valve A had to be closed. However, it was believed by the workforce that despite the operating instructions, closing B had an effect similar to closing A and in fact produced less disruption to downstream production.
   Possible cause: violation as a result of mistaken information and an informal company culture to concentrate on production rather than safety goals (wrong intention).

Source: Adapted from CCPS (1994).

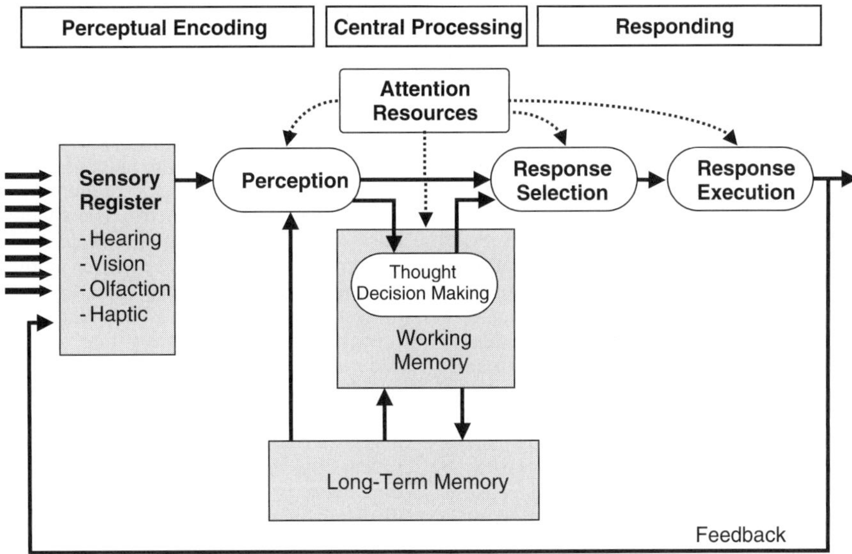

Figure 2 Generic model of human information processing. (Adapted from Wickens et al., 2004.)

on the designs of workplaces, operating procedures, training programs, job aids, and communication protocols, and can produce excessive workload demands by imposing multiple conflicting and shifting performance objectives and by exerting pressure to meet production goals, often at the expense of safety considerations. How work cultures (the cultures associated with those people responsible for producing products and services) become established is a complex phenomenon, and though numerous factors can play a role, the strongest influence is most likely to be the organizational climate (Section 9). Problematic work cultures are very resistant to change, and their remediation usually requires multiple interventions at both the management and operator levels over extended periods of time.

Although the human factors and ergonomics, cognitive engineering, and sociotechnical perspectives appear to suggest different approaches for predicting and analyzing human error, the study of human error will often require the collective consideration of these different perspectives. Human capabilities and limitations from a human factors and ergonomics perspective provide the fundamental basis for pursuing more rigorous cognitive engineering analyses of human error. Similarly, the cognitive engineering perspective, in its requirements for more detailed analyses of work contexts, would be remiss to exclude sociotechnical considerations (Chapter 10).

3.2 Human Fallibility

3.2.1 Human Information Processing

The basis for many human errors derives from fundamental limitations that exist in the human's sensory, cognitive, and motor processes (Chapter 5). These limitations are best understood by considering a generic model of human information processing (Wickens et al., 2004) that conceptualizes the existence of various *processing resources* for handling the flow and transformation of information (Figure 2).

According to this model, sensory information received by the body's various receptor cells gets stored in a system of sensory registers that has an enormous storage capacity. However, this information is available for further processing only briefly. Through the process of selective attention, subsets of this vast collection of information become designated for further processing in an early stage of information processing known as perception. Here, information can become meaningful through comparison with information in long-term memory (LTM), which may result in a response or the need for further processing in a short-term memory store referred to as working memory (WM). A good deal of our conscious effort is dedicated to WM activities such as visualizing, planning, evaluating, conceptualizing, and making decisions, and much of this WM activity depends on information that can be accessed from LTM. Rehearsal of information in WM enables it to be encoded into LTM; otherwise, it decays rapidly. WM also has relatively severe capacity constraints governing the amount of information that can be kept active. The current contention is that within WM there are separate storage systems for accommodating visual information in an analog spatial form or verbal information in an acoustical form, and an attentional control system for coordinating these two storage systems. Ultimately, the results of WM/LTM analysis can lead to a response (e.g., a motor action or decision), or to the revision of thoughts. Note that although this sequence of information processing is depicted in Figure 2 as flowing from left to right, in principle it can begin anywhere.

With the exception of the system of sensory registers and LTM the processing resources in this model may require *attention*. Often thought of as mental effort, attention is conceptualized here as a finite and flexible internal energy source under conscious control whose intensity can be modulated over time. Although attention can be distributed among the information-processing resources, fundamental limitations in attention constrain the capacities of these resources—that is, there is only so much information that can undergo perceptual coding or WM analysis. Focusing attention on one of these resources will, in many cases, handicap other resources. Thus if a North American rents a car with a manual transmission in Great Britain, the experience of driving on the left-side of the road may require substantial allocation of attention to perceptual processing in order to avoid collisions with other drivers, perhaps at the expense of being able to smoothly navigate the stick shift (which is now located to the left of the driver), or at the expense of using WM resources to keep adequate track of one's route. Whatever attention is allocated to WM may be needed for working out the cognitive spatial transformations required for executing left-hand and right-hand turns.

Attention may also be focused almost exclusively on WM, as often occurs during intense problem solving or when planning activities. The ability to divide attention, which is the basis for time sharing, is often observed in people who may have learned to rapidly shift attention between tasks. This skill may require knowledge of the temporal and knowledge demands of the tasks and the possibility for one or more of the tasks having become automated in the sense that very little attention is needed for their performance. Various dichotomies within the information-processing system have been proposed, for example, between the visual and auditory modalities and between early (perceptual) versus later (central and response) processing (Figure 2), to account for how people are able, in time-sharing situations, to more effectively utilize their processing capacities (Wickens, 1984).

Many design implications arise from the errors that human sensory and motor limitations can cause or contribute to. Indeed, human factors studies are often preoccupied with deriving design guidelines for minimizing such errors. Knowledge concerning human limitations in contrast sensitivity, hearing, bandwidth in motor movement, and in sensing tactile feedback can be used to design visual displays, auditory alarms, manual control systems, and protective clothing (such as gloves that are worn in surgery) that are less likely to produce errors in detection and response. Much of the focus on human error, however, is on the role that cognitive processing plays. Even seemingly simple situations involving errors in visual processing may in fact be rooted in much more complex information processing as illustrated in the following example.

3.2.2 Example: Medication Error

Consider the following prescription medication error, which actually occurred. A physician opted to change the order for 50 mg of a leukemia drug to 25 mg by putting a line through the zero and inserting a "2" in front of the "5." The resulting dose was perceived by the pharmacist as 250 mg and led to the death of a 14-year-old boy. The line that was meant to indicate a cross-out was not centered and turned out to be much closer to the right side of the circle (due to *psychomotor variability*; see Figure 1); thus, it could easily have been construed as just a badly written zero. Also, when one considers that perception relies on both bottom-up processing (where the stimulus pattern is decomposed into features) and top-down processing (where context and thus expectations are used for recognition), the possibility that a digit was crossed out may have countered expectations (i.e., it does not usually occur).

If one were to presume that the pharmacist had a high workload (and thus diminished resources for processing the prescription) and a relative lack of experience or knowledge concerning dosage ranges for this drug, it is easy to understand how this error can come about. The dynamics of the error can be put into a more complete perspective when potential barriers are considered, such as an automatic checking system that could have screened the order for a potentially harmful dosage or interactions with other drugs, or a procedure that would have required the physician to rewrite any order that had been altered. Even if these barriers were in place, which was not the case, there is a high likelihood that they would be bypassed. In fact, if such a procedure were to be imposed on physicians, routine violations would be expected given the contexts within which many physicians work.

3.2.3 Long-Term Memory and Its Implications for Human Error

LTM has been described as a parallel distributed architecture that is being reconfigured continuously through selective activation and inhibition of massively interconnected neuronal units (Rumelhart and McClelland, 1986). These reconfiguration processes occur within distinct modules that are responsible for different representations of information, such as mental images or sentence syntax. In the process of adapting to new stimuli or thoughts, the complex interactions between neuronal units that are produced give rise to the generalizations and rules that are so critical to human performance. With regard to the forms of knowledge stored in LTM, we usually distinguish between the general knowledge we have about the world, referred to as *semantic memory*, and knowledge about events, referred to as *episodic memory*.

When items of information, such as visual images, sounds, and thoughts based on existing knowledge, are processed in WM at the same time, they become associated with each other in LTM. The retrieval of this information from LTM will then depend on the strength of the individual items as well as the strengths of their associations with other items. Increased frequency and recency of activation are assumed to promote stronger (i.e., more stable) memory traces, which are otherwise subject to negative exponential decays.

Much of our basic knowledge about things can be thought of as being stored in the form of *semantic networks* that are implemented through parallel distributed architectures. Other knowledge representation schemes commonly invoked in the human factors literature are schemas and mental models. *Schemas* typically represent knowledge organized about a concept or topic. When they reflect processes or systems for which there are relationships between inputs and outputs that the human can mentally visualize and experiment with (i.e., "run," like a simulation program), the schemas are often referred to as *mental models*. The organization of knowledge in LTM as schemas or mental models is also likely based on semantic networks.

The constraints associated with LTM architecture can provide many insights into human fallibility and how this fallibility can interact with situational contexts to produce errors. For example, implicit in the existence of parallel associative networks is the ability to recall both items of information and patterns (i.e., associations) of information based on partial matching of this information with the contents of memory. Because the contexts within which humans operate often produce what Reason (1990) has termed *cognitive underspecification*, the implication is that at some point in the processing of information the specification of information may be incomplete. It may be incomplete due to perceptual processing constraints, WM constraints, or LTM (i.e., knowledge) limitations, or due to external constraints, as when there is little information available on the medical history of a patient undergoing emergency treatment or when piping and instrumentation diagrams have not been updated. LTM organization can overcome these limitations by retrieving some items of information that provide a match to the inputs, and thus enable an entire rule, by previous association with other items of information in LTM, to be activated. Unfortunately, that rule may not be appropriate for the particular situation. Similarly, for instance in the case of a radiologist who has recently encountered a large number of tumors of a particular type, the increased activation levels that are likely to be associated with this diagnosis may result in a greater tendency for arriving at this diagnosis in future situations.

3.2.4 Information Processing and Decision-Making Errors

Human decision making that is not guided by normative prescriptive models (Chapter 8) is an activity fraught with fallibility, especially in complex dynamic environments. As illustrated in Figure 3, human limitations in decision making can arise from a number of information-processing considerations (Figure 2) that directly or indirectly implicate LTM (Wickens et al., 2004). For example, if the information the human opts to select for WM activity, which may be guided by past experiences, is fuzzy or incomplete, intensive interpretation or integration of this information may be needed. Also, any hypotheses that the decision maker generates regarding this information will be highly dependent on information that can be retrieved from LTM, and their evaluation could require searching for additional information. Although any hypothesis for which adequate support is found can become the basis for an action, the possible candidate actions that would need to be evaluated in WM would first need to be retrieved from LTM. In addition, the possible outcomes associated with each action, the estimates of the likelihoods of

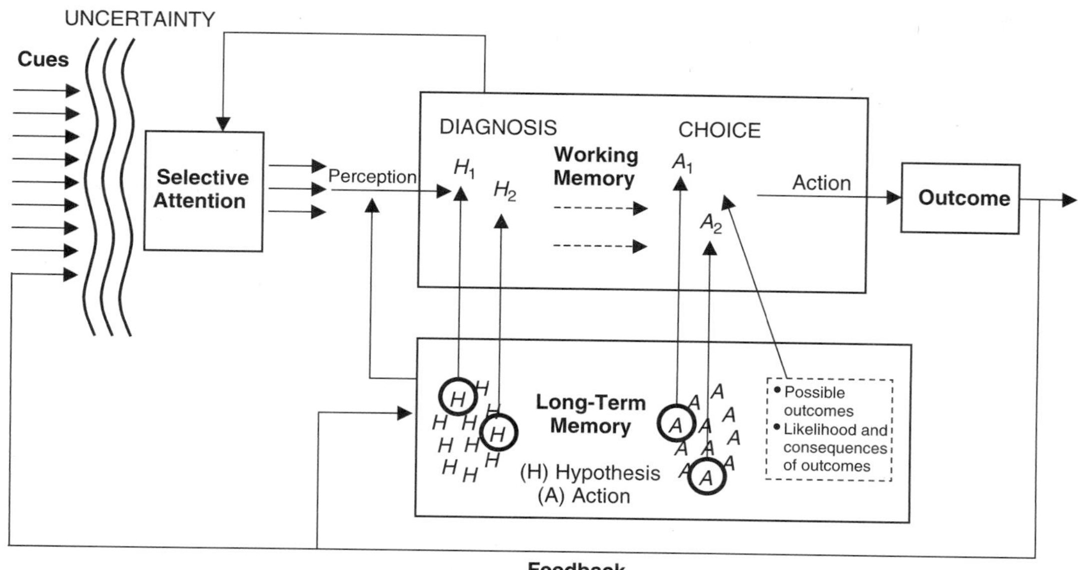

Figure 3 Information-processing model of decision making. (Adapted from Wickens et al., 2004.)

these outcomes, and the negative and positive implications of these actions would also require retrieval from LTM.

From an information-processing perspective, there are numerous factors that could constrain this decision-making process, particularly factors that could influence the amount or quality of information brought into WM and the retrieval of information from LTM. These constraints often lead to shortcuts in decision making, such as *satisficing* (Simon, 1966), whereby people opt for choices that are good enough for their purposes and adopt strategies for sampling information that they perceive to be most relevant. In general, the human's natural tendency to minimize cognitive effort (Sharit, 2003) opens the door to a wide variety of shortcuts or heuristics that are efficient and usually effective in negotiating environmental complexity, but under the right coincidence of circumstances can lead to ineffective choices or actions that become designated as errors. For example, with respect to the cues of information that we perceive, there is a tendency to overweight cues occurring earlier than later in time or that change over time. WM will only allow for a limited number of possible hypotheses, actions, or outcomes of actions to be evaluated, and LTM architecture will accommodate these limitations by making information that has been considered more frequently or recently (the "availability" heuristic) more readily available and by enabling its partial-matching capabilities to classify cues as more representative of a hypothesis than may be warranted. Many other heuristics (Wickens et al., 2004), such as *confirmation bias* (the tendency to consider confirming and not disconfirming evidence when evaluating hypotheses), *cognitive fixation* (remaining fixated on initial hypotheses and underutilizing subsequent information), and the tendency to judge an event as likely if its features are representative of its category (e.g., judging a person as having a particular occupation based on the person's appearance even though the likelihood of having that occupation is extremely low) derive primarily from a conservation of cognitive effort.

An enormous investment by the human in WM activities (i.e., an extensive commitment to functioning in an attentional mode) would be required to expose the biases that these heuristics can potentially induce. It is important to note, however, that to exclude the possibility that a human's situational assessments are in fact rational, explanations of human judgments and behaviors on the basis of cognitive biases require a sound understanding of the specific context (Fraser et al., 1992).

3.2.5 Levels of Human Performance and Dispositions for Errors

Considerations related to LTM architecture enable many different types of human errors to be accounted for by a few powerful principles. This few-to-many mapping between underlying memory mechanisms and different error types will be influenced by the nature of human performance, particularly on how information-processing resources (Figure 2) are used, other aspects of human fallibility (Section 3.2.6), and situational variables. A framework that distinguishes between skill-based, rule-based, and knowledge-based levels of performance—Rasmussen's *SRK framework*—emphasizes fundamentally different approaches to processing information and is thus particularly appealing for understanding the role of human performance in analyzing and predicting different types of human errors (Rasmussen, 1986).

Activities performed at the skill-based level are highly practiced routines that require little conscious attention. Referring to Figure 2, these activities map perception directly to actions, bypassing WM. Following an intention for action that could originate in WM or from environmental cues, the responses associated with the intended activity are so well integrated with the activity's sensory features that they are elicited in the form of highly automatic routines. Given the frequent repetitions of consistent mappings from sensory features to motor responses, the meaning imposed on perception by LTM can be thought of as hardwired to the human's motor response system.

The rule-based level of performance makes use of rules that have been established in LTM based on past experiences. WM is now a factor, as rules (of the if–then type) or schemas may be brought into play following the assessment of a situation or problem. More attention is thus required at this level of performance, and the partial matching characteristics of LTM can prove critical. When stored rules are not effective, as is often the case when new or challenging problems arise, the human is usually forced to devise plans that involve exploring and testing hypotheses, and must continuously refine the results of these efforts into a mental model or representation that can provide a satisfactory solution. At this knowledge-based level of performance heavy demands on information-processing resources are exacted, especially on WM, and performance is vulnerable to LTM architectural constraints to the extent that WM is dependent on LTM for problem solving.

In reality, many of the meaningful tasks that people perform represent mixtures of skill, rule, and knowledge-based levels of performance. Although performance at the skill-based level results in a significant economy in cognitive effort, the reduction in resources of attention comes at a risk. For example, consider a task other than the one that is intended that contains features that are similar to those of the intended task. If the alternative activity is frequently performed and therefore associated with skill-based automatic response patterns, all that is needed is a context that can distract the human from the intention and allow the human to be "captured" by the alternative (incorrect) task. This situation represents example 4 in Table 1 in the case of an inadvertent closure of a valve. In other situations the capture by a skill-based routine may result in the exclusion of an activity. For example, suppose that task A is performed infrequently and task B is performed routinely at the skill-based level. If the initial steps are identical for both tasks but task A requires an additional step, this step is likely to be omitted during execution of the task. Untimely interruptions are often the basis for omissions at the skill-based level of performance. In some circumstances, interruptions or moments of inactivity during skill-based routines may instigate thinking about where one is in the sequence of steps. By directing attention to routines that are not designed to be examined, steps could be performed out of sequence (reversal errors) or be repeated (Reason, 1990).

Many of the errors that occur at the rule-based level involve inappropriate matching of either external cues or internally generated information with the conditional components of rules stored in LTM. Generally, conditional components of rules that have been satisfied on a frequent basis or that appear to closely match prevailing conditions are more likely to be activated. The prediction of errors at this level of performance will thus require knowing what other rules the human might consider, thus necessitating detailed knowledge not only about the task but also about the process (e.g., training or experience) by which the person acquired rule-based knowledge. Mistakes in applying rules generally involve the misapplication of rules with proven success or the application of *bad rules* (Reason, 1990). Mistakes in applying rules with proven success often occur when *first exceptions* are encountered. Consider the case of an endoscopist who relies on indirect visual information when performing a colonoscopy. Based on past experiences and available knowledge, the sighting of an anatomical landmark during the performance of this procedure may be interpreted to mean that the instrument is situated at a particular location within the colon, when in fact the presence of an anatomical deformity in this patient may render the physician's interpretation as incorrect (Cao and Milgram, 2000). These first exception errors often result in the decomposition of general rules into more specific rule forms and reflect the acquisition of expertise. General rules, however, usually have higher activation levels in LTM given their increased likelihood of encounter, and under contextual conditions involving high workload and time constraints, they will be the ones more likely to be invoked. Rule-based mistakes that occur by applying bad (e.g., inadvisable) rules are also not uncommon, as when a person who is motivated to achieve high production values associates particular work conditions with the opportunity for implementing shortcuts in operations.

At the knowledge-based level of performance, when needed associations or schemas are not available in LTM, control shifts primarily to intensive WM activities. This level of performance is often associated with large degrees of freedom that characterize how a human "moves through the problem space," and suggests a much greater repertory of behavioral responses and corresponding expressions of error. Contextual factors that include task characteristics and personal factors that include emotional state, risk attitude, and confidence in intuitive abilities can play a significant role in shaping the error modes, making these types of errors much harder to predict. It is at this level of performance that we observe undue weights given to perceptually salient cues or early data, confirmation bias, use of the availability and representative heuristics (especially for assessing relationships between causes and effects), underestimation and overestimation of the likelihood of events in response to observed data, vagabonding (darting from issue to issue, often not even realizing that issues are being revisited, with essentially no effective movement through the problem space), and encysting (overattention to a few details at the expense of other, perhaps more relevant information).

3.2.6 Other Aspects of Human Fallibility

There are many facets to human fallibility, and all have the potential to contribute to human error. For example, personality traits that reflect dispositions toward confidence, conscientiousness, and perseverance could influence both the possibility for errors and the nature of their expression at both the rule- and knowledge-based levels of performance, especially under stress. Overconfidence can lead to risk-taking behaviors and has been implicated as a contributory factor in a number of accidents.

Sleep deprivation and fatigue are forms of human fallibility whose manifestations are often regarded as contextual factors. In fact, in the maritime and commercial aviation industries, these conditions are often attributed to company or regulatory agency rules governing hours of operation and rest time. The effects of fatigue may be to regress skilled performers to the level of unskilled performers (CCPS, 1994) through widespread degradation of abilities that include decision making and judgment, memory, reaction time, and vigilance. NASA has determined that about 20% of incidents reported to its Aviation Safety Reporting System (Section 6.3), which asks pilots to report problems anonymously, are fatigue-related (Kaye, 1999a). On numerous occasions pilots have been found to fall asleep at the controls, although they usually wake up in time to make the landing.

Another facet of human fallibility with important implications for human error is *situation awareness* (Chapter 20), which refers to a person's understanding or mental model of the immediate environment (Endsley, 1995). As in the case of fatigue, situation awareness represents an aspect of human fallibility that can be heavily influenced by contextual factors. In principle, any factor that could disrupt a human's ability to acquire or perceive relevant data concerning the elements in the environment, or compromise one's ability to understand the importance of that data and relate the data to events that may be unfolding in the near future, presumably can degrade situation awareness. Comprehending the importance of the various types of information in the environment also implies the need for temporal awareness—the need to be aware of how much time tasks require and how much time is available for their performance (Grosjean and

Terrier, 1999). Thus, potentially many factors related to both human fallibility and context can influence situation awareness. Increased knowledge or expertise should allow for better overall assessments of situations, especially under conditions of high workload and time constraints, by enabling elements of the problem and their relationships to be identified and considered in ways that would be difficult for those who are less familiar with the problem. In contrast, poor display designs that make integration of data difficult can easily impair the process of assessing situations. In operations involving teamwork, situation awareness can become disrupted by virtue of the confusion created by the presence of too many persons being involved in activities.

Finally, numerous affective factors can corrupt a human's information-processing capabilities and thereby predispose the human to error. Personal crises could lead to distractions, and emotionally loaded information can lead to the substitution of relevant information with "information trash." Similarly, a human's susceptibility to panic reactions and fear can impair information-processing activities critical to human performance.

3.3 Context

Human actions are embedded in contexts and can only be described meaningfully in reference to the details of the context that accompanied and produced them (Dekker, 2005). The possibility for human fallibility to result in human error as well as the expression of that error will thus depend on the context in which task activities occur. Although the notion of a context is often taken as obvious, it is not easy to define, leading to commonly encountered alternative expressions, such as *scenario*, *situation*, *situational context*, *situational details*, *contextual features*, *contextual dynamics*, *contextual factors*, and *work context*. Designers of advanced computing applications often speak in terms of providing functionalities that are responsive to various user contexts. Building on a definition of context proposed by Dey (2001) in the domain of context-aware computer applications, *context* is defined as any information that can be used to characterize the situation of a person, place, or object, as well as the dynamic interactions among these entities. This definition of context would regard a process such as training as an entity derived from these interactions and would also encompass information concerning how situations are developing and the human's responses to these situations.

Figure 1 reveals some representative contextual factors. In this depiction, the presumption is that higher-order *context-shaping factors* can influence contextual factors that are more directly linked to human performance. Contexts ultimately derive from the characterization of these factors and their interactions. Analysis of the interplay of human fallibility and context as a basis for understanding human error will be beneficial to the extent that relevant contextual factors can be identified and analyzed in detail.

A number of quantitative approaches to human error assessment (Section 5) employ concepts that are related to context. For example, several of these approaches use *performance-shaping factors* (PSFs) to either modify the probability estimate assigned to an activity performed in error (Swain and Guttmann, 1983) or as the basis for the estimation of human error (Embrey et al., 1984). Any environmental, individual, organizational, or task-related factor that could influence human performance can, in principle, qualify as a PSF; thus PSFs appear to be related to contextual factors. These approaches, however, by virtue of emphasizing probabilities as opposed to possibilities for error, assume additive effects of PSFs on human performance rather than interactive effects. In contrast, implicit to the concept of a context is the interactive complexity among contextual factors. A sociotechnical method for quantifying human error referred to as *STAHR* (Phillips et al., 1990) is somewhat more consistent with the concept of context than approaches based on PSFs. This method utilizes a hierarchical network of influence diagrams to represent the effects of direct influences on human error, such as time pressure and quality of training, as well as the effects of less direct influences, such as organizational and policy issues, which project their influences through the more direct factors. However, while STAHR imposes a hierarchical constraint on influences, the concept of context implicit to Figure 1 imposes no such constraint, thus enabling influences to be represented as an unconstrained network (Figure 4).

Generally, the emphasis on predicting the possibility for error as opposed to the probability of error relaxes the assessments required of contextual factors. In making these assessments, some of the possible considerations could include the extent to which a contextual factor is present (i.e., the level of activation of a network node) and the extent to which it can influence other factors (i.e., the level of activation of a network arc), as illustrated in Figure 4. Temporal characteristics underlying these influences could also be included. Also, as conceptualized in Figure 1, contextual factors can be refined to any degree of detail, and practitioners and analysts would need to determine for specific task domains of interest the appropriate level of contextual analysis. For example, the introduction of new technology into activities involving teamwork (Section 7.3) would require the characterization of each person's role with respect to the technology as well as analysis of how team communication may become altered as a consequence of these new roles. Links to other contextual factors come to mind immediately. The creation of new tasks may result in fragmented jobs that impose higher workload demands and less reliable mental models, due to the difficulty in forming meaningful associations in memory. These factors, in turn, can affect adversely communication among team members. New training protocols that do not anticipate many of these influences may further predispose the human to error by directing attention away from important cues.

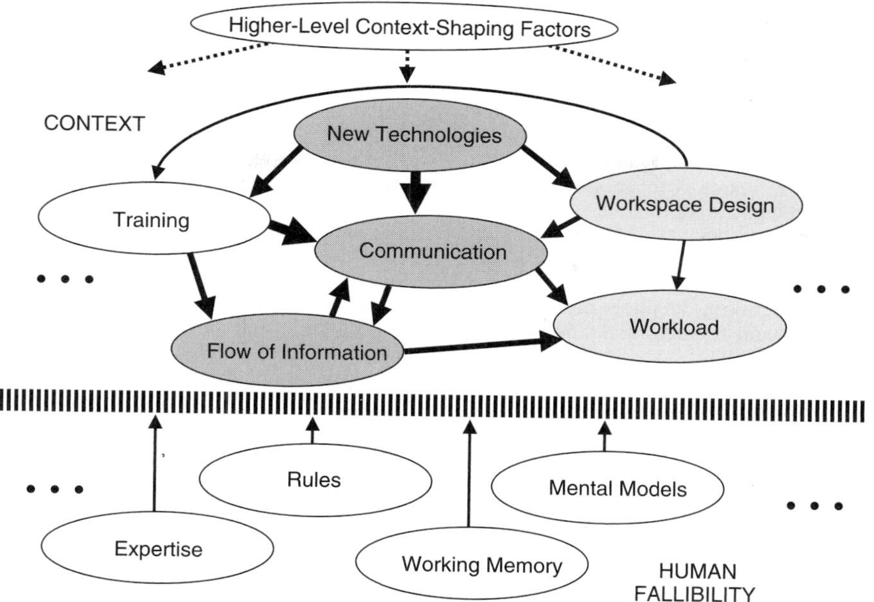

Figure 4 Influences between contextual factors representing part of a relevant work context, and their potential interplay with representative human fallibility factors. Activation levels of contextual factors are denoted by different degrees of shading and their degrees of influence are denoted by arrow widths. Temporal characteristics associated with these influences could also be included. Human fallibility factors can affect contextual factors as well as their influences. Influences among human fallibility factors are not depicted.

In some models of accident causation, the concept of a triggering event is used to draw attention to a "spark," such as a distraction or a pipe break, which sets off a chain of events (including human errors) that can ultimately lead to an accident. As defined here, a trigger represents just another contextual factor. Some triggering events, such as the random failure of a pump, may not necessarily have any contextual factors influencing it, whereas other triggers (e.g., a disruption in a work process resulting from a late discovery that a needed tool is absent) could very well be influenced by other contextual factors. Investigations of accidents (Section 10) are often aimed at exposing multiple chains of causal events and identifying critical paths whose disruption could have prevented the accident. Similarly, there may be nodes or paths within the network of contextual factors (Figure 4) that may be more responsive to human fallibility, and closer examination of these nodes and their links may inform the analyst of strategies for reducing human error or adverse events.

Finally, the possibility also exists for describing larger-scale work domain contexts that are capable of bringing about adverse outcomes through their interplay with human fallibility. In this regard, the views of Perrow (1999), which constitute a *system theory of accidents*, have received considerable attention. According to Perrow, the structural analysis of any system, whether technological, social, or political, reveals two loosely related concepts or dimensions, interactive complexity and coupling, whose sets of attributes govern the potential for adverse consequences. *Interactive complexity* can be categorized as either *complex* or *linear* and applies to all possible system components, including people, materials, procedures, equipment, design, and the environment. The relatively stronger presence of features such as reduced proximity in the spacing of system components, increased interconnectivity of subsystems, the potential for unintended or unfamiliar feedback loops, the existence of multiple and interacting controls (which can be administrative as well as technological), the presence of information that tends to be more indirect and incomplete, and the inability to easily substitute people in task activities predispose systems toward being complex as opposed to linear. Complex interactions are more likely to be produced by complex systems than linear systems, and because these interactions tend to be less perceptible and comprehensible the human's responses to problems that occur in complex systems can often further increase the system's interactive complexity.

Most systems can also be characterized by their degree of coupling. Tightly coupled systems are much less tolerant of delays in system processes than are loosely coupled systems and are much more invariant to materials and operational sequences. Although each type of system has both advantages and disadvantages, loosely coupled systems provide more opportunities for recovery from events with potentially adverse consequences, often through creative, flexible, and adaptive responses by people. To compensate for the fewer opportunities for recovery provided by tightly coupled systems, these systems generally require more built-in safety devices and redundancy than do loosely coupled systems.

Although Perrow's account of technological disasters focuses on the properties of systems themselves rather than human error associated with design, operation, or management of these systems, many of the catastrophic accidents chronicled by Perrow do in fact concern interactions between technological, human factors, organizational, and sociocultural systems, and technical systems are in their own right economic, social, and political constructs. Thus, despite the virtue in his theory of dispelling such accidents as having resulted from human error, his model has been criticized for its marginalization of factors at the root of technological accidents (Evan and Manion, 2002). These criticisms, however, do not preclude the possibility of augmenting Perrow's model with additional perspectives on system processes that would endow it with the capability for providing a reasonably compelling basis for predisposing the human to error.

3.4 Barriers

Various methods exist for building in barriers to human error. For example, computer-interactive systems can force the user to correct an invalid entry prior to proceeding, provide warnings about actions that are potentially error inducing, and employ self-correction algorithms that attempt to infer the user's intentions. Unfortunately, each of these methods can also be breached, depending on the context in which it is used. Forcing functions can initiate a process of backtracking by the user that can lead to total confusion and thus more opportunity for error (Reason, 1990), and warnings can be ignored under high workloads.

The facilitation of errors by computer-interactive systems was found to occur in a study by Koppel et al. (2005) on the use of hospital-computerized physician order-entry (CPOE) systems, contradicting widely held views that these systems significantly reduce medication prescribing errors. In this study, errors were grouped into two categories: (1) information errors arising from the fragmentation of data and the failure to integrate information across the various hospital information systems, and (2) human–machine interface flaws that fail to adequately consider the practitioner's behaviors in response to the constraints of the hospital's organizational work structure. An example of an error related to the first category is when the physician orders new medications or modifies existing medications. If current doses are not first discontinued, the medications may actually become increased or decreased, or be added on as duplicative or conflicting medication. Detection of these errors is hindered by flaws in the interface that may require 20 screens for viewing a single patient's medications. Complex organizational systems such as hospitals can make it extremely difficult for designers to anticipate the many contexts and associated problems that can arise from interactions with the systems that they design (Section 7.3.1). Although it may make more sense to have systems such as CPOEs monitored by practitioners and other workers for their error-inducing potential rather than have designers attempt to anticipate all the contexts associated with the use of these systems, this imposes the added burden of ensuring that mechanisms are in place for collecting the appropriate data, communicating this information to designers, and validating that the appropriate interventions have been incorporated.

Many of the electronic information devices (including CPOEs) that are currently in use in complex systems such as health care were implemented under the assumption that they would decrease the likelihood of human error. However, the benefits of reducing or even eliminating the possibility for certain types of errors often come at the risk of new errors, exemplifying how the introduction of barriers can create new windows of opportunity for errors through the alteration of existing contexts (Section 3.1). For example, in hospital systems the reliance on information in electronic form can disturb critical communication flows and is less likely than face-to-face communication to provide the cues and other information necessary for constructing appropriate models of patient problems.

One of the most frequently used barriers in industry—the written work procedure—is also one that is highly vulnerable to violation. Many of the procedures designed for high-hazard operations include warnings, contingencies (information on when and how to "back out" when dangerous conditions arise during operations), and other supporting features. To avoid the recurrence of past incidents, these procedures are updated continuously. Consequently, they grow in size and complexity to the point where they can contribute to information overload, increasing the possibility of missing or confusing important information (Reason, 1997). Procedures that disrupt the momentum of human actions are especially vulnerable to violation.

Humans themselves are quite adept at detecting and correcting many of the skill-based errors they make and are thus often relied upon to serve as barriers. Self-correction, however, implies two conditions: that the human depart from automated processing, even if only momentarily, and that the human invest attentional resources periodically to check whether the intentions are being met and that cues are available to alert one to deviation from intention (Reason, 1990). This would apply to both slips and omissions of actions. Redundancy in the form of cues presented in multiple modalities is a simple and very effective way of increasing a person's likelihood of detecting and correcting these types of errors. This strategy is illustrated in the case of the ampoule-swap error in hospital operating rooms (Levy et al., 2000). Many drug solutions are contained in ampoules that do not vary much in size and shape, often contain clear liquid solutions, and have few distinguishing features. If an anesthesiologist uses the wrong ampoule to fill a syringe and inadvertently "swaps in" a risky drug such as potassium chloride, serious consequences could ensue. Contextual factors such as fatigue and distractions make it unreasonable to expect medical providers to invest the attentional resources necessary for averting these types of errors. Moreover, the use of warning signs on bins that store ampoules containing "risky solutions" are poor solutions to this problem, as they require that the human maintain *knowledge in the head*—specifically, in WM—thus making this information vulnerable to memory loss resulting from delays or distractions between retrieving the ampoule and preparing the solution. The more reliable solution that was suggested by these investigators was to provide tactile cues on both the storage bins and the ampoules. For example, wrapping a rubber band around the ampoule following its removal from the bin provides an alerting cue in the form of tactile feedback prior to loading the ampoule into the syringe.

Not surprisingly, the human's error detection abilities are greatly reduced at the knowledge-based level of performance. Error detection in these more complex situations will depend on discovering that the wrong goal has been selected or recognizing that one's movement through the problem space is not consistent with the goal. In this regard, strategic errors (e.g., in goal definition) are expected to be much harder to discover than tactical errors (e.g., in choosing which subsystem to diagnose). Human error detection and recovery at the knowledge-based level of performance may in fact represent a highly evolved form of expertise. Interestingly, whereas knowledge-based errors decrease with increased expertise, skill-based errors increase. Also, experienced workers, as compared to beginners, tend to disregard a larger number of errors that have no work-related consequences, suggesting that with expertise comes the ability to apply higher-order criteria for regulating the work system, thus enabling the allocation of attention to errors to occur on a more selective basis (Amalberti, 2001).

A very common barrier to human error is having other people available for error detection. As with hardware components, human redundancy will usually lead to more reliable systems. However, successful human redundancy often requires that the other people be external to the operational situation, and thus possibly less subject to tendencies such as cognitive fixation. In a study of 99 simulated emergency scenarios involving nuclear power plant crews, Woods (1984) found that none of the errors involving diagnosis of the system state were detected by the operators who made them and that only other people were able to detect a number of them. In contrast, half the errors categorized as slips (i.e., errors in execution of correct intentions) were detected by the operators who made them. These results also suggest that team members can often be subject to the same error-producing tendencies as individuals.

Barriers to human error need not always be present by design. As implied in Perrow's system theory of accidents (Section 3.3), a complex mixture of a system's properties can produce conditions that are conducive to human error as well as to its detection and correction. This phenomenon is routinely demonstrated in large-scale hospital systems where one encounters an assortment of patient problem scenarios, a variety of health care services, complex flows of patient information across various media on a continual 24-hour basis, and a large variability in the skill levels of health care providers, who must often perform under conditions of overload and fatigue while being subjected to various administrative constraints. The complex interactions that arise under these circumstances provide multiple opportunities for human error, arising from missed or misunderstood information or confusion in following treatment protocols. Fortunately, there usually exist multiple layers of redundancy in the form of alternative materials (e.g., medications and equipment), treatment schedules, and health care workers to thwart the serious propagation of many of these errors. Thus, despite a number of constraints that exist in hospital systems, particularly in the provision of critical care, these systems are sufficiently loosely coupled to overcome many of the risks that arise in patient care, including those that are generated by virtue of discontinuities or gaps in treatment (Cook et al., 2000). However, even if adverse consequences are indeed averted in many of these cases, one must acknowledge the possibility that the quality of patient care may become significantly compromised in the process.

Finally, there is always the possibility that the perceived presence of barriers such as intelligent sensing systems and corrective devices may actually increase a person's risk-taking behavior. Adjusting risk-taking behavior to maintain a constant level of risk is in line with *risk-homeostasis theory* (Wilde, 1982). These adjustments presume that humans are reasonably good at estimating the magnitude of risk, which generally does not appear to be the case. Nonetheless, a disturbing implication of this theory is the possibility that interventions by organizations directed at improving the safety climate could, instead, result in work cultures that promote attitudes that are not conducive to safe operations.

3.5 Example: Wrong-Site Surgery

Wrong-site errors in health care encompass surgical procedures performed on a wrong part of the body, wrong side of the body, wrong person, or at the wrong level of a correctly identified anatomical site. The Joint Commission on Accreditation of Healthcare Organizations (JCAHO) considers wrong-site surgeries to be *sentinel events* that require immediate investigation and response. As of March 2000, JCAHO has reported wrong-site surgery to be the fourth most commonly reported sentinel event, following patient suicide, medication error, and operative or postoperative complications. It seems inconceivable that this type of error, which carries potentially devastating consequences, could become a common occurrence in organizations comprised of so many highly trained practitioners. While human fallibility, as always, plays a fundamental role, its interplay with contextual factors and existing barriers suggests that these errors are more complex than they appear.

A common factor in wrong-site surgery is the involvement of multiple surgeons on a case. Each of the various physicians has a relatively narrow focus of attention (e.g., the cardiologist is focused on whether the heart can withstand surgery), which decreases the likelihood that the patient will be surrounded by health care providers who are knowledgeable about the case and thus limits the benefits of human redundancy. Another factor in wrong-site surgery is the need to perform multiple procedures during a single trip to the operating room. This factor provides the necessary distractions for a slip. The likelihood that a distraction could result in an unintended action is increased to the extent that the surgeon has "frequently" or "recently" performed surgeries at the unintended site or when patient care is transferred to another surgeon. Fatigue, sleep deprivation, and unusual patient characteristics such as massive obesity (which could alter the positioning of the patient) are also capable of promoting unintended actions by disrupting the surgeon's focused attention.

Presurgical procedures and problems with the way that team members communicate during an operation can also contribute to the occurrence of wrong-site errors. Ideally, an entire team should be required to verify that the correct patient and the correct limb have been prepared for surgery. However, when the surgical team fails to review the patient record or image data in the time period immediately prior to the surgery, memory concerning the correct surgical site can become flawed. Incomplete or inaccurate communication among surgical team members can also occur when some team members are excluded from participating in the site verification process, team members exchange roles during the day of surgery, or when the entire team depends exclusively on the surgeon to identify the surgical site (the latter often occurs in work cultures that accept the surgeon's decision as final). Many of these communication problems become magnified under time constraints stemming from pressure from hospital administrators to speed things up.

A tactic that has recently received considerable attention is marking the operative site and involving the patient in the process. However, even this seemingly straightforward policy can be problematic. If surgeons were to employ their own marking techniques, such as "No" on the wrong limb or "Yes" on the proper site, confusion may occur to the point of increasing the likelihood of wrong-site surgery. Standardization is thus critical, and the recommended procedure is for the surgeon to initialize the operative site. This barrier alone, however, is insufficient. For example, if the marked site is draped out of the surgeon's field of view and the surgeon does not recall whether the site was or was not marked, the possibility for error still exists. Thus, a verification checklist should also be in place that includes all documents referencing the intended procedure and site, informed consent, and direct observation of the marked operative site. Strict reliance on x-rays or the patient's chart can prove inadequate in cases where the data are incorrect or associated with the wrong patient, and patient involvement is not always possible, depending on the patient's condition.

Violation of these barriers is not uncommon. Some surgeons see signing as a waste of time and a practice that could contaminate the operative site. They are insistent that wrong-site surgery errors would not happen to them and that the focus of the medical profession should be on ridding itself of incompetent surgeons rather than instituting wide-reaching programs (Prager, 1998). This attitude, however, is not surprising in a profession that has relied largely on people avoiding mistakes rather than creating systems to minimize them. It also reflects a very traditional perspective to human error whereby the responsibility or blame for errors is placed solely on those who committed them and suggests that a culture shift among surgeons may be needed. To utilize the surgeon's time more efficiently, for hospitalized patients implementation might involve the operating surgeon initializing the intended operative site at the time consent is obtained, thus requiring that the physician be present during consent. The JCAHO has constructed a universal protocol for eliminating wrong-site surgery which ensures that the surgical site is marked while the patient is conscious and that there is a final pause and verification among all surgical team members to ensure that everyone is in agreement with the procedure. This protocol became effective in July 2004 for all JCAHO-accredited hospitals.

4 ERROR TAXONOMIES AND PREDICTING HUMAN ERROR

Many areas of scientific investigation use classification systems or taxonomies as a way of organizing knowledge about a subject matter. In the case of human error, the taxonomies that have been proposed have theoretical as well as practical value. The taxonomies that emphasize observable behaviors are primarily of practical value. They can be used retrospectively to gather data on trends that point to weaknesses in design, training, and operations, as well as prospectively, in conjunction with detailed analyses of tasks and situational contexts, to predict possible errors and to suggest countermeasures for detecting, minimizing, or eliminating these errors. Human error taxonomies can also be directed at specific tasks or operations. For example, a taxonomy could be developed for the purpose of characterizing all the various observable ways that a particular task can be performed incorrectly, analogous to the use of failure mode and effects analysis (Kumamoto and Henley, 1996) to identify a component's failure modes and their corresponding causes and effects. In the health care industry, the diversity of medical procedures and the variety of circumstances under which these procedures are performed may, in fact, call for highly specific error taxonomies.

For more cognitively complex tasks, it may be possible to classify errors according to stages of information processing (Figure 2), thereby differentiating errors related to perception from errors related to failures in working memory. However, many of these errors of cognition can only be inferred from assumptions concerning the human's goals and observed behaviors, and to some extent from contextual factors. The characterization of performance as skill-, rule-, or knowledge-based (Section 3.2.5) has proven particularly useful in thinking about the ways in which information-processing failures can arise, in light of the distinctions in information-processing activities that are presumed to occur at each of these levels. Generally, taxonomies that focus on the cognitive or causal end of the error spectrum have the ability to propose types of errors that might occur under various circumstances and thus can shape or augment our understanding of human limitations in information processing.

A very simple error taxonomy that bears a long history (Sanders and McCormick, 1993) differentiates errors of omission (forgetting to do something) from errors of commission (doing something incorrectly). Errors of commission are often further categorized into errors related to sequence, timing, substitution, and actions not included in a person's current plans (Hollnagel, 1993). *Sequence errors* include actions that are repeated (which may result in restarting a process) or are reversed (which may result in jumping ahead in a sequence). *Timing errors* refer to actions that do not occur when they are required; thus they may occur prematurely or after some delay. *Substitution errors* refer to single actions or sets of actions that are performed in place of the expected action or action set. Errors involving the inclusion of additional actions are referred to as *intrusions* when they are capable of disrupting the planned sequence of actions. Disruptions can lead to *capture* by the sequence, *branching* to an incorrect sequence, or *overshooting* the action sequence beyond the satisfaction of its objective.

Figure 5 and Tables 2 to 4 illustrate several other error taxonomies. The flowchart in Figure 5 classifies different types of human errors that can occur under skill-, rule-, and knowledge-based levels of performance. This flowchart seeks to answer questions concerning how an error occurred. Similar flowcharts are provided by the author to address the more preliminary issue in the causal chain (i.e., why an error occurred) as well as the external manifestation of the error (i.e., what type of error occurred). Reason's (1990) taxonomy (Table 2) also exploits the distinctions among skill-, rule-, and knowledge-based levels of performance, but draws attention to how error modes related to skill-based slips and lapses differ from error modes related to rule- and knowledge-based mistakes. The taxonomies presented in Tables 3 and 4 demonstrate various schemes for classifying errors based on stages of information processing.

In addition to their usefulness for analyzing accidents for root causes (Section 10.2), error taxonomies that emphasize cognitive or causal factors have predictive value as well. Predicting human error, however, is a difficult matter. It may indeed be possible to construct highly controlled experimental tasks that "trap" people into particular types of skill-based slips and lapses and some forms of rule-based mistakes. However, the multidimensional complexity surrounding actual work situations and the uncertainty associated with the human's goals, intentions, and emotional and attentional states introduce many layers of guesswork into the process of establishing reliable mappings between human fallibility and context. In 1991, Senders and Moray stated: "To understand and predict errors ... usually requires a detailed task analysis" (p. 60). Nothing has changed since to diminish the validity of this assertion. In fact, the current emphasis on *cognitive task analysis* (CTA) techniques and our greater understanding of mechanisms underlying human error have probably made the process of predicting human error more laborious than ever, as it should be. Expectations of shortcuts are unreasonable; error prediction by its very nature should be a tedious process and will often be influenced by the choice of taxonomy.

Task analysis (TA), which is fundamental to error prediction, describes the human's involvement with a system in terms of task requirements, actions, and cognitive processes (Chapter 14). It can be used to provide a broad overview of task requirements (that are often useful during the preliminary stages of product design) or a highly detailed description of activities. These descriptions could include time constraints and activity time lines; sequential dependencies among activities; alternative plans for performing an operation; contingencies that may arise during the course of activities and options for handling these contingencies; the feedback available at each step of the process; characterizations of information flow between different subsystems; and descriptions of displays, controls, training, and interactions with other people. Tabular formats are often used to illustrate the various relationships between these factors and task activities. Many different TA methods exist (Kirwan and Ainsworth, 1992; Luczak, 1997; Shepherd, 2000) and identifying an appropriate method for a particular problem or work domain can be critical.

In CTA, the interest is in determining how the human conceptualizes tasks, recognizes critical information and patterns of cues, assesses situations, makes discriminations, and uses strategies for solving problems, forming judgments, and making decisions. Successful application of CTA for enhancing system performance will depend on a concurrent understanding of the cognitive processes underlying human performance in the work domain and the constraints on cognitive processing the work domain imposes (Vicente, 1999). In developing new systems, meeting this objective may require multiple, coordinated approaches. As Potter et al. (1998) have noted: "No one approach can capture the richness required for a comprehensive, insightful CTA" (p. 395).

As with TA, many different CTA techniques are presently available (Hollnagel, 2003). TA and CTA, however, should not be viewed as mutually exclusive enterprises—in fact, the case could be made that TA methods that incorporate CTA represent "good" task analyses. As anticipating the precise time and

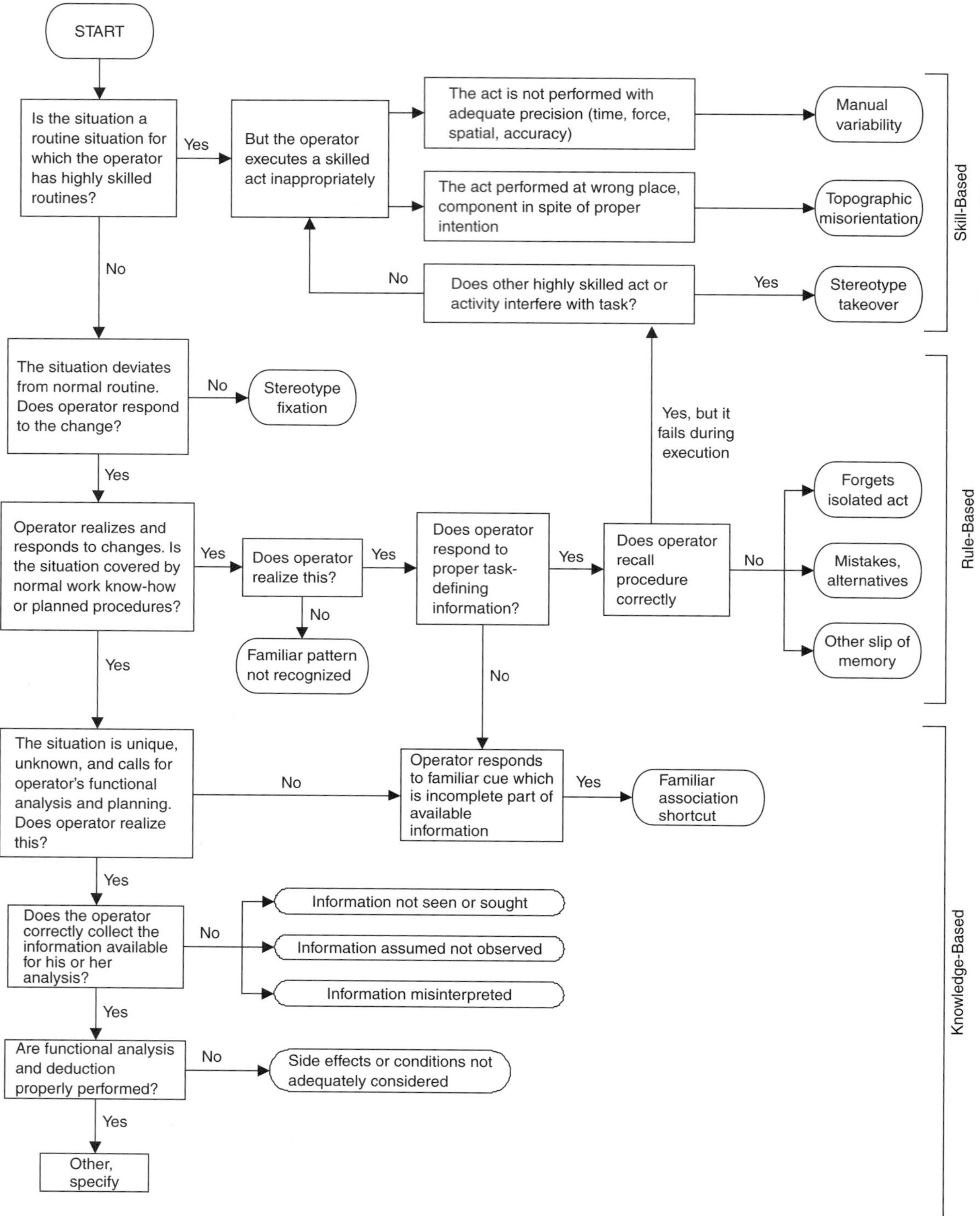

Figure 5 Decision flow diagram for analyzing an event into one of 13 types of human error. (From Rasmussen, 1982; copyright 1982, with permission from Elsevier.)

Table 2 Human Error Modes Associated with Rasmussen's SRK Framework

Skill-Based Performance	
Inattention	Overattention
Double-capture slips Omissions following interruptions Reduced intentionality Perceptual confusions Interference errors	Omissions Repetitions Reversals
Rule-Based Performance	
Misapplication of Good Rules	Application of Bad Rules
First exceptions Countersigns and nonsigns Informational overload Rule strength General rules Redundancy Rigidity	Encoding deficiencies Action deficiencies Wrong rules Inelegant rules Inadvisable rules
Knowledge-Based Performance	
Selectivity Workspace limitations Out of sight, out of mind Confirmation bias Overconfidence Biased reviewing Illusory correlation Halo effects Problems with causality	Problems with complexity Problems with delayed feedback Insufficient consideration of processes in time Difficulties with exponential developments Thinking in causal series and not causal nets Thematic vagabonding Encysting

Source: Reason (1990).

mode of error is generally unrealistic, the use of TA techniques should be directed at uncovering the possibility for errors and prioritizing these possibilities. Given what we can surmise about human fallibility, the contexts within which human activities occur, and the barriers that may be in place, the relevant questions are then as follows: What kinds of actions by people are possible or even reasonable that would, by one's definition, constitute errors? What are the possible consequences of these errors? What kinds of barriers do these errors and their consequences call for? Depending on whether the analysis is to be applied to a product or process that is still in the conceptual stages, to a newly implemented process, or to an existing process, broad applications of TA techniques that may include mock-ups, walkthroughs, simulations, interviews, and direct observations are needed to identify the relevant contextual elements. In-depth task analyses that incorporate CTA techniques could then provide the details necessary for evaluating the various possibilities for interplay between context and human fallibility (Sharit, 1998).

Table 3 External Error Modes Classified According to Stages of Human Information Processing

1. Activation/detection
   - 1.1 Fails to detect signal/cue
   - 1.2 Incomplete/partial detection
   - 1.3 Ignore signal
   - 1.4 Signal absent
   - 1.5 Fails to detect deterioration of situation
2. Observation/data collection
   - 2.1 Insufficient information gathered
   - 2.2 Confusing information gathered
   - 2.3 Monitoring/observation omitted
3. Identification of system state
   - 3.1 Plant-state-identification failure
   - 3.2 Incomplete-state identification
   - 3.3 Incorrect-state identification
4. Interpretation
   - 4.1 Incorrect interpretation
   - 4.2 Incomplete interpretation
   - 4.3 Problem solving (other)
5. Evaluation
   - 5.1 Judgment error
   - 5.2 Problem-solving error (evaluation)
   - 5.3 Fails to define criteria
   - 5.4 Fails to carry out evaluation
6. Goal selection and task definition
   - 6.1 Fails to define goal/task
   - 6.2 Defines incomplete goal/task
   - 6.3 Defines incorrect or inappropriate goal/task
7. Procedure selection
   - 7.1 Selects wrong procedure
   - 7.2 Procedure inadequately formulated/shortcut invoked
   - 7.3 Procedure contains rule violation
   - 7.4 Fails to select or identify procedure
8. Procedure execution
   - 8.1 Too early/late
   - 8.2 Too much/little
   - 8.3 Wrong sequence
   - 8.4 Repeated action
   - 8.5 Substitution/intrusion error
   - 8.6 Orientation/misalignment error
   - 8.7 Right action on wrong object
   - 8.8 Wrong action on right object
   - 8.9 Check omitted
   - 8.10 Check fails/wrong check
   - 8.11 Check mistimed
   - 8.12 Communication error
   - 8.13 Act performed wrongly
   - 8.14 Part of act performed
   - 8.15 Forgets isolated act at end of task
   - 8.16 Accidental timing with other event/circumstance
   - 8.17 Latent error prevents execution
   - 8.18 Action omitted
   - 8.19 Information not obtained/transmitted
   - 8.20 Wrong information obtained/transmitted
   - 8.21 Other

Source: Kirwan (1994).

Even when applied at relatively superficial levels, TA techniques are well suited for identifying mismatches between demands imposed by the work

Table 4 Human Error Classification Scheme

1. Observation of system state
    - Improper rechecking of correct readings
    - Erroneous interpretation of correct readings
    - Incorrect readings of appropriate state variables
    - Failure to observe sufficient number of variables
    - Observation of inappropriate state variables
    - Failure to observe any state variables
2. Choice of hypothesis
    - Hypotheses could not cause the values of the state variables observed
    - Much more likely causes should be considered first
    - Very costly place to start
    - Hypothesis does not functionally relate to the variables observed
3. Testing of hypothesis
    - Stopped before reaching a conclusion
    - Reached wrong conclusion
    - Considered and discarded correct conclusion
    - Hypothesis not tested
4. Choice of goal
    - Insufficient specification of goal
    - Choice of counterproductive or nonproductive goal
    - Goal not chosen
5. Choice of procedure
    - Choice would not fully achieve goal
    - Choice would achieve incorrect goal
    - Choice unnecessary for achieving goal
    - Procedure not chosen
6. Execution of procedure
    - Required stop omitted
    - Unnecessary repetition of required step
    - Unnecessary step added
    - Steps executed in wrong order
    - Step executed too early or too late
    - Control in wrong position or range
    - Stopped before procedure complete
    - Unrelated inappropriate step executed

Source: Rouse and Rouse (1983).

context and the human's capabilities for meeting these demands. Although hypothesizing specific error forms will become more difficult at this level of analysis, windows of opportunity for error still can be readily exposed that, in and of themselves, can suggest countermeasures capable of reducing risk potential. For example, these analyses may determine that there is insufficient time to input information accurately into a computer-based documentation system, that the design of displays is likely to evoke control responses that are contraindicated, or that sources of information on which high-risk decisions are based contain incomplete or ambiguous information. This coarser approach to predicting errors or error-inducing conditions that derives from analyzing demand-capability mismatches can also highlight contextual and cognitive considerations that can form the basis for a more focused application of TA and CTA techniques.

Table 5 depicts a portion of a type of TA known as a *hierarchical task analysis* (HTA) that was developed

Table 5 Part of a Hierarchical Task Analysis Associated with Filling a Chlorine Tanker

0. Fill tanker with chlorine.
   Plan: Do tasks 1 to 5 in order.
1. Park tanker and check documents (not analyzed).
2. Prepare tanker for filling.
   Plan: Do 2.1 or 2.2 in any order, then do 2.3 to 2.5 in order.
   - 2.1 Verify tanker is empty.
     Plan: Do in order:
       - 2.1.1 Open test valve.
       - 2.1.2 Test for Cl_2.
       - 2.1.3 Close test valve.
   - 2.2 Check weight of tanker.
   - 2.3 Enter tanker target weight.
   - 2.4 Prepare fill line.
     Plan: Do in order:
       - 2.4.1 Vent and purge line.
       - 2.4.2 Ensure main Cl_2 valve is closed.
   - 2.5 Connect main Cl_2 fill line.
3. Initiate and monitor tanker filling operation.
   Plan: Do in order:
   - 3.1 Initiate filling operation.
     Plan: Do in order:
       - 3.1.1 Open supply line valves.
       - 3.1.2 Ensure tanker is filling with chlorine.
   - 3.2 Monitor tanker filling operation.
     Plan: Do 3.2.1, do 3.2.2 every 20 minutes; on initial weight alarm, do 3.2.3 and 3.2.4; on final weight alarm, do 3.2.5 and 3.2.6.
       - 3.2.1 Remain within earshot while tanker is filling.
       - 3.2.2 Check road tanker.
       - 3.2.3 Attend tanker during last filling of 2 or 3 tons.
       - 3.2.4 Cancel initial weight alarm and remain at controls.
       - 3.2.5 Cancel final weight alarm.
       - 3.2.6 Close supply valve A when target weight is reached.
4. Terminate filling and release tanker.
   - 4.1 Stop filling operation.
     Plan: Do in order:
       - 4.1.1 Close supply valve B.
       - 4.1.2 Clear lines.
       - 4.1.3 Close tanker valve.
   - 4.2 Disconnect tanker.
     Plan: Repeat 4.2.1 five times, then do 4.2.2 to 4.2.4 in order.
       - 4.2.1 Vent and purge lines.
       - 4.2.2 Remove instrument air from valves.
       - 4.2.3 Secure blocking device on valves.
       - 4.2.4 Break tanker connections.
   - 4.3 Store hoses.
   - 4.4 Secure tanker.
     Plan: Do in order:
       - 4.4.1 Check valves for leakage.
       - 4.4.2 Secure log-in nuts.
       - 4.4.3 Close and secure dome.
   - 4.5 Secure panel (not analyzed).
5. Document and report (not analyzed).

Source: CCPS (1994). Copyright 1994 by the American Institute of Chemical Engineers, and reproduced by permission of AIChE.

for analyzing the task of filling a storage tank with chlorine from a tank truck. The primary purpose of this TA was to identify potential human errors that could contribute to a major flammable release resulting either from a spill during unloading of the truck or from a tank rupture. Table 6 illustrates the use of this HTA for predicting external error modes. The error taxonomy shown in Table 3 can easily be adapted for predicting the types of errors listed in Table 6. This taxonomy can also be linked to more underlying psychological mechanisms, allowing errors with identical or similar external manifestations to be distinguished and thus adding considerable depth to the understanding of potential errors predicted from the TA. As discussed in Section 5.4, this ability not only results in more accurate quantification of error data but also provides the basis for more effective error-reduction strategies. An example of such a scheme is the *human error identification in systems technique* (HEIST), which classifies external error modes according to the eight stages of human information processing listed in Table 3. The first column in a HEIST table consists of a code whose initial letter(s) refers to one of these eight stages. The next letter in the code refers to one of six general PSFs: time (T), interface (I), training/experience/familiarity (E), procedures (P), task organization (O), and task complexity (C). The external error modes are then linked to underlying psychological error mechanisms (PEMs). Many of these mechanisms are consistent with the failure modes that appear in Reason's error taxonomy (Table 2).

Table 7 presents an extract from a HEIST table corresponding to two of the eight stages of human information processing listed in Table 3: activation/detection and observation/data collection. For these two stages of information processing, more detailed explanations of the PEMs listed in the HEIST table may be found in Table 8. A complete HEIST table and the corresponding listing of PEMs can be found in Kirwan (1994).

On a final note, task analysts contending with complex systems will often need to consider various properties of the wider system or subsystem in which human activities take place. As Shepherd (2000) has stated: "Any task analysis method which purports to serve practical ends needs to be carried out beneath a general umbrella of systems thinking" (p. 11). There are a variety of ways in which systems can be characterized or decomposed (Sharit, 1997), and for any particular system these various descriptions could lead to the consideration of different activities for analysis as well as different strategies for performing these analyses.

5 QUANTIFYING HUMAN ERROR

5.1 Historical Antecedents

Quantifying human error presumes that a probability can be attached to its occurrence. Is this a realistic endeavor? An objective assignment of probabilities to events requires that *human error probability* (HEP) be defined as a ratio of the number of observed occurrences of the error to the number of opportunities for that error to occur. Based on this definition, it can be argued that with the exception of routine skill-based activities, estimates of HEPs are not easily attainable or likely to be accurate. Assuming that most organizations would be more interested in gauging the possibility for human error and understanding its causality and consequences, the more compelling question is: Why do we need a quantitative estimate?

The catalyst behind quantification of human error was the mandate for industries involved in high-hazard operations to perform *probabilistic risk analyses* (PRAs). Most industries that carry out such assessments, such as the chemical processing and nuclear power industries, are concerned about hazards arising from interactions among various system events, including hardware and software failures, environmental anomalies, and human errors that are capable of producing injuries, fatalities, disruptions to production, and plant and environmental damage. The two primary hazard analysis techniques that have become associated with PRAs are *fault tree* (FT) analysis and *event tree* (ET) analysis. The starting point for each of these methods is an undesirable event. Other hazard analysis techniques (CCPS, 1992) or methods based on expert opinion are often used to identify these events. FTs utilize Boolean logic models to depict the relationships among hardware, human, and environmental events that can lead to the undesirable *top event*. When FTs are used as a quantitative method, *basic events* (for which no further analysis of the cause is carried out) are assigned probabilities or occurrence rates, which are then propagated into a probability or rate measure associated with the top event (Dhillon and Singh, 1981). The contributions of each of the singular events to the top event can also be computed, making this technique very suitable for cost–benefit analyses that can be used as a basis for specifying design interventions. As a qualitative analysis tool, FTs can identify the various combinations of events (or cut sets) that could lead to the top event; for many applications this information is sufficiently revealing for satisfying safety objectives.

Whereas a FT represents a deductive, top-down decomposition of an undesirable event (such as a loss in electrical power), an ET corresponds to an inductive analysis that determines how this undesirable event can propagate. These trees are thus capable of depicting the various sequences of events that can become triggered by the initiating event, as well as the risks associated with each of these sequences. Figure 6 illustrates a simple ET consisting of two operator actions and two safety systems. When ETs are constructed to address only sequences of human actions in response to the initiating event, the ET is sometimes referred to as an *operator action event tree* (OAET). In OAETs, each branch of the tree represents either a success or an HEP associated with the required actions specified along the column headings. These trees can easily accommodate paths signifying recovery from previous errors. In many

Table 6 Human Errors and Error Reduction Recommendations for the HTA in Table 5[a]

Error Type	Error Description	Recovery	Consequences and Comments	Error Reduction Recommendations: Procedures	Training	Equipment
Wrong information obtained	Wrong weight is entered.	On check	Alarm does not sound before tanker overfills.	Validate target weight independently.	Ensure that operator double-checks data entered; record values in checklist.	Provide automatic setting of weight alarms from unladen weight; install computerized logging system and built-in checks on tanker reg. no. and unladen weight linked to warning system; display differences.
Check omitted	Tanker is not monitored while filling.	On initial weight alarm	Alarm will alert the operator if set correctly. Equipment fault (e.g., leaks not detected early and remedial action delayed).	Provide secondary task involving other personnel; supervisor checks operation periodically.	Stress importance of regular checks for safety.	Provide automatic log-in procedure.
	Operator fails to attend.	On step 3.2.5	If alarm is not detected within 10 minutes, tanker will overfill.	Ensure work schedule allows operator to do this without pressure.	Illustrate consequences of not attending.	Repeat alarm in secondary area; provide automatic interlock to terminate loading if alarm is not acknowledged; provide visual indication of alarm.
	Final weight alarm is taken as initial weight alarm.	No recovery	Tanker overfills.	Note differences between the sound of the two alarms in checklist.	Alert operators during training about differences in sounds of alarms.	Use completely different tones for initial and final weight alarms.
	Tanker valve is not closed.	On step 4.2.1	Failure to close tanker valve would result in pressure not being detected during the pressure check in step 4.2.1.	Perform independent check on action; use checklist.	Ensure that operator is aware of consequences of failure.	Valve position indicator would reduce probability of error.
Operation omitted, operation incomplete	Lines are not fully purged.	On step 4.2.4	Failure of operator to detect pressure in lines could lead to leak when tanker connections are broken	Specify a procedure to indicate how to check if fully purged.	Ensure that training covers symptoms of pressure in line.	Line pressure indicator used at controls, interlock device on line pressure.
Operation omitted	Locking nuts are left unsecured.	None	Failure to secure locking nuts could result in leakage during transportation.	Use checklist.	Stress safety implication of training.	Locking nuts gives tactile feedback when secure.

Source: Adapted from CCPS (1994). Copyright 1994 by the American Institute of Chemical Engineers, and reproduced by permission of AIChE.
[a]Possible errors derive from Table 3.

Table 7 Extract from a HEIST Table

Code	Error-Identifier Prompt	External Error Mode	System Cause/Psychological Error-Mechanism	Error-Reduction Guidelines
AT1	Does the signal occur at the appropriate time? Could it be delayed?	Action omitted or performed either too early or too late	Signal timing deficiency, failure of prospective memory	Alter system configuration to present signal appropriately; generate hard copy to aid prospective memory; repeat signal until action has occurred.
AI1	Could the signal source fail?	Action omitted or performed too late	Signal failure	Use diverse/redundant signal sources; use a higher-reliability signal system; give training and ensure that procedures incorporate investigation checks on "no signal."
AI2	Can the signal be perceived as unreliable?	Action omitted	Signal ignored	Use diverse signal sources; ensure higher signal reliability; retrain if signal is more reliable than it is perceived to be.
AI3	Is the signal a strong one, and is it in a prominent location? Could the signal be confused with another?	Action omitted, or performed too late, or wrong act performed	Signal-detection failure	Prioritize signals; place signals in primary (and unobscured) location; use diverse signals; use multiple-signal coding; give training in signal priorities; make procedures cross-reference the relevant signals; increase signal intensity.
AI4	Does the signal rely on oral communication?	Action omitted or performed too late	Communication failure, lapse of memory	Provide physical backup/substitute signal; build required communications requirements into procedures.
AE1	Is the signal very rare?	Action omitted or performed too late	Signal ignored (false alarm), stereotype fixation	Give training for low-frequency events; ensure diversity of signals; prioritize signals into a hierarchy of several levels.
AE2	Does the operator understand the significance of the signal?	Action omitted or performed too late	Inadequate mental model	Training and procedures should be amended to ensure that significance is understood.
AP1	Are procedures clear about action following the signal or the previous step, or when to start the task?	Action omitted or performed either too early or too late	Incorrect mental model	Procedures must be rendered accurate, or at least made more precise; give training if judgment is required on when to act.
AO1	Does activation rely on prospective memory (i.e., remembering to do something at a future time, with no specific cue or signal at that later time)?	Action omitted or performed either too late or too early	Prospective memory failure	Proceduralize task, noting calling conditions, timings of actions, etc.; utilize an interlock system preventing task from occurring at undesirable times; provide a later cue; emphasize this aspect during training.
AO2	Will the operator have other duties to perform concurrently? Are there likely to be distractions? Could the operator become incapacitated?	Action omitted or performed too late	Lapse of memory, memory failure, signal-detection failure	Training should prioritize signal importance; improve task organization for crew; use memory aids; use a recurring signal; consider automation; utilize flexible crewing.
AO3	Will the operator have a very high or low workload?	Action omitted or performed either too late or too early	Lapse of memory, other memory failure, signal-detection failure	Improve task and crew organization; use a recurring signal; consider automation; utilize flexible crewing; enhance signal salience.

(continued overleaf)

Table 7 (*continued*)

Code	Error-Identifier Prompt	External Error Mode	System Cause/Psychological Error-Mechanism	Error-Reduction Guidelines
AO4	Will it be clear who must respond?	Action omitted or performed too late	Crew-coordination failure	Emphasize task responsibility in training and task allocation among crew members; utilize team training.
AC1	Is the signal highly complex?	Action omitted, or wrong act performed either too late or too early	Cognitive overload, inadequate mental model	Simplify signal; automate system response; give adequate training in the nature of the signal; provide online, automated, diagnostic support; develop procedures that allow rapid analysis of the signal (e.g., use of flowcharts).
AC2	Is the signal in conflict with the current diagnostic mindset?	Action omitted or wrong act performed	Confirmation bias, signal ignored	Procedures should emphasize disconfirming as well as confirmatory signals; utilize a shift technical advisory in the shift structure; carry out problem-solving training and team training; utilize diverse signals; implement automation.
AC3	Could the signal be seen as part of a different signal set? Or is, in fact, the signal part of a series of signals to which the operator needs to respond?	Action performed too early or wrong act performed	Familiar-association shortcut/ stereotype takeover	Training and procedures could involve display of signals embedded within mimics or other representations showing their true contexts or range of possible contexts; use fault-symptom matrix aids; etc.
OT1	Could the information or check occur at the wrong time?	Failure to act, or action performed either too late or too early, or wrong act performed	Inadequate mental model/ inexperience/ crew coordination failure	Procedure and training should specify the priority and timing of checks; present key information centrally; utilize trend displays and predictor displays if possible; implement team training.
OI1	Could important information be missing due to instrument failure?	Action omitted or performed either too late or too early, or wrong act performed	Signal failure	Use diverse signal sources; maintain backup power supplies for signals; have periodic manual checks; procedures should specify action to be taken in event of signal failure; engineer automatic protection/action; use a higher-reliability system.
OI2	Could information sources be erroneous?	Action omitted or performed either too late or too early, or wrong act performed	Erroneous signal	Use diverse signal sources; procedures should specify cross-checking; design system-self-integrity monitoring; use higher-reliability signals.
OI3	Could the operator select an incorrect but similar information source?	Action omitted or performed either too late or too early, or wrong act performed	Mistakes alternatives, spatial misorientation, topographic misorientation	Ensure unique coding of displays, cross-referenced in procedures; enhance discriminability via coding; improve training.
OI4	Is an information source accessed only via oral communication?	Action omitted or performed either too late or too early, or wrong act performed	Communication failure	Use diverse signals from hardwired or softwired displays; ensure backup human corroboration; design communication protocols.
OI5	Are any information sources ambiguous?	Action omitted or performed either too late or too early, or wrong act performed	Misinterpretation, mistakes alternatives	Use task-based displays; design symptom-based diagnostic aids; utilize diverse information sources; ensure clarity of information displayed; utilize alarm conditioning.

(*continued overleaf*)

Table 7 (*continued*)

Code	Error-Identifier Prompt	External Error Mode	System Cause/Psychological Error-Mechanism	Error-Reduction Guidelines
OI6	Is an information source difficult or time-consuming to access?	Action omitted or performed too late, or wrong act performed	Information assumed	Centralize key data; enhance data access; provide training on importance of verification of signals; enhance procedures.
OI7	Is there an abundance of information in the scenario, some of which is irrelevant, or a large part of which is redundant?	Action omitted or performed too late	Information overload	Prioritize information displays (especially alarms); utilize overview mimics (VDU or hardwired); put training and procedural emphasis on data-collection priorities and data management.
OE1	Could the operator focus on key indication(s) related to a potential event while ignoring other information sources?	Action omitted or performed too late, or wrong act performed	Confirmation bias, tunnel vision	Provide training in diagnostic skills; enhance procedural structuring of diagnosis, emphasizing checks on disconfirming evidence; implement a staff-technical-advisor role; present overview mimics of key parameters showing whether system integrity is improving or worsening or adequate.
OE2	Could the operator interrogate too many information sources for too long, so that progress toward stating identification or action is not achieved?	Action omitted or performed too late	Thematic vagabonding, risk-recognition failure, inadequate mental model	Provide training in fault diagnosis; provide team training; put procedural emphasis on required data collection time frames; implement high-level indicators (alarms) of system-integrity deterioration.
OE3	Could the operator fail to realize the need to check a particular source? Is there an adequate cue prompting the operator?	Action omitted or performed either too late or too early, or wrong act performed	Need for information not prompted, prospective memory failure	Provide procedural guidance on checks required, training, use of memory aids, use of attention-gaining devices (false alarms, central displays, and messages)
OE4	Could the operator terminate the data collection/observation early?	Action omitted or performed either too early or too late, or wrong act performed	Overconfidence, inadequate mental model, incorrect mental model, familiar-association shortcut	Provide training in diagnostic procedures and verification; provide procedural specification of required checks, etc.; implement a shift-technical-advisor role.
OE5	Could the operator fail to recognize that special circumstances apply?	Action omitted or performed either too late or too early, or wrong act performed	Failure to consider special circumstances, slip of memory, inadequate mental model	Ensure training for, as well as procedural noting of, special circumstance; STA; give local warnings in the interface displays/controls.
OP1	Could the operator fail to follow the procedures entirely?	Action omitted or wrong act performed	Rule violation, risk-recognition failure, production–safety conflict, safety-culture deficiency	Provide training in use of procedures; involve operator in development and verification of procedures.

(*continued overleaf*)

Table 7 ***(continued)***

Code	Error-Identifier Prompt	External Error Mode	System Cause/Psychological Error-Mechanism	Error-Reduction Guidelines
OP2	Could the operator forget one or more items in the procedures?	Action omitted or performed either too early or too late, or wrong act performed	Forget isolated act, slip of memory, place-losing error	Ensure an ergonomic procedure design; utilize tick-off sheets, place keeping aids, etc.; provide team training to emphasize checking by other team member(s).
OO1 (AO2)	Will the operator have other duties to perform concurrently? Are there likely to be distractions? Could the operator become incapacitated?	Action omitted or performed too late	Lapse of memory, memory failure, signal-detection failure	Training should prioritize signal importances; develop better task organization for crew; use memory aids; use a recurring signal; consider automation; use flexible crewing.
OO2 (AO3)	Will the operator have a very high or low workload?	Action omitted or performed either too late or too early	Lapse of memory, other memory failure, signal-detection failure	Establish better task and crew organization; utilize a recurring signal; consider automation; use flexible crewing; enhance signal salience.
OO3 (AO4)	Will it be clear who must respond?	Action omitted or performed too late	Crew-coordination failure	Improve training and task allocation among crew; provide team training.
OO4	Could information collected fail to be transmitted effectively across shift-handover boundaries?	Failure to act, or wrong action performed, or action performed either too late or too early, or an error of quality (too little or too much)	Crew-coordination failure	Develop robust shift-handover procedures; training; provide team training across shift boundaries; develop robust and auditable data-recording systems (logs).
OC1	Does the scenario involve multiple events, thus causing a high level of complexity or a high workload?	Failure to act, or wrong action performed, or action performed either too early or too late	Cognitive overload	Provide emergency-response training; design crash-shutdown facilities; use flexible crewing strategies; implement shift-technical-advisor role; develop emergency operating procedures able to deal with multiple transients; engineer automatic information recording (trends, logs, printouts); generate decision/diagnostic support facilities.

Source: Adapted from Kirwan (1994).

PRA applications, FTs and ETs are combined—each major column of the ET can represent a top event whose failure probability can be computed through the evaluation of a corresponding FT model (Figure 7).

Quantitative solutions to FTs or ETs that address only machine or material components are ultimately dictated by well-documented mathematical methods for computing component reliability, either in terms of the probability that the component or subsystem functions normally each time it is used or in terms of the probability that the component will not fail during some prescribed time of use (Kapur and Lamberson, 1977). The realization that human interaction with other system components, including other humans, may have a marked effect on the outcomes of PRAs required developing methods for assessing human reliability, thus establishing the field of *human reliability analysis* (HRA). A variety of methods of HRA are currently available that range from relatively quick assessment procedures to those that involve detailed analyses (Kirwan, 1994). Almost all these methods rely on the idea of PSFs, discussed earlier (Section 3.3); the methods differ, however, in how PSFs are used to generate HEPs for various activities. To illustrate the different approaches to deriving HEPs that these methods can take, two

Table 8 Psychological Error Mechanisms for Two of the Stages of Information Processing Presented in Table 7

Activation/Detection

1. *Vigilance failure:* lapse of attention. Ergonomic design of interface to allow provision of effective attention-gaining measures; supervision and checking; task-organization optimization, so that the operators are not inactive for long periods and are not isolated.
2. *Cognitive/stimulus overload:* too many signals present for the operator to cope with. Prioritization of signals (e.g., high-, medium-, and low-level alarms); overview displays; decision-support systems; simplification of signals; flowchart procedures; simulator training; automation.
3. *Stereotype fixation:* operator fails to realize that situation has deviated from norm. Training and procedural emphasis on range of possible symptoms/causes; fault-symptom matrix as a job aid; decision support system; shift technical advisor/supervision.
4. *Signal unreliable:* operator treats signal as false due to its unreliability. Improved signal reliability; diversity of signals; increased level of tolerance on the part of the system, or delay in effects of error, which allows error detection and correction (decreases "coupling"); training in consequences associated with incorrect false-alarm diagnosis.
5. *Signal absent:* signal absent due to a maintenance/calibration failure or a hardware/software error. Provide signal; redundancy/diversity in signaling-design approach; procedures/training to allow operator to recognize when signal is absent.
6. *Signal-discrimination failure:* operator fails to realize that the signal is different. Improved ergonomics in the interface design; enhanced training and procedural support in the area of signal differentiation; supervision checking.

Observation/Data Collection

7. *Attention failure:* lapse of attention.
8. *Multiple signal coding:* enhanced alarm salience; improved task organization with respect to backup crew and rest pauses.
9. *Inaccurate recall:* operator remembers data incorrectly (usually, quantitative data). Nonreliance on memorized data, which would necessitate better interface design — as data are received, they can either be acted on while still present on a display (controls and displays are co-located) or at least be logged onto a "scratch pad"; sufficient displays for presenting all information necessary for a decision/action simultaneously; printer usage; training in nonreliance on memorized data.
10. *Confirmation bias:* operator only selects data that confirm given hypothesis and ignores other disconfirming data sources. Problem-solving training; team training [including training in the need to question decisions, and in the ability of the team leader(s) to take constructive criticism]; shift technical advisor (diverse, highly qualified operator who can "stand back" and consider alternative diagnoses), functional procedures: high-level information displays; simulator training; high-level alarms for system-integrity degradation; automatic protection.
11. *Thematic vagabonding:* operator flits from datum to datum, never actually collating it meaningfully. Problem-solving training; team training; simulator training; functional-procedure specification for decision-timing requirements; high-level alarms for system-integrity degradation.
12. *Encystment:* operator focuses exclusively on only one data source. Problem-solving training; team training [including training in the need to question decisions and in the ability of the team leader(s) to take constructive criticism]; shift technical advisor; functional procedures; high-level information displays; simulator training; high-level alarms for system-integrity degradation.
13. *Stereotype fixation revisited:* need for information is not prompted by either memory or procedures. Emergency procedure enhancements, and emphasis of key symptoms and indicators to be checked; team training; problem-solving training; alarm reprioritization; simulator training.
14. *Crew-functioning problem:* allocation of responsibility or priorities is unclear, with the result that data collection/observation fails.
15. *Cognitive/stimulus overload:* operator too busy, or being bombarded by signals, with the result that effective data collection/observation fails. See item 2.

Source: Kirwan (1994).

well-known techniques that were originally developed for application in the nuclear power industry will be described.

5.2 THERP

The *technique for human error rate prediction*, generally referred to as THERP, is detailed in a work by Swain and Guttmann (1983) sponsored by the U.S. Nuclear Regulatory Commission. Its methodology is driven by decomposition: Human tasks are first decomposed into clearly separable actions or subtasks; HEP estimates are then assigned to each of these actions; and finally, these HEPs are aggregated to derive probabilities of task failure, which reflect human reliability.

THERP is a highly systematic procedure. Its initial steps are directed at establishing which work activities will require emphasis and the time and skill requirements and concerns for human error associated with these activities. Factors related to error detection and the potential for error recovery are also determined. The results of these efforts are represented by a type of event tree referred to as a *probability tree*. Each relevant subtask in a probability

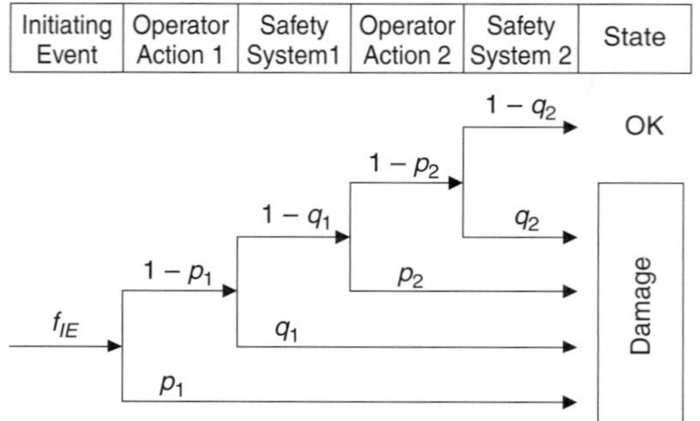

Figure 6 Event tree, where f_{IE} represents the probability of the initiating event, p_1 and p_2 represent human error probabilities associated with two operator actions, and q_1 and q_2 represent system failure probabilities for two safety systems. Typically, the "damage" consequence in the last column is stated more specifically in terms of different accident possibilities. Also, depending on the "level" of the risk analysis, the last column could be extended to reflect the different consequences of the various possible accidents. (From Kumamoto and Henley, 1996; © 2004 IEEE.)

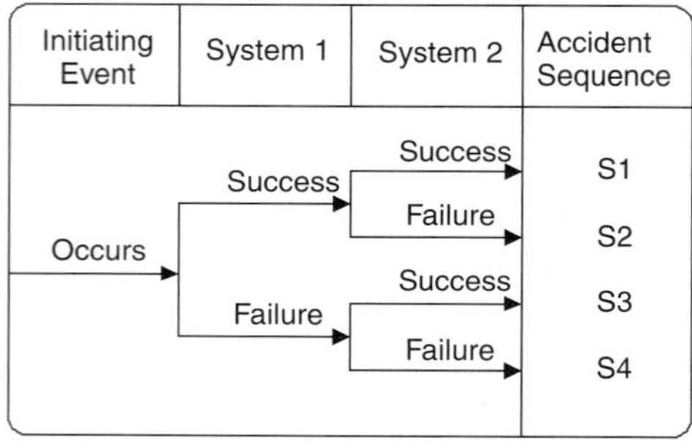

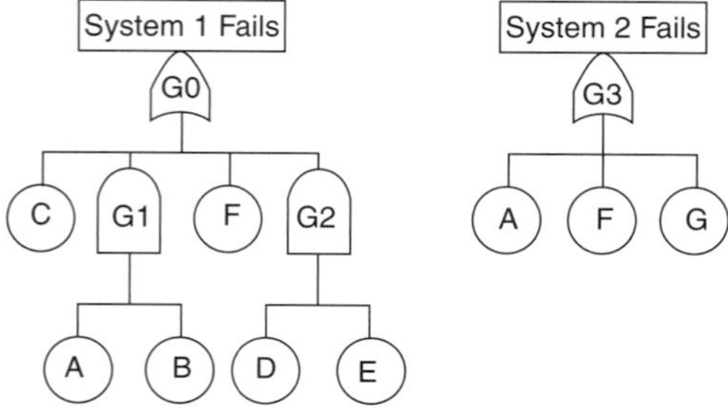

Figure 7 Coupling of event trees and fault trees. The probabilities of failure associated with systems 1 and 2 in the event tree would be derived from the two corresponding fault tees. (From Kumamoto and Henley, 1996; © 2004 IEEE.)

tree is characterized by two limbs, representing either successful or unsuccessful performance (Figure 8).

The next set of steps in THERP constitutes the quantitative assessment stage. First, HEPs are assigned to each of the limbs of the tree corresponding to incorrect performance. These probabilities, referred to as *nominal HEPs*, in theory are presumed to represent medians of lognormal probability distributions. Associated with each nominal HEP are *upper and lower uncertainty bounds* (UCBs), which reflect the variance associated with any given error distribution. The square root of the ratio of the upper to the lower UCB defines the *error factor* (the value selected for this factor will depend on the variability believed to be associated with the probability distribution for that error). Swain and Guttmann (1983) provide values of nominal HEPs and their corresponding error factors for a variety of nuclear power plant tasks. For some tasks the nominal HEPs that are provided refer to *joint HEPs* because it is the performance of a team rather than that of an individual worker that is being evaluated. Generally, the absence of existing hard data from the operations of interest will require that nominal HEPs be derived from other sources, which include (1) expert judgment elicited through techniques such as direct numerical estimation or paired comparisons (Swain and Guttmann, 1983; Kirwan, 1994), (2) simulators (Gertman and Blackman, 1994), and (3) data from jobs similar in psychological content to the operations of interest.

To account for more specific individual-, environmental-, and task-related influences on performance, nominal HEPs are subjected to a series of refinements. First, nominal HEPs are modified based on the influence of PSFs, resulting in *basic HEPs* (BHEPs). In some cases, guidelines are provided in tables indicating the direction and extent of influence of particular PSFs on nominal HEPs; for example, adjustments that are to be made in nominal HEPs due to the influence of the PSF of stress are provided as a function of type of task and worker experience. Next, a nonlinear dependency model is incorporated which considers positive dependencies that exist between adjacent limbs of the tree, resulting in *conditional HEPs* (CHEPs). In a positive dependency model, failure on a subtask increases the probability of failure on the following subtask, and successful performance of a subtask decreases the probability of failure in performing the subsequent task element. Instances of negative dependence can be accounted for but require the discretion of the analyst. In the case of positive dependence, THERP provides equations for modifying BHEPs to CHEPs based on the extent to which the analyst believes dependencies exist.

At this point, success and failure probabilities are computed for the entire task. Various approaches to these computations can be taken. The most straightforward approach is to multiply the individual CHEPs associated with each path on the tree leading to failure, sum these individual failure probabilities to arrive at the probability of failure for the total task, and then assign UCBs to this probability. More complex approaches to these computations take into account the variability associated with the combinations of

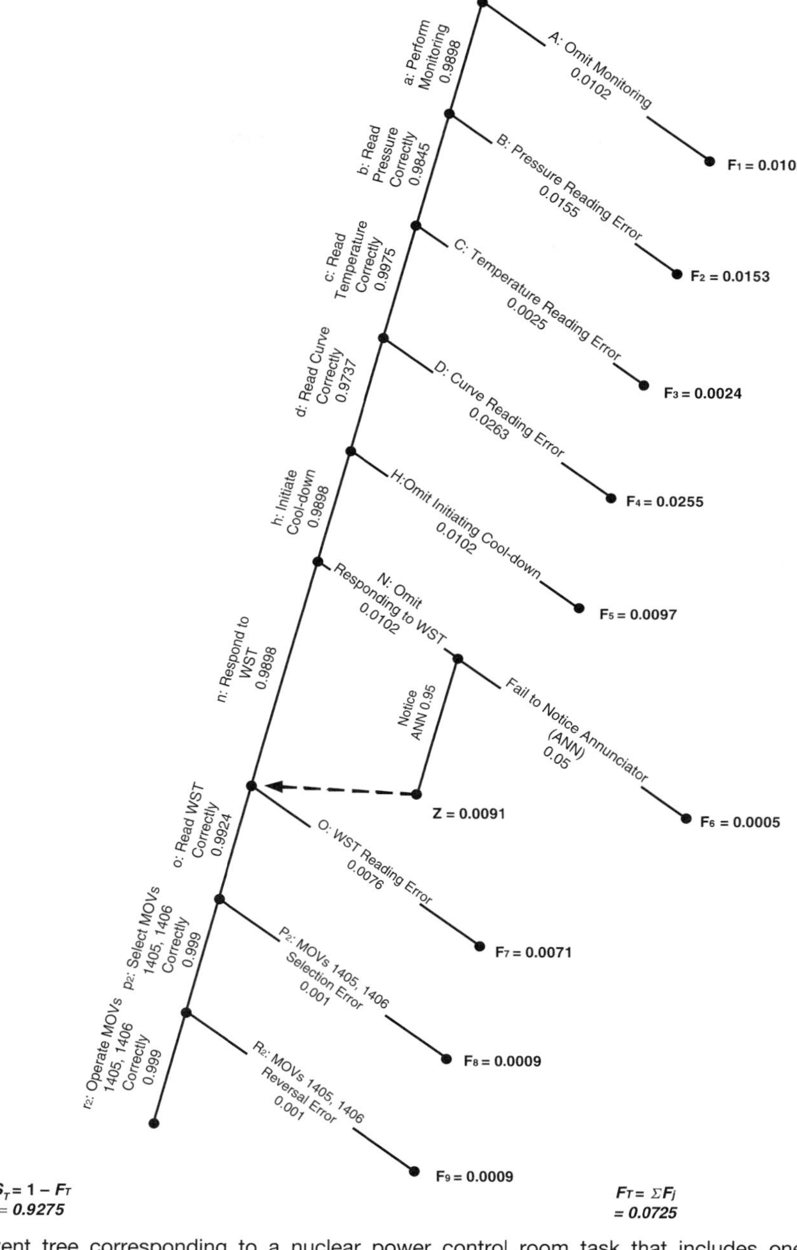

Figure 8 HRA event tree corresponding to a nuclear power control room task that includes one recovery factor. (From Kumamoto and Henley, 1996; © 2004 IEEE.)

events comprising the probability tree (Swain and Guttmann, 1983).

The final steps of THERP consider the ways in which errors can be recovered and the kinds of design interventions that can have the greatest impact on task success probability. Common recovery factors include the presence of annunciators that can alert the operator to the occurrence of an error, co-workers potentially capable of catching or discovering (in time) a fellow worker's errors, and various types of scheduled walk-through inspections. As with conventional ETs, these recovery paths can easily be represented in HRA probability trees (Figure 8). In the case of annunciators or inspectors, the relevant failure limb is extended into two additional limbs: one failure limb and one success limb. The probability that the human responds successfully to the annunciator or that the inspector spots the operator's error is then fed back into the success path of the original tree. In the case of recovery by fellow team members, BHEPs are modified to CHEPs by considering the degree of dependency between the operator and one or more fellow workers who are in a position to notice the error. The effects of recovery factors can be determined by repeating the computations for total task failure.

In addition to considering error recovery factors, the analyst can choose to perform sensitivity analysis. One approach to sensitivity analysis is to identify the most probable errors on the tree, propose design modifications corresponding to those task elements, estimate the degree to which the corresponding HEPs would become reduced by virtue of these modifications, and evaluate the effect of these design interventions on the computation of the total task failure probability. The final step in THERP is to incorporate the results of the HRA into system risk assessments such as PRAs.

An obvious deficiency of THERP is its inability to handle human errors that have a more complex cognitive basis. Despite attempts to embellish THERP [e.g., through "sneak analysis" methods that may enable the analyst to identify decision making errors (Hahn and deVries, 1991)], THERP's underlying emphasis on decomposition and subsequent aggregation of individual actions has been questioned. For example, Hollnagel (1993) has argued that human reliability cannot be accounted for by considering "each action on its own" but rather, by considering "actions as a whole sequence," and has developed an alternative approach to HRA based on a modeling framework for predicting cognitive reliability that can also be used to support system risk assessments (Hollnagel, 1998). Although THERP's inability to adequately address more cognitively complex tasks and the underlying causality of human error tends to cast it as shallow, the insights concerning system operations acquired through THERP's attention to detail ultimately are likely to make it more useful than the numbers it makes available to quantitative risk assessments such as PRAs. In this respect, THERP shares many of the characteristics of PRAs: The quantitative products provided by PRAs are often considered to be less important than the ability for these risk assessments to identify deficiencies in design, provide a better understanding of interdependencies among systems and operations, and offer insights for improving procedures and operator training (Kumamoto and Henley, 1996).

5.3 SLIM–MAUD

The *success likelihood index methodology* (SLIM) represents another procedure for deriving HEPs (Embrey et al., 1984). In contrast to THERP, SLIM allows the analyst to focus on any human action or task. Consequently, this method can provide inputs into PRAs at various system levels; that is, the HEPs can reflect relatively low-level actions that cannot be further decomposed, as well as more broadly defined actions that encompass many of these lower-level actions. This increased flexibility, however, comes at the expense of a greatly reduced emphasis on task analysis and an increased reliance on subjective assessments.

SLIM assumes that the probability that a human will carry out a particular task or action successfully depends on the combined effects of a number of relevant PSFs. For each action under consideration, task domain experts are required to identify the relevant set of PSFs; assess the relative importance (or weights) of each of these PSFs with respect to the likelihood of some potential error mode associated with the action; and independent of this assessment, rate how good or bad each PSF actually is. Relative importance weights for the PSFs are derived by asking each analyst to assign a weight of 100 to the most important PSF, and then assign weights ranging from 0 to 100 to each of the remaining PSFs based on the importance of these PSFs relative to the one assigned the value of 100. Normalized weights are derived by dividing each weight by the sum of the weights for all the PSFs. The judges then rate each PSF on each action or task, with the lowest scale value indicating that the PSF is as poor as it is likely to be under real operating conditions, and the highest scale value indicating that the PSF is as good as it is likely to be in terms of promoting successful task performance. The likelihood of success for each human action is determined by summing the product of the normalized weights and ratings for each PSF, resulting in numbers (SLIs) that represent a scale of success likelihood.

The SLIs are useful in their own right. For example, if the actions under consideration represent alternative modes of response in an emergency scenario, the analyst may be interested in determining which types of responses are least or most likely to succeed. However, for the purpose of conducting PRAs, SLIM converts the SLIs to HEPs. An estimate of the HEP is derived using the following relationship:

$$\text{probability of success} = a \times \text{SLI} + b$$

where HEP is 1− the probability of success. To derive the two constants in this equation, the probabilities of success must be available for at least two tasks taken from the cluster of tasks for which the relevant set of PSFs was identified. However, methods exist for deriving HEPs even if information on such "reference" tasks is not available. Methods also exist for deriving upper and lower uncertainty bounds for these HEPs, which PRAs typically require.

Multiattribute utility decomposition (MAUD) provides a user-friendly computer-interactive environment for implementing SLIM. This feature ensures that many of the assumptions that are critical to the theoretical underpinnings of SLIM are met. For example, MAUD can determine if the ratings for the various PSFs by a given analyst are independent of one another and whether the relative importance weights elicited for the PSFs are consistent with the analyst's preferences. In addition, MAUD provides procedures for assisting the expert in identifying the relevant PSFs. Further details concerning SLIM–MAUD are provided in Embrey et al. (1984) and Kirwan (1994).

5.4 Human Error Data

As indicated in the discussion of THERP, fundamental data on HEPs can come from a variety of sources. Ideally, HEP data should derive from the relevant operating experience or at least from similar industrial experiences. However, as Kirwan (1994) notes, a number of problems are associated with collecting this type of quantitative HEP data. For example, many workers will be reluctant to report errors due to the threat of reprisals, and mechanisms for investigating errors are often nonexistent.

Even if these problems could be overcome, there are still other issues to contend with concerning the collection of useful HEP data. One problem is that errors that do not lead to a violation of a company's technical specifications or that are recovered almost immediately will probably not be reported. Also, data on errors associated with very low probability events, as in the execution of recovery procedures following an accident, may not be sufficiently available to produce reliable estimates and thus often require simulator studies for their generation. Finally, error reports are usually confined to the observable manifestations of an error (the external error modes). Without knowledge of the underlying cognitive processes or psychological mechanisms, errors that are in fact dissimilar (Table 1) may be aggregated. This would not only corrupt the HEP data but could also compromise error-reduction strategies.

In a study covering over 70 incidents in the British nuclear industry, it was possible to compile data on external error modes, PSFs, and psychological error mechanisms, and to derive 34 different HEPs (Kirwan et al., 1990), suggesting the possibility for collecting reasonably accurate operational-experience human error data. More typically, HEPs are derived from other sources, including expert judgments, laboratory experiments, and simulator studies. Table 9 presents examples of HEP data from several of these sources. Additional data on HEPs that include upper and lower uncertainty bounds and the effects of PSFs on nominal HEPs may be found in Swain and Guttmann (1983) and Gertman and Blackman (1994). More recently, Kirwan (1999) has reported on the construction of a HEP database in the UK referred to as CORE-DATA (computerized operator reliability and error database) for supporting HRA activities. CORE-DATA currently contains a large number of HEPs; its long-term objective is to apply its data to new industrial contexts through the development of extrapolation rules.

6 INCIDENT REPORTING SYSTEMS

6.1 Design, Data Collection, and Management Considerations

Information systems allow extensive data to be collected on incidents, accidents, and human errors, and thus afford excellent opportunities for organizations to learn. The distinction between accidents and incidents varies among authors and government regulatory agencies. Generally, accidents imply injury to persons or reasonable damage to property, whereas incidents usually involve the creation of hazardous conditions that if not recovered could lead to an accident. *Accidents* and *adverse events* are terms that are often used interchangeably, as are *incident*, *near miss*, and *close call*.

Capturing information on near misses is particularly advantageous. Depending on the work domain, near misses may occur hundreds of times more often than adverse events. If near misses are regarded as events that did not result in accidents by virtue of chance factors alone, the contexts surrounding near misses should be highly predictive of accidents. The reporting of near misses, especially in the form of short event descriptions or detailed anecdotal reports, would then provide a potentially rich set of data that could be used as a basis for proactive interventions. Moreover, fewer barriers exist in reporting them (Barach and Small, 2000). However, to anticipate hazardous scenarios and provide the proactive accident prevention function necessary for enabling organizations to improve continuously, *incident reporting systems* (IRSs) must be capable of identifying the underlying causes of the reported events.

The role of management is critical to the successful development and implementation of an IRS (CCPS, 1994). Management not only allocates the resources for developing and maintaining the system but can also influence the development of work cultures that may be resistive to the deployment of IRSs. In particular, organizations that have instituted "blame cultures" (Reason, 1997) are unlikely to advocate IRSs that emphasize underlying causes of errors, and workers in these organizations are unlikely to volunteer information to these systems. Ultimately, management's attitudes concerning human error causation will be reflected in the data that will be collected. The adoption of a system-induced perspective on human error that is consistent with Figure 1 would imply the need for an information system that emphasizes the collection of data on possible causal factors, including organizational and management policies responsible for creating the latent conditions for errors. Data on near misses would be viewed as indispensable for providing early warnings about how the interplay between human fallibility and situational contexts can penetrate barriers. System-based perspectives to human error are also conducive to a dynamic approach to data collection—if the methodology is proving inadequate in accounting for or anticipating human error, it will probably be modified (Figure 9).

Worker acceptance of an IRS that relies on voluntary reporting entails that the organization meet

Table 9 Examples of HEP Data Derived from Various Sources

Error	Probability
Data from Operational Plants	
1. Invalid address keyed into process-control computer This error occurred in a computer-controlled-batch chemical plant. When a valve sticks, or another malfunction occurs, the operator goes through a sequence on the computer which includes entering an address code for the component to be manipulated. The operator could, however, enter the wrong address (i.e., either an address for which there is no item, or the address for the wrong item); the HEP reflects the sum of these two alternative errors. There is a plant mimic available, prompt feedback is given of control actions, and the task occurs in normal operations.	0.007
2. Precision error: incorrect setting of chemical interface pressure In this event, an interface-pressure setting was set incorrectly, allowing an aqueous solution to pass into the stock tank, where it subsequently crystallized — which has a highly serious consequence. The error was caused largely by the failure on the part of the operator to be precise enough when setting the equipment.	0.03
3. Welders worked on wrong line Welders at a chemical plant worked on a vent line by mistake and holed a pipe.	0.04
4. Erroneous discharge of contaminants into the sea In this event, material was discharged into the sea erroneously, partly due to a communications failure across two shifts.	0.0007
5. Fuel-handling machine moved while still attached to a static fuel tank In this event, the fuel-handling machine in question, resembling a large overhead crane but with a very limited view, from the crane cab, of the flasks it carries, was moved by the operator while it was still in fact attached to a flask via flexible hoses, thus rupturing the hoses. This accident was in part caused by a communication failure across a shift break.	0.0005
6. Critical safety system not properly restored following maintenance In this event, a U.S. boiling-water-reactor (BWR) core-spray-pump system was left in an incorrect line-up configuration after testing. Testing is done by the operator in the CCR five times per year on five similar systems. This particular error occurred on the control switches on the CCR panels. The consequences are serious, since the effect is to disable a backup safety system.	0.0006
7. Operator works on wrong pump In this event, an operator on the plant was instructed to work on a pump in the west part of the plant but instead worked erroneously on the identical east plant.	0.03
8. Wrong fuel container moved In a supervised and heavily logged operation, the wrong fuel container was moved via the crane. The operator in the crane cab did not have a direct view of the containers but could only see them via a CCTV facility. The operator was, however, in communications with local operators who could see the containers directly.	0.0007
Data Derived from Ergonomics Experiments	
9. Human-recall performance with digital displays A six-digit sequence was presented for 2.6 seconds. The subject then had to write down the digit sequence in the intervening 10 seconds before the next sequence was presented. Seventy-two slides of six-digit sequences were shown to each subject. The error in question involved not writing down the correct sequence.	0.03
10. Inspectors' level of accuracy in spotting soldering defects in a complex system In a study of the capabilities of quality-control inspectors, the inspectors examined a complex unit with 1500 wires soldered to various terminals over a 3-hour period. Thirty defects had been placed in each unit, and the inspectors had to find all these defects, which were similar to the kinds of defects that they would find or look for every day.	0.2
11. Typing performance Each touch-typist in this experiment was instructed to type out a 1000-character piece of text as fast as possibie without exceeding an error rate of 1%.	0.01
12. Network problem solving: a premature diagnosis Subjects were required to find the faulty component in a network of AND units. If two units feed into another unit, and one or both of the first two units are unhealthy, the third unit will read "unhealthy." However, the operator can only see which units are healthy or unhealthy at the end of the line of connected units, although he can also see how all the units are interconnected. Thus, the correct diagnosis involves determining which unit is unhealthy and is affecting the other units (only one unit is unhealthy in each network) and requires the operator to perform a type of fault diagnosis known as *backward chaining*. This task is very similar to a fault diagnosis for electrical maintenance panels. The number of units per network ranged from 16 to 24, always with four main "lines" leading to four final	0.07

Table 9 ***(continued)***

Error	Probability
output states (healthy or unhealthy). A *premature* diagnosis implied that the operator identified the faulty unit without first having carried out enough tests conclusively to determine which unit was faulty (irrespective of whether the premature diagnosis was correct or not: the task cannot afford the operator to make premature guesses).	
13. Failure to carry out a one-step calculation correctly	0.01
14. Failure to carry out a seven- to 13-step calculation correctly	0.27
Simulator-Derived Data	
15. Emergency manual trip in a nuclear control room Prior to a fault appearing, the operator would be occupied with normal operations in a simulated control room. Initially, when a fault appeared, the operator was expected to try to control the fault, but it quickly became apparent that this was not possible, the operator was required instead to shut down (trip) the plant. The faults in question comprised a control-rod runout, a blower failure, a gas-temperature rise, and a coolant-flow fault. Tripping the plant required a single pushbutton activation. The fault rate in this scenario was 10 signals per hour (normally, it would have been on the order of 1 in 10,000 hours). The operator had only 30 to 90 seconds to respond by tripping the reactor, during which time the operator would have had to detect and diagnose the problem and then take action almost immediately.	0.2
16. Omission of a procedural step in a nuclear control room This HEP is based on a number of different scenarios, which were faced by shift teams in a full-scope nuclear power plant (NPP) simulation in the United States. The shift teams, all of whose members were being recertified as NPP operators, were required to deal with a number of emergency scenarios.	0.03
17. Selection of wrong control (discrimination by label only) This HEP, which was derived from a number of NPP simulator scenarios, was based on 20 incorrect (unrecovered) selections from out of a total of 11,490 opportunities for control selection.	0.002
18. Selection of wrong control (functionally grouped) As above, but this time the HEP is based on only four unrecovered errors out of 27,055 opportunities for error.	0.0002
19. Equipment turned in wrong direction As above, based on the unrecovered errors again, and with equipment that does not violate a population stereotype (i.e., with normal, expected turning conventions).	0.00002

Source: Kirwan (1994).

three requirements: exact a minimal use of blame; ensure freedom from the threat of reprisals, and provide feedback indicating that the system is being used to affect positive changes that can benefit all stakeholders. Accordingly, workers would probably not report the occurrence of accidental damage to an unforgiving management and would discontinue voluntarily offering information on near misses if insights gained from intervention strategies are not shared (CCPS, 1994). It is therefore essential that reporters of information perceive IRSs as error management or learning tools and not as disciplinary instruments.

In addition to these fundamental requirements, two other issues need to be considered. First, consistent with user-centered design principles (Nielsen, 1995), potential users of the system should be involved in its design and implementation as they would with any newly designed (or redesigned) product, although for very large populations of potential users this may not be practical. Second, effective training is critical to the system's usefulness and usability. When human errors, near misses, or incidents occur, the people who are responsible for their reporting and investigation need to be capable of addressing in detail all considerations related to human fallibility, context, and barriers that affect the incident. Thus, training may be required for recognizing that an incident has in fact occurred and for providing full descriptions of the event. Training would also be necessary for ensuring that these data are input correctly into the information system and for verifying that the system's knowledge base adequately supports the representation of this information. Analysts would need training on applying the system's tools, including the use of any modeling frameworks for analyzing causality of human error and on interpreting the results of these application tools. They would also need training on generating summary reports and recommendations and on making modifications to the system's database and inferential tools if the input data imply the need for such adjustments. Access control would also need to be addressed. For each category of system user (e.g., manager, human factors analyst, employee) a *reading authority* (who is allowed to retrieve information from the system) and a *writing authority* (who is allowed to update the database) need to be specified.

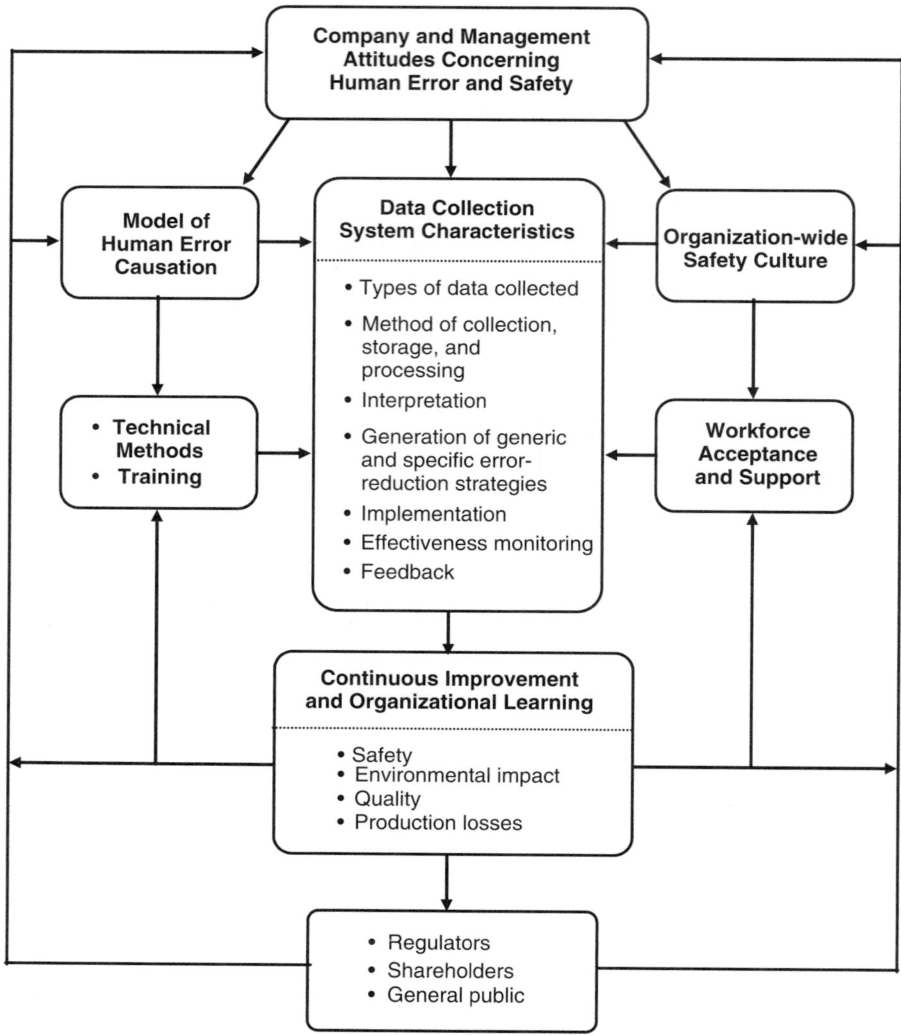

Figure 9 Data collection system for error management. (Adapted from CCPS, 1994. Copyright 1994 by the American Institute of Chemical Engineers, and reproduced by permission of AIChE.)

Data for input into IRSs can be of two types: quantitative data, which lend themselves more easily to coding and classification, and qualitative data in the form of free-text descriptions. Kjellén (2000) has specified the basic requirements for a safety information system in terms of data collection, distribution and presentation of information, and overall information system attributes. To meet data collection requirements, the input data need to be reliable (if the analysis were to be repeated, it should produce similar results), accurate, and provide adequate coverage (e.g., on organizational and human factors/ergonomics issues) needed for exercising efficient control. Foremost in the distribution and presentation of information is the need for relevant information. *Relevance* will depend on how the system will be used. If the objective is to analyze statistics on accidents in order to assess trends, a limited set of data on each accident or near miss would be sufficient and the nature of these data can often be specified in advance. However, suppose that the user is interested in querying the system regarding the degree to which new technology and communication issues have been joint factors in incidents involving errors of omission. In this case, the relevance will be decided by the coverage. Generally, the inability to derive satisfactory answers to specific questions will signal the need for modifications of the system.

In addition to relevance, the information should be comprehensible and easy to survey; otherwise, its use will be restricted to highly trained analysts, prompting high-level management to view the system with suspicion. Overall, the information system should promote involvement between management and employees, thus fostering organizational learning. Finally, the system should be cost-efficient. As in most cost–benefit analyses, costs will be much easier to assess than benefits. Investment, operations, and maintenance costs are relatively straightforward to determine, as are potential benefits resulting from cost reductions associated with the handling, storing, and distribution of various safety-related documents. Benefits associated with reductions in adverse outcomes such as accidents, production delays, and reduced qualities are generally much more difficult to assess.

In searching the database, the user may restrict the search to events that meet criteria defined on one of the standard four (nominal, ordinal, interval, or ratio) scales of measurement (e.g., find all near misses involving workers with less than six months of experience) or to events that include keywords in free-text descriptions (e.g., find all near misses of radiation overexposure that resulted in disruptions to production schedules). Data entered and coded based on standard forms of measurement are relatively easy to manage, whereas data that have been documented and stored in unstructured free-text descriptions may require intelligent software agents for analysis and interpretation. All information searches, however, afford the possibility for type I errors (wanted data that are not found) and type II errors (unwanted data identified as hits).

6.2 Historical Antecedent

An idea related to IRSs, that of the modest suggestion box, has been around for hundreds of years. Both IRSs and suggestion programs are tools designed to capture problem-related data from interested parties regarding the operations of an organization, and are deployed by organizations in order to learn about and improve themselves. One of the earliest suggestion programs, implemented by the British Navy in 1770 (Robinson and Stern, 1998), was motivated by the recognition that persons within the organization should have a way of speaking out without fear of reprisals. The first suggestion box was implemented in the Scottish firm William Denny & Brothers in 1880, and the first U.S. company to implement a companywide suggestion program was National Cash Register in 1892. The suggestion program gained rapid acceptance following World War II, when it was adapted by quality initiatives to meet various objectives, such as safety (Turrell, 2002). At the Toyota Motor Corporation, the suggestion program is part of the Kaizen or "continuous improvement" approach to manufacturing and represents an extremely important feature of the Toyota production system. Implemented in 1951, it took nine years to achieve a 20% participation rate. In 1999, data from the Toyota Motor manufacturing plant in Kentucky indicated that 5048 of 7800 employees contributed 151,327 ideas into the system and that nearly all were implemented (Leech, 2004), resulting in $41.5 million in savings. Unquestionably, the benefits that can potentially be accrued from constructive use of worker feedback can have a powerful impact on an organization's effectiveness and are the basis for the appeal of IRSs in industry.

6.3 The Aviation Safety Reporting System

The Aviation Safety Reporting System (ASRS) was developed in 1976 by the Federal Aviation Administration (FAA) in conjunction with the National Aeronautics and Space Administration (NASA). Many significant improvements in aviation practices have since been attributed to the ASRS, and these improvements have largely accounted for the promotion and development of IRSs in other work domains (Table 10).

The ASRS's mission is threefold: to identify deficiencies and discrepancies in the National Aviation System (NAS), to support policy formulation and planning for the NAS, and to collect human performance data and strengthen research in the aviation domain. All pilots, air traffic controllers, flight attendants, mechanics, ground personnel, and other personnel associated with aviation operations can submit confidential reports if they have been involved in or observed any incident or situation that could have a potential effect on aviation safety. Preaddressed postage-free report forms are available online and are submitted to the ASRS via the U.S. Postal Service. However, unlike other systems, the ASRS presently is not equipped to handle online submissions of information. The ASRS database can be queried by accessing its Internet site (http://asrs.arc.nasa.gov), and is also available on CD-ROM.

ASRS reports are processed in two stages by a group of analysts composed of experienced pilots and air traffic controllers. In the first stage, each report is read by at least two analysts who identify incidents and situations requiring immediate attention. Alerting messages are then drafted and sent to the appropriate group. In the second stage, analysts classify the reports and assess causes of the incident. Their analyses and the information contained in the reports are then incorporated into the ASRS database. The database consists of the narratives submitted by each reporter and coded information that is used for information retrieval and statistical analysis procedures.

Several provisions exist for disseminating ASRS outputs. These include alerting messages that are sent out in response to immediate and hazardous situations, the *CALLBACK* safety bulletin, which is a monthly publication containing excerpts of incident report narratives and added comments (Figure 10), and the ASRS *Directline*, which is published to meet the needs of airline operators and flight crews. In addition, in response to database search requests, ASRS staff communicates with the FAA and the National Transportation Safety Board (NTSB) on an institutional level in support of various tasks, such as accident investigations, and conducts and publishes research related primarily to human performance issues.

Table 10 Attributes of Incident Reporting Systems

Reporting System	Ownership	Regulatory	Mandatory	Voluntary	Anonymous	Confidential	Narrative	Immunity	Threshold	Feedback
Aviation safety reporting system	Federally funded, administered by NASA	Yes	No	Yes	After filed	Yes	Yes	Yes	All nonaccidents	Yes (*Callback*)
Aviation safety airways program	American Airlines	No	No	Yes	No	Yes	Yes	No	All noncrashes	Yes
Airline Pilots Association	FAA in with private pilot association	No	No	Yes	No	Yes	Yes	No	All incidents	Yes
British Airways safety information system										
Air safety report	British Airways	No	Yes	No	No	Yes	Yes	No	Safety-related events	Yes (*Flywise*)
Confidential human factors reporting program	British Airways	No	No	Yes	No	Yes	No, but can expand	No	Human factor data	Yes
Special event search and master analysis	British Airways	Yes	Yes	No	Yes	Yes	N/A	Yes	Monitors fight data records	Yes
Human factors failure analysis classification system	U.S. navy and marines	Yes	Yes	No	No	No	Yes	No	All crashes	Yes
NASA	Federal	Yes	Yes	No	No	Yes	Yes	No	All safety events	Yes
Prevention and recovery information system for monitoring and analysis	Institutional	No	No	Yes	Yes	Yes	Yes	No	Accidents and near misses	Yes
Human factors information systems	Federal with private input (INPO)	Yes	No	Yes	No	Yes	Yes	Yes	Human factor issues related to nuclear safety	Yes
NRC allegations systems process	Federal	Yes	No	Yes	No	Yes	Yes	Yes	All safety concerns	Yes
Diagnostic misadministration reports-regulatory information distribution system	Federal, nuclear regulatory control	Yes	Yes	No	No, patient ID is needed	No	Yes	?	All misadministration	Yes

Source: Barach and small (2000).

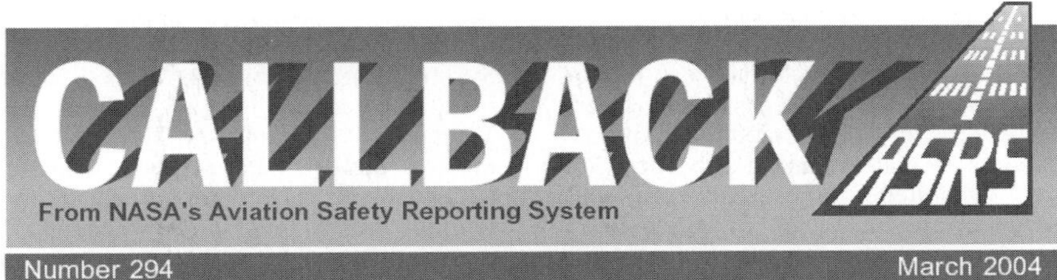

Caution: Clear Weather Ahead

Restricted visibility. Micro Burst. Icing. Embedded cells. SIGMET. **No matter what your affiliation with aviation, certain meteorological terms can evoke a sense of apprehension, even anxiety. But eventually spring arrives, better weather prevails, and forecasts feature a more benign vocabulary. *Clear. Light and variable. High pressure. CAVU.* Welcome words signal that it's time to relax. Up to a point. If** there are any benefits associated with flight operations

landing checklist for the C172N doesn't say "flaps" and because of the lack of normal pattern procedure, I landed with only 20 degrees of flaps. I was wondering why speed did not decrease. I floated and made a long touchdown. I immediately applied brakes but could not stop in the remaining runway. I ran off the runway and the airplane sustained some damage.... I could not understand why I was floating. I could have, and should have, gone around, but somehow I didn't. It was a beautiful day. I was not mentally alert.

Visual Mindset Challenge

Unrestricted visibility led this MD80 Captain to believe that everyone could see what he could see.

■ *We were cleared to taxi down Runway 18R, exit at Taxiway W6 and give way to another aircraft on Taxiway W. As we cleared the runway...tower cleared a commuter for takeoff from Runway 18R. I realized at that time that although we were clear of the runway, part of our aircraft would be over the hold short line in order to keep Taxiway W clear. As another aircraft cleared Taxiway W6 I was unsure if the tower wanted us to go north or south on "W" and by then a B737 was cleared to take off on Runway 18R. After takeoff the B737 pilot called the tower to report that [our aircraft] might want to pull a little further off the runway next time. I mistakenly believed the tower was watching the situation and wouldn't clear anyone for takeoff until we had time to clear. I should have immediately called the tower to let him know we might not be clear of the runway. The great weather created a mindset that all parties could see what was going on....*

Figure 10 Excerpt of an incident report narrative from the *CALLBACK* bulletin.

Many major U.S. airlines have their own internal programs for tracking human errors, especially among pilots, and these programs usually rely on some form of IRS. The Aviation Safety Action Partnership (ASAP), an American Airlines program, is somewhat unique in that it collaborates with the FAA to determine how pilot errors should be handled. The event-review team includes an FAA official, company managers, and representatives from the Allied Pilots Association (the pilots' union). Typically, if the FAA determines that a pilot error has occurred, it issues a citation with penalties ranging from warning letters to license revocation. However, American Airlines pilots who file ASAP reports are assured that the FAA will exact no punishment, or less severe punishment, as long as the error was unintentional. It has been estimated that without ASAP the FAA would be aware of fewer than 1% of the errors of American Airline pilots (Kaye, 1999b).

In contrast to these fairly conventional industry-specific IRSs, United Airlines has adopted a more sophisticated approach to dealing with pilot error.

Its flight operations quality assurance program uses optical recorders in most of its daily flights to reduce pilot error by capturing a pilot's every move electronically (Kaye, 1999c). These disks are later analyzed by computer, and if something wrong, dangerous, or outside normal operating procedure is identified, a team of 10 United Airlines pilots examines the problem and determines a course of action. For example, if a proficiency issue is identified, the team can authorize training for that pilot. The optical recorders could also be connected to operating systems other than the cockpit. For instance, when linked to its maintenance systems, it enabled United Airlines to discover that some internal engine parts were cracking from too much heat. Electronic monitoring presumes relinquishing privacy; thus, acceptance of the program will require that workers acknowledge the possibility that more of their mistakes can become corrected.

6.4 Medical Incident Reporting Systems

The medical industry is currently struggling with what has been termed an epidemic of adverse events stemming from medical error. (This industry defines an adverse event as an injury or death resulting from medical management, and medical error as the failure of a planned action to be completed as intended.) Taking a cue from other complex high-risk industries such as nuclear power and chemical processing, the health care industry is increasingly considering, developing, and deploying IRSs to deal with patient safety concerns and related issues. Despite acknowledgment by the Institute of Medicine that there are an enormous number of preventable injuries to patients (Kohn et al., 1999), implementing IRSs in the health care industry has lagged behind other industries and for good reason. Compared to other industries, the health care industry interacts with the public on a highly personal basis, and protecting reports on near misses, incidents, and accidents is likely to be met with resistance from a public that especially in the United States, is entrenched in a culture of litigation and that is seeing an increasing part of their income being allocated to health care costs (Section 1.1). Collecting reports on medical error that are anonymous may not appease the public—amnesty of unsafe acts that lead to near misses and adverse events would probably not go over very well. Interestingly, medical IRSs have been successful in gaining acceptance in Australia and New Zealand, where legal protection for those who report events has been enforced (Rosenthal et al., 2001).

Not surprisingly, standardization of definitions of errors, near misses, and adverse events in the medical industry, which is fundamental to the industry's ability to gather information, learn about patient safety, and institute intervention strategies, has been difficult to establish. It is also questionable whether a true safety culture that supports IRSs exists in the medical industry. Despite these issues, there have been a number of successful implementations of IRSs in the health care industry, in particular in transfusion medicine, intensive care, anesthesia, occupational medicine, and pharmacy. One example is the Veteran's Administration Patient Safety Reporting System (PSRS), which developed out of an agreement in 2000 between NASA and the Department of Veteran's Affairs (VA). The PSRS allows all VA medical facility staff to report voluntarily any events and concerns related to patient safety confidentially without being subject to reprisals. The types of events that can be reported include close calls (i.e., near misses), unexpected situations involving death, physical or psychological injury of a patient or employee, and lessons learned related to patient safety. Ultimately, the information is made available through alerts, publications such as the Patient Safety Bulletin, and research studies. Although still in use, the PSRS now serves as a complement to a more recent reporting system being operated by the VA that utilizes a root-cause analysis methodology (Section 10.2) for analyzing adverse events and near misses, and provides strategies for decreasing the likelihood of the event's reoccurrence.

Another example of a medical IRS is the Anesthesia Critical Incident Reporting System (CIRS) operated by the Department of Anesthesia at the University of Basel in Switzerland. Using a Web-based interface, contributors worldwide can anonymously report information on incidents in anesthesia practice and review information collected on those incidents. The CIRS IRS defines a *reportable event* as "an event under anesthetic care which has the potential to lead to an undesired outcome if left to progress." Contributors can also report events resulting from team interactions. The design of this system was based on the experiences of the Australian AIMS study, another influential IRS for reporting anesthesia incidents.

As of this writing, the U.S. Senate has proposed a bill that would set up a confidential, voluntary system for reporting medical errors in hospitals without fear of litigation. The goal of the bill, which is pending committee review and action, is to encourage health care providers to report errors so they can be analyzed by patient safety organizations for the purpose of producing better procedures and safety protocols that could improve the quality of care. Notably, in his statement supporting the passage of this bill, Donald Palmisano, the immediate past president of the American Medical Association, stated that "the Aviation Safety Reporting System serves as a successful model for this system."

6.5 Limitations of Incident Reporting Systems

Some IRSs, by virtue of their inability to cope with the vast number of incidents in their databases, have apparently become "victims of their own success" (Johnson, 2002). The Federal Aviation Administration's (FAA's) ASRS and the Food and Drug Administration's MedWatch Reporting System (designed to gather data on regulated, marketed medical products, including prescription drugs, specialized nutritional products, and medical devices) both contain over a half a million incidents. Because their database technologies were not designed to manage this magnitude of data, users who query these systems are having trouble extracting useful information and often fail to identify important cases. This is particularly true of the many IRSs that rely on *relational database* technology. In these systems, each incident is stored as a record and incident identifiers are used to link similar records in response to user queries. Relational database techniques, however, do not adapt well to changes in the nature of incident reporting or in the models of incident causation. Also, different organizations in the same industry tend to classify events differently, which reduces the benefits of drawing on the experiences of IRSs across different organizations. It can also be extremely difficult for people who were not involved in the coding and classification process to develop appropriate queries (Johnson, 2002).

Problems with IRSs can also arise when large numbers of reports on minor incidents are stored. These database systems may then begin to drift toward reporting information on quasi-incidents and precursors of quasi-incidents, which may not necessarily provide the IRS with increased predictive capability (Amalberti, 2001). As stated by Amalberti: "The result is a bloated and costly reporting system with not necessarily better predictability, but where everything can be found; this system is chronically diverted from its true calling (safety) to serve literary or technical causes. When a specific point needs to be proved, it is (always) possible to find confirming elements in these extra-large databases" (p. 113). There is, however, a counterargument to this view: that in the absence of a sufficient number of true incidents, the judicious examination of quasi-incidents may reveal vulnerabilities within the system that would normally be concealed. In this regard, exploiting the potential of quasi-incidents in IRSs suggests the possibility for a proactive capability that may indeed reflect the existence of a highly evolved safety culture.

There is a drift of a different sort that would be advantageous to catalog, but unfortunately is not amenable to capture by the current state-of-the-art in incident reporting. These drifts reflect the various adaptations by an organization's constituents to the external pressures and conflicting goals to which they are continuously subjected (Dekker, 2005). *Drifting into failure* may occur, for instance, when a worker confronts increasingly scarce resources while under pressure to meet higher production standards. If the adaptive responses by the worker to these demands gradually become absorbed into the organization's definition of normal work operations (Section 8), work contexts that may be linked to system failures are unlikely to be reported. The intricate, incremental, and transparent nature of the adaptive processes underlying these drifts is manifest at both the horizontal and vertical levels of an organization. Left unchecked, aggregation of these drifts seals an organization's fate by effectively excluding the possibility for proactive risk management solutions. In the case of the accident in Bhopal (Casey, 1993), these drifts were personified at all levels of the responsible organization. Although IRSs can, in theory, monitor these types of drifts, to do so these systems may need to be driven by new models of organizational dynamics and armed with new levels of intelligence (Dekker, 2005).

A much more fundamental problem with IRSs is the difficulty in assuring anonymity to reporters, especially in smaller organizations. Although most IRSs are confidential, anonymity is more conducive to obtaining disclosures of incidents. Unfortunately, anonymity precludes the possibility for follow-up interviews, which are often necessary for clarifying reported information (Reason, 1997).

Being able to follow up interviews, however, does not always resolve problems contained in reports. Gaps in time between the submission of a report and the elicitation of additional contextual information can result in important details being forgotten or confused, especially if one considers the many forms of bias that can affect eyewitness testimony (Table 11). Biases that can affect reporters of incidents can also affect the teams of people (i.e., analysts) that large-scale IRSs often employ to analyze and classify the reports. For example, there is evidence that persons who have received previous training in human factors are more likely to diagnose human factors issues in incident reports than persons who have not received this type of training (Lekberg, 1997).

Variability among analysts can also derive from the confusion that arises when IRSs employ classification schemes for incidents that are based on detailed taxonomies. Difficulty in discriminating between the various terms in the taxonomy may result in low recall systems, whereby some analysts fail to identify potentially similar incidents. In general, concerns associated with interanalyst reliability stemming from bias and differences in analysts' abilities can impede an organization's ability to learn. More specifically, limitations in analysts' abilities to interpret causal events reduces the capability for organizations to draw important conclusions from incidents, and analyst bias can lead to organizations using IRSs for supporting existing preconceptions concerning human error and safety. As alluded to earlier, training all analysts to the same standard, although a resource-intensive proposition for large organizations, is necessary for minimizing variability associated with inferring causality.

Although there are no software solutions to all these problems, a number of recommendations discussed by Johnson (2002) deserve consideration. For example, for IRSs that are confidential but not anonymous, computer-assisted interviewing techniques can mitigate some of the problems associated with follow-up elicitations of contextual details from reporters. By relying on frames and scripts that are selected in response to information from the user, these techniques can ensure that particular questions are asked in particular situations, thus reducing interanalyst biases stemming from the use of different interview approaches. The success of these approaches, however, depends on ensuring that the dialogue is appropriate for the situation that the reporter is being asked to address. Information-retrieval engines that are the basis for Web search also offer promise, due

Table 11 Forms of Eyewitness Testimony Biases in Reporting

- *Confidence bias:* arises when witnesses unwittingly place the greatest store in their colleagues who express the greatest confidence in their view of an incident. Previous work into eyewitness testimonies and expert judgments has shown that it may be better to place greatest trust in those who do not exhibit this form of overconfidence (Johnson, 2003).
- *Hindsight bias:* arises when witnesses criticize individuals and groups on the basis of information that may not have been available at the time of an incident.
- *Judgment bias:* arises when witnesses perceive the need to reach conclusions about the cause of an incident. The quality of the analysis is less important than the need to make a decision.
- *Political bias:* arises when a judgment or hypothesis from a high-status member commands influence because others respect that status rather than the value of the judgment itself. This can be paraphrased as "pressure from above."
- *Sponsor bias:* arises when a witness testimony can affect indirectly the prosperity or reputation of the organization they manage or for which they are responsible. This can be paraphrased as "pressure from below."
- *Professional bias:* arises when witnesses may be excluded from the society of their colleagues if they submit a report. This can be paraphrased as "pressure from beside."
- *Recognition bias:* arises when witnesses have a limited vocabulary of causal factors. They actively attempt to make any incident "fit" with one of those factors, irrespective of the complexity of the circumstances that characterize the incident.
- *Confirmation bias:* arises when witnesses attempt to make their evidence confirm an initial hypothesis.
- *Frequency bias:* occurs when witnesses become familiar with particular causal factors because they are observed most often. Any subsequent incident is therefore likely to be classified according to one of these common categories irrespective of whether an incident is actually caused by those factors.
- *Recency bias:* occurs when a witness is heavily influenced by previous incidents.
- *Weapon bias:* occurs when witnesses become fixated on the more "sensational" causes of an incident. For example, they may focus on the driver behavior that led to a collision rather than the failure of a safety belt to prevent injury to the driver.

Source: Adapted from Johnson (2002).

to their flexibility in exploiting semantic information about the relationships between terms or phrases that are contained in a user's query and in the reports. In some instances, these search techniques have been integrated with relational databases in order to capitalize on fields previously encoded into the database. However, the integration of these techniques cannot assure users that their queries will find similar incidents (i.e., the precision may be low), or as the large results lists that are typically generated from Web-based searches imply, return almost every report in the system (i.e., recall may be too high). Alternatives to relational databases and information retrieval techniques that have been suggested include conversational case-based reasoning, where the user must answer a number of questions in order to obtain information concerning incidents of interest. The possibility also exists for determining differences among analysts in the patterns of their searches, and thus insights into their potential biases, by tracing their interactions with these systems (Johnson, 2003).

Finally, a very different type of concern with IRSs arises when these systems are used as a basis for quantitative human error applications. In these situations, the voluntary nature of the reporting may invalidate the data that are used for deriving error likelihoods (Thomas and Helmreich, 2002). From a probabilistic risk assessment (Section 5.1) and risk management perspective, this issue can undermine decisions regarding allocating resources for resolving human errors: Which errors do you attempt to remediate if it is unclear how often the errors are occurring?

7 AUTOMATION AND HUMAN ERROR

7.1 Human Factors Considerations in Automation

Innovations in technology will always occur and will bring with them new ways of performing tasks and doing work. Whether the technology completely eliminates the need for the human to perform a task or results in new ways of performing tasks through automation of selective task functions, the human's tasks will probably become reconfigured (Chapter 60). The human is especially vulnerable when adapting to new technology. During this period, knowledge concerning the technology and the impact it may have when integrated into task activities is relatively unsophisticated and biases deriving from previous work routines are still influential.

Automating tasks or system functions by replacing the human's sensing, planning, decision making, or manual activities with computer-based technology often requires making allocation of function decisions—that is, deciding which functions to assign to the human and which to delegate to automatic control (Sharit, 1997). Because these decisions ultimately can have an impact on the propensity for human error, consideration may also need to be given to the level of automation to be incorporated into the system (Parasuraman et al., 2000; Kaber and Endsley, 2004). Higher levels imply that automation will assume greater autonomy in decision making and control. The primary concern with technology-centered systems is that they deprive themselves of the benefits deriving from the human's ability to anticipate, search for, and discern relevant data based on the current context; make generalizations and inferences based on past experience; and modify activities based on changing constraints. Determining the optimal level of automation, however, is a daunting task for the designer.

While levels of automation somewhere between the lowest and highest levels may be the most effective way to exploit the combined capabilities of both the automation and the human, identifying an ideal level of automation is complicated by the need also to account for the consequences of human error and system failures (Moray et al., 2000).

Many of the direct benefits of automation are accompanied by indirect benefits in the form of error reduction. For example, the traffic alert and collision avoidance system in aviation that assesses airspace for nearby traffic and warns the pilot if there is a potential for collision can overcome human sensory limitations, and robotic assembly cells in manufacturing can minimize fatigue-induced human errors. Generally, reducing human physical and cognitive workload enables the human to attend to other higher-level cognitive activities, such as the adoption of strategies for improving system performance. Reckless design strategies, however, that automate functions based solely on technical feasibility can often lead to a number of problems (Bainbridge, 1987). For instance, manual and cognitive skills that are no longer used due to the presence of automation will deteriorate, jeopardizing the system during times when human intervention is required. Situations requiring rapid diagnosis that rely on the human having available or being able quickly to construct an appropriate mental model will thus impose higher WM demands on humans who are no longer actively involved in system operations. The human may also need to allocate significant attention to monitoring the automation, which is a task humans do not perform well. These problems are due largely to the capability for automation to insulate the human from the process, and are best handled through training that emphasizes ample hands-on simulation exercises encompassing varied scenarios. The important lesson learned is that "disinvolvement can create more work rather than less, and produce a greater error potential" (Dekker, 2005, p. 165).

Automation can also be clumsy for the human to interact with, making it difficult to program, monitor, or verify, especially during periods of high workload. A possible consequence of clumsy automation is that it "tunes out small errors and creates opportunities for larger ones" (Weiner, 1985) by virtue of its complex connections to, and control of important systems. Automation has also been associated with *mode errors*, a type of mistake in which the human acts based on the assumption that the system is in a particular mode of operation (either because the available data support this premise or because the human instructed the system to adopt that mode) when in fact it is in a different mode. In these situations, unanticipated consequences may result if the system remains capable of accommodating the human's actions. The tendency for a system to mask its operational mode represents just one of the many ways that automation can disrupt situation awareness.

More generally, when the logic governing the automation is complex and not fully understood by the human, the actions taken by automatic systems may appear confusing. In these situations, the human's tendency for partial matching and biased assessments (Section 3.2) could lead to the use of an inappropriate rule for explaining the behavior of the system—a mistake that in the face of properly functioning automation could have adverse consequences. These forms of human–automation interaction have been examined in detail in flight deck operations in the cockpit and have been termed *automation surprises* (Woods et al., 1997). Training that allows the human to explore the various functions of the automation under a wide range of system or device states can help reduce some of these problems. However, it is also essential that designers work with users of automation to ensure that the user is informed about what the automation is doing and the basis for why it is doing it. In the past, slips and mistakes by flight crews tended to be errors of commission. With automation, errors of omission have become more common, whereby problems are not perceived and corrective interventions are not made in a timely fashion.

Another important consideration is *mistrust* of automation, which can develop when the performance of automatic systems or subsystems is perceived to be unreliable or uncertain (Lee and Moray, 1994). Lee and See (2004) have defined *trust* as an attitude or expectancy regarding the likelihood that someone or something will help the person achieve his or her goal in situations characterized by uncertainty and vulnerability. As these authors have pointed out, many parallels exist between the trust that we gain in other people and the trust we acquire in complex technology, and as in our interactions with other people, we tend to rely on automation we trust and reject automation we do not trust. Mistrust of automation can provide new opportunities for errors, as when the human decides to assume manual control of a system or decision-making responsibilities that may be ill-advised under the current conditions.

Like many decisions people make, the decision to rely on automation can be strongly influenced by emotions. Consequently, even if the automation is performing well, the person's trust in it may become undermined if its responses are not consistent with expectations (Rasmussen et al., 1994). Mistrust of automation can also lead to its disuse, which impedes the development of knowledge concerning the system's capabilities and thus further increases the tendency for mistrust and human error. Overreliance on automation can also lead to errors in those unlikely but still possible circumstances in which the automation is malfunctioning, or when it encounters inputs or situations unanticipated in its design that the human believes it was programmed to handle.

Lee and See (2004) have developed a conceptual model of the processes governing trust and its effect on reliance that is based on a dynamic interaction among the following factors: the human, organizational, cultural, and work contexts; the automation; and the human–automation interface. As a framework for guiding the creation of appropriate trust in automation, their model suggests that the algorithms governing the automation need to be made more transparent to the user, that the interface should provide information regarding the capabilities of the automation in a format that is easily understandable, and that training should address the varieties of situations that can affect the capabilities of the automation.

Organizational and work culture influences also need to be considered. If automation is imposed on workers, especially in the absence of a good rationale regarding its purpose or how human–automation interaction may enhance the work experience or improve the potential for job enrichment, the integrity and meaningfulness of work may become threatened, resulting in work cultures that promote unproductive and possibly dangerous behavioral strategies. Finally, as the work of Cao and Taylor (2004) described below suggests, the adverse effects that interacting with complex technology can have on team communication may require the need to address the concept of *meta-trust*, the trust people have that other people's trust in automation is appropriate (Lee and See, 2004).

7.2 Examples of Human Error in Commercial Aviation

The cockpits of commercial airliners contain numerous automated systems. Central among these systems is the flight management system (FMS). The FMS can be programmed to follow an assigned flight plan route, allowing a plane to navigate itself to a series of checkpoints and providing the estimated time and distance to these checkpoints. It can also determine speed and power settings that optimize fuel consumption, prevent the plane from descending below an altitude restriction, and display navigational information. Working in conjunction with the FMS is the autopilot, which allows the plane to assume and maintain a specific heading, level off at an assigned altitude, or climb or descend at a specific rate, and an auto-throttle system, which sets the throttles for specific airspeeds. In addition, the traffic alert and collision avoidance system notifies pilots about potential collisions with other aircraft and provides instructions on how to avoid that aircraft, the stormscope warns pilots when severe weather lies ahead, and the wind shear system allows pilots to detect wind shear during takeoff and approach to landing. The pilot can also employ an automatic landing system. These automatic systems have the potential to reduce pilot workload significantly and thus enhance safety. However, they can perform so many functions that pilots can lose sight of where they are or what tasks they need to perform. Some examples of these situations are discussed below.

In 1998 the pilots of a Boeing 757 failed to notice that the auto-throttle system had disengaged. The pilots sensed a slight vibration, and after detecting a dangerously low airspeed, the captain correctly attributed the vibration to a loss of lift by the wings. To regain the required airspeed, the throttles were advanced and a slow descent was initiated. However, upon descent the aircraft nearly collided with another plane and both planes needed to be instructed to adopt new courses. The captain claimed that no warning had been provided to alert the crew that the automatic throttle system had disengaged.

In the aftermath of the crash of the American Airlines flight 965 near Cali, Columbia, in 1995, the FAA's human factors team suggested that pilots might not know how to interpret computer system information. The pilots of that flight accepted an offer to land on a different runway, forcing them to rush their descent. In the process, they incorrectly programmed their FMS to direct their plane to Bogota, which was off course by more than 30 miles, and ultimately flew into a 9000-foot mountain.

On a normal approach into Nagoya, Japan, in 1994, the first officer of a China Airlines Airbus A-300 hit the wrong switch on the autopilot, sending the plane into an emergency climb. The throttles increased automatically and the nose pitched up. As the pilots reduced power and tried to push the nose down, the flightdeck computers became even more determined to make the plane climb. The nose rose to 53 degrees, and despite adding full power, the airspeed dropped to 90 mph, which was too slow to maintain the plane in the air. The aircraft crashed tail first into the ground near the runway.

In 1998 a Boeing 737 bound for Denver was instructed by air traffic controllers to descend quickly to 19,000 feet to avoid an oncoming plane. The captain attempted to use the FMS to execute the descent, but the system did not respond quickly enough. Following a second order by air traffic controllers, the captain opted to turn off the FMS and assume manual control, resulting in a near miss with the other plane. The captain attributed his "error" to reliance on automation.

7.3 Adapting to Automation and New Technology

7.3.1 Designer Error

As is the case with user performance of various types of products, the performance of designers will also depend on the operational contexts in which they are working and will be susceptible to many of the same forms of errors (Smith and Geddes, 2003). Working against designers is the increased specialization and heterogeneity of work domains, which is making it exceedingly difficult for them to anticipate the effects on users of introducing automation and new technologies. Nonetheless, errors resulting from user interactions with new technologies are now often attributed to designers. Designer errors could arise from inadequate or incorrect knowledge about the application area (i.e., a failure for designers to anticipate important scenarios) or the inability to anticipate how the product will influence user performance (i.e., insufficient understanding by designers).

In reality, designers' conceptualizations are nothing more than initial hypotheses concerning the

*This section is adapted from Kaye (1999a).

collaborative relationship between their technological product and the human. Accordingly, their beliefs regarding this relationship need to be gradually shaped by data that are based on actual human interaction with these technologies, including the transformations in work experiences that these interactions produce (Dekker, 2005). However, as Dekker notes, in practice the validation and verification studies by designers are usually limited, providing results that may be informative but "hardly about the processes of transformation (different work, new cognitive and coordination demands) and adaptation (novel work strategies, tailoring of the technology) that will determine the sources of a system's success and potential for failure once it has been fielded" (p. 164). In the study on computerized physician order-entry systems discussed in Section 3.4, many of the errors that were identified were probably rooted in constraints of these kinds that were imposed on the design process.

Although designers have a reasonable number of choices available to them that can translate into different technical, social, and emotional experiences for users, like users they themselves are under the influence of sociocultural (Evan and Manion, 2002) and organizational factors (Figure 1). For example, the reward structure of the organization, an emphasis on rapid completion of projects, and the insulation of designers from the consequences of their design decisions can induce designers to give less consideration to factors related to ease of operation and even safety (Perrow, 1983). Although these circumstances would appear to shift the attribution of user errors from designers to management, designer errors and management errors both represent types of *latent errors* that are responsible for creating the preconditions for user errors (Reason, 1990). Perrow (1999) contends that a major deficiency in the design process is the inability of designers and management to appreciate human fallibility by failing to take into account relevant information that could be supplied by human factors and ergonomics specialists. This concern is given serious consideration in user-centered design practices (Nielsen, 1995). However, in some highly technical systems where designers may still be viewing their products as closed systems governed by perfect logic, this issue remains unresolved. The way the FAA has approached this problem has been through recommendations to manufacturers that they make displays and controls easier to use and that they develop a better understanding of pilot vulnerabilities to complex environments. For example, in Boeing's modern air fleets, all controls and throttles provide visual and tactile feedback to pilots—thus the control column that a pilot normally pulls back to initiate climb will move back on its own when a plane is climbing on autopilot.

7.3.2 The Keyhole Property, Task Tailoring, and System Tailoring

Much of our core human factors knowledge concerning human adaptation to new technology in complex systems is derived from experiences in the nuclear power and aviation industries. These industries were forced to address the consequences of imposing on their workers major transformations in the way that system data were presented. In nuclear power control rooms, the banks of hardwired displays were replaced by one or a few computer-based display screens, and in cockpits the analog single-function single displays were replaced by sophisticated software-driven electronic integrated displays. These changes drastically altered the human's visual–spatial landscape and offered a wide variety of schemes for representing, integrating, and customizing data. For those experienced operators who were used to having the entire data world available to them at a glance, adapting to the new technology was far from straightforward. The mental models and strategies that were developed based on having all system state information available simultaneously were not likely to be as successful when applied to these newly designed environments, making these operators more predisposed to errors than were their less experienced counterparts.

In complex work domains such as health care that require the human to cope with a potentially enormous number of different task contexts, anticipating the user's adaptation to new technology can become so difficult for designers that they themselves, like the practitioners who will use their products, can be expected to conform to strategies of minimizing cognitive effort. Instead of designing systems with operational contexts in mind, a cognitively less taxing solution is to identify and make available all possible information that the user may require but to place the burden on the user to search for, extract, or configure the information as the situation demands. These designer strategies are often manifest as computer mediums that exhibit the *keyhole property*, whereby the size of the available viewports (e.g., windows) is very small relative to the number of data displays that potentially could be examined (Woods and Watts, 1997). Unfortunately, this approach to design makes it more likely that the user can "get lost in the large space of possibilities" and makes it difficult to find the right data at the right time as activities change and unfold.

In a study by Cook and Woods (1996) on adapting to new technology in the domain of cardiac anesthesia, physiological monitoring equipment dedicated to cardiothoracic surgery was upgraded from separate devices to a computer system that integrated the functions of four devices onto a single color display. However, the flexibilities that the new technology provided in display options and display customization also created the need for physicians to direct attention to interacting with the patient monitoring system. By virtue of the keyhole property there were now new interface management tasks to contend with. These tasks derived in part from the need to access highly interrelated data serially, thus potentially degrading the accuracy and efficiency of the mental models the physicians required for making patient intervention decisions. New interface management tasks also included the need to declutter displays periodically to avoid obscuring data channels that required monitoring. This requirement resulted from collapsing into a single device the data world previously made available by the multi-instrument configuration.

To cope with these potentially overloading situations, physicians were observed to tailor both the computer-based system (*system tailoring*) and their own cognitive strategies (*task tailoring*). For example, the physicians discovered that the default blood pressure display configuration for the three blood pressures that were routinely displayed was unsuitable—the waveforms and numeric values (derived from digital processing) changed too slowly and eliminated important quantitative information. Rather than exploit the system's flexibility, the physicians simplified the system by constraining the display of data into a fixed spatially dedicated default organization. This required substantial effort, initially to force the preferred display configuration prior to the initiation of a case, then to ensure that this configuration is maintained in the event that the computer system performs automatic window management functions. To tailor their tasks, they planned their interactions with the device to coincide with self-paced periods of low criticality, and developed stereotypical routines to avoid getting lost in the complex menu structures rather than exploiting the system's flexibility. In the face of circumstances incompatible with task-tailoring strategies, which are bound to occur in this complex work domain, the physicians had no choice but to confront the complexity of the device, thus diverting information-processing resources from the patient management function (Cook and Woods, 1996). This irony of automation, whereby the burden of interacting with the technology tends to occur during those situations when the human can least afford to divert attentional resources, is also found in aviation. As noted, automation in cockpits can potentially reduce workload by allowing complete flight paths to be programmed through keyboards. Changes in the flight path, however, require that pilots divert their attention to the numerous keystrokes that need to be input to the keyboard, and these changes tend to occur during takeoff or descent—the phases of flight containing the highest risk and that can least accommodate increases in pilot workload (Strauch, 2002).

Task tailoring reflects a fundamental human adaptive process. Thus, humans should be expected to shape new technology to bridge gaps in their knowledge of the technology and fulfill task demands. The concern with task tailoring is that it can create new cognitive burdens, especially when the human is most vulnerable to demands on attention, and mask the real effects of technology change in terms of its capability for providing new opportunities for human error (Dekker, 2005).

7.3.3 Effects of New Technology on Team Communication

Cao and Taylor (2004) recently examined the effects of introducing a remote surgical robot on communication among the operating room (OR) team members. Understanding the potential for human errors brought about from interactions among team members in the face of this new technology requires closely examining contextual factors such as communication, teamwork, flow of information, work culture, uncertainty, and overload (Figure 1). In their study, a framework referred to as *common ground* (Clark and Schaefer, 1989) was used to analyze communication for two cholecystectomy procedures that were performed by the same surgeon: one using conventional laparoscopic instruments and the other using a robotic surgical system. Common ground represents a person's knowledge or assumptions about what other people in the communication setting know. It can be established through past and present experiences in communicating with particular individuals, the knowledge or assumptions one has about those individuals, and general background information. High levels of common ground would thus be expected to result in more efficient and accurate communication.

In the OR theater, common ground can become influenced by a number of factors. For instance, the surgeon's expectations for responses by team members may depend on the roles (such as nurse, technician, or anesthesiologist) that those persons play. Other factors that can affect the level of common ground include familiarity with team members, which is often undermined in the OR due to rotation of surgical teams, and familiarity with the procedure. When new technology is introduced, all these factors conspire to erode common ground and thus potentially compromise patient safety. Roles may change, people become less familiar with their roles, the procedures for using the new technology are less familiar, and expectations for responses from communication partners becomes more uncertain. Misunderstandings can propagate through team members in unpredictable ways, ultimately leading to new forms of errors.

The introduction of a remote master–slave surgical robot into the OR necessitates a physical barrier, and what Cao and Taylor (2004) observed was that the surgeon, now removed from the surgical site, had to rely almost exclusively on video images from this remote surgical site. Instead of receiving a full range of sensory information from the visual, auditory, haptic, and olfactory senses, the surgeon had to contend with a "restricted field of view and limited depth information from a frequently poor vantage point" (p. 310) and increased uncertainty regarding the status of the remote system. These changes potentially overload the surgeon's visual system and create more opportunities for decision-making errors, due to gaps in the information that is being received. Also, in addition to the need for obtaining information on patient status and the progress of the procedure, the surgeon had to cope with information-processing demands deriving from the need to access information about the status of the robotic manipulator. Thus, to ensure effective coordination of the procedure, the surgeon was now responsible for verbally distributing more information to the OR team members than with conventional laparoscopic surgery.

Overall, significantly more communication within the OR team was observed under robotic surgery conditions than with conventional laparoscopic surgery. Moreover, the communication patterns were haphazard, which increased the team member's uncertainty concerning when information and what information should be distributed or requested and thereby the potential for human error resulting from miscommunication and lack of communication. Use of different terminologies in referring to the robotic system and startup confusion contributed to the lack of common ground. Although training on the use of this technology was provided to these surgical team members, the findings suggested the need for training to attain common ground. This could possibly be achieved through the use of rules or an *information visualization system* that could facilitate the development of a shared mental model among the team members (Stout et al., 1999).

8 HUMAN ERROR IN MAINTENANCE ACTIVITIES

To function effectively, almost all systems require maintenance. Most organizations require both scheduled (preventive) maintenance and unscheduled (active) maintenance. Whereas unscheduled maintenance is required when systems or components fail, preventive maintenance attempts to anticipate failures and thereby minimize system unavailability. Frequent scheduled maintenance can be costly, and organizations often seek to balance these costs against the risks of equipment failures. Lost in this equation, however, is a possible "irony of maintenance"—that an increased frequency in scheduled maintenance may actually increase system risk by providing more opportunities for human interaction with the system (Reason, 1997). This increase in risk is more likely if assembly rather than disassembly operations are called for, as the comparatively fewer constraints associated with assembly operations makes these activities much more susceptible to various errors, such as identifying the wrong component, applying inappropriate force, or omitting an assembly step.

Maintenance environments are notorious for breakdowns in communication, often in the form of implicit assumptions or ambiguity in instructions that go unconfirmed (Reason and Hobbs, 2003). When operations extend over shifts and involve unfamiliar people, these breakdowns in communication can propagate into catastrophic accidents, as was the case in the explosion aboard the Piper Alpha oil and gas platform in the North Sea (Reason and Hobbs, 2003) and the crash of ValueJet flight 592 (Strauch, 2002). Incoming shift workers are particularly vulnerable to errors following commencement of their task activities, especially if maintenance personnel in the outgoing shift conclude their work at an untimely point in the procedure and fail to brief incoming shift workers adequately as to the operational context about to be confronted (Sharit, 1998). In these cases, incoming shift workers are placed in the difficult position of needing to invest considerable attentional resources almost immediately in order to avoid an incident or accident.

Many preventive maintenance activities initially involve searching for flaws prior to applying corrective procedures, and these search processes are often subject to various expectancies that could lead to errors. For example, if faults or flaws are seldom encountered, the likelihood of missing such targets will increase; if they are encountered frequently, properly functioning equipment may be disassembled. Maintenance workers are also often required to work in restricted spaces that are error inducing by virtue of the physical and cognitive constraints that these work conditions impose (Reynolds-Mozrall et al., 2000).

Flawed partnerships between maintenance workers and troubleshooting equipment can also give rise to errors. As with other types of automation or aiding devices, troubleshooting aids can compensate for human limitations and extend human capabilities when designed appropriately. However, these devices are often opaque and may be misused or disregarded (Parasuraman and Riley, 1997), depending on the worker's self-confidence, prior experiences with the aid, and knowledge of co-worker attitudes toward the device. For instance, if the logic underlying the software of an expert troubleshooting system is inaccessible, the user may not trust the recommendations or explanations given by the device (Section 7.1) and therefore choose not to replace a component that the device has identified as faulty.

Errors resulting from interruptions are particularly prevalent in maintenance environments. Interruptions due to the need to assist a co-worker or following the discovery that the work procedure called for the wrong tool or equipment generally require the worker to leave the scene of operations, and the most likely error in these types of situations is an omission. In fact, memory lapses probably constitute the most common errors in maintenance, suggesting the need for incorporating good reminders (Table 12). Reason and Hobbs (2003) emphasize the need for mental readiness and mental rehearsal as ways that maintenance workers can inoculate themselves against errors that could arise from interruptions, time pressure, communication, and unfamiliar situations that may arise.

Written work procedures are pervasive in maintenance operations, and there may be numerous problems with the design of these procedures that can predispose their users to errors (Drury, 1998). Violations of these procedures are also relatively common, and management has been known to consider such violations as causes and contributors of adverse events—a belief that is both simplistic and unrealistic. The assumptions that go into the design of procedures are typically based on normative models of work operations. However, the actual contexts under which real work takes place are often very different from those that the designers of the procedures have envisioned or were willing to acknowledge. To the followers of the procedures, who must negotiate

Table 12 Characteristics of Good Reminders

Universal Criteria

- *Conspicuous*. It should be able to attract the person's attention at the critical time.
- *Contiguous*. It should be located as closely as possible in both time and distance to the to-be-remembered (TBR) task step.
- *Context*. It should provide sufficient information about when and where the TBR step should be carried out.
- *Content*. It should inform the person about what has to be done.
- *Check*. It should allow the person to check off the number of discrete actions or items that should be included in correct performance of the task.

Secondary Criteria

- *Comprehensive*. It should work effectively for a wide range of TBR steps.
- *Compel*. It should (when warranted or possible) block further progress until a necessary prior step has been completed.
- *Confirm*. It should help the person to establish that the necessary steps have been completed. In other words, it should continue to exist and be visible for some time after the performance of the step has passed.
- *Conclude*. It should be readily removable once the time for the action and its checking have passed.

Source: Adapted from Reason (1997).

their tasks while being subjected to limited resources, conflicting goals, and pressures from various sources, the cognitive process of transforming procedures into actions is likely to expose incomplete and ambiguous specifications that at best appear only loosely related to the actual circumstances (Dekker, 2005). A worker's ability to adapt (and thereby violate) these procedures successfully may in fact be lauded by management and garner respect from fellow workers. However, if these violations happen to become linked to accidents, management would probably deny the existence of any unspoken approval of these informal activities and retreat to the official doctrine: Safety can result only if workers follow procedures.

As indicated by Dekker (2005), skilled workers who attempt to adapt procedures to the situation face a double bind: "If rote rule following persists in the face of cues that suggest procedures should be adapted, this may lead to unsafe outcomes. People can get blamed for their inflexibility, their application of rules without sensitivity to context. If adaptations to unanticipated conditions are attempted without complete knowledge of circumstance or certainty of outcome, unsafe results may occur too. In this case, people get blamed for their deviations, their nonadherence" (p. 140). Dekker suggests that organizations monitor (Section 6.5) and understand the basis for the gaps between procedures and practice and develop ways of supporting the cognitive skill of applying procedures successfully across different situations by enhancing workers' judgments of when and how to adapt.

9 ORGANIZATIONAL AND WORK GROUP CULTURES

As with people who live in the same regions or share similar religious beliefs, members of groups within companies, such as maintenance workers, control room operators, or workers involved in transporting goods can also embody beliefs and practices that reflect their shared values. These various work group cultures can be influenced by select individuals who choose to impose their views on subordinates, as well as by the norms that characterize the entire organization. Although cultural factors associated with the organization are generally assumed to be responsible for the norms adopted by work group cultures, in reality organizational culture can have varying degrees of influence on the development and behavior of any particular work group culture.

Strauch (2002) has noted that cultural factors "can make the difference between effective and erroneous performance" (p. 111), and identified two cultural antecedents to error: acceptance of authority and identification with the group. In Hofstede's (1991) analysis of the influence of company cultures on behaviors among individuals, identification with the group was termed *individualism–collectivism*, and acceptance of authority was referred to as *power distance*. Whereas individually oriented people place personal goals ahead of organizational goals, collectivist-oriented persons tend to identify with the company (or work group), so more of the responsibility for errors that they commit would be deflected onto the company. These distinctions thus underlie attitudes that can possibly affect the degree to which workers prepare mentally for potential errors (Section 11).

Power distance refers to the differences in power that employees perceive between themselves and subordinates and superiors. In cultures with high power distance, subordinates are less likely to point out or comment to others about errors committed by superiors as compared to workers in company cultures with low power distance. Although differences in power distance tend to be associated with different countries, this factor can have a considerable impact in ethnically diverse organizations that have become commonplace in many Western societies. Thus, in a Canadian hospital a nurse originating from and trained in the Philippines, a country with a relatively high power distance score, may be less willing than her Canadian counterpart to question a possibly incorrect medication order by a physician. Cultures in which workers tend to defer to authority can also suppress the organization's capability for learning. For example, workers may be less willing to make suggestions that can improve training programs or operational procedures (Section 6.2).

A third cultural factor identified by Hofstede, *uncertainty avoidance*, refers to the willingness or ability to deal with uncertainty. This factor also has implications for human error. For example, workers in cultures that are low in uncertainty avoidance are probably more likely to invoke performance at the knowledge-based level (Section 3.2.5) in response to novel or unanticipated situations for which rules are not available.

Can companies with good cultures be differentiated from those with bad cultures? High-reliability organizations (Section 11) that anticipate errors and encourage safety at the expense of production, that have effective error-reporting mechanisms without fear of reprisals, and that maintain channels of communication across all levels of the company's operations generally reflect good cultures. Questionable hiring practices, poor economic incentives, inflexible and outmoded training programs, the absence of incident reporting systems and meaningful accident investigation mechanisms, managerial instability, and the promotion of atmospheres that discourage communication between superiors and subordinates are likely to produce poor organizational and work group cultures.

Errors associated with maintenance operations can often be traced to organizational culture. This was clearly the case in the crash of ValueJet flight 592 into the Florida Everglades in 1996 just minutes after takeoff. The crash occurred following an intense fire in the airplane's cargo compartment that made its way into the cabin and overcame the crew (Strauch, 2002). Unexpended and unprotected canisters of oxygen generators, which can inadvertently generate oxygen and heat and consequently ignite adjacent materials, had somehow managed to become placed onto the aircraft. Although most of the errors that were uncovered by the investigation were associated with maintenance technicians at SabreTech—the maintenance facility contracted by ValueJet to overhaul several of its aircraft—these errors were attributed to practices at SabreTech that reflected organizational failures. Specifically, the absence of information on the work cards concerning the removal of oxygen canisters from two ValueJet airplanes that were being overhauled led to the failure by maintenance personnel to lock or expend the generators. There was also a lack of communication across shifts concerning the hazards associated with the oxygen generators; although some technicians who had removed the canisters from the other aircraft knew of the hazards, others did not. In addition, procedures for briefing incoming and outgoing shift workers concerning hazardous materials and for tracking tasks performed during shifts were not in place. Finally, parts needed to secure the generators were unavailable, and none of the workers in shipping and receiving, who were ultimately responsible for placing the canisters on the airplane, was aware of the hazards.

Relevant to this discussion was the finding that the majority of the technicians that removed oxygen canisters from ValueJet airplanes as part of the overhaul of these aircraft were not SabreTech personnel but contractor personnel. In the absence of an adequately informed organizational culture, it comes as no surprise that management would be oblivious to the implications of outsourcing on worker communication and task performance. Further arguments concerning the importance of organizational culture for system safety can be found in Reason (1997) and Vicente (2004).

9.1 The *Columbia* Accident

The *Columbia* space shuttle accident in 2003 exposed a failed organizational culture. The physical cause of the accident was a breach in the thermal protection system on the leading edge of *Columbia*'s left wing about 82 seconds after the launch. This breach was caused by a piece of insulating foam that separated from the external tank in an area where the orbiter attaches to the external tank. The *Columbia* Accident Investigation Board's (2003) report stated that "NASA's organizational culture had as much to do with this accident as foam did," that "only significant structural changes to NASA's organizational culture will enable it to succeed," and that NASA's current organization "has not demonstrated the characteristics of a learning organization" (p. 12).

To some extent NASA's culture was shaped by compromises with political administrations that were required to gain approval for the space shuttle program. These compromises imposed competing budgetary and mission requirements that resulted in a "remarkably capable and resilient vehicle" but one that was "less than optimal for manned flights" and "that never met any of its original requirements for reliability, cost, ease of turnaround, maintainability, or, regrettably, safety" (p. 11). The organizational failures are almost too numerous to document: unwillingness to trade off scheduling and production pressures for safety; shifting management systems and a lack of integrated management across program elements; reliance on past success as a basis for engineering practice rather than on dependable engineering data and rigorous testing; the existence of organizational barriers that compromised communication of critical safety information and discouraged differences of opinion; and the emergence of an informal command and decision-making apparatus that operated outside the organization's norms. According to the *Columbia* Accident Investigation Board, deficiencies in communication, both up and down the shuttle program's hierarchy, were a foundation for the *Columbia* accident.

These failures were largely responsible for missed opportunities, blocked or ineffective communication, and flawed analysis by management during *Columbia*'s final flight that hindered the possibility of a challenging but conceivable rescue of the crew by launching the *Atlantis*, another space shuttle craft, to rendezvous with *Columbia*. The accident investigation board concluded: "Some Space Shuttle Program managers failed to fulfill the implicit contract to do whatever is possible to ensure the safety of the crew. In fact, their management techniques unknowingly imposed barriers that kept at bay both engineering concerns and dissenting views, and ultimately helped create 'blind spots' that prevented them from seeing the danger the foam strike posed" (p. 170). Essentially, the position adopted by managers concerning whether the debris strike created a safety-of-flight issue placed the burden on engineers to prove that the system was unsafe.

Numerous deficiencies were also found with the Problem Reporting and Corrective Action database, a critical information system that provided data on any nonconformances. In addition to being too time consuming and cumbersome, it was also incomplete. For example, only foam strikes that were considered in-flight anomalies were added to this database, which masked the extent of this problem.

Finally, what is particularly disturbing was the failure of the shuttle program to detect the foam trend and appreciate the danger that it presented. Shuttle managers discarded warning signs from previous foam strikes and normalized their occurrences. In so doing, they desensitized the program to the dangers of foam strikes and compromised the flight readiness process. Many workers at NASA knew of the problem. However, in the absence of an effective mechanism for communicating these "incidents" (Section 6) proactive approaches for identifying and mitigating risks were unlikely to be in place. In particular, a proactive perspective to risk identification and management could have resulted in a better understanding of the risk of thermal protection damage from foam strikes, tests being performed on the resilience of the reinforced carbon–carbon panels, and either the elimination of external tank foam loss or its mitigation through the use of redundant layers of protection.

10 INVESTIGATING HUMAN ERROR IN ACCIDENTS AND INCIDENTS

10.1 Causality and Hindsight Bias

Investigations of human error are generally performed as part of accident and incident investigations (Chapter 41). In conducting these investigations, the most fundamental issue is the attribution of causality to incident and accident events. Currently, there are a variety of techniques that investigators can choose from to assist them in performing causal analysis (Johnson, 2003).

A related issue concerns the level of detail required for establishing causality (Senders and Moray, 1991). At one level of analysis a cause can be an interruption; at a different level of analysis the cause can be attributed to a set of competing neural activation patterns that result in an action slip. Dilemmas regarding the appropriate level of causal analysis are usually resolved by considering the requirements of the investigative analysis. Generally, analysts can be expected to employ heuristics such as satisficing (Section 3.2), whereby decisions and judgments are made that appear good enough for the purposes of the investigation. Investigators also need to be aware of the possible cognitive biases that reporters of incidents and accidents may be harboring (Table 11). These same biases can also play a part in how witnesses attribute blame and thus in how they perceive relationships between causes and effects.

What makes determining causes of accidents especially problematic for investigators is that they typically work with discrete fragments of information derived from decomposing continuous and interacting sequences of events. This ultimately leads to various distortions in the true occurrence of events (Woods, 1993). A further complication is *hindsight bias*, which derives from the tendency to judge the quality of a process based on whether positive or negative outcomes ensued (Fischhoff, 1975; Christoffersen and Woods, 1999). Because accident investigators usually have knowledge about negative outcomes, through hindsight they can look back and identify all the failed behaviors that are consistent with these outcomes (Section 1.1). It is then highly probable that a causal sequence offering a crisp explanation for the incident will unfold to the investigator. A foray through Casey's (1993) reconstructed accounts of a number of high-profile accidents attributed to human error would probably transform many in the lay public into hindsight experts.

Although hindsight bias can assume a number of forms (Dekker, 2001), they all derive from the tendency to treat actions in isolation and thus distort the context in which the actions took place. The pervasiveness of the hindsight bias has led Dekker (2005) to suggest the intriguing possibility that it may actually be serving an adaptive function—that is, the hindsight bias is not so much about explaining what happened as it is about future survival, which would necessarily require the decontextualization of past failures into a "linear series of binary choices." The true bias thus derives from the belief that the oversimplification of rich contexts into a series of clearly defined choices will increase the likelihood of coping with complexity successfully in the future. However, in reality, by obstructing efforts at establishing cause and effect, the hindsight bias actually jeopardizes the ability to learn from accidents (Woods et al., 1994) and thus the ability to predict or prevent future failures.

10.2 Methods for Investigating Human Error

Investigations of human error can be pursued using either informal approaches or methods that are much more systematic or specialized. Strauch's (2002) approach, which reflects a relatively broad and informal perspective to this problem, emphasizes antecedent factors (e.g., equipment, operator, maintenance, and cultural factors), data collection and analysis issues, and factors such as situation awareness and automation, all of which are interwoven with case studies. The more specialized methods for investigating human error are generally referred to as accident or incident analysis techniques, although in some cases these tools can also be used to assess the potential risks associated with systems or work processes. Examples of techniques used exclusively for investigating accidents include change analysis (Kepner and Tregoe, 1981) and the sequentially timed events plotting procedure (Hendrick and Benner, 1987).

Change analysis techniques are based on the well-documented general relationship between change and increased risk. These techniques make use of accident-free reference bases to identify systematically changes or differences associated with the accident or incident situation. A simple worksheet is usually all that is required for exploring potential changes contributory to adverse outcomes. Listed in the rows of the first column of this worksheet are factors that are stated in terms of questions regarding who, where, what, when, how, and why with respect to task factors, working conditions, initiating events, and management control factors. The next columns, respectively, address each of these factors in terms of the present (accident or incident) situation, prior (accident-free) situation, a comparison of these two situations in order to identify changes or differences, and a list of the differences. Finally, differences are analyzed for their effect on the accident or incident in terms of both their independent and interactive contributions. When used in conjunction with TA or CTA methods, change analysis can be applied both retrospectively to facilitate the identification of underlying causes of human error, and proactively to predict adverse consequences by investigating potential problems associated with proposed changes in normal or stable functioning systems.

Another well-known technique, the management oversight and risk tree (MORT), relies on a logic diagram for investigating the various factors contributing to an accident (Johnson, 1980). Factors considered by MORT include lines of responsibility, barriers to unwanted energy sources, and management factors. By reasoning backward through a sequence of contributory factors, responding "yes" and "no" to questions along the way, and through the availability of accompanying text that aids the analyst in judging whether a factor is adequate or less than adequate, MORT assists the analyst in detecting omissions, oversights, or defects, and may be especially useful for identifying organizational root causes (Gertman and Blackman, 1994).

The accident investigation method that has recently been given the most attention is *root-cause analysis* (RCA). This method usually refers to the formal application of a root cause decision tree diagram for investigating why a particular event occurred. In the SOURCE (seeking out the underlying root causes of events) method of performing RCA (ABS Consulting Group, 1998), *root causes* are defined as "the most basic causes that can reasonably be identified, which management has control to fix and for which effective recommendations for preventing recurrence can be generated" (p. 2). The major steps in the SOURCE RCA process are illustrated in Figure 11. The first step, data collection, represents the most time-consuming step of the process. Although data collection is assumed to occur throughout the analysis, data from relatively unstable sources, such as people and certain types of physical data, need to be collected as soon as possible. The interviewing technique employed by the investigator will probably be the most critical factor in determining the effectiveness of data gathering (Strauch, 2002).

The next step, *causal factor charting*, utilizes a causal factor chart to describe in sequence the events leading up to and following the incident, as well as the conditions surrounding these events. A skeletal causal chart is generated based on the initial data collected; this chart is then modified progressively as data accumulate. Other elements in addition to those illustrated in Figure 11 can be incorporated into causal factor charts, including presumptive events, presumptive conditions, presumptive causal factors, and items of note. A number of principles, guidelines, and procedures are offered for supporting the causal factor charting process (ABS Consulting Group, 1998).

The third step of this process involves the use of a tree diagram called a *root-cause map* to identify the underlying reasons for each causal factor identified during causal factor charting. For each causal factor the analyst must determine which top-level node in the map is most applicable. Based on this decision, the analyst then works down through the lower

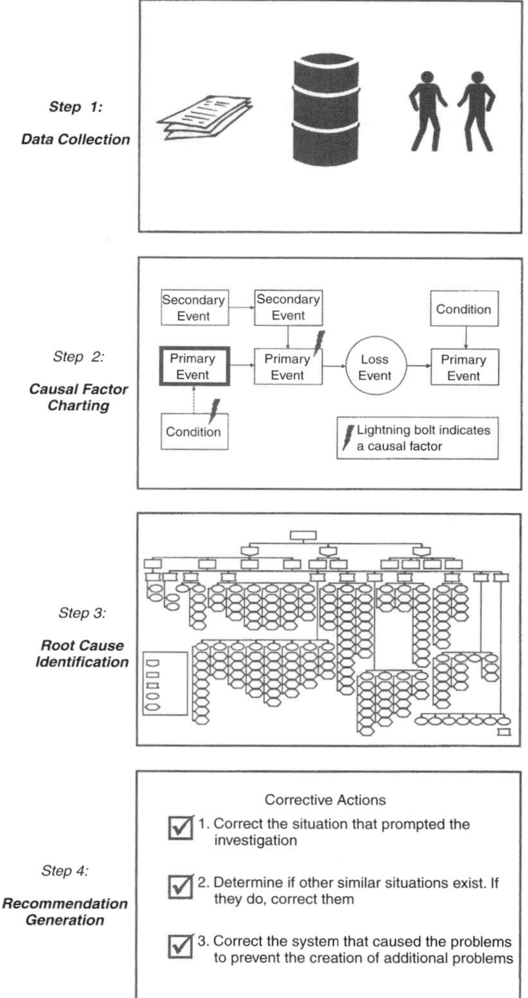

Figure 11 Root-cause analysis method. (From ABS Consulting Group, 1998.)

(more specific) levels of the map, selecting the most applicable node at each level. The three upper-level nodes of the map correspond to equipment failures, personnel failures, and other failures; however, only the first two categories are analyzed for root causes. At the second level these three nodes are subdivided into 10 problem category nodes. Examples of these categories are equipment design problem, equipment misuse, contract employee, natural phenomena, and sabotage or horseplay. The third level of the map consists of nodes corresponding to 11 major root-cause categories; examples of these categories include procedures, human factors engineering, training, and communications. In the transition from the second to the third level, the map allows for a number of points of intersection between equipment failures and personnel failures, thus allowing all failures to be traced back to some type of human error. At the fourth level of the map these categories become subdivided into near root causes, which in turn are subdivided at the bottom level into a detailed set of root causes. To aid the investigator in using the root-cause map, examples for each node are provided in terms of typical issues and typical recommendations.

At this point in the process, a root-cause summary table can be generated that links each causal factor in the chart with one or more paths in the map that terminate at root causes and to recommendations that address each of these root causes. These tables then form the basis for investigative reports that comprise the final step of RCA.

The availability of a systematic method for performing incident and accident investigations within a high-risk organization will increase an organization's potential for learning, improvement, and development of positive work cultures. However, as with IRSs these benefits are anticipated only when these investigations are not used as a basis for reprisals and when workers are informed about and involved in the investigative process.

11 TOWARD MINIMIZATION OF HUMAN ERROR AND THE CREATION OF SAFE SYSTEMS

Human error is a complex phenomenon. Recent evidence from neuroimaging studies has linked an *error negativity*, an event-related brain potential probably originating from the anterior cingulate cortex, to the detection by individuals of action slips, errors of choice, and other errors (Nieuwenhuis et al., 2001; Holroyd and Coles, 2002), possibly signifying the existence of a neurophysiological basis for a preconscious action-monitoring system. However, suggestions that these kinds of findings may offer possibilities for predicting human errors in real-time operations (Parasuraman, 2003) are probably overstated. Event-related brain potentials may provide insight into attentional preparedness and awareness of response conflicts, but the complex interplay of factors responsible for human error (Section 3.1) takes these discoveries out of contention as meaningful explanatory devices.

Although managers often speak in terms of the need for eliminating human error, this goal is neither desirable nor reasonable. The benefits that derive from the realization that errors have been committed should not be readily dismissed; they play a critical role in human adaptability, creativity, and the manifestation of expertise. The elimination of human error is also inconceivable if only because human fallibility will always exist. Tampering with human fallibility, for example by increasing the capabilities of working memory and attention, would probably facilitate the design and production of new and more complex systems, and ultimately, new and unanticipated opportunities for human error. More realistically, the natural evolution of knowledge and society should translate into the emergence of new systems, new forms of interaction among people and devices, and new sociopolitical and organizational cultures that will, in turn, provide new opportunities for enabling human fallibility.

However, in no way should these suppositions detract from the goal of human error reduction, especially in complex high-risk systems. As a start, system hardware and software need to be made more reliable, better partnerships between humans and automation need to be established, barriers that are effective in providing detection and absorption of errors without adversely affecting contextual and cognitive constraints need to be put in place, and incident-reporting systems that enable organizations to learn and anticipate, especially when errors become less frequent and thus deprive analysts with the opportunity for preparing and coping with their effects, need to become more ubiquitous.

Organizations also need to consider the impact that various economic incentives may have on shaping work behaviors (Moray, 2000) and the adoption of strategies and processes for implementing features that have come to be associated with *high-reliability organizations* (HROs) (Rochlin et al., 1987; Roberts, 1990). By incorporating fundamental characteristics of HROs, particularly the development of cultures of reliability that anticipate and plan for unexpected events, try to monitor and understand the gap between work procedures and practice (Dekker, 2005), and place value in organizational learning, the adverse consequences of interactive complexity and tight coupling that Perrow's theory predicts (Section 3.3) can largely be countered.

In addition, methods for describing work contexts and for determining and assessing the perceptions and assessments that workers make in response to these contexts, as well as rigorous TA and CTA techniques for determining the possible ways that fallible humans can become ensnared by these situations, need to be investigated, implemented, and continuously evaluated in order to strengthen the predictive capabilities of human error models. These methods also need to be integrated into the conceptual, development, and testing stages of the design process to better inform designers (of both products and work procedures) about the potential effects of design decisions, thus bridging the gap between the knowledge and intentions of the designer and the needs and goals of the user.

Problems created by poor designs and management policies traditionally have been dumped on training departments (CCPS, 1994). Instead of using training to compensate for these problems, it should be given a proactive role in minimizing, detecting, and recovering errors. This can be achieved through innovative training methods that emphasize management of task activities under uncertainty and time constraints; integrate user-centered design principles for establishing performance support needs (such as the need for planning aids); give consideration to the kinds of cues that are necessary for developing situation awareness (Endsley et al., 2003) and for interpreting common-cause and common-mode system failures; and utilize simulation methods effectively for providing extensive exposure to a wide variety of contexts. By including provisions in training for imparting mental preparedness, people will be better able to anticipate the anomalies they might encounter and the errors they might make, and to develop error-detection skills (Reason and Hobbs, 2003).

Although worker selection (Chapter 17) is a potentially explosive issue, it can be used to exploit individual variability in behavioral tendencies and cognitive capabilities and thus provide better human–system fits (Damos, 1995). Bierly and Spender (1995) have documented the extraordinary safety record of the U.S nuclear navy and attributed it in part to a culture that insisted on careful selection of people who were highly intelligent, very motivated, and who were then thoroughly trained and held personally accountable for their tasks. These characteristics created the work culture context for communications that could: be carried out under conditions of high risk and high stress; flow rapidly either top-down or bottom-up through the chain of command; and encompass information about mistakes, whether technical, operational, or administrative, without fear of reprisals.

However, perhaps the greatest challenge in reducing human error is managing these error-management processes (Reason and Hobbs, 2003)—defense strategies need to be aggregated coherently (Amalberti, 2001). Too often these types of error-reduction enterprises, innovative as they may be, remain isolated or hidden from each other. This needs to change—all programs that can influence error management need to be managed as a unified synergistic entity.

REFERENCES

ABS Consulting Group (1998), *Root Cause Analysis Handbook: A Guide to Effective Incident Investigation*, Government Institutes Division, Knoxville, TN.

Amalberti, R. (2001), "The Paradoxes of Almost Totally Safe Transportation System," *Safety Science*, Vol. 3, pp. 109–126.

Bainbridge, L. (1987), "Ironies of Automation," in *New Technology and Human Error*, J. Rasmussen, K. Duncan, and J. Leplat, Eds., Wiley, New York, pp. 273–276.

Barach, P., and Small, S. (2000), "Reporting and Preventing Medical Mishaps: Lessons from Non-medical Near Miss Reporting Systems," *British Medical Journal*, Vol. 320, pp. 759–763.

Bierly, P. E., and Spender, J. C. (1995), "Culture and High Reliability Organizations: The Case of the Nuclear Submarine," *Journal of Management*, Vol. 21, pp. 639–656.

Cao, C. G. L., and Milgram, P. (2000), "Disorientation in Minimal Access Surgery: A Case Study," in *Proceedings of the IEA 2000/HFES 2000 Congress*, San Diego, CA, Vol. 4, pp. 169–172.

Cao, C. G. L., and Taylor, H. (2004), "Effects of New Technology on the Operating Room Team," in *Work with Computing Systems, 2004*, H. M. Khalid, M. G. Helander, and A. W. Yeo, Eds., Damai Sciences, Kuala Lumpur, Malaysia, pp. 309–312.

Casey, S. (1993), *Set Phasers on Stun and Other True Tales of Design, Technology, and Human Error*, Aegean Park Press, Santa Barbara, CA.

CCPS (1992), *Guidelines for Hazard Evaluation Procedures, with Worked Examples*, 2nd ed., Center for Chemical Process Safety, American Institute of Chemical Engineers, New York.

CCPS (1994), *Guidelines for Preventing Human Error in Process Safety*, Center for Chemical Process Safety, American Institute of Chemical Engineers, New York.

Christoffersen, K., and Woods, D. D. (1999), "How Complex Human–Machine Systems Fail: Putting 'Human Error' in Context," in *The Occupational Ergonomics Handbook*, W. Karwowski and W. S. Marras, Eds., CRC Press, Boca Raton, FL, pp. 585–600.

Clark, H. H., and Schaefer, E. F. (1989), "Contributing to Discourse," *Cognitive Science*, Vol. 13, pp. 259–294.

Columbia Accident Investigation Board (2003), *Report Volume 1*, U.S. Government Printing Office, Washington, DC.

Cook, R. I., and Woods, D. D. (1996), "Adapting to New Technology in the Operating Room," *Human Factors*, Vol. 38, pp. 593–611.

Cook, R. I., Render, M., and Woods, D. D. (2000), "Gaps in the Continuity of Care and Progress on Patient Safety," *British Medical Journal*, Vol. 320, pp. 791–794.

Cullen, D. J., Bates, D. W., Small, S. D., Cooper, J. B., Nemeskal, A. R., and Leape, L. L. (1995), "The Incident Reporting System Does Not Detect Adverse Drug Events: A Problem for Quality Improvement," *Joint Commission Journal on Quality Improvement*, Vol. 21, pp. 541–548.

Damos, D. (1995), "Issues in Pilot Selection," in *Proceedings of the 8th International Symposium on Aviation Psychology*, Ohio State University, Columbus, OH, pp. 1365–1368.

Dekker, S. W. A. (2001), "The Disembodiment of Data in the Analysis of Human Factors Accidents," *Human Factors and Aerospace Safety*, Vol. 1, pp. 39–58.

Dekker, S. W. A. (2005), *Ten Questions About Human Error: A New View of Human Factors and System Safety*, Lawrence Erlbaum Associates, Mahwah, NJ.

Dey, A. K. (2001), "Understanding and Using Context," *Personal and Ubiquitous Computing*, Vol. 5, pp. 4–7.

Dhillon, B. S., and Singh, C. (1981), *Engineering Reliability: New Technologies and Applications*, Wiley, New York.

Drury, C. G. (1998), "Human Factors in Aviation Maintenance," in *Handbook of Aviation Human Factors*, D. J. Garland, J. A. Wise, and V. D. Hopkin, Eds., Lawrence Erlbaum Associates, Mahwah, NJ, pp. 591–606.

Embrey, D. E., Humphreys, P., Rosa, E. A., Kirwan, B., and Rea, K. (1984), *SLIM–MAUD: An Approach to Assessing Human Error Probabilities Using Structured Expert Judgment*, NUREG/CR-3518, U.S. Nuclear Regulatory Commission, Washington, DC.

Endsley, M. R. (1995), "Toward a Theory of Situation Awareness," *Human Factors*, Vol. 37, pp. 32–64.

Endsley, M. R., Bolté, B., and Jones, D. G. (2003), *Designing for Situation Awareness: An Approach to User-Centred Design*, CRC Press, Boca Raton, FL.

Evan, W. M., and Manion, M. (2002), *Minding the Machines: Preventing Technological Disasters*, Prentice-Hall, Upper Saddle River, NJ.

Fischhoff, B. (1975), "Hindsight–Foresight: The Effect of Outcome Knowledge on Judgment Under Uncertainty," *Journal of Experimental Psychology: Human Perception and Performance*, Vol. 1, pp. 278–299.

Fraser, J. M., Smith, P. J., and Smith, J. W. (1992), "A Catalog of Errors," *International Journal of Man–Machine Systems*, Vol. 37, pp. 265–307.

Gertman, D. I., and Blackman, H. S. (1994), *Human Reliability and Safety Analysis Data Handbook*, Wiley, New York.

Grosjean, V., and Terrier, P. (1999), "Temporal Awareness: Pivotal in Performance?" *Ergonomics*, Vol. 42, pp. 443–456.

Hahn, A. H., and deVries J. A., II (1991), "Identification of Human Errors of Commission Using Sneak Analysis," in *Proceedings of the Human Factors Society 35th Annual Meeting*, pp. 1080–1084.

Helander, M. G. (1997), "The Human Factors Profession," in *Handbook of Human Factors and Ergonomics*, 2nd ed., G. Salvendy, Ed., Wiley, New York, pp. 3–16.

Hendrick, K., and Benner, L., Jr. (1987), *Investigating Accidents with STEP*, Marcel Dekker, New York.

Hofstede, G. (1991), *Cultures and Organizations: Software of the Mind*, McGraw-Hill, New York.

Hollnagel, E. (1993), *Human Reliability Analysis: Context and Control*, Academic Press, London.

Hollnagel, E. (1998), *Cognitive Reliability and Error Analysis Method*, Elsevier Science, New York.

Hollnagel, E., Ed. (2003), *Handbook of Cognitive Task Design*, Lawrence Erlbaum Associates, Mahwah, NJ.

Holroyd, C. B., and Coles, M. G. H. (2002), "The Neural Basis of Human Error Processing: Reinforcement Learning, Dopamine, and the Error-Related Negativity," *Psychological Review*, Vol. 109, pp. 679–709.

Johnson, W. G. (1980), *MORT Safety Assurance Systems*, Marcel Dekker, New York.

Johnson, C. (2002), "Software Tools to Support Incident Reporting in Safety-Critical Systems," *Safety Science*, Vol. 40, pp. 765–780.

Johnson, C. (2003), *Failure in Safety-Critical Systems: A Handbook of Accident and Incident Reporting*, University of Glasgow Press, Glasgow.

Kaber, D. B., and Endsley, M. R. (2004), "The Effects of Level of Automation and Adaptive Automation on Human Performance, Situation Awareness and Workload in a Dynamic Control Task," *Theoretical Issues in Ergonomics Science*, Vol. 4, pp. 113–153.

Kapur, K. C., and Lamberson, L. R. (1977), *Reliability in Engineering and Design*, Wiley, New York.

Kaye, K. (1999a), "Automated Flying Harbors Hidden Perils," *South Florida Sun-Sentinel*, September 27.

Kaye, K. (1999b), "United Has Eye in the Sky: Optical Recorders Check Crews," *South Florida Sun-Sentinel*, September 27.

Kaye, K. (1999c), "Program Urges Error Reporting, Mitigates Penalty," *South Florida Sun-Sentinel*, September 27.

Kepner, C. H., and Tregoe, B. B. (1981), *The New Rational Manager*, Kepner-Tregoe Inc., Princeton, NJ.

Kirwan, B. (1994), *A Guide to Practical Human Reliability Assessment*, Taylor & Francis, London.

Kirwan, B. (1999), "Some Developments in Human Reliability Assessment," in *The Occupational Ergonomics Handbook*, W. Karwowski and W. S. Marras, Eds., CRC Press, Boca Raton, FL, pp. 643–666.

Kirwan, B., and Ainsworth, L. K. (1992), *Guide to Task Analysis*, Taylor & Francis, London.

Kirwan, B., Martin, B. R., Rycraft, H., and Smith, A. (1990), "Human Error Data Collection and Data Generation," *International Journal of Quality and Reliability Management*, Vol. 7.4, pp. 34–66.

Kjellén, U. (2000), *Prevention of Accidents Through Experience Feedback*, Taylor & Francis, London.

Kohn, L. T., Corrigan, J. M., and Donaldson, M. S., Eds. (1999), *To Err Is Human: Building a Safer Health System*, National Academy Press, Washington, DC.

Koppel, R., Metlay, J. P., Cohen, A., Abaluck, B., Localio, A. R., Kimmel, S., and Strom, B. L. (2005), "Role of Computerized Physician Order Entry Systems in Facilitating Medication Errors," *Journal of the American Medical Association*, Vol. 293, pp. 1197–1203.

Kumamoto, H., and Henley, E. J. (1996), *Probabilistic Risk Assessment and Management for Engineers and Scientists*, 2nd ed., IEEE Press, Piscataway, NJ.

Leape, L. L., Brennan, T. A., Laird, N. M., Lawthers, A. G., Localio, A. R., Barnes, B. A., Hebert, L., Newhouse, J. P., Weiler, P. C., and Hiatt, H. H. (1991), "The Nature of Adverse Events in Hospitalized Patients: Results from the Harvard Medical Practice Study II," *New England Journal of Medicine*, Vol. 324, pp. 377–384.

Lee, J. D., and Moray, N. (1994), "Trust, Self-Confidence, and Operators' Adaptation to Automation," *International Journal of Human–Computer Studies*, Vol. 40, pp. 153–184.

Lee, J. D., and See, K. A. (2004), "Trust in Automation: Designing for Appropriate Reliance," *Human Factors*, Vol. 46, pp. 50–80.

Leech, D. S. (2004), "Learning in a Lean System," Defense Acquisition University Publication, Department of Defense, retrieved May 12, 2004, from http://acc.dau.mil/.

Lekberg, A. (1997), "Different Approaches to Accident Investigation: How the Analyst Makes the Difference," in *Proceedings of the 15th International Systems Safety Conference*, Sterling, VA, International Systems Safety Society, pp. 178–193.

Levy, J., Gopher, D., and Donchin, Y. (2002), "An Analysis of Work Activity in the Operating Room: Applying Psychological Theory to Lower the Likelihood of Human Error," in *Proceedings of the Human Factors and Ergonomics Society 46th Annual Meeting*, Human Factors and Ergonomics Society, Santa Monica, CA, pp. 1457–1461.

Luczak, H. (1997), "Task Analysis," in *Handbook of Human Factors and Ergonomics*, 2nd ed., G. Salvendy, Ed., Wiley, New York, pp. 340–416.

Moray, N. (2000), "Culture, Politics and Ergonomics," *Ergonomics*, Vol. 43, pp. 858–868.

Moray, N., Inagaki, T., and Itoh, M. (2000), "Adaptive Automation, Trust, and Self-Confidence in Fault Management of Time-Critical Tasks," *Journal of Experimental Psychology: Applied*, Vol. 6, pp. 44–58.

Nielsen, J. (1995), *Usability Engineering*, Academic Press, San Diego, CA.

Nieuwenhuis, S. N., Ridderinkhof, K. R., Blom, J., Band, G. P. H., and Kok, A. (2001), "Error Related Brain Potentials Are Differentially Related to Awareness of Response Errors: Evidence from an Antisaccade Task," *Psychophysiology*, Vol. 38, pp. 752–760.

Norman, D. A. (1981), "Categorization of Action Slips," *Psychological Review*, Vol. 88, pp. 1–15.

Parasuraman, R. (2003), "Neuroergonomics: Research and Practice," *Theoretical Issues in Ergonomics Science*, Vol. 4, pp. 5–20.

Parasuraman, R., and Riley, V. (1997), "Humans and Automation: Use, Misuse, Disuse, and Abuse," *Human Factors*, Vol. 39, pp. 230–253.

Parasuraman, R., Sheridan, T. B., and Wickens, C. D. (2000), "A Model for Types and Levels of Human Interaction with Automation," *IEEE Transactions on Systems Man, and Cybernetics, Part A: Systems and Humans*, Vol. 30, pp. 276–297.

Perrow, C. (1983), "The Organizational Context of Human Factors Engineering," *Administrative Science Quarterly*, Vol. 27, pp. 521–541.

Perrow, C. (1999), *Normal Accidents: Living with High-Risk Technologies*, Princeton University Press, Princeton, NJ.

Phillips, L. D., Embrey, D. E., Humphreys, P., and Selby, D. L. (1990), "A Sociotechnical Approach to Assessing Human Reliability," in *Influence Diagrams, Belief Nets and Decision Making: Their Influence on Safety and Reliability*, R. M. Oliver and J. A. Smith, Eds., Wiley, New York.

Potter, S. S., Roth, E. M., Woods, D. D., and Elm, W. C. (1998), "A Framework for Integrating Cognitive Task Analysis into the System Development Process," in *Proceedings of the Human Factors and Ergonomics Society 42nd Annual Meeting*, Human Factors and Ergonomics Society, Santa Monica, CA, pp. 395–399.

Prager, L. O. (1998), "Sign Here," *American Medical News*, Vol. 41, October 12, pp. 13–14.

Rasmussen, J. (1982), "Human Errors: A Taxonomy for Describing Human Malfunction in Industrial Installations," *Journal of Occupational Accidents*, Vol. 4, pp. 311–333.

Rasmussen, J. (1986), *Information Processing and Human–Machine Interaction: An Approach to Cognitive Engineering*, Elsevier, New York.

Rasmussen, J., Pejterson, A. M., and Goodstein, L. P. (1994), *Cognitive Systems Engineering*, Wiley, New York.

Reason, J. (1990), *Human Error*, Cambridge University Press, New York.

Reason, J. (1997), *Managing the Risks of Organizational Accidents*, Ashgate, Aldershot, Hampshire, England.

Reason, J., and Hobbs, A. (2003), *Managing Maintenance Error: A Practical Guide*, Ashgate, Aldershot, Hampshire, England.

Reynolds-Mozrall, J., Drury, C. G., Sharit, J., and Cerny, F. (2000), "The Effects of Whole-Body Restriction on Task Performance," *Ergonomics*, Vol. 43, pp. 1805–1823.

Roberts, K. H. (1990), "Some Characteristics of One Type of High Reliability Organization," *Organization Science*, Vol. 1, pp. 160–176.

Robinson, A. G., and Stern, S. (1998), *Corporate Creativity*, Berrett-Koehler, San Francisco.

Rochlin, G., La Porte, T. D., and Roberts, K. H. (1987), "The Self-Designing High Reliability Organization: Aircraft Carrier Flight Operations at Sea," *Naval War College Review*, Vol. 40, pp. 76–90.

Rosenthal, J., Booth, M., and Barry, A. (2001), "Cost Implications of State Medical Error Reporting Programs: A Briefing Paper," National Academy for State Health Policy, Portland, ME.

Rouse, W. B., and Rouse, S. (1983), "Analysis and Classification of Human Error," *IEEE Transactions on Systems, Man, and Cybernetics*, Vol. SMC-13, pp. 539–549.

Rumelhart, D. E., and McClelland, J. L., Eds. (1986), *Parallel Distributed Processing: Explorations in the Microstructure of Cognition*, Vol. 1, *Foundations*, MIT Press, Cambridge, MA.

Sanders, M. S., and McCormick, E. J. (1993), *Human Factors in Engineering and Design*, 7th ed., McGraw-Hill, New York.

Senders, J. W., and Moray, N. P. (1991), *Human Error: Cause, Prediction, and Reduction*, Lawrence Erlbaum Associates, Mahwah, NJ.

Sharit, J. (1997), "Allocation of Functions," in *Handbook of Human Factors and Ergonomics*, 2nd ed., G. Salvendy, Ed., Wiley, New York, pp. 301–339.

Sharit, J. (1998), "Applying Human and System Reliability Analysis to the Design and Analysis of Written Procedures in High-Risk Industries," *Human Factors and Ergonomics in Manufacturing*, Vol. 8, pp. 265–281.

Sharit, J. (2003), "Perspectives on Computer Aiding in Cognitive Work Domains: Toward Predictions of Effectiveness and Use," *Ergonomics*, Vol. 46, pp. 126–140.

Shepherd, A. (2000), *Hierarchical Task Analysis*, Taylor & Francis, London.

Simon, H. A. (1966), *Models of Man: Social and Rational*, Wiley, New York.

Smith, P. J., and Geddes, N. D. (2003), "A Cognitive Systems Engineering Approach to the Design of Decision Support Systems," in *The Human–Computer Interaction Handbook: Fundamentals, Evolving Technologies, and Emerging Applications*, J. A. Jacko and A. Sears, Eds., Lawrence Erlbaum Associates, Mahwah, NJ.

Stout, R. M., Cannon-Bowers, J. A., Salas, E., and Milanovich, D. M. (1999), "Planning, Shared Mental Models, and Coordinated Performance: An Empirical Link Is Established," *Human Factors*, Vol. 41, pp. 61–71.

Strauch, B. (2002), *Investigating Human Error: Incidents, Accidents, and Complex Systems*, Ashgate, Aldershot, Hampshire, England.

Swain, A. D., and Guttmann, H. E. (1983), *Handbook of Human Reliability Analysis with Emphasis on Nuclear Power Plant Applications*, NUREG/CR-1278, U.S. Nuclear Regulatory Commission, Washington, DC.

Thomas, E. J., and Helmreich, R. L. (2002), "Will Airline Safety Models Work in Medicine?" in *Medical Error: What Do We Know? What Do We Do?* M. M. Rosenthal and K. M. Sutcliffe, Eds., Jossey-Bass, San Francisco, pp. 217–234.

Turrell, M. (2002), "Idea Management and the Suggestion Box," White Paper, Imaginatik Research, retrieved May

12, 2004, from http://www.imaginatik.com/web/nsf/docs/idea_reports_imaginatik.

Vicente, K. J. (1999), *Cognitive Work Analysis: Toward Safe, Productive, and Healthy Computer-Based Work*, Lawrence Erlbaum Associates, Mahwah, NJ.

Vicente, K. J. (2004), *The Human Factor: Revolutionizing the Way People Live with Technology*, Routledge, New York.

Weiner, E. L. (1985), "Beyond the Sterile Cockpit," *Human Factors*, Vol. 27, pp. 75–90.

Wickens, C. D. (1984), "Processing Resources in Attention," in *Varieties of Attention*, R. Parasuraman and R. Davies, Eds., Academic Press, New York, pp. 63–101.

Wickens, C. D., Liu, Y., Becker, S. E. G., and Lee, J. D. (2004), *An Introduction to Human Factors Engineering*, 2nd ed., Prentice-Hall, Upper Saddle River, NJ.

Wilde, G. J. S. (1982), "The Theory of Risk Homeostasis: Implications for Safety and Health," *Risk Analysis*, Vol. 2, pp. 209–225.

Woods, D. D. (1984), "Some Results on Operator Performance in Emergency Events," *Institute of Chemical Engineers Symposium Series*, Vol. 90, pp. 21–31.

Woods, D. D. (1993), "Process Tracing Methods for the Study of Cognition Outside the Experimental Psychology Laboratory," in *Decision Making in Action: Models and Methods*, G. Klein, R. Calderwood, and J. Orasanu, Eds., Ablex, Norwood, NJ, pp. 227–251.

Woods, D. D., and Watts, J. C. (1997), "How Not to Navigate Through Too Many Displays," in *Handbook of Human–Computer Interaction*, 2nd ed., M. Helander, T. K. Landauer, and P. Prabhu, Eds., Elsevier Science, New York, pp. 617–650.

Woods, D. D., Johannesen, L. J., Cook, R. I., and Sarter, N. B. (1994), "Behind Human Error: Cognitive Systems, Computers, and Hindsight," CSERIAC State-of-the Art-Report, Crew Systems Ergonomics Information Analysis Center, Wright-Patterson Air Force Base, OH.

Woods, D. D., Sarter, N. B., and Billings, C. E. (1997), "Automation Surprises," in *Handbook of Human Factors and Ergonomics*, 2nd ed., G. Salvendy, Ed., Wiley, New York, pp. 1926–1943.

CHAPTER 28

ERGONOMICS OF WORK SYSTEMS*

Stephen M. Popkin and Heidi D. Howarth
U.S. Department of Transportation
Cambridge, Massachusetts

Donald I. Tepas
University of Connecticut
Stores, Connecticut

1 INTRODUCTION

The exclusion or misapplication of ergonomic work system principles has been estimated to cost the U.S. economy upward of $60 billion a year in lost work time and lower productivity (Mapes, 1990). Although no current comprehensive statistics are available, it is generally held that there has been a worldwide increase in the number of people who do not work regular and/or fixed diurnal hours. This is true of the United States, which saw a doubling of the number of people working flexible hours in the professional services sector (28.8%) (BLS, 2002) between 1991 and 2001. Other countries, especially developing countries,

*Portions of this work were supported by the U.S. Federal Railroad Administration Human Factors Program. Simon Folkard, Johannes Gärtner, and David Nash provided useful comments on previous drafts. The views of the authors do not purport to reflect the position of the Federal Railroad Administration or the U.S. Department of Transportation.

have seen huge increases in the number of round-the-clock manufacturing operations. Mexico, for instance, had a 30% annual growth in manufacturing in their *maquiladora* zones between 1988 and 1993. Given the number of people affected by working nontraditional shift schedules, it is important for those managing such shiftwork operations to apply ergonomic principles and operations analyses when developing shift systems.

There are three basic types of shiftwork operations, defined by the nature of the work being performed and its associated staffing requirements. The first is service-based operations, which includes, for instance, health care providers, security and emergency response personnel, and recreation and entertainment establishments. The degree to which these services are required depends on the needs of the society. These needs may vary by time of day, day of week, season, and so on, thereby requiring proportionally greater or fewer staff at any given time. The second type of operation is process-based, in which the nature of the work requires continuous or nearly continuous utilization and staffing in order to regain capital expenditure and minimize downtime. Examples of process operations include chemical and oil refineries, textile weavers, and paper milling. These operations may run around the clock only on weekdays, or seven days a week, depending on the demand for the product and the expense of staffing weekends versus restarting the process. The third type of shiftwork is operations-based and covers activities that are considered valuable by the company to run 24 hours, although not required as such by society or the nature of the work being performed. Examples of this type of operation include banking and finance, construction, and shipping. Each of these three types of operations calls for different work system solutions that incorporate one or more nontraditional work schedules.

As perhaps the most researched and potentially problematic alternate shift schedule option, night work has been practiced routinely at least since Roman times (Scherrer, 1981). The relatively recent introduction of artificial light and the need to realize profits from large capital investments has undoubtedly played an important role in making night work all the more practical and necessary. Additionally, there are many features of the contemporary workplace that have further increased the need for work at all hours of the day, such as communication and travel across multiple time zones to support the demands of international trade; development of new manufacturing processes that cannot be performed without continuous operation; the popularity of agile manufacturing and just-in-time methods; investment in automation and robotics that require around-the-clock operation; and growth in the demand for support services to respond to these factors. What were once considered optional and nonusual work hours have clearly become a required ingredient for successful competition in the global workplace, both in the United States and worldwide.

In many work environments, both large and small, work schedule design continues to be the domain of managers who may not have a background in ergonomics and view this job task as a collateral duty rather than a critical element on which he or she will be evaluated. Consequently, many of the resulting schedules are born of considerations other than what has been established empirically, including the continuance of traditional practices and fulfilling economic and practical concerns (e.g., production demands, collective bargaining agreements), the outcomes of which may prove deleterious to worker performance, safety, and health (Rosa, 1991; Costa, 1996; Horne and Reyner, 1998). Available data do, in fact, suggest that in most workplaces the origin, justification, and impact of the work schedule used is unknown (Tepas, 1994), as are many of the legislative guidelines that affect work schedules.

As it currently stands, regulations that do exist in the United States to govern work hours treat all times of day as equal and interchangeable. For example, within the hours of service (HOS) regulations for commercial motor vehicle operators, acceptable limits for time spent driving are based solely upon the number of hours worked relative to the number of hours of off time, not whether driving occurs at night versus during the day. This translates into millions of workers who are probably assigned to hundreds of different work schedules of unknown ergonomic merit. The magnitude of this problem is further complicated by the rapid rise of workers participating in both formal and informal flexible schedules (BLS, 2002). Although in some cases this type of arrangement can result in higher levels of job satisfaction (Baltes et al., 1999), the reasons why, for example, men and women utilize alternative work schedules may differ (Sharpe et al., 2002). Additionally, whether or not such arrangements ultimately benefit the organization probably depends on what business outcomes are deemed most important. For example, a meta-analysis conducted by Baltes et al. (1999) found that flexible work schedule arrangements were more likely to be associated with a positive impact on absenteeism rates than on productivity levels.

Ultimately, due to the nature of such work arrangements, workers on alternate shift schedules have an increased probability of experiencing negative outcomes, as well as diminished on-the-job performance associated with their work schedules. Not surprisingly, performance and effectiveness are constant concerns of employers, for instance, in the commercial transportation industry, where suboptimal work schedule practices have been linked to accidents (NTSB, 1989). This is a substantial problem that led the U.S. National Transportation Safety Board (NTSB) to request that all government transportation agencies review their hours of service laws and regulations (NTSB, 1990). In 1995, responding to the NTSB's recommendation, Congress required the Federal Motor Carrier Safety Association (FMCSA) to reform its HOS regulations in an effort to reduce accidents and incidents related to both operator fatigue and decrements in operator vigilance, each of which may place drivers and the general public at risk. As a result, in 2003 FMCSA issued the first significant revision to the HOS regulations in more than 60 years. These new regulations were due for compliance by January 2004 but were revoked in July 2004 by the U.S. Court of Appeals primarily because of their lack of consideration for driver health (*Public citizen et al. v. Federal Motor Carrier Safety Administration*, 2004). Clearly, ergonomic work scheduling is an area that is not without political and judicial controversy and consequences.

Alternatively, from a science perspective, shiftwork research had initially focused its efforts on elucidating the problems associated with the use and unfettered spread of nontraditional work schedules. As a result, ergonomic design efforts aimed at decreasing or eliminating the practice of night work and other "undesirable" work scheduling practices are being implemented in several countries, including France and Germany. However, demands of the current, global, 24/7 society are ever-increasing, and with technology continually speeding up various processes, one must now view increased implementations of alternative work schedules as a requisite part of the contemporary workplace. Those responsible must, consequently, aim to develop work arrangements that are sensitive to human limitations pertaining to working at night and for long periods of time. Despite the fact that humans are diurnal animals and subject to fatigue, the demands of a global economy require schedule designs and methods that expand, rather than contract, work hours. It is quite clear that amateur evaluations of work schedules with regard to health and safety are often faulty; hence, expert assessment is needed to design work schedules with consideration to both prevention and evaluation of outcomes that may negatively affect the worker and/or organization.

2 HISTORICAL BACKGROUND

Although people have been employed on alternate work schedules for centuries, prior to 1950 the research literature on the effects of working various shift schedules is limited. Indeed, in the United States, most modern labor laws governing work hours have their origins before this date, and as such are not based on scientific data. Legal constraints on night work in the United States have always been quite modest compared to those in other nations, and unlike many industrialized countries, the U.S. government has never ratified any of the International Labour Organisation (ILO) conventions regulating night and shiftwork.

Interest in scientific work schedule research has grown steadily since World War II. One might argue that the first modern study of shiftworkers in North America was published in 1965 (Mott et al., 1965) and the first U.S. meeting dedicated to shiftwork problems was conducted in 1975 by the National Institute of Occupational Safety and Health (NIOSH) (Rentos and Shepard, 1976). European and Asian support for work schedule research appears to be more significant and long-standing. In 1966, European investigators founded what is now known as the Scientific Committee on Night and Shiftwork as part of the International Commission on Occupational Health (ICOH) (Wedderburn and Tepas, in press). In subsequent years, this ICOH committee has sponsored and published 16 international symposia and the *Shiftwork International Newsletter* (Tepas, 2003). These symposia are the only dedicated, regular shiftwork research meetings in the world, although periodic, industry-specific meetings on work scheduling and related issues have become more commonplace (*Proceedings of the 5th International Conference on Fatigue and Transportation*, 2003). Still today, the number of European and Asian publications and meetings on work schedule issues surpass those from North America. In fact, the 2003 Santos Symposium in Brazil marked only the second time the International Symposium on Night and Shiftwork was held in the Americas.

As much of the work schedule research has been conducted in industrialized Asia and Europe, there continues to be a need for scientific investigation, especially in other countries, given cultural and regulatory differences. These factors define the possible number of solutions pertaining to hours of work that are considered acceptable and feasible within a society. Overall, research findings have demonstrated this need by exploring the multivariate nature and impact of work systems operating within the context of social and cultural variables. Although there is much to be learned from the research conducted on the variety of work system practices in other nations, it is also clear that work systems should not be imported without careful consideration and evaluation. The changes introduced by the current technological revolution make the search for better physiological and sociological work systems a global challenge. As such, it is appropriate to note that simple universal solutions to work schedule problems should not be expected and that work schedule practices will continue to vary based on country, industry, and cultural norms (Ong and Kogi, 1990; U.S. Congress, Office of Technology Assessment, 1991).

3 WORK SYSTEM DEFINITIONS AND REPRESENTATIONS

Given the immense diversity of work system practices, it is reasonable to suggest that globally, thousands of different work systems are being used. This is not surprising or undesirable given the multivariate nature of work systems and the social, cultural, and economic contexts within which they operate. The availability of a wide range of possible solutions to work system problems increases the options available to the ergonomic expert, but with a proliferation of options, confusion may also arise if terms are not used consistently or if they are applied differentially. For this reason, development of a consistent, precise, and science-bound terminology and representation method is essential to understanding, discussing, and applying these diverse work systems in a reliable manner.

Although the use of work system terminology is often consistent within a given workplace, shop-floor terms do vary from workplace to workplace, country to country, and expert to expert (Gärtner et al., 2003). Unfortunately, these differences in usage often lead to confusion and promote misinterpretation of the literature. The following three examples demonstrate these differences and the ensuing confusion. First, in the United States, to some the term *swing shift* refers to any worker or group of workers who are employed on a schedule where the time of day one works changes. For others, the same term is used only to indicate the afternoon–evening (second) shift. Second, U.S. workers often refer to the *night shift* as the *third shift*. However, this shift is also sometimes called the first, graveyard, or lobster shift. Finally, most workers in the United States consider a work system that changes its hours once per week to be an example of *rapid* or *fast* shift rotation, whereas many Europeans would reserve the designation "rapid" or "continental" for rotating schedules that change work hours several times within a week.

As the standardization of work system-related terminology has yet to be acted upon in a formal manner, this leaves the potential for misunderstanding the literature in which these types of decisions should be grounded. Technical reports and journal articles routinely obfuscate critical issues by failing to fully define, describe, or comprehend the work system terms and schedules to which they refer. It is for this reason that we chose to operationally define these terms and provide the reader with a method to represent and visualize complex work system information in a way that clearly highlights its salient features (Gärtner et al., 2004).

Where feasible, to promote consistency, these definitions are very similar to those used in the second edition of this handbook (Tepas et al., 1997); however, terms are updated where necessary, given the evolving state of the work system terminology in the United States. Furthermore, they are employed throughout the remainder of this chapter wherever possible, with the exception of cases where authors' operational definitions were inconsistent with our own.

As a caution, rather than assuming that the terms used in this chapter are consistent with the readers' understanding, we suggest thorough and careful study of the definitions and how they are reflected in the work system representation format. This will help to ensure an accurate understanding of subsequent sections of the chapter. Moreover, the authors approve of, recommend, and encourage use of this terminology and representation format by other authors in future scientific publications and presentations until more formal definitions are provided through a standards-setting process.

Table 1 Work System Definitions

Shift: the hours of a given day that a person or a group of people is scheduled to be at the workplace.

Off time: the hours of a given day that a person or a group of people is not normally required to be at the workplace; often includes commuting time.

Schedule: the sequence of consecutive shifts and off time assigned to a particular person or a group of people as their usual work assignment.

Permanent shift: a schedule for a person or a group of people that does not normally require work on more than one type of shift (i.e., the time of day of work is constant).

Rotating shifts: a schedule that normally requires a person or a group of people to work more than one type of shift (i.e., the time of day when work is performed changes regularly).

Relief shift: a schedule comprised of extra staffing required to cover for absenteeism and other on-site work demands.

Flexible hours: a schedule that splits the workday into two types of time: core time, when the employee must be performing work-related duties, and flexible time, which is a set of hours flanking the core hours that the employee can choose to work within in order to fulfill the required shift length. It is used almost exclusively in nonmanufacturing organizations because of its difficulty to implement in line or crew-based operations (Baltes et al., 1999).

Basic pattern: the minimum number of days on and days off required to complete the specific sequence of shifts and off time constituting a given schedule (i.e., the number of days until a schedule begins to repeat).

Locked/unlocked: determined by whether or not the basic pattern repeats at seven days or a multiple of seven days. A locked schedule has a basic pattern of seven days or multiples of seven days. An unlocked schedule is any schedule that does not exist as a multiple of seven days.

Full cycle: the minimum number of days required to arrive at a point where the basic pattern of a schedule begins to repeat on the same day of the week.

Workweek: defined by the Fair Labor Standards Act as based on a work month of 168 hours (24 hours per day; seven days per week), which can start at any point in time within a given week as long as it repeats each week thereafter.

Payweek: the interval by which the company has decided or agreed to pay its employees.

Work system: all of the schedule(s) implemented in a given workplace to meet the real or perceived requirements of a given plant, process, or service, up to seven-day coverage.

3.1 Work System Definitions

Work system definitions are presented in Table 1. The basic unit of every work system is a *shift*; more specifically, the time of day a worker is required or chooses (in the case of flexible hours) to be at the workplace to perform job-related activities. By definition, all workers who are scheduled to be at a workplace on a repeated basis are shiftworkers. However, given markedly different effects on operations and workers' lives, those working exclusively daytime hours during the workweek are considered *dayworkers*, and those working all other schedules are considered to be *shiftworkers*. Each worker is due to report on a schedule for the shifts to which he or she has been assigned. This schedule includes *off time*, which consists of a number of hours (often 24 hours or more) when the worker is not expected to be at the workplace. If work time is not scheduled, such as with on-call operations, there is no formal work system. However, any workplace that employs more than one person needs some means of ensuring that all work requirements are met, whether they be formal or informal.

When a person works at the same time and day on a consistent basis, work hours and schedules are termed *permanent*. On the other hand, for a worker who is scheduled to be on the job over different hours on specified days, yet still following planned, nonflexible hours, the schedule is termed *rotating*. Flexible and on-call operations add a degree of variability to work times. Flexible working times are still planned schedules, typically providing the worker with a several-hour window within which they may set their work start and end times. On-call schedules, on the other hand, are unplanned work assignments that do not afford the worker much flexibility in shift start or end time. Rather, he or she must be available to report to work within a specified amount of time when summoned for duty.

An important and often ignored component of a work schedule is its de facto scheduling of a worker's off time. A schedule with a variable pattern of shift start and end times necessarily influences the available time that a worker has to plan his or her personal activities. The easier it is to understand work/off-time patterns, the easier it is for the worker to plan the other aspects of his or her life. For this reason, the terms *basic pattern* and *full cycle* have been devised. The number of days required for a worker on a given schedule to complete a single, repetitive sequence of shifts and off time is the basic pattern of a schedule. When there is variability in the day(s) of the week that are worked, the number of days required for the basic pattern to begin to repeat on the *same* days of the week is the full cycle of a schedule. For many schedules, the basic pattern is equal to the length of the full cycle, but for others, the full cycle is longer (never shorter) than the basic pattern. Basic patterns of seven days or multiples of seven days are considered *locked*, whereas basic patterns that are not in multiples of seven are considered *unlocked*. Because of the shorter duration or regularity of the schedule pattern, locked schedules tend to be easier for the worker to use when planning off time.

The variety of work systems that might be designed and used is considerable, since one work system may include many different work schedules. For example, a given work system may incorporate work schedules involving both permanent and rotating hours for various shifts, with basic patterns and full cycles of differing lengths. Specific examples are presented in subsequent text and graphical representations.

3.2 Shift Definitions

Table 2 provides operational definitions for the various shift types that are presented in this chapter. These terms are common U.S. designations and are referred to in the remaining text. However, for ease and accuracy of comparison and understanding, alternative terminology is included in parentheses after each definition, where applicable. As shown in the table, different work systems necessitate different numbers and/or arrangements of shifts, the most basic being two- and three-shift systems. Additionally, many work systems require less straightforward arrangements and incorporate irregular and/or split shifts.

Table 2 Work Shift Definitions

Three-Shift Systems

First shift: a work period of about eight hours' duration that generally falls between the hours of 0600 and 1700 (also known as the *morning* or *day* shift).

Second shift: a work period of about eight hours' duration that generally falls between the hours of 1500 and 0100 (also known as the *afternoon–evening* or *swing* shift).

Third shift: a work period of about eight hours' duration that generally falls between the hours of 2200 and 0700 (also known as the *night* or *graveyard* shift).[a]

Two-Shift Systems

Day shift: a work period of about 10 hours' or more duration that generally falls between the hours of 0600 and 2200.

Night shift: a work period of about 10 hours' or more duration that generally falls between the hours of 2200 and 0600.

Other Work Shift Classifications

Split shift: any work period that is regularly scheduled to include two or more work periods of less than seven hours, separated by more than one hour away from work, on the same workday.

Irregular shifts: scheduled work periods that vary their shift starting time and duration in an inconsistent or unpredictable way.

On-call shifts: unscheduled work periods that vary their shift starting time and duration in an inconsistent way.

Non-workday: any calendar day in which only off time is scheduled.

[a]ILO: C171 night work convention, 1990: "(a) the term *night work* means all work which is performed during a period of not less than seven consecutive hours, including the interval from midnight to 5 a.m., to be determined by the competent authority after consulting the most representative organizations of employers and workers or by collective agreements"; http://www.ilo.org/ilolex/english/recdisp1.htm.

As part of a three-shift system, the third shift is defined based on the convention specified by the ILO for a night shift (ILO, 1990). The remaining shift definitions use the ILO specifications for night shift hours as a model and appear to be fairly characteristic of workplace practices in the United States. To avoid confusion with regard to overtime work and to facilitate a logical distinction between two- and three-shift systems, the current definitions limit use of the terms *day* and *night* to shifts that are significantly longer than eight hours. Users are advised to provide additional operational specifications to the current terminology whenever an exact fit with these definitions is not possible.

It is important to note that the definitions presented do not include the term *crew*. When workers are employed on exactly the same schedule and work the same hours, they are often termed a crew. This terminology can be problematic, because it leads one to assume that the crew is in fact one group of people working together in an interactive and cooperative manner. In a large operation, however, it is the case that persons working identical schedules may never meet or interact. It is also possible that members of a crew may be assigned and scheduled to work together, but for some reason they are unable to function as a team.

3.3 Work System Characteristics

When a formal system is used to schedule workers' shifts and off time, the term *work system* is best used to describe all the shift schedules practiced in the given workplace. The five general categories of work system operations are defined in Table 3. A work system that includes schedules for most seven-day-a-week operations is referred to as a *continuous operation*. A *discontinuous operation* runs less than seven days a week; in most cases this is a system that typically excludes Sunday and/or Saturday work. When a continuous or discontinuous operation regularly uses shifts longer than eight hours, they are said to utilize *compressed operations*. This category assumes that the standard workweek is somewhere around 40 hours and that the worker can complete his or her workweek in less than five days. When there is a predictable, periodic need for a discontinuous operation to run seven days a week, the operations may select a *semicontinuous* operation to account for these extra hours. Finally, continuous operations necessitated by unusual events and requiring prolonged periods of performance are sometimes termed *sustained operations*. This category often involves people "performing at close to a nonstop rate for as long as they can" (Krueger, 1989, p. 129).

Table 3 Primary Work System Operations

Discontinuous operations: a work system that does not employ around-the-clock and/or seven-day-a-week scheduling. These systems use schedules that often do not require a person or a group of people to work on weekends (Saturday and/or Sunday).

Continuous operations: a work system that employs around-the-clock and seven-day-a-week scheduling. These systems use schedules that normally require a person or a group of people to work some weekends.

Semicontinuous operations: a work system that is usually discontinuous, but at company discretion, may provide planned, optional weekend work to its employees; often used by companies that have variable demand for product or are affected by seasonal constraints.

Compressed operations: a system that employs schedules that normally include shifts of more than eight hours in length resulting in a workweek of less than five full shifts.

Sustained operations: a system that allows for shifts of more than 12 hours in length. These longer shifts are not usually scheduled, for they often require the worker to perform for as long as he or she is able.

Table 4 Symbols for Work System Representation Tables

Symbol	Meaning	Symbol	Meaning
1	First shift	S1	A designated schedule for a given work system
2	Second shift	S2	Another schedule for the same work system
3	Third shift	S3	Another schedule for the same work system
D	Day shift	S4	Another schedule for the same work system
N	Night shift	M1	The first 28 days of a given designated schedule
S	Split shift	M2	Days 29 to 56 of the same schedule (follows M1)
•	Non-workday	M3	Days 57 to 84 of the same schedule (follows M2)
M	Monday	M4	Days 85 to 112 of the same schedule (follows M3)
T	Tuesday		
W	Wednesday		
R	Thursday		
F	Friday		
S	Saturday		
K	Sunday		

3.4 Work System Representation Method

Based on the definitions and terminology provided in Sections 3.2 and 3.3, specific work systems can be presented schematically using a variety of software and paper-and-pencil tools and methodologies. A relatively simple and straightforward representation method, similar to that used in the prior version of this chapter (Tepas et al., 1997), has been updated for inclusion here. Table 4 contains the symbols and associated meanings for this representation method, and an introduction and example appear in Table 5. This table illustrates a work system in which all workers are employed on a single, traditional, fixed daytime work schedule for a period of 16 consecutive weeks. Work time for all employees is limited to eight-hour daytime periods on weekdays. A discontinuous operation with permanent hours is depicted: only one work shift is used, all workers are assigned to the same schedule, and the schedule is identical every week. In this simple but traditional example, the basic pattern and the full cycle are both seven days, as the same days of the week are worked every week.

3.5 Work Systems Examples

The following sections present a variety of work system shift combinations using this work schedule representation method. Readers who are familiar with the various work systems may wish to skip ahead to Section 4.

Table 5 Work System with Permanent Hours and Discontinuous Operations: Single Schedule[a]

	M	T	W	R	F	S	K	M	T	W	R	F	S	K	M	T	W	R	F	S	K	M	T	W	R	F	S	K
S1M1	1	1	1	1	1	•	•	1	1	1	1	1	•	•	1	1	1	1	1	•	•	1	1	1	1	1	•	•
S1M2	1	1	1	1	1	•	•	1	1	1	1	1	•	•	1	1	1	1	1	•	•	1	1	1	1	1	•	•
S1M3	1	1	1	1	1	•	•	1	1	1	1	1	•	•	1	1	1	1	1	•	•	1	1	1	1	1	•	•
S1M4	1	1	1	1	1	•	•	1	1	1	1	1	•	•	1	1	1	1	1	•	•	1	1	1	1	1	•	•

[a]Basic pattern (locked)/full cycle = seven days. Notation symbols for this and subsequent work system tables are given in Table 4.

3.5.1 Discontinuous Operations

Table 6 provides an example of a discontinuous operation with permanent hours over a period of four consecutive weeks. In this case, workers are assigned to one of three work shifts, work hours do not vary within a given schedule, and all work shifts occur on weekdays. Again here the basic pattern and the full cycle are seven days, as the same days are worked each week. The system in Table 7 is also limited to discontinuous operations, however, in this case, it is rotating, not permanent. Three schedules are used, where the table shows the work system over four consecutive weeks. Each of these schedules involves work on three different shifts and begins with five consecutive workdays on a specific shift, followed by two non-workdays, then work on another shift. Thus, rotation occurs once every seven days and in each case, proceeds from first to second to third shift before a return to the first shift. This is referred to as *forward* rotation representing a *phase delay* within the basic pattern, as the shift start time is later on each succeeding week up to the return. Again the basic pattern and the full cycle are of equal length, but in this case they are 21 days in duration. The work system shown in Table 8 is identical to that in Table 7, except that the direction of rotation is *backward*, representing a *phase advance* for the worker. Table 9 provides an example of how rotation rate can vary in length. This work system is the same as that found in Table 7; however, in this case the rotation is extended to every 28 days. The basic pattern and full cycle remain equal, but are increased to 84 days.

3.5.2 Continuous Operations

Continuous operations require around-the-clock, seven-day-a-week staffing and scheduling. Because these work systems involve covering more hours per week than are found in discontinuous operations, they usually include more than three work schedules. Differences in basic pattern and full cycle length are also likely. In practice, systems with rotating hours appear to be more common for continuous operations. This may be related to the fact that although permanent shifts are feasible for continuous work systems they usually require more schedules than a comparable rotating system to provide full shift coverage.

Table 10 shows a European work system, the *continental rota*, which is a continuous-operation, forward-rotating system. Four schedules are required for this system, where the table depicts four consecutive weeks for each, and both the basic pattern and full cycles are equal to 28 days. Another European work system is shown in Table 11. This continuous, forward-rotating system is known as the *metropolitan rota*, more concisely referred to as the 2–2–2–2 (two first shifts, two second shifts, two third shifts, and two non-workdays). The table shows eight consecutive workweeks for each

Table 6 Work System with Permanent Hours and Discontinuous Operations: Three Schedules[a]

	M	T	W	R	F	S	K	M	T	W	R	F	S	K	M	T	W	R	F	S	K	M	T	W	R	F	S	K
S1M1	1	1	1	1	1	•	•	1	1	1	1	1	•	•	1	1	1	1	1	•	•	1	1	1	1	1	•	•
S2M1	2	2	2	2	2	•	•	2	2	2	2	2	•	•	2	2	2	2	2	•	•	2	2	2	2	2	•	•
S3M1	3	3	3	3	3	•	•	3	3	3	3	3	•	•	3	3	3	3	3	•	•	3	3	3	3	3	•	•

[a]Basic pattern (locked)/full cycle = seven days.

Table 7 Work System with Rotating Hours and Discontinuous Operations: Three Schedules and Forward Rotation[a]

	M	T	W	R	F	S	K	M	T	W	R	F	S	K	M	T	W	R	F	S	K	M	T	W	R	F	S	K
S1M1	1	1	1	1	1	•	•	2	2	2	2	2	•	•	3	3	3	3	3	•	•	1	1	1	1	1	•	•
S2M1	2	2	2	2	2	•	•	3	3	3	3	3	•	•	1	1	1	1	1	•	•	2	2	2	2	2	•	•
S3M1	3	3	3	3	3	•	•	1	1	1	1	1	•	•	2	2	2	2	2	•	•	3	3	3	3	3	•	•

[a]Basic pattern (locked)/full cycle = 21 days.

Table 8 Work System with Rotating Hours and Discontinuous Operations: Three Schedules and Backward Rotation[a]

	M	T	W	R	F	S	K	M	T	W	R	F	S	K	M	T	W	R	F	S	K	M	T	W	R	F	S	K
S1M1	1	1	1	1	1	•	•	3	3	3	3	3	•	•	2	2	2	2	2	•	•	1	1	1	1	1	•	•
S2M1	2	2	2	2	2	•	•	1	1	1	1	1	•	•	3	3	3	3	3	•	•	2	2	2	2	2	•	•
S3M1	3	3	3	3	3	•	•	2	2	2	2	2	•	•	1	1	1	1	1	•	•	3	3	3	3	3	•	•

[a]Basic pattern (locked)/full cycle = 21 days.

Table 9 Work System with Slowly Rotating Hours and Discontinuous Operations: Three Schedules and Forward Rotation[a]

	M	T	W	R	F	S	K	M	T	W	R	F	S	K	M	T	W	R	F	S	K	M	T	W	R	F	S	K
S1M1	1	1	1	1	1	•	•	1	1	1	1	1	•	•	1	1	1	1	1	•	•	1	1	1	1	1	•	•
S2M1	2	2	2	2	2	•	•	2	2	2	2	2	•	•	2	2	2	2	2	•	•	2	2	2	2	2	•	•
S3M1	3	3	3	3	3	•	•	3	3	3	3	3	•	•	3	3	3	3	3	•	•	3	3	3	3	3	•	•
S1M2	2	2	2	2	2	•	•	2	2	2	2	2	•	•	2	2	2	2	2	•	•	2	2	2	2	2	•	•
S2M2	3	3	3	3	3	•	•	3	3	3	3	3	•	•	3	3	3	3	3	•	•	3	3	3	3	3	•	•
S3M2	1	1	1	1	1	•	•	1	1	1	1	1	•	•	1	1	1	1	1	•	•	1	1	1	1	1	•	•
S1M3	3	3	3	3	3	•	•	3	3	3	3	3	•	•	3	3	3	3	3	•	•	3	3	3	3	3	•	•
S2M3	1	1	1	1	1	•	•	1	1	1	1	1	•	•	1	1	1	1	1	•	•	1	1	1	1	1	•	•
S3M3	2	2	2	2	2	•	•	2	2	2	2	2	•	•	2	2	2	2	2	•	•	2	2	2	2	2	•	•
S1M4	1	1	1	1	1	•	•	1	1	1	1	1	•	•	1	1	1	1	1	•	•	1	1	1	1	1	•	•
S2M4	2	2	2	2	2	•	•	2	2	2	2	2	•	•	2	2	2	2	2	•	•	2	2	2	2	2	•	•
S3M4	3	3	3	3	3	•	•	3	3	3	3	3	•	•	3	3	3	3	3	•	•	3	3	3	3	3	•	•

[a]Basic pattern (locked)/full cycle = 84 days.

Table 10 Work System with Rotating Hours and Continuous Operations (Continental Rota): Four Schedules and Forward Rotation[a]

	M	T	W	R	F	S	K	M	T	W	R	F	S	K	M	T	W	R	F	S	K	M	T	W	R	F	S	K
S1M1	1	1	2	2	3	3	3	•	•	1	1	2	2	2	3	3	•	•	1	1	1	2	2	3	3	•	•	•
S2M1	2	2	3	3	•	•	•	1	1	2	2	3	3	3	•	•	1	1	2	2	2	3	3	•	•	1	1	1
S3M1	3	3	•	•	1	1	1	2	2	3	3	•	•	•	1	1	2	2	3	3	3	•	•	1	1	2	2	2
S4M1	•	•	1	1	2	2	2	3	3	•	•	1	1	1	2	2	3	3	•	•	•	1	1	2	2	3	3	3

[a]Basic pattern (locked)/full cycle = 28 days.

Table 11 Work System with Rotating Hours and Continuous Operations (Metropolitan Rota): Four Schedules and Forward Rotation[a]

	M	T	W	R	F	S	K	M	T	W	R	F	S	K	M	T	W	R	F	S	K	M	T	W	R	F	S	K
S1M1	1	1	2	2	3	3	•	•	1	1	2	2	3	3	•	•	1	1	2	2	3	3	•	•	1	1	2	2
S2M1	2	2	3	3	•	•	1	1	2	2	3	3	•	•	1	1	2	2	3	3	•	•	1	1	2	2	3	3
S3M1	3	3	•	•	1	1	2	2	3	3	•	•	1	1	2	2	3	3	•	•	1	1	2	2	3	3	•	•
S4M1	•	•	1	1	2	2	3	3	•	•	1	1	2	2	3	3	•	•	1	1	2	2	3	3	•	•	1	1
S1M2	3	3	•	•	1	1	2	2	3	3	•	•	1	1	2	2	3	3	•	•	1	1	2	2	3	3	•	•
S2M2	•	•	1	1	2	2	3	3	•	•	1	1	2	2	3	3	•	•	1	1	2	2	3	3	•	•	1	1
S3M2	1	1	2	2	3	3	•	•	1	1	2	2	3	3	•	•	1	1	2	2	3	3	•	•	1	1	2	2
S4M2	2	2	3	3	•	•	1	1	2	2	3	3	•	•	1	1	2	2	3	3	•	•	1	1	2	2	3	3

[a]Basic pattern = eight days (unlocked); full cycle = 56 days.

of the four schedules required. For each of these schedules the basic pattern is eight days and the full cycle is 56 days.

Shifts of greater than eight hours in duration are sometimes used with continuous work systems. In some cases, the system may be termed a *compressed operation*, in that a workweek often includes fewer than five full shifts. In other cases, such as the scheduling of residents at hospitals, compressed operations are simply a means for scheduling long working hours, in that work time exceeds 40 hours per workweek. Table 12 provides an example of a compressed operation, referred to as the 4–4, or the 4–4–4–4 (four day shifts, four non-workdays, four night shifts, four non-workdays). The table shows 16 consecutive workweeks for each of the four required schedules. For each schedule the basic pattern is 16 days and the full cycle is 112 days. A second compressed operations work system is shown in Table 13. This system is sometimes referred to as the 3–3 or the 3–3–3–3 and also requires four basic schedules. Twelve consecutive workweeks for each of the schedules are presented. For each of these schedules, the basic pattern is 12 days and the full cycle is 84 days. The work system shown in Table 14 is known as EOWEO, an acronym for "every other weekend off." This compressed operations work system is said to have originated in the United States, and it results in an equal distribution of full weekend non-workdays for all workers. The system features two three-day weekend non-workdays every month for each work schedule. Again, four basic schedules with rotating work hours are used, and four workweeks for each schedule are shown. In this case, the basic pattern and the full cycle are both 28 days.

3.5.3 Irregular Operations

It is difficult to define *irregular operations* because of the many ways an operation can vary if it is not "regular." Some formal work systems include schedules that alter their shift start time and duration in an inconsistent but predictable manner in order to meet operational demands and constraints. Scheduling these shifts in advance allows employees to plan and prepare for upcoming work episodes.

Work schedules can also vary shift start time and shift duration in an unpredictable manner, which

Table 12 Work System with Rotating Hours and Continuous Operations (Compressed 4–4–4–4): Four Schedules and Forward Rotation[a]

	M	T	W	R	F	S	K	M	T	W	R	F	S	K	M	T	W	R	F	S	K	M	T	W	R	F	S	K
S1M1	D	D	D	D	•	•	•	•	N	N	N	N	•	•	•	•	D	D	D	D	•	•	•	•	N	N	N	N
S2M1	N	N	N	N	•	•	•	•	D	D	D	D	•	•	•	•	N	N	N	N	•	•	•	•	D	D	D	D
S3M1	•	•	•	•	D	D	D	D	•	•	•	•	N	N	N	N	•	•	•	•	D	D	D	D	•	•	•	•
S4M1	•	•	•	•	N	N	N	N	•	•	•	•	D	D	D	D	•	•	•	•	N	N	N	N	•	•	•	•
S1M2	•	•	•	•	D	D	D	D	•	•	•	•	N	N	N	N	•	•	•	•	D	D	D	D	•	•	•	•
S2M2	•	•	•	•	N	N	N	N	•	•	•	•	D	D	D	D	•	•	•	•	N	N	N	N	•	•	•	•
S3M2	N	N	N	N	•	•	•	•	D	D	D	D	•	•	•	•	N	N	N	N	•	•	•	•	D	D	D	D
S4M2	D	D	D	D	•	•	•	•	N	N	N	N	•	•	•	•	D	D	D	D	•	•	•	•	N	N	N	N
S1M3	N	N	N	N	•	•	•	•	D	D	D	D	•	•	•	•	N	N	N	N	•	•	•	•	D	D	D	D
S2M3	D	D	D	D	•	•	•	•	N	N	N	N	•	•	•	•	D	D	D	D	•	•	•	•	N	N	N	N
S3M3	•	•	•	•	N	N	N	N	•	•	•	•	D	D	D	D	•	•	•	•	N	N	N	N	•	•	•	•
S4M3	•	•	•	•	D	D	D	D	•	•	•	•	N	N	N	N	•	•	•	•	D	D	D	D	•	•	•	•
S1M4	•	•	•	•	N	N	N	N	•	•	•	•	D	D	D	D	•	•	•	•	N	N	N	N	•	•	•	•
S2M4	•	•	•	•	D	D	D	D	•	•	•	•	N	N	N	N	•	•	•	•	D	D	D	D	•	•	•	•
S3M4	D	D	D	D	•	•	•	•	N	N	N	N	•	•	•	•	D	D	D	D	•	•	•	•	N	N	N	N
S4M4	N	N	N	N	•	•	•	•	D	D	D	D	•	•	•	•	N	N	N	N	•	•	•	•	D	D	D	D

[a]Basic pattern (unlocked) = 16 days; full cycle = 112 days.

Table 13 Work System with Rotating Hours and Continuous Operations (Compressed 3–3–3–3): Four Schedules and Forward Rotation[a]

	M	T	W	R	F	S	K	M	T	W	R	F	S	K	M	T	W	R	F	S	K	M	T	W	R	F	S	K
S1M1	D	D	D	•	•	•	N	N	N	•	•	•	D	D	D	•	•	•	N	N	N	•	•	•	D	D	D	•
S2M1	•	•	•	D	D	D	•	•	•	N	N	N	•	•	•	D	D	D	•	•	•	N	N	N	•	•	•	D
S3M1	N	N	N	•	•	•	D	D	D	•	•	•	N	N	N	•	•	•	D	D	D	•	•	•	N	N	N	•
S4M1	•	•	•	N	N	N	•	•	•	D	D	D	•	•	•	N	N	N	•	•	•	D	D	D	•	•	•	N
S1M2	•	•	D	D	D	•	•	•	N	N	N	•	•	•	D	D	D	•	•	•	N	N	N	•	•	•	D	D
S2M2	N	N	•	•	•	D	D	D	•	•	•	N	N	N	•	•	•	D	D	D	•	•	•	N	N	N	•	•
S3M2	•	•	N	N	N	•	•	•	D	D	D	•	•	•	N	N	N	•	•	•	D	D	D	•	•	•	N	N
S4M2	D	D	•	•	•	N	N	N	•	•	•	D	D	D	•	•	•	N	N	N	•	•	•	D	D	D	•	•
S1M3	N	•	•	•	D	D	D	•	•	•	N	N	N	•	•	•	D	D	D	•	•	•	N	N	N	•	•	•
S2M3	•	N	N	N	•	•	•	D	D	D	•	•	•	N	N	N	•	•	•	D	D	D	•	•	•	N	N	N
S3M3	D	•	•	•	N	N	N	•	•	•	D	D	D	•	•	•	N	N	N	•	•	•	D	D	D	•	•	•
S4M3	•	D	D	D	•	•	•	N	N	N	•	•	•	D	D	D	•	•	•	N	N	N	•	•	•	D	D	D

[a]Basic pattern (unlocked) = 12 days; full cycle = 84 days.

Table 14 Work System with Rotating Hours and Continuous Operations (EOWEO): Four Schedules and Forward Rotation[a]

	M	T	W	R	F	S	K	M	T	W	R	F	S	K	M	T	W	R	F	S	K	M	T	W	R	F	S	K
S1M1	D	D	•	•	N	N	N	•	•	D	D	•	•	•	N	N	•	•	D	D	D	•	•	N	N	•	•	•
S2M1	•	•	D	D	•	•	•	N	N	•	•	D	D	D	•	•	N	N	•	•	•	D	D	•	•	N	N	N
S3M1	N	N	•	•	D	D	D	•	•	N	N	•	•	•	D	D	•	•	N	N	N	•	•	D	D	•	•	•
S4M1	•	•	N	N	•	•	•	D	D	•	•	N	N	N	•	•	D	D	•	•	•	N	N	•	•	D	D	D

[a]Basic pattern (locked)/full cycle = 28 days.

Table 15 On-Call Work System[a]

	M	T	W	R	F	S	K	M	T	W	R	F	S	K	M	T	W	R	F	S	K	M	T	W	R	F	S	K
S1M1				D	1	3	•	1	2	1	•	D	2	•	2	N	N	•	D	N	•	1	•	1	1	N	3	•
S1M2	N	•	D	•	N	2	•	•	1	1	•	3	N	•	N	N	2	•	3	•	W	•	X	N	N	N	•	3
S1M3	2	1	•	D	3	•	D	2	•	Y	•	•	•	•	3	•	Z	2	•	•	•	2	•	D	•	N	N	•
S1M4	N	•	3																									

Source: NTSB (1989).
[a]Actual work schedule of a locomotive engineer for 90 days leading to a fatal train collision. W, worked two shifts: one from 0000 to 0800 and another from 1600 to 0030; X, worked two shifts: one from 0100 to 0400 and another from 1345 to 0400; Y, worked two shifts: one from 0045 to 0830 and another 1630 to 0045; Z, worked two shifts: one from 0130 to 0830 and another 1830 to 0500.

makes it quite difficult, if not impossible, for the person to determine in advance exactly when and how long he or she will work. Table 15 provides an example of such on-call operations. It presents the work schedule for a locomotive engineer whose fatal crash has been attributed in part to irregular and unpredictable work hours (NTSB, 1989). The actual work hours of the engineer have been adapted into the representation method used in this chapter.

Despite the fact that irregular and on-call operations are in common use and considered of higher risk with regard to the probability of accidents, there is no agreed-upon or standard way to measure the variability of such work systems. In general, although it is reasonable to conclude that highly variable work systems are easily identified by the layperson, this is nevertheless difficult to quantify. Moreover, it is difficult to predict the degree to which such a system constitutes a health or safety risk unless, at a minimum, adequate rest time and education are provided. For this reason, at present, risk assessment of irregular and on-call work schedules requires expert advice and analysis.

3.5.4 Mixed Operations

The work systems discussed thus far represent only a small number of the total that are possible. With the exception of irregular operations, the work systems presented contain schedules that are similar in shift length and are either rotating or permanent. More often, it is the case that work schedules are mixed within one system. These may include both permanent and rotating shifts of different lengths, part- and full-time workers, as well as other alternative work systems practices.

Table 16 provides an example of a mixed operation with continuous workweeks that uses split, eight-, and 12-hour shifts, as well as full- and part-time workers. Each of the six schedules utilizes permanent shifts. New to this example is the *split shift*, which contains a period of off time in between two periods of work within one shift. This results in work start and end times that are extended and thus do not fall neatly within previously discussed shift designations. For each of the schedules, the basic pattern and the full cycle are seven days in length. A variation of this schedule, not shown, would involve 12-hour rotating shifts, a further mixing of operations. A final example of this type of work system is provided in Table 17, depicting continuous operations in which each of the four schedules employ eight-hour shifts, where three are permanent and one rotates.

3.6 Discussion

The taxonomy proposed in this section is intended to facilitate communication between researchers and practitioners. The importance of using such a means of classification to clearly and accurately describe the temporal characteristics of work systems cannot be overstated. Failure to do so may unintentionally obfuscate significant variables and make it much more difficult to evaluate and compare systems. Further, a unified methodology allows for more rapid transfer of information, thereby increasing the overall knowledge base for understanding the effects of shiftwork. Nevertheless, do note that this is not an evaluation methodology, as proper evaluation of a work system includes many variables that are not obvious from these charts.

The examples presented here show how to apply a representation methodology to work systems and

Table 16 Work System with Permanent Hours, Split Shifts, and Continuous Operations (Weekend Warrior System): Six Schedules[a]

	M	T	W	R	F	S	K	M	T	W	R	F	S	K	M	T	W	R	F	S	K	M	T	W	R	F	S	K
S1M1	1	1	1	1	1	•	•	1	1	1	1	1	•	•	1	1	1	1	1	•	•	1	1	1	1	1	•	•
S2M1	S	S	S	S	S	•	•	S	S	S	S	S	•	•	S	S	S	S	S	•	•	S	S	S	S	S	•	•
S3M1	2	2	2	2	2	•	•	2	2	2	2	2	•	•	2	2	2	2	2	•	•	2	2	2	2	2	•	•
S4M1	3	3	3	3	3	•	•	3	3	3	3	3	•	•	3	3	3	3	3	•	•	3	3	3	3	3	•	•
S5M1	•	•	•	•	•	D	D	•	•	•	•	•	D	D	•	•	•	•	•	D	D	•	•	•	•	•	D	D
S6M1	•	•	•	•	•	N	N	•	•	•	•	•	N	N	•	•	•	•	•	N	N	•	•	•	•	•	N	N

[a]For all schedules, basic pattern (locked)/full cycle = seven days.

Table 17 Work System with Permanent and Rotating Hours, Continuous Operations (Hybrid System): Four Schedules[a]

	M	T	W	R	F	S	K	M	T	W	R	F	S	K	M	T	W	R	F	S	K	M	T	W	R	F	S	K
S1M1	1	1	1	1	1	1	1	•	•	1	1	1	1	1	1	1	•	•	1	1	1	1	1	1	1	•	•	•
S2M1	•	•	2	2	2	2	2	2	2	•	•	2	2	2	2	2	2	2	•	•	2	2	2	2	2	2	2	2
S3M1	3	3	•	•	3	3	3	3	3	3	3	•	•	3	3	3	3	3	3	3	•	•	3	3	3	3	3	3
S4M1	2	2	3	3	•	•	•	1	1	2	2	3	3	3	•	•	1	1	2	2	2	3	3	•	•	1	1	1

[a]S1, S2, and S3 have permanent hours, and S4 is the "continental rota." For all schedules, basic pattern (locked)/full cycle = 28 days.

provide points of discussion that demonstrate the variety and complexity of scheduling options. In no sense should they be viewed as a comprehensive list of all the work systems available or in use. Furthermore, experience and the research literature both suggest that the merit of a specific work system is relative and depends upon the interaction among human factors variables and operational conditions. This is a key point, as using an inappropriate system can lead to diminished worker productivity, safety, and health. It also follows that a specific work system may prove to be appropriate in one workplace at a given time and inappropriate in another place or time. The complexity of the myriad of variables involved suggests that a universal and general ranking of all work systems with regard to their absolute merit is not possible or useful. In subsequent sections of this chapter, we review human factors variables and operational parameters that should be considered in the selection of a work schedule, as well as how one might evaluate a work system.

4 WORK SYSTEM DESIGN CONSIDERATIONS

Traditionally, night work and extended hours have been viewed as difficult and optional duty, with a few exceptions, such as nurses and doctors working to monitor and care for sick people 24 hours a day, petroleum refineries that cannot be easily turned off, and search and rescue operations, as time is always a critical factor if victims are to survive. In many cases, seniority rights and career choices were further means for workers to avoid undesirable schedules, and historically, acute fatigue was perceived as the only significant human factors problem, with the recruitment and retention of high-quality workers as the major personnel problem. However, scientific research on the impact of shiftwork and the changes in work associated with technological developments make these traditional viewpoints an oversimplification of the situation.

In the contemporary workplace, there are numerous human factors concerns to be considered, of which fatigue is one. This is not surprising, as many new technologies and services require around-the-clock operation and certain methods, such as agile manufacturing, often demand nontraditional work hours. Moreover, the increased focus on international finance and intelligence gathering has greatly expanded the number and extent of 24-hour operations to include more white-collar positions. In a very real sense, mastery of continuous and sustained operations can increase profits and mission success if the associated human factors pitfalls can be avoided. Unfortunately, many managers continue to assume that the traditional approach to addressing shiftwork and extended operations is adequate.

This is indeed not the case, as research during the past 60 years has led to major changes in how we think about work schedule issues. Studies have clearly confirmed that biology and performance vary as a function of time of day (U.S. Congress, Office of Technology Assessment, 1991; Jewett, 1997) and that these variations are not the same as time on task variations (Paley and Tepas, 1994a; Smith et al., 1998). Night work has been associated with both acute fatigue and chronic, negative health and performance effects (Duchon and Keran, 1990; Dirkx, 1993). Significant interactions between off-the-job and on-the-job behavior (e.g., Thierry and Meijman, 1994; Khaleque, 1999) have been confirmed, and over time, the demography of the shiftworking population itself has changed. As a result, a contemporary human factors perspective of work schedules must take a total systems approach. In this section we provide a review of the factors that must be considered.

4.1 Circadian Variation

Numerous laboratory and field studies have demonstrated that people have an internal biological timekeeping system. This internal system is referred to as an *endogenous clock, oscillator*, or *biological clock*. It has a natural cycle that was previously thought to run about 25 hours (Wever, 1979); however, it is more likely to be just over 24 hours in duration (Czeisler et al., 1999). Under usual societal conditions, this clock is reset daily so that it has a cycle time of exactly 24 hours in normal, healthy people, and it is therefore termed *circadian* (from the Latin *circa*, about, and *dies*, a day). Physical and social time cues, such as sunlight or mealtimes, called *zeitgebers* (German for "time giver"), are responsible for these fairly modest and normal resettings. Conversely, research indicates that the human circadian system is typically resistant to sudden large changes in routine and is quite stable (Barton and Folkard, 1993). For example, people placed in isolation for weeks without a clock but with total freedom to determine their activity schedule continue to live on a self-selected routine based on a circadian day of about 25 hours (Wever, 1979; Czeisler et al., 1999).

The circadian system produces concomitant changes in most physiological and behavioral variables. Physiological variables peak at different times of day, but under stable conditions, they are in relative synchrony with one another (Comperatore and Krueger, 1990). Both common sense and theory suggest that good health requires maintenance of this normal synchrony of the behavior rest–activity cycle (including work) with circadian biological variation. Under these conditions, circadian rhythms are entrained, or synchronized, with body temperature or blood serum melatonin level often used as a measure of this variation (Monk and Embrey, 1981; Laakso et al., 1991). Figure 1 is a simplified model of human circadian variation over time of day. As is shown in this figure, body temperature reaches a minimum (trough) in the early morning, between 0300 and 0500, and a maximum (peak) about 12 to 15 hours later. There is also literature to suggest a brief dip in this rhythm between 1300 and 1500, termed the *postlunch* or *postprandial dip* (Hayashi and Hori, 1998).

Work schedules that are limited to daytime activity with nighttime rest maintain and promote circadian rhythm synchrony, thereby sustaining the entrainment of alertness and vigilance to body temperature variation (i.e., decreased body temperature is related to decrements in alertness and vigilance) (Folkard and Monk, 1979; Monk and Embrey, 1981). In this case, work acts as a facilitating *zeitgeber*. Since research indicates that many circadian biological systems are quite resilient to abrupt changes in routine, synchrony of circadian systems with alternate work schedules, such as the third shift, may not be easily attained. As modeled in Figure 2, experts have proposed that exposure to some work schedules can result in a decrease in the amplitude of the circadian variations (function A) or circadian rhythm desynchronization (function B). Either or both of these changes in the synchronization of the circadian systems are indicative of an intolerance to shiftwork (Reinberg et al., 1984). In fact, both laboratory and field studies of shiftworkers suggest that over 10 days of uninterrupted nighttime work may be needed before full circadian adjustment occurs (Knauth and Rutenfranz, 1976; Åkerstedt, 1977; Dirkx, 1993; Totterdell et al., 1995).

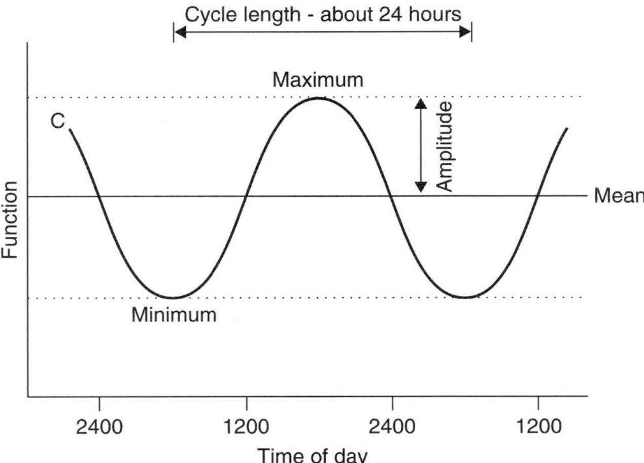

Figure 1 Simplified model of human circadian variation as measured by body temperature. Oral temperature is found to reach a minimum in the early morning and a maximum around 12 hours later. Alertness ratings and performance are often entrained to these circadian variations. (From Tepas et al., 1997.)

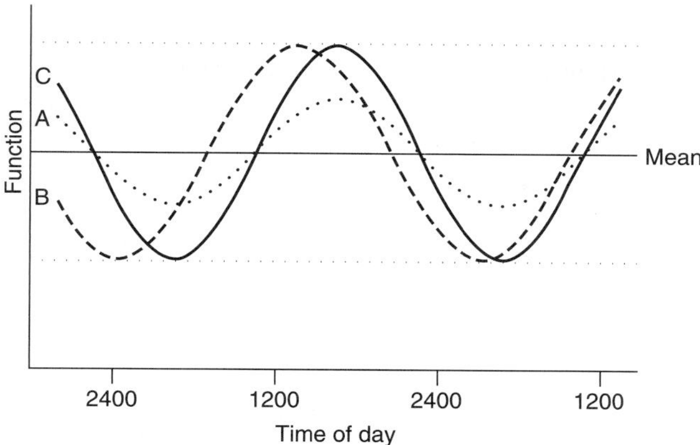

Figure 2 Simplified model of changes in circadian systems that may be associated with exposure to some shiftwork systems. Function C is the usual human circadian variation, as presented in Figure 1. Function A is an example of a decrease in the amplitude of the circadian variation, and function B is an example of circadian rhythm desynchronization. (From Tepas et al., 1997.)

In addition, work schedules that include work on the third shift require job performance during times around the workers' circadian trough. Along with the postprandial dip, these times have been termed *zones of extra vulnerability*, wherein circadian variations make peak performance unlikely and may increase the risk of accidents (Smith et al., 1995). Taken from these viewpoints, regular work at chronobiologically inappropriate times and the resulting impact on circadian rhythm variation are both seen as hazards for ergonomic work schedule designers to factor in when planning new work systems. Job workload, safety criticality, environment, and staffing levels, among other variables, all need to be evaluated and considered when work is performed regularly during these zones. For this reason, an occupational health and safety approach may be advisable, in which working during nighttime hours is considered exposure to a hazard, and thus a safety plan is drafted regarding how this hazard will be controlled.

As mentioned above, research has indicated that 10 or more days of night work may be required for adjustment to occur. This is not a prescription for longer periods of continuous night work, however. Because night-shift workers often revert to daytime schedules on non-workdays (Tepas and Carvalhais, 1990), circadian desynchronization is possible unless a nighttime regime is maintained on both work and non-workdays. In essence, the need for 10 days suggests that full circadian adjustment is an unlikely outcome for these workers, making this an important issue for work system design. Although there do appear to be individual differences in biological regulation that are related to shiftwork tolerance (Härmä, 1993), it is unlikely that these differences can or should be used as criteria to select workers for nontraditional shifts. Rather, it is reasonable to suggest that a particular person's ability to tolerate working shifts is multifactorial and not well understood. It also appears to be as much a function of preference and social variables as it is a function of physiology.

4.2 Light Exposure

Given that circadian desynchronization is a common outcome of shiftwork, it should come as no surprise that during the last 20 years there has been considerable interest in the use of light to control or change circadian variation. Animal research has clearly demonstrated that light functions as a powerful *zeitgeber*, and, if appropriately timed, brief exposure to light can result in fairly abrupt and major changes in circadian rhythms. More recent laboratory and field studies indicate that similar changes are also reproducible in humans (Czeisler et al., 1990; Eastman and Miescke, 1990; Crowley et al., 2003). These results demonstrate the possibility that applying specific light exposure techniques may facilitate adjustment to shiftwork through circadian rhythm reentrainment. However, research has yet to demonstrate a consistent set of recommendations regarding the optimal light levels required for reliable impact on human circadian rhythms (Rosa et al., 1990; Martin and Eastman, 1998). For example, lighting parameters used have tended to vary from study to study with regard to intensity, wavelength, temporal placement, and exposure duration, making interpretation of results difficult.

Nevertheless, if the systematic research of Eastman and her colleagues (Eastman et al., 1994; Crowley et al., 2003) is reliable and representative, circadian change requires bright light at night, dark sunglasses during the day, an anchored sleep time soon after shift completion, and several days of optimal exposure. Taken together, these finings suggest that the conditions required for light exposure to have a positive impact on work schedules may be difficult to obtain and utilize for practical commercial applications (Czeisler et al., 1990; Martin and Eastman, 1998). However, as a well-lighted workplace serves, if nothing else, to improve visibility, there is probably little reason to suggest that the light exposure schemes used in these studies could have a negative impact on job performance or health.

4.3 Sleep

Research has clearly shown that sleep is a robust benchmark for the study and evaluation of work systems. The impact of work hours on sleep is manifest in at least two ways. Initially, there is an acute and immediate physiological response to a sudden alteration of when a person is awake and asleep. Examples of this type of alteration include transmeridian jet travel and changing on to or off of a night shift (Samel et al., 1997). These examples are distinguished by the fact that transmeridian travel is an acute, transitory event from which recovery occurs over a period of time. However, in the case of shiftwork, every episode of working during the nighttime may be considered similar in disruption to flying from Los Angeles to Beijing. Exposure to five night shifts in a row would therefore be similar to five consecutive transmeridian flights. It is easy to understand how this type of schedule might impede recovery or adaptation, as there is continual disruption to the sleep-wake cycle.

Shiftwork research findings on the acute effects of working during nighttime hours indicate that the problems that initially manifest include subjective insomnia-like complaints and changes in polysomnographic sleep-stage sequencing, suggesting a disturbed sleep (Torsvall et al., 1989). Over time, a chronic response associated with extended exposure to work schedules that conflict with sleeping during nighttime hours may develop. At this point, getting enough sleep is a regular difficulty for the worker, and his or her polysomnographic record begins to resemble that of a person who is sleep deprived. It is this reduction in workday sleep length, and the resulting chronic loss of sleep, that troubles many experienced shiftworkers (Tepas and Carvalhais, 1990).

If this lost sleep is not recovered, there is a carryover effect and a sleep debt accumulates. Continued exposure to working nighttime hours can consequently result in *accumulated sleep debt* (ASD), a state of chronic sleep deprivation for the worker. It is assumed that as ASD increases, worker health and safety risks increase concomitantly (Tepas and Mahan, 1989).

Figure 3 provides a good example of the sleep lengths many studies have reported for experienced shiftworkers: on workdays second-shift workers sleep the most, third-shift workers sleep the least, and first-shift workers report sleep lengths that are somewhere in between. For all three groups of experienced permanent shiftworkers, sleep on non-workdays is significantly longer than workday sleep (Tepas and Carvalhais, 1990; Kaliterna et al., 1993). However, reports of non-workday sleep durations do not differ significantly from each other. A study on transit bus operators working split shifts found similar results, where the mean reported workday sleep length totaled 386 minutes, compared to 489 minutes for non-workday sleep reports (Howarth, 2003). In this sample, shorter workday sleep was probably due to early morning shift starts. A reduction in workday sleep duration, together with small, positive correlations between workday and non-workday sleep length (Tepas and Carvalhais, 1990; Howarth, 2003), suggest that it is unlikely that non-workday sleep serves to make up lost workday sleep in permanent shiftworkers. If non-workday sleep were indeed compensating for reductions in workday sleep duration, the direction of the correlation would instead have been negative. Moreover, in studies of rotating shiftworkers, similar and sometimes greater night-shift workday sleep reductions are evident (Tepas et al., 1981b; Paley and Tepas, 1994a). Therefore, among third-shift workers, there is little evidence to support the notion that one can easily recover from ASD.

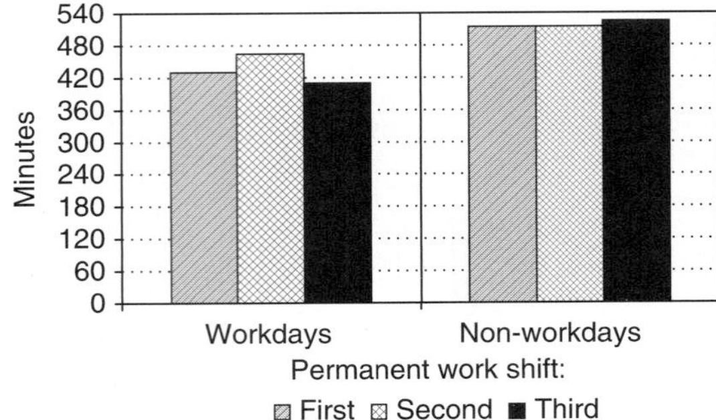

Figure 3 Mean workday and non-workday sleep length in minutes for permanent discontinuous hourly shiftworkers on eight-hour shifts. For all three shift groups, workday sleep is significantly shorter than non-workday sleep, with third-shift workers sleeping the least during the workweek. Differences between shifts for non-workday sleep are not significant. Data graphed from a survey of 1262 workers, as reported by Tepas and Carvalhais (1990).

There is also literature suggesting that ASD persists and is difficult to avoid. Figure 4 depicts data from a longitudinal study of rotating shiftworkers (Gersten, 1987) who were surveyed three times over a six-year period. The reduction in sleep associated with third-shift work at the time of the first survey was still present after six years. As there is also little evidence that workers who prefer the night shift are self-selected, naturally short sleepers (Tepas and Mahan, 1989), it is reasonable to suggest that a major problem for experienced night workers is getting enough sleep.

In that vein, it should be noted that shiftwork has recently been added to the International Classification of Sleep Disorders (ICSD-9), under heading 307.45, *Shift Work Sleep Disorder*, with symptoms ranging from transitory insomnia to excessive waketime sleepiness. Despite qualifying as a medical disorder, the authors caution the reader against concluding that all shiftworkers are sick or will become sick as a result of working shifts. This is clearly not the case, even

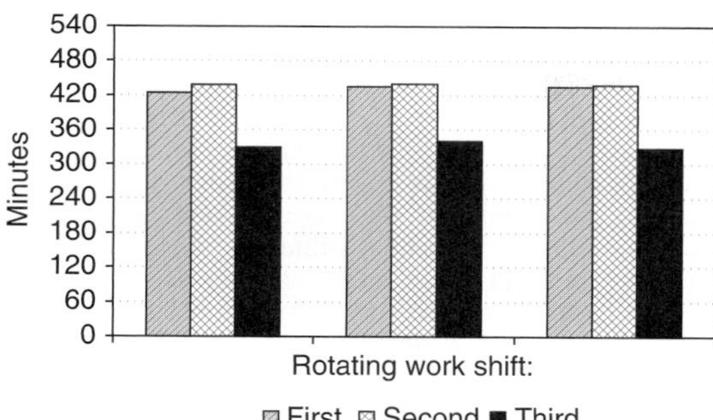

Figure 4 Mean workday sleep length in minutes for experienced rotating shiftworkers. The same shiftworkers were surveyed three times over a six-year period. For each survey, the workday sleep length was significantly shorter on the third shift. Changes over time were not statistically significant. Data graphed from survey results reported by Gersten (1987).

though certain shiftwork systems may have features that can affect aspects of health. For example, it is known that human growth hormone is secreted almost exclusively during nighttime sleep, a process that is obviously disrupted in shiftworkers who work during this time. It is also the case that there are nighttime metabolic changes affecting how food is processed and stored, which when combined with a nutritionally poor diet, may lead to weight gain and increased risk of other health outcomes, such as obstructive sleep apnea. As health maintenance is complex, involving a constellation of variables, it is therefore difficult to causally link to shiftwork schedules or the work itself. Nevertheless, it is evident that the problems experienced by those working shifts are more than a simple reduction in duration and/or a change in the timing of sleep.

4.4 Napping

Many studies have reported that night and rotating workers have a higher incidence of napping than those working other shifts (Tepas 1981a, 1985; Chan et al., 1989). However, despite the presence of nappers in the workforce, this activity remains neither fully understood nor well defined, largely because it is complex and multidimensional. It should not be surprising, then, that there is no consensus in the literature regarding the utility of shiftworkers using this strategy as a coping mechanism. This is probably a result of the fact that although napping has the potential for positive effects (i.e., reduction of *acute* sleepiness and sleep loss), there may also be negative outcomes, such as the reduction of main sleep period duration (Chan et al., 1989; Torsvall et al., 1989; Rosa, 1993). The potential for varied outcomes makes selecting a napping strategy contingent upon taking a systems approach that considers workplace variables. It is important to note that increased napping behavior may be indicative of poorly designed work or the inability of the worker to otherwise cope with an assigned schedule. In practice, strategies for dealing with shiftwork are often selected and combined without considering systems issues, and therefore workers may include napping as a compensatory measure when in reality a schedule redesign is necessary. This section discusses three napping strategies that should be regarded by practitioners: cultural, prophylactic, and replacement.

4.4.1 Cultural Napping

Cultures, including corporate cultures, differ significantly in the degree to which they promote or discourage napping. Many cultures have incorporated the nap as accepted or expected behavior. The siesta of hot climates is a good example; one investigator found that over 7.5% of the population he studied took naps on a regular basis (Taub, 1971). In Japan, napping on the job during work hours is a long-standing, widely practiced, and accepted tradition (Matsumoto et al., 1982).

In the U.S., however, sleeping on the job at any time is typically considered an infraction of workplace rules that may lead to employee dismissal, although changes in this prohibition are slowly appearing. In some cases, railroads that had unilaterally been dismissing employees who napped have now instituted policy allowing napping in certain situations, such as when a train is stopped at a siding (Sherry, 2000). In another example, there exist transit bus operations that provide a napping facility for employees to use during off times or breaks in split shifts (TRB, 2002).

Explanations for the culturalization of naps are often "sermons" by "believers" and therefore very difficult to objectively evaluate. Some propose that sleep during the heat of the day in tropical climates is a way of avoiding the hazards of heat stress, while others posit that napping assures quality work by reducing employee sleepiness and fatigue. In contrast, empirical research on cultural napping does support the existence of *recreational* nappers, who habitually nap for pleasure and differ from those who take naps for other reasons (Evans et al., 1977). Regardless of individual beliefs, a good systems approach to shiftwork should take into account variations in cultural napping practices.

4.4.2 Prophylactic Napping

Undoubtedly, some workers take naps to prepare themselves for an anticipated future period of sleep loss. This is sometimes referred to as "putting sleep in the bank." Åkerstedt et al. (1989) reported that longer nap durations were associated with less subjective fatigue reported over the shift; similarly, Carskadon and Dement (1981) noted that napping while on certain restricted sleep schedules may ameliorate further decline in subjective feelings of sleepiness. However, in some cases, laboratory studies of naps taken before a period of total sleep deprivation appear to yield only modest improvements in performance (Dinges et al., 1987). Additionally, there is evidence that naps may have a long-term carryover effect (Gillberg, 1984), as well as a stronger or weaker inertial effect, depending on factors such as duration and/or placement (Jewett et al., 1999). It should be noted that, problematically, there is no consistency within the literature pertaining to nap length and nap placement with regard to time of day or proximity to a work period (Sallinen et al., 2003).

Anecdotal discussions with rotating shiftworkers suggest that some of them, in fact, adopt a quite different strategy than the use of prophylactic napping. Instead, these workers may sleep deprive themselves before switching to the night shift, in an effort to promote "good" sleep after their first night on that schedule. How prevalent this strategy is remains unknown. A late-evening nap before a first night shift has the potential for a positive and immediate effect on work performance a few hours later. However, the long-term impact on third-shift workers of prophylactic napping remains open to question.

4.4.3 Replacement Napping

A replacement napping strategy is intended to make up for previously lost sleep. Such napping behavior may occur either during work hours, commonly referred to as *maintenance napping*, or outside of work. Maintenance napping may be both necessary and helpful when work schedules call for sustained performance or for schedules that result in chronic circadian desynchronization. In these cases, napping may improve alertness by anchoring future sleep and thereby maintaining circadian rhythm synchronization (Naitoh and Angus, 1989; Dinges et al., 1991). In the United States, where sleeping on the job is most often prohibited, elevated levels of shiftworker napping are presumed to be efforts at replacing lost sleep. However, the reader is cautioned that just as napping is not a good way to cure insomnia, frequent napping by permanent third-shift workers should also probably be avoided due to its potential effect on primary sleep periods. As an example of this, problems falling or staying asleep were more likely to be reported by third-shift workers who napped frequently (Härmä et al., 1989; Tepas, 1993; Matsumoto and Harada, 1994). Additionally, the extent to which the effects of sleep inertia (i.e., how long it takes to wake up and perform well) might be problematic for a worker, especially in the case of maintenance napping, must be considered before concluding that this type of strategy is warranted (Naitoh et al., 1993). Although such napping is often perceived by workers as a good recovery effort, its use should be approached with caution. It has both the potential to produce negative and immediate performance decrements due to sleep inertia (Jewett et al., 1999), as well as lead to long-term complaints about sleep problems.

4.5 Fatigue

In many traditional viewpoints, fatigue is considered a product of "doing something" for a long period of time and is measured simply as the amount of time spent performing or being exposed to a task. In the workplace, this approach equates fatigue approximately to the length of any given work shift. There is considerable evidence, however, indicating that fatigue is, in fact, not a singular variable; rather, it is a complex, multidimensional construct. This assertion does not discount the importance of time on task as a contributor to the development of fatigue, but it does preclude the notion that fatigue is equivalent to time on task.

A complete discussion of fatigue is beyond the scope of this chapter. However, in conjunction with time on task, three facets of this construct deserve mention, as each is particularly relevant to the process of assessing and evaluating work schedules and demonstrating that time on task cannot solely account for fatigue outcomes. These are sleepiness and tiredness, health, and time of day.

4.5.1 Sleepiness and Tiredness

Johnson and Naitoh (1974) suggested that the only certainty about sleep loss is an increase in sleepiness, while at the same time noting that sleep deprivation increased feelings of fatigue. Corroborating this relationship, Kribbs and Dinges (1994) pointed out that fatigue resulted from inadequate rest and was thereby related to sleepiness and being tired. As such, one can reasonably argue that changes in sleepiness and tiredness occur in a similar fashion to changes in fatigue. Research has shown, however, that how physically and mentally tired a worker feels is often not correlated with the length of the workday (Tepas and Popkin, 1994), supporting the idea that time on task is not the singular, causal factor contributing to fatigue outcomes. In fact, as depicted on the right panel in Figure 5, a majority of workers on all shifts reported feeling sleepy at work, and significantly more night workers responded this way. Because each of the shifts was eight hours in duration, reductions in sleep length, and not simply time on task, probably contributed to these findings.

4.5.2 Subjective Health

A number of health conditions are associated with fatigue. A discussion of these ailments (e.g., insomnia

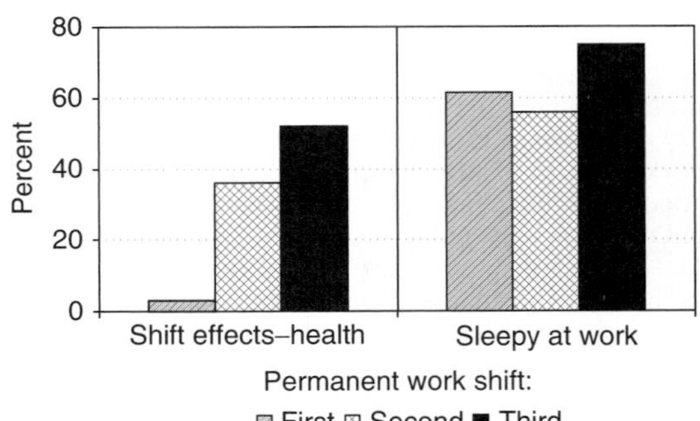

Figure 5 Percentage of workers responding to two survey items, one asking whether they feel that their health would improve if they worked another schedule (left panel), and the other asking about feeling sleepy at work (right panel). These data are from a study of 1490 hourly workers on permanent discontinuous shiftwork (Tepas et al., 1985). For both items, the rate is significantly higher for third-shift workers.

and sleep apnea) is beyond the scope of this chapter; however, it is generally held that increased levels of fatigue are associated with deteriorating health. Accordingly, changes in subjective health ratings may indeed be outcomes of chronic exposure to fatigue-inducing work systems, especially if appropriate treatment and/or time to recover are lacking. As depicted in Figure 5, when workers were asked whether they thought that their health would improve if they worked a different schedule, third-shift workers responded affirmatively at a significantly higher rate than those on the first or second shift. These data should not be interpreted to mean that night workers experience more health problems; however, they do suggest a link between subjective health and schedule worked. Other research has demonstrated that subjective reports of well-being and mood fluctuate with work schedule variables, further challenging the historical, singular notion of fatigue (Paley and Tepas, 1994a).

4.5.3 Time of Day

More so than sleepiness and tiredness or subjective health, time of day directly confronts the assumptions of a time-on-task-based definition of fatigue. Figure 6 superimposes sleep and work times for first-, second-, and third-shift workers on the model of circadian variation presented earlier in the chapter. More than 90% of a large sample of survey data from permanent and rotating shiftworkers in the United States fit this model (Tepas et al., 1985). As depicted in the case of rotating shiftworkers, changing shifts not only alters work and sleep times, but also the order in which these activities occur. Additionally, this figure shows how biologically based circadian variation in alertness is differentially influenced by the work and sleep behavior of shiftworkers. In particular, the way in which this daily variation interacts with time on task is a function of shift timing or, more specifically, time of day.

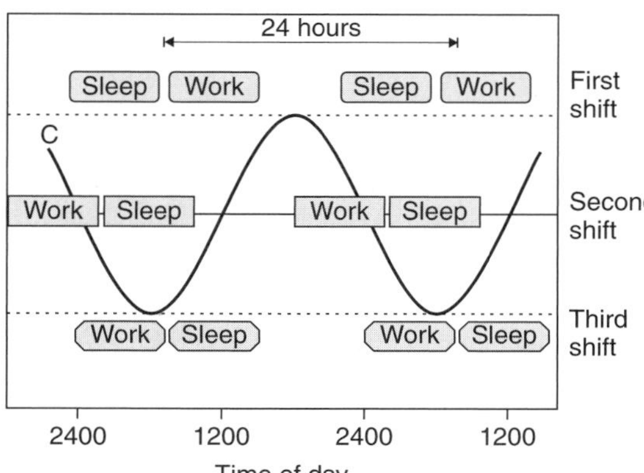

Figure 6 Model of when people work and sleep while on shiftwork, superimposed on the circadian model presented in Figures 1 and 2. The model shown in this figure fits greater than 90% of a total 3150 participants in two studies. These samples included workers on rotating and permanent shiftwork, and both hourly and salary workers. (From Tepas et al., 1997.)

With regard to experiencing sleepiness at work, shift differences have been reported, as shown on the right panel in Figure 5 (Tepas et al., 1985). One explanation of this finding is that ASD (described in Section 4.3) was a contributing factor and was most severe for those working at night. This is likely; however, the influence of time of day on reported sleepiness must also be considered. As body temperature drops, sleepiness increases, such that greater levels would be expected at 0300 as compared to 1100. Consequently, an eight-hour shift starting at midnight is unlikely to generate equal amounts of fatigue as an eight-hour shift starting at 0900. Overall, it must be emphasized that persons responsible for the assessment and/or design of work schedules and systems fully consider the complex interaction effects among the variables discussed in this section.

As a final point, it should be mentioned that the past 10 years has seen an increase in research and modeling efforts aimed at better understanding the construct of fatigue and its effects on biological and performance outcomes (Mallis et al., 2004). Most of this research frames fatigue as a combination of temporal placement of an event on the circadian rhythm, and the duration of the preceding sleep and current waking periods. Moreover, the authors of these research initiatives attest that a majority of the variance in performance and reported sleepiness can be accounted for by careful manipulation of these few variables. This is, however, likely to be an oversimplification, as noted by Tepas and Price (2001). These researchers presented a much more multivariate approach to address the outcomes related to chronic exposure to fatigue-inducing work schedules including such items as health impacts and shiftwork intolerance. Although operations managers and workers may prefer to focus on acute needs and impacts, it is important they also understand that chronic fatigue is multidimensional. Ultimately, if not addressed, outcomes including health problems, attrition, and other, though perhaps less-often considered, bottom-line effects (e.g., staffing issues, hiring, training) may result.

4.6 Social Variables

Historically, many employers have ignored the impact of social variables, including off-time activity, on the workplace performance and well-being of their workers. In recent years, a number of factors have gradually led more employers to be concerned with off-time activity for its potential effects on worker health and performance. These factors, all of which significantly escalate the cost of doing business, include the increased cost of employer-paid health care and provisions to care for children/dependents; progressively more workplace diversity in age, gender, and race; increasing costs of performance errors in the automated workplace; and a higher rate of worker turnover. Although employers have no true ability to control or influence social factors that are related to the shifts their employees work, they can and should strive to provide a well-designed work system and supporting activities that make shiftwork more usable for and agreeable to the majority of their workforce.

Shiftwork has the potential to have a significant positive or negative impact on worker off-time and social activity (Gordon et al., 1981; Thierry and Meijman, 1994; Khaleque, 1999), and the resulting off-time behavior may, in turn, have a significant effect on work satisfaction and performance (Thierry and Jansen, 1984). A reexamination of Figure 6 helps illustrate this when you consider that many workers are *job bound*. That is, they must be at work at a specific time and remain there for a given period. Sleep is also (typically) a required daily activity but allows workers some flexibility in determining *when* and *how long* they will sleep. The model presented in this figure shows the self-selected practices of most shiftworkers, choices that are tempered by social as well as biological factors. As depicted, the sleep of first-shift workers is truncated to the extent that they must report for work each morning after rising from their major sleep period. Alternatively, workers on the second or third shift are less likely to have to truncate sleep for the purposes of reporting to work, as their shifts start later in the day. For third-shift workers, who sleep after work, wakeup time is not at all job bound. However, these workers report the shortest sleep periods and consequently probably experience the greatest ASD (Tepas and Carvalhais, 1990). Since an estimated 85% of these third-shift workers use an alarm or some other device to awaken (Tepas, 1982), it is reasonable to argue that social variables play a role in their reduced sleep length. That is, the third-shift reductions in sleep duration are probably more than simply a function of endogenous circadian variables. Rather, it is probable that most third-shift workers cut their sleep short for reasons including wanting to spend time with family or friends in the late afternoon and early evening, or to meet the mandatory demands of everyday life.

In addition to differences in sleep-length duration, the home, family, and social lives of workers have been shown to vary dramatically with work shift schedule. As depicted in Figure 7, data from 1490 workers on permanent, discontinuous shifts indicated that approximately 60% of families of both second- and third-shift workers preferred that their loved one work a different schedule (Tepas et al., 1985). This is in contrast to only a very small percentage of families of first-shift workers. Within the same data set, worker social life may also be viewed as potentially disrupted, as significantly fewer second- and third-shift workers reported their friends working the same shift as them. Additionally, cross-sectional (Tepas et al., 1985) and longitudinal (Gersten, 1987) data from two studies each suggested that working the third shift may increase the divorce rate, as did results from a subsample of couples who took part in a national survey (Presser, 2003).

Importantly, not all relationships between work system and social variables are negative. Second- and third-shift workers in the Tepas et al. (1985) sample appeared, in fact, to have *less* difficulty seeing a doctor about their health, and some divorced second- and third-shift workers who were single parents reported that their work schedule made child care easier. It is important to recognize that numerous social variables interact to determine worker suitability for a given work schedule. This, in turn, suggests that good work system design requires consideration of social variables in addition to workplace, biological, and economic variables.

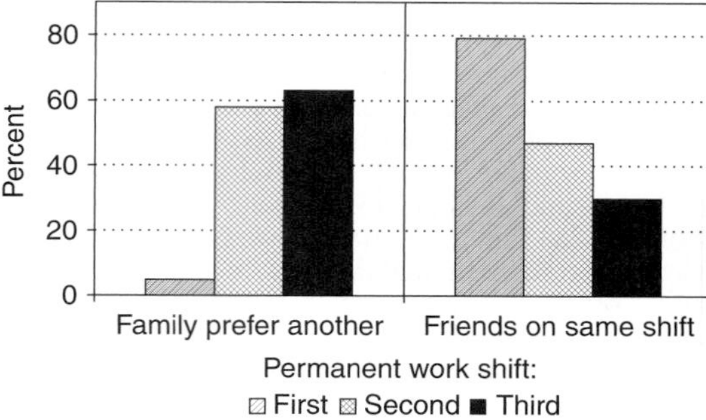

Figure 7 Percentage of workers responding "yes" to two survey items, one asking whether their family preferred that they work another schedule (left panel) and the other asking whether their friends worked the same shift as they do (right panel). These data are from 1490 hourly workers on permanent discontinuous shifts, as also shown in Figure 5. The shift difference is statistically significant for both graphs.

4.7 Diet and Exercise

The timing of eating and drinking, as Figure 6 illustrates, cannot be the same for all work shifts. For example, third-shift workers sleep during the hours around noon. When compared to those on the first shift, these workers report that they eat fewer meals, have poorer appetites, are less satisfied with their eating habits, and eat at different times of the day (Tepas, 1990a). These differences raise questions regarding the degree to which nutrition may vary with shift, and if eating and drinking can be manipulated to improve worker shift adjustment.

European studies have failed to find any differences in the overall nutrient intake of shiftworkers compared to control groups (Romon-Rousseaux et al., 1986; Lennernas, 1993). However, there is support for a difference in the circadian distribution of food intake and caffeine consumption that is unique to those working at night. Lennernas et al. (1995) evidenced a reduction in intake during the night shift itself, but found no differences among shifts when intake was aggregated over a 24-hour period. Additionally, when controlling for age and other demographic variables, shift differences in caffeine and alcohol intake appear to be modest or nonexistent (Tepas, 1990a; Lennernas, 1993). It is noteworthy that cultural variation in food consumption and habits make interpretation and the generalizability of findings in this area difficult to reconcile (Takagi, 1972; Cervinka et al., 1984).

Several researchers have suggested the use of special diets (Ehret, 1981; Wurtman, 1986); however, the merit of such eating plans is questionable, as they appear to be based mostly on experimentation with animals and have not been tested satisfactorily with shiftworkers (Tepas, 1990a). In one case where a field study was performed, shiftworker food intake was manipulated, but no significant changes in mood or alertness resulted (Romon-Rousseaux et al., 1987). Although many consultants advise shiftworkers on diet issues, Popkin (1994) reported data suggesting that this sort of information, while well received, does not appear to lead to any major behavioral changes. However, as basic laboratory research indicates that meals (Smith, 1988), caffeine (Walsh et al., 1990), and alcohol (Wilson et al., 1956) may result in fairly immediate and persistent postconsumption behavioral effects, it seems reasonable to suggest that the dietary intake of shiftworkers is an area worthy of additional research and recommendations. Factors such as optimal food choices, quantities, and how to time intake all appear relevant to shiftworker lifestyle.

A further issue to highlight with regard to diet is the difficulty shiftworkers may have in obtaining healthy meal choices as related to the time of day of work and/or the nature of the job. For example, for night-shift workers who do not bring their food to work, meal options may be severely limited, as most restaurants or workplace eating areas are probably not open 24 hours (excepting fast food, an often less-than-healthy option). Also, for certain jobs, such as transit bus operators, actual meal breaks to purchase and eat food may, in reality, not exist. Such workers often do not have long break periods, if any, to obtain food between trips on their routes. If they do have a scheduled layover that is long enough to step out, it is also likely that company policy prohibits them from leaving a vehicle unattended. Equally problematic, if an operator brings along his or her own food, it is often the case that eating while driving is forbidden by companies, although this appears nevertheless to be practiced. This ultimately leaves many transit operators with limited food options; consequently, many resort to unhealthy choices that are quick and found in company vending machines. Even though there appears to be a trend toward stocking such machines with more produce, microwaveable meals, and nutritionally sound snacks, the fact remains that shiftworkers continue to have few healthy food options to choose from and minimal or no true break times during which to eat.

Given the various health ailments that appear to be related to working shifts (see Section 4.9), research has in recent years also sought to determine whether shiftworkers are less physically fit than day workers. In a study using the General Health Questionnaire paired with physical fitness tests, it was evidenced that shiftworkers had lower fitness levels than those of their day-working counterparts (Yamada et al., 2001). Similarly, Ishizaki et al. (2004) conducted a cross-sectional study of Japanese workers to assess the relationship between body mass index and waist-to-hip ratio. Their findings indicated a deleterious relationship with shiftwork for both of these indexes. Finally, in a study on shiftwork tolerance and physical fitness levels, Härmä (1996) determined that, among other benefits, being fit appears to increase shiftwork tolerance. Nevertheless, many shiftworkers and dayworkers alike struggle to integrate regular exercise into their lifestyles.

Additional research has incorporated exercise or physical training into routines as a means of promoting circadian adjustment to shiftwork. Generally, findings suggest that appropriately timed exercise or physical training may facilitate circadian phase shifts and consequently promote adaptation to night-shift work (Härmä et al., 1988; Eastman et al., 1995). Results such as these are encouraging but must be interpreted with caution if other relevant individual-difference factors have not been controlled statistically or otherwise.

4.8 Individual Differences

A persistent notion for many years has been the idea that some people are especially suited for night work. With this notion comes the temptation to try to select the "right" workers and/or to prohibit the "wrong" workers from performing shiftwork. Some of the factors that have been evaluated for use in this process include age, gender, personality, and shift preference (discussed in Section 4.11). Despite attempts to demonstrate otherwise, in a review of individual differences in shiftwork adjustment, Monk and Folkard (1985) noted that "it seems unlikely that we will ever reach a position where we can distinguish between people who are suited and those who are not suited to [shiftwork] in general" (p. 237). This conclusion continues to be appropriate (Kaliterna et al., 1995).

4.8.1 Age

Although individual difference variables are not a feasible criterion for selecting workers for shiftwork, research does suggest a number of ways that such variables are relevant to work system design. For example, several studies demonstrated that older shiftworkers have particular problems associated with night shiftwork, as they were found to report shorter sleep durations and increased ill health compared to their younger counterparts (e.g., Tepas et al., 1993; Bourdouxhe et al., 1999). These results suggest that work system designs requiring extensive work at night may not be appropriate for older workers; however, age findings must always be interpreted with caution, as the general population also ages and changes with regard to health and sleep. Therefore, rather than advocate age restrictions or requirements for shiftwork, employers should instead be sensitive to and plan for the possibility of age effects. Work systems that are initially deemed acceptable may indeed become unacceptable or unhealthy as an organization's workforce ages. Thus, there is a constant need for evaluation of schedules over time with regard to work force demographics.

4.8.2 Gender

Investigating gender differences in shiftwork tolerance presents a challenge for different reasons. Historically, there existed specific prohibitions against employing women on the night shift. As these provisions have gradually been removed, more and more women have entered into shiftwork, and gender differences have consequently become an issue. Research findings appear to be somewhat inconclusive and are probably complicated by gender roles within many societies that have traditionally placed the primary burden of child care and other household responsibilities on women. In one study, Oginska et al. (1993) compared age- and occupation-matched male and female shiftworkers on various shiftwork tolerance outcomes. Results indicated that women slept less, reported more sleep disturbances and increased drowsiness during work, and suffered more health symptoms related to shiftwork intolerance (see Section 4.9) than did their male counterparts. By contrast, in other research, analysis of data from three additional samples of workers all suggested that the impact of working the night shift may be less on women than on men (Dekker and Tepas, 1990). As societal viewpoints evolve further with regard to women's roles, domestically and in shiftwork, this research area will continue to merit study.

4.8.3 Morningness–Eveningness

There has been considerable study of the concept *morningness–eveningness*, as coined by Horne and Ostberg (1976). Research in this area is driven by the notion that individual preferences full on a continuum between evening types (*owls*) and morning types (*larks*), and that people may be assigned to work shifts on the basis of where they lie on this continuum. In an effort to further the development of this construct, Smith et al. (1989) completed a psychometric analysis of three morningness–eveningness questionnaires, and Brown (1993) used their results to produce a scale suitable for assessment of industrial workers. In general, however, studies have found that distributions of scale scores on this dimension are nearly normal and certainly not dichotomous. In fact, very few workers are clearly identifiable as evening types; consequently, it would be difficult to assemble a large number of them for night-shift work (Mahan et al., 1987). The possibility of designing a work system based on morningness–eveningness principles seems therefore inappropriate, although practical utility may still exist for use of this concept as a means to assess persons who appear particularly unable to cope with working shifts.

4.9 Health Status

Several published studies on shiftworkers have concluded that shiftwork does not affect mortality adversely (Taylor and Pocock, 1972; Bøggild et al., 1999), and Poole et al. (1992) asserted that there are no long-term ill effects of shiftwork. More recent views suggest that working at night or on shift systems is most strongly related to medical disorders, including gastrointestinal disease, coronary heart disease, and compromised pregnancy outcomes (Knutsson, 2003). Epidemiological studies of shiftworkers are not simple to conduct, however, and there are many contributing factors to consider in addition to shift, which is why mixed findings are not surprising (for a review, see Costa, 2003).

Nonetheless, there is significant reason for concern (Waterhouse et al., 1993). Some European and Japanese studies (e.g., Koller et al., 1978; Angersbach et al., 1980; Morikawa et al., 2001) have shown that shiftworkers are sick and/or absent from work more than dayworkers. Also, it is noted that health problems in shiftworkers are particularly difficult to fully explain and understand, since many of those who cannot tolerate shiftwork drop out. This appears to result in a population of shiftwork "survivors" who are healthier than both their day-working counterparts and those who left shiftwork, as this is often due to illness.

Over the last 15 years, evidence has accumulated from Scandinavian and European research indicating that work environment conditions contribute to the etiology of cardiovascular disease (CVD) (Knutsson, 2003). In a review of 17 such studies, a 40% excess risk for CVD was found for shiftworkers compared with dayworkers (Bøggild et al., 1999). Although a strong link appears to be present, this research should not be considered conclusive in a global sense, since cultural factors and shift design features cannot be generalized and are difficult to isolate. Research also exists suggesting that generally, gastrointestinal disorders are more common in shiftworkers than in dayworkers, and specifically, that shiftwork and peptic ulcer disease are related (Knutsson, 2003). These latter findings probably maintain their merit, despite the fact that much of this research dates back to the 1970s, when diagnosis was often not verified through the use of endoscopy or x-ray procedures (Knutsson, 2003). Finally, with regard to women, studies undertaken largely outside the United States have evidenced an association among shiftwork and various compromised pregnancy outcomes, such as miscarriage, low birth weight, and preterm birth (Mamelle et al., 1984; Axelsson, 1989; Xu et al., 1994).

Taken together, despite the caveats associated with epidemiological research, these studies suggest that when ergonomic design and evaluation of shiftwork systems is preformed, issues related to worker health should not be ignored. Additionally, the fact that some drug susceptibility and illness symptoms vary with time of day presents another reason for paying specific attention to shiftworker health (Reinberg, 1974). Here, a conservative and precautionary position is warranted, as persons dependent on drugs and/or suffering from disease may be at additional risk if they are required to work during the nighttime.

In the United States, given the current obesity epidemic and growing concern about the health of Americans (NIH, 2004; DHHS, 2004), employers are becoming acutely aware of the costs associated with an unhealthy lifestyle. Under the Americans with Disabilities Act of 1990, however, it may be effectively impossible to deny shiftwork to unhealthy but otherwise capable persons. For this reason, good ergonomic work system design is all the more important as a means of controlling employer health care costs and absenteeism. In Germany and other countries, special periodic health checks for employed workers performing nighttime work are recommended and practiced (Rutenfranz, 1982). As this is not currently the case in the United States, workers must seek out such care themselves. At a minimum, it may be beneficial for people to inform their physicians that they are shiftworkers, and company doctors should be made aware of any plans to change or install shiftwork operations.

In the United States, permanent nighttime work is most frequently assigned to new employees by default, since the premium pay for this work is often small, and experienced workers may use their seniority to bid out when possible. In practice, this can result in a situation where those who perform work at night are the youngest, fittest, and most recently hired. Alternatively, in many other industrialized nations, premium pay for this work is high and older, more experienced employees may use their tenure instead to obtain this work. It is therefore not unreasonable to argue that low or no premium pay for third-shift work may be a good preventive measure, in that it discourages older, more susceptible workers from exposing themselves to an operational hazard.

4.10 Accidents and Injuries

The economic benefit of a shiftwork operation is diminished when accidents and injuries occur at the workplace. Although large-scale incidents such as the *Exxon Valdez* grounding and Three Mile Island highlight the potential dangers of poor work systems and nighttime work, a more in-depth examination of the risks associated with shiftwork in general is in order. Folkard and Tucker (2003) explored the issue of relative risk associated with both rotating and permanent shiftwork operations on a number of dimensions. These included looking across each shift, over successive night shifts, and with regard to shift length and intervals between at-work rest breaks. As might be expected, relative risk was shown to increase from first to third shift, across four successive night shifts, and with increasing time intervals between rest breaks. A large increase in relative risk was also found for shifts that were longer than nine hours in length.

Although each of these findings has implications for work system and job design (e.g., the number of breaks and length of time between them), they require qualification. In most industries, risk is not constant across or within shifts; rather, there are transient risk factors that may increase relative risk, if only for a brief period. A transient risk analysis of an operation provides information on these vulnerabilities, based on which, models may be developed and subsequently applied to work schedule analysis. However, it is important to note that the success of any given model assumes good-quality data. Such data are often problematic to obtain and/or interpret, as the number of accidents and injuries recorded in any given workplace vary with respect to individual company rules and compliance exhibited by the staff and managers. As a starting point, given the potential for increased risk of accidents and injuries, shiftwork operations should strive to develop a positive safety culture with a well-thought-out and implemented injury data surveillance system. Only then will evaluation of the impacts of a work system and its subsequent redesign (as necessary) be possible.

Beyond basic components of the work system itself, there is literature to suggest that injury is related to shiftworker feelings of fatigue and ability to recover (Swaen et al., 2003), as well as age (Ringenbach and Jacobs, 1995) and circadian rhythm and social factors (Monk et al., 1996). Furthermore, an often-overlooked aspect of shiftwork accident and injury rates is the commute to and from work. Rarely is it the case that police or accident investigators report work schedules of those involved in noncommercial vehicle roadway accidents, when logically, this time spent commuting adds to a person's workday, albeit not at the workplace. This is unfortunate, as at least one study has found a relationship between commute duration and the rate and intensity of feeling sleepy at work (Popkin et al., 2001).

As a final step, once data have been collected and a solid understanding of accident and injury rate at an operation has been established, managers should weigh the various economic trade-offs in determining appropriate risk-cutoff criteria for the organization. It is important that these criteria be developed by a team that is comprised of all stakeholders prior to implementing a work system redesign. The current HOS for flight crews, for example, are considered by some to be potentially inadequate. However, a change in these regulations and the ensuing alterations in work schedules that would be necessitated are believed to be too costly for the industry to bear, despite any potential gains (Goode, 2003).

4.11 Worker Preferences

Employee acceptability ratings of different work systems vary significantly from one workplace to another (Tepas et al., 1985). Moreover, it has been found that individual experiences with work schedules within a workplace influence preferences (Tepas, 1990b). Data also indicate that workers may judge a work schedule as acceptable while nevertheless perceiving that it results in a negative impact on quality control, workplace safety, and health (Colligan and Tepas, 1986). Numerous studies in the field of industrial and organizational psychology have chronicled findings indicating that knowledge of how workers perceive problems and issues may be just as important as objective facts about a situation. This point should not be overlooked when assessing, evaluating, or considering a change in the work system of a given workplace.

There is substantial merit in regarding employees as experts, given their direct experience with shiftwork. For this reason, workers can be considered a rich source of data when redesigning a work system. Systematic surveys, interviews, record studies, and/or group discussions are all good ways of gathering information. Most of these methods are also an available means of incorporating the workforce into work system assessment, evaluation, and selection. As an added benefit, these methods provide a way to test the accuracy of management and labor perceptions of worker and workplace shift preferences and problems. Viewing workers as the foremost experts serves to minimize the possibility that outside consultants or zealous managers overlook important workplace variables.

As always, care must be taken to ensure that any information gathered from workers is reliable, valid, and representative. In most cases, this requires the use of an outside, third-party expert to fully and believably facilitate unbiased data collection, anonymous responding, and informed consent. To help ensure worker cooperation in the collection of this information and instating changes in work systems, it is advisable to inform workers in advance that they will have access to group data and project reports. Finally, it should be noted that many attitudinal measurements have little meaning unless they can be compared to equivalent norms or standards gathered from other workplaces, a process referred to as *benchmarking*.

4.12 Performance

Kleitman pioneered the study of circadian performance rhythms with research conducted in laboratories, caves, and submarines, all environments with minimal or nonexistent time cues. Using a number of fairly simple tasks (such as sorting playing cards), he concluded that performance varies with time of day and body temperature (Kleitman and Jackson, 1950). Following Kleitman's lead, many investigators took the position that body temperature was a reliable and simple way to index circadian changes in performance, including job performance.

As noted previously, just as physiological variables are circadian by nature but peak at different times of the day, so do various performance measures. Under many conditions, alertness and body temperature both appear to be at their lowest ebb in the early morning hours and continue to show a strong link throughout the day (Folkard and Monk, 1979). However, this parallelism between alertness and body temperature does not necessarily carry forward to indicate a causal link with job performance. For example, Monk and Embrey (1981) demonstrated that some tasks follow a performance pattern that differs from that of alertness. In their research, performance on a low-level memory task was shown to peak at a much different time of day than performance on a high-level memory task. Additionally, laboratory studies that produced a dissociation between physiological measures, similarly found differentially desynchronized performance rhythms (Fröberg, 1979; Folkard et al., 1983; Monk et al., 1983).

As measured in the actual workplace, job performance has been difficult to study, and therefore surrogates, such as accidents and injuries (see Section 4.10), must often be used. It is likely that once job performance can be measured in real time, as is emerging in several commercial transportation modes, we will begin to better understand the direct effects of work schedules on performance and company bottom lines. Performance-based work system management may then be a more realistic possibility.

Although circadian variation in task performance, especially where a high degree of vigilance is required, has been demonstrated in the literature (Dinges, 1992, 1995), predicting performance efficiency solely using the circadian temperature rhythm is ill advised due to the influence of task type. These findings are important from a work systems viewpoint, as they suggest that interactions with assorted work schedules may result in different time-of-day effects for various tasks. Thus, the simple generalization that workplace performance is poorest in the early morning hours must be limited and qualified. It is, in fact, the case that some work schedules may result in the highest level of performance for one type of task in the early morning hours, at the same time yielding the lowest level of performance for another task (Folkard and Monk, 1979).

In selecting or evaluating a work system, one must therefore consider not only time-of-day variation in performance, but also the types of job duties that are required. Moreover, it should be recognized that chronic reductions in sleep length, as associated with permanent third-shift workers, are related to persistent decrements in worker performance on some tasks (Tepas et al., 1981c). This combination of factors makes an ergonomic assessment much more complex and emphasizes the need for experts in the evaluation of work systems.

4.13 Recovery Time

Once exposed to night and shiftwork operations, the worker becomes at risk for ASD as well as suffering health and performance decrements (Tepas and Carvalhais, 1990; Moline et al., 1992). Initially, rates of performance recovery after working the night shift or being sleep deprived appear to be rather rapid (i.e., a matter of several days to eliminate acute effects, according to Totterdell et al., 1995). Similarly, a series of studies conducted at Walter Reed Army Institute for Research found that it takes more than four days to fully recover from two weeks of restricted sleep. Taken together, these results suggest that recovery is indeed possible and fairly rapid when exposure to sleep deprivation is brief, but that extended exposure, such as working shifts, quickly diminishes the ability to recover. This emphasizes the point that if recovery is to be expected, days off are indeed an invaluable part of any shift system and should not be considered "spare capacity" by either the employee or manager. Nevertheless, it is readily acknowledged that worker off time cannot be regulated and that certain persons may actually need to recover from their non-workdays more than they need to recover from work time.

In contrast, literature also suggests that the long-term exposure to shiftwork may have a certain degree of positive impact. For example, Duchon and Keran (1990) found that the longer shiftworkers remained on the night shift, the better their mood, subjective health, and performance, and the less time that was spent in bed searching for sleep (Dirkx, 1993). Dirkx (1993) also noted that the longer a person remained on a particular shift, the more likely it was that useful cognitive coping strategies were employed. It is therefore not unreasonable to offer that the longer one is exposed to shiftwork, the more readily the body is able to adapt and recover some level of performance. However, as noted above, chronic or extended exposure to such a stressor eventually precludes effective recovery from occurring (Tepas et al., 1993).

Other long-term effects on health and well-being that are not addressed as easily may also exist. Colligan et al. (1979) found that shiftworkers, especially those on rotating shifts, took more sick leave than dayworkers. Furthermore, according to the Austrian model of chronic shiftwork effects, workers are exposed to many situational and biological factors that wear on them over time. As a result, the successful worker is projected to be able to engage in shiftwork for about 40 years, after which disease and disorders will manifest at a rate making it impossible for him or her to continue (Tepas et al., 1993). This supported Moline et al. (1992), who suggested that advancing age tends to be related to more hours of wakefulness, increased sleepiness and fatigue, and decreased feelings of alertness and well-being when exposed to displaced sleep, such as that resulting from working nights.

In terms of worker health and performance, recovery appears highly complex and dependent on both the characteristics of the work system and the particular person exposed to it. Although more research is clearly needed to tease out the differential effects of these variables on off time, it is important to understand that (1) shiftwork is a stress agent that may affect the body negatively, and (2) off time is built into the work system to offer the employee time to recover. Use of employee off-duty time to fill holes in shift coverage should therefore only be employed as a brief, temporary stop-gap measure in special situations and should not be considered a solution to deficient staffing.

4.14 Discussion

Advances in our understanding of work systems have repeatedly confirmed the complexity of the human factors and operational variables that must be considered in the ergonomic assessment, design, and evaluation of such systems. For example, circadian variation in the endogenous biological clock is linked with sleep behavior and variation in alertness levels. These three factors interact with each other and with other variables, such as work task performance, mental and physical fatigue, individual difference variables, and aspects of worker napping and social behavior, to name a few. Given the complex nature of these interactions, it is legitimate to conclude that there is no single or universal answer to the question of what is the best work schedule.

Rather, one should consider what is an appropriate work system for a specific job, given a particular group of workers, and existing operational and cultural constraints. Therein lie buried a wide range of variables that must not be overlooked, yet we are not at present able to address using personnel selection methods. Workplace lore perpetuates the idea that there is a category of workers who thrive when they are employed on a regular basis at hours other than those of the first shift. For the present, this myth remains unsubstantiated. Even if support for this idea should evidence itself in the future, it is unlikely that people with these special characteristics or abilities exist in large enough numbers to meet the expanding needs of the current technological revolution demanding continuous operations. Therefore, the design and evaluation of work systems should be approached as a systems-based usability study, requiring expert operations analysis and ergonomic assessment, selection, and evaluation if it is to succeed.

5 WORK SYSTEM VARIABLES

Analyzing, selecting, and evaluating a work system requires consideration of the human factors variables discussed in the preceding section, as well as a number of key design and implementation topics. In this section we discuss briefly some of the more prominent variables.

5.1 Rotating versus Permanent Shifts

In practice, there is little support for the proposal that permanent shiftworkers can fully adapt to working evening or nighttime hours. Examining main sleep period length, Figures 3 and 4 support this conclusion. Statistical analysis of the time-of-day data for sleep onset and offset as related to Figure 3 is even more revealing. Like the sleep length data, non-workday differences between the three shift schedules with regard to sleep onset and offset time were negligible or nonexistent. That is, on non-workdays, most shiftworkers, including those on the third shift, chose to sleep at night and be active during the day (Tepas and Carvalhais, 1990).

As such, it may be reasonable to think of permanent third-shift workers as rotating shiftworkers, since their sleep-wake patterns typically change on days off. Society is daytime oriented, and shiftworkers seemingly maintain this orientation during their non-workdays to avoid social desynchronization and to be able to participate in family and community activities. These benchmark findings indicate that experienced shiftworkers may regularly engage in a diurnal lifestyle on non-workdays. This lifestyle should not, however, be considered typically diurnal, as it fluctuates whenever a person is not bound by their work schedule, rotating from nocturnal to diurnal and back.

Given this, if one views nighttime work hours as a hazard to be minimized, it becomes appropriate to argue in favor of rotating hours. In support of this position, rotating hours may be considered a practical method for spreading nighttime work hours across a greater portion of the workforce, thereby minimizing exposure for any individual worker. This assumes, however, that the probable hazards associated with rotating hours are fewer than those associated with permanent third-shift work, and that the exposure affects all workers equally. A more moderate approach suggests that rotating hours and nighttime work each have their own hazards and benefits, and that by weighing the positive and negative aspects of various schedules, an option that best matches the requirements of the job, considers ergonomic factors, and addresses the needs of the worker group may be determined.

5.2 Rotation Rate

Experts remain divided as to what is an appropriate work system rotation rate. In addition, there appear to be cultural differences regarding what is perceived as acceptable. For example, in the United States, a work system such as that shown in Table 7 is considered an acceptable but fast rotation rate, while the system in Table 10 is often deemed quite unacceptable. In Europe, on the other hand, a work system such as the one in Table 7 is not thought of as a fast rotation and is considered by many to be less acceptable than the system in Table 10. Generally, in Europe, if a worker rotates in some manner through first, second, and third shifts in about a week, the rotation is termed *fast*. In the United States, this arrangement is sometimes referred to as *swift* rotation.

In his review of the shift rotation-rate literature, Wilkinson (1992) determined that "with the possible exception of personal convenience, permanent or slowly rotating night shift systems emerge superior to rapidly rotating ones as an alternative to weekly rotation" (p. 1441). Responding to Wilkinson, Wedderburn (1992) argued in favor of rapidly rotating systems, stating that "a strong case for the use of permanent night shift can only be made by being selective in sources, and narrow in the range of outcomes considered" (p. 1447). In a second response, Folkard (1992) expressed the opinion that permanent shift systems may result in better performance, whereas swiftly rotating systems are perhaps more likely to make individual workers happy. In his view, a work system that is neither permanent, nor swiftly rotating, such as that presented in Table 7, is worse than either of these alternatives. Furthermore, Williamson and Sanderson (1986) provided some data to support Folkard's contention about happy workers. When holding shift length constant, they found evidence for improved subjective job satisfaction, health, and well-being following an increase in rotation speed.

Both Folkard (1992) and Monk (1986) present positions which suggest that the search for a single ideal rotation rate is somewhat like the search for one best and universal shiftwork system. As no single best choice may ever emerge, it is recommended that the ergonomic principles discussed in this chapter are used to determine a best solution in light of the particular characteristics of a given situation. Because a direct, controlled, and comprehensive workplace comparison of swift, fast, slow rotation, and permanent shift systems has never been performed, additional research continues to be necessary. Findings from such a study would help to further define the advantages and disadvantages that each of these approaches undoubtedly has, but must be conducted without the expectation that a single best solution will be uncovered. As a final point, though worker personal satisfaction is an important consideration when designing a new work system, it should not be the sole focus of such an effort. Sometimes what may make a worker happy initially is not scientifically prudent or healthy on a long-term basis.

5.3 Direction of Rotation

Many rotating shift schedules allow the option of forward or backward rotation. Table 7 is an example of forward shift rotation, and Table 8 shows the option of backward rotation using the same basic work system. As most field studies that change rotation direction also simultaneously alter other work system variables, this makes it impossible to determine what impact the change in rotation has had on system operation.

Studies conducted by Czeisler et al. (1982) provide a good example of this limitation. Their circadian rhythm laboratory research indicates that the biological clock has a forward-moving tendency that makes forward rotation easier for the human body to cope with. Common sense also leads experts to argue for forward rotation, since it tends to be easier to fall asleep when delaying one's bedtime than when advancing one's bedtime (i.e., going to bed earlier). This result is explained both by the relationship between the location of the extension of wakefulness relative to one's biological clock and the additional pressure to sleep that has built up as a result of increased hours of wakefulness. In a study of employees working on the same schedule in a single factory but rotating in opposite directions, Lavie et al. (1992) collected actigraphic data that supported the claim that forward rotation is preferable in certain situations. Workers may nevertheless argue in favor of backward rotation, as it tends to provide them with a longer off-time periods.

In another examination of this issue, Barton and Folkard (1993) surveyed 261 shiftworkers from various areas of the service sector using the Standard Shiftwork Index. These participants were subsequently segregated into three categories for purposes of comparison: phase-delayed, phase-advanced without quick return, and phase-advanced with quick return (i.e., only eight hours off from ending the day shift to beginning the night shift). When controlling for age and experience, the data showed poorer physical and psychological health, need for more sleep, more social and domestic disruption, and lower job satisfaction for advancing versus delaying shift systems. However, when comparing the two advancing systems separately, it became evident that the quick return was responsible for the majority of these decrements. Only sleep disturbance remained as a significant factor associated with the phase-advanced without-quick-returns group. Thus, although it appears that forward rotation may facilitate improved health and alertness in comparison to backward rotation, much work remains to understand the relationship between rotation rate and on-the-job safety and performance.

5.4 Length of Shift

After decades of effort by organized labor to reduce the length of the workday, there is now a popular move toward increasing it. In some cases, there is an interest in moving to 12-hour workdays, which result in a shorter workweek and more consecutive days off. A thorough review of the literature on these compressed workweeks has been published by Duchon and Smith (1993), who concluded that compressed workweeks exhibit both positive and negative effects. These authors recommend that "in industries, where accidents are a serious concern, special measures and evaluation in the use of extended workdays be considered" (p. 37).

The importance of taking a systems view when considering the potential merits of a long workday is demonstrated by two research efforts involving firefighters. Table 18 shows the compressed work schedule of a group of U.S. firefighters studied by Paley et al. (1998). In this sample, the day shift was 10 hours long and the night shift was 14 hours long, sleep at work was permitted when duty allowed, and the work system was judged as acceptable for that particular site. Knauth et al. (1995) studied German firefighters who also worked 14-hour-long night shifts as operators in a control room, answering calls, making decisions, and dispatching equipment. In contrast to the Paley et al. (1998) findings, the work schedule was determined to be unacceptable for this group. Although the recommendations of these two studies of firefighters working 14-hour night shifts were opposing, they are not at odds with each other. Rather, each was entirely appropriate, having resulted from proper consideration of ergonomic factors, including differences between the two groups in task assignments and work rules.

Worker preferences regarding shift length are another area where findings are sometimes inconsistent. Survey data in one study revealed both that 12-hour workdays were acceptable to workers and that a typical workday length of eight hours was most preferred (Tepas, 1990b). Mixed findings regarding shift length have also been revealed in field studies where objective data were gathered. For example, Rosa et al. (1989) demonstrated a negative impact of 12-hour shifts on the performance of control room operators in a continuous processing plant. In a second analysis of the same data, Lewis and Swain (1986) dismissed the significance of Rosa et al.'s (1989) findings and recommended the adoption of extended workdays for similar operations. However, in a 3.5-year follow-up study, Rosa (1991) was able to duplicate his original findings, thereby showing long-term persistent decrements in performance and alertness attributable to working 12-hour shifts.

Inconsistencies such as those discussed above raise a warning flag. The combination of investigators and employees eager to incorporate innovative work schedules may inadvertently lead to overlooking problems and the premature adoption of new work systems. For those organizations contemplating compressed workweeks, Duchon and Smith (1993) properly suggest that "since no a priori predictions from prior research can be made with certainty about the probable consequences of introducing 10- and 12-hour shifts into particular work groups, rigorous evaluation should be made on a continual basis" (p. 48).

One final point to consider regarding shift length is the use of overtime work to make up for inadequate staffing, thereby extending employee work hours. Although scheduled overtime shifts (e.g., special event staffing needs) remain rare, in some occupations at least, there tend to be more informal, regular, anticipated overtime needs and expectations by both workers and managers. A recent study on railroad dispatchers found that over half of the participants worked eight or more hours of overtime per week, 27% of whom reported that they were expected to work these hours (Popkin et al., 2001). Even though not formally a part of this work system, extended shifts

Table 18 Work System with Rotating Hours and Continuous Operations (Compressed Schedule): Four Schedules[a]

	M	T	W	R	F	S	K	M	T	W	R	F	S	K	M	T	W	R	F	S	K	M	T	W	R	F	S	K
S1M1	D	D	N	N	•	•	•	•	D	D	N	N	•	•	•	•	D	D	N	N	•	•	•	•	D	D	N	N
S2M1	N	N	•	•	•	•	D	D	N	N	•	•	•	•	D	D	N	N	•	•	•	•	D	D	N	N	•	•
S3M1	•	•	D	D	N	N	•	•	•	•	D	D	N	N	•	•	•	•	D	D	N	N	•	•	•	•	D	D
S4M1	•	•	•	•	D	D	N	N	•	•	•	•	D	D	N	N	•	•	•	•	D	D	N	N	•	•	•	•
S1M2	•	•	•	•	D	D	N	N	•	•	•	•	D	D	N	N	•	•	•	•	D	D	N	N	•	•	•	•
S2M2	•	•	D	D	N	N	•	•	•	•	D	D	N	N	•	•	•	•	D	D	N	N	•	•	•	•	D	D
S3M2	N	N	•	•	•	•	D	D	N	N	•	•	•	•	D	D	N	N	•	•	•	•	D	D	N	N	•	•
S4M2	D	D	N	N	•	•	•	•	D	D	N	N	•	•	•	•	D	D	N	N	•	•	•	•	D	D	N	N

[a]Basic cycle = eight days; major cycle (unlocked) = 56 days.

were nevertheless quite common. As the ability of a work system to function properly is, in part, related to appropriate staffing levels and the maintenance of employee off time, recurrent and expected use of overtime is a signal that staffing levels should be reviewed.

5.5 Basic Pattern and Full Cycle Length

In a previous section, it was discussed how shift schedules can vary in their basic pattern and full cycle length. As a general rule, a locked basic pattern and a short full cycle are preferred over unlocked basic patterns or longer full cycles. Two concepts are relevant here. First, a locked basic pattern makes short-term off-time planning easier for the worker, since the schedule repeats every seven (or multiple of seven) days. This weekly component of schedule predictability, as shown in the continental rota in Table 10, provides the benefit of a short and easily trackable locked basic pattern.

Second, a shorter full cycle also makes it easier to track a shift schedule and plan ahead. That is, it facilitates understanding of the work system and when a worker or group of workers is to report for duty. If schedule characteristics are easily grasped, relatively long-term off-time planning is simplified for both the employee and managers, who must plan coverage for workers on leave, for instance. The work system shown in Table 6 is a good example of such a schedule. At the other extreme, the work system shown in Table 12 is an example of a shift schedule with a long full cycle that makes planning ahead more difficult. The metropolitan rota in Table 11, given its short, unlocked basic pattern paired with a long full cycle, is an example of a work system that facilitates tracking, but makes planning ahead difficult.

Many work systems incorporate shift schedules with both unlocked, basic cycles and relatively long full cycles. Table 13 provides an example of a system of this sort. To overcome the problems inherent in such a system, managers sometimes issue workers a year-long pocket card or calendar with days off marked to facilitate personal planning. Although the value of such a card is evident, use of a work system with a shorter cycle length is often a better solution to the problem. In most cases it appears that a shift system that makes it difficult for a worker to track and plan ahead also makes it difficult for a manager to schedule adequate personnel on a regular basis.

In conclusion, it must be pointed out that empirical data in support of the commonsense ideas noted in this section are hard to come by. Moreover, it may be true that these apply more to the problems associated with use of a new work system than to a person's ability to cope with a system once it is in practice for a period of time. It has, in fact, been argued that most shiftworkers can eventually become experts at planning around any work system (Wedderburn, 1981).

5.6 Weekend Work

By definition, continuous work schedules require that someone work on weekends. This may be approached in a number of ways. Table 12 is an example of a system in which some weekend work is required by all workers. Another approach assigns some workers to shift schedules that only work weekends; these workers, often referred to as "weekend warriors," are depicted in Table 16. Finally, some approaches combine shift schedules that are limited to weekdays with other shift schedules that include both weekdays and weekends.

As our society is not only daytime but also weekend oriented, schedules that allow all workers some weekend days off are clearly preferable. Important family, religious, and other social events are almost always associated with weekends, and for some workers, Saturday non-workdays may be more important than Sunday non-workdays (Wedderburn, 1981). A work system such as EOWEO (Table 14) is attractive to some because it provides all workers with three-day weekends, twice every four weeks on a regular and predictable basis. At the other extreme, a system such as that shown in Table 16, which requires workers on certain shift schedules to work every weekend, may be particularly problematic because it requires a commitment to work that is in conflict with community values. This may lead to increases in absenteeism and/or turnover (Tepas et al., 1986).

5.7 Countermeasures

Acknowledging that shiftwork, and night work in particular, carries with it the risk of performance decrements and feelings of sleepiness, workers and managers have devised various strategies for coping with shiftwork operations. Proper management of sleep time (i.e., sleep hygiene) has been shown to be the best countermeasure for use in combating sleepiness (Zarcone, 2000). This entails properly anchoring and respecting the sleep time of the shiftworker (Minors and Waterhouse, 1983), as well as having access to a sleeping environment that is conducive to sleep. An operation might include providing its workforce with this type of information when it educates them about a new or existing work system.

If anchored sleep is not possible, as is the case with irregular and rotating work systems, shiftworkers often rely on napping to regain lost sleep (Åkerstedt and Torsvall, 1985). In fact, this strategy is one of the countermeasures more frequently used and studied, and has been demonstrated as effective if employed strategically (Rosa, 1993; Rosekind et al., 1996). Another common and scientifically based means of combating fatigue is the consumption of caffeine while at work (Bonnet and Arand, 1994; Horne and Reyner, 1996; Reyner and Horne, 1997). In their research, Horne and Reyner point out that the timing and combination of napping with caffeine may also be effective, especially in environments where on-the-job napping is allowed.

Other countermeasure approaches require precise timing and a physician's guidance to be used properly. These include the use of melatonin or light therapy to reentrain one's circadian rhythm to a new shift schedule (Arendt et al., 2000; Cajochen et al., 2000) and sedatives and stimulants to either promote sleep or remain alert (Mitler and Aldrich, 2000; Roehrs and Roth, 2000). In general, regular and/or frequent use of any of the foregoing countermeasures should be approached with caution. Their actual or perceived need may be indicative of a poor work system or other health problems.

Similarly, operations-situated countermeasures, often labeled fatigue monitoring technologies (Horberry et al., 2001) or fitness-for-duty screeners, should be avoided. It is likely that such technologies have not been empirically validated or calibrated to local operations, may be easily defeated by the worker, may be used to determine liability, and/or patch a poorly designed or implemented work system, rather than providing any true value or protection to the worker or operations.

5.8 Education Programs

It is not uncommon for workers to be inserted into new work systems with little preparation. Often, workers change schedules, or work systems are altered, with little advance planning or notice. Instant work system changes or unilateral work system implementations are not likely to be well received and may add to operational costs by promoting absenteeism, tardiness, and worker turnover. Educational programs may minimize the disturbance and cost of such change. In principle, these programs can take a variety of formats, such as lectures, group discussion, videotapes, and special literature. In practice, many efforts at this are informational but lack the rigor in content development and evaluation of student knowledge that is typically gained through a formal education approach (Tepas, 1993).

Educational programs for shiftworkers often have two objectives. First, a program explains the work system that is to be instated to those affected by it at the workplace. This includes a full description of the work system, an explanation of why it was selected, discussion of what is expected from the system and workforce, and details of the implementation process. Doing so ensures that workers know when they are supposed to be at work, thereby helping to minimize absences and decrease complaints. Second, an educational program may provide workers with advice and information on what they, their employers, and those who they live with might do at work, during off time, and on non-workdays to overcome the potential problems or hazards associated with the particular work system. Topics that address this include operational risk factors, such as those relating to sleep, diet, and circadian aspects of work schedules, as well as strategies for dealing with shiftwork in general (e.g., fatigue countermeasures).

The use and distribution of educational programs and informational material appears to be growing. For example, the U.S. Federal Transit Administration sponsored the development of a report, the *Toolbox for Transit Operator Fatigue* (TRB, 2002), for use by transit agencies as a basis for developing their own fatigue management programs. This publication includes discussion of the current scientific understanding of fatigue, as well as guidelines for the development of educational programs and specific materials that may be incorporated into such programs. Additional information on the design of successful training programs may be found in Chapter 18.

One important precaution with regard to educational programs and informational materials is to guard against their use as substitutes for needed improvements to work systems or as a means of placing the responsibility on workers to find their own solutions to problems related to working shifts. Also, although available commercial programs may appear to have face validity, little is known regarding whether they actually succeed in changing worker behavior or improve coping. Nevertheless, the popularity of such programs has been noted by Popkin (1994). Survey data collected from shiftworkers who were provided with a tailored educational program showed that over 90% of the workers would recommend the program to others. However, related to the foregoing point regarding behavioral change, these workers reported that they made little use of the strategies presented within that program.

5.9 Discussion

In this section we discussed briefly some of the key shiftwork issues that the human factors and ergonomics professional should address when evaluating or designing a work system. Most of these issues are complex and interactive, and thus one must approach them using a systems perspective. A comprehensive review of each is beyond the scope of this chapter, and it is likely that more will be revealed with future research. Currently, in many cases, experts remain divided in their opinions as to how a potential hazard is best addressed. The fact remains, however, that an appropriate work system solution at one location may be an inappropriate solution at another, and laboratory research recommendations may not be relevant or practical in the field. There is no "magic bullet" answer, as all work system solutions require compromise given the complexity of the interacting variables and operational constraints. Nevertheless, employing ergonomic strategies in the design, implementation, and evaluation of work systems will help ensure the best possible result.

6 WORK SYSTEM DESIGN AND IMPLEMENTATION

So far, this chapter has focused on terminology and the human factors dimensions of work systems. In the remaining section we provide an outline of the broader domains, evaluation methods, and systems concerns that ergonomists and operations analysts should consider when examining the hours of work in most organizations. This is not meant to be a step-by-step "how-to" guide, but rather, a means to provide the reader with concepts and variables that should be considered in whatever implementation and evaluation processes he or she chooses to follow.

6.1 Assessment of Existing Work Systems

Some of the domains that should be evaluated when assessing the net value of a given work system are listed in Table 19. This is not intended to be a comprehensive list for all workplaces, but it does include the major factors to consider, where applicable. It should immediately be recognized that each of the items on the list contains a human element that interacts with many of the other factors associated with a given work system.

Table 19 Checklist of Work System Domains

Community work systems practices: What work systems are practiced by other operations in the immediate community? How well do they seem to work? How accepting is the community of these work systems?

Corporate work system practices: What work systems are practiced at other operations within the corporation? How well are they working?

Plant work system history: What work systems have been practiced in this plant? When were they practiced? How well did they work? Why were they changed?

Local job market: What is the past, current, and future status of the local job market? What kinds of skills and expertise are or are not available in the workforce?

Organized labor: What does the union contract say about hours of work? Are adequate mechanisms in place to allow input and comments from union representatives?

Health and safety: What local and national occupational health and safety regulations apply to this installation? Are any environmental regulations relevant?

Plant personnel records: Do these records contain demographic, health, and work performance information that might be helpful? Is shift assignment information in the files? Are the files in an accessible and updatable digital format?

Local utility costs: Do local utility costs vary with time of day or day of the week? How do utility costs vary with volume used?

Legal requirements: What local and national work hour and wage legislation apply to this installation? What is their impact?

Maintenance requirements: How is equipment maintenance handled, and does it have restrictions? How is building maintenance handled?

Supervision and personnel requirements: Does the work system practiced interact with the personnel and supervision requirements of this installation? Is the staffing level appropriate?

Shipping and receiving: Are adequate shipping and receiving services available? Do costs and services vary with time of day or day of week?

Support services: Are medical, food, transportation, and other installation and community support services adequate?

Productivity and performance indicants: What objective measures of productivity and performance are used at this installation, and do they allow for adequate evaluation of shift schedule differences?

Worker participation: Are adequate mechanisms in place to allow representative and cooperative input and comment from all workers?

Demand for product: What are the characteristics of the demand for the goods or services produced by this installation? What is the anticipated future market like? Is it a stable market, or is it subject to cyclic changes?

For example, an equipment maintenance program is required for the long-term satisfactory operation of a continuous work system. In some cases, the only way to handle this preventive maintenance is with the purchase of duplicate equipment so that production can run continuously. For other work systems, it is possible to build in down time so that maintenance can be performed. With the latter solution, maintenance workers are usually faced with an irregular or on-call schedule, and although production time is lost, additional equipment is not required. However, it may also be possible to complete the repairs during regular work hours, with minimal production loss and little equipment duplication. A good assessment effort requires that all reasonable maintenance alternatives be considered and reviewed through a systematic process. Because people perform the maintenance and because maintenance tasks may have an inflexible schedule of their own, each of these alternatives has a financial cost as well as a human factor to consider in the assessment.

The checklist in Table 19 should be reviewed before electing to keep or change an existing work system. Note that financial cost, a factor always salient to organizational decision-makers, is associated with the human element that is present for most of these items. Moreover, many of the domains require skills beyond those of the ergonomist, or data that are not typically made available to an outside expert. Thus, by necessity, the review and assessment process requires a team of relevant persons, many of whom are employees of the workplace being evaluated. A team effort such as this increases workplace participation in the assessment and often serves as a good way to educate its participants with regard to the complexity of making work system decisions. Furthermore, including multiple stakeholders helps foster a better decision-making and buy-in process, whereby the end result is more likely to be agreed upon by all sides and subsequently be adopted and successful. Given unique corporate cultures and inherent complexities in the makeup of workforces, it is important to tailor this approach to individual operations.

6.2 Designing New Work Systems

The process described in Section 6.1 is the foundation for evaluating an existing work system, as well as for designing a new one. Methods such as surveys, interviews, focus groups, and retrieving archival company records may be used to gather data. (For a more detailed examination of information gathering, see Chapters 2 and 44.) As the first step in an iterative design process, examination of each domain listed in Table 19 will provide preliminary information needed to select potential work system arrangements. Thereafter, once a number of possibilities have been proposed, a second iteration should begin, reviewing in more detail and precision the financial and human costs associated with each system under consideration. A third iteration of this process may result in the final design for a new work system. An iterative approach is vital as a means of assurance that the best possible selection is made. It is all too easy to choose a work system prematurely based on its anticipated financial rewards or worker preferences, perhaps overlooking long-term human costs that may negate these benefits. Worker opinions, while important, should not alone drive the decision process, for like company management, employees may prioritize maximizing their own financial rewards or off time, unaware of the complex nature in which a work system may affect job safety, performance, and quality of life.

The design and work system selection process should always be couched within an evaluation framework. Table 20 presents a practical work system development scenario that incorporates evaluation. This methodology uses a survey for assessment purposes and provides a forum for the discussion of significant issues. Other methods may be more appropriate in some situations, and the use of control groups is desirable, where practical. In any case, the primary importance of gaining individual worker cooperation in assessing human factors variables cannot be overstated. Workers and managers must understand the work system changes that are being considered, agree on an implementation strategy, and be willing to participate and cooperate with each other. Although one may not be able to provide every worker with all of the information needed to allow him or her to elect a solution, experts can and should work diligently to gather survey data from every possible worker. Only through this effort will managers be provided with a fair and accurate view of the human element in their operation for use as a means of helping them make a final decision. If a survey is to provide this function, adequate worker response is required. Rates of 90% or higher are realistic and to be expected when confidentiality is practiced and workers are offered participation in the survey design (Tepas et al., 1985).

Table 20 Work System Development Scenario

Step 1: Develop a business case for making work system change that is salient to all affected stakeholders.
Step 2: Review this methodology with all parties, revise the methodology if needed, and agree to participate in implementation of the chosen methodology.
Step 3: Develop a survey to collect worker self-reports of their demographics, work system history, perceptions, preferences, and health status. Offer the survey to all workers for their voluntary, anonymous, and confidential responses.
Step 4: Analyze the survey data and develop work system recommendations grounded in human factors and ergonomics. Present recommendations to all parties.
Step 5: Perform a financial and technical evaluation of all work system recommendations, and develop a final recommendation. Gain acceptance by all parties of this work system recommendation and establish a schedule for implementation.
Step 6: If a new work system is recommended and agreed to, define criteria for use in a future evaluation of the success or failure of the new system. Agree on a schedule for implementation of the new work system. Select a time at which an evaluation of the new work system will be made.
Step 7: Resolve implementation problems between old and new systems with all stakeholders (e.g., workweek, vacations, holidays, overtime, pay).
Step 8: Develop and implement an instructional program for all workers. This is aimed primarily at teaching workers how to use the new work system as well as how they might best cope with it.
Step 9: If a new work system is being installed, implement the work system following the agreed-upon schedule.
Step 10: Following installation of the new work system, schedule an evaluation of that system, and subsequently initiate development of a future plan of action.

Table 21 Assessment Elements

- Deciding whether to evaluate
- Defining the evaluation problem
- Designing the evaluation
- Collecting information
- Analyzing information
- Reporting the evaluation
- Costing the evaluation
- Contracting the evaluation
- Managing the evaluation
- Staffing the evaluation

6.3 Implementing Work Systems

The importance of evaluation, both before and after implementation of a new work system, cannot be overly stressed. Evaluation is itself a long-standing discipline that grew out of the field of education. A thorough, proper evaluation will likely require the inclusion of someone who is schooled in this specialty on the implementation team. Proper evaluation will also take considerable time and financial resources. Table 21 presents the Joint Committee on Standards for Educational Evaluation's (1994) list of assessment elements for conducting a full-scale evaluation. Those interested in learning more are encouraged to obtain this source for further details, as well as to read Chapter 44.

Obviously, errors in the financial evaluation of such items as utilities and maintenance can be disastrous. Of equal importance, however, are costly human factors errors that may lead to accidents, increased health care costs, high turnover rates, and labor problems (Imberman, 1983; Duchon and Smith, 1993; Monk et al., 1996; Oginski et al., 2000). Unfortunately, it is easy to overlook these potential human factors-related losses, since they take time to develop and are sometimes masked by other changes in the workplace. For this reason, a preimplementation evaluation plan is a crucial part of a good work system evaluation, in that the information obtained provides the ergonomist with baseline data for later comparison. The use of a repeated measures design is also an important option to consider when there is no control group.

As noted earlier and in Table 20, an instructional training program is a necessary step in the implementation of a new work system. Workers and managers alike must understand the new system; in particular, what is expected of them, and how they might cope with these changes. Many training programs offer instruction and information in the name of education, though without regard for actual worker needs. Data collected during the evaluation and design of the system being implemented should therefore include a thorough needs assessment to assist in the development of a truly educational program, since the success of training may not be outwardly obvious (Popkin and Coplen, 1995). Although managers often provide workers with extensive training regarding job skills, many ignore the fact that learning how to cope with shiftwork is also a job skill and in some countries is a shared responsibility between employee and employer. All too often, managers get caught up in a rush to implement a new work system quickly and ignore this important step. Adopting a strong evaluation framework from the start will help protect the project from this type of oversight.

6.4 Pitfalls to Be Avoided

As has been highlighted throughout this chapter, the range of work schedules and systems available for potential implementation is immense. These systems include a host of interacting biological, psychological, social, and industrial variables. Given the current level of empirical understanding of these interactions, it should be obvious that work system design is currently as much art as it is science. Table 22 provides a number of human factors that work schedule evaluators and designers should consider for their potential as problem areas. The factors listed are quite general, and therefore should not be considered problematic for all forms of shiftwork.

Age and gender provide two examples of factors that may not be the pitfalls that many believe them to be. There is evidence to support both the proposition that older workers find it difficult to tolerate the third shift (Tepas et al., 1993), just as there are findings demonstrating that older workers have no more difficulty with 12-hour shifts than do their younger counterparts (Keran et al., 1994). Paley and Tepas (1994b) also presented data that failed to show differences in a number of outcome variables between older and younger firefighters who worked a rotating schedule that allowed for sleeping at the workplace. With regard to gender, ILO conventions formerly limited third-shift work by women (Kogi and Thurman, 1990). Existing research suggests, however, that the impact of the third shift on the sleep of women may actually be less than the impact on men (Dekker and Tepas, 1990).

The importance of and need for a systematic evaluation of each work has been stated throughout this chapter. A good starting point for such an evaluation is found in Table 22. As is true in all areas of ergonomic investigation, care must be taken to avoid the traditional pitfalls of investigator bias and expectancy, Hawthorne effects, and halo errors, among others. Since many of the hazards of shiftwork are associated with long-term chronic effects, special effort should be made to ensure that a data surveillance system is in place to capture agreed-upon operational measures and leading indicators, and that the evaluation schedule allows sufficient time for chronic symptoms to develop, appear, and be measured. Although no definitive data are available, it is suggested that a new work system be fully implemented for at least six months before any attempt is made to measure chronic long-term effects.

With consideration of factors such as those noted in Table 22, the BEST network, a team of distinguished European shiftwork experts, developed a set of guidelines for shiftworkers (Wedderburn, 1991). Modified and updated to incorporate an American viewpoint, Table 23 provides the U.S. version of the BEST list of key considerations in the design of a shift system. For the most part, these are factors associated with work systems and the work itself that are likely to cause shiftwork coping problems. This material is thorough; however, it should not be viewed as a list of absolute prohibitions. More realistically, taken together, Tables 22 and 23 should be viewed as helpful aids for spotting potential trouble areas.

6.5 Regulatory and Policy Considerations

The construction of work systems must be performed within the context of applicable governmental and company laws, regulations, policies and contractual obligations. The work schedules of commercial transportation system operators, for example, are often limited by hours-of-service regulations. In the United States, most of these regulations were written decades

Table 22 Human Factors That Are Likely to Cause Shiftwork Coping Problems

- Age
- Gender
- Marital status
- Young children at home
- Moonlighting
- Workload at home and on the job
- Circadian biological clock
- Sleep disorders
- Frequent napping
- Physical illness and disease
- Psychological characteristics
- Addictions
- Constipation, pain, chronic medical conditions
- Social rhythms, roles, and responsibilities
- Work/rest schedule experience
- Religious beliefs
- Physical and mental demands of work
- Perceived risk of exposure and errors

Table 23 Key Considerations in the Design of a Shift System[a]

1. ***Number of consecutive night shifts***
   - Permanent night shifts can be hazardous and require special consideration and limitation.
   - Although open to debate, current research seems to point in favor of fewer (two to four) consecutive night shifts worked.
2. ***The scheduling of work***
   - The length of a work shift should depend on the types of tasks being performed and the workload.
   - An excessive number of consecutive workdays (more than eight) should be scheduled with caution, and screening workers for physical and mental stress may be necessary.
   - There is no firm evidence supporting either direction of rotation; however, there seems to be some advantage to rotating schedules in a forward direction.
3. ***The scheduling of off time***
   - The purpose of off time is to allow workers to recover from the fatiguing effects of work, to sleep, and to socialize. Off time ideally should be scheduled to allow all three activities to occur.
   - The amount of off time should increase concomitantly as the number of consecutive workdays and the length of the work shifts increases.
   - A shift system should aim to maximize the amount of off time between work shifts.
   - Travel time to and from work must be considered when scheduling off time between work shifts.
   - For social reasons, shift systems should aim to include some weekend off time for all workers.
4. ***Considerations for time of day***
   - Do not treat all hours of the day in the same manner. Acceptable daytime work shift lengths may prove to be excessive during the night.
   - Maximize the opportunity for workers to sleep during the night. Early morning shifts, for example, tend to lead to reductions in sleep length.
5. ***Consideration of psychosocial variables***
   - Keep the rotation schedules as regular as possible. It is preferable that a worker need not check the schedule each day to figure out whether he or she must work.
   - Whenever possible, allow for flexibility. This may include organizational designs such as flextime or simply allowing workers to exchange shifts with co-workers.
   - Consider workload, both mental and physical.
   - Minimize changes in shift schedules.

[a]These guidelines are an adapted and modified version of those published by the European Foundation for the Improvement of Living and Working Conditions (Wedderburn, 1991).

ago, when little was known about the impact of shiftwork on employee safety and performance. Inasmuch as their objective is clearly to protect the public by preventing accidents, for the most part, HOS regulations continue to be unbound by the results of systematic scientific research. Most would agree that some regulation is warranted, but the degree to which current regulations prevent accidents is not clear. Additionally, as discussed in Section 1, the fact that the NTSB's (1990) "Most Wanted Transportation Safety Improvements" list had required the study and revision of HOS in all modes of transportation is an indication that those regulations were not deemed adequate.

This is not altogether surprising, as HOS regulations in the United States continue to hold a number of faulty assumptions, having been based on a model of work performance that originated prior to the current empirical understanding of biological clocks and the expansion of our knowledge about the dynamics of work, rest, and recovery. Among other things, these assumptions maintain that all hours of the day may be treated as equal and interchangeable; fatigue is simply a function of the number of consecutive hours worked, and hours worked and recovery time are linked exclusively; acute exposure to nighttime work hours is a more serious issue than chronic exposure; and off time is most likely to be spent resting/sleeping. Science does not support these oversimplified viewpoints, however, and it is not currently possible to legislate what an employee does during his or her time away from work. In actuality, even dayworkers may develop problems that are associated with shiftwork if adequate rest is not obtainable during off time. Furthermore, such regulations mean little if they are not actively enforced and regularly evaluated and updated. Although important and necessary, changing HOS regulations is likely to be politically charged, due in part to the expense of redesigning work systems and operating practices to be consistent with new requirements and laws.

There are at least two other national approaches to regulating the working time of employees with transportation jobs that are considered safety-critical. Australia chose to eliminate its HOS regulations in favor of a nonprescriptive approach, as fostered through its Occupational Health and Safety Act. A key principle of the act is the requisite *duty of care* that employers have to provide a safe place of work for employees. Shiftwork is considered similar to other environmental hazards, and therefore a hazard control plan must be developed and approved by the government. Another major feature of the Australian approach is referred to as the *chain of responsibility*. Each being a "link" in the chain, the company, worker, and (where applicable) the consigner may all be held liable if the hazard control plan is not followed, the stipulations of which may include aspects related to the placement and duration of employee sleep time. While this approach provides managers with significant flexibility in developing work schedules, it remains unknown whether this means of regulating employee working time does in fact benefit the worker and/or his or her on-the-job safety. This is especially noteworthy, as many of the tools used to evaluate work systems and their implementation as described in such plans have not yet been fully validated for this use.

The European Union represents a third option regarding how to regulate hours of work. Using an iterative approach over years, a set of Working Time Directives exist that specify in great detail what hours and shift patterns are permissible for transportation operators. These directives are by design more consistent with the body of scientific literature discussed throughout this chapter, and thus may be considered more equitable than the rather arbitrary regulations in the United States. However, this means of regulation, like that used in the United States, is rigid and as such may preclude the use of an otherwise advantageous work system. It is also presumed within these regulations that the worker utilizes his or her off time appropriately for recovery, which may not always be the case.

It is important to note that none of these three approaches has yet been formally evaluated for its impact on worker safety and health or the consistency with which it is enforced. Until this is done and a valid approach has been identified and deployed, regulatory solutions may simply be regarded as political artifacts that must be considered when developing a new scheduling system. Furthermore, as the deficiencies of the existing regulations have not gone unnoticed by the U.S. Court of Appeals, it is vital that the deployment of workers in the U.S. transportation industry be made with proper vigilance and caution. For additional discussion of transportation issues, see Chapter 59.

As a final consideration regarding regulatory impact on work systems, in the United States there is no governmental restriction of pay differentials for less desirable work schedules, which makes this an often-negotiated point during collective bargaining. Operations are allowed to establish premium pay rates for such shifts, presumably as compensation for the inconvenience and hardship they impose or to ensure an adequate staffing level. Unfortunately, this approach has the potential to alter a work system that ideally has its roots in occupational health and safety to one that is shaped at least in part by collective bargaining. In such operations, if work is bid by seniority, this may result in the most difficult shifts being left for the junior staff. Alternately, the most senior employees may choose to work these shifts, possibly for higher pay, as well as having the first option for overtime work. In conclusion, there are a variety of factors, including those political in nature (within organizations, as well as governmental) that must be considered both when developing a new work system and when trying to construct or redesign laws, regulations, and policy regarding employee hours of work.

6.6 Work Systems Tools

Designing new work systems and assigning people to specific slots within a work schedule have two important dimensions in common; both take considerable time and patience. Consequently, there are a number of ongoing efforts to provide helpful and structured guidance, services, and tools that strive to make this effort easier. As assigning people to particular work schedules is a large and repetitive task, the development of computer-assisted scheduling (CAS) methods is a logical step to assist in meeting this need. This concept is not new, dating back at least 30 years (Tepas, 1985). Unfortunately, however, such tools were often designed by persons lacking in training and/or understanding regarding the human factors aspects of work systems. Expert work system designers, on the other hand, experienced the opposite dilemma, in that they understood the relevant ergonomic principles and criteria but struggled with the actual design of a computer-assisted tool (Nachreiner et al., Gärtner, 2001). At present, we are not aware of any published research addressing the reliability and validity of these tools for use in work system design and evaluation. Until this is done, one should take great care when using the output from such tools in making work systems decisions.

The current state of the art in computer work system design and analysis has overcome some of its initial problems and use of such software is becoming increasingly popular. Typically, the user is allowed to visualize and compare multiple work schedule and staffing solutions based on a rule set including operating, regulatory, and collective bargaining constraints, as well as ergonomic criteria, such as those found on Table 23. Figure 8 provides output from one such software product, in which the on-duty/off time of an irregularly scheduled locomotive engineer is depicted along with summary statistics (Gärtner et al., 2004). Visualizations, such as the raster plot of the work periods shown in the figure, in addition to the descriptive statistics provided regarding salient work schedule properties, give the user a powerful tool to explore the merit of a particular work schedule and system (Coplen and Popkin, 1995). It should be noted, however, that the goal of such a tool is not to produce a single "best" schedule, as every schedule has merits and drawbacks. Rather, CAS tools are valuable in that they provide a work-scheduling professional with information beyond that available using traditional paper-and-pencil methods.

Other tools and resources, some of which have been incorporated into computer software, are also available to persons who must make decisions about work schedules. One such tool category is fatigue modeling. The aim of these models is to understand the relationship between human performance, sleepiness and/or fatigue, time of day, and time awake. However, existing models are in no way able to capture all the intricacies required in designing a work system, nor to date have any been shown to be reliable or valid (Dinges, 2004; Van Dongen, 2004). This is not to say, however, that they do not have value; rather, the user must thoroughly understand the model's theoretical and empirical underpinnings and elect to use its output in a way that is consistent with this understanding (Raslear and Coplen, 2004).

Additionally, during the past decade there has been sizable growth in the number of consulting firms that specialize in work schedule design and fatigue management. Despite current business trends in the United

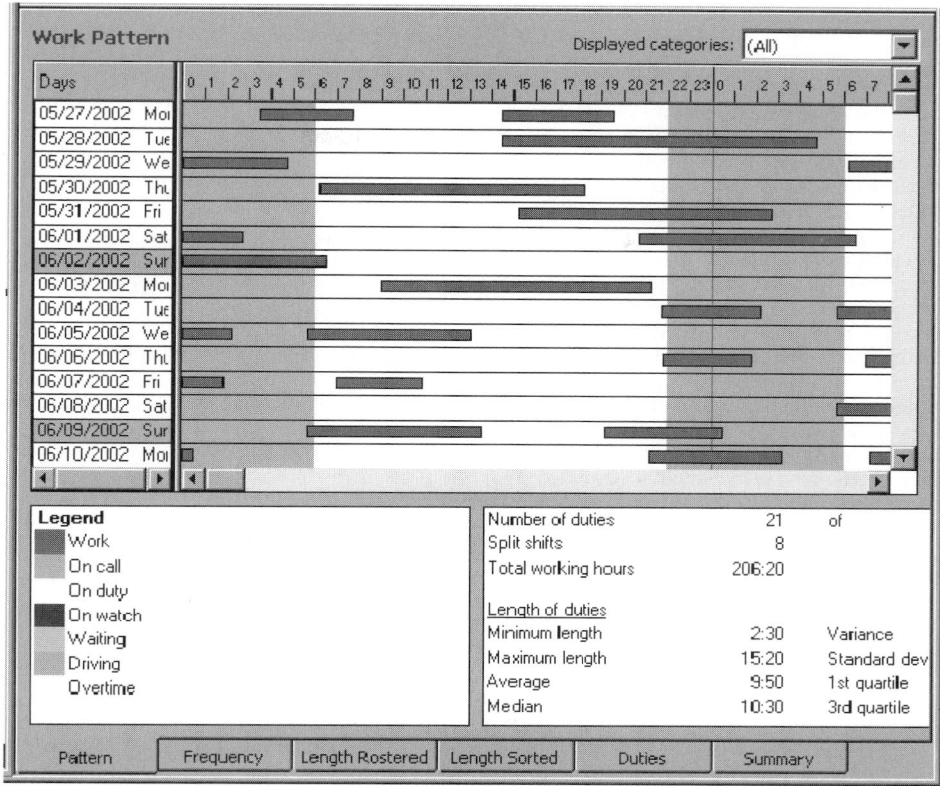

Figure 8 Graphical and statistical analysis made possible through the current generation of work schedule evaluation software. In this case, work schedule data are presented from a locomotive engineer involved in the 1986 Hinton train accident in Alberta, Canada (Smiley, 1990). The engineer was working an on-call schedule, as can easily be seen through both graphical and statistical presentations of the high variability in work start times and durations.

States to outsource, larger companies are encouraged to consider, where possible, maintaining in-house capabilities for the ergonomic design, implementation, and evaluation of work systems. The importance of constantly and proactively iterating this process has been stressed throughout the chapter. In some cases, the ability and commitment of an organization to do so regularly and make requisite changes promptly on its own may yield a better bottom line than reliance on an outside firm to provide this service. When such a firm is brought in only occasionally, it is often in a reactive fashion to fix a significant problem, and therefore a work system redesign is likely to take place without obtaining all the necessary information or stakeholder buy-in regardless of the process used. In the end, it is nevertheless reasonable to argue that a certain competitive advantage is present for companies that are able to instate and maintain proper ergonomic work systems, whether this is accomplished in-house or by using consultants.

As a final consideration regarding work scheduling tools for organizations, some attempts have been made to improve at least general awareness of the effects of work schedules on fatigue through the use of a competency-based approach to training those responsible for scheduling shiftworkers. Although proposals do exist for the development of a unified approach to address this deficiency with a set of tools and training, to date in the United States this need remains unmet (Tepas, 1999, 2003).

7 CONCLUSIONS

Shiftwork is not a new idea, and it involves a significant segment of the workforce in industrially developed and developing countries. For decades, the complexities of work schedule design were not considered or understood, and therefore the problems faced by shiftworkers drew minimal attention. In recent years, advances in chronobiology and ergonomics have generated findings that allow the arguable proposal that shiftwork can place workers at risk. Shiftwork *does* change people, and many of these changes do not go away, even, and especially, with long-term exposure to such work systems. Research has also demonstrated that scheduling work hours and minimizing the associated risk inherent in certain work systems is a complex task due to a host of interacting variables.

We are now well into the information technology revolution, and computer programs are scheduling work and people; however, this has not necessarily served to make the task any less complex. Frequently, such activities continue to be performed without fully addressing the impact of the resulting hours of work on the human, although applications are becoming more sophisticated in this regard. The *misapplication hypothesis* proposes that the misapplication of computers serves to decrease productivity (Tepas, 1994) and may become reality if new technology simply creates additional jobs that are characterized by long work hours, irregular shift periods, and/or inappropriate work systems. Our existing knowledge does not yet, and may never, allow off-the-shelf delivery of work systems with a proven positive impact. However, scientific research has now established a methodology to design ergonomically-appropriate work systems and to evaluate their implementation, and as such, it is incumbent upon us to use it.

As the current technology revolution continues, the time demands of international business, the development of new continuous process methods, the need for additional and more sophisticated security services, and the use of just-in-time techniques all make it quite clear that there will be a continued need for around-the-clock operations. More people working increasingly different types of shiftwork is a likely result of these developments. Just as contemporary research has demonstrated the risks of shiftwork, good work system management has the potential for providing organizations with benefits that will protect the health of workers and make operations safe and productive. It has been said that mastery of the clock was the key to the Industrial Revolution. Similarly, mastery of the biological clock may be the key to the demands generated by the technology revolution that is now under way.

REFERENCES

Åkerstedt, T. (1977), "Inversion of the Sleep Wakefulness Pattern: Effects on Circadian Variations in Psychophysiological Activation," *Ergonomics*, Vol. 20, pp. 459–474.

Åkerstedt, T., and Torsvall, L. (1985), "Napping in Shift Work," *Sleep*, Vol. 8, pp. 105–109.

Åkerstedt, T., Torsvall, L., and Gillberg, M. (1989), "Shift Work and Napping," in *Sleep and Alertness: Chronobiological, Behavioral, and Medical Aspects of Napping*, D. F. Dinges and R. J. Broughton, Eds., Raven Press, New York, pp. 205–220.

Angersbach, D., Knauth, P., Loskant, H., Karvonen, M. J., Undeutsch, K., and Rutenfranz, J. (1980), "A Retrospective Cohort Study Comparing Complaints and Diseases in Day and Shiftworkers," *International Archives of Occupational and Environmental Health*, Vol. 45, pp. 127–140.

Arendt, J., Stone, B., and Skene, D. (2000), "Jet Lag and Sleep Disruption," in *Principles and Practice of Sleep Medicine*, M. H. Kryger, T. Roth, and W. C. Dement, Eds., W.B. Saunders, Philadelphia, pp. 591–599.

Axelsson, G., Rylander, R., and Molin, I. (1989), "Outcome of Pregnancy in Relation to Irregular and Inconvenient Work Schedules," *British Journal of Industrial Medicine*, Vol. 46, pp. 393–398.

Baltes, B. B., Briggs, T. E., Huff, J. W., Wright, J. A., and Neuman, G. A. (1999), "Flexible and Compressed Workweek Schedules: A Meta-analysis of Their Effects on Work-Related Criteria," *Journal of Applied Psychology*, Vol. 84, No. 4, pp. 496–513.

Barton, J., and Folkard, S. (1993), "Advancing Versus Delaying Shift Systems," *Ergonomics*, Vol. 35, pp. 59–64.

BLS (Bureau of Labor Statistics) (2002), "Workers on Flexible and Shift Schedules 2001 Summary," retrieved September 2004, from ftp://ftp.bls.gov/pub/news.release/flex.txt.

Bøggild, H., Saudicani, P., Hein, H. O., and Gyntelberg, F. (1999), "Shift Work, Social Class, and Ischaemic Heart Disease in Middle Aged and Elderly Men: A 22 Year Follow up in the Copenhagen Male Study," *Occupational and Environmental Medicine*, Vol. 56, pp. 640–645.

Bonnet, M. H., and Arand, D. L. (1994), "The Use of Prophylactic Naps and Caffeine to Maintain Performance During a Continuous Operation," *Ergonomics*, Vol. 37, pp. 1009–1020.

Bourdouxhe, M., Quéinnec, Y., Granger, D., Baril, R., Guertin, S., Massicotte, P., et al. (1999), "Aging and Shiftwork: The Effects of 20 Years of Rotating 12-Hour Shifts Among Petroleum Refinery Operators," *Experimental Aging Research*, Vol. 25, pp. 323–329.

Brown, F. M. (1993), "Psychometric Equivalence of an Improved Basic Language Morningness (BALM) Scale Using Industrial Populations Within Comparisons," *Ergonomics*, Vol. 36, pp. 191–197.

Cajochen, C., Zeitzer, J, Czeisler, C., and Dijk, D. (2000), "Dose–Response Relationship for Light Intensity and Ocular and Electroencephalographic Correlates of Human Alertness," *Behavioural Brain Research*, Vol. 115, pp. 75–83.

Carskadon, M., and Dement, W. (1981), "Cumulative Effects of Sleep Restriction on Daytime Sleepiness," *Psychophysiology*, Vol. 18, No. 2, pp. 107–113.

Cervinka, R., Kundi, M., Koller, M., Haider, M., and Arnhof, J. (1984), "Shift Related Nutrition Problems," in *Psychological Approaches to Night and Shift Work: International Research Papers*, A. A. I. Wedderburn and P. A. Smith, Eds., Heriot-Watt University, Edinburgh, pp. 14.1–14.18.

Chan, O., Phoon, W., Gan, S., and Ngui, S. (1989), "Sleep–Wake Patterns and Subjective Sleep Quality of Day and Night Workers: Interaction Between Napping and Main Sleep Episodes," *Sleep*, Vol. 12, No. 5, pp. 439–448.

Colligan, M. J., and Tepas, D. I. (1986), "The Stress of Hours of Work," *American Industrial Hygiene Association Journal*, Vol. 47, pp. 686–695.

Colligan, M., Frockt, I., and Tasto, D. (1979), "Frequency of Sickness Absences and Work Site Clinic Visits Among Nurses as a Function of Shift," *Applied Ergonomics*, Vol. 10, pp. 79–85.

Comperatore, C. A., and Krueger, G. P. (1990), "Circadian Rhythm Desynchronosis, Jet Lag, Shift Lag, and Coping Strategies," in *Shiftwork*, A. J. Scott, Ed., Hanley and Belfus, Philadelphia.

Coplen, M. K., and Popkin, S. M. (1995), "Raster Plot Representations of Shiftwork Data," poster session presented at the 12th International Symposium of Night and Shiftwork, Ledyard, CT.

Costa, G. (1996), "The Impact of Shift and Night Work on Health," *Applied Ergonomics*, Vol. 27, pp. 9–16.

Costa, G. (2003), "Factors Influencing Health of Workers and Tolerance to Shift Work," *Theoretical Issues in Ergonomics Science*, Vol. 4, pp. 263–288.

Crowley, S. J, Lee, C., Tseng, C. Y., Fogg, L. F., and Eastman, C. (2003), "Combinations of Bright Light, Scheduled Dark, Sunglasses, and Melatonin to Facilitate Circadian Entrainment to Night Shift Work," *Journal of Biological Rhythms*, Vol. 18, No. 6, pp. 513–523.

Czeisler, C. A., Moore-Ede, M. C., and Coleman, R. M. (1982), "Rotating Shift Work Schedules That Disrupt Sleep Are Improved by Applying Circadian Principles," *Science*, Vol. 217, pp. 460–462.

Czeisler, C. A., Johnson, M. P., Duffy, J. F., Brown, E. N., Ronda, J. M., and Kronauer, R. E. (1990), "Exposure to Bright Light and Darkness to Treat Physiological Maladaptation to Night Work," *New England Journal of Medicine*, Vol. 322, pp. 1153–1159.

Czeisler, C. A., Duffy J. F., Shanahan T. L., Brown E. N., Mitchell J. F., Rimmer D. W., et al. (1999), "Stability, Precision, and Near-24-Hour Period of the Human Circadian Pacemaker," *Science*, Vol. 284, No. 5423, pp. 2177–2181.

Dekker, D. K., and Tepas, D. I. (1990), "Gender Differences in Permanent Shiftworker Sleep Behavior," in *Shiftwork: Health, Sleep and Performance*, G. Costa, G. Cesana, K. Kogi, and A. Wedderburn, Eds., Peter Lang, Frankfurt-am-Main, Germany, pp. 77–82.

DHHS (U.S. Department of Health and Human Services) (n.d.), *Smallstep.gov Website*, retrieved September 2004, from http://smallstep.gov/.

Dinges, D. (1992), "Adult Napping and Its Effects on the Ability to Function," in *Why We Nap*, C. Stampi, Ed., Birkhauser, Boston, pp. 118–134.

Dinges, D. (1995), "An Overview of Sleepiness and Accidents," *Journal of Sleep Research*, Vol. 4, pp. 4–14.

Dinges, D. (2004), "Critical Research Issues in Development of Biomathematical Models of Fatigue and Performance," *Aviation, Space, and Environmental Medicine*, Vol. 75, No. 3, Pt. II, pp. A181–A191.

Dinges, D. F., Orne, M. E., Whitehouse, W. G., and Orne, E. C. (1987), "Temporal Placement of a Nap for Alertness: Contributions of Circadian Phase and Prior Wakefulness," *Sleep*, Vol. 10, pp. 313–329.

Dinges, D. F., Connell, L. J., Rosekind, M. R., Gillen, K. A., Kribbs, N. B., and Graeber, R. C. (1991), "Effects of Cockpit Naps and 24-hr Layovers on Sleep Debt in Long-Haul Transmeridian Flight Crews," *Sleep Research*, Vol. 20, p. 406.

Dirkx, J. (1993), "Adaptation to Permanent Night Work: The Number of Consecutive Work Nights and Motivated Choice," *Ergonomics*, Vol. 36, pp. 29–36.

Duchon, J., and Keran, C. (1990), "The Adjustment to a Slowly Rotating Shift Schedule: Are Two Weeks Better Than One? *Proceedings of the Human Factors Society 34th Annual Meeting*, Orlando, FL, pp. 899–903.

Duchon, J. C., and Smith, T. J. (1993), "Extended Workdays and Safety," *International Journal of Industrial Ergonomics*, Vol. 11, pp. 37–50.

Eastman, C. I., and Miescke, K. J. (1990), "Entrainment of Circadian Rhythms with 26-h Bright Light and Sleep–Wake Schedules," *American Journal of Physiology*, Vol. 259, pp. R1189–R1197.

Eastman, C. I., Stewart, K. T., Mahoney, M. P., Liu, L., and Fogg, L. F. (1994), "Dark Goggles and Bright Light Improve Circadian Rhythm Adaptation to Night-Shift Work," *Sleep*, Vol. 17, pp. 535–543.

Eastman, C., Hoese, E., Youngstedt, S., and Liu, L. (1995), "Phase-Shifting Human Circadian Rhythms with Exercise During the Night Shift," *Physiology and Behavior*, Vol. 58, No. 6, pp. 1287–1291.

Ehret, C. F. (1981), "New Approaches to Chronohygiene for the Shift Worker in the Nuclear Power Industry," in *Night and Shift Work: Biological and Social Aspects*, A. Reinberg, N. Vieux, and P. Andlauer, Eds., Pergamon Press, Oxford.

Evans, F. J., Cook, M. P., Cohen, H. D., Orne, E. C., and Orne, M. T. (1977), "Appetitive and Replacement Naps: EEG and Behavior," *Science*, Vol. 197, pp. 687–689.

Folkard, S. (1992), "Is There a 'Best Compromise' Shift System?" *Ergonomics*, Vol. 35, pp. 1453–1463.

Folkard, S., and Monk, T. H. (1979), "Shiftwork and Performance," *Human Factors*, Vol. 21, pp. 483–492.

Folkard, S., and Tucker, P. (2003), "Shiftwork, Safety and Productivity," *Occupational Medicine*, Vol. 53, pp. 95–101.

Folkard, S., Wever, R. A., and Wildgruber, C. M. (1983), "Multi-oscillatory Control of Circadian Rhythms in Human Performance," *Nature*, Vol. 305, pp. 223–226.

Fröberg, J. E. (1979), *Performance in Tasks Differing in Memory Load and Its Relationship with Habitual Activity Phase and Body Temperature*, FOA Report C-52002-H6, Forsvarets Forskningsanstalt, Research Institute of National Defense, Stockholm.

Gärtner, J. (2001), "Interactive Computer Aided Shift Scheduling," *Journal of Human Ergology*, Vol. 30, pp. 21–26.

Gärtner J., Åkerstedt, T., Folkard, S., Nachreiner, F., and Popkin, S. (2003), "Actually, Is This a Night Shift? Are Shiftwork Design-Recommendations Becoming Meaningless Due to Fuzzy Terms? in *Proceedings of the 16th International Symposium on Night and Shiftwork*, Santos, Brazil, F. M. Fischer, L. Rotenberg, and C. R. de Castro Moreno, Eds., Vol. 20, No. 2, p. 73.

Gärtner J., Popkin, S., Leitner, W., Wahl, S., Åkerstedt, T., and Folkard, S. (2004), "Analyzing Irregular Working Hours: Lessons Learned in the Development of RAS 1.0: The Representation and Analysis Software," *Chronobiology International.*

Gersten, A. H. (1987), "Adaptation in Rotating Shift Workers: A Six Year Follow-up Study," unpublished doctoral dissertation, Illinois Institute of Technology, Chicago.

Gillberg, M. (1984), "The Effects of Two Alternative Timings of a One-Hour Nap on Early Morning Performance," *Biological Psychology*, Vol. 19, pp. 45–54.

Goode, J. (2003), "Are Pilots at Risk of Accidents Due to Fatigue?" *Journal of Safety Research*, Vol. 34, pp. 309–313.

Gordon, G. H., McGill, W. L., and Maltese, J. W. (1981), "Home and Community Life of a Sample of Shift Workers," in *Biological Rhythms, Sleep and Shift Work*, L. C. Johnson, D. I. Tepas, W. P. Colquhoun, and M. J. Colligan, Eds., Spectrum, New York.

Härmä, M. (1993), "Individual Differences in Tolerance to Shiftwork: A Review," *Ergonomics*, Vol. 36, pp. 101–109.

Härmä, M. (1996), "Ageing, Physical Fitness and Shiftwork Tolerance," *Applied Ergonomics*, Vol. 27, No. 1, pp. 25–29.

Härmä, M., Ilmarinen, J., Knauth, P., and Rutenfranz, J. (1988), "Physical Training Intervention in Female Shift Workers, II: The Effects of Intervention on the Circadian Rhythms of Alertness, Short-Term Memory, and Body Temperature," *Ergonomics*, Vol. 31, No. 1, pp. 51–63.

Härmä, M., Knauth, P., and Ilmarinen, J. (1989), "Daytime Napping and Its Effects on Alertness and Short-Term Memory Performance in Shiftworkers," *International Archives of Occupational and Environmental Health*, Vol. 61, pp. 341–345.

Hayashi, M., and Hori T. (1998), "The Effects of a 20-min. Nap Before Post-lunch Dip," *Psychiatry and Clinical Neuroscience*, Vol. 52, pp. 203–204.

Horberry, T., Hartley, L., Mabbott, N., and Krueger, G. (2001), "Fatigue Detection Technologies for Trucks and Commercial Vehicles: Possibilities and Potential Pitfalls," *Business Briefing: Global Truck and Commercial Vehicle Technology*, January, pp. 58–63.

Horne, J. A., and Ostberg, 0. (1976), "A Self-Assessment Questionnaire to Determine Morningness–Eveningness in Human Circadian Rhythms," *International Journal of Chronobiology*, Vol. 4, pp. 97–110.

Horne, J. A., and Reyner, L. A. (1996), "Counteracting Driver Sleepiness: Effects of Napping, Caffeine, and Placebo," *Psychophysiology*, Vol. 33, pp. 306–309.

Horne, J., and Reyner, L. (1998), "Vehicle Accidents Related to Sleep: A Review," *Occupational and Environmental Medicine*, Vol. 56, pp. 289–294.

Howarth, H. D. (2003), "An Investigation of Sleep and Fatigue in Transit Bus Operators on Different Work Schedules" (doctoral dissertation, University of Connecticut, Storrs, CT, 2002), *Dissertation Abstracts International*, Vol. 63, No. 11-A, p. 4118.

ILO (International Labour Organisation) (1990), "C171 Night Work Convention, 1990," retrieved September 2004, from http://www.ilo.org/ilolex/english/convdisp1.htm.

Imberman, W. (1983), "Who Strikes—and Why?" *Harvard Business Review*, Vol. 61, pp. 18–28.

Ishizaki, M., Morikawa, Y., Nakagawa, H., Honda, R., Kawakami, N., and Haratani, T. (2004), "The Influence of Work Characteristics on Body Mass Index and Waist to Hip Ratio in Japanese Employees," *Industrial Health*, Vol. 42, No. 1, pp. 41–49.

Jewett, M. (1997), "Models of Circadian and Homeostatic Regulation of Human Performance and Alertness," unpublished doctoral dissertation, Harvard University, Cambridge, MA.

Jewett, M., Wyatt, J., and Ritz-De Cecco, A. (1999), "Time Course of Sleep Inertia Dissipation in Human Performance and Alertness," *Journal of Sleep Research*, Vol. 8, pp. 1–8.

Johnson, L. C., and Naitoh, P. (1974), *The Operational Consequences of Sleep Deprivation and Sleep Deficit*, AGARD-AG-193, NATO, Technical Editing and Reproduction, London.

Joint Committee on Standards for Educational Evaluation, James R. Sanders, Chair (1994), *Program Evaluation Standards*, 2nd ed., Sage Publications, Thousand Oaks, CA.

Kaliterna, L., Vidaček, S., Radošević-Vidaček, B., and Prizmić, Z. (1993), "The Reliability and Stability of Various Individual Difference and Tolerance to Shiftwork Measures," *Ergonomics*, Vol. 36, pp. 183–189.

Kaliterna, L., Vidaček, S., Prizmić, Z., and Radošević-Vidaček, B. (1995), "Is Tolerance to Shiftwork Predictable from Individual Difference Measures?" *Work and Stress*, Vol. 9, pp. 140–147.

Keran, C. M., Duchon, J. C., and Smith, T. J. (1994), "Older Workers and Longer Work Days: Are They Compatible?" *International Journal of Industrial Ergonomics*, Vol. 13, pp. 113–123.

Khaleque, A. (1999), "Sleep Deficiency and Quality of Life of Shift Workers," *Social Indicators Research*, Vol. 46, No. 2, pp. 181–189.

Kleitman, N., and Jackson, D. P. (1950), "Body Temperature and Performance Under Different Routines," *Journal of Applied Physiology*, Vol. 3, pp. 309–328.

Knauth, P., and Rutenfranz, J. (1976), "Experimental Studies of Permanent Night and Rapidly Rotating Shift Systems," *International Archives of Occupational and Environmental Health*, Vol. 37, pp. 125–137.

Knauth, P., Keller, J., Schindele, G., and Totterdell, P. (1995), "A Fourteen-Hour Night Shift in the Control Room of a Fire-Brigade," *Work and Stress*, Vol. 9, pp. 176–186.

Knutsson, A. (2003), "Health Disorders and Shift Workers," *Occupational Medicine*, Vol. 53, pp. 103–108.

Kogi, K., and Thurman, J. E. (1990), "Development of New International Standards on Night Work," in *Shiftwork: Health, Sleep and Performance*, G. Costa, G. Cesana, K. Kogi, and A. Wedderburn, Eds., Peter Lang, Frankfurt-am-Main, Germany, pp. 19–24.

Koller, M., Kundi, M., and Cervinka, R. (1978), "Field Studies of Shift Work in an Austrian Oil Refinery, I: Health and Psychosocial Wellbeing of Workers Who Drop Out of Shiftwork," *Ergonomics*, Vol. 21, pp. 835–847.

Kribbs, N. B., and Dinges, D. (1994), "Vigilance Decrement and Sleepiness," in *Sleep Onset: Normal and Abnormal Processes*, R. D. Ogilvie and J. R. Harsh, Eds., American Psychological Association, Washington, DC, pp. 113–125.

Krueger, G. P. (1989), "Sustained Work, Fatigue, Sleep Loss and Performance: A Review of the Issues," *Work and Stress*, Vol. 3, pp. 129–141.

Laakso, M., Porkka-Heiskanen, T., Stenberg, D., and Alila, A. (1991), "Interindividual Differences in the Responses of Serum and Salivary Melatonin to Bright Light," in *Role of Melatonin and Pineal Peptides in Neuroimmunomodulation*, F. Fraschini and R. J. Reter, Eds., Plenum Press, New York, pp. 307–311.

Lavie, P., Tzischinsky, O., Epstein, R., and Zomer, J. (1992), "Sleep–Wake Cycle in Shift Workers on a 'Clockwise' and 'Counter-clockwise' Rotation System," *Israel Journal of Medical Sciences*, Vol. 28, pp. 636–644.

Lennernas, M. (1993), "Nutrition and Shift Work: The Effect of Work Hours on Dietary Intake, Meal Patterns and Nutritional Status Parameters," *Comprehensive Summaries of Uppsala Dissertations from the Faculty of Medicine*, No. 402, Almqvist & Wiksell International, Stockholm.

Lennernas, M., Hambraeus, L., and Åkerstedt, T. (1995), "Shift Related Dietary Intake in Day and Shift Workers," *Appetite*, Vol. 25, No. 3, pp. 253–265.

Lewis, P. M., and Swain, D. J. (1986), "Evaluation of a 12-Hour Day Shift Schedule," in *Proceedings of the Human Factors Society 30th Annual Meeting*, pp. 885–889.

Mahan, R. P., Tepas, D. I., and Carvalhais, A. B. (1987), "Morningness–Eveningness Norms for Industrial Shift

Workers," in *Contemporary Advances in Shiftwork Research: Theoretical and Practical Aspects in the Late Eighties*, A. Oginski, J. Pokorski, and J. Rutenfranz, Eds., Medical Academy, Krakow, pp. 429–433.

Mallis, M. M., Mejdal, S., Nguyen, T. T., and Dinges, D. F. (2004), "Summary of the Key Features of Seven Biomathematical Models of Human Fatigue and Performance," *Aviation, Space, and Environmental Medicine*, Vol. 75, No. 3, Pt. II, pp. A4–A14.

Mamelle, N., Laumon, B., and Lazar, P. (1984), "Prematurity and Occupational Activity During Pregnancy," *American Journal of Epidemiology*, Vol. 119, pp. 309–322.

Mapes, G. (1990), "Beating the Clock," *Wall Street Journal,* April 10, pp. A1, A6.

Martin, S. K., and Eastman, C. I. (1998), "Medium-Intensity Light Produces Circadian Rhythm Adaptation to Simulated Night-Shift Work," *Sleep*, Vol. 21, No. 2, pp. 154–165.

Matsumoto, K., and Harada, M. (1994), "The Effect of Night-Time Naps on Recovery from Fatigue Following Night Work," *Ergonomics*, Vol. 37, No. 5, pp. 899–907.

Matsumoto, K., Matsui, T., Kawamori, M., and Kogi, K. (1982), "Effects of Nighttime Naps on Sleep Patterns of Shiftworkers," *Journal of Human Ergology*, Vol. 11 (Suppl.), pp. 279–289.

Minors, D., and Waterhouse, J. (1983), "Does 'Anchor Sleep' Entrain Circadian Rhythms? Evidence from Constant Routine Studies," *Journal of Physiology*, Vol. 345, pp. 451–467.

Mitler, M., and Aldrich, M. (2000), "Stimulants: Efficacy and Adverse Effects," in *Principles and Practice of Sleep Medicine*, M. H. Kryger, T. Roth, and W. C. Dement, Eds., W.B. Saunders, Philadelphia, pp. 429–440.

Moline, M., Pollak, C., Monk, T., Lester, L., Wagner, D., Zendell, S., et al. (1992), "Age-Related Differences in Recovery from Simulated Jet Lag," *Sleep*, Vol. 15, No. 1, pp. 28–40.

Monk, T. H. (1986), "Advantages and Disadvantages of Rapidly Rotating Shift Schedules: A Circadian Viewpoint," *Human Factors*, Vol. 38, pp. 553–557.

Monk, T. H., and Embrey, D. E. (1981), "A Field Study of Circadian Rhythms in Actual and Interpolated Task Performance," in *Night and Shift Work: Biological and Social Aspects*, A. Reinberg, N. Vieux, and P. Andlauer, Eds., Pergamon Press, Oxford.

Monk, T. H., and Folkard, S. (1985), "Individual Differences in Shiftwork Adjustment," in *Hours of Work: Temporal Factors in Work-Scheduling*, S. Folkard and T. H. Monk, Eds., Wiley, New York, pp. 227–237.

Monk, T. H., Weitzman, E. D., Fookson, J. E., Moline, M. L., Kronauer, R. E., and Gander, P. H. (1983), "Task Variables Determine Which Biological Clock Controls Circadian Rhythms in Human Performance," *Nature*, Vol. 304, pp. 543–545.

Monk, T., Folkard, S., and Wedderburn, A. (1996), "Maintaining Safety and High Performance on Shiftwork," *Applied Ergonomics*, Vol. 27, No. 1, pp. 17–23.

Morikawa, Y., Miura, K., Ishizaki, M., Nakagawa, H., Kido, T., Naruse, Y., et al. (2001), "Sickness Absence and Shift Work Among Japanese Factory Workers," *Journal of Human Ergology*, Vol. 30, No. 1–2, pp. 393–398.

Mott, P. E., Mann, F. C., McLoughlin, Q., and Warwick, D. P. (1965), *Shift Work: The Social, Psychological and Physical Consequences*, University of Michigan Press, Ann Arbor, MI.

Nachreiner, F., Grzech-Sukalo, H., Qin, L., Moehlmann, D., and Will, W. (1995), "Computer-Aided Design of Shift Schedules for Public Transport Operations," *Shiftwork International Newsletter*, Vol. 12, p. 46.

Naitoh, P., and Angus, R. G. (1989), "Napping and Human Functioning During Prolonged Work," in *Sleep and Alertness: Chronobiological, Behavioral, and Medical Aspects of Napping*, D. F. Dinges and R. J. Broughton, Eds., Raven Press, New York.

Naitoh, P., Kelly, T., and Babkoff, H. (1993), "Sleep Inertia: Best Time Not to Wake Up?" *Chronobiology International*, Vol. 10, pp. 109–118.

NIH (National Institutes of Health) (2004), "NIH Releases Research Strategy to Fight Obesity Epidemic," posted August 24, retrieved September, 2004, from http://www.nih.gov/news/pr/aug2004/niddk-24.htm.

NTSB (1989), "Head-End Collision of Consolidated Rail Corporation Freight Trains UBT-506 and TV-61 near Thompsontown, Pennsylvania, January 14, 1988," *Railroad Accident Report*, NTSB/RAR-89/02, National Transportation Safety Board, Washington, DC.

NTSB (1990), "NTSB Adopts First List of 'Most Wanted' Safety Items," *Safety Information*, SB 90 48/5299A, National Transportation Safety Board, Washington, DC.

Oginska, H., Pokorski, J., and Oginski, A. (1993), "Gender, Ageing, and Shiftwork Intolerance," *Ergonomics*, Vol. 36, No. 1–3, pp. 161–168.

Oginski, A., Oginska, H., Pokorski, J., Kmita, W., and Gozdziela, R. (2000), "Internal and External Factors Influencing Time-Related Injury Risk in Continuous Shift Work," *International Journal of Occupational Safety and Ergonomics*, Vol. 6, No. 3, pp. 405–421.

Ong, C. N., and Kogi, K. (1990), "Shiftwork in Developing Countries: Current Issues and Trends," in *Shiftwork*, A. J. Scott, Ed., Hanley & Belfus, Philadelphia, pp 417–428.

Paley, M. J., and Tepas, D. I. (1994a), "Fatigue and the Shiftworker: Firefighters Working on a Rotating Shift Schedule," *Human Factors*, Vol. 36, pp. 269–284.

Paley, M. J., and Tepas, D. I. (1994b), "The Effect of Tenure on Fire Fighter Adjustment to Shiftwork," presented at the 23rd International Congress of Applied Psychology, Madrid, Spain.

Paley, M. J., Price, J. M., and Tepas, D. I. (1998), "The Impact of a Change in Rotating Shift Schedules: A Comparison of the Effects of 8, 10 and 14 Hour Work Shifts," *International Journal of Industrial Ergonomics*, Vol. 21, No. 3–4, pp. 293–305.

Pilcher, J. J., Teichman, H. M., Popkin, S. M., Hildebrand, K. R., and Coplen, M. K. (2004), "Effect of Day Length on Sleep Habits and Subjective On-Duty Alertness in Irregular Work Schedules," *Transportation Research Record*, Vol. 1865, pp. 72–79.

Poole, C. J. M., Wright, A. D., and Nattrass, M. (1992), "Control of Diabetes Mellitus in Shift Workers," *British Journal of Industrial Medicine*, Vol. 49, pp. 513–515.

Popkin, S. M. (1994), "An Evaluation of the Impact of an Educational Program for Freight Locomotive Engineers on Irregular Work Schedules," in *Proceedings of the 12th Triennial Congress of the International Ergonomics Association*, Vol. 5, pp. 33–35.

Popkin, S. M., and Coplen, M. K. (1995), "Effects of an Educational Program for Shiftworkers and Their Spouses," presented at the 12th International Symposium of Night and Shiftwork, Ledyard, CT.

Popkin, S., Gertler, J., and Reinach, S. (2001), *Preliminary Examination of Railroad Dispatcher Workload, Stress, and Fatigue*, DOT/FRA/ORD-01/08, U.S. Department of Transportation, Washington, DC.

Presser, H. B. (2003), *Working in a 24/7 Economy: Challenges for American Families*, Russell Sage Foundation, New York.

Proceedings of the 5th International Conference on Fatigue and Transportation (2003), Murdoch University, Fremantle, Western Australia.

Public citizen et al. *v. Federal Motor Carrier Safety Administration* (2004), U.S. Court of Appeals, p. 22.

Raslear, T. G., and Coplen, M. (2004), "Fatigue Models as Practical Tools: Diagnostic Accuracy and Decision Thresholds," *Aviation, Space, and Environmental Medicine*, Vol. 75, No. 3, Pt. II, pp. A168–A172.

Reinberg, A. (1974), "Chronopharmacology in Man," in *Chronobiological Aspects of Endocrinology*, J. Aschoff, F. Ceresa, and F. Halberg, Eds., F.K. Schattauer, Stuttgart, Germany, pp. 157–185.

Reinberg, A., Andlauer, P., DePrins, J., Malbecq, W., Vieux, N., and Bourdeleau, P. (1984), "Desynchronization of the Oral Temperature Circadian Rhythm and Intolerance to Shift Work," *Nature*, Vol. 308, pp. 272–274.

Rentos, P. G., and Shepard, R. D., Eds. (1976), *Shift Work and Health: A Symposium*, HEW Publication NIOSH 76–203, U.S. Government Printing Office, Washington, DC.

Reyner, L. A., and Horne, J. A. (1997), "Suppression of Sleepiness in Drivers: Combination of Caffeine with a Short Nap," *Psychophysiology*, Vol. 34, pp. 721–725.

Ringenbach, K., and Jacobs, R. (1995), "Injuries and Aging Workers," *Journal of Safety Research*, Vol. 26, pp. 169–176.

Roehrs, T., and Roth, T. (2000), "Hypnotics: Efficacy and Adverse Effects," in *Principles and Practice of Sleep Medicine*, M. H. Kryger, T. Roth, and W. C. Dement, Eds., W.B. Saunders, Philadelphia, pp. 429–440.

Romon-Rousseaux, M., Beuscart, R., Thuilliez, J. C., Frimat, P., and Furon, D. (1986), "Influence of Different Shift Schedules on Eating Behavior and Weight Gain in Edible-Oil Refinery Workers," in *Night and Shiftwork: Long Term Effects and Their Prevention*, M. Haider, M. Koller, and R. Cervinka, Eds., Peter Lang, Frankfurt-am-Main, Germany, pp. 433–440.

Romon-Rousseaux, M., Lancry, A., Poulet, I., Frimat, P., and Furon, D. (1987), "Effect of Protein and Carbohydrate Snacks on Alertness During the Night," in *Contemporary Advances in Shiftwork Research: Theoretical and Practical Aspects in the Late Eighties*, A. Oginski, J. Pokorski, and J. Rutenfranz, Eds., Medical Academy, Krakow, pp. 133–141.

Rosa, R. R. (1991), "Performance, Alertness and Sleep After *3.5* Years of 12 h Shifts: A Follow-up Study," *Work and Stress*, Vol. 5, pp. 107–116.

Rosa, R. R. (1993), "Napping at Home and Alertness on the Job in Rotating Shift Workers," *Sleep*, Vol. 16, pp. 727–735.

Rosa, R. R., Colligan, M. J., and Lewis, P. (1989), "Extended Workdays: Effects of 8-Hour and 12-Hour Rotating Shift Schedules on Performance, Subjective Alertness, Sleep Patterns, and Psychosocial Variables," *Work and Stress*, Vol. 3, pp. 21–32.

Rosa, R. R., Bonnet, M. H., Bootzin, R. R., Eastman, C. I., Monk, T., Penn, P. E., et al. (1990), "Intervention Factors for Promoting Adjustment to Nightwork and Shiftwork," in *Shiftwork*, A. J. Scott, Ed., Hanley & Belfus, Philadelphia, pp. 391–414.

Rosekind, M. R., Gander, P. H., Gregory, K. B., Smith, R. M., Miller, D. L., Oyung, R., et al. (1996), "Managing Fatigue in Operational Settings, I: Physiological Considerations and Countermeasures," *Behavioral Medicine*, Vol. 21, pp. 157–165.

Rutenfranz, J. (1982), "Occupational Health Measures of Night- and Shiftworkers," *Journal of Human Ergology*, Vol. 11 (Suppl.), pp. 67–86.

Sallinen, M., Härmä, M., Mutanen, P., Ranta, R., Virkkala, J., and Müller, K. (2003), "Sleep–Wake Rhythm in an Irregular Shift System," *Journal of Sleep Research*, Vol. 12, No. 2, pp. 103–112.

Samel, A., Wegmann, H., and Vejvoda, M. (1997), "Aircrew Fatigue in Long-Haul Operations," *Accident Analysis and Prevention*, Vol. 29, pp. 439–452.

Scherrer, J. (1981), "Man's Work and Circadian Rhythm Through the Ages," in *Night and Shift Work: Biological and Social Aspects*, A. Reinberg, N. Vieux, and P. Andlauer, Eds., Pergamon Press, Oxford.

Sharpe, D. L., Hermsen, J. M., and Billings, J. (2002), "Gender Differences in Use of Alternative Full-Time Work Arrangements by Married Workers," *Family and Consumer Sciences Research Journal*, Vol. 31, No. 1, pp. 78–111.

Sherry, P. (2000), *Fatigue Countermeasures in the Railroad Industry: Past and Current Developments*, University of Denver, Intermodal Transportation Institute, Counseling Psychology Program, Denver, CO.

Smiley, A. (1990), "The Hinton Rail Crash," *Accident Analysis and Prevention*, Vol. 22, pp. 443–455.

Smith, A. (1988), "Effects of Meals on Memory and Attention," in *Practical Aspects of Memory: Current Research and Issues*, Vol. 2, *Clinical and Educational Implications*, M. M. Gruneberg, P. E. Morris, and R. N. Sykes, Eds., Wiley, Chichester, West Sussex, England, pp. 477–482.

Smith, C. S., Reilly, C., and Midkiff, A. (1989), "Evaluation of Three Circadian Rhythm Questionnaires with Suggestions for an Improved Measure of Morningness," *Journal of Applied Psychology*, Vol. 74, pp. 728–738.

Smith, L., Folkard, S., and Macdonald, I. (1995), "Zones of Extra Vulnerability," *Shiftwork International Newsletter*, Vol. 12, p. 58.

Smith, L., Folkard, S., Tucker, P., and Macdonald, I. (1998), "Work Shift Duration: A Review Comparing Eight Hour and 12 Hour Shift Systems," *Occupational and Environmental Medicine*, Vol. 55, pp. 217–229.

Swaen, G., Van Amelsvoort, L., Bultmann, U., and Kant, I. (2003), "Fatigue as a Risk Factor for Being Injured in an Occupational Accident: Results from the Maastricht Cohort Study, *Occupational and Environmental Medicine*, Vol. 60 (Suppl. 1), pp. i88–92.

Takagi, K. (1972), "Influence of Shift Work on Time and Frequency of Meal Taking," *Journal of Human Ergology*, Vol. 1, pp. 195–205.

Taub, J. M. (1971), "The Sleep–Wakefulness Cycle in Mexican Adults," *Journal of Cross-Cultural Psychology*, Vol. 44, pp. 353–362.

Taylor, P. J., and Pocock, S. J. (1972), "Mortality of Shift and Day Workers 1956–68," *British Journal of Industrial Medicine*, Vol. 29, pp. 201–207.

Tepas, D. I. (1982), "Adaptation to Shiftwork: Fact or Fallacy?" *Journal of Human Ergology*, Vol. 11 (Suppl.), pp. 1–12.

Tepas, D. I. (1985), "Flexitime, Compressed Workweeks and Other Alternative Work Schedules," in *Hours of Work*, S. Folkard and T. H. Monk, Eds., Wiley, New York, pp. 147–162.

Tepas, D. I. (1990a), "Do Eating and Drinking Habits Interact with Work Schedule Variables?" *Work and Stress*, Vol. 4, pp. 203–211.

Tepas, D. I. (1990b), "Condensed Working Hours: Questions and Issues," in *Shiftwork: Health, Sleep and Performance*, G. Costa, G. Cesana, K. Kogi, and A. Wedderburn, Eds., Peter Lang, Frankfurt-am-Main, Germany, pp. 271–282.

Tepas, D. I. (1993), "Educational Programs for Shiftworkers, Their Families, and Prospective Shiftworkers," *Ergonomics*, Vol. 36, pp. 199–209.

Tepas, D. I. (1994), "Technological Innovation and the Management of Alertness and Fatigue in the Workplace," *Human Performance*, Vol. 7, pp. 165–180.

Tepas, D. I. (1999), "Work Shift Usability Testing," in *The Occupational Ergonomics Handbook*, W. Karwowski and W. S. Marras, Eds., CRC Press, Boca Raton, FL, pp. 1741–1758.

Tepas, D. I. (2003), "Workware Decision Support Systems: A Comprehensive Methodological Approach to Work-Scheduling Problems," *Theoretical Issues in Ergonomic Science*, Vol. 4, No. 3–4, pp. 319–326.

Tepas, D. I., and Carvalhais, A. B. (1990), "Sleep Patterns of Shiftworkers," *Occupational Medicine: State of the Art Reviews*, Vol. 5, No. 2, pp. 199–208.

Tepas, D. I., and Mahan, R. P. (1989), "The Many Meanings of Sleep," *Work and Stress*, Vol. 3, pp. 93–102.

Tepas, D. I., and Popkin, S. M. (1994), "Duration-of-Work-day and Time-of-Day as Predictors of Workday Sleep Length and End of Workday Ratings of Being Tired or Tense," paper presented at the 11th International Symposium on Night and Shiftwork, La Trobe University, Melbourne, Australia.

Tepas, D. I., and Price, J. M. (2001), "What Is Stress and What Is Fatigue? In *Stress, Workload, and Fatigue*, P. A. Hancock and P. A. Desmond, Eds., Lawrence Erlbaum Associates, Mahwah, NJ, pp. 607–622.

Tepas, D. I., Armstrong, D. R., Byrnes, E. A., and Canning, P. M. (1981a), "Napping and Sleep Extension Among Shift Workers," *Sleep Research*, Vol. 10, p. 300.

Tepas, D. I., Walsh, J. K., and Armstrong, D. (1981b), "Comprehensive Study of the Sleep of Shift Workers," in *Biological Rhythms, Sleep, and Shift Work*, L. C. Johnson, D. I. Tepas, W. P. Colquhoun, and M. J. Colligan, Eds., Spectrum, New York.

Tepas, D. I., Walsh, J. K., Moss, P. D., and Armstrong, D. (1981c), "Polysomnographic Correlates of Shift Worker Performance in the Laboratory," in *Night and Shift Work: Biological and Social Aspects*, A. Reinberg, N. Vieux, and P. Andlauer, Eds., Pergamon Press, Oxford.

Tepas, D. I., Armstrong, D. R., Carlson, M. L., Duchon, J. C., Gersten, A. H., and Lezotte, D. V. (1985), "Changing Industry to Continuous Operations: Different Strokes for Different Plants," *Behavior Research Methods, Instruments and Computers*, Vol. 17, pp. 670–676.

Tepas, D. I., Carlson, M. L., Duchon, J. C., Gersten, A., and Mahan, R. P. (1986), "Moving a Plant from Discontinuous to Continuous Operation Using Weekend Work Shifts," in *Night and Shiftwork: Long-Term Effects and Their Prevention*, M. Haider, M. Koller, and R. Cervinka, Eds., Peter Lang, Frankfurt-am-Main, Germany, pp. 379–386.

Tepas, D. I., Duchon, J. C., and Gersten, A. H. (1993), "Shiftwork and the Older Worker," *Experimental Aging Research*, Vol. 19, pp. 295–320.

Tepas, D. I., Paley, M. J., and Popkin, S. M. (1997), "Work Schedules and Sustained Performance," in *Handbook of Human Factors and Ergonomics*, 2nd ed., G. Salvendy, Ed., Wiley, New York, pp. 1021–1058.

Thierry, H., and Jansen, B. (1984), "Work and Working Time," in *Handbook of Work and Organizational Psychology*, J. D. D. Drenth, H. Thierry, P. J. Willems, and C. J. deWolff, Eds., Wiley, New York, pp. 597–642.

Thierry, H., and Meijman, T. (1994), "Time and Behavior at Work," in *Handbook of Industrial and Organizational Psychology*, H. C. Triandis, M. D. Dunnette, and L. M. Hough, Eds., Consulting Psychologists Press, Palo Alto, CA, pp. 341–414.

Torsvall, L., Åkerstedt, T., Gillander, K., and Knutsson, A. (1989), "Sleep on the Night Shift: 24-Hour EEG Monitoring of Spontaneous Sleep/Wake Behavior," *Psychophysiology*, Vol. 26, No. 3, pp. 352–358.

Totterdell, P., Spelten, E., Smith, L., Barton, J., and Folkard, S. (1995), "Recovery from Work Shifts: How Long Does It Take?" *Journal of Applied Psychology*, Vol. 80, pp. 43–57.

TRB (2002), *Toolbox for Transit Operator Fatigue*, Report 81, Transit Cooperative Research Program, Transportation Research Board, Washington, DC.

U.S. Congress, Office of Technology Assessment (1991), *Biological Rhythms: Implications for the Workers*, OTA-BA-463, U.S. Government Printing Office, Washington, DC.

Van Dongen, H. P. A. (2004), "Comparison of Mathematical Model Predictions to Experimental Data of Fatigue and Performance," *Aviation, Space, and Environmental Medicine*, Vol. 75, No. 3, Pt. II, pp. A15–36.

Walsh, J. K., Muehlbach, M. J., Humm, T. M., Dickins, Q. S., Sugerman, J. L., and Schweitzer, P. K. (1990), "Effect of Caffeine on Physiological Sleep Tendency and Ability to Sustain Wakefulness at Night," *Psychopharmacology*, Vol. 101, pp. 271–273.

Waterhouse, J. M., Folkard, S., and Minors, D. S. (1993), "Effects of a Change in Shift-Work on Health," *Occupational Medicine*, Vol. 43, p. 167.

Wedderburn, A. A. I. (1981), "How Important Are the Social Effects of Shiftwork?" in *Biological Rhythms, Sleep, and Shift Work*, L. C. Johnson, D. I. Tepas, W. P. Colquhoun, and M. J. Colligan, Eds., Spectrum, New York.

Wedderburn, A. A. I. (1991), "Guidelines for Shiftworkers," *Bulletin of European Shiftwork Topics, No. 3*, European Foundation for the Improvement of Living and Working Conditions, Dublin, pp. 257–269.

Wedderburn, A. A. I. (1992), "How Fast Should the Night Shift Rotate? A Rejoinder," *Ergonomics*, Vol. 35, pp. 1447–1451.

Wedderburn, A., and Tepas, D. (in press), "Working Time," in *International Encyclopedia of Ergonomics and Human Factors*, 2nd ed., W. Karwowski, Ed., CRC Press, Boca Raton, FL.

Wever, R. A. (1979), *The Circadian System of Man*, Springer-Verlag, Berlin.

Wilkinson, R. T. (1992), "How Fast Should the Night Shift Rotate?" *Ergonomics*, Vol. 35, pp. 1425–1446.

Williamson, A. M., and Sanderson, J. W. (1986), "Changing the Speed of Shift Rotation: A Field Study," *Ergonomics*, Vol. 29, pp. 1085–1096.

Wilson, R. H. L., Newman, E. J., and Newman, H. W. (1956), "Diurnal Variations in Rate of Alcohol Metabolism," *Journal of Applied Physiology*, Vol. 8, No. 5, pp. 556–558.

Wurtman, J. J. (1986), *Managing Your Mind and Mood Through Food*, Rawson Associates, New York.

Xu, X., Ding, M., Li, B., and Christiani, D. C. (1994), "Association of Rotation Shiftwork with Preterm Births and Low Birth Weight Among Never Smoking Women Textile Workers in China," *Occupational and Environmental Medicine*, Vol. 51, pp. 470–474.

Yamada, Y., Kameda, M., Noborisaka, Y., Suzuki, H., Honda, M., and Yamada, S. (2001), "Comparisons of Psychosomatic Health and Unhealthy Behaviors Between Cleanroom Workers in a 12-Hour Shift and Those in an 8-Hour Shift," *Journal of Human Ergology*, Vol. 30, No. 1–2, pp. 399–403.

Zarcone, V. (2000), "Sleep Hygiene," in *Principles and Practice of Sleep Medicine*, M. H. Kryger, T. Roth, and W. C. Dement, Eds., W.B. Saunders, Philadelphia, pp. 657–662.

CHAPTER 29

PSYCHOSOCIAL APPROACH TO OCCUPATIONAL HEALTH

Mika Kivimäki and Kari Lindström
Finnish Institute of Occupational Health and University of Helsinki
Helsinki, Finland

1 INTRODUCTION

The aim of this chapter is to provide ergonomic and human factor practitioners and occupational health professionals with an understanding of psychosocial factors at work, how these factors can be identified, and how and why they can influence health and well-being. We have therefore provided an overview on major theories, defined core concepts, reviewed seminal research findings, and described frequently used assessment methods with respect to psychosocial factors at work. We also discuss evidence-based interventions that have been implemented to modify psychosocial factors at work to avoid adverse health effects and promote well-being.

2 THEORIES AND CONCEPTS

In Western countries with advanced occupational safety legislation, traditional occupational hazards, such as exposures to toxic chemicals, cold, and noise, account for only a part of the effect of work on health. Common trends in modern worklife include global competition, organizational changes such as downsizing and mergers, the growing proportion of the workforce with various kinds of temporary work arrangements, and a growing number of dual-career families. Such trends and many other characteristics of modern work are likely to influence the health and well-being of employees through psychosocial rather than physicochemical factors.

Not surprisingly, a psychosocial paradigm, in addition to the conventional physicochemical approach, forms an essential part of contemporary occupational health research. Psychosocial factors at work refer to the aspects of work design, organization, and management, and their social and organizational contexts, that have the potential to cause harm or influence employee health and well-being favorably (Cox and Griffiths, 1996). A central mediating mechanism between psychosocial factors at work and ill heath involves work stress.

2.1 Biological Model

The biomedical research approach to stress reactions has played a central role in determining the mechanisms that mediate between psychosocial factors and health. The foundation for the biology of stress was developed by Cannon (1929) and Selye (1936, 1956), who studied the autonomic nervous and neuroendocrine systems. Stress was described as a set of physiological reactions called the *general adaptation syndrome* (GAS), which was thought to be activated in a nonspecific, stereotyped form by any environmental challenge (Selye, 1956). The syndrome is characterized by the mobilization of energy resources for fight and flight. The syndrome proceeds through the following three stages: the alarm reaction, when the body responds to an environmental challenge; resistance (or adaptation), when the body attempts to restore itself; and if the stressful condition continues, exhaustion with the risk of stress-related disorders and diseases.

The recognition that the physiological reactions to stress cannot only protect, but also damage, the body has inspired further conceptual development in biological stress research. According to a recent view, the neuroendocrine, autonomic nervous, and immune systems are all mediators of attempts to adapt challenges in daily life (McEwen, 1998). This physiological accommodation to challenge is referred to as *allostasis*. It is adaptive in the short run, but prolonged overactivity of stress systems and inefficiently managed allostatic responses may cause wear and tear that can play a role in, for example, cardiovascular disease, infection, and accelerated aging.

2.2 Cognitive Model

Emphasis on the cognitive aspects of stress originated from attempts to understand why the same environmental challenges or stressors did not induce similar stress reactions in all people. One pioneer in the field was Lazarus (1966), who claimed that appraisal processes largely explain individual differences in responses: A stressor does not have an effect unless it is recognized and assessed by the person. Appraisal processes partially determine the quality and intensity of the reactions of stressors.

According to Lazarus, stress appraisal includes two phases. In the primary phase, a stressor is detected and its potential harmfulness is assessed. Stress perceptions relate to harm, threat, and challenge. *Harm* refers to psychological damage that has already taken place (e.g., loss of job). *Threat* is the anticipation of harm that has not yet taken place but may happen (e.g., job insecurity). *Challenge* results from potentially harmful demands that one feels confident of avoiding through the mobilization of resources. The secondary appraisal phase includes the assessment of the resources available to confront a stressor. Resources are then activated and manifested as coping strategies to counteract the stressors (problem-focused coping) or to alleviate stress reactions (emotion-focused coping).

2.3 Models Linking the Person and the Environment

Since the launching of Lazarus's model, a fundamental assumption in many stress theories has been that the nature and consequences of stress are determined by both the person and the environment. As an example, the *person–environment fit theory* proposes that stress arises from misfit between a person and the environment (Edwards et al., 1998). *Stress* is defined as a subjective appraisal that indicates that supplies are insufficient to fulfill the person's needs. *Supplies* refer to extrinsic and intrinsic resources and rewards (e.g., money, food, shelter, social involvement, and the opportunity to achieve) that may fulfill a person's needs. The misfit may also involve that between demands and abilities.

The *cybernetic theory of organizational stress* complements the person–environment fit model by putting more emphasis on the mechanism of reducing the misfit and achieving a balance between the person and the environment (Cummings and Cooper, 1979). The basic premise in the cybernetic stress theory is that deviations from a goal direct a person's behavior rather than predetermined internal mechanisms that aim blindly. A corresponding principle is commonly applied in other disciplines, such as biology and physics. According to the cybernetic stress theory, people appraise a situation as stressful when there is a mismatch between their actual and preferred states (i.e., discrepancy between the person and the environment). Such feedback information induces strain, guides the choice of adjustment processes, and stimulates coping.

A further specification of the stress-producing interaction between the person and the environment involves a cycle consisting of the following four basic elements (McGrath and Beehr, 1990): (1) a situation (i.e., a potentially stress-inducing event or condition), (2) the perceived situation (i.e., a person's interpretation of the objective situation), (3) the response selection (i.e., a person's choice of responses to cope with the situation), and (4) the actual coping behavior. As the last step can affect the actual situation, it may initiate a new cycle. The four elements of the cycle are connected by the following four processes: the appraisal process (how the situation is interpreted by the person—between elements 1 and 2); the choice process (what coping responses and strategies are chosen—between elements 2 and 3); the performance process (execution of coping—between elements 3 and 4); and the outcome process (how the coping executed affects the situation—between elements 4 and 1).

3 WORK ORGANIZATION AND JOB DESIGN

An important aim of research on psychosocial factors at work is to formulate theoretical concepts at a level of generalization that allows for their identification in a wide range of occupations. Stressful characteristics refer to the psychosocial aspects of worklife that produce intense, recurrent, and long-lasting stress experiences, at least in a substantial proportion of those exposed. Conceptualizations of psychosocial factors cover job characteristics, characteristics of work organization and social relations, as well as work contexts and cultures. These conceptualizations overlap and are not independent.

3.1 Job Characteristics

The model for psychosocial factors at work most often cited and most widely tested is the two-dimensional *demand–control model* conceptualized by Karasek and Theorell (1990). This model proposes that employees who do not have enough job control to meet their job demands are in a *job strain situation* which, if prolonged, increases the risk of stress-related diseases. Thus, adverse stress reactions are hypothesized to occur when a worker's decision latitude is low in a task and the psychological demands of the job are high. Job control (or decision latitude) refers to both socially predetermined control over detailed aspects of task performance (e.g., pace, quantity of work, policies and procedures, time of breaks, and scheduled hours) and skill discretion (i.e., control over the use of skills by the worker).

In the demand–control model, the cross-tabulation of demands and job control create three other conditions in addition to job strain. These conditions are related to a lower health burden and include active jobs, passive jobs, and low-strain jobs. *Active jobs* refer to high demand–high control situations. If the demands are not overwhelming, such situations are associated with learning and growth. The employee makes choices about how to cope with new stressors. If effective, the choices will contribute to the employee's coping repertoires and expand the range of solutions to environmental challenges. There may also be spillover from work to other life domains, as research suggests that workers with active jobs are the most active in their leisure time and political activity. Low demands combined with low control, called *passive jobs*, are assumed to cause an unmotivating job setting, which leads to gradual loss of previously acquired skills. *Low-strain jobs* are those with low demands but high decision latitude. An expanded version of the demand–control model adds social support to the model as a third component (Johnson and Hall, 1988). The highest risk of illness is assumed to relate to *iso-strain jobs*, characterized by high demands, low job control, and low social support.

More recent developments in research on psychosocial factors at work have broadened the view from proximal work characteristics to cover aspects of the person and the labor market context. A promising example is the *effort–reward imbalance model* of Siegrist (1996). This model maintains that an experienced imbalance between high effort spent at work and low reward received is particularly stressful, as this imbalance violates core expectations about reciprocity and adequate exchange at work.

Not only high demands and challenges at work, but also certain coping strategies (e.g., overcommitment and high personal need for control) and heavy obligations in private life (e.g., heavy debts) may contribute to a high expenditure of effort. Low rewards can be related to small financial compensation from work, low esteem (e.g., lack of help or acceptance by supervisors and colleagues), and poor career opportunities (no promotion prospects, job insecurity, and status inconsistency).

According to Siegrist (1996), at least three situations increase the likelihood of a continued health-damaging imbalance between efforts and rewards: (1) an employee perceives having no alternative choice in the labor market, (2) an employee accepts imbalance for a certain time for strategic reasons (e.g., due to an expectation that it will improve chances for career promotion and related rewards at a later stage), and (3) a person's excessive work-related overcommitment and high need for approval may lead to continued high efforts despite the low rewards received.

Research on quality of worklife (QWL) shares elements with the demand–control and effort–reward imbalance models, but with its main focus on outcomes such as well-being and job satisfaction, it is less health oriented. According to QWL principles, the criteria in job design should deal with the physical work environment, compensation systems, institutional rights and decisions, job content, internal and external social relations, and career development (Davis and Wacker, 1982). In a job design context, the main interest focuses on the following characteristics: variety of tasks and task identity, feedback from job, perceived and individual autonomy, and self-regulation.

3.2 Management and Supervisory Systems

Management and supervisory styles are important targets in occupational health research. A hierarchical authorative management style has been found to be associated with poor employee well-being (Smith et al., 1992), whereas participatory leadership is more advantageous (Lawler, 1986). Recent findings suggest that management characterized by high organizational justice has special occupational health relevance. Organizational justice involves both a procedural and a relational component (Moorman, 1991). The former indicates whether decision-making procedures include input from affected parties, are applied consistently, suppress bias, are accurate, are correctable, and are ethical (called *procedural justice*). The latter element refers to whether supervisors treat workers politely and considerately (called *relational justice*). Low organizational justice has been associated with job dissatisfaction, retaliation, and lower work commitment (Moorman, 1991; Dailey and Kirk, 1992; Shapiro and Brett, 1993). In contrast, high organizational justice has been related to better employee health both cross-sectionally (Elovainio et al., 2002) and longitudinally (Kivimäki et al., 2003a; Kivimäki et al., in press).

3.3 Roles and Interpersonal Relations

According to Kahn (1973), the three basic types of problems in organizational role setting are role ambiguity, role conflict, and role overload. *Role ambiguity* refers to a lack of clarity regarding role expectations and uncertainty concerning the outcomes of one's role performance. Role ambiguity may imply poor goal clarity (an employee is unaware of the outputs expected) and poor process clarity (although goals are clear, the means for achieving them may not be) (Sawyer, 1992). *Role conflict* reflects incompatible role expectations placed on an employee. *Role overload* is defined as a scope that is too wide with too many different kinds of expectations.

More recently, *workplace bullying* has been recognized as a serious health and safety problem in work life. The term refers to a situation in which one or more persons are subjected to persistent and repetitive negative acts by one or more co-workers, supervisors, or subordinates, and the person or persons feel unable to defend him/herself or themselves (Vartia, 2003). Physical bullying is rare, but other forms abound. Rayner and Hoel (1997) presented the following five categories of potential intimidating behavior: threats to professional status, threats to personal standing, isolation, overwork, and destabilization. Workplace bullying has been associated with increased absence rates among the victims and also among other employees working in units where bullying occurs (Kivimäki et al., 2000a).

Interpersonal relations in terms of social networks and social support from colleagues and supervisors can also represent a significant health resource in workplaces. The forms of social support include at least (1) emotional support through caregiving and affectionate concern, (2) appraisal support through evaluative feedback and affirmation, (3) informal support through suggestions or guidance, and (4) instrumental support through organized opportunities (House, 1981). In addition to a direct effect on health, social support has been hypothesized as buffering part of the adverse effects of stressors on health by changing one's perception of the stressor, by calming the stress-related neuroendocrine response, or by being linked with better tangible aid in a crisis.

3.4 Organizational Climate, Culture, and Organizational Restructuring

Several features of the organizational context in which employees work are subsumed under the concepts of organizational climate and culture. *Climate* is broadly defined as perceptions of organizational practices reported by people who work in a specific organizational setting (Rousseau, 1998). *Operationalizations* tap aspects such as communication, conflict, leadership, and reward emphasis (e.g., reward or punishment orientation). *Culture* constitutes the values, norms, and ways of behaving that members of the organization share. At least the following five elements of organizational culture have been identified: (1) unconscious beliefs that shape member's interpretations, called *fundamental assumptions*; (2) *values* (i.e., preferences for certain outcomes over others); (3) *behavioral norms*, referring to beliefs regarding appropriate and inappropriate types of behavior; (4) *patterns of behavior*, meaning observable recurrent practices; and (5) *artifacts*, defined as symbols and objects used to express cultural messages (e.g., mission statements and logos). There is a large body of research on the link between organizational climate and employee well-being, but the corresponding links for organizational culture have been studied less frequently.

The associations between actual structural changes in organizations (e.g., personnel reduction or downsizing and mergers) and employee health and well-being have drawn the attention of occupational health researchers. Since the recessions that hit most industrialized countries in the 1990s, evidence has accumulated on the health risks of survivors of corporate downsizing. One of the first studies in the field was conducted among the municipal employees of a town called Raisio, in Finland. The Raisio study found that the risk of health problems, as indicated by medically certified sickness absence and other indicators of health, was at least twice as great after major downsizing than after no downsizing (Vahtera et al., 1997; Kivimäki et al., 2000b). Half of this excess risk was attributable to an elevated level of work stress after major downsizing. Adverse effects on the health of the survivors of downsizing, including an elevated risk of fatal cardiovascular disease, have since been reported in several other studies (Quinlan et al., 2001; Vahtera et al., 2004a). Some studies suggest that not only downsizing, but also repeated exposure to rapid personnel expansion (possibly related to the centralization of functions) predict increased health problems, as indicated by rates of long-term sickness absence and hospitalization among employees (Westerlund et al., 2004).

3.5 Work–Life Interface

Interrelationships between work life and private life have been recognized in modern occupational health literature. The increasing numbers of dual-earner families and single parents and the evolving 24-hour society are likely to increase problems in combining work and family. Work–family conflict is defined as a form of role conflict in which the role pressures from work are incompatible with the demands of family life or in which employees have insufficient energy or time or both to perform work and family roles successfully (Greenhaus and Beutell, 1985; Grandey and Cropanzano, 1999).

A seminal study by Gardell (1982) reported that the effect of monotony caused by repetitive mechanized work carried over into private life in the form of general passivity. Other research has suggested that conflict between work and family is associated with job dissatisfaction, burnout, depression, and life and marital dissatisfaction (Allen et al., 2000). Worktime arrangements (e.g., day work vs. shift work, number of workhours per week, flexible workhours, level of worktime autonomy) have been found to be associated with work–family interference and employee health (Ala-Mursula et al., 2004; Jansen et al., 2004.) and are therefore a promising target for intervention. There is asymmetry in the permeable boundaries between the domains of work and family in that the effects of family-to-work interference on employee health may be less than the effects related to work-to-family interference.

4 RESPONSES TO PSYCHOSOCIAL FACTORS AT WORK

In this section we deal with the health relatedness of psychosocial factors at work, although these factors can also be regarded as contributors to work motivation, job satisfaction, and work performance. We begin with a description of the common physiological changes attributed to a stressful psychosocial work environment. This description is followed by an overview of the responses in the psychological and behavioral domains.

4.1 Physiological Changes

The central components of the physiological stress system are located in the phylogenetically oldest parts of the brain, the hypothalamus and the brain stem (McEwen, 1998). Activation of the stress system helps the body to overcome the influence of short-term physical stressors and therefore postpones all functions that may be irrelevant to immediate survival, such as digestion and growth. Although its precise nature varies according to the stressor, the function of a physiological stress response is essentially to prepare for, or to maintain, physical exertion through cognitive arousal, sensory vigilance, bronchodilation, tachycardia, raised blood pressure, elevated hemoconcentration, and energy mobilization.

Cardiovascular functioning is regulated by various factors, of which the autonomic nervous system and the hypothalamic–pituitary–adrenal cortex (HPA) axis are known to be activated during stress response. Activation of these systems releases circulating catecholamines (epinephrine and norepinephrine) and glucocorticoids (e.g., cortisol). Prolonged exposure to an increased secretion of these stress hormones can result in pathophysiologic consequences. According to the neuroendocrine hypothesis, chronic stress drives susceptible persons toward the *metabolic syndrome*, characterized by abnormalities such as glucose intolerance, insulin resistance, lipoprotein disturbances (e.g., a high concentration of low-density lipoprotein cholesterol), reduced fibrinolysis, and often, central obesity (Brunner, 2002).

The *immune response* is a series of events carried out by immune cells as a result of the body's exposure to foreign material or trauma. Blunting of the immune response may lead to an increased susceptibility to infectious diseases and malignancies, whereas an overactive immune response is associated with autoimmune and inflammatory disease. Studies on chronic stress suggest that prolonged stress exposure is associated with an increased susceptibility to the viral-induced common cold (Cohen et al., 1991), with a weaker antibody and viral-specific T-lymphocyte response to influenza vaccinations (Glaser et al., 1998) and with a slowing of wound healing (Kiecolt-Glaser et al., 1995). In contrast, short-term acute stress may enhance immunity and thus be beneficial. Psychosocial factors at work are also related to other biological systems (e.g., by triggering physiological mechanisms that increase muscle tension that may, in the long run, contribute to the development or intensification of musculoskeletal symptoms) (Bongers et al., 2002).

4.2 Health Risk Behavior

Psychosocial factors at work have been suggested to affect health-related behavior. Several studies have reported that a higher prevalence or intensity of smoking is associated with low job control, high job demands, and job strain (Pieper et al., 1989; Green and Johnson, 1990; Niedhammer et al., 1998b; Johnson et al., 1999). In some studies, work stress has been associated negatively with smoking cessation (Caplan et al., 1975). Although activation of the sympathetic nervous system in the fight-or-flight response is known to suppress digestive processes, evidence indicates that an adverse psychosocial environment may be associated with an increase in the amount eaten, particularly of sweet and fatty foods (Wardle and Gibson, 2002) and be predictive of weight gain (Kivimäki et al., 2002). The prospective Whitehall II study of British civil servants found a poor psychosocial work environment, as indicated by high effort–reward imbalance, to be associated with alcohol dependence among men but not among women (Head et al., 2004). Behavioral changes manifested in stressful work situations may reflect efforts to cope, or they may develop as part of the overall symptom pattern.

4.3 Burnout and Job Satisfaction

In the field of psychology, the multidimensional construct of burnout has been developed to describe a long-term stress syndrome (Maslach, 1998). The syndrome develops as a result of the chronic depletion of resources and is characterized by the following three clusters of symptoms: (1) exhaustion (emotional, intellectual, physical), (2) depersonalization (emotional detachment and cynicism), and (3) lowered professional accomplishment, self-efficacy, and self-esteem. An exhausted employee feels used up and drained and lacks enough energy to face another day or another person in need. *Depersonalization* is the interpersonal dimension of burnout and refers to a negative or excessively detached response to other people, often accompanied by a loss of idealism. Although it is supposed to be self-protective, there is a risk that detachment will turn into dehumanization. *Reduced personal accomplishment* refers to a decline in the sense of self-efficacy.

The potential of psychosocial factors contributing to positive outcomes at work has been studied extensively in relation to job satisfaction, especially in organizational psychology (Judge et al., 2002). Overall job satisfaction is characterized as a general positive attitude toward work. Aspects of job satisfaction include satisfaction with pay, employment security, social relationships at work, management, and personal growth.

5 OCCUPATIONAL STRESS AND DISEASE

There is evidence on the relationships between psychosocial factors and diseases with major public health relevance: that is, coronary heart disease, musculoskeletal disorders, and depression. These diseases make a significant contribution to the burden of disease of total populations and define a substantial part of potentially work-related suffering. Psychosocial factors at work may also be associated with several other morbidities and conditions.

5.1 Coronary Heart Disease

Coronary heart disease (CHD) is the leading single cause of death among men in many industrialized countries, and it is an important cause of death among women. Two aspects of the association between psychosocial factors at work and CHD have received particular attention among researchers. The first is the question of whether psychosocial factors are causally related to CHD incidence or to newly diagnosed CHD (the role of psychosocial actors in the etiology of CHD). A second question involves the effect of psychosocial factors on survival or prognosis among people with CHD. Despite a large body of research on these issues, both questions remain open to debate.

Several plausible mechanisms have been identified through which psychosocial factors may have an impact on the risk of CHD (McEwen, 1998; Brunner, 2002). They include overactivation and dysregulation of the autonomic nervous system and the HPA axis; both, if prolonged, are assumed to increase disease risk, to disrupt existing disease processes, and to act as triggers of acute events such as heart attack. In addition to these direct biological effects, psychosocial factors at work may influence CHD risk indirectly through health risk behavior (e.g., by increasing smoking intensity or the likelihood of sedentary lifestyle and an unhealthy diet). Moreover, adverse psychosocial factors at work (e.g., low social support) may adversely affect help-seeking behavior.

A systematic review of prospective studies published up to 2001 reported five papers with strong support for the effect of psychosocial work factors on CHD (Kuper et al., 2002a). Another five papers were either moderately supportive or supportive for only a subset of the population, a particular outcome, or a particular exposure. Three papers found a lack of a clear association between psychosocial factors at work and CHD. Since publication of the systematic review, two new studies have reported an association between job strain, effort–reward imbalance, and CHD (Kivimäki et al., 2002; Kuper et al., 2002b; Kuper and Marmot, 2003). Well-controlled, large-scale intervention studies are lacking.

Demonstrating a causal relationship between psychosocial factors at work and CHD is particularly challenging, due to the possibilities for confounding and the difficulties with measurements. CHD takes decades to develop, and in addition to adulthood risk factors, it is associated with a large variety of other factors, such as intrauterine development, growth and health in childhood, and lifetime socioeconomic circumstances. Psychosocial factors at work are not randomly distributed, and they do not necessarily remain stable over time. Thus, there is obviously a risk of obtaining both false positive and false negative findings.

5.2 Musculoskeletal Disorders

Musculoskeletal problems, particularly back disorders, are very common complaints in Western societies, and they cause a great deal of trouble for people and considerable sickness costs to society. Several potential pathophysiological pathways link psychosocial work factors with musculoskeletal disorders (Marras et al., 2000; Bongers et al., 2002). High demands can increase muscle tension and muscle coactivation and decrease micropauses in muscle activity and therefore lead to extra loading of the musculoskeletal system. In addition, high stress can hamper the ability to reduce physiological activation to resting levels after work and thus affect recovery adversely. The responses of stress systems can also lead to increased sensitization to pain stimuli.

Low job control has been associated with an increased rate of musculoskeletal sickness absence and an increased risk of hospitalization due to musculoskeletal disorders (Kivimäki et al., 2001; Kaila-Kangas et al., 2004). According to a systematic review with a high-quality filter, high job stress and high job demands in particular are associated with shoulder, upper arm, and wrist problems (Bongers et al., 1993, 2002). However, due to the limitations of the studies covered by the review, firm conclusions on the etiological and prognostic role of psychosocial work factors in musculoskeletal disorders are not yet possible.

5.3 Depression

Depression has large derogatory effects on a person's quality of life and functioning and is likely to lower productivity, cause absenteeism, and increase substance use and accidents (Lecrubier, 2000; Goldberg and Steury, 2001). According to the Global Burden of Disease Study, depression will account for 15% of the disease burden throughout the world by 2020 (Murray and Lopez, 1997). Several prospective studies suggest a relationship between psychosocial factors at work (e.g., high demands, low job control, effort–reward imbalance, low organizational justice) and subsequent depression or other mental health disorders (Stansfeld et al., 1997, 1999; Niedhammer et al., 1998a; Paterniti et al., 2002; Kivimäki et al., 2003a). However, methodological limitations in the measurement of exposure or outcome or both limit the strength of this evidence. Use of self-reports in assessing psychosocial factors at work is particularly problematic in relation to mental health studies, as it is difficult to eliminate the possibility of reporting bias due to subclinical mental health problems.

6 METHODS OF MEASUREMENT

Psychosocial factors at work share features that make their measurement challenging (Cox and Griffiths, 1996). They are usually not visible in the way that physical and chemical hazards are, nor can they be assessed directly. They are often related to and, in part, determined by social relationships, the way other employees behave, and the way behavior is managed. All these features increase the risk of measuring the psychosocial work environment imprecisely.

Several methodological challenges are also evident in respect to causal associations between psychosocial factors and health outcomes. First, psychosocial factors are often associated with other health risk factors, and it is therefore difficult to determine whether psychosocial factors would independently predict health or whether their associations with disease are noncausal and redundant with respect to other etiological factors. Demonstrating the biological pathways that link psychosocial factors to specific diseases and demonstrating outcome specificity (i.e., the absence of associations between psychosocial factors and nonstress-related diseases) might strengthen the evidence. Second, health effects are expected to result only after long-term exposure to an adverse psychosocial work environment, but repeated measurements over time to identify such exposures are often impractical. Third, health problems may influence psychosocial factors (reverse causation) and cause selective sample attrition (healthy worker effect), and both complicate the interpretation of findings. Fourth, randomized controlled trials represent the strongest study design but are seldom feasible in relation to psychosocial factors at work for ethical and practical reasons. Therefore, methodologically strongest evidence is almost impossible to accumulate.

6.1 Self-Report Measures

Psychosocial factors at work are typically operationalized using self-report measurements. In such instances, job incumbents respond to a questionnaire or undergo a structural interview. Many questionnaire measures have been developed according to good practice in psychometrics (for some of them, computer programs for data analysis are also available) (Elo, 1994). One difficulty is to determine the extent to which the responses to these measurements reflect job characteristics and to the extent to which they reflect subjective appraisal processes that may even vary between people working in the same psychosocial work environment. Using complementary assessment strategies, such as inferred measurements, may help to overcome this problem.

Coping strategies are determined primarily with questionnaires. For example, the ways of coping checklist includes items covering problem-focused coping and emotion-focused coping (Folkman and Lazarus, 1980). A checklist to measure the appraisal of work stress situations is also available (Dewe, 1992). In addition to these context-specific measures, some indicators of coping styles have been developed to reveal coping as a traitlike construct.

In research on burnout, a standard measurement tool is the Maslach burnout inventory (MBI), which assesses all three dimensions of the syndrome (Maslach and Jackson, 1998). Occupation-specific versions of this measure have been developed, one for people working in human services, another for use with employees in educational settings, and still another applicable also to occupations that are not people-oriented (Maslach, 1998). Emotional reactions to work, both negative and positive, and job satisfaction are important outcomes to measure for an understanding of responses to psychosocial factors at work. Emotional reactions can be assessed using questionnaires, state-trait measures, checklists, projective tests, and biological measures.

It is noteworthy that the use of self-reported measures to indicate both psychosocial factors at work and the responses to these factors makes a study open to common-method variance bias. For example, people's different response styles, negative affectivity, and other factors affecting both exposure and outcome may inflate the relationships artificially.

6.2 Inferred and Archival Measures

In assessments of psychosocial factors at work, a strategy that complements employee self-reports is the use of ecological (inferred) measures that are not based directly on subjective evaluations. Instead of individual scores from the responses to self-assessment questionnaires, studies have used occupation mean scores or work-unit mean scores. In this measurement strategy, the mean score is applied to all members of the occupation or the work unit. The use of mean scores eliminates individual-related variation in the assessment and effectively reduces the risk of bias due to common-method variance, but the mean scores are insensitive to true variation in psychosocial factors within an occupation or work unit. Measurements inferring non-self-report of psychosocial factors include, for example, expert ratings and observation.

Archival data such as sickness absence records, hospitalization, and other morbidity records and mortality data are used increasingly to assess health-related outcomes in studies of psychosocial factors at work. One advantage is the minimization of potential recall and response biases attributable to self-reported indicators of health. Routinely collected register-based measures are also practical in assessments of health changes across multiple time points, and they minimize problems related to common-method variance.

In many countries, administrative hospital discharge data contain information about dates of hospitalizations and coded diagnoses that allow researchers to determine cases of serious illness and medical conditions. Linking these records and information from mortality records to the data is a standard procedure in occupational health research in the field of medicine and epidemiology.

Sickness absence records are collected routinely in many workplaces. Certified absences are also a by-product of the medical care process. There is evidence to suggest that medically certified recorded sickness absences (but not self-certified absences) can accurately reflect the health of working populations, at least when health is understood in terms of physical and social functioning. Medically certified sick leaves are a strong correlate of morbidity and diseases (Marmot et al., 1995), and they are highly predictive of disability retirement (Borg et al., 2001; Gjesdal and Bratberg, 2002; Kivimäki et al., 2004) and overall and cause-specific mortality (Kivimäki et al., 2003a; Vahtera et al., 2004b). Illness episodes that do not lead to hospital admission are typically unavailable in health measures based on morbidity records, but they are nevertheless seen in absence figures. Obviously, the recording of absences can be inaccurate in some organizations, and such organizations are therefore not appropriate targets for studies using sickness absence as the outcome measure. Some employees take sick leave without actual illness or are ill without sick leave. These factors form a greater validity problem with the use of self-certified absences than with the use of long-term medically certified absences.

6.3 Physiological Measures of Stress Response

Physiological methods can be used to assess potential responses to psychosocial factors at work. Measurements, such as heart rate, heart rate variability, blood pressure, respiratory rate and electromyography, can be taken over a continuous time period. As these measures are associated with a physiological system that can be considered fast acting in response to stress stimuli, they have the potential to identify episodic responses to stress.

Biochemical measures can be obtained from various fluids of the body, such as urine, blood, sweat, and saliva. Often-measured biochemicals include catecholamines, glucocorticoids, triglycerides, and lipids. In contrast to, for example, heart rate and respiratory rate measurements, practical restrictions on the collection procedures used for biochemical measures necessarily dictate that they reflect responses to stress over fairly long periods. As a result, biochemical measurements are relatively insensitive with respect to identifying effects of specific events. It is important to exercise control in the types and quantities of food, liquid, and drugs ingested by a worker prior to and during the biochemical measurement process, as they may bias findings.

CHD is a particularly relevant outcome in research on psychosocial factors. Established CHD is actually the end stage of a prolonged pathogenic process. Abnormalities in arterial and myocardial structures and functions precede, by some decades, the onset of clinically manifest CHD (i.e., acute myocardial infarction, unstable or stable angina, sudden cardiac death, heart failure). Noninvasive measurements of some of these abnormalities are currently possible. Structural changes in the arteries include the thickening of arterial walls (atherosclerosis). Repeated ultrasonic measurements of the carotid intima-media thickness are able to determine the progression of arterial wall thickening. In atherosclerosis, arteries also become stiffer (the term *sclerosis* denotes a hardening of the arteries). This process causes abnormalities in endothelial function, and it is possible to assess these abnormalities using ultrasonic measurements. Electrocardiography, chest radiography, echocardiography, and magnetic resonance imaging have been used to assess increased left ventricular mass of the myocardium, another correlate of CHD.

7 INDIVIDUAL- AND GROUP-ORIENTED INTERVENTION

At least three types of intervention can be carried out to tackle psychosocial stressors and stress reactions at work. The individual- and group-oriented approaches are usually secondary and aim at reversing or slowing the progression of ill health. Tertiary forms are also initiated to treat health disorders. The purpose of primary prevention is to change or develop job and organizational characteristics in order to improve employee health and well-being (Murphy and Sauter, 2004). Table 1 summarizes the categories of methods used to reduce individual stress. They are discussed in more detail in the following sections.

7.1 Stress Management and Coping Training

Interventions employing stress management and coping training fall into the category of tertiary prevention, aimed at modifying a person's reactions to perceived stressful situations or events or his or her behavior in stressful situations. Stress management is often targeted more at individuals by using, for example, muscle relaxation and meditation, although group- and organization-focused forms of interventions are on the increase (Giga et al., 2003; Murphy and Sauter, 2004). Evaluation of these interventions has been carried out in a short-term time frame. A somewhat different approach was applied in a Dutch intervention, in which the employees met during nine weekly meetings to read inspirational workplace stories, comment on these stories, and handle their own strain. Their coping resources, cognitive–rational coping, state of mind, confidence, and home–work balance improved (Horan, 2003).

A meta-analysis has been carried out on the following four types of interventions: cognitive–behavioral intervention, relaxation techniques, multimodal programs, and organization-focused interventions. The first three intervention types, which are usually called *stress management intervention*, had a moderate or small effect on complaints, psychological resources and responses, and perceived quality of work life. The organizational interventions had no effects (van der Klink et al., 2001).

Rahe et al. (2002) studied a workplace stress management program aimed at reducing illness and the use of health services. Even though the intensity of the intervention varied between the groups, each group

Table 1 Individual-Level Means to Deal with Occupational Stress

Intervention	Mechanism of Stress Control	References
Stress management, including relaxation and biofeedback	Modify perceptions or behavior or physiological reactions.	Roskies et al. (1989)
Cognitive or behavioral methods	Modify perceptions and behavior.	Bramson (1985), Meichenbaum (1985), Bunce and West (1996), van der Klink et al. (2001), Murphy and Sauter (2004)
Debriefing	Institute cognitive restructuring.	Braverman (1992)
Health education and lifestyle changes	Modify behavior.	Ivancevich and Matteson (1988)
Career development	Enhance technical skills, knowledge, and competence.	Kram (1985), Senge (1990), Russell (1991), Aryee and Tan (1992), Nonaka and Takeuchi (1995)

reported significant improvement in stress, anxiety, and coping throughout the year. The most intensive intervention groups displayed a more rapid reduction in negative responses to stress as well as less need for health services. However, doubts have been expressed about stress management intervention because it may fail to recognize the wider organizational context in which the behavior takes place (Dewe and O'Driscoll, 2002).

7.2 Cognitive–Behavioral Approach

The cognitive–behavioral approach involves a variety of strategies that are meant to change the appraisal of the stressful situation and the psychological responses to the situation. Some of the guiding principles of these strategies are (1) a person's responses to the environment are a reaction to his or her own cognitive interpretations of it; (2) cognitions (thoughts), emotions, and behavior are causally related to each other; and (3) a person's expectations, beliefs, and attributions predict negative consequences (Murphy, 1988). These types of intervention can be primarily classified as secondary, in which the perception of stress and behavior is modified. According to a meta-analysis of the benefits of interventions on work-related stress, cognitive–behavioral interventions were more effective than the other types (relaxation techniques, multimodal interventions, and organization-focused interventions) (van der Klink et al., 2001). Cognitive–behavioral interventions have been effective especially when improving the perceived quality of life, enhancing psychological resources, and reducing complaints are the goals. Often, however, a combination of methods is used in a group-level approach.

The stress inoculation training developed by Meichenbaum (1985) involves the steps of preparation, training in skills, and training for application. In the stage of preparation, the clients are guided to recognize the connections of their maladaptive thoughts and beliefs with their maladaptive emotional reactions. The second step aims at developing skills through advice on how to confront stressors and apply efficient coping strategies. In the third phase, clients practice the new skills. Stressful stimuli can be used at this step to increase the ecological validity of the situation and to "desensitize" workers.

A comparison between cognitive–behavioral stress management aimed at changing the perception of stress and so-called innovative interventions focused also on the causes of stress and aimed at developing innovative response to stress, and individual action plans showed that the process variables in the groups were important for solutions that could be maintained. This finding may reflect the importance of contextualizing of the solutions (Bunce and West, 1996).

Research indicates that the key factors in the implementation of occupational stress intervention are the creation of a social climate conducive to learning from failure, the provision of opportunities for multilevel participation in designing the intervention, and an awareness of tacit negative behavior that possibly undermines the objectives of the intervention. It is also crucial to define the roles and responsibilities of the participants clearly before and after the intervention (Nytrø et al., 2000). The cognitive–behavioral approach is participatory and can focus also on the individual–organization interface (Landsbergis and Vivona-Vaughan, 1995).

7.3 Debriefing

Debriefing is planned as a program for treating traumatic psychological experiences immediately to avoid posttraumatic stress disorders and prevent the development of other long-term problems resulting from untreated trauma. The method is applicable in accident situations, near-accidents and injuries, violence at work, catastrophes, and any other situation involving acute dramatic crisis. A debriefing program is comprised of an initial session with the victim and the counselor, preferably within 1 to 2 hours of the crisis event. Rescue personnel can also be involved in the session, which serves as a forum for the immediate exchange of the experiences of those involved. Instructions for effective coping can also be given in this session.

The main debriefing sessions are organized one to two days after the event, when people have already started to recover. The program follows a systematic plan (Braverman, 1992). Similar types of interventions can be organized in cases of an organizational crisis, such as downsizing, reorganizations, and mergers. In these instances a company-wide program with multiple—not only individual—approaches would be recommended.

Critical incident stress debriefing (CISD) and critical incident stress management (CISM) have been somewhat poorly defined (Devilly and Cotton, 2003). Meta-analysis data suggest that CISD may be noxious, and generic psychological debriefing is probably ineffective. More emphasis should be placed on screening and providing early intervention for those who may develop pathological reactions.

7.4 Health Education and Lifestyle Changes

Health education is a classic method for preventing health problems by changing health habits and lifestyle. Quitting smoking, weight control, and physical exercise are typical topics of health education. In more comprehensive programs, instructions are also given for such topics as improving sleeping habits, nutrition, alcohol consumption, substance abuse (drug testing procedures), and social activities (Ivancevich and Matteson, 1988). Health education is often carried out in the form of campaigns called *wellness programs*. When they are available, they also belong to the regular preventive functions of occupational health service (Laitinen et al., 2002). Health education has been focused on work-related and general risk factors (e.g., musculoskeletal disorders, cardiovascular disease, cancer, HIV, and AIDS). Both traditional and team-oriented health education have been used.

7.5 Career and Skill Development

Career development interventions are designed to meet the needs of people in various career stages, of various ages, of both genders, and of the unemployed. Such intervention can take the form of self-assessment tools, individual counseling, information services, and assessment and development programs (Russell, 1991). Rapid technological and structural changes in companies have illustrated the need for developing employees' skills. At the organization level, self-assessment tools for personnel, such as career workbooks, can guide workers in determining their strengths and weaknesses and in identifying job and career opportunities. Such tools have been used as the basis for career planning workshops at the organization level. An assessment center can be used for improving the understanding of one's own skills and goals. Training programs have been initiated on the basis of these ideas. Mentoring programs also help to form closer relationships between junior and senior co-workers. These relationships provide opportunities for career development and social support, as well as friendship (Kram, 1985). Recently, e-mentoring has also been used. One recent model of life-span development in adult career advancement is the model of *selective optimization with compensation* (SOC). It is suited for conceptualizing career development and work performance throughout a person's life span (Vondracek and Porfeli, 2002).

Skill development is positively related to career or job commitment (Aryee and Tan, 1992). Skill development is dependent on individual motivation or outside incentives or pressures, especially when job demands and the work organization are changing. The learning organization model refers to an organization that emphasizes continuous learning and development at the individual, group, and organization levels. Tacit knowledge and the model of Nonaka and Takeuchi (1995) are important for enriching the learning organization model. Back-to-work group intervention models have been applied to manage the transition from school to work life, to support the transition to retirement, and to help unemployed people return to work (Vinokur et al., 2000).

Traditionally, interventions to prevent career stagnation have been suggested as a part of career development programs because they are both a means with which to fight periods of frustration and an opportunity to find new challenges or to reappraise one's life goals (Weiner et al., 1992). The creation of new personal and organizational goals is necessary. During midlife, people generally become more concerned with decreasing job opportunities and their changing work-role identity, although they still have a rather strong need for career advancement (Buunk and Janssen, 1992). The maintenance of work ability (MWA) is a Finnish model applied by occupational health services to maintain and promote the work ability of employees and working groups throughout working age (Huuskonen et al., 1999).

8 IMPROVING WORK AND WORK ORGANIZATIONS

Improving work and work organizations is usually a practically oriented activity supported by consultants and other specialists from inside or outside the organization. The theoretical background is often a loose combination of several theories about organizations and human behavior. The intervention traditions and models are strongly dependent on the prevailing social values. The goal of a specific developmental intervention at work obviously influences the choice of the type of intervention. The interventions are mainly primary and focus either on the functioning of the organization, its productivity, profit, and quality of production and services, or on promoting workers' health, aspirations, and competence through job redesign, and improvements in the work and its characteristics from the human point of view. Table 2 summarizes the categories of organizational-level intervention approaches for stress reduction. They are discussed in more detail in the following sections.

8.1 Employee Assistance Programs

Employee assistance programs (EAPs) are strategies developed in companies and carried out at the company level to help employees to overcome problems

Table 2 Organizational Means to Deal with Occupational Stress

Intervention	Mechanism of Stress Control	References
EAPs (partly)	Provide coping resources.	Philips and Mushinski (1992)
Healthy work organization	Create a healthy and productive work organization.	Cox and Howarth (1990), Cooper and Cartwright (1994), Lindström et al. (2000), Murphy and Sauter (2000)
Ergonomic improvements	Provide a comfortable physical work environment.	Konz (1990), Westgaard and Winkel (1996); see also Chapters 22 and 31
Job redesign	Modify tasks, responsibilities, relationships, or technology.	Hackman and Oldham (1980), Lawler (1986), Smith and Carayon (1995); see also Chapter 16
Organizational change	Manage organization restructuring for greater psychological well-being.	Nadler (1988), Schweiger and Dennis (1991); see also Chapters 16 and 19

that hamper organizational behavior and work performance. The program ideology was first developed for interventions against alcohol abuse and to help alcoholic employees. Drug abusers were soon included in the programs as well. Positive experiences have been found in the form of both personal and financial savings. As a result, the same basic model was adopted for interventions concerning many other problems of employees, including work-related stress, mental health and personal problems, and difficulties in the family.

The programs have been modified further, and sometimes they now also include assistance for financial and legal problems. The EAP is a referral program designed to help a clinician or counselor assess the situation and make a referral. It is not meant to involve long-term support or therapy. The feasibility of such programs has been shown in the successful prevention of problems met by the personnel administration, in the saving of occupational health care costs, and in the improvement of productivity. See Philips and Mushinski (1992) for a review.

A meta-analysis of employee health management programs (EHMPs) showed that voluntary general programs were unrelated to job performance and negatively related to absenteeism, but the effects on absenteeism wane when the program is not voluntary. The programs were also minimally related to job satisfaction and slightly associate with turnover. Therefore, their benefits may lie in helping employees deal with their health problems (DeGroot and Kiker, 2003). In the evaluation of two workplace wellness programs, the successful case showed a relationship between the time employees spent on sick leave and improved fitness (Watson and Gauthier, 2003).

The question is, then, whether the programs should be broader and cover more areas than alcohol and substance abuse and employees' individual problems. These programs are typically carried out in the United States, although similar trials have been initiated in Europe. However, in Europe they generally overlap somewhat with preventive occupational health care. The programs have been evaluated, but the results are difficult to interpret because the content and process have been documented only vaguely. For evaluation purposes, the practices should be described in more detail.

EAPs have, in some cases, been enlarged and can be called *workplace wellness programs* (e.g., in Australia). Their evaluation is difficult, especially because the utilization rates across the organization vary according to their consistency (Csiernik, 2003). Best-practice guidelines have been proposed for EAP policies. High program utilization has been related to the existence of written policies, their broad distribution, an adequate level of staffing, provision of training for supervisors, and high client confidentiality (Weiss, 2003). It has been suggested that organization development (OD) activities should be included in EAPs, and some trials have been carried out (Beard, 2000).

8.2 Promoting Healthy and Productive Work Organizations

The creation of a healthy work organization has a dual focus, in that both the workers' well-being and organizational performance are involved (Murphy and Sauter, 2000). The organizational approach to health (Cox and Howarth, 1990) emphasizes the relationship between organizational factors and workers' health. Cox and Howarth see organizational health as the capability of an organization both to function effectively in relation to various environmental factors and to respond to environmental changes.

As prerequisites for a healthy work organization, Cooper and Cartwright (1994) list factors that have usually been considered related to occupational stress. On the intervention side, they underline the central role of occupational stress and its management at both the individual and organizational levels. Regular stress audits, employing the use of job and organizational stress screening, should be used as a tool in the maintenance of organizational health. When carrying out such audits, the companies and workplaces should have organization-directed strategies.

A value-based organization using the healthy company model involves strategies for profit making and the valuation of people (Rosen, 1992). Management plays a key role in the implementation of these strategies, which are open communication and employee involvement, learning and renewal, valued diversity, institutional fairness, equitable rewards and recognition, and general economic security. A people-centered technology, a health-enhancing environment, and meaningful work are all important. The balance between work and family life is also emphasized.

The occupational health approach applied by the National Institute for Occupational Health and Safety (NIOSH) (United States) and the Finnish Institute of Occupational Health (FIOH) with respect to a healthy work organization has grounded the value base of these organizations clearly in the health and well-being of people. These models for analysis and intervention are based on job and organizational stress, followed by individual- and organization-oriented interventions (Murphy and Sauter, 2000; Lindström, 2001). This approach to occupational health serves as a good basis when socially sustainable development in an organization is the goal of job design or organizational development.

The empirical results from small and medium-size enterprises have shown that organizational interventions aimed at improving supervisory and management practices, as well as mutual collaboration within the organization, improve both the well-being of the personnel and the productivity of the company (Lindström et al., 2000). Good continuous improvement practices and leadership support were also related to both productivity and well-being.

8.3 Improving Environmental and Ergonomic Conditions

The reader is referred to Chapters 15, 16, and 19 for explicit details on how to implement ergonomic measurements, evaluation procedures, and improvements. *Ergonomics* is the science of fitting the environment and activities to the capabilities, dimensions, and needs of people. In this section we have a specific interest in ergonomics since it is applied to adapt work conditions to the physical, psychological, and social nature of the person. Good ergonomic practice recommends that all aspects of the work system be included in job redesign improvements. From an ergonomic perspective, the design of the environment, workstations, tasks, work organization, and tools or technology should reasonably accommodate the employees' capacities, dimensions, strengths, and skills.

Strong management support for an ergonomic program is crucial for the success of ergonomic intervention providing psychosocial benefits. The participation of employees in the planning and implementation phases is crucial. Both the involvement of managers and supervisors in the ergonomic program and the involvement of employees and their willingness to participate in defining problems, proposing solutions, and improving work practices are of paramount importance, as is the application of engineering improvements to reduce biomechanical and physiological loads (Ketola et al., 2002).

8.4 Strategy for Job Redesign

The reader is referred to Chapters 15 and 16 for details on how to maximize job design considerations. In this section we define a strategy for redesigning jobs to reduce job stress, but it does not elaborate on the specific characteristics of the job elements needed to achieve proper job design, since they have been described elsewhere. Two models explain how job characteristics are related to the well-being and health of employees. The demand–control model (Karasek and Theorell, 1990) emphasizes the importance of the combination of work demands and job control. Stress symptoms and increased CHD morbidity can be a result of high job demands combined with low job control or low social support. The other model describes the imbalance between job demands and rewards as a risk factor for elevated stress reactions and cardiovascular disease (Siegrist, 1996).

We must, however, point out that no job provides complete psychological satisfaction and is also free of all stress. In any redesign process there are trade-offs among specific improvements and the achievement of the best overall job design solution. In these trade-offs one needs to think about how best to balance the various needs to achieve a solution that will have the greatest positive benefit for both employee health and productivity.

Smith and Sainfort (1989) proposed a balance among various elements of the work "system" (Figure 1). The essence of this approach is to reduce the negative health consequences caused by stress by balancing the various elements of the work system to reduce unwanted loading. Proper work organization and job design can best be achieved by providing each work system element with characteristics that meet recognized criteria for proper physical loads, work cycles, processes, and job contents and that suit individual physiological and psychological needs. The best designs eliminate all sources of stress, job dissatisfaction, and discomfort and even promote well-being and satisfaction and increase competence.

For instance, a proper workload can be established by using appropriate methods of work analysis (Konz, 1990), whereas aspects of job control can be developed

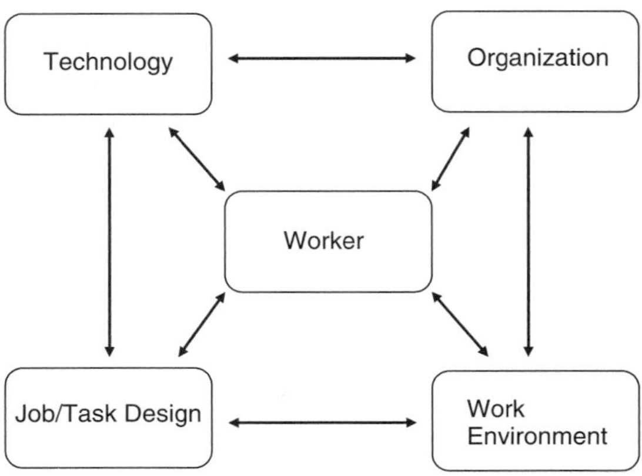

Figure 1 Model of the work system. (From Smith and Sainfort, 1989; copyright 1989, with permission from Elsevier.)

through the use of worker participation (Lawler, 1986) (see Chapter 19). The negative influence of inadequate skill to use new technology can be offset by increased employee training. Similarly, the adverse influence of low job content, repetition, and boredom can be balanced by an organizational supervisory structure that promotes employee involvement and control over tasks and job enlargement that introduces task variety. Organizational structure can be adapted to enrich jobs to provide support for the individual, such as increased staff or shared responsibilities or increased financial resources.

8.5 Organizational Change Management

The management of change depends on the type of change, which can be future- or past-oriented. Organizational changes can occur through gradual development or strategic actions (Nadler, 1988). The change processes, however, are now more often structural ones, meaning that paradigmatic and cultural changes also occur. Change can also cover only narrow or technical issues (e.g., the implementation of more advanced information technology, a new service or product).

Planned changes and developmental intervention processes should follow the developmental cycle. They start with problem or goal definition and the committing and informing of all interest groups within the workplace. They then proceed to an analysis of the current situation, based, for example, on survey results, and subsequently, to joint planning of the intervention. The implementation of the intervention requires clear goals, subgoals, steps, a time schedule, duties for people, and ways to overcome obstacles and problems during the process. The process seldom proceeds as planned; instead, it needs reevaluation and checks of the methods, and even the goals, during the process. The processes and the outcome should be evaluated. This evaluation helps the organization learn about its experiences so that it is more prepared for the next project. This step is usually skipped, however, because, very often, new changes and projects are already being implemented before the previous changes and projects have come to an end.

A major cause of worker resistance to new work activities is the appearance of changes "out of the blue." For the successful implementation of changes in work processes and the subsequent enhancement of worker health, performance, and satisfaction, it is of importance that organizations have a transition policy that includes worker participation in all stages of the change process. In other words, workers should participate in the planning, then in the selection of equipment, and finally in the daily operation of the work system.

The most difficult change situations are those involving the downsizing of activities or the merger of earlier independent units or organizations. Such changes usually create insecurity about the future and provoke anxiety among personnel (Kilpatrick et al., 1991). Open communication has proved to be necessary and positive from the workers' point of view during the whole structural change process (Schweiger and Dennis, 1991). A large-scale downsizing of work units has led, in the long run, to the increased morbidity and mortality among the survivors (Kivimäki et al., 2000b, 2001; Vahtera et al., 2004a). Increased long-term absenteeism due to downsizing has also led to the loss of the savings from the personnel downsizing (Vahtera et al., 1997).

Mergers are common in both the public and private sectors. Insufficient preparation and a lack of supervisory information and support during the transition from the old organization to the new create elevated job stress and lower well-being (Lindström et al., 2003). Management of the change and employee participation are therefore important.

9 SUMMARY

Evidence has accumulated on the psychosocial work factors that are relevant in terms of employee health and well-being. Several organizational behavior approaches can be applied to diminish or eliminate the harmful effects of poor workplace design that lead to psychological distress. Examples of these approaches include removing the sources of stress (stressors) through job redesign, balancing the positive and negative aspects of work to provide an acceptable level of stress, developing a corporate culture that promotes self-worth and productive employment (a healthy organization), providing organizational resources for dealing with the consequences of stress (employee assistance programs), and enhancing employee skills and career development. In addition, personal means can be used to deal with stress: for example, relaxation methods, lifestyle improvements, exercise, psychotherapy, cognitive restructuring, and social support.

If the psychosocial aspects of work design are properly developed and implemented, employee health and performance can benefit, and a healthy enterprise will emerge. If job stress is not controlled properly, it can lead to reduced job satisfaction, poor health, and low morale among employees, as well as to reduced employee and corporate performance.

REFERENCES

Ala-Mursula, L., Vahtera, J., Pentti, J., and Kivimäki, M. (2004), "Effect of Employee Worktime Control on Health: Prospective Cohort Study," *Occupational and Environmental Medicine*, Vol. 61, pp. 239–246.

Allen, T. D., Herst, D. E. L., Bruck, C. S., and Sutton, M. (2000), "Consequences Associated with Work-to-Family Conflict: A Review and Agenda for Future Research," *Journal of Occupational Health Psychology*, No. 5, pp. 278–308.

Aryee, S., and Tan, K. (1992), "Antecedents and Outcomes of Career Commitment," *Journal of Vocational Behavior*, Vol. 40, pp. 288–305.

Beard, M. (2000), "Organizational Development: An EAP Approach," *Employee Assistance Quarterly*, Vol. 16, No. 1–2, pp. 117–140.

Bongers, P. M., de Winter, C. R., Kompier, M. A., and Hildebrandt, V. H. (1993), "Psychosocial Factors at

Work and Musculoskeletal Disease," *Scandinavian Journal of Work, Environment and Health*, Vol. 19, pp. 297–312.

Bongers, P. M., Kremer, A. M., and ter Laak, J. (2002), "Are Psychosocial Factors Risk Factors for Symptoms and Signs of the Shoulder, Elbow, or Hand/Wrist? A Review of the Epidemiological Literature," *American Journal of Industrial Medicine*, Vol. 41, pp. 32–48.

Borg, K., Hensing, G., and Alexanderson, K. (2001), "Predictive Factors for Disability Pension: An 11-Year Follow Up of Young Persons on Sick Leave Due to Neck, Shoulder, or Back Diagnoses," *Scandinavian Journal of Public Health*, Vol. 29, pp. 104–112.

Bramson, R. M. (1985), "Toward Effective Coping: The Basic Steps," in *Stress and Coping: An Anthology*, 2nd ed., A. Monat, and R. S. Lazarus, Eds., Columbia University Press, New York, pp. 356–370.

Braverman, M. (1992), "Post Trauma Crisis Intervention in the Workplace," in *Stress and Well-Being at Work: Assessments and Interventions for Occupational Mental Health*, J. C. Quick, L. R. Murphy, and J. J. Hurrell, Jr., Eds., American Psychological Association, Washington, DC, pp. 299–316.

Brunner, E. (2002), "Stress Mechanisms in Coronary Heart Disease," in *Stress and the Heart: Psychosocial Pathways to Coronary Heart Disease*, S. A. Stansfeld and M. G. Marmot, Eds., BMJ Books, London, pp. 181–199.

Bunce, D., and West, M. A. (1996), "Stress Management and Innovation Interventions at Work," *Human Relations*, Vol. 49, No. 2, pp. 209–231.

Buunk, B. P., and Janssen, P. P. M. (1992), "Relative Deprivation, Career Issues, and Mental Health Among Men in Mid-life," *Journal of Vocational Behavior*, Vol. 40, pp. 338–350.

Cannon, W. B. (1929), *Bodily Changes in Pain, Hunger, Fear and Rage: An Account of Recent Researches in the Function of Emotional Excitement*, Appleton, New York.

Caplan, R. D., Cobb, S., and French, Jr., J. R. (1975), "Relationships of Cessation of Smoking with Job Stress, Personality, and Social Support," *Journal of Applied Psychology*, Vol. 60, pp. 211–219.

Cohen, S., Tyrrell, D. A., and Smith, A. P. (1991). "Psychological Stress and Susceptibility to the Common Cold," *New England Journal of Medicine*, Vol. 325, pp. 606–612.

Cooper, C. L., and Cartwright, S. (1994), "Healthy Mind; Healthy Organization: A Proactive Approach to Occupational Stress," *Human Relations*, Vol. 47, No. 4, pp. 455–471.

Cox, T., and Griffiths, A. (1996), "Assessment of Psychosocial Hazards at Work," in *Handbook of Work and Health Psychology*, M. J. Schabracq, J. A. M. Winnubst, and C. L. Cooper, Eds., Wiley, Chichester, West Sussex, England, pp. 127–146.

Cox, T., and Howarth, I. (1990), "Organizational Health, Culture and Helping," *Work and Stress*, Vol. 4, No. 2, pp. 107–110.

Csiernik, R. (2003), "Employee Assistance Program Utilization: Developing a Comprehensive Scorecard," *Employee Assistance Quarterly*, Vol. 18, No. 3, pp. 45–60.

Cummings, T. G. and Cooper, C. L. (1979). "A Cybernetic Framework for Studying Occupational Stress," *Human Relations*, Vol. 32, pp. 395–418.

Dailey, R. C., and Kirk, D. J. (1992), "Distributive and Procedural Justice as Antecedents of Job Dissatisfaction and Intent to Turnover," *Human Relations*, Vol. 45, pp. 305–317.

Davis, L. E., and Wacker, G. J. (1982), "Job Design," in *Handbook of Industrial Engineering*, G. Salvendy, Ed., Wiley, New York, Sections 2.5.1–2.5.31.

DeGroot, T., and Kiker, D. S. (2003), "A Meta-analysis of the Non-monetary Effects of Employee Health Management Programs," *Human Resource Management*, Vol. 42, No. 1, pp. 53–69.

Devilly, G. J., and Cotton, P. (2003), "Psychological Debriefing and the Workplace: Defining a Concept, Controversies and Guidelines for Intervention," *Australian Psychologist*, Vol. 38, No. 2, pp. 144–150.

Dewe, P. J. (1992), "Applying the Concept of Appraisal to Work Stressors: Some Explanatory Analysis," *Human Relations*, Vol. 45, pp. 143–164.

Dewe, P., and O'Driscoll, M. (2002), "Stress Management Interventions: What Do Managers Actually Do? *Personnel Review*, Vol. 31, No. 2, pp. 143–165.

Edwards, J. R., Caplan, R. D., and van Harrison, R. (1998), "Person–Environment Fit Theory," in *Theories of Organizational Stress*, C. L. Cooper, Ed., Oxford University Press, Oxford, pp. 28–67.

Elo, A.-L. (1994), "Assessment of Mental Stress Factors at Work," in *Occupational Medicine*, O. B. Dickerson and E. P. Ovarth, Eds., C.V. Mosby, St. Louis, MO, pp. 945–959.

Elovainio, M., Kivimäki, M., and Vahtera, J. (2002), "Organizational Justice: Evidence of a New Psychosocial Predictor of Health," *American Journal of Public Health*, Vol. 92, pp. 105–108.

Folkman, S., and Lazarus, R. (1980), "An Analysis of Coping in a Middle Aged Community Sample," *Journal of Health and Social Behaviour*, Vol. 21, pp. 219–239.

Gardell, B. (1982). "Scandinavian Research on Stress in Working Life," *International Journal of Health Services*, Vol. 12, pp. 31–41.

Giga, S. I., Noblet, A. J., Faragher, B., and Cooper, C. L. (2003), "The UK Perspective: A Review of Research on Organisational Stress Management Interventions," *Australian Psychologist*, Vol. 38, No. 2, pp. 158–164.

Gjesdal, S., and Bratberg, E. (2002), "The Role of Gender in Long-Term Sickness Absence and Transition to Permanent Disability Benefits: Results from a Multiregister Based, Prospective Study in Norway, 1990–1995," *European Journal of Public Health*, Vol. 12, pp. 180–186.

Glaser, R., Kiecolt-Glaser, J. K., Malarkey, W. B., and Sheridan, J. F. (1998), "The Influence of Psychological Stress on the Immune Response to Vaccines," *Annals of the New York Academy of Science*, Vol. 840, pp. 649–655.

Goldberg, R. J., and Steury, S. (2001), "Depression in the Work Place: Costs and Barriers to Treatment," *Psychiatric Services*, Vol. 52, No. 12, pp. 1639–1643.

Grandey, A. A., and Cropanzano, R. (1999), "The Conservation of Resources Model Applied to Work–Family Conflict and Strain," *Journal of Vocational Behavior*, Vol. 54, pp. 350–370.

Green, K. L., and Johnson, J. V. (1990), "The Effects of Psychosocial Work Organization on Patterns of Cigarette Smoking Among Male Chemical Plant Employees," *American Journal of Public Health*, Vol. 80, pp. 1368–1371.

Greenhaus, J. H., and Beutell, N. J. (1985), "Sources and Conflict Between Work and Family Roles," *Academy of Management Review*, Vol. 10, pp. 76–88.

Hackman, J. R., and Oldham, G. R. (1980), *Work Redesign*, Addison-Wesley, Reading, MA.

Head, J., Stansfeld, S. A., and Siegrist, J. (2004), "The Psychosocial Work Environment and Alcohol Dependence: A Prospective Study," *Occupational and Environmental Medicine*, Vol. 61, pp. 219–224.

Horan, A. P. (2003), "An Effective Workplace Stress Management Intervention: Chicken Soup for the Soul at Work™ Employee Groups," *Work: Journal of Prevention, Assessment and Rehabilitation*, Vol. 18, No. 1, pp. 3–13.

House, J. S. (1981), *Work Stress and Social Support*, Addison-Wesley, Reading, MA.

Hurrell, J. J., Jr., and Murphy, L. R. (1996), "Occupational Stress Intervention," *American Journal of Industrial Medicine*, Vol. 29, pp. 338–341.

Huuskonen, M., Bergström, M., Lindström, K., and Rantanen, J. (1999), "Strategies to Promote Well-Being in Small Enterprises," *American Journal of Industrial Medicine*, Suppl. 1, pp. 89–92.

Ivancevich, J. M., and Matteson, M. T. (1988), "Promoting the Individual's Health and Well-Being," in *Causes, Coping and Consequences of Stress at Work*, C. L. Cooper and R. Payne, Eds., Wiley, Chichester, West Sussex, England, pp. 267–299.

Jansen, N. W. H., Kant, I., Nijhuis, F. J. N., Swaen, G. M. H., and Kristensen, T. S. (2004), "Impact of Worktime Arrangements on Work–Home Interference Among Dutch Employees," *Scandinavian Journal of Work, Environment and Health*, Vol. 30, pp. 139–148.

Johnson, J., and Hall, E. (1988), "Job Strain, Work Place Social Support, and Cardiovascular Disease: A Cross-Sectional Study of a Random Sample of the Swedish Working Population," *American Journal of Public Health*, Vol. 7, pp. 1336–1342.

Jonsson, D., Rosengren, A., Dotevall, A., et al. (1999), "Job Control, Demands and Social Support at Work in Relation to Cardiovascular Risk Factors in MONICA 1995, Gotebörg," *Journal of Cardiovascular Risk*, No. 6, pp. 379–385.

Judge, T. A., Parker, S. K., Colbert, A. E., Heller, D., and Ilies, R. (2002), "Job Satisfaction: A Cross-Cultural Review," in *Handbook of Industrial, Work and Organizational Psychology*, Vol. 2, *Organizational Psychology*, N. Anderson and D. S. Ones, Eds., Sage, Thousand Oaks, CA, pp. 25–52.

Kahn, R. L. (1973), "Conflict, Ambiguity, and Overload: Three Elements in Job Stress," *Occupational Health*, No. 3, pp. 2–9.

Kaila-Kangas, L., Kivimäki, M., Riihimäki, H., Luukkonen, R., Kirjonen, J., and Leino-Arjas, P. (2004), "Psychosocial Factors at Work as Predictors of Hospitalization for Back Disorders: A 28-Year Follow-up of Industrial Employees," *Spine*, Vol. 29, No. 16.

Karasek, R., and Theorell, T. (1990), *Healthy Work*, Basic Books, New York.

Ketola, R., Toivonen, R., Häkkänen, M., Luukkonen, R., Takala, E.-P., and Viikari-Juntura, E. (2002), "Effects of Ergonomic Intervention in Work with Video Display Units," *Scandinavian Journal of Work, Environment and Health*, Vol. 28, No. 1, pp. 18–24.

Kiecolt-Glaser, J. K., Marucha, P. T., Malarkey, W. B., and Glaser, R. (1995), "Slowing of Wound Healing by Psychological Stress," *Lancet*, Vol. 346, pp. 1194–1196.

Kilpatrick, A. O., Johnson, J. A., and Jones, J. K. (1991), "Organizational Downsizing in Hospitals: Consideration for Management Development," *Journal of Management Development*, Vol. 10, pp. 44–52.

Kivimäki, M., Elovainio, M., and Vahtera, J. (2000a), "Workplace Bullying and Sickness Absence in Hospital Staff," *Occupational and Environmental Medicine*, Vol. 57, pp. 656–660.

Kivimäki, M., Vahtera, J., Pentti, J., and Ferrie, J. E. (2000b), "Factors Underlying the Effect of Organisational Downsizing on Health of Employees: A Longitudinal Cohort Study," *British Medical Journal*, Vol. 320, pp. 971–975.

Kivimäki, M., Vahtera, J., Ferrie, J. E., Hemingway, H., and Pentti, J. (2001), "Organisational Downsizing and Musculoskeletal Problems in Employees: A Prospective Study," *Occupational and Environmental Medicine*, Vol. 58, pp. 811–817.

Kivimäki, M., Leino-Arjas, P., Luukkonen, R., Riihimäki, H., Vahtera, J., and Kirjonen, J. (2002), "Work Stress and Risk of Cardiovascular Mortality: Prospective Cohort Study of Industrial Employees," *British Medical Journal*, Vol. 325, pp. 857–860.

Kivimäki, M., Elovainio, M., Vahtera, J., and Ferrie, J. E. (2003a), "Organisational Justice and Health of Employees: Prospective Cohort Study," *Occupational and Environmental Medicine*, Vol. 60, pp. 27–34.

Kivimäki, M., Head, J., Ferrie, J. E., Shipley, M., Vahtera, J., and Marmot, M. G. (2003b), "Sickness Absence as a Global Measure of Health: Evidence from Mortality in the Whitehall II Prospective Cohort Study," *British Medical Journal*, Vol. 327, pp. 364–368.

Kivimäki, M., Forma, P., Wikström, J., Halmeenmäki, T., Pentti, J., Elovainio, M., and Vahtera, J. (2004), "Sickness Absence as a Risk Marker of Future Disability Pension: The 10-Town Study," *Journal of Epidemiology and Community Health*, Vol. 58, pp. 710–711.

Kivimäki, M., Ferrie, J. E., Brunner, E., Head, J., Shipley, M. J., Vahtera, J., & Marmot, M. G. (in press). "Justice at Work and Reduced Risk of Coronary Heart Disease Among Employees: The Whitehall II Study," *Archives of Internal Medicine*.

Konz, S. (1990), *Work Design: Industrial Ergonomics*, 3rd ed., Publishing Horizons, Worthington, OH.

Kram, K. E. (1985), *Mentoring at Work*, Scott, Foresman, Glenview, IL.

Kramer, R. M., and Tyler, T. R. (1996), *Trust in Organizations: Frontiers of Theory and Research*, Sage, London.

Kuper, H. and Marmot, M. (2003). "Job Strain, Job Demands, Decision Latitude, and Risk of Coronary Heart Disease Within the Whitehall II Study," *Journal of Epidemiology and Community Health*, Vol. 57, pp. 147–153.

Kuper, H., Marmot, M., and Hemingway, H. (2002a), "Systematic Review of Prospective Cohort Studies of Psychosocial Factors in the Etiology and Prognosis of Coronary Heart Disease," *Seminars in Vascular Medicine*, No. 2, pp. 267–314.

Kuper, H., Singh-Manoux, A., Siegrist, J., and Marmot, M. (2002b), "When Reciprocity Fails: Effort–Reward Imbalance in Relation to Coronary Heart Disease and Health Functioning Within the Whitehall II Study," *Occupational and Environmental Medicine*, Vol. 59, pp. 777–784.

Laitinen, J., Ek, E., and Sovio, U. (2002), "Stress-Related Eating and Drinking Behavior and Body Mass Index and Predictors of This Behavior," *Preventive Medicine*, Vol. 34, pp. 29–39.

Landsbergis, P. A., and Vivona-Vaughan, E. (1995), "Evaluation of an Occupational Stress Intervention in a Public Agency," *Journal of Organizational Behavior*, Vol. 16, pp. 29–48.

Lawler, E. E., III (1986), *High Involvement Management*, Jossey-Bass, San Francisco.

Lazarus, R. S. (1966), *Psychological Stress and the Coping Process*, McGraw-Hill, New York.

Lecrubier, Y. (2000), "Depressive Illness and Disability," *European Neuropsychopharmacology*, Vol. 10, Suppl. 4, pp. S439–S443.

Lindström, K. (2001), "Work Organizations: Health and Productivity Issues," in *International Encyclopedia of Ergonomics and Human Factors*, Vol. III, W. Karwowski, Ed., Taylor & Francis, New York, pp. 1608–1611.

Lindström, K., Schrey, K., Ahonen, G., and Kaleva, S. (2000), "The Effects of Promoting Organizational Health on Worker Well-Being and Organizational Effectiveness in Small and Medium-Sized Enterprises," in *Healthy and Productive Work*, L. R. Murphy and C. L. Cooper, Eds., Taylor & Francis, New York, pp. 83–104.

Lindström, K., Turpeinen, M., and Kinnunen, J. (2003), "Effects of Data System Changes on Job Characteristics and Well-Being of Hospital Personnel: A Longitudinal Study," in *Human-Centred Computing: Cognitive, Social and Ergonomic Aspects, Proceedings of HCI International 2003, 2nd International Conference on Universal Access in Human–Computer Interaction*, Crete, D. Harris, V. Duffy, M. Smith, and C. Stephanidis, Eds., Lawrence Erlbaum Associates, London, pp. 88–92.

Marmot, M., Feeney, A., Shipley, M., North, F., and Syme, S. L. (1995), "Sickness Absence as a Measure of Health Status and Functioning: From the UK Whitehall II Study," *Journal of Epidemiology and Community Health*, Vol. 49, pp. 124–130.

Marras, W. S., Davis, K. G., Heaney, C. A., et al. (2000), "The Influence of Psychosocial Stress, Gender, and Personality on Mechanical Loading of the Lumbar Spine," *Spine*, Vol. 25, No. 23, pp. 3045–3054.

Maslach, C. (1998), "A Multidimensional Theory of Burnout," in *Theories of Organizational Stress*, C. L. Cooper, Ed., Oxford University Press, Oxford, pp. 68–87.

Maslach, C., and Jackson, S. E. (1998), *The Maslach Burnout Inventory*, research ed., Consulting Psychologist Press, Palo Alto, CA.

McEwen, B. S. (1998), "Protective and Damaging Effects of Stress Mediators," *New England Journal of Medicine*, Vol. 338, pp. 171–179.

McGrath, J. E. and Beehr, T. A. (1990). "Time and Stress Response: Some Temporal Issues in the Conceptualization and Measurement of Stress," *Stress Medicine*, Vol. 6, pp. 93–104.

Meichenbaum, D. (1985), *Stress Inoculation Training*, Allyn & Bacon, Boston.

Moorman, R. H. (1991), "Relationship Between Organizational Justice and Organizational Citizenship Behaviors: Does Fairness Perception Influence Employee Citizenship?" *Journal of Applied Psychology*, Vol. 76, pp. 845–855.

Murphy, L. R. (1988), "Workplace Interventions for Stress Reduction and Prevention," in *Causes, Coping and Consequences of Stress at Work*, C. L. Cooper and R. Payne, Eds., Wiley, Chichester, West Sussex, England, pp. 301–339.

Murphy, L. R., and Sauter, S. L. (2000), "Models of Healthy Work Organizations," in *Healthy and Productive Work: An International Perspective*, L. R. Murphy and C. L. Cooper, Eds., Taylor & Francis, New York, pp. 1–11.

Murphy, L. R., and Sauter, S. L. (2004), "Work Organization Interventions: State of Knowledge and Future Directions," *Social and Preventive Medicine*, Vol. 49, pp. 79–86.

Murray, C. J., and Lopez, A. D. (1997), "Alternative Projections of Mortality and Disability by Cause, 1990–2020: Global Burden of Disease Study," *Lancet*, Vol. 349, pp. 1498–1504.

Nadler, D. A. (1988), "Organizational Frame Bending: Types of Change in the Complex Organization," in *Corporate Transformation: Revitalizing Organizations for a Competitive World*, R. H. Kilmann and T. J. Covin, Eds., Jossey-Bass, San Francisco, pp. 66–83.

Niedhammer, I., Goldberg, M., Leclerc, A., Bugel, I., and David, S. (1998a), "Psychosocial Factors at Work and Subsequent Depressive Symptoms in the Gazel Cohort," *Scandinavian Journal of Work Environment Health*, Vol. 24, pp. 197–205.

Niedhammer, I., Goldberg, M., Leclerc, A., et al. (1998b), "Psychosocial Work Environment and Cardiovascular Risk Factors in an Occupational Cohort in France," *Journal of Epidemiology and Community Health*, Vol. 52, pp. 93–100.

Nonaka, I., and Takeuchi, H. (1995), *The Knowledge-Creating Company: How Japanese Companies Create the Dynamics of Innovation*, Oxford University Press, New York.

Nytrø, K., Saksvik, P. Ø., Mikkelsen, A., Bohle, P., and Quinlan, M. (2000), "An Appraisal of Key Factors in the Implementation of Occupational Stress Interventions," *Work and Stress*, Vol. 14, No. 3, pp. 213–225.

Paterniti, S., Niedhammer, I., Lang, T., and Consoli, S. M. (2002), "Psychosocial Factors at Work, Personality Traits and Depressive Symptoms: Longitudinal Results from the GAZEL Study," *British Journal of Psychiatry*, Vol. 181, pp. 111–117.

Philips, K. (1989), "Psychophysiological Consequences of Behavioural Choice in Aversive Situations," in *Stress, Personal Control and Health*, A. Steptoe and A. Appels, Eds., Wiley, Chichester, West Sussex, England, pp. 239–256.

Philips, S. B., and Mushinski, M. H. (1992), "Configuring an Employee Assistance Program to Fit the Corporation's Structure: One Company's Design," in *Stress and Well-Being at Work: Assessments and Interventions for Occupational Mental Health*, J. C. Quick, L. R. Murphy and J. J. Hurrell, Jr., Eds., American Psychological Association, Washington, DC, pp. 317–328.

Pieper, C., LaCroix, A. Z., and Karasek, R. A. (1989), "The Relation of Psychosocial Dimensions of Work with Coronary Heart Disease Risk Factors: A Meta-analysis of Five United States Data Bases," *American Journal of Epidemiology*, Vol. 129, pp. 483–494.

Quinlan, M., Mayhew, C., and Bohle, P. (2001), "The Global Expansion of Precarious Employment, Work Disorganization, and Consequences for Occupational Health: A Review of Recent Research," *International Journal of Health Services*, Vol. 31, pp. 335–414.

Rahe, R. H., Taylor, C. B., Tolles, R. L., Newhall, L. M., Veach, T. L., and Bryson, S. (2002), "A Novel Stress and Coping Workplace Program Reduces Illness and Healthcare Utilization," *Psychosomatic Medicine*, Vol. 64, No. 2, pp. 278–286.

Rayner, C., and Hoel, H. (1997), "A Summary Review of Literature Relating to Workplace Bullying," *Journal of Applied Social Psychology*, No. 7, pp. 181–191.

Rosen, R. H. (1992), *The Healthy Company*, Jeremy P. Tarcher/Perigee, Los Angeles.

Roskies, E., Seraganian, P., Oseasohn, R., Smilga, C., Martin, N., and Hanley, J. A. (1989), "Treatment of Psychological Stress Responses in Healthy Type A Men," in *Advances in the Investigation of Psychological Stress*, R. W. J. Neufeld, Ed., Wiley, New York, pp. 284–304.

Rousseau, D. M. (1998), "Organizational Climate and Culture," in *Encyclopaedia of Occupational Health and Safety*, 4th ed., Vol. II, Part V, *Psychosocial and Organizational Factors, 34.36*, J. M. Stellman, Ed., International Labour Office, Geneva.

Russell, J. E. A. (1991), "Career Development Interventions in Organizations," *Journal of Vocational Behavior*, Vol. 38, pp. 237–287.

Sawyer, J. (1992), "Goal and Process Clarity: Specification of Multiple Constructs of Role Ambiguity and a Structural Equation Mode of Their Antecedents and Consequences," *Journal of Applied Psychology*, Vol. 77, pp. 130–142.

Schweiger, D. M., and Dennis, A. S. (1991), "Communication with Employees Following the Merger: A Longitudinal Held Experiment," *Academy of Management Journal*, Vol. 34, pp. 110–135.

Selye, H. (1936), "A Syndrome Produced by Diverse Nocuous Agents," *Nature*, Vol. 138, p. 32.

Selye, H. (1956), "What Is Stress?" *Metabolism*, Vol. 5, p. 525.

Senge, P. (1990), *The Fifth Discipline: The Art and Practice of the Learning Organization*, Century Business, London.

Shapiro, D. L., and Brett, J. M. (1993), "Comparing Three Processes Underlying Judgements of Procedural Justice: A Field Study of Mediation and Arbitration," *Journal of Personality and Social Psychology*, Vol. 65, pp. 1167–1177.

Siegrist, J. (1996), "Adverse Health Effects of High-Effort/Low-Reward Conditions," *Journal of Occupational Health Psychology*, No. 1, pp. 27–41.

Smith, M. J., and Carayon, P. (1995), "New Technology, Automation and Work Organization: Stress Problems and Improved Technology Implementation Strategies," *International Journal of Human Factors in Manufacturing*, No. 5, pp. 99–116.

Smith, M. J., and Sainfort, P. C. (1989), "A Balance Theory of Job Design for Stress Reduction," *International Journal of Industrial Ergonomics*, Vol. 4, No. 1, pp. 67–79.

Smith, M. J., Carayon, P., Sanders, K. J., Lim, S.-Y., and LeGrande, D. (1982), "Electronic Monitoring, Job Design and Worker Stress," *Applied Ergonomics*, Vol. 23, pp. 17–27.

Smith, M. J., Carayon, P., Sanders, K. J., Lim, S.-Y., and LeGrande, D. (1992), "Electronic monitoring, job design and worker stress," *Applied Ergonomics*, Vol. 23., pp. 17–27.

Stansfeld, S. A., Rael, E. G. S., Head, J., et al. (1997), "Social Support and Psychiatric Sickness Absence: A Prospective Study of British Civil Servants," *Psychological Medicine*, Vol. 27, pp. 35–48.

Stansfeld, S. A., Fuhrer, R., Shipley, M. J., et al. (1999), "Work Characteristics Predict Psychiatric Disorders: Prospective Results from the Whitehall II Study," *Occupational and Environmental Medicine*, Vol. 56, pp. 302–307.

Tsutsumi, A., Kayaba, K., Yoshimura, M., et al. (2003), "Association Between Job Characteristics and Health Behaviors in Japanese Rural Workers," *International Journal of Behavioral Medicine*, Vol. 10, pp. 125–142.

Vahtera, J., Kivimäki, M., and Pentti, J. (1997), "Effect of Organisational Downsizing on Health of Employees," *Lancet*, Vol. 350, pp. 1124–1128.

Vahtera, J., Kivimäki, M., Pentti, J., Linna, A., Virtanen, M., and Ferrie, J. E. (2004a), "Organisational Downsizing, Sickness Absence and Mortality: The 10-Town Prospective Cohort Study," *British Medical Journal*, Vol. 328, pp. 555–557.

Vahtera, J., Pentti, J., and Kivimäki, M. (2004b), "Sickness Absence as a Predictor of Mortality Among Male and Female Employees," *Journal of Epidemiology and Community Health*, Vol. 58, pp. 321–326.

van der Klink, J. J. L., Blonk, R. W. B., Schene, A. H., and van Dijk, F. J. H. (2001), "The Benefits of Interventions for Work-Related Stress," *American Journal of Public Health*, Vol. 91, No. 2, pp. 270–276.

Vartia, M. (2003), *Workplace Bullying: A Study on the Work Environment, Well-Being and Health*, *People and Work Research Reports* 56, Finnish Institute of Occupational Health, Helsinki.

Vinokur, A., Schul, Y., Vuori, J., and Price, R. (2000), "Two Years After a Job Loss: Long-Term Impact of the JOBS Program on Reemployment and Mental Health," *Journal of Occupational Health Psychology*, Vol. 5, No. 1, pp. 32–47.

Vondracek, F. W., and Porfeli, E. J. (2002), "Life-Span Developmental Perspectives on Adult Career Development: Recent Advances," in *Adult Career Development: Concepts, Issues and Practices*, 3rd ed., S. G. Niles, Ed., National Career Development Association, Columbus, OH, pp. 20–38.

Wardle, J., and Gibson, E. L. (2002), "Impact of Stress on Diet: Processes and Implications," in *Stress and the Heart: Psychosocial Pathways to Coronary Heart Disease*, S. A. Stansfeld and M. G. Marmot, Eds., BMJ Books, London, pp. 124–149.

Watson, W., and Gauthier, J. (2003), "The Viability of Organizational Wellness Programs: An Examination of Promotion and Results," *Journal of Applied Social Psychology*, Vol. 33, No. 6, pp. 1297–1312.

Weiner, A., Remer, R., and Remer, P. (1992), "Career Plateauing: Implications for Career Development Specialists," *Journal of Career Development*, Vol. 19, No. 1, pp. 37–48.

Weiss, R. M. (2003). "Effects of Program Characteristics on EAP utilization," *Employee Assistance Quarterly*, Vol. 18, 61–70.

Wells, K. B., Stewart, A., Hays, R. D., Burnam, M. A., Rogers, W., Daniels, M., Berry, S., Greenfield, S., and Ware, J. (1989), "The Functioning and Well-Being of Depressed Patients," *Journal of the American Medical Association*, Vol. 262, No. 7, pp. 914–919.

Westerlund, H., Ferrie, J., Hagberg, J., Jeding, K., Oxenstierna, G., and Theorell, T. (2004), "Workplace Expansion, Long-Term Sickness Absence, and Hospital Admission," *Lancet*, Vol. 363, pp. 1193–1197.

Westgaard, R. H., and Winkel, J. (1996), "Guidelines for Occupational Musculoskeletal Load as a Basis for Intervention: A Critical Review," *Applied Ergonomics*, Vol. 27, No. 2, pp. 79–88.

CHAPTER 30
MANUAL MATERIALS HANDLING

David Rodrick* and Waldemar Karwowski
University of Louisville
Louisville, Kentucky

1 INTRODUCTION

1.1 MMH and Low Back Disorders

With continuing technological progress, developments in the field of logistics and distribution, and implementation of automation in industrial processes, numerous *manual materials handling* (MMH) tasks are still carried out every day in businesses around the world. These tasks include unaided lifting, lowering, carrying, pushing, pulling, and holding activities. Musculoskeletal disorders (MSDs) of the back, which include lower back disorders (LBDs) and lower back pain (LBP), are among the leading causes of occupational injury and disability in industrialized countries. In 1988, of 127 million active workers, 22.4 million (17.6%) reported "back pain for a week or more" in the 12 months preceding the National Health Interview Survey (NHIS). Work-related back pain, caused by injury or repeated activities, accounted for approximately 53% of self-reported causes (Park et al., 1997). In 1993, back disorders accounted for 27% of all nonfatal occupational injuries and illnesses that involved sick leave. Approximately 100 million workdays are lost each year due to back disabilities. LBP disables 5.4 million Americans per year and it is the most frequent reason for filing a workers' compensation claim. The annual incidence rate is estimated to be around 1 to 2% (Hashemi et al., 1997). According to the U.S. Bureau of Labor Statistics, in 2000 there were more than 255,774 overexertion back injuries resulting in lost workdays (BLS, 2000).

The total compensable cost for LBP in the United States was estimated to be \$11.1 billion in 1986. The average cost of a worker's compensation claim for LBP was \$8300, about twice the average cost for all other claims combined. Medical costs account for about one-third of the total costs, with the remaining expenditures going to wage and related costs (Webster and Snook, 1994). In 1993, the cost for the compensation coverage in the United States was reported to be approximately \$57 billion, with about 60% going for wage replacement alone (Schmulowitz, 1995). According to Hashemi et al. (1997), the costliest LBP claims (10%) at a large insurance company in 1992 were responsible for a large percentage of the total costs (86%). Seven percent of the claims with a greater

*Present address: Florida State University, Tallahassee, Florida.

than one year length of disability accounted for 75% of the costs and 84.2% of total disability days (Hashemi et al., 1997).

Recent evidence show that work-related musculoskeletal disorders (WMSDs), such as LBDs, account for more than one-third of all the occupational injuries and illness in the United States (BLS, 2001). Leigh et al. (1997) estimated that the total cost of occupational low back pain in the United States was $49.2 billion in 1992, and Murphy and Volinn (1999) estimated the costs for workers' compensation claims alone for LBD in the United States to be $8.8 billion in 1995. It is believed that work-related LBDs are triggered by a complex process. This complex process involves the interaction of physical work factors (e.g., biomechanical factors), nonphysical work factors (e.g., psychosocial stress and work organization factors), and individual characteristics (e.g., prior medical history, strength capabilities, and personality traits). Other psychosocial factors, such as perceived effort and work satisfaction or dissatisfaction, may also play a role in the development and/or reporting of LBDs (Bigos et al., 1991; Kerr et al., 2001).

1.2 Epidemiology and Etiology of Low Back Pain

Incidences of LBP are common among the general population. About 85% of the general population develops LBP at least once in their lifetimes (Von Korff et al., 1988). About one-third of American workers are employed in jobs that may significantly increase their risk of developing or aggravating back disorders and disabilities. Since 1989, the National Institute for Occupational Safety and Health (NIOSH) has listed musculoskeletal diseases as a leading priority for research and disease prevention efforts in the United States. It called particular attention to the need for instituting treatment and preventive research programs related to *traumatogen workplace hazards* (NIOSH, 1996).

A comprehensive review of the epidemiological data on musculoskeletal disorders by NIOSH (1997) and the National Research Council (NRC, 2001), concluded that "strong evidence" exists for the following occupational risk factors of LBDs: lifting, forceful movement, and whole-body vibration, and to a lesser degree for heavy physical work and awkward posture (Burdorf and Sorock, 1997; NIOSH, 1997). Since the exact anatomical cause of LBP cannot be identifiable, etiological research has concentrated on identifying and quantifying several personal and work-related factors (Frymoyer and Gordon, 1989; Riihimäki, 1991). Studies found that the prevalence was highest between the ages of 35 and 55 years and there were no significant gender differences. In addition to the specific occupational risk factors for back problems, nonwork-related risk factors such as smoking, sedentary lifestyles, education, and some psychosocial factors were also identified as important risk factors (Pope et al., 1991; Erdil and Dickerson, 1997). Other demographic and anthropometric factors, such as gender, height, weight, exercise, and marital status, however, appeared to be less important factors associated with back disorders in occupational populations (Burdorf and Sorock, 1997).

As reported in Cole and Grimshaw (2003), many epidemiological studies consistently reported LBP as the most frequent cause of activity limitation among those aged less than 45 years (McCoy et al., 1997; Pool-Goudzwaard et al., 1998) and the third most frequent beyond this age (Liemohn et al., 1988; Sullivan, 1989). Lee et al. (2001) identified LBP as an acute pain in the lumbar spinal region, localized below the belt line and above the gluteal sulcus, occurring intermittently or continuously over a period of two days or more. Previous studies found that those who suffer from LBP experience numbness, tingling, or radiating pain down through the buttocks and the lower extremities due to soft tissue damage (King, 1993; Waddell, 1996). Some research has suggested that the onset of LBP typically occurs early in the third decade of life (Frymoyer and Nachemsom, 1991), while other studies have suggested that the incidence of LBP tends to increase in early adolescence, possibly coinciding with puberty (Burton et al., 1996; Leboeuf-Yde and Kyvik, 1998). In their studies, Dempsey et al. (1997) and Sjolie and Ljunggren (2001) have established that the early incidents of LBP experienced in the adolescent years are related to those experienced in adult life. Consequently, these studies suggest that a better understanding of the etiology of LBP could be established if future research efforts concentrated on juvenile LBP rather than on the previously established LBP of adults (Sjolie and Ljunggren, 2001).

Several other studies reported that the incidence of LBP increases with age, before reaching a plateau or declining around 60 to 65 years of age (Burdorf and Sorock, 1997; Dempsey et al., 1997; Lee et al., 2001). On the other hand, Frymoyer and Nachemsom (1991) hypothesized that the degenerative changes attributable to osteoporosis tend to cause females to experience LBP more frequently as they increase in age. According to Lee et al. (2001), the pain experienced as a result of LBP is more prolonged and disabling later in life, irrespective of gender. Despite the large amount of scientific research that has been performed in this area, our understanding of the nature of LBP is still to some extent obscure, due to the vast uncertainties that surround its etiology (Ekholm et al., 1982; Liemohn et al., 1988; King, 1993; de Looze et al., 1998; Pool-Goudzwaard et al., 1998; Levangie, 1999). However, due to the widespread prevalence and the socioeconomic impact that it has on society, Frymoyer et al. (1980) urged that an important aim of the studies should be defining the common risk factors associated with the development of LBP.

Adams et al. (1986) reported that LBDs can develop as a result of a person's personal characteristics and lifestyle. However, previous investigations have also identified several mechanical risk factors to LBP, and the most common of these is manual materials handling (Goel et al., 1985; de Looze et al., 1994; Sparto et al., 1997). Macfarlane et al. (1997) reported that the inherent risk associated with MMH is greater for females, and several studies have found that the risk increases when a movement involves excessive forward flexion (Esola et al., 1996; Granata and Wilson, 2001; Sjolie and Ljunggren, 2001) or spinal torsion (Hakkanen et al., 1997; Macfarlane et al., 1997; van Dieen et al., 1999). Still, some other studies have theorized that during the performance of these tasks, the lumbar spine is exposed to excessive compressive loads that have the potential to hasten degenerative changes in the intervertebral disks or lead to fracturing of the intervertebral endplates (Cheng et al., 1998; Innes and Straker, 1999; van Dieen and Kingma, 1999; Marras et al., 2001).

1.3 Risk Factors for Low Back Disorder and Pain

As reported earlier by Ayoub et al. (1997), in a study by Riihimäki (1991) various work- and individual-related risk factors have been stated to be associated with LBP and LBDs. However, Ayoub et al. (1997) concluded that precise knowledge about the extent to which these factors are etiologic and the extent to which they are symptom-precipitating or symptom-aggravating is still limited. An important review of epidemiological studies on risk factors of LBP using five comprehensive publications on LBP was made by Hildebrandt (1987). The study found a total of 24 work-related factors that were regarded by at least one of the sources reviewed as risk indicators of LBP. These risk factors include the following categories (Ayoub et al., 1997): (1) general (heavy physical work, work postures in general), (2) static workload (static work postures in general, prolonged sitting, standing or stooping, reaching, no variation in work posture), (3) dynamic work load [heavy manual handling, lifting (heavy or frequent, unexpected heavy, infrequent, torque), carrying, forward flexion of trunk, rotation of trunk, pushing/pulling], (4) work environment (vibration, jolt, slipping/falling), and (5) work content (monotony, repetitive work, work dissatisfaction).

Riihimäki (1991) reduced the list of factors above to "generally accepted" risk factors (i.e., referenced by at least three literature sources): (1) general (heavy physical work), (2) static workload (prolonged sitting), (3) dynamic workload (heavy manual handling, heavy lifting, frequent lifting, trunk rotation, and pushing/pulling), and (4) work environment (vibrations). Furthermore, the study also reported a total of 55 individual factors cited by at least one of the sources as a risk indicator of LBP. Ayoub et al. (1997) grouped these individual factors into the following categories: (1) constitutional factors [age, gender, weight, back muscle strength (absolute and relative), fitness, back mobility, genetic factors], (2) postural–structural factors, (3) medical factors, (4) psychosocial factors, and (5) demographic factors. However, Ayoub et al. (1997) included only the following risk factors as "generally accepted" factors, which are: (1) constitutional factors [(a) age, (b) relative muscle strength, (c) physical fitness], (2) medical factors [(a) back complaints in the past, (b) psychosocial factors (not specified)], and (3) other factors (work experience).

1.3.1 Work-Related Risk Factors

Several epidemiological studies have reported that the type of work involved in an occupation is closely associated with the risk of suffering from an LBD (Bernard, 1997; Burdorf and Sorock 1997). More specifically, according to Bigos et al. (1986) and Spengler et al. (1986), manual materials handling activities dominate occupationally related LBD risk. Previous retrospective studies on industrial injuries have identified MMH as the most common cause of LBD. These studies estimated that lifting and MMH account for 50 to 75% of all back injuries (Bigos et al., 1986; Spengler et al., 1986; Snook, 1989). Nachemsom (1975) assumed that back pain is discogenic and has a mechanical origin. In their study, by examining the functional spinal units of 86 cadavers with a known nature of work and LBD history, Videman et al. (1990) confirmed the notion that LBD risk was associated with physically heavy work, such as unaided MMH tasks. The study found increased degeneration in the spines of those persons who had performed physically heavy work. This finding supports the fact that occupationally related LBDs are often associated with spine loading. Even though studies have also identified psychosocial (Bigos et al., 1986) association to back pain, a recent prospective study by Krause et al. (1998) showed that biomechanical factors and psychosocial factors could independently predict LBD. However, most researchers feel that biomechanical risk factors must be present before psychosocial factors are able to play a role. Thus, biomechanical loading is believed to be the primary injury pathway for occupationally related LBDs.

Several investigations by Marras et al. (1993, 1995, 2000) and Norman et al. (1998) documented exposure to a host of the physical workplace conditions and their association with risk. These surveillance studies were able to describe the physical workplace parameters for lifting tasks, such as origin height, destination height, frequency of lifting, lift asymmetry, weight of object lifted, load moment, and their association with risk. Further, these studies also documented the association between trunk kinematics and LBD risk. Studies by Marras developed a multiple logistic regression model based on these studies that would predict how risk changes as the combination of these workplace features and trunk kinematic characteristics change (Marras et al., 1993, 1995, 2000). Similar findings were reported by Norman et al. (1998) in an independent study. To summarize, these studies demonstrated that the better the physical job demands are described, the better the association with risk of LBD. These risk models have documented the levels at which each of five variables such as lifting rate, trunk twisting velocity, trunk lateral velocity, load moment, and sagittal flexion angle become more adverse.

1.3.2 Psychosocial and Individual Risk Factors

Studies have shown that several personal or individual factors also play a role in increasing the risk of LBDs. Age appeared to have some relationship to the occurrence of LBD. Frequency of LBP symptoms has been observed to reach a maximum between 35 and 55 years of age, whereas lost time increases with age (Andersson, 1981).

Anthropometric characteristics of persons with LBP have also been studied extensively. However, those studies could not established any such strong correlation between back pain and stature, weight, body build, and low back pain, although tallness has been associated with a greater-than-average risk of back pain in certain investigations (Kelsey, 1975; Heliovaara, 1987). In the study by Heliövaara (1987), the relative risk was found to be 2.3 in men (over 179 cm) and 3.7 in women (over 169 cm) compared to those who were at least 10 cm shorter. Chaffin and Park (1973) found strength to be an important indicator of LBP risk. It was found in their study that back pain incidence was three times more among those workers who did not have the strength to perform a job.

Cady et al. (1979, 1985) found that personal fitness is correlated with less chronic LBP. Nevertheless, it is generally believed that fitness is a difficult concept to quantify. Battie et al. (1989) argued that neither aerobic capacity nor trunk mobility was a good indicator of future back pain. Another individual factor, smoking, was found to be associated with a greater risk of low back disorder (Kelsey, 1975; Kelsey et al., 1984; Deyo and Bass, 1989). Deyo and Bass (1989) and Scott et al. (1999) found a dose–response relationship between smoking and LBP. However, it can also be argued that whereas smoking may not be a risk factor in itself, it could certainly be identified as a confounding factor.

A study by Allread (2000) showed that mismatches between personality characteristics and job demands could be associated with greater rates of musculoskeletal disorders. Recent studies found that interpersonal communication (Marras et al., 2000) and mental stress due to pacing and precision of work (Davis et al., 2002) could influence muscle activities and result in greater spinal loading and eventually the onset of LBP. These studies showed that individual factors such as personality and gender could result in variations in muscle recruitment patters (coactivity) and differences in spine loadings (Marras et al., 2000; Davis et al., 2002). Another recent study showed that previous history of back injury significantly increased trunk muscle coactivity and spine loading (Marras et al., 2001). Finally, another very recent study (Roppnen et al., 2004) investigated the role of genetics and environment in lifting force trunk endurance. The study found that genetics had a dominant role over environment in isokinetic lifting, while environmental factors play the dominant role in isometric trunk extensor endurance.

1.4 Unification of Risk Factors

From previous sections it can be concluded that workplace factors and individual factors combine to affect LBD risk. It is generally believed that one has limited control over the manipulation of individual factors in the control of risk. However, to appreciate the risk attributable to work, one must understand the contribution of the individual factors to overall LBD risk as well. One must, therefore, work toward understanding how the various dimensions of risk combine to influence the overall risk of LBD.

The National Research Council (NRC, 2001) has recently proposed a model that helps explain the interrelationship between the various factors leading to LBD. Figure 1 depicts this interrelationship schematically. The model focuses on a potential causal pathway to musculoskeletal disorders related to a load–tolerance relationship of human tissue. The biomechanical system responds to workplace and individual conditions through the systematic recruitment of muscles that result in body movements and the application of forces outside the body needed for task/job performance. The muscle co-contractions also result in subsequent loading of the structures within the torso. If the internal loading due to task demands exceeds the tolerance of a tissue within the torso, there is potential to stimulate pain receptors (nociceptors) and/or cause structural damage (vertebral endplate damage, disk damage, muscle strain, etc), which can lead to an injury.

As discussed by Marras (2004), the sequence of events above can be influenced by individual factors as well as workplace factors (see Figure 1). Since both workplace and individual factors associated with risk have a common pathway to injury, their roles should be considered in the causality of LBD (NRC, 2001). Individual factors (such as age, conditioning, personality, etc.) can modify the coactivity response of the trunk muscles, affect the tissue tolerance (through altered tolerance or adaptation), and/or influence the outcomes of pain or injury. Workplace factors can also influence this injury pathway. Physical loads imposed by the work demands (influenced by workplace design), organizational factors (such as pacing or overtime), and social context (such as interpersonal relationships) have all been shown to influence the recruitment of trunk muscles (Schultz et al., 1987; Marras et al., 1993, 1999, 2000). These factors can also influence tissue tolerance and pain perception (Cavanaugh et al., 1997; Siddall and Cousins, 1997; Yingling and McGill, 1999; Callaghan et al., 2001).

2 MMH CAPACITY DESIGN

People who perform heavy physical work are subjected not only to forces and stresses from the immediate physical environment, but also to mechanical forces generated from within the body. As a result of these forces and stresses, a strain is produced on the worker's musculoskeletal system as well as on other systems, such as the cardiopulmonary system (Ayoub and Mital, 1989). The philosophy of ergonomics is to design work systems where the demands of the job are within the capacities of the workforce. Ayoub et al. (1997) proposed a generic manual materials handling model (Figure 2) that shows the MMH work system,

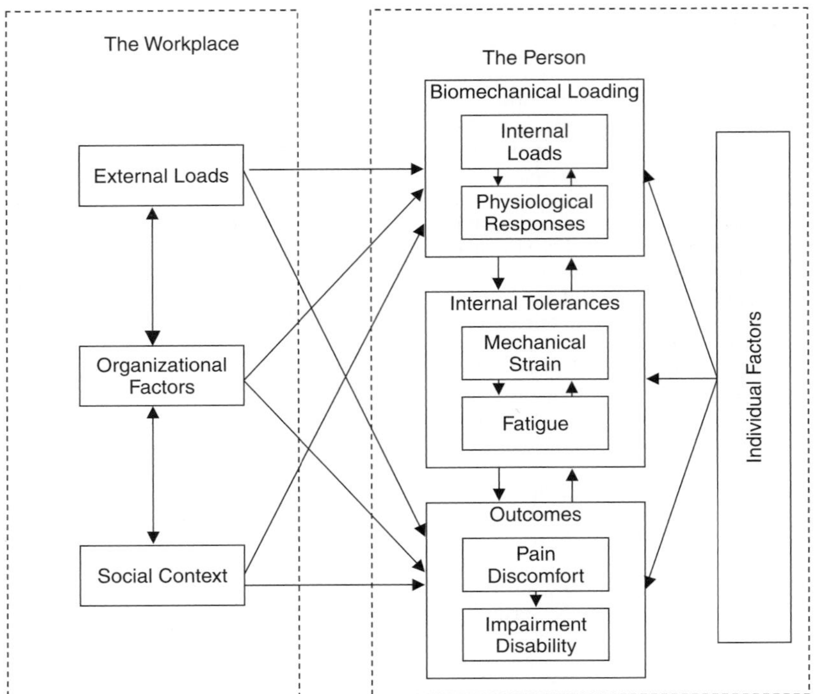

Figure 1 Conceptual model of risk factors and lower back pain. (From NRC, 2001.)

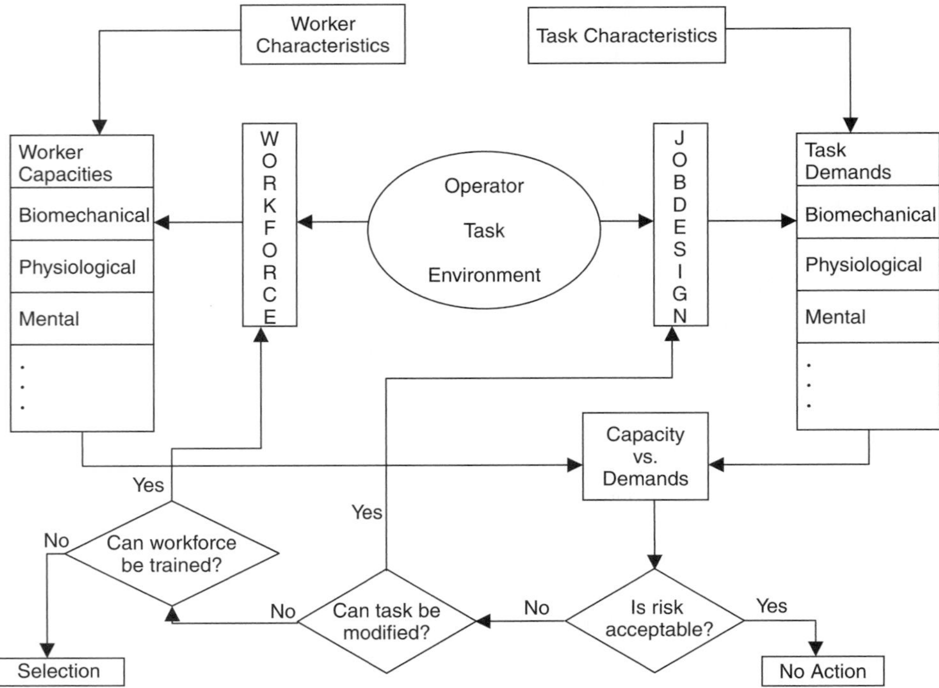

Figure 2 Manual materials handling model. (From Ayoub et al., 1997.)

its components, and the focus of ergonomics on the accommodation of a large percent of the population. According to the model, the work system identifies the risks involved in a particular work situation and the ability to control these risks through the modification of task designs. As can be seen from this model, several capacities, such as the biomechanical capacities of the musculoskeletal system and the physiological capacities of the cardiopulmonary system, must be contrasted with the biomechanical and physiological demands of the tasks. These capacities should be the focus when considering MMH tasks.

Dempsey (1998) proposed a systems approach to defining the task demands/worker capacity ratio for MMH systems (Figure 3). Dempsey (1998) defined task demands in terms of organizational, material, task/workplace, and environmental characteristics. Worker capacity was defined in terms of personal characteristics and biomechanical, physiological, and psychological capacities. According to this approach, the ratio of task demands to worker capacity results in the occurrence of potential undesirable outcomes, such as fatigue, discomfort, and injury. Similarly, the ratio influences factors such as productivity and quality. This systems approach proposed the outcomes to be interrelated. For instance, fatigue may have detrimental effects on productivity or discomfort on performance. Thus, according to Dempsey (1998), the overall goal of the MMH task design should be optimization of the system by considering the summation of positive and negative effects. Accordingly, unilateral concern on only one outcome, such as injuries or productivity, will not necessarily optimize system performance. In some cases there will be concomitant improvements of outcomes even when the focus is only on one outcome. For example, reducing fatigue through proper work–rest schedules can enhance overall productivity.

In the context of MMH, several approaches have been proposed and used by different investigators to establish safe handling limits: (1) the psychophysical approach, (2) the physiological approach (3) the biomechanical approach, and (4) the cognitive engineering approach. It is quite clear that these approaches to MMH design criteria are different and appear to be unrelated. The psychophysical approach relies on the worker's perceived exertion, to quantify his or her tolerance level, thereby establishing the maximum acceptable weights or forces (MAW/F) for different MMH activities [e.g., maximum acceptable weight of lift (MAWL), maximum frequency to lift]. The

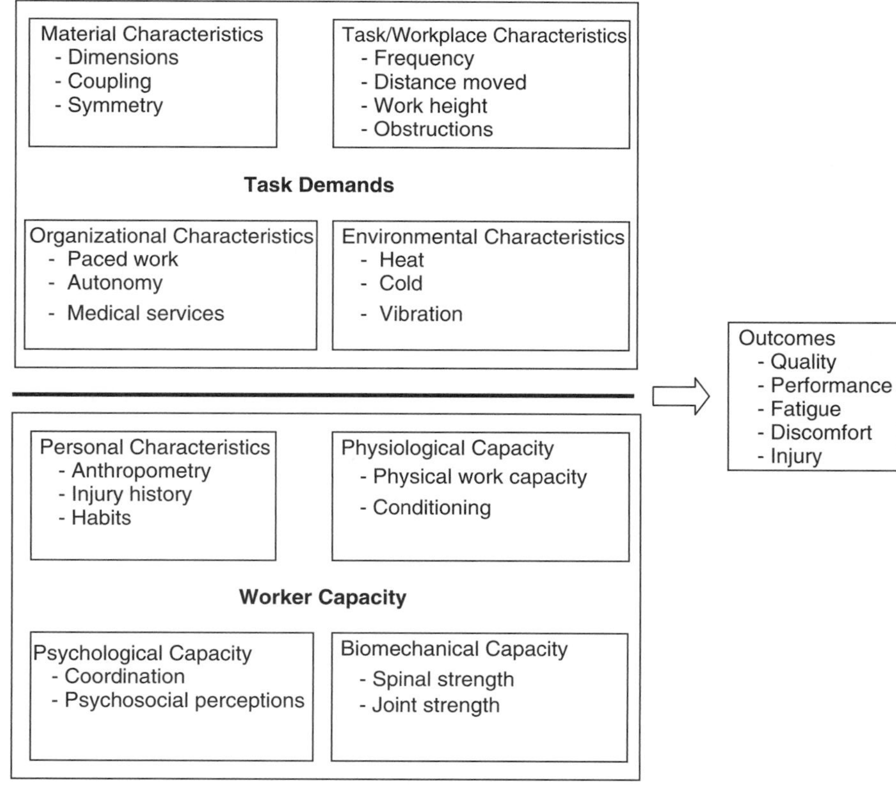

Figure 3 Factors influencing the task demands/worker ratio model. (From Dempsey, 1998; with permission from Taylor & Francis Ltd., http://www.tandf.co.uk/journals.)

biomechanical approach focuses on the establishment of tissue tolerance limits of the body, especially the spine (e.g., compressive and shear force limits tolerated by the lumbar spine). The physiological approach focuses on the physiological responses of the body to physical work; therefore, this approach relates metabolic and circulatory costs of performing MMH activities to workers' physiological capacity. According to Dempsey (1998), to define acceptable task demands relative to worker capacity, researchers from various fields have developed criteria. Primarily, these criteria are based on the principles of biomechanics, physiology, and psychophysics. Karwowski et al. (1999) proposed for a cognitive engineering approach where cognitive abilities of people who perform MMH tasks were considered. Epidemiological results have also been used to develop criteria, but historically, epidemiology has been used to evaluate various criteria or to examine the etiology of injuries associated with MMH. Each of the various classes of criteria or approaches is reviewed critically in the following sections.

2.1 Psychophysical Approach

The primary goal of the psychophysical approach is to design tasks that are "acceptable" to the majority of workers who perform a physical task. The psychophysical approach requires workers to adjust the weight, force, or frequency of a particular task to maximum acceptable values. Borg (1962) and Eisler (1962) found that the perception of both muscular effort and force follow the psychophysical function, where a worker's sensation magnitude (ψ) about the stimulus grows as a power function of the stimulus intensity (ϕ). Stevens (1975) reported the relationship between the strength of the sensation (ψ) and the intensity of its physical stimulus (ϕ) by the power function:

$$\psi = k\phi^n$$

where ψ is the strength of sensation, k a constant, ϕ the intensity of physical stimulus, and n the slope of the line that represents the power function when plotted on log-log coordinates. In the psychophysical approach, databases are usually generated in laboratories for use in the field, or models are developed that predict psychophysical values from task and worker variables.

2.1.1 Limitations

The psychophysical approach has been utilized extensively in MMH research for years. At the present time, there are extensive databases on lifting and lowering (Snook and Ciriello, 1991), as well as various multipliers and correction factors (Mital et al., 1993) to extend the range of applicability of the data. However, psychophysical databases on pushing, pulling, holding, carrying, and one-handed handling are less extensive.

One of the methodological issues in application of the psychophysical approach to design limits in manual lifting tasks is perception of load heaviness. Karwowski and Yates (1986) investigated the reliability of the psychophysical approach to manual lifting of liquids. Karwowski (1991) also studied the psychophysical acceptability and perception of load heaviness by females. In subsequent studies, Karwowski et al. (1992) discussed the problem of discriminability of load heaviness and its implications for the selection of the maximum acceptable, and the maximum safe loads of lift (Karwowski, 1996).

One potential limitation of psychophysical data that most researchers are critical about is the use of data generated in short trials for a longer period of work (such as for eight hours or more of work). There remains uncertainty about the relationship between psychophysical limits and physiological criteria. Several previous studies had indicated that psychophysical assessments determined in short periods (i.e., 20 to 25 minutes) are acceptable for eight hours for low-frequency tasks, but not for moderate- to high-frequency tasks (Mital, 1983; Ciriello et al., 1990; Fernandez et al., 1991).

As it is quite clear, psychophysical determinations in short periods do not allow workers to adjust the weight as will be required by the physiological demand of performing the task for eight hours. Ciriello et al. (1990) stated that psychophysical data collected in short periods are valid for task frequencies of up to 4.3 per minute. Dempsey (1998) recommended that caution must therefore be used when using psychophysical data to design tasks with moderate to high frequencies. The psychophysical approach will result in violation of physiological guidelines if short periods are used for selecting the acceptable values for moderate- to high-frequency tasks.

Like the other approaches, one shortcoming of the psychophysical approach is the lack of sufficient epidemiological data to support this approach. As Dempsey (1998) pointed out, an inherent assumption required by the psychophysical approach is that the data represent values that would protect workers from injuries, that is, the values selected by subjects in laboratories represent loads that will not induce injury in real settings. This assumption can only be validated epidemiologically. Studies done by Snook (1978) and Liles et al. (1984) are worth mentioning in this context. Snook (1978) found that 25% of jobs investigated involved tasks acceptable to less than 75% of the population, but that these jobs resulted in one-half of the back injuries recorded. Liles et al. (1984) found that the rate of back injuries rose rapidly at a job severity index (JSI) value of 1.5. JSI is a time- and frequency-weighted ratio of task demands to operator capacity, where operator capacity is predicted with the models of Ayoub et al. (1978) (discussed in later section), which use strength and anthropometric variables to predict MAWL. Although both studies support use of the psychophysical approach as a valid job design criterion, Dempsey (1998) proposed that further validation is needed.

2.1.2 Existing Databases

Numerous psychophysical design databases are available for various types of MMH tasks (e.g., lifting, carrying, lowering, pushing, and pulling). Snook (1978) reported a large database for designing lifting, lowering, pushing, pulling, and carrying tasks for both males and females. This database is a compilation of a series of psychophysical experiments conducted by Snook and colleagues (Snook, 1965, 1971; Snook and Irvine, 1966, 1967, 1968, 1969; Snook et al., 1970; Snook and Ciriello, 1974a, b; Ciriello and Snook, 1978). Snook (1978) also reported a field study that showed that one-fourth of industrial tasks examined were acceptable to less than 75% of the workforce, but that these jobs accounted for one-half of the back injuries. This study provided support for use of the psychophysical approach in industry and resulted in the recommendation that jobs accommodate at least 75% of the population. Ayoub et al. (1978) reported a database that is applicable to six ranges of lifts and frequencies of between two and eight lifts per minute. This study also provided regression equations to predict individual psychophysical lifting capacity from strength and anthropometric variables.

Mital (1984a) provided a database applicable to males and females, three ranges of lift, three box sizes, and frequencies between one and 12 lifts per minute for eight-hour work shifts. A total of 74 experienced subjects participated in the study. The data were adjusted, as Mital (1983), in a previous study, found that psychophysical determinations made in short periods (i.e., 20 to 25 minutes) need to be adjusted for longer work periods. Mital (1984b) presented a database similar to Mital (1984a) except that the data were applicable for 12-hour work shifts. Again, the data were adjusted using Mital's (1983) results.

Snook and Ciriello (1991) provided an update to the database reported by Snook (1978). The revisions were based on the findings of further studies by Ciriello and Snook (1983) and Ciriello et al. (1990). The database was revised to reflect additional data and a wider range of conditions for some tasks. Specific details of the differences are given in Snook and Ciriello (1991).

Using Kim's (1990) approach, Mital et al. (1993), developed a database that simultaneously satisfies the biomechanical, psychophysical, and physiological approaches. Mital's database also took epidemiological results into account. The study utilized the following criteria to modify Snook and Ciriello's (1991): (1) spinal compression limits of 2689 and 3920 N for females and males, respectively [following epidemiological results of Liles et al. (1984)], (2) intraabdominal pressure limit of 90 mmHg, and (3) an energy expenditure criterion of 21 to 23% of treadmill aerobic capacity or 28 to 29% of bicycle aerobic capacity. Furthermore, Mital et al. (1993) also presented maximum acceptable frequencies of lift for lifting various loads with one hand, mean acceptable holding times for numerous postures and five different load levels, and maximum acceptable forces and weights of lift for MMH tasks performed in numerous unusual postures. Finally, Mital et al. (1993) presented various corrective multipliers to adjust lifting data. Multipliers for work duration, limited headroom, asymmetrical lifting, load asymmetry, couplings, load placement clearance, and heat stress were also provided. Such multipliers extend the range of conditions under which the data can be applied.

Some research efforts were made to establish databases for specific materials handling tasks that are not considered as "standard" MMH tasks, such as mining and maintenance activities. Smith et al. (1992) reported the results of a series of psychophysical studies to determine acceptable weights for one- and two-handed lifting and lowering tasks performed in unusual postures. Many of the tasks are applicable to maintenance work, as they include activities such as lifting while lying on one' s side, lifting while kneeling, and so on. Overall, 99 tasks were studied. The data from these studies were also presented by Mital et al. (1993), along with illustrations of the tasks.

Gallagher (1991) investigated the effects of stooping vs. kneeling, symmetry, and vertical lift distance under restricted headroom conditions to address materials handling talks in the mining industry. Gallagher concluded that MMH tasks in low-seam coal mines should be designed in accordance with lifting capacity in the kneeling posture. In another study, Gallagher and Hamrick (1992) utilized psychophysical data for three specific items commonly handled in underground coal mines, such as rock dust bags, ventilation stopping blocks, and crib blocks. The results indicated that miners are often required to lift loads that exceed psychophysical limits.

Dempsey (1998) pointed out that using such available data replaces conducting a study for every work task and group of workers. Tables provided by the various investigators can be used to estimate the MAW/F for a range of job conditions and work populations. The databases provided in tabular format often make allowances for certain task, workplace, and/or worker characteristics. Use of these databases begins with the determination of the various characteristics with which the database is stratified. Using this information, the value in the tables that applies to the particular work situation can be used as the permissible limit.

This approach can also be used to assess the limits for combination of MMH tasks. For example, Snook and Ciriello (1991) suggest calculating the MAW/F for all individual components for the combination using the same frequency, and then using the lowest or the critical task MAW/F for the entire combination. However, it would be possible to determine the MAW/F for common combination tasks as investigated by Jiang et al. (1986) and Taboun and Dutta (1989).

2.2 Physiological Approach

The physiological approach is concerned with the physiological responses of the body to the physical task that is being performed. During the performance of work, physiological changes take place within the body. Changes in work methods, performance level, or certain environmental factors are usually reflected in the stress levels of the worker and may be evaluated by physiological methods (Dempsey, 1998). The basis of the physiological approach to risk assessment is the comparison of the physiological responses of the body to the stress of performing a task with levels of permissible physiological limits. As reported in Ayoub et al. (1997), many physiological studies of MMH tended to concentrate on whole-body indicators of fatigue, such as heart rate, energy expenditure, blood lactate, or oxygen consumption as a result of the workload. To determine the limit recommended for energy expenditure, Mital et al. (1993) urged that two basic questions be addressed: (1) what the upper limit of VO_2 consumption is as a percentage of aerobic capacity, and (2) which aerobic capacity should be used to express this percentage. The types of aerobic capacity used for setting a criterion can certainly affect the appropriateness of the criterion. Aerobic capacity is very specific to the task with which the capacity is determined. The most frequently used aerobic capacities are those determined by treadmill and bicycle ergometer, and usually, each results in a different value.

In several studies, aerobic capacity is determined for specific MMH tasks. For instance, Khalil et al. (1985) determined aerobic capacity for various lifting tasks using a submaximal technique by either holding frequency constant and increasing the load or holding the load constant and increasing frequency. For the determinations where load was held constant and frequency varied, aerobic capacity increased up to a frequency of 5 lifts per minute, then leveled off. For the trials where frequency was held constant and load was varied, aerobic capacity increased as the load constant increased. The aerobic capacities for lifting were between 56.66 and 90.79% of the bicycle aerobic capacities. In another study, Kim (1990) found that lifting aerobic capacities were between 41.3 and 85.43% of bicycle aerobic capacities. Thus, bicycle aerobic capacities were determined too high for MMH tasks. As mentioned earlier, due to the complexity of aerobic capacity for MMH tasks and conflicting results, many researchers have chosen to use the simpler treadmill or bicycle aerobic capacities.

The fundamental strategy of physiological approach is that once an aerobic capacity is determined, the percentage of the capacity at which the criterion will be set must be selected. The most commonly selected value has been 33% of the maximum aerobic capacity for an eight-hour day (NIOSH, 1981; Garg et al., 1992), which is assumed to be appropriate for the worker population. Generally, a limit of 1 liter of O_2 per minute of uptake has been used, which corresponds to about 5 kcal/min energy expenditure. This value is related to 33% of the aerobic capacity of the "average" person. Waters et al. (1993) based their lifting equation on 50th percentile 40-year-old female data, which resulted in a maximum energy expenditure of 3.1 kcal/min for an eight-hour day. Thus, Dempsey (1998) pointed out fairly wide differences in different guidelines. Accordingly, if the aerobic capacity used to set MMH limits overestimates the MMH task aerobic capacity, the criterion selected may be too high. To confirm this notion, Asfour et al. (1986) found that the 1.0 L/min criterion was too high compared to 33% of lifting aerobic capacity. As Dempsey (1998) argued, the appropriateness of various energy expenditure criteria is somewhat unknown, as few researchers have used task-specific aerobic capacities.

In the study by Bhambhani et al. (1997) on men it was reported that (1) the metabolic cost of loaded walking was significantly higher during the 20-kg load compared with the 15-kg load, (2) there were no significant differences in the cardiovascular responses measured (i.e., cardiac output, heart rate, stroke volume, systolic blood pressure, and diastolic blood pressure between the two load-carriage walks), (3) the peripheral oxygen extraction was not significantly different between loads, and (4) the rate of perceived exertion during the 20-kg load-carriage walk was significantly higher than that observed during the 15-kg load-carriage walk. Bhambhani and Maikala (2000) later found that the trend in these responses was similar in men and women.

Craig et al. (1998) investigated the relationship among injury occurrence, aerobic capacity ($VO_{2\,max}$), and body composition for high-frequency MMH tasks. The study found a significant relationship between injury occurrence and relative aerobic capacity and body composition (percentage of body fat). The study also found geographical location to be a significant factor for the occurrence of injury.

2.2.1 Predictive Models

Studies involving physiological response to manual handling, especially measurement of oxygen consumption during a physical activity, are difficult to conduct, due partly to the nature of measurement devices used. To avoid these difficulties, numerous researchers have developed regression models that predict oxygen consumption from personal, task, and workplace variables. The most comprehensive and flexible set of models is that of Garg et al. (1978). Other models have been developed for more specific MMH tasks. An assumption of the models of Garg et al. (1978) is that the net metabolic cost of an activity is the sum of the metabolic costs of the individual components. However, this assumption was not confirmed by several studies (Genaidy et al., 1985; Taboun and Dutta, 1989). The absolute errors associated with this assumption were found to be between 19 and 36.8% for lifting and between 27.1 and 44.8% for lowering tasks (Genaidy et al., 1985). In their study, Taboun and Dutta (1989) found that the additivity assumption errors associated with a combination lifting and carrying task ranged between −25.25 and 60.36%. Thus, use of these models is associated with what can be significant errors due to the incorrect additivity assumption. A critical issue with respect to predictive models is the accuracy and generalizability of the models.

Another important issue pointed out by Dempsey (1998) is that most of the models are based on limited sample sizes, and typically, the models are derived from data collected from student subjects. Kim (1990) developed models for predicting oxygen consumption for three ranges of lifts and compared the prediction accuracy of his and other models using the data from various studies. The study found that error rates ranged between 5.56 and 27.7%. Kim (1990) also used the various models to estimate the load corresponding to a given oxygen consumption (inverse prediction), as was done by Waters et al. (1993). It was found that the absolute errors were as high as 339%, while all errors exceeded 20%. The author subsequently recommended against using such regression models based on the study findings.

2.2.2 Epidemiological Limitations

One limitation of physiological criteria is the lack of demonstrated relationships between physiological load and injury rates (Dempsey, 1998). Various studies have hypothesized about the potential for increased risk of LBD associated with cardiovascular demands and fatigue (Brown, 1975; Garg and Saxena, 1979). However, there is no compelling epidemiological evidence to support such hypotheses. According to Leamon (1994a), it is unlikely that oxygen uptake criteria will be able to differentiate the LBD propensity of different work designs. As Dempsey (1998) urged, there is a need for studies to relate physiological stresses to various outcomes, such as fatigue or injury, or other appropriate metrics, such as productivity or performance. Physiological criteria may be more important for preventing fatigue and discomfort than for preventing compensable injuries. In the context of maximizing the performance of the MMH system, physiological criteria are essential (Dempsey, 1998).

2.3 Biomechanical Approach

The primary biomechanical criteria used are based on spinal compression and the maximum voluntary torque capabilities of the various major joints involved in performing a MMH task. Regardless of the criterion used, there are inadequacies in the data that are used to develop these criteria, as well as inadequacies in the methods used to estimate values in field situations. However, in the last two decades, numerous studies were conducted to address various biomechanical issues, ranging from determination of spinal compression limits to dynamic modeling. In the following sections we provide useful information based on recent research findings.

2.3.1 Spinal Compression Limits

One of the factors influencing the validity of using L5/S1 compression as a design criterion is the validity of using cadaver data to specify in vivo spinal compression limits (Dempsey, 1998). The first issue is whether the in vivo spinal response to compression is the same as experimentally prepared in vitro spine specimens, which often have a portion or all of the supporting ligaments and musculature removed. It is unlikely that the in vivo spine responds to stresses in the same manner as in vitro specimens, raising doubts concerning the applicability of cadaver data. Adams (1995) provides a review of studies that have addressed issues such as temperature differences, thawing effects on spinal specimens, specimen fixation, the testing environment, and related factors that influence the applicability of using in vitro data.

The second issue related to cadaver data is whether or not the loads applied to in vitro specimens are representative of the loads encountered by the in vivo spine. There is a growing interest on cumulative loading of the spine (discussed in the following section), and the phenomenon of cumulative loading on the spine may not be captured by cadaver studies. To address this issue, Brinckmann et al. (1988) studied the compressive strength of the lumbar vertebrae subjected to cyclical loading. It was found that the probability of failure increased as the number of loadings increased. If LBD were related to repeated trauma to the spine over months or even years, even studies similar to that of Brinckmann et al. (1988) may be inadequate, as it would be difficult to simulate the spinal loadings experienced by workers in various occupations. The mechanical properties of the intervertebral disks depend on both the loading history and the applied load (Adams et al., 1996), indicating the need for realistic loading patterns in in-vitro studies. The viscoelastic load-bearing properties of the intervertebral disks would seem to accentuate the need for realistic loading of experimental specimens (Dempsey, 1998). In fact, according to Iatridis et al. (1996), the viscoelastic characteristics of the nucleus pulposus depend on the loading conditions. In the study it was hypothesized that injury outcomes (e.g. fracture versus degeneration) under different loading speeds may be related to the highly rate-sensitive mechanical behaviors of the nucleus.

The third issue about the use of cadaver data is that cadaver studies have primarily emphasized compression due to axial loading while ignoring shear forces due to nonaxial loading, and torsional stress (Dempsey, 1998). It is speculated that most MMH tasks impose such stresses, as evidenced by such factors as twisting and bending. The focus on compression forces in the investigations may have hindered researchers from examining the influence of other types of loadings on the occurrence and severity of LBD. Adams (1995) argued that since the spine is typically subjected to complex loadings comprising combinations of shear, compression, bending, and torsion, complex loadings should be applied to specimens in the laboratory.

2.3.2 Cumulative Spinal Loading

Recent research has focused on measurement and analysis of cumulative spinal loading and associated occurrence of LBDs due to manual materials handling tasks. Different studies have used different methods to estimate spinal loading. In their studies, Norman and colleagues (Norman et al., 1998; Daynard et al., 2001; Kerr et al., 2001) used motion analysis to derive the peak static spinal load and multiplied these values by the number of repeats and the duration of each task to yield a shift exposure. Jäger at al. (2000) in their DOLLY study utilized a postural sampling approach and a biomechanical model to calculate load for an entire shift period. Following the load calculation, they extrapolated the data to yield results for longer periods. Three weighting systems were utilized in the study for force relative to exposure time to calculate shift dose. The Mainz–Dortmund dose model (MDD) was based only on those instances when the lumbar disk compression force exceeded 3200 N.

In another study, Callaghan et al. (2001) analyzed lifting tasks using five documenting approaches and compared them to a "gold standard," the biomechanical model outputs for the entire lifting cycle (sampling rate of 30 Hz). They concluded that there were significant errors (average error range between task and subject was 27 to 69%) for four of the approaches that used discrete measures to represent the time-varying cyclic exposure. Sullivan et al. (2002) examined inter- and intraobserver reliability of calculating lumbar spine loads. The results showed that compression and moment demonstrated the highest reliability in comparison to joint and reaction shear forces.

In their study, Seidler et al. (2001, 2003) investigated the relationship between lumbar spine disease and cumulative occupational exposure to lifting or carrying and to working postures with extreme forward bending using a case–control study design. They employed structured questionnaires/interviews to extract data on postures and loads and used one of the three algorithms described by Jäger et al. (2000) for cumulative compressive loading. Stuebbe et al. (2002) used a work sampling approach to collect postural data for compressive force calculation. The cumulative load was computed as the sum across all samples.

2.3.3 Trunk Motion Characteristics

Trunk motion characteristics have been identified as important risk factors in MMH operations in the production industry (Marras et al., 1993, 1995). The notion in recent research is that trunk velocity characteristics within each of the three planes (i.e., sagittal, coronal, and transverse) of motion are often more predictive of LBD than position, range of motion (ROM), or acceleration (Marras et al., 1995). Furthermore, in a three-dimensional model of loads on the lumbar spine, Marras and Sommerich (1991a, b) showed that spine compression increased directly with trunk velocity and that shear forces increased when the trunk became more asymmetric.

In recent years, awareness of the role of biomechanical stress on cumulative pathogenesis of LBDs has led to attempts to quantify the effects of repeated physical loading associated with MMH tasks (Keyserling et al., 1987; Kumar, 1990; Seidler et al., 2001; Stuebbe et al., 2002). It has been demonstrated in clinical studies that repeated exposure to loads well below the threshold for causing traumatic, instantaneous injury can result in the eventual fatigue failure of the vertebral endplate (Brinckmann et al., 1988). Adverse effects of repetitive loading on spinal structures have been shown in in-vitro studies (Adams and Hutton, 1982; Hansson et al., 1987; Callaghan and McGill, 2001) and radiographic investigations (Brinckmann et al., 1988). Marras (1998) has proposed a time-varying load-tolerance model, which is depicted in Figure 4.

According to Marras (1998), the time-varying load tolerance model assumes that the tissue tolerance decreases due to repeated loading, thereby decreasing the safety margin and lowering the load threshold for back injury, compared to the traditional model of biomechanical risk that assumes a constant tissue tolerance level. This response and the rate of decrease can be attributed to the accumulated loading that a tissue has experienced and the way in which the force, repetition, and postural factors have interacted during exposure. Since working life is typically comprised of a complex set of individual and interactive risk factors that act repeatedly over an extended period of time, a methodology to assess the cumulative, aggregate risk incurred by the biological system may be helpful in predicting the overall risk of LBD due to repeated spinal loading. Recently, the role of cumulative physical workload as a risk factor for lumbar spine disorders has been the subject of increased focus (Seidler et al., 2001, 2003; Callaghan, 2004).

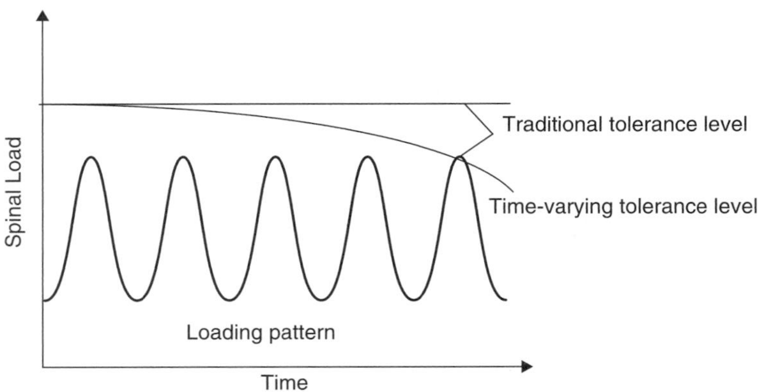

Figure 4 Time-varying load-tolerance model. (From Marras, 1998; with permission from Taylor & Francis Ltd., http://www.tandf.co.uk/journals.)

2.3.4 Biomechanical Modeling

According to Ayoub et al. (1997), the existing biomechanical models can be divided into two- and three-dimensional models as well as static and dynamic. All of these models have the goal of estimating stresses imposed on body segments, joints, and the lumbar spine. The models developed by Chaffin (1969), Martin and Chaffin (1972), Park and Chaffin (1974), and Garg and Chaffin (1975) are all static in nature. Static models assume that the lifting action is performed quite slowly and smoothly such that forces caused by acceleration can be neglected. This implies the fact that the effects of inertia are not included in static models. Ayoub et al. (1980) argued that this assumption is invalid and can seriously affect the estimation of forces exerted on the body segment. In all occupational biomechanical models, the human body is modeled as a system of rigid links of fixed length, mass, and center of gravity.

Dynamic models such as those developed by Fisher (1967), El-Bassoussi (1974), Ayoub and El-Bassoussi (1976), Kromodihardjo and Mital (1986, 1987), and Chen and Ayoub (1988) provide data for analysis in the form of the time–displacement relationships of the body segments (kinematic analysis) and the forces and torques involved (kinetic analysis). The static models have been popular and applied more widely than dynamic models for the following reasons: (1) static models require simpler logic and task data than dynamic models, and (2) most static models compare the stresses produced by a manual task with allowable stresses (static strength data, which are readily available). For the application and future research purpose, in this chapter we discuss two widely used models: static whole-body kinematic models and dynamic whole-body biomechanical models.

Static Whole-Body Kinematic Models Chaffin and his associates at the University of Michigan developed a static three-dimensional kinematic model of the musculoskeletal system that can be used to evaluate biomechanical responses to whole-body exertions such as lifting, pushing, and pulling (Chaffin and Baker, 1970; Garg and Chaffin, 1975; Chaffin and Andersson, 1991; Chaffin et al., 1999). This model has generally been used for two purposes: (1) to compare the strength demands of a task to the strength capabilities of the workforce to estimate the percentage of adult males and females that are capable of performing the task, and (2) to predict compressive forces acting at the L5/S1 spinal disk during static exertions. The model has been used extensively to evaluate whole-body tasks that are performed at normal (nonjerky) movement speeds on an infrequent basis (typically less than once every five minutes). Because the model does not consider the effects of fatigue, it is believed not to be appropriate for highly repetitive tasks (psychophysical or metabolic job analysis tools are preferred) or highly dynamic motions (dynamic models are preferred). Despite these limitations, the model has been used to predict biomechanical responses to strenuous exertions associated with common manual handling tasks. Figure 5 illustrates a typical outcome of the modeling software.

Dynamic Whole-Body Biomechanical Models Dynamic biomechanical models for evaluating whole-body materials handling activities have been developed by several investigators (Leskinen et al., 1983; McGill and Norman, 1985; Jäger and Luttmann, 1989; Marras and Sommerich, 1991a, b). As is generally believed, these dynamic models are inherently more complex than static models. In addition to considering external forces acting on the body (kinetics) and posture, these models consider the effects of motion dynamics, including velocity and acceleration (kinematics). The fundamental assumption is that due to inertia, acceleration or deceleration of body segments and any load in the hands requires the application of additional force, as stated by Newton's second law ($F = ma$). Because the body is composed of multiple links and multiple joint centers, dynamic models require high-frequency measurements of many reference points to determine instantaneous locations, velocities, and accelerations of model components. For this reason, dynamic models are often restricted to laboratory environments where accelerometers, goniometers, and/or motion analysis equipment can be used to collect reliable data. Despite this limitation, dynamic models provide important insights into the additional biomechanical strain imposed by rapid motions (Marras, 2000).

In their study, Jäger and Luttmann (1989) evaluated the effects of movement dynamics and other task factors on selected indices of biomechanical strain on the lower back by using a whole-body model (the Dortmunder model). The basic representation of the skeletal system in the Dortmunder model was similar to the Michigan model described earlier. However, the in Jäger and Luttmann (1989) model, lumbar spine is depicted as a system of five joints representing the five lumbar intervertebral disks (compared to the Michigan model, which used only the L5/S1 joint in the lower back region). The Dortmunder model was used to estimate the moment at the L5/S1 joint, compressive forces at the L5/S1 disk, and shear forces at the L5/S1 disk under various static and dynamic task conditions using 50th percentile male anthropometry during symmetric, sagittal plane lifting. The static analysis produced results consistent with the Michigan model. All three outcome measures increased as the weight in the hands increased from a no-load condition to a 50-kg load. These outcomes also increased as the horizontal location of the load in front of the L5/S1 increased. When the hands were empty, it was found that increased trunk flexion angle (forward bending) caused increases in L5/S1 moment, compression, and shear force.

Dynamic analysis with the Dortmunder model demonstrated the significance of velocity and acceleration. The task of raising objects of three weights from floor to elbow height was simulated under three conditions, such as slow, medium, and fast lifting. Under

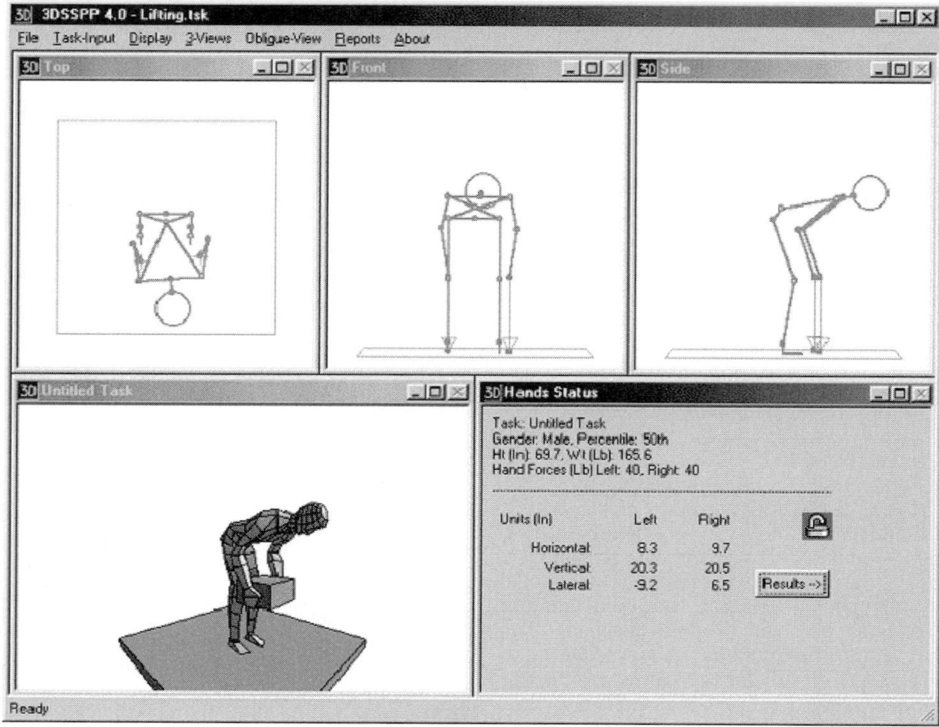

Figure 5 Output of a lifting task using 3DSSPP strength modeling.

the slow condition, compressive forces at the L5/S1 were similar to static conditions. At medium and high speeds, peak compressive forces were about 20 and 50% higher, respectively, than under static conditions.

Jäger and Luttmann (1989) also simulated a jerking motion, assuming that all of the upward acceleration was completed in 0.1 second. Under these conditions, the peak L5/S1 compression for lifting a 20-kg load from the floor was approximately 8000 N, which was more than double the peak load under static conditions and considerably higher than compressive loads shown to cause mechanical damage in cadaver tissues.

Marras and Sommerich (1991a, b) developed a three-dimensional dynamic biomechanical model for evaluating loads on the lumbar spine during lifting. Instead of using a single equivalent trunk extensor muscle, this model extended earlier work by Schultz and Andersson (1981) by including 10 functional muscle groups in the lower back. By considering multiple muscle groups, Marras and Sommerich were able to evaluate the effects of co-contraction and asymmetric postures on spinal stresses during lifting.

The model developed by Marras and colleagues was used to evaluate the effects of trunk velocity, torque output, and posture symmetry on biomechanical loads on the spine. During these exertions, EMG measured muscle activity in 10 trunk muscles (Figure 6) and was used as input to the model

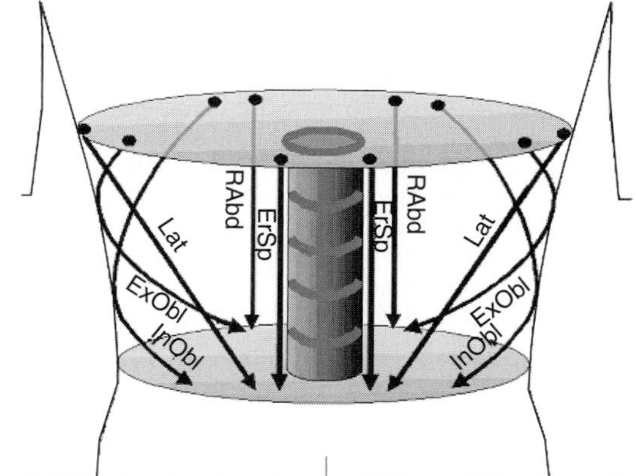

Figure 6 Ten trunk muscles used in EMG-assisted dynamic biomechanical modeling. (From Marras and Granata, 1997.)

along with subject anthropometry and trunk kinetics. The model calculated compression, shear, and torsion loading in the lumbar spine. It was found that under symmetric conditions, peak compression at the L5/S1 increased with velocity and trunk torque output. On the contrary, under asymmetric conditions, peak compression was level over the range of velocities tested and approximately 25% lower than under similar symmetric conditions. Furthermore, peak anterior/posterior shear forces were greater under symmetric conditions and increased with the magnitude of trunk torque output. Lateral shear force was approximately 40 N under asymmetric conditions compared to about 10 N under equivalent symmetric conditions.

In subsequent follow-up cross-sectional epidemiological studies, Marras et al. (1993, 1995) used historical medical records of low-back injuries to classify over 400 cyclical jobs as either high risk or low risk. Dynamic trunk motions were measured for each job using a triaxial goniometer system [lumbar motion monitor (LMM)] to document three-dimensional angular position, velocity, and acceleration of the lumbar spine while workers performed their jobs. In addition, basic biomechanical variables (weight lifted, lift frequency, posture, etc.) were determined for each job. Logistic regression was used to identify the following five risk factors that distinguished between high- and low-risk jobs: lifting frequency, load moment, trunk lateral velocity, trunk twisting velocity, and trunk sagittal angle. The results found odds ratio for five variables combined as 10.7.

In another study, Allread et al. (1996) used the lumbar motion monitor in a laboratory study to evaluate the effects of lift symmetry and to compare one- vs. two-handed lift technique. The study included one- and two-handed lifts ranging between 0° (in the sagittal plane) and 135°. Trunk motion characteristics associated with increased risk of back injury (Marras et al., 1993) were all higher with one-handed lifts, and velocities and accelerations increased substantially with the angle of asymmetry. Figure 7 illustrates the three-dimensional kinematic output of LMM using the EMG-assisted model.

Kim and Chung (1995) conducted a laboratory study of the EMG activity of the lower back during dynamic lifting. Eight healthy males participated in four two-hour trials, lifting and lowering weights between floor and knuckle height. The study investigated the effects of normalized lifting frequency, heavy lifting, and symmetric/asymmetric postures. This study found that muscle activity (normalized EMG) was significantly higher during heavy lifting and during asymmetric lifting, while muscle fatigue during the two-hour lifting task (measured by a decrease in the mean power frequency of the EMG signal) was significantly greater during asymmetric and frequent lifting.

It is generally agreed that static models are not only easier to develop but are easier and typically cheaper to use since no kinematic data are required. Dynamic models are more difficult to develop and use, but they provide more accurate and realistic estimates of stresses. Thus, there are certainly trade-offs between static and dynamic models. According to Dempsey (1998), owing to the inaccuracies associated with various aspects of biomechanical calculations, the most appropriate use of biomechanical model estimates is a relative comparison of alternative task designs rather than using values in an absolute sense. Rather than, for example, determining if stresses associated with a particular task exceed some criterion, the stresses

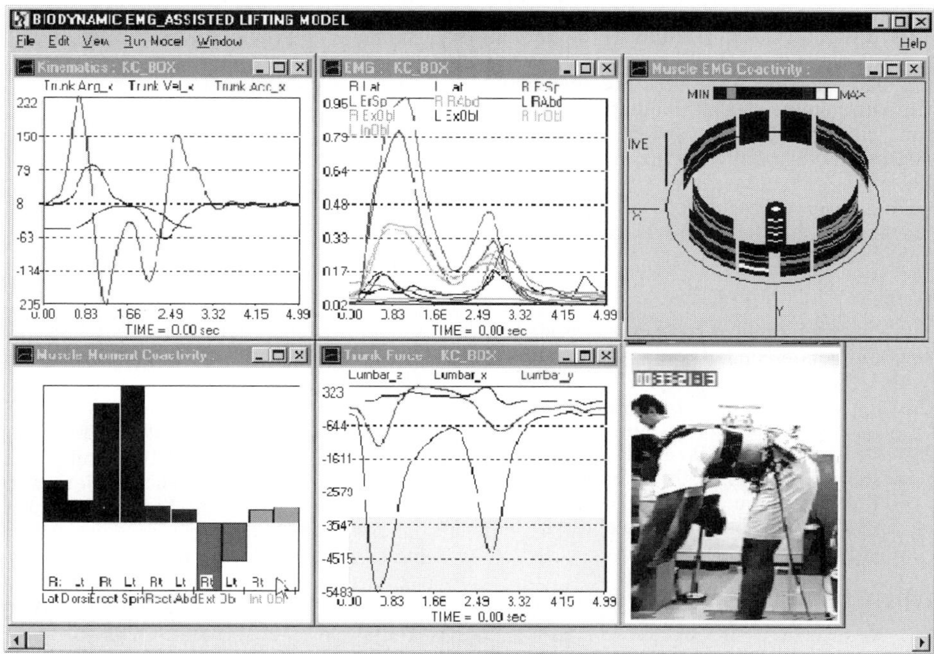

Figure 7 EMG-assisted model used in a Windows environment. (From Marras, 2000.)

are calculated for multiple task designs and the least stressful design is adopted. Although there are errors in the absolute values, the effect of the constancy of the errors across conditions allows for a relative comparison.

2.3.5 Epidemiological Evidence of Biomechanical Criteria

The most critical question concerning the use of L5/S1 compression is whether L5/S1 compression is a criterion capable of reducing LBD. As Dempsey (1998) proposed earlier, there is a need to correlate in vitro experimental specimens with in vivo workers. Certainly in vitro damage could be documented with imaging techniques to determine if the nature and severity of damage assumed to be work related is similar to damage predicted by experimental studies of cadaver specimens. Indeed, Leamon's (1994a) review of the epidemiological support for using L5/S1 compression as a criterion for job design questions the use of this criterion. Subsequently, Leamon (1994b) argued that disability may be largely dependent on factors unrelated to effects on the L5/S1 joint and that such factors account for the lack of a predictable relationship between L5/S1 compression and disability. The primary focus of the use of L5/S1 compression as a design criterion is the prevention of mechanical damage to the vertebrae and/or intervertebral disks.

Emphasizing these mechanical failures can contribute significantly to loss prevention, provided that work exposure to mechanical stresses is related to the failures. In their study relevant to this issue, Jensen et al. (1994) studied magnetic resonance images (MRIs) from persons free from LBP and found a considerable number of structural abnormalities. It was concluded that disk bulges or protrusions found in the MRIs of LBP patients might frequently be coincidental. This study added to the growing literature which shows that anatomical abnormalities are common in people without LBP (Deyo, 1994). The findings of the study are very critical when attempts are made to relate physical exposures in the workplace to structural spinal damage. LBP incorrectly attributed to mechanical damage would be a serious confounding factor (Dempsey, 1998).

2.4 Comparison of Design Criteria

Both practitioners and researchers have concerns about the differences in recommended weight limits that result from the biomechanical, physiological, and psychophysical approaches. Each approach utilizes a unique criterion to generate a load limit assumed to be safe, and each approach is valid only under certain conditions; for example, the biomechanical approach is valid for low-frequency tasks, whereas the physiological approach is most applicable for high-frequency tasks (Mital et al., 1993).

2.4.1 Psychophysical versus Biomechanical Approaches

Earlier we noted that the psychophysical approach violates biomechanical criteria for low-frequency tasks. In one study, Nicholson (1989) computed lumbosacral compression estimates for psychophysical values presented by Snook (1978). The study found that a 3400-N limit was exceeded only for floor-to-knuckle lifting. The limit was exceeded by about 10%. However, the study used a static model and it was assumed, using results in the literature, that the static model would underestimate dynamic forces by about 40%. When 40% was added, most of Snook's values resulted in spinal compression values exceeded the 3400-N limit, except those tasks in which the load was lifted close to the body. Chaffin and Page (1994) criticized maximum acceptable weights of lift because some values violated a spinal compression value of 3400 N. The 3400-N value is a biomechanical criterion selected to protect 99% of males and 75% of females (NIOSH, 1981). Chaffin and Page (1994) examined psychophysical values from Snook and Ciriello (1991) that accommodate 99% of males to the 3400-N limit, and these values produced spinal compressions below the 3400-N limit.

Mital et al. (1993) modified Snook and Ciriello's (1991) database to comply with spinal compression criteria of 2689 and 3920 N for females and males, respectively. Thus, to date, there is not complete consensus as to what spinal compression criterion should be used. The degree to which psychophysical values violate biomechanical criteria will obviously depend on the specific criterion chosen (Dempsey, 1998).

2.4.2 Psychophysical versus Physiological Approaches

The psychophysical approach may suggest handling limits that result in energy expenditure criteria being violated. A primary concern is whether or not workloads selected in short periods (20 to 25 minutes) are valid for eight hours of work (in some cases, more than eight hours). Subjects may not be able to adequately project the physiological burden of a given workload to an eight-hour day. Snook and Ciriello (1991) italicized values in their database that violate energy expenditure limits. As expected, the violations occur for the higher-frequency tasks. Mital et al. (1993) modified some values in this database to comply with energy expenditure limits.

Several studies have been done to investigate the conflicts between the psychophysical and physiological approaches. The approach has typically been to have subjects select workloads during short periods and then determine if the subjects would prefer to lower the load when performing the task for longer periods (four to 12 hours). Legg and Myles (1981) had soldiers select psychophysical loads for lifting from the floor to a height equal to 40% of their stature at a frequency of 2.5 lifts per minute. The soldiers were given 20 minutes to select the load. Legg and Myles then had the soldiers lift the loads for eight hours. All subjects were able to lift the selected loads for eight hours without "metabolic or cardiovascular evidence of fatigue." This study supports the validity of psychophysically selected loads for a frequency of 2.5 lifts per minute.

Mital (1983) had the subjects select loads in 25-minute adjustment periods to be lifted at frequencies between 1 and 12 lifts per minute. Subjects were instructed to select loads they could handle for eight and 12 hours. Subjects were then asked to lift the loads for 12 hours, and were allowed to adjust the loads. It was found that on average, males decreased the loads at approximately 3.4% per hour and females decreased the loads by 2% per hour. It was concluded that the psychophysical approach overestimates handling capacity when shorter periods are used to select acceptable workloads. However, it should be noted that Mital (1983) did not analyze the data for each frequency, but aggregated the results.

Karwowski and Yates (1986) utilized female subjects who selected weights of lift for frequencies of 1, 3, 6, and 12 lifts per minute that they felt they could handle for eight hours. The study found that after four hours of lifting, the weights selected by the subjects for frequencies of 1, 3, and 6 lifts per minute were not significantly different from the weights selected at 30 minutes. For 12 lifts per minute the mean weight selected at four hours was 23% less than the mean weight selected at 30 minutes. It was concluded that psychophysical adjustments made in short periods may not be valid for longer work periods when the frequency of lifting is above 6 lifts per minute.

Ciriello et al. (1990) found that weights and forces selected for lifting, lowering, pushing, pulling and carrying by subjects in 40-minute adjustment periods were not significantly different from the weights and forces selected after four hours. The highest frequency used was 4.3 handling tasks per minute. Thus, Ciriello et al. (1990) concluded that weights and forces selected in short periods are valid for longer periods for frequencies at or below 4.3 per minute.

Fernandez et al. (1991) examined the validity of psychophysical maximum acceptable weight of lift (MAWL) values selected in 25-minute periods for eight hours of work and frequencies of 2 and 8 lifts per minute. The study involved two experimental conditions that allowed the subjects to decrease the MAWL selected in the 25-minute periods during the eight-hour task, and the other condition determined the number of subjects who were willing to lift the MAWL selected in the 25-minute periods for eight hours at 2 lifts per minute. The study found that MAWL decreased to 88% of the value selected in the 25-minute period, whereas at 8 lifts per minute, MAWL decreased by 83%. Furthermore, all subjects were willing to lift the MAWL selected in the short period for eight hours at a frequency of 2 lifts per minute, but at 8 lifts per minute, only three of the 12 subjects were willing to lift for eight hours. Thus, the study confirmed the suggestion that psychophysically determined MAWL for a short period of time cannot be applicable for a longer period of time for high-frequency lifting tasks.

2.5 Cognitive Engineering Approach

Karwowski (1996) argued that assumptions made for the psychophysical approach to setting limits in manual handling tasks have never been examined fully with sufficient scientific scrutiny to ensure acceptance of the research community. Furthermore, Karwowski et al. (1999) pointed out that the classical psychophysical approach cannot take into account the cognitive processes that govern human judgment of load heaviness and load acceptability. In this point of view, Karwowski et al. (1999) argued that in addition to immediate perception of exertion, an assessment measure of the acceptable physical workload should also rely on the subjects' cognitive judgment. They reported results of two laboratory experiments. In the first experiment, the classical concept of maximum acceptable weight of lift was compared to the alternative concept of maximum safe weight of lift (MSWL). The study found a significant difference in maximum acceptable and safe weight of lifting where acceptable weight was higher than the safe weight. Further, the study also found a significant difference in box color with respect to weight selection (weight selected higher for a white box than for a black box). Comparison of the rate of perceived exertion (RPE) showed that subjects worked at lower levels of perceived physical exertion when specifically instructed to pay attention to their own safety.

In the second experiment, the linguistic magnitude estimation (LME) method (Karwowski, 1991) was used to model mathematically the human assessment of four categories of load heaviness, such as *acceptable*, *safe*, *not-too-heavy*, and *too-heavy loads* for continuous lifting. The study also investigated the concept of the load indifference in an assessment of load heaviness, its acceptability, and its safety. The study found a nonlinear relationship between load heaviness and weight lifted, the sensation of acceptability and weight lifted, and the sensation of safety and weight lifted. In light of the findings, Karwowski et al. (1999) argued that at least some of the results of the classical psychophysical experiments, which are to indicate what is acceptable to the human subjects, may not be suitable as direct estimates for load limitation and for design purposes for lifting tasks.

Karwowski et al. (1999) suggested that human judgment of load heaviness should be modeled using the cognitive engineering approach, focusing on a person's cognitive processes, which underlie the development of acceptability criteria for lifted loads and the assessment of safety over the course of load adjustments and selection. It was argued further that this unexplored area of research should lead to greater understanding of human capacities and limitations in manual lifting tasks.

3 TOOLS AND DATABASES FOR TASK DESIGN

3.1 Human Capacity Data for Lifting

The design data presented in this section are based on the database of Snook and Ciriello (1991). These data represent several decades of research that has been conducted at the Liberty Mutual Research Center for Safety and Health, and is based on an industrial subject pool. For a detailed discussion of the data collection protocols as well as additional data, the reader is referred to Snook and Ciriello (1991), Smith et al. (1992), and Mital et al. (1993).

Various studies showed that the psychophysical approach tends to result in higher recommended loads than the biomechanical and physiological approaches at low and high frequencies, respectively. Mital et al. (1993) modified Snook and Ciriello's (1991) lifting data so that the data satisfied biomechanical and physiological criteria. Thus, the data presented an attempt to satisfy the biomechanical, psychophysical, and physiological approaches simultaneously.

Table 1 provides Snook and Ciriello's (1991) two-handed lifting data for males and females as modified by Mital et al. (1993). The data were modified so that a job severity index (discussed in Section 4.1) value of 1.5 is not exceeded, which corresponds to 27.24 kg. In a similar way, a spinal compression value of 30% was used for the biomechanical criterion, which corresponds to a maximum load of 27.24 kg for males and 20 kg for females. The database also took into consideration the physiological criterion of energy expenditure. The limits selected were 4 kcal/min for males and 3 kcal/min for females for an eight-hour working day (Mital et al., 1993). Mital et al. (1993) also utilized seven different modifiers to modify the table values: (1) duration of the activity, (2) quality of coupling, (3) asymmetrical lifting, (4) load asymmetry, (5) head room, (6) load placement clearance, and (7) heat stress.

The design data for maximal acceptable weights for two-handed pushing/pulling tasks, maximal acceptable weights for carrying tasks, and maximal acceptable holding times can be found in Snook and Ciriello (1991) and Mital et al. (1993). The maximal acceptable weights for manual handling in unusual postures are presented by Smith et al. (1992).

3.2 NIOSH Lifting Equation

In 1981, the National Institute for Occupational Safety and Health published the *Work Practices Guide for Manual Lifting*. This guide applies to symmetrical and smooth lifting of moderate-width objects in the sagittal plane only (no twisting) using good couplings (secure handholds and low slip potential at the floor), with unrestricted posture and favorable temperature conditions. NIOSH (1981) recommendations were based on two levels of hazard: the action limit (AL) and maximum permissible limits (MPLs). The AL is based on a biomechanical criterion of 3400-N spinal compression, a physiological criterion of 3.5 kcal/min, and a psychophysical criterion that the load be acceptable to 75% of women and 99% of men. The MPL is based on a biomechanical criterion of 6400-N spinal compression, a physiological criterion of 5 kcal/min, and psychophysical data indicating that the load would be acceptable to only 1% of women and 25% of men. In 1991, the equation was revised to extend the range of conditions over which the equation applies.

3.3 Revised NIOSH Lifting Equation

As discussed by Karwowski and Rodrick (2001), the 1991 revised NIOSH lifting equation (RNLE) introduced the concept of the *recommended weight limit* (RWL), which was designed to protect 90% of the mixed (male/female) industrial working population against LBP (Waters et al., 1993). The RNLE is based on three main components: (1) the standard lifting location, (2) load constant, and (3) risk factor multipliers. The standard lifting location (SLL) serves as the three-dimensional reference point for evaluating the parameters defining the worker's lifting posture. The SLL for the 1981 Guide was defined as a vertical height of 75 cm and a horizontal distance of 25 cm with respect to the midpoint between the ankles. The 25 cm is the minimum horizontal distance in lifting that does not interfere with the front of the body (Garg and Badger, 1986; Garg, 1989).

The load constant (LC) refers to a maximum weight value for the SLL. For the revised equation, the LC was reduced from 40 kg to 23 kg. The reduction in the load constant was driven, in part, by the need to increase the 1981 horizontal displacement value from a 15- to a 25-cm displacement. Table 2 shows definitions of the relevant terms utilized by the 1991 equation. The RWL is the product of the load constant and six multipliers:

$$\text{RWL (kg)} = \text{LC} \times \text{HM} \times \text{VM} \times \text{DM} \times \text{AM} \times \text{FM} \times \text{CM}$$

The multipliers (M) are defined in terms of the related risk factors, including the horizontal location (HM), vertical location (VM), vertical travel distance (DM), asymmetry angle (AM), frequency of lift (FM), and coupling (CM). The multipliers for frequency and coupling are defined using relevant tables. In addition to lifting frequency, the work duration and vertical distance factors are used to compute the frequency multiplier (see Table 3). Table 4 shows the coupling multiplier (CM), and Table 5 provides information about the coupling classification.

The horizontal location (H) is measured from the midpoint of the line joining the inner ankle bones to a point projected on the floor directly below the midpoint of the hand grasps (i.e., load center). If significant control is required at the destination (i.e., precision placement), H should be measured at both the origin and destination of the lift. This procedure is required if there is a need to (1) regrasp the load near the destination of the lift, (2) hold the object at the destination momentarily, or (3) position or guide the load at the destination. If the distance is less than 10 inches (25 cm), H should be set to 10 inches (25 cm). The vertical location (V) is defined as the vertical height of the hands above the floor and is measured vertically from the floor to the midpoint between the hand grasps, as defined by the large middle knuckle. The vertical location is limited by the floor surface and the upper limit of vertical reach for lifting (i.e., 70 inches or 175 cm).

Table 1 Recommended Weight (kg) of Lift for Male (Female) Industrial Workers for Two-Handed Symmetrical Lifting for Eight Hours

Container Size (cm)		Frequency of Lift															
		$\frac{1}{8}$ h		$\frac{1}{30}$ min		$\frac{1}{5}$ min		1/min		4/min		8/min		12/min		16/min	
Width	Length																
Floor to 80-cm Height																	
75	90	17	(12)	14	(9)	14	(8)	11	(7)	9	(7)	7[b]	(6)	6	(5)	4.5	(4)
	75	24	(14)	21	(11)	20	(10)	16	(9)	13	(9)	10.5[b]	(8)	9	(7)	7	(6)
	50	27[a]	(17)	27[a]	(13)	27	(12)	22	(11)	17	(10)	14[b]	(9)	12	(8)	9.5	(7)
	25	27[a]	(20[a])	27[a]	(15)	27[a]	(14)	27[a]	(13)	21	(12)	17.5[b]	(11)	15	(9)	12	(7)
	10	27[a]	(20[a])	27[a]	(17)	27[a]	(16)	27[a]	(14)	25	(14)	20.5[b]	(13)	18	(11)	14.5	(9)
49	90	20	(13)[a]	17	(9)	16	(8)	13	(8)	10	(8)	7[b]	(7)	7	(6)	6.5	(5)
	75	27[a]	(16)	24	(12)	24	(10)	19	(10)	14	(9)	10[b]	(8)	10[b]	(7)	9	(6)
	50	27[a]	(19)	27[a]	(14)	27[a]	(13)	26	(12)	19	(11)	15[b]	(10)	12.5[b]	(9)	10[b]	(8)
	25	27[a]	(20[a])	27[a]	(17)	27[a]	(15)	27[a]	(14)	24	(13)	18.5	(11)	15[b]	(10)	12[b]	(8)
	10	27[a]	(20[a])	27[a]	(19)	27[a]	(17)	27[a]	(15)	28	(15)	22	(13)	17.5[b]	(11)	15[b]	(9)
34	90	23	(15)[a]	19	(11)	19	(10)	15	(9)	11	(9)	7[b]	(8)	7[b]	(7)	6.5	(7)
	75	27[a]	(19)	27[a]	(14)	27[a]	(13)	22	(12)	17	(11)	10[b]	(9)	10[b]	(8)	9.5	(7)
	50	27[a]	(20[a])	27[a]	(17)	27[a]	(16)	27[a]	(14)	22	(13)	15[b]	(11)	14[b]	(10)	12[b]	(8)
	25	27[a]	(20[a])	27[a]	(20[a])	27[a]	(18)	27[a]	(17)	27[a]	(15)	20[b]	(13)	17[b]	(12)	14[b]	(10)
	10	27[a]	(20[a])	27[a]	(20[a])	27[a]	(20[a])	27[a]	(19)	27[a]	(18)	25[b]	(15)	21[b]	(13)	15[b]	(11)
Floor to 132-cm Height																	
75	90	15	(10)	13	(7.5)	13	(6.5)	10	(6)	8	(6)	6[b]	(5)	6	(4)	4	(3)
	75	22	(12)	20	(9)	19	(8)	14.5	(7.5)	12	(7.5)	10[b]	(6.5)	9	(6)	7	(5)
	50	27[a]	(14)	25	(11)	24	(10)	20	(9)	15	(8)	13	(7.5)	11	(6.5)	9	(6)
	25	27[a]	(17)	27[a]	(12.5)	27[a]	(11.5)	24.5	(11)	18	(10)	15	(9)	12	(7.5)	11	(6.5)
	10	27[a]	(19)	27[a]	(14)	27[a]	(13)	27[a]	(11.5)	22	(11.5)	19	(11)	16	(9)	13	(8)
49	90	18	(11)	16	(7.5)	15	(6.5)	12.5	(6.5)	9	(6.5)	6[b]	(6)	6[b]	(5)	5[b]	(4)
	75	27	(13)	22.5	(10)	22.5	(8)	18	(8)	14	(7.5)	10[b]	(6.5)	9[b]	(6)	8[b]	(5)
	50	27[a]	(16)	27[a]	(11.5)	27[a]	(11)	24	(10)	18	(9)	14[b]	(8)	12[b]	(7.5)	10[b]	(6.5)
	25	27[a]	(17)	27[a]	(14)	27[a]	(12.5)	27[a]	(11.5)	22	(11)	18	(9.5)	14[b]	(8)	11[b]	(7)
	10	27[a]	(19)	27[a]	(16)	27[a]	(14)	27[a]	(12.5)	27	(12.5)	21	(11)	17[b]	(9)	14[b]	(7.5)
34	90	22	(12.5)	18	(9)	18	(8)	14	(7.5)	11	(7.5)	6[b]	(6.5)	6[b]	(6)	5[b]	(5)
	75	27[a]	(16)	26	(11.5)	25	(11)	21	(10)	16	(9)	10[b]	(8)	9[b]	(6.5)	8	(5.5)
	50	27[a]	(19)	27[a]	(14)	27[a]	(13)	27[a]	(11.5)	22	(11)	14[b]	(9.5)	12[b]	(8)	10[b]	(7)
	25	27[a]	(20[a])	27[a]	(17)	27[a]	(15)	27[a]	(14)	27	(12.5)	20[b]	(11)	14[b]	(10)	11[b]	(9)
	10	27[a]	(20[a])	27[a]	(19)	27[a]	(17)	27[a]	(16)	27[a]	(15)	21	(13)	17[b]	(11)	14[b]	(9)
Floor to 183-cm Height																	
75	90	15	(9)	12	(6)	12	(6)	9.5	(5)	8	(5)	6[b]	(4.5)	5	(4)	3	(3)
	75	21	(11)	18	(8)	17	(7)	14	(7)	11	(7)	9[b]	(6)	8	(5)	6	(4.5)
	50	27[a]	(12.5)	24	(10)	23	(9)	19	(8)	15	(7)	12	(7)	10	(6)	8	(5.5)
	25	27[a]	(15)	27[a]	(11)	27[a]	(10)	24	(10)	18	(9)	14	(8)	12	(7)	9	(6)
	10	27[a]	(17)	27[a]	(12.5)	27[a]	(12)	27[a]	(10)	22	(10)	18	(10)	15	(8)	12	(7)
49	90	17	(10)	15	(7)	14	(6)	11	(6)	9	(6)	6[b]	(5.5)	6[b]	(4.5)	4[b]	(3.5)
	75	24	(12)	21	(9)	21	(7)	16	(7)	12	(7)	9[b]	(6)	9[b]	(5)	7[b]	(4.5)
	50	27[a]	(14)	27[a]	(10)	27[a]	(10)	22	(9)	16	(8)	14[b]	(7)	12[b]	(7)	10[b]	(6)
	25	27[a]	(15)	27[a]	(12)	27[a]	(11)	27[a]	(10)	20	(10)	17	(8.5)	14[b]	(7)	11[b]	(6.5)
	10	27[a]	(17)	27[a]	(14)	27[a]	(12)	27[a]	(11)	23	(11)	20	(10)	17[b]	(8)	14[b]	(7)
34	90	20	(11)	16	(8)	16	(7)	13	(7)	9	(7)	6[b]	(6)	6[b]	(5)	4[b]	(4.5)
	75	27[a]	(14)	24	(10)	24	(10)	19	(9)	15	(8)	9[b]	(7)	9[b]	(6)	7[b]	(5)
	50	27[a]	(17)	27[a]	(12)	27[a]	(12)	26	(10)	19	(10)	14[b]	(8.5)	12[b]	(7)	10[b]	(6)
	25	27[a]	(20)	27[a]	(15)	27[a]	(13.5)	27[a]	(12)	23	(11)	20[b]	(10)	14[b]	(9)	11[b]	(8)
	10	27[a]	(20[a])	27[a]	(17)	27[a]	(15)	27[a]	(14)	27[a]	(13.5)	24	(12)	17[b]	(10)	14[b]	(8)

Table 1 ***(continued)***

Container Size (cm)		Frequency of Lift															
Width	Length	$\frac{1}{8}$ h		$\frac{1}{30}$ min		$\frac{1}{5}$ min		1/min		4/min		8/min		12/min		16/min	
		80- to 132-cm Height															
75	90	19	(13)	18	(11)	16	(10)	15	(9)	13	(8)	7[b]	(6)	6[b]	(6)	5[b]	(5)
	75	25	(15)	23	(13)	21	(12)	20	(11)	17	(9)	8[b]	(7)	8[b]	(7)	7[b]	(6)
	50	27[a]	(17)	27[a]	(15)	26	(14)	25	(13)	21	(11)	12[b]	(9)	11[b]	(9)	9[b]	(8)
	25	27[a]	(20)	27[a]	(17)	27[a]	(16)	27[a]	(14)	26	(12)	17[b]	(11)	13[b]	(10)	12[b]	(9)
	10	27[a]	(20[a])	27[a]	(19)	27[a]	(17)	27[a]	(16)	27[a]	(14)	23	(12.5)	20	(11)	16[b]	(9.5)
49	90	19	(13)	18	(11)	16	(10)	15	(9)	13	(8)	7[b]	(6[b])	6[b]	(6)	5[b]	(5)
	75	25	(15)	23	(13)	21	(12)	20	(11)	17	(9)	8[b]	(7[b])	8[b]	(7)	7[b]	(6)
	50	27[a]	(17)	27[a]	(15)	26	(14)	25	(13)	21	(11)	12[b]	(9[b])	11[b]	(9)	9[b]	(8)
	25	27[a]	(20)	27[a]	(17)	27[a]	(16)	27[a]	(14)	26	(12)	17[b]	(11)	13[b]	(10)	12[b]	(9)
	10	27[a]	(20[a])	27[a]	(19)	27[a]	(17)	27[a]	(16)	27[a]	(14)	23	(12.5)	20	(11)	16[b]	(9.5)
34	90	22	(14)	20	(12)	18	(11)	17	(10)	14	(9)	7[b]	(7[b])	6[b]	(6.5[b])	5[b]	(6.5[b])
	75	27[a]	(17)	26	(14)	23	(13)	22	(12)	18	(11)	8[b]	(8.5[b])	8[b]	(8.5[b])	7[b]	(8)
	50	27[a]	(19)	27[a]	(17)	27[a]	(15)	27[a]	(14)	23	(13)	12[b]	(11[b])	11[b]	(10)	9[b]	(8.5)
	25	27[a]	(20[a])	27[a]	(19)	27[a]	(17)	27[a]	(16)	27	(14)	17[b]	(13.5)	13[b]	(11.5[b])	12[b]	(11)
	10	27[a]	(20[a])	27[a]	(20[a])	27[a]	(19)	27[a]	(18)	27[a]	(16)	24	(14.5)	21	(13)	16[b]	(11.5)
		80- to 183-cm Height															
75	90	16	(11)	15	(9.5)	13	(9)	12	(8)	11	(7)	7[b]	(5[b])	6[b]	(5)	5[b]	(4.5)
	75	22	(13)	20	(11)	18	(10.5)	17	(9.5)	15	(8)	8[b]	(6[b])	8[b]	(6)	6	(5)
	50	27[a]	(15)	25	(13)	23	(12)	21	(11)	19	(10)	12[b]	(8[b])	11[b]	(8)	8	(7)
	25	27[a]	(17.5)	27[a]	(15)	27	(14)	26	(12)	23	(10.5)	17[b]	(10)	13[b]	(9)	11	(8)
	10	27[a]	(19)	27[a]	(17)	27[a]	(15)	27[a]	(14)	27	(12)	22	(11)	18	(10)	13	(8)
49	90	16	(11)	15	(9.5)	13	(9)	12	(8)	11	(7)	7[b]	(6[b])	6[b]	(5)	5[b]	(4.5)
	75	22	(13)	20	(11)	18	(10.5)	17	(9.5)	15	(8)	8[b]	(6[b])	8[b]	(6)	6	(5)
	50	27[a]	(15)	25	(13)	23	(12)	21	(11)	19	(10)	12[b]	(8[b])	11[b]	(8)	8	(7)
	25	27[a]	(17.5)	27[a]	(15)	27	(14)	26	(12)	23	(10.5)	17[b]	(10)	13[b]	(9)	11	(8)
	10	27[a]	(19)	27[a]	(17)	27[a]	(15)	27[a]	(14)	27	(12)	22	(11)	18	(10)	13	(8)
34	90	18	(12)	17	(10.5)	15	(10)	14	(9)	12	(8)	7[b]	(6[b])	6[b]	(6[b])	5[b]	(6[b])
	75	24	(15)	22	(12)	20	(11)	19	(10.5)	16	(10)	8[b]	(7.5[b])	8[b]	(7.5[b])	7	(7)
	50	27[a]	(17)	27[a]	(15)	25	(13)	24	(12)	20	(11)	12[b]	(10[b])	11[b]	(9)	9	(7.5)
	25	27[a]	(19)	27[a]	(17)	27[a]	(15)	27[a]	(14)	24	(12)	20	(11)	16	(10[b])	12	(10)
	10	27[a]	(20[a])	27[a]	(19)	27[a]	(17)	27[a]	(16)	27[a]	(14)	22	(13)	18	(11)	13	(10)
		132- to 183-cm Height															
75	90	15	(9)	14	(8)	12	(7)	12	(7)	9	(7)	7[b]	(5[b])	6	(4[b])	4	(3[b])
	75	20	(11)	18	(9)	15	(9)	15	(8)	12	(8)	9[b]	(6[b])	8	(5[b])	6	(4[b])
	50	25	(13)	23	(11)	20	(10)	19	(9)	16	(9)	12[b]	(8)	10	(7)	7	(6)
	25	27[a]	(14)	27	(12)	25	(11)	23	(10)	19	(10)	15[b]	(9)	12[b]	(8)	10	(7)
	10	27[a]	(16)	27[a]	(14)	27[a]	(13)	27	(12)	22	(11)	17[b]	(10)	13[b]	(9)	12	(8)
49	90	18	(10)	16	(9)	14	(8)	14	(7)	11	(7)	7[b]	(5[b])	7	(4[b])	5	(3[b])
	75	23	(12)	21	(10)	19	(9)	18	(9)	14	(8)	9[b]	(6[b])	8[b]	(5[b])	6	(4[b])
	50	27[a]	(14)	27	(12)	24	(11)	23	(10)	18	(9)	12[b]	(8)	10[b]	(7)	9	(6)
	25	27[a]	(15)	27[a]	(13)	27[a]	(12)	27[a]	(11)	21	(10)	15[b]	(9)	12[b]	(8)	10	(7)
	10	27[a]	(17)	27[a]	(15)	27[a]	(14)	27[a]	(13)	25	(11)	17[b]	(10)	13[b]	(9)	11	(8)
34	90	20	(12)	18	(11)	17	(10)	16	(9)	13	(8)	7[b]	(6[b])	6[b]	(6[b])	5	(6[b])
	75	26	(14)	24	(12)	22	(11)	21	(11)	17	(9)	9[b]	(7[b])	8[b]	(7[b])	8	(7[b])
	50	27[a]	(17)	27[a]	(14)	27[a]	(13)	26	(12)	21	(11)	12[b]	(9[b])	11[b]	(9)	10	(8)
	25	27[a]	(19)	27[a]	(16)	27[a]	(15)	27[a]	(14)	25	(12)	15[b]	(11)	14[b]	(10)	13	(9)
	10	27[a]	(20[a])	27[a]	(18)	27[a]	(16)	27[a]	(15)	27[a]	(14)	17[b]	(12[b])	16[b]	(11)	15	(9.5)

Source: Adapted from Mital et al. (1993).
[a]Weight limited by biomechanical design criterion (3930 N spinal compression for males, 2689 N for females).
[b]Weight limited by physiological design criterion (4 kcal/min for males, 3 kcal/min for females).

Table 2 Terms of the 1991 NIOSH Equation

Multiplier	Formula[a]
Load constant	LC = 23 kg
Horizontal	$HM = 25/H$
Vertical	$VM = 1 - (0.003\|V - 75\|)$
Distance	$DM = 0.82 + 4.5/D$
Asymmetry	$AM = 1 - 0.0032A$
Frequency	FM (see Table 3)
Coupling	CM (see Table 4)

Source: After Waters et al. (1993).

[a] *H*, horizontal distance of the hands from the midpoint of the ankles, measured at the origin and destination of the lift (cm); *V*, vertical distance of the hands from the floor, measured at the origin and destination of the lift (cm); *D*, vertical travel distance between the origin and destination of the lift (cm); *A*, angle of asymmetry (angular displacement of the load from the sagittal plane), measured at the origin and destination of the lift (degrees); *F*, average frequency of lift (lifts/minute), *C*, load coupling, the degree that appropriate handles, devices, or lifting surfaces are present to assist lifting and reduce the possibility of dropping the load.

Table 3 Frequency Multipliers for the 1991 Lifting Equation

	Continuous Work Duration					
	<8 h		<2 h		<1 h	
Frequency Lifts (min)	$V < 75$	$V > 75$	$V < 75$	$V > 75$	$V < 75$	$V > 75$
0.2	0.85	0.85	0.95	0.95	1.00	1.00
0.5	0.81	0.81	0.92	0.92	0.97	0.97
1	0.75	0.75	0.88	0.88	0.94	0.94
2	0.65	0.65	0.84	0.84	0.91	0.91
3	0.55	0.55	0.79	0.79	0.88	0.88
4	0.45	0.45	0.72	0.72	0.84	0.84
5	0.35	0.35	0.60	0.60	0.80	0.80
6	0.27	0.27	0.50	0.50	0.75	0.75
7	0.22	0.22	0.42	0.42	0.70	0.70
8	0.18	0.18	0.35	0.35	0.60	0.60
9	0	0.15	0.30	0.30	0.52	0.52
10	0	0.13	0.26	0.26	0.45	0.45
11	0	0	0	0.23	0.41	0.41
12	0	0	0	0.21	0.37	0.37
13	0	0	0	0	0	0.34
14	0	0	0	0	0	0.31
15	0	0	0	0	0	0.28

Source: After Waters et al. (1993).

Table 4 Coupling Multipliers for the 1991 Lifting Equation

Couplings	$V < 75$ cm	$V \geq 75$ cm
Good	1.00	1.00
Fair	0.95	1.00
Poor	0.90	0.90

Source: After Waters et al. (1993).

Table 5 Coupling Classification

- Good coupling
  1. For containers of optimal design, such as some boxes and crates, a "good" hand-to-object coupling would be defined as handles or handhold cutouts of optimal design.
  2. For loose parts or irregular objects, which are not usually containerized, such as castings, stock, and supply materials, a "good" hand-to-object coupling would be defined as a comfortable grip in which the hand can easily be wrapped around the object.
- Fair coupling
  1. For containers of optimal design, a "fair" hand-to-object coupling would be defined as handles or handhold cutouts of less than optimal design.
  2. For containers of optimal design with no handles or handhold cutouts or for loose parts or irregular objects, a "fair" hand-to-object coupling is defined as a grip in which the hand can be flexed about 90°.
- Poor coupling
  1. Containers of less than optimal design or loose parts or irregular objects that are bulky, hard to handle, or have sharp edges.
  2. Lifting nonrigid bags (i.e., bags that sag in the middle).

Source: After Waters et al. (1994).

Notes:

1. An optimal handle design is 0.75 to 1.5 inches (1.9 to 3.8 cm) in diameter, ≥4.5 inches (11.5 cm) in length, has 2 inches (5 cm) of clearance, is cylindrical in shape, and has a smooth, nonslip surface.
2. An optimal handhold cutout has the following approximate characteristics: ≥1.5 inches (3.8 cm) in height, 4.5 inches (11.5 cm) in length, a semioval shape, ≥ has 2 inches (5 cm) of clearance, a smooth nonslip surface, and a container thickness of ≥0.25 inch (0.60 cm) (e.g., double-thickness cardboard).
3. An optimal container design is ≤16 inches (40 cm) in frontal length, ≤12 inches (30 cm) in height, and has a smooth, nonslip surface.
4. A worker should be capable of clamping the fingers at nearly 90° under the container, such as is required when lifting a cardboard box from the floor.
5. A container is considered less than optimal if it has a frontal length >16 inches (40 cm), a height >12 inches (30 cm), rough or slippery surfaces, sharp edges, an asymmetric center of mass, unstable contents, or requires the use of gloves. A loose object is considered bulky if the load cannot be balanced easily between the hand grasps.
6. A worker should be able to wrap the hand comfortably around the object without causing excessive wrist deviations or awkward postures, and the grip should not require excessive force.

The vertical travel distance variable (D) is defined as the vertical travel distance of the hands between the origin and destination of the lift. For lifting tasks, D

can be computed by subtracting the vertical location (V) at the origin of the lift from the corresponding V at the destination of the lift. For lowering tasks, D is equal to V at the origin minus V at the destination. The variable (D) is assumed to be at least 10 inches (25 cm), and no greater than 70 inches (175 cm). If the vertical travel distance is less than 10 inches (25 cm), D should be set to 10 inches (25 cm).

The asymmetry angle A is limited to the range 0 to 135°. If $A > 135°$, AM is set equal to zero, which results in a RWL of 0. The asymmetry multiplier (AM) is $1 - 0.0032A$. The AM has a maximum value of 1.0 when the load is lifted directly in front of the body and a minimum value of 0.57 at 135° of asymmetry.

The frequency multiplier (FM) is defined by (1) the number of lifts per minute (frequency), (2) the amount of time engaged in the lifting activity (duration), and (3) the vertical height of the lift from the floor. Lifting frequency (F) refers to the average number of lifts made per minute as measured over a 15-minute period. Lifting duration is classified into three categories: short, moderate, and long duration. These categories are based on the pattern of continuous work- and recovery-time (i.e., light work) periods.

A continuous work-time period is defined as a period of uninterrupted work. Recovery time is defined as the duration of light work activity following a period of continuous lifting. Short duration defines lifting tasks that have a work duration of one hour or less, followed by a recovery time equal to 1.2 times the work time. Moderate duration defines lifting tasks that have a duration of more than one hour but not more than two hours, followed by a recovery period of at least 0.3 times the work time. Long duration defines lifting tasks that have a duration between two and eight hours, with standard industrial rest allowances (e.g., morning, lunch, and afternoon rest breaks).

The lifting index (LI) provides a relative estimate of the physical stress associated with a manual lifting job and is equal to the load weight divided by the RWL. According to Waters et al. (1994), the RWL and LI can be used to guide ergonomic design in several ways:

1. The individual multipliers can be used to identify specific job-related problems. The general redesign guidelines related to specific multipliers are shown in Table 6.
2. The RWL can be used to guide the redesign of existing manual lifting jobs or to design new manual lifting jobs.
3. The LI can be used to estimate the relative magnitude of physical stress for a task or job. The greater the LI, the smaller the fraction of workers capable of sustaining the level of activity safely.
4. The LI can be used to prioritize ergonomic redesign. A series of suspected hazardous jobs could be rank ordered according to the LI, and a control strategy could be developed according to the rank ordering (i.e., jobs with lifting indices about 1.0 or higher would benefit the most from redesign).

Table 6 General Design/Redesign Suggestions for Manual Lifting Tasks[a]

If HM is less than 1.0	Bring the load closer to the worker by removing any horizontal barriers or reducing the size of the object. Lifts near the floor should be avoided; if unavoidable, the object should fit easily between the legs.
If VM is less than 1.0	Raise/lower the origin/destination of the lift. Avoid lifting near the floor or above the shoulders.
If DM is less than 1.0	Reduce the vertical distance between the origin and the destination of the lift.
If AM is less than 1.0	Move the origin and destination of the lift closer together to reduce the angle of twist, or move the origin and destination farther apart to force the worker to turn the feet and stp, rather than twist, the body.
If FM is less than 1.0	Reduce the lifting frequency rate, reduce the lifting duration, or provide longer recovery periods (i.e., light work period).
If CM is less than 1.0	Improve the hand-to-object coupling by providing optimal containers with handles or handhold cutouts, or improve the handholds for irregular objects.
If the RWL at the destination is less than at the origin	Eliminate the need for significant control of the object at the destination by redesigning the job or modifying the container/object characteristics.

[a]As recommended by Waters et al. (1994).

Finally, it should be noted that the RNLE of the 1991 equation should not be used if any of the following conditions occur:

- Lifting/lowering with one hand
- Lifting/lowering for over eight hours
- Lifting/lowering while seated or kneeling
- Lifting/lowering in a restricted work space
- Lifting/lowering unstable objects
- Lifting/lowering while carrying, pushing or pulling
- Lifting/lowering with wheelbarrows or shovels

- Lifting/lowering with high speed motion (faster than about 30 inches/second)
- Lifting/lowering with unreasonable foot/floor coupling (<0.4 coefficient of friction between the sole and the floor)
- Lifting/lowering in an unfavorable environment (i.e., temperature significantly outside 66 to 79°F (19 to 26°C) range; relative humidity outside the range 35 to 50%

3.3.1 Computer Simulation

As discussed by Karwowski and Rodrick (2001), practical implications of the 1991 revised lifting equation for industry can be determined when applying a realistic range of values for the risk factors (Karwowski, 1992). Karwowski and Gaddie (1995) simulated the RNLE of 1991 under a broad range of conditions using SLAM II (Pritsker, 1986), a *simulation language for alternative modeling*, as the product of the six independent factor multipliers represented as attributes of an entity flowing through the network. For this purpose, probability distributions for all the relevant risk factors were chosen such that they were representative of the real industrial workplaces (Ciriello et al., 1990; Brokaw, 1992; Karwowski and Brokaw, 1992; Marras et al., 1993). Except for the vertical travel distance factor, coupling and asymmetry multipliers, all factors were defined using either normal or lognormal distributions. For all the factors defined as having lognormal distributions, the procedure was developed to adjust for the required range of real values whenever necessary. The SLAM II computer simulation was run for a total of 100,000 trials (i.e., randomly selected scenarios which realistically define the industrial tasks). Descriptive statistical data were collected for all the input (lifting) factors, the respective multipliers, and the resulting recommended weight limits. The input factor distributions were examined to verify the intended distributions.

For all lifting conditions examined, the distribution of recommended weight limit values had a mean of 7.22 kg and a standard deviation of 2.09 kg. In 95% of all cases, the RWL was at or below the value of 10.5 kg, or about 23.1 lb. In 99.5% of all cases the RWL value was at or below 12.5 kg (27.5 lb). This implies that when the LI is set to 1.0 for task design or evaluation purposes, only 0.5% of the (simulated) industrial lifting tasks would have RWL values greater than 12.5 kg. Taking into account the lifting task duration, in 99.5% of the simulated cases, the RWL values were equal to or lower than 13.0 kg (28.6 lb) for up to one hour of lifting task exposure, 12.5 kg (27.5 lb) for less than two hours of exposure, and 10.5 kg (23.1 lb) for lifting over an eight-hour shift.

From a practical point of view, these values define simple and straightforward lifting limits [i.e., the threshold RWL values (TRWL) that can be used by practitioners for the purpose of immediate and easy-to-perform risk assessment of manual lifting tasks performed in industry]. For example, if the TRWL value of 27.5Ylb is exceeded, a more thorough examination of the tasks identified and evaluation of the physical capacity of the exposed workers should be performed. Since the RWL values do not exceed the acceptable lifting capabilities of 99% of male workers and 75% of female workers, this means protecting about 90% of the industrial workers if there is a 50:50 split between males and females.

3.3.2 Validation

In recent years, several studies were conducted to validate the revised NIOSH lifting equation. Potvin and Bent (1997) conducted a study to determine the 1991 NIOSH equation horizontal distances associated with the three different box widths and lift starting heights. The study found that horizontal distance was positively related to box width and that there was a significant interaction effect of box width and starting height on horizontal distances. Waters et al. (1998) studied the accuracy of measurement of the revised NIOSH equation. The study used 27 nonergonomists who evaluated a simulated lifting task eight weeks after a one-day training session. It was found that small interobserver variability exists, especially for horizontal distance. However, measurements of coupling and asymmetric variables were found to be least accurate.

In an epidemiological study, Wang et al. (1998) evaluated the relationship between low-back discomfort ratings and use of the revised NIOSH lifting guide to assess the risk of manual materials handling tasks. The study found that 42 of the 97 jobs analyzed had a recommended weight limit of zero, which was attributed to either a horizontal distance or a lifting frequency that exceeded the bounds of the NIOSH lifting index. Apparently, it was concluded that the limits for horizontal distance and maximum allowable frequency were too stringent to accommodate many existing MMH jobs. For the remaining 55 jobs, the significant positive correlation obtained between the lifting index and the severity of low back discomfort suggests that the lifting index is reliable in assessing the potential risk of low back injury in MMH.

Waters et al. (1999) conducted a cross-sectional study to determine the correlation between the prevalence of low back pain and exposure to manual lifting stressors measured with the lifting index. A logistic regression analysis showed that as the lifting index increased from 1.0 to 3.0, the odds of low back pain increased, with a peak and statistically significant odds ratio occurring in the $2 < \text{lifting index} < 3$ category (odds ratio = 2.45). However, for jobs with a lifting index higher than 3.0, the odds ratio was lower (odds ratio = 1.45).

Dempsey et al. (2001) conducted a laboratory experiment to investigate the accuracy of NIOSH equation parameter measurements made by eight subjects following a four-hour training session. The study measured five individual tasks: two single tasks and three parts of a multiple-component simulated palletizing operation. Significant differences between reference parameter measurements and average measurements made by subjects were found. The sensitivity analysis showed frequency and horizontal location as the most important parameters. These parameters also tended to have the highest measurement errors.

In a cross-cultural study, Yeung et al. (2001) documented how workers from different cultures and with different physical characteristics evaluate physical effort in industrial lifting activities. The results showed that personal knowledge and expertise can be used successfully to evaluate the physical effort required for the performance of various lifting tasks. More specifically, there were no differences in cognitive reasoning patterns in evaluating the physical effort between workers in Hong Kong and the United States. The participants from the two different cultures both rated the weight of load as the most important variable in evaluating lifting activities.

In a recent usability study, Dempsey (2002) found that qualitative results from training sessions indicated that frequency, asymmetry, and duration were the parameters that required relatively longer instruction periods and resulted in the most questions. The variable nature of lifting/lowering demands found in many jobs resulted in difficulty applying the NIOSH lifting equation. Approximately 35% of 1103 lifting and lowering tasks had at least one parameter outside acceptable ranges, while a majority of workers (62.8%) reported other manual handling tasks that contradict assumptions made in development of the equation.

3.3.3 Beyond the NIOSH Lifting Equation

Hidalgo et al. (1997) developed a comprehensive lifting model (CLM) for the evaluation and design of manual tasks extending the 1991 NIOSH lifting equation. The model utilizes two stages with 11 task-related, personal, and environmental variables. In addition to revised NIOSH equation variables, this model incorporated various percentages of working population, gender, wide range of duration of lifting and lifting frequency, heat stress effects, and physiological and biomechanical tolerance limits. The model was developed as a data-driven model instead of including any human judgment. In the first stage, the model was built using the psychophysical data. In the second stage, discounting factors of various variables were tested and adjusted using the physiological and biomechanical data. Two lifting indices are proposed to evaluate lifting tasks for a group of workers [relative lifting safety index (RLSI)] and for an individual worker [personal lifting safety index (PLSI)].

Shoaf et al. (1997) developed a set of mathematical models for manual lowering, pushing, pulling, and carrying activities that would result in establishing load capacity limits to protect the lower back against occupational LBDs. The study proposed separate mathematical modeling of lowering, pushing, pulling, and carrying. To establish safe guidelines, psychophysical, biomechanical, and physiological data were used for each mathematical modeling, and both task and personal factors were considered in the models' development. The study found load capacity values lower than limits established previously for the same activities. The study also validated the hypothesis of Karwowski (1983) and Karwowski and Ayoub (1984) that the combination of physiological and biomechanical stresses leads to the overall measure of task acceptability in terms of the psychophysical responses observed.

4 PREVENTION AND REDUCTION OF LOWER BACK DISORDERS

The application of ergonomic principles to the design of MMH tasks is one of the most effective approaches to controlling the incidence and severity of LBDs. The goal of ergonomic job design is to reduce the ratio of task demands to worker capability to an acceptable level. The application of ergonomic principles to task and workplace design reduces stresses permanently. Such changes are preferable to altering other aspects of the MMH system, such as work practices. For example, worker training may be ineffective if practices trained are not reinforced and refreshed (Kroemer, 1992), whereas altering the workplace is a lasting physical intervention.

In cases where elimination of the need for MMH is physically or economically unfeasible, stresses must be reduced by decreasing job demands and minimizing stressful body movements by altering the task, workplace, and/or the material being handled. Figure 11 provides a sample of techniques to achieve these goals. In addition to the engineering controls, there are also techniques for job (re)design based on the relationships between job stress and injury occurrence. Each of the techniques can be used to prioritize job interventions in terms of injury risk associated with different jobs.

4.1 Job Severity Index

As discussed by Ayoub et al. (1997), the *job severity index* (JSI) is a time- and frequency-weighted ratio of worker capacity to job demands. For this purpose, the worker capacity is predicted with the models developed by Ayoub et al. (1978), which used isometric strength and anthropometric data to predict psychophysical lifting capacity. JSI can be calculated by the formula

$$\text{JSI} = \sum_{i=1}^{n} \frac{\text{hours}_i \times \text{days}_i}{\text{hours}_t \times \text{days}_t} \sum_{j=1}^{m_i} \left(\frac{F_j}{F_i} \times \frac{\text{WT}_J}{\text{CAP}_j} \right)$$

where

n = number of task groups
hours_i = exposure hours/day for group i
days_i = exposure days/week for group i
hours_t = total hours/day for job
days_t = total days/week for job
m_i = number of tasks in group i
WT_j = maximum required weight of lift for task j
CAP_j = adjusted capacity of the person working at task j
F_j = lifting frequency for task j
F_i = total lifting frequency for group i

Thus,

$$F_i = \sum_{j=1}^{m_i} F_j$$

The job severity index (JSI) can be determined using the following steps:

1. Depending on the task parameters and other relevant data, a regression model is developed. Table 7 provides the estimated regression coefficients used to predict individual worker capacity (CAPj). The equations provide predictions of MAWL plus body weight based on the strength and anthropometric variables given in the table. Body weight is subtracted from the prediction to obtain a person's predicted MAWL.
2. Once MAWL is predicted, the capacity is adjusted for frequency using the appropriate equation in Table 8 with respect to task parameters and gender.
3. Finally, the frequency-adjusted MAWL is adjusted for the dimension of the box in the sagittal plane using the appropriate equation in Table 9 with respect to task parameters and gender.

In the JSI formula, the denominator for the incidence and severity rates is 100 full-time employees, which is 200,000 exposure hours. Therefore, JSI can be reduced to a desirable level by increasing worker capacity (e.g., selecting a worker with higher capacity) or altering task and job parameters to reduce JSI to an acceptable level. Liles et al. (1984) performed a field study to determine the relationship between JSI and the incidence and severity of LBDs. A total of 453 subjects were included in the study. The results of the field study indicated that both incidence and severity of recordable back injuries rose rapidly at values of JSI greater than 1.5.

Ayoub et al. (1997) discussed the following example, which illustrates calculations required for the determination of JSI based on the task parameters in Table 10. The job involves one task group with two lifting tasks. The first task is a floor-to-knuckle height (F-K) lift, and the second task is a floor-to-shoulder height (F-S) lift. The first step involves determining the capacity of the person performing the job. The job is performed by a 30-year-old female weighing 59 kg with the following strength and anthropometric

Table 7 Estimated Regression Coefficients Used to Predict MAWL Plus Body Weight (kg) for Males and Females

Lifting Range[a]	Constant Term	Gender Code Coefficient[b]	Weight Code Coefficient[c]	Arm Strength (kg) Coefficient	Age Coefficient	Shoulder Height (cm) Coefficient	Back Strength (kg) Coefficient	Abdominal Depth (cm) Coefficient	Dynamic Endurance (min) Coefficient
F-K	−32.733	−12.852	10.996	0.143	−0.251	0.556	0.056	2.229	0.797
F-S	−65.958	−7.332	5.410	0.185	−0.271	0.652	0.077	2.936	1.183
F-R	−18.718	−8.824	7.337	0.210	−0.382	0.344	0.068	2.821	0.647
K-S	−25.020	−8.370	5.307	0.265	−0.275	0.348	0.105	2.853	0.642
K-R	−35.921	−8.581	7.835	0.297	−0.226	0.495	0.018	2.338	0.962
S-R	−16.982	−8.883	9.232	0.096	−0.269	0.402	0.099	2.146	0.494

Source: Adapted from Ayoub et al. (1978).
[a]F, floor; K, knuckle height; S, shoulder height; R, reach height.
[b]0 for males, 1 for females.
[c]0 if body weight is ≤ median weight, 1 if body weight is > median weight, where the median weight is 61.2 kg for females and 77.1 kg for males.

Table 8 Equations to Adjust MAWL Predictions for Frequency[a]

Range of Lift	Frequency of Lift (FY) (lifts/min) $0.1 < FY < 1.0$	Equation	$1.0 \leq FY \leq 12.0$	Equation
		Male Capacity		
F-K, F-S, F-R	$CAP \times FY^{-0.184697}$	(1)	$CAP - 0.91(FY - 1)$	(3)
K-S, K-R, S-R	$CAP \times FY^{-0.138650}$	(2)	Use equation (3)	
		Female Capacity		
F-K, F-S, F-R	$CAP \times FY^{-0.187818}$	(4)	$CAP - 0.5(FY - 1)$	(6)
K-S, K-R, S-R	$CAP \times FY^{-0.156150}$	(5)	Use equation (6)	

Source: Adapted from Ayoub et al. (1983).
[a]CAP, capacity as determined from Table 7.

Table 9 Equations to Correct Frequency-Adjusted MAWL Predictions for Box Size[a]

	Box Size (BX) in Sagittal Plane (cm)			
Range of Lift	$31 \leq BX \leq 46$	Equation	$BX \geq 46$	Equation
Male Capacity				
F-K, F-S, F-R	$CAP_{FA} + 0.30(46 - BX)$	(1)	$CAP_{FA} + 0.14(46 - BX)$	(3)
K-S, K-R, S-R	$CAP_{FA} + 0.20(46 - BX)$	(2)	Use equation (3)	
Female Capacity				
F-K, F-S, F-R	$CAP_{FA} + 0.20(46 - BX)$	(4)	$CAP_{FA} + 0.07(46 - BX)$	(6)
K-S, K-R, S-R	$CAP_{FA} + 0.10(46 - BX)$	(5)	$CAP_{FA} + 0.04(46 - BX)$	(7)

Source: Adapted from Ayoub et al. (1983).
[a]CAP_{FA}, frequency-adjusted capacity as determined from Table 8.

Table 10 Task Parameters for JSI Example

Parameter	Task 1	Task 2
$hours_i$	8	8
$days_i$	5	5
$hours_t$	8	8
$days_t$	5	5
WT_j (kg)	20	15
Box size[a] (cm)	40	35
F_j	5	2

[a]Dimension in sagittal plane.

measurements: arm strength 20 kg, shoulder height 132 cm, back strength 51 kg, abdominal depth 20 cm, and dynamic endurance 2.62 minutes. Using the coefficients in Table 7, the person's MAWL plus body weight (MAWLBW) prediction for the F-K task is

$$\begin{aligned}\text{MAWLBW} &= -32.733 - 12.852(1) + 10.996(0)\\ &\quad + 0.143(20) - 0.251(30) + 0.556(132)\\ &\quad + 0.056(51) + 2.229(20) + 0.797(2.62)\end{aligned}$$

MAWLBW = 72.66 kg; thus, the subject's predicted MAWL for the F-K lift is 13.66 kg. The next step involves correcting the MAWL value for frequency using equation (6) in Table 8. This correction reduces the MAWL to 11.66 kg. Finally, the frequency-adjusted MAWL is corrected for box size using equation (4) from Table 9, resulting in a final value of 12.86 kg. The person's MAWL plus body weight (MAWLBW) prediction for the F-S task is

$$\begin{aligned}\text{MAWLBW} &= -65.958 - 7.332(1) + 5.410(0)\\ &\quad + 0.185(20) - 0.271(30) + 0.652(132)\\ &\quad + 0.077(51) + 2.936(20) + 1.183(2.62)\end{aligned}$$

MAWLBW = 74.1 kg; thus, the subject's predicted MAWL for the F-K lift is 15.1 kg. The frequency-adjusted MAWL value, using equation (6) in Table 8, is 14.6 kg. The frequency-adjusted MAWL is corrected for box size using equation (4) from Table 9, resulting in a final value of 16.8 kg. Once all capacities have been estimated, the JSI value for the job can be calculated. Using the equation presented earlier, the JSI for the job is

$$\begin{aligned}\text{JSI} &= \left(\frac{8}{8}\right)\left(\frac{5}{5}\right)\left[\left(\frac{5}{7}\right)\left(\frac{20.0}{12.86}\right) + \left(\frac{2}{7}\right)\left(\frac{15.0}{16.8}\right)\right]\\ &= 1.37\end{aligned}$$

The JSI value of 1.37, even though it is below 1.5, indicates that the job is a candidate for redesign. Since the JSI value for task 1 is 1.56, this task should be redesigned. Potential solutions outlined by Ayoub et al. (1997) include eliminating the need to handle the material manually, providing the material to the operator at knuckle height, reducing the frequency of handling, reducing the weight of the load, and as a last resort, selecting an operator with a higher capacity. In a recent study by Tharmmaphornphilas and Norman (2004), JSI was utilized as a tool for job rotation to reduce worker fatigue and injury. This study utilized an integer programming method to minimize the maximum JSI values.

As discussed by Karwowski and Rodrick (2001), Chaffin and Park (1973) investigated the efficacy of using the lifting strength rating (LSR) to predict the risk of LBP. LSR was defined as the ratio of maximum load lifted on a job to the maximum isometric strength of a "large/strong" man in a position similar to the task. The follow-up of 411 people lasted one year, which indicated that an LSR value greater than 0.2 should be considered potentially harmful. It was also found that jobs with an LSR value between 0.9 and 1.0 had an overall incidence rate of LBP approximately two times higher than jobs with LSR values between 0.2 and 0.8, and approximately four times higher than jobs with LSR values below 0.2. Chaffin (1974) reported the relationship between the ratio of the weight of load lifted for the task with the highest LSR value to the mean isometric strengths of employees. Thus, instead of using the isometric strength of a large/strong man, actual strength values of employees were used. Incident rates were approximately three times greater when the stated ratio exceeded 1.0 than when it did not. Chaffin et al. (1978) also used the job strength rating

(JSR), the ratio of the maximum strength requirements of the job to the average task-specific isometric strengths of all workers in a job. The relationship between JSR and the incidence rate of low back disorders appeared to be fairly linear. The relationship between JSR and severity rate was not linear, but values greater than 1 resulted in considerably higher severity rates than values less than 1. Like JSI, JSR can be used either to select employees for stressful tasks or to prioritize redesign requirements.

4.2 Task or Workplace Redesign

According to Ayoub et al. (1997), the application of ergonomic principles to the design of MMH tasks is one of the most effective approaches to controlling the incidence and severity of LBDs. The primary goal of ergonomic task design is to reduce the amount of task demands to an acceptable level with respect to worker capability. In addition to that, the application of ergonomic principles to task and workplace design reduces stresses permanently.

Ayoub et al. (1997) proposed a generic ergonomic model to task (re)design, which is summarized in Figure 8. According to this model, the optimal solution is to eliminate the need to handle materials manually. This can be achieved in either of two ways: by implementing mechanical aids or by redesigning the work area layout so that all materials are located on the same level. The model also suggests that in cases where elimination of the need for MMH is physically or economically unfeasible, stresses must be reduced by decreasing job demands and minimizing stressful body movements by altering the task, workplace, and/or the material being handled.

As pointed out by Snook (2004), workplaces must be designed for people with low back pain as well as

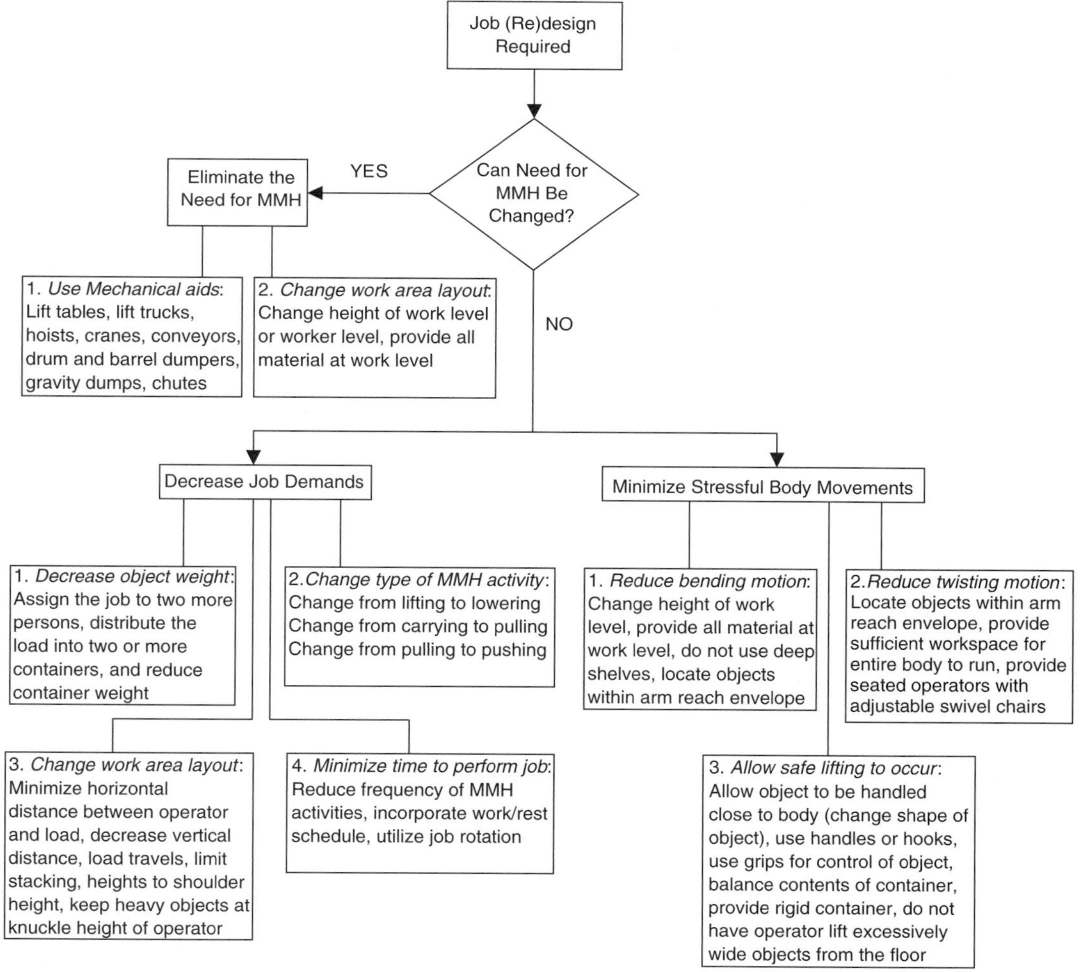

Figure 8 Generic MMH task re(design) model. (From Ayoub et al.,1997.)

for people without low back pain. Workers with low back pain have difficulty bending forward and difficulty handling objects when excessive forward reaching is involved. The basic ergonomic principle is to get things up off the floor. Objects can often be handled with little stress if excessive bending and forward reaching are avoided. Manual handling of tasks should also be designed according to accepted guidelines for maximum weights and forces (Waters et al., 1993). In a randomized clinical trial of 130 workers who had been absent from work for six weeks because of back pain, Loisell et al. (1997) investigated the effects of an occupational intervention that included ergonomic redesign and time-limited light duties (the Sherbrooke model). The study found that the median time of regular work was 67 days for workers with occupational intervention, compared to 131 days for workers without occupational intervention.

Lemstra and Olszynski (2003) conducted a prospective study on a company that changed from standard care to an occupational management approach. The new approach included worker rotation schedules, reduced lifting loads, and ergonomic redesign of tasks. Following the intervention, the study found that the total days lost per 100,000 hours worked dropped from 60.9 days for standard care to 1.1 days for the occupational management approach. Another large prospective study by Marras et al. (2000) demonstrated that the introduction of lift tables and lift aids such as overhead pulley systems and vacuum hoists reduced the incidence rate of reported LBDs significantly in repetitive manual handling jobs.

4.3 Proactive Return-to-Work Program

Previous studies indicate that there is strong evidence that the longer a worker is off work with low back pain, the lower their chances of ever returning to work (e.g., Hashemi et al., 1997; Waddell and Burton, 2000). Carter and Birrell (2000) noted "once a worker is off work for 4–12 weeks, they have a 10–40% risk (depending on the setting) of still being off work at one year; after 1–2 years absence it is unlikely they will return to any form of work in the foreseeable future, irrespective of further treatment." The preceding statistics emphasize the importance of providing modified, alternative, or part-time work as a means of returning the disabled employee to the job as quickly as possible (Snook, 2004). This statement is supported by the findings of Krause et al. (1998), Carter and Birrell (2000), and Waddell and Burton (2000).

4.4 Communication

Carter and Birrell (2000) and Waddell and Burton (2000) also pointed out that there is moderate evidence that communication, cooperation, and commonly agreed upon goals between the worker with low back pain, the occupational health team, supervisors, management, and primary health care professionals is fundamental for improvement in clinical and occupational health management and outcomes.

4.5 Selection of Health Care Providers

As proposed by Snook (2004), management should select health care providers who diagnose and treat low back pain according to accepted guidelines. In a recent Australian study, McQuirk et al. (2001) demonstrated the effectiveness of evidence-based treatment guidelines. In this study, 13 special urban and rural clinics were established with trained medical practitioners who treated LBP patients according to evidence-based guidelines. The study found a significantly greater reduction in pain, significantly fewer patients requiring continuing care, and a significantly greater proportion of patients fully recovered at 12 months for patients treated according to the guidelines.

4.6 Management Commitment

Carter and Birrell (2000) and Waddell and Burton (2000) reported that there is a general consensus, but only limited scientific evidence, that workplace organizational and/or management strategies may reduce absenteeism and duration of work loss. These authors argued that high job satisfaction and good industrial relations are the most important organizational characteristics associated with low disability and sickness absence rates attributed to low back pain.

4.7 Supervisor Training

Snook (2004) argued that in general, supervisors do not respond well when workers experience low back pain. Previous studies conducted by Fitzler and Berger (1982, 1983) consisted of conservative, supportive, in-house treatment of low back pain while keeping the employee on the job. In those studies, a significant part of the program was the training given to management, supervisors, and workers on the true nature of low back pain and the positive acceptance of low back pain when it occurs. The studies reported that over a three-year period, the number of lost-time LBP cases was reduced by 32%, and the cost of LBP compensation claims was reduced by over 90%. A disability management program for supervisors was reported by McLellan et al. (2001). As part of the program, the supervisors were given a 1.5-hour training session to reinforce a proactive and supportive response to work-related musculoskeletal symptoms and injuries among employees. Results showed that 38.5% of supervisors reported decreases in lost time within their department.

4.8 Coordination with Organized Labor

It is believed traditionally that labor unions are opposed to a worker's early return to work after a disabling episode of low back pain. Various studies (e.g., Spitzer, 1993; Malmivaara et al., 1995) reported that an employee is entitled to time off for even a minor episode, despite sound medical evidence that activity and work will hasten healing. Snook (2004) proposes that organized labor must be involved in the planning and execution of a disability reduction program. There must be agreement between labor and management on what constitutes the best interests of employees.

5 MMH RESEARCH NEEDS

As advocated by Dempsey (1998), the most critical need in MMH research, common to all three approaches, is in the area of epidemiological studies that provide evaluation and verification of the various limits-setting criteria. The outlines of the research needs in MMH are presented in Table 11. These research directions are based on the limitations of the three criteria used for MMH tasks design. It is believed that these studies will serve to advance the knowledge of the etiology of LBD as well as providing information on valid interventions to reduce the incidence and severity of MMH-related injuries. Dempsey and Westfall (1997) advocate the development of explicit risk models (ERMs). ERMs are models that provide a prediction of injury risk given a vector of worker, task, and workplace variables. These models may clarify the etiology of LBD as well as providing a design tool to practitioners attempting to prioritize ergonomic interventions. The benefit of such models cannot be understated.

After examining the reliability and validity of estimation methods and practicality for field measurement, Waters et al. (in press), proposed future research directions for cumulative spinal loading as exposure to lower back injury due to MMH tasks as follows:

1. The relationship between questionnaire- and video-based methods should be studied to determine the appropriate correction factors to assist in the reporting of valid integrated spinal loading measures.
2. Spinal loading estimation methods should be developed for individual- and population-based studies. With respect to biomechanical modeling, Waters et al. (in press) urged for novel methods for the integrated force time and speed of lift in individual-based studies. New methods should be established on the basis of total mechanical work required during a work shift for population-based studies.
3. Different methods of integration should be validated with respect to lower back outcomes. To attain this, investigations should be carried out on existing algorithms as well as to propose new ones.
4. Different types of spinal parameters should be examined to determine the most sensitive measure in detecting lower back outcomes.
5. A dose limit should be established with respect to lower back outcomes. In this regard, limits should be defined as a function of the degree of repetitiveness of the task during a work shift.

It can be concluded that prospective studies should be used to reveal the contribution of psychosocial factors to the occurrence of low back pain due to MMH tasks. From the discussion of the published literature presented above, it is quite evident that many studies that considered both work-related and psychosocial factors could not distinguish the true contributions of each of these factors. Furthermore, many of those studies were retrospective cross-sectional in nature. Thus, they could have been contaminated by the recall bias of the respondents. One advantage of the prospective epidemiological study (e.g., a cohort study that documents the details of the exposures before the disorder/injury/pain is reported) would be that the

Table 11 Summary of Primary MMH Research Needs with Respect to Design Criteria

Biomechanical Criteria	Epidemiological Criteria	Physiological Criteria	Psychophysical Criteria
Establish if in vitro spine failures are similar in nature to in vivo damage	Validation and comparison of the various criteria	A more thorough understanding of MMH aerobic capacities	Epidemiological verification
Determination of the validity of L5/S1 compression as a criterion	Development of multivariate statistical models that explicitly predict the risk of LBD given a risk vector	Predictive models (for VO_2) based on large samples	Verification of physiological acceptability
Determination of the role of cumulative loading in the development of low-back injuries		Determination of relationship between physiological load and various outcomes measures such as fatigue and injury	Predictive models based on dynamic strength
A better understanding of the effects of shear and torsional loadings on the occurrence of LBD		Determination of the percentage of MMH aerobic capacity that industrial workers find acceptable as a means of establishing physiological criteria	
Data on dynamic capabilities of major joints			

temporal relationship between exposure and outcomes could be established.

6 MMH TRAINING AND STANDARDS

6.1 Training Programs

Kroemer et al. (1994) reported that about 50% of injured persons return to work within one week after an incidence of low back pain. Recent evidence shows that as the duration of absence from work increases, the likelihood of successful return diminishes (Hashemi et al., 1997; Carter and Birrell, 2000; Waddell and Burton, 2000). Snook (2004) argued that the main deterrents to returning to work after low back injury include those related to the worker (malingering, illness behavior, etc.), the management (lack of follow-up, no work modification), the union (rigid work rules), the practitioner (inappropriate or ineffective treatment), or the lawyer (lump-sum settlements instead of rehabilitation). Ayoub et al. (1997) also pointed out that in many cases the employment conditions for these workers are subject to specific medical work restrictions.

Personnel training in "safe" lifting techniques has been advocated and practiced for many years as the means to reduce the incidence (i.e., frequency and severity rates) of LBDs (Ayoub et al., 1997). From all the research evidence, there are two fundamental rules that the trainers/practitioners tend to adhere to while lifting: (1) keep the load as close to the body as possible, and (2) avoid twisting. Another most popular lifting technique is the straight-back/bent knees method, which advocates keeping the back straight and bending the knees and lifting using the legs (Ayoub et al., 1997). Kroemer et al. (1994) proposed the kinetic lift, a variation of the preceding technique, with a predefined posture of the body. Considering all these principles, Ayoub et al. (1997) proposed the generic manual handling guidelines summarized in Table 12.

Table 12 Guidelines for Proper Manual Lifting

Things to follow:

1. Design manual lifting (and lowering) out of the task and workplace. If it needs to be done by a person, perform it between knuckle and shoulder height.
2. Be in good physical shape. If not used to lifting and vigorous exercise, do not attempt to do difficult lifting or lowering tasks.
3. Think before acting. Place material conveniently. Make sure that sufficient space is cleared. Have handling aids available.
4. Get a good grip on the load. Test the weight before trying to move it. If it is too bulky or heavy, get a mechanical lifting aid, or somebody else or help, or both.
5. Get the load close to the body. Place the feet close to the load. Stand in a stable position, have the feet point in the direction of movement.
6. Involve primarily straightening of the legs in lifting.

Things to avoid:

1. Do NOT twist the back, or bend sideways.
2. Do NOT lift or lower, push or pull, awkwardly.
3. Do NOT hesitate to get help, either mechanical or by another person.
4. Do NOT lift or lower with arms extended.
5. Do NOT continue heaving when the load is too heavy.

Source: After Kroemer et al. (1995) and Ayoub et al. (1997).

6.2 Effects of Training

Based on the relationship between spinal characteristics and injury mechanism, it can be assumed that materials handling techniques used by workers influence the likelihood of developing LBP/LBD. However, based on an extensive review of over 100 published studies, Kroemer et al. (1994) suggested that there was strong evidence to conclude that training in safe lifting techniques was not an effective program for prevention of LBDs. Indeed, many other studies that investigated the effectiveness of lifting training programs on prevention of LBDs in the workplace failed to show significant reductions in LBDs or effectiveness of the training program (Daltroy et al., 1997; van Poppel et al., 1998). Previously, Snook et al. (1980) showed that the number of work-related LBDs in companies that used the training programs in "safe lifting" did not differ statistically from those that did not use such programs. Recently, Lavender et al. (2002) pointed out that two primary factors lead to an unsuccessful training program: (1) method of training (whether the training involves a teaching of intellectual concepts or a motor skill), and (2) beliefs of the trainees about the behavior learned (whether the learned behavior will be effective in the work environment or will be detrimental to productivity demands).

6.3 Standards

The International Organization for Standards (ISO) is a worldwide standards provider in many areas. ISO also has a specific manual materials standard, ISO 11228. Part 1 of the standard, ISO 11228-1:2003, describes limits for manual lifting and carrying considerating the intensity, frequency, and duration of the task. The limits recommended can be used in the assessment of several task variables and in health risk evaluations for the working population (Dickinson, 1995). This standard does not include holding objects (without walking), pushing or pulling objects, lifting with one hand, manual handling while seated, or lifting by two or more people. Holding, pushing, and pulling objects are included in other parts (Parts 2 and 3) of ISO 11228, which are currently at the stage of committee drafts. ISO/TS 20646-1:2004 presents guidelines for application of various ergonomics standards related to local muscular workload (LMWL) and specifies activities required to reduce LMWL in workplaces.

7 CONCLUSIONS

The Bureau of Labor Statistics data for 1992–1995 indicated a declining trend in injuries and illnesses requiring days away from work. For instance, the incidence rate of overexertion (due to manual lifting) declined from 52.1 per 10,000 workers in 1992 to 41.1 in 1995. According to NIOSH (1997), there may be several reasons for these declines, which include secular trends in reporting injury, socioeconomic trends, awareness, and work improvements due to effective prevention and intervention programs. According to Liberty Mutual (2004), the frequency of all serious work-related injuries fell 6% between 2000 and 2001. Since there were fewer, but more expensive, serious work-related injuries in 2001, the total cost of injuries did not decline despite the drop in frequency. It should also be noted that Liberty Mutual's (2004) index of leading occupational injuries shows that the leading cause of workplace injuries in 2001 was an overexertion, accounting for 27.3% of all injuries. The top three injury causes (overexertion, falls on same level, and bodily reaction) were the fastest growing of all injury causes, representing 50.1% of the total costs (i.e., about $23 billion a year or $450 million a week). The injuries noted above include a large number of disabling injuries to the lower back due to either cumulative exposure to manual handling of loads over a long period of time or to isolated incidents of overexertion when handling heavy objects.

Although lower back disorders due to manual lifting tasks in industry may not be fully preventable at present, many different strategies have been developed and can be used to reduce the potential for disabilities resulting from the onset of lower back pain in the workplace. There is sufficient evidence in the subject literature (NIOSH, 1997; Karwowski and Marras, 2003; NRC, 2001; Snook, 2004) to suggest that approaches based on the science of ergonomics, as discussed in this chapter, should help in this quest.

REFERENCES

Adams, M. A. (1995), "Mechanical Testing of the Spine: An Appraisal of Methodology, Results, and Conclusions," *Spine*, Vol. 20, pp. 2151–2156.

Adams, M. A., and Hutton, W. C. (1982), "Prolapsed Intervertebral Disc: A Hyperflexion Injury," *Spine*, Vol. 7, pp. 184–191.

Adams, M. A., McMillan, D. W., Green, T. P., and Dolan, P. (1986), "Sustained Loading Generates Stress Concentrations in Lumbar Intervertebral Discs," *Spine*, Vol. 21, pp. 434–438.

Allread, W. G. (2000), "An Investigation of the Relationship Between Personality and Risk Factors for Musculoskeletal Disorders," unpublished dissertation, Ohio State University, Columbus, OH.

Allread, W. G., Marras, W. S., and Parnianpour, M. (1996), "Trunk Kinematics of One-Handed Lifting and the Effects of Asymmetry and Load Weight," *Ergonomics*, Vol. 39, pp. 322–334.

Andersson, G. B. J. (1981), "Epidemiologic Aspects on Low Back Pain in Industry," *Spine*, Vol. 6, pp. 53–60.

Asfour, S. S., Genaidy, A. M., Khalil, T. M., and Muthuswamy, S. (1986), "Physiological Responses to Static, Dynamic and Combined Work," *American Industrial Hygiene Association Journal*, Vol. 47, No. 12, pp. 798–802.

Ayoub, M. M., and El-Bassoussi, M. M. (1976), "Dynamic Biomechanical Model for Sagittal Lifting Activities," in *Proceedings of the 6th Congress of the International Ergonomics Association*, pp. 355–359.

Ayoub, M. M., and Mital, A. (1989), *Manual of Materials Handling*, Taylor & Francis, London.

Ayoub, M. M., Bethea, N. J., Deivanayaga, M. S., Asfour, S. S., Bakken, G. M., Liles, D., Mital, A., and Sherif, M. (1978), *Determination and Modelling of Lifting Capacity, Final Report*, HEW (NIOSH) Grant 5R010H-000545-02, National Institute of Occupational Safety and Health, Washington, DC.

Ayoub, M. M., Mital, A., Bakken, G. M., Asfour, S. S., and Bethea, N. J. (1980), "Development of Strength and Capacity Norms for Manual Materials Handling Activities: The State of the Art," *Human Factors*, Vol. 22, No. 3, pp. 271–283.

Ayoub, M. M., Dempsey, P. G., and Karwowski, W. (1997), "Manual Materials Handling," in *Handbook of Human Factors and Ergonomics*, 2nd ed., G. Salvendy, Ed., Wiley, New York, pp. 1085–1123.

Battie, M. C., Bigos, S. J., Fisher, L. D., Hansson, T. H., Jones, M. E., and Wortley, M. D. (1989), "Isometric Lifting Strength as a Predictor of Industrial Back Pain Reports," *Spine*, Vol. 14, No. 8, pp. 851–856.

Bernard, B. P., Ed. (1997), *Musculoskeletal Disorders and Workplace Factors: A Critical Review of Epidemiological Evidence for Work-Related Musculoskeletal Disorders of the Neck, Upper Extremity, and Low Back*, U.S. Department of Health and Human Services, Public Health Service, Centers for Disease Control, National Institute for Occupational Safety and Health, Washington, DC.

Bhambhani, Y., and Maikala, R. (2000), "Gender Differences During Treadmill Walking with Graded Loads: Biomechanical and Physiological Comparisons," *European Journal of Applied Physiology*, Vol. 81, pp. 75–83.

Bhambhani, Y., Buckley, S., and Maikala, R. (1997), "Physiological and Biomechanical Responses During Treadmill Walking with Graded Loads," *European Journal of Applied Physiology*, Vol. 76, pp. 544–551.

Bigos, S. J., Spengler, D. M., Martin, N. A., Zeh, J., Fisher, L., Nachemsom, A., and Wang, M. H. (1986), "Back Injuries in Industry: A Retrospective Study, II: Injury Factors," *Spine*, Vol. 11, pp. 246–251.

Bigos, S. J., Battié, M. C., Spengler, D. M., Fisher, L. D., Fordyce, W. E., Hansson, T. H, Nachemsom, A. L., and Wortley, M. D. (1991), "A Prospective Study of Work Perceptions and Psychosocial Factors Affecting the Report of Back Injury," *Spine*, Vol. 16, pp. 1–6.

BLS (Bureau of Labor Statistics) (2000), http.bls.gov. oshhome.htm.

Borg, G. A. V. (1962), *Physical Performance and Perceived Exertion*, Gleerup, Lund, Sweden.

Brinckmann, P., Biggemann, M., and Hilweg, D. (1988), "Fatigue Fracture of Human Lumbar Vertebrae," *Clinical Biomechanics (Bristol, Avon)*, Vol. 3, Suppl. 1, pp. S1–S23.

Brokaw, N. (1992), "Implications of the Revised NIOSH Lifting Guide of 1991: A Field Study," unpublished M.S. thesis, Department of Industrial Engineering, University of Louisville, Louisville, KY.

Brown, J. R. (1975), "Factors Contributing to the Development of Low-Back Pain in Industrial Workers," *American Industrial Hygiene Association Journal*, Vol. 36, pp. 26–31.

Burdorf, A., and Sorock, G. (1997), "Positive and Negative Evidence of Risk Factors for Back Disorders," *Scandinavian Journal of Work, Environment and Health*, Vol. 23, pp. 243–256.

Burton, A. K., Symonds, T. L., Zinzen, E., et al. (1996), "Is Ergonomics Intervention Alone Sufficient to Limit Musculoskeletal Problems in Nurses," *Occupational Medicine*, Vol. 47, pp. 25–32.

Cady, L. D., Bischoff, D. P., O'Connell, E. R., Thomas, P. C., and Allan, J. H. (1979), "Strength and Fitness and Subsequent Back Injuries in Firefighters," *Journal of Occupational Medicine*, Vol. 21, pp. 269–272.

Cady, L. D., Thomas, P. C., and Karwasky, R. J. (1985), "Program for Increasing Health and Physical-Fitness of Fire Fighters," *Journal of Occupational and Environmental Medicine*, Vol. 27, No. 2, pp. 110–114.

Callaghan, J. P. (2004), "Cumulative Spine Loading: From Basic Science to Application," in *The Occupational Ergonomics Handbook*, 2nd ed., W. Karwowski and W. S. Marras, Eds., CRC Press, Boca Raton, FL.

Callaghan, J. P., and McGill, S. M. (2001), "Intervertebral Disc Herniation: Studies on a Porcine Model Exposed to Highly Repetitive Flexion/Extension Motion with Compressive Force," *Clinical Biomechanics (Bristol, Avon)*, Vol. 16, pp. 28–37.

Callaghan, J. P., Salewytsch, A. J., and Andrews, D. M. (2001), "An Evaluation of Predictive Methods for Estimating Cumulative Spinal Loading," *Ergonomics*, Vol. 44, pp. 825–837.

Carter, J. T. and Birrell, L. N., Eds. (2000), *Occupational Health Guidelines for the Management of Low Back Pain*. Faculty of Occupational Medicine, London.

Cavanaugh, J. M., Ozaktay, A. C., Yamashita, T., et al. (1997), "Mechanism of Low Back Pain: A Neurophysiologic and Neuroanatomic Study," *Clinical Orthopaedics and Related Research*, Vol. 335, pp. 166–180.

Chaffin, D. B. (1969), "A Computerized Biomechanical Model: Development of and Use in Studying Gross Body Actions," *Journal of Biomechanics*, Vol. 2, pp. 429–441.

Chaffin, D. B. (1974), "Human Strength Capability and Low-Back Pain," *Journal of Occupational Medicine*, Vol. 16, No. 4, pp. 248–254.

Chaffin, D. B., and Andersson, G. B. J. (1991), *Occupational Biomechanics*, Wiley, New York.

Chaffin, D. B., and Baker, W. H. (1970), "A Biomechanical Model for Analysis of Symmetric Sagittal Plane Lifting," *AIIE Transactions*, Vol. 2, pp. 16–27.

Chaffin, D. B., and Page, G. B. (1994), "Postural Effects on Biomechanical and Psychophysical Weight-Lifting Limits," *Ergonomics*, Vol. 37, No. 4, pp. 663–676.

Chaffin, D. B., and Park, K. S. (1973), "A Longitudinal Study of Low-Back Pain as Associated with Occupational Weight Lifting Factors," *American Industrial Hygiene Association Journal*, Vol. 34, pp. 513–525.

Chaffin, D. B., Herrin, G. D., and Keyserling, W. M. (1978), "Pre-employment Strength Testing," *Journal of Occupational Medicine*, Vol. 20, No. 6, pp. 403–408.

Chaffin, D. B., Andersson, G. B. J., and Martin, B. J. (1999), *Occupational Biomechanics*, Wiley, New York.

Chen, H. C., and Ayoub, M. M. (1988), "Dynamic Biomechanical Model for Asymmetrical Lifting," in *Trends in Ergonomics/Human Factors*, Vol. V, F. Aghazadeh, Ed., Elsevier, Amsterdam, pp. 879–886.

Cheng, C. K., Chen, H. H., Kuo, H. H., Lee, C. L., Chen, W. J., and Liu, C. L. (1998). "A Three-Dimensional Mathematical Model for Predicting Spinal Joint Force Distribution During Manual Lifting," *Clinical Biomechanics*, 13, Suppl. 1, pp. S59–S64.

Ciriello, V. M., and Snook, S. H. (1978), "The Effects of Size, Distance, Height, and Frequency on Manual Handling Performance," in *Proceedings of the Human Factors Society 22nd Annual Meeting*, Human Factors Society, Santa Monica, CA, pp. 318–322.

Ciriello, V. M. and Snook, S. H. (1983). "A Study of Size, Distance, Height, and Frequency Effects on Manual Handling Tasks," *Human Factors*, Vol. 25, No. 5, pp. 473–483.

Ciriello, V. M., and Snook, S. H. (1999), "Survey of Manual Handling Tasks," *International Journal of Industrial Ergonomics*, Vol. 23, pp. 149–156.

Ciriello, V. M., Snook, S. H., Blick, A. C., and Wilkinson, P. L. (1990), "The Effects of Task Duration on Psychophysically-Determined Maximum Acceptable Weights and Forces," *Ergonomics*, Vol. 33, No. 2, pp. 187–200.

Ciriello, V. M., Snook, S. H., and Hughes, G. J. (1993), "Further Studies of Psychophysically Determined Maximum Acceptable Weights and Forces," *Human Factors*, Vol. 35, pp. 175–186.

Cole, M. H., and Grimshaw, P. N. (2003), "Low Back Pain and Lifting: A Review of Epidemiology and Aetiology," *Work*, Vol. 21, pp. 173–184.

Craig, B. N., Congleton, J. J., Kerk, C. J., Lawler, J. M., and McSweeney, K. P. (1998), "Correlation of Injury Occurrence Data with Estimated Maximal Aerobic Capacity and Body Composition in a High-Frequency Manual Materials Handling Task," *American Industrial Hygiene Association Journal*, Vol. 59, pp. 25–33.

Daltroy, L. H., et al. (1997), "A Controlled Trial of an Educational Program to Prevent Low Back Injuries," *New England Journal of Medicine*, Vol. 337, pp. 322–328.

Davis, K. G., Marras, W. S., Heaney, C. A., Waters, T. R., and Gupta, P. (2002), "The Impact of Mental Processing and Pacing on Spine Loading," *Spine*, Vol. 27, No. 23, pp. 2645–2653.

Daynard, D., Yassi, A., Cooper, J. E., Tate, R., Norman, R., and Wells, R. (2001), "Biomechanical Analysis of Peak and Cumulative Spinal Loads During Simulated Patient-Handling Activities: A Substudy of a Randomized Controlled Trial to Prevent Lift and Transfer Injury of Health Care Workers," *Applied Ergonomics*, Vol. 32, pp. 199–214.

de Looze, M. P., Kingma, I., Thunnissen, W., vanWijk, M. J., and Toussaint, H. M. (1994), "The Evaluation of a Practical Biomechanical Model Estimating Lumbar Moments in Occupational Activities," *Ergonomics*, Vol. 37, No. 9, pp. 1495–1502.

de Looze, M. P., Dolan, P., Kingma, I., and Baten, C. T. M. (1998), "Does An Asymmetric Straddle-Legged Lifting Movement Reduce the Low-Back Load?" *Human Movement Science*, Vol. 17, pp. 243–259.

Dempsey, P. G. (1998), "A Critical Review of Biomechanical, Epidemiological, Physiological and Psychosocial Criteria for Designing Manual Materials Handling Tasks," *Ergonomics*, Vol. 41, No. 1, pp. 73–88.

Dempsey, P. G. (2002), "Usability of the Revised NIOSH Lifting Equation," *Ergonomics*, Vol. 45, No. 12, pp. 817–828.

Dempsey, P. G., and Westfall, P. H. (1997), "Developing Explicit Risk Models for Predicting Low-Back Disability: A Statistical Perspective," *International Journal of Industrial Ergonomics*, Vol. 19, pp. 483–497.

Dempsey, P. G., Burdorf, A., and Webster, B. S. (1997), "The Influence of Personal Variables on Work-Related Low-Back Disorders and Implications for Future Research," *Journal of Occupational and Environmental Medicine*, Vol. 39, No. 8, pp. 748–759.

Dempsey, P. G., Burdorf, A., Fathallah, F., Sorock, G. S., and Hashemi, L. (2001), "Influence of Measurement Accuracy on the Application of the 1991 NIOSH Equation," *Applied Ergonomics*, Vol. 32, No. 1, pp. 91–99.

Deyo, R. A. (1994), "Magnetic Resonance Imaging of the Spine: Terrific Test or Tar Baby?" *New England Journal of Medicine*, Vol. 331, pp. 115–116.

Deyo, R. A., and Bass, J. E. (1989), "Lifestyle and Low-Back-Pain: The Influence of Smoking and Obesity," *Spine*, Vol. 14, No. 5, pp. 501–506.

Dickinson, C. E. (1995), "Proposed Manual Handling International and European Standards," *Applied Ergonomics*, Vol. 26, No. 4, pp. 265–270.

Eisler, H. (1962), "Subjective Scale of Force for a Large Muscle Group," *Journal of Experimental Psychology*, Vol. 64, No. 3, pp. 253–257.

Ekholm, J., Arborelius, U. P., and Nemeth, G. (1982), "The Load on the Lumbo-Sacral Joint and Trunk Muscle Activity During Lifting," *Ergonomics*, Vol. 25, No. 2, pp. 145–161.

El-Bassoussi, M. M. (1974), "A Biomechanical Dynamic Model for Lifting in the Sagittal Plane," Ph.D. dissertation, Texas Tech University, Lubbock, TX.

Erdil, M., and Dickerson, O. B., Eds. (1997), *Cumulative Trauma Disorders: Prevention, Evaluation and Treatment*, Van Nostrand Reinhold, New York.

Esola, M. A., McClure, P. W., Kelley, F. G., and Siegler, S. (1996), "Analysis of Lumbar Spine and Hip Motion During Forward Bending in Subjects with and Without a History of Low Back Pain," *Spine*, Vol. 21, pp. 71–78.

Fernandez, J. E., Ayoub, M. M., and Smith, J. L. (1991), "Psychophysical Lifting Capacity over Extended Periods," *Ergonomics*, Vol. 34, No. 1, pp. 23–32.

Fisher, B. O. (1967), "Analysis of Spinal Stresses During Lifting," M.S. thesis, University of Michigan, Ann Arbor, MI.

Fitzler, S. L., and Berger, R. A. (1982), "Attitudinal Change: The Chelsea Back Program," *Occupational Health and Safety*, Vol. 51, pp. 24–26.

Fitzler, S. L. and Berger, R. A. (1983), "Chelsea Back Program: One Year Later," *Occupational Health and Safety*, Vol. 51, pp. 52–54.

Frymoyer, J. W., and Gordon, S. L., Eds. (1989), *New Perspectives on Low Back Pain: Proceedings of the American Academy of Orthopedic Surgeons Symposium*, Chicago.

Frymoyer, J. W., and Nachemsom, A. (1991), "Natural History of Low Back Disorders", in *The Adult Spine: Principles and Practice*, J. W. Frymoyer, T. B. Ducker, N. M. Hadler, J. P. Kostuik, J. N. Weinstein, and T. S. Whitecloud, Eds., Raven Press, New York, pp. 1537–1550.

Frymoyer, J. W., Pope, M. H., Costanza, M. C., Rosen, J. C., Goggin, J. E., and Wilder, D. G. (1980), "Epidemiologic Studies of Low-Back Pain," *Spine*, Vol. 5, No. 5, pp. 419–423.

Gallagher, S. (1991), "Acceptable Weights and Physiological Costs of Performing Combined Manual Handling Tasks in Restricted Postures," *Ergonomics*, Vol. 34, No. 7, pp. 939–952.

Gallagher, S., and Hamrick, C. A. (1992), "Acceptable Workloads for Three Common Mining Materials," *Ergonomics*, Vol. 35, pp. 1013–1031.

Garg, A. (1989), "An Evaluation of the NIOSH Guidelines for Manual Lifting with Special Reference to Horizontal Distance," *American Industrial Hygiene Association Journal*, Vol. 50, No. 3, pp. 157–164.

Garg, A., and Badger, D. (1986), "Maximum Acceptable Weights and Maximum Voluntary Strength for Asymmetric Lifting," *Ergonomics*, Vol. 29, No. 7, pp. 879–892.

Garg, A., and Chaffin, D. B. (1975), "A Biomechanical Computerized Simulation of Human Strength," *IIE Transactions*, Vol. 14, No. 4, pp. 272–281.

Garg, A., and Saxena, U. (1979), "Effects of Lifting Frequency and Technique on Physical Fatigue with Special Reference to Psychophysical Methodology and Metabolic Rate," *American Industrial Hygiene Association Journal*, Vol. 40, pp. 894–903.

Garg, A., Chaffin, D. B., and Herrin, G. D. (1978), "Prediction of Metabolic Rates for Manual Materials Handling Jobs," *American Industrial Hygiene Association Journal*, Vol. 39, No. 8, pp. 661–675.

Garg, A., Rodgers, S. H., and Yates, J. W. (1992), "The Physiological Basis for Manual Lifting," in *Advances in Industrial Ergonomics and Safety*, S. Kumar, Ed., Taylor & Francis, London, pp. 867–874.

Genaidy, A. M., Asfour, S. S., Khalil, T. M., and Waly, S. M. (1985), "Physiological Issues in Manual Materials Handling," in *Trends in Ergonomics/Human Factors*, Vol. II, R. E. Eberts and C. G. Eberts, Eds., North-Holland, Amsterdam, pp. 571–576.

Goel, V. K., Fromknecht, S. J., Nishiyama, K., Weinstein, J., and Liu, Y. K. (1985), "The Role of Lumbar Spinal Elements in Flexion," *Spine*, Vol. 10, No. 6, pp. 516–523.

Granata, K. P., and Wilson, S. E. (2001), "Trunk Posture and Spinal Stability," *Clinical Biomechanics*, Vol. 16, pp. 650–659.

Hakkanen, M., Viikari-Juntura, E., and Takala, E. (1997), "Effects of Changes in Work Methods on Musculoskeletal Load: An Intervention Study in the Trailer Assembly," *Applied Ergonomics*, Vol. 28, No. 2, pp. 99–108.

Hansson, T. H., Keller, T. S., and Spengler, D. M. (1987), "Mechanical Behavior of the Lumbar Spine, II: Fatigue Strength During Dynamic Compressive Loading," *Journal of Orthopaedic Research*, Vol. 5, No. 9, pp. 479–487.

Hashemi, L., Webster, B. S., Clancy, E. A., and Volinn, E. (1997), "Length of Disability and Cost of Workers' Compensation Low Back Claims," *Journal of Occupational and Environmental Medicine*, Vol. 39, pp. 937–945.

Heliövaara, M. (1987), "Body Height, Obesity, and Risk of Herniated Lumbar Intervertebral Disc," *Spine*, Vol. 12, pp. 469–472.

Hidalgo, J., Genaidy, A., Karwowski, W., Christensen, D., Huston, R., and Stambough, J. (1997), "A Comprehensive Lifting Model: Beyond the NIOSH Lifting Equation," *Ergonomics*, Vol. 40, No. 9, pp. 916–927.

Iatridis, J. C., Weidenbaum, M., Setton, L. A., and Mow, V. C. (1996), "Is the Nucleus Pulposus a Solid or a

Fluid? Mechanical Behaviors of the Human Intervertebral Disc," *Spine*, Vol. 10, pp. 1174–1184.

Innes, E., and Straker, L. (1999), "Validity of Work-Related Assessments," *Work*, Vol. 13, No. 2, pp. 125–152.

Jäger, M., and Luttmann, A. (1989), "Biomechanical Analysis and Assessment of Lumbar Stress During Load Lifting Using a Dynamic 19-Segment Human Model," *Ergonomics*, Vol. 32, No. 1, pp. 93–112.

Jäger, M., Jordan, C., Luttmann, A., Laurig, W. and DOLLY Group (2000), "Evaluation and Assessment of Lumbar Load During Total Shifts for Occupational Manual Materials Handling Jobs Within the Dortmund Lumbar Load Study: DOLLY," *International Journal of Industrial Ergonomics*, Vol. 25, pp. 553–571.

Jensen, M. C., Brant-Zawadzki, M. N., Obuchowski, N., Modic, M. T., Malkasian, D., and Ross, J. S. (1994), "Magnetic Resonance Imaging of the Lumbar Spine in People Without Back Pain," *New England Journal of Medicine*, Vol. 331, pp. 69–73.

Jiang, B. C., Smith, J. L., and Ayoub, M. M. (1986), "Psychophysical Modelling of Manual Materials Handling Capacities Using Isoinertial Strength Variables," *Human Factors*, Vol. 28, No. 6, pp. 691–702.

Karwowski, W. (1983), "A Pilot Study of the Interaction between Physiological, Biomechanical and Psychological Stresses Involved in Manual Lifting Tasks," in *Proceedings of the Ergonomics Society Conference*, Taylor & Francis, Cambridge, pp. 95–100.

Karwowski, W. (1991), "Psychophysical Acceptability and Perception of Load Heaviness by Females," *Ergonomics*, Vol. 34, No. 4, pp. 487–496.

Karwowski, W. (1992), "Comments on the Assumption of Multiplicity of Risk Factors in the Draft Revisions to NIOSH Lifting Guide," in *Advances in Industrial Ergonomics and Safety IV*, S. Kumar Ed., Taylor & Francis, London.

Karwowski, W. (1996), "Maximum Safe Weight of Lift: A New Paradigm for Setting Design Limits in Manual Lifting Tasks Based on the Psychophysical Approach," in *Proceedings of the Human Factors and Ergonomics Society 40th Annual Meeting*, Santa Monica, CA, pp. 614–618.

Karwowski, W., and Ayoub, M. M. (1984), "Fuzzy Modeling of Stresses in Manual Lifting Tasks," *Ergonomics*, Vol. 27, No. 6, pp. 641–649.

Karwowski, W., and Brokaw, N. (1992), "Implications of the Proposed Revisions in a Draft of the Revised NIOSH Lifting Guide (1991) for Job Redesign: A Field Study," in *Proceedings of the 36th Annual Meeting of the Human Factors Society*, Atlanta, GA, pp. 659–663.

Karwowski, W., and Gaddie, P. R. (1995), "Simulation of the 1991 Revised NIOSH Manual Lifting Equation," in *Proceedings of the Human Factors and Ergonomics Society Annual Meeting*, Santa Monica, CA, pp. 699–701.

Karwowski, W., and Marras, W. S., Eds. (2003), *Occupational Ergonomics: Principles of Work Design*, CRC Press, Boca Raton, FL.

Karwowski, W., and Rodrick, D. (2001), "Physical Tasks: Analysis, Design and Operation," in *Handbook of Industrial Engineering*, 3rd ed., G. Salvendy, Ed., Wiley, New York, pp. 1041–1110.

Karwowski, W., and Yates, J. W. (1986), "Reliability of the Psychophysical Approach to Manual Lifting of Liquids by Females," *Ergonomics*, Vol. 29, No. 2, pp. 237–248.

Karwowski, W., Shumate, C., Yates, J. W., and Pongpatana, N. (1992), "Discriminability of Load Heaviness in Manual Lifting: Implications for the Psychophysical Approach," *Ergonomics*, Vol. 35, No. 7–8, pp. 729–744.

Karwowski, W., Lee, W. G, Jamaldin, B., Gaddie, P., and R. Jang (1999), "Beyond Psychophysics: A Need for Cognitive Modeling Approach to Setting Limits in Manual Lifting Tasks," *Ergonomics*, Vol. 42, No. 1, pp. 40–60.

Kelsey, J. L. (1975), "An Epidemiological Study of the Relationship Between Occupations and Acute Herniated Lumbar Intervertebral Discs," *International Journal of Epidemiology*, Vol. 4, pp. 197–205.

Kelsey, J. L., Githens, P. B., White, A. A., Holford, T. R., Walter, S. D., O'Connor, T., Ostfeld, A. M., Weil, U., Southwick, W. O., and Calogero, J. A. (1984), "An Epidemiologic Study of Lifting and Twisting on the Job and Risk for Acute Prolapsed Lumbar Intervertebral Disc," *Journal of Orthopaedic Research*, Vol. 2 No. 1, pp. 61–66.

Kerr, M. S., Frank, J. W., Shannon, H. S., Norman, R. W. K., Wells, R. P., Patrick Neumann, W., Bombardier, C., and the Ontario Universities Back Pain Study Group (2001), "Biomechanical and Psychosocial Risk Factors for Low Back Pain at Work," *American Journal of Public Health*, Vol. 91, pp. 1069–1075.

Keyserling, W. M., Fine, L. J., and Punnett, L. (1987), "Computer-Aided Analysis of Trunk and Shoulder Posture," in *Musculoskeletal Disorders at Work*, P. Buckle, Ed., Taylor & Francis, London, pp. 83–96.

Khalil, T. M., Genaidy, A. M., Asfour, S. S., and Vinciguerra, T. (1985), "Physiological Limits in Lifting," *American Industrial Hygiene Association Journal*, Vol. 46, pp. 220–224.

Kim, H.-K. (1990), "Development of a Model for Combined Ergonomic Approaches in Manual Materials Handling Tasks," Ph.D. dissertation, Texas Tech University, Lubbock, TX.

Kim, S. H., and Chung, M. K. (1995), "Effects of Posture, Weight, and Frequency on Trunk Muscular Activity and Fatigue During Repetitive Lifting Tasks," *Ergonomics*, Vol. 38, No. 5, pp. 853–863.

King, A. I. (1993), "Injury to the Thoraco-Lumbar Spine and Pelvis," in *Accidental Injury: Biomechanics and Prevention*, A. M. Nahum and J. W. Melvin, Eds., Springer-Verlag, New York, pp. 429–459.

Krause, N., Dasinger, L. K., and Neuhauser, F. (1998), "Modified Work and Return to Work: A Review of the Literature," *Journal of Occupational Rehabilitation*, Vol. 8, No. 2, pp. 113–139.

Kroemer, K. H. E. (1992), "Personnel Training for Safer Material Handling," *Ergonomics*, Vol. 35, No. 9, pp. 1119–1134.

Kroemer, K., Kroemer, H., and Kroemer-Elbert, K. (1994), *Ergonomics: How to Design for Ease and Efficiency*, Prentice Hall, Englewood Cliffs, NJ.

Kromodihardjo, S., and Mital, A. (1986), "Kinetic Analysis of Manual Lifting Activities, I: Development of a Three Dimensional Computer Model," *International Journal of Industrial Ergonomics*, Vol. 1, pp. 77–90.

Kromodihardjo, S., and Mital, A. (1987), "Biomechanical Analysis of Manual Lifting Tasks," *Journal of Biomechanical Engineering*, Vol. 109, pp. 132–138.

Kumar, S. (1990), "Cumulative Load as a Risk Factor for Back Pain," *Spine*, Vol. 15, pp. 1311–1316.

Lavender, S. A., Lorenz, E., and Andersson, G. B. J. (2002), "Training in Lifting: Do Good Lifting Techniques Adversely Affect Case-Handling Times?" *Professional Safety*, December, pp. 30–35.

Leamon, T. B. (1994a), "L5/S1: So Who Is Counting? *International Journal of Industrial Ergonomics*, Vol. 13, pp. 259–265.

Leamon, T. B. (1994b), "Research to Reality: A Critical Review of the Validity of Various Criteria for the Prevention of Occupationally Induced Low Back Pain Disability," *Ergonomics*, Vol. 37, No. 12, pp. 1959–1974.

Leboeuf-Yde, C., and Kyvik, K. O. (1998), "At What Age Does Low-Back Pain Become a Common Problem?" *Spine*, Vol. 23, pp. 228–234.

Lee, P., Helewa, A., Goldsmith, C. H., Smythe, H. A., and Stitt, L. W. (2001), "Low Back Pain: Prevalence and Risk Factors in an Industrial Setting," *Journal of Rheumatology*, Vol. 28, No. 2, pp. 346–351.

Legg, S. J., and Myles, W. S. (1981), "Maximum Acceptable Repetitive Lifting Workloads for an 8 Hour Work-Day Using Psychophysical and Subjective Rating Methods," *Ergonomics*, Vol. 24, No. 12, pp. 907–916.

Leigh, P., Markowitz, S., Fahs, M., Shin, P., and Landrigan, P. (1997), "Occupational Injury and Illness in the United States: Estimated Costs, Morbidity, and Mortality," *Archives of Internal Medicine*, Vol. 157, pp. 1557–1568.

Lemstra, M., and Olszynski, W. P. (2003), "The Effectiveness of Standard Care, Early Intervention, and Occupational Management in Worker's Compensation Claims," *Spine*, Vol. 28, No. 3, pp. 299–304.

Leskinen, T. P. J., Salhammar, H. R., and Kuorinka, A. A. (1983), "The Effect of Inertial Factors on Spinal Stress When Lifting," *Engineering in Medicine*, Vol. 12, No. 2, pp. 87–89.

Levangie, P. K. (1999), "The Association Between Static Pelvic Asymmetry and Low Back Pain," *Spine*, Vol. 24, No. 12, pp. 1234–1242.

Liberty Mutual (2004), "Liberty Mutual Workplace Safety Index of Leading Occupational Injuries," http://www.libertymutual.com/omapps/ContentServer?cid=1029415781973&pagename=ResearchCenter/Page/StandardOrange&c=Page.

Liemohn, W., Snodgrass, L. B., and Sharpe, G. L. (1988), "Unresolved Controversies in Back Management: A Review," *Journal of Orthopaedic and Sports Physical Therapy*, Vol. 9, No. 7, pp. 239–244.

Liles, D. H., Deivanayagam, S., Ayoub, M. M., and Mahajan, P. (1984), "A Job Severity Index for the Evaluation and Control of Lifting Injury," *Human Factors*, Vol. 26, pp. 683–693.

Loisell, P., Abenhaim, L., Durand, P., Esdaile, J. M., Suissa, S., Simard, L., Simard, R., Turcotte, J., and Lemaire, J. (1997), "A Population-Based, Randomized Clinical Trial on Back Pain Management," *Spine*, Vol. 22, No. 24, pp. 2911–2918.

Macfarlane, G. J., Thomas, E., Papageorgiou, A. C., Croft, P. R., Jayson, M. I. V., and Silman, A. J. (1997), "Employment and Physical Work Activities as Predictors of Future Low Back Pain," *Spine*, Vol. 22, No. 10, pp. 1143–1149.

Malmivaara, A., Hakkinen, U., Aro, T., Heinrichs, M.-L., Koskenniemi, L., Kuosma, E., Lappi, S., Paloheimo, R., Servo, C., Vaaranen, V., and Hernberg, S. (1995), "The Treatment of Acute Lowback Pain: Bed Rest, Exercises, or Ordinary Activity?" *New England Journal of Medicine*, Vol. 332, No. 6, pp. 351–355.

Marras, W. S. (1998), "Occupational Biomechanics," in *Occupational Ergonomics Handbook*, W. Karwowski and W. S. Marras, Eds., CRC Press, Boca Raton, FL.

Marras, W. S. (2000). "Occupational Low Back Disorder Causation and Control," *Ergonomics*, Vol. 43, No. 7, pp. 880–902.

Marras, W. S. (2001). "Spine Biomechanics, Government Regulation, and Prevention of Occupational Low Back Pain," *Spine Journal: Official Journal of the North American Spine Society*, Vol. 1, No. 3, pp. 163–165.

Marras, W. S. (2004), "State-of-the Art Research Perspectives on Musculoskeletal Disorder Causation and Control: The Need for an Integrated Understanding of Risk," *Journal of Electromyographys and Kinesiology*, Vol. 14, No. 1, pp. 1–5.

Marras, W. S., and Granata, K. P. (1997), "The Development of an EMG-Assisted Model to Assess Spine Loading During Whole-Body Free-Dynamic Lifting," *Journal of Electromyography and Kinesiology*, Vol. 7, pp. 259–268.

Marras, W. S., and Sommerich, C. M. (1991a), "A Three-Dimensional Motion Model of Loads on the Lumbar Spine, I: Model Structure," *Human Factors*, Vol. 33, No. 2, pp. 123–137.

Marras, W. S., and Sommerich, C. M. (1991b), "A Three-Dimensional Motion Model of Loads on the Lumbar Spine, II: Model Validation," *Human Factors*, Vol. 33, No. 2, pp. 139–149.

Marras, W. S., Lavender, S. A., Leurgans, S. E., Rajulu, S. L., Allread, W. G., Fathallah, F. A., and Ferguson, S. A. (1993), "The Role of Dynamic Three-Dimensional Trunk Motion in Occupationally-Related Low Back Disorders," *Spine*, Vol. 18, pp. 617–628.

Marras, W. S., Lavender, S. A., Leurgans, S., Fathallah F., Allread, W. G., Ferguson, S. A., and Rajulu, S. (1995), "Biomechanical Risk Factors for Occupationally Related Low Back Disorder Risk," *Ergonomics*, Vol. 38, No. 2, pp. 377–410.

Marras, W. S., Ferguson, S. A., Gupta, P., Bose, S., Parnianpour, M., Kim, J., and Crowell, R. R. (1999), "The Quantification of Low Back Disorder Using Motion Measures: Methodology and Validation," *Spine*, Vol. 24, No. 20, pp. 2091–2100.

Marras, W. S., Allread, W. G., Burr, D. L., and Fathallah, F. A. (2000), "Prospective Validation of a Low-Back Disorder Risk Model and Assessment of Ergonomic Interventions Associated with Manual Materials Handling Tasks," *Ergonomics*, Vol. 43, No. 11, pp. 1866–1886.

Marras, W. S., Davis, K. G., Ferguson, S. A., Lucas, B. R., and Gupta, P. (2001), "Spine Loading Characteristics of Patients with Low Back Pain Compared with Asymptomatic Individuals," *Spine*, Vol. 26, No. 23, pp. 2566–2574.

Martin, J. B., and Chaffin, D. B. (1972), "Biomechanical Computerized Simulation of Human Strength in Sagittal Plane Activities," *AIIE Transactions*, Vol. 4, No. 1, pp. 19–28.

McCoy, C. E., Hadjipavlou, A. G., Overman, T., Necessary, J. T., and Wolf, C. (1997), "Work-Related Low Back Injuries Caused by Unusual Circumstances," *Journal of Orthopaedic and Sports Physical Therapy*, Vol. 26, No. 5, pp. 260–265.

McGill, S. M., and Norman, R. W. (1985), "Dynamically and Statically Determined Low Back Moments During Lifting," *Journal of Biomechanics*, Vol. 18, No. 12, pp. 877–885.

McLellan, R. K., Pransky, G., and Shaw, W. S. (2001), "Disability Management Training for Supervisors: A Pilot Intervention Program," *Journal of Occupational Rehabilitation*, Vol. 11, No. 1, pp. 33–41.

McQuirk, B., King, W., Govind, J., Lowry, J., and Bogduk, N. (2001), "Safety, Efficacy, and Cost Effectiveness of Evidence-Based Guidelines for the Management of Acute Lowback Pain in Primary Care," *Spine*, Vol. 26, No. 23, pp. 2615–2622.

Mital, A. (1983), "The Psychophysical Approach in Manual Lifting: A Verification Study," *Human Factors*, Vol. 25, No. 5, pp. 485–491.

Mital, A. (1984a), "Comprehensive Maximum Acceptable Weight of Lift Database for Regular 8-Hour Work Shifts," *Ergonomics*, Vol. 27, No. 11, pp. 1127–1138.

Mital, A. (1984b), Revised lifting capability data base for the industrial workforce, in Proceedings of the Human Factors Society 28th Annual Meeting, Santa Monica, CA: Human Factors Society, pp. 581–585.

Mital, A., Nicholson, A. S., and Ayoub, M. M. (1993), *A Guide to Manual Materials Handling*, Taylor & Francis, London.

Murphy, P. L., and Volinn, E. (1999), "Is Occupational Low Back Pain on the Rise?" *Spine*, Vol. 24, pp. 691–697.

Nachemsom, A. (1976), "The Lumbar Spine, An Orthopedic Challenge," *Spine*, Vol. 1, pp. 59–71.

Nicholson, A. S. (1989), "A Comparative Study of Methods for Establishing Load Handling Capabilities," *Ergonomics*, Vol. 32, pp. 1125–1144.

NIOSH (1981), *Work Practices Guide for Manual Lifting*, Technical Report 81–122, U.S. Government Printing Office, Washington, DC.

NIOSH (1996), *National Occupational Research Agenda*. U.S. Department of Health and Human Services, Public Health Service, Center for Disease Control and Prevention, National Institute for Occupational Safety and Health, Washington, DC, April.

NIOSH (1997), "Low Back and Musculoskeletal Disorders: Evidence for Work-Relatedness," in *Musculoskeletal Disorders (MSDs) and Workplace Factors/TOC*, B. P. Bernard, Ed., National Institute for Occupational Safety and Health, Cincinnati, OH, July.

Norman, R., Wells, R., Neumann, P., Frank, J., Shannon, H., Kerr, M., and Ontario Universities Back Pain Study (OUBPS) Group (1998), "A Comparison of Peak vs Cumulative Physical Work Exposure Risk Factors for the Reporting of Back Pain in the Automotive Industry," *Clinical Biomechanics*, Vol. 13, pp. 561–573.

NRC (2001) *Musculoskeletal Disorders and the Workplace: Low Back and Upper Extremities*, National Research Council, National Academy Press, Washington, DC.

Park, K. S., and Chaffin, D. B. (1974), "A Biomechanical Evaluation of Two Methods of Manual Load Lifting," *AIIE Transactions*, Vol. 6, No. 2, pp. 105–113.

Park, H. C., Wagener, D. K., and Parsons, V. L. (1997), "Back Pain Among U.S. Workers: Comparison of Workers Attributes According to Self-Reported Causes of Back Pain," *International Journal of Occupational and Environmental Health*, Vol. 3, pp. 37–44.

Pool-Goudzwaard, A. L., Vleeming, A., Stoeckart, R., Snijders, C. J., and Mens, J. M. A. (1998), "Insufficient Lumbopelvic Stability: A Clinical, Anatomical and Biomechanical Approach to 'A Specific' Low Back Pain," *Manual Therapy*, Vol. 3, No. 1, pp. 12–20.

Pope, M. H., Wilder, D. G., and Krag, M. H. (1991), "Biomechanics of the Lumbar Spine," in *The Adult Spine: Principles and Practice*, J. W. Frymoyer, T. B. Ducker, N. M. Hadler, J. P. Kostuik, J. Weinstein, and T. S. Whitecloud, Eds., Raven Press, New York, pp. 1487–1501.

Potvin, J. R., and Bent, L. R. (1997), "NIOSH Equation Horizontal Distances Associated with the Liberty Mutual (Snook) Lifting Table Box Widths," *Ergonomics*, Vol. 40, No. 6, pp. 650–655.

Pritsker, A. A. B. (1986), *Introduction to Simulation and SLAM II*, 3rd ed., Wiley, New York.

Riihimäki, H. (1991), "Low-Back Pain, Its Origin and Risk Indicators," *Scandinavian Journal of Work, Environment and Health*, Vol. 17, pp. 81–90.

Roppnen, A., Levalahti, E., Videman, T., Kaprio, J., and Battie, M. (2004), "The Role of Genetics and Environment in Lifting Force and Isometric Trunk Extensor Endurance," *Physical Therapy*, Vol. 84, No. 7, pp. 608–621.

Schmulowitz, J. (1995), "Workers' Compensation: Coverage, Benefits, and Costs," *Social Security Bulletin*, Vol. 58, pp. 51–57.

Schultz, A. B., and Andersson, G. B. J. (1981), "Analysis of Loads on the Lumbar Spine," *Spine*, Vol. 6, No. 1, pp. 76–82.

Schultz, A., Cromwell, R., et al. (1987), "Lumbar Trunk Muscle Use in Standing Isometric Heavy Exertions," *Journal of Orthopaedic Research*, Vol. 5, No. 3, pp. 320–329.

Scott, S. C., Goldberg, M. S., Mayo, N. E., Stock, S. R., and Poitras, B. (1999), "The Association Between Cigarette Smoking and Back Pain in Adults," *Spine*, Vol. 24, pp. 1090–1098.

Seidler, A., Bolm-Audorff, U., Heiskel, H., Henkel, N., Roth-Kuver, B., Kaiser, U., Bickeböller, R., Willingstorfer, W. J., Beck, W., and Elsner, G. (2001), "The Role of Cumulative Physical Work Load in Lumbar Spine Disease: Risk Factors for Lumbar Osteochondrosis and Spondylosis Associated with Chronic Complaints," *Occupational and Environmental Medicine*, Vol. 58, pp. 735–746.

Seidler, A., Bolm-Audorff, U., Siol, T., Henkel, N., Fuchs, C., Schug, H., Leheta, F., Marquardt, G., Schmitt, E., Ulrich, P. T., Beck, W., Missalla, A., and Elsner, G. (2003), "Occupational Risk Factors for Symptomatic Lumbar Disc Herniation; A Case–Control Study," *Occupational and Environmental Medicine*, Vol. 60, pp. 821–830.

Shoaf, C., Genaidy, A., Karwowski, W., Waters, T., and Christensen, D. (1997), "Comprehensive Manual Handling Limits for Lowering, Pushing, Pulling and Carrying Activities," *Ergonomics*, Vol. 40, No. 11, pp. 1183–1200.

Siddall, P. J., and Cousins, M. J. (1997), "Spinal Pain Mechanisms," *Spine*, Vol. 22, No. 1, pp. 98–104.

Sjolie, A. N., and Ljunggren, A. E. (2001), "The Significance of High Lumbar Mobility and Low Lumbar Strength for Current and Future Low Back Pain in Adolescents," *Spine*, Vol. 26, No. 23, pp. 2629–2636.

Smith, J. L., Ayoub, M. M., and McDaniel, J. W. (1992), "Manual Materials Handling Capabilities in Non-standard Postures," *Ergonomics*, Vol. 35, pp. 807–831.

Snook, S. H. (1965), "Group Work Capacity: A Technique for Evaluating Physical Tasks in Terms of Fatigue," unpublished report, Liberty Mutual Insurance Company, Hopkinton, MA.

Snook, S. H. (1971), "The Effects of Age and Physique on Continuous Work Capacity," *Human Factors*, Vol. 17, pp. 467–479.

Snook, S. H. (1978), "The Design of Manual Handling Tasks," *Ergonomics*, Vol. 21, pp. 963–985.

Snook, S. H. (1989), "The Control of Lowback Disability: The Role of Management," in *Manual Material Handling: Understanding and Preventing Back Trauma*, J. D. McGlothlin, T. G. Bobick, and K. H. E. Kroemer, Eds., American Industrial Hygiene Association, Fairfax, VA, pp. 97–101.

Snook, S. H. (2004), "Work-Related Low Back Pain: Secondary Intervention," *Journal of Electromyography and Kinesiology*, Vol. 14, pp. 153–160.

Snook, S. H., and Ciriello, V. M. (1974a), "Maximum Weights and Work Loads Acceptable to Female Workers," *Journal of Occupational Medicine*, Vol. 16, pp. 527–534.

Snook, S. H., and Ciriello, V. M. (1974b), "The Effects of Heat Stress on Manual Lifting Tasks," *American Industrial Hygiene Association Journal*, Vol. 35, pp. 681–685.

Snook, S. H. and Ciriello, V. M. (1991). The design of manual handling tasks: revised tables of maximum acceptable weights and forces, *Ergonomics*, 34(9), 1197–1213.

Snook, S. H., and Irvine, C. H. (1966), "The Evaluation of Physical Tasks In industry," *American Industrial Hygiene Association Journal*, Vol. 27, pp. 228–233.

Snook, S. H., and Irvine, C. H. (1967), "Maximum Acceptable Weight of Lift," *American Industrial Hygiene Association Journal*, Vol. 28, pp. 322–329.

Snook, S. H., and Irvine, C. H. (1968), "Maximum Frequency of Lift Acceptable to Male Industrial Workers," *American Industrial Hygiene Association Journal*, Vol. 29, pp. 531–536.

Snook, S. H., and Irvine, C. H. (1969), "Psychophysical Studies of Physiological Fatigue Criteria," *Human Factors*, Vol. 11, pp. 291–300.

Snook, S. H., Irvine, C. H., and Bass, S. F. (1970), "Maximum Weights and Work Loads Acceptable to Male Industrial Workers," *American Industrial Hygiene Association Journal*, Vol. 31, pp. 579–586.

Snook, S. H., Campanelli, R. A., and Ford, R. J. (1980), *A Study of Back Injuries at Pratt-Whitney Aircraft*, Liberty Mutual Insurance Company Research Center, Hopkinton, MA.

Sparto, P. J., Parnianpour, M., Reinsel, T. E., and Simon, S. (1997), "The Effect of Fatigue on Multijoint Kinematics, Coordination, and Postural Stability During a Repetitive Lifting Test," *Journal of Orthopaedic and Sports Physical Therapy*, Vol. 25, No. 1, pp. 3–12.

Spengler, D. M. J., Bigos, S. J., Martin, N. A., Zeh, J., Fisher, L., and Nachemsom, A. (1986), "Back Injuries in Industry: A Retrospective Study," *Spine*, Vol. 11, pp. 241–256.

Spitzer, W. O. (1993), "Lowback Pain in the Workplace: Attainable Benefits Not Attained," *British Journal of Industrial Medicine*, Vol. 50, pp. 385–388.

Stevens, S. S. (1975), *Psychophysics: Introduction to Its Perceptual, Neural, Social Prospects*, Wiley, New York.

Stuebbe, P., Genaidy, A., Karwowski, W., Young, G. K., and Alhemood, A. (2002), "The Relationships Between Biomechanical and Postural Stresses, Musculoskeletal Injury Rates, and Perceived Body Discomfort Experienced by Industrial Workers: A Field Study," *International Journal of Occupational Safety and Ergonomics*, Vol. 8, No. 2, pp. 259–280.

Sullivan, M. S. (1989), "Back Support Mechanisms During Manual Lifting," *Physical Therapy*, Vol. 69, pp. 38–49.

Sullivan, D., Bryden, P., and Callaghan, J. P. (2002), "Inter- and Intra-observer Reliability of Calculating Cumulative Lumbar Spine Loads," *Ergonomics*, Vol. 45, pp. 788–797.

Taboun, S. M., and Dutta, S. P. (1989), "Energy Cost Models for Combined Lifting and Carrying Tasks," *International Journal of Industrial Ergonomics*, Vol. 4, No. 1, pp. 1–17.

Tharmmaphornphilas, W., and Norman, B. A. (2004), "A Quantitative Method for Determining Proper Job Rotation Intervals," *Annals of Operations Research*, Vol. 128, pp. 251–266.

van Dieen, J. H., and Kingma, I. (1999), "Total Trunk Muscle Force and Spinal Compression Are Lower in Asymmetric Moments as Compared to Pure Extension Moments," *Journal of Biomechanics*, Vol. 32, pp. 681–687.

van Dieen, J. H., Hoozemans, M. J. M., and Toussaint, H. M. (1999), "Stoop or Squat: A Review of Biomechanical Studies on Lifting Techniques," *Clinical Biomechanics*, Vol. 14, No. 10, pp. 685–696.

van Poppel, M. N., et al. (1998), "Lumbar Supports and Education for the Prevention of Low Back Pain in Industry: A Randomized Controlled Trial," *Journal of the American Medical Association*, Vol. 279, pp. 1789–1794.

Videman, T., Nurminen, M., and Troup, J. D. (1990), "Lumbar Spinal Pathology in Cadaveric Material in Relation to History of Back Pain, Occupation, and Physical Loading," *Spine*, Vol. 15, pp. 728–740.

Von Korff, M., Dworkin, S. F., LeResche, L., and Kruger, A. (1988), "An Epidemiologic Comparison of Pain Complaints," *Pain*, Vol. 32, pp. 173–183.

Waddell, G. (1996), "Low Back Pain: A Twentieth Century Health Care Enigma," *Spine*, Vol. 21, No. 24, pp. 2820–2825.

Waddell, G., and Burton, A. K. (2000), *Occupational Health Guidelines for the Management of Low Back Pain at Work: Evidence Review*, Faculty of Occupational Medicine, London.

Wang, M. J., Garg, A., Chang, Y. C., Shih, Y. C., Yeh, W. Y., and Lee, C. L. (1998), "The Relationship Between Low Back Discomfort Ratings and the NIOSH Lifting Index," *Human Factors*, Vol. 40, No. 3, pp. 509–515.

Waters, T. R. (2004), "National Efforts to Identify Research Issues Related to Prevention of Work-Related Musculoskeletal Disorders," *Journal of Electromyography and Kinesiology*, Vol. 14, pp. 7–12.

Waters, T. R., Putz-Anderson, V., Garg, A., and Fine, L. J. (1993), "Revised NIOSH Equation for the Design and Evaluation of Manual Lifting Tasks," *Ergonomics*, Vol. 36, No. 7, pp. 749–776.

Waters, T. R., Putz-Anderson, V., and Garg, A. (1994), *Applications Manual for the Revised NIOSH Lifting Equation*, National Institute for Occupational Safety and Health, Cincinnati, OH.

Waters, T. R., Putz-Anderson, V., and Baron, S. (1998), "Methods for Assessing the Physical Demands of Manual Lifting: A Review and Case Study from Warehousing," *American Industrial Hygiene Association Journal*, Vol. 59, No. 12, pp. 871–881.

Waters, T. R., Baron, S., Piacitelli, L. A., Anderson, V. P., Skov, T., Haring-Sweeney, M., Wall, D. K., and Fine, L. J. (1999), "Evaluation of the Revised NIOSH Lifting Equation: A Cross-Sectional Epidemiologic Study," *Spine*, Vol. 24, pp. 386–395.

Waters, T., Yeung, S., Genaidy, A., Callaghan, J., Barriera-Viruet, H., and Welge, J. (in press), "Cumulative Spinal Loading Exposure Methods for Manual Material Handling Tasks, 1: Is Cumulative Spinal Loading Associated with Lower Back Disorders?" *Theoretical Issues in Ergonomic Science*.

Webster, B. S., and Snook, S. H. (1994), "The Cost of 1989 Workers' Compensation Low Back Pain Claims," *Spine*, Vol. 19, pp. 111–116.

Yeung, S. S., Genaidy, A., Karwowski, W., Huston, R., and Beltran, J. (2001), "Assessment of Manual Lifting Activities Using Worker Expertise: A Comparison of Two Workers Population," *Asian Journal of Ergonomics* Vol. 2, pp. 11–24.

Yingling, V. R., and McGill, S. M. (1999), "Anterior Shear of Spinal Motion Segments: Kinematics, Kinetics, and Resultant Injuries Observed in a Porcine Model," *Spine*, Vol. 24, No. 18, pp. 1882–1889.

CHAPTER 31

WORK-RELATED UPPER EXTREMITY MUSCULOSKELETAL DISORDERS

Carolyn M. Sommerich and W. S. Marras
The Ohio State University
Columbus, Ohio

Waldemar Karwowski
University of Louisville
Louisville, Kentucky

1 INTRODUCTION

1.1 Extent of the Problem

Estimates of the economic burden due to work-related musculoskeletal disorders (WMSD) are as high as $45 to 54 billion in the United States annually (NRC, 2001). In 1996, the National Institute for Occupational Safety and Health unveiled the National Occupational Research Agenda (NORA), which was designed to address "the broadly recognized need to focus research in the areas with the highest likelihood of reducing the still significant toll of workplace injury and illness" (NIOSH, 1996). Musculoskeletal disorders were designated as one of 21 NORA priority research areas. Musculoskeletal disorders include non-traumatic injuries to the back, trunk, upper extremity, neck, and lower extremity. For 2002, the U.S. Bureau of Labor Statistics reported 487,915 cases of occupational MSDs (BLS, 2002, Table 11), accounting for one-third of the 1,436,194 lost-time occupational injuries and illnesses recorded for that year (excluding farms with fewer than 11 employees). The BLS defines

MSDs as "cases where the nature of injury is: sprains, tears; back pain, hurt back; soreness, pain, hurt, except back; carpal tunnel syndrome; hernia; or musculoskeletal system and connective tissue diseases and disorders and when the event or exposure leading to the injury or illness is: bodily reaction/bending, climbing, crawling, reaching, twisting; overexertion; or repetition. Cases of Raynaud's phenomenon, tarsal tunnel syndrome, and herniated spinal discs are not included. Although these cases may be considered MSD's, the survey classifies these cases in categories that also include non-MSD cases" (BLS, 2002, Table 11, footnote 2). Of those MSDs, 50.4% (246,103 cases) affected the back, 10.8% (52,641 cases) affected the shoulder, and 15.1% (73,707 cases) affected the upper extremity (fingers, hand, or wrist). Although lower than occupational back-related MSDs in number of incident annual cases, upper extremity MSDs (UEMSDs) are costly, due to the extent of the lost workdays typically associated with each case. Overall, in 2002 the median number of lost workdays for a lost-time occupational injury or illness was seven days. By contrast, the median lost workdays for some specific upper extremity MSDs were: bursitis, nine days; ganglion cyst, 13 days; tendonitis, 15 days; synovitis, 24 days; carpal tunnel syndrome, 30 days; and tenosynovitis, 48 days (BLS, 2002, Table R67). Work-related upper extremity musculoskeletal disorders (WUEDs) are not unique to the U.S. workforce. Buckle and Devereux (2002) reported that 5.4 million workdays are lost annually in the UK due to work-related upper limb disorder; about 22 workdays (about one month) are lost for each case. The associated economic burden, based on employers' costs for an average injury, was estimated to be £ 1.25 billion.

1.2 Definitions

According to Hagberg et al. (1995), the umbrella term *work-related musculoskeletal disorders* (WMSDs) defines those disorders and diseases of the musculoskeletal system that have a proven or hypothetical work-related causal component. Musculoskeletal *disorders* are pathological entities in which the functions of the musculoskeletal system are disturbed or abnormal, whereas *diseases* are defined pathological entities with observable impairments in body configuration and function. Although WUEDs are a heterogeneous group of disorders, and the current state of knowledge does not allow for a general description of the course of these disorders, it is possible nevertheless to identify a group of *generic risk factors*, including biomechanical factors, such as static and dynamic loading on the body and posture, cognitive demands, and organizational and psychosocial factors, for which there is evidence of work relatedness and a higher risk of developing WUEDs.

According to Hagberg et al. (1995), such generic risk factors, which typically interact and accumulate to form cascading cycles, are assumed to be directly responsible for pathophysiological phenomena, which depend on location, intensity, temporal variation, duration, and repetitiveness of the generic risk factors. It was also proposed that both insufficient and excessive loading on the musculoskeletal system have deleterious effects and that the pathophysiological process is dependent on a person's characteristics with respect to body responses, coping mechanisms, and adaptation to risk factors. The generic risk factors, workplace design features, and pathophysiological phenomena are parts of the generic model for WUED prevention proposed by Armstrong et al. (1993).

1.3 Cumulative Trauma Disorders of the Upper Extremity

In the United States, the term most often used for WUEDs is *cumulative trauma disorders* (CTDs) of the upper extremity. According to Putz-Anderson (1993), CTDs can be defined by combining the separate meanings for each word. *Cumulative* indicates that these disorders develop gradually over periods of time as a result of repeated stresses. The cumulative concept is based on the assumption that each repetition of an activity produces some trauma or wear and tear on the tissues and joints of the particular body part. The term *trauma* indicates bodily injury from mechanical stresses, whereas *disorders* refers to physical ailments. The definition above also stipulates a simple cause-and-effect model for CTD development. According to such a model, since the human body needs sufficient intervals of rest time between episodes of repeated strains to repair itself, if the recovery time is insufficient, combined with a high repetition of forceful and awkward postures, the worker is at higher risk of developing a CTD. In the context of the generic model for prevention proposed by Armstrong et al. (1993), the definition above is oriented primarily toward biomechanical risk factors for WUEDs and therefore is incomplete.

According to Ranney et al. (1995), the evidence that chronic musculoskeletal disorders of the upper extremities are work related is growing rapidly. Several recent comprehensive reviews, examining multiple sources of evidence and data, have concluded that there is sufficient evidence to link WUED to workplace exposures (Bernard, 1997; Viikari-Juntura and Silverstein, 1999; NRC, 2001; Buckle and Devereux, 2002). Armstrong et al. (1993) concluded that presently it is not possible to define the dose–response relationships and exposure limits for the WUED problems. To establish the work relatedness of these disorders, both the quantification of exposures involved in work and a determination of health outcomes, including details of the specific disorders (Luopajarvi et al., 1979; Moore et al., 1991; Stock, 1991; Hagberg, 1992), are needed. Also, more detailed medical diagnoses are required for choosing appropriate exposure measures as well as for structuring treatment, screening, and prevention programs (Ranney et al., 1995).

2 CONCEPTS AND CHARACTERISTICS

2.1 Epidemiology

According to the World Health Organization (WHO, 1985), an *occupational disease* is a disease for which there is a direct cause–effect relationship between hazard and disease (e.g., silica-silicosis). Work-related diseases (WRDs) are defined as multifactorial when the work environment and the performance of work contribute significantly to the causation of disease (WHO, 1985). Work-related diseases can be partially caused by adverse work conditions. However, personal characteristics, environmental factors, and sociocultural factors are also recognized as risk factors for these diseases.

As reviewed by Armstrong et al. (1993) and summarized by Hagberg et al. (1995), Bernard (1997), and NRC (2001), the evidence of work relatedness of musculoskeletal disorders is established by the pattern of evidence supplied through numerous epidemiologic studies conducted over the last 35 years of research in the field. It was also noted that the incidence and prevalence of musculoskeletal disorders in the reference populations of the studies were low but not zero, indicating that there are non-work-related causes of these disorders as well. Such variables as cultural differences and psychosocial and economic factors, which may influence one's perception and tolerance of pain and consequently affect the willingness to report musculoskeletal problems, may have a significant impact on the progressions from disorder to work disability (WHO, 1985; Leino, 1989). Descriptions of common musculoskeletal disorders and related job activities were summarized by Kroemer et al. (1994) and are shown in Table 1.

2.2 Evidence of Work Relatedness of Upper Extremity Disorders

The WRDs of the upper extremity include, among others, carpal tunnel syndrome, tendonitis, ganglionitis, tenosynovitis, bursitis, and epicondylitis (Putz-Anderson, 1993). Workers employed in construction, food preparation, clerical and computer work, product fabrication, and mining are at a high risk of developing WUEDs. NIOSH (1977) reported that 15 to 20% of workers in these jobs are at a potential risk of WUEDs. Although the occurrence of WUEDs at work has been well documented (Hagberg et al., 1995), because of the high complexity of the problem, there is a lack of clear understanding of the cause–effect relationship characteristics for these disorders, which may prevent implementation of the effective control measures. The problem may be confounded by poor management–labor relationships and lack of willingness to talk openly to each other about the potential problems and how to solve them for the fear of legal litigation, including claims of unfair labor practices.

2.2.1 Definition of Risk Factors

Risk factors are defined as variables that are believed to be related to the probability of a person's developing a disease or disorder (Kleinbaum et al., 1982). Hagberg et al. (1995) classified the generic risk factors for development of WMSDs by considering their explanatory value, biological plausibility, and the relation to work environment: These generic risk factors are (1) fit, reach, and see; (2) musculoskeletal load; (3) static load; (4) postures; (5) cold, vibration, and mechanical stresses; (6) task in invariability; (7) cognitive demands; and (8) organizational and psychosocial work characteristics. These WMSD risk factors are present at varying levels for different jobs and tasks. They are assumed to interact and to have an accumulative effect, forming the cascading cycles described by Armstrong et al. (1993), the extent and severity of which depends on their intensity, duration, and so on, meaning that the mere presence of a risk factor does not necessarily suggest that an exposed worker is at excessive risk of injury (Armstrong et al., 1993).

2.2.2 Biomechanical Factors

In 1986, Armstrong et al. (1986) identified the following categories of *biomechanical risk factors* for development of WUEDs: (1) forceful exertions and motions; (2) repetitive exertions and motions; (3) extreme postures of the shoulder (elbow above midtorso or reaching down and behind), forearm (inward or outward rotation with a bent wrist), wrist (palmar flexion or full extensions), and hand (pinching); (4) mechanical stress concentrations over the base of the palm, on the palmar surface of the fingers, and on the sides of the fingers; (5) duration of exertions, postures, and motions; (6) effects of hand–arm vibration; (7) exposure to a cold environment; (8) insufficient rest or break time; and (9) the use of gloves. In addition, wrist angular flexion–extension acceleration was also determined to be a potential risk factor for hand–wrist cumulative trauma disorders under conditions of dynamic industrial tasks (Marras and Schoenmarklin, 1993; Schoenmarklin et al., 1994). More recently, comprehensive reviews of epidemiological and experimental studies of work-related MSDs have concluded that there is sufficient evidence of associations between physical exposures in the workplace and MSDs (Bernard, 1997; Viikari-Juntura and Silverstein, 1999; NRC, 2001; Buckle and Devereux, 2002). Conclusions from the reviews by Bernard (1997) and NRC (2001) are summarized in Tables 2 and 3, respectively. Tables 4 and 5 present a summary of work-related postural risk factors for wrist and shoulder disorders.

As discussed by Armstrong et al. (1993), poor design of tools with respect to weight, shape, and size can impose extreme wrist positions and high forces on a worker's musculoskeletal system. For example, holding a heavier object requires increased power grip and high tension in the finger flexor tendons, which causes increased pressure in the carpal tunnel. Furthermore, a task that induces hand and arm vibration causes an involuntary increase in power grip through a reflex of the strength receptors. Vibration can also cause protein leakage from the blood vessels in the nerve trunks and result in edema and increased pressure in the nerve trunks and therefore can also result in edema and increased pressure in the nerve (Lundborg et al., 1990).

Several millions of workers in occupations such as vehicle operation are intermittently exposed every year to hand–arm vibration that significantly stresses the musculoskeletal system (Haber, 1971). Hand–arm vibration syndrome (HAVS), which is characterized

Table 1 Common WMSDs

Disorder[a]	Description	Typical Job Activities
Carpal tunnel syndrome (writer's cramp, neuritis, median neuritis) (N)	The result of compression of the median nerve in the carpal tunnel of the wrist. This tunnel is an opening under the carpal ligament on the palmar side of the carpal bones. Through this tunnel pass the median nerve and the finger flexor tendons. Thickening of the tendon sheaths increases the volume of tissue in the tunnel thereby increasing pressure on the median nerve. The tunnel volume is also reduced if the wrist is flexed or extended, or ulnarly or radially pivoted.	Buffing, grinding, polishing, sanding, assembly work, typing, keying, cashiering, playing musical instruments, surgery, packing, housekeeping, cooking, butchering, hand washing, scrubbing, hammering
Cubital tunnel syndrome (N)	Compression of the ulnar nerve below the notch of the elbow. Tingling, numbness, or pain radiating into ring or little fingers.	Resting forearm near elbow on a hard surface and/or sharp edge, also when reaching over obstruction
DeQuervain's syndrome (or disease) (T)	A special case of tenosynovitis which occurs in the abductor and extensor tendons of the thumb, where they share a common sheath. This condition often results from combined forceful gripping and hand twisting, as in wringing clothes.	Buffing, grinding, polishing, sanding, pushing, pressing, sawing, cutting, surgery, butchering, use of pliers, "turning" control such as on a motorcycle, inserting screws in holes, forceful hand wringing
Epicondylitis ("tennis elbow")	Tendons attaching to the epicondyle (the lateral protrusion at the distal end of the humerus bone) become irritated. This condition is often the result of impacts of jerky throwing motions, repeated supination and pronation of the forearm, and forceful wrist extension movements. The condition is well known among tennis players, pitchers, bowlers, and people hammering. A similar irritation of the tendon attachments on the inside off the elbow is called medical epicondylitis, also known as "golfer's elbow."	Turning screws, small parts assembly, hammering, meat cutting, playing musical instruments, playing tennis, pitching, bowling
Ganglion (T)	A tendon sheath swelling that is filled with synovial fluid, or a cystic tumor at the tendon sheath or a joint membrane. The affected area swells up and causes a bump under the skin, often on the dorsal or radial side of the wrist. (Since it was in the past occasionally smashed by striking with a Bible or heavy book, it was also called a "Bible bump.")	Buffing, grinding, polishing, sanding, pushing, pressing, sawing, cutting, surgery, butchering, use of pliers, "turning" control such as on a motorcycle, inserting screws in holes, forceful hand wringing
Neck tension syndrome (M)	An irritation of the levator scapulae and the trapezius muscle of the neck, commonly occurring after repeated or sustained work.	Belt conveyor assembly, typing, keying, small parts assembly, packing, load carrying in hand or on shoulder, overhead work
Pronator (teres) syndrome	Result of compression of the median nerve in the distal third of the forearm, often where it passes through the two heads of the pronator teres muscle in the forearm; common with strenuous flexion of elbow and wrist.	Soldering, buffing, grinding, polishing, sanding

Table 1 (*continued*)

Disorder[a]	Description	Typical Job Activities
Shoulder tendonitis (rotator cuff syndrome or tendonitis, supraspinatus tendonitis, subacromial bursitis, subdeltoid bursitis, partial tear of the rotator cuff) (T)	The rotator cuff consists of four muscles and their tendons that fuse over the shoulder joint. They medially and laterally rotate the arm and help to abduct it. The rotator cuff tendons must pass through a small bony passage between the humerus and the acromion with a bursa as cushion. Irritation and swelling of the tendon or of the bursa are often caused by continuous muscle and tendon effort to keep the arm elevated.	Punch press operations, overhead assembly, overhead welding, overhead painting, overhead auto repair, belt conveyor assembly work, packing, storing, construction work, postal "letter carrying," reaching, lifting, carrying load on shoulder
Tendonitis (T)	An inflammation of a tendon. Often associated with repeated tension, motion, bending, being in contact with a hard surface, vibration. The tendon becomes thickened, bumpy, and irregular in its surface. Tendon fibers may be frayed or torn apart. In tendons without sheaths, such as the biceps tendon, the injured area may calcify.	Punch press operation, assembly work, wiring, packaging, core making, use of pliers
Tenosynovitis (tendosynovitis, tendovaginitis) (T)	Inflammation of the synovial sheaths. The sheath swells. Consequently, movement of the tendon with the sheath is impeded and painful. The tendon surfaces can become irritated, rough, and bumpy. If the inflamed sheath presses progressively onto the tendon, the condition is called stenosing tendosynovitis. "DeQuervain's syndrome" (see there) is a special case occurring at the thumb; the "trigger finger" (see there) condition occurs in flexors of the fingers.	Buffing, grinding, polishing, sanding, pushing, pressing, sawing, cutting, surgery, butchering, use of pliers, "turning" control such as on a motorcycle, inserting screws in holes, forceful hand wringing
Thoracic outlet syndrome (neurovascular compression syndrome, cervicobrachial disorder, brachial plexus neuritis, costoclavicular syndrome, hyperabduction syndrome) (V,N)	A disorder resulting from compression of the nerves and blood vessels of the brachial plexus between clavicle and first and second ribs. If this neurovascular bundle is compressed by the pectoralis minor muscle, blood flow to and from the arm is reduced. This ischemic condition makes the arm numb and limits muscular activities.	Buffing, grinding, polishing, sanding, overhead assembly, overhead welding, overhead painting, overhead auto repair, typing, keying, cashiering, wiring, playing musical instruments, surgery, truck driving, stacking, material handling, postal "letter carrying," carrying heavy loads with extended arms
Trigger finger (or thumb) (T)	A special case of tenosynovitis (see there) where the tendon forms a nodule and becomes nearly locked, so that its forced movement is not smooth but in a snapping or jerking manner. This is a special case of stenosing tenosynovitis crepitans, a condition usually found with the digit flexors at the A1 ligament.	Operating trigger finger, using hand tools that have sharp edges pressing into the tissue or whose handles are too far apart for the user's hand so that the end segments of the fingers are flexed while the middle segments are straight
Ulnar artery aneurysm (V, N)	Weakening of a section of the wall of ulnar artery as it passes through the Guyon tunnel in the wrist; often from pounding or pushing with the heel of the hand. The resulting "bubble" presses on the ulnar nerve in the Guyon tunnel.	Assembly work

(*continued overleaf*)

Table 1 (*continued*)

Disorder[a]	Description	Typical Job Activities
Ulnar nerve entrapment (Guyon tunnel syndrome) (N)	Results from the entrapment of the ulnar nerve as it passes through the Guyon tunnel in the wrist. It can occur from prolonged flexion and extension of the wrist and repeated pressure on the hypothenar eminence of the palm.	Playing musical instruments, carpentering, brick laying, use of pliers, soldering, hammering
White finger ("dead finger," Raynaud's syndrome, vibration syndrome) (V)	Stems from insufficient blood supply bringing about noticeable blanching. Finger turns cold or numb and tingles, and sensation and control of finger movement may be lost. The condition results from closure of the digit's arteries caused by vasospasms triggered by vibrations. A common cause in continued forceful gripping of vibrating tools particularly in a cold environment.	Chain sawing, jackhammering, use of vibrating tool, sanding, paint scraping, using vibrating tool too small for the hand, often in a cold environment

Source: Adapted from Kroemer et al. (1994).
[a]Type of disorder: N, nerve; M, muscle; V, vessel; T, tendon.

Table 2 Evidence for Causal Relationship between Physical Work Factors and MSDs

Body Part	Risk Factor	Strong Evidence	Evidence	Insufficient Evidence	Evidence of No Effect
Neck and neck/shoulder	Repetition		×		
	Force		×		
	Posture	×			
	Vibration			×	
Shoulder	Repetition		×		
	Force			×	
	Posture		×		
	Vibration			×	
Elbow	Repetition			×	
	Force		×		
	Posture			×	
	Combination	×			
Hand/wrist					
Carpal tunnel syndrome	Repetition		×		
	Force		×		
	Posture			×	
	Vibration		×		
	Combination	×			
Tendinitis	Repetition		×		
	Force		×		
	Posture		×		
	Combination	×			
Hand–arm vibration syndrome	Vibration	×			

Source: Bernard (1997).

by intermittent numbness and blanching of the fingers with reduced sensitivity to heat, cold, and pain, affects up to 90% of workers in occupations such as chipping, grinding, and chain-sawing (Wasserman et al., 1974; Taylor and Pelmear, 1976). HAVS is caused primarily by vibration of a part or parts of the body of which the main sources are handheld power tools such as chain saws and jackhammers.

2.2.3 Work Organization Factors

Work organization and environmental factors may also play a significant role in development of

Table 3 Summary of Epidemiological Studies Reviewed by NRC (2001) for Evidence of Association between Work-Related Physical Exposures and WUED

Focus of Study	Number of Studies Providing Risk Estimates	Number of Studies That Found Significant Positive Associations Between Physical Risk Factor Exposure and WUED
Risk factor		
Manual materials handling	28	24
Repetition	8	4
Force	3	2
Repetition and force	2	2
Repetition and cold	1	1
Vibration	32	26
Disorder		
Carpal tunnel syndrome	18[a]	12
Hand–arm vibration syndrome	13	12

[a]Nine studies provided 18 risk estimators.

Table 4 Postural Risk Factors Reported in the Literature for the Wrist

Risk Factor	Results: Outcome and Details
Wrist flexion	CTS, exposure of 20 to 40 hours per week; increased median nerve stresses (pressure); increased finger flexor muscle activation for grasping; median nerve compression by flexor tendons.
Wrist extension	Median nerve compression by flexor tendons; CTS, exposure of 20 to 40 hours per week; increased intracarpal tunnel pressure for extreme extension of 90°.
Wrist ulnar deviation	Exposure response effect found: If deviation greater than 20°, increased pain and pathological findings.
Deviated wrist positions	Workers with carpal tunnel syndrome used these postures more often.
Hand manipulations	More than 1500 to 2000 manipulations per hour led to tenosynovitis.
Wrist motion	1276 flexion extension motions led to fatigue; higher wrist accelerations, and velocities in high-risk wrist WMSD jobs.

Source: Adapted from Kuorinka and Forcier (1995).

WUEDs. *Work organization* is defined as "the objective nature of the work process. It deals with the way in which work is structured, supervised, and processed" (Hagberg et al., 1995). The mechanisms by which work organizational factors can modify the risk for WUEDs include modifying the extent of exposure to other risk factors (physical and environmental) and modifying a person's stress response, thereby increasing the risk associated with a given level of exposure. Specific work organization factors that have been shown to fall into at least one of these categories include (but are not limited to) the following: (1) wage incentives, (2) machine-paced work, (3) workplace conflicts of many types, (4) absence of worker decision latitude, (5) time pressures and work overload, and (6) unaccustomed work during training periods or after returning from long-term leave. As discussed by Hagberg et al. (1995), the organizational context in which work is carried out has major influences on a worker's physical and psychological stress and health. The work organization defines the level of work output required (work standards), the work process (how the work is carried out), the work cycle (work–rest regimens), the social structure, and the nature of supervision.

2.2.4 Psychosocial Work Factors

Psychosocial work factors are defined as "the subjective aspects of work organization and how they are perceived by workers and managers" (Hagberg et al., 1995). Factors commonly investigated include job dissatisfaction and perceptions of workload, supervisor, and co-worker support, job control, monotony of work, job clarity, and interactions with clients. Recent reviews have concluded that high perceived job stress and nonwork-related stress (worry, tension, distress) are consistently linked to WUEDs (NRC, 2001; Bongers et al., 2002). Bernard (1997) and NRC (2001) also found high job demands/workload to be linked to WUED in the majority of studies they reviewed that considered it. The NRC (2001) review concluded that the evidence was insufficient for linking WUED and low decision latitude, social support, or limited rest-break opportunities. MacDonald et al. (2001) presented evidence of high degrees of correlation between some physical and psychosocial work factors. This makes studying these factors more difficult, requiring collection of information on both types of factors and use of sophisticated data analysis techniques in order to draw correct conclusions about any associations between risk factors and WUED. However, it also means that the risk factors

Table 5 Postural Risk Factors Reported in the Literature for the Shoulder

Risk Factor	Results: Outcome and Details
More than 60° abduction or flexion for more than one hour per day	Acute shoulder and neck pain
Less than 15° median upper arm flexion and 10° abduction for continuous work with low loads	Increased sick leave resulting from musculoskeletal problems
Abduction greater than 30°	Rapid fatigue at greater abduction angles
Abduction greater than 45°	Rapid fatigue at 90°
Shoulder forward flexion of 30°, abduction greater than 30°	Hyperabduction syndrome with compression of blood vessels
Hands no greater than 35° above shoulder level	Impairment of blood flow in the supraspinatus muscle
Upper arm flexion or abduction of 90°	Onset of local muscle fatigue
Hands at or above shoulder height	Electromyographic signs of local muscle fatigue in less than one minute
Repetitive shoulder flexion	Tendonitis and other shoulder disorders
Repetitive shoulder abduction or flexion	Acute fatigue
Postures invoking static shoulder loads	Neck–shoulder symptoms negatively related to movement rate
Arm elevation	Tendinitis and other shoulder disorders
Shoulder elevation	Pain
Shoulder elevation and upper arm abduction	Neck–shoulder symptoms
Abduction and forward flexion invoking static shoulder loads	Neck–shoulder symptoms, shoulder pain and sick leave resulting from musculoskeletal problems
Overhead reaching and lifting	Pain

Source: Adapted from Kuorinka and Forcier (1995).

may be linked through work organization elements, and that by thoughtfully addressing those elements, exposures to both physical and psychosocial risk factors may be reduced.

2.2.5 Individual Factors

Individual characteristics of a worker, including anthropometry, health, sex, and age, may alter the way in which work is performed and may affect a worker's capacity for or tolerance of exposure to physical or other risk factors. In particular, Lundberg (2002) and Treaster and Burr (2004) determined that women were at greater risk than men for WUED development. Treaster and Burr (2004) found this to hold, even after accounting for confounders, such as age, and for work factor exposure. They suggested a number of reasons why this might occur, including differences in exposures due to mismatches between female workers and their workstations, tools, strength requirements of tasks, and so on (anthropometry); psychosocial or psychological factors, which may be the result of differences in job status (e.g., many women's jobs may tend to have less autonomy); or the perceived need to work harder to prove one's self in a male-dominated workplace or profession; additional psychological pressure or workload due to responsibilities outside work (care of home, children, or aging parents); or biological differences, possibly related to the effects of sex hormones on soft tissues (Hart et al., 1998).

2.2.6 Basic Classification of Disorders

Since most manual work requires the active use of the arms and hands, the structures of the upper extremity are particularly vulnerable to soft tissue injury. WUEDs are typically associated with repetitive manual tasks with forceful exertions, such as those performed at assembly lines, or when using hand tools, computer keyboards, and other devices or operating machinery. These tasks impose repeated stresses to the upper body, i.e., muscles, tendons, ligaments, nerve tissues, and neurovascular structures. WEUD may be classified by the type of tissue that is primarily affected. Table 6 lists a number of upper extremity MSDs by the tissue that is primarily affected.

Generally, the greater the exposure to a single risk factor or combination of factors, the greater the risk of a WMSD. Furthermore, as the number of risk factors present increases, so does the risk of injury. The interaction between risk factors is more likely to have a multiplicative rather than an additive effect. Evidence for this can be found in the investigation of the effects of repetition and force exposure by Silverstein et al. (1986, 1987). However, risk factors may pose minimal risk of injury if sufficient exposure does not occur or if sufficient recovery time is provided. It is known that changes in the levels of risk factors will result in changes in the risk of WUEDs. Therefore, a reduction in WUEDs risk factors should also reduce the risk for WMSDs.

2.3 Physical Assessments of Workers

Ranney et al. (1995) performed precise physical assessments of workers in highly repetitive jobs as part of a cross-sectional study to assess the association between musculoskeletal disorders and a set of work-related risk factors. A total of 146 female workers employed in five different industries (garment and

Table 6 Upper Extremity MSDs, Classified by Tissue That Is Primarily Affected

Tendon-Related Disorders	Nerve-Related Disorders	Muscle-Related Disorders	Circulatory/Vascular Disorders	Joint-Related Disorders	Bursa-Related Disorders
Paratenonitis, peritendinitis, tendinitis, tendinosis, tenosynovitis	Carpal tunnel syndrome	Muscle strain	Hypothenar hammer syndrome	Osteoarthritis	Bursitis
Epicondylitis	Cubital tunnel syndrome	Myofascial pain, trigger points, myositis, fibromyalgia, fibrositis	Raynaud's syndrome		
Stenosing tenosynovitis (DeQuervain's disease; trigger finger)	Guyon canal syndrome				
Dupuytren's contracture	Radial tunnel syndrome				
Ganglion cyst	Thoracic outlet syndrome, digital neuritis				

Source: Almekinders (1998) and Buckle and Devereux (2002).
Notes (based on Almekinders (1998), NRC (2001); refer also to Figure 5):

- *Paratenonitis:* involves tendon sheath; para-, meso, and epitenon.
- *Peritendinitis:* involves para-, meso, and epitenon (may also involve tendon sheath, depending on reference).
- *Tendinitis:* involves tendon, endotenon.
- *Tendinosis* (insertional or midsubstance): involves tendon, endotenon, tendon–bone junction; refers to situation where there are degenerative changes within the tendon without evidence of inflammatory cells.
- *Tenosynovitis:* involves tendon sheath: para-, meso-, and epitenon; refers to inflammation of the involved tissue(s). Stenosing tenosynovitis occurs when tendon gliding is restricted due to thickening of the sheath or tendon.

automotive trim sewing, electronic assembly, metal parts assembly, supermarket cashiering, and packaging) were examined for the presence of potential work-related musculoskeletal disorders. The prerequisites for selection of industries and tasks within these industries were (l) the existence of repetitive work using the upper limb, (2) at least five to 10 female workers performing the same repetitive job, (3) a range of jobs from light to demanding, (4) minimal job rotation, (5) no major change in the plant for at least one year, and (6) the support from both union or employee group and management.

The study showed that 54% of the workers had evidence of musculoskeletal disorders in the upper extremities that were judged as potentially work related. Many workers had multiple problems, and many were affected bilaterally (33% of workers). Muscle pain and tenderness was the largest problem in both the neck–shoulder area (31%) and in forearm–hand musculature (23%). Most forearm muscle problems were found on the extensor side. Carpal tunnel syndrome was the most common form of disorder, with 16 workers affected (seven people affected bilaterally). DeQuervain's tenosynovitis and wrist flexor tendinitis were the most commonly found tendon disorders in the distal forearm (12 workers affected for each diagnosis). In view of the study results, it was concluded that muscle tissue is highly vulnerable to overuse; that the stressors that affect muscle tissue, such as static loading, should be studied in the forearm as well as in the shoulder; and that exposure should be evaluated bilaterally. Finally, the predominance of forearm muscle and epicondyle disorders on the extensor side was linked to the dual role of these muscles for supporting the hands against gravity plus postural stability during grasping.

The criteria for establishing the work site diagnosis for various WMSDs are shown in Tables 7 and 8. Studies that include physical assessment to identify cases of a disorder are often viewed as more rigorous than those that rely exclusively on subjective recall in response to questions concerning musculoskeletal discomfort or disorder. However, there are a variety of ways in which physical examinations can be conducted and interpreted. When studying a disorder, adhering to a specific definition of the disorder is a key factor in classifying study participants as cases or noncases, and thereafter, determining the degree of association between the disorder and the risk factor(s) being studied. Recognizing the importance of this, criteria are put forth in order that researchers

Table 7 Minimal Clinical Criteria for Establishing Work Site Diagnoses for Work-Related Muscle or Tendon Disorders

Disorder	Symptoms	Examination
Neck myalgia	Pain in one or both sides of the neck increased by neck movement	Tender over paravertebral neck muscles
Trapezius myalgia	Pain on top of shoulder increased by shoulder elevation	Tender top of shoulder or medial border of scapula
Scapulothoracic pain syndrome[a]	Pain in scapular region increased by scapular movement	Tender over rib angles, 2, 3, 4, 5, and/or 6
Rotator cuff tendonitis[b]	Pain in deltoid area or front of shoulder increased by glenohumeral movement	Rotator cuff tenderness[c]
Triceps tendonitis	Elbow pain increased by elbow movement	Tender triceps tendon
Arm myalgia	Pain in muscle(s) of the forearm	Tenderness in a specific muscle of the arm
Epicondylitis/tendonitis[d]	Pain localized to lateral or medial aspect of elbow	Tenderness of lateral or medial epicondyle localized to this area or to soft tissues attached for a distance of 1.5 cm
Forearm myalgia[e]	Pain in the proximal half of the forearm (extensor or flexor aspect)	Tenderness in a specific muscle in the proximal half of the forearm (extensor or flexor aspect) more than 1.5 cm distal to the condyle
Wrist tendonitis[f]	Pain on the extensor or flexor surface of the wrist	Tenderness is localized to specific tendons and is not found over bony prominences
Extensor finger tendonitis[f,g]	Pain on the extensor surface of the hand	Tenderness is localized to specific tendons and is not found over bony prominences
Flexor finger tendonitis[f,g]	Pain on the flexor aspect of the hand or distal forearm	Pain on resisted finger flexion localized to area of tendon
Tenosynovitis (finger/thumb)	Clicking or catching of affected digit on movement; may be pain or a lump in the palm	Demonstration of these complaints, tenderness anterior to metacarpal of affected digit
Tenosynovitis, DeQuervain's	Pain on the radial aspect of wrist	Tenderness over first tendon compartment and positive Finkelstein's test
Intrinsic hand myalgia	Pain in muscles of the hand	Tenderness in a specific muscle in the hand

Source: Adapted from Ranney et al. (1995).

[a]Crepitation on circumduction of the shoulder.
[b]Positive impingement test.
[c]Frozen shoulder excluded.
[d]Positive Mills's test or reverse Mills's test (lateral or medial epicondylitis).
[e]Pain localized to the muscle belly of the muscle being stressed during resisted activity.
[f]Pain localized to tendon being stressed during resisted activity.
[g]Only diagnosed moderate or severe. Classification of severity of muscle/tendon problems: *mild*, above criteria met; *moderate*, pain persists more than two hours after cessation or work but is gone after a night's sleep, or tenderness plus pain on resisted activity if localized in an anatomically correct manner, or see notes a, b, and d to f; *severe*, pain not completely relieved by a night's sleep.

might begin to use common methods of classifying study of patients/participants. Consensus criteria for conducting physical assessments for classification of carpal tunnel syndrome in epidemiological studies are provided by Rempel et al. (1998). Sluiter et al. (2001) provided criteria for identifying 11 specific UEMSDs, including CTS, and a four-step approach for determining work relatedness of a disorder once it is identified. Although these two groups took an expert consensus approach to criteria development, Helliwell et al. (2003) took a statistical approach, relying on multivariate modeling to identify the most discriminating symptoms and signs for classifying six different UEMSDs. Authors of the latter two documents also address "nonspecific" upper extremity disorders as well.

3 ANATOMY OF THE UPPER EXTREMITY

The anatomy of the upper extremity provides for great functionality but also puts certain soft tissue components at risk of damage from repeated or sustained compressive, shear, and/or tensile loading.

Table 8 Minimal Clinical Criteria for Establishing Work Site Diagnoses for Work-Related Neuritis

Disorder	Symptoms	Examination
Carpal tunnel syndrome	Numbness and/or tingling in thumb, index, and/or midfinger with particular wrist postures and/or at night	Positive Phalen's test or Tinel's sign present over the median nerve at the wrist
Scalenus anticus syndrome	Numbness and/or tingling on the preaxial border of the upper lip	Tender scalene muscles with positive Adson's or Wright's test
Cervical neuritis	Pain, numbness, or tingling following a dermatomal pattern in the upper limb	Clinical evidence of intrinsic neck pathology
Lateral antebrachial neuritis	Lateral forearm pain, numbness, and tingling	Tenderness of coracobrachialis origin and reproduction of symptoms on palpation here or by resisted coracobrachialis activity
Pronator syndrome	Pain, numbness, and tingling in the median nerve distribution distal to the elbow	Tenderness of pronator teres or superficial finger flexor muscle, with tingling in the median nerve distribution on resisted activation of same
Cubital tunnel syndrome	Numbness and tingling distal to elbow in ulnar nerve distribution	Tender over ulnar nerve with positive Tinel's sign and/or elbow flexion test
Ulnar tunnel syndrome	Numbness and tingling in ulnar nerve distribution in the hand distal to the wrist	Positive Tinel's sign over the ulnar nerve at the wrist
Wartenberg's syndrome	Numbness and/or tingling in distribution of the superficial radial nerve	Positive Tinel's sign on tapping over the radial sensory nerve
Digital neuritis	Numbness or tingling in the fingers	Positive Tinel's sign on tapping over digital nerves

Source: Adapted from Ranney et al. (1995).

The shoulder complex joins the upper extremity to the axial skeleton. It provides the greatest range of motion of all the body's joints, yet this comes with several associated costs, including reduced joint stability and potential for entrapment of various soft tissues when the arm is elevated or loaded. All handheld loads pass through the shoulder joint, but their effects are magnified increasingly the farther in the transverse plane the hands are located away from the shoulder. The elbow is a simpler joint than the shoulder, yet can also be a sight of nerve entrapment. The hand is small in dimension, yet is capable of producing large amounts of force. In addition, the hand is capable of configuring itself in a variety of orientations and can generate force with either the whole hand in a power grip or with combinations of fingers in opposition to the thumb as in a pinch grip. It is this very flexibility in capability that makes the upper extremity susceptible to cumulative trauma disorders. Refer to Figures 1 to 3 for some views of the anatomy of the upper extremity.

3.1 Anatomy of the Hand

The anatomy of the hand is illustrated in Figure 1. To achieve a variety of functions, the hand is constructed so that it contains numerous small muscles which facilitate fine, precise positioning of the hand and fingers, but few power-producing muscles. One of the only power-producing muscles in the hand is a group of three muscles that form the thenar group, which flex, abduct, and position the thumb for opposition. Strong grasping is produced by extrinsic finger flexor muscles that are located in the forearm. Force is transmitted to

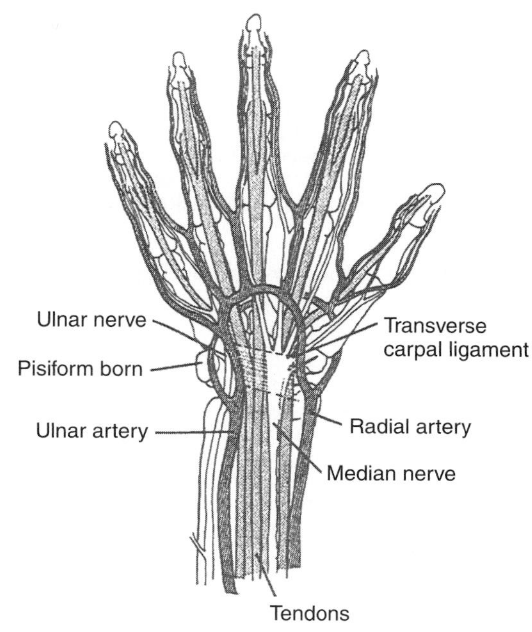

Figure 1 Anatomy of the hand.

the fingers through a network of long tendons (tendons attach muscles to bone). These tendons pass from the muscles in the forearm through the wrist (and through the carpal canal), through the hand, and to the fingers. These tendons are secured at various points along this path with ligaments that keep the tendons in close proximity to the bones. The transverse carpal ligament forms the palmar boundary of the carpal tunnel, and the carpal bones form the remainder of the boundary. The tendons of the upper extremity are also encased in a sheath, which assists in the sliding of the tendons that occurs in concert with muscle contraction and prevents the tendons from sliding directly over the carpal bones or the transverse carpal ligament. A common sheath envelops the nine tendons passing through the carpal tunnel. For the fingers to generate force, a great deal of tension must be passed through these tendons. Since there are various possible combinations of tendons experiencing tension depending on the configuration of the fingers, type of grip, and the grip force required, many of these tendons experience friction. This frictional component can be exacerbated by several factors, including position of the wrist and fingers, motion of the wrist and fingers, and insufficient rest periods. The extrinsic finger flexor muscles are paired with a set of extrinsic finger extensors on the dorsal side of the forearm and hand.

Other structures in the hand are also important to the development of cumulative trauma in the distal portion of upper extremity. As shown in Figure 1, two major blood vessels pass through the hand. Both the radial artery and the ulnar artery provide the tissues and structures of the hand with a blood supply. One of the key structures of the hand that is often involved with cumulative trauma experiences is the nerve structure. The median nerve enters the lower arm and passes through the carpal canal. Once it passes through the carpal canal, the median nerve becomes superficial at the base of the wrist and then branches off to serve the thumb, index finger, middle finger, and the radial side of the ring finger. This nerve also serves the palmar surface of the hand connected to these fingers as well as the dorsal portion up to the first two knuckles on the fingers mentioned above, as well as the thumb up to the first knuckle.

3.2 Anatomy of the Forearm and Elbow

Located on the humerus, the medial epicondyle is the attachment site for the primary wrist flexor muscles and the extrinsic finger flexor muscles; the lateral epicondyle is the attachment site for the primary wrist extensor muscles and extrinsic finger extensor muscles. Repeated activation of either of these groups of muscles has been associated with development of epicondylitis at the relevant epicondyle. At the elbow, the ulnar nerve passes between the olecranon process of the ulna and the medial epicondyle. The space is referred to as the cubital tunnel. Cubital tunnel syndrome may develop as a result of compressive loading of the ulnar nerve as it passes through that tunnel when the elbow is flexed, either repeatedly or over a sustained period. Direct pressure can also be applied to the nerve when part of the body's weight is supported by the elbows, depending on the shallowness of the space. The extrinsic finger extensor muscles, located on the posterior side of the forearm, are a common site of discomfort, as are their tendons and the tendons of the thumb's extrinsic extensor and abductor muscles. An illustration of the anterior view of the forearm appears in Figure 2.

3.3 Anatomy of the Shoulder

The glenohumeral joint, where the head of the humerus partially contacts the glenoid fossa of the scapula, is what most people think of when the term *shoulder* is used. However, there are four articulations that make up the shoulder complex; the other three are the acromioclavicular, sternoclavicular, and scapulothoracic joints. The four joints work in a coordinated manner to provide the wide range of motion possible in a healthy shoulder. The shoulder complex is also composed of approximately 16 muscles and numerous ligaments. The extensive range of motion at the glenohumeral joint is afforded, in part, because of the minimal contact made between the humerus and scapula. The connection is secured by a group of muscles referred to as the *rotator cuff* (teres minor, infraspinatus, supraspinatus, and subscapularis muscles). They create a variety of torques about the joint and also help protect against subluxation (incomplete dislocation). The supraspinatus tendon is thought to be particularly susceptible to injury (tears and tendinitis) for three reasons: (1) It may have an avascular zone near its insertion, (2) it is placed under significant tension when the arm is elevated, and (3) it passes through a confined space above the humeral head and below the acromion. Scapular anatomy also

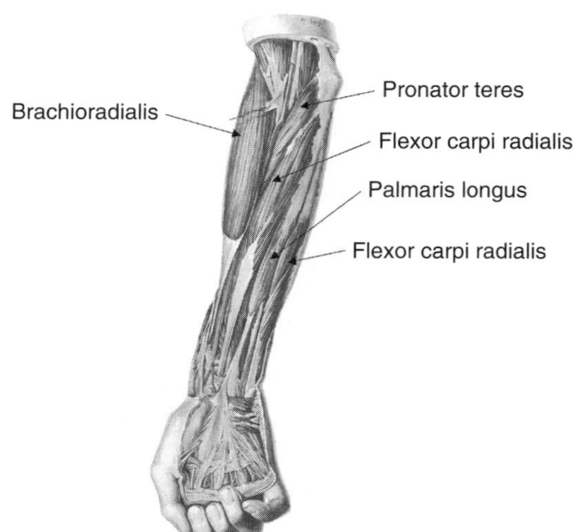

Figure 2 Superficial muscles of the anterior forearm. The two sets of extrinsic muscles that flex the fingers are part of the intermediate and deep layers of anterior forearm muscles. (From LifeART, 2000.)

makes the supraspinatus tendon prone to compression between the humeral head and coracoacromial arch. The coracoacromial arch is formed by the acromion of the scapula and the coracoacromial ligament, which joins the coracoid process and acromion. The arch is a structure above the supraspinatus that can apply a compression force to the tendon if the humeral head migrates superiorly (Soslowsky et al., 1994). The subdeltoid and subacromial bursas are subject to compressive forces when the humerus is elevated, as is the tendon of the long head of the biceps. That tendon is also subject to frictional forces as it moves relative to the humerus within the bicipital groove, when the humerus is elevated.

Other structures within the shoulder complex are also important to the development of cumulative trauma in the proximal portion of upper extremity. These include the brachial plexus (the anterior primary rami of the last four cervical spinal nerves and the first thoracic nerve, which go on to form the radial, median, and ulnar nerves in the upper extremity), the subclavian artery, and the subclavian vein. These structures all pass over the first cervical rib, in close proximity to it. The artery and plexus pass in between the anterior and medial scalene muscles (which attach to the first cervical rib), and the vein lies anterior to the anterior scalene muscle, which separates the vein from the artery. The plexus, artery, and nerve all pass underneath the pectoralis minor muscle. These structures can be compressed by the muscles or bones in proximity to them when the humerus is elevated or when the shoulders are loaded directly (such as when wearing a backpack) or indirectly (as when holding a load in the hands). Additionally, the muscles of the shoulder, particularly trapezius, infraspinatus, supraspinatus, and levator scapula, are also common sites of pain and tenderness (Norregaard et al., 1998). The anterior view of the shoulder, from MRI, is illustrated in Figure 3.

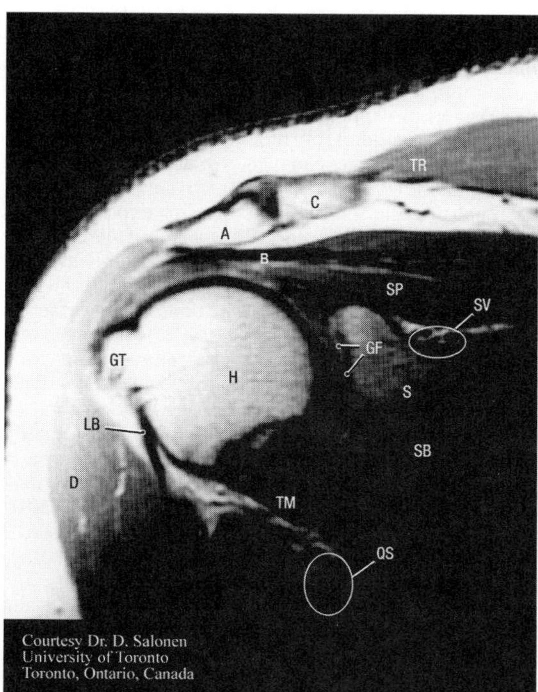

A	Acrominon	SP	Supraspinatus muscle
C	Clavicle	D	Deltoid muscle
H	Head of humerus	TM	Teres minor
GT	Greater tuberosity on humerus	LB	Tendon of the long head of the biceps muscle
TR	Trapezius muscle	B	Subdeltoid bursa

Figure 3 Frontal plane view of the shoulder. (Modified from LifeART, 2000.)

4 CAUSATION MODELS FOR DEVELOPMENT OF DISORDERS

4.1 A Conceptual Model

Armstrong et al. (1993) developed a conceptual model for the pathogenesis of work-related musculoskeletal disorders which is not specific to any particular disorder. The model is based on the set of four cascading and interacting state variables of exposure, dose, capacity, and response, which are measures of the system state at any given time. The response at one level can act as dose at the next level (see Figure 4). Furthermore, it is assumed that a response to one or more doses can diminish or increase the capacity for responding to successive doses. This conceptual model for development of WUEDs reflects the multifactorial nature of these disorders and the complex nature of the interactions among exposure, dose, capacity, and response variables. The model also reflects the complexity of interactions among the physiological, mechanical, individual, and psychosocial risk factors.

In the model proposed, *exposure* refers to the external factors (i.e., work requirements) that produce the internal dose (i.e., tissue loads and metabolic demands and factors). Workplace organization and hand tool design characteristics are examples of such external factors, which can determine work postures and define loads on the affected tissues or the velocity of muscular contractions. *Dose* is defined by a set of mechanical, physiological, or psychological factors that in some way disturb an internal state of the affected worker. Mechanical disturbance factors may include tissue forces and deformations produced as a result of exertion or movement of the body. Physiological disturbances are such factors as consumption of metabolic substrates, or tissue damage, whereas psychological disturbance factors are those related to, for example, anxiety about work or inadequate social support.

Changes in the states of variables of the worker are defined as *responses*. A response is an effect of the dose caused by exposure. The model also allows for a given response to constitute a new dose, which then produces a secondary response (called the *tertiary response*). For example, hand exertion can cause elastic deformation of tendons and changes in

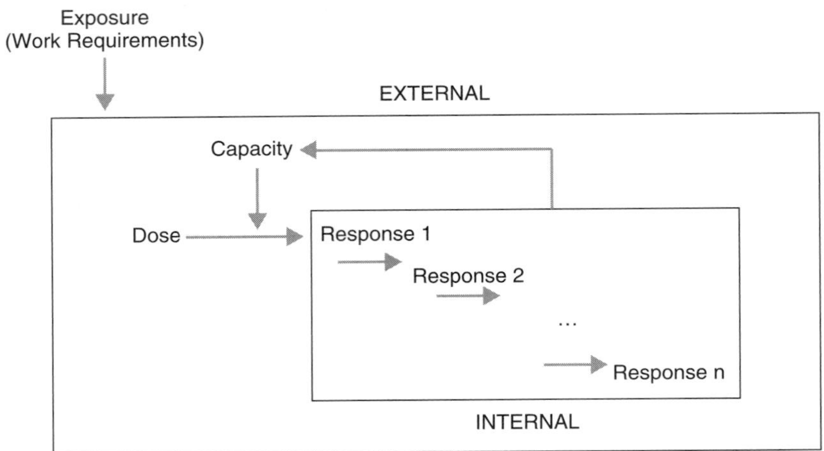

Figure 4 Conceptual model for development of WMSDs proposed by Armstrong et al. (1993).

tissue composition and/or shape, which in turn may result in hand discomfort (Armstrong et al., 1993). The dose–response time relationship implies that the effect of a dose can be immediate or the response may be delayed for a long period of time.

The proposed model stipulates that system changes (responses) can also result in either increased dose tolerance (adaptation) or reduced dose tolerance lowering the system capacity. *Capacity* is defined as the worker's ability (physical or psychological) to resist system destabilization resulting from various doses. Whereas capacity can be reduced or enhanced by previous doses and responses, it is assumed that most people are able to adapt to certain types and levels of physical activity. Table 9 shows characterization of WMSDs with respect to exposure–dose relationship, the worker's capacity, and the response model proposed by Armstrong et al. (1993).

The main purpose of the *dose–response model* is to account for the factors and processes that result in WMSDs to specify acceptable limits with respect to work design parameters for a given person. The exposure, dose, response, and capacity variables need to be measured and quantified. Exposure can be measured using the job title or job classification, questionnaires on possible risk factors, job checklists, or direct measurements. Dose can be measured by estimating muscle forces and joint positions. Worker capacity can be measured using anthropometry, muscle strength, and psychological characteristics. The model proposed should be useful in the design of studies on the etiology and pathomechanisms of WMSDs. The model should also complement epidemiological studies that focus on associations between the physical workload, psychological demands, and environmental risk factors of work at one end, and the manifestations of symptoms, diseases, or disabilities at the other.

4.2 Pathomechanical and Pathophysiological Models

As discussed previously, many epidemiological studies are consistent in their finding of *statistically significant associations* between workplace exposures to various physical risk factors and the incidence and/or prevalence of upper extremity MSDs in workers. Experimental studies on humans (*in vivo*) and investigations that utilize cadavers (*in vitro*) can provide *somewhat more direct evidence* to support hypotheses regarding the internal responses to exposure doses of which Armstrong et al. (1993) wrote. However, the types of experimental studies that can establish *direct causal links* are those based on animal models, where the exposure dose and potential confounding factors can be strictly controlled, and the effects measured over time, to provide a view of the natural history of a disorder's progression. A number of reviews have been published recently, which piece together information from these various types of studies in order to examine, from all sides, the patterns of evidence in support of workplace physical exposures as causes of musculoskeletal disorders in workers. These include NRC (2001) and Barr and Barbe (2002), which reviewed studies concerning the mechanobiology and pathophysiology of tendons, muscles, peripheral nerves, and other tissues that are involved in MSDs; Buckle and Devereux (2002), who conducted a similar review for the European Commission; Visser and van Dieen (in press), who focused on upper extremity muscle disorders; and Viikari-Juntura and Silverstein (1999), who focused on carpal tunnel syndrome. A sampling of key details and excerpts from their conclusions are provided herein. Readers are encouraged to read the full reviews and original research studies for a deeper appreciation of the strength of the evidence provided by these studies.

Table 9 Characterization of Work-Related Musculoskeletal Disorders in General and Muscle, Tendon, and Nerve Disorders in Particular According to Sets of Cascading Exposure and Response Variables as Conceptualized in the Model

Exposure Dose	Worker's Capacity	Response
Musculoskeletal system Work load Work location Work frequency	Body size and shape Physiological state Psychological state	Joint position Muscle force Muscle length Muscle velocity Frequency
Muscle disorders Muscle force Muscle velocity Frequency Duration	Muscle mass Muscle anatomy Fiber type and composition Enzyme concentration Energy stores Capillary density	Membrane permeability Ion flow Membrane action potentials Energy turnover (metabolism), muscle enzymes, and energy stores Intramuscular pressure Ion imbalances Reduced substrates Increased metabolites and water Increase in blood pressure, heart rate, cardiac output, muscle blood flow Muscle fatigue Pain Free radicals Membrane damage Z-disk ruptures Afferent activation
Tendon disorders Muscle force Muscle length Muscle velocity Frequency Joint position Compartment pressure	Anthropometry Tendon anatomy Vascularity Synovial tissue	Stress Strain (elastic and viscous) Microruptures Necrosis Inflammation Fibrosis Adhesions Swelling Pain
Nerve disorders Muscle force Muscle length Muscle velocity Frequency Joint position Compartment pressure	Anthropometry Nerve anatomy Electrolyte status Basal compartment pressure	Stress Strain Ruptures in perineural tissues Protein leakage Ruptures in perineural tissue Protein leakage in nerve trunks Edema Increased pressure Impaired blood flow Numbness, tingling, conduction block Nerve action potentials

Source: Adapted from Armstrong et al. (1993).

4.2.1 Tendons

Tendons are a complex composite material consisting of collagen fibrils embedded in a matrix of proteoglycans. A schematic representation of tendon structure is provided in Figure 5.

Biomechanics Tendons transmit tensile loads between the muscles and the bones to which they are attached. However, they are also subjected to compressive and frictional/shear loads from adjacent structures (bone, other tendons, tendon sheaths, muscle, etc.).

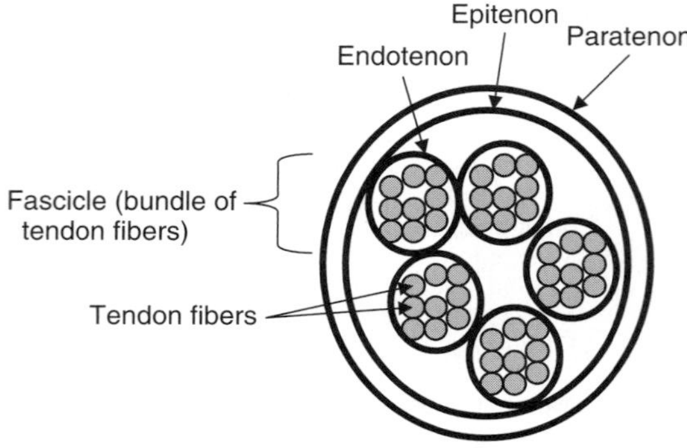

Figure 5 Tendon structure. The paratenon is essentially doubled for those tendons that are surrounded by a synovial sheath (not shown in the figure). The mesotenon (not shown in the figure) are synovial layers that pass from the tendon to the wall of the sheath.

Estimates of these loads have been made by modeling tendons as mechanical pulley systems (Armstrong and Chaffin, 1979). When loads on tendons are excessive (in magnitude, duration, repetition, or in some combination of these), damage is thought to occur. Moore (2002) presented a biomechanical model, which incorporated the pulley tendon model, to express the relative importance of load duration and repetition in the development of tendon entrapment at the dorsal wrist compartments. The most common of these is de Quervain's disease (stenosing tenosynovitis), which involves the tendons of the abductor pollicis longus and extensor pollicis brevis (two extrinsic thumb-moving muscles).

Pathophysiology and Pathomechanics The NRC (2001) review of the mechanobiology of tendons concluded that "basic science studies support the conclusion that repetitive motion or overuse loading can cause chronic injury to tendon tissues." The review noted that fibrocartilaginous tissue is found within tendons, where they wrap around bone (e.g., the long head of biceps tendon within the bicipital groove of the humerus). The review also cited animal studies in which excessive repetitive loading was shown to result in degenerative changes to tendons and result in edema, increased numbers of capillaries, fibrosis, and other changes to the paratenon (the loose connective tissue that surrounds tendons that are not surrounded by a sheath). These changes are similar to those associated with peritendinitis and tendinosis in humans. Recently, Barbe et al. (2003) found evidence of tendon fraying at the muscle–tendon junction within the reaching limb of rats that performed a voluntary, low-force, repetitive reaching task for six to eight weeks. They also found progressive increases, over the course of the study, in the number of infiltrating macrophages in tendons and other tissues, in both forward limbs, but more so in the reaching limb. [*Macrophages* are large cells that possess the property of ingesting bacteria, foreign particles, and other cells. *Infiltrating macrophages* are typically found at sites of inflammation. *Inflammation* is a fundamental pathologic process the occurs in response to an injury or abnormal stimulation caused by a physical, chemical, or biologic agent (McDonough, 1994).] Effects were also seen in tendons that were not directly related to the reaching task, indicating not just a local effect but also a systemic effect of the repetitive hand/paw-use intensive task.

4.2.2 Muscle

Muscle is a composite structure made up of muscle cells, organized networks of nerves and blood vessels, and extracellular connective matrix. Cells are fused together to form each muscle fiber, the basic structural element of skeletal muscle. Muscle is unique among the tissues in the body, for its ability to contract in response to a stimulus from a motoneuron. A *motor unit* is made up of a single alpha motoneuron and the muscle fibers it innervates. There are essentially three types of fibers or motor units. All the fibers within a motor unit are of the same type. Type I fibers are small and are recruited first and at low levels of contraction. They have a high capillary density and are fatigue resistant. Type II fibers are larger and are recruited later, when more force is required. They have low capillary density and are the most fatiguable of the three types. Type IIA have a mix of properties of the other two types. Muscles usually contain all three types in varying proportions, based on the role of the muscle as well as the construction of the person.

Biomechanics Muscle fiber damage can occur when external loads exceed the tolerance of the active contractile components and the passive connective tissue. Nonfatiguing muscle exertions require energy and oxygen, supplied by the blood. The flow of blood can be reduced when a muscle contracts if the intramuscular pressure increases beyond about 30 mmHg (capillary closing pressure). Järvholm et al. (1988, 1991) demonstrated reductions in blood flow in rotator cuff muscles as a function of arm position (abduction or flexion) and weight held in the hand (0 to 2 kg). Even mild elevations, 30° of abduction combined with 45° of flexion, increased intramuscular pressure to 70 mmHg. Contraction of the supraspinatus to only 10% of maximum capability increased intramuscular pressure to 50 mmHg. Elevated intramuscular pressure may lead to localized muscle fatigue or more serious outcomes. Studies have shown that intramuscular pressure sustained at 30 mmHg for eight hours resulted in muscle fiber atrophy, splitting, necrosis, and other damage (NRC, 2001).

Pathophysiology and Pathomechanics The NRC (2001) review of the mechanobiology of skeletal muscle concluded that "the scientific studies reviewed support the conclusion that repetitive mechanical strain exceeding tolerance limits . . . results in chronic skeletal muscle injury." In addition to finding changes in tendon, Barbe et al. (2003) also found infiltrating macrophages in the muscles of the reaching and nonreaching limb of test rats which performed the repetitive, low-force reaching task. Muscles in the paw and the distal forearm were affected, as were muscles that were involved in the task only indirectly (forearm extensors, upper arm and shoulder muscles) or not at all (tibial muscles). Heat shock protein cells increased, first in the intrinsic hand muscles and then in the distal forelimb flexor muscles. Heat shock proteins (HSPs) have a protective role in the cell. Cells increase their production of HSPs when they experience acute or chronic stress. Precursors that stimulate HSP production include inflammation, ischemia, and nerve crush as well as other stimulating factors.

Visser and van Dieen (in press) reviewed several hypotheses concerning the pathogenesis of work-related upper extremity muscle disorders and concluded that no complete proof existed in the literature for any of them but that some of them were likely to interact or follow one another in a downward spiral of damage:

> It appears that the selective and sustained motor unit recruitment [of small type I fibers] in combination with homeostatic disturbances possibly due to limitations in blood supply and metabolite removal offers a plausible basis for the pathogenesis of muscle disorder in low intensity tasks. The bulk of the findings from the biopsy studies reviewed, which indicate mitochondrial dysfunction of type I fibers in myalgic muscles, could also be accounted for by such a mechanism .... As a response to the release of metabolites in the muscle the circulation increases. Sympathetic activation (stress) might lead to a reduction of circulation and an increase of muscle activation. Sustained exposure can result in an accumulation of metabolites, stimulating nociceptors. This process can be enhanced in subjects with relatively large type I fibers and low capillarization, which paradoxically may have developed as an adaptation to the exposure. Nociceptor activation can disturb the proprioception and thereby the motor control most likely leading to further increased disturbance of muscle homeostasis. In addition, in the long run a reduction of the pain threshold and an increase of pain sensitivity can develop. It is worth noting that initial nociceptor stimulation may be a response to metabolite accumulation and not to tissue damage.

4.2.3 Peripheral Nerves

Peripheral nerves are composed of nerve fibers, connective tissue, and blood vessels. A nerve fiber is a long process that extends from a nerve cell body. The term *nerve* refers to a bundle of axons, some of which send information from the spinal cord to the periphery (e.g., muscles, vessels) and some of which send information from peripheral tissues (e.g., skin, muscle, tendon) to the spinal cord. Figure 6 illustrates a single nerve cell (motor neuron) and a nerve.

Biomechanics Peripheral nerves are well vascularized, in order to supply energy needs for impulse transmission and axonal transport (transportation of nutrition and waste products within a nerve cell). If pressure is elevated within the nerve (due to edema or external compression), blood flow is reduced, and as a result, impulse transmission and axonal transport can be slowed or disrupted. Structural damage may also occur.

Pathophysiology and Pathomechanics Numerous studies of nerve function have identified threshold levels of pressure, ranging from 20 to 40 mmHg, above which nerve function, including circulation, axonal transport, and impulse conduction, is compromised (Rydevik et al., 1981; Szabo and Gelberman, 1987). Damage to nerves comes in the form of loss of myelin (Mackinnon et al., 1984), large myelinated fibers (Hargens et al., 1979), and damage to small unmyelinated fibers (Hargens et al., 1979). The physiological and histological effects of pressure on a nerve depend on the amount of pressure, how it is applied (dispersed or focal), and duration of application. Vibration can also cause damage to peripheral nerves, including breakdown of myelin, interstitial and

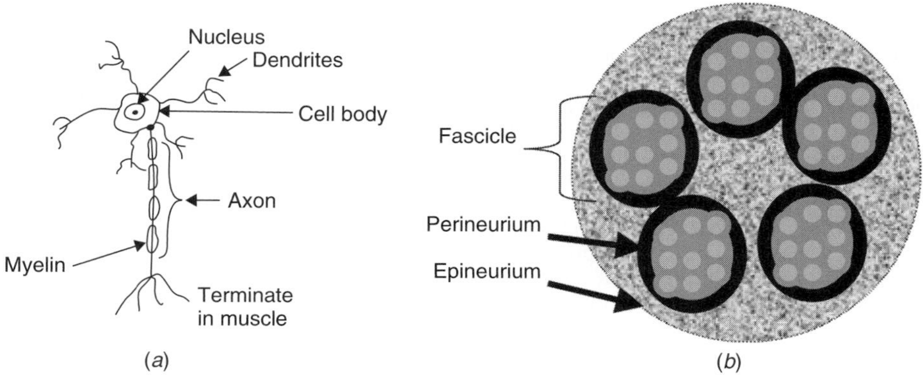

Figure 6 (*a*) Neuron; and (*b*) nerve.

perineural fibrosis, and axonal loss, based on evidence from biopsies from humans exposed to vibrating hand tools and empirical animal studies (NRC, 2001).

4.2.4 Carpal Tunnel Syndrome

Carpal tunnel syndrome (CTS) is the most commonly encountered peripheral neuropathy (Falkiner and Myers, 2002; Werner and Andary, 2002). The term describes a constellation of symptoms that result from localized compression of the median nerve within the carpal tunnel. The carpal tunnel is a fibro-osseous canal created by the carpal bones (floor and walls) and the flexor retinaculum (ceiling). The contents include the median nerve, the eight extrinsic flexor tendons of digits 2 to 5 [flexor digitorum superficialis and profundus (FDS and FDP), respectively], and tendons of the flexor pollicis longus (FPL) and flexor carpi radialis (FCR) (the latter is somewhat separated from the other contents, in its own subtunnel). Part of the volume of the carpal tunnel is also taken up by the synovial sheaths that surround the tendons.

The contents of the tunnel are not static. The nerve moves transversely within the tunnel, and relative to the tendons, when the extrinsic finger flexors are activated isometrically (Nakamichi and Tachibana, 1992), when the fingers are flexed or extended (Ham et al., 1996), and when the wrist is flexed or extended (Zeiss et al., 1989), and change in the nerve's location seems to be somewhat variable from one person to the next. The extrinsic finger flexor tendons slide proximally with finger flexion and distally with finger extension. Further, in some people lumbrical muscles may enter the tunnel during pinching (Ditmars, 1993) and appear consistently to enter the tunnel with full finger flexion (Siegel et al., 1995; Ham et al., 1996). Distal fibers of the FDS and FDP may enter during wrist extension (Keir and Bach, 2000). The cross-sectional area of the tunnel also changes, being smaller when the fingers are fully extended than when flexed (Ham et al., 1996). The area of the tunnel is reduced in wrist flexion and extension compared with a neutral posture (Keir, 2001).

Biomechanics Soft tissues appear to be subjected to mechanical stress within the carpal tunnel. The median nerve takes on an oval or somewhat flattened shape within the carpal tunnel (Robbins, 1963; Zeiss et al., 1989). In a study of cadaver hands (donor health history unknown), Armstrong et al. (1984) observed increased subsynovial and adjacent connective tissue densities, increased synovial hyperplasia, and arteriole and venule muscular hypertrophy, with changes being most pronounced near the distal wrist crease. The authors suggested that the types of changes, similar to those seen in CTS biopsy specimens, and extent to which their severity corresponded to their proximity to the distal wrist crease, indicated that repeated flexion and extension at the wrist imposes mechanical stress on the tissues that cross that joint and that their alterations were in direct response to that stress. They also suggested that highly repetitive use of the wrist might bring about alterations that would be severe enough to elicit CTS symptoms. In patients undergoing CTS release surgery, Schuind et al. (1990) found fibrous hyperplasia of the flexor tendon synovium and increased amounts of collagen fibers (irregular and disorganized). More advanced lesions contained necrotic areas, as well. The authors concluded that these histologic lesions were "typical of a connective tissue undergoing degeneration under repeated mechanical stresses."

Pathophysiology and Pathomechanics The median nerve can be damaged by direct force (damaging myelin and other structural components) and by elevated hydrostatic pressure (ischemic response). Schuind et al. (1990) hypothesized that following an initial mechanical stress upon the flexor tendon synovium, a vicious cycle develops in which changes in synovium increase frictional loads on the tendons as they move back and forth through the tunnel, which causes further irritation to the synovium. Effects on the median nerve are increased pressure, due to a reduction of free volume within the carpal tunnel, and possibly restrictions on the nerve's freedom to move within the tunnel during wrist movement (contact stress). This is consistent with the cascade model of work-related upper-limb musculoskeletal disorders proposed by Armstrong et al. (1993). Another cascade model theorizes that elevated pressure within the carpal tunnel leads to ischemia in the tendons, sheaths, and median nerve. This is followed by tissue swelling, which further increases the pressure in the tunnel and can result in physiological and histological changes in the median nerve (Lluch, 1992).

Studies of nerve function have identified threshold levels of pressure, ranging from 20 to 40 mmHg, above which nerve function, including circulation, axonal transport, and impulse conduction, is compromised (Rydevik et al., 1981; Szabo and Gelberman, 1987). Short-term laboratory-based studies have also shown that carpal tunnel pressure (CTP) in healthy people exceeds these levels with nonneutral wrist postures (Keir et al., 1998b) or in using the hand in ordinary ways, such as pressing with the fingertip or pinching (Keir et al., 1998a) or typing (Sommerich et al., 1996). Even with hands inactive and wrists in a neutral posture, CTP is elevated in people with CTS (Rojviroj et al., 1990). Cyclic movement of the wrist was shown to induce sustained, elevated CTP in patients with early or intermediate CTS (Szabo and Chidgey, 1989).

Viikari-Juntura and Silverstein (1999) examined the pattern of evidence of the role of physical factors in the development of CTS, by reviewing studies from various areas: epidemiological, experimental, cadaver, and animal. One of the consistent and unifying threads they found was the connection (manifestation) of these external, physical factors (posture, force, repetition, and external pressure) as carpal tunnel pressure (CTP), or as having an effect on CTP. They concluded that there is sufficient evidence that duration, frequency, or intensity of exposure to forceful repetitive work and extreme wrist postures is likely to be related to the occurrence of CTS in working populations. Recent work by Clark et al. (2003, 2004) provides additional insight into the effects of performing highly repetitive tasks, in rats performing voluntary hand/paw use-intensive tasks. Effects of a low-force, repetitive reaching task included increased numbers of infiltrating macrophages (a sign of inflammation) in the median nerve, with a greater increase in the reach than in the nonreach limb; myelin degradation; and fibrosis, particularly in the epineurium at the wrist and just distal to the carpal ligament (associated with increased compression) (Clark et al., 2003). Effects of a high-force, repetitive grasping task included infiltrating macrophages in all connective tissue associated with the nerve and increased collagen type I (fibrosis) in perineurium, epineurium, and surrounding tissues. For both of these effects, there was no difference between the reach and nonreach limbs (Clark et al., 2004). These bilateral effects are important, because bilateral presentation of CTS in workers is taken by some to be an indication that work is not the cause of the condition.

4.3 Causal Models Summary

Evidence of the role of physical factors in the development of upper-extremity MSDs, from epidemiological, experimental, cadaver, and animal studies, seems to support the conceptual model of Armstrong et al. (1993). Additional animal models, to replicate and extend the work of Clark and colleagues, will serve to clarify and begin to quantify the relationships between external dose and internal response. Eventually, such work will lead to quantitative, preventive models that can be used to reduce risk of WUED in human workers.

5 QUANTITATIVE MODELS FOR CONTROL OF DISORDERS

5.1 Challenges

Today, only a few quantitative models that are based on the physiological, biomechanical, or psychophysical data and that relate the specific job risk factors for musculoskeletal disorders to increased risk of developing such disorders have been developed. As discussed by Moore and Garg (1995), this is mainly because (1) the dose–response (cause–effect) relationships are not well understood; (2) measurement of some task variables, such as force and even posture, is difficult in an industrial setting; and (3) the number of task variables is very large. However, it is generally recognized that biomechanical risk factors, such as force, repetition, posture, recovery time, duration of exposure, static muscular work, use of the hand as a tool, and type of grasp, are important for explaining the causation mechanism of WUEDs (Armstrong and Lifshiz, 1987; Keyserling et al., 1993). Given the foregoing knowledge, even though limited in scope and subject to epidemiological validation, a few methodologies that allow discrimination between safe and hazardous jobs in terms of workers being at increased risk of developing the WUEDs have been developed and reported in the subject literature. Some of the quantitative data and models for evaluation and prevention of WUEDs available today are described below.

5.2 Maximum Acceptable Forces for Repetitive Wrist Motions for Females

Snook et al. (1995) utilized the psychophysical methodology to determine the maximum acceptable forces for various types and frequencies of repetitive wrist motion, including (1) flexion motion with a power grip (handle diameter, 40 mm; handle length, 135 mm); (2) flexion motion with a pinch grip (handle thickness, 5 mm; handle length, 55 mm); and (3) extension motion with a power grip (handle diameter, 40 mm; handle length, 135 mm). Subjects were instructed to work as if they were on an incentive basis, getting paid for the amount of work that they performed. They were asked to work as hard as they could (i.e., against as much resistance as they could) without developing unusual discomfort in the hands, wrists, or forearms.

Fifteen women worked seven hours each day, two days per week, for 40 days in the first experiment. Repetition rates of 2, 5, 10, 15 and 20 motions per minute were used with each flexion and extension task. Maximum acceptable torques were determined for the various motions, grips, and repetition rates without dramatic changes in wrist strength, tactile sensitivity, or number of symptoms. Fourteen women worked in the second experiment, performing a wrist flexion motion (power grip) 15 times per minute, seven hours per day, five days per week, for 23 days. In addition to the four dependent variables, which included the maximum acceptable torque, maximum isometric wrist strength, tactile sensitivity, and symptoms, performance errors and duration of force were measured. The most common health symptom reported was muscle soreness (55.3%), located mostly in the hand and wrist (51.8%). Numbness in the palmar side of the fingers and thumb (69.1%) and stiffness on the dorsal (back) side of the fingers and thumb (30.6%) were also reported. The number of symptoms increased consistently as the day progressed. The number of symptoms reported was two to three times higher after the seventh hour of work than after the first hour of work (similar to the two to four times higher rate in the two days per week exposure). Symptoms reports were 4.1 times higher at the end of the day than at the beginning of the day before testing began.

The maximum acceptable torque determined during the five days per week exposure was 36% lower than the task performed only two days per week. Based on assumption that maximum acceptable torque decreases 36.3% for the other repetition rates used during the two days per week exposure, and using the adjusted means and coefficients of variation from the two days per week exposure, the maximum acceptable torques were estimated for different repetition of wrist flexion (power grip) and different percentages of the population. Torques were then converted into forces by dividing each torque by the average length of the handle lever (0.081 m). The estimated maximum acceptable forces for female wrist flexion (power grip) are shown in Table 10. Tables 11 and 12 show the estimated maximum acceptable forces for female wrist flexion (pinch grip) and wrist extension (power grip), respectively. The torques were converted into forces by dividing by 0.081 m for the power grip and 0.123 m for the pinch grip.

5.3 Hand Activity Level

In 2002 the American Congress of Governmental Industrial Hygienists (ACGIH) adopted hand activity level within the section on ergonomics in its annual publication of threshold limit values (TLVs). This annual publication contains recommendations and guidelines "to assist in the control of potential workplace health hazards" (ACGIH, 2002). The TLV considers the dual exposures of average hand activity level and peak hand force for monotask jobs performed for four or more hours per day. Peak force is normalized to the strength capability of the workforce for the activity being assessed. Combinations of force and activity either fall within an acceptable range (below the action limit line), which means that nearly all workers could be exposed repeatedly without adverse health effects, a midrange that is delineated by the action limit line below and the threshold limit line above in which some workers might be at risk, and a region that exceeds the TLV in which jobs with similar characteristics have been shown to be associated with elevated risk of UEMSDs. The TLV does not specifically account for contact stress, nonneutral postures, cold, gloves, or vibration, and ACGIH urges use of professional judgment when modifying factors such as these are also present in the task. More specific information about the HAL can be found at http://www.acgih.org/ and other locations on the Web.

Table 10 Maximum Acceptable Forces for Female Wrist Flexion (Power Grip) (Newtons)

Percentage of Population	Repetition Rate				
	2/min	5/min	10/min	15/min	20/min
90	14.9	14.9	13.5	12.0	10.2
75	23.2	23.2	20.9	18.6	15.8
50	32.3	32.3	29.0	26.0	22.1
25	41.5	41.5	37.2	33.5	28.4
10	49.8	49.8	44.6	40.1	34.0

Source: Adapted from Snook et al. (1995).

Table 11 Maximum Acceptable Forces for Female Wrist Flexion (Pinch Grip) (Newtons)

Percentage of Population	Repetition Rate				
	2/min	5/min	10/min	15/min	20/min
90	9.2	8.5	7.4	7.4	6.0
75	14.2	13.2	11.5	11.5	9.3
50	19.8	18.4	16.0	16.0	12.9
25	25.4	23.6	20.6	20.6	16.6
10	30.5	28.3	24.6	24.6	19.8

Source: Adapted from Snook et al. (1995).

Table 12 Maximum Acceptable Forces for Female Wrist Extension (Power Grip) (Newtons)

Percentage of Population	Repetition Rate				
	2/min	5/min	10/min	15/min	20/min
90	8.8	8.8	7.8	6.9	5.4
75	13.6	13.6	12.1	10.9	8.5
50	18.9	18.9	16.8	15.1	11.9
25	24.2	24.2	21.5	19.3	15.2
10	29.0	29.0	25.8	23.2	18.3

Source: Adapted from Snook et al. (1995).

5.4 Strain Index

5.4.1 Model Structure

Moore and Garg (1995) developed a self-styled semi-quantitative job analysis methodology for identifying industrial jobs associated with distal upper extremity disorders (elbow, forearm, wrist, and hand). An existing body of knowledge and theory of the physiology, biomechanics, and epidemiology of distal upper extremity disorders was used for that purpose. The following major principles were derived from the physiological model of localized muscle fatigue:

1. The primary task variables are intensity of exertion, duration of exertion, and duration of recovery.
2. *Intensity of exertion* refers to the force required to perform a task one time and is characterized as a percentage of maximal strength.
3. *Duration of exertion* describes how long an exertion is applied. The sum of duration of exertion and duration of recovery is the cycle time of one exertional cycle.
4. Wrist posture, type of grasp, and speed of work are considered by means of their effects of maximal strength.
5. The relationship between strain on the body (endurance time) and intensity of exertion is nonlinear.

The following were major principles derived from the biomechanical model of the viscoelastic properties of components of a muscle–tendon unit:

1. The primary task variables for the viscoelastic properties are intensity and duration of exertion, duration of recovery, number of exertions, wrist posture, and speed of work.
2. The primary task variables for intrinsic compression are intensity of exertion and nonneutral wrist posture.

3. The relationship between strain on the body and intensity of effort is nonlinear.

Finally, the major principles derived from the epidemiological literature and used for the purpose of model development were as follows:

1. The primary task variables associated with an increased prevalence or incidence of distal upper extremity disorders are intensity of exertion (force), repetition rate, and percentage of recovery time per cycle.
2. Intensity of exertion is the most important task variable related to disorders of the muscle–tendon unit.
3. Wrist posture may not be an independent risk factor because it may contribute to an increased incidence of distal upper extremity disorders when combined with intensity of exertion.
4. The roles of other task variables have not been clearly established epidemiologically.

Moore and Garg (1995) compared exposure factors for jobs associated with WUEDs to jobs without prevalence of such disorders. They found that the intensity of exertion, estimated as a percentage of maximal strength and adjusted for wrist posture and speed of work, was the major discriminating factor. The relationship between the incidence rate for distal upper extremity disorder and job risk factors was defined as follows:

$$\frac{\mathrm{IR} = 30 \times F^2}{\mathrm{RT}^{0.6}} \quad (1)$$

where IR is the incidence rate (per 100 workers per year), F the intensity of exertion (% of maximum strength), and RT the recovery time (% of cycle time).

5.4.2 Elements of the Strain Index

The strain index (SI) proposed by Moore and Garg (1995) is the product of six multipliers that correspond to six task variables: (1) intensity of exertion, (2) duration of exertion, (3) exertions per minute, (4) hand–wrist posture, (5) speed of work, and (6) duration of task per day. An ordinal rating is assigned for each of the variables according to the exposure data. The ratings that are applied to model variables are presented in Table 13. The multipliers for each task variable related to these ratings are shown in Table 14. The strain index score is defined as follows:

$$\begin{aligned}\mathrm{SI} = {} & (\text{intensity of exertion multiplier}) \\ & \times (\text{duration of exertion multiplier}) \qquad (2) \\ & \times (\text{exertions per minute multiplier}) \\ & \times (\text{posture multiplier}) \\ & \times (\text{speed of work multiplier}) \\ & \times (\text{duration per day multiplier})\end{aligned}$$

Intensity of exertion, the most critical variable of the SI, is defined as the percentage of maximum strength required to perform a task once. The intensity of exertion is estimated by an observer using verbal descriptors (see Table 13) and assigned corresponding rating values (1, 2, 3, 4, or 5). The multiplier values (Table 14) are defined based on the rating score raised to a power of 1.6 to reflect the nonlinear nature of the relationship between intensity of exertion and manifestations of strain according to the psychophysical theory. The multipliers for other task variables are modifiers to the intensity of the exertion multiplier.

Duration of exertion is defined as the percentage of time that an exertion is applied per cycle. The terms *cycle* and *cycle time* refer to the exertional cycle and average exertional cycle time, respectively. The duration of recovery per cycle is equal to the exertional cycle time minus the duration of exertion per cycle. The duration of exertion is the average duration of exertion per exertional cycle (calculated by dividing all durations of a series of exertions by the number of exertions observed). The percentage duration of exertion is calculated by dividing the average duration of exertion per cycle by the average exertional cycle time, then multiplying the result by 100:

$$\%\ \text{duration of exertion} = \frac{\text{average duration of exertion per cycle}}{\text{Average exertional cycle time}} \times 100 \quad (3)$$

The percentage duration of exertion calculated is compared to the ranges in Table 13 and assigned the appropriate rating. The corresponding multipliers are identified using Table 14.

Efforts per minute is the number of exertions per minute (e.g., repetitiveness) and is synonymous with frequency. Efforts per minute are measured by counting the number of exertions that occur during a representative observation period (as described for determining the average exertional cycle time). The results measured are compared to the ranges shown in Table 13 and given the corresponding ratings. The multipliers are defined in Table 14.

Posture refers to the anatomical position of the wrist or hand relative to neutral position and can be rated qualitatively using verbal anchors. As shown in Table 14, posture has four relevant ratings. Postures that are "very good" or "good" are essentially neutral and have multipliers of 1.0. As hand or wrist postures progressively deviate beyond the neutral range to extremes, they are graded as "fair," "bad," and "very bad."

Speed of work refers to perceived pace of the task or job and can be estimated subjectively. Once a verbal anchor is selected, a rating is assigned according to Table 13. *Duration of task per day* is defined as a total time that a task is performed per day. As such, this variable reflects the beneficial effects of task diversity such as job rotations and the adverse effects of prolonged activity such as overtime. Duration of

Table 13 Rating Criteria for the Strain Index

Rating	Intensity of Exertion	Duration of Exertion (% of Cycle)	Efforts per Minute	Hand–Wrist Posture	Speed of Work	Duration per Day (h)
1	Light	<10	<4	Very good	Very slow	≤1
2	Somewhat hard	10–29	4–8	Good	Slow	1–2
3	Hard	30–49	9–14	Fair	Fair	2–4
4	Very hard	50–79	15–19	Bad	Fast	4–8
5	Near maximal	≥ 80	≥ 20	Very bad	Very fast	≥ 8

Source: Adapted from Moore and Garg (1995).

Table 14 Multipliers for the Strain Index

Rating	Intensity of Exertion	Duration of Exertion (% of Cycle)	Efforts per Minute	Hand–Wrist Posture	Speed of Work	Duration per Day (h)
1	1	0.5	0.5	1.0	1.0	0.25
2	3	1.0	1.0	1.0	1.0	0.50
3	6	1.5	1.5	1.5	1.0	0.75
4	9	2.0	2.0	2.0	1.5	1.00
5	13	3.0[a]	3.0	3.0	2.0	1.50

Source: Adapted from Moore and Garg (1995).

[a]If duration of exertion is 100%, the efforts/minute multiplier should be set to 3.0.

task per day is measured in hours and assigned a rating according to Table 13.

Application of the strain index involves five steps: (1) collecting data, (2) assigning rating values, (3) determining multipliers, (4) calculating the SI score, and (5) interpreting the results. The values of intensity of exertion, wrist posture, and speed of work can be estimated using the verbal descriptors in Table 13. The values of percentage duration of exertion per cycle, efforts per minute, and duration per day are based on measurements and counts. These values are then compared to the appropriate column in Table 14 and assigned a rating. The SI multipliers are determined from Table 14. Table 15 shows the numerical example for calculating the strain index.

5.4.3 Application

In a preliminary test of the ability of the strain index to distinguish between jobs that were high or low risk for distal UEMSDs, the SI was found to have a sensitivity of 0.92 (able to identify correctly 11 of 12 known positive jobs) and a specificity of 1.0 (able to identify correctly 13 of 13 known negative jobs), when 25 jobs in a pork processing facility were assessed. *Negative* is used in the medical sense to indicate lack of a finding. In other words, negative jobs were those that had not been associated with distal upper extremity MSDs. From that study came the suggestion of using an SI score of 5 as a threshold to distinguish between safe and hazardous jobs. Rucker and Moore (2002) performed another assessment of the SI, this time on 28 jobs from two different manufacturing companies. The sensitivity, specificity, positive predictive value, and negative predictive value were 1.00, 0.91, 0.75, and 1.00, respectively, providing additional evidence of the validity of the tool for assessment of single tasks.

5.4.4 Limitations

The proposed strain index methodology aims to discriminate between jobs that expose workers to risk

Table 15 Example to Demonstrate the Procedure for Calculating the SI Score

	Intensity of Exertion	Duration of Exertion	Efforts per Minute	Posture	Speed of Work	Duration per Day (h)
Exposure dose	Somewhat hard	60%	12	Fair	Fair	4–8
Rating	2	4	3	3	3	4
Multiplier	3.0	2.0	1.5	1.5	1.0	1.0
	SI Score = 3.0×2.0×1.5×1.5×1.0×1.0 = 13.5					

Source: Adapted from Moore and Garg (1995).

factors (task variables) that cause WUEDs vs. jobs that do not. However, according to Moore and Garg (1995), the strain index is not designed to identify jobs associated with an increased risk of any specific disorder. It is anticipated that jobs identified by the strain index to be in the high-risk category will exhibit higher levels of WUEDs among workers who currently perform or historically performed those jobs believed to be hazardous. Finally, the authors caution that large-scale studies are needed to validate and update the methodology proposed. The strain index has the following primary limitations in terms of its application:

1. It is designed primarily to predict distal upper extremity disorders involving muscle–tendon units and CTS rather than all UEMSDs [e.g., hand–arm vibration syndrome (HAVS), ganglion cysts, osteoarthritis].
2. The strain index has not been developed to predict disorders beyond the distal upper extremity, such as disorders of the shoulder, shoulder girdle, neck, or back.
3. No method has been developed for using the strain index to assess multiple tasks.

6 ERGONOMICS EFFORTS TO CONTROL DISORDERS

6.1 Strategies for Prevention of Musculoskeletal Injuries

Facing the growing challenges of musculoskeletal injuries in the contemporary workplace, the *Proposed National Strategies for the Prevention of Leading Work-Related Diseases and Injuries* (NIOSH, 1986) identified environmental hazards and human biological hazards among the four main factors that contribute to human diseases. *Environmental hazards* to the musculoskeletal system associated with work were described as workplace traumatogens (i.e., a source of biomechanical stress from job demands that exceed the worker's strength or endurance, such as heavy lifting or repetitive and forceful manual exertions). Traumatogens can be measured by determining the frequency, magnitude, and direction of forces imposed on the body in relation to posture and the point of application. *Human biological factors* include the anthropometric or innate attributes that influence a worker's capacity for performing a job safely. Examples include the worker's physical size, strength, range of motion, and work endurance. These factors account partially for variability in performance capability in the population and the potential for a mismatch between the worker and job that can be addressed by applying ergonomic principles of work design. To reduce the extent of work-related musculoskeletal injuries, progress in four methodological areas was expected (NIOSH, 1986): (1) identifying the biomechanical hazards accurately, (2) developing effective health promotion and hazard control interventions, (3) changing management concepts and operational policies with respect to expected work performance, and (4) devising strategies for disseminating knowledge on control technology and promoting their application through incentives.

With the issuance of the *National Occupational Research Agenda* in 1996 (NIOSH, 1996), NIOSH built upon these previous strategies by including musculoskeletal disorders and several related areas (e.g., organization of work, indoor environment, special populations at risk, exposure assessment methods, intervention effectiveness research, risk assessment methods, and surveillance research methods) in its list of 21 declared priority areas, defined as areas with the highest likelihood of reducing workplace injuries/illnesses. Consistent with NIOSH's vision are the recommendations from the NRC (2001), which included encouraging the institution or extension of ergonomic and other preventive, science-based strategies and identifying areas that need further research (improved tools for exposure assessment, improved measures of outcome and case definition for use in epidemiologic and intervention studies, further quantification of exposure-outcome relationships).

6.2 Applying Ergonomic Principles and Processes

Ergonomic job design (and redesign) efforts focus on fitting characteristics of the job to capabilities of workers. In simple terms, this can be accomplished, for example, by reducing excessive strength requirements and exposure to vibration, improving the design of hand tools and work layouts, designing out unnatural postures at work, or addressing the problem of work–rest requirements for jobs with high production rates. From the occupational safety and health perspective, the current state of ergonomics knowledge should allow for management of WUEDs in order to minimize human suffering, potential for disability, and the related worker compensation costs. Application of ergonomics can help to (1) identify working conditions under which the WUEDs might occur, (2) develop engineering design measures aimed at elimination or reduction of the known job risk factors, and (3) identify the affected worker population and target it for early medical and work intervention efforts.

Workplace and work design–related risk factors, which often overlap, typically involve a combination of poor work methods, inadequate workstations and hand tools, and high production demands. A *risk factor* is defined as an attribute or exposure that increases the probability of the disease or disorder (Putz-Anderson, 1993). As discussed before, the biomechanical risk factors for WUEDs include repetitive and sustained exertions, awkward postures, and application of high mechanical forces. Vibration and cold environments may also accelerate the development of WUEDs. Tools that can be used to identify the potential for development of WUEDs include plant walk-throughs and/or more detailed work-methods analyses. Checklists, analytical tools [such as the strain index (Moore and Garg, 1995) or HAL (ACGIH, 2002)], and/or expertise of the analyst are utilized to identify undesirable work site conditions or worker activities that can contribute to injury.

Since job redesign decisions may require some design trade-offs (Putz-Anderson, 1993), the ergonomic intervention process should follow these steps: (1) perform a thorough job analysis to determine the nature of specific problems, making sure to identify the root causes for the problem; (2) evaluate and select the most appropriate intervention(s), based on the root cause and other factors relevant to the particular circumstance; (3) develop and apply conservative treatment (implement the intervention), on a limited scale if possible; (4) monitor progress to ensure that the intervention has the intended effect and no adverse consequences; and (5) adjust or refine the intervention as needed.

6.3 Administrative and Engineering Controls

The control of WUEDs requires consideration of the following aspects of this complex problem: (1) WUEDs diagnosis, (2) treatment, (3) rehabilitation and return to work, (4) WUEDs surveillance, (5) surveillance and control of risk factors at the micro- and macro levels, (6) training and education, and (7) management and leadership with regard to WUEDs-related organizational and social aspects (Hagberg et al., 1995). The specific recommendations for prevention of WUEDs can be classified as being either primarily administrative (i.e., focusing on personnel solutions) or engineering (i.e., focusing on redesigning tools, workstations, and jobs) (Putz-Anderson, 1993). In general, administrative controls are those actions taken by the management that are intended to limit the potentially harmful effects of a physically stressful job on individual workers. Administrative controls, which are focused on the workers, are modifications of existing personnel functions such as worker training, job rotation, and matching employees to job assignments. A summary of selected ergonomics measures that aim to control the incidence of WUEDs is shown in Table 16.

With respect to biomechanical risk factors, prevention and control efforts for WUEDs should be directed toward fulfilling several recommendations based on ergonomics principles for workplace design, work methods, and work organization. As discussed by Putz-Anderson (1993), these may include, for example, the following recommendations: (1) permit several different working postures; (2) place controls, tools, and materials between waist and shoulder heights for ease of reach and operation; (3) use jigs and fixtures for holding purposes; (4) resequence jobs to reduce the repetition; (5) automate highly repetitive operations; (6) allow self-pacing of work whenever feasible; and (7) allow frequent (voluntary and mandatory) rest breaks.

Furthermore, with respect to hand tools used at work, the following general work design guidelines are provided:

1. Make sure that the center of gravity of the tool is located close to the body and the tool is balanced.
2. Use power tools to reduce the force and repetition required.
3. Consider redesigning the straight tool handle; bend it as necessary to preserve the neutral posture of the wrist.
4. Use tools with pistol grip or straight grips, respectively, where the axis in use is horizontal or vertical (or when the direction of force is perpendicular to the workplace).
5. Avoid tools that require working with a flexed wrist and extended arm at the same time, or tools that call for the flexion of distal phalanges (last joints) of the fingers.
6. Minimize the tool weight; suspend all tools heavier than 20 N (or 2 kg of force) by a counterbalancing harness.
7. Align the tool's center of gravity with the center of the grasping hand.
8. Use special-purpose tools that facilitate fitting the task to the worker (avoid standard off-the-shelf tools for specific repetitive operations).
9. Design tools so that workers can use them with either hand.
10. Use a power grip where power is needed, and a precision grip for precise tasks.
11. The handles and grips should be cylindrical or oval with a diameter between 3.0 and 4.5 cm (for precise operations the diameter recommended is from 0.5 to 1.2 cm).
12. The minimum handle length should be 10.0 cm, whereas a 11.5- to 12.0-cm handle is preferable.
13. A handle span of 7.5 to 8.0 cm can be used by male and female workers for plier-type handles.
14. Triggers on power tools should be at least 5.1 cm wide, allowing their activation by two or three fingers.
15. Avoid form-fitting handles that cannot be adjusted easily.
16. Provide handles that are nonporous, nonslip, and nonconductive (thermally and electrically).

6.4 Ergonomics Programs for Prevention

An important component of WUED management efforts is development of well-structured and comprehensive ergonomics programs. According to Alexander and Orr (1992), the basic components of such a program should include the following: (1) health and risk factor surveillance, (2) job analysis and improvement, (3) medical management, (4) training, and (5) program evaluation. An excellent program should include participation of all levels of management; medical, safety, and health personnel; labor unions; engineering; facility planners; and workers, and contain the following elements:

1. Routine (monthly or quarterly) reviews of the OSHA log (injury records) for patterns of

Table 16 Ergonomic Measures to Control Common WMSDs

Disorder	Avoid in General	Avoid in Particular	Recommendation	Design Considerations
Carpal tunnel syndrome	Rapid, often repeated finger movements, wrist deviation	Dorsal and palmar flexion, pinch grip, vibrations between 10 and 60 Hz		Workplace design
Cubital tunnel syndrome	Resting forearm on sharp edge or hard surface			Workplace design
DeQuervain's syndrome	Combined forceful gripping and hard twisting			Workplace design
Epicondylitis	"Bad tennis backhand"	Dorsiflexion, pronation		Workplace design
Pronator syndrome	Forearm pronation	Rapid and forceful pronation, strong elbow and wrist flexion	Use large muscles but infrequently and for short time	Design of work object
Shoulder tendonitis, rotator cuff syndrome	Arm elevation	Arm abduction, elbow elevation	Let wrists be in line with forearm	Design of job task
Tendonitis	Often repeated movements, particularly with force exertion; hard surface in contact with skin, vibrations	Frequent motions of digits, wrists, forearm shoulder	Let shoulder and upper arm be relaxed	Design of hand tools ("bend tool, not the wrist")
Tenosynovitis, DeQuervain's syndrome, ganglion	Finger flexion, wrist deviation	Ulnar deviation dorsal and palmar flexion, radial deviation with firm grip	Let forearms be horizontal or more declined	Design for round corners, use pad
Thoracic outlet syndrome	Arm elevation, carrying loads	Shoulder flexion, arm hyperextension		Design of work object placement
Trigger finger or thumb	Digit flexion	Flexion of distal phalanx alone		Workplace design
Ulnar artery aneurysm	Pounding and pushing with heel of the hand			Workplace design
Ulnar nerve entrapment	Wrist flexion and extension	Wrist flexion and extension, pressure of hypothenar eminence		Workplace design
White finger, vibration syndrome	Vibrations, tight grip, cold exposure	Vibrations between 40 and 125 Hz		Workplace design
Neck tension syndrome	Static head posture	Prolonged static head–neck posture	Alternate head–neck postures	Workplace design

Source: Adapted from Kroemer et al. (1994).

injury and illness (dedicated computer programs can be used to identify problem areas).

2. Workplace audits for ergonomic problems that are a routine part of the organization's culture (more than one audit annually for each operating area), and timely interventions as a response to the problems identified.
3. A knowledge by management and workers regarding the list of most critical problems (i.e., jobs with the job title clearly identified).
4. Application of both engineering solutions and administrative controls, with engineering solution treated as the long-term solutions.
5. Awareness of ergonomic considerations by design engineering that utilizes them in new or reengineered designs. (People are an important design consideration.)
6. Frequent refresher training in ergonomics, including short courses and seminars for site-appointed "ergonomists."

6.5 Employer Benefits from Ergonomic Programs

In 1997, the U.S. General Accounting Office issued a report in response to a charge to (1) identify core elements of effective ergonomics programs and describe how these are operationalized within companies, (2) examine whether or not such programs have proven beneficial to companies and employees where they have been implemented, and (3) address the implications for employers who have not adopted such programs (GAO, 1997). The core elements they identified were consistent with those listed above. The parts of the report that are particularly interesting are the five case studies that are included, which provide details of the ergonomic program experiences of five different companies, in different sectors of industry, and how each tailored the generic components of a successful program to fit their circumstances, company culture, and so on. Examples of specific benefits realized by the companies are also provided and include reductions in numbers of lost workdays, workers' compensation costs (total and per case), and MSD incidence rates. Methods for supporting the case for ergonomics based on economics are provided by a number of authors (Andersson, 1992a, b; Simpson and Mason, 1995; Oxenburgh, 1997).

On a smaller scale, specific interventions may require cost–benefit analysis to justify them if a substantial initial investment is required. Seeley and Marklin (2003) employed the cost–benefit analysis method described by Rouse and Boff (1997) to calculate the expected benefit to an electric utility company as a result of replacing the manual cutters and presses their linemen used with battery-powered models. Based on the quantification of expected benefits (reductions in medical and workers' compensation costs due to upper extremity MSDs, costs to replace injured workers, training costs for replacement workers, and additional medical expenses for employees who postpone reporting injuries until they become severe), they determined a payback period of only four months for the $300,000 cost of the new tools. Details of their methodology are provided in Seeley and Marklin (2003).

6.6 Surveillance

6.6.1 Surveillance System

To evaluate the extent of WUEDs in a working population, a surveillance system should be used. *Surveillance* refers to the ongoing systematic collection, analysis, and interpretation of health and exposure data. Relevant to this chapter, this refers to the process of describing and monitoring work-related MSD occurrence. Surveillance is used to determine which jobs need further evaluation and where ergonomic interventions may be warranted. Surveillance data are used to determine the need for occupational safety and health action and to plan, implement, and evaluate ergonomic interventions and programs (Klaucke et al., 1988). Health and hazard (job risk factor) surveillance provides employers and employees with a means of evaluating WUEDs and workplace ergonomic risk factors systematically by monitoring trends over time. This information can also be of benefit for planning, implementing, and evaluating ergonomic interventions.

Although the climate for standards making has cooled in the United States, the final draft of the standard for management of WMSDs from the ANSI Z365 Committee is still a valuable source of information on the elements of a management program, including the surveillance component (ANSI Z365 Committee, 2002). The draft standard describes surveillance as including (1) review and analysis of existing records on worker injury and illness (OSHA 300 logs, company medical records, etc.), (2) worker reports concerning MSD symptoms or potential risk factors in the workplace, and (3) job surveys (cursory or screening-level reviews of jobs conducted to identify potential risk factors and the degree of risk they might pose to workers). The goal of surveillance, and of ergonomics programs in general, is to reduce or eliminate MSD risk factor exposure through both reactive (after MSDs or their symptoms develop in workers) and proactive (identify hazards before workers who are exposed to the hazard develop a problem) approaches.

6.6.2 Worker Health Data

Analysis of existing records will be used to estimate the potential magnitude of the problem in the workplace. The number of employees in each job, department, or similar population needs to be determined first. Then the incidence rates can be calculated on the basis of hours worked:

$$\text{incidence (new case) rate (IR)} = \frac{\text{no. of new cases during time} \times 200,000}{\text{work hours during time}} \quad (4)$$

where *time* refers to the time period of interest, typically one year, and *work hours* refers to the total number of hours worked by all employees in the group for which the rate is being calculated. The incidence rate is equivalent to the number of new cases per 100 full-time workers (assuming that each works 40 hours per week and 50 weeks per year). Workplace-wide incidence rates (IRs) can be calculated for all WUEDs classified by body location for each department, process, or type of job. If specific work hours are not readily available, the number of full-time equivalent employees in each area multiplied by 2000 hours can be used to estimate the denominator. Another important calculation is that of severity, used to describe the number of lost workdays. One way to calculate severity is to substitute the number of lost workdays for the number of new cases in equation (4). Another would be to examine the number of lost workdays per case. *Prevalence* refers to the number of existing cases relative to the number of employees in the group. These numbers can be compared within a company among different departments to see where problems exist. They can also be compared with data from the Bureau of Labor

Statistics, to provide a sense of where a company's statistics are relative to those of its industry or business sector. That information can be found on the BLS Web site (http://www.bls.gov/iif/oshcdnew.htm). [Note that BLS incidence rate data are provided per 10,000 workers, not per 100 as in equation (4).]

In addition to making use of existing records, information about current symptoms can be sought through use of employee surveys. These usually provide employees with diagrams and other means by which to indicate where they are experiencing symptoms, as well as the intensity and frequency of occurrence. In their chapter on surveillance, Hagberg et al. (1995) provide some examples of symptom surveys. Once a symptom survey has been conducted, the employer must be prepared to follow up with job analysis if problems are identified through the symptom survey.

6.6.3 Job Surveys

Job surveys are performed to identify specific jobs and processes that may put employees at risk of developing WMSDs. Conduct surveys of all jobs, a representative sample, or jobs that have been identified as potential problems through some other method (such as jobs with excessive turnover or absenteeism, or when a substantial change is made to a job). Job surveys may include walk-throughs; conversations with employees, supervisors, and/or company health personnel; use of checklists; or other basic methods.

6.6.4 Data Collection Instruments

The surveillance system aims to link the occurrence of WMSDs to work-related risk factors. Ideally, the surveillance should make it possible to identify workplace risk factors before symptoms develop. Surveillance data collection instruments can be passive or active in nature (Hagberg et al., 1995). A summary of active and passive surveillance methods are listed in Table 17. The passive surveillance process relies on information collected from existing databases and records (e.g., company dispensary logs, insurance records, workers' compensation records, accident reports, and absentee records) to identify the WRMD cases and patterns and potential problem jobs. Passive surveillance records are often useful in helping to determine the frequency with which active surveillance tools should be used and the interventions required or in assessing the effectiveness of ergonomics programs. In addition, brief job analysis or physical demand analysis to assess the suitability of a job for the return to work of an injured worker can also be used for passive risk factor surveillance.

Active surveillance uses specifically designed tools and information, such as checklists and job analysis. As shown in Table 18, there can be both health active surveillance and risk factor active surveillance. Since most musculoskeletal disorders produce some symptoms of pain or discomfort, health questionnaires are useful in identifying new or incipient problems as well as for assessing the effectiveness of medical interventions and ergonomic controls. In addition to symptom questionnaires, medical interviews and examination can also be used in active health surveillance (Table 19).

Table 17 Passive and Active Surveillance Methods

Passive Surveillance[a]	Active Surveillance
Information source and method already exist and are usually designed for other administrative purposes	Information source and method specifically designed for surveillance
Relatively inexpensive	Modest to quite expensive
Usually requires additional coding of information for the purpose [e.g., surrogate(s) of exposure, such as job titles]	Since tools are "tailor made," includes at least job title information and other data considered important by surveillance analyst; will include data for linking of information between risk factor and WMSD data
Examples: health and safety logs, medical department logs, workers' compensation data, early retirement, medical insurance, absenteeism and transfer records, accident reports, product quality, productivity	*Examples:* for WMSD surveillance: confidential questionnaires without personal identifiers, questionnaire interviews, physical examinations; for risk factor surveillance: workplace walk-throughs, job checklists, postural discomfort surveys

Source: Adapted from Kuorinka and Forcier (1995).
[a]Used mostly for health surveillance since, in practice, no existing records have been used to obtain information on risk factors associated with WMSDs.

Table 18 Summary of Tools Used in Surveillance

Approach	Tools
Health	
Passive	Existing records
Active, level 1	Symptoms surveys or questionnaires (self or group administered)
Active, level 2	Health-related interviews and/or brief physical exams
Risk factors	
Active, level 1	Quick checklists of risk factors
Active, level 2	In-depth job analysis

Source: Adapted from Kuorinka and Forcier (1995).

Table 19 Examples of Tools for WMSD Surveillance

Focus of Surveillance	Methods of Surveillance	
	Passive	Active
Health (WMSDs)	Company dispensary logs	Checklists
	Insurance records	Questionnaires
	Workers' compensation records	Interviews
	Accident reports	Physical exams
	Transfer requests	
	Absentee records	
	Grievances	
Workplace risk factors (associated with WMSDs)	Not really used for WMSD risk factor yet[a]	Checklists
		Questionnaires
		Job analysis

Source: Adapted from Kuorinka and Forcier (1995).

[a]The use of surrogate measures for exposure (e.g., job title of firm's department) could be viewed as "passive surveillance."

6.6.5 Analysis and Interpretation of Data

The surveillance data can be analyzed and interpreted to study possible associations between the WMSD surveillance data and the risk factor surveillance data (Hagberg et al., 1995). The two principal goals of the analysis are (1) to help identify patterns in the data that reflect differences between jobs or departments, and (2) to target and evaluate intervention strategies. This analysis can be done on the number of existing WMSD cases (cross-sectional analysis) or during a specific period of time on the number of new WMSD cases, in a retrospective and prospective fashion (retrospective and prospective analysis).

One of the simplest ways to assess the association between risk factors and WMSDs is to calculate the odds ratios (see Table 20). For this example, the prevalence data obtained in health surveillance are linked with the data obtained in risk factor surveillance. In the example shown in Table 20 (for more details, see Hagberg et al., 1995), one risk factor is selected at a time (e.g., overhead work for more than four hours). Using the data obtained in surveillance, the following numbers of employees are counted:

- Employees with WMSDs exposed to more than four hours of overhead work (15 workers)
- Employees with WMSDs not exposed to more than four hours of overhead work (15 workers)
- Employees without WMSDs exposed to more than four hours of overhead work (25 workers)
- Employees without WMSDs not exposed to more than four hours of overhead work (85 workers)

Table 20 Examples of Odds Ratio Calculations for a Firm of 140 Employees

Risk Factor (e.g., Overhead Work for More Than Four Hours) Is:	WMSDs are[a]: Present	Not Present	Total
Present	15 (A)	25 (B)	40 (A + B)
Not present	15 (C)	85 (D)	100 (C + D)
Total	30 (A + C)	110 (B + D)	140 (*N*)

Source: Adapted from Kuorinka and Forcier (1995).

[a]Number in each cell indicates the count of employees with or without WMSD and the risk factor. Odds ratio (OR) $= (A \times D)/(B \times C) = (15 \times 85)/(25 \times 15) = 3.4$.

The overall prevalence for the company is 30/140 or 21.4%. The prevalence for those exposed to the risk factor is 37.5% (15/40) compared with 15.0% (15/100) for those not exposed. The risk of having a WMSD depending on exposure to the risk factor, the odds ratio, can be calculated using the number of existing cases of WMSD (prevalence). In the example above, those exposed to the risk factor have 3.4 times the odds of having the WMSD than those not exposed to the risk factor. An odds ratio of greater than 1 indicates an elevated risk. Such ratios can be monitored over time to assess the effectiveness of the ergonomics program in reducing the risk of WMSD, and a variety of statistical tests can be used to assess the patterns seen in the data.

6.7 Procedures for Job Analysis

Detailed job analysis typically consists of analyzing the job at the element or micro level. Job surveys, on the other hand, can be used for establishing work relatedness, for prioritizing jobs for further analysis, or for proactive risk factor surveillance. Job analysis involves breaking down the job into component actions, measuring and quantifying risk factors, and identifying the problems and conditions contributing to each risk factor. Tools that might be employed to perform a job analysis include videotape, tape measure, scale to weigh tools and parts, stopwatch to measure exposure duration, and possibly more sophisticated tools (electrogoniometers to measure wrist joint posture and motion; electromyographic equipment to assess muscle activity; vibration analysis equipment for assessing a powered hand tool's vibration characteristics). Exposures are characterized by magnitude, duration, and rate of repetition. Work organization factors, such as number of hours in a shift (8, 10, 12 hours), job rotation schedule if applicable, and pay system (hourly, incentive, etc.). These data may be examined relative to existing research findings, regarding levels of exposure associated with elevated risk, or they may be used as input to tools such as the strain index (Moore and Garg, 1995),

HAL (ACGIH, 2002), or others, to determine the degree of risk posed by the hazards (risk factors). The Web site of the Washington State Department of Labor and Industries provides assessment tools as well (http://www.lni.wa.gov/Safety/Topics/Ergonomics/ServicesResources/Tools/default.asp).

The job analysis should be performed at a sufficient level of detail to identify potential work-related risk factors associated with WMSDs and include the following steps: (1) collection of pertinent information about the job (number of employees on the job, which jobs precede and follow it, cycle time, tools used, etc.), (2) interview of a representative sample of workers, (3) breakdown of the job into tasks or elements, (4) description of the component actions of each task or element, (5) measurement and quantification of WMSD risk factors, (6) identification of the risk factors for each task element, (7) identification of the problems contributing to risk factors (root cause analysis), and (8) summary of the problems and needs for intervention for the job. If intervention is required, once it has been developed, with input from workers and others, and put in place, follow-up assessments should be performed to ensure that the intervention is effective in dealing with the former problem and that no new problems are inadvertently introduced with the intervention. It is also important to document all of these steps within the analysis process in order to track progress, provide justification for changes, and share information with others about successful interventions.

6.8 Medical Management

6.8.1 Basic Activities

The primary objective of medical management in occupational health and safety programs is the prevention of work-related disorders and injuries (Hagberg et al., 1995). The specific goals of occupational health programs relevant to prevention of musculoskeletal disorders were specified by the AMA (1972) as follows: (1) protecting employees against health and safety hazards in their work situation; (2) evaluating workers' physical, mental, and emotional capacity before job placement; (3) ensuring that employees can perform the work with an acceptable degree of efficiency and without endangering their own health and safety or that of others; (4) ensuring adequate medical care and rehabilitation for the occupationally ill or injured; and (5) encouraging and assisting with measures for personal health maintenance, including the acquisition of a personal physician whenever possible.

Medical management of WUEDs includes medical diagnosis, treatment, rehabilitation and return to work, and work hardening (Karwowski and Kasdan, 1988). In addition to these activities, medical management should also be involved in both passive and active health surveillance, job-skills training programs, and ergonomic task force activities (Hagberg et al., 1995). As discussed in Section 6.5.4, the use of injury reports for health surveillance purposes is a form of passive health surveillance. The effective passive health surveillance requires data that have a high sensitivity for WUEDs. Injury reports should be followed up by workplace visits and an evaluation. In a population of workers, or in a specific job category where there is a high risk of WUEDs, it may be necessary to perform the active health surveillance (i.e., periodic medical evaluations to identify workers in the early stages of a disease) and to target these workers for early secondary prevention efforts (i.e., medical treatment).

6.8.2 Medical Treatment

In general, the medical treatment efforts for WUEDs in the acute phase are similar to the treatments used for non-work-related disorders. As discussed by Hagberg et al. (1995), the general therapeutic objectives for WUEDs should include the following: (1) promotion of rest for the anatomical structures affected, (2) diminished spasms and inflammation, (3) reduction of pain, (4) increase in strength and endurance, (5) increase in range of motion, (6) alteration of mechanical and neurological structures, (7) increase in functional and physical work capacity, and (8) modification of work content and social environment.

6.8.3 Rehabilitation, Return to Work, and Work-Hardening Programs

A program that promotes healing and helps an injured worker to return to work, and specifies appropriate job-placement conditions based on different job tasks and work requirements, is called *occupational rehabilitation.* Since the injury may not always have only a physical basis, psychosocial (at work and outside work) and psychological disability aspects are essential parts of the rehabilitation process. According to the Commission on Accreditation of Rehabilitation Facilities (CARF, 1989), a *work-hardening program* is a highly structured, goal-oriented, individualized treatment program designed to maximize a person's ability to return to work. Such a program uses a set of conditioning tasks that are graded progressively in quest to improve biomechanical, neuromuscular, cardiovascular, and psychosocial functions with real or simulated work activities.

6.8.4 Preemployment and Preplacement Screening

Although there is no scientific evidence that screening can predict the development of WUEDs, preemployment and preplacement screening may be an important part of medical management activities (Hagberg et al., 1995). According to the American College of Occupational and Environmental Medicine, the Committee on Occupational Medical Practice (ACOM, 1990), screening refers to the application of at least one test (or examination) to workers in order to identify apparently healthy workers who are at high risk of developing a specific WUED from those workers who are not. Although the screening tests are not diagnostic, preemployment screening and examination are typically performed before any offer of employment can be made. On the other hand, a preplacement screening process is an examination of an employee who has already received an offer of employment and addresses a question of employee placement in a specific job.

7 SUMMARY

7.1 Balancing the Work System for Ergonomics Benefits

As pointed out by Hagberg et al. (1995), there are no perfect jobs or perfect workplaces that are free of all work-related hazards and provide ideal psychosocial conditions for complete satisfaction for all employees. Therefore, one must consider the trade-offs between competing needs for ergonomic improvements at the workplace and establish a basis for identifying the most critical workplace characteristics for design or redesign. Such trade-offs between the biomechanical factors, personal factors, and work organizational factors, including work stress, coping strategies, and organizational practices, require one to balance various ergonomic needs to achieve the solution that will have the greatest benefits for employee health and productivity.

The balance theory-based model proposed by Smith and Sainfort (1989) takes a systems approach by focusing on the interactions between the worker, including the physical characteristics, perceptions, personality and work behavior; the physical and social environments; and the organizational structure that defines the nature and level of worker involvement, interaction, control, and supervision. The capabilities of technologies available to a worker to perform a specific job affect task performance and the worker's skills and knowledge needed for their effective use. Task requirements affect the required skills and knowledge of the worker. Both the tasks and technologies affect the content of the job and physical demands. The balance theory-based model can be used to establish relationships between interacting elements such as job demands, job design factors, and ergonomic loads. Demands that are placed on the worker create loads that can be healthy or harmful. Harmful loads may lead to physical and psychological stress responses that can produce adverse health effects such as WUEDs. It should be noted that a number of personal considerations may also contribute to the physical and psychological effects. These include the strength and health of the worker, previous musculoskeletal or nerve injury, personality, perceptual-motor skills and abilities, physical conditioning, prior experience and learning, motives, goals, and needs and intelligence.

7.2 Ergonomics Guidelines

The expected benefits of managing WUEDs in industry are improved productivity and quality of work products, enhanced safety and health of the employees, higher employee morale, and accommodation of people with various degrees of physical abilities. Strategies for managing the WUEDs at work should focus on prevention efforts and should include, at the plant level, employee education, ergonomic job redesign, and other early intervention efforts, including engineering design technologies such as workplace reengineering and active and passive surveillance. At the macro level, management of the WUEDs should aim to provide adequate occupational health care provisions, legislation, and industry-wide standardization.

Already widely recognized in Europe (Wilson, 1994), ergonomics has to be seen as a vital component of the value-adding activities of a company. Even in strictly financial terms, the benefits of an ergonomics management program will outweigh the costs of the program. A company must be prepared to engage in a participative culture and to utilize participative techniques. The ergonomics-related problems and consequent interventions should go beyond engineering solutions and must include design for manufacturability, total quality management, work organization, workplace redesign, and worker training. Only then will the promise of ergonomics in managing the WUEDs at work be fulfilled.

In the absence of generally applicable guidelines and criteria on minimizing and/or optimizing risk factor exposure, two complementary approaches have merit for the prevention of WUEDs: (1) general guidelines that describe in general terms the principles and policies to be adopted in preventing WUEDs, and (2) specific guidelines that aim at the design and redesign of work and tasks that are known in detail (Hagberg et al., 1995). Since the specific guidelines draw on both scientific knowledge and the collective industrial experience, they may be much more detailed and often contain quantitative data.

Most of the current guidelines for control of the biomechanical risk factors for WUEDs at work aim to (1) reduce the exposure to highly repetitive and stereotyped movements, (2) reduce excessive force levels, and (3) reduce the need for sustained postures. For example, to control the extent of force required to perform a task, one should (1) reduce the force required through tool and fixture redesign, (2) distribute the application of force, or (3) increase the mechanical advantage of the (muscle) lever system. Because of neurophysiological needs of the working muscles, adequate rest pauses (determined based on the scientific knowledge of the physiology of muscular fatigue and recovery) should be scheduled to provide relief for the most active muscles used on the job. Furthermore, reduction in task repetition can be achieved, for example, by (1) task enlargement (increasing variety of tasks to perform), (2) increase in the job cycle time, and (3) work mechanization and automation.

Finally, it should be noted that many of the recommendations offered by ergonomics may be difficult to implement in practice without full understanding of the production processes, plant layouts, or quality requirements, and total commitment from all management levels and workers of the company. This is because many of the guidelines are not specific, and define what to avoid (e.g., avoid high contact forces and static loading, avoid extreme or awkward joint positions, avoid repetitive finger action, avoid tool vibration) but do not define how to avoid these risk factors. In view of the above, involvement of professional ergonomists (i.e., those who are certified by the Board of Certification in Professional Ergonomics), along with engineering personnel and production workers in a truly participative manner, is critical to the success of ergonomic intervention efforts. Furthermore, ergonomics must be treated with the same level of attention and significance as other business functions of the plant (e.g., quality management control) and be accepted as the cost of doing business rather than add-on activity requiring action only when problems arise.

REFERENCES

ACGIH (2002), *TLVs and BEIs*, American Congress of Government Industrial Hygienists, Cincinnati, OH.

ACOM (1990), "Preplacement/Preemployment Physical Examinations," *Journal of Occupational Medicine*, Vol. 32, pp. 295–299.

Alexander, D. C., and Orr, G. B. (1992), "The Evaluation of Occupational Ergonomics Programs," in *Proceedings of the Human Factors Society 36th Annual Meeting*, Santa Monica, CA, pp. 697–701.

Almekinders, L. C. (1998), "Tendinitis and Other Chronic Tendinopathies," *Journal of the American Academy of Orthopaedic Surgeons*, Vol. 6, No. 3, pp. 157–164.

AMA (1972), *Scope, Objectives, and Functions of Occupational Health Programs*, American Medical Association, Chicago.

Andersson, E. R. (1992a), "Economic Evaluation of Ergonomic Solutions, I: Guidelines for the Practitioner," *International Journal of Industrial Ergonomics*, Vol. 10, pp. 161–171.

Andersson, E. R. (1992b), "Economic Evaluation of Ergonomic Solutions, II: The Scientific Basis," *International Journal of Industrial Ergonomics*, Vol. 10, pp. 173–178.

ANSI Z365 Committee (2002), "Management of Work-Related Musculoskeletal Disorders," American National Standards Institute/National Safety Council, retrieved 2004, from http://www.nsc.org/ehc/z365.htm.

Armstrong, T. J., and Chaffin, D. B. (1979), "Some Biomechanical Aspects of the Carpal Tunnel," *Journal of Biomechanics*, Vol. 12, pp. 567–570.

Armstrong, T. J., and Lifshiz, Y. (1987), "Evaluation and Design of Jobs for Control of Cumulative Trauma Disorders," in *Ergonomic Interventions to Prevent Musculoskeletal Injuries in Industry*, Lewis Publishers, Chelsea, England, pp. 73–85.

Armstrong, T. J., Castelli, W. A., Evans, F. G., and Diaz-Perez, R. (1984), "Some Histological Changes in Carpal Tunnel Contents and Their Biomechanical Implications," *Journal of Occupational Medicine*, Vol. 26, pp. 197–201.

Armstrong, T. J., Radwin, R. G., Hansen, D. J., and Kennedy, K. W. (1986), "Repetitive Trauma Disorders: Job Evaluation and Design," *Human Factors*, Vol. 28, pp. 325–336.

Armstrong, T. J., Buckle, P., Fine, L. J., Hagberg, M., Jonsson, B., Kilbom, A., Kuorinka, I. A., Silverstein, B. A., Sjogaard, G., and Viikari-Juntura, E. R. (1993), "A Conceptual Model for Work-Related Neck and Upper-Limb Musculoskeletal Disorders," *Scandinavian Journal of Work, Environment and Health*, Vol. 19, pp. 73–84.

Barbe, M. F., Barr, A. E., Gorzelany, I., Amin, M., Gaughan, J. P., and Safadi, F. F. (2003), "Chronic Repetitive Reaching and Grasping Results in Decreased Motor Performance and Widespread Tissue Responses in a Rat Model of MSD," *Journal of Orthopaedic Research*, Vol. 21, pp. 167–176.

Barr, A. E., and Barbe, M. F. (2002), "Pathophysiological Tissue Changes Associated with Repetitive Movement: A Review of the Evidence," *Physical Therapy*, Vol. 82, pp. 173–187.

Bernard, B. P. (1997), *Musculoskeletal Disorders and Workplace Factors*, DHHS (NIOSH) Publication 97-141, U.S. Department of Health and Human Services, Public Health Service, Centers for Disease Control and Prevention, National Institute for Occupational Safety and Health, Washington, DC.

BLS (2002), *Bureau of Labor Statistics: Annual Survey of Occupational Injuries and Illnesses*, U.S. Department of Labor, retrieved 2004, from http://www.bls.gov/iif/oshcdnew.htm.

Bongers, P. M., Kremer, A. M., and ter Laak, J. (2002), "Are Psychosocial Factors, Risk Factors for Symptoms and Signs of the Shoulder, Elbow, or Hand/Wrist? A Review of the Epidemiological Literature," *American Journal of Industrial Medicine*, Vol. 41, pp. 315–342.

Buckle, P. W., and Devereux, J. J. (2002), "The Nature of Work-Related Neck and Upper Limb Musculoskeletal Disorders," *Applied Ergonomics*, Vol. 33, pp. 207–217.

CARF (1989), *Standards Manual for Organizations Serving People with Disabilities*, Commission on Accreditation of Rehabilitation Facilities, Tucson, AZ.

Clark, B. D., Barr, A. E., Safadi, F. F., Beitman, L., Al-Shatti, T., Amin, M., Gaughan, J. P., and Barbe, M. F. (2003), "Median Nerve Trauma in a Rat Model of Work-Related Musculoskeletal Disorder," *Journal of Neurotrauma*, Vol. 20, pp. 681–695.

Clark, B. D., Al-Shatti, T. A., Barr, A. E., Amin, M., and Barbe, M. F. (2004), "Performance of a High-Repetition, High-Force Task Induces Carpal Tunnel Syndrome in Rats," *Journal of Orthopaedic and Sports Physical Therapy*, Vol. 34, pp. 244–253.

Ditmars, D. M., Jr. (1993), "Patterns of Carpal Tunnel Syndrome," *Hand Clinics*, Vol. 9, pp. 241–252.

Falkiner, S., and Myers, S. (2002), "When Exactly Can Carpal Tunnel Syndrome Be Considered Work-Related?" *ANZ Journal of Surgery*, Vol. 72, pp. 204–209.

GAO (1997), *Worker Protection: Private Sector Ergonomics Programs Yield Positive Results*, GAO/HEHS-97-163, U.S. General Accounting Office, Washington, DC.

Haber, L. D. (1971), "Disabling Effects of Chronic Disease and Impairment," *Journal of Chronic Disease*, Vol. 24, pp. 269–487.

Hagberg, M. (1992), "Exposure Variables in Ergonomic Epidemiology," *American Journal of Industrial Medicine*, Vol. 21, pp. 91–100.

Hagberg, M., Silverstein, B., Wells, R., Smith, M. J., Hendrick, H. W., Carayon, P., and Perusse, M. (1995), in *Work Related Musculoskeletal Disorders (WMSDs): A Reference Book for Prevention*, I. Kourinka and L. Forcier, Eds., Taylor & Francis, London.

Ham, S. J., Kolkman, W. F. A., Heeres, J., den Boer, J. A., and Vierhout, P. A. M. (1996), "Changes in the Carpal Tunnel Due to Action of the Flexor Tendons: Visualization with Magnetic Resonance Imaging," *Journal of Hand Surgery [American Volume]*, Vol. 21, pp. 997–1003.

Hargens, A. R., Romine, J. S., Sipe, J. C., Evans, K. L., Mubarak, S. J., and Akeson, W. H. (1979), "Peripheral Nerve-Conduction Block by High Muscle-Compartment Pressure," *Journal of Bone and Joint Surgery*, Vol. 61-A, pp. 192–200.

Hart, D. A., Archambault, J. M., Kydd, A., Reno, C., Frank, C. B., and Herzog, W. (1998), "Gender and Neurogenic Variables in Tendon Biology and Repetitive Motion Disorders," *Clinical Orthopaedics*, June (351), pp. 44–56.

Helliwell, P. S., Bennett, R. M., Littlejohn, G., Muirden, K. D., and Wigley, R. D. (2003), "Towards Epidemiological Criteria for Soft-Tissue Disorders of the Arm," *Occupational Medicine (London)*, Vol. 53, pp. 313–319.

Järvholm, U., Palmerud, G., Styf, J., Herberts, P., and Kadefors, R. (1988), "Intramuscular Pressure in the Supraspinatus Muscle," *Journal of Orthopaedic Research*, Vol. 6, pp. 230–238.

Järvholm, U., Palmerud, G., Karlsson, D., Herberts, P., and Kadefors, R. (1991), "Intramuscular Pressure and Electromyography in Four Shoulder Muscles," *Journal of Orthopaedic Research*, Vol. 9, pp. 609–619.

Karwowski, W., and Kasdan, M. L. (1988), "The Partnership of Ergonomics and Medical Intervention in Rehabilitation of Workers with Cumulative Trauma Disorders of the Hand," in *Ergonomics in Rehabilitation*, A. Mital and W. Karwowski, Eds., Taylor & Francis, London, pp. 35–53.

Keir, P. J. (2001), "Magnetic Resonance Imaging as a Research Tool for Biomechanical Studies of the Wrist," *Seminars in Musculoskeletal Radiology*, Vol. 5, pp. 241–250.

Keir, P. J., and Bach, J. M. (2000), "Flexor Muscle Incursion into the Carpal Tunnel: A Mechanism for Increased Carpal Tunnel Pressure? *Clinical Biomechanics (Bristol, Avon)*, Vol. 15, pp. 301–305.

Keir, P., Bach, J. M., and Rempel, D. (1998a), "Fingertip Loading and Carpal Tunnel Pressure: Differences Between a Pinching and a Pressing Task," *Journal of Orthopaedic Research*, Vol. 16, pp. 112–115.

Keir, P. J., Bach, J. M., and Rempel, D. M. (1998b), "Effects of Finger Posture on Carpal Tunnel Pressure During Wrist Motion," *Journal of Hand Surgery [American Volume]*, Vol. 23, pp. 1004–1009.

Keyserling, W. M., Stetson, D. S., Silverstein, B. A., and Brouwer, M. L. (1993), "A Checklist for Evaluating Ergonomic Risk Factors Associated with Upper Extremity Cumulative Trauma Disorders," *Ergonomics*, Vol. 36, pp. 807–831.

Klaucke, D. N., Buehler, J. W., Thacker, S. B., Parrish, R. G., Trowbridge, R. L., and Berkelman, R. L. (1988), "Guidelines for Evaluating Surveillance Systems," *Morbidity and Mortality Weekly Report*, Vol. 37, Suppl. 5, pp. 1–18.

Kleinbaum, D. G., Kupper, L. L., and Morgenstern, H. (1982), *Epidemiologic Research*, Van Nostrand Reinhold, New York.

Kroemer, K., Kroemer, H., and Kroemer-Elbert, K. (1994), *Ergonomics: How to Design for Ease and Efficiency*, Eds., Prentice Hall, Englewood Cliffs, NJ.

Kuorinka, I., and Forcier, L. Eds. (1995), *Work Related Musculoskeletal Disorders: A Reference Book for Prevention*, Taylor & Francis, London.

Leino, P. (1989), "Symptoms of Stress Predict Musculoskeletal Disorders," *Journal of Epidemiology and Community Health*, Vol. 43, pp. 293–300.

LifeArt (2000), "Grant's Atlas 10th Edition Images" [computer file], Lippincott Williams & Wilkins, Baltimore.

Lluch, A. L. (1992), "Thickening of the Synovium of the Digital Flexor Tendons: Cause or Consequence of the Carpal Tunnel Syndrome?" *Journal of Hand Surgery [British Volume]*, Vol. 17, pp. 209–212.

Lundberg, U. (2002), "Psychophysiology of Work: Stress, Gender, Endocrine Response, and Work-Related Upper Extremity Disorders," *American Journal of Industrial Medicine*, Vol. 41, pp. 383–392.

Lundborg, G., Dahlin, L. B., Hansson, H. A., Kanje, M., and Necking, L. E. (1990), "Vibration Exposure and Peripheral Nerve Fiber Damage," *Journal of Hand Surgery [American Volume]*, Vol. 15, pp. 346–351.

Luopajarvi, T., Kuorinka, I., Virolainen, M., and Holmberg, M. (1979), "Prevalence of Tenosynovitis and Other Injuries of the Upper Extremities in Repetitive Work," *Scandinavian Journal of Work, Environment and Health*, Vol. 5, Suppl. 3, pp. 48–55.

MacDonald, L. A., Karasek, R. A., Punnett, L., and Scharf, T. (2001), "Covariation Between Workplace Physical and Psychosocial Stressors: Evidence and Implications for Occupational Health Research and Prevention," *Ergonomics*, Vol. 44, pp. 696–718.

Mackinnon, S. E., Dellon, A. L., Hudson, A. R., and Hunter, D. A. (1984), "Chronic Nerve Compression: An Experimental Model in the Rat," *Annals of Plastic Surgery*, Vol. 13, pp. 112–120.

Marras, W. S., and Schoenmarklin, R. W. (1993), "Wrist Motions in Industry," *Ergonomics*, Vol. 36, pp. 341–351.

McDonough, J. T., Jr., Ed. (1994), *Stedman's Concise Medical Dictionary*, Williams & Wilkins, Baltimore.

Moore, A., Wells, R., and Ranney, D. (1991), "Quantifying Exposure in Occupational Manual Tasks with Cumulative Trauma Disorder Potential," *Ergonomics*, Vol. 34, pp. 1433–1453.

Moore, J. S. (2002), "Biomechanical Models for the Pathogenesis of Specific Distal Upper Extremity Disorders," *American Journal of Industrial Medicine*, Vol. 41, pp. 353–369.

Moore, J. S., and Garg, A. (1995), "The Strain Index: A Proposed Method to Analyze Jobs for Risk of Distal Upper Extremity Disorders," *American Industrial Hygiene Association Journal*, Vol. 56, pp. 443–458.

Nakamichi, K., and Tachibana, S. (1992), "Transverse Sliding of the Median Nerve Beneath the Flexor Retinaculum," *Journal of Hand Surgery [British Volume]*, Vol. 17, pp. 213–216.

NIOSH (1977), *National Occupational Hazard Survey, 1972–1974*, Publication 78-114, National Institute for Occupational Safety and Health, Washington, DC.

NIOSH (1986), *Proposed National Strategies for the Prevention of Leading Work-Related Diseases and Injuries, Part 1*, PB87-114740, National Institute for Occupational Safety and Health, Washington, DC.

NIOSH (1996), *National Occupational Research Agenda*, National Institute for Occupational Safety and Health, retrieved July 18, 2004, from http://www2a.cdc.gov/NORA/default.html.

Norregaard, J., Jacobsen, S., and Kristensen, J. H. (1998), "A Narrative Review on Classification of Pain Conditions of the Upper Extremities," *Scandinavian Journal of Rehabilitation Medicine*, Vol. 31, pp. 153–164.

NRC (2001), *Musculoskeletal Disorders and the Workplace*, National Academy of Sciences, Washington, DC.

Oxenburgh, M. S. (1997), "Cost–Benefit Analysis of Ergonomics Programs," *American Industrial Hygiene Association Journal*, Vol. 58, pp. 150–156.

Putz-Anderson, V. (1993), *Cumulative Trauma Disorders: A Manual for Musculoskeletal Diseases of the Upper Limbs*, Taylor & Francis, London.

Ranney, D., Wells, R., and Moore, A. (1995), "Upper Limb Musculoskeletal Disorders in Highly Repetitive Industries: Precise Anatomical Physical Findings," *Ergonomics*, Vol. 38, pp. 1408–1423.

Rempel, D., Evanoff, B., Amadio, P. C., de Krom, M., Franklin, G., Franzblau, A., Gray, R., Gerr, F., Hagberg, M., Hales, T., Katz, J. N., and Pransky, G. (1998), "Consensus Criteria for the Classification of Carpal Tunnel Syndrome in Epidemiologic Studies," *American Journal of Public Health*, Vol. 88, pp. 1447–1451.

Robbins, H. (1963), "Anatomical Study of the Median Nerve in the Carpal Tunnel and Etiologies of the Carpal-Tunnel Syndrome," *Journal of Bone and Joint Surgery*, Vol. 45-A, pp. 953–966.

Rojviroj, S., Sirichativapee, W., Kowsuwon, W., Wongwiwattananon, J., Tamnanthong, N., and Jeeravipoolvarn, P. (1990), "Pressures in the Carpal Tunnel: A Comparison Between Patients with Carpal Tunnel Syndrome and Normal Subjects," *Journal of Bone and Joint Surgery [Br.]*, Vol. 72, pp. 516–518.

Rouse, W. B., and Boff, K. R. (1997), "Assessing Cost/Benefits of Human Factors," in *Handbook of Human Factors and Ergonomics*, 2nd ed., G. Salvendy, Ed., Wiley, New York, pp. 1617–1633.

Rucker, N., and Moore, J. S. (2002), "Predictive Validity of the Strain Index in Manufacturing Facilities," *Applied Occupational and Environmental Hygiene*, Vol. 17, pp. 63–73.

Rydevik, B., Lundborg, G., and Bagge, U. (1981), "Effects of Graded Compression on Intraneural Blood Flow," *Journal of Hand Surgery*, Vol. 6, pp. 3–12.

Schoenmarklin, R. W., Marras, W. S., and Leurgans, S. E. (1994), "Industrial Wrist Motions and Incidence of Hand/Wrist Cumulative Trauma Disorders," *Ergonomics*, Vol. 37, pp. 1449–1459.

Schuind, F., Ventura, M., and Pasteels, J. L. (1990), "Idiopathic Carpal Tunnel Syndrome: Histologic Study of Flexor Tendon Synovium," *Journal of Hand Surgery [American Volume]*, Vol. 15, pp. 497–503.

Seeley, P. A., and Marklin, R. W. (2003), "Business Case for Implementing Two Ergonomic Interventions at an Electric Power Utility," *Applied Ergonomics*, Vol. 34, pp. 429–439.

Siegel, D. B., Kuzman, G., and Eakins, D. (1995), "Anatomic Investigation of the Role of the Lumbrical Muscles in Carpal Tunnel Syndrome," *Journal of Hand Surgery [American Volume]*, Vol. 20, pp. 860–863.

Silverstein, B. A., Fine, L. J., and Armstrong, T. J. (1986), "Hand Wrist Cumulative Trauma Disorders in Industry," *British Journal of Industrial Medicine*, Vol. 43, pp. 779–784.

Silverstein, B. A., Fine, L. J., and Armstrong, T. J. (1987), "Occupational Factors and Carpal Tunnel Syndrome," *American Journal of Industrial Medicine*, Vol. 11, pp. 343–358.

Simpson, G., and Mason, S. (1995), "Economic Analysis in Ergonomics," in *Evaluation of Human Work*, 2nd ed., J. R. Wilson and E. N. Corlett, Eds., Taylor & Francis, London, pp. 1017–1037.

Sluiter, J. K., Rest, K. M., and Frings-Dresen, M. H. (2001), "Criteria Document for Evaluating the Work-Relatedness of Upper-Extremity Musculoskeletal Disorders," *Scandinavian Journal of Work, Environment and Health*, Vol. 27, pp. 1–102.

Smith, M. J., and Sainfort, P. C. (1989), "A Balance Theory of Job Design for Stress Reduction," *International Journal of Industrial Ergonomics*, Vol. 4, pp. 67–79.

Snook, S. H., Vaillancourt, D. R., Ciriello, V. M., and Webster, B. S. (1995), "Psychophysical Studies of Repetitive Wrist Flexion and Extension," *Ergonomics*, Vol. 38, pp. 1488–1507.

Sommerich, C. M., Marras, W. S., and Parnianpour, M. (1996), "A Quantitative Description of Typing Biomechanics," *Journal of Occupational Rehabilitation*, Vol. 6, pp. 33–55.

Soslowsky, L. J., An, C. H., Johnston, S. P., and Carpenter, J. E. (1994), "Geometric and Mechanical Properties of the Coracoacromial Ligament and Their Relationship to Rotator Cuff Disease," *Clinical Orthopaedics*, Vol. 304, July, pp. 10–17.

Stock, S. R. (1991), "Workplace Ergonomic Factors and the Development of Musculoskeletal Disorders of the Neck and Upper Limbs: A Meta-analysis," *American Journal of Industrial Medicine*, Vol. 19, pp. 87–107.

Szabo, R. M., and Chidgey, L. K. (1989), "Stress Carpal Tunnel Pressures in Patients with Carpal Tunnel Syndrome and Normal Patients," *Journal of Hand Surgery*, Vol. 14A, pp. 624–627.

Szabo, R. M., and Gelberman, R. H. (1987), "The Pathophysiology of Nerve Entrapment Syndromes," *Journal of Hand Surgery*, Vol. 12A, pp. 880–884.

Taylor, W., and Pelmear, P. L. (1976), "Raynaud's Phenomenon of Occupational Origin: An Epidemiological Survey," *Acta Chirurgica Scandinavica. Supplementum*, Vol. 465, pp. 27–32.

Treaster, D. E., and Burr, D. (2004), "Gender Differences in Prevalence of Upper Extremity Musculoskeletal Disorders," *Ergonomics*, Vol. 47, pp. 495–526.

Viikari-Juntura, E., and Silverstein, B. (1999), "Role of Physical Load Factors in Carpal Tunnel Syndrome," *Scandinavian Journal of Work, Environment and Health*, Vol. 25, pp. 163–185.

Visser, B., and van Dieen, J. H. (in press), "Pathophysiology of Upper Extremity Muscle Disorders," *Journal of Electromyography and Kinesiology*.

Wasserman, D. E., Badger, D. W., Doyle, T. E., and Margoilies, L. (1974), "Industrial Vibration: An Overview," *ASSE Journal*, Vol. 19, pp. 38–42.

Werner, R. A., and Andary, M. (2002), "Carpal Tunnel Syndrome: Pathophysiology and Clinical Neurophysiology," *Clinical Neurophysiology*, Vol. 113, pp. 1373–1381.

WHO (1985), *Identification and Control of Work-Released Diseases*, Technical Report 174, World Health Organization, Geneva.

Wilson, J. R. (1994), "Devolving Ergonomics: The Key to Ergonomics Management Programs," *Ergonomics*, Vol. 37, pp. 579–594.

Zeiss, J., Skie, M., Ebraheim, N., and Jackson, W. T. (1989), "Anatomic Relations Between the Median Nerve and Flexor Tendons in the Carpal Tunnel: MR Evaluation in Normal Volunteers," *American Journal of Roentgenology*, Vol. 153, pp. 533–536.

CHAPTER 32

WARNINGS AND HAZARD COMMUNICATIONS

Michael S. Wogalter
North Carolina State University
Raleigh, North Carolina

Kenneth R. Laughery
Rice University
Houston, Texas

1 INTRODUCTION

Safety communications, such as warnings, are used to inform people about hazards and to provide instructions so as to avoid or minimize undesirable consequences. Warnings may be used to address environmental hazards as well as hazards associated with the use of products. In the United States, interest in warnings is also related to litigation concerns. The adequacy of warnings has become a prevalent issue in product liability and personal injury litigation. According to the Restatement of Torts (second) and to the Theory of Strict Liability, if a product needs a warning and it is absent or defective, the product is defective (see Madden, 1999).

Regulations, standards, and guidelines as to when and how to warn have been developed more extensively in the last three decades. Also, there has been a substantial increase in research activity on the topic during this time. Human factors specialists, or ergonomists, have played a major role in the research and the technical literature that has resulted.

In this chapter we review some of the major concepts and findings regarding factors that influence warning effectiveness. Most of the review is presented in the context of a communication–human information processing (C-HIP) model. The model is not only useful for organizing research findings, but it also provides a predictive and investigative tool. Following the presentation of the model and the review of major concepts and findings, a number of suggestions and recommendations for designing warnings is presented.

1.1 Hazard Control Hierarchy

In the United States and many other parts of the world, product manufacturers are responsible for providing safe products. To meet this responsibility, they should undertake a hazard analysis (e.g., fault tree, failure modes, critical incident) and examine pertinent databases to determine what hazards the product may pose in foreseeable use and misuse. Once hazards are identified, the next step is to determine how the hazards can be controlled. It should be recognized that warnings are usually not the first choice for controlling hazards and promoting safety. Rather, it is one tool that designers and manufacturers may use. Compared to other methods for protecting people and property, warnings have relatively limited reliability. Even the best warnings are not always 100% reliable or effective. The classic *hazard control hierarchy*, or a variant of it, is frequently a part of the analysis (Sanders and McCormick, 1993). This hierarchy defines a sequence of approaches in order of preference for dealing with hazards. The basic sequence is first to design the hazard out, second to guard, and third to warn. The first preference, the notion of eliminating the hazard through alternative design, is generally the best. If a type of poison can be removed from a product (e.g., lead in paint) and a safer substitute used, then the reformulation should be adopted. Beveling a sharp edge would eliminate or lessen a cutting hazard, and so on. Frequently, it is not possible to eliminate all hazards and still have the product function as intended.

The second line of defense is guarding; its purpose is to prevent contact between people and the hazard. There are several forms of guarding. Personal protective equipment, road barricades, and a lock on an electrical box are examples of physical barriers. Designing tasks in such a way to keep people away from a hazard is an example of a procedural guard. The "dead-man" switch on a lawn mower that shuts off the rotor when the handle is released is one such example, and requiring a physician's prescription to buy certain drugs is another. However, guarding, like hazard elimination, is not always a feasible solution.

The third line of defense is to warn. Warnings are the third priority in this sequence because, as mentioned above, they are not always reliable. Depending on the circumstances, the person at risk may not see or hear a warning, may not understand it, may not believe it, or may not be motivated to comply. Influencing behavior is sometimes difficult, and seldom foolproof. An implication of the hazard-control priority sequence is that warnings are not a substitute for good design or guarding. Indeed, where appropriate, warnings should be viewed as a supplement, not a substitute, to the other approaches to safety (Lehto and Salvendy, 1995).

In addition to the three-part hierarchy, other approaches may be effective in dealing with hazards. Generally, they fall into the same category as warnings in that they mostly involve communications that are intended to influence behavior. Training and personnel selection are examples. Another approach that includes elements similar to procedural guarding and warnings is supervisory control. These three approaches are applicable primarily to hazards in work environments.

1.2 Purposes of Warnings

Warnings have several purposes. First, warnings are a method for communicating important safety or safety-related information to a target audience who can then make better, more informed decisions regarding safety issues. Second, warnings are ultimately intended to reduce or prevent health problems, workplace accidents, personal injury, and property damage. To accomplish this, warnings are intended to influence or modify people's behavior in ways that will improve safety. Third, warnings may serve as a reminder, to call into awareness the hazard that may otherwise be latent in long-term memory.

There are two additional points to be noted regarding the purpose of warnings. First, warnings are a means of shifting or assigning responsibility for safety to people in the system (e.g., the product user, the worker) in situations where hazards cannot be designed out or guarded adequately. This is not to say that people do not have safety responsibilities independent of warnings; of course they do. Rather, a purpose of warnings is to provide the information necessary to enable them to carry out such responsibilities. The second point concerns people's right to know. The notion is that even in situations where the likelihood of warnings being effective may not be high, people have the right to be informed about safety problems confronting them. Obviously, this aspect of warnings is more of a personal, societal, and legal concern than a human factors issue, and although it is not addressed further in this chapter, it is a matter that is related to the overall purposes of warnings.

2 WHAT, WHO, WHEN, AND WHERE TO WARN

What to Warn Warnings are a form of safety communications. There are many kinds of warnings. Warnings can be in the form of signs, labels, product inserts and manuals, tags, audio and video tapes, face-to-face verbal statements, and so on. Printed warnings are generally text and graphics. Auditory warnings may be verbal and/or nonverbal. In this chapter we describe factors that are generally applicable to all types of warnings, although the examples given are geared mostly toward visual warnings associated with products. There are three kinds or categories of information to be included in warning-message text: hazard information, consequences information, and instructions. Each of these categories is addressed in subsequent sections.

Who to Warn Persons at risk are to be warned. The general principle regarding who should be warned is that it should include everyone who may be exposed to the hazard (who are at risk) and everyone who may be able to do something about it.

Warnings may be directed to a very specific audience. For example, warnings about toxic shock syndrome from the use of tampons would be directed primarily to women of childbearing age. On the other hand, warnings may be intended for the general public, such as an electric shock warning on the consumer appliances.

Although warnings are usually directed at end users, they may also be directed at intermediaries such as physicians who prescribe medications and job supervisors who make decisions about workplace safety. Who is to be warned is obviously a factor in the design of warnings used. Warnings for surgical equipment such as a laser scalpel can be written at higher levels of technical verbiage, because it can be assumed that users have had education and training to understand the language. This point relates to the concept of learned intermediary, which is frequently used in medical communications about prescription drugs. The warnings directed to physicians can be much more complex than would be appropriate for end users (patients), because physicians have had extensive training relevant to the use of drugs by their patients. This notion carries with it the assumption that end users may be less capable of understanding at least some of the warning material. A similar situation is communicating warnings about products that are hazardous to children. Here the learned intermediary is the caretaker.

Consumer products, however, are generally intended for a much wider group of people, sometimes the entire population. In such cases, the warning designer must be sensitive to the capabilities of a wide range of users. A general principle in warnings is that they should be written to take into account the lowest level of abilities training and experience in the target population. Some persons are color blind. Therefore, color should not be the only indicative cue for a hazard; there should other cues (i.e., have redundancy). Some persons, such as older adults, have visual decrements (e.g., presbyopia), which results in reduced acuity, decreased contrast discrimination, and increased glare, all of which suggests that larger print should be used for textual warnings than is commonly employed on product labels. Some persons cannot hear well; thus, auditory warnings need to be distinctly discriminable (in loudness and tonal quality) from background noise. These are just a sample of considerations based on sensory capabilities. Other considerations are cognitive. People may be limited in their ability to understand technical information. Consider that the U.S. population now includes many Spanish speakers. Lim and Wogalter (2003) found that although many English language users in the United States believe that immigrants should learn English, most acknowledge that safety communications might also need to be in Spanish.

When and Where to Warn The placement of a warning in time and location is important. The warning should be available when and where it is needed. Having read a warning on a previous occasion does not mean that it will be remembered or that it will transfer to the current situation. In general, the preferred location for a product warning is on the product, but this location may not always be possible. Space constraints or the nature of the product (e.g., a small clamp) may impose such limits. Determination of potential locations for the warning generally requires a task analysis (Frantz and Rhoades, 1993). Later in the chapter some solutions to these problems are offered.

3 COMMUNICATION–HUMAN INFORMATION PROCESSING MODEL

In this section a theoretical context is presented that will serve as an organizing framework for reviewing some of the major concepts and findings regarding factors that influence warning effectiveness. Specifically, a communication–human information processing (C-HIP) model is described. First, a few comments about communications and human information processing.

Communications Warnings are a form of safety communications. Communication models were around for most of the twentieth century (Lasswell, 1948; Shannon and Weaver, 1949). A typical, very basic model shows a sequence starting with a source, who encodes a message into a channel that is transmitted to a receiver, who receives a decoded version of that message. Noise may enter into the system at several points in the sequence, reducing the correspondence between the message sent and the one received. The warning sender may be a product manufacturer, government agency, employer, or other. The receiver is the user of the product, the worker, or any other person at risk. The message, of course, is the safety information to be communicated. The medium refers to the channels or routes through which information gets to the receiver from the sender. Understanding and improving these components of a safety communication system increases the probability that the message will be conveyed successfully.

However, the communication of warnings is seldom as simple as implied by a sequential communication model. Frequently, more than one medium or channel may be available and/or involved; multiple messages in different formats and/or containing different information may be called for, and the receiver or target audience may include different subgroups with varying characteristics. An example of such a warning situation would occur when a product with associated hazards is being used in a work environment. Figure 1 illustrates a communication model that might be applicable. It shows the distribution of safety information from several entities to the receiver and that feedback may influence the kind of safety information given. It also shows that in addition to the sender (manufacturer) and receiver (end user), other people or entities may be involved, such as distributors and employers. Further, each of these entities may be both receivers and senders of safety information. There are also more routes through which warnings may travel, such as from the manufacturer to the distributor to the employer to the user, from the manufacturer to the employer to the user, or directly from the manufacturer to the user (as on a product label). The warnings may take various forms. One example is safety rules that an

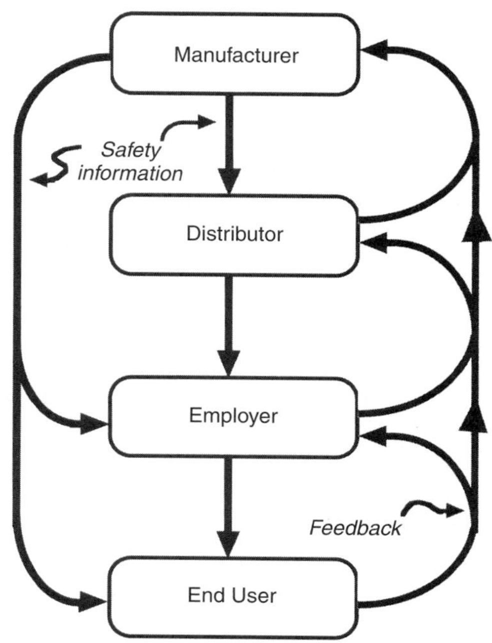

Figure 1 Distribution of safety information and feedback.

employer sets to govern the behavior of employees. By collecting feedback information, the warnings can be adjusted to meet criteria such as comprehension. Thus, warnings or warning systems may be much more complex than just a sign or label. Warning systems as a concept is discussed in more detail later in the chapter.

Human Information Processing Cognition is a core area of psychology that is concerned with mental processes such as attention and memory. Since the 1960s, much of the theoretical work has been described in terms of stages of processing. Numerous models have been developed and tested, with recent versions becoming more complex. Central to this activity has been the notion of stages. In the next section, a model that incorporates some basic stages of mental processing is described. It can be viewed as an elaboration of the communication model's receiver stage.

C-HIP Model The communications–human information processing model (Wogalter et al., 1999a) is a framework for showing stages of information flow from a source to a receiver, who in turn may cognitively process the information subsequently to produce compliance behavior. The model is displayed in Figure 2. The conceptual stages of source, channel, and receiver are taken from a very simple communication model. The receiver stage is divided into several human information processing substages prior to carrying out the compliance behavior. These substages are attention switch, attention maintenance, comprehension, attitudes and beliefs, and motivation.

At each stage of the model, warning information is processed and, if successful at that stage, "flows through" to the next stage. If processing at a stage is unsuccessful, it can produce a bottleneck, blocking the flow from getting to the next stage. This is the weak-link-in-the-chain phenomenon. If all the stages are successful, the process ends in behavior (compliance). Although processing of the warning might not make it all of the way to the last stage, it still may be effective at influencing earlier stages. For example, a warning might positively influence comprehension but not change behavior. Such a warning cannot be said to be totally "ineffective," since it produces better understanding and potentially can lead to better, more informed decisions. However, it is ineffective in the sense that it does not curtail unsafe behavior.

The C-HIP model can be particularly useful describing the factors that influence warning effectiveness. It also can be helpful in diagnosing and understanding warning failures and inadequacies. If a source (or sender) does not issue a warning, no information will be transmitted and thus nothing will be communicated to the receiver. Even if a warning is issued by the source, it will not be effective if the channel or transmission medium is poorly matched with the message, the receiver, or the environment. Each of the processing stages within the receiver can also produce a bottleneck, preventing further processing. The receiver might not notice the warning and thus not be directly affected. Even if the warning is noticed, the individual may not maintain attention to the warning to encode the information. If the receiver encodes the details of the warning, it still may not be understood. If understood, it still might not be believed; and so on.

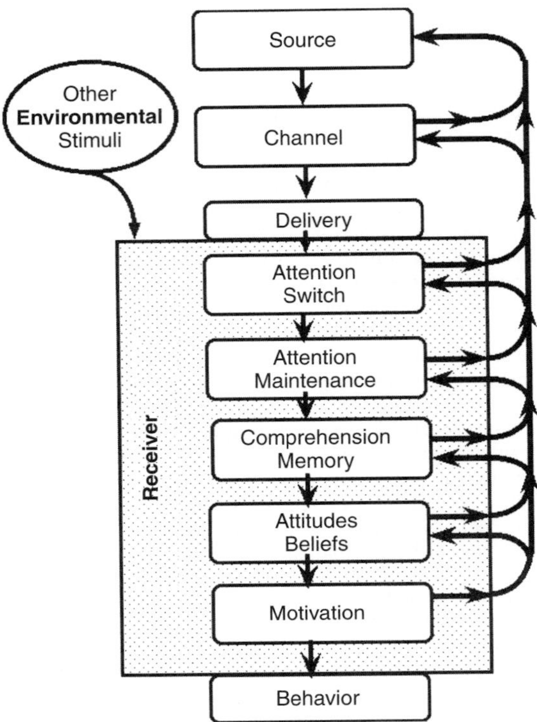

Figure 2 Communication–human information processing (C-HIP) model.

Although the processing described above is linear, there are feedback loops from later stages to earlier stages, as illustrated in Figure 2. For example, when a warning stimulus becomes habituated from repeated exposures over time, attention is less likely to be allocated to the warning on subsequent occasions. Here, memory (as part of the comprehension stage) affects an earlier attention stage of processing. Another example is that some people might not believe that a product or situation is hazardous, and as a consequence, not look for a warning. A third example is that the person may not understand the warning and therefore switch attention to read it again. These nonlinear effects between the stages resulting from feedback show how later stages influence earlier stages in ongoing cognitive processing.

In the sections that follow, each stage of the C-HIP model is described with some of the factors that influence it. The purpose is to assist in analyzing how or why warnings may fail or, conversely, what they have to accomplish to succeed. In many respects the model is similar to the information-processing models employed by others (Lehto and Miller, 1986; Lehto and Papastavrou, 1993; Rogers et al., 2000). The model given here is somewhat different from the ones given in Wogalter et al. (1999a) and Wogalter and Laughery (2005). Two main differences are the inclusion of other environmental influences, and the need to deliver the message to the receiver. The purpose of including these additional elements is to emphasize their impact on the warning processes. An additional minor change is that attention is explicitly split into two distinct stages of switch and maintenance. Table 1 gives a summary of some of the primary considerations associated with successful processing at each stage.

3.1 Source

The source is the originator or initial transmitter of the warning information. The source can be a person(s) or an organization (e.g., company or government). Research shows that differences in the perceived characteristics of the source can influence people's beliefs about the credibility and relevance of the warning (Wogalter et al., 1999b). Information from a reliable, expert source (e.g., the Surgeon General, the U.S. Food and Drug Administration) is given greater credibility, particularly when the expertise is relevant (e.g., the American Medical Association for a health-related warning) (Wogalter et al., 1999b). An important aspect that is discussed in more detail later is that a warning attributed to an expert source may aid in changing erroneous beliefs and attitudes that the receiver may have.

A critical role of the source is to determine if there is a need for a warning and if so, what should be warned. This decision typically hinges on the outcome of hazard analyses that determine foreseeable ways injuries could occur.

There are several principles or rules that guide when a warning should be employed: (1) a significant hazard exists; (2) the hazard, consequences, and appropriate safe modes of behavior are not known to the persons at risk; (3) the hazards are not open and obvious (i.e., the appearance of the product or environment does not communicate them); and (4) a reminder is needed to assure awareness of the hazard at the proper time. There are other considerations in deciding what to warn about, such as the likelihood of an undesirable event and the severity of the potential outcomes.

Assuming that the product or environment is determined to need a warning, one or more channels of communication must be used to reach the receiver.

3.2 Channel

The channel is the way or medium in which information is transmitted from the source to one or more receivers. Warnings can be presented on product labels, on posters, in brochures, as part of audio-video presentations, given orally, and so on. Most commonly, warnings are sent via the visual (printed text warnings and pictorial symbols) and auditory (alarm tones, live voice, and voice recordings) modalities as opposed to the other senses. There are exceptions: an odor added to flammable gases such as propane makes use of the olfactory sense, and a pilot's control stick that is designed to vibrate when the aircraft begins to stall makes use of the tactile and kinesthetic senses.

3.2.1 Media and Modality

There are two basic dimensions of the channel. The first concerns the media in which the information is embedded. The second dimension of the channel is the sensory modality used to capture the information by the receiver. Media and modalities are closely tied. Some studies have examined whether presentation of a language-based warning is more effective when presented in the visual (text) vs. the auditory (speech) modality. The results are conflicting (although generally either one is better than no presentation whatsoever). Some cognitive research (Penney, 1989) suggests that longer, more complex messages may be better presented visually and shorter messages auditorily. The auditory modality is better for attracting attention (a stage described below). However, auditory presentation can be less effective than visual presentation, particularly for processing lengthy, complex messages because (1) it is primarily temporal/sequential in nature, (2) its processing speed is slower, and (3) the ability to review previously presented material is often not possible. These characteristics tend to overload working memory (or maintenance attention, to be discussed later).

3.2.2 Multiple Methods and Redundancy

Research has generally found that presenting warnings in two modalities is better than one modality. Thus, a video-based warning is better if the words are shown on the screen while the same information is given orally. This method provides redundancy. If a person is not watching the screen, people can still hear it. If the person is blind or deaf, the information is available in the other modality. A similar concept for media is described in the next section.

3.2.3 Warning System

The idea that a warning is only a sign or a portion of a label is too narrow a view of how such safety information gets transmitted. Warning systems for a particular environment or product may consist of a number of components. In the context of the communication model presented in Figure 1, the components may include a variety of media and messages.

A warning system for an over-the-counter (OTC) pharmaceutical product such as a multisymptom cold medication may consist of several components: a printed statement on the box, a printed statement on the bottle, and a printed package insert. In addition, there may be text and/or speech warnings in television advertisements about the product. A warning system for pneumatic tools regarding the hazard of long-term vibration exposure causing damage to the nervous and vascular systems of the hand (vibration-induced white

Table 1 Methods and influences of the Communication–Human Information Processing (C-HIP) Stages

C-HIP Stage	Methods and Influences
Source	Determines that hazard is not designed out or guarded. Credible, expert.
Channel	Visual (signs, labels, tags, inserts, product manuals, video, etc.). Auditory (simple and complex nonverbal; voice; live or recorded). Other senses: vibration, smell, pain. Generally, transmission in more than one modality is better.
Delivery	Make sure that message gets to target audience(s).
Receiver	Consider demographics of target audiences (e.g., older adults, illiterates, cultural and language differences, persons with sensory impairments).
Attention switch	Should be high salience (conspicuous/prominent) in cluttered and noisy environments. Visual: high contrast, large. Presence of pictorial symbols aids noticeability. Auditory: louder and distinguishable from surround. Present when and where needed (placed proximal in time and space). Avoid habituation by changing stimulus.
Attention maintenance	Enables message encoding by examining/reading or listening. Visual: legible font and symbols, high-contrast aesthetic formatting, brevity. Auditory: intelligible voice, distinguishable from other sounds.
Comprehension and memory	Enables informed judgment. Understandable message provides necessary information to avoid hazard. Try to relate information to knowledge already in users' heads. Explicitness enables elaborative rehearsal and storage of information. Pictorials can benefit understanding and substitute for some wording; may be useful for certain demographic groups. At subsequent exposures, warning can cue or remind user of information. Comprehension testing needed to determine whether warning communicates intended/needed information.
Beliefs/attitudes	Familiarity reduces perceived hazard and warning processing. Persuasive argument and excellent warning design needed when beliefs are seriously discrepant with truth. May influence receiver's earlier stages.
Motivation	Energizes person to carry out next stage. Low cost (time, effort, money) facilitates compliance. Perceived high cost increases likelihood of noncompliance. Benefited by warning explicitness and perceived injury severity. Affected by social influence, time stress, mental workload.
Behavior	Carrying out safe behavior that does not result in injury or property damage.

finger) might consist of a number of components. Examples include warnings embossed on the tool, a removable tag attached to the product when new, accompanying sheets or a stapled manual, and printing on the box. In addition, manufacturers might provide employers with supplemental materials such as videos and posters to assist in employee training sessions. Organizations, including government agencies and consumer and trade groups, could provide additional materials. With the growing use of the Internet, information may be made available on Web sites. Another example would be warnings for a solvent used in a work environment for cleaning parts. Here the components might include printed on-product labels, printed flyers that accompany the product, statements in advertisements about the product, verbal statements from the salesperson to the purchasing agent, and material safety data sheets provided to the employer.

The components of a warning system may not be identical in terms of content or purpose. For example, some components may be intended to capture attention and direct the person to another component, where more information is presented. Similarly, different components may be intended for different target audiences. In the example of the solvent above, the label on the product container may be intended for everyone associated with the use of the product, including the end user, whereas the information in the material safety data sheet (MSDS) may be directed more to fire personnel or to an industrial toxicologist or safety engineer working for the employer (Smith-Jackson and Wogalter, in press).

3.2.4 Direct and Indirect Communications

The distinction between direct and indirect effects of warnings concerns the routes by which information gets to the target person. A direct effect occurs as a result of the person being exposed directly to the warning. That is, he or she reads or hears the warning directly. But warnings can also accomplish their purposes when delivered indirectly. One example is the woman who did not read the warnings about Toxic Shock Syndrome on a tampon box, but learns about the hazard in a conversation with her neighbor. The employer or physician who reads warnings and then communicates the information verbally to employees or patients is another example. Moreover, the print and broadcast news media may present information that is given in warning labels. The point is that a warning put out by a manufacturer may have utility even if the consumer or user is not exposed to the warning directly.

An example of where an indirect effect was considered in the design of a product warning concerned a herbicide used in agricultural settings. Given that significant numbers of farmworkers in parts of the United States read Spanish but not English, there was reason to put the warning in both languages. However, there are sometimes space constraints on product containers. One suggested strategy was to include a short statement on the label in Spanish indicating that the product was hazardous and that the user should get someone translate the rest of the label before using the product.

There are situations where we rely on indirect communications to transmit warning information. Employers and physicians are examples already noted; adults who have responsibility for the safety of children are another important category. In the design of warning systems, empowering indirect warnings could enhance the spread of warning information to relevant targets.

3.3 Delivery

Although the source may try to disseminate warnings in one or more channels, the warnings might not reach some of the targets at risk. For example, a safety brochure that is developed and produced by a governmental agency that is never distributed is not very helpful. Purchasers of used products are at risk because the manufacturer's product manual is frequently not available or is not transferred to new owners at resale (Rhoades et al., 1991; Wogalter et al., 1998b). Without the manual, the user may not know what the correct and incorrect uses of the product are or what the maintenance schedule is, both of which could affect safety. Williamson (2006) describe problems associated with communicating warnings on the flash-fire hazard associated with burning plastic-based insulation. Although some warnings accompany bulk lots of insulation when shipped from the manufacturer/distributor to job sites and some technical warnings may be seen by architects and high-level supervisors, the warnings infrequently make it downstream to construction workers who may be working with or around the product. The point here is that although a warning may be put out by a source (through some channel), it may have limited utility if it does not reach the targets at risk either directly or indirectly.

3.4 Receiver

In this section we focus on the receiver; that is, the person(s) or target audience to whom the warning is directed. As noted earlier, the primary theoretical context for presenting this analysis is an information-processing model. This model with respect to the receiver, shown in Figure 2, defines a sequence of processing stages through which warning information flows. By examining each of the stages and the factors that influence success or failure at each stage, a better understanding of how warnings should be designed and whether they are likely to be effective can be attained.

For a warning to communicate information and influence behavior effectively, attention must be switched to it and then maintained long enough for the receiver to extract the necessary information. Next, the warning must be understood and must concur with the receiver's existing beliefs and attitudes. If there is disagreement, the warning must be sufficiently persuasive to evoke an attitude change toward agreement. Finally, the warning must motivate the receiver to perform proper compliance behavior. The next several sections are organized around these stages of information processing.

3.4.1 Attention

One of the goals of a warning is to capture attention and then hold it long enough for the contents to be processed. In the following sections we address these two attention issues.

Switch of Attention The first stage in the human-information-processing portion of the C-HIP model concerns the switch of attention. An effective warning must initially attract attention. Often, this attraction must occur in environments that also have other stimuli competing for attention.

For a warning to capture attention, it must first be available to the recipient. As noted earlier, warning messages that do not arrive at the end user will not have direct effects. Assuming that the warning is present, it needs to be sufficiently salient (conspicuous or prominent) to capture attention. Warnings typically have to compete for attention, and several design factors influence how well they compete.

Size and Contrast Bigger is generally better. Increased print size and contrast against the background have been shown to benefit subsequent recall (Barlow and Wogalter, 1993). Young and

Wogalter (1990) found that print warnings with highlighting and bigger, bolder print led to higher comprehension of and memory for owner's manual warnings.

Context plays an important role with regard to size effects on salience. It is not just the size of the warning that is important, but also its size relative to other information in the display. A bold warning on a product label where there are other informational items in larger print is less likely to be viewed than those larger items.

For some products, the available surface area on which warnings can be printed is limited. This is particularly true for small product containers such as pharmaceuticals. Methods available to increase the surface area for print warnings include adding tags or peel-off labels (Barlow and Wogalter, 1991; Wogalter et al., 1999d). Another method is to put some minimum critical information on a primary label and direct the user to additional warning information in a secondary source, such as in a well-designed owner's manual or package insert. Wogalter et al. (1995) have shown that such a procedure can be effective.

Color Although there are some problems with the use of color, such as color blindness, fading, and lack of contrast with certain other colors, people are generally strongly in favor of the use of color. A colored signal word attracts attention more effectively than one that is black like the rest of the print (e.g., Laughery et al., 1993b). The ANSI Z535 (2002) standard relies on color in the signal word panel to attract attention.

Pictorial Symbols Pictorial symbols can be useful for capturing attention (Jaynes and Boles, 1990; Young and Wogalter, 1990; Laughery et al., 1993a; Kalsher et al., 1996; Bzostek and Wogalter, 1999). One general symbol that attracts attention is the alert icon (triangle enclosing an exclamation point) (Laughery et al., 1993a) that is found in the signal word panel in ANSI (2002) Z535-style warnings.

Placement A general principle is that warnings located close to the hazard, both physically and in time, will increase the likelihood of a proper attention switch (Frantz and Rhodes, 1993; Wogalter et al., 1995). A warning on the battery of a car regarding a hydrogen gas explosion hazard is much more likely to be effective than a warning in the car owner's manual. A verbal warning given two days ago before a farmworker uses a hazardous pesticide is less likely to be remembered and effective than one given immediately prior to using the product.

A warning, even a good one, that is located in a out-of-view location reduces its likely effectiveness drastically. In general, placement of warnings directly on a hazardous product is preferred (Wogalter et al., 1987). However, there are several factors to be considered in warning placement. One is visibility: A warning should be placed so that users are likely to see it (Frantz and Rhoades, 1993). For example, a warning on one side of a tall rolling cart (with a high center of gravity) may not be seen if the user does not examine that side of the cart before use. People generally do not read owner's manuals of cars they rent; thus, if not given some better way of warning about the particulars of the vehicle, such as a special stickers or a quick-tip chart, drivers will not be aware of important safety information. Manufacturers need to consider how their product may be used so they can be better prepared to select proper locations for warnings. In general, warnings should be located near other information that will be needed to perform a task. Task analyses are likely to be beneficial here.

With most languages, people tend to scan printed material left to right and top to bottom. Thus, warnings should be located near the top or to the left and not be buried in the middle or at the bottom. Wogalter et al. (1987) showed that warnings in a set of instructions for mixing chemicals were more likely to be noticed and complied with if placed before the task instructions than if following them.

Related to the concern about warning locations, however, is the fact that at times practical considerations limit the options. A small container such as on some over-the-counter medications may simply not have room for all of the information that should go into the warning. Some options for this problem are discussed later.

Formatting Another factor that can influence attention is formatting. Visual warnings that are formatted to be aesthetically pleasing, with plenty of white space and coherent information groupings (Hartley, 1994), are more likely to attract attention (Wogalter and Vigilante, 2003). If a warning contains large amounts of text, people may decide that too much effort is required to read it, and direct their attention to something else.

Repeated Exposure A related issue is that repeated and long-term exposure to a warning may result in a loss of attention-capturing ability (Wogalter and Laughery, 1996). This habituation can occur over time, even with well-designed warnings. Where feasible, changing a warning's format or content can slow the habituation process (Wogalter and Brelsford, 1994). We discuss habituation in more detail in a later section.

Other Environmental Stimuli Other stimuli in the environment may compete with the warning for attention capture. Other stimuli may include the presence of other persons, various objects that comprise the context, and tasks the person is performing. Thus, the warning must stand out from the background (i.e., be salient or conspicuous) in order to be noticed. This factor is particularly important because people typically do not actively seek hazard and warning information. Usually, people are focused on the tasks they are trying to accomplish. Safety considerations that may be important to a person are simply not always on one's mind. Hence, the warning needs to be conspicuous.

Auditory Warnings Auditory warnings are frequently used to attract attention. Auditory signals are omnidirectional, so the receiver does not have to be looking at a particular location to be alerted. Like print warnings, their success on the attention criterion is largely a matter of salience. Auditory warnings should be more intense and distinctively different from expected background noise. Often, auditory warnings are used in conjunction with visual warnings, with the auditory warning serving to call attention to the need to read or examine a visual/written warning that contains specific information.

Maintenance of Attention People may notice the presence of a warning but not stop to examine it. A warning that is noticed but fails to maintain attention long enough for its content to be encoded is of little direct value. Attention must be maintained on the message for some length of time to extract meaning from the material (Wogalter and Leonard, 1999). During this process, the information is encoded or assimilated with existing knowledge in memory.

With brief warnings the message information may be acquired very quickly, sometimes as fast as a glance. For longer warnings to maintain attention, they need to have qualities that generate interest and do not require considerable effort. Some of the same design features that facilitate the switch of attention also help to maintain attention. For example, large print not only attracts attention, but also increases legibility, thus making reading less effortful and more likely.

Legibility If the warning has very small print, it may not be legible, making it difficult to read. Some persons may not be able to read it even with visual correction, and some who might be able to read it with some effort will not. Older adults with age-related vision problems are a particular concern (Wogalter and Vigilante, 2003). Distance and environmental conditions such as fog, smoke, and veiling glare can affect legibility.

Sanders and McCormick (1993) give data on the legibility of fonts developed for military applications. Legibility of type can be affected by numerous factors, including choice of font, stroke width, letter compression and distance between them, case, resolution, and justification. Although there is not much research to support a clear preference for the use of certain fonts in warnings over others, the general recommendation is to use relatively plain, familiar fonts. It is sometimes recommended that a serif font such as Times or Times Roman be used for small-sized text and a sanserif font such as Universe or Helvetica be used in applications requiring headline type sizes. The ANSI (2002) Z535.2 and Z535.4 standard documents have a chart of print size and expected reading distances in good and degraded conditions.

Contrast and color is another consideration. Black on white or the reverse has the highest contrast, but legibility can be adequate with other combinations, such as black print on yellow and white print on red. The selection of color should also be governed by the context in which the warning is presented (Young, 1991). One would not want to put a red and white warning on a largely red background.

Formatting People are more likely to maintain attention if a warning is "readable" with respect to layout. Visual warnings formatted to be aesthetically pleasing are more likely to hold attention (and thus be examined and the information extracted) than is a single chunk of dense text (Wogalter and Vigilante, 2003). Formatting can show the organization of the warning material, making it easier to assimilate or accommodate into memory. In general, the use of generous white space and bulleted lists are preferred to long, dense paragraphs (Desaulniers, 1987; Wogalter and Post, 1989). Although aesthetically pleasing at a distance, full justification (the straight alignment of the beginning and ending words at both margins) is more difficult to read than "ragged right" (justification only of the left margin), where the spacing between letters and words is consistent.

Pictorial Symbols Interest is also facilitated by the presence of well-designed pictorial symbols. Further, research indicates that people prefer warnings that have a pictorial symbol to warnings without one (Young et al., 1995; Kalsher et al., 1996).

Auditory Warnings Simple nonverbal auditory warnings are generally used as alert (attention-getting) signals. Frequently, these signals carry very little information other than an attention-switch cue. After the alert is given, the visual modality is generally used to access further information (Sorkin, 1987; Sanders and McCormick, 1993).

3.4.2 Comprehension

Comprehension concerns one's ability to grasp the meaning of a warning. Some comprehension may derive from subjective understanding such as its hazard connotation, and some from more direct understanding of the language and the symbols used.

Hazard Connotation The idea of hazard connotation is that certain aspects of the warning may convey some level or degree of hazard. It is an overall perception of risk, a subjective understanding of the danger conveyed by the warning components. A similar type of connoted hazard was shown in research by Wogalter et al. (1997) for various container types.

In the United States, current standards such as ANSI (2002) Z535 and guidelines (e.g., Westinghouse Electric Corporation, 1981; FMC Corporation, 1985) recommend that warning signs and labels contain a signal word panel that includes one of the terms DANGER, WARNING, or CAUTION. According to ANSI Z535, these terms are intended to denote decreasing levels of hazard, respectively. Figure 3 shows two ANSI-type warning signal word panels. According to ANSI Z535, the DANGER panel should be used for hazards where serious injury or death *will* occur if warning compliance behavior is not followed, such as around high-voltage electrical circuits. The WARNING panel is used when serious injury *might* occur, such as severe chemical burns or exposure to highly flammable gases. The CAUTION panel is used when less severe personal injuries or damage to

DANGER

CAUTION

Figure 3 Two signal word panels including alert symbol and color. Note that the DANGER panel is white print on a red background, and the CAUTION is black print on a yellow background. Not shown is the WARNING panel, which is black print on an orange background.

property might occur, such as getting hands caught in operating equipment. Research shows that laypersons often fail to differentiate between CAUTION and WARNING, although both are interpreted as connoting lower levels of hazard than DANGER (e.g., Wogalter and Silver, 1995). The term NOTICE is intended for messages that are important but do not relate to injuries. The term DEADLY, which has been shown in several research studies to connote hazard significantly above DANGER, has not been adopted by ANSI, yet might be considered for hazards that are significantly above those using the term DANGER.

Different characteristics of sounds can lead to different hazard connotations. Sounds that are more intense, of higher frequency, or have rises in pitch and/or faster beats can cue greater perceived hazard urgency (Edworthy et al., 1991). The same effects are true with voice (Barzegar and Wogalter, 1998; Hollander and Wogalter, 2000; Weedon et al., 2000; Hellier et al., 2002).

According the ANSI standard, the signal words DANGER, WARNING, AND CAUTION are to be accompanied by specific colors (red, orange, and yellow, respectively). This assignment provides redundancy. However, the colors for WARNING (orange) and CAUTION (yellow) are not readily distinguished with regard to hazard connotation, although red (for DANGER) generally has a significantly higher hazard connotation than the other two colors (e.g., Chapanis, 1994). Color can also be used to change the hazard connotation. For example, the signal word DANGER with the color orange connotes less hazard than the same term with the color red.

Competence There are many dimensions of receiver competence that may be relevant to the design of warnings. For example, sensory deficits might be a factor in the ability of some special target audiences to be directly influenced by a warning. A blind person would not be able to receive a written warning, nor would a deaf person receive an auditory warning.

At the opposite end of the sequence of events is behavior. If special equipment is required to comply with the warning, it must be available or obtainable. If special skills are required, they must be present in the receiver population. It is not difficult to find examples of warnings that violate considerations of people's limitations. One example is the common warning instruction on containers of solvents: "Avoid breathing fumes." This instruction can be difficult to carry out because users may not see or smell the vapors and appropriate respirator equipment may not be available.

Three characteristics of receivers related to cognitive competence are important in warning design: technical knowledge, language, and reading ability. The communication of hazards associated with medications, chemicals, and mechanical devices is often technical in nature. If the target audience does not have the relevant technical competence, the warning is not likely to be successful. The level or levels of knowledge and understanding of the audience must be taken into account. This point is discussed further in a later section.

The issue of language is straightforward, and it is increasingly important. Subgroups in the United States speak and read languages other than English, such as Spanish. As trade becomes more international, requirements for warnings to be directed to non-English readers will increase. Ways of dealing with this problem include warnings stated in multiple languages and the use of pictorials.

Reading ability is another target audience characteristic where its importance is obvious. Yet, high reading levels, such as a grade 12, are not uncommon for warnings intended for those with lower reading abilities. In general, reading level should be as low as feasible. For general target audiences, the reading level might need to be in the range grade 4 to 6. Clearly, if comprehension of a warning is to be achieved, reading levels must be consistent with reading abilities of receivers. There are readability formulas or indices based on word frequency of use, length of words, number of words in statements, and so on, that are used to estimate reading grade level (Duffy, 1985). These formulas have limitations, such as being notorious for giving inaccurate estimates. However, they can be useful as a preliminary guide to achieving a warning that will be understood. A discussion of reading level measures and their application to the design of instructions and warnings may be found in Duffy (1985).

An additional point on reading ability concerns illiteracy. There are estimates that over 16 million functionally illiterate adults exist in the U.S. population. If so, successfully communicating warnings may require more than simply keeping reading levels to a minimum. Although simple solutions to this problem do not exist, pictorials, speech warnings, special training programs, and so on, may be important components of warning systems for such populations.

Message Content The content of the warning message should include information about the hazard, the consequences of the hazard, and instructions on how to avoid the hazard.

Hazard Information The point of giving hazard information is to tell the target audience what the safety problem is (i.e., what can go wrong). Generally, this information is specific to the environment or product. Examples are:

Toxic Vapors
Slippery Floor
High Voltage

A general principle is that the hazard should be spelled out in the warning. However, there are exceptions when the hazard is (1) general knowledge, (2) known from previous experience, or (3) "open and obvious" (the latter is a concept that is described in more detail in a subsequent section). Where these conditions do not exist, hazard information is an important part of the warning (Wogalter et al., 1987).

Consequences Consequences information concerns the nature of the injury, illness, or property damage that could result from a hazard. Hazard information and consequence information are usually closely linked in the sense that one leads to the other, or stating it in the reverse, one is the outcome of the other. In warnings, statements regarding these two elements should generally be sequenced; an example is

Toxic Vapor
Severe Lung Damage

For purposes of getting and holding the receiver's attention, however, there are situations where it is desirable to put consequences information near the beginning of the warning (just after the icon and signal word) in larger and bolder print (Young et al., 1995). This is particularly true for severe consequences such as death, paralysis, or severe lung damage. Hence, the hazard and consequence statements above might be better presented as

Severe Lung Damage
Toxic Vapor

There are also occasions or situations when the hazard information is presented and understood, so it may not be necessary to state the consequences in the warning. This point is related to the open and obvious aspects of hazards. For example, a sign indicating "Slippery Floor" probably does not need to include a consequence statement "You Could Fall." It is reasonable to assume that people will correctly infer the appropriate consequence. Although it is desirable to keep warnings as brief as possible (the brevity criterion is discussed in a later section), there is a potential problem with omitting consequence information; specifically, people may not make the correct inference regarding injury, illness, or property damage outcomes. Thus, it is important in designing

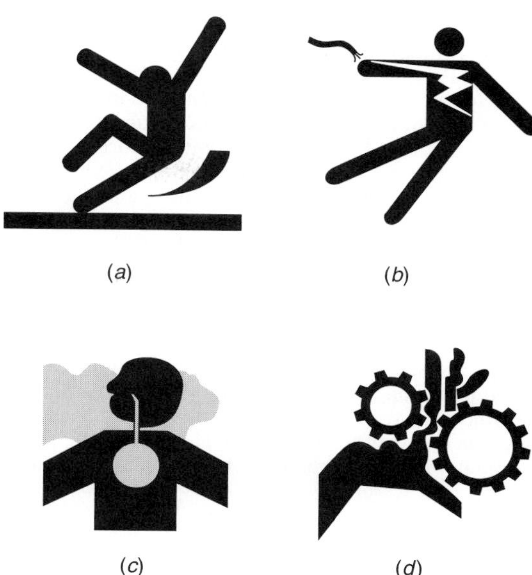

Figure 4 Pictorials conveying hazard information: (*a*) slippery floor; (*b*) electricity; (*c*) toxic gas; (*d*) pinch point.

warnings to assess, if necessary, whether people will infer the consequences correctly.

A common shortcoming of warnings is that consequences information is not explicit; that is, it does not provide important specific details. The statement "May be hazardous to your health" in the context of a toxic vapor hazard does not tell the receiver whether he or she may develop a minor cough or suffer severe lung damage (or some other outcome). This issue is discussed in Section 4.3. The point is that knowing about severe consequences can be a motivational factor in attending to and complying with the warning message, a consideration discussed further in Section 3.4.4.

Pictorials can also be used to communicate consequence information. Some pictorials (e.g., for a slippery floor hazard) convey both hazard and consequence information without it being stated directly. Examples are shown in Figure 4.

Instructions In addition to getting people's attention and telling them what the hazard and potential consequences are, warnings should instruct people about what to do or not do. Typically, but not always, instructions in a warning follow the hazard and consequence information. An example of an instructional statement that might go with the hazard and consequence statements above is

Severe Lung Damage
Toxic Vapors
Must Use Respirator Type 1234

This instruction assumes, of course, that the receiver will know what a Type 1234 respirator is and have access to one.

Pictorials can be used to communicate instructions. Figure 5 shows examples of instructional information used in warnings. Note that some pictorials use a prohibition symbol, a circle containing the pictorial with a slash through it. Both the circle and slash are usually red, although sometimes they are black.

Sometimes a distinction is made between warnings and instructions. Warnings are communications about safety; instructions may or may not concern safety. "Keep off the grass" is an instruction that generally has nothing to do with safety (unless the grass is infested with fire ants, in which case the statement alone clearly would not be an adequate warning). When instructions are concerned with safety information or safe behavior, they can be viewed as part of a warning. In short, warnings include instructions, but not all instructions are part of a warning.

Explicitness An important design principle relevant to warning comprehension is explicitness (Laughery et al., 1993a). Explicit messages contain information that is sufficiently clear and detailed to permit the receiver to understand at an appropriate level the nature of the hazard, the consequences, and the instructions. The key here is the word *appropriate*. A classic example is: "Use with adequate ventilation." Does this statement mean open a window, use a fan, or something much more technical in terms of volume of airflow per unit time? Obviously, the instruction is not clear. Warnings are frequently not detailed or specific enough. However, sometimes, as stated earlier, technical details are not necessary and may be detrimental. The following two examples are warnings with hazard, consequence, and instructional statements that are not sufficiently explicit.

Dangerous Environment
Health Hazard
Use Precaution

Mechanical Hazard
You Could Be Injured
Exercise Care

Alternatives to the above that would be considered more explicit and appropriate are:

Severe Lung Damage
Toxic Vapors
Must Use Respirator Type 1234

Pinch Point Hazard — Moving Rollers
Hand May Be Severely Crushed or Amputated
Do Not Operate Without Guard X In Place

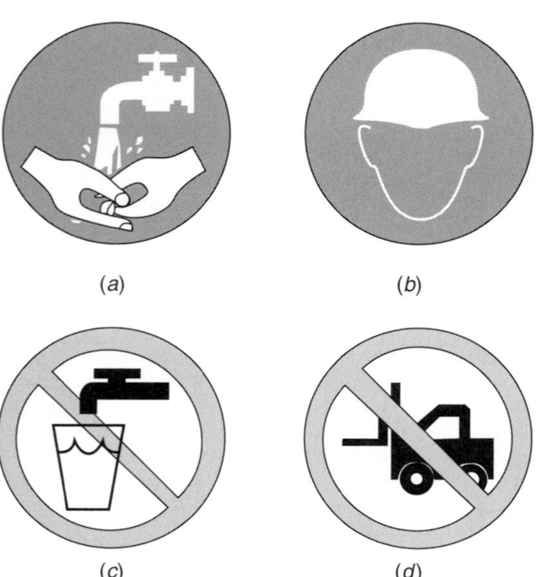

Figure 5 Pictorials conveying instructional information: (*a*) wash hands; (*b*) wear hard hat; (*c*) do not drink water; (*d*) no forklifts in area.

Pictorial Symbols Pictorial symbols are used to communicate hazard-related information, often in conjunction with the printed text message. Guidelines such as ANSI (1991) and FMC Corporation (1985) place considerable emphasis on the use of safety symbols. Pictorials are particularly useful in helping to increase comprehension (Lerner and Collins, 1980; Collins, 1983; Zwaga and Easterby, 1984; Boersema and Zwaga, 1989; Laux et al., 1989; Wolff and Wogalter, 1993, 1998; Dewar, 1999). They can contribute to understanding when illiterates or non-English readers are part of the target audience. Also, they can be useful where there are time constraints, such as traveling on a highway, because well-designed pictorials can cue large amounts of knowledge in a glance.

Although pictorials can assist in the comprehension of warning information, comprehension is also a primary concern or criterion for pictorials. In some pictorials, the depiction directly represents the information or object being communicated and will be understood if the person recognizes the intended depiction. Figure 6 shows two examples of direct representation. One shows both a hazard and consequences by depicting a raging fire, and the other shows both the hazard and the instructions, depicting the need for an eye shield. In other pictorials, the symbol may be recognized but its meaning has to be learned. People may recognize a skull and crossbones, but the fact that it represents a poison hazard would have to be learned. Some pictorials are completely abstract, such as the symbols for "do not enter" and biohazard shown in Figure 7, and must be learned to be understood. As

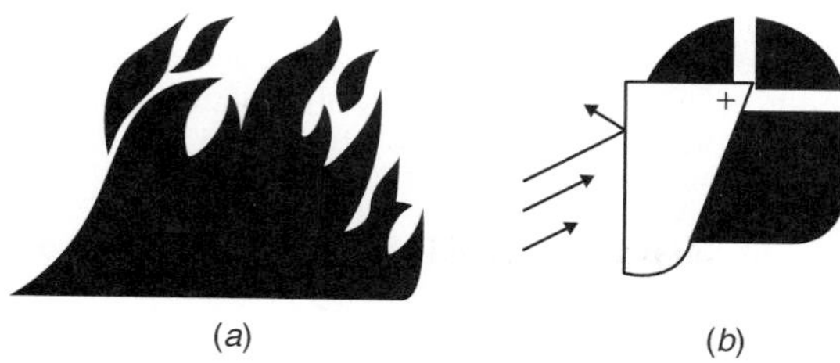

Figure 6 Pictorials showing a direct representation: (*a*) raging fire; (*b*) wear eye shield.

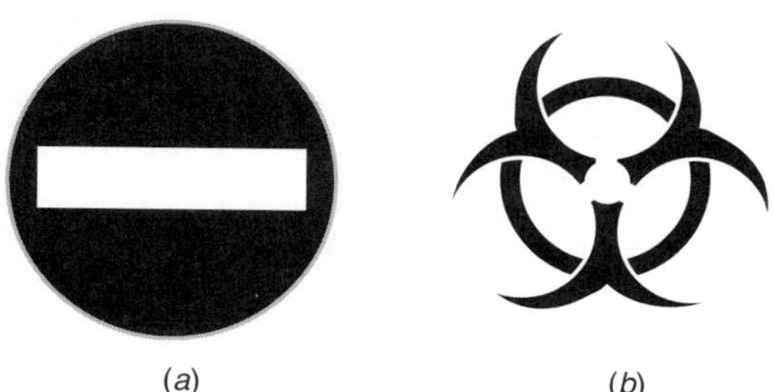

Figure 7 Pictorials that can be recognized after learning: (*a*) do not enter; (*b*) biohazard.

a general principle, pictorials containing symbols or pictures that directly represent the information are preferred, especially for general target audiences.

What is an acceptable level of comprehension for pictorials? This question has been addressed in the ANSI (2002) Z535 standard, which suggests a goal of 85% comprehension by the target audience. Two criteria seem relevant here. The first is simply that the pictorial should be designed to accomplish the highest level of comprehension attainable. If 85% cannot be achieved, it may still be useful, depending on the alternatives. A second criterion is that the pictorial not be misinterpreted or communicate incorrect information. According to the ANSI (2002) Z535 standard, an acceptable symbol must have less than 5% critical confusions (opposite meaning or a meaning that would produce unsafe behavior). Wogalter (1999) cites an example of a misinterpretation of a pictorial that was part of a warning for the drug Acutane. This drug is used for severe acne but causes birth defects in babies of women taking the drug during pregnancy. The pictorial shows a side-view outline shape of a pregnant woman within a circle–slash negation sign. The intended meaning of the pictorial is that women should not take the drug if they are pregnant. However, some women interpreted the symbol incorrectly to mean that the drug might help in preventing pregnancy.

Habituation Repeated exposure to a warning over time may result in its being less effective in attracting attention. Even a well-designed warning incorporating the features described in this chapter will eventually become habituated if encountered repeatedly. Although the result implies that the person has learned the information, sometimes the warning may become habituated with only partial knowledge. Although habituation is a problem, warnings with the features described in this chapter are more likely to slow down the habituation process than warnings without the features. Although there are no easy solutions to this problem, one approach that may have some utility is to have warnings that vary from time to time. Rotational warnings such as on cigarette packages in the United States were intended to serve this purpose. However, these warnings have not changed in content or appearance in several decades, and regular smokers have probably habituated to them.

Memory and Experience There are several ways to enhance safety knowledge. Employer training, mentioned earlier, is one method. Experience is another way that people may acquire safety knowledge. "Learning the hard way" by having experienced an incident or personally knowing someone who did can certainly result in such knowledge. However, such experiences are not good experiences to have (!), and they do not necessarily produce accurate perceptions of risk. We discuss this topic in more detail in Section 3.4.3.

Warnings as Reminders Although people may have knowledge of a hazard, they may not be aware of it at the time they are at risk. In short, there is a distinction between awareness and knowledge. This distinction is analogous to the short- and long-term memory distinction in cognitive psychology. Short-term or working memory is sometimes thought of as consciousness, whereas long-term memory is one's knowledge of the world. The point is that people may have information or experience in their overall knowledge base, but at a given time, it is not part of what they are thinking about. It is not enough to say that people know something. Rather, it is important that people be aware of (be thinking about) the relevant information at the critical time. No one knew better than the three-fingered punch press operators of the 1920s that their hand should not be under the piston when it stroked, but such incidents continued to occur. The solution was an engineering control, requiring guards to be in place for the press to punch. Thus, the distinction between knowledge and awareness has implications for the role of warnings as reminders, where their presence may cue information in long-term memory so as to bring forth related, previously dormant knowledge into conscious awareness.

There are several circumstances in which warning reminders are useful and/or needed. Some of the more noteworthy are when (1) a hazardous situation or product (that is not open and obvious) is encountered infrequently, and forgetting may be a factor; (2) distractions occur during the performance of a task or the use of a product (e.g., environmental stimuli); and (3) heavy task loads exceed attentional capacity to access related knowledge (high mental workload and involvement in the task).

When warnings are intended only to function as reminders, it generally is not necessary to provide the same amount of information that would normally be required. Here the emphasis should be more on noticeability, getting the person's attention. The key point in considering the need for reminder warnings is to keep in mind that hazard knowledge on the part of a target audience (e.g., through prior training) does not guarantee that the relevant knowledge will be available when needed.

"Open and Obvious" A source of information about dangers is the situation or product itself. In U.S. law there is a concept of "open and obvious." This concept means that the appearance of a situation or product or the manner in which it functions may communicate the nature of the safety problem. For example, it is apparent to all but the youngest children that a knife can cut. The hazard and consequence of a fall from a height in a construction setting is open and obvious. Of course, many hazardous situations are not open and obvious, such as those associated with many types of chemical hazards.

Technical Information Many warnings require an appreciation of technical information for full and complete understanding of the material. The chemical content of a toxic material, the maximum safe level of a substance in the atmosphere in parts per million (ppm), and the biological reaction to exposure to a substance are examples. Although there are circumstances where it is appropriate to communicate such information (e.g., to the toxicologist on the staff of a chemical plant or the physician prescribing medicine), as a general rule it is neither necessary nor useful to communicate such information to a general target audience. Indeed, it may be counterproductive in the sense that encountering such information may result in the receiver not attending to the remainder of the message. The end user of the toxic material typically does not need to know technical chemical information, such as its density in the atmosphere. Rather, he or she needs to be informed that the substance is toxic, what it can do in the way of injury or illness, and how to use it safely. Where there are multiple groups within the target audience (the toxicologist and the employee, the physician and the patient, the parent and the child), different components of the warning system can and often should be used to communicate to the different groups.

Auditory Warnings Besides simple auditory alerts described in Section 3.4.1, auditory warnings may be used for the specific purpose of conveying particular meanings. These auditory warnings may be nonverbal (different distinguishable sounds to cue different things) or verbal (speech/voice).

Nonverbal Warnings Nonverbal auditory warnings can be divided further into simple and complex. Such simple warnings were mentioned in the context of attention switch. Complex nonverbal signals are composed of sounds of differing (sometimes dynamically) in amplitude, frequency, and temporal pattern. Their purpose is to communicate different levels or types of hazards. They can transmit more information than simple auditory warnings, but the listener must know what the auditory signal means. Training must be given to associate its meaning. Only a limited number of different complex signals should be used, usually not more than a few, because people are limited in discriminating and remembering them (Cooper, 1977; Banks and Boone, 1981).

Voice Warnings Complex warning messages can also be transmitted via voice (speech). In recent years, voice chips and digitized sound processors have been developed making voice warnings feasible for a wide range of novel approaches and applications. Under certain circumstances, voice warnings can be more effective than printed signs in transmitting information (Wogalter and Young, 1991; Wogalter et al., 1993a). Some problems are, however, inherently associated with voice warnings. Transmitting speech messages requires longer durations than simple auditory warnings or reading an equivalent message. Comprehension can also be a problem with complex voice messages. To be effective, voice messages should be intelligible and brief. Nevertheless, this medium for communicating safety information would appear to have considerable potential.

3.4.3 Beliefs and Attitudes

If a warning captures and maintains attention successfully and is understood, it still might fail to elicit safety behavior, due to discrepant beliefs and attitudes held by the receiver. Beliefs refer to a person's knowledge of a topic that is accepted as true. Attitudes are similar to beliefs but have greater emotional involvement (DeJoy, 1999). According to the C-HIP model, a warning will be processed successfully at this stage if it concurs with the receiver's current beliefs and attitudes. The warning message will tend to reinforce what the receiver already knows and in the process will tend to make those beliefs and attitudes stronger and more resistant to change. If, however, the warning information does not concur with the receiver's existing beliefs and attitudes, for it to be effective, the beliefs and attitudes must be altered by the warning. The warning must be salient and the message must be strong and persuasive to override preexisting beliefs and motivate compliance.

People's experiences with a situation or product can result in their believing that it is safer than it is. It can also be a problem when people believe that their own abilities or competence will enable them to overcome the hazard, such as the young adult male who believes that he can safely do a shallow dive into the shallow end of a swimming pool.

Risk Perception One of the important factors in whether people will read and comply with warnings is their perception of the level of hazard and consequences associated with the situation or product. The greater the perceived level of hazard and consequences, the more responsive people will be to warnings (Wogalter et al., 1991, 1993b). Persons who do not perceive a product as being hazardous are less likely to notice or read an associated warning (Wogalter et al., 1991; Wogalter, 1993b). Perceived hazard is also closely related to the expected injury's severity level. The greater the potential injury, the more hazardous the product is perceived (Wogalter et al., 1991). Even if the warning is read and understood, compliance may be minimal if the level of hazard is believed to be low.

Familiarity Familiarity beliefs are formed from past similar experience where at least some relevant information has been acquired and stored in memory. Familiarity may produce a belief that everything that needs to be known about a product or situation is already known (Wogalter et al., 1991, 1993b). A person who is familiar with a piece of equipment might assume that a new, similar piece of equipment operates in the same way (which may not be true), thus reducing the likelihood that a warning would be read. Numerous studies have explored the effects of people's familiarity/experience with a product on how they respond to warnings associated with the product. Results indicate that the more familiar people are with a product, the less likely they are to look for, notice, or read a warning (Godfrey et al., 1983; Godfrey and Laughery, 1984; LaRue and Cohen, 1987; Otsubo, 1988; Wogalter et al., 1991). Some research has also examined the effects of familiarity on compliance (Goldhaber and deTurck, 1988; Otsubo, 1988). The results have shown that greater familiarity is associated with a lower likelihood to comply with warnings.

This notion "familiarity breeds contempt," however, should not be overemphasized for at least two reasons. First, people more familiar with a situation or product may have more knowledge about the hazards and consequences as well as an understanding about how to avoid them Second, people in situations or using products more frequently are exposed to the warnings more often, which increases the opportunity to be influenced by them. Nevertheless, where familiarity is a factor, it should be realized that stronger warnings or perhaps other efforts will be required. Clearly, then, products that are used repetitively or used in highly familiar environments pose special warning challenges.

Prior experience can be influential in other ways. Having experienced some form of injury or having personal knowledge of someone else being injured has been shown to result in overestimating the degree of danger associated with some situation or product. Similarly, the lack of such experiences may lead to underestimating such dangers or not thinking about them at all (Wogalter et al., 1991, 1993b).

A related point concerns the problem of overestimating what people know. Experts in a domain may be so facile with certain knowledge that they fail to realize that nonexperts do not have similar skills and knowledge. To the extent that it is assumed incorrectly that people have certain information and knowledge, there may be a tendency to provide inadequate warnings. Thus, an important part of job, environment, and product design is to take into account the target audience's understanding and knowledge of hazards and their consequences. Further discussion of this issue may be found in Laughery (1993).

3.4.4 Motivation

Even if people believe a warning, they may not comply. Motivation is tied very closely to the response end of the process leading to behavior. In some respects it is energizing to a person to carry out an activity that they might not otherwise do. Among the most influential factors for motivation with respect to warnings is the cost of compliance and the cost of noncompliance (severity of the potential injury, illness, or property damage). If the warning calls for actions that are inconvenient, time consuming, or costly, there is a likelihood that it will not be effective unless the consequences of noncompliance are perceived as being highly undesirable.

Cost of Compliance The cost associated with compliance can be a strong motivator. Generally, compliance with a warning requires that people take some action. Usually, there are costs associated with taking action. The cost of compliance may include time, effort, or even money to carry out the behavior instructed by the warning. When people perceive the costs of compliance to be greater than the benefits, they are less likely to perform the safety behavior. This problem is commonly encountered in warnings where the instructions given are inconvenient, difficult, or occasionally impossible to carry out. "Do not breathe vapors" clearly cannot be accomplished by stopping breathing. "Always have two or more persons to lift" is not possible if no one else is around. "Wear rubber gloves when handling this product" may be inconvenient if the user does not have them and the hardware store is two miles away.

Thus, the requirement to expend extra time or effort can reduce motivation to comply with a warning (Dingus et al., 1991; Wogalter et al., 1987, 1989). One way of reducing the cost of compliance is to make the directed behavior easier to perform. For example, if hand protection is required when using a product, gloves might accompany the product. The general rule is that safe use of a product should be as simple, easy, and convenient as possible.

The costs of noncompliance with a warning can also have a powerful influence on compliance motivation. This effect is particularly true when the possible consequences of the hazards are severe. As already discussed, possible injuries associated with noncompliance should be explicitly stated in the warning (Laughery et al., 1993a). Explicit injury-outcome statements such as "Can cause liver disease—a condition that almost always leads to death" provide reasons for complying and are preferred to general, nonexplicit statements such as "Can lead to serious illness." In a sense, compliance decisions can be viewed in part as a trade-off between the perceived cost of noncompliance and the perceived cost of compliance.

Severity of Consequences An issue related to the costs of noncompliance is the severity of consequences. Perceived severity of injury is tied intimately to risk perception, as discussed in Section 3.4.3. Severity of injury is a major factor in reported willingness to comply with warnings. People's notions of hazardousness are based almost entirely on the seriousness of the potential outcome (Wogalter et al., 1991, 1993b). Further, people do not readily consider the likelihood or probability of such events in their hazard-related judgments (Wogalter and Barlow, 1990; Young et al., 1990, 1992). These findings emphasize the importance of clear, explicit consequences information in warnings. Such information can be critical to people's risk perception and their evaluation of trade-offs between cost of compliance and cost of noncompliance.

Social Influence and Stress Another motivator of warning compliance is social influence. Research (Wogalter et al., 1989) has shown that if people see others comply with a warning, they are more likely to comply themselves. Similarly, seeing others not comply lessens the likelihood of compliance. Social influence is an external factor with respect to warnings in that it is not part of the warning design.

Other factors that influence motivation to comply with a warning are time stress (Wogalter et al., 1998a) and mental workload (Wogalter and Usher, 1999). In high stress and high workload situations, competing activities distribute away some of the cognitive resources available for processing warning information and carrying out compliance behavior.

3.4.5 Behavior

The last stage of the sequential process is to carry out the warning-directed safe behavior. Determining what people will do in the context of a warning is a very desirable measure of its effectiveness. Behavioral compliance research shows that warnings can change behavior (e.g., Laughery et al., 1994; Cox et al., 1997; Wogalter et al., 2001). The main issue is in contemporary research is to determine the factors and conditions that underlie whether or not a warning will be effective in producing compliance. Silver and Braun (1999) have reviewed published research that has measured compliance with warnings under various conditions. Many researchers have used intentions to comply because of the difficulty in measuring behavior under conditions or circumstances that enable conclusions to be drawn. Wogalter and Dingus (1999) showed that indirect measures may also be useful where a residual outcome of the behavior is examined (e.g., whether a pair of protective gloves have been used, according to its stretch marks).

3.4.6 Summary and Benefits of C-HIP

The foregoing review of factors influencing warning effectiveness was organized around the C-HIP model. This model divides the processing of warning information into separate stages that must be completed successfully for compliance behavior to occur. A bottleneck at any given stage can inhibit processing at subsequent stages. Table 1 summarizes some of factors that influence the processing at each stage.

The basic C-HIP model can be a valuable tool in developing and evaluating warnings. Identifying potential processing bottlenecks can be useful in determining why a warning may or may not be successful. The model, in conjunction with empirical data obtained in various types of testing, can identify specific deficiencies in the warning system. Suppose that a manufacturer finds that a critical warning on their product label is not working to prevent accidents. The first reaction to solving the compliance problem might be to increase the size of the sign so that more people are likely to see it. But noticing the sign (the attention switch stage) might not be the problem. Potentially, user testing could show that all users report having seen the warning (attention switch stage), having read the warning (attention maintenance stage), having understood the warning (comprehension and memory stage), and having believed the message (the beliefs and attitudes stage). Thus, the problem with the manufacturer's warning in this case is likely to be at the motivation stage—users may not be complying because they believe the cost of complying with the warning (e.g., wearing uncomfortable personal protection equipment) outweighs the perceived belief about getting injured by not wearing the equipment. By using the model as an investigative tool, one can determine the specific causes of a warning's failure and not waste resources trying to fix the wrong aspect of the warning design.

For the practitioner, the model has utility in determining the adequacy and potential effectiveness of a warning. To the extent that a warning fails to meet various design criteria, the model can be a basis for judging adequacy. The lack of signal words, color, and pictorials or a poor location can be a basis for judging its adequacy regarding attention. A high reading level, the use of technical terminology, or the omission of critical information may be the basis of a warning's comprehension inadequacy. Failure to provide explicit consequences information in circumstances where the outcome of noncompliance may be catastrophic is inconsistent with adequacy criteria regarding motivation. Considerations such as these can be useful in formulating opinions regarding why a warning was not successful.

3.4.7 Demographic Factors

The sections above have provided a review of major concepts and findings organized on the basis of the C-HIP model. There are also relevant demographic characteristics of receivers. Receivers differ, and such differences must be considered in warning design. Laughery and Brelsford (1991) discussed a number of relevant dimensions along which intended receivers may differ. Several such factors have already been discussed, including experience and competence.

A number of studies have shown that gender and age may be related to how people respond to warnings. With regard to gender, results suggest a tendency for women to be more likely than men to look for and read warnings (Godfrey et al., 1983; LaRue and Cohen, 1987; Young et al., 1989). Similarly, some research results show that women are more likely to comply with warnings (Viscussi et al., 1986; Goldhaber and deTurck, 1988; Desaulniers, 1987). However, many other studies either do not report or do not find a gender difference.

Although results regarding age are mixed, there is a trend that people older than 40 are more likely to take precautions in response to warnings (Desaulniers, 1987). On the other hand, some research (Wogalter et al., 1999d; Wogalter and Vigilante, 2003) has shown that older adults have more difficulty reading small print on product labels than do younger adults. Other research (Easterby and Hakiel, 1981; Collins and Lerner, 1982; Ringseis and Caird, 1995) has shown that older subjects had lower levels of comprehension than that of younger adults for safety signs involving pictorials. Results such as these suggest that older people may be more influenced by warnings, but legibility and comprehension need to be considered.

Other potentially important demographics include locus of control (Laux and Brelsford, 1989; Donner, 1991) and self-efficacy (Lust et al., 1993). Persons who believe that they can control their destiny and/or who are less confident in a situation or task are more likely to read available warnings than persons who believe that fate controls their lives and/or who are more confident in a situation or task. When designing warnings for the general population, it may not be possible to address all of the needs of different people with a single warning; thus, a multimethod systems approach may be needed to meet the needs of the varying target audience.

4 DESIGNING FOR APPLICATION

It is important to design warning systems that will maximize their effectiveness. In this section we consider basic guidelines and principles to assist in the design and production of warnings.

4.1 Standards

A starting point in designing warnings is to consider guidelines such as the American National Standards Institute's Z535 (ANSI, 2002). This five-part series includes descriptions of safety colors, signs, symbols, labels, and tags. According to these guidelines, printed warnings should possess four textual components: (1) a signal word panel such as DANGER, WARNING, or CAUTION (with corresponding red, orange, or yellow color) and an alert symbol (triangle enclosing an exclamation mark) to attract attention to the warning and connote levels of hazard, (2) a hazard statement that describes the nature of the hazard briefly, (3) a description of the possible consequences associated with noncompliance, and (4) instructions for how to avoid the hazard. Research indicates that each of these four components benefit warning efficacy. There may be exceptions when one (or more) of the component information items is clear or redundant from the other statements (Wogalter et al., 1987; Young et al., 1995) or from the presence of a pictorial symbol. Pictorial symbols can provide information on the hazard, consequences, or appropriate (or inappropriate) behavior and so can be used in lieu of some of the component text. The symbols should meet certain comprehension criteria to be acceptable for use as a warning by itself. Both the ANSI (2002) Z535.3 and ISO (2001) 9186 symbol standards provide guidelines and methods to assess symbol comprehension.

4.2 Potential Warning Components

Use of standards and guidelines only may not always produce an effective warning. In Table 2 we present a checklist of factors that should be considered in designing warnings. These factors are based not only on standards and guidelines but also on empirical research. Although not an exhaustive list, the table contains a set of factors that the warning literature indicates should be considered in warning design. Thus, one means of assessing a warning's effectiveness is simply to determine the extent to which the design meets appropriate criteria, such as those given in Table 2. With respect to attention, if no signal word is used, no color is employed, the print is small, the message is embedded in other types of information, and so on, then the effectiveness of the warning may be questioned. With respect to comprehension, if the reading level is high, technical language is used, or the statements are vague and not explicit, the warning may not be interpreted as intended.

Implementation of specific factors may also depend on situational-specific considerations such as target audience knowledge and/or characteristics of the product. For example, not all of the textual components in the table are necessary if members of the target audience are aware of the procedures needed to avoid injury at the appropriate time.

4.3 Principles

In addition to the factors in Table 2, there are other principles or guidelines that should be kept in mind when designing warnings or warning systems. These principles are described in the following sections.

4.3.1 Principle 1: Be Brief and Complete

As a general rule, warnings should be as brief as possible. Two separate statements should not be included if one will do, such as in the slippery floor example cited earlier. Longer warnings or those with nonessential information are less likely to be read, and they may be more difficult to understand. Obviously, this criterion should not be interpreted as a license to omit important information. The brevity criterion conflicts to some extent with the explicitness criterion. Being explicit about every hazard could result in very long warnings. A way to find a "happy medium" between brevity and completeness is discussed in Section 4.3.2.

Table 2 Warning Design Guidelines

Warning Component	Design Guidelines
Signal words	DANGER — Indicates immediately hazardous situation that will result in death or serious injury if not avoided; use only in extreme situations. Use white print on a red background (ANSI Z535). WARNING — Indicates a potentially hazardous situation that may result in death or serious injury if not avoided. Use black print on an orange background. CAUTION — Indicates a potentially hazardous situation that may result in minor or moderate injury. Use black print on a yellow background. NOTICE — Indicates important nonhazard information. Use white print on a blue background. Although not in ANSI Z535, the term DEADLY connotes higher levels of hazard than DANGER. On the left side of the signal word is the alert symbol (triangle surrounding an exclamation mark).
Format	Text should be high contrast, preferably black print on white or yellow background, or vice versa. Left-justify text. Consistently position component elements. Orient messages to read from left to right. Each statement starts on it own line. Use white space or bullet points to separate statements or sets of statements. Give priority to the most important warning statements (e.g., position at the top).
Wording	Use as little text as necessary to convey the message clearly. Give information about the hazard, instructions on how to avoid the hazard, and consequences of failing to comply. Be explicit. Tell exactly what the hazard is, what the consequences are, what to do or not do. Use short statements rather than long, complicated sentences. Use concrete rather than abstract wording. Use short, familiar words. Use active rather than passive voice. Remove unnecessary connector words (e.g. prepositions, articles). Avoid using words or statements that might have multiple interpretations. Avoid using abbreviations unless they have been tested on the user population. Use multiple languages when necessary.
Pictorials symbols	When used alone, symbols should have at least 85% comprehension scores, with no more than 5% critical confusions (opposite or very wrong answers). Pictorials not passing a comprehension test should be accompanied by words, but critical confusions should still be avoided. Use bold shapes. Avoid including irrelevant details. Prohibition (circle slash) should not obscure critical elements of symbol. Should be legible under degraded conditions (e.g., distance, size, abrasion).
Font	Text should be legible enough to be seen by the intended audience and expected viewing distance. Use mixed-case letters. Avoid using all caps except for signal words or for specific emphasis. Use sanserif fonts (Arial, Helvetica, etc.) for signal words and larger text. Use serif (Times, Times New Roman, etc.) fonts for smaller text. Use plain, familiar, nonfancy font.
Other	Locate/position so presentation is where it will be seen or heard. Test to assure that message fulfills C-HIP stages in Table 1.

A concept related to completeness is overwarning. The term *overwarning* is sometimes used to label the extent to which our world is filled with warnings. The negative is that people may not attend to them or may become highly selective, attending only to some. The notion is that if warnings were to be put on everything, people would tune them out. Although this notion has face validity, there has been few empirical data assessing the limits implied. Nevertheless, overwarning may be a valid concern, and unnecessary warnings should be avoided. Prioritization, discussed in the next section, is a useful approach in dealing with lengthier warnings for products and equipment that have multiple hazards.

4.3.2 Principle 2: Prioritize

Prioritization concerns what hazards to warn about and emphasize when multiple hazards exist. How are priorities defined in deciding what to include or delete, how to sequence items, or how much relative emphasis to give them? The criteria overlap the rules about what and when to warn. According to Vigilante and Wogalter (1997a, b), considerations include:

1. *Likelihood.* The more likely it is that an undesirable event will occur, the greater the priority that it should be warned.
2. *Severity.* The more severe the potential consequences of a hazard, the greater the priority that it should be warned. If a chemical product poses a skin contact hazard, a higher priority would be given to a severe chemical burn consequence than if it were a minor rash.
3. *Known (or not known) to target population.* If the hazard is already known and understood or if it is open and obvious, warnings may not be needed (except as a possible reminder).
4. *Importance.* Is it important for people to know? In most cases, people want the opportunity to know about risks. Some hazards may be more important to people than others.
5. *Practicality.* There are occasions when limited space (a small label) or limited time (a television commercial) does not permit all hazards to be addressed in a single component of the warning system and still have a readable label.

As a general rule, unknown and important hazards leading to more severe consequences and/or those more likely to occur should have higher priority than less severe or less likely hazards. Higher-priority warnings should be placed first on the product label, and if not practical to place them all there, the lower-priority ones might go on other warning system components, such as package inserts or manuals.

4.3.3 Principle 3: Know the Receiver

Gather information and data about relevant receiver characteristics. This task may require time, effort, and money; but without it, the warning designer and ultimately the receiver will be at a serious disadvantage.

4.3.4 Principle 4: Design for a Low-End Receiver

When there is variability in the target population, which is almost always the case (especially when the audience is the general public), design for the low-end extreme. Safety communications should not be written at the level of the average or median percentile person in the target audience. Such warnings will present comprehension problems for people at lower competence, experience, and knowledge levels.

4.3.5 Principle 5: Employ a Warning System

When the target audience consists of subgroups that differ on relevant dimensions, or when they may be involved under different conditions, consider employing a warning system that includes different components. Do not assume that everything will be accomplished with a single warning.

4.3.6 Principle 6: Design for Durability

Warnings should be designed to last as long as needed. There are circumstances in which durability is typically not a problem. A product off the shelf of a drugstore that will be consumed immediately and completely is an example. On the other hand, products with a long life, such as cars and lawn mowers, may present a challenge. Similarly, situations where warnings are exposed to weather, such as on construction sites, or to extensive handling, such as on some containers, may pose durability problems.

4.3.7 Principle 7: Test the Warning

In addition to considering design criteria, it is frequently necessary to carry out some sort of testing to evaluate a particular warning or several prototype warnings. This approach may entail using small groups of people to give ideas for improvement and/or formal assessments involving larger numbers of people giving independent evaluations. Of course, the sample should be representative of the target audience.

To assess attention, a warning could be placed on a product and people could be asked to carry out a relevant task using the product to determine if they look at or notice it. Regarding comprehension, conducting studies to assess the extent to which a warning is understood probably has one of the best cost–benefit ratios of any procedure in the warnings design process. Relative to behavioral studies, comprehension can be assessed easily, quickly, and at low cost. Well-established methodologies involving memory tests, open-ended response tests, interviews, and so on, are applicable. Such studies can be exceptionally valuable in determining what information in the warning was or was not understood as well as what might be done in the way of redesign to increase the level of comprehension. Studies can also be carried out to determine the extent to which members of the target audience accept the warning information as true and, where appropriate, believe it to be applicable to them (beliefs and attitudes). Negative results on these dimensions would indicate that the warning lacks sufficient persuasiveness. Motivation can be assessed by obtaining measures of compliance intentions. Although such intention measures will generally reflect higher levels than will actual compliance, they can be useful for determining whether or not the warning is likely to be effective as well as for comparing warnings to determine which would probably be more effective.

Although behavioral compliance studies are generally difficult to execute, in situations where negative consequences of an ineffective warning are high, the effort may be warranted. Sometimes behavioral intentions are measured as a proxy because of the relative ease in collecting such data and/or the difficulty (including ethical considerations) in collecting behavioral data.

Studies carried out to evaluate the potential effectiveness of a warning must, of course, incorporate appropriate principles of research design. The selection of subjects to be representative of the target population, avoiding confounding by extraneous variables, and guarding against contamination by expected outcomes are a few of the more salient factors that must be considered. For a more complete discussion of approaches to evaluating warning effectiveness, see Frantz et al. (1999), Wogalter and Dingus (1999), Wogalter et al. (1999c), and Young and Lovvoll (1999).

5 SUMMARY AND CONCLUSIONS

Warning design and effectiveness are comprised of many factors and considerations. In this chapter we have presented an overview of the current status of issues, including research, guidelines, and criteria for designing warnings. Approaches to dealing with environmental or product hazards are generally prioritized such that one first tries to solve the problem by design, then by guarding, then by warning. Thus, in the domain of safety, warnings are viewed as a third but important line of defense.

Warnings can properly be viewed as communications, whose purposes include informing and influencing the behavior of people. Warnings are not simply signs or labels. They can include a variety of media through which various types of information get communicated to a broad spectrum of people. The use of various media or channels and an understanding of the characteristics of the receivers or target audiences to whom warnings are directed are important in the design of effective warnings. The concept of a warning system with multiple components or channels for communication to a variety of receivers is central in this regard.

The design of warnings can and should be viewed as an integral part of systems design. Too often, it is carried out after the environment or product design is essentially completed, a kind of afterthought phenomenon. Importantly, warnings cannot and should not be expected to serve as a cure for bad design.

In this chapter we have covered the C-HIP model that included several processing stages based on communication theory and human information-processing theory. As part of this discussion, relevant factors influential at each stage were presented. In addition, guidelines and principles for warning design in application were presented.

Determining whether or not a warning will influence behavior is usually a difficult assignment. In addition to the ethical problems of exposing people to hazards, actual field studies testing warnings are likely to be time consuming and costly. Certainly, where feasible, such studies are desirable. Also, although laboratory or other controlled simulations of warning situations can be useful in assessing behavioral effects, such approaches leave open questions of generalizability. Studies that examine the effects of warnings on attention, comprehension, beliefs and attitudes, and motivation to comply can be valuable as part of the process of designing and assessing warnings. Such studies can help in isolating why a warning is not effective. A behavioral study which shows that people do not comply with a warning may not tell us if it failed because it was not noticed, or because it was not understood, or because it was not believed, or because it was unable to motivate. Studies employing attention, comprehension, risk perception, or behavioral intention measures can provide information that, in turn, can be useful in developing alternative warning designs that are effective (e.g., Wogalter and Young, 1994).

The issue of warning effectiveness has received a great deal of attention in recent years, especially the means by which effectiveness is assessed. Several criteria can be employed in assessing warnings, including whether they capture and maintain attention, are understood, are consistent with or capable of modifying beliefs and attitudes, motivate people to comply, and result in people behaving safely. The assessment of warning effectiveness employing approaches such as these can and should be part of the warning design process.

REFERENCES

ANSI (1991), Z535, *Accredited Standards for Color, Safety Signs, Symbols, Labels, and Tags*. Parts 1 to 5. National Electrical Manufacturers Association, Arlington VA.

ANSI (2002), Z535 *Accredited Standards for Color Signs Symbols, Labels and Tags*, National Electrical Manufacturers Association, Arlington, VA.

Banks, W. W., and Boone, M. P. (1981), *Nuclear Control Room Enunciators: Problems and Recommendations*, NUREG/CR-2147, National Technical Information Service, Springfield, VA.

Barlow, T., and Wogalter, M. S. (1991), "Increasing the Surface Area on Small Product Containers to Facilitate Communication of Label Information and Warnings," in *Proceedings of Interface '91*, Human Factors Society, Santa Monica, CA, pp. 88–93.

Barlow, T., and Wogalter, M. S. (1993), "Alcoholic Beverage Warnings in Magazine and Television Advertisements," *Journal of Consumer Research*, Vol. 20, pp. 147–155.

Barzegar, R. S., and Wogalter, M. S. (1998), "Intended Carefulness for Voiced Warning Signal Words," *Proceedings of the Human Factors and Ergonomics Society*, Vol. 42, pp. 1068–1072.

Boersema, T., and Zwaga, H. J. G. (1989), "Selecting Comprehensible Warning Symbols for Swimming Pool Slides," *Proceedings of the Human Factors Society*, Vol. 33, pp. 994–998.

Bzostek, J. A., and Wogalter, M. S. (1999), "Measuring Visual Search Time for a Product Warning Label as a Function of Icon, Color, Column, and Vertical Placement," *Proceedings of the Human Factors and Ergonomics Society*, Vol. 43, pp. 888–892.

Chapanis, A. (1994), "Hazards Associated with Three Signal Words and Four Colours on Warning Signs," *Ergonomics*, Vol. 37, pp. 265–275.

Collins, B. L. (1983), "Evaluation of Mine-Safety Symbols," *Proceedings of the Human Factors Society*, Vol. 27, pp. 947–949.

Collins, B. L., and Lerner, N. D. (1982), "Assessment of Fire-Safety Symbols," *Human Factors*, Vol. 24, pp. 75–84.

Cooper, G. E. (1977), *A Survey of the Status and Philosophies Relating to Cockpit Warning Systems*, NASA-CR-152071, NASA Ames Research Center, Moffett Field, CA.

Cox, E. P., III, Wogalter, M. S.,Stokes, S. L., and Murff, E. J. T. (1997), "Do Product Warnings Increase Safe Behavior? A Meta-analysis," *Journal of Public Policy and Marketing,* Vol. 16, pp. 195–204.

DeJoy, D. M. (1999), "Beliefs and Attitudes," in *Warnings and Risk Communication,* M. S. Wogalter, D. M. DeJoy, and K. R. Laughery, Eds., Taylor & Francis, London, pp. 183–219.

Desaulniers, D. R. (1987), "Layout, Organization, and the Effectiveness of Consumer Product Warnings," *Proceedings of the Human Factors Society,* Vol. 31, pp. 56–60.

Dewar, R., (1999), "Design and Evaluation of Graphic Symbols," in *Visual Information for Everyday Use: Design and Research Perspectives,* H. J. G. Zwaga, T. Boersema, and H. C. M. Hoonhout, Eds., Taylor & Francis, London, pp. 285–303.

Dingus, T. A., Hathaway, J. A., and Hunn, B. P. (1991), "A Most Critical Warning Variable: Two Demonstrations of the Powerful Effects of Cost on Warning Compliance," *Proceedings of the Human Factors Society,* Vol. 35, pp. 1034–1038.

Donner, K. A. (1991), "Prediction of Safety Behaviors from Locus of Control Statements," in *Interface '91, Proceedings of the 7th Symposium on Human Factors and Industrial Design in Consumer Products*, Human Factors Society, Santa Monica, CA, pp. 94–98.

Duffy, T. M. (1985), "Readability Formulas: What's the Use?" in *Designing Usable Texts,* T. M. Duffy and R. Waller, Eds., Academic Press, Orlando, FL.

Easterby, R. S., and Hakiel, S. R. (1981), "The Comprehension of Pictorially Presented Messages," *Applied Ergonomics,* Vol. 12, pp. 143–152.

Edworthy, J., Loxley, S., and Dennis, I. (1991), "Improving Auditory Warning Design: Relationship Between Warning Sound Parameters and Perceived Urgency," *Human Factors*, Vol. 33, pp. 205–231.

FMC Corporation (1985), *Product Safety Sign and Label System*, FMC Corporation, Santa Clara, CA.

Frantz, J. P., and Rhoades, T. P. (1993), "A Task Analytic Approach to the Temporal Placement of Product Warnings," *Human Factors*, Vol. 35, pp. 719–730.

Frantz, J. P., Rhoades, T. P., and Lehto, M. R. (1999), "Practical Considerations Regarding the Design and Evaluation of Product Warnings," in *Warnings and Risk Communication,* M. S. Wogalter, D. M. DeJoy, and K. R. Laughery, Eds., Taylor & Francis, London, pp. 291–311.

Godfrey, S. S., and Laughery, K. R. (1984), "The Biasing Effect of Familiarity on Consumer's Awareness of Hazard," *Proceedings of the Human Factors Society,* Vol. 28, pp. 483–486.

Godfrey, S. S., Allender, L., Laughery, K. R., and Smith, V. L. (1983), "Warning Messages: Will the Consumer Bother to Look?" *Proceedings of the Human Factors Society,* Vol. 27, pp. 950–954.

Goldhaber, G. M., and deTurck, M. A. (1988), "Effects of Consumer's Familiarity with a Product on Attention and Compliance with Warnings," *Journal of Products Liability*, Vol. 11, pp. 29–37.

Hartley, J. (1994), *Designing Instructional Text*, 3rd ed., Kogan Page, London, and Nichols, East Brunswick, NJ.

Hellier, E., Edworthy, J., Weedon, B., Walters, K., and Adams, A. (2002), "The Perceived Urgency of Speech Warnings: Semantics Versus Acoustics," *Human Factors*, Vol. 44, pp. 1–17.

Hollander, T. D., and Wogalter, M. S. (2000), "Connoted Hazard of Voice Warning Signal Words: An Examination of Auditory Components," *Proceedings of the International Ergonomics Association and the Human Factors and Ergonomics Society Congress*, Vol. 44, No. 3, pp. 702–705.

ISO (2001), *Graphical Symbols: Test Methods for Judged Comprehensibility and for Comprehension*, ISO 9186, International Organization for Standards, Geneva.

Jaynes, L. S., and Boles, D. B. (1990), "The Effects of Symbols on Warning Compliance," *Proceedings of the Human Factors Society,* Vol. 34, pp. 984–987.

Kalsher, M. J., Wogalter, M. S., and Racicot, B. M. (1996), "Pharmaceutical Container Labels and Warnings: Preference and Perceived Readability of Alternative Designs and Pictorials," *International Journal of Industrial Ergonomics*, Vol. 18, pp. 83–90.

LaRue, C., and Cohen, H. (1987), "Factors Influencing Consumer's Perceptions of Warning: An Examination of the Differences Between Male and Female Consumers," *Proceedings of the Human Factors Society*, Vol. 31, pp. 610–614.

Lasswell, H. D. (1948), "The Structure and Function of Communication in Society," in *The Communication of Ideas*, L. Bryson, Ed., Wiley, New York.

Laughery, K. R. (1993), Everybody Knows: Or Do They?" *Ergonomics in Design*, July, pp. 8–13.

Laughery, K. R., and Brelsford, J. W. (1991), "Receiver Characteristics in Safety Communications," *Proceedings of the Human Factors Society,* Vol. 35, pp. 1068–1072.

Laughery, K. R., Vaubel, K. P., Young, S. L., Brelsford, J. W., and Rowe, A. L. (1993a), "Explicitness of Consequence Information in Warning," *Safety Science*, Vol. 16, pp. 597–613.

Laughery, K. R., Young, S. L., Vaubel, K. P., and Brelsford, J. W. (1993b), "The Noticeability of Warnings on Alcoholic Beverage Containers," *Journal of Public Policy and Marketing*, Vol. 12, pp. 38–56.

Laughery, K. R., Wogalter, M. S., and Young, S. L., Eds. (1994), *Human Factors Perspectives on Warnings: Selections from Human Factors and Ergonomics Society Annual Meetings, 1980–1993*, Human Factors and Ergonomics Society, Santa Monica, CA.

Laux, L., and Brelsford, J. W. (1989), "Locus of Control, Risk Perception, and Precautionary Behavior," in *Proceedings of Interface 89: 6th Symposium on Human Factors and Industrial Design in Consumer Products,* Human Factors Society, Santa Monica, CA, pp. 121–124.

Laux, L. F., Mayer, D. L., and Thompson, N. B. (1989), "Usefulness of Symbols and Pictorials to Communicate Hazard Information," in *Proceedings of the Interface 89: 6th Symposium on Human Factors and Industrial Design in Consumer Products,* Human Factors Society, Santa Monica, CA, pp. 79–93.

Lehto, M. R., and Miller, J. M. (1986), *Warnings*, Vol. 1, *Fundamentals, Design and Evaluation Methodologies*, Fuller Technical Publications, Ann Arbor, MI.

Lehto, M. R., and Papastavrou, J. D. (1993), "Models of the Warning Process: Important Implications Towards Effectiveness," *Safety Science*, Vol. 16, pp. 569–595.

Lehto, M. R., and Salvendy, G. (1995), "Warnings: A Supplement Not a Substitute for Other Approaches to Safety," *Ergonomics*, Vol. 38, pp. 2155–2163.

Lerner, N. D., and Collins, B. L. (1980), *The Assessment of Safety Symbol Understandability by Different Testing*

Methods, PB81-185647, National Bureau of Standards, Washington, DC.

Lim, R. W., and Wogalter, M. S. (2003), "Beliefs About Bilingual Labels on Consumer Products," *Proceedings of the Human Factors and Ergonomics Society*, Vol. 47, pp. 839–843.

Lust, J. A., Celuch, K. G., and Showers, L. S. (1993), "A Note on Issues Concerning the Measurement of Self-Efficacy," *Journal of Applied Social Psychology*, Vol. 23, pp. 1426–1434.

Madden, M. S. (1999), "The Law Related to Warnings," in *Warnings and Risk Communication,* M. S. Wogalter, D. M. DeJoy, and K. R. Laughery, Eds., Taylor & Francis, London, pp. 315–329.

Otsubo, S. M. (1988), "A Behavioral Study of Warning Labels for Consumer Products: Perceived Danger and Use of Pictographs," *Proceedings of the Human Factors Society,* Vol. 32, pp. 536–540.

Penney, C. G. (1989), "Modality Effects and the Structure of Short-Term Verbal Memory," *Memory and Cognition*, Vol. 17, pp. 398–422.

Rhoades, T. P., Frantz, J. P., and Hopp, K. M. (1991), "Product Information: Is It Transferred to the Second Owner of a Product?" in *Proceedings of Interface '91*, Human Factors Society, Santa Monica, CA, pp. 100–104.

Ringseis, E. L., and Caird, J. K. (1995), "The Comprehensibility and Legibility of Twenty Pharmaceutical Warning Pictograms," *Proceedings of the Human Factors and Ergonomics Society*, Vol. 39, pp. 974–978.

Rogers, W. A., Lamson, N., and Rouseau, G. K. (2000), "Warning Research: An Integrative Perspective," *Human Factors*, Vol. 42, pp. 102–139.

Sanders, M. S., and McCormick, E. J. (1993), *Human Factors in Engineering and Design*, 7th ed., McGraw-Hill, New York.

Shannon, C. E., and Weaver, W. (1949), *The Mathematical Theory of Communication,* University of Illinois Press, Urbana, IL

Silver, N. C., and Braun, C. C. (1999), "Behavior," in *Warnings and Risk Communication,* M. S. Wogalter, D. M. DeJoy, and K. R. Laughery, Eds., Taylor & Francis, London, pp. 245–262.

Smith-Jackson, T. L., and Wogalter, M. S. (in press), "Application of Mental Models Approach to MSDS Design," *Theoretical Issues in Ergonomics Science*.

Sorkin, R. D. (1987), "Design of Auditory and Tactile Displays," in *Handbook of Human Factors,* G. Salvendy, Ed., Wiley, New York.

Vigilante, W. J., Jr., and Wogalter, M. S. (1997a), "On the Prioritization of Safety Warnings in Product Manuals," *International Journal of Industrial Ergonomics*, Vol. 20, pp. 277–285.

Vigilante, W. J., Jr,, and Wogalter, M. S. (1997b), "The Preferred Order of Over-the-Counter (OTC) Pharmaceutical Label Components," *Drug Information Journal*, Vol. 31, pp. 973–988.

Viscussi, W. K., Magat, W. A., and Huber, J. (1986), "Informational Regulation of Consumer Health Risks: An Empirical Evaluation of Hazard Warnings," *Rand Journal of Economics*, Vol. 17, pp. 351–365.

Weedon, B., Hellier, E., Edworthy, J., and Walters, K. (2000), "Perceived Urgency in Speech Warnings," *Proceedings of the International Ergonomics Association and the Human Factors and Ergonomics Society Congress,* Vol. 44, No. 3, pp. 690–693.

Westinghouse Electric Corporation (1981), *Product Safety Label Handbook*, Westinghouse Printing Division, Trafford, PA.

Williamson, R. B. (2006), "Warnings for Fires Safety," in *Handbook of Warnings*, M. S. Wogalter, Ed., Lawrence Erlbaum Associates, Mahwah, NJ.

Wogalter, M. S. (1999), "Factors Influencing the Effectiveness of Warnings," in *Visual Information for Everyday Use: Design and Research Perspectives,* H. J. G. Zwaga, T. Boersema, and H. C. M. Hoonhout, Eds., Taylor & Francis, London, pp. 93–110.

Wogalter, M. S., and Barlow, T. (1990), "Injury Likelihood and Severity in Warnings," *Proceedings of the Human Factors Society,* Vol. 34, pp. 580–583.

Wogalter, M. S., and Brelsford, J. W. (1994), "Incidental Exposure to Rotating Warnings on Alcoholic Beverage Labels," *Proceedings of the Human Factors and Ergonomics Society*, Vol. 38, pp. 374–378.

Wogalter, M. S., and Dingus, T. A. (1999), "Methodological Techniques for Evaluating Behavioral Intentions and Compliance," in *Warnings and Risk Communication,* M. S. Wogalter, D. M. DeJoy, and K. R. Laughery, Eds., Taylor & Francis, London, pp. 53–82.

Wogalter, M. S., and Laughery, K. R. (1996), "WARNING: Sign and Label Effectiveness," *Current Directions in Psychology*, Vol. 5, pp. 33–37.

Wogalter, M. S., and Laughery, K. R. (2005), "Effectiveness of Consumer Product Warnings: Design and Forensic Considerations," in *Handbook of Human Factors in Litigation*, I. Noy and W. Karwowski, Eds., Taylor & Francis, London.

Wogalter, M. S., and Leonard, S. D. (1999), "Attention Capture and Maintenance," in *Warnings and Risk Communication,* M. S. Wogalter, D. M. DeJoy, and K. R. Laughery, Eds., Taylor & Francis, London, pp. 123–148.

Wogalter, M. S., and Post, M. P. (1989), "Printed Computer Instructions: The Effects of Screen Pictographs and Text Format on Task Performance," in *Proceedings of Interface' 89,* Human Factors Society, Santa Monica, CA, pp. 133–138.

Wogalter, M. S., and Silver, N. C. (1995), "Warning Signal Words: Connoted Strength and Understandability by Children, Elders, and Non-native English Speakers," *Ergonomics*, Vol. 38, pp. 2188–2206.

Wogalter, M. S., and Usher, M. (1999), "Effects of Concurrent Cognitive Task Loading on Warning Compliance Behavior," *Proceedings of the Human Factors and Ergonomics Society*, Vol. 43, pp. 106–110.

Wogalter, M. S., and Vigilante, W. J., Jr. (2003), "Effects of Label Format on Knowledge Acquisition and Perceived Readability by Younger and Older Adults," *Ergonomics*, Vol. 46, pp. 327–344.

Wogalter, M. S., and Young, S. L. (1991), "Behavioural Compliance to Voice and Print Warnings," *Ergonomics*, Vol. 34, pp. 79–89.

Wogalter, M. S., and Young, S. L. (1994), "Enhancing Warning Compliance Through Alternative Product Label Designs," *Applied Ergonomics*, Vol. 25, pp. 53–57.

Wogalter, M. S., Godfrey, S. S., Fontenelle, G. A., Desaulniers, D. R., Rothstein, P. R., and Laughery, K. R. (1987), "Effectiveness of Warnings," *Human Factors*, Vol. 29, pp. 599–612.

Wogalter, M. S., Allison, S. T., and McKenna, N. (1989), "Effects of Cost and Social Influence on Warning Compliance," *Human Factors*, Vol. 31, pp. 133–140.

Wogalter, M. S., Brelsford, J. W., Desaulniers, D. R., and Laughery, K. R. (1991), "Consumer Product Warnings: The Role of Hazard Perception," *Journal of Safety Research*, Vol. 22, pp. 71–82.

Wogalter, M. S., Kalsher, J. J., and Racicot, B. (1993a), "Behavioral Compliance with Warnings: Effects of Voice, Context and Location," *Safety Science*, Vol. 16, pp. 637–654.

Wogalter, M. S., Brems, D. J., and Martin, E. G. (1993b), "Risk Perception of Common Consumer Products: Judgments of Accident Frequency and Precautionary Intent," *Journal of Safety Research*, Vol. 24, pp. 97–106.

Wogalter, M. S., Barlow, T., and Murphy, S. (1995), "Compliance to Owner's Manual Warnings: Influence of Familiarity and the Task-Relevant Placement of a Supplemental Directive," *Ergonomics*, Vol. 38, pp. 1081–1091.

Wogalter, M. S., Laughery, K. R., and Barfield, D. A. (1997). "Effect of Container Shape on Hazard Perceptions," *Proceedings of the Human Factors and Ergonomics Society*, Vol. 41, pp. 390–394.

Wogalter, M. S., Magurno, A. B., Rashid, R., and Klein, K. W. (1998a), "The Influence of Time Stress and Location on Behavioral Compliance," *Safety Science,* Vol. 29, pp. 143–158.

Wogalter, M. S., Vigilante, W. J., and Baneth, R. C. (1998b), "Availability of Operator Manuals for Used Consumer Products," *Applied Ergonomics*, Vol. 29, pp. 193–200.

Wogalter, M. S., DeJoy, D. M., and Laughery, K. R. (1999a), "Organizing Framework: A Consolidated Communication–Human Information Processing (C-HIP) Model," in *Warnings and Risk Communication,* M. S. Wogalter, D. M. DeJoy, and K. R. Laughery, Eds., Taylor & Francis, London, pp. 15–24.

Wogalter, M. S., Kalsher, M. J., and Rashid, R. (1999b), "Effect of Signal Word and Source Attribution on Judgments of Warning Credibility and Compliance Likelihood," *International Journal of Industrial Ergonomics*, Vol. 24, pp. 185–192.

Wogalter, M. S., Conzola, V. C., and Vigilante, W. J. (1999c), "Applying Usability Engineering Principles to the Design and Testing of Warning Messages," *Proceedings of the Human Factors and Ergonomics Society*, Vol. 43, pp. 921–925.

Wogalter, M. S., Magurno, A. B., Dietrich, D., and Scott, K. (1999d), "Enhancing Information Acquisition for Over-the-Counter Medications by Making Better Use of Container Surface Space," *Experimental Aging Research*, Vol. 25, pp. 27–48.

Wogalter, M. S., Young, S. L., and Laughery, K. R., Eds. (2001), *Human Factors Perspectives on Warnings*, Vol. 2, *Selections from Human Factors and Ergonomics Society Annual Meetings, 1993–2000*, Human Factors and Ergonomics Society, Santa Monica, CA.

Wolff, J. S., and Wogalter, M. S. (1993), "Test and Development of Pharmaceutical Pictorials," in *Proceedings of Interface '93*, Human Factors and Ergonomics Society, Santa Monica, CA, pp. 187–192.

Wolff, J. S., and Wogalter, M. S. (1998), "Comprehension of Pictorial Symbols: Effects of Context and Test Method," *Human Factors*, Vol. 40, pp. 173–186.

Young, S. L. (1991), "Increasing the Noticeability of Warnings: Effects of Pictorial, Color, Signal Icon and Border," *Proceedings of the Human Factors Society*, Vol. 34, pp. 580–584.

Young, S. L., and Lovvoll, D. R. (1999), "Intermediate Processing: Assessment of Eye Movement, Subjective Response and Memory," in *Warnings and Risk Communication,* M. S. Wogalter, D. M. DeJoy, and K. R. Laughery, Eds., Taylor & Francis, London, pp. 27–51.

Young, S. L., and Wogalter, M. S. (1990), "Comprehension and Memory of Instruction Manual Warnings: Conspicuous Print and Pictorial Icons," *Human Factors,* Vol. 32, pp. 637–649.

Young, S. L., Martin, E. G., and Wogalter, M. S. (1989), "Gender Differences in Consumer Product Hazard Perceptions," in *Proceedings of Interface '89*, Human Factors and Ergonomics Society, Santa Monica, CA, pp. 73–78.

Young, S. L., Brelsford, J. W., and Wogalter, M. S. (1990), "Judgments of Hazard, Risk and Danger: Do They Differ?" *Proceedings of the Human Factors Society*, Vol. 34, pp. 503–507.

Young, S. L., Wogalter, J. S., and Brelsford, J. D. (1992), "Relative Contribution of Likelihood and Severity of Injury to Risk Perceptions," *Proceedings of the Human Factors Society*, Vol. 36, pp. 1014–1018.

Young, S. L., Wogalter, M. S., Laughery, K. R., Magurno, A., and Lovvoll, D. (1995), "Relative Order and Space Allocation of Message Components in Hazard Warning Signs," *Proceedings of the Human Factors and Ergonomics Society*, Vol. 39, pp. 969–973.

Zwaga, H. J. G., and Easterby, R. S. (1984), "Developing Effective Symbols or Public Information," in *Information Design: The Design and Evaluation of Signs and Printed Material,* R. S. Easterby and H. J. G. Zwaga, Eds., Wiley, New York.

CHAPTER 33

USE OF PERSONAL PROTECTIVE EQUIPMENT IN THE WORKPLACE

Carolyn K. Bensel
U.S. Army Natick Soldier Center
Natick, Massachusetts

William R. Santee
U.S. Army Research Institute of Environmental Medicine
Natick, Massachusetts

1 INTRODUCTION

Personal protective equipment (PPE) is a general term encompassing a wide variety of clothing and equipment that is worn to protect against hazards to personal life and health. Examples of PPE include gloves, foot protection, eye and face protection, protective hearing devices, hard hats, respirators, and full body suits. Some PPE items, such as safety shoes, are barely distinguishable from regular work attire. Others are ensembles that completely encapsulate the body, such as the chemical protective outfit pictured in Figure 1. A commonality among different types of PPE is that these items are the last line of defense. Depending on the hazard, the outcome of failure of the protective equipment ranges from increased probability of injury to immediate threat to life. Therefore, in a program of hazard control, use of PPE is a last resort.

Figure 1 Military-style chemical protective uniform, with a partially permeable overgarment, rubber gloves and overboots, and a negative-pressure respirator. (Courtesy of the Soldier Systems Center.)

2 HAZARD CONTROLS

Recognition of the hazard, definition of its extent and severity, and identification of persons at risk is the vital first step toward worker safety. The information gathered in this process is also the foundation for actions that will be taken to minimize exposure. Management of safety problems, although varying with the nature of the hazard, consists of three approaches or levels of control. Listed in order of decreasing desirability, these are engineering controls, administrative controls, and personal protective equipment use.

2.1 Engineering Controls

The most direct approach of removing the hazard by changing the relevant equipment, materials, or work process is not always feasible. However, it may be possible to isolate the source of the hazardous condition, placing it away from people. Locating the source so as to minimize the number of workers who are exposed requires space. Where space is constrained, a physical barrier enclosing the source is an option. A large number of hazardous conditions are dealt with in this way. Sound-absorbing or vibration-damping materials are used in noise reduction. Guards attached to machines prevent contact with rotating machinery, falling objects, high-voltage power sources, and escaping liquids. Local exhaust ventilation is used to remove airborne contaminants. Integral to engineering control of hazardous conditions is machine maintenance. The equipment itself, as well as safety devices installed on it, must be kept in good working order to prevent exposure to the hazard and creation of additional safety problems. Engineering controls also include provision of equipment and facilities needed to carry out good housekeeping practices and to dispose of waste properly.

2.2 Administrative Controls

Support by the highest levels of management is at the core of effective administrative controls of hazards. Only with this support will awareness and implementation of sound safety practices spread throughout an organization. Training programs are one method of communicating management emphasis while providing education regarding the hazards that exist in the workplace and how to deal with them. Training programs should be tailored to all the various levels in the organization. Information on relevant government regulations, legal considerations, and costs would be of interest to an audience comprised of supervisors. Workers and technical staff need to know specific requirements and prohibitions. Screening of potential employees is another method available to organizations. People with a predisposition for health problems that could be brought on by exposure to the particular hazards in the workplace and those with existing conditions that could be exacerbated by exposure are thereby identified before being assigned to jobs that put their health at risk. For those employees who must be exposed to the hazard, a work schedule of planned rotation through the environment for a few hours at a time reduces exposure time and increases recovery time. Monitoring the environment and monitoring the workers operating in it are additional methods that an organization can employ for maintenance of the health and safety of personnel. These are means of ensuring that the status of the hazard has not changed, precipitating an increased risk to personnel.

2.3 Personal Protective Equipment

From the perspectives of both the employer and the worker, use of PPE is the least desirable of the three approaches for managing exposure to workplace hazards. The employer's responsibilities extend to selecting and acquiring the appropriate equipment, training workers in its use, ensuring that it is worn properly and when required, and maintaining and replacing the equipment as needed. Workers may abuse the equipment, compromising its protective integrity, or get a false sense of security and engage in risky practices. A most common occurrence is that workers avoid wearing PPE. There are numerous reasons for this, including physiological discomfort, interference with performance of job activities, and low motivation. In this chapter we emphasize the role of PPE, particularly chemical protective items, in producing heat stress and imposing mechanical and psychological burdens on the user that can negatively affect job performance. Actions that can be taken to reduce some of the deleterious effects of PPE are also presented.

3 THERMAL ENVIRONMENT, PHYSIOLOGY, AND PROTECTIVE CLOTHING

The thermal environment, human physiology, and clothing variables all influence thermal stress. The body's core temperature for normal functioning is $37 \pm 1^\circ$C. Thermoregulatory mechanisms maintain a stable core temperature. Thermoregulation is accomplished by the exchange of heat between the body and the surrounding environment. Clothing modifies the heat exchange. Humans typically use this function of clothing to advantage and dress for the immediate climatic conditions. Requirements to wear personal protective clothing, though, supersede dressing appropriately for the climate.

3.1 Thermal Environment

The environmental parameters that determine the potential for heat exchange are air temperature (T_a), which is also referred to as dry-bulb temperature; air movement (V) or wind speed; humidity; and nonionizing radiation. With regard to nonionizing radiation, ultraviolet light (UV) affects human tissue but has little impact on the body's thermal balance. Longer wavelengths of the electromagnetic spectrum, those in the visible and near-infrared (IR) range, may affect thermal balance significantly. Outdoors, the body may be exposed to the direct rays of solar radiation as well as to diffuse sources. Indoors, radiation emanates from hot or cold surfaces or a point source. Radiant energy is quantified by calculating mean radiant temperature ($\overline{T}_r$). One method for determining $\overline{T}_r$ uses a temperature sensor inside a black copper sphere, which measures black globe temperature (T_g). Black globe temperature (°C), air velocity (m/s), and air temperature (°C), are entered in the following formula to obtain $\overline{T}_r$:

$$\overline{T}_r = [(T_g + 273)^4 + (2.5 \times 10^8)V^{0.6}(T_g - T_a)]^{1/4} - 273 \quad °C$$

There are a number of expressions for *humidity*, the amount of water vapor in a given quantity of air. These include water vapor pressure, dew-point temperature, absolute humidity, and relative humidity. *Relative humidity* (RH) is the ratio between the actual moisture content and the saturation value at the prevailing temperature, expressed as a percentage. When the air is saturated with water vapor, the RH is 100%. The value of RH depends on the air temperature. The temperature at which saturation occurs is the *dew-point temperature* (T_{dp}). The wet-bulb temperature (T_{wb}) varies with humidity and equals the dry-bulb temperature when the RH is 100%. The natural wet-bulb temperature (T_{nwb}) is obtained by exposing a thermometer fitted with a wetted cotton wick to the natural, or prevailing, air movement. Information on instruments for measuring the environment and work-site placement of the instruments is available from a number of sources (Santee et al., 1994; ISO, 1998; Parsons, 2002; ASHRAE, 2004).

3.2 Physiology

A dual mechanism, triggered by peripheral temperature sensors and controlled centrally by the hypothalamus, regulates heat exchange to maintain thermal balance. The exchange of heat between the body and the environment is accomplished through conduction, convection, radiation, and evaporation. The potential for heat exchange is determined primarily by the thermal environment and clothing. Conduction, convection, and radiation are termed *dry* or *sensible heat exchange*; they do not involve evaporation of water from the body surface, and exchange can be measured or "sensed" as differences in temperature.

Evaporation is termed *wet* or *insensible heat exchange*; it results from a water phase change and is not reflected directly by a change in temperature. Water released by sweat glands absorbs body heat to change from liquid to vapor. Water vapor pressure, the local concentration of water in the air, exerts pressure in proportion to the amount of water present. The concentration of water, as reflected by the water vapor pressure, determines the direction of any exchange of water vapor. The vapor passes into the environment if the vapor pressure on the skin is higher than the vapor pressure in the surrounding environment. When not all the water released can be absorbed by the environment, liquid sweat accumulates on the skin. Further increases in sweat production do not result in heat loss because rather than being evaporated, the sweat drips from the skin. Water vapor, and heat, are also lost by breathing.

3.2.1 Thermal Balance

The internal temperature of the body is a summation of the surface heat exchanges and internal heat production, with the remaining energy being either work or heat storage. The relationship among these terms is expressed in the *heat balance equation*:

$$\pm S = M - (\pm W_k) \pm K \pm C \pm R \pm E \qquad \text{W/m}^2$$

where S is the body heat storage, M the internal heat production or energy metabolism, W_k the mechanical work (+ when energy leaves the body, − when gravity works against the body), K the conduction (gain = +, loss = −), C the convective heat exchange (gain = +, loss = −), R the radiative heat exchange (gain = +, loss = −), E the evaporative heat exchange (condensation = +, loss = −).

If the body maintains thermal balance, S is zero. The heat storage term is negative when there is a net loss of heat from the body to the environment and positive when there is heat gain. Body temperature decreases with negative heat storage and increases with positive heat storage. In warm environments, the evaporation required to compensate for net dry heat gain (E_{req}) is calculated as

$$E_{req} = M \pm C \pm R \qquad \text{W/m}^2$$

If the required evaporative cooling rate is not impaired by the humidity in the environment, body temperature is in thermal equilibrium. Ambient vapor pressure and air velocity impose a maximum evaporative capacity (E_{max}), which is high in dry environments and much lower in humid ones. The proportion of sweat-wetted skin surface area needed to eliminate the required amount of heat from the body by evaporation is estimated as the ratio E_{req}/E_{max}. With values of the ratio between 0.20 and 0.40, marked sensations of discomfort are experienced. Sweat begins to drip rather than to evaporate at 0.70. Increases in values of the ratio to above 0.80 result in limited tolerance. The value equals 1.00 for totally wetted skin.

3.2.2 Measurement of Physiological Variables

Thermal stress imposed by the environment is a combination of factors that affect adversely the potential for thermoregulatory heat transfer between the person and the environment. Heat stress can be divided into compensated and uncompensated heat stress. *Compensated heat stress* exists when heat loss to the environment occurs at a rate in balance with heat production so that the body maintains thermal balance. *Uncompensated heat stress* occurs when evaporative cooling requirements exceed the environment's evaporative cooling capacity, resulting in a gain in body heat storage. A number of physiological parameters are used to assess the strain on the body's thermoregulatory mechanisms resulting from thermal stress. Five of the parameters are described here.

Core temperature, the temperature of deep, central areas of the body, is often represented by rectal temperature (t_{re}) or esophageal temperature. *Skin temperature* is measured at the skin surface and may vary greatly over the body's surface. Therefore, measurements are taken at a number of points, and the weighted points are summed to get the mean surface temperature ($\bar{t}_{sk}$). A number of different computational schemes are used (Bensel and Santee, 1997). *Sweat rate*, the rate of water loss through the skin, is most easily measured by weighing a person at regular intervals and calculating the change in body weight after adjusting for food and water intake. Water lost from the lungs during breathing, termed respiratory water loss, is often estimated as a percentage of total water loss.

Metabolic heat production can be quantified by measuring *oxygen consumption*. One liter of oxygen consumed is approximately equal to production of 20 kJ of heat. The rates of oxygen consumption and heat production are determined by measuring the oxygen concentration and volume of expired air. Metabolic rate is often expressed in terms of body weight (W/kg) or body surface area (W/m^2). Body surface area can be obtained from tables or calculated if height and weight are known (DuBois and DuBois, 1916; Sendroy and Collison, 1960). When measurement of oxygen consumption is not feasible, extensive tables of estimates from past research can be consulted (Durnin and Passmore, 1967). Estimates should be employed with caution because there are large inter- and intraindividual variations in metabolic rates. *Heart rate* (HR) may be measured with a "sports watch" heart rate monitor that transmits to a wrist-mounted display from sensors on a chest band. For continuous monitoring, heart rate is recorded as an electrocardiogram from electrodes affixed to the chest. Heart rate is approximated by measuring pulse rate, which can be taken by palpation of the carotid artery at the throat or the radial artery at the wrist.

3.3 Protective Clothing

Clothing is the body's outer shell relative to the environment. Relative to the body, clothing establishes the immediate microenvironment to which the body is exposed. There are a number of clothing properties that affect heat exchange between the body and the environment. The theory underlying measurement of clothing properties is presented in a number of sources (ASHRAE, 2001; Parsons, 2002).

3.3.1 Resistance to Passage of Dry Heat

Resistance to heat transfer by convection and radiation is combined into one general clothing property, *insulation*, expressed in an arbitrary unit, the clo. In SI units, 1 clo equals 0.155°C · m^2/W. One clo unit approximates the insulation of a business suit, plus shirt, underclothing, and the air layers between them. A boundary layer of air at the clothing surface provides some resistance to heat exchange. The insulation of this air layer (I_a) is added to the intrinsic insulation provided by the clothing and air trapped between the clothing and the skin surface (I_{cl}) to obtain total insulation (I_T).

Electronically heated devices have been developed to measure insulation. One device, the guarded flat-plate, is used to test flat pieces of fabric (ASTM, 1985). Insulation of finished garments and clothing ensembles is measured on thermal manikins, which are full-sized human models (ASTM, 1999b). Table 1 contains clo values obtained on thermal manikins for a variety of clothing. Lists of clo values for an extensive array of clothing and ensembles are presented in a number of reference sources (e.g., McCullough et al., 1985; ISO, 1995; ASHRAE, 2001).

3.3.2 Resistance to Evaporation

If water vapor passes completely through the clothing to the external environment, heat is transferred from the body to the environment. If water vapor recondenses on the skin or within the clothing, heat is not lost to the environment. Resistance of clothing to evaporation is expressed by the water vapor permeability index (i_m), a dimensionless index (Woodcock,

Table 1 Insulation and Water Vapor Permeability for Selected Torso Clothing

Clothing	I_T (clo)	i_m	i_m/I_T	I_{cl} (clo)
Sweatpants, sweatshirt	1.35	0.45	0.33	0.74
Trousers, long-sleeved shirt	1.21	0.45	0.37	0.61
Knee-length skirt, long-sleeved blouse, slip, panty hose	1.22	—	—	0.67
Long-sleeved coveralls, T-shirt	1.30	—	—	0.72
Insulated coveralls, thermal underwear	1.94	0.39	0.20	1.37
Two-piece work uniform	1.34	0.41	0.31	
Tyvek coverall and hood over two-piece work uniform	1.67	0.27	0.16	
Chemical protective overgarment (charcoal in foam) over two-piece work uniform	2.17	0.28	0.13	
Chemical protective ensemble, including overgarment (charcoal in foam), respirator, hood, gloves, and overboots	2.44	0.30	0.12	
Temperate-zone winter clothing	3.20	0.40	0.13	
Arctic winter clothing	4.30	0.43	0.10	

Source: Data from Goldman (1988) and ASHRAE (2001).

1962). Methods for measuring permeability are modifications of those for measuring insulation. The theoretical value of i_m can range from 0 for completely moisture-impermeable clothing to a maximum of 1 for completely permeable clothing.

The maximum potential for evaporative heat transfer through the clothing to the environment is a function of the ratio of the permeability index to the total insulation. This ratio approximates the percentage of the maximum evaporative potential for a given environment that may be realized when wearing specified clothing. For a nude man who is sweating, the ratio is about 0.6. Table 1 includes the values of the permeability index and the ratio i_m/I_T for a variety of clothing items.

3.3.3 Wind Resistance

Even at low speeds, wind disrupts the air layer at the clothing surface, affecting ambient insulation. At higher speeds, wind may also pass into fabrics, disturbing air trapped within the clothing and affecting intrinsic insulation. The decrease in intrinsic insulation depends largely on fabric characteristics. Tightly woven materials function as *wind breakers*, whereas loose, open weaves are affected more adversely by wind penetration. Thermal manikin testing to determine insulation values is normally done at several wind speeds to acquire data on the effect of wind on I_T. There are also estimated adjustments for wind penetration in the form of temperature decrements that should be subtracted from the actual thermometer reading to obtain still air temperatures corrected for wind speed (Burton and Edholm, 1955; Bensel and Santee, 1997). The adjustments should be applied when determining clo requirements for various ambient temperatures and metabolic rates.

3.3.4 Clothing Surface Area

Heat exchange between the body and the environment takes place at the surface of the skin and the clothing. As the outer clothing surface area increases, the total amount of heat transferred increases. Consequently, wearing thicker and thicker clothing, which increases outer surface area, contributes proportionally less and less to protection against heat loss in a cold environment. The clothing area factor (f_{cl}) is a dimensionless unit obtained by dividing the surface area of the clothed body by the nude body surface area.

3.3.5 Additional Clothing Variables

Thickness is so closely correlated with insulation that measuring thickness is a way to estimate insulation. The rule of thumb is that 1 cm of thickness equals an I_T of 1.58 clo, or 1 clo equals a thickness of 0.62 cm. In high-temperature environments, weight is a critical variable insofar as an increase in clothing weight generally represents an increase in insulation and thermal stress. Clothing designs sometimes incorporate openings for air circulation, and body motion creates a pumping action that expands and contracts air spaces within the clothing. Air circulation can greatly modify convective heat transfer within clothing and have a substantial impact on insulation.

3.4 Relating the Thermal Environment, Physiology, and Protective Clothing

Maintenance of health and safety requires a means for predicting the thermal strain that a person will experience given the status of the environmental, physiological, and clothing variables. There are a number of indices for assessing the thermal environment, some of which are used in national and international guidelines (OSHA, 1999b; Parsons, 1999). Table 2 presents summaries of some of the indices. The *wet-bulb globe temperature* (WBGT) *index* is a simple, empirical method for assessing hot environments (ISO, 1989; OSHA, 1999b). Indoors, and outdoors in the absence of a solar load, the WBGT index is computed from the following formula:

$$\text{WBGT} = 0.7T_{\text{nwb}} + 0.3T_g \qquad {}^\circ\text{C}$$

Outdoors, with a solar load, air temperature must also be measured and the formula is then as follows:

$$\text{WBGT} = 0.7T_{\text{nwb}} + 0.2T_g + 0.1T_a \qquad {}^\circ\text{C}$$

Threshold limits recommended by the Occupational Safety and Health Administration (OSHA, 1999b) for heat-acclimatized persons are presented in Table 3. The limits were set to avoid rises in core temperature above 38°C and assume that one layer of clothing is being worn. With the WBGT index, wear of protective clothing is assumed to have the same effect as exposure to a hotter environment. The OSHA recommendations suggest that WBGT readings be increased by 2°C when coveralls are worn, by 4°C when winter work clothing is used, and by 6°C for permeable clothing that presents a barrier to the passage of water. The U.S. Army and Air Force recommend that the WBGT index value be increased by 3°C in humid environments when a ballistic protective vest is used. When Army and Air Force personnel are wearing partially impermeable ensembles for protection against chemical, biological, and nuclear hazards, the recommended increase in the WBGT index value is 6°C for light work and 12°C for moderate and heavy work levels (U.S. Department of the Army and Air Force, 2003).

The *physiological strain index* (PSI) is unlike the other thermal strain indices listed in Table 2, insofar as the inputs are actual physiological measurements made on a person rather than measurements made of the thermal environment. The PSI was constructed as a simple means for directly evaluating heat strain in situations in which a person is exposed to conditions posing an immediate risk of heat illness, such as high environmental temperatures or wearing of protective clothing with high insulative and low moisture vapor permeability properties. The PSI can be used at any time during a person's exposure to extreme conditions (Moran et al., 1998). The PSI

Table 2 Thermal Strain Indices

Index	References	Inputs	Comments
		Indices for Heat Exposure	
Wet-bulb globe temperature (WBGT)	Yaglou and Minard (1957), NIOSH (1986), ISO (1989)	T_a, T_{nwb}, T_g	WBGT requires simple input and output calculations; not recommended for conditions of high humidity.
Heat stress index (HSI)	Belding and Hatch (1955), ISO (2004)	T_a, T_{wb}, T_g or $\overline{T}_r$, V, M	HSI is the ratio of evaporative heat loss required to maintain a constant body temperature (E_{req}) to the maximum amount of sweat that can be evaporated under the given climatic conditions (E_{max}). Required sweat rate index (S_{req}), a further development of HSI, is used in ISO 7933.
Oxford index (WD)	Leithead and Lind (1964)	T_a, T_{wb}	WD can be used to determine tolerance times.
Physiological strain index (PSI)	Moran et al. (1998)	t_{re}, HR	PSI describes physiological strain on a scale of 0 to 10; assumes maximal acceptable increases in t_{re} and HR are 3°C and 120 beats/min, respectively.
		Equivalent Temperature Indices	
Effective temperature	Houghten and Yaglou (1923)	T_a, T_{wb}, V	ET relates actual conditions to an equivalent, calm, saturated environment; overemphasizes effects of humidity in cool and neutral conditions and underemphasizes its effects in warm conditions.
New effective temperature (ET*)	Gagge et al. (1971), ASHRAE (2001)	T_a, $\overline{T}_r$, V, P_a, i_m, ω, M	ET* was developed to replace ET; includes skin wettedness (ω) and water vapor pressure (P_a) parameters in calculating temperature of an environment at 50% RH that results in equivalent total heat loss from the skin as in the actual environment.

Table 3 OSHA Recommended WBGT Levels (°C) for Heat Stress Exposure for Acclimatized, Lightly Clothed Workers[a]

	Workload		
Hourly Work–Rest Cycle	Light (<230*W*)	Moderate (230–400*W*)	Heavy (>400*W*)
Continuous work	30.0	26.7	25.0
75% work/25% rest	30.6	28.0	25.9
50% work/50% rest	31.4	29.4	27.9
25% work/75% rest	32.2	31.1	30.0

Source: OSHA (1999b).

[a]Limits are for a "standard" worker weighing 70 kg with a 1.8-m^2 body surface area. The WBGT of the resting area is assumed to approximate the WBGT of the workplace.

assumes that the maximal acceptable increase in t_{re} is 3°C, based on a maximal change from 36.5°C to 39.5°C. The maximal acceptable increase in HR is taken as 120, based on a maximal change from 60 to 180 beats/min. Use of the PSI requires that simultaneous readings of rectal temperature (t_{re0}) and heart rate (HR_0) be taken at the beginning of an exposure period. Monitoring of a person's physiological status is done by subsequently taking simultaneous readings of rectal temperature (t_{ret}) and heart rate (HR_t) and applying the following formula:

$$\text{PSI} = 5(t_{ret} - t_{re0})(39.5 - t_{re0})^{-1} + 5(\text{HR}_t - \text{HR}_0)(180 - \text{HR}_0)^{-1}$$

The PSI values are scaled to a range of 0 to 10, and descriptors of strain are assigned to the scale: 0–2, no/little strain; 3–4, low; 5–6, moderate; 7–8, high; 9–10, very high strain.

Equivalent temperature indices are based on the assumption that the net effect of the thermal environment can be characterized in temperature units. The *effective temperature* (ET) *scale* was developed from research in which subjects compared thermal sensations experienced in still, saturated air (100% RH), set at a given temperature, with sensations produced under other combinations of temperature, humidity, and wind speed. Conditions producing the same subjective impressions of warmth or cold were assigned the same ET value. The ET scale is best suited to warm environments, where radiation effects are minimal, light clothing is being worn, and people are sedentary. Furthermore, different climates with the same ET values do not result in the same tolerance times, rectal and skin temperatures, or sweat rates. The "*new*" *effective temperature scale* (ET*) was developed to address the limitations of the original ET scale (Gagge et al., 1971; ASHRAE, 2001). More variables are entered into the calculation of ET* to obtain a more accurate prediction of thermal strain than those of the original ET yields.

Computer-based mathematical models for predicting physiological responses to extreme temperature environments are becoming increasingly sophisticated and are likely to be used more widely in the future than the thermal strain indices. The majority of thermal models are applicable to heat. A heat strain model that has been validated for military personnel in both laboratory and field studies is the U.S. Army Research Institute of Environmental Medicine (USARIEM) model (Pandolf et al., 1986). The model predicts how long an activity can be sustained before a preselected level of heat casualties will occur and estimates water consumption requirements. The model also calculates a work–rest cycle that will allow an activity to be sustained indefinitely while holding heat casualties below the level selected.

4 PHYSIOLOGICAL IMPLICATIONS OF PPE USE

When the human body overheats, blood vessels dilate and heart rate increases to carry warm blood to the skin, where heat is lost to the environment by radiation and convection. The sweat glands are also stimulated to release water for evaporative cooling. Radiative heat loss is a direct function of surface temperature, so that the higher the skin temperature, the greater the loss of thermal radiation. When air temperature is higher than skin temperature, the direction of convection is reversed and the body may gain heat. Under such conditions, the primary mechanism for heat loss is evaporation of sweat. If the water vapor is not transferred to the environment, there is no evaporative heat loss and liquid sweat accumulates on the skin.

4.1 Thermal and Respiratory Strain

In warm temperatures, the added insulation of clothing hampers evaporation of sweat. Wearing impermeable or partially permeable PPE seriously compromises attaining the required evaporative cooling, and work durations must be curtailed to avoid heat casualties. Table 4 is the output of a USARIEM thermal model, indicating suggested maximal work times, work–rest cycles, and drinking water requirements when wearing an impermeable suit, full-facepiece respirator, impermeable gloves, and impermeable overboots and when wearing regular clothing. Even at 15°C, activities must be limited if heat casualties are to be avoided with the protective outfit.

Items of PPE tend to be heavy and bulky. For example, firefighting PPE, including turnout gear, gloves, hood, helmet, respirator, and self-contained breathing apparatus (SCBA), weighs between 20 and 30 kg. The additional weight and bulk of PPE contribute to an increased metabolic cost of working while wearing protective clothing and equipment. Comparison of the energy cost of walking in normal indoor clothing and in firefighting gear with SCBA yielded an increase in energy cost of 34% with the firefighting gear, which was attributable primarily to carrying the weight of the SCBA (Raven et al., 1979). Estimated increases in metabolic rate during walking are presented in Table 5 for a number of PPE items. With the increased metabolic rate, there is increased heat production, further contributing to the thermal strain associated with PPE use.

Without heat loss through evaporation, there may be reduced sweat production, additional increases in core temperature, and eventually, heat illness. A rectal temperature of 42°C is often lethal. Less extreme rectal temperatures are associated with a variety of heat illnesses (Table 6). Aside from the influence of the environment and clothing, the incidence of heat stress is affected by intra- and interindividual factors, which are listed in Table 7.

Like other items of PPE, respirators hamper sweat evaporation. In addition, they impose breathing resistance and a resulting respiratory load on the wearer. The more resistant the inlets and outlets of a respirator are to inspiration and expiration of air, the shorter the time a person can perform physical work before experiencing labored breathing (dyspnea) and subsequently being unable to continue working. Airflow resistance of respirators is usually measured as pressure, in mm H_2O or pascal, when a steady airflow is applied at 85 L/min. The data in Table 8 illustrate the effect that inspiratory resistance has on the capacity for heavy work. Both inspiratory and expiratory resistances affect the respiratory load, although inspiratory resistance has a greater impact on the ability to carry out physical activities at heavy work rates.

When a respirator is worn, not all of the air expired with each breath passes out of the respirator to the environment; some is retained within the respirator. The retained air comprises the external dead-space volume. On the next inspiration, a large proportion of the carbon dioxide–enriched air in the dead space is inhaled. The increase in inhaled CO_2 increases ventilation, measured as the volume of air moved out of the lungs over a one-minute period. With increased ventilation, there is an increase in energy expenditure and a concomitant increase in body heat production, contributing further to thermal strain.

4.2 Reducing Thermal and Respiratory Strain

Interjecting scheduled rest periods throughout a work activity will decrease the amount of heat produced for the amount of time that PPE is worn by reducing

Table 4 USARIEM Thermal Model Outputs for a Two-Piece Work Uniform and a Fully Enclosed Impermeable Suit for Chemical Handling[a]

	Work Uniform				Impermeable Chemical Suit			
T_a (°C)	Rest (105*W*)	Light (250*W*)	Moderate (425*W*)	Heavy (600*W*)	Rest (105*W*)	Light (250*W*)	Moderate (425*W*)	Heavy (600*W*)
				Maximum Work Time (min)				
15	NL	NL	NL	NL	NL	NL	81	43
20	NL	NL	NL	171	NL	NL	63	38
25	NL	NL	NL	110	NL	NL	52	33
30	NL	NL	NL	78	NL	96	44	28
35	NL	NL	111	55	NL	67	37	23
40	NL	NL	56	37	154	53	32	18
45	152	55	34	21	84	44	26	14
				Recommended Hourly Work–Rest Cycle (min)				
15	60/0	60/0	60/0	60/0	60/0	60/0	30/30	18/42
20	60/0	60/0	60/0	42/18	60/0	60/0	22/38	13/47
25	60/0	60/0	60/0	36/24	60/0	60/0	11/49	5/55
30	60/0	60/0	60/0	30/30	60/0	0/60	0/60	0/60
35	60/0	60/0	33/27	21/39	60/0	0/60	0/60	0/60
40	60/0	60/0	11/49	5/55	0/60	0/60	0/60	0/60
45	0/60	0/60	0/60	0/60	0/60	0/60	0/60	0/60
				Recommended Water Consumption (L/h)				
15	0.2	0.2	0.6	1.0	0.2	0.6	1.6	2.1
20	0.2	0.4	0.8	1.2	0.2	1.0	2.0	2.1
25	0.2	0.6	1.1	1.5	0.5	1.4	2.1	2.1
30	0.5	0.9	1.3	1.8	1.0	1.9	2.1	2.1
35	0.7	1.2	1.7	2.1	1.5	2.1	2.1	2.1
40	1.1	1.6	2.1	2.1	2.1	2.1	2.1	2.1
45	1.9	2.1	2.1	2.1	2.1	2.1	2.1	2.1

[a]NL, No limit. A person can sustain work for at least four hours in the heat category specified. This model is a developmental, analytical tool that has not been determined to be completely safe for use in making decisions that could affect the health and safety of personnel. Results apply only to the specified population and worker physical condition: persons fully hydrated, acclimatized for 15 days, average male soldier (height = 175.1 cm, weight = 77.1 kg), light casualty rate (<5%). Environmental parameters: RH = 50%, wind speed = 2 m/s, full sun.

Table 5 Estimates of Increases in Metabolic Rate for PPE Items

Item	Increase in Metabolic Rate (*W*)
Safety shoes/ankle-height boots	18
High safety boots	36
Full-facepiece respirator	72
SCBA	108
Light chemical protective coverall	36
Chemical protective ensemble with hood, gloves, and boots	90
Firefighting turnout gear with helmet, trousers, gloves, and boots	135

Source: Adapted from Hanson (1999).

internal heat production (M) and allowing heat to be transferred to the environment. It is extremely beneficial if the protective clothing can be removed during rest breaks. This promotes heat loss and allows more total work to be accomplished. As indicated in Table 7, acclimatizing to the environment is among the factors that contribute positively to maintenance of thermal regulation in heat. However, when PPE is worn, the heat stress imposed may negate the benefits of a worker's acclimatization, raising the risk of heat illness (Goldman, 1988). Being well hydrated before beginning work and replacing fluids whenever possible during work are critical in maximizing heat tolerance while wearing PPE.

Another means of reducing thermal strain associated with PPE use is addressed in the OSHA (1989) regulation on hazardous waste operations and emergency response to release of hazardous substances. The approach presented is to decrease or increase the level of protection as the degree of hazard warrants, thereby minimizing the occasions on which people are overprotected and at increased risk for heat illness (OSHA, 1999a). Four protection levels, referred to as the *EPA levels of protection*, are defined by OSHA (Table 9).

Table 6 Symptoms, Etiology, and Treatment of Heat Illnesses

Illness	Symptoms	Etiology	Treatment
Heat exhaustion (In the event of collapse or unconsciousness, medical attention should be sought)	Clammy, moist skin	Dehydration	Rest lying down in cool, shaded area.
	Headache, dizziness, nausea, fainting, collapse	Circulatory strain due to profusion of blood in the skin	If person is conscious, provide drinking water; do not use salt.
	Heart rate over 160 to 180 beats/min; t_{re}: 37.5–38.5°C	Reduced blood flow to the brain	Loosen/remove clothing. Splash cold water on body; massage arms and legs; if unconscious, treat for heat stroke; do not return to hot environment until after overnight rest.
Dehydration exhaustion	Fatigue, nausea, headache, fainting, collapse, small volume of urine	Dehydration resulting from sweating	Provide drinking water; do not use salt; do not return to hot environment until after overnight rest; carbohydrate–electrolytic replacement liquids (sports drinks) are effective during recovery.
Heat cramps	Painful spasms of arm, leg, and abdominal muscles	Loss of body salt in sweat	Massage affected area.
	Heavy sweating and extreme thirst	Drinking of large volume of water dilutes electrolytes; water enters muscles, resulting in spasms	Take adequate amount of salt with meals; loosen clothing.
Heat rash	Tiny raised red vesicles on skin	Plugging of sweat gland ducts with prolonged exposure of skin to heat, humidity, sweat	Apply mild drying lotion.
Heat stroke (Life-threatening medical emergency requiring immediate medical attention)	Headache, dizziness, confusion, diarrhea, vomiting, coma, convulsions; onset may be sudden	Failure of body's temperature-regulating mechanisms	Move to shaded area, remove clothing, wrap in wet sheet, pour chilled water on body and fan vigorously; avoid overcooling.
	Hot, dry skin that is red, mottled, or cyanotic	Reduced blood flow to brain and other vital organs	Treat shock once temperature is lowered.
	t_{re}: 40.5°C or higher; usually fatal at t_{re} of 42°C	Dehydration and sustained exertion in heat	Clear all vomit from nose and mouth.

Source: Data from Leithead and Lind (1964), OSHA (1999b), U.S. Department of the Army and Air Force (2003).

The U.S. Army also uses a flexible system for protection of troops against chemical agents. The Army system of *mission-oriented protective posture* (MOPP) entails the wearing of protective clothing consistent with the chemical threat (Table 10).

Activities are under way continually to identify lighter-weight materials for use in PPE. There are extensive national and international resources available through the World Wide Web for being kept apprised of the latest technologies in protective items and current standards and regulations pertaining to PPE use. Government sites are prime sources of information, as are organizations that are responsible for preparation of standards, such as the International Organization for Standardization (www.iso.org), the National Fire Protection Association (www.nfpa.org),

Table 7 Factors Affecting the Occurrence of Heat Stress

- *Hydration state*. Hypohydration results in lower sweat production and an increase in core temperature.
- *Acclimatization*. Repeated heat exposure leads to earlier onset of sweating, a higher sustained sweat rate, and lower core temperature and heart rate.
- *Age*. Sweating mechanism and circulatory system become less responsive with age, and there is high level of skin blood flow, possibly due to impaired thermoregulatory mechanism.
- *Physical fitness*. Exercise that increases maximal aerobic capacity improves thermoregulatory responses in the heat.
- *Subcutaneous fat*. Subcutaneous fat provides an insulative barrier, reducing transfer of heat from muscles to skin.
- *Gender*. Although studies indicate that sweating and vasodilation occur at higher core temperatures in women than in men, when controlled for fitness and menstrual phase, gender differences in the follicular phase are questionable. Women in the luteal phase have significantly higher core temperatures, which may affect thermoregulatory responses.
- *Body size*. Leaner persons are at an advantage because they have a larger ratio of surface area to body mass and thus greater capacity to dissipate heat.
- *Diet*. Regular consumption of a balanced diet serves to replace salt and other electrolytes lost in sweat, maintaining sweating efficiency.
- *Previous heat illness*. Previous occurrence of heat stroke increases susceptibility to subsequent heat illness.
- *Drugs and alcohol*. Use interferes with the functioning of the central and peripheral nervous system, negatively affecting heat tolerance.

Table 8 Exercise Duration and Ventilation While Exercising Wearing Respirators That Differed in Inspiratory Resistance[a]

Inspiratory Resistance (kPa)	Exercise Duration (min)	Ventilation (L/min)
Control (minimal resistance)	17.6	86.8
0.20	15.1	84.7
0.29	13.5	78.3
0.39	9.8	73.9
0.49	5.2	56.1

Source: Caretti and Whitley (1998).

[a]Data are means for six men and three women who exercised to exhaustion on a treadmill set at a grade and velocity to elicit 80% of their maximal aerobic capacity. Respirator resistance pressures were measured at a steady airflow rate of 85 L/min.

and the European Committee for Standardization (www.cenorm.be).

The wear of cooling devices during work periods is another means for countering the stress of the microenvironments associated with PPE. Some success has been achieved in lowering the temperature of skin under a respirator by wetting absorbent material placed on the outside surface of the respirator and relying on evaporation (Fox and DuBois, 1993). Systems for body cooling include vests that hold ice or frozen gel, wetted overgarments, and more complex devices, requiring power sources and heat sinks, that deliver conditioned air or liquid to whole body garments, vests, or caps. Auxiliary cooling extends tolerance times, but work–rest cycles are still needed (Pandolf et al., 1995). The devices vary in effectiveness, and carrying the weight of portable power supplies can negate the benefits of cooling. Selection of the most appropriate system depends on environmental parameters, workload, worker mobility requirements, and availability of resources such as power.

Methods can also be employed for body cooling during rest breaks, as opposed to cooling during work periods. Air and liquid cooling during breaks have proven effective in lowering rectal and skin temperatures and increasing the number of work periods that can be completed by people wearing chemical protective ensembles, even though the protective ensemble is kept sealed during the break period (Constable et al., 1994). Firefighters sometimes have the opportunity to remove pieces of their protective gear during breaks. Submersion of the forearms in cool water (17.4°C) during breaks or delivery of a fine mist propelled by a fan has resulted in lower rectal and skin temperatures and has extended tolerance times over those of the passive cooling experienced simply by removing firefighting gear (Selkirk et al., 2004).

5 PERFORMANCE IN PPE

There are many elements comprising human performance. The categories most relevant to PPE use are tasks that are primarily manual or entail basic sensorimotor processes; intellectual activities requiring mental processing of information, referred to as *cognitive tasks*; and subjective behaviors, behaviors that relate to a person's mood, attitudes, or feelings.

5.1 Sensorimotor Performance

Although physiological stress is a major concern with PPE use, the protective items themselves, independent of the thermal burden they represent, degrade workers' sensory and sensorimotor processes (Bensel, 1997). Thus, PPE can be said to impose a mechanical burden on the user. Lenses of eyewear fog, obscuring vision. Due to lens dimensions, full-facepiece respirators can restrict the visual field substantially compared with no respirator. This is illustrated in Figure 2 for respirators that had a separate lens for each eye. Wear of respirators and hoods also affects speech intelligibility. In a study of U.S. Army chemical protective ensembles, it was found that listeners without respirators and hoods made almost three times more errors identifying simple words when the person speaking was wearing a respirator and hood than when

Table 9 EPA Levels of Protection

Basic Components	Optional Components	Protection Provided	Conditions for Use
	Level A		
Completely encapsulating, vapor-protective suit Pressure-demand, full-facepiece SCBA or positive pressure–supplied air respirator with escape SCBA Inner chemical-resistant gloves Chemical-resistant safety boots Two-way radio communication system	Auxiliary cooling system Outer gloves Hard hat	Highest available level of respiratory, skin, and eye protection from solid, liquid, and gaseous chemicals	Chemicals have been identified and pose high level of hazards to respiratory system, skin, and eyes. Substances are present with known or suspected skin toxicity or carcinogenity. Operations must be conducted in confined or poorly ventilated areas.
	Level B		
Liquid splash protection suit Pressure-demand, full-facepiece SCBA or positive pressure–supplied air respirator with escape SCBA Inner chemical-resistant gloves Chemical-resistant safety boots Hard hat Two-way radio communication system	Auxiliary cooling system Outer gloves	Same level of respiratory protection as level A, but less skin protection; liquid splash protection, but no protection against chemical vapors or gases	Chemicals have been identified and do not require high level of skin protection. Initial site surveys are required until higher levels of hazards are identified. Primary hazards associated with site entry are from liquid and not vapor contact.
	Level C (Not Acceptable for Chemical Emergency Response)		
Support function, chemical-resistant garment Full facepiece, air-purifying, canister-equipped respirator Chemical-resistant gloves Chemical-resistant safety boots Hard hat Two-way radio communication system	Face shield Escape SCBA	Same level of skin protection as level B, but lower level of respiratory protection; liquid splash protection, but no protection against chemical vapors or gases	Contact with chemicals on site will not affect the skin. Air contaminants have been identified and concentrations measured. A respirator canister is available that can remove the contaminant. The site and its hazards have been completely characterized.
	Level D (Not Acceptable for Chemical Emergency Response)		
Coveralls Safety boots/shoes Safety glasses or chemical splash goggles	Face shield Escape SCBA Gloves	No respiratory protection, minimal skin protection	The atmosphere contains no known hazards. Work activities preclude splashes, immersion, potential for inhalation, or direct contact with hazard chemicals.

Source: OSHA (1999a).

the speaker was wearing neither. Listeners had even more difficulty understanding spoken words when they themselves were wearing a respirator and hood (Bensel et al., 1987).

Measurements made of range of motion about body joints indicate that one or more of the chemical protective items comprising a complete ensemble can impose limitations on the extent of movement possible (Bensel et al., 1987). Using U.S. Army chemical protective clothing, it was found that head flexion and rotation were restricted in the complete ensemble compared to movements in a normal military duty uniform or in the protective overgarment alone (Table 11). Simple movements were also performed while wearing various combinations of chemical protective items; the respirator was the item that most restricted these movements.

Table 10 U.S. Army MOPP Levels

MOPP Level	Overgarment	Overboots	Respirator	Gloves
0	Readily available	Readily available	Carried by soldier	Readily available
1	Worn; can be left open for ventilation	Carried by soldier	Carried by soldier	Carried by soldier
2	Worn; can be left open for ventilation	Worn	Carried by soldier	Carried by soldier
3	Worn; can be left open for ventilation	Worn	Worn	Carried by soldier
4	Worn; respirator hood and all garment openings securely closed	Worn	Worn	Worn

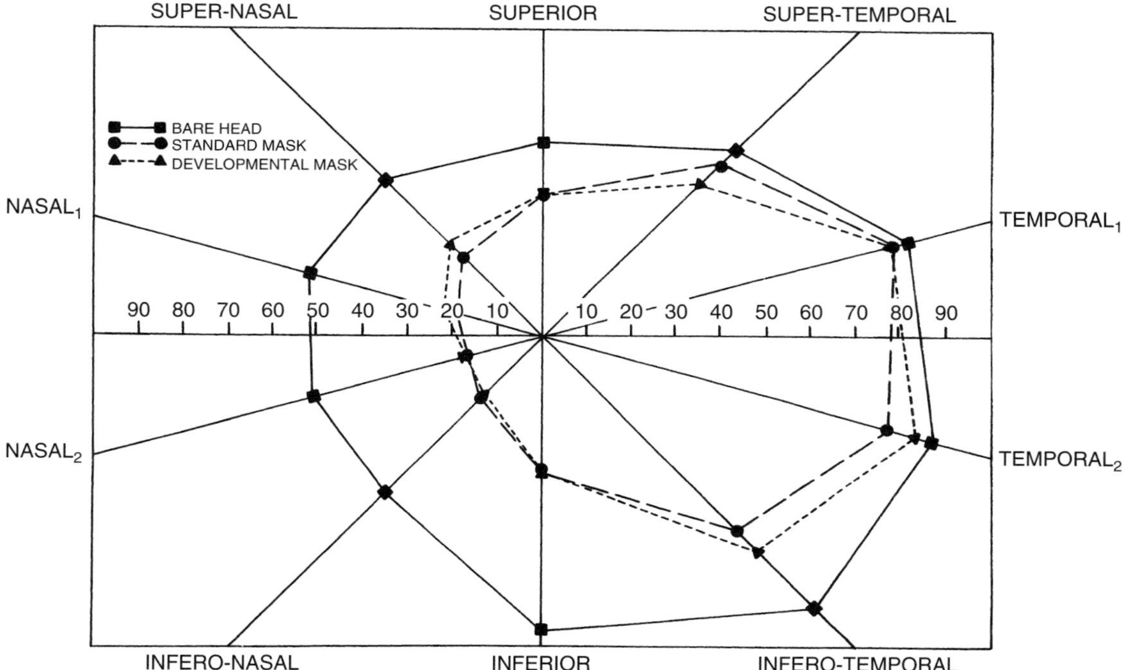

Figure 2 Perimetric measurements made monocularly of the visual field for two respirators and the bare head. (From Bensel et al., 1992.)

Table 11 Mean Scores on Tests for Three Clothing Configurations[a]

Test	Work Clothes	Chemical Protective Overgarment	Chemical Protective Ensemble
Ventral–dorsal head flexion (deg)	141_A	139_A	120_B
Lateral head rotation (deg)	156_A	148_A	106_B
O'Connor finger dexterity test (s)	74_A	73_A	106_B
Pursuit rotor time on target (s)	104_A	100_{AB}	95_B

Source: Bensel (1997).

[a]For each test, means with different subscripts differed significantly ($p < 0.05$) in Tukey post hoc comparisons.

Time to complete a test of fine finger dexterity, the *O'Connor finger dexterity test*, was also substantially longer when the complete chemical protective outfit, including the respirator, was worn compared to times recorded when the normal duty uniform or the overgarment alone was worn (Table 11). The protective item contributing most to poor dexterity performance was the handwear; the respirator also had an effect, but to a lesser extent. The combined effect of the gloves and the respirator was greater than the sum of the effects when each was worn individually (Bensel et al., 1987). There was a similar finding on a test of visual–motor coordination, the *pursuit rotor* (Table 11). Time on target decreased slightly when either the respirator or gloves were used relative to use of the overgarment alone. Wearing both the respirator and the gloves resulted in performance decrements greater than the sum of the decrements associated with each item alone.

The effects of the mechanical burden that PPE can impose include longer times to complete work activities and failure to undertake some tasks. Workers also become frustrated trying to perform tasks and may remove the PPE to accomplish them. For example, protective gloves may be put aside to perform activities requiring fine dexterous manipulations.

5.2 Cognitive Performance

Cognitive tasks involve a minimum of physical effort and large mental or perceptual motor components. Tasks categorized as those involving a cognitive component include reaction time, mathematical processing, and storage and retrieval of information from memory. These are considered to be simple cognitive tasks. More complex cognitive tasks include responding to signals occurring infrequently, monitoring and responding to inputs from two or more sources simultaneously, and operating vehicles or machines.

The effects of PPE on cognitive performance are generally ascribed to the high temperature of the microenvironment. However, unlike the relationship between environmental and physiological parameters in determining physical tolerance to heat, there is no clear-cut relationship on which to base predictions of the effects on cognitive performance of heat associated with PPE wear. Indeed, heat may produce an increment, a decrement, or no effect on cognitive performance, depending on exposure duration, task skill level, acclimatization status, and the nature of the task (Ramsey, 1995).

One interpretation of heat effects, based on dividing cognitive tasks into simple tasks and complex ones, indicates that tasks in the former category will be minimally affected and performance may even improve, whereas a decrease in performance levels of tasks in the latter category will begin in the range 30 to 33°C WBGT (Ramsey, 1995; Pilcher et al., 2002). Another interpretation, drawing on the inverted-U shaped curve to relate arousal to performance, is that there is an optimal arousal level for a particular task type. Thus, heat may increase arousal level, facilitating performance on one task and impairing it on another (Poulton, 1976). A third interpretation maintains that cognitive performance is degraded when core temperature changes and that when core temperature is static, there may be an improvement in performance (Hancock, 1986). These interpretations are helpful in explaining extant research findings, but there is no basis on which to make specific predictions regarding the extent to which heat will affect cognitive performance.

5.3 Subjective Behaviors

Use of PPE can elicit negative psychological reactions. In situations in which thermal strain is experienced due to the PPE, users' responses to the heat can include drowsiness, restlessness, dizziness, and hyperirritability. The hazards present in the situations in which workers are operating can also evoke fears and anxieties indicative of psychological stress, despite the protection afforded by the clothing and equipment being worn. A taxonomy of symptoms of psychological stress is presented in Table 12. Particularly when wearing PPE that encapsulates the body completely, workers may react negatively to the PPE itself and display symptoms of psychological stress. Feelings of isolation and adverse reactions due to diminished sensory input (e.g., vision, hearing, smell) are prominently associated with the use of encapsulating PPE.

Full-facepiece respirators have been implicated as the items most responsible for many of the negative subjective reactions to PPE. Field and laboratory studies of chemical protective respirator use conducted

Table 12 Symptoms of Psychological Stress

Emotional	
Aggression	Irritability
Anxiety	Loneliness
Apathy	Low self-esteem
Depression	Moodiness
Emotional outbursts	Negativism
Excitability	Nervousness
Frustration	Restlessness
Guilt	Tension
Physical	
Breathing difficulties	Inadequate intake of food and liquids
Diarrhea	Muscle cramps
Dry mouth	Nausea
Fatigue	Numbness or tingling in extremities
Frequent urination	Rapid heart rate
Headaches	Sweating
Hot and cold spells	Vomiting
Mental	
Denial	One-track thinking
Forgetfulness	Rationalization
Impulsivity	Regressive reactions
Nightmares	Risk taking

Source: U.S. Army Combined Arms Combat Developments Activity (1986).

by the U.S. military services have established that toxic agents need not be present in the environment and that harmless agent simulants need not be in use for psychological symptoms to appear when military personnel are training with respirators (Bensel, 1994). One series of studies consisted of one- to two-hour training exercises which did not involve the presence or threat of toxic agents or simulants. As many as 69% of participants reported experiencing symptoms such as panic, anxiety, rapid breathing, claustrophobia, and dizziness. Some persons removed their respirators and ceased their training activities; a number manifested indications of hyperventilation.

Complaints of respiratory distress and evidence of hyperventilatory symptoms occur relatively frequently when respirators are used. Symptoms of hyperventilation are listed in Table 13. *Hyperventilation* has been defined as a state in which there is an increased amount of air entering the lungs, resulting in reduction of carbon dioxide tension. Hyperventilation can be produced by increasing the frequency or the depth of respiration or both. The immediate effects are reflected in cerebral circulation, with a slowing of delivery of blood to the brain. There are organic causes of hyperventilation, but episodes can occur as a result of anxiety and psychological stress (Morgan, 1983). One of the implications of a hyperventilatory state is a decrement in cognitive performance, particularly on tasks with a perceptual motor component. The immediate step to resolve a case of hyperventilation is to remove the person from the hazardous environment so that the respirator can be removed.

For workplace productivity to suffer, negative psychological reactions to PPE use need not be as extreme as hyperventilatory episodes. If persons are experiencing considerable discomfort or psychological stress, decrements in their performance are likely, and the length of time they can work in PPE will be limited. In releasing a revised standard for respiratory protection, OSHA (1998) estimated that 5 million workers in 1.3 million workplaces fall under the rule, using respirators regularly or intermittently in the course of their work. Some percentage of these people will suffer negative psychological reactions when wearing a respirator, and as a result, productivity will suffer. Thus, considering only this one item of PPE, the effects of PPE wear on performance in workplaces can be substantial.

5.4 Maximizing Performance

Actions taken to reduce thermal and respiratory strain will also reduce performance decrements associated with PPE use. There are, as well, a number of additional factors influencing performance that can be addressed. In selecting PPE for a workplace, the type and level of protection required relative to the hazards are generally the primary considerations. Expanding selection criteria to include the comfort and functioning of the intended users can ease the burden that PPE imposes on sensory and sensorimotor processes and on the psychological well-being of the user. Guidance on integrating these considerations into the selection process is available from several sources (ASTM, 1999a, 2000; OSHA, 1999c).

It is critical for job performance and the integrity of the protection from hazards that PPE fit properly. The sizes available relative to the body dimensions of the workers that must be outfitted should be part of the PPE selection process. Once PPE is on hand, attention must be paid to achieving a proper fit for each person. There are references to consult regarding fitting processes (ASTM, 1996, 1999a, 2002; OSHA, 1998, 1999a, c). A proper fit is needed even in completely encapsulating garments (Figure 3). Also, attention

Table 13 Symptoms of Hyperventilation

General	***Cardiovascular***	***Neurologic***
Fatigue	Palpitations	Dizziness
Weakness	Tachycardia	Lightheadedness
Exhaustion	Precordial pain	Paresthesia
	Reynaud's phenomenon	Tetany
		Disturbance of consciousness or vision
Respiratory	***Gastrointestinal***	***Musculoskeletal***
Shortness of breath	Globus hystericus	Muscle pain, cramps
Chest pain	Epigastric pain	Tremors
Dry mouth	Aerophagia	Stiffness
Yawning		
	Psychological	
	Tension	
	Anxiety	
	Insomnia	
	Nightmares	

Source: Missri and Alexander (1978).

Figure 3 Self-contained suit for protection against toxicological agents. (Courtesy of the Soldier Systems Center.)

must be given to achieving a proper fit in handwear and footwear, as just about every job function requires mobility and some degree of dexterous manipulation of objects.

There are specific and extensive procedures available from the OSHA (1998) to apply to the fitting of respirators. The OSHA regulations also recognize that some people are not able to use a respirator because of medical conditions or negative psychological reactions. A formal medical evaluation process is required to make a determination as to whether an employee can be permitted to use a respirator and therefore work at sites where respiratory protection is required.

The most important and most neglected element of minimizing performance decrements associated with PPE wear is training. Effective training programs include the following introductory information: an explanation of the hazards for which PPE is being worn; symptoms that indicate exposure to the hazard; limitations in the protection afforded by PPE; and proper PPE storage, maintenance, and inspection procedures. Also important as introductory training for use of some PPE is provision of methods for decontamination.

Background knowledge regarding PPE is important, but it is only the beginning of an effective training program. Training should include a phase in which workers use PPE in a safe environment. It has been found effective to acclimate workers to PPE over a number of days, progressing gradually from less than one hour of wear to continuous wear for the durations expected on the job. Execution of work-related tasks and rehearsal of decontamination procedures, if required, can then be introduced. All users of PPE, but particularly those who use it only intermittently, also need to maintain proficiency through regularly scheduled training.

Effective training programs in the use of PPE minimize performance decrements in several ways. There is a physiological benefit of training to perform work-related tasks in PPE because people learn to pace their physical activities and to implement prescribed work–rest cycles and water consumption guidance. The negative impacts of sensorimotor impairments can also be reduced as people learn to modify work procedures and improvise ways to perform tasks while wearing PPE. Anxiety levels also tend to decrease with repeated use of PPE, presumably because the person learns that it is possible to work in PPE for a period of time despite the discomfort and restrictions. Finally, training increases users' confidence in PPE and confidence in their abilities to perform well while wearing it.

REFERENCES

ASHRAE (2001), *2001 ASHRAE Handbook: Fundamentals,* American Society of Heating, Refrigerating, and Air-Conditioning Engineers, Atlanta, GA.

ASHRAE (2004), *Thermal Environmental Conditions for Human Occupancy*, Standard 55, American Society of Heating, Refrigerating, and Air-Conditioning Engineers, Atlanta, GA.

ASTM (1985), *Standard Test Method for Thermal Transmittance of Textile Materials*, ASTM D 1518-85, American Society for Testing and Materials, West Conshohocken, PA.

ASTM (1996), *Standard Practice for Body Measurements and Sizing of Fire and Rescue Services Uniforms and Other Thermal Hazard Protective Clothing*, ASTM F 1731-96, American Society for Testing and Materials, West Conshohocken, PA.

ASTM (1999a), *Standard Practices for Qualitatively Evaluating the Comfort, Fit, Function, and Integrity of Chemical-Protective Suit Ensembles*, ASTM F 1154-99a, American Society for Testing and Materials, West Conshohocken, PA.

ASTM (1999b), *Standard Test Method for Measuring the Thermal Insulation of Clothing Using a Heated Manikin*, ASTM F 1291-99, American Society for Testing and Materials, West Conshohocken, PA.

ASTM (2000), *Standard Test Method for Evaluation of Glove Effects on Wearer Hand Dexterity Using a Modified Pegboard Test*, ASTM F 2010-00, American Society for Testing and Materials, West Conshohocken, PA.

ASTM (2002), *Standard Terminology Relating to Body Dimensions for Apparel Sizing*, ASTM D 5219-02, American Society for Testing and Materials, West Conshohocken, PA.

Belding, H. S., and Hatch, T. F. (1955), "Index for Evaluating Heat Stress in Terms of Resulting Physiological Strain," *Heating, Piping and Air Conditioning,* Vol. 27, pp. 129–136.

Bensel, C. K. (1994), *Psychological Aspects of Nuclear, Biological, and Chemical Protection of Military Personnel*, TTCP/SGU/94/007, The Technical Cooperation Program, Washington, DC.

Bensel, C. K. (1997), "Soldier Performance and Functionality: Impact of Chemical Protective Clothing," *Military Psychology,* Vol. 9, pp. 287–300.

Bensel, C. K., Teixeira, R. A., and Kaplan, D. B. (1987), *The Effects of U.S. Army Chemical Protective Clothing on Speech Intelligibility, Visual Field, Body Mobility, and Psychomotor Coordination of Men*, NATICK/TR-87/037, U.S. Army Natick Research, Development and Engineering Center, Natick, MA.

Bensel, C. K., Teixeira, R. A., and Kaplan, D. B. (1992), *The Effects of Two U.S. Army Chemical Protective Clothing Systems on Speech Intelligibility, Visual Field, Body Mobility, and Psychomotor Coordination of Men*, NATICK/TR-92/025, U.S. Army Natick Research, Development and Engineering Center, Natick, MA.

Bensel, C. K., and Santee, W. R. (1997), "Climate and Clothing," in *Handbook of Human Factors and Ergonomics*, 2nd ed., G. Salvendy, Ed., Wiley, New York.

Burton, A. C., and Edholm, O. G. (1955), *Man in a Cold Environment,* Edward Arnold, London.

Caretti, D. M., and Whitley, J. A. (1998), "Exercise Performance During Inspiratory Resistance Breathing Under Exhaustive Constant Load Work," *Ergonomics,* Vol. 41, pp. 501–511.

Constable, S. H., Bishop, P. A., Nunneley, S. A., and Chen, T. (1994), "Intermittent Microclimate Cooling During Rest Increases Work Capacity and Reduces Heat Stress," *Ergonomics,* Vol. 37, pp. 277–285.

DuBois, D., and DuBois, E. F. (1916), "A Formula to Approximate Surface Area if Height and Weight Be Known," *Archives of Internal Medicine,* Vol. 17, pp. 863–871.

Durnin, J. V. G. A., and Passmore, R. (1967), *Energy, Work, and Leisure,* William Heinemann, London.

Fox, S. H., and DuBois, A. B. (1993), "The Effect of Evaporative Cooling of Respiratory Protective Devices on Skin Temperature, Thermal Sensation, and Comfort," *American Industrial Hygiene Association Journal,* Vol. 54, pp. 705–710.

Gagge, A. P., Stolwijk, J. A. J., and Nishi, Y. (1971), "An Effective Temperature Scale Based on a Simple Model of Human Regulatory Response," *ASHRAE Transactions,* Vol. 77, pp. 247–262.

Goldman, R. F. (1988), "Standards for Human Exposure to Heat," in *Environmental Ergonomics: Sustaining Human Performance in Harsh Environments,* I. B. Mekjavic, E. W. Banister, and J. B. Morrison, Eds., Taylor & Francis, London.

Hancock, P. A. (1986), "Sustained Attention Under Thermal Stress," *Psychological Bulletin,* Vol. 99, pp. 263–281.

Hanson, M. A. (1999), "Development of a Draft British Standard: The Assessment of Heat Strain for Workers Wearing Personal Protective Equipment," *Annals of Occupational Hygiene,* Vol. 43, pp. 309–319.

Houghten, F. C., and Yaglou, C. P. (1923), "Determining Lines of Equal Comfort," *Transactions of the American Society of Heating and Ventilating Engineers,* Vol. 29, pp. 163–176.

ISO (1989), *Hot Environments: Estimation of the Heat Stress on Working Man, Based on the WBGT-Index*, ISO 7243, International Organization for Standardization, Geneva, Switzerland.

ISO (1995), *Ergonomics of the Thermal Environment: Estimation of the Thermal Insulation and Evaporative Resistance of a Clothing Ensemble*, ISO 9920, International Organization for Standardization, Geneva, Switzerland.

ISO (1998), *Ergonomics of the Thermal Environment: Instruments for Measuring Physical Quantities*, ISO 7726, International Organization for Standardization, Geneva, Switzerland.

ISO (2004), *Ergonomics of the Thermal Environment: Analytical Determination and Interpretation of Heat Stress Using Calculation of the Predicted Heat Strain*, ISO 7933, International Organization for Standardization, Geneva, Switzerland.

Leithead, C. S., and Lind, A. R. (1964), *Heat Stress and Heat Disorders*, F.A. Davis, Philadelphia.

McCullough, E. A., Jones, B. W., and Huck, J. (1985), "A Comprehensive Data Base for Estimating Clothing Insulation," *ASHRAE Transactions*, Vol. 91, pp. 29–47.

Missri, J. C., and Alexander, S. (1978), "Hyperventilation Syndrome: A Brief Review," *Journal of the American Medical Association*, Vol. 240, pp. 2093–2096.

Moran, D. S., Shitzer, A., and Pandolf, K. B. (1998), "A Physiological Strain Index to Evaluate Heat Stress," *American Journal of Physiology*, Vol. 275, pp. R129–R134.

Morgan, W. P. (1983), "Hyperventilation Syndrome: A Review," *American Industrial Hygiene Association Journal*, Vol. 44, pp. 685–689.

NIOSH (1986), *Criteria for a Recommended Standard ... Occupational Exposure to Hot Environments*, U.S. Government Printing Office, Washington, DC.

OSHA (1989), "OSHA Hazardous Waste Operations and Emergency Response Regulation," 29 CFR 1910.120, http://www.osha.gov/pls/oshaweb.

OSHA (1998), "OSHA Respiratory Protection: Final Rule," 29 CFR 1910.134 with Appendices, http://www.osha.gov/pls/oshaweb.

OSHA (1999a), "Chemical Protective Clothing," TED 01-00-015, Sect. VIII, Ch. 1, *OSHA Technical Manual*, http://www.osha.gov/dts/osta/otm/otm_viii/otm_viii_1.html.

OSHA (1999b), "Heat Stress," TED 01-00-015, Sect. III, Ch. 4, *OSHA Technical Manual*, http://www.osha.gov/dts/osta/otm/otm_iii/otm_iii_4.html.

OSHA (1999c), "Respiratory Protection," TED 01-00-015, Sect. VIII, Ch. 2, *OSHA Technical Manual*, http://www.osha.gov/dts/osta/otm/otm_viii/otm_viii_2.html.

Pandolf, K. B., Stroschein, L. L., Drolet, R. R., Gonzalez, R. R., and Sawka, M. N. (1986), "Prediction Modeling of Physiological Responses and Human Performance in the Heat," *Computers in Biological Medicine*, Vol. 16, pp. 319–329.

Pandolf, K. B., Gonzalez, J. A., Sawka, M. N., et al. (1995), *Tri-Service Perspectives on Microclimate Cooling of Protective Clothing in the Heat*, T95-10, U.S. Army Research Institute of Environmental Medicine, Natick, MA.

Parsons, K. C. (1999), "International Standards for the Assessment of the Risk of Thermal Strain on Clothed Workers in Hot Environments," *Annals of Occupational Hygiene*, Vol. 43, pp. 297–308.

Parsons, K. C. (2002), *Human Thermal Environments*, 2nd ed., Taylor & Francis, London.

Pilcher, J. J., Nadler, E., and Busch, C. (2002), "Effects of Hot and Cold Temperature Exposure on Performance: A Meta-analytic Review," *Ergonomics,* Vol. 45, pp. 682–698.

Poulton, E. C. (1976), "Arousing Environmental Stresses Can Improve Performance, Whatever People Say," *Aviation, Space, and Environmental Medicine*, Vol. 47, pp. 1193–1204.

Ramsey, J. D. (1995), "Task Performance in Heat: A Review," *Ergonomics*, Vol. 38, pp. 154–165.

Raven, P. B., Dodson, A. T., and Davis, T. O. (1979), "The Physiological Consequences of Wearing Industrial Respirators: A Review, " *American Industrial Hygiene Association Journal*, Vol. 40, pp. 517–534.

Santee, W. R., Matthew, W. T., and Blanchard, L. A. (1994), "Effects of Meteorological Parameters on Adequate Evaluation of the Thermal Environment," *Journal of Thermal Biology*, Vol. 19, pp. 187–198.

Selkirk, G. A., McLellan, T. M., and Wong, J. (2004), "Active Versus Passive Cooling During Work in Warm Environments While Wearing Firefighting Protective Clothing," *Journal of Occupational and Environmental Hygiene*, Vol. 1, pp. 521–531.

Sendroy, J., and Collison, H. A. (1960), "Nomogram for Determination of Human Body Surface Area from Height and Weight," *Journal of Applied Physiology*, Vol. 15, pp. 958–959.

U.S. Army Combined Arms Combat Developments Activity (1986), *Extended Operations in Contaminated Areas*, Field Circular FC 50-12, Author, Fort Leavenworth, KS.

U.S. Department of the Army and Air Force (2003), *Heat Stress Control and Casualty Management*, TB MED 507/AFPAM 48-152(I), Headquarters, Washington, DC.

Woodcock, A. H. (1962), "Moisture Transfer in Textile Systems, Part I," *Textile Research Journal*, Vol. 32, pp. 628–633.

Yaglou, C. P., and Minard, D. (1957), "Control of Heat Casualties in Military Training Centers," *American Medical Association Archives of Industrial Health,* Vol. 16, pp. 302–316.

CHAPTER 34
HUMAN SPACE FLIGHT

Barbara Woolford
NASA Johnson Space Center
Houston, Texas

Frances Mount
NASA Johnson Space Center
Houston, Texas

1 UNIQUE FACTORS IN SPACE FLIGHT

The first human space flight, in the early 1960s, was aimed primarily at determining whether humans could indeed survive and function in microgravity. Would eating and sleeping be possible? What mental and physical tasks could be performed? Subsequent programs increased the complexity of the tasks the crew performed. Table 1 summarizes the history of U.S. space flight, showing the projects, their dates, crew sizes, and mission durations. With over 40 years of experience with human space flight, the emphasis now is on how to design space vehicles, habitats, and missions to produce the greatest returns to human knowledge. What are the roles of humans in space flight in low Earth orbit, on the moon, and in exploring Mars?

1.1 Gravity

The most obvious factor specific to space flight is gravity. Orbiting Earth, crews experience free-fall, or microgravity. This affects all aspects of life and requires special considerations when designing habitat, equipment, tools, and procedures. During launch and entry, crews experience hypergravity for short periods of time. Extensive research and experience with high-performance aircraft has provided great understanding of these environments, and indeed, the tasks to be performed are similar to aviation tasks. On the surface of the moon and Mars, gravity is substantially lower than on Earth but is definitely sufficient to allow designing habitats, equipment, and tasks analogously to those on Earth.

1.2 Mission Constraints

Accommodations for humans in space are constrained by the three major mission drivers: mass, volume, and power, each of which drives the cost of a mission. Mass and volume determine the size of the launch vehicle directly; they limit consumables such as air, water, and propellant; and they affect crew size and the types of activities the crew performs. Power is a limiting factor for a space vehicle. All environmental features—atmosphere, temperature, lighting—require power to be maintained. Power can be generated from batteries, fuel cells, or solar panels. Each of these

Table 1 U.S.-Crewed Space Programs to Date

Program	Dates	U.S. Crew Size	Mission Length
Mercury	1961–1963	1	Up to 34 hours
Gemini	1961–1962	2	Up to 6 days
Apollo	1968–1972	3	Up to 12.5 days
Skylab	1973	3	Up to 84 days
Apollo–Soyuz (ASTP)	1975	3	Up to 9 days
Space Transportation System (STS)	1981–current	2–10	3–17 days
Shuttle–Mir	1995–1998	2 Russian, 1 U.S.	Up to 6 months
International Space Station (ISS)	2000–current	2–6 (including international partners)	Approx. 6 months

Source: Date from NASA (2003, 2004b).

sources requires lifting mass and volume from Earth, driving mission cost.

1.3 Mission Duration

The habitability and human factors requirements for space flight are driven by mission duration. The Space Transportation System (STS) was designed for missions on the order of two weeks—analogous to a camping trip. With Mir and the International Space Station (ISS), mission durations of six months became standard, requiring far more concern for habitability and for crew efficiency, training, and sustenance. As NASA begins to plan for a mission to the Mars surface, with travel times on the order of six months each way and a possible surface stay of 18 months, it must address providing all support and services to crew members: health maintenance, training, recreation, food, clothing, and so on.

1.4 Communications

To date, the model for space exploration has had a very small crew—from a maximum of seven or eight on a shuttle flight, to just two people on the ISS—supported by a very large group of scientific and engineering experts on the ground. The crew and ground personnel are linked through the Mission Control Center (MCC). This model has been essential because such a small crew cannot be expert in all the critical subsystems on board. There are too few people to understand the subsystems in sufficient detail to operate and maintain them under nominal circumstances, let alone when malfunctions occur. But this model depends on rapid two-way communications. Video and audio transmissions allow the MCC to see and hear the crew, and to transmit questions and procedures in a short enough time to be responsive to time-critical events. Even on the lunar surface, communications lags are on the order of seconds. But with a mission to Mars, the nature of communications and the roles of the ground and flight crews will be reexamined to consider a delay of 20 minutes each way.

1.5 Crew Time

Crew time is becoming recognized as another mission driver. The size of the crew affects mass and volume requirements directly. Designing equipment and procedures to maximize returns from crew time is beginning to be considered in the earliest stages of mission planning. Detailed studies of how crew time was actually used during Skylab (Bond, 1977) showed that approximately one-third of the crew time was spent in sleep, one-third in other forms of self-sustenance such as hygiene, exercise, eating, and recreation, and one-third was actually devoted to operating the spacecraft and scientific experiments.

2 ANTHROPOMETRY AND BIOMECHANICS

2.1 Changes in Posture and Body Size

In a microgravity environment the body changes. Immediately on reaching free-fall, the body assumes a neutral posture quite different from standing or sitting postures on Earth. The neck, shoulders, elbows, hips, and knees all flex somewhat, and the shoulders also abduct and rotate with a large intersubject variability. The result affects a crew member's line of sight, height, and reach envelope. The range of postures observed on one Shuttle mission is shown in Figure 1. Table 2 gives the joint angles. Figure 2 illustrates reach envelopes based on a typical posture for a 95th percentile crew member. After a short while, on the order of hours, the body height changes due to spinal elongation. Height increases about 3% during the first day or so in microgravity. The distribution of body fluids also changes. Greater amounts of body fluids move to the head and torso, affecting hand size, facial appearance, the voice, and perhaps the sense of smell. Space suits and gloves, which must fit snugly, must accommodate changes in hand size and stature.

2.2 Changes in Strength

Changes in strength over time in microgravity have been a focus of research because of the direct effect on the ability to perform physical tasks. Jaweed (1994) reports significant (10 to 20%) decreases between the preflight and postflight strength in the antigravity muscles (back and legs) after as few as five to 10 days on orbit. This, taken with the loss of bone mass observed (Schneider et al., 1994), indicates that countermeasures must be taken for long-duration flights and that tasks that can be performed early in flight might be more difficult or dangerous after an extended

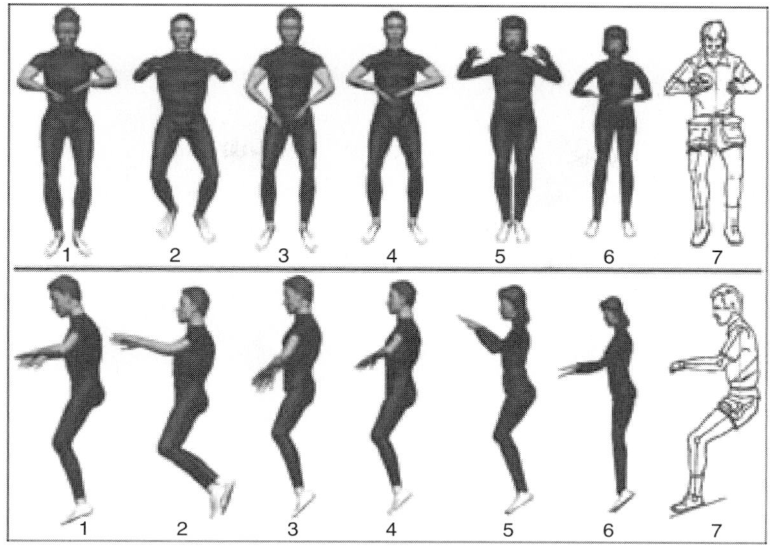

Figure 1 Neutral postures in microgravity. Bodies 1 to 6 are actual crew members. Body 7 is a composite posture based on Skylab data.

Table 2 Crew Microgravity Posture Measurements (deg)[a]

Anthropometric Measurement Joint Angles[b]	Skylab Composite Left–Right	Crew 1 Left–Right	Crew 2 Left–Right	Crew 3 Left–Right	Crew 4 Left–Right	Crew 5 Left–Right	Crew 6 Left–Right
Hip flexion	50	33	33–29	33	33	29	12
Hip abduction	18.5	6.5–5.5	20–16	13–17.5	15.5–16	3.5–4.5	4–9
Knee flexion	50	50	83–87	50	50	44	11–12
Ankle plantar extension	21	6–7	15–14.5	29–30	27–24	16–14	35–41
Waist flexion	0	13	0	1	0	0	2
Neck flexion	24	16	18	16	5	7	16
Left neck lateral bend	0	0	0	3	0	0	0
Shoulder flexion	36	49–46	67–64	29	33–35	60–57	36
Shoulder abduction	50	32–33	26–26.5	27–29	40.5	24–45	23–36
Medial shoulder rotation	86.6	58–61	45.5–41	71–77	74.5–74	25.5–26.5	50–48
Elbow flexion	90	78	45–53	61–57	94–91	78–80	51–64
Wrist extension	0	0	3–0	0	0	0	0
Wrist ulnar bend	0	0	0	0	0–9	0–3	0
Forearm pronation	N/A	N/A	26	20–N/A	N/A–2	16–N/A	N/A–5
Forearm supination	30	7–10	N/A	N/A–30	15–N/A	N/A–4	14–N/A
Finger flexion	0	42	60	30	21–57	55–47	25–35

[a]Crews 1 to 6 correspond to the body positions shown in Figure 1. Skylab composite corresponds to illustration 7.
[b]Angles are based on an upright stature coordinate system.

time in microgravity. The most common countermeasure for strength loss is exercise, particularly of the legs and back. Typical equipment includes bicycle ergometers and treadmills. When designing spacecraft, volume must be allowed for equipment storage and deployment. Significant periods of crew time, on the order of an hour per day per person, must be reserved for exercise. Design and location of equipment must address isolation of vibration and noise.

3 ENVIRONMENTAL FACTORS

3.1 Human Factors in a Closed Environment

NASA strives to close the spacecraft environment in the sense that every effort is made to recycle air and water rather than to carry replacement oxygen and water on a mission. This greatly affects design of the habitat and equipment. Materials must not release compounds that are difficult to remove from the atmosphere; this eliminates a variety of plastics

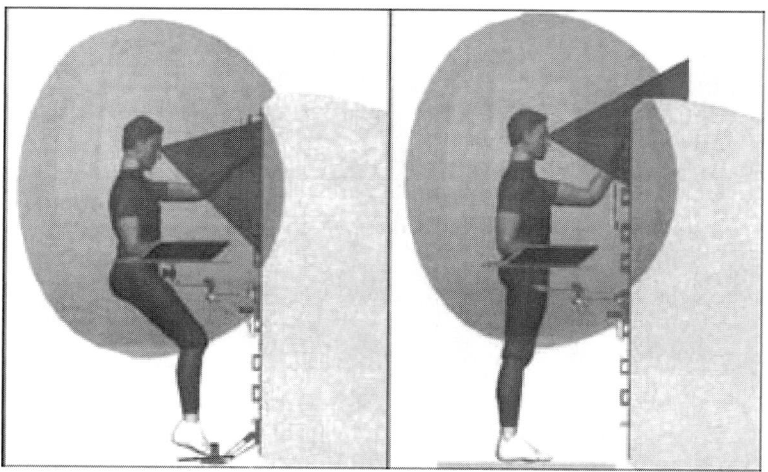

Figure 2 A 95th percentile stature crew member is shown in a neutral body posture (left) and standing vertically (right). The elliptical gray shading indicates the reach envelope. The darker gray cone indicates the viewing area.

and certain types of finishes for other materials. Materials must be compatible with cleaning materials and biocides that are safe for the environment; they must be incompatible with flourishing colonies of bacteria and mold.

3.2 Atmosphere

Crew members in a system must be provided with an environment to enable them to survive and function as a system component in space. An artificial atmosphere of suitable composition and pressure is the most immediate need. It supplies the oxygen their blood must absorb and the pressure their body fluids require. Humans can survive in a wide range of atmospheric compositions and pressures. Atmospheres deemed sufficient for human survival are constrained by the following considerations:

- There must be sufficient total pressure to prevent the vaporization of body fluids.
- There must be free oxygen at sufficient partial pressure for adequate respiration.
- Oxygen partial pressure must not be so great as to induce oxygen toxicity.
- For a long duration (in excess of two weeks), some physiologically inert gas must be provided to prevent atelactasis.
- All other atmospheric constituents must be physiologically inert or of low enough concentration to preclude toxic effects.
- The breathing atmosphere composition should have minimal flame or explosive hazard.

Mission planning must take the foregoing considerations for atmospheric conditions and balance them with the constraints of the mission: length of mission, mission objectives, requirement for pre-breathe (for extravehicular activity), research requirements for the mission, and equipment in the vehicle.

3.3 Water

In addition to the obvious need for drinking water, water is required for a variety of other uses, including personal use, hygiene, and housekeeping. If plants are to be grown during the mission, that is an additional water requirement. Typical water requirements for drinking, hygiene, and washing for each crew member are 2.84 to 5.16 kg per person per day for standard operational mode (NASA, 1995). A crew depends on water that is clean and safe. The use of water that is reclaimed and stored depends on its quality.

Water management systems changed with the design of the space vehicles and life support requirements of each program. During early Mercury, Gemini, and Skylab missions, water was filled up in tanks, built into the vehicle before launch, and carried into space. However, during the Apollo missions, the water source came from the fuel cells; fuel cells convert hydrogen and oxygen to generate power with water as the by-product. This marked a major breakthrough in the water management technology because water tanks did not have to be prefilled before the launch. The shuttle orbiter uses four 168-lb-capacity steel tanks. The potable water source comes from the fuel cell by-product, water (NASA, 2004a). The excess water from the shuttle is used to meet the water requirements of the ISS under normal mission configuration.

3.4 Noise

Noise can affect human physiology and health in a number of ways (Wheelwright et al., 1994). From the perspective of human factors, noise can affect performance by interfering with communications, interfering with sleep, and causing annoyance. The SpaceHab is a modular laboratory that fits in the cargo bay of the Space Shuttle. In an assessment of the SpaceHab-1 mission (STS-57), Mount et al. (1994) found that although the measured noise levels did not generally exceed the levels permitted for the shuttle flight deck or middeck, noise levels were substantially above design limits for the SpaceHab. This is probably because of the number and nature of experiments and equipment that were located there. However, most crew members required earplugs during sleep, even though they slept in the shuttle. Crew members principally used the intercom rather than unaided voice to communicate, even when in the same area, and reported difficulty in concentration and noise-induced headaches and fatigue.

Large space vehicles present a significant acoustics challenge because of obvious difficulties with controlling a number of connected, operating modules with payloads and equipment to perform vehicle functions and experiments, sustaining crew, and keeping them in good physical condition. Modules have equipment such as fans, pumps, compressors, avionics, and other noise-producing hardware or systems to serve their functional and life support needs. Payload racks with operating equipment create continuous or intermittent noises, or a combination of both. Payload rack contributions to the total on-orbit noise can be and has been shown to be significant. The crew exercises on a treadmill and with other conditioning devices that generate noise. Communications between crew and ground, which are raised to communicate over the background environment, adds to the overall crew noise exposure. The crew members have to work and live in the resulting acoustic environment. The acoustics challenge is further complicated by the fact that there are numerous suppliers of modules, hardware, and payloads from across and outside the United States (Goodman, 2003).

The Mir audible noise measurement experiment was designed to characterize the Mir internal environment background noise levels during the docked period of STS-74 with Mir. The NC 50 curve was exceeded at all measurement locations except the *Kvant 2* airlock compartment. During the docked time period, Mir science and exercise activity was low. Overall, the crew's subjective impression of the Mir acoustic environment was favorable; however, some hearing loss was noted at the end of the mission (NASA, 1996).

The ISS is a complicated and sophisticated machine. ISS hardware is divided into categories, including the module (or spacecraft), government-furnished equipment, and payloads (science experiments). These different categories of hardware are governed by different requirements. Acoustic noise emissions verification is performed through actual test measurements of the hardware to the greatest extent possible. However, in some instances a fully integrated end item is not available due to schedule mismatches or physical limitations to the hardware configuration, or the payload may be delivered to ISS and placed in a rack already onboard. An acoustic test-correlated analytical model is used to predict overall noise levels in this case so that crew safety can be ensured. Remedial actions are performed to quiet hardware when necessary (Allen and Goodman, 2003).

Flight data was taken in the ISS in 2003. When first flown, before the payloads were factored in, the U.S. laboratory exceeded the NC-50 module requirement. The requirement was waived predicated on planned modification of three hardware items: the pump package assembly, the carbon dioxide removal assembly, and the medium-rate outage recorder. Two of these items have been modified, and the third modification is under assessment. With the addition of the payloads and science equipment, the U.S. lab is reasonably close to the total module systems requirement of NC-52, except that the noise level has been higher in the aft end of the module. The node and airlock are shown to be at acceptable levels. Measurements taken in the Russian modules—functional cargo block, service module, and docking compartment—exceed specification limits.

Waivers have been granted with the intent to implement modifications as soon as feasible. Noise levels have improved but are still excessively high. The acoustic levels (measured in ground testing) of other ISS international partner modules are expected to be acceptable (Goodman, 2003).

3.5 Lighting

Lighting is essential to performing virtually every task in space. When windows are present and unshuttered, the typical 90-minute low Earth orbit of the shuttle or station cause problems with time for eyes to adapt to the rapid disappearance of sunlight. In the study by Mount et al. (1994), the most frequent report of lighting problems was that sunlight made electronic displays and video monitors difficult or impossible to read. However, some activities, such as remote manipulator operations, require out-the-window viewing, and Earth watching is a favorite crew activity in any spare time. Wheelwright et al. (1994) and the Man–Systems Integration Standards (NASA, 1995) provide tables and guidelines for illumination levels for various intravehicular and extravehicular tasks.

Two critical tasks requiring vision of external targets are docking the shuttle to the ISS and using remote manipulators to position space-suited crew members or large structural components. In low Earth orbit, there is a change from light to dark every 90 minutes. In vacuum, shadows are much sharper than in an atmosphere, where water vapor, dust particles, and other airborne particles scatter light. To ensure adequate light, tasks may be scheduled to be performed in those parts of the orbit when the combination of sunlight and artificial light are predicted to provide adequate contrast and visibility (Bowen, 2004). NASA developed software that models realistic images of complex environments. Measured data are used to develop models of shuttle and station artificial light. Natural lighting, such as sun and Earth shine, are also incorporated into the lighting analyses. By incorporating the measured reflectance of each material into the lighting models, an accurate calculation of the amount of light entering a camera can be made. Using this calculated light distribution with the model of the shuttle cameras, camera images can be simulated accurately. Use of these lighting images are essential to predict available lighting during space operations requiring camera viewing, such as the assembly of ISS components. In preparing for a shuttle visit to ISS, mission planners simulate the lighting environment for critical tasks at 1-minute intervals.

3.6 Dust and Debris

Debris and dust in the orbiter crew compartment of early shuttle missions created crew health concerns and physiological discomfort and was the cause of some equipment malfunctions. Debris from orbiters during flight and processing was analyzed, quantified, and evaluated to determine its source. Selected ground support equipment and some orbiter hardware were redesigned to preclude or reduce particularization/debris generation. New filters and access ports for cleaning were developed and added to most air-cooled avionics boxes. Most steps to reduce debris were completed before flight STS-26, in 1988. After these improvements were made, there was improved crew compartment habitability and less potential for equipment malfunction (Goodman, 1992).

For future lunar/Mars exploration missions, the problem of dust in these environments is recognized. However, our knowledge at this time is limited as to the specifics of the dust. We have some data from previous lunar missions and are supplementing it with derived data. Derived data from our limited, but growing knowledge of Mars is forming a basis of our need for requirements for dust abatement. The dust will cause a serious problem for extravehicular activity (EVA) suits and equipment used external to the vehicle. There is also a concern for dust in the vehicle habitation area. Dust inside the vehicle could increase crew time due to more frequent filter changes and other chores to remove dust from equipment. Basic habitability could also be affected if the dust were to accumulate on display screens and cooking equipment.

4 HABITABILITY AND ARCHITECTURE

4.1 Architecture

Habitability as a discipline is concerned with providing a space vehicle that within some understandably necessary size restraints, provides a comfortable, functionally efficient habitat that will support mixed crews living and working together for the duration of the mission. Attention must be given to the morale, comfort, and health of crews with differing backgrounds, cultures, and physical size. Architectural design of crew interfacing elements should be comfortable for the extremes of any crew population. The *habitability architecture* design concerns are mainly the fixed architectural elements such as (1) the geometric arrangements of compartments, (2) passageways and traffic paths, (3) windows, (4) color, (5) workstations, (6) off-duty areas, (7) stowage, and (8) lighting (NASA, 1983).

4.1.1 Compartments

The success of an extended mission on a space vehicle depends on the crew being an integral part of the interior design. The focus of any vehicle design should be crew-centered. The arrangement and design of any habitable compartment should take into account the possibility of a subsystem failure or damage that could require quick, efficient evacuation. The actual vehicle arrangement depends on the specific program's goals and definition. Based on space flight history, configuration should take into account the following:

- Sleeping and private areas should be separate from traffic paths and noise generators.
- Areas that are to be used by more than one crew member at a time should be arranged to avoid bottlenecks. These are areas such as the galley, workstations, and waste management systems.
- Traffic flow analysis should be done for crew tasks and activities.
- Switches should be located in proximity of associated equipment.
- Adequate electrical outlets should be provided to reduce the use of extension power cords and the resulting "spaghetti all over".
- A dedicated desk/work area should be provided for general paperwork associated with vehicle keeping.

Skylab experience has shown that crew members were able to operate equipment easily from any orientation. Basically, a crew member established a local orientation based on himself or herself and proceeded without difficulty. However, it was also shown that crew could much more easily orient themselves in a room with equipment oriented with consistent up and down directions. An inconsistent zero-g orientation of one module caused orientation problems that were time consuming. The conclusion is that a common plane for visual reference should be designated throughout each module.

Habitable volume is defined as free, pressurized volume, excluding the space required for equipment, fixtures, furniture, and so on. It does not include "nooks and crannies" (i.e., spaces too small for human access). Total volume requirements depend on the specific program goals of the particular mission. Volume requirements for specific workstations have to be calculated after determination of the tasks required at the workstation and number of crew involved (NASA, 1983).

4.1.2 Passageways and Traffic Paths

A *passageway* is defined as a pass-through area between two nonadjacent compartments. Passageways shall be kept free of sharp and protruding objects. Skylab crew members liked the large "ship-type" doorways. They found round hatches to be much less satisfactory.

Traffic paths consist of three types. *Emergency paths* are those used for crew passage to emergency equipment such as oxygen bottle or mask, firefighting equipment, pressure controls, and escape hatches. *Primary paths* are those used for personnel and equipment transfer between major habitable compartments, or between a compartment and a workstation or off-duty area. *Secondary paths* provide access behind equipment, between equipment and structural members, and around workstations. All traffic paths can be superimposed to form a total traffic pattern, which in conjunction with detailed task analysis, can be used to determine the most efficient placement of mobility aids. This traffic pattern and task analysis must also be used to design out potential bottlenecks in a space vehicle.

To be avoided are the bottlenecks experienced on Skylab missions. They were: insufficient passage room in areas with workstations, too much activity in one place (e.g., conflicting placement of shower and tool kit), and the inability to use the waste management equipment if there was someone using the handwashing equipment (NASA, 1983).

4.1.3 Windows

All habitable volumes should include windows that are adequate for terrestrial and celestial references. Windows are necessary for observation of scientific phenomena, monitoring of EVA, observation of the vehicle exterior, photography, and general viewing. Sufficient window locations should always be provided to view Earth, for both Earth observation experiments and crew recreation and well-being.

All viewing windows and the area adjacent to them should be considered a crew workstation. Sufficient work space and restraint equipment should be provided at view ports for one or more crew members to perform assigned tasks. A window should be installed in the pressure hatch that allows the flight crew to observe the EVA crew in the airlock. Windows that are to be utilized for special photography and scientific experiments must be designed with an aperture size that is compatible with the equipment and tasks specified for that location. Space flights have shown window gazing to be the prime off-duty activity for crew members. Window viewing has been a treasured pasttime on all missions to date.

The design of viewing windows should not impose difficult housekeeping tasks on the crew. Cleaning equipment should be provided for removal of fingerprints and other stains that may accumulate. The equipment must be compatible with the coating(s) on the window and not scratch or affect the optical quality of the window or disturb any surface coating.

Each window should have a sufficiently clear area around it to permit any body position for viewing. A positive means of defogging the windows should be provided. All window covers and/or shutters shall be operated by a device that is easy for any crew member to use. All viewing windows should be provided with a crew-operated, opaque sunshade located within the interior of the spacecraft that is capable of restricting all sunlight from entering the habitable compartments (NASA, 1983).

4.1.4 Color

Color should be used to provide visual stimulation for the vehicle occupants and to create different moods for relieving the monotony of prolonged confinement. Factors required in color planning are room volume, function, architecture materials, safety, and required color coding. As the Skylab mission grew in length, the interior color scheme became less acceptable. The crew of the 84-day mission felt that the color scheme was too drab and suggested that accent colors should be used more extensively. Color coding should be used as a supplement to nomenclature to enhance discrimination and to assist the crew in rapid identification of functions. Coding of EVA equipment should be used with colors that will not deteriorate from solar exposure. All EVA handrails should be a standard color. The color should have a high contrast ratio with the background (NASA, 1983).

4.1.5 Workstations

A *workstation* is defined as any location in the space vehicle where a dedicated task or activity is performed exclusive of the recreation, personal maintenance, and sleep areas. Tasks and activities include vehicle stabilization and control, systems management, experiments, science, and maintenance (equipment repair). With any workstation, analysis should be done to determine the tasks, operator activities, tools, and equipment necessary for each workstation. To make efficient use of space, multiuse workstation can be considered.

All necessary equipment, tools, restraints, lights, and power outlets should be provided at each workstation. Adequate space should be provided for the crew to perform the assigned tasks efficiently and safely. Where possible, workstations and associated equipment should be standardized throughout the entire vehicle to aid in the efficiency of tasks. Part of the workstation analysis should cover adjacent workstations and any impact that might arise from two crew members working at adjacent workstations at the same time. An analysis of traffic flow should be completed to determine placement of a workstation without bottlenecks.

Flight experience has shown that anything “usable” will be used as a kickoff point or as a grabbing point to change direction of travel. All workstations should be planned to limit inadvertent control activation and/or deactivation by passing crew members. A restraint system should be incorporated into a workstation design with compatibility to the task to be done (NASA, 1983).

4.1.6 Off-Duty Areas

There should be a dedicated area for off-duty activities, with a minimum space for the entire crew. This allows for socialization. Stowage areas should be provided in a dedicated recreation area and in the personal space area for items to be used during recreation activity and off-duty time (NASA, 1983). There has been agreement from crewmembers on U.S. missions and also from crew during analog studies that they do not like to have the same table used for dining as well as a maintenance bench and/or as a biology work area (Mount, 2002).

4.1.7 Stowage

Stowage space must be provided. For efficient use the space should be near the stations where the stowed items will be used. A method should be provided for locating stowed equipment and supplies. This is extremely important for a mission like the International Space Station, where crews are changed out periodically, but large quantities of the stowed equipment and supplies stay (NASA, 1983).

4.1.8 Ambient Lighting

For the most part, lighting follows the same requirements as for an Earth structure, but spacecraft hardware designers face a few human factors challenges not usually encountered in earthbound environments. In general, design of any space vehicle must take into account the constraints of power and weight limitations. This has an impact on the number of lights and their specifications. General lighting for all vehicles designed and built in the U.S. space program have been fluorescent luminaires. New types of lighting are being considered, such as LEDs (light-emitting diodes). Fluorescent lighting has to be sealed to contain the mercury in case of breakage. The use of fixed luminaires for general illumination within the relatively small habitable volume of a spacecraft implies that an astronaut may frequently find one or more of these light sources in her or his field of view as she or he floats in microgravity. This creates potential direct glare sources.

Additionally, many astronauts are old enough to have experienced typical symptoms of presbyopia. The loss of the full range of accommodation in their viewing close and distant objects is often simply compensated for by their use of corrective eyeglasses or contact lenses. These means are not available to an astronaut during extravehicular activities in a spacesuit, however. The dry, low-pressure, high-oxygen content environment within the spacesuit precludes the use of contact lenses, and the helmet does not provide adequate interior space for eyeglasses. If the helmet were roomy enough to allow eyeglasses to be worn, it is likely that internal light reflections between the lenses of the eyeglasses and the interior of the faceplate would prove problematic. This means that when planning an EVA task, lack of eyeglasses and light levels must be taken into account (Bowen, 2004).

While in low Earth orbit there is a change from light to dark every 90 minutes. This affects the EVA task planning, due to the changes in light and shadows. The Graphics Research and Analysis Facility at Johnson Space Center uses an accurate lighting model to produce realistic images of this complex, ever-changing environment. Measured data are used to develop models of shuttle and station artificial lights along with the natural lighting from sun and Earth shine. This information is incorporated into the task analysis for EVA tasks (Maida, 2002).

4.2 Considerations for Self-Sustenance

The spacecraft must be designed to provide for all aspects of life. For long-duration missions, private compartments are used for sleep and certain personal activities, such as recreational reading or communicating with family and friends. Since the sleep compartment is the single location in which the crew member spends the most time, it has been found to be most effective to shield the compartment heavily against radiation.

4.2.1 Sleep

An individual sleep compartment should be provided for each crew member. The private sleeping accommodations should have a privacy curtain, partitions, and stowage lockers. Each sleep area should be located as far as possible from noise, activity, and public area. Since there is no up or down in weightlessness, the position of the body did not matter during sleep (Figure 3). Some astronauts have been bothered by an effect known as *head nod*. If the head is not secure when fully relaxed during sleep, the head develops a nodding motion. Astronauts can secure the sleep restraint (sleeping bag) to limit this nod. Skylab sleep restraints were similar to sleeping bags with neck holes and arm slits. Straps were on the front and back so the crew member could be tightened for a steady, snug position. The Space Shuttle missions sometimes split the crew into two shifts to enable around-the-clock science. Figure 4 shows the compartments provided for the off-duty crew's sleep. Figure 5 illustrates a nonstandard sleep bag.

4.2.2 Food

Since the first food was consumed in orbit in 1962, improvements and developments have been made and are continuing to be made in the food systems for manned space flight. The food system for the Mercury flights was limited in scope and purpose. Food was used in most cases to obtain general information on the effects of null gravity on food ingestion and digestion and to determine types of food and packaging for longer-duration space flights. Food for Mercury flights consisted of purees in aluminum tubes, coated tubes, and rehydratables.

The Gemini food system began with an all dehydrated food system that provided four meals per day per crewman. This was later changed to three meals per day and a wider variety of food was supplied. The food consisted of bite-size cubes with an expanded variety, and rehydratable foods which included beverages, pudding, soups, fruits, and vegetables. The initial Apollo food system was based on the dehydrated system used for Gemini; however, greater attention was focused on astronaut preference. The availability of hot water increased the selection of foods and enhanced the palatability. The thermostabilized food in a flexible pouch, fresh bread, canned fruit and puddings, and frozen sandwiches for

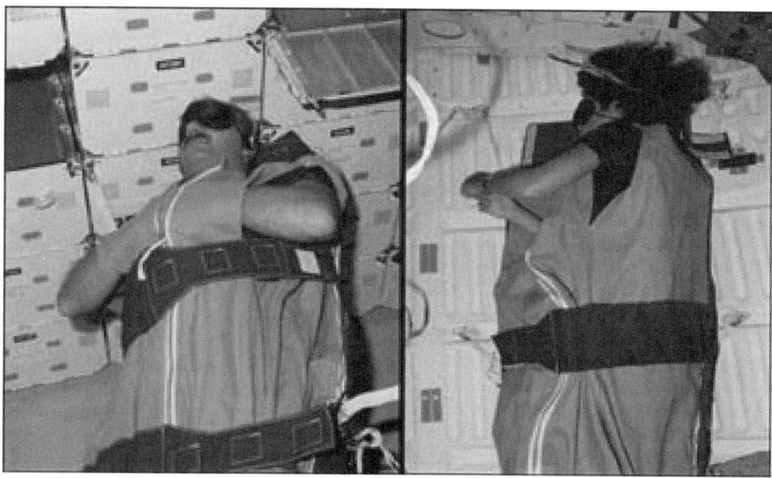

Figure 3 Two examples of using the shuttle sleep restraint while sleeping in the shuttle middeck.

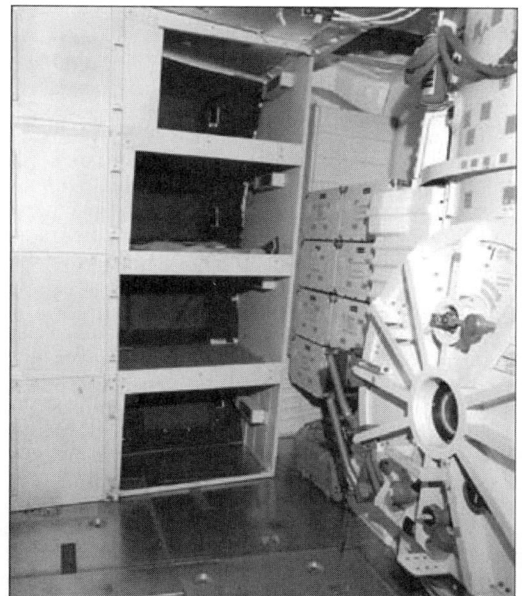

Figure 4 Shuttle sleep compartments are used when a crew is working multiple shifts during a flight.

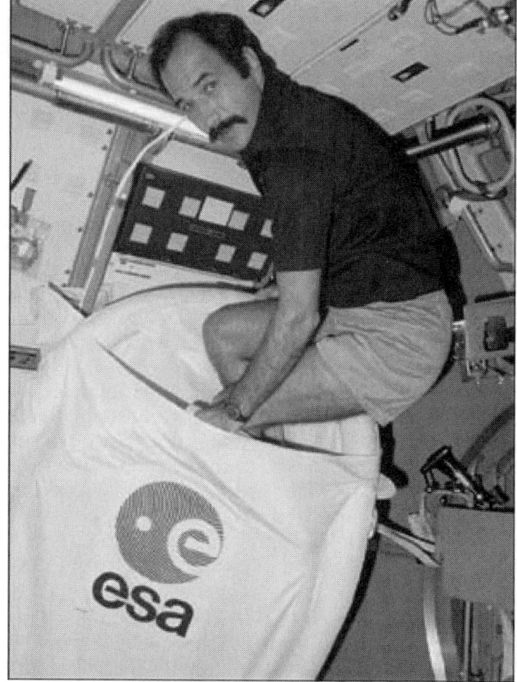

Figure 5 Unique sleep restraint used by an ESA crew member.

launch day were some of the items introduced on Apollo. Results from Apollo proved that food could be consumed from an open container using normal utensils in microgravity.

A completely new food system was designed for the Skylab program. The new system was required because (1) the food was launched with the orbiting laboratory and would be exposed to unusual environmental extremes and long-term on-orbit storage; (2) the metabolic studies on board required precise intakes of several nutrients; (3) all water had to be launched, so rehydratables offered no weight advantage; and (4) refrigerators, freezers, and food warmers would be available. To meet the long-shelf-life requirement, all Skylab foods were packaged in full-panel pull-out aluminum cans. Cabin pressure required that the aluminum cans be "over-canned" in canisters to withstand pressure variances. This resulted in the rehydratables being packaged in three containers: a plastic pouch, a can, and a canister. Beverages were packaged in a polyethylene collapsible container which expanded on reconstitution. Menus for Skylab were repeated every six days.

The Apollo–Soyuz test program (ASTP) food system maximized menu variety and incorporated the most acceptable food and packages developed for Apollo and Skylab with the mission constraints (i.e., no freezer or food warmer, limited weight and volume, and limited supply of hot water). An ASTP type of food system with some modification was used for the first four shuttle flights. A carry-on portable food warmer was used to heat foods to serving temperature since no hot water was available. The shuttle food system packages were introduced on STS-3 and STS-4 in test meals, and the fifth mission was all packaged in the shuttle container. The container is an injection-molded rigid base with a thermoformed flexible lid and a dehydration port. The package functions as a container for rehydratable foods as well as beverages. The shuttle package was designed in conjunction with the galley, which provides a meal assembly and preparation area, some food storage, hot and cold water, and a forced-air convection oven for warming food.

The design of the shuttle package significantly reduced the production process and eliminated numerous failure points. Most of the package production steps are automated or semiautomated. The shuttle food system uses a portable food tray for assembly and consumption of meals. This concept was first tested on ASTP and has been used since. At the present time, research is ongoing to look into advanced technology for future food systems for lunar and/or Mars long-term missions. This includes the growing of crops on board a space vehicle.

Historically, food weight for U.S. space food systems has been dependent on the water supply. When fuel cells are used (as on the shuttle), water as a by-product is available for use with the food. When solar panels are used, all water has to be launched, and the advantages of dehydrated foods are diminished. The weight varies from a completely dehydrated system of 1.7 lb to the Skylab system with frozen foods and a weight of 4.2 lb per person per day. Water is a necessity and must be provided either in the food or if the food is dehydrated, then added back prior to consumption. When dehydrated food is used, the original flavor is rarely attained when hydrated. The type of food as classified by the method of preservation also influences the weight of the food system, the palatability, and the preparation. Table 3 shows the support elements or activities associated with each classification of preservation. Compared to dehydrated foods, which require six support elements, most natural-form foods, such as cookies and dried fruits, only require opening the package.

The shuttle food system was designed specifically for typical shuttle missions, which are usually around seven to 10 days in length. Although the system works well for shuttle, it would be deficient in several areas for longer missions. The primary packaging material will not protect the food if stored unrefrigerated for long periods. In addition, the foods are packaged in single-service containers, which is an inefficient method storage for long-duration missions. Although the shuttle packages weigh only 1 pound per person per day, this may be prohibitive for extended missions. The shuttle single-service packages also generate a considerable amount of trash, which could pose a problem over an extended period.

Table 3 Types of Food and Associated Equipment Required

Food Type	Equipment Required
Natural form	Equipment to open package (if necessary)
Thermostabilized, irradiated	Method of heating Utensils Serving tray or dish Cleanup equipment
Frozen	Freezers Method of heating Utensils Serving tray or dish Cleanup equipment
Dehydrated (freeze dried, spray dried, etc.)	Hot and cold water Method to open package Method of heating Utensils Serving tray or dish Cleanup equipment

Source: NASA (1983).

The Skylab food system was packaged and designed for long-duration missions; however, little of the technology would be transferable, due to the uniqueness of the metabolic studies that directed the food system design. Additionally, the Skylab food system was overpackaged and would be problematic on missions that are weight and volume critical.

The International Space Station menu composition is an extension of the menu system established for the space Shuttle/Mir Phase 1 program, which consisted of 50% Russian and 50% American foods. For Mir, meals A and C were provided by Russia, and the United States supplied meals B and D. Meal D was not considered a meal, but was a snack that could be eaten anytime during the day. Experience on Mir indicated that having Russian food for meals A and C resulted in little or no U.S. breakfast items. Conversely, cosmonauts did not get their usual snack items since the United States supplied these. A unique system that alternates these combinations every other day is used on ISS. Now the United States provides breakfast and lunch on one day and dinner and the snack the next day, with Russia providing the other meals on the same rotation. Meal D is now called a snack and can be eaten anytime during the day. The menu format for shuttle/Mir was a six-day cycle, whereas ISS is currently utilizing an eight-day cycle with plans to expand to a 10 to 12-day menu cycle in the future. The percentage of thermostabilized foods in the U.S. menu has constantly increased for the ISS food program. This is due to a higher preference for these items by crew members (Kloeris and Bourland, 2003).

The next possible step after ISS is long-duration manned space flights beyond low Earth orbit. The duration of these missions may be as long as 2.5 years and will probably include a stay on a lunar or planetary surface. The primary goal of the food system in these long-duration exploratory missions is to provide the crew with a palatable, nutritious, and safe food system and minimize volume, mass, and waste. The paramount importance of the food system in a long-duration manned exploration mission should not be underestimated. During long-duration space missions, several physiological effects may occur, including weight loss, fluid shifts, dehydration, constipation, electrolyte imbalance, calcium loss, potassium loss, decreased red blood cell mass, and space motion sickness. The menu will provide the crew with changes in the nutrient levels that may be required due to the longer-duration mission.

The acceptability of the food system is of much greater importance, due to the longer-mission durations and the partial energy intake that is often observed in space flight. The decreased energy intake might significantly compromise the survival of the crew. The food system will initially emphasize technologies for space-vehicle application (ISS and shuttle) and then slowly focus on technologies toward tasks that support exploration. As the food system is developed, it must continually integrate and determine the impact on the air recovery, water recovery, biomass production, solid waste management, and thermal control systems. The needs and constraints of the other life-support elements must be balanced with the food system to provide a well-integrated life-support system for long-duration space missions. The food system will need to consider the availability of power, volume, and water availability as the entire food system is developed (Perchonok and Bourland, 2002).

4.2.3 Personal Hygiene

Managing personal waste and cleaning the skin and hair are problematic because of the lack of gravity and the cost of lifting water to orbit. Except for Skylab, dedicated volumes for various activities have been very limited. Early bodily waste management systems can be described succinctly as "baggies." Since Skylab, there have been a variety of suction-based toilets for collecting fecal matter and urine. The principal systems for personal hygiene for each major spacecraft are described below.

Skylab Personal hygiene for the Skylab crew members was supported in the waste management compartment (WMC). The WMC included a fecal–urine collector, a hand washer, stowage for personal hygiene items and kits, and a drying station. There was also a shower aboard the Skylab. Pressurized water flow combined with a suction device to collect the water caused the water to flow "down." It was considered a pleasant experience but was very time consuming, about 45 minutes from start to finish. This included cleanup activity.

Mir The Mir personal hygiene subsystem consisted of toilets for body waste management, hand washing units, a shower, and personal hygiene kits. For the last two years the shower was on board, and it was used as an air shower (sauna). It was removed to make way for other required equipment.

Shuttle For washing, the shuttle crew is provided with a personal hygiene system hose located in the waste collection system (WCS) compartment. Water is squirted onto a washcloth using the hose. Some crew prefer to use the hygiene port provided at the galley because it provides hot water. The hose for the galley hygiene port is long enough to be extended to the WCS for cleansing and grooming. The crew is provided with no-rinse body bath and no-rinse shampoo.

ISS The Russian segment is generally the same as for Mir, without a shower. In the U.S. segment the personal hygiene subsystem provides a WMC. Wet wipes and towels are used from the Russian segment. Occasionally, ISS crew members have rigged up a bathing device for their use. There are differing opinions on the results (Mohanty, 2001).

4.2.4 Exercise

Exercise regimens prescribed for space missions have required gradually longer and more frequent periods of exercise, particularly as the length of mission has increased. On the first prolonged (18-day) Soviet manned flight, *Soyuz 9*, physical exercises were performed by the cosmonauts for two one-hour periods each day. In subsequent 24-day flights, 2.5 hours of exercise per day was employed, including walking and running on a treadmill. By 1975, the standard program involved three exercise periods per day, with a variety of equipment, for a total of 2.5 hours, with the selection of exercises on the fourth day being optional. Over the three missions of the Skylab program a similar increase in exercise quantity was imposed, although the total amounts were less than those used by the Soviets. On the last manned Skylab mission, a treadmill was provided, which allowed more vigorous exercise.

Throughout the Skylab missions, successive improvements were seen in postflight leg strength and volume changes, orthostatic tolerance and recovery time, and cardiac output and stroke volume, even though each mission lasted four weeks longer than the last. *Skylab 4*, was an 84-day mission. Results of exercise on Soviet missions have shown a similar pattern of reduced physiological deconditioning in response to more strenuous exercise programs (NASA, 1982).

The exercise requirement for ISS is 2.5 hours daily with 1.0 hours for aerobic exercise (cycle ergometry or treadmill locomotion) and 1.5 hours for resistive exercise conditioning. Each time segment includes 15 minutes for setup and 15 minutes for set-down of equipment. Usually, astronauts exercise six days a week, with day 7 as active rest (the astronauts can exercise if they want to). They usually start exercise conditioning after space motion sickness has resolved and all transfer of payload has occurred. The Russians do not start exercise countermeasures until flight day 30. The shuttle requirements are different and depend on mission length and crew member roles. They apply only to use of the cycle ergometer.

4.2.5 Recreation

With any space vehicle design for a long-term mission, an area for recreation should be designated to provide for social interaction, Earth viewing, games, videotape viewing, music, and active and passive participatory activities. A quiet area should be provided for a crew member to read, listen to music, and write.

4.3 Vehicle Maintenance

With the exception of Skylab and ISS, in-flight maintenance provisions and planning on U.S. space programs have not been supported by definitive program requirements. The Skylab mission acknowledged a substantive role for maintenance to achieve mission objectives. The wisdom of this decision was validated by the major repair and maintenance tasks required during the brief lifetime of the program. The shuttle program was to have no in-flight maintenance, with all maintenance tasks planned to be done on the ground. Over the life of the program this has changed, due to the necessity of preventive maintenance, even on the short missions, and unanticipated problems (Mount, 1989).

On-orbit maintenance was recognized as an essential consideration within the International Space Station program (NASA, 2004c). A three-tiered maintenance concept was adopted that is similar to that employed by military organizations. The primary mode of on-orbit maintenance was designated as organizational maintenance and consisted primarily of removal and replacement of orbital replaceable units (ORUs) (comparable to line replaceable units in military applications). This was supplemented by in situ maintenance for systems that did not lend themselves to the modular ORU design approach, such as utility lines and secondary structure. The option was retained for intermediate-level maintenance, which would consist of on-orbit repair of ORUs. Intermediate-level maintenance has been employed to a limited extent in applications such as replacement of circuit cards within avionics ORUs. Crew member training for maintenance has focused on the development of general skills and on types of maintenance tasks. However, extensive training on highly specific actions is done in some specific instances.

Future missions will be challenged by their extended duration, limited or no resupply opportunities once the mission has begun, and extended round-trip communication times (Watson et al., 2003). These factors will require such missions to be almost entirely self-sufficient. An additional constraint will be the need to carefully control and minimize the mass and volume of equipment and supplies used to support maintenance activities. It is expected that maintenance will be performed at the level of piece parts, so that the required replacement parts will be as small as possible. However, performing maintenance at this level carries significant implications from multiple perspectives.

First, hardware must be designed to enable crew members to perform the required maintenance. Not only must the equipment be accessible but it must also be possible for units to be disassembled as necessary to enable piece-part replacement. Additionally, commonality and standardization of piece parts must be imposed to obtain mass and volume benefits. If not, the number of unique piece-parts could be so great as to negate any potential benefit. This maintenance concept will also require more extensive diagnostic capabilities than have been used heretofore in space. Every effort should be made to incorporate these capabilities within the systems themselves, to minimize the amount of stand-alone test equipment that is required. Preparation of all potential maintenance procedures in advance will probably be prohibitively expensive, so means must be available to provide crew members with necessary information and guidance when needed. An attractive concept would be capable of automatically generating needed procedures based on input from diagnostic systems and from hardware design information stored onboard. Finally, maintenance at this level will require the ability to perform quality assurance tests (Watson, 2004).

Future missions will probably require operations in multiple gravitational environments, including the microgravity environment of Earth-orbit or in-space transit, lunar gravity (approximately $0.17g$), and Martian gravity (approximately $0.38g$). Design for maintenance must take these environments into account. For example, a microgravity environment offers three-dimensional freedom of motion, facilitating access to all areas within a spacecraft volume. However, a microgravity environment introduces significant challenges from the standpoint of reacting forces that must typically be applied during maintenance tasks. Fractional-g environments will restrict mobility and access to some degree (e.g., restricting access to hardware in overhead locations) but will facilitate the application of forces by crew members. Another subtle advantage to working in fractional-g environments is that unrestrained parts and tools remain where placed and do not tend to float away and become lost(Watson, 2004).

With longer missions maintenance must be planned and all contingencies must be anticipated. Simple maintenance tasks take on great complexities when in microgravity. What might be considered a simple task on Earth, such as using a slot-head screwdriver, could be impossible in space. Automation is being developed to save crew time and increase productivity, but we need to know all the ramifications when the automation (and robotics) breaks down (Mount, 1989). As automated capabilities become increasingly prominent in maintenance operations, the potential for their failure and appropriate fallback positions must be considered. Tasks and hardware for which robotic intervention is planned should retain manual intervention as a backup capability. Designs should not preclude manual troubleshooting even if embedded diagnostics are planned. Interchangeability of hardware within and among spacecraft should be a key design objective (Watson, 2004).

Considerations to be given for support of maintenance in space fall into four categories (Mount, 1989):

1. Crew provisions
   - Crew interface at appropriate sites
   - Personnel and equipment restraints
   - Access (both physical and visual)
   - Work envelope (volume)
   - Tools and task support equipment
   - Procedural and reference data
   - Suits and protective equipment
2. Hardware
   - Design for maintainability
   - Redundancy in design
   - Materials
   - Fasteners
   - Connectors
   - Mounts
   - Structural interfaces
   - Sensors/instrumentation
   - Piece parts/orbital replacement units (ORUs)
3. Software
   - Architecture (subelement compatibility, maintainability, reconfigurability)
   - Automation, robotics, and artificial intelligence
   - Fault detection, isolation, and recovery support
   - Integrated computer-assisted training support
   - Inventory control and management
4. Supporting disciplines and processes
   - Safety, reliability, maintainability, and quality assurance
   - Configuration management (control, documentation, accounting)

4.4 Restraints

Launch and reentry require significant structural strength; loads of up to 5g are experienced in nominal conditions. But once in orbit, the microgravity environment enables objects to be held in place with very little force; hook and loop fasteners dot the surfaces.

Figure 6 Operating the remote manipulator system requires a stable restraint carefully adjusted.

On the other hand, some force must be provided to hold anything in place. Restraints are needed for both personnel and equipment in microgravity. The most common restraint for crew members is a foot restraint. In a location where a person will be working for extended periods of time, platforms can be used that tilt to accommodate a neutral posture, with the feet angled down, and with height adjustments.

Tasks of various durations requiring various degrees of force or dexterity require different types of restraints. Short, easy tasks can often be performed with toes stuck under a handle or one hand on a handhold. Tasks such as attaching a module to the ISS using the remote manipulator system, which take many hours and a high degree of hand–eye coordination, require a restraint such as that shown in Figure 6. This restraint provides support for the feet and thighs. Another example of restraints is shown in Figure 7, illustrating use of existing hardware for a temporary restraint.

5 SLEEP AND CIRCADIAN RHYTHM

5.1 Sleep Shifting and Light

Circadian and sleep components, two physiological processes, interact in a dynamic manner to regulate changes in alertness, performance, and timing of sleep. Light can aid in shifting circadian rhythms to an earlier or later time within the biological day. Also,

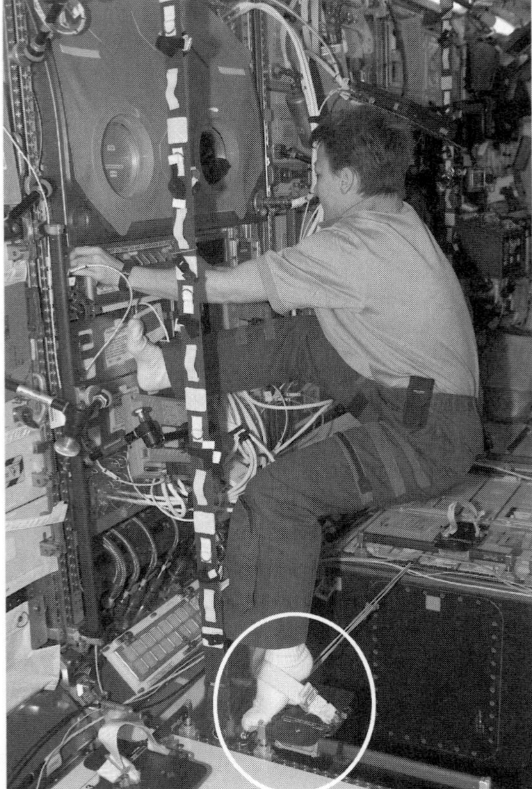

Figure 7 Brief activities can be performed with simpler restraints.

use of bright light during nighttime can result in significant improvement in performance and alertness levels (Campbell and Dawson, 1990). NASA currently uses light treatment to help crew members adapt their circadian system prior to missions, allowing the astronauts to be physiologically alert when critical tasks are required (Czeisler, 1999). The timed use of bright light to facilitate circadian phase shifts was effective in the STS-35 mission, the first mission requiring both dual shifts and a night launch. Subjective reports indicated that crew members were able to obtain better-quality sleep during the day and remain more alert during the night after using bright light exposure to facilitate their schedule inversion prior to the launch dates (Czeisler et al., 1999). Astronauts in space are exposed to variable light levels, due to the non-24-hour orbital cycle (day/night) of space operations, such as the 90-minute orbital cycle of the shuttle. Additionally, light levels in the space environment can be variable. Field data have shown that light levels aboard spacecraft can be as low as 10 lux during the highest-activity portions of the day and as high as 79,433 lux on the flight deck (Dijk et al., 2001). The Soviets recommended 400 to 500 lux of full-spectrum light for work on spacecraft, and results demonstrated an improvement in performance when the location of lights on *Salyut-7* was changed to maximize lighting (Bluth, 1984).

Around-the-clock operational tasks often require splitting crews into two separate shifts, which requires that half the crew invert their sleep–wake cycles. A procedure called *slam shifting*, which involves abrupt shifts of up to 12 hours, is now used to align the sleep–wake schedules of shuttle and ISS crews upon docking. Staggered sleep schedules on an eight-day mission did not work, since the crew tended to retain ground-based work–rest cycles and the schedules resulted in increased fatigue and irritability. On a one-year flight, where sleep times for docking operations were shifted by 4.5 to 5.0 hours a total of 14 times, asthenia, end-of-day fatigue, and sleep disruptions were documented (Grigor'yev et al., 1990).

Current astronaut crew scheduling guidelines allow for astronauts' schedules to be lengthened by no more than two hours (phase delay) and shortened by no more than 30 minutes (phase advance) within a given day (NASA, 1992). Schedules can be lengthened only if there is an operational requirement. For example, if the shuttle is going to dock with ISS during a time that the ISS crew is scheduled to be sleeping, operations would require the ISS crew to shift to a new schedule in the days preceding, in order to be awake and alert for the docking (Mallis and DeRoshia, 2003).

5.2 Mars Day Circadian Entrainment

With NASA's continuing support of a manned mission to Mars, the effects of a Mars light–dark cycle must be investigated to determine a person's ability to adapt to a Mars cycle and its impact on physiological alertness. The Martian day, otherwise known as a sol, is about 39 minutes longer than an Earth day (a sol period is 24.6 hours). Although this period length is well within the circadian range of entrainment according to previous studies conducted in relatively bright light (23 to 27 hours) (Aschoff and Wever, 1981), preliminary laboratory results have suggested that in dim-light conditions, such as found indoors, humans cannot reliably entrain to a 24.6-hour Mars sol. People differ as to their circadian rhythm, and the 25% of the population who have periods shorter than 24 hours will have the greatest challenges acclimatizing to a Mars sol (Mallis and DeRoshia, 2003).

6 ASTRONAUT SELECTION

Today's astronauts come from an international pool of candidates, including the European Space Agency, the Russian program, and the U.S. program. Each country selects its own astronaut candidates according to its own criteria. Here we discuss the U.S. experience.

Planning the first astronaut selection in 1958, Allen O. Gamble, one of the psychologists on the medical team, realized that there needed to be some job or task analyses—a difficult challenge, as no one had ever flown in space before. He listed the duties of the first astronauts as the following (Link, 1965):

1. To survive; that is, to demonstrate the ability of humans to fly in space and return safely

2. To perform; that is, to demonstrate the human capacity to act usefully under conditions of space flight
3. To serve as backup for automatic controls and instrumentation; that is, to add reliability to the system
4. To serve as scientific observers; that is, to go beyond what instruments and satellites can observe and report
5. To serve as engineering observers and, acting as true test pilots, to improve the flight system and its components

Since the late 1950s there has always been some system of psychological selection, although there have been many changes in criteria and procedure. Originally, psychological assessment was extensive, requiring 30 hours of psychological testing, plus interviews and evaluation by a team made up of a psychiatrist, an industrial–organizational psychologist, and management. In the 1960s the Lovelace clinic tested several women; 25 female pilots completed the same psychological evaluations as those given the males chosen for the Mercury project. Of these, 13 of them enrolled in an unofficial astronaut training program; none were declared official astronaut candidates.

From 1958 through 1969, astronaut selection occurred at least four more times. Since applicants already had extensive, often hazardous, flight experience, criteria emphasized emotional stability, motivation and energy, self-concept, and quality of interpersonal relationships. Psychological testing now required only 6.5 hours, and the clinical evaluation was primarily psychiatric rather than psychological. This shift toward clinical content paralleled a shift away from research, reducing the data available for systematic scientific selection into astronaut selection. By 1983, Jones and Annes (1983) could write: "Presently, no psychological testing is done." Instead, the evaluation consisted of two consulting psychiatrists who separately interviewed each candidate for two hours. This screening, although completed by expert aviation psychiatrists, did not have specific and objective criteria by which to rate each candidate.

After a hiatus of nine years, in 1978, astronaut selection began again for the Space Shuttle program, including nonaviators, scientists, and women. It was not until the 1980s that NASA hired its own psychiatrist and soon thereafter, a psychologist, to work in the operational arena. From 1988 through 1990, a newly established in-house group met to improve the selection process. This Working Group on Psychiatric and Psychological Selection of Astronauts distinguished between the roles of psychology and psychiatry and rewrote NASA psychiatric standards to include disqualifying psychiatric disorders based on the then current American Psychiatric Association's *Diagnostic and Statistical Manual*.

Holland (1999) notes that by 1989, clinical testing had returned, giving some objective data to be used by the psychiatrists, but it was still a medical model. By the 1994–1995 selection cycle, nonmedical evaluations based on industrial–organizational principles and techniques were added to the clinical and medical models. Based on these organizational studies, Galarza and Holland (1999) have listed the critical psychological proficiencies needed for space flight: "mental/emotional stability, ability to perform under stressful conditions, group living skills, teamwork skills, ability to cope with prolonged family separations, motivation, judgment/decision making, conscientiousness, communication skills, leadership capability." Currently, astronaut candidates complete an extensive battery of tests and undergo several hours of interview to determine their suitability. Interviewers undergo training and extensive review of their work.

7 CONCLUSIONS

After 40 years of human space flight we have gathered great quantities of information dealing with the crew and their interfaces. With a new mission in front of us, going beyond low Earth orbit, we must learn more about the challenges of long-term missions. We must gather much more data from ISS missions. Additionally, we must take advantage of analogs that are consistent with the perceived challenges of long-term missions and glean what we can to augment our knowledge base.

ACKNOWLEDGMENTS

The authors wish to acknowledge the substantial contributions provided by Melissa Mallis of Ames Research Center to Section 5, and of Edna Fiedler of the National Space Biomedical Research Institute to Section 6.

REFERENCES

Allen, C. S., and Goodman, J. R. (2003), "Preparing for Flight: The Process of Assessing the ISS Acoustic Environment," in *Proceedings of NOISE-CON 2003*, Cleveland, OH.

Aschoff, J., and Wever, R. (1981), "The Circadian System of Man," in *Handbook of Behavioral Neurobiology*, Vol. 4, *Biological Rhythms*, Plenum Press, New York, pp. 311–331.

Bluth, B. J. (1984), Human Systems Interfaces for Space Stations," in *AIAA/NASA Space Systems Technology Conference Proceedings*, Costa Mesa, CA, National Aeronautics and Space Administration, Washington, DC, pp. 40–49.

Bond, R. L. (1977), *Application of Skylab Workday Analysis to Future Programs*, JSC 12856, Johnson Space Center, Houston, TX.

Bowen, C. (2004), personal communication.

Campbell, S. S., and Dawson, D. (1990), "Enhancement of Nighttime Alertness and Performance with Bright Ambient Light," *Physiological Behavior*, No. 48, pp. 317–320.

Czeisler, C. A. (1999), *Circadian Entrainment, Sleep–Wake Regulation and Neurobehavioral Performance During Extended Duration Space Flight*, Report 20000029467, National Aeronautics and Space Administration Center for AeroSpace Information, Moffett Field, CA.

Czeisler, C. A., Dijk, D., and Neri, D. F. (1999), "Ambient Light Intensity, Actigraphy, Sleep and Respiration, Circadian Temperature and Melatonin Rhythms and Daytime Performance of Crew Members During Space Flight on STS-90 and STS-95 Missions" [abstract], in *Proceedings of the First Biennial Space Biomedical Investigators' Workshop*, League City, TX, National Aeronautics and Space Administration, Houston, TX.

Dijk, D. J., Neri, D. F., and Wyatt, J. K. (2001), "Sleep, Performance, Circadian Rhythms, and Light–Dark Cycles During Two Space Shuttle Flights," *American Journal of Physiology: Regulatory, Integrative, and Comparative Physiology*, Vol. 281, pp. R1647–R1664.

Galarza, L., and Holland, A. W. (1999), "Critical Astronaut Proficiencies Required for Long-Duration Space Flight," in *Proceedings of the 29th International Conference on Environmental Systems*, ICES 1999CD Paper 1999-01-2096, Society of Automotive Engineers, Denver, CO.

Goodman, J. R. (1992), "Space Shuttle Crew Compartment Debris/Contamination," ICES Paper 921345, in *Proceedings of the 22nd International Conference on Environmental Systems*, Seattle, WA.

Goodman, J. R. (2003), "International Space Station Acoustics," in *Proceedings of NOISE-CON 2003*, Cleveland, OH.

Grigor'yev, A. I., Bugrov, S. A., and Bogomolov, V. V. (1990), "Review of the Major Medical Results of the 1-Year Flight on Space Station 'Mir,'" *Kosmichesleaya Biologiya i Aviakosmichesleaya Meditsina*, Vol. 24, No. 5, pp. 3–10.

Holland, A. W. (1999), "Psychology of Spaceflight," in *Human Spaceflight: Mission Analysis and Design*, W. J. Larson and L. K. Pranke, Eds., McGraw-Hill, New York.

Jahweed, M. M. (1994), "Muscle Structure and Function," in *Space Physiology and Medicine*, 3rd ed., A. E. Nicogossian, C. L. Huntoon, and S. L. Pool, Eds., Lea and Febiger, Philadelphia, PA.

Jones, D. R., and Annes, C. A. (1983), "The Evolution and Present Status of Mental Health Standards for Selection of USAF Candidates for Space Missions," *Aviation, Space, and Environmental Medicine*, Vol. 54, No. 8, pp. 730–734.

Kloeris, V. L., and Bourland, C. T. (2003), "The Food System for the International Space Station: The First Five Increments," in *Proceedings of the 33rd International Conference on Environmental Systems*, Society of Automotive Engineers, Denver, Co.

Link, M. M. (1965), *Space Medicine in Project Mercury*, Special Publication 4003, National Aeronautics and Space Administration, Washington, DC.

Maida, J. (2002), "Graphics Research and Analysis Facility (GRAF) Brochure," Johnson Space Center, Houston, TX.

Mallis, M. M., and DeRoshia, C. S. (2003), "Circadian Rhythms, Sleep, and Performance in Space: A Focus of NASA Research," in *Special Supplement of Aviation, Space and Environmental Medicine*, Proceedings of the New Directions in Behavioral Health: Integrating Research and Application Conference, December 2–3.

Mohanty, S. (2001), *Design Concepts for Zero-G Whole Body Cleansing on ISS Alpha*, Part II, *Individual Design Project*, NASA CR-2001-208931, National Aeronautics and Space Administration, Washington, DC.

Mount, F. E. (1989), "Session Summary: Maintenance," in *Proceedings of the Manned System: A Human Factors Symposium and Workshop*, American Astronautical Society, Houston, TX.

Mount, F. E. (2002), "Habitability: An Evaluation," in *Isolation: NASA Experiments in Closed-Environment Living*, H. W. Lane, R. L. Sauer, and D. L. Feeback, Eds., Univelt, San Diego, CA, pp. 87–116.

Mount, F. E., Adam, S., McKay, T., Whitmore, M., Merced-Moore, D., Holden, T., Wheelwright, C., Koros, A., O'Neal, M., Toole, J., and Wolf, S. (1994), *Human Factors Assessments of the STS-57 SpaceHab 1 Mission*, NASA TM-104802, National Aeronautics and Space Administration, Washington, DC.

NASA (1982), *Space Physiology and Medicine*, NASA SP-447, National Aeronautics and Space Administration, Washington, DC.

NASA (1983), *Crew Interface Panel Space Station Habitability Requirements Document*, JSC 19517, National Aeronautics and Space Administration, Johnson Space Center, Houston, TX.

NASA (1992), *Appendix K of the Space Shuttle Crew Procedures Management Plan*, Mission Operations Directorate, Operations Division, National Aeronautics and Space Administration, Johnson Space Center, Houston, TX.

NASA (1995), *Man–Systems Integration Standards*, NASA STD-3000, National Aeronautics and Space Administration, Washington, DC.

NASA (1996), *Risk Mitigation Experiment 1305: Mir Audible Noise Measurement*, JSC 27431, National Aeronautics and Space Administration, Johnson Space Center, Houston, TX.

NASA, (2003), *Astronaut Fact Book*, NP-2003-07-008JSC, National Aeronautics and Space Administration, Johnson Space Center, Houston, TX.

NASA (2004a), "National Aeronautics and Space Administration, Research Overview," http://lsda.jsc.nasa.gov/scripts/cf/glass_.cfm?term=Life_Support_Systems.

NASA (2004b), "National Aeronautics and Space Administration, History," http//spaceflight.nasa.gov/history/shuttle-mir/h-tl-text-main.htm.

NASA, (2004c), *SSP 41000AR, ISS System Specification*, National Aeronautics and Space Administration, Johnson Space Center, Houston, TX.

Perchonok, M., and Bourland, C. (2002), *NASA Food Systems: Past, Present, and Future*, National Space Biomedical Research Institute, Houston, TX.

Schneider, V. S., LeBlanc, A. D., and Taggard, L. C. (1994), "Bone and Mineral Metabolism," In *Space Physiology and Medicine*, 3rd ed., A. E. Nicogossian, C. L. Huntoon, and S. L. Pool, Eds., Lea and Febiger, Philadelphia, PA.

Watson, J. K. (2004), personal communication.

Watson, J. K., Ivins, M. S., Robbins, W. W., Van Cise, E. A., et al. (2003), *Supportability Concepts for Long-Duration Human Exploration Missions*, AIAA Paper 2003–6240, AIAA Space 2003 Conference, Houston, TX.

Wheelwright, C. D., Lengel, R. D., and Koros, A. S. (1994), "Noise, Vibration and Illumination," in *Space Biology and Medicine*, Vol. II, *Life Support and Habitability*, American Institute of Aeronautics and Astronautics, Washington, DC.

CHAPTER 35

CHEMICAL, DUST, BIOLOGICAL, AND ELECTROMAGNETIC RADIATION HAZARDS

Danuta Koradecka, Małgorzata Pośniak, Elżbieta Jankowska, Jolanta Skowroń, and Jolanta Karpowicz
Central Institute for Labour Protection–National Research Institute
Warsaw, Poland

1 HAZARDOUS CHEMICAL AGENTS IN THE WORKING ENVIRONMENT

Chemical agents are one of most common hazardous agents in the working environment. Occupational exposure to these agents exists in every workplace, in particular in the chemical industry but also in other branches of the economy: in offices and hospitals and in agriculture. These agents are dangerous both to human health and to the environment.

Human reaction to chemical compounds depends primarily on the dose absorbed, but also on many other agents: physical and chemical properties; methods of entry; duration of exposure; genetic, immunologic, and endocrinologic factors; age; personal habits; medication; previous exposures; and external parameters (e.g., temperature, humidity, airflow). The effects of exposure to chemical compounds may be irritation or sensitizing of the skin or mucous membranes. Exposure may also cause various changes in such human systems as the central nervous system, the liver, the digestive tract, and the kidneys, and the intensity may be acute or chronic. Long-term exposure effects include cancer and mutagenic diseases.

According to the Chemical Abstracts Service, which tracks substances cited in the scientific literature, there are approximately 16 million chemicals in the world. About 100,000 of them are used commonly in the working environment: from the chemical industry to hairdressing establishments to construction sites and offices to agriculture. A wide variety of chemical agents are also used in hospitals, as anesthetic, cytostatic, and other agents. These days the use of chemical agents is very popular, not only in those types of workplaces but also in domestic, educational, and recreational activities (e.g., in cleaning products, adhesives, and cosmetics). As a result, hazards arising from the use of chemical agents may exist in many workplaces and services, as well as during such daily activities as cooking, washing, cleaning, and gardening.

The toxicological properties of two-thirds of the chemicals that are in common use have not been investigated fully. Of all the registered chemicals with known toxicological effects, about 4000 are classified as carcinogenic and 3000 as allergenic agents (http://osha.eu.int/ew2003). Exposure to chemical agents present in the working environment may be the cause of occupational risk either as a result of direct contact with the human body or as a result of producing some form of energy (e.g., through chemical reactions) which can have hazardous effects on human health. For a chemical agent to harm humans directly, its molecules must come into contact with some part of the body. The effects of such contact may be short term—burning of the skin, irritation of the respiratory tract, or chemical asphyxia—or long term—cancer or asbestosis.

Many chemical agents are combustible and thus can cause fires or explosions. Fires and explosions at the workplace can pose a serious threat to workers and to material assets, especially if appropriate emergency measures are not taken, and they almost always badly damage a company's assets.

According to EU directives (EC, 1989, 1998) and regional laws, employers are legally obligated to provide workers with information about occupational health and safety risks. Occupational risk assessment is the overall process of estimating the magnitude of risk and deciding whether or not that risk is either tolerable or acceptable according to EN standards. The process of risk assessment is a systematic examination of all aspects of work to consider what can cause injury or harm, whether hazards can be eliminated, and if not, what preventive or protective measures are or should be in place to control the risks.

1.1 Definitions

- *Chemical agent*: any chemical element or compound, on its own or admixed, as it occurs in the natural state or as produced, used, or released, including release as waste, by any work activity, whether or not produced intentionally, and whether or not placed on the market (EC, 1998).
- *Dangerous substance*: any liquid, gas, or solid that poses a toxicological risk to employees' health, as well as one with hazardous properties arising from its physical and chemical nature.
- *Hazardous chemical agent*: any chemical agent that meets the criteria for classification as explosive, oxidizing, extremely flammable, highly flammable, flammable, very toxic, toxic, harmful, corrosive, irritant, sensitizing, carcinogenic, mutagenic, toxic to reproduction, and dangerous for the environment.
- *Hazard*: the intrinsic capacity of a chemical agent to cause harm (EC, 1998).
- *Risk*: the likelihood that the potential for harm will be attained under the conditions of use and/or exposure (EC, 1998).
- *Exposure to chemical agents*: any work situation in which a chemical agent is present and a worker comes into contact with the agent, normally via inhalation or through the skin (CEN, 1995a).
- *Threshold limit value–time-weighted average (TLV-TWA)*: the time-weighted average concentration for a conventional eight-hour workday and a 40-hour workweek, to which it is believed that nearly all workers may be repeatedly exposed, day after day for a working lifetime, without adverse effect (ACGIH, 2005a).
- *Threshold limit value–short-term exposure limit (TLV-STEL)*: a 15-minute TWA exposure that should not be exceeded at any time during a workday even if the eight-hour TWA is within the TLV-TWA. Exposures above the TLV-TWA up to the TLV-STEL should be less than 15 minutes, should occur less than four times per day, and there should be at least 60 minutes between successive exposures in this range. An averaging period other than 15 minutes may be recommended when this is warranted by observed biological effects (ACGIH, 2005a).
- *Threshold limit value–ceiling (TLV-C)*: the concentration that should not be exceeded during any part of the working exposure (ACGIH, 2005a).

1.2 Methods for Measuring Concentration of Hazardous Chemical Agents in Workplace Air

The analytical procedures for investigating chemical air pollutants in the workplace should ensure that results are obtained with the required level of reliability. The requirements of ISO/IEC 17025 (ISO, 2005) should be implemented by all occupational hygiene laboratories, especially in the range of aspects connected with a quality system, using calibrated equipment and validated methods of measurement. The staff of each laboratory has to possess relevant experience and training for carrying out analysis of chemical agents in workplace air. Participation in interlaboratory comparisons or proficiency-testing programs is the best procedure for monitoring the competence of OHS laboratories.

Methods for determining concentration of chemical compounds in workplace air which normally have been used for occupational exposure assessment contain two separate stages: air sampling and analysis (Cohen and Hering, 1995). The purpose of air sampling is to characterize air quality for occupational exposure assessment. Sufficient samples must be collected for reliable estimation of chemical compounds concentration equal to at least one-tenth the TLV value. For sampling gases and vapors, instantaneous or integrated samples may be collected in the breathing zone of a worker. Passive dosimeters or battery-operated personal pumps connected with absorbers or absorbers for gas and vapors and with filters for particular matter sampling are especially useful for evaluating occupational exposure to chemical agents.

Sampling must be carried out following the indications for the method of measurement selected. Transport and storage or safekeeping of samples are very important parts of the air sampling procedure. Special care must be taken to ensure appropriate conditions for samples. The next step, analysis in the laboratory, includes the process of sample preparation and quantification using suitable analytical techniques: gas chromatography, high-performance liquid or thin-layer chromatography, or various kinds of spectrophotometry (e.g., UV, IR, AAS).

General requirements for procedures for the measurement of chemical agents in workplace air have been included in standard EN 482 (CEN, 1994). Validation, the most important requirement, provides information on the functional characteristics of the method and ensures a high degree of confidence in both the method and the results obtained. The validating requirements include, among others, a limit value; maximum accuracy and precision values, which must be achieved under laboratory conditions similar to real conditions; and possible environmental influences. The validation of a measurement procedure is established as a result of systematic laboratory investigations confirming that the characteristics of the procedure comply with the specifications relating to the intended use of the analytical results. The overall uncertainty in these types of measurements must be $\leq$ 30% for the range 0.5 to 2 TLV and $\leq$ 50% for the range 0.1 to 0.5 TLV, as indicated in EN 482.

According to the type of determining chemical agent, the type of air sampling, and the equipment, various requirements have to be considered. Personal sampling devices for all procedures with active sampling systems should comply with requirements of standards EN 1232 (CEN, 1997b) (pumps for flows to 5 L/min) and EN 12919 (CEN, 1999) (pumps for flows in excess of 5 L/min). Methods for determining gases and vapors adsorbed on various solid sorbents will also comply with the requirements in EN 1076 (CEN, 1997a), or in the case of passive samplers, with ISO 16107 (ISO, 1999) and EN 838 (CEN, 1995b). Methods for determining chemical agents present in workplace air in the shape of particles and which require size selectors during sampling must comply with standards EN 481 (CEN, 1993) and EN 13205 (CEN, 2001a). Chemical agents present as mixtures of particles and vapors in the atmosphere should be determined using methods that comply with the requirements recommended by standard EN 13936 (CEN, 2001b).

A suitable measurement procedure of particular chemical agents in workplace air should be applied. They are prepared in a standard form, with known validation protocols and accessible validation reports. Methods developed and published by competent institutions in the range of air purity protection in the workplace are preferred: for example, those issued by the National Institute of Occupational Safety and Health (NIOSH, 1994), Occupational Safety and Health Administration (OSHA, 2001), and Health and Safety Executive (HSE, n.d.).

Determination of chemical agents in workplace air may also be carried out in accordance to the recommended procedures specified in international and European standards. For example, the analytical procedure for measuring metals and metalloids should comply with the requirements of standards ISO 15202-1 (ISO, 2000), ISO 15202-2 (ISO, 2001a), and EN 13890 (CEN, 2002). Guidance for air sampling and analysis of volatile organic compounds, including hydrocarbons, halogenated hydrocarbons, esters, glycol ethers, ketenes, and alcohols, is presented in ISO 16200-1 (ISO, 2001b) and ISO 16200-2 (ISO, 2001c).

Frequently, more than one published method is available for specific chemical agents. Differences in those methods may exist due to the various analytical techniques used, differences in the limit values established in different countries, and changes in limit values over time. Guidance for assessing exposure by inhalation of chemical agents for comparison with limit values and the measurement strategy are described in standard EN 689 (CEN, 1995a).

1.3 Assessment of Risks Posed by Hazardous Chemical Agents in the Workplace

Risk evaluation is fundamentally a process of information and investigation of the hazardous properties of chemical agents present and the conditions under which people work with them, in order to determine the risks present, the persons exposed and any possible harm that may occur (including the possible existence of individual susceptibility), and finally, evaluating the possibility that such harm may, in fact, occur.

Occupational risk assessment posed by chemical agents is a very complicated process that includes at least the following five steps:

1. Identification of chemical agents used in the workplace and those generated by processes such as vulcanization and thermodestruction processes in the plastic and foundry industries
2. Collection of data about the hazardous properties of these substances
3. Assessment of exposure to the substances identified
4. Estimation of the risks, taking into account the reliability and adequacy of existing preventive or precautionary measures
5. Development of an action plan with suitable preventive measures, if necessary

During assessment of occupational risk posed by chemical agents, risk should be evaluated with criteria generally used for industrial hygiene:

1. Hazardous properties of chemical agents, in particular information on safety data sheets
2. Occupational exposure limit values or legally established biological limit values
3. Modes of exposure (skin, inhalation, ingestion)
4. Duration of exposure
5. Working conditions with regard to the agents identified, including quantities
6. Conclusions from health surveillance studies, if available

1.3.1 Identification of Risk Factors

Risk posed by hazardous chemical agents arises both during direct contact with the human body and through the action of the energy involved in a chemical reaction such as a fire or explosion. The first step in occupational risk assessment is for all chemical agents to be identified. Also, all risk factors that may cause the following risks should be described:

1. Risk of fire and/or explosion
2. Risk generated by hazardous chemical reactions
3. Risk due to inhalation of all chemical compounds
4. Risk as a result of chemical compound absorption through the skin and contact with the skin or mucous membranes
5. Risk due to ingestion and penetration through the parenteral route
6. Risk due to inadequate chemical installations

The preparation of a list of all chemical agents in the workplace and evaluation of work processes and procedures are the main steps in identifying potential risk. The list of agents should include primary products, impurities, intermediates, final products, reaction products, and by-products. When determining workplace factors, the following should be reviewed: job functions, work techniques, production processes, safety procedures, emission sources, duration of exposure, ventilation, and other prevention measures (CEN, 1995a). Frequently, detailed investigation using appropriate analytical techniques should be carried out to identify risk due to inhalation. Gathering information about the hazardous properties of all chemical agents identified is the next step in risk assessment.

1.3.2 Sources of Information on Hazardous Chemical Agents

Chemical agents present in the workplace may pose risks to the health or safety of workers in consideration of their hazardous physicochemical or toxicological properties, the temperature or pressure at which they occur in the workplace, their capacity to displace atmospheric oxygen from the workplace, and in the manner in which they act on human health in the workplace. To determine the capacity of chemical agents present in the workplace to pose risks, the hazardous properties of these agents and the way in which they are used or are present must be determined. Information on hazardous properties of chemical agents present in the workplace can be obtained from safety data sheets, occupational exposure limit values, and other sources.

Safety Data Sheets for Chemical Products In the case of harmful substances used in the production process, storage and handling data obtained from their material safety data sheets (MSDSs) can be used for their identification. Those sheets contain complex information about dangerous properties of individual chemical substances, the type and scale of risk they pose, and the rules of conduct related to them. The use of chemical substances that do not have a chemical safety data sheet is inadmissible. The purpose of the MSDSs is to provide professional users with effective and adequate information about the hazard posed to human health, safety, and the environment, to enable them to assess the possible risks posed to workers by the use of these agents. The information included in safety data sheets should meet the requirements set in Directive 98/24/EC (EC, 1998) and should be prepared by a competent person in accordance with the standard ISO 11014 (ISO, 1994).

Occupational Exposure Limit Values and Biological Limit Values Occupational exposure limit values and biological limit values are specific reference parameters used in assessing risks due to exposure to chemical agents in the workplace. Limit values may be of two types, depending on whether they have been established solely taking into account health criteria or whether they include viability criteria. In the first case, they form references to ensure workers' health. In the second case, which includes limits of genotoxic agents, they form references for the level of risk that must not be exceeded at any time. Lists of limit values must clearly distinguish between the two types of value.

Other Sources Frequently, information of interest on chemicals can be obtained from other sources, such as regulations of ADR, RID, or ICAO-TI, or various articles and monographs based on existing scientific and technical information.

1.3.3 Assessment of Exposure

Assessment of occupational exposure to chemical agents provides information of the type, intensity, length, and frequency of occurrence of dangerous substances in the workplace, including the combined effects. The results of determining chemical agents' concentration in workplace air are the basis of the exposure evaluation procedure. The guidance and strategy described in EN 689 (CEN, 1995a) are recommended when assessing exposure to hazardous chemical agents by inhalation, and there are limit values for them. In accordance with the requirements of this standard for measuring the concentration of chemical compounds, personal air sampling in the worker's breathing area should be used. The results of these measurements should ensure a correct eight-hour exposure as well as long- and short-term exposure. Classification of exposure into one of three categories—acceptable, tolerable, or unacceptable—concludes the quantitative procedure of assessing occupational exposure to chemical agents.

1.3.4 Risk Estimation

The main criteria of occupational risks due to exposure to hazardous chemical agents are occupational limit values (TLVs). These values are intended for use in the practice of industrial hygiene as guidelines or recommendations in the control of potential workplace health hazards and for no other use (e.g., in the evaluation or control of community air pollution nuisances; in estimating the toxic potential of continuous, uninterrupted exposures or other extended work periods; as proof or disproof of an existing disease or physical condition).

To estimate risk due to inhalation of chemical agents, a three-level scale may be accepted. Application of a scale of risk assessment according to which risk can be defined as low, medium, or high allows easy comparison of the assessment with the regulations in force. For chemical agents (except carcinogenic agents), risk is assessed as low if the worker's exposure does not exceed 0.5 TLV, as medium if it is higher than 0.5 TLV but does not exceed that value, and as high if the health standards are not met. Risk connected with the presence of a carcinogenic agent in workplace air, even if exposure is lower than a TLV, is always assessed as high (Pośniak and Skowroń, 2000).

The evaluation of risks arising from the capacity of hazardous chemical agents to give rise to accidents (in particular, fires, explosions, or other hazardous chemical reactions) covers (1) hazards arising from the physical and chemical nature of the chemical agents; (2) risk factors identified in storage, transport, and use; and (3) the estimated consequences in the event of occurrence. There are complex methodologies, such as HAZOP, fault trees, and event trees, for evaluating risks of this type which we do not deal with in greater detail because they are universally known and applied. These methodologies should be used in accordance with the following criteria: (1) they should be used when the consequences of the occurrence of the risk might be very serious, in terms of human and material or environmental loss, both to the company itself or its vicinity; (2) they require a thorough knowledge of the installations; (3) their application normally requires the involvement of a work team, guaranteeing a thorough knowledge of various areas (process, instrumentation, maintenance, prevention, engineering, etc.); and (4) given the severity of the possible consequences, it is normal to focus the analysis on the maximum loss to which an accident may give rise.

1.3.5 Principles of Risk Prevention

According to the main requirements of Council Directive 98/24/EC (EC, 1998), employees should enforce all prevention measures for eliminating or reducing risks to the health and safety of workers who do work involving hazardous chemical agents. The general principles for prevention must be followed whenever hazardous chemical agents exist in the working environment. Occupational risks at work involving these agents must be eliminated or reduced to a minimum by:

1. Designing work processes and engineering controls to avoid or minimize the release of hazardous agents into the workplace (e.g., the design of adequate collective protection equipment against gases, vapors, and aerosols; the provision of suitable equipment and materials that ensure the health and safety of workers)
2. Where possible, eliminating hazardous chemical agents or substituting nonhazardous alternatives
3. Reducing to a minimum the number of workers and the duration of time to which they are exposed to dangerous chemical agents, as well as their quantity
4. Appropriate hygienic measures
5. Suitable working procedures, especially with arrangements for safe handling, storage, and transport of hazardous chemical agents

Where exposure cannot be prevented by these measures, personal protective measures can be followed, including the use of appropriate personal protective equipment.

2 DUST

Dust—solid particles suspended in inhaled air—is one of the most common hazards in human working and living environments. The effects of human exposure to dust can cause mechanical injury of mucosa or skin, allergic diseases, pneumoconiosis, and cancer. A reduction in the risk of occupational diseases resulting from exposure to dust is thus one of most important problems connected with assuring occupational safety. Protecting workers against the adverse effects of dust requires (1) determination of the type and basic parameters of dust emitted to the working environment, (2) estimation of workers' exposure to the harmful effect of dust present in the working environment, (3) risk assessment of workers exposed to dust, and (4) application of suitable collective protective equipment against dust, which makes it possible to eliminate dust from workplace air, but if it is not possible, use of adequate personal protective equipment.

2.1 Definitions

- *Total airborne particles*: all particles surrounded by air in a given volume of air. Because all measuring instruments are size selective to some extent, it is often impossible to measure the total airborne particle concentration (CEN, 1993; ISO, 1995).
- *Inhalable fraction*: mass fraction of total airborne particles that is inhaled through the nose and mouth. The inhalable fraction depends on the speed and direction of the air movement, the rate of breathing, and other factors (CEN, 1993; ISO, 1995).
- *Thoracic fraction*: mass fraction of inhaled particles that penetrate beyond the larynx (CEN, 1993; ISO, 1995).
- *Respirable fraction*: mass fraction of inhaled particles that penetrate to the unciliated airways (CEN, 1993; ISO, 1995).

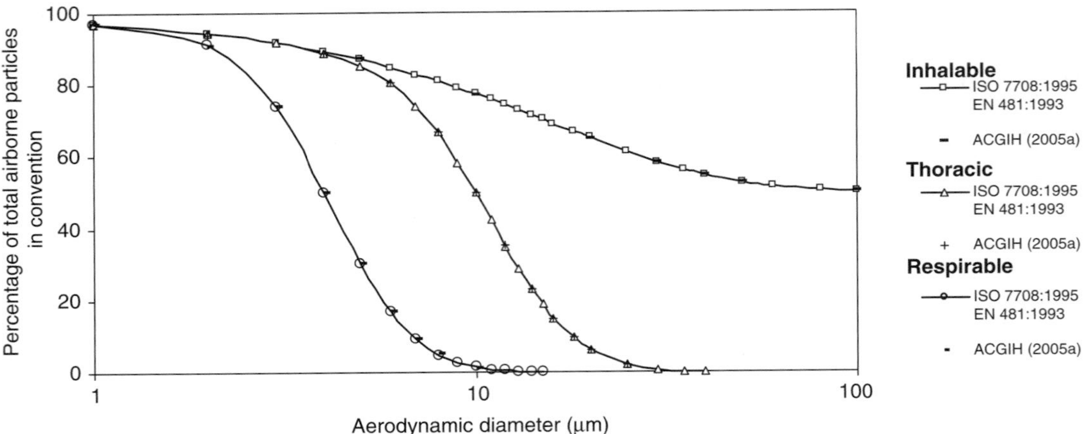

Figure 1 Inhalable, thoracic, and respirable conventions as percentages of total airborne particles.

- *Particle aerodynamic diameter*: the diameter of a sphere of density 1 g/cm^3 with the same terminal velocity due to gravitational force in calm air as the particle, under the prevailing conditions of temperature, pressure, and relative humidity (CEN, 1993; ISO, 1995).
- *Threshold limit value–time-weighted average (TLV-TWA)*: the time-weighted average concentration for a conventional eight-hour workday and a 40-hour workweek, to which it is believed that nearly all workers may be repeatedly exposed, day after day for a working lifetime, without adverse effects (ACGIH, 2005a).
- *Dust concentration*: mass (number) of solid particles in the unit volume of gas, in mg/m^3 (number of particles/m^3).
- *Air filter*: a device designed to remove suspended particles from the air flowing through it.
- *Collective protection equipment (CPE) against dust*: technical devices used in machinery and systems of buildings to protect people against dust (local and general ventilation systems equipment with air filters).
- *Personal protective equipment (PPE)*: products used by workers to protect against dust.

2.2 Sources and Properties of Dust and Particle Size-Selective Sampling Criteria

Properties of dust emitted to the working environment depend strongly on the type of technological process being utilized. Parameters such as mass (number) concentration of dust, particle size and shape, chemical composition, and crystal structure are the basis for estimating exposure to harmful effects, risk assessment, and selection of CPE or PPE. Size distribution of particles emitted at workplaces depends on their formation. Particles generated in a technological process are usually suspended in the air (total airborne particles) as a polydisperse aerosol. The number of particles inhaled by workers from total airborne particles depends on the properties of particles, the speed and direction of air movement near the body, the breathing rate, and on whether breathing is through the nose or mouth (CEN, 1993, ISO, 1995). In the standards there are definitions of sampling conventions for particle-size fractions that are to be used in assessing the possible health effects resulting from inhalation of airborne particles in the workplace. Figure 1 presents the inhalable, thoracic, and respirable conventions as percentages of total airborne particles according to standards from CEN (1993), ISO (1995), ACGIH (2005).

For dust, the American Conference of Governmental Industrial Hygienists (ACGIH, 2005) has recommended particle size-selective TLVs in three forms: (1) inhalable particulate mass TLVs (IPM-TLVs) for those materials that are hazardous when deposited anywhere in the respiratory tract, (2) thoracic particulate mass TLVs (TPM-TLVs) for those materials that are hazardous when deposited anywhere within the lung airways and the gas-exchange region, and (3) respirable particulate mass TLVs (RPM-TLVs) for those materials that are hazardous when deposited in the gas-exchange region.

2.3 Methods for Measuring Dust, Estimation of Exposure to Dust, and Risk Assessment

General requirements for the performance of procedures for the measurement of dust in workplace atmospheres are described in standard EN 482 (CEN, 1994). The test methods for the measuring procedures are given in this standard in general terms. As test methods depend on the specific measuring procedures or devices, detailed test methods are specified in specific standards (see also Baron and Willeke, 2001). Performance requirements given in the standard include unambiguity, selectivity, overall uncertainty (a combination of precision and bias) for minimum specified measuring ranges, averaging time, and so on. Fulfilling these requirements is very important during the sampling process and for dust sampling instruments. Methods for testing dust sampling instruments under prescribed laboratory conditions, and performance requirements that are specific to dust sampling instruments, are specified in standard EN 13205 (CEN, 2001).

Assessing occupational exposure to air contaminants in a representative way is a challenging task. Guidance for the assessment of exposure by inhalation of dust for comparison with limit values and measurement strategy is described in standard EN 689 (CEN, 1995a). Risk assessment is a difficult process which includes (1) identification of the type of dust present in the working environment, (2) determination of dust concentration for specified size-selection fraction and chemical composition of dusts and calculation of exposure index for dust, (3) carrying out an estimation of workers' exposure to the harmful effects of dust, (4) risk assessment of workers exposed to dust, and (5) determination of acceptable risk. For risk assessment, various methods and scales can be used, but all parameters of the working environment that can have an adverse effect because of exposure to dust always have to be taken into consideration.

2.4 Principles of Risk Prevention

2.4.1 Medical Prevention

Medical procedures in relation to workers exposed to dust consist of a medical examination to determine a worker's suitability for employment in a variety of workplaces and periodic exams during employment to ensure avoidance of occupational diseases.

2.4.2 Technical Prevention

Elimination or reduction of dust spreading through the work environment can be done by using various types of collective protection equipment, which has priority over the use of personal protective equipment. This equipment includes general mechanical ventilation devices as well as local ventilation equipped with air filters (ACGIH, 2004). The aim of ventilation, which is a continuous or periodic air exchange in the working environment, is (1) improvement of the condition and composition of air in the working environment according to hygienist requirements (protection of human health) and technological requirements (necessity to obtain products with a predetermined property), and (2) regulation of air parameters in the workrooms, such as concentration of pollution, temperature, humidity, and velocity and direction of air movement.

In work areas in which considerable amounts of dust are emitted into individual workplaces, the best solution is airtight sealing of technological processes, which means tight encasing of the dust emission region. If this is not possible, partial encasing or local ventilation installation is used. In workrooms, local ventilation equipment must be used together with general ventilation systems. In systems for general ventilation and in local ventilation devices, the elements primarily responsible for the quality of air supply or carry-off from work areas are cleaning systems (one- or multistage) equipped with air filters. When the use of collective protection equipment does not ensure the required air purity in workrooms, personal protective equipment adequate for the type of dust present in the working environment must be used.

3 BIOLOGICAL AGENTS

Biological agents are organisms or toxins that have illness-producing effects on people, livestock, and crops. They are found in many sectors of employment but are rarely visible, so workers are not always able to appreciate the risk they pose. Biological agents include bacteria, viruses, fungi (yeasts and molds), parasites, and other microorganisms and their associated toxins. They have the ability to affect human health adversely in a variety of ways, ranging from relatively mild, allergic reactions to serious conditions, even death. These organisms are ubiquitous in the natural environment; they are found in water, soil, plants, and animals. Because many microbes reproduce rapidly and require minimal resources for survival, they are a potential danger in a wide variety of occupational settings.

Whenever people are in contact with natural or organic materials such as soil, clay, plant materials (hay, straw, cotton, etc.), substances of animal origin (wool, hair, etc.), food, organic dust (e.g., flour, paper dust, from animals), waste, wastewater, and blood and other body fluids, they may be exposed to biological agents. Microorganisms can enter the human body via damaged skin or mucous membranes. They can be inhaled or swallowed, leading to infections of the upper respiratory tract or the digestive system.

3.1 Definitions

There are many definitions of biological agents. In European legislation (EC, 2000) biological agents are "micro-organisms (microbiological entity, cellular or noncellular, capable of replication or transferring genetic material), including those which have been genetically modified, cell cultures (the *in vitro* growth of cells derived from multicellular organisms) and human endoparasites, which may be able to provoke any infections, allergy or toxicity." This definition does not include animal and plant toxins, exoparasites, allergens, or toxins produced by microorganisms (endotoxin, mycotoxin, glucans) (Dutkiewicz and Górny, 2002).

According to ACGIH (2005a), "biologically derived airborne contaminants include bioaerosols (airborne particles composed of or derived from living organisms) and volatile organic compounds that organisms release. Bioaerosols include microorganisms (i.e., culturable, nonculturable, and dead microorganisms) and fragments, toxins, and particulate waste products from all varieties of living things." ACGIH has developed and separately published guidance on the assessment, control, remediation, and prevention of biologically derived contamination in indoor environments.

Indoor biological contamination is defined as the presence of (1) biologically derived aerosols, gases, and vapors of a kind and concentration likely to cause disease or predispose people to disease; (2) inappropriate concentrations of outdoor bioaerosols, especially in buildings designed to prevent their entry; or (3) indoor microbial growth and remnants of biological growth that may become aerosolized and to which people may be exposed. The term *biological agents* refers to a substance of biological origin that is capable of producing an adverse effect (e.g., an infection or hypersensitivity, irritant, inflammatory, or other response).

3.2 Classification of Biological Agents

Biological agents can cause three types of diseases: (1) infections caused by parasites, viruses, or bacteria; (2) allergies initiated by exposure to mold, organic dust such as flour dust and animal dander, enzymes, and mites; and (3) poisoning or toxic effects.

Biological agents usually are classified into four risk categories according to their potential to cause diseases and the possibilities of prevention and treatment: (1) biological agents that are unlikely to cause human disease, (2) biological agents that can cause human disease and might be a hazard to workers but for which effective prophylaxis or treatment is usually available, (3) biological agents that can cause severe human disease and present a serious hazard to workers but for which effective prophylaxis or treatment is usually available, and (4) biological agents that can cause severe human disease and pose a serious hazard to workers but for which no effective prophylaxis or treatment is available.

3.3 Measurement of Airborne Microorganisms

Measurements could aim to locate sources emitting microorganisms, to measure a worker's daily or work shift exposure, to identify peaks in exposure, to test the efficiency of control measures, or to control actions taken to diminish the exposure. The following measurement options can be used to measure microorganisms and endotoxin: (1) microbial cells by direct counting (the total number); (2) microbial cells and cell aggregates by culturing on agar media (the culturable number); (3) cellular components of microorganisms from viable, nonviable, or disintegrated microorganisms (e.g., constituents of cell structure that may also have inflammatory properties, such as endotoxin and glucans); (4) primary metabolites (e.g., ATP) that may serve as markers of microorganisms or of their vital activity; and (5) secondary metabolites (e.g., mycotoxins) that may be found in microorganisms and other particles in the aerosol. Sampling of aerosols of microbiological origin should be made in accordance with the principles of sampling to assess workers' exposure to other substances hazardous to health. Static or personal exposure to bioaerosols can be intermittent and of short duration, and it can be related to specific work activities. The sampler used must have known and documented sampling efficiency (e.g., be capable of sampling total microorganisms, viable microorganisms, or microbial components). The methods used for analyzing the sample should be selected according to the type of microorganism or microbial component (cultivation methods, microscopic methods, endotoxin–LAL method) (CEN, 2000).

Nonviable microorganisms are not living organisms; as such, they are not capable of reproduction. The bioaerosol is collected on a "greased" surface or membrane filter. The microorganisms are then enumerated and identified using microscopy, classical microbiology, molecular biological, or immunochemical techniques. When sampling for culturable bacteria and fungi, the bioaerosol is generally collected by impaction onto the surface of a broad-spectrum solid medium (agar), filtration through a membrane filter, or impingement into an isotonic liquid medium (water-based). Organisms collected by impaction onto an agar surface may be incubated for a short time, replica-plated (transferred) onto selective or differential media, and incubated at various temperatures for identification and enumeration of microorganisms. Impingement collection fluids are plated directly on agar, serially diluted, and plated, or the entire volume of fluid is filtered through a membrane filter. The filter is then placed on an agar surface and all colonies may be replica-plated. Culturable microorganisms may be identified or classified by using microscopy, classical microbiology, or molecular biology techniques, such as restriction fragment length polymorphic (RFLP) analysis. Classical microbiology techniques include observation of growth characteristics; cellular or spore morphology; simple and differential staining; and biochemical, physiological, and nutritional testing for culturable bacteria. Analytical techniques that may be applied to both nonviable and viable microorganisms, but which do not distinguish among them, include polymerase chain reaction (PCR) and enzyme-linked immunosorbent assay (ELISA). Such methods may be used to identify specific microorganisms and to locate areas of contamination (Jensen and Schafer, 1998; Skowroń and Gołofit-Szymczak, 2004).

When microbial numbers have been derived from counting colonies grown on agar plates, the results are expressed as colony-forming units per cubic meter (CFU/m^3). When microbial numbers have been derived from microscopic counting, the results are expressed as the total number of microorganisms per cubic meter of air sampled. Endotoxin is expressed as endotoxin units per cubic meter (EU/m^3) of air sampled and related to the reference standard used. There are currently no occupational exposure limits for biological contaminants. The essential difference between biological agents and other hazardous substances is in their ability to reproduce. A small amount of a microorganism may grow considerably in a very short time under favorable conditions.

Bioaerosols may contain different microorganisms and/or different components from these. Microorganisms may be classified in different taxonomic groups, such as gram-positive and gram-negative bacteria, actinomycetes, fungi, protozoa, algae, and viruses. These may be further classified to genus or species level. Immunologic reactions (e.g., allergic and/or toxic reactions) can result from exposure to microorganisms irrespective of their viability. Exposure to biological agents very often leads to adverse health effects in susceptible persons. Elaboration of values for biologically derived airborne contaminants seems to be necessary to prevent harmful exposure in occupational and nonoccupational environments, to ensure reliability of measurement methods and proper interpretation of the results.

ACGIH (2005a) actively solicits information, comments, and data in the form of peer-reviewed literature on health effects associated with bioaerosol exposures in occupational and related environments that may help the Bioaerosols Committee evaluate the potential for proposing exposure guidelines for selected biologically derived airborne contaminants: gram-negative bacterial endotoxin, (1–3) beta, D-glucan. The Scandinavian CFU-oriented projects for occupational exposure limit (OEL) suggest 5 to 10×10^3 CFU/m^3 for total microorganisms, 1×10^3 CFU/m^3 for gram-negative bacteria, and $1 \div 2 \times 10^2$ ng/m^3 for endotoxin (Malmros et al., 1992). The project prepared in Poland suggests 1×10^5 CFU/m^3 for total microorganisms, 5×10^4 CFU/m^3 for fungi, 2×10^4 CFU/m^3 for gram-negative bacteria and thermophilic actinomycetes, with a reduction of the values by half if the respirable fraction equals or exceeds 50% of the total count, and 2×10^2 ng/m^3 for endotoxin. This project is based on the fact that at a continuous exposure to microbial concentrations above 10^5 CFU/m^3, work-related respiratory disorders in workers are very common. The presence in indoor air of microorganisms from risk groups 3 and 4, independent of the concentration, should always be inadmissible and result in prevention actions (Dutkiewicz, 1997; Górny and Dutkiewicz, 2002).

3.4 Risk Assessment and Prevention

Employers have a duty to provide and maintain for employees, as far as practicable, a working environment that is safe and without risks to health. This includes providing a safe system of work, information, training, supervision, and where appropriate, personal protective equipment. The employer must identify hazards in the workplace and, if practicable, eliminate these hazards. If this is not practicable, the employer must take measures to control the hazard and reduce the risk to workers. When a work activity involves a deliberate, intentional use of biological agents, such as cultivating a microorganism in a microbiological laboratory or using it in food production, the biological agent will be known, can be monitored more easily, and measures can be taken to prevent exposure. When the occurrence of biological agents is an unintentional consequence of the work (e.g., in waste sorting or in agricultural activities), the assessment of risks to which workers are exposed will be more difficult. In any case, for some of the activities involved, information on exposures and protection measures is available.

If exposure is not avoidable, it should be kept to a minimum by limiting the number of workers who are exposed and reducing their exposure time. The control measures introduced in a workplace must be tailored to the working processes. The employer also has the responsibility to provide workers with information and training to enable them to recognize the hazards and to follow safe working procedures. The measures that need to be taken to eliminate or reduce the risks to workers will depend on the particular biohazard, but there are a number of common actions that can be implemented. Many biological agents are communicated via air, such as exhaled bacteria or toxins of mold grains. The production of aerosols and dusts should be avoided in the manufacturing process and during cleaning and/or maintenance; good housekeeping, hygienic working procedures, and use of relevant warning signs are key elements of safe and healthy working conditions. Many microorganisms have developed mechanisms to survive or resist heat, dehydration, or radiation: for example, by producing spores. The workplace must develop decontamination measures for waste, equipment, and clothing and appropriate hygienic measures for workers, as well as proper instructions for safe disposal of waste, emergency procedures, and first aid. In some cases preventive measures may include vaccinations of workers most at risk.

4 ELECTROMAGNETIC FIELDS AND RADIATION HAZARDS

Electrical appliances and wireless telecommunication systems are very common sources of electromagnetic fields and radiation (EMF). Occupational use of various intended or unintended EMF sources results in a significant population of highly or permanently exposed workers, mainly among health care staff and industrial workers. The general public is usually exposed to low-level EMF. Workers' exposure is significantly different from that of the general public, because of its higher level, longer duration, shorter distance between human body and sources, significant complexities of frequencies and modulation (which do not exist in the public environment), and simultaneous exposure to various chemical and physical agents. In this section we focus on specific topics concerning occupational EMF exposure assessment: exposure characteristics of various worker groups and measurement methods as well as safety guidelines and methods for exposure reduction.

4.1 Electromagnetic Fields and Radiation in the Environment

4.1.1 Basic Properties

The electromagnetic field is a manifestation of electrical forces associated with electric charges or current. Electric field, $\boldsymbol{E}$, is associated with the presence of electric charge and exerts forces on another electric charge. The magnetic field, $\boldsymbol{H}$, is the result of the physical movement of an electric charge (electric current), and it exerts physical forces on electric charges but only when such charges are in motion. Static or time-varying magnetic, electric, and electromagnetic fields or radiation (e.g., high-frequency electromagnetic fields) are marked together as EMF.

Time-varying EMF is characterized by its frequency and wavelength. Wavelength and frequency are in close relationship:

$$f = c/\lambda \tag{1}$$

where f is the frequency expressed in hertz, c is the velocity of light (= 300,000,000 m/s), and λ is the wavelength expressed in meters. The wavelength of EMF can be estimated quickly from the following formula:

$$\lambda = 300/f \tag{2}$$

where λ is expressed in meters and f is expressed in MHz.

EMFs are part of the radiation spectrum (see Figure 2) of frequencies up to 300 GHz. The radiation of a higher frequency is infrared (IR). The wavelength of EMFs is longer than 1 mm (see Figure 3). Fields or radiations of frequency in the range of 300 MHz to 300 GHz are called *microwaves* (MW). This radiation has a wavelength sufficiently short for a practical use of waveguides for its transmission and reception. Electromagnetic radiation of any frequency, which is useful for telecommunication, is called *radio-frequency* (RF) *radiation*. It is possible to find various definitions of RF; for example, the frequency range defined by the Institute of Electrical and Electronics Engineers is 300 Hz to 300 GHz (IEEE, 2004). EMFs of the frequency below 300 Hz are called extremely low frequency (ELF) fields. Power frequency is the frequency of currents used for electrical power

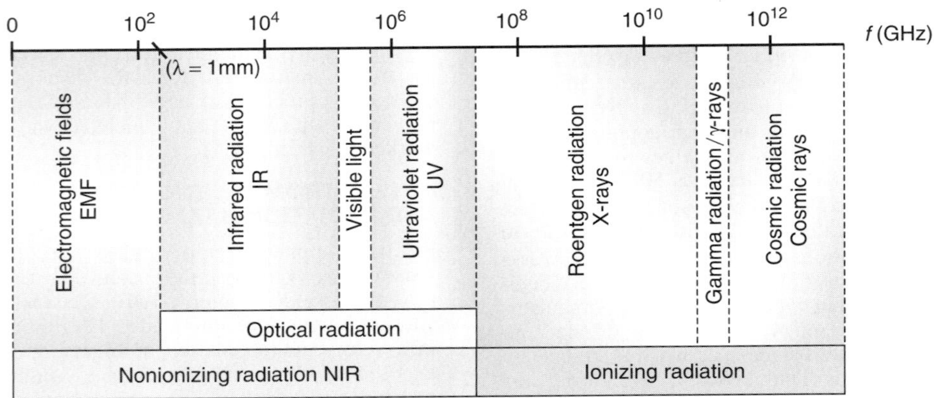

Figure 2 Electromagnetic radiation spectrum.

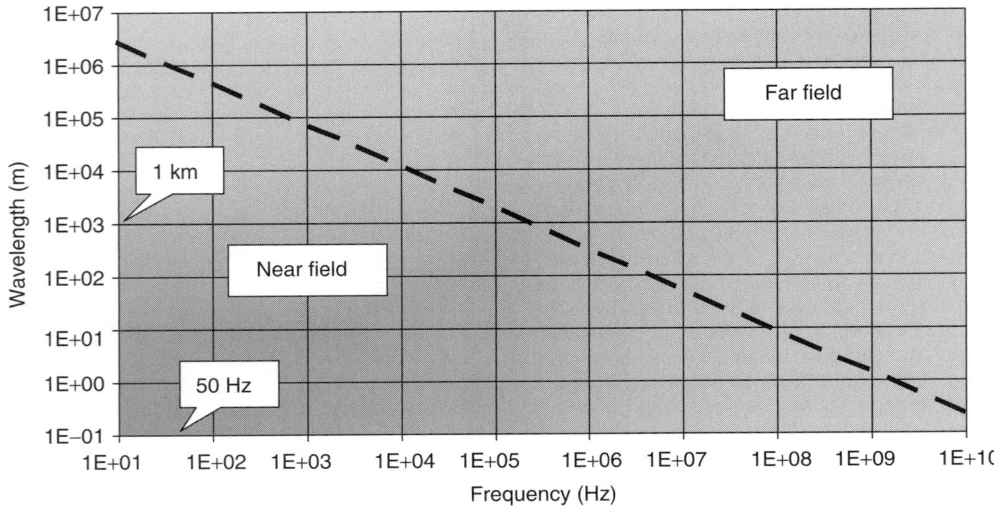

Figure 3 EMF wavelengths.

Table 1 Frequency Bands

Band	Wavelength	Frequency
EHF (extremely high frequency)	1 mm–1 cm	300–30 GHz
SHF (superhigh frequency)	1 cm–10 cm	30–3 GHz
UHF (ultrahigh frequency)	10 cm–1 m	3–0.3 GHz
VHF (very high frequency)	1 m–10 m	300–30 MHz
HF (high frequency)	10 m–100 m	30–3 MHz
MF (medium frequency)	100 m–1000 m	3–0.3 MHz
LF (low frequency)	1 km–10 km	300–30 kHz
VLF (very low frequency)	10 km–100 km	30–3 kHz

Source: UNEP/WHO/IRPA (1993).

transmitting and of EMFs produced by it (50 Hz in Europe, 60 Hz in North America). It is possible to find various definitions of frequency bands (e.g., definitions used for telecommunication purposes), see Table 1.

EMFs are part of nonionizing radiation (NIR). This means that EMFs are characterized by energy per photon lower than about 12 eV (corresponding to wavelengths greater than 100 nm and frequencies lower than 3×10^{15} Hz) and do not normally have sufficient energy to produce ionization in matter. Electric and magnetic field strengths are vector quantities, characterized by the magnitude and direction of the vector. In a three-dimensional orthogonal coordinate system, a vector is represented by three orthogonal components defining its direction and its magnitude (module), expressed by a square root of the sum of all squared components (see Figure 4).

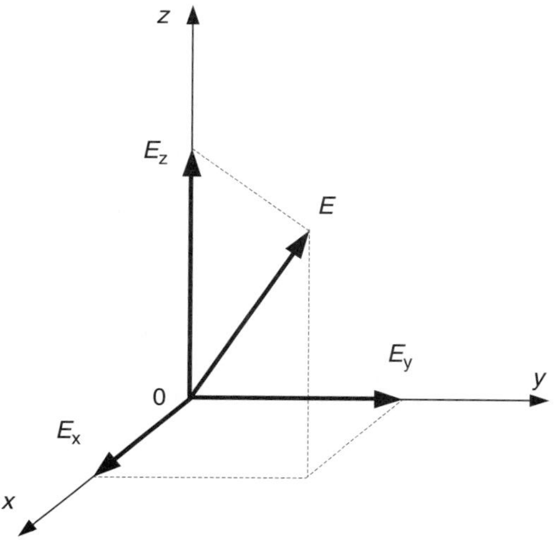

Figure 4 Representation of vector quantities.

EMF's behavior and interaction with the elements of the environment (including the human body and the medium in which there is EMF propagation) are defined by *Maxwell's equations*:

$$\text{rot}\, \boldsymbol{E} = -\frac{\partial \boldsymbol{B}}{\partial t} \tag{3}$$

$$\text{rot}\, \boldsymbol{H} = \boldsymbol{J} + \frac{\partial \boldsymbol{D}}{\partial t} \tag{4}$$

$$\text{div}\, \boldsymbol{D} = \rho \tag{5}$$

$$\text{div}\, \boldsymbol{B} = 0 \tag{6}$$

$$\text{div}\, \boldsymbol{J} = -\frac{\partial \rho}{\partial t} \tag{7}$$

$$\boldsymbol{D} = \varepsilon \boldsymbol{E} \tag{8}$$

$$\boldsymbol{B} = \mu \boldsymbol{H} \tag{9}$$

$$\boldsymbol{J} = \sigma(\boldsymbol{E} + v \times \boldsymbol{B}) \tag{10}$$

where $\boldsymbol{E}$ is the electric field strength, $\boldsymbol{B}$ the magnetic flux density, t the time, $\boldsymbol{H}$ the magnetic field strength, $\boldsymbol{J}$ the current density, $\boldsymbol{D}$ the electric flux density, ρ the volume density of charge, ε the permittivity, μ the magnetic permeability, σ the conductivity, and v the velocity.

Electric field strength, $\boldsymbol{E}$, is expressed in volts per meter (V/m), and magnetic field strength, $\boldsymbol{H}$, is expressed in amperes per meter (A/m). A magnetic field can be specified equivalently as magnetic flux density, $\boldsymbol{B}$, and expressed in tesla (T) or gauss (G). Magnetic permeability in a vacuum and in the air, as well as in nonmagnetic (including biological) materials has the value $\mu = 4\pi \times 10^{-7}$ H/m. For practical use, the estimated values of magnetic field strength and magnetic flux density corresponding to each other can be found according to Table 2.

Physical properties of EMF depend on the distance from the source. In the area called the far field, the plane-wave model (see Figure 5) correctly represents EMF propagation:

1. Wavefronts have a planar geometry.
2. $\boldsymbol{E}$ and $\boldsymbol{H}$ vectors and the direction of propagation are mutually perpendicular.
3. The phase of the $\boldsymbol{E}$ and $\boldsymbol{H}$ fields is the same.
4. The ratio of E/H amplitudes (impedance) is constant throughout space [$E/H = 377$ ohms (Ω) in free space].

Table 2 Corresponding Values of Magnetic Flux Density and Magnetic Field Strength in a Vacuum and in the Air, as Well as in Nonmagnetic (Including Biological) Materials

		H (A/m)	***B*** (μT)	***B*** (mG)
Magnetic field strength	***H*** (A/m)	1	1.25	12.5
Magnetic flux density	***B*** (μT)	0.8	1	10
	B (mG)	0.08	0.1	1

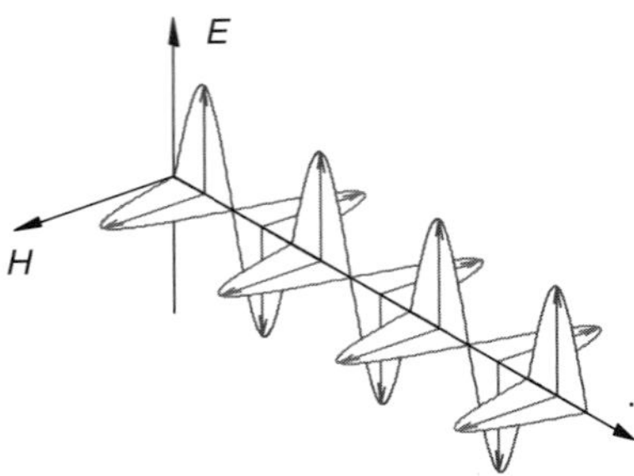

Figure 5 EMF in the far-field area.

The characteristic of EMFs in the near-field area is more complicated:

1. The maxima and minima of ***E*** and ***H*** fields do not occur at the same points along the direction of propagation.
2. EMFs' structure may be highly inhomogeneous.
3. There may be substantial variations from the plane-wave impedance of 377 Ω (both an almost pure E and H field are possible).

It is possible to find various definitions of the distance from the EMF source where EMF can be taken as the far field. This distance depends on the source's geometry and frequency, usually near field passing into far field at the distance from the source roughly of 1 to 3 EMF wavelengths.

4.1.2 Quantities and Units

Electric, magnetic, electromagnetic, and dosimetric quantities and corresponding SI units are as follows:

H magnetic field strength [ampere per meter (A/m)]
B magnetic flux density [tesla (T)]
E electric field strength [volt per meter (V/m)]
D electric flux density [coulomb per square meter (C/m^2)]
S power density [watt per square meter (W/m^2)]
U voltage [volt (V)]
I current [ampere (A)]
J current density [ampere per square meter (A/m^2)]
σ conductivity [siemens per meter (S/m)]
ρ volume density of charge [coulomb per cubic meter (C/m^3)]
λ wavelength [meter (m)]
μ magnetic permeability [henry per meter (H/m)]
μ_0 magnetic permeability in a vacuum or in air ($4\pi \times 10^{-7}$ H/m),
μ_r relative magnetic permeability (in the air = 1)
ε permittivity [farad per meter (F/m)]
Z_0 impedance of free space ($120\pi\Omega \approx 377\ \Omega$)
f frequency [hertz (Hz)]
v velocity [meter per second (m/s)]
t time [second (s)]
SA specific energy absorption [joule per kilogram (J/kg)]
SAR specific energy absorption rate [watt per kilogram (W/kg)]

Submultiple and multiple units are as follows:

Prefix to Unit	Symbol	Submultiple or Multiple Meaning	
nano	n	$\times 10^{-9}$	($\times$0.000000001)
micro	μ	$\times 10^{-6}$	($\times$0.000001)
milli	m	$\times 10^{-3}$	($\times$0.001)
—	—	$\times 10^{0}$	($\times$1)
kilo	k	$\times 10^{3}$	($\times$1000)
mega	M	$\times 10^{6}$	($\times$1000000)
giga	G	$\times 10^{9}$	($\times$1000000000)

4.1.3 EMF Sources

EMFs exist permanently in the general public and occupational environment (UNEP/WHO/IRPA, 1993; ICNIRP, 1998). Significant exposure of workers is caused by many industrial appliances, especially of high power consumption. Electrical power production and distribution result in occupational exposure to 50/60 Hz electric and magnetic fields in the vicinity of high-voltage power lines and substations, power generators, and so on. Various industrial processes require EMF heating of elements produced. Induction heaters and sealers used for these processes are sources of a high levels of worker exposure (mainly 27 MHz). Broadcasting antennas, mobile phone base stations, and radar installations usually produce significant EMF exposure of the technical staff (from various RF bands). Health care staff can be exposed to a high level of EMFs while operating MRI devices (static magnetic fields), physiotherapy diathermy (27 MHz), and electrosurgery devices (300 kHz to 2 MHz). Cables supplying welding electrodes are sources of significant magnetic fields of various frequencies (from static to RF). Besides the examples mentioned, numerous EMF sources are operated in the working environment, such as NMR spectrometers, antitheft devices, supply cables, and electric vehicles and railways. The variability of frequency characteristics of a particular type of device is significant and should be checked very carefully during the opening stage of an assessment of workers' exposure.

4.2 Electromagnetic Fields and Radiation Interaction with the Human Body and Technical Structures

EMF interactions with the environment result in various energetic processes. The nature of particular effects depends primarily on the EMF frequency as well as geometrical and electrical properties (permeability and conductivity) of the exposed structures. The direct effects of EMF exposure result from direct interaction of fields with the human body; indirect effects involve interactions with an object at a different electric potential from the body.

4.2.1 Direct Hazards

Three basic mechanisms were established for various direct interactions between living matter and time-varying electric and magnetic fields (UNEP/WHO/IRPA, 1993; IEEE, 2005).

1. *Low-frequency electric field interaction with the human body.* The interaction of time-varying electric fields with the human body results in a flow of electric current, the polarization of bound charge (formation of electric dipoles), and the reorientation of electric dipoles already present in tissue. The relative magnitudes of these different effects depend on conductivity and permittivity, which vary with the type of body tissue (see Figure 6) and depend on the frequency of the applied field (Reilly, 1998; www.emfdosimetry.org). The distribution of induced currents depends on exposure conditions, on the size and shape of the body, and on the body's position in the field.

2. *Low-frequency magnetic field interaction with the human body.* The physical interaction of time-varying magnetic fields with the human body results in induced electric fields and circulating electric currents (eddy currents) in the body. The magnitudes of the induced field and the current density are proportional to the radius of the loop, the electrical conductivity of the tissue, and the rate of change and magnitude of the magnetic flux density (Reilly, 1998). The distribution of the current induced in any part of the body depends on the distribution of the conductivity of the tissue.

3. *Absorption of energy from electromagnetic fields.* EMF exposure at frequencies above about 100 kHz can lead to significant absorption of energy and temperature increases. EMF can be divided into four frequency ranges of various energy absorption by the human body (Durney et al., 1986):

   (a) About 100 kHz–20 MHz, at which absorption in the trunk decreases rapidly with decreasing frequency, and significant absorption may occur in the neck and legs.
   (b) About 20–300 MHz, at which relatively high absorption can occur in the entire body, and to even higher values if partial body (e.g., head) resonances are considered.
   (c) About 300 MHz–several GHz, at which there is significant local, nonuniform absorption.
   (d) Above about 10 GHz, at which energy absorption occurs primarily at the body surface.

Thermal and Nonthermal Effects of a High Level of Human Body Exposure As a consequence of the above-mentioned interactions of EMF with the human body, various thermal and nonthermal effects of a high level of human body exposure can occur during (or immediately after) exposure (called short-term effects). The effects of a high level of EMF exposure are well investigated and they are taken as the established human mechanism (ICNIRP, 1998; IEEE, 2002, 2005). Thermal effects occurring as a consequence of EMFs' energy absorption are dominant at frequencies above 100 kHz. SAR is the widely adopted dosimetric measure for thermal effect. SAR was defined as the rate at which energy is absorbed per unit mass of body tissue and is expressed in watts per kilogram (W/kg). Whole-body average SAR and local SAR values are used to evaluate and limit excessive energy deposition in the whole body or in small parts of the body. Exposure to low-frequency electric and magnetic fields normally results in negligible energy absorption and no measurable temperature rise in the body.

In the case of low-frequency EMFs, exposure effects are related primarily to membrane polarization (Reilly, 1998; IEEE, 2002): the alteration of the cellular membrane's natural resting potential by the electric field induced in the body. Depolarization of nerve and muscle membranes can lead to their excitation, herein referred to as *electrostimulation*, which can result in (1) aversive or painful stimulation of sensory or motor neurons, (2) muscle excitation that may lead to injury while potentially hazardous activities are being performed, (3) excitation of neurons or direct alteration of synaptic activity within the brain, and cardiac excitation.

In the case of exposure to magnetic fields of the frequency below 1 Hz and static magnetic fields, there can be magnetohydrodynamic effects, which apply to forces on moving charges in fluids. A magnetohydrodynamic effect can influence the blood flow. It the case of a high level of exposure, which is possible in the occupational environment only, the following results of exposure can occur (IEEE, 2005): (1) localized RF heating effects, which can be seen only in association with high-power industrial uses of RF or medical applications; (2) surface heating effects, which can come from accidental significant exposure from open waveguides for high-powered GHz sources and the potential use of microwave-based nonlethal weapons for crowd control; (3) microwave hearing effects, which consist of the perception of a barely audible click, buzz, or hiss, from pulsed radar signals in a very quiet environment; and (4) whole-body heating effects, which produce discomfort due to absorbed RF energy, require sustained application of high (e.g., kW) RF power usually associated with climbing on energized broadcast antenna towers and working close to unshielded RF heaters or sealers.

Possible Effects of a Low Level of Human Body Exposure The possibility of adverse effects of long-term or chronic exposure to low levels of EMFs remains controversial. Mechanisms of such interactions are usually classified as proposed (ICNIRP, 2001; IARC, 2002) and mentioned in connection with hypotheses related to cancer, reproductive or immune effects, nervous system effects, and so on. Although

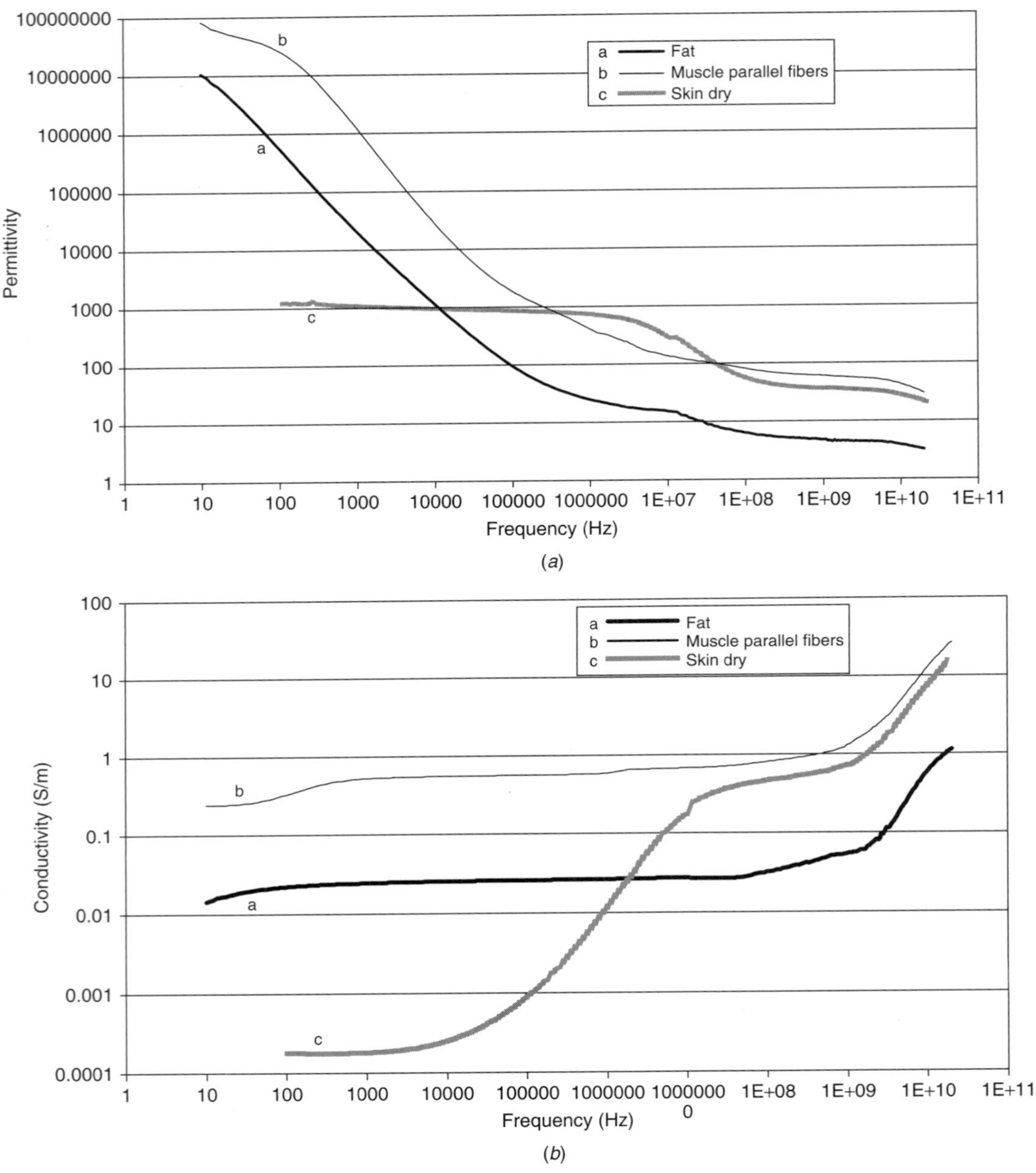

Figure 6 (*a*) Permittivity and (*b*) conductivity of selected human tissues. (Composed from www.emfdosimetry.org, 2004.)

these mechanisms cannot be dismissed as irrelevant, progress in research should be monitored for further updating of EMF limitation.

4.2.2 Indirect Hazards

All conductive objects are able to receive EMF energy. When EMF exposure results from the human body being in the proximity of various conductive structures, the following indirect hazards should be taken into consideration (ICNIRP, 1998; IEEE, 2002, 2005): (1) contact currents that result when the human body comes into contact with an object at a different electric potential (i.e., when either the body or the object is charged by an electric field) or when the human body and conductive objects create a closed loop exposed to the magnetic field; (2) coupling of EMF to medical devices worn by, or implanted in, a person; and (3) transient discharges (sparks), which can occur when a person and a conducting object exposed to a strong static or time-varying electric field come into close proximity.

Contact and Induced Currents Contact currents of a magnitude high enough to produce their perception and pain could occur in the area of EMF lower than limits of exposure, which are intended for protection against direct hazards. This is possible when conductive structures of significant size are exposed to EMF. Such effects could initiate hazardous situations and accidents for workers involved in professional activities in the proximity of exposed structures.

In a frequency range of approximately 100 kHz to 110 MHz, shocks and burns can result either from a person touching an ungrounded metal object that has acquired a charge in a field or from contact between a charged person and a grounded metal object. RF shocks and burns from passively energized conductors (reradiating structures) can be seen in high-powered RF environments at MF, HF, and VHF frequencies, such as broadcast sites, and when long conductors, such as a hoisting cable of a tall crane, are in the vicinity (e.g., 1000 m) of AM radio broadcasting antennas. Contact currents can also occur in the proximity of high-voltage power lines, as a result of a different electric potential of grounded and insulated objects (exposed to a strong electric field) or as a result of creating by the conducting objects a closed loop of significant area (receiving magnetic field).

Electromagnetic Influence on Electronic Devices (Including Medical Implants) Induced voltages and currents, which occur in electrically conductive structures (among others in metallic objects and the human body), spark discharges, and static magnetic fields can influence sensitive electronic devices. Significant changes in the performance of exposed devices, and even their damage is possible, depending on the magnitude and frequency of EMF and the geometrical size of conductive structures receiving EMF energy. It is possible that a particular electronic device is much more sensitive to EMF exposure than healthy people are, and that compliance with safety guidelines referring to workers' and the general public's exposure may not preclude interference with electronic devices and medical devices such as cardiac pacemakers and defibrillators and cochlear implants. Detailed recommendations concerning the immunity of electronic devices to EMF exposure results are included in standards on electromagnetic compatibility (EMC). The most common recommendations concerning the problems described are as follows:

1. People with cardiac pacemakers should avoid exposure to static magnetic fields higher than 0.5 mT, magnetic fields of power frequency (50/60 Hz) higher than 0.1 mT, and electric fields of power frequency higher than 1 kV/m.
2. People with some ferromagnetic implants or electrically activated devices (other than cardiac pacemakers) may be affected by fields above a few mT.
3. In the area where magnetic flux densities exceed 3 mT, precautions should be taken to prevent hazards from flying metallic objects.
4. Analog watches, credit cards, magnetic tapes, computer disks, precise electronic devices, and so on, may be affected adversely by exposure to 1 mT.

Explosive Hazard Sparks, which can occur when the human body (or any other object) and a conducting object exposed to a strong static or time-varying electric field come into close proximity can be the reason for an explosion in the environment where there are explosive chemicals or dust. Induced currents, which can occur in the cables of electro-explosive installations (detonators) localized in the vicinity of AM or FM broadcasting installations can produce enough heat to start an explosion.

4.3 Methods of Electromagnetic Field and Radiation Assessment

The internationally accepted EMF exposure limitations of widest use in the world have been published by ICNIRP (1998), IEEE (2002, 2005), and EC (2004). Various terms and exposure limits have been proposed by them, but the general approach consists of many common ideas. The main principle is to protect exposed population at least against known adverse health effects. Separate limitations concern workers and members of the general public (lower limits). These limitations are based on short-term, immediate health effects such as stimulation of peripheral nerves and muscles, shocks and burns caused by touching conducting objects, and elevated tissue temperatures resulting from absorption of energy during exposure to EMF. Potential long-term effects of exposure, such as an increased risk of cancer, other possible adverse health effects, or suggested hypersensitivity of some people to EMF exposure have not yet been taken into consideration when establishing exposure limitation. However, the International Agency for Research on Cancer recently classified low-frequency magnetic fields as potentially carcinogenic (IARC, 2002).

National regulations of various countries refer directly or indirectly to the above-mentioned international limitations. In some countries additional "precautionary measures" were introduced into national regulations. They took into consideration the possibility of long-term effects and they established a more restrictive limitation of EMF exposure concerning long duration of exposure (e.g., in living areas or in the case of whole-shift duration of exposure). Two steps toward exposure limitations were introduced by international bodies: (1) limitations concerning *internal measures* of exposure effects occurring inside an exposed body, determining maximum admissible exposure conditions; and (2) limitations concerning *external measures* of exposure level, determining environmental conditions of exposure that require special attention (e.g., inspection measurements, exposure evaluation, workers' training).

Internal measures cannot be measured directly in the real environment. External measures are intended for practical use; they can be measured directly at workplaces. Theoretical analyses or measurements (in general complementary) can accomplish EMF exposure assessment. If possible, theoretical analyses should be done prior to taking measurements.

4.3.1 Internal Measures

The following internal measures were used in international EMF exposure limitation documents:

- Current density (***J***), in the frequency range up to 10 MHz
- In situ electric field (***E***) in the frequency range up to 3350 Hz
- Specific energy absorption rate (SAR) in the frequency range 100 kHz to 10 GHz
- Specific energy absorption (SA) for pulsed fields in the frequency range 300 MHz to 10 GHz

The set of these physical quantities defines minimum safety restrictions on EMF exposure and is usually termed *basic restrictions* (IEEE, ICNIRP) or *exposure limit values* (EC) (see Tables 3 to 5).

4.3.2 External Measures

External measures are provided for practical exposure assessment to determine whether the basic restrictions are likely to be exceeded. Some external measures are derived from relevant internal measures using measurement and/or computational techniques, and some address perception and adverse indirect effects of exposure to EMF (ICNIRP, 1998; IEEE, 2002, 2005). The derived quantities of external measures are electric field strength (***E***), magnetic field strength (***H***), magnetic flux density (***B***), power density (***S***), and contact and induced currents flowing through the limbs (***I***). Among the external measures, only power density in the air, outside the body, also belongs to the basic restrictions, in the frequency range 10 to 300 GHz, although it can be measured readily. This is so because in the relevant frequency range, EMF depth of penetration is so small that in practice any internal effects of exposure are observed. The values of external measures established by various international or national limitations differ slightly (see, e.g., comparison presented in Figures 7 and 8 and Tables 6 and 7).

Compliance with the permissible values established for external measures' quantities will ensure compliance with the relevant internal measures. If the measured or calculated value of a particular external measure exceeds its permissible level, it does not necessarily follow that the internal measures will also be exceeded. Whenever the level of an external measure is exceeded, it is necessary to test compliance with the relevant internal measure and to determine whether additional protective measures are necessary.

Table 3 Internal Measures for EMF Exposure: Permissible Values for Current Density Caused by Time-Varying Electric and Magnetic Fields

Limitation	Frequency Range	Current Density for Head and Trunk [mA/m² (rms)] Occupational Exposure	General Public Exposure
ICNIRP (1998) (basic restriction)	Up to 1 Hz	40	8
	1–4 Hz	40/f	8/f
	4 Hz–1 kHz	10	2
	1 kHz–10 MHz	f/100	f/500
EC (2004) (exposure limit value)	Up to 1 Hz	40	
	1–4 Hz	40/f	
	4 Hz–1 kHz	10	
	1 kHz–10 MHz	f/100	

Table 4 Internal Measures for EMF Exposure: Permissible Values for In Situ Electric Field Caused by Time-Varying Electric and Magnetic Fields[a]

Exposed Tissue	f_c (Hz)	Electric Field Strength, E_0 [(V/m) rms] Occupational Exposure	General Public Exposure
Brain	20	0.0177	0.00589
Heart	167	0.943	0.943
Arms distal from the elbows and legs distal from the knees	3350	2.1	2.1
Other tissues	3350	2.1	0.701

Source: [IEEE (2002)] (basic restriction).
[a] $E_i = E_0$ for $f \le f_c$, $E_i = E_0(f/f_c)$ for $f \ge f_c$, where E_i is the maximum allowed induced *in situ* electric field and E_0 is the rheobase *in situ* electric field.

4.3.3 Electromagnetic Fields and Radiation Measurements

EMFs are usually measured with handheld devices equipped with various sensors (Durney et al., 1986; IEEE, 2002, 2005): dipole antennas (for an E field), loop antennas (for time-varying B and H fields or high-frequency E fields), or Hall probes (for static magnetic or low-frequency B fields). Power density, S, in a far-field area can be calculated from the results of E field measurements. In the near-field area, E and H components have to be measured independently. A shielded loop antenna is required for measurements of a high-frequency H field. Broadband rms (root-mean-square/effective value) meters as well as a selective meter can be used, depending on the frequency composition of measured fields. EMF sensors can also

Table 5 Internal Measures for EMF Exposure: Permissible Values for SAR Caused by Time-Varying Electric and Magnetic Fields

		SAR (W/kg)					
		Occupational Exposure			General Public Exposure		
Limitation	Frequency Range	Whole-Body Average	Localized (Head and Trunk)	Localized (Limbs)	Whole-Body Average	Localized (Head and Trunk)	Localized (Limbs)
IEEE (2005) (basic restriction)	100 kHz–3 GHz	0.4	10	20	0.08	2	4
ICNIRP (1998) (basic restriction)	100 kHz–10 GHz	0.4	10	20	0.08	2	4
EC (2004) (exposure limit value)	100 kHz–10 GHz	0.4	10	20			

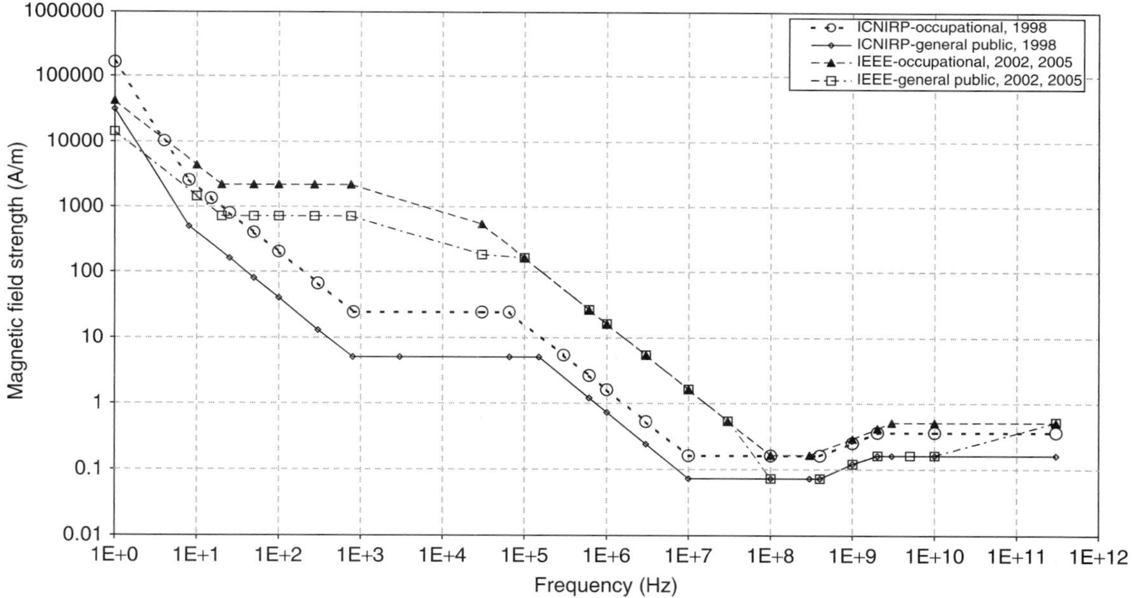

Figure 7 Comparison of MPE (IEEE, 2005) and reference levels (ICNIRP, 1998) concerning occupational and general public exposure to magnetic fields.

be used with spectrum analyzers. EMF meters can be equipped with flat-response wide-band antennas or *shaped response antennas* fitted to the frequency characteristics of permissible electric or magnetic field strength levels. In the case of EMFs composed of components of various frequencies, a shaped response antenna or frequency analysis of a measured field is required to allow estimation of the exposure factor. The definitions of exposure factors depend on the EMF frequency range. The EMF of a nonuniform spatial distribution can be spatially averaged over human body volume. Time averaging is always used (e.g., 6-minute averaging of E and H fields from the frequency range 100 kHz to 10 GHz, adopted by international recommendations) (ICNIRP, 1998; IEEE, 2005).

4.3.4 Numerical Methods for Computer-Aided Electromagnetic Hazards Assessment

Numerical methods can be used successfully for calculating internal electric field strength ($\boldsymbol{E}_{\text{in}}$), induced current density ($\boldsymbol{J}$), specific energy absorption (SA), or specific energy absorption rate (SAR) distribution

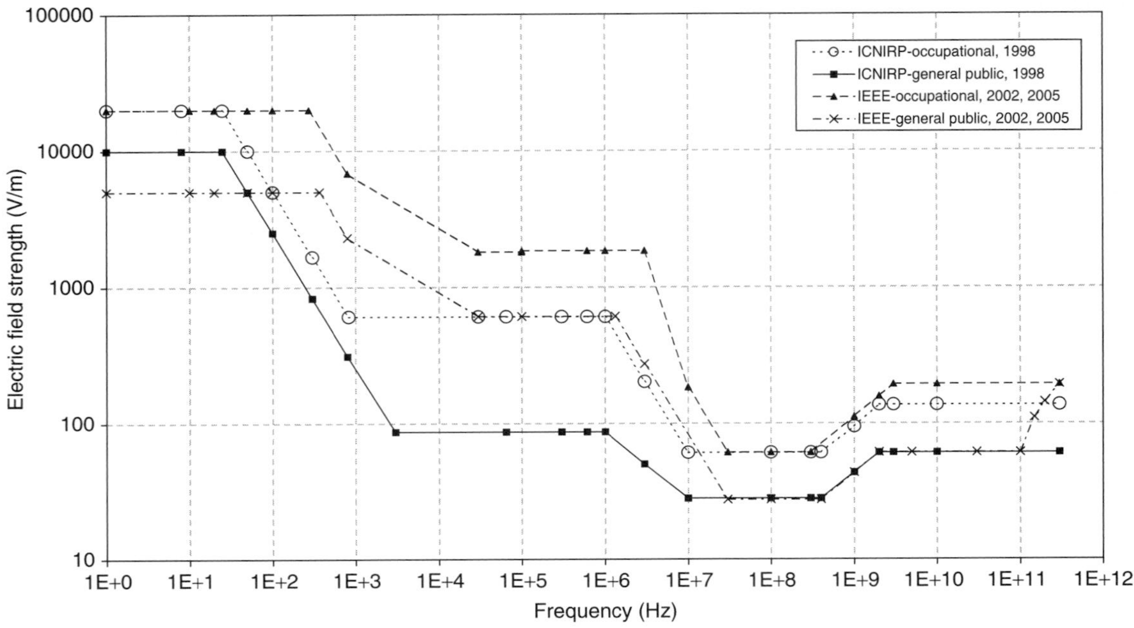

Figure 8 Comparison of MPE (IEEE, 2005) and reference levels (ICNIRP, 1998) concerning occupational and general public exposure to electric fields.

Table 6 External Measures for EMF Exposure: Permissible Values for Induced Current Caused by Time-Varying Electric and Magnetic Fields[a]

		Maximum Current (mA)			
		Occupational Exposure		General Public Exposure	
Limitation	Frequency Range	Through Both Feet	Through Each Foot	Through Both Feet	Through Each Foot
IEEE (2002, 2005) (maximum permissible exposure)	Up to 0.003 MHz	6.0	3.0	2.7	1.35
	0.003–0.1 MHz	2000*f*	1000*f*	900*f*	450*f*
	0.1–110 MHz	200	100	90	45
ICNIRP (1998) (reference level)	10–110 MHz		100		45
EC (2004) (action value)	10–110 MHz		100		

[a]*f* is frequency in MHz.

in humans or animals exposed to EMFs. These quantities can be calculated using anatomically and electrically realistic models of the body which have a high degree of anatomical resolution, and specialized computational methods, (ICNIRP, 1998; IEEE, 2005).

Because of the electrical inhomogeneity of the body, current densities should be calculated as averages over a cross section of 1 cm^2 perpendicular to the current direction. Localized SAR averaging mass is any 10 g of contiguous tissue. This 10 g of tissue is intended to be a mass of contiguous tissue with nearly homogeneous electrical properties. A simple geometry such as cubic tissue mass can be used, provided that the dosimetric quantities calculated will be taken as conservative values for the purpose of comparison with exposure guidelines. The two numerical methods FDTD (finite-difference time domain) and FEM (finite-element method) are in most common use.

4.4 Electromagnetic Hazards Elimination and Reduction

Appropriate protective measures must be implemented when exposure exceeds permissible levels in the workplace (ICNIRP, 1998; EC, 2004; IEEE, 2005; ILO, 1994). It should be noted that assessment of workers' exposure does not apply to the level of the emission of EMFs from machinery and devices. If analysis and/or measurement concerning an EMF emitted indicate high

Table 7 External Measures for EMF Exposure: Permissible Values for Contact Current Caused by Time-Varying Electric and Magnetic Fields[a]

Limitation	Frequency Range	Maximum Current (mA) Occupational Exposure	General Public Exposure
IEEE (2002, 2005) (maximum permissible exposure)	Up to 3 kHz	1.5	0.5
	3–100 kHz	0.50*f*	0.167*f*
	0.1–100 MHz	50	16.7
ICNIRP (1998) (reference level)	To 2.5 kHz	1.0	0.5
	2.5–100 kHz	0.4*f*	0.2*f*
	100 kHz–110 MHz	40	20
EC (2004) (action value)	Up to 2.5 kHz	1.0	
	2.5–100 kHz	0.4*f*	
	100 kHz–110 MHz	40	

[a]*f* is frequency in kHz.

levels, this does not necessarily imply that people will be exposed to excessive EMF levels at the workplace. The first step in protective measures should be an analysis of the possibility of EMF emission reduction (e.g., a good safety design of the EMF source, active or passive EMF shielding, interlocks or similar health protection mechanisms, manipulators and automatization of operation, minimalization of the dimensions of the EMF source). Additionally, employers should introduce administrative controls, such as limitations on access, the use of audible and visible warnings in areas of high levels of EMFs, and medical surveillance for exposed workers. Personal protection measures such as the use of protective clothing, should be taken when no other protective measures can be implemented successfully.

REFERENCES

ACGIH (2005a), *Threshold Limit Values for Chemical Substances and Physical Agents and Biological Exposure Indices*, American Conference of Governmental Industrial Hygienists, Cincinnati, OH.

ACGIH (2005b), *Industrial Ventilation: A Manual of Recommended Practice*, 25th ed., American Conference of Governmental Industrial Hygienists, Cincinnati, OH.

Baron, P. A., and Willeke, K., Eds., (2001). *Aerosol Measurement: Principles, Techniques, and Application*, Wiley, New York.

CEN (1993), *Workplace Atmospheres: Size Fraction Definitions for Measurement of Airborne Particles*, EN 481:1993, European Committee for Standardization, Brussels, Belgium.

CEN (1994), *Workplace Atmospheres: General Requirements for the Performance of Procedures for Measurement of Chemical Agents*, EN 482:1994, European Committee for Standardization, Brussels, Belgium.

CEN (1995a), *Workplace Atmospheres: Guidance for the Assessment of Exposure by Inhalation to Chemical Agents for Comparison with Limit Values and Measurement Strategy*, EN 689:1995, European Committee for Standardization, Brussels, Belgium.

CEN (1995b), *Workplace Atmospheres: Diffusive Samplers for the Determination of Gases and Vapours—Requirements and Test Methods*, EN 838:1995, European Committee for Standardization, Brussels, Belgium.

CEN (1997a), *Workplace Atmospheres: Pumped Sorbets Tubes for the Determination of Gases and Vapours—Requirements and Test Methods*, EN 1076:1997, European Committee for Standardization, Brussels, Belgium.

CEN (1997b), *Workplace Atmospheres: Pumps for Personal Sampling of Chemical Agents—Requirements and Test Methods*, EN 1232:1997, European Committee for Standardization, Brussels, Belgium.

CEN (1999), *Workplace Atmospheres: Pumps for the Sampling of Chemical Agents with a Volume Flow Rate of over 5 l/min—Requirements and Test Methods*, EN 12919:1999, European Committee for Standardization, Brussels, Belgium.

CEN (2000), *Workplace Atmospheres: Guidelines for Measurement of Airborne Micro-organisms and Endotoxin*, EN 13098:2000, European Committee for Standardization, Brussels, Belgium.

CEN (2001a), *Workplace Atmospheres: Assessment of Performance of Instruments for Measurement of Airborne Particle Concentrations*, EN 13205:2001, European Committee for Standardization, Brussels, Belgium.

CEN (2001b), *Workplace Atmospheres: Measurement of Chemical Agents Present as Mixtures of Airborne Particles and Vapour—Requirements and Test Methods*, EN 13936:2001, European Committee for Standardization, Brussels, Belgium.

CEN (2002), *Workplace Atmospheres: Procedures for Measuring Metals and Metalloids in Airborne Particles—Requirements and Test Methods*, EN 13890:2002, European Committee for Standardization, Brussels, Belgium.

Cohen, B. S., and Hering, S. V. (1995), *Air Sampling Instruments for Evaluation of Atmospheric Contaminants*, 8th ed., American Conference of Governmental Industrial Hygienists, Cincinnati, OH.

Durney, C. H., Massoudi, H., and Iskander, M. F. (1986), *Radiofrequency Radiation Dosimetry Handbook*, 4th ed., USAFSAM-TR-85-73, USAF School of Aerospace Medicine, Brooks Air Force Base, TX.

Dutkiewicz, J. (1997), "Bacteria and Fungi in Organic Dust as Potential Health Hazard," *Annals of Agricultural Environmental Medicine*, Vol. 4, pp. 11–16.

Dutkiewicz, J., and Górny, R. L. (2002), "Biological Factors Hazardous to Human Health: Classification and Criteria of Exposure Assessment," *Medycyna Pracy*, Vol. 53, No. 1, pp. 29–39 (in Polish).

EC (1989), "Directive 89/391/EEC of 12 June 1989 on the Introduction of Measures to Encourage Improvements in the Safety and Health of Workers at Work," *Official Journal of the European Communities*," L-183, June 29.

EC (1998), "Council Directive 98/24/EC of 7 April 1998 on the Protection of the Health and Safety of Workers from the Risks Related to Chemical Agents at Work" (fourteenth individual directive within the meaning of Article 16(1) of Directive 89/391/EEC), *Official Journal of the European Communities*, L-131/11-23, May 5.

EC (2000), "Directive 2000/54/EC of the European Parliament and of the Council of 18 September 2000 on the Protection of Workers from Risk Related to Exposure to

Biological Agents in the Work," *Official Journal of the European Communities*, L-262/21-45, October 17.

EC (2004), "Directive 2004/40/EC of the European Parliament and of the Council on the Minimum Health and Safety Requirements Regarding the Exposure of Workers to the Risks Arising from Physical Agents (Electromagnetic Fields)" (18th individual directive within the meaning of Article 16(1) of Directive 89/391/EEC), *Official Journal of the European Communities*, L-184, May 18.

Górny, R. L., and Dutkiewicz, J. (2002), "Bacterial and Fungal Aerosols in Indoor Environment in Central and Eastern European Countries," *Annals of Agricultural Environmental Medicine*, Vol. 9, pp. 17–23.

HSE (n.d.), *Methods for the Determination of Hazardous Substances*, Health and Safety Executive, Occupational Medicine and Hygiene Laboratory, London; http/www.hse.gov.uk/pubns/mdhs.

HSE (1997), *Monitoring strategies for toxic substances*, Health and Safety Executive Books, London; http://www.hsebooks.com.

IARC (International Agency for Research on Cancer) (2002), "Non-ionizing Radiation, 1: Static and Extremely Low-Frequency (ELF) Electric and Magnetic Fields," *IARC Monographs 80*, IARC Press, Lyon, France, p. 429.

ICNIRP (International Commission on Non-ionizing Radiation Protection) (1998), "Guidelines for Limiting Exposure to Time-Varying Electric, Magnetic, and Electromagnetic Fields (up to 300 GHz)," *Health Physics,* Vol. 74, No. 4, April, pp. 494–522.

ICNIRP (2001), International Commission on Non-ionizing Radiation Protection Standing Committee on Epidemiology: A. Ahlbom, E. Cardis, A. Green, M. Linet, D. Savitz and A. Swerdlow, "Review of the Epidemiological Literature on EMF and Health," *Environmental Health Perspective*, Vol. 109, Suppl. 6, pp. 911–933.

IEEE (2002), *Standard for Safety Levels with Respect to Human Exposure to Radio Frequency Electromagnetic Fields, 0 to 3 kHz*, Standard C95.6, Institute of Electrical and Electronics Engineers, New York.

IEEE (2005), *Standard for Safety Levels with Respect to Human Exposure to Radio Frequency Electromagnetic Fields, 3 kHz to 300 GHz*, Standard C95.1, Institute of Electrical and Electronics Engineers, New York.

ILO (1994), *Protection of Workers from Power Frequency Electric and Magnetic Field, Occupational Safety and Health Series*, No. 69, International Labour Organization, Geneva, Switzerland.

ISO (1994), *Safety Data Sheets for Chemical Products*, ISO 11014:1994, International Organization for Standardization, Geneva, Switzerland.

ISO (1995), *Air Quality: Particle Size Fraction Definitions for Health-Related Sampling*, ISO 7708:1995, International Organization for Standardization, Geneva, Switzerland.

ISO (1999), *Workplace Atmospheres: Protocol for Evaluating the Performance of Diffusive Samplers*, ISO 16107:1999, International Organization for Standardization, Geneva, Switzerland.

ISO (2000), *Workplace Air: Determination of Metals and Metalloids in Airborne Particulate Matter by Inductively Coupled Plasma Atomic Emission Spectrometry*, Part 1, *Sampling*, ISO 15202-1:2000, International Organization for Standardization, Geneva, Switzerland.

ISO (2001a), *Workplace Air: Determination of Metals and Metalloids in Airborne Particulate Matter by Inductively Coupled Plasma Atomic Emission Spectrometry*, Part 2, *Sample Preparation*, ISO 15202-2:2001, International Organization for Standardization, Geneva, Switzerland.

ISO (2001b), *Workplace Air Quality: Sampling and Analysis of Volatile Organic Compounds by Solvent Desorption/Gas Chromatography*, Part 1, *Pumped Sampling Method*, ISO 16200-1:2001, International Organization for Standardization, Geneva, Switzerland.

ISO (2001c), *Workplace Air Quality: Sampling and Analysis of Volatile Organic Compounds by Solvent Desorption/Gas Chromatography*, Part 2, *Diffusive Sampling Method*, ISO 16200-2:2001, International Organization for Standardization, Geneva, Switzerland.

ISO (2005), *General Requirement for the Competence of Testing and Calibration Laboratories*, ISO/IEC 17025:2005, International Organization for Standardization, Geneva, Switzerland.

Jensen, P. A., and Schafer, M. P. (1998), "Sampling and Characterization of Bioaerosols," retrieved October 13, 2005, from http:/www.cdc.gov/niosh/nmam/pdfs/chapter-j.pdf.

Malmros, P., Sigsgaard, T., and Bach, B. (1992), "Occupational Health Problems Due to Garbage Sorting," *Waste Management Research*, Vol. 10, pp. 227–234.

NIOSH (1994), *Manual of Analytical Methods*, 4th ed., National Institute for Occupational Safety and Health, Cincinnati, OH.

OSHA (2001), *Analytical Methods Manual*, Occupational Safety and Health Administration, Washington, DC.

Pośniak, M., and Skowroń, J. (2000), "Polish System of Assessing Occupational Risk Posed by Chemical Compounds," *International Journal of Occupational Safety and Ergonomics*, Special Issue, pp. 103–109

Reilly, P. J. (1998), *Applied Bioelectricity: From Electrical Stimulation to Electropathology*, Springer-Verlag, New York.

Skowroń, J., and Gołofit-Szymczak, M. (2004), "Microbiological Air Pollution in the Working Environment: Sources, Types and Monitoring," *Bromatology and Toxicological Chemistry*, Vol. 37, No. 1, pp. 91–98 (in Polish).

UNEP/WHO/IRPA (U.N. Environment Programme/World Health Organization/International Radiation Protection Association) (1993), *Electromagnetic Fields (300 Hz –300 GHz)*, Environmental Health Criteria 137, World Health Organization, Geneva, Switzerland.

PART 6
PERFORMANCE MODELING

CHAPTER 36

MODELING HUMAN PERFORMANCE IN COMPLEX SYSTEMS

K. Ronald Laughery, Jr., Christian Lebiere, and Susan Archer
Micro Analysis and Design, Inc.
Boulder, Colorado

1 INTRODUCTION

Over the past few decades, human factors and ergonomics practitioners have been called upon increasingly early in the system design and development process. Early inputs from all disciplines result in better and more integrated designs, as well as lower costs, than if one or more disciplines is solely in charge, finds out late in the development stage that changes are required, and then calls upon the expertise of the other disciplines. Our goal as human factors and ergonomics practitioners should be to provide substantive and well-supported input regarding the human(s), his or her interaction(s) with the system, and the resulting total performance. Total performance includes a number of converging measures, including task latency, type and probability of errors, quality of performance, and workload measures. Furthermore, we should be prepared to provide this input from the earliest stages of system concept development and then throughout the entire system or product life cycle.

To meet this challenge, many human factors and ergonomics tools and technologies have evolved over the years to support early analysis and design. Two specific types of technologies are design guidance (e.g., Boff et al., 1986; O'Hara et al., 1995) and high-fidelity rapid prototyping of user interfaces (e.g., Dahl et al., 1995). Design guidance technologies, either in the form of handbooks or computerized decision support systems, put selected portions of the human factors and ergonomics knowledge base at the fingertips of the designer, often in a form tailored to a particular problem, such as nuclear power plant design or Unix computer interface design. However, design guides have the shortcoming that they do not often provide methods for making quantitative trade-offs in *system* performance as a function of design. For example, design guides may tell us that a high-resolution color display will be better than a black-and-white display, and they may even tell us the value in terms of increased response time and reduced error rates. However, this type of guidance will rarely provide good insight into the value of this improved element of the human's performance to the *overall system's* performance. As such, design guidance has limited value for providing concrete input to *system-level performance prediction*.

Rapid prototyping, on the other hand, supports analysis of how a specific design and task allocation will affect human and system-level performance. The disadvantage of prototyping, as with all human-subject experimentation, is that it can be slow and costly. In particular, prototypes of hardware-based systems, such as aircraft and machinery, are very expensive to develop, particularly at early design stages when there are many widely divergent design concepts. Despite the expense, hardware and software prototyping are important tools for the human factors practitioner, and their use is growing in virtually every application area.

Although these technologies are valuable to the human factors practitioner, what is often needed is an integrating methodology that can extrapolate from the base of human factors and ergonomics data, as reflected in design guides and the literature, to support system-level performance predictions as a function of design alternatives. This methodology should also bind with rapid prototyping and experimentation in a mutually supportive and iterative way. As has become the case in many engineering disciplines, a prime candidate for this integrating methodology is computer modeling and simulation.

Computer modeling of human behavior and performance is not a new endeavor. Computer models of complex cognitive behavior have been around for over 20 years (e.g., Newell and Simon, 1972; Card et al., 1983) and tools for computer modeling of task-level performance have been available since the 1970s (e.g., Wortman et al., 1978). However, three trends have changed appreciably in the past decade to promote the use of computer modeling and simulation of human performance as a standard tool for the practitioner. First are the rapid increase in computer power and the associated development of easier-to-use modeling tools. People with an interest in predicting human performance through simulation can select from a variety of computer-based tools (for a comprehensive list of these tools, see McMillan et al., 1989). Second is the increased focus by the research community on the development of *predictive* models of human performance rather than simply descriptive models. For example, the GOMS model (Gray et al., 1993) represents the integration of research results into a model for making predictions of how humans will perform in a realistic task environment. Another example is the research in cognitive workload that has been represented as computer algorithms (e.g., McCracken and Aldrich, 1984; Farmer et al., 1995). Given a description of the tasks and equipment with which humans are engaged, these algorithms support assessment of when workload-related performance problems are likely to occur, and often include identification of the quantitative impact of those problems on overall system performance (Hahler et al., 1991). These algorithms are particularly useful when embedded as key components in computer simulation models of the tasks and the environment. Third is the integration of those algorithms into cognitive architectures that integrate cognition, perception, and action into a single computational framework that can be applied to a broad range of tasks, from basic laboratory experiments used to validate the architectural mechanisms to predict operator performance on complex practical tasks (Gray et al., 1997).

Perhaps the most powerful aspect of computer modeling and simulation is that it provides a method through which the human factors and ergonomics team can "step up to the table" with the other engineering disciplines, which also rely increasingly on quantitative computer models. What we discuss in this chapter are the methods through which the human factors and ergonomics community can contribute early to system design trade-off decisions.

1.1 Chapter Objectives

In this chapter we discuss some existing computer tools for modeling and simulating human–system performance. It is intended to provide the reader with an understanding of the types of human factors and ergonomics issues that can be addressed with modeling and simulation and some of the tools that are now available to assist the human factors and ergonomics specialist in conducting model-based analyses, and an appreciation of the level of expertise and effort that will be required to use these technologies. We begin with two caveats. The first is that we are not yet at a point where computer modeling of human behavior allows sufficiently accurate predictions that no other analysis method (e.g., prototyping) is likely to be needed. In the early stages of system concept development, high-level modeling of human–system interaction may be all that is possible. As the system moves through the design process, human factors and ergonomics designers will often want to augment modeling and simulation predictions with prototyping and experimentation. In addition to providing high-fidelity system performance data, these data can be used to constrain, enhance, and refine the models. This concept of human performance modeling supporting and being supported by experimentation with human subjects is represented in Figure 1. In essence, simulation provides the human factors and ergonomics practitioner with a means of extending the knowledge base of human factors and of amplifying the effectiveness of limited experimentation.

The second caveat is that the technologies discussed here are evolving rapidly. We can be certain that every tool discussed is undergoing constant change, and that new modeling tools are being developed. We are discussing computer-based tools, and we expect the pace of change in these tools to mirror the pace in other software tools, such as word processors, spreadsheets,

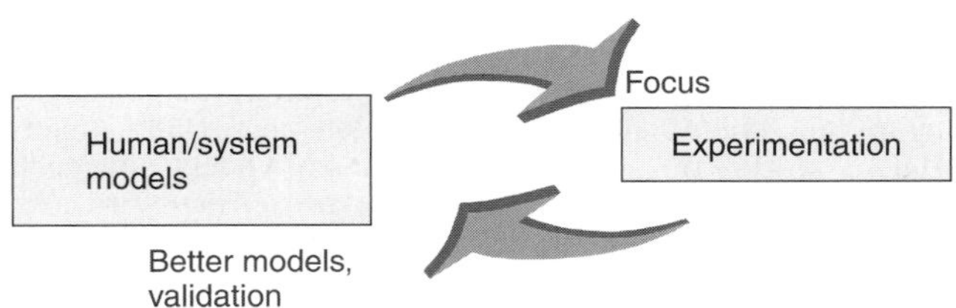

Figure 1 Synergy between modeling and experimentation.

presentation and productivity tools, and Internet-based applications. These detailed discussions of several of the modeling tools are included to facilitate better understanding of human performance modeling tools. We encourage the reader to follow citations in this chapter to assess the current state of any tool. Most of these modeling tools have large, active user communities that maintain Web sites to provide introductory tutorials, software downloads, validated models, and published papers. These resources are invaluable both for the experienced modeler trying to stay abreast of recent developments and the novice user attempting to get up to speed on a new technology.

2 QUESTIONS ADDRESSED BY HUMAN PERFORMANCE MODELS

Below are a few classes of problems to which human–system modeling has been applied:

- How long will it take a human or team of humans to perform a set of tasks as a function of system design, task allocation, and individual capabilities?
- What are the trade-offs in performance for different combinations of design, task allocation, and individual capability selections?
- What are the workload demands on the human as a function of system design and automation?
- How will human performance and resulting system performance change as the demands of the environment change?
- How many people are required on a team to ensure safe, successful performance?
- How should tasks be allocated to optimize performance?
- How will environmental stressors such as heat, cold, or the use of drugs affect human–system performance?

The list above is a sample rather than an exhaustive list. The tools we discuss in this chapter are inherently flexible, and we consistently discover that these tools can be used to solve problems that the tool developers never conceived. To assess the potential of simulation to answer questions, in every potential human performance modeling project we should first determine the specific questions that the project is trying to answer. Then we can conduct a critical assessment of what is important in the human–machine system being modeled. This will define the required content and fidelity of the model. The questions that should be considered about the system include:

1. *Human performance representation.* What time or duration of performance is important? How is human performance initiated, and what resolution of behavior is required? What aspects of human performance, including task management, load management, and goal management, are expected? How much is known and constrained about the knowledge and strategies that human users bring to bear on this task?
2. *Equipment representation.* What equipment is used to accomplish the task? To what level of functional and physical description can and should equipment be represented? Is it operable by more than one human or system component?
3. *Interface requirements.* What information needs to be conveyed to the humans, and when? Is transformation of information required? How often is information updated and monitored?
4. *Control requirements.* What processes need to be controlled by the human, and to what level of resolution? How much attention is required by the human to perform control changes?
5. *Logical and physical constraints.* How is performance supported through equipment operability and procedural sequences? What alarms and alerts should be represented?
6. *Simulation driver.* What makes the system function? The occurrence of well-defined events (e.g., a procedure), the passage of time (e.g., the control of a vehicle), or a hybrid of both?

By defining the purpose of the model and then answering the questions above, the human factors practitioner will get a sense of what is important in the system and therefore what may need to be represented in a model. In using human performance models, perhaps the most significant task of the human factors practitioner will be to determine what aspects of the human–machine system to include in the model and what to leave out. Many modeling studies have failed because of the inclusion of too many factors that although a part of human–system performance, were not system performance drivers. Consequently, the models become overly complex and expensive to develop. In our experience, it is better to begin with a model with too few aspects of the system represented and then add to it than to begin a modeling project by trying to model everything. The first approach may succeed, whereas the second is often doomed.

Additionally, the human factors practitioner should consider the measures of effectiveness of the system that the model should be designed to predict. In building the model, it is important to remember that the goal will be to predict measures of human performance that will affect system performance. Therefore, a clear definition of what is important to performance is necessary. The following aspects of performance measures should be considered:

1. *Success criteria.* What operational success measures are important to the system? Can these be stated in relative terms, or must they be measured in absolute terms?
2. *Range of performance to be studied.* What experimental variables are to be explored by the model? How important is it to establish a range of performance for each experimental

condition as a function of the stochastic (i.e., random) behavior of the system?

By asking the foregoing questions prior to beginning a modeling project, the human factors practitioner can develop a better sense of what is important in the system in terms of both aspects that drive system performance and the measures of effectiveness that are truly of interest. Then, and only then, can a human performance modeling project begin with a reasonable hope of success.

In the remainder of this chapter we discuss two classes of modeling tools for human performance simulation, then report on recent efforts to unify those two complementary classes in order to leverage their strengths and alleviate their shortcomings. After discussing each class of modeling tool, we provide specific examples of a modeling tool and then provide case studies about how these tools have been used in answering real human performance questions.

3 CLASSES OF SIMULATION MODELS

Human performance can be highly complex and involve many types of processes and behavior. Over the years many models have been developed that predict sensory processes (e.g., Gawron et al., 1983), aspects of human cognition (e.g., Newell, 1990), and human motor response (e.g., Fitts's law). The current literature in the areas of cognitive engineering, error analysis, and human–computer interaction contains many models, descriptions, methodologies, metaphors, and functional analogies. However, in this chapter, we are not focusing on the models of these individual elements of human behavior but rather, on models that can be used to describe human performance in systems. These human–system performance models typically include some of these elemental behavioral models as components, but provide a structural framework that allows them to be integrated with each other and put in the context of human performance of tasks in systems.

We separate the world of human–system performance models into two general categories that can be described as reductionist models and first-principle models. *Reductionist models* use human–system task sequences as the primary organizing structure, as shown in Figure 2. The individual models of human behavior for each task or task element are connected to this task sequencing structure. We refer to it as reductionist because the process of modeling human behavior involves taking the larger aspects of human–system behavior (e.g., "perform the mission") and then reducing them successively to smaller elements of behavior (e.g., "perform the function," "perform the tasks"). This continues until a level of decomposition is reached at which reasonable estimates of human performance for the task elements can be made. One can also think of this as a top-down approach to modeling human–system performance. The example of this type of modeling that we use in this chapter is *task network modeling*, where the basis of the human–system model is a task analysis.

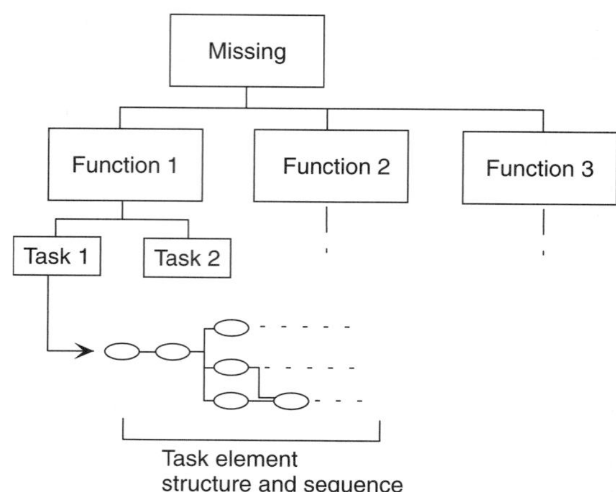

Figure 2 Reductionist models of human performance.

First-principle models of human behavior are structured around an organizing framework that represents the underlying goals, principles, and mechanisms of human performance (Figure 3). Tools that support first-principle modeling of human behavior have structures embedded in them that represent elemental aspects of human performance. For example, these models might directly represent processes such as goal-seeking behavior, task scheduling, sensation and perception, cognition, and motor output. In turn, those processes might invoke fundamental actions such as shifts of attention, memory retrieval, and conflict resolution among competing courses of action. To use tools that support first-principle modeling, one must describe how the system and environment interacts with the human processes being modeled. In this chapter we focus on the Adaptive Control of Thought–Rational (ACT-R) cognitive architecture (Anderson and Lebiere, 1998).

It is worth noting that these two modeling strategies are not mutually exclusive and, in fact, can be mutually supportive in any given modeling project. Often, when one is modeling using a reductionist approach, one needs models of basic human behavior to represent behavioral phenomena accurately and therefore must draw on elements of first-principle models. Alternatively, when one is modeling human–system performance using a first-principle approach, some aspects of human–system performance and interrelationships between tasks may be more easily defined using a reductionist approach. Both classes of model have been used to model individual and team performance. It is also worth noting that recent advances in human performance modeling tool development are blurring the distinctions between these two classes (e.g., Hoagland et al., 2001). Increased emphasis on interoperability between models has caused researchers and developers to focus on integrating reductionist and first-principle models. In the final section of this chapter we present one such attempt at integrating the ACT-R cognitive

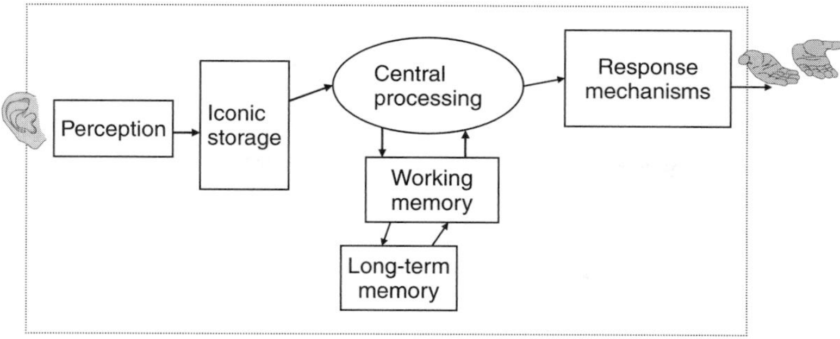

Figure 3 First-principle models of human performance.

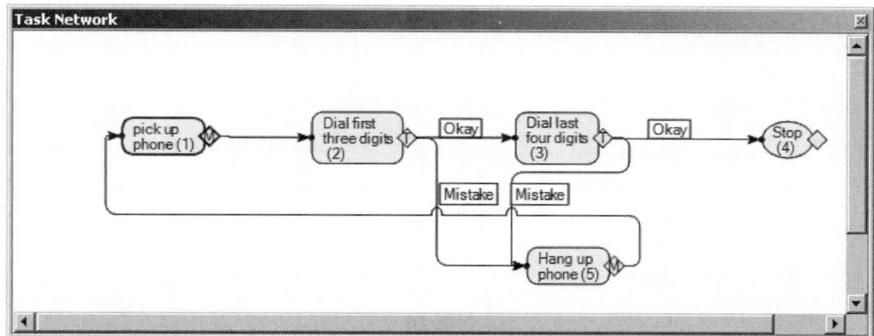

Figure 4 Task network model representing a human dialing a telephone.

architecture with the Improved Performance Research Integration Tool (IMPRINT).

4 REDUCTIONIST APPROACH: TASK NETWORK MODELING

One technology that has proven useful for predicting human–system performance is *task network modeling*. In a task network model, human performance is decomposed into tasks. The fidelity of this decomposition can be selective, with some functions being decomposed several levels and others just one or two. This is, in human factors engineering terms, the task analysis. The sequence of tasks is defined by constructing a *task network*. This concept is illustrated in Figure 4, which presents a sample task network for dialing a telephone.

Task network modeling is an approach to modeling human performance in complex systems that has evolved for several reasons. First, it is a reasonable means for extending the human factors staple: the task analysis. Task analyses organized by task sequence are the basis for the task network model. Second, task network models can include sophisticated submodels of the system hardware and software to create a closed-loop representation of relevant aspects of the human–machine system. Third, task network modeling is relatively easy to use and understand. Recent advancements in task network modeling technology have made this technology more accessible to human factors practitioners. Finally, task network modeling can provide efficient, valid, and useful input to many types of issues. With a task network model, the human factors engineer can examine a design (e.g., control panel redesign) and address questions such as "How much longer will it take to perform this procedure?" and "Will there be an increase in the error rate?" Generally, task network models can be developed in less time and with substantially less effort than would be required if a prototype were developed and human subjects used. However, as stated before, for revolutionary designs, modeling may not alleviate the need for empirical data collection.

Task network models of human performance have been subjected to validation studies with favorable results (e.g., Lawless et al., 1995; Engh et al., 1998). However, as with any modeling approach, the real level at which validation must be considered is with respect to a particular model, not with respect to the general approach.

4.1 Components of a Task Network Model

To represent complex, dynamic human–system behavior, many aspects of the system may need to be modeled in addition to simply task lists and sequence. In this section we use the task network modeling tool *Micro Saint Sharp* as an example. The basic ingredient of a Micro Saint Sharp task network model is

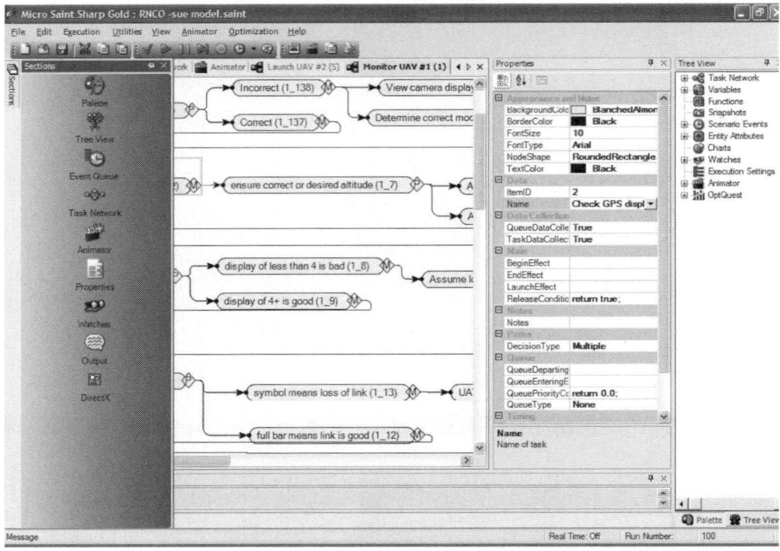

Figure 5 Main window in Micro Saint Sharp for task network construction and viewing.

the task analysis as represented by a network or series of networks. The level of system decomposition (i.e., how finely we decompose the tasks) and the amount of the system that is simulated depends on the particular problem. For example, in a power plant model, one can create separate networks for each of the operators and one for the power plant itself. Although the networks may be independent, performance of the tasks can be interrelated through shared variables. The relationships among different components of the system, represented by different segments of the network, can then communicate through changes in these shared variables. For example, when an operator manipulates a control, this may initiate an "open valve" task in a network representing the plant. This could ripple through to a network representing other operators and subsystems and their response to the open valve. This basic task network is built in Micro Saint Sharp via a point-and-click drawing palette. Through this environment, the user creates a network as shown in Figure 5. Networks can be embedded within networks, allowing for hierarchical construction. In addition, the shape of the nodes on the diagram can be chosen to represent specific types of activity.

To reflect complex task behavior and interrelationships, more detailed characteristics of the tasks need to be defined. By double-clicking on a task, the user opens up the task description window, as shown in Figure 6. Below are descriptions of each of the items on the tabs in this window.

- *Task number.* This value is an arbitrary number for task referencing.
- *Task name.* This parameter contains a text string used to identify the task.
- *Time distribution.* Micro Saint Sharp conducts Monte Carlo simulations with task performance times sampled from a distribution as defined by this option (e.g., normal, beta, exponential).
- *Mean time.* This parameter defines average task performance time for this task. This can be a number, equation, or algorithm, as can all values in the fields described below.
- *Standard deviation.* This value contains the standard deviation of the task performance time, assuming that the user has chosen a distribution that is parameterized by a standard deviation.
- *Release condition.* Data in this field determine when a task begins executing. For example, a condition stating that this task will not start before an operator is available might be represented by a release condition such as the following:

$$\text{operator} >= 1;$$

In other words, for the task to begin, at least one operator must be available. If all operators are busy, the value of the variable "operator" would equal zero until a task is completed, at which time an operator becomes available. This task would wait until the condition was true before beginning execution, which would probably occur as a result of the operator completing the task that he or she is currently performing.

- *Beginning effect.* This field permits the user to define how the system will change as a result of the commencement of this task. For example, if this task used an operator that other tasks might need, we could set the following condition to show that the operator is unavailable while he

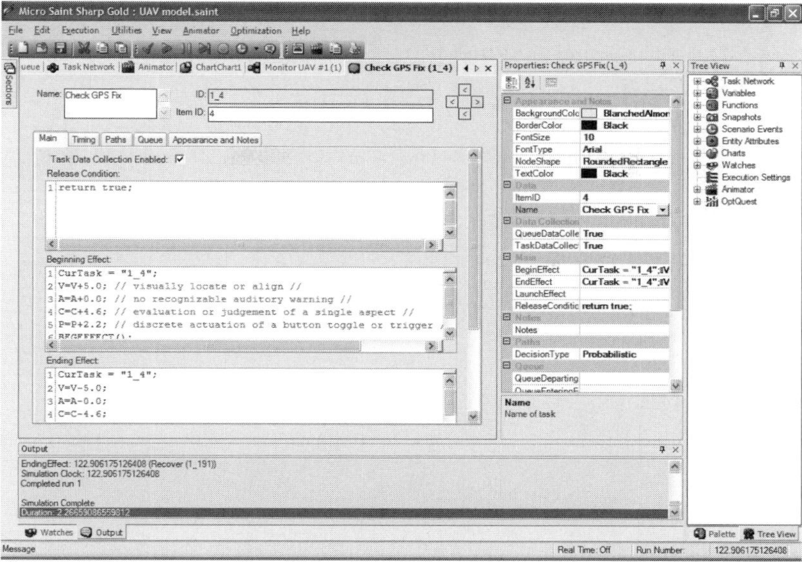

Figure 6 User interface in Micro Saint Sharp for providing input on a task.

or she performed this task:

$$\text{operator} = \text{operator} - 1;$$

Assignment and modification of variables in beginning effects are one principal way in which tasks are interrelated.

- *Launch effect.* This data element is similar to a task beginning effect but is used to launch high-resolution animation of the task.
- *Ending effect.* This field contains the definition of how the system will change as a result of the completion of this task. From the previous example, when this task was complete and the operator became available, we could set the ending effect as follows:

$$\text{operator} = \text{operator} + 1;$$

at which point another task waiting for an operator to become available could begin. Ending effects are another important way in which tasks can be interrelated through the assignment and modification of variables.

Another notable aspect of the task network diagram window shown in Figure 5 is the diamond-shaped icon that follows every task. This icon encapsulates data that describe the paths and the associated logic that will be executed when this task is completed. Often, this logic represents a human decision-making process. In that case, the branches align to potential courses of action that the modeled human could select. To define the decision logic, the Micro Saint Sharp user would use the "Paths" tab on the task description dialogue, as shown in Figure 7. There are three general types of decisions to model:

- *Probabilistic.* In probabilistic decisions, the human will begin one of several tasks based on a random draw weighted by the probabilistic branch value. These weightings can be dynamically calculated to represent the current context of the decision. For example, this decision type might be used to represent human error likelihoods and would be connected to the subsequent tasks that would be performed.
- *Tactical.* In tactical decisions, the human will begin one of several tasks based on the branch with the highest "value." This could be used to model the many types of rule-based decisions that humans make, as illustrated in Figure 7.
- *Multiple.* This would be used to begin several tasks at the completion of this task, such as when one human issues a command that begins other crew members' activities.

The fields on Figure 7 labeled "decision code" represent the values associated with each branch. The values can be numbers, expressions, or complicated algorithms defining the probability (for probabilistic branches) or the desirability (for tactical and multiple branches) of taking a particular branch in the network. Again, any value on this screen can be not simply numbers but also variables, algebraic expressions, logical expressions, or groups of algebraic and logical expressions that would, essentially, form a subroutine. As the model executes, Micro Saint Sharp includes a parser that evaluates the expressions included in the branching logic when it is encountered in the task

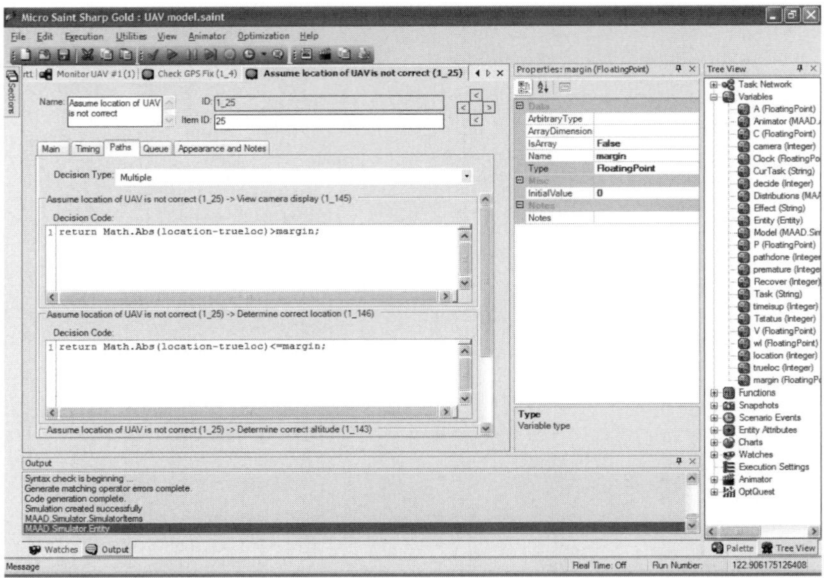

Figure 7 User interface in Micro Saint Sharp for defining task branching decision logic.

network flow. This results in a dynamic network in which the flow through the tasks can be controlled with variables that represent equipment state, scenario context, or the task loading of the humans in the system, to name a few examples. It is the power of this parser that provides many task network models with the ability to address complex problems.

There are other aspects of task network model development. Some items define a simulation scenario, defining continuous processes within the model, and defining queues in front of tasks. Further details of these features can be obtained from the Micro Saint Sharp *User's Guide* (Micro Analysis and Design, 2004). As a model is being developed and debugged, the user can execute the model to test it and collect data. The user can rearrange, open, and close a variety of windows to represent a variety of display modes providing differing levels of information during execution. The simulation speed can also be controlled, to include pausing after every simulated task. Typically, during execution the user will display the task network on the screen, and tasks that are currently executing will be highlighted. In this mode, the analyst can get a very clear picture of what events are occurring in what sequence in the model, greatly aiding debugging. Additionally, an animator mode is available. In this mode, the user can draw a graphical representation of the system. Changes on the graphical background can be tied to the task flow, providing a powerful method to communicate the model's findings to stakeholders. Figure 8 presents a sample display during model animation. Once a model is executed and data are collected, the analyst has a number of alternatives for data analysis. The data created during a model execution can be reviewed within Micro Saint Sharp or can be exported to statistical and graphics packages for postprocessing.

As stated before, the basis for task network models of human performance is the mainstay of human engineering analysis, the task analysis. Much of the information discussed above is generally included in the task analysis. Task network modeling, however, greatly increases the power of task analysis since the ability to simulate a task network with a computer permits *prediction* of human performance rather than simply the *description* of human performance that a task analysis provides. What may not be as apparent, however, is the power of task network modeling as a means of modeling human performance in *systems*. Simply by describing the systems activities in this step-by-step manner, complex models of the system can be developed where the human's interaction with the system can be represented in a closed-loop manner. The preceding discussion, in addition to being an introduction to the concepts, is also intended to support the argument that task network modeling is a mature technology ready for application in a wide range of problem domains.

4.2 Task Network Model of a Process Control Operator

This simple hypothetical example illustrates how many of the basic concepts of task network modeling can be applied to studying human performance in a process control environment. It is intended to illustrate many of the concepts described above. The simple human task that we want to model is of an operator responding to an annunciator. The procedure requires that the operator compare readings on two meters. Based on

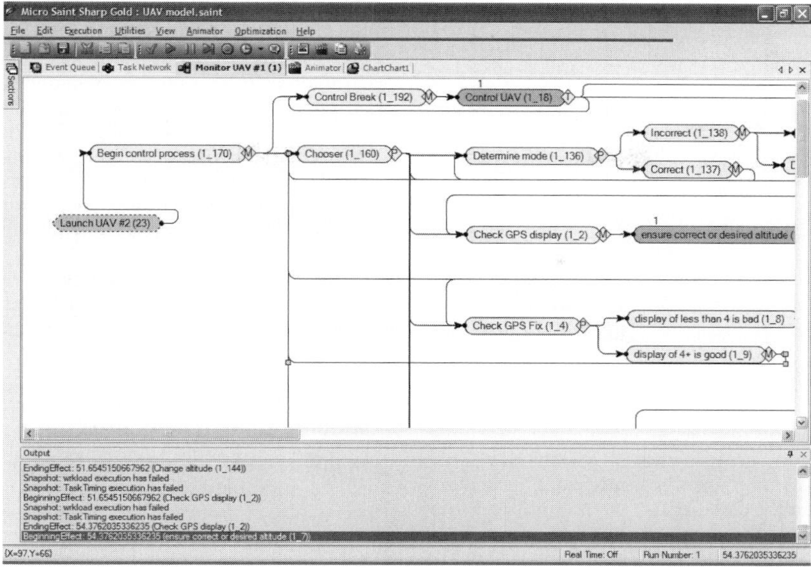

Figure 8 Task network animation during model execution in Micro Saint Sharp.

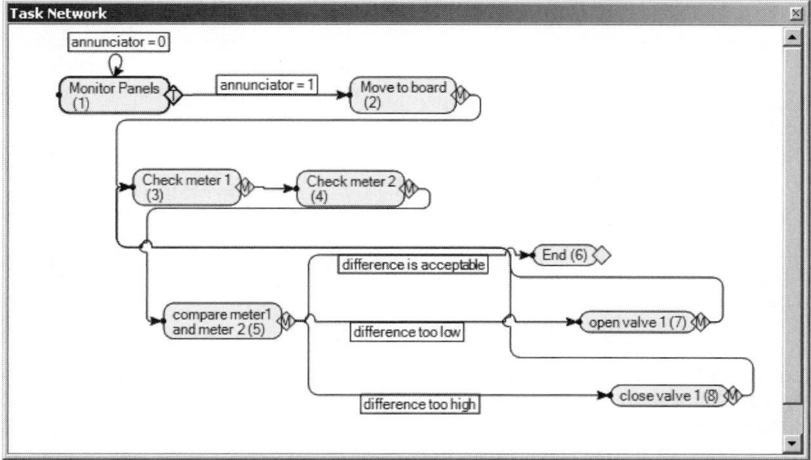

Figure 9 Task network model of a process control operator responding to an annunciator.

the relative values of these readings, the operator must either open or close a valve until the values on the two meters are nearly the same. The task network in Figure 9 represents the operator activities for this model. Also, to allow the study of the effects of different plant dynamics (e.g., control lags), a simple one-node model of the line in which the valve is being opened is included in Figure 10.

The operator portion of the model will run the "monitor panels" task until the values of the variables "meter1" and "meter2" are different. The simulation could begin with these values being equal and then precipitate a change in values based on what is referred to as a *scenario event* (e.g., an event representing the effects of a line break on a plant state). This event could be as simple as

$$\text{meter1} = \text{meter1} + 2.0;$$

or as complex as an expression defining the change in the meter as a function of line-break size, flow rates, and so on. An issue that consistently arises in model construction is how complex the plant system model should be. If the problem under study

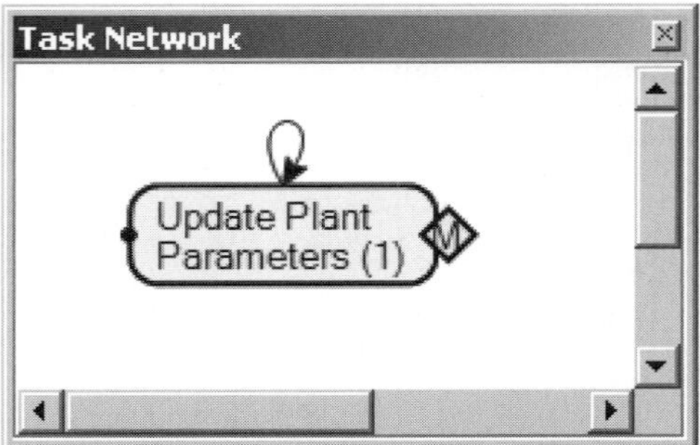

Figure 10 Simple one-node model of the plant integrated with the detailed operator model.

is purely operator performance, simple models will usually suffice. However, if overall plant behavior is of interest, the models of plant dynamics, such as meter values, are more important. Again, we recommend the "start simple" approach whenever possible.

When the transient occurs and the values of meter1 and meter2 start to diverge, the annunciator signal will trigger. This annunciator would be triggered in the plant portion of the model by a task-ending effect such as

if meter1 <> meter2 then annunciator = 1;

Once the plant model sets the value of the variable "annunciator" to 1, the operator will begin to move to the appropriate board. Then the operator will continue through a loop to check the values for meter1 and meter2 and either open valve1, close valve1, or make no change. The determination of whether to make a control input is determined by the difference in values between the two meters. If the value is less than the acceptable threshold, the operator would open the valve further. If the value is greater than the threshold, the operator would close the valve. This opening and closing of the valve would be represented by changes in the value of the variable valve1 as a task-ending effect of the tasks open valve1 and close valve1. In this simple model, operators do not consider rates of change in values for meter1 and therefore would get into an operator-induced oscillation if there were any response lag. A more sophisticated operator model could use rates of change in the value for meter1 in deciding whether to open or close valves.

Again, this is a very small model reflecting simple operator activity on one control via a review of two displays. However, it illustrates how large models of operator teams looking at numerous controls and manipulating many displays could be built via the same building blocks used in this model. The central concepts of a task network and shared variable reflecting human–system dynamics remain the same.

Given a task network model of a process control operator in a "current" control room, how might the model be modified to address human-centered design questions? Some examples are (1) modifying task times based on changes in the time required to access a new display; (2) modifying task times and accuracies based on changes in the content and format of displays; (3) changing task sequence, eliminating tasks, and/or adding tasks based on changes in plant procedures; (4) changing allocation of tasks and ensuing task sequence based on reallocation of tasks among operators; and, (5) changing task times and accuracies based on stressors such as sleep loss or the effects of circadian rhythm. This is not intended as a definitive list of all the ways that these models may be used to study design or operations concepts but should illustrate how these models can be used to address design and operational issues.

4.3 Use of Task Network Modeling to Address Specific Design Concerns

In this section we examine two case studies in the use of task network simulation for studying human performance issues. The first case study explores how task network modeling can be used to assess task allocation issues in a cognitively demanding environment. The second example explores how task network modeling has been used to extend laboratory and field research on human performance under stress to new task environments. We should state clearly that these examples are intended to be representative of the types of issues that task network modeling can address as well as approaches to modeling human performance with respect to these issues. *They are not intended to be comprehensive with respect to either the issues that might be addressed or the possible techniques that the human factors practitioner might apply.* Simulation modeling is a technology whose application leaves much room for creativity on the part of the human factors practitioner with respect to application areas and methods. These two case studies are representative.

4.3.1 Crew Workload Evaluation

Perhaps the greatest contributor to human error in many systems is the extensive workload placed on the human operator. The inability of the operator to cope effectively with all of his or her information and responsibilities contributes to many accidents and inefficiencies. In recognition of this problem, new automation technologies have been introduced to reduce workload during periods of high stress. Some of these technologies are in the form of enhanced controls and displays, some are in the form of tools that "push" information to the operator and alert the operator in order to focus attention, and still others consist of adaptive tools that "take over" tasks when they sense that the operator is overloaded. Unfortunately, these technical solutions often introduce new tasks to be performed that affect the visual, auditory, and/or psychomotor workload of the operators.

Recently, new concepts in crew coordination have focused on better management of human workload. This area shows tremendous promise and is benefiting from efforts of human factors researchers. However, their efforts are hindered because there are limited opportunities to examine empirically the performance of different combinations of equipment and crew composition in a realistic scenario or context. Additionally, high workload is not typically caused by a single task but by situations in which multiple tasks must be performed or managed simultaneously. It is not simply the quantity of tasks that can lead to overload, but also depends on the composition of those tasks. For example, two cognitive tasks being performed in parallel are much more effortful than a simple motor task and an oral communication task being performed together. The occurrence of these situations will not typically be discovered through normal human engineering task analysis or subjective workload analysis until there is a system to be tested. That is often too late to influence design. To rectify this problem, there has been a significant amount of recent research and development aimed at human workload *prediction* models. Predictive models allow the designers of a system to estimate operator workload *without human subject experimentation*. From this and other research, a solid theoretical basis for human workload prediction has evolved as is described in Wickens (1984).

In this section we discuss a study using task network modeling to predict the impact of task allocation on human workload. Although these examples are posed in the context of the design of a military system, the same techniques have been used in nonmilitary applications such as process control and user–computer interface design.

4.3.2 Modeling the Workload of a Future Command and Control Process

The Army command and control (C2) community is concerned with how new information technology and organizational changes projected for tomorrow's battlefield will affect soldier tasks and workload. To address this concern, an effort was undertaken to model soldier performance under current and future operational conditions. In this way, the impact of performance differences could be quantitatively assessed so that equipment and doctrine design could be influenced in a timely and effective manner.

In one C2 project, the primary concern was to determine how tasks should be allocated and automated such that a C2 team could evaluate all the relevant data and make decisions within an environment with particularly high time pressure. Specifically, the effort was to address the following key questions:

- How many crew members do you need?
- How do you divide tasks among jobs?
- How does decision authority flow?
- Can the crew meet decision timeline requirements?
- Is needed information usable and accessible?

Task network modeling was used to study crew member, task, and scenario combinations in order to examine these questions. Figure 11 shows the top-level diagram of the task network. Essentially, the crew members receive and monitor information about the system and the environment until an event occurs that pushes them out of the 10000 and 20000 networks into either a series of planning tasks or a series of evaluation, decision, direction, and execution tasks. The purpose of the planning task is to update tactical battle plans based on new information received from the system or the environment. Receipt of new intelligence data about the enemy's intention or capability is an example of an event that would cause crew members to undertake planning tasks. Similarly, receipt of information from the system about resource limitations might trigger the crew members to proceed down the alternative path (through evaluate to execute). Specifically, limited resources might cause crew members to evaluate whether the engagement is proceeding appropriately (30000), decide how to adjust system parameters (40000), direct the appropriate response to the correct level of command (50000), and then execute

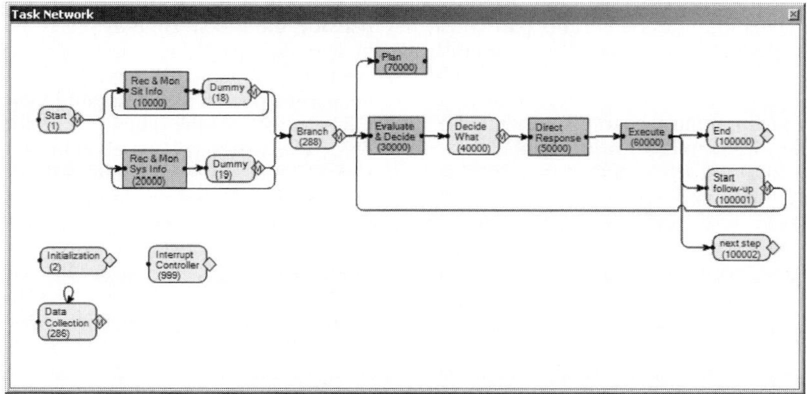

Figure 11 Upper-level task network.

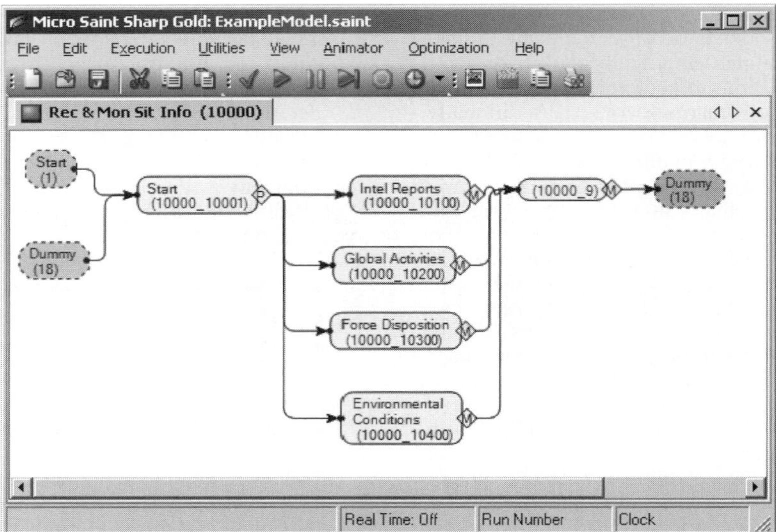

Figure 12 Second-level task network.

the order (60000). Upon completion, crew members would return to monitoring the system and situation.

Each rectangle in the task network shown in Figure 11 actually consists of a network of tasks. An example of the tasks that belong to network 10000 are shown in Figure 12. As described in Figure 7, the tasks in network 10000 are linked by probabilistic and tactical decisions. Each of the tasks in the C2 task network is associated with several items of human performance data:

- *Task performance time.* These data consist of a mean, standard deviation, and distribution. The data were collected from a combination of three sources: (1) human factors literature (e.g., Fitt's law), (2) empirical studies during operator-in-the-loop simulator exercises, and (3) subject matter experts.
- *Branching logic.* Although the task network indicates a general process flow, this particular model was designed to respond to scenario events. Because of that design decision, each task includes logic to determine the following task. For example, if the scenario is very intense and multiple target tracks are available, crew members would follow a different task flow than if they were performing routine system checks.
- *Release rules.* Logic controlling the number and types of parallel tasks each crew member can perform is contained in each task's release condition.

Since one purpose of the model was to examine various task allocation strategies, the model was designed to incorporate several measures of crew member workload. The basis of this technique is an assumption that excessive human workload is not usually caused by one particular task required of the operator. Rather, the human having to perform several tasks simultaneously leads to overload. Since the factors that cause this type of workload are intricately linked to these dynamic aspects of the human's task requirements, task network modeling provides a good basis for studying how task allocation and sequencing can affect operator workload.

However, task network modeling is not inherently a model of human workload. The only relevant output common to all task network models is the time required to perform a set of tasks and the sequence in which the tasks are performed. Time information alone would suffice for some workload evaluation techniques, such as Siegel and Wolf (1969), whereby workload is estimated by comparing the time available to perform a group of tasks to the time required to perform the tasks. Time available is driven by system performance needs, and time required can be computed with a task network model. However, it has long been recognized that this simplistic analysis misses many aspects of the human's tasks that influence both perceived workload and ensuing performance. At the very least, this approach misses the fact that some pairs of tasks can be performed in combinations better than other pairs of tasks.

One of the most promising theories of operator workload, which is consistent with task network modeling, is the multiple resource theory proposed by Wickens (e.g., Wickens et al., 1983). Simply stated, the multiple resource theory suggests that humans have several different resources that can be tapped simultaneously and with varying levels of interresource conflict and competition. Depending on the nature of

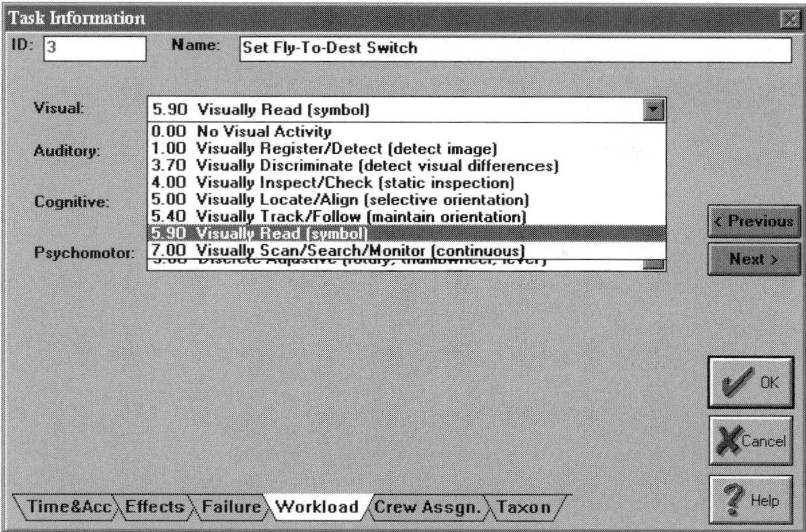

Figure 13 Visual workload scale.

the information-processing tasks required of a human, these resources would have to process information sequentially (if different tasks require the same types of resources) or possibly in parallel (if different tasks required different types of resources). There are many versions of this multiple resource theory in workload literature (e.g., McCracken and Aldrich, 1984; Archer and Adkins, 1999). In this chapter we provide a discussion of the underlying methodology of the basic theory.

Multiple resource workload theory is implemented in a task model in a fairly straightforward manner. First, each task in the task network is characterized by the workload demand required in each human resource, often referred to as a *workload channel*. Examples of commonly used channels include auditory, visual, cognitive, and psychomotor. Particular implementations of the theory vary in the channels that are included and the fidelity with which each channel is measured (high, medium, low vs. 7-point scale). As an example, the scale for visual demand is presented in Figure 13.

Similar scales have been developed for the auditory, cognitive, and psychomotor channels. Using this approach, each operator task can be characterized as requiring some amount of each of the four types of resources, as represented by a value between 1 and 7. All operator tasks can be analyzed with respect to these demands and values assigned accordingly. In performing a set of tasks pursuant to a common goal (e.g., engage an enemy target), crew members frequently must perform several tasks simultaneously, or at least nearly so. For example, they may be required to monitor a communication network while visually searching a display for target track. Given this, the workload literature indicates that the crew member may either accept the increased workload (with some risk of performance degrading) or begin dumping tasks perceived as less important. To factor these two issues into task network simulations, two approaches can be incorporated: (1) evaluate combined operator workload demands for tasks that are being performed concurrently, and/or (2) determine when the operator would begin dumping tasks due to overload.

During a task network simulation, the model of the crew may indicate that they are required to perform several tasks simultaneously. The task network model evaluates total attentional demands for each human resource (e.g., visual, auditory, psychomotor, and cognitive) by combining the attentional demands across all tasks that are being performed simultaneously. This combination leads to an overall workload demand score for each crew member.

To implement this approach in Micro Saint Sharp, the task beginning effect can be used to increment variables that represent the current workload score in each resource. Then, while the tasks are being performed, these variables track attentional demands. When the tasks are completed, the task ending effects can decrement the values of these variables accordingly. Therefore, if these workload variables were recorded and then plotted as the model runs, the output would look something as shown in Figure 14. This result can be used to identify points of high workload throughout the scenario being modeled. The human factors practitioner can then review the tasks that led to the points of high workload and determine whether they should be reallocated or redesigned in order to alleviate the peak. This is a common approach to modeling workload.

Once the task networks were verified with knowledgeable crew members, they became part of the human factors team's analytical test bed. Figure 15

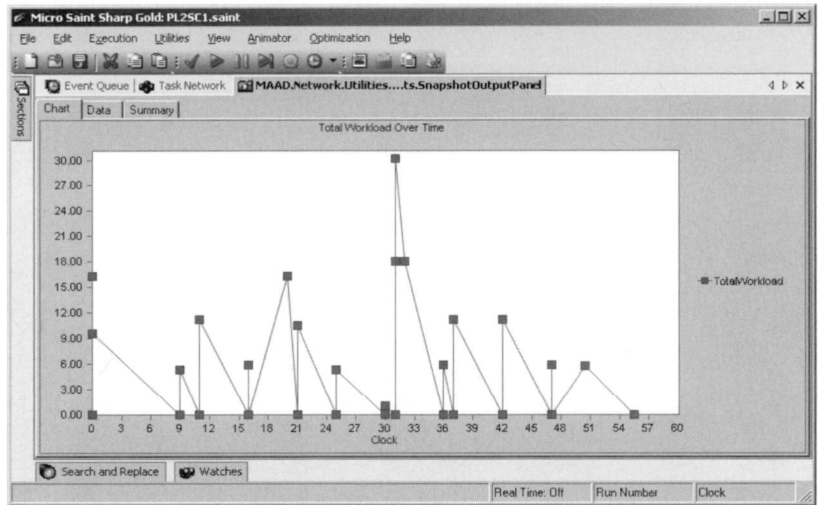

Figure 14 Workload output from a task network model.

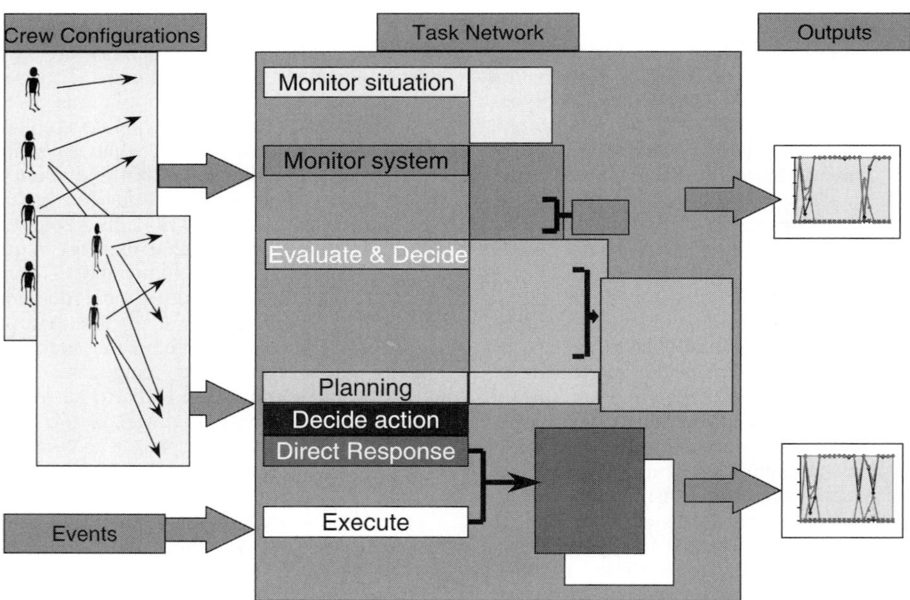

Figure 15 Overall method for examining workload in a complex system.

shows the overall method that can be used to examine aspects of crew member performance across a wide variety of operational scenarios and crew configuration concepts. The center of this diagram, labeled the task network, represents the tasks that the crew performs. The network itself, representing the flow of the tasks, does not change between model runs. Rather, the model has been parameterized so that an event scenario stimulates the network. The left side of the diagram illustrates the types of data that are used to drive the task network model. In this case, those data include crew configurations, or allocations of tasks to different crew members and automation devices, as well as scenario events. The scenario events represent an externally generated time-ordered list of the events that trigger the crew members to perform tasks in the task network. The right side of Figure 15 represents the types of outputs that can be produced from this

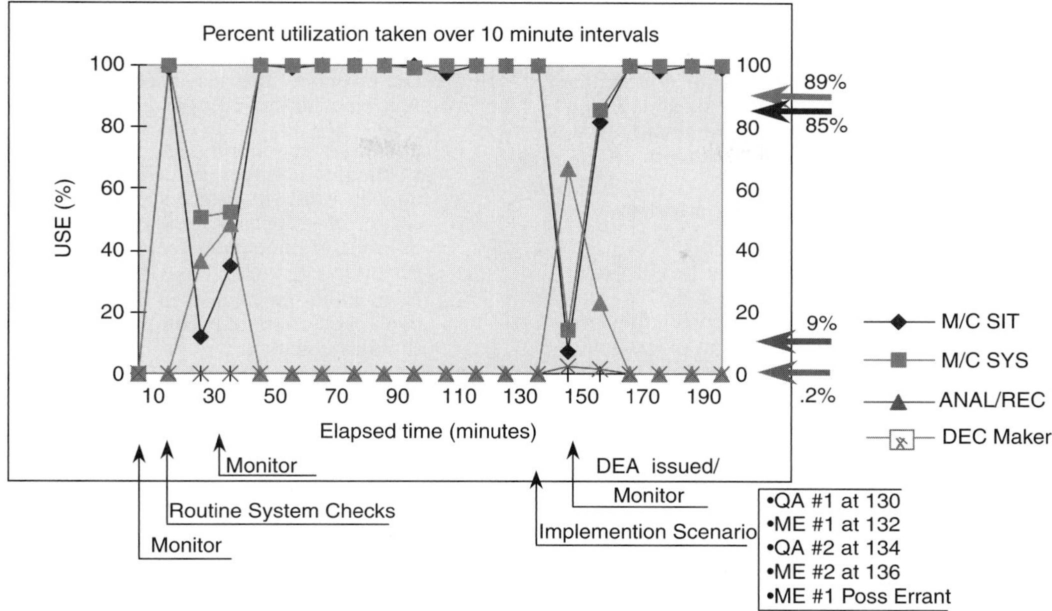

Figure 16 Model predictions of operator utilization over time.

task network model. One of the primary outputs is a crew member workload graph, such as that shown in Figure 14. Another is operator utilization, as shown in Figure 16.

4.3.3 Extensions to Other Environments

The workload analysis methodology described above has recently been developed into a stand-alone task network modeling tool by the Army Research Laboratory (ARL) Human Research and Engineering Directorate (HRED) as part of the Improved Performance Research Integration Tool (IMPRINT) (Archer and Adkins, 1999). IMPRINT integrates task network modeling software with features that specifically support the multiple resource theory of workload discussed above. It provides the human factors practitioner with an environment that supports the analysis of task assignment to crew members based on four factors:

1. *Workload of crew members.* Tasks should be assigned to minimize the amount of time that crew members will spend in situations of excessive workload.
2. *Time performance requirements.* Tasks must be assigned and sequenced so that they are completed within the available time. This consideration is essential since time constraints often will drive the need to perform several tasks simultaneously.
3. *Likelihood of successful performance and consequences of failure.* Tasks must be assigned and sequenced so that they can be completed within a specified accuracy measure.
4. *Access to controls and displays.* Tasks cannot be assigned to crew members that do not have access to the necessary controls and displays.

Of course, there are numerous theoretical questions regarding this simplistic approach to assessing workload in an operational environment. However, even the use of this simple approach has been shown to provide useful insight during design. For example, in a study conducted by the Army (Allender, 1995), a three-man crew design was evaluated using a task network model. The three-man model was constructed using data from a prototype four-man system. From this model-based analysis, the three-man design was found to be unworkable. Later, human subjects experimentation verified that the model's workload predictions were sufficiently accurate to point the design team in a valid direction.

IMPRINT also includes built-in constructs for simulating *workload management* strategies that operators would employ to accommodate points of high operator workload (Plott, 1995). The ultimate result of simulating the workload management strategies is that the operator task network being modeled is *dynamic.* In other words, the task sequence, operator assignments, and individual task performance may change in response to excessive operator workload as the task network model executes. These changes may be as simple as one operator handing tasks off to another operator to reduce workload to an acceptable level or as complex as the operator beginning to time-share tasks in order to complete all the tasks assigned, potentially with associated task performance penalties. Ultimately, the tool provides an estimate of system-level performance as a result of these realistic workload management strategies. This innovation in modeling provides greater fidelity in efforts that model human behavior in the context of system performance, particularly in high workload environments such as complex system control and management.

4.3.4 Extending Research Findings to New Task Environments

Task network modeling was used by LaVine et al. (1995) to extend laboratory data and field data collected on one set of human tasks to predicting performance on similar tasks. The problem of extending laboratory or field human performance data to other tasks has plagued the human engineering community for years. We know intuitively that human performance data can be used to predict performance for similar tasks. However, it is often the case that the task whose performance we want to predict is similar in some ways but different in others. The approach described below uses a skill taxonomy to quantify task similarity, and therefore provides a means for determining how other tasks will be affected when exposed to a common stressor on human performance. Once functional relationships are defined between a skill type and a stressor, task network modeling is used to determine the effect of the stressor on performance of a complex task that uses many of these skills simultaneously.

The specific approach below is being used by the U.S. Army to predict crew performance degradation as a function of a variety of stressors. It is not intended to represent a universally acceptable taxonomy for simulating human response to stress. The selection of the best taxonomy would depend on the particular tasks and stressors being studied. What this example is intended to illustrate is another way that task network modeling can be used to predict human performance by *making a series of reasonable assumptions* that can be played together in a model for the purpose of making predictions that would be impossible to make otherwise. The methodology for predicting human performance degradation as a function of stressors consists of three parts: (1) a *taxonomy* for classifying tasks according to basic human skills, (2) *degradation functions* for each skill type for each stressor, and (3) *task network models* for the human-based system whose performance is being predicted. Conceptually, either laboratory or field data can be used to develop links between a human performance stressor (e.g., heat, fatigue) and basic human skills. By selecting a skills taxonomy that is sufficiently discriminating to make this assumption reasonable, one can assume that the effects of the stressor on all tasks involving the skill will be approximately the same. The links between the level of a stressor (e.g., fatigue) and resulting skill performance (e.g., the expected task time increase from fatigue) are defined mathematically as the degradation function. The task network model is the means for linking these back to complex human–system performance.

Taxonomy The basic premise behind the taxonomy is that the tasks that humans perform can be broken down into basic human skills or atomic tasks (Roth, 1992). The taxonomy that was used by Roth consists of five skill types described by Roth as follows:

1. *Attention:* the ability to attend actively to a stimulus complex for extended periods of time in order to detect specified changes or classes of changes that indicate the occurrence of some phenomenon that is critical to task performance.
2. *Perception:* the ability to detect and categorize specific stimulus patterns embedded in a stimulus complex.
3. *Psychomotor skill:* the ability to maintain one or more characteristics of a situation within a set of defined conditions over a period of time, either by direct manipulation, or by manipulating controls that cause changes in the characteristics.
4. *Physical skill:* the ability to accomplish sustained, effortful muscular work.
5. *Cognitive skill:* the ability to apply concepts and rules to information from the environment and from memory in order to select or generate a course of action or a plan (includes communicating the course of action or plan to others).

These five skills covered most of the tasks that were of interest to the Army for this study and still provided a manageable number of categories for an analyst to use.

Degradation Functions The degradation functions quantitatively link skill performance to the level of a stressor. The degradation functions can be developed from any data source, including standard test batteries or actual human tasks. Through statistical analysis, one can build skill degradation functions for each taxon. These functions map the performance decrement expected on a skill based on the parameters of the performance-shaping factor (e.g., time since sleep). An example of these functions is presented in Figure 17.

Incorporating the Degradation Functions into Task Network Models to Predict Overall Human–System Performance Degradation The key to making this approach useful to predicting complex human performance is the task network model of the new task. In the task network model of the human's activities, all tasks are defined with respect to the percentage of each skill required from the taxonomy. For example, the following are ratings for tasks faced by a console operator responding to telephone contacts:

Detect ring	50% attention, 50% perception
Select menu item using a mouse	40% attention, 60% psychomotor
Interpret customer's request for information	100% cognitive

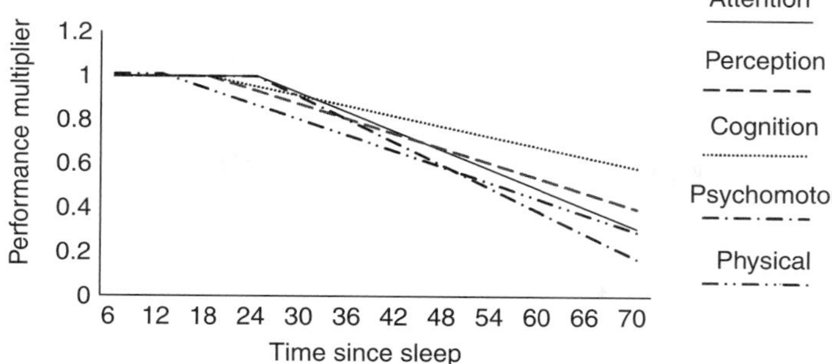

Figure 17 Performance degradation functions associated with each of the human skills from the taxonomy.

In building the task network model, mathematical expressions can be developed that degrade a specific task's performance through an arithmetic weighting of skill degradation multipliers that are derived from the degradation functions. For example, if the fatigue parameter was "time since sleep" and the value of that parameter was "36 hours since sleep," the task time performance multipliers would be as follows in the example above:

Attention performance multiplier	0.82
Perception performance multiplier	0.808
Cognition performance multiplier	0.856
Psychomotor performance multiplier	0.784
Physical performance multiplier	0.727

Based on these multipliers and the task weightings above, the specific task effects would be:

- Detect ring (50% attention, 50% perception)

$$\text{task multiplier} = 0.5 \times 0.82 + 0.5 \times 0.808 = 0.814$$

- Select menu item using a mouse (40% attention, 60% psychomotor)

$$\text{task multiplier} = 0.4 \times 0.82 + 0.6 \times 0.784 = 0.7984$$

- Interpret customer's request for information (100% cognitive)

$$\text{task multiplier} = 0.856$$

In a model of the complex tasks being examined by LaVine et al. (1995), the task networks consisted of several dozen or even several hundred tasks. Through the approach described above, each task in a model exhibited a unique response to a stressor depending on the particular skills that it required. The task network model then provided the means for relating the individual task performance to overall human–system performance as a function of stressor level (e.g., the time to perform a complex series of tasks involving decision making and error correction). Through this type of analysis, LaVine et al. were able to develop curves such as that shown in Figure 18 relating human performance to a stressor. These relationships would have been virtually impossible to develop experimentally.

Again, there were a number of simplifying assumptions that were made in this research. However, by being willing to accept these assumptions, LaVine et al. were able to characterize how complex human–system performance would be affected by a variety of stressors over a wide range in a relatively short time. As such, they were able to estimate the effects of stressors that would have otherwise been pure guesswork.

4.4.4 Summary

Once again, the above are intended to serve as examples, not a catalog of problems or approaches that are appropriate for task network modeling. Task network modeling is an approach to extend task and systems analysis to make predictions of human–system performance. The creative human factors and ergonomics practitioner will find many other useful applications and approaches.

5 FIRST-PRINCIPLE APPROACH: ADAPTIVE CONTROL OF THOUGHT–RATIONAL COGNITIVE ARCHITECTURE

The other fundamental approach to modeling human performance is based on the mechanisms that underlie and cause human behavior. Since this approach is based on fundamental principles of the human and his or her interaction with the system and environment, we have designated them as *first-principle models*. By integrating these models with models of the system and environment, the human factors specialist can predict the full behavior of large-scale interactive human–machine systems. The ACT-R cognitive architecture (Anderson and Lebiere, 1998) is a production system theory that models the steps of cognition by a sequence of production rules that fire to coordinate retrieval of information from the environment and from

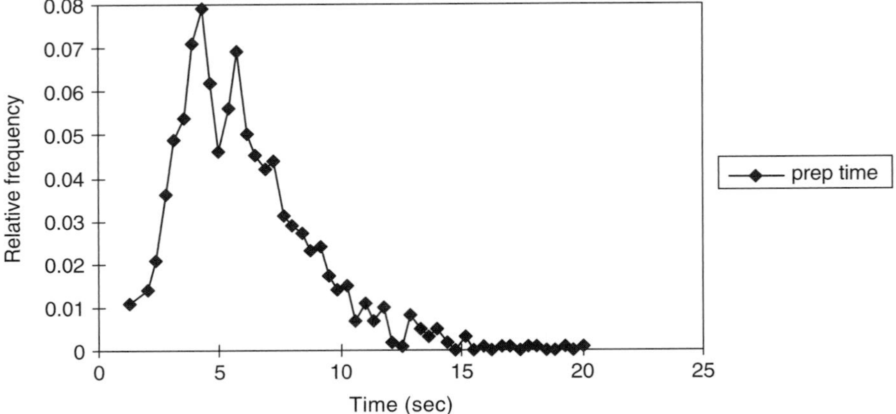

Figure 18 Frequency distribution of expected human performance as a function of time since sleep that was derived using task network modeling.

memory. It is a cognitive architecture that can be used to model a wide range of human cognition. It has been used to model tasks from memory retrieval (Anderson et al., 1998) to visual search (Anderson et al., 1997). The range of models developed, from those purely concerned with internal cognition to those focused on perception and action, makes ACT-R a plausible candidate to model complex tasks involving the interaction of one (or more) human operator with complex systems with the goal of evaluating the design of those systems. In all domains, ACT-R is distinguished by the detail and fidelity with which it models human cognition. It makes claims about what occurs cognitively every few hundred milliseconds in performance of a task. ACT-R is situated at a level of aggregation above those of basic brain processes (targeted by other modeling approaches, such as neural networks) but considerably below such complex tasks as air-traffic control. The new version of the theory has been designed to be more relevant to tasks that require deploying significant bodies of knowledge under conditions of time pressure and high information-processing demand. This is because of the increased concern with the temporal structure of cognition and with the coordination of perception, cognition, and action.

5.1 ACT-R

ACT-R is a unified architecture of cognition developed over the last 30 years at Carnegie Mellon University. At a fine-grained scale it has accounted for hundreds of phenomena from the cognitive psychology and human factors literature. The most recent version, ACT-R 5.0 (Anderson et al., in press), is a modular architecture composed of interacting modules for declarative memory, perceptual systems such as vision and audition modules, and motor systems such as manual and speech modules, all synchronized through a central production system (see Figure 19). This modular view of cognition is a reflection both of functional constraints and of recent advances in neuroscience concerning the localization of brain functions. ACT-R is also a hybrid system that combines a tractable symbolic level that enables the easy specification of complex cognitive functions, with a subsymbolic level that tunes itself to the statistical structure of the environment to provide the graded characteristics of cognition such as adaptivity, robustness, and stochasticity.

The central part of the architecture is the production module. A production can match the contents of any combination of buffers, including the goal buffer, which holds the current context and intentions; the retrieval buffer, which holds the most recent chunk retrieved from declarative memory; the visual and auditory buffers, which hold the current sensory information; and the manual and vocal buffers, which hold the current state of the motor and speech module. The highest-rated matching production is selected to effect a change in one or more buffers, which in turn trigger an action in the corresponding module(s). This can be an external action (e.g., movement) or an internal action (e.g., requesting information from memory). Retrieval from memory is initiated by a production specifying a pattern for matching in declarative memory. Each chunk competes for retrieval, with the most active chunk being selected and returned in the retrieval buffer. The activation of a chunk is a function of its past frequency and recency of use, the degree to which it matches the pattern requested, plus stochastic noise. Those factors confer memory retrievals, and behavior in general, desirable "soft" properties such as adaptivity to changing circumstances, generalization to similar situations, and variability (Anderson and Lebiere, 1998).

The current goal is a central concept in ACT-R, which as a result provides strong support for goal-directed behavior. However, the most recent version of the architecture is less goal-focused than its predecessors by allowing productions to match

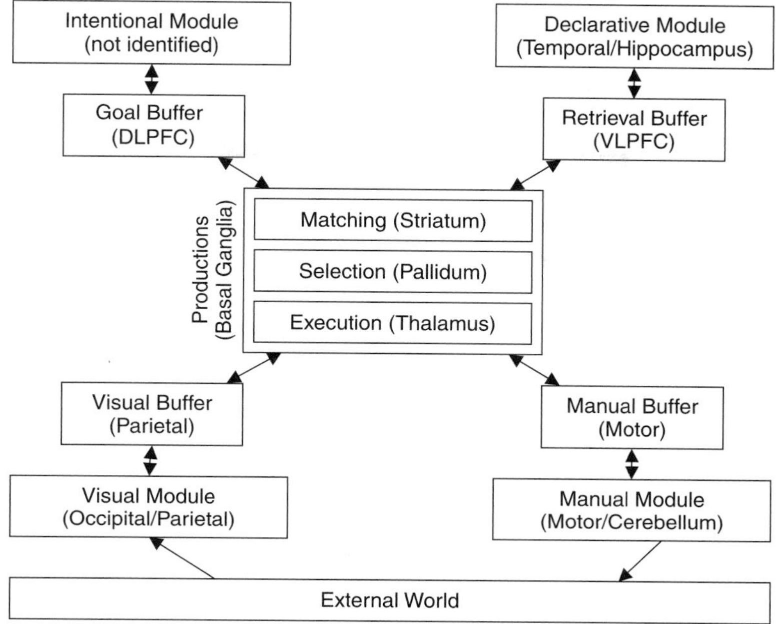

Figure 19 Modular view of ACT-R cognitive architecture.

to any source of information, including the current goal, information retrieved from declarative memory, objects in the focus of attention of the perceptual modules, and the state-of-the-action modules. The content of many of those buffers, especially the perceptual buffers, might have changed not as a function of an internal request but as a result of an external event happening, perhaps unexpectedly, in the outside world. This emphasis on asynchronous pattern matching of a wide variety of information sources better enables ACT-R to operate and react efficiently in a dynamic fast-changing world through flexible goal-directed behavior which gives equal weight to internal and external sources of information.

There are three main distinctions in the ACT-R architecture. First, there is the procedural–declarative distinction that specifies two types of knowledge structures: chunks for representing declarative knowledge and productions for representing procedural knowledge. Second, there is the symbolic level, which contains the declarative and procedural knowledge, and the subsymbolic level of neural activation processes that determine the speed and success of access to chunks and productions. Finally, there is a distinction between the performance processes by which the symbolic and subsymbolic layers map onto behavior and the learning processes by which these layers change with experience.

Human cognition can be characterized as having two principal components: (1) the knowledge and procedures codified through specific training within the domain, and (2) the natural cognitive abilities that manifest themselves in tasks as diverse as memory, reasoning, planning, and learning. The fundamental advantage of an integrated architecture like ACT-R is that it provides a framework for modeling basic human cognition and integrating it with specific symbolic domain knowledge of the type specified by domain experts (e.g., rules specifying what to do in a given condition, a type of knowledge particularly well suited for representation as production rules). However, performance described by symbolic knowledge is mediated by parameters at the subsymbolic level that determine the availability and applicability of symbolic knowledge. Those parameters underlie ACT-R's theory of memory, providing effects such as decay, priming, and strengthening and make cognition adaptive, stochastic, and approximate, capable of generalization to new situations and robustness in the face of uncertainty. They also can account for the limitations of human performance, such as latencies to perform tasks and errors that can originate from a number of sources. Finally, they provide a basis for representing individual differences such as those in working memory capacity, attentional focus, motivation, and psychomotor speed as well as the impact of external behavior moderators such as fatigue (Lovett et al., 1999; Taatgen, 2001) through continuous variations of those subsymbolic architectural parameters that affect performance in complex tasks.

Because they influence quantitative predictions of performance so fundamentally, we describe in some more detail the subsymbolic level in which continuously varying quantities are processed, often in parallel, to produce much of the qualitative structure of human cognition. These subsymbolic quantities participate in neural-like activation processes that determine the speed and success of access to chunks in declarative memory as well as the conflict resolution among production rules. ACT-R also has a set of learning processes that can modify these subsymbolic quantities. Formally, activation reflects the log posterior odds that a chunk is relevant in a particular situation. The *activation* A_i of a chunk i is computed as the sum of its base-level activation B_i plus its context activation:

$$A_i = B_i + \sum_j W_j S_{ji}$$

In determining the context activation, W_j designates the attentional weight given the focus element j. An element j is in the focus, or in context, if it is part of the current goal chunk (i.e., the value of one of the goal chunk's slots). S_{ji} stands for the strength of association from element j to chunk i. ACT-R assumes that there is a limited capacity of source activation and that each goal element emits an equal amount of activation. Source activation capacity is typically assumed to be 1 (i.e., if there are n source elements in the current focus each receives a source activation of $1/n$). The associative strength S_{ji} between an activation source j and a chunk i is a measure of how often i was needed (i.e., retrieved in a production) when chunk j was in the context. Associative strengths provide an estimate of the log likelihood ratio measure of how much the presence of a cue j in a goal slot increases the probability that a particular chunk i is needed for retrieval to instantiate a production. The *base-level activation* of a chunk is learned by an architectural mechanism to reflect the past history of use of a chunk i:

$$B_i = \ln \sum_{j=1}^{n} t_j^{-d} \approx \ln \frac{nL^{-d}}{1-d}$$

where t_j stands for the time elapsed since the jth reference to chunk i,d is the memory decay rate, and L denotes the lifetime of a chunk (i.e., the time since its creation). As Anderson and Schooler (1991) have shown, this equation produces the power law of forgetting (Rubin and Wenzel, 1990) as well as the power law of learning (Newell and Rosenbloom, 1981). When retrieving a chunk to instantiate a production, ACT-R selects the chunk with the highest activation A_i. However, some stochasticity is introduced in the system by adding Gaussian noise of mean zero and standard deviation σ to the activation A_i of each chunk. In order to be retrieved, the activation of a chunk needs to reach a fixed retrieval threshold τ that limits the accessibility of declarative elements. If the Gaussian noise is approximated with a sigmoid distribution, the *probability* P of chunk i *to be retrieved* by a production is

$$P = \frac{1}{1 + e^{-(A_i - \tau)/s}}$$

where $s = \sqrt{3}\sigma/\pi$. The activation of a chunk i is related directly to the latency of its retrieval by a production p. Formally, *retrieval time* T_{ip} is an exponentially decreasing function of the chunk's activation A_i:

$$T_{ip} = Fe^{-fA_i}$$

where F is a time scaling factor. In addition to the latencies for chunk retrieval as given by the retrieval time equation, the total time of selecting and applying a production is determined by executing the actions of a production's action part, whereby a value of 50 ms is typically assumed for elementary internal actions. External actions, such as pressing a key, usually have a longer latency determined by the ACT-R/PM perceptual-motor module (Byrne and Anderson, 2001). In summary, subsymbolic activation processes in ACT-R make a chunk active to the degree that past experience and the present context (as given by the current goal) indicate that it is useful at this particular moment.

Just as subsymbolic activation processes control which chunk is retrieved from declarative memory, the process of selecting which production to fire at each cycle, known as conflict resolution, is also determined by subsymbolic quantities called utility that are associated with each production. The utility, or *expected gain*, E of a production is defined as

$$E = PG - C$$

where G is the value of the goal to which the production applies, and P and C are estimates of the goal's probability of being completed successfully and the expected cost in time until that completion, respectively, after this production fires. Just as for retrieval, *conflict resolution* is a stochastic process through the injection of noise in each production's utility, leading to a probability of selecting a production i given by

$$p(i) = \frac{e^{E_i/t}}{\sum_j e^{Ej/t}}$$

where $t = \sqrt{6}\sigma/\pi$. Just as for the base-level activation, a production's probability of success and cost are learned to reflect the past history of use of that production, specifically the past number of times that that production led to success or failure of the goal to which it applied, and the subsequent cost that resulted, as specified by

$$P = \frac{\text{successes}}{\text{successes} + \text{failures}}$$

$$C = \frac{\sum \text{costs}}{\text{successes} + \text{failures}}$$

Costs are defined in terms of the time to lead to a resolution of the current goal. Thus, the more or less successful a production is in leading to a solution to the goal and the more or less efficient that solution is, the more or less likely that the production is to be selected in the future. Similar computations are at work in other modules, such as the perceptual-motor modules. Especially important are the parameters controlling the time course of processing as one attempts to execute a complex action or as one shifts visual attention to encode a new stimulus (Byrne and Anderson, 2001). ACT-R can predict not only direct quantitative measures of performance such as latency and probability of errors, but from the same mechanistic basis can also arise more global, indirect measures of performance, such as cognitive workload. Although ACT-R has traditionally shied away from such meta-awareness measures and concentrated on matching directly measurable data such as external actions, response times, and eye movements, it is by no means incapable of doing so. For the purpose of the task described below, Lebiere (2001) proposed a measure of cognitive workload in ACT-R grounded in the central concept of unit task (Card et al., 1983). *Workload* is defined as the ratio of time spent in critical unit tasks to the total time spent on task. *Critical unit tasks* are defined as tasks that involve actions, such as a goal to respond to a request for action with a number of mouse clicks, or tasks that involve some type of pressure, such as a goal to scan a display result from the detection of an event onset. The ratio is scaled to fit the particular measurement scale used in the self-assessment report. Lebiere (2001) describes possible elaborations of this basic measure.

5.2 AMBR

In this section we describe in some detail the constraints and requirements of the process of developing an ACT-R model for a task of moderate complexity and the range of quantitative predictions that one can expect from such a model. The task is a synthetic air-traffic control simulation that was developed for the agent-based modeling of behavior representation (AMBR) comparison (Pew and Gluck, in press) that arose from a report (Pew and Mavor, 1998) that highlighted the need for more robust, realistic human performance models (HPMs) for use in simulations for training and system acquisition

The AMBR project was designed to advance the state of the art in cognitive and behavioral modeling, especially models of integrative performance, requiring the coordination of memory, learning, multitasking, interruption handling, and perceptual and motor systems in order to scale more effectively to real-world environments. The program provided a structure to gather human performance data and evaluate the accuracy and predictiveness of the models. The AMBR program was organized as a series of comparisons among alternative modeling approaches including ACT-R but also the Air Force Research Laboratory's DCOG (Eggleston et al., 2001), CHI Systems, Inc.'s COGNET/iGEN (Zachary et al., 2001), and George Mason University's EASE (Chong, 2001).

The task designed to elicit the desired behaviors is a synthetic air-traffic control simulation. This domain requires a controller to manage one sector of airspace, especially the transition of aircraft into and out of the sector. Scenarios can vary the number, speed, altitude, and type of aircraft requesting access to the sector and can be complicated by having them arrive from multiple directions and adjoining sectors. This is a rich enough infrastructure to create a variety of scenarios having variable task load levels and varying levels of planning complexity. Figure 20 displays a screen shot

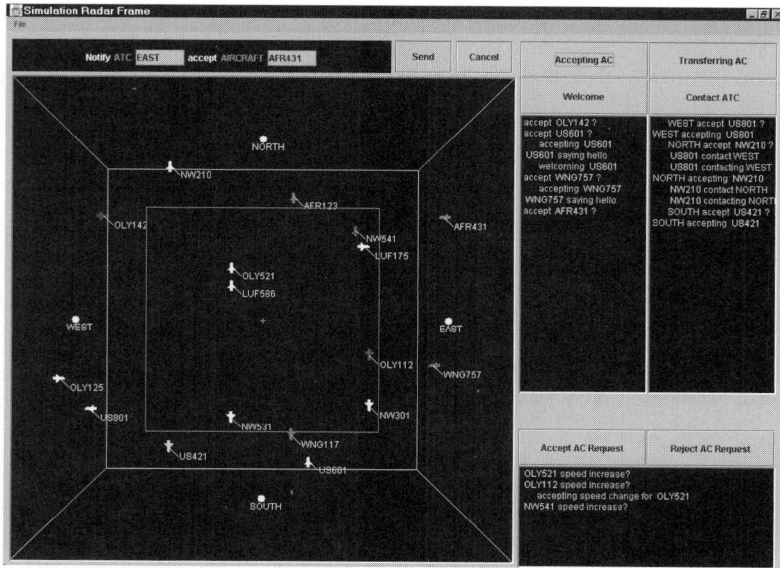

Figure 20 Screen shot of the AMBR simulation.

of the simulation. The main part of the screen on the left contains a graphical representation of the entire airspace, with the part controlled by the human or model agent contained in the central yellow square. The rest of the airspace is divided by the yellow lines in four regions, north, east, south, and west, each managed by a separate controller. At any point during the simulation a number of airplanes (the exact number being a parameter controlling the difficulty of the task) are present in the airspace, flying through the central region or entering or exiting it. The task of the central controller is to exchange messages with the airplanes (each tagged with its identifying code, e.g., UAL344) and neighboring controllers to manage their traversal of its airspace. Those messages are displayed in the text windows on the right of the screen, with each window dedicated to a specific message category. The top left window concerns messages sent when a plane is entering the central controller's region, while the top right window concerns messages sent when a place is exiting the central region. Both windows include messages exchanged between controllers as well as messages between the central controller and the plane itself. The bottom window concerns messages from and to planes requesting a speed increase, which should be granted unless that plane is overtaking another plane, which is the only airspace conflict that this simplified task allows.

A single event involves a number of messages being exchanged, all of which are appended to the relevant text window. For example, in the case of a plane about to enter the central region, a message requesting permission to enter will first be sent to the central controller from the controller of the neighboring region from which the plane originates. The central controller must reply to the other controller in a timely manner to accept the plane, then contact the plane to welcome it to the airspace. Those two cannot be performed in immediate succession, but instead, require waiting for the first party contacted (in this case the other controller) to reply before taking the final action. This delay allows for the interleaving of unit tasks but also requires the maintenance of the currently incomplete tasks in working memory. Messages from other tasks can arrive when a task is being processed, thus requiring some search of the text window to identify the messages relevant to a task. A message is composed by clicking a button above the relevant text window (e.g., accepting AC), then clicking in the graphical window on the intended recipient (e.g., another controller) and optionally the target of the message (i.e., a plane, unless it is the intended recipient, in which case this is omitted), then the send button above the graphic window. The message being composed is displayed at the top left of the display in a text window.

To measure performance on the task objectively, penalties were assessed for a variety of failures to act in a timely manner. To evaluate the impact of system design, a decision support condition contrasted with a support condition were implemented to dissociate two aspects of multitasking behavior. In the standard condition, subjects had to parse the messages printed in the text windows on the right side of the screen to determine which planes needed attention and which functions needed to be performed on them. In the assisted condition, planes that require assistance were color-coded in the graphical display on the left side of the screen according to the task that needed to be performed (green for accept, blue for welcome, orange for transfer, yellow for contact, magenta for speed change, and red for holding). This helped the subjects track visually which tasks needed to be attended to and removed any necessity to parse the text windows on the left, a complex and time-consuming task. Therefore, it dissociated the maintenance and updating of the queue of to-be-attended tasks from the resolution of conflicts between high-priority tasks. Two sets of scenarios were created: One set was provided to the developers as a model on which to base their designs, and another set was reserved to be used at the time of the competitive validation (i.e., the fly-off). Human performance data on the first set of scenarios were provided to the developers to fine-tune their model. The data from the second set of scenarios were withheld until after the fly-off for comparison with the model performance. The range of behavior requirements of both sets had the same scope, but the ways in which those behaviors were exercised were not identical, to test the robustness and predictiveness of the models.

5.3 Model Development

If it is to justify its structural costs, a cognitive architecture should facilitate the development of a model in several ways. It should limit the space of possible models to those that can be expressed concisely in its language and work well with its built-in mechanisms. It should provide for significant transfer from models of similar tasks, either directly in the form of code or more generally in the form of design patterns and techniques. Finally, it should provide learning mechanisms that allow the modeler to specify in the model only the structure of the task and let the architecture learn the details of the task in the same way that human cognition constantly adapts to the structure of its environment. These architectural advantages not only reduce the amount of knowledge engineering required and the number of trial-and-error development cycles, providing significant savings in time and labor, but also improve the predictiveness of the final model. If the "natural" model (derived a priori from the structure of the task, the constraints of the architecture, and the guidelines from previous models of related tasks) provides a good fit to the empirical data, one can be more confident that it will generalize to unforeseen scenarios and circumstances than if it is the result of post hoc knowledge engineering and data analysis. That is the approach that we adopted in developing a model of this task, and indeed more generally, our design and use of the ACT-R architecture.

Of course, in domains involving a large body of expertise, it makes sense to encode in the cognitive model the accepted knowledge of the field. But in synthetic tasks or in tasks involving new system design, specific established knowledge is usually inexistent or inaccessible. Thus, we did not try to reverse-engineer the subjects' strategies but instead, tried to develop the simplest and most natural model for the architecture. We organized the model around a few goal types with their associated productions. Goal types correspond closely to the unit tasks in human–computer interaction (Card et al., 1983) as well as to the tasks in task network models (e.g., Allender et al., 1995). Five goal types, called color-goal, text-goal, scan-text, scan-screen, and process, were defined, together with a total of 36 very simple productions. Goals were simple and would hold just a few elements, such as the aircraft currently being handled together with related information such as its position and the action to be performed, in accordance with architectural constraints. Overall, such model development need not take more than a few days. Two basic modes of human interaction with the simulation were defined: one in which the operator had to rely mostly on text messages scrolling in windows to identify events that required action (the text condition), and one in which aircraft on the radar screen that required action would turn a color corresponding to the action (the color condition). The simulation also had three speeds (low, medium, and high) that controlled how much time the subjects would have (10, 7.5, and 5 minutes, respectively) to perform a given number of actions.

The goal type color-goal was the top goal for the color condition. Five productions were defined that applied to that goal. They scanned the radar screen continuously, identified an aircraft that had turned color, mapped the color into the required action by relying on five simple memory chunks encoding the instructions that the subjects were given regarding the color-action mappings, then created a goal to perform the given action on the aircraft. The goal-type process executed the sequence of mouse clicks required to perform the action. Twelve productions were defined to handle the five possible actions. This required clicking on a button identifying the action, then on the aircraft, then perhaps on a neighboring controller, then finally on the send button.

As expected, the text condition was both more difficult for the subjects and slightly more complicated for the model. The goal type text-goal was the top goal for the text condition. Four productions were defined to cycle through the three text windows and the radar screen looking for aircraft requiring action by creating goals of type scan-text and scan-screen, respectively. A goal of type scan-text would handle the scanning of a single text window for a new message from another controller requesting action. A production was defined to scan the window systematically for such a message. If one was found, another production would attempt to retrieve a memory of handling such a request. Memories for such requests would be created automatically by the architecture when the corresponding goal was completed, but their availability was subject to their subsymbolic parameters, which were in turn subject to decay as well as reinforcement. If no memory could be retrieved, the window would be scanned for another message, indicating completion. If none could be found, a process goal would be created to perform the action requested. Note that this is the same goal as in the color condition. A key component of the model was an additional production that would detect the onset of a new message in another window and interrupt the current goal to scan that window instead. This allowed the model to be sensitive to new events and handle them promptly. Scanning the radar screen was accomplished in a similar manner by goals of type scan-screen and their eight associated productions.

Finally, all the architectural parameters that control the performance of the simulation were left at their default values provided by previous models. A key aspect of our methodology, which is also pervasive in ACT-R modeling, is the use of Monte Carlo simulations to reproduce not only the aggregate subject data (such as the mean performance or response time) but also the variation that is a fundamental part of human cognition. Especially when evaluating system design, it is essential not only to capture an idealized usage scenario but as broad a range of performance as possible. In that view, the model doesn't represent an ideal or even average subject, but instead, each model run is meant to be equivalent to a subject run, in all its variability and unpredictiveness. For that to happen, it is essential that the model not be merely a deterministic symbolic system but be able to exhibit meaningful nondeterminism. To that end, randomness is incorporated in every part of ACT-R's subsymbolic level, including chunk activations, which control their probability and latency of retrieval; production utilities, which control their probability of selections; and production efforts, which control the time that they spent executing.

Moreover, as has been found in other ACT-R models (e.g., Lerch et al., 1999), that randomness is amplified in the interaction of the model with a dynamic environment: Even small differences in the timing of execution might mean missing a critical deadline, which results in an error condition, which requires immediate attention, which might cause another missed deadline, and so on. To model the variation as well as the mean of subject performance, the model was always run as many times as there were subject runs. For that to be a practical strategy of model development, it is essential that the model run very fast, ideally significantly faster than real time. Our model ran up to five times faster than real time on a desktop PC, making it possible to run a full batch of 48 scenarios in about an hour and a half, enabling a relatively quick cycle of model development.

5.4 Modeling Results

Because the variability in performance between runs, even of the same subject, is a fundamental characteristic of this task, we ran as many model runs as there were subject runs. Figure 21 compares the mean performance in terms of penalty points for subjects and model for color (left three bars) and text (right three bars) condition by increasing workload level. The

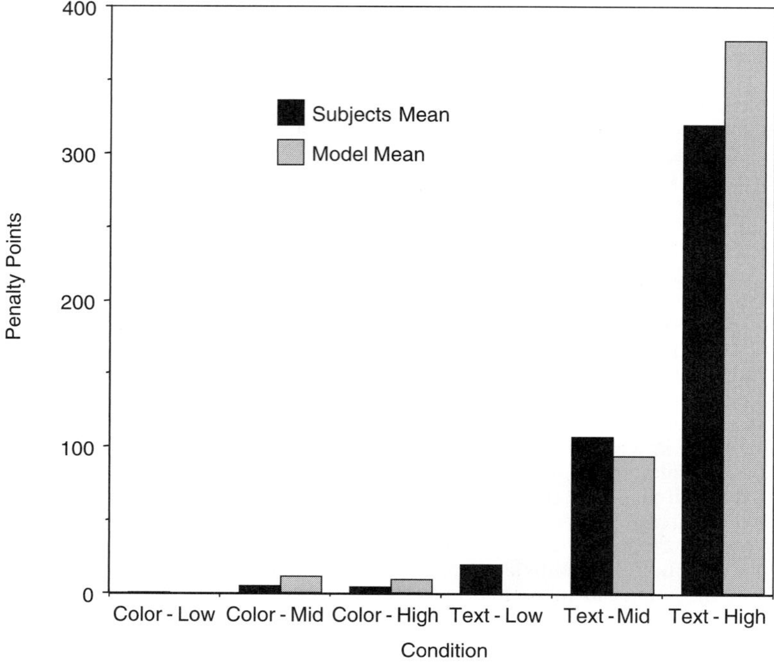

Figure 21 Mean performance as a function of workload and system design.

model matches the data quite well, including the strong effects of color vs. text condition and of workload for the unaided (text) condition.

Because ACT-R includes stochasticity in chunk retrieval, production selection, and perceptual/motor actions, and because that stochasticity is amplified by the interaction with a highly dynamic simulation, it can reproduce a large part of the variability in human performance, as indicated by Figure 22 which plots the individual subject and model runs for the two conditions that generated a significant percentage of errors (text condition in medium and high workload). The range of performance in the medium-workload condition is reproduced almost perfectly other than for two outliers, and a significant portion of the range in the high condition is also reproduced, albeit shifted slightly too upward. It should be noted that each model run is the result of an identical model that differs from another only in its run-time stochasticity. The model neither learns from trial to trial nor is modified to take into account individual differences.

The model reproduces not only the subject performance in terms of total penalty points, but also matches well to the detailed subject profile in terms of penalties accumulated under eight different error categories, as plotted in Figure 23. It should be emphasized that those errors were not engineered in the model, but instead, resulted directly from the limitations of the cognitive architecture applied to a demanding, fast-paced dynamic task.

The model also fits the mean response times (RTs) for each condition, as shown in Figure 24, which plots the detailed pattern of latencies to perform a required action for each condition and number of intervening events (i.e., number of planes requiring action between the time of a given plane requiring action and the time the action is actually performed). The model predicts very accurately the degradation of RT as more events compete for attention, including the somewhat counterintuitive exponential (note that RT is plotted on a log scale) increase in RT as a function of number of events rather than a more straightforwardly linear increase. The differences in RT between conditions are primarily a function of the time taken by the perceptual processes of scanning radar screen and text windows.

Finally, the model reproduces the subjects' answers to the self-reporting workload test administered after each trial. As shown in Figure 25, the simple definition of workload described in Section 5.3 captures the main workload effects, specifically effects of display condition and schedule speed. The latter effect results from reducing the total time to execute the task (i.e., the denominator) while keeping the total number of events (roughly corresponding to the numerator) constant, thereby increasing the ratio. The former effect results from adding to the process tasks the message scanning tasks resulting from onset detection in the text condition, thus increasing the numerator while keeping the denominator constant, thereby increasing the ratio as well. Another quantitative effect

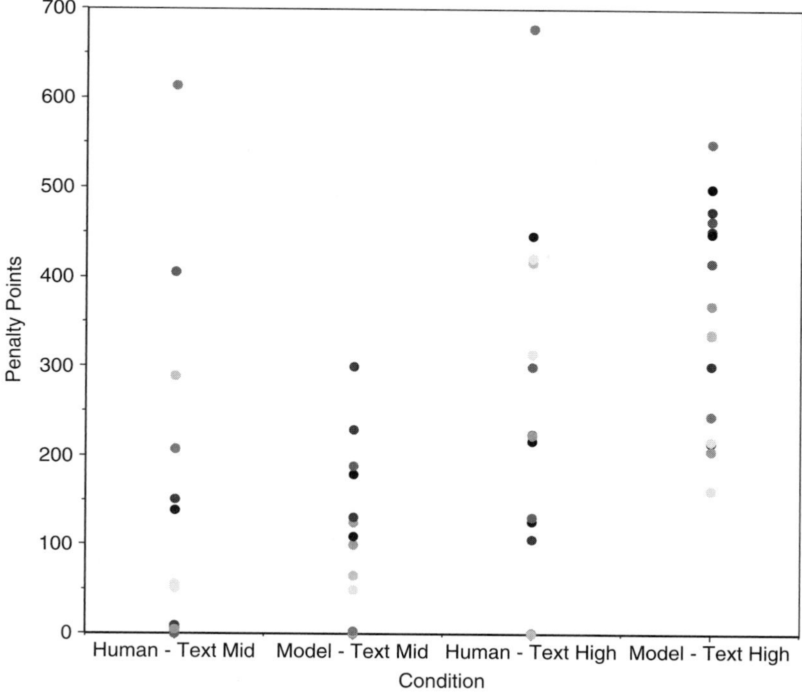

Figure 22 Mean performance as a function of workload and system design.

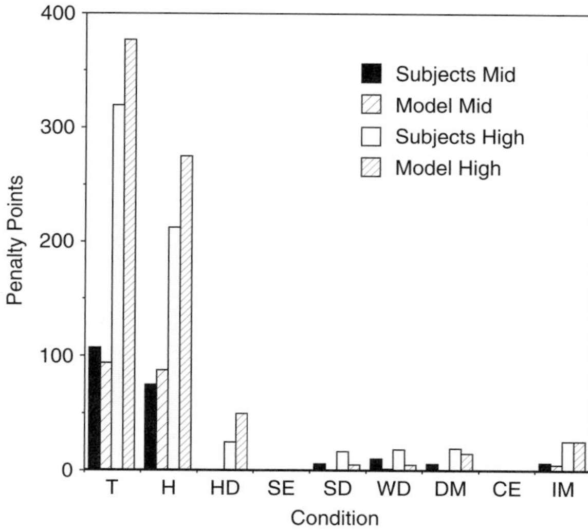

Figure 23 Penalty points for a variety of error categories.

that is reproduced is the higher rate of impact of schedule speed in the text condition (and the related fact that workload in the slowest text condition is higher than workload in the fastest color condition). This is primarily a result of task embedding [i.e., the fact that a process task can be (and often is) a subgoal of another critical unit task (scanning a message window following the detection of an onset in that window)], thus making the time spent in the inner critical task count twice.

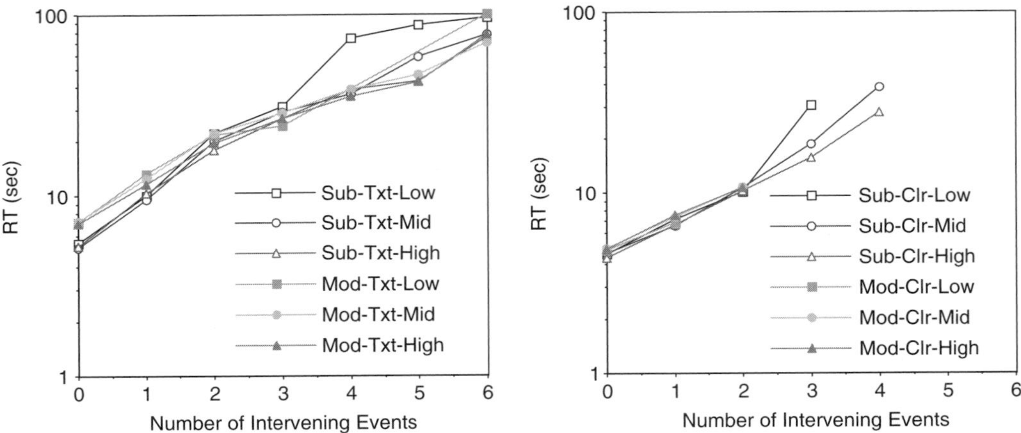

Figure 24 Response time as a function of intervening events.

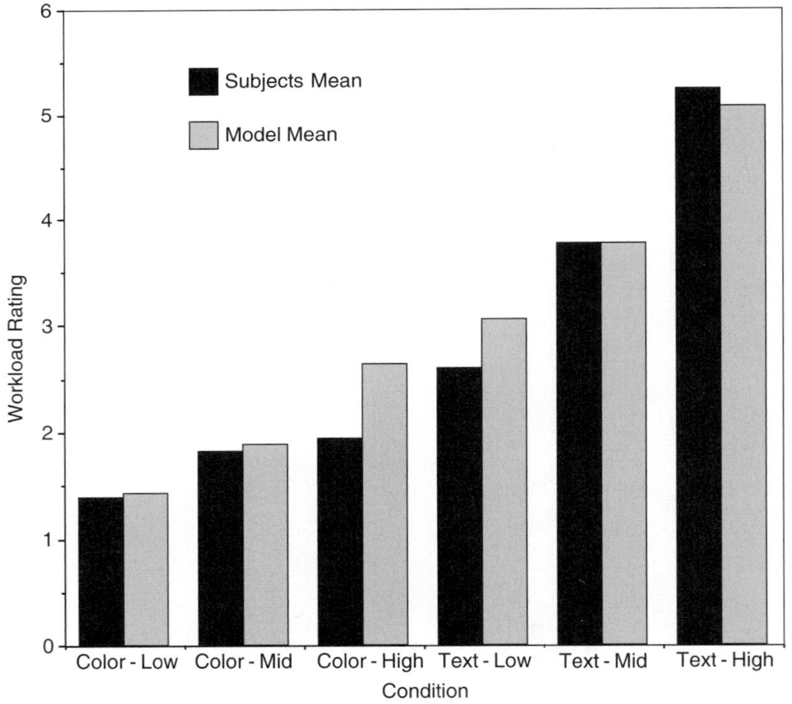

Figure 25 Workload levels for various speed and display conditions.

Lebiere (in press) reports the results of a second phase of the AMBR comparison in which we model had to learn how to categorize airplanes properly based on a simple pass–fail feedback. This model is similar to the one described here but leverages even more extensively the subsymbolic aspects of the architecture, especially the learning equations described in the ACT-R introductory section, to perform the learning task as a constrained component of the entire task. In summary, the advantages of this model are that it is relatively simple, required almost no parameter tuning or knowledge engineering, provides a close fit to both the mean and variance of a wide range of subject performance measures as well as workload estimates, and suggests a straightforward account of multitasking behavior within the existing constraints of the ACT-R architecture.

6 INTEGRATION OF THE APPROACHES

Because ACT-R and IMPRINT were targeted at different behavioral levels, they complement each other perfectly. IMPRINT is focused on the task level, how high-level functions break down into smaller-scale tasks, and the logic by which those tasks follow each other to accomplish those functions. ACT-R is targeted at the "atomic" level of thought, the individual cognitive, perceptual, and motor acts that take place at the subsecond level. As shown in Figure 19 and in the previous example, the current goal is a central concept in ACT-R which corresponds directly to the concept of unit task. At each cycle, a production will be chosen that best applies to the goal, knowledge might be retrieved from declarative memory and perceptual and motor actions taken. Those cycles will repeat until the current goal is solved, at which point it is popped and another one is selected. The ACT-R theory specifies in detail the performance and learning that takes place at each cycle within a specific goal, but has comparatively little to say about the selection of those goals. Since goals in ACT-R correspond closely to tasks in IMPRINT, that weakness matches IMPRINT's strength perfectly. Conversely, since IMPRINT requires the characteristics of each task to be specified as part of the model, ACT-R can be used to generate those detailed characteristics in a psychologically plausible way without requiring extensive data collection. Thus, an integrated ACT-R/IMPRINT is structured along as pictured in Figure 26.

An IMPRINT model specifies the network of tasks used to accomplish the functions targeted by the model (e.g., landing a plane and taxiing safely to the gate). The network specifies how higher-order functions are decomposed into tasks and the logic by which these tasks are composed together. As input, it takes the distribution of times to complete the task and the accuracy with which the task is completed. It can also take as input the workload generated by each task. Additional inputs include events generated by the simulation environment. Finally, a number of additional general parameters, such as personnel characteristics, level of training, and familiarity and environmental stressors can be specified. IMPRINT specifies the performance function by which these parameters modulate human performance. The outputs include mission performance data such as time and accuracy, as well as aggregate workload data.

An ACT-R model specifies the knowledge structures, such as declarative chunks and production rules, that constitute the user knowledge relevant to the tasks targeted by the model. It also specifies the goal structures reflecting the task structure and the architectural and prior knowledge parameters that modulate the model's performance. For each goal on which ACT-R is focused (i.e., made the current goal), it generates a series of subsecond cognitive, perceptual, and motor actions. The result of those actions is the total time to accomplish the goal, as well as how the goal was accomplished, including any error that might result. Errors in ACT-R originate from a broad range of sources. They include memory failures, including the failure to retrieve a needed piece of information or the retrieval of the wrong piece of information; choice failures, including the selection of the wrong production rule; and attentional failures, such as the failure to detect the salient piece of information by the perceptual modules. Although those errors could arise

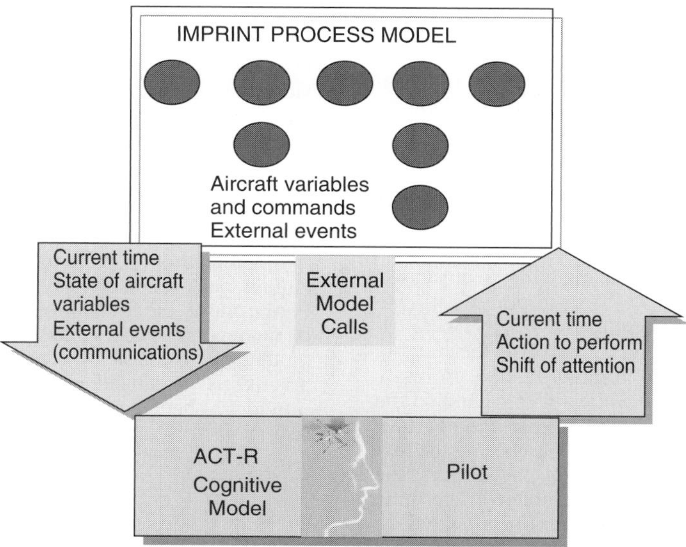

Figure 26 Integrated ACT-R/IMPRINT model.

because of faulty symbolic knowledge (either declarative or procedural), it is often not the case, especially in domains that involve highly trained crews. More often, those errors occur because the subsymbolic parameters associated with chunks or productions do not allow the model to access them reliably or quickly enough to be deployed in the proper situation.

Moreover, because those parameters vary stochastically and their effect is amplified by the interaction with a dynamic environment, those times and errors will not be deterministic but will vary with each execution, as is the case for human operators. Thus, the ACT-R model for a particular goal can be run whenever IMPRINT selects the corresponding task to generate the time and error distribution for that task in a manner that reflects the myriad cognitive, perceptual, and motor factors that enter into the actual performance of the task. As seen in the previous example, ACT-R can also generate workload estimates for each goal that reflect the cognitive demands of the actions taken to perform that particular subtask, then pass those estimates to IMPRINT, which can then combine them into global workload estimates for the entire task.

6.1 Sample Applications

As a practical application of the IMPRINT and ACT-R integration, a complex and dynamic task was selected for a modeling effort. Researchers with the National Aeronautics and Space Administration (NASA) were interested in developing models of pilot navigation while taxiing from a runway to a gate. Research on pilot surface operations had shown that pilots can commit numerous errors during taxi procedures (Hooey and Foyle, 2001). NASA was hoping to reduce the number and scope of pilot error during surface operations by using information displays that would improve the pilots' overall situation awareness.

NASA researchers provided the IMPRINT and ACT-R modeling teams with data describing pilot procedures during prelanding and surface taxi operations. These data included videotapes of pilots in the NASA Ames Advanced Concept Flight Simulator (ACFS), which is a simulated cockpit capable of duplicating pilot taxiing operations. A detailed, scaled map of Chicago's O'Hare airport was also provided, which included runway signage. Other types of documentation were provided to give the IMPRINT and ACT-R modeling team the information necessary to duplicate runway taxiing behavior by pilots.

The IMPRINT and ACT-R modeling teams used the scaled map of Chicago's O'Hare airport to estimate the time between runway taxi turns. IMPRINT handled the higher-level, task-oriented parts of the taxiing and landing operations (i.e., turning, talking on radio, looking at instrumentation), while ACT-R handled the more cognitive and decision-making parts of the task (i.e., remembering where to turn, remembering the taxi route). By using the scaled map of the airport, the IMPRINT and ACT-R teams were able to determine the amount of time between each taxi turn (based on an estimated plane speed that was correlated with the simulated speeds from the videotape data) and then use those data to estimate the decay rate for the list of memory elements (i.e., runway names) that the pilot would have to remember.

Using this integrated architecture allowed the team to represent a complex, dynamic task and by exploiting each architecture's strengths, the modeling process was enhanced and streamlined. The resulting model could account for a broad range of possible taxiing errors within a constrained first-principle framework, as was the case for the stand-alone AMBR model, but in addition benefited by integration with the task network model, which provided a convenient task-based organizing framework to minimize the authoring requirements for the cognitive model as well as to provide a high-productivity tool to simulate the environment and aircraft with which the cognitive model interacts.

Craig et al. (2002) performed a similar integration of ACT-R into the combat automation requirements tool (CART)* model (Brett et al., 2002), a task network model used in the acquisition process of the joint strike fighter. The task to be performed was target acquisition, more specifically, management of the shoot list, which allows a pilot to select potential targets to be identified by high-resolution radar. Using a methodology similar to that described above, specific subtasks were identified for which additional cognitive fidelity was required and reimplemented in the form of ACT-R goals and associated production rules. ACT-R then interacted with the CART model, providing plausible performance for cognitive subtasks such as prioritizing targets and recalling items identified previously.

7 SUMMARY

In this chapter we have reviewed the need for simulating performance of complex human-based systems as an integral part of system design, development, testing, and life-cycle support. We have also defined two fundamentally different approaches to modeling human performance, a reductionist approach and a first-principle approach. Additionally, we have provided detailed examples of two modeling environments that typify these two approaches along with representative case studies. Finally, we described an integrated tool that attempts to leverage the advantages of both approaches into an efficient and principled modeling package.

As we have stated and demonstrated repeatedly throughout this chapter, the technology for modeling human performance in systems is evolving rapidly. Furthermore, the breadth of questions being addressed by models is expanding constantly. Necessity being the mother of invention, we encourage the human factors practitioner to consider how computer simulation can provide a better and more cost-effective basis for human factors analysis and in turn stimulate further developments in modeling and simulation tools to better serve their needs.

*The reader should note that the CART model capabilities are now subsumed into the IMPRINT tool.

REFERENCES

Allender, L., Kelley, T., Salvi, L., Headley, D. B., Promisel, D., Mitchell, D., Richer, C., and Feng, T. (1995), "Verification, Validation, and Accreditation of a Soldier-System Modeling Tool," in *Proceedings of the 39th Human Factors and Ergonomics Society Meeting*, October 9–13, San Diego, CA, Human Factors and Ergonomics Society, Santa Monica, CA.

Allender, L. (1995), personal communication, December.

Anderson, J. R., and Lebiere, C. (1998), *The Atomic Components of Thought*, Lawrence Erlbaum Associates, Mahwah, NJ.

Anderson, J. R., and Schooler, L. J. (1991), "Reflections of the Environment in Memory," *Psychological Science*, Vol. 2, pp. 396–408.

Anderson, J. R., Matessa, M., and Lebiere, C. (1997), "ACT-R: A Theory of Higher Level Cognition and Its Relation to Visual Attention," *Human–Computer Interaction*, Vol. 12, No. 4, pp. 439–462.

Anderson, J. R., Bothell, D., Lebiere, C., and Matessa, M. (1998), "An Integrated Theory of List Memory," *Journal of Memory and Language*, Vol. 38, pp. 341–380.

Anderson, J. R., Bothell, D., Byrne, M. D., Douglas, S., Lebiere, C., and Qin, Y. (in press), "An Integrated Theory of the Mind," *Psychological Review*.

Archer, S. G., and Adkins, R. (1999), *Improved Performance Research Integration Tool (IMPRINT) Analysis Guide*, Army Research Laboratory Technical Report, Aberdeen Proving Ground, MD.

Boff, K. R., Kaufman, L., and Thomas, J. P. (1986). *Handbook of Perception and Cognition*, Wiley, New York.

Brett, B. E., Doyal, J. A., Malek, D. A., Martin, E. A., Hoagland, D. G., and Anesgart, M. N. (2002), *The Combat Automation Requirements Testbed (CART) Task 5 Interim Report: Modeling a Strike Fighter Pilot Conducting a Time Critical Target Mission*, AFRL-HE-WP-TR-2002-0018, Wright-Patterson Air Force Base, OH.

Byrne, M. D., and Anderson, J. R. (2001), "Serial Modules in Parallel: The Psychological Refractory Period and Perfect Time-Sharing," *Psychological Review*, Vol. 108, pp. 847–869.

Card, S. K., Moran, T. P., and Newell, A. (1983), *The Psychology of Human–Computer Interaction*, Lawrence Erlbaum Associates, Mahwah, NJ.

Chong, R. (2001), "Low-Level Behavioral Modeling and the HLA: An EPIC-Soar Model of an Enroute Air-Traffic Control Task," in *Proceedings of the 10th Conference on Computer Generated Forces and Behavior Representation*, Norfolk, VA.

Craig, K., Doyal, J., Brett, B., Lebiere, C., Biefeld, E., and Martin, E. (2002), "Development of a Hybrid Model of Tactical Fighter Pilot Behavior Using IMPRINT Task Network Model and ACT-R," in *Proceedings of the 11th Conference on Computer Generated Forces and Behavior Representation*, Orlando, FL.

Dahl, S. G., Allender, L., Kelley, T., and Adkins, R. (1995), "Transitioning Software to the Windows Environment: Challenges and Innovations," in *Proceedings of the 1995 Human Factors and Ergonomics Society Meeting*, Human Factors and Ergonomics Society, Santa Monica, CA, October.

Eggleston, R. G., Young, M. J., and McCreight, K. L. (2001), "Modeling Human Work Through Distributed Cognition," in *Proceedings of the 10th Annual Conference on Computer Generated Forces and Behavior Representation*, Norfolk, VA.

Engh, T., Yow, A., and Walters, B. (1998), *An Evaluation of Discrete Event Simulation for Use in Operator and Crew Performance Evaluation*, report to the Nuclear Regulatory Commission, NRC, Washington, DC.

Farmer, E. W., Belyavin, A. J., Jordan, C. S., Bunting, A. J., Tattershall, A. J., and Jones, D. M. (1995). *Predictive Workload Assessment: Final Report*, DRA/AS/MMI/CR95100/, Defence Research Agency, Farnborough, Hampshire, England, March.

Gawron, V. J., Laughery, K. R., Jorgensen, C. C., and Polito, J. (1983), "A Computer Simulation of Visual Detection Performance Derived from Published Data," in *Proceedings of the Ohio State University Aviation Psychology Symposium*, Columbus, OH, April.

Gray, W. D., Young, R. M., and Kirschenbaum, S. S. (1997), Special Issue: Cognitive Architectures and Human–Computer Interaction, *Human–Computer Interaction*, Vol. 12, No. 4.

Hahler, B., Dahl, S., Laughery, R., Lockett, J., and Thein, B. (1991). "CREWCUT: A Tool for Modeling the Effects of High Workload on Human Performance," presented at the 35th Annual Human Factors Society Meeting, San Francisco.

Hoagland, D. G., Martin, E. A., Anesgart, M., Brett, B. S., Lavine, N., and Archer, S. G. (2001). "Representing Goal-Oriented Human Performance in Constructive Simulations: Validation of a Model Performing Complex Time-Critical-Target Missions," in *Proceedings of the Simulation Interoperability Workshop*, Orlando, FL, Spring.

Hooey, B. L., and Foyle, D. C. (2001), "A Post-hoc Analysis of Navigation Errors During Surface Operations: Identification of Contributing Factors and Mitigating Strategies," in *Proceedings of the 11th Symposium on Aviation Psychology*, Ohio State University, Columbus, OH.

LaVine, N. D., Peters, S. D., and Laughery, K. R. (1995). *A Methodology for Predicting and Applying Human Response to Environmental Stressors*, Micro Analysis and Design, Inc., Boulder, CO, December.

Lawless, M. L., Laughery, K. R., and Persensky, J. J. (1995), *Micro Saint to Predict Performance in a Nuclear Power Plant Control Room: A Test of Validity and Feasibility*, NUREG/CR-6159, Nuclear Regulatory Commission, Washington, DC, August.

Lebiere, C. (2001), "A Theory-Based Model of Cognitive Workload and Its Applications," in *Proceedings of the 2001 Interservice/Industry Training, Simulation and Education Conference (I/ITSEC)*, NDIA, Arlington, VA.

Lebiere, C. (in press), "Constrained Functionality: Application of the ACT-R Cognitive Architecture to the AMBR Modeling Comparison," in *Modeling Human Behavior with Integrated Cognitive Architectures: Comparison, Evaluation, and Validation*, R. W. Pew and K. A. Gluck, Eds., Lawrence Erlbaum Associates, Mahwah, NJ.

Lerch, F. J., Gonzalez, C., and Lebiere, C. (1999), "Learning Under High Cognitive Workload," in *Proceedings of the 21st Conference of the Cognitive Science Society*, Lawrence Erlbaum Associates, Mahwah, NJ, pp. 302–307.

Lovett, M. C., Reder, L. M., and Lebiere, C. (1999), "Modeling Working Memory in a Unified Architecture: An ACT-R Perspective," in *Models of Working Memory*,

A. Miyake and P. Shah, Eds., Cambridge University Press, Cambridge, MA.

McCracken, J. H., and Aldrich, T. B. (1984), *Analysis of Selected LHX Mission Functions: Implications for Operator Workload and System Automation Goals*, Technical Note ASI 479-024-84(B) prepared by Anacapa Sciences, Inc., June.

McMillan, G. R., Beevis, D., Salas, E., Strub, M. H., Sutton, R., and Van Breda, L. (1989). *Applications of Human Performance Models to System Design*, Plenum Press, New York.

Micro Analysis and Design (2004). *Micro Sharp User's Guide*, Micro Analysis and Design, Inc., Boulder, CO.

Newell, A., (1990), *Unified Theories of Cognition*, Harvard University Press, Cambridge, MA.

Newell, A., and Rosenbloom, P. S. (1981), "Mechanisms of Skill Acquisition and the Power Law of Practice," in *Cognitive Skills and Their Acquisition*, J. R. Anderson, Ed., Lawrence Erlbaum Associates, Mahwah, NJ, pp. 1–56.

Newell, A., and Simon, H. A. (1972), *Human Problem Solving*, Prentice-Hall, Englewood Cliffs, NJ.

O'Hara, J. M., Brown, W. S., Stubler, W. F., Wachtel, J. A., and Persensky, J. J. (1995), *Human–System Interface Design Review Guideline: Draft Report for Comment*, NUREG-0700 Rev. 1, U.S. Nuclear Regulatory Commission, Washington, DC.

Pew, R. W., and Gluck, K. A. (in press). *Modeling Human Behavior with Integrated Cognitive Architectures: Comparison, Evaluation, and Validation*, Lawrence Erlbaum Associates, Mahwah, NJ.

Pew, R. W., and Mavor, A. S. (1998), *Modeling Human and Organizational Behavior: Application to Military Simulations*, National Academy Press, Washington, DC.

Plott, B. (1995), *Software User's Manual for WinCrew, the Windows-Based Workload and Task Analysis Tool*, U.S. Army Research Laboratory, Aberdeen Proving Ground, MD.

Roth, J. T. (1992), *Reliability and Validity Assessment of a Taxonomy for Predicting Relative Stressor Effects on Human Task Performance*, Technical Report 5060-1 prepared under contract DNA001-90-C-0139, Micro Analysis and Design, Inc., Boulder, CO, July.

Rubin, D. C., and Wenzel, A. E. (1990), "One Hundred Years of Forgetting: A Quantitative Description of Retention," *Psychological Review*, Vol. 103, pp. 734–760.

Siegel, A. I., and Wolf, J. A. (1969), *Man–Machine Simulation Models*, Wiley-Interscience, New York.

Taatgen, N. A. (2001), "A Model of Individual Differences in Learning Air Traffic Control," in *Proceedings of the 4th International Conference on Cognitive Modeling*, Lawrence Erlbaum Associates, Mahwah, NJ, pp. 211–216.

Wickens, C. D. (1984), *Engineering Psychology and Human Performance*, Charles E. Merrill, Columbus, OH.

Wickens, C. D., Sandry, D. L., and Vidulich, M. (1983). "Compatibility and Resource Competition Between Modalities of Input, Central Processing, and Output," *Human Factors*, Vol. 25.

Wortman, D. B., Duket, S. D., Seifert, D. J., Hann, R. L., and Chubb, A. P. (1978), *Simulation Using SAINT: A User-Oriented Instruction Manual*, AMRL-TR-77–61, Aerospace Medical Research Laboratory, Wright-Patterson Air Force Base, OH, July.

Zachary, W., Santarelli, T., Ryder, J., Stokes, J., and Scolaro, D. (2001), "Developing a Multi-tasking Cognitive Agent Using the COGNET/iGEN Integrative Architecture," in *Proceedings of the 10th Annual Conference on Computer Generated Forces and Behavior Representation*, Norfolk, VA.

CHAPTER 37

MATHEMATICAL MODELS IN ENGINEERING PSYCHOLOGY: OPTIMIZING PERFORMANCE

Donald L. Fisher
University of Massachusetts Amherst
Amherst, Massachusetts

Richard Schweickert
Purdue University
West Lafayette, Indiana

Colin G. Drury
University at Buffalo: State University of New York
Buffalo, New York

1 INTRODUCTION

Mathematical models of human behavior serve essential scientific and applied functions in human factors and ergonomics, and have long done so. Their bearing on the advance of theoretical behavioral science (e.g., Townsend and Ashby, 1983) is well documented, and their utility in more applied areas such as biomechanics is well established (e.g., Chaffin et al., 1999). However, their utility for the engineering psychologist is often overlooked or little discussed. In fact, there is a general sense that the engineering psychologist cannot yet use mathematical models to design an actual interface. In a recent issue of *Human Factors*, where a special section was devoted to the discussion of quantitative formal models of human performance, this sense is clearly articulated in one of the articles: "An aim of human factors research is to have models that allow for the advance of user-friendly environments. This is still a distant dream because existing models are not yet sufficiently sophisticated" (Jax et al., 2003).

Clearly, there is no model or technique that can handle all situations. But we believe that there do exist models that have played and can continue to play a central role in the design of user-friendly environments (Fisher, 1993). Unfortunately, they lie scattered throughout a broad and varied literature, one not always easily accessible to many readers. Moreover, designers aspire to consider the entire task and its environment, but available models are likely to be targeted for particular aspects of the situation, thereby not coming to the attention of the design community. Finally, many of the models that could potentially be useful in the design process are not extended in that direction, perhaps because the optimization techniques needed to make this transformation are part of a field, operations research, which generally does not overlap

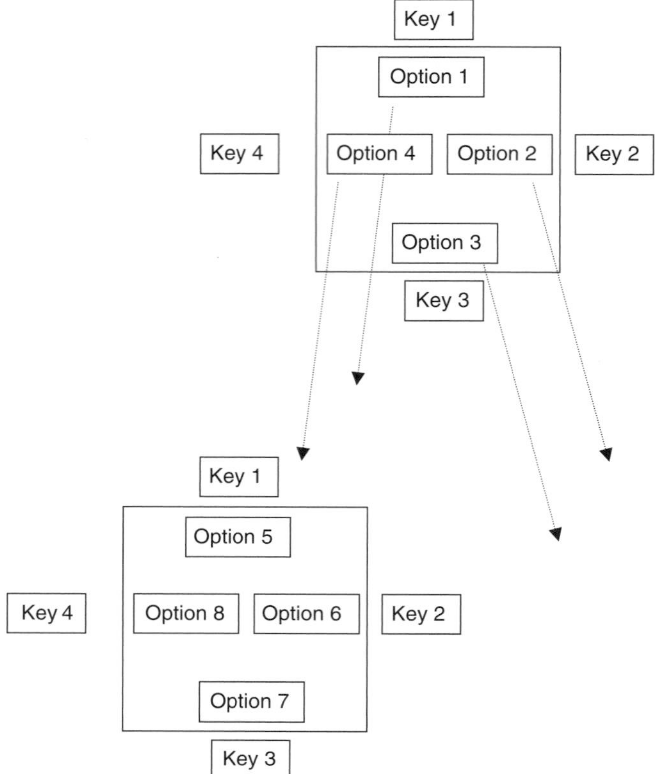

Figure 1 ATM arrangement of keys and options.

with engineering psychology. One of the purposes of a chapter like this is to bring to the attention of the more general reader models and optimization techniques that do allow for the creation of real, user-friendly environments.

Design lies at the center of the link between theory (the scientific model) and practice (the application or environment). It is useful to have an overall understanding of this link at the start of the chapter since this understanding makes clear the more limited but still essential role that mathematical models play in the design process, one that many models can now play. The interface between the human operator and the environment is the design element, natural or built. This design element can be as simple as a kitchen utensil or as complex as the displays in a nuclear power plant control room. The design element is frequently constrained, but within those constraints can take on an infinite or very large number of different configurations. For example, consider just the location of the keys on an ATM. Suppose that there are four keys positioned on the ATM, one key near each edge of a small square display window and each key accessing a different option (Figure 1). Suppose that any option can be assigned to any key. And suppose that the menu hierarchy is three levels deep, with four options in each display (only two levels are displayed in Figure 1). Given this configuration, at the bottom level of the hierarchy there are 16 menus each with four options, or a total of 64 different terminal options. Then there are over 1 quadrillion possible arrangements of the options in each display (4! arrangements in each display, so 24^{1+4+16} arrangements overall).

The link from theory to practice through design is now easily made. Clearly, it is not possible to evaluate all of the different arrangements (designs) experimentally. This is where mathematical models have a critical role, in engineering in general and, more specifically, in human factors engineering (Byrne and Gray, 2003). They can be used to predict performance for each different configuration of an interface and in this case, for each different arrangement of the menu options. By itself, this may be all that is required. One can simply iterate through all possibilities and identify the one or ones that optimize performance. However, when the number of different configurations gets too large, or is infinite, it is necessary either to derive the optimal solution analytically or to use methods that can approximate it. This is where the knowledge of optimization techniques becomes critical and it is why in this chapter some attention is given to such techniques.

Formally, we can treat optimization as finding the maximum (or minimum) of an objective function: for example, a weighted sum of key variables:

$$F(x) = \sum_{i=1}^{n} c_i x_i \tag{1}$$

where the variables are x_i and the weights on each variable are c_i. Of course, more complex and more general objective functions beyond the simple linear sum of equation (1) are possible. Generally, it is not possible to choose arbitrary values of the variables x_i, as these are constrained. In the ATM menu design example, we must choose arrangements that include one and only one placement of each of the four keys. In general, there will be a set of constraints, for example:

$$\sum_{i=1}^{n} a_i x_i \leq b_j \qquad (1 \leq j \leq m) \tag{2}$$

where each of m linear combinations of the variables x_i and weights a_i must be less than some constant b_j. Such optimization problems have been studied extensively in operations research, and solution procedures have been developed for many classes of problems: for example, linear programming (as above) or integer programming, where the x_i values are constrained to integers. Not all mathematical models are optimization models, but many can be used in this way, as the remainder of this chapter shows.

Although we focus on the use of mathematical models to optimize performance, we do not do so exclusively, for such models have a broader practical utility than simply the optimal design of an interface. First, the parameters of such models can indicate something about a quantity such as the relative speed of latent processes, a quantity that would be important if, say, one group has been exposed to a toxin and another group has not been so exposed (Smith and Langolf, 1981) or if, say, younger and older adults are being compared performing a particular task (Salthouse and Somberg, 1982). Second, models also have an important role to play before implementing an interface and incurring all the expenses that go along with such an implementation. Specifically, they can be used to estimate whether the interface will perform as desired (Gray et al., 1993). Third, models have still another, perhaps surprising role to play. They can identify situations where the intuitively most obvious course of action leads to paradoxical results (Meyer and Bitan, 2002). Fourth, because a mathematical model makes explicit the variables and parameters considered, the discipline of mathematical modeling also forces users to make explicit these variables and relationships, thus providing a solid foundation for the testing, extension, or perhaps ultimate negation of the model. Somewhat paradoxically, there is an art to mathematical modeling, and that art is to abstract from the complex, real system of humans and devices those aspects most necessary for an accurate, yet economical prediction of performance. Of course, models have a theoretical purpose as well and we will necessarily make clear their utility in that context as we go forward. Specifically, they are useful not only because they can provide testable predictions of complex theories (e.g., Sternberg, 1975), but also because one can determine in certain cases whether models that appear on the surface as different indeed are such (Townsend, 1972).

The selection of examples below is necessarily limited by space, but also by a desire to give readers enough background material to understand how mathematical models, together with appropriate optimization techniques, can have a radical impact on design. Having said this, an attempt has been made to be as broad as possible within the proscribed confines, giving to readers some sense of the models that currently are being used to assist in the design of interfaces as diverse as workstations, variable message signs, menu hierarchies, intelligent tutors, and warnings, among others.

Finally, we should end the introduction with something of a disclaimer. We recognize that some of the most creative design lies outside the scope of the procedures considered in this chapter. For example, consider the design of menu hierarchies that was discussed briefly above. Recent efforts to improve performance include the compression of speech (Sharit et al., 2003), and with cell phones and other technologies with very small display windows, the presentation of a portion of the hierarchy rather than just the current command (Tang, 2001). These more qualitative changes can and often do bring about much larger changes than can be realized through the modeling and optimization techniques described below.

2 GENERAL MATHEMATICAL MODELS

To begin, we want to describe some general tools that can be brought to bear on the design of an optimal interface. Perhaps the class of tools used most frequently by engineering psychologists are networks that represent the latent processes involved in the performance of a task. Once the arrangement of the cognitive processes is understood, quantitative tools exist to estimate the response time. However, these tools often need to be augmented. For example, the output of an encoding process might not be perfect, changing from one trial to the next even though the objective stimulus did not change. The output of a decision process might change based on the existing payoff matrix; or the time it takes to execute a response might change as a function of the number of possible responses. The tools needed to handle these and other more complex variations on behavior are discussed below.

2.1 Task Analysis and Activity Network Models

One can describe human performance in a task by starting with a general model of the human as an operator, identifying the components of the model critical for the task under consideration, and then using the model so constrained to predict performance. Introductions to the main general models for this purpose are discussed in Anderson (1993) and Byrne and Anderson (1998) for ACT-R, Newell (1990) for SOAR, and Meyer and Kieras (1997a, b) for EPIC. It is often more direct to start by modeling the cognitive, perceptual, and motor activities in the specific task that the operator is performing. This approach is congruous with the state of knowledge in experimental psychology. A psychologist may study memory, but the actual experiments will be conducted on a task such as recognition of an object. At this time, much knowledge in experimental psychology is organized as knowledge about the latent activities required to perform specific tasks such as searching a display or drawing a figure. A component of interest, such as perception, is emphasized, but the smallest unit of study is the task. This organizational scheme may be a temporary phase, or if tasks turn out to be natural fundamental units, it may be permanent. In any case, most contemporary models are not for the entire human system, nor are they for isolated components. Most models are for a component within a task, even if presented in the literature as a model for the component alone. The implication is that if a model does not yet exist for performance in a new situation, it is unlikely that one can be made simply by snapping together existing models of components. A new model will usually need to be developed for the components in their new context.

2.1.1 Activity Networks

For context, the modeler needs a functioning model for the entire task. The handiest model for a task is often an activity network (e.g., Elmaghraby, 1977; Pritsker, 1979). An activity network indicates the arrangement of activities in the task, some following one upon another and some going on concurrently (see Figure 2). Each vertex v_i in the network represents an activity a_j, and an arrow from one vertex to another indicates the order in which the activities must be performed. A path from a vertex v_1 to a vertex v_2, going along the arrows in the proper directions, indicates that the activity represented at vertex v_1 precedes the activity represented at vertex v_2. Two activities a and b are called *sequential* if either a precedes b or b precedes a. For example, in Figure 2 the activity "call begins," represented at vertex v_1, precedes the activity "system response," represented at vertex v_2. Two activities are called *concurrent* if and only if they are not sequential. For example, "system response" and "listen to beep" are concurrent. (Note that two activities are called concurrent if they could in principle be carried out simultaneously, even if it happens that one of them is finished before the other starts.)

The duration of the task depends on the durations of the individual activities. Suppose that all the activities in the network must be completed for the task to be completed (there are other possibilities, of course). An activity network of this form is called a *critical path network*. The task will be completed when and only when all the activities on the longest path through the network are completed. The longest path through the network is called the *critical path*, and the duration of the task is the sum of the durations of all the activities on the critical path. For example, in Figure 2 the activity durations are listed immediately below the activity inside the vertex (assume that these durations are constant). The duration of the activities along the top path sums to 8; the durations of the activities along the bottom path sums to 11. Thus, the task duration is equal to 11. If the durations of the activities are constants, then to reduce the task completion time of a critical path network, one must identify the activities on the critical path and minimize their durations. Shortening the duration of an activity not on the critical path has no effect.

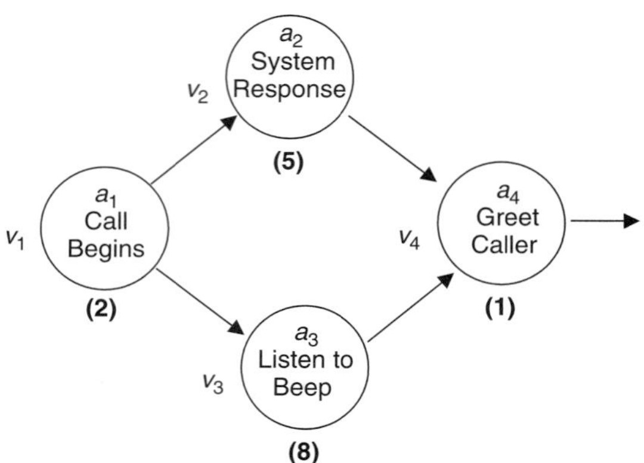

Figure 2 Activity network.

Application: Workstation Design

When new workstations were designed for operators in a New York telephone company, the hope was that with faster displays and fewer keystrokes, the time per call would drop. But before the new workstations were installed, an attempt was made to determine whether this would actually be the case. Gray et al. (1993) modeled the way the new workstations would be used, extrapolating from videotapes of operators using the old workstations. The modelers used a technique called CPM-GOMS (John and Newell, 1989; John, 1990) to construct networks for the activities in phone calls. The technique is an extension of the GOMS technique for task analysis in terms of goals, operators, methods, and selection rules (Card et al., 1983). The modeling predicted that the time per call would actually increase, and an increase was indeed found in later data from the new workstations.

What the modeling revealed was that activities that were performed more quickly with the new workstations would not shorten by much the overall time to complete the task, because the faster activities would be going on concurrently with other slow activities. On the other hand, despite the need for fewer keystrokes with the new workstations, some new keystrokes would occur when there were no concurrent slow activities, so the time for the new keystrokes increased the call completion time. Here is an example where a model can be used not only to predict whether an interface will perform as desired, but also why, if such is not the case, the performance will be less than desired. Given the right lead time, this in turn can suggest how to redesign the interface so that performance improves before it is implemented.

2.1.2 OP Diagrams

It was assumed above when discussing activity networks that the durations of the processes were constant. In practice, durations of activities are random variables, so one is interested in the probability an activity is on the critical path (i.e., its *criticality*). With random activity durations, calculating the mean and variance of the task completion time is usually intractable, so simulations are carried out with programs such as MATLAB or programs especially designed for activity networks, such as MICRO-SAINT. When durations have exponential or gamma distributions, exact formulas can be found with an OP (order-of-processing) diagram (Fisher and Goldstein, 1983). For example, if we let T_i be the duration of activity a_i, then for the activity network in Figure 2 we want to find the following expected value: $E[\max\{T_1 + T_2 + T_4, T_1 + T_3 + T_4\}]$. It is this expectation that can be computed easily if the foregoing conditions are met and the task is represented in an OP diagram, a discussion to which we now turn.

Figure 3 shows the beginning part of an OP diagram for a driver reading an electronic variable message sign that presents words one at a time. Each individual word is displayed, perceptually encoded, and comprehended. At any given time, a certain subset of activities will be executing; for example, the comprehension of the first word might go on simultaneously with the displaying of the second word. Such a set of activities which are executing simultaneously defines a state, and each possible state is represented by a vertex in the OP diagram. In the OP diagram in Figure 3, $w1$ denotes the display of the first word, $e1$ its encoding, $c1$ its comprehension, and so on. In the state represented by the first vertex (labeled s_1), the first word is displayed ($w1$) and the driver is encoding it ($e1$). It is assumed that the durations of these activities are continuous random variables. Thus, the probability that the two activities finish at the same point in time is zero, so such an event does not need to be represented in the OP diagram. In this case, one of these activities will finish first (not both), and when it does, the state is exited. The driver may finish encoding the first word before its display ends. In that case, in the next state (s_2) the first word is still being displayed, and the driver is comprehending it. The transition from the first state to this next state is indicated by an arrow labeled with the activity whose completion leads to this next state, in this case, $e1$. The other way to exit the first state is for the display of the first word to finish before the driver has encoded it. The OP diagram indicates that in this case, the driver does not complete the encoding of the first word ($e1$ appears in parentheses on the arc exiting from the state), and fails at reading the sign.

Every critical path network can be converted to an OP diagram (software for this is available from the first author). In addition, situations such as two activities that must go on one at a time, but in any order, can be represented in an OP diagram but not in a critical path network. After an OP diagram has been constructed, it can be used to calculate quantities such as the mean and variance of the time to complete the task and the probabilities of the task completing in various ways, such as success or failure at reading a variable message sign (Fisher and Goldstein, 1983; Goldstein and Fisher, 1991, 1992). In this case, if we let T_i be the duration of activity a_i, the expected time to complete the task is no longer the expectation of the maximum of the path durations. Instead, it is now a probability mixture of conditional expectations, where each conditional expectation is the time on average it takes to complete all activities along a path. For example, if the top path through the OP diagram were taken, and it consisted of only those states listed (s_1, s_2, s_3, s_5), we would want to compute the following conditional expectation: $E[T_1 + T_2 + T_3 + T_5 | path(s_1, s_2, s_3, s_5)]$. We would need to weight this by the probability that path (s_1, s_2, s_3, s_5) is taken and then do the same thing for all other paths in the OP diagram. Equations for the calculations may be found in Fisher and Goldstein (1983).

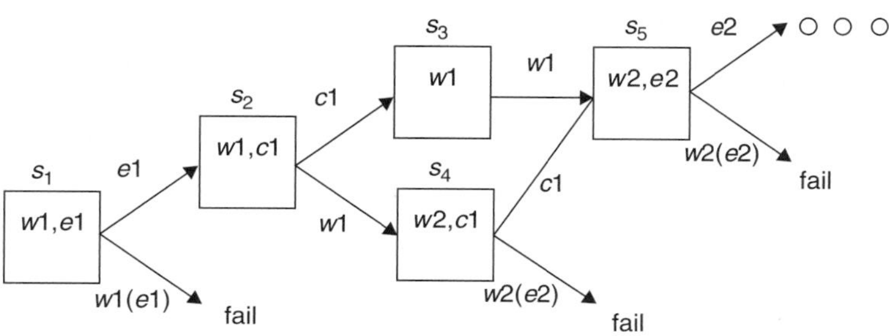

Figure 3 OP diagram.

2.1.3 Identifying Activity Durations

Several assumptions have been made above. First, it has been assumed that the durations of the activities are known if calculations or simulations are needed. Durations are ideally found by observation, as in Gray et al. (1993). Durations of certain common activities are available in the literature (some references are in Schweickert et al., 2003). Another source is expert opinion. However, it is often difficult for an expert to produce an accurate estimate of the duration of an activity. Schweickert et al. (2003) proposed that it may be more natural for an expert to produce a rank ordering of the differences in durations between pairs of activities. For example, in a telephone operator's task, an expert may judge that the difference in durations between greeting a customer and pressing the enter key is less than the difference in durations between entering a credit card number and listening to a beep. When the judgments are entered into a multidimensional scaling program (e.g., Shepard, 1962; Kruskal 1964), scale values for activity durations are produced. If the actual mean and variance of the duration of at least one activity is known for calibration, the scale values can be converted to estimates of mean activity durations. The estimates are, of course, only approximations. But Schweickert et al. (2003) found that they can lead to excellent predictions of the criticality and the product of criticality and duration for individual activities in simulations. (Activity durations were assumed to have gamma distributions.)

2.1.4 Identifying the Arrangement of Activities

A second assumption has been made. Specifically, it has been assumed that the arrangement of the activities is known. The way the activities are arranged in a task is ordinarily found by observation, discussion with operators, and inference. Sometimes an activity network has not been constructed, but a task analysis has been carried out, and the resulting diagram can be readily be converted to an activity network (Schweickert et al., 2003; see also Anderson, 1993). In many information-processing tasks, activities are unobservable mental processes such as perceiving and deciding. An analyst may be able to infer their existence or ask about the operator's knowledge of them. Another approach is based on experimentation, using the technique of selective influence for activity networks. With the technique, the activity arrangement is obtained by manipulating experimental factors, such as the number of elements in a display, with the intention of using each factor to influence selectively the duration of a single activity, such as a visual search. The effects of the factors on task completion time and their interactions provide the information needed to construct a critical path network or to show that no such network is possible for the factors used (Schweickert, 1978; Schweickert and Townsend, 1989; Schweickert et al., Goldstein, 1992). The key idea is that when two activities are influenced selectively, patterns in task completion times will be different depending on whether the pair of activities is sequential or concurrent. With this information for pairs of activities, a network can be constructed. After an activity network for the task has been constructed, simulations or calculations can be used to model effects of changes such as aging (Fisher and Glaser, 1996) or equipment modification.

Application: Variable Message Signs

Above, the use of an OP diagram to model a driver reading a variable message sign was discussed (Figure 3). The OP diagram can be used to predict the probability that a driver reads each of the words in the message, and therefore understands the signs. In practice, it is often possible to present the message one, two, or three times in the legibility zone and, even, to vary the duration of each page in a multiple-page message (a message so long that it cannot be presented in its entirety on a single page of a variable message sign). It is by no means clear exactly how long each page should be displayed to maximize the probability that drivers understand the variable message sign when the message is displayed one or more times in the legibility zone. It is clear from recent research that displaying the message more than once increases the likelihood that drivers will understand the message (Dutta et al., 2005). The next step is to use an OP diagram to find the page durations that maximize the probability that drivers understand the message. In this case one would need to add an additional process to the OP diagram, which reflects the time that drivers had to read the message. This will be a function of the driver's speed and the distance from the sign at which the message on the sign first becomes legible. For any given setting of the parameters, one can easily compute the probability that the words on both pages are understood. The optimal page durations can then be estimated by iterating through the space of possible page durations.

2.2 Signal Detection Theory

Task analysis is a powerful methodology for understanding what a person does (task description) and interpreting how a person performs (task analysis). A basic distinction in task analysis is often made between resource-limited and data-limited tasks (Norman and Bobrow, 1975). In a *resource-limited task*, such as the ATM menu tasks noted earlier, performance improves as more resources (time in this case) are devoted to the task. If we try to rush the sequence of button pushes, errors are more likely. All of the tasks that were described in Section 2.1 were resource-limited. In contrast, *data-limited tasks* do not show increased performance with more resources. These tasks are limited by the quality of the incoming data, so that no matter how many processing resources are employed, the performance (e.g., detecting or recognizing a signal) does not improve. Another way to characterize the two types of tasks is as follows: A resource-limited task exhibits a speed–accuracy trade-off (SATO), whereas a data-limited task does not. An example of a data-limited task would be trying to hear an important news broadcast on a radio at the limit of reception. If the signal-to-noise ratio is too low, trying to analyze more intensely what was heard will hardly make the signal more recognizable. These data-limited tasks have often been modeled by signal detection theory (Green and Swets, 1966), a discussion to which we now turn.

In a signal detection task, accuracy is the dependent variable. A subject must decide whether or not a weak signal is present. When a signal is presented, it produces neural activation, but the same signal does not always produce the same amount of activation. To make matters worse, an amount of activation usually produced by a signal can sometimes be produced in the absence of a signal by background noise from the environment (or from the nervous system itself). Error-free performance is not possible under these circumstances. According to signal detection theory (SDT), the best that one can do is set a particular amount of activation, call it x_c, as a criterion. If the amount of activation present exceeds the criterion, one decides that a signal is present; otherwise, one decides that noise alone was present. The result is a 2×2 classification of events:

1. If a signal is present and the observer says that a signal is present, the event is called a *hit.*
2. If a signal is present and the observer says that a signal is not present, the event is called a *miss*.
3. If a signal is not present, and the observer says that a signal was present, the event is called a *false alarm*.
4. If a signal is not present, and the observer says that a signal is not present, the event is called a *correct rejection*.

In the prototypical version of signal detection theory, the activation produced by a signal is normally distributed with mean μ_s, and the activation produced by noise alone is normally distributed with mean μ_n. The variance of the two distributions is assumed to be the same, σ^2. The more intense the signal, the greater the mean activation produced by it. The greater the difference between the mean of the signal distribution and that of the noise distribution, the more sensitive the observer will be. A measure of sensitivity is d':

$$d' = \frac{\mu_s - \mu_n}{\sigma}$$

The means and variances of the activation distributions, and hence d', are assumed to be influenced by characteristics of the signal and noise but to be beyond the control of the observer.

What is under control of the observer is the location of the criterion. The location of the criterion is frequently specified, not by the value of x_c, but by the ratio of the density functions of the signal, $f_s(x_c)$, and noise, $f_n(x_c)$, distributions evaluated at the criterion x_c. This ratio is often called the *response bias*. A measure of the response bias is β, defined as follows:

$$\beta = \frac{f_s(x_c)}{f_n(x_c)}$$

To estimate d' and β from data, one ordinarily assumes that the signal and noise distributions are normal, with equal variance, so these quantities are said to be *parametric*.

Signal detection theory is a good example of a mathematical model that can be used as an optimization model, directly influencing practice. The two parameters, d' and β, describe the actual performance of the task when the distributions of signal and noise are given. But we can go further and use mathematical optimization techniques to find where a person *should* place their criterion (optimum β) rather than just where a person *does* place it. To do this, we develop an objective function: for example, the long-term expected payoff over many trials. To optimize performance, the observer should adjust the criterion, taking the probability of a signal into account as well as the costs and benefits of the various correct responses and errors (see, e.g., Macmillan and Creelman, 1991). The constraint set is implicit in the model of signal and noise given above.

Specifically, if one can assign values to hits (V_{H}), misses (V_{M}), correct rejections (V_{CR}), and false alarms (V_{FA}), and if one knows the probability of a signal, $p(s)$ and, by extension, the probability of noise, $p(n)$, the criterion x_c which maximizes the expected gain can easily be found by knowing the optimal β where β_{op} is defined as follows:

$$\beta_{\text{op}} = \frac{p(n)}{p(s)} \frac{V_{\text{CR}} + V_{\text{FA}}}{V_{\text{H}} + V_{\text{M}}} \tag{3}$$

Note that this optimum is independent of the actual distributions of signal and noise.

Above, we have talked about signal detection theory outside the context of activity networks and more general OP diagrams. We now want to show how one can easily and immediately incorporate the elements of signal detection theory into the framework of OP diagrams. Suppose that a signal is presented and a response obtained. Then one might have only three processes: encoding, decision, and response. However, when a signal is presented, there are two different responses. Correspondingly, when noise is presented, there are two different responses. To model the task, one will need two different OP diagrams, one used when the signal is presented and one used when noise is presented. Consider just the case when the signal is presented. The state in which the decision process completes will now have two transitions associated with that completion, one indicating that the subject responds that the signal is present and one indicating that the subject responds that the signal is absent. The probabilities of these transitions are, respectively, the probability of a hit and the probability of a miss obtained from signal detection theory.

Application: Inspection Tasks I

An obvious application of signal detection theory is to data-limited tasks such as matching a paint color to a sample, or listening to a car engine to detect a maladjusted tappet. These are inspection tasks and are considered in more detail in Sections 6.1 to 6.3.

2.3 Information Theory

Above, we described tasks in which one uses knowledge about the operation of the encoding and decision processes to assign values to the parameters in an OP diagram, thereby increasing the overall power of these diagrams. Here, we want to describe tasks in which one uses knowledge about the operation of the response selection processes, again to assign values to the parameters in an OP diagram. Specifically, we want to know how response times will vary as a function of the number of different responses that can be made in a given task. As an example, consider an in-vehicle collision warning system. One could potentially warn drivers not only that a collision was going to occur, but where (in general) that collision was going to occur. As someone concerned with the design of such a system, one would like to know whether drivers will respond most quickly if they are warned that the collision will occur somewhere in front or somewhere behind the middle of the vehicle. Or, instead, should one warn the driver that the collision will occur in front, in back, to the left, or to the right? And, of course, more complex schemes are possible. To understand what needs to be done, some appreciation is needed for the role that information theory can play in this decision.

Generally speaking, one might assume that it is difficult, if not impossible, to decide on a common definition of information, let alone quantify that definition. Yet this is exactly what was done so elegantly by Shannon and Weaver (1949). Briefly, imagine two events, one very likely and one very unlikely. Suppose that the first event is: "The sun will rise tomorrow." The second event is: "The winning lottery number tomorrow will be 978654133." Most people would agree that there is much more information in the second message than in the first. So, let's take as a basic axiom the following: If the probability $p(x_i)$ of a message x_i is greater than the probability $p(x_j)$ of a message x_j, the information $I(x_i)$ in message x_i is less than the information $I(x_j)$ in message x_j. Rather surprisingly, together with a few other reasonable axioms, it follows necessarily that the information in a message can be written as follows:

$$I(x_i) = -\log p(x_i)$$

Assume that there are n messages in a set X of messages. Then one is frequently interested in the expected information in the message set, what is often called the *uncertainty*:

$$U(X) = E[I(X)] = \sum_{i=1}^{n} I(x_i) \cdot p(x_i)$$

$$= \sum_{i=1}^{n} [-\log p(x_i)] \cdot p(x_i)$$

Now, taking this one step further, imagine that we have a set S of n stimuli and a set R of n responses. For the sake of simplicity, assume that r_i is the correct response to stimulus s_i. Then we can ask how much information in the stimulus set is transmitted by the responses. In theory, anything between none of the information and all of the information could be transmitted. If a subject always gives the correct response, all of the information in the stimuli is transmitted by the responses. However, suppose that the probability that a subject gives response r_i to stimulus s_i is equal to chance ($1/n$) and, in fact, $p(r_j) = 1/n$, $j = 1, \ldots, n$. Then, none of the information in the stimuli is contained in the responses since the response that is made is entirely independent of the stimulus that is presented. A measure consistent with these intuitions, defined as the *information transmitted*, is easy to develop and is defined as follows:

$$T(S, R) = U(S) + U(R) - U(S, R)$$

where $U(S, R) = -\sum_{i=1}^{n} \sum_{j=1}^{n} p(s_i, r_j) \log p(s_i, r_j)$.

In light of these developments, Hick (1952) asked the following question: Would the response time in a task be related to the information transmitted? To answer this question, he ran an experiment in which the number n of stimuli shown to participants varied across conditions. Each stimulus was associated with a unique response. The probability $p(s_i)$ that a particular stimulus could occur was simply set to $1/n$. He found a linear relation between the information transmitted and the response time:

$$RT(n) = a + bT_{n+1}(S, R) \qquad (4)$$

The interpretation of this relation is as follows. Suppose that the number n of stimuli were a power of 2 ($n = 2^k$, k a positive integer) and subjects always responded correctly. Then the information transmitted (assuming no errors) can be shown to be equal to k, which is the minimum number of binary decisions it takes to identify one of n stimuli (again, $n = 2^k$). However, the reader will note that the information transmitted is indexed by $n + 1$, not n as one might expect. Hick argued that the respondent making the decision might be choosing among $n + 1$ responses, n of which were associated with a particular stimulus and one of which was associated with the absence of a stimulus. Hyman (1953) observed that the number of responses and information transmitted were well correlated. To determine whether it was the information, not the number of responses, which was controlling response time, he covaried the two and found that the ordering of the response times was consistent with the measure of the information transmitted, not the number of responses.

Predictions of response times from information theory are easily demonstrable in laboratory tasks with a relatively small number of alternatives, perhaps 16 or less (4 bits of information). However, they can be extended successfully to tasks with much higher levels of information per stimulus. For example, Bishu and Drury (1988) showed a good fit of information theory to complex surface wiring tasks in the communications industry, with information per stimulus up to 30 bits.

Finally, and as above with signal detection theory, we want to bring the discussion back to the more general framework of OP diagrams, if only briefly. In a task in which there are no errors (or few) and the number of stimuli in the message set varies across blocks of trials, a simple serial model can be used to represent the latent encoding, response selection, and response execution processes in an OP diagram or other network. The distribution of the duration of the response selection process is one of the parameters in the model. The mean of the distribution can be determined for each number n of stimuli in the message set using information theory. From equation (1) it follows directly that this mean is equal to bT_{n+1} (S,R). The sum of the means of the encoding and response execution processes is a.

Application: In-Vehicle Collision Warning Systems

At the start of this subsection, brief mention was made of in-vehicle collision warning systems. The utility of information theory for the design of such systems can now be made more clear. Suppose that participants are asked to indicate as quickly as possible from which direction an alarm has sounded. The first question is whether response times increase as the number of locations from which a warning can sound increases. In a recent experiment this number was varied between two and five (Wallace and Fisher, 1998). Response times increased linearly as a function of the information transmitted, going up by almost 50% when the uncertainty in the stimulus set was largest. Of course, although times are longer with more alarm locations, subjects know better where to focus their attention when the location of the collision is delineated more clearly. Thus, although with few alarm locations drivers may quickly be able to determine that the collision is in front or in back of them, they will not know where more precisely to look. Additional time will be needed by the driver to find the object with which a collision is imminent. A complete model of the time that it takes a driver to locate the source of a collision will require not only a model of how quickly the driver can locate the general area of concern, but additionally, a model of how quickly within that general area of concern the driver can find the actual object that is creating the collision risk. There is every reason to believe that such a model can be constructed based on related models of visual search (e.g., Arani et al., 1984) and could be used to identify the number of warning locations that will minimize the time it takes drivers to respond appropriately to a threat.

Application: Mail Sorting

Mail sorting represents another instantiation of an information-theoretic model. The sorter needs the address of an envelope and sorts it into the correct slot from hundreds of slots in a mail route, each slot representing one address. Hoffmann et al. (1993) used information theory to predict mail sorting times for Australian Post. In fact, such a model can be manipulated mathematically. Drury (1993) studied a mail-sorting system where part of the incoming mail stream was sorted into the correct order automatically. Ordered mail restricts the choices that are available [e.g., if the previous mail piece went into slot i out of n, the information to be processed in the next piece would be a choice between $(n - i)$ alternatives]. This formed the basis of predicted savings in mail sorting time from preordering of the mail.

2.4 Other Tools

We have described just several of many different tools that might be used to model the latent cognitive processes that govern the performance of participants in both laboratory and field settings. Many of these other tools, including queuing networks (Liu, 1996; Liu et al., 2004), associative networks (Anderson and Bower, 1973), connectionist networks (Rumelhart and McClelland, 1986), and shortest-route networks have their equivalent as OP diagrams and have been discussed elsewhere (Rouse, 1980). Other more general models, ACT-R (Anderson, 1983; Byrne and Anderson, 1998), SOAR (Newell, 1990), and EPIC (Meyer and Kieras, 1997a, b), about which we spoke earlier, can also easily be incorporated in the OP network framework.

3 VISUAL AND MEMORY SEARCH

The range of applications of task analyses and activity networks to human factors problems broadly defined as requiring cognition is way too large to even begin to catalog, let alone cover in some detail. However, we felt that there is one area that stands centrally in human factors, has an extensive history of mathematical modeling (Townsend and Ashby, 1983), and continues to be a critical element throughout many of the advances in human factors. In particular, we want to begin the discussion of mathematical modeling with a review of the early attempts to quantify the search and scanning processes. Almost every one of us is involved daily in one form of search or another, not the least of which is the search for a particular option on an ATM or PC, or the search through voice mail. Below, models of visual and memory search are discussed and applications of these models are detailed.

3.1 Visual Search

The visual search through a display can be as simple as the scanning of an array of symbols presented to a person seated in front of a computer or as complex as the scanning of traffic visible to a driver who may be looking for a particular license plate number in a sea of cars. At the heart of all models of visual scanning are four latent cognitive processes: an encoding e of the information to which attention is being paid; a comparison c of the encoded information with the target; a decision to end the search since the target is present and respond p that such is the case; or a decision to end the search since no target is present and respond a that such is the case.

In theory, there are at least four different ways that one might scan a visual display for a target, the most obvious being a serial scan that terminates when the target is identified (*serial, self-terminating*). If there were multiple targets, the scan could not terminate until all stimuli in the display had been identified (*serial, exhaustive*). In some cases, one could also imagine users scanning the display in parallel, either stopping when the target was identified (*parallel, self-terminating*) or continuing until all stimuli had been scanned (*parallel, exhaustive*). It is straightforward to represent the architecture for each of the models and predict the response time when the durations of the latent processes are constant. However, it becomes more difficult to represent the architecture and predict the moments of the response time distributions when the durations of the latent processes are random variables, especially as constraints are added, say, to the number of decision processes in a parallel model that can be ongoing simultaneously.

To give the reader a sense for how the modeling is undertaken, a simple derivation will be made of the expected time that it takes a person to find a target, when the search is a serial, self-terminating one and the display consists of an array of symbols (say, letters). Let E_i represent the time to encode the ith symbol scanned in the display, whatever it is. Let C_i represent the time to compare the ith symbol scanned with the target. The subscript i will be dropped here because it is assumed that the distributions of these times are identical. Let Y represent the time to respond that the target is present and A represent the time to respond that a target is absent. Let F be an indicator random variable that is set to i if the target is identified as the ith symbol scanned in the search. Finally, let $P(F = i)$ be the probability that the target is in the ith location. Then the expected time, $E[T(\text{present})]$, to find a target when there are n symbols can be written as the weighted sum of the time on average, $E[T(\text{present})|F = i]$, that it takes to find the target in each of the i different positions:

$$E[T(\text{present})] = \sum_{i=1}^{n} E[T(\text{present})|F = i]P(F = i)$$

If the target is in the ith position, there are i encoding operations and i comparisons plus one decision to respond that the target is present. Let e, c, y, and a represent, respectively, the expected encoding, comparison, target present, and target absent process durations (these symbols also are used to label the processes; the meaning will always be clear from the context). Assuming that the target is equally likely to be in any one of the n positions [i.e., $P(F = i) = 1/n$], we find

$$\begin{aligned} E[T(\text{present})] &= \sum_{i=1}^{n} \frac{ie + ic + y}{n} \\ &= \frac{(n+1)(e+c)}{2} + y \end{aligned} \tag{5}$$

The expected time to scan the display when the target is absent is simply $E[T(\text{absent})] = n(e + c) + a$. We can rewrite the expected time to scan the display when the target is present and absent as a linear function of the number of items in the display, as indicated by

$$\begin{aligned} E[T(\text{present})] &= y + \frac{e+c}{2} + \frac{(e+c)n}{2} \\ E[T(\text{absent})] &= a + (e+c)n \end{aligned} \tag{6}$$

Note that the slope, $(e + c)/2$, of the linear function relating the expected target present response time to the number of stimuli in the display is half the size of the slope, $e + c$, relating the expected target absent response time to the number of stimuli in the display. This is just one of many examples where a mathematical model makes predictions that can easily be tested with actual data. Note that as long as the assumption that the target is equally likely to be in any one of the n positions on each scan is valid, it matters not in what order or orders the stimuli are scanned.

Applications: Menu Hierarchies

In the introduction to this chapter, we explained the role that mathematical models and optimization techniques can play in the design process by referring to the construction of the optimal menu hierarchy for an ATM. We now want to continue this discussion, specifying here and in detail the exact quantitative procedures that one can use to design the optimal menu hierarchy, not for an ATM, but for a PC. Users today interact with menu hierarchies constantly, whether on their cell phones, at their ATMs, or on their computers at home and at work. Regardless of the technology, it is still the case that the underlying structure of the menu hierarchy has a large impact on the time it takes users to access information at the terminal nodes. It has been shown that given some very simple assumptions, one can identify the structure of a particular hierarchy that minimizes the time on average that it takes users to access the information in that hierarchy. There are two cases. In the first case, the number of menus at each level in the hierarchy is equal to twice the number in the superordinate level, and the number of options in each menu is identical across all menus (Lee and MacGregor, 1985). In the second

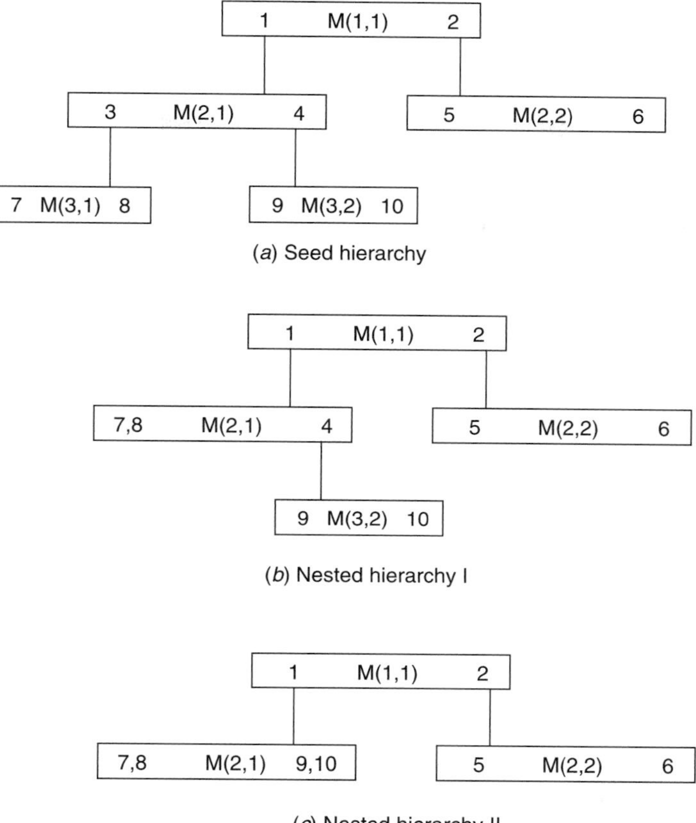

(*a*) Seed hierarchy

(*b*) Nested hierarchy I

(*c*) Nested hierarchy II

Figure 4 Menu hierarchy.

case, there are no constraints on the initial structure of the hierarchy (Fisher et al., 1990). It is second case that we want to consider here.

Suppose that users scan serially through the options in a menu, stopping when they identify the option that leads them to the appropriate next level. Then, it is easy enough to derive expressions for how long on average it will take users to identify a terminal option. What is not so obvious is what alternative structures one should consider. For example, take the menu in Figure 4*a*. Call this the *seed hierarchy*. It is the menu that a design team might produce, one that is most detailed. Five menus are labeled, beginning at the top, M(1,1), and so on. There are two options in each of the five menus. For example, in menu M(2,1) there are two options, options 3 and 4. There are three *terminal menus*, M(3,1), M(3,2), and M(2,2), and six *terminal options* (7, 8, 9, 10, 5, and 6) in these menus. Ideally, one would like to examine the complete space of semantically well-defined hierarchies that can lead to the retrieval of information from the six terminal options. However, there is currently no way automatically of generating all such semantically well-defined hierarchies. Still, there is something one can do that makes it possible to identify a large subset of the semantically well-defined hierarchies. Specifically, suppose that a *nested hierarchy* is defined as one that is formed from the seed hierarchy by replacing one or more options in a menu with all of the terminal options that come beneath it. Examples include the nested hierarchies in Figure 4*b* and *c*. There are 6 nested hierarchies that can be formed from the one seed hierarchy. In a slightly more complex example, assume that there are 64 terminal options, two options in the top-level menu, two options in each of the second-level menus, and so on down to the 64 terminal options in each of the 32 sixth-level menus. Then it can easily be shown that there are over 1 million different nested hierarchies, each of which is semantically well defined. Experimentally, it would be impossible to search this space exhaustively. However, a recursive algorithm can easily be implemented which identifies the hierarchy that minimizes the expected terminal option access time (Fisher et al., 1990). It is an example of the application of dynamic programming, one of the optimization techniques to which reference was made earlier.

Very briefly, the time on average that it takes to find a terminal option from a nonterminal menu can be written recursively as the time on average that it takes to find the option in the current menu that leads to the terminal option plus a probability mixture of the time on average it takes to find the terminal option from each of the menus that can be reached from the current menu. This recursive formula is then implemented in computer code. Each menu in the hierarchy is represented on the computer as a node in a linked list, with the link from each option in a menu pointing to the menu which can be reached from that option. To search the entire space of nested hierarchies, one compares the time on average it takes to reach a terminal option from the current menu in the seed hierarchy with the time on average that it takes to reach a terminal option when all terminal options are nested in the current menu. If the time on average it takes to reach the terminal option from the current menu with the nested terminal options is shorter than the time on average that it takes to reach the terminal option with the options left unnested, one can replace the seed hierarchy with the nested hierarchy so constructed. In this way, the search space is reduced dramatically. Here is an example where optimization techniques, not just mathematical models, play a critical role in the design process.

3.2 Memory Search

Above, when discussing visual search, we focused on a serial, self-terminating model. Interestingly, in a memory search task, it is not a serial, self-terminating model of the scanning process that best explains the results (Sternberg, 1966, 1975). Briefly, in a memory search task, a participant is given a list of stimuli to memorize (say, four digits). The number of digits in the memory set is referred to as the *memory set size*. After memorizing the digits, the experimental trial begins. A probe digit is then displayed. The participant must indicate whether the probe is in the memory set. Response time is graphed as a function of the memory set size. The best fitting lines relating the response times to memory set size for the case where the target is and is not present are often roughly parallel. The serial, self-terminating model cannot explain such parallelism, as can easily be seen from equation (6), if, among other things, the assumption that the distributions of the process durations associated with each item in the memory set do not depend on the identity of the item is generalized across memory sets of different sizes. However, a serial, exhaustive model can easily explain these results (Sternberg, 1966; Townsend, 1972). Note that in memory search, unlike visual search, only the probe digit needs to be encoded since all of the items in the memory set have already been encoded. Thus, the formula for memory search will include only one encoding.

This might be seen as the rather tidy end to the puzzle of how it is that items in memory are scanned. However, the resolution depends on a number of critical assumptions, one of which, as we just stated, is that the distributions of the first and second comparison times in the serial, self-terminating model are identical. To see that this assumption is required, we need to refer to OP diagrams again and show that a parallel exhaustive model can mimic a serial, self-terminating one. The parallel exhaustive model is represented in Figure 5. To begin, imagine that there are two items in the memory set and one target on the screen that matches one of the items in the memory set. Then the target is first encoded (state s_1 in Figure 5). If the search were a parallel exhaustive one, the two items in the memory set could be compared in parallel with the target (s_2). In this example, let c_i represent the comparison time of the ith memory set item with the target. If c_1 finished before c_2, execution of c_2 would continue by itself (s_3), and finally, the participant would respond (s_5). Alternatively, if c_2 finished before c_1, execution of c_1 would continue by itself (s_4) and again the participant would respond (s_5). Note that because the search is exhaustive, both items in the memory set need to be compared before the response is executed. The rather surprising finding here (Townsend, 1972) is that the expected response time with this serial, exhaustive model can mimic the expected response time with a serial, self-terminating model. To see this, imagine a serial, self-terminating model where the first comparison is relatively fast (the system is not fatigued) and the second is relatively slow. In particular, the distribution of the first comparison time in the serial, self-terminating model is equal to the distribution of the time spent in state s_2 and the distribution of the comparison time of the second (first) stimulus in memory with the target when it is processed last is equal to the distribution of the time spent in state s_3 (s_4). Finally, imagine that in the serial, self-terminating model the probability that the comparison of the target with the second item in memory occurs after the comparison of the target with the first item in memory is equal to this same probability in the parallel, exhaustive model (i.e., the probability of the path represented by the transition from s_2 to s_3). Then the expected response times of the serial, exhaustive and serial, self-terminating models will be identical. Fortunately, more detailed tests exist which can resolve the mimicking (Schweickert, 1978; Townsend and Wenger, 2004). This example very nicely points out the importance of quantifying models wherever possible, thereby reducing needless testing

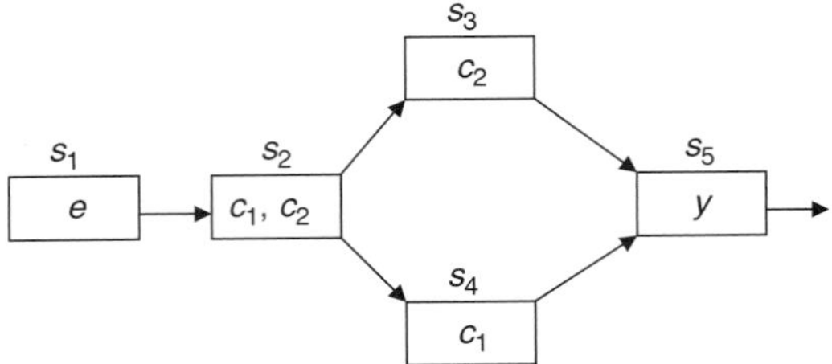

Figure 5 OP diagram: memory search.

or exploration of alternatives that turn out not to be identifiably different.

Application: Toxicology

Perhaps one of the more elucidating applications of memory search models was developed by Smith and Langolf (1983). They asked a simple question: Could low levels of mercury exposure produce effects on the speed of processing in people who were otherwise asymptomatic? To test this hypothesis, they had chlor-alkali workers exposed to different levels of mercury perform a simple memory search task, estimating for each person the comparison time (among other quantities). They then regressed the estimated comparison time for each person on the corresponding level of exposure of that person to mercury. There were significant effects of the mercury level on comparison times. From a practical standpoint, the most heavily exposed workers had an increase of 100% in their scanning times, suggesting a serious reduction in short-term memory capacity (Baddeley, 1992). Interestingly, this model developed for purely theoretical purposes has become some 30 years later a useful tool for uncovering neurobehavioral toxicity (Fiedler et al., 1996).

4 VIGILANCE

When signals occur rarely, a common finding is that observers tend to miss more signals after some time on task. This is called a *vigilance decrement*. Often, there is a decrease in false alarms as well; that is, there is a decline in the total number of reports of a signal, correct or incorrect (see Davies and Parasuraman, 1982, and See et al., 1995, for reviews). It is natural to use signal detection theory to determine whether as time goes by, observers become less sensitive to signals, or less prone to say "signal," or both. Results are complicated, but as a rough guide, Parasuraman and Davies (1977) found that sensitivity (measured, e.g., by d') tends to decline when one stimulus occurs at a time (so identification relies on memory) and stimuli occur relatively frequently. The criterion (measured, e.g., by x_c or β) tends to increase when stimuli are presented simultaneously (in a same or different task) or stimuli occur rarely.

As described earlier, one reason that signal detection theory is useful is that it divides the observer's processing into encoding and decision, and provides separate measures characterizing each, d' and β. Variables such as the probability of a signal are predicted, and often found in data, to significantly change the value of the decision parameter β, which represents the observer's choice of where to put the criterion, but not to significantly change the value of the encoding parameter d', which represents the observer's sensitivity to the signal. However, it is readily acknowledged that the assumptions underlying the calculation and interpretation of d' and β are not met exactly.

To investigate alternatives, See et al., (1997) compared the performance in vigilance tasks of two measures of sensitivity and five measures of response bias, with an emphasis on the latter. (For the formulas used, see See et al., 1997.) The two measures of sensitivity were the parametric d', which was discussed above, and a widely used nonparametric analog, A'. The authors report that the two measures were functionally equivalent. They were highly correlated over subjects, each declined with time on task, and neither was influenced by signal probability or payoff scheme, factors thought to influence the response bias rather than the sensitivity. Thus, d' seemed to function as expected.

However, such was not the case for response bias. For response bias, two parametric measures β and c, and three nonparametric measures, B'', B'_H, and B''_D, were compared. The authors report that β did not perform well; in particular, it was the least sensitive measure to signal probability and payoff scheme. The best-performing measure was c (Ingham, 1970),

$$c = \frac{x_c - (\mu_s + \mu_c)/2}{\sigma}$$

(The measure c is the z score for the criterion, x_c, but calculated with respect to a mean halfway between the means of the signal and noise distributions.) Of the nonparametric measures, the best performing was B''_D (Donaldson, 1992),

$$B''_D = \frac{(1-H)(1-F) - HF}{(1-H)(1-F) + HF}$$

where H is the probability of a hit and F is the probability of a false alarm. Both c and B''_D were sensitive to signal probability and payoff scheme, both indicated a predicted change in bias over time, and both functioned well even when performance was at chance level. The data of See et al. (1997) satisfied a test for normal distributions with equal variance, and it would be useful to know whether their conclusions hold in other situations.

Recently, Balakrishnan (1998a, b) has argued that violations of the assumptions underlying the use of parameters d' and β could be causing misleading interpretations of data, but these pass unnoticed because of the apparent robustness of the measures under signal probability or payoff manipulations. Signal detection theory assumes that the distributions of activation produced by a signal or by noise are not influenced by the probability of a signal. This may be true for the distribution of a single sampled amount of neural activity, such as when signals are rare. But neural activation may be extended over locations and over time when signals are frequent. In this case, the observer may sample several amounts of activation. The probability distribution of a sample depends on the sample size. Hence, a change in the number of amounts of activation sampled would be a change in encoding produced by a variable (the probability of a signal) thought to influence only the decision. (Note that if several samples are taken and the results combined to produce a better decision, we have moved from a data-limited to a resource-limited task. How the data are combined determines how the discriminability changes with the number of samples, or more generally with the time over which samples are taken. This would be another example of a speed–accuracy trade-off.) Whether this change is registered as a change in d' or β or both would depend on the forms of the underlying distributions and the way the size of the sample of observations is determined. To avoid such problems, Balakrishnan (1998a) proposes that distribution-free measures of response bias be used, and he has developed new measures based on confidence ratings. Analysts using signal detection theory often ask observers to give a number indicating their confidence that a particular stimulus was a signal or noise. For example, response 1 would indicate very low confidence that the stimulus was a signal, and response 8 would indicate very high confidence that the stimulus was a signal. Note that with this procedure, the analyst does not know what the observer would actually say if asked whether the stimulus was signal or noise. Balakrishnan proposes modifying the procedure slightly, so the observer produces a judgment about signal or noise together with a confidence rating that the judgment is correct. For example, response 1 would indicate a judgment of noise, with very high confidence, and response 5 would indicate a judgment of noise, with very low confidence. Similarly, response 6 would indicate a judgment of signal, with very low confidence, and response 10 would indicate a judgment of signal, with very high confidence.

The modified procedure leads to Balakrishnan's (1998a) distribution-free measures of response bias. An overall measure of bias is Ω, the total of the amount of bias at each rating having a bias (not all ratings have a bias). For example, suppose that the rating 5 stands for "with low confidence, the stimulus is noise" and the rating 6 stands for "with low confidence, the stimulus is signal." For simplicity, suppose that the probability of signal equals the probability of noise, equals $\frac{1}{2}$. Suppose that when 5 is used, $\frac{3}{4}$ of the time it is used for noise and $\frac{1}{4}$ of the time it is used for signal. Then there is no bias at rating 5. It is intended to indicate noise, and it usually does. It contributes nothing to Ω. But suppose that when 5 is used, $\frac{1}{4}$ of the time it is used for noise and $\frac{3}{4}$ of the time it is used for signal. Then there is a bias at rating 5. It contributes to Ω. Again, Ω is summed over only those ratings that have bias.

The amount of bias at rating 5, when it has a bias, is the total proportion of trials for which the rating 5 was used. Suppose that there were 200 noise trials and 200 signal trials, 400 total. Suppose that rating 5 was used six times for noise stimuli and 30 times for signal stimuli. The amount of bias contributed by rating 5 is $36/400 = 0.09$. An overall measure of bias is Ω, the total of the amount of bias at each rating having a bias.

Using the measure of bias Ω in a vigilance task, Balakrishnan (1998b) found no evidence of bias in the decision rule for either relatively frequent or relatively rare signals (i.e., values of Ω were close to 0 for probability of a signal 0.5 and 0.1). On the other hand, there was evidence for increased sensitivity for more frequent signals, indicated by increased A'. In other words, signal frequency appears to influence the encoding rather than the decision, the opposite of the usual signal detection theory interpretation. For rare signals the hypothesis of equal variances for signal and noise distributions can be rejected for the data of Balakrishnan (1998b). Instead, the variance is large for rare signals. The performance measures themselves do not indicate how the processing was done, but Balakrishnan points out that the results are consistent with models in which the subject samples repeatedly from the stimulus, rather than just once.

Application: Inspection Tasks II

In any particular application, the payoffs are maximized when the criterion is set as indicated in equation (3), provided that the assumptions of signal detection theory are met. If the assumptions are met, then to improve performance, one would train observers to locate their criteria appropriately. However, if, contrary to assumption, signal frequency influences encoding by changing the size of the sample the observer takes from the stimulus, then to improve performance training would emphasize taking time for adequate observation of the stimuli.

5 INSPECTION

Whether unaided manual or partially or fully automated, test and inspection tasks abound. The most obvious examples are from manufacturing quality control with products ranging in complexity from apples to Apple computers. Other examples come from checkout procedures in spacecraft, from aircraft structural inspection, from airport security, and even from inspection of restaurants for compliance with hygiene regulations. What these inspection examples have in common is that their input is an item whose state is unknown (apple, restaurant) and their output is the same item whose state has been determined. Unfortunately, not all state determinations are perfect: there are inspection errors, as noted in Section 2.2. In the current section, our interest is in mathematical models of inspection and their use in performance optimization. For a chapter-length treatment of test and inspection, see Drury (2001).

As with any modeling activity the starting point should be a task analysis (see Section 2.1). Many detailed task analyses of inspection have been performed, resulting in a generic function-level model (e.g., Drury, 2001): initiate, present, search, decision, and response. Task analyses of specific inspection tasks go much deeper than this to provide insights and best practices (e.g., Drury, 1999). Here, however, we restrict ourselves to what are usually seen as the most difficult functions: search and decision. These are often the functions having the lowest reliability (Drury et al., 1997) and taking the greatest time to complete. Each has several useful mathematical models (e.g., visual search theory, Section 3.1) and opportunities for optimization making them appropriate instances for the current chapter. In addition, the models can be combined to give a more integrated view of the entire inspection task and can be used where parts of the task are automated. Both of these extensions are considered after models of search and decision are given individually in the context of inspection tasks.

5.1 Visual and Memory Search: Alternative Model

The visual search networks in Section 3.1 were developed to model rapid search processes, where each distracter is compared with the target (or target set) until a match is found. In contrast, inspection tasks often involve large and complex objects (e.g., circuit boards or aircraft internal structures) when the search process takes place much more slowly (many seconds or even minutes) and there are no distracters as such, just other elements that may or may not have a defect (e.g., an IC chip placed backward or a crack in aircraft structure). Here a model of search as a sequence of eye fixations appears more appropriate. Such models have existed for many years and are based on the following facts:

1. Visual information is available only when the eye is stationary or tracking a moving object. These fixations typically take between 0.2 and 1.0 s and the rapid saccadic movements between fixations preclude visual information intake.
2. In a single fixation, the probability of detecting a target falls off with angle between the target and the optic axis. This means that a target is detectable only (with a given probability) in an area around the optic axis known as the *visual lobe*. Note that the visual lobe is not the fovea: Lobe size depends on target–background contrast and can range from subfoveal size for very difficult targets to almost the entire visual field of view for extremely easy targets.

A major preoccupation of visual search modelers has been how successive fixations are chosen from the visual field to perform the search task. The general consensus is that the fixation sequence arises partly from top-down factors (e.g., a predetermined search sequence based on the inspector's experience) and partly from bottom-up factors (e.g., a potential target at the periphery of vision, leading the next saccade to fixate that point). Models of this type are available (e.g., Wolfe, 1994).

If we treat the bottom-up information as essentially an "end game" to confirm a target, the top-down aspect has typically been modeled as either a random process or a systematic process (Morawski et al., 1980). A random process is characterized as having no memory for previous fixations, while a systematic process assumes perfect memory. In fact, a more general model of partial memory was devised by Arani et al. (1984), with the random and systematic models as special cases. They showed that memory has to be almost perfect to invalidate a random model, collaborating many studies that have fitted both models to the data and found adequate fits for the random model.

The random model assumes that each fixation i is chosen randomly from a set of possible fixations (i.e., sampling with replacement). From this model it is easy to derive the cumulative search time distribution [i.e., the probability $P(t)$ that the target will be located at or before time t] as

$$P(t) = 1 - e^{-\lambda t}$$

Here, the parameter λ is a constant incorporating lobe size and fixation duration information. In fact, both the mean and the standard deviation of this exponential distribution are equal to $1/\lambda$. If the model is valid (and it typically is), a useful deduction from the model is that the time to detect a target will be extremely variable. A typical cumulative probability distribution of a random model is shown in Figure 6.

Note that the process has diminishing returns in that the expected gain from continued search decreases. This is the basis for an optimization model of stopping time in search. The model was developed first in Tsao

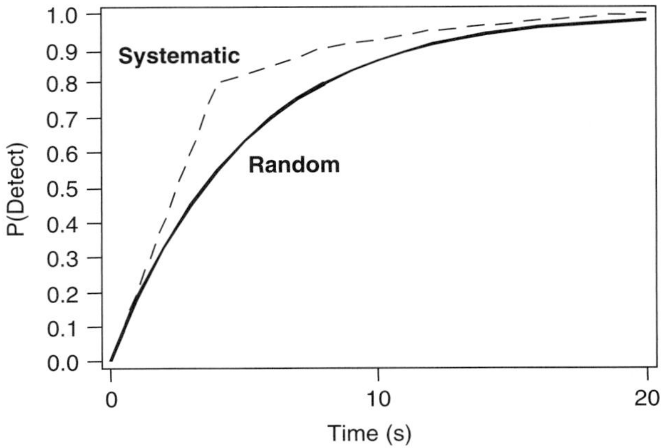

Figure 6 Cumulative probability distribution of a random model.

et al. (1979) and developed much more fully in Chi and Drury (1998). If the value of detecting a target is v, the probability of an item having a target is p, and the cost of a unit of inspection time is k, we can find the three outcomes of a search task with the probability and value of each: (1) the target is present and found, (2) the target is present and not found, and (3) the target is not present. No false alarm is logically possible for a search task: either the target is found or it is not. If we multiply probability by value and sum over all outcomes, we have the expected value of the search task. This gives, after simplification;

$$\text{value expected} = -kt + vp(1 - e^{-\lambda t})$$

Setting the first derivative with respect to t at zero to maximize expected value gives

$$k = vp\lambda e^{-\lambda t_{\text{opt}}} \qquad t_{\text{opt}} = \frac{1}{\lambda} \ln \frac{\lambda vp}{k}$$

Note that t_{opt} increases when p is high, v is high, and k is low. Thus, a longer time should be spent inspecting each area where (1) there is a greater prior probability of a defect, (2) there is a greater value to finding a defect, and (3) there is a lower cost of the inspection.

This describes a possible optimum behavior in a search task where a single target can occur. It has been verified by Drury and Chi (1995) as a reasonable model of actual inspection behavior. For a slightly different model, where the operator covers the search field in overlapping fixations, an equivalent optimization model was devised by Bavejo et al. (1996). That model described human field of view movements rather than the (untested) eye movements, very well.

Extensions have been made to search models to cover multiple instances of a target (Drury and Hong, 2000) and multiple instances of multiple target types (Hong and Drury, 2002). Our optimization model of search can also be applied to multiple targets of the same or different types. Hong (2002) derived and validated such a model against human search time data with good results.

5.2 Decision Model: SDT Revisited

After completing the search function, the inspector will now have either (1) not found a target, in which case the item is by definition good; or (2) have found one (or even more than one) target that needs to be assessed for acceptance or rejection against a standard. The decision process (2) thus arises only when a potential target (an *indication* in nondestructive inspection terminology) has been found. At this point, any model that has two states of the item (defect, no defect) and two decision outcomes (accept, reject) is appropriate. SDT (Section 2.2) certainly meets this criterion, and has a long history of application to inspection tasks. For example, Drury and Addison (1973) found that increased feedback in a glass inspection task raised the discriminability d', and the benefit persisted over many months of measurement. In the 1970s and 1980s it became quite fashionable to apply SDT to inspection: for example, the studies reported in the book by Drury and Fox (1975). Legitimate warnings were raised about the use of a parametric form of SDT: for example, the assumption of normal distributions of equal variance (Megaw, 1979).

Optimization aspects of the SDT model, for example, the choice of optimum criterion β_{opt}, are directly testable. In noninspection decision tasks, the general finding is that β_{opt} does not change as rapidly as it should with changes in the cost–value structure or changes in the a priori probability of a defect. This has become known as the *sluggish beta* phenomenon. It arises if the decision maker does not take all of the available information into account in reaching a decision (i.e., the human as a degraded optimizer). This really calls into question whether any model of the human as a maximizer of expected value has any validity, and eventually led to much more general models of the human as a satisficer rather than an optimizer, most famously the model of Newell and Simon (1972). However, if we treat the human as a degraded optimizer in the decision aspects of an inspection task (Chi and Drury, 2001), reasonable agreement with performance data is found. The warning is raised, however, that people may not be optimizers in the mathematically strict sense in inspection decisions, perhaps because they do not deal well with low-probability events (e.g., finding a bomb in an airline passenger's bag) or large costs (e.g., the tragedy of losing an airline to terrorism).

5.3 Reintegration of Search and Decision

Inspection tasks typically include both search and decision components. For example, Drury (2002) showed that a single unified model (essentially the generic functions, initiate, present, search, decision, and response, listed above) described all of the various airport security inspection processes. Some processes have no search (e.g., listening to a car engine for a loose tappet), whereas some have no decision (e.g., inspection of jet engine hubs for cracks where any crack must cause rejection). These are, however, the exception. The integration of models such as visual search and SDT involves issues beyond each individual model.

An early integration example was Drury (1973), who collected many studies of the "speed effect" in inspection [i.e., the speed–accuracy trade-off (SATO)]. With visual search following a random cumulative model [e.g., equation (5)] and SDT providing the ultimate levels of type 1 and type 2 error when search is complete, the SATO model shows the probability of a hit increasing over time with diminishing returns to reach some ultimate value less than 1.0. The probability of a false alarm starts at zero for very short inspection times and gradually increases with inspection time, but leveling at a different value. This model fitted much of the available SATO data and was interpreted as evidence for a search-plus-decision model. Later, Chi and Drury (2001) tested optimization aspects of this model against human performance in a task of inspecting circuit boards, again finding good agreement.

The second use of an integrated model is in diagnosis of inspection error. At its simplest the search-plus-decision model shows that search alone cannot produce false alarms. Hence, if there is only a search process, false alarms will be extremely rare, as was shown to be the case by Drury and Forsman (1996). Where both search and decision occur, it is possible to separate the errors from the two functions with some additional effort. In a task of inspecting jet engine bearings, Drury and Sinclair (1983) used the fact that search was visual whereas decision was tactile to differentiate between search errors and decision errors. They found that search performance was poor, but consistent across inspectors, whereas decision performance varied widely among inspectors. Their analysis led to the development of a successful training–retraining program for these inspectors (Drury and Kleiner, 1990). In another study of aircraft structural inspection, Drury et al. (1997) were able to use videotape analysis to separate the two functions, with findings very similar to those of Drury and Sinclair.

Finally, an integration of the two models has implications for analysis of the results of all inspection tasks. When SDT was first used for vigilance tasks, search was not recognized. However, many of the misses in inspection arise from the search process, not from a bias toward acceptance in the decision process. Hence, interpreting overall inspection results in terms of SDT is erroneous unless no search is involved. This is what Wiener (1975) called *jumping off the d-prime end*, and it is a legitimate criticism of early inspection modeling work (e.g., Drury and Addison, 1973). Unfortunately, this misinterpretation still happens. Again, the moral is that there is nothing as valuable as a good model, nor as misleading as an inappropriate model.

Application: Automated Inspection

If we can model the human inspector with reasonable success, how can we extend this modeling to situations where human and automation perform inspection tasks jointly? There are now excellent automated alternatives to some aspects of inspection; see, for example, the detailed review in Drury (2000). How can we incorporate human and automation models into overall inspection models to derive appropriate levels of automation (cf. Parasuraman et al., 2000)? Part of the problem is that most papers on automated inspection denigrate human roles, emphasize the wonders of algorithms, and often provide data only on probability of detection, ignoring false alarms.

The most obvious first step to explain the integration of human and automated inspection is to compare their relative merits directly. Drury and Sinclair (1983) examined an automated system for jet engine bearing inspection using the same measures of performance as were used for human inspectors, in this case ROC curves. Their conclusion was that neither human nor automation was particularly effective, leading to recommendations to improve the automated system and to the subsequent human training program (Drury and Kleiner, 1990).

A more comprehensive step was taken by Hou et al. (1993), who examined search and decision separately for human and automation. They use an SDT measure of discriminability (A) as well as inspection speed to compare human and algorithmic alternatives for each function. Their conclusion was that both purely automated systems, as well as the unaided manual system, were inferior to hybrid human–algorithm systems for circuit board inspection. We may be able to compare different human and automation hybrids by direct measurement, but true a priori allocation of function will come only when we can predict from models of the alternatives which hybrid systems to build and test.

6 DUAL TASKS

Above, we have been describing tasks in which a person needs to make just one response in a given situation, that response depending on the stimulus ensemble presented to the person. However, the real world sometimes provides people with much more challenging situations. For example, suppose that a sign tells a driver to change lanes, and then a car cuts in front. The responses are to turn and to brake. Typically, the time to respond to the second stimulus is greater than it would be if it had been presented alone. This and many other findings can be explained by a model of *dual-task* performance originally proposed by Davis (1957) and studied extensively since (e.g., McCann and Johnston, 1992; Pashler, 1998).

The model is shown in Figure 7 Stimulus $s1$ is presented, followed after a brief interval by stimulus $s2$, with required responses $r1$ and $r2$. Each stimulus requires perceptual processing, denoted $a1$ or $a2$ as appropriate, cognitive processing, denoted $b1$ or $b2$, and motor preparation processing, denoted $c1$ or $c2$. The interval between the stimuli is denoted SOA (stimulus-onset asynchrony). Perceptual and motor processing of either stimulus can go on concurrently with any other processing. But cognitive processing (response selection) for the two stimuli is sequential, so process $b1$ must be completed before process $b2$ can start. The delay in responding to the second stimulus is due to $b2$ waiting for $b1$ to finish.

One can see from the bar chart in Figure 7 that the processes can be represented as activities in a

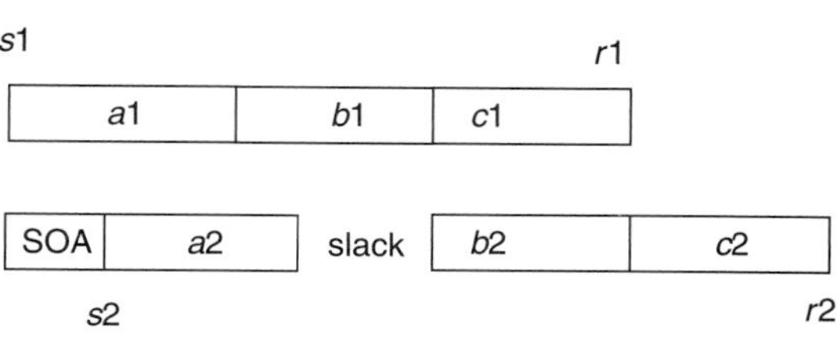

Figure 7 Dual processing network.

critical path network, so knowledge of the process arrangement can be obtained through the method of selective influence for activity networks (Schweickert, 1978; Schweickert and Townsend, 1989; Schweickert et al., 1992). Analysis of selective influence for this particular model is sometimes called *locus-of-slack logic* (e.g., McCann and Johnston, 1992). Much is known about where factors such as stimulus quality, display size, arithmetic difficulty, and so on, have their effects. Results are too extensive to summarize here, but a review is given by Pashler (1998). Recent reviews of the methodology are given by Logan (2002) and Townsend and Wenger (2004).

The model assumes that only one response selection process goes on at a time, and it is worth taking a moment to consider why this might be true. Many models of the response selection process have been proposed. The two major ones are accumulators and random walks. With each, a stimulus is presumed to transmit information over time. Each incoming piece of information is classified as favoring one of the possible responses. In an accumulator, information favoring a response increases the activation of that response. In a random walk, information favoring a response increases the net activation of that response and decreases the net activation of every other possible response. In both models, for each possible response a criterion has been set. As soon as the activation for some response reaches its criterion, that response is made.

For a given stimulus, every possible response is made with some probability and there is a probability that no response is made. The behavioral data consist of those probabilities, together with the probability distributions of the response times. In evaluating the models it is helpful to realize that an accumulator model can always be constructed to account perfectly for the behavioral data (Dzhafarov, 1993). Hence, goodness of fit is important, but of more importance is whether experimental factors can be explained coherently. On the whole for these models, results are as one would expect, for example, stimulus quality ordinarily influences the rate at which activation increases and payoffs ordinarily influence the criterion values. For a review, see Luce (1986).

A single neuron begins firing when the algebraic sum of the excitation and inhibition reaching it exceeds a threshold, so the resources needed for an accumulator or a random walk in itself seem small. But with either an accumulator or a random walk, the system must be set by (1) selecting the possible responses, (2) forming temporary associations between anticipated incoming bits of information and the responses they favor for the situation, and (3) setting the values of the response criteria. The resources for assembling and maintaining the settings may be considerable. The resources needed lead to Welford's (1967) single-channel hypothesis that only one decision (i.e., response selection) is made at a time.

Many have considered whether the hypothesis is wrong, and it is possible that the two response selection processes go on concurrently, at least sometimes. In the EPIC model (Meyer and Kieras, 1997a, b) a person has the option of executing two response selection processes concurrently and without interference. Most models allowing concurrent response selection assume that it is slower when concurrent (e.g., Navon and Miller, 2002; Tombu and Jolicoeur, 2003).

Application: Scheduling Stimulus Presentations

For timing the information presented in displays, one useful finding is that a delay in the onset of the second stimulus need not delay the response to the second stimulus, or might only delay it by a small amount (e.g., Smith, 1969). In other words, the perceptual processing of the second stimulus may not be on the critical path to the second response, and so may have slack. It is also useful to note that there is evidence that humans are able to control the order of the cognitive processes; that is, they can control whether $b1$ is executed before $b2$, or vice versa (Ehrenstein et al., 1997).

A relevant finding from scheduling theory is that if a process is faster when executed alone than when executed concurrently with another process, and if the goal is to minimize the average of the times at which each task is finished, with respect to the same time zero starting point, the optimal schedule is usually to allocate all the capacity to one process and then allocate it all to the other process (i.e., to schedule the processes sequentially rather than concurrently) (Conway et al., 1967). To see this, suppose that processes a and b each take 1 unit of time if executed alone, and 2 units of time if executed concurrently. Suppose that they are both ready to start at time 0. The average of their completion times with respect to a time zero starting point is 1.5 if they are sequential and 2 if they are concurrent. In other words, humans may schedule response selection processes sequentially not because they must, but because it is optimal.

If there is a choice as to when to present stimuli in a display, there is an apparent advantage to displaying the second stimulus soon. If, as illustrated, perception of the second stimulus is not on the critical path (note the presence of slack), having the second stimulus appear early would seem to do no harm. Further, if response selection for the second stimulus is not delayed, having the second stimulus available as soon as possible would be helpful. There are two potential problems with this analysis. First, as soon as a stimulus is presented, irrelevant information from it may be sent to processes for the other stimulus (crosstalk) (see, e.g., Hommel, 1998). This may lead to increased response times and errors. Second, presenting the stimuli close together in time may lead to parallel processing, which can be more inefficient if, as noted above, the goal is to minimize the average completion time of two processes, measured from the same starting point.

7 PSYCHOMOTOR PROCESSES

As noted above, processing time in a task can roughly be categorized as perceptual, cognitive, and motor, and the largest of these is often motor time. The study of movement is interdisciplinary, but as with sensation, the physics of the system is an integral part of the modeling. For an introduction to models, see Jagacinski and Flach (2003), and for a discussion of controversies, see *Controversies in Neuroscience, I: Movement Control* (Editors, 1992).

One of the simplest models to lead to a reasonable approximation of human movement is the mass, spring, and damper in Figure 8 [see Crossman and Goodeve (1963) for an early presentation]. A force $F(t)$ is applied over time by an agonist muscle to a limb of mass m. A restoring force is produced by an antagonist muscle represented by a spring. (The agonist could be represented by a spring as well.) Another force is damping due, say, to friction from sliding across a mouse pad or to an internal source such as the antagonist muscle itself or a joint. (The damper is illustrated as a piston in a cylinder filled with oil.) We consider a horizontal movement rather than a rotation through an angle; an analysis of a rotation would not be very different.

Let the position of the limb at time t be x, with position 0 at time 0. By Hooke's law, the spring produces a force proportional to its displacement from its equilibrium position, q. The direction of the force depends on whether the spring is stretched or contracted. The direction is opposite to the direction of the spring's displacement from its equilibrium position, so the force due to the spring is $-k(x-q)$. Note that if the agonist is represented by a spring producing force $-k_1(x-q_1)$, and the antagonist is represented by a spring producing force $-k_2(x-q_2)$, the sum of the two forces is $-(k_1+k_2)[x-(k_1q_1+k_2q_2)/(k_1+k_2)]$. That is, the model with two springs can be replaced by an equivalent model with a single

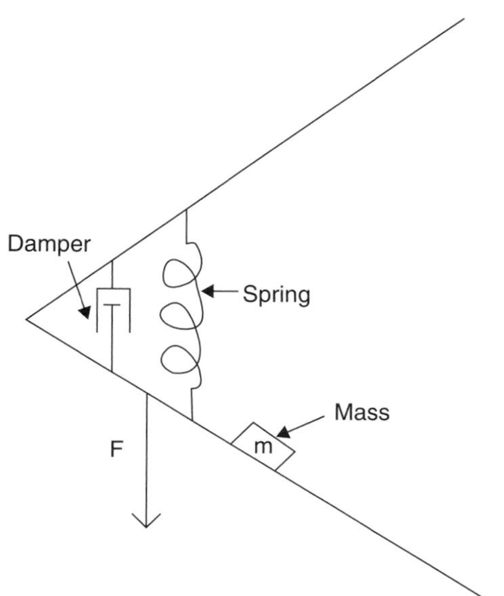

Figure 8 Mass, spring, and damper.

spring. The force due to the damper is proportional to the velocity but in direction opposite to it, that is, $-bdx(t)/dt$. By Newton's second law, the sum of all the forces equals the mass times the acceleration. That is,

$$m\frac{d^2x}{dt^2} = F(t) - k(x-q) - \frac{b\,dx}{dt} \tag{7}$$

where $F(t)$ is the force applied by the agonist muscle and $-k(x-q)$ is the force applied by the antagonist muscle. Most movements are more complicated, and the model would include components such as gravity and multiple joints.

A model of motion is useful not only for describing the motion but also for considering how the motion is controlled. For optimal control (i.e., producing an input that maximizes some objective function), general principles are found, for example, in Kirk (1970) and Hogan (1988). For biological systems, two main ways of controlling a movement have been proposed. Suppose that the goal is to produce a position A of the limb. The first way is for the system to estimate and then apply the force $F(t)$ needed to produce position A. The second way is not ordinarily available in a physical system, where the characteristics of the spring and damper are fixed. But in a biological system the stiffness and other features of the muscles can be changed. Hence to produce movement, the equilibrium position of the spring can be set directly to that needed to produce the limb position goal (Feldman, 1966). With both methods, later corrections can be made based on feedback, although these are more often considered with the first method.

To be useful, a model must produce known findings about human movement. One of the main results is Fitts's law. Suppose that a person moves a limb to a target of width W, centered at a distance A away from the starting position. The time to make the movement is well approximated by

$$MT = c + d\log_2\frac{2A}{W} \tag{8}$$

where c and d are free parameters (Fitts, 1954). The quantity $\log_2(2A/W)$ is called the *index of difficulty*. The relation was first fit to data for people moving a stylus back and forth continually between two targets, each of width W, with the centers of the targets separated by distance A. But it fits well for many other situations, for example, for moving a finger to a calculator button or a mouse pointer to a target (Card et al., 1983).

The simple mass, spring, and damper model leads to Fitts's law approximately when a force is applied to the limb (Langolf et al., 1976; for review, see Jagacinski and Flach, 2003). We illustrate the method of presetting the equilibrium position by giving a similar derivation leading to Fitts's law as an approximation for short movement times. Suppose that the goal is to move the limb from starting position 0 to position A. With this method of control, the force $F(t)$ in equation (7) is 0. Then the solution to equation (7) is (e.g., Resnick and Halliday, 1963)

$$x - q = Ce^{-bt/2m}\cos(\omega' t + \delta)$$

where $\omega' = \sqrt{k/m - (b/2m)^2}$, and C and δ are parameters depending on the initial conditions. It is straightforward to check that this is a solution by taking first and second derivatives. When the motion is bounded, all solutions have this form for the underdamped case (when b is small). In that case, the movement is a damped oscillation. Through trigonometric identities, the solution can be expressed in various equivalent ways.

Suppose that at time 0, the initial position x is 0, and the initial velocity dx/dt is 0. Then it is straightforward to see that $C = -q/\cos\delta$, δ is the angle whose tangent is $-b/2m\omega'$, and $\cos\delta = \sqrt{1 - b^2/4mk}$.

To set the limb position goal to A, set the equilibrium position q to A. With these parameters, the position x of the limb at time t is

$$x = A - Ae^{-bt/2m}\frac{\cos(\omega' t + \delta)}{\cos\delta}$$

According to the model, the limb will oscillate, passing back and forth over the target position A, and stopping at it at time infinity. (Oscillations in a movement can often be viewed when moving a mouse pointer to a target on a computer screen.) We are interested in the time at which the limb moves into the target interval and does not leave it. As an approximation, let us say that this happens when the envelope of the oscillations is within the target interval. Consider the lower boundary of the envelope (the result is the same if we consider the upper boundary). It reaches the target interval when

$$\mathrm{A} - \frac{W}{2} = A - \frac{Ae - bt/2m}{\cos\delta}$$

Then

$$\frac{W}{2} = \frac{Ae - bt/2m}{\cos\delta}$$

$$\ln\frac{W}{2A} = -\frac{bt}{2m} + \ln(\cos\delta)$$

$$t = \frac{2m}{b}\ln\frac{2A}{W} + \frac{2m}{b}\ln(\cos\delta)$$

The log can be changed from base e to base 2 by multiplying by $\ln 2$, and the result is in the form of Fitts's law.

We mention one more regularity for checking a working model. In the task for which Fitts's law applies, a person is given a target's center position and width and produces a movement time. In a slightly different task, a person is given a target position A and a movement time goal MT and produces a movement to a position. The standard deviation of the movement positions produced, W_e, is well approximated by Schmidt's law,

$$W_e = \frac{cMT}{A}$$

where c is a free parameter (Schmidt et al., 1979). The two tasks are similar, as are the two laws, and a model leading to each as a special case has been proposed by Meyer et al. (1990).

The basic mass, spring, and damper model described above fails to predict some aspects of movement accurately (see, e.g., Langolf et al., 1976; Jagacinski et al., 1980), so there are many variations of the basic model. Controlling a movement with a single brief initial step function force predicts movements more asymmetrical than are found in data, and better fits are produced by assuming an initial accelerating force and a final decelerating force (the bang-bang model; see Jagacinski and Flach, 2003). The equilibrium point hypothesis has trouble explaining fast movements. De Lussanet et al. (2002) propose as an improvement controlling the movement by moving the equilibrium point to its goal position at a constant velocity rather than in a jump. Some evidence that the equilibrium point moves is provided in an experiment by Bizzi et al. (1992).

It is difficult to differentiate between the hypothesized methods of control (i.e., control by directly producing forces and control by setting the equilibrium point). One difficulty is that when changes in the system are observed, it is difficult to establish that they occurred for the purpose of control. Naturally, for a complicated system, different methods of control are probably used in different situations, as proposed by Schmidt and McGown (1980).

Fitts's law describes terminal aiming tasks, where the path to the target is unimportant but hitting the target at the terminal point of the aiming task is vital. This task is self-paced, and in fact, Fitts's law describes a speed–accuracy trade-off. A rather different form of self-paced movement is a path control task where the operator must move along a path without exceeding lateral boundaries. Examples are walking along a narrow corridor, driving along a narrow road, or even sewing along a seam of fixed width. An early study of line drawing along fixed width paths (Drury, 1971) derived a model based on the operator as an intermittently acting servomechanism. At any instant, the operator finds himself or herself at some point across the allowed width and must choose how to make an open-loop movement during the next sampling interval. Drury et al. (1987) and Montazer et al. (1988) modeled this task as one of choosing a direction (angle θ to centerline) and a distance (R in direction θ) for the next movement. They assumed that the objective function was to maximize the distance traveled along the path ($R\cos\theta$) while minimizing the probability of going outside the path boundaries. The

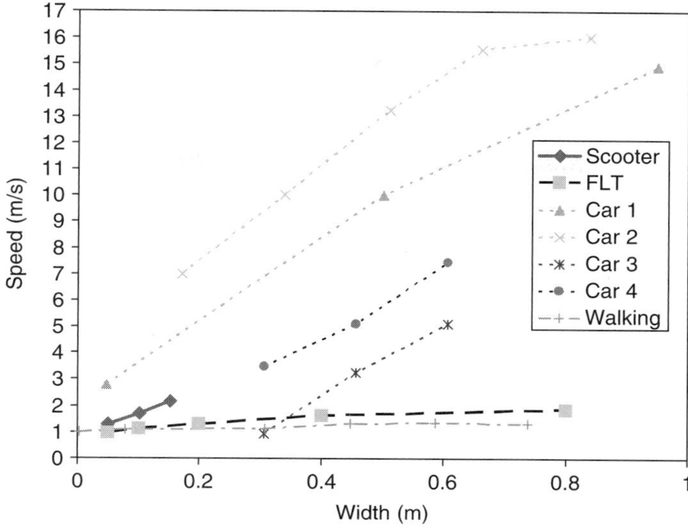

Figure 9 Relation between speed and target width.

constraint set was derived from models of the buildup of lateral error in blind movements (e.g., Beggs and Howarth, 1970). The optimization model gave the same formulation as Drury's original (1971) model, namely

$$\text{speed} = \text{constant} \times \text{path width}$$

The model was derived and validated for movements on straight and circular courses. For very large widths, the speed is limited by other factors, such as a speed limit on a highway, so the linear relationship will eventually flatten out at high widths. Figure 9 shows speed–width relationships found by a number of authors for many different vehicles, including unpublished data on a personal "scooter" with side-by-side wheels. As in other examples, the optimization model provides a good description of performance.

Application: Manual Assembly

Many workers perform manual assembly tasks when their hands are located above their shoulders or far out from the sides of their body. Although this is not the preferred location (Konz, 1967), it is still a situation common in the workforce (Wiker et al., 1989). Some such tasks resemble a repetitive Fitts's tapping task. In fact, it has been estimated that anywhere between 40 and 80% of the cycle times in typical manual assembly tasks are due to the move and positioning elements required to perform the task (Arberg, 1963). To study such elements by themselves, Wiker et al. asked subjects to move a tool back and forth, repeatedly, between two holes. The tool was shaped like a small hand drill and contained a stylus that the subject had to position in the center of the hole. In this task it was possible to adjust, among other things, the movement amplitude (the distance between the two holes, let this be labeled A), direction (horizontal or vertical), target hole diameter ($2W$), positioning tolerance (the distance T between the outer perimeter of a pin being placed in the center of a round hole and the edge of the hole), task duration, tool mass (m, kg), duty cycle (number of repetitions per second, N), and hand elevation (height of arm above shoulder height, E). Wiker et al. (1989) found that there was a lawful relation between these factors and the movement and positioning time (MT) [an elaboration of Fitts's law first proposed by Hoffman (1981; cited in Chung, 1983)]:

$$\begin{aligned} MT(\text{ms}) &= 106 + (68 + 0.015E + 0.00034EmN)\text{IDM} \\ &\quad + (24 + 0.001EN)\text{IDP} \end{aligned} \tag{9}$$

where the index of move, IDM, and index of position, IDP, are defined as follows:

$$\text{IDM} = \log_2 \frac{A}{2W} \qquad \text{IDP} = \log_2 \frac{2W}{T}$$

[The relation between equations (9) and (8) is immediately apparent given the definition of IDM, which was referred to above in a different context as the index of difficulty.] As an example of the changes in the movement and positioning times observed as a function of changes in the independent variables, Wiker et al. (1989) find that the move and positioning times increased, respectively, by 15.3 and 26.5% when the arm went from 15° below shoulder height to 60° above shoulder height. The equation above is particularly useful if, say, one wants to identify the optimal elevation, assuming that workers of many different heights will be performing the job.

8 TRAINING, EDUCATION, AND INSTRUCTIONAL SYSTEMS

Training continues to be of central importance in human factors in both the private and public sectors. However, quantitative models that could be used in training, and more broadly in education, have remained elusive for the most part. This is changing radically. Below we discuss some of the very earliest work and then segue to a discussion of some more current research.

8.1 Paired Associate Models

Much was done in the 1950s and 1960s with paired associate learning (Estes, 1950; Bush and Mosteller, 1951). In that work, and in much that followed (Bower, 1961), an attempt was made to predict the rate at which the associations between stimuli and responses were learned. In a typical paradigm, participants would be given a stimulus, say a letter, and be asked to produce the correct response, say a particular digit. A number of factors were varied, including the total number of stimuli in the list, the time between repetitions of the same stimulus, and the time between the last trial and an evaluation of performance.

In the simplest case, the one we describe here, the stimulus–response pair is assumed to be in one of two states, either learned (C, or conditioned) or not learned ($\overline{C}$, not conditioned). Let c be the probability that it is learned on any one trial. Let g be the probability P(correct) that the participant guesses the correct answer, assuming that the stimulus–response association is not learned; assume that this probability is 1 if the association is learned. Then the model can be presented formally as a two-state Markov chain where:

	Trial $n+1$			
Trial n	C	$\overline{C}$		P(correct)
C	1	0	C	1
$\overline{C}$	c	$1-c$	$\overline{C}$	g

It can easily be shown that the probability that a participant responds correctly after n trials is equal to:

$$P(\text{correct}) = 1 - (1-g)(1-c)^n$$

Now, suppose that one wanted to maximize the joint probability that each of the stimuli had been learned after n training trials, where n is greater than the number of paired associates. Then Karush and Dear (1966) have derived the optimal training strategy. It is so simple that it can easily be described in a sentence or two. One must first train exactly once all stimulus–response pairs. One then needs to keep track of an index for each stimulus–response pair: That index is set to the zero if the response on the preceding trial was incorrect; it is set to the number of successive correct responses if the response on the preceding trial was correct. One then selects to train on the next trial the stimulus with the smallest such index. Unfortunately, simple two-state models cannot explain some of the results. In fact, Katsikopoulos and Fisher (2001) have shown that if one is going to explain several of the most critical results, one will need two Markov chains, one applied on each trial in which an association is trained and one applied on each trial in which an association is not trained. Moreover, the chains will need, at a minimum, four states. This has made the analytic derivation of the optimal global training strategy all but impossible. Instead, optimization is performed over a relatively small horizon, or simulations are used to approximate the best training schedule.

Application: Morse Code

The applications of the work in paired-associate learning are relatively few, but still potentially significant. In the military, soldiers continue to be trained in the use of Morse code as a backup. The associations between the stimuli, "dots" and "dashes," and the responses, letters and digits, must be learned. The training takes a very long time, so the military was interested in learning whether something could be done to reduce the training time (Fisher and Townsend, 1993). For an application such as this, the assumptions required to identify the optimal order in which to train the stimuli as set forth by Karush and Dear (1966) are reasonably well satisfied, and so their method can be applied with very little additional cost (since the training was already being done on a computer and the stimuli and responses recorded). Other applications might well include the learning of simple multiplication facts (the multiplicands are the stimulus and the product is the response) or the vocabulary in a foreign language (the foreign word is the stimulus and the English word equivalent is the response).

8.2 Complex Skill Acquisition

A more complex model of learning is needed than the paired-associate model described above if one is going to explain the majority of learning that goes on in the classroom and elsewhere. Anderson (1983) has developed just such a model, one that is widely used and constantly under development, its most recent version being referred to as ACT-R (Anderson et al., 2002). Very briefly, the process starts with a task analysis, often using the GOMS procedures mentioned above (Card et al., 1983). Once a task is decomposed into increasingly better defined goals, a working model is built from two general types of knowledge: declarative and procedural knowledge.

Declarative knowledge is based on the instructions that the participant receives when performing a task and on the task analysis. These are stored in a semantic network. For example, a student will learn that the sum of three angles in a triangle must equal 180°. *Procedural knowledge* is stored as production rules,

Table 1 Production Rules and Compilation

Rule 1:	IF	the goal is to add three numbers, n_1, n_2, and n_3,
	THEN	send a retrieval request to declarative memory for the sum $s_{1+2} = n_1 + n_2$ of the first two numbers.
Rule 2:	IF	the goal is to add three numbers AND the sum s_{1+2} of the first two numbers is retrieved,
	THEN	send a retrieval request to declarative memory for the sum of (a) the first two numbers, s_{1+2}, and (b) the third number n_3; label this last sum s_{1+2+3}.
Rule 3:	IF	the goal is to add three numbers AND the sum s_{1+2+3} of the first three numbers is retrieved,
	THEN	the answer is the retrieved sum, s_{1+2+3}.
Rule 1&2:	IF	the goal is to add the two numbers 3 and 5 together with a third number, n_3,
	THEN	send a request to declarative memory for the sum of 8 and n_3.

rules that take actions consistent with the declarative component, including retrieving information from memory, focusing attention on a certain area of the display, pressing a key in response to a stimulus, and so on. For example, the three rules in Table 1, described in more detail in Taatgen and Lee (2003), are used to add three numbers (see rules 1, 2 and 3). There are various hard constraints on both the general form that productions can take and the execution of those productions. Most relevant here are two such constraints. First, only one item at a time can be retrieved from declarative memory. Second, the production rules must fire in sequence.

From the standpoint of learning, the question that must be addressed is how participants over time become faster when performing complex tasks. In general terms, skill acquisition has been broken down into three stages: the cognitive, associative, and autonomous stages (Fitts, 1964). The answer within the framework of ACT-R has varied over time. Most recently, Taatgen and Anderson (2002) have suggested that with enough use, two rules that occur one after one another can be combined in such a way that the retrieval request from declarative memory in the first rule is no longer assumed to be necessary. Thus, only one production rule is necessary, a rule consisting of the IF part of the first rule and the THEN part of the second rule. An example is given in Table 1 in rule 1&2. A relatively time-consuming step has now been saved, in particular the retrieval from declarative memory of the sum of the first two numbers. This is identified by them as *production compilation* and is consistent with their results in several different experiments.

Application: Training Air-Traffic Controllers

Taatgen and Lee (2003) used ACT-R to model the improvement in performance over time of participants (college undergraduates) asked to perform the Kanfer–Ackerman air-traffic controller task (Ackerman, 1988; Ackerman and Kanfer, 1994). The task is a complex one and can be decomposed hierarchically into the unit task level (e.g., land a plane on the runway), the functional level (e.g., find a runway to land), and the keystroke level (e.g., press a particular key). Taatgen and Lee identified the declarative knowledge and task-independent procedural rules needed to perform each task at each level. In addition, they identified formally under what conditions declarative knowledge and task-independent production rules could be combined into task-specific production rules through the mechanism that was identified above as production compilation. Using values for parameters obtained outside the Kanfer–Ackerman air-traffic controller task, they predicted for each of the first 10 trials how performance at the unit task level, functional level, and keystroke level would vary. They find that qualitatively their model fits the results very nicely. This is an example where the mathematical model could be used to limit the selection of interfaces to examine more completely, perhaps in an experiment, assuming that too large a number were available initially to evaluate fully.

9 WARNINGS

A warning device can be characterized in terms of signal detection theory. Consider a smoke alarm that goes off when the concentration of certain particles in the air exceeds a critical value. The critical value corresponds to the response bias. The standardized difference between the mean concentration of particles when there is a fire and when there is not corresponds to the sensitivity of the device. In most environments the operator is not in a position to change the frequency of hits (there is a fire and the warning device indicates such), misses, false alarms (there is no fire and the warning device indicates that there is such), and correct rejections of the device itself. However, in some environments the operator can, and probably will, actually want to alter the frequency of false alarms (without tinkering with the warning system mechanics).

For example, consider nurses working in an intensive care unit (ICU). Such nurses will want to reduce the number of times that the warning sounds. This will reduce the number of times that a true emergency is present, given that the warning sounded, as well as the number of times that no emergency is present, given that the warning sounded. Meyer and Bitan (2002) realized that this will in turn reduce the information in the warning. At the extreme, if the operator is always able to take actions that prevent the warning from sounding in a true emergency, only false alarms will be generated. It is known that the operator's response time is influenced by the frequency of false alarms produced by the warning system; most important in this context, response time decreases when the frequency of false alarms increases (Getty et al., 1995).

Table 2 Positive Predictive Value

Device's Response	System State		Device's Response	System State	
	F	N		F	N
f	90	10	f	9	10
n	10	90	n	1	90
	$\text{PPV} = \dfrac{90}{90+10} = 0.9$			$\text{PPV} = \dfrac{9}{9+10} = 0.47$	

This raises the question of whether a warning device is more valuable or less valuable for better operators (i.e., operators who have reduced the number of times that an emergency situation occurs).

Because the answer depends on the combined effect of several quantities, this question would be difficult to answer without a model. Using signal detection theory, Meyer and Bitan (2002) calculated the predictive value of warnings as a function of the probability of a system failure. The positive predictive value is the probability that there actually is a failure given that the device produces an alarm. Let F denote an actual failure of the system and f denote that an alarm is given. Then the positive predictive value, PPV, is $P(F|f) = \text{hits}/(\text{hits} + \text{false alarms})$. Analogously, the negative predictive value, NPV, is the probability the system is normal given that the device does not produce an alarm. Let N denote the normal state of the system, and let n denote that no warning is given. Then the negative predictive value is $P(N|n)$. It is clear from the information in Table 2 that the positive predictive value decreases as the operator reduces the number of instances of system failure from 100 (left side) to 10 (right side). The negative predictive value can easily be shown to increase here.

Given the above, Meyer and Bitan (2002) asked how one might index the overall effect on the operator of changes in these two indices, the positive and negative predictive values, that were heading in opposite directions. They suggested several ways that one might combine this information, one of which is to use information theory and, in particular, to compute the information transmitted by the warning system, something we have already discussed above. Meyer and Bitan (2002) considered three hypothetical warning devices with different values of d' and $\beta = 1$ (i.e., neutral bias). For each, they found that the negative predictive value increased slightly as the probability of actual failure decreased from 0.200 to 0.001. However, the positive predictive value decreased and was considerably lower at the low end of this range than at the high end, and so was the information transmitted. In short, the diagnostic value of an alarm is worse for better operators.

Application: Intensive Care Units

The conclusion above was borne out in an experiment on a simulated intensive care nurse's workstation, with an imperfect device warning that a patient needed attention (Meyer and Bitan, 2002). Performance of participants improved over time in the experiment. But the positive predictive value of an alarm decreased and the negative predictive value increased in such a way that the information transmitted by the alarm decreased as they became better operators. Unfortunately, the behavioral implications of these findings are not immediately clear. On the one hand, it is known that as the informative value of a warning goes down, operators are less likely to take action. On the other hand, it is clearly beneficial to reduce the number of situations in which a warning is required. Here is a situation where a mathematical model of a system has uncovered a problem that would probably not have been recognized. However, by itself it cannot be used to solve the problem. More information is needed about the performance of operators in such complex situations.

10 SUMMARY

We set out to show the broad range of uses to which mathematical models can be put in the design of the interface between the human user and the environment. Most centrally, we set out to show that such models not only can be used to design an interface, but can actually be used to optimize the interface. We gave several such examples, including (1) the design of optimal menu hierarchies, (2) the design of optimal inspection schedules, and (3) the design of optimal training sequences. Of course, as we noted above, and as we made clear throughout the chapter, mathematical models have a broader use than simply that of optimizing the interface. We gave several examples of these other uses as well, including (1) the prediction of performance with a particular interface to determine whether it will function as desired and therefore should be considered for implementation, (2) the identification of the effects of neurotoxic agents on the speed of latent processes, and (3) the determination of whether two apparently different models actually make different predictions.

We are hopeful that mathematical modeling will continue to play an important and increasing role in the design of the interface when such is appropriate. There are several indications that such will be the case, including a recent special issue of *Human Factors* devoted entirely to mathematical models as well as the formation just this year within the Human Factors and Ergonomics Society of a technical group whose interests focus on modeling. And, of course, we hope that this chapter will motivate others in the research community to think more broadly about how they too might apply one or more of the many modeling techniques described herein to the design of an interface.

ACKNOWLEDGMENTS

We have benefited greatly from comments by Jerry Balakrishnan and Richard J. Jagacinski and want to thank them for their help.

REFERENCES

Ackerman, P. L. (1988), "Determinants of Individual Differences During Skill Acquisition: Cognitive Abilities and Information Processing," *Journal of Experimental Psychology: General*, Vol. 117, pp. 288–318.

Ackerman, P. L., and Kanfer, R. (1994), *Kanfer–Ackerman Air Traffic Controller Task© CD_ROM Database, Data Collection Program and Playback Program*, Office of Naval Research, Cognitive Science Program, Arlington, VA.

Anderson, J. A. (1983), *The Architecture of Cognition*, Harvard University Press, Cambridge, MA.

Anderson, J. R. (1993), *Rules of the Mind*, Lawrence Erlbaum Associates, Mahwah, NJ.

Anderson, J. R., and Bower, G. H. (1973), *Human Associative Memory*, V.H. Winston & Sons, Washington DC.

Anderson, J. A., Bothell, D., Byrne, M., and Lebiere, C. (2002), "An Integrated Theory of Mind," retrieved from http://act-r.psy.cmu.edu/papers/403/IntegratedTheory.pdf.

Arani, T., Karwan, M. H., and Drury, C. G. (1984), "A Variable-Memory Model of Visual Search," *Human Factors*, Vol. 26, pp. 631–639.

Arberg, U. (1963), *Frequency of Occurrence of Basic MTM Motions, Research Report 1*, Svenska MTM-Foreningen, Stockholm, Sweden.

Baddeley, A. D. (1992), "Is Working Memory Working?" *Quarterly Journal of Experimental Psychology*, Vol. 44A, No. 1, pp. 1–31.

Balakrishnan, J. D. (1998a), "Some More Sensitive Measures of Sensitivity and Response Bias," *Psychological Methods*, Vol. 3, pp. 68–90.

Balakrishnan, J. D. (1998b), "Measures and Interpretations of Vigilance Performance: Evidence Against the Detection Criterion," *Human Factors*, Vol. 40, pp. 601–623.

Bavejo, A., Drury, C. G., Karwan, M., and Malone, D. M. (1996), "Derivation and Test of an Optimum Overlapping-Lobe Model of Visual Search," *IEEE Transactions on Systems, Man, and Cybernetics*, Vol. 28, pp. 161–168.

Beggs, W. D. A., and Howarth, C. I. (1970), "Movement Control in a Repetitive Motor Task," *Nature*, Vol. 225, pp. 752–753.

Bishu, R. R., and Drury, C. G. (1988), "Information Processing in Assembly Tasks: A Case Study," *Applied Ergonomics*, Vol. 19, No. 2, pp. 90–98.

Bizzi, E., Hogan, N., and Mussa-Ivaldi, F. A. (1992), "Does the Nervous System Use Equilibrium-Point Control to Guide Single and Multiple Joint Movements?" *Behavioral and Brain Sciences*, Vol. 15, pp. 603–611.

Bower, G. H. (1961), "Application of a Model to Paired-Associate Learning," *Psychometrika*, Vol. 26, pp. 255–280.

Bush, R. R., and Mosteller, F. A. (1951), "A Mathematical Model for Simple Learning," *Psychological Review*, Vol. 58, pp. 505–512.

Byrne, M. D., and Anderson, J. R. (1998), "Perception and Action," in *The Atomic Components of Thought*, J. R. Anderson and C. Lebiere, Eds., Lawrence Erlbaum Associates, Mahwah, NJ, pp. 167–200.

Byrne, M. D., and Gray, W. D. (2003), "Returning Human Factors to an Engineering Discipline: Expanding the Science Base Through a New Generation of Quantitative Methods" (Preface to the Special Section), *Human Factors*, Vol. 45, pp. 1–4.

Card, S. K., Moran, T. P., and Newell, A. (1983), *The Psychology of Human–Computer Interaction*, Lawrence Erlbaum Associates, Mahwah, NJ.

Chaffin, D. B., Andersson, G. B. J., and Martin, B. J. (1999), *Occupational Biomechanics*, 3rd ed., Wiley, New York.

Chi, C.-F., and Drury, C. G. (1998), "Do People Choose an Optimal Response Criterion in an Inspection Task?" *IIE Transactions*, Vol. 30, pp. 257–266.

Chi, C.-F., and Drury, C. (2001), "Limits to Human Optimization in Inspection Performance," *International Journal of Systems Science*, Vol. 32, No. 6, pp. 689–701.

Chung, C. S. (1983), "A Study of Human Arm Movements with Applications to Predetermined Motion Time Systems," unpublished master's thesis, University of Melbourne, Australia.

Conway, R. W., Maxwell, W. L., and Miller, L. W. (1967), *Theory of Scheduling*, Addison-Wesley, Reading, MA.

Crossman, E. R. F. W., and Goodeve, P. J. (1963), "Feedback Control of Hand-Movement and Fitts' Law," presented at the meeting of the Experimental Psychology Society, Oxford, July; *Quarterly Journal of Experimental Psychology* (1983), Vol. 35A, pp. 251–278.

Davies, D. R., and Parasuraman, R. (1982), *The Psychology of Vigilance*, Academic Press, London.

Davis, R. (1957), "The Human Operator as a Single Channel Information System," *Quarterly Journal of Experimental Psychology*, Vol. 9, pp. 119–129.

de Lussanet, M. H. E., Smeets, J. B. J., and Brenner, E. (2002), "Relative Damping Improves Linear Mass-Spring Models of Goal-Directed Movements," *Human Movement Science*, Vol. 21, pp. 85–100.

Donaldson, W. (1992), "Measuring Recognition Memory," *Journal of Experimental Psychology: General*, Vol. 121, pp. 275–277.

Drury, C. G. (1971), "Movements with Lateral Constraint," *Ergonomics*, Vol. 14, pp. 293–305.

Drury, C. G. (1973), "The Effect of Speed of Working on Industrial Inspection Accuracy," *Applied Ergonomics*, Vol. 4, pp. 2–7.

Drury, C. G. (1993), "A Note on Fitts' Law and Assembly Order," *Ergonomics*, Vol. 36, No. 7, pp. 801–806.

Drury, C. G. (1999), "Human Factors Good Practices in Fluorescent Penetrant Inspection," in *Human Factors in Aviation Maintenance: Phase Nine, Progress Report*, DOT/FAA/AM-99/xx, National Technical Information Service, Springfield, VA.

Drury, C. G. (2000), "Global Quality: Linking Ergonomics and Production," *International Journal of Production Research*, Vol. 38, No. 17, pp. 4007–4018.

Drury, C. G. (2001), "Human Factors and Automation in Test and Inspection," in *Handbook of Industrial Engineering*, 3rd ed., G. Salvendy, Ed., Wiley, New York, pp. 1887–1920.

Drury, C. G. (2002), "A Unified Model of Security Inspection," in *Proceedings of the Aviation Security Technical Symposium*, Federal Aviation Administration, Atlantic City, NJ.

Drury, C. G., and Addison, J. L. (1973), "An Industrial Study of the Effects of Feedback and Fault Density on Inspection Performance," *Ergonomics*, Vol. 16, pp. 159–169–.

Drury, C. G., and Chi, C.-F. (1995), "A Test of Economic Models of Stopping Policy in Visual Search," *IIE Transactions*, Vol. 27, pp. 392–393.

Drury, C. G., and Forsman, D. R. (1996), "Measurement of the Speed Accuracy Operating Characteristic for Visual Search," *Ergonomics*, Vol. 39, No. 1, pp. 41–45.

Drury, C. G., and Fox, J. G., Eds. (1975), *Human Reliability in Quality Control*, Taylor & Francis, London.

Drury, C. G., and Hong, S.-K. (2000), "Generalizing from Single Target Search to Multiple Target Search," *Theoretical Issues in Ergonomics Science*, Vol. 1, No. 4, pp. 303–314.

Drury, C. G., and Kleiner, B. M. (1990), "Training in Industrial Environments," in *Proceedings of the Human Factors Association of Canada Meeting*, Ottawa, Ontario, Canada, pp. 99–108.

Drury, C. G., and Sinclair, M. A. (1983), "Human and Machine Performance in an Inspection Task," *Human Factors*, Vol. 25, pp. 391–400.

Drury, C. G., Montazer, M. A., and Karwan, M. H. (1987), "Self-Paced Path Control as an Optimization Task," *IEEE Transactions*, Vol. SMC-17.3, pp. 455–464.

Drury, C. G., Spencer, F. W., and Schurman, D. (1997), "Measuring Human Detection Performance in Aircraft Visual Inspection," in *Proceedings of the 41st Annual Human Factors and Ergonomics Society Meeting*, Albuquerque, NM.

Dutta, A., Fisher, D. L., and Noyce, D. A. (2005), "Use of a Driving Simulator to Evaluate and Optimize Factors Affecting Understandability of Variable Message Signs," *Transportation Research F*, Vol. 7, pp. 209–227.

Dzhafarov, E. N. (1993), "Grice-Representability of Response Time Distribution Families," *Psychometrika*, Vol. 58, pp. 281–314.

Editors (1992), "Controversies in Neuroscience I: Movement Control" (Special Issue), *Behavioral and Brain Sciences*, Vol. 15, No. 4.

Ehrenstein, A., Schweickert, R., Choi, S., and Proctor, R. W. (1997), "Scheduling Processes in Working Memory: Instructions Control the Order of Memory Search and Mental Arithmetic," *Quarterly Journal of Experimental Psychology*, Vol. 50A, pp. 766–802.

Elmaghraby, S. E. (1977), *Activity Networks: Project Planning and Control by Network Models*, Wiley, New York.

Estes, W. K. (1959), "Toward a Statistical Theory of Learning," *Psychological Review*, Vol. 57, pp. 94–107.

Feldman, A. G. (1966), "Functional Tuning of the Nervous System During Control of Movement or Maintenance of a Steady Posture, III: Mechanographic Analysis of the Execution by Man of the Simplest Motor Task," *Biophysics*, Vol. 11, pp. 766–775.

Fiedler, N., Feldman, R. G., Jacobson, J., Rahill, A., and Wetherell, A. (1996), "The Assessment of Neurobehavioral Toxicity: SGOMSEC Joint Report," *Environmental Health Perspectives*, Vol. 104, Suppl. 2, pp. 179–191.

Fisher, D. L. (1993), "Optimal Performance Engineering," *Human Factors*, Vol. 35, pp. 115–140.

Fisher, D. L., and Glaser, R. (1996), "Molar and Latent Models of Cognitive Slowing: Implications for Aging, Dementia, Depression, Development and Intelligence," *Psychonomic Bulletin and Review*, Vol. 3, pp. 458–480.

Fisher, D. L., and Goldstein, W. M. (1983), "Stochastic PERT Networks as Models of Cognition: Derivation of the Mean, Variance, and Distribution of Reaction Time Using Order-of-Processing (OP) Diagrams," *Journal of Mathematical Psychology*, Vol. 27, pp. 121–115.

Fisher, D. L., and Townsend, J. T. (1993), *Models of Morse Code Skill Acquisition: Simulation and Analysis*, U.S. Army Research Institute, Training Systems Research Division, Automated Instructional Systems Technical Area, Alexandria, VA, February.

Fisher, D. L., Yungkurth, E., and Moss, S. (1990), "Optimal Menu Hierarchy Design: Syntax and Semantics," *Human Factors*, Vol. 32, No. 6, pp. 665–683.

Fitts, P. M. (1954), "The Information Capacity of the Human Motor System in Controlling the Amplitude of Movement," *Journal of Experimental Psychology*, Vol. 47, pp. 381–391.

Fitts, P. M. (1964), "Perceptual–Motor Skill Learning," in *Categories of Human Learning*, A. W. Melton, Ed., Academic Press, New York, pp. 243–285.

Getty, D. J., Swets, J. A., Pickett, R. M., and Gonthier, D. (1995), "System Operator Response to Warnings of Danger: A Laboratory Investigation of the Predictive Value of a Warning on Human Response Time," *Journal of Experimental Psychology: Applied*, Vol. 1, pp. 19–33.

Goldstein, W. M., and Fisher, D. L. (1991), "Stochastic Networks as Models of Cognition: Derivation of Response Time Distributions Using the Order-of-Processing Method," *Journal of Mathematical Psychology*, Vol. 35, pp. 214–241.

Goldstein, W. M., and Fisher, D. L. (1992), "Stochastic Networks as Models of Cognition: Deriving Predictions for Resource-Constrained Mental Processing," *Journal of Mathematical Psychology*, Vol. 36, pp. 129–145.

Gray, W. D., John, B. E., and Atwood, M. E. (1993), "Project Ernestine: Validating a GOMS Analysis for Predicting and Explaining Real-World Task Performance," *Human–Computer Interaction*, Vol. 8, pp. 237–309.

Green, D. M., and Swets, J. A. (1966), *Signal Detection Theory and Psychophysics*, Wiley, New York.

Hick, W. E. (1952), "On the Rate of Information Gain," *Quarterly Journal of Experimental Psychology*, Vol. 4, pp. 11–26.

Hogan, N. (1988), "Physical Systems Theory and Controlled Manipulation," in *Natural Computation*, W. Richards, Ed., MIT Press, Cambridge, MA, pp. 430–442.

Hoffmann, E. R., Macdonald, W. A., and Almond, G. A. (1993), "Quantification of the Cognitive Difficulty of Mail Sorting," *International Journal of Industrial Ergonomics*, Vol. 11, No. 2, April, pp. 83–98.

Hommel, B. (1998), "Automatic Stimulus–Response Translation in Dual-Task Performance," *Journal of Experimental Psychology: Human Perception and Performance*, Vol. 24, pp. 1368–1394.

Hong, S. K. (2002), "Human Performance in Visual Search for Multiple Targets," unpublished doctoral dissertation, State University of New York at Buffalo, Buffalo, NY.

Hong, S. K., and Drury, C. G. (2002), "Sensitivity and Validity of Visual Search Models for Multiple Targets," *Theoretical Issues in Ergonomic Science*, Vol. 3.1, pp. 85–110.

Hou, T. S., Lin, L., and Drury, C. G. (1993), "An Empirical Study of Hybrid Inspection Systems and Allocation of Inspection Function," *International Journal of Human Factors in Manufacturing*, Vol. 3, pp. 351–367.

Hyman, R. (1953), "Stimulus Information as a Determinant of Reaction Time," *Journal of Experimental Psychology*, Vol. 45, pp. 188–196.

Ingham, J. G. (1970), "Individual Differences in Signal Detection," *Acta Psychologica*, Vol. 34, pp. 39–50.

Jagacinski, R. J., and Flach, J. M. (2003), *Control Theory for Humans: Quantitative Approaches to Modeling Performance*, Lawrence Erlbaum Associates, Mahwah, NJ.

Jagacinski, R. J., Repperger, D. W., Moran, M. S., Ward, S. L., and Glass, B. (1980), "Fitts' Law and the Microstructure of Rapid Discrete Movements," *Journal of Experimental Psychology: Human Perception and Performance*, Vol. 6, pp. 309–320.

Jax, S. A., Rosenbaum, D. A., Vaughan, J., and Meulenbroek, R. G. J. (2003), "Computational Motor Control and Human Factors: Modeling Movements in Real and Possible Environments," *Human Factors*, Vol. 45, pp. 5–27.

John, B. E. (1990), "Extensions of GOMS Analyses to Expert Performance Requiring Perception of Dynamic Visual and Auditory Information," in *Proceedings of the CHI '90 Conference on Human Factors in Computing Systems*, Association for Computing Machinery, New York, pp. 107–115.

John, B. E., and Newell, A. (1989), "Toward an Engineering Model of Stimulus–Response Compatibility," in *Stimulus–Response Compatibility: An Integrated Perspective*, R. W. Proctor and T. G. Reeve, Eds., Elsevier, New York, pp. 427–479.

Karush, W., and Dear, R. E. (1966), "Optimal Stimulus Presentation Strategy for a Stimulus Sampling Model of Learning," *Journal of Experimental Psychology*, Vol. 3, pp. 19–47.

Katsikopoulos, K., and Fisher, D. L. (2001), "Formal Requirements of Markov State Models for Paired Associate Learning," *Journal of Mathematical Psychology*, Vol. 45, No. 2, pp. 324–333.

Kirk, D. E. (1970), *Optimal Control Theory: An Introduction*, Prentice-Hall, Englewood Cliffs, NJ.

Konz, S. (1967), "Design of Work Stations," *Journal of Industrial Engineering*, Vol. 18, pp. 413–423.

Kruskal, J. B. (1964), "Multidimensional Scaling by Optimizing Goodness of Fit to a Nonmetric Hypothesis," *Psychometrika*, Vol. 29, pp. 1–27.

Langolf, G. D., Chaffin, D. B., and Foulke, J. A. (1976), "An Investigation of Fitts' Law Using a Wide Range of Movement Amplitudes," *Journal of Motor Behavior*, Vol. 8, pp. 113–128.

Lee, E., and MacGregor, J. (1985), "Minimizing User Search Time in Menu Retrieval Systems," *Human Factors*, Vol. 27, pp. 157–162.

Liu, Y. (1996), "Queueing Network Modeling of Elementary Mental Processes," *Psychological Review*, Vol. 103, pp. 116–136.

Liu, Y., Feyen, R., and Tsimhoni, O. (2004), *Queueing Network-Model Human Processor (QN-MHP): A Computational Architecture for Multitask Performance* Report 04–05, Department of Industrial and Operations Engineering, University of Michigan, Ann Arbor, MI.

Logan, G. D. (2002), "Parallel and Serial Processing," in *Stevens' Handbook of Experimental Psychology: Methodology in Experimental Psychology*, 3rd ed., J. Wixted, Ed., Wiley, New York, Vol. 4, pp. 271–300.

Luce, R. D. (1986), *Response Times: Their Role in Inferring Elementary Mental Organization*, Oxford University Press, New York.

Macmillan, N. A., and Creelman, C. D. (1991), *Detection Theory: A User's Guide*, Cambridge University Press, Cambridge.

McCann, R. S., and Johnston, J. C. (1992), "Locus of the Single-Channel Bottleneck in Dual-Task Interference," *Journal of Experimental Psychology: Human Perception and Performance*, Vol. 18, pp. 471–484.

Megaw, E. D. (1979), "Factors Affecting Visual Inspection Accuracy," *Applied Ergonomics,* Vol. 10, No. 1, pp. 27–32.

Meyer, J., and Bitan, Y. (2002), "Why Better Operators Receive Worse Warnings," *Human Factors*, Vol. 44, pp. 343–353.

Meyer, D. E., and Kieras, D. E. (1997a), "A Computational Theory of Executive Cognitive Processes and Multiple-Task Performance, 1: Basic Mechanisms," *Psychological Review*, Vol. 104, pp. 3–65.

Meyer, D. E., and Kieras, D. E. (1997b), "A Computational Theory of Executive Cognitive Processes and Multiple-Task Performance, 2: Accounts of Psychological Refractory-Period Phenomena," *Psychological Review*, Vol. 104, pp. 749–791.

Meyer, D. E., Smith, J. E. K., Kornblum, S., Abrams, R. A., and Wright, C. E. (1990), "Speed–Accuracy Tradeoffs in Aimed Movements: Toward a Theory of Rapid Voluntary Action," in *Attention and Performance XIII*, M. Jeannerod, Ed., Lawrence Erlbaum Associates, Mahwah, NJ, pp. 173–226.

Montazer, M. A., Drury, C. G., and Karwan, M. H. (1989), "An Optimization Model of Self-Paced Tracking on Circular Courses," *IEE Transactions*, Vol. SMC 18.6, pp. 908–916.

Morawski, T., Drury, C. G., and Karwan, M. H. (1980), "Predicting Search Performance for Multiple Targets," *Human Factors*, Vol. 22, No. 6, pp. 707–718.

Navon, D., and Miller, J. (2002), "Queuing or Sharing? A Critical Evaluation of the Single-Bottleneck Notion," *Cognitive Psychology*, Vol. 44, pp. 193–251.

Newell, A. (1990), *Unified Theories of Cognition*, Harvard University Press, Cambridge, MA.

Newell, A., and Simon, H. A. (1972), *Human Problem Solving*, Prentice-Hall, Englewood Cliffs, NJ.

Norman, D. A., and Bobrow, D. G. (1975), "On Data-Limited and Resource Limited Processes," *Cognitive Psychology*, Vol. 7, pp. 44–64.

Parasuraman, R., and Davies, D. R. (1977), "A Taxonomic Analysis of Vigilance Performance," in *Vigilance: Theory, Operational Performance, and Physiological Correlates*, R. R. Mackie, Ed., Plenum Press, New York, pp. 559–574.

Parasuraman, R., Sheridan, T. B., and Wickens, C. D. (2000), "A Model for Types and Levels of Human Interaction with Automation," *IEEE Transactions on Systems, Man and Cybernetics, Part A: Systems and Humans*, Vol. 30, pp. 286–297.

Pashler, H. (1998), *The Psychology of Attention*, MIT Press, Cambridge, MA.

Pritsker, A. A. B. (1979), *Modeling and Analysis Using Q-GERT Networks*, 2nd ed., Wiley, New York.

Resnick, R., and Halliday, D. (1963), *Physics: For Students of Science and Engineering*, Wiley, New York.

Rouse, W. B. (1980), *Systems Engineering Models of Human–Machine Interaction*, North-Holland, New York.

Rumelhart, D. E. and McClelland, J. L. (1986), *Parallel Distributed Processing*, Vol. 1, *Explorations in the Microstructure of Cognition*, MIT Press, Cambridge, MA.

Salthouse, T. A., and Somberg, B. L. (1982), "Isolating the Age Deficit in Speeded Performance," *Journal of Gerontology*, Vol. 37, pp. 349–357.

Schmidt, R. A., and McGown, C. (1980), "Terminal Accuracy of Unexpectedly Loaded Rapid Movements: Evidence for a Mass-Spring Mechanism in Programming," *Journal of Motor Behavior*, Vol. 12, pp. 149–161.

Schmidt, R. A., Zelaznik, N. H., Hawkins, B., Frank, J. S., and Quinn, J. T., Jr. (1979), "Motor Output Variability: A Theory for the Accuracy of Rapid Motor Acts," *Psychological Review*, Vol. 86, pp. 415–451.

Schweickert, R. (1978), "A Critical Path Generalization of the Additive Factor Method: Analysis of a Stroop Task," *Journal of Mathematical Psychology*, Vol. 18, pp. 105–139.

Schweickert, R., and Townsend, J. T. (1989), "A Trichotomy: Interactions of Factors Prolonging Sequential and Concurrent Mental Processes in Stochastic Discrete Mental (PERT) Networks," *Journal of Mathematical Psychology*, Vol. 33, pp. 328–347.

Schweickert, R., Fisher, D. L., and Goldstein, W. (1992), *General Latent Network Theory: Structural and Quantitative Analysis of Networks of Cognitive Processes*, Technical Report 92-1, Mathematical Psychology Program, Purdue University, West Lafayette, IN.

Schweickert, R., Fisher, D. L., and Proctor, R. (2003), "Steps Towards Building Mathematical and Computer Models from Cognitive Task Analysis," *Human Factors*, Vol. 45, pp. 77–103.

See, I. E., Howe, S. R., Warm, J. S., and Dember, W. N. (1995), "Meta-analysis of the Sensitivity Decrement in Vigilance," *Psychological Bulletin*, Vol. 117, pp. 230–249.

See, J. E., Warm, J. S., Dember, W. N., and Howe, S. R. (1997), "Vigilance and Signal Detection Theory: An Empirical Evaluation of Five Measures of Response Bias," *Human Factors*, Vol. 39, pp. 14–29.

Shannon, C. E., and Weaver, W. (1949), *The Mathematical Theory of Communications*, University of Illinois Press, Urbana, IL.

Sharit, J., Czaja, S. J., Nair, S., and Lee, C. C. (2003), "Effects of Age, Speech Rate, and Environmental Support in Using Telephone Voice Menu Systems," *Human Factors*, Vol. 45, pp. 234–251.

Shepard, R. N. (1962), "The Analysis of Proximities: Multidimensional Scaling with an Unknown Distance Function, I," *Psychometrika*, Vol. 27, pp. 125–140.

Smith, M. C. (1969), "The Effect of Varying Information on the Psychological Refractory Period, in *Attention and Performance II*, W. G. Koster, Ed., *Acta Psychologica*, Vol. 30, pp. 220–231.

Smith, P. J., and Langolf, G. D. (1981), "The Use of Sternberg's Memory Scanning Paradigm in Assessing Effects of Chemical Exposure," *Human Factors*, Vol. 23, pp. 701–708.

Sternberg, S. (1966), "High-Speed Scanning in Human Memory," *Science*, Vol. 153, pp. 652–654.

Sternberg, S. (1975), "Memory-Scanning: New Findings and Current Controversies," *Quarterly Journal of Experimental Psychology*, Vol. 27, pp. 1–32.

Taatgen, N. A., and Anderson, J. R. (2002), "Why Do Children Learn to Say 'Broke'? A Model of the Past Tense Without Feedback," *Cognition*, Vol. 86, pp. 123–155.

Taatgen, N. A. and Lee, F. J. (2003), "Production Compilation: A Simple Mechanism to Model Complex Skill Acquisition," *Human Factors*, Vol. 45, pp. 61–76.

Tang, K. E. (2001), "Menu Design with Visual Momentum for Compact Smart Products," *Human Factors*, Vol. 43, pp. 267–277.

Tombu, M., and Jolicoeur, P. (2003), "A Central Capacity Sharing Model of Dual-Task Performance," *Journal of Experimental Psychology: Human Perception and Performance*, Vol. 29, pp. 3–18.

Townsend, J. T. (1972), "Some Results Concerning the Identifiability of Parallel and Serial Processes," *British Journal of Mathematical and Statistical Psychology*, Vol. 25, pp. 168–199.

Townsend, J. T., and Ashby, F. G. (1983), *The Stochastic Modeling of Elementary Psychological Processes*, Cambridge University Press, Cambridge.

Townsend, J. T., and Wenger, M. J. (2004), "The Serial–Parallel Dilemma: A Case Study in a Linkage of Theory and Method," *Psychonomic Bulletin and Review*, Vol. 11, pp. 391–418.

Tsao, Y. C., Drury, C. G., and Morawski, T. B. (1979), "Human Performance in Sampling Inspection," *Human Factors,* Vol. 21, No. 1, pp. 99–105.

Wallace, J. S., and Fisher, D. L. (1998), "Sound Localization: Theory and Practice," *Human Factors*, Vol. 40, pp. 50–68.

Welford, A. T. (1967), "Single-Channel Operation in the Brain," *Acta Psychologica*, Vol. 27, pp. 5–22.

Wiener, E. L. (1975), "Individual and Group Differences in Inspection," in *Human Reliability in Quality Control*, C. G. Drury and J. G. Fox, Eds, Taylor & Francis, London, pp. 19–30.

Wiker, S. F., Langolf, G. D., and Chaffin, D. B. (1989), "Arm Posture and Human Movement Capability," *Human Factors*, Vol. 31, pp. 421–441.

Wolfe, J. M. (1994), "Guided Search 2.0: A Revised Model of Visual Search," *Psychonomic Bulletin and Review*, Vol. 1, No. 2, pp. 202–238.

CHAPTER 38

SUPERVISORY CONTROL

Thomas B. Sheridan
Massachusetts Institute of Technology
Cambridge, Massachusetts

1 DEFINING SUPERVISORY CONTROL

This chapter is intended as a tutorial on supervisory control. It is not a comprehensive or even-handed review of the literature in human–robot interaction, monitoring, diagnosis of failures, human error, mental workload, or other closely related topics. Sheridan (1992, 2000), Mouloua and Parasuraman (1994), Mouloua and Koonce (1997), Sarter and Amalberti (2000), and Degani (2004) cover these aspects more fully.

The term *supervisory control* is derived from the close analogy between the characteristics of a supervisor's interaction with subordinate human staff members and a person's interaction with "intelligent" automated subsystems. A supervisor of people gives general directives that are understood and translated into detailed actions by staff members. In turn, staff members aggregate and transform detailed information about process results into summary form for the supervisor. The degree of intelligence of staff members determines the ability to delegate, and the level of involvement of their supervisor in the process. Automated subsystems permit the same sort of interaction to occur between a human supervisor and the process (Ferrell and Sheridan, 1967; Sheridan et al. 1983a). Supervisory control behavior is interpreted to apply broadly to vehicle control (aircraft and spacecraft, ships, and undersea vehicles), continuous process control (oil, chemicals, power generation), and robots and discrete tasks (manufacturing, space, undersea, mining).

In the strictest sense, the term *supervisory control* indicates that one or more human operators are setting initial conditions for intermittently adjusting and receiving information from a computer that itself closes a control loop (i.e., interconnects) through external sensors, effectors, and the task environment. In a broader sense, supervisory control means interaction with a computer to transform data to produce integrated (chunked) displays or to retransform operator commands to generate detailed control actions. Figure 1 compares supervisory control with direct manual control (*a*) and full automatic control (*e*). Diagrams (*c*) and (*d*) characterize supervisory control in the strict formal sense; diagram (*b*) characterizes supervisory control in the latter (broader) sense.

The essential difference between these two characterizations of supervisory control is that in the first and stricter definition the computer can act on new information independent of and with only blanket authorization and adjustment from the supervisor; that is, the computer implements discrete sets of instructions by itself, closing the loop through the environment. In the second definition the computer's detailed implementations are *open loop*; that is, feedback from the task has no effect on computer control of the task except through the human operator. The two situations may appear similar to the supervisor, since he or she always sees and acts through the computer (analogous to a staff) and therefore may not know whether it is acting open loop or closed loop in its fine behavior. In either case the computer may function principally on the efferent or motor side to implement the supervisor's commands (e.g., do some part of the task entirely and leave other parts to the human, or provide some control compensation to ease all of the task for the human). Alternatively, the computer may function principally on the display side (e.g., to integrate and interpret incoming information from below or to give advice to the supervisor as to what to do next as an "expert system"). Or it may work equally on the efferent and afferent sides.

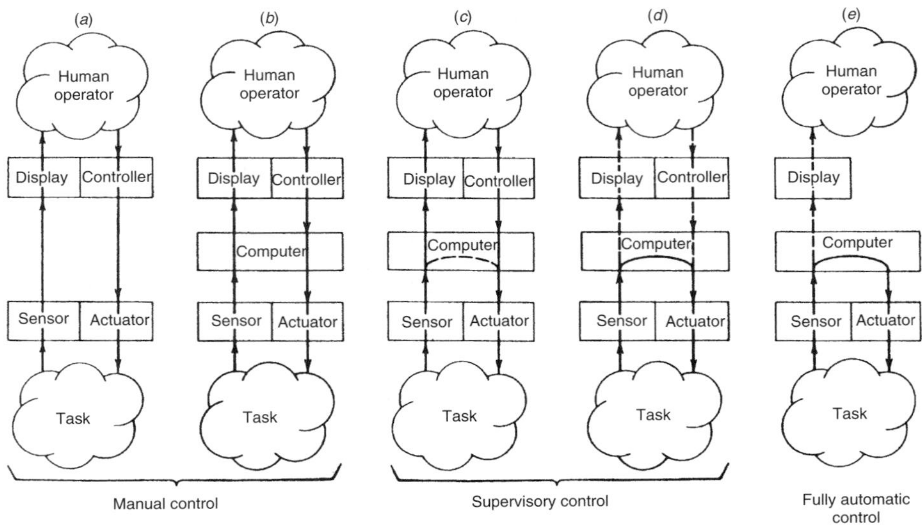

Figure 1 Supervisory control as related to direct manual control and full automation.

2 SUPERVISORY CONTROL IN VARIOUS TECHNOLOGICAL SYSTEMS

Supervisory control is emerging rapidly in many industrial, military, medical, and other contexts, although this form of human interaction with technology is relatively little recognized or understood in a formal way. From the pyramid-building pharaohs of Egypt through all of the history of technology, there surely has been concern about how best to extend the capabilities of human workers. Early in the present century, against the backdrop of the newly mechanized production line, Taylor's scientific management (Taylor, 1922) catalyzed a formal intellectual consciousness about the human factors involved. Taylor intended a new interest in the sensorimotor aspects of human performance. What he did not intend was the subsequent criticism of his essentially mechanistic approach that it was dehumanizing.

The 1940s and 1950s saw *human factors* (*ergonomics* in Europe) emerge, first in essentially empirical "knobs and dials" form, concentrating on the human–machine interface itself. This was supported over the next decade by the theoretical underpinnings of man–machine systems (Sheridan and Ferrell, 1974). Such theories included control, information, signal detection, and decision theories originally developed for application to physical systems but now applied explicitly to the human operator. As contrasted with human factors engineering at the interface, human–machine systems analysis considers characteristics of the entire causal "loop" of decision, communication, control, and feedback—through the operator's physical environment and back again to the human.

From the late 1950s the computer began to intervene in the causal loop: electronic compensation and stability augmentation for control of aircraft and similar systems, electronic filtering of signal patterns in noise, and electronic generation of simple displays. It was obvious that if vehicular or industrial systems were equipped with sensors that could be read by computers, and by motors that could be driven by computer, then even though the overall system was still very much human controlled, control loops between those sensors and motors could be closed automatically. Thus, the chemical plant operator was relieved of keeping the tank at a given level or the temperature at a reference; he or she needed only to set in that desired level or temperature signal from time to time. So, too, after the autopilot was developed for the aircraft, the human pilot needed only to set in the desired altitude to heading; an automatic system would strive to achieve this reference, with the pilot monitoring to ensure that the aircraft did in fact go where desired.

The automatic building elevator, of course, has been in place for many years, and is certainly one of the first implementations of supervisory control. Recently, developers of new systems for word processing and handling of business information (i.e., without the need to control any mechanical processes) have begun thinking along supervisory control lines.

The full generality of the idea of supervisory control came to the author and his colleagues (Sheridan, 1960; Ferrell and Sheridan, 1967) as part of research on how people on Earth might control vehicles on the moon through three-dimensional round-trip time delays (imposed by the speed of light). Under such constraint, remote control of lunar roving vehicles or manipulators was shown to be possible only by performing in "move-and-wait" fashion. This means that the operator can commit only to a small incremental movement open loop, that is, without feedback (which actually is as large a movement as is reasonable without risking collision or other error), then stopping and waiting one delay period for feedback to "catch up," then repeating the process in steps until the task is completed.*

It was shown that if instead of the human operator remaining within the control loop, he or she communicates a goal state relative to the remote environment, and if the remote system incorporates the capability to measure proximity to this goal state, the achievement of this goal state can be turned over to remote subordinate control system for implementation. In this case there is no delay in the control loop implementing the task, and thus there is no instability.

There necessarily remains, of course, a delay in the supervisory loop. This delay in the supervisor's confirmation of desired results is acceptable as long as (1) the subgoal is a sufficiently large "bite" of the task, (2) the unpredictable aspects of the remote environment are not changing too rapidly (i.e., disturbance bandwidth is low), and (3) the subordinate automatic systems is trustworthy.

If these conditions obtain and as computers gradually become more capable both in hardware and software (and as "machine intelligence" finally makes its real if modest appearance), it is evident that telemetry transmission delay is in no way a prerequisite to the usefulness of supervisory control. The incremental goal specified by the human operator need not be simply a new steady-state reference for a servomechanism (as in resetting a thermostat) in one or even several dimensions (e.g., resetting both temperature and humidity, or commanding a manipulator endpoint to move to a new position, including three translations and three rotations relative to its initial position). Each new goal statement can be the specification of an entire trajectory of movements (as the performance of a dance or a symphony) together with programmed branching conditions (what to do in case of a fall or a broken violin string, or how to respond contingent on audience applause).

In other words, the incremental goal statement is a program of instructions in the full sense of a computer program, which make the human supervisor an intermittent real-time computer programmer, acting relative to the subordinate computer much the same as a teacher or parent or boss behaves relative to a student or a child or subordinate worker. The size and complexity of each new program is necessarily a function of how much the computer can (be trusted to) cope with in one bite, which in turn depends on the computer's own sophistication (knowledge base) and the complexity (uncertainty) of the task.

*Attempts to drive or manipulate continuously only produce instability, as simple control theory predicts (i.e., where loop gains exceeds unity at a frequency such that the loop time delay is one half-cycle, instead of errors being nulled out, they are only reinforced). Performing remote manipulation with delayed force feedback was shown by Ferrell (1966) to be essentially impossible since forces at unexpected times act as significant disturbances to produce instability. At least the visual feedback can be ignored by the operator.

Supervisory control has been embodied in various forms and in various industries usually without being called that. (More likely, each developer or vendor has its own cute acronym emphasizing how "smart" and easy it is to use the new product.) Aircraft autopilots are now "layered," meaning that the pilot can select among various forms and levels of control. At the lowest level the pilot can set in a new heading or rate of climb. Or he or she can program a sequence of heading changes at various waypoints, or a sequence of climb rates initiated at various altitudes. Or program the inertial guidance system to take the aircraft to (within a fraction of a mile of) a distant city. Given the existence of certain ground-based equipment, the pilot can program an automatic landing on a given runway, and so on. Sheridan (2002) reviews how such automation is creeping into the aircraft flight deck. Sarter and Amalberti (2000) describe the modern flight management system in some detail.

Supervisory control of a simpler sort is now evident in the cruise control system of current automobiles and trucks and is being upgraded in the form of "advanced" or "intelligent" cruise control, wherein a radar detector controls speed to maintain a safe distance behind a leading vehicle. Because of the telecommunication time delay, supervisory control is really the only acceptable approach to controlling planetary rovers and other space vehicles from the Earth (a Mars rover, for example, experiences a 30-minute time delay for telecommunications).

Modern chemical and nuclear plants can similarly be programmed to perform heating, mixing, and various other processes according to a time line, and including various sensor-based conditions for shutting down or otherwise aborting the operation. Nandi and Ruhe (2002) describe the use of supervisory control in sintering furnaces. Seiji et al. (2001) provide an extensive review of modern supervisory control in nuclear power plants.

In modern hospital operating rooms, intensive care units, and ordinary patient wards there are numerous supervisory control systems at work. The modern anesthesiology workstation is a good example. Drugs in liquid or gaseous form are pumped into the patient at rates programmed by the anesthesiologist and by sensors monitoring patient respiration heart rate and other variables.

More and more a multiplicity of computers are used in a supervisory control system, as shown in Figure 2. One typically large computer is in the control room to generate displays and interpret commands. We call this a *human-interactive computer* (HIC), part of a *human-interactive system* (HIS). It in turn forwards that command to various microprocessors that actually close individual control loops through their own associated sensors and effectors. We call the latter *task-interactive computers* (TICs), each part of its own *task-interactive system* (TIS).

The examples cited above characterize the first or stricter definition of supervisory control previously given (Figure 1*c* and *d*), where the computer, once programmed, makes use of its own artificial sensors

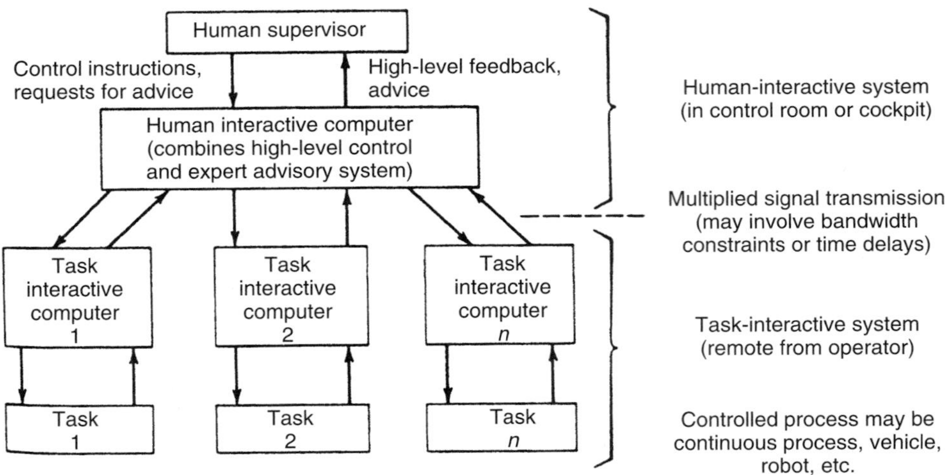

Figure 2 Hierarchical nature of supervisory control.

to ensure completion of the tasks assigned. Many familiar systems, such as automatic washing machines, dryers, dishwashers, or stoves, once programmed, perform their operations open loop; that is, there is no measurement or knowledge of results. If the task can be performed in such open-loop fashion, and if the human supervisor can anticipate the task conditions and is good at selecting the right open-loop program, there is no reason not to employ this approach. To the human supervisor, whether the lower-level implementation is open or closed loop is often opaque and/or of no concern; the only concern is whether the goal is achieved satisfactorily. For example, a programmable microwave oven without the temperature sensor in place operates open loop, whereas the same oven with the temperature sensor operates closed loop. To the human supervisor or programmer, they look the same.

A very important aspect of supervisory control is the ability of the computer to "package" information for visual display to the human operator, including data from many sources; from the past, present, or even predicted future; and presented in words, graphs, symbols, pictures, or some combination. Ubiquitous examples of such integrated displays are in aircraft and air-traffic control, chemical and power plants, and various other industrial or military settings too numerous to review here. General interest in supervisory control became evident in the mid-1970s (Edwards and Lees, 1974; Sheridan and Johannsen, 1976; Wiener and Curry, 1980; Sheridan and Hennessy, 1984) and continues to grow.

3 SUPERVISORY ROLES, LOCI, AND LEVELS OF HUMANS AND COMPUTERS

The human supervisor's *roles* are (1) *planning* off-line what task to do and how to do it; (2) teaching (or programming) the computer what was planned; (3) *monitoring* the automatic action online to make sure that all is going as planned and to detect failures; (4) *intervening*, which means the supervisor takes over control after the desired goal state has been reached satisfactorily, or interrupts the automatic control in emergencies to specify a new goal state and reprogram a new procedure; and (5) *learning* from experience so as to do better in the future. These are usually time-sequential steps in task performance.

We may view these steps as being within three nested loops, as shown in Figure 3. The innermost loop, monitoring, closes on itself; that is, evidence of something interesting or completion of one part of cycle of monitoring strategy leads to more investigation and monitoring. We might include minor online tuning of the process as part of monitoring. The middle loop closes from intervening back to teaching; that is, human intervention usually leads to programming of a new goal state in the process. The outer loop closes from learning back to planning; intelligent planning for the next subtask is usually not possible without learning from the last one.

The three supervisory loops operate at different time scales relative to one another. Revisions in fine-scale monitoring behavior take place at brief intervals. New programs are generated at somewhat longer intervals. Revisions in significant task planning occur only at still longer intervals. These differences in time scale further justify Figure 3.

For each of the five roles or stages of the supervisory process there are three *loci* of human function in a physiological sense: *sensory* functions (accessing displays, observing, perceiving), *cognitive* functions internal to the supervisor (evaluating the situation, accessing memory, making decisions), and *response* functions. *S.C.R.* are the classic designators to differentiate these functional elements of causation through the operator.

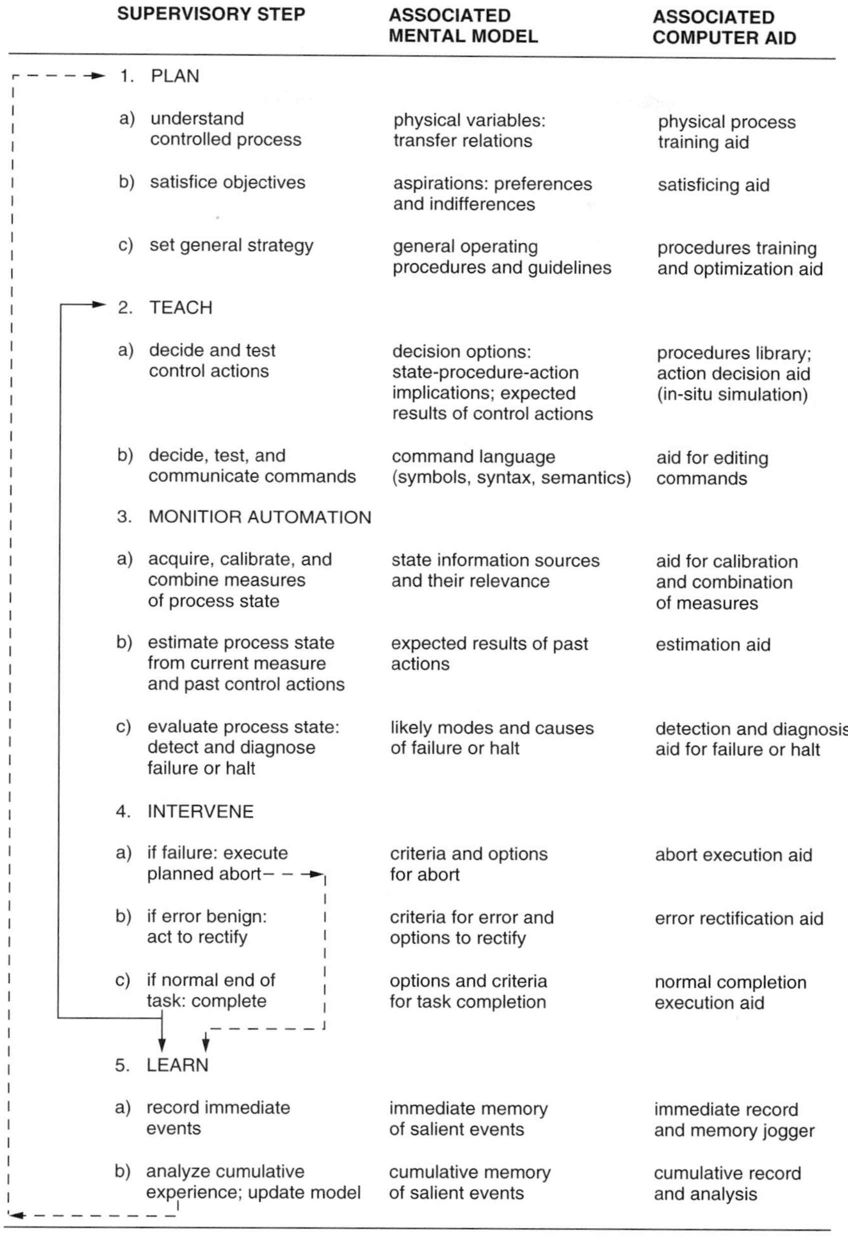

SUPERVISORY STEP	ASSOCIATED MENTAL MODEL	ASSOCIATED COMPUTER AID
1. PLAN		
a) understand controlled process	physical variables: transfer relations	physical process training aid
b) satisfice objectives	aspirations: preferences and indifferences	satisficing aid
c) set general strategy	general operating procedures and guidelines	procedures training and optimization aid
2. TEACH		
a) decide and test control actions	decision options: state-procedure-action implications; expected results of control actions	procedures library; action decision aid (in-situ simulation)
b) decide, test, and communicate commands	command language (symbols, syntax, semantics)	aid for editing commands
3. MONITIOR AUTOMATION		
a) acquire, calibrate, and combine measures of process state	state information sources and their relevance	aid for calibration and combination of measures
b) estimate process state from current measure and past control actions	expected results of past actions	estimation aid
c) evaluate process state: detect and diagnose failure or halt	likely modes and causes of failure or halt	detection and diagnosis aid for failure or halt
4. INTERVENE		
a) if failure: execute planned abort	criteria and options for abort	abort execution aid
b) if error benign: act to rectify	criteria for error and options to rectify	error rectification aid
c) if normal end of task: complete	options and criteria for task completion	normal completion execution aid
5. LEARN		
a) record immediate events	immediate memory of salient events	immediate record and memory jogger
b) analyze cumulative experience; update model	cumulative memory of salient events	cumulative record and analysis

Figure 3 Functional and temporal nesting of supervisory roles.

Finally, we may appeal to the *levels* of behavior introduced by Rasmussen (1976); *skill-based* behavior (continuous, typically well-learned, sensory-motor behavior analogous to what can be expected from a servomechanism); *rule-based* behavior (what an "artificially intelligent" computer can do in recognizing a pattern of stimuli, then triggering an "if–then" algorithm to execute an appropriate response); and finally, *knowledge-based* behavior ("high-level" situation assessment and evaluation, consideration of alternative actions in light of various goals, decision and scheduling of implementation—a form of behavior that machines are not now good at).

We consider the foregoing three meta-characteristics of supervision: *role, loci*, and *level*, as three independent dimensions of such behavior. The human-interactive computer (HIC) is conceived to be a large enough computer to communicate in a human-friendly way, using near-natural language, good graphics, and so on. This includes being able to accept and interpret commands and to give the supervisor useful feedback. The HIC should be able to recognize patterns in data sent up to it from below and decide on appropriate algorithms for response, which it sends down as instructions. Eventually, the HIC should be able to run "what would happen if . . ." simulations and be able to give useful advice from a knowledge base, that is, include an expert system.

The HIC, located near the supervisor in a control room or cockpit, may communicate across a barrier of time or space with a multiplicity of task-interactive computers (TICs), which probably are microprocessors distributed throughout the plant or vehicle. The latter are usually coupled intimately with artificial sensors and actuators, in order to deal in low-level language and to close relatively tight control loops with objects and events in the physical world.

The human supervisor can be expected to communicate with the HIC intermittently in information "chunks" (alphanumeric sentences, video pages, etc.) while the task communicates with the TIC continuously in computer words at the highest possible bit rates. The availability of these computer aids means that the human supervisor, while retraining the knowledge-based behavior function, is likely to download some of the rule-based programs and almost all of the skill-based programs into the HIC. The HIC, in turn, should download a few of the rule-based programs, and most of the skill-based programs, to the appropriate TICs.

Figure 4 presents the functions of Figure 3 in the form of a flowchart. Each supervisory function is shown above, and the (usually multiple) automated subsystems of the TIC are shown below. Normally,

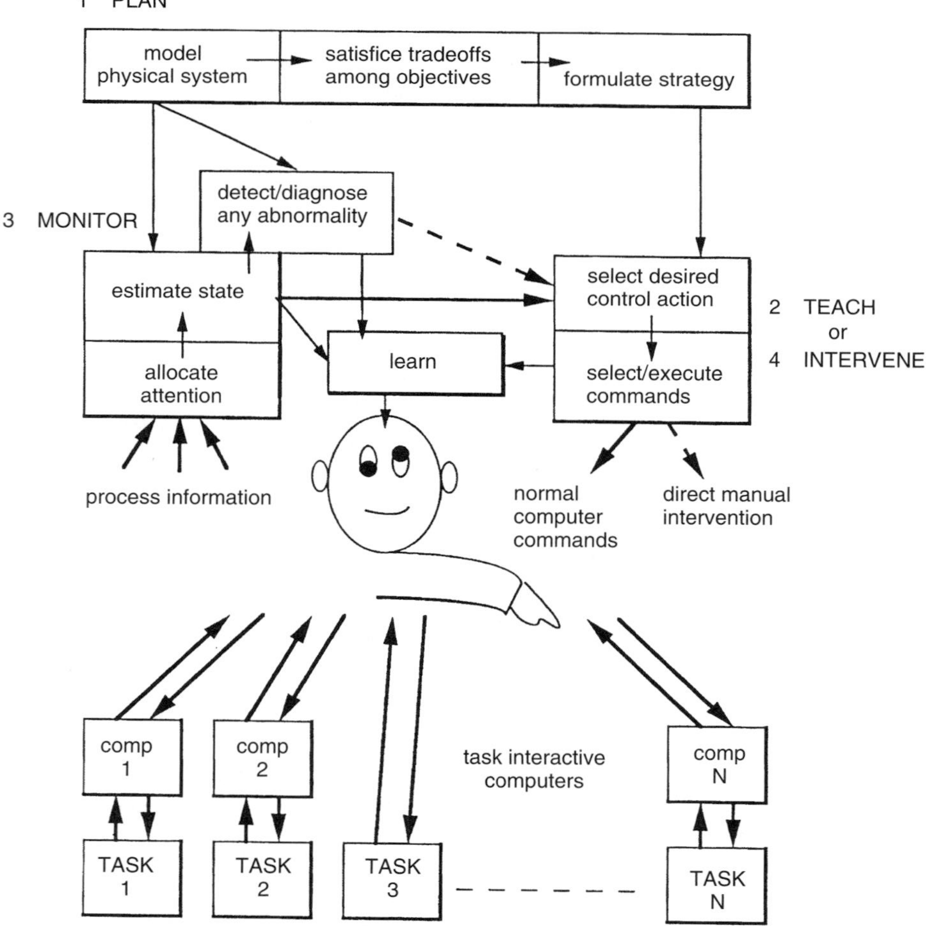

Figure 4 Flowchart of supervisor functions (including both mental models and decision aids). (From Sheridan, 1992.)

for any given task, the planning and learning roles are performed off-line relative to the online human-mediated and automatic operations of the other parts of system, and therefore are shown at the top with light lines connecting them to the rest of the system. Teaching precedes monitoring on the first cycle, but thereafter follows monitoring and intervening (as necessary) within the intermediate loop. The inner-loop monitoring role is carried out within the "estimate state" and "allocate attention" boxes.

Allocation of functions between the human and the machine need not be fixed. There have been numerous papers discussing the potential for dynamic allocation—where the allocation changes as a function of the flow of demands and the workload of the two entities (see, e.g., the February 2000 Special Issue on Function Allocation of the *International Journal of Human–Computer Studies*). In the sections that follow the various supervisory roles are discussed in more detail, bringing in examples of research problems and prototype systems to aid the supervisor in these roles.

4 PLANNING AND LEARNING: COMPUTER REPRESENTATION OF RELEVANT KNOWLEDGE

The first and fifth supervisory roles described previously, planning and learning, may be considered together since they are similar activities in many ways. Essentially, in planning the supervisor asks "What would happen if ...?" questions of the accumulated knowledge base and considers what the implications are for hypothetical control decisions. In learning, the supervisor asks "What did happen?" questions of the database for the more recent subtasks and considers whether the initial assumptions and final control decisions were appropriate.

The designer of an automatic control system or manual control system must ask: "What variables do I wish to make do what, subject to what constraints and what criteria?" The planning role in supervisory control requires that the same kinds of questions be answered, because in a sense, the supervisor is redesigning an automatic control system each time that he or she programs a new task and goal state. Absolute constraints on time, tools, and other resources available need to be clear, as do the criteria of trade-off among time, dollars and resources spent, accuracy, and risk of failure.

Just as computer simulation figures into planning, it also figures into supervisory control—the difference being that such simulation may more likely be subjected to time stress in supervisory control. Simulation requires acquiring some idea of how the process (system to be controlled) works, that is, a set of equations relating the various controllable variables, the various uncontrollable but measurable variables (disturbances), and the degree of unpredictability (noise) on measured system response variables. This is a common representation of knowledge. Given measured inputs and outputs, there are well-established means to infer the equations if the processes are approximately linear and differentiable.

Once such a model is in place, the supervisor can posit hypothetical inputs and observe what the outputs would be. Also, one may use such a process model as an "observer" (in the sense of modern control theory). Namely, when control signals are put into both the model and actual processes, and the model parameters are then trimmed to force certain model outputs to conform to corresponding actual process outputs that can be measured (Figure 5), other process outputs that are inconvenient to measure may be estimated ("observed") from the model. Just as this is a theoretical prerequisite to optimal automatic control of physical systems, so it is likely to be a useful practice to aid humans in supervisory control (Sheridan, 1984a).

A different type of knowledge representation is that used by the artificial intelligence (AI) community. Here knowledge is usually couched in the form of if–then logical statements called *production rules*, semantic association networks, and similar forms. The input to a simulated program usually represents in cardinal numbers a hypothetical physical input to a simulated physical system. In contrast, the input to the AI knowledge base can be a question about relationships for given data or a question about data for given relationships. This can be in less restrictive

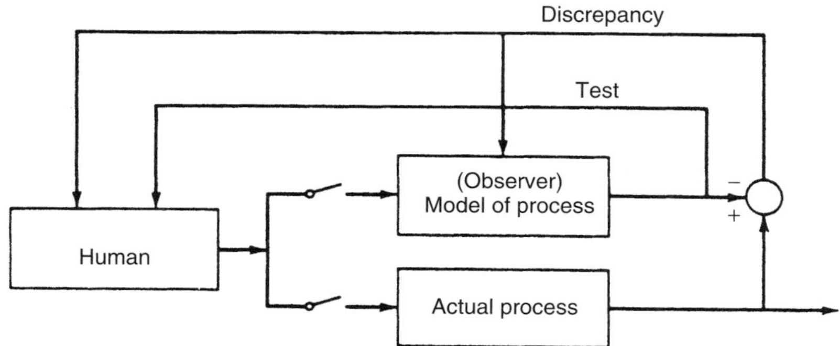

Figure 5 Use of computer-based observer as an aid to supervisor.

ordinal form (e.g., networks of diadic relations) or in nominal form (e.g., lists).

Currently, there is great interest in how best to transfer expertise from the human brain (knowledge representation, mental model) into the corresponding representation or model within the computer, how best to transfer it back, and when to depend on each of those sources of information. This research on mental models has a lively life of its own (Falzon, 1982; Gentner and Stevens, 1983; Rouse and Morris, 1984; Sheridan, 1984a; Moray, 1997) quite independent of supervisory control.

An important aspect of planning is visualization. The now rather sophisticated tool of computer simulation, when augmented by computer graphics, enables remarkable visualization possibilities. When further augmented by human interactive devices such as head-mounted visual and auditory displays and high-bandwidth force-reflecting haptics (mechanical arms), the operator can be made to feel present in a virtual world, as has been popularized by the oxymoron *virtual reality*. Of course, the idea of virtual reality is not new. The original idea of Edwin Link's first flight simulators (developed early in the 1940s) was to make the pilot trainee feel as if he or she were flying a real aircraft. First they were instrument panels only, then a realistic out-the-window view was created by flying a servo-driven video camera over a scale model, and finally, computer graphics were used to create the out-the-window images. Now all commercial airlines and military services routinely train with computer-display, full-instrument, moving-platform flight simulators. Similar technology has been applied to ship, automobile, and spacecraft control. The salient point for the present discussion is that the new simulation capabilities now permit visualization of alternative plans as well as better understanding of complex state information in situ, during monitoring. That same technology, of course, can be used to convey a sense of presence in an environment that is not simulated but is quite real and merely remote—communicated via closed-circuit video with cameras slaved to the observer's head.

Supervisory aiding in planning of the moves of a telerobot are illustrated by the work of Park (1991). His computer-graphic simulation let a supervisor try out moves of a telerobot arm before committing to the actual move. He assumed that for some obstacles the positions and orientation were already known and represented in a computer model. The user commanded each straight-line move to a subgoal point in three-space by designating a point on the floor or the lowest horizontal surface (such as a tabletop) by moving a cursor to that point (say, *A* in Figure 6*a*) and clicking, then lifting the cursor by an amount corresponding to the desired height of the subgoal point (say, *A*) above that floor point and observing on the graphic model a blue vertical line being generated from the floor point to the subgoal point in space. This process was repeated for successful subgoal points (say, *B* and *C*). Using the computer display, the user could view the resulting trajectory model from any desired perspective (although the "real" environment could be viewed only from the perspective provided by the video camera's location). Either of two collision-avoidance algorithms could be invoked: a detection algorithm that indicated where on some object a collision occurred as the arm was moved from one point to another, or an automatic avoidance algorithm that found (and drew on the computer screen) a minimum-length, no-collision trajectory from the starting point to the new subgoal point. Park's aiding scheme also allowed new observed objects to be added to the model by graphically "flying" them into geometric correspondence with the model display. Another aid was to generate *virtual objects* for any portion of the environment in the umbral region (not visible) after two video views (Figure 6*b*). In this case the virtual objects were treated in the same way in the model and in the collision-avoidance algorithms as the visible objects. Experiments with this technique showed that it was easy to use and that it avoided collisions.

At the extreme of time desynchronization is recording an entire task on a simulator, then sending it to the telerobot for reproduction. This might be workable when one is confident that the simulation matches the reality of the telerobot and its environment, or when small differences would not matter (e.g., in programming telerobots for entertainment). Doing this would certainly make it possible to edit the robot's maneuvers until one was satisfied before committing them to the actual operation. Machida et al. (1988) demonstrated such a technique by which commands from a master–slave manipulator could be edited much as one edits material on a videotape recorder or a word processor. Once a continuous sequence of movements had been recorded, it could be played back either forward or in reverse at any time rate. It could be interrupted for overwrite or insert operations. Their experimental system also incorporated computer-based checks for mechanical interference between the robot arm and the environment.

A number of planning aids are manifest in modern air-traffic control. Computers are used to show the expected arrival of aircraft at airports and the gaps between them. This helps the human controller to command minor changes in aircraft speed or flight path to smooth the flow. The center TRACON automation system (CTAS) assists in providing an optimal schedule and three-dimensional spacing. Other systems use radar data to project ahead and alert the controller to potential conflicts (violation of aircraft separation standards) (Wickens et al., 1997).

5 TEACHING THE COMPUTER

Teaching or programming a task, including a goal state and a procedure for achieving it, and including constraints and criteria, can be formidable or quite easy, depending on the command hardware and software. By *command hardware* is meant the way in which human response (hand, foot, or voice) is converted to physical signals to the computer. Command hardware can be either analogic or symbolic. *Analogic* means that there is a spatial or temporal isomorphism among

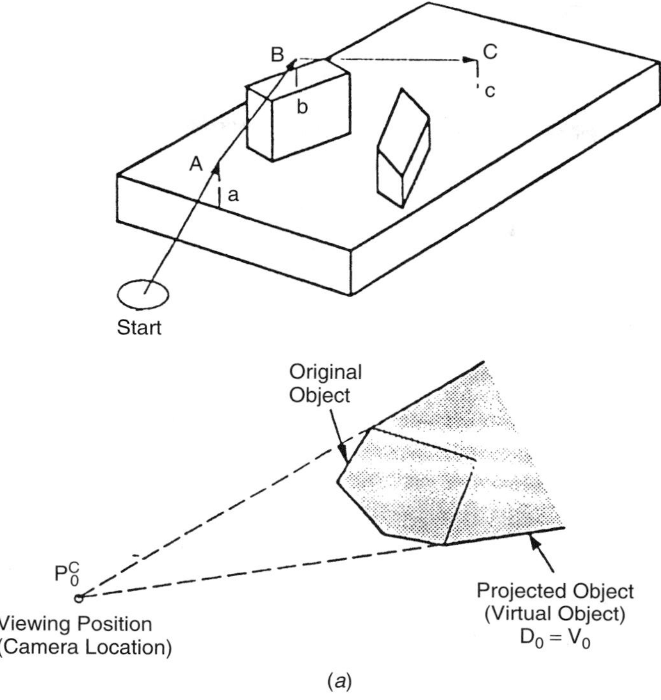

(*a*)

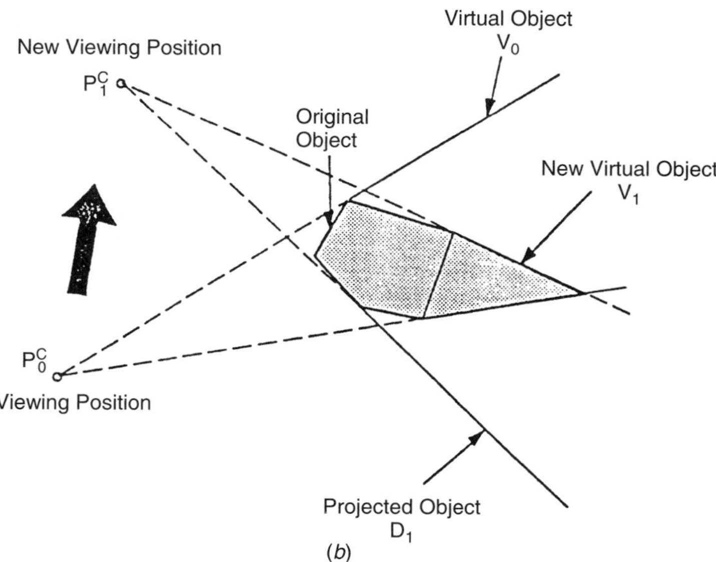

(*b*)

Figure 6 Park's display of computer aid for obstacle avoidance: (*a*) human specification of subgoal points on graphic model; (*b*) generation of virtual obstacles for a single viewing position (above) and a pair of viewing positions (below). (From Park, 1991.)

human response, semantic meaning, and/or feedback display. For example, moving a control up rapidly to increase the magnitude of a variable quickly, which causes a display indicator to move up quickly, would be a proper analogic correspondence.

Symbolic command, by contrast, is accomplished by depressing one or a unique series of keys (as typing words on a typewriter), or uttering one or a series of sounds (as in speaking a sentence), each of which has a distinguishable meaning. For symbolic commands a particular series or concatenation of such responses has a different meaning from other concatenations. Spatial or temporal correspondence to the meaning or desired result is not a requisite. Sometimes analogic and symbolic can be combined: for example, where up–down keys are both labeled and positioned accordingly.

It is natural for people to intermix analogic and symbolic commands or even to use them simultaneously. Typical industrial robots are taught by a combination of grabbing hold and leading the endpoint of the manipulator around in space relative to the workpiece, at the same time using a switch box on a cable (a *teach pendant*) to key in codes for start, stop, speed, and so on, between various reference positions. This happens, for example, when a person talks and points at the same time, or plays the piano and conducts a choir with his or her head or free hand.

Supervisory command systems have been developed for mechanical manipulators that utilize both analogic and symbolic interfaces with the supervisor and that enable teaching to be both rapid and available in terms of high-level language. Brooks (1979) developed such a system, which he called SUPERMAN, which allows the supervisor to use a master arm to identify objects and demonstrate elemental motions. He showed that even without time delay for certain commands, which refer to predefined location, supervisory control that included both teaching and execution took less time and had fewer errors than manual control.

Yoerger (1982) developed a more extensive and robust supervisory command system that enables a variety of arm–hand motions to be demonstrated, defined, called on, and combined under other commands. In one set of experiments, Yoerger compared three different procedures for teaching a robot arm to perform a continuous seam weld along a complex curved workpiece. The end effector (welding tool) had to keep 1 in. away and retain an orientation perpendicular to the curved surface to be welded and move at constant speed. Yoerger tested his subjects in three command (teaching) modes. The first mode was for the human teacher to move the master (with slave following in master–slave correspondence) relative to the workpiece in the desired trajectory. The computer would memorize the trajectory and then cause the slave end effector to repeat the trajectory exactly. The second mode was for the human teacher to move the master (and slave) to each of a series of positions, pressing a key to identify each. The human would then key in additional information specifying the parameters of a curve to be fit through these points and the speed at which it was to be executed, and the computer would then be called upon for execution. The third mode was to use the master–slave manipulator to contact and trace along the workpiece itself, to provide the computer with the knowledge of the location and orientation of the surfaces to be welded. Then, using the typewriter keyboard, the human teacher would specify the positions and orientations of the end effector *relative* to the workpiece. The computer could then execute the task instructions relative to the geometric references given.

Identifying the geometry of the workpiece analogically, and then giving symbolic instructions relative to it, proved the constant winner. The reasons for this advantage apparently are the same as for Brook's results described previously, provided of course that the time spent in the teaching loop is sufficiently short.

Teaching airplane autopilots is a good example of the reaching role in supervisory control. Modern airplanes can now adjust their throttle, pitch, and yaw damping characteristics automatically. They can take off and climb to altitude autonomously, or fly to a given latitude and longitude, and can maintain altitude and direction despite wind disturbances. They can approach and land automatically in zero-visibility conditions. To do these tasks, airplanes make use of artificial sensors, motors, and computers, programmed in supervisory fashion by pilots and ground controllers. In this sense airplanes are telerobots in the hands of their pilot teachers. In the aviation world the supervising pilot is called a flight manager. Figure 7 provides a metaphoric summary of pilot information requirements for performing this task.

New aviation technology (Billings, 1991; Sarter and Woods, 1994; Hopkin, 1995; Sheridan, 2002) includes TCAS (traffic alert and collision avoidance system), ARTS (automated radar terminal system), SSR (secondary surveillance radar), and ILS/MLS (instrument- or machine-aided landing systems). The "glass cockpit" emerged with Boeing's 757 and 767, in which integrated computer-graphic CRT, LED, and LCD displays integrate information heretofore presented on separate displays. These have replaced the older multiple independent mechanical flight instruments and have permitted simplification of the instrument panel. Autopilots have been provided with multiple control modes, e.g., for going to and holding a new altitude, flying to a set of latitude–longitude coordinates, or making an automatic landing when the airport has the supporting equipment. In the Airbus A320 a primary flight mode is fly-by-wire through miniature sidekicks, in dramatic contrast to the old control yokes. In the cockpit, computer-generated and computer-based expert systems give the pilot advice on engine conditions, how to save fuel, and other topics. Performance management systems are now available to optimize fuel and time.

The flight management system (FMS) is the aircraft embodiment of the human-interactive computer discussed previously and currently is where the supervisory teaching is done. The typical FMS has a CRT

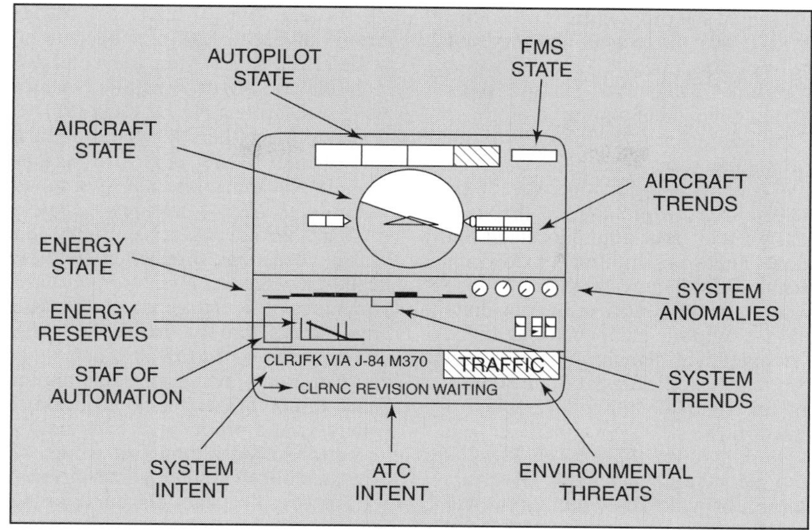

Figure 7 Pilot information requirements. (Courtesy of C. Billings, NASA.)

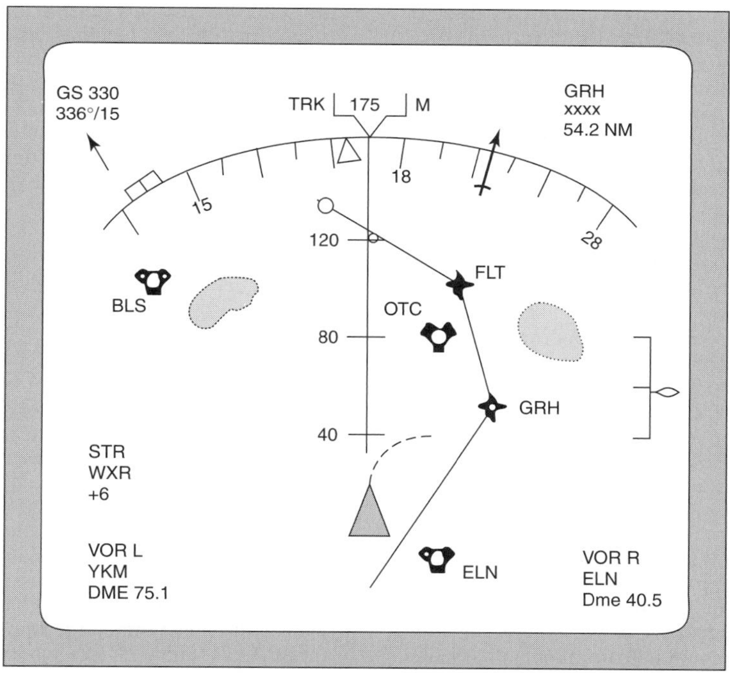

Figure 8 Computer-generated map display showing weather, flight plan route, and navigation aids. (Courtesy of C. Billings, NASA.)

display and both generic and dedicated keysets. More than 1000 modules provide maps for terrain and navigational aids, procedures, and synoptic diagrams of various electrical and hydraulic subsystems. A proposed electronic map that would show planned flight route, weather, and other navigational aids is illustrated in Figure 8. When the pilot enters a certain flight plan, the FMS can visualize the trajectory automatically and call attention to any waypoints that appear to be erroneous on the basis of a set of reasonable assumptions.

(This might have prevented the programmed trajectory that allegedly took KAL 007 into Soviet territory and resulted in its being shot down.)

The problem of authority is one of the most difficult (Boehm et al., 1983). Popular mythology is that the pilot is (or should be) in charge at all times. But when a human turns control over to an automatic system, it is the exception that she or he can do something else for awhile (as in the case of setting one's alarm clock and going to sleep). It is also recognized that there are limited windows of opportunity for escaping from the automation (once you get on an elevator you can get off only at discrete floor levels). People are seldom inclined to "pull the plug" unless they receive clear signals indicating that such action must be taken, and unless circumstances make it convenient for them to do so. Examples of some current debates follow.

1. Should there be certain states, or a certain envelope of conditions, for which the automation will simply seize control from the pilot? In the MD11 it is impossible to exceed critical boundaries of speed and attitude that will bring the aircraft into stall or other unsafe flight regimes. The pilot can approach the boundaries of the safe flight envelope only by exerting much more than the normal force on the control stick.
2. Should the computer deviate from a programmed flight plan automatically if critical unanticipated circumstances arise? The MD11 will deviate from its programmed plan if it detects wind shear.
3. If the pilot programs certain maneuvers ahead of time, should the aircraft execute these automatically at the designated time or location, or should the pilot be called upon to provide further concurrence or approval? The A320 will not initiate a programmed descent unless it is reconfirmed by the pilot at the required time.
4. In the case of a subsystem abnormality, should the affected subsystem be reconfigured automatically, with after-the-fact display of what has failed and what has been done about it? Or should the automation wait to reconfigure until after the pilot has learned about the abnormality, perhaps been given some advice on the options, and had a chance to take initiative? The MD11 goes a long way in automatic fault remediation.

Along with the advance of computer science in natural language understanding, it will be important to learn how to cope with the "fuzziness" (Zadeh, 1984; Kosko, 1992) inherent in the way that people think about, and therefore communicate about, their tasks. That is, both memorized "rules" and typed or spoken messages would by nature be sentences consisting of fuzzy terms. As an example, a fuzzy rule for driving a car might be: "If your car is going *fast* and if the car ahead is *very* close or *moderately* close and going *slow*, brake." The italicized terms are fuzzy sets, which may be defined with varying degrees of *membership* over a range of numerical values of speed and distance. Given a number of statements such as the one above, and given membership functions for each fuzzy term over the physical variables, the "relative truth" of each of several control actions (e.g., brake, accelerate, coast) can be determined. Buharali and Sheridan (1982) demonstrated that a computer could be taught to drive a car by giving rules repeatedly, where the computer thereby would come to "know what it didn't know" with regard to various combinations of conditions and could ask the supervisor–teacher for additional rules to cover its "domains of ignorance."

The big advantages of fuzzy logic, fuzzy control, fuzzy decision support systems, and so on, are that the rules can be quite arbitrary, and how the fuzzy variables map onto physical variables of the real works can be as "crisp" or as "soft" as one wishes for a particular context—much as normal human communication. Some applied mathematicians and computer scientists accustomed to closed-form analysis and proof have been negative about "fuzzy" because so far, although it works well, "fuzzy" is not so tractable to mathematical theorems and proofs.

To complete this discourse on supervisor teaching of computer automation, it is important to emphasize that simple and ideal command-and-feedback patterns are not to be expected as systems get more complex. In interactions between a human supervisor and his or her subordinates, or a teacher and the students, it can be expected that the teaching process will not be a one-way communication. Some feedback will be necessary to indicate whether the message is understood or to convey a request for clarification on some aspect of the instructions. Further, when the subordinate or student does finally act on the instruction, the supervisor may not understand from the immediate feedback what the subordinate has done and may ask for further details. This is illustrated in Figure 9 by the light arrows, where the bold arrows characterize the conventional direction of information in feedback control.

6 MONITORING OF DISPLAYS AND DETECTION OF FAILURES

The human supervisor monitors the automated execution of the task to ensure proper control (Parasuraman, 1987). This includes intermittent adjustment or trimming if the process performance remains within satisfactory limits. It also includes detection of if and when it goes outside limits, and the ability to diagnose failures or other abnormalities. The subject of failure detection in human–machine systems has received considerable attention (Rasmussen and Rouse, 1981). Moray (1986) regards such failure detection and diagnosis as the most important human supervisory role. I prefer the view that all five supervisory roles are essential and that no one can be placed above the others.

The supervisory controller tends to be removed from full and immediate knowledge about the controlled process. The physical processes that he or she must monitor tend to be large in number and distributed widely in space (e.g., around a ship or plant). The physical variables may not be immediately sensible by him or her (e.g., steam flow and pressure) and may be computed from remote measurements on other variables. Sitting in the control room or cockpit, the

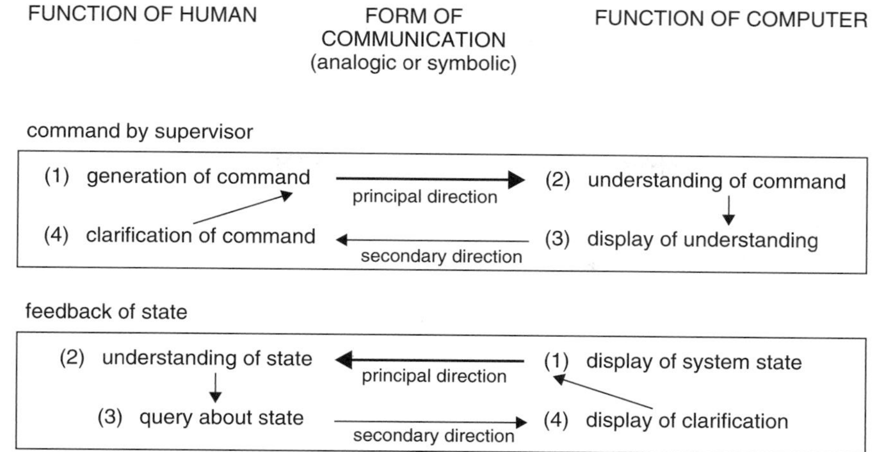

Figure 9 Intermediate feedback in command and display. Heavy arrows indicate the conventional understanding of functions. Light arrows indicate critical additional functions that tend to be neglected. (From Sheridan, 1992.)

supervisor is dependent on various artificial displays to give feedback of results as well as knowledge of new reference inputs or disturbances. These factors greatly affect how he or she detects and diagnoses abnormalities in the process, but whether removal from active participation in the control loop makes it harder (Ephrath and Young, 1981) or easier (Curry and Ephrath, 1977) remains an open question. Curry and Nagel (1974), Niemala and Krendel (1974), Gai and Curry (1978), and Wickens and Kessell (1979, 1981) have studied various psychophysical aspects of this problem.

In traditional control rooms and cockpits the tendency has been to provide the human supervisor with an individual and independent display of each variable, and for a large fraction of these to provide a separate additional alarm display that lights up when the corresponding variable reaches or exceeds some value. Thus, modern aircraft may easily have over 1000 displays and modern chemical or power plants 5000 displays. In the writer's experience, in one nuclear plant training simulator, during the first minute of a "loss of coolant accident," 500 displays were shown to have changed in a significant way, and in the second minute, 800 more.

Clearly, no human being can cope with so much information coming simultaneously from so may seemingly disconnected sources. Just as clearly, such signals in any real operating system actually are highly correlated. In real-life situations in which we move among people, animals, plants, or buildings our eyes, ears, and other senses easily take in and comprehend vast amounts of information just as much as in the power plant. Our genetic makeup and experience enable us to integrate the bits of information from different parts of the retina and from different senses from one instant to the next, presumably because the information is correlated. We say we "perceive patterns" but do not pretend to understand how. In any case the challenge is to design displays in technological systems to somehow integrate the information to enable the human operator to perceive patterns in time and space and across the senses. As with teaching (command), the forms of display may be either analogic (e.g., diagrams, plots) or symbolic (e.g., alphanumerics) or some combination.

In the nuclear power industry the *safety parameter display system* (SPDS) is now required of all plants in some form. The idea of the SPDS is to select a small number (e.g., six to 10) of variables that tell the most about plant safety status, and to display them in integrated fashion, such that by a glance the human operator can see whether something is abnormal, if so what, and to what relative degree. Figure 10 shows an example of an SPDS. Figure 10 gives the high-level or overview display (a single computer "page"). If the operator wishes more detailed information about one variable or subsystem, he or she can page down (select lower levels), as in Figure 10*b*. These can be diagrams having lines or symbols that change color or flash to indicate changed status, and alphanumerics to give quantitative or more detailed status. These can also be bar graphs or cross plots, or integrated in other forms. One novel technique is the *Chernoff face* (Figure 10), in which the shapes of eyes, ears, nose, and mouth differ systematically to indicate different values of variables, the idea being that facial patterns are easily perceived. Allegedly, the Nuclear Regulatory Commission, fearful that some enterprising designer might employ this technique before it was proven, formally forbade it as an acceptable SPDS.

Since detection and diagnosis of system failure is a critical task for the supervisor, computer aiding by the HIC in comparing, computing, and displaying has great potential. Various techniques have been proposed for doing this. One such technique (Tsach et al., 1983) compares key measurements continuously from the plant to corresponding variables of an online computer

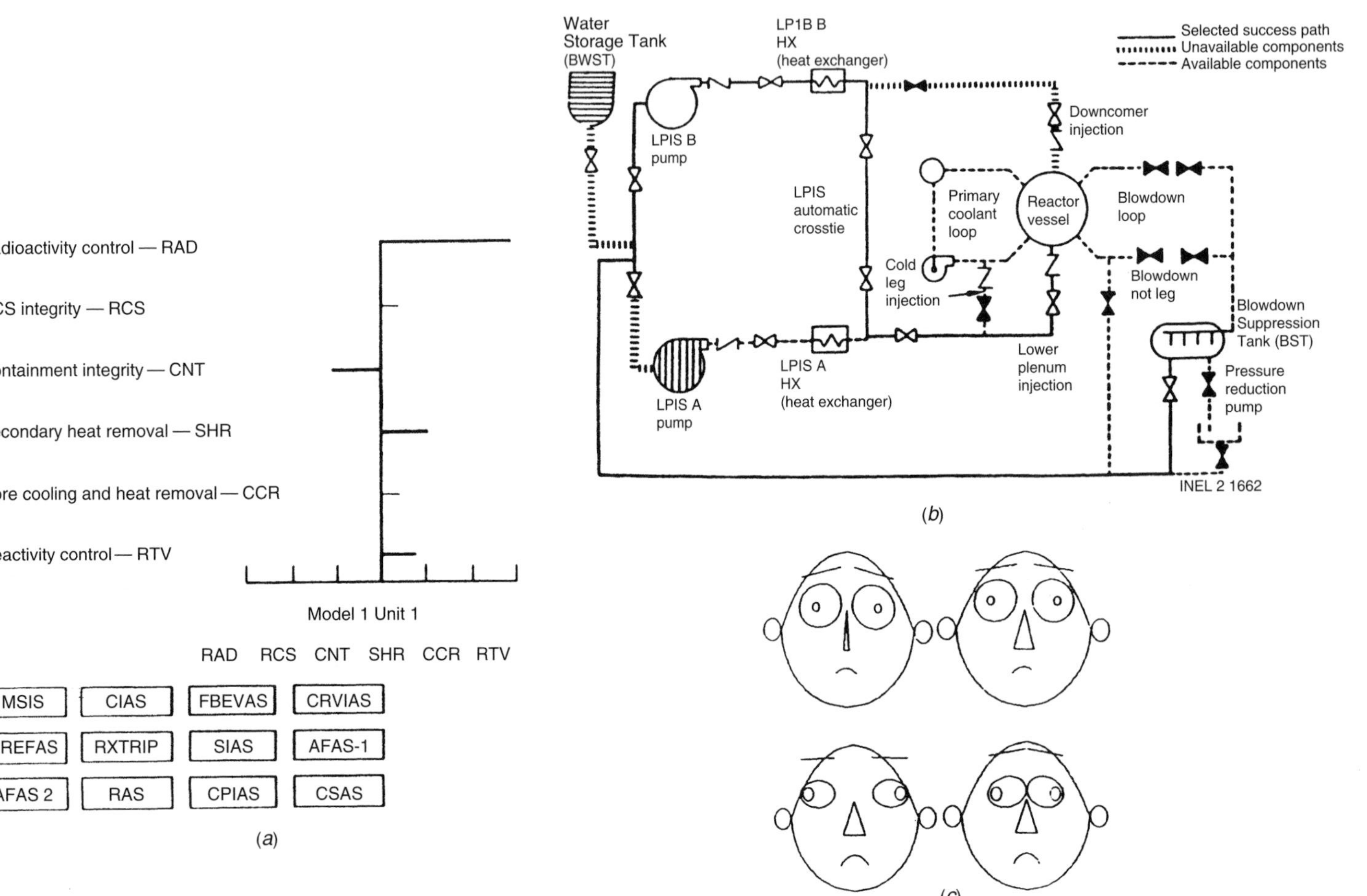

Figure 10 Safety parameter display system for a nuclear power plant.

model; then a computer-graphic display focuses the operator's attention on the discrepancies that indicate abnormality. Figure 11 shows one type of iconic display developed for this system: a polygon whose vertices indicate the degree to which each variable (of one subsystem in this case) is below or above a normal range (torus). The display therefore "points" to corresponding discrepancies between the measured and model variables as they evolve in time.

As noted previously (Figure 5), an important potential of the HIC is for modeling the controlled process. Such a model may then be used to generate a display of observed state variables that cannot be seen or measured directly. Another use is to run the model in fast time to predict the future, given of course that the model is calibrated to reality at the beginning of each such predictive run. A third use, now being developed for application to remote control of manipulators and vehicles in space, helps the human operator cope with telemetry time delays (as shown in Figure 12, wherein video feedback in necessarily delayed by at least several seconds). By sending control signals to a computer model as a basis for superposing the corresponding graphic model on the video, the graphic model will "lead" the video picture and indicate what the video will do several seconds hence. This has been shown to speed up the execution of simple manipulation tasks by 70 to 80% (Noyes and Sheridan, 1984).

Computer-driven displays for monitoring are gradually finding their way into older technical systems such as railway systems. One example is for high-speed trains. A train driver's main job is to control speed given that his or her controlled process has very large momentum [it takes up to 3 km to stop a display recently proposed for the driver's (locomotive engineer's) cab in high-speed train traveling at 300 km per hour, even under emergency braking]. Currently, speed signals are being moved from wayside indicators into the cab itself, but realistically, the operator cannot see ahead for more than 1 km, usually not even that, especially at night. This means that if a truck is stalled at a grade crossing, the operator cannot receive any information about it in time to stop. Most speed constraints are set by curves in the track, grade crossings, or population densities, which are fixed and can be learned by the operator. But because of track maintenance, rock slides, snow, and so on, there may be other speed constraints that are not so easily anticipated. For

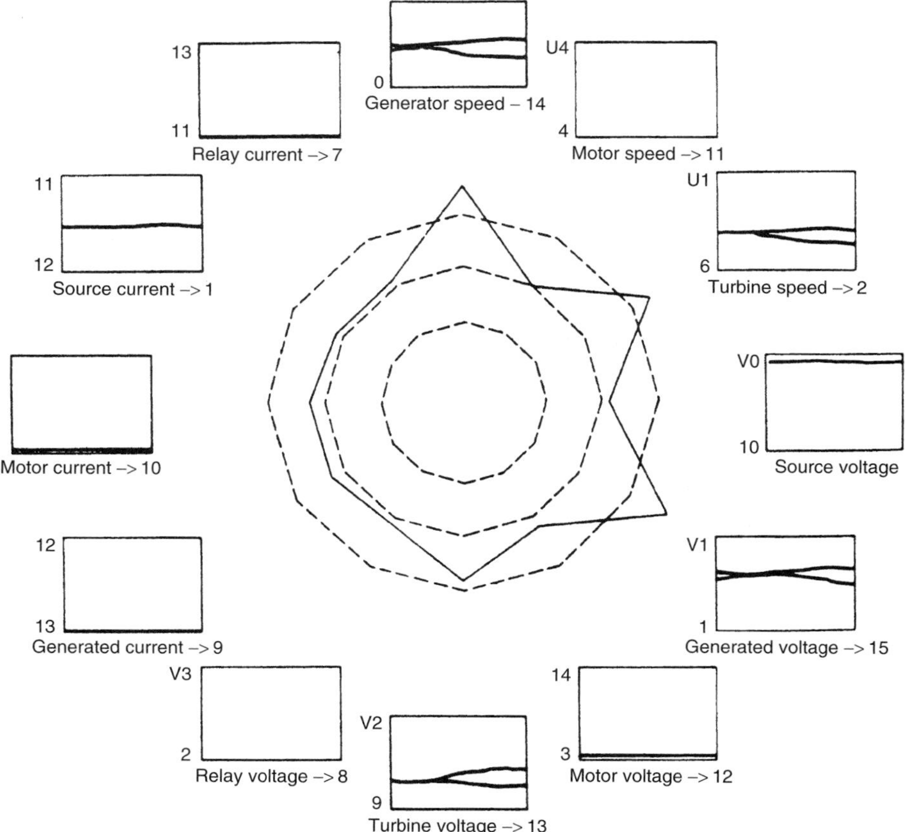

Figure 11 Computer display of the Tsach–Sheridan–Buharali system for failure detection and location in process control.

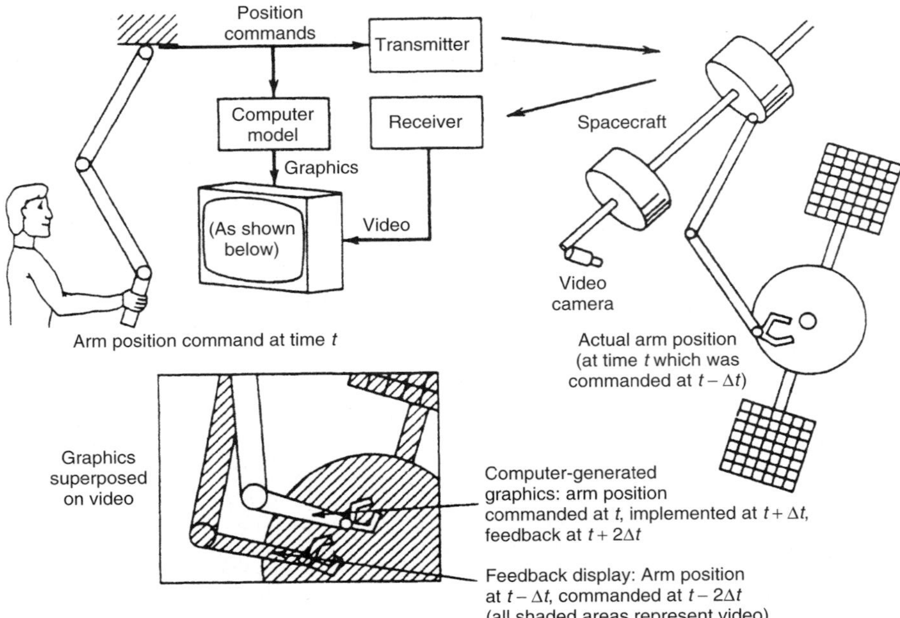

Figure 12 Predictor display for delayed telemanipulator.

these reasons we developed a computer-based display that previews the track for several kilometers, showing curves, speed limits, and other features, both fixed and changing (Figure 13). It also shows prediction curves of how speed will change as a function of track distance ahead (according to a dynamic model) if the current throttle setting is maintained, as well as predication curves for maximum service braking and maximum emergency braking (a different system). Finally, it indicates the continuous throttle settings that will get the train to the next station on schedule (assuming that the train is at or near schedule currently and the winds are known), meeting the known speed limits, and under these constraints minimizing fuel. The latter is akin to the flight director in an aircraft. This system was tested in a dynamic human-in-the-loop train simulator with a number of trained driver subjects. It improved performance significantly over driving with the conventional displays (Askey, 1995).

Advances in computer graphics, as driven by the computer game industry, film animation and special effects, and other simulations (virtual reality) has meant that computer displays are theoretically limitless in what they can display: dynamically at high resolution, in color, and on a head-mounted display if that is called for. The challenge is then for the display designer: What will provide the most effective interaction with the human supervisor?

A final aspect of supervisory monitoring and display concerns format adaptivity—the ability to change the format and/or the logic of the displays as a function of the situation. Displays in aerospace and industrial systems now have fixed formats (e.g., the labels, scales, and ranges are designed into the display). Alarms have fixed set points. However, future computer-generated displays even for the same variables may be different at various mission stages or in various conditions. Thus, formats may differ for aircraft takeoff, landing, and on-route travel, and be different for plant startup, full-capacity operation, and emergency shutdown. Some alarms have no meaning or may be expected to go off when certain equipment is being tested or taken out of service. In such a case adaptive formatted alarms may be suppressed or the set points changed automatically to correspond to the operating mode. Future displays and alarms could also be formatted or adjusted to the personal desires of the supervisor, to provide any time scale, degree of resolution, and so on, necessary at the time. Ideally, some future displays could adapt based on a running model of how the human supervisor's perception was being enhanced.

A currently popular research challenge is to measure highway vehicle driver task workload and whether driver's use of the potential in-vehicle information distracters, such as cell phone, radio, navigation system, and so on, should be prohibited during busy demands of traffic (Boer, 2000; Llaneras, 2000; Lee et al., 2002). This would make the driver interfaces adaptive.

There are hazards, of course, in allowing emergency displays to be too flexible, to the point where they cause errors rather than preventing them. *Mode errors*, where the operator believes that he or she is operating

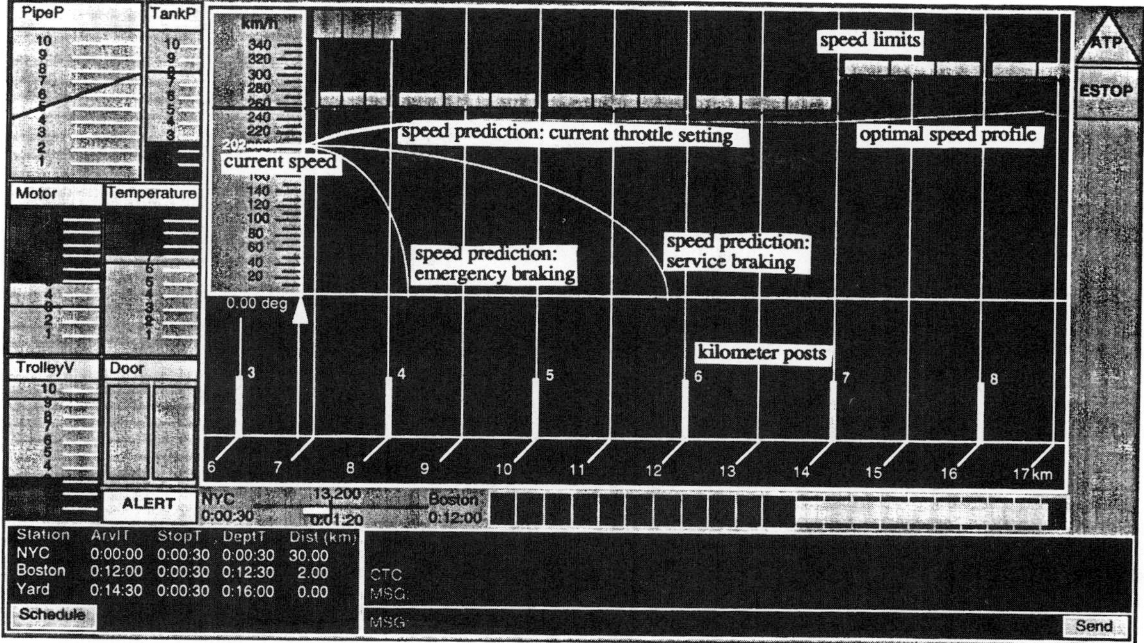

Figure 13 Proposed preview, predictor, and advisory display for drivers of high-speed trains. (From Askey, 1995.)

in one mode but actually is operating in a different mode, can be dangerous. An example of where flexibility in monitoring displays went awry was in an aircraft accident that occurred in Europe several years ago. In this instance the pilot could ask to have either descent rate (thousands of feet per minute) or descent angle (degrees) presented, and depending on how the model control panel had been set, the number was indicated by two digits displayed at the same location. In this case the pilot forgot which mode he had requested (although that information was also displayed, but at a different location). The result was a misreading and a tragic crash.

7 INTERVENING AND HUMAN RELIABILITY

Sarter and Woods (2000) and Wiener (1988) write about *automation surprises*, the tendency of automatic systems to catch the human supervisor off-guard such that the human thinks: What is the automation doing now? What will it do next? How did I get into this mode? Why did it do this? How do I stop the machine from doing this? Why won't it do what I want?

The challenge of surprise is a great one, and there are not easy answers. Computers do what they have been programmed to do, which is not always what the user intended. User education—toward better understanding of how the system works—is one remedy. Another is to provide error messages that are couched in a language understandable to the operator (not in the jargon of the computer programmer, so familiar to all users of computers). Generally, the solution lies in some form of feedback—to lead the human in making a mild or radical intervention, as appropriate.

The supervisor decides to intervene when the computer has completed its task and must be retaught for the next task, when the computer has run into difficulty and requests of the supervisor a decision as to which way to go, or when the supervisor decides to stop automatic action because he or she judges that system performance is not satisfactory. Intervention is a problem that really has not received as much attention as teaching and monitoring. Yet systems are being planned in which the supervisory operator is expected to receive advice from a computer-based system about remote events and within seconds decide whether to accept the computer's advice (in which case the response is commanded automatically), or reject the advice and generate his or her own commands (in effect, intervene in an otherwise automatic chain of events).

One such system is a state-of-the-art traffic management system currently operating in Boston's Central Artery/Tunnel, the primary traffic system through the city, completed recently at a cost of $14 billion. Magnetic, optical, infrared, and smoke detectors signal sudden changes in traffic density (accident), and fire and other emergencies, indicating to the operator where in the system the problem is and displaying hopefully appropriate video pictures (from one or several of 400 cameras). The operator can accept the computer's advice (to call up fire, police, or tow trucks; turn on fire extinguishers; modify electronic signals; etc.) or devise his or her own commands. The operator can change the programmed execution after it has begun, but at considerable time or other costs. He or she can also add or subtract commands, and so on. The aim is to provide quicker, safer emergency response.

It is at the intervention stage that human error most reveals itself. Errors in learning from past experience, planning, teaching, and monitoring will surely exist. Many of these are likely to be corrected as the supervisor notes them "during the doing." It is after the automatic system is functioning and the supervisor is monitoring intermittently that those human errors make a difference and where it is therefore critical that the human supervisor intervene in time and take appropriate action when something goes wrong. Thus, the intervention stage is where human error is most manifest.

If human error is not caught by the supervisor, it is perpetuated slavishly by the computer, much as happened to the *Sorcerer's Apprentice*. For this reason supervisory control may be said to be especially sensitive to human error. Several factors affect the supervisor's decision to intervene and/or his or her success in doing so.

1. *Trade-off between collecting more data and taking action in time.* The more data collected from the more sources, the more reliable is the decision of what, if anything, is wrong, and what to do about it. Weighed against this is that if the supervisor waits too long, the situation will probably get worse, and corrective action may be too late. Formally, the optimization of this decision is called the *optional stopping problem.*

2. *Risk taking.* The supervisor may operate from either risk-averse criteria such as minimax (minimize the worst outcome that could happen) or more risk-neutral criteria such as expected value (maximize the subjectively expected gain). Depending on the criterion, the design of a supervisory control system may be very different in complexity and cost.

3. *Mental workload.* This problem is aggravated by supervisory control. When a supervisory control system is operating well in the automatic mode, the supervisor may have little concern. When there is a failure and sudden intervention is required, the mental workload may be considerably higher than in direct manual control, where in the latter case the operator is already participating actively in the control loop. In the former case the supervisor may have to undergo a sudden change from initial inattention, moving physically and mentally to acquire information and learn what is going on, then making a decision on how to cope. Quite likely this will be a rapid transient from very little to very high mental workload.

Mental workload can be at issue in any human operation. The topic is reviewed by Moray (1979, 1982), Williges and Wierwille (1979), and Hart and Sheridan (1984). Ruffle-Smith (1979) studied pilot errors and fault detection under heavy cognitive workload in a realistic flight simulator and found that crews made approximately one error every 5 minutes.

Although the subject of human error is currently of great interest, there is no consensus on either a taxonomy or a theory of causality of errors. One common error taxonomy relates to locus of behavior: sensory, memory, decision, or motor. Another useful distinction is between errors of omission and those of commission. A third is between slips (correct intentions that inadvertently are not executed) and mistakes (intentions that are executed but that lead to failure).

In supervisory control there are several problems of human error worth particular mention. One is the type of slip called *capture*. This occurs when the intended task requires a deviation from a well-rehearsed (behaviorally) and well-programmed (in the computer) procedure. Somehow, habit, augmented by other cues from the computer, seems to capture behavior and drive it on to the next (unintended) step in the well-rehearsed and computer-reinforced routine.

A second supervisory error, important in both planning and failure diagnosis, results from the human tendency to seek confirmatory evidence for a single hypothesis currently being entertained (Gaines, 1976). It would be better if the supervisor could keep in mind a number of alternative hypotheses and let both positive and negative evidence contribute symmetrically in accordance with the theory of Bayesian updating (Sheridan and Ferrell, 1974). Norman (1981), Reason and Mycielska (1982), Rasmussen (1982), and Rouse and Rouse (1983) provide reviews of human error research from their different perspectives.

Theoretically, anything that can be specified in an algorithm can be given over to the computer, so that the reason the human supervisor is present is to add novelty and creativity: precisely those ingredients that cannot be prespecified. This means, in effect, that the best or most correct human behavior cannot be prespecified and that variation from precise procedure must not always be viewed as errant noise. The human supervisor, by the nature of his or her function, must be allowed room by the system design for what may be called *trial and error* (Sheridan, 1983).

What training should the human supervisory controller receive to do a good job at detecting failures and intervening to avoid errors? As the supervisor's task becomes more cognitive, is the answer to provide training in theory and general principles? Curiously, the literature seems to provide a negative answer (Duncan, 1981). In fact, Moray (1986) in his review concludes that "there seems to be no case in the literature where training in the theory underlying a complex system has produced a dramatic change in fault detection or diagnosis." Rouse (1985) similarly concludes "that the evidence [e.g., Morris and Rouse (1985)] does not support a conclusion ... that diagnosis of the unfamiliar requires theory and understanding of system principles." Apparently, frequent hands-on experience in a simulator (i.e., with simulated failures) is the best way to enable a supervisor to retain an accurate mental model of a process.

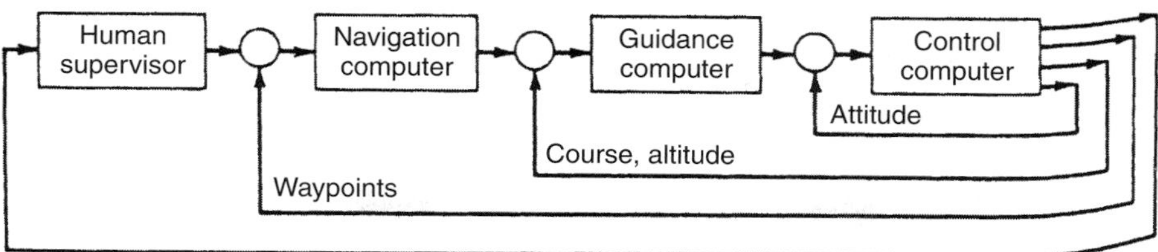

Figure 14 Nested control loops of aerospace vehicle.

8 MODELING SUPERVISORY CONTROL

For 35 years various models of supervisory control have been proposed. Most of these have been models of particular aspects of supervisory control, not apparently claiming to model all or even very many aspects of it. The simplest model of supervisory control might be that of nested control loops (Figure 14), where one or more inner loops are automatic and the outer one is manual. In aerospace vehicles the innermost of four nested loops is typically called "control," the next "guidance," and the next "navigation," each having a set point determined by the next outer loop. Hess and McNally (1997) have shown how conventional manual control models can be extended to such multiloop situations. The outer loop in this generic aerospace vehicle includes the human operator, who, given mission goals, programs in the destination. In driving a car the functions of navigation, guidance, and control are all done by a person and can be seen to correspond roughly to knowledge-based, skill-based, and rule-based behavior.

Figure 15 is a qualitative functional model of supervisory control, showing the various cause–effect loops or relationships among elements of the system, and emphasizing the symmetry of the system as viewed from top and bottom (human, task) of the hierarchy. Figure 16 is an abbreviated version of Rasmussen's qualitative model referred to in Section 6, showing in particular the nesting of skill-based, rule-based, and knowledge-based behavioral loops. Figure 17 extends Rasmussen's model to show various interactions with computer aids having comparable levels of intelligence.

One problem of interest to the supervisor is how often he or she should sample the input, how often he or she should update a control setting, or both, particularly if there is a cost incurred each time he or she does so. Given assumptions on the magnitude distribution and autocorrelation of inputs, a utility function for the value of performance resulting from a particular input and particular control action in combination, and a discrete cost of sampling, Sheridan (1970a) showed how an optimal sampling strategy could be derived to maximize expected gain. Sheridan (1976) suggested a framework for how a supervisor equipped with a variety of sensing options and a variety of motor options could try various combinations of these in *thought experiments* or simulations with an *internal model* of the controlled process and utility function. Using Bayes' theorem, it is shown how expected utility is maximized.

One problem the supervisor faces is allocating attention between different tasks, where each time that he or she switches tasks there is a time penalty in transfer, typically different for different tasks, and possibly involving uses of different software procedures, different equipment, and even bodily transportation of himself or herself to different locations. Given relative worths for time spent attending to various tasks, it has been shown (Sheridan, 1970b) that dynamic programming enables the optimal allocation strategy to be established. Moray et al. (1982) applied this model to deciding whether human or computer should control various variables at each succeeding moment. For simpler experimental conditions, the model fit the experimental data (subjects acted like utility maximizers), but as task conditions became complex, apparently it did not. Wood and Sheridan (1982) did a similar study where supervisors could select among alternative automatic machines (differing in both rental cost and productivity) to do assigned tasks or do the tasks themselves. Results showed the supervisors to be suboptimal, paying too much attention to costs and too little to productivity, and in some cases using the automation when they could have done the tasks more efficiently manually. Govindaraj and Rouse (1981) modeled the supervisor's decisions to divert attention from a continuous task to perform or monitor a discrete task.

Rouse (1977) utilized a queueing theory approach to model whether from moment to moment a task should be assigned to a computer or to the operator. The allocation criterion was to minimize service time under costs constraints. Results suggested that human–computer "misunderstanding" of one another degraded efficiency more than limited computer speed. In a related flight simulation study, Chu and Rouse (1979) had a computer perform those tasks that had waited in the queue beyond a certain time. Chu et al. (1980) extended this idea to have the computer learn the pilot's priorities and later make suggestions when the pilot was under stress.

Tulga and Sheridan (1980) and later Pattipatti et al. (1983) utilized a model of allocation of attention among multiple task demands, a task displayed on the computer screen to the subject as is represented in Figure 18. Instead of being stationary, these demands appear at random times (not being known until they appear), exist for given periods of time, then disappear at the end of that time with no more opportunity to gain

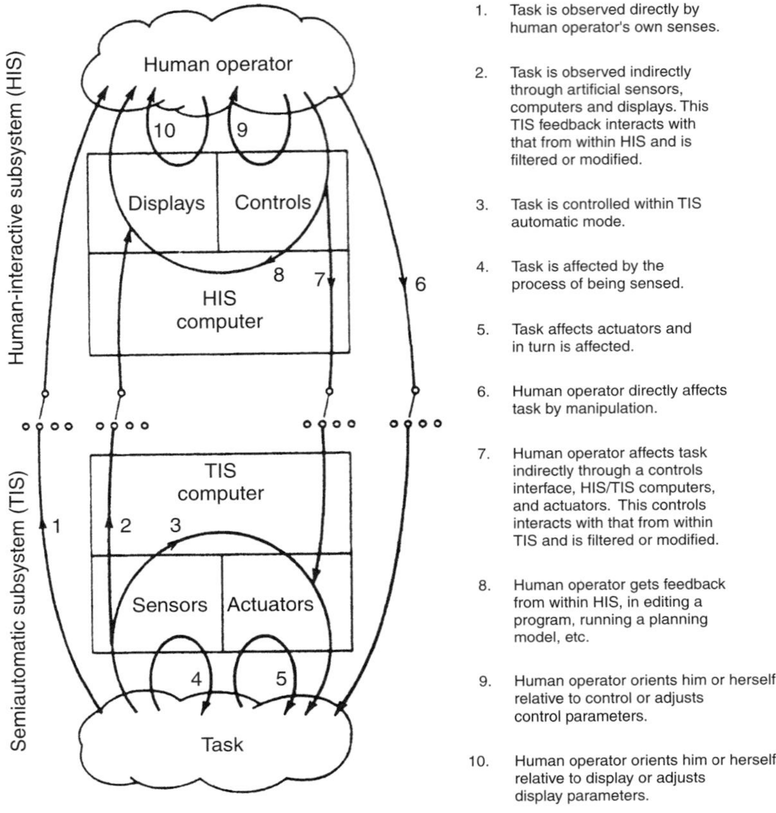

Human operator
Human-interactive subsystem (HIS)
10
9
Displays
Controls
8
7
6
HIS computer
TIS computer
Semiautomatic subsystem (TIS)
1
2
3
Sensors
Actuators
4
5
Task
1. Task is observed directly by human operator's own senses.
2. Task is observed indirectly through artificial sensors, computers and displays. This TIS feedback interacts with that from within HIS and is filtered or modified.
3. Task is controlled within TIS automatic mode.
4. Task is affected by the process of being sensed.
5. Task affects actuators and in turn is affected.
6. Human operator directly affects task by manipulation.
7. Human operator affects task indirectly through a controls interface, HIS/TIS computers, and actuators. This controls interacts with that from within TIS and is filtered or modified.
8. Human operator gets feedback from within HIS, in editing a program, running a planning model, etc.
9. Human operator orients him or herself relative to control or adjusts control parameters.
10. Human operator orients him or herself relative to display or adjusts display parameters.

Figure 15 Multiloop model of supervisory control. (From Sheridan, 1984b.)

anything by attending to them, While available, they take differing amounts of time to complete and have differing rewards for completion, which information may be available after they appear and before they are "worked on." The human decision maker in this task need not allocate attention in same temporal order in which the task demands become known, nor in the same order in which their deadline will occur. Instead, he or she may attend first to that task which has the highest payoff or takes the least time, and/or may plan ahead a few moves so as to maximize gains. The Tulga–Sheridan experimental results suggest that subjects approach optimal behavior, which, when heavily loaded (i.e., there are more opportunities than he or she can possibly cope with), simply amounts to selecting the task with highest payoff regardless of time to deadline. These subjects also reported that their sense of subjective workload was greatest when by arduous planning they could barely keep up with all tasks presented. When still more tasks came at them and they had to select which they could do and which they had to off-load, subjective workload decreased.

Using as a "front end" some attention allocation mechanisms similar to those of the Tulga–Sheridan and Pattipatti–Kleinman–Ephrath models, Baron et al. (1980) extended the Baron and Kleinman optimal model and called it PROCRU (Figure 19). It was built originally to model crew selection and implementation of procedures in aircraft approach and landing. Optimum decision and control algorithms maximize expected gain for given nominal procedure requests from the ground, aerodynamic disturbances, vehicle dynamics, and objective function.

There are a number of questions that researchers and designers of supervisory control systems must cope with. Among these are (1) how much autonomy is appropriate for the TIC, (2) how much the TIC and the HIC should tell the human supervisor, and (3) how responsibilities should be allocated among the TIC, HIC, and supervisor (Johannsen, 1981).

The famous Yogi Berra allegedly counseled: "Never make predictions, especially about the future!" Nevertheless, it is ethically mandatory that we predict as best we can. However, recent decades have seen a shift away from monolithic, computationally predictive models toward frameworks or categorizations of models, each of which may be quite simple—involving elementary control laws, a few heuristics, or pattern recognition rules. Thus, as knowledge and understanding of supervisory control has grown, along with its

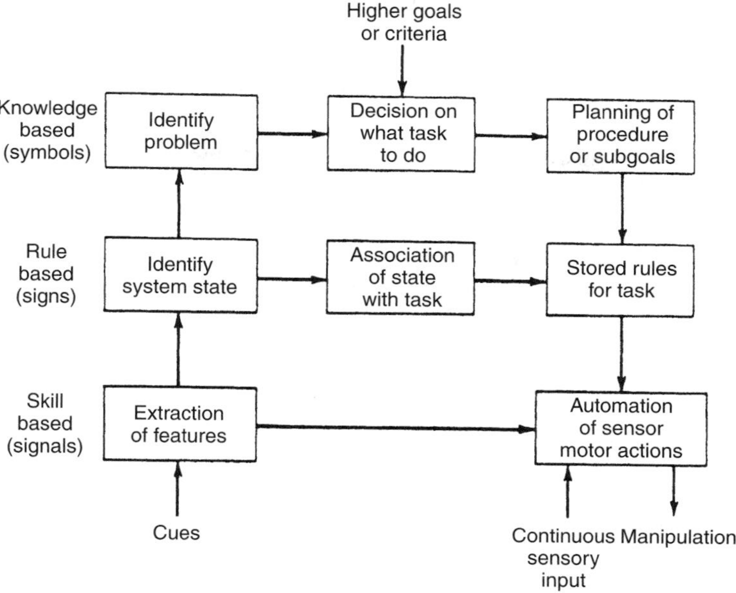

Figure 16 Simplification of Rasmussen's qualitative model of human behavior.

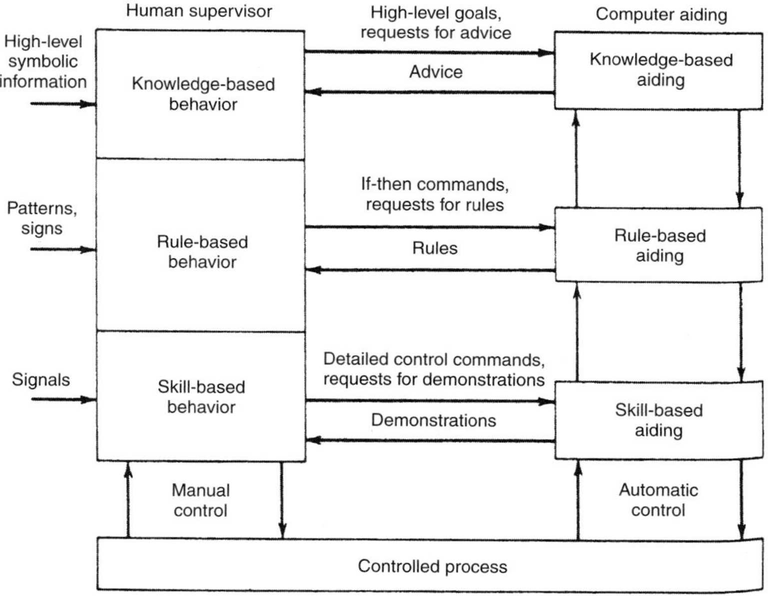

Figure 17 Supervisor interactions with computer decision aids at knowledge, rule, and skill levels.

complexity, researchers have come to realize that they need not and cannot be held to comprehensive predictive models, desirable as they may be.

The most difficult, and it might even be said impossible, aspect of supervisory control to model is that of setting in goals, conditions, and values. Even though overall goals may be given to an actual system (or given in an experiment), how those are translated into subgoals and conditional statements remains elusive. The same is true for communicating values (criteria, coefficients of utility, etc.). Although this act of evaluation remains the *sine qua non* of why

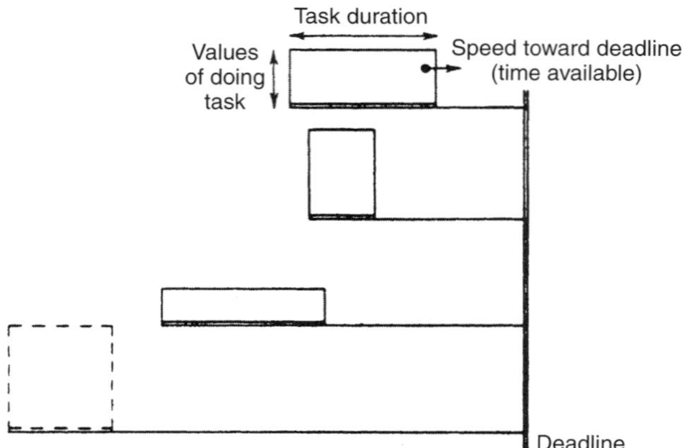

Figure 18 Multitask computer display used in the Tulga–Sheridan experiment.

human participation in system control must remain, there is little prospect for mathematical modeling of this aspect in the near future.

9 SOCIAL IMPLICATIONS AND THE FUTURE OF SUPERVISORY CONTROL

One near certainty is that as technology of computers, sensors, and displays improves, supervisory control will become more prevalent. This should occur in two ways: (1) a greater number of semiautomated tasks will be controlled by a single supervisor (a greater number of TICs will be connected to a single HIC), and (2) the sophistication of cognitive aids, including expert systems for planning, teaching, monitoring, failure detection, and learning, will increase and include more of what we now call knowledge-based behavior in the HIC.

The World Wide Web has enabled easy worldwide communication (for those properly equipped). One aspect of that communication that up to now has hardly become manifest is the ability to exercise remote control. A number of experimental demonstrations have been performed on controlling robots between continents, but delayed feedback still poses a difficulty for continuous control, so supervisory control clearly has an advantage here. In the future we should see many more applications of moderate- and long-distance remote control.

Concurrently, understanding by the layperson (including those of both corporate and government bureaucracies) should come to understand the potential of supervisory control much better. At the present time the layperson tends to see automation as "all or none," where a system is controlled either manually or automatically, with nothing in between. In robotized factories the media tend to focus on the robots, with little mention of design, installation, programming, monitoring, fault detection and diagnosis, maintenance, and various learning functions that are performed by people. In the space program the same is true; options are seen to be either "automated," "astronaut in EVA," or "astronaut or ground controlling telemanipulator" without much appreciation for the potential of supervisory control.

In considering the future of supervisory control relative to various degrees of automation, and to the complexity or unpredictability of task situations to be dealt with, a representation such as Figure 20 comes to mind. The meaning of the four extremes of this rectangle are quite identifiable. Supervisory control may be considered to be a frontier (line) advancing gradually toward the upper right-hand corner.

For obvious reasons, the tendency, has been to automate what is easiest and to leave the rest to the human. This has sometimes been called the *technological imperative*. From one perspective this dignifies the human contribution; from another it may lead to a hodge-podge of partial automation, making the remaining human tasks less coherent and more complex than need be, resulting in overall degradation of system performance (Bainbridge, 1983; Parsons, 1985).

As discussed previously, supervisory control may involve varying degrees of computer aiding on the afferent or sensing/analyzing side, as well as on the efferent or control execution side. Table 1 suggests a scale of degrees of automation that separates the afferent (sensing) from the efferent (taking action). It breaks the afferent functions into components dealing with (1) experience, (2) sensing present data, (3) interpreting present data, and (4) formulating action alternatives. This *degrees of automation* idea, originally presented in Sheridan and Verplank (1978), has been picked up and used by others in various ways. Parasuraman et al. (2000) added the idea that the successive stages of information acquisition, information analysis, action decision, and action implementation are usually automated to different degrees; the best degree of automation is seldom the same at the various stages.

Human-centered automation has become a popular phrase, and therefore it is important to comment on

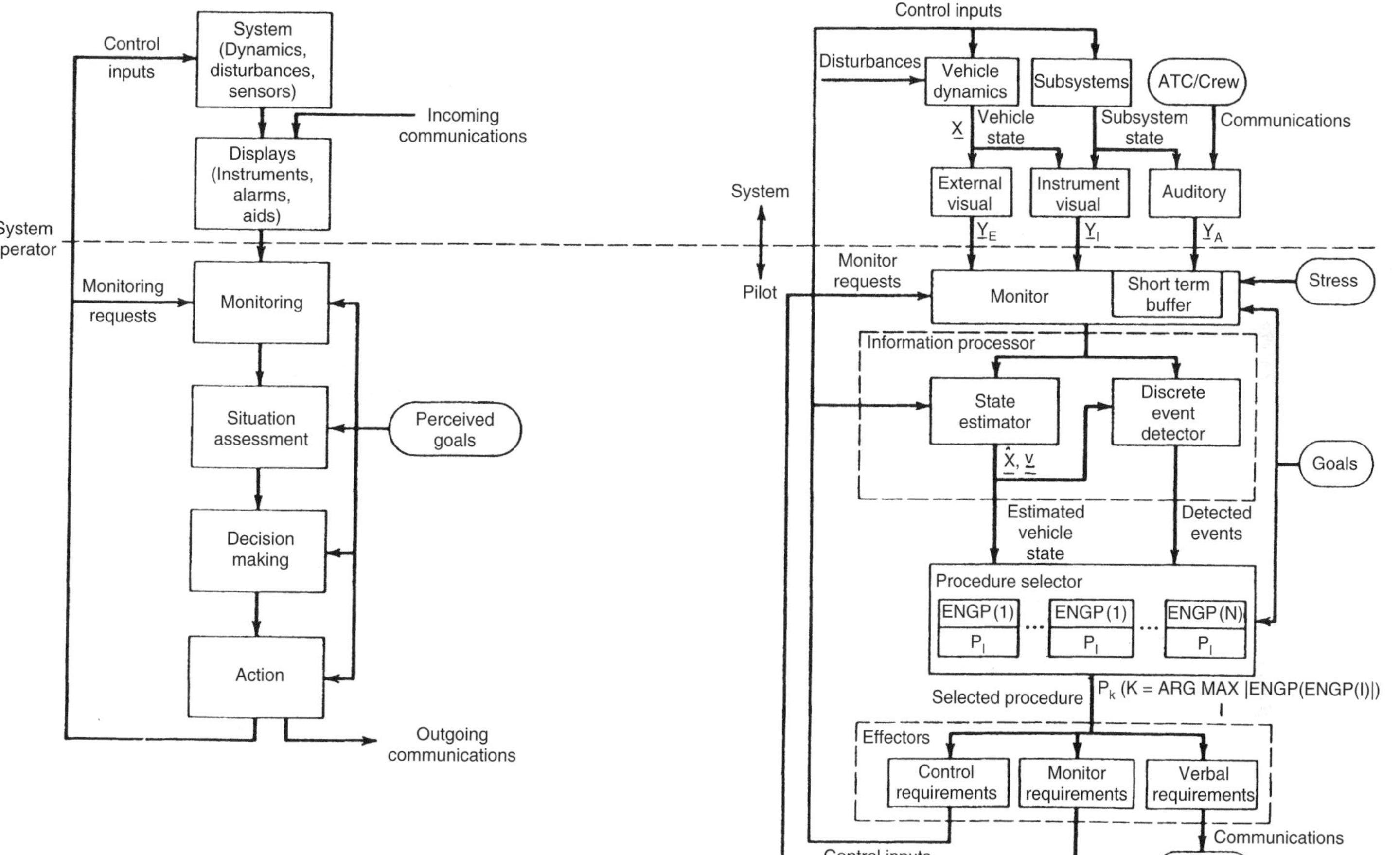

Figure 19 PROCRU model of Baron et al. (1980).

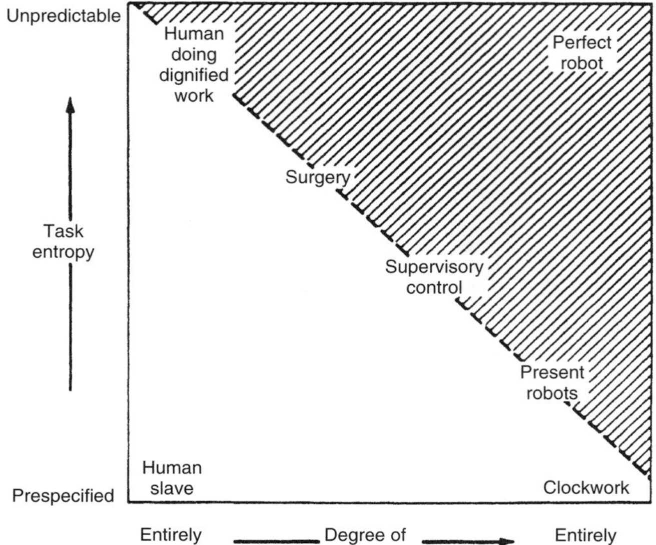

Figure 20 Combinations of human and computer control to achieve tasks at various levels of difficulty.

its alternative meanings. Below are 10 alternative meanings (stated in italics) that the author has gleaned from current literature. In every case the meaning must be qualified, as is done by the one or two sentences following each particular meaning of the phrase.

1. *Allocate to the human the tasks best suited to the human, allocate to the automation the tasks best suited to it.* Yes, but for some tasks it really is easier to do them manually than to initialize the automation to do them. And at the other end of the spectrum are tasks that require so much skill or art or creativity that it simply is not possible to program a computer to do them.

2. *Keep the human operator in the decision and control loop.* That is a good idea provided that the control tasks are of appropriate bandwidth, attentional demand, and so on.

3. *Maintain the human operator as the final authority over the automation.* Realistically, this is not always the safest solution. It depends on the task context. In nuclear plants, for example, there are safety functions that cannot be entrusted to the human operator and cannot be overridden by him or her. Examples have been given previously in the case of aircraft automation.

4. *Make the human operator's job easier, more enjoyable, or more satisfying through friendly automation.* That is fine if operator ease and enjoyment are the primary considerations, and if ease and enjoyment necessarily correlate with operator responsibility and system performance, but often these conditions are not the case.

5. *Empower the human operator to the greatest extent possible through automation.* Again one must remember that operator empowerment is not the same as system performance. Maybe the designer knows best. Don't encourage megalomaniacal operators.

6. *Support trust by the human operator.* Trust of the automation by the operator is often a good thing, but not always. Too much trust is just as bad as not enough trust.

7. *Give the operator computer-based information about everything that he or she should want to know.* We now have many examples of where too much information can overwhelm the operator, to the point where performance breaks down and even when the operator originally wanted "all" the information.

8. *Engineer the automation to reduce human error and keep response variability to the minimum.* This, unfortunately, is a simplistic view of human error. Taken literally it reduces the operator to an automaton, a robot. Modest levels of error and response variability enhance learning (Darwin's *requisite variety*).

9. *Make the operator a supervisor of subordinate automatic control system(s).* Although this is a chapter on supervisory control, it must be noted that for some tasks direct manual control may be best.

10. *Achieve the best combination of human and automatic control, where best defined by explicit system objectives.* Again, in some ideal case, where objectives can reliably be reduced to mathematics, this would be just fine. Unfortunately, automatic judgment of what is good and bad in a particular situation is seldom possible, even for a machine programmed with the best available algorithms or heuristics. Fortunately, judgment of what is good and bad in a particular situation is almost the essence of what it is to be human.

Table 1 Degrees of Automation

	Degree of Compute Aiding in Acquiring and Analyzing Data			
	Past Data	Current Data	Situation Assessment	Action Alternatives
0. Human does it all (within capability).	Measures and stores relevant past data	Observes relevant current data	Assess current system state or situation relative to goals	Conjures up various action alternatives within resource constraints
1. Computer acquires all relevant data.	Stores all relevant past data	Measures all relevant current data	Estimates current state or situation	Determines all relevant action alternatives
2. Computer displays all relevant data to human.	Displays relevant past data, trends, and so on	Displays relevant current data, relates it to past	Displays current state or situation	Displays all relevant action alternatives
3. Computer selects and displays a narrow set of data to human.	Selects and displays a narrow set of past data	Selects and displays a narrow set of current data	Assesses and displays current state or situation relative to assumed goals	Determines and displays a few most salient action alternatives
4. Computer selects and displays a single recommendation with justification.	Selects and displays only those past data sufficient to support recommendations	Selects and displays only that current data are sufficient to support recommendations	Assesses and displays only limited information sufficient to specify current state relative to assumed goals	Determines and displays a single recommendation for action

	Degree of Computer Aiding in Implementation		
	Selection of Action Alternative	Conditions Accompanying Implementation	
0. Human does it all (within capability).	(a) Decides on action independent of computer. (b) Selects an action proposed by computer.	Implements action if, when, and how he or she decides.	
1. Computer implements, with human say over if, when, or how.	(a) What human selects. (b) What computer recommended	(a) Only if and when human approves (b) By a deadline if human does not stop it in time	(a) Done jointly with human participation (b) Done by computer only
2. Computer implements independent of human.	What computer recommended.	(a) And necessarily tells human after the fact what it did (b) And tells human after the fact what it did only if human asks (c) And tells human after the fact what it did if it thinks he or she should be told	

The bottom line is that proper use of automation depends upon context, which in turn depends upon designer and operator judgment.

I have written elsewhere about the long-term social implications of supervisory control (Sheridan, 1980; Sheridan et al. 1983b). My concerns are reviewed here very briefly:

1. *Unemployment.* This is the factor most often considered. More supervisory control means more efficiency, less direct control, fewer jobs.

2. *Desocialization.* Although cockpits and control rooms now require two- to three-person teams, the trend is toward fewer people per team, and eventually one person will be adequate in most installations. Thus, cognitive interaction with computers will replace that with other people. As supervisory control systems are interconnected, the computer will mediate more and more interpersonal contact.

3. *Remoteness from the product.* Supervisory control removes people from hand-on interaction with the workpiece or other product. They become not only separated in space but also desynchronized in time. Their functions or actions no longer correspond to how the product itself is being handled or processed mechanically.

4. *Deskilling.* minu1.11pt Skilled workers "promoted" to supervisory controller may resent the transition because of fear that when and if called on to take over and do the job manually, they may not be able to. They also feel loss of professional identity built up over an entire working life.

5. *Intimidation by higher stakes.* Supervisory control will encourage larger aggregations of equipment,

higher speeds, greater complexity, higher costs of capital, and probably greater economic risk if something goes wrong and the supervisor does not take the appropriate action.

6. *Discomfort in the assumption of power.* The human supervisor will be forced to assume more and more ultimate responsibilities. Depending on one's personality, this could lead to insensitivity to detail, anxiety about being up to the job requirements, or arrogance.

7. *Technological illiteracy.* Supervisory controllers may lack the technological understanding of how the computer does what it does. They may come to resent this and resent the elite class who do understand.

8. *Mystification.* Human supervisors of computer-based systems could become mystified about the power of the computer, even seeing it as a kind of magic or "big brother" authority figure.

9. *Sense of not being productive.* Although the efficiency and mechanical productivity of a new supervisory control system may far exceed that of an earlier manually controlled system that a given person has experienced, that person may come to feel no longer productive as a human being.

10. *Eventual abandonment of responsibility.* As a result of the factors described previously, supervisors may eventually feel that they are no longer responsible for what happens; the computers are.

These 10 potential negatives may be summarized with a single word: alienation. In short, if human supervisors of the new breed of computer-based systems are not given sufficient familiarization with and feedback from the task, sufficient sense of retaining their old skills, or ways of finding identity in new ones, they may well come to feel alienated. They must be trained to feel comfortable with their new responsibility, must come to understand what the computer does and not be mystified, and must realize that they are ultimately in charge of setting the goals and criteria by which the system operates. If these principles of human factors are incorporated into the design, selection, training, and management, supervisory control has a positive future.

10 CONCLUSIONS

Computer technology, both hard and soft, is driving the human operator to become a supervisor (planner, teacher, monitor, and learner) of automation and an intervener within the automated control loop for abnormal situations. A number of definitions, models, and problems have been discussed. There is little or no present consensus that any one of these models characterizes in a satisfactory way all or even very much of supervisory control with sufficient predictive capability to entrust to the designer of such systems. It seems that for the immediate future we are destined to run breathless behind the lead of technology, trying our best to catch up.

REFERENCES

Askey, S. (1995), "Design and Evaluation of Decision Aids for Control of High Speed Trains: Experiments and Model," Ph.D. dissertation, MIT, Cambridge, MA.

Bainbridge, L. (1983), "Ironies of Automation," *Automatica*, Vol. 19, pp. 775–779.

Baron, S., Zacharias, G., Muraldiharan, R., and Lancraft, R. (1980), "PROCRU: A Model for Analyzing Flight Crew Procedures in Approach to Landing," in *Proceedings of the 16th Annual Conference on Manual Control*, MIT, Cambridge, MA, pp. 488–520.

Billings, C. S. (1991), *Human-Centered Aircraft Automation: A Concept and Guideline*, NASA TM-103885, NASA Ames Research Center, Moffett Field, CA.

Boehm-Davis, D., Curry, R., Wiener, E., and Harrison, R. (1983), "Human Factors of Flight Deck Automation: Report on a NASA-Industry Workshop," *Ergonomics*, Vol. 26, pp. 953–961.

Boer, E. R. (2000), "Behavioral Entropy as an Index of Workload," in *Proceedings of the 44th Annual Meeting of the Human Factors and Ergonomics Society*, San Diego, CA, July 30–August 4.

Brooks, T. L. (1979), "SUPERMAN: A system for Supervisory Manipulation and the Study of Human–Computer Interactions," S.M. thesis, MIT, Cambridge, MA.

Buharali, A., and Sheridan, T. B. (1982), "Fuzzy Set Aids for Telling a Computer How to Decide," in *Proceedings of the IEEE International Conference on Cybernetics and Society*, Seattle, WA, pp. 643–647.

Chu, Y., and Rouse, W. B. (1979), "Adaptive Allocation of Decision Making Responsibility Between Human and Computer in Multi-task Situations," *IEEE Transactions on Systems, Man and Cybernetics*, Vol. 9, pp. 769–778.

Chu, Y. Y., Steeb, R., and Freedy, A. (1980), *Analysis and Modeling of Information Handling Tasks in Supervisory Control of Advanced Aircraft*, PATR-1080-80-6, Perceptronics, Woodlands, CA.

Curry, R. E., and Ephrath, A. R. (1977), "Monitoring and Control of Unreliable Systems," in *Monitoring Behavior and Supervisory Control*, T. B. Sheridan and G. Johannsen, Eds., Plenum Press, New York, pp. 193–203.

Curry, R. E., and Nagel, D. C. (1974), "Decision Behavior Changing Signal Strengths," *Journal of Mathematical Psychology*, Vol. 14, pp. 1–24.

Degani, A. (2004), *Taming HAL: Designing Interfaces Beyond 2001*, Palgrave, New York.

Duncan, K. D. (1981), "Training for Fault Diagnosis in Industrial Process Plants," in *Human Detection and Diagnosis of System Failures*, J. Rasmussen and W. B. Rouse, Eds., Plenum Press, New York, pp. 553–524.

Edwards, E., and Lees, F. (1974), *The Human Operator in Process Control*, Taylor & Francis, London.

Ephrath, A. R., and Young, L. R. (1981), "Monitoring vs. Man-in-the-Loop Detection of Aircraft Control Failures," in *Human Detection and Diagnosis of System Failures*, J. Rasmussen and W. B. Rouse, Eds., Plenum Press, New York, pp. 143–154.

Falzon, P. (1982), "Display Structures: Computability with the Operator's Mental Representation and Reasoning Processes," in *Proceedings of the 2nd Annual Conference on Human Decision Making and Manual Control*, pp. 297–305.

Ferrell, W. R. (1966), "Delayed Force Feedback," *Human Factors*, October, pp. 449–455.

Ferrell, W. R., and Sheridan, T. B. (1967), "Supervisory Control of Remote Manipulation," *IEEE Spectrum*, Vol. 4, No. 10, pp. 81–88.

Gai, E. G., and Curry R. E. (1978), "Preservation Effects in Detection Tasks with Correlated Decision Intervals," *IEEE Transactions on Systems, Man and Cybernetics*, Vol. 8, pp. 93–110.

Gaines, B. R. (1976), "On the Complexity of Casual Models," *IEEE Transactions on Systems, Man and Cybernetics*, Vol. 6, pp. 56–59.

Gentner, D., and Stevens, A. L., Eds. (1983), *Mental Models*, Lawrence Erlbaum Associates, Mahwah, NJ.

Govindaraj, T., and Rouse, W. B. (1981), "Modeling the Human Controller in Environments That Include Continuous and Discrete Tasks," *IEEE Transactions on Systems, Man and Cybernetics*, Vol. 11, pp. 411–417.

Hart, S. G., and Sheridan, T. B. (1984), "Pilot Workload, Performance, and Aircraft Control Automation," in *Proceedings of the AGARD Symposium on Human Factors Considerations in High Performance Aircraft*.

Hess, R. A., and McNally, B. D. (1997), "Automation Effect in a Multi-loop Manual Control System," *IEEE Transactions on Systems, Man Cybernetics*, Vol. SMC-16, No. 1, pp. 111–121.

Hopkin, V. D. (1995), *Human Factors in Air Traffic Control*, Taylor & Francis, London.

Johannsen, G. (1981), "Fault Management and Supervisory Control of Decentralized Systems," in *Human Detection and Diagnosis of System Failures*, J. Rasmussen and W. B. Rouse, Eds., Plenum Press, New York.

Kim, S. (1997), "Theory of Human Intervention and Design of Human–Computer Interfaces in Supervisory Control: Application to Traffic Incident Management," Ph.D. dissertation, MIT, Cambridge, MA.

Kosko, B. (1992), *Neural Networks and Fuzzy Systems*, Prentice Hall, Englewood Ciffs, NJ.

Lee, J. D., McGehee, D., Brown, T. L., and Reyes, M. (2002), "Collision Warning Timing, Driver Distraction, and Driver Response to Imminent Rear End Collision in a High Fidelity Driving Simulator," *Human Factors*, Vol. 44, No. 2, pp. 314–334.

Llaneras, R. E. (2000), "NHTSA Driver Distraction Internet Forum," www.nrd.nhtsa.dot.gov/departments/nrd-13/DriverDistraction.html.

Machida, K., Toda, Y., Iwata, T., Kawachi, M., and Nakamura, T. (1988), "Development of a Graphic Simulator Augmented Teleoperator System for Space Applications," in *Proceedings of the 1988 AIAA Conference on Guidance, Navigation, and Control*, Part I, pp. 358–364.

Moray, N., Ed. (1979), *Mental Workload: Its Theory and Measurement*, Plenum Press, New York.

Moray, N. (1982), "Subjective Mental Workload," *Human Factors*, Vol. 24, pp. 25–40.

Moray, N. (1986), "Monitoring Behavior and Supervisory Control," in *Handbook of Perception and Human Performance*, Vol. 2, K. Boff, L. Kaufman, and J. P. Thomas, Eds., Wiley, New York.

Moray, N. (1997), "Models of Models of... Mental Models," in *Perspectives on the Human Controller*, T. Sheridan and T. van Lunteren, Eds., Lawrence Erlbaum Associates, Mahwah, NJ.

Moray, N., Sanderson, P., Shiff, B., Jackson, R., Kennedy, S., and Ting, L. (1982), "A Model and Experiment for the Allocation of Man and Computer in Supervisory Control," in *Proceedings of the IEEE International Conference on Cybernetics and Society*, pp. 354–358.

Morris, N. M., and Rouse, W. B. (1985), "The Effects of Type of Knowledge upon Human Problem Solving in a Process Control Task," *IEEE Transactions on Systems, Man and Cybernetics*, Vol. 15, pp. 698–707.

Mouloua, M., and Koonce, J., Eds. (1997), *Human–Automation Interaction*, Lawrence Erlbaum Associates, Mahwah, NJ.

Mouloua, M., and Parasuraman, R., Eds. (1994), *Human Performance in Automated Systems: Recent Research and Trends*, Lawrence Erlbaum Associates, Mahwah, NJ.

Nandi, H., and Ruhe, W. (2002), "On Line Modeling and New Generation of Supervisory Control System for Sintering Furnaces," Comp US Controls, Inc., Indiana, PA.

Niemala, R., and Krendel, E. S. (1974), "Detection of a Change in Plant Dynamics of a Man–Machine System," in *Proceedings of the 10th Annual Conference on Manual Control*, pp. 97–112.

Norman, D. A. (1981), "Categorization of Action Slips," *Psychological Review*, Vol. 88, pp. 1–15.

Noyes, M. V., and Sheridan, T. B. (1984), "A Novel Predictor for Telemanipulation Through a Time Delay," in *Proceedings of the Annual Conference on Manual Control*, NASA Ames Research Center, Moffett Field, CA.

Parasuraman, R. (1987), "Human–Computer Monitoring," *Human Factors*, Vol. 29, pp. 695–706.

Parasuraman, R., Sheridan, T. B., and Wickens, C. D. (2000), "A Model for Types and Levels of Human Interaction with Automation," *IEEE Transactions on Systems, Man and Cybernetics,* Vol. 30, No. 3, pp. 286–297.

Park, J. H. (1991), "Supervisory Control of Robot Manipulators for Gross Motions," Ph.D. dissertation, MIT, Cambridge, MA, August.

Parsons, H. M. (1985), "Automation and the Individual: Comprehensive and Comparative Views," *Human Factors*, Vol. 27, No. 1, pp. 99–111.

Pattipatti, K. R., Kleinman, D. L., and Ephrath, A. R. (1983), "A Dynamic Decision Model of Human Task Selection Performance," *IEEE Transactions on Systems, Man and Cybernetics*, Vol. 13, pp. 145–166.

Rasmussen, J. (1976), "Outlines of a Hybrid Model of the Process Plant Operator," in *Monitoring Behavior and Supervisory Control*, T. B. Sheridan and G. Johannsen, Eds., Plenum Press, New York.

Rasmussen, J. (1982), "Human Errors: A Taxonomy for Describing Human Malfunction in Industrial Installations," *Journal of Occupational Accidents*, Vol. 4, pp. 311–335.

Rassmussen, J., and Rouse, W. B., Eds. (1981), *Human Detection and Diagnosis of System Failures*, Plenum Press, New York.

Reason, J. T., and Mycielska, K. (1982), *Absent Minded? The Psychology of Mental Lapses and Everyday Errors*, Prentice-Hall, Englewood Cliffs, NJ.

Rouse, W. B. (1977), "Human–Computer Interaction in Multi-task Situations," *IEEE Transactions on System, Man and Cybernetics*, Vol. 7, No. 5, pp. 384–392.

Rouse, W. B. (1985), "Supervisory Control and Display Systems," in *Human Productivity Enhancement*, J. Zeidner, Ed., Praeger, New York.

Rouse, W. B., and Morris, N. M. (1984), *On Looking into the Black Box: Prospects and Limits in the Search for Mental Models*, Search Technology, Norcross, GA.

Rouse, W. B., and Rouse, S. H. (1983), "Analysis and Classification of Human Error," *IEEE Transactions on System, Man and Cybernetics*, Vol. 13, No. 4, pp. 539–599.

Ruffel-Smith, H. P. A. (1979), *A Simulator Study of the Interaction of Pilot Workload with Error, Vigilance and Decisions*, NASA TM-78482, NASA Ames Research Center, Moffett Field, CA.

Sarter, N. B., and Amalberti, R., Eds. (2000), *Cognitive Engineering in the Aviation Domain*, Lawrence Erlbaum Associates, Mahwah, NJ.

Sarter, N., and Woods, D. D. (1994), "Decomposing Automation: Autonomy, Authority, Observability and Perceived Animacy," in *Human Performance in Automated Systems: Recent Research and Trends*, M. Mouloua and R. Parasuraman, Eds., Lawrence Erlbaum Associates, Mahwah, NJ.

Sarter, N. B., and Woods, D. D. (2000), "Learning from Automation: Surprises and 'Going Sour' Accidents," in *Cognitive Engineering in the Aviation Domain*, N. B. Sarter and R. Amalberti, Eds., Lawrence Erlbaum Associates, Mahwah, NJ, p. 327.

Seiji, T., Ohga, Y., and Koyama, M. (2001), "Advanced Supervisory Control Systems for Nuclear Power Plants," *Hitachi Review*, Vol. 50, No. 3.

Sheridan, T. B. (1960), "Human Metacontrol," in *Proceedings of the Annual Conference on Manual Control*, Wright-Patterson Air Force Base, OH.

Sheridan, T. B. (1970a), "On How Often the Supervisor Should Sample," *IEEE Transactions on Systems Science and Cybernetics*, Vol. 6, pp. 140–145.

Sheridan, T. B. (1970b), "Optimum Allocation of Personal Presence," *IEEE Transactions on Human Factors in Electronics*, Vol. 10, pp. 242–249.

Sheridan, T. B. (1976), "Toward a General Model of Supervisory Control," in *Monitoring Behavior and Supervisory Control*, T. B. Sheridan and G. Johannsen, Eds., Plenum Press, New York.

Sheridan, T. B. (1980), "Computer Control and Human Alienation," *Technology Review*, Vol. 83, October, pp. 60–73.

Sheridan, T. B. (1983), "Measuring, Modeling and Augmenting Reliability of Man–Machine Systems," *Automatica*, Vol. 19.

Sheridan, T. B. (1984a), *Interaction of Human Cognitive Models and Computer-Based Models in Supervisory Control*, Man–Machine Systems Laboratory, MIT, Cambridge, MA.

Sheridan, T. B. (1984b), "Supervisory Control of Remote Manipulators, Vehicles and Dynamic Processes," in *Advances in Man–Machine Systems Research*, W. B. Rouse, Ed., Vol. 1, JAI Press, New York.

Sheridan, T. B. (1992), *Telerobotics, Automation, and Human Supervisory Control*, MIT Press, Cambridge, MA.

Sheridan, T. B. (2002), *Humans and Automation*, Wiley, Hoboken, NJ.

Sheridan, T. B., and Ferrell, W. R. (1974), *Man–Machine Systems*, MIT Press, Cambridge, MA.

Sheridan, T. B., and Hennessy, R. T., Eds. (1984), *Research and Modeling of Supervisory Control Behavior*, National Research Council, Committee on Human Factors, National Academy Press, Washington, DC.

Sheridan, T. B., and Johannsen, G., Eds. (1976), *Monitoring Behavior and Supervisory Control*, Plenum Press, New York.

Sheridan, T. B., and Verplank, W. L. (1978), *Human and Computer Control of Undersea Teleoperators*, Man–Machine Systems Laboratory, MIT, Cambridge, MA.

Sheridan, T. B., Fischoff, B., Posner, M., and Pew, R. W. (1983a), "Supervisory Control Systems," in *Research Needs in Human Factors*, National Academy Press, Washington, DC.

Sheridan, T. B., Vamos, T., and Aida, S. (1983b), "Adapting Automation to Man, Culture and Society," *Automatica*, Vol. 19, No. 6, pp. 605–612.

Taylor, F. W. (1922), *Principles of Scientific Management*, Harper Brothers, New York.

Tsach, U., Sheridan, T. B., and Buharali, A. (1983), "Failure Detection and Location in Process Control: Integrating a New Model-Based Technique with Other Methods," in *Proceedings of the American Control Conference*, San Francisco, June, pp. 22–24.

Tulga, M. K., and Sheridan, T. B. (1980), "Dynamic Decisions and Workload in Multi-task Supervisory Control," *IEEE Transactions on Systems, Man and Cybernetics*, Vol. 10, No. 5, pp. 217–231.

Wickens, C. D., and Kessell, C. (1979), "The Effects of Participatory Model and Task Workload on the Detection of Dynamic System Failures," *IEEE Transactions on Systems, Man and Cybernetics*, Vol. 9, pp. 24–34.

Wickens, C. D., and Kessell, C. (1981), "Failure Detection in Dynamic Systems," in *Human Detection and Diagnosis of System Failures*, J. Rasmussen and W. B. Rouse, Eds., Plenum Press, New York.

Wickens, C., Mavor, A., and McGee, J. (1997), *Flight to the Future*, National Academy Press, Washington, DC.

Wiener, E. L. (1988), "Cockpit Automation," in *Human Factors in Aviation*, E. L. Wiener and D. Nagel, Eds., Academic Press, San Antonio, TX, pp. 433–461.

Wiener, E. L., and Curry, R. E. (1980), "Flight Deck Automatic: Promises and Problems," *Ergonomics*, Vol. 23, pp. 995–1011.

Williges, R. C., and Wierwille, W. W. (1979), "Behavioral Measures of Aircrew Mental Workload," *Human Factors*, Vol. 21, pp. 549–574.

Wood, W., and Sheridan, T. B. (1982), "The Use of Machine Aids in Dynamic Multi-task Environments: A Comparison of an Optimal Model to Human Behavior," in *Proceedings of the IEEE International Conference on Cybernetics and Society*, Seattle, WA, pp. 668–672.

Yoerger, D. (1982), "Supervisory Control of Underwater Telemanipulators: Design and Experiment, Ph.D. dissertation, MIT, Cambridge, MA.

Zadeh, L. A. (1984), "Making Computers Think Like People," *IEEE Spectrum*, Vol. 21, pp. 26–32.

CHAPTER 39

DIGITAL HUMAN MODELING FOR CAE APPLICATIONS

Anders Sundin*
National Institute for Working Life
Göteborg, Sweden

Roland Örtengren
Chalmers University of Technology
Göteborg, Sweden

1 INTRODUCTION

Global competition and rapidly changing customer demands have resulted in great changes in production methods and the configuration of manufacturing systems. Products and production systems are rapidly and constantly changing to meet these demands, resulting in an increased number of product versions to be presented with a shorter market life. Traditional product design and production planning are insufficient to cope with these highly dynamic product development cycles variations. Against this background, virtual product and production development is becoming more important. This is especially true for vehicle manufacturing companies, where, for example, Paulin (2002) has found a significant decrease regarding costs for prototype built as well as improved verification time due to a virtual verification process. Within product design there are several areas that traditionally have used computer software [e.g., in the 1980s when computer-aided design (CAD) software promoted efficiency in the design process]. More recently, new software opened up possibilities for three-dimensional (3D) design and solid modeling, also enabling functional analysis and simulation of products and systems. The term *computer-aided engineering* (CAE) was introduced in the engineering world, and software was used for analysis and simulation (e.g., in design and structural analysis). Today, CAE includes engineering disciplines and software such as CAD, material selection/utilization, structural analysis and optimization, dynamic and kinematics mechanical system analysis, computational fluid dynamics, manufacturing simulation, and safety/occupant simulation. During the past decade, the use of CAE software has both developed and increased in the production area [e.g., within materials flow analysis, production engineering

*Present address: WM-data Caran AB, Göteborg, Sweden, anders.sundin@caran.com

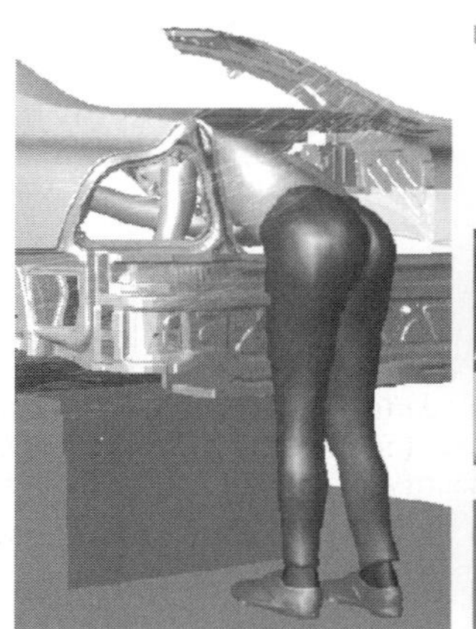
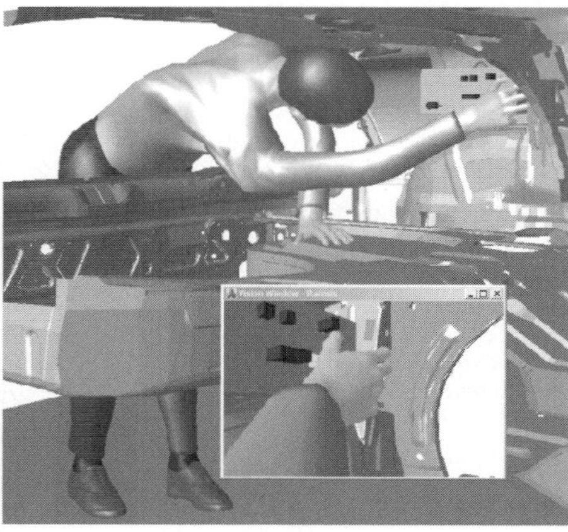

Figure 1 Typical computer manikin application (eM-Human Advanced/RAMSIS): analysis of a preliminary assembly position at the Volvo S60 Road Traffic Information (RTI) unit. (Courtesy of Volvo Car Corporation.)

(assembly sequences etc.), robot simulation, and manual assembly]. Here, most commonly, 3D computer models are used in visualization and simulation prior to two-dimensional (2D) representations. However, the introduction of computer support for visualization of layouts and solutions in workplaces design was slow to develop. Even today paper drawings or sketches, often in 2D form, are still used.

Almost all products to be developed will be used by human beings. Workers of different genders, heights, and so on, also populate most workplaces and production systems. All these individuals and their needs differ when acting as users, consumers, or workers, thus leading to the obvious demand for products, workplaces, and so on, to be designed to meet those needs, or as Porter et al. (2004) puts it: "There are ethical, legislative and financial reasons why products and services should be designed, wherever appropriate and possible, for the widest range of consumer ages, shapes, sizes, needs, preferences and abilities." This obvious fact has highlighted the need for the use of digital humans in the development of products and workplaces; thus, during the past decade, digital representations of humans have become more and more common. The development of digital human modeling (DHM) tools began in the 1960s and has gained momentum in recent years, leading to an improvement in the content, usefulness, and problem-solving ability of the software.

DHM software is used primarily for industrial purposes, but also for consumer and/or user issues. The software has a range of applications in virtual environments, but the main aim is to provide anthropometric support: analysis, measurement, and evaluation in the design of products and production, virtually analyzing reach, vision, and so on, before any physical objects are built (Figure 1).

Besides the capacity and usefulness of the tool itself, the actual utilization of the software is equally important. There are many aspects and prerequisites that must be considered in order to adapt and use the tool in a correct way (e.g., competences, processes, communication, interpretation, and documentation of results). Methodologies and simulation procedures are important, as are the methodologies for the integration of ergonomics simulation into the design and production planning processes. But above all, the use of digital human modeling software requires knowledge of both technical and human factors.

This chapter covers a description of the area of DHM, including a short historical review of how digital human models are built and their primary use. The focus is mainly on physical digital human modeling, not on cognitive and performance digital human modeling. Important aspects regarding context and organization are covered and we also provide users with advice and guidance in the implementation and practical use of the models in an organization. Finally, the advantages and limitation are touched upon.

2 DIGITAL HUMAN MODELING

The term *digital human modeling* is used throughout the chapter. But what does it involve? In the attempt to represent the complex human being digitally, functions being modeled include both physical and cognitive performance human aspects. Generally, a model is a copy or an image of something, often a miniature representation of a physical object, but it can equally well be used to replicate a function or a process. The purpose of a model is to replace the reality with a cheaper or simpler form so that the consequences of additions or manipulations of reality can be studied before any actions are decided or undertaken. A model can be an image or a physical reproduction of the real object, but it can also be a set of rules or equations that can be used to derive or calculate the response to certain input conditions. As a verb, *model* refers to the procedure of building, creating, or designing such models. In this chapter the term *model* is used both as a noun and as a verb. To help the reader recognize

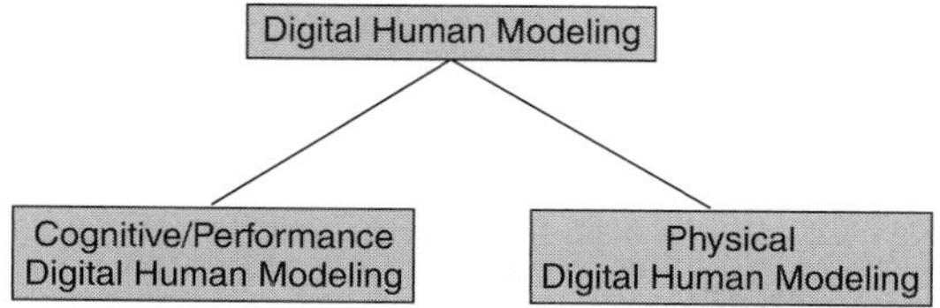

Figure 2 Division of the area of digital human modeling into two classes, cognitive/performance and physical digital human modeling.

and understand the differences in computer tools in the area of digital human modeling, the area has been divided into two major classes, cognitive/performance and physical digital human modeling (Figure 2).

2.1 Cognitive and Performance DHM

Cognitive and performance digital human modeling includes the modeling and simulation of the cognitive and performance aspects of the human being [e.g., modeling of human–computer interaction (HCI), man–machine interaction (MMI), behavioral realism, artificial intelligence and interactivity of synthetics agents*]. Aspects such as employee qualifications and skills as well as the modeling and evaluation of psychological stress are included. The simulation of cognitive performance is considered more complex than the simulation of physical performance. Target areas for this type of modeling and simulation are cognitive and performance-related fields such as emergency management, game development, law enforcement, aviation, the armed forces, and the manufacturing industry. Examples of software are Micro Saint (Micro Analysis and Design, 2004), NASA TLX (NASA, 2004), Computer-Aided Systems Human Engineering (CASHE) (Ergon, 2004b), and the Performance Visualization System (PVS), which can be used to study human workload, workplace design, and integrated human–system performance. Another software is REBA, which models and evaluates psychological stress (e.g., tiredness, monotony, and wear-out) (e.g., Pohlandt et al., 2003). Some of these modeling tools are based on logistics and manufacturing simulation tools. Compared to the area of physical modeling, the area of cognitive and performance modeling is not as well known or developed, due primarily to its complexity and abstract nature. In this chapter we do not describe the area of cognitive and performance digital human modeling any further, but instead, focus on the modeling and simulation of physical attributes. Some references for further information on cognitive and performance modeling are Pew and Mavor (1998), Zülch and Vollstedt (2000), HSIAC (2004), and the University of California Research Center for Virtual Environments and Behavior (RECVEB) (2004).

Agent is a virtual human figure representation controlled by a computer program. An agent is not to be mistaken for an a*vatar*, that is, a virtual human figure representation controlled by a live person.

2.2 Physical DHM

Physical digital human modeling encompasses the traditional domains of ergonomics: applied physiology (biomechanics in ergonomics and the ergonomics of working postures) and occupational ergonomics (work-related musculoskeletal disorders) (Eklund et al., 1997, division of ergonomics) as well as energy expenditure, vision, and safety. The human models reproduce either the anatomical shape and structure of the human body or human physiological reactions and performance under certain conditions. The models are usually portrayed in the form of linked segments and are used to identify ergonomic problems such as the effects of heavy loads on the human musculoskeletal system or difficulties in reaching or seeing objects, including the physical accommodation of large variations in people's size or strength when operating the products or systems being designed. Because of the variety of human shapes and behaviors, human models can take many different forms. Human models have been created and used since ancient times, but digital models have only existed since the advent of the digital computer. Two-dimensional and especially 3D articulated digital models have now largely replaced the previous 2D human cardboard models or templates used by designers.

In this context, two types of human models are of particular interest. The first is the biomechanical model, which makes it possible to calculate forces and moments in the joints and other body parts during exertion of work or sporting activities. These models have made it possible to understand how loads affect the body structures and how injury can result. An evaluation of such models as tools for applied research is given by Delleman et al. (1992). The available software uses either only basic graphic visualization, or none at all, which is often beneficial in the given situation. Examples of digital human models for biomechanics use specifically are 3DSSPP and Ergowatch, including 4D Watbak,† NIOSH, and Snook tools (University of Waterloo, 2004b), ErgoIntelligence Manual Materials Handling (NexGen Ergonomics, 2004a), ergoSHAPE (Launis and Lehtelä, 1992; FIOH, 2004), MADYMO (TNO, 2004), which is an engineering tool for the design and optimization of occupant safety systems, and ABBA (Ergon, 2004a), workplace and stress analysis software.

Development of the second type, the 2- or 3D structural models, are now called *computer manikins*, started in the 1960s. Software examples are Jack (UGS, 2004), SAFEWORK (Safework, 2004), RAMSIS (Human solutions, 2004), and ManneQuinPro

†4D WATBAK is a biomechanical modeling tool that calculates acute and cumulative loads on the major body joints, particularly the lumbar spine region. It can be used to estimate the risk of injury associated with a variety of occupational actions, including pushing, pulling, lifting, lowering, holding, and carrying (University of Waterloo, 2004a).

(NexGen Ergonomics 2004b). The programs available have manikins that are more humanlike in appearance and movement, use graphic visualization, and act in a computer-generated environment (i.e., environment built in the manikin software itself or in other software and then imported into the computer manikin software). Chaffin (2001) and Landau (2000) have provided reviews and descriptions of this area as well as of how this technology is being used in practice.

3 DEVELOPMENT AND GROWTH OF DHM

Human modeling software has developed from different fields: for example, human factors engineering and ergonomics consulting firms, robotics development, automotive and aerospace engineering, university research, military use, game development, and virtual reality software companies. In line with other engineering software development, human modeling software has also moved from a 2D to a 3D world, where it is possible to interact in a realistic three-dimensional virtual environment.

A large number of manikin models have been developed. For reviews, see Aune and Jürgens (1989), Landau (2000), and Chaffin (2001). Early human models to be mentioned were the First Man (later Boeman), developed for the Boeing Aircraft Company, and COMBIMAN (later CrewChief), developed for the USAF Aerospace Medical Research Laboratory. Many programs have appeared on the market [e.g., SAMMIE (System for Aiding Man–Machine Interaction Evaluation) developed in the same era as that of the aforementioned First Man and COMBIMAN, Safework, Jack, ErgoMan, RobcadMan, Anthropos, Deneb/Ergo, RAMSIS, McDonnell Douglas Human-Modeling System (MDHMS), APOLINEX, ManneQuinPro, and dV/Manikin].

Important early work was carried out in the United States by Don Chaffin at the University of Michigan (e.g., Chaffin, 1986), and included the 2D and 3D Static Strength Prediction Program, 2D- and 3DSSPP (University of Michigan, 2004a) and the HUMOSIM laboratory for movement studies (University of Michigan, 2004b), and by Norman Badler at the University of Pennsylvania, who developed the Jack computer manikin software (e.g., Badler, 1993, 2000), which forms an important foundation of today's human modeling. Badler also considered behavioral aspects. Others worthy of mention are Nadia Magnenat-Thalmann and Daniel Thalmann, who improved the realism and autonomy of the models during the 1980s [e.g., the early animations Dream flight, 1982; Rendez-vous in Montreal, 1987; Marilyn in Geneva, 1995 (Magnenat-Thalmann and Thalmann, 1991)], and Heiner Bubb, who developed the software RAMSIS together with specific car applications (e.g., Artl and Bubb, 1999).

Parallel to the development of manikins for industrial use, military authorities continuously commissioned research and funded the development of simulation tools and digital humans, digital humans that more or less acted as agents (i.e., independent of human control). A recent example of this is the virtual soldier called Santos, which is a digital human environment being developed by the Virtual Soldier Research (VSR) program at the University of Iowa (Virtual Soldier Research, 2004). Besides basic manikin features (e.g., a biomechanical model with dynamic simulation, including joint and torque analysis, vision and reach analysis, etc.), Santos also has features such as fatigue and endurance evaluation, volitional cognition and situation awareness, behavior representation and prediction, crowd simulation, object avoidance, and clothing simulation. Apart from the functions mentioned, which are suitable for military simulations, its appearance is also promising, with realistic skin deformation and contracting muscles. Santos is an example of a digital human model represented in both the areas, physical and cognitive/performance digital human modeling, as described in Sections 2.1 and 2.2.

During the last few years, a major change has taken place in the area of human modeling software and the map of the available commercial software in this field is complex, due primarily to mergers and takeovers. Some of the above-mentioned software is still available independently, although some systems have now merged. One reason for these mergers is the trend to incorporate human modeling and ergonomics analysis in design and production system simulation software in order to offer more and more complete CAE software packages. Another reason is that the interest in human modeling and ergonomics has increased, leading to a growing commercial market for developers of manikins and CAE software. Larger companies providing simulation solutions, including human models, are, for example, Tecnomatix, providing eM-Engineer, including AnyMan and RAMSIS; Dassault Systemes/DELMIA, providing the V5 platform, including V5 Human (Safework Pro); and UGS, providing E-factory, including Jack.

Terms and spelling have also changed over the past years, from several different spellings and terms (e.g., *computer-aided ergonomics system, virtual human, digital human, human model, mannequin, manequin*, and *mannikin*) to the most common names today, *human model* and *computer manikin*, or just *manikin* (supported by CEN, 2003). This is thus a simplification and refinement of the concept formation.

Extensive research and development efforts have been devoted to DHM over the years. Initially, it was necessary to focus on the software itself and how the digital human could be represented and realized. At first there were only stick figures, but these were later developed into more complex representations with many segments and joints. However, even in the early days there were also thoughts on how representation and motions affect realism. An example is the work made of Chris Landreth in the 1980s (animator at Alias/Wavefront responsible for films such as *The End* and the more recent *Bingo*), in which he explained how realism was not the same as likeness. *Realism* is when a human model physically resembles a human but is revealed due to its nonhumanlike behavior. *Likeness* means that any strange creature (e.g., a space alien) can behave like a human in motions, body language, and expressions, thus projecting an air of reality, despite its nonhumanlike appearance. To create a humanlike digital human model, it is important to aim for both realism and likeness. Efforts have been made by all software developers to create increasingly realistic human models. Good examples are the Anthropos manikin, which had an attractive and smooth appearance at an early stage and the recent Santos.

There have also been extensive developments inside the shell and under the skin of digital human models. Early models were simple, but later developed into useful analysis tools. Good examples of manikins that had an advanced structure and features at an early stage are Jack, RAMSIS, and Safework. Usually, the level of analysis available is adequate for most users. They offer a wide range of possibilities, but as a result, sometimes difficult to learn and use. Many of the analysis models included have been taken from a "paper version" and incorporated into the software. This transfer of old but accurate models into a digital tool has not always been an optimal solution. The original of, for example, RULA* (rapid upper limb analysis) was used by ergonomists in a traditional manner. Incorporation of such analysis models into DHM software leads to the risk that the user may employ the model inaccurately, not knowing its limitations, background, and so on. However, used properly, these models and tools can provide good support for ergonomics work (McAtamney and Corlett, 1993). Other analysis tools and modules incorporated in DHM software are, for example, the NIOSH† lifting equation (NIOSH, 2004), the Burandt–Schultetus analysis method,‡ and OWAS§ (Owako Working Posture Analysis System) (Tampere University of Technology, 2004).

Although industrial applications were the driving force behind the first designs, most manikins have been developed for the working environment or biomechanical applications. Researchers at institutes and university departments have developed manikins using various types of software, resulting in different solutions to the problem of how to represent the morphology and biomechanics of the human body in a computer program. Recently, software development has been taken over by large consultancy or software companies, due partly to the growing market interest and partly due to the high cost of software development, maintenance, and support. The computer manikin software available today is very advanced compared to the early versions and thus also very expensive. Computer manikins have been specially developed for use in the following branches/areas, although this list does not claim to be complete:

- Engineering (design in CAD applications)
- Aircraft industry and government agencies (design, dynamics)
- Car industry (design, crash testing, assembly)
- Textile industry (clothing design)
- Robotics (movement control)
- Medicine (rehabilitation, orthopedics, surgery)
- Sports sciences (movement studies, dance notation)
- Computer graphics (film animation, commercials)
- Workplace design
- Maintenance engineering
- Hazardous environments
- Microgravity
- Education and learning

The advanced DHM tools available today have functionality meeting most needs. Many companies use them frequently, and their use has spread and reached new users in different disciplines via other forms of CAE software, which has led to other demands. Besides the DHM tool functions, the importance of their practical application in an organization has risen. Surveys and research show that the focus of research and development needs to be shifted from functions and structure of DHM tools toward their use and how to handle them. In other words, DHM tools are adequate in terms of their functionality, but improvements and guidelines are needed on implementation and integration in a company or organization, on education, data handling and documentation, validation of the simulation process, dissemination of results, and so on (Figure 3). This shift has already started. Examples of recent studies focusing on the process within human simulation are Green (2000), Dukic et al. (2003), and Sundin and Sjöberg (2003).

4 DHM IN PRODUCT AND PRODUCTION SYSTEM DEVELOPMENT

In product development, as well as in workplace and production system design, the use of computerized visualization has increased in line with the rapid development of computerized design tools. However, the upsurge in the development of analysis, simulation, and visualization tools (i.e., CAE tools) has not only paved the way for increased efficiency and quality but also for a deeper understanding of the design process. Due to the development of CAE tools, the use

*RULA is a survey method for the investigation of work-related upper limb disorders. It assesses workers' exposure to musculoskeletal loading and allows screening of a large number of operators.

†The U.S. National Institute for Occupational Safety and Health (NIOSH) lifting equations consist of ergonomic standard methods by which lifting and carrying tasks can be evaluated in an effort to reduce the prevalence of lifting-related low back pain. They provide load limits for lifting and lowering actions.

‡The Burandt–Schultetus analysis method calculates the maximum permissible force.

§OWAS is a method for the evaluation of postural load during work and is based on a simple and systematic classification of work postures combined with observations of work tasks.

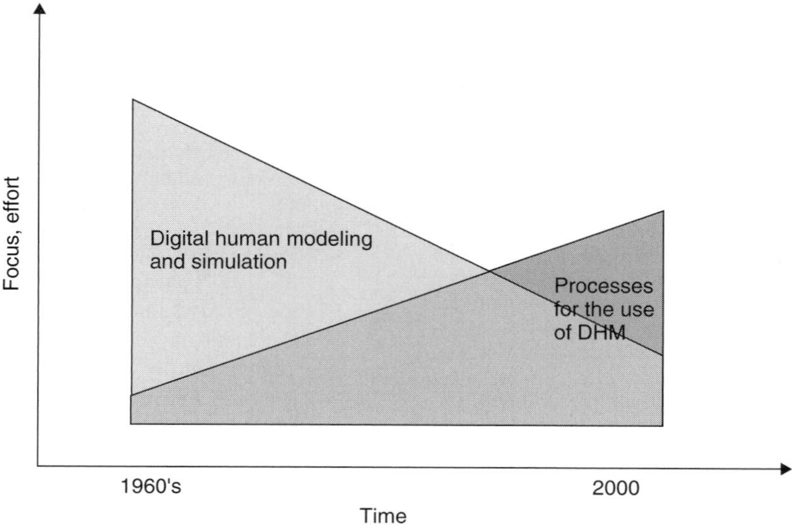

Figure 3 Shift toward usage. During recent years the focus and efforts of the research and development on digital human modeling have shifted toward how to use them rather than how to build them, due to the fact that previous development focused on the functionality of the software at the expense of how an organization or company should best employ it.

of computerized ergonomics tools has become more common.

Figure 4 illustrates the use of visualization and ergonomics simulation in the product life cycle (PLC). PLC is described schematically with linear activities, but in reality each activity is more or less carried out in parallel. It is also commonly recognized that the costs associated with the entire PLC increase the later simulations and corrections are carried out. Although visualization is in widespread use in all phases of PLC today, ergonomics simulation is not. Ergonomics simulation is mainly in use early in the PLC, but also increasing in the later phases, especially in service and maintenance applications.

DHM tools permit verification of whether or not a task is ergonomically acceptable by predicting and analyzing reach, clearance, fit, loads, and line of sight, and thus, in every step of the product realization process, can help predict how well the dimensions of a product or production system will fit the body dimensions of future users. As almost all products are used or handled by people in various ways, digital human modeling tools are of great importance today. The same is also true for workplaces and production systems where products are manufactured (i.e., people are involved in production, assembly, etc). All these people have different needs, whether acting as users, consumers, or workers, thus leading to the obvious demand that products and workplaces should be designed to meet those needs. As product realization of today, including both product and production development, involves more and more digital tools, there is thus a great need for functional, accurate, and user-friendly digital human modeling tools.

DHM tools can be applied in a wide range of areas within product realization, such as product design, crash testing, workplace design, industrial engineering (e.g., packaging, assembly path analysis, assembly), and maintenance. The tools are most advantageous when no physical environment is available (when no product, prototype, or mock-up is built) or when the environment is inaccessible or hazardous (spacecraft, nuclear power plants, etc). However, the tools are also useful when an environment exists (e.g., testing changes in a workplace) or a complement to physical attributes, as identified by Sundin et al. (2003), among others. In a case study aimed at productivity and ergonomic improvements at an early stage in the design of a new bus chassis, they identified benefits despite the availability of physical prototypes. For example, fast digital redesign and testing could be made in a rapid development process, and tests could be made when the chassis was repositioned and rotated, which is not easily done with heavy physical prototypes. The product realization tools available today are used primarily for sophisticated anthropometric and biomechanical analyses, where they fulfill their purpose adequately, despite lacking some features. They are less suitable for cognitive analysis, which has also been less in demand by users until recently. However, the need for, as well as interest in, such analyses is on the increase, and there are other tools available and under development for that purpose (see Section 2.1).

The evaluation of manual work and manual operation of plant and machinery is commonplace in both product and production system development. In addition to postural aspects, visibility is crucial in, for

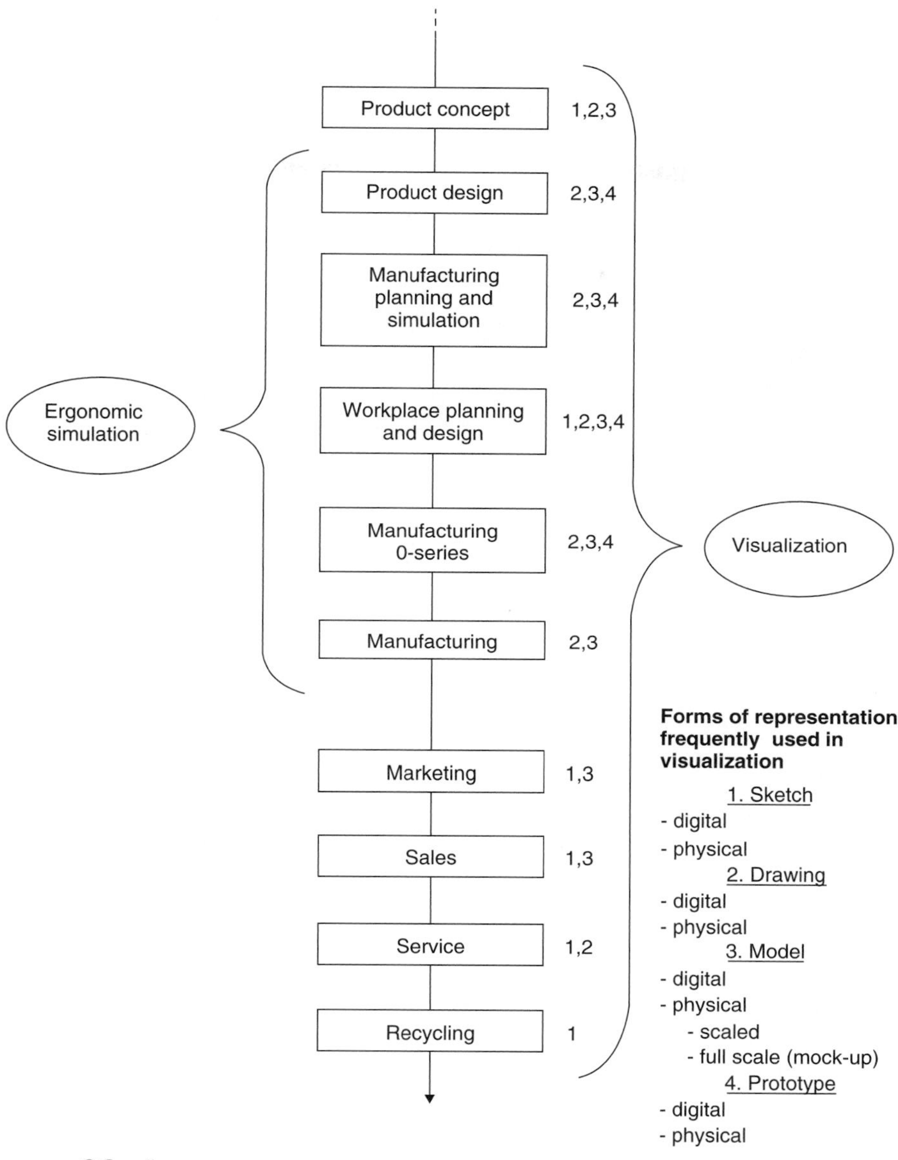

Figure 4 Main use of ergonomic simulation and visualization and its various representational forms in the product life cycle. (From Sundin, 2001.)

example, the operation of machines and controls in vehicles, and in assembly, maintenance, and supervision. In computer manikin software, vision modules are included composed by a field-of-view (FOV) function and a function allowing the user to see through the eyes of the manikin (Figure 5). However, the visibility functions available today have limitations, as pointed out by Hudelmaier (2002) and Dukic (2003) and described further in Section 5.3.2.

The manikin software available at present is limited to cases with a single worker. As assembly processes often require several workers performing independent and simultaneous work and in some cases acting together in synchronized tasks, there is a need for developments in this area. Some software can populate the virtual environment with several manikins and there have also been attempts to provide multimanikin coordination and communication (e.g., Toma et al.,

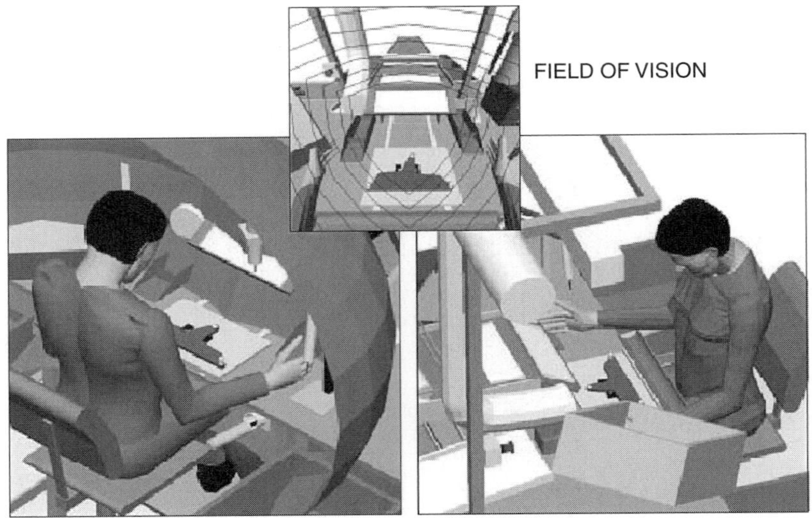

Figure 5 Analysis functions (ENVISION/ERGO) Top: the field of vision, where lines represent the range of vision from close to peripheral. To the left: a reach envelope depicting points that can be reached by the right hand, where all equipment and details within the envelope can be reached comfortably without the need to lean. (Courtesy of Delmia.)

2003) in which independent or synchronized tasks are carried out in the same work cell where the field of view module is used to coordinate several manikins.

5 PHYSICAL DHM AND COMPUTER MANIKINS

5.1 Anthropometrics and Anthropometrics Design

It is the designer's job to design products in such a way that they fit the variability of the target population. Data on the size and shape of the human body are provided in anthropometrics tables, handbooks, and standards (e.g., Van Cott and Kinkade, 1972; Pheasant, 2001; and European and International CEN and ISO standards). The tables give percentiles, sometimes also means and standard deviations, of standardized body measurements such as stature, eye height, elbow height, height when seated, arm reach, and circumference which can be used by designers in adapting equipment and workplaces to the users. The anthropometric measurements in the tables are given independently of each other, which means that no information on correlation between the variables is provided. Therefore, it becomes difficult to base the design on more than one parameter, as it is not possible to tell how large a percentile of the target population the design will accommodate. A design should fit the variability of the population from a 5th percentile woman to a 95th percentile man. The critical dimensions are those concerning clearance and reach, and in most situations, the products should be adjustable. Important dimensions in emergency situations should be based on the 1st and 99th percentiles.

When the design is to be adapted to two or more body measurements at the same time, different approaches are possible depending on which data are available. If only tabular data without correlations are available, one possibility is to combine the same percentile values, although combinations of percentiles can lead to mistakes, as pointed out by Haslegrave (1986). Another possibility, which is not correct either, is to start with 50th percentile values for all measurements of interest. One parameter is then adjusted to the percentile of interest and the others are adjusted by scaling. In both cases it must be remembered that persons having these measurements do not really exist and it remains unclear how large a share of the population is accommodated.

Quite another approach is to use combinations of selected percentiles to generate body types for the fit analysis. The 5th, 50th, and the 95th percentiles correspond to a small, medium, and tall person, respectively. Corpulence can be handled in the same way, thus including thin, average, and high corpulence as well as the relation between length of leg and trunk (e.g., short legs and long trunk, medium legs and medium trunk, and long legs and short trunk). A selection of combinations can be used in the analysis of the design. These problems are also discussed by Dai et al. (2003), who describe three methods: the single-parameter method, the multiple-parameter method, and the boundary manikin method. An illustrative discussion on the variability of anthropometric measurements and proportions for human modeling has been presented by Greil and Jürgens (2000). Porter et al. (2004) have suggested using a database and saving the information about each person as a complete data set, used together with another program, named HADRIAN (Human Anthropometric Data Requirements Investigation and Analysis). HADRIAN is a

CAD tool that integrates the database of persons, including their anthropometry, mobility/capability, disability, coping strategies, and a wealth of background data, with a task analysis tool.

Most anthropometric databases are based on surveys that were carried out a long time ago; thus, the information is no longer up to date and does not reflect the anthropometrics of current populations. Therefore, new surveys have been initiated for both general use and for specific purposes (e.g., the CAESAR study, the aim of which is to establish new databases based on body scanning of large numbers of American and European men and women) (Robinette et al., 2002). A survey based on scanning is also being planned in Sweden, where car manufacturers and the clothing industry are the interested parties. Data from these new surveys are stored in databases where all values pertaining to one person are kept together. This leads to entirely new possibilities to analyze data and create representative body types for equipment design (e.g., Rioux et al., 2004; Robinette, 2004), now referred to as *3D anthropometry*.

5.2 Biomechanics

A computer manikin consists of a number of segments, which are linked through articulated joints. The number of segments is a compromise between the high number required to obtain high accuracy and the low number necessary for ease of handling. As illustrated in Figure 6, the design can usually be viewed in different ways. The individual segments are rigid and have defined physical properties (mass, dimensions, point of gravity coordinates, moments of inertia, etc.). The joints are designed to resemble human joints, and although simpler, they have a similar range of motion restrictions. The surface is represented by a wire frame that is programmed to have a certain flexibility so that it can stretch and shrink, forming a smooth surface when the segments move in relation to each other. When the structure and the physical properties are defined, a set of static or dynamic equations can be established to form a biomechanical model that allows calculation of the external as well as internal forces and torques acting on or between the body segments. This is used in the evaluation modules of most computer manikin software. In principle, it is possible to design very detailed biomechanical models in which individual muscles and ligaments are represented. Since the number of variables increase faster than the number of equations to calculate them, an indeterminate condition arises. To some extent this can be rectified by applying additional criteria, such as that for load sharing between muscles. But even then, the degree of confidence in the accuracy of the results depends on the possibilities of validating the biomechanical models and the results calculated. This should be performed in specially designed experiments in which the calculated results are compared to measurements, something very difficult to achieve without hurting or injuring the subjects. Due to the problem of validation, it is vital that the scientific basis of the biomechanical models incorporated in the computer manikin software is specified accurately.

Simulating human postures and movements is a very difficult problem, owing to the complexity of the human musculoskeletal system. Problems have arisen in this area due to the fact that biomechanical modeling does not permit inclusion of the large number of DOF associated with various joints. The determination of various joint angles when a working posture has to be modified is also problematic. Models involving many joints are difficult to handle, especially if accuracy is important. Thus, posture and motion prediction

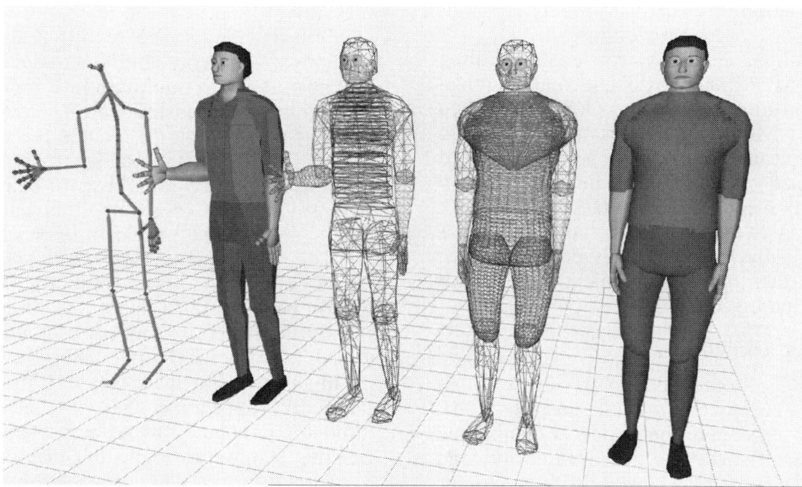

Figure 6 Representations of a computer manikin (Jack); the linked segments are visible on the extreme left, in the wire frame third from left and in the shaded frame second from right. (Courtesy of H. Sjöberg, Chalmers University of Technology.)

methods that can support realistic ergonomics analysis are the most important function in digital human modeling and simulation. However, although most human motion prediction methods take anthropometry into consideration; they nevertheless fail to deal adequately with the task at hand.

At present there are different ways of achieving accurate posture and motion prediction. There are two central schools of thought regarding posture prediction. The first uses anthropometrical data, collected from experiments using human subjects, which are statistically analyzed to form a predictive model for posture, such as various regression models. This is referred to as *empirical statistical modeling*. The second school of thought often uses biomechanics and inverse kinematics as a predictive tool for postures that have been estimated, as opposed to observed, as being a likely posture for a task.

The first school, which uses the above-mentioned motion capture technique, includes Thalmann and Thalmann (1990), Chaffin et al. (1999), Park et al. (2002), Wang (2002), and Faraway (2003). In their recent work, Park et al. (2003) start from a user scenario and search for *root motions* in a motion database, a database consisting of human motions collected by means of motion capture that closely matches the new scenario. Thereafter, the root motions are divided into groups, and one motion from each group is chosen and adapted to meet the requirements of the input scenario, thus providing the movement for the new scenario based on similar real human movements. In this approach, the predicted movements for the digital human are based on real human movements, which appears to be the correct way to solve the problem. However, the drawback of such an approach is the need for a quantitative and qualitative database or collection of real human movements. Artl and Bubb (1999) employed a somewhat different approach consisting of precalculated target-directed movements, by means of a combination of statistical probability models with dynamic movement restrictions.

The second school includes, for example, Jung et al. (1995) and Mi (2004), who use mathematical modeling of human motion based on well-established kinematics theories. Mi employs a task-based approach simulating human reaction to and performance of different tasks, based on the fact that humans perform differently in response to various tasks. Each task comprises a number of human performance measures that are mathematically represented by cost functions. Cost functions are then optimized subject to a number of constraints, including joint limits.

5.3 Ergonomics Analysis

The purpose of an ergonomic evaluation of a designed product or piece of equipment, a work space, or a task is to ascertain to what degree the design is compatible with human characteristics, abilities, limitations, and needs. An evaluation can address different topics and areas: physical, physiological, psychological, and social. As mentioned above, in their present state, computer manikins are best suited for the evaluation of physical and to some extent physiological conditions. According to Chaffin (2001), the focus is mostly on fit, clearance, and line-of-sight issues, followed by human motion analysis, which is less common, due to its less humanlike results. An evaluation can be made by using methods with different degrees of sophistication, but evaluations based on the judgment of an experienced ergonomist should not be disregarded since they are important when more formal methods are either too specific or lacking. Computer manikin software allows the modeling and evaluation of human activities by means of an evaluation scheme. The 3D representation should permit an experienced ergonomist to assess whether or not requirements concerning space, form, reach (see Figure 5), distance, access, and posture have been fulfilled. Vision can also be evaluated based on the judgment of an experienced ergonomist. When appropriate models are available, the evaluation can also include static and dynamic biomechanical load analysis, and results can be compared to values recommended for acceptable load and either be approved or disapproved. In principle, it should be possible to evaluate short-term effects (fatigue, pain) as well as long-term effects (overload injuries, repetitive strain injuries). Other ergonomic aspects can be evaluated by means of either judgment/assessment or the application of a suitable method, provided that they can be visualized or otherwise represented by computer manikin software.

Although the computer manikins available on the market today are similar, they each have specific uses (e.g., design and testing of car interiors and visualization, animation and simulation of manufacturing systems, and materials handling). According to a survey from 1996 in Chaffin (2001), which listed the most desirable features in future human simulation software, designers' preferences are (1) being able to include different anthropometric data sets and population subgroups; (2) being able to include a variety of clothing, gloves, and helmets; (3) being able to predict the strength and endurance of different populations in a given task; (4) being able to simulate realistic motions and postures with minimum task input descriptions, in both physically constrained and unconstrained conditions; (5) being able to include the provision of hand grip, strength, and visual sight lines with and without mirrors and obstructions; (6) being able to include the provision of task time line analyses; (7) being able to include the provision of reach and fit analyses for a variety of conditions; and (8) achieving seamless acceptance of input–output commands and data, and/or being able to apply the program within various CAD systems commonly used when designing and specifying products, tools, and workstations.

In general, the necessity for manikins to have a humanlike appearance has not been a central issue, although it appears to be important to achieve successful communication with other departments, managers, and so on. Thus, the ongoing increase in collaboration between departments, companies, and countries in global product development highlights the need for more humanlike computer manikins. Lämkull

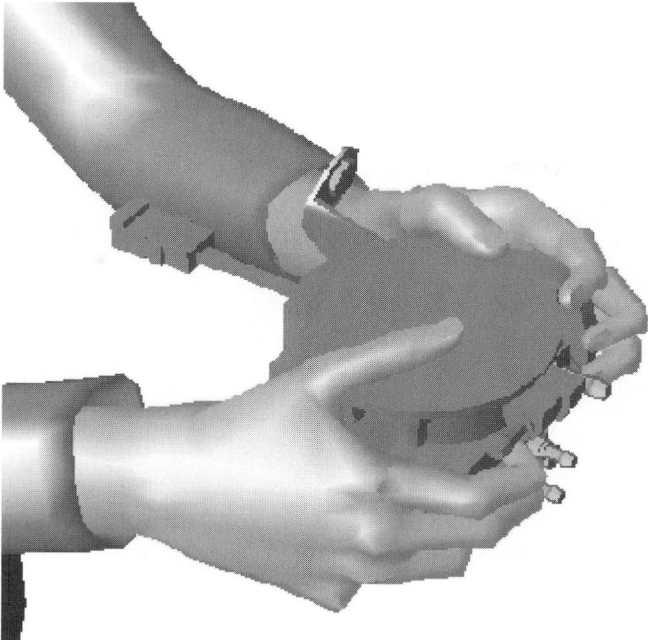

Figure 7 To facilitate analysis, auto grasp (RAMSIS) is a function where the fingers close automatically around an object. Similar function is body posture wizards that quickly identify a body position, which usually acts as a starting point. (Courtesy of J. Nordling, Tecnomatix.)

et al. (2005) investigated how the appearance of four manikins affected body posture assessment. Manufacturing engineering managers, simulation engineers, and ergonomists took part in the study. Results show that different appearance modes affect the ergonomic judgment. A more realistic looking manikin is rated higher.

5.3.1 Manipulation Strategies

When using computer manikins in ergonomics analysis, posture and movements can be controlled in different ways. The simplest but most tedious method is to control each degree of freedom (DOF) by means of the mouse or keyboard. To create an animation, a series of postures has to be produced and assembled as film frames. The distance in time between frames can be quite large, as it is possible to calculate the postures for the frames in between through interpolation. Because of the time it takes to make postural adjustments, various more or less automated means have been developed (e.g., for posture and grasp) (Figure 7).

Motion capture is another manipulation method, where the recorded movements of a real person control the segments of the computer manikin (Figure 8). Thus, the manikin reproduces the movements of a real person in real time (see, e.g., Speyer et al., 2002). Sensors are put on landmarks on a real human (i.e., on the forehead, elbows, knees, etc.). Each marker is located in space (position and sometimes also orientation) and connected to the system via cables or an infrared (IR) technique. The body sensors can also be complemented with a glove for detailed registration of hand and finger movements. An advantage of motion capture is its generation of very realistic movements. On the other hand, to ensure that the recorded situation is sufficiently realistic, good-quality physical equipment is required.

Another way of controlling movement is to develop a programming language for *scripted animation*. The user writes a script similar to a programming language. The script can be low level, where the user writes coordinate data for points where different objects will pass at a certain time. A high-level language allows complex movement specification such as sit, stand up, or reach for an object. These movement-related commands are then translated into coordinate time data.

5.3.2 Vision Analysis

Vision is perhaps the most important sense when carrying out different manual tasks. The vision is used to give information about the body's position and orientation in relation to objects that are to be handled or manipulated and to targets that are to be reached through loco-motor activities. Although certain tasks can be carried out without visual control, it is normally necessary to work under full visual control. Therefore, it is also necessary to evaluate visual demands when designing equipment and workplaces. In computer manikin software, vision modules are included composed by a field-of-view (FOV) function, often representing a person's field of vision as view

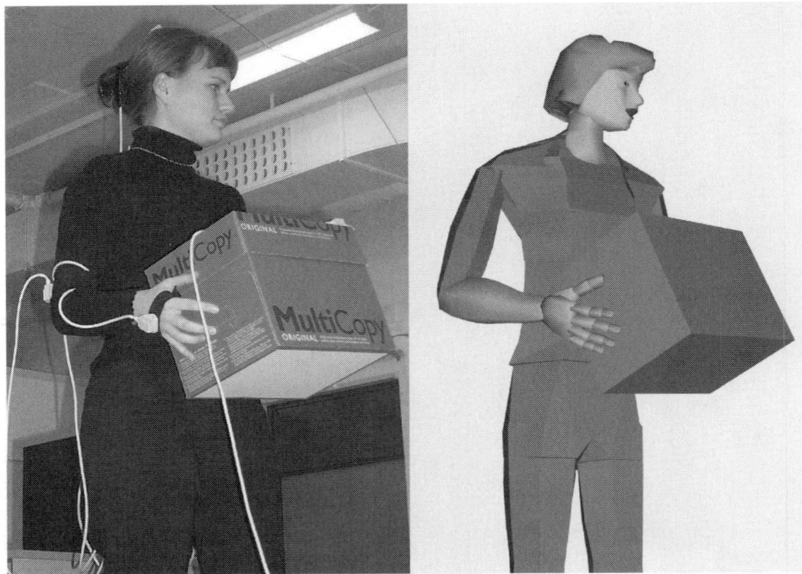

Figure 8 Motion capture system Ascension MotionStar/Jack attached to a person whose movements drive the computer manikin. (Courtesy of H. Sjöberg, Chalmers University of Technology.)

cones, view angles, or lines of sight (see Table 1), and a function allowing the user to see through the eyes of the manikin (see Figure 5).

By comparing the knowledge about the human visual system and the computer manikin software available on the market, limitations can be identified [e.g., fields of view do not consider and represent clearly the different properties of the different fields of view (sharp vision in foveal area and blurred vision in peripheral area)]. The visual fields of the computer manikin are represented differently in different computer manikins. Table 1 shows how the field of view is indicated in Ramsis, Jack, and Safework. The opening angles and the level of acuity of the central, middle, and outer areas differ from one computer manikin to another. The optimum area can be considered as the cones area of human eye and the maximum area as the rods area in the retina.

By means of the vision module, it is possible to show lines of sight from the manikin's eyes and central viewing fields to a certain angle. In most software it is possible to see through the manikin's eyes, monocular or binocular, display the viewpoints, and get direct impressions of which objects in the viewing field are seen and which are obscured. In today's computer manikin programs the algorithms display the field of view dynamically so that the field of view changes and follows the position of the manikin's head. Possibilities for vision through a mirror can also be provided. By means of a mirror module, it is possible to define at some location and then get a display of the mirror picture. Using this, the effect of body size on the field of view in the mirror can be evaluated and the need for seat adjustment determined.

5.4 Standardization

Computer manikins have also been subject to standardization. At present three organizations are working on computer manikin standards. Organized by the Society of Automotive Engineers International (SAE), the G-13 group is developing standards to facilitate human factors assessment in equipment design. The group consists of industrial scientists and designers, who also sponsor the annual Digital Human Modeling for Design and Engineering Symposium. The European Standards Organization CEN and the International Standards Organization (ISO) have several working groups devoted to the development of standards for ergonomics, biomechanics, anthropometry, and computer manikins, and the organizations are working together on some standards. The first standard on computer manikins, EN ISO 15536 (*Ergonomics: Computer Manikins and Body Templates*, Part 1, *General Requirements*), was issued jointly by CEN and ISO. The standard does not specify how computer manikins should be designed in view of the fact that the development is not deemed to have reached the necessary level. Instead, the standard sets requirements for specification of computer manikins and the connected anthropometric databases. Other standards on computer manikins and standards relevant for the use of computer manikins in design of equipment and workplaces are under development.

6 CONTEXTUAL AND ORGANIZATIONAL ASPECTS OF DHM

Digital human modeling and simulation provide a powerful tool for the analyst. However, knowing

Table 1 Vision Module Characteristics for Three Computer Manikin Software Packages

Computer Manikin	Vision Module Characteristics	Examples from the Software
Ramsis Version 3.7	Three areas of vision and acuity are defined: • *Sharp sight area*: opening angle ± 5° • *Optimum sight area*: opening angle ± 15° • *Maximum sight area*: opening angle ± 50°	
Jack Version 4.0	It is possible to define view cones, one for each eye: view cones length (default value 200 cm), view cones angle (default value 40°). Angles and length have no limitations. Visual fields defined by: • *Eye point*: both, right, or left • *Type*: achromatic, green, blue, yellow, red, and blind spot (achromatic field represents the peripheral vision)	
Safework Version 5.11	Four types of vision are defined: binocular, ambinocular (total field of vision seen simultaneously by both eyes), monocular left, monocular right, and stereo. For each type of vision one can define the field-of-view characteristics (horizontal monocular, horizontal ambinocular, vertical top, and vertical bottom angles). Visual characteristics are displayed as peripheral cones, central cones, blind spot cones and central spot cones	

Source: After Dukic (2003).

that these tools are complex with many functions and possibilities also places high demands on the user. Another central aspect is how these simulations are processed and organized within the organization, as pointed out by Peacock et al. (2001). There are four main aspects that should be highlighted, further elaborated in the following sections: (1) requirements on competence and contextual knowledge, (2) how the software is implemented and integrated in the organization, (3) what formal processes and methods are employed, and (4) how the organization uses the visualization techniques in a participative and collaborative environment.

6.1 Competence and Contextual Knowledge

Competence and contextual knowledge are vital factors for successful human simulation. Rönnäng et al. (2003) discuss one of these aspects: how users with different backgrounds interpret the ergonomics simulation. In their study they asked one group of production engineers and one group of ergonomists to analyze a workplace presented in the form of a computer manikin simulation. The results differed between the groups on, for example, the problems identified and how to solve them (see Section 7). In Dukic et al. (2002, 2003) the contextual knowledge is considered very important for successful ergonomics simulation (i.e., the simulation engineer requires some contextual knowledge of the area that he or she simulates, while the ergonomist needs some technical experience, including familiarity with the software used). The latter is also relevant for engineers, as they must know what can and cannot be simulated and analyzed with a DHM tool. Lack of certain competencies can sometimes be compensated by using a participatory process. See Section 6.4 for more information on participatory processes and methods.

6.2 Implementation and Integration

The initial problem posed by human simulation tools is to implement and integrate them in a company to ensure their accurate and efficient use. If a tool is bought without a predetermined purpose and without introductory courses and training of personnel, there may be problems (e.g., ramp-up may take a long time, due to lack of training), and if only one person is assigned to its operation, difficulties may arise when he or she is absent. There should be a plan for integrating the DHM software with other existing systems [e.g., databases and product data management (PDM) systems]. In a larger corporation, there may be a need for an investigation of the type of human simulation tools used in different departments or different countries. Blomé et al. (2003) give an example of such a survey, which investigated tools, information flow, and procedures for human simulations between departments at Saab Automobile. The results highlighted the need for better adaptation of the tools to the organization, and vice versa, in addition to the necessity of integrating working procedures and methodology to ensure efficient use.

6.3 Processes and Methods

It is becoming more and more obvious that the main barrier to more cost-effective and correct use of digital human modeling is the lack of work procedures and processes. Methods and instructions are often lacking on, for example, how to initiate a simulation (i.e., correct instructions from the person who requests the simulation), how to carry out the simulation (e.g., construction of the virtual environment or choice of posture or manikin type), and how to document and communicate the results.

The goals of or reason for a simulation can vary but are often not clearly expressed by those who request a simulation from a simulation engineer. Common goals, expressed in Dukic et al. (2002), are (1) to identify a problem, (2) to illustrate a problem, (3) to find a solution, and (4) for learning purposes. If the goals or reasons are not stated or documented, simulation tasks and results may be useless. The reason for having structured methods supporting the actual analysis made by the simulation engineer is to facilitate validation of the results. Thus, the person who carries out the analysis will be less critical (i.e. the results should be the same independent of the person who does the job). Moreover, the level of traceability and reusability of previous simulations is low. A simulation engineer usually saves his or her work on his or her own disk area and sometimes also deletes the simulations after some time. This creates problems for colleagues and makes it impossible to reuse the results in future projects. If results after all are documented and saved, there is often information missing, such as why a simulation was made, why a solution was taken, and the discussion and decisions that led to this solution.

Today, developers of DHM tools, researchers, and users in industry are striving to find effective and synergetic ways to handle these problems. Important approaches on a holistic level are both the seamless integration of different software packages and increased use of process management software, where company product and process data are gathered and used jointly by different persons within a corporation, irrespective of geographic location. Geyer and Rösch (2002) as well as Schneider (2003) give examples where the functions include product assembly sequencing, process and production planning, production performance analysis, and workplace analysis. All relevant and updated information, including the bill of materials, parts geometries, tasks to be carried out, and workplace configuration, is stored and used for different purposes. For example, a manufacturing engineer can design the layout of an assembly line together with a conceptual design of equipment, while an ergonomist can optimize the work system. In such a system, standard postures and movement sequences can be stored together with the planning data, accelerating the process of simulating human operations. A challenge in achieving an effective process management system is to establish a suitable level of geometric detail of visualization for different purposes. A high level of detail is needed for precise measurements of an object, resulting in a large file for the system to handle, whereas the level of detail of objects used for a dynamic collision check, for example, can be lower, thus enhancing system performance.

Another important step is to structure the simulation itself in addition to the pre- and postsimulation steps (i.e., steps related to the actual simulation), where different persons are involved (e.g., simulation engineers, production engineers, and designers). Such protocols and databases have been developed previously (e.g., as described in Green, 2000; Sundin et al., 2002b); Wartenberg et al., 2002; (Hanson et al., 2004). In the latter example, a Web-based system aimed at formalizing the simulation process for global use within a corporation was developed. This system consists of three major phases: background/order, method, and results/discussion. In background/order, the person who requests the simulation (e.g., a production engineer) describes the aim and purpose together with the desired output and results. In this phase, the person who requests the simulation collaborates closely with the simulation engineer. This step results in the formal order. The following step, method, formalizes actual use of the tool and is decided and handled exclusively by the simulation engineer before and during the simulation or, if necessary, together with the person who requests the simulation. At this stage the types of manikins and the environment files are specified, together with a detailed description of the task, and its limitations, to be performed by the digital human in the simulation. A hierarchical task analysis can be carried out as a support, which is also recommended by Sundin et al. (2002). The final section, results/discussion, documents a presentation and discussion of the results. Information from all three steps is stored in a searchable and printable database. Protocols and databases are a great help for handling a simulation and are sometimes also a necessity in large simulation projects to prevent a number of faults and misuses, an aspect also described by Ziolek and Nebel (2003). Moreover, protocols and databases can have positive effects, such as speeding up the learning process for new tool users, reducing the differences in results between users, and spreading an awareness of ergonomics within a company, in addition to saving and keeping track of ergonomics problems and solutions over time, thus making the organization less independent on specific employees.

6.4 Participatory Ergonomics and Collaborative Design

The participatory ergonomics method is a way of improving the final result in, for example, workplace or product design by involving people who are familiar with the work process or use the workplace or product. The use of participatory ergonomics has evolved from different cultures and backgrounds. Cases with participatory approaches were used in Europe during the early 1980s. The approaches were, however, defined differently in different cases (Wilson and Haines, 1997).

One reason for involving people who are familiar with the work process or use the workplace or product is that they have unique knowledge and experience, which is not normally taken into account. When interviewing ergonomics consultants about their interventions toward reduced work-related musculoskeletal disorders, Whysall et al. (2004) noted that little attention was paid to employees' knowledge and attitudes. Central to the participatory ergonomics work is the creation of a group or team made up of participants with different backgrounds (e.g., workers, production engineers, production managers, physiotherapists, and perhaps also researchers). The involvement also give rise to solution ownership within the group, leading to a commitment to change, which usually means that the ensuing changes are carried out more quickly and easily. Furthermore, it can be regarded as a learning experience, as it promotes an interest in and understanding of ergonomics (Wilson and Haines, 1998).

When using participatory ergonomics there are some difficulties to be aware of. It may be a process difficult to initiate and support [e.g., due to (mid) management resistance as well as lack of time and motivation of those involved]. Moreover, consensus does not necessarily lead to the best solution. Participatory ergonomics is also seen as a slower and more complex process, as well as in some cases giving rise to unrealistic expectations: for example, in terms of what and how much will be changed (Wilson and Haines, 1998). Furthermore, both the commitment of those involved and the understanding of the proposed solutions are important in participatory ergonomics. Kuorinka (1997) holds that conventional tools, tables, drawings, CAD, and so on, are too complex and may not be applicable in a participatory context. Instead, he suggests that a hands-on approach is needed, as some participants find abstract and conceptual issues difficult to understand, especially at the start of the process. An opposing view is that the visualization and simulation tools available today within the area of CAE in general, and in the area of digital human modeling in particular, is most suitable for boosting the use of participatory ergonomics, thus eliminating the grounds for criticism. This view is put forward by Söderman (1998), Sundin (2001), Dukic et al. (2002), and Johansson (2004).

It should also be added that both what is understood as participatory ergonomics and how it is implemented may vary somewhat. An important framework for the definition and use of participatory ergonomics has been proposed by Wilson and Haines (1998); complements have also been suggested (Sundin, 2001). The framework has now been validated and consists of the following nine dimensions: permanence, involvement, level of influence, decision making, mix of participants, requirement, focus, remit, and role of ergonomics specialist (Haines et al., 2002).

Traditionally, participatory ergonomics has focused on the production process and the workers' involvement in the development of new workplaces or the improvement of existing ones. However, there are few studies available that focus on how to implement participatory ergonomics in the early steps of the product life cycle (PLC), which includes all phases from product concept to recycling (see Section 4), and on the involvement of product designers. With the aim of achieving more effective and ergonomic production systems, a new concept of participatory ergonomics has been proposed by Sundin (2001), participatory ergonomics design (PED). PED was introduced was to encourage the use of participatory ergonomics in the product design phase, which is not normally the case today (Figure 9).

Collaboration in product development projects (i.e., collaborative engineering), includes all parts of the product development process, such as brainstorming, concept generation, sharing of CAD models, and detailed design work. Increased collaboration, both within and between companies (e.g., customer–suppliers) in different locations or countries, is of great interest in terms of the sharing and transfer of information, where the information transfer includes computer models. Collaboration can take place over time but also interactively in real time, in addition to sharing the same CAD model. CAE offers possibilities, via a database, to easily share and use updated CAD models in the performance of different simulation studies at various locations as well as allowing departments other than engineering departments access to product data. Via Web browsers, for example, it is possible for management or financial departments to view and interact with a product in 3D at an early stage, thus giving them access to a simplified product model based on the same CAD data as used by the engineering department at the same point in time.

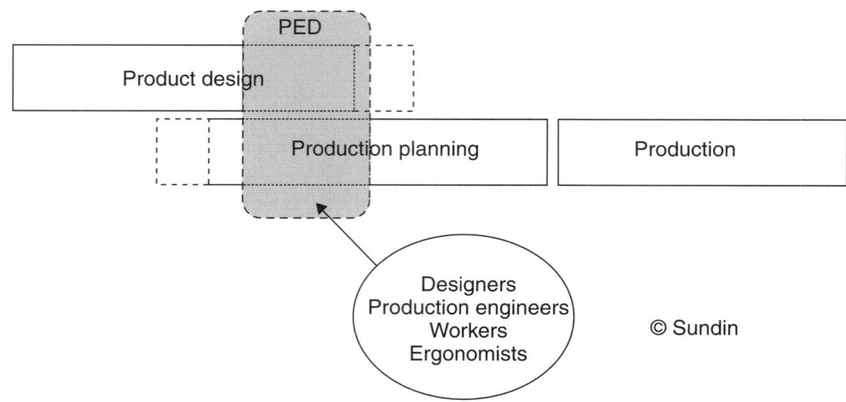

Figure 9 Participatory ergonomics design in the product development process. (From Sundin, 2001.)

When sharing complex product knowledge in collaborative product development, 3D representations function as critical knowledge representations of especially tacit knowledge (Yap et al., 2003). However, Paulin (2002) has found that the use of virtual prototypes obstruct knowledge creation and transfer, especially regarding incoming and outgoing knowledge transfer and justification of proposed solutions. The possibility of improving collaborative engineering and design with the increased use of DHM is discussed by Chaffin (2001), who suggests that this would better enhance human–hardware ergonomics assessments at an early stage in the product life cycle. More information on collaborative engineering can be found in Larsson (2002), Editorial (2003), Francis and Avijit (2003), and Fagerström (2004).

7 CASE STUDIES

The aim of this section is to refer the reader to some of the DHM work, primarily to recent studies where today's available DHM software has been used. Conferences are an important source of information in this fast-developing field. The International Digital Human Modeling for Design and Engineering Symposium (DHMS), held annually by the International Society of Automotive Engineers, is of major importance. Other key conferences are the International Computer Aided Ergonomics and Safety (CAES) conference, the International Conference on Human Aspects of Advanced Manufacturing Agility and Hybrid Automation (HAAMAHA), and the International Conference on Work with Computing Systems (WWCS).

Recent successful case studies using DHM have been presented by Kaiser et al. (2003), who used ANTHROPOS ErgoMAX for reach and vision evaluation of alternative solutions in design of a semiautomatic production line, and Kaiser and Klein (2003), who used RAMSIS posture and vision evaluation in the concept and design phases of construction machines. An additional example is presented by Ambrose et al. (2003), where an incident and risk analysis of roof bolter operators was carried out using the Jack manikin. This analysis was performed to complement investigation reports and laboratory experiments, where the use of human subjects was not feasible due to safety and ethical aspects. In Jimmerson's (2001) study, the conclusion was that despite the limitations of the digital human model, its use in automobile door assembly provided significant opportunities to shorten and improve the product and design process and that the benefits realized justified the cost and time spent to develop the simulation. In addition, numerous physical prototypes that would otherwise have been necessary answering clearance and reach concerns could be avoided.

An example of operator workstation design based on population data only is provided by Cerney et al. (2002), who used 3D-scanned population data in addition to measured data, but without the presence of a human figure. Results were visualized in a fully immersive virtual reality environment. Schrader et al. (2002) describe an application where mixed reality was used to increase haptic feedback. A physical mock-up of a driver's car interior was combined with a virtual reality head-mounted display, a motion capture system, and a RAMSIS computer manikin. This setup allowed a test driver, represented by the RAMSIS model, to experience the car interior in the form of a mock-up in driving scenarios. A similar application was presented by Kasch et al. (2002), where the VR-AHTROPOS manikin was used in combination with a fully immersive virtual reality environment in the development of aircraft cabins, with focus on passenger and crew processes, as well as maintenance aspects. Kozycki and Gordon (2002) also used manikins in the design of helicopter crew stations. They generated *boundary manikins* from a population database, using a method that reduces the size of the accommodation envelope from a large to a smaller number of body dimensions, thus dispensing with a large proportion of the original variation, which is critical for the design of accommodation. The models, which were based on the Jack manikin, were also adjusted for bulky clothing and equipment. Posture data were collected to better identify postures that accurately reflected those of actual pilots. See Dai et al. (2003) for a discussion of boundary manikins.

As already mentioned, vision is one important function that is analyzed by means of DHM tools, as well as being much in demand from designers, especially in the area of car design. However, there is a need for improvements. Work in this area is presented by Hudelmaier (2002), who overcame the fundamental weakness of having the eye point center as the basic measuring point, as the center point is merely fictional, based on the average of the statistical distribution of the eye points of the population. A system called ARGUS has been developed which when combined with the RAMSIS manikin software, enhances the current vision analysis methods and also includes binocular analysis instead of the more common monocular. Additional work based on case and laboratory studies is presented by Dukic (2003), who describes a visual demand model in which the characteristics of the task, operator, and environment influence the visual requirements for performing a task, resulting in the need for specific methods.

Studies by Sundin et al. (2001, 2002) and Wartenberg et al. (2002) reveal the importance of considering methodological aspects of DHM. A human factors analysis was carried out in the preliminary design phase of Cupola, a European Space Agency (ESA) module for manned space flights to the International Space Station (ISS). Jack manikin software was employed at an early stage of the design process prior to the production of any flight hardware. Different methods were used in support of the Jack analyses: hierarchical task analysis, a file exchange protocol, and a relational database. The Cupola study highlighted the fact that successful use of computer manikins depends on the specific project or context in which the manikin is used as well as on the methodological approach and preparatory work. Dukic et al. (2002) presented a case study that further explored the processes and methods as well as the role of the user in DHM. The study covers the role of computer manikins in the evaluation of assembly tasks played by computer manikins at Volvo Car Corporation during the development of a sport utility vehicle (SUV). Results show that the simulations failed to reveal a number of problems, due primarily to the lack of communication between manufacturing engineers, simulation engineers, and the design department, but also as a result of inadequate training, lack of clarity when ordering a simulation, insufficient participation during simulations, and defective documentation. It was stated that knowledge of the potential of the simulation tools was lacking (e.g., that the tools allowed analysis of the field of view). All of the foregoing aspects influenced the virtual analysis work process, illustrated in Figure 10.

In a case study presented by Rönnäng et al. (2003), a rather novel question is given: How does a computer manikin user's background and knowledge influence the results of ergonomics simulation? Subjects taking part in the study were production engineers and ergonomists. A manual task comprising production and ergonomics problems was used. The task was simulated by means of the Jack computer manikin, and the animated simulation was shown to the test subjects. Using basic functions such as rotating, zooming, and still pictures, the subjects were asked to detect problems and make recommendations. The study revealed that there are differences in how production engineers and ergonomists interpret results: Engineers focused on technical issues and practical solutions, whereas ergonomists highlighted psychosocial aspects using a holistic perspective. These differences show that guidelines are needed for the utilization and interpretation of DHM programs in participative processes. Jimmerson (2001) exemplified the need

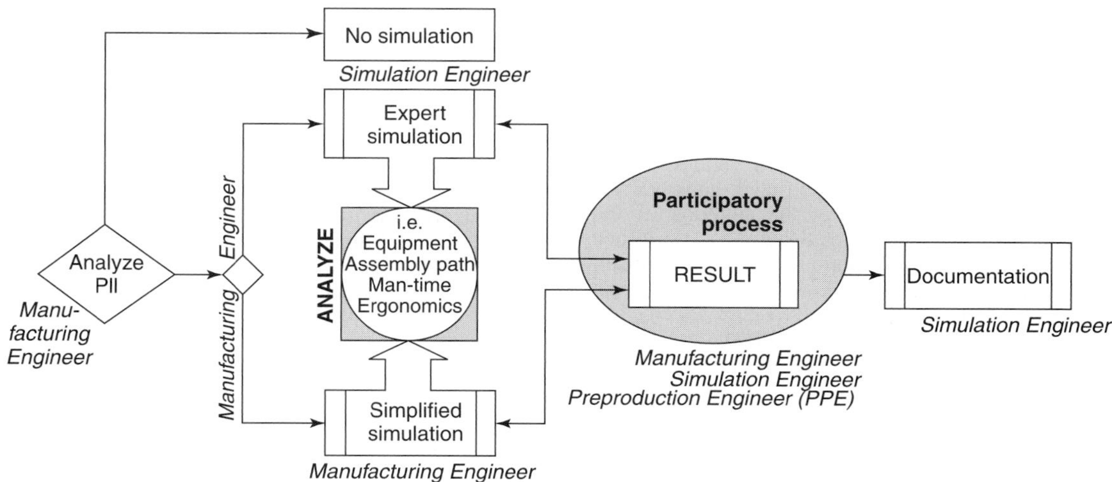

Figure 10 Car producer's process description of process and inspection instructions investigated in the case study. (From Dukic et al., 2002.)

to simplify DHM procedures in collaborative engineering and illustrated the importance of improved collaboration between different groups of people within a company. Although still undeveloped, the use of DHM in collaborative engineering has been facilitated, thanks to the greater emphasis on combining engineering and ergonomics software. In addition, the participatory ergonomics approach, as an aspect of collaborative engineering, is facilitated by this development, as illustrated in several case studies by Sundin (2001).

8 GENERAL BENEFITS OF DHM

Throughout this chapter we have been describing the area of DHM. In this and the following section we summarize benefits and disadvantages of DHM. First, when combined with other CAE solutions, DHM has significantly reduced the time and costs involved in product realization and in some ways has dramatically changed the attitude toward human factors as well as how products and production systems are designed to fit humans. DHM is most advantageous when no physical environment is available (when no product, prototype, or mock-up is built) or when the environment is inaccessible or hazardous (spacecraft, nuclear power plants, etc.), although it also offers advantages even when physical environments exist.

Use of DHM tools has also had positive effects on communication and collaboration between actors, by facilitating common understanding of a problem or solution. In addition to the enhancement of perception achieved by the use of computerized visualization of products and workplaces, another benefit mentioned by Andreoni and Pedotti (2001) is that virtual design allows ergonomists to become involved at an earlier stage of the design process instead of having to wait until a physical mock-up is built. This trend has also created a new and positive role for the ergonomist, which, due to its difference, is sometimes perceived as daunting.

As mentioned previously, one major obstacle to the efficient use of DHM tools is the slow positioning of the manikin by means of keyboard and mouse. Motion capture systems are one way of coping with this and lead to more efficient use of DHM. Such a system makes it possible to register and scale the human in a virtual environment and subsequently to control, record, and replay the motions of the computer manikin (see also Section 5.3.1). When combining the benefits of motion capture with human modeling software and virtual reality, the experience of an environment can be improved dramatically and a user can become totally immersed in the virtual environment and "act" as a manikin. As a result, a designer can capture the experience of various people in his or her target user group: for example, becoming aware of how it feels for a small woman to manage in a workplace designed for average-sized men, or for a big man to drive a small sports car. This is one example of situations that have been very difficult to experience before the introduction of modern DHM tools.

9 LIMITATIONS AND PROBLEMS OF DHM

In general, working with virtual instead of physical objects has both advantages and limitations. As exemplified by Paulin (2002), the use of virtual prototypes increases the possibility of finding geometrical problems, but decreases finding problems related to tactile sensations. The experience of DHM tools differs among users. Some, either new users or those unfamiliar with ergonomics, consider the tools difficult both to use and when translating and understanding analysis results, leading to a risk of inappropriate use (e.g., acceptance of hazardous postures). Other, more experienced users, especially from the automotive and the aerospace industries, experience the tool as inadequate (e.g., lacking in functions) and are therefore obliged to modify them to meet their own requirements.

A clear and informative summary of DHM has been presented by Chaffin (2001), several points of which are worth mentioning. First, the positioning of the manikin is still a major problem, as it is both difficult and time consuming; thus, there is a great need for robust and simple posture and motion prediction capabilities. Second, the inaccuracy of positions and movements can sometimes make a huge difference and is closely related to the first point. All computer manikins differ from how real humans move and position their body parts. A study carried out by Oudenhuijzen and Zehner (2000) assessed the anthropometrics of DHM software in terms of accuracy. Live subjects were measured and compared to CAD measurements on manikins (BHMS, COMBIMAN, Jack, RAMSIS, and Safework) modeled as the subjects. Results showed that accuracy not only differed between human subjects and manikins, but also between manikins. As all developers cooperated in this verification study, the results have hopefully led to improved models.

As highlighted by Peacock et al. (2001), there are few tools that address the consequences of repetitive motion; thus, simulation analysis results are not necessarily directly related to injury. However, in the 4D Watbak (University of Waterloo, 2004) software, included in the Ergowatch package, research methods used in epidemiological studies of low back pain have been incorporated in the software. The system increases the capability of current biomechanical modeling approaches by calculating shift-long cumulative loading on the spine as well as the peak hand and spine load forces. Epidemiological evidence is also used to provide insight into the risk of low back injury in the presence of multiple proven risk factors (Neumann et al., 1999)

Inaccurate human models or results from analyses can create problems, problems that users are sometimes unaware of. Verification and validation of digital human models and computer manikins are thus necessary to improve the analysis and its results. Several studies have been published on this subject (e.g., Conradi and Alexander, 2002; Bartels and Kwitowski, 2003; Doi and Haslegrave, 2003; Parkingson et al., 2003). The integration of old but accurate models [e.g., rapid upper limb analysis (RULA)] into a digital tool is another problem area. Incorporation of such analysis models into DHM software leads to the risk that the user may employ the model inaccurately, being unaware of its limitations, background, and so on. Problems in this area include how the users scale anthropometric models, described in Section 5.1.

Yet another well-known difficulty is that too much time is required for file conversions and on the building of the virtual environment. This is due primarily to software compatibility problems, as described by Dukic et al. (2002). In addition to the designers' preferences listed in Section 5.3, industrial engineers want to be able to assess pressure when applying load (e.g., in an assembly situation), as well as the possibility of creating simulation reports automatically.

The above-mentioned drawbacks are associated with embedded functions or models that still lack optimum solutions, central to the accurate and correct use of DHM tools. Other drawbacks that have recently arisen are related to the implementation and use of DHM tools, especially in organizations unfamiliar with DHM or becoming advanced users. The main problems are deficiencies in how simulations are requested, structured in terms of execution, and documented. These problems are discussed in more detail in Sections 6.2 and 6.3. Furthermore, long-time suppliers of product components and production equipment to companies using DHM tools (e.g., the aerospace industry) need to be abreast of the rapid development in the use of virtual tools in order to survive, due either to market forces or to formal customer contracts. They should be familiar with the computer systems and software and be able to deliver 3D models in addition to physical products.

10 GUIDELINES FOR DHM IN CAE APPLICATIONS

The following section is an attempt to provide a user with some advice in the form of eight items to consider before buying or implementing a DHM tool.

1. *Type of user.* The type of company (small or large company or a noncommercial organization) and the degree of knowledge about CAE tools will influence the choice of software. One important question is what type of applications you intend to use the DHM tool for and how frequently. This leads to the question of whether you should buy your own software or simply hire consultants. Three categories of users were defined, based on a survey carried out in Sweden (Sundin and Sjöberg, 2002a): (1) users with their own software, (2) users without software but who hired consultants, and (3) the consultants themselves. All three categories have different aims and different approaches in their use of DHM and important aspects need to be considered for each group. For example, if hiring consultants, the company may obtain extra value during the project, including not only ergonomics knowledge but also other engineering know-how. On the other hand, much of the knowledge and experience gained during the project more or less departs with the consultants.

2. *Choice of digital human modeling tools.* The level of investment depends on your requirements (i.e., how advanced the DHM tool needs to be). Digital human models such as 3DSSPP and Ergowatch are the most suitable software for biomechanical calculations. For more complex vision and reach assessments and a possibility to interact and act in a virtual environment, computer manikins are the obvious choice. These are, however, more expensive than the aforementioned, by roughly a factor of 100.

3. *Target users and training.* Who are the target users, what is their level of training, and what skills have they got? This question also affects the choice of software. Experience of CAD programs is a major advantage when using computer manikins. If the user does not possess such experience, an introductory course is vital to ensure effective use of the software. It is also important that the user be able to start using the software immediately after introductory course. If the user is an ergonomist, it might be necessary to devote extra effort to the technical aspects, such as file handling or coordinate system. If the user is an engineer who is unfamiliar with ergonomics, he or she should receive training in the area.

4. *Implementation and integration into an organization.* In addition to dedicated personnel and training, a plan is needed for the integration of DHM software with existing company systems (e.g., databases and product data management systems). In a larger corporation there may be a need to investigate what types of human simulation tools are already in use in different departments or different branches based in other locations or countries. There is also a need to establish methods and processes to facilitate the use of DHM, much as instructions on how to initiate a simulation, how to carry out the simulation (e.g., how to build the virtual environment or choose a manikin type or posture), and how to communicate and document the results. Information on the software and its capabilities and on agreed descriptions of processes and methods must be spread throughout the organization.

5. *Building the virtual environment.* Some software uses a human model only and offers no possibility for modeling of an environment. Other software allows one both to model environments in the program itself and to import environment files from other programs. It is well known that handling and conversions of geometry files between programs can cause problems (i.e., it is often time consuming and difficult), result in poor or changed geometries, and can cause loss of important data. Information is invariably lost due to approximations in the geometries in the conversions from CAD programs to a computer manikin program. This fact also limits the possibility to convert back to a CAD program from a computer manikin program. To save time, the best way is to import environment files (e.g., machines and other equipment) in an original format compatible with the computer manikin program. However, factory equipment objects are often built in 2D CAD, which is unsuitable for use in a 3D computer manikin program, resulting in the need

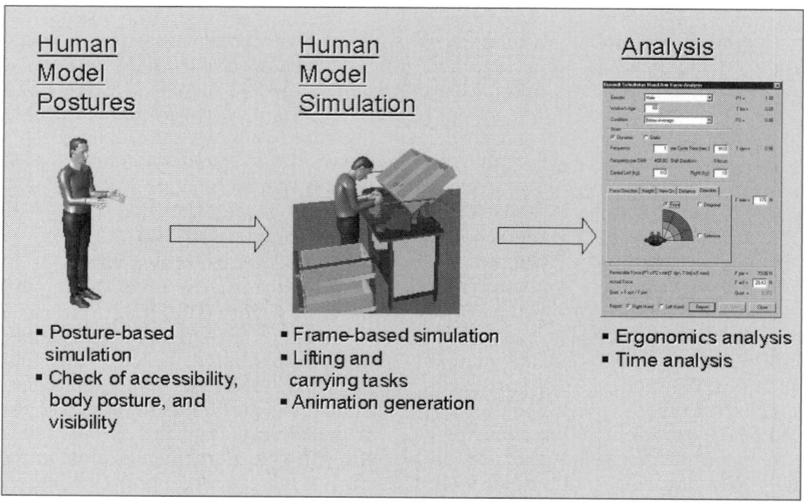

Figure 11 Stepwise working process for DHM ergonomics analysis. (Based on J. Nordling, Tecnomatix.)

to build factory environments from scratch, which is time consuming. A holistic product management process considerably facilitates the building of such an environment. See Section 6.3 for a description of product management processes.

6. *Ergonomics analysis.* In the ergonomics analysis phase the support processes are vital for efficiency and quality. Methodology documents or guidelines should be formulated on the topics in this section to enhance efficiency and quality in simulation. Further details and references on this topic may be found in Section 6.3.

To carry out an analysis, its aim and purpose should first be defined. Is it to identify a problem, to find a solution, or to illustrate a problem? When using computer manikin programs the analyst specifies and adds the type of manikin, specifies the task to be performed by the manikin, and considers restrictions after the environment has been built but before the actual simulation or analysis starts. Thereafter, analysis begins by manipulating and moving the manikin and objects, using, for example, predefined postures to identify a realistic posture (Figure 11). Some computer manikin programs have automatic generators for posture and hand grasping to facilitate manipulation. When a posture is identified or an animation is created and accepted as valid by the analyst and/or others, different analysis tools embedded in the software can be used. The most common tools and methods are (1) reach envelopes for arms; (2) vision analysis; (3) static body posture analysis, such as OWAS, RULA, and 3DSSPP; (4) strain analysis using the Burand–Schultetus model or a low-back analysis model; and (5) repetitive task analysis using the Niosh lifting equation model.

To achieve enhanced results, a participatory approach during the ergonomics analysis phase is advantageous. This approach gathers different competencies, including end users, all of whom contribute their own specific knowledge (e.g., ergonomists, operators, and designers). See Section 6.4 for more information.

7. *Handling of results and documentation.* Results of the analysis, together with comments and reflections from the working process, should be documented during and after the analysis. Results can consist of screenshot pictures, animations or films, written text, and so on, and should be documented as simulation reports and saved. This can be simplified with the help of templates, or preferably, saving the data in a database where it can be reused, thus ensuring that the right version is always available. The information can later also be used as input for work instructions (Figure 12). Besides documenting the results, it is important to (1) document the reason for and aim of the simulation, (2) document reflections and comments during the work process, (3) use an information and documentation system designed for ease of access in everyday work, (4) use a dynamic database system so that the most up-to-date versions of files are used, and (5) provide easy access to specifications, standards, and regulations. With a carefully planned documentation strategy, it will be easier to communicate, share, and spread information within a project and to transfer information and knowledge to future projects. It is also advantageous to have quick links to company ergonomics standards, or even if possible, to include these in the DHM software.

8. *Continuity of use.* Continuity is important for efficient use of the software. One problem is ensuring continuity in cases where there are few simulation tasks over time due to project cycles or being a small company with little need for simulation, which gives rise to periods where the simulation engineer does not use the software at all, whereby his or her skills

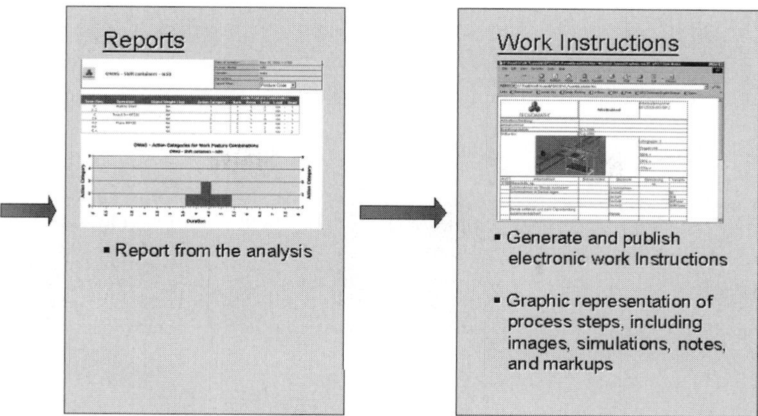

Figure 12 Continuation of the stepwise working process in Figure 11. In these steps simulation reports and work instructions are created based on ergonomics simulation data. (Based on J. Nordling, Tecnomatix.)

decline. A solution can be to share simulation software capacity and skills in a *simulation pool*, which is owned jointly and used by several departments or companies. Another critical situation is when only one or a few staff members have mastered the software. If they are absent or leave the company, knowledge can be lost.

11 CONCLUSIONS AND THE FUTURE OF DHM

In the attempt to compile the quite big and sprawling area of digital human modeling, connected to many disciplines such as engineering, computer science, and physiology, some concluding thoughts are as follows. As products and production systems are changing rapidly and constantly to meet the demands of global competition, virtual product and production development is becoming more important, leading to an increase in the use of CAE software together with DHM tools that has gained momentum particularly in recent years. Successful use of DHM requires awareness of its limitations and deficiencies as well as support for its strength. DHM will not succeed in all applications but has the potential to play an important role renewing, and in some part revolutionizing, how human aspects are considered in product realization as well the role of engineers and ergonomists. The positive effects are prevailing, and organizations using DHM see great benefits in cost reduction, time saving, and products and workplaces better fitted to humans. However, limitations are also reported, such as poor posture and motion prediction capabilities, inaccurate human models, and lack of procedures and process descriptions. In this setting, actual utilization of the software is as important as the capacity and usefulness of the tool itself. Thus, DHM hold challenges for theorists as well as practitioners regarding theory, models, validity, methods, and so on.

Earlier visions of DHM have been shared publicly. One given by Biferno just a few years ago is starting to come true. This vision gave an overall view of the role of human modeling and the relation to a company's infrastructure, and the aim was to give human modeling analyses a place as a core process in a company. Biferno also outlined a vision where human modeling becomes standard validation and benchmarking methods and has the function of embedded ergonomic training and technical support, a vision in some part fulfilled (Biferno, 2000). Badler's (2000) vision on future human models is also coming true in some parts, especially the anticipation of a better integration with other engineering tools (e.g., PDM and CAE models) and multiple levels of details to be able to use a smooth transition between views of a simulation. Parts of the vision, not being fulfilled as quickly, are smarter models, including autonomy with action skills, decision making, "thoughts," and reaction processes based on incoming knowledge about the outside world.

Vibration comfort is a central issue in automotive design. Future DHM tools will naturally contain improved analysis methods for comfort and vibration assessments. Work in this direction has already been developed and integrated (e.g., in the manikin RAMSIS; Pankoke et al., 2002; Schmale et al., 2002; Verver and van Hoof, 2002). Another issue playing an important role in the future is how DHM and other tools can be used for the assessment of user friendliness and affect (i.e., emotions and feelings toward a product or interface). In addition, today it is only possible to assess individual work tasks. There is, however, a need to assess the loads on an operator over an entire day and the total work load on a complete assembly line for single or multiple workers, and thus to be able to simulate changes over time.

12 DEFINITIONS RELATED TO HUMAN MODELING IN CAE APPLICATIONS

- *Animation* is a sequence of pictures that when played together give a perception of movement. According to the *Random House Compact*

Unabridged Dictionary, animation is described as "make alive, to give motion to."

- *CAD* (computer-aided design) refers to the design of products by means of a computer program. According to Groover (1984), CAD can be described as "any activity that involves the use of the computer to create or modify an engineering design."
- *CAE* (computer-aided engineering) employs software tools for supporting engineering work.
- *Computer manikin* is a 2D or 3D computer representation of the human body, based on anthropometric measurements, link and joint structure, and movement characteristics (CEN, 2003), which is incorporated in software and acts in a digital environment.
- *Computer manikin software* is computer modeling software consisting of a computer manikin, tools for controlling and manipulating the manikin (e.g., posture, anthropometric measurements), functions for mimicking human characteristics and behavior (e.g., biomechanical, strength, movements), and means to position the manikin in relation to the computer model of the physical environment (CEN, 2003).
- *Computerized visualization* is the "computer-supported creation and presentation of e.g. a workplace, product or solution" (Majchrak et al., 1987).
- *Digital human models* are general software models of the human structure or function which can be used to predict the human response to a given set of or changes in input conditions. They are often used for biomechanical analysis of a single load situation. Available software does not normally use graphic visualization, or includes only simple forms of visualization.
- *Digital mock-up* is a digital model of a product or workplace used for the testing of space demands, special features, or appearance.
- *Ergonomics simulation* is when a human and his or her activity are virtually analyzed and simulated. Visualization, analysis, and animation are the most common forms of ergonomics simulation (i.e., the human is visualized or sometimes moves in an animation while the user himself or herself or embedded biomechanical models carry out the analysis).
- *Mock-up* is a basic full-scale physical model of a product or workplace used to test space demands and special features to demonstrate appearance and tactile qualities. These models are usually produced in plastic, clay, or wood and despite their realistic appearance have few, or in many cases, no real functions.
- *Product realization* includes both product and production system development and involves the complete chain from concept generation and product design, production system, and workplace design to manufacturing.
- *Prototype* is a specially built product that is real in design, function, and appearance but not in production method.
- *Simulation*, according to the Swedish National Encyclopedia, is "to represent a system by another for the purpose of studying its dynamic behavior or to refine the operation of the system in a laboratory setting. The reason for the simulation can be that the system is too complex, inaccessible or too expensive or dangerous." "Modern simulation work with mathematical models, comprising e.g. fundamental laws of science, geometrical descriptions and material characteristics."
- *Tool* is a term used for a computer simulation program. A tool is something that is used as support in an activity or task. The Swedish National Encyclopedia says that "a tool is a thing used to process something by hand." However, according to the *Encyclopaedia Britannica*, it is "something [such as an instrument or apparatus] used in performing an operation or necessary in the practice of a vocation or profession."
- *Virtual reality* is seen as the most advanced means of visualizing environments or results from different areas. To qualify as a VR system, a system must be able to respond to user actions, have real-time 3D graphics, and give the user a sense of immersion; all three of these characteristics must be fulfilled (Pimentel and Teixeira, 1995).

ACKNOWLEDGMENTS

We are grateful to all colleagues, both those involved in DHM and others, who have contributed valuable inputs to and comments on this chapter. We also thank industrial partners for their contribution in projects where important knowledge building has been gained as well for their direct contributions to the chapter. Finally, we send our thanks to Gullvi and Margaretha for their support toward the writing of this chapter.

REFERENCES

Ambrose, D. H., Bartels, J., Kwitowski, A., Helinski, R. F., and Gallagher, S. (2003), "Machine Injury Prediction by Simulation Using Human Models," in *Proceedings of the Digital Human Modelling Conference,* August, Montreal, Quebec, Canada, 2003-01-2190, CD-ROM, Society of Automotive Engineers International, Warrendale, PA.

Andreoni, G., and Pedotti, A. (2001), "Virtual Workplace Design," in *International Encyclopedia of Ergonomics and Human Factors*, W. Karwowski, Ed., Taylor & Francis, New York, Vol. 2, pp. 971–974.

Artl, F., and Bubb, H. (1999), "Simulation of Target Directed Movements Within the CAD-Manmodel RAMSIS," in *Proceedings of the SAE Digital Human Modeling Conference*, The Hague, The Netherlands.

Aune, I. A., and Jürgens, H. W. (1989), "Ergonomische Studien: Computermodelle des menschlichen Körpers," in *Aufträgsstudie für das Bundesamt für Wehrtechnik und Beschaffung*, Bericht 27, ISSN 0933–4092, Bundesamt für Wehrtechnik und Beschaffung, Koblenz, Germany.

Badler, N. (1993), "Computer Graphics Animation and Control," in *Simulating Humans*, Oxford University Press, New York.

Badler, N. (2000), "What's Next for Digital Human Models," keynote, in *Proceedings of the SAE International Digital Human Modeling for Design and Engineering 2000 International Conference and Exposition,* June 6–8, Dearborn, MI.

Bartels, J. R., and Kwitowski, A. (2003), "Verification of a Roof Bolter Simulation Model," in *Proceedings of the Digital Human Modelling Conference,* August, Montreal, Quebec, Canada, 2003-01-2217, CD-ROM, Society of Automotive Engineers International, Warrendale, PA.

Biferno, M. (2000), "Application of Human Modeling Technology in Industry: Using a Systems Approach," keynote, in *Proceedings of the SAE International Digital Human Modeling for Design and Engineering 2000 International Conference and Exposition,* June 6–8, Dearborn, MI.

Blomé, M., Dukic, T., Hanson, L., and Högberg, D. (2003), "Simulation of Human–Vehicle Interaction in Vehicle Design at Saab Automobile: Present and Future," *Proceedings of the Digital Human Modelling Conference,* August, Montreal, Quebec, Canada, 2003-01-2192, CD-ROM, Society of Automotive Engineers International, Warrendale, PA.

CEN (2003), *Ergonomics: Computer Manikins and Body Templates, Part 1: General Requirements*, CEN/TC 122, ISO/FDIS 15536-1:2003, manuscript of the final draft, European Standard prEN ISO 15536-1, European Committee for Standardization, Brussels, Belgium.

Cerney, M. M., Duncan, J. R., and Vance, J. M. (2002), "Using Population Data and Immersive Virtual Reality for Ergonomic Design of Operator Workstations," in *Proceedings of the Digital Human Modeling Conference*, Munich, June, VDI/SAE, Ed., VDI Verlag, Düsseldorf, Germany, pp. 107–120.

Chaffin, D. B. (1986), "Computerized Models for Occupational Biomechanics," *Scandinavian Journal of Occupational Health, Safety and Biomechanics*, Vol. 10.

Chaffin, D. B., Ed. (2001), *Digital Human Modeling for Vehicle and Workplace Design*, Society of Automotive Engineers, Warrendale, PA.

Chaffin, D. B., Faraway, J., and Zhang, X. (1999), "Simulating Reach Motions," in *Proceedings of the SAE Digital Human Modeling Conference*, The Hague, The Netherlands.

Conradi, J., and Alexander, T. (2002), "Interrelation Between Empirical and Motion Simulation Data of a Digital Human Model," in *Proceedings of the SAE Digital Human Modeling Conference*, Munich, June, SAE/VDI, Ed., VDI Verlag, Düsseldorf, Germany, pp. 357–365.

Dai, J., Teng, Y., and Oriet, L. (2003), "Application of Multiparameter and Boundary Mannequin Techniques in Automotive Assembly Process," in *Proceedings of the Digital Human Modelling Conference,* August, Montreal, Quebec, Canada, 2003-01-2198, CD-ROM, Society of Automotive Engineers International, Warrendale, PA.

Delleman, N. J., Drost, M. R., and Huson, A. (1992), "Value of Biomechanical Macromodels as Suitable Tools for the Prevention of Work-Related Low Back Problems," *Clinical Biomechanics*, Vol. 7, No. 3, pp. 138–148.

Doi, M. A. C., and Haslegrave, C. M. (2003), "Evaluation of JACK for Investigating Postural Behavior at Sewing Machine Workstations," in *Proceedings of the Digital Human Modelling Conference,* August, Montreal, Quebec, Canada, 2003-01-2218; CD-ROM, Society of Automotive Engineers International, Warrendale, PA.

Dukic, T. (2003), "Applying Computer Manikins to Evaluate the Visual Demands in Manual Tasks," Licentiate thesis, Chalmers University of Technology, Göteborg, Sweden.

Dukic, T., Rönnäng, M., Örtengren, R., Christmansson, M., and Davidsson, A. J. (2002), "Virtual Evaluation of Human Factors for Assembly Line Work: A Case Study in an Automotive Industry," in *Proceedings of the SAE Digital Human Modeling Conference*, Munich, June, SAE/VDI, Ed., VDI Verlag, Düsseldorf, Germany, pp. 129–150.

Dukic, T., Rönnäng, M., Christmansson, M., Falck, A.-C., Sjöberg, H., Sundin, A., Wartenberg, C., and Örtengren, R. (2003), "The Use of Virtual Tools in the Development of Production Systems at Volvo Cars: A Case Study of the Development of the Production System for P28/XC90, Final Assembly" (in Swedish), Chalmers Tekniska Högskola, Göteborg, Sweden.

Editorial (2003), Special Issue on Knowledge Sharing in Collaborative Design Environment, *Computers in Industry*, Vol. 52, pp. 1–3.

Eklund, J., Cederqvist, T., and Chen, F., Eds. (1997), *Physical Ergonomics*, Division of Industrial Ergonomics, Linköping University, Linköping, Sweden.

Ergon (2004a), "ABBA (Arbeitsplatz Begehungs- und Belastungs Analyse), Technische Universität Darmstadt," retrieved April 21, 2004, from http://www.tu-darmstadt.de, http://www.ergon-gmbh.com/web/ergon_en/products/software.htm, Ergon Verlag, Stuttgart, Germany.

Ergon (2004b), "Computer Aided Systems Human Engineering (CASHE): Performance Visualization System (PVS)," retrieved April 20, 2004, from http://iac.dtic.mil/hsiac/Products.htm#CASHE, Ergon Verlag, Stuttgart, Germany.

Fagerström, B. (2004), *Managing Distributed Product Development: An Information and Knowledge Perspective*, Engineering and Industrial Design, Division of Product and Production Development, Chalmers University of Technology, Göteborg, Sweden.

Faraway, J. (2003), "Data-Based Motion Prediction," in *Proceedings of the Digital Human Modelling Conference*, August, Montreal, Quebec, Canada, 2003-01-2229, CD-ROM, Society of Automotive Engineers International, Warrendale, PA.

FIOH (2004), "ergoSHAPE," retrieved April 20, 2004, from http://www.ttl.fi/Internet/Suomi/Aihesivut/Ergonomia/Tyokalut/ergoshape.htm, Finnish Institute of Occupational Health, Helsinki, Finland.

Francis, E. H. T., and Avijit, R. (2003), "CyberCAD: A Collaborative Approach in 3D-CAD Technology in a Multimedia-Supported Environment," *Computers in Industry*, Vol. 52, pp. 127–145.

Geyer, M., and Rösch, B. (2002), "3D-Human Simulation and Manufacturing Process Management," in *Proceedings of the Digital Human Modeling Conference*, Munich, June, VDI/SAE, Ed., VDI Verlag, Düsseldorf, Germany, pp. 121–128.

Green, R. F. (2000), "A Generic Process for Human Model Analysis," in *Proceedings of the SAE International Digital Human Modeling for Design and Engineering 2000 International Conference and Exposition*, June 6–8, Dearborn, MI, 2000-01-2167.

Greil, H., and Jürgens, H. W. (2000), "Variability of Dimensions and Proportions in Adults or How to Use Classic Anthropometry in Man Modeling," in *Ergonomic Software Tools in Product and Production Design: A Review of Recent Developments in Human Modeling and Other Design Aids*, K. Landau, Ed., Ergon Verlag, Stuttgart, Germany, pp. 7–27.

Groover, M. P. (1984), *CAD/CAM Computer Aided Design and Manufacturing*, Prentice-Hall, Englewood Cliffs, NJ.

Haines, H., Wilson, J. R., Vink, P., and Koningsveld, E. (2002), "Validating a Framework for Participatory Ergonomics (the PEF)," *Ergonomics*, Vol. 45, No. 4, pp. 309–327.

Hanson, L., Blomé, M., Dukic, T., and Högberg, D. (2004), "Cinide and Documentation System to Support Digital Human Modeling Applications," *International Journal of Industrial Ergonomics*, accepted for publication.

Haslegrave, C. N. (1986), "Characterizing the Anthropometric Extremes of the Population," *Ergonomics*, Vol. 29, pp. 281–301.

HSIAC (2004), retrieved April 21, 2004, from http://iac.dtic.mil/hsiac, Human Systems Information Analysis Center, Wright-Patterson Air Force Base, Dayton, OH.

Hudelmaier, J. (2002), "Analysing Driver's View in Motor Vehicles," in *Proceedings of the SAE Digital Human Modeling Conference*, Munich, June, SAE/VDI, Ed., VDI Verlag, Düsseldorf, Germany, pp. 167–182.

Human Solutions (2004), "RAMSIS," retrieved June 21, 2004, from http://www.ramsis.de/, Human Solutions, Kaiserslautern, Germany.

Jimmerson, G. D. (2001), "Digital Human Modeling for Improved Product and Process Feasibility Studies," in *Digital Human Modeling for Vehicle and Workplace Design*, D. B. Chaffin, Ed., Society of Automotive Engineers, Warrendale, PA, pp. 127–135.

Johansson, P. (2004), *On the Use of Visualisation Tools in Product Design: Exploring Possibilities and Problems of Virtual Reality Techniques*, Department for Mechanical Engineering, Linköping University, Linköping, Sweden.

Jung, E. S., Kee, D., and Chung, M. K. (1995), "Upper Body Reach Posture Prediction for Ergonomic Evaluation Models," *International Journal of Industrial Ergonomics*, Vol. 16, pp. 95–107.

Kaiser, R., and Klein, T. (2003), "Optimizing Working Conditions in Construction Machines Using a CAD-Based 3D Man Model," in *Proceedings of the Digital Human Modelling Conference*, August, Montreal, Quebec, Canada, 2003-01-2183, CD-ROM, Society of Automotive Engineers International, Warrendale, PA.

Kaiser, R., Kuhn, P., and Anskohl, M. (2003), "Ergonomics Workplace Design Using a 3D Man Model," in *Proceedings of the Digital Human Modelling Conference*, August, Montreal, Quebec, Canada, 2003-01-2191, CD-ROM, Society of Automotive Engineers International, Warrendale, PA.

Kasch, D., Bauch, A. and Schmitt, D. (2002), "See, Test and Fly in the Aircraft Cabin Before the First Seat Is Ordered," in *Proceedings of the SAE Digital Human Modeling Conference*, Munich, June, SAE/VDI, Ed., VDI Verlag, Düsseldorf, Germany, pp. 183–189.

Kozycki, R., and Gordon, C. C. (2002), "Applying Human Figure Modeling Tools to the RAH-66 Comanche Crewstation Design," in *Proceedings of the SAE Digital Human Modeling Conference*, Munich, June, VDI/SAE, Ed., VDI Verlag, Düsseldorf, Germany, pp. 191–200.

Kuorinka, I. (1997), "Tools and Means of Implementing Participatory Ergonomics," *International Journal of Industrial Ergonomics*, Vol. 19, pp. 267–270.

Lämkull, D., Hanson, L., and Örtengren, R. (2005), "Human Models' Appearance Impact on Observers' Ergonomic Assessment," in *Proceedings of the Digital Human Modeling for Design and Engineering Symposium*, June 14–16, Iowa City, IA, 2005-01-2722; CD-ROM, Society of Automotive Engineers International, Warrendale, PA.

Landau, K., Ed. (2000), *Ergonomic Software Tools in Product and Production Design: A Review of Recent Developments in Human Modeling and Other Design Aids*, Ergon Verlag, Stuttgart, Germany.

Larsson, A. (2002), *Socio-technical Aspects of Distributed Collaborative Engineering*, Department of Applied Physics and Mechanical Engineering, Division of Computer Aided Design, LuleÅUniversity of Technology, LuleÅ, Sweden.

Launis, M., and Lehtelä, J. (1992), "ergoSHAPE: A Design Oriented Ergonomic Tool for AutoCAD," in *Computer Applications in Ergonomics, Occupational Safety and Health*, M. Mattila and W. Karwowski, Eds., North-Holland, Amsterdam, pp. 121–128.

Magnenat-Thalmann, N., and Thalmann, D. (1991), "Complex Models for Visualizing Humans," *IEEE Computer Graphics and Applications*, Vol. 11, pp. 32–44.

Majchrak, A., Chang, T.-C., Barfield, W., and Salvendy, G., Eds. (1987), *Human Aspects of Computer-Aided Design*, Taylor & Francis, Philadelphia.

McAtamney, L., and Corlett, N. (1993), "RULA: A Survey Method for the Investigations of Work-Related Upper Limb Disorders," *Applied Ergonomics*, Vol. 24, No. 2, pp. 91–99.

Mi, Z. (2004), "Task Based Prediction of Upper Body Motion," doctoral dissertation, http://www.engineering.uiowa.edu/amalek/VSR/ZanMi2004.pdf, University of Iowa, Ames, IA.

Micro Analysis and Design (2004), "Micro Saint," retrieved April 20, 2004, from http://www.maad.com/index.pl/micro_saint and http://www.adeptscience.co.uk/products/mathsim/microsaint/tour.html, Micro Analysis and Design, Boulder, CO.

NASA (2004), "NASA Task Load Index (TLX)," retrieved April 20, 2004, from http://iac.dtic.mil/hsiac/Products.htm, NASA Ames Research Center, Moffett Field, CA.

Neumann, W. P., Wells, R. P. and Norman, R. W. (1999), "4D WATBAK: Adapting Research Tools and Epidemiological Findings to Software for Easy Application by Industrial Personnel," in *Proceedings of the International Conference on Computer-Aided Ergonomics and Safety*, Barcelona, Spain.

NexGen Ergonomics (2004a), "ErgoIntelligence Manual Materials Handling," retrieved April 20, 2004, from http://www.nexgenergo.com/ NexGen Ergonomics, Quebec, Canada.

NexGen Ergonomics (2004b), "ManneQuinPRO," retrieved April 21, 2004, from http://www.nexgenergo.com/, NexGen Ergonomics, Quebec, Canada.

NIOSH (2004), "Lifting Equation," DHHS (NIOSH) Publication 94–110, retrieved April 27, 2004, from http://www.

cdc.gov/niosh/94-110.html, National Institute for Occupational Safety and Health, Atlanta, GA.

Oudenhuijzen, A. J. K., and Zehner, G. (2000), "Digital Human Modelling Systems: A Procedure for Verification and Validation Using the F-16 Crew Station," in *Proceedings of the SAE International Digital Human Modeling for Design and Engineering 2000 International Conference and Exposition*, June 6–8, Dearborn, MI, 2000-01-2159, SAE International, Washington, DC.

Pankoke, S., Balzulat, J., and Wölfel, H. P. (2002), "Vibrational Comfort with CASIMIR and RAMSIS Using a Dynamic Finite-Element Model of the Human Body," in *Proceedings of the SAE Digital Human Modeling Conference*, Munich, June, SAE/VDI, Ed., VDI Verlag, Düsseldorf, Germany, pp. 493–504.

Park, W., Chaffin, D., and Martin, B. (2002), "Memory-Based Motion Simulation," in *Proceedings of the SAE Digital Human Modeling Conference*, Munich, June, VDI/SAE, Ed., VDI Verlag, Düsseldorf, Germany, pp. 255–270.

Park, W., Chaffin, D., Rider, K., and Martin, B. (2003), "Simulating Complex Manual Handling Motions via Motion Modification: Performance Evaluation of Motion Modification Algorithm," in *Proceedings of the Digital Human Modelling Conference*, August, Montreal, Quebec, Canada, 2003-01-2227; CD-ROM, Society of Automotive Engineers International, Warrendale, PA.

Parkingson, M., Reed, M. P., and Klinkenberger, A. (2003), "Assessing the Validity of Kinematically Generated Reach Envelopes for Simulations of Vehicle Operators," in *Proceedings of the Digital Human Modelling Conference*, August, Montreal, Quebec, Canada, 2003-01-2216, CD-ROM, Society of Automotive Engineers International, Warrendale, PA.

Paulin, D. (2002), *Virtual Verification: Impact of a New York Method on the Final Verification Process*, Department of Operations Management and Work Organization, Chalmers University of Technology, Göteborg, Sweden.

Peacock, B., Reed, H., and Fox, R. (2001), "Ergonomics Analysis of Sheet-Metal Handling," in *Digital Human Modeling for Vehicle and Workplace Design*, D. B. Chaffin, Ed., Society of Automotive Engineers, Warrendale, PA, pp. 113–126.

Pew, R. W., and Mavor, A. S. (1998), *Modeling Human and Organizational Behavior: Application to Military Simulations*, National Academy Press, Washington, DC.

Pheasant, S. (2001), *Bodyspace: Anthropometry, Ergonomics and the Design of Work*, Taylor & Francis, London.

Pimentel, K., and Teixeira, K. (1995), *Virtual Reality Through the New Looking Glass*, McGraw-Hill, New York.

Pohlandt, A., Debitz, U., Schulze, F., and Richter, P. (2003), "A Tool for Human-Centred Job Design," in *Human Aspects in Production Management: Proceedings of the IFIP WG 5.7 Working Conference on Human Aspects in Production Management*, Vol. 1, *Esim-European Series in Industrial Management*, G. Zülch, S. Stowasser, and H. S. Jagdev, Eds., Shaker Verlag, Aachen, Germany, Vol. 5, pp. 90–98.

Porter, M. J., Case, K., Marshall, R., Gyi, D., and Sims neé Oliver, R. (2004), " 'Beyond Jack and Jill': Designing for Individuals Using HADRIAN," *International Journal of Industrial Ergonomics*, Vol. 33, pp. 249–264.

Rioux, M., Abdali, O., Victor, H., and Paquet, E. (2004), "Exploring Anthropometric Data Through Cluster Analysis," in *Proceedings of the Digital Human Modeling for Design and Engineering Symposium*, 2004-01-218841, Society of Automotive Engineers, Warrendale, PA.

Robinette, K. (2004), "An Alternative 3-D Shape Descriptor for Database Mining," in *Proceedings of the Digital Human Modeling for Design and Engineering Symposium*, 2004-01-2185, Society of Automotive Engineers, Warrendale, PA.

Robinette, K., Blackwell, S., Daanen, H., Boehmer, M., Fleming, S., Kelly, S., and Brill, T. (2002), *Civilian American and European Surface Anthropometry Resource (CAESAR): Final Report*, Vol. 1, *Summary*, AFRL-HE-WP-TR-2002-0169, Air Force Research Laboratory, Wright Patterson, Air Force Base, OH.

Rönnäng, M., Örtengren, R., and Dukic, T. (2003), "It's in the Eye of the Beholder: Who Should Be the User of Computer Manikin Tools?" in *Proceedings of the Digital Human Modelling Conference*, August, Montreal, Quebec, Canada, 2003-01-2196, CD-ROM, Society of Automotive Engineers International, Warrendale, PA.

Safework (2004), retrieved June 30, 2004, from http://www.safework.com/safework_pro/sw_pro.html, Safework, Inc., Montreal, Quebec Canada.

Schmale, G., Stelzle, W., Kreienfeld, T., Wolf, C.-D., Härtel, T. and Jödicke, R. (2002), "COSYMAN: A Simulation Tool for Optimization of Seating Comfort in Cars," in *Proceedings of the SAE Digital Human Modeling Conference*, Munich, June, SAE/VDI, Ed., VDI Verlag, Düsseldorf, Germany, pp. 301–311.

Schneider, F.-J. (2003), "The Virtual Factory at Opel," in *Human Aspects in Production Management: Proceedings of the IFIP WG 5.7 Working Conference on Human Aspects in Production Management*, Vol. 1, *Esim-European Series in Industrial Management*, G. Zülch, S. Stowasser, and H. S. Jagdev, Eds., Vol. 5, Shaker Verlag, Aachen, Germany, pp. 114–119.

Schrader, K., Remlinger, W., and Meier, M. (2002), "Mixed Reality with RAMSIS," in *Proceedings of the Digital Human Modeling Conference*, Munich, June, SAE/VDI, Ed., VDI Verlag, Düsseldorf, Germany, pp. 151–166.

Söderman, M. (1998), *Product Representations: Understanding the Product in the Design Process*, Chalmers University of Technology, Göteborg, Sweden.

Speyer, H., Wirsching, H.-J., Devolder, S., and van Raemdonk, R. (2002), "ALERT: Motion Capturing Using a Suit," in *Proceedings of the SAE Digital Human Modeling Conference*, Munich, June, SAE/VDI, Ed., VDI Verlag, Düsseldorf, Germany, pp. 357–365.

Sundin, A. (2001), "Participatory Ergonomics in Product Development and Workplace Design, Supported by Computerised Visualisation and Human Modelling," doctoral dissertation, Department of Human Factors Engineering, Chalmers University of Technology, Göteborg, Sweden.

Sundin, A., and Sjöberg, H. (2002a), "The Use of Computer Manikins Used in Sweden," in *Proceedings of the 34th Annual Congress of the Nordic Ergonomics Society, Humans in Complex Environments*, October, 1–3, KolmÅrden, D. Caldenfors, J. Eklund, and L. Kiviloog, Eds., Linköping University, Linköping, Sweden, pp. 745–750.

Sundin, A., and Sjöberg, H. (2003), "The Adoption of Virtual Visualisation and Simulation of Human Aspects in Product Development and Manufacturing Design," in *Proceedings of the 8th International Conference on Human Aspects of Advanced Manufacturing: Agility and Hybrid Automation*, May 26–30, Rome, S. Bagnara, A. Rizzo,

S. Pozzi, F. Rizzo, and L. Save, Eds., National Research Council of Italy, Institute of Cognitive Science and Technologies, Rome, pp. 287–292.

Sundin, A., Örtengren, R., and Sjöberg, H. (2001), "Proactive Human Factors Engineering Analysis in Space Station Design Using the Computer Manikin Jack," in *SAE Transactions: Journal of Passenger Cars—Mechanical Systems*, paper awarded to be among the most outstanding technical research in that field 2000, 2000-01-2166, pp. 2237–2242, Society of Automotive Engineers International, Warrendale, PA.

Sundin, A., Törner, M., Rönnäng, M., Dukic, T., Wartenberg, C., Sjöberg, H., Örtengren, R., and Frid, J. (2002b), "Prerequisites for Extensive Computer Manikin Analysis: An Example with Hierachical Task Analysis, File Exchange Protocol and a Relational Database," in *SAE Transactions: Journal of Passenger Cars—Mechanical Systems*, paper awarded to be among the most outstanding technical research in that field 2001, 2001-01-2101, Society of Automotive Engineers International, Warrendale, PA.

Sundin, A., Christmansson, M., and Larsson, M. (2003), "A Different Perspective in Participatory Ergonomics in Product Development Improves Assembly Work in the Automotive Industry," *International Journal of Industrial Ergonomics*, Vol. 33, pp. 1–14.

Tampere University of Technology (2004), "Owako Working Posture Analysis System (OWAS)," retrieved April 27, 2004, from http://turva.me.tut.fi/owas/, Tampere University of Technology, Tampere, Sweden.

Thalmann, M., and Thalmann, D. (1990), *Computer Animation Theory and Practice*, 2nd rev. ed., Springer-Verlag, New York.

TNO–Nederlandse Organisatie voor toegepast-natuurwetenschappelijk (2004), "MADYMO (Mathematical Dynamical Models)," retrieved April 20, 2004, from http://www.automotive.tno.nl/smartsite.dws?id=537, The Netherlands Organisation for Applied Scientific Research Delft, The Netherlands.

Toma, L., Teng, Y., and Oriet, L. (2003), "Multi-mannequin Coordination and Communication in Digital Workcells," in *Proceedings of the Digital Human Modelling Conference*, August, Montreal, Quebec, Canada, 2003-01-2197, CD-ROM, Society of Automotive Engineers International, Warrendale, PA.

UGS (2004), "Jack," retrieved June 21, 2004, from http://www.ugs.com/products/efactory/jack/.

University of California (2004), retrieved April 8, 2004, from http://www.psych.ucsb.edu/research/recveb/research_area_socialpsych.htm, Research Center for Virtual Environments and Behavior, University of California, Santa Barbara, CA.

University of Michigan (2004a), "3D Static Strength Prediction Program ™ (3DSSPP)," retrieved April 20, 2004, from http://www.engin.umich.edu/dept/ioe/3DSSPP/index.html, Center for Ergonomics, University of Michigan, Ann Arbor, MI

University of Michigan (2004b), "Human Motion Simulation (HUMOSIM) at the Center for Ergonomics, University of Michigan," retrieved April 21, 2004, from http://www.engin.umich.edu/dept/ioe/HUMOSIM/index.html, University of Michigan, Ann Arbor, MI.

University of Waterloo (2004a), "4D WATBAK," retrieved April 21, 2004, from http://www.ahs.uwaterloo.ca/~escs/rwtools.html, Faculty of Applied Health Sciences, Department of Kinesiology, University of Waterloo, Waterloo, Ontario, Canada.

University of Waterloo (2004b), "Ergowatch," retrieved April 21, 2004, from http://www.ahs.uwaterloo.ca/~escs/rwtools.html, Faculty of Applied Health Sciences, Department of Kinesiology, University of Waterloo, Waterloo, Ontario, Canada.

Van Cott, H. P., and Kinkade, R. G. (1972), *Human Engineering Guide to Equipment Design*, American Institute for Research, Washington, DC.

Verver, M. M., and van Hoof, J. (2002), "Vibration Analysis with MADYMO Human Models," in *Proceedings of the SAE Digital Human Modeling Conference*, Munich, June, SAE/VDI, Ed., VDI Verlag, Düsseldorf, Germany, pp. 447–456.

Virtual Soldier Research (2004), "Santos, Virtual Soldier," retrieved April 21, 2004, from http://www.digital-humans.org/main.htm, Center for Computer Aided Design, University of Iowa, Ames, IA.

Wang, X. (2002), "Prediction of Lower-Limb Movements of Clutch Pedal Operation from an Existing Motion Database," in *Proceedings of the SAE Digital Human Modeling Conference*, Munich, June, SAE/VDI, Ed., VDI Verlag, Düsseldorf, Germany, pp. 271–284.

Wartenberg, C., Frid, J., and Törner, M. (2002), "Relational Database as a Tool in Industrial Design: Experience from a Human Factors Engineering Analysis of a Module for Manned Space-Flight," *International Journal of Industrial Ergonomics*, Vol. 30, pp. 371–385.

Whysall, Z. J., Haslam, R. A., and Haslam, C. (2004), "Processes, Barriers, and Outcomes Described by Ergonomics Consultants in Preventing Work-Related Musculoskeletal Disorders," *Applied Ergonomics*, Vol. 35, pp. 343–351.

Wilson, J. R., and Haines, H. (1997), "Participatory Ergonomics," in *Handbook of Human Factors and Ergonomics*, 2nd ed., G. Salvendy, Ed., Wiley, New York, pp. 490–513.

Wilson, J. R., and Haines, H. M. (1998), *Development of a Framework for Participatory Ergonomics*, Health and Safety Executive, HSE Books, Norwich, England.

Yap, A. Y., Ngwenyama, O., and Osei-Bryson, K.-M. (2003), "Levering Knowledge Representation, Usage, and Interpretation to Help Reengineer the Product Development Life Cycle: Visual Computing and Tacit Dimensions of Product Development," *Computers in Industry*, Vol. 51, pp. 89–110.

Ziolek, S., and Nebel, K. J. (2003), "Human Modelling: Controlling Misuse and Misinterpretation," in *Proceedings of the Digital Human Modelling Conference*, August, Montreal, Quebec, Canada. 2003-01-2178, CD-ROM, Society of Automotive Engineers International, Warrendale, PA.

Zülch, G., and Vollstedt, T. (2000), "Personnel-Integrated and Personnel-Oriented Simulation: A New Guideline of the German Association of Engineers," in *Information and Communication Technology (ICT) in Logistics and Production Management*, J. O. Strandhagen and E. Alfnes, Eds., Department of Production and Quality Engineering, Norwegian University of Science and Technology, Trondheim, Norway, pp. 185–192.

CHAPTER 40
VIRTUAL ENVIRONMENTS

Kay M. Stanney
University of Central Florida
Orlando, Florida

Joseph Cohn
Naval Research Laboratory
Washington, D.C.

1 INTRODUCTION

Until recently, the notion of virtual environments (VEs) becoming a mainstream technology was little more than a pipe dream. Typical VEs required specialized hardware and software to run and highly skilled software and hardware engineers to develop and maintain. A far cry from such technology utopias as *Star Trek's* Holodeck, these systems remained little more than experimental prototypes, one-of-a kind testbeds which under even the slightest scientific exploration often generated far more questions than they answered. Yet, after nearly four decades of research and development, VEs have finally begun to realize their potential. This maturation owes much to advancements in the computer industry which, proceeding at a Moore's Law rate, have evolved the simple desktop PC into a powerful graphical rendering system. As technology barriers have been shattered, so have many of the human factors related challenges that were typically associated with earlier technology-limited systems. Now, more than ever, it is possible to contemplate the likelihood that such systems will soon be supporting a range of applications, from education to entertainment.

These changes, positive though they may be, mandate a review of the current state of the art in VE technology. This chapter endeavors to answer this call by providing design and implementation strategies, discussing health and safety concerns and potential countermeasures, and presenting the latest in VE usability engineering approaches. Current efforts in a number of application domains are reviewed. The chapter should enable readers to better specify design and implementation requirements for VE applications and prepare them to use this advancing technology in a manner that maximizes the benefits while minimizing health and safety concerns.

2 SYSTEM REQUIREMENTS

Virtual environments provide multisensory computer-generated experiences, which are driven by the hardware and software used to generate the virtual world (see Figure 1). The hardware interfaces consist primarily of (1) multisensory display devices used to present information that renders the virtual experience, (2) tracking devices used to identify head and limb position and orientation, and (3) interaction techniques that allow users to navigate through and interact with the virtual world. The software interfaces include (1) modeling software used to generate VEs, (2) autonomous agents that inhabit VEs, and (3) communication networks used to support multiuser virtual environments.

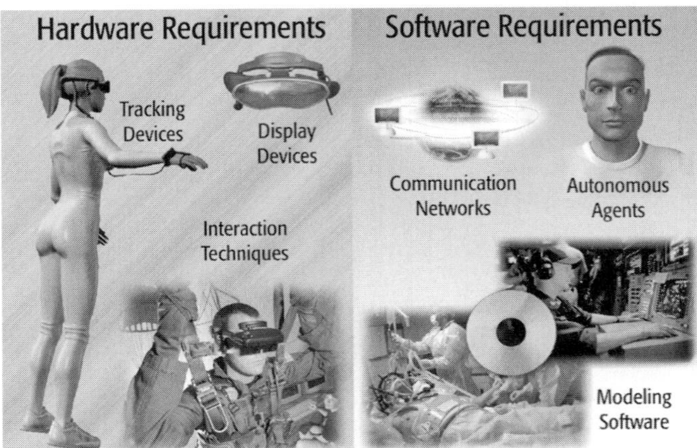

Figure 1 Hardware and software requirements for virtual environment generation. (Courtesy of Branka Wedell.)

2.1 Hardware Requirements

Virtual environments require very large physical memories, high-speed processors, high-bandwidth mass storage capacity, and high-speed interface ports for interaction devices (Durlach and Mavor, 1995). The memory bandwidth problem is being assuaged by SDRAM, which incorporates new features that allow it to keep pace with bus speeds as high as 100 (peak bandwidth 800 MB/s) to 133 MHz (peak bandwidth 1.1 GB/s), with promises of 400 MHz and higher-speed memory in the near term (Stanney and Zyda, 2002). The latest processors to surpass the gigahertz barrier include AMD's Athlon and Opteron; Hewlett-Packard's 8700; IBM's Power4; Intel's Itanium, "Xeon," and Pentium IV, the latter of which has hyperthreading (HT) technology, which improves processor multitasking performance and responsiveness; and Sun's UltraSPARC IV, which marks the first milestone in Sun's chip multithreading. The future looks even brighter, with promises of massive parallelism in computing via the realization of molecular feature size limits in integrated circuits (Appenzeller, 2000). Graphics processing units (GPUs) will also boost computing power substantially, potentially on the order of 1000% (Blachford, 2004; Buck et al., 2004). With the rapidly advancing ability to generate complex and large-scale virtual worlds, hardware advances in multimodal input–output (I/O) devices, tracking systems, and interaction techniques are needed to support generation of increasingly engaging virtual worlds.

2.1.1 Multimodal I/Os

Virtual environments can provide sensorially rich experiences that engage multiple senses. The eyes can be stimulated by enticing visuals, the ears by three-dimensional (3D) sound, the body by tactile gloves or vests, and the nose by olfactory stimulation. Many different types of peripheral devices are available to provide these sensory inputs and technologies for integrating them into a realistic multimodal experience continue to advance.

Visual Displays In terms of visual displays, the main differentiating factor that often sets VE apart from other interactive solutions is the provision of stereo viewing (although not all VEs are in stereo), which can be provided in low- to high-end technology solutions. Low-technology stereo viewing methods, include anaglyph methods, where a viewer wears glasses with distinct color-polarized filters; parallel or cross-eyed methods, in which right and left images are displayed adjacently (in parallel or crossed), requiring the viewer to actively fuse the separate images into one stereo image; parallax stereogram methods, in which an image is made by interleaving columns of two images from a left- and right-eye perspective image of a 3D scene; polarization methods, in which the images for the left and right eyes are projected on a plane through two orthogonal linearly polarizing filters (i.e., the right image is polarized horizontally; the left is polarized vertically) and glasses with polarization filters are donned to see the 3D effect; Pulfrich methods, in which an image of a scene moves sideways across the viewer's field of view and one eye is covered by a dark filter so that the darkened image reaches the brain later, causing stereo disparity; and shutter glass methods, in which images for the right and left eyes are displayed in quick alternating sequence and special shutter glasses are worn that "close" the right or left eye at the correct time (Halle, 1997; Taylor, 2004; Vince, 2004). All of these low-technology solutions are limited in terms of their resolution, the maximum number of views that they can display, and clunky implementation; they can also be associated with pseudoscopic images (i.e., the depth of an object can appear to flip inside out).

Spatially immersive displays (SIDs) surround viewers physically, often in a room-sized display, with panoramic large field-of-view imagery generally projected via fixed front or rear projection display units (Majumder, 1999). The views provided by SIDS are generally responsive to a user's point of view, orientation, and actions. Examples of this type of display include the Cave Automated Virtual Environment (CAVE) (Cruz-Neira et al., 1993), ImmersaDesk, PowerWall, Infinity Wall, and Vision-Dome (Majumder, 1999). Issues with SIDS include a stereo view that is correct for only one or a few viewers, noticeable overlaps between adjacent projections, and image warp on curved screens (Taylor, 2004).

There are several desktop stereo display options. One of the earliest desktop displays was the Responsive Workbench, which projected stereoscopic images onto a horizontal tabletop display surface via a projector-and-mirrors system that was viewed through shutter glasses (Agrawala et al., 1997). More recent efforts include the Totally Active WorkSpace (TAWS), which is a CAVE-like structure where the user works on a glass desk surface (Johnson et al., 1999). Another is the Personal Penta Panel (P3), which is an open box made out of five flat panels that users place their tracked head and hands into and are presented with a surround stereo view. There is also the Personal Augmented Reality Immersive System (PARIS), which is a desktop augmented reality device. A newcomer is the Interactive Real-time Imaging Solutions, IRIS-3D (http://www.iris3d.com/), which generates high-resolution auto-stereoscopic visualizations from a desktop environment through a dual-channel 3D projection display system. These desktop display systems have advantages over SIDS because they are smaller, easier to configure in terms of mounting cameras and microphones, easier to integrate with gesture and haptic devices, and more readily provide access to conventional interaction devices such as mice, joysticks, and keyboards. Issues with such displays include stereo that is only accurate for one viewer and a limited display volume (Taylor, 2004).

In terms of visual displays, the one that has received the greatest attention, in both hype and disdain, is almost certainly the head-mounted display (HMD). One benefit of HMDs is their compact size, as an HMD when coupled with a head tracker can be used to provide a similar visual experience as a multitude of bulky displays associated with SID and desktop solutions. There are three main types of HMDs: monocular (i.e., one display source), biocular (i.e., two displays with separate display and optics paths that show one image), and binocular (i.e., stereoscopic viewing via two image generators) (Vince, 2004). HMDs generally use CRT or LCD image sources, with CRTs typically providing higher resolution, although at the cost of increased weight. When coupled with tracking devices, HMDs can be used to present 3D visual scenes that are updated as a user moves his or her head about a virtual world. Although this often provides an engaging experience, due to poor optics, sensorial mismatches, and slow update rates, these devices are also often associated with adverse effects such as eyestrain and nausea (May and Badcock, 2002). Further, although HMDs have come down substantially in weight, rendering them more suitable for extended wear, they are still hindered by cumbersome designs, obstructive tethers, suboptimal resolution, and insufficient field of view. These adverse effects may be the reason behind why, in a recent review of HMD devices, over 30% had been discontinued by their manufacturers (Bungert, 2004). Of the HMDs available, there are several low- to mid-cost models that are relatively lightweight (approximately 39 to 1000 g) and provide a horizontal field of view (30 to 40° per eye) and resolution (180 K to 2.4 M pixels/LCD) exceeding predecessor systems.

Another option in visual displays includes volumetric displays that fill a volume of space with a "floating" image (Halle, 1997). With volumetric displays, imagery can be seen from a multitude of viewing angles, generally without the need for goggles. A recent example of this type of display is the Perspecta Spatial 3D, which projects images onto a rotating flat screen inside a glass dome (Favalora et al., 2003). Issues with volumetric displays include low resolution and the tendency for transparent images to lose interposition cues (Taylor, 2004). Also, view-independent shading of objects is not possible with volumetric displays, and current solutions do not exhibit arbitrary occlusion by inter position of objects (Halle, 1997).

There are thus many options for a VE visual display (e.g., low-technology stereo imaging, SIDs, desktop stereo displays, HMDs, volumetric displays). VE developers must determine which is most appropriate for their application context (see Table 1). The way of the future seems to be wearable computer displays (e.g., Microvision's Nomad Augmented Vision System, MicroOptical), which can incorporate miniature LCDs directly into conventional eyeglasses (Lieberman, 1999; Chinthammit et al., 2003). If designed effectively, these devices should eliminate the tethers and awkwardness of current designs while enlarging field of view and enhancing resolution.

Audio Displays One has "to understand that also sound is half of the experience of a total movie-going experience" (Stephen Spielberg; statement made at an endowment reception for the USC film school, October 2002; Waterman, 2002). The same could be said for virtual environments. Unfortunately, in terms of VEs, the development of audio displays has lagged considerably behind video displays. Shilling (2002) suggests that "even though the medium is non-interactive, a movie with a properly designed audio track represents a better example of an auditory virtual environment than most high-end simulations developed by industry and the military."

Audio can be presented via spatialized or non-spatialized displays. Just as stereo visual displays are a defining factor for VE systems, so are "interactive" spatialized audio displays. The 3D Working Group of the Interactive Audio Special Interest Group, which is supervised by the MIDI Manufacturers Association, (1998, p. 9) defines interactive 3D audio as that which allows for "on-the-fly positioning of sounds anywhere in a three-dimensional space

Table 1 Strengths, Weaknesses, and Potential Applications of Various Visual Displays

Visual Display Type	Strengths	Weaknesses	Potential Applications
Low-technology stereo viewing methods	Can be provided in a wide range of display sizes; produce photo realistic images	Limited resolution; limited maximum number of views; clunky implementation; potential for pseudoscopic images	Art and museum exhibits; medical imaging; online applications
Spatially immersive displays	Physically surround viewers (i.e., unencumbered viewers); wide field-of-view imagery; high resolution; views generally responsive to a user's point of view, orientation, and actions; group viewing	Stereo view correct for only one or a few viewers; noticeable overlaps between adjacent projections; image warp on curved screens	Collaborative design or exploration; architecture; collaborative telepresence or teleconferencing
Desktop stereo displays	Small and easy to configure; easy to integrate with gesture and haptic devices; readily provide access to conventional interaction devices	Stereo only accurate for one viewer; limited display volume	Command and control; product design; information visualization and exploration; medical training, imaging, and exploration; entertainment; virtual classrooms
Head-mounted displays	Compact size; provide a similar visual experience as a multitude of bulky displays; can be highly immersive	Poor optics; sensorial mismatches; slow update rates; may cause eyestrain and nausea; cumbersome designs; obstructive tethers; suboptimal resolution; insufficient field of view	Hands-free operations; augmented reality (i.e., 3D mapping of graphical images onto real images); medical training, imaging, and exploration; remote maintenance; entertainment
Volumetric displays	Image visible from a wide range of viewpoints; provides sense of ocular accommodation	View-independent shading of objects not possible; current solutions do not exhibit arbitrary occlusion of one part of an image volume by another; low resolution; tendency for transparent images to lose interposition cues	Command and control; product design; information visualization and exploration; medical imaging and exploration; entertainment; in general, nonphoto realistic applications

surrounding a listener." Thus, it is 3D positioning and interactivity (i.e., sounds created on the fly based on user input) that differentiate spatialized sound. VRSonic's SoundScape3D (http://www.vrsonic.com/), Aureal's A3D Interactive (http://www.a3d.com/), and AuSIM3D (http://ausim3d.com/) are examples of positional 3D audio technology.

Research in spatial audio display has traditionally focused on the technological problems of spatialization; not much attention has been paid to the challenges of sound design for spatial audio environments. The tools that are currently being used for creating these environments are largely based on traditional sound design methods developed for film, television, and music production. Unfortunately, these methods cannot adequately represent the dynamism and spatial complexity of the virtual soundscape. Recently, however, there have been promising developments in new sound modeling paradigms and sound design principles that will hopefully lead to a new generation of tools for designing effective spatial audio environments (Hahn et al., 1998; Fouad and Ballas, 2000; Fouad, 2004). For example, VRSonic's ViBe technology is a new modeling paradigm for sound that captures the complex dynamic behaviors of real-world sounds; it allows sound designers to create behavioral audio models that encapsulate the sonic behaviors of objects.

Spatialized audio can be instrumental in helping a person pinpoint the direction and distance of a target object (e.g., adversary's position in a training scenario; direction of fire), monitor multiple streams of auditory information (e.g., command

and control), monitor information about events outside one's field of view (e.g., air-traffic control), aid night operations, direct visual attention, and enhance activities that benefit from high levels of engagement (e.g., training, entertainment) (Shilling and Shinn-Cunningham, 2002; Nelson and Bolia, 2003).

Developers must decide if sounds should be presented via headphones or speakers. For nonspatialized audio, most audio characteristics (e.g., timbre, relative volume) are generally considered to be equivalent whether projected via headphones or speakers. This is not so for spatialized audio, in which the information must generally be projected through headphones in order to exploit subtle nuances. However, spatialized audio headphones tend to be more expensive than speakers, and they also tend to require customization to an individual listener for optimum effect; while speaker systems have a larger footprint, are sensitive to room acoustics, and the listener must be placed in the "sweet spot" (i.e., ideal listening position) for the proper 3D audio effect to be realized (Shilling, 2002).

With true 3D audio, a sound can be placed in any location, right or left, up or down, near or far (Begault, 1994). Systems that produce true 3D audio use a head-related transfer function (HRTF) to represent the manner in which sound sources change as a listener moves his or her head, which can be specified with knowledge of the source position as well as the position and orientation of the listener's head (Butler, 1987; Cohen, 1992). The HRTF is dependent on the physiological makeup of the listener's ear (i.e., the pinna does a nonlinear fitting job in the HRTF). Recent advances have allowed for the development of personalized HRTF functions; however, currently, these functions still require a significant amount of calibration time and fail to provide adequate cues for front-to-back or up-down localization (Crystal Rivers Engineering, 1995). These functions also have yet to include reverberation or echoes, which would cause auditory cues to seem more realistic and provide robust relative source distance information (Shilling and Shinn-Cunningham, 2002). Such acoustic characteristics are instead simulated, for example, by causing reverberation as a function of space geometry and surface reflectivity (Funkhouser and Sequin, 1993). With this technique, heuristics and simplified acoustic models of sound are used to provide basic environmental effects due to such factors as room size and material composition.

Ideally, a more generalized HRTF could be developed that would be applicable to a multitude of users. This may be possible by using a best-fit HRTF selection process in which one finds the nearest matching HRTF in a database of candidate HRTFs by comparing the physiological characteristics of stored HRTFs to those of a target user (Algazi et al., 2001).

Haptic Displays Haptic interaction involves all aspects of touch and movement of the hand or body segments as they apply to computer interaction. Haptic displays can be used to support two-way communications between humans and interactive systems, enabling bidirectional interaction between a user and his or her surroundings (Hale and Stanney, 2004). In general, haptic displays are effective at alerting people to critical tasks (e.g., warning), providing a spatial frame of reference within one's personal space, and supporting hand–eye coordination tasks. Texture cues, such as those conveyed via vibrations or varying pressures, are effective as simple alerts and may speed reaction time and aid performance in degraded visual conditions (Massimino and Sheridan, 1993; Akamatsu, 1994; Biggs and Srinivasan, 2002; Mulgund et al., 2002). Kinesthetic devices are advantageous when tasks involve hand–eye coordination (e.g., object manipulation), where haptic sensing and feedback are key to performance. Currently available haptic interaction devices include static displays (i.e., convey deformability or Braille); vibrotactile, electrotactile, and pneumatic displays (i.e., convey tactile sensations such as surface texture and geometry, surface slip, surface temperature); force feedback systems (i.e., convey object position and movement distances); and exoskeleton systems (i.e., enhance object interaction and weight discrimination) (Hale and Stanney, 2004).

Vestibular Displays The vestibular system, which is located in the inner ear and consists of the otolith organs and three semicircular canals arranged at right angles to one another, serves to balance the body (Howard, 1986). While the otolith organs detect primarily linear accelerations, each of the semicircular canals detects angular motion [i.e., rotation (roll, pitch, or yaw) along a given axis]. When movement along a plane occurs, displacement of hair follicles within the semicircular canals is interpreted by the brain as acceleration. The "vestibular system can be exploited to create, prevent, or modify acceleration perceptions" (Lawson et al., 2002, p. 137). For example, by simulating acceleration cues, a person can be psychologically transported from his or her veridical location, such as sitting in a chair in front of a computer, to a simulated location, such as the cockpit of a moving airplane.

While vestibular cues can be stimulated via many different techniques, three of the most promising methods are physical motion of the user (e.g., motion platforms), wide field-of-view visual displays that induce vection (i.e., an illusion of self-motion) and locomotion devices that induce illusions of self-motion without physical displacement of the user through space (e.g., walking in place, treadmills, pedaling, foot platforms) (Hettinger, 2002; Hollerbach, 2002; Lawson et al., 2002). Of these options, motion platforms are probably the most advanced. Motion platforms are generally characterized via their range of motion/degrees of freedom and actuator type (Isdale, 2000). In terms of range of motion,

motion platforms can move a person in many combinations of translational (i.e., surge-longitudinal motion, sway-lateral motion, heave-vertical motion) and rotational (i.e., roll, pitch, yaw) degrees of freedom (DOF). A single-DOF translational motion system might provide a vibration sensation via a "seat shaker." A common six-DOF configuration is a hexapod, which consists of a frame with six or more extendable struts (actuators) connecting a fixed base to a movable platform. In terms of actuators, electrical actuators are quiet and relatively maintenance free; however, they are not very responsive and they cannot hold the same load as can hydraulic or pneumatic systems. Hydraulic and pneumatic systems are smoother, stronger, and more accurate; however, they require compressors, which may be noisy. Servos are expensive and difficult to program.

Although other multimodal interactions are possible (e.g., gustatory, olfactory), there has been limited research and development beyond the primary three interaction modes (i.e., visual, auditory, haptic), although efforts have been made to support advances in olfactory interaction (Jones et al., 2004).

2.1.2 Tracking Systems

Tracking systems allow determination of a user's head or limb position and orientation, or the location of handheld devices, to allow interaction with virtual objects and traversal through 3D computer-generated worlds (Foxlin, 2002). Tracking is what allows the visual scene in a VE to coincide with a user's point of view, thereby providing an egocentric real-time perspective. Tracking systems must be coupled carefully with the visual scene, however, to avoid unacceptable lags (Kalawsky, 1993), which often manifest as a set of symptoms collectively termed *cybersickness* (Kennedy et al., 1997). Advances in tracking technology have been realized in terms of drift-corrected gyroscopic orientation trackers, outside-in optical tracking for motion capture, and laser scanners (Foxlin, 2002). The future of tracking technology is probably the hybrid tracking system, with an acoustic-inertial hybrid on the market (see http://www.isense.com/products/) and several others in research labs (e.g., magnetic-inertial, optical-inertial, and optical-magnetic). In addition, ultrawideband radio technology holds promise for an improved method of omnidirectional point-to-point ranging.

Tracking technology also allows for gesture recognition, in which human position and movement are tracked and interpreted to recognize semantically meaningful gestures (Turk, 2002). Tracking devices that are worn (e.g., gloves, bodysuits) are currently more advanced than passive techniques (e.g., camera, sensors), yet the latter hold much promise for the future, as they are more powerful and less obtrusive than those that must be worn.

2.1.3 Interaction Techniques

Although one may think of joysticks and gloves when considering VE interaction devices, there are many techniques that can be used to support interaction with and traversal through a virtual environment. Interaction devices support traversal, pointing and selection of virtual objects, tool use (e.g., through force and torque feedback), tactile interaction (e.g., through haptic devices), and environmental stimuli (e.g., temperature, humidity) (Bullinger et al., 2001).

Supporting traversal throughout a VE via motion interfaces is of primary importance (Hollerbach, 2002). Motion interfaces are categorized as either active (i.e., locomotion) or passive (i.e., transportation). Active motion interfaces require self-propulsion to move about a virtual environment (e.g., treadmill, pedaling device, foot platforms) (Templeman et al., 1999). Passive motion interfaces transport users within a VE without significant user exertion (e.g., inertial motion, as in a flight simulator, or noninertial motion, such as in the use of a joystick or gloves). The utility, functionality, cost, and safety of active motion interfaces beyond traditional options (e.g., joysticks) have yet to be proven. In addition, beyond physical training, concrete applications for active interfaces have yet to be clearly delineated.

Another interaction option is speech control. Speaker-independent continuous speech recognition systems are currently commercially available (Huang, 1998). For these systems to provide effective interaction, however, additional advances are needed in acoustic and language modeling algorithms to improve the accuracy, usability, and efficiency of spoken language understanding.

Gesture interaction allows users to interact through nonverbal commands conveyed via physical movement of the fingers, hands, arms, head, face, or other body limbs (Turk, 2002). Gestures can be used to specify and control objects of interest, direct navigation, manipulate the environment, and issue meaningful commands. To support natural and intuitive interaction, a variety of interaction techniques can be coupled. Combining speech interaction with nonverbal gestures and motion interfaces can provide a means of interaction that closely captures real-world communications.

2.2 Software Requirements

Software development of VE systems has progressed tremendously, from proprietary and arcane systems, to development kits that run on multiple platforms (i.e., general-purpose operating systems to workstations) (Pountain, 1996). Virtual environment system components are becoming modular and distributed, thereby allowing VE databases (i.e., editors used to design, build, and maintain virtual worlds) to run independently of visualizer and other multimodal interfaces via network links. Standard APIs (Application Program Interfaces) (e.g., OpenGL, Direct3D, Mesa) allow multimodal components to be hardware-independent. Virtual environment programming languages are advancing, with APIs, libraries, and particularly, scripting languages allowing nonprogrammers to develop virtual worlds (Stanney and Zyda, 2002). Advances are also being made in modeling of autonomous agents and communication networks used to support multiuser virtual environments.

2.2.1 Modeling

A VE consists of a set of geometry, the spatial relationships between the geometry and the user, and the change in geometry invoked by user actions or the passage of time (Kessler, 2002). Generally, modeling starts with building geometry components (e.g., graphical objects, sensors, viewpoints, animation sequences) (Kalawsky, 1993). These are often converted from CAD data. These components are then imported into the VE modeling environment and rendered when appropriate sensors are triggered. Color, surface textures, and behaviors are applied during rendering. Programmers control the events in a VE by writing task functions, which become associated with imported components.

A number of 3D modeling languages and toolkits are available that provide intuitive interfaces and run on multiple platforms and renderers (e.g., 3D Studio Max, AC3D Modeler, AccuRender, ACIS 3D, Ashlar-vellum Argon/Xenon/Cobalt, Carrara, CINEMA 4D, DX Studio, EON Studio, MultiGen Creator and Vega, RenderWare, solidThinking) (Ultimate 3D Links, 2003). In addition, there are scene management engines (e.g., R3vis Corporation's OpenRM Scene Graph is an API that provides cross-platform scene management and rendering services) that allow programmers to work at a higher level, defining characteristics and behaviors for more holistic concepts (Kershner, 2002; Menzies, 2002). There have also been advances in photo-realistic rendering tools (e.g., Electric Image Animation System), which are evolving toward full-featured physics-based global illumination rendering systems (Heirich and Arvo, 1997; Merritt and Bacon, 1997). Taken together, these advances in software modeling allow for generation of complex and realistic VEs that can run on a variety of platforms, permitting access to VE applications by both small- and large-scale application development budgets.

2.2.2 Autonomous Agents

Autonomous agents are synthetic or virtual human entities that possess some degree of autonomy, social ability, reactivity, and proactiveness (Allbeck and Badler, 2002). They can have many forms (e.g., human, animal), which are rendered at various levels of detail and style, from cartoonish to physiologically accurate models. Such agents are a key component of many VE applications involving interaction with other entities, such as adversaries, instructors, or partners (Stanney and Zyda, 2002).

There has been significant research and development in modeling embodied autonomous agents. As with object geometry, agents are generally modeled off-line and then rendered during real-time interaction. Although the required level of detail varies, modeling of hair and skin generally adds realism to an agent's appearance (Allbeck and Badler, 2002). There are a few toolkits available to support agent development, with one of the most notable offered by Boston Dynamics, Inc. (BDI) (http:///www.bdi.com/), a spin-off from the MIT Artificial Intelligence Laboratory. BDI's products allow VE developers to work directly in a 3D database, interactively specifying an agent's physical behaviors, as well as emotional and psychological dimensions. The resulting agents move realistically, respond to simple commands, and travel about a VE as directed. Another option is ArchVision's (http://www.archvision.com/) 3D Rich Photorealistic Content (RPC) People.

2.2.3 Networks

Distributed networks allow multiple users at diverse locations to interact within the same virtual environment. Improvements in communications networks are required to allow realization of such shared experiences in which users, objects, processes, and autonomous agents from diverse locations interactively collaborate (Durlach and Mavor, 1995). Yet the foundation for such collaboration has been built within the Next Generation Internet (NGI). The NGI initiative (http://www.ngi.gov/) aimed to connect a number of universities and national labs at speeds 100 to 1000 times faster than the 1996 Internet in order to experiment with collaborative-networking technologies, such as high-quality videoconferencing and audio and video streams. The NGI's successor, the Large Scale Networking (LSN) Coordinating Group (http://www.hpcc.gov/iwg/lsn.html), aims to develop technologies and services that enable wireless, optical, mobile, and wireline communications and enhance networking software, with a goal of achieving the terabit per second objective, which proved elusive to the NGI.

In addition, Internet2 is using existing networks [e.g., NSF's vBNS (very high-speed Backbone Network Service)] to determine the transport designs necessary to carry real-time multimedia data at high speed (http://apps.internet2.edu/). Distributed VE applications can leverage the special capabilities (i.e., high bandwidth, low latency, low jitter) of these advancing network technologies to provide shared virtual worlds (Singhal and Zyda, 1999).

3 DESIGN AND IMPLEMENTATION STRATEGIES

Many conventional HCI techniques can be used to design and implement VE systems; however, there are unique cognitive, content, products liability, and usage protocol considerations that must be addressed.

3.1 Cognitive Aspects

The fundamental objective of VE systems is to provide multimodal interaction or, when sensory modalities are missing, perceptual illusions that support human information processing in pursuit of a VE application's goals, which could range from training to entertainment. Ancillary yet fundamental to this goal is to minimize cognitive obstacles, such as navigational difficulties, that could render a VE application's goals inaccessible.

3.1.1 Multimodal Interaction Design

Virtual environments are designed to provide users with direct manipulative and intuitive interaction with multisensory stimulation (Bullinger et al., 2001). The number of sensory modalities stimulated and the quality of this multisensory interaction are critical to the realism and potential effectiveness of VE systems (Popescu et al., 2002). Yet there is currently a limited understanding of how to effectively provide such sensorial parallelism (Burdea, 1996); however, Stanney et al. (2004) have provided a set of preliminary cross-modal integration rules. These rules consider aspects of multimodal interaction, including temporal and spatial coincidence, working memory capacity, intersensory facilitation effects, congruency, and inverse effectiveness. When multimodal sensory information is provided to users it is essential to consider such rules governing the integration of multiple sources of sensory feedback. People have adapted their perception-action systems to "expect" a particular type of information flow in the real world; VEs run the risk of breaking these perception–action couplings if either the full range of sensory stimuli are not supported or if they are supported in a manner that is not contiguous with real-world expectations. Such pitfalls can be avoided through consideration of the coordination between sensing and user command and the transposition of senses in the feedback loop. Specifically, command coordination considers user input as primarily monomodal and feedback to the user as multimodal. Designers need to consider which input modalities are most appropriate to support execution of a given task within the VE, if there is any need for redundant user input, and whether or not users can effectively handle such parallel input (Stanney et al., 1998a, 2004).

A limiting factor in supporting multimodal sensory stimulation in VEs is the current state of interface technologies. With the exception of the visual modality, current levels of technology simply cannot even begin to reproduce virtually those sensations, such as haptics and audition, which users expect in the real world. One solution to current technological short comings, *sensorial transposition*, occurs when a user receives feedback through senses other than those expected, which may occur because a command coordination scheme has substituted available sensory feedback for those that cannot be generated within a virtual environment. Sensorial substitution schemes may be one for one (e.g., visual for force) or more complex (e.g., visual for force and auditory; visual and auditory for force). If designed effectively, command coordination and sensory substitution schemes should provide multimodal interaction that allows for better user control of the virtual environment. On the other hand, if designed poorly, these solutions may in fact exacerbate interaction problems.

3.1.2 Perceptual Illusions

When sensorial transpositions are used, there is an opportunity for perceptual illusions to occur. With perceptual illusions, certain perceptual qualities perceived by one sensory system are influenced by another sensory system (e.g., "feel" a squeeze when you *see* your hand grabbing a virtual object). Such illusions could simplify and reduce the cost of VE development efforts (Storms, 2002). For example, when attending to a high-quality visual image coupled with a low-quality auditory display, auditory–visual cross-modal perception allows for an increase in the perceived quality of the visual image. Thus, in this case, if the visual image is the focus of the task, there may be no need to use a high-quality auditory display. Unfortunately, little is known about how to leverage these phenomena to reduce development costs while enhancing one's experience in a virtual environment. Perhaps the one exception is vection (i.e., a compelling illusion of self-motion throughout a virtual world), which is known to be enhanced via a number of display factors, including a wide field of view and high spatial frequency content (Hettinger, 2002). Other such illusions exist (e.g., visual dominance) and could be leveraged similarly if perceptual and cognitive design principles are identified that can be used to trigger and capitalize on these illusory phenomena.

3.1.3 Navigation and Wayfinding

Effective multimodal interaction design and use of perceptual illusions can be impeded if navigational complexities arise. Navigation is the aggregate of wayfinding (i.e., cognitive planning of one's route) and the physical movement that allows travel throughout a virtual environment (Darken and Peterson, 2002). Since VEs typically lack the ability to support the full complement of sensory cues that users expect, it is likely that the ability to develop robust mental models of the spatial layout of a VE is reduced, and users can become disoriented or lost. For example, if a VE system lacks vestibular input from which to derive heading, such as that provided via an HMD, heading disorientation would be anticipated. Consequently, a number of tools and techniques have been developed to aid wayfinding in virtual worlds, including maps, landmarks, trails, and direction finding. These tools can be used to display current position, current orientation (e.g., compass), log movements (e.g., "breadcrumb" trails), demonstrate or access the surround (e.g., maps, binoculars), or provide guided movement (e.g., signs, landmarks) (Chen and Stanney, 1999). Darken and Peterson (2002) provide a number of principles concerning how best to use these tools. In addition, Stanney (2005) have provided preliminary technology specification guidelines for training spatial knowledge acquisition. If applied effectively to VEs, these principles should lead to reduced disorientation and enhanced wayfinding in large-scale virtual environments.

3.2 Content Development

Content development is concerned with the design and construction of virtual objects and synthetic environments that support a VE experience (Isdale et al., 2002). Although this medium can leverage existing HCI design principles for content preparation (Proctor et al., 2002), VEs have unique design challenges that arise due to the demands of real-time, multimodal, collaborative interaction. In fact, content designers are just starting to appreciate and determine what it means to create a full sensory experience with user control of both point of view and narrative development. For example, aesthetics is thought to be a product of agency (e.g., pleasure of being), narrative potential, presence and co-presence (e.g., existing in and sharing the virtual experience), as well as transformation (e.g., assuming another persona) (Murray, 1998; Church, 1999). In general, content development should be about stimulating perceptions (e.g., sureties, surprises), as well as contemplation over the nature of being (Fencott, 1999; Isdale et al., 2002).

Existing design techniques, for example from computer games and theme park development, can be used to support VE content development. Game development techniques that can be leveraged in VE content development include but are not limited to providing a clear sense of purpose, emotional objectives, perceptual realism, intuitive interfaces, multiple solution paths, challenges, a balance of anxiety and reward, and an almost unconscious flow of interaction (Isdale et al., 2002). From theme park design, content development suggestions include having a story that provides an all-encompassing theme and thus the "rules" that guide VE design, providing location and purpose, using cause-and-effect to lead users to their own conclusions, and anchoring users in the familiar (Carson, 2000a, b). Although these suggestions provide guidelines for VE content development, considerable creativity is still an essential component of the process. Isdale et al. (2002) suggest that the challenges of VE content development highlight the need for art to complement technology.

3.3 Products Liability

Those who implement VE systems must be cognizant of potential products liability concerns. Exposure to a VE system often produces unwanted side effects that could render users incapable of functioning effectively upon return to the real world. These adverse effects may include nausea and vomiting, postural instability, visual disturbances, and profound drowsiness (Stanney et al., 1998b). As users subsequently take on their normal routines unaware of these lingering effects, their safety and well-being may be compromised. If a VE product occasions such problems, liability of VE developers or system administrators could range from simple accountability (i.e., reporting what happened) to full legal liability (i.e., paying compensation for damages) (Kennedy and Stanney, 1996a; Kennedy et al., 2002). To minimize their liability, manufacturers and corporate users should design systems and provide usage protocols to minimize risks, warn users about potential aftereffects, monitor users during exposure, assess users' risk, and debrief users upon postexposure.

3.4 Usage Protocols

To minimize products liability concerns, VE usage protocols should be carefully designed. Adverse responses to exposure vary directly with the stimulus intensity of the VE and susceptibility of the person exposed (Stanney et al., 2002). To minimize VE stimulus intensity, the following should be considered:

- Are system lags and latencies minimized and stable?
- Are frame rates optimized?
- Is the interpupilary distance (IPD) of the visual display adjustable?
- Is a large field of view causing excessive vection, such that the spatial frequency content of the visual scene should be reduced?
- Is multimodal feedback integrated such that sensory conflicts are minimized?

Self-report has been found useful in gauging individual susceptibility. In particular, the Motion History Questionnaire (MHQ) (Kennedy and Graybiel, 1965) has been found effective in determining susceptibility to motion sickness associated with VE exposure (Kennedy et al., 2001). The MHQ assesses susceptibility based on past occurrences of sickness in inertial environments. Those persons determined to be susceptible to motion sickness can be expected to experience more than twice the level of adverse effects to VE exposure as that of nonsusceptible persons (Stanney et al., 2003). These people should thus be carefully monitored during and after VE exposure.

Regardless of the strength of the stimulus or the susceptibility of the user, following a systematic usage protocol can minimize the adverse effects associated with VE exposure. During VE exposure, the room should be arranged such that there is adequate airflow and comfortable thermal conditions, as sweating often precedes an emetic response (Stanney et al., 2002). All persons, regardless of their susceptibility, should be educated about the potential risks of VE exposure, including the potential for nausea, malaise, disorientation, headache, dizziness, vertigo, eyestrain, drowsiness, fatigue, pallor, sweating, increased salivation, and vomiting. Users should be prepared for the transition into the HMD by informing them that there will be a perceptual adjustment period. To minimize fatigue, which can exacerbate adverse effects, all equipment should be adjusted to ensure comfortable fit and unobstructed movement. Particularly for strong VE stimuli, initial exposure should be short (e.g., 10 minutes), and reexposure should be prohibited for two to five days. All users should be monitored during exposure and red flags (e.g., profuse sweating, burping, verbal frustration, restricted head or body movement) should be attended to prudently. Users demonstrating any of these behaviors should be observed closely, as they may experience an emetic response, and extra care should be taken with these persons postexposure. Some people may be unsteady upon postexposure and may need assistance when initially standing up after exposure. After exposure the well-being of users should be assessed, for which a derivative of the field sobriety test can be used (Kennedy and Stanney, 1996b).

Depending on the strength of the VE stimulus, the amount of time after exposure that users must remain on premises before driving or participating in other such high-risk activities should be determined. People should be informed that upon postexposure they may experience disturbed visual functioning, visual flashbacks, and unstable locomotor and postural control for prolonged periods following exposure.

4 HEALTH AND SAFETY ISSUES

The health and safety issues associated with VE exposure complicate usage protocols and lead to products liability concerns. It is thus essential to understand these issues if one is going to utilize VE technology. There are both physiological and psychological risks associated with VE exposure, the former being related primarily to sickness and aftereffects and the latter being concerned primarily with the social impact.

4.1 Cybersickness, Adaptation, and Aftereffects

Motion sickness-like symptoms and other aftereffects (i.e., balance disturbances, visual stress, altered hand–eye coordination) are unwanted by-products of VE exposure. The sickness related to VE systems is commonly referred to as *cybersickness* (McCauley and Sharkey, 1992). Some of the most common symptoms exhibited include dizziness, drowsiness, headache, nausea, fatigue, and general malaise (Kennedy et al., 1993). More than 80% of users will experience some level of disturbance, with approximately 12% ceasing exposure prematurely due to this adversity (Stanney et al., 2003). Of those who drop out, approximately 10% can be expected to have an emetic response (i.e., vomit); however, only 1 to 2% of all users will have such a response. These adverse effects are known to increase in incidence and intensity with prolonged exposure duration (Kennedy et al., 2000). Although most users will experience some level of adverse effects, symptoms vary substantially from one person to another as well as from one system to another (Kennedy and Fowlkes, 1992). These effects can be assess via the Simulator Sickness Questionnaire (Kennedy et al., 1993), with values above 20 requiring due caution (i.e., warn and observe users) (Stanney et al., 2002).

To overcome such adverse effects, people generally undergo physiological adaptation during VE exposure. This adaptation is the natural and automatic response to an intersensorily imperfect VE and is elicited due to the plasticity of the human nervous system (Welch, 1978). Due to technological flaws (e.g., slow update rate, sluggish trackers), users of VE systems may be confronted with one or more intersensory discordances (e.g., visual lag, a disparity between seen and felt limb position). To perform effectively in the VE, they must compensate for these discordances by adapting their psychomotor behavior or visual functioning. Once interaction with a VE is discontinued, these compensations persist for some time after exposure, leading to aftereffects.

Once VE exposure ceases and users return to their natural environment, they are probably unaware that interaction with the VE has potentially changed their ability to interact effectively with their normal physical environment (Stanney and Kennedy, 1998). Several different kinds of aftereffects may persist for prolonged periods following VE exposure (Welch, 1997). For example, hand–eye coordination can be degraded via perceptual-motor disturbances (Rolland et al., 1995; Kennedy et al., 1997), postural sway can arise (Kennedy and Stanney, 1996b), as can changes in the vestibulo-ocular reflex (VOR), i.e., one's ability to stabilize an image on the retina (Draper et al., 1997). The implications of these aftereffects are that (1) VE exposure duration may need to be minimized, (2) highly susceptible persons or those from clinical populations (e.g., those prone to seizures) may need to avoid or be banned from exposure, (3) users should be closely monitored during VE exposure, and (4) users' activities should be closely monitored for a considerable period of time upon VE postexposure to avoid personal injury or harm.

4.2 Social Impact

Like its ancestors (e.g., television, computers), virtual environment technology has the potential for negative social implications through misuse and abuse (Kallman, 1993). Yet violence in VE is nearly inevitable, as evidenced by the violent content of popular video games. Such animated violence is a known favorite over the portrayal of more benign emotions such as cooperation, friendship, or love (Sheridan, 1993). The concern is that users who engage in what seems like harmless violence in the virtual world may become desensitized to violence and mimic this behavior in the look-alike real world.

Currently, it is not clear whether or not such violent behavior will result from VE exposure, but early research is not reassuring. Calvert and Tan (1994) found VE exposure to increase significantly the physiological arousal and aggressive thoughts of young adults. Perhaps more disconcerting was that neither aggressive thoughts nor hostile feelings were found to decrease due to VE exposure, thus providing no support for catharsis. Such increased negative stimulation may subsequently be channeled into real-world activities. The ultimate concern is that VE immersion may potentially be a more powerful perceptual experience than past, less interactive technologies, thereby increasing the negative social impact of this technology (Calvert, 2002). A proactive approach is needed which weighs the risks and potential consequences associated with VE exposure against the benefits. Waiting for the onset of harmful social consequences should not be tolerated.

5 VIRTUAL ENVIRONMENT USABILITY ENGINEERING

Most VE user interfaces are fundamentally different from traditional graphical user interfaces, with unique I/O devices, perspectives, and physiological interactions. Thus, when developers and usability practitioners attempt to apply traditional usability engineering methods to the evaluation of VE systems, they find few if any that are particularly well suited to these environments (for notable exceptions, see Gabbard, 1997; Stanney et al., 2000; Hix and Gabbard, 2002). There is a need to modify and optimize available techniques to meet the needs of VE usability engineering, as well as to better characterize factors unique to VE usability, including sense of presence and VE ergonomics.

5.1 Usability Techniques

Assessment of usability for VE systems must go beyond traditional approaches, which are concerned with the determination of effectiveness, efficiency, and user satisfaction. Evaluators must consider whether multimodal input and output is optimally presented and integrated, navigation is supported to allow the VE to be readily traversed, object manipulation is intuitive and simple, content is immersive and engaging, and the system design optimizes comfort while minimizing sickness and aftereffects. It is an impressive task to ensure that all of these criteria are met.

Gabbard (1997) developed a taxonomy of VE usability characteristics that can serve as a foundation for identifying and evaluating usability criteria particularly relevant to VE systems. Stanney et al. (2000) used this taxonomy as the foundation on which to develop an automated system, MAUVE (Multi-criteria Assessment of Usability for Virtual Environments), which assesses VE usability in terms of how effectively each of the following are designed: navigation, user movement, object selection and manipulation, visual output, auditory output, haptic output, presence, immersion, comfort, sickness, and aftereffects. MAUVE can be used to support expert evaluations of VE systems, similar to the manner in which traditional heuristic evaluations are conducted. Due to such issues as cybersickness and aftereffects, it is essential to use these or other techniques to ensure the usability of VE systems, not only to avoid rendering them ineffective but also to ensure that they are not hazardous to users.

5.2 Sense of Presence

One of the usability criteria unique to VE systems is sense of presence. Virtual environments have the unique advantage of leveraging people's imaginative ability to "transport" themselves psychologically to another place, one that may not exit in reality (Sadowski and Stanney, 2002). To support such transportation, VEs provide physical separation from the real world by immersing users in the virtual world, say via an HMD, then impart sensorial sensations via multimodal feedback that would naturally be present in the real environment. Focus on generating such presence is one of the primary characteristics distinguishing VEs from other means of displaying information.

Presence has been defined as the subjective perception of being immersed in and surrounded by a virtual world rather than the physical world that one is situated in currently (Stanney et al., 1998b). Virtual environments that engender a high degree of presence are thought to be more enjoyable, effective, and well received by users (Sadowski and Stanney, 2002). To enhance presence, designers of VE systems should spread detail around a scene; let user interaction determine when to reveal important aspects; maintain a natural and realistic, yet simple appearance; and utilize textures, colors, shapes, sounds, and other features to enhance realism (Kaur, 1999). To generate the feeling of immersion within the environment, designers should isolate users from the physical environment (use of an HMD may be sufficient), provide content that involves users in an enticing situation supported by an encompassing stimulus stream, provide natural modes of interaction and movement control, and utilize design features that enhance vection (Stanney et al., 2000). Presence can be assessed via Witmer and Singer's (1998) presence questionnaire or techniques used by Slater and Steed (2000), as well as a number of other means (Sadowski and Stanney, 2002).

5.3 Virtual Environment Ergonomics

Ergonomics, which focuses on fitting a product or system to the anthropometric, musculoskeletal, cardiovascular, and psychomotor properties of users, is an essential element of VE system design (McCauley-Bell, 2002). Supporting user comfort while donning cumbersome HMDs or unwieldy peripheral devices is an ergonomics challenge of paramount importance because discomfort could supersede any other sensations (e.g., presence, immersion). If a VE produces discomfort, participants may limit their exposure time or possibly avoid repeat exposure. Overall physical discomfort should thus be minimized, while user safety is maximized.

Ergonomic concerns affecting comfort include visual discomfort resulting from visual displays with improper depth cues, poor contrast and illumination, or improperly set IPDs (Stanney et al., 2000). Physical discomfort can be driven by restrictive tethers, awkward interaction devices, or heavy, awkward, and constraining visual displays. To enhance the ergonomics of VE systems, several factors should be considered, including the following (McCauley-Bell, 2002):

- Is operator movement inhibited by the location, weight, or window of reach of interaction devices or HMDs?
- Does layout and arrangement of interaction devices and HMDs support efficient and comfortable movement?
- Is any limb overburdened by heavy interaction devices or HMDs?
- Do interaction devices require awkward and prolonged postures?
- If a seat is provided, does it support user movement, and is it of the right height with adequate back support?

- If active motion interfaces are provided (e.g., treadmills), are they adjustable to ensure fit to the anthropometrics of users?
- Are the noise and sound levels within ergonomic guidelines, and do they support user immersion?

6 APPLICATION DOMAINS

The application of many of the technologies and strategies outlined throughout this chapter may best illustrated by a brief survey of two distinct, yet complementary domains. The first involves applying VE tools to the challenging problem of providing training. The second involves exploiting the novelty factor of VE technologies to open new frontiers in entertainment. Each domain stands to inform the other both in terms of development strategies leading to successful system design and in lessons-learned.

6.1 Training Applications

6.1.1 Military Applications

The U.S. military, possibly the largest, most unified training organization in the world, has focused on modeling and simulation in general, and virtual environments in particular, as a means to resolve many of the training deficits that result from the rigors of military life. Such challenges include, but are not limited to, sustaining skills and knowledge sets acquired in a schoolhouse setting during prolonged periods of deployment far from instructional tools, the acquisition of new skills and knowledge in the absence of the schoolhouse/instructor-based training framework, and the provision of a large-scale environment within which large numbers of people may participate in simulated, distributed training exercises that closely match the real-world environment. Although live training exercises may, at first blush, seem to provide the most realism, in fact, as training exercises become more closely aligned to real-world conditions, a series of factors, such as cost, logistics, and safety, move to the forefront and make such training events unwieldy to organize and difficult to guide. Due to their promise of reduced hardware footprint, software reusability, and networkability, VE systems, combined with projected continued decreases in cost, provide an ideal solution to these challenges.

These characteristics make VE-based training systems ideally suited for supporting training in conditions in which space, technical expertise, and time are limited, a situation in which today's U.S. military continues to find itself. As the pace with which military operations continues to quicken, and as the length of time for which military personnel are deployed continues to increase, the U.S. military will need to find technology solutions that enable the provision of training to individuals and teams deployed or located aboard ships around the world. Such deployed locations are, of course, far from schoolhouses, which typically supply trained instructors, and cannot support the large contingent of technicians that current legacy-type simulations often require for support and maintenance.

The Office of Naval Research has developed a comprehensive Virtual Technologies and Environments (VIRTE) program which addresses many of these challenges in multiple stages (Cohn et al., 2003; Muller et al., 2003). Each stage involves two components, the first focusing on identifying technology solutions and the second emphasizing a rigorous evaluation process grounded in solid human factors practice. Ultimately, each system is validated through a unique demonstration of real-world performance enhancement through exposure to VEs and has its own unique evaluative metrics. For the first stage, these metrics included a heavy reliance on usability engineering principles to guide system development (Stanney et al., 2000; Becker et al., 2004). For the second, which moves away from the single person/single platform paradigm to a multiple people/single platform paradigm, these metrics will shift to more advanced analyses of the minimum ecological validity that individual user interfaces must have (Grant and Magee, 1998; Templeman et al., 1999), as well as to the development of a team performance evaluation metrics suite. For the final stage, the paradigm will shift again to multiple people/multiple platforms, and metrics will evolve to focus on both collective training, an advanced form of team training, and on an evaluation of the benefits of integrating performance-enhancing tools within virtual environments. The intent with this overall approach is to demonstrate a spiral development cycle not only for the production of actual systems but also in the collective understanding of what VE systems are (and are not) capable of accomplishing. Currently, three vehicle systems, the Virtual Environment Landing Craft Air Cushion (VE LCAC) (Schaffer et al., 2003), the Virtual Environment Expeditionary Fighting Vehicle (VE EFV), and the Virtual Environment Helicopter Navigation Training System (VE HELO) (Milham et al., 2004) have been prototyped, validated, and delivered to their respective military customers (Muller et al., 2003).

6.1.2 Medical Applications

Until quite recently, the notion of training for complex surgical procedures, let alone the idea of training students using any tool other than cadavers may have been met with incredulity. However, along with a trend toward using anatomical manikins to provide medical students and skilled surgeons with opportunities to hone their skills, advances in technology have combined to make VE tools more commonplace within the medical training domain (Satava and Jones, 2002; Mantovani et al., 2003; Sheehan, 2004). The critical challenge lies in making the leap from cadaver—which clearly supports a wide range of training applications—to manikin—which still retains many features common to cadavers, such as accurate placement and design of organs—to virtual—which relies on advanced technologies to mimic many of the sensorial feedback cues critical to training delicate surgical skills.

A 1995 National Institutes of Standards and Technologies report (Moline, 1995) provides some insight into the wide range of medical applications for which VE is suited, as well as identifying some of the challenges this technology must resolve to become fully effective for medical training. Some of the earliest medical VE successes came in the area of psychotherapy, where VE technology has been used to treat acrophobia, fear of flying, arachnophobia, and other psychological disorders (Strickland et al., 1997; North et al., 2002). Other applications, such as remote surgery via telepresence and visualization of large-scale databases (Mantovani et al., 2003), are also easily realized using current technology. However, some applications, such as simulating actual medical procedures, push VE to its current limits, requiring significant advances in the development of both the technology and underlying mathematical formulations. For example, if the training objective is to provide visual familiarization of a particular pathology, one need only have advanced visual presentation capabilities coupled with a high-end graphics rendering system (Zajtchuk and Satava, 1997). On the other hand, if the training objective is to provide actual surgical training, which requires significantly advanced algorithms describing the interactions between the trainee, their surgical equipment, and the organs upon which they will be acting, one must have access not only to high-end visual rendering and display systems, but also advanced tactile feedback simulations, high-definition models of the region on which they will be operating, and as yet unexplored feedback cues (Basdogan et al., 2001).

6.1.3 Other Training Applications

The training applications described thus far are highly specialized. Consequently, the technology requirements underlying their development are high end, comparatively high cost, and cater to a select user population. For VE tools to become more commonplace, without requiring a long-term and expensive research and development base, it is likely that current technologies will need to be assessed in terms of the types of training they can support, a twist on the more common approach of defining training requirements and then developing system requirements that support them (Sticha et al., 2002). Potential applications range from providing visually rich, immersive representations of concrete (Yair et al., 2001) and abstract (Trindage et al., 2002) scientific and mathematical principles, to providing training *en masse* (McComas et al., 2002).

As technologies for cueing multiple sensory modalities become more affordable, it is likely that current VE training applications will benefit by supporting increased levels of experiential learning, such as moving from simple visual displays to those that provide seamless cueing across the visual, haptic, and aural modalities. Ultimately, it is possible to imagine a scenario in which virtual schoolhouses, which are currently little more than integrated or networked Web-based sites, will allow unprecedented levels of collective interaction between fully immersed participants in real time. Advanced multimodal interaction may be necessary to realize gains in such educational VE applications because early educational VE projects have yet to demonstrate any significant gains (Moshell and Hughes, 2002). This is probably due not only to the current complexity and novelty of the technology, but also to the need for instructional design techniques particularly well suited to educational VE applications. Efforts in this area should lead the way to new and dynamic approaches in experiential learning.

6.2 Entertainment Applications

Virtual environments are inherently technocentric, whereas training is inherently human-centric (Cohn, 2003). The evolution of most multimedia tools, of which VE is arguable the latest incarnation, is guided by an outlook that emphasizes continuously surveying the state of the art and then determining how best to integrate any and all relevant components (Mayer, 2001). VE is no different, in that more often than not, it is the user who must adapt to the technology rather than the technology that must be adapted to the user. The result of this approach is a fundamental failure to focus on developing technologies that best match desired training goals and objectives. Consequently, many training applications, such as those described in Section 6.1, may in fact be further ahead of our grasp than suggested by individual success stories.

Nevertheless, one application that is admittedly technocentric and that is not only driven by technology advancements, but often provides the impetus necessary to make industry-wide paradigm shifts, is entertainment (Badiqué et al., 2002). Current VE entertainment applications include games, online communities, location-based entertainment, theme parks, and other venues. From interactive arcades to cybercafés, the entertainment industry has leveraged the unique characteristics of the VE medium, providing dynamic and exciting experiences. Such applications tend to set the pace, both technological and in terms of content, for advances in VE technology in general. Loosely put, the goal of such entertainment technologies is to provide consumers with experiences that they would not ordinarily encounter in their typical daily activities. Consequently, the goals and objectives of entertainment-based systems are quite broadly defined (and, sometimes, not necessarily defined at all!), shaped primarily through user feedback vis-à-vis factors such as ease of use, engagement level, interest level, and so on. These factors are molded to a large extent by popular culture. By attending to these global trends, developers can determine what technologies exist—or need to be created—to support current user demand.

Earlier sections in this chapter laid out a range of technologies, such as autonomous agents, human–computer interfaces, and content development tools, and their associated challenges. An estimation of the level of success that gaming systems, which in fact are one commonplace form of VE along the virtuality continuum (Milgram and Kishino, 1994), have had in pushing technocentric solutions on consumers despite these challenges is evidenced by the degree to which these devices and their associated interfaces have virtually swept the marketplace. Measured against many of the criteria that a system developer might use to evaluate systems, these games suffer from interfaces that have poor ecological validity (e.g., using a game pad to locomote and to manipulate objects within the environment), minimally intelligent agents (e.g., teammates and opponents in sports games), and little, if any, ability to modify scenario content above and beyond the capabilities provided as part of the development kit incorporated within the game. However, measured against criteria that are entertainment specific, such as engagement level, these technologies appear to have significant appeal. Moreover, factors such as nonmodifiable scenarios and low-intelligence agents, which would be a liability from a training perspective, are dealt with in a manner uniquely tailored to the consumer industry: New versions, with different scenarios and different agent behaviors, are released at intervals determined by overall market interest.

7 CONCLUSIONS

There is little doubt that the arena in which the development of VE technologies plays out has significantly matured over that of only a decade ago. Advances in a range of technologies, from graphics rendering engines, to the display systems that support them, to new technologies supporting unprecedented levels of multimodal cueing, have merged to provide the possibility of an integrated VE technology that is literally light years beyond anything Mortin Heilig could have imagined when he introduced his Sensorama Simulator in 1960. In parallel, our current understanding of how to utilize these technologies to maximize their benefits across a range of applications, as well as how to ensure that human factors best practices are adhered to throughout the development cycle, has also progressed. Stone (2002) suggests that the technological revolution over the close of the preceding millennium has set a foundation that should ensure the presence of VE applications for at least the next two decades.

At the same, though, it is imperative that the VE community not ignore the sobering fact that just as VE tools stand poised to provide a wide range of unique service, so have other multimodal technologies stood at similar ground, only to fall far short of their promise, in part due to a failure of system developers to fully appreciate the role of humans (Mayer, 2001). If there is a glimmer of hope, it is to be found in the simple fact that new areas of research have arisen whose goals are to better understand the intersection between humans and the computer systems they use.

ACKNOWLEDGMENTS

The authors thank Hesham Fouad for contributions to Section 2.1.1. This material is based on work supported in part by the National Science Foundation (NSF) under grant IRI-9624968; the Office of Naval Research (ONR) under grants N00014-98-1-0642, N00014-03-C-0194, and N00014-04-C-0024; Navair Orlando under contract N61339-99-C-0098; and the National Aeronautics and Space Administration (NASA) under grant NAS9-19453. Any opinions, findings, and conclusions or recommendations expressed in this material are those of the authors and do not necessarily reflect the views or the endorsement of the NSF, ONR, Navair, or NASA. This chapter represents an updated version of K. M. Stanney, "Virtual Environments," in *Handbook of Human–Computer Interaction*, J. Jacko and A. Sears, Eds., Lawrence Erlbaum Associates, Mahwah, NJ, pp. 621– 634.

REFERENCES

Agrawala, M., Beers, A. C., McDowall, I., Fröhlich, B., Bolas, M., and Hanrahan, P. (1997), "The Two-User Responsive Workbench: Support for Collaboration Through Individual Views of a Shared Space, in *SIGGRAPH '97*, ACM Press, New York, pp. 327–332.

Akamatsu, M. (1994), "Touch with a Mouse: A Mouse Type Interface Device with Tactile and Force Display," in *Proceedings of the 3rd IEEE International Workshop on Robot and Human Communication*, Nagoya, Japan, July 18–20, pp. 140–144.

Algazi, V. R., Duda, R. O., Thompson, D. M., and Avendano, C. (2001), "The CIPIC HRTF Database," in *Proceedings of the 2001 IEEE Workshop on Applications of Signal Processing to Audio and Electroacoustics*, Mohonk Mountain House, New Paltz, NY, October 21–24, pp. 99–102.

Allbeck, J. M., and Badler, N. I. (2002), "Embodied Autonomous Agents," in *Handbook of Virtual Environments: Design, Implementation, and Applications*, K. M. Stanney, Ed., Lawrence Erlbaum Associates, Mahwah, NJ, pp. 313–332.

Appenzeller, T. (2000), "The Chemistry of Computing: Computers Made of Molecule-Size Parts Could Build Themselves," *U.S. News & World Report*, May 1, http://www.usnews.com/usnews/issue/000501/chips.htm.

Badiqué, E., Cavazza, M., Klinker, G., Mair, G., Sweeney, T., Thalmann, D., and Thalmann, N. M. (2002), "Entertainment Applications of Virtual Environments," in *Handbook of Virtual Environments: Design, Implementation, and Applications*, K. M. Stanney, Ed., Lawrence Erlbaum Associates, Mahwah, NJ, pp. 1143–1166.

Basdogan, C., Ho, C., and Srinivasan, M. A. (2001), "Virtual Environments for Medical Training: Graphical and Haptic Simulation of Common Bile Duct Exploration," *IEEE/ASME Transactions on Mechatronics*, Vol. 6, No. 4, pp. 267–285.

Becker, W., Cohn, J. V., Lackey, S., and Allen, R. C. (2004), "Applying Human Centric Design Methodology to Training System Development and Validation," in *Proceedings of the 2004 IMAGE Conference*, Scottsdale, AZ, July 11–15.

Begault, D. (1994), *3-D Sound for Virtual Reality and Multimedia*, Academic Press, San Diego, CA.

Biggs, S. J., and Srinivasan, M. A. (2002), "Haptic Interfaces," in *Handbook of Virtual Environments: Design, Implementation, and Applications*, K. M. Stanney, Ed., Lawrence Erlbaum Associates, Mahwah, NJ, pp. 93–115.

Blachford, N. (2004), "The Future of Computing, 4: The Next Generation," *CS News: Exploring the Future of Computing*, http://www.osnews.com/story.php?news_id=6089.

Buck, I., Foley, T., Horn, D., Sugerman, J., Fatahalian, K., Houston, M., and Hanrahan, P. (2004), "Brook for GPUs: Stream Computing on Graphics Hardware," in *Proceedings of SIGGRAPH 2004*, Los Angeles, August 8–12.

Bullinger, H.-J., Breining, R., and Braun, M. (2001), "Virtual Reality for Industrial Engineering: Applications for Immersive Virtual Environments," in *Handbook of Industrial Engineering: Technology and Operations Management*, 3rd ed., G. Salvendy, Ed., Wiley, New York, pp. 2496–2520.

Bungert, C. (2004), "HMD/Headset/VR-Helmet Comparison Chart," http://www.stereo3d.com/hmd.htm.

Burdea, G. (1996), *Force and Touch Feedback for Virtual Reality*, Wiley, New York.

Butler, R. A. (1987), "An Analysis of the Monaural Displacement of Sound in Space," *Perception and Psychophysics*, Vol. 41, pp. 1–7.

Calvert, S. L. (2002), "The Social Impact of Virtual Environment Technology," in *Handbook of Virtual Environments: Design, Implementation, and Applications*, K. M. Stanney, Ed., Lawrence Erlbaum Associates, Mahwah, NJ, pp. 663–680.

Calvert, S. L., and Tan, S. L. (1994), "Impact of Virtual Reality on Young Adult's Physiological Arousal and Aggressive Thoughts: Interaction Versus Observation," *Journal of Applied Developmental Psychology*, Vol. 15, pp. 125–139.

Carson, D. (2000a), "Environmental Storytelling, 1: Creating Immersive 3D Worlds Using Lessons Learned from the Theme Park Industry," http://www.gamasutra.com/features/20000301/carson_01.htm.

Carson, D. (2000b), "Environmental Storytelling, 2: Bringing Theme Park Environment Design Techniques Lessons to the Virtual World," http://www.gamasutra.com/features/20000405/carson_01.htm.

Chen, J. L., and Stanney, K. M. (1999), "A Theoretical Model of Wayfinding in Virtual Environments: Proposed Strategies for Navigational Aiding," *Presence: Teleoperators and Virtual Environments*, Vol. 8, No. 6, pp. 671–685.

Chinthammit, W., Seibel, E. J., and Furness, T. A., III (2003), "A Shared-Aperture Tracking Display for Augmented Reality," *Presence: Teleoperators and Virtual Environments*, Vol. 12, No. 1, pp. 1–18.

Church, D. (1999), "Formal Abstract Design Tools," *Games Developer Magazine*, August, pp. 44–50; http://www.gamasutra.com/features/19990716/design_tools_02.htm.

Cohen, M. (1992), "Integrating Graphic and Audio Windows," *Presence: Teleoperators and Virtual Environments*, Vol. 1, No. 4, pp. 468–481.

Cohn, J. V. (2003), "Exploiting Human Information Processing to Enhance Virtual Environment Training," in *Proceedings of 111th Annual American Psychological Association Conference*, Toronto, Ontario, Canada, August 7–10.

Cohn, J. V., Schmorrow, D., Nicholson, D., Templeman, J., and Muller, P. (2003), "Virtual Technologies and Environments for Expeditionary Warfare Training," in *Proceedings of the NATO Human Factors and Medicine Symposium on Advanced Technologies for Military Training*, Genoa, Italy, October 15–17.

Cruz-Neira, C., Sandin, D. J., and DeFanti, T. A. (1993), "Surround-Screen Projection-Based Virtual Reality: The Design and Implementation of the CAVE," *ACM Computer Graphics*, Vol. 27, No. 2, pp. 135–142.

Crystal Rivers Engineering (1995), *Snapshot: HRTF Measurement System*, CRE, Groveland, CA.

Darken, R. P., and Peterson, B. (2002), "Spatial Orientation, Wayfinding, and Representation," in *Handbook of Virtual Environments: Design, Implementation, and Applications*, K. M. Stanney, Ed., Lawrence Erlbaum Associates, Mahwah, NJ, pp. 493–518.

Draper, M. H., Prothero, J. D., and Viirre, E. S. (1997), "Physiological Adaptations to Virtual Interfaces: Results of Initial Explorations," in *Proceedings of the Human Factors and Ergonomics Society 41st Annual Meeting*, Human Factors and Ergonomics Society, Santa Monica, CA, p. 1393.

Durlach, B. N. I., and Mavor, A. S. (1995), *Virtual Reality: Scientific and Technological Challenges*, National Academy Press, Washington, DC.

Favalora, G. E., Napoli, J., Hall, D. M., Dorval, R. K., Giovinco, M. G., Richmond, M. J., and Chun, W. S. (2003), "100 Million-Voxel Volumetric Display," in *SPIE Proceedings of the 16th Annual International Symposium on Aerospace/Defense Sensing, Simulation, and Controls*, 4297, pp. 227–235.

Fencott, C. (1999), "Content and Creativity in Virtual Environment Design," in *Proceedings of Virtual Systems and Multimedia '99*, University of Abertay, Dundee, Scotland, September 1–3, pp. 308–317.

Fouad, H. (2004), "Ambient Synthesis with Random Sound Fields," in *Audio Anecdotes: Tools, Tips, and Techniques for Digital Audio*, K. Greenebaum, Ed., A.K. Peters, Natick, MA.

Fouad, H., and Ballas, J. (2000), "An Extensible Toolkit for Creating Virtual Sonic Environments," in *International Conference on Auditory Displays, ICAD 2000*, Atlanta, GA, April 2–5.

Foxlin, E. (2002), "Motion Tracking Requirements and Technologies," in *Handbook of Virtual Environments: Design, Implementation, and Applications*, K. M. Stanney, Ed., Lawrence Erlbaum Associates, Mahwah, NJ, pp. 163–210.

Funkhouser, T., and Sequin, C. (1993), "Adaptive Display Algorithm for Interactive Frame Rates During Visualization of Complex Virtual Environments," in *Proceedings of SIGGRAPH'93*, ACM Press Books, New York, pp. 247–254.

Gabbard, J. L. (1997), "Taxonomy of Usability Characteristics in Virtual Environments," Final Report to the Office of Naval Research, http://iwb.sv.vt.edu/publications.

Grant, S. C., and Magee, L. E. (1998), "Contributions of Proprioception to Navigation in Virtual Environments," *Human Factors: Journal of the Human Factors Society*, Vol. 40, No. 3, pp. 489–497.

Hahn, J. K., Fouad, H., Gritz, L., and Lee, J. W. (1998), "Integrating Sounds and Motions in Virtual Environments," *Presence: Teleoperators and Virtual Environments*, Vol. 7, No. 1, pp. 67–77.

Hale, K. S., and Stanney, K. M. (2004), "Deriving Haptic Design Guidelines from Human Physiological, Psychophysical, and Neurological Foundation," *IEEE Computer Graphics and Applications*, Vol. 24, No. 2, pp. 33–39.

Halle, M. (1997), "Autostereoscopic Displays and Computer Graphics," *Computer Graphics*, Vol. 31, No. 2, May, pp. 58–62.

Heirich, A., and Arvo, J. (1997), "Scalable Monte Carlo Image Synthesis," *Parallel Computing* (Special Issue on Parallel Graphics and Visualization), Vol. 23, No. 7, pp. 845–859.

Hettinger, L. J. (2002), "Illusory Self-Motion in Virtual Environments," in *Handbook of Virtual Environments: Design, Implementation, and Applications*, K. M. Stanney, Ed., Lawrence Erlbaum Associates, Mahwah, NJ, pp. 471–491.

Hix, D., and Gabbard, J. L. (2002), "Usability Engineering of Virtual Environments," in *Handbook of Virtual Environments: Design, Implementation, and Applications*, K. M. Stanney, Ed., Lawrence Erlbaum Associates, Mahwah, NJ, pp. 681–699.

Hollerbach, J. M. (2002), "Locomotion Interfaces," in *Handbook of Virtual Environments: Design, Implementation, and Applications*, K. M. Stanney, Ed., Lawrence Erlbaum Associates, Mahwah, NJ, pp. 239–254.

Howard, I. P. (1986), "The Vestibular System," in *Handbook of Perception and Human Performance*, K. R. Boff, L. Kaufman, and J. P. Thomas, Eds., Wiley, New York, pp. 11–1 to 11–30.

Huang, X. D. (1998), "Spoken Language Technology Research at Microsoft," *16th ICA and 135th ASA '98*, Seattle, WA, June 20–26.

Isdale, J. (2000), "Motion Simulation," *VR News: April Tech Review*, http://vr.isdale.com/vrTechReviews/Motion Simulation_April2000.htm

Isdale, J., Fencott, C., Heim, M., and Daly, L. (2002), "Content Design for Virtual Environments," in *Handbook of Virtual Environments: Design, Implementation, and Applications*, K. M. Stanney, Ed., Lawrence Erlbaum Associates, Mahwah, NJ, pp. 519–532.

Johnson, A., Leigh, J., DeFanti, T., Sandin, D., Brown, M., and Dawe, G. (1999), "Next-Generation tele-immersive Devices for Desktop Trans-oceanic Collaboration," in *Proceedings of the IS&T/SPIE Conference on Visual Communications and Image Processing '99*, San Jose, CA, January 23–29, pp. 1420–1429.

Jones, L., Bowers, C. A., Washburn, D., Cortes, A., and Vijaya Satya, R. (2004), "The Effect of Olfaction on Immersion into Virtual Environments," in *Human Performance, Situation Awareness and Automation: Current Research and Trends*, D. A. Vincenzi, M. Mouloua, and P. A. Hancock, Eds., Lawrence Erlbaum Associates, Mahwah, NJ, Vol. 2, pp. 282–285.

Kalawsky, R. S. (1993), *The Science of Virtual Reality and Virtual Environments*, Addison-Wesley, Wokingham, Berkshire, England.

Kallman, E. A. (1993), "Ethical Evaluation: A Necessary Element in Virtual Environment Research," *Presence: Teleoperators and Virtual Environments*, Vol. 2, No. 2, pp. 143–146.

Kaur, K. (1999), "Designing Virtual Environments for Usability," unpublished doctoral dissertation, City University, London.

Kessler, G. D. (2002), "Virtual Environment Models," in *Handbook of Virtual Environments: Design, Implementation, and Applications*, K. M. Stanney, Ed., Lawrence Erlbaum Associates, Mahwah, NJ, pp. 255–276.

Kennedy, R. S., and Fowlkes, J. E. (1992), "Simulator Sickness Is Polygenic and Polysymptomatic: Implications for Research," *International Journal of Aviation Psychology*, Vol. 2, No. 1, pp. 23–38.

Kennedy, R. S., and Graybiel, A. (1965), *The Dial Test: A Standardized Procedure for the Experimental Production of Canal Sickness Symptomatology in a Rotating Environment*, Report 113, NSAM 930, Naval School of Aerospace Medicine, Pensacola, FL.

Kennedy, R. S., and Stanney, K. M. (1996a), "Virtual Reality Systems and Products Liability," *Journal of Medicine and Virtual Reality*, Vol. 1, No. 2, pp. 60–64.

Kennedy, R. S., and Stanney, K. M. (1996b), "Postural Instability Induced by Virtual Reality Exposure: Development of a Certification Protocol," *International Journal of Human–Computer Interaction*, Vol. 8, No. 1, pp. 25–47.

Kennedy, R. S., Lane, N. E., Berbaum, K. S., and Lilienthal, M. G. (1993), "Simulator Sickness Questionnaire: An Enhanced Method for Quantifying Simulator Sickness," *International Journal of Aviation Psychology*, Vol. 3, No. 3, pp. 203–220.

Kennedy, R. S., Stanney, K. M., Ordy, J. M., and Dunlap, W. P. (1997), "Virtual Reality Effects Produced by Head-Mounted Display (HMD) on Human Eye–Hand Coordination, Postural Equilibrium, and Symptoms of Cybersickness," *Society for Neuroscience Abstracts*, Vol. 23, p. 772.

Kennedy, R. S., Stanney, K. M., and Dunlap, W. P. (2000), "Duration and Exposure to Virtual Environments: Sickness Curves During and Across Sessions," *Presence: Teleoperators and Virtual Environments*, Vol. 9, No. 5, pp. 466–475.

Kennedy, R. S., Lane, N. E., Grizzard, M. C., Stanney, K. M., Kingdon, K., Lanham, S., and Harm, D. L. (2001), "Use of a Motion History Questionnaire to Predict Simulator Sickness," presented at the Driving Simulation Conference 2001, Sophia-Antipolis (Nice), France, September 5–7.

Kennedy, R. S., Kennedy, K. E., and Bartlett, K. M. (2002), "Virtual Environments and Products Liability," in *Handbook of Virtual Environments: Design, Implementation, and Applications*, K. M., Stanney, Ed., Lawrence Erlbaum Associates, Mahwah, NJ, pp. 543–553.

Kershner, J. (2002), "Object-oriented scene management," *GameDev.net*, http://www.gamedev.net/reference/articles/article1812.asp.

Lawson, B. D., Sides, S. A., and Hickinbotham, K. A. (2002), "User Requirements for Perceiving Body Acceleration," in *Handbook of Virtual Environments: Design, Implementation, and Applications*, K. M. Stanney, Ed., Lawrence Erlbaum Associates, Mahwah, NJ, pp. 135–161.

Lieberman, D. (1999), "Computer Display Clips onto Eyeglasses," *Technology News*, http://www.techweb.com/wire/story/TWB19990422S0003.

Majumder, A. (1999), "Intensity Seamlessness in Multiprojector Multisurface Displays," Technical Report, http://www.cs.unc.edu/~majumder/, Department of Computer Science, University of North Carolina, Chapel Hill, NC.

Mantovani, F., Castelnuovo, G., Gaggioli, A., and Riva, G. (2003), "Virtual Reality Training for Health-Care Professionals," *CyberPsychology and Behavior*, Vol. 6, No. 4, pp. 389–395.

Massimino, M., and Sheridan, T. (1993), "Sensory Substitution for Force Feedback in Teleoperation," *Presence: Teleoperators and Virtual Environments*, Vol. 2, No. 4, pp. 344–352.

May, J. G., and Badcock, D. R. (2002), "Vision and Virtual Environments," in *Handbook of Virtual Environments: Design, Implementation, and Applications*,

K. M. Stanney, Ed., Lawrence Erlbaum Associates, Mahwah, NJ, pp. 29–63.

Mayer, R. E. (2001), *Multi-media Learning*, Cambridge University Press, Cambridge.

McCauley, M. E., and Sharkey, T. J. (1992), "Cybersickness: Perception of Self-Motion in Virtual Environments," *Presence: Teleoperators and Virtual Environments,* Vol. 1, No. 3, pp. 311–318.

McCauley-Bell, P. R. (2002), "Ergonomics in Virtual Environments," in *Handbook of Virtual Environments: Design, Implementation, and Applications*, K. M. Stanney, Ed., Lawrence Erlbaum Associates, Mahwah, NJ, pp. 807–826.

McComas, J., MacKay, M., and Pivik, J. (2002), "Effectiveness of Virtual Reality for Teaching Pedestrian Safety," *CyberPsychology and Behavior*, Vol. 5, No. 3, pp. 185–190.

Menzies, D. (2002), "Scene Management for Modelled Audio Objects in Interactive Worlds," in *Proceedings of the 8th International Conference on Auditory Displays*, Kyoto, Japan, July 2–5.

Merritt, E. A., and Bacon, D. J. (1997), "Raster3D: Photorealistic Molecular Graphics," *Methods in Enzymology*, Vol. 277, pp. 505–524.

Milgram P., and Kishino, F. (1994), "A Taxonomy of Mixed Reality Visual Displays," *IEICE Transactions on Information and Systems* (Special Issue on Networked Reality), E77–D(12), pp. 1321–1329.

Milham, L., Kingdon-Hale, K., Stanney, K., Graeber, D., and Cohn, J. (2004). "Methodology and Preliminary Findings of a Virtual Environment Operational Transfer of Training Study," poster presented at the *2005 Annual American Psychological Association Conference*, Honolulu, Hawaii, July 28–August 1.

Moline, J. (1995), "Virtual Environments for Health Care," White Paper for the Advanced Technology Program (ATP), National Institute of Standards and Technology, http://www.itl.nist.gov/iaui/ovrt/projects/health/vr-envir.htm.

Moshell, J. M., and Hughes, C. E. (2002), "Virtual Environments as a Tool for Academic Learning," in *Handbook of Virtual Environments: Design, Implementation, and Applications*, K. M. Stanney, Ed., Lawrence Erlbaum Associates, Mahwah, NJ, pp. 893–910.

Mulgund, S., Stokes, J., Turieo, M., and Devine, M. (2002), *Human/machine interface modalities for soldier systems technologies*, Final Report 71950–00, TIAX, Cambridge, MA.

Muller, P., Cohn, J. V., and Nicholson, D. (2003), "Developing and Evaluating Advanced Technologies for Military Simulation and Training," in *Proceedings of the 25th Annual Interservice/Industry Training, Simulation and Education Conference*, Orlando, FL, December 1–4.

Murray, J. H. (1998), *Hamlet on the Holodeck: The Future of Narrative in Cyberspace*, MIT Press, Cambridge, MA; http://web.mit.edu/jhmurray/www/HOH.html.

Nelson, T., and Bolia, R. (2003), "Evaluating the Effectiveness of Spatial Audio Displays in a Simulated Airborne Command and Control Task," in *Proceedings of the Human Factors and Ergonomics Society 47th Annual Meeting*, Denver, CO, October 13–17, pp. 202–206.

North, M. M., North, S. M., and Coble, J. R. (2002), "Virtual Reality Therapy: An Effective Treatment for Psychological Disorders," in *Handbook of Virtual Environments: Design, Implementation, and Applications*, K. M. Stanney, Ed., Lawrence Erlbaum Associates, Mahwah, NJ, pp. 1065–1078.

Popescu, G. V., Burdea, G. C., and Trefftz, H. (2002), "Multimodal Interaction Modeling," in *Handbook of Virtual Environments: Design, Implementation, and Applications*, K. M. Stanney, Ed., Lawrence Erlbaum Associates, Mahwah, NJ, pp. 435–454.

Pountain, D. (1996), "VR Meets Reality: Virtual Reality Strengthens the Link Between People and Computers in Mainstream Applications," *Byte Magazine*, July, http://www.byte.com/art/9607/sec7/art5.htm.

Proctor, R. W., Vu, K.-P. L., Salvendy, G., Degen, H., Fang, X., Flach, J. M., Gott, S. P., Herrmann, D., Krömker, H., Lightner, N. J., Lubin, K., Najjar, L., Reeves, L., Rudorfer, A., Stanney, K., Stephanidis, C., Strybel, T. Z., Vaughan, M., Wang, H., Weber, H., Yang, Y., and Zhu, W. (2002), "Content Preparation and Management for Web Design: Eliciting, Structuring, Searching, and Displaying Information," *International Journal of Human–Computer Interaction*, Vol. 14, pp. 25–92.

Rolland, J. P., Biocca, F. A., Barlow, T., and Kancherla, A. (1995), "Quantification of Adaptation to Virtual-Eye Location in See-thru Head-Mounted Displays," *Proceedings of the IEEE Virtual Reality Annual International Symposium '95*, IEEE Computer Society Press, Los Alamitos, CA, pp. 56–66.

Sadowski, W., and Stanney, K. (2002). "Presence in Virtual Environments," in *Handbook of Virtual Environments: Design, Implementation, and Applications*, K. M. Stanney, Ed., Lawrence Erlbaum Associates, Mahwah, NJ, pp. 791–806.

Satava, R. M., and Jones, S. B. (2002), "Medical Applications of Virtual Environments," in *Handbook of Virtual Environments: Design, Implementation, and Applications*, K. M. Stanney, Ed., Lawrence Erlbaum Associates, Mahwah, NJ, pp. 937–957.

Schaffer, R., Cullen, S., Cohn, J. V., and Stanney, K. (2003), "A Personal LCAC Simulator Supporting a Hierarchy of Training Requirements," in *Proceedings of the 25th Annual Interservice/Industry Training, Simulation and Education Conference*, Orlando, FL, December 1–4.

Sheehan, C. (2004), "Ambulances May Get Virtual Doctor," Associated Press, http://www.boston.com/business/technology/articles/2004/12/24/ambulances_may_get_virtual_doctors/.

Sheridan, T. B. (1993), "My Anxieties About Virtual Environments," *Presence: Teleoperators and Virtual Environments*, Vol. 2, No. 2, pp. 141–142.

Shilling, R. (2002), "Entertainment Industry Sound Design Techniques to Improve Presence and Training Performance in VE," *MSIAC's M&S Journal Online*, Winter, Vol. 4, No. 2, http://www.msiac.dmso.mil/journal/FA02/shil41.html.

Shilling, R. D., and Shinn-Cunningham, B. (2002), "Virtual Auditory Displays," in *Handbook of Virtual Environments: Design, Implementation, and Applications*, K. M. Stanney, Ed., Lawrence Erlbaum Associates, Mahwah, NJ, pp. 65–92.

Singhal, S., and Zyda, M. (1999), *Networked Virtual Environments: Design and Implementation*, SIGGRAPH Series, ACM Press Books, New York.

Slater, M., and Steed, A. (2000), "A Virtual Presence Counter," *Presence: Teleoperators and Virtual Environments*, Vol. 9, No. 5, pp. 413–434.

Stanney, K. M., and Kennedy, R. S. (1998), "Aftereffects from Virtual Environment Exposure: How Long Do

They Last? in *Proceedings of the 42nd Annual Human Factors and Ergonomics Society Meeting*, Chicago, October 5–9, pp. 1476–1480.

Stanney, K. M., and Zyda, M. (2002), "Virtual Environments in the 21st Century," in *Handbook of Virtual Environments: Design, Implementation, and Applications*, K. M. Stanney, Ed., Lawrence Erlbaum Associates, Mahwah, NJ, pp. 1–14.

Stanney, K. M., Mourant, R., and Kennedy, R. S. (1998a), "Human Factors Issues in Virtual Environments: A Review of the Literature," *Presence: Teleoperators and Virtual Environments*, Vol. 7, No. 4, pp. 327–351.

Stanney, K. M., Salvendy, G., Deisigner, J., DiZio, P., Ellis, S., Ellison, E., Fogleman, G., Gallimore, J., Hettinger, L., Kennedy, R., Lackner, J., Lawson, B., Maida, J., Mead, A., Mon-Williams, M., Newman, D., Piantanida, T., Reeves, L., Riedel, O., Singer, M., Stoffregen, T., Wann, J., Welch, R., Wilson, J., and Witmer, B. (1998b), "Aftereffects and Sense of Presence in Virtual Environments: Formulation of a Research and Development Agenda" (report sponsored by the Life Sciences Division at NASA Headquarters), *International Journal of Human–Computer Interaction*, Vol. 10, No. 2, pp. 135–187.

Stanney, K. M., Mollaghasemi, M., and Reeves, L. (2000), *Development of MAUVE, the Multi-criteria Assessment of Usability for Virtual Environments System*, Final Report, Contract N61339-99-C-0098, Naval Air Warfare Center, Training Systems Division, Orlando, FL.

Stanney, K. M., Kennedy, R. S., and Kingdon, K. (2002), "Virtual Environments Usage Protocols," in *Handbook of Virtual Environments: Design, Implementation, and Applications*, K. M. Stanney, Ed., Lawrence Erlbaum Associates, Mahwah, NJ, pp. 721–730.

Stanney, K. M., Kingdon, K., Nahmens, I., and Kennedy, R. S. (2003), "What to Expect from Immersive Virtual Environment Exposure: Influences of Gender, Body Mass Index, and Past Experience," *Human Factors*, Vol. 45, No. 3, pp. 504–522.

Stanney, K., Samman, S., Reeves, L., Hale, K., Buff, W., Bowers, C., Goldiez, B., Nicholson, D., and Lackey, S. (2004), "A Paradigm Shift in Interactive Computing: Deriving Multimodal Design Principles from Behavioral and Neurological Foundations," *International Journal of Human–Computer Interaction*, Vol. 17, No. 2, pp. 229–257.

Stanney, K. M., Cohn, J., Milham, L., Hale, K., Darken, R., and Sullivan, J. (2005), "Deriving Training Strategies for Spatial Knowledge Acquisition from Behavioral, Cognitive, and Neural Foundations," manuscript under review.

Sticha, P. J., Campbell, R. C., and Knerr, C. M. (2002), *Individual and Collective Training in Live, Virtual and Constructive Environments: Training Concepts for Virtual Environments*, Study Report 2002-05, U.S. Army Research Institute for the Behavioral and Social Sciences, Alexandria, VA.

Stone, R. J. (2002), "Applications of Virtual Environments: An Overview," in *Handbook of Virtual Environments: Design, Implementation, and Applications*, K. M. Stanney, Ed., Lawrence Erlbaum Associates, Mahwah, NJ, pp. 827–856.

Storms, R. L. (2002), "Auditory–Visual Cross-Modality Interaction and Illusions," in *Handbook of Virtual Environments: Design, Implementation, and Applications*, K. M. Stanney, Ed., Lawrence Erlbaum Associates, Mahwah, NJ, pp. 455–469.

Strickland, D., Hodges, L., North, M., and Weghorst, S. (1997), "Overcoming Phobias by Virtual Exposure," *Communications of the ACM*, Vol. 40, No. 8, pp. 34–39.

Taylor, H. (2004), "3D Vision and Displays," *Multimedia Technology Class Notes*, http://www.macs.hw.ac.uk/cs/online/9ig2/4/index.htm.

Templeman, J. N., Denbrook, P. S., and Sibert, L. E. (1999), "Virtual Locomotion: Walking in Place Through Virtual Environments," *Presence: Teleoperators and Virtual Environments*, Vol. 8, No. 6, pp. 598–617.

3D Working Group of the Interactive Audio Special Interest Group (1998), *3D Audio Rendering and Evaluation Guidelines*, Version 1.0, MIDI Manufacturers Association, Los Angeles, http://www.iasig.org/pubs/3dl1v1.pdf.

Trindage, J., Fiolhais, C., and Almeida, L. (2002), "Science Learning in Virtual Environments: A Descriptive Study," *British Journal of Educational Technology*, Vol. 33, No. 4, pp. 471–488.

Turk, M. (2002), "Gesture Recognition," in *Handbook of Virtual Environments: Design, Implementation, and Applications*, K. M. Stanney, Ed., Lawrence Erlbaum Associates, Mahwah, NJ, pp. 223–237.

Ultimate 3D Links (2003), "3D Modeling and Animation Software," http://www.3dlinks.com/oldsite/software_modcom.cfm.

Vince, J. (2004), *Introduction to Virtual Reality*, 2nd ed., Springer-Verlag, Berlin.

Waterman, S. (2002), "Steven Spielberg Always and Forever Entertainment," *Entertainment News*, http://www.manmademultimedia.com/magazine/news/ent/index6.html.

Welch, R. B. (1978), *Perceptual Modification: Adapting to Altered Sensory Environments*, Academic Press, New York.

Welch, R. B. (1997), "The Presence of Aftereffects," in *Design of Computing Systems: Cognitive Considerations*, G. Salvendy, M. Smith, and R. Koubek, Eds., Elsevier Science, Amsterdam, The Netherlands, pp. 273–276.

Witmer, B., and Singer, M. (1998), "Measuring Presence in Virtual Environments: A Presence Questionnaire," *Presence: Teleoperators and Virtual Environments*, Vol. 7, No. 3, pp. 225–240.

Yair, Y., Mintz, R., and Litvak, S. (2001), "3D-Virtual Reality in Science Education: An Implication for Astronomy Teaching," *Journal of Computers in Mathematics and Science Teaching*, Vol. 20, No. 3, pp. 293–305.

Zajtchuk, R., and Satava, R. M. (1997), "Medical Applications of Virtual Reality," *Communications of the ACM*, Vol. 40, No. 1, pp. 63–64.

PART 7

EVALUATION

CHAPTER 41
ACCIDENT AND INCIDENT INVESTIGATION

Patrick G. Dempsey
Liberty Mutual Research Institute for Safety
Hopkinton, Massachusetts

1 INTRODUCTION

Of all the applications and benefits of human factors found in a comprehensive handbook, determining how a system led to an accident, minor or catastrophic, is among the least proactive and enjoyable. However, experience is often very instructive, even if the result is determination of what went wrong, and not right. The systems philosophy of human factors is particularly well suited to investigating accidents and incidents to determine the causative influences so that corrective actions can be taken. An example of this is commercial air travel, where the accident rate per miles flown has decreased substantially over time, in part due to the corrective actions implemented after accidents. The same can be said of highway safety, as death rates per mile driven continue to decrease.

The investigation of accidents can take on a variety of forms, and investigations vary in depth from little or no investigation to multiyear investigations costing millions of dollars. Perhaps the most rudimentary of these approaches are summaries of administrative databases used for insurance or regulatory purposes. Beyer (1928) used accident statistics to provide a sense of the seriousness of the burden of occupational injuries and illnesses, in some cases categorizing data due to the loss sources. Although Beyer's (1928) text (and the preceding first and second editions) may have been one of the first to address industrial safety, the topic of accident investigation was not covered explicitly. However, numerous safety innovations contained within clearly resulted from lessons learned from previous accidents.

At the Silver Jubilee Congress of the National Safety Council in October 1938, S. E. Whiting used a simulated confined-entry scenario of workers entering tanks to demonstrate how carbon dioxide in confined spaces can lead to death (Whiting, 1939). Although these were conceptually the opposite of an accident investigation, a key characteristic of the scenarios is that they went beyond descriptions of hardware, including safety equipment. The scenarios alluded to organizational factors such as inaction by bystanders because of concerns over who had decision authority, and cognitive considerations such as flawed mental models of what caused workers to become unconscious. The complexities and interactions between system components was apparent in the descriptions. The realistic nature of the simulations suggest that actual cases were used to develop the scenarios.

Within a few decades, accident prevention and investigation began to take on a more sophisticated and systemic view of causation. Heinrich's (1959) accident sequence depicted by dominos acknowledged that the cause of accidents was multifactorial and involved a sequence of factors. The three factors requisite for an accident were ancestry and social environment, fault of the person, and unsafe act and/or mechanical or physical hazard. Although these components may seem dated relative to contemporary theories and philosophies of accident causation, a broader interpretation of ancestry and social environment to include organizational culture and personal factors would be consistent with more current thought. Unsafe acts continue to occur, although the other components would be replaced by more systemic views, such as how a particular system (humans, environment, and hardware) can lead to errors that ultimately culminate in an incident or accident. Similarly, a more contemporary approach would be to determine what the organizational factors are that increase the propensity for unsafe acts.

On the morning of February 1, 2003, the Space Shuttle *Columbia*, launched by the U.S. National Aeronautics and Space Administration (NASA) a few weeks prior on January 16, broke apart during reentry to the Earth's atmosphere, leading to the loss of the lives of the seven astronauts aboard. A piece of insulating foam that was part of the thermal protection system had separated 81.9 seconds after launch on January 16. When superheated air penetrated the leading-edge insulation, which eventually melted the aluminum structure of the left wing, the weakening of the structure and increasing aerodynamic forces led to loss of control, wing failure, and eventually, to breakup of the *Columbia* (Columbia Accident Investigation Board, 2003).

The Columbia Accident Investigation Board (CAIB) oversaw an accident investigation process that involved a staff over 120 persons in conjunction with 400 NASA engineers lasting nearly seven months. Over 25,000 searchers worked on the ground to collect debris from the spacecraft (CAIB, 2003). Although the physical cause of the *Columbia* accident was attributed to the piece of insulating foam that separated shortly after launch, the CAIB's conclusions regarding the chain of events that led to the disaster reached much further back in time than the foam separation. Examples of the findings included the conclusion that NASA's safety culture had become reactive, complacent, and too optimistic. During the mission and after the foam strike was known, managers resisted new information; thus, communication within the organization was inhibited. Because there were numerous foam incidents during previous missions that did not result in problems, managers were conditioned to believe that foam strikes were maintenance issues to be solved after landing (CAIB, 2003). These and the other extensive findings illustrate just how intricately accident investigations sometimes need to be conducted, and how much accident investigation has matured from earlier forms. This particular case also illustrates many of the concepts to be discussed, ranging from the more traditional fault tree analysis, which is more oriented toward hardware, to investigating organizational influences.

The goal of this chapter is to provide an overview of the numerous approaches to accident and incident investigation in occupational settings, covering a range of approaches that range from using administrative databases to more complex systems approaches. More attention will be given to human-centered approaches, as a mismatch between human capabilities and the demands posed by systems is often a key component of accidents. The type of investigation selected will vary depending on the frequency and severity of the incident(s) in question.

2 BASIC PRINCIPLES OF ACCIDENT INVESTIGATION

Although organizations should always strive to prevent accidents, a key component of a safety program is having the resources and procedures ready to respond to an accident if an accident should occur. Prior to an accident, it is critical that factors such as authority for the investigation be established. All of the accident investigation methodologies discussed later rely on details of the accident; thus, collecting the information in a timely and thorough manner is critical. Many organizations have accident report and investigation forms specifically developed to collect information about factors such as the demographics of the injured person(s), location, and ambient conditions at the time of the accident. All employees should know to whom accidents should be reported, in addition to procedures for calling for emergency response from fire departments or emergency medical technicians when necessary.

An important question that often arises relates to who should conduct the investigation. Vincoli (1994) argues that the manager responsible for the employees involved should lead the investigation, as it is his or her responsibility to ensure the safety of the employees. The reasons for this suggestion is because management can, among other items, marshal resources for the investigation and obtain organizational support, define and implement corrective actions, resolve conflicts, and make employees aware of the outcome as well as conduct follow-up procedures to ensure that corrective actions have been taken (Vincoli, 1994). This does not exclude others from contributing, however, and site safety professionals or safety consultants can and should contribute to the process. Depending on the severity, others, such as insurance company representatives or government investigators, may become involved. For example, the Occupational Safety and Health Administration (OSHA) in the United States must be notified whenever a fatality occurs. OSHA collects information on the individual (demographics, experience, nature of injury, etc.), accident data (measurements, photos, workplace layout, etc.), information on the equipment or process being used (machine type, manufacturer, model, warning devices, etc.), information on potential witnesses, and information on whether the employer has an active safety program and whether or not the program or any of its components address the type of accident that has occurred.

The information collected should provide a sufficient description of the system, including the personnel, the machine/equipment being used, and the task. It is important to try to ascertain what goal the behavior carried out at the time of the accident was directed at, especially if a nonroutine task was being conducted or the operator was troubleshooting. The importance of the task is addressed in more detail in Section 3.3.

An important component of the investigation is recording the scene through photography or videography. Vincoli (1994) provides a comprehensive overview of what the accident photography kit should include. Aside from the camera, lenses, film, and so on, scales, rulers, and a perspective grid are recommended. A digital camera with a liquid crystal display has the advantage of being able to preview the pictures to ensure that the required information is present in the photographs. This can also ensure that the photographs are properly exposed, and geometric issues such as potential parallax problems can be assessed.

3 EPIDEMIOLOGIC APPROACHES

Accident investigation is often thought of only as something conducted for catastrophic events, yet the combined analysis of similar accidents, even if the accidents are somewhat minor, can be a powerful means of determining underlying causes of a class of accidents. Epidemiology, which is the study of the determinants and distribution of health-related states in populations (Last, 1995), is not typically discussed in the context of accident investigation. However, numerous principles of epidemiology are relevant to the study of common classes of accidents, such as injuries caused by a particular object (e.g., a chainsaw) or associated with a particular task (e.g., meat cutting). Several approaches to the study of injuries resulting from accidents are discussed to provide an overview of how epidemiology can contribute to understanding the causes of accidents. It should be noted that all of the methods discussed are not appropriate for investigating single accidents or incidents but rather are cases where a reasonable sample size is available for study. Although this may seem to be a disadvantage, a clear advantage is that the results of these studies can often be implemented and prevent multiple future incidents or accidents.

3.1 Administrative Data

Administrative data collected for other purposes, such as paying Workers' Compensation claims or tracking injuries and illnesses to satisfy regulatory requirements (e.g., OSHA Log of Work-Related Injuries and Illnesses—Form 300 used in the United States) have valuable information on the potential factors associated with the injuries and illnesses recorded. Often, these data have short narratives of the accident scenario and may have administrative codes describing the nature of injury (e.g., contusion, laceration, fracture) and cause (e.g., struck by/against, fall on same level).

As an example of the use of administrative data, Dempsey and Hashemi (1999) analyzed a large sample of Workers' Compensation claims attributed to manual materials handling to determine if the claims suggested specific areas for intervention or future research. Although the majority of the claims were due to musculoskeletal overexertion, a number of high-cost traumatic injuries were uncovered that would be better analyzed through one of the more detailed techniques for rarer events discussed later in the chapter. A large number of acute traumatic injuries to the upper extremity, such as lacerations and contusions, also suggested the need for more widespread use of basic personal protective equipment such as gloves.

Another example of a study that utilized administrative data was the analysis of motor vehicle crashes in construction work zones by Sorock et al. (1996). The accident narrative fields from insurance claims were used as the basis of the study. Over 3600 claims were analyzed by categorizing the claims according to pre-crash activity (stopping, merging, cutting off, reversing, pre-crash error) and crash types (rear-end impact, hitting large object, hitting small object, side impact, overturning). Of the pre-crash actions classified, stopping was most common, although the group of claims associated with various pre-crash errors (e.g. lost control, asleep, failure to yield) had the largest mean and median costs.

One method worth mentioning is the discontinued American National Standards Institute (ANSI) standard, ANSI Z16.2, *Information Management for Occupational Safety and Health* (previously titled *Method of Recording Basic Facts Relating to the Nature and Occurrence of Work Injuries*). This method required collecting items such as nature of injury, part of body, source of injury, agency of accident, and unsafe act, which could then be coded using the coding system provided. Although these are not administrative data per se, older safety texts often mention using the coded data similarly to the suggestions for using Workers' Compensation claims that were discussed earlier.

A final note on the limitations of administrative data should be made. Although these data can provide insights into accidents, the data are often of variable quality, particularly narratives. Since the data are not always for the purposes of gaining an understanding of accident causation, information will be missing. There may also be coding errors, especially for body part and nature-of-injury determinations. Often, the people coding these data do not have medical training; thus, the codes selected may represent the closest to what the coder believes to be correct. Little information is usually available to assess validity. More active approaches discussed throughout the remainder of the chapter can be utilized to overcome these limitations.

3.2 Case-Crossover Methodology

The case-crossover method is a study design used to investigate transient risk factors for discrete outcomes such as occupational injury. The method has been used to investigate risk factors for myocardial infarction (Mittleman et al., 1993), and more recently to investigate cell phone use and motor vehicle collisions (Redelmeier and Tibshirani, 1997). The method is based on determining what was different in the time period immediately prior to an accident that is different from "normal" conditions. The hazard period investigated varies with the nature and duration of the exposure being investigated (e.g., an object falling has a very short hazard period, whereas a sedating antihistamine may have effects for several hours) (Sorock et al., 2001a). For some types of accidents there can be multiple risk factors; thus, the longest period needs to be considered. A key advantage of this approach is that subjects act as their own controls, avoiding the difficulties posed by, for example, finding suitable control subjects for a case–control study. Case–control studies can present difficulties when worker populations are limited, and the cases and controls need to be matched by demographics and exposures.

An example of the case-crossover method applied to occupational problems is the study of acute traumatic hand injury (Sorock et al., 2001b, 2004; Lombardi et al., 2003). Workers with acute traumatic hand injuries seeking treatment at several participating clinics were asked to participate in a telephone interview following treatment, preferably within a day of the accident. Volunteers were called and interviewed about eight transient exposures prior to their accident: using a machine, tool, or work material that performed differently than usual; performing an unusual task; using an unusual work method; being distracted or rushed; feeling ill; working overtime; and glove use at the time. The most important risks suggested were using a machine, tool, or work material that performed differently than usual, unusual work methods, and performing unusual tasks. This methodology is appropriate to use for similar types of accidents, and additional studies are currently under way examining other classes of occupational traumatic injuries, such as eye injuries.

3.3 Scenario Analysis

Drury and Brill's (1983) scenario analysis is a task analysis–based approach to accident investigation, developed for investigating consumer product accidents. The intention was to go beyond traditional accident investigation techniques to incorporate consideration of the task in addition to characteristics of the person, equipment, and environment. Since task analysis is the basis, uncovering the mismatch between the task demands and the limitations of the human body subsystem of interest is the goal. Although more descriptive than the analytical case-crossover method, the advantage is that the method allows for more open-ended data collection and is more rooted in ergonomics and human performance. Although the method is not rooted in epidemiologic methods, there is a clear link, with the goal being to understand the underlying risk factors for accidents.

The scenario analysis approach is based on classifying accidents into hazard patterns (or scenarios), including a description of the victim, product, environment, and task. This approach was considered useful if no more than six hazard patterns describe more 90% of the in-depth investigations. Once the generic hazard patterns are developed, a questionnaire to collect information is then developed. In addition to the information on the victim (e.g., age, sex, weight, body part injured), environment (e.g., indoor versus outdoor, lighting, weather conditions), and product (e.g., type, make, shape, weight), detailed information on task performance is collected. This includes information on the action intended prior to the accident; at the moment the task could not be completed as intended; at the moment the victim took a new, perhaps corrective action but before the injury occurred; and at the moment of the injury. The relationship between task demands and operator capacity can be assessed at each stage. The actual interview is somewhat longer than the interview typically used in conjunction with the case-crossover method, but this allows for more in-depth information to be collected.

In summary, Drury and Brill's (1983) approach uses archival data to generate hazard patterns, which then form the basis of data collection for future incidents. This is somewhat analogous to the case-crossover method discussed above in that knowledge of prior accidents or risk factors is necessary to formulate the interview for injured workers. The case-crossover method has the advantage of being more analytic, whereas the scenario analysis technique is more capable of uncovering human factors issues that lead to accidents and injury. Both methods have proven quite successful in gaining an understanding of the risk factors for numerous types of accidents.

4 SYSTEMS SAFETY TECHNIQUES

There are a number of systems safety techniques that can be utilized for proactive investigations of potential risks in a system to maximize reliability as well as for retrospective accident investigations. These methods sometimes encourage concentrating on hardware failures, but are nevertheless useful components of accident investigations. Several of these techniques have been in existence for some time and have been refined considerably, the most common of which are discussed next.

4.1 Fault Tree Analysis

Fault tree analysis was developed at Bell Laboratories at the request of the U.S. Air Force due to concern over potential catastrophes associated with the Minuteman missile system being developed by Boeing (Hammer, 1985). Many accident investigators find the method particularly useful because it utilizes deductive logic (Vincoli, 1994). The original intention was to develop a method that allowed probabilities of different potential sequences culminating in an accident to be estimated. If an accident probability is available, a risk assessment can be performed by multiplying the probabilities of various undesired events by their predicted costs. The question of the value of human life is often the most difficult question to answer, as this approach requires a common denominator across predictions.

Fault tree analysis is conducted through the use of Boolean logic. The top of the fault tree, in the shape of a rectangle, represents the end effect under investigation, such as an accident. It should be noted that a safe state can be used as the top event to delineate the factors that need to occur to have a safe system. Symbols are then used to represent the different logic operators, including AND gates and OR gates. All possible sequences of events are then mapped out, and as the procedure becomes more complex, the shape of a tree sometimes becomes apparent. The events or system states that need to occur before failure are mapped out. If probabilities for each logic gate are available or can be estimated, probabilities for different branches can be estimated.

Seven different fault trees were constructed during investigation of the *Columbia* tragedy discussed earlier (CAIB, 2003). The number of elements in the fault trees ranged from 3 to 883, the latter illustrating how complex fault trees are when systems are complex.

4.2 Management Oversight and Risk Tree

The management oversight and risk tree (MORT) technique has similarities to fault tree analysis in that it also uses Boolean logic in a graphical format. MORT can be used to assess the adequacy of safety program elements, or it can be used retrospectively to investigate accidents, in particular the management components that may have contributed to failure, for example by creating conditions conducive to being complacent about safety or failure to correct previous safety issues. Rather than simply diagramming a physical system, MORT includes systems issues such as hazard review processes, assumed risks, and safety program review. An extensive discussion of MORT, including examples of completed trees, is provided by Johnson (1973).

5 SWISS CHEESE AND THE HUMAN FACTORS ANALYSIS AND CLASSIFICATION SYSTEM

Reason (1997) used a "Swiss cheese" metaphor to illustrate that the culmination of events in damage to humans or assets depends on failures at different levels. Holes in different layers of defenses that could allow penetration of accident trajectories led to the Swiss cheese metaphor. The holes are the result of what Reason calls either active failures or latent conditions. Reason (1997) defines *active failures* to be unsafe acts by personnel that are likely to have a direct impact on the safety of the system. *Latent conditions* arise from strategic and other top-level decisions (e.g., poor design, undetected manufacturing defects) that spread through organizations and affect the corporate culture. The CAIB (2003) report discussed earlier is an excellent reference for those wishing an in-depth view of latent failures identified during the *Columbia* investigation. Unfortunately, many industrial safety programs often fail to address or investigate these issues, due to an acute focus on procedures, maintenance, and similar factors.

Figure 1 illustrates several of the concepts associated with the Swiss cheese model. The box at the top illustrates the layered defenses concept and the notion that the accident trajectory can be stopped by different defenses, or alternatively, that different components of an organization need to be coordinated to prevent accidents. The triangle is the system that produces the accident, comprised of the personnel (unsafe acts), the workplace, and organizational factors. Causation is bottom-up, whereas the accident investigation process is top-down.

The Human Factors and Analysis and Classification System (HFACS) (Shappell and Wiegmann, 2001) puts into operational terms the concepts of active and latent failures from Reason's (1997) Swiss cheese model. Wiegmann and Shappell (2003) provide a comprehensive overview of HFACS, including illustrative case studies of previous accident investigations. HFACS was developed and refined by analyzing accident reports. Although the approach is discussed within the aviation context, some adaptation provides an excellent method that can be applied beyond aviation.

The "unsafe acts" portion of HFACS is broken down into errors and violation. Errors are further broken down into skill-based, decision, and perceptual errors. Regardless of the context in which this taxonomy is used, human factors principles will be critical to understanding the human capabilities or limitations that contributed to the error or errors that led to the accident. Violations are classified as routine or exceptional, with exceptional representing more egregious violations.

The "preconditions for unsafe acts" component of HFACS describes the environmental factors, conditions of operators, and personnel factors that can lead to increased propensity for unsafe acts. Wiegmann and Shappell (2003) provide a broad range of examples for each of these. The "conditions of operators" concept has drawn interest for many occupations over the years, especially in the area of testing operators, particularly in the transportation industry, for what has been termed fitness for duty.

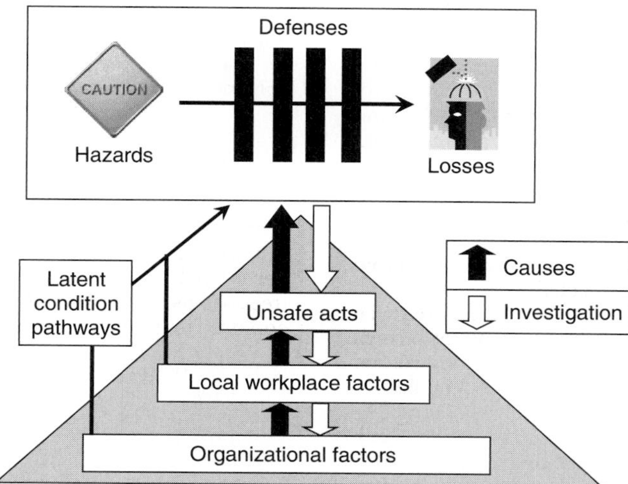

Figure 1 Stages in the development and investigation of an organizational accident. (Adapted from Reason, 1997.)

HFACS includes an unsafe supervision component broken down into inadequate supervision, planned inappropriate operations, failure to correct problems, and supervisory violations. Thus, management accountability and culpability is a critical component of HFACS. These factors become particularly important for nonroutine work such as construction, where planning a job and the safety measures taken by supervisors and employees are critical, and the lack of such planning has led to accidents.

The last component of HFACS is that of organizational influences, which includes resource management, organizational climate, and organizational process. This is perhaps the most difficult component to describe during an investigation. To truly understand factors such as organizational customs, communication in an organization, and procedural influences, a great deal of data gathering through interviews may be required. The CAIB (2003) report is an excellent example of a comprehensive set of findings regarding organizational influences. The interested reader is strongly encouraged to consult the case studies provided by Wiegmann and Shappell (2003) for in-depth information on application of this taxonomic system for investigating accidents.

6 CONCLUSIONS

Accident investigation can take on many forms, ranging from an analysis of administrative data to investigations of single incidents that last for months. Regardless of the nature or severity of the accident, gaining an understanding of the human capabilities and limitations contributing to the accident is often the key to understanding causation. Even in cases where "hardware faults" were seemingly the cause, there are causative contributions influenced by humans.

One of the best ways to explore different approaches to accident investigation and to gain an understanding of the scope and depth of different investigations is through published case studies. An interesting set of case studies centered around human factors is presented by Casey (1993), with human factors violations highlighted for a range of accident types, from the Bhopal disaster to a medical context. A case study describing how three cosmonauts from the former Soviet Union died in 1971 due to a pressure equalization valve that could not be turned quickly enough following depressurization is presented. Although the valve was intended to be used under such situations, operation had not been tested under the extreme conditions where physical capabilities were greatly reduced. Although the particular physical capabilities required were not difficult to understand or even to define empirically, the underlying issue was why no one had the foresight to test the system under operational conditions (or more realistically, simulated conditions). More in-depth cases studies are also provided, including an analysis of the Bhopal Union Carbide disaster in India in 1984 which killed more than 2500 residents. Perhaps most distressing about many of these case studies is that the actions that would have prevented many of the disasters were not excessively burdensome or technically infeasible.

Kletz (2001) provides an illustrative set of in-depth case studies, many of which are rather famous due to their severity, including the Three-Mile Island and Chernobyl nuclear disasters and the King's Cross railway station fire. The case studies include detailed overviews, in some cases engineering drawings of the relevant physical systems or components. Recommendations for prevention or mitigation at each stage of the accidents are provided, as are comprehensive references for additional sources of information. One advantage of studying such well-known disasters is that there is often detailed information that has been gathered from witnesses, physical evidence, and so on. Although most accident investigations will not reach this level of scope and depth, these case studies provide one of the best means of becoming familiar with accident investigation techniques and the wide array of people, equipment, procedures, and organizational characteristics that often need to be considered during an accident investigation.

REFERENCES

Beyer, D. S. (1928), *Industrial Accident Prevention*, 3rd ed., Houghton Mifflin, Boston.

CAIB (2003), *Columbia Accident Investigation Board Report*, Vol. I, National Aeronautics and Space Administration and U.S. Government Printing Office, Washington, DC.

Casey, S. M. (1993), *Set Phasers on Stun and Other True Tales of Design, Technology, and Human Error*, Aegean Publishing, Santa Barbara, CA.

Dempsey, P. G., and Hashemi, L. (1999), "Analysis of Workers' Compensation Claims Associated with Manual Materials Handling," *Ergonomics*, Vol. 42, No. 1, pp. 183–195.

Drury, C. G., and Brill, M. (1983), "Human Factors in Consumer Product Accident Investigation," *Human Factors*, Vol. 25, pp. 329–342.

Hammer, W. (1985), *Occupational Safety Management and Engineering*, 3rd ed., Prentice-Hall, Englewood Cliffs, NJ.

Heinrich, H. W. (1959), *Industrial Accident Prevention: A Scientific Approach*, 4th ed., McGraw-Hill, New York.

Johnson, W. G. (1973), *MORT: The Management Oversight and Risk Tree*, SAN 821-2, U.S. Atomic Energy Commission and U.S. Government Printing Office, Washington, DC.

Kletz, T. A. (2001), *Learning from Accidents*, 3rd ed., Gulf Professional Publishing, Oxford.

Last, J. M. (1995), *A Dictionary of Epidemiology*, 3rd ed., Oxford University Press, New York.

Lombardi, D. A., Sorock, G. S., Hauser, R. B., Nasca, P. C., Eisen, E. A., Herrick, R. F., and Mittleman, M. A. (2003), "Temporal Factors and the Prevalence of Transient Exposures at the Time of an Occupational Traumatic Hand Injury," *Journal of Occupational and Environmental Medicine*, Vol. 45, pp. 832–840.

Mittleman, M. A., Maclure, M., Tofler, G. H., Sherwood, J. B., Goldberg, R. J., and Muller, J. E. (1993), "Triggering of Acute Myocardial Infarction by Heavy Physical Exertion," *New England Journal of Medicine*, Vol. 329, pp. 1677–1683.

Reason, J. (1997), *Managing the Risks of Organizational Accidents*, Ashgate Publishing, Aldershot, Hampshire, England.

Redelmeier, D. A., and Tibshirani, R. J. (1997), "Association Between Cellular-Telephone Calls and Motor Vehicle Collisions," *New England Journal of Medicine*, Vol. 336, No. 7, pp. 453–458.

Shappell, S., and Wiegmann, D. (2001), "Applying Reason: The Human Factors Analysis and Classification System," *Human Factors and Aerospace Safety*, Vol. 1, pp. 59–86.

Sorock, G. S., Ranney, T. A., and Lehto, M. R. (1996), "Motor Vehicle Crashes in Roadway Construction Workzones: An Analysis Using Narrative Text from Insurance Claims," *Accident Analysis and Prevention*, Vol. 28, No. 1, pp. 131–138.

Sorock, G. S., Lombardi, D. A., Gabel, C. L., Smith, G. S., and Mittleman, M. A. (2001a), "Case-Crossover Studies of Occupational Trauma: Methodological Caveats," *Injury Prevention*, Vol. 7, Suppl. I, pp. i38–i42.

Sorock, G. S., Lombardi, D. A., Hauser, R. B., Eisen, E. A., Herrick, R. F., and Mittleman, M. A. (2001b), "A Case-Crossover Study of Occupational Traumatic Hand Injury: Methods and Initial Findings," *American Journal of Industrial Medicine*, Vol. 39, pp. 171–179.

Sorock, G. S., Lombardi, D. A., Peng, D. K., Hauser, R. B., Eisen, E. A., Herrick, R. F., and Mittleman, M. A. (2004), "Glove Use and the Relative Risk of Acute Hand Injury: A Case-Crossover Study," *Journal of Occupational and Environmental Hygiene*, Vol. 1, pp. 182–190.

Vincoli, J. W. (1994), *Basic Guide to Accident Investigation and Loss Control*, Wiley, New York.

Whiting, S. E. (1939), "Men Who Go Down to Their Death in Tanks," demonstration at the Silver Jubilee of the National Safety Council, October 10–14, 1938, Liberty Mutual Insurance Company, Boston.

Wiegmann, D. A., and Shappell, S. A. (2003), *A Human Error Approach to Aviation Accident Analysis: The Human Factors Analysis and Classification System*, Ashgate Publishing, Aldershot, Hampshire, England.

CHAPTER 42

HUMAN FACTORS AND ERGONOMICS AUDITS

Colin G. Drury
University at Buffalo: State University of New York
Buffalo, New York

1 PURPOSE

This chapter has two interrelated aims. First, we examine the human–machine tasks of inspecting, checking, and auditing, to provide understandings that can guide work design, equipment design, and job aid development. Second, we apply this knowledge to inspecting, checking, and auditing of human factors/ergonomics aspects of human–machine systems. As part of this aim, we provide a detailed review and worked example of recent human factors/ergonomics audit programs. Throughout the chapter examples are given from a wide range of domains, from product usability audits, through aviation preflight checklists, to inspection of products for quality assurance.

2 INSPECTING, CHECKING, AND AUDITING

The idea of inspecting is as old as civilization. In the Sumerian epic *Gilgamesh*, almost 5000 years old, the narrator invites the reader to examine the quality of the walls of Uruk, built by Gilgamesh, the king:

> Look at its wall which gleams like copper (?),
>
> inspect its inner wall, the likes of which no one can equal!
>
> Take hold of the threshold stone—it dates from ancient times!
>
> Go close to the Eanna Temple, the residence of Ishtar,
>
> such as no later king or man ever equaled!
>
> Go up on the wall of Uruk and walk around,
>
> examine its foundation, inspect its brickwork thoroughly.
>
> Is not (even the core of) the brick structure made of kiln-fired brick,
>
> and did not the Seven Sages themselves lay out its plans?" (Gilgamesh, Tablet 1)

The essence of the examining function is all there already. Bodily senses (look, take hold of) are used to compare the existing item (wall) with some implied or actual standard (e.g., kiln-dried brick). Inspection can have more formal definitions (e.g., in dictionaries and quality control texts), but a simple definition from Drury (2002) provides a reasonable modern start: "The [test and] inspection system determines the suitability of a product or process to fulfill its intended function, within given parameters of accuracy, cost and timeliness."

Inspection is thus a decision function, for management, for the general public, and for an individual. Should the process be stopped before it produces more defects? Does this café meet local standards of cleanliness, so that people can eat there safely? Is this aircraft safe for me to fly in? Does this box of strawberries contain any bad fruit? The most basic decision is a go/no-go decision: Does this item or process fulfill its intended function, or not? In practice, the amount of inspection, checking, and auditing needed to reach such a simple conclusion may be considerable. For example, a national safety board may need considerable evidence after a major accident to determine that the system (aircraft, train, ferry, spacecraft) is fit to resume normal operations. As noted above, most inspection decisions are simpler than this.

In principle, inspection can be done by the producer or consumer of the goods or services directly, but this is not always satisfactory. Thus, the person machining a part can determine whether or not it meets standards, or the ultimate end user can inspect the part. Much of the quality revolution from the 1970s onward has been concerned with pushing such decisions back along the production system to ensure that decisions are made at the source (e.g., Evans and Lindsay, 1993; Drury,

1997), to prevent errors from propagating through a production system. Indeed, the preferred solution is to inspect the *process* rather than the *product* to ensure that defects are extremely unlikely to be produced. That is the aim of in-process statistical process control (e.g., Devor et al., 1992). Unlike our example of the box of strawberries, most ultimate consumers are not equipped to be able to inspect or check the complex devices and processes they use in daily living. In an agrarian economy, a farmer could examine a spade produced for him by the local blacksmith (e.g., Jones, 1981) with enough skill and visual/haptic information to reach a reasonable conclusion about whether the design and construction quality met his requirements. In our more specialized economy, consumers cannot make valid quality judgments on automobiles, computers, or even the safety of a local chemical plant because they lack both the specialized knowledge required and access to the inner workings of the product or system.

In such cases, customers rely on the judgment of professional inspectors, checkers, and auditors to make informed decisions. This reliance raises questions of honesty, trust, competence, and human–machine system design. From a sociotechnical systems perspective, inspectors must be seen as independent of producers of goods and services or their findings will not be accepted. For example, in civil aviation the U.S. Federal Aviation Administration (FAA) decrees that airline inspectors charged with checking airworthiness are kept organizationally independent of aviation maintenance technicians, who perform repairs, adjustments, and replacements. This independence can lead to role ambiguity and conflict in their job. For example, McKenzie (1958) noted that inspection is always of people: The inspector is judging the work of others by examining the outputs. Early work by McKenzie (1958), Thomas and Seaborne (1961), and Jamieson (1966), established that social pressures are an important part of the inspector's job, and that inspectors can at times change their behavior and performance in response to such pressures. The author worked in one factory where the customer returned shipments of product when they could not be used right away (e.g., during a strike) by finding a defect in the shipment. Later, when the customer needed the product, the factory repackaged the same shipments, often receiving complements on their improved quality!

3 INSPECTION AND HUMAN FACTORS/ERGONOMICS

As noted above, the study of inspection and inspectors is at least 50 years old, although most concentrated on inspection of products in a manufacturing setting. There are many other types, as noted by Drury (2002):

- *Regulatory inspection*: to ensure that regulated industries meet or exceed regulatory norms. Examples are review of restaurants against local service codes, fire safely inspection of buildings, and safety inspection of workplaces.
- *Maintenance*: to detect failure arising during the service life of a product. This failure detection function can be seen in inspection of road and rail bridges for structural determination or of civil airliners for stress cracks or corrosion.
- *Security*: to detect items deliberately concealed. These may be firearms or bombs carried onto aircraft, drugs smuggled across borders, or camouflaged targets in aerial photographs. They can also be suspicious happenings on a real-time video monitor at a security station. Law enforcement has many examples of searching crime sites for evidence.
- *Design review*: to detect discrepancies or problems with new designs. Examples are the checking of building drawings for building code violations, of chemical plant blueprints for possible safety problems, or of new restaurant designs for health code violations.
- *Functionality testing*: to detect lack of functionality in a completed system. This functional inspection can often include problem diagnosis, as with checks of avionics equipment in aircraft. Often functional inspection is particularly dangerous and costly, as in test flying aircraft or checking out procedures for a chemical process.

All of these inspection applications have much in common. Indeed, Drury (2003) has proposed a unified model of inspection in the security domain and shows how it can apply to all security inspection systems. This model was hardly new; it was based on function analytical models developed earlier by Drury (1978), Sinclair (1984), and Wang and Drury (1989). A recent incarnation can be seen in Jiang et al. (2003). Table 1 gives a generic functional breakdown of inspection, showing the major functions of inspection with the correct outcomes and errors arising from each function.

The reader is referred directly to Drury (2002) for a detailed consideration of inspection, including automated inspection and test, and a design methodology for "design for inspectability." In the current chapter, only an overview is provided to help understand the context for both checklists and audits. Each function is considered in the order given in Table 1. A single example of safety inspection of a workplace will be used to illustrate each function.

1. *Setup*. In this function, the inspection system is prepared for use. Needed tools, equipment, and supplies are procured, procedures are available to aid the inspector, and the inspector has been trained to perform the task correctly. For a workplace safety inspection, there will be some equipment needing calibration (e.g., psychrometer, air sampling systems), a written procedure in the form of a checklist or computer program (e.g., Wilkins et al., 1997), and safety inspectors who have undertaken training and often, certification. Certification is one way in which the competence and independence of inspectors is maintained, leading presumably to a higher degree

Table 1 Generic Function, Outcome, and Error Analysis of Test and Inspection

Function	Outcome	Errors
Setup	Inspection system functional, calibrated correctly, and capable	1.1. Incorrect equipment 1.2. Nonworking equipment 1.3. Incorrect calibration 1.4. Incorrect or inadequate system knowledge
Present	Item (or process) presented to inspection system	2.1. Wrong item presented 2.2. Item mispresented 2.3. Item damaged by presentation
Search	Indications of all possible nonconformities detected, located	3.1. Indication missed 3.2. False indication detected 3.3. Indication mislocated 3.4. Indication forgotten before decision
Decision	All indications located by search measured correctly and classified correct outcome decision reached	4.1. Indication measured incorrectly 4.2. Indication classified incorrectly 4.3. Wrong outcome decision 4.4. Indication not processed
Respond	Action specified by outcome taken correctly	5.1. Nonconforming action taken on conforming item 5.2. Conforming action taken on nonconforming item

Source: Drury (2002).

of public trust in the inspection process. The SETUP function places demands on the regulatory system to provide the needed antecedents of effective inspection (i.e., sufficient resources).

2. *Present.* Here the inspector (suitably equipped) and the entity to be inspected come together so that inspection can take place. The narrator of *Gilgamesh* urges the reader to "... go up to the wall of Uruk and walk around..." Often, this function is purely mechanical: The manufacturing inspector has products arrive and depart on a conveyor, the safety inspector accesses all areas of a factory that could contain indications of an unsafe condition, and the FAA inspector goes to the file cabinets where maintenance records are kept to check that maintenance was performed and signed off correctly. The managerial implications for PRESENT are that inspectors must know the places they need to examine, and the organization being inspected must provide open and timely access to such sites. In safety inspection by a regulatory agency, there is often special legal provision for sites to be made available with little or no prior notice, to prevent concealment of safety concerns known to management.

3. *Search.* Here we arrive at the first of the two most important and error-prone functions of inspection. Search is typically a sequential serial process during which the entire item to be inspected is brought under scrutiny piece by piece. The most obvious form of search is visual, and this has a long history of study, going back to the earliest days of human factors (Blackwell, 1946; Lamar, 1960). People (i.e., inspectors) search an area by successive visual fixations where the eye remains essentially stationary. What can be detected during a single fixation is a function of target and background characteristics (e.g., Overington, 1973) with considerable research (e.g., Treisman, 1986) on combinations of target and background conditions favoring rapid, parallel ("preattentive") search. Mathematical models (e.g., Morawski et al., 1980; Wolfe, 1994) of the visual search process emphasize two features:

(a) Over what area around the fixation point is the target detectable in a single fixation ("visual lobe")?

(b) How are successive fixations sequenced to achieve coverage of the entire area?

Visual lobe models (e.g., Engel, 1971; Eriksen, 1990) determine how much area can be covered in a single fixation, and hence the number of fixations required for coverage. The time to search an item completely varies directly with the number of fixations required, and hence the reciprocal of lobe area (Morawski et al., 1980). Successive fixations are determined partly by top-down strategy and partly by bottom-up features of the currently fixated area (Wolfe, 1994). Top-down strategy has been modeled (e.g., Hong and Drury, 2002) as sequential, random, or a mixture (e.g., Arani et al., 1984). It is determined in part by the inspector's knowledge and expectations of where targets are likely. In an inspection context, a key derivation from search models is the *stopping time*, when the inspector decides that enough time has been spent on one item and moves to the next item. Stopping time chosen by inspectors accords quite well with the predictions of optimum stopping models (e.g., Chi and Drury, 1995; Baveja et al., 1996). Stopping time is a physical manifestation of how many resources the inspector (or the system giving inspection instructions) is willing to devote to each item inspected.

Visual search is not the only form of search important in inspection; there is also procedural search, where an inspector goes through a list of places that require examination. Procedural search is used extensively in aviation checklists (see later) where, for example, a preflight inspection of a general aviation aircraft follows a written procedure requiring examination of control surfaces, tires, fuel, structural joints, and so on. For each item on the checklist, the inspector is trained on what defects to look for. An example is fuel, where a small sample of fuel is drawn from a low point in the fuel line to check for water contamination, with water drops appearing as spheres in the fuel sample.

In a safety inspection, the inspector will have a list of key items/areas to search for indications of lack of safety. Inspectors will use their senses to examine each item in this procedural search, often requiring visual search within the procedural search. For example, inspectors must check that guarding is present on moving machinery, a largely procedural task. When they check safety records, such as the OSHA log in the United States, a visual search is required to determine whether all fields have been filled in and signed correctly. Note that in both these examples, considerable knowledge and skill is required to understand what would be an indication of a safety violation.

As noted in Table 1, the successful outcome of search is something detected that *could* be a defect. This is known in the nondestructive inspection (NDI) community as an *indication*. Subsequent inspection functions are concerned with how to deal with each indication. Note that if search fails, the indication is missed and subsequent functions cannot proceed. At least in visual inspection there is considerable evidence (Drury and Sinclair, 1983; Drury et al., 1997) that the search function is quite error prone, with only about 50% of defects ever being located as indications.

4. *Decision.* This is the function in which the indication is judged against a standard to determine whether it is a true defect. If inspection is about decision, this function represents the essence of inspection. Decision requires human (or machine-aided) judgment against a standard, so a standard must be prespecified. Standards in manufacturing inspection can come from physical properties (hardness, conductivity, surface finish) that can be measured with appropriate guaging. The decision in such cases of what the SQC community calls *variables inspection* can be automated quite simply and is thus rarely an appropriate human task. Not all measurements and standards can be implemented so simply. How do we quantify blemishes in the surface finish of automobile paint work (Lloyd et al., 2000) or corrosion areas on an aircraft fuselage (Wenner and Drury, 1996). These examples of *attributes inspection* require a complex, typically human judgment. Often, signal detection theory (SDT) has been used as a model of this part of the inspection process (e.g., Drury and Sinclair, 1993; Chi and Drury, 1995) although at times it has been misapplied to the overall inspection process, hence lumping search errors with wrong decisions to accept a true defect (e.g., Drury and Addison, 1973). SDT suggests a separation of decision difficulty (discriminability) from bias in reporting/not reporting defects (criteria), thus providing a useful link to different remedial actions, depending on whether discriminability or criterion needs changing. For example, Drury et al. (1997) found that the decision function of inspection was extremely inconsistent between inspectors, implying the need for better training and job aids in aircraft inspection.

For safety inspection, discriminability represents the decision difficulty. This can be very easy when the implied standard is zero (e.g., *any* missing machine guard is a violation) or more difficult (e.g., how much untidiness represents "poor housekeeping"). The decision criterion reflects the willingness to report. From SDT this is a function of the a priori probability of a defect and the relative costs of the two errors:

- *Miss*: not reporting a true defect
- *False alarm*: reporting an indication that was not a true defect

In safety inspection, the pressures on the inspector can be high. A false alarm can be used by the factory being inspected to bring disrepute on a regulatory agency and on the actual inspector. A miss can lead to an accident, injury, or even a Bhlopal-like disaster. Similar pressures exist for aircraft inspectors, security operators, and even the intelligence community. The decision to report becomes even more difficult (from SDT) when the defect found is extremely rare. A recent example is the failure to find a crack in the titanium hub of a jet engine that caused the accident in 1997 at Penascola, Florida. The inspector, despite years of experience, had never encountered a crack in a titanium hub before. More details on mathematical approaches to the decision function may be found in Drury (2002).

5. *Response.* When the decision has been made, the action chosen must be taken. In a manufacturing context, this can be a simple as removing a defective item from the production process or as complex as stopping the process to diagnose the "root cause" of the defect being produced. In more general contexts, the action is often written (e.g., a repair order for a crack in an aircraft structure, written warning of unsanitary conditions in a restaurant, or a safety citation to company management for a missing machine guard).

Although response is a relatively mechanical function, it can be subject to errors. Aircraft inspectors who intend to write up all defects at the end of inspection can forget some defects, surgeons can mark the wrong leg for amputation, the safety citation can be incomplete. This is one function where computer-based automation can help by making response simple and immediate. For example, Drury et al. (2000) developed a computer-based task card system for aircraft inspection that allowed easy generation of repair orders. Even more could have been done with drop-down menus for fault type (crack, corrosion, etc.) or a search function for the appropriate reference in the structural repair manual.

Throughout this short treatment of inspection we have seen where and why good human factors/ergonomics practices can reduce error potential. There are also overarching considerations of job design and automation that can have a great impact on errors [e.g., the hybrid automation studies of Hou et al. (1993) and Jiang et al. (2003)]. Again, details may be found in Drury (2002).

The audit and checking activities associated with nonmanufacturing applications can now be placed in a suitable context.

4 CHECKING AND CHECKLISTS

When inspection is too complex to be carried out by an inspector unaided by procedural notes, a job aid is required to lead the inspector through the task. The nonmanufacturing examples given already (e.g., aircraft maintenance, safety inspectors) are typical of those requiring and using job aids. The simplest job aid for any procedural task (e.g., preparation for landing an aircraft) is the checklist. All pilots carry checklists for many complex procedures when a sequence of actions must be performed in a standard order. In addition to the landing preparation noted above, there are checklists for preflight inspection, startup/taxi, pretakeoff, climb, cruise, postlanding, and engine shutdown. These are typically short laminated paper lists in general aviation, or computer-based lists for corporate and passenger jets. Glider pilots use laminated checklists but also use mnemonics, such as "STALLS" for prelanding, where they might not have time to consult a written checklist. Safety inspectors in industry typically use a written checklist of several pages.

If the form of a checklist (paper, computer, mnemonic) can vary, so can the content and structure. Most checklists are used as memory aids for well-practiced tasks, so that they are structured as lists of commands, each of which is relatively tense: "switch both magnetos on: open fuel cock: prime engine for 5 seconds." The user is expected to know which are the magneto switches and which way they move for "on." Users are also expected to understand *why* each action is required so that they have some strategy for recovering from malfunctions (e.g., one magneto not working correctly).

In contrast, more detailed procedures are used for inspection and maintenance of aircraft and spacecraft. These procedures will spell out each step in detail, often with part numbers, numerical settings (e.g., tightening torque) and include warnings and cautions as well as a rationale for the overall procedure. With a computer-based system, detailed procedures can also be viewed as checklists: for example, when the procedure is being repeated after a short interval. Drury et al. (2000) used simple hypertext links to move between the checklist steps and the more detailed procedures in their program for inspection workcards. Much human factors design and evaluation has gone into the physical design of such procedures (e.g., Patel et al. 1994; Chervak et al., 1996; Drury et al., 2000), so in the remainder of this section we concentrate on classic checklists, as they are most often encountered in nonmanufacturing inspection and audit.

Checklists have their limitations, though. The cogent arguments put forward by Easterby (1967) provide a good early summary of these limitations in the context of design checklists, and most are still valid today. Checklists are only of use as an aid to designers of systems at the earliest stages of the process. By concentrating on simple questions, often requiring yes or no answers, some checklists may reduce human factors to a simple stimulus–response system rather than encouraging conceptual thinking. Easterby quotes Miller (1967): "I still find that many people who should know better seem to expect magic from analytic and descriptive procedures. They expect that formats can be filled in by dunces and lead to inspired insights.... We should find opportunity to exorcise this nonsense" (Easterby, 1967, p. 554).

Easterby finds that checklists can have a helpful structure, but often have vague questions, make nonspecified assumptions, and lack quantitative detail. Checklists are seen as appropriate for some parts of ergonomics analysis (as opposed to synthesis) and are even more appropriate to aid operators (not ergonomists) in following procedural steps. Clearly, we should be careful, even 30 years on, to heed these warnings. Many checklists are developed, and many of these published, that contain design elements fully justifying such criticisms.

Most formal studies of checklist use have been in an aviation context, both in maintenance (Pearl and Drury, 1995) and in preflight inspection (Ockerman and Pritchett, 1998, 2000a,b). They have also found widespread use in the flight operations side of aviation, with detailed analysis by Degani and Wiener (1990). The Degani and Wiener (1990) study laid the basis for much subsequent work on checklists. They analyzed incident reports from NTSB and ASRS (NASA's Aviation Safety Reporting System), finding that the main checklist errors resulted from overlooking items following interruptions or distractions, particularly when working under time pressure or toward the end of the working day. Their recommended countermeasures were use of a "challenge and response" operating philosophy, grouping several items together and using a logical flow pattern. In particular, they advocated a "geographical" sequence of steps, good formatting/typography, and that "operators should keep checklists as short as possible to minimize interruptions." They also reviewed then-current technologies that could assist checklist use. An earlier study of checklists for circuit board inspection (Goldberg and Gibson, 1986) also found that a logically organized checklist outperformed a randomly organized one.

Patel et al. (1993) found that during the initial inspection of an aircraft on arrival at maintenance, the sequence of tasks in the checklist or workcard did not match the sequence of tasks that aircraft maintenance technicians typically followed. In a related study, Pearl and Drury (1995), questionnaires and videotapes showed that mechanics tended to sequence their tasks using spatial cues on the airplane rather than the order specified on their workcard. The study also revealed that aviation maintenance technicians who performed low-level inspections used spatial locations of tasks to sequence them. In addition, many aircraft mechanics rarely used the checklist and viewed it as an only guide for inexperienced mechanics. Experienced inspectors felt that they had acquired sufficient skill to perform the inspection task using their memory and referred to the checklist only occasionally.

A more recent series of investigations (Ockerman and Pritchett, 1998, 2000a,b) have examined the relationship between the medium (paper vs. wearable computers) on which the procedure was displayed, the presentations of procedure context, overreliance, and inspection performance for a preflight inspection task. The studies found that inspection performance could be influenced by the presence of procedure context information presented with procedures. Their 1998 study also observed that one-third of the participants used their memory and not the task guidance system to perform the preflight inspection. They observed that in some sessions the subjects performed the task from memory and consulted the checklist only to see if anything was forgotten, echoing the Pearl and Drury (1995) findings from maintenance.

Computer applications for checklists were also advocated by Degani and Wiener (1990) and tested in an aviation maintenance/inspection environment by Drury et al. (2000). The latter study measured the impact of a hypertext-based computer program on the usability of work documentation in maintenance and inspection. Based on data collected in 1992–1993 at an airline partner, they concluded that computer-based inspection job aids were effective, although much of their effectiveness was attributed to good job aid design rather than computerization per se. Their task was one of detailed inspection of aircraft structures, and used a checklist only as a top-level job aid, with more detailed instructions and data available via hyperlinks.

Major findings of all of these studies should be applicable to audit checklists, despite their somewhat different domains within aviation. Checklists are good job performance aids for repetitive tasks. They involve little explanation of detail or rationale for the sequence of operations, being mainly reminders of the correct sequence, often with facilities for marking (the "check" in "checklist") each item as it is performed to minimize the effect of interruptions or distractions (Degani and Wiener, 1990). The findings include (1) a geographical sequence is probably best; (2) good design principles should be followed; (3) technology can improve checklists; and (4) checklists are not always used, with reliance often placed on memory. The latter finding was reinforced by Wenner and Drury (2000), who note that some people did not read or follow very explicit instructions for performing a task.

Recently, two closely related studies of checklists were performed using a simulated repetitive aircraft inspection task with engineering student participants. The first (Larock, 2000; Larock and Drury, 2003) measured the effects of checklist layout and the number of sign-offs when a task was repeated eight times on eight days. The second (Pai, 2003) examined the use of computer-based checklists under the best design conditions found by Larock and Drury (2003), using the same task repeated over six days.

The first study compared functionally ordered and spatially ordered checklists, and also whether each of the 108 items had to be signed off individually or whether they were signed off in 37 logically related subsets. As expected, spatial ordering was better than functional ordering for both accuracy and speed. The number of sign-offs and the checklist layout interacted for sequence errors, where the best combination was spatial ordering and signing off in 37 groups. Over the course of eight daily trials, participants became faster at the task but tended to develop a spatial strategy for either checklist. The second study used this combination to test the efficacy of various computer implementations of the original paper checklist. A personal digital assistant (Palm-Pilot) was used with its built-in application of the "to-do list." A more user-friendly program was written specifically for the task studied and was implemented on the PDA and on a laptop computer. The conclusion was that the three computer implementations did not differ from each other, and all gave better speed and accuracy than those for the paper-based checklist.

These studies reinforce conclusions 1 to 4 noted above and showed that checklist behavior is not merely an artifact of using aviation professionals as subjects. Checklists emerge as a powerful tool, but one that needs careful human factors design to reach its maximum performance. Outside of aviation operations, there is little evidence that checklists are designed with these human factors findings in mind.

To develop checklists for auditing safety, ergonomics, or human factors, the design principles above should be followed. In addition, checklists need to be validated with actual users to ensure that their content, structure, and format do indeed lead to reliable performance.

5 AUDITING WITH SPECIFIC APPLICATION TO HUMAN FACTORS

When we audit an entity, we perform an examination of it. Dictionaries typically emphasize official examinations of (financial) accounts, reflecting the accounting origin of the term. Accounting texts go further: for example, "testing and checking the records of an enterprise to be certain that acceptable policies and practices have been consistently followed" (Carson and Carlson, 1977, p. 2). In the human factors field, the term is broadened to include nonfinancial entities, but remains faithful to the concepts of checking, acceptable policies/practices, and consistency.

As with inspecting, auditing is mentioned in antiquity, at least in current translations. "[H]e who does not pay the fine annually shall owe ten times the sum, which the treasurer of the goddess shall exact; and if he fails in doing so, let him be answerable and give an account of the money at his audit" (Plato, *Laws*, Book VI).

Human factors audits can be applied, as can human factors itself, to both products and processes. Both applications have much in common, as any *process* can be considered as a *product* of a design procedure, but in this section we emphasize process audits because product evaluation is covered in detail in Chapter 50. Product usability audits have their own history (e.g., Malde, 1992), which is best accessed through the product design and evaluation literature (e.g., McClelland, 1990).

5.1 Need for Auditing Human Factors

Human factors or ergonomics programs have become a permanent feature of many companies, with typical examples shown in Alexander and Pulat (1985). As with any other function, human factors/ergonomics needs tools to measure its effectiveness. Earlier, when human factors operated through individual projects, evaluation could take place on a project-by-project basis. Thus, the interventions to improve apparel sewing workplaces described by Drury and Wick (1984) could be evaluated to show changes in productivity and reductions in cumulative trauma disorder causal factors. Similarly, Hasslequist (1981) showed productivity, quality, safety, and job satisfaction following human factors interventions in a computer-component assembly line. In both cases, the objectives of the intervention were used to establish appropriate measures for the evaluation.

Ergonomics/human factors, however, is no longer confined to operating in a project mode. Increasingly, the establishment of a permanent function within an industry has meant that ergonomics is more closely related to the strategic objectives of the company. As Drury et al. (1989) have observed, this development requires measurement methodologies that also operate at the strategic level. For example, as a human factors group becomes more involved in strategic decisions about identifying and choosing the projects it performs, evaluation of the individual projects is less revealing. All projects performed could have a positive impact, but the group could still have achieved more with a more astute choice of projects. It could conceivably have had a more beneficial impact on the company's strategic objectives by stopping all projects for a period to concentrate on training the management, workforce, and engineering staff to make more use of ergonomics.

Such changes in the structure of the ergonomics/human factors profession indeed demand different evaluation methodologies. A powerful network of individuals, for example, who can, and do, call for human factors input in a timely manner can help an enterprise more than a number of individually successful project outcomes. Audit programs are one of the ways in which such evaluations can be made, allowing a company to focus its human factors resources most effectively. They can also be used in a prospective, rather than retrospective, manner to help quantify the needs of the company for ergonomics/human factors. Finally, they can be used to determine which divisions, plants, departments, or even product lines are in most need of ergonomics input.

5.2 Design Requirements for Audit Systems

Returning to the definition of an audit, the emphasis is on checking, acceptable policies, and consistency. The aim is to provide a fair representation of the business for use by third parties. A typical audit by a certified public accountant would comprise the following steps (adapted from Koli, 1994):

1. *Diagnostic investigation.* Describe the business and highlight areas requiring increased care and high risk.
2. *Test for transaction.* Trace samples of transactions grouped by major area and evaluate.
3. *Test of balances.* Analyze content.
4. *Formation of opinion.* Communicate judgment in an audit report.

Such a procedure can also form a logical basis for human factors audits. The first step chooses the areas of study, the second samples the system, the third analyses these samples, and the final step produces an audit report. These define the broad issues in human factors audit design:

1. *How to sample the system.* How many samples are to be used, and how are they distributed across the system?
2. *What to sample.* What specific factors are to be measured, from biomechanical to organizational?
3. *How to evaluate the sample.* What standards, good practices, or ergonomic principles are to be used for comparison?
4. *How to communicate the results.* What techniques are to be used for summarizing the findings, and how far can separate findings be combined?

A suitable audit system needs to address all of these issues, but some overriding design requirements must first be specified.

5.2.1 Breadth, Depth, and Application Time

Ideally, an audit system would be broad enough to cover any task in any industry, would provide highly detailed analysis and recommendations, and would be applied rapidly. Unfortunately, the three variables of breadth, depth, and application time are likely to trade off in a practical system. Thus, a thermal audit (Parsons, 1992) sacrifices breadth to provide considerable depth based on the heat balance equation but requires measurement of seven variables. Some can be obtained rapidly (air temperature, relative humidity), but some take longer (clothing insulation value, metabolic rate). Conversely, structured interviews with participants in an ergonomics program (Drury, 1990a) can be broad and rapid, but quite deficient in depth.

At the level of audit instruments such as questionnaires or checklists, there are comprehensive surveys such as the Position Analysis Questionnaire (McCormick, 1979); the Arbeitswissenschaftliche Erhebungsverfahren zur Tätikgkeitsanalyse (AET) (Rohmert and Landau, 1989), which takes two to three hours to complete; or the simpler Work Analysis Checklist (Pulat, 1992). Alternatively, there are simple single-page checklists such as the Ergonomics-Working Position-Sitting Checklist (SHARE, 1990), which can be completed in a few minutes. Analysis and reporting can range in depth from merely tabulating the number of ergonomic standards violated, to expert systems that provide prescriptive interventions (Ayoub and Mital, 1989).

Most methodologies fall between the various extremes given above, but the goal of an audit system with an optimum trade-off between breadth, depth, and time is probably not realizable. A better practical course would be to select several instruments and use them together to provide the specific breadth and depth required for a particular application.

5.2.2 Use of Standards

The human factors/ergonomics profession has many standards and good practice recommendations. These differ by country (ANSI, BSI, DIN), although commonality is increasing through joint standards such as those of the International Standards Organisation (ISO). Some standards are quantitative, such as heights for school furniture (BSI, 1980), sizes of characters or a VDT screen (ANSI/HFS-100), and occupational exposure to noise. Other standards are more general in nature, particularly those which involve management actions to prevent or alleviate problems, such as the OSHA guidelines for meatpacking plants (OSHA, 1990). Generally, standards are more likely to exist for simple tasks and environmental stressors and are hardly to be expected for the complex cognitive activities with which human factors predictions increasingly deal. Where standards exist, they can represent unequivocal elements of audit procedures, as a workplace that does not meet these standards is in a position of legal violation. A human factors program that tolerates such legal exposure should clearly be held accountable in any audit.

Merely meeting legal requirements, however, is an insufficient test of the quality of ergonomics/human factors efforts. Many legal requirements are arbitrary or outdated: for example, weight limits for manual materials handing in some countries. Additionally, other aspects of a job with high ergonomic importance may not be covered by standards: for example, the presence of multiple stressors, work in restricted spaces resulting in awkward postures, or highly repetitive upper extremity motions. Finally, there are many "human factors good practices" that are not the subject of legal standards. Examples are the NIOSH lifting equation (Waters et al., 1993), the Illuminating Engineering Society codes (IES, 1993), or the zones of thermal comfort defined by ASHRAE (1989) or Fanger (1970). In some cases, standards are available in a different jurisdiction from that being audited. As an example, the military standard MIL-1472D (DoD, 1989) provides detailed standards for control and display design that are equally appropriate to process controls in manufacturing industry but have no legal weight there.

In the legal sense, standards are a particularly reactive phenomenon. It may take many years (any many injuries and accidents) before a standard is found necessary and agreed upon. The NIOSH lifting equation referenced above addresses a back injury problem that is far from new, yet it still has no legal force. Standards for upper extremity cumulative trauma disorder prevention have lagged disease incidence by many years. Perhaps because of busy legislative agendas, we cannot expect rapid legal reaction unless a highly visible major disaster occurs. Human factors problems are both chronic and acute, so that legislation based on acute problems as the sole basis for auditing is unlikely ever to be effective.

Despite the lack of legislation covering many human factors concerns, standards and other instantiations of good practice do have a place in ergonomics audits. Where they exist, they can be incorporated into an audit system without becoming the only criterion. Thus, noise levels in the United States have a legal limit of 90 dBA for hearing protection purposes. But at levels far below this, noise can disrupt communications (Jones and Broadbent, 1987) and distract from task performance. An audit procedure can assess the noise on multiple criteria (i.e. on hearing protection and on communication interruptions), with the former criterion used on all jobs and the latter only where verbal communication is an issue.

If standards and other good practices are used in a human factors audit, they provide a quantitative basis for decision making. Measurement reliability can be high and validity self-evident for legal standards. However, it is good practice in auditing to record only the measurement used, and not its relationship to the standard, which can be established later. This removes any temptation by the analyst to "bend" the measurement to reach a predetermined conclusion. Illumination measurements, for example, can vary considerably over a workspace, so that the audit question:

Work surface illumination > 750 lux ☐ **yes** ☐ **no**

could be answered legitimately either way for some workspaces by choice of sampling point. Such temptation can be removed, for example, by the following audit question:

Illumination at four points on workstation:

☐ ☐ ☐ ☐ **lux**

Later analysis can establish whether, for example, the mean exceeds 750 lux or whether any of the four points fall below this level.

It is also possible to provide later analyses that combine the effects of several simple checklist responses, as in Parsons' (1992) thermal audit, where no single measure would exceed good practice even though the overall result would be cumulative heat stress.

5.2.3 Evaluation of an Audit System

For a methodology to be of value, it must demonstrate validity, reliability, sensitivity, and usability. Most texts that cover measurement theory treat these aspects in detail (e.g., Kerlinger 1964). Shorter treatments are found within human factors methodology texts (e.g., Drury, 1990b; Osburn, 1987).

Validity is the extent to which a methodology measures the phenomenon of interest. Does our ergonomics audit program indeed measure the quality of ergonomics in the plant? It is possible to measure validity in a number of ways, but ultimately all are open to argument. For example, if we do not know the "true" value of the "quality of ergonomics" in a plant, how can we validate our ergonomics audit program? Broadly, there are three ways in which validation can be tested.

Content validity is perhaps the simplest but least convincing measure. If each of the items of our measurement device display the correct content, validity is established. Theoretically, if we could list all of the possible measures of a phenomenon, content validity would describe how well our measurement device samples these possible measures. In practice, it is assessed by having experts in the field judge each item for how well its content represents the phenomenon studied. Thus, the heat balance equation would be judged by most thermal physiologists to have a content that well represents the thermal load on an operator. Not all aspects are as easily validated!

Concurrent (or *prediction*) *validity* has the most immediate practical impact. It measures empirically how well the output of the measurement device correlates with the phenomenon of interest. Of course, we must have an independent measure of the phenomenon of interest, which raises difficulties. To continue our example, if we used the heat balance equation to assess the thermal load on operators, there should be a high correlation between this and other measures of the effects of thermal load. Perhaps measures such as frequency of temperature complaints or of heat disorders: heat stroke, hyperthermia, hypothermia, and so on. In practice, however, measuring such correlations would be contaminated by, for example, the propensity to report temperature problems or individual acclimatization to heat. Overall outputs from a human factors audit (if such overall outputs have any useful meaning) should correlate with other measures of ergonomic inadequacy, such as injuries, turnover, quality measures, or productivity. Alternatively, we can ask how well the audit findings agree with independent assessments of qualified human factors engineers (Keyserling et al., 1992; Koli et al., 1993) and thus validate against one interpretation of current good practice.

Finally, there is *construct validity*. This is concerned with inferences made from scores, evaluated by considering all empirical evidence and models. Thus, a model may predict that one of the variables being measured should have a particular relationship to another variable not in the measurement device. Confirming this relationship empirically would help validate the particular construct underlying our measured variable. Note that different parts of an overall measurement device can have their construct validity tested in different ways. Thus, in a board human factors audit, the thermal load could differentiate between groups of operators who do and do not suffer from thermal complaints. In the same audit a measure of difficulty in a target aiming task could be validated against Fitts's law. Other ways to assess construct validity are those that analyze clusters or factors within a group of measures. Different workplaces audited on a variety of measures, and the scores, which are then subjected to factor analysis, should show an interpretable, logical structure in the factors derived. This method has been used on large databases for job-evaluation-oriented systems such as McCormick's position analysis questionnaire (PAQ) (McCormick, 1979).

Reliability refers to how well a measurement device can repeat a measurement on the same sample unit. Classically, if a measurement X is assumed to be composed of a true value X_t and a random measurement error X_e, then

$$X = X_t + X_e$$

For uncorrelated X_t and X_e, taking variances gives

$$\text{Variance}(X) = \text{Variance}(X_t) + \text{Variance}(X_e)$$

or

$$\text{V}(X) = \text{V}(X_t) + \text{V}(X_e)$$

We can define the reliability of the measurement as the fraction of measurement variance accounted for by true measurement variance:

$$\text{reliability} = \frac{\text{V}(X_t)}{\text{V}(X_t) + \text{V}(X_e)}$$

Typically, reliability is measured by correlating the scores obtained through repeated measurements. In an audit instrument, this is often done by having two (or more) auditors use the instrument on the same set of workplaces. The square of the correlation coefficient between the scores (either overall scores, or separately for each logical construct) is then the reliability. Thus, PAQ was found to have an overall reliability of 0.79, tested using 62 jobs and two trained analysts (McCormick, 1979).

Sensitivity defines how well a measurement device differentiates between entities. Does an audit system for human–computer interaction find a difference between software generally acknowledged to be "good" and "bad"? If not, perhaps the audit system lacks sensitivity, although of course there may truly be no difference between the systems except blind prejudice. Sensitivity can be affected adversely by poor reliability, which increases the variability in a measurement relative to a fixed difference between entities (i.e., gives a poor signal-to-noise ratio). Low sensitivity can also come from a floor or ceiling effect. These arise where almost all of the measurements cluster at a high or low limit. For example, if an audit question on the visual environment was

Does illumination exceed 10 lux? ☐ **yes** ☐ **no**

almost all workplaces could answer "yes" (although the author has found a number that could not meet even this low criterion). Conversely, a floor effect would be a very high threshold for illuminance. Sensitivity can arise too when validity is in question. Thus, heart rate is a valid indicator of heat stress but not of cold stress. Hence, exposure to various degrees of cold stress would be measured only insensitively by heart rate.

Usability refers to the auditor's ease of use of the audit system. Good human factors principles should be followed: for example, document design guidelines in constructing checklists (Wright and Barnard, 1975; Patel et al., 1993). If the instrument does not have good usability, it will be used less often and may even show reduced reliability due to auditors' errors.

5.3 Audit System Design

As outlined in Section 2, the audit system must choose a sample, measure the sample, evaluate it, and communicate the results. In this section we approach these issues systematically.

An audit system is not just a checklist; it is a methodology that often includes the technique of a checklist. The distinction needs to be made between methodology and techniques. Almost three decades ago, Easterby (1967) used Bainbridge and Beishon's (1964) definitions:

- *Methodology*: a principle for defining the necessary procedures
- *Technique*: a means to execute a procedural step

Easterby notes that a technique may be applicable in more than one methodology.

5.3.1 Sampling Scheme

In any sampling, we must define the unit of sampling, the sampling frame and the sample choice technique. For a human factors audit the unit of sampling is not as self-evident as it appears. From a job evaluation viewpoint (e.g., McCormick, 1979), the natural unit is the job that is composed of a number of tasks. From a medical viewpoint the unit would be the individual. Human factors studies focus on the task/operator/machine/environment (TOME) system (Drury, 1992a and b) or, equivalently, the software/hardware/environment/liveware (SHEL) system (ICAO, 1989). Thus, from a strictly human factors viewpoint, the specific combination of TOME can become the sampling unit for an audit program.

Unfortunately, this simple view does not cover all the situations for which an audit program may be needed. Although it works well for the rather repetitive tasks performed at a single workplace typical of much manufacturing and service industry, it cannot suffice when these conditions do not hold. One relaxation is to remove the stipulation of a particular incumbent, allowing for jobs that require frequent rotation of tasks. This means that the results for one task will depend on the incumbent chosen, or that several tasks will need to be combined if an individual operator is of interest. A second relaxation is that the same operator may move to different workplaces, thus changing environment as well as task. This is typical of maintenance activities, where a mechanic may perform any one of a repertory of hundreds of tasks, rarely repeating the same task. Here, the rational sampling unit is the task, which is observed for a particular operator at a particular machine in a particular environment. Examples of audits of repetitive tasks (Mir, 1982; Drury, 1990a) and maintenance tasks (Chervak and Drury, 1995) are given later to illustrate these different approaches.

Definition of the sampling frame, once the sampling unit is settled, is more straightforward. Whether the frame covers a department, a plant, a division, or an entire company, enumeration of all sampling units is possible at least theoretically. All workplaces, or jobs, or individuals can in principle be listed, although in practice the list may never be up to date in an agile industry where change is the normal state of affairs. Individuals can be listed from personnel records, tasks from work orders or planning documents, workplaces from plant layout plans. A greater challenge, perhaps, is to decide whether indeed the entire plant really is the focus of the audit. Do we include office jobs or just production? What about managers, foremen, part-time janitors, and so on? A good human factors program would see all of these tasks or people as worthy of study, but in practice they may have had different levels of ergonomic effort expended upon them. Should some tasks or groups be excluded from the audit merely because most participants agree that they have few pressing human factors problems? These are issues that need to be decided explicitly before the audit sampling begins.

Choice of the sample from the sampling frame is well covered in sociology texts. Within human factors it typically arises in the context of survey design (Sinclair, 1990). To make statistical inferences from the sample to the population (specifically to the sampling frame), our sampling procedure must allow the laws of probability to be applied. The sampling methods used most often are described here.

Random Sampling Each unit within the sampling frame is equally likely to be chosen for the sample. This is the simplest and most robust method, but it may not be the most efficient. Where subgroups of interest (strata) exist and these subgroups are not equally represented in the sampling frame, one collects unnecessary information on the most populous subgroups and insufficient information on the least populous. This is because our ability to estimate a population statistic from a sample depends on the absolute sample size and not, in most practical cases, on the population size. As a corollary, if subgroups are of no interest, random sampling loses nothing in efficiency.

Stratified Random Sampling Each unit within a particular stratum of the sampling frame is equally likely to be chosen for the sample. With stratified random sampling we can make valid inferences about each of the strata. By weighting the statistics to reflect the size of the strata within the sampling frame, we can also obtain population inferences. This is often the preferred auditing sampling method, as, for example, we would wish to distinguish between different classes of tasks in our audits: production, warehouse, office, management, maintenance, security, and so on. In this way our audit interpretation could give more useful information concerning where ergonomics is being used appropriately.

Cluster Sampling Clusters of units within the sampling frame are selected, followed by random or nonrandom selection within clusters. Examples of clusters would be the selection of particular production lines within a plant (Drury, 1990a), or selection of "representative" plants within a company or division. The difference between cluster and stratified sampling is that in cluster sampling only a subset of possible units within the sampling frame is selected, whereas in stratified sampling, all of the sampling frame is used, as each unit must belong to one stratum. Because clusters are not randomly selected, the overall sample results will not reflect population values, so that statistical inference is not possible. If units are chosen randomly within each cluster, statistical inference within each cluster is possible. For example, if three production lines are chosen as clusters and workplaces sampled randomly within each, the clusters can be regarded as fixed levels of a factor and the data subjected to analysis of variance to determine whether there are significant differences between levels of that factor. What is sacrificed in cluster sampling is the ability to make *population* statements. Continuing this example, we could state that the lighting in line A is better than in line B or C, but still not be able to make statistically valid statements about the plant as a whole.

5.3.2 Date Collection Instrument

So far we have assumed that the instrument used to collect the data from the sample is based on measured data where appropriate. Although this is true of many audit instruments, this is not the only way to collect audit data. There have been interviews with participants (Drury, 1990a), interviews and group meetings to locate potential errors (Fox, 1992), and use of archival data such as injury of quality records (Mir, 1982). All have potential uses with, as remarked earlier, a judicious range of methods often providing the appropriate composite audit system.

One consideration regarding audit technique design and use is the extent of computer involvement. Computers are now inexpensive, portable, and powerful, so that they can be used to assist data collection, data verification, data reduction, and data analysis (Drury, 1990a). With the advent of more intelligent interfaces, checklist questions can be answered from mouse clicks on buttons, or selection from menus, as well as the more usual keyboard entry. Data verification can take place at entry time by checking for out-of-limits data, or odd data, such as the ratio of luminance to illuminance implying a reflectivity greater than 100%. In addition, branching in checklists can be made easier, with only valid follow-on questions highlighted. The "checklist user's manual" can be built into the checklist software using context-sensitive help facilities, as in the EEAM checklist (Chervak and Drury, 1995). Computers can, of course, be used for data reduction (e.g., finding the insulation value of clothing from a clothing inventory), data analysis, and results presentation.

Having made the case for computer use, some precautions are in order. Computers are still bulkier than simple pencil-and-paper checklists. Computer reliability is not perfect, so that inadvertent data loss is still a real possibility. Finally, software and hardware dates much more rapidly than hard copy, so that results stored safely on the latest media may be unreadable 10 years later. How many of us can still read punched cards or 8-inch floppy disks? In contrast, hard copy records are still available from before the start of the common era.

Checklists and Surveys as Audit Tools For many practitioners the proof of the effectiveness of an ergonomics effort lies in the ergonomic quality of the work systems it produces. A plant or office with appropriate human–machine function allocation, well-designed workplaces, comfortable environment, adequate placement/training, and inherently satisfying jobs almost by definition has been well served by human factors. Such a facility may not have human factors specialists, just good designers of environment, training, organization, and so on, working independently, but this would generally be a rare occurrence. Thus, a checklist to measure such inherently ergonomic qualities has great appeal as part of an audit system. We have covered the design aspects of checklists in general, so we concentrate here on their use in the context of human factors/ergonomics audits.

Such checklists are almost as old as the discipline. An early paper by Burger and deJong (1964) lists four earlier checklists for ergonomic job analysis before going on to develop their own. Theirs was commissioned by the International Ergonomics Association in 1961 and is usually known as the IEA checklist. It was

Table 2 IEA Checklist Structure and Typical Questions

Structure

Load: 1. Mean 2. Peaks (intensity, frequency, duration)		A Worker	B Environment	C Working Method, Tools, Machines
I. Physical load	1. Dynamic			
	2. Static			
II. Perceptual load	1. Perception			
	2. Selection, decision			
	3. Control of movement			
III. Mental load	1. Individual			
	2. Group			

Typical Question

I/B. Physical Load/Environment	2.1. Physiological Criteria
	1. Climate: high and low temperatures
	1. Are these extreme enough to affect comfort or efficiency?
	2. If so, is there any remedy?
	3. To what extent is working capacity adversely affected?
	4. Do personnel have to be specially selected for work in this particular environment?

based in part on one developed at the Philips Health Centre by G. J. Fortuin and provided in detail in Burger and deJong's paper.

Like any other questionnaire, a checklist needs to have both a helpful overall structure and well-constructed questions. It should also be proven reliable, valid, sensitive, and usable, although precious few meet all these criteria. In the remainder of this section, a selection of checklists is presented as typical of (reasonably) good practice. Emphasis will be on objective, structure, and question design. Note that checklists are not the only approach possible. Westwater and Johnson (1995) compared them with expert evaluation and empirical user testing in evaluating PDA design. They concluded that user-based evaluations led to more insights for this evaluation.

1. *IEA checklist.* The IEA checklist (Burger and deJong, 1964) was designed for ergonomic job analysis over a wide range of jobs. It uses the concept of functional load to give a logical framework relating physical load, perceptual load, and mental load to the worker, the environment, and working methods/tools/machines. Within each cell (or subcell, e.g., physical load could be static or dynamic) the load was assessed on different criteria, such as force, time, distance, occupational medical, and psychological criteria. Table 2 shows the structure and typical questions. Dirken (1969) modified the IEA checklist to improve the questions and methods of recording. He found that it could be applied in a median time of 60 minutes per workstation. No data are given on evaluation of the IEA checklist, but its structure has been so influential that it included here for more than historical interest.

2. *Position Analysis Questionnaire.* The PAQ is a structured job analysis questionnaire using 187 worker-oriented elements to characterize the human behaviors involved in jobs (McCormick et al., 1969). The PAQ is structured into six divisions, with the first three representing the classic experimental psychology approach (information input, mental process, work output) and the next a broader sociotechnical view (relationships with other persons, job context, other job characteristics). Table 3 shows these major divisions, examples of job elements in each, and the rating scales employed for response.

Construct validity was tested by factor analyses of databases containing 3700 and 2200 jobs, which established 45 factors. Thirty-two of these fit neatly into the original six-division framework, with the remaining 13 being classified as "overall dimensions." Further proof of construct validity was based on 76 human attributes derived from the PAQ, rated by industrial psychologists and the ratings subjected to principal components analysis to develop dimensions "which had reasonably similar attribute profiles" (McCormick, 1979, p. 204). As noted earlier, interreliability, was 0.79, based on another sample of 62 jobs.

The PAQ covers many of the elements of concern to human factors engineers and has indeed much influenced subsequent instruments, such as AET. With

Table 3 PAQ Structure and Scales

Structure

Division	Definition	Examples of Questions
1. Information input	Where and how does the worker get the information he uses in performing his job?	1. Use of written materials 2. Near-visual differentiation
2. Mental processes	What reasoning, decision making, planning, and information-processing activities are involved in performing the job?	1. Level of reasoning in problem solving 2. Coding–decoding
3. Work output	What physical activities does the worker perform, and what tools or devices does he use?	1. Use of keyboard devices 2. Assembling–unassembling
4. Relationships with other persons	What relationships with other people are required in performing the job?	1. Instructing 2. Contacts with public or customers
5. Job context	In what physical or social contexts is the work performed?	1. High temperature 2. Interpersonal; conflict situations
6. Other job characteristics	What activities, conditions, or characteristics other than those described above are relevant to the job?	1. Specified work pace 2. Amount of job structure

Scales

Types of Scales		Scale Values	
Code	Type of Rating	Rating	Definition
U	Extent of use	N	Does not apply
I	Importance of the job	1	Very minor
T	Amount of time	2	Low
P	Possibility of occurrence	3	Average
A	Applicability (yes/no only)	4	High
S	Special code	5	Extreme

Source: McCormick (1979).

good reliability and useful (although perhaps dated) construct validity, it is still a useful instrument if the natural unit of sampling is the job. The exclusive reliance on rating scales applied by the analyst goes rather against current practice of comparison of measurements against standards or good practices.

3. *AET (Arbeitswissenschaftliche Erhebungsverfahren zur Tätikgkeitsanalyse).* The AET has been published in German (Landau and Rohmert, 1981) and later in English (Rohmert and Landau, 1983). It is the job analysis subsystem of a comprehensive system of work studies. It covers "the analysis of individual components of man-at-work systems as well as the description and scaling of their interdependencies" (Rohmert and Landau, 1983, pp. 9–10). As with all good techniques, it starts from a model of the system (REFA, 1971, referenced in Wagner, 1989), to which is added Rohmert's stress–strain concept. The latter sees strain as being caused by the intensity and duration of stresses impinging on an operator's individual characteristics. It is seen as useful in the analysis of requirements and work design, organization in industry, personnel management, and vocational counseling and research.

AET itself was developed over many years, using PAQ as an initial starting point. Table 4 shows the structure of the survey instrument with typical questions and rating scales. Note the similarity between AET's job demands analysis and the first three categories of the PAQ, and between the scales used in AET and PAQ (Table 3).

Measurements of validity and reliability of AET are discussed by H. Luczak in an appendix to Landau and Rohmert (1981), although no numerical values are given. Cluster analysis of 99 AET records produced groupings that supported the AET constructs. Seeber et al. (1989) used AET along with two other work analysis methods in 170 workplaces. They found that AET provided the most differentiating aspects (suggesting sensitivity). They also measured postural complaints and showed that only the AET groupings for 152 female workers found significant differences between complaint levels, thus helping establish construct validity.

Like PAQ before it, AET has been used on many thousands of jobs, mainly in Europe. A sizable database is maintained that can be used for both norming of new jobs analyzed, and analysis to test research hypotheses. It remains a most useful instrument for work analysis.

4. *Ergonomics Audit Program* (Mir, 1982; Drury, 1990a). This program was developed at the request

Table 4 AET Structure and Scales

Structure

Part	Major Divisions	Sections
A. Work systems analysis	1. Work objects	1.1. Material work objects 1.2. Energy as work object 1.3. Information as work object 1.4. Man, animals. plants as work objects
	2. Equipment	2.1. Working equipment 2.2. Other equipment
	3. Work environment	3.1. Physical environment 3.2. Organizational and social environment 3.3. Principles and methods of remuneration
B. Task analysis	1. Tasks relating to material work objects 2. Tasks relating to abstract work objects 3. Man-related tasks 4. Number and repetitiveness of tasks	
C. Job demand analysis	1. Demands on perception	1.1. Mode of perception 1.2. Absolute/relative evaluation of perceived information 1.3. Accuracy of perception
	2. Demands for decision	2.1. Complexity of decisions 2.2. Pressure of time 2.3. Required knowledge
	3. Demands for response/activity	3.1. Body postures 3.2. Static work 3.3. Heavy muscular work 3.4. Light muscular work, active light work 3.5. Strenuousness and frequency of movements

Scales

Types of Scales		Scale Values	
Code	Type of Rating	Duration Value	Definition
A	Does this apply?	0	Very infrequent
F	Frequency	1	Less than 10% of shift time
S	Significance	2	Less than 30% of shift time
D	Duration	3	30% to 60% of shift time
		4	More than 60% of shift time
		5	Almost continuously during whole shift

of a multinational corporation to be able to audit its various divisions and plants as ergonomics programs were being instituted. The system developed was a methodology of which the workplace survey was one technique. Overall, the methodology used archival data or outcome measures (injury reports, personnel records, productivity) and critical incidents to rank-order departments within a plant. A cluster sampling of these departments gives either the ones with the highest need (if the aim is to focus ergonomic effort) or a sample representative of the plant (if the objective is an audit). The workplace survey is then performed on the sampled departments.

The workplace survey was designed based on ergonomic aspects derived from a task/operator/machine/environment model of the person at work. Each aspect formed a section of the audit, and sections could be omitted if there were clearly not relevant (e.g., manual materials handling aspects for data-entry clerks). Questions within each section were based on standards, guidelines, and models, such as the NIOSH (1981) lifting equation, ASHRAE's (1990) *Handbook of Fundamentals* for thermal aspects, and Givoni and Goldman's (1972) model for predicting heart rate. Table 5 shows the major sections and typical questions.

Data were entered into the computer program and a rule-based logic evaluated each section to provide messages to the user in the form of either a "section shows no ergonomic problems" message:

MESSAGE
Results from analysis of auditory aspects:
Everything OK in this section.

or discrepancies from a single input:

Table 5 Workplace Survey Structure and Typical Questions

Section	Major Classification	Examples of Questions
1. Visual aspects		Nature of task? Illuminance at task (midfield, outer field)?
2. Auditory aspects		Noise level (dBA)? Main source of noise?
3. Thermal aspects		Strong radiant sources present? Wet bulb temperature? (Clothing inventory)
4. Instruments, controls, displays	Standing vs. seated Displays Labeling Coding Scales, dials, counters Control–display relationships Controls	Are controls mounted between 30 and 70 in.? Signals for crucial visual checks? Are trade names deleted? Color codes same for control and display? All numbers upright on fixed scales? Grouping by sequence or subsystem? Emergency button diameter > 0.75 in.?
5. Design of workplaces	Desks Chairs Posture	Seat to underside of desk > 6.7 in.? Height easily adjustable to 15–21 in.? Upper arms vertical?
6. Manual materials handling	NIOSH (1981) lifting guide	Task, H, V, D, F
7. Energy expenditure		Cycle time? Object weight? Type of work?
8. Assembly/repetitive aspects		Seated, standing, or both? If heavy work, is bench 6–16 in. below elbow height?
9. Inspection aspects		Number of fault types? Training time until unsupervised?

MESSAGE
Seats should be padded, covered with nonslip materials, and have the front edge rounded.

or discrepancies based on the integration of several inputs:

MESSAGE
The total metabolic workload is 174 watts.
Intrinsic clothing insulation is 0.56 clo.
Initial rectal temperature is predicted to be 36.0°C.
Final rectal temperature is predicted to be 37.1°C.

Counts of discrepancies were used to evaluate departments by ergonomics aspect, while the messages were used to alert company personnel to potential design changes. The latter use of the output as a training device for nonergonomic personnel was seen as desirable in a multinational company rapidly expanding its ergonomics program.

Reliability and validity have not been assessed, although the checklist has been used in a number of industries (Drury, 1990a). The workplace survey has been included here because despite its lack of measured reliability and validity, it shows the relationship between audit as methodology and checklist as technique.

5. *ERGO, EEAM, and ERNAP* (Koli et al., 1993; Chervak and Drury, 1995). These checklists are both part of complete audit systems for different aspects of civil aircraft hangar activities. They were developed for the Federal Aviation Administration to provide tools for assessing human factors in aircraft inspection (ERGO) and maintenance (EEAM) activities, respectively. Inspection and maintenance activities are nonrepetitive in nature, controlled by task cards issued to technicians at the start of each shift. Thus, the sampling unit is the task card, not the workplace, which is highly variable between task cards. Their structure was based on extensive task analyses of inspection and maintenance tasks, which led to generic function descriptions of both types of work (Drury et al., 1990). Both systems have sampling schemes and checklists. Both are computer-based with initial data collection on either hard copy or direct into a portable computer. Recently, both have been combined into a single program (ERNAP) distributed by the FAA's Office of Aviation Medicine. The structure of ERNAP and typical questions are given in Table 6.

As in Mir's ergonomics audit program, the ERNAP, the checklist is again modular, and the software allows formation of data files, selection of required modules, analysis after data entry is completed, and printing of audit reports. Similarly, the ERGO, EEAM, and ERNAP instruments use quantitative or yes/no questions comparing the value entered with standards and good practice guides. Each takes about 30 minutes per task. Output is in the form of an audit report for each workplace, similar to the messages given

Table 6 ERNAP Structure and Typical Questions

Audit Phase	Major Classification	Examples of Questions
I. Premaintenance	Documentation	Is feedforward information on faults given?
	Communication	Is shift change documented?
	Visual characteristics	If fluorescent bulbs are used, does flicker exist?
	Electric/pneumatic equipment	Do pushbuttons prevent slipping of fingers?
	Access equipment	Do ladders lave nonskid surfaces on landings?
II. Maintenance	Documentation	Does inspector sign off workcard after each task?
	Communication	Explicit verbal instructions from supervisor?
	Task lighting	Light levels in four zones during task (fc)?
	Thermal issues	Wet-bulb temperature in hanger bay (°C)?
	Operator perception	Satisfied with summer thermal environment?
	Auditory issues	Noise levels at five times during task (dBA)?
	Electrical and pneumatic	Are controls easily differentiated by touch?
	Access equipment	Is correct access equipment available?
	Hand tools	Does the tool handle end in the palm?
	Force measurements	What force is being applied (kg)?
	Manual material handling	Does task require pushing or pulling forces?
	Vibration	What is total duration of exposure on this shift?
	Repetitive motion	Does the task require flexion of the wrist?
	Access	How often was access equipment repositioned?
	Posture	How often were following postures adopted?
	Safety	Is inspection area cleaned adequately for inspection?
	Hazardous material	Were hazardous materials signed out and in?
III. Postmaintenance	Buyback	Are discrepancy worksheets readable?

by Mir's workplace survey, but in narrative form. Output in this form was chosen for compatibility with existing performance and compliance audits used by the aviation maintenance community.

Reliability of a first version of ERGO was measured by comparing the output of two auditors on three tasks. Significant differences were found at $p < 0.05$ on all three tasks, showing a lack of interrater reliability. Analysis of these differences showed them to be due largely to errors on questions requiring auditor judgment. When such questions were replaced with more quantitative questions, the two auditors had no significant disagreements on a later test. Validity was measured using concurrent validation against six Ph.D. human factors engineers who were asked to list all ergonomic issues on a power plant inspection task. The checklist found more ergonomic issues than the human factors engineers. Only a small number of issues were raised by the engineers that were missed by the checklist. For the EEAM checklist, again an initial version was tested for reliability with two auditors and achieved the same outcome for only 85% of the questions. A modified version was tested and the reliability was considered satisfactory with 93% agreement. Validity was again tested against four human factors engineers; this time the checklist found significantly more ergonomic issues than the engineers, without missing any of the issues they raised.

The ERNAP audits have been included here to provide examples of a checklist embedded in an audit system where the workplace is *not* the sampling unit. They show that nonrepetitive tasks can be audited in a valid and reliable manner. In addition, they demonstrate how domain-specific audits can be designed to take advantage of human factors analyses already made in the domain.

6. *Upper Extremity Checklist* (Keyserling et al., 1993). As its name suggests, this checklist is narrowly focused on biomechanical stresses to the upper extremities that could lead to cumulative trauma disorders (CTDs). It does not claim to be a full-spectrum analysis tool but is included here as a good example of a special- purpose checklist that has been carefully constructed and validated. The checklist (Table 7) was designed for use by management and labor to fulfill a requirement in the OSHA guidelines for meatpacking plants. The aim is to screen jobs rapidly for harmful exposures rather than to provide a diagnostic tool. Questions were designed based on the biomechanical literature, structured into six sections. Scoring was based on simple presence or absence of a condition or on a three-level duration score. As shown in Table 7, the two or three levels were scored as 0, $\surd$ or *, depending on the stress rating built into the questionnaire. These symbols represented insignificant, moderate, or substantial exposures. A total score could be obtained by summing moderate and substantial exposures.

The upper extremity checklist was designed to be bias toward false positives (i.e., to be very sensitive). It was validated against detailed analyses of 51 jobs by an ergonomics expert. Each section (except the first, which recorded only dominant hand) was considered as giving a positive screening if at least one * rating was recorded. Across the various sections, there was reasonable agreement between checklist users and the expert analysis, with the checklist being generally more sensitive, as was its aim. The original reference

Table 7 Upper Extremity Checklist: Structure and Scoring

Structure

Major Section	Examples of Questions
Worker information	Which hand is dominant?
Repetitiveness	Repetitive use of the hands and wrists? If "yes," then: Is cycle <30 s? Repeated for >50% cycle?
Mechanical stress	Do hard or sharp objects put localized pressure on: Back or side of fingers? Palm or base of hand?
Force	Lift, carry, push, or pull objects >4.5 kg? If gloves worn, do they hinder gripping?
Posture	Is pinch grip used? Is there wrist deviation?
Tools, handheld objects and equipment	Is vibration transmitted to the operator's hand? Does cold exhaust air blow on the hand or wrist?

Scoring Scheme

Question	Scoring		
Is there wrist deviation?	No	Some	> 33% cycle
	0	√	*

Overall evaluation:

total score = number of √ + number of *

shows the findings of the checklist applied to 335 manufacturing and warehouse jobs.

As a special-purpose technique in an area of high current visibility for human factors, the upper extremity checklist has proven validity, can be used by those with minimal ergonomics training for screening jobs, and takes only a few minutes per workstation. The same team has also developed and validated a legs, trunk, and neck job screening procedure along similar lines (Keyserling et al., 1992).

7. *Ergonomic Checkpoints.* The workplace improvement in small enterprises (WISE) methodology (Kogi, 1994) was developed by the International Ergonomics Association (IEA) and the International Labour Office (ILO) to provide cost-effective solutions for smaller organizations. It consists of a training program and a checklist of potential low-cost improvements. This checklist, called *ergonomics checkpoints*, can be used both as an aid to discovery of solutions and as an audit tool for workplaces within an enterprise.

The 128-point checklist has now been published (Kogi and Kuorinka, 1995). It covers the nine areas shown in Table 8. Each item is a statement rather than a question and is called a *checkpoint*. For each checkpoint there are four sections, also shown in Table 8. There is no scoring system as such; rather, each checkpoint becomes a point of evaluation of each workplace for which it is appropriate. Note that each checkpoint also covers why that improvement is important and a description of the core issues underlying it. Both of these help the move from rule-based reasoning to knowledge-based reasoning as nonergonomists continue to use the checklist. A similar idea was embodied in the Mir (1982) ergonomic checklist.

8. *Other Checklists.* The sample of successful audit checklists above has been presented in some detail to provide the reader with their philosophy, structure, and sample questions. Rather then continue in the same vein, other interesting checklists are outlined in Table 9. Each entry shows the domain, the types of issues addressed, the size or time taken in use, and whether validity and reliability have been measured. Most textbooks now provide checklists, and a few of these are cited. No claim is made that Table 9 is comprehensive; rather, it is a sampling with references so that readers can find a suitable match to their needs. The first nine entries in the table are conveniently co-located in Landau and Rohmert (1989). Many of their reliability and validity studies are reported in this publication. The next entries are results of the Commission of European Communities fifth ECSC program, reported in Berchem-Simon (1993). Others are from texts and original references. The author has not personally used all of these checklists, so cannot endorse them specifically. Also, omission of a checklist from this table implies nothing about its usefulness.

Other Data Collection Methods Not all data come from checklists and questionnaires. We can audit a human factors program using outcome measures alone. However, outcome measures such as injuries, quality, and productivity are nonspecific to human factors: Many other external variables can affect them. An obvious example is changes in the reporting threshold for injuries, which can lead to sudden apparent increases and decreases in the safety of a department or plant. Additionally, injuries are (or should be) extremely rare events. Thus, to obtain enough data to perform meaningful statistical analysis may require aggregation over many disparate locations and/or time periods. In ergonomics audits, such outcome measures are perhaps best left for long-term validation or for use in selecting cluster samples.

Besides outcome measures, interviews represent a possible data collection method. Whether directed or not (e.g., Sinclair, 1990), they can produce critical incidents, human factors examples, or networks of communication (e.g., Drury, 1990a) that have value as part of an audit procedure. Interviews are used routinely as part of design audit procedures in large-scale operations such as nuclear power plants (Kirwan, 1989) or naval systems (Malone et al., 1988).

A novel interview-based audit system was proposed by Fox (1992) based on methods developed by British

Table 8 Ergonomic Checkpoints

Structure of the Checklist

Major Section	Typical Checkpoints
Materials handling	Clear and mark transport ways.
Handtools	Provide handholds, grips, or good holding points for all packages and containers.
Productive machine safety	Use jigs and fixtures to make machine operations stable, safe, and efficient.
Improving workstation design	Adjust working height for each worker at elbow level or slightly below it.
Lighting	Provide local lights for precision or inspection work.
Premises	Ensure safe wiring connections for equipment and lights.
Control of hazards	Use feeding and ejection devices to keep hands away from dangerous parts of machinery.
Welfare facilities	Provide and maintain good changing, washing, and sanitary facilities to keep good hygiene and tidiness.
Work organization	Inform workers frequently about the results of their work.
Structure of Each Checkpoint	
Why?	Reasons why improvements are important
How?	Description of several actions each of which can contribute to improvement
Some More Hints	Additional points which are useful for attaining the improvement
Points to Remember	Brief description of the core element of the checkpoint

Source: K. Kogi, private communication, November 13, 1995.

Table 9 Selection of Published Checklists

Name	Reference	Coverage	Reliability	Validity
TBS	Hacker et al. (1983)	Mainly mental work		Vs. AET
VERA	Volpert et al. (1983)	Mainly mental work		Vs. AET
RNUR	RNUR (1976)	Mainly physical work		
LEST	Guélaud(1975)	Mainly physical work		
AVISEM	AVISEM (1977)	Mainly physical work		
GESIM	GESIM (1988)	Mainly physical work		
RHIA	Leitner and greiner (1987)	Task hindrances, stress	0.53–0.79	Vs. many
MAS	Groth (1989)	Open structure, derived from AET		Vs. AET
JL and HA	Mattila and Kivi (1989)	Mental, physical work, hazards	0.87–0.95	
	Bolijn (1993)	Physical work for women	Tested	
	Panter (1993)	Load handling		
	Portillo Sosa (1993)	VDT standards		
Work analysis	Pulat (1992)	Mental and physical work		
Thermal audit	Parsons (1992)	Thermal audit from heat balance		Content
WAS	Yoshida and Ogawa (1991)	Workplace and environment	Tested	Vs. expert
Ergonomics	SHARE (1990)	Short workplace checklists		
	Cakir et al. (1980)	VDT checklist		

Source: First nine from Landau and Rohmert (1989), next three from Berchem-Simon (1993)

Coal (reported Simpson, 1994). Here an error-based approach was taken, using interviews and archival records to obtain a sampling of actual and possible errors. These were then classified using Reason's (1990) active/latent failure scheme and orthogonally by Rasmussen's (1987) skill-, rule-, knowledge-based framework. Each active error is thus a conjunction of skill/mistake/violation with skill/rule/knowledge. Within each conjunction, performance-shaping factors can be deduced and sources of management intervention listed. This methodology has been used in a number of mining-related studies (see Section 5.4.2).

5.3.3 Data Analysis and Presentation

Human factors as a discipline cover a wide range of topics from workbench height to function allocation in automated systems. An audit program can only hope to abstract and present a part of this range. With our consideration of sampling systems and data collection devices we have seen different ways in which an unbiased abstraction can be aided. At this stage the data consist of large numbers of responses to large numbers of checklist items or detailed interview findings. How can, or should, these data be treated for best interpretation?

Here there are two opposing viewpoints: One is that the data are best summarized across sample units but not across topics. This is typically the way the human factors professional community treats the data, giving summaries in published papers of the distribution of responses to individual items on the checklist. In this way, findings can be more explicit: for example, that the lighting is an area that needs ergonomics effort, or that the seating is generally poor. Adding together lighting and seating discrepancies is seen as perhaps obscuring the findings rather than assisting in their interpretation.

The opposite viewpoint, in many ways, is taken by the business community. For some, an overall figure of merit is a natural outcome of a human factors audit. With such a figure in hand, the relative needs of different divisions, plants, or departments can be assessed in terms of ergonomic and engineering effort required. Thus, resources can be distributed rationally from a management level. This view is heard by those who work in manufacturing and service industries who ask after an audit "How did we do?" and expect a very brief answer. The proliferation of the spreadsheet, with its ability to sum and average rows and columns of data, has encouraged people to do just that with audit results. Repeated audits fit naturally into this view, as they can become the basis for monthly, quarterly, or annual graphs of ergonomic performance.

Neither view alone is entirely defensible. Of course, summing lighting and seating needs produces a result that is logically indefensible and that does not help diagnosis. But equally, decisions must be made concerning optimum use of limited resources. The human factors auditor, having chosen an unbiased sampling scheme and collected data on (presumably) the correct issues, is perhaps in an excellent position to assist in such management decisions. But so, too, are other stakeholders, primarily the workforce.

Audits are not, however, the only use of some of the data collection tools. For example, the Keyserling et al. (1993) upper extremity checklist was developed specifically as a screening tool. Its objective was to find which jobs/workplaces are in need of detailed ergonomic study. In such cases, summing across issues for a total score has an operational meaning (i.e., that a particular workplace needs ergonomic help).

Where interpretation is made at a deeper level than just a single number, a variety of presentation devices have been used. These must show scores (percent of workplaces, distribution of sound pressure levels, etc.) separately but so as to highlight broader patterns. Much is now known about separate versus integrated displays and emergent features (e.g., Wickens, 1992, pp. 121–122), but the traditional profiles and spider's web charts are still the most usual presentation forms. Thus, Wagner (1989) shows the AVISEM profile for a steel industry job before and after automation. The nine different issues ("rating factors") are connected by lines to show emergent shapes for the old and the new jobs. Landau and Rohmert's (1981) original book on AET shows many other examples of profiles. Klimer et al. (1989) present a spider web diagram to show how three work structures influenced 10 issues from the AET analysis. Mattila and Kivi (1989) present their data on the job load and hazard analysis system applied to the building industry in the form of a table. For six occupations, the rating on five different loads/hazards is presented as symbols of different sizes within the cells of the table.

There is little that is novel in the presentation of audit results: Practitioners tend to use the standard tabular or graphical tools. But audit results are inherently multidimensional, so that some thought is needed if the reader is to be helped toward an informed comprehension of the audit's outcome.

5.4 Audit Systems in Practice

Almost any of the audit programs and checklists referred to in previous sections give examples of their use in practice. Only two examples will be given here, as others are readily accessible. These examples were chosen as they represent quite different approaches to auditing.

5.4.1 Auditing a Decentralized Business

From 1992 to 1996, a major U.S.-based apparel manufacturer had run an ergonomics program, aimed primarily at the reduction of workforce injuries in backs and upper extremities. As detailed in Drury et al. (1999), the company during that time comprised nine divisions and employed about 45,000 workers. Of particular interest was the fact that the divisions enjoyed great autonomy, with only a small corporate headquarters with a single executive responsible for all risk management activities. The company had grown through mergers and acquisitions, meaning that different divisions had different degrees of vertical integration. Hence, core functions such as sewing, pressing, and distribution were common to most divisions, while some also included weaving, dyeing, and embroidery. In addition, the products and fabrics presented quite different ergonomic challenges, from delicate undergarments, through heavy jeans, to knitted garments and even luggage.

The ergonomics program was similarly diverse. It started with a corporate launch by the highest-level executives, then was rolled out to the divisions and to individual plants. The pace of change was widely variable. All divisions were given a standard set of workplace analysis and modification tools (based on Drury and Wick, 1984) but were encouraged to develop their own solutions to problems in a way appropriate to their specific needs.

Evaluation took place continuously, with regular meetings between representatives of plants and divisions to present results of before-and-after workplace studies. However, there was a need for a broader audit of the entire corporation aimed at understanding how much had been achieved for the multimillion-dollar investment, where the program was strong or weak, and what program needs were emerging for the future.

During 1995, a team of auditors visited all nine divisions and a total of 12 plants spread across eight divisions. This was three years after the initial corporate launch and about two years after the start of shop-floor implementation.

A three-part audit methodology was used. First, a workplace survey was developed based on elements of the program itself, supplemented by direct comparisons to ergonomics standards and good practices. Table 10 shows this 50-item survey form, with data added for the percentage of "yes" answers where the responses were not measures or scale values. The workplace survey was given at a total of 157 workplaces across the 12 plants. Second, a user survey (Table 11) was used in an interview format with 66 consumers of ergonomics, typically plant managers, production managers, human resource managers, or their equivalent at the division level, usually vice presidents. Finally, a total of 27 providers of ergonomics services were given a similar provider survey (Table 12) interview. Providers were mainly engineers, with three human resources specialists and one line supervisor. From these three audit methods the corporation wished to provide a time snapshot of how effectively the current ergonomics programs was meeting their needs for reduction of injury costs. While the workplace survey measured how well ergonomics was being implemented at the workplace, the user and provider surveys provided data on the roles of the decision makers beyond the workplace.

Detailed audit results are provided in Drury et al. (1999), so only examples and overall conclusions are covered in this chapter. Workplaces showed some evidence of good ergonomic practice, with generally satisfactory thermal, visual, and auditory environments. There were some significant differences ($p < 0.05$) between workplace types rather than between divisions or plants; for example, better lighting (> 700 lux) was associated with inspection and sewing. Also, higher thermal load was associated with laundries and machine load/unload. Overall, 83% of workplaces met the ASHRAE (1990) summer comfort zone criteria. As shown in Table 13, the main ergonomics problem areas were in poor posture and manual materials handling. Where operators were seated (only 33% of all workplaces), seating was relatively good. In fact, many of the workforce had been supplied with well-designed chairs as part of the ergonomics program.

To obtain a broad perspective, the three general factors at the end of Table 10 were analyzed. Apart from cycle time (W48), the questions related to workers having seen the corporate ergonomics video (W49) and having experienced a workplace or methods change (W50). Both should have received a "yes" response if the ergonomics program were reaching the entire workforce. In fact, both showed highly significant differences between plants ($X_8^2 = 92.0$, $p < 0.001$ and $X_8^2 = 22.2$, $p < 0.02$, respectively). Some of these differences were due to two divisions lagging in ergonomics implementation, but even beyond this were large between-plant differences. Overall, 62% of the workforce had seen the ergonomics video, a reasonable value but one with wide variance between plants and divisions. Also, 38% of workplaces had experienced some change, usually ergonomics-related, a respectable figure after only two to three years of the program.

From the user and provider surveys, an enhanced picture emerged. Again, there was variability between divisions and plants, but 94% of the users defined ergonomics as fitting the job to the operator rather than training or medical management of injuries. Most users had requested an ergonomic intervention within the past two months, but other "users" had never in fact used ergonomics.

The solutions employed ranged widely, with a predominance of job aids such as chairs or standing pads. Other frequent categories were policy changes (e.g., rest breaks, rotation, box weight reduction) and workplace adjustment to the individual operator. There were few uses of personal aids (e.g., splints) or referrals to physicians as ergonomic solutions. Changes to the workplace clearly predominated over changes to the individual, although a strong medical management program was in place when required. When questioned about ergonomics results, all mentioned safety (or workplace comfort or ease of use), but some also mentioned others. Cost or productivity benefits were the next most common response, with a few additional ones relating to employee relations, absence/turnover, or job satisfaction. Significantly, only one respondent mentioned quality.

The major user concern at the plant level was time devoted to ergonomics by providers. At the corporate level, the need was seen for more rapid job analysis methods and corporate policies (e.g., on back belts or "good" chairs). Overall, 94% of users made positive comments about the ergonomics program.

Ergonomics providers were almost always trained in the corporate or division training seminars, usually near the start of the program. Providers' chief concern was for the amount of time and resources they could spend on ergonomics activities. Typically, ergonomics was only one job responsibility among many. Hence, broad programs, such as new chairs or back belts, were supported enthusiastically, as they gave the maximum perceived impact for the time devoted. Other solutions presented included job aids, workplace redesign (e.g., moving from seated to standing jobs for long-seam sewing), automation, rest breaks, job rotation, packaging changes, and medical management. Specific needs were seen in the area of corporate or supplier help in obtaining standard equipment solutions, and of more division-specific training. As with users, the practitioners enjoyed their ergonomics activity and thought it worthwhile.

Recommendations arising from this audit were that the program was reasonably effective at present but had some long-term needs. The corporation sees itself as an industry leader and wants to move beyond a relatively superficial level of ergonomics application. To do this will require more time resources for job analysis and change implementation. Corporate help

Table 10 Ergonomics Audit: Workplace Survey with Overall Data

Number: Division: Plant: Job Type:

	Yes	No	Factor
1. Postural aspects			
W1	68%		Frequent extreme motions of back, neck, shoulders, wrists
W2	66%		Elbows raised or unsupported more than 50% of time
W3	22%		Upper limbs contact nonrounded edges
W4	73%		Gripping with fingers
W5	36%		Knee/foot controls
1.1. Seated			
W6	12%		Leg clearance restricted
W7	21%		Feet unsupported/legs slope down
W8	17%		Chair/table restricts thighs
W9	22%		Back unsupported
W10	37%		Chair height not adjustable easily
1.2. Standing			
W11	3%		Control requires weight on one foot more than 50% time
W12	37%		Standing surface hard
W13	92%		Work surface height not adjustable easily
1.3. Hand tools			
W14	77%		Tools require hand/wrist bending
W15	9%		Tools vibrate
W16	63%		Restricted to one handed use
W17	39%		Tool handle ends in palm
W18	20%		Tool handle has nonrounded edges
W19	56%		Tool uses only two or three fingers
W20	9%		Requires continuous or high force
W21	41%		Tool held continuously in one hand
2. Vibration			
W22	14%		Vibration reaches body from any source
3. Manual materials handling			
W23	40%		More than five moves per minute
W24	36%		Loads unbalanced
W25	14%		Lift above head
W26	28%		Lift off floor
W27	83%		Reach with arms
W28	78%		Twisting
W29	60%		Bending trunk
W30	3%		Floor wet or slippery
W31	0%		Floor in poor condition
W32	17%		Area obstructs task
W33	4%		Protective clothing unavailable
W34	2%		Handles used
4. Visual aspects			
W35			Task nature: 1, rough; 2, moderate; 3, fine; 4, very fine
W36			Glare/reflection: 0, none; 1, noticeable; 2, severe
W37			Color contrast: 0, none; 1, noticeable; 2, severe
W38			Luminance contrast: 0, none; 1, noticeable; 2, severe
W39			Task illuminance (foot candles)
W40	69%		Luminance: task > midfield > outerfield = yes
5. Thermal aspects			
W41			Dry-bulb temperature (°F)
W42			Relative humidity (%)
W43			Airspeed: 1, just perceptible; 2, noticeable; 3, severe
W44			Metabolic cost
W45			Clothing (clo value)
6. Auditory aspects			
W46			Maximum sound pressure level (dBA)
W47			Noise sources: 1, m/c; 2, other m/c; 3, general; 4, other
7. General factors			
W48			Primary cycle time (seconds)
W49	62%		Seen ergonomics video
W50	38%		Any ergonomics changes to workplace or methods

Table 11 Ergonomics Audit: User Survey

Number: Division: Plant: Job Type:

U1. What is ergonomics?
U2. Who do you call to do ergonomics?
U3. When did you last ask them to do ergonomics?
U4. Describe what they did.
U5. Who else should we talk to about ergonomics?
U6. General comments on ergonomics.

Table 12 Ergonomics Audit: Provider Survey

Number: Division: Plant: Job Type:

P1. What do you do?
P2. How do you get contacted to do ergonomics?
P3. When were you last asked to do ergonomics?
P4. Describe what you did.
P5. How long have you been doing ergonomics?
P6. How were you trained in ergonomics?
P7. What Percent of your time is spent on ergonomics?
P8. Where do you go for more detailed ergonomics help?
P9. What ergonomics implementation problems have you had?
P10. How well are you regarded by management?
P11. How well are you regarded by workforce?
P12. General comments on ergonomics.

could also be provided in developing more rapid analysis methods, standardized video-based training programs, and more standardized solutions to recurring ergonomics problems. Many of these changes have since been implemented.

On another level, the audit was a useful reminder to the company of the fact that it had incurred most of the up-front costs of a corporate ergonomics program and was now beginning to reap the benefits. Indeed, by 1996, corporate injury costs and rates had decreased by about 20% per year after peaking in 1993. Clearly, the ergonomics program was not the only intervention during this period, but it was seen by management as the major contributor to improvement. Even on the narrow basis of cost savings, the ergonomics program was a success for the corporation.

5.4.2 Error Reduction at a Colliery

In a two-year project reported by Fox (1992) and Simpson (1994), the human error audit described in Section 5.3.2 was applied to two colliery haulage systems. The results of the first study are presented here. In both systems, data collection focused on potential errors and the performance shaping factors (PSFs) that can influence these errors. Data were collected by "observation, discussion and measurement within the framework of the broader man–machine systems and checklist of PSFs," taking some 30 to 40 shifts at each site. The entire haulage system from surface operations to delivery at the coal face was covered.

The first study found 40 active failures (i.e., direct error precursors) and nine latent failures (i.e., dormant states predisposing the system to later errors). Four broad classes of active failures were (1) errors associated with locomotive maintenance (7 errors) (e.g., fitting incorrect thermal cutoffs), (2) errors associated with locomotive operation (10 errors) (e.g., locos not returned to service bay for a 24-hour check), (3) errors associated with loads and load security (7 errors), (e.g., failure to use spacer wagons between overhanging loads), (4) errors associated with the design/operation of the haulage route (10 errors), (e.g., continued use despite potentially unsafe track), plus (5) a small miscellaneous category.

The latent failures were (Fox, 1992) (1) quality assurance in supplying companies, (2) supply-ordering procedures within the colliery, (3) locomotive design, (4) surface "makeup" of supplies, (5) lack of equipment at specific points, (6) training, (7) attitudes to safety, and (8) the safety inspection/reporting/action procedures. As an example of item 3, locomotive design, the control positions were not consistent across the locomotive fleet, despite all originating from the same manufacturer. Using the slip/mistake/violation categorization, each potential error could be classified so that the preferred source of action (intervention) could be specified.

This audit led to the formation of two teams, one to tackle locomotive design issues and the other for safety reporting and action. As a result of team activities, many ergonomic actions were implemented. These included management actions to ensure a uniform wagon fleet, autonomous inspection/repair teams for tracks, and multifunctional teams for safety initiatives.

The outcome was that the accident rate dropped from 35.40 per 100,000 person-shifts to 8.03 in one year. This brought the colliery from worst in the regional group of 15 collieries to best in the group, and indeed in the UK. In addition, personnel indicators, such as industrial relations climate and absence rates, improved.

6 CONCLUSIONS

In this chapter we have arrived at human factors audits through a context of inspection and checklist design. It should be obvious by now that checklists are a subset of audits, which are in turn a subset of inspection. Within the context of inspection, we have seen that all inspections follow a short logical sequence of functions and that each function has considerable scope for model-based and empirical design to improve the human factors and system performance. Nonmanufacturing applications have been emphasized, with the focus on processes and broader systems rather than on repetitively produced products.

Inspecting, checking, and auditing are interesting, as they all have human factors design aspects but can all be applied to both the processes being audited and to the auditing process itself. Whether inspecting nonmanufacturing items, or checking items on a checklist, or performing an audit, there is prescriptive advice on how to develop or choose a system that accords with human factors good practices.

Table 13 Responses to Ergonomics User

Question and Concern	Corporate Mgt.	Corporate Staff	Plant Mgt.	Plant Staff
1. What is ergonomics?				
1.1. Fitting job to operator	1	6	10	5
1.2. Fitting operator to job	0	6	0	0
2. Who do you call on to get ergonomics work done?				
2.1. Plant ergonomics people	0	3	3	2
2.2. Division ergonomics people	0	4	5	2
2.3. Personnel department	3	0	0	0
2.4. Engineering department	1	8	6	11
2.5. We do it ourselves	0	2	1	0
2.6. College interns	0	0	4	2
2.7. Vendors	0	0	0	1
2.8. Everyone	0	1	0	0
2.9. Operators	0	1	0	0
2.10. University faculty	0	0	1	0
2.11. Safety	0	1	0	0
3. When did you last ask them for help?				
3.1. Never	0	4	2	0
3.2. Sometimes/infrequently	2	0	1	0
3.3. 1 year or more ago	0	1	4	0
3.4. 1 month or so ago	0	0	2	0
3.5. Less than 1 month ago	1	0	3	4
5. Who else should we talk to about ergonomics?				
5.1. Engineers	0	0	3	2
5.2. Operators	1	1	2	0
5.3. Everyone	0	0	2	0
6. General ergonomics comments				
6.1. Ergonomics concerns				
6.11. Workplace design for safety/ease/stress/fatigue	2	5	13	5
6.12. Workplace design for cost savings/productivity	1	0	2	1
6.13. Workplace design for worker satisfaction	1	1	0	1
6.14. Environment design	2	1	3	0
6.15. The problem of finishing early	0	0	1	1
6.16. The seniority/bumping problem	0	3	1	0
6.2. Ergonomics program concerns				
6.21. Level of reporting of ergonomics	0	1	7	0
6.22. Communication/who does ergonomics	7	1	4	0
6.23. Stability/staffing of ergonomics	0	0	10	4
6.24. General evaluation of ergonomics				
Positive	1	3	3	4
Negative	4	10	10	3
6.25. Lack of financial support for ergonomics	0	0	1	0
6.26. Lack of priority for ergonomics	2	2	1	4
6.27. Lack of awareness of ergonomics	2	1	6	1

REFERENCES

Alexander, D. C., and Pulat, B. M. (1985), *Industrial Ergonomics: A Practitioner's Guide*, Industrial Engineering and Management Press, Atlanta, GA.

Arani, T., Karwan, M. H., and Drury, C. G. (1984), "A Variable-Memory Model of Visual Search," *Human Factors*, Vol. 26, No. 6, pp. 631–639.

ASHRAE (1989), "Physiological Principles, Comfort and Health," Chapter 8 in *Fundamentals Handbook*, American Society of Heating, Refrigerating, and Air-Conditioning Engineers, Atlanta, GA.

ASHRAE (1990), *Handbook of Fundamentals*, American Society of Heating, Refrigerating, and Air-Conditioning Engineers, Atlanta, GA.

AVISEM (1977), *Techniques d'amélioration des conditions de travail dans l'industrie*, Editions Hommes et Techniques, Suresnes, France.

Ayoub, M. M., and Mital, A. (1989), *Manual Materials Handling*, Taylor & Francis, London.

Bainbridge, L., and Beishon, R. J. (1964), "The Place of Checklists in Ergonomic Job Analysis," in *Proceedings of the 2nd I.E.A. Congress*, Dortmund, Ergonomics Congress Proceedings Supplement.

Baveja, A., Drury, C. G., Karwan, M., and Malone, D. M. (1996), "Derivation and Test of an Optimum Overlapping-Lobe Model of Visual Search," *IEEE Transactions on Systems, Man, and Cybernetics,* Vol. 28, pp. 161–168.

Berchem-Simon, O., Ed. (1993), *Ergonomics Action in the Steel Industry*, EUR 14832 EN, Commission of the European Communities, Luxembourg.

Blackwell, H. R. (1946), "Contrast Thresholds of the Human Eye," *Journal of the Optical Society of America*, Vol. 36, pp. 624–643.

Bolijn, A. J. (1993), *Research into the Employability of Women in Production and Maintenance Jobs in Steelworks*, Commission of the European Communities, Luxembourg.

BSI (1980), "Educational furniture," *Specification for Functional Dimensions, Identification and Finish of Chairs and Tables for Educational Institutions*, BS5873-1: 1980, British Standards Institution, London.

Burger, G. C. E., and de Jong, J. R. (1964), "Evaluation of Work and Working Environment in Ergonomic Terms," *Aspects of Ergonomic Job Analysis*, Vol. 7 pp. 185–201.

Cakir, A., Hart, D. M., and Stewart, T. F. M. (1980), *Visual Display Terminals*, Wiley, New York.

Carson, A. B., and Carlson, A. E. (1977), *Secretarial Accounting,* 10th ed., South-Western, Cincinnati, OH.

Chervak, S., and Drury, C. G. (1995), "Simplified English Validation," in *Human Factors in Aviation Maintenance: Phase 6 Progress Report*, DOT/FAA/AM-95/xx, Federal Aviation Administration, Office of Aviation Medicine, National Technical Information Service, Springfield, VA.

Chervak, S., Drury, C. G., and Ouellette, J. L. (1996), "Simplified English for Aircraft Workcards," in *Proceedings of the Human Factors and Ergonomics Society 39th Annual Meeting*, pp. 303–307.

Chi, C.-F., and Drury, C. G. (1995), "A Test of Economic Models of Stopping Policy in Visual Search," *IIE Transactions*, Vol. 27, pp. 392–393.

Degani, A., and Wiener, E. L. (1990), *Human Factors of Flight-Deck Checklists: The Normal Checklist,* Contractor Report 177548, NASA Ames Research Center, Moffett Field, CA.

Devor, R. E., Change, T.-H., and Sutherland, J. W. (1992), *Statistical Quality Design and Control*, Prentice-Hall, Englewood Cliffs, NJ.

Dirken, J. M. (1969), "An Ergonomics Checklist Analysis of Printing Machines, *ILO* (Geneve), Vol. 2, pp. 903–913.

Drury, C. G. (1978), "Integration of Human Factors Models into Statistical Quality Control," *Human Factors*, Vol. 20, No. 5, pp. 561–572.

Drury, C. G. (1990a), "The Ergonomics Audit," in *Contemporary Ergonomics*, E. J. Lovesey, Ed., Taylor & Francis, London, pp. 400–405.

Drury, C. G. (1990b), "Computerized Data Collection in Ergonomics," in *Evaluation of Human Work,* J. R. Wilson and E. N. Corlett, Eds., Taylor & Francis, London, pp. 200–214.

Drury, C. G. (1992a), "Inspection Performance," in *Handbook of Industrial Engineering,* G. Salvendy, Ed., Wiley, New York, pp. 2282–2314.

Drury, C. G. (1992b), "Design for Inspectability," in *Design for Manufacturability: A Systems Approach to Concurrent Engineering and Ergonomics*, M. H. Helander and M. Nagamachi, Eds., Taylor & Francis, London, pp. 204–216.

Drury, C. G. (1997), "Ergonomics Audits: Why and How," in *Proceedings of the 13th Triennial Congress of the International Ergonomics Association'97*, Tampere, Finland, Vol. 2, pp. 284–286.

Drury, C. G. (2002), "A Unified Model of Security Inspection," in *Proceedings of the Aviation Security Technical Symposium,* Federal Aviation Administration, Atlantic City, NJ.

Drury, C. G. (2003), "Service Quality and Human Factors," *AI and Society,* Vol. 17, No. 2, April–May, pp. 78–96.

Drury, C. G., and Addison, J. L. (1973), "An Industrial Study of the Effects of Feedback and Fault Density on Inspection Performance," *Ergonomics*, Vol. 16, pp. 159–169.

Drury, C. G., and Sinclair, M. A. (1983), "Human and Machine Performance in an Inspection Task," *Human Factors,* Vol. 25, No. 4, pp. 391–400.

Drury, C. G., and Wick, J. (1984), "Ergonomic Applications in the Shoe Industry," in *Proceedings of the International Conference on Occupational Ergonomics,* Vol. 1, pp. 489–483.

Drury, C. G., Kleiner, B. M. and Zahorjan, J. (1989), "How Can Manufacturing Human Factors Help Save a Company: Intervention at High and Low Levels," in *Proceedings of the Human Factors Society 33rd Annual Meeting*, Denver, CO, pp. 687–689.

Drury, C. G., Prabhu, P., and Gramopadhye, A. (1990), "Task Analysis of Aircraft Inspection Activities: Methods and Findings," in *Proceedings of the Human Factors Society 34th Annual Meeting*, Santa Monica, CA, pp. 1181–1185.

Drury, C. G., Spencer, F. W., and Schurman, D. (1997), "Measuring Human Detection Performance in Aircraft Visual Inspection," in *Proceedings of the 41st Annual Human Factors and Ergonomics Society Meeting,* Albuquerque, NM.

Drury, C. G., Broderick, R. L., Weidman, C. H. and Reynolds Mozrall, J. L. (1999), "A Corporate-Wide Ergonomics Programme: Implementation and Evaluation," *Ergonomics*, Vol. 42, No. 1, pp. 208–228.

Drury, C. G., Patel, S. C., and Prabhu, P. V. (2000), "Relative, advantage of Portable Computer-Based Workcards for Aircraft Inspection," *International Journal of Industrial Ergonomics*, Vol. 26, pp. 163–176.

Easterby, R. S. (1967), "Ergonomics Checklists: An Appraisal," *Ergonomics*, Vol. 10, No. 5, pp. 548–556.

Engel, F. L. (1971), "Visual Conspicuity, Directed Attention, and Retinal Locus," *Visual Research*, Vol. 11, pp. 563–576.

Eriksen, C. W. (1990), "Attentional Search of the Visual Field," in *Visual Search,* D. Brogan, Ed., Taylor & Francis, London, pp. 3–19.

Evans, J. R., and Lindsay, W. (1993), *The Management and Control of Quality*, West Publishing Company, Minneapolis/St. Paul, MN.

Fanger, P. O. (1970), *Thermal Comfort, Analyses and Applications in Environmental Engineering*, Danish Technical Press, Copenhagen, Denmark.

Fox, J. G. (1992), "The Ergonomics Audit as an Everyday Factor in Safe and Efficient Working," *Progress in Coal, Steel and Related Social Research*, pp. 10–14.

Givoni, B., and Goldman, R. F. (1972), "Predicting Rectal Temperature Response to Work, Environment, and Clothing," *Journal of Applied Physiology*, Vol. 32, No. 6, pp. 812–822.

Goldberg, J. H. and Gibson, D. C. (1986), "The Effects of Training Method and Type of Checklist upon Visual Inspection Accuracy," in *Proceedings of the Annual International Industrial Ergonomics and Safety Conference*, Louisville, KY, North-Holland, Amsterdam.

Groth, K. M. (1989), "The Modular Work Analysis System (MAS)," in *Recent Developments in Job Analysis, Proceedings of the International Symposium on Job Analysis* Taylor & Francis, New York, pp. 253–261.

Guélaud, F., Beauchesne, M.-N., Gautrat, J. and Roustang., G. (1975), "Pour une analyse des conditions de travail ouvrier dans l'entreprise," in *Recherche du Laboratoire d'Economie et de Sociologie du Travail C.M.R.S. Aix-en-Provence*, 3rd ed., Librairie Armand Colin, Paris.

Hacker, W., Iwanowa, A., and Richter, P. (1983), *Tätigkeitsbewertungssystem* Paper presented at the Psychodiagnostisches Zentrum, Berlin (OST).

Hasslequist, R. J. (1981), "Increasing Manufacturing Productivity Using Human Factors Principles," in *Proceedings of the Human Factors Society 25th Annual Meeting*, Santa Monica, CA, pp. 204–206.

Hong, S.-K., and Drury, C. G. (2002), "Sensitivity and Validity of Visual Search Models for Multiple Targets," *Theoretical Issues in Ergonomic Science*, Vol. 3, No. 1, pp. 1–26.

Hou, T.-S., Lin, L., and Drury, C. G. (1993), "An Empirical Study of Hybrid Inspection Systems and Allocation of Inspection Function," *International Journal of Human Factors in Manufacturing*, Vol. 3, pp. 351–367.

ICAO (1989), *Human Factors Digest No. 1 Fundamental Human Factors Concepts,* Circular 216-AN/131, International Civil Aviation Organization, Montreal, Canada.

IES (1993), *Lighting Handbook, Reference and Application,* 8th ed., Illuminating Engineering Society of North America, New York.

Jamieson, G. H. (1966), "Inspection in the Telecommunications Industry: A Field Study of Age and Other Performance Variables," *Ergonomics*, Vol. 9, pp. 297–303.

Jiang, X., Melloy, B. J., and Gramopadhye, A. (2003), "Evaluation of Best System Performance: Human, Automated and Hybrid Inspection Systems," *Human Factors and Ergonomics in Manufacturing,* Vol. 13, No. 2, pp. 137–152.

Jones, J. C. (1981), *Design Methods: Seeds of Human Factors*, Wiley, New York.

Jones, D. M., and Broadbent, D. E. (1987), "Noise," in *Handbook of Human Factors Engineering,* G. Salvendy, Ed., Wiley, New York.

Kerlinger, F. N. (1964), *Foundations of Behavioral Research*, Holt, Rinehart and Winston, New York.

Keyserling, W. M., Brouwer, M., and Silverstein, B. A. (1992), "A Checklist for Evaluation of Ergonomic Risk Factors Resulting from Awkward Postures of the Legs, Trunk and Neck," *International Journal of Industrial Ergonomics*, Vol. 9, No. 4, pp. 283–301.

Keyserling, W. M., Stetson, D. S., Silverstein, B. A., and Brouwer, M. L. (1993), "A Checklist for Evaluating Ergonomic Risk Factors Associated with Upper Extremity Cumulative Trauma Disorders," *Ergonomics*, Vol. 36, No. 7, pp. 807–831.

Kirwan, B. (1989), "A Human Factors and Human Reliability Programme for the Design of a Large UK Nuclear Chemical Plant," in *Proceedings of the Human Factors Society 33rd Annual Meeting*, Denver, CO, pp. 1009–1013.

Klimer, F., Kylian, H., Schmidt, K.-H., and Rutenfranz, J. (1989), "Work Analysis and Load Components in an Automobile Plant After the Implementation of New Technologies," in *Recent Developments in Job Analysis*, K. Landau and W. Rohmert, Eds., Taylor & Francis, New York, pp. 331–340.

Kogi, K. (1994), "Introduction to WISE (Work Improvement in Small Enterprises) Methodology and Workplace Improvements Achieved by the Methodology in Asis," in *Proceedings of the 12th Triennial Congress of the International Ergonomics Association*, Vol. 5, Human Factors Association of Canada, Toronto, Ontario, Canada, pp. 141–143.

Kogi, K., and Kuorinka, I., Eds. (1995), *Ergonomic Checkpoints*, ILO Publications, Geneva, Switzerland.

Koli, S. T. (1994), "Ergonomic Audit for Non-repetitive Task," unpublished M.S. thesis, State University of New York at Buffalo, Buffalo, NY.

Koli, S., Drury, C. G., Cuneo, J., and Lofgren, J. (1993), "Ergonomic Audit for Visual Inspection of Aircraft," in *Human Factors in Aviation Maintenance: Phase Four, Progress Report*, DOT/FAA/AM-93/xx, National Technical Information Service, Springfield, VA.

Lamar, E. S. (1960), "Operational Background and Physical Considerations Relative to Visual Search Problems," in *Visual Search Techniques*, A. Morris and P. E. Horne, Eds., National Research Council, Washington, DC.

Landau, K., and Rohmert, W. (1981), *Fallbeispiele zur Arbeitsanalyse*, Hans Huber, Bern, Switzerland.

Landau, K., and Rohmert, W., Eds. (1989), *Recent Developments in Job Analysis: Proceedings of the International Symposium on Job Analysis*, University of Hohenheim, March 14–15, Taylor & Francis, New York.

Larock, B. (2000), "The Effects of Spatial and Functional Job-Aid Checklists on a Frequently Performed Aircraft Maintenance Task," unpublished Master's thesis, State University of New York at Buffalo, Buffalo, NY.

Larock, B., and Drury, C. G. (2003), "Repetitive Inspection with Checklists: Design and Performance," in *Proceedings of the Ergonomics Society Annual Conference*, Edinburgh Scotland, April 15–17.

Leitner, K., and Greiner, B. (1987), "Assessment of Job Stress: The RHIA Instrument," in *Recent Developments in Job Analysis*, in *Proceedings of the International Symposium on Job Analysis*, K. Landau and W. Rohmert, Eds., University of Hohenheim, March 14–15, 1989, Taylor & Francis, New York.

Lloyd, C. J., Boyce, P., Ferzacca, N., Eklund, N., and He, Y. (2000), "Paint Inspection Lighting: Optimization of Lamp Width and Spacing," *Journal of the Illuminating Engineering Society,* Vol. 28, pp. 99–102

Malde, B. (1992), "What Price Usability Audits? The Introduction of Electronic Mail into a User Organization," *Behaviour and Information Technology,* Vol. 11, No. 6, pp. 345–353.

Malone, T. B., Baker, C. C., and Permenter, K. E. (1988), "Human Engineering in the Naval Sea Systems Command," in *Proceedings of the Human Factors Society 32nd Annual Meeting*, Anaheim, CA, Vol. 2, pp. 1104–1107.

Mattila, M., and Kivi, P. (1989), "Job Load and Hazard Analysis: A Method for Hazard Screening and Evaluation," in *Recent Developments in Job Analysis: Proceedings of the International Symposium on Job Analysis*, K. Landau and W. Rohmert, Eds., University of Hohenheim, March 14–15, Taylor & Francis, New York, pp. 179–186.

McClelland, I. (1990), "Product Assessment and User Trials," in *Evaluation of Human Work*, J. R. Wilson and E. N. Corlett, Eds., Taylor & Francis New York, pp. 218–247.

McCormick, E. J. (1979), *Job Analysis: Methods and Applications,* AMACOM, New York.

McCormick, W. T., Mecham, R. C., and Jeanneret, P. R. (1969), *The Development and Background of the Position Analysis Questionnaire,* Occupational Research Center, Purdue University, Lafayette, IN.

McKenzie, R. M. (1958), "On the Accuracy of Inspectors," *Ergonomics*, Vol. 1, No. 3, pp. 258–272.

Miller, R. B. (1967), "Task Taxonomy: Science or Art," in *The Human Operator in Complex Systems*, W. T. Singleton, R. S. Easterby, and D. Whitfield, Eds., Taylor & Francis, London.

Mir, A. H. (1982), "Development of Ergonomic Audit System and Training Scheme," unpublished M.S. thesis, State University of New York at Buffalo, Buffalo, NY.

Morawski, T., Drury, C. G., and Karwan, M. H. (1980), "Predicting Search Performance for Multiple Targets," *Human Factors*, Vol. 22, No. 6, pp. 707–718.

NIOSH (1981), *Work Practices Guide for Manual Lifting*, DHEW-NIOSH Publication 81–122, National Institute for Occupational Safety and Health, Cincinnati, OH.

Ockerman, J. J., and Pritchett, A. R. (1998), "Preliminary Investigation of Wearable Computers for Task Guidance in Aviation Inspection," presented at the 2nd International Symposium on Wearable Computers, Pittsburgh, PA.

Ockerman, J. J., and Pritchett, A. R. (2000a), "Reducing Over-Reliance on Task-Guidance Systems," in *Proceedings of the Human Factors and Ergonomics Society 44th Annual Meeting*, San Diego, CA.

Ockerman, J. J., and Pritchett, A. R. (2000b), "Wearable Computers for Aiding Workers in Appropriate Procedure Following," unpublished paper submitted to the International Symposium on Wearable Computers.

Osburn, H. G. (1987), "Personnel Selection," in G. Salvendy, Ed., *Handbook of Human Factors*, Wiley, New York, pp. 911–933.

OSHA (1990), *Ergonomics Program Management Guidelines for Meatpacking Plants*, OSHA-3121, Occupational Safety and Health Administration, U.S. Department of Labor, Washington, DC.

Overington, I. (1973), *Vision and Acquisition*, Pentech Press, London.

Pai, S. (2003), "Effectiveness of Technological Aids in Checklist Use," unpublished Master's thesis, State University of New York at Buffalo, Buffalo, NY.

Panter, W. (1993), "Biomechanical Damage Risk in the Handling of Working Materials and Tools—Analysis, Possible Approaches and Model schemes," in O. Berchem-Simon (Ed.), *Ergonomics Action in the Steel Industry EUR 14832 EN*, Commission of the European Communities, Luxembourg.

Parsons, K. C. (1992), "The Thermal Audit: A Fundamental Stage in the Ergonomics Assessment of Thermal Environment," in *Contemporary Ergonomics 1992*, E. J. Lovesey Ed., Taylor & Francis, London, pp. 85–90.

Patel, S., Drury, C. G., and Prabhu, P. (1993), "Design and Usability Evaluation of Work Control Documentation," in *Proceedings of the Human Factors and Ergonomics Society 37th Annual Meeting*, Seattle, WA, pp. 1156–1160.

Patel, S., Drury, C. G., and Lofgren, J. (1994), "Design of Workcards for Aircraft Inspection," *Applied Ergonomics*, Vol. 25, No. 5, pp. 283–293.

Pearl, A., and Drury, C. G. (1995), "Improving the Reliability of Maintenance Checklists," in *Human Factors in Aviation Maintenance, Phase Four, Progress Report*, DOT/FAA/AM-93/xx, National Technical Information Service, Springfield, VA.

Portillo Sosa, J. (1993), "Design of a Computer Programme for the Detection and Treatment of Ergonomic Factors at Workplaces in the Steel Industry," in O, Berchem-Simon (Ed.), *Ergonomics Action in the Steel Industry EUR 14832 EN*, Commission of the European Communities, Luxembourg; pp. 421–427.

Pulat, B. M. (1992), *Fundamentals of Industrial Ergonomics,* Prentice-Hall, Englewood Cliffs, NJ.

Rasmussen, J. (1987), "Reasons, Causes and Human error," in *New Technology and Human Error*, J. Rasmussen, K. Duncan, and J. Leplat, Eds., Wiley, New York, pp. 293–301.

Reason, J. (1990), *Human Error*, Cambridge University Press, New York.

Régie Nationale des Usine Renault (RNUR) (1976), "Les profils de ostes: Méthode d'analyse des conditions de travail," in *Collection Hommes et Savoir*, Masson, Sirtés, Paris.

Rohmert, W., and Landau, K. (1983), *A New Technique for Job Analysis*, Taylor & Francis, London.

Rohmert, W., and Landau, K. (1989), "Introduction to Job Analysis," *A New Technique for Job Analysis*, Taylor & Francis, London, Part 1, pp. 7–22.

Seeber, A., Schmidt, K.-H., Kierswelter, E., and Rutenfranz, J. (1989), "On the Application of AET, TBS and VERA to Discriminate Between Work Demands at Repetitive Short Cycle Tasks," in *Recent Developments in Job Analysis*, K. Landau and W. Rohmert, Eds., Taylor & Francis, New York, pp. 25–32.

SHARE (1990), *Inspecting the Workplace,* SHARE Information Booklet, Occupational Health and Safety Authority, Canberra, Australia.

Simpson, G. C. (1994), "Ergonomic Aspects in Improvement of Safe and Efficient Work in Shafts," in *Ergonomics Action in Mining*, Eur 14831, Commission of the European Communities, Luxembourg, pp. 245–256.

Sinclair, M. A. (1984), "Ergonomics of Quality Control," workshop document, International Conference on Occupational Ergonomics, Toronto, Ontario, Canada.

Sinclair, M. A. (1990), "Subjective Assessment," in *Evaluation of Human Work*, J. R. Wilson and E. N. Corlett, Eds., Taylor & Francis, London pp. 58–88.

Thomas, L. F., and Seaborne, A. E. M. (1961), "The Sociotechnical Context of Industrial Inspection," *Occupational Psychology*, Vol. 35, pp. 36–43.

Treisman, A. (1986), "Features and Objects in Visual Processing," *Scientific American*, Vol. 255, pp. 114B–125.

Volpert,U.S. DoD (1989), *Human Engineering Design Criteria for Military Systems, Equipment and Facilities*, MIL-STD-1472D, Department of Defense, Washington, DC.

Wagner, R. (1989), "Standard Methods Used in French-Speaking Countries for Workplace Analysis," in *Recent Developments in Job Analysis*, K. Landau and W. Rohmert, Eds., Taylor & Francis, New York, pp. 33–42.

Wang, M. J., and Drury, C. G. (1989), "A Method of Evaluating Inspector's Performance Differences and Job

Requirements," *Applied Ergonomics*, Vol. 20, No. 3, pp. 181–190.

Waters, T. R., Putz-Anderson, V., Garg, A., and Fine, L. J. (1993), "Revised NIOSH Equation for the Design and Evaluation of Manual Lifting Tasks," *Rapid Communications, Ergonomics, 1993*, Vol. 36, No. 7, pp. 748–776.

Wenner, C., and Drury, C. G. (1996), "Active and Latent Failures in Aircraft Ground Damage Incidents," in *Proceedings of the Human Factors and Ergonomic Society 39th Annual Meeting*, pp. 796–799.

Wenner, C., and Drury, C. G. (2000), "Analyzing Human Error in Aircraft Ground Damage Incidents," *International Journal of Industrial Ergonomics,* Vol. 26, No. 2, pp. 177–199.

Westwater, M. G., and Johnson, G. I. (1995), "Comparing Heuristic, User-Centred and Checklist-Based Evaluation Approaches," in *Contemporary Ergonomics 1995*, S. A. Robertson, Ed., Taylor & Francis, London, pp. 538–543.

Wickens, C. D. (1992), *Engineering Psychology and Human Performance*, 2nd ed., Harper-Collins, New York.

Wilkins, J. R., Bean, T. L, Mitchell, G. L., Crawford, J. M., and Eicher, L. C. (1997), "Development and Application of a Pen-Based Computer Program for Direct Entry of Agricultural Hazard Data," *Applied Occupational and Environmental Hygiene*, Vol. 12, No. 2, pp. 105–110.

Wolfe, J. M. (1994), "Guided Search II: A Revised Model of Visual Search," *Psychonomic Bulletin and Review 1994*, Vol. 1, No. 2, pp. 202–238.

Wright, P., and Barnard, P. (1975), "Just Fill in This Form: A Review for Designers," *Applied Ergonomics*, Vol. 6, pp. 213–220.

Yoshida, H., and Ogawa, K. (1991), "Workplace Assessment Guideline—Checking Your Workplace," in W. Karwowski and J. W. Yates (Eds.), *Advances in Industrial Ergonomics and Safety III*, Taylor & Francis, London, pp. 23–28.

CHAPTER 43

COST–BENEFIT ANALYSIS OF HUMAN SYSTEMS INVESTMENTS

William B. Rouse
Georgia Institute of Technology
Atlanta, Georgia

Kenneth R. Boff
Air Force Research Laboratory
Wright-Patterson Air Force Base, Ohio

1 INTRODUCTION

The past decade has been a period of very serious scrutiny of the activities of most enterprises. Business processes have been reengineered and enterprises have been downsized or, more popularly, right-sized. Every aspect of an enterprise must now provide value to customers, earn revenues based on this value, and pay its share of costs. Aspects of an enterprise that do not satisfy these criteria are targeted for elimination.

This philosophy seems quite reasonable and straightforward. However, implementation of this philosophy becomes rather difficult when the "value" provided is indirect and abstract. When anticipated benefits are not readily measurable in monetary units and affect things amenable to monetary measurement only indirectly, it can be very difficult to assess the worth of investments in such benefits. There is a wealth of examples of such situations. With any reasonable annual discount rate, the tangible discounted cash flow of benefits from investments in libraries and education, for example, would be so small as to make it very difficult to justify societal investments in these institutions and activities. Of course, we feel quite justified arguing for such investments. Thus, there obviously must be more involved in such an analysis than just discounted cash flow.

In this chapter we address types of human factors and ergonomics investments that have these intangible characteristics in addition to more tangible attributes. One type is research and development (R&D). This type of investment is often made for the purpose of creating long-term value. It will certainly require years and may take decades before returns are fully realized. It is easy to see how R&D can be difficult to justify in terms of impacts on, for instance, this year's sales and profits, or current operational readiness.

Another type of investment with these intangible characteristics involves products and services that enhance human effectiveness. This includes selection, training, system design, job design, organizational development, health and safety, and in general, the wide range of things done to assure and enhance the effectiveness of people in organizations ranging from businesses to military units. In particular, investments focused on increasing human potential rather than direct job performance outputs are much more difficult to justify than those with near-term financial returns (Rouse et al., 1997b).

In this chapter we also address the complex interaction of these two types of investments: namely, R&D investments in human effectiveness. This is done by building on previous efforts—by the authors and many others—addressing the two elements of this interaction. Investing in R&D to enhance human effectiveness presents a confluence of difficulties related to representing and quantifying benefits, as well as attributing costs. Nevertheless, there is a widely shared sense that such investments are socially and economically important. It is difficult, however, to justify particular projects on the basis of such perceptions.

A primary difficulty involves the trade-off between the relatively short-term payoffs of direct improvements in job performance and the inherently long-term benefits of R&D efforts aimed at enhancing human effectiveness. Short-term investments usually involve less uncertainty and fewer risks. In contrast, revolutionary high-payoff innovations usually emerge from much earlier R&D investments. Thus, small, certain, near-term returns compete with large, uncertain, long-term, and potentially very substantial returns. The methodology presented in this chapter enables addressing both types of investments.

In general, several issues underlie the difficulties of justifying the aforementioned types of long-term investments. As just noted, a fundamental issue concerns the associated uncertainties. Not only are the magnitudes and timing of returns uncertain—the very nature and characteristics of returns are uncertain. With R&D investments, for instance, the eventual payoffs from investments are almost always greater for unanticipated applications than for the originally envisioned applications (Burke, 1996). Further, organizations that make the original investments are often unable to take advantage of the eventual returns from R&D (Christensen, 1997).

These findings raise concerns about whether or not the outcomes of R&D will actually be employed. Newly emerging technologies and competitors' initiatives may diminish the value of the outcomes. We assert that R&D should be viewed as the means for creating technology options that address the contingent needs of an enterprise (Rouse and Boff, 2001, 2003a, 2004). The notion of options, which we formalize in the next section, implies that deployment of the outcome of R&D is contingent on the situation at hand when the decision to exercise an option must be made.

Another central issue relates to the preponderance of intangible outcomes for these types of investments. For example, investments in training may enhance leadership skills of managers or commanders. Investments in organizational development can improve the cohesiveness of mental models of management teams or command teams and enhance the shared nature of these models. However, it is difficult to capture fully such impacts in terms of tangible bottom-line metrics.

It is important to differentiate between intangible outcomes and those that are tangible but difficult to translate into monetary benefits or costs. For example, an investment might decrease pollution, which is very tangible, but it may be difficult to translate this projected reduction to estimated economic gain. This is a mainstream issue in economics and not unique to cost–benefit analyses.

A further issue concerns cost–benefit analyses across multiple stakeholders. Most companies' stakeholders include customers, shareholders, employees, suppliers, and communities, among others. Government agencies often have quite diverse sociopolitical constituencies who benefit—or stand to lose benefits—in a myriad of ways, depending on investment decisions. For example, government-sponsored market research may be part of a regional economic development plan or part of a broader political agenda focused on creating jobs. In general, diverse constituencies are quite likely to attempt to influence decisions in a variety of ways. These situations raise many basic questions relative to the importance of benefits and costs for the various stakeholders.

Yet another issue concerns the difference between assessing cost–benefits and predicting cost–benefits. It is certainly valuable to know whether past investments were justified. However, it would be substantially more valuable to be able to predict whether anticipated investments will later provide benefits that justify the initial investments. Of course, the limits of our abilities to predict outcomes are not unique to cost–benefit analysis.

The types of investment problems addressed in this chapter are rife with many uncertainties, intangibles, and stakeholders and the associated unpredictability. These issues are explored in this chapter in the context of alternative frameworks for performing cost–benefit analyses. This leads to clear conclusions about how best to handle these types of investments methodologically. Application of the resulting methodology is then illustrated in the context of three investment problems involving technologies for aiding, training, and assuring the health and safety of personnel in military systems.

2 COST–BENEFIT FRAMEWORKS

There are a variety of frameworks for scrutinizing and justifying investments:

- *Cost–benefit analysis*: methods for estimating and evaluating time sequences of costs and benefits associated with alternative courses of action
- *Cost–effectiveness analysis*: methods for estimating and evaluating time sequences of costs and multiattribute benefits to assure that the greatest benefits accrue for given costs
- *Life-cycle costing*: methods for estimating and evaluating costs of acquisition, operation, and retirement of alternative solutions over their total cycles of life

- *Affordability analysis*: methods for estimating and evaluating life-cycle costs compared to expected acquisition, operations, and maintenance budgets over the total life cycle of an alternative investments
- *Return on investment analysis*: methods for projecting the ratio, expressed as a percentage, of anticipated free cash flow to planned resource investments

In this chapter we focus on cost–benefit in a broad sense, including many aspects of other approaches. For more traditional treatments of cost–benefit analysis, as well as worked examples, see Layard and Glaister (1994) and Gramlich (1997). We should, at the outset, contrast approaches to analysis of investments with those for managing investments. R&D funnels, multistage decision processes, and so on are intended to assess progress and evaluate the attractiveness of continued investment. Reviews of these constructs can be found in Cooper et al. (1998b) and Rouse and Boff (2001).

Cost–benefit analyses are very straightforward when one considers fixed monetary investments made now to earn a known future stream of monetary returns over some time period. Things get much more complicated, however, when investments occur over time, some of which may be discretionary, and when returns are uncertain. Further complications arise when one must consider multiple stakeholders' preferences regarding risks and rewards. Additional complexity is added when returns are indirect and intangible rather than purely monetary. These complications and complexity are more common than are situations where the straightforward cost–benefit analyses are applicable. In this section we discuss alternative frameworks for addressing cost–benefit analyses and compare these alternatives relative to their abilities to address the issues considered in Section 1.

2.1 Traditional Economic Analysis

The time value of money is the central concept in this traditional approach. Resources invested now are worth more than the same amounts gained later. This is due to the costs of the investment capital that must be paid, or forgone, while waiting for subsequent returns on the investment. The time value of money is represented by discounting the cash flows produced by the investment to reflect the interest that would, in effect at least, have to be paid on the capital borrowed to finance the investment.

The following equations summarize the basic calculations of the discounted cash flow model:

$$\text{NPV} = \sum_{i=0}^{N} \frac{r_i - c_i}{(1+\text{DR})^i} \tag{1}$$

$$\text{IRR} = \text{DR such that} \sum_{i=0}^{N} \frac{r_i - c_i}{(1+\text{DR})^i} = 0 \tag{2}$$

$$\text{CBR} = \frac{\sum_{i=0}^{N} c_i/(1+\text{DR})^i}{\sum_{i=0}^{N} r_i/(1+\text{DR})^i} \tag{3}$$

Given projections of costs, $c_i, i = 0, 1, \ldots, N$, and returns, $r_i, i = 0, 1, \ldots, N$, the calculations of net present value (NPV), internal rate of return (IRR), or cost–benefit ratio (CBR) are quite straightforward elements of financial management (Brigham and Gapenski, 1988). The only subtlety is choosing a discount rate, DR, to reflect the current value of future returns decreasing as the time until those returns will be realized increases. It is quite possible for DR to change with time, possibly reflecting expected increases in interest rates in the future. Equations (1) to (3) must be modified appropriately for time-varying discount rates.

The metrics in equations (1) to (3) are interpreted as follows: NPV reflects the amount that one should be willing to pay now for benefits received in the future. These future benefits are discounted by the interest paid now to receive these later benefits. IRR, in contrast, is the value of DR if NPV is zero. This metric makes it possible to compare alternative investments by forcing the NPV of each investment to zero. Note that this assumes a fixed interest rate and reinvestment of intermediate returns at the internal rate of return. CBR simply reflects the discounted cash outflows divided by the discounted cash inflows, or benefits.

2.2 Multiattribute Utility Models

Cost–benefit calculations become more complicated when benefits are not readily transformable to economic terms. Benefits such as safety, quality of life, and aesthetic value are very difficult to translate into strictly monetary values. Multiattribute utility models provide a means for dealing with situations involving mixtures of economic and noneconomic attributes.

Let cost attribute i at time j be denoted by $c_{ij}, i = 1, 2, \ldots, L$ and $j = 0, 1, \ldots, N$, and benefit attribute i and time j be denoted by $b_{ij}, i = 1, 2, \ldots, M$ and $j = 0, 1, \ldots, N$. The values of these costs and benefits are transformed to common utility scales using $u(\mathbf{c}_{ij})$ and $u(\mathbf{b}_{ij})$. These utility functions serve as inputs to the overall utility calculation at time j as shown by (Keeney and Raiffa, 1976)

$$U(\mathbf{c}_j, \mathbf{b}_j) = U[u(c_{1j}), u(c_{2j}), \ldots, u(c_{Lj}), u(b_{1j}), u(b_{2j}), \ldots, u(b_{Mj})] \tag{4}$$

which provides the basis for an overall calculation across time using

$$U(\mathbf{C}, \mathbf{B}) = U[U(\mathbf{c}_1, \mathbf{b}_1), U(\mathbf{c}_2, \mathbf{b}_2), \ldots, U(\mathbf{c}_N, \mathbf{b}_N)] \tag{5}$$

Note that the time value of benefits depicted in equations (1) to (3) is included in equations (4) and (5) by dealing with the time value of costs and returns explicitly and separately from uncertainty.

An alternative approach involves assessing utility functions for discounted costs and benefits, possibly discounted as represented in equations (1) to (3). With this approach, streams of costs and benefits are collapsed across time before the values are transformed to utility scales. The validity of this simpler approach depends on the extent to which people's preferences for discounted costs and benefits reflect their true preferences.

The mappings from c_{ij} and b_{ij} to $u(c_{ij})$ and $u(b_{ij})$, respectively, enable dealing with the subjectivity of preferences for noneconomic benefits. In other words, utility theory enables one to quantify and compare things that are often perceived as difficult to objectify. Unfortunately, models based on utility theory do not always reflect the ways in which human decision making actually works.

Subjective expected utility (SEU) theory reflects these human tendencies. Thus, to the extent that one accepts that perceptions are reality, one needs to consider the SEU point of view when one makes expected utility calculations. In fact, one should consider making these calculations using both objective and subjective probabilities to gain an understanding of the sensitivity of the results to perceptual differences.

Once one admits the subjective, one needs to address the issue of whose perceptions are considered. Most decisions involve multiple stakeholders: in other words, people who hold a stake in the outcome of a decision. It is, therefore, common for multiple stakeholders to influence a decision. Consequently, the cost–benefit calculation needs to take into account multiple sets of preferences. The result is a group utility model (Keeney and Raiffa, 1976; Kirkwood, 1979)

$$U = U[U_1(\mathbf{C}, \mathbf{B}), U_2(\mathbf{C}, \mathbf{B}), \ldots, U_K(\mathbf{C}, \mathbf{B})] \quad (6)$$

where K is the number of stakeholders.

Formulation of such a model requires that two important issues be resolved. First, mappings from attributes to utilities must enable comparisons across stakeholders. In other words, one has to assume that $u = 0.8$, for example, implies the same value gained or lost for all stakeholders, although the mapping from attribute to utility may vary for each stakeholder. Thus, all stakeholders may, for instance, have different needs or desires for safety and hence different utility functions. They may also have different time horizons within which they expect benefits: For example, stakeholders of different generations, some perhaps not yet born, have different time horizons within which they expect to receive benefits. However, once the mapping from attributes to utility is performed and utility metrics are determined, one has to assume that these metrics can be compared quantitatively.

The second important issue concerns the relative importance of stakeholders. Equation (6) implies that the overall utility attached to each stakeholder's utility can differ. For example, it is often the case that primary stakeholders' preferences receive more weight than the preferences of secondary stakeholders. The difficulty of this issue is obvious. Who decides? Is there a super stakeholder, for instance? Do the groups of stakeholders, or their representatives, simply vote on who gets how much weight? Such a procedure has its own theoretical problems that cannot be addressed here.

Beyond these two more theoretical issues, there are substantial practical issues associated with determining the functional forms of $u(c_{ij})$ and $u(b_{ij})$ and the parameters within these functional relationships. This is also true for the higher-level forms represented by equations (4) to (6). As the number of stakeholders (K), cost attributes (L), benefit attributes (M), and time periods (N) increases, these practical assessment problems can be quite daunting.

2.3 Option Pricing Theory

Many investment decisions are not made all at once. Instead, initial investments are made to create the potential for possible future and usually larger investments involving much greater benefits than are probable for the initial investments. For example, investments in R&D are often made to create the intellectual property and capabilities that will support or provide the opportunity subsequently to decide whether or not to invest in launching new products or services. These launch decisions are contingent on R&D reducing uncertainties and risks, as well as further market information being gained in the interim between the R&D investment decision and possible launch decision. In this way, R&D investments amount to purchasing options to make future investments and earn subsequent returns. These options, of course, may or may not be exercised.

Boer (1998, 1999), Luehrman (1998), and Amram and Kulatilaka (1999) advocate using option pricing theory to analyze investments involving such contingent downstream decisions. Option pricing theory focuses on establishing the value of an option to make an investment decision, in an uncertain environment, at a later date. Developing option-based models begins with consideration of the effects sought by the investment and the capabilities needed to provide these effects. In the private sector, desired effects are usually profits, perhaps expressed as earnings per share, and needed capabilities are typically competitive market offerings. Options can relate to which technologies are deployed and/or which market segments are targeted. Purchasing options may involve R&D investments, alliances, mergers, acquisitions, and so on. Exercising options involves deciding which technologies will be deployed in which markets and investing accordingly.

In the public sector, effects are usually couched in terms of provision of some public good, such as defense. More specific effects might be expressed in terms of measures of surveillance and reconnaissance coverage, for instance. Capabilities would then be defined as alternative means for providing the desired effects. Options in this example might relate to technologies that could enable the capabilities for providing these effects. Attractive options would be those that could provide given effects at lower costs of development, acquisition, and/or operations.

Option-based valuations are economic valuations. Various financial projections are needed as input to option calculations, including (1) investment to "purchase" option, including timing; (2) investment to "exercise" option, including timing; (3) free cash flow (profits and/or cost savings) resulting from exercise; and (4) volatility of cash flow, typically expressed as a percentage. The analyses needed to create these projections are often substantial. For situations where cash flows are solely cost savings, it is particularly important to define credible baselines against which savings are estimated. Such baselines should be choices that would actually be made were the options of interest not available.

The models employed for option-based valuations were initially developed for valuation of financial instruments (Black and Scholes, 1973). For example, an option might provide the right to buy shares of stock at a predetermined price some time in the future. Valuation concerns what such an option is worth. This depends, obviously, on the likelihood that the stock price will be greater than the predetermined price associated with the option. More specifically, the value of the option equals the discounted expected value of the stock at maturity, conditional on the stock price at maturity exceeding the exercise price, minus the discounted exercise price, all times the probability that at maturity the stock price is greater than the exercise price (Smithson, 1998). Net option value equals the option value calculated in this manner minus the cost of purchasing the option.

Thus, there are net present values embedded in the determination of net option values. However, in addition, there is explicit representation of the fact that one will not exercise an option at maturity if the current market share price is less than or equal to the exercise price. As mentioned earlier, sources such as Luenberger (1997), Boer (1998, 1999), Luehrman (1998), Smithson (1998), and Amram and Kulatilaka (1999) provide a wealth of illustrations of how option values are calculated for a range of models.

It is important to note that the options addressed in this chapter are usually termed "real" options in the sense that the investments associated with these options are usually intended to create tangible assets rather than purely financial assets. Application of financially derived models to nonfinancial investments often raises the issue of the extent to which assumptions from financial markets are valid in the domains of nonfinancial investments. This concern is usually addressed with sensitivity analysis.

The assumptions underlying the option-pricing model and the estimates used as input data for the model are usually subject to much uncertainty. This uncertainty should be reflected in option valuations calculated. Therefore, what is needed is a probability distribution of valuations rather than solely a point estimate. This probability distribution can be generated using Monte Carlo simulation to vary model and input variables systematically using assumed distributions of parameter/data variations. The *Technology Investment Advisor* (Rouse et al., 2000) is an example of a tool available to support these types of sensitivity analyses. These analyses enable consideration of options in terms of both returns and risks. Interesting "what if?" scenarios can be explored. A question that we have encountered frequently when performing these analyses is: How bad can it get and have this decision still make sense? This question reflects a desire to understand thoroughly the decision being entertained, not just to get better numbers.

The option value resulting from the formulation above is premised totally on the assumption that waiting does not preempt deciding later. In other words, the assumption is that the decision to exercise an option cannot be preempted by somebody else deciding earlier. In typical situations where other actors (e.g., competitors) can affect possible returns, it is common to represent their impact in terms of changes in projected cash flows (Amram and Kulatilaka, 1999). In many cases, competitors acting first will decrease potential cash flows that will decrease the option value. It is often possible to construct alternative competitive scenarios and determine an optimal exercise date.

A central attraction of this model is the explicit recognition that the purpose of an investment now (i.e., purchasing an option) is to assure the option to make a subsequent investment later (i.e., exercise the option). Thus, for example, one invests in creating new technologies for the option of later incorporating these technologies in product and service lines. The significance of the contingent nature of this decision makes an option-pricing model a much better fit than a traditional discounted cash flow model. However, not all long-term investment decisions have substantial contingent elements. For example, one may invest in training and development to have the option later of selecting among talented managers for elevation to executive positions. There are minimal investments associated with exercising such options—almost all of the investment occurs up front. Thus, option-pricing models are not useful for such decisions.

2.4 Knowledge Capital Approach

Tangible assets and financial assets usually yield returns that are important elements of a company's overall earnings. It is often the case, however, that earnings far exceed what might be expected from these "hard" assets. For example, companies in the software, biotechnology, and pharmaceutical industries typically have much higher earnings than companies with similar hard assets in the aerospace, appliance, and automobile industries, to name just a few. It can be argued that these higher earnings are due to greater knowledge capital among software companies, and so on. However, since knowledge capital does not appear on financial statements, it is very difficult to identify and, better yet, project knowledge earnings.

Mintz (1998) summarizes a method developed by Baruch Lev for estimating knowledge capital and earnings. This article in *CFO* drew sufficient attention to be discussed in *The Economist* (1999) and reviewed by Strassman (1999). In general, both reviews applauded the progress represented by Mintz's article, but also noted the shortcomings of his proposed metrics. The key, Mintz and Lev argue, is to partition earnings into knowledge earnings and hard asset earnings. This is accomplished by first projecting normalized annual earnings from an average of three past years and estimates for three future years using readily available information:

$$\begin{aligned}\text{knowledge earnings} &= \text{normalized annual earnings}\\ &\quad - \text{earnings from tangible assets}\\ &\quad - \text{earnings from financial assets} \qquad (7)\end{aligned}$$

Earnings from tangible and financial assets are calculated from reported asset values using industry averages of 7% and 4.5% for tangible and financial assets, respectively. Knowledge capital is then estimated by dividing knowledge earnings by a knowledge capital discount rate:

$$\text{knowledge capital} = \frac{\text{knowledge earnings}}{\text{knowledge capital discount rate}} \qquad (8)$$

Based on an analysis of several knowledge-intensive industries, Mintz and Lev use 10.5% for this discount rate.

Using this approach to calculating knowledge capital, Mintz compares 20 pharmaceutical companies to 27 chemical companies. He determines, for example, a knowledge capital/book value ratio of 2.45 for pharmaceutical companies and 1.42 for chemical companies. Similarly, the market value/book value ratio is 8.85 for pharmaceutical companies and 3.53 for chemical companies. Considering this correlation between knowledge capital and market value, Strassman (1999) points out that Mintz's estimates do not fully explain the full excess of market values over book values.

The key issue within this overall approach is being able to partition earnings. Whereas earnings from financial assets should be readily identifiable, the distinction between tangible and knowledge assets is problematic. Further, using industry average return rates to attribute earnings to tangible assets does not allow for the significant possibility of tangible assets having little or no earnings potential. Finally, of course, simply attributing all earnings "left over" to knowledge assets amounts to giving knowledge assets credit for everything that cannot be explained by traditional financial methods.

Nevertheless, the knowledge capital construct appears to have potential application to investments involving, for example, R&D or training and development. The purpose of these two types of investments seems obviously to be that of increasing knowledge capital. Further, companies that make investments for this purpose do seem to create more knowledge capital. The key for cost–benefit analyses is being able to project investment returns in terms of knowledge capital, and, in turn, project earnings and separate these earnings into knowledge earnings and hard earnings. Further, one needs to be able to do this for specific investment opportunities, not just the company as a whole.

2.5 Comparison of the Frameworks

Table 1 provides a comparison of the four frameworks just reviewed. It is important to note that this assessment is not really an apples-to-apples comparison. Multiattribute utility theory provides much more of

Table 1 Comparison of Cost–Benefit Frameworks

	Framework			
Issue	Traditional Economic Analysis	Multiattribute Utility Models	Option Pricing Theory	Knowledge Capital Approach
Representation of uncertainties	Focuses on expected revenues and costs without consideration of variances.	Probabilistic uncertainties and stakeholders preferences regarding uncertainties are central to models.	Volatility of returns is a central construct within this model.	Focuses on actual and expected earnings without consideration of variances.
Intangible vs. tangible outcomes	All outcomes must be converted to monetary units.	Preferences regarding intangible outcomes can be incorporated.	All outcomes must be converted to monetary units.	All outcomes must be converted to monetary units.
Multiple stakeholders in cost–benefits	One-dimensional nature of costs and benefits implies one stakeholder.	Formulations for multiple stakeholders are available and limitations are understood.	One-dimensional nature of costs and benefits implies one stakeholder.	One-dimensional nature of costs and benefits implies one stakeholder.
Assessing vs. projecting cost–benefits	Depends on abilities to project monetary costs and benefits.	Depends on abilities to project attributes of utility functions.	Depends on abilities to project monetary costs and benefits.	Difficult to project effects of particular investments.

a general framework than the other three approaches that emphasize financial metrics. Nevertheless, these four approaches represent the dominant alternatives.

Traditional economic analyses are clearly the most narrow. However, in situations where they apply, these analyses are powerful and useful. Most of the investment situations addressed in this chapter do not fit these narrow characteristics. For example, if R&D investments in human effectiveness are viewed within a traditional framework, with typical discount rates, no one would ever invest anything in such R&D. But people do make such investments and, thus, there must be more to it than just NPV, IRR, and CBR. [In fact, Cooper et al. (1998b) have found that companies relying solely on financial metrics for R&D investment decisions tend to be the poorest performers of R&D in terms of subsequent market success.] One view is that R&D reduces uncertainty and buys time before committing very substantial resources to productization, process development, and so on. Option pricing theory seems to be a natural extension of traditional methods to enable handling these complications. As noted earlier, several authors have advocated this approach for analyses of R&D investments (e.g., Boer, 1998, 1999; Lint and Pennings, 1998).

The knowledge capital approach provides another, less mathematical way of capturing the impact of R&D investments in human effectiveness. The difficulty with this approach, which is probably inherent to its origins in accounting and finance, is that it does not address the potential impact of alternative investments. Instead, it serves to report the overall enterprise score after the game.

Multiattribute utility models can, in principle, address the full range of complications and complexity discussed thus far. Admittedly, the ability to create a rigorous multi-attribute utility model depends on the availability of substantial amounts of information regarding stakeholders' preference spaces, probability density functions, etc. However, in the absence of such information, a much more qualitative approach can be quite useful, as is discussed later in this chapter. The value of the multiattribute utility approach also depends on being able to compare overall utilities of alternative investments, which, in turn, depends on being able to compare different stakeholders' utilities of the alternatives. This ability to transform a complex, multidimensional comparison into a scalar comparison is laden with assumptions. The saving grace of the approach, in this regard, is that it makes these assumptions quite explicit and, hence, open to testing. This does not, of course, guarantee that they will be tested.

Expected utility calculations serve to show how one alternative is better than another, rather than providing absolute scores. Thus, differences in expected utilities among alternatives are usually more interesting than the absolute numbers. In fact, the dialogue among stakeholders that is often associated with trying to understand the sources of expected utility differences can provide crucial insights into the true nature of differences among alternatives.

Overall, one must conclude that multiattribute utility models provide the most generalizable approach. This is supported by the fact that multiattribute models can incorporate metrics such as NPV, option value, and knowledge capital as attributes within the overall model; indeed, the special case of one stakeholder, linear utility functions, and NPV as the sole attribute is equivalent to the traditional financial analysis. Different stakeholders' preferences for these metrics can then be assessed and appropriate weightings determined. Thus, use of multiattribute models does not preclude also taking advantage of the other approaches: The four approaches can therefore be viewed as complementary rather than competing. For these reasons, the multiattribute approach is carried forward in the remainder of the chapter.

3 COST–BENEFIT METHODOLOGY

Cost–benefit analysis should always be pursued in the context of particular decisions to be addressed. A valuable construct for facilitating an understanding of the context of an analysis is the value chain from investments to returns. More specifically, it is quite helpful to consider the value chain from investments (or costs), to products, to benefits, to stakeholders, to utility of benefits, to willingness to pay, and finally, to returns on investments. This value chain can be depicted as:

investments (costs)	to	resulting products over time
products over time	to	benefits of products over time
benefits over time	to	range of stakeholders in benefits
range of stakeholders	to	utility of benefits to each stakeholder
utility to stakeholders	to	willingness to pay for utility gained
willingness to pay	to	returns to investors

The process starts with investments that result, or will result, in particular products over time. Products need not be end products—they might be knowledge, skills, or technologies. These products yield benefits, also over time. A variety of people—or stakeholders—have a stake in these benefits. These benefits provide some level of utility to each stakeholder. The utility perceived—or anticipated—by each stakeholder affects their willingness to pay for these benefits. Their willingness to pay affects their "purchase" behaviors, which result in returns for investors.

The central methodological question concerns how one can predict the inputs and outputs of each element of this value chain. This question is addressed elsewhere in some detail for R&D management (Rouse et al., 1997a; Rouse and Boff, 1998) and for human effectiveness (Rouse and Boff, 2003a). Briefly, a variety of models have been developed for addressing this need for prediction. These models are very interesting and offer much potential. However, they suffer from a central shortcoming. With few exceptions, there is an almost overwhelming lack of data for estimating model parameters, as well as a frequent lack of adequate input data. Use of data from baselines can help, but the validity of these baselines depends on new systems and products being very much like their predecessors. Overall, the paucity of data dictates development of a more qualitative methodology whose usefulness is not determined totally by the availability of hard data. We outline such a methodology in the remainder of this section.

As indicated in our earlier comparison of four frameworks for addressing cost–benefit analysis, the most broadly applicable of these alternatives are multiattribute utility models. In the remainder of this section we describe a seven-step methodology:

1. Identify the stakeholders in alternative investments.
2. Define the benefits and costs of alternatives in terms of attributes.
3. Determine the utility functions for attributes (benefits and costs).
4. Decide how the utility functions should be combined across stakeholders.
5. Assess the parameters within the utility models.
6. Forecast the levels of the attributes (benefits and costs).
7. Calculate the expected utility of alternative investments.

It is important to note that this methodology is, by no means, novel and builds upon works by many others related to multiattribute analysis (e.g., Keeney and Raiffa, 1976; Sage, 1977; Hammond et al., 1998; Matheson and Matheson, 1998; Sage and Armstrong, 2000).

3.1 Step 1: Identify the Stakeholders

The first step involves identifying the stakeholders who are of concern relative to the investments being entertained. Usually, this includes all the people in the value chain summarized earlier. This might include, for example, those who will provide the resources that will enable a solution, those who will create the solution, those who will implement the solution, and those who will benefit from the solution.

3.2 Step 2: Define the Benefit and Cost Attributes

The next step involves defining the benefits and costs involved from the perspective of each stakeholder. These benefits and costs define the attributes of interest to the stakeholders. Usually, a hierarchy of benefits and costs emerges, with more abstract concepts at the top (e.g., feasibility, acceptability, and validity; Rouse, 1991) and concrete measurable attributes at the bottom.

3.3 Step 3: Determine the Stakeholders' Utility Functions

The value that stakeholders attach to these attributes is defined by stakeholders' utility functions. The utility functions enable mapping disparate benefits and costs to a common scale. A variety of techniques are available for assessing utility functions (Keeney and Raiffa, 1976).

3.4 Step 4: Determine the Utility Functions across Stakeholders

Next, one determines how utility functions should be combined across stakeholders. At the very least, this involves assigning relative weights to different stakeholders' utilities. Other considerations, such as desires for parity, can make the ways in which utilities are combined more complicated. For example, equation (5) may require interaction terms to assure that all stakeholders gain some utility.

3.5 Step 5: Assess the Parameters of the Utility Functions

The next step focuses on assessing parameters within the utility models. For example, utility functions that include diminishing or accelerating increments of utility for each increment of benefit or cost involve rate parameters that must be estimated. As another instance, estimates of the weights for multistakeholder utility functions have to be estimated. Fortunately, there are a variety of standard methods for making such estimates.

3.6 Step 6: Forecast the Levels of the Attributes

With the cost–benefit model fully defined, one must next forecast levels of attributes or, in other words, benefits and costs. Thus, for each alternative investment, one must forecast the stream of benefits and costs that will result if this investment is made. Quite often, these forecasts involve probability density functions rather than point forecasts. Utility theory models can easily incorporate the impact of such uncertainties on stakeholders' risk aversions. On the other hand, information on probability density functions may not be available, or may be prohibitively expensive. In these situations, beliefs of stakeholders and subject matter experts can be employed, perhaps coupled with sensitivity analysis (see step 7), to determine where additional data collection may be warranted.

3.7 Step 7: Calculate the Expected Utilities

The final step involves calculating the expected utility of each alternative investment. These calculations are performed using specific forms of equations (4) to (6). This step also involves using sensitivity analysis to assess, for example, the extent to which the rank ordering of alternatives, by overall utility, changes as parameters and attribute levels of the model are varied.

3.8 Use of the Methodology

Some elements of the cost–benefit methodology just outlined are more difficult than others. The overall calculations are quite straightforward. The validity of the resulting numbers depends, of course, on stakeholders and attributes having been identified appropriately. It depends further on the quality of the inputs to the calculations. These inputs include estimates of model parameters and forecasts of attribute levels. As indicated earlier, the quality of these estimates is often compromised by lack of available data. Perhaps the most difficult data collection problems relate to situations where the impacts of investments are both uncertain and very much delayed. In such situations, it is not clear which data should be collected and when they should be collected.

A recurring question concerns the importance that should be assigned to differences in expected utility results. If alternative A yields $U(\mathrm{A}) = 0.648$ and alternative B yields $U(\mathrm{B}) = 0.553$, is A really that much better than B? In fact, is either set of utilities sufficiently great to justify an investment? These questions are best addressed by considering past investments. For successful past investments, what would their expected utilities have been at the time of the investment decisions? Similarly, for unsuccessful past investments, what were their expected utilities at the time? Such comparisons often yield substantial insights.

Of course, the issue is not always A versus B. Quite often the primary question concerns which alternatives belong in the portfolio of investments and which do not. Portfolio management is a fairly well-developed aspect of new product development (e.g., Gill et al., 1996; Cooper et al., 1998a). Well-known and recent books on R&D/technology strategy pay significant attention to portfolio selection and management (e.g., Roussel et al., 1991; Matheson and Matheson, 1998; Boer, 1999; Allen, 2000). In fact, the conceptual underpinnings of option pricing theory are based on notions of market portfolios (Amram and Kulatilaka, 1999).

Most portfolio management methods rely on some scoring or ranking mechanism to decide which investments will be included in the portfolio. Expected utility is a quite reasonable approach to creating such scores or ranks. This is particularly useful if sensitivity analysis has been used to explore the basis and validity of differences among alternatives interactively. A more sophisticated view of portfolio management considers interactions among alternatives in the sense that synergies between two alternatives may make both of them more attractive (Boer, 1999; Allen, 2000). Also, correlated risks between two alternatives may make both of them less attractive. A good portfolio has an appropriate balance of synergies and risks.

In principle at least, the notions of portfolio synergy and risk can be handled within multiattribute utility models. This can be addressed by adding attributes that are characteristics of multiple rather than individual alternatives. In fact, such additional attributes might be used to characterize the entire portfolio. An important limitation of this approach is the probable significant increase in the complexity of the overall problem formulation. Indeed, this is an issue in general when multiattribute utility models are elaborated to better represent problem complexities.

Beyond these technical issues, it is useful to consider how this cost–benefit methodology should affect decision making. To a very great extent, the purpose of this methodology is to get the right people to have the right types of discussions and debates on the right issues at the right time. If this happens, the value of people's insights from exploring the multiattribute model usually far outweighs the importance of any particular numbers.

The practical implications of this conclusion are quite simple. Very often, decision making happens within working groups that view large-screen computer-generated displays of the investment problem formulation and results as they emerge. Such groups perform sensitivity analyses to determine the critical assumptions or attribute values that are causing some alternatives to be more highly rated or ranked than others. They use "what if?" analyses to explore new alternatives, especially hybrid alternatives.

This approach to investment decision making helps to decrease substantially the impact of limited data being available. Groups quickly determine which elements of the myriad of unknowns really matter—where more data are needed, and where more data, regardless of results, would not affect decisions. A robust problem formulation that can be manipulated, redesigned, and tested for sanity provides a good way for decision-making groups to reach defensible conclusions with some level of confidence and comfort.

4 THREE EXAMPLES

Human effectiveness concerns enhancing people's direct performance (aiding), improving their potential to perform (training), and assuring their availability to perform (health and safety). These are central issues in human systems integration. Investments in human effectiveness also have the potential of increasing returns on other investments by, for example, enabling people to take full advantage of new technologies.

Three examples of aiding, training, and health and safety investments are discussed in this section: VCATS (aiding), DMT (training), and PTOX (health and safety). These examples focus on enhancing human effectiveness and human systems integration in military systems—particularly, Air Force systems. The applicability of these technologies and the relevance of the following analysis of the impacts of these technologies to other military services and to nonmilitary problems should also be readily apparent.

4.1 Visually Coupled Targeting and Acquisition System

The Visually Coupled Targeting and Acquisition System (VCATS) provides aid to military aircraft pilots. VCATS includes a helmet-mounted tracker and display (HMT/D), associated signal processing sensor/transducer hardware, interchangeable panoramic night vision goggle with head-up display (PNVG-HUD), and extensive upgrades to the aircraft's operational flight program software (Rastikis, 1998). VCATS enables the pilot to cue and be cued by on- and off-board systems, sensors, and weapons, as well as to be coupled spatially and temporally with the control processes implemented with the HMT/D and PNVG-HUD. The system is particularly effective in helping pilots to cue weapons and sensors to targets, to maintain "ownship" formation situation awareness, and to avoid threats through provision of a real-time, three-dimensional portrayal of the pilots' tactical and global battlefield status. In general, VCATS enables pilots to acquire targets and threats faster. This results in improvements in terms of (1) how far, (2) how quickly, and (3) how long—for both initial contacts and countermeasures.

To a great extent, the case for advanced development has already been made for VCATS, and current support is substantial. However, the transition from advanced development to production involves assuring that the options created by VCATS and validated by combat pilots are exercised. The case has also been argued for ongoing investments in basic research and exploratory development to assure that VCATS has future technology options, particularly for migration to multirole fighter aircraft. The maturity of the program should help in making this case in terms of benefits already demonstrated. However, in the current budget climate, there is also substantial risk that VCATS research may be viewed as essentially "done." This raises the potential for negative decisions regarding further investments.

4.2 Distributed Mission Training

Distributed mission training (DMT) involves aircraft, virtual simulators, and constructive models that, collectively, provide opportunities for military pilots to gain experiences deemed important to their performance proficiency relative to anticipated mission requirements (Andrews, 2000). The desired training experiences are determined from competencies identified as needed to fulfill mission requirements. These competency requirements are translated to training requirements stated in terms of types and durations of experiences deemed sufficient to gain competency.

The case to be made for DMT involves investments to address research issues and technology upgrades of near-term capabilities. The primary options-oriented argument is that investments in R&D in DMT will create contingent possibilities for cost savings in training due to reduced use of actual aircraft. More specifically, if exercised, DMT options will provide cash flows of savings that justify the investments needed to field this family of technologies.

A much more subtle options-oriented argument concerns the training experiences provided by DMT that could not otherwise be obtained. Clearly, the opportunity to have relevant training experiences must be better than not having these experiences. The option therefore relates to proficiency versus. possible lack of proficiency. As straightforward as this may seem, it quickly encounters the difficulty of projecting mission impacts—and the value of these impacts—of not having proficient personnel. One possible approach to quantifying these benefits is to project the costs of using real aircraft to gain the desired proficiences. Although these costs are likely to be prohibitive—and thus never would be seriously considered—they nevertheless characterize the benefits of DMT.

4.3 Predictive Toxicology

Predictive toxicology (PTOX) is concerned with projecting the impacts on humans from exposure to operational chemicals (individual and mixtures). The impact can be characterized in terms of the possibility of performance decrement and consequent loss of force effectiveness, possible military and civilian casualties, and potential long-term health impacts. Also of concern are the impacts of countermeasures relative to sustaining immediate performance and minimizing long-term health impacts (OSTP, 1998).

The case to be made involves investment in basic research and exploratory development programs, with longer-term investment in an advanced development program to create deployable predictive toxicology capabilities. The requisite R&D involves developing and evaluating models for predicting performance and health impacts of operational chemicals. Advanced development will focus on field sensing and prediction, termed *deployment toxicology*. The nature of the necessary models is strongly affected by the real-time requirements imposed by deployment.

4.4 Applying the Methodology

The remainder of this section addresses primarily steps 1 to 4 of the cost–benefit methodology in the context of these three examples related to human effectiveness aspects of human systems integration. These steps constitute the framing steps of the methodology rather than the calculation steps. Appropriate framing of cost–benefit analyses is critical to subsequent calculations being meaningful and useful.

4.4.1 Step 1: Identify the Stakeholders

This step involves identifying people (usually, types of people) and organizations that have a stake in costs and benefits. All three of the examples involve three classes of stakeholders: warfighters, developers, and the public. A key issue concerns the relative importance of these three types of stakeholders. Some would argue that warfighter preferences dominate decisions. Others recognize the strong role that developers and their constituencies play in procurement decisions. Yet another argument is that the dominating factor is value to the public, with the other stakeholders being secondary in importance.

Warfighters as stakeholders include military personnel in general, especially for PTOX. Warfighters of particular importance include aircraft pilots, personnel who support flight operations, and military commanders. Developers as stakeholders include companies and their constituencies (e.g., stockholders, employees, and communities). Several agents, including Congress, the executive functions within the military services, and the military procurement establishment, represent the public's interests. Pilots and other military personnel are users of the technologies of interest, developers are the providers, and the public's agents are the customers for these technologies. There are obvious trade-offs across the interests of users, providers, and customers.

4.4.2 Step 2: Define the Benefit and Cost Attributes

Benefits and costs tend to fall in general classes. Example benefits for military organizations and contractors include:

- *Enhanced impact*: increased lethality, survivability, and availability
- *Enhanced operability*: decreased response time and increased throughput
- *Enhanced design*: new techniques and larger pool of experienced people
- *Increased opportunities*: new tactics and countermeasures

Example cost attributes applicable to military procurement include:

- *Investment costs*: capital investments and R&D costs
- *Recurring costs*: operating and G&A costs
- *Time costs*: time from development to fielding to competent use
- *Opportunity costs*: other cost–benefits forgone

These general classes of benefits and costs can be translated into specific benefit and cost attributes for the three classes of stakeholders in VCATS, DMT, and PTOX. Benefits for warfighters (users) include enhanced performance (e.g., response time), confidence in performance, and health and safety—in varying combinations for the three examples. Costs for these stakeholders include learning time and changing their ways of doing things to assure compatibility between new and legacy technologies.

Benefits for companies and their constituencies (providers) include R&D funds received, subsequent intellectual property created, and competitive advantages that result. Also important are jobs and economic impacts on the community. Direct costs include bid and proposal costs as well as opportunity costs. Less direct costs include, for instance, economic development resources and incentives provided to the companies by their communities.

The primary benefit sought by the public's agents (customer) is mission performance/dollar. It can easily be argued for all three examples that mission performance is increased. Unfortunately, it is difficult to attach a value to this increase. For example, what is the value of being able to generate 5% more sorties per time period? The answer depends on whether more sorties are needed.

Few would argue with the importance of meeting mission requirements successfully. However, if the types of innovations represented by these examples enable exceeding mission requirements, what are such increases worth? This is a politically sensitive question. If better performance is of substantive value, why wasn't this level of performance specified in the original requirements?

A good way to avoid this difficulty is to take mission requirements as a given and determine how much money could be saved in meeting these requirements by adopting the technologies in question. For example, could requirements be met with fewer aircraft, pilots, and support personnel? As shown in Table 2, the cost savings due to these decreases can be viewed as benefits of the technologies. It also might be possible for VCATS, DMT, or PTOX to enable meeting mission requirements with less capable systems rather than just fewer systems. This possibility provides substantial opportunities for increased benefits due to these technologies.

Note that this philosophy amounts to trying to provide a given level of defense for the least

Table 2 Public Benefits–Costs for Three Examples

	VCATS	DMT	PTOX
Benefits	Fewer aircraft and associated personnel to meet mission requirements due to better performance and fewer aircraft losses	Fewer aircraft and associated personnel to meet mission requirements due to better performance, fewer aircraft losses, and fewer aircraft for training	Fewer personnel to meet mission requirements and decreased medical costs due to fewer people affected by toxic materials, fewer people lost to toxic effects, fewer people to care for people affected, and decreased downstream medical costs
Costs	Initial investment (option price) for proposed R&D costs and later, contingent investment (exercise price) for subsequent fielding of technology	Initial investment (option price) for proposed R&D costs and later, contingent investment (exercise price) for subsequent fielding of technology	Initial investment (option price) for proposed R&D costs and later, contingent investment (exercise price) for subsequent fielding of technology

investment. Another approach might be to attempt to provide the most defense per investment dollar. However, this immediately begs the question of how much defense is enough. Unlike the business world, where value is defined by the marketplace and hence can provide a basis for optimization [see Nevins and Winner (1999) for a good example], there is no widely agreed-upon approach to measuring military value and optimizing accordingly.

The rationale for the benefits indicated in Table 2 for each of the three examples includes the following points:

1. VCATS enables pilots to compete with threats, increase the number of wins versus losses, and counter threats (e.g., missiles) in ways that they could not do otherwise. Consequently, it must be possible to meet *fixed* mission requirements with fewer aircraft and associated infrastructure. These benefits can be translated into financial returns in terms of cost avoidance.

2. DMT provides opportunities to practice behaviors that would not otherwise be practiced, for the most part due to the costs of practice. This decreases the probability of not performing acceptably given inadequate training. DMT also provides training experiences that would not otherwise be possible. For example, in the DMT environment, pilot "kills" actually disappear. In contrast, field exercises often "reuse" kills because of the costs of getting adversaries into the exercise in the first place.

3. PTOX enables larger proportions of deployed forces to be fully functional and less dependent on medical surveillance or medication, and permits earlier intervention, before the onset of problems. In principle, this should enable reducing the size of deployed force, which is critical for increasingly likely expeditionary military missions (Fuchs et al., 1997). PTOX also provides cost avoidance due to downstream health impacts. The ability to predict the body burden of toxicity during deployment should enable removing personnel from risk once the burden is approaching predetermined limits. These capabilities are also likely to be very important for nonmilitary operations such as disaster cleanup.

It is *not* essential that the savings indicated in Table 2 actually occur. For example, it may be that the number of aircraft is not decreased, perhaps due to factors far beyond the scope of these analyses. Nevertheless, one can attribute to these technologies the benefits of having provided opportunities to meet mission requirements in less costly ways. Technologies that provide such opportunities are valuable—the extent of this value is the extent of the opportunities for savings.

This argument puts all three examples on common ground. The benefits of all alternative technologies can be expressed as reduced costs to meet requirements. From an options pricing perspective, these savings can be viewed as free cash flow returned on investments in these technologies. The *option price* comprises the R&D costs. The *exercise price* represents the subsequent costs of fielding the technologies. Thus, assuming that costs savings can be projected (albeit with substantial volatility), the option values of investing in these technologies can be calculated.

4.4.3 Step 3: Determine the Stakeholders' Utility Functions

Different stakeholders' preferences over the benefit and cost attributes will vary substantially with specific situations. However, there is a small family of functional relationships that captures most, if not all, expressed preferences (Keeney and Raiffa, 1976). Thus, although context-specific tailoring is needed, it can be performed within a prescribed (and preprogrammed) set of functions, both within and across stakeholders. Similarly, alternative parameter choices can be prescribed in terms of choices of weightings.

An important aspect of cost–benefit analyses, as advocated in this chapter, is the likely nonlinear nature of utility functions. In particular, diminishing returns and aspiration levels tend to be central to stakeholders' *preference spaces*. In other words, whereas linear functions imply that incremental increases (or decreases) of attributes always yield the same incremental changes in utility, nonlinear functions lead to shifting preferences as attributes increase (or decrease). Figure 1 portrays a range of example utility functions. Parts (*a*) and (*d*) illustrate linear relationships, (*b*) and (*e*) show accelerating relationships, and (*c*) and (*f*) show diminishing relationships.

To illustrate how these types of relationships can be employed to represent the preferences of users, providers, and customers, the general forms of each type of stakeholders' utility function are shown:

$$U_{\text{user}} = U[u(\text{performance}), u(\text{confidence}), u(\text{cost of change})] \quad (9)$$

$$U_{\text{provider}} = U[u(\text{resources}), u(\text{advantage}), u(\text{cost of pursuit})] \quad (10)$$

$$U_{\text{customer}} = u(\text{option value}) \quad (11)$$

where, as noted earlier, users are concerned primarily about impacts of investments on their performance, their confidence in their performance, and the costs of changing their ways of performing; providers are concerned with the investment resources supplied to develop the technologies in question, the competitive advantages created by the intellectual property created, and the costs of pursuing the investment opportunities; and finally, customers are focused on the financial attractiveness of the investments as reflected in the option values of the alternatives, which are based on projected cash flows (i.e., costs savings), volatility of cash flows, magnitudes of investments required, and time periods until returns are realized.

Considering the elements of equations (9) to (11), the appropriate functional forms from Figure 1 are likely to be as follows:

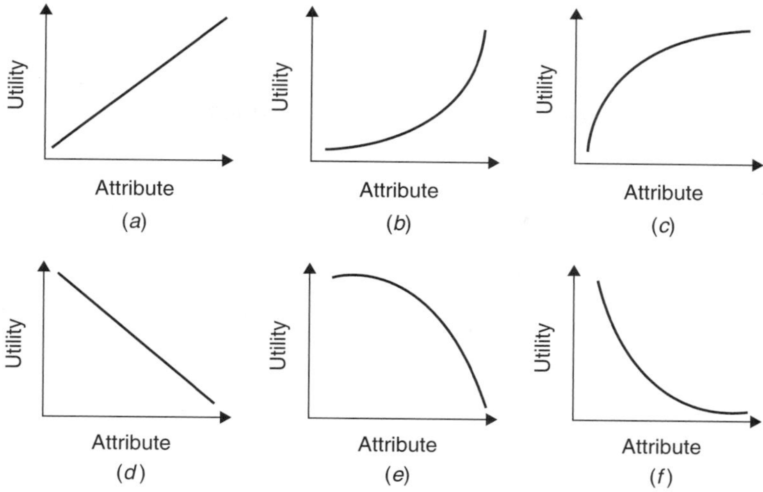

Figure 1 Utility functions: (*a*) more is better; (*b*) accelerating returns; (*c*) diminishing returns; (*d*) less is better; (*e*) accelerating decline; (*f*) diminishing decline.

- u(performance) is an accelerating returns function (Figure 1*b*):
  - VCATS is least concave since relatively modest performance improvements are of substantial utility.
  - DMT is moderately concave since training on otherwise untrained tasks must produce substantial improvements to yield high utility.
  - PTOX is most concave since major decreases in performance risk are needed to assure high utility increases of personnel availability.
- u(confidence) is a linear function (Figure 1*a*) since greater confidence is always better, but there are unlikely to be significant thresholds.
- u(cost of change) is an accelerating decline function (Figure 1*e*) since low to moderate costs are easily sustained, whereas larger costs present difficulties.
- u(resources) is an accelerating returns function (Figure 1*b*) since moderate to large resources are needed to make opportunities attractive.
- u(advantage) is a linear function (Figure 1*a*) since greater advantage is always better, but there are unlikely to be significant thresholds.
- u(cost of pursuit) is an accelerating decline function (Figure 1*e*) since low to moderate costs are easily sustained, whereas larger costs present difficulties.
- u(option value) is a linear function (Figure 1*a*) since customers will inherently gain the expected value across a large number of investments.

It is important to note the importance of this last assumption. If customers' (i.e., the public's) utility function were not linear, it would be necessary to entertain assessing the specific form of their function. Unlike users and providers, the public is not so easily identified and interviewed.

With the identification of the stakeholders (step 1) and framing of the cost–benefit attributes (step 2), the process of determining the form of stakeholders' utility functions (step 3) can draw upon considerable standard "machinery" of decision analysis. The specific versions of the functional forms discussed above are likely to vary with VCATS, DMT, and PTOX. However, the overall formulation chosen is quite general.

4.4.4 Step 4: Determine the Utility Functions across Stakeholders

Another important aspect of the utility functions is their typical lack of alignment across stakeholders. Specifically, either different stakeholders care about different things, or possibly they care about the same things in different ways. For example, customers may be very price-sensitive, whereas users, who seldom pay prices themselves, are usually much more concerned with effects on their job performance. For the types of investment problems considered in this chapter, preferences typically differ across time horizons and across people with vested interests in different investment opportunities. Thus far in the formulation of the three examples, the stakeholders do not have attributes in common. Nevertheless, they are likely to have competing preferences, since, for example, the alternative providing the greatest performance impact may not have the largest option value.

Differing preferences across stakeholders are often driving forces in pursuing cost–benefit analyses. These differing preferences can be aggregated, and traded off, by formulating a composite utility function such as

$$U = U[U_{\text{user}}, U_{\text{provider}}, U_{\text{customer}}] \quad (12)$$

Often, equation (12) will be linear in form with weights assigned to component utility functions to reflect the relative importance of stakeholders. Slightly more complicated are multilinear forms, which include products of component functions (e.g., $U_{\text{user}} \times U_{\text{customer}}$). Multilinear formulations tend to assure that all stakeholders gain nonzero utility because, otherwise, zero in either term in a product yields zero overall.

Considering trade-offs across stakeholders, it is important to note that the formulation of the analysis can often be usefully expanded to include a broader set of stakeholders. These additional stakeholders may include other entities that will benefit by advances of the technologies in question, although they may have little or no stake in the immediate application for the technology. It is also quite possible that stakeholders such as "the public" have multiple interests (e.g., military effectiveness and public safety from toxic risks).

Broadening the analysis in this way is likely to have differing impacts on the assessment for the three examples due to the natures of the technologies and issues being pursued. The three examples differ in this regard in the following ways: (1) VCATS addresses a rather esoteric set of issues from the public's perspective; (2) DMT addresses an issue with broad general support from the public but narrower specific constituencies; and (3) PTOX addresses strong cross-cutting health and safety issues of substantial concern to the public. These differences suggest that PTOX would gain a larger ΔU value than DMT, and DMT would in turn gain a larger ΔU value than VCATS, by broadening the number of stakeholders and issues. Quite simply, the spin-off benefits of PTOX are likely to be perceived as much greater by a larger number of stakeholders.

However, if the formulation is further broadened to consider the likelihood that the desired technologies will emerge elsewhere if investments are not made in these efforts, the ΔU impacts are likely to be the opposite. PTOX research and development are being pursued by several agencies. DMT has broad applicability for both military and nonmilitary applications and consequently is being pursued by other parties. VCATS, in contrast, is highly specialized and is unlikely to emerge from other sources.

These two possibilities for broadening the formulation, in terms of stakeholders and issues, clearly illustrate the substantial impact of the way in which cost–benefit assessments are framed. If the framing is too focused, important spin-off benefits may not be included. On the other hand, framing the analysis too broadly may raise issues that are difficult to quantify, even roughly, and include stakeholders whose preferences are difficult to assess. Of course, many modeling efforts face such difficulties (Sage and Rouse, 1999).

4.4.5 Steps 5–7: Calculate the Overall Cost–Benefit

The remaining steps of the cost–benefit methodology involve assessing parameters of utility functions, forecasting levels of attributes, and calculating expected utilities. Performing these steps obviously depends on having data on stakeholders' preferences and projected/targeted attribute levels. Discussion of such data is well beyond the scope of this chapter, and in light of the nature of the examples, it would be difficult to publish the requisite data.

The needed data can, in many instances, be quite difficult to compile. It can be particularly difficult to relate returns on human effectiveness investments to organizational impacts. Relationships between human and organizational performance are needed. These relationships should answer the following types of questions:

- How do improvements in human performance (e.g., via aiding) translate to increased organizational impacts? Specifically, how does a 2-second improvement in pilot response time due to VCATS affect mission performance?
- How do improvements in the human potential to perform (e.g., via training) translate to actual performance and consequent increased organizational impacts? Specifically, how does increased practice via DMT affect subsequent performance and, in turn, translate to improved mission performance?
- How do improvements in human availability to perform (e.g., via health and safety) translate to actual performance and consequent increased organizational impacts? Specifically, how does prevention of toxic exposure due to PTOX affect immediate unit performance and thereby affect mission performance?

These can be difficult questions. However, they are not inherently cost–benefit questions. Instead, they are fundamental system design questions (Sage and Rouse, 1999). If answers are possible, cost–benefit analyses are more straightforward.

For the VCATS, DMT, and PTOX examples, it may be possible to translate human performance improvements to organizational impacts via mission models. Such models are typically used to determine, for example, the "logistics footprint" needed to support a targeted sortie generation rate or, as another illustration, the combat wins and losses likely with competing defensive measures and countermeasures. Such models can be used, perhaps with extensions, to project the impacts of faster responses due to VCATS, improved task performance due to DMT, and increased personnel availability due to PTOX. It is important to note, however, that even if such projections are not available, the multiattribute methodology presented here can still be employed. However, the validity of cost–benefit assessments and predictions will then depend on subjective perceptions of attribute levels and the relative importance of attributes. Any limitations of this more subjective approach reflect underlying limitations of knowledge rather than inherent limitations of the methodology.

Once $U[U_{user}, U_{provider}, U_{customer}]$ is fully specified, both functionally and in terms of parameters of these functions, one is in position to project attribute levels (e.g., option values), to calculate the expected utility of the alternative investments (e.g., VCATS, DMT, and PTOX), and to perform sensitivity analyses. This provides the basis for making investment decisions. There are several ways that these cost–benefit assessments can be used to inform decision making.

The most common way of using expected utility cost–benefit assessments is to rank order alternative investments in terms of decreasing $U[U_{user}, U_{provider}, U_{customer}]$ and then allocate investment resources from highest ranked to lowest ranked until resources are exhausted. This approach allows the possibility of alternatives, with mediocre $U_{customer}$ making the cut by having substantial U_{user} and $U_{provider}$. To avoid this possibility, one can rank order by $U[U_{user}, U_{provider}, U_{customer}]$ all alternatives with $U_{customer} > U_{co}$, which implies a minimum acceptable option value.

If resources are relatively unconstrained, one can invest in all alternatives for which $U_{user} > U_{uo}$, $U_{provider} > U_{po}$, and $U_{customer} > U_{co}$. This reflects situations where all stakeholders prefer investment to no investment. Of course, one can also rank order these alternatives by $U[U_{user}, U_{provider}, U_{customer}]$ to determine priorities for investment. However, if resources are truly unconstrained, this rank ordering will not change the resulting investment decisions.

4.5 Summary

The three examples discussed in this section have portrayed a cross section of human effectiveness investments to enhance human systems integration, ranging from aiding to training to health and safety investments. The discussion has shown how this range of investment alternatives can be fully addressed with an overarching multiattribute utility, multistakeholder cost–benefit formulation. The stakeholder classes of user, provider, and customer are broadly applicable. The classes of attributes discussed also have broad applicability.

These examples have also served to illustrate the merits of a hybrid approach. In particular, option value theory has been used to define the issue of primary interest to customers—assuring that investments make financial sense—and this issue has then been incorporated into the overall multiattribute formulation. This enabled including in the formulation a substantial degree of objective rigor as well as important subjective attributes and perceptions. As a result, rigor is not sacrificed, but instead is balanced with broader, less quantifiable considerations.

It is useful to note that the knowledge capital construct was not employed in the formulation for these three examples, despite the intuitive appeal of the notion that investments in human effectiveness increase knowledge capital (Davenport, 1999). Although the formulation reported here could have included increases in knowledge capital as possible benefits, there is no basis for predicting such impacts. Subjective estimates could, of course, be employed. However, this construct is not defined with sufficient crispness to expect reliable estimates from subject matter experts.

The discussion of these examples of human effectiveness investments have served to illustrate the value of an overall cost–benefit formulation. The generality of this formulation allows it to be applied to analyses of a wide variety of human system integration investment decisions. The types of information needed to support such analyses are defined by this formulation. Although the availability of information remains a potential difficulty, this formulation nevertheless substantially ameliorates the typical problems of comparing ad hoc analyses of competing investments. Also of great importance, this formulation enables cross-stakeholder comparisons and trade-offs that for the lack of a suitable methodology are usually ignored or resolved in ad hoc manners.

5 CONCLUSIONS

It is difficult to make the case for investments in long-term investments that will provide highly uncertain and intangible returns. In this chapter we have reviewed alternative ways to characterize such investments and presented an overall methodology that incorporates many of the advantages of these alternatives. This methodology has been illustrated in the context of R&D investments in the human effectiveness aspects of human systems integration.

Central to the cost–benefit analysis methodology presented is a multiattribute, multistakeholder formulation. This formulation includes nonlinear preference spaces that are not necessarily aligned across stakeholders. The nonlinearities and lack of alignment provide ample opportunities for interesting trade-offs.

It is important to stress the applicability of this methodology to nearer-term human effectiveness investments, which may or may not involve R&D. Although the time frame will certainly affect choices of attributes—for instance, option values may not be meaningful for near-term investments—the overall cost–benefit methodology remains unchanged. This chapter focused on long-term R&D investments because such analyses are the most difficult to frame and perform.

It is also useful to indicate that cost–benefit analysis, as broadly conceptualized in this chapter, can be a central element in assessment activities related to life-cycle costing (e.g., affordability) and program/contract management (e.g., earned value management) (EVM, 2000). For the former, attributes reflecting life-cycle costs can easily be incorporated. For the latter, costs and benefits can be tracked and compared to original projections. This does, of course, require that benefits be attributable to ongoing processes and not just outcomes.

When coupled with appropriate methods and tools for predicting attribute levels (Sage and Rouse, 1999), this cost–benefit methodology can enable cost–benefit predictions and thereby support investment decision making. Using attributes such as option values and potentially knowledge capital can make it possible to translate the intuitive appeal of R&D and human effectiveness investments into more tangible measures of value.

Note also that the methodology includes many of the elements necessary to developing a business case for human effectiveness investments. Markets (stakeholders), revenues (benefits), and costs are central issues in business case development and in this methodology. However, this methodology also supports the valuation of investments with broader constituencies (e.g., the public) and ranges of issues (e.g., jobs created) than are typically considered in business cases.

Finally, we have also found that use of the methodology presented here provides indirect advantages in terms of causing decision-making groups to clarify and challenge underlying assumptions. This helps decision makers avoid being trapped by common delusions that would mislead them relative to probable cost–benefits (Rouse, 1998).

REFERENCES

Allen, M. S. (2000), *Business Portfolio Management: Valuation, Risk Assessment, and EVA Strategies*, Wiley, New York.

Amram, M., and Kulatilaka, N. (1999), *Real Options: Managing Strategic Investment in an Uncertain World*, Harvard Business School Press, Boston.

Andrews, D. H. (2000), "Distributed Mission Training," in *International Encyclopedia of Ergonomics and Human Factors*, W. Karwowski, Ed., Taylor & Francis, Philadelphia.

Black, F., and Scholes, M. (1973), "The Pricing of Options and Corporate Liabilities," *Journal of Political Economy*, Vol. 81, No. 4, pp. 637–659.

Boer, F. P. (1998), "Traps, Pitfalls, and Snares in the Valuation of Technology," *Research Technology Management*, September–October, pp. 45–54.

Boer, F. P. (1999), *The Valuation of Technology: Business and Financial Issues in R&D*, Wiley, New York.

Brigham, E. F., and Gapenski, L. C. (1988), *Financial Management: Theory and Practice*, Dryden Press, Chicago.

Burke, J. (1996), *The Pinball Effect: How Renaissance Water Gardens Made the Carburetor Possible and Other Journeys Through Knowledge*, Little, Brown, Boston.

Christensen, C. M. (1997), *The Innovator's Dilemma: When New Technologies Cause Great Firms to Fail*, Harvard Business School Press, Boston.

Cooper, R. G., Edgett, S. J., and Kleinschmidt, E. J. (1998a), *Portfolio Management for New Products*, Addison-Wesley, Reading, MA.

Cooper, R. G., Edgett, S. J., and Kleinschmidt, E. J. (1998b), "Best Practices for Managing R&D Portfolios," *Research Technology Management*, Vol. 41, No. 4, pp. 20–33.

Davenport, T. O. (1999), *Human Capital: What It Is and Why People Invest It*, Jossey-Bass, San Francisco.

Economist (1999), "A Price on the Priceless: Measuring Intangible Assets," *The Economist*, June 12, pp. 61–62.

EVM (Earned Value Management Center) (2000), http://evms.dcmdw.dla.mil/.

Fuchs, R., McCarthy, J., Corder, J., Rankine, R., Miller, W., and Gawron, V. (1997), *United States Air Force Expeditionary Forces*, Air Force Scientific Advisory Board, Washington, DC.

Gill, B., Nelson, B., and Spring, S. (1996), "Seven Steps to New Product Development," in *The PDMA Handbook of New Product Development*, M. D. Rosenau, Jr., Ed., Wiley, New York.

Gramlich, E. M. (1997), *A Guide to Cost–Benefit Analysis*, Waveland Press, Prospect Heights, IL.

Hammond, J. S., Keeney, R. L., and Raiffa, H. (1998), *Smart Choices: A Practical Guide to Making Better Decisions*, Harvard Business School Press, Boston.

Keeney, R. L., and Raiffa, H. (1976), *Decisions with Multiple Objectives: Preferences and Value Tradeoffs*, Wiley, New York.

Kirkwood, C. W. (1979), "Pareto Optimality and Equity in Social Decision Analysis," *IEEE Transactions on Systems, Man, and Cybernetics*, Vol. 9, No. 2, pp. 89–91.

Layard, R., and Glaister, S., Eds. (1994), *Cost–Benefit Analysis*, Cambridge University Press, Cambridge.

Lint, O., and Pennings, E. (1998), "R&D As an Option on Market Introduction," *R&D Management*, Vol. 28, No. 4, pp. 279–287.

Luehrman, T. A. (1998), "Investment Opportunities as Real Options," *Harvard Business Review*, July–August, pp. 51–67.

Luenberger, D. G. (1997), *Investment Science*, Oxford University Press, Oxford.

Matheson, D., and Matheson, J. (1998), *The Smart Organization: Creating Value Through Strategic R&D*, Harvard Business School Press, Boston.

Mintz, S. L. (1998), "A Better Approach to Estimating Knowledge Capital," *CFO*, February, pp. 29–37.

Nevins, J. L., and Winner, R. I. (1999), *Ford Motor Company's Investment Efficiency Initiative: A Case Study*, Paper P-3311, Institute for Defense Analyses, Alexandria, VA, April.

OSTP (1998), *A National Obligation: Planning for Health Preparedness for and Readjustment of the Military, Veterans, and Their Families After Future Deployments, Presidential Review Directive* 5, Office of Science and Technology Policy, Washington, DC, August.

Rastikis, L. (1998), "Human-Centered Design Project Revolutionizes Air Combat," *CSERIAC Gateway*, Vol. 9, No. 1, pp. 1–6.

Rouse, W. B. (1991), *Design for Success: A Human-Centered Approach to Designing Successful Products and Systems*, Wiley, New York.

Rouse, W. B. (1998), *Don't Jump to Solutions: Thirteen Delusions That Undermine Strategic Thinking*, Jossey-Bass, San Francisco.

Rouse, W. B., and Boff, K. R. (2001), "Strategies for Value: Quality, Productivity, and Innovation in R&D/Technology Organizations," *Systems Engineering*, Vol. 4, No. 2, pp. 87–106.

Rouse, W. B., and Boff, K. R. (2003a), "Value Streams in Science and Technology: A Case Study of Value Creation and Intelligent Tutoring Systems," *Systems Engineering*, Vol. 6, No. 2, pp. 76–91.

Rouse, W. B., and Boff, K. R. (2003b), "Cost/Benefit Analysis for Human Systems Integration: Assessing and Trading Off Economic and Non-economic Impacts of HSI," in *Handbook of Human Systems Integration*, H. R. Booher, Ed., Wiley, New York.

Rouse, W. B., and Boff, K. R. (2004), "Value-Centered R&D Organizations: Ten Principles for Characterizing, Assessing and Managing Value," *Systems Engineering*, Vol. 7, No. 2, pp. 167–185.

Rouse, W. B., Boff, K. R., and Thomas, B. G. S. (1997a), "Assessing Cost/Benefits of R&D Investments," *IEEE Transactions on Systems, Man, and Cybernetics, Part A*, Vol. 27, No. 4, pp. 389–401.

Rouse, W. B., Kober, N., and Mavor, A., Eds. (1997b), *The Case for Human Factors in Industry and Government*, National Academy Press, Washington, DC.

Rouse, W. B., Howard, C. W., Carns, W. E., and Prendergast, E. J. (2000), "Technology Investment Advisor: An Options-Based Approach to Technology Strategy," *Information • Knowledge • Systems Management*, Vol. 2, No. 1, pp. 63–81.

Roussel, P. A., Saad, K. N., and Erickson, T. J. (1991), *Third Generation R&D: Managing the Link to Corporate Strategy*," Harvard Business School Press, Cambridge, MA.

Sage, A. P. (1977), *Systems Methodology for Large-Scale Systems*, McGraw-Hill, New York.

Sage, A. P., and Armstrong, J. (2000), *An Introduction to Systems Engineering*, Wiley, New York.

Sage, A. P., and Rouse, W. B., Eds. (1999), *Handbook of Systems Engineering and Management*, Wiley, New York.

Smithson, C. W. (1998), *Managing Financial Risk: A Guide to Derivative Products, Financial Engineering, and Value Maximization*," McGraw-Hill, New York.

Strassman, P. A. (1999), "Does Knowledge Capital Explain Market/Book Valuations?" *Knowledge Management*, September; www.strassman.com/pubs/km.

CHAPTER 44
METHODS OF EVALUATING OUTCOMES

Paula J. Edwards, François Sainfort, Thitima Kongnakorn, and Julie A. Jacko
Georgia Institute of Technology
Atlanta, Georgia

1 INTRODUCTION

An underlying goal of all human factors research is to produce research that is compelling and meaningful. Meister (2004) suggests that this is accomplished in part by selecting appropriate research objectives and through careful experimental design. In Chapter 11, designing an experiment to collect reliable, valid data relevant to examining the identified research objectives was discussed. In this chapter we focus on another aspect of sound experimental design: selecting appropriate measures and outcomes to collect during an experiment and selecting appropriate methods of analyzing the experimental outcomes and drawing conclusions from them.

As demonstrated in this chapter, selecting the appropriate data outcomes to collect and the appropriate methods of analyzing the outcomes go hand in hand. Certain characteristics of the data collected, such as the measurement type, drive which evaluation methods may be used to analyze the data, which in turn influence the types of conclusions that can be drawn from the results of that analysis. Other characteristics of the data, such as level of objectivity and specificity, affect the conclusions that can be made from the analysis and how or where those conclusions can be applied.

To help human factors researchers select the appropriate data outcomes and analysis methods for their research, we define characteristics of various outcome data and describe analysis and evaluation methods frequently used to analyze those outcomes. We begin by characterizing outcomes along a number of dimensions and discussing the implications these characteristics have both on selecting appropriate outcome measures and selecting appropriate evaluation method(s). Next, a variety of methods of evaluating both structured and unstructured outcomes data are described. For structured data, statistical analysis methods frequently used in human factors are presented. For each method, the purpose, assumptions, methods, and results are described as well as guidelines for interpreting the results and drawing conclusions. Next, several methods of analyzing unstructured data, such as content analysis, are presented. The chapter concludes with an example of applying methods of analyzing structured and unstructured data to the analysis of survey data. Applying this knowledge of outcome characteristics and evaluations methods should enable human factors researchers to produce research outcomes and conclusions that provide compelling and meaningful insight into the field of human factors.

2 TYPES OF OUTCOMES

The types of outcomes data that are produced in human factors research are as varied as the humans they strive to measure and analyze. To choose the appropriate data to collect to support the research objectives, we must first understand the nature and characteristics of the data. Outcomes data and measures can be classified along a variety of dimensions. The dimensions most relevant to selecting appropriate outcomes and analysis methods are (1) level of structure, (2) level of objectivity, (3) specificity, (4) measurement type, and (5) multiplicity.

2.1 Level of Structure

In the dictionary (Merriam-Webster, 2004), *structure* is defined as "something arranged in a definite pattern of organization." Depending on the research methods used, the resulting data may be structured, unstructured, or a combination of the two. The level of structure is one of the most significant factors driving the methods appropriate for analyzing your data.

Based on this definition of structure, *unstructured data* are not arranged in a definite pattern. Unstructured data include descriptions, observation notes, answers to open-ended questions, video and audio recordings, and pictures. Field studies typically generate large amounts of detailed data that reflect the richness of the work being observed (Wixon, 1995). This is usually unstructured data in the form of notes on observed actions and recordings of observed activities. Whereas the unstructured, or qualitative, data resulting from these observations contain invaluable detail on the activity being observed, the raw, unstructured data generally use more subjective methods of analyzing outcomes. Additionally, because unstructured data are so rich in detail, it can be more difficult to present findings clearly with this type of data. Because of their disadvantages, researchers often use unstructured data to produce structured data, as in the case of coding or classifying types of activities observed, or create figures and tables to summarize and communicate the details in a more structured manner. These methods of analyzing unstructured data are addressed in more detail in Section 5.

Structured data, conversely, has a definite pattern in the way it is collected and stored. For example, consider a survey that asks: What factor is most important to reducing errors? If the question were open-ended, the answer would be unstructured, since the subject could write in anything that comes to mind. On the other hand, if the question were multiple choice, the answer would have structure since all participants would have to choose from a limited number of available choices. Structured data typically lend themselves to quantification and are often referred to as *quantitative data*. Other examples of structured data include category classifications, rating scales, counts of events, and times/durations. When the data are structured, a wider range of analysis methods are available for evaluating the outcomes. In the case of structured data, the data items are sometimes also referred to as *measures*.

2.2 Level of Objectivity

Objectivity is "expressing or dealing with facts or conditions as perceived without distortion by personal feelings, prejudices, or interpretations" (Merriam-Webster, 2004). When it comes to human factors outcomes, the level of objectivity is a continuum with objective and subjective at either extreme and considerable gray area in between. At one end of the spectrum we have *objective data*, which are recorded "without the aid or expression of the subject whose performance is being recorded" (Meister, 2004, p. 81). For purely objective data, task performance is recorded manually or automatically, with minimal human involvement in the measurement. For example, time to complete a task, missed targets, height, and other physical measures are objective. *Subjective data*, on the other hand, are based on the subject or experimenter's opinions, values, and interpretations. Subjective measures rely on human perception, cognition, judgment, and experience (Wickens et al., 1998). Examples of subjective measures include responses from interviews/questionnaires, verbal reports of activities or thought processes, self-ratings, and other personal judgments. To illustrate the continuum of objectivity, consider verbal reports collected during an experiment. Concurrent verbal reports collected using a think-aloud protocol, although subjective, are more objective than retrospective reports, which are more objective than subjects' explanations of their behavior (Ericsson, 1998).

So which is better, objective or subjective? The answer, of course, depends on the research objectives. Although the level of objectivity has little effect on which methods may be used to analyze the data, it has a great effect on the validity and the generalizability of the research and the conclusions that can be drawn from the research. In human factors, the subjects' subjective opinion of their performance, confidence, or workload is sometimes of as much or more interest than their objective performance on the task. In many cases, it is useful for human factors researchers to collect *both* subjective and objective measures. If the results of both the objective and subjective measures support a particular conclusion, there is greater confidence in the validity of that conclusion. Therefore, when selecting outcome measures, researchers should consider whether objective measures, subjective measures, or a combination are best suited to supporting their research objectives and the conclusions they hope to draw from their research.

2.2.1 Preference-Based Measures

Within the set of subjective measures is a subset of measures, referred to as preference-based measures, which indicate a subject's likes or dislikes based on his or her experience and values. Survey questions that ask what a subject likes, prefers, or values, or that ask the subject to rate value or importance of items, are examples of preference-based measures. Similarly, comparisons and choices among options presented to subjects also capture preferences.

These measures are worth distinguishing from other subjective measures because they are notoriously difficult to measure reliably. This is caused in part by the nature of values and preferences and the uncertainty present in applying values to different sets of options and choices. People tend not to know their preferences in an unfamiliar situation, especially those with lasting consequences, and even when they are known, those preferences tend to be labile (Fischhoff et al., 1988). Often, preferences are "constructed" as people learn about or experience options. When preferences are known, they are extremely difficult to measure. Research has shown that people tend to construct their preferences during the process of elicitation, and that the method used to elicit their preferences can affect their final expressed preferences (Slovic, 1995). For example, studies in health care have shown that the use of different methods of eliciting preferences (Chapman and Elstein, 2000), physician's explanations of treatment alternatives (Mazur and Hickam, 1994), and framing of information about alternatives (positive, negative, or neutral) (Llewellyn-Thomas et al., 1995) affected patient's stated treatment preferences. Therefore, when preference-based subjective measures are of interest, special care should be taken during experimental design to ensure the validity of these measures. The researcher must take steps to ensure that the way the preference-based measures are collected does not bias the results.

2.3 Specificity

Specificity indicates whether a measure refers to a particular task, industry, or situation, as opposed to being applicable to a variety of areas. Again, this characteristic is a continuum ranging from specific to generic. *Specific measures* are tailored to measuring a phenomenon of interest but have little generalizability to other phenomena. At the other end of the scale, *generic measures* are generalizable to a variety of tasks or situations. However, it can be quite difficult to define generic measures that are sensitive enough to capture the phenomenon of interest in a variety of situations that may be quite different. For example, in human–computer interaction (HCI), target highlight time (THT) is a measure used to quantify the salience of feedback received from the interface as the user completes a drag-and-drop task. THT, a specific measure for drag-and-drop tasks, is much more sensitive than total task time to measuring reaction time to feedback because task time is strongly affected by a number of other factors. Task time is a generic measure, and although it is less sensitive in this case, it has the advantage of being generalizable to other computer tasks. So task time could be used to compare drag-and-drop performance to performance on a different computer task, such as point-and-click.

Another example relates to measuring health outcomes resulting from various treatments. A generic measure would capture a range of health status dimensions (e.g., physical function, mental health, social function). On the other hand, a disease-specific measure will zoom in on specific symptoms and implications related to a disease of interest (e.g., lower back pain).

The specificity of a measure affects the conclusions drawn from the data. If very specific measures are used, the results may not generalize to a broad enough range of situations, thereby limiting the scope and applicability of the conclusions. Conversely, if the measures are too broad, they may not be sensitive enough to detect the phenomenon of interest through the statistical analysis methods used. Therefore, it is crucial to reconcile the specificity of the measures with the research objectives as the study measures are chosen. In some cases, researchers choose to collect a combination of specific and generic measures. In selecting generic measures, it is recommended that researchers look at industry standard metrics and measures used in related research to enable comparability across studies.

2.4 Measurement Type

In structured data, the measurement type determines to a large degree the statistical methods that may be used to analyze the data. The measurement type indicates the amount of information the measure contains with respect to the value being measured (Sheskin, 1997). The four measurement types, in order from least to greatest information provided, are nominal/categorical, ordinal/rank order, interval, and ratio. These categories are defined as follows (Sheskin, 1997; Argyrous, 2000):

1. *Nominal/categorical*: indicates the category to which an item belongs. The category may be represented as a number or text, but even if the category is represented as a number, it cannot be manipulated mathematically in a meaningful way. In human factors, nominal data would include gender, race, and part number. For each of these measures, it is not possible to add or rank order them, since the name/number is used solely for identification purposes.

2. *Ordinal/rank order*: indicates rank orders using numbers but does not provide information on the magnitude of difference between two ranks. An example of ordinal data from human factors is participant rankings of the preferred tool or method. If the participant is asked to rank three proposed tools, from the one they prefer most to the one they prefer least, those rankings (1 through 3) do not indicate if the participant strongly prefers the first ranked system to the second, or only slightly prefers the first to the second.

3. *Interval*: indicates both order and magnitude of difference between values by providing a number along a scale that has intervals of equal distance between equal values on the scale. This means that a difference of 10 between two values represents the same magnitude of change regardless of whether the initial value was small or large. However, interval scales arbitrarily assign a zero score rather than having a true zero. For example, consider a question that asks participants to rate on a scale from 1 to 10 how hard they had to work physically to complete a task. A 1-point increase in "work" is the same 1-point increase in work whether the initial rating is 3 or 7. However, there is no true measurement of zero work; 1 on the scale was selected arbitrarily to represent an extremely low workload.

Table 1 Analyses Appropriate for Each Measurement Type

Measurement Type	Compare Counts/ Frequencies	Greater Than/ Less Than Comparisons	Ratio Comparisons	Mathematical Operations	Statistical Analysis
Nominal	×				
Ordinal	×	×			
Interval	×	×		×	×
Ratio	×	×	×	×	×

4. *Ratio*: like interval measures, provides a number along a scale that indicates both order and magnitude of difference between values. Unlike interval measures, however, ratio measures have a true zero point. Because there is a true zero point, it is possible to compare scores by taking their ratios. For example, age has a true zero point, so one can meaningfully say that a person who is 40 is twice as old as someone who is 20. In contrast, it is not appropriate to compare ratios of work in the previous example. Since the true zero point for work is unknown, saying that a task with a work rating of 10 is twice as much work as a task with work rating of 5 is inaccurate.

The measurement type has an enormous influence on the methods that can be used to analyze the data. Table 1 indicates which type of analysis is appropriate for each measurement type. Several of these analysis methods are discussed in further detail in Section 4.

2.5 Dimensionality

Another important characteristic of structured outcome data is the degree of dimensionality. Some outcomes can be captured directly with one overall measure (e.g., the time required to perform a given task, or how one person feels about his or her own health at a point in time as measured by a global 5-point Likert scale with five possible answers: excellent, very good, good, fair, or poor). However, many outcomes need to be captured through a number of dimensions. In the case of multiple dimensions, a complex measurement task often then consists of appropriately aggregating the various dimensions into one overall quantitative measure of the outcome of interest. For example, although performance on a job clearly contains many different aspects, one may wish to combine the performance on different aspects into an overall assessment of global performance. To represent overall performance adequately, one needs to understand the potential relationships between the various elements.

Many scales are constructed based on different items and thus appear to be inherently multidimensional. However, it is important to differentiate between a scale and an index. A *scale* typically is comprised of multiple items whose values are caused by an underlying construct (or latent variable) (Bollen, 1989). On the other hand, an *index* consists of several cause indicators or individual variables that together determine, or at least strongly relate to or influence, the level of the construct of interest (Devellis, 1991). Thus, in a scale, the items that comprise the scale typically correlate to each other, and multiplicity of items increases the overall reliability of the scale. On the other hand, for an index, constituent variables may be independent of each other (e.g., physical function versus social function, both important cause indicators of overall health). Creating an index may or may not be important in terms of analyzing and understanding outcomes. In terms of analyzing multiple outcomes, however, an index is desirable if one overall metric is desired to represent the outcomes. Furthermore, an index allows for unidimensional analytical approaches to be used, whereas multidimensional outcomes typically require multivariate analysis methods.

2.6 Summary: Selecting Appropriate Outcomes Data

As demonstrated in the preceding sections, a variety of outcome characteristics influence the methods used to analyze outcomes and the conclusions that can be drawn from those outcomes. Therefore, selecting the appropriate outcomes data to collect begins with the definition of the research objectives. However, it is important to consider the research objectives in broad terms when developing a data collection and analysis plan for human factors research. Instead of just thinking in terms of the phenomenon to be studied or the intervention to be evaluated, also think about the goals for communicating the outcomes before determining which outcomes data to collect. If these objectives and goals are not well established up front, it is unlikely that you will by chance collect data that supports an ill-defined research objective.

Once the research objectives are established and well understood, the next step is to select what data outcomes to collect. It is always useful to begin by looking at the domain literature to identify frequently used data and measures and any gaps in those outcomes. Using measures consistent with other research is useful in that it is a prerequisite for having results that are comparable with other studies. However, do not be afraid to bridge any gaps that exist in the literature by creating new measures. For example, there may be a need for a new, more sensitive measure of a phenomenon of interest or a more generalizable measure that enables comparisons across multiple related tasks.

Next consider the types of conclusions that should be drawn to support the research objectives. What level of objectivity is required? What specificity? What analysis method(s) will enable drawing those conclusions? Do the required analysis methods impose any restrictions on the structure or measurement type of the outcomes data? All of these questions should be answered to develop a data collection and analysis plan for the research study. Addressing these topics up front helps ensure that the outcomes data collected will be valid and credible and will support the research objectives identified. Of course, this does not guarantee that the outcomes will always produce the expected results—human factors data are always full of surprises!

3 MEASUREMENT OF OUTCOMES

Measurement is a fundamental activity of science. As Krantz et al. (1971, p. 1) explain: "When measuring some attribute of a class of objects or events, we associate numbers (or other familiar mathematical entities, such as vectors) with the objects in such a way that the properties of the attribute are faithfully represented as numerical properties." Although this process can be relatively straightforward for physical measures such as length or density, it can be very difficult for psychosociological constructs, such as stress or health status. Whether one measure is created or multiple measures are used, fundamental psychometric properties need to be tested properly before using the measurement system created. These fundamental properties are described briefly in the next section. In addition, we describe briefly methods that can be used when multidimensional outcomes need to be aggregated into an overall scale or an index.

3.1 Psychometric Properties

There are two overall fundamental properties in measurement: reliability and validity. Ghiselli et al. (1981) consider reliability a fundamental issue in psychosociological measurement and in the context of developing scales, define it as the proportion of variance attributable to the true score of the latent variable. Although several terminologies exist, there are essentially three types of validity in scale development: content validity, criterion-related validity, and construct validity. *Content validity* refers to the extent to which a set of items selected in a scale covers the content domain. *Criterion-related validity* refers to the extent to which the scale created relates to an existing criterion or "gold standard." Finally, *construct validity* refers to the extent to which a scale "behaves" as it is expected to, according to theoretical relationships with existing constructs, where these relationships have been formulated prior to the development of the scale. A number of techniques exist to ascertain the reliability and validity of scales (see, e.g., Devellis, 1991).

3.2 Multidimensional Outcomes

As mentioned above, many (complex) evaluation problems are by nature multidimensional and require the construction of a scale or an index. Scales and indices differ in fundamental ways and require very different techniques for their development. Devellis (1991) provides a structured eight-step process as a guideline for scale development:

1. Determine clearly what to measure.
2. Generate an item pool.
3. Determine the format for measurement.
4. Have the initial pool reviewed by experts.
5. Consider the inclusion of validation items.
6. Administer the items to a development sample.
7. Evaluate the items.
8. Optimize the scale length.

As part of the final step, data reduction techniques for creating scales are commonly used and are described in greater detail in Section 4.3.

As opposed to creating a scale, multiattribute utility theory (MAUT) can be applied directly to create an index. Edwards and Newman (1982, p. 10) distinguish "four different classes of reasons for evaluations: curiosity, monitoring, fine tuning, and programmatic choice. ... These reasons for evaluation share two common characteristics that make MAUT applicable to them all. The first is that, implicitly or explicitly, all require comparison of something with something else. ... The second characteristic is that [entities to be evaluated] virtually always have multiple objectives." Thus, MAUT is applicable to many situations where multiple outcomes need to be aggregated into an overall index.

These multiple objectives lead to the identification of what Keeney and Raiffa (1976) call *evaluators* or *attributes* and purport to describe completely the consequences of any of the possible actions or entities to be evaluated. According to Keeney and Raiffa, each attribute itself must be comprehensive and measurable, and the set of attributes describing the consequences must be complete, operational, decomposable, nonredundant, and minimal (p. 50). Often, these attributes can be structured meaningfully into a hierarchy. On top of the hierarchy is the all-inclusive objective, which indicates the reason for being interested in the problem in the first place. Figure 1 illustrates such a hierarchy in the context of creating an index comprising multiple outcomes to measure overall health.

MAUT provides a way of aggregating the information describing each entity on the multiple attributes into a summary measure or index. As described by Edwards and Newman (1982, p. 79), "the goal of MAUT is to come up with one number for each [entity] of evaluation, expressing in highly concentrated form how well that [entity] does on all evaluative dimensions. But whether that much compression is appropriate depends very much on the purpose of the evaluation." MAUT is widely used to combine multiple outcomes for at least three reasons. First, it is very appealing and convenient to have a summary index. It is especially useful to have a summary index when the objective of the evaluation is to monitor changes, to compare alternatives, or to assign another quantity

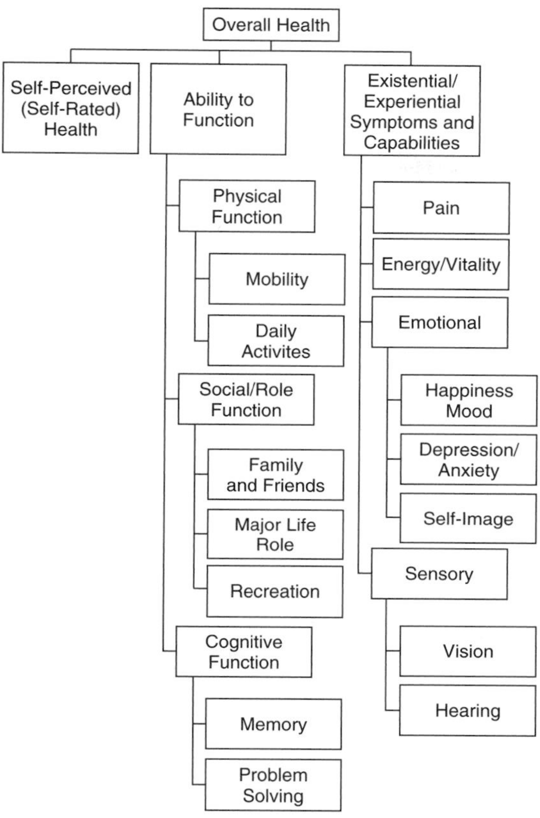

Figure 1 Hierarchy of attributes to measure overall health. [Reprinted with permission from Fryback (1998), 1998 by the National Academy of Sciences, courtesy of the National Academies Press Washington, DC.]

in proportion to the index score (e.g., assigning monetary rewards as a function of performance). Second, as pinpointed by von Winterfeldt and Edwards (1986), in practice, the additive and multiplicative models are the only workable ones; they are relatively simple and therefore easily accepted forms of aggregation. Third, MAUT is grounded in theory: The standard decomposition theorem for additive and multiplicative utility functions is seen as a determinant, to provide theoretical justification for the development of such evaluative models.

4 ANALYSIS OF STRUCTURED OUTCOMES DATA

Both unstructured and structured outcomes data result from human factors research. However, our discussion of analysis methods will begin by focusing on analyzing structured outcomes data, or measures, since the results from these analytical methods are frequently the primary focus of research results presented in the human factors literature. There are a variety of statistical and graphical methods for exploring and analyzing structured data. In this section we review methods that are frequently used by human factors researchers. The methods are grouped according to the objective of the analysis. For each method we discuss when it is appropriate to use the method, the type of results produced by the method, and how to draw conclusions from the results. Delving into the statistical details of these methods is beyond the scope of this chapter, so suggested statistics reference(s) are also provided for each method. The decision model presented in Figure 2 is provided to help researchers decide which statistical tests are appropriate based on their analysis objectives and characteristics of the data.

4.1 Exploring Your Data

As Box et al. (1978) point out, when doing statistical analysis on outcomes data, it is important not to forget what you know about the subject matter in your field. One way to build subject matter knowledge is to explore your data prior to completing any statistical tests. Human factors data can be quite different from data found in other domains. One of the distinguishing characteristics of human factors data is that it tends to be very noisy. This is especially true when the population being studied is very heterogeneous. Because the people being studied vary in physical and mental capabilities, expertise, and other factors, their performance on the same task will naturally vary. To compound this, even the same person acting under varying environmental factors may perform differently. The noise inherent in human factors data can reduce the power of statistical methods to detect effects. This is why it is very important to get to know your data before you start running statistical tests. By exploring and getting to know your data, you develop an initial understanding of potential occurrences and trends that can be used to validate and spot potential problems in the statistical analysis.

4.1.1 Exploring the Data Distribution

There are several graphical techniques useful for getting to know your data. For example, box plots (Figure 3) and histograms (Figure 4) may be used to get an overall feel for the distribution of the data. Both of these types of plots illustrate the minimum value, maximum value, and overall distribution and variability of the data. Additionally, a box plot indicates quartiles and the median value and highlights data points that are outliers or extreme values.

Identifying extreme values and outliers is important, as these data points are far outside the expected range of values, which is determined based on the variability of the data, measured by the standard deviation. These data points should be examined since they may be caused by a data collection or related error. For example, the value may be the result of a data entry error. In some cases, an unexpected event may have occurred during the trial, causing the data to be invalid. For example, in a task that is being timed, if the subject starts the task, then pauses to ask the experimenter a question, the task time for that trial may be artificially inflated and may need to be excluded from the data

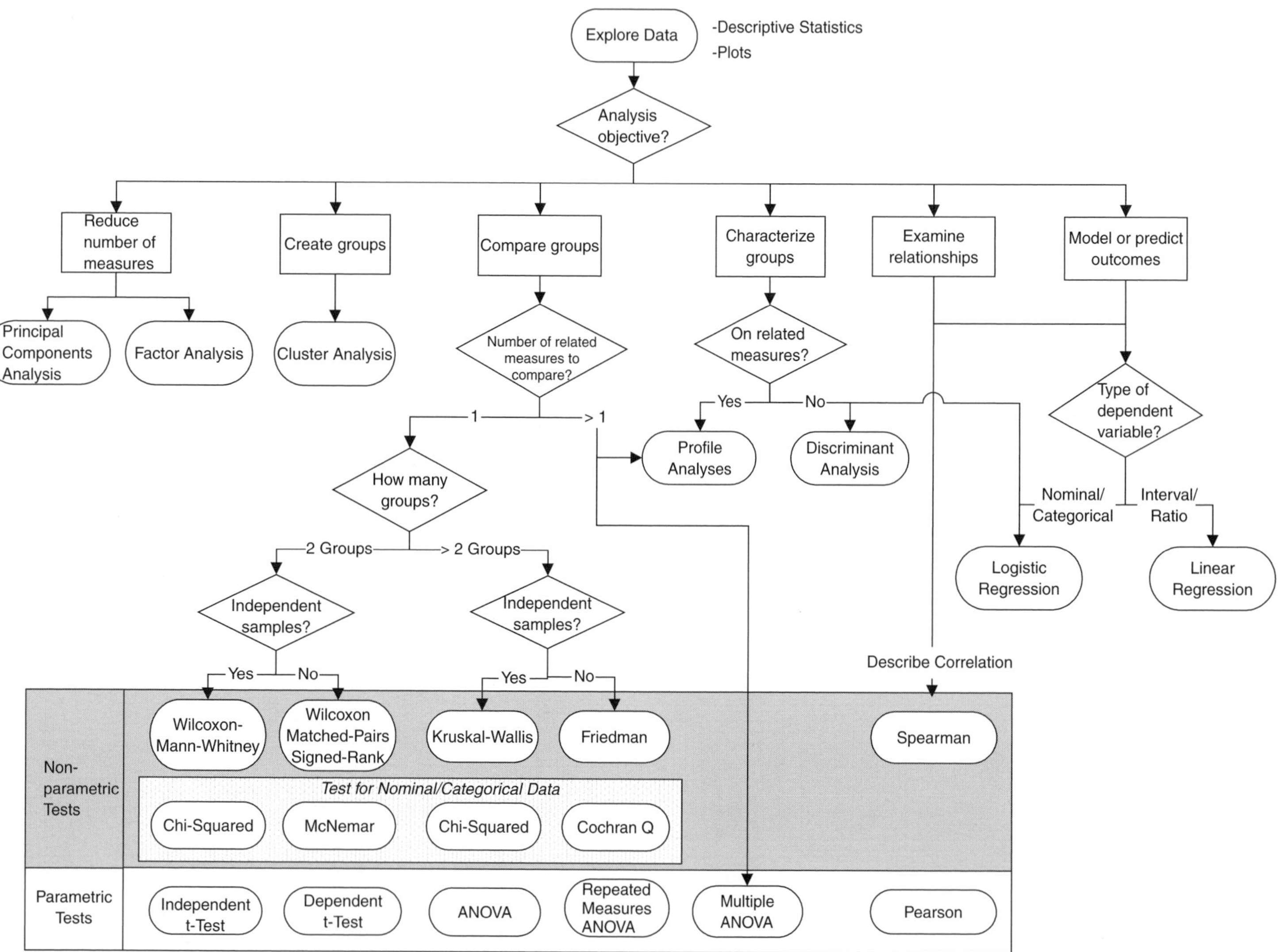

Figure 2 Decision model for analyzing structured measures.

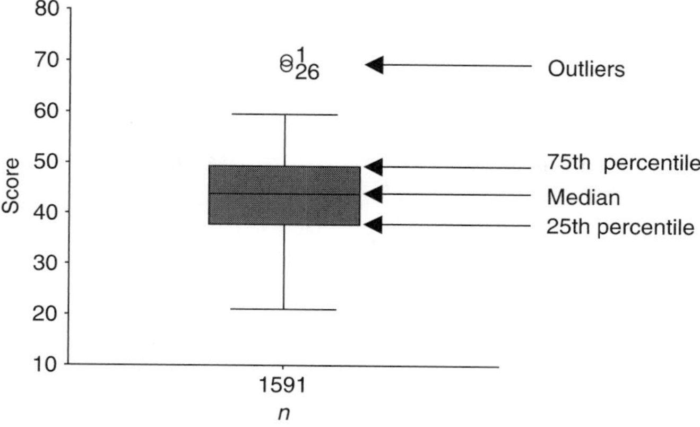

Figure 3 Box plot.

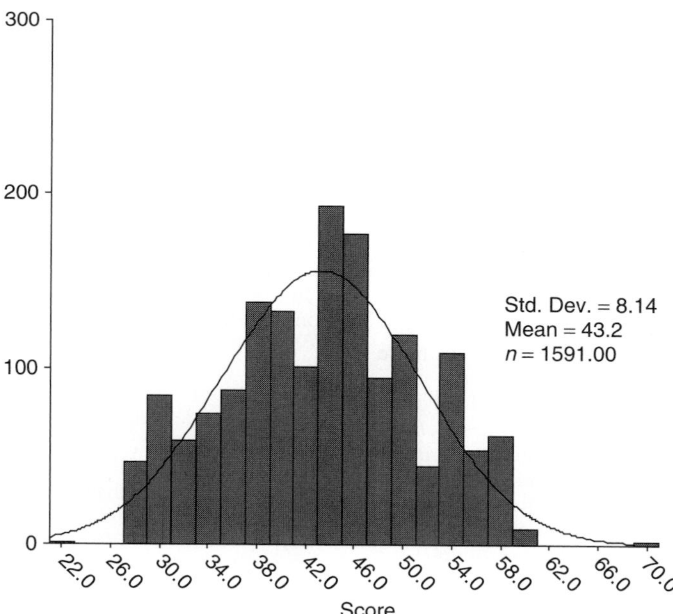

Figure 4 Histogram.

analysis. In another example, if all of the extreme values and outliers are from the same participant, there may be some characteristic of that participant that is causing the person to perform very differently from the others. For example, one participant may have more experience with the experimental task, causing the person to perform much better than the other participants. The researcher needs to be aware of the cause of this difference and make an educated decision about whether or not it is appropriate to include the participant in the data analysis. If participant(s) or trials are excluded from analysis, the reason for excluding them should be presented when reporting the study findings, to ensure that correct conclusions are drawn from the study.

4.1.2 Descriptive Statistics

Descriptive statistics are used to convey information about the central tendency and dispersion of the data set. One frequently reported measure of central tendency is the *mean*, or average, value. However, one must be careful when reporting means because extreme values, which should be identified in your data exploration, can distort the mean. For example, if a researcher wanted to examine the net worth of a group of 10 people selected at random and Warren Buffet happened to be one of those people, Warren Buffet's net worth (in the billions) would distort the mean, making it in the hundreds of millions even if the net worth of all the others in the group was less than $100,000. Because the mean is vulnerable to this distortion, it is often useful to look at the median and mode as well. The *median* is the value that splits the data set in half, with 50% of the values greater than that value and 50% less than that value. To find the median, the values are arranged in order from least to greatest; the middle value is the median. If there are an even number of values, the median is the average of the two middle values. The *mode*, on the other hand, is the value that occurs most frequently.

In addition to examining measures of central tendency, examine the dispersion of the data as well. Dispersion is important for two primary reasons. First, dispersion, or variability, affects the power of statistical tests to detect significant effects, which is discussed in more detail later in the chapter. Second, in human factors, researchers are sometimes interested in reducing variability to gain more consistent performance over time. As a measure of dispersion, researchers frequently report the *variance* (σ^2) or *standard deviation* (σ). These two values are directly related, as the standard deviation is the square root of the variance. The variance is calculated based on the square of the difference between each value and the mean and therefore increases as more values are a greater distance from the mean. Statistical packages and spreadsheet programs calculate this value, so the mathematical equation is not provided here. However, any statistics book provides the mathematical definitions of these measures.

When discussing dispersion, it is also often worthwhile to examine the *range* of the values, defined by the *minimum* and *maximum values*, and *percentiles*. The X percentile is defined as the value at which $X\%$ of the values fall at or below that value. The range and quartiles (the 25th, 50th, and 75th percentiles) are depicted in a box plot, as discussed in Section 4.1.1.

References for further details: Sheskin (1997), Field (2000)

4.2 Statistical Analysis Methods

A variety of statistical analysis methods can be used to make inferences about structured outcomes measures. The purpose of these statistical methods is to help you understand and draw conclusions from the outcomes measures. The appropriate statistical test to use depends on the analysis objective and on characteristics of the data. The reader is again referred to the decision model in Figure 2 for assistance in choosing the appropriate method(s). The methods discussed in this section may be used to accomplish the following objectives:

- *Comparing groups*: t-test, Wilcoxon–Mann–Whitney, Wilcoxon matched-pairs signed-rank, chi-squared, ANOVA, repeated measures ANOVA, Kruskal–Wallis, Friedman
- *Characterizing groups*: profile analyses, discriminant analysis
- *Creating groups*: cluster analysis
- *Describing and modeling relationships*: correlation, linear regression, logistic regression

Since this is a book on human factors and ergonomics, not statistics, formulas and detailed mathematical explanations are not provided for these statistical methods. Since most statistical packages provide functions that automate these calculations, the focus is on understanding conceptually how each method works, when it is appropriate to use it, and how to understand and interpret the results in order to draw conclusions. Before delving into the methods, several general statistical concepts need to be reviewed.

4.2.1 General Statistical Concepts

In inferential statistical methods, the researcher seeks to determine whether or not phenomena observed in the data are caused by random variation in the data. If it is not caused by random variation, we can make inferences about the nature and cause of those phenomena. In making these inferences, several general statistical concepts play a role both in designing experiments and in interpreting and communicating results. In this section we review these concepts in the context of an example experiment in which a researcher wants to compare the time to complete a task using the current method to the time using a proposed new method. The researcher collects the data and computes the mean time to complete the task under each method. The new method has a lower mean time, but how can they be sure that the lower time is actually caused by using the new method instead of just by chance?

Type I and II Errors Inferential statistical methods typically frame the research question as a hypothesis, which is then tested to determine whether or not it is true. In our example, the researcher's hypothesis would be that there is a difference between the mean times of the two methods. When testing a hypothesis, researchers are vulnerable to making two types of errors. First, if the researcher concludes that the hypothesis is true when the result is actually caused by chance, he or she has made a *type I error*. In our example, the researcher would make a type I error if he or she concluded that the new method reduced the time to complete the task, when in reality the lower mean time was only caused by chance variation in the data. To avoid these errors, researchers typically limit the type I error rate (α). In human factors research, α is usually set at 5%, which means that type I errors will occur at most in 5 of every 100 tests.

In the second type of error, *type II error*, the researcher concludes that the result was caused by chance when, in reality, the hypothesis is true. If our researcher concluded that there was no difference in the times for the two methods when the proposed method is actually faster, they would commit a type II error. The likelihood of making a type II error (β) is an indicator of the *power* ($1 - \beta$) of the test, its ability to detect a difference when one actually exists. The type I error rate (α), the sample size (n), the mean, and the variance of the data determine the power of the test (Neter et al., 1996). For researchers designing experiments, understanding this relationship between α and the power of the test is crucial. Since the industry practice is to control α at no more than 0.05, the only way to increase the power of a test is to collect more data, increasing n, or control the experiment to reduce variability in the data, which can be difficult with human factors data, as mentioned in the discussion on experimental design in Chapter 11. Some statistical packages, such as MiniTab, provide calculators that estimate the sample size needed to achieve the desired power given an estimate of the variance and mean. When designing an experiment, it is highly recommended that you estimate the power to ensure that enough data are collected to achieve the desired power for statistical testing.

Experimentwise Error In situations where more than two groups are being examined or several, possibly related, dependent measures are being examined, researchers need to be careful to manage the *experimentwise error* rate to ensure the validity of their results. Experimentwise error is the combined type I error rate for all the statistical tests being performed. Take a simple example. A researcher is comparing the task time of three experimental groups. If the researcher uses the t-test to compare each test to the other tests, three t-tests are completed, comparing group 1 to 2, 1 to 3, and 2 to 3. If the researcher sets the type I error rate (α) to 0.05, each test has a 0.95 chance that there will be no type I errors. Since there are three tests, each of which is assumed to be independent, the experimentwise probability that no type I error is $0.95 \times 0.95 \times 0.95 = 0.857$. This means that the experimentwise error rate $(1 - 0.857)$ is 0.143. In other words, 14.3% of the time there will be a type I error!

However, statisticians are aware of this phenomenon, so tests are available that account for it and control the experimentwise error to the desired α level. For example, this is why ANOVA is used to compare more than two groups instead of multiple t-tests, as demonstrated in the example. ANOVA examines all of the groups together to determine whether any of them differ significantly at the α level of experimentwise error. Once it is established that there is a significant difference at the experiment level, post hoc comparisons are completed to determine which pairs are different. These post hoc comparisons adjust to account for the number of comparisons being performed.

***p*-Values** Researchers generally provide p-values when reporting the results of statistical analyses. The p-value indicates the likelihood of making a type I error. In other words, the p-value is the probability that the researcher will conclude that the hypothesis is true when the result is actually caused by chance. To continue the task time comparison example, if the statistical test resulted in $p = 0.25$, it would indicate that there is a 25% chance that the difference in the mean task time is simply caused by random variation in the data. This is clearly higher than the 5% α threshold that is generally accepted, so in this case, the researcher would conclude that there is not evidence to support the conclusion that there is a difference in the task times. This means that either there truly is no difference in the task times or the test did not have enough power to detect the difference, in which case the researcher could collect additional data to increase the power of the test. On the other hand, if the p-value were 0.03, it would indicate only a 3% chance that the difference in times is due to chance, and the researcher could conclude that the new process affects the task time.

Confidence Intervals When comparing groups, the p-value indicates whether or not there is a difference between the groups but gives no indication of the magnitude of this difference. Confidence intervals (CIs) provide this information, making the results of the study easier for practitioners to interpret and apply. CIs indicate the magnitude for a value such as the mean or the mean difference between two groups. Because the mean is calculated from a sample of data, it merely provides an estimate for the actual population mean. Consequently, with a different sample of data, the estimate of the mean, although close to the first mean, is unlikely to be exactly the same. A CI provides a range in which the actual mean falls. The confidence interval is for a certain α level, typically $\alpha = 0.05$, and is referred to as the $(1 - \alpha)$-level CI. The interval is calculated using the mean and the variance (Wu and Hamada, 2000). In a 95% confidence interval, there is a 95% chance that the actual mean falls within the upper and lower bound of the interval. How does knowing this help us?

In our example, the researcher might want to understand the magnitude of the difference in task time between the current method and the method proposed. If the 95% CI for the difference in task time ($\text{time}_{\text{current}} - \text{time}_{\text{proposed}}$) was found to be 33 to 45 seconds, it would indicate that the proposed process saves 33 to 45 seconds on the task.

Statistical versus Practical Significance Just because an analysis indicates that a result is statistically significant does not mean that it has practical significance as well. For example, in the CI presented previously, we reported that the time savings achieved by using the proposed method is 33 to 45 seconds. If the task normally takes 60 minutes to complete, this time savings amounts to a mere 1% saving, so it may not be cost-effective to change to the new method. On the other hand, if the task normally takes 5 minutes, the time savings amounts to an 11 to 15% savings, a

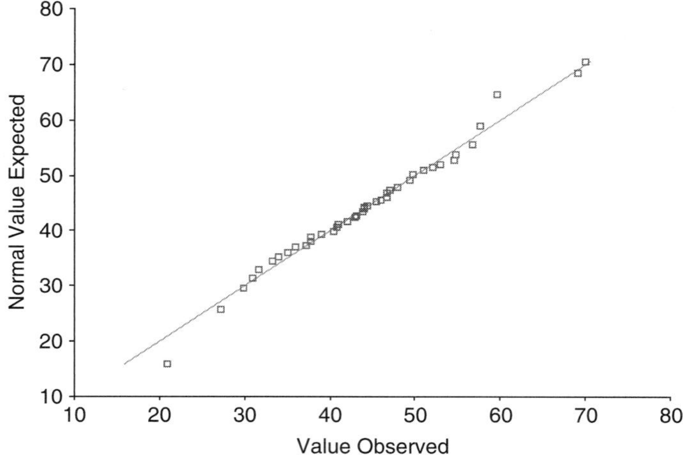

Figure 5 Normal probability plot (normally distributed data).

much more practically significant result. This illustrates why it is important for researchers to understand and report practical significance as well as statistical significance. Reporting practical significance makes it easier for people less familiar with statistics and human factors methods to understand and apply the results, increasing the use and impact of the results.

Parametric and Nonparametric Tests In inferential statistics, many tests are classified as parametric or nonparametric. This distinction is made because *parametric tests* are based on certain assumptions about characteristics of the distribution of the data, whereas *nonparametric tests* make no such assumptions about the distribution of the data (Sheskin, 1997; Sprent and Smeeton, 2001). For example, when comparing two independent samples, the t-test, a parametric test, or the Mann–Whitney U-test (a nonparametric test) may be used to analyze the data. Which should the researcher choose?

In general, parametric tests are appropriate for interval and ratio data, whereas nonparametric tests are used for categorical/nominal and/or ordinal/rank order data. With interval and ratio data, it is always preferable to use parametric tests because they have more statistical power. However, when using parametric tests, it is important to verify that the underlying assumptions of the test are met. Although many of these tests are robust enough to handle some departures from the assumptions (Sheskin, 1997; Newton and Rudestam, 1999), large departures from the assumptions may make the test inappropriate for use with the given data set. For example, the t-test assumes that the data being analyzed are characterized by a normal distribution (normality assumption) and that the variance of the underlying population is homogeneous (homogeneity of variance assumption). In the discussion of each test, the assumptions are listed and, if appropriate, ways to validate those assumptions. However, since many parametric tests assume that the data or the error between the data and a model of the data are normally distributed, it is worth recalling the normal distribution. Figures 3 and 4 are a box plot and histogram, respectively, of data that resemble a normal distribution. The reference line in the histogram displays how data with a normal distribution would be shaped—with a large number of values in the middle near the mean and progressively fewer occurrences as you move farther away from the mean. In the box plot, the line representing the median is located in the center of the box and the box and lines are fairly balanced. All of these things indicate that the data resemble a normal distribution. To confirm this, examine the normal probability (Q–Q) plot for these data (Figure 5). Since most of the data points fall along a straight line, this also indicates that the data resemble a normal distribution. In this case, we could conclude that the assumption of normality is met.

When significant violations of the parametric test assumptions are observed, the researcher has two alternatives. One alternative is to transform the measure (y) using one of the power transformations such as $\log(y)$ or $1/y$, so that the transformed measure meets the test assumptions. However, when using a transformation, researchers should take care to ensure that the results of the analysis are interpretable. This means that the results of the analysis on the transformed measure must still be meaningful (e.g., if $y =$ death rate, then $1/y =$ survival rate and the $1/y$ transformation has an interpretable meaning). Also, the researcher must take care to present the results in a manner that is clear, even to those with limited statistical knowledge. Many statistics books, (e.g., Neter et al., 1996; Newton and Rudestam, 1999; Wu and Hamada, 2000 provide details on how to select the appropriate data transformation. Refer to one of these books for more information on this topic.

The second alternative when a parametric test is not appropriate is to use an equivalent nonparametric test. In our t-test example, if the data set significantly violated the normality and homogeneity of variance assumptions, the researcher could instead use the Mann–Whitney test to compare the two groups. This test is based strictly on the rank order of the data points, so it makes no underlying assumptions about the distribution of the data or its variance. However, because it is based on the ranks, it sacrifices the additional information provided in the interval/ratio data, making it less powerful. This trade-off between the power provided by parametric tests and the absence of data distribution assumptions in nonparametric tests is crucial for researchers when selecting the appropriate test for their data.

Note that there is some debate in the applied statistics community over whether a parametric or a nonparametric test should be used when there are departures from the parametric test assumptions. However, as Sheskin (1997) demonstrates with examples in his book, frequently when both a parametric test and its nonparametric counterpart are applied to the same data set, they result in the same or similar conclusions. Therefore, the prudent researcher should use the decision process illustrated in Figure 6 when trying to decide whether a parametric or a nonparametric test is appropriate.

4.2.2 Comparing and Creating Groups

Now that the review of general statistical concepts is complete, we can focus on the methods used to accomplish the researcher's analysis goals. In human factors research, researchers are frequently interested in examining groups of participants, items, or events. For preexisting groups, the researcher may want to compare groups on some dependent measure of interest or characterize those groups based on a number of different measures. In other cases, the researcher may be interested in creating new groups of related participants, items, or events, to develop a better understanding of relationships and patterns in the data.

Of these three general analysis goals related to groups, comparing two or more experimental groups on a dependent measure of interest is the one most frequently seen in human factors research. For example, a researcher may be interested in examining worker efficiency, measured by task time, when using each of several available tools. After designing the experiment and collecting the data, as described in Chapter 11, the researcher must use the appropriate statistical test to analyze the data and draw conclusions about the influence of each tool on task time. In the first part of this section we review statistical tests used to compare two groups (e.g., t-test, Mann–Whitney U-test) and to compare two or more groups (e.g., ANOVA, Friedman test).

After the review of methods for comparing groups, two methods are presented for characterizing groups. The first is profile analyses, which characterize each group on a set of related measures. Profile analyses may also be used to compare the profiles of each group to determine whether or not they differ. The second method is discriminant analysis, which examines

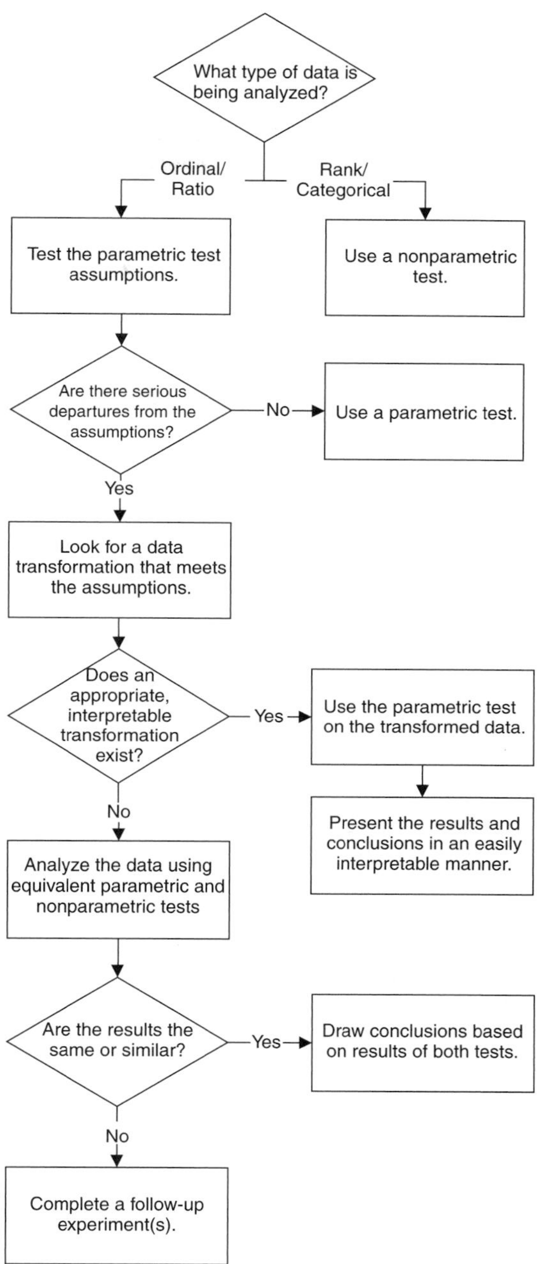

Figure 6 Decision model for choosing between use of a parametric or a nonparametric test.

independent measures for items in each group to determine which measures can be used to separate the groups. These identified discriminant measures are then used to define rules for classifying or predicting which group a new item will belong to given its discriminant measures. The section concludes with cluster analysis, an exploratory method used to create groups. Cluster analysis examines characteristics of individual items and creates groups such that items in each group are highly similar to each other and highly different from items in other groups.

Comparing Two Groups The tests presented here are used to determine if the variable of interest differs between two groups. The appropriate test to use depends on three factors. First, are the two groups dependent or independent? Referring to the experimental design concepts in Chapter 11, within-subjects (repeated measures) designs are dependent because each subject receives each experimental condition. This means that the values for each condition are related since they come from the same participant. In contrast, in between-subjects designs different sets of randomly selected participants receive each experimental condition and are assumed to be independent of each other. The second factor in selecting the appropriate test is whether or not the data adequately meet the assumptions of the test, as described in the "Assumptions" section for each test. The third factor is measurement type of the dependent measure. Table 2 identifies these factors and the appropriate test for each. Note that since the McNemar test is not used frequently compared to the other tests, it is not discussed in detail in this chapter. However, more information about this test may be found in Sheskin (1997) or Sprent and Smeeton (2001).

***t*-Test** The t-test is a parametric test used to compare means of a dependent measure for two groups. There are two forms of the test: (1) the *dependent* or *paired t-test*, used when the same subjects received both experimental conditions; and (2) the *independent t-test*, used when different subjects are assigned to each experimental condition. Both tests are used to examine the question: Do the means of the dependent measure differ in the two populations represented by the groups? If the difference between the groups is small, it may only be caused by chance variation in the data, but if the difference is large, we may conclude that there is, indeed, a significant difference between the two groups. The primary difference in the two tests is how they examine the values.

Table 2 Tests for Comparing Two Groups

	Parametric:	Nonparametric	
	Interval/ Ratio	Ordinal/ Rank	Categorical
Dependent groups	Dependent (paired) t-test	Wilcoxon signed rank test	McNemar test
Independent groups	Independent t-test	Mann-Whitney U-test	Chi-squared test

Assumptions The assumptions of both forms of the t-test are that (1) the dependent variable of each underlying population is normally distributed, and (2) the two populations have equal population variances (i.e., homogeneity of variance). The independent t-test also assumes that the each group is selected randomly from the population it represents (i.e., the two groups are independent of each other). Note that the assumption of homogeneity of variance can be tested using Hartley's $F_{\max}$ test for homogeneity of variance.

Methods and Results Both forms of the t-test calculate a test (t) statistic that is used to determine the p-value. Conceptually, the t-statistic is the ratio of the difference in the means to the standard error of the difference. The difference in the two tests is that the independent t-test examines the means of each group [mean(Y_1) − mean(Y_2)], while the dependent t-test examines the mean difference of the pairs of results for each participant [mean$(Y_{i1} - Y_{i2})$, where i = participant number]. When completing a t-test using a statistical package, the results typically provide the t-statistic and the associated p-value. The p-value indicates the likelihood that the difference in the means occurred by chance. Therefore, the smaller the p-value is, the greater evidence the test provides that the two groups are indeed different.

Drawing Conclusions In human factors literature, p-values less than or equal to 0.05 are usually considered significant. Therefore, if the p-value is 0.05 or less, the researcher can conclude that evidence indicates that there is a significant difference in the dependent variable between the two groups (experimental conditions). However the p-value does not indicate the size of the difference between the groups; therefore, it is recommended that the researcher also consider and report the size of the difference when drawing conclusions. This can be done by calculating a 95% CI on the difference between the population means. Again, most statistical packages calculate the CI, so it is quite easy to include this additional information in the results. This helps ensure that the results have practical as well as statistical significance.

References for further details: Sheskin (1997), Lomax (2001)

Wilcoxon–Mann–Whitney Test The Wilcoxon–Mann–Whitney test is a nonparametric test used to examine differences between groups on ordinal/rank data where the two groups are independent. This test is approximately the same as the Mann–Whitney U-test, which yields comparable results using slightly different computations. This test examines the question: Do the two groups have different median values (i.e., do the mean ranks for the two groups differ)? If the underlying population for the two groups is the same, we would expect both groups to have a similar distribution of ranks from low to middle to high. Conversely, if the underlying populations are different, we would expect one group to be concentrated in the low ranks and the other in the high ranks.

Assumptions This test assumes that each group is selected randomly from the population it represents (i.e., the two groups are independent of each other). It also assumes that the two groups come from populations with similarly shaped distributions, although no assumptions are made about what that shape is.

Methods and Results In this test, the results from the two groups are combined, the data are sorted from least to greatest, and the data are ranked overall. Then the sum of the ranks for each group is calculated. Most statistical packages calculate automatically the p-value by comparing the sum of the ranks for the group with the smallest sum of ranks to the relevant critical value. If the p-value is not provided, it can be determined by comparing the smaller sum of ranks to critical values, typically provided in the appendix of statistics books.

Drawing Conclusions As with the t-test, the p-value indicates the likelihood that the observed differences in rank sums are from chance as opposed to being due to differences between the underlying populations for each group. A significant p-value (e.g., $p \leq 0.05$) provides evidence that the median of one group is lower than the median of the other group. In other words, the group with the lower rank sum has lower values of the dependent variable than those of the other group.

References for further details: Sheskin (1997), Argyrous (2000), Sprent and Smeeton (2001)

Wilcoxon Matched-Pairs Signed-Rank Test The Wilcoxon matched-pairs signed-rank test is a nonparametric test used to examine differences between groups when the two groups are *dependent*. This is in contrast to the Wilcoxon–Mann–Whitney test, which is used to examine two *independent* groups. The Wilcoxon matched-pairs signed-rank test is the nonparametric equivalent of the dependent (paired) t-test. The test is based on the ranks of the differences between scores for each participant. The test examines the question: Is the median value for the difference between the two scores equal to zero (i.e., does the experimental condition cause a difference in the scores)?

Assumptions This test assumes that participants are selected randomly from the population represented and that each subject receives both experimental conditions. It also assumes that original scores for each participant are ordinal/ratio data and that the distribution of scores for each experimental condition comes from populations with similarly shaped distributions, although no assumptions are made about what that shape is.

Methods and Results In this test, the ranks are generated based on the difference between the ordinal/ratio score obtained in each experimental condition for each participant. Therefore, the original ordinal/ratio score is needed. First the difference for each participant is calculated ($Y_{1i} - Y_{2i}$). Then the absolute value of the difference scores are ranked in order from least to greatest. These difference ranks are split into two groups: the positive ranks, which are those ranks where the difference score was positive ($Y_{1i} - Y_{2i} > 0$), and the negative ranks, where the difference score was negative ($Y_{1i} - Y_{2i} < 0$). Next the sum of the positive ranks is calculated and compared to the sum of the negative ranks. The smaller of the two rank sums is used as the Wilcoxon's T-statistic, which is compared to the expected value of the rank sum if there was no difference between the groups. When using a statistical package to complete this test, the researcher is shielded from these calculations. The package usually calculates and reports the positive and negative rank sums, Wilcoxon's T-statistic, and the p-value.

Drawing Conclusions As with the Wilcoxon–Mann–Whitney test, the p-value indicates the likelihood that the difference observed between the smallest rank sum observed (Wilcoxon's T) and the rank sum expected are by chance, as opposed to being due to differences between the scores in each group. A significant p-value (e.g., $p \leq 0.05$) provides evidence that the median difference in scores is not zero. To determine whether experimental condition 1 or 2 resulted in higher differences, we must examine the positive and negative rank sums. If the difference scores were calculated as $Y_{1i} - Y_{2i}$, a larger positive rank sum indicates that condition 1 resulted in higher scores than condition 2. This is logical since a larger positive rank sum indicates that more participants had a positive difference in scores ($Y_{1i} - Y_{2i}$) and/or that the absolute value of the positive differences was larger than those observed for participants with negative differences.

References for further details: Sheskin (1997), Argyrous (2000), Sprent and Smeeton (2001)

Chi-Squared Test The chi-squared test is a nonparametric test used to examine differences between groups using nominal/categorical data. For example, it could be used to determine if there were significant differences in the number of males and females in two experimental groups. This test can be used to examine two or more groups.

Comparing More Than Two Groups The tests presented here are used to determine if the dependent measure differs between more than two groups of interest. Similar to the case in which two groups are being examined, the appropriate test to use depends on the answers to four questions: (1) Are the groups dependent or independent? (2) Do the data adequately meet the assumptions of the parametric test? (3) What is the measurement type of the dependent measure? (4) Are multiple, correlated measures being compared across groups? Table 3 identifies these factors and the appropriate test for each. Note that since the Cochran Q test is not used frequently compared to the other tests, it is not discussed in detail in this chapter. However, more information about this

Table 3 Tests for Comparing More Than Two Groups

	Parametric: Interval/ Ratio	Nonparametric Ordinal/ Rank	Categorical
Dependent groups	Repeated measures ANOVA test	Kruskal–Wallis test	Cochran Q test
Independent groups	ANOVA test	Friedman test	Chi-squared test
Multiple, correlated measures	Multiple ANOVA test		

test may be found in Sheskin (1997) or Sprent and Smeeton (2001).

ANOVA Test The analysis of variance (ANOVA) test is used to determine if two or more independent groups differ on an interval/ratio dependent measure. This test answers the question: Is there a difference in the mean for at least two of the groups? ANOVA is closely related to the t-test and is preferred for examining more than two groups because it controls the experimentwise error rate. A variety of ANOVA procedures exist to support analyzing a variety of experimental designs, such as those with two grouping factors or using a mixed design. For simplicity, we address only one-factor ANOVA here since the concepts used in this procedure extend to more advanced ANOVA procedures.

Assumptions ANOVA assumes that the data being analyzed comprise a randomly selected sample. In the model on which the ANOVA is based, the error terms are normally distributed with mean zero. The variances of the data in each group are approximately equal (homogeneity of variance). Note that ANOVA is based on the general linear model, so these assumptions are the same as many of the assumptions for linear regression. See "Linear Regression" in Section 4.2.3 for more details on testing these assumptions.

Methods and Results ANOVA determines if the mean is equal for all the groups. To compare the groups, an ANOVA table is constructed which breaks down the sources of the variation in the dependent measure. For the one-factor ANOVA, there are two possible sources of variation: the independent measure used for grouping or random variation. The ratio of the variation accounted for by the independent measure [mean square treatment (MSTr)] to the random variation [mean square error (MSE)] is the F-statistic. The F-statistic is compared to a threshold value to determine the likelihood (p-value) that the differences in the means of the groups are due to random variation or actual differences in the mean of the underlying population. Most statistical packages report the full ANOVA table, including the breakdown of the variation (in terms of sum of squares, degrees of freedom, and the mean square for each source of variation), the F-statistic, and the p-value.

Results from the ANOVA F-test indicate only whether or not there is a difference between at least two of the groups. To determine which groups are actually different, post hoc tests, or paired comparisons, must also be performed. However, completing these multiple comparisons increases the experimentwise error rate (α). Several post hoc comparison methods exist that are designed to control the experimentwise error rate. Two commonly used methods are the Bonferroni and the Tukey. In both of these methods, the t-test is used to compare the means for each pair of groups. To control the error rate, the critical value to which the t-statistic is compared is adjusted to be more stringent and to account for the number of comparisons being made. Most statistical packages allow you to choose which post hoc method to use. The Bonferroni method is more conservative—it is more robust than other methods, but as a result has less power to detect differences. The Tukey method is more sensitive but is not as robust when departures from test assumptions occur. When using statistical packages to complete paired comparisons, of interest in the output results are the mean difference between the groups, the p-value, and the 95% CI on the mean difference.

Because ANOVA analysis is based on the general linear model, it is important to complete a residual analysis as part of this procedure to ensure that the data meet all of the assumptions of the model. See "Linear Regression" in Section 4.2.3 for more details on how to validate these assumptions.

Drawing Conclusions To draw conclusions, first look at the p-value for the ANOVA F-test. If the p-value is sufficiently low, we may conclude that there is a difference among the groups and examine the results of the post hoc tests to determine which groups differ and the direction of that difference. When examining post hoc test results, first look at the p-values for each pair to identify which groups have means that differ significantly. Once these have been identified, examine the 95% CI on the mean difference to determine the direction of the difference based on the sign of the mean [e.g., positive (A − B) indicates group A > group B] and the magnitude of the difference.

References for further details: Neter et al. (1996), Sheskin (1997), Argyrous (2000), Wu and Hamada (2000)

Repeated Measures ANOVA Test The repeated measures, or within-subjects, ANOVA test is used to determine if two or more dependent groups differ on an interval/ratio dependent variable of interest. This test is used to analyze data from repeated measures and blocked experimental designs where each participant receives each experimental condition. This test answers the question: Is there a difference in the mean for at least two of the experimental conditions? The repeated measures ANOVA is related conceptually to the dependent t-test and is closely related to the ANOVA used to examine multiple, independent groups.

Assumptions This test assumes that the data are a randomly selected sample. In the ANOVA model on which the test is based, the error terms are normally distributed with mean zero. In contrast to the ANOVA test, the repeated measures ANOVA assumes sphericity instead of homogeneity of variance. Sphericity is a more complex concept related to the underlying variance and covariance of the populations being examined. For more information on this assumption, see Sheskin (1997).

Methods and Results Like ANOVA, this test examines whether the means for all groups are equal. The primary difference in the repeated measures ANOVA is that it acknowledges and accounts for the variation between participants, in effect comparing each participant's performance across groups. To do this, this method decomposes the variation in the dependent variable into three possible sources of variation: (1) the independent measure used for grouping, (2) the participant completing the trial, or (3) random variation. In this case, two F-statistics may be calculated, one for the grouping measure (as in ANOVA) and one for the blocking variable, the participant. As with ANOVA, most statistical packages report the full ANOVA table, including the breakdown of the variation (in terms of sum of squares, degrees of freedom, and the mean square for each source of variation), the F-statistics, and the p-value. Upon completing the repeated measures ANOVA, a post hoc test should be completed using the methods described in the section on ANOVA to examine the differences between groups in more detail.

Drawing Conclusions Drawing conclusions for repeated measures ANOVA is essentially the same as that for ANOVA, so refer to that section for more detail. The primary difference is that in addition to drawing conclusions about differences in the dependent measure related to the grouping measure, you can also draw conclusions about whether or not the participants varied significantly on the dependent measure. If there are differences among participants, further investigation may be warranted to try to determine if another characteristic of the participant is at the root of these differences.

References for further details: Neter et al. (1996), Sheskin (1997), Wu and Hamada (2000), Lomax (2001)

Multivariate ANOVA (MANOVA) Test MANOVA is conceptually similar to ANOVA, except that MANOVA examines differences among groups on a set of correlated dependent measures. MANOVA takes steps to manage the experimentwise error by not only accounting for the number of groups being compared, as in ANOVA, but also the number of dependent measures being examined. If the MANOVA test is significant, it indicates that there is a significant difference among the groups on at least one of the measures. Once this conclusion is made, use ANOVAs and the appropriate post hoc tests to analyze each of the dependent measures individually to determine which dependent measures vary and between which groups those measures vary. Due to the complexity of this test, the reader is referred to the references for further details for additional information regarding specific assumptions, methods, and results of MANOVA.

References for further details: Johnson and Wichern (1988), Johnson (1998), Field (2000)

Kruskal–Wallis Test The Kruskal–Wallis test is the nonparametric equivalent of ANOVA. It is appropriate for ordinal data and examines differences in ranks between independent groups. It is extension of the Wilcoxon–Mann–Whitney test and is used to analyze more than two groups. As in the Wilcoxon–Mann–Whitney test, this test examines the question: Do the mean ranks for the groups differ? If there were true differences between the groups, we would expect one or more groups to be concentrated in the low ranks and the others to be concentrated in the high ranks.

Assumptions Kruskal–Wallis assumes that the groups represent a random sample and are independent of each other. It also assumes that the groups come from populations with similarly shaped distributions, although no assumptions are made about what that shape is.

Methods and Results As in the Wilcoxon–Mann–Whitney test, the data from all groups are combined and sorted from least to greatest, and the data are ranked overall. Then the sum of the ranks for each group is calculated. The test statistics, H, is calculated based on these sums. Statistical packages calculate the H statistic and associated p-value by comparing H to the relevant critical value. As with the ANOVA test, a significant p-value for this test only indicates that two or more groups vary in their median rank. To determine which groups differ, pairwise comparisons must be completed. Kruskal–Wallis uses the Wilcoxon–Mann–Whitney test to compare each pair of groups. To control experimentwise error (α), the Bonferroni method is typically used to adjust α. For this method, divide the target α by the number of comparisons (j). In other words, if your target is $\alpha = 0.05$ and you are completing four paired comparisons, the target α' for each comparison is $0.05/4 = 0.0125$. This means that p-values greater than 0.0125 would be rejected in the paired comparisons.

Drawing Conclusions If the p-value for the Kruskal–Wallis test is sufficiently low, we conclude that there is a difference among the groups and examine the results of the post hoc tests to determine which groups differ. In the post hoc tests, if any pairs differ significantly (i.e., they have a p value below the

Bonferroni revised threshold, α'), we may conclude that those pairs of groups differ. To determine the directionality of the difference, compare the sum of the ranks for each group. The group with the higher rank sum can be inferred to have higher values of the dependent measure than those of the other group.

References for further details: Sheskin (1997), Lomax (2001), Sprent and Smeeton (2001)

Friedman Test The Friedman test is the nonparametric equivalent of the repeated measures ANOVA. Like the Kruskal–Wallis test, the Friedman test uses ordinal data to examine whether the mean ranks differ for the groups. However, the Friedman test is used for dependent groups, such as those found in repeated measures designs.

Assumptions This test assumes that the data being analyzed comprise a randomly selected sample.

Methods and Results To analyze differences among the groups, the Friedman test examines the rank of each group within a participant's results. This is in contrast to the Kruskal–Wallis test, which assigns an overall rank to pooled scores for all groups. Therefore, the Friedman test results in a set of rankings for each participant. After the ranks are assigned, the ranks for each group are summed. These rank sums are then used to calculate the Friedman test statistic (χ_r^2). When this test is run using statistical packages, the test statistic and the associated p-value are typically reported. As in the previous tests examining multiple groups, if the Friedman test is significant, post hoc comparison tests are required to determine which pairs of groups differ. One method of accomplishing this is to use the Wilcoxon matched-pairs signed-ranks test to compare each pair of groups. To control the experimentwise error (α), the Bonferroni method is typically used to adjust α ($\alpha' = \alpha$/number of tests), as described in the Kruskal–Wallis test.

Drawing Conclusions The first step in drawing conclusions is to examine the p-value of the Friedman test. If p is small (e.g., $p < 0.05$), we may conclude that there is a difference among the groups. If there is a difference in the groups, examine the post hoc tests results to determine which groups differ. In the paired comparisons, identify any pairs that differ significantly (i.e., they have a p-value less than the Bonferroni revised threshold, α'). For the pairs that differ, compare the sum of the ranks for each group in the pair to determine which group has higher values. The group with the higher rank sum can be inferred to have higher values of the dependent measure than those of the other group.

References for further details: Sheskin (1997), Lomax (2001), Sprent and Smeeton (2001)

Chi-Squared Test The chi-squared (χ^2) test is a nonparametric test used to examine differences between groups on a nominal/categorical measure. The test is typically employed in association with a contingency or crosstab table. A contingency table is an $r \times c$ table where there is a row for each of the r groups and a column for each of the c possible values of the nominal measure being examined. Each cell represents the frequency with which the given row–column pair occurred. The chi-squared test examines whether or not there is a relationship between the group and the nominal measure. For example, is the number of females the same for all groups, or do a disproportionate proportion of women fall into a particular group?

Assumptions This test assumes that categorical data are used and that the $r \times c$ categories are mutually exclusive. That is, a participant/object will occur in only one cell in the contingency table. It also assumes that the data are from a random sample. An additional assumption is that the expected frequency of each cell is at least 1 and that no more than 20% of the cells have an expected value of 5 or less.

Methods and Results The chi-squared test compares the expected frequency of a category to the frequency of that category observed for each of the $r \times c$ categories in the table. The expected frequency is calculated based on the assumption that there is no relationship between the grouping measure and the nominal measure, so it is based on the proportion of overall participants who fall into that particular group. So in our example of examining gender differences among the groups, if you have 80 participants who are evenly divided among four experimental groups, we would expect each group to be 50% female, which translates to an expected frequency of 10 females [20 participants per group $\times$ 0.5 (expected percentage of females)]. The χ^2 statistic is calculated based on the difference between the frequency observed and the frequency expected. When using a statistical package to complete this test, results typically include the contingency table with observed and expected frequencies (or proportions), the χ^2 statistic, degrees of freedom, and the p-value. *Note:* The degrees of freedom are calculated as $(r - 1)(c - 1)$.

Drawing Conclusions Again, we begin drawing conclusions by examining the p-value. If p is sufficiently small, evidence indicates that there is a significant difference between the expected and observed frequencies in the cells. This, in turn indicates that there is a relationship between the grouping measure and the nominal measure. However p does not provide an indication of how they are related. Which cells differ from the expected frequencies? To determine this, most sources recommend completing paired comparisons in the form of 2×2 contingency tables for the subsets of the larger $r \times c$ table. As mentioned previously, these multiple comparisons can inflate the experimentwise error rate, α. To control the error rate in the chi-squared test, we again use the Bonferroni method, which divides the target α by the number of comparisons (j).

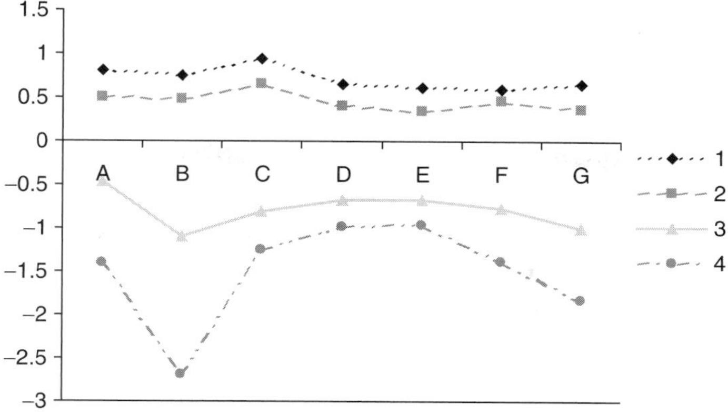

Figure 7 Profile plot for four groups.

Now we know which cells differ, but still cannot draw any conclusions regarding the magnitude of the differences, which indicate the strength of the relationship. There are a number of measures of association or correlation that indicate the size of the effect that can be examined. These include the phi coefficient, for 2 × 2 tables only, and Cramer's phi, for tables larger than 2 × 2. Phi coefficients range from −1 to 1, with the absolute value indicating the strength of the effect. Cramer's phi ranges from 0 to 1 with a similar interpretation. Cohen (1988) suggests the following recommendations for interpreting the size of an effect:

- *Small effect:* between 0.10 and 0.30
- *Medium effect:* between 0.30 and 0.50
- *Large effect:* greater than or equal to 0.50

References for further details: Sheskin (1997), Argyrous (2000), Lomax (2001)

Characterizing and Creating Groups In some cases we are interested in understanding the characteristics of known groups or in using the characteristics of items to create new groups. Several approaches are available for accomplishing either of these goals. To characterize a group based on a number of correlated measures, profile analyses are appropriate. On the other hand, discriminant analysis can be used to examine groups on a number of independent measures to determine which measures most distinguish groups from each other. These discriminating characteristics can then be used to classify or predict which group new items might fall into. When there are only two groups, logistic regression is sometimes used to predict classification of items into the two groups. Sometimes, however, we are simply interested in creating groups of items that are similar to each other on a number of characteristics. In this case, cluster analysis proves useful. In the following section we describe profile analyses, discriminant analysis, and cluster analysis. Logistic regression is described in Section 4.2.3.

Profile Analyses Profile analyses are used to examine the means of several measures for two or more groups. The measures are typically correlated. For example, they may be the results of several tests given to a participant. The profile for each group is a plot of the mean of each measure (see Figure 7). Profile analyses attempt to answer the following questions: Are the profiles for two groups parallel (e.g., is the difference in the means the same for all measures)? If they are parallel, are they coincident (e.g., is the mean for both groups the same)? Do the profiles show any trends?

Assumptions Profile analyses assume that each group's measures are independent of those of the other groups and normally distributed. It also assumes that the measures being used to create the profile use the same scale/units.

Methods and Results Profile analyses uses a number of statistical tests, based on those discussed previously, to compare the difference in the mean values of the two groups for each measure. The primary difference between the two is that the statistical methods used in profile analysis take into account the number of measures being examined and their covariance, similar to MANOVA. To determine whether or not two profiles are parallel, the T^2-statistic is used to test if the difference between two groups in the means for each measure is the same.

Drawing Conclusions Looking at the profiles in Figure 7, we would intuitively suspect that the profiles for groups 1 and 2 could be parallel. If the T^2-statistics for these groups produced a sufficiently small p-value, we could conclude that this is indeed the case.

Reference for further details: Johnson and Wichern (1988)

Discriminant Analysis Discriminant analysis is used to explore features of items (e.g., participants, events) in existing groups to determine which features differentiate items in one group from the other group(s). In many cases, these discriminant features, represented by measures, can be used to develop rules for classifying new items into one of the existing groups. For example, a researcher who was examining an assembly task might group participants in a study into two groups: those who completed the task successfully and those who did not. This researcher could use discriminant analysis to determine which participant and environmental characteristics differentiate those who completed the task successfully from those who did not. These discriminant features could, in turn, be used to develop rules for predicting whether or not the task will be completed successfully in certain situations. Discriminant analysis develops rules used to predict which group a given item will belong to given a set of measures describing that item. In general, the goals of discriminant analysis are (1) to identify and describe the features most distinctly separate items in known groups, and (2) to derive rules used to optimally sort or classify new items into one of the given groups. In this section we examine the simple case of discriminating between two groups, although the method can be used for more than two groups.

Assumptions Discriminant analysis assumes that the populations of the groups being examined have a multivariate normal distribution with equal covariance.

Methods and Results The first step in discriminant analysis is to define a set of discriminant rules based on the means and variance–covariance of the measures for the two groups being examined. Although in many cases they result in the same rules, four different methods may be used to develop these rules: the likelihood rule, the linear discriminant function rule (Fisher's method), the Mahalanobis distance rule, and the posterior probability rule. The linear discriminant function rule (Fisher's method) is the best known. This method finds a linear transformation (Y) of the measures (X's) such that the Y's of the two groups are separated as much as possible. The discriminant rule compares the Y of the new item to the midpoint between the two groups. If the value is greater than the midpoint, it is assigned to the group with larger values of Y, and vice versa.

Once the discriminant rules are established, it is important to assess the adequacy of these rules by estimating the probability of classifying an item correctly. As researchers, we want to choose the rule that classifies items correctly a sufficiently high portion of the time. After all, a rule that only works 50% of the time isn't very useful; we might as well flip a coin to assign an item to a group. Of the three methods that may be used, the simplest is to use resubstitution estimates. For this method, create predicted classifications for the original data based on the discriminant rules and compare how frequently the prediction is correct. The method is problematic because it tends to overestimate the probability of correct classification. For large data sets a second method of assessing adequacy of the rules is to use holdout data. In this method, a subset of the data available is withheld when the discriminant rule is created. The discriminant rule is used to predict classifications for the holdout group; then the percentage of correct classifications is examined. For smaller data sets, this method should not be used since all of the data are required to create the best possible discriminant rule. The most accurate method is known as cross-validation, jackknifing, or Lachenbruch's holdout method. This method is more complex than the other two, so refer to one of the references for further details on how to complete this method of estimating the likelihood of a correct classification.

Drawing Conclusions The first step in drawing conclusions is to examine the probability of classifying an item correctly. If this percentage is sufficiently high considering the type of data and the purpose for which the discriminant analysis is intended, we may conclude that the discriminant rule is an adequate predictor for classification. Next, examine the coefficient/weight of each measure in the discriminant rule. The measures with higher coefficients factor more heavily in attempting to separate the two groups; therefore, those measures are better for discriminating between groups.

References for further details: Johnson and Wichern (1988), Johnson (1998)

Cluster Analysis Clustering is an exploratory method used to examine a set of ungrouped items (e.g., participants, events) and to identify any naturally occurring clusters, or groups, of items. This is in contrast to discriminant analysis, in which there were preexisting groups and the goal was to define rules for classifying new items into those preexisting groups. In cluster analysis we start with a set of items (e.g., participants) and a number of descriptive measures related to those items (e.g., a variety of behavioral measures related to task performance). The primary goal is to identify groups of items (participants) that are highly similar to each other on the descriptive measures (behavior), but highly different from the items in the other groups. In other words, we want to minimize the distance between items within a group and to maximize the distance between groups.

Once these groups are defined, the methods identified previously for comparing groups can be used to examine differences between the groups on the descriptive measures used to create the groupings. Although this can provide interesting insights, it is often more interesting to examine the groups for differences on measures that were not used to create the groupings. For example, if behavioral measures were used to group participants, it might be interesting to examine the resulting groups for differences in demographic variables such as age, education, and experience level. Note that cluster analysis is exploratory in nature and is more of an art than a science. There are a number of arbitrary judgments that the researcher must make, such as the appropriate clustering method and number of clusters. Therefore, validating that the resulting groups are meaningful and useful is crucial to successful cluster analysis.

Methods and Results Cluster analysis uses one of several available algorithms to identify groups of items based on a number of similarity and/or dissimilarity measures. Before using an algorithm, always explore the data using scatterplots and other graphical representations to see if there are any apparent natural groupings. If obvious groupings are observed, they can be used to validate the number of clusters and assignment of items to those clusters after the clustering algorithm is run. They may also be helpful for selecting the appropriate clustering algorithm. There are two primary classes of algorithms for identifying clusters in a data set: nonhierarchical and hierarchical. Nonhierarchical methods begin with a given number of clusters and an initial set of seed points around which the clusters will be built. The disadvantage of these methods is that the researcher must make an initial guess at the number of groups and location of the initial seed points. If the data are already thoroughly understood, it may be possible to set these effectively, but frequently this is not the case. Therefore, hierarchical clustering algorithms are more widely recommended.

Frequently used hierarchical clustering algorithms begin with each item in a single group and merge the closest groups successively. These algorithms, known as *agglomerative hierarchical methods*, attempt to merge the most similar items into a group while keeping the groups dissimilar. When using these methods, several different linkage methods can be used to build the clusters, and these methods vary in how they measure dissimilarity, the distance between two clusters. Two frequently used methods are nearest neighbor and furthest neighbor. In nearest neighbor, at each step the two closest items/clusters are merged and the dissimilarity between two clusters is measured as the distance between the two closest members. Both methods result in a series of merged clusters, and it is up to the researcher to decide which number of clusters is appropriate. This can be done by examining a dendrogram, or hierarchical tree diagram (Figure 8). The dendrogram depicts each successive merging step through the lines used to join items (cases). The horizontal distance of the line indicates the distance

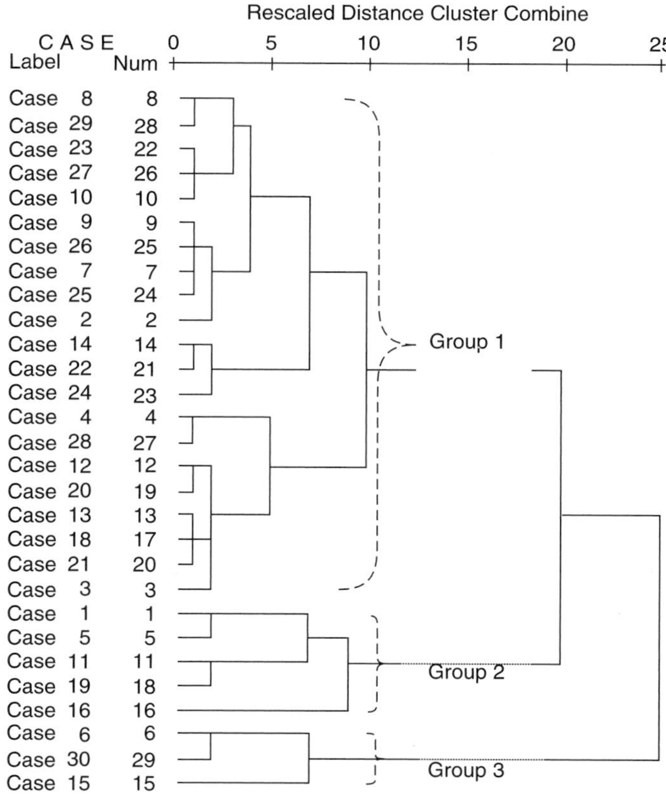

Figure 8 Dendrogram from cluster analysis.

between the two merged clusters. Therefore, we want to choose the number of clusters that results in the smallest horizontal lines connecting the members, indicating that the members are quite close, and the largest horizontal lines connecting the identified clusters, indicating the clusters are far apart. In Figure 8 it appears that selecting three clusters will accomplish this goal.

If a dendrogram does not present an obvious choice of groups, it is sometimes useful to try a different linkage method to see if it will produce more distinct groups. As this process demonstrates, the appropriate linkage method and number of groups is a subjective judgment made by the researcher. This is the reason that cluster analysis is considered exploratory in nature. However, there are statistical tests that can also be used to validate the selected number of clusters. See the references identified at the end of this section for details on these procedures. Once the appropriate number of clusters is identified, each item is assigned to a cluster and further analysis can be performed on each cluster (group).

Drawing Conclusions Before drawing conclusions as to the results of a cluster analysis, we first determine if the clusters are meaningful and useful. Examining the dendrogram gives a good indication of whether or not the groups created are distinct. If they are not distinct, it is unlikely that the groups will be useful. Another way is to examine two- and three-dimensional scatterplots of the measures used to create the groups, using a different symbol or color to represent items in each group. If the items form distinct visual clusters, it is an indication that the grouping is good. After confirming visually that the groups are distinct, it is also useful to use the statistical methods described previously to compare the groups on the measures used to create the groups. This provides statistical evidence that the groups indeed differ from each other on these measures. Once we are confident that the groups are meaningful, it is often interesting to compare the groups on measures that were not used to create the groups. This can sometimes give insight into complex relationships between measures used in cluster analysis and other measures that might lead to interesting directions for further research and analysis.

References for further details: Johnson and Wichern (1988), Johnson (1998)

4.2.3 Examining and Modeling Relationships

In previous sections we have focused on examining and in some cases, creating groups. However, frequently, researchers are interested in examining relationships between certain measures, independent of specific groups. For example, a researcher might want to examine the relationship that several independent measures have with a dependent measure. Alternatively, we might want to examine relationships among independent variables to determine if some redundant variables can be excluded from future experimental designs and/or analysis. Although a number of advanced techniques for modeling performance are presented in Part VI, the simple modeling methods presented here focus on understanding and characterizing relationships among a number of measures. In fact, the insights gained through these simple modeling methods can inform the development of more complex models.

Correlation Using Pearson and Spearman Coefficients Correlation is used to examine the relationship between two ordinal and/or interval/ratio measures. Correlation measures are used to examine the question: What is the degree of relationship or association between two measures? Correlation analysis is used to determine the strength of the relationship between the two variables—it does not imply causality in the relationship. Two commonly used measures of correlation are the Pearson product-moment coefficient, a parametric measure used for interval/ratio data, and the Spearman rank-order coefficient, a nonparametric measure used for ordinal data.

Assumptions Both the Pearson and Spearman tests assume that the data used comprise a random sample. The Pearson coefficient additionally assumes that both variables use an interval/ratio scale and that the two variables have a bivariate normal distribution. In a bivariate normal distribution, both variables and the linear combination of the variables are normally distributed. The latter half of the assumption means that the Pearson coefficient assumes that the relationship between the two variables is linear.

Methods and Results The Pearson coefficient examines the degree to which a linear relationship exists between the two measures of interest. The coefficient (r) ranges from -1 to 1 and the absolute value of r ($|r|$) is an indicator of the strength of the relationship. The closer $|r|$ is to 1, the stronger the relationship between the two variables. The sign of r indicates the direction of the linear relationship, with positive values indicating a direct relationship and negative values indicating an inverse relationship. r^2 represents the proportion of the variance in one measure accounted for by the other measure. When using the Pearson coefficient, create a scatterplot of the two variables to make sure that the assumption of a linear relationship holds. If a curvilinear relationship exists, r will be 0 even if a relationship between the two variables does actually exist. Additionally, large sample sizes are best for this analysis. When small sample sizes are used, several factors, such as the presence of outliers and restrictions on the range of one of the variables, can distort the value of r. For this reason, in experiments with a small number of observations, even if r is large, the p-value may indicate that the correlation is not significant. Conversely, in an experiment with a large sample, the p-value may indicate a significant correlation even though r is relatively small.

In contrast to the Pearson method, Spearman's rank-order coefficient determines the degree to which a monotonic relationship exists between the two variables rather than a linear relationship. The Spearman coefficient accomplishes this by examining the relationship between two ordinal variables by analyzing the ranks of the two variables. Each participant is ranked on each variable, and the coefficient (r_S) is calculated based on the difference in the ranks for each participant. Once r_S is calculated, the interpretation of the correlation coefficient and p-value are the same as the interpretation using the Pearson coefficient (r).

Drawing Conclusions To draw conclusions, first look at the p-value and correlation coefficient (r). If p is large, it does not necessarily indicate that no relationship exists since, as mentioned previously, p is sensitive to the number of observations. For example, when using the Pearson coefficient, if p is large, r is large, and the number of observations is small, it might be advisable to collect more data. The additional data would make r less vulnerable to distortions and reduce the threshold used to calculate p. In the case where p is small enough to be declared significant, you must still examine r to determine the strength of the relationship. As mentioned previously, Cohen (1988) provides suggestions for interpreting the size of a correlation effect based on r: small effect ($0.10 \leq r < 0.30$), medium effect ($0.30 \leq r < 0.50$), and large effect ($r \geq 0.50$). For an example of correlation analysis, see analysis 1 in the case study presented in Figure 14.

References for further details: Neter et al. (1996), Sheskin (1997), Lomax (2001)

Linear Regression Linear regression is used to examine the effect that certain measures (predictors) have on a dependent measure when the dependent measure consists of interval/ratio data. Linear regression models predict the value of the dependent measure based on the values of predictors. A linear regression explores two primary questions: (1) What effect do the predictors examined have on the dependent measure? (2) Given a set of values for the predictors, what is the value predicted for the dependent measure? For example, we might use linear regression to examine the factors that influence task completion time for a computer task. The measures used as predictors may be of any measurement type, but nominal data must be recoded to use a series of 0–1 indicators, where 1 indicates that a given category is present. For example, if job type is a predictor with three possible values (administrative, managerial, engineering), create three new variables, one for each job type, and assign a 1 to the variable representing the participant's job type and 0 to the remaining two variables. These three variables should be used in the analysis instead of the original nominal variable. There are two forms of linear regression: simple linear regression, which uses only one predictor in the model, and multiple linear regression, which includes multiple predictors.

Assumptions Linear regression is based on a general linear model and therefore must meet the assumptions of that model. The first assumption is that the relationship between the predictors and the dependent measure is linear. This assumption can be validated by examining a scatterplot of the data and ensuring that the plot resembles a straight line. If the scatterplot between a predictor and the dependent measure is curvilinear and resembles an upright or inverted U, linear regression may still be used if both linear and quadratic components (X and X^2) are included in the model. Additionally, multiple linear regression assumes that the predictors are *not* highly correlated. When two or more predictors are correlated, multicollinearity exists and this causes the regression model to be imprecise. There are several tests for multicollinearity; see Neter et al. (1996) for more details. When multicollinearity exists, drop redundant predictors from the model or use the data reduction techniques discussed in Section 4.3 to develop a set of independent predictors based on the predictors correlated.

The next set of assumptions is related to the error terms, or residuals, of the model. The residuals are calculated by subtracting the observed value of the dependent measure from the value predicted. The linear model assumes that residuals are independent and randomly distributed. It also assumes that at each value of a predictor (X), the error terms are normally distributed with a mean of zero and that for all values of X, the variance of the error terms is the same (homogeneity of variance). To check several of these assumptions, examine scatterplots of the residuals. If there is a pattern to the residuals, it indicates that they are not random. This usually means that the model is a bad fit. This can be caused if predictors are missing from the model or if the underlying relationship between the dependent variable and one or more of the variables is not linear. If the distribution of points is not centered around zero, it indicates that the mean of the errors is not zero and that the model is consistently overestimating (mean < 0) or underestimating (mean > 0) the value predicted. If the scatterplot is funnel shaped, as illustrated in Figure 9, it indicates that the error variances change with the value of X, and the homogeneity of variance assumption is violated. In contrast, Figure 10 depicts a good residual plot that supports the assumptions that errors are randomly distributed with mean zero and with homogeneous variance. When the homogeneity of variance assumption is violated, the violation can often be resolved by using a transformation of the dependent variable, as discussed in "Parametric and Nonparametric Tests" in Section 4.2.1.

To test the assumption that the error terms are normally distributed, create a normal probability plot of the residuals. If the points on this plot are in a straight line, the distribution of residuals is close to a normal distribution, as illustrated in Figure 11. When running the analysis, make sure that the option to create these plots is selected so that the statistical package will generate the residual graphs automatically as part of the linear regression analysis.

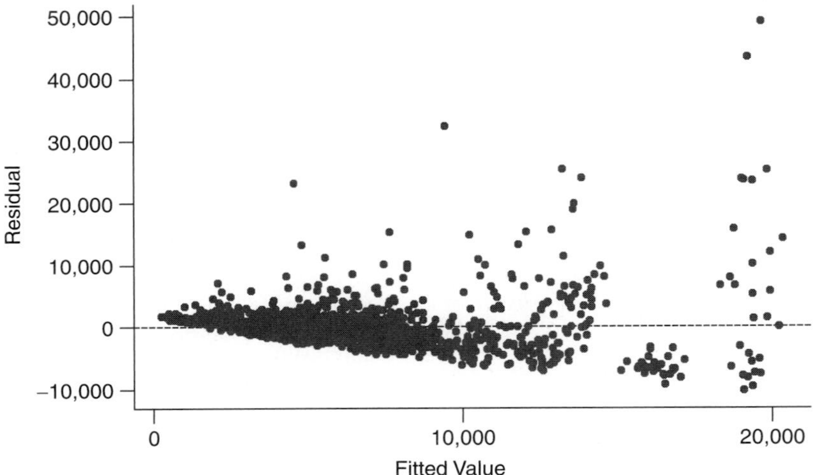

Figure 9 Residual plot where the homogeneity of variance assumption is violated.

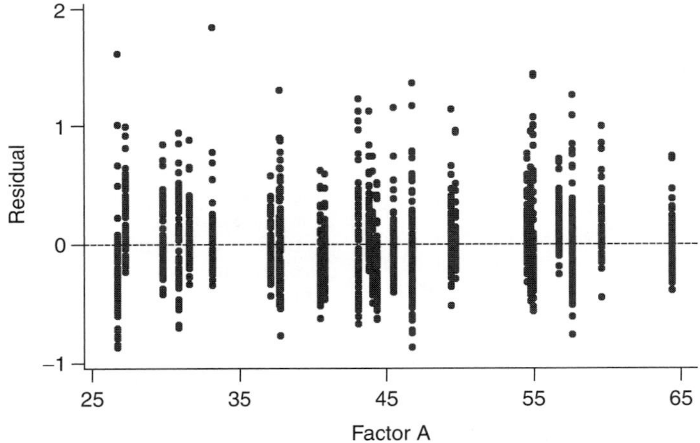

Figure 10 Good residual plot that meets the error distribution assumptions.

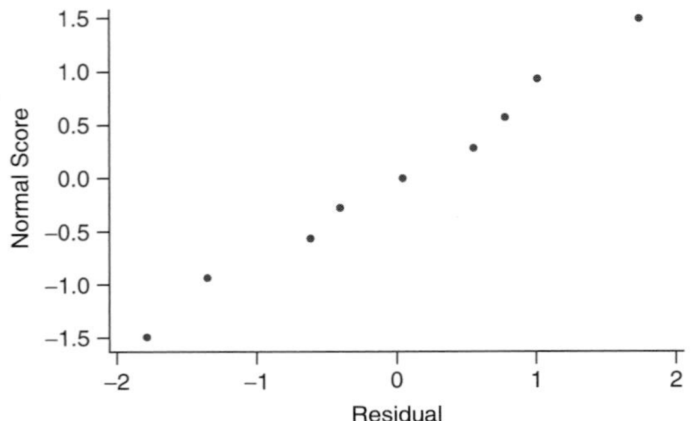

Figure 11 Normal probability plot of residuals when error terms are normally distributed.

Methods and Results Linear regression attempts to model the relationship between the dependent measure (Y) and a number of predictor measures or factors (X_i). Multiple regression generates a model in the following format: $Y = \beta_0 + \beta_1 X_1 + \cdots + \beta_i X_i$. In linear regression, the goal is to develop a parsimonious model with sufficient explanatory power. In other words, we want to gain the strongest match between the model and the data with the least number of predictors. A model with fewer predictors is preferred because it is simpler to interpret and use. In general, the Pareto principle applies to model building—a small number of factors account for a large portion of the variability. Therefore, we can often reduce the size of the model without sacrificing a significant portion of its explanatory power.

Several methods may be used to build a model from the predictors, including forward elimination, backward elimination, stepwise selection, and best subsets. Stepwise selection and best subsets are generally the preferred methods. In stepwise selection the predictor that contributes most to the explanatory model is added to the model. This continues, and as additional predictors are added, the predictors added previously are checked to make sure that they are still significant given the addition of the new predictor. The process ends when all of the predictors in the model are significant and all predictors excluded from the model are not significant. In the best subsets method of model selection, all possible model combinations are examined and compared using Mallow's C_p statistic, which indicates the explanatory power of the model adjusted to account for the number of variables. This adjustment factor ensures that more parsimonious models will be favored over larger models when the explanatory power of both models is similar. When comparing C_p statistics, choose a model with a C_p value that is low and close to 1 plus the number of predictors.

After the model has been selected, run the linear regression for the model. The results of the regression include the coefficient and p-value for each predictor and an r^2 and adjusted r^2 for the model. These two values are indicators of the proportion (0 to 100%) of the variation in the data that is explained by the model, so large values of r^2 are desirable. The adjusted r^2 adjusts the r^2 value to penalize models with a larger number of variables; therefore, it is the preferred measure of model fit.

Drawing Conclusions After selecting a model and running the linear regression, first examine the p-values for each predictor included in the model to ensure that all predictors are statistically significant. Look at the r^2 and adjusted r^2 values to ensure that the model's explanatory power is sufficient for the intended use of the model. "Sufficient" will depend on the intended use. For example, in a situation where the goal is to tightly control the dependent measure within a certain range of tolerances, a higher explanatory power may be required. Next, examine the residual plots to ensure that none of the assumptions have been violated. Also, in cases where the explanatory power of the model is low, examining plots of the residuals versus the possible predictor variables may provide insight into how to improve the model. A pattern in one of these plots might indicate that the linear model is not adequately capturing an underlying relationship. For example, if the residuals make a U-shaped pattern, including a quadratic component (X_i^2) in the model might improve the fit.

After validating that the model meets the necessary assumptions and has sufficient explanatory power, the model can be interpreted and applied. The coefficient for each predictor indicates the size and direction of the effect that a predictor has on the dependent measure. Be careful when comparing the coefficients of several predictors to understand their relative contribution to the dependent measure. Differences in the scales used to measure those factors can skew the comparison. For example, if predictors A and B are in the model of dependent measure C and the range of values for these measures are 1 to 5, 100 to 1000, and 0 to 1000, respectively, A's coefficient may be larger than B's simply because of the difference in scale. To make comparisons of the size of the effect of several factors when different scales are used, it may be useful to run the regression using standardized values of the predictors. However, interpretation of the coefficient in terms of the effect of an incremental change in a given factor can be more complex and less generalizable (Lomax, 2001), since the standardized value is based on the variance of the specific sample.

Note that when drawing conclusions about the effects of predictors, it is best to limit them to the range of values examined for each predictor. It can be dangerous to extrapolate these trends to values far outside the range observed in the experiment since the underlying relationship between the predictor and the dependent measure may differ when the predictor is at levels extremely different from those examined.

References for further details: Neter et al. (1996), Wu and Hamada (2000), Lomax (2001)

Logistic Regression A logistic regression is used to examine the effect that certain measures (predictors) have on a dependent measure when the dependent measure is nominal/categorical. Frequently, the dependent measure is binary (e.g., yes/no, true/false, 0/1). Conceptually, logistic regression is similar to linear regression except that instead of modeling and predicting the value of the dependent measure, logistic regression models the likelihood that the dependent measure will have a certain value (e.g., yes). Logistic regression explores two primary questions: (1) What effect do the predictors have on the probability that the event represented by the dependent variable will occur? (2) Given a set of values for the predictors, what is the probability that the event will occur? For example, a logistic regression could examine the likelihood that a worker will complete a task successfully given a certain set of tools and worker characteristics.

Assumptions Logistic regression assumes that the dependent variable is categorical/nominal.

Methods and Results The logistic regression analyzes the data to provide a model of how each predictor (X_i) affects the probability (π) that the event (e.g., Y = "yes") will occur. The model takes the format

$$\pi = \frac{e^{\beta_0+\beta_1 X_1+\cdots+\beta_i X_i}}{1 + e^{\beta_0+\beta_1 X_1+\cdots+\beta_i X_i}}$$

where X_i is the value of the ith predictor and β_i is the coefficient representing the effect that the ith predictor has on the likelihood (π) that the event will occur. Note that the equation representing the power to which e is raised is the same as the equation used in a linear regression. As in linear regression, the goal in building a logistic regression is to create a parsimonious model with sufficient explanatory power. Therefore, the first step is to determine which of the available predictors to include in the model. Although several methods of variable selection are available, the backward stepwise method is used most frequently (Neter et al., 1996; Field, 2000). This method begins by including all predictors in the model, and at each step removes the predictor that has the smallest effect on model fit. This continues until all the predictors in the model meet an established threshold of significance, typically $p \leq 0.05$. Once the model is developed, examine the model's chi-squared statistic, derived from the likelihood ratio, to ensure that the model has an adequate fit to the data.

When using a statistical package to complete a logistic regression, the model output includes the estimated coefficient and odds ratio for each factor and a p-value indicating the level of significance of that factor as well as the model chi-squared statistics. It also reports a classification table that reports the number of cases in which the value predicted matched the value observed and the percentage of correct classifications. It is also useful to have the package save the predicted probability that the event will occur and the predicted value of the dependent measure, since these can both be used to examine how well the model fits the data. The predicted value is determined by comparing the predicted probability to an established threshold (usually, 0.50). If the probability is greater than the threshold, the predicted value is "yes," the event will occur; otherwise, it is "no."

Drawing Conclusions The first step in drawing conclusions is to determine if the model has an adequate fit. If the model chi-squared statistics has a sufficiently small p-value and the percentage of correct classifications is high, we can conclude that the model has an adequate fit. Unfortunately, in logistic regression there is no value equivalent to the r^2 used in linear regression to estimate the degree of fit between the model and the data. However, when the response being modeled is binary (yes/no or 0/1), an ROC curve may be used to estimate the degree of fit. An ROC curve (Figure 12) indicates how well the model predicts the dependent measure by comparing the true positive rate (i.e., predicted = yes and actual = yes) to the false positive rate (i.e., predicted = yes and actual = no). If chance were used to predict the dependent variable, the prediction would be correct on average 50% of the time. The diagonal line in the graph indicates this chance fit between the predicted and actual values. The area under the ROC curve represents the fit of the model. Therefore, if there is substantially more area under the ROC curve than under the diagonal line, the fit is good. A statistical test is used to test whether the area under the curve is significantly different from 50% and provide a 95% CI on the area. The closer the area is to 1 (100%), the better the model fit.

Once the model is deemed adequate, the results of the model may be interpreted. The model may be used for prediction by plugging in a set of values for the factors in the model to determine the likelihood that the event will occur given those values. Alternatively, the model coefficients for each predictor may be examined to determine that predictor's effect size. Unfortunately, this is not as straightforward as effect estimation in linear regression, since in logistic regression, a unit increase in X_i multiplies the odds of the event by $e^{\beta i}$. Therefore, to interpret the effect, we examine the odds ratio ($e^{\beta i}$), which is the ratio of the probability of the event with a unit increase in X_i to the probability of the event without the unit increase in X_i. The odds ratio should be compared to a baseline of 1 (100%). So an odds ratio of 1.25 indicates a 25% increase in the event's likelihood, and an odds ratio of 5 would indicate a fivefold increase in the odds. Conversely, an odds ratio of less than 1 would indicate that an increase in X_i reduces the likelihood of the event. By examining the odds ratio for each factor in the model, we can identify those factors that have the greatest positive and negative impact on the dependent variable being examined.

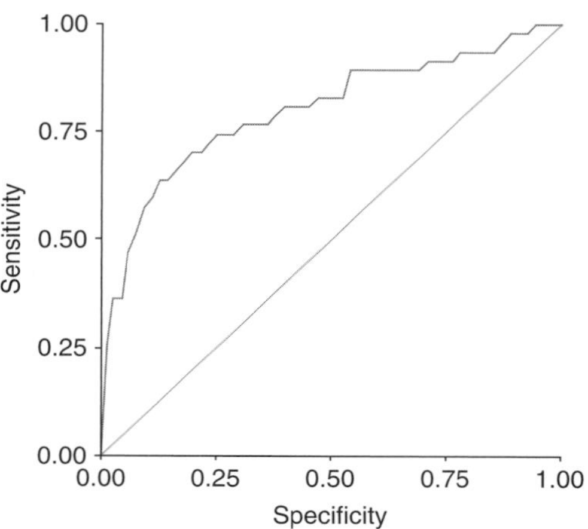

Figure 12 ROC curve for a logistic regression.

References for further details: Neter et al. (1996), Johnson (1998), Field (2000)

4.3 Data Reduction Techniques

Several of the methods described in Section 4.2 assume that the measures being examined (e.g., predictors in linear regression, dependent measures in MANOVA) are independent of each other. However, frequently in human factors research a number of the collected measures are correlated but capture different aspects of a given phenomenon. This is often the case in complex concepts such as decision satisfaction, where it is difficult for one measure to capture every facet of the concept. When examining these correlated measures independently, it can be difficult to isolate the independent components of the phenomenon that are driving the underlying variation in those measures.

When this is the case, data reduction techniques can be useful to isolate those independent components that explain the variation, thereby reducing the number of measures to be analyzed and aiding in interpretation of the data.

For example, data reduction techniques are often used to examine questionnaire data, as demonstrated in a study by Sainfort and Booske (2000) which examined post-decision satisfaction. See their case study in Figure 14. In this study, after completing a decision task, participants were asked to complete a questionnaire related to their satisfaction with their decision. The questionnaire presented 10 statements that addressed several aspects of decision satisfaction (e.g., "My decision is sound," "More information would help") and asked the participant to rate each statement on a scale from 1 (strongly disagree) to 5 (strongly agree). Obviously, the answers to a number of these statements are expected to be correlated. By examining the correlations among the measures and using factor analysis to isolate the underlying constructs affecting the variation in answers, Sainfort and Booske were able to identify four underlying dimensions of decision satisfaction: self-efficacy, satisfaction with choice, usability of information, and adequacy of information. Thus, in this example, factor analysis enabled the researchers to reduce a set of 10 highly correlated measures to four relatively independent measures and as a result, gained insight into the underlying structure of the highly abstract concept of post-decision satisfaction.

In this section we examine two data reduction techniques: principal component analysis and factor analysis. These two methods are conceptually very similar but differ in the mathematics used to accomplish the results. However, research has demonstrated that the two methods yield highly similar results, especially when sample sizes are large (Fava and Velicer, 1992). In general, sample sizes of at least 160 are recommended, but in some cases larger sample sizes (e.g., 300 or more) are necessary to ensure stability of the components (Guadagnoli and Velicer, 1988). These two data reduction techniques are often not an end in and of themselves. Typically, the new component/factor variables created as a result of this analysis are used as input for other analysis techniques discussed in previous sections, such as regression or group comparisons.

4.3.1 Principal Component Analysis

The primary objectives of principal component analysis (PCA) are to reduce the dimensionality of a data set by discovering the true dimensionality of the data and to identify new and meaningful underlying variables that represent the true dimensionality (Johnson, 1998). PCA is typically used as an exploratory technique to help researchers understand the data, especially the correlation structure of the data. PCA results in a set of new variables, principal components, that are uncorrelated and account for as much of the variability in the original data as possible.

Methods and Results The focus of PCA is to explain the variability in the measures through the components identified. To accomplish this, PCA produces an orthogonal transformation of the measures into a number of principal components, which constitute a linear combination of the measures being examined and depend solely on the covariance of those measures. The method begins by identifying the component that accounts for the largest portion of the variation, as indicated by the eigenvalue. This continues, with each component accounting for less of the overall variation.

One of the first steps in PCA is determining the appropriate number of components. In other words, how many "true" dimensions are represented by the data? There are several methods for determining this number, although it is more of an art than a science. The first method is to establish a minimum threshold, usually 1, for a component's eigenvalue. Alternatively, we could examine a scree plot, which plots the eigenvalues of each component as illustrated in Figure 13. In the scree plot, look for the cutoff at which the eigenvalues level off or create an elbow. All components to the right of the elbow should be eliminated, since they add little additional explanatory power. Based on this method, for the results in Figure 13, we would keep components 1 and 2 and eliminate the others. The third method for selecting the appropriate number of components is to establish a threshold for the amount of variability in the original data that is accounted for by the principal components and then select the minimum number of components that meet this threshold. For example, if a researcher wanted to account for 80% of the variation in the original data, and the first four components accounted for 50, 25, 10, and 8% of the variation, respectively, the researcher would use the first three components, which account for $50 + 25 + 10 = 85\%$ of the variation.

Once the principal components are identified and derived, the factor loadings are reported for each component. The ith principal component (C_i) for a given item is calculated based on the factor loading (β_i) and value (X_{ij}, where j represents the item) of each measure that loads on that component: $C_{ij} = \beta_1 X_{1j} + \beta_2 X_{2j} + \cdots + \beta_i X_{ij}$. The absolute value of the factor loading indicates the weight of that measure in calculating the component. Absolute values close to 1 are very important to the component. The sign of the factor loading indicates the direction of the relationship between the component and original measure; a positive value indicates that an increase in the measure will increase the value of the component score. These factor loadings are used to calculate principal component scores for each item based on the values of the original measures. After these values are calculated, they can be used for subsequent analysis. Frequently, principal components are used as inputs for cluster analysis, discriminant analysis, and multiple linear regression.

Drawing Conclusions Although many researchers use the results of PCA as input to additional

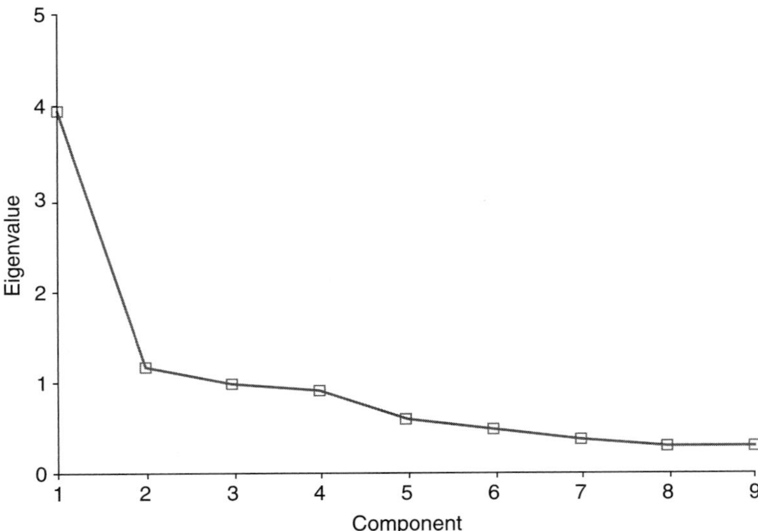

Figure 13 Scree plot in principal component analysis.

analyses, there are several interesting conclusions that can be made from the results of PCA themselves. These conclusions center around understanding the true dimensionality of the data. For example, if the percentage of variability accounted for by the components selected is sufficiently large, the number of components identified can be interpreted as the number of true dimensions in the construct being measured by the data. By examining the factor loadings for each measure that contributes to a component, we can determine if that component, which represents a dimension of the data, has any interpretable meaning. It is easiest to interpret the components when a measure has a high factor loading for one component but not on any of the others, and when the combination of measures that have a high loading are conceptually or logically related in some way. For example, if we completed a PCA on a number of measures that contribute to productivity and one component had high factor loadings for a worker's education level, number of training certifications, and number of years on the job, we might interpret this component as the worker's knowledge level. Note that in many cases, the resulting principal components may not be interpretable. When this is the case, it is often worthwhile to complete a factor analysis on the data set as well, especially since functionality in the current generation of statistical packages makes running either analysis relatively quick and easy.

References for further details: Johnson and Wichern (1988), Johnson (1998)

4.3.2 Factor Analysis

Conceptually, factor analysis (FA) and PCA are quite similar. Both methods examine the true dimensionality for a large set of correlated measures. However unlike PCA, which does not depend on an underlying model, FA depends on a reasonable statistical model. Also, FA is focused on explaining the covariance and correlation among the measures, which is in contrast to PCA, which focuses on explaining the variability of the measures. These are subtle distinctions from a conceptual standpoint, but they affect the mathematics used to develop the components, or factors, that result from each method. In PCA, the components identified are linear combinations of the measures. In contrast, in FA, the measures are linear combinations of the factors identified. In most cases, the results of the two methods are highly similar, but sometimes FA results are more interpretable as a result of factor rotation, as discussed next. As with PCA, the output of FA is a new set of uncorrelated measures or underlying factors.

Methods and Results Similar to PCA, FA analyzes the variance or covariance of the measures being examined to identify factors (components) that account for most of the variability in those measures. However, in FA, several different methods can be used to estimate the factors. Guidelines as to which method is best are relatively vague, but two frequently used methods are principal factor, which is similar to PCA, and maximum likelihood, which should only be used to analyze multivariate normal data. As in PCA, the researcher must determine the number of factors to include in the analysis. The criteria used in PCA, which are used to examine the eigenvalues, are typically used as a starting point, although additional guidelines can be used. For example, Johnson (1998) identifies minimizing Akaike's information criterion (AIC) or Schwarz's Bayesian criterion (SBC) as objective rules that can be used to determine the ideal number of factors.

In factor analysis, the set of factors, and their factor loadings generated through the initial analysis are not unique. By multiplying the resulting factor loadings by an orthogonal matrix, the factor loadings can be rotated in space. What exactly does this mean? We won't dwell on the mathematics, but conceptually, rotating the factor loadings obtains a mathematically equivalent set of factor loadings that may be easier to interpret. In other words, we as researchers can look for the rotated set of factor loadings that can be interpreted most meaningfully. Note that some statisticians and researchers view this as a criticism of FA, since exactly which rotation is "meaningful" is open to interpretation; however, many people view this as an advantage since it enables the resulting factors to be more easily interpreted and applied (Johnson, 1998).

Several methods are available for rotating the initial factor loadings. Factor rotation methods have two goals. The primary goal is to rotate the factor loadings for each factor so that all measures either load heavily on that factor (i.e., have an absolute value near 1) or load very little on it (i.e., have an absolute value near zero). The secondary goal is for each measure to load heavily on only one factor. Meeting these goals typically simplifies interpretation since the measures that weigh heavily on a factor are isolated and each measure (ideally) contributes significantly to only one factor. There are two classes of rotation methods: orthogonal rotation methods, which maintain the independence between the identified factors, and oblique rotation methods, which are appropriate only when the factors are not assumed to be independent. In general, orthogonal methods are recommended over oblique rotation methods unless there are theoretical grounds for assuming that the factors are not independent. Of the orthogonal rotation methods available, varimax is used most frequently

Drawing Conclusions The conclusions drawn from FA are almost identical to those drawn in PCA: Researchers may use the resulting factors from FA as input to additional analyses and/or draw conclusions about the true dimensionality of the data. The number of components identified can be interpreted as the number of true dimensions in the construct being measured by the data, and by examining the factor loadings for each factor, we can determine if that factor has any interpretable meaning. Because FA uses factor rotation to ensure that factors have high factor loadings for a subset of measures and low factor loadings for the remaining measures, meaningful interpretation of the factors is usually easier than in PCA. (For example, see the results highlighted in the case study in Figure 14.) However, because of the variety of methods that can be used to determine the number of factors, estimate the factors, and rotate the factors to find meaningful factor loadings, factor analysis is fairly subjective. As such, be prepared to support your conclusions with tests to ensure validity and reliability. When presenting your conclusions, provide sound theoretical support from related work in the literature. Furthermore, validity can be examined by comparing the resulting factors to related outcome measures not used in the FA. Reliability can be examined by completing the same analysis on a similar data set and comparing the number of factors and factor loadings that result from each data set. See the case study in Figure 14 for an example of how the researchers established the validity and reliability of their results.

References for further details: Johnson and Wichern (1988), Johnson (1998)

5 ANALYSIS OF UNSTRUCTURED OUTCOMES DATA

It is apparent that a wide variety of methods are available to analyze structured outcomes. But what can be done with all of the unstructured data that are generated? Unstructured outcomes are generated from a number of research methods. For example, unstructured outcomes are the results of open-ended interview or survey questions, observations from the field, observations during an experimental task, documents, and transcripts from video recording and think-aloud and other verbal protocols. These data are rich in detail, and unlocking the themes and trends from these data can provide invaluable insight into the topic being researched.

Unfortunately, unlocking these secrets can be quite challenging, especially when dealing with large amounts of data. Researchers usually begin by pulling out the data that are relevant, where relevance is determined by the nature of the data collected and the research questions to be answered (Wixon, 1995). Once the relevant data are identified, several techniques can be used to develop and communicate findings from unstructured data. In the following sections we discuss methods that can be used to accomplish these analysis objectives for unstructured data: (1) impose structure on unstructured data using coding and classification, (2) provide an aggregate view of the topic using figures and tables, and (3) provide a detailed record of the topic through documentation.

5.1 Content Analysis and Coding

Content analysis, or coding, is used by researchers to understand and summarize unstructured data. In content analysis, qualitative unstructured data are examined systematically to identify key themes and subsequently, to classify events, observations, and answers into one or more of those themes or categories. Although there are subtle distinctions between coding and content analysis, we avoid those semantic discussions for now and focus on methods of completing these types of analysis, which are, for the most part, the same. Content analysis and coding are used in a variety of fields, especially in market research and the social sciences, where more descriptive research methods are frequently used. In human factors, coding has been used, for example, (*text continues on page 1179*)

Research Objectives: Given the proliferation of decision aids used to help individuals make complex, value-laden decisions, there is a need for reliable, objective methods for evaluating the effectiveness of these decision aids. In this study, Sainfort and Booske (2000) develop the decision-attitude scale, which is used to assess the decision maker's satisfaction after making a decision.

Data Collection Method: Participants examined information on health insurance plans using a computer-based decision aid. Participants selected a plan at two points in time: first after viewing a minimal amount of information about the plans, and again after having the opportunity to view extensive information about each plan. At each decision point, participants completed the decision attitude survey consisting of ten statements related to various aspects of decision satisfaction. Participants rated whether or not they agreed with each statement using a scale from 1 (strongly disagree) to 5 (strongly agree).

Analysis 1: Examine relationships among decision satisfaction answers
Test: Pearson's Correlation
Results: Correlation coefficients

		1	2	3	4	5	6	7	8	9
1	I had no problem using the information	1.00								
2	I am comfortable with my decision	0.50	1.00							
3	The information was easy to understand	0.64	0.39	1.00						
4	I wish someone else had made the decision for me	0.10	0.07	0.09	1.00					
5	It was difficult to make a choice	0.40	0.52	0.43	0.13	1.00				
6	I am satisfied with my decision	0.54	0.70	0.46	0.07	0.50	1.00			
7	My decision is sound	0.41	0.57	0.40	0.14	0.49	0.61	1.00		
8	More information would help	0.18	0.38	0.31	0.05	0.33	0.37	0.37	1.00	
9	My decision is the right one for my situation	0.29	0.46	0.35	0.12	0.40	0.49	0.57	0.40	1.00
10	Consulting someone else would have been helpful	0.20	0.30	0.37	0.04	0.31	0.33	0.22	0.49	0.32

Conclusions: Examining the correlation coefficients among the responses to each statement, most of the responses are correlated, with the exception of statement #4, which appears to be relatively independent of the other responses. Because of the correlations among the response, the researchers' next step was to use factor analysis in order to gain insight into the underlying factors driving the variation in these responses.

Analysis 2: Exploring underlying factors of the correlated responses
Test: Factor Analysis. Response #4 was excluded from analysis since the correlation indicated it is independent. To ensure the reliability of the scale, the factor analysis was performed on the responses obtained at the first decision point. A second factor analysis was completed using the responses obtained at the second decision point to ensure that comparable factors and factor loadings resulted from both data sets.
Results: Three factors, accounting for 71% of the total variation, were identified. As an example, the factor loadings for factor 1 are provided below, sorted on the factor loadings. Factor 1 had an Eigenvalue of 4.37 and accounted for 48.6% of the variation.

Response	*Loadings on Factor 1*
7. My decision is sound	0.828
2. I am comfortable with my decision	0.741
9. My decision is the right one for my situation	0.729
6. I am satisfied with my decision	0.721
5. It was difficult to make a choice	0.574
8. More information would help	0.364

Response	*Loadings on Factor 1*
1. I had no problem using the information	0.312
3. The information was easy to understand	0.197
10. Consulting someone else would have been useful	0.082

Conclusions: Based on the results, the first five responses (items 7, 2, 9, 6, & 5) factor most heavily on factor 1 (loading >0.55). These items also had much lower loadings on the other two factors. The remaining four factors, all had low loadings on this factor (loading <0.4) and high loadings on one of the two other factors (loading >0.75 for one of the remaining factors and loading <0.3 on the remaining factor). Considering the items that loaded heavily on factor 1, the researchers concluded that factor 1 represents 'Satisfaction with Choice'. The remaining two factors had similar interpretability, representing 'Usability of Information' and 'Adequacy of Information'.

Note: In this study, a number of tests were completed to ensure the validity of the identified factors in the decision satisfaction scale. In the interest of brevity, only the results from one test are highlighted here.
Test: Paired t-test
Results: Paired t-tests examined difference on each of the factors between the time of the first decision point and the second:

Factor	*t-statistic*	*p-value*
Satisfaction with Choice	−3.461	<0.01
Adequacy of Information	−9.619	<0.01
Usability of Information	−1.608	0.11

Conclusions: Because additional information was made available to participants after the first decision point, intuitively, one would expect participants to be more satisfied with their choice and the adequacy of the information at the second decision point Results indicate that participants were indeed more satisfied with their choice (p<0.01) and the adequacy of the information (p<0.01) at the second decision point. This is indicates that the scale is sensitive enough to detect the differences in these two aspects of decision satisfaction. Usability of information had only a marginal difference (p=0.11), which, again, makes intuitive sense because though more information was presented between decision points one and two, a similar presentation format was used, so information usability would not necessarily improve. These results led the researchers to conclude that the scale had adequate discriminant ability to detect changes over time attributed to the presentation of additional information.

Figure 14 Case study: measuring post-decision satisfaction.

in incident and accident analysis and to provide structure to outcomes from verbal protocols such as think-aloud and other unstructured outcomes gathered during a research study.

When completing content analysis, the goal is to apply the coding scheme to the data objectively and systematically. The coding scheme is, in effect, the set of categories or themes to which an event/observation/answer may belong and the rules used to classify a given item into the appropriate category or theme. If the coding scheme is not applied consistently, the reliability of the coded data, and as a result the validity of the research results, is questionable. Therefore, close attention must be paid to the process used to code the data to ensure the reliability of the results.

5.1.1 Coding Process

The following steps comprise the coding process. For purposes of simplicity, an example of coding responses to an open-ended survey question is used. However, the same method could be applied to coding verbal protocols, observation notes, or other forms of unstructured data. Throughout this process, be on the lookout to avoid systematic biases that can result from inconsistent use of the coding scheme.

1. Develop the coding scheme.
   a. If there is an appropriate existing coding scheme that has been validated and used in the literature for the content being analyzed, use that coding scheme to improve the external validity and comparability of the results. For example, coding and classification schemes are available for analyzing certain types of communication. If an appropriate existing scheme is available, skip to step 2.
   b. To develop a specific reliable coding scheme, first examine the responses from a representative sample of the surveys. If there are only a small number of surveys, examine all of the responses.
   c. Based on knowledge of the actual responses and research objectives, identify a set of categories for classifying responses. Design the categories such that minimal information is lost is the coding. Also, design the categories to be as distinct from each other as possible to reduce confusion in classification.

   d. Establish definite rules and criteria for assigning a response to a category. Provide examples to clarify complex rules, especially when coding requires a judgment on the part of the coder. (For example, assigning a rating of "level of understanding" based on a participant's response requires more coder judgment than that needed for coding a fact-based response such as assigning a job category based on the participant's job title.) Defining clear rules for assigning responses to categories is *crucial* to ensuring intercoder reliability.
2. Establish a coding protocol.
   a. Determine the number of coders. At least two coders should be used so that intercoder reliability can be assessed. For large data sets, more than two coders may be needed.
   b. Recruit the coders. If possible, people with experience in content analysis and with relevant research content knowledge should be used.
   c. Define the coding protocol, which should include how many responses each coder will code, how questions regarding coding of questionable responses will be resolved, how coding will be validated, and how differences in coding will be resolved.
3. Test and revise the coding scheme.
   a. Using a small sample of responses, preferably not the same sample as that used to establish the initial coding scheme, test the coding scheme. If any categories appear too broad or narrow, adjust the coding scheme as needed. Also, if any of the classification rules are unclear, revise them as needed to ensure that they will be applied consistently.
4. Train the coders on the coding scheme and coding protocol. Allow them to practice with examples and to ensure that they have an adequate understanding of the coding scheme.
5. Code the responses.
   a. Assign the appropriate number of responses to each coder. For small samples, have each coder code all responses. For larger samples, make sure that a representative sample of the responses is coded by multiple coders so that intercoder reliability can be assessed.
   b. The primary researcher or a coding supervisor should resolve any questions that arise during the coding. Any clarifications should be communicated to *all* coders, so that coding will be consistent among all coders.
   c. The primary researcher or a coding supervisor should spot-check a sample of the responses to ensure that they are coded properly. Any problems identified in the coding should be resolved as quickly as possible to prevent the need to recode responses.
6. Assess intercoder reliability using one of the methods described in Section 5.1.3.

5.1.2 Improving Intercoder Reliability

In their study of coder variability, Kalton and Stowell (1979) make several observations and recommendations intended to reduce the amount of variability in coding among coders. Considering and applying these findings when designing and executing coding schemes can improve intercoder reliability:

- When constructing coding frames, take into account the actual responses so that those responses can be more readily mapped to the codes.
- When constructing coding frames, limit the use of catch-all codes (e.g., "other"), since these codes tend to be applied unreliably. Instead, try to use more clear-cut codes whenever possible.
- Be aware that fact-based codings, which are more objective, tend to have higher reliability than judgmental codings, which require more interpretation on the coder's part.
- Use training and strict supervision to ensure that coding frames are applied uniformly.
- Especially when judgmental codings are required, the researcher should stay closely involved with the coding operations (training, spot-checking coded data, etc.) to ensure that the codes are being interpreted and used as intended to support the research objective.

5.1.3 Assessing Intercoder Reliability

Intercoder reliability is the degree of match between the assigned codes of two different coders who apply the coding scheme independently to the same set of responses. Even when researchers take great care to ensure that there is adequate intercoder reliability, they should always assess and report intercoder reliability to ensure the credibility of their results. Unfortunately, many researchers omit this step, probably because there is relatively little agreement on the best way to assess intercoder reliability. Frequently, researchers report the percentage of agreement among coders. However, the one thing that most content analysis researchers agree on is that this is not the best approach. Instead, estimates of intercoder reliability should include adjustments that correct for chance agreement among coders (Hughes and Garrett, 1990; Grayson and Rust, 2001). Consider, for example, a case in which two coders are coding data that can fall into one of two categories. By random chance, those two coders will agree 50% of the time. Therefore, the assessment of reliability needs to indicate that agreement between coders is better than agreement that would be achieved by random chance.

A number of methods for assessing intercoder reliability have been presented in the content analysis literature. All of these measures range from 0 to 1, with values closer to 1 indicating high intercoder reliability. Cronbach's alpha is an intraclass correlation measure frequently used to measure agreement using the ratio of the true score variance to the sum of the true score variance and the error variance. Other methods used frequently include Scott's pi, Krippendorff's alpha, and Cohen's kappa. These three measures are conceptually similar: They are calculated by taking the difference between the agreement observed between coders and the agreement expected using random chance and adjusted based on the agreement expected using random chance (the chance correction). They differ in how they calculate the chance correction. Cohen's kappa is sometimes criticized because in certain cases the maximum possible value of kappa is less than 1. Due to the assumptions made in Scott's pi and Krippendorff's alpha, they are appropriate only in cases where intercoder bias is assumed to be negligible.

Once the appropriate method of calculating intercoder reliability is selected, there are several aspects of intercoder reliability that should be examined, even though the overall reliability is the only thing that is usually reported. First, if there are more than two coders, examine the reliability of each coder. If one coder's reliability is significantly less than the others, it indicates that they may have been using the coding scheme inconsistently with the other coders. Next, examine the reliability of each code. This will highlight any codes that are used less consistently than the others. This may indicate that the rules for assigning responses to this category need to be clarified or the category itself needs to be redefined. For an even deeper assurance of coder reliability, examine the percentages of coders assigning responses to each code. This helps pinpoint where the discrepancies are that reduce the reliability of specific coders or codes. For example, is coder A assigning a disproportionate amount of responses to the "other" category? Is coder B underutilizing category 2? By identifying these discrepancies, appropriate steps can be taken to revise the coding scheme or provide additional training to the coders to improve the overall reliability of the results.

5.1.4 Using the Results

Once the coding is completed, the results are structured data, usually in the form of frequencies of occurrence for each category. These results in and of themselves may provide the answers to the research questions of interest. Otherwise, these structured data derived from the unstructured data may be used as input for one or more of the structured analysis methods described previously. In either case, when reporting results based on the coded data, be sure to report an overview of the coding method used and the intercoder reliability achieved so that other researchers can be assured of the validity of the results and conclusions.

References for further details: Krippendorff (1981), Hughes and Garrett (1990), Oppenheim (1992)

5.2 Figures and Tables

Summarizing unstructured data in an informative figure or table can also be an invaluable tool for communicating findings gleaned from unstructured data. In some cases, figures and tables are more effective than text for communicating certain types of information. They are most effective when combined with written documentation that complements (but does not duplicate) the content in the figure or table. Take, for example, the decision model for selecting the appropriate structured analysis method presented in Figure 2. This figure does a much better job of conveying in a clear and concise manner the methods presented in the chapter, their relationships, and the factors that go into selecting the appropriate method than a written description of that information could. The text in the chapter supplements the figure by providing additional details on each of the decision points and methods identified in the figure. The figure, in effect, provides readers with a map of the content presented, helping them understand the big picture up front before delving into the details of a specific method. Using a figure instead of a text description to present the decision factors and relationships between goals and methods shortens the chapter. Also, presenting the big picture of the content up front makes it easier for readers to understand the detailed material as they read it. Both of these factors combined (hopefully!) make the chapter easier to read.

The appropriate type of figure or table to use will depend on the type of unstructured data being summarized and the purpose and audience of the document (or presentation) in which the figure or table will be included. A wide variety of figures and tables can be observed in the human factors and other literature. Examples of the several useful types include:

- *Maps.* Maps of physical space are obviously effective for conveying information relative to existing and proposed workspaces. They can also be effective for conveying paths that worker and/or materials must follow in the course of a given task.
- *Concept maps.* Concept maps are a physical representation of relationships between ideas or concepts in a knowledge space. In a concept map, concepts that are closely related conceptually are placed close to each other on the content map, and vice versa. Concept maps, in effect, present the relationship between abstract concepts in a physical manner.
- *Flowcharts.* Flowcharts have many obvious uses, including conveying processes or a series of tasks, communicating information/material flows, and others.
- *Decision models.* Decision models can be used to map out the steps in a decision process. They include decision points, decision factors, and related activities (e.g., data gathering) that are part of the overall decision process. In situations where multiple people are involved in

a decision, the decision model may also indicate who is responsible at each point in the process.

- *Organization charts*. Organization charts graphically illustrate members of an organization (people or suborganizations) and how they are related. These can be quite useful for depicting divisions of responsibility, chains of command, and communication channels.
- *Hierarchies*. Hierarchies are useful for presenting the breakdown of high-level concepts or groupings into lower-level concepts or grouping. For example, task hierarchies are used to break high-level tasks into the steps required to complete that task.
- *Time lines*. Time lines are useful for presenting a series of events that took place during research. They can also be effective at presenting a historical perspective on a particular domain of research or results of related research over time.
- *Literature tables*. Literature tables are useful to summarize concisely the relevant literature related to a particular research study or meta-analysis.
- *Matrixes*. Matrixes are useful for summarizing data along a small number of dimensions. For example, a matrix could present communication patterns and topics by presenting people along two axes and using a symbol in the row–column square to indicate frequency or topic of communication between the two people.

It is left to the reader to determine which type of figure or table will best convey the information they have to present. However, when developing these tables and figures, think like a human factors researcher and design them to be easy to perceive and use. Use the following guidelines (adapted from Gillan et al., 1998) to present quantitative data in papers, to improve the readability and usability of figures and tables.

- Design your figure or table to support your readers' cognitive tasks.
- Make sure that all labels and text are readable. Use a sufficiently large font size and high contrast between the background and text (e.g., white background, black text).
- Use clear, concise wording. Long words and phrases can clutter the display and make it difficult to read. Also, more concise wording enables the use of larger font sizes.
- Make all colors and symbols used to convey information clearly discernible from each other and convey their meaning clearly through a legend or key (if there are many symbols) or labeling (if there are only two or three symbols).
- Use symbols, colors, and other meaningful features consistently within a figure or table and if possible, across all figures and tables.
- Consider how readers will perceive the figure or table.
  - Make the main point visible at first glance.
  - Attract the readers' attention to the most important features.
  - Use graphical techniques (line thickness, color, etc.) for emphasis.
- Eliminate clutter to improve visual searching of the figure or table.
- Place related items close to each other so that relationships between them are more readily apparent.
- Whenever possible, use structures and presentation methods that are familiar to the user. For example, in flowcharts, adhere to the generally accepted meanings associated with different shapes.

5.3 Documentation

Research results are presented in document(s) that detail a study's findings. When large volumes of unstructured data are involved, documentation may be the only way to communicate and present those findings. However, especially in the case of unstructured data, it can be difficult to determine the appropriate scope, level of detail, and organization for documenting the findings. When writing, keep in mind that the goal is not to write the largest document possible, including every detail encountered in the unstructured research outcomes. Instead, as my fifth-grade English teacher instructed, the goal is to make it long enough to cover the subject, but short enough to be interesting. The last thing you want to do is spend weeks writing a detailed work analysis, only to have the 4-inch-thick monster end up as someone's doorstop. "Brevity is the soul of wit," and it saves you and your readers' time and increases the readability of your document. So the primary question becomes how you can present results based on unstructured outcomes clearly and concisely.

There are a number of technical writing guides that help provide the answer to that question. Many of these guides (e.g., Gerson and Gerson, 1997; Reep, 1997) provide guidelines for specific types of documents, so we will not delve into those details here. However, it is useful to review more general principles and methods for good technical writing. These principles are helpful in determining which unstructured outcomes to focus on in the report and how to present those outcomes.

1. *Know the audience*. To communicate effectively with the audience through the document, you must first identify and understand the audience. A variety of audience characteristics should affect how your present the material. For example, it is important to consider the audience's subject knowledge or level of expertise (Gerson and Gerson, 1997; Reep, 1997). Readers who are experts will probably need less detailed explanations of concepts and definitions than novice or lay readers. Also consider the reader's motivation or purpose for reading the document. This

will have a strong influence on the appropriate scope, level of detail, and structure of the document.

2. *Define the objective(s) of the document.* Before writing the first word, define the objective(s) of your document. Beginning with these objectives in mind, it will be easier to organize and present the information needed to achieve those objectives. Without a well-defined audience and objectives, you may find yourself mired in a series of extensive reorganizations and revisions. In the case where there are multiple objectives, consider carefully if all of the objectives identified can be met in one document. If any of the objectives require presenting significantly different information or target drastically different audiences, it may be more effective to address those objectives in separate documents. Although writing two documents may take more time, it will be time well spent if it ensures that the documents will be read and your objectives achieved.

3. *Keep in mind general objectives in technical writing.* Gerson and Gerson (1997) recommend that writers strive for clarity, conciseness, accuracy, and organization in their writing. Writing for *clarity* helps ensure that the audience understands what they have read. To achieve clarity, provide answers to anticipated questions, provide specific details, and use terms that are easily understood by the audience. Writing *concisely* is beneficial, as it saves time for both the author and reader, but it can also improve comprehension of the material. Conciseness can be achieved by limiting sentence length, omitting redundancies, and avoiding wordy phrases. *Accuracy* entails ensuring that the document is grammatically, factually, and textually correct. Accuracy can usually be achieved by thorough proofreading. Be especially careful to check figures, equations, and references. The *organization* of the document is crucial to communicating effectively. Although the appropriate organization will depend on the purpose and content of the document, be sure that the information is organized and presented logically. One way to achieve this is to start by constructing an outline.

4. *Construct an outline.* An outline is a useful tool for improving the clarity, conciseness, and organization of the document. Whether informal or formal, outlines usually begin as a list of major topics and subtopics that will be addressed in the document. The list is transformed into an outline as the author(s) reorders the topics until they are presented in a logical order that facilitates understanding the material and achieving the stated document objectives. As Reep (1997) points out, developing an outline enables the author to see the document structure before he or she begins writing. It also enables the author to focus on presenting and explaining information as they write the rough draft, as opposed to organizing and writing at the same time, which often results in poor organization and redundant content presentation. When multiple authors are working on the same document, creating an outline has added benefits. By creating an outline, the authors discuss and agree to the overall content and structure before they begin writing. Then each author can draft his or her sections of the document in parallel, knowing how they fit into the overall content and structure. This greatly simplifies merging those sections later in the writing process.

5. *Use figures and tables.* When appropriate, use figures and tables, as discussed earlier, to convey information. In many cases, figures and tables can convey complex information or provide a big picture overview more clearly and concisely than can a written description. Gerson and Gerson (1997) provide useful criteria for using figures and tables in documents. Good figures and tables:

- Are integrated with the text (i.e., text explains graphic, and vice versa)
- Add to the material conveyed in the text but are not redundant with the text
- Communicate important information that would be difficult to obtain easily in longer text
- Do not include details that would detract from the information conveyed
- Are located close to the text that refers to them (preferably on the same page)
- Are appropriately sized
- Are readable
- Are labeled correctly with legends, headings, and titles
- Use a style consistent with other figures and tables in the document
- Are well-conceived and executed

6. *Use appendixes to convey supplementary information.* Use appendixes to provide useful additional detail that only some readers will need or that is too detailed for inclusion in the main body of the document. For example, an appendix might include highly technical information or examples of surveys. Also, when extensive statistical analysis has been completed, it can be useful to highlight the significant results in the main document and include the full statistical results in an appendix in order to improve the clarity and conciseness of the main document.

6 ANALYZING SURVEYS

The results of surveys are a special case in analyzing both structured and unstructured outcomes data for human factors research. As shown in Chapter 11, surveys may vary greatly in their purpose, content, and structure, and great care must be taken in their development to ensure that the survey gathers data that are reliable and valid. In this chapter we do not revisit those concepts. Instead, our focus is on applying the analysis techniques discussed previously to analyze survey results in a way which ensures that the results and conclusions are valid.

In terms of format, survey questions can be either structured or unstructured (i.e., open-ended). In structured questions, answers may be nominal/categorical, ordinal, or interval/ratio. As demonstrated earlier, the measurement type of the answer has a strong influence on which analysis methods can be used to examine the data and find answers to the study's research questions. Therefore, we recommend considering the research objectives and the data analysis methods that support meeting those objectives when designing survey questions. This ensures that the format of the question results in answers that are of the appropriate measurement type to support the analysis methods that will result in achieving your research objectives.

6.1 Validating the Data

Before analyzing survey results, it is important to review the responses to ensure that there are no gaps in the data. One of the steps in validating the data is to check the overall response rate to ensure that it is sufficiently high. This is not as much of an issue for surveys administered by researchers in the course of lab-based research where response rates should be 100%. However, in mail or telephone surveys, response rates are very important. If a large number of potential participants do not respond or there is a pattern to the participants that do not respond, it can inadvertently bias the survey results. If this is the case, do not even bother analyzing the results. Unfortunately, this means that you should reexamine your survey and data collection procedures and modify them as needed to increase your response rate; then begin collecting data again.

If the overall response rate is adequate, review the individual questions to determine if there are missing answers. If a participant did not answer a large portion of the questions, it could indicate that he or she did not understand the survey or questions. If this is the case, it may make sense to remove that person's entire survey from the results. In general, if a participant is missing data for only one or two questions, it is all right to leave the data in for analysis that does not involve that question. Of course, if the missing data are crucial to the analysis, the participant's entire survey may need to be excluded. For example, if a survey is examining gender differences in communication and a participant omits his or her gender, that data is useless. If a large number of participants omit answers to the same question, it could indicate that the question was confusing or that participants were uncomfortable answering the question. In this case, again, the research must judge whether on not it is valid to include this question in the analysis. Whatever the cause for the missing data, it is crucial for the researcher to be aware that the data are missing and of the implications the missing data have or can have on the results. It is up to the researcher to determine the appropriate course of action to ensure that the missing data do not compromise the results of the survey analysis.

6.2 Analysis of Unstructured Answers

Although unstructured questions are typically less time consuming to write, their answers are more time consuming to analyze. The appropriate analysis of unstructured answers, of course, depends on the purpose of the question and research. Some researchers, interested in general information gathering only, may be able to get away with simply reading and summarizing the answers. Unfortunately, a more rigorous method is often required to summarize the answers in a structured manner so that statistical methods can be used to summarize and analyze the results. For this more rigorous analysis of answers, coding is required.

Note that coding for survey answers can be more difficult than coding a researcher's field notes and other unstructured data captured by the researchers involved in a study. One reason for this increased difficulty is that answers from different participants are often not directly comparable (Alreck and Settle, 1985). When unstructured data are gathered from researchers, they (hopefully) share a common understanding of the research objectives and intent and a relatively similar vocabulary. However, survey participants frequently do (and should) vary widely. Participants may use different vocabularies, have different meanings for similar words, or have different understanding of the question's intent. Additionally, their answers may be vague and the researcher is often unable follow up with a participant to get clarification on an answer. (Also, given the amount of time that has elapsed between when the survey is administered and when a clarification is requested, the validity of the clarification may be called into question.) Therefore, when coding unstructured answers to survey questions, it is vital to take special care in developing a coding scheme and assessing intercoder reliability as recommended in Section 5.1.

For unstructured questions, the purpose of coding the answers is often to capture common themes. For example, if the question was "What did you like best about this interface?", the researcher might want to identify common concepts or themes regarding what participants liked best about the interface. Coding the data based on these concepts results in nominal/categorical measure(s). Using these nominal measures, the researcher could count the frequency with which each concept was mentioned and examine relationships between these frequencies and demographic or other measures collected through the research study. In other words, coding the unstructured answers imposes structure on those data so that the researcher can use analysis methods for structured data to gain further insights from the data.

6.3 Analysis of Structured Answers

When analyzing structured answers from surveys, a number of the analysis methods described previously in this chapter may be used. However, there are some special considerations that must be taken in analyzing survey answers. In this section we address some of those considerations, as well as mapping some of the general measurement and statistical concepts discussed previously as to their use for survey analysis.

6.3.1 Measurement Types

At the beginning of this chapter, several measurement types were defined. The answers to structured

Table 4 Example Question Formats

Question Format	Example	Measurement Type
Multiple choice	What is your job category? (a) Sales (b) Administrative (c) Engineering (d) Management (e) Other	Nominal/categorical
Single response item	Have you completed XYZ certification?	Nominal/categorical
Order/rank items	Rank the three tools based on which is easiest to use. (1 = easiest, 3 = hardest)	Ordinal
Likert scale	1 Strongly agree, 2 Agree, 3 Neutral, 4 Disagree, 5 Strongly disagree	Interval
Frequency scale	1 Always, 2 Often, 3 Sometimes, 4 Seldom, 5 Never	Interval

survey questions can be classified using these measurement types, which determine the appropriate analysis methods. Table 4 identifies the measurement type for several common survey question formats. Note that in multiple-choice questions, where the participant can select multiple items, for data analysis purposes it is sometimes easier to treat each possible answer as a separate single response (yes/no) item and/or to create a measure that indicates the number of responses. For example, if a question asks users to mark all the types of computer applications they use, code each application as a separate answer (e.g., word processing (Y/N), spreadsheet (Y/N), etc.). This enables comparing use among application types and comparing application use to other measures collected in the study (e.g., task performance, participant preferences). Additionally, comparing the number of application types used (i.e., a count of the application types used) to other participant responses and/or task performance can also provide interesting insights. In this example, the number of application types used can measure the breadth of computer experience, which can be a crucial covariate in human–computer interaction research.

6.3.2 Exploring and Describing the Data

As in any analysis of structured outcomes, the first step is exploring and describing the data (answers). The methods described in this chapter apply here. Begin by graphing the answers individually through histograms and box plots to examine the distribution, range, and variability of answers. Next, describe the data using the mean, median, percentiles, minimum, and maximum. Note that the appropriate descriptors will depend on the research objectives (e.g., measuring central tendency vs. range and distribution of values) and the type of data. For example, with ranking data, it may make more sense to examine percentages of responses (e.g., 60% ranked this item 1, 10% ranked it 2, and 30% ranked it 3) instead of mean ranks (what does a mean rank of 1.7 mean?).

After examining each answer individually, the next step is to explore possible relationships between answers. Here, again, scatterplots are useful graphical means of examining possible relationships. Also use correlation to examine and describe relationships among ordinal and interval/ratio answers. Use contingency tables (crosstabs) and the chi-squared test to examine relationships among nominal/categorical answers. When correlations exist, they often provide interesting insight into the relationship between two answers, but keep in mind that correlations do not necessarily imply causality. These correlations are, however, crucial to determining the appropriate way to analyze and make inferences from the data.

6.3.3 Making Inferences from the Data

Once we have gotten familiar with the structured data, we can use the more advanced statistical methods presented in this chapter to make inferences about the results. However, because of the nature of survey data, special care must be taken when using these methods. Survey data require special considerations for two reasons: (1) since surveys generally include many questions, statistical tests may be used many times to examine the results for different questions/sets of questions; and (2) the answers in many surveys, especially surveys dealing with user preferences and opinions, are frequently correlated.

Recall our earlier discussion on experimentwise error. Because we are performing statistical tests on a number of potentially correlated answers, there is a danger that the experimentwise error in the survey analysis will be inflated. Although many researchers overlook experimentwise error, researchers are encouraged to consider this and, whenever possible, use statistical methods and analysis approaches that manage this error appropriately. For example, in cases where answers to questions are logically and statically correlated, MANOVA tests should be used to examine group differences instead of only doing separate ANOVA analyses for each question. MANOVA tests are designed to control experimentwise error and provide additional insights into group difference on combinations of answers.

Aside from considerations on experimentwise error, the structured data analysis methods presented earlier can be used as described. Refer once more to Figure 2 for assistance in selecting the appropriate methods based on the analysis objectives and data characteristics of the answers. Although the appropriate analyses will depend on the content of the survey collected and the context of the research design, the following examples are presented to generate ideas on how these methods could be used:

- *Compare demographic groups on survey answers.* For example, use gender or age group to group participants; then compare each group's answers using the appropriate method [e.g., *t*-test for gender (two groups) or ANOVA for age group (more than two groups)].
- *Use demographic and other data to model a preference-related response.* For example, in an experiment examining two interfaces, demographic characteristics, previous experience measures, and performance measures could be used to create a logistic regression model to model the likelihood that a participant will prefer one interface to the other.
- *Use correlated preference or opinion responses to develop a new measurement for a complex concept.* For example, the research presented in the case study in Figure 14 used factor analysis to examine questions that addressed various aspects of decision satisfaction and to decompose those answers into factors that capture four underlying dimensions of decision satisfaction.

7 CONCLUSIONS

As demonstrated in this chapter, a number of factors influence which outcomes data to gather in human factors research and how to analyze and draw conclusions from those outcomes. We first described the characteristics of outcomes data and how those characteristics affect how the outcomes may be analyzed and the types of conclusions that can be draw from the analysis and outcomes. Next, we examined a number of methods for analyzing both structured outcomes, often using statistical methods, and unstructured outcomes, which require less concrete analysis methods. The discussion of both the outcome characteristics and analysis methods has stressed how both of these relate to the research objectives through the types of conclusions they support. By applying this information, human factors researchers can ensure that the outcomes data they collect and the analysis methods they use produce reliable, valid results that support their research goals.

REFERENCES

Alreck, P. L., and Settle, R. B. (1985), *The Survey Research Handbook*, Richard D. Irwin, Homewood, IL.

Argyrous, G. (2000), *Statistics for Social and Health Research with a Guide to SPSS*, Sage Publications, London.

Bollen, K. A. (1989), *Structural Equations with Latent Variables*, Wiley, New York.

Box, G. E. P., Hunter, W. G., and Hunter, J. S. (1978), *Statistics for Experimenters: An Introduction to Design, Data Analysis, and Model Building*, Wiley, New York.

Chapman, G. B., and Elstein, A. S. (2000), "Cognitive Processes and Biases in Medical Decision Making," in *Decision Making in Health Care*, G. B. Chapman and F. A. Sonnenberg, Eds., Cambridge University Press, Cambridge, pp. 183–210.

Cohen, J. (1988), *Statistical Power Analysis for the Behavioral Sciences*, 2nd ed., Lawrence Erlbaum Associates, Mahwah, NJ.

Devellis, R. F. (1991), *Scale Development: Theory and Applications*, Sage Publications, Newbury Park, CA.

Edwards, W., and Newman, J. R. (1982), *Multiattribute Evaluation*, Sage Publications, Beverly Hills, CA.

Ericsson, K. A. (1998), "Protocol Analysis," in *A Companion to Cognitive Science*, W. Bechtel and G. Graham, Eds., Blackwell, Oxford, pp. 425–432.

Fava, J. L., and Velicer, W. F. (1992), "An Empirical Comparison of Factor, Image, Component, and Scale Score," *Multivariate Behavioral Research*, Vol. 27, No. 3, pp. 301–322.

Field, A. (2000), *Discovering Statistics Using SPSS for Windows*, Sage Publications, London.

Fischhoff, B., Slovic, P., and Lichtenstein, S. (1988), "Knowing What You Want: Measuring Labile Values," in *Decision Making: Descriptive, Normative and Prescriptive Interactions*, D. E. Bell, H. Raiffa, and A. Tversky, Eds., Cambridge University Press, Cambridge, pp. 398–421.

Fryback, D. G. (1998), "Methodological Issues in Measuring Health Status and Health-Related Quality of Life for Population Health Measures: A Brief Overview of the 'HALY' Family of Measures," in *Summarizing Population Health: Directions for the Development and Application of Population Metrics*, M. J. Field and M. R. Gold, Eds., National Academy Press, Washington, DC.

Gerson, S. J., and Gerson, S. M. (1997), *Technical Writing: Process and Product*, 2nd ed., Prentice Hall, Upper Saddle River, NJ.

Ghiselli, E. E., Campbell, J. P., and Zedeck, S. (1981), *Measurement Theory for the Behavioral Sciences*, W.H. Freeman, San Francisco.

Gillan, D. J., Wickens, C. D., Hollands, J. G., and Carswell, C. M. (1998), "Guidelines for Presenting Quantitative Data in HFES Publications," *Human Factors*, Vol. 40, No. 1, pp. 28–41.

Grayson, K., and Rust, R. (2001), "Interrater Reliability," *Journal of Consumer Psychology*, Vol. 10, No. 1–2, pp. 71–73.

Guadagnoli, E., and Velicer, W. F. (1988), "Relation of Sample Size to the Stability of Component Patterns," *Psychological Bulletin*, Vol. 103, No. 2, pp. 265–275.

Hughes, M. A., and Garrett, D. E. (1990), "Intercoder Reliability Estimation Approaches in Marketing: A Generalizability Theory Framework for Quantitative Data," *Journal of Marketing Research*, Vol. 27, No., pp. 185–195.

Johnson, D. E. (1998), *Applied Multivariate Methods for Data Analysts*, Duxbury Press, Pacific Grove, CA.

Johnson, R. A., and Wichern, D. W. (1988), *Applied Multivariate Statistical Analysis*, 2nd ed., Prentice Hall, Englewood Cliffs, NJ.

Kalton, G., and Stowell, R. (1979), "A Study of Coder Variability," *Applied Statistics*, Vol. 28, No. 3, pp. 276–289.

Keeney, R. L., and Raiffa, H. (1976), *Decisions with Multiple Objectives*, Wiley, New York.

Krantz, D. H., Luce, R. D., Suppes, P., and Tversky, A. (1971), *Foundations of Measurement*, Vol. 1, *Additive and Polynomial Representations*, Academic Press, San Diego, CA.

Krippendorff, K. (1981), *Content Analysis*, Sage Publications, London.

Llewellyn-Thomas, H. A., McGreal, M. J., and Thiel, E. C. (1995), "Cancer Patients' Decision Making and Trial-Entry Preferences: The Effects of 'Framing' Information About Short-Term Toxicity and Long-Term Survival," *Medical Decision Making*, Vol. 15, No. 1, pp. 4–12.

Lomax, R. G. (2001), *An Introduction to Statistical Concepts for Educational and Behavioral Sciences*, Lawrence Erlbaum Associates, Mahwah, NJ.

Mazur, D. J., and Hickam, D. H. (1994), "The Effect of Physicians' Explanations on Patients' Treatment Preferences," *Medical Decision Making*, Vol. 14, No. 3, pp. 255–258.

Meister, D. (2004), *Conceptual Foundations of Human Factors Measurement*, Lawrence Erlbaum Associates, Mahwah, NJ.

Merriam-Webster (2004), "Merriam-Webster Online," retrieved April 14, 2004, from http://www.webster.com/cgi-bin/dictionary.

Neter, J., Kutner, M. H., Nachtsheim, C. J., and Wasserman, W. (1996), *Applied Linear Statistical Models*, 4th ed., WCB McGraw-Hill, Boston.

Newton, R. R., and Rudestam, K. E. (1999), *Your Statistical Consultant*, Sage Publications, Thousand Oaks, CA.

Oppenheim, A. N. (1992), *Questionnaire Design, Interviewing, and Attitude Measurement*, Pinter Publishers, London.

Reep, D. C. (1997), *Technical Writing: Principles, Strategies, and Readings*, 3rd ed., Allyn & Bacon, Needham Heights, MA.

Sainfort, F. C., and Booske, B. C. (2000), "Measuring Post-decision Satisfaction," *Medical Decision Making*, Vol. 20, No. 1, pp. 51–61.

Sheskin, D. J. (1997), *Handbook of Parametric and Nonparametric Statistical Procedures*, CRC Press, Boca Raton, FL.

Slovic, P. (1995), "The Construction of Preference," *American Psychologist*, Vol. 50, No. 5, pp. 364–371.

Sprent, P., and Smeeton, N. C. (2001), *Applied Nonparametric Statistical Methods*, 3rd ed., Chapman & Hall/CRC Press, Boca Raton, FL.

von Winterfeldt, D., and Edwards, W. (1986), *Decision Analysis and Behavioral Research*, Cambridge University Press, Cambridge.

Wickens, C. D., Gordon, and Liu, Y. (1998), *An Introduction to Human Factors Engineering*, Prentice Hall, Upper Saddle River, NJ.

Wixon, D. (1995), "Qualitative Research Methods in Design and Development," *ACM Interactions*, October, pp. 19–24.

Wu, C. F. J., and Hamada, M. (2000), *Experiments: Planning, Analysis, and Parameter Design Optimization*, Wiley, New York.

PART 8

HUMAN–COMPUTER INTERACTION

CHAPTER 45
VISUAL DISPLAYS

Kevin B. Bennett, Allen L. Nagy, and John M. Flach
Wright State University
Dayton, Ohio

1 INTRODUCTION

Advances in computer science and artificial intelligence currently provide new forms of computational power with the potential to support human problem solving. One use of this computational power is to provide an expert system or an automatic assistant that provides advice to the human operator at the appropriate times. For example, there has been some progress in the use of production systems and neural networks as the drivers for decision support. An alternative, complementary use is to integrate information graphically (or more generally, "perceptibly"). Here computational power is used to create and manipulate representations of the target world rather than to create autonomous machine problem solvers. Perhaps the most general term that has been applied to this endeavor is *representation aiding* (Zachary, 1986; Woods and Roth, 1988; Woods, 1991). Representation aiding offers a unique opportunity to improve overall performance of human–machine systems. The technologies needed to produce computer graphics are mature, and when designed properly, representation aids maintain the flexibility of the human in the loop and improve the capability of the overall system to respond to unforeseen circumstances. The challenge in providing effective representation aids centers around how best to use these technological capabilities to support human decision making and problem solving. Although this chapter focuses on concepts and principles of design to meet this challenge, we see representation aiding and machine problem solvers as complementary tools in the designer's tool chest and we expect that for very complex systems both approaches will be necessary.

In contrast to most other treatments of display design, we did not provide a "cookbook" of detailed guidelines and recommendations (primarily because they tend to be conflicting and difficult to apply). Instead, we chose to describe a set of general heuristics for display design. Because these heuristics are necessarily abstract, we have made the discussion more concrete by illustrating them within the context of a simple domain. We describe how the heuristics apply to that domain and annotate our written descriptions with concrete graphic examples. Our goal is to transfer functional knowledge of display design to practitioners.

We begin our discussion with a description of basic physiological, perceptual, and technological considerations in display design. These considerations are the foundation for display design and represent the baseline conditions that must be met for a display to be effective. We next consider four alternative approaches to display design. Each approach emphasizes a different conceptual aspect of the display design puzzle, and each approach has both strengths and weaknesses. A fifth approach is outlined; this approach draws from the strengths of the earlier approaches and incorporates new considerations that are particularly relevant to the design of displays for complex, dynamic domains. We discuss some analytical tools of the approach and illustrate their use in determining the various types of information that are required for a simple domain. We describe alternative displays and discuss how each display provides a specific mapping that emphasizes certain aspects of the domain but deemphasizes or even eliminates other aspects. We end the chapter by considering the limitations of our discussion and examples and additional challenges for display design.

2 PHYSIOLOGICAL, PERCEPTUAL, AND TECHNOLOGICAL CONSIDERATIONS

In this section we consider fundamental aspects of the visual system and visual perception that are relevant for display design. Information on the surface of a display is most often represented by a difference in perceived brightness or a difference in perceived color between the information-carrying stimuli and the background of the display field. This section is concerned primarily with the detection and perceived appearance of these differences. Although this chapter is focused primarily on emissive displays, it is useful to begin by discussing some of the differences between reflective and emissive displays and the implications of these differences for visual perception. Emissive displays, such as the cathode ray tube (CRT), generate the light that is used to produce text, symbols, or pictures that carry information. Reflective displays such as road signs, pages in a textbook, and the speedometer in an automobile do not produce any light, but reflect some portion of the light that falls on them. Although emissive displays are much more versatile and flexible in some respects, it is probably safe to say that the use of reflective displays to present information was, and still is, far more common. With regard to the visual system and visual perception, there are some fundamental differences between reflective and emissive displays. We begin by examining properties of achromatic, or colorless, displays that illustrate these differences and later in this section, take up chromatic displays.

2.1 Reflective Displays

The surface of a reflective display reflects some portion of the light energy that falls on it in many different directions. The percentage of light reflected, known as the *reflectance* of the surface, and the dependence of this percentage on the wavelength of the light, known as the *spectral reflectance function* of the surface, are determined by the physical properties of the surface (Nassau, 1983). We begin by discussing surfaces with flat spectral reflectance functions that reflect approximately the same percentage of light for all wavelengths. Images are placed on the surface by changing the properties of the surface in local regions. For example, suppose that a printer for a personal computer deposits black ink on a gray page so as to form text. The gray page reflects a percentage, perhaps 50%, of the light energy at each wavelength falling on it. The ink deposited on the page appears very dark because it reflects only a small percentage, for example, 5%, of the light energy falling on it. Suppose that an observer views this page tacked to the wall painted uniformly white so that the surface of the wall has a reflectance of 90%.

The reflectance of surfaces varies with the angle of incidence of the illumination and the angle at which the reflectance is measured. Reflectances of surfaces can be described with two components, a specular component and a diffuse component (Shafer, 1985; Hunter and Herold, 1987). The specular component is mirrorlike, in that a large proportion of the light is reflected off at an angle equal to the angle of incidence. The diffuse component is characterized by light reflected off in all directions. Shiny surfaces such as mirrors have a large specular component and a small diffuse component, whereas matte surfaces such as a velvet cloth have a large diffuse component and a small specular component. For simplicity we ignore these complexities here. Figure 1*a* illustrates idealized spectral reflectance curves for the page, the ink, and the wall. Real spectral reflectance curves would only approximate flat curves. Surfaces with flat curves are neutral in the sense that they do not change the spectral quality of the light that falls on them.

To characterize the light reflected back from the surface, we need to know something about the light falling on the surface. A typical spectrum for sunlight is shown in Figure 1*b*, where the relative energy is plotted as a function of wavelength. This spectrum is referred to as typical because the spectrum for sunlight varies with time of day, time of year, latitude, and atmospheric conditions. Not all of the energy in sunlight is effective in generating a visual response. Some wavelengths of light are more likely than others to be absorbed by the receptors in the eye, the rods and cones. A function describing the relative effectiveness of different wavelengths for photopic or cone vision (Figure 2) was standardized by the CIE in 1924 (Wyszecki and Stiles, 1982). This function, known as the *photopic luminosity function*, has served as a standard in science and industry ever since.

A similar function for scotopic or rod vision was standardized in 1951 (Wyszecki and Stiles, 1982). Since most displays are viewed under photopic conditions, we concentrate on cone vision here. To

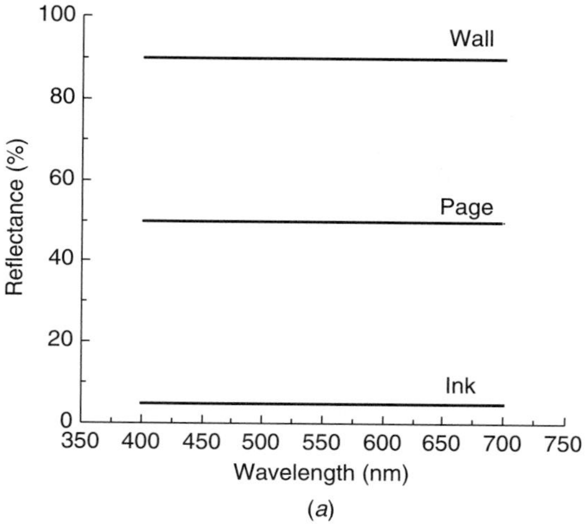

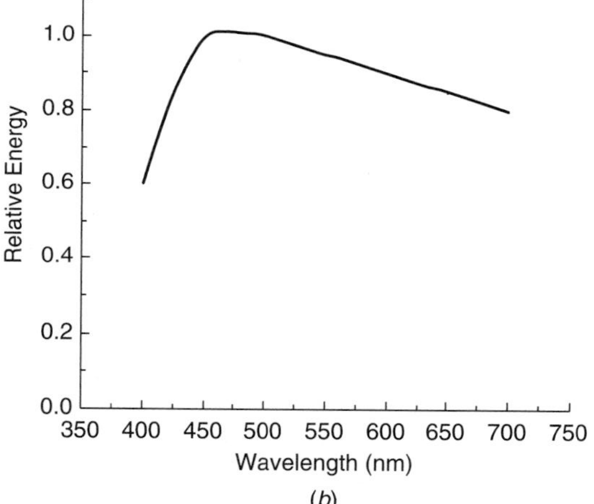

Figure 1 (*a*) Idealized spectral reflectance curves for the ink, the page, and the wall in the example described in the text; (*b*) relative energy at each wavelength in sunlight.

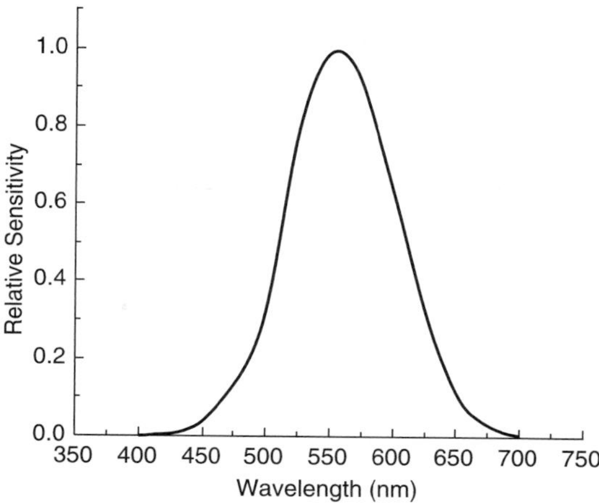

Figure 2 CIE 1924 photopic luminosity curve.

get a measure of the visual effectiveness of the light energy from the sun, we multiply the energies at each wavelength in Figure 1*b* by the value of the photopic luminosity function at that wavelength. The sum or integral of these weighted energies, multiplied by a constant to convert the energy units to a convenient unit of visual effectiveness, is known as the *luminance* of the source. A commonly used unit for luminance today is the candela per square meter (Cd/m^2).

For our purposes, the more important measure is the amount of light that actually falls on the wall, the page, and the ink. This quantity is known as *illuminance*, the amount of visually effective light that actually falls on a surface in space. We assume that the wall is evenly illuminated so that this measure is the same across the wall, the text, and the page. A common unit of illuminance is the lux. The measurement of luminance, and the related quantity illuminance, is itself a complex topic, and many different types of units are used in measuring light. [For discussions of light measurement, see Grum and Bartleson (1980) and Wyszecki and Stiles, (1982).] To find the amount of visually effective light reflected from the surface, we multiply the reflectance at each wavelength times the illuminance provided by the sunlight at each wavelength. Alternatively, we could measure directly the amount of visually effective light reflected in a particular direction using a device called a photometer. [For a discussion of devices for measuring light, see Post (1992).]

An important property of reflective displays, such as our page of printed text mounted on the wall, is that the physical contrast between the text and the page, or the page and the wall, does not vary with the amount of light falling on them as long as all of the surfaces are illuminated at the same level. The term *physical contrast* is used to refer to the difference in the light reflected from two regions of a scene. The physical contrast of a stimulus on a background is often defined as the *contrast ratio*, $\Delta L/L$, the difference between the light reflected from the stimulus and the background divided by the background level. In our example the physical contrast between the text and the page could be specified as the difference in the amounts of light reflected by the ink and by the page divided by the amount of light reflected by the page. Note that as the amount of light falling on the wall is changed, the physical contrast ratios calculated for the text and the page, the text and the wall, and the page and the wall will remain constant (Figure 3). The reader can demonstrate this by setting up the contrast ratios and demonstrating that the light level, which appears in both the numerator and the

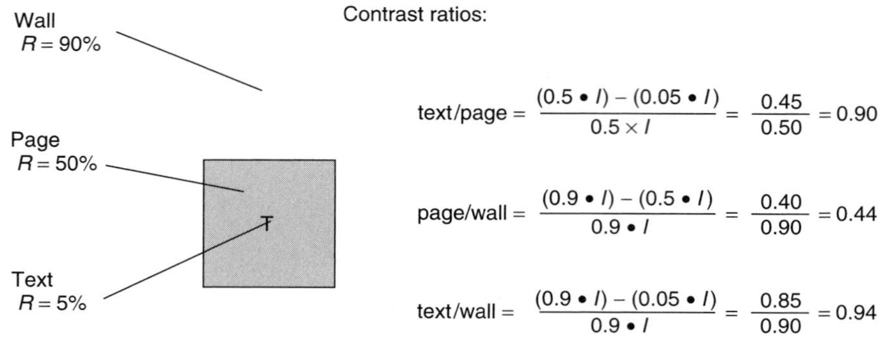

Figure 3 Calculation of contrast ratios for the page, the text, and the wall. The values of *R* indicate the reflectances of the three surfaces in the figure. *I* represents the illumination level, which is identical for all three surfaces in the figure and therefore cancels out of the equations.

denominator of the contrast ratio, will cancel out, and the contrast ratios are determined by the reflectances alone.

The human visual system appears to have evolved to take advantage of the reflective properties of surfaces. One of the earliest relationships established in the study of visual perception is that the intensity difference between a stimulus and a background necessary for detection of the stimulus is a constant proportion of the intensity of the background field. This rule, known as *Weber's law*, is often written in equation form as $\Delta I = kI$. Here ΔI refers to the difference between the intensity of the stimulus and the intensity of the background, k is the proportionality constant or the Weber fraction, and I is the intensity of the background field. Weber's law indicates that the visual system becomes less sensitive to differences between the stimulus and the background as the intensity of the background field increases. That is, in order to keep the stimulus detectable, the difference between the stimulus and the background must be increased as the background is increased. Notice, however, that if we rearrange Weber's law by dividing both sides of the equation by I, we get $(\Delta I/I) = k$.

At threshold, the difference between the intensities of the stimulus and the background (ΔI) divided by the background intensity (I) is constant. This is exactly the situation for the reflective displays described above. It means that if the text on a page is detectable at any light level, it will remain detectable as the light level is changed. A somewhat different form of Weber's law also applies to the discrimination of two stimuli presented on a background. In this case, at threshold the difference in the contrasts between the two stimuli relative to the background must be a constant proportion of one of the contrasts (Whittle, 1986, 1992; Nagy and Kamholz, 1995). Thus, for reflective displays, if two stimuli at different contrast levels on a background are discriminable from each other, they will remain discriminable as the illumination level is changed. It is well known that Weber's law is only approximately true and that it breaks down under many conditions, perhaps most important, when the light levels involved are low and approach absolute threshold. However, the change in the sensitivity implied by Weber's law is an important property of the visual system. It is a component of another property of the visual system known as *lightness constancy*, which refers to the fact that the visual system operates in such a manner as to keep the perceived appearance of reflective objects approximately constant under changing illumination levels. That is, the wall, the page, and the text in our example appear white, gray, and black, respectively, whether they are viewed outdoors under intense sunlight or indoors under dim illumination. Lightness constancy depends on many other factors in addition to the change in sensitivity indicated by Weber's law and has been a topic of intense interest in the last couple of decades (Gilchrist et al., 1983; Adelson, 1993).

2.2 Emissive Displays

We will use a CRT as an example of an emissive display. CRTs generate light by shooting beams of electrons at substances called *phosphors* which are painted on the screen of the CRT. When the electrons hit a point on the screen, light energy is given off by the phosphor at that point. The intensity of the light given off can be changed by varying the strength of the beam of electrons directed at the point. Images are created on the screen by varying the intensity of the electron beam hitting different points on the screen. The physical contrast between different regions of the screen can be defined in the same manner as for reflective displays.

Suppose that we mount the CRT on the white wall and use it to generate a page of dark text on a gray page. Suppose also that we adjust the CRT so that the page gives off 50 units of light and the text gives off 5 units of light. The physical contrast ratio between the text and the page would be 0.90, as it was for the reflective display (see Figure 3). Suppose that the white wall is illuminated initially so that 90 units of light are reflected from it. Also suppose for the moment that the surface of the CRT reflects none of this light. In this case the contrast ratios between the three surfaces would be the same as in our first example with the reflective page of text, and we might expect that the CRT display would look very similar to the reflective display.

Note what happens as the illumination falling on the wall is increased, however. The intensity of the light reflected from it increases, but the intensities of the lights from the text and the page on the CRT do not change. The contrast ratio between the text and the page on the CRT remains constant, but the contrast ratios between the page and the wall, and the text and the wall, increase. Thus, we might expect the appearances of the text and the page to change considerably as the light level falling on the wall is changed. If we regard the text and the page as individual incremental stimuli against the large background provided by the wall, Weber's law suggests that their discriminability will decrease as the light reflected from the wall increases. The decrease in discriminability occurs because the difference in contrast ratios decreases with increasing light level. In this case the decrease in the sensitivity of the visual system with increasing background light level reduces the ability to detect the difference between the text and the page which remains constant.

Any light that is reflected from the glass face of the CRT will reduce the discriminability of the text on the page even further, because it will be reflected from both the region containing the dark text and the region containing the page. The reflected light actually reduces the physical contrast between the text and the page and makes them even less discriminable. Thus, emissive displays behave quite differently than reflective displays in natural environments. These differences do not present much of a problem when emissive displays are placed in a constant environment such as an office illuminated by a fixed light source. However, when emissive displays are placed in natural environments in which the illumination level may vary by a factor of a million or more, the problems caused by the varying contrast ratios are evident. For example, this problem occurs when emissive displays are used in aircraft. The detectability and the appearance of elements within the display may vary dramatically. To keep the appearance of the text and the page constant, the light levels given off by the CRT must be adjusted in accord with the change in the illumination of the wall.

2.3 Factors Affecting Perceived Contrast

Besides the physical contrast, there are many other factors, such as adaptive state, location in the visual field, eye movements, and the interpretation of the perceived illuminant, which affect the perceived contrast of a stimulus against a background. One of the most important of these factors is stimulus size. In the last few decades this problem has been investigated very successfully with an approach based on Fourier analysis. [For extensive reviews and applications, see Ginsburg (1986), Olzack and Thomas (1986), Graham (1989), DeValois and DeValois (1990), Pavel and Ahumada (1997), and Makous (2003).] Fourier analysis suggests that any pattern of light and dark on the retina can be described as a sum of sinusoidal components of different frequency and amplitude. The application of this idea to visual perception involves measuring an observer's sensitivity to a number of sinusoidal patterns of different spatial frequency (Figure 4). These repetitive spatial patterns of light and dark are known as *gratings*.

Spatial frequency is essentially a measure of the size of the bars in the pattern. The spatial frequency of the pattern is defined as the number of cycles that occur in 1° of visual angle. As spatial frequency increases, there are more cycles per degree of visual angle and the bars become smaller. Visual angle is used as the unit of size because it gives a measure of the size of the image on the retina (e.g., a book 12 in. long makes a larger image on the retina when it is held up close to the eye than when it is held far away). To get a measure of the size of an image on the retina, the distance between an object and the observer's eye must be considered. Thus, the visual angle subtended by an object is defined as twice the arctan of the height/2 divided by the distance (Figure 5).

Sensitivity is measured by finding the physical contrast level at which a given pattern of light and dark is just detectable. To give a measure of sensitivity, the reciprocal of the threshold is calculated by dividing 1 by the threshold contrast. The measure of physical contrast typically used in this approach is slightly different from the contrast ratio described above and is called the *Michelson contrast*. It is defined as $L_{\max} - L_{\min}$ divided by $L_{\max} + L_{\min}$, where $L_{\max}$ is defined as the maximum luminance level in the pattern and $L_{\min}$ is defined as the minimum luminance in the pattern. The curve described by plotting contrast sensitivity against the spatial frequency of the grating pattern is called the *contrast sensitivity function*.

A typical contrast sensitivity function for photopic or cone vision obtained from a human observer is shown in the Figure 6. The curve shows that when spatial frequency is low (i.e., the bars are large), the sensitivity to contrast is low. As spatial frequency is increased, the sensitivity increases up to spatial frequencies of about 5 to 10 cycles per degree. With further increases in spatial frequency (i.e., smaller and smaller bars), sensitivity falls off rapidly until at a spatial frequency of approximately 50 cycles per degree, a grating of 100% contrast (the highest physical contrast obtainable) is not visible. Spatial patterns of even greater frequency also are not visible. Thus, very fine patterns are visible only if the spatial frequency is below 50 cycles per degree and they are very high contrast.

Over the last few decades many physical factors, such as overall light level, number of cycles present in the pattern, and the location of the pattern in the visual field, have been shown to affect the contrast sensitivity function. The shape of the curve as well as the overall sensitivity can vary considerably. The shape and height of the curve are affected by several components within

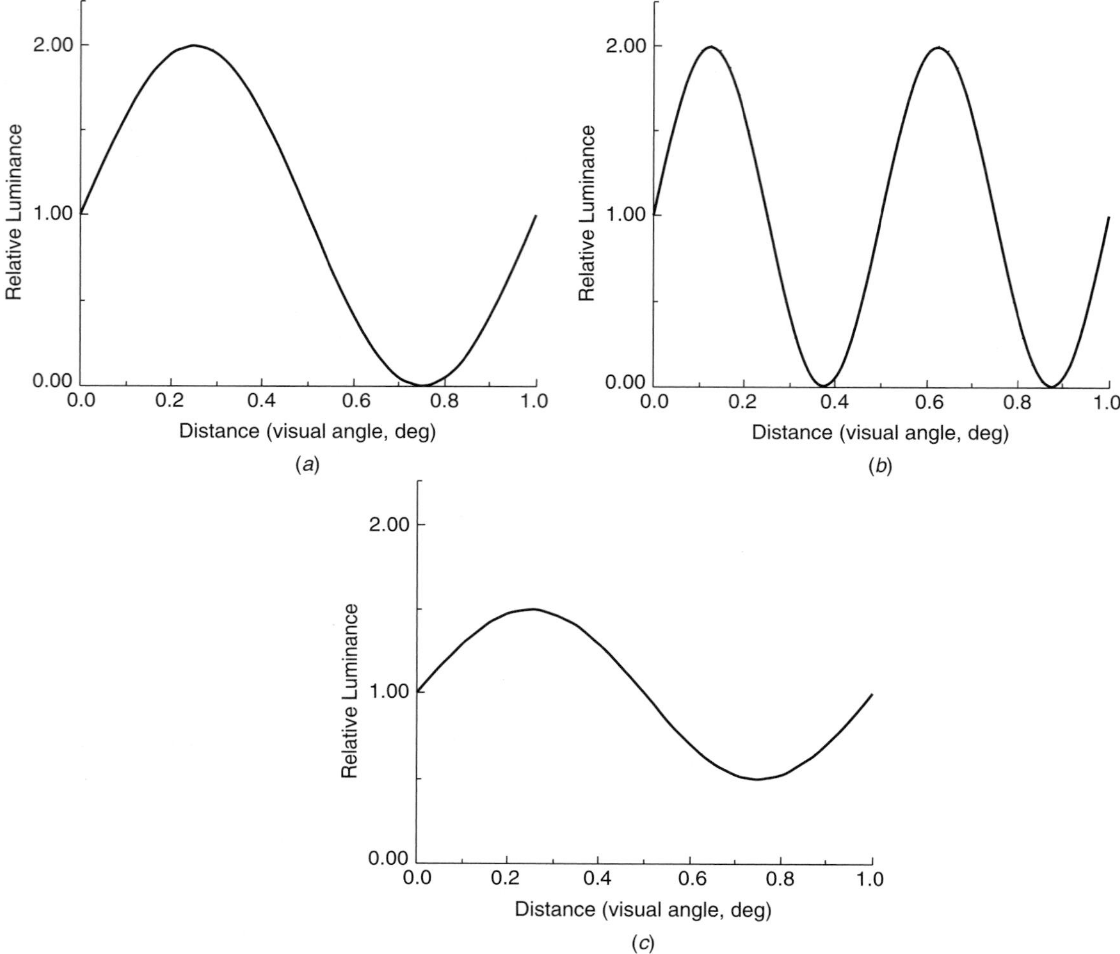

Figure 4 Variation in luminance for sinusoidal patterns: (*a*) spatial frequency of 1 cycle/degree at a contrast of 100%; (*b*) spatial frequency of 2 cycles/degree at a contrast of 100%; (*c*) spatial frequency of 1 cycle/degree at a contrast of 50%.

the visual system that play a role in determining the contrast sensitivity function. For example, the optics of the eye, the lens and cornea, which form an image of the pattern on the retina, influence the contrast sensitivity function, because they do not form a perfect image of the external pattern on the retina. A good introductory treatment of the optics of the eye is given by Millodot (1982). The distribution of rods and cones on the retina also plays a role in determining the contrast sensitivity function. The rods and cones absorb light and initiate neural signals in the visual system. Thus, their size and the distances between them have some effect on the contrast sensitivity function. A good introduction to the sampling properties of rods and cones is given by Wandell (1995). The way the rods and cones are connected to the neurons that carry signals out of the eye also plays a role in determining the contrast sensitivity function, because many receptors are connected to each neuron. Psychophysical evidence suggests that the visual system may be organized into approximately five to seven neural channels, each sensitive to a different band of spatial frequencies (Olzack and Thomas, 1986). Thus, the contrast sensitivity function is the result of many factors which have been studied intensely over the last few decades. Nevertheless, it is a very useful and fundamental description of the ability of a human observer to detect contrast in patterns of different size. For example, recent studies suggest that the recognition of text may be mediated by the same mechanisms that mediate the contrast sensitivity function (Alexander et al., 1994; Solomon and Pelli, 1994).

The perceived contrast of patterns that are well above threshold is not simply related to the contrast sensitivity function (Cannon and Fullenkamp, 1991). That is, if we measure the threshold contrast for

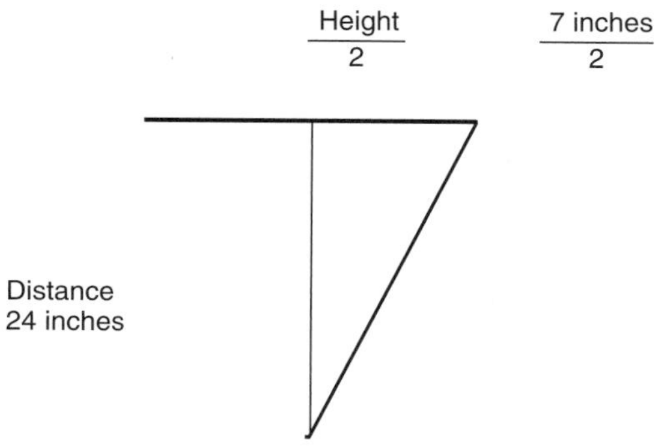

Figure 5 Calculation of the visual angle.

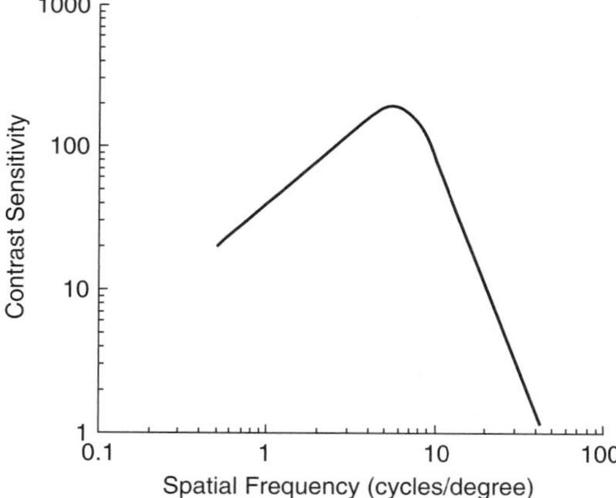

Figure 6 Typical plot of a contrast sensitivity function for a human observer. (Based on data from DeValois and DeValois, 1990.)

sinusoidal patterns at a number of different spatial frequencies and then increase the physical contrast of all of these patterns so that the contrast for each one is five times the threshold contrast, the patterns will not appear to have equal contrasts. This is similar to the situation in audition where equal-loudness curves for tones of different frequencies do not have the same shape as the audibility curve, a plot of threshold as a function of frequency, and change shape as the loudness level is raised. Thus, the contrast sensitivity function can be used to predict whether a pattern of a given spatial frequency is visible, but it cannot be used to predict accurately the perceived contrast of patterns that are well above threshold. For example, if a display designer wants to equate the perceived contrast of patterns of different size that are well above threshold, the contrast sensitivity function cannot be used to do this accurately.

The notions of visual angle, spatial frequency, and contrast sensitivity that were introduced briefly above are very useful in thinking about both reflective and emissive displays. Here we concentrate on emissive displays. Consider a standard CRT display that is 9.5 in. wide and 7 in. high. Assume that this CRT has 640 columns of pixels, each containing 480 rows (standard 640×480 resolution). If the observer views this display from a distance of 2 ft, the screen subtends about 22.4° horizontally and 16.6° vertically (see Figure 5), and each pixel subtends about 0.035°. If we want to make patterns of light and dark bars on the screen, we might want to know the highest spatial frequency that can be represented. If we make alternate pixels black and white, we need two pixels to make one cycle, which will subtend 0.07°. Thus, the highest spatial frequency that can be represented accurately will be 1/0.07, or slightly over 14 cycles per degree.

Looking back at our representative contrast sensitivity function, we see that this frequency is well below the upper limit of approximately 50 cycles per degree. Looking at the vertical axis, we find that the sensitivity at 14 cycles per degree is approximately 30. For an observer to detect this pattern on the screen, we can determine that the Michelson contrast will have to be approximately 1/30 or 3.3%. These calculations also tell us something else. Patterns with spatial frequencies higher than 14 cycles per degree just cannot be represented accurately on the monitor. Thus, if we want to view an image with a lot of fine details at high spatial frequencies, such as a digitized photograph that subtends 9.5 by 7 in., spatial frequencies greater than 14 cycles per degree that were visible when the original photograph was viewed from a distance of 2 ft will not be represented accurately on the monitor if they are composed of spatial frequencies above 14 cycles per degree.

One solution to this problem is to use a monitor with higher resolution or smaller pixels. For example, if we could pack 1280×960 pixels into the same 9.5×7 in. screen, patterns with spatial frequencies up to nearly 29 cycles per degree could be represented. To make a display that matches the upper limit on the resolution of the visual system we would need to pack about 2240×1660 pixels into the display. A 9.5×7 CRT with this resolution would permit the presentation of patterns with spatial frequencies up to 50 cycles per degree at a viewing distance of 2 ft. This would be very difficult to accomplish with the present technology, making the display and the computer hardware that drives it very expensive.

It is also possible to portray patterns with spatial frequencies greater than 14 cycles per degree on the original CRT by moving the observer farther away so that each pixel subtends a smaller visual angle. The drawback to this approach is that the entire display field now subtends a smaller portion of the field of view. For example, if we move the observer back to a distance of about 4 ft, patterns with spatial frequencies up to nearly 29 cycles per degree could be portrayed on the screen. This example helps to illustrate a fundamental trade-off in emissive displays, the trade-off between field of view and resolution. With a fixed number of pixels, this trade-off is always present in an emissive display. If the pixels are spread over a larger viewing area, the resolution will be poor. If they are packed into a smaller viewing area, the resolution will improve but the field of view will decrease.

The resolution of an emissive display may be limited either by the display itself or by the hardware that drives it: that is, the video card in a computer or the signals generated on a television cable. The detail in an image, or the spatial frequencies that can be portrayed, and the field of view that is visible will be limited by this resolution and the size of the screen.

2.4 Color

Although black-and-white pictures carry much of the information in the real world, they do not carry information about color. Color in images is certainly important for aesthetic reasons, but in addition to the aesthetic qualities it brings to an image color serves two important basic functions (Boynton, 1990). First, chromatic contrast between two regions in an image can add to the luminance contrast between these regions to make the difference between the regions much more noticeable, especially when the luminance contrast is small. Second, since color is perceived to be a property of an object (although, in fact, it also depends on illumination, as we will see), it is useful in identifying objects, searching for them, or grouping them. Boynton (1990) regards the second function of color, which he describes as related to categorical perception, as the more important one. It is probably because of these categorical properties that color is often used as a coding device and as a means of segregating information in visual displays (see Widdel and Post, 1992).

Several excellent treatments of the basics of human color vision and the science of specifying colors for applications are available (e.g., Wyszecki and Stiles, 1982; Pokorny and Smith, 1986; Travis, 1991; Post, 1992, 1997; Kaiser and Boynton, 1997; Gegenfurtner and Sharpe, 1999; Nagy, 2003), so a very brief review will be given here. Normal human color vision depends on the presence of three types of cone receptors in the retina. These cones differ in the type of light-absorbing pigment contained in them. One of these pigments absorbs best, meaning the greatest percentage of the light falling on it, in the short-wavelength region of the spectrum; hence the cone containing it is referred to as the *S cone*. The second pigment absorbs best in the middle of the spectrum, and the cone containing it is referred to as the *M cone*. The third pigment absorbs best at slightly longer wavelengths than the M pigment and the cone containing it is referred to as the *L cone*.

The differences in the signals generated in these cones by a given light provide some information about the spectral content of the light. For example, a light source that gives off more energy in the long-wavelength portion of the spectrum than in the middle- or short-wavelength regions would tend to stimulate the L cones more than the other two cone types. On the other hand, a light source that gives off more energy in the short-wavelength region would tend to stimulate the S cones more than the other two types. The differences in the stimulation of the cone types serve as a means for discriminating between the lights, and result in the perception of color.

Since there are only three types of cones, normal human color vision is said to be *three-dimensional* or *trichromatic*. Furthermore, since there are only three signals from different types of cones in the visual system, it follows that only three numbers are needed to specify the perceptual quality of a color. Much effort has gone into developing systems of specifying colors with three numbers such that they represent the perceptual qualities of the stimulus in useful ways. The fact that only three numbers are needed to specify the chromatic quality of a stimulus also means that there are many physically different stimuli that stimulate the three cones in the same way and thus appear to be the same color. Stimuli that are physically different but appear to be the same are called *metamers*.

Consider the reflective display example given above. Suppose that we print the text on our gray page using red ink rather than black ink. The ink appears red because it tends to absorb short- and middle-wavelength light that falls on it, while reflecting long-wavelength light. A spectral reflectance curve showing the percentage of light reflected as a function of wavelength for red ink might look like the curve shown in Figure 7. To get the light reflected back from the ink, we multiply the reflectance at each wavelength times the energy at each wavelength. To calculate the luminance of this light, we would weight the reflected energy at each wavelength by the photopic luminosity function and integrate or sum over the entire curve as we did for achromatic stimuli above. However, the

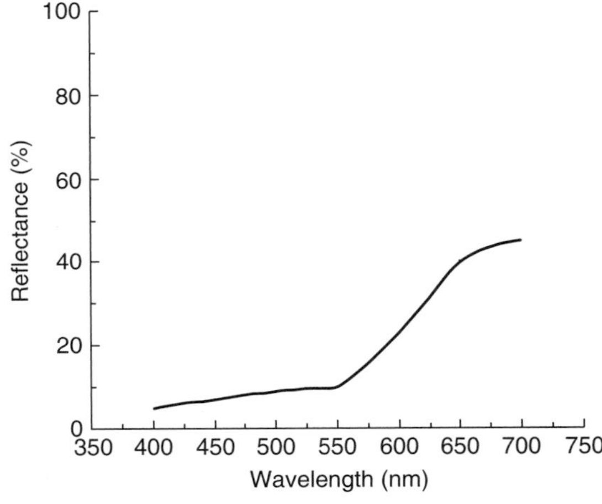

Figure 7 Spectral reflectance curve for red ink.

text appears to differ from the gray page and the white wall in color as well as in lightness. To characterize this difference, we would like some means of measuring the colors of the text, the page, and the wall. The most widely used system for doing this is based on the CIE 1931 chromaticity diagram. This diagram is based on color matches of normal human observers. A good introduction to the color-matching experiment and the development of chromaticity diagrams can be found in Kaiser and Boynton (1997). In the color-matching task, observers were asked to adjust the intensities of three primary lights that were mixed together in a single stimulus field so as to match the colors of a wide variety of other lights presented in another stimulus field. The CIE chromaticity diagram uses three numbers (related to the intensities of the primaries needed to make a match in the color-matching experiment) to represent the color or, more specifically, the chromaticity of a stimulus. These numbers are called the *chromaticity coordinates* of the color and are referred to as x, y, and z. The color-matching data were normalized so that the values of these three chromaticity coordinates add up to 1 for any real color. As a result, only two of the chromaticity coordinates need to be given to specify a color, because the third can always be obtained by subtracting the sum of the other two from 1. Therefore, all colors can be represented in a two-dimensional diagram such as the CIE 1931 diagram shown in Figure 8, where only x and y are plotted. Many measuring instruments have been developed and are commercially available for measuring the CIE coordinates of a color. [See Post (1992) for some discussion of these.]

The chromaticity coordinates specify the chromatic properties of a color but do not specify its appearance, because the appearance of the color can change with many viewing conditions that do not change its chromaticity coordinates. For example, the size of the stimulus, in terms of visual angle, can affect the color appearance even though the chromaticity coordinates of the ink used to make it do not change (Poirson and Wandell, 1993). This is a severe limitation on the meaning and usefulness of the CIE chromaticity diagram. One would like to have a system in which the appearance of the color is specified, but this is a very difficult problem that has not yet been solved. Nevertheless, the specification of colors in the chromaticity diagram is still very useful, because any two stimuli with the same chromaticity coordinates will appear to be identical in color when viewed under the same conditions. What the chromaticity coordinates specify is how to make a color that will appear the same as a given sample under the same viewing conditions.

The chromaticity coordinates of a reflective display change with the chromaticity of the light used to illuminate it. The change occurs because the amount of light reflected back from an object at each wavelength depends in part on the amount of light falling on it. Therefore, when the chromaticities of objects, or dyes, or paints are specified, they are usually given with reference to a standard light source. [For a discussion of standardized light sources, see Wyszecki and Stiles (1982).] One might expect that the change in the chromaticity coordinates accompanying a change in the light source would change the color appearance of a reflective display. Such changes in light source are actually quite common. As noted above, the spectral quality of daylight changes with time of day, atmospheric conditions, season, and location on Earth. A large variety of artificial light sources are commercially available, and these can differ considerably in the spectral quality of the light given off. However, these changes do not generally result in large changes in the appearances of objects, because mechanisms within the visual system act to maintain a constant color appearance despite these changes in illumination. Color constancy has generally been shown to be less than perfect (Arend and Reeves, 1986; Brainard and Wandell, 1992). However, it appears to work well enough to prevent confusing changes in the appearance of reflective objects. The visual mechanisms mediating color constancy have been of intense interest over the past few decades (D'Zmura and Lennie, 1986; Maloney and Wandell, 1986). Selective adaptation within the three cone mechanisms is thought to be one of the major mechanisms mediating color constancy (Worthy and Brill, 1986) much as the change in sensitivity described by Weber's law plays a role in lightness constancy.

Although mechanisms of color constancy work to maintain a constant appearance in reflective displays, they actually work against the maintenance of a constant appearance in emissive displays, much as mechanisms of lightness constancy worked against the constant appearance of black-and-white emissive displays. Color CRTs take advantage of the fact that human color vision is trichromatic by using only three different phosphors. Each phosphor emits light of a

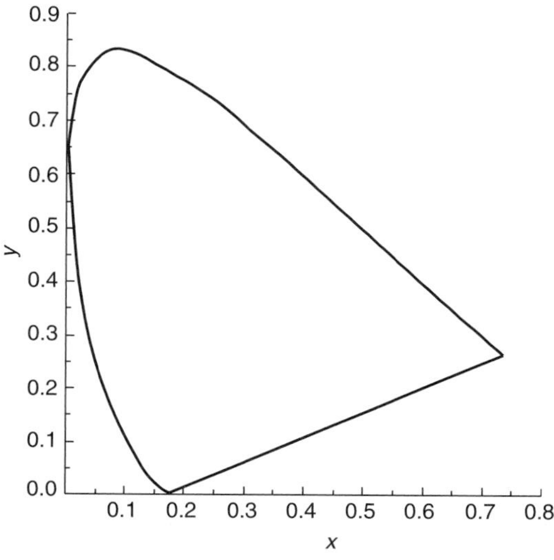

Figure 8 CIE 1931 chromaticity diagram. (Based on data from Wyszecki and Stiles, 1982.)

different color when it is stimulated. The light from the three phosphors is mixed together in different proportions to give all other colors, including white.

The chromaticity of a color produced on an emissive display does not change with changes in the illumination of the surroundings. Thus, the mechanisms of color constancy, activated by changes in the illumination of the surroundings, introduce changes in the appearance of these chromaticities, which may be quite noticeable to the observer. Under some conditions these changes in appearance may be large enough to cause some confusion in identifying objects on the basis of color.

2.4.1 Factors Affecting Perceived Color Contrast

Much as the perception of achromatic contrast is affected by many factors, color contrast is affected by many factors, such as light level, adaptive state, location in the visual field, and stimulus size. The spatial frequency approach has also been applied to the detection of color contrast. It is possible to produce grating patterns which vary sinusoidally in color, with little or no variation in luminance. The color contrast between the bars of the grating required for detection of the pattern can be measured as a function of the spatial frequency (Kelly, 1974; Noorlander and Koenderink, 1983; Mullen, 1985; Sekiguchi et al., 1993). Typical results for red/green and yellow/blue gratings are shown in Figure 9. Comparison of the results for chromatic patterns with those shown for luminance patterns reveals clear differences. Sensitivity to color contrast is high at low spatial frequencies but begins to fall off dramatically at rather low spatial frequencies compared to luminance contrast. Above spatial frequencies of approximately 12 cycles per degree color contrast is not detectable even at the highest color contrasts producible. Thus, chromatic contrast information is limited to fairly low spatial frequencies, or large patterns, as compared to luminance contrast information. Within this range of spatial frequencies, the color appearance of the bars of a pattern that is well above threshold is also affected by spatial frequency (Poirson and Wandell, 1993). As the spatial frequency of the pattern is increased, the apparent color contrast between the bars is reduced. Thus, the detectability of color contrast and the color appearance of stimuli is affected dramatically by stimulus size.

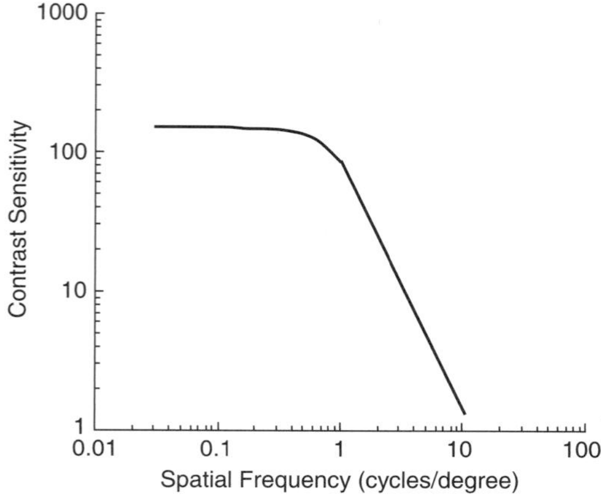

Figure 9 Typical plot of contrast sensitivity for isoluminant chromatic gratings. (Based on data from Mullen, 1985.)

3 FOUR ALTERNATIVE APPROACHES TO DISPLAY DESIGN

Earlier we discussed physiological, perceptual, and technological considerations in designing visual displays. This has been the traditional focus for human factors research: to design displays that are legible. For example, the knowledge that a user will be seated a particular distance from a particular type of display under a particular set of ambient lighting conditions can be used to determine the appropriate size and luminance contrast that will be necessary for the characters to be seen. Thus, the previous considerations provide us with an understanding of the baseline conditions of display design that must be met (are necessary) for a person to use a display.

Although these considerations are necessary for the design of effective displays, they are not sufficient. Compliance with these considerations will make the *data* required to complete domain tasks available but may not provide the *information* necessary to support an observer in decision making and action. Woods (1991) makes an important distinction between design for data availability and design for information extraction. Designs that consider only data availability often impose unnecessary burdens on the user: to collect relevant data, to maintain these data in memory, and to integrate these data mentally to arrive at a decision. These mental activities require extensive knowledge, tax-limited cognitive resources (attention, short-term memory), and therefore increase the probability of poor decision making and errors.

Our discussion of design for information extraction will begin with a consideration of four broadly defined approaches to display design. Each approach is complementary in the sense that it approaches the display design problem from different conceptual perspectives (i.e., graphical arts, psychophysical, attention-based, and problem solving/decision making).

3.1 Aesthetic Approach

Tufte (1983, 1990) reviews the design of displays from an aesthetic, graphic arts perspective. Tufte (1983) describes principles of design for data graphics or statistical graphics which are designed expressly to present quantitative data. One principle is the *data-ink ratio*, a measurement of the relative salience of data vs. nondata elements in a graph. It is computed by determining the amount of ink that is used to convey the data and dividing this number by the total amount of ink that is used in the graphic. A higher data-ink ratio (a maximum of 1.0) represents the more effective presentation of information. A second measure of graphical efficiency is *data density*. Data density is computed by determining the number of data points represented in the graphic and dividing this number by the total area of the graphic. The higher the data density, the more effective the graphic. Other principles include eliminating graphical elements that interact (e.g., moire vibration), eliminating irrelevant graphical structures (e.g., containers and decorations), and aesthetics (e.g., effective labels, proportion, and scale).

The two versions of a statistical graphic that are shown in Figure 10 illustrate several of Tufte's principles. The version in Figure 10*a* is poorly designed; the version in Figure 10*b* is more effectively designed. In Figure 10*b* the irrelevant data container (the box) that surrounds the graph in Figure 10*a* has been eliminated. In addition, several other nondata graphical structures have been removed (grid lines). In fact, these grid lines are made conspicuous by their absence in Figure 10*b*. Together, these manipulations produce both a higher data-ink ratio and a higher data density for the version in Figure 10*b*. In Figure 10*a* the striped patterns on the bar graphs produce an unsettling moire vibration and have been replaced in Figure 10*b* with gray-scale patterns. In addition, the bar graphs in Figure 10*b* have been visually segregated by spatial separation. Finally, the three-dimensional perspective in Figure 10*a* complicates visual comparisons and has been removed in Figure 10*b*.

Tufte (1990) broadens the scope of these principles and techniques by considering nonquantitative displays as well. Topics that are discussed include micro/macro designs (the integration of global and local visual information), layering and separation (the visual stratification of different categories of information), small multiples (repetitive graphs that show the relationship between variables across time, or across a series of variables), color (appropriate and inappropriate use of), and narratives of space and time (graphics that preserve or illustrate spatial relations or relationships over time). The following quotations (Tufte, 1990) summarize many of the key principles:

- It is not how much information there is, but rather, how effectively it is arranged. (p. 50)
- Clutter and confusion are failures of design, not attributes of information. (p. 51)
- Detail cumulates into larger coherent structures... Simplicity of reading derives from the context of detailed and complex information, properly arranged. A most unconventional design strategy is revealed: to clarify, add detail. (p. 37)
- Micro/macro designs enforce both local and global comparisons and, at the same time, avoid the disruption of context switching. All told, exactly what is needed for reasoning about information. (p. 50)
- Among the most powerful devices for reducing noise and enriching the content of displays is the technique of layering and separation, visually stratifying various aspects of the data.... What matters—inevitably, unrelentingly—is the proper relationship among information layers. These visual relationships must be in relevant proportion and in harmony to the substance of the ideas, evidence, and data displayed. (pp. 53–54)

This final principle, layering and separation, is graphically illustrated in Figure 10*c* and *d*. These two versions of the same display vary widely in terms of the visual stratification of the information that they contain. In Figure 10*c* all of the graphical elements are at the same level of visual prominence; in Figure 10*d* there are at least three levels of visual prominence. The lowest layer of visual prominence is associated with the nondata elements of the display. The various display grids have thinner, dashed lines, and their labels have also been reduced in size and made thinner. The medium layer of perceptual salience is associated with the individual variables. The graphical forms that represent each variable have been gray-scale coded, which contributes to separating these data from the nondata elements. Similarly, the lines representing the system goals (G_1 and G_2) have been made bolder and dashed. In addition, the labels and digital values that correspond to the individual variables are larger and bolder than their nondata counterparts. Finally, the highest level of visual prominence has been reserved for those graphical elements, which represent higher-level system properties (e.g., the bold lines that connect the bar graphs). The visual stratification could have been enhanced further through the use of color. The techniques of layering and separation will facilitate an observer's ability to locate and extract information.

To summarize, Tufte (1983, 1990) addresses the problem of presenting three-dimensional, multivariate data on flat, two-dimensional surfaces (focusing primarily on static, printed material) very admirably. He attacks the problem from a largely aesthetic perspective and provides numerous examples of both good and bad display design that clearly illustrate the associated design principles. Although there are critical aspects of dynamic display design for complex domains that are not considered, the principles can be generalized.

3.2 Psychophysical Approach

Cleveland and his colleagues have also developed principles for the design of statistical graphics. However, in contrast to the aesthetic conceptual perspective of Tufte, Cleveland has used a psychophysical approach. As an introduction, consider the following quotation (Cleveland, 1985, p. 229):

> When a graph is constructed, quantitative and categorical information is encoded by symbols, geometry, and color. Graphical perception is the visual

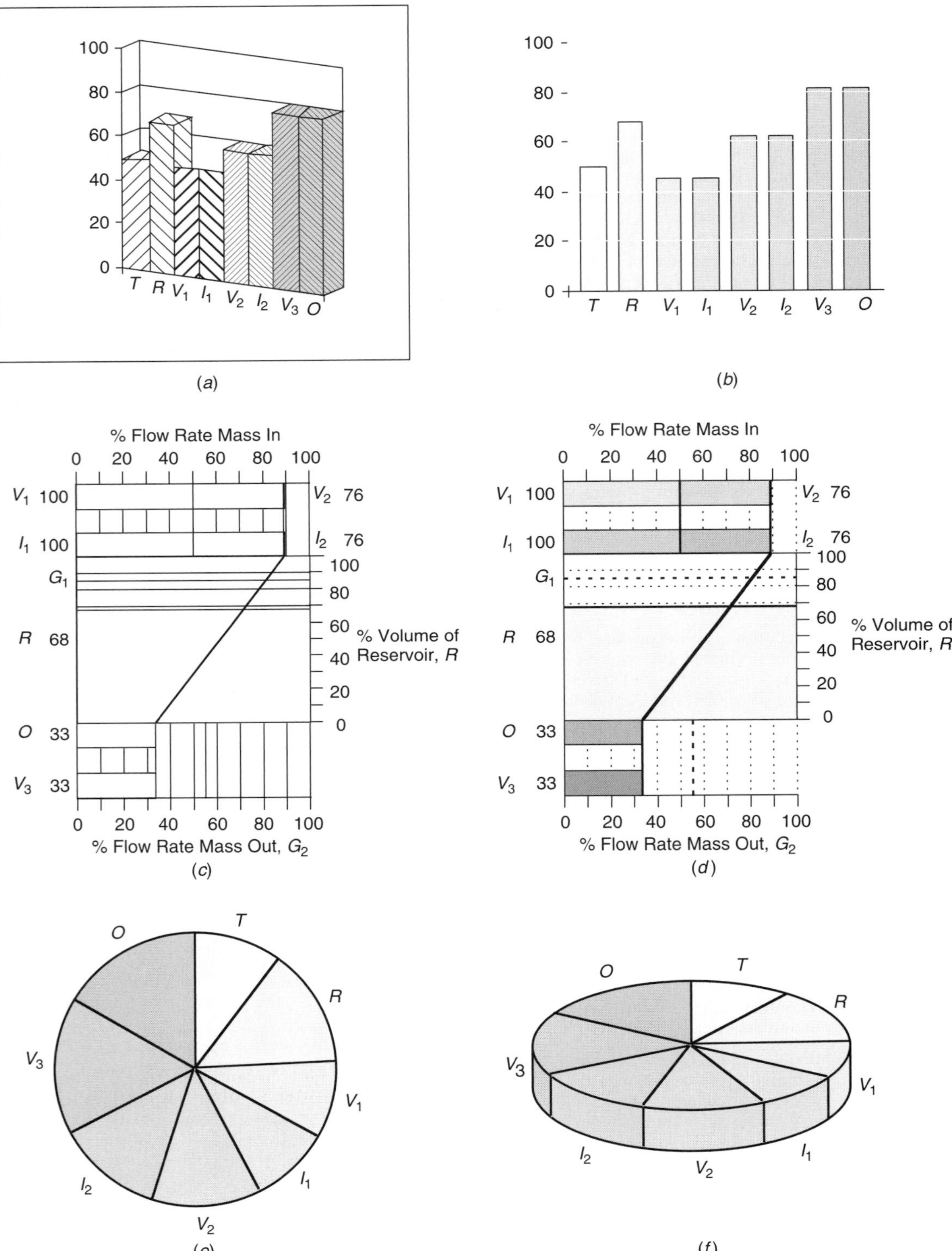

Figure 10 Six alternative mappings. Parts (*a*) and (*b*) represent alternative versions of a separable (bar graph) graphical format that provide a less effective (*a*) and a more effective (*b*) mapping. Parts (*c*) and (*d*) represent alternative versions of a configural display format that provide a less (*c*) and a more (*d*) effective mapping, due primarily to layering and separation. Parts (*e*) and (*f*) represent the least effective mappings.

decoding of this encoded information. Graphical perception is the vital link, the raison d'etre, of the graph. No matter how intelligent the choice of information, no matter how ingenious the encoding of the information, and no matter how technologically impressive the production, a graph is a failure if the visual decoding fails. To have a scientific basis for graphing data, graphical perception must be understood. Informed decisions about how to encode data must be based on knowledge of the visual decoding process.

In their efforts to understand graphical perception Cleveland and his colleagues have considered how psychophysical laws (e.g., Weber's law, Stevens' law) are relevant to the design of graphic displays. For example, psychophysical studies using magnitude estimation have found that judgments of length are less biased than judgments of area or volume. Therefore, visual decoding should be more effective if data have been encoded into a format that requires length discriminations as opposed to area or volume discriminations. Cleveland and his colleagues have tested this, and similar intuitions, empirically. Their experimental approach was to take the same quantitative information, to provide alternative encodings of this quantitative information (graphs that required different "elementary graphical–perception tasks"), and to test observers' ability to extract the information.

The results of these experiments provided a rank ordering of performance on basic graphical perception tasks: position along a common scale, position along identical, nonaligned scales, length, angle/slope, area, volume, and color hue/color saturation/density (ordered from best to worst performance) (Cleveland, 1985, p. 254). Guidelines for display design were developed based on these rankings. Specifically, graphical encodings should be chosen that require the highest-ranking graphical perceptual task of the observer during the visual decoding process. For example, consider the three graphs illustrated in Figure 10*b*, *e*, and *f*. For decoding information contained in Figure 10*b*, an observer is required to judge position along a common scale (in this case, the vertical extent of the various bar graphs). For Figure 10*e* the observer is required to judge angles and/or area. Finally, to decode the information in Figure 10*f*, the observer is required to judge volume (note that because of the three-dimensional representation, angles and area are no longer valid cues). According to the rankings, Cleveland and his colleagues would therefore predict that performance would be best with the bar chart, intermediate with the pie chart, and worst with the three-dimensional pie chart.

3.3 Attention-Based Approach

A third perspective on display design is to consider the problem in terms of visual attention and object perception. From this conceptual perspective, designers have a number of interface resources at their disposal for encoding information in graphical displays (e.g., chromatic contrast, luminance contrast, the integration of individual variables into geometrical objects, and animation). A great deal of basic research has attempted to identify the factors that control the distribution of attention to visual stimuli. The results have important theoretical and practical implications for display design.

Understanding these implications requires a brief consideration of the continuum of attention demands that operators might face in complex, dynamic domains. At one end of the attention continuum are tasks that require selective responses to specific elements in the display ("focused" tasks). This might refer to a response contingent on the height of a single bar in a bar graph or on the position of a pointer on a radial display. At the opposite end of this continuum are tasks that require the distribution of attention across many features that must be considered together to choose an appropriate response ("integration" tasks). For example, the response might be contingent on the relative position of numerous bars within a bar graph. Thus, tasks can be characterized in terms of the relative demands for selective attention to respond to specific features with specific actions and distributed or divided attention in which multiple display elements must be considered together to choose the appropriate actions.

Attention-based approaches to display design have examined how the design of visual representations can help to meet the cognitive load posed by this continuum of attention demands. Garner (Garner, 1970, 1974; Garner and Felfoldy, 1970) and Pomerantz (Pomerantz et al., 1977; Pomerantz, 1986; Pomerantz and Pristach, 1989) have used the speeded classification task to examine the dimensional structure of stimuli. Carswell and Wickens (1990) have generalized these results by investigating perceptual dimensions that are representative of those found in visual displays. Three qualitatively different relationships between stimulus dimensions have been proposed: separable, integral, and configural (Pomerantz, 1986).

Separable Dimensions A separable relationship is defined by a lack of interaction among stimulus dimensions. Each dimension retains its unique perceptual identity within the context of the other dimension. Observers can attend selectively to an individual dimension and ignore variations in the irrelevant dimension. On the other hand, no new properties emerge as a result of the interaction among dimensions. Thus, performance suffers when both dimensions must be considered to make a discrimination. This pattern of results suggests that separable dimensions are processed independently. An example of separable dimensions are color and shape: The perception of color does not influence the perception of shape, and vice versa.

Integral Dimensions An integral relationship is defined by a strong interaction among dimensions such that the unique perceptual identities of individual dimensions are lost. Integral stimulus dimensions are processed in a highly interdependent fashion: a change in one dimension necessarily produces changes in the second dimension. In their discussion of two integral stimulus dimensions, Garner and Felfoldy (1970, p. 237) state that "in order for one dimension to exist, a level on the other must be specified." As a result of this highly interdependent processing, a redundancy gain occurs. However, focusing attention on the individual stimulus dimensions becomes very difficult, and performance suffers when attention to one (selective attention) or both (divided attention) dimensions are required. An example of an integral stimulus is perceived color: it is a function of both hue and brightness.

Configural Dimensions A configural relationship refers to an intermediate level of interaction between perceptual dimensions. Each dimension maintains its unique perceptual identity, but new properties are also created as a consequence of the interaction between them. These properties have been referred to as *emergent features*. Using parentheses as our graphic elements will allow us to demonstrate several examples of emergent features. Depending on the orientation, a pair of parentheses can have the emergent features of vertical symmetry, () and) (, or parallelism,)) and ((. Pomerantz and Pristach (1989, p. 636) state that "emergent features may be global (i.e., not localized to any particular position within the figure), such as symmetry or closure, or they may be local, such as vertices that result from intersections of line segments." There are two significant aspects of performance with configural dimensions. First, relative to integral and separable stimulus dimensions, there is a smaller divided attention cost, suggesting that performance can be enhanced when both dimensions must be considered to make a discrimination. The second noteworthy aspect of this pattern of results is that there is an *apparent* failure of selective attention. Bennett and Flach (1992) discuss why this failure may be apparent and not inherent; Bennett and Walters (2001) investigate design strategies to overcome potential costs.

3.3.1 Proximity Compatibility Principle

Wickens and his colleagues (e.g., Wickens and Carswell, 1995) have applied the results of the visual attention research to the problem of display design. Their principle of *proximity compatibility* emphasizes the relationship between task demands and the graphical form of a display. *Perceptual proximity* (display proximity) refers to the perceptual similarity between information sources in a display. Perceptual proximity can be defined along several dimensions, including (1) spatial proximity (e.g., physical distance—near or far), (2) chromatic proximity (e.g., the same or different colors), (3) physical dimensions (e.g., information is encoded using the same or different physical dimensions—length vs. volume), (4) perceptual code (e.g., digital vs. analog), and (5) geometric form (e.g., object vs. separate displays). For example, when individual variables are mapped into a closed geometric form, the display is high in display proximity; when each variable has its own unique representation (e.g., a bar graph), the display is low in proximity.

Processing proximity (mental proximity) refers to the continuum of attentional demands, that is, to the extent to which information from the various sources in a display must be (or need not be) considered together to accomplish a particular task. There are three major categories of processing proximity: integrative processing, nonintegrative processing, and independent processing. Information from different sources must be explicitly combined in integrative processing, and this represents a high level of processing proximity. Integrative processing includes both computational processing (involving numerical operations) and Boolean processing (involving logical operations). Nonintegrative processing represents an intermediate level of processing proximity and involves "some other features of similarity instead of (or in addition to) their need for combination" (Wickens and Carswell, 1995, p. 476). Examples include (1) metric similarity (similarity of units), (2) statistical similarity (extent of covariation), (3) functional similarity (semantic relatedness), (4) processing similarity (similarity of computational procedures), and (5) temporal similarity (temporal proximity). Finally, *independent processing* refers to the situation where different information sources need not be considered together (in fact, one information source is independent of another).

Briefly stated, the principle of proximity compatibility maintains that the *display proximity* should match the *task proximity*. Performance on integrated tasks (high mental proximity) is predicted to be facilitated by displays that have high perceptual proximity (e.g., object display). Similarly, performance on focused tasks (low mental proximity) is predicted to be facilitated by displays that have low perceptual proximity (e.g., bar graph displays).

Implications Researchers continue to investigate the potential trade-offs between display type [object (Figure 10*d*) vs. separate (Figure 10*b*)] and task type (integrated vs. focused). Initially, a straightforward trade-off was predicted: Object displays would produce superior performance for integration tasks, whereas separable displays would produce superior performance for focused tasks. In general, laboratory research comparing performance differences between object and separate displays when integration tasks must be completed has revealed significant advantages for object displays (Bennett and Flach, 1992). However, there is a general consensus that these performance advantages are not attributable to objectness per se (Sanderson et al., 1989; Buttigieg and Sanderson, 1991; Bennett and Flach, 1992; Bennett et al., 1993; Wickens and Carswell, 1995). Instead, the quality of performance at integration tasks is dependent on the quality of the mapping between the emergent features produced by a display and the inherent data relationships that exist in the domain (this point is discussed at length in subsequent sections).

There is much less consensus on the second major prediction regarding the potential costs for configural displays (relative to separable displays) when individual variables must be considered. We believe that a single display may support performance at both integration and focused attention tasks (Bennett and Flach, 1992). The attention and object perception literature (in particular, the principle of configurality) leaves open the possibility that a single geometric display may be designed to support performance for both distributed and focused attention tasks. One way to consider objects is as a set of hierarchical features (including elemental features, configural features, and global features) that vary in their relative salience. For example, Treisman (1986, p. 35.54) observed that "if an object is complex, the perceptual description we form may be hierarchically structured, with global entities defined by subordinate elements and subordinate elements related to each other by the global description." Observers may focus attention at various levels in the hierarchy at their discretion, and in particular, there may be no inherent cost associated with focusing attention on elemental features.

From a practical standpoint, Bennett and Walters (2001) found that potential costs associated with the extraction of low-level data from configural displays can be essentially eliminated by annotating the analog display with digital information. Both integrated (manual control and fault detection) and focused (quantitative estimates of individual process variables) tasks were administered. Four display design techniques (i.e., bar graphs/extenders, scale markers/scale grids, color coding/layering/separation, and digital values) were applied alone, and in combination, to a configural display. The composite display (incorporating all four techniques) produced very good performance for both task types. Bennett and Walters (p. 431) concluded that "participants could select and use the specific design features in the composite display that were appropriate for tasks at each boundary [focused and integrated].... The results represent progress toward a fundamental display design goal: single graphical displays capable of supporting performance at multiple tasks." The design technique primarily responsible for improved performance at focused tasks was the presence of digital values; see Calcaterra and Bennett (2003) for an investigation of different strategies for the annotation of configural displays involving alternative placements of digital values.

3.4 Problem-Solving and Decision-Making Approach

The fourth perspective on display design to be discussed is problem solving and decision making. Recently, there has been an increased appreciation for the creativity and insight that experts bring to human–machine systems. Under normal operating conditions, a person is perhaps best characterized as a decision maker. Depending on the perceived outcomes associated with different courses of action, the amount of evidence that a decision maker requires to choose a particular option will vary. In models of decision making, this is called a *decision criterion*. Under abnormal or unanticipated operating conditions, a person is characterized most appropriately as a creative problem solver. The cause of the abnormality must be diagnosed, and steps must be taken to correct the abnormality (i.e., an appropriate course of action must be determined). This involves monitoring and controlling system resources, selecting between alternatives, revising diagnoses and goals, determining the validity of data, overriding automatic processes, and coordinating the activities of other people. Thus, the literature on reasoning, problem solving, and decision making has important insights for display design.

There is a vast literature on problem solving, ranging from the seminal work of the Gestaltists (e.g., Wertheimer, 1959), to the paradigmatic contributions of Newell and Simon (1972), to contemporary approaches. For the Gestalt psychologists, perception and cognition (more specifically, problem solving) were intimately intertwined. The key to successful problem solving was viewed as the formation of an appropriate gestalt, or representation, that revealed the "structural truths" of a problem. For example, Wertheimer (1959, p. 235) states that "thinking consists in envisaging, realizing structural features and structural requirements...." The importance of a representation is still a key consideration today; it is probably not an overstatement to conclude that the primary lesson to be learned from the problem-solving literature is that the representation of a problem has a profound influence on the ease or difficulty of its solution.

Historically, decision research has focused on developing models that describe the generation of multiple alternatives (potentially, all alternatives), evaluation (ranking) of these alternatives, and selection of the most appropriate alternative. By and large, perception was ignored. In contrast, recent developments in decision research, stimulated by research on naturalistic decision making (e.g., Klein et al., 1993), has begun to give more consideration to the generation of alternatives in the context of dynamic demands for action. Experts are viewed as generating and evaluating a few "good" alternatives. The emphasis is on recognition (e.g., how this is problem similar, or dissimilar, to problems encountered before). As a result, perception plays a dominant role. This change in emphasis has increased awareness of perceptual processes and dynamic action constraints in decision making.

These trends have, either directly or indirectly, led researchers in interface design to focus on the representation problem. Perhaps the first explicit realization of the power of graphic displays to facilitate understanding was the STEAMER project (Hollan et al., 1984, 1987), an interactive inspectable training system. STEAMER provided alternative conceptual perspectives: "conceptual fidelity" of a propulsion engineering system through the use of analogical representations. In addition, the current approach to the design of human–computer interfaces (direct manipulation) (Hutchins et al., 1986; Shneiderman, 1986, 1993) can be viewed as an outgrowth of this general approach. More recently, scientific visualization (the role of diagrams and representation in discovery and invention) is being investigated vigorously (Brodie et al., 1992; Earnshaw and Wiseman, 1992). Thus, the challenge for display design from this perspective is to provide appropriate representations that support humans in their problem-solving endeavors.

4 REPRESENTATION-AIDING APPROACH TO DISPLAY DESIGN

It should be noted that in the aesthetic, psychophysical, and attention-based approaches, little consideration is given to a domain behind the display. It was not necessary for us to describe the "problem" behind the displays shown in Figure 10. However, the correspondence between the visual structure in a representation and the constraints in a problem is fundamental to the problem-solving and decision-making approaches. Recently, a number of research groups have recognized that effective interfaces depend on both the mapping from human to display (the coherence problem) and the mapping from display to a work domain or problem space (the correspondence problem). Terms used to articulate this recognition include direct perception (Moray et al., 1994), ecological interface design (Rasmussen and Vicente, 1989; Vicente, 1991, 1999; Burns and Hajdukiewicz, 2004), representational design (Woods, 1991), or semantic mapping (Bennett and Flach, 1992).

Woods and Roth (1988) have illustrated the problem of interface design in terms of a *cognitive triad* that we illustrate in Figure 11. The three components of the triad are (1) the cognitive demands produced by the domain of interest, (2) the resources of the cognitive agent(s) available to meet those demands, and (3) the representation of the domain through which the agent experiences and interacts with the domain (the interface).

Each of these three components contributes a set of constraints that will influence the effectiveness (and/or the pleasurableness) of the interaction. A particular domain will introduce a particular set of constraints (e.g., tasks, goals, limits) that will determine the nature of the work to be completed. Another set of constraints are introduced by the cognitive agent (human, machine) that completes the work. For a human agent this will include a specific set of cognition/perception/action capabilities and limitations. The functionality/design of the interface introduces a third set of constraints: Particular characteristics of the interface will introduce cognitive demands that will vary in terms of the nature and amount of cognitive resources that are required. These three sources of constraints are independent but mutually interactive and mutually constraining. The effectiveness of graphical decision support will ultimately depend upon the quality of *very specific sets of mappings* between these constraints. [See Bennett and Walters (2001) for further discussion of these points for display design.] Thus, the focus of our approach is not on information-processing characteristics, graphical forms, events, trajectories, tasks, or procedures per se. Instead, the focus is on the quality of the mappings between the person, the interface, and the domain (labeled coherence and correspondence in Figure 11).

4.1 Correspondence Problem: The Semantics of Work

Correspondence refers to the issue of content: What information should be present in the interface in order to meet the cognitive demands of the work domain? Correspondence is defined neither by the domain itself nor the interface itself: It is a property that arises from the interaction of the two. Thus, in Figure 11, correspondence is represented by the labeled arrows that connect the domain and the interface. One convenient way to conceptualize correspondence is as the quality of the mapping between the interface and the workspace, where these mappings can vary in terms of the degree of specificity (consistency, invariance, or correspondence). As we will demonstrate, within this mapping there can be a one-to-one correspondence, a many-to-one, a one-to-many, or a many-to-many mapping between the information that exists in the interface and the structure within the workspace.

4.1.1 Rasmussen's Abstraction Hierarchy

Addressing the issue of correspondence requires a deep understanding and explicit description of the "semantics" of a work domain. Rasmussen's (1986) abstraction hierarchy is a theoretical framework for describing domain semantics in terms of a nested hierarchy of functional constraints (including goals, physical laws, regulations, organizational/structural constraints, equipment constraints, and temporal/spatial constraints). One way to think about the abstraction hierarchy is that it provides structured categories of information (i.e., the alternative conceptual perspectives) that a person must consider in the course of accomplishing system goals. Consider the following passage from Rasmussen (1986, p. 21):

> During emergency and major disturbances, an important control decision is to set up priorities by selecting the level of abstraction at which the task should be initially considered. In general, the highest priority will be related to the highest level of abstraction. First, judge overall consequences of the disturbances for the system function and safety in order to see whether the mode of operation should be switched to a safer state (e.g., standby or emergency shutdown). Next, consider whether the situation can be counteracted by reconfiguration to use alternative functions and resources. This is a judgment at a lower level of function and equipment. Finally, the root cause of the disturbance is sought to determine how it can be corrected. This involves a search at the level of physical functioning of parts and components. Generally, this search for the physical disturbance is of lowest priority (in aviation, keep flying—don't look for the lost light bulb!).

Thus, in complex domains, situation awareness requires the operator to understand the process at

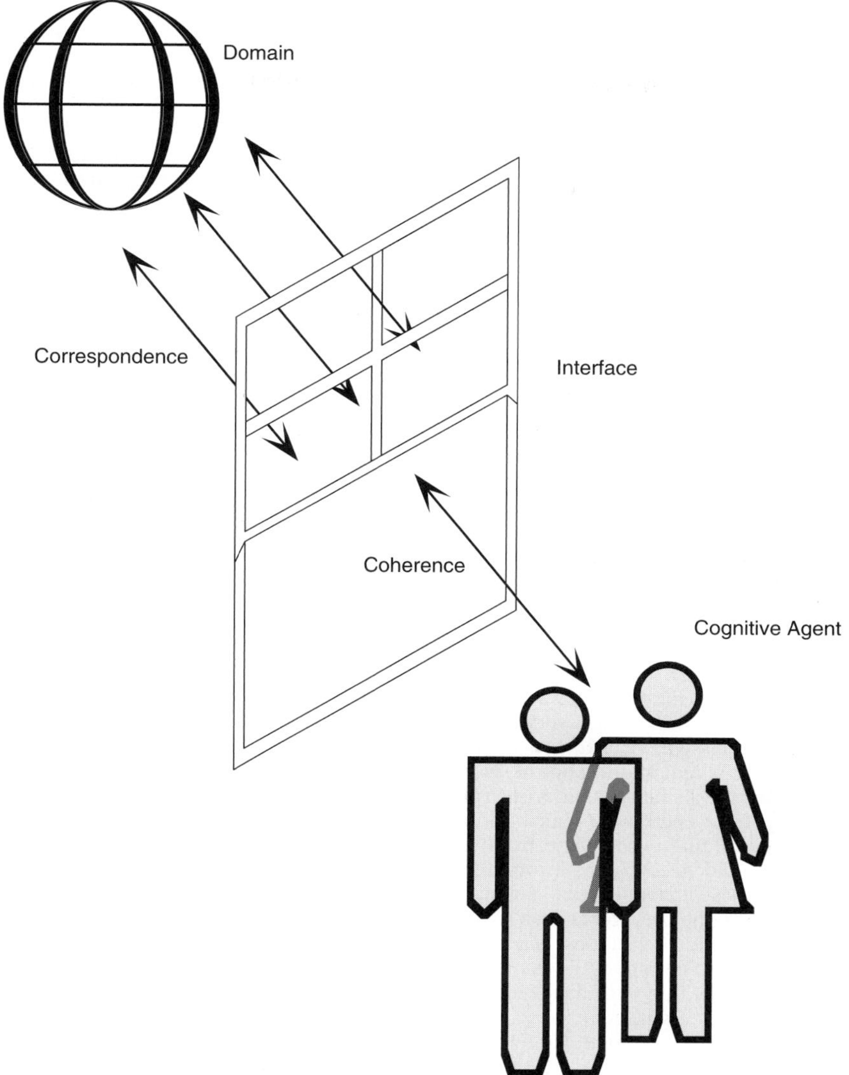

Figure 11 Cognitive triad: a cognitive systems engineering perspective. Any domain produces cognitive demands that must be met by the cognitive agents interacting with (or controlling) the domain. The cognitive agent possesses cognitive resources that must be used to meet these demands. The interface is the medium (or representation) through which the cognitive agent views and controls the domain. The effectiveness of an interface is determined by both correspondence and coherence.

different levels of abstraction. Further, the operator must be able to understand constraints at one level of abstraction in terms of constraints at other levels. The correspondence question asks whether the hierarchy of constraints that define a work domain are reflected in the interface.

4.2 Coherence Problem: The Syntax of Form

Coherence refers to the mapping from the representation to the human perceiver. Here the focus is on the visual properties of the representation. What distinctions within the representation are discriminable to the human operator? How do the graphical elements fit together or coalesce within the representation? Is each element distinct or separable? Are the elements absorbed within an integral whole, thus losing their individual distinctness? Or do the elements combine to produce configural or global properties? Are some elements or properties of the representation more or less salient than other elements or properties?

In general, coherence addresses the question of how the various elements within a representation compete for attentional and cognitive resources. Just as work domains can be characterized in terms of a nested hierarchy of constraints, complex visual representations can be perceived as a hierarchy of nested structures, with local elements combining to produce more global patterns or symmetries.

4.3 Mapping Problem

In human–machine systems, a display is a representation of an underlying domain, and the user's tasks are defined by that domain rather than by the visual characteristics of the display itself. Thus, whether or not a display will be effective is be determined by both correspondence and coherence. More specifically, the effectiveness of the display is determined by the quality of the mapping among the agent, interface, and domain.

The constraints that characterize a particular work domain will have a substantial impact on the type of representations that will provide effective support. Rasmussen et al. (1994) have analyzed the various types of constraints that characterize different work domains and have developed a continuum for classification. At one end of the continuum are domains in which the unfolding events arise from the laws of nature (e.g., process control). An example of such a law is the conservation of mass: If more mass is flowing into a reservoir than out of a reservoir, the level of fluid that it contains will rise. In these "law-driven" domains, the user is required to control, monitor, and compensate for the demands that arise from the domain. At the opposite end of the spectrum are domains in which the unfolding events arise from the user's intentions, goals, and needs (e.g., information search and retrieval). In these "intent-driven" domains, the demands are created by the user rather than by the domain. The domain structure is more loosely coupled (e.g., attributes that differentiate among books of fiction) and the process of searching and identifying the appropriate information (e.g., a particular book of fiction to read) is ultimately user-dependent. Note that an understanding of the domain structure is still critical to the development of effective decision support (Flach et al., 2005).

The design strategies and techniques that are required to develop effective interfaces for these two categories of domains are quite different. In law-driven domains, the constraints of the system (e.g., physical, functional, and goal-related structure) are the primary consideration. Display design involves the development of abstract geometrical forms that reflect these inherent constraints. A simple example is using an axis in a graph to represent time. In configural representations the geometrical display constraints will generally take the form of symmetries: equality (e.g., length, angle, area), parallel lines, collinearity, or reflection. In addition, Gestalt properties of closure and good form are useful. These display constraints will produce the emergent features that were discussed in Section 3.3, each particular representation that is chosen will produce a different set of display constraints, defined by the spatiotemporal structure (the visual appearance of the display over time).

The core problem in implementing effective configural displays for law-driven work domains is to provide visual representations that are perceived as accurate reflections of the abstract domain constraints: Are the critical domain constraints reflected appropriately in the geometrical constraints in the display? Are breaks in the domain constraints (e.g., abnormal or emergency conditions) reflected by breaks in the geometrical constraints (e.g., emergent features such as nonequality, nonparallelism, nonclosure, bad form)? Only when this occurs will the cognitive agent be able to obtain meaning about the underlying domain in an effective fashion.

One source of ideas for configural displays is the graphical representations that engineers use to make design decisions. For example, Beltracchi (1987, 1989) (see also Moray et al., 1994; Rasmussen et al., 1994) has designed a configural display for controlling the process of steam generation in nuclear power plants based on the temperature–entropy graphic used to evaluate thermodynamic engines (Rankine cycle display). Effective representation aiding for law-driven domains allows trained operators to use high-capacity perceptual and motor skills to monitor and control the system as opposed to limited capacity resources (e.g., working memory).

A very different design strategy is required for intent-driven domains, where the needs and goals of the user are the driving force in the unfolding interaction. Relative to law-driven domains, agents working in intent-driven domains will interact with the system more sporadically, will have far less training and experience, and will possess more diverse sets of skills or knowledge. Under these circumstances the appropriate interface design strategy is to use metaphors and icons.

Metaphorical representations use spatial or symbolic relations from other, more familiar work domains. They are designed to relate the functioning of the system and the requirements for interaction to concepts and activities with which the majority of potential agents will already be familiar. Ultimately, the goal is to enhance the transfer of skills from one domain to another. Perhaps the most obvious example is the "desktop" metaphor that is used in personal computer systems. Another example is the BookHouse metaphor, developed by Pejtersen (Goodstein and Pejtersen, 1989; Pejtersen, 1992) to facilitate library information retrieval. Rasmussen et al. (1994, pp. 289–291) describe the metaphor and its justification:

> The use of the BookHouse metaphor serves to give an invariant structure to the knowledge base. ... Since no overall goals or priorities can be embedded in the system, but depend on the particular user, a global structure of the knowledge base reflects subsets relevant to the categories of users having different needs and represented by

different rooms in the house. ... This gives a structure for the navigation that is easily learned and remembered by the user. ... The user "walks" through rooms with different arrangements of books and people. ... It gives a familiar context for the identification of tools to use for the operational actions to be taken. It exploits the flexible display capabilities of computers to relate both information in and about the data base, as well as the various means for communicating with the data base to a location in a virtual space. ... This approach supports the user's memory of where in the BookHouse the various options and information items are located. It facilitates the navigation of the user so that items can be remembered in given physical locations that one can then retraverse in order to retrieve a given item and/or freely browse in order to gain an overview.

In addition to metaphors and configural displays, there is a third type of representation: analogical. This type of representation is appropriate when the constraints of the work domain are fundamentally spatial. For example, STEAMER used an analogical representation of the spatial layout of the feedwater system to show the connections among component processes. Also, the standard flight display for representing pitch and roll (attitude) is an analog to the spatial relations between the aircraft and the horizon. In general, where the domain constraints themselves are naturally spatial, designers should consider whether the interface might provide a direct analog of these constraints.

Whether analogical, configural, metaphorical, or combined representations are used, the key to successful design is the quality of the mapping. The visual salience of the information in the display must reflect the relative importance of that information in terms of the work domain. For analogical displays the spatial analogs must scale appropriately to the real task constraints. For configural displays the geometric symmetries must correspond to higher-order constraints on the process. For metaphorical displays, the intuitions and skills elicited by the representational domain must map appropriately to the target domain.

5 EXAMPLE-BASED TUTORIAL OF THE REPRESENTATION-AIDING APPROACH

The concepts and principles of display design that have been introduced thus far include correspondence, coherence, process constraints, display constraints, and the mappings between process and display constraints. These concepts and principles are necessarily abstract, and for them to be useful for display design they must be presented in a clear and unambiguous fashion. In this section we provide a tutorial that illustrates these concepts and principles through a series of concrete examples.

We begin with an analysis of a law-driven domain: a simple system from the domain of process control. The goal is to provide a description of the associated process constraints. We then consider various types of displays that could be devised for the system. The goal is to consider the alternative mappings between process constraints and geometric (display) constraints that are provided by each representation: in particular, the implications for correspondence and coherence. The representations are chosen to illustrate the continuum of visual forms from separable, through configural, to integral geometries. We then examine one representation in greater detail and discuss the implications of this mapping for normal and abnormal operating conditions. We end the section with a set of practical guidelines for display design.

5.1 Simple Domain from Process Control

The process is a generic one that might be found in process control, and it is represented graphically in the lower portion of Figure 12. There is a reservoir (or tank, represented by the large rectangle in the middle of the figure) that is filled with a fluid (e.g., coolant). The volume, or level, of the reservoir (R) is represented by the filled portion of the rectangle. Fluid can enter the reservoir through the two pipes and valves located above the reservoir; fluid can leave the reservoir through the pipe and valve located below the reservoir. We categorize the information in this simple process using a simple distinction in which the term *low-level data* refers to local constraints or elemental state variables that might be measured by a specific sensor. The term *higher-level properties* will be used to refer to more global constraints that reflect relations or interactions among multiple variables.

Low-Level Data (Process Variables) There are two goals associated with this simple process. First, there is a goal (G_1) associated with R, the level of the reservoir. The reservoir should be maintained at a relatively high level to ensure that sufficient resources are available to meet long-term increases in demanded output flow rate (O). The second goal (G_2) refers to the specific rate of output flow that must be maintained to meet an external demand. These goals are achieved and maintained by adjusting three valves (V_1, V_2, and V_3) that regulate flow through the system (I_1, I_2, and O). Thus, this simple process is associated with a number of process variables that can be measured directly: these low-level data are listed in the upper, left-hand portion of Figure 12 (V_1, V_2, V_3, I_1, I_2, O, G_1, G_2, and R).

High-Level Properties (Process Constraints) In addition, there are relationships between these process variables that must be considered when controlling the process (see the upper, right-hand portion of Figure 12). The most important high-level properties are goal-related: Does the actual reservoir volume level (R) match the goal of the system (G_1)$-K_5$? Does the actual system output flow rate (O) match the flow rate that is required (G_2)$-K_6$? Even for this simple process, some of the constraints or (high-level properties) are fairly complex. For example, an important property of the system is mass balance, which is determined

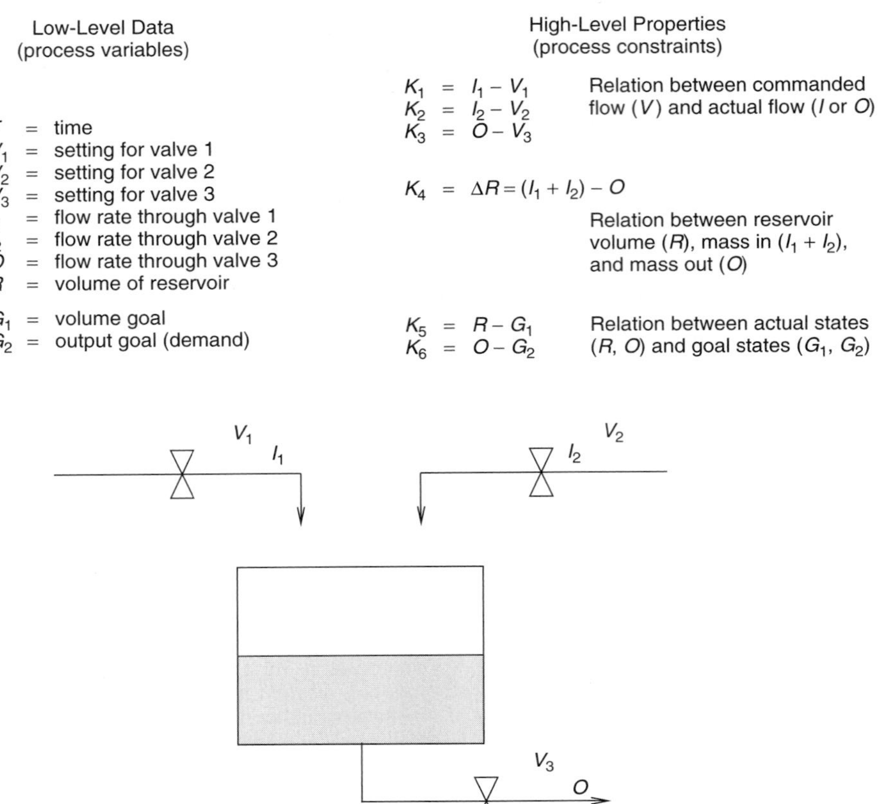

Figure 12 Simple domain from process control that has a reservoir for storing mass, two input streams that increase the volume of mass in the reservoir, and a single output stream that decreases the volume. The low-level data (the measured domain variables), the high-level properties (constraints that arise from the interaction of these variables and the physical design), and the domain goals (requirements that must be met for the system to be functioning properly) are listed.

by comparing the mass leaving the reservoir (O, the output flow rate) to mass entering the reservoir (the combined input flow rates of I_1 and I_2). This relationship determines the direction and the rate of change for the volume inside the reservoir (ΔR). For example, if mass in and mass out are equal, the mass is balanced, ΔR will equal 0.00, and R will remain constant.

Controlling even this simple process will depend on a consideration of both high-level properties and low-level data. As the earlier example indicates, decisions about process goals (e.g., maintaining a sufficient level of reservoir volume) generally require consideration of relationships between variables (is there a net inflow, a net outflow, or is mass balanced?) as well as the values of the individual variables themselves (what is the current reservoir volume?).

5.1.1 Abstraction Hierarchy Analysis

The constraints of the simple process in Figure 12 will be characterized in terms of the abstraction hierarchy. Typically, the hierarchy has five separate levels of description, ranging from the physical form of a domain to the higher-level purposes it serves. The highest level of constraints refers to the *functional purpose* or design goals for the system. For our simple process these are constraints K_5 and K_6. For example, consider the relationship between R and G_1. When the actual reservoir volume (R) equals the goal reservoir volume (G_1), the difference between these two values will assume a constant value (0.00). This process constraint is represented by the equation associated with the higher-level property K_5 in Figure 12. For an actual work domain, the associated values (costs and benefits) underlying these particular goals might be considered. The *abstract functions* or physical laws that govern system behavior are another important source of constraints. In our example, K_4 reflects the law of conservation of mass. Change of mass in the reservoir (ΔR) should be determined by the difference between the residual mass in ($I_1 + I_2$) and the mass out (O). K_1, K_2, and K_3 represent similar constraints associated with the mass flow. Flow is proportional to valve setting (this assumes a constant-pressure head). Further constraints arise as a result of the *generalized function* (sources, storage, sink). In our example, there are two sources: a single store and a single sink. The physical processes behind each general function represents another source of constraint, *physical function*. In this case there are two feedwater streams, a single output stream, and a reservoir for storage. Similarly, the moment-to-moment values of each variable (T, V_1, V_2, V_3, I_1, I_2, O, and R) should be considered at the level of physical function. Finally, the level of *physical form* provides information concerning the physical configuration of the system, including information related to causal connections, length of pipes, position of valves on pipes, and size of the reservoir. All of these constraints will be satisfied if the process is being controlled in a proper fashion.

To summarize, an abstraction hierarchy analysis provides information about the hierarchically nested constraints that constitute the semantics of a domain and therefore defines the information that must be present in the interface for a person to perform successfully. The product of this analysis (interrelated categories of information) provides a structured framework for display development, as we will demonstrate shortly. It should be emphasized that this analysis and description is independent of the interface and therefore differs from traditional task analysis. Although space limitations do not permit a complete discussion, we view abstraction hierarchy analysis and task analysis (traditional or cognitive) as complementary processes that are necessary for the development of effective displays.

5.2 Coherence and Correspondence: Alternative Mappings

In this section we provide six examples that illustrate alternative mappings between domain semantics and representations (displays) for our simple process (Figure 13). The discussion is organized in terms of the distinction between integral, configural, and separable dimensions that was outlined in Section 3.3. One goal is to illustrate what these terms, originally coined in the attention literature, mean in the context of display design for complex systems. A second goal is to focus on the quality of the mapping that each display provides, especially with respect to the ability of each display to convey information at various levels of abstraction (see Sections 4.1.1 and 5.1.1). To illustrate the quality of the mapping explicitly, we have provided a summary listing (at the right of each display in Figure 13) that sorts the associated process constraints into two categories (P and D). Process constraints that are represented directly in the display (i.e., which can be "seen") have been placed in the P category (*P*erceived). Process constraints that are not represented directly, and must be computed or inferred, are placed in the D category (*D*erived). Process constraints that are related to physical structure are represented by the theta symbol (ϕ); process constraints related to the functional structure are represented by the symbol ($\int$).

Separable Displays Figure 13*a* represents a separable display that contains a single display for each individual process variable present. Each display is represented in the figure by a circle, but no special significance should be attached to the symbology: The circles could represent digital displays, bar graphs, and so on. For example, four instantiations of this display are shown in Figure 10*a*, *b*, *e*, and *f*. In Figure 10*a* and *b* the display constraints are the relative heights of the bars in response to changes in the underlying variables.

In terms of the abstraction hierarchy, the class of displays represented by Figure 13*a* provides information only at the level of physical function: individual variables are represented directly. Thus, there is not likely to be a selective attention cost for low-level data. However, there is likely to be a divided attention cost, because the observer must derive the high-level properties. To do so, the observer must have an internalized model of the functional purpose, the abstract functions, the general functional organization, and the physical process. For example, to determine the direction (and cause) of ΔR would require detailed internal knowledge about the process, since no information about physical relationships (ϕ') or functional properties ($\int$) is present in the display.

Simply adding information about high-level properties does not change the separable nature of the display. In Figure 13*b* a second separable display has been illustrated. In this display the high-level properties (constraints) have been calculated and are displayed directly, including information related to functional purpose (K_5 and K_6) and abstract function (K_1, K_2, K_3, and K_4). This does off-load some of the calculational requirements (e.g., ΔR). However, there is still a divided attention cost. Even though the high-level properties have been calculated and incorporated into the display, the relationships among and between levels of information in the abstraction hierarchy are still not apparent. The underlying cause of a particular system state still must be derived from the separate information that is displayed. Thus, although some low-level integration is accomplished in the display, the burden for understanding the causal structure still rests in the observer's stored knowledge.

Configural Displays The first configural display, illustrated in Figure 13*c*, provides a direct representation of much of the low-level data that are present in the display in Figure 13*a*. However, it also provides additional information that is critical to completing domain tasks: information about the physical structure of the system (ϕ'). This "mimic" display format was introduced in STEAMER (Hollan et al., 1984), and issues in the animation of these formats have been investigated more recently (Bennett, 1993; Bennett and Madigan, 1994; Bennett and Nagy, 1996; Bennett and Malek, 2000).

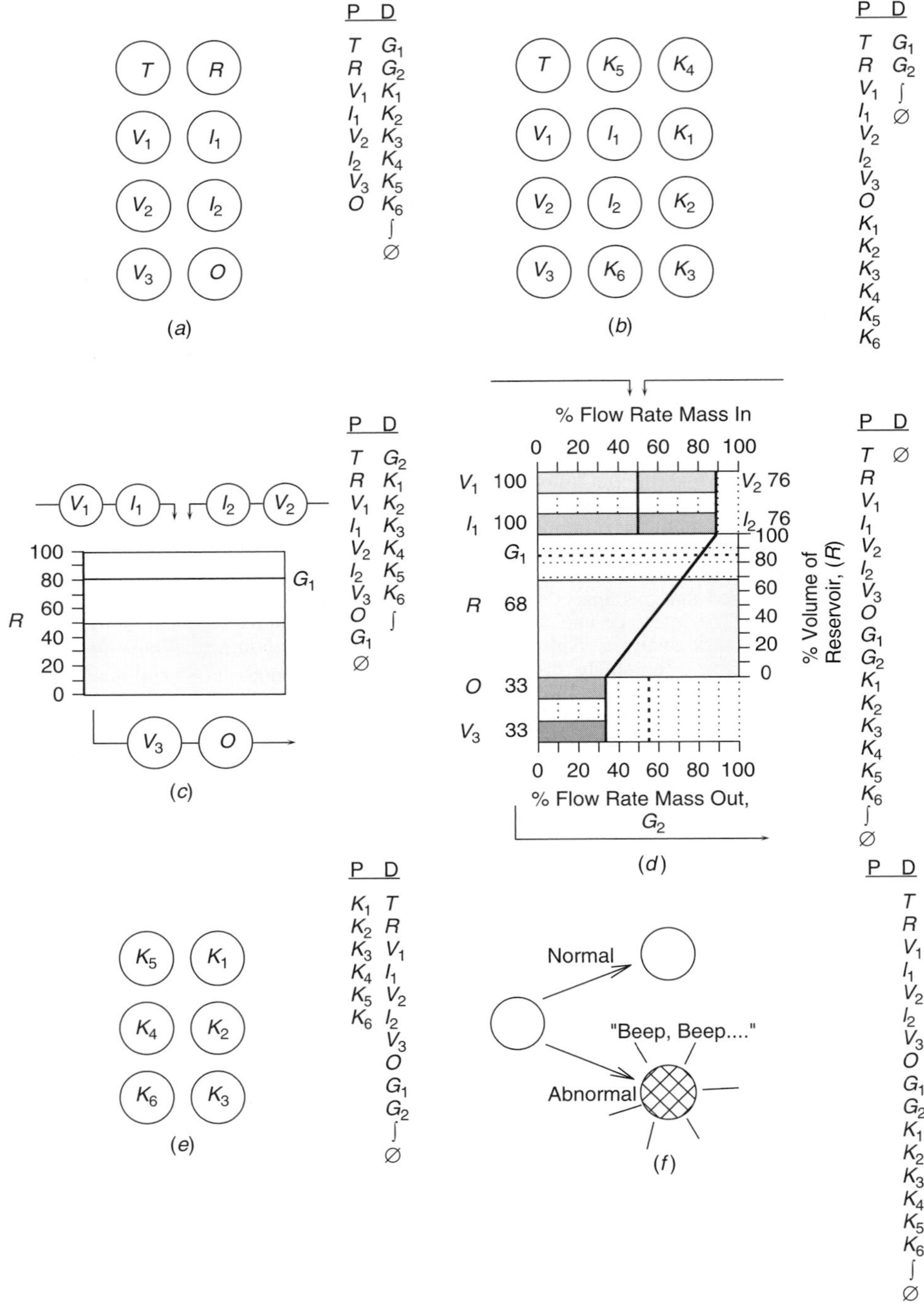

Figure 13 Six alternative mappings for the domain constraints described in Figure 12. The circles represent generic separable displays, which could be bar graphs, pie charts, or digital displays. The data and properties outlined in Figure 12 have been placed in two categories for each mapping: P for data that can be perceived directly from the display and D for data that must be derived from the display by the observer. Parts (*a*) and (*b*) represent separable mappings, (*c*) and (*d*) represent configural mappings, and (*e*) and (*f*) represent integral mappings. These mappings illustrate how the terms *separable, configural*, and *integral* have different meanings when applied to display design (as opposed to their meaning in the attention literature).

The mimic display is an excellent format for representing the generalized functions in the process. It has many of the properties of a functional flow diagram or flowchart. The elements can represent physical processes (e.g., feedwater streams), and by appropriately scaling the diagram, relations at the level of physical form can be represented (e.g., relative positions of valves). Also, the moment-to-moment values of the process variables can easily be integrated within this representation. This display not only includes information with respect to generalized function, physical function, and physical form, but the organization provides a visible model illustrating the relations across these levels of abstraction. This visual model allows the observer to "see" some of the logical constraints that link the low-level data. Thus, the current value of I_2 can be seen in the context of its physical function (feedwater stream 2) and its generalized function (source of mass); in fact, its relation to the functional purpose in terms of G_1 is also readily apparent from the representation.

Just as in the displays listed in Figure 13*a* and *b*, there is not likely to be a cost in selective attention with respect to the low-level data. However, although information about physical structure illustrates the causal factors that determine higher-level system constraints, the burden of computing these constraints (e.g., determining mass balance) rests with the observer. Thus, what is missing in the mimic display is information about abstract function (information about the physical laws that govern normal operation).

The second configural display, illustrated in Figure 13*d*, is slightly more complex [the logic is similar to that of Vicente (1991)] and will be described in detail before discussing the quality of the mapping that it provides. The valve settings V_1 and V_2 are represented as back-to-back horizontal bar graphs that increase or decrease in horizontal extent with changes in settings. The measured flow rates (I_1 and I_2) have the same configuration of graphical elements and are located below the valve settings in the display. The horizontal bar graphs depicting valve settings and flow rates for a particular pipe (e.g., V_1 and I_1) are connected with a bold vertical line (in Figure 13*d* both of the lines are perpendicular because the settings and flow rates are equal in both input streams). The volume of the reservoir (R) is represented by a bold horizontal line and as the filled portion of the rectangle inside the reservoir. The value of R can be read from the scale on the right side of the display and the associated digital value on the left; in Figure 13*d* the value of R is 68. The associated reservoir volume goal (G_1) is represented by the bold horizontal dashed line (approximately 85). The flow rate of the mass leaving the reservoir is represented by the horizontal bar graph labeled O at the bottom of the display; the corresponding valve setting is represented by the bar graph labeled V_3. These two bar graphs are also connected by a bold vertical line. The mass output goal (G_2) is represented by the bold vertical dashed line (approximately 55). The relationship between mass in ($I_1 + I_2$) and mass out (O) is highlighted by the bold angled line that connects the corresponding bar graphs.

Unlike the displays discussed previously, this configural display integrates information from all levels of the abstraction hierarchy in a single representation, making extensive use of the geometrical constraints of equality, parallel lines, and collinearity. The general functions are related through a funnel metaphor with input (source) at the top, storage in the center, and output (sink) on the bottom. The abstract functions are related using equality and the resulting collinearity across the bar graphs. For example, the constraints on mass flow (K_1, K_2, K_3) are represented in terms of equality of the horizontal extent of the bars labeled V_1/I_1, V_2/I_2, and V_3/O. In addition, the constraints relating rate of volume change and mass balance (K_4) are represented by the horizontal extent of $I_1 + I_2$ relative to the horizontal extent of O, and these relationships are highlighted by the bold line connecting these bars. Thus, the mass balance is represented by the symmetry between the input bar graphs and the output bar graphs; the slant of the line connecting them should be proportional to rate of change of mass in the reservoir. Constraints at the level of functional purpose are illustrated by the difference between the goal and the relevant variable. For example, the constraint on mass inventory (K_5) is shown using the relative position between the hatched area representing volume within the reservoir and the bold horizontal dashed line representing the goal level G_1.

Although not a direct physical analog, this configural display preserves important physical relations from the process (e.g., volume and filling). In addition, it provides a direct visual representation of the process constraints and connects these constraints so as to make the functional logic of the process visible within the geometric form. As a result, performance for both selective (focused) and divided (integration) tasks is likely to be facilitated substantially.

Integral Displays Figure 13*e* shows an integral mapping in which each of the process constraints are shown directly, providing information at the higher levels of abstraction. However, the low-level data must be derived. In addition, there is absolutely no information about the functional processes behind the display, and therefore the display does not aid the observer in relating the higher-level constraints to the physical variables. Because there would normally be a many-to-one mapping from physical variables to the higher-order constraints, it would be impossible for the observer to recover information from this display at lower levels of abstraction.

Figure 13*f* shows the logical extreme of this continuum. In this display the process variables and constraints are integrated into a single "bit" of information that indicates whether or not the process is working properly (all constraints are at their designed value). It should be obvious that although these displays may have no divided attention costs, they do have selective attention costs and provide little support for problem solving when the system fails.

Summary This section has focused on issues related to the quality of mapping between process constraints and display constraints. Even the simple domain that we chose for illustrative purposes has a nested structure of domain constraints: There are multiple constraints that are organized hierarchically both within and between levels of abstraction. The six alternative displays achieved various degrees of success in mapping these constraints. The principle of correspondence is illustrated by the fact that these formats differ in terms of the amount of information about the underlying domain that is present. The display in Figure 13*f* has the lowest degree of correspondence; the displays in Figure 13*b* and *d* have the highest degree of correspondence. These two displays are roughly equivalent in correspondence, with the exception of the two goals that are present in Figure 13*d* but absent in Figure 13*b*. Although these two displays are roughly equivalent in correspondence, it should be clear from the prior discussion that they are definitely not equivalent in terms of coherence. Figure 13*d* allows a person to perceive information concerning the physical structure, functional structure, and hierarchically nested constraints in the domain directly, a capability that is not supported by the format in Figure 13*b*. The coherence of Figure 13*d* will be explored in greater detail in the following section. This section has also illustrated the duality of meaning for the terms *integral, configural*, and *separable*. In attention, these terms refer to the relationship between perceptual dimensions, as described in Section 3.3; in display design, these terms refer more appropriately to the nature of the mapping between the domain and the representation.

5.3 Representation Aiding: Normal and Abnormal Operating Conditions

In Section 10 we outlined differences in correspondence and coherence that resulted from six alternative mappings for our simple domain. In this section we explore issues related to coherence in greater detail, focusing on Figure 13*d* and the implications of the mapping for performance under both normal and abnormal, or emergency operating conditions. To begin, we discuss the facilitating role that graphical constraints representing information in the abstraction hierarchy (in particular, abstract function—the physical laws that govern normal operation) can play under normal conditions. Properly designed configural displays will provide a powerful representation for control: breaks in the domain constraints will generally be seen as breaks in display constraints (e.g., nonsymmetries) and will suggest appropriate control inputs. This information is, perhaps, even more important for detecting faults (e.g., a leak). The possibility that these types of displays can change the fundamental nature of the behavior that is required on the part of the operator will also be entertained. Finally, the implications for the reduction of errors (more likely to occur under abnormal or emergency conditions) will be discussed.

The mapping between domain constraints and geometrical constraints that is provided in the configural display shown in Figure 13*d* provides a powerful representation for control under normal operating conditions. In Figure 14*a* the display is shown with values for system variables indicating that all constraints are satisfied. The figure indicates that the flow rate is larger for the first mass input valve (I_1, V_1) than for the second (I_2, V_2) but that the two flow rates added together match the flow rate of the mass output valve (O, V_3). In addition, the two system goals (G_1 and G_2) are being fulfilled.

In contrast, Figure 14*b–d* illustrate failures to achieve system goals. In these displays not only is the violation of the goal easily seen, but each system variable is seen in the context of the control requirements. Thus, in Figure 14*b* it is apparent that the K_5 constraint is not being met (the actual level of the reservoir is higher than the goal). It is also apparent that the K_4 constraint is broken. The orientation of the line connecting mass in ($I_1 + I_2$) and the mass out (O) utilizes the funnel metaphor to indicate that a positive net inflow for mass exists (mass in is greater than mass out). In essence, the deviation in orientation of this line from perpendicular is an emergent feature corresponding to the size of the difference. Under these circumstances control input is required immediately: An adjustment at valve 1 and/or valve 2 will be needed to avoid overflow from the reservoir. The observer can see these valves in the context of the two system goals; the representation makes it clear that these are the appropriate control inputs to make. For example, although adjusting valve 3 from 54 to a value greater than 70 would also cause the reservoir volume to drop, it is an inappropriate control input because goal 2 would then be violated.

In Figure 14*c* the situation is exactly the same, with one exception: There is a negative net inflow for mass, as indicated by the reversed orientation of the connecting line. Under these circumstances the operator can see that no immediate control input is required. Because mass in is less than mass out, the reservoir volume is falling, and this is exactly what is required to meet the G_1 reservoir volume goal. Of course, a control input will be required at some point in the future (mass will need to be balanced when the reservoir level approaches the goal). Similarly, in Figure 14*d* the observer can see that the K_5 and K_6 constraints are broken and that an adjustment to valve 3 (a decrease in output) is needed to meet the output requirements (G_2) and the volume goal (G_1).

Thus, in complex dynamic domains it is the pattern of relationships between variables, as reflected in the geometric constraints, that determines the significance of the data that are presented. It is this pattern that ultimately provides the basis for action, even when the action hinges on the value of an individual variable. When properly designed, configural displays will directly reflect these critical data relationships and suggest the appropriate control input.

A similar logic applies for operational support under abnormal or emergency conditions. As in Figure 14, Figure 15*a* represents a configuration with all system constraints being met. In Figure 15*b* the first

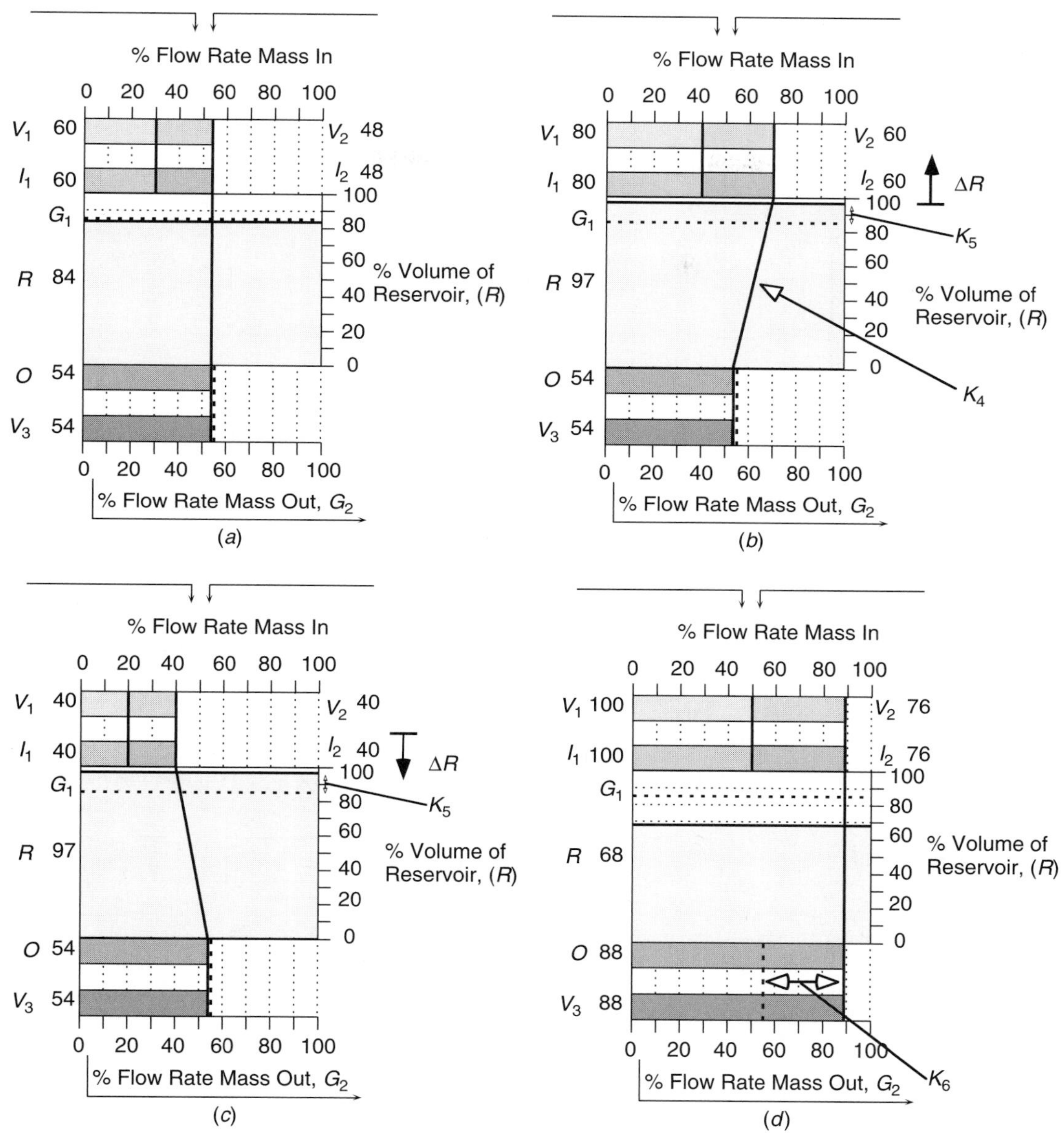

Figure 14 Mapping between the domain constraints (data, properties, goals) and the geometric constraints (visual properties of the display, including emergent features such as symmetry and parallelism) under relatively normal operating conditions.

constraint (K_1) is broken. There are two aspects of the display geometry, indicating that the flow rate (I_1) does not match the commanded flow or valve setting (V_1). First, the horizontal extent of the two bar graphs in the top left portion of the display are not equal, and this relationship is emphasized by the bold line connecting the two graphs (similar to the connecting line for mass balance). There are a number of potential causes for this discrepancy, which include (1) a leak in the valve, (2) a leak in the pipe prior to the point at which the flow rate is measured, or (3) an obstruction in the pipe. In contrast, the fact that the line connecting V_2 and I_2 is not perpendicular (but is parallel to the first connector line) does not indicate that the K_2 constraint is broken. Instead, this is an indication that the commanded and actual mass flows in the second mass input stream are equal (and therefore that the discrepancy is isolated in the first mass input stream).

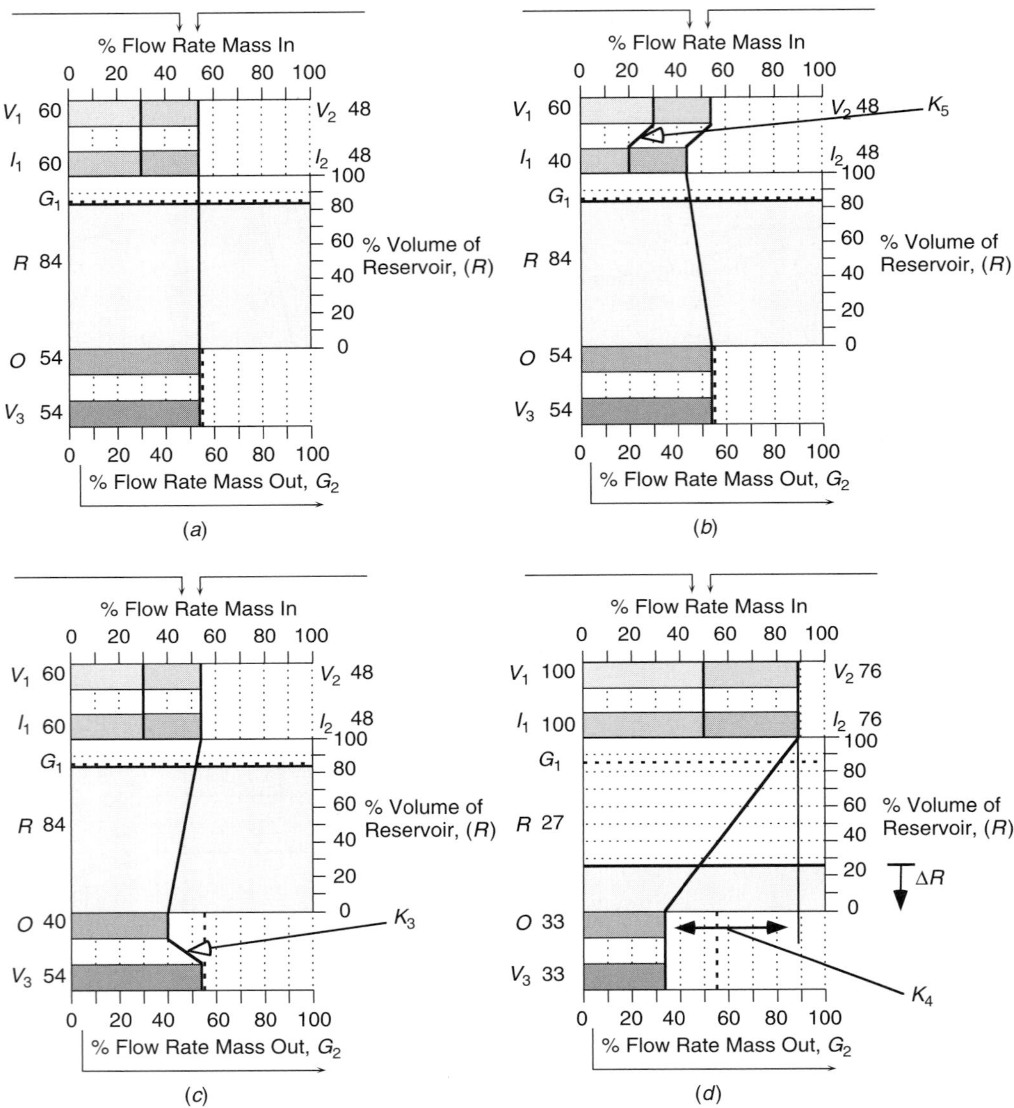

Figure 15 Mapping between the domain constraints (data, properties, goals) and the geometric constraints (visual properties of the display, including emergent features such as symmetry and parallelism) under abnormal or emergency operating conditions.

A similar mapping between geometrical constraints and domain constraints represents a fault in the K_3 constraint, as illustrated in Figure 15*c*.

Figure 15*d* illustrates changes in the visual display (breaks in the geometrical constraints) that are associated with a fault in the system (a break in the mass balance constraint, K_4). In this example, there is a positive net inflow of mass, which is normally associated with an increase in the volume of the reservoir (again, suggested by the funnel metaphor). However, in this case the mass inventory is falling, as we have indicated in the diagram by the downward-pointing arrow located near the ΔR symbol (this is difficult to represent in a static diagram but would be seen clearly on a dynamic display). Again, there are several potential explanations for this fault. The most likely explanation is that there is a leak in the reservoir itself; however, there could be a leak in the pipe between the reservoir and the point at which the flow measurement is taken. It should be noted that while the nature of the fault can be seen (e.g., leak or blockage in feedwater line) this representation would not be very helpful in physically locating the leak within the plant (e.g., locating valve 1).

These examples illustrate that properly designed displays can change the fundamental type of behavior that is required of an operator under both normal and abnormal operating conditions. With separable displays (e.g., the separable configurations illustrated in Figure 13) the operators are required to engage in knowledge-based behaviors: they must rely on internal models of system structure and function (and therefore use limited capacity resources—working memory) to detect, diagnose, and correct faults. As a result, the potential for errors is increased dramatically. In contrast, properly designed configural displays present externalized models of system structure and function through geometric constraints. This allows operators to utilize skill-based behaviors (e.g., visual perception and pattern recognition) that do not require limited capacity resources. As a result, the potential for errors will be decreased dramatically. As Rasmussen and Vicente (1989) have noted, changing the required behavior from knowledge-based behavior to rule- or skill-based behavior is a goal for display design.

Properly designed configural displays will also reduce the possibility of *underspecified action errors* (Rasmussen and Vicente, 1989). In complex, dynamic domains people can form incorrect hypotheses about the nature of the existing problem if they do not consider the relevant subsets of data (Woods, 1988). Observers may focus on these incorrect hypotheses and ignore disconfirming evidence, showing a kind of tunnel vision (Moray, 1981). Observers may also exhibit *cognitive hysteresis* and fail to revise hypotheses as the nature of the problem changes over time (Lewis and Norman, 1986). Configural displays that reflect the semantics of a domain directly can reduce the probability of these types of errors by forcing an observer to consider relevant subsets of data.

One final point needs to be made in closing this section. It is important to note that even though multiple variables may testify with regard to higher-level domain properties, configural displays are *not* always the appropriate design solution. Ultimately, the design decision hinges on the relationships and interactions between variables. Consider the case of a mobile army commander engaged in tactical battlefield operations (Talcott et al., in press). There are five combat resources that are the primary determinants of the higher-level property of combat power: tanks, personnel carriers, ammunition, fuel, and personnel. The key design criterion is that these resources are essentially independent; there is no physical law or causal relationship that explicitly defines higher-order patterns between variables. It is true that the combat resources can be correlated: A unit engaged in an intense offensive battle might suffer equipment and personnel losses while expending substantial amounts of fuel and ammo. However, any number of factors can change the relationship between them. For example, ammunition is likely to be expended more quickly than fuel in a defensive mission.

The use of a highly configural representation, such as a polar graphic display (Woods et al., 1981) for the five combat resources is a tempting, but inappropriate, design choice. This display would produce numerous, salient, and hierarchically nested emergent features that highlighted the relationships and interactions between combat resources. This produces a poor mapping between display and domain: the display constraints (i.e., the particular asymmetries and other distortions of the polar graphic) would not uniquely specify underlying domain constraints (i.e., higher-level domain properties arising from the interaction of variables). Many, if not most of the emergent features produced by the display would be essentially meaningless and would therefore need to be ignored (an extremely difficult, if not impossible task for actors to accomplish). Talcott et al. (in press) chose a separable format similar to that portrayed in Figure 10*b*. This representation maintains the independence of the individual combat resources while still providing limited configurality (Sanderson et al., 1989).

5.4 Practical Guidelines

In conclusion, we believe that the application of this approach to display design will improve overall human machine performance through the development of configural displays that support normal control as well as fault detection, diagnosis, and repair. The potential for errors will be decreased dramatically, because the critical information for control is represented directly in the interface. This, in turn, dramatically reduces the requirement for knowledge-based reasoning on the basis of internalized models. To summarize, we offer three general heuristics for graphic display design.

1. Each relevant process variable should be represented by a distinct element within the display. If precise information about this variable is desirable, a reference scale or supplemental digital information should be provided.
2. The display elements should be organized so that the emergent properties (symmetries, closure, parallelism) that arise from their interaction correspond to higher-order constraints within the process. Thus, when process constraints are broken (i.e., a fault occurs), the corresponding geometric constraints are also broken (the display symmetry is broken).
3. The symmetries within the display should be nested (from global to local) in a way that reflects the hierarchical structure of the process. High-order process constraints (e.g., at the level of functional purpose or abstract function) should be reflected in global display symmetries; lower-order process constraints (e.g., functional organization) should be reflected in local display symmetries.

6 CHALLENGES OF COMPLEX SYSTEMS

The simple process described above is convenient for a tutorial introduction to some of the important decisions that must be made when designing a graphical representation. However, this example greatly underestimates the complexity seen in many advanced human–technological systems (e.g., nuclear power, air traffic control, advanced tactical aviation, command and control centers for managing military and space operations, minimally invasive and remote surgery). These systems typically have multiple modes of operation (each with different constraints and boundary conditions) and require multiple windows into the process. In these systems, the goal remains the same, to make the real constraints of the work process (at all levels of abstraction) visible to the human operator. The designer must still address the problems of correspondence (so that all relevant process constraints are represented in the interface) and coherence (so that the representation is comprehensible to the human operator). For these complex systems, however, it will not be possible to achieve both correspondence and coherence with a single graphic display. Thus, the added problem of navigating through multiple views (i.e., windows, screens, pages) must be addressed.

A principal threat to these complex systems is *mode error* (Woods, 1984), which occurs when the operator loses track of dynamic changes in the operating constraints governing a process. The operator responds to one set of constraints (i.e., mode) when a different set of constraints is, in fact, governing the process. The design challenge is to coordinate the multiple windows necessary for complete representation with the changing operational modes. In simple terms, the problem is how the interface can be designed to ensure that the appropriate window is always coupled with the appropriate mode: that is, to ensure that the important information is salient at the appropriate times.

Two classes of solutions might be considered for dealing with the navigation problems that typically lead to mode errors, computational and graphical. Computational solutions or adaptive interfaces include an inference engine that manages the representation automatically. This computational engine adjusts the representation automatically based on inferences about the state of the system and the state of the operator. Projects such as the "pilot's associate" are examples of attempts to design automatic systems to aid operators to navigate through the representations associated with a complex work domain. However, the focus of this chapter is on graphical solutions. For this reason we use the remaining space to consider briefly some graphical approaches to this problem.

Woods (1984) introduced the term *visual momentum* to refer to the cognitive costs associated with switching from one reference frame to another. If visual momentum is high, the cost of switching views is low. In this case, the new display is consistent with expectations created by the prior display. If visual momentum is low, there is a high cost of switching. That is, the new display is not consistent with expectations and the cognitive system must effectively recalibrate before information can be extracted from the new display. To ensure high visual momentum, the design of each graphical display must be considered relative to the other displays that operators may be using. Are the graphical conventions (e.g., coordinates, scales, directions, motions, colors, S-R mappings) used in one display consistent with those in another?

A graphical device that Woods (1984) has suggested to increase visual momentum is the use of *landmarks*, graphical elements that provide an orientation point that relates one display to another. Just as a tall building or mountain that is visible from many different parts of the landscape might help a person to orient to the geography, graphical landmarks can be designed with the objective of aiding the operator to orient within the functional landscape of the work domain. For example, Aretz (1991) used a shaded wedge within an electronic map display as a landmark to specify the region within the map that corresponded to the head-up forward view of the pilot.

Another graphical device to help operators navigate across multiple display pages is a map or overview display. This display can be implemented as a separate window or as an embedded landmark in all windows. This overview might use a flow diagram or hierarchical tree structure to show functional links among the multiple display pages.

The concrete examples we have provided involve law-driven work domains and configural displays. We provide one brief description of an intent-driven domain and a metaphorical display. The BookHouse interface designed by Goodstein and Pejtersen (1989) uses a spatial metaphor in which rooms in a "house" are set up for various categories of users. This spatial metaphor allows the operator to apply natural abilities for navigating in three-dimensional spaces to the task of navigating in the more abstract space of a library database. For a more detailed treatment of intent-driven domains and metaphorical displays, consider Flach et al. (2005), who discuss ecological interface design for the Web.

In the BookHouse, the three-dimensional space is implemented in a two-dimensional display. Virtual reality systems now offer the possibility for effective three-dimensional representations. With these systems, designers have the opportunity to maximize the transfer of natural human ability to orient and navigate in three-dimensional environments to more abstract environments, and to combine natural three-dimensional representations with imagery obtained by advanced sensor systems. For example, virtual displays for minimally invasive surgery are being designed that integrate the three-dimensional image of a patient's anatomy with information obtained by MRI scans and other advanced imaging technologies. Thus, virtual three-dimensional spatial metaphors might provide another technique for integrating complex information from distributed sensors into a coherent representation.

The central theme of this chapter is that problem solving can be critically influenced by the nature of visual representations. Building effective representations requires designers to go beyond the simple psychophysical questions of data availability to the more complex questions of information availability, where information refers to the specification of domain constraints and boundary conditions. This specification depends both on the mapping from display to human (i.e., coherence) and that from display to domain (i.e., correspondence).

ACKNOWLEDGMENTS

The authors would like to thank Brian Tsou for discussions and comments on earlier drafts and the helpful comments provided by reviewers. Funding in support of this work was provided to Kevin Bennett by the Ohio Board of Regents (Wright State University Research Challenge Grant 662613). John Flach was partially supported during the preparation of this manuscript by a grant from the Air Force Office of Scientific Research. Opinions expressed are those of the authors and do not represent an official position of any of the supporting agencies.

REFERENCES

Adelson, E. H. (1993), "Perceptual Organization and the Judgement of Brightness," *Science*, Vol. 262, pp. 2042–2044.

Alexander, K. R., Xie, W., and Derlacki, D. J. (1994), "Spatial Frequency Characteristics of Letter Identification," *Journal of the Optical Society of America, Series A*, Vol. 11, pp. 2375–2382.

Arend, L. E., and Reeves, A. (1986), "Simultaneous Color Constancy," *Journal of the Optical Society of America, Series A*, Vol. 3, pp. 1743–1751.

Aretz, A. J. (1991), "The Design of Electronic Map Displays," *Human Factors*, Vol. 33, No. 1, pp. 85–101.

Beltracchi, L. (1987), "A Direct Manipulation Interface for Heat Engines Based upon the Rankine Cycle," *IEEE Transactions on Systems, Man, and Cybernetics*, Vol. 17, No. 3, pp. 478–487.

Beltracchi, L. (1989), "Energy, Mass, Model-Based Displays, and Memory Recall," *IEEE Transactions on Nuclear Science*, Vol. 36, No. 3, pp. 1367–1382.

Bennett, K. B. (1993), "Encoding Apparent Motion in Animated Mimic Displays," *Human Factors*, Vol. 35, No. 4, pp. 673–691.

Bennett, K. B., and Flach, J. M. (1992), "Graphical Displays: Implications for Divided Attention, Focused Attention, and Problem Solving," *Human Factors*, Vol. 34, No. 5, pp. 513–533.

Bennett, K. B., and Madigan, E. (1994), "Contours and Borders in Animated Mimic Displays," *International Journal of Human–Computer Interaction*, Vol. 6, No. 1, pp. 47–64.

Bennett, K. B., and Malek, D. A. (2000), "Evaluation of Alternative Waveforms for Animated Mimic Displays," *Human Factors*, Vol. 42, No. 3, pp. 432–450.

Bennett, K. B., and Nagy, A. L. (1996), "Spatial and Temporal Considerations in Animated Mimic Displays," *Displays,* Vol. 17, No. 1, pp. 1–14.

Bennett, K. B., and Walters, B. (2001), "Configural Display Design Techniques Considered at Multiple Levels of Evaluation," *Human Factors*, Vol. 43, No. 3, pp. 415–434.

Bennett, K. B., Toms, M. L., and Woods, D. D. (1993), "Emergent Features and Configural Elements: Designing More Effective Configural Displays," *Human Factors*, Vol. 35, No. 1, pp. 71–97.

Boynton, R. M. (1990), "Human Color Perception," in *Science of Vision*, K. N. Leibovic, Ed., Springer-Verlag, New York, pp. 211–253.

Brainard, D. H., and Wandell, B. A. (1992), "Asymmetric Color-Matching: How Color Appearance Depends on the Illuminant," *Journal of the Optical Society of America, Series A*, Vol. 9, pp. 1433–1448.

Brodie, K. W., Carpenter, L. A., Earnshaw, R. A., Gallop J. R., Hubbold, R. J., Mumford, A. M., Osland, C. D., and Quarendon, P., Eds. (1992), *Scientific Visualization: Techniques and Applications*, Springer-Verlag, Berlin.

Burns, C. M., and Hajdukiewicz, J. R. (2004), *Ecological Interface Design*, CRC Press, Boca Raton, FL.

Buttigieg, M. A., and Sanderson, P. M. (1991), "Emergent Features in Visual Display Design for Two Types of Failure Detection Tasks," *Human Factors*, Vol. 33, No. 6, pp. 631–651.

Calcaterra, J. A., and Bennett, K. B. (2003), "The Placement of Digital Values in Configural Displays," *Displays*, Vol. 24, No. 2, pp. 85–96.

Cannon, M. W., and Fullenkamp, S. C. (1991), "A Transducer Model for Contrast Perception," *Vision Research*, Vol. 31, pp. 983–998.

Carswell, C. M., and Wickens, C. D. (1990), "The Perceptual Interaction of Graphical Attributes: Configurality, Stimulus Homogeneity, and Object Integration," *Perception and Psychophysics*, Vol. 47, pp. 157–168.

Cleveland, W. S. (1985), *The Elements of Graphing Data*, Wadsworth, Belmont, CA.

DeValois, R. L., and DeValois, K. K. (1990), *Spatial Vision*, Oxford University Press, New York.

D'Zmura, M., and Lennie, P. (1986), "Mechanisms of Color Constancy," *Journal of the Optical Society of America, Series A*, Vol. 3, pp. 1662–1672.

Earnshaw, R. A., and Wiseman, N., Eds. (1992), *An Introductory Guide to Scientific Visualization*, Springer-Verlag, Berlin.

Flach, J. M., Bennett, K. B., Stappers, P. J., and Saakes, D. P. (2005), "Searching for Meaning in Complex Databases: An Ecological Perspective," in *Handbook of Human Factors in Web Design*, R. Proctor and K. L. Vu, Eds., Lawrence Erlbaum Associates, Mahwah, NJ, pp. 408–423.

Garner, W. R. (1970), "The Stimulus in Information Processing," *American Psychologist*, Vol. 25, pp. 350–358.

Garner, W. R. (1974), *The Processing of Information and Structure*, Lawrence Erlbaum Associates, Mahwah, NJ.

Garner, W. R., and Felfoldy, G. L. (1970), "Integrality of Stimulus Dimensions in Various Types of Information Processing," *Cognitive Psychology*, Vol. 1, pp. 225–241.

Gegenfurtner, K. R., and Sharpe, L. T. (1999), *Color Vision: From Genes to Perception*, Cambridge University Press, Cambridge.

Gilchrist, A., Delman, S., and Jacobson, A. (1983), "The Classification and Identification of Edges as Critical to the Perception of Reflectance and Illumination," *Perception and Psychophysiology*, Vol. 33, pp. 425–436.

Ginsburg, A. (1986), "Spatial Filtering and Visual Form Perception," in *Handbook of Human Perception and Performance*, K. R. Boff, L. Kaufmann, and J. P. Thomas, Eds., Wiley, New York, pp. 34/1–34/41.

Goodstein, L. P., and Pejtersen, A. M. (1989), *The Book House System: Functionality and Evaluation*, Riso-M-2793, Riso National Laboratory, Roskilde, Denmark.

Graham, N. V. S. (1989), *Visual Pattern Analyzers*, Oxford University Press, New York.

Grum, F., and Bartleson, C. J. (1980), *Optical Radiation Measurements*, Vol. 2, *Colorimetry*, Academic Press, New York.

Hollan, J. D., Hutchins, E. L., and Weitzman, L. (1984), "Steamer: An Interactive Inspectable Simulation-Based Training System," *AI*, Summer, pp. 15–27.

Hollan, J. D., Hutchins, E. L., McCandless, T. P., Rosenstein, M., and Weitzman, L. (1987), "Graphical Interfaces for Simulation," in *Advances in Man–Machine Systems Research*, W. B. Rouse, Ed., JAI Press, Greenwich, CT, Vol. 3, pp. 129–163.

Hunter, R. S., and Herold, R. W. (1987), *The Measurement of Appearance*, Wiley, New York.

Hutchins, E. L., Hollan, J. D., and Norman, D. A. (1986), "Direct Manipulation Interfaces," in *User Centered System Design*, D. A. Norman, and S. W. Draper, Eds., Lawrence Earlbaum Associates, Mahwah, NJ, pp. 87–124.

Kaiser, P. K., and Boynton, R. M. (1997), *Human Color Vision*, Optical Society of America, Washington, DC.

Kelly, D. H. (1974), "Spatio-Temporal Frequency Characteristics of Color-Vision Mechanisms," *Journal of the Optical Society of American, Series A*, Vol. 64, pp. 983–990.

Klein, G. A., Orasanu, J., and Zsambok, C. E., Eds. (1993), *Decision Making in Action: Models and Methods*, Ablex Publishing, Norwood, NJ.

Lewis, C., and Norman, D. A. (1986), "Designing for Error," in *User Centered System Design*, D. A. Norman and S. W. Draper, Eds., Lawrence Earlbaum Associates, Mahwah, NJ.

Makous, W. (2003), "Threshold and Suprathreshold Spatiotemporal Contrast Sensitivity," in *Encyclopedia of Optical Engineering*, Marcel Dekker, New York, pp. 2828–2850.

Maloney, L. T., and Wandell, B. A. (1986), "Color Constancy: A Method for Recovering Surface Reflectance," *Journal of the Optical Society of America, Series A*, Vol. 3, pp. 29–33.

Millodot, M. (1982), "Image Formation in the Eye," in *The Senses*, H. Barlow and J. D. Mollen, Eds., Cambridge University Press, New York, pp. 46–61.

Moray, N. (1981), "The Role of Attention in the Detection of Errors and the Diagnosis of Failures in Man–Machine Systems," In *Human Detection and Diagnosis of System Failures*, J. Rasmussen and W. B. Rouse, Eds., Plenum Press, New York.

Moray, N., Lee, J., Vicente, K. J., Jones, B. G., and Rasmussen, J. (1994), "A Direct Perception Interface for Nuclear Power Plants," in *Proceedings of the Human Factors and Ergonomics Society 38th Annual Meeting*, Human Factors and Ergonomics Society, Santa Monica, CA, pp. 481–485.

Mullen, K. (1985), "The Contrast Sensitivity of Human Color Vision to Red–Green and Blue–Yellow Chromatic Gratings," *Journal of Physiology*, Vol. 359, pp. 381–400.

Nagy, A. L. (2003), "Visual Search: Color and Display Issues," in *Encyclopedia of Optical Engineering*, Marcel Dekker, New York, pp. 2955–2962.

Nagy, A. L., and Kamholz, D. (1995), "Luminance Discrimination, Color Contrast, and Multiple Mechanisms," *Vision Research*, Vol. 35, pp. 2147–2155.

Nassau, K. (1983), *The Physics and Chemistry of Color*, Wiley, New York.

Newell, A., and Simon, H. A. (1972), *Human Problem Solving*, Prentice Hall, Englewood Cliffs, NJ.

Noorlander, C., and Koenderink, J. J. (1983), "Spatial and Temporal Discrimination Ellipsoids in Colour Space," *Journal of the Optical Society of America, Series A*, Vol. 73, pp. 1533–1543.

Olzack, L., and Thomas, J. (1986), "Seeing Spatial Patterns," in *Handbook of Human Perception and Performance*, K. R. Boff, L. Kaufmann, and J. P. Thomas, Eds., Wiley, New York, pp. 7/1–7/56.

Pavel, M., and Ahumada, A. J. (1997), "Model Based Optimization of Display Systems," in *Handbook of Human–Computer Interaction*, 2nd ed. M. G. Helander, T. K. Landauer, and P. V. Prabhu, Eds., Elsevier, New York, pp. 65–86.

Pejtersen, A. M. (1992), "The BookHouse: An Icon Based Database System for Fiction Retrieval in Public Libraries," in *The Marketing of Library and Information Services*, Vol. 2, B. Cronin, Ed., Aslib, London, pp. 572–591.

Poirson, A. B., and Wandell, B. A. (1993), "The Appearance of Colored Patterns," *Journal of the Optical Society of America, Series A*, Vol. 12, pp. 2458–2471.

Pokorny, J., and Smith, V. (1986), "Colorimetry and Color Discrimination," in *Handbook of Human Perception and Performance*, K. R. Boff, L. Kaufmann, and J. P. Thomas, Eds., Wiley, New York, pp. 8/1–8/51.

Pomerantz, J. R. (1986), "Visual Form Perception: An Overview," in *Pattern Recognition by Humans and Machines*, Vol. 2, *Visual Perception*, H. C. Nusbaum and E. C. Schwab, Eds., Academic Press, Orlando, FL, pp. 1–30.

Pomerantz, J. R., and Pristach, E. A. (1989), "Emergent Features, Attention, and Perceptual Glue in Visual Form Perception," *Journal of Experimental Psychology: Human Perception and Performance*, Vol. 15, No. 4, pp. 635–649.

Pomerantz, J. R., Sager, L. C., and Stoever, R. J. (1977), "Perception of Wholes and of Their Component Parts: Some Configural Superiority Effects," *Journal of Experimental Psychology: Human Perception and Performance*, Vol. 3, pp. 422–435.

Post, D. (1992), "Colorimetric Measurement, Calibration, and Characterization of Self-luminous Displays," in *Color in Electronic Displays*, H. Widdel and D. L. Post, Eds., Plenum Press, New York, pp. 299–312.

Post, D. (1997), "Color and Human Computer Interaction," in *Handbook of Human–Computer Interaction*, 2nd ed., M. G. Helander, T. K. Landauer, and P. V. Prabhu, Eds., Elsevier, New York, pp. 573–616.

Rasmussen, J. (1986), *Information Processing and Human–Machine Interaction: An Approach to Cognitive Engineering*, Elsevier, New York.

Rasmussen, J., and Vicente, K. (1989), "Coping with Human Errors Through System Design: Implications for Ecological Interface Design," *International Journal of Man–Machine Studies*, Vol. 31, pp. 517–534.

Rasmussen, J., Pejtersen, A. M., and Goodstein, L. P. (1994), *Cognitive Systems Engineering*, Wiley, New York.

Sanderson, P. M., Flach, J. M., Buttigieg, M. A., and Casey, E. J. (1989), "Object Displays Do Not Always Support Better Integrated Task Performance," *Human Factors*, Vol. 31, No. 2, pp. 183–198.

Sekiguchi, N., Williams, D. R., and Brainard, D. H. (1993), "Aberration Free Measurements of the Visibility of Isoluminant Gratings," *Journal of the Optical Society of America, Series A*, Vol. 10, pp. 2105–2117.

Shafer, S. A. (1985), "Using Color to Separate Reflection Components," *Color Research and Applications*, Vol. 10, pp. 210–218.

Shneiderman, B. (1986), *Designing the User Interface*, Addison-Wesley, Reading, MA.

Shneiderman, B., Ed. (1993), *Sparks of Innovation in Human–Computer Interaction*, Ablex, Norwood, NJ.

Solomon, J. A., and Pelli, D. G. (1994), "The Visual Filter Mediating Letter Identification," *Nature (London)*, Vol. 369, pp. 395–397.

Talcott, C. P., Bennett, K. B., Martinez, S. G., Shattuck, L., and Stansifer, C. (in press), "Perception-Action Icons: An Interface Design Strategy for Intermediate Domains," *Human Factors*.

Travis, D. (1991), *Color Displays: Theory and Practice*, Academic Press, London, UK.

Treisman, A. M. (1986), "Properties, Parts, and Objects," in *Handbook of Perception and Human Performance*, K. Boff, L. Kaufmann, and J. Thomas, Eds., Wiley, New York, pp. 35/1–35/70.

Tufte, E. R. (1983), *The Visual Display of Quantitative Information*, Graphics Press, Chesire, CT.

Tufte, E. R. (1990), *Envisioning Information*, Graphics Press, Chesire, CT.

Vicente, K. J. (1991), *Supporting Knowledge Based Behavior Through Ecological Interface Design*, Technical Report EPRL-91-1, Engineering Psychology Research Laboratory and Aviation Research Laboratory, University of Illinois, Urbana–Champaign, IL.

Vicente, K. J. (1999), *Cognitive Work Analysis: Toward Safe, Productive, and Healthy Computer-Based Work*, Lawrence Erlbaum Associates, Mahwah, NJ.

Wandell, B. A. (1995), *Foundations of Vision*, Sinauer Associates, Sunderland, MA.

Wertheimer, M. (1959), *Productive Thinking*, Harper & Row, New York.

Whittle, P. (1986), "Increments and Decrements: Luminance Discrimination," *Vision Research*, Vol. 26, pp. 1677–1692.

Whittle, P. (1992), "Brightness, Discriminability, and the 'Crispening Effect,'" *Vision Research*, Vol. 32, pp. 1493–1508.

Wickens, C. D., and Carswell, C. M. (1995), "The Proximity Compatibility Principle: Its Psychological Foundation and Its Relevance to Display Design," *Human Factors*, Vol. 37, pp. 473–494.

Widdel, H., and Post, D. L. (1992), *Color in Electronic Displays*, Plenum Press, New York.

Woods, D. D. (1984), "Visual Momentum: A Concept to Improve the Cognitive Coupling of Person and Computer," *International Journal of Man–Machine Studies*, Vol. 21, pp. 229–244.

Woods, D. D. (1988), "Coping with Complexity: The Psychology of Human Behavior in Complex Systems," in *Mental Models, Tasks and Errors: A Collection of Essays to Celebrate Jens Rasmussen's 60th Birthday*, L. P. Goodstein, H. B. Andersen, and S. E. Olsen, Eds., Taylor & Francis, New York.

Woods, D. D. (1991), "The Cognitive Engineering of Problem Representations," in *Human–Computer Interaction and Complex Systems*, G. R. S. Weir and J. L. Alty, Eds., Academic Press, London, pp. 169–188.

Woods, D. D., and Roth, E. M. (1988), "Cognitive Systems Engineering," in *Handbook of Human–Computer Interaction*, M. Helander (Ed.), Elsevier Science, Amsterdam, pp. 1–41.

Woods, D. D., Wise, J. A., and Hanes, L. F. (1981), "An Evaluation of Nuclear Power Plant Safety Parameter Display Systems," in *Proceedings of the Human Factors Society 25th Annual Meeting*, Human Factors and Ergonomics Society, Santa Monica, CA, pp. 110–114.

Worthy, J. A., and Brill, M. H. (1986), "Heuristic Analysis of Color Constancy," *Journal of the Optical Society of America, Series A*, Vol. 3, pp. 1708–1712.

Wyszecki, G., and Stiles, W. S. (1982), *Color Science*, 2nd ed., Wiley, New York.

Zachary, W. (1986), "A Cognitively Based Functional Taxonomy of Decision Support Techniques," *Human–Computer Interaction*, Vol. 2, pp. 25–63.

CHAPTER 46

INFORMATION VISUALIZATION

Chris North
Virginia Polytechnic Institute and State University
Blacksburg, Virginia

1 INTRODUCTION

The information revolution is changing the way that many people live and think. Vast quantities and diverse types of information are being generated, stored, and disseminated, raising serious issues about how to make such information usable. The need to understand and extract knowledge from stored information is becoming a ubiquitous task. As examples, in everyday life people must sort through a variety of personal information, such as e-mail communications, schedules, news, finances, and computer directories. Students can access countless digital libraries of educational materials. Online shoppers must make decisions among dozens of alternative products, models, vendors, and prices. New disciplines such as bioinformatics are leading the revolution in information-intensive science, using high-throughput data collection technologies such as microarrays and online data repositories. Government intelligence analysts must sift through massive collections of information gathered on a daily basis from sensor networks and other sources.

Information visualization has evolved as an approach to make large quantities of complex information intelligible. An information visualization is a visual user interface to information, with the goal of providing users with information *insight* (Spence, 2001). The basic method is to generate interactive visual representations of the information that exploit the perceptual capabilities of the human visual system and the interactive capabilities of the cognitive problem-solving loop (Ware, 2004).

The goal of this chapter is to highlight the critical high-level design issues in the information visualization design process. Lower-level details of visual display and human perception can be found elsewhere in this book. Other aspects of the design process that apply to the design of user interfaces in general, such as evaluation methods, are also covered in other chapters. While other major references focus on the "what" and "why" of information visualization (Card et al., 1999; Chen, 1999; Wickens and Hollands, 2000; Spence, 2001; Ware, 2004; Shneiderman and Plaisant, 2005), here we emphasize the "how."

1.1 Insight

Human vision contains millions of photoreceptors and is capable of rapid parallel processing and pattern recognition (Ware, 2004). The impressive bandwidth of vision as a mode of communication leads to the efficient transfer of data from digital storage to human mind. Yet a more important benefit is the human ability to reason visually about the data and extract higher-level knowledge, or insight, beyond simple data transfer (Card et al., 1999). This enables people to infer mental models of the real phenomena represented by the data.

	A	B	C	D	E	F	G	H
1	Name	State	Population	% College Grad	Income per Capita	Median Rent	Employees	Retail Sales
2	Aleutians East	AK	2,305	12.9	20,114	518	1,410	6,693
3	Aleutians West	AK	5,259	14.8	20,298	428	4,321	24,233
4	Anchorage	AK	251,335	26.9	26,619	528	93,037	2,612,640
5	Bethel	AK	15,525	13.1	15,493	460	2,269	59,687
6	Bristol Bay	AK	1,023	18.9	29,699	464	392	9,653
7	Dillingham	AK	4,360	15.3	22,616	518	1,060	23,077
8	Fairbanks North Star	AK	83,374	25.2	19,149	471	17,746	702,135
9	Haines	AK	2,181	17.6	26,413	405	371	15,154
10	Juneau	AK	29,378	30.7	26,066	587	8,369	289,620
11	Kenai Peninsula	AK	46,151	17.9	22,761	410	8,934	329,438
12	Ketchikan Gateway	AK	14,422	20.2	28,789	533	5,583	152,834

(*a*)

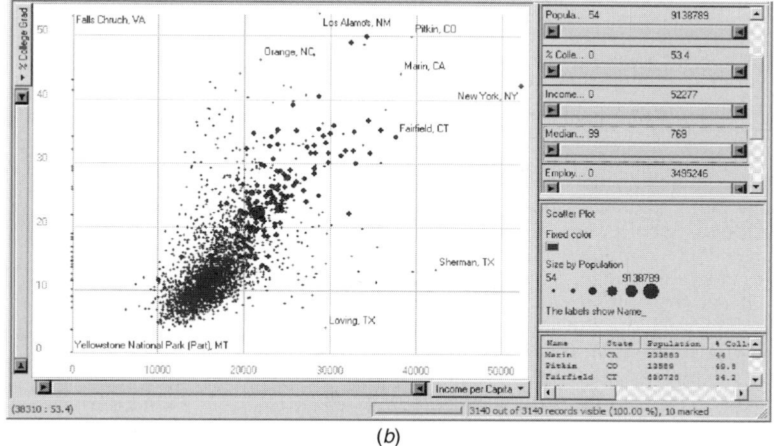

(*b*)

Figure 1 Census demographics data set of 3140 U.S. counties shown in (*a*) spreadsheet form and (*b*) scatterplot form using Spotfire. The plot shows counties by income per capita vs. percentage of adult population that has college degree, with dots sized by population and labels for some outliers. The plot is interactive and reveals details of a county when selecting its dot, in the text subwindow at the lower right. Dots can be filtered by other county attributes, such as median rent, using the dynamic query slider widgets on the right. [From Ahlberg and Wistrand (1995), courtesy of Spotfire.]

For example, Figure 1 demonstrates the mapping of a database of census demographics. From the visual representation, one can readily recognize the approximate proportional relationship between education and income, various outliers, such as New York, NY, and the predominance of large population counties in high income and education. These insights are not explicitly stored within the data set but are inferred through visual pattern recognition. These insights are not so readily identifiable from the textual representation. Clearly, the design of the visual representation is important. A poorly designed visualization can hide insight, or even mislead with incorrect insight.

Visualization can enable a broad range of information insight, and several such insights are listed below (see also Wehrend and Lewis, 1990; Zhou and Feiner, 1998; Wickens and Hollands, 2000; Shneiderman and Plaisant, 2005). The first two are simplistic and can be supported readily by textual or query-based user interfaces such as spreadsheets or search forms, because they are precise and have solutions consisting of a single data entity. However, the latter are more complex and are well supported by visualization. These involve open-ended questions with complex answers that require seeing the whole. A strength of visualization is the capacity for discovery, the recognition of new insights unexpected by the users and potentially unforeseen by the visualization designers.

Simple insights:

- *Summaries*: minimum, maximum, average, percentages
- *Find*: known item search

Complex insights:

- *Patterns*: distributions, trends, frequencies, structures
- *Outliers*: exceptions
- *Relationships*: correlations, multiway interactions
- *Tradeoffs*: balance, combined minimum/maximum
- *Comparisons*: choices (1 : 1), context (1 : M), sets (M : N)
- *Clusters*: groups, similarities
- *Paths*: distance, multiple connections, decompositions
- *Anomalies*: data errors

1.2 Design

Like any user interface, effective information visualizations are difficult to design. Fundamentally, information visualizations make *abstract information* perceptible. Abstract information has no inherent perceptual form, as in the case of databases or computer directories. Hence, there are no natural constraints on the types of visual representations that creativity can produce for abstract information, and the possibilities are limitless. As a result, there is significant challenge, excitement, and opportunity both in creating novel visual representations and in identifying the most effective representations among the endless possibilities. In contrast, *scientific visualization* (Rosenblum et al., 1994) typically emphasizes the visualization of data that represent physical three-dimensional phenomena, which offer some natural constraints on visual representations and focus the challenges on realism.

The two most challenging characteristics of information that make designing effective information visualizations difficult are (1) *complexity* (supporting diverse abstract information that may have multiple interrelated data types and structures), and (2) *scalability* (supporting very large quantities of information). Because of these characteristics, visual representations alone are not sufficient, and interactive techniques must also be designed. Although the principles of static graphs and illustrations are fundamental to visualization design (Cleveland, 1993; Gillan et al., 1998; Wilkinson, 1999; Tufte, 2001), new human–computer interaction issues related to these two challenging characteristics come to the forefront in information visualization design.

As an overview, the visualization design process involves iterative requirements analysis, design, and evaluation (e.g., Rosson and Carroll, 2001). In the requirements analysis phase, it is important to identify the two primary inputs to design: the characteristics of the information to be visualized and the types of insights that the visualization should enable. Characteristics of the information include the data schema, underlying structures, and quantity. Since the number of data attributes and desired insights may be large, identifying a prioritization of attributes and insights will be helpful in balancing design trade-offs. Other elements of requirements analysis include broader user tasks, users' domain knowledge, data semantics, and computer system requirements. In the design phase, major design decisions (presented in this chapter) include the visual mapping of the information, the representation of information structures, visual overview strategies, navigation strategies, and interaction techniques.

The evaluation phase must be considered continually during the design process (Plaisant, 2004). A claims analysis identifies the positive and negative impacts of a visualization design's features on its insight capability and seeks to overcome or balance these trade-offs through iterative design (Rosson and Carroll, 2001). Begin with analytic evaluations to determine if designs meet requirements, such as scalability to data quantity and appropriateness for producing desired insights. In later iterations, empirical evaluations involving users should be undertaken, such as the "wizard of Oz" technique, usability testing, or controlled experiments (Chen and Yu, 2000; Tory and Möller, 2004a). Desired insights identified in requirements analysis should be implemented as benchmark user tasks in the empirical evaluations. Alternatively, since benchmark tasks often overly constrain the testing to simplistic insights that discount the discovery aspect of visualization, the insight-based methodology (Saraiya et al., 2004) attempts to measure the insight generated by visualizations by using an open-ended experimental protocol without benchmark tasks. In the following sections we highlight the major design decisions in the information visualization design process.

2 VISUALIZATION PIPELINE

The *visualization pipeline* is the computational process of converting information into a visual form with which users can interact (Card et al., 1999) (Figure 2). The first step is to transform raw information into a well-organized canonical data format. The resulting format typically consists of a data set containing a set of data entities each of which has associated data attribute values. Various data-processing steps can be used to manipulate the data as needed. Derived data, such as data mining or clustering results, can be very useful for assisting in insight generation (Fayyad et al., 2001). The second step, the heart of the visualization process, is to map the data set into visual form. The visual form contains visual glyphs that correspond to the data set entities. The third step embeds this visual form into views, which display the visual form on screen and provide various view transformations, such as navigation. The view is then presented to the user through the human visual system. Users interpret the view to (partially) reconstruct the underlying

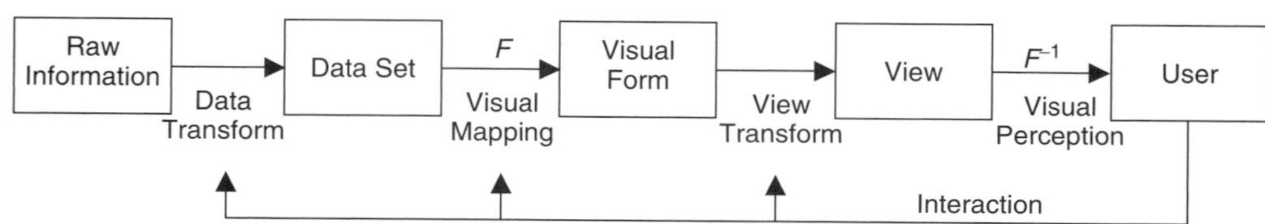

Figure 2 Visualization pipeline, converting information into interactive visual representations. (Adapted from Card et al., 1999.)

information. Finally, users can interact with any of the steps in the pipeline to alter the resulting visualization and make further interpretations. This entire pipeline comprises an information visualization.

2.1 Visual Mapping

The visual mapping at the second step is the heart of visualization and must be designed carefully. The goal is to communicate information from computer to human. The medium of communication is a visual representation of the information. The data set is mapped computationally into visual form by some function F, which takes the data set as input and generates the visual representation as output. Then, when the visual representation is communicated to users, they must cognitively reverse the visual mapping by inverting function F to decode the information from the visual representation. It is yet unclear how, when, and to what degree F^{-1} is cognitively applied in the perceptual process, and a variety of models exist (Ware, 2004). Although some cognitive reasoning operates on the visual representation itself, eventually meaning must be decoded. Nonetheless, this visual communication process implies four important characteristics of the visual mapping function F:

1. *Computable.* F is a mathematical function that can be computed by some algorithm. Although there is significant room for creativity in the design of these functions, execution of the functions must be algorithmic.

2. *Invertible.* It must be possible to use F^{-1}, the inverse of mapping function F, to reconstruct the data from the visual representation to a desired degree of accuracy. If this is not possible, the visualization will be ambiguous, misleading, or not interpretable.

3. *Communicable.* F (or preferably F^{-1}) must be known by the user to decode the visual representation. It must be communicated with the visualization or already known by the user through prior experience. In usability terms, this is a learnability issue.

4. *Cognizable.* F^{-1} should minimize cognitive load for decoding the visual representation. This is a human perception and performance issue.

The visual mapping step is accomplished by two substeps (Card et al., 1999) (Figure 3). First, each data entity is mapped into a visual *glyph.* The vocabulary of possible glyphs consists primarily of points (dots, simple shapes), lines (segments, curves, paths), regions (polygons, areas, volumes), and icons (symbols, pictures). Second, attribute values of each data entity are mapped onto *visual properties* of the entity's glyph. Common visual properties of glyphs include spatial position (x, y, z), size (length, area, volume), color (gray scale, hue, intensity), orientation (angle, slope, unit vector), and shape. Other visual properties include texture, motion, blink, density, and transparency. For example, in Figure 1, U.S. counties are mapped to circular points. Income and education

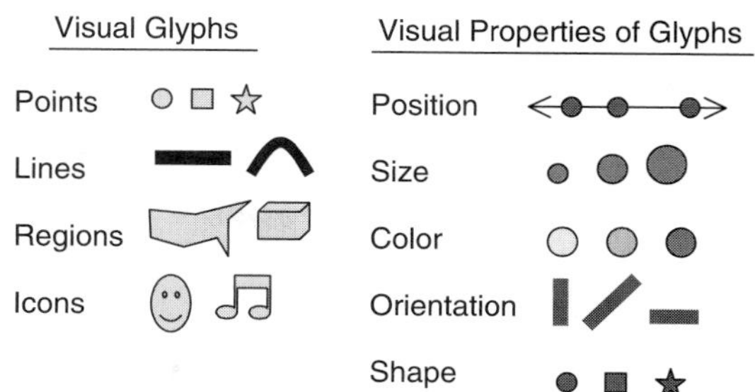

Figure 3 Vocabulary of glyphs and some visual properties of glyphs. (Adapted from Card et al., 1999.)

levels of each county are mapped to the horizontal and vertical position of the point, respectively, and population value is mapped to the size of the point.

2.2 Visual Properties

In general, data attributes should be prioritized according to the problem requirements and desired insights. The prioritization can then be applied to map the higher-priority data attributes to the most effective visual properties. Spatial position properties are the most effective and should be reserved to lay out the data set in the visual representation according to the most important data attributes.

The remaining visual properties, called *retinal properties* (Bertin, 1983), can be used next. The effectiveness of these properties are determined by many interdependent factors, including preattentive processing (Healey et al., 1996), perceptual independence (separability) (Ware, 2004), data type (quantitative, ordinal, categorical) (Card et al., 1999), polarity (greater than, less than) (Ware, 2004), task (Carswell, 1992; Wickens and Hollands, 2000), and attention (Chewar et al., 2002). Commonly accepted orderings of the effectiveness of these attributes are based on empirical evidence (e.g., Cleveland and McGill, 1984; Nowell et al., 2002), as well as experience (Bertin, 1983; Mackinlay, 1986). The order shown in Figure 3 is intended for quantitative data. For categorical data, color and shape become more predominant. Some visualization design systems attempt to use such rules to generate effective mappings automatically [e.g., Apt (Mackinlay, 1986) and Sage (Roth et al., 1994)].

Finally, for any remaining attributes, interaction techniques can be applied. In general, direct visual mapping of information is most effective for rapid insight; interaction techniques require slower physical actions by the user to reveal insights. When mapping additional data attributes would overly clutter the visual representation and reduce comprehension of more important attributes, interaction techniques can be used instead. Interaction techniques enable users to alter the visual mapping function or other stages of the visualization pipeline, based on additional attributes. By viewing the resulting changes in the visual representation, users can infer additional information about those attributes. For example, the *dynamic queries* technique provides interactive query widgets for other attributes and can be used to dynamically filter the entity glyphs (Ahlberg and Wistrand, 1995) (Figure 1).

3 INFORMATION STRUCTURE

The visual mapping process provides an initial starting point for visualization design, but more advanced methods are needed as data complexity increases. Identifying underlying structures within the target information helps to further guide the design process. These structures provide high-level organization to a data set and often provide guidance for the design of appropriate visualizations. Since these structures are likely to be very important to users' mental models of the information, they are typically mapped to the spatial position attributes and form the primary layout of the visualization. In general, there are four common classes of information structures (adapted from Shneiderman (1996), Card et al. (1999), and Spence (2001)), as discussed in Sections 3.1–3.4. These are not strict or mutually exclusive classifications, but useful guidelines.

3.1 Tabular Structure

Tables consist of rows (entities) and columns (attributes). This is often referred to as *multidimensional* or *multivariate data*, because each attribute defines a dimension of the data space within which each entity identifies a single point. Examples include databases and spreadsheet tables, such as the census data in Figure 1. Visualizations of tables that contain a small number of attributes can be designed relatively easily using the visual mapping process. However, such visualizations lack scalability to many attributes, due to the limited number of nonconflicting visual properties to choose from. To address this problem, a variety of creative methods have been developed for tables of many attributes. Primarily, these involve the use of more complex glyphs and spatial layouts.

TableLens (Rao and Card, 1994) (Figure 4*a*) preserves the tabular spreadsheet visual representation but converts cells to horizontal bar glyphs with cell values mapped to bar length. This exploits the length property, which is excellent for encoding quantitative data. Also, since the bars are very thin, many values can be packed onto the screen, providing an excellent overview of a large data set. TableLens encodes each data entity (row) with multiple glyphs (bars), one glyph for each of the entity's attribute values (columns). Interactively selecting a set of rows will expand them to reveal the detailed data values in textual form. Users can vertically sort the table by any attribute. By spatially arranging the data according to one attribute, distributions and relationships to other attributes can be seen. However, it is perceptually difficult to relate two nonsorted attributes. Hence, users must sort each attribute interactively to explore all potential relationships. The *proximity compatibility principle* (Wickens and Hollands, 2000) predicts that representations that use a single glyph per data entity, such as in scatterplots (Figure 1), would be better than TableLens for recognizing relationships between attributes. But as indicated previously, such representations are more limited in scalability. Herein is the trade-off: TableLens provides an overview of many attributes with reasonable capability for relationship insights, whereas scatterplots provide excellent insight on relationships but only for the two attributes mapped to x and y (and potentially a small number of other attributes using color, size, etc.).

To analyze the scalability of TableLens, consider an approximate screen resolution of 1000×1000 pixels. If each bar glyph is only 1 pixel thick, 1000 data entities (rows) can be shown. Columns will need to be approximately 50 to 100 pixels wide to enable reasonable visual discretion of quantitative data such as percentages, leading to 10 to 20 attributes being visible. Hence, this approach can display a tabular data set containing 1000 data entities and 20 attributes. Much larger data sets can be explored in TableLens by using its aggregation and interactive navigation (e.g., scrolling) strategies, but only 1000×20 values can be perceived simultaneously. For data sets larger than the screen size, TableLens aggregates adjacent rows by showing averages or minimum and maximum values.

The Cartesian coordinate system uses orthogonal axes to visually map two or three attributes of a tabular data set to space. However, orthogonal axes fundamentally limit scalability of attributes. As an alternative, Parallel Coordinates (Inselberg, 1997) (Figure 4*b*) displays attribute axes as parallel vertical lines. Each data entity is mapped to a polyline that connects the entity's attribute values on each attribute axis. Hence, attributes are mapped to the vertical position of the respective vertices of the polyline. Users can recognize clusters of similar entities and relationships between adjacent attributes. Patterns of crossing lines between adjacent axes indicate an inverse relationship between those two attributes, and noncrossing lines indicate a proportional relationship. To combat occlusion and clutter, selecting entities interactively highlights their polylines across all axes. A scalability analysis of Parallel Coordinates would give a result similar to that of TableLens. Other arrangements of axes include radial (Kandogan, 2000) and circumferential (Miller, 2004).

Use of more complex iconic glyphs includes star plots (Chambers et al., 1983), which map data attributes to the length of the radial needles emanating from a star icon, and Chernoff faces (Chernoff, 1973), which attempts to exploit the human ability to rapidly recognize facial features and expressions. Although these iconic methods do not scale up very well, they are useful for combining with other information structures (such as networks) because they leave the spatial position properties available for other uses (Ward, 2002).

The reverse approach is to simplify the glyphs and visual representations by splitting the attributes up into separate views. For example, four attributes can be displayed using the x and y axes of two scatterplots. An interactive technique called *brushing and linking* relates the two plots (Becker and Cleveland, 1987).

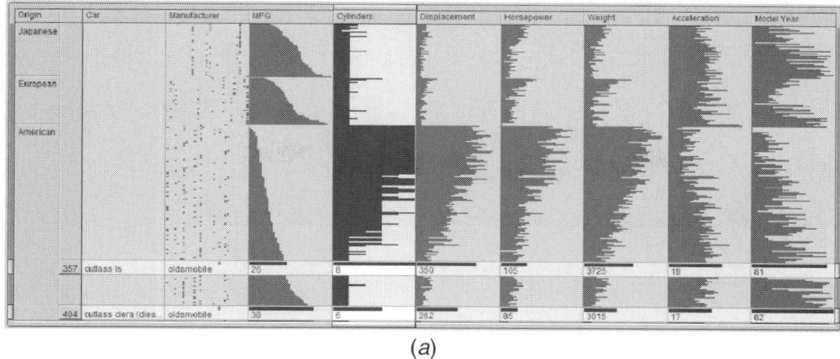

(*a*)

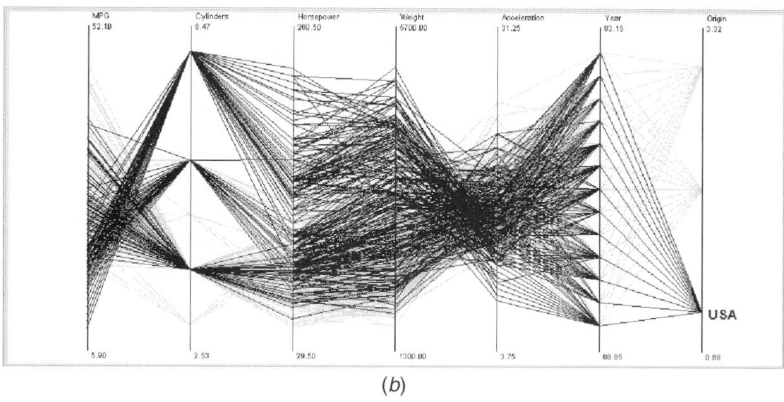

(*b*)

Figure 4 (*a*) InXight TableLens; (*b*) XmdvTool's Parallel Coordinates. Both show the same data set about automobiles. Clearly, the United States dominates the market in large, powerful, gas-guzzling cars. Two outliers, cars with large engines yet high MPG, are highlighted in TableLens, revealing their detailed values. TableLens sorts the data by one attribute, whereas Parallel Coordinates sorts all the attributes simultaneously. [(*a*) From Rao and Card (1994), courtesy of InXight); (*b*) from Ward (1994), courtesy of Matthew Ward.]

When users select glyphs in one plot (*brushing*), the corresponding glyphs for the same underlying data entities are highlighted in the other plot (*linking*). Scatterplot matrixes take this to the extreme, showing plots for all combinations of attribute pairs (Cleveland, 1993). The brushing-and-linking technique can also be used to view a single data set in several different visual representations at the same time, such as plots and geographic maps, to relate the various contexts (Roth et al., 1996; North et al., 2002) (see Section 6.2).

3.2 Spatial and Temporal Structures

Spatial and temporal structures have a strong one-, two-, or three-dimensional component in which navigation is likely to be required. One-dimensional (1D) examples include time lines, music, video streams, lists, linear documents, and slide shows. Two-dimensional (2D) examples are road maps, satellite images, photographs, and blueprints. Three-dimensional (3D) examples are MRI and CT medical scans, CAD/CAM architectural plans, and virtual environments. Continuous functions, including those with domains greater than three dimensions, also fall in this category (Tory and Möller, 2004b). These spatial and temporal structures are the most natural for mapping onto spatial displays.

For example, the Music Animation Machine (Malinowski, 2004) (Figure 5*a*) provides a simple time line representation of music that scrolls as the music plays. Notes are represented as horizontal bars, with vertical position representing pitch, horizontal position representing timing, length indicating duration, and color indicating other attributes, such as instrument, timbre, or hand (in the case of piano). Similarly, LifeLines (Plaisant et al., 1996) represents events in a person's medical history, but in a more compact form, with zooming for navigation. For time lines that contain periodic cycles, such as calendars, visual spirals can be used to proximate the cycles while maintaining a continuous line (Carlis and Konstan, 1998). Streaming video can be viewed as a 3D video cube (2D frames + 1D time) (Elliott and Davenport, 1994).

In 3D data spaces, the main challenge is viewing the interior of the 3D structure beyond the exterior surface when occlusion is problematic. Architectural walk-through applications (polygonal data)

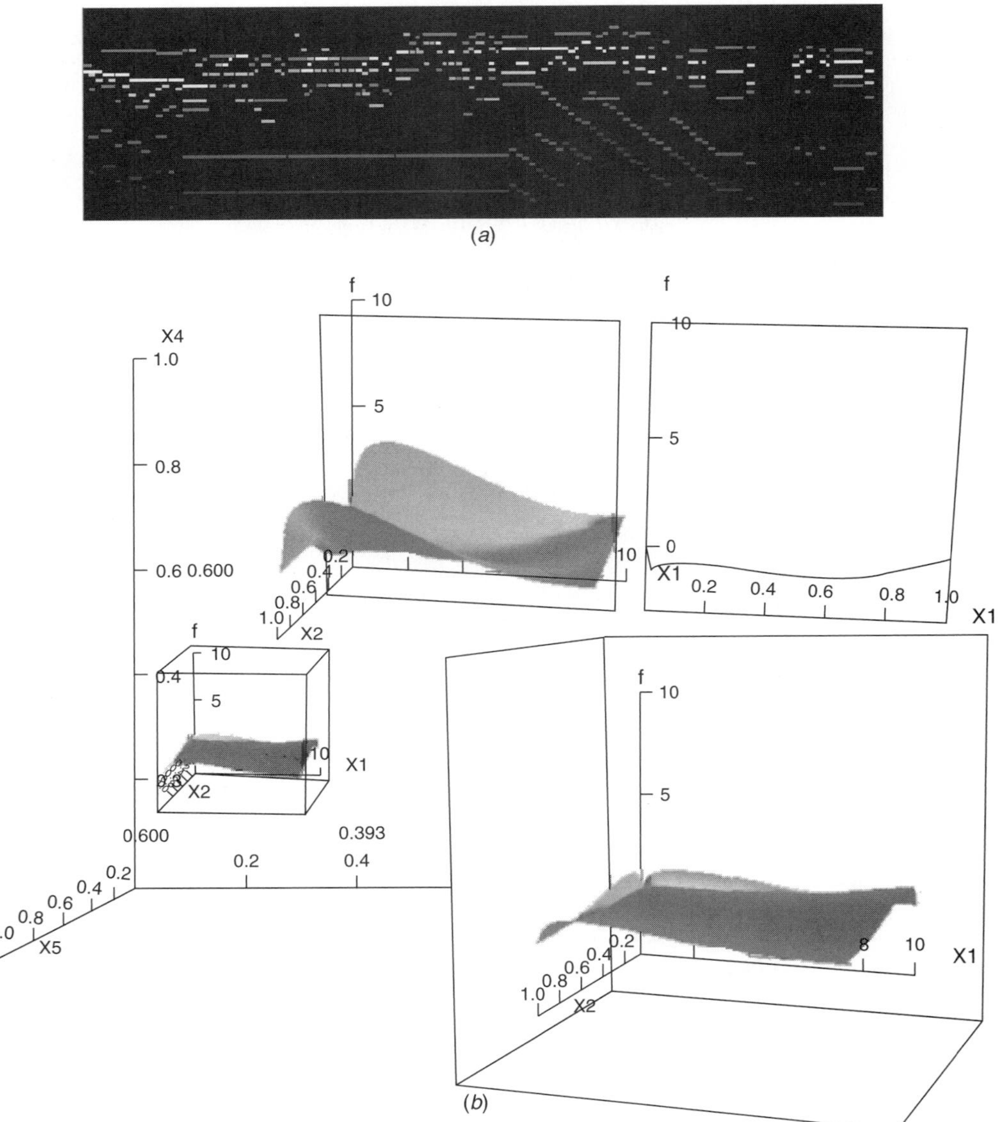

Figure 5 (*a*) Music Animation Machine showing a portion of Bach's Brandenburg Concerto No. 4, third movement; (*b*) Worlds within Worlds nests inner coordinate frames within outer frames to decompose hyperdimensional spaces. [(*a*) From Malinowski (2004), courtesy of Stephen Malinowski; (*b*) from Beshers and Feiner (1993), © 1993 IEEE.]

typically use first-person perspective projection, with six-degree-of-freedom navigation for a lifelike experience (Stoakley et al., 1995). For medical imagery (volumetric data), strategies include slicing and transparency. For example, the Visible Human Explorer presents 2D slices that can be animated through the 3D body (North et al., 1996). In 3D volume rendering, transparency can give users x-ray vision into the space by adjusting the opacity of various materials within the space through interactive control of the visual transfer function (Kniss et al., 2001).

Hyperdimensional continuous spaces must somehow be reduced to three or fewer dimensions for display. Worlds within Worlds (Beshers and Feiner, 1993) (Figure 5*b*) displays subspaces of hyperdimensional functions by nesting a 3D coordinate frame within another 3D coordinate frame. The location of the origin (0,0,0) of the inner frame within the outer frame determines the values of the outer frame dimensions used to generate the subspace for the inner frame. By interactively sliding the inner frame around the inside of the outer frame, the full space can be explored. Repeated nesting can enable greater numbers of dimensions. Other methods include hierarchical axes (Mihalisin et al., 1991) and slicing (van Wijk and van Liere, 1993).

3.3 Tree and Network Structure

Tree and network structures contain specific *connections* between individual entities. In graph theory terms, a network consists of a set of vertices (entities) connected by a set of edges (connections), which can be either directed or undirected. Like data entities, connections can also contain attributes. Examples include communications networks, literature citations, or Web hyperlinks between pages. Tree structures are a special subset of networks that are distinct and common enough in digital information to warrant separate treatment. Trees have a hierarchical structure that connects data entities by parent–child connections. To be a tree, each child entity should have only one parent. Examples include computer file directories, menu systems, organization charts, and taxonomies such as the Dewey decimal system. Other useful variants of tree structures exist, such as multitrees (Furnas and Zacks, 1994) and polyarchies (Robertson et al., 2002). New types of insights involve understanding the connection structure, such as the breadth or depth of the tree. The primary challenge for visualization is the spatial layout of the network or tree to reveal the structure of the connections. The secondary challenge is to visualize data attributes of the entities and connections.

3.3.1 Trees

For tree structures, two primary approaches exist for representing parent–child connections visually: *link* and *containment*. The *link* approach uses node–link diagrams. Entities are mapped to visual nodes, and connections are mapped to visual links between the nodes. Alternative spatial layouts of node–link diagrams include nested–indented, as in Windows Explorer or Mac Finder; top–down or left–right, as in SpaceTree (Grosjean et al., 2002); radial, as in Hyperbolic Tree (Lamping et al., 1995); or 3D ConeTrees, which combines radial and left–right (Robertson et al., 1993) (Figure 6*a*). These systems emphasize the display of a single data attribute as a text label on the nodes. Node–link diagrams tend to be space consuming, due to the amount of white space needed within each of these spatial layouts, making it difficult to get beyond 100 or 1000 nodes visible. Since large tree structures cannot be displayed completely on the screen, each layout requires interactive navigation. The focus + context technique (Section 5.3) is a natural match for tree navigation, enabling users to drill down within an individual branch of focus in the tree while maintaining the context of the path to the root and siblings. Hyperbolic Tree shrinks the size of nodes near the periphery to pack more nodes on the display. Three-dimensional approaches such as ConeTrees exploit the third dimension for additional space to lay out the tree, but due to occlusion, it is unclear whether that extra space is useful. The most important factor in three-dimensional designs is the interactive navigation (Wiss and Carr, 1999). Simple six-degree-of-freedom camera movement through a static three-dimensional scene is clearly not effective in these structures. ConeTrees employs a much more efficient interaction technique, cascading rotation of the three-dimensional cones to bring the desired child nodes to the front.

The *containment* approach for tree layout is exemplified by Treemap (Johnson and Shneiderman, 1991) (Figure 6*b*). Child nodes, represented as rectangles, are contained visually within their parent nodes as in Venn diagrams. Treemap is space-filling, to maximize the use of every available pixel, and scales easily to 10,000 entities. Data attributes are mapped to retinal properties of the node rectangles, such as size and color. Hence, Treemap emphasizes the visualization of nontextual attributes. In dense Treemaps, not enough space is left for textual node labels. Nodes can be arranged within their parent node according to a variety of algorithms. The original Treemap used a slice-and-dice algorithm. It was simple but tended to generate rectangles with many different aspect ratios, some square and some long narrow, which are difficult to compare visually. Newer algorithms generate squarified Treemaps (Bederson et al., 2002). SunBurst (Stasko et al., 2000) offers a radial version of the containment approach based on the stacked pie chart. In comparison to Treemap, SunBurst can improve learnability for novices but reduces scalability because the number of leaf nodes is limited by one-dimensional circumferential space rather than the full two-dimensional area available to Treemap.

3.3.2 Networks

For visualizing networks, the node–link approach is dominant. Many algorithms have been devised to lay out network diagrams spatially (Herman et al., 2000) and increasingly are tuned to specific types of networks (Figure 7*a*). Designs must consider network features such as number of nodes and links, directedness of links, node degree, any common patterns within the network structure, and attributes of nodes and links that should be visible. Links can be drawn as straight lines, arcs, or orthogonal polylines. Algorithms can apply aesthetic constraints such as minimizing link crossings, minimizing link lengths, and maximizing symmetries (Ware et al., 2002). In general, the goal is to lay out the network to reveal hidden network patterns and avoid the "bowl of spaghetti" phenomenon. Some common layout algorithms include circular, layering, concentric layering, force directed, and clustering. SeeNet (Becker et al., 1995) arranges communications nodes according to geographical position and raises links off the surface as three-dimensional arcs. Arc properties such as color and line thickness are used to represent communications type and bandwidth. H3 (Munzner, 1998) arranges nodes inside a three-dimensional sphere that can be rotated and navigated similar to Hyperbolic Tree. Navigation can be

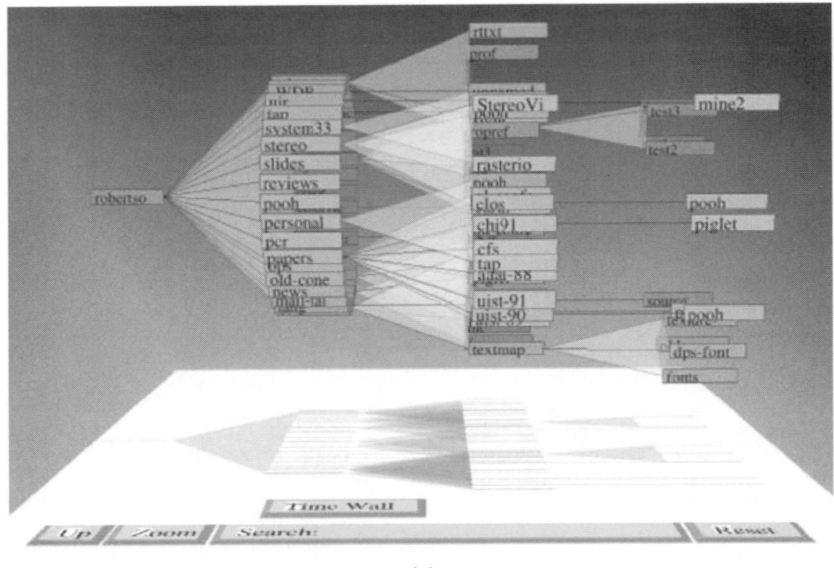

(*a*)

(*b*)

Figure 6 (*a*) ConeTrees; (*b*) SequoiaView Treemap, showing directory structures on a computer system. ConeTrees emphasizes the tree structure and node labels, whereas Treemap emphasizes node attributes such as file size and type. [(*a*) From Robertson et al. (1993), © 1993 ACM, Inc., used by permission; (*b*) from van Wijk and van de Wetering (1999), courtesy of Jarke van Wijk.]

used to reduce network complexity by representing the network from the perspective of one node in focus (Andrews et al., 1995).

A different approach is to visualize the network as an adjacency matrix (Hetzler et al., 1998) (Figure 7*b*). An adjacency matrix emphasizes the connections instead of the nodes, mapping each potential connection to a cell in the matrix. Finally, it is also possible to reduce networks to trees using minimum spanning trees or hierarchical aggregation (Feiner, 1988), thereby enabling the use of tree visualization methods.

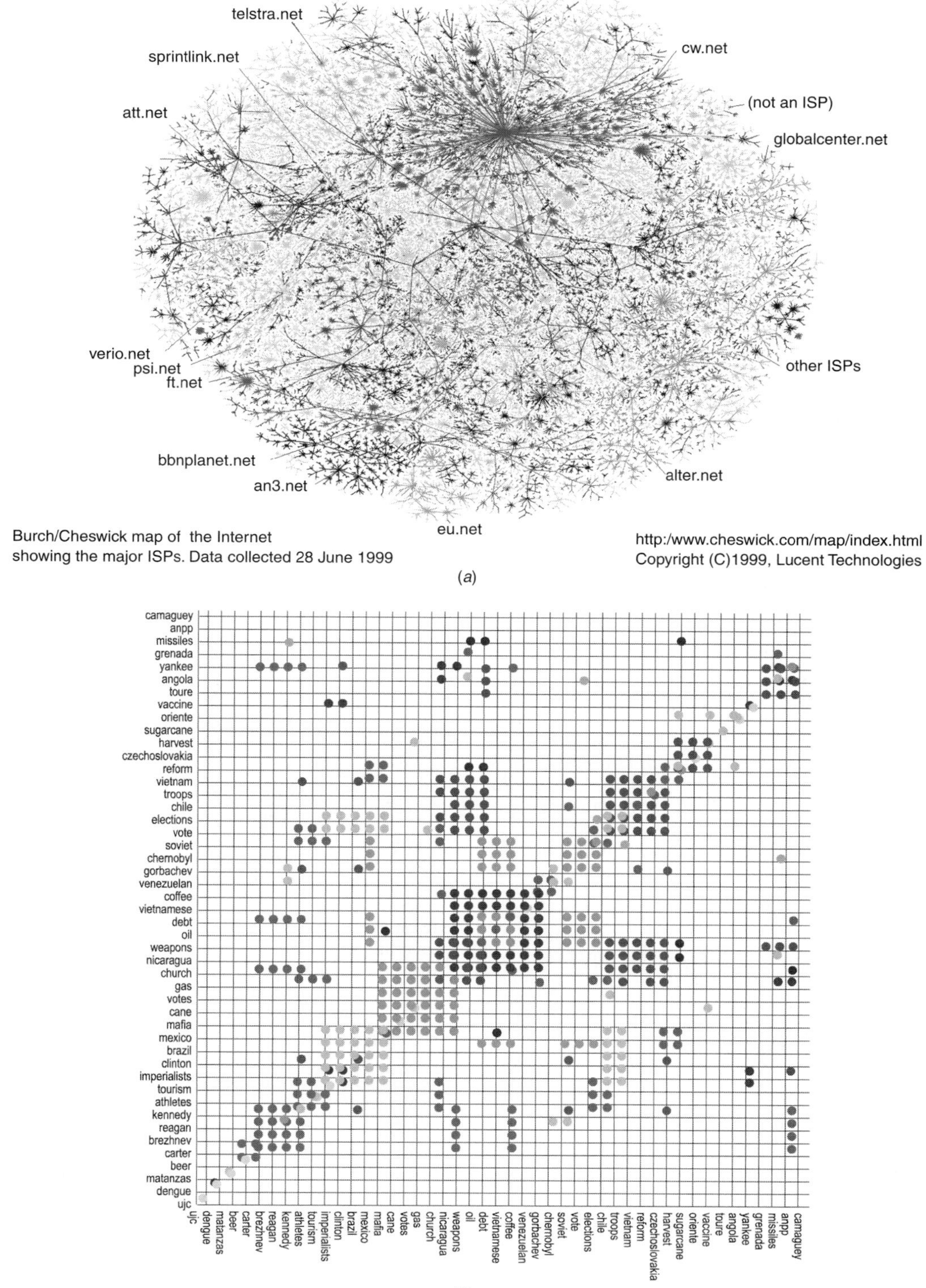

Figure 7 (a) Map of the Internet color-coded by major ISPs; (b) Connex adjacency matrix representation of a relationship network. [(a) From Cheswick, (1998), courtesy of William Cheswick; (b) from Hetzler et al. (1998), courtesy of Pacific Northwest National Laboratory.]

3.4 Text and Document Collection Structure

This structure consists of arbitrary collections of documents, often text. Examples include digital libraries, news archives, digital image repositories, and software code. Of the four types of information structure, this type is the least structured and hence can be the most difficult to design visualizations for. Text is particularly challenging to map to visual form, because it is not obvious how text can be the input to a mapping function as described in Section 2.1. Mapping functions must take advantage of the minimal structure and other characteristics of text to generate useful data for computing visual representations. External structures of text or document collections such as tables of contents (tree structure), meta-data (tabular structure), or citations (network structure) are categorized as other types of information structures and were discussed earlier. The emphasis of the structure described in this section is on the full text or the documents themselves. Solutions range from the macro scale (overview of large collections) to the micro scale (a single document fragment).

A major class of text visualizations focuses on providing semantic maps of large document collections based on document topics. Generally, the goal is to cluster documents spatially in the visualization such that similar documents (documents containing similar content) are near each other and dissimilar documents are distant. This creates a map of the document space based on the metaphor of a physical library in which books are arranged carefully by topic. Similarity between documents can be measured in many ways and is generally the domain of *information retrieval* (Baeza-Yates and Ribiero-Neto, 1999). A common method is to compare the frequency of occurrence of dictionary words or phrases between the two documents. Densities of document clusters can then be analyzed to extract topic keywords for labeling the map. Document Galaxies (Wise et al., 1995) and Kohonen self-organizing maps (Lin, 1992) map individual documents to tiny dots that are clustered by text content. Selecting a dot from the map reveals a document summary or opens the full document. ThemeView (Wise et al., 1995) (Figure 8*a*) emphasizes the documents' topics, creating a three-dimensional terrain landscape representing the themes in the collection. Themes are mapped to terrain landscapes of mountains, with theme strength mapped to mountain height. Mountains that are adjacent or joined indicate the presence of documents that span both themes.

The keyword query approach provides a more focused map, based on keywords specified by the user. VIBE (Korfhage, 1995) visualizes how documents relate to the keywords. It spreads the user's keywords around the periphery of the display. Then, document dots are mapped into the space according to their strength of match to each keyword, using a spring-based attraction model. TileBars (Hearst, 1995) inverts the map, showing how the keywords relate to each document. Document hits are listed as in a normal textual search engine, but each document has a tilebar that shows the density of the keywords in each section of the document.

Finally, documents can be arranged by the users themselves or by some default order. Miniature representations of the documents can be displayed to promote browsing by content. Web Book and Forager (Card et al., 1996) collects favorite Web pages in a virtual three-dimensional book that users can flip through quickly and scan visually. Books can be arranged on a virtual bookshelf. With DataMountain (Robertson et al., 1998), users arrange images of favorite Web pages or photos on an inclined plane (Figure 8*b*), taking advantage of spatial memory for recall. At the lowest level of text visualization, SeeSoft (Eick et al., 1992) visualizes the text of software code using a miniaturized representation. It displays each line of text as a tiny colored line segment (more on SeeSoft in Section 4.2).

3.5 Combining Multiple Structures

Frequently in real-world applications, information involves complex combinations of multiple information structures. Furthermore, information of one structure type could be massaged computationally into a different structure type to offer new ways to conceptualize the information. For example, a single e-commerce Web site may consist of a text document collection of product pages which also contains a network structure of hyperlinks, is organized by a tree-structured site map, and has tabular meta-data about product prices and page accesses. A visualization designer should consider each of these separate structures as a potential visual index into the underlying product information.

A frequent insight goal in these situations is to relate the various structures. However, designing a visual representation that effectively combines multiple structures is difficult. Since a structure typically consumes the primary spatial portion of the visual mapping, combining multiple structures in a single mapping can result in conflict. A primary decision is whether to attempt to combine them or to separate the structures into multiple views (Baldonado et al., 2000). Multiple views simplify the design, since each structure can use its most optimal mapping independently. The structures can then be related by interactive linking between the views (see Section 6.2). Linking is useful for querying one structure with respect to another. However, because interactive linking reveals only a small number of associations at a time, users must mentally integrate the relationships between the structures over time and can easily miss interesting associations. On the other hand, integrating two structures into a single view typically requires that one structure be used as the spatial basis, while the other is dismantled and embedded within that space. This enables a clear representation of how the second structure depends on the first, but clarity of the second structure can be lost. Another potential solution is animating or morphing between the two structures (Robertson et al., 2002).

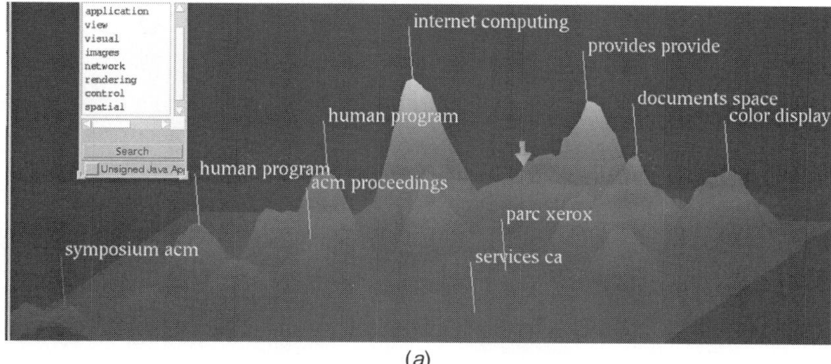

(*a*)

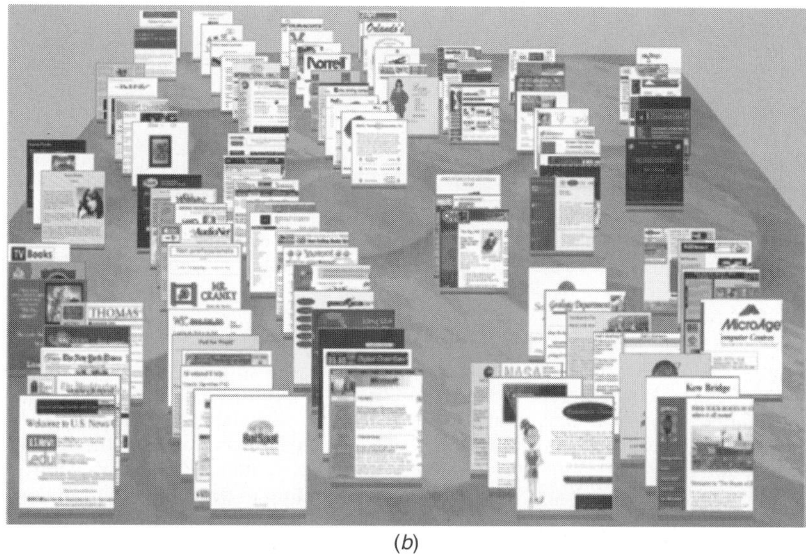

(*b*)

Figure 8 (*a*) ThemeView represents common themes in a large document collection as labeled mountains in a landscape. Probing reveals specific documents. (*b*) DataMountain lets users visually organize Web favorites, documents, or digital photos for later recall. [(*a*) From Wise et al. (1995), courtesy of Pacific Northwest National Laboratory; (*b*) from Robertson et al. (1998), © 1998 ACM, Inc., used by permission.]

An example is PathSim (Polys et al., 2004) (Figure 9), an information-rich virtual environment for biology simulation which combines three-dimensional spatial structure of human anatomy with tabular structure of data collected on viral infection within anatomical components. In this design the tabular data are visually embedded directly within the three-dimensional anatomy as small manipulable visualizations adjacent to their corresponding anatomical components. The tabular structure is dismantled to associate portions of the data set visually with components in the three-dimensional scene. Although this supports the task of understanding the effects in each anatomical component, it does not enable a single overview of all tabular results. To overcome this, a heads-up display is included according to the multiple-views approach. It shows aggregated tabular information as a summary of what is visible in the entire scene as users navigate in the three-dimensional anatomy.

4 OVERVIEW STRATEGIES

Designing methods for the visual representation of very large quantities of information is one of the fundamental problems in visualization research. As information quantity increases, it becomes more difficult to pack all the information visually on the available screen space. There simply are not enough pixels. Even if there were enough pixels, including all the details in a single display might make it appear visually cluttered. In general, a naive visualization design would be to consume the entire display with the full detail of only a few of the data entities, and thereby limit the display to a relatively small portion of the full data set. This is analogous to peering

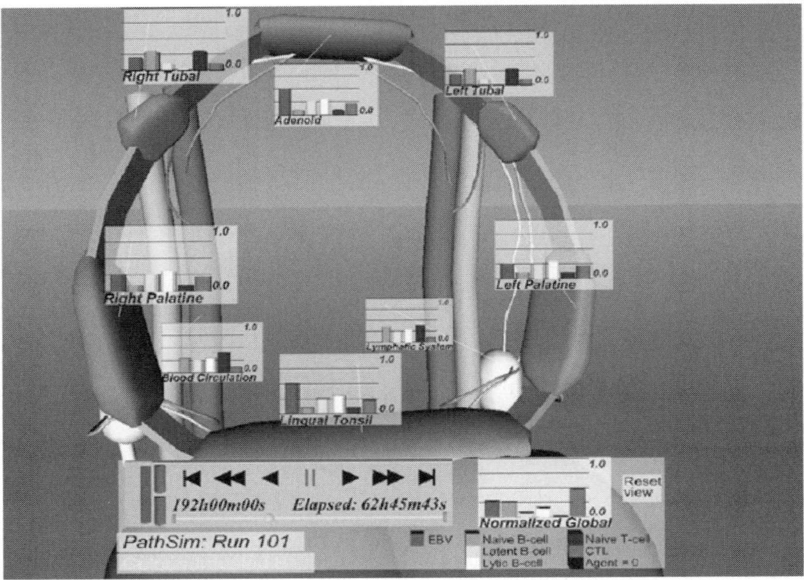

Figure 9 PathSim reveals simulated viral infection data within the human anatomy, combining tabular structure with three-dimensional spatial structure. Zooming in or out navigates to lower or higher levels of anatomical structure, and shows correspondingly lesser or greater levels of aggregation of tabular data. Here, we see the effects of an Epstein–Barr virus infection of the tonsils. (From Polys et al., 2004.)

into a vast room through a tiny keyhole and is called the *keyhole problem*. For example, the spreadsheet in Figure 1 shows the detailed numerical data, but the scrolling window reveals only about 40 rows at a time.

To support visualization of very large information spaces, Shneiderman suggests the design mantra "*overview first, zoom and filter, then details on demand*" (Shneiderman and Plaisant, 2005). The solution to the keyhole problem is to start users with a broad overview of the full information space, sacrificing information details. Then provide interaction mechanisms that enable users to zoom in on desired information and filter out anything not of interest. Finally, quickly retrieve and display detailed information about individual data entities when selected by the user. There are several advantages to providing an initial visual overview of the information:

- An overview supports the formation of mental models of the information space.
- It reveals what information is present or not present.
- It reveals relationships between the parts of the information, providing broader insights.
- It enables direct access and navigation to parts of the information simply by selecting them from the overview.
- It encourages exploration.

Empirical evidence confirms that the use of visual overviews results in improved user performance in various information-seeking tasks [some studies are listed in North (2001) and Hornbæk et al. (2002)]. In general, visualization designers should seek to pack as much information into the overview as cleanly as possible. A major design decision is choosing which information to percolate up to the overview and which information to bury in the lower detail levels that can only be reached through user interaction. This is somewhat analogous to choosing which products to show in the store window. Ideally, an overview should provide some "scent" of all the detailed information hiding beneath it (Pirolli and Card, 1999).

To create overviews that attempt to pack a large data set onto a relatively small screen, there are two possible approaches in the visual mapping process: (1) reducing the quantity of data in the data set before the mapping is applied, or (2) reducing the physical size of the visual glyphs created in the mapping.

4.1 Reducing Data Quantity

One method for reducing the data quantity while maintaining reasonable representation of the original data is *aggregation*. Aggregation groups entities within the data set, creating a new data set with fewer total entities. Each aggregate becomes an entity itself, temporarily replacing the need for all entities within the aggregate. For example, a histogram applies aggregation to represent data distribution on one attribute (Spence, 2001).

When using aggregation, the first design decision is choosing which entities should be grouped together. Entities can be grouped by common attribute values (Stolte et al., 2002) or by more advanced methods such as clustering algorithms (Yang et al., 2003) or nearest neighbors. The next decision is determining the new attribute values of the aggregates. Ideally, aggregates' values should be representative of the member entities contained. Statistical summaries such as mean, minimum, maximum, and count are commonly used. Aggregation can be iteratively applied to generate tree structures of groups and subgroups (Conklin et al., 2002). The final decision is the visual representation of the aggregates, which, ideally, reveals some hint of their contents. Aggregate Towers (Rayson, 1999) (Figure 10*a*) groups entities spatially if they overlap on a map. The aggregates are shown as towers whose height represents the number of entities in the aggregate. Zooming out of the map causes further aggregation as needed, and zooming in on the map segregates towers until no towers are needed. XmdvTool (Yang et al., 2003) (Figure 10*b*) clusters tabular data in a parallel coordinates plot. The extent of the contents of each aggregate is revealed by a glowing shadow that emanates from the aggregate's polyline.

Aggregation can also be used to group data attributes together. Dimensionality reduction methods reduce the number of data attributes in large multidimensional tabular data sets so that they can be visualized more easily (Rencher, 2002). The reduced set of attributes should approximately capture the main trends found in the full set of attributes. For example, *principal components analysis* projects the data entities onto a subspace of the original data space that best preserves the variance in the data. *Multidimensional scaling* uses measures of similarity between entities,

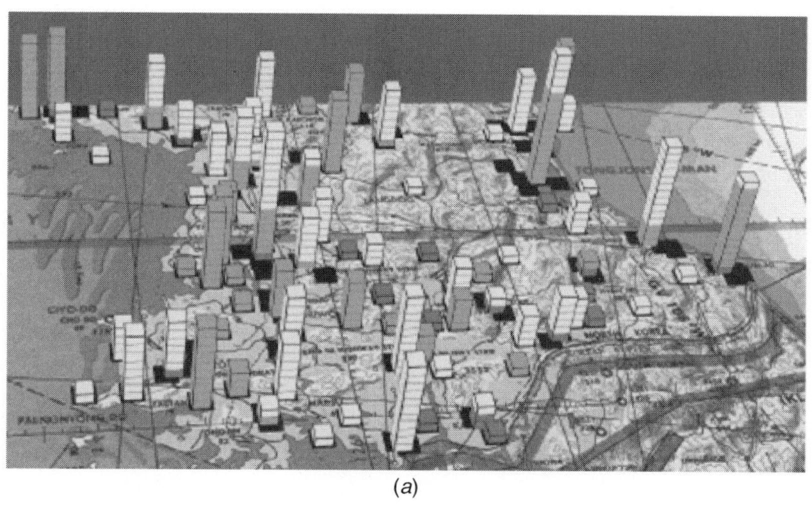

(*a*)

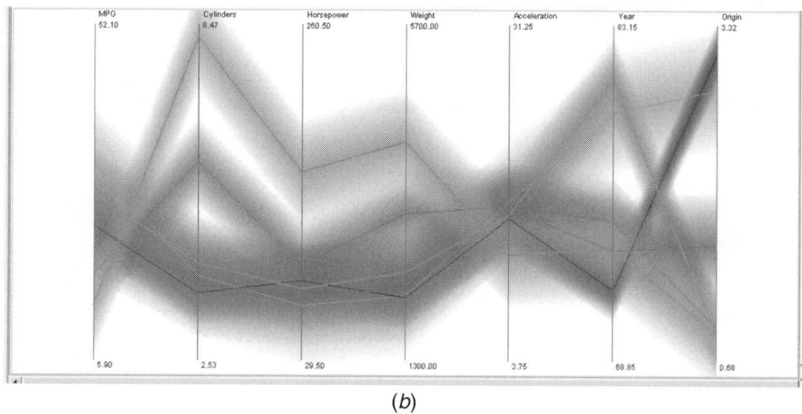

(*b*)

Figure 10 (*a*) Aggregate Towers stacks military units that spatially overlap on a map. The black footprints show the spatial coverage of each tower. Zooming in separates the units. (*b*) XmdvTool's Parallel Coordinates clusters the data from Figure 4*b* into six entities to reduce clutter. Translucent shading reveals the approximate spread of each cluster. [(*a*) From Rayson (1999), © 1999 IEEE; (*b*) from Yang et al., (2003), courtesy of Matthew Ward.]

based on their many attribute values, to compute a one-, two-, or three-dimensional map that groups similar entities spatially.

Filtering can also be used to reduce data quantity. VIDA (Woodruff et al., 1998) selects a representative subset of data entities based on data density and entity importance. Spotfire (Ahlberg and Wistrand, 1995) relegates less important data attributes to interactive methods such as dynamic queries, eliminating them from the input to the initial visual mapping function. Tree-structured information is easily reduced simply by filtering deeper levels of a tree to visualize the upper levels as an overview.

4.2 Miniaturizing Visual Glyphs

Alternatively, emphasis can be placed on miniaturization of the visual glyphs generated by the visual mapping process. Tufte argues for increased data density in visual displays by maximizing the data per unit area of screen space and maximizing the data/ink ratio (Tufte, 2001). A higher data/ink ratio is accomplished by minimizing the quantity of "ink" required for each visual glyph, and eliminating chart junk that wastes ink on unimportant nondata elements.

SeeSoft (Eick et al., 1992) (Figure 11*a*) provides an overview of textual software code using miniaturization. Similar to TableLens (Figure 4*a*), each line of code is reduced to a single line segment of colored pixels whose length is proportional to the number of characters in the line of code. In this way, large software projects of up to 50,000 lines of code can be viewed in a single screen. Color coding can be used to reveal other attributes of the lines of code, such as which programmer wrote it, whether it has been tested, or the amount of CPU time required to execute the line (code profiling). Pixel Bar Charts (Keim et al., 2002) reduces the size of visual glyphs for tabular data to a single pixel, colored by one attribute and ordered on the display by another attribute. The Information Mural (Jerding and Stasko, 1998) (Figure 11*b*) takes miniaturization to the subpixel level. When many glyphs overlap and occlude each other, Mural visualizes the density of the glyphs like an x-ray image.

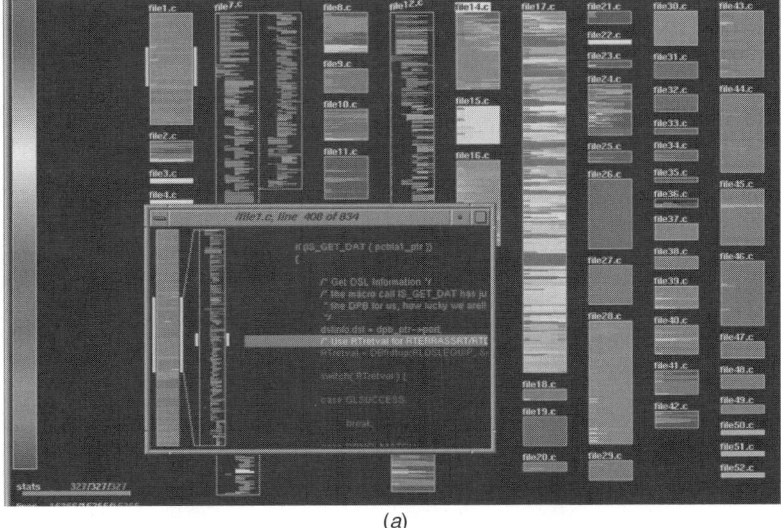

(*a*)

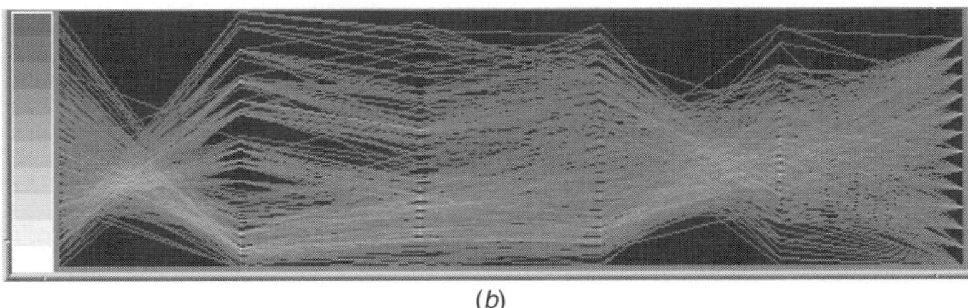

(*b*)

Figure 11 (*a*) SeeSoft provides a miniaturized visual overview of software code, shading each line of code by an attribute such as date authored; (*b*) Information Mural shows the density of a parallel coordinates plot, enabling users to see hidden patterns that would otherwise be occluded within dense clutter as in Figure 4*b*. [(*a*) From Eick et al. (1992), courtesy of Stephen Eick; (*b*) from Jerding and Stasko (1998), courtesy of John Stasko.]

5 NAVIGATION STRATEGIES

After employing an overview strategy to provide a broad view of a large information space, the next design concern is that of navigation. Interactive methods are needed to support navigation between the broad overview and the details of the information. To support this need, three primary navigation design strategies have evolved: *zoom + pan*, *overview + detail*, and *focus + context*. These strategies reside at the third stage of the visualization pipeline, view transformation (see Figure 2).

These navigation strategies should be contrasted with the naive strategy called *detail-only*. Detail-only is the baseline strategy that does not employ an overview. It provides only the detail-level view of a portion of the information space (e.g., the spreadsheet in Figure 1*a*). Users can navigate by scrolling or panning to access the rest of the information space. In general, the detail-only strategy should be avoided. The principal disadvantage is disorientation due to lack of overview, leaving the user lost in the information space and wondering: Where am I? Where do I want to go? How do I get there?

5.1 Zoom + Pan

Zoomable visualizations begin with the overview and then enable users to zoom dynamically into the information space to reach details of interest. Users can zoom back out to return to the overview, and zoom in again to another portion of detail. Users can also pan across the space without zooming out. Zooming can be a smooth continuous navigation through the space as in Pad + + (Bederson et al., 1996) (Figure 12) or can be used to drill down through discrete levels of scale as in Treemaps (Johnson and Shneiderman, 1991). Although the zooming strategy provides an overview, disorientation remains when zooming in. It is easy for users to become lost in the space when zooming in and panning, since the overview is no longer present.

5.2 Overview + Detail

Overview + detail uses multiple views to display an overview and a detail view simultaneously. A field-of-view indicator in the overview indicates the location of the detail view within the information space. The views are linked such that manipulating the field of view in the overview causes the detail view to navigate accordingly. Similarly, when users navigate directly in the detail view, the field of view updates to provide location feedback. This strategy is commonly found in various map and image browsing software (Plaisant et al., 1995). In SeeSoft (Eick et al., 1992), the miniaturized overview of text operates as a scrollbar for a detailed view of the actual text (Figure 11*a*, center). A zoom factor of 30 : 1 between overview and detailed view is the usability limit for navigating two-dimensional images, but intermediate views can be chained to reach higher total zoom factors (Plaisant et al., 1995). In navigating three-dimensional worlds, Worlds in Miniature (Stoakley et al., 1995) provides a small three-dimensional overview map attached to a virtual glove (Figure 13, bottom center) to help orient users within the world. Overview + detail preserves overview to avoid disorientation in the detailed view but suffers from a visual discontinuity between the overview and detailed views. Ideally, the detail view should not overlap or occlude the overview, but

Figure 12 Zooming sequence in Pad++ from a Web page to an embedded folder to an embedded text file. [From Bederson et al. (1996), courtesy of Ben Bederson.]

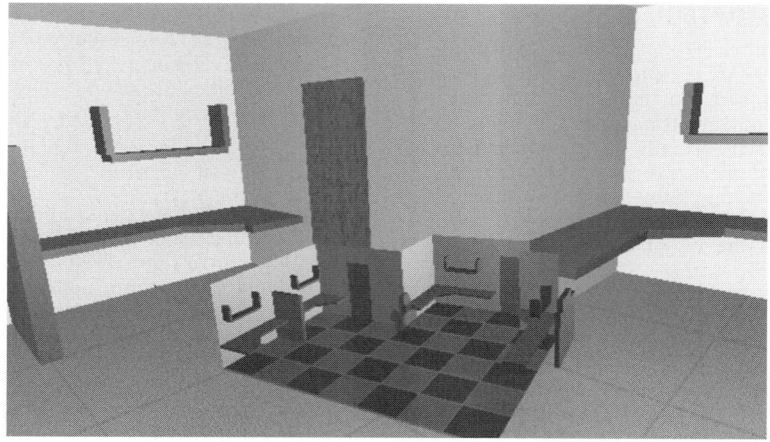

Figure 13 Worlds in Miniature uses overview + detail for navigating three-dimensional virtual environments [From Stoakley et al. (1995), © 1995 ACM, Inc., used by permission.]

pop-ups such as tooltips and magnifying glasses are reasonable for small amounts of detailed information.

5.3 Focus + Context

Focus + context expands a focus region directly within the overview context. The focus is enlarged and magnified to provide detailed information for that portion of the information space. Users can navigate simply by sliding the focus across the overview to reveal details for other portions of the space. To make room for the expanded focus region, the surrounding overview must be pushed back partially by distorting or warping the overview. For this reason, this strategy is sometimes referred to as fisheye (Furnas, 1986) or distortion-oriented (Leung and Apperley, 1994) displays. Without distortion, the magnified region would occlude the adjacent context like a magnifying glass. Since the near context is the most important part of the context, the magnifying glass effect is undesirable, and distortion is required to preserve the overview. In general, the focal point is magnified the most, and the degree of magnification decreases with distance from the focal point. Careful design based on a variety of metaphors can help to minimize the negative effects of the distortion.

Several variants of the focus + context strategy have been developed for navigating one- and two-dimensional spaces, including:

- *Bifocal* (Spence, 2001): uses two distinct levels of magnification, such as TableLens (Rao and Card, 1994) (Figure 4*a*)
- *Perspective*: wraps information on three-dimensional angled surfaces, such as Perspective Wall (Robertson et al., 1993) (Figure 14*a*)
- *Wide-angle*: creates a classic visual fisheye effect, such as Hyperbolic Tree (Lamping et al., 1995)
- *Nonlinear*: uses more complex magnification functions to create a magnified bubble effect (Keahey and Robertson, 1996) (Figure 14*b*)

As an alternative to spatial distortion, focus + context screens (Baudisch et al., 2002) offer *resolution distortion*, which may provide a better match to the human visual system. Fisheyes have also been developed for navigating three-dimensional spaces (Carpendale et al., 1997). The focus + context strategy offers continuity of detail within overview context but suffers from disorientation caused by dynamic distortion.

Although studies have repeatedly shown advantages of these three strategies over the detail-only strategy, comparisons among the three are inconclusive and depend strongly on the specifics of the individual designs, data domains, and user tasks (e.g., Hornbæk et al., 2002). An analytic summary follows:

Zoom + pan:
- \+ Screen space efficient
- \+ Infinite scalability
- − Lose overview when zooming in
- − Slower navigation

Overview + detail:
- \+ Stable overview
- \+ Scalable; chained views; multiple overviews or foci
- − Visual disconnect between views; back and forth glancing
- − Views compete for screen space; smaller overview

Focus + context:
- \+ Detail visually connected to surrounding context
- − Limited scalability; typically under 10 : 1 zoom factor
- − Distortion; unstable overview

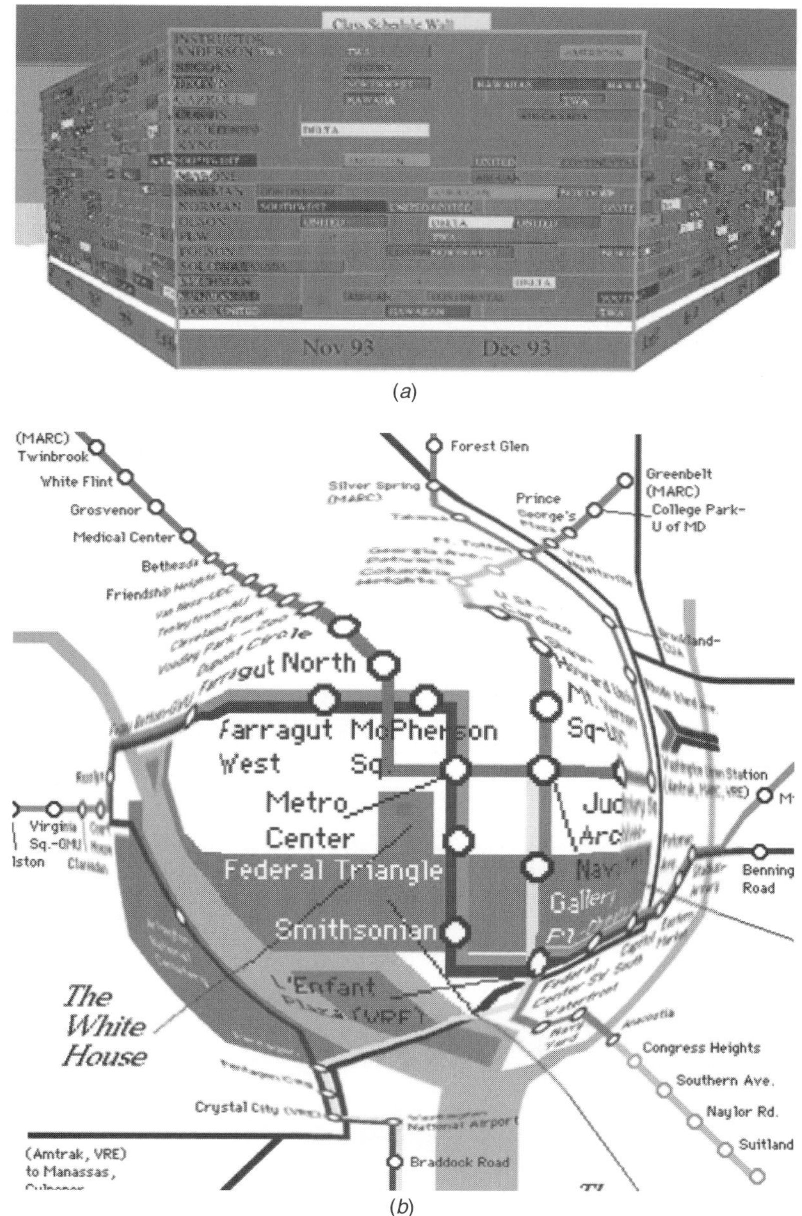

(*a*)

(*b*)

Figure 14 (*a*) Perspective Wall wraps a one-dimensional time line around a bent wall. The front portion provides detailed information, and perspective sidewalls provide overview context. (*b*) Nonlinear magnification can create this bubble effect, which magnifies the focus region, squeezes the near context, and maintains an otherwise stable far context. [(*a*) From Robertson et al. (1993), courtesy of PARC; (*b*) from Keahey and Robertson (1996), © 1996 IEEE.]

6 INTERACTION STRATEGIES

Interaction strategies support further scalability and complexity of visualized information. Although it is preferable to map all data onto the display visually in a form that effectively reveals all desired insights without interaction, this is generally impossible for data of even modest complexity. Interaction strategies overcome this limitation by enabling users to explore additional mappings and insights interactively over time. Many interactive techniques exist. A few major categories of interaction strategies should be considered in every visualization design.

6.1 Selecting

The most fundamental need in visualization is interactive selection of individual data entities or subsets of data entities. Users select entities to identify data that is of interest to them. This is useful for many reasons, including viewing detailed information about the entities (details on demand), highlighting entities that are obscured or occluded in a crowded display, grouping a set of related entities, or extracting entities for future use.

In general, there are two possible criteria by which users can specify selections. First, users can select data entities directly. Direct manipulation visualizations enable users to select entities in a visualization directly using a variety of techniques (Wills, 1996), such as pointing at individual entities' glyphs (as in Figure 4*a*), lassoing a group of glyphs, and so on. Second, users can select data entities indirectly through selection criteria on information structures (Section 3). For example, XmdvTool (Ward, 1994) enables users to make selections in tabular data in parallel coordinates by specifying range criteria on data attributes. In Figure 4*b*, all American-made cars are highlighted by selecting the USA range on the Origin axis. Other structure-based selection techniques include selecting an entire branch in a tree structure, selecting a path in a network structure, or selecting a ThemeView mountain (Wise et al., 1995) in a document collection structure. Selection techniques should be designed to enable users to easily select entities, select additional entities into the current selection, remove entities from the current selection, and clear the selection. Selecting is sometimes called *brushing*, because it is like painting glyphs with a special type of paintbrush that behaves according to the selection technique.

6.2 Linking

Linking is useful to relate information interactively among multiple views (Baldonado et al., 2000; North, 2001). Information can be mapped differently into separate views to reveal different perspectives or different portions of the information. The most common form of linking is called *brushing and linking* (Becker and Cleveland, 1987). Interactive selections of entities in one view are propagated to other views to automatically highlight corresponding entities, enabling users to recognize relationships. This strategy enables users to take advantage simultaneously of the different strengths of different visual representations. This is particularly useful for relating between different information structures (Figure 15), essentially using

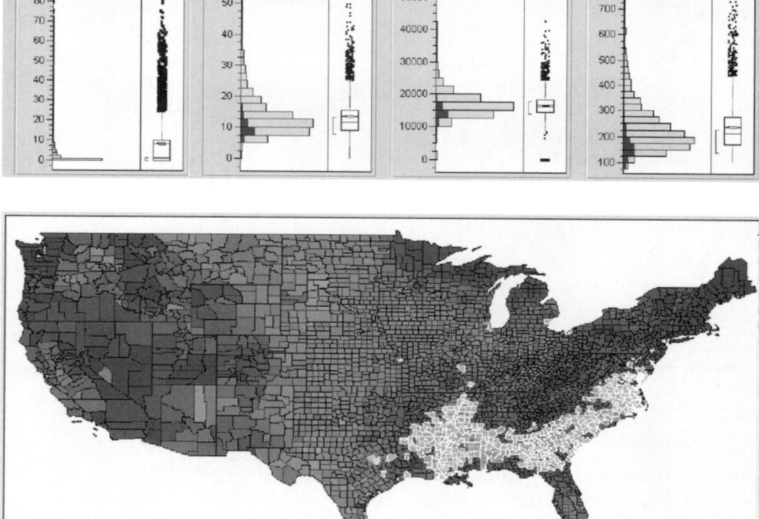

Figure 15 Interactive brushing and linking between histogram plots (top) and a geographic map (bottom) of a census counties data set. Histograms (tabular information structure) show data distributions for four county attributes: percent population black, percent population with college degree, income per capita, and median rent. The map (two-dimensional spatial information structure) is colored by "percent farmland." Selecting the counties that are more than 25% African American in the first histogram also highlights those counties in the other views. The highlighting in the map reveals that those counties are clustered in the southern and southeastern regions. Selecting counties in the map would similarly highlight them in the histograms. [The histograms are generated by JMP (SAS, 2004), and map by ArcView (ESRI, 2004). They are linked by Snap-Together Visualization (North et al., 2002).]

one structure to query another. Users can select entities according to criteria in one structure, which then shows the distribution of those entities within the other structure. Although linking is commonly used to relate two views of the same data set in a one-to-one fashion, it can also be used to relate entities across many-to-many database relationships for more complex scenarios. Linking also helps users coordinate multiple views during navigation or other interactive operations, such as synchronized scrolling. Tools such as Snap-Together Visualization (North et al., 2002) enable users to mix and match a wide variety of views to produce customized combinations of linked views.

6.3 Filtering

Interactive filtering enables users to dynamically reduce information quantity in the display and focus in on information of interest. *Dynamic queries* (Ahlberg and Wistrand, 1995) apply direct manipulation principles to querying attribute values. Visual widgets such as the range slider (Figure 1*b*, right) enable users to adjust query parameters rapidly and view filtered results in the visualization in real time. The widgets also provide a visual representation of the current query parameters. Because of the rapid feedback, dynamic query filters can be used not just to reduce information quantity but also to explore relationships between mapped attributes and query attributes. For example, in Figure 1, filtering with the query slider for "unemployment" to eliminate the low-unemployment counties from the display reveals that counties with high unemployment are all in the low-income and low-education area of the plot. The rapid query feedback also eliminates the difficulty of zero-hit or megahit query results, because users can quickly adjust the query parameters until a desirable number of hits is acquired. For example, by further filtering on "unemployment" in Figure 1, users find that there are 11 counties with an unemployment rate over 20%, most of which are located near the border with Mexico. Dynamic queries are the inverse of brushing; brushing highlights selected data, while dynamic queries elide nonselected (filtered) data.

Magic Lenses (Fishkin and Stone, 1995) offer a spatially localized form of filter. For more advanced queries involving complex combinations of Boolean operations, metaphors such as Filter Flow (Young and Shneiderman, 1993) enable users to construct virtual pipelines of filters.

6.4 Rearranging and Remapping

Since a single mapping of information to visual form may not be adequate, it is straightforward to enable users to customize the mapping or choose among several mappings. Since the spatial layout is the most salient visual mapping, rearranging the spatial layout of the information is the most potent for generating different insights. For example, TableLens (Rao and Card, 1994) (Figure 4*a*) can spatially rearrange its view by choosing a different attribute to sort by, and Parallel Coordinates (Inselberg, 1997) (Figure 4*b*) can rearrange the left-to-right order of its axes. This enables users to explore relationships among attributes.

In general, any part of the mapping process throughout the visualization pipeline can be under user control. For example, Spotfire (Ahlberg and Wistrand, 1995) users can customize the scatterplot view (Figure 1*b*) by choosing data attributes to map to various visual properties, such as x, y, color, and size. It also provides a variety of visual representations to choose from, including heat maps (colored spreadsheets), parallel coordinates, histograms, and pie and bar charts. Visage (Roth et al., 1996) emphasizes a technique called *data-centric interaction*, in which users can select data entities directly and drag them to different views to display them in new ways. At the extreme are systems such as Sage and SageBrush (Roth et al., 1994) that let users design new visual mappings for a data set using a set of basic primitives as described in Section 2. Sage can also automatically generate certain visual mappings for a given data set and task using a rule-based expert system.

7 THE FUTURE

Information visualization is a relatively young field (e.g., the *IEEE Symposium on Information Visualization* started in 1995). Significant further research is needed on new visual mappings, overview strategies, interaction and navigation strategies, evaluation methods, and guidelines. A few critical areas of need that should be explored in the foreseeable future include:

1. *Visualization of massive heterogenous data.* Recent developments in intelligence analysis and homeland security require new abilities to analyze terabytes of textual, voice, and video data in unstructured collections. Bioinformatics is driving the need for new methods to visualize megadimensional tabular data sets, containing thousands or millions of data attributes and huge networks.
2. *Integrating visualization with the analysis context.* Visualization is not an independent task but must be integrated with data management, information retrieval, statistical analysis, data mining (Shneiderman, 2002), decision support, task management, and content authoring and publishing.
3. *High-resolution visualization.* New large high-resolution display technologies can fundamentally impact interactive visualization (Funkhouser and Li, 2000). New visualization strategies must be devised to expand the limits of visualization by effectively taking advantage of significantly increased quantity of pixels and larger physical screen sizes.
4. *Visualization evaluation.* To support the iterative design of increasingly advanced visualizations, new evaluation methods are needed to identify and measure the long-term effect of visualizations on high-level insight generation and information analysis.

As researchers explore future visualization innovations and practitioners apply visualization design principles to new domains, the proliferation of effective information visualizations will lead to widespread improvements in the usability of information and to increased generation of valuable insight.

REFERENCES

Ahlberg, C., and Wistrand, E. (1995), "IVEE: An Information Visualization and Exploration Environment," in *Proceedings of the IEEE Symposium on Information Visualization*, Atlanta, GA, pp. 66–73; http://www.spotfire.com/.

Andrews, K., Kappe, F., and Maurer, H. (1995), "Hyper-G and Harmony: Towards the Next Generation of Networked Information Technology," in Companion Proceedings of the ACM *Conference on Human Factors in Computing Systems*, Denver, CO, pp. 33–34.

Baeza-Yates, R., and Ribiero-Neto, B. (1999), *Modern Information Retrieval*, Addison-Wesley, Reading, MA.

Baldonado, M., Woodruff, A., and Kuchinsky, A. (2000), "Guidelines for Using Multiple Views in Information Visualization," in *Proceedings of the ACM Conference on Advanced Visual Interfaces*, Palermo, Italy, pp. 110–119.

Baudisch, P., Good, N., Bellotti, V., and Schraedley, P. (2002), "Keeping Things in Context: A Comparative Evaluation of Focus Plus Context Screens, Overviews, and Zooming," in *Proceedings of the ACM Conference on Human Factors in Computing Systems*, Minneapolis, MN, pp. 259–266.

Becker, R., and Cleveland, W. (1987), "Brushing Scatterplots," *Technometrics*, Vol. 29, No. 2, pp. 127–142.

Becker, R. A., Eick, S. G., and Wilks, A. R. (1995), "Visualizing Network Data," *IEEE Transactions on Visualization and Computer Graphics*, Vol. 1, No. 1, pp. 16–28.

Bederson, B., Hollan, J., Perlin, K., Meyer, J., Bacon, D., and Furnas, G. (1996), "Pad++: A Zoomable Graphical Sketchpad for Exploring Alternate Interface Physics," *Journal of Visual Languages and Computing*, Vol. 7, No. 1, pp. 3–32.

Bederson, B., Shneiderman, B., and Wattenberg, M. (2002), "Ordered and Quantum Treemaps: Making Effective Use of 2D Space to Display Hierarchies," *ACM Transactions on Graphics*, Vol. 21, No. 4, October, pp. 833–854.

Bertin, J. B. (1983), *Semiology of Graphics: Diagrams, Networks, Maps*, translated by W. J. Berg, University of Wisconsin Press, Madison, WI.

Beshers, C., and Feiner, S. (1993), "AutoVisual: Rule-Based Design of Interactive Multivariate Visualizations," *IEEE Computer Graphics and Applications*, Vol. 13, No. 4, pp. 41–49.

Card, S., Robertson, G., and York, W. (1996), "The WebBook and the Web Forager: An Information Workspace for the World-Wide Web," in *Proceedings of the ACM Conference on Human Factors in Computing Systems*, Vancouver, British Columbia, Canada, p. 111.

Card, S., Mackinlay, J., and Shneiderman, B. (1999), *Readings in Information Visualization: Using Vision to Think*, Morgan Kaufmann, San Francisco.

Carlis, J., and Konstan, J. (1998), "Interactive Visualization of Serial Periodic Data," in *Proceedings of ACM Symposium on User Interface Software and Technology* San Francisco, pp. 29–38.

Carpendale, M. S. T., Cowperthwaite, D. J., and Fracchia, F. D. (1997), "Extending Distortion Viewing from 2D to 3D," *IEEE Computer Graphics and Applications*, Vol. 17, No. 4, pp. 42–51.

Carswell, C. (1992), "Reading Graphs: Interactions of Processing Requirements and Stimulus Structure," in *Precepts, Concepts, and Categories: The Representation and Processing of Information*, B. Burns, Ed., Elsevier Science, Amsterdam, pp. 605–645.

Chambers, J., Cleveland, W., Kleiner, B., and Tukey, P. (1983), *Graphical Methods for Data Analysis*, Wadsworth, Belmont, CA.

Chen, C. (1999), *Information Visualisation and Virtual Environments*, Springer-Verlag, London.

Chen, C., and Yu, Y. (2000), "Empirical Studies of Information Visualization: A Meta-analysis," *International Journal of Human–Computer Studies*, Vol. 53, No. 5, pp. 851–866.

Chernoff, H. (1973), "The Use of Faces to Represent Points in k-Dimensional Space Graphically," *Journal of the American Statistical Association*, Vol. 68, pp. 361–368.

Cheswick, B. (1998), "The Scenic Route," *Wired*, Vol. 6, No. 12; Also, Internet Mapping Project, http://research.lumeta.com/ches/map/.

Chewar, C. M., McCrickard, D. S., Ndiwalana, A., North, C., Pryor, J., and Tessendorf, D. (2002), "Secondary Task Display Attributes: Optimizing Visualizations for Cognitive Task Suitability and Interference Avoidance," in *Proceedings of the Symposium on Data Visualisation*, Barcelona, Spain, pp. 165–171.

Cleveland, W. (1993), *Visualizing Data*, Hobart Press, Summit, NJ.

Cleveland, W. S., and McGill, R. (1984), "Graphical Perception: Theory, Experimentation, and Application to the Development of Graphical Methods," *Journal of the American Statistical Association*, Vol. 79, No. 387, pp. 531–554.

Conklin, N., Prabhakar, S., and North, C. (2002), "Multiple Foci Drill-Down Through Tuple and Attribute Aggregation Polyarchies in Tabular Data," in *Proceedings of the IEEE Symposium on Information Visualization*, Boston, pp. 131–134.

Eick, S. G., Steffen, J. L., and Sumner, E. E., Jr. (1992), "SeeSoft: A Tool for Visualizing Line Oriented Software Statistics," *IEEE Transactions on Software Engineering*, Vol. 18, No. 11, pp. 957–968.

Elliott, E., and Davenport, G. (1994), "Video Streamer," in *ACM Conference Companion on Human Factors in Computing Systems*, Boston, pp. 65–68.

ESRI (2004), *ArcView Desktop GIS*, Environmental Systems Research Institute, Redlands, CA, http://www.esri.com/.

Fayyad, U. M., Grinstein, G., Wierse, A., and Fayyad, U. (2001), *Information Visualization in Data Mining and Knowledge Discovery*, Morgan Kaufmann, San Francisco.

Feiner, S. (1988), "Seeing the Forest for the Trees: Hierarchical Displays of Hypertext Structures," in *Proceedings of the ACM Conference on Office Information Systems*, Palo Alto, CA, pp. 205–212.

Fishkin, K., and Stone, M. (1995), "Enhanced Dynamic Queries via Movable Filters," in *Proceedings of the ACM Conference on Human Factors in Computing Systems*, Denver, CO, pp. 415–420.

Funkhouser, T., and Li, K. (2000), "Onto the Wall: Large Displays," Special Issue, *IEEE Computer Graphics and Applications*, Vol. 20, No. 4.

Furnas, G. (1986), "Generalized Fisheye Views," in *Proceedings of the ACM Conference on Human Factors in Computing Systems*, Boston, pp. 16–23.

Furnas, G., and Zacks, J. (1994), "Multitrees: Enriching and Reusing Hierarchical Structure," in *Proceedings of the ACM Conference on Human Factors in Computing Systems*, Boston, pp. 330–336.

Gillan, D., Wickens, C., Hollands, J., and Carswell, C. (1998), "Guidelines for Presenting Quantitative Data in HFES Publications," *Human Factors*, Vol. 36, pp. 419–440.

Grosjean, J., Plaisant, C., and Bederson, B. (2002), "Space-Tree: Supporting Exploration in Large Node Link Tree, Design Evolution and Empirical Evaluation," in *Proceedings of the IEEE Symposium on Information Visualization*, Boston, pp. 57–64.

Healey, C. G., Booth, K. S., and Enns, J. T. (1996), "High-Speed Visual Estimation Using Preattentive Processing," *ACM Transactions on Human–Computer Interaction*, Vol. 3, No. 2, pp. 107–135.

Hearst, M. (1995), "TileBars: Visualization of Term Distribution Information in Full Text Information Access," in *Proceedings of the ACM Conference on Human Factors in Computing Systems*, Denver, CO, pp. 59–66.

Herman, I., Melancon, G., and Marshall, M. S. (2000), "Graph Visualization and Navigation in Information Visualization: A Survey," *IEEE Transactions on Visualization and Computer Graphics*, Vol. 6, No. 1, pp. 24–43.

Hetzler, B., Harris, W. M., Havre, S., and Whitney, P. (1998), "Visualizing the Full Spectrum of Document Relationships," in *Structures and Relations in Knowledge Organization: Proceedings of the 5th International ISKO Conference*, Ergon Verlag, Würzburg, Germany, pp. 168–175.

Hornbæk, K., Bederson, B., and Plaisant, C. (2002), "Navigation Patterns and Usability of Zoomable User Interfaces with and Without an Overview," *ACM Transactions on Computer–Human Interaction*, Vol. 9, No. 4, pp. 362–389.

Inselberg, A. (1997), "Multidimensional Detective," in *Proceedings of the IEEE Symposium on Information Visualization*, Phoenix, AZ, pp. 100–107.

Jerding, D. F., and Stasko, J. T. (1998), "The Information Mural: A Technique for Displaying and Navigating Large Information Spaces," *IEEE Transactions on Visualization and Computer Graphics*, Vol. 4, No. 3, pp. 257–271.

Johnson, B., and Shneiderman, B. (1991), "Treemaps: A Space-Filling Approach to the Visualization of Hierarchical Information Structures," in *Proceedings of the 2nd International IEEE Visualization Conference*, San Diego, CA, pp. 284–291.

Kandogan, E. (2000), "Star Coordinates: A Multi-dimensional Visualization Technique with Uniform Treatment of Dimensions," in *Extended Proceedings of the IEEE Symposium on Information Visualization*, Salt Lake City, UT, pp. 9–12.

Keahey, T. A., and Robertson, E. (1996), "Techniques for Non-linear Magnification Transformations," in *Proceedings of the IEEE Symposium on Information Visualization*, San Francisco, pp. 38–45.

Keim, D., Hao, M., Dayal, U., and Hsu, M. (2002), "Pixel Bar Charts: A Visualization Technique for Very Large Multi-attribute Data Sets," *Information Visualization*, Vol. 1, No. 1, pp. 20–34.

Kniss, J., Kindlmann, G., and Hansen, C. (2001), "Interactive Volume Rendering Using Multidimensional Transfer Functions and Direct Manipulation Widgets," in *Proceedings of the IEEE Conference on Visualization*, San Diego, CA, pp. 255–262.

Korfhage, R. (1995), "VIBE: Visual Information Browsing Environment," in *Proceedings of the ACM Conference on Research and Development in Information Retrieval*, Seattle, WA, p. 363.

Lamping, J., Rao, R., and Pirolli, P. (1995), "A Focus + Context Technique Based on Hyperbolic Geometry for Visualizing Large Hierarchies," in *Proceedings of the ACM Conference on Human Factors in Computing Systems*, Denver, CO, pp. 401–408.

Leung, Y. K., and Apperley, M. D. (1994), "A Review and Taxonomy of Distortion-Oriented Presentation Techniques," *ACM Transactions on Computer–Human Interaction*, Vol. 1, No. 2, pp. 126–160.

Lin, X. (1992), "Visualization for the Document Space," in *Proceedings of the IEEE Conference on Visualization*, Boston, pp. 274–281.

Mackinlay, J. (1986), "Automating the Design of Graphical Presentations of Relational Information," *ACM Transactions on Graphics*, Vol. 5, No. 2, pp. 110–141.

Malinowski, S. (2004), "Music animation machine," http://www.musanim.com/.

Mihalisin, T., Timlin, J., and Schwegler, J. (1991), "Visualizing Multivariate Functions, Data, and Distributions," *IEEE Computer Graphics and Applications*, Vol. 11, No. 3, pp. 28–35.

Miller, J. (2004), "Daisy analytics," http://www.daisy.co.uk/.

Munzner, T. (1998), "Exploring Large Graphs in 3D Hyperbolic Space," *IEEE Computer Graphics and Applications*, Vol. 18, No. 4, pp. 18–23.

North, C. (2001), "Multiple Views and Tight Coupling in Visualization: A Language, Taxonomy, and System, in *Proceedings of the CSREA CISST Workshop on Fundamental Issues in Visualization*, Las Vegas, NV, pp. 626–632.

North, C., Shneiderman, B., and Plaisant, C. (1996), "User Controlled Overviews of an Image Library: A Case Study of the Visible Human," in *Proceedings of the ACM Digital Libraries Conference*, Bethesda, MD, pp. 74–82.

North, C., Conklin, N., and Saini, V. (2002), "Visualization Schemas for Flexible Information Visualization," in *Proceedings of the IEEE Symposium on Information Visualization*, Boston, pp. 15–22.

Nowell, L., Schulman, R., and Hix, D. (2002), "Graphical Encoding for Information Visualization: An Empirical Study," *Proceedings of the IEEE Symposium on Information Visualization*, Boston, pp. 43–50.

Pirolli, P., and Card, S. (1999), "Information Foraging," *Psychology Review*, Vol. 106, No. 4, pp. 643–675.

Plaisant, C. (2004), "The Challenge of Information Visualization Evaluation," in *Proceedings of the ACM Conference on Advanced Visual Interfaces*, Gallipoli, Italy, pp. 109–116.

Plaisant, C., Carr, D., and Shneiderman, B. (1995), "Image Browsers: Taxonomy, Guidelines, and Informal Specifications," *IEEE Software*, Vol. 12, No. 2, March, pp. 21–32.

Plaisant, C., Milash, B., Rose, A., Widoff, S., and Shneiderman, B. (1996), "LifeLines: Visualizing Personal Histories," in *Proceedings of the ACM Conference on Human Factors in Computing Systems*, Vancouver, British Columbia, Canada, pp. 221–227.

Polys, N., Bowman, D., North, C., Laubenbacher, R., and Duca, K. (2004), "PathSim Visualizer: An Information-Rich Virtual Environment Framework for Systems Biology," in *Proceedings of the ACM Web3D Symposium*, Monterey, CA, pp. 7–14.

Rao, R., and Card, S. (1994), "The TableLens: Merging Graphical and Symbolic Representations in an Interactive Focus + Context Visualization for Tabular Information," in *Proceedings of the ACM Conference on Human Factors in Computing Systems*, Boston, pp. 318–322; http://www.tablelens.com.

Rayson, R. (1999), "Aggregate Towers: Scale Sensitive Visualization and Decluttering of Geospatial Data," in *Proceedings of the IEEE Symposium on Information Visualization*, San Francisco, pp. 92–99.

Rencher, A. (2002), *Methods of Multivariate Analysis*, 2nd ed., Wiley, New York.

Robertson, G., Card, S., and Mackinlay, J. (1993), "Information Visualization Using 3D Interactive Animation," *Communications of the ACM*, Vol. 36, No. 4, pp. 57–71.

Robertson, G., Czerwinski, M. Larson, K., Robbins, D., Thiel, D., and van Dantzich, M. (1998), "Data Mountain: Using Spatial Memory for Document Management," in *Proceedings of the 11th Annual ACM Symposium on User Interface Software and Technology*, San Francisco, pp. 153–162.

Robertson, G., Cameron, K., Czerwinski, M., and Robbins, R. (2002), "Polyarchy Visualization: Visualizing Multiple Intersecting Hierarchies," in *Proceedings of the ACM Conference on Human Factors in Computing Systems*, Minneapolis, MN, pp. 423–430.

Rosenblum, L., Earnshaw, R., Encarnacao, J., Hagen, H., Kaufman, A., Klimenko, S., Nielson G., Post F., and Thalmann, D. (1994), *Scientific Visualization: Advances and Challenges*, Academic Press, San Diego, CA, in association with the IEEE Computer Society, Los Alamitos, CA.

Rosson, M. B., and Carroll, J. (2001), *Usability Engineering: Scenario-Based Development of Human–Computer Interaction*, Morgan Kaufmann, San Francisco.

Roth, S. F., Kolojejchick, J., Mattis, J., and Goldstein, J. (1994), "Interactive Graphic Design Using Automatic Presentation Knowledge," in *Proceedings of the ACM Conference on Human Factors in Computing Systems*, Boston, pp. 112–117.

Roth, S. F., Lucas, P., Senn, J. A., Gomberg, C. C., Burks, M. B., Stroffolino, P. J., Kolojejchick, J. A., and Dunmire, C. (1996), "Visage: A User Interface Environment for Exploring Information," in *Proceedings of the IEEE Symposium on Information Visualization*, San Francisco, pp. 3–12.

Saraiya, P., North, C., and Duca, K., (2004), "An Evaluation of Microarray Visualization Tools for Biological Insight," in *Proceedings of the IEEE Symposium on Information Visualization*, Austin, TX, pp. 1–8.

SAS (2004), *JMP: The Statistical Discovery Software*, SAS Institute, Cary, NC; http://www.jmp.com/.

Shneiderman, B. (1996), "The Eyes Have It: A Task by Data Type Taxonomy for Information Visualization, in *Proceedings of the IEEE Symposium on Visual Languages*, Boulder, CO, pp. 336–343.

Shneiderman, B. (2002), "Inventing Discovery Tools: Combining Information Visualization with Data Mining," *Information Visualization*, Vol. 1, No. 1, pp. 5–12.

Shneiderman, B., and Plaisant, C. (2005), *Designing the User Interface: Strategies for Effective Human–Computer Interaction*, 4th ed., Addison-Wesley, Reading, MA.

Spence, R. (2001), *Information Visualization*, Addison-Wesley, Reading, MA.

Stasko, J., Catrambone, R., Guzdial, M., and Mcdonald, K. (2000), "An Evaluation of Space-Filling Information Visualizations for Depicting Hierarchical Structures," *International Journal of Human–Computer Studies*, Vol. 53, No. 5, pp. 663–694.

Stoakley, R., Conway, M. J., and Pausch, R. (1995), "Virtual Reality on a WIM: Interactive Worlds in Miniature," in *Proceedings of the ACM Conference on Human Factors in Computing Systems*, Denver, CO, pp. 265–272.

Stolte, C., Tang, D., and Hanrahan, P. (2002), "Polaris: A System for Query, Analysis and Visualization of Multidimensional Relational Databases," *IEEE Transactions on Visualization and Computer Graphics*, Vol. 8, No. 1, pp. 52–65.

Tory, M., and Möller, T. (2004a), "Human Factors in Visualization Research," *IEEE Transactions on Visualization and Computer Graphics*, Vol. 10, No. 1, pp. 72–84.

Tory, M., and Möller, T. (2004b), "Rethinking Visualization: A High-Level Taxonomy," in *Proceedings of the IEEE Symposium on Information Visualization*, Austin, TX, pp. 151–158.

Tufte, E. (2001), *The Visual Display of Quantitative Information*, 2nd ed., Graphics Press, Cheshire, CT.

van Wijk, J., and van de Wetering, H. (1999), "Cushion Treemaps: Visualization of Hierarchical Information," in *Proceedings of the IEEE Symposium on Information Visualization*, San Francisco, pp. 73–80; http://www.win.tue.nl/sequoiaview/.

van Wijk, J., and van Liere, R. (1993), "HyperSlice: Visualization of Scalar Functions of Many Variables," in *Proceedings of the IEEE Visualization Conference*, San Jose, CA, pp. 119–125.

Ward, M. (1994), "XmdvTool: Integrating Multiple Methods for Visualizing Multivariate Data," in *Proceedings of the IEEE Visualization Conference*, Washington DC pp. 326–333; http://davis.wpi.edu/~xmdv/.

Ward, M. (2002), "A Taxonomy of Glyph Placement Strategies for Multidimensional Data Visualization," *Information Visualization*, Vol. 1, No. 3, pp. 194–210.

Ware, C. (2004), *Information Visualization: Perception for Design*, Morgan Kaufmann, San Francisco.

Ware, C., Purchase, H., Colpoys, L., and McGill, M. (2002), "Cognitive Measurements of Graph Aesthetics," *Information Visualization*, Vol. 1, No. 2, pp. 103–110.

Wehrend, S., and Lewis, C. (1990), "A Problem-Oriented Classification of Visualization Techniques," in *Proceedings of the IEEE Visualization Conference*, San Francisco, pp. 139–143.

Wickens, C., and Hollands, J. (2000), *Engineering Psychology and Human Performance*, Prentice Hall, Upper Saddle River, NJ.

Wilkinson, L. (1999), *The Grammar of Graphics*, Springer-Verlag, New York.

Wills, G. (1996), "Selection: 524,288 Ways to Say 'This Is Interesting,'" in *Proceedings of the IEEE Symposium on Information Visualization*, San Francisco, CA, pp. 54–60.

Wise, J., Thomas, J., Pennock, K., Lantrip, D., Pottier, M., Schur, A., and Crow, V. (1995), "Visualizing the Non-visual: Spatial Analysis and Interaction with Information from Text Documents," in *Proceedings of the IEEE Symposium on Information Visualization*, Atlanta, GA, pp. 51–58.

Wiss, U., and Carr, D. A. (1999), "An Empirical Study of Task Support in 3D Information Visualizations," in *Proceedings of the IEEE International Conference on Information Visualization*, London, pp. 392–399.

Woodruff, A., Landay, J., and Stonebraker, M. (1998), "Constant Information Density in Zoomable Interfaces,", in *Proceedings of the Conference on Advanced Visual Interfaces*, L'Aquila, Italy, pp. 57–65.

Yang, J., Ward, M., and Rundensteiner, E. (2003), "Interactive Hierarchical Displays: A General Framework for Visualization and Exploration of Large Multivariate Data Sets," *Computers and Graphics Journal*, Vol. 27, No. 2, pp. 265–283.

Young, D., and Shneiderman, B. (1993), "A Graphical Filter/Flow Representation of Boolean Queries: A Prototype Implementation and Evaluation," *Journal of the American Society of Information Science*, Vol. 44, No. 6, pp. 327–339.

Zhou, M., and Feiner, S. (1998), "Visual Task Characterization for Automated Visual Discourse Synthesis," in *Proceedings of the ACM Conference on Human Factors in Computing Systems*, Los Angeles, pp. 392–399.

CHAPTER 47
ONLINE COMMUNITIES

Chadia Abras
Goucher College
Baltimore, Maryland

1 INTRODUCTION

The concept of community today is a group of people that are united by culture but who do not have to live in the same neighborhood (Wellman and Gulia, 1999). The word *community* evokes different meanings for different people, and to define it has become a complicated task. Online communities add yet another dimension to an already complicated definition, one of virtuality and elusiveness. An online community includes some characteristics found in face-to-face communities; in addition, it has to contend with the technological aspects of communications, and with the absence of face-to-face contact. Etzioni and Etzioni (1999, p. 241) define community as "having two attributes: First, it is a web of affect-laden relationships that encompasses a group of individuals—relationships that criss-cross and reinforce one another, rather than simply a chain of one-on-one relationships ... referred to as bonding. Second, a community requires a measure of commitment to a set of shared values, mores, meanings, and a shared historical identity—in short, a culture."

The term *online community* evokes a warm feeling in some people; in others it can lead to disturbing thoughts, due to the existence of hate groups and groups plotting terrible crimes (Preece, 2000, 2001a). Fernback (1997) asserts that community is easily defined by individuals but becomes complex and nondefinable in academic settings. A community encompasses a "material and symbolic dimension" (Davies, 1997, p. 39), and a sense of self in relation to others, interdependence, and belonging (Davies, 1997). A community does not need to have physical boundaries; scholars define it as being a social network rather than defining it in terms of space (Wellman and Gulia, 1999). Communities do not have to be bound together by space; they can be separated geographically and still be considered a community (Etzioni and Etzioni, 1999). A community in cyberspace emerges from the interaction between technology and human beings. It is computer-mediated communication (CMC) that allows human relationships and communication to occur in cyberspace. Therefore, a virtual community in cyberspace exists when enough people come together (Rheingold, 1996) united by a common need or desire. Computer-linked communities may cover the globe, with several locations around the world, but with the members all meeting in one place, a URL address on the Internet. These communities are different from face-to-face gathering, because physical location is irrelevant and the participants could choose to be invisible. Another aspect of online community is that it costs relatively little to socialize on the Internet (Sproull and Faraj, 1997). Rheingold (1993) envisioned that virtual communities would be more nurturing and that people would feel safer in creating close ties, since the lack of physical presence in cyberspace eliminates the distrust and fear of others that may occur in traditional community building. However, he maintains that roots in virtual communities are shallow; members can join easily or leave without questions being asked. The departure is not noticeable by the other members and therefore is less painful (Q. Jones, 1997). Virtual communities are also social networks without physical boundaries (Rheingold, 1993; S. Jones, 1995). These social networks have loose, more permeable boundaries, interaction is with diverse people, and hierarchies are no longer as well defined as in face-to-face interactions. The connection is more to a group than to individuals, and the groups have overlapping boundaries (Wellman et al., 2002).

Hillery (1955) analyzed 94 definitions of community in sociology and found that only one element was common to all: They deal with people. However, 69 of the definitions agree that in addition to people a community should include social interactions and common ties (Hamman, 2001). Sociologists have been trying to define community for many years, and the definitions provided are always changing to fit the ever-changing nature of a dynamic society (Wellman, 1982). Online communities are no exception; in the short time they have existed, several definitions have evolved and new definitions are always emerging. Rheingold (1996) sees a virtual community as possessing all the desires, needs, and behaviors of face-to-face communities; people meet others if they fulfill a certain need or if they have the same purpose. They also need to share emotions, discuss ideas, feud, fall in love, and find friends. However, in this virtual setting the attributes change, location is no longer an issue, physical presence is not required, and the mode of communication is textual and lacks a visual presence. It is not a communication between human and machine, but rather, communication between humans separated by time and geography, through the aid of technology. The *where* in an online community becomes a *node* or a *URL*, where people gather virtually and connection comes from the sharing of ideas among them (S. Jones, 1998).

According to Schuler (1996), community is a *web* of social relations. "Ideally, the web of community is a unity, a cohesive force that is supportive, builds relationships, and encourages tolerance"(p. 32). Many aspects of everyday life come into play when defining community, some of which apply to online communities and some of which do not. Online communities have aspects that are particular to the virtual world and that are not present in face-to-face communities. In 1997 in a workshop held at the ACM CHI Conference, participants identified several core attributes that should be included in the definition of an online community (Whittaker et al., 1997; Preece, 2000):

- Members have a shared goal, interest, need, or activity that provides the primary reason for belonging to the community.
- Members engage in repeated and active participation that could lead to intense interactions and shared activities.
- Participants often develop strong emotional ties.
- Members have access to shared resources.
- Policies govern the behavior online.
- Reciprocity is important.
- Members share the same social conventions, language, and protocols.

Several noncore attributes were also identified:

- The roles that people play in the community and their reputation in those roles should be considered.
- Members should be aware of membership boundaries and the group identity.
- Initiation criteria should be required for joining.
- Community history and long duration of existence should be tracked.
- Notable events and rituals should be recorded.
- Shared physical environment should be created.
- Voluntary membership should be allowed.

What was absent or not evident in the brainstorming session is the notion of presence, whether virtual or real. Somehow this was not as important in early definitions of online communities but discussed in detail later, especially in the worlds of MUDs (multiuser dungeons, domains) and MOO (MUD object oriented), where choosing a gender is part of representation (Bruckman, 1996; Reid, 1996; Turkle, 1999).

Online community researchers can draw on earlier definitions from sociology to define community (Wellman and Gulia, 1999), but more research is needed before one can draw a more complete definition of online communities. The type of community is to be considered, whether it is educational, business oriented, a community of practice, or a gaming community. From these attributes, Preece (2000, p. 10) proposes a working definition for an online community:

- *People* who interact socially, satisfy their own needs, and perform special roles.
- A shared *purpose*, such as interests, need, information exchange, or service.
- *Policies* that guide people's interactions, such as tacit assumptions, rituals, protocols, rules, and laws.
- *Computer systems* that support and mediate social interaction and facilitate a sense of togetherness.

Preece identifies these four elements as high-level criteria but also acknowledges that many low- or moderate-level criteria could be included or understood in this working definition. Communities are complex, and defining them in detail will always remain elusive. As in sociology, a definition of online community will take many years and several debates before eventually shaping up into an inclusive definition (Wellman, 1982; Preece, 2000).

What is apparent from all the definitions presented above is that community includes a group of people who share a common interest and who are connected by ties (weak or strong). A group of people alone does not constitute community; therefore, online communities need an aspect of communication present on the site, and some type of connection to individual members or to the group. There are many types of online communities, such as:

- *Guest book communities.* A tool lets users post a passage about who they are.

- *Listserver communities.* Participants subscribe to a common list.
- *Chat communities.* Synchronous chat is possible.
- *Forum communities or discussion board communities.* Users post messages in a forum.
- *UseNet news groups.* Users conduct a collection of discussions on various topics.
- *Immersive graphic environments.* These are synchronous active environments with three-dimensional graphics (Kollock and Smith, 1999; Preece and Maloney-Krichmar, 2002).
- *Wiki communities.* These provide a collaborative hypertext social environment with an emphasis on easy access to and modification of information. A wiki is a Web page in progress that can be modified by any of the participants—an asset recognized in professional groups, where ideas and projects are constantly evolving and forming.
- *Web log communities.* These are social in nature and rely on software that emphasizes easy communication and exchange. They constitute a kind of travelogue, documenting, muses, anecdotes, thoughts, opinions, and activities.

There are online communities related to almost any topic: health, wealth, hobbies, religion, politics, culture, science, education, business, sports, and games. The social configurations in community can be local, such as families, or global, such as organizations, in which collectivity describes the social cohesion needed to sustain the community over time. Such mechanisms can include solidarity, commitments, and common interests (Wenger, 1998a).

2 CREATING AND DESIGNING SUCCESSFUL COMMUNITIES

In this section, several views on how to build successful online communities are discussed. Researchers do not always agree on the most important method of community building. Depending on their area of expertise, they can have a business-oriented or academic-oriented approach. Regardless of the approach taken, building social capital is a primary concern for all since without it a community is not sustainable.

2.1 Community Developers and Designers' Approach

Earlier recommendations on design principles for online communities go to 1984, with Axelrod's (1984) three principles, which included more social than aesthetic recommendations. He recommended that the developer should arrange it so that the participants could meet again, they should be able to recognize each other, and they should have information about each other's behavior in the group. Ostrom (1990) added more specific recommendations, such as that group boundaries be clearly defined, rules to govern the community be included, behavior be monitored, and sanctions be employed when necessary. Goodwin (1994) proposed nine principles that are essential for making online communities work. They include the use of software that promotes communication and rules regarding length of postings and moderators' behavior. He suggests that postings not be limited in size and that it is the moderator's responsibility to recruit talkative people to populate the discussion area. He also recommends that the moderator not get involved in disputes but that misbehavior be addressed. Some researchers advocate building a history of the community to give the feeling of continuity (Kollock, 1998a).

Professional designers and developers agree that doing your homework before designing a community is vital to its success. At the heart of the preliminary research is understanding your audience: For whom are you designing? This research can be done through questionnaires and focus groups (Kim, 2000; Powazek, 2002).

Some researchers see the purpose as the most important aspect that can lead to the success of a community (Kim, 2000; Preece, 2000; Abras et al., 2003). They also stress that communities evolve and that the designer should be ready to modify the purpose to fit the criteria of an evolving community. The designer should define and describe the type of community being built and therefore must understand the target audience. One basic principle in Kim's (2000) design is derived from Maslow's hierarchy of needs: Maslow states that people need to fulfill basic needs before fulfilling higher-order needs, which he calls self-actualization. In the case of online communities, the basic or physiological need is system access, the ability to maintain one's identity and participate in a Web community (Kim, 2000, p. 9), and a higher-order need of self-actualization would be the ability to take on a community role that develops skills and opens up new opportunities (Kim, 2000, p. 9).

Powazek (2002) sees the content as the most important aspect for the success of an online community, since it draws people to the site and keeps them coming back. Content should be focused in order to attract and keep an audience (Andrews, 2002). The next step to consider is design, which should be flexible, simple, readable, beautiful, and geared toward the audience that one is trying to reach (Powazek, 2002). Design does not have to contain flashy graphics for a community to succeed (Kollock, 1998a). Powazek (2002) asserts that most of all, one should design for the community by including what he terms community features, such as discussion boards and chat areas. To foster communication, the discussion area should be an equal part of the site and not a side button.

Policies should be well defined and enforced to maintain stability in the community (Kim, 2000; Preece, 2000; Powazek, 2002). Members' opinions, ideas, and desires should be integrated at the site as the community grows; but the site should be well moderated to exclude low-quality material (Kim, 2000; Powazek, 2002). One should choose well-designed software for communication (Kim, 2000), but to be in control of the system and to maintain a constant design throughout the site, it is advisable to program or design the software instead of purchasing or downloading it (Powazek, 2002).

Including the option to use smilies, avatars, and photographs is a good strategy, because it helps the members construct an identity and it makes them feel more personally involved (Kim, 2000; Preece, 2000). Kim (2000) encourages the construction of social identity by including a profile for each member, which is one way of building trust in the community, since the members have an identity for each member with whom they are communicating. Intimacy is another key ingredient in community building, since it fosters trust, respect, and honesty (Powazek, 2002). Another avenue that fosters trust is to program an acceptable level of scarcity and risk, since they encourage the formation of groups to manage the risk (Kollock, 1998a). Trust is a key ingredient in building social capital, which is the glue that holds communities together. It is a social resource that helps sustain a community (Preece, 2002). Social capital encourages the members of a community to collaborate, it fosters interactivity (Putnam, 1995), and it helps society run smoothly (Uslaner, 2000). One way of building social capital online is to ensure that technology serves human needs. For the technology to serve this need, it has to be made available to all people regardless of income, social class, gender, or education. One way of ensuring this process is to encourage the production of low-cost hardware and software (Preece, 2002, p. 37; Quan-Haase and Wellman, 2003).

In the end, the online community needs to be constantly nurtured for it to grow and continue to thrive. Online community building is constant, hard work and requires a great deal of effort on the developer's part (Kim, 2000; Preece, 2000; Powazek, 2002).

2.2 Participatory Community-Centered Development Approach

Preece (2000) states that like all interaction designs, developing online communities involves technology, but in the case of online communities, sociability is essential. It is also paramount to put the user at the center of development, by using a participatory community-centered development approach. In this approach, the community is consulted and involved in the design right from the start. It is a participatory design that strongly advocates including the participants in the design (Muller, 1992; Schuler and Namioka, 1993; Preece, 2000), and it is modeled after Norman's (1986) user-centered design, which focuses on the user rather than the technology (Figure 1).

Each step of the design is followed by an evaluation period and redesign of the community according to the findings. Step 1 in the design involves understanding the needs of the community. Unlike the designers' approach, where assessing the users' needs involves asking what they want, in this participatory community-development approach, the developer must ask specifically what kind of discussion space they want, what their Internet habits are, and how they actually read and send messages. To be able to assess the users' needs, the developer must understand who the members of the community are and assess its purpose.

Step 2 addresses selecting the technology for communication, such as discussion boards or chat

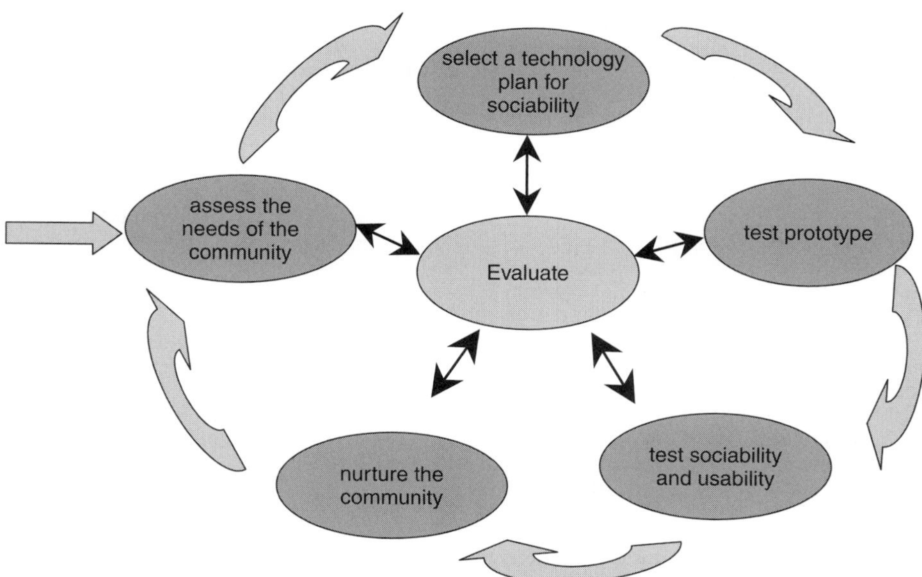

Figure 1 Participatory community-centered development model. (From Preece, 2000.)

areas, and tailoring the software to fit the needs of the intended community. At the same time, the designer is planning for sociability through planning policies and the social structure. Step 3 requires testing the prototype with all the intended software included. In this step, iterative evaluation is done through many tests involving several small tasks. Step 4 is more detailed and relies on in-depth usability testing of Web design, consistency of language, fonts, and navigation. At this stage sociability testing involves the policies and refining the purpose.

Step 5, the most important, is an ongoing process as long as the online community exists. The community has to be well advertised and the members welcomed, supported, and nurtured to ensure that they will come back. Moderators and leaders assume and refine their roles. Evaluation and change in the design should reflect the dynamic nature of the community. This process is very important throughout a community's life. The software provides the place and structure in which a group of people can evolve to create a community. Community development is iterative by nature, and this development should reflect an evolving community (Preece, 2000, pp. 209–211).

There are many types of online communities; in this chapter, three are discussed in detail: (1) online communities of interest, (2) online communities of practice, and (3) e-learning online communities.

3 ONLINE COMMUNITIES OF INTEREST

Communities are at the basis of every stable society; humans are social beings who form communities that are essential for their emotional and material well-being (Rheingold, 1993; Wellman and Gulia, 1999; Preece, 2000). Communities bring people together who are alike, or who share a common interest, goal, or problem. In this age of the Internet, and for the last 20 years, online communities are becoming a new form of social gathering. The advent of digital communication technology, especially the World Wide Web, has made this new form of community possible. These online communities are becoming an increasingly important source of support and resources. In today's fast-paced lifestyle, people often seek support in the most convenient and easiest way possible, and that is online.

Rheingold (1996, p. 413) states that: "Virtual communities emerged from a surprising intersection of humanity and technology." The first online communication emerged with the inception of e-mail in 1972, developed by ARPAnet, a government defense agency. Online communities did not appear until 1975 with the emergence of listservers, which allowed a group of people to communicate via e-mail (Preece et al., 2003). However, bulletin boards, which have existed from around the same time as listservers, allow users to communicate through a Web site without the need for e-mail. Both technologies are what constitutes communication online and are at the heart of online communities, but online communities are much more. Licklider and Taylor predicted in 1968 that in 20 years communities would be thriving online. They described them as "communities not of common location, but of common interest" (Rheingold, 1996). In 1991, the World Wide Web (WWW), developed by Tim Berners-Lee, was released by CERN (Communauté Européenne des Recherches Nucléaires); this event facilitated the widespread use of online communities. From the early listservers and bulletin boards, communities online evolved into more complex gatherings of people with the same needs. Rheingold (1996) talks about his life on the WELL (Whole Earth 'Lectronic Link), a virtual community that he describes as being a neighborhood café where people can discuss any subject that interests them. As most people, he originally sought this community for human contact, and over the years, he felt a sense of place. The evolution of these communities has prompted researchers to study social networks. Instead of interactions between individuals online, the researchers are studying group interactions and computer-mediated theories such as social influence, critical mass, and social information. Social network analysis concentrates on social actors. The researchers have to consider relations between members, ties (weak or strong), and the degree of complexity of these ties (Garton et al., 1997).

Furthermore, in studies of online communities, authors such as Markham (1998) provide several accounts of people she has interviewed online. An example is Terri, who seems to have accepted online communication as valid and embraced her new community as real. To her it is a different experience but as valid as off-line communication. Cherny's (1999) ElseMOO, a virtual world spun from LambdaMoo, is a real community. In it, she describes how the daily interactions, communication, conversation, turn taking, interactivity, and back channels take on a different form in this text-based, virtual environment. In contrast, Hakken (1999) sees more of a futuristic online world, with *cyborgs* rather than humans populating it. Cyborgs in cyberspace are entities that are different from the entities that populate real space. They have more advanced information technology (AIT) into their bodies, which makes them hybrids between humans and technology. In cyberspace, the cyborg possesses a *cyborgic* characteristic rather than a biological one; he refers to them as *creolized* and *multilocal*.

Online communities are proliferating in cyberspace; many communities of interest are more and more popular with Internet users who wish to share their interests with others online. Communities of support, especially health communities, are among the most successful, since people are able to seek support, help, and advice without having to leave the comfort of their homes. Gaming communities are among the first to become popular online, and of course, one cannot ignore hate groups, which are a minor part of online communities.

3.1 Types of Online Communities

Online communities of Interest could deal with many topics, areas of interest, and hobbies, such as academic, health, religious interests, and cooking, among others. Online health communities are the most numerous and are becoming very popular; many are support communities, and some are information-retrieval communities established by medical associations. In these communities it is important for the users to feel secure about telling their stories without fear of private information being used by others. They also need to be sure that the information provided can be trusted. They need a very well-established support system and support network from the group. Members need to be heard and need access to the community at all times (Preece and Ghozati, 1998b, 2000). The people in online communities are looking for information as well as empathy from other group members (Preece and Ghozati, 1998a).

For online health communities to function, the designer has to concentrate on some issues that are common to all. The source of information has to be specified, in order for the patient to assess its validity. In a study of several of these communities, researchers have found that some of the information given is false (Culver et al., 1997) because the advise given is from lived experiences and not medical expertise (Farnham et al., 2002). It was found that the members of these communities also need to be guaranteed privacy and to be protected from exploitation and flaming. Eighty-nine percent (89%) of people surveyed in online health groups expressed concern about the site selling their private information, and 85% expressed concern that their insurance company might raise their premium if they find out what health sites they visit (Fox and Rainie, 2000).

These members also need empathy (Preece, 1999), which could be facilitated by community software, and ultimately, they need to meet people who can sympathize with them (Preece, 2000). In a study conducted on HutchWorld on a support online community for cancer patients, the researchers found that Internet access greatly enhanced the lives of many patients. However, the synchronous part of the site did not succeed since this form of communication requires a large number of participants to sustain its existence, and therefore it failed to reach critical mass, but the asynchronous support system was successful (Farnham et al., 2002).

3.2 Determining Success in Online Communities

The numbers of online communities are estimated to be in the thousands (Preece and Ghozati, 1998a). Creating a safe, private, and successful environment will foster better trust and therefore social capital needed for the sustenance of such communities. The research suggests characteristics that make these communities successful, such as a strong purpose, empathy, and trust. Schuler (1996) states that technology could be the answer, the medical field is looking to make medical information accessible, convenient, comprehensible, timely, nonthreatening, anonymous, and controlled by users. User control in this instance refers to a sense of control over the process of recovery. Some areas of concern in the success of online communities are discussed next.

3.2.1 Interactivity and Focus

Interactivity is usually assessed by thread depth (Rafaeli and Sudweeks, 1997; Cummings et al., 2002). However, thread depth could also indicate the type of community; for example, in empathic communities the threads are broad and shallow, and in education communities many are narrow and deep. The amount of on-topic discussions indicates how well the group is focused (Preece, 2001b).

3.2.2 Reciprocity and Lurking

Reciprocity is also a concern in a community. If too many members take and few give back, the community can fail (Kollock, 1998b). In general, in a face-to-face community, whatever is given is expected to be repaid. In online communities the formula changes; according to Wellman and Gulia (1999), the exchanges are between people who have weak ties and who may never have met face to face, and therefore they are less likely to give or reciprocate support to others. However, there is evidence that support and reciprocity do exist online even between individuals who are networked by weak ties. Rheingold (1993) gives evidence that reciprocity does exist online, but it might have a different form than in face to face. In a way, the person you help might not be the one who reciprocates; it can be any other member in the group. Each act given by a person is seen by the entire group, and this engenders an image of group reciprocity (Rheingold, 1993). The group acts as one large entity ready to help anyone who is in need. Reciprocity and attachments in online communities are probably directed to individuals but are seen by the general group, which gives the feeling of group reciprocity. Another issue is the number of members needed to sustain an online community and what number of lurkers is still acceptable in a successful community. In large communities, having 90% of the members as lurkers will not affect its success, whereas in smaller communities, a high number of lurkers could lead to its demise (Nonnecke and Preece, 2000a,b).

3.2.3 Social Presence

Visualization to support social presence is an area of concern in online community research. Erickson and Kellogg (2000) compare visibility to a glass window in a door, it makes socially significant information visible; it supports awareness and holds people accountable for their actions. Donath (2002), on the other hand, takes a semantic approach to visualization. She discusses the *PeopleGarden* approach. Users are represented by flowers; different colors represent first postings, replies, and lurking. In using the semantic approach, colors can indicate provocateurs and leaders. Smith and Fiore (2001) use the Netscan project to document the use of the Vector Markup Language (Web browser–based vector) in Usenet newsgroups that include a set of visualization components that depict how each message is represented. The Netscan project allows participants to develop a better sense of other players and helps them form cooperative relationships (Smith, 2002).

3.2.4 Trust

One important issue in online communities is trust, which has been classified into at least three types: security of credits cards and online monetary transactions, trust of people's actions, and trust of what people say (Preece, 2001b). As Olson and Olson (2000) assert, trust is inherent in some people; in some situations trust is needed even if some risk is taken, and in other cases trust is inferred by other people's behavior. Therefore, all these issues have to be taken into account when assessing and measuring trust online. Some companies can inspire trust simply because they are known and people came to know that they can trust information and monetary transactions that come from them (Resnick et al., 2000). In general, in e-commerce communities, guidelines are required for developers to enable them to build trust online (Shneiderman, 2000). It is much harder to establish and therefore measure trust between individuals in an online community than in e-commerce communities, yet there is an indication that trust is more freely given online than had been expected (Feng et al., 2003).

3.2.5 Identity

The Internet lacks social richness; hence, the person with whom one is communicating can be anonymous. We are unaware of gender, race, social class, age, and education in online identities (Wellman and Gulia, 1999). Online people can present a different persona than in off-line communication, which leads to diversity in online communities. Online persons are diverse in terms of the real self, but probably more homogeneous as to the online self.

In a virtual world, physical presence takes on a different perspective and becomes a problem to those whose traditional view of community comes from contact and interaction face to face. Physical presence face to face is a point of reference that determines part of a person's identity. Online physical presence is either nonexistent or manifests itself through words, symbols, photos, or avatars. These online personas can be problematic, especially in gaming communities such as MUDs and MOOs, where the persona of a player can evolve and morph to fit the online emotional evolution of the player (Reid, 1996). Online people can take on any gender, race, or physical appearance they choose; they can remain hidden with only the text they project and the imagination of the other participants to construct a physical presence of them. Class and race take on different definitions according to what the participant wishes to project. Depending on the community, status could be determined by level of participation, the role the participant plays in the community, or the level of control that he or she has reached, such as wizard or magician in gaming communities (Reid, 1996). Gender is elusive in online communities, and in some instances it is more advantageous to project a male image to avoid harassment or to avoid being ignored (Bruckman, 1996; Reid, 1996). People need to know the gender of the person they are communicating with in order to have a point of reference, even though they know that the gender presented might not be what that person is in real life (Turkle, 1999).

As stated earlier, Kim (2000) sees building a social identity as one step toward building trust and one way is to create a members' bank (a record of members' personal information) in which information about members is stored. This information is part of the members' social identity as seen by the group. Furthermore, this personal identity should evolve into a social one: how long the individuals have been members, how often they post, and how active they are in the daily activities of the community. The recorded social activity of the members cultivates a social identity and therefore engenders trust.

3.2.6 Privacy

Privacy is another important issue that is closely linked to trust. The members need to know that the community is not going to exploit personal or credit information without their consent. Most Web-based communities post their privacy information on the Web. Needless to say, posting a privacy statement is useless unless it is backed by action. Privacy policies need to be reinforced for the members to trust that any personal information given is not going to be exploited (Kim, 2000; Preece, 2000). In education communities, privacy takes on a different role: The members need to know that their ideas are not going to be exploited and that they can ask and answer questions without being judged (Preece, 2000). Furthermore, the members need to know that their true identity can remain hidden if they choose it to be. In health communities privacy is very important because medical problems are very personal and members need reassurance that their medical information will not be used, sold, or exploited in any way (Preece, 2000).

3.2.7 Safety

Online communities have a darker side as well, as is the case in physical communities. Dibbell (1996) introduced Mr. Bungle, a virtual rapist and the harsh realities of the LambdaMOO (MUD, object-oriented) environment. Even though the rape occurred in cyberspace, the author describes the victim as being distraught and outraged, as if the incident has happened in the physical world. In this virtual world, characters move without a physical body, yet they possess a *psychic double* that could be hurt as easily. Female characters are often besieged with unwanted attention and sexual advances in online communities, especially in the synchronous world of MUDs and MOOs (Bruckman, 1996). To make members feel more at ease, which will make them come back to the community, moderators need to reprimand people who misbehave.

4 ONLINE COMMUNITIES OF PRACTICE

A community of practice (CoP) is a group of professionals who come together in pursuit of solutions to shared goals and interests. In this pursuit they employ the same practices, share the same ideas, work with the same tools, and use the same language (Wenger, 1998a; McDermott, 1999; Preece, 2004). Wenger (1998a) describes three important dimensions to the community of practice: (1) a domain of knowledge that is common to the members of the group; (2) a community of people bound together into a social entity in which they interact and build relationships and trust; and (3) a practice in which the members develop a shared repertoire, resources, tools, and build an accumulated knowledge of the community (Allee, 2000). To these three dimensions in an online CoP, a fourth dimension should be added: (4) the software, tools, and technologies that facilitate the group's interactions, stores their knowledge, and disseminate it in a logical, comprehensible manner. These communities can exist face to face, online, or as is always the case, as a combination of both environments. As networked communities, they could be locally networked (i.e., Intranet) or virtual (McDermott, 1999; Preece, 2004). Originally, the term *community of practice* was used to describe any group of persons who come together in pursuit of a solution to a common problem or a common business practice. However, today, the term is used almost exclusively to describe a group of practitioners in a professional organization or government agency (Preece, 2004). CoPs develop for many reasons; some come together because of best practices, some create guidelines, some are created for knowledge repositories, and some meet to brainstorm solutions for common problems in the company (McDermott, 1999). For these communities to be effective, they usually do not cross hierarchical boundaries, and they function best when members are at the same level of decision making (Snyder and de Souza Briggs, 2003; Preece, 2004).

Online CoPs participants usually have met before coming online, come together due to a shared practice in a common business interest, and are bound by informal relationships (Lave and Wenger, 1991; Wenger, 1991). The members of these communities come together because of an interest in applying and learning a common practice (Snyder, 1997). Wenger (1998a,(p. 4)) stresses that learning is a social participation where individuals are "active participants in the *practices* of social communities." Within these groups, they construct identities in relations to other members of their community. In this model, action and belonging are closely related and they determine individual behavior, identity, and perception. The components of a social theory of learning include the following (Wenger, 1998a, p. 5):

1. *Meaning:* a way of talking about our (changing) ability—individually and collectively—to experience our life and the world as meaningful
2. *Practice:* a way of talking about a shared historical and social recourses, frameworks, and perspectives that can sustain mutual engagement in action
3. *Community:* a way of talking about the shared historical configurations in which our enterprises are defined as worth pursuing and our participation is recognizable as competence
4. *Identity:* a way of talking about how learning changes who we are and creates personal histories of becoming in the context of our communities

In the next section, the relationship between knowledge and practice and identity and practice are discussed specifically in the context of online CoPs.

4.1 Knowledge and Practice

Knowledge management in organizations has shifted from sharing documents and comparing data to sharing ideas and understanding the logic used by other colleagues. Many companies have recently started using the Internet and other technologies to link professionals worldwide (McDermott, 2001). Knowledge is a key resource that could give a company an edge in a competitive business. However, disseminating it in practice is still very little understood (Wenger, 1998b). Knowledge could be tacit, which consists of opinions and beliefs; therefore, this type of knowledge stresses more bias, prejudice, and assumptions, making it harder to define. On the other hand, explicit knowledge is more factual and subscribes to the Western world of scientific paradigm (Preece, 2004). Up to recently, most knowledge disseminated has been explicit, although recently, companies have started to recognize the value of tacit knowledge in developing relationships between members, and in keeping the information flow alive. However, no matter how useful, if not handled appropriately tacit knowledge could result in the creation of *information junkyards*, since people are unwilling to sift through huge file cabinets looking for a resource that might or might not be there (McDermott, 2001). Both tacit and explicit knowledge are vital in knowledge management, since explicit knowledge is needed for disseminating facts, and tacit knowledge is needed for disseminating, opinions, thoughts, and expertise (Nonaka and Takeuchi, 1995; Snyder, 1997; McDermott, 2001; Wenger etal., 2002; Preece, 2004). Tools for retrieving explicit knowledge abound, but what is needed are tools for retrieving and managing tacit knowledge.

For knowledge to be disseminated appropriately, management tools must allow the user to organize the information in a logical, controllable fashion. These tools could facilitate the flow of information within an Intranet store, therefore making better use of the raw information already available while sifting, abstracting, and helping to share new information. Such tools could be ontology based, with user profile models for each user. One such tool is OntoShare, which has the capability of extracting key words from Web pages and other resources and then sharing them with appropriate users in a community of practice according to their interests as specified in their user profiles (Davies et al., 2003). Another tool described by Tennison and Shadbolt (1998), APECKS, a personal ontology server, keeps enriched Web resources active by creating a "living archive," preventing the services from falling into disuse (Domingue et al., 2001).

4.1.1 Trust and Identity

For knowledge to be shared, trust between the members of the community is paramount. Without trust, people will not communicate with each other, and therefore knowledge will not be shared. Fostering trust is closely tied to developing a trustworthy identity online. Identity as stated by Wenger (1998a) is an integral part of a social theory of learning and therefore is deeply rooted in practice. Since developing a practice is dependent on the formation of a community, whose members interact while negotiating each other's identities. This identity could be (1) lived (it involves participation and reification), (2) negotiated (it is ongoing and negotiated), (3) social (as part of a community, it gains a social character), (4) a learning process (it incorporates past and future into the meaning of the present), and (5) a local–global interplay between activities and globalization (Wenger, 1998a, p. 163).

Fostering identities is crucial in building trust in a community and eventually, participation. Trust building is based on personal trust, as well as on understanding each other, which develops from a deep insight of understanding each other's individual practice, thoughts, opinions, and behavior (Wenger et al., 2002).

Trust is needed to build social capital, which leads to behavioral change that results in greater knowledge sharing, which eventually influences organizational performance (Lesser and Storck, 2001). Lesser and Storck argue that CoPs are able to influence organizational performance through the development of social capital by creating connections among members who may or not be co-located and by fostering relationships that build trust among practitioners, and therefore CoPs "serve as generators of social capital" (Lesser and Storck, 2001, p. 4), trust, increase in knowledge sharing, and eventually, increase in organizational performance.

4.1.2 Organizational Performance and CoPs

Collaborative technologies are increasingly becoming a part of companies in disseminating knowledge management. Some of these technologies could be listservers, MOO collaborative environments, as well as special-purpose collaborative environments, such as the Answer Garden, an asynchronous conversation application (Ackerman, 1998). These collaborative environments are thought to increase interaction and exchange of knowledge among co-workers (Millen and Fontaine, 2003). As investment into these environments increases, there is a need for research to formalize the amount of benefit a company is receiving as a return on its investment. In a multicompany study, Millen et al. (2002) were able to create three distinct groupings of benefits: (1) personal (increase in personal skill and know-how), (2) community (increase trust among members, sharing of expertise and resources), and (3) organizational (improved sales, customer satisfaction, and decrease of employee turnover among other benefits). If they want to improve performance, organizations need to view CoPs as an important element in the life of the company. IBM, one of many companies that foster the creation of CoPs within its company's borders, boasts the creation of 60 knowledge networks or CoPs worldwide (Gongola and Rizzuto, 2001).

5 E-LEARNING ONLINE COMMUNITIES

The primary function of education online communities is to support students' learning, and the community could be in the form of a syllabus online, with a discussion board for interaction, or a class conducted totally online, without face-to-face interaction. The latter type can be supported via educational software such as WebCT, or Blackboard, or the instructor can set up a Web site and link it to a discussion board. Other, more elaborate forms of these education communities would be a Diversity University MOO community to teach programming (Preece, 2000), or more specifically, to teach programming to young children (Bruckman, 1998), and Virtual Universities, which are designed to simulate the physical world (Hiltz, 1994). Other forms can be communities of support for educators to exchange ideas, which could be a type of community of practice since it allows educators to discuss techniques and business practices (Wenger, 1998a) or to foster communication and collaboration among graduate students (Maloney-Krichmar et al., 2002). Yet another form of these communities is popular on college campuses: a made-to-order community that the campus can purchase at a price. Such communities include Campus Pipeline and MyBytes. They promise an instant campus complete with chat areas, e-mail, and other features. These products have not been successful for building community, since all they provide is information retrieval (Clark, 2001). Another example is the Cyberschool project started on the Virginia Tech campus in 1994, which has been a success with students, since there is a noticeable improvement in their performance, but is less popular with faculty, since it added a major workload to already full schedules (Luke, 2001).

Education communities are *communities of learning*; in this concept, community and education are linked, with the community helping to support education, and education in turn supporting the community through resource sharing. At the heart of this link between the two is cooperative learning, which is essential in a learning environment (Schuler, 1996). Communities of practice rely heavily on learning, and in itself, learning is a social structure. Learning in these communities is an "engine of practice, and practice is the history of that learning" (Wenger, 1998a, p. 96).

E-learning communities have specific needs that should be addressed to foster communication and interactivity. There is a need for resources to communicate with students, professors, complete assignments, and check grades. In addition, careful guidance by faculty members and feedback is paramount in the success of these communities. As in the case of face-to-face classes, enjoyment is important to foster learning (Alessi and Trollip, 1991; Preece, 2000). Some of these needs could be organized in external and internal factors as presented below.

5.1 External Factors

Many external factors are listed in the literature cited, but they all clustered around two categories: Course management and course delivery.

5.1.1 Course Management

Course management is an important aspect of student success. Students often complained about delayed responses in handing out assignments, no response to e-mail, no feedback on postings or grades, and ambiguity in instructions concerning assignments (Hara, 1998). The lack of response from the teacher can often lead to a feeling of isolation. The student is then left wondering: "Well, how am I doing in the class?"

A gap between the teacher and the student widens with telecourses (Brenner, 1997, p. 4).

To close the gap, the teacher has to make an extra effort to give students feedback on all aspects of their performance in class. If material is not given promptly, it will create frustration and lead to low performance (Malan et al., 1991).

5.1.2 Course Delivery

Brenner (1997) includes most of the factors associated with successes in distant learning under the heading of psychological factors. He states that for the delivery of a successful course, the teacher has to consider the *learner interface with technology* as well as distance education delivery methods. The use of technology is one strategy that the instructor can use to minimize the separation between teacher and student (Simonson, 1997). However, the use of certain types of technology, such as the use of two-way video, is not effective unless the class size is small and manageable. If the class is large, it lowers the enjoyment and therefore affects performance (Surgue et al., 1999). If technology does not work properly, it creates frustrations and therefore results in lower performance. If the student is not competent with computer use, or if there are too many links in assignments, the results were often less than optimal (Hara, 1998). To improve performance, teachers must accommodate to the learning styles of their students. In general, students prefer to learn through the visual sense (Gordon, 1995), which makes the Internet a perfect tool for visual learners.

5.2 Internal Factors

Students' attitude toward the course stemmed mostly from internal affects. Shuemer points out that "the learner must have the psychological will power to direct the learning process.... The distance learner must possess a high-degree of self-discipline, self-organization, and self-planning" (Brenner, 1997, p. 4).

5.2.1 Motivation

Students with lower abilities when entering the class usually were less enthusiastic about it, and their performances were often linked directly to their attitudes. They often performed less well than students with more positive attitudes (Surgue et al., 1997). Often, anxieties about a course and course satisfaction were also linked to performance; however, field-dependent and field-independent personalities had the same performance in online courses (Brenner, 1997). Field-dependent people are oriented toward social activities. They like to be around people, whereas field-independent people prefer more solitary activities (Brenner, 1997). Brenner refuted the notion that people who need to be around others cannot succeed in online settings.

Motivation is sometimes affected by external factors. It was found that students who had no access to a certain program performed better than students who were on campus. Since the first group had no way other than online of getting the information, they were highly motivated. On the other hand, the group that was on campus had no motivation to succeed, since face-to-face contact was an available alternative (Lia-Hoagberg et al., 1999).

5.2.2 Discipline

To succeed, students have to pace their study activities and make a continual effort to keep up with the schedule. They should not attempt multiple tasks at once, since it is daunting, and they should give themselves a specific time to sit at the computer to accomplish the task (Mory et al., 1998).

When asked, several students who dropped out of a course indicated that the reason for their failure was lack of time, the course was too time consuming, they had personal study problems, or they had a time management problem (Morgan and Tam, 1999).

In the end it is left up to the teacher to provide structure in the course. It is advisable to assist students in choosing the course that best fits their needs. When they are in a course, the teacher should aid in their progress and should assist them in managing their time. Helping them with study skills is just as crucial as delivering content. The teacher should have realistic expectations and should not try to include a full course load that worked face to face, since online time is slightly different (Malan et al., 1991).

The online environment lends itself to a constructivist approach to learning since it is student-centered and tends to encourage students to think independently through the construction of meaning (Berge, 1998; Preece, 2000). More important, it is more of a social constructivism, since it includes a community of learners, who are learning through social interaction, by using their knowledge of the world as a tool for learning (Vygotsky, 1962; Berge, 1998; Preece,

2000). Berge (1998, p. 22) defines social constructivism as one that "marries a constructivist's strong focus on creating understanding through requiring learners to explain, elaborate, and defend the position they hold to others. Participants interact by interpreting, evaluating and critiquing peers' comments and by sharing information."

In education, to define success in e-learning, research has concentrated on software and hardware that best delivers instruction in an online classroom (R. Jones, 1989; Kies et al., 1997). Also, it concentrates on teachers' methods of instruction or course delivery (Bednar and Charles, 1999; Kirby, 1999), course design, course management, and instructor training (Everett, 1998; Kitchen and McDougall, 1999). In more recent studies, these communities are referred to as online education communities instead of DE classes. These studies present failures and successes but never fully study the issue from a design and implementation perspective: How does one design and implement a successful education online community (Clark, 2001; Luke, 2001)? There is a need for more specific guidelines for designing and implementing an online education community.

To be able to understand the mechanisms of e-learning, one has to have a better understanding of its nature. Moore and Kearsley (1996, p. 2) define it as a "planned learning that normally occurs in a different place from teaching and as a result requires special techniques of course design, special instructional techniques, special methods of communication by electronic and other technology, as well as special organizational and administrative arrangements."

By definition, e-learning is a passive form of instruction where the student and teacher are separated by time and place. The modes of designing, transmitting, and retrieving information have changed. They could all occur synchronously or asynchronously. The teacher and the student no longer interact face to face, but rather, through a computer screen. Conversation alone is no longer the only source of instruction; the students have at their disposal all the resources of the Internet and all the teacher has to offer. The mode of input, delivery, and retrieval are presented in Figure 2.

The nature of the e-learning teacher is no longer one of lecturer but is that of a mediator and guide. She no longer needs to interact directly with students, but as in conventional education, distant learners need to have sufficient interaction with the instructor to allow exchange of ideas (Moore and Kearsley, 1996). E-learning allows students to attend the institution of their choice even though they are separated from it geographically. The Internet and e-learning make it possible as well to have access to better resources and to have a more individualized instruction. Designing an e-learning course is an involved process that requires a team of technicians, course developers, and teachers. One important aspect of design is the delivery of audio and video files to enhance synchronous interaction between student and teacher. The delivery of audio and video files via the Internet, called *teleconferencing*, involves (Moore and Kearsley, 1996; Steiner, 1999):

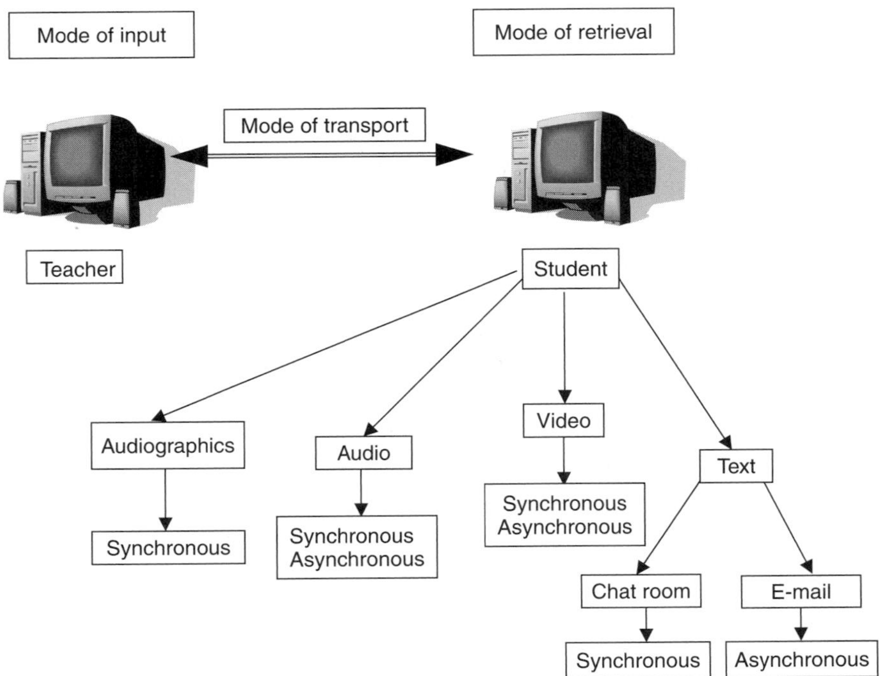

Figure 2 E-learning and information retrieval.

- *Audioconferencing.* The most common form of interaction, this is a connection that requires a bridge to connect more than two people at the same time.
- *Audiographics.* These send visual images to support instruction by voice. This form of teleconferencing requires a camera on each end in order to activate the video aspect of the communication.
- *Two-way videoconferencing.* This can be delivered via satellite, cable, slow scan, or compressed video. The slow scan requires a telephone line and the compressed video requires digital telephone lines and a codec.
- *Computer conferencing.* This can be done via electronic mail. It can be synchronous, but for the most part it is asynchronous

Videoconferencing might be the replacement of face-to-face instruction since it can deliver a full view of the participants setting and can deliver a clear voice (Kies et al., 1997), but the price can place it out of reach for most students unless the institute sets up a place for them to meet with the teacher at the institute's expense. This setting will eliminate the free scheduling and the one-on-one feature that the e-learner is seeking. Audiographics, audioconferencing, and computer conferencing are three forms of communication most used in e-learning. The software is very important in delivering the teacher's message, but the student must be considered as one of the important elements in the schema.

The student must be a self-directed learner; she has to depend on self-discipline and be self-motivated to accomplish the task at hand. If, on the other hand, the student is dependent on emotional support from the teacher, she will probably not be a good candidate in this form of isolated learning (Lewis et al., 1995). Whether or not the student is dependent on the teacher for support, the teacher should always give encouraging feedback. She should also assign group work using audioconferencing to encourage teamwork and to encourage the students to practice the target language at the same time. During a group activity, the teacher should always be present as a mediator and observer (Rohfeld and Hiemstra, 1995) but should refrain from taking over the conversation. The e-learner is usually an adult between the ages of 25 and 30. Adult learners in general need to define course content for themselves. They have a fear of failure and they are more likely to thrive in an isolated environment, where they have control over the material learned (Moore and Kearsley, 1996).

Many of the paralinguistic and social cues relied upon to make inferences about context and communication are virtually nonexistent in online communication. Howell-Richardson (1995) observed that pragmatic constraints are reduced in an online setting, and stimulation to participants is absent. Verano (1988) cited a research in which the participants had to watch a video. When the material was presented in an interactive fashion, the participants did better at recalling details, and their understanding was enhanced. Verano also remarked that both the audio and video parts of the message are needed; otherwise, the learner will not understand the gist of the message. Video is to the spoken language as the book and the newspaper are to the written language. It gives permanency to speech, which is otherwise ephemeral. It provides a paradigm for imitation and a basis for analysis. It is mainly a topic stimulus (Coleman, 1991).

It is apparent that the advantages of e-learning are numerous; the learner could study at her own pace, go over a point slowly, and skip the points that she understood, which is not possible in a regular classroom setting. One could also accommodate several learning preferences online. The disadvantages are just as numerous. The main concern is communication; even with teleconferencing, communication is less than optimal. Group work suffers because of time and distance, and most of all, the social aspect is absent online. It is an isolating experience; the student cannot relate to others and their problems. The learner is unable to relate his or her weaknesses to similar ones in the classroom (Cowan, 1995).

6 CONCLUSIONS

Online communities are constantly evolving and changing to accommodate the nature of the participants. As access to the Internet becomes more available, the number of online communities is going to accelerate, and the issues of trust, safety, and empathy are going to be more apparent. Community developers often do not know what makes a community thrive, and others fail to reach their target audience. A good place to start is by asking participants what their needs are, to design for them, and to evolve and change constantly to accommodate their needs and desires (Preece, 2000). Considering the point of view of the member is not a new notion and has been advocated by many researchers in user-centered designs (Norman, 1986; Shneiderman, 1998; Lazar, 2001).

Social presence is another concern for many researchers in online communities since it can drive users to further explore the use of media and to move beyond the limits of the physical presence. Lee and Nass (2003) found that manipulating paralinguistic cues resulted in the feeling of stronger social presence. In their study, they found that people responded more favorably to an extrovert personality, which in turn fostered a greater sense of social presence. Earlier work by Erickson et al. (2002) concentrated on designing the *babble system*, which supports synchronous and asynchronous text-based conversations in order to foster communication. The aim was to create a time line of social proxy that leaves a trace of people presence online. Participants in the conversation are able to see who is online, when they entered, and when they left. The idea is to make online presence visible, which in turn allows participants to draw inferences, and in turn, helps shape their collective activity. Work by Donath (2002) supports the idea of social presence and its importance in creating connections and supporting conversation online. In later work, Erickson (2003) reported on five years of research in designing visualization for social activity. In this last report, he reevaluates six claims on how to represent people in computational contexts. The claims state that visualization should be the same for everyone, because it makes people accountable for their actions, and therefore fosters trust and eventually helps build social capital. However, trust is a complex concept, with multiple dimensions, and the designer's goal is to establish it and keep it functioning under difficult conditions, because once shaken, it is extremely difficult to rebuild (Shneiderman, 2002).

As ubiquitous computing is becoming more popular, online communities will also have to contend with smaller devices, and Web pages will have to be adaptable to accommodate various sizes (Golub, 2002). Weiser (1991) created ubicomp and introduced the availability of information in an instant. He suggested devices that would support infrastructure and new models of interaction that allow widespread access of information. Weiser's vision includes natural interfaces that facilitate a variety of communication; context-aware applications capable of adapting due to information sensed from the physical environment, and universal access to all sorts of live experiences, where the interaction with computers is more akin to the rich interaction with people.

REFERENCES

Abras, C., Maloney-Krichmar, D., and Preece, J. (2003), "Evaluating an Online Academic Community: 'Purpose' Is the Key," presented at HCI International, Crete, Greece, June 22–27.

Ackerman, M. S. (1998), "Augmenting Organizational Memory: A Field Study of Answer Garden," *ACM Transactions on Information Systems*, Vol. 16, No. 3, pp. 203–224.

Alessi, S. M., and Trollip, S. R. (1991), *Computer-Based Instruction: Methods and Development*, 2nd ed., Prentice Hall, Englewood Cliffs, NJ.

Allee, V. (2000), "Knowledge Networks and Communities of Practice," *Journal of the Organization Development Network*, Vol. 32, No. 4; http://www.odnetwork.org.

Andrews, D. (2002), "Audience-Specific Online Community Design," *Communications of the ACM*, Vol. 45, No. 4, pp. 64–68.

Axelrod, R. (1984), *The Evolution of Cooperation*, Basic Books, New York.

Bednar, A. K., and Charles, M. T. (1999), "A Constructivist Approach for Introducing Pre-service Teachers to Educational Technology: Online and Classroom Education," presented at the Society for Information Technology and Teacher Education International Conference, San Antonio, TX.

Berge, Z. L. (1998), "Conceptual Frameworks in Distance Training and Education," in *Distance Training: How Innovative Organizations Are Using Technology to Maximize Learning and Meet Business Objectives*, D. A. Schreiber and Z. L. Berge, Eds., Jossey-Bass, San Francisco, pp. 19–36.

Brenner, J. (1997), "An Analysis of Students' Cognitive Styles in Asynchronous Distance Education Courses," *Inquiry*, Vol. 1, No. 2, pp. 37–44.

Bruckman, A. (1996), "Gender Swapping on the Internet," in *High Noon on the Electronic Frontier*, P. Ludlow, Ed., MIT University Press, Cambridge, MA, pp. 317–325.

Bruckman, A. (1998), "Community Support for Constructionist Learning: Computer Supported Collaborative Work," *Journal of Collaborative Computing*, Vol. 7, pp. 47–86.

Cherny, L. (1999), *Conversation and Community: Chat in a Virtual World*, CSLI Publications, Stanford, CA.

Clark, N. (2001), "Education, Communication, and Consumption: Piping in the Academic Community," in *Online Communities: Commerce, Community Action, and the Virtual University*, C. Werry and M. Mowbray, Eds., Prentice Hall, Upper Saddle River, NJ, pp. 129–151.

Coleman, J. A. (1991), "Interactive Multimedia," in *Computers as a Tool in Language Teaching*, W. Brierley and I. R. Kemble, Eds., Ellis Horwood, New York, pp. 87–111.

Cowan, J. (1995), "The Advantages and Disadvantages of Distance Education," in *Distance Education for Language Teachers*, R. Howard and I. McGrath, Eds., Multilingual Matters, Clevedon, Somerset, England, pp. 14–20.

Culver, J. D., Gerr, F., and Frumkin, H. (1997), "Medical Information on the Internet," *Journal of General Internal Medicine*, Vol. 47, pp. 466–470.

Cummings, J., Butler, B., and Kraut, R. (2002), "The Quality of Online Social Relationships," *Communications of the ACM*, Vol. 45, No. 7, pp. 103–108.

Davies, M. (1997), "Fragmented by Technology: A Community in Cyberspace," *Interpersonal Computing and Technology: An Electronic Journal for the 21st Century*, Vol. 5, pp. 1–2.

Davies, J., Duke, A., and Sure, Y. (2003), "OntoShare: A Knowledge Management Environment for Virtual Communities of Practice," in *Proceedings of the International Conference on Knowledge Capture*, Sanibel Island, FL, pp. 20–27.

Dibbell, J. (1996), "A Rape in Cyberspace: Or How an Evil Clown, a Haitian Trickster Spirit, Two Wizards, and a Cast of Dozens Turned Database into a Society," in *High Noon on the Electronic Frontier*, P. Ludlow, Ed., MIT University Press, Cambridge, MA, pp. 375–395.

Domingue, J., et al. (2001), "Supporting Ontology Driven Document Enrichment Within Communities of Practice," in *Proceedings of the International Conference on Knowledge Capture*, Victoria, British Columbia, Canada, pp. 30–37.

Donath, J. (2002), "A Semantic Approach to Visualizing Online Conversations," *Communications of the ACM*, Vol. 45, No. 4, pp. 45–49.

Erickson, T. (2003), "Designing Visualizations of Social Activity: Six Claims," presented at CHI 2003, April 5–10. New Horizons, Ft. Lauderdale, FL.

Erickson, T., and Kellogg, W. A. (2000), "Social Translucence: An Approach to Designing Systems That Support Social Processes," *ACM Transactions on Computer–Human Interaction*, Vol. 7, No. 1, pp. 59–83.

Erickson, T., Halverson, C., Kellogg, W. A., and Wolf, T. (2002), "Social Translucence: Designing Social Infrastructures That Make Collective Activity Visible," *Communications of the ACM*, Vol. 45, No. 4, pp. 40–44.

Etzioni, A., and Etzioni, O. (1999), "Face-to-Face and Computer-Mediated Communities: A Comparative Analysis," *Information Society*, Vol. 15, No. 4, pp. 241–248.

Everett, D. R. (1998), "Taking Instruction Online: The Art of Delivery," presented at the Society for Information Technology and Teacher Education International Conference, Washington, DC.

Farnham, S., Cheng, L., Stone, L., Zaner-Godsey, M., Hibbeln, C., Syrjala, K., Clark, A. M., and Abrams, J. (2002), "HutchWorld: Clinical Study of Computer-Mediated Social Support for Cancer Patients and Their Caregivers," presented at CHI 2002, Changing the World, Changing Ourselves, Minneapolis, MN, April 20–25.

Feng, J., Lazar, J., and Preece, J. (2003), "Interpersonal Trust and Empathy Online: A Fragile Relationship," Extended Abstracts (CD-ROM), CHI 2003, Fort Lauderdale, FL.

Fernback, J. (1997), "The Individual Within the Collective: Virtual Ideology and the Realization of Collective Principles," in *Virtual Culture: Identity and Communications*, S. Jones, Ed., Sage Publications, Thousands Oaks, CA, pp. 36–54.

Fox, S., and Rainie, L. (2000), *The Online Health Care Revolution: How the Web Helps Americans Take Better Care of Themselves*," Pew Internet and American Life Project, Washington, DC.

Garton, L., Haythornthwaite, C., and Wellman, B. (1997), "Studying Online Social Networks," retrieved October 11, 2002, from http://www.ascuscorg/jcmc/vol3/issue1/garton.html.

Golub, E. (2002), "A Brief History of the Net," retrieved March 15, 2002, from http://www.cs.umd.edu/~egolub/abhotn.shtml.

Gongola, P., and Rizzuto, C. R. (2001), "Evolving Communities of Practice: IBM Global Services Experience," *IBM Systems Journal*, Vol. 40, No. 4; http://www.research.ibm.com/journal/sj/404/gongola.html.

Goodwin, M. (1994), "Nine Principles for Making Virtual Communities Work," *Wired*, Vol. 2, No. 6, pp. 72–73.

Gordon, H. R. D. (1995), *Description of the Productivity and Learning Style Preferences of On- and Off-Campus Distance Education Participants at Marshall University*, Productivity Environmental Preference Survey, Technical Report HE-028-777, Marshall University, Huntington, WV.

Hakken, D. (1999), *Cyborgs @ Cyberspace? An Ethnographer Looks to the Future*, Routledge, New York.

Hamman, R. B. (2001), "Computer Networks Linking Network Communities," in *Online Communities: Commerce, Community Action, and the Virtual University* C. Werry and M. Mowbray, Eds., Prentice Hall, Upper Saddle River, NJ.

Hara, N. (1998), *Student's Perspectives in a Web-Based Distance Education Course*, Report HE-031-714, Mid-Western Educational Research Association, Chicago: (ERIC Document Reproduction Service ED 426 633).

Hillery, G. J. (1955), "Definitions of Community: Areas of Agreement," *Rural Sociology*, Vol. 20, pp. 11–123.

Hiltz, S. R. (1994), *The Virtual Classroom: Learning Without Limits via Computer Networks*, Ablex Publishing, Norwood, NJ.

Howell-Richardson, C. (1995), "Interaction Across Computer-Conferencing," in *Distance Education for Language Teachers*, R. Howard and I. McGrath, Eds., Multilingual Matters, Clevedon, Somersetshire, England, pp. 118–132.

Jones, Q. (1997), "Virtual-Communities, Virtual-Settlements and Cyber-archaeology: A Theoretical Outline," *Journal of Computer Mediated Communication* [On-line], Vol. 3, No. 3. http://jcmc.indiana.edu/vol3/issue3/.

Jones, R. L. (1989), "Using Interactive Audio With Computer-Assisted Language Learning," in *Modern Technology in Foreign Language Education: Applications and Projects*, W. F. Smith, Ed., National Textbook Company, Lincolnwood, IL, pp. 219–226.

Jones, S. (1995), "Understanding Community in the Information Age," in *Cybersociety: Computer-Mediated Communication and Community*, S. Jones, Ed., Sage Publications, Thousand Oaks, CA, pp. 10–33.

Jones, S. (1998), "Information, Internet, and Community: Notes Toward an Understanding of Community in the Information Age," in *Cybersociety: Revisiting Computer-Mediated Communication and Community*," S. Jones, Ed., Sage Publications, Thousand Oaks, CA.

Kies, J. K., Williges, R. C., and Rosson, M. B. (1997), "Evaluating Desktop Video Conferencing for Distance Learning," *Computer Education*, Vol. 28, pp. 79–91.

Kim, A. J. (2000), *Community Building on the Web: Secret Strategies for Successful Online Communities,"* Peachpit Press, Berkeley, CA.

Kirby, E. (1999), "Building Interaction in Online and Distance Education Courses," presented at the Society for Information Technology and Teacher Education International Conference, San Antonio, TX.

Kitchen, D., and McDougall, D. (1999), "Collaborative Learning on the Internet," *Journal of Educational Technology Systems*, Vol. 27, No. 3, pp. 245–258.

Kollock, P. (1998a), "Design Principles in Online Communities," *PC Update*, Vol. 15, pp. 58–60.

Kollock, P. (1998b), "The Economies of Online Cooperation: Gifts and Public Goods in Cyberspace," in *Communities in Cyberspace*, M. Smith and P. Kollock, Eds., Routledge, London, pp. 220–239.

Kollock, P., and Smith, M. (1999), "Communities in Cyberspace," in *Communities in Cyberspace*, M. A. Smith and P. Kollock, Eds., Routledge, London, pp. 3–25.

Lave, J., and Wenger, E. (1991), *Situated Learning: Legitimate Peripheral Participation*, Cambridge University Press, Cambridge.

Lazar, J. (2001), *User-Centered Web Development*, Jones & Bartlett Computer Science, Boston.

Lee, K. M., and Nass, C. (2003), "Designing Social Presence of Social Actors in Human–Computer Interaction," presented at CHI 2003, April 5–10, New Horizons, Ft. Lauderdale, FL.

Lesser, E. L., and Storck, J. (2001), "Communities of Practice and Organizational Performance," *IBM Systems Journal*, Vol. 40, No. 4, http://www.research.ibm.com/journal/sj/404/lesser.html.

Lewis, J., Whittaker, J., and Julian, J. (1995), "Distance Education for the 21st Century: The Future of National and International Telecomputing Networks in Distance Education, in *Computer Mediated Communication and the Online Classroom*, Vol. 3, *Distance Learning*, Z. L. Berge & M. P. Collins, Eds., Hampton Press, Cresskill, NJ.

Lia-Hoagberg, B., Vellenga, B., and Miller, M. (1999), "A Partnership Model of Distance Education: Student's Perceptions of Correctedness and Professionalization," *Journal of Professional Nursing*, Vol. 15, No. 2, pp. 116–122.

Luke, T. (2001), "Building a Virtual University: Working Realities from the Virginia Tech Cyberschool," in *Online Communities: Commerce, Community Action, and the Virtual University*, C. Werry and M. Mowbray, Eds., Prentice Hall, Upper Saddle River, NJ, pp. 153–174.

Malan, R. F., Rigby, D. S., and Glines, L. J. (1991), "Support Services for the Independent Study Student," in *The Foundations of American Distance Education*, B. L. Watkins and S. J. Wright, Eds., Kendall/Hunt, Dubuque, IA, pp. 159–172.

Maloney-Krichmar, D., Abras, C., and Preece, J. (2002), "Revitalizing an Online Community," presented at the 2002 International Symposium on Technology and Society: Social Implications of Information and Communication Technology, Raleigh, NC, June 6–8.

Markham, A. N. (1998), *Life Online: Researching Real Experience in Virtual Space*, Altamira, Walnut Creek, CA.

McDermott, R. (1999), "Nurturing Three Dimensional Communities of Practice: How to Get the Most Out of Human Networks," *Knowledge Management Review*, http://www.co-i-l.com/coil/knowledge-garden/cop/dimensional.shtml.

McDermott, R. (2001), "Knowing in Community: 10 Critical Success Factors in Building Communities of Practice," *Leveraging Knowledge*, http://www.co-i-l.com/coil/knowledge-garden/cop/knowing.shtml.

Millen, R. D., and Fontaine, M. A. (2003), "Improving Individual and Organizational Performance Through Communities of Practice," in *Proceedings of the 2003 International ACM SIGGROUP Conference on Supporting Group Work*, Sanibel Island, FL, pp. 205–211.

Millen, R. D., Fontaine, M. A., and Muller, M. J. (2002), "Understanding the Benefit and Costs of Communities of Practice," *Communications of the ACM*, Vol. 45, No. 4, pp. 69–73.

Moore, M. G., and Kearsley, G. (1996), *Distance Education: A Systems View*, Wadsworth, Belmont, CA.

Morgan, C. K., and Tam, M. (1999), "Unraveling the Complexities of Distance Education Student Attrition," *Distance Education*, Vol. 20, No. 1, pp. 96–108.

Mory, E. H., Gambill, L. E., and Browning, J. B. (1998), *Instruction on the Web: The Online Student's Perspective*, Report IR-018-809, Society for Information Technology and Teacher Education International Conference, Washington, DC (ERIC Document Reproduction Service ED 421 090).

Muller, M. J. (1992), "Retrospective on a Year of Participatory Design Using the PICTIVE Technique," presented at the CHI 1992 Conference on Human Factors in Computing Systems, Monterey, CA.

Nonaka, I., and Takeuchi, H. (1995), *The Knowledge Creating Company*, Oxford University Press, New York.

Nonnecke, B., and Preece, J. (2000a), "Lurker Demographics: Counting the Silent," presented at the CHI 2000 Conference on Human Factors in Computing Systems, The Hague, The Netherlands.

Nonnecke, B., and Preece, J. (2000b), "Persistence and Lurkers: A Pilot Study," presented at the HICSS-33 IEEE Computer Society Conference, Maui, HI.

Norman, D. A. (1986), "Cognitive Engineering," in *User-Centered Systems Design*, D. Norman and S. Draper, Eds., Lawrence Erlbaum Associates, Mahwah, NJ.

Olson, J. S., and Olson, G. M. (2000), "i2i Trust in E-Commerce," *Communications of the ACM*, Vol. 43, pp. 41–44.

Ostrom, E. (1990), *Governing the Commons: The Evolution of Institutions for Collective Action*, Cambridge University Press, New York.

Powazek, D. (2002), *Design for Community: The Art of Connecting Real People in Virtual Places*, New Riders, Indianapolis, IN.

Preece, J. (1999), "Empathic Communities: Reaching Out across the Web," *Interactions Magazine*, Vol. 2, issue 2, pp. 32–43.

Preece, J. (2000), *Online Communities: Designing Usability, Supporting Sociability*, Wiley, Chichester, West Sussex, England.

Preece, J. (2001a), "Designing Usability, Supporting Sociability: Questions Participants Ask About Online Communities," presented at the Human–Computer Interaction Conference, Interact'01, Tokyo.

Preece, J. (2001b), "Sociability and Usability in Online Communities: Determining and Measuring Success," *Behaviour and Information Technology*, Vol. 20, No. 5, pp. 347–356.

Preece, J. (2004), "Etiquette, Empathy and Trust in Communities of Practice: Stepping-Stones to Social Capital," *Journal of Universal Computer Science*, Vol. 10, No. 3; http://www.jucs.org.

Preece, J., and Ghozati, K. (1998a), "In Search of Empathy Online: A Review of 100 Online Communities," presented at the 1998 Association for Information Systems Americas Conference, Baltimore, MD.

Preece, J., and Ghozati, K. (1998b), "Offering Support and Sharing Information: A Study of Empathy in a Bulletin Board Community," presented at the Conference on Computer Virtual Environments, Manchester, Lancashire, England.

Preece, J., and Ghozati, K. (2000), "Experiencing Empathy Online," in *The Internet and Health Communication: Experience and Expectations*, R. Rice and J. Katz, Eds., Sage, Thousand Oaks, CA, pp. 237–260.

Preece, J., and Maloney-Krichmar, D. (2002), "Online Communities: Focusing on Sociability and Usability," in *Handbook of Human–Computer Interaction*, J. Jacko and A. Sears, Eds., Lawrence Erlbaum Associates, Mahwah, NJ, pp. 596–620.

Preece, J., Maloney-Krichmar, D., and Abras, C. (2003), "History of Emergence of Online Communities," in *Encyclopedia of Community*, B. Wellman, Ed., Sage, Thousand Oaks, CA, pp. 1023–1027.

Putnam, R. D. (1995), "Bowling Alone: America's, Declining Capital," *Journal of Democracy*, Vol. 6, No. 1, pp. 65–78.

Quan-Haase, A., and Wellman, B. (2004), "How Does the Internet Affect Social Capital," in *IT and Social Capital*, M. Huysman and V. Wulf, Eds., MIT Press, Cambridge, MA, pp. 151–176.

Rafaeli, S., and Sudweeks, F. (1997), "Networked Interactivity," *Journal of Computer-Mediated Communication*, Vol. 2, No. 4, http://jcmc.indiana.edu/vol2/issue4/.

Reid, E. (1996), "Text-Based Virtual Realities: Identity and the Cyborg Body," in *High Noon on the Electronic Frontier*, P. Ludlow, Ed., MIT University Press, Cambridge, MA, pp. 327–345.

Resnick, P., Zeckhauser, R., Friedman, E., and Kuwabara, K. (2000), "Reputation Systems," *Communications of the ACM*, Vol. 43, No. 12, pp. 45–48.

Rheingold, H. (1993), *The Virtual Community: Homesteading on the Electronic Frontier*, Addison-Wesley, Reading, MA.

Rheingold, H. (1996), "A Slice of My Life in My Virtual Community," in *High Noon on the Electronic Frontier*, P. Ludlow, Ed., MIT Press, Cambridge, MA, pp. 413–436.
Rohfeld, R. W., and Hiemstra, R. (1995), "Moderating Discussions in Electronic Classrooms," in *Computer Mediated Communication and the Online Classroom*, Vol. 3, *Distance Learning*. Z. L. Berge and M. P. Collins, Eds., Hampton Press, Cresskill, NJ.
Schuler, D. (1996), *New Community Networks: Wired for Change*, Addison-Wesley, Reading, MA.
Schuler, D., and Namioka, A. (1993), *Participatory Design: Principles and Practices*, Lawrence Erlbaum Associates, Mahwah, NJ.
Shneiderman, B. (1998), *Designing the User Interface: Strategies for Effective Human–Computer Interaction*, 3rd ed., Addison-Wesley, Reading, MA.
Shneiderman, B. (2000), "Designing Trust into Online Experiences," *Communications of the ACM*, Vol. 43, No. 12, pp. 57–59.
Shneiderman, B. (2002), *Leonardo's Laptop: Human Needs and the New Computing Technologies,* MIT Press, Cambridge, MA.
Simonson, M. (1997), "Distance Education: Does Anyone Really Want to Learn at a Distance?" *Contemporary Education*, Vol. 68, No. 2, pp. 104–107.
Smith, M. (2002), "Tools for Navigating Large Social Cyberspaces," *Communications of the ACM*, Vol. 45, No. 4, pp. 51–55.
Smith, M. A., and Fiore, A. T. (2001), "Visualization Components for Persistent Conversations," presented at CHI 2001, Seattle, WA.
Snyder, W. M. (1997), "Communities of Practice: Combining Organizational Learning and Strategy Insights to Create a Bridge to the 21st Century," http://www.co-i-l.com/coil/knowledge-garden/cop/cols.shtml.
Snyder, W. M., and de Souza Briggs, X. (2003), *Communities of Practice: A New Tool for Government Managers/Collaboration Series*, IBM Center for Business and Government, online http://www.businessofgovernment.org/pdfs/snyder_report.pdf
Sproull, L., and Faraj, S. (1997), "Atheism, Sex, and Databases: The Net as a Social Technology," in *Culture of the Internet*, S. B. Kiesler, Ed., Lawrence Earlbaum Associates, Mahwah, NJ, pp. 35–51.
Steiner, V. (1999), "What Is Distance Education?" http://www.wested.org/tie/dlrn/distance.html.
Surgue, B., Rietz, T., and Hansen, S. (1999), "Distance Learning: Relationships Among Class Size, Instructor Location, Student Perceptions, and Performance," *Performance Improvement Quarterly*, Vol. 12, No. 3, pp. 4–57.
Tennison, J., and Shadbolt, N. R. (1998), "APECKS: A Tool to Support Living Ontologies," in *Proceedings of the 11th Banff Knowledge Acquisition Workshop,* Banff, Alberta, Canada, April 18–23.
Turkle, S. (1999), "Tinysex and Gender Trouble," *IEEE Technology and Society Magazine*, Vol. 4, Winter, pp. 8–20.
Uslaner, E. M. (2000), "Social Capital and the Net," *Communications of the ACM*, Vol. 43, No. 12, pp. 60–64.
Verano, M., (1988), "USAFA Interactive Study in Spanish," in *Modern Technology in Foreign Language Education: Applications and Projects*, W. F. Smith, Ed., National Textbook Company, Lincolnwood, IL, pp. 249–256.
Vygotsky, L. (1962), *Thought and Language*, MIT Press, Cambridge, MA.
Weiser, M. (1991), "The Computer for the 21st Century," *Scientific American*, Vol. 265, No. 3, pp. 94–104.
Wellman, B. (1982), "Studying Personal Communities," in *Social Structure and Network Analysis*, P. M. N. Lin, Ed., Sage Publications, Beverly Hills, CA.
Wellman, B., and Gulia, M. (1999), "Virtual Communities as Communities: Net Surfers Don't Ride Alone," in *Communities in Cyberspace*, P. Kollock and M. Smith, Eds., Routledge, Berkeley, CA.
Wellman, B., Boase, J., and Chen, W. (2002), "The Networked Nature of Community: Online and Offline," *IT & Society*, Vol. 1, No. 1, pp. 151–165.
Wenger, E. (1991), "Communities of Practice: Where Learning Happens," *Benchmark,* Fall, pp. 82–84.
Wenger, E. (1998a), *Communities of Practice: Learning, Meaning, and Identity*, Cambridge University Press, Cambridge.
Wenger, E. (June 1998b), "Communities of Practice: Learning as a Social System, *Systems Thinker,* http://www.co-i-l.com/coil/knowledge-garden/cop/lss.shtml.
Wenger, E., McDermott, R., and Snyder, W. M. (2002), *Cultivating Communities of Practice: A Guide to Managing Knowledge*, Harvard Business School Press, Boston.
Whittaker, S., Issacs, E., and O'Day, V. (1997), "Widening the Net: Workshop Report on the Theory and Practice of Physical and Network Communities," *SIGCHI Bulletin*, Vol. 29, No. 3, pp. 27–30.

CHAPTER 48

HUMAN FACTORS AND INFORMATION SECURITY

E. Eugene Schultz
High Tower Software
Aliso Viejo, California

1 OVERVIEW

In this chapter we delve into the relationship between information security (sometimes called *computer security*) and usability design. The goals of *information security* are to protect the confidentiality, integrity, and availability of systems, information, applications, and network devices, as well as to prevent repudiation (untruthful denial) of electronically based transactions. Security-related breaches manifest themselves in a variety of forms, including intrusions into systems, worm and virus infections, misuse, denial of service, integrity compromises in systems and/or data, scams, hoaxes, and many others. A taxonomy of the major security-related tasks that people must perform includes the following tasks: identification and authentication, assurance of integrity, confidentiality, availability, and system integrity, and intrusion detection. *Usability* flaws in a number of corresponding areas—password selection, third-party authentication, file access control, Web server configuration, firewall configuration, encryption of sensitive information, electronic commerce transactions, auditing and logging, and intrusion detection—are analyzed. These flaws are almost certainly also linked to the most costly form of security-related incident—damage and disruption to systems and data due to insiders. Employees and contractors who are disgruntled may, for example, be less motivated than other users to overcome usability hurdles in computing systems, something that may escalate damage and/or disruption considerably. Better default parameters in operating system and application software and the availability of settings that produce pervasive changes in the security of systems and applications, thus obviating the need to interact with systems many times or with more difficulty to tighten these settings, would go a long way in addressing the usability problems discussed in this chapter.

2 INTRODUCTION

Using computers is a way of life almost everywhere around the world. Without computers, life in virtually every country would be radically different. Although appreciating how different life would be without computers is easy, many people fail to appreciate what happens if computers are unreliable for a variety of reasons. In some cases, such as when computers are used to regulate energy flow within buildings, computer unreliability might not radically disrupt computing activity because people could in these circumstances simply take manual control of functions normally performed by computers. But in other cases, such as in air traffic control systems, the ramifications of computers becoming unreliable can potentially be considerably more draconian.

Computers become unreliable for a variety of well-known reasons: electrical failure, damage due to water or fire, and hardware and software flaws ("bugs") that threaten the normal operation of computers, to name a few. Strangely, until relatively recently, people have been largely unaware of the many security-related reasons for unreliability in computing systems, such as remote access by unauthorized users or the execution of malicious programs, even though security threats may be more costly and disruptive than others, such as power outages.

Protecting computers, the information residing in them and sent over networks, applications, and network components such as routers and switches against security-related threats falls within the purview of the field of information security, also known as computer security (Garfinkle et al., 2003). Included in the goals of information security are ensuring confidentiality of information; protecting the integrity of systems, data, and applications; ensuring that systems, data, and applications are available when needed; and ensuring that anyone who initiates an electronic transaction cannot repudiate or deny having done so afterward. Although once an obscure area, information security has become increasingly important over the last 10 to 15 years as the number, magnitude, and impact of security breaches have grown. According to the Computer Emergency Response Team Coordination Center (CERT/CC) (2004), the number of security breaches reported has increased dramatically in recent years. Although the amount of financial loss caused by many security breaches is relatively small (i.e., a few thousand dollars), some security breaches cause losses of millions of dollars. Federal and state laws that require the implementation of security measures to protect certain types of information, such as personal information that could be used in identity theft, and that have prescribed penalties for computer-related actions such as gaining unauthorized access to systems have also greatly contributed to the growth of information security in numerous countries.

3 SECURITY BREACHES

The nature of security breaches varies considerably. The most common types of security breaches include:

1. *Intrusions into systems and network devices.* These are commonly known as *hacker attacks*. In most of these attacks, perpetrators break into user or system administrator's accounts using passwords captured ("sniffed") as they are entered during local logins or as they traverse networks during remote logins. *Brute force attacks*, in which attackers run programs that enter one password after another until one finally succeeds, are another variation of this type of attack. Alternatively, many attackers run programs that attempt to exploit vulnerabilities in systems and applications to gain unauthorized access. Once intruders break into a system, they often engage in activities such as reading users' files and e-mail and planting Trojan horse programs that allow them once again to gain access to the victim system later.

2. *Virus and worm infections.* Viruses are self-replicating programs that spread because of user actions, whereas worms are self-replicating programs that spread independently of user actions (Schultz and Shumway, 2001). The fact that worms work independent of users makes them particularly troublesome; in 2003, for example, the MSBlaster worm and its variants infected well over a half million PCs connected to the Internet (Symantec Corporation, 2003).

3. *Misuse and subversion by trusted individuals, such as employees and contractors.* Misuse and subversion are a less common but in many cases the most costly category of security breaches. These types of malfeasance are often due to motives such as greed or revenge.

4. *Denial of service attacks.* Among the most common of all security breaches, these are intended to shut down or disrupt computing activities. They also are among the most costly of all because of organizations' dependence upon computing services.

5. *Integrity compromises.* Integrity compromises occur when perpetrators place malicious programs in systems they have accessed or modify files within these systems. Web page defacements are the most widely known type of integrity compromise, although unauthorized modification of system files and critical data such as financial data are generally much more costly than is typical.

6. *Scams.* Scams are schemes in which e-mail, electronic messaging, Web sites, or chat rooms are used to con unsuspecting users out of something (usually, money). *Phishing*, currently the most common type of scam, entails sending e-mail that threatens people such as bank customers with disruption of services if recipients do not enter personal and/or financial information in a form on a Web page that appears to belong to a legitimate company. Other scams offer recipients of messages that appear to come from people such as deposed African political figures a large commission in return for helping transfer what is described as millions of dollars to the United States. The catch is that recipients must first send a sum of money to a designated address as a "measure of good faith."

7. *Hoaxes.* In hoaxes, bogus information is disseminated electronically. For example, certain network postings falsely claim that Windows operating systems contain routines that covertly glean data stored in these systems for the National Security Agency (NSA).

8. *Spamming.* Spam is unwanted e-mail ("junk mail") and pop-up messages. Although sending e-mail does not in and of itself constitute a security breach, the fact that the overwhelming majority of spam has a sender address that has been falsified does. Spam is thus, in effect, often a type of repudiation attack. Furthermore, spam now constitutes such a large proportion of the e-mail that users receive that organizations often lose large amounts of money each year from lost productivity (because each user must read and then delete each spam message).

Despite the growing importance of information security, system administrators and users often resist using measures that improve security. User resistance toward systems with which users must interact is a well-known phenomenon (Turnage, 1990), as is the fact that systems with poor user interaction methods lead to greater user resistance than do other systems (e.g., Markus, 1983; Al-Ghatani and King, 1999). Although usability design in systems is generally less than optimal, poor usability design abounds in computing systems and devices designed to improve information security, as explained shortly. The weaknesses in this design may be "the straw that broke the camel's back," in that measures used to raise security too often create usability barriers that cause people to neglect or abandon them, leaving their systems, applications, data, and network devices vulnerable to all types of attacks.

4 TAXONOMY OF INFORMATION SECURITY TASKS

Analyzing the tasks that system administrators and users must perform in securing systems, applications, and data is a good starting point for examining usability issues in information security. Schultz et al.'s (2001) taxonomy of security tasks provides an analysis of six major security-related tasks:

1. *Identification and authentication. Identification* means proving one's identity. *Authentication*, very similar in meaning to identification, means proving one's identity for the purpose of accessing a system or network. The most common type of identification and authentication task is entering a password, although many other identification and authentication methods [such as inserting a smart card and then entering a short PIN (personal identification number)] exist. Effective identification and authentication help prevent perpetrators from masquerading as other users.
2. *Protecting data integrity.* Although numerous data integrity protection methods exist, the most commonly used method is setting file system permissions to prevent all but very few users from being able to change, replace, or delete files and directories. Software for detecting changes in files and directories (often called *tripwire software*) is also becoming used more frequently. Data integrity protection methods help to prevent unauthorized deletion of and/or changes in data.
3. *Protecting data confidentiality.* Setting file permissions appropriately is also the most commonly used method of protecting data confidentiality. Controlling against privilege escalation in systems by drastically limiting the number of superusers is another data confidentiality assurance method because superusers can read every file on their system, regardless of what permissions have been set. Data confidentiality methods help prevent unauthorized disclosure and/or possession of information.
4. *Ensuring data availability.* Assuring that data and the applications that use them are available is another critical information security task. Tasks that help achieve this goal include making system and data backups as well as using other measures, such as fault-tolerant storage systems.* These tasks help in guarding against unauthorized deletion of or denying access to information and the programs that use them.
5. *Ensuring system integrity.* The methods discussed previously that are used to protect data integrity are used in ensuring system integrity. Additionally, installing patches for vulnerabilities in systems and applications helps prevent unauthorized modification of system files. Inspecting system files for unauthorized changes is still another often-used method. These measures help to prevent unauthorized deletion and/or changes in system files.
6. *Intrusion detection.* Intrusion detection means identifying attacks. Intrusion detection means discovering attacks that have occurred and their outcomes (in terms of their success). Although intrusion detection is not really a security countermeasure per se in that it does not help directly in preventing attacks, it is nevertheless a much used measure in that it enables technical experts to quickly identity and thwart attacks that are under way, thereby minimizing their impact. The most basic form of intrusion detection is inspecting system audit logs, an arduous task at best. Intrusion detection is often performed by special software and hardware; even so, user interaction tasks are necessary.

Now that the basic kinds of tasks that must be performed in information security have been introduced, we'll explore the types of usability hurdles in information security and the impact they have.

5 FLAWS IN USABILITY DESIGN

As mentioned earlier, the area of information security abounds with examples of poor usability design. We'll look at the usability design of password-based authentication, third-party authentication, file access control methods, Web configuration, firewall configuration, encryption of sensitive information, electronic commerce transactions, auditing and logging, and intrusion detection.

5.1 Password Entry

Previous work by Proctor et al. (2000) demonstrates that entering a user name–password combination to log in to a system or network is not at all difficult from a human factors perspective. The fact that users become highly practiced over time in entering passwords when they see the appropriate prompt helps overcome the few usability hurdles that this task poses. A task analysis of generic password-based log ins shows that users must engage in only a few relatively simple actions:

1. Visually sight the dialog box and the prompts and input field within (see Figure 1).
2. Use the pointing device to align the cursor/pointer to the correct location.
3. Home both hands at the keyboard.
4. Recall the password.
5. Enter the password by pressing the appropriate keystroke sequence.
6. Click on <OK> or press the <ENTER> key.

*RAID, the redundant array of independent drives, is the most frequently used fault tolerance solution. RAID distributes data across multiple drives; if any drive fails, the data will thus be available on another.

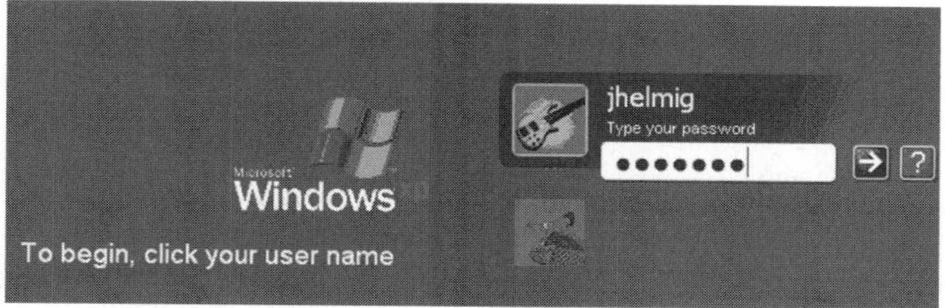

Figure 1 Log-on screen in Windows XP.

Although entry of a log-in name–password sequence is relatively easy for users, there is a more difficult human factors problem of which few users are aware. Hackers and password cracking tools have become so efficient that passwords that users normally choose can be compromised in a very short time (Skoudis, 2004). Users can choose stronger (more difficult-to-guess) passwords, but doing so makes them more difficult to remember (Proctor et al. 2002). For example, "password" would be an easy-to-remember password, but it would be trivial to guess or crack. "6f*2Sl&," which has just as many characters as "password," would be much more difficult to guess or crack, but few users would be able to remember this password. Many users are not aware of what differentiates good passwords from ones that are easy to guess or crack; even if they know how to create good passwords, they nevertheless often choose weak ones (such as combinations of characters that include their user account names) because of the additional effort needed to create good passwords (Bishop and Klein, 1995). To compensate for the inability to remember more-difficult-to-remember passwords, users could write them down, but doing so would enable anyone who found the slips of paper or whatever else on which the passwords are written to use them to gain unauthorized access to the users' accounts.

Proctor et al. (2002) conducted studies testing the effects of proactive constraints, that is, limiting the types of passwords that users could create, and password length on the ability to crack passwords. As expected, longer passwords were more difficult to crack than were shorter ones. They found that proactive constraints produced different effects for shorter passwords than for longer passwords, however. When passwords were shorter (five characters in length), however, constraints elevated resistance to cracking more than when passwords were longer (eight characters in length). Constraints and password length together required the most effort on the part of users and produced only slightly better resistance to password cracking than did password length alone, suggesting that requiring longer passwords but not imposing additional constraints represents a good trade-off point between usability and security.

5.2 Third-Party Authentication Tasks

Many information security experts believe that passwords are now too dangerous to use and that other forms of authentication must supplant password-based authentication. Third-party authentication is any form of authentication that is not built into operating systems but that is, instead, provided through vendor products or methods. Password-based authentication is built into virtually every commercial off-the-shelf operating system, but smart card–based authentication, one of the most common types of third-party authentication, is not. Smart cards are actually miniature computers with chips that are built-into an object such as a plastic card (Corcoran, 2000). To authenticate using smart cards, users must insert a smart card into a smart card reader, which is normally attached to a keyboard. Proctor et al. (2000) performed a task analysis on a typical smart card–based user interaction sequence. They found that to authenticate using smart cards users had to:

1. Visually sight a prompt on the display terminal that directs the user to proceed with the smart card authentication process.
2. Visually sight the smart card.
3. Use several fingers and the thumb to grasp the smart card.
4. Visually sight the smart card reader.
5. Use the hand and arm to move the smart card toward the smart card reader until it is in close proximity.
6. Rotate the hand until the smart card is at the proper angle to be inserted.
7. Insert the smart card until it fits inside the smart card reader.
8. Visually sight the display terminal (or listen for auditory feedback) for confirmation that the smart card was read successfully.
9. Grasp the smart card in several fingers and the thumb, moving the hand and finger away from the smart card reader.
10. Confirm that smart card data have been processed properly and are valid.
11. Use the hand to place the smart card on a surface, such as a table surface.
12. Let go of the smart card with the fingers and thumb.
13. Read a prompt that begins the "normal" log-in name–password entry sequence.

14. Home the hand on the keyboard.
15. Engage in the steps normally required for a password-based log-in but enter the PIN instead.

The task sequence for this generalized smart card–based authentication task sequence involves many steps that are not included in password-based logins, showing that the former type of authentication presents additional levels of difficulty for users. Although the exact tasks vary in different smart card implementations, one thing seems clear—users who are accustomed to password-based authentication are likely to resist smart card–based authentication because a greater amount of work and more opportunity for error are inherent in the latter. Given the far greater strength of smart card—vs. password-based authentication, this is unfortunate. This is just one of the many cases in information security in which security and usability are orthogonal to each other. Other forms of third-party authentication, such as biometric authentication, authentication based on fingerprints, retinal patterns, facial shape, and so on, are available, but do they require less work on the part of users? A task analysis on user interaction with a generic fingerprint-based biometric authentication device showed that this method involved substantially more steps than does password-based authentication but somewhat fewer steps than smart card–based authentication (Proctor et al., 2000). The usability limitations in third-party authentication thus are not limited to task sequences involving smart card–based authentication.

5.3 File Access Control

File access control mechanisms are used not only to protect information from unauthorized disclosure and possession, but also from unauthorized modification and deletion. Virtually every modern operating system includes a file system that offers permissions such read, write, and execute to control access to files and directories by users and/or groups. Write access is particularly potentially dangerous; it generally allows not only modification of content, but also deletion of files and/or directories. Setting file and directory permissions is, however, far from optimal from a usability perspective. One of the problems is that an excessive number of user interaction task steps is often required. Consider, for example, the following interaction steps required for changing the permissions of a file or directory in Windows NT (Schultz et al., 2001). To change the permissions for a single user, the system administrator or owner of a file must:

1. Inspect the desktop visually to find the icon for the Windows Explorer or visually sight the Windows and E keys on the keyboard.
2. Bring up the Windows Explorer by double-clicking on the desktop icon or pressing the Windows and E keys simultaneously.
3. Use a pointing device to scroll through groups of icons for files and directories until sighting the desired one.
4. Right-click on the selected icon using the pointing device.
5. Click on Properties using the pointing device.
6. Visually scan the tabs at the top of the new screen that appears to find the Security tab.
7. Click on Security.
8. Of the options that appear, click on Permissions.
9. Click on Add.
10. Scroll through the list of groups and users.
11. Click on the Show Users box.
12. Scroll down to the desired user name.
13. Click on the user name to highlight.
14. Scroll through Type of Access.
15. Highlight the desired type of access (e.g., Full Control, Change, Read) and release the selection button on the pointing device.
16. Click on OK on three different screens.

Worse yet, the standard Windows NT file permissions are only the beginning when it comes to securing files and directories because they are in effect high-level permissions intended mainly for the convenience of system administrators and users. Many *individual* or *advanced permissions* are also available to allow for more granular file and directory access control. In Windows NT there are seven such permissions: Read, Write, Execute, Delete, Change Permissions, Take Ownership, and Full Control. In Windows 2000 there are 14 such permissions: Traverse Folder/Execute File, List Folder/Read Data, Read Attributes, Read Extended Attributes, Create Files/Write Data, Create Folders/Append Data, Write Attributes, Write Extended Attributes, Delete Subfolders and Files, Delete, Read Permissions, Change Permissions, Take Ownership, and Synchronize. To add or delete any of these additional permissions, additional user interaction steps, almost none of which are intuitive to novice users, must be performed.

Various methods of changing file permissions without having to use a graphical user interface (GUI) also exist. These alternative methods may involve considerably fewer steps on the part of the user, but require the entry of commands and flags in a syntax that is extremely unforgiving. For example, in Windows NT, 2000, XP, and Server 2003, someone must enter the following command to change file permissions on a file named Payroll in a folder named Finance for a user (named Brown in this example) from change (modify) to read-only:

```
cacls D:\Finance\Payroll /E /R Brown:C /G brown:R
```

In Linux and Unix systems users must enter the following command to remove Read and Write access from the group that owns the file "foo" in user Brown's home directory as well as from world (other):

```
chmod og-rw /home/Brown/foo
```

Experienced system administrators would have little trouble entering any of these commands to change file permissions, but only because of prolonged practice. Less experienced system administrators and users would experience extreme difficulty entering these commands correctly because of the nonintuitive nature of the command primitives and syntax.

5.4 Configuring Web Servers

The universal popularity of the World Wide Web (WWW) makes optimizing usability design in Web servers a necessity. Web sites face enormous competition, to the point that user interaction with the majority of them must be satisfactory to users if users are going to be willing to visit (and, in particular, revisit) them. The main usability limitations in the Web arena thus instead involve Webmasters' interactions with Web servers. Interaction sequences involved in setting up and administering Web servers tend to be extremely nonintuitive. For example, consider how to deny default access to directories in Apache Web servers (Schultz, 2002a). The Webmaster must insert the following directives (commandlike objects) in the server's configuration:

```
<Directory />
Order Deny,Allow
Deny from all
</Directory>
```

The Webmaster must next make exceptions by specially permitting access to the directories intended for Web access by users. For example, the following directives allow access to the/usr/users/*/public_html and/usr/local/httpd directories:

```
<Directory /usr/users/*/public_html>
Order Deny,Allow
Allow from all
</Directory>
<Directory /usr/local/httpd>
Order Deny,Allow
Allow from all
</Directory>
```

Although experienced Webmasters can deal readily with Apache directories, novices are not so fortunate. Directives have a difficult syntax, one that once again illustrates the usability problems that plague information security. Ironically, Microsoft's Internet Information Server (IIS) Web server is considerably easier to use because of well-designed control panels that obviate the need for recalling complex syntactic conventions such as Apache directives. At the same time, however, IIS Webmasters must often go through four to six levels of menus to get to a particular function even though menu depth produces a longer menu navigation time than does menu breadth (Schultz and Curran, 1986).

5.5 Configuring Firewalls

Firewalls are barriers between one network and another that are put in place to insulate one network from attacks from another (Cheswick et al., 2003). Implementing well-designed and well-maintained firewalls is one of the most important security measures that an organization can put in place. Firewalls vary considerably in their functionality; some do little more than block certain kinds of incoming packets bound for certain IP addresses and/or ports, whereas others analyze packets very thoroughly to determine whether or not they constitute desirable input to applications before sending them on via an entirely new connection that they create.

Regardless of the functionality, most commercially available firewalls have one thing in common: poor usability design. Consider, for example, the following access control entries in a Cisco PIX firewall:

```
#access-list acl_out permit tcp any any eq telnet
#access-list acl_out deny tcp any any
#access-list acl_out deny udp any any
#access-list acl_in permit tcp any host 128.13.23.9 eq
   ftp
#access-list acl_in permit tcp any host 128.13.23.9 eq
netbios-ssn
```

Each of these entries controls packet traffic in a unique manner. For example, the topmost entry says in effect that all telnet packets (i.e., packets sent in connection with the telnet service that allows one system to connect to another) are allowed to go outbound from the network in which PIX is placed, regardless of where they originated and where they are being sent. The second and third rules say that all other outbound traffic based on the tcp (transmission control protocol) and udp (user datagram protocol) protocols is blocked, again independently of the source or destination. The fourth rule says that all inbound ftp (file transfer protocol) packets bound for IP address 128.13.23.9 are allowed. The fifth and final rule says that all inbound NetBIOS packets, packets often used in connection with Windows network sessions, bound for the same system are also allowed.

In this case, one might expect that a very experienced firewall administrator would readily understand each of these rules, although this might not be true for a less experienced firewall administrator. Even if these assumptions are true, however, this might not make as much difference when it comes to making errors in firewall configurations as one might expect. Wool (2004) has conducted studies that have showed that firewall administrators often make errors in firewall rules that leave internal networks exposed to attacks that well-configured firewalls would block. Wool asserts that these errors are due to the fact that firewall interfaces deal with directionality of packets,

Keys	Validity	Trust	Size	Description
Ann Campi <acampi@nai.com>			2048/1024	DH/DSS public k
Anne Hutton <ahutton@lbl.gov>			1024	RSA public key
Bill Blanke <wjb@pgp.com>			4096/1024	DH/DSS public k
Bill Marlow <bill@marlow1.com>			2048/1024	DH/DSS public k
Chanda Groom <chanda_groom@nai.c...			2048/1024	DH/DSS public k
Christopher Jay Manders <CJManders...			2048/1024	DH/DSS public k
Clinton W. Kreitner <clintkreitner@aol....			2048/1024	DH/DSS public k
Damon Gallaty <dgal@pgp.com>			3072/1024	DH/DSS public k
David Sockol <davidsockol@emagined....			2048/1024	DH/DSS public k
Dennis Szerszen <szerszen@bellsouth...			2048/1024	DH/DSS public k

Figure 2 PGP public key list display.

that is, whether they are inbound or outbound, differently from how firewall administrators think about traffic flow through firewalls. Worse yet, some vendors fail to provide explanations of directionality in documentation provided with firewalls. This incongruity, according to Wool, results in poor usability that leaves firewall administrators confused and error-prone when they configure firewalls.

The fact that the order of rules within an access control list (ACL) is extremely important in determining exactly how the rules work is another critical human factors consideration, as is the likelihood that the ACL will be extremely long, sometimes as much as 6000 (or even more) entries. Furthermore, several firewalls, many versions of PIX included, do not allow firewall administrators to edit the ACLs directly. Instead, they must add new rules that affect the existing ACL list. This makes the job of obtaining exactly the right rule set in the correct order even more difficult.

5.6 Encrypting Messages Sent over the Network

Anyone who expects the content of any message sent across a network or any information sent during a network session to be safe from unauthorized reading is badly deluded. Attackers often use hardware or software to capture the content of all packets going over the network, thereby enabling them to glean not only cleartext passwords, but also potentially valuable information such as credit card numbers. *Encryption*, which means systematically scrambling the content of a cleartext message using a key and then applying a key to unscramble it (Schneier, 1998), provides a potentially very strong type of protection against unauthorized reading or possession of messages and network session content.

Despite the many advantages of encryption, the use of encryption by everyday users is rare, once again because of associated usability problems. Whitten and Tygar (1999) showed how deficient the usability design of a well-known encryption program, PGP (Pretty Good Privacy), is. PGP can be used to encrypt the content of messages and files sent across the network. The fact that free versions of this software exist and that PGP can run on a variety of operating systems makes it a good candidate for widespread use. To use this tool to send an encrypted message to another user, users of Windows systems must double-click on an icon (a gray PGP padlock) on the desktop and choose a menu selection called Encrypt to encrypt the contents of a message (or possibly the contents of the clipboard). Once the user decides which to encrypt, the screen shown in Figure 2 appears.

If the user wants to encrypt a message sent to someone such as someone in Figure 1, the user must first scroll to that user's name and then confirm that the person has the same type of encryption (such as DH/DSS public key encryption) that the originator of the message has. Very few users understand enough about encryption to make this decision. The user must next double-click on the other user's name, making the name and key information appear in a window below the list of names. If the user clicks only once, the error dialog box shown in Figure 3 appears.

The error message provides no meaningful feedback to users. Once the user double-clicks on the name of another user, an additional set of nonintuitive choices appears related to whether the user wants a Secure Viewer and/or Conventional Encryption. If the user checks both of the radio buttons representing these choices, another error message that provides no meaningful feedback appears. If the user manages to click on the correct radio buttons, the message content looks like the one shown in Figure 4. The user can (finally) send the message. The fact that even experienced PGP users often have trouble using PGP and that adding, deleting, and generating keys needed for encryption

Figure 3 PGP error dialog box.

```
-----BEGIN PGP MESSAGE-----
Version: PGPfreeware 6.5.8 for non-commercial use <http://www.pgp.com>

qANQR1DBwE4DH9zEKh/Jxd0QA/0WR5pJGWQ8Nao5mp88JroWivxGe8/P4ruqBZqk
HfHCrp/G5Z5kEFB4OEQTkqcGhX5/7/tNnfPuEiTj83DrYh7y6tEkMjF6Wjqt0p6I
RcyEMpHqVeuarfPaYk1hIoB26g8p8eqt2XURXnoYw/Kp0nj/PLEO1eYMQlkb4jht
xczSIAP/cABA8Vz3ggjeDeKQnNMz8UojVhxsa8H2X6a0aWLIn480/eDK5DySEVG
Qf0cE/anygdtlkrCYmcXcGPTopqvPhuI5BR2hLCXQG1KU5W/6z5Wjmavl5hDfueM
1Jm0895IGy7qCeySIKtBGCqGdAi4jsKh9I7NgGxzFfEu8zDmbr7JwLDllf1MC8mr
YrYzODfrG5SO4x9ZkPnDyE2hmYdbgMSgTZBacubwy2PgGJqPCgwv58+tnJAF0c4h
UYTp2DR7tBZNdP78FV3uM7BCUEVYV6RhJE6zauyjd3nlBp4kTNZxEN2MO6Hn//j1
Qoakp2xCa1pRGeBNh5talRI1Xf3mLFjHeUSIdNx3IgPsBE5PcX7bO+yrKUlLeMCD
xNpFuEpi+qqQKYG1lAMNstX5bSUw9rpaQmwk/8M2nv9FIttuRLXGpEgSghCojRZQ
QLj5KT9jLVXxeZ1sBlntC6hDvXMrWKoDxVrVfp+9EoxXLUIFzBJXk++CUPhfGIrw
1ndU1NSiS95BTlIZ5Lwa70XMwZ+Q1ZdEHTYpPogXkYRjCC/3R6v6j/kV8iBlnCK5
bpnlXI5ghzWA5gP6pBDAMr4MJ4WRHjpJDQIAFcgJh3713xdy3oeA9BNe4E3kxLSg
9paoEes1gClA9s8oMgLlqwEnySriAZnmxA==
=GJxi
-----END PGP MESSAGE-----
```


Figure 4 PGP-encrypted message.

and decryption involve additional long and conceptually different interaction sequences attests further to the usability problems inherent in the use of PGP.

Not all encryption is as difficult to use as is PGP. Users of Windows 2000 and XP systems can, for example, encrypt the contents of any file of which they are the owner by performing the following interaction steps:

1. Bring up the Windows Explorer.
2. Sight the icon for the to-be-encrypted file.
3. Right-click on the icon to Properties.
4. Click on Advanced.
5. Click on Encrypt Contents to Secure Data.
6. Click on Apply.
7. Click on Encrypt the File Only.
8. Click on OK twice.

At this point all might seem well as far as the user goes, but there is a serious problem of which most users are unlikely to be aware. Although the file will be encrypted transparently whenever the user closes it and it will also be decrypted transparently whenever the user opens it, if something happens to the user's key—if it should be deleted or corrupted—the contents of the file will be unrecoverable because they cannot be decrypted. As a precaution, the user (or the system administrator acting on behalf of the user) needs to engage in additional tasks related to making an "escrow" key for file recovery purposes. Doing so is not a trivial task from a human–computer interaction standpoint, another case in point for the conclusion that information security and usability requirements are often opposed to each other.

5.7 Electronic Commerce Transactions

Electronic commerce transactions require secrecy, integrity, and nonrepudibility more than anything else. Many ways of achieving secrecy exist, but secrecy alone is not enough in many such transactions. Several corporations created a special protocol, the Secure Electronic Transaction (SET) protocol, to address all three needs at once by encrypting all traffic generated in connection with transactions, using strong user authentication, confirming credit card numbers, and approving each transaction. SET does not merely encrypt network traffic, however; it also keeps personal information obtained from merchants as well as the specific types of purchases made from financial institutions that process the transactions.

Despite the inherent goodness of SET from an information security perspective, SET's usability design weaknesses have made it an extremely unpopular protocol with those who use it. As Schultz (in press) points out, to start a SET transaction, each customer must request and then fill in entries in an electronic wallet or digital certificate, a kind of electronic credit card that contains information about a customer and that customer's credentials. The issuing institution (normally, a bank or credit card company) provides copies of certificates issued to third-party merchants. These certificates contain the public keys* of both the merchant and the issuing institution. The customer initiates a transaction, causing the customer's Web browser to receive and validate the merchant's certificate. The browser uses the merchant's public key to encrypt a message related to the transaction and the issuing institution's public key to encrypt the payment information. This information, as well as information that uniquely links payment to this particular transaction, is sent to the issuing institution and the merchant.

*A public key is one half of a public–private key pair in which encryption of data is performed using one key, and decryption is performed using the other. This type of encryption is often called *public key encryption.*

The merchant confirms the identity of the customer by verifying the customer's digital signature* contained within the customer's certificate. Next, the merchant transmits the order message to the issuing institution. The order message contains the issuing institution's public key, customer payment information, and the merchant's certificate. The issuing institution confirms the identity of the merchant and verifies the message itself. The issuing institution verifies the payment portion of the message and then digitally signs and sends authorization back to the merchant, who can then supply the goods or services specified in the customer's order.

If you are confused at this point, you can readily understand how those who use SET often feel. Many of the major steps in a SET transaction can be broken down into multiple individual user interaction tasks. Many of the steps involving the merchant and issuing institution are automated, however, so it is the customer who is faced with the majority of the human–computer interaction tasks, many of which are not very intuitive. SET shows once again that information security and usability design are very often orthogonal to each other.

5.8 Auditing and Logging

Auditing and logging in operating systems enables system administrators to examine what users and applications have done, something that can lead to corrective action such as disabling accounts of malicious users as well as modifying information security policy. Accessing audit logs is not generally very difficult in most operating systems. In Unix and Linux, for example, the root user (the superuser) needs only to enter the who and last commands to discover who is currently logged in and the log-in and log-out times of each user, respectively. A major exception is in Novell NetWare. To view Netware audit reports for volume auditing, the system administrator must enter AUDITCON and then select the Change current server to choose the appropriate service. From the AUDITCON main menu, the system administrator must select Change current volume to designate the volume of interestand, then select Auditor volume log-in from the AUDITCON main menu. The Enter volume password input box will appear. The system administrator must enter the auditing password for the chosen volume and then press <ENTER> and then select "Auditing reports," a selection within the Available audit options menu that will be displayed. The difficulty of performing these tasks is self-explanatory.

The generally more difficult part from a human–computer interaction standpoint is configuring logging. Unix and Linux system logging (syslog) is controlled by configuring a file, /etc/syslog.conf. There are eight priorities of logging: emerg (highest), alert, crit, err, warning, notice, info, debug (lowest). There are also seven types of logable messages, each concerning a different part of or function within the system: kernel, user, mail, daemon, auth, lpr, and local. syslog messages can be sent to one or more of the following: the system console, a central log server, and/or to a file within the system in which syslog has been enabled.

The entries in an /etc/syslog.conf file in a Unix system are as follows:

```
*.crit;kern.debug;auth.info    /dev/console
*.alert;user.notice            root
auth.debug                     /var/adm/authlog
mail.notice                    /var/adm/maillog
```

The first line specifies that any type of event that has a priority of critical or higher† will be sent to the terminal (console) on which the event occurred. Any kernel-related event with a priority of debug or higher as well as any authentication-related event with a priority of information or higher will also be sent to the terminal. The second line entries cause any event with a priority of alert or higher to be sent to the root (superuser) account in the system in which this event has occurred; additionally, any user-related event with a priority of notice or higher will also be sent to root. The third line specifies that any authentication-related event will be sent to a local file, /var/adm/authlog. The fourth line causes all mail-related events with a priority of notice or higher to be sent to the mail log, /var/adm/maillog. Regardless of any specific entry, there is nothing in the format of the /etc/syslog.conf file that makes configuring system logging straightforward.

The built-in graphical user interface in Windows 2000, XP, and Server 2003 makes configuring Security Logging (Auditing) in these systems somewhat more intuitive. Nevertheless, the number of user interaction steps involves is still excessive. On a Windows XP Professional workstation, for example, one must:

1. Go from Start to Control Panel.
2. Double-click on Administrative Tools.
3. Double-click on Local Security Policy.
4. Enumerate the Security Settings container.
5. Enumerate the Local Policies container.
6. Click on Audit Policy.
7. For each setting or change desired, double-click on the name of the audit category (Audit account logon events, Audit policy changes, Audit privilege use, and so on).
8. Click on Success and/or Failure.
9. Click on Apply.
10. Click on OK.

*A *digital signature* is a cryptographic method used to uniquely identify each person using public key encryption as well as other cryptographic methods. Digital signatures help protect against repudiation.

†This is not necessarily true, however. In some flavors of Unix, choosing a priority of debug results in only debug-related events being logged.

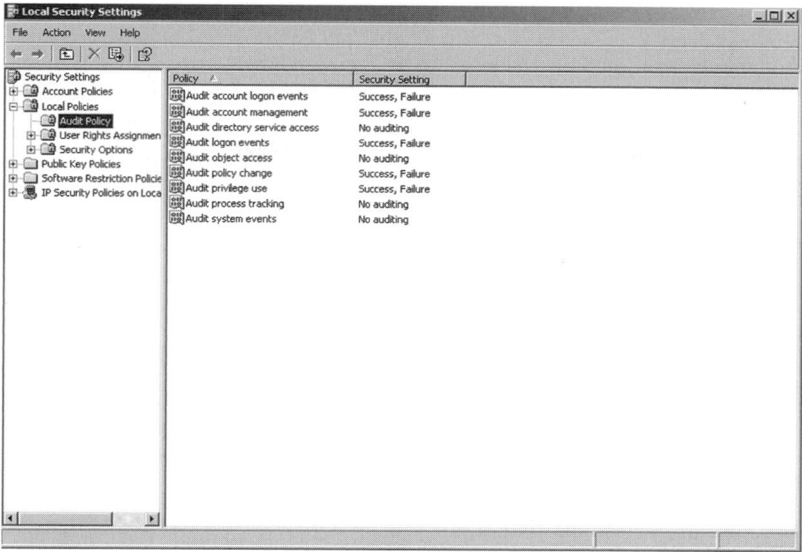

Figure 5 Audit Policy screen in Windows XP.

11. Repeat steps 7 to 10 for each additional audit category.

Figure 5 shows the Audit Policy configuration screen. Even though configuring Security Logging in Windows systems is more intuitive than in Unix and Linux, the number of interaction steps, especially when multiple audit categories must be selected, is fairly large.

5.9 Intrusion Detection Monitoring

Intrusion detection means identifying security breaches that occur. Intrusion detection has become an important component of most organizations' information security program. Intrusion detection enables technical staff to identify and respond readily to incidents, thereby minimizing the amount of financial and other types of loss (Endorf et al., 2004).

Intrusion detection systems (IDSs) automate the process of detecting intrusions, thereby increasing proficiency and reducing the number of personnel needed.

Although most IDSs are not all that difficult to configure, reading the output of these systems can be quite a challenge. Consider, for example, the following output from Snort, the most widely used IDS today (Caswell and Foster, 2003):

```
[**] SCAN-SYN FIN [**]
11/02-16:01:36.792199 109.10.0.1:21 ->
  16.16.90.1:21
TCP TTL:24 TOS: 0x0 ID:39426
**SF**** Seq: 0x27896E4 Ack: 0xB35C4BD Win:
  0x404
```

Even a proficient technical person would have difficulty understanding what this output means without careful study of Snort documentation. Following is another example of Snort output:

```
[**] [1:1959:1] RPC portmap request NFS UDP [**]
[Classification: Decode of an RPC Query] [Priority:
  2]
08/14-04:12:43.991442 109.10.0.1:46637 ->
16.16.90.1:111
UDP TTL:250 TOS:0x0 ID:38580 IpLen:20
  DgmLen:84 DF
Len: 56
```

Snort is only one of many IDSs. Bro (see ftp://ftp.ee.lbl.gov/.vp-bro-pub-0.7a90.tar.gz), another IDS, also nicely illustrates the problem of poor usability design in information security with output such as the following:

```
Nov 16 03:08:39 AddressDropped dropping address
spock.bcc.com.pl (ftp)
Nov 16 03:15:01 WeirdActivity 218.73.102.106/1039
> nsx/dns: repeated_SYN_with_ack
Nov 16 03:31:23 AddressDropped low port trolling
a213-22-132-227.netcabo.pt 258/tcp
Nov 16 04:50:44 AddressDropped dropping address
12.31.179.246 (4000/tcp)
Nov 16 06:25:23 SensitivePortmapperAccess rpc:
cs4/917 > guacamole.cchem.berkeley.edu/portmap
pm_dump: (done)
Nov 16 06:30:48 SensitivePortmapperAccess rpc:
jackal.icir.org/1721 > arg/portmap pm_dump: (nil)
Nov 16 06:30:49 AddressScan 66.243.211.244 has
scanned 10000 hosts (445/tcp)
```

```
Nov 16 06:30:50 PortScan 218.204.91.85 has
scanned 50 ports of siblys.dhcp
Nov 16 06:30:50 AddressDropped dropping address
216.101.181.5 (4000/tcp)
Nov 16 06:30:50 SensitiveConnection hot: neutrino
200b > 147.8.137.149/telnet 463b 14.2s "root"
Nov 16 06:30:50 WeirdActivity p508c7fc5.dip.t-
dialin.net -> 131.243.3.162:
excessively_large_fragment
Nov 16 06:30:50 SensitiveConnection hot:
p508d918a.dip.t-dialin.net 0b }2 muaddib/IRC ?b
0.6s inbound IRC
Nov 16 06:30:50 OutboundTFTP outbound TFTP:
sip000d28083467.dhcp -> inoc-dba.pch.net
Nov 16 06:30:52 SensitiveConnection hot:
198.128.27.21 560b > 208.254.3.160/https 4202b
0.5s <IRC source sites>
Nov 16 06:30:53 WormPhoneHome worm
  phone-home
signature mcr-88-4 -> 218.146.108.51/9900
```

The output of Bro almost seems to be designed to be confusing to everyone but people who thoroughly understand how the system works. Bro's user interface is not at all atypical of today's IDSs.

6 IMPLICATIONS

Failure to pay sufficient attention to effective usability design in the information security arena has caused a plethora of dire consequences, the most apparent of which is failure to implement measures needed to defend systems and networks against attack. If too much effort, confusion, error, and/or frustration results from attempting to engage in security-related tasks, users will simply refrain from engaging in these tasks or will perform them inadequately. Consequently, stronger forms of authentication than password-based authentication will not be implemented, firewalls will not be inadequately configured, file system permissions will allow too much access by too many users, sensitive data and passwords will traverse networks in cleartext, operating system patches will never get installed, auditing will not be enabled or will be configured inadequately, intrusion detection data will be ignored, and so on.

Why does the world of information security repeatedly turn its proverbial back on the principles of effective human–computer interaction? The "Sherman M51 Tank" analogy may help in understanding what may be happening. Over a half century ago the Sherman M51 tank represented a major advance in warfare from the standpoint of the firepower it delivered, but poor human factors design greatly hampered tank crews' ability to operate it. The weaponry-related advantages apparently outweighed the human factors–related disadvantages, as judged by the many M51 tanks that were built and deployed. In information security the same kinds of trade-offs apply. When one considers the advantages of third-party authentication, such as smart card–based authentication compared to normal password-based authentication, information security professionals will readily endorse third-party authentication. At the same time, however, users and system administrators (and, in particular, managers) may not understand just how advantageous smart card–based authentication is compared to the usability disadvantages, leading them to favor conventional (password-based) authentication.

Additional negative consequences of failing to consider usability design also need to be considered. Vendors of products designed to improve security are in general not exactly reaping record profits. Vendors could considerably boost their sales by redesigning the user interfaces to their products.

Perhaps most important in the information security arena, however, is the potential relevance of the usability problems documented in this chapter to the insider threat. Security breaches instigated by insiders—employees, contractors, and consultants—account for far more financial and other forms of loss than any other source (Schultz, 2002b). User resistance to interaction tasks with poor usability design is well documented. This resistance surfaces in a variety of ways, including passive behavior, verbal behavior, hesitance to continue in interacting with computers, loss of attention to tasks, and many others (Martinko et al., 1996). A study by the Information Security Forum (http://www.securityforum.org) in 2000 shows that inadequate security behavior of staff members rather than poor security measures per se account for as much as 80% of all security-related loss. Furthermore, even when staff members realize that security controls have been put in place with sound justification, they quickly reject controls that are ineffective, inefficient, and ambiguous (Leach, 2003). Poor usability design thus appears to be closely linked to insider attacks, internal misuse, and insider error that result in massive losses. Although the link between usability design and insider-related loss is currently indirect, empirical studies on these issues will in time provide more definitive results.

7 SOLUTIONS

Applying well-accepted principles of usability design in the information security arena is the most obvious solution to the problems presented in this chapter. Table 1 lists the problems presented in this chapter and possible usability solutions for each. A good high-level approach to an effective solution is to assume that most people who have security needs are not very aware of exactly what these needs are and what must be done to meet them. First and foremost, operating systems and applications need to have more secure settings right out of the box. Vendors typically use default settings that cause the least disruption to users rather than providing settings that raise security to at least an acceptable minimum. The unfortunate result is higher susceptibility to attacks. Allowing options in user interfaces that set security to a desired level without requiring that users know all the individual settings and what they mean is an excellent solution. A good example of how this can be done is the security options for the Microsoft Internet Explorer (IE), a widely used Web browser, as shown in Figure 6.

Table 1 Usability Problems in Information Security and Possible Solutions

Task	Usability Problems	Possible Solutions
Password entry	Stronger passwords require more effort to create, are more difficult to remember.	Use an alternative form of authentication that does not require users to create and remember authentication credentials.
Third-party authentication	Task involves excessive number of user interaction steps.	Design more efficient interaction sequences.
Setting file access controls	The number and difficulty of user interaction steps are often overwhelming to users.	Design more efficient and simpler interaction sequences.
Configuring Web server security parameters	Syntax for changing parameters is often nonintuitive; interaction with menus may involve an excessive number of levels.	Syntax should be made more intuitive; menus should have greater breadth, not depth.
Configuring firewall security parameters	Syntax for changing parameters is often nonintuitive.	Syntax should be made more intuitive.
Message encryption	User interaction may involve nonintuitive steps; may involve an excessive number of steps.	Design more efficient and simpler user interaction steps.
Electronic commerce transactions	User interaction may involve nonintuitive steps; number of steps may be excessive; error messages may be confusing.	Design more efficient and simpler user interaction steps; error messages should be made clearer and more informative.
Configuring auditing and logging	Syntax for setting auditing and logging parameters is often nonintuitive; number of steps may be excessive.	Syntax should be made more intuitive; design more efficient user interaction steps.
Reading intrusion detection output	Output may be cryptic and poorly formatted.	Output should be easier to interpret and should be formatted in a manner that facilitates recognition of important data.

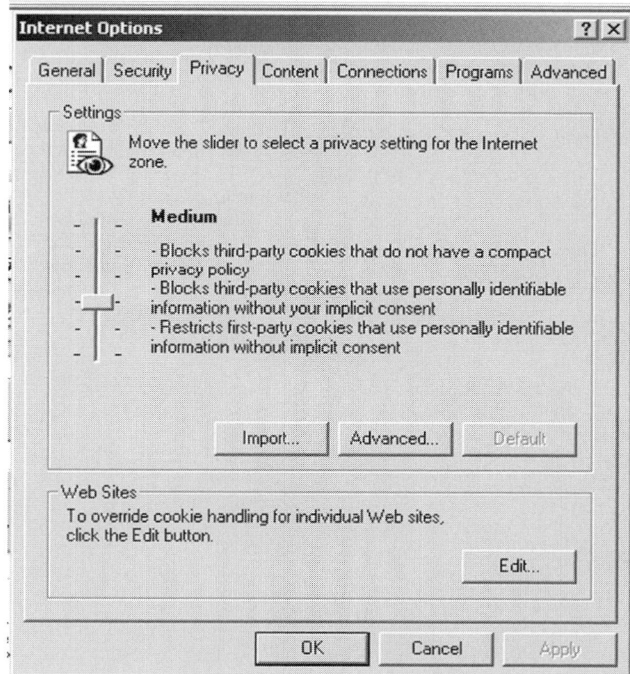

Figure 6 Dialog box for setting privacy level in Windows Internet Explorer.

A slide bar allows users to choose privacy levels varying from high to medium to low, thereby precluding the need to navigate to and choose settings from an excessive number of screens. In Figure 6 the user has chosen a level of security that is slightly below medium. The result is that third-party cookies (objects used to keep information about users in Web transactions) from Web sites that do not have a defined privacy policy will be blocked. The IE user interface in this example also is conducive to explorability—users can choose privacy levels without being locked in to a particular choice. Users may not fully understand the choices they explore and/or choose, but they will be better off than the way things are with the current user interfaces in the information security arena.

REFERENCES

Al-Ghatani, S. S., and King, M. (1999), "Attitudes, Satisfaction and Usage: Factors Contributing to Each in the Acceptance of Information Technology," *Behaviour and Information Technology*, Vol. 18, pp. 277–297.

Bishop, M., and Klein, D. V. (1995), "Improving System Security via Proactive Password Checking," *Computers and Security*, Vol. 14, pp. 233–249.

Caswell, B., and Foster, J. C. (2003), *Snort 2.0 Intrusion Detection*, Syngress, Rockland, MA.

CERT/CC (Computer Emergency Response Team Coordination Center) (2004), "2004 E-Crime Watch Survey shows significant increase in electronic crimes," http://www.cert.org/about/ecrime.html.

Cheswick, B., Bellovin, S., and Rubin, A. (2003), *Firewalls and Internet Security: Repelling the Wily Hacker*, 2nd ed., Addison-Wesley, Reading, MA.

Corcoran, D. (2000), "Security-Related Exposures and Solutions in Smartcards," *Information Security Bulletin*, Vol. 5, No. 9, pp. 13–22.

Endorf, C., Schultz, E. E., and Mellander, J. (2004), *Intrusion Detection and Prevention*, McGraw-Hill, New York.

Garfinkle, S., Spafford, G., and Schwartz, A. (2003), *Practical Unix and Internet Security*, 3rd ed., O'Reilly & Associates, Sebastopol, CA.

Leach, J. (2003), "Improving User Security Behavior," *Computers and Security*, Vol. 22, No. 8, pp. 685–692.

Markus, M. L. (1983), "Power, Politics, and MIS Implementation," *Communications of the ACM*, Vol. 26, pp. 430–444.

Martinko, M. J., Henry, J. W., and Zmud, R. W. (1996), "An Attributional Explanation of Individual Resistance to the Introduction of Information Technologies in the Workplace," *Behaviour and Information Technology*, Vol. 15, pp. 313–330.

Proctor, R. W., Lien, M. C., Salvendy, G., and Schultz, E. E. (2000), "A Task Analysis of Usability in Third-Party Authentication," *Information Security Bulletin*, Vol. 5, pp. 49–56.

Proctor, R. W., Lien, M. C., Vu, K.-P. L, Schultz, E. E., and Salvendy, G. (2002), "Improving Computer Security for Authentication of Users: Influence of Proactive Password Restrictions," *Behavior Research Methods, Instruments and Computers*, Vol. 34, pp. 163–169.

Schneier, B. (1998), *Applied Cryptography*, 2nd ed., Wiley, New York.

Schultz, E. E. (in press), "Web Security and Privacy," in *Human Factors in Web Design*, R. W. Proctor and K.-P. L Vu, Eds., Lawrence Erlbaum Associates, Mahwah, NJ.

Schultz, E. E. (2002a), "Guidelines for Securing Apache Web Servers," *Network Security*, December, pp. 8–14.

Schultz, E. E. (2002b), "A Framework for Understanding and Predicting Insider Attacks," *Computers and Security*, Vol. 21, No. 6, pp. 526–531.

Schultz, E. E. and Curran, P. S. (1986), "Menu Structure and Ordering of Menu Selections: Independent or Interactive Effects?" *SIGCHI Bulletin*, Vol. 18, pp. 69–71.

Schultz, E. E. and Shumway, R. (2001), *Incident Response: A Strategic Guide for Handling Security Incidents*, New Riders, Indianapolis, IN.

Schultz, E. E., Proctor, R. W., Lien, M. C., and Salvendy, G. (2001), "Usability and Security: An Appraisal of Usability Issues in Information Security Methods," *Computers and Security*, Vol. 20, pp. 620–634.

Skoudis, E. (2004), *Malware: Fighting Malicious Code*, Prentice Hall, Upper Saddle River, NJ.

Symantec Corporation (2003), "W32.Blaster.Worm," retrieved August 29, 2003, from http://securityresponse.symantec.com/avcenter/venc/data/w32.blaster.worm.html.

Turnage, J. J. (1990), "The Challenge of New Workplace Technology for Psychology," *American Psychologist*, Vol. 45, pp. 171–178.

Whitten, A., and Tygar, J. D. (1999), "Why Johnny Can't Encrypt: A Usability Evaluation of PGP 5.0, in *Proceedings of the 8th USENIX Security Symposium*, Usenix Association, Berkeley, CA.

Wool, A. (2004), "A Quantitative Study of Firewall Configuration Errors," *IEEE Computer*, Vol. 37, No. 6, pp. 62–67.

CHAPTER 49

USABILITY TESTING

James R. Lewis
International Business Machines Corporation Software Group
Boca Raton, Florida

1 INTRODUCTION

Usability testing is an essential skill for usability practitioners—professionals whose primary goal is to provide guidance to product developers for the purpose of improving the ease of use of their products. It is by no means the *only* skill with which usability practitioners must have proficiency, but it is an important one. A recent survey of experienced usability practitioners (Vredenburg et al., 2002) indicated that usability testing is a very frequently used method, second only to the use of iterative design.

One goal of this chapter is to provide an introduction to the practice of usability testing. This includes some discussion of the concept of usability and the history of usability testing, various goals of usability testing, and running usability tests. A second goal is to cover more advanced topics, such as sample size estimation for usability tests, computation of confidence intervals, and the use of standardized usability questionnaires.

2 THE BASICS

2.1 What Is Usability?

The term *usability* came into general use in the early 1980s. Related terms from that time were *user friendliness* and *ease of use*, which *usability* has since displaced in professional and technical writing on the topic (Bevan et al., 1991). The earliest publication (of which I am aware) to include the word *usability* in its title was Bennett (1979).

It is the nature of language that words come into use with fluid definitions. Ten years after the first use of the term *usability*, Shackel (1990) wrote, "one of the most important issues is that there is, as yet, no generally agreed definition of usability and its measurement" (p. 31). As recently as 1998, Gray and Salzman stated: "Attempts to derive a clear and crisp definition of usability can be aptly compared to attempts to nail a blob of Jell-O to the wall" (p. 242).

There are several reasons why it has been so difficult to define usability. Usability is not a property of a person or thing. There is no thermometer-like instrument that can provide an absolute measurement of the usability of a product (Dumas, 2003). Usability is an emergent property that depends on the interactions among users, products, tasks, and environments.

Introducing a theme that will reappear in several parts of this chapter, there are two major conceptions of usability. These dual conceptions have contributed to the difficulty of achieving a single agreed-upon definition. One conception is that the primary focus of usability should be on measurements related to the accomplishment of global task goals (summative, or measurement-based evaluation). The other conception is that practitioners should focus on the detection and elimination of usability problems (formative, or diagnostic evaluation).

The first conception has led to a variety of similar definitions of usability, some embodied in current standards (which, to date, have emphasized summative evaluation). For example:

- "The current MUSiC definition of usability is: the ease of use and acceptability of a system or product for a particular class of users carrying out specific tasks in a specific environment; where 'ease of use' affects user performance and satisfaction, and 'acceptability' affects whether or not the product is used" (Bevan et al., 1991, p. 652).
- Usability is the "extent to which a product can be used by specified users to achieve specified goals with effectiveness, efficiency and satisfaction in a specified context of use" (ISO, 1998, p. 2; ANSI, 2001, p. 3).
- "To be useful, usability has to be specific. It must refer to particular tasks, particular environments and particular users" (Alty, 1992, p. 105).

One of the earliest formative definitions of usability (ease of use) is from Chapanis (1981, p. 3):

> Although it is not easy to measure "ease of use," it is easy to measure difficulties that people have in using something. Difficulties and errors can be identified, classified, counted, and measured. So my premise is that ease of use is inversely proportional to the number and severity of difficulties people have in using software. There are, of course, other measures that have been used to assess ease of use, but I think the weight of the evidence will support the conclusion that these other dependent measures are correlated with the number and severity of difficulties.

Practitioners in industrial settings generally use both conceptualizations of usability during iterative design. Any iterative method must include a stopping rule to prevent infinite iterations. In the real world, resource constraints and deadlines can dictate the stopping rule (although this rule is valid only if there is a reasonable expectation that undiscovered problems will not lead to drastic consequences). In an ideal setting, the first conception of usability can act as a stopping rule for the second. Setting aside, for now, the question of where quantitative goals come from, the goals associated with the first conception of usability can define when to stop the iterative process of the discovery and resolution of usability problems. This combination is not a new concept. In one of the earliest published descriptions of iterative design, Al-Awar et al. (1981, p. 31) wrote: "Our methodology is strictly empirical. You write a program, test it on the target population, find out what's wrong with it, and revise it. The cycle of test–rewrite is repeated over and over until a satisfactory level of performance is reached. Revisions are based on the performance, that is, the difficulties typical users have in going through the program."

2.2 What Is Usability Testing?

Imagine the two following scenarios.

Scenario 1 Mr. Smith is sitting next to Mr. Jones, watching him work with a high-fidelity prototype of a Web browser for personal digital assistants (PDAs). Mr. Jones is the third person that Mr. Smith has watched performing these tasks with this version of the prototype. Mr. Smith is not constantly reminding Mr. Jones to talk while he works, but is counting on his proximity to Mr. Jones to encourage verbal expressions when Mr. Jones encounters any difficulty in accomplishing his current task. Mr. Smith takes written notes whenever this happens, and also takes notes whenever he observes Mr. Jones faltering in his use of the application (e.g., exploring menus in search of a desired function). Later that day he will use his notes to develop problem reports and, in consultation with the development team, will work on recommendations for product changes that should eliminate or reduce the impact of the reported problems. When a new version of the prototype is ready, he will resume testing.

Scenario 2 Dr. White is watching Mr. Adams work with a new version of a word-processing application. Mr. Adams is working alone in a test cell that looks almost exactly like an office, except for the large mirror on one wall and the two video cameras overhead. He has access to a telephone and a number to call if he encounters a difficulty that he cannot overcome. If he places such a call, Dr. White will answer and provide help modeled on the types of help provided at the company's call centers. Dr. White can see Mr. Adams through the one-way glass as she coordinates the test. She has one assistant working the video cameras for maximum effectiveness and another who is taking time-stamped notes on a computer (coordinated with the video time stamps) as different members of the team notice and describe different aspects of Mr. Adams's task performance. Software monitors Mr. Adams's computer, recording all keystrokes and mouse movements. Later that day, Dr. White and her associates will put together a summary of the task performance measurements for the tested version of the application, noting where the performance measurements do not meet the test criteria. They will also create a prioritized list of problems and recommendations, along with video clips that illustrate key problems, for presentation to the development team at their weekly status meeting.

Both of these scenarios provide examples of usability testing. In scenario 1 the emphasis is completely on usability problem discovery and resolution (formative, or diagnostic evaluation). In scenario 2 the primary emphasis is on task performance measurement (summative, or measurement-focused evaluation), but there is also an effort to record and present usability problems to the product developers. Dr. White's team knows that they cannot determine if they've met the usability performance goals by examining a list of problems, but they also know that they cannot provide appropriate guidance to product development if they present only a list of global task measurements. The problems observed in the use of an application provide

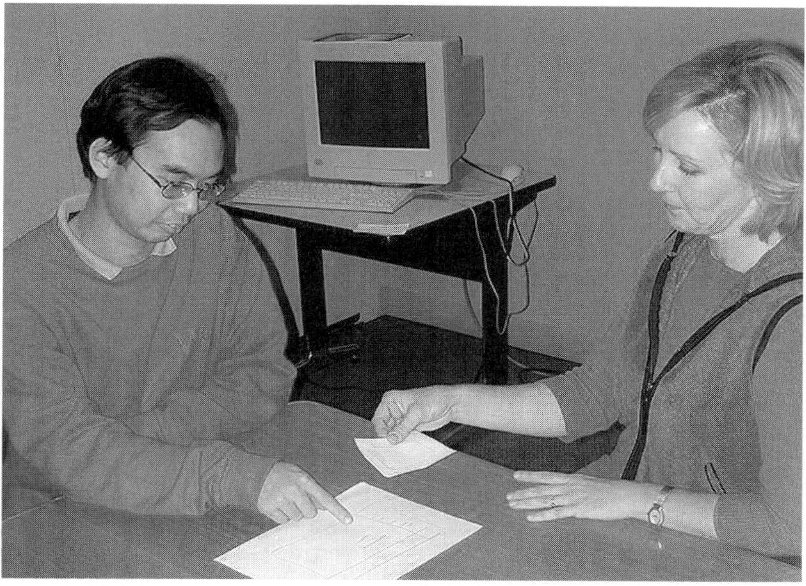

Figure 1 Practitioner and participant engaging in an informal usability test with a pencil-and-paper prototype. (Courtesy of IBM.)

important clues for redesigning the product (Chapanis, 1981; Norman, 1983). Furthermore, as J. Karat (1997, p. 693) observed: "The identification of usability problems in a prototype user interface (UI) is not the end goal of any evaluation. The end goal is a redesigned system that meets the usability objectives set for the system such that users are able to achieve their goals and are satisfied with the product."

These scenarios also illustrate the defining properties of a usability test. During a usability test, one or more observers watch one or more participants perform specified tasks with the product in a specified test environment (compare this with the ISO/ANSI definition of usability presented earlier in this chapter). This is what makes usability testing different from other user-centered design (UCD) methods. In interviews (including the group interview known as a focus group), participants do not perform worklike tasks. Usability inspection methods (such as expert evaluations and heuristic evaluations) also do not include the observation of users or potential users performing worklike tasks. The same is true of techniques such as surveys and card sorting. Field studies (including contextual inquiry) can involve the observation of users performing work-related tasks in target environments but restrict the control that practitioners have over the target tasks and environments. Note that this is not necessarily a bad thing, but it is a defining difference between usability testing and field (ethnographic) studies.

This definition of usability testing permits a wide range of variation in technique (Wildman, 1995). Usability tests can be very informal (as in scenario 1) or very formal (as in scenario 2). The observer might sit next to the participant, watch through a one-way glass, or watch the on-screen behavior of a participant who is performing specified tasks at a location halfway around the world. Usability tests can be think-aloud (TA) tests, in which observers train participants to talk about what they're doing at each step of task completion and prompt participants to continue talking if they stop. Observers might watch one participant at a time, or might watch participants work in pairs. Practitioners can apply usability testing to the evaluation of low-fidelity prototypes (see Figure 1), Wizard of Oz (WOZ) prototypes (Kelley, 1985), high-fidelity prototypes, products under development, predecessor products, or competitive products.

2.2.1 Where Did Usability Testing Come From?

The roots of usability testing lie firmly in the experimental methods of psychology (in particular, cognitive and applied psychology) and human factors engineering, and are strongly tied to the concept of iterative design. In a traditional experiment, the experimenter draws up a careful plan of study that includes the exact number of participants that the experimenter will expose to the different experimental treatments. The participants are members of the population to which the experimenter wants to generalize the results. The experimenter provides instructions and debriefs the participant, but at no time during a traditional experimental session does the experimenter interact with the participant (unless this interaction is part of the experimental treatment). The more formative (diagnostic, focused on problem discovery) the focus of a usability test, the less it is like a traditional experiment (although the requirements for sampling from a legitimate population of users, tasks, and environments still apply). Conversely, the more summative (focused on measurement) a usability test is, the more it should resemble the mechanics of a traditional experiment. Many of the principles of psychological experimentation that exist to protect experimenters from threats to reliability and validity (e.g., the control of demand characteristics) carry over into usability testing (Wenger and Spyridakis, 1989; Holleran, 1991).

As far as I can tell, the earliest accounts of iterative usability testing applied to product design came from Alphonse Chapanis and his students (Al-Awar et al., 1981; Chapanis, 1981; Kelley, 1984) and had an almost immediate influence on product development practices at IBM (Kennedy, 1982; J. Lewis, 1982) and other companies, notably Xerox (D. Smith et al., 1982) and Apple (Williams, 1983). Shortly thereafter, John Gould and his associates at the IBM T. J. Watson Research Center began publishing influential papers on usability testing and iterative design (Gould and Boies, 1983; Gould and C. Lewis, 1984; Gould et al., 1987; Gould, 1988).

The driving force that separated iterative usability testing from the standard protocols of experimental psychology was the need to modify early product designs as rapidly as possible (as opposed to the scientific goal of developing and testing competing theoretical hypotheses). As Al-Awar et al. (1981, p. 33) reported: "Although this procedure [iterative usability test, redesign, and retest] may seem unsystematic and unstructured, our experience has been that there is a surprising amount of consistency in what subjects report. Difficulties are not random or whimsical. They do form patterns."

When difficulties of use become apparent during the early stages of iterative design, it is hard to justify continuing to ask test participants to perform the test tasks. There are ethical concerns with intentionally frustrating participants who are using a product with known flaws that the design team can and will correct. There are economic concerns with the time wasted by watching participants who are encountering and recovering from known error-producing situations. Furthermore, any delay in updating the product delays the potential discovery of problems associated with the update or problems whose discovery was blocked by the presence of the known flaws. For these reasons, the earlier you are in the design cycle, the more rapidly you should iterate the cycles of test and design.

2.2.2 Is Usability Testing Effective?

The widespread use of usability testing is evidence that practitioners believe that usability testing is effective. Unfortunately, there are fields in which practitioners' belief in the effectiveness of their methods does not appear to be warranted by those outside the field (e.g., the use of projective techniques such as the Rorschach test in psychotherapy) (Lilienfeld et al., 2000). In our own field, a number of recently published papers have questioned the reliability of usability problem discovery (Molich et al., 1998, 2004; Kessner et al., 2001).

The common finding in these studies has been that observers (either individually or in teams across usability laboratories) who evaluated the same product produced markedly different sets of discovered problems. Molich et al. (1998) had four independent usability laboratories carry out inexpensive usability tests of a software application for new users. The four teams reported 141 different problems, with only one problem common among all four teams. Molich et al. (1998) attributed this inconsistency to variability in the approaches taken by the teams (task scenarios, level of problem reporting). Kessner et al. (2001) had six professional usability teams independently test an early prototype of a dialog box. None of the problems were detected by every team, and 18 problems were described by one team only. Molich et al. (2004) assessed the consistency of usability testing across nine independent organizations that evaluated the same Web site. They documented considerable variability in methodologies, resources applied, and problems reported. The total number of reported problems was 310, with only two problems reported by six or more organizations, and 232 problems uniquely reported. "Our main conclusion is that our simple assumption that we are all doing the same and getting the same results in a usability test is plainly wrong" (Molich et al., 2004, p. 65).

This is important and disturbing research, but there is a clear need for much more research in this area. A particularly important goal of future research should be to reconcile these studies with the documented reality of usability improvement achieved through iterative application of usability testing. For example, a limitation of research that stops with the comparison of problem lists is that it is not possible to assess the magnitude of the usability improvement (if any) that would result from product redesigns based on design recommendations derived from the problem lists (Wixon, 2003). When comparing problem lists from many labs, one aberrant set of results can have an extreme effect on measurements of consistency across labs, and the more labs that are involved, the more likely this is to happen.

The results of these studies (Kessner et al., 2001; Molich et al., 1998, 2004) stand in stark contrast to the published studies in which iterative usability tests (sometimes in combination with other UCD methods) have led to significantly improved products (Al-Awar et al., 1981; Kennedy, 1982; J. Lewis, 1982, 1996b; Kelley, 1984; Gould et al., 1987; Bailey et al., 1992; Bailey, 1993; Ruthford and Ramey, 2000). For example, in a paper describing their experiences in product development, Marshall et al. (1990, p. 243) stated: "Human factors work can be reliable—different human factors engineers, using different human factors techniques at different stages of a product's development, identified many of the same potential usability defects." Published cost–benefit analyses (Bias and Mayhew, 1994) have demonstrated the value of usability engineering processes that include usability testing, with cost–benefit ratios ranging from 1 : 2 for smaller projects to 1 : 100 for larger projects (C. Karat, 1997).

Most of the papers that describe the success of iterative usability testing are case studies (such as Marshall et al., 1990), but a few have described designed experiments. R. Bailey et al. (1992) compared two user interfaces derived from the same base interface: one modified via heuristic evaluation and the other modified via iterative usability testing (three iterations, five participants per iteration). They conducted this experiment with two interfaces, one character-based and the other a graphical user interface (GUI), with the same basic outcomes. The number of changes indicated by usability testing was much smaller than the number indicated by heuristic evaluation, but user performance was the same with both final versions of the interface. All designs after the first iteration produced faster performance than, and for the character-based interface were preferred to, the original design. The time to complete the performance testing was about the same as that required for the completion of multireviewer heuristic evaluations.

G. Bailey (1993) provided additional experimental evidence that iterative design based on usability tests leads to measurable improvements in the usability of an application. In the experiment, he studied the designs of eight designers, four with at least four years of professional experience in interface design and four with at least five years of professional experience in computer programming. All designers used a prototyping tool to create a recipes application (eight applications in all). In the first wave of testing, Bailey videotaped participants performing tasks with the prototypes, three different participants per prototype. Each designer reviewed the videotapes of the people using his or her prototype, and used the observations to redesign his or her application. This process continued until each designer indicated that it was not possible to improve his or her application. All designers stopped after three to five iterations. Comparison of the first and last iterations indicated significant improvement in measurements such as number of tasks completed, task completion times, and repeated serious errors.

In conclusion, the results of the studies of Molich et al. (1998, 2004) and similar studies show that usability practitioners must conduct their usability tests as carefully as possible, document their methods completely, and show proper caution when interpreting their results. On the other hand, as Landauer stated (1997, p. 204): "There is ample evidence that expanded task analysis and formative evaluation can, and almost always do, bring substantial improvements in the effectiveness and desirability of systems." This is echoed by Desurvire et al. (1992, p. 98): "It is generally agreed that usability testing in both field and laboratory is far and above the best method for acquiring data on usability."

2.3 Goals of Usability Testing

The fundamental goal of usability testing is to help developers produce more usable products. The two conceptions of usability testing (formative and summative) lead to differences in the specification of goals in much the same way that they contribute to differences in fundamental definitions of usability (diagnostic problem discovery and measurement). Rubin (1994, p. 26) expressed the formative goal as follows: "The overall goal of usability testing is to identify and rectify usability deficiencies existing in computer-based and electronic equipment and their accompanying support materials prior to release." Dumas and Redish (1999, p. 11) provided a more summative goal: "A key component of usability engineering is setting specific, quantitative, usability goals for the product early in the process and then designing to meet those goals."

These goals are not in direct conflict, but they do suggest different focuses that can lead to differences in practice. For example, a focus on measurement typically leads to more formal testing (less interaction between observers and participants), whereas a focus on problem discovery typically leads to less formal testing (more interaction between observers and participants). In addition to the distinction between diagnostic problem discovery and measurement tests, there are two common types of measurement tests: comparison against objectives and comparison of products.

2.3.1 Problem Discovery Test

The primary activity in diagnostic problem discovery tests is the discovery, prioritization, and resolution of usability problems. The number of participants in each iteration of testing should be fairly small, but the overall test plan should be for multiple iterations, each with some variation in participants and tasks. When the focus is on problem discovery and resolution, the assumption is that more global measures of user performance and satisfaction will take care of themselves (Chapanis, 1981). The measurements associated with problem-discovery tests are focused on prioritizing problems and include frequency of occurrence in the test, likelihood of occurrence during normal usage (taking into account the anticipated usage of the part of the product in which the problem occurred), and magnitude of impact on the participants who experienced the problem. Because the focus is not on precise measurement of the performance or attitudes of participants, problem-discovery studies tend to be informal, with a considerable amount of interaction between observers and participants. Some typical stopping rules for iterations are a preplanned number of iterations or a specific problem-discovery goal, such as "Identify 90% of the problems available for discovery for these types of participants, this set of tasks, and these conditions of use." See the section below on sample size estimation and adequacy for more detailed information on setting and using these types of problem-discovery objectives.

2.3.2 Measurement Test Type I: Comparison against Quantitative Objectives

Studies that have a primary focus of comparison against quantitative objectives include two fundamental activities. The first is the development of the usability objectives. The second is iterative testing to determine if the product under test has met the objectives. A third activity (which can take place during iterative testing) is the enumeration and description of usability problems, but this activity is secondary to the collection of precise measurements.

The first step in developing quantitative usability objectives is to determine the appropriate variables to measure. As part of the work done for the European MUSiC (Measuring the Usability of Systems in Context) project, Rengger (1991), produced a list of potential usability measurements based on 87 papers out of a survey of 500 papers. He excluded purely diagnostic studies and also excluded papers if they did not provide measurements for the combined performance of a user and a system. He categorized the measurements into four classes:

- *Class 1*: goal achievement indicators (such as success rate and accuracy)
- *Class 2*: work rate indicators (such as speed and efficiency)
- *Class 3*: operability indicators (such as error rate and function usage)
- *Class 4*: knowledge acquisition indicators (such as learnability and learning rate)

In a later discussion of the MUSiC measures, Macleod et al. (1997) described measures of effectiveness (the level of correctness and completeness of goal achievement in context) and efficiency (effectiveness related to cost of performance, typically the effectiveness measure divided by task completion time). Optional measures were of productive time and unproductive time, with unproductive time consisting of help actions, search actions, and snag (negation, canceled, or rejected) actions.

Their (Macleod et al., 1997) description of the measures of effectiveness and efficiency seem to have influenced the objectives expressed in ISO 9241-11 (ISO, 1998, p. iv): "The objective of designing and evaluating visual display terminals for usability is to enable users to achieve goals and meet needs in a particular context of use. ISO 9241-11 explains the benefits of measuring usability in terms of user performance and satisfaction. These are measured by the extent to which the intended goals of use are achieved, the resources that have to be expended to achieve the intended goals, and the extent to which the user finds the use of the product acceptable."

In practice [and as recommended by ANSI (2001)], the fundamental global measurements for usability tasks are successful task completion rates (for a measure of effectiveness), mean task completion times (for a measure of efficiency), and mean participant satisfaction ratings (either collected on a task-by-task basis or at the end of a test session; see Section 3.3 for more information on measuring participant satisfaction). There are many other measurements that practitioners could consider (Nielsen, 1997; Dumas and Redish, 1999), including but not limited to (1) the number of tasks completed within a specified time limit, (2) the number of wrong menu choices, (3) the number of user errors, and (4) the number of repeated errors (same user committing the same error more than once).

After determining the appropriate measurements, the next step is to set the goals. Ideally, the goals should have an objective basis and shared acceptance across the various stakeholders, such as marketing, development, and test groups (J. Lewis, 1982). The best objective basis for measurement goals is data from previous usability studies of predecessor or competitive products. For maximum generalizability, the historical data should come from studies of similar types of participants completing the same tasks under the same conditions (Chapanis, 1988). If this information is not available, an alternative is for the test designer to recommend objective goals and to negotiate with the other stakeholders to arrive at a set of shared goals.

"Defining usability objectives (and standards) isn't easy, especially when you're beginning a usability program. However, you're not restricted to the first objective you set. The important thing is to establish some specific objectives immediately, so that you can measure improvement. If the objectives turn out to be unrealistic or inappropriate, you can revise them" (Rosenbaum, 1989, p. 211). Such revisions, however, should take place only in the early stages of gaining experience and taking initial measurements with a product. It is important not to change reasonable goals to accommodate an unusable product.

When setting usability goals, it is usually better to set goals that make reference to an average (mean) of a measurement than to a percentile. For example, set an objective such as "The mean time to complete task 1 will be less than 5 minutes" rather than "95% of participants will complete task 1 in less than 10 minutes." The statistical reason for this is that sample means drawn from a continuous distribution are less variable than sample medians (the 50th percentile of a sample), and measurements made away from the center of a distribution (e.g., measurements made to attempt to characterize the value of the 95th percentile) are even more variable (Blalock, 1972). Cordes (1993) conducted a Monte Carlo study comparing means and medians as measurements of central tendency for time-on-task scores, and determined that the mean should be the preferred metric for usability studies (unless there is missing data due to participants failing to complete tasks, in which case the mean from the study will underestimate the population mean).

A practical reason to avoid percentile goals is that the goal can imply a sample size requirement that is unnecessarily large. For example, you can't measure accurately at the 95th percentile unless there are at least 20 measurements (in fact, there must be many more than 20 measurements for accurate measurement). For more details, see Section 3.1.

An exception to this is the specification of successful task completions (or any other measurement that is based on counting events), which necessarily requires a percentile goal, usually set at or near 100% (unless there are historical data that indicate an acceptable lower level for a specific test). If 10 out of 10 participants complete a task successfully, the observed completion rate is 100%, but a 90% exact binomial confidence interval for this result ranges from 74 to 100%. In other words, even perfect performance for 10 participants with this type of measure leaves open the possibility (with 90% confidence) that the true completion rate could be as low as 75%. See Section 3.2.2 for more information on computing and using this information in usability tests.

After the usability goals have been established, the next step is to collect data to determine if the product has met its goals. Representative participants perform the target tasks in the specified environment as test observers record the target measurements and identify, to the extent possible within the constraints of a more formal testing protocol, details about any usability problems that occur. The usability team conducting the test provides information about goal achievement and prioritized problems to the development team, and a decision is made regarding whether or not there is sufficient evidence that the product has met its objectives. The ideal stopping rule for measurement-based iterations is to continue testing until the product has met its goals.

When there are only a few goals, it is reasonable to expect to achieve all of them. When there are many goals (e.g., five objectives per task multiplied by 10 tasks, for a total of 50 objectives), it is more difficult to determine when to declare success and to stop testing. Thus, it is sometimes necessary to specify a meta-objective of the percentage of goals to achieve.

Despite the reluctance of some usability practitioners to conduct statistical tests to quantitatively assess the strength of the available evidence regarding whether or not a product has achieved a particular goal, the best practice is to conduct such tests. The best approach is to conduct multiple t-tests or nonparametric analogs of t-tests (J. Lewis, 1993) because this gives practitioners the level of detail that they require. There is a well-known prohibition against doing this because it can lead investigators to mistakenly accept as real that some differences that are due to chance [technically, alpha (α) inflation]. On the other hand, if this is the required level of information, it is an appropriate method (Abelson, 1995). Furthermore, the practice of avoiding alpha inflation is a concern more related to scientific hypothesis testing than to usability testing (Wickens, 1998), although usability practitioners should be aware of its existence and take it into account when interpreting their statistical results. For example, if you compare two products by conducting 50 t-tests with alpha set to 0.10, and only five (10%) of the t-tests are significant (have a p value below 0.10), you should question whether or not to use those results as evidence of the superiority of one product over the other. On the other hand, if substantially more than five of the t-tests are significant, you can be more confident that the differences indicated are real.

In addition to (or as an alternative to) conducting multiple t-tests, practitioners should compute confidence intervals for their measurements. This applies to the measurements made for the purpose of establishing test criteria (such as measurements made on predecessor versions of the target product or competitive products) and to the measurements made when testing the product under development. See Section 3.2 for more details.

2.3.3 Measurement Test Type II: Comparison of Products

The second type of measurement test is to conduct usability tests for the purpose of direct comparison of one product with another. As long as there is only one measurement that decision makers plan to consider, a standard t-test (ideally, in combination with the computation of confidence intervals) will suffice for the purpose of determining which product is superior.

If decision makers care about multiple dependent measures, standard multivariate statistical procedures (such as MANOVA or discriminant analysis) are not often helpful in guiding a decision about which of two products has superior usability. The statistical reason for this is that multivariate statistical procedures depend on the computation of centroids (a weighted average of multiple dependent measures) using a least-squares linear model that maximizes the difference between the centroids of the two products (Cliff, 1987). If the directions of the measurements are inconsistent (e.g., a high task completion rate is desirable, but a high mean task completion time is not), the resulting centroids are uninterpretable for the purpose of usability comparison. In some cases it is possible to recompute variables so they have consistent directions (e.g., recomputing task completion rates as task failure rates). If this is not possible, another approach is to convert measurements to ranks (J. Lewis, 1991a) or standardized (Z) scores (Jeff Sauro, personal communication, March 1, 2004) for the purpose of principled combination of different types of measurements.

To help consumers compare the usability of different products, the American National Standards Institute (ANSI) has published the Common Industry Format (CIF) for usability test reports (ANSI, 2001). Originally developed at the National Institute of Standards and Technology (NIST), this test format requires measurement of effectiveness (accuracy and completeness—completion rates, errors, assists), efficiency (resources expended in relation to accuracy and completeness—task completion time), and satisfaction (freedom from discomfort, positive attitude toward use of the product—using any of a number of standardized satisfaction questionnaires). It also requires a complete description of participants and tasks.

Morse (2000) reviewed the NIST IUSR project conducted to pilot test the CIF. The purpose of the CIF is to make it easier for purchasers to compare the usability of different products. The pilot study ran into problems, such as inability to find a suitable software product for both supplier and consumer, reluctance to share information, and uncertainty about how to design a good usability study. To date, there has been little if any use (at least, no published use) of the CIF for its intended purpose.

2.4 Variations on a Theme: Other Types of Usability Tests

2.4.1 Think Aloud

In a standard, formal usability test, test participants perform tasks without necessarily speaking as they work. The defining characteristic of a think-aloud (TA) study is the instruction to participants to talk about what they are doing as they do it (in other words, to produce verbal reports). If participants stop talking (as commonly happens when they become very engaged in a task), they are prompted to resume talking.

The most common theoretical justification for the use of TA is from the work in cognitive psychology (specifically, human problem solving) of Ericsson and Simon (1980). Responding to a review by Nisbett and Wilson (1977) that described various ways in which verbal reports were unreliable, Ericsson and Simon provided evidence that certain kinds of verbal reports could produce reliable data. They stated that reliable verbalizations are those that participants produce during task performance that do not require additional cognitive processing beyond the processing required for task performance and verbalization.

Some discussions of usability testing hold that the best practice in usability testing is to use the TA method in all usability testing. For example, Dumas (2003) encouraged the use of TA because (1) TA tests are more productive for finding usability problems (Virzi et al., 1993), and (2) thinking aloud does not affect user ratings or performance (Bowers and Snyder, 1990). As the references indicate, there is some evidence in support of these statements, but the evidence is mixed.

Earlier prohibitions against the use of TA in measurement-based tests assumed that thinking aloud would cause slower task performance. Bowers and Snyder (1990), however, found no measurable task performance or preference differences between a test group that thought aloud and one that didn't. Surprisingly, there are some experiments in which the investigators reported better task performance when participants were thinking aloud. Berry and Broadbent (1990) provided evidence that the process of thinking aloud invoked cognitive processes that improved rather than degraded performance, but only if people were given (1) verbal instructions on how to perform the task, and (2) the requirement to justify each action aloud. Wright and Converse (1992) compared silent with TA usability testing protocols. The results indicated that the think-aloud group committed fewer errors and completed tasks faster than the silent group, and the difference between the groups increased as a function of task difficulty.

Regarding the theoretical justification for and typical practice of TA, Boren and Ramey (2000) noted that TA practice in usability testing often does not conform to the theoretical basis most often cited for it (Ericsson and Simon, 1980). "If practitioners do not uniformly apply the same techniques in conducing thinking-aloud protocols, it becomes difficult to compare results between studies" (Boren and Ramey, 2000, p. 261). In a review of publications of TA tests and field observations of practitioners running TA tests, they reported inconsistency in explanations to participants about how to think aloud, practice periods, styles of reminding participants to think aloud, prompting intervals, and styles of intervention. They suggest that rather than basing current practice on Ericsson and Simon, a better basis would be speech communication theory, with clearly defined communicative roles for the participant (in the role of domain expert or valued customer, making the participant the primary speaker) and the usability practitioner (the learner or listener, thus a secondary speaker).

Based on this alternative perspective for the justification of TA, Boren and Ramey (2000) have provided guidance for many situations that are not relevant in a cognitive psychology experiment, but are in usability tests. For example, they recommend that usability practitioners running a TA test should continually use acknowledgment tokens that do not take speakership away from the participant, such as "mm hm?" and "uh-huh?" (with the interrogative intonation) to encourage the participant to keep talking. In normal communication, silence (as recommended by the Ericsson and Simon protocols) is not a nonresponse—the speaker interprets it in a primarily negative way as indicating aloofness or condescension. They avoided providing precise statements about how frequently to provide acknowledgments or somewhat more explicit reminders (such as "And now...?") because the best cues come from the participants. Practitioners need to be sensitive to these cues as they run the test.

The evidence indicates that relative to silent participation, TA can affect task performance. If the primary purpose of the test is problem discovery, TA appears to have advantages over completely silent task completion. If the primary purpose of the test is task performance measurement, the use of TA is somewhat more complicated. As long as all the tasks in the planned comparisons were completed under the same conditions, performance comparisons should be legitimate. The use of TA almost certainly prevents generalization of task performance outside the TA task, but there are many other factors that make it difficult to generalize specific task performance data collected in usability studies.

For example, Cordes (2001) demonstrated that participants assume that the tasks they are asked to perform in usability tests are possible (the "I know it can be done or you wouldn't have asked me to do it" bias). Manipulations that bring this assumption into doubt can have a strong effect on quantitative usability performance measures, such as increasing the percentage of participants who give up on a task. If uncontrolled, this bias makes performance measures from usability studies unlikely to be representative of real-world performance when users are uncertain as to whether the product they are using can support the desired tasks.

2.4.2 Multiple Simultaneous Participants

Another way to encourage participants to talk during task completion is to have them work together (Wildman, 1995). This strategy is similar to TA in its strengths and limitations. Hackman and Biers (1992) compared three think-aloud methods: thinking aloud alone (Single), thinking aloud in the presence of an observer (Observer), and verbalizations occurring in a two-person team (Team). They found no significant differences in performance or subjective measures. The Team condition produced more statements of value to designers than the other two conditions, but this was probably due to the differing number of participants producing statements in the different conditions. There were three groups, with 10 participants per group for Single and Observer, and 20 participants (10 two-person teams) for the Team condition. "The major result was that the team gave significantly more verbalizations of high value to designers and spent more time making high value comments. Although this can be reduced to the fact that the team spoke more overall and that there are two people talking rather than one, this finding is not trivial" (Hackman and Biers, 1992, p. 1208).

2.4.3 Remote Evaluation

Recent advances in the technology of collaborative software have made it easier to conduct remote software tests (tests in which the usability practitioner and the test participant are in different locations). This can be an economical alternative to bringing one or more users into a lab for face-to-face user testing. A participant in a remote location can view the contents of the practitioner's screen, and in a typical system the practitioner can decide whether the participant can control the desktop. System performance is typically slower than that of a local test session.

Some of the advantages of remote testing are (1) access to participants who would otherwise be unable to participate (international, special needs, etc.), (2) the capability for participants to work in familiar surroundings, and (3) no need for either party to install or download additional software. Some of the disadvantages are (1) potential uncontrolled disruptions in the participant's workplace, (2) lack of visual feedback from the participant, and (3) the possibility of compromised security if the participant takes screen captures of confidential material. Despite these disadvantages, McFadden et al. (2002) reported data that indicated that remote testing was effective at improving product designs and that the test results were comparable to the results obtained with more traditional testing.

2.5 Usability Laboratories

A typical usability laboratory test suite is a set of soundproofed rooms with a participant area and observer area separated by a one-way glass and with video cameras and microphones to capture the user experience (Marshall et al., 1990; Nielsen, 1997), possibly with an executive viewing area behind the primary observers' area. The advantages of this type of usability facility are quick setup, a place where designers can see people interacting with their products, videos to provide a historical record and backup for observers, and a professional appearance that raises awareness of usability and reassures customers about commitment to usability. In a survey of usability laboratories, Nielsen (1994) reported a median floor space of 63 m^2 (678 ft^2) for the observer room and 13 m^2 (144 ft^2) for test rooms. This type of laboratory (see Figure 2) is especially important if practitioners plan to conduct formal, summative usability tests.

If the practitioner focus is on formative, diagnostic problem discovery, this type of laboratory is not essential (although it is still convenient). "It is possible to convert a regular office temporarily into a usability laboratory, and it is possible to perform usability testing with no more equipment than a notepad" (Nielsen, 1997, p. 1561). Making an even stronger statement against the perceived requirement for formal laboratories, Landauer (1997, p. 204) stated: "Many usability practitioners have demanded greater resources and more elaborate procedures than are strictly needed for effective guidance—such as expensive usability labs rather than natural settings for test and observations, time consuming videotaping and analysis where observation and note-taking would serve as well, and large groups of participants to achieve statistical significance when qualitative naturalistic observation of task goals and situations, or of disastrous interface or functionality flaws, would be more to the point."

2.6 Test Roles

There are several ways to categorize the roles that testers need to play in the preparation and execution of a usability test (Rubin, 1994; Dumas and Redish, 1999). Most test teams will not have a person assigned to each role, and most tests (especially informal problem discovery tests) do not require every role. The actual distribution of skills across a team might vary from these roles, but the standard roles help to organize the skills needed for effective usability testing.

2.6.1 Test Administrator

The test administrator is the usability test team leader. He or she designs the usability study, including the specification of the initial conditions for a test session and the codes to use for data logging. The test administrator's duties include conducting reviews with the rest of the test team, leading in the analysis of data, and putting together the final presentation or report. People in this role should have a solid understanding of the basics of usability engineering, ability to tolerate ambiguity, flexibility (knowing when to deviate from the plan), and good communication skills.

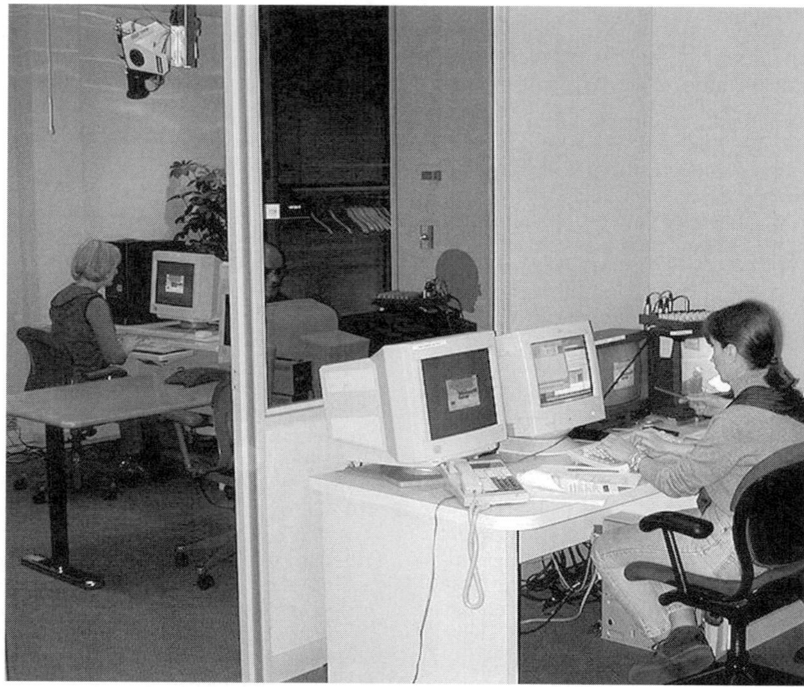

Figure 2 Usability laboratory. (Courtesy of IBM.)

2.6.2 Briefer

The briefer is the person who interacts with the participants (briefing them at the start of the test, communicating with them as required during the test, and debriefing them at the end of the test sessions). On many teams, the same person takes the roles of administrator and briefer. In a think-aloud study, the briefer has the responsibility to keep the participant talking. The briefer needs to have sufficient familiarity with the product to be able to decide what to tell participants when they ask questions. People in this role need to be comfortable interacting with people, and need to be able to restrict their interactions to those that are consistent with the purposes of the test without any negative treatment of the participants.

2.6.3 Camera Operator

The camera operator is responsible for running the audiovisual equipment during the test. He or she must be skilled in the setup and operation of the equipment and must be able to take directions quickly when it is necessary to change the focus of the camera (e.g., from the keyboard to the user manual).

2.6.4 Data Recorder

The video record is useful as a data backup when things start happening quickly during the test and as a source for video examples when documenting usability problems. The primary data source for a usability study, however, is the notes that the data recorder takes during a test session. There just isn't time to take notes from a more leisurely examination of the video record. Also, the camera doesn't necessarily catch the important action at every moment of a usability study.

For informal studies, the equipment used to record data might be nothing more than a notepad and pencil. Alternatively, the data recorder might use data-logging software to take coded notes (often time-stamped, possibly synchronized with the video). Before the test begins, the data recorder needs to prepare the data-logging software with the category codes defined by the test administrator. Taking notes with data-logging software is a very demanding skill, so the test administrator does not usually assign additional tasks to the person taking this role.

2.6.5 Help Desk Operator

The help desk operator takes calls from the participant if the user experiences enough difficulty to place the call. The operator should have some familiarity with the call-center procedures followed by the company that has designed the product under test and must also have skills similar to those of the briefer.

2.6.6 Product Expert

The product expert maintains the product and offers technical guidance during the test. The product expert must have sufficient knowledge of the product to recover quickly from product failures and to help the other team members understand the system's actions during the test.

2.6.7 Statistician

A statistician has expertise in measurement and the statistical analysis of data. Practitioners with an educational background in experimental psychology typically have sufficient expertise to take the role of statistician for a usability test team. Informal tests rarely require the services of a statistician, but the team needs a statistician to extract the maximum amount of information from the data gathered during a formal test (especially if the purpose of the formal test was to compare two products using a battery of measurements).

2.7 Planning the Test

One of the first activities that a test administrator must undertake is to develop a test plan. To do this, the administrator must understand the purpose of the product, the parts of the product that are ready for test, the types of people who will use the product, what they are likely to use the product for, and in what settings.

2.7.1 Purpose of the Test

At the highest level, is the primary purpose of the test to identify usability problems or to gather usability measurements? The answer to this question provides guidance as to whether the most appropriate test is formal or informal, think-aloud or silent, problem discovery or quantitative measurement. After addressing this question, the next task is to define any more specific test objectives. For example, an objective for an interactive voice response system (IVR) might be to assess whether participants can accomplish key tasks without encountering significant problems. If data are available from a previous study of a similar IVR, an alternative objective might be to determine whether participants can complete key tasks reliably faster with the new IVR than they did with the previous IVR. Most usability tests will include several objectives.

If a key objective of the test is to compare two products, an important decision is whether the test will be within-subjects or between-subjects. In a within-subjects test, every participant works with both products, with half of the participants using one product first and the other half using the other product first (a technique known as *counterbalancing*). In a between-subjects study, the test groups are completely independent. In general, a within-subjects test leads to more precise measurement of product differences (requiring a smaller number of participants for equal precision, due primarily to the reduction in variability that occurs when each participant acts as his or her own control) and the opportunity to get direct subjective product comparisons from participants. For a within-subjects test to be feasible, both products must be available and set up for use in the lab at the same time, and the amount of time needed to complete tasks with both products must not be excessive. If a within-subjects test is not possible, a between-subjects test is a perfectly valid alternative. Note that the statistical analyses appropriate for these two types of tests are different.

2.7.2 Participants

To determine who will participate in the test, the administrator needs to obtain or develop a user profile. A user profile is sometimes available from the marketing group, the product's functional specification, or other product planning documentation. It is important to keep in mind that the focus of a usability test is the end user of a product, not the expected product purchaser (unless the product will be purchased by end users). The most important participant characteristic is that the participant is representative of the population of end users to whom the administrator wants to generalize the results of the test. Practitioners can obtain participants from employment agencies, internal sources if the participants meet the requirements of the user profile (but avoiding internal test groups), market research firms, existing customers, colleges, newspaper ads, and user groups.

To define representativeness, it is important to specify the characteristics that members of the target population share but are not characteristic of nonmembers. The administrator must do this for the target population at large and any defined subgroups. Within group definition constraints, administrators should seek heterogeneity in the final sample to maximize the generalizability of the results (Chapanis, 1988; Landauer, 1997) and to maximize the likelihood of problem discovery. It is true that performance measurements made with a homogeneous sample will almost always have greater precision than measurements made with a heterogeneous sample, but the cost of that increased precision is limited generalizability. This raises the issue of how to define homogeneity and heterogeneity of participants. After all, at the highest level of categorization, we are all humans, with similar general capabilities and limitations (physical and cognitive). At the other end of the spectrum, we are all individuals—no two alike.

One of the most important defining characteristics for a group in a usability test is specific relevant experience, both with the product and in the domain of interest (work experience, general product experience, specific product experience, experience with the product under test, and experience with similar products). One common categorization scheme is to consider people with less than three months' experience as novices, with more than a year of experience as expert, and those in between as intermediate (Dumas and Redish, 1999). Other individual differences that practitioners routinely track and attempt to vary are education level, age, and gender.

When acquiring participants, how can practitioners define the similarity between the participants they can acquire and the target population? An initial step is to develop a taxonomy of the variables that affect human performance (where performance should include the behaviors of indicating preference and other choice behaviors). Gawron et al. (1989) produced a human performance taxonomy during the development of a

Table 1 Results of Aykin and Aykin (1991) Review of Individual Differences in HCI

Individual Difference	Measurement Method	Crossed Interactions
Level of experience	Various methods	No
Jungian personality types	Myers–Briggs type of indicator	No
Field dependence/independence	Embedded figures test	Yes; field-dependent participants preferred organized sequential item number search mode, but field-independent subjects preferred the less organized keyword search mode (Fowler et al., 1985)
Locus of control	Levenson test	No
Imagery	Individual differences questionnaire	No
Spatial ability	VZ-2	No
Type A/type B personality	Jenkins activity survey	No
Ambiguity tolerance	Ambiguity tolerance scale	No
Gender	Unspecified	No
Age	Unspecified	No
Other (reading speed and comprehension, intelligence, mathematical ability)	Unspecified	No

human performance expert system. They reviewed existing taxonomies and filled in some missing pieces. They structured the taxonomy as having three top levels: environment, subject (person), and task. The resulting taxonomy took up 12 pages in their paper and covered many areas that would normally not concern a usability practitioner working in the field of computer system usability (e.g., ambient vapor pressure, gravity, acceleration). Some of the key human variables in the Gawron et al. (1989) taxonomy that could affect human performance with computer systems are:

- *Physical characteristics*
  - Age
  - Agility
  - Handedness
  - Voice
  - Fatigue
  - Gender
  - Body and body part size
- *Mental state*
  - Attention span
  - Use of drugs (both prescription and illicit)
  - Long-term memory (includes previous experience)
  - Short-term memory
  - Personality traits
  - Work schedule
- *Senses*
  - Auditory acuity
  - Tone perception
  - Tactual
  - Visual accommodation
  - Visual acuity
  - Color perception

These variables can guide practitioners as they attempt to describe how participants and target populations are similar or different. The Gawron et al. (1989) taxonomy, however, does not provide much detail with regard to some individual differences that other researchers have hypothesized to affect human performance or preference with respect to the use of computer systems: personality traits and computer-specific experience.

Aykin and Aykin (1991) performed a comprehensive review of the published studies to that date that involved individual differences in human–computer interaction (HCI). Table 1 lists the individual differences that they found in published HCI studies, the method used to measure the individual difference, and whether there was any indication from the literature that manipulation of that individual difference led to a crossed interaction.

In statistical terminology, an interaction occurs whenever an experimental treatment has a different magnitude of effect depending on the level of a different, independent experimental treatment. A crossed interaction occurs when the magnitudes have different signs, indicating reversed directions of effects. As an example of an uncrossed interaction, consider the effect of turning off the lights on the typing throughput of blind and sighted typists. The performance of the sighted typists would probably be worse, but the presence or absence of light should not affect the performance of the blind typists. As an extreme example of a crossed interaction, consider the effect of language on task completion for people fluent only in French or English. When reading French text, French speakers would outperform English speakers, and vice versa.

For any of these individual differences, the lack of evidence for crossed interactions could be due to a paucity of research involving the individual difference or could reflect the probability that individual differences will not typically cause crossed interactions in

HCI. In general, a change made to support a problem experienced by a person with a particular individual difference will either help other users or simply not affect their performance.

For example, John Black (personal communication, 1988) cited the difficulty that field-dependent users had working with one-line editors at the time (decades ago) when that was the typical user interface to a mainframe computer. Switching to full-screen editing resulted in a performance improvement for both field-dependent and field-independent users—an uncrossed interaction because both types of users improved, with the performance of field-dependent users becoming equal to (thus improving more than) that of field-independent users. Landauer (1997) cites another example of this, in which Greene et al. (1986) found that young people with high scores on logical reasoning tests could master database query languages such as SQL with little training, but older or less able people could hardly ever master these languages. They also determined that an alternative way of forming queries, selecting rows from a truth table, allowed almost everyone to make correct specification of queries, independent of their abilities. Because this redesign improved the performance of less able users without diminishing the performance of the more able, it was an uncrossed interaction. In a more recent study, Palmquist and Kim (2000) found that field dependence affected the search performance of novices using a Web browser (with field-independent users searching more efficiently), but did not affect the performance of more experienced users.

If there is a reason to suspect that an individual difference will lead to a crossed interaction as a function of interface design, it could make sense to invest the time (which can be considerable) to categorize users according to these dimensions. Another situation in which it could make sense to invest the time in categorization by individual difference would be if there were reasons to believe that a change in interface would greatly help one or more groups without adversely affecting other groups. (This is a strategy that one can employ when developing hypotheses about ways to improve user interfaces.) It always makes sense to keep track of user characteristics when categorization is easy (e.g., age or gender). Another potential use of these types of variables is as covariates (used to reduce estimates of variability) in advanced statistical analyses (Cliff, 1987).

Aykin and Aykin (1991) reported effects of users' levels of experience, but did not report any crossed interactions related to this individual difference. They did report that interface differences tended to affect the performance of novices but had little effect on the performance of experts. It appears that behavioral differences related to user interfaces (Aykin and Aykin, 1991) and cognitive style (Palmquist and Kim, 2000) tend to fade with practice. Nonetheless, user experience has been one of the few individual differences to receive considerable attention in HCI research (Fisher, 1991; Mayer, 1997; Miller et al., 1997; B. Smith et al., 1999).

According to Mayer (1997), relative to novices, experts have (1) better knowledge of syntax, (2) an integrated conceptual model of the system, (3) more categories for more types of routines, and (4) higher-level plans.

Fisher (1991) emphasized the importance of discriminating between computer experience (which he placed on a novice-experienced dimension) and domain expertise (which he placed on a naive-expert dimension). LaLomia and Sidowski (1990) reviewed the scales and questionnaires developed to assess computer satisfaction, literacy, and aptitudes. None of the instruments they surveyed specifically addressed measurement of computer experience. Miller et al. (1997) published the Windows Computer Experience Questionnaire (WCEQ), an instrument specifically designed to measure a person's experience with Windows 3.1. The questionnaire took about 5 minutes to complete and was reliable (coefficient $\alpha = 0.74$; test–retest correlation $= 0.97$). They found that their questionnaire was sensitive to three experiential factors: general Windows experience, advanced Windows experience, and instruction. B. Smith et al. (1999) distinguished between subjective and objective computer experience. The paper was relatively theoretical and "challenges researchers to devise a reliable and valid measure" (p. 239) for subjective computer experience, but did not offer one.

One user characteristic not addressed in any of the literature cited is one that becomes very important when designing products for international use: cultural characteristics. For example, in adapting an interface for use by members of another country, it is extremely important that all text be translated accurately. It is also important to be sensitive to the possibility that these types of individual differences might be more likely than others to result in crossed interactions.

For comparison studies, having multiple groups (e.g., males and females or experts and novices) allows the assessment of potential interactions that might otherwise go unnoticed. Ultimately, the decision for one or multiple groups must be based on expert judgment and a few guidelines. For example, practitioners should consider sampling from different groups if they have reason to believe:

- There are potential and important differences among groups on key measures (Dickens, 1987).
- There are potential interactions as a function of group (Aykin and Aykin, 1991).
- The variability of key measures differs as a function of group.
- The cost of sampling differs significantly from group to group.

Gordon and Langmaid (1988) recommended the following approach to defining groups:

1. Write down all the important variables.
2. If necessary, prioritize the list.

3. Design an ideal sample.
4. Apply common sense to collapse cells.

For example, suppose that a practitioner starts with 24 cells, based on the factorial combination of six demographic locations, two levels of experience, and the two levels of gender. The practitioner should ask himself or herself whether there is a high likelihood of learning anything new and important after completing the first few cells, or whether additional testing would be wasteful. Can one learn just as much from having one or a few cells that are homogeneous within cells and heterogeneous between cells with respect to an important variable, but are heterogeneous within cells with regard to other, less important variables? For example, a practitioner might plan to (1) include equal numbers of males and females over and under 40 years of age in each cell, (2) have separate cells for novice and experienced users, and (3) drop intermediate users from the test. The resulting design requires testing only two cells (groups), but a design that did not combine genders and age groups in the cells would have required eight cells.

The final issue is the number of participants to include in the test. According to Dumas and Redish (1999), typical usability tests have 6 to 12 participants divided among two to three subgroups. For any given test, the required sample size depends on the number of subgroups, available resources (time/money), and the purpose of the test (e.g., precise measurement or problem discovery). It also depends on whether a study is single-shot (needing a larger sample size) or iterative (needing a smaller sample size per iteration, building up the total sample size over iterations). For more detailed treatment of this topic, see Section 3.1.

2.7.3 Test Task Scenarios

As with participants, the most important consideration for test tasks is that they are representative of the types of tasks that real users will perform with the product. For any product, there will be a core set of tasks that anyone using the product will perform. People who use barbecue grills use them to cook. People who use desktop speech dictation products use them to produce text. For usability tests, these are the most important tasks to test.

After defining these core tasks, the next step is to list any more peripheral tasks that the test should cover. If a barbecue grill has an external burner for heating pans, it might make sense to include a task that requires participants to work with that burner. If in addition to the basic vocabulary in a speech dictation system, the program allows users to enable additional special topic vocabularies such as cooking or sports, it might make sense to devise a task that requires participants to activate and use one of these topics. Practitioners should avoid frivolous or humorous tasks because what is humorous to one person might be offensive to another.

From the list of test tasks, create scenarios of use (with specific goals) that require participants to perform the identified tasks. Critical tasks can appear in more than one scenario. For repeated tasks, vary the task details to increase the generalizability of the results. When testing relatively complex systems, some scenarios should stay within specific parts of the system (e.g., typing and formatting a document) and others should require the use of different parts of the system (e.g., creating a figure using a spreadsheet program, adding it to the document, attaching the document to a note, and sending it to a specified recipient).

The complete specification of a scenario should include several items. It is important to document (but not to share with the participant) the required initial conditions so it will be easy to determine before a test session starts if the system is ready. The written description of the scenario (presented to the participant) should state what the participant is trying to achieve and why (the motivation), keeping the description of the scenario as short as possible to keep the test session moving quickly. The scenario should end with an instruction for the action the participant should take upon finishing the task (to make it easier to measure task completion times). The descriptions of the scenario's tasks should not typically provide step-by-step instructions on how to complete the task, but should include details (e.g., actual names and data) rather than general statements. The order in which participants complete scenarios should reflect the way in which users would typically work and with the importance of the scenario, with important scenarios done first unless there are other less important scenarios that produce outputs that the important scenario requires as an initial condition. Not all participants need to receive the same scenarios, especially if there are different groups under study. The tasks performed by administrators of a Web system that manages subscriptions will be different from the tasks performed by users who are requesting subscriptions.

Here are some examples of scenarios:

- Frank Smith's business telephone number has changed to (896) 555-1234. Please change the appropriate address book entry so you have this new phone number available when you need it. When you have finished, please say "I'm done."
- You've just found out that you need to cancel a car reservation that you made for next Wednesday. Please call the system that you used to make the reservation (1-888-555-1234) and cancel it. When you have finished, please hang up the phone and say "I'm done."

2.7.4 Procedure

The test plan should include a description of the procedures to follow when conducting a test session. Most test sessions include an introduction, task performance, posttask activities, and debriefing.

A common structure for the introduction is for the briefer (review Section 2.6) to start with the purpose of the test, emphasizing that its goal is to improve the product, not to test the participant. Participation is voluntary, and the participant can stop at any time without penalty. The briefer should inform the participant that all test results will be confidential. The participant should be aware of any planned audio or video recording. Finally, the briefer should provide any special instructions (e.g., think-aloud instructions) and answer any other questions that the participant might have.

The participant should then complete any preliminary questionnaires and forms, such as a background questionnaire, an informed consent form (including consent for any recording, if applicable), and if necessary, a confidential disclosure form. If the participant will be using a workstation, the briefer should help the participant make any necessary adjustments (unless, of course, the purpose of the test is to evaluate workstation adjustability). Finally, the participant should complete any prerequisite training. This can be especially important if the goal of the study is to investigate usability after some period of use (ease of use) rather than immediate usability (ease of learning).

The procedure section should indicate the order in which participants will complete task scenarios. For each participant, start with the first task scenario assigned and complete additional scenarios until the participant finishes (or runs out of time). The procedure section should specify when and how to interact with participants, according to the type of study. This section should also indicate when it is permissible to provide assistance to participants if they encounter difficulties in task performance.

Normally, practitioners should avoid offering assistance unless the participant is visibly distressed. When participants initially request help at a given step in a task, refer them to documentation or other supporting materials if available. If that doesn't help, provide the minimal assistance required to keep the participant moving forward in the task, note the assistance, and score the task as failed. When participants ask questions, try to avoid direct answers, instead turning their attention back to the task and encouraging them to take whatever action seems right at that time. When asking questions of participants, it is important to avoid biasing the participant's response. Try to avoid the use of loaded adjectives and adverbs in questions (Dumas and Redish, 1999). Instead of asking if a task was easy, ask the participant to describe what it was like performing the task. Give a short satisfaction questionnaire (such as the ASQ; see Section 3.3 for details) at the end of each scenario.

After participants have finished the assigned scenarios, it is common to have them complete a final questionnaire, usually a standard questionnaire and any additional items required to cover other test- or product-specific issues. For standardized questionnaires, ISO lists the SUMI (Software Usability Measurement Inventory) (Kirakowski and Corbett, 1993; Kirakowski, 1996) and PSSUQ (Post-Study System Usability Questionnaire) (J. Lewis, 1995, 2002). In addition to the SUMI and PSSUQ, ANSI lists the QUIS (Questionnaire for User Interaction Satisfaction) (Chin et al., 1988) and SUS (System Usability Scale) (Brooke, 1996) as widely used questionnaires. After completing the final questionnaire, the briefer should debrief the participant. Toward the end of debriefing, the briefer should tell the participant that the test session has turned up several opportunities for product improvement (this is almost always true) and thank the participant for his or her contribution to product improvement. Finally, the briefer should discuss any questions that the participant has about the test session, and then take care of any remaining activities, such as completing time cards. If any deception has been employed in the test (which is rare but can happen legitimately when conducting certain types of simulations), the briefer must inform the participant.

2.7.5 Pilot Testing

Practitioners should always plan for a pilot test before running a usability test. A usability test is a designed artifact, and like any other designed artifact needs at least some usability testing to find problems in the test procedures and materials. A common strategy is to have an initial walkthrough with a member of the usability test team or some other convenient participant. After making the appropriate adjustments, the next pilot participant should be a more representative participant. If there are no changes made to the design of the usability test after running this participant, the second pilot participant can become the first real participant (but this is rare). Pilot testing should continue until the test procedures and materials have become stable.

2.7.6 Number of Iterations

It is better to run one usability test than not to run any at all. On the other hand, "usability testing is most powerful and most effective when implemented as part of an iterative product development process" (Rubin, 1994, p. 30). Ideally, usability testing should begin early and occur repeatedly throughout the development cycle. When development cycles are short, it is a common practice to run, at a minimum, exploratory usability tests on prototypes at the beginning of a project, to run a usability test on an early version of the product during the later part of functional testing, and then to run another during system testing. Once the final version of the product is available, some organizations run an additional usability test focused on the measurement of usability performance benchmarks. At this stage of development, it is too late to apply information about any problems discovered during the usability test to the soon-to-be-released version of the product, but the information can be useful as early input to a follow-on product if the organization plans to develop another version of the product.

2.7.7 Ethical Treatment of Test Participants

Usability testing always involves human participants, so usability practitioners must be aware of professional practices in the ethical treatment of test participants. Practitioners with professional education in experimental psychology are usually familiar with the guidelines of the American Psychology

Association (APA; see http://www.apa.org/ethics/), and those with training in human factors engineering are usually familiar with the guidelines of the Human Factors and Ergonomics Society (HFES) (see http://www.hfes.org/About/Code.html). It is particularly important (Dumas, 2003) to be aware of the concepts of informed consent (participants are aware of what will happen during the test, agree to participate, and can leave the test at any time without penalty) and minimal risk (participating in the test does not place participants at any greater risk of harm or discomfort than situations normally encountered in daily life). Most usability tests are consistent with guidelines for informed consent and minimal risk. Only the test administrator should be able to match a participant's name and data, and the names of test participants should be confidential. Anyone interacting with a participant in a usability test has a responsibility to treat the participant with respect.

Usability practitioners rarely use deception in usability tests. One technique in which there is potential use of deception is the Wizard of Oz (WOZ) method (originally, the OZ Paradigm) (Kelley, 1985; see also http://www.musicman.net/oz.html). In a test using the WOZ method, a human (the Wizard) plays the part of the system, remotely controlling what the participant sees happen in response to the participant's manipulations. This method is particularly effective in early tests of speech recognition interactive voice response (IVR) systems because all the Wizard needs is a script and a phone (Sadowski, 2001). Often, there is no compelling reason to deceive participants, so they know that the system they are working with is remotely controlled by another person for the purpose of early evaluation. If there is a compelling need for deception (e.g., to manage the participant's expectations and encourage natural behaviors), this deception must be revealed to the participant during debriefing.

2.8 Reporting Results

There are two broad classes of usability test results, problem reports and quantitative measurements. It is possible for a test report to contain one type exclusively (e.g., the ANSI Common Industry Format has no provision for reporting problems), but most usability test reports will contain both types of results.

2.8.1 Describing Usability Problems

"We broadly define a usability defect as: Anything in the product that prevents a target user from achieving a target task with reasonable effort and within a reasonable time. ... Finding usability problems is relatively easy. However, it is much harder to agree on their importance, their causes and the changes that should be made to eliminate them (the fixes)" (Marshall et al., 1990, p. 245).

The best way to describe usability problems depends on the purpose of the descriptions. For usability practitioners, the goal should be to describe problems in such a way that the description leads logically to one or more potential interventions (recommendations). Ideally, the problem description should also include some indication of the importance of fixing the problem (most often referred to as problem severity). For more scientific investigations, there can be value in higher levels of problem description (Keenan et al., 1999), but developers rarely care about these levels of description. They just want to know what they need to do to make things better while also managing the cost (both monetary and time) of interventions (Gray and Salzman, 1998).

The problem description scheme of C. Lewis and Norman (1986) has both scientific and practical merit because their problem description categories indicate, at least roughly, an appropriate intervention. They stated (p. 413) that "although we do not believe it possible to design systems in which people do not make errors, we do believe that much can be done to minimize the incidence of error, to maximize the discovery of the error, and to make it easier to recover from the error." They separated errors into mistakes (errors due to incorrect intention) and slips (errors due to appropriate intention but incorrect action), further breaking slips down into mode errors (which indicate a need for better feedback or elimination of the mode), capture errors (which indicate a need for better feedback), and description errors (which indicate a need for better design consistency). In one study using this type of problem categorization, Prümper et al. (1992) found that expertise did not affect the raw number of errors made by participants in their study, but experts handled errors much more quickly than novices. The types of errors that experts made were different from those made by novices, with experts' errors occurring primarily at the level of slips rather than mistakes (knowledge errors).

Using an approach similar to that of C. Lewis and Norman (1986), Rasmussen (1986), described three levels of errors: skill-based, rule-based, and knowledge-based. Two relatively new classification schemes are Structured Usability Problem Extraction, or SUPEX (Cockton and Lavery, 1999), and the User Action Framework, or UAF (Andre et al., 2000). The UAF requires a series of decisions, starting with an interaction cycle (planning, physical actions, assessment) based on the work of Norman (1986). Most classifications require four or five decisions, with interrater reliability [as measured with kappa (κ)] highest at the first step ($\kappa = 0.978$) but remaining high through the fourth and fifth steps ($\kappa > 0.7$).

Whether any of these classification schemes will see widespread use by usability practitioners is still unknown. There is considerable pressure on practitioners to produce results and recommendations as quickly as possible. Even if these classification schemes see little use by practitioners, effective problem classification is a very important problem to solve as usability researchers strive to compare and improve usability testing methods.

2.8.2 Crafting Design Recommendations from Problem Descriptions

As indicated by the title of this section, the development of recommendations from problem descriptions is a craft rather than a rote procedure. A well-written problem description will often strongly imply an intervention, but it is also often the case that there might be several ways to attack a problem. It can be helpful for practitioners to discuss problems and potential interventions with the other members of their team and to get input from other stakeholders as necessary (especially, the developers of the product). This is especially important if the practitioner has observed problems but is uncertain as to the appropriate level of description of the problem.

For example, suppose that you have written a problem description about a missing Help button in a software application. This could be a problem with the overall design of the software or might be a problem isolated to one screen. You might be able to determine this by inspecting other screens in the software, but it could be faster to check with one of the developers.

The first recommendations to consider should be for interventions that will have the widest impact on the product. "Global changes affect everything and need to be considered first" (Rubin, 1994, p. 285). After addressing global problems, continue working through the problem list until there is at least one recommendation for each problem. For each problem, start with interventions that would eliminate the problem, then follow, if necessary, with other less drastic (less expensive, more likely to be implemented) interventions that would reduce the severity of the remaining usability problem. When different interventions involve different trade-offs, it is important to communicate this clearly in the recommendations. This approach can lead to two tiers of recommendations: those that will happen for the version of the product currently under development (short-term) and those that will happen for a future version of the product (long-term).

2.8.3 Prioritizing Problems

Because usability tests can reveal more problems than there are resources to address, it is important to have some means for prioritizing problems. There are two approaches to prioritization that have appeared in the usability testing literature: (1) judgment-driven (Virzi, 1992) and (2) data-driven (J. Lewis et al., 1990; Rubin, 1994; Dumas and Redish, 1999). The bases for judgment-driven prioritizations are the ratings of stakeholders in the project (such as usability practitioners and developers). The bases for data-driven prioritizations are the data associated with the problems, such as frequency, impact, ease of correction, and likelihood of usage of the portion of the product that was in use when the problem occurred. Of these, the most common measurements are frequency and impact (sometimes referred to as severity, although, strictly speaking, severity should include the effect of all of the types of data considered for prioritization). In a study of the two approaches to prioritization, Hassenzahl (2000) found a lack of correspondence between data-driven and judgment-driven severity estimates. This suggests that the preferred approach should be data-driven.

The usual method for measuring the frequency of occurrence of a problem is to divide the number of occurrences within participants by the number of participants. A common method (Rubin, 1994; Dumas and Redish, 1999) for assessing the impact of a problem is to assign impact scores according to whether the problem (1) prevents task completion, (2) causes a significant delay or frustration, (3) has a relatively minor effect on task performance, or (4) is a suggestion. This is similar to the scheme of J. Lewis et al. (1990), in which the impact levels were (1) scenario failure or irretrievable data loss (e.g., the participant required assistance to get past the problem or caused the participant to believe the scenario to be properly completed when it wasn't), (2) considerable recovery effort (recovery took more than 1 minute or the participant repeatedly experienced the problem within a scenario), (3) minor recovery effort (the problem occurred only once within a scenario with recovery time at or under 1 minute), or (4) inefficiency (a problem not meeting any of the other criteria).

When considering multiple types of data in a prioritization process, it is necessary to combine the data in some way. A graphical approach is to create a problem grid with frequency on one axis and impact on the other (see Figure 3). High-frequency, high-impact problems would receive treatment before low-frequency, low-impact problems. The relative treatment of high-frequency, low-impact problems and low-frequency, high-impact problems depends on practitioner judgment.

An alternative approach is to combine the data arithmetically. Rubin (1994) described a procedure for combining four levels of impact (using the criteria described above with 4 assigned to the most serious

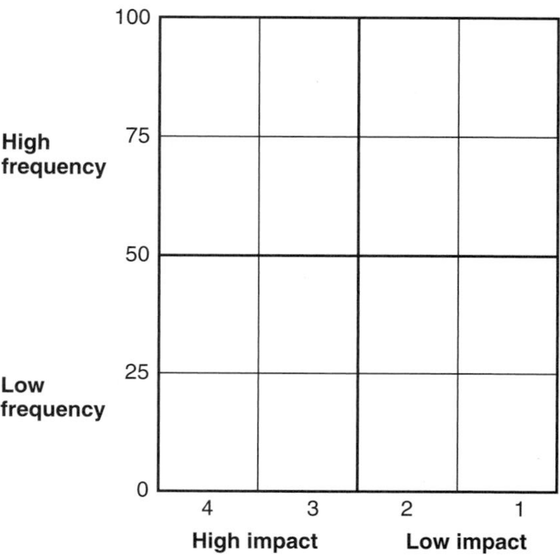

Figure 3 Sample problem grid.

level) with four levels of frequency (4: frequency $\geq$ 90%; 3: 51 to 89%; 2: 11 to 50%; 1: $\leq$ 10%) by adding the scores. For example, if a problem had an observed frequency of occurrence of 80% and had a minor effect on performance, its priority would be 5 (a frequency rating of 3 plus an impact rating of 2). With this approach, priority scores can range from a low of 2 to a high of 8. If information is available about the likelihood that a user would work with the part of the product that enables the problem, this information would be used to adjust the frequency rating. Continuing the example, if the expectation is that only 10% of users would encounter the problem, the priority would be 3 (a frequency rating of 1 for the 10% $\times$ 80%, or an 8% likelihood of occurrence plus an impact rating of 2).

A similar strategy is to multiply the observed percentage frequency of occurrence by the impact score. The range of priorities depends on the values assigned to each impact level. Assigning 10 to the most serious impact level leads to a maximum priority (severity) score of 1000 (which can optionally be divided by 10 to create a scale that ranges from 1 to 100). Appropriate values for the remaining three impact categories depend on practitioner judgment, but a reasonable set is 5, 3, and 1. Using those values, the problem with an observed frequency of occurrence of 80% and a minor effect on performance would have a priority of 24 (80 $\times$ 3/10). It is possible to extend this method to account for likelihood of use using the same procedure as that described by Rubin (1994), which in the example resulted in modifying the frequency measurement from 80% to 8%. Another way to extend the method is to categorize the likelihood of use with a set of categories such as very high likelihood (assigned a score of 10), high likelihood (assigned a score of 5), moderate likelihood (assigned a score of 3), and low likelihood (assigned a score of 1), and multiplying all three scores to get the final priority (severity) score (then optionally divide by 100 to create a scale that ranges from 1 to 100). Continuing the previous example with the assumption that the task in which the problem occurred has a high likelihood of occurrence, the problem's priority would be 12 (5 $\times$ 240/100). In most cases, applying the different data-driven prioritization schemes to the same set of problems should result in a very similar prioritization (but there has been no research published on this topic).

2.8.4 Working with Quantitative Measurements

The most common use of quantitative measurements is to characterize performance and preference variables by computing means, standard deviations, and ideally, confidence intervals. Practitioners use these results to compare observed to target measurements when targets are available. When targets are not available, the results can still be informative, for example, for use as future target measurements or as relatively gross diagnostic indicators.

The failure to meet targets is an obvious diagnostic cue. A less obvious cue is an unusually large standard deviation. Landauer (1997) describes a case in which the times to record an order were highly variable. The cause for the excessive variability was that a required phone number was sometimes, but not always, available, which turned out to be an easy problem to fix. Because the means and standard deviations of time scores tend to correlate, one way to detect an unusually large variance is to compute the coefficient of variation by dividing the standard deviation by the mean (Jeff Sauro, personal communication, April 26, 2004) or the normalized performance ratio by dividing the mean by the standard deviation (Moffat, 1990). Large coefficients of variation (or, correspondingly, small normalized performance ratios) are potentially indicative of the presence of usability problems.

3 ADVANCED TOPICS

This section covers more advanced topics in usability testing, including sample size estimation for problem discovery and measurement tests (both comparative and parameter estimation), confidence intervals based on t-scores and binomial confidence intervals, and standardized usability questionnaires. This chapter contains a considerable amount of information about statistical topics because statistical methods do not typically receive much attention in chapters on usability testing, and properly practiced, these techniques can be very valuable. On the other hand, practitioners should keep in mind that the most important factors that lead to successful usability evaluation are the appropriate selection of participants and tasks. No statistical analysis can repair a study in which you watch the wrong people doing the wrong activities.

3.1 Sample Size Estimation

The purpose of this section is to discuss the principles of sample size estimation for three types of usability test: population parameter estimation, comparative (also referred to as experimental), and problem discovery (also referred to as diagnostic, observational, or formative). This section assumes some knowledge of introductory applied statistics, so if you're not comfortable with terms such as *mean, variance, standard deviation*, p, *t-score*, and *Z-score*, refer to an introductory statistics text such as Walpole (1976) for definitions of these and other fundamental terms.

Sample size estimation requires a blend of mathematics and judgment. The computations are straightforward, and it is possible to make reasoned judgments (e.g., judgments about expected costs and precision requirements) for those values that the mathematics cannot determine.

3.1.1 Sample Size Estimation for Parameter Estimation and Comparative Studies

Traditional sample size estimation for population parameter estimation and comparative studies depends on having an estimate of the variance of the dependent measure(s) of interest and an idea of how precise (the magnitude of the critical difference and the statistical confidence level) the measurement must be (Walpole, 1976). Once you have that, the rest is mathematical mechanics (typically, using the formula for the t statistic).

You can (1) get an estimate of variance from previous studies using the same method (same or similar tasks and measures), (2) run a quick pilot study to get the estimate (e.g., piloting with four participants should suffice to provide an initial estimate of variability), or (3) set the critical difference your are trying to detect to some fraction of the standard deviation (Diamond, 1981). (See the following examples for more details about these different methods.)

Certainly, people prefer precise measurement to imprecise measurement, but all other things being equal, the more precise a measurement is, the more it will cost, and running more participants than necessary is wasteful of resources (Kraemer and Thiemann, 1987). The process of carrying out sample size estimation can also lead usability practitioners and their management to a realistic determination of how much precision they really need to make their required decisions.

Alreck and Settle (1985) recommend using a 'what if' approach to help decision makers determine their required precision. Start by asking the decision maker what would happen if the average value from the study was off the true value by 1%. Usually, the response would be that a difference that small wouldn't matter. Then ask what would happen if the measurement were off by 5%. Continue until you determine the magnitude of the critical difference. Then start the process again, this time pinning down the required level of statistical confidence. Note that statistically unsophisticated decision makers are likely to start out by expecting 100% confidence (which is only possible by sampling every unit in the population). Presenting them with the sample sizes required to achieve different levels of confidence can help them settle in on a more realistic confidence level.

Example 1: Parameter Estimation Given Estimate of Variability and Realistic Criteria This example illustrates the process of computing the sample size requirement for the estimation of a population parameter given an existing estimate of variability and realistic measurement criteria. For speech recognition applications, the recognition accuracy is an important value to track due to the adverse effects misrecognitions have on product usability. Thus, part of the process of evaluating the usability of a speech recognition product is estimating its accuracy. For this example, suppose that:

- Recognition variability (variance) from a previous similar evaluation = 6.35
- Critical difference (d) = 2.5%
- Desired level of confidence: 90%.

The appropriate procedure for estimating a population parameter is to construct a confidence interval (Bradley, 1976). To determine the upper and lower limits of a confidence interval, add to and subtract the following from the observed mean:

$$d = \text{SEM} \times t_{\text{crit}} \qquad (1)$$

where SEM is the standard error of the mean (the standard deviation, S, divided by the square root of the sample size, n) and t_{crit} is the t-value associated with the desired level of confidence (found in a t-table, available in most statistics texts). Setting the critical difference to 2.5 is the same as saying that the value of SEM $\times\ t_{\text{crit}}$ should be equal to 2.5. In other words, you don't want the upper or lower bound of the confidence interval to be more than 2.5 percentage points away from the observed mean, for a confidence interval width equal to 5.0.

Calculating the SEM depends on knowing the sample size, and the value of t_{crit} also depends on the sample size, but you don't know the sample size yet. Iterate using the following method.

1. Start with the Z-score for the desired level of confidence in place of t_{crit}. For 90% confidence, this is 1.645. (By the way, if you actually *know* the true variability for the measurement rather than just having an estimate, you're done at this point because it's appropriate to use the Z-score rather than a t-score. However, you almost never know the true variability, but must work with estimates.)
2. Algebraic manipulations based on the formula SEM $\times Z = d$ results in $n = Z^2S^2/d^2$, which for this example is $n = (1.645^2)(6.35)/2.5^2$, which equals 2.7. Always round sample size estimates up to the next whole number, so this initial estimate is 3.
3. Now you need to adjust the estimate by replacing the Z-score with the t-score for a sample size of 3. For this estimate, the degrees of freedom (df) to use when looking up the value in a t table is $n - 1$, or 2. This is important because the value of Z will always be smaller than the appropriate value of t, making the initial estimate smaller than it should be. For this example, t_{crit} is 2.92.
4. Recalculating for n using 2.92 in place of 1.645 produces 8.66, which rounds up to 9.
5. Because the appropriate value of t_{crit} is now a little smaller than 2.92 (because the estimated sample size is now larger, with $9 - 1$, or 8, degrees of freedom), recalculate n again, using t_{crit} equal to 1.860. The new value for n is 3.515, which rounds up to 4.
6. Stop iterating when you get the same value for n on two iterations or you begin cycling between two values for n, in which case you should choose the larger value. Table 2 shows the full set of iterations for this example, which ends by estimating the appropriate sample size as 5. Note that there is nothing in these computations that makes reference to the size of

Table 2 Full Set of Iterations for Example 1

	Initial	1	2	3	4	5
t_{crit}	1.645	2.92	1.86	2.353	2.015	2.132
t^2_{crit}	2.71	8.53	3.46	5.54	4.06	4.55
S^2	6.35	6.35	6.35	6.35	6.35	6.35
d	2.5	2.5	2.5	2.5	2.5	2.5
Estimated n	2.7493	8.663	3.515	5.6252	4.1252	4.618
Rounded up	3	9	4	6	5	5
df	2	8	3	5	4	4

the population. Unless the size of the sample is a significant percentage of the total population under study (which is rare, but correctable using a finite population correction), the size of the population is irrelevant. Alreck and Settle (1985) explain this with a soup-tasting analogy. Suppose that you're cooking soup in a one-quart saucepan and want to test if it's hot enough. You would stir it thoroughly, then taste a teaspoonful. If it were a two-quart saucepan, you'd follow the same procedure—stir thoroughly, then taste a teaspoonful.

Diamond (1981) points out that you can usually get by with an initial estimate and one iteration because most researchers don't mind having a sample size that's a little larger than necessary. If the cost of each sample is high, though, it makes sense to iterate until reaching one of the stopping criteria. Note that the initial estimate establishes the lower bound for the sample size (3 in this example), and the first iteration establishes the upper bound (9 in this example).

Example 2: Parameter Estimation Given Estimate of Variability and Unrealistic Criteria The measurement criteria in Example 1 were reasonable—90% confidence that the interval (limited to a total length of 5%) contains the true mean. This example shows what happens when the measurement criteria are less realistic, illustrating the potential cost associated with high confidence and high measurement precision. Suppose that the measurement criteria for the situation described in Example 1 were less realistic, with:

- Recognition variability from a previous similar evaluation = 6.35
- Critical difference (d) = 0.5%
- Desired level of confidence: 99%

In that case, the initial Z-score would be 2.576, and the initial estimate of n would be:

$$n = \frac{(2.576^2)(4.9)}{0.5^2} = 168.549 \tag{2}$$

which rounds up to 169. Recalculating n with t_{crit} equal to 2.605 (t with 168 degrees of freedom) results in n equal to 172.37, which rounds up to 173. (Rather then continuing to iterate, note that the final value for the sample size must lie between 169 and 173.) There might be some industrial environments in which usability investigators would consider 169 to 173 participants a reasonable and practical sample size, but they are rare. (On the other hand, collecting data from this number of participants or more in a mailed survey is common.)

Example 3: Parameter Estimation Given No Estimate of Variability For both Examples 1 and 2, it doesn't matter if the estimate of variability came from a previous study or a quick pilot study. Suppose, however, that you don't have any idea what the measurement variability is, and it's too expensive to run a pilot study to get an initial estimate. This example illustrates a technique (from Diamond, 1981) for getting around this problem. To do this, though, you need to give up the definition of the critical difference (d) in terms of the variable of interest and replace it with a definition in terms of a fraction of the standard deviation.

In this example, the measurement variance is unknown. To get started, the testers have decided that with 90% confidence, they do not want d to exceed half the value of the standard deviation. The measurement criteria are:

- Recognition variability from a previous similar evaluation = N/A
- Critical difference (d) = $0.5S$
- Desired level of confidence: 90%

The initial sample size estimate is

$$n = \frac{1.645^2 S^2}{(0.5S)^2} = \frac{(1.645^2)}{(0.5)^2} = 10.824 \tag{3}$$

which rounds up to 11. The result of the first iteration, replacing 1.645 with t_{crit} for 10 degrees of freedom (1.812), results in a sample size estimation of 13.13, which rounds up to 14. The appropriate sample size is therefore somewhere between 11 and 14, with the final estimate determined by completing the full set of iterations.

Example 4: Comparing a Parameter to a Criterion For an example comparing a measured parameter to a criterion value, suppose that you have a product requirement that installation should take no more than 30 minutes. In a preliminary evaluation, participants needed an average of 45 minutes to complete installation. Development has fixed a number of usability problems found in that preliminary study, so you're ready to measure installation time again, using the following measurement criteria:

- Performance variability from the previous evaluation = 10.0
- Critical difference (d) = 3 minutes
- Desired level of confidence: 90%

The interpretation of these measurement criteria is that you want to be 90% confident that you can detect a difference as small as 3 minutes between the mean of the data gathered in the test and the criterion you're trying to beat. In other words, the installation will pass if the observed mean time is 27 minutes or less, because the sample size should guarantee an upper limit to the confidence interval that is no more than 3 minutes above the mean (as long as the observed variance is less than or equal to the initial estimate of the variance). The procedure for determining the sample size in this situation is the same as that of Example 1, shown in Table 3. The outcome of these iterations is a sample size requirement of 6 because the sample size estimates begin cycling between 5 and 6.

Example 5: Sample Size for a Paired t-Test When you obtain two comparable measurements from each participant in a test (a within-subjects design), you can assess the results using a paired t-test. Another name for a *paired t-test* is a *difference score t-test*, because the measurements of concern are the mean and standard deviation of the set of difference scores rather than the raw scores. Suppose that you plan to obtain recognition accuracy scores from participants who have dictated test texts into your product under development and a competitor's product [following all the appropriate experimental design procedures such as counterbalancing the order of presentation of products to participants; see a text such as Myers (1979) for guidance in experimental design], using the following criteria:

- Difference score variability from a previous evaluation = 5.0
- Critical difference (d) = 2%
- Desired level of confidence: 90%

This situation is similar to that of Example 4 because the typical goal of a difference scores t-test is to determine if the average difference between scores is statistically significantly different from 0. Thus, the usability criterion in this case is 0, and you want to be 90% confident that if the true difference between system accuracies is 2% or more, you will be able to detect it because the confidence interval for the difference scores will not contain 0. Table 4 shows the

Table 3 Full Set of Iterations for Example 4

	Initial	1	2	3	4
t_{crit}	1.645	2.353	1.943	2.132	2.015
t^2_{crit}	2.706	5.537	3.775	4.545	4.060
s^2	10	10	10	10	10
d	3	3	3	3	3
d^2	9	9	9	9	9
Estimated n	3.006694	6.151788	4.194721	5.050471	4.511361
Rounded up	4	7	5	6	5
df	3	6	4	5	4

Table 4 Full Set of Iterations for Example 5

	Initial	1	2	3	4
t_{crit}	1.645	2.353	1.943	2.132	2.015
t^2_{crit}	2.706	5.537	3.775	4.545	4.060
s^2	5	5	5	5	5
d	2	2	2	2	2
d^2	4	4	4	4	4
Estimated n	3.382531	6.920761	4.719061	5.68178	5.075281
Rounded up	4	7	5	6	6
df	3	6	4	5	5

iterations for this situation, leading to a sample size estimate of 6.

Example 6: Sample Size for a Two-Groups t-Test Up to this point, the examples have all involved one group of scores and have been amenable to similar treatment. If you have a situation in which you plan to compare scores from two independent groups, things get a little more complicated. For one thing, you now have two sample sizes to consider, one for each group.

To simplify things in this example, assume that the groups are essentially equal (especially with regard to performance variability), which should be the case if the groups contain participants from a single population who have received random assignment to treatment conditions. In this case it is reasonable to believe that the sample size for both groups will be equal, which simplifies things. For this situation, the formula for the initial estimate of the sample size for each group is

$$n = \frac{2Z^2S^2}{d^2} \tag{4}$$

Note that this is the similar to the formula presented in Example 1, with the numerator multiplied by 2. After getting the initial estimate, begin iterating using the appropriate value for t_{crit} in place of Z. For example, suppose that you needed to conduct the experiment described in Example 5 with independent groups of participants, keeping the measurement criteria the same:

- Estimate of variability from a previous evaluation = 5.0
- Critical difference (d) = 2%
- Desired level of confidence: 90%

In that case, iterations would converge on a sample size of nine participants per group, for a total sample size of 18, as shown in Table 5.

This illustrates the well-known measurement efficiency of experiments that produce difference scores (within-subjects designs) relative to experiments involving independent groups (between-subjects designs). For the same measurement precision, the estimated sample size for Example 5 was six participants, one-third the sample size requirement estimated for this example.

Doing this type of analysis gets more complicated if you have reason to believe that the groups are different, especially with regard to variability of performance. In that case you would want to have a larger sample size for the group with greater performance variability in an attempt to obtain more equal precision of measurement for each group. Advanced market research texts (such as Brown, 1980) provide sample size formulas for these situations.

Table 5 Full Set of Iterations for Example 6

	Initial	1	2	3
t_{crit}	1.645	1.943	1.833	1.86
t_{crit}^2	2.706	3.775	3.360	3.460
s^2	5	5	5	5
d	2	2	2	2
d^2	4	4	4	4
Estimated n	6.765	9.438	8.400	8.649
Rounded up	7	10	9	9
df	6	9	8	8

Example 7: Making Power Explicit in the Sample Size Formula The power of a procedure is not an issue when estimating the value of a parameter, but it is an issue when testing a hypothesis (as in Example 6). In traditional hypothesis testing, there is a null (H_0) and an alternative (H_a) hypothesis. The typical null hypothesis is that there is no difference between groups, and the typical alternative hypothesis is that the difference is greater than zero. When the alternative hypothesis is that the difference is nonzero, the test is two-tailed because you can reject the null hypothesis with either a sufficiently positive or a sufficiently negative outcome. If you have reason to believe that you can predict the direction of the outcome, or if an outcome in only one direction is meaningful, you can construct an alternative hypothesis that considers only a sufficiently positive or a sufficiently negative outcome (a one-tailed test). For more information, see an introductory statistics text (such as Walpole, 1976).

When you test a hypothesis (e.g., that the difference in recognition accuracy between two competitive dictation products is nonzero), there are two ways to make a correct decision and two ways to be wrong, as shown in Table 6. Strictly speaking, you never accept the null hypothesis, because the failure to acquire sufficient evidence to reject the null hypothesis could be due to (1) no significant difference between groups, or (2) a sample size too small to detect an existing difference. Rather than accepting the null hypothesis, you fail to reject it.

Returning to Table 6, the two ways to be right are (1) to fail to reject the null hypothesis (H_0) when it is true, or (2) to reject the null hypothesis when it is false. The two ways to be wrong are (1) to reject the null hypothesis when it is true (type I error), or (2) to fail to reject the null hypothesis when it is false (type II error). Table 7 shows the relationship between these concepts and their corresponding statistical testing terms.

Table 6 Possible Outcomes of a Hypothesis Test

	Reality	
Decision	H_0 Is True	H_0 Is False
Insufficient evidence to reject H_0	Fail to reject H_0	Type II error
Sufficient evidence to reject H_0	Type I error	Reject H_0

Table 7 Statistical Testing Terms

Statistical Concept	Testing Term
Acceptable probability of a type I error	α
Acceptable probability of a type II error	β
Confidence	$1-\alpha$
Power	$1-\beta$

The formula presented in Example 6 for an initial sample size estimate was

$$n = \frac{2Z^2S^2}{d^2} \qquad (5)$$

In Example 6, the Z-score was set for 90% confidence (which means that $\alpha = 0.10$). To take power into account in this formula, you need to add another Z-score to the formula, the Z-score associated with the desired power of the test (as defined in Table 7). Thus, the formula becomes

$$n = \frac{2(Z_\alpha + Z_\beta)^2S^2}{d^2} \qquad (6)$$

So, what was the value for power in Example 6? When beta (β) equals 0.5 (in other words, when the power is 50%), the value of z_β is 0, so z_β disappears from the formula. Thus, in this example the implicit power was 50%. Suppose that you want to increase the power of the test to 80%, reducing β to 0.2.

- Estimate of variability from a previous evaluation = 5.0
- Critical difference (d) = 2%
- Desired level of confidence: 90% ($Z_\alpha = 1.645$)
- Desired power: 80% ($Z_\beta = 1.282$)

With this change, the iterations converge on a sample size of 24 participants per group, for a total sample size of 48, as shown in Table 8. To achieve the stated goal for power results in a considerably larger sample size. Note that the stated power of a test is relative to the critical difference—the smallest effect worth finding. Either increasing the value of the critical difference or reducing the power of a test will result in a smaller required sample size.

Table 8 Full Set of Iterations for Example 7

	Initial	1	2	3
$t(\alpha)$	1.645	1.721	1.714	1.717
$t(\beta)$	1.282	1.323	1.319	1.321
t(total)	2.927	3.044	3.033	3.038
$t(\text{total})^2$	8.567	9.266	9.199	9.229
S^2	5	5	5	5
d	2	2	2	2
d^2	4	4	4	4
Estimated n	21.418	23.165	22.998	23.074
Rounded up	22	24	23	24
df	21	23	22	23

Appropriate Statistical Criteria for Industrial Testing In scientific publishing, the usual criterion for statistical significance is to set the permissible type I error (α) equal to 0.05. This is equivalent to seeking to have 95% confidence that the effect is real rather than random and is focused on controlling the type I error (the likelihood that you decide that an effect is real when it's random). There is no corresponding scientific recommendation for the type II error (β, the likelihood that you will conclude an effect is random when it's real), although some suggest setting it to 0.20 (Diamond, 1981). The rationale behind the emphasis on controlling the type I error is that it is better to delay the introduction of good information into the scientific database (a type II error) than to let erroneous information in (a type I error).

In industrial evaluation, the appropriate values for type I and II errors depend on the demands of the situation, whether the cost of a type I or II error would be more damaging to the organization. Because we are often resource-constrained, especially with regard to making timely decisions to compete in dynamic marketplaces, this paper has used measurement criteria (such as 90% confidence rather than 95% confidence and fairly large values for d) that seek a greater balance between type I and II errors than is typical in work designed to result in scientific publications. Nielsen (1997) has suggested that 80% confidence is appropriate for practical development purposes. For an excellent discussion of this topic for usability researchers, see Wickens (1998). For other technical issues and perspectives, see Landauer (1997).

Another way to look at the issue is to ask the question, "Am I typically interested in small high-variability effects or large low-variability effects?" The correct answer depends on the situation, but in usability testing, the emphasis is on the detection of large low-variability effects (either large performance effects or frequently occurring problems). You shouldn't need a large sample to verify the existence of large low-variability effects. Some writers equate sample size with population coverage, but this isn't true. A small sample size drawn from the right population provides better coverage than a large sample size drawn from the wrong population. The statistics involved in computing confidence intervals from small samples compensate for the potentially smaller variance in the small sample by forcing the confidence interval to be wider than that for a larger sample (specifically, the value of t is greater when samples are smaller).

Coming from a different tradition than usability research, many market research texts provide rules of thumb recommending large sample sizes. For example, Aaker and Day (1986) recommend a minimum of 100 per group, with 20 to 50 for subgroups. For national surveys with many subgroup analyses, the typical total sample size is 2500 (Sudman, 1976). These rules of thumb do not make any formal contact with statistical theory and may in fact be excessive, depending on the goals of the study. Other market researchers (e.g., Banks, 1965) do promote a careful evaluation of the goals of a study.

> It is urged that instead of a policy of setting uniform requirements for type I and II errors, regardless of the economic consequences of the various decisions to be made from experimental data, a much more flexible approach be adopted. After all, if a researcher sets himself a policy of always choosing the apparently most effective of a group of alternative treatments on the basis of data from unbiased surveys or experiments and pursues this policy consistently, he will find that in the long run he will be better off than if he chose any other policy. This fact would hold even if none of the differences involved were statistically significant according to our usual standards or even at probability levels of 20 or 30 percent. (Banks, 1965, p. 252)

Finally, Alreck and Settle (1985) provide an excellent summary of the factors indicating appropriate use of large and small samples. Use a large sample size when:

1. Decisions based on the data will have very serious or costly consequences.
2. The sponsors (decision makers) demand a high level of confidence.
3. The important measures have high variance.
4. Analyses will require dividing the total sample into small subsamples.
5. Increasing the sample size has a negligible effect on the cost and timing of the study.
6. Time and resources are available to cover the cost of data collection.

Use a small sample size when:

1. The data will determine few major commitments or decisions.
2. The sponsors (decision makers) require only rough estimates.
3. The important measures have low variance.
4. Analyses will use the entire sample or just a few relatively large subsamples.
5. Costs increase dramatically with sample size.
6. Budget constraints or time limitations limit the amount of data you can collect.

Tips on Reducing Variance Because measurement variance is such an important factor in sample size estimation for these types of studies, it generally makes sense to attempt to manage variance (although in some situations, such management is out of a practitioner's control). Here are some ways to reduce variance:

- Make sure that participants understand what they are supposed to do in the study. Unless potential participant confusion is part of the evaluation (and it sometimes is), it can only add to measurement variance.
- One way to accomplish this is through practice trials that allow participants to get used to the experimental situation without unduly revealing study-relevant information.
- If appropriate, use expert rather than novice participants. Almost by definition, expertise implies reduced performance variability (increased automaticity) (Mayer, 1997). With regard to reducing variance, the farther up the learning curve, the better.
- A corollary of this is that if you need to include both expert and novice users, you should be able to get equal measurement precision for both groups with unequal sample sizes (fewer experts required than novices—which is good, because experts are typically harder than novices to recruit as participants).
- If appropriate, study simple rather than complex tasks.
- Use data transformations for measurements that typically exhibit correlations between means and variances or standard deviations. For example, frequency counts often have proportional means and variances (treated with the square-root transformation), and time scores often have proportional means and standard deviations (treated with the logarithmic transformation) (Myers, 1979).
- For comparative studies, use within-subjects designs rather than between-subjects designs whenever possible.
- Keep user groups as homogeneous as possible (but although this reduces variability, it can, simultaneously, pose a threat to a study's external validity if the test group is more homogenous than the population under study) (Campbell and Stanley, 1963).

Keep in mind that it is reasonable to use these tips only when their use does not adversely affect the validity and generalizability of the study. Having a valid and generalizable study is far more important than reducing variability.

Tips for Estimating Unknown Variance Parasuraman (1986) described a method for estimating variability if you have an idea about the largest and smallest values for a population of measurements but don't have the information you need to actually calculate the variability. Estimate the standard deviation (the square root of the variability) by dividing the difference between the largest and smallest values by 6.

This technique assumes that the population distribution is normal and then takes advantage of the fact that 99% of a normal distribution will lie in the range of plus or minus three standard deviations of the mean.

Nielsen (1997) surveyed 36 published usability studies and found that the mean standard deviation for measures of expert performance was 33% of the mean value of the usability measure (in other words, if the mean completion time was 100 seconds, the mean standard deviation was about 33 seconds). For novice-user learning the mean standard deviation was 46%, and for measures of error rates the value was 59%.

Churchill (1991) provided a list of typical variances for data obtained from rating scales. Because the number of items in the scale affects the possible variance (with more items leading to more variance), the table takes the number of items into account. For five-point scales, the typical variance is 1.2 to 2.0; for seven-point scales it is 2.4 to 4.0; and for 10-point scales it is 3.0 to 7.0. Because data obtained using rating scales tends to have a more uniform than normal distribution, he advises using a number nearer the high end of the listed range when estimating sample sizes.

Measurement theorists who agree with Steven's (1951) principle of invariance might yell "foul" at this point because they believe that it is not permissible to calculate averages or variances from rating scale data. There is considerable controversy on this point (see, e.g., Lord, 1953; Nunnally, 1978; Harris, 1985). Data reported by J. Lewis (1993) indicate that taking averages and conducting t-tests on multipoint rating data provides far more interpretable and consistent results than the alternative of taking medians and conducting Mann–Whitney U-tests. When you make claims about the meaning of the outcomes of your statistical tests you do have to be careful not to act as if rating scale data are interval data rather than ordinal data. An average rating of 4 might be better than an average rating of 2, but you can't claim that it is twice as good (a ratio claim), nor can you claim that the difference between 4 and 2 is equal to the difference between 4 and 6 (an interval claim).

3.1.2 Sample Size Estimation for Problem-Discovery (Formative) Studies

"Having collected data from a few test subjects—and initially a few are all you need—you are ready for a revision of the text" (Al-Awar et al., 1981, p. 34). "This research does not mean that all of the *possible* problems with a product appear with 5 or 10 participants, but most of the problems that are going to show up with one sample of tasks and one group of participants will occur early" (Dumas, 2003, p. 1098).

Although these types of general guidelines have been helpful, it is possible to use more precise methods to estimate sample size requirements for problem-discovery usability tests. Estimating sample sizes for tests that have the primary purpose of discovering the problems in an interface depends on having an estimate of p, characterized as the average likelihood of problem occurrence or, alternatively, the problem-discovery rate. As with comparative studies, this estimate can come from previous studies using the same method and similar system under evaluation or can come from a pilot study. For standard scenario-based usability studies, the literature contains large-sample examples that show p ranging from 0.16 to 0.42 (J. Lewis, 1994). For heuristic evaluations, the reported value of p from large-sample studies ranges from 0.22 to 0.60 (Nielsen and Molich, 1990).

When estimating p from a small sample, it is important to adjust its initially estimated value because a small-sample estimate of p (e.g., fewer than 20 participants) has a bias that results in potentially substantial overestimation of its value (Hertzum and Jacobsen, 2003). A series of Monte Carlo experiments (J. Lewis, 2001a) have demonstrated that a formula combining Good–Turing discounting with a normalization procedure provides a reasonably accurate adjustment of initial estimates of $p(p_{\text{est}})$, even when the sample size for that initial estimate has as few as two participants (preferably four participants, though, because the variability of estimates of p is greater for smaller samples) (J. Lewis, 2001a; Faulkner, 2003). This formula for the adjustment of p is

$$p_{\text{adj}} = \frac{1}{2}\left[\left(p_{\text{est}} - \frac{1}{n}\right)\left(1 - \frac{1}{n}\right)\right] + \frac{1}{2}\left(\frac{p_{\text{est}}}{1 + \text{GT}_{\text{adj}}}\right) \qquad (7)$$

where GT_{adj} is the Good–Turing adjustment to probability space (which is the proportion of the number of problems that occurred once divided by the total number of different problems). The $p_{\text{est}}/(1 + \text{GT}_{\text{adj}})$ component in the equation produces the Good–Turing adjusted estimate of p by dividing the observed, unadjusted estimate of $p(p_{\text{est}})$ by the Good–Turing adjustment to probability space. The $(p_{\text{est}} - 1/n)(1 - 1/n)$ component in the equation produces the normalized estimate of p from the observed, unadjusted estimate of p and n (the sample size used to estimate p). The reason for averaging these two different estimates is that the Good–Turing estimator tends to overestimate the true value of p, and the normalization tends to underestimate it. For more details and experimental data supporting the use of this formula for estimates of p based on sample sizes from 2 to 10 participants, see J. Lewis (2001a).

Adjusting the Initial Estimate of p Because this is a new procedure, this section contains a detailed illustration of the steps used to adjust an initial estimate of p. To start with, organize the problem-discovery data in a table (e.g., Table 9) that shows which participants experienced which problems. With four participants and eight observed problems, there are 32 cells in the table. The total number of problem occurrences is 16, so the initial estimate of $p(p_{\text{est}})$ is 0.50 (16/32). Note that averaging the proportion of problem occurrence across participants or across problems also equals 0.50.

To apply the Good–Turing adjustment, count the number of problems that occurred with only one participant. In Table 9 this happened for three problems (problems 4, 6, and 8) out of the eight unique

Table 9 Hypothetical Results for a Problem-Discovery Usability Study

	Problem									
Participant	1	2	3	4	5	6	7	8	Count	Proportion
1	×		×		×		×	×	5	0.63
2	×	×			×		×		4	0.50
3	×		×	×	×				4	0.50
4	×	×				×			3	0.38
Count	4	2	2	1	3	1	2	1	16	
Proportion	1.00	0.50	0.50	0.25	0.75	0.25	0.50	0.25		0.50

problems listed in the table. Thus, the value of GT_{adj} is 0.375 ($\frac{3}{8}$), and the value of $p_{\text{est}}/(1 + \text{GT}_{\text{adj}})$ is 0.36 (0.5/1.375).

To apply the normalization adjustment, start by computing $1/n$, which in Table 9 is 0.25 ($\frac{1}{4}$). The value of $(p_{\text{est}} - 1/n)(1 - 1/n)$ is 0.19 [(0.25)(0.75)].

The average of the two adjustments produces p_{adj}, which in this example equals 0.28 ((0.36 + 0.19)/2). In this example, the adjusted estimate of p is almost half of the initial estimate.

Using the Adjusted Estimate of p Once you have an appropriate (adjusted) estimate for p, you can use the formula $1 - (1 - p)^n$ [derivable from both the binomial probability formula (J. Lewis, 1982, 1994) and the Poisson probability formula (Nielsen and Landauer, 1993)] for various values of n from, say, 1 to 20, to generate the curve of diminishing returns expected as a function of sample size. It is possible to get even more sophisticated, taking into account the fixed and variable costs of the evaluation (especially the variable costs associated with the study of additional participants) to estimate when running an additional participant will result in costs that exceed the value of the additional problems discovered (J. Lewis, 1994).

The Monte Carlo experiments reported in J. Lewis (2001a) demonstrated that an effective strategy for planning the sample size for a usability study is first to establish a problem-discovery goal (e.g., 90% or 95%). Run the first two participants and based on those results, calculate the adjusted value of p using equation (7). This provides an early indication of the probable sample size required, which might estimate the final sample size exactly or, more likely, underestimate by one or two participants (but will provide an early estimate of the required sample size). Collect data from two more participants (for a total of four). Recalculate the adjusted estimate of p using equation (7) and project the required sample size using $1 - (1 - p)^n$. The estimated sample size requirement based on data from four participants will generally be highly accurate, allowing accurate planning for the remainder of the study. Practitioners should do this even if they have calculated a preliminary estimate of the required sample size from an adjusted value for p obtained from a previous study.

Figure 4 shows the discovery rates predicted for problems of differing likelihoods of observation during a usability study. Several independent studies have verified that these types of predictions fit observed data very closely for both usability and heuristic evaluations (Nielsen and Molich, 1990; Virzi, 1990, 1992; Wright and Monk, 1991; Nielsen and Landauer, 1993; J. Lewis, 1994). Furthermore, the predictions work both for predicting the discovery of individual problems with a given probability of detection and for modeling the discovery of members of sets of problems with a given mean probability of detection (J. Lewis, 1994). For usability studies, the sample size is the number of participants. For heuristic evaluations, the sample size is the number of evaluators.

Table 10 shows problem detection sample size requirements as a function of problem detection probability and the cumulative likelihood of detecting the problem at least once during the study. The sample size required for detecting the problem twice during a study appears in parentheses. To use this information to establish a usability sample size, you need to determine three things:

1. What is the average likelihood of problem detection probability (p)? This plays a role similar to the role of variance in the previous examples. If you don't know this value (from previous studies or a pilot study), you need to decide on the lowest problem detection probability that you want to (or have the resources to) tackle. The smaller this number, the larger the required sample size.
2. What proportion of the problems that exist at that level do you need (or have the resources) to discover during the study (in other words, the cumulative likelihood of problem detection)? The larger this number, the larger the required sample size.
3. Are you willing to take single occurrences of problems seriously or must problems appear at least twice before receiving consideration? Requiring two occurrences results in a larger sample size.

For values of p or problem-discovery goals that are outside tabled values, you can use the following formula [derived algebraically from Goal $= 1 - (1 - p)^n$] to compute directly the sample size required for a given problem-discovery goal (taking single occurrences of problems seriously) and value of p:

$$n = \frac{\log(1 - \text{goal})}{\log(1 - p)} \tag{8}$$

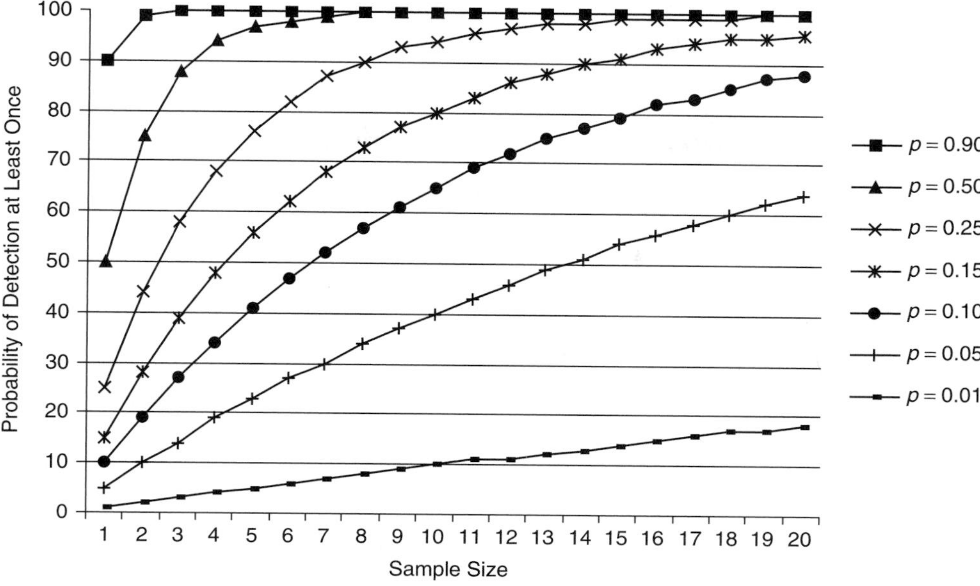

Figure 4 Predicted discovery as a function of problem likelihood.

Table 10 Sample Size Requirements for Problem Discovery (Formative) Studies

Problem Occurrence Probability	Cumulative Likelihood of Detecting the Problem at Least Once (Twice)					
	0.50	0.75	0.85	0.90	0.95	0.99
0.01	69 (168)	138 (269)	189 (337)	230 (388)	299 (473)	459 (662)
0.05	14 (34)	28 (53)	37 (67)	45 (77)	59 (93)	90 (130)
0.10	7 (17)	14 (27)	19 (33)	22 (38)	29 (46)	44 (64)
0.15	5 (11)	9 (18)	12 (22)	15 (25)	19 (30)	29 (42)
0.25	3 (7)	5 (10)	7 (13)	9 (15)	11 (18)	17 (24)
0.50	1 (3)	2 (5)	3 (6)	4 (7)	5 (8)	7 (11)
0.90	1 (2)	1 (2)	1 (3)	1 (3)	2 (3)	2 (4)

In the example from Table 9, the adjusted value of p was 0.28. Suppose that the practitioner decided that the appropriate problem-discovery goal was to find 97% of the discoverable problems. The computed value of n is 10.6 (log(0.03)/log(0.72), or $-1.522/-0.143$). The practitioner can either round the sample size up to 11 or adjust the problem-discovery goal down to 96.3% $[1-(1-0.28)^{10}]$.

J. Lewis (1994) created a return-on-investment (ROI) model to investigate appropriate cumulative problem detection goals. It turned out that the appropriate goal depended on the average problem detection probability in the evaluation, the same value that has a key role in determining the sample size. The model indicated that if the expected value of p was small (say, around 0.10), practitioners should plan to discover about 86% of the problems. If the expected value of p was larger (say, around 0.25 or 0.50), practitioners should plan to discover about 98% of the problems. For expected values of p between 0.10 and 0.25, practitioners should interpolate between 87 and 97% to determine an appropriate goal for the percentage of problems to discover.

The cost of an undiscovered problem had a strong effect on the magnitude of the maximum ROI, but, contrary to expectation, it had only a minor effect on sample size at maximum ROI. (J. Lewis, 1994). Usability practitioners should be aware of these costs in their settings and their effect on ROI (Boehm, 1981), but these costs have relatively little effect on the appropriate sample size for a usability study.

In summary, there is compelling evidence that the law of diminishing returns, based on the cumulative binomial probability formula, applies to problem-discovery studies. To use this formula to determine an appropriate sample size, practitioners must form an idea about the expected value of p (the average likelihood of problem detection) for the study and the percentage of problems that the study should uncover. Practitioners can use the ROI model from J. Lewis (1994) or their own ROI formulas to estimate an appropriate goal for the percentage of problems to discover and can examine data from their own or published usability studies to get an initial estimate of p (which published studies to date indicate can range at least from 0.16 to 0.60). With these two estimates, practitioners can use Table 10 (or, for computations outside tabled values, the appropriate equations) to estimate appropriate sample sizes for their usability studies.

It is interesting to speculate that a new product that has not yet undergone any usability evaluation is likely to have a higher p than an established product that has gone through several development iterations (including usability testing). This suggests that it is easier (takes fewer participants) to improve a completely new product than to improve an existing product (as long as that existing product has benefited from previous usability evaluation). This is related to the idea that usability testing is a hill-climbing procedure, in which the results of a usability test are applied to a product to push its usability up the hill. The higher up the hill you go, the more difficult it becomes to go higher, because you have already weeded out the problems that were easy to find and fix.

Practitioners who wait to see a problem at least twice before giving it serious consideration can see from Table 10 the sample size implications of this strategy. Certainly, all other things being equal, it is more important to correct a problem that occurs frequently than one that occurs infrequently. However, it is unrealistic to assume that the frequency of detection of a problem is the only criterion to consider in the analysis of usability problems. The best strategy is to consider problem frequency and other problem data (such as severity and likelihood of use) simultaneously to determine which problems are most important to correct rather than establishing a cutoff rule such as “fix every problem that appears two or more times.”

Note that in contrast to the results reported by Virzi (1992), the results reported by J. Lewis (1994) did not indicate any consistent relationship between problem frequency and impact (severity). It is possible that this difference was due to the difference in the methods used to assess severity [judgment-driven in Virzi (1992); data-driven in J. Lewis (1994)]. Thus, the safest strategy is for practitioners to assume independence of frequency and impact until further research resolves the discrepancy between the outcomes of these studies.

It is important for practitioners to consider the risks as well as the gains when using small samples for usability studies. Although the diminishing returns for inclusion of additional participants strongly suggest that the most efficient approach is to run a small sample (especially if p is high, if the study will be iterative, and if undiscovered problems will not have dangerous or expensive outcomes), human factors engineers and other usability practitioners must not become complacent regarding the risk of failing to detect low-frequency but important problems.

One could argue that the true number of possible usability problems in any interface is essentially infinite, with an essentially infinite number of problems with nonzero probabilities that are extremely close to zero. For the purposes of determining sample size, the p we are really dealing with is the p that represents the number of discovered problems divided by the number of discoverable problems, where the definition of a discoverable problem is vague but almost certainly constrained by details of the experimental setting, such as the studied scenarios and tasks and the skill of the observer(s). Despite this vagueness and some recent criticism of the use of p to model problem discovery (Caulton, 2001; Woolrych and Cockton, 2001), these techniques seem to work reasonably well in practice (Turner et al., in press).

Examples of Sample Size Estimation for Problem-Discovery (Formative) Studies This section contains several examples illustrating the use of Table 10 as an aid in selecting an appropriate sample size for a problem-discovery study.

A. Given the following problem-discovery criteria:
   - Detect problems with an average probability of 0.25
   - Minimum number of detections required: 1
   - Planned proportion to discover: 0.90

The appropriate sample size is nine participants.

B. Given the same discovery criteria, except that the practitioner requires problems to be detected twice before receiving serious attention:
   - Detect problems with an average probability of 0.25
   - Minimum number of detections required: 2
   - Planned proportion to discover: 0.90

The appropriate sample size would be 15 participants.

C. Returning to requiring a single detection, but increasing the planned proportion to discover to .99:
   - Detect problems with an average probability of 0.25
   - Minimum number of detections required: 1
   - Planned proportion to discover: 0.99

The appropriate sample size would be 17 participants.

D. Given the following extremely stringent discovery criteria:
   - Detect problems with an average probability of 0.01

- Minimum number of detections required: 1
- Planned proportion to discover: 0.99

The sample size required would be 459 participants (an unrealistic requirement in most settings, implying unrealistic study goals).

Note that there is no requirement to run the entire planned sample through the usability study before reporting clear problems to development and getting those problems fixed before continuing. These required sample sizes are total sample sizes, not sample sizes per iteration. The following testing strategy promotes efficient iterative problem discovery studies and is similar to strategies published by a number of usability specialists (Rosenbaum, 1989; Bailey et al., 1992; Jeffries and Desurvire, 1992; Nielsen, 1993; Kantner and Rosenbaum, 1997; Macleod et al., 1997; Fu et al., 2002).

1. Start with an expert (heuristic) evaluation or one-participant pilot study to uncover the obvious problems. Correct as many of these problems as possible before starting the iterative cycles with step 2. List all unresolved problems and carry them to step 2.
2. Watch a small sample of participants (e.g., three or four) use the system. Record all observed usability problems. Calculate an adjusted estimate of p based on these results and reestimate the required sample size.
3. Redesign based on the problems discovered. Focus on fixing high-frequency and high-impact problems. Fix as many of the remaining problems as possible. Record any outstanding problems so they can remain open for all following iterations.
4. Continue iterating until you have reached your sample size goal (or must stop for any other reason, such as running out of time).
5. Record any outstanding problems remaining at the end of testing and carry them over to the next product for which they are applicable.

This strategy blends the benefits of large and small sample studies. During each iteration, you observe only three or four participants before redesigning the system. Therefore, you can quickly identify and correct the most frequent problems (which means that you waste less time watching the next set of participants encounter problems that you already know about). With five iterations, for example, the total sample size would be 15 to 20 participants. With several iterations you will identify and correct many less–frequent problems because you record and track the uncorrected problems through all iterations.

Note that using this sort of iterative procedure affects estimates of p as you go along. The value of p in the system you end with should generally be lower than the p you started with (as long as the process of fixing problems doesn't create as many other problems). For this reason it's a good idea to recompute the adjusted value of p after each iteration.

Evaluating Sample Size Effectiveness Given Fixed n Suppose that you know you have time to run only a limited number of participants, are willing to treat a single occurrence of a problem seriously, and want to determine what you can expect to get out of a problem-discovery study with that number of participants. If that number were six, for example, examination of Table 10 indicates:

- You are almost certain to detect problems that have a 0.90 likelihood of occurrence (it only takes two participants to have a 99% cumulative likelihood of seeing the problem at least once).
- You are almost certain (between 95 and 99% likely) to detect problems that have a 0.50 likelihood of occurrence (for this likelihood of occurrence, the sample size required at 95% is 5, and at 99% is 7).
- You've got a reasonable chance (about 80% likely) of detecting problems that have a 0.25 likelihood of occurrence (for this likelihood of occurrence, the required sample size at 75% is 5, and at 85% is 7).
- You have a little better than even odds of detecting problems that have a 0.15 likelihood of occurrence (the required sample size at 50% is 5).
- You have a little less than even odds of detecting problems that have a 0.10 likelihood of occurrence (the required sample size at 50% is 7).
- You are not likely to detect many of the problems that have a likelihood of occurrence of 0.05 or 0.01 (for these likelihoods of occurrence, the sample size required at 50% is 14 and 69 respectively).

This analysis illustrates that although a problem-discovery study with a sample size of six participants will typically not discover problems with very low likelihoods of occurrence, the study is almost certainly worth conducting. Applying this procedure to a number of different sample sizes produces Table 11. The cells in Table 11 are the probability of having a problem with a specified occurrence probability happen at least once during a usability study with the given sample size.

Estimating the Number of Problems Available for Discovery Another approach to assessing sample size effectiveness is to estimate the number of undiscovered problems. Returning to the situation illustrated in Table 9, the adjusted estimate of p is 0.28 with four participants and eight unique problems. The estimated proportion of problems discovered with those four participants is 0.73 $[1 - (1 - 0.28)^4]$. If eight problems are about 73% of the total number of problems available for discovery, the total number of

Table 11 Likelihood of Discovering Problems of Probability p at Least Once in a Study with Sample Size n

	Sample Size, n						
Problem Occurrence Probability, p	3	6	9	12	15	18	21
0.01	0.03	0.06	0.09	0.11	0.14	0.17	0.19
0.05	0.14	0.26	0.37	0.46	0.54	0.60	0.66
0.10	0.27	0.47	0.61	0.72	0.79	0.85	0.89
0.15	0.39	0.62	0.77	0.86	0.91	0.95	0.97
0.25	0.58	0.82	0.92	0.97	0.99	0.99	1.00
0.50	0.88	0.98	1.00	1.00	1.00	1.00	1.00
0.90	1.00	1.00	1.00	1.00	1.00	1.00	1.00

problems available for discovery (given the constraints of the testing situation) is about 11 (8/0.73). Thus, there appear to be about three undiscovered problems. With an estimate of only three undiscovered problems, the sample size of four is approaching adequacy.

Contrast this with the MACERR study described in J. Lewis (2001a), which had an estimated value of p of 0.16 with 15 participants and 145 unique problems. For this study, the estimated proportion of discovered problems at the end of the test was 0.927 $[1 - (1 - 0.16)^{15}]$. The estimate of the total number of problems available for discovery was about 156 (145/0.927). With about 11 problems remaining available for discovery, it might be wise to run a few more participants.

On the other hand, with an estimated 92.7% of problems available for discovery extracted from the problem-discovery space defined by the test conditions, it might make more sense to make changes to the test conditions (in particular, to make reasonable changes to the tasks) to create additional opportunities for problem discovery. This is one of many areas in which practitioners need to exercise professional judgment, using the available tables and formulas to guide that judgment.

Some Tips on Managing p Because p (the average likelihood of problem discovery) is such an important factor in sample size estimation for usability tests, it generally makes sense to attempt to manage it (although in some situations, such management is out of a practitioner's control). Here are some ways to increase p:

- Use highly skilled observers for usability studies.
- Use multiple observers rather than a single observer (Hertzum and Jacobsen, 2003).
- Focus evaluation on new products with newly designed interfaces rather than older, more refined interfaces.
- Study less-skilled participants in usability studies (as long as they are appropriate participants).
- Make the user sample as heterogeneous as possible, within the bounds of the population to which you plan to generalize the results.
- Make the task sample as heterogeneous as possible.
- Emphasize complex rather than simple tasks.
- For heuristic evaluations, use examiners with usability and application-domain expertise (double experts) (Nielsen, 1992).
- For heuristic evaluations, if you must make a trade-off between having a single evaluator spend a lot of time examining an interface vs. having more examiners spend less time each examining an interface, choose the latter option (Dumas et al., 1995; Virzi, 1997).

Note that some of the tips for increasing p are the opposite of those that reduce measurement variability.

3.1.3 Sample Sizes for Nontraditional Areas of Usability Evaluation

Nontraditional areas of usability evaluation include activities such as the evaluation of visual design and marketing materials. As with traditional areas of evaluation, the first step is to determine if the evaluation is comparative/parameter estimation or problem discovery.

Part of the problem with nontraditional areas is that there is less information regarding the values of the variables needed to estimate sample sizes. Another issue is whether these areas are focused inherently on detecting more subtle effects than is the norm in usability testing, which has a focus on large low-variability effects (and correspondingly small sample size requirements). Determining this requires the involvement of someone with domain expertise in these nontraditional areas. It seems, however, that even these nontraditional areas would benefit from focusing on the discovery of large low-variability effects. Only if there were a business case which held that investment in a study to detect small, highly variable effects would ultimately pay for itself should you conduct such a study.

For example, in *The Survey Research Handbook*, Alreck and Settle (1985) point out that the reason that survey samples rarely contain fewer than several hundred respondents is due to the cost structure of surveys. The fixed costs of the survey include activities such as determining information requirements, identifying survey topics, selecting a data collection method, writing questions, choosing scales, composing the questionnaire, and so on. For this type of research, the additional or marginal cost of including hundreds of additional respondents can be very small relative to the fixed costs. Contrast this with the cost (or feasibility) of adding participants to a usability study in which there might be as little as a week or two between the availability of testable software and the deadline for affecting the product, with resources limiting the observation of participants to one at a time and the test scenarios requiring two days to complete. The potentially high cost of observing participants in usability tests is one reason why usability researchers have devoted considerable attention to sample size estimation, despite some assertions that sample size estimation is relatively unimportant (Wixon, 2003).

"Since the numbers don't know where they came from, they always behave just the same way, regardless" (Lord, 1953, p. 751). What potentially differs for nontraditional areas of usability evaluation isn't the behavior of numbers or statistical procedures, but the researchers' goals and economic realities.

3.2 Confidence Intervals

A major trend in modern statistical evaluation has been a reduced focus on hypothesis testing and a move toward more informative analyses such as effect sizes and confidence intervals (Landauer, 1997). For most applied usability work, confidence intervals are more useful than effect sizes because they have the same units of measurement as the variables from which they are computed. Even when confidence intervals are very wide, they can still be informative, so practitioners should routinely report confidence intervals for their measurements. Although 95% confidence is a commonly used level, confidence as low as 80% will often be appropriate for applied usability measurements (Nielsen, 1997).

3.2.1 Intervals Based on *t*-Scores

Formulas for the computation of confidence intervals based on t-scores are algebraically equivalent to those used to estimate required sample sizes for measurement-based usability tests, but isolate the critical difference (d) instead of the sample size (n):

$$d = \text{SEM} \times t_{\text{crit}} \quad (9)$$

where SEM is the standard error of the mean (the standard deviation, S, divided by the square root of the sample size, n) and t_{crit} is the t-value associated with the desired level of confidence (found in a t-table, available in most statistics texts). (Practitioners who are concerned about departures from normality can perform a logarithmic transformation on their raw data before computing the confidence interval, then transform the data back to report the mean and confidence interval limits.)

For example, suppose that a task in a usability test with seven participants has an average completion time of 5.4 minutes with a standard deviation of 2.2 minutes. The SEM is 0.83 ($2.2/7^{1/2}$). For 90% confidence and 6 ($n - 1$) degrees of freedom, the tabled value of t is 1.943. The computed value of d is 1.6 [(0.83)(1.943)], so the 90% confidence interval is 5.4 ± 1.6 minutes.

As a second example, suppose that the results of a within-subjects test of the time required for two installation procedures showed that the mean of the difference scores (version A minus version B) was 2 minutes with a standard deviation of 2 minutes for a sample size of eight participants. The SEM is 0.71 ($2/8^{1/2}$). For 95% confidence and 7 ($n - 1$) degrees of freedom, the tabled value of t is 2.365. The computed value of d is 1.7 [(0.71)(2.365)], so the 95% confidence interval is 2.0 ± 1.7 minutes (ranging from 0.3 to 3.7 minutes). Because the confidence interval does not contain 0, this interval indicates that with α of 0.05 (where α is 1 minus the confidence expressed as a proportion rather than a percentage) you should reject the null hypothesis of no difference. The evidence indicates that version A takes longer than version B. The major advantage of a confidence interval over a significance test is that you also know with 95% confidence that the magnitude of the difference is probably no less then 0.3 minute and no greater than 3.7 minutes. If the versions are otherwise equal, version B is the clear winner. If the cost of version B is greater than the cost of version A (e.g., due to a need to license a new technology for version B), the decision about which version to implement is more difficult but is certainly aided by having an estimate of the upper and lower limits of the difference between the two versions.

3.2.2 Binomial Confidence Intervals

As discussed above, confidence intervals constructed around a mean can be very useful. Many usability measurements, however, are proportions or percentages computed from count data rather than means. For example, a usability defect rate for a specific problem is the proportion computed by dividing the number of participants who experience the problem divided by the total number of participants.

The statistical term for a study designed to estimate proportions is a *binomial experiment*, because a given problem either will or will not occur for each trial (participant) in the experiment. For example, a participant either will or will not install an option correctly. The point estimate of the defect rate is the observed proportion of failures (p). However, the likelihood is very small that the point estimate from a study is exactly the same as the true percentage of failures, especially if the sample size is small (Walpole, 1976). To compensate for this, you can calculate interval estimates that have a known likelihood of containing the true proportion (Steele and Torrie, 1960). You can use these binomial confidence intervals to describe the proportion of usability defects effectively, often with only a small sample (J. Lewis, 1996a). Cordes and Lentz (1986) and J. Lewis (1996a) provided BASIC programs for the computation of exact binomial confidence intervals. There are similar programs available at the Web site of the Southwest Oncology Group Statistical Center (SOGSC, 2004), the GraphPad Web site (GraphPad, 2004), and the Measuring Usability Web site (Sauro, 2004).

Some programs (Cordes and Lentz, 1986; Lewis, 1996a; SOGSC, 2004) produce binomial confidence intervals that always contain the exact binomial confidence interval. Other programs (GraphPad, 2004; Sauro, 2004) also produce a new type of interval called *approximate* binomial confidence intervals (Agresti and Coull, 1998). Exact and approximate binomial confidence intervals differ in a number of ways. An exact binomial confidence interval guarantees that the actual confidence is equal to or greater than the nominal confidence. An approximate interval guarantees that the average of the actual confidence in the long run will be equal to the nominal confidence, but for any specific test, the actual confidence could be lower than the nominal confidence. On the other hand, approximate binomial confidence intervals tend to be narrower than exact intervals. When sample sizes are large ($n > 100$), the two types of intervals are virtually indistinguishable. When sample sizes are small though, there can be a considerable difference in the width of the intervals, especially when the observed proportion is close to 0 or 1. The exact interval often has an actual confidence closer to 99% when the nominal confidence is 95%, making it too conservative.

Monte Carlo studies that have compared exact and approximate binomial confidence intervals using standard statistical distributions (Agresti and Coull, 1998) and data from usability studies (Sauro and Lewis, 2005) generally support the use of approximate rather than exact binomial confidence intervals. When the actual confidence of an approximate binomial confidence interval is below the nominal level, the actual level tends to be close to the nominal level. For example, Agresti and Coull (1998) found that the actual level for 95% approximate binomial confidence intervals using the adjusted-Wald method was never less than 89%.

> In forming a 95% confidence interval, is it better to use an approach that guarantees that the actual coverage probabilities are *at least* .95 yet typically achieves coverage probabilities of about .98 or .99, or an approach giving narrower intervals for which the actual coverage probability could be less than .95 but is usually quite *close* to .95? For most applications, we would prefer the latter (Agresti and Coull, 1998, p. 125).

This conclusion, that using approximate binomial confidence intervals will tend to produce superior decisions relative to the use of exact intervals, seems to apply to usability test data (Sauro and Lewis, 2005). If, however, it is critical for a specific test to achieve or exceed the nominal level of confidence, then it is reasonable to use an exact binomial confidence interval.

When using binomial confidence intervals, note that if the failure rate is fairly high, you do not need a very large sample to acquire convincing evidence of failure. In the first evaluation of a wordless graphic instruction (J. Lewis and Pallo, 1991), 9 of 11 installations (82%) were incorrect. The exact 90% binomial confidence interval for this outcome ranged from 0.53 to 0.97. This interval allowed us to argue that without intervention, the failure rate for installation would be at least 53% (and more likely closer to the observed 82%).

This suggests that a reasonable strategy for binomial experiments is to start with a small sample size and record the number of failures. From these results, compute a confidence interval. If the lower limit of the confidence interval indicates an unacceptably high failure rate, stop testing. Otherwise, continue testing and evaluating in increments until you reach a specified level of precision or you reach the maximum sample size allowed for the study.

This method can rapidly demonstrate with a small sample that a usability defect is unacceptably high if the criterion is low and the true defect rate is high. Although the confidence interval will be wide (50 percentage points in the graphic symbols example), the lower limit of the interval may be clearly unacceptable. When the true defect rate is low or the criterion is high, this procedure may not work without a large sample size. The decision to continue sampling or to stop the study should be determined by a reasonable business case that balances the cost of continued data collection against the potential cost of allowing defects to go uncorrected.

You cannot use this procedure with small samples to prove that a success rate is acceptably high. With small samples, even if the defect percentage observed is zero or close to 0%, the interval will be wide, so it will probably include defect percentages that are unacceptable. For example, suppose that you have run five participants through a task and all five have completed the task successfully. The 90% confidence interval on the percentage of defects for these results ranges from 0 to 45%, with a 45% defect rate almost certainly unacceptable. If you had 50 out of 50 successful task completions, the 90% binomial confidence interval would range from 0 to 6%, which would indicate a greater likelihood of the true defect rate being close to 0%. The moral of the story is that it is relatively easy to prove (requires a small sample) that a product is unacceptable, but it is difficult to prove (requires a large sample) that a product is acceptable.

3.3 Standardized Usability Questionnaires

Standardized satisfaction measures offer many advantages to the usability practitioner. Specifically, standardized measurements provide objectivity, replicability, quantification, economy, communication, and scientific generalization (Nunnally, 1978). The first published standardized usability questionnaires appeared in the late 1980s (Chin et al., 1988; Kirakowski and

Dillon, 1988). Questionnaires focused on the measurement of computer satisfaction preceded these questionnaires (e.g., the Gallagher Value of MIS Reports Scale and the Hatcher and Diebert Computer Acceptance Scale) (see LaLomia and Sidowski, 1990, for a review), but these questionnaires were not applicable to scenario-based usability tests.

The most widely used standardized usability questionnaires are the Questionnaire for User Interaction Satisfaction (QUIS) (Chin et al., 1988), the Software Usability Measurement Inventory (SUMI) (Kirakowski and Corbett, 1993; Kirakowski, 1996), the Post-Study System Usability Questionnaire (PSSUQ) (J. Lewis, 1992, 1995, 2002), and the Software Usability Scale (SUS) (Brooke, 1996). The most common application of these questionnaires is at the end of a test (after completing a series of test scenarios). The After-Scenario Questionnaire (ASQ) (J. Lewis, 1991b) is a short three-item questionnaire designed for administration immediately following the completion of a test scenario. The ASQ takes less than a minute to complete. The longer standard questionnaires typically have completion times of less than 10 minutes (Dumas, 2003).

The primary measures of standardized questionnaire quality are reliability (consistency of measurement) and validity (measurement of the intended attribute) (Nunnally, 1978). There are several ways to assess reliability, including test–retest and split-half reliability. The most common method for the assessment of reliability is coefficient α, a measurement of internal consistency. Coefficient α can range from 0 (no reliability) to 1 (perfect reliability). Measures that can affect a person's future, such as IQ tests or college entrance exams should have a minimum reliability of 0.90 (preferably, reliability greater than 0.95). For other research or evaluation, measurement reliability in the range of 0.70 to 0.80 is acceptable (Nunnally, 1978; Landauer, 1997).

A questionnaire's validity is the extent to which it measures what it claims to measure. Researchers commonly use the Pearson correlation coefficient to assess criterion-related validity (the relationship between the measure of interest and a different concurrent or predictive measure). These correlations do not have to be large to provide evidence of validity. For example, personnel selection instruments with validities as low as 0.30 or 0.40 can be large enough to justify their use (Nunnally, 1978). Another approach to validity is content validity, typically assessed through the use of factor analysis (which also helps questionnaire developers discover or confirm clusters of related items that can form reasonable subscales).

Regarding the appropriate number of scale steps, it is true that more scale steps are better than fewer scales steps, but with rapidly diminishing returns. The reliability of individual items is a monotonically increasing function of the number of steps (Nunnally, 1978). As the number of scale steps increase from 2 to 20, the increase in reliability is very rapid at first, but tends to level off at about 7. After 11 steps there is little gain in reliability from increasing the number. The number of steps in an item is very important for measurements based on a single item but is less important when computing measurements over a number of items (as in the computation of an overall or subscale score).

3.3.1 QUIS

The QUIS (Shneiderman, 1987; Chin et al., 1988; see also http://lap.umd.edu/QUIS/) is a product of the Human–Computer Interaction Lab at the University of Maryland. Its use requires the purchase of a license. Chin et al. (1988) evaluated several early versions of the QUIS (Versions 3 through 5). They reported an overall reliability (coefficient α) of 0.94 but did not report any subscale reliability.

The QUIS is currently at Version 7. This version includes demographic questions, an overall measure of system satisfaction, and 11 specific interface factors. The QUIS is available in two lengths, short (26 items) and long (71 items). The items are nine-point scales anchored with opposing adjective phrases (such as "confusing" and "clear" for the item "messages which appear on screen").

3.3.2 CUSI and SUMI

The Human Factors Research Group (HFRG) at University College Cork published their first standardized questionnaire, the Computer Usability Satisfaction Inventory (CUSI), in 1988 (Kirakowski and Dillon, 1988). The CUSI was a 22-item questionnaire containing two subscales: affect and competence. Its overall reliability was 0.94, with 0.91 for affect and 0.89 for competence.

The HFRG replaced the CUSI with the SUMI (Kirakowski and Corbett, 1993; Kirakowski, 1996), a questionnaire that has six subscales: global, efficiency, affect, helpfulness, control, and learnability. Its 50 items are statements (such as "The instructions and prompts are helpful") to which participants indicate that they agree, are undecided, or disagree. The SUMI has undergone a significant amount of psychometric development and evaluation to arrive at its current form. The results of studies (McSweeney, 1992; Wiethoff et al., 1992) with results that included significant main effects of system, SUMI scales, and their interaction, support its validity. The reported reliabilities of the six subscales (measured with coefficient α) are:

- Global: 0.92
- Efficiency: 0.81
- Affect: 0.85
- Helpfulness: 0.83
- Control: 0.71
- Learnability: 0.82

One of the greatest strengths of the SUMI is the database of results that is available for the construction of interpretive norms. This makes it possible for practitioners to compare their results with those of similar products (as long as there are similar products in the database). Another strength is that the SUMI is available in different languages (such as UK English, American English, Italian, Spanish, French, German, Dutch, Greek, and Swedish). Like the QUIS, practitioners planning to use SUMI must purchase a license for its use (which includes questionnaires and scoring software). For an additional fee, a trained psychometrician at the HFRG will score the results and produce a report.

3.3.3 SUS

Usability practitioners at Digital Equipment Corporation (DEC) developed the SUS in the mid-1980s (Dumas, 2003). The 10 five-point items of the SUS provide a unidimensional (no subscales) usability measurement that ranges from 0 to 100. In the first published account of the SUS, Brooke (1996) stated that the SUS was robust, reliable, and valid but did not publish the specific reliability or validity measurements. With regard to validity, "it correlates well with other subjective measures of usability (e.g., the general usability subscale of the SUMI)" (Brooke, 1996, p. 194). DEC has copyrighted the SUS, but according to Brooke (1996, p. 194), "the only prerequisite for its use is that any published report should acknowledge the source of the measure."

3.3.4 PSSUQ and CSUQ

The PSSUQ is a questionnaire designed for the purpose of assessing users' perceived satisfaction with their computer systems. It has its origin in an internal IBM project called SUMS (System Usability MetricS), headed by Suzanne Henry in the late 1980s. A team of human factors engineers and usability specialists working on SUMS created a pool of seven-point scale items based on the work of Whiteside et al. (1988) and from that pool selected 18 items to use in the first version of the PSSUQ (J. Lewis, 1992). Each item was worded positively, with the scale anchors "strongly agree" at the first scale position (1) and "strongly disagree" at the last scale position (7). A "not applicable" (NA) choice and a comment area were available for each item [see J. Lewis (1995) for examples of the appearance of the items].

The development of the Computer System Usability Questionnaire (CSUQ) followed the development of the first version of the PSSUQ. Its items are identical to those of the PSSUQ except that their wording is appropriate for use in field settings or surveys rather than in a scenario-based usability test, making it, essentially, an alternative form of the PSSUQ. For a discussion of CSUQ research and comparison of the PSSUQ and CSUQ items, see J. Lewis (1995).

An unrelated series of IBM investigations into customer perception of usability revealed a common set of five usability characteristics associated with usability by several different user groups (Doug Antonelli, personal communication, January 5, 1991). The 18-item version of the PSSUQ addressed four of these five characteristics (quick completion of work, ease of learning, high-quality documentation and online information, and functional adequacy), but did not address the fifth (rapid acquisition of productivity). The second version of the PSSUQ (J. Lewis, 1995) included an additional item to address this characteristic, bringing the total number of items up to 19.

J. Lewis (2002) conducted a psychometric evaluation of the PSSUQ using data from several years of usability studies (primarily studies of speech dictation systems, but including studies of other types of applications). The results of a factor analysis on these data were consistent with earlier factor analyses (J. Lewis, 1992, 1995) used to define three PSSUQ subscales: system usefulness (SysUse), information quality (InfoQual), and interface quality (IntQual). Estimates of reliability were also consistent with those of earlier studies. Analyses of variance indicated that variables such as the specific study, developer, state of development, type of product, and type of evaluation significantly affected PSSUQ scores. Other variables, such as gender and completeness of responses to the questionnaire, did not. Norms derived from the new data correlated strongly with norms derived from earlier studies.

Significant correlation analyses indicated scale validity (J. Lewis, 1995). For a sample of 22 participants who completed all PSSUQ and ASQ items in a usability study (Lewis et al., 1990), the overall PSSUQ score correlated highly with the sum of the ASQ ratings that participants gave after completing each scenario [$r(20) = 0.80$, $p = 0.0001$]. The overall PSSUQ score correlated significantly with the percentage of successful scenario completions [$r(29) = -0.40$, $p = 0.026$]. SysUse [$r(36) = -0.40$, $p = 0.006$] and IntQual [$r(35) = -0.29$, $p = 0.08$] also correlated with the percentage of successful scenario completions.

One potential criticism of the PSSUQ has been that some items seemed redundant and that this redundancy might inflate estimates of reliability. J. Lewis (2002) investigated the effect of removing three items from the second version of the PSSUQ (items 3, 5, and 13). With these items removed, the reliability of the overall PSSUQ score (using coefficient α) was 0.94 (remaining very high), and the reliabilities of the three subscales were:

- SysUse: 0.90
- InfoQual: 0.91
- IntQual: 0.83

All of the reliabilities exceeded 0.80, indicating sufficient reliability to be valuable as usability measurements (Anastasi, 1976; Landauer, 1997). Thus, the third (and current) version of the PSSUQ has 16 seven-point scale items (see Table 12 for the items and their normative scores).

Note that the scale construction is such that lower scores are better than higher scores and that the means of the items and scales all fall below the scale midpoint of 4. With the exception of item 7 ("The system gave error messages that clearly told me how to fix problems"), the upper limits of the confidence intervals are below 4. This shows that practitioners should not use the scale midpoint exclusively as a reference from which they would judge participants' perceptions of usability. Rather, they should also use the norms shown in Table 12 (and comparison with these norms

Table 12 PSSUQ Version 3 Items, Scales, and Normative Scores (99% Confidence Intervals)[a]

Item/Scale	Item Text/Scale Scoring Rule	Norm (99% CI) Lower Limit	Mean	Upper Limit
Q1	Overall, I am satisfied with how easy it is to use this system.	2.60	2.85	3.09
Q2	It was simple to use this system.	2.45	2.69	2.93
Q3	I was able to complete the tasks and scenarios quickly using this system.	2.86	3.16	3.45
Q4	I felt comfortable using this system.	2.40	2.66	2.91
Q5	It was easy to learn to use this system.	2.07	2.27	2.48
Q6	I believe I could become productive quickly using this system.	2.54	2.86	3.17
Q7	The system gave error messages that clearly told me how to fix problems.	3.36	3.70	4.05
Q8	Whenever I made a mistake using the system, I could recover easily and quickly.	2.93	3.21	3.49
Q9	The information (such as on-line help, on-screen messages and other documentation) provided with this system was clear.	2.65	2.96	3.27
Q10	It was easy to find the information I needed.	2.79	3.09	3.38
Q11	The information was effective in helping me complete the tasks and scenarios.	2.46	2.74	3.01
Q12	The organization of information on the system screens was clear.	2.41	2.66	2.92
Q13	The interface[b] of this system was pleasant.	2.06	2.28	2.49
Q14	I liked using the interface of this system.	2.18	2.42	2.66
Q15	This system has all the functions and capabilities I expect it to have.	2.51	2.79	3.07
Q16	Overall, I am satisfied with this system.	2.55	2.82	3.09
SysUse	Average items 1 through 6.	2.57	2.80	3.02
InfoQual	Average items 7 through 12.	2.79	3.02	3.24
IntQual	Average items 13 through 15.	2.28	2.49	2.71
Overall	Average items 1 through 16.	2.62	2.82	3.02

Source: J. Lewis (2002).

[a] SysUse, system usefulness; InfoQual, information quality; IntQual, interface quality; CI, confidence interval. Scores can range from 1 (strongly agree) to 7 (strongly disagree), with lower scores better than higher scores.

[b] The "interface" includes those items that you use to interact with the system. For example, some components of the interface are the keyboard, the mouse, the microphone, and the screens (including their graphics and language).

is probably more meaningful than comparison with the scale midpoint).

The way that item 7 stands out from the others indicates:

- It should not surprise practitioners if they find this in their own data.
- It is a difficult task to provide usable error messages throughout a product.
- It may well be worth the effort to focus on providing usable error messages.
- If practitioners find the mean for this item to be equal to or less than the mean of the other items in InfoQual (assuming that they are in line with the norms), they have been successful in creating better-than-average error messages.

The consistent pattern of relatively poor ratings for InfoQual versus IntQual [seen across all the studies; for details and complete normative data, see J. Lewis (2002)] suggests that practitioners who find this pattern in their data should not conclude that they have poor documentation or a great interface. Suppose, however, that this pattern appeared in the first iteration of a usability evaluation and the developers decided to emphasize improvement to the quality of their information. Any subsequent decline in the difference between InfoQual and IntQual would be evidence of a successful intervention.

Another potential criticism of the PSSUQ is that the items do not follow the typical convention of varying the tone of the items so that half of the items elicit agreement and the other half elicit disagreement. The rationale for the decision to align the items consistently was to make it as easy as possible for participants to complete the questionnaire. With consistent item alignment, the proper way to mark responses on the items is clearer, potentially reducing response errors due to participant confusion. Also, the use of negatively worded items can produce a number of undesirable effects (Barnette, 2000; Ibrahim, 2001), including problems with internal consistency and factor structure. The setting in which balancing the tone of the items is likely to be of greatest value is when participants do not have a high degree of motivation for providing reasonable and honest responses (e.g., in clinical and educational settings). Obtaining reasonable and honest responses is rarely a problem in most usability testing settings.

Additional key findings and conclusions from J. Lewis (2002) were:

- There was no evidence of response styles (especially, no evidence of extreme response style) in the PSSUQ data.
- Because there is a possibility of extreme response and acquiescence response styles in cross-cultural research (Grimm and Church, 1999; Baumgartner and Steenkamp, 2001; Clarke, 2001; van de Vijver and Leung, 2001), practitioners should avoid using questionnaires for cross-cultural comparison unless that use has been validated. Other types of group comparisons with the PSSUQ are valid because any effect of response style should cancel out across experimental conditions.
- Scale scores from incomplete PSSUQs were indistinguishable from those computed from complete PSSUQs. These data do not provide information concerning how many items a participant might ignore and still produce reliable scale scores. They do suggest that, in practice, participants typically complete enough items to produce reliable scale scores.

The similarity of psychometric properties across the various versions of the PSSUQ, despite the passage of time and differences in the types of systems studied, provide evidence of significant generalizability for the questionnaire, supporting its use by practitioners for measuring participant satisfaction with the usability of tested systems. Due to its generalizability, practitioners can confidently use the PSSUQ when evaluating different types of products and at different times during the development process. The PSSUQ can be especially useful in competitive evaluations (for an example, see J. Lewis, 1996b) or when tracking changes in usability as a function of design changes made during development. Practitioners and researchers are free to use the PSSUQ and CSUQ (no license fees), but anyone using them should cite the source.

3.3.5 ASQ

The ASQ (J. Lewis, 1991b, 1995) is an extremely short questionnaire (three 7-point scale items using the same format as the PSSUQ). The items address three important aspects of user satisfaction with system usability: ease of task completion ("Overall, I am satisfied with the ease of completing the tasks in this scenario"), time to complete a task ("Overall, I am satisfied with the amount of time it took to complete the tasks in this scenario"), and adequacy of support information ["Overall, I am satisfied with the support information (on-line help, messages, documentation) when completing tasks"]. The overall ASQ score is the average of responses to these three items.

Because the questionnaire is short, it takes very little time for participants to complete, an important practical consideration for usability studies. Measurements of ASQ reliability (using coefficient α) have ranged from 0.90 to 0.96 (J. Lewis, 1995). A significant correlation between ASQ scores and successful scenario completion [$r(46) = -0.40$, $p < 0.01$) in J. Lewis et al. (1990; analysis reported in J. Lewis, 1995) provided evidence of concurrent validity. Like the PSSUQ and CSUQ, the ASQ is available for free use by practitioners and researchers, but anyone using the ASQ should cite the source.

4 WRAPPING UP

4.1 Getting More Information about Usability Testing

This chapter has provided fundamental and some advanced information about usability testing, but there is only so much that you can cover in a single chapter. For additional chapter-length treatments of the basics of usability testing, see Nielsen (1997) and Dumas (2003). There are also two well-known books devoted to the topic of usability testing. Dumas and Redish (1999) is one of these book-length treatments of usability testing. The content and references are somewhat dated. The 1999 copyright date is a bit misleading, as the body of the book has not changed since its 1993 edition. The 1999 edition does include a new preface and some updated reading recommendations and provides excellent coverage of the fundamentals of usability testing.

The other well-known usability testing book is Rubin (1994). Like Dumas and Redish (1999), the content and references are 10 years out of date. It, too, covers the fundamentals of usability testing (which haven't really changed for 20 years) very well and contains many useful samples of a variety of testing-related forms and documents.

For late-breaking developments in usability research and practice, there are a number of annual conferences that have usability evaluation as a significant portion of their content. Companies making a sincere effort in the professional development of their usability practitioners should ensure that their personnel have access to the proceedings of these conferences and should support attendance at one or more of these conferences at least every few years. These major conferences are:

- Usability Professionals Association (http://www.upassoc.org/)
- Human–Computer Interaction International (http://www.hci-international.org/)
- ACM Special Interest Group in Computer–Human Interaction (http://www.acm.org/sigchi/)
- Human Factors and Ergonomics Society (http://hfes.org/)
- INTERACT (held every two years; see, e.g., http://www.interact2005.org/)

4.2 Research Challenge: Improved Understanding of Usability Problem Detection

The recently published papers that have questioned the reliability of usability problem discovery (Molich et al., 1998, 2004; Kessner et al., 2001; Hertzum and Jacobsen, 2003) have raised a number of questions. Dumas (2003, p. 1112) responded: "It is not clear why there is so little overlap in problems. Are slight variations in method the cause? Are the problems really the same but just described differently? We look to further research to sort out the possibilities." Hertzum and Jacobsen (2003) noted the following as potential causes of lack of reliability across evaluations: (1) vague goal analyses that lead to variability in task scenarios, (2) vague evaluation procedures, and (3) vague problem criteria that lead to acceptance of anything as a usability problem.

Developing a better understanding of why these studies produced their results, which are so at odds with the apparent success of usability testing (Al-Awar et al., 1981; Kennedy, 1982; J. Lewis, 1982, 1996b; Gould et al., 1987; Kelley, 1984; Marshall et al., 1990; Bailey et al., 1992; Bailey, 1993; Ruthford and Ramey, 2000), should be one of the top usability research efforts of the coming decade. An improved understanding might provide guidance about how or whether practitioners should change the way they conduct usability tests. One of the most important components of this research effort should be to investigate the cognitive mechanisms that underlie the detection of usability problems. Even after over 20 years of professional practice with usability methods, there is still no general consensus on the boundaries of what constitutes a usability problem or on the appropriate levels of description of usability problems (J. Lewis, 2001b). Doctoral candidates should take note—this is a topic rich with theoretical and practical consequences!

For example, it seems reasonable that the task of the observer in a usability study involves classical signal-detection issues (Massaro, 1975). The observer monitors participant behavior, and at any given moment must decide whether that observed behavior is indicative of a usability problem. Thus, there are two ways for an observer to make correct decisions (rejecting nonproblem behaviors correctly, identifying problem behaviors correctly) and two ways to make incorrect decisions (identifying nonproblem behaviors as indicative of a usability problem, failing to identify problem behaviors as indicative of a usability problem). In signal detection terms, the names for these right and wrong decisions are correct rejection, hit, false alarm, and miss. The rates for these types of decisions depend independently on both the skill and the bias of the observer. Applying signal detection theory to the assessment of the skill and bias of usability test observers is a potentially rich, but to date untapped, area of research, with potential application for both selection and training of observers.

4.3 Usability Testing: Yesterday, Today, and Tomorrow

It seems clear that usability testing (both summative and formative) is here to stay, and that its general form will remain similar to the forms that emerged in the late 1970s and early 1980s. The last 25 years have seen the introduction of more usability evaluation techniques and some consensus (and some continuing debate) on the conditions under which to use the various techniques (of which usability testing is a major one). In the last 15 years, usability researchers have made significant progress in the areas of standardized usability questionnaires and sample size estimation for formative usability tests. As we look to the future, usability practitioners should monitor the continuing research that will almost certainly take place in developing a better understanding of the cognitive mechanisms of usability problem discovery because such an understanding has the potential to increase the reliability of usability testing.

In the meantime, practitioners will continue to perform usability tests, exercising professional judgment as required. Usability testing is not a perfect usability evaluation method in the sense that it does not guarantee the discovery of all possible usability problems, but it doesn't have to be perfect to be useful and effective. It is, however, important to understand its strengths, limitations, and current best practices to ensure its proper (most effective) use.

ACKNOWLEDGMENTS

I want to thank Gavriel Salvendy for giving me the opportunity and encouragement to write this chapter. I also want to express my deepest appreciation to the colleagues who took the time to review the first draft of the chapter under a very tight deadline: Patrick Commarford, Richard Cordes, Barbara Millet, Jeff Sauro, and Wallace Sadowski.

REFERENCES

Aaker, D. A., and Day, G. S. (1986), *Marketing Research*, Wiley, New York.

Abelson, R. P. (1995), *Statistics as Principled Argument*, Lawrence Erlbaum Associates, Mahwah, NJ.

Agresti, A., and Coull, B. (1998), "Approximate Is Better Than 'Exact' for Interval Estimation of Binomial Proportions," *The American Statistician*, Vol. 52, pp. 119–126.

Al-Awar, J., Chapanis, A., and Ford, R. (1981), "Tutorials for the First-Time Computer User," *IEEE Transactions on Professional Communication*, Vol. 24, pp. 30–37.

Alreck, P. L., and Settle, R. B. (1985), *The Survey Research Handbook*, Richard D. Irwin, Homewood, IL.

Alty, J. L. (1992), "Can We Measure Usability?" in *Proceedings of Advanced Information Systems*, Learned Information, London, pp. 95–106.

Anastasi, A. (1976), *Psychological Testing*, Macmillan, New York.

Andre, T. S., Belz, S. M, McCreary, F. A., and Hartson, H. R. (2000), "Testing a Framework for Reliable Classification of Usability Problems," in *Proceedings of the IEA 2000/HFES 2000 Congress*, Human Factors and Ergonomics Society, Santa Monica, CA, pp. 573–576.

ANSI (2001), *Common Industry Format for Usability Test Reports*, ANSI-NCITS 354-2001, American National Standards Institute, Washington, DC.

Aykin, N. M., and Aykin, T. (1991), "Individual Differences in Human–Computer Interaction," *Computers and Industrial Engineering*, Vol. 20, pp. 373–379.

Bailey, G. (1993), "Iterative Methodology and Designer Training in Human–Computer Interface Design," in

INTERCHI '93 Conference Proceedings, ACM, New York, pp. 198–205.

Bailey, R. W., Allan, R. W., and Raiello, P. (1992), "Usability Testing vs. Heuristic Evaluation: A Head to Head Comparison," in *Proceedings of the Human Factors and Ergonomics Society 36th Annual Meeting*, Human Factors and Ergonomics Society, Santa Monica, CA, pp. 409–413.

Banks, S. (1965), *Experimentation in Marketing*, McGraw-Hill, New York.

Barnette, J. J. (2000), "Effects of Stem and Likert Response Option Reversals on Survey Internal Consistency: If You Feel the Need, There Is a Better Alternative to Using Those Negatively Worded Stems," *Educational and Psychological Measurement*, Vol. 60, pp. 361–370.

Baumgartner, H., and Steenkamp, J. B. E. M. (2001), "Response Styles in Marketing Research: A Cross-National Investigation," *Journal of Marketing Research*, Vol. 38, pp. 143–156.

Bennett, J. L. (1979), "The Commercial Impact of Usability in Interactive Systems," *Infotech State of the Art Report: Man/Computer Communication*, Vol. 2, pp. 289–297.

Berry, D. C., and Broadbent, D. E. (1990), "The Role of Instruction and Verbalization in Improving Performance on Complex Search Tasks," *Behaviour and Information Technology*, Vol. 9, pp. 175–190.

Bevan, N., Kirakowski, J., and Maissel, J. (1991), "What Is Usability?" in *Human Aspects in Computing, Design and Use of Interactive Systems and Work with Terminals, Proceedings of the 4th International Conference on Human–Computer Interaction*, H. J. Bullinger, Ed., Elsevier Science, Stuttgart, Germany, pp. 651–655.

Bias, R. G., and Mayhew, D. J. (1994), *Cost-Justifying Usability*, Academic Press, Boston.

Blalock, H. M. (1972), *Social Statistics*, McGraw-Hill, New York.

Boehm, B. W. (1981), *Software Engineering Economics*, Prentice-Hall, Englewood Cliffs, NJ.

Boren, T., and Ramey, J. (2000), "Thinking Aloud: Reconciling Theory and Practice," *IEEE Transactions on Professional Communications*, Vol. 43, pp. 261–278.

Bowers, V., and Snyder, H. (1990), "Concurrent Versus Retrospective Verbal Protocols for Comparing Window Usability," in *Proceedings of the Human Factors Society 34th Annual Meeting*, Human Factors Society, Santa Monica, CA, pp. 1270–1274.

Bradley, J. V. (1976), *Probability; Decision; Statistics*, Prentice-Hall, Englewood Cliffs, NJ.

Brooke, J. (1996), "SUS: A 'Quick and Dirty' Usability Scale," in, *Usability Evaluation in Industry*, P. Jordan, B. Thomas, and B. Weerdmeester, Eds., Taylor & Francis, London, pp. 189–194.

Brown, F. E. (1980), *Marketing Research: A Structure for Decision Making*, Addison-Wesley, Reading, MA.

Campbell, D. T., and Stanley, J. C. (1963), *Experimental and Quasi-experimental Designs for Research*, Rand McNally, Chicago.

Caulton, D. A. (2001), "Relaxing the Homogeneity Assumption in Usability Testing," *Behaviour and Information Technology*, Vol. 20, pp. 1–7.

Chapanis, A. (1981), "Evaluating Ease of Use," unpublished manuscript prepared for IBM, available from J. R. Lewis.

Chapanis, A. (1988), "Some Generalizations About Generalization," *Human Factors*, Vol. 30, pp. 253–267.

Chin, J. P., Diehl, V. A., and Norman, K. L. (1988), "Development of an Instrument Measuring User Satisfaction of the Human–Computer Interface," in *CHI '88 Conference Proceedings: Human Factors in Computing Systems*, E. Soloway, D. Frye, and S. B. Sheppard, Eds., ACM, Washington, DC, pp. 213–218.

Churchill, G. A., Jr. (1991), *Marketing Research: Methodological Foundations*, Dryden Press, Ft. Worth, TX.

Clarke, I. (2001), "Extreme Response Style in Cross-Cultural Research," *International Marketing Review*, Vol. 18, pp. 301–324.

Cliff, N. (1987), *Analyzing Multivariate Data*, Harcourt Brace Jovanovich, San Diego, CA.

Cockton, G., and Lavery, D. (1999), "A Framework for Usability Problem Extraction," in *Human Computer Interaction, INTERACT '99*, IOS Press, Amsterdam, pp. 344–352.

Cordes, R. E. (1993), "The Effects of Running Fewer Subjects on Time-on-Task Measures," *Journal of Human–Computer Interaction*, Vol. 5, pp. 393–403.

Cordes, R. E. (2001), "Task-Selection Bias: A Case for User-Defined Tasks," *Journal of Human–Computer Interaction*, Vol. 13, pp. 411–419.

Cordes, R. E., and Lentz, J. L. (1986), *Usability-Claims Evaluation Using a Binomial Test*, Technical Report 82.0267, International Business Machines Corporation, Tucson, AZ.

Desurvire, H. W., Kondziela, J. M., and Atwood, M. E. (1992), "What Is Gained and Lost When Using Evaluation Methods Other Than Empirical Testing," in *Proceedings of HCI '92: People and Computers VII*, A. Monk, D. Diaper, and M. D. Harrison, Eds., Cambridge University Press, New York, pp. 89–102.

Diamond, W. J. (1981), *Practical Experiment Designs for Engineers and Scientists*, Lifetime Learning Publications, Belmont, CA.

Dickens, J. (1987), "The Fresh Cream Cakes Market: The Use of Qualitative Research as Part of a Consumer Research Programme," in *Applied Marketing and Social Research*, U. Bradley, Ed., Wiley, New York, pp. 23–68.

Dumas, J. S. (2003), "User-Based Evaluations," in *The Human–Computer Interaction Handbook*, J. A. Jacko and A. Sears, Eds., Lawrence Erlbaum Associates, Mahwah, NJ, pp. 1093–1117.

Dumas, J., and Redish, J. C. (1999), *A Practical Guide to Usability Testing*, Intellect, Portland, OR.

Dumas, J., Sorce, J., and Virzi, R. (1995), Expert Reviews: How Many Experts Is Enough? in *Proceedings of the Human Factors and Ergonomics Society 39th Annual Meeting*, Human Factors and Ergonomics Society, Santa Monica, CA, pp. 228–232.

Ericsson, K. A., and Simon, H. A. (1980), "Verbal Reports as Data," *Psychological Review*, Vol. 87, pp. 215–251.

Faulkner, L. (2003), "Beyond the Five-User Assumption: Benefits of Increased Sample Sizes in Usability Testing," *Behavior Research Methods, Instruments, and Computers*, Vol. 35, pp. 379–383.

Fisher, J. (1991), "Defining the Novice User," *Behaviour and Information Technology*, Vol. 10, pp. 437–441.

Fowler, C. J. H., Macaulay, L. A., and Fowler, J. F. (1985), "The Relationship between Cognitive Style and Dialogue Style: An Exploratory Study," In *People and Computers: Designing the Interface*, P. Johnson and S. Cook, Eds., Cambridge University Press, Cambridge, UK, pp. 186–198.

Fu, L., Salvendy, G., and Turley, L. (2002), "Effectiveness of User Testing and Heuristic Evaluation as a Function of Performance Classification," *Behaviour and Information Technology*, Vol. 21, pp. 137–143.

Gawron, V. J., Drury, C. G., Czaja, S. J., and Wilkins, D. M. (1989), "A Taxonomy of Independent Variables Affecting Human Performance," *International Journal of Man–Machine* Studies, Vol. 31, pp. 643–672.

Gordon, W., and Langmaid, R. (1988), *Qualitative Market Research: A Practitioner's and Buyer's Guide*, Gower Publishing, Aldershot, Hampshire, England.

Gould, J. D. (1988), "How to Design Usable Systems," in *Handbook of Human–Computer Interaction*, M. Helander, Ed., North-Holland, Amsterdam, pp. 757–789.

Gould, J. D., and Boies, S. J. (1983), "Human Factors Challenges in Creating a Principal Support Office System: The Speech Filing System Approach," *ACM Transactions on Information Systems*, Vol. 1, pp. 273–298.

Gould, J. D., and Lewis, C. (1984), *Designing for Usability: Key Principles and What Designers Think*, Technical Report RC-10317, International Business Machines Corporation, Yorktown Heights, NY.

Gould, J. D., Boies, S. J., Levy, S., Richards, J. T., and Schoonard, J. (1987), "The 1984 Olympic Message System: A Test of Behavioral Principles of System Design," *Communications of the ACM*, Vol. 30, pp. 758–769.

GraphPad. (2004), "Confidence Interval of a Proportion or Count," http://graphpad.com/quickcalcs/ConfInterval1.cfm.

Gray, W. D., and Salzman, M. C. (1998), "Damaged Merchandise? A Review of Experiments That Compare Usability Evaluation Methods," *Human–Computer Interaction*, Vol. 13, pp. 203–261.

Greene, S. L., Gomez, L. M., and Devlin, S. J. (1986), "A Cognitive Analysis of Database Query Production," in *Proceedings of the 30th Annual Meeting of the Human Factors Society*, Human Factors Society, Santa Monica, CA, pp. 9–13.

Grimm, S. D., and Church, A. T. (1999), "A Cross-Cultural Study of Response Biases in Personality Measures," *Journal of Research in Personality*, Vol. 33, pp. 415–441.

Hackman, G. S., and Biers, D. W. (1992), "Team Usability Testing: Are Two Heads Better Than One? in *Proceedings of the Human Factors and Ergonomics Society 36th Annual Meeting*, Human Factors and Ergonomics Society, Santa Monica, CA, pp. 1205–1209.

Harris, R. J. (1985), *A Primer of Multivariate Statistics*, Academic Press, Orlando, FL.

Hassenzahl. M. (2000), "Prioritizing Usability Problems: Data Driven and Judgement Driven Severity Estimates," *Behaviour and Information Technology*, Vol. 19, pp. 29–42.

Hertzum, M., and Jacobsen, N. J. (2003), "The Evaluator Effect: A Chilling Fact About Usability Evaluation Methods," *Journal of Human–Computer Interaction*, Vol. 15, pp. 183–204.

Holleran, P. A. (1991), "A Methodological Note on Pitfalls in Usability Testing," *Behaviour and Information Technology*, Vol. 10, pp. 345–357.

Ibrahim, A. M. (2001), "Differential Responding to Positive and Negative Items: The Case of a Negative Item in a Questionnaire for Course and Faculty Evaluation," *Psychological Reports*, Vol. 88, pp. 497–500.

ISO (1998), *Ergonomic Requirements for Office Work with Visual Display Terminals (VDTs)*, Part 11, *Guidance on Usability*, ISO 9241-11:1998(E), ISO, Geneva, Switzerland:

Jeffries, R., and Desurvire, H. (1992), "Usability Testing vs. Heuristic Evaluation: Was There a Contest? *SIGCHI Bulletin*, Vol. 24, pp. 39–41.

Kantner, L., and Rosenbaum, S. (1997), "Usability Studies of WWW Sites: Heuristic Evaluation vs. Laboratory Testing," in *Proceedings of SIGDOC 1997*, Salt Lake City, UT, pp. 153–160.

Karat, C. (1997), "Cost-Justifying Usability Engineering in the Software Life Cycle," in *Handbook of Human–Computer Interaction*, 2nd ed., M. Helander, T. K. Landauer, and P. Prabhu, Eds., Elsevier, Amsterdam, The Netherlands, pp. 767–778.

Karat, J. (1997), "User-Centered Software Evaluation Methodologies," in *Handbook of Human–Computer Interaction*, 2nd ed., M. Helander, T. K. Landauer, and P. Prabhu, Eds., Elsevier, Amsterdam, pp. 689–704.

Keenan, S. L., Hartson, H. R., Kafura, D. G., and Schulman, R. S. (1999), "The Usability Problem Taxonomy: A Framework for Classification and Analysis," *Empirical Software Engineering*, Vol. 1, pp. 71–104.

Kelley, J. F. (1984), "An Iterative Design Methodology for User-Friendly Natural Language Office Information Applications," *ACM Transactions on Information Systems*, Vol. 2, pp. 26–41.

Kelley, J. F. (1985), "CAL: A Natural Language Program Developed with the OZ Paradigm: Implications for Supercomputing Systems," in *Proceedings of the First International Conference on Supercomputing Systems*, ACM, New York, pp. 238–248.

Kennedy, P. J. (1982), "Development and Testing of the Operator Training Package for a Small Computer System," in *Proceedings of the Human Factors Society 26th Annual Meeting*, Human Factors Society, Santa Monica, CA, pp. 715–717.

Kessner, M., Wood, J., Dillon, R. F., and West, R. L. (2001), "On the Reliability of Usability Testing," in *Conference on Human Factors in Computing Systems: CHI 2001 Extended Abstracts*, J. Jacko and A. Sears, Eds., ACM Press, Seattle, WA, pp. 97–98.

Kirakowski, J. (1996), "The Software Usability Measurement Inventory: Background and Usage," in *Usability Evaluation in Industry*, P. Jordan, B. Thomas, and B. Weerdmeester, Eds., Taylor & Francis, London, pp. 169–178 see also http://www.ucc.ie/hfrg/questionnaires/sumi/index.html.

Kirakowski, J., and Corbett, M. (1993), "SUMI: The Software Usability Measurement Inventory," *British Journal of Educational Technology*, Vol. 24, pp. 210–212.

Kirakowski, J., and Dillon, A. (1988), *The Computer User Satisfaction Inventory (CUSI): Manual and Scoring Key*, Human Factors Research Group, University College of Cork, Cork, Ireland.

Kraemer, H. C., and Thiemann, S. (1987), *How Many Subjects? Statistical Power Analysis in Research*, Sage Publications, Newbury Park, CA.

LaLomia, M. J., and Sidowski, J. B. (1990), "Measurements of Computer Satisfaction, Literacy, and Aptitudes: A Review," *International Journal of Human–Computer Interaction*, Vol. 2, pp. 231–253.

Landauer, T. K. (1997), "Behavioral Research Methods in Human–Computer Interaction," in *Handbook of Human–Computer Interaction*, 2nd ed., M. Helander, T. K. Landauer, and P. Prabhu, Eds., Elsevier, Amsterdam, pp. 203–227.

Lewis, C., and Norman, D. (1986), "Designing for Error," in *User Centered System Design: New Perspectives on Human–Computer Interaction*, D. A. Norman and S. W. Draper, Eds., Lawrence Erlbaum Associates, Mahwah, NJ, pp. 411–432.

Lewis, J. R. (1982), "Testing Small System Customer Set-up," in *Proceedings of the Human Factors Society 26th Annual Meeting*, Human Factors Society, Santa Monica, CA, pp. 718–720.

Lewis, J. R. (1991a), "A Rank-Based Method for the Usability Comparison of Competing Products," in *Proceedings of the Human Factors Society 35th Annual Meeting*, Human Factors Society, Santa Monica, CA, pp. 1312–1316.

Lewis, J. R. (1991b), "Psychometric Evaluation of an After-Scenario Questionnaire for Computer Usability Studies: The ASQ," *SIGCHI Bulletin*, Vol. 23, pp. 78–81.

Lewis, J. R. (1992), "Psychometric Evaluation of the Post-study System Usability Questionnaire: The PSSUQ," in *Proceedings of the Human Factors Society 36th Annual Meeting*, Human Factors Society, Santa Monica, CA, pp. 1259–1263.

Lewis, J. R. (1993), "Multipoint Scales: Mean and Median Differences and Observed Significance Levels," *International Journal of Human–Computer Interaction*, Vol. 5, pp. 382–392.

Lewis, J. R. (1994), "Sample Sizes for Usability Studies: Additional Considerations," *Human Factors*, Vol. 36, pp. 368–378.

Lewis, J. R. (1995), "IBM Computer Usability Satisfaction Questionnaires: Psychometric Evaluation and Instructions for Use," *International Journal of Human–Computer Interaction*, Vol. 7, pp. 57–78.

Lewis, J. R. (1996a), "Binomial Confidence Intervals for Small Sample Usability Studies," in *Advances in Applied Ergonomics: Proceedings of the 1st International Conference on Applied Ergonomics, ICAE '96*, G. Salvendy and A. Ozok, Eds., Istanbul, Turkey, pp. 732–737.

Lewis, J. R. (1996b), "Reaping the Benefits of Modern Usability Evaluation: The Simon Story," In *Advances in Applied Ergonomics: Proceedings of the 1st International Conference on Applied Ergonomics, ICAE '96*, G. Salvendy and A. Ozok, Eds., Istanbul, Turkey, pp. 752–757.

Lewis, J. R. (2001a), "Evaluation of Procedures for Adjusting Problem-Discovery Rates Estimated from Small Samples," *Journal of Human–Computer Interaction*, Vol. 13, pp. 445–479.

Lewis, J. R. (2001b), "Introduction: Current Issues in Usability Evaluation," *International Journal of Human–Computer Interaction*, Vol. 13, pp. 343–349.

Lewis, J. R. (2002), "Psychometric Evaluation of the PSSUQ Using Data from Five Years of Usability Studies," *International Journal of Human–Computer Interaction*, Vol. 14, pp. 463–488.

Lewis, J. R., and Pallo, S. (1991), *Evaluation of Graphic Symbols for Phone and Line*, Technical Report 54.572, International Business Machines Corporation, Boca Raton, FL.

Lewis, J. R., Henry, S. C., and Mack, R. L. (1990), "Integrated Office Software Benchmarks: A Case Study," in *Proceedings of the 3rd IFIP Conference on Human–Computer Interaction, INTERACT '90*, D. Diaper et al., Eds., Elsevier Science, Cambridge, pp. 337–343.

Lilienfeld, S. O., Wood, J. M., and Garb, H. N. (2000), "The Scientific Status of Projective Techniques," *Psychological Science in the Public Interest*, Vol. 1, pp. 27–66.

Lord, F. M. (1953), "On the Statistical Treatment of Football Numbers," *American Psychologist*, Vol. 8, pp. 750–751.

Macleod, M., Bowden, R., Bevan, N., and Curson, I. (1997), "The MUSiC Performance Measurement Method," *Behaviour and Information Technology*, Vol. 16, pp. 279–293.

Marshall, C., Brendan, M., and Prail, A. (1990), "Usability of Product X: Lessons from a Real Product," *Behaviour and Information Technology*, Vol. 9, pp. 243–253.

Massaro, D. W. (1975), *Experimental Psychology and Information Processing*, Rand McNally, Chicago.

Mayer, R. E. (1997), "From Novice to Expert," In M. G. Helander, T. K. Landauer, and P. V. Prabhu, Eds., *Handbook of Human–Computer Interaction*, 2nd ed., Elsevier, Amsterdam, pp. 781–795.

McFadden, E., Hager, D. R., Elie, C. J., and Blackwell, J. M. (2002), "Remote Usability Evaluation: Overview and Case Studies," *International Journal of Human–Computer Interaction*, Vol. 14, pp. 489–502.

McSweeney, R. (1992), "SUMI: A Psychometric Approach to Software Evaluation," unpublished M.A. (Qual.) thesis in applied psychology, University College of Cork, Cork, Ireland.

Miller, L. A., Stanney, K. M., and Wooten, W. (1997), "Development and Evaluation of the Windows Computer Experience Questionnaire (WCEQ)," *International Journal of Human–Computer Interaction*, Vol. 9, pp. 201–212.

Moffat, B. (1990), "Normalized Performance Ratio: A Measure of the Degree to Which a Man–Machine Interface Accomplishes Its Operational Objective," *International Journal of Man–Machine Studies*, Vol. 32, pp. 21–108.

Molich, R., Bevan, N., Curson, I., Butler, S., Kindlund, E., Miller, D., and Kirakowski, J. (1998), "Comparative Evaluation of Usability Tests," in *Usability Professionals Association Annual Conference Proceedings*, Usability Professionals Association, Washington, DC, pp. 189–200.

Molich, R., Ede, M. R., Kaasgaard, K., and Karyukin, B. (2004), "Comparative Usability Evaluation," *Behaviour and Information Technology*, Vol. 23, pp. 65–74.

Morse, E. L. (2000), "The IUSR Project and the Common Industry Reporting Format," in *Proceedings of the ACM Conference on Universal Usability*, Arlington, VA, pp. 155–156.

Myers, J. L. (1979), *Fundamentals of Experimental Design*, Allyn & Bacon, Boston.

Nielsen, J. (1992), "Finding Usability Problems Through Heuristic Evaluation," in *CHI '92 ACM Conference Proceedings*, Monterey, CA, pp. 373–380.

Nielsen, J. (1993), *Usability Engineering*, Academic Press, San Diego, CA.

Nielsen, J. (1994), "Usability Laboratories," *Behaviour and Information Technology*, Vol. 13, pp. 3–8.

Nielsen, J. (1997), "Usability Testing," in *Handbook of Human Factors and Ergonomics*, 2nd ed., G. Salvendy, Ed., Wiley, New York.

Nielsen, J., and Landauer, T. K. (1993), "A Mathematical Model of the Finding of Usability Problems," in *Proceedings of the ACM INTERCHI'93 Conference*, Amsterdam. pp. 206–213.

Nielsen, J., and Molich, R. (1990), "Heuristic Evaluation of User Interfaces," in *Conference Proceedings on Human*

Factors in Computing Systems, CHI '90*, ACM, New York, pp. 249–256.
Nisbett, R. E., and Wilson, T. D. (1977), "Telling More Than We Can Know: Verbal Reports on Mental Processes," *Psychological Review*, Vol. 84, pp. 231–259.
Norman, D. A. (1983), "Design Rules Based on Analyses of Human Error," *Communications of the ACM*, Vol. 26, pp. 254–258.
Norman, D. (1986), "Cognitive Engineering," in *User Centered System Design: New Perspectives on Human–Computer Interaction*, D. A. Norman and S. W. Draper, Eds., Lawrence Erlbaum Associates, Mahwah, NJ, pp. 31–61.
Nunnally, J. C. (1978), *Psychometric Theory*, McGraw-Hill, New York.
Palmquist, R. A., and Kim, K. S. (2000), "Cognitive Style and On-line Database Search Experience as Predictors of Web Search Performance," *Journal of the American Society for Information Science*, Vol. 51, pp. 558–566.
Parasuraman, A. (1986), "Nonprobability Sampling Methods," in *Marketing Research*, Addison-Wesley, Reading, MA, pp. 498–516.
Prümper, J., Zapf, D., Brodbeck, F. C., and Frese, M. (1992), "Some Surprising Differences Between Novice and Expert Errors in Computerized Office Work," *Behaviour and Information Technology*, Vol. 11, pp. 319–328.
Rasmussen, J. (1986), *Information Processing and Human–Machine Interaction: An Approach to Cognitive Engineering*, Elsevier, New York.
Rengger, R. (1991), "Indicators of Usability Based on Performance," in *Human Aspects in Computing, Design and Use of Interactive Systems and Work with Terminals: Proceedings of the 4th International Conference on Human–Computer Interaction*, H. J. Bullinger, Ed., Elsevier Science, Stuttgart, Germany, pp. 656–660.
Rosenbaum, S. (1989), "Usability Evaluations Versus Usability Testing: When and Why?" *IEEE Transactions on Professional Communication*, Vol. 32, pp. 210–216.
Rubin, J. (1994), *Handbook of Usability Testing: How to Plan, Design, and Conduct Effective Tests*, Wiley, New York.
Ruthford, M. A., and Ramey, J. A. (2000), "Design Response to Usability Test Findings: A Case Study Based on Artifacts and Interviews," *IEEE International Professional Communication Conference*, IEEE Press, Piscataway, NJ, pp. 315–323.
Sadowski, W. J. (2001), "Capabilities and Limitations of Wizard of Oz Evaluations of Speech User Interfaces," in *Proceedings of HCI International 2001: Usability Evaluation and Interface Design*, Lawrence Erlbaum Associates, Mahwah, NJ, pp. 139–143.
Sauro, J. (2004), "Restoring Confidence in Usability Results," http://www.measuringusability.com/conf_intervals.htm.
Sauro, J., and Lewis, J. R. (2005), "Estimating Completion Rates from Small Samples Using Binomial Confidence Intervals: Comparisons and Recommendations," In *Proceedings of the Human Factors and Ergonomics Society 49th Annual Meeting*, Human Factors and Ergonomics Society, Santa Monica, CA.
Shackel, B. (1990), "Human Factors and Usability," in J. Preece and L. Keller, Eds., *Human–Computer Interaction: Selected Readings*, Prentice Hall International, Hemel Hempstead, Hertfordshire, England, pp. 27–41.
Shneiderman, B. (1987), *Designing the User Interface: Strategies for Effective Human–Computer Interaction*, Addison-Wesley, Reading, MA.
Smith, B., Caputi, P., Crittenden, N., Jayasuriya, R., and Rawstorne, P. (1999), "A Review of the Construct of Computer Experience," *Computers in Human Behavior*, Vol. 15, pp. 227–242.
Smith, D. C., Irby, C., Kimball, R., Verplank, B., and Harlem, E. (1982), "Designing the Star User Interface," *Byte*, Vol. 7, No. 4, pp. 242–282.
SOGSC (Southwest Oncology Group Statistical Center) (2004), "Binomial Confidence Interval," http://www.swogstat.org/stat/public/binomial_conf.htm.
Steele, R. G. D., and Torrie, J. H. (1960), *Principles and Procedures of Statistics*, McGraw-Hill, New York.
Stevens, S. S. (1951), "Mathematics, Measurement, and Psychophysics," in *Handbook of Experimental Psychology*, S. S. Stevens, Ed., Wiley, New York, pp. 1–49.
Sudman, S. (1976), *Applied Sampling*, Academic Press, New York.
Turner, C. W., Lewis, J. R., and Nielsen, J. (in press), "Determining Usability Test Sample Size, in *The International Encyclopedia of Ergonomics and Human Factors*.
van de Vijver, F. J. R., and Leung, K. (2001), "Personality in Cultural Context: Methodological Issues," *Journal of Personality*, Vol. 69, pp. 1007–1031.
Virzi, R. A. (1990), "Streamlining the Design Process: Running Fewer Subjects," in *Proceedings of the Human Factors Society 34th Annual Meeting*, Human Factors Society, Santa Monica, CA, pp. 291–294.
Virzi, R. A. (1992), "Refining the Test Phase of Usability Evaluation: How Many Subjects Is Enough?" *Human Factors*, Vol. 34, pp. 457–468.
Virzi, R. A. (1997). "Usability Inspection Methods," in *Handbook of Human–Computer Interaction*, 2nd ed., M. G. Helander, T. K. Landauer, and P. V. Prabhu, Eds., Elsevier, Amsterdam, pp. 705–715.
Virzi, R. A., Sorce, J. F., and Herbert, L. B. (1993), "A Comparison of Three Usability Evaluation Methods: Heuristic, Think-Aloud, and Performance Testing," in *Proceedings of the Human Factors and Ergonomics Society 37th Annual Meeting*, Human Factors and Ergonomics Society, Santa Monica, CA, pp. 309–313.
Vredenburg, K., Mao, J. Y., Smith, P. W., and Carey, T. (2002), "A Survey of User Centered Design Practice," in *Proceedings of CHI 2002*, ACM, Minneapolis, MN, pp. 471–478.
Walpole, R. E. (1976), *Elementary Statistical Concepts*, Macmillan, New York.
Wenger, M. J., and Spyridakis, J. H. (1989), "The Relevance of Reliability and Validity to Usability Testing," *IEEE Transactions on Professional Communication*, Vol. 32, pp. 265–271.
Whiteside, J., Bennett, J., and Holtzblatt, K. (1988), "Usability Engineering: Our Experience and Evolution," in *Handbook of Human–Computer Interaction*, M. Helander, Ed., North-Holland, Amsterdam, pp. 791–817.
Wickens, C. D. (1998), "Commonsense Statistics," *Ergonomics in Design*, Vol. 6, No. 4, pp. 18–22.
Wiethoff, M., Arnold, A., and Houwing, E. (1992), *Measures of Cognitive Workload*, MUSiC ESPRIT Project 5429 document code TUD/M3/TD/2.
Wildman, D. (1995), "Getting the Most from Paired-User Testing," *Interactions*, Vol. 2, No. 3, pp. 21–27.
Williams, G. (1983), "The Lisa Computer System," *Byte*, Vol. 8, No. 2, pp. 33–50.

Wixon, D. (2003), "Evaluating Usability Methods: Why the Current Literature Fails the Practitioner," *Interactions*, Vol. 10, No. 4, pp. 28–34.

Woolrych, A., and Cockton, G. (2001), "Why and When Five Test Users Aren't Enough," in *Proceedings of IHM–HCI 2001 Conference*, J. Vanderdonckt, A. Blandford, and A. Derycke, Eds., Cépadèus Éditions, Toulouse, France, Vol. 2, pp. 105–108.

Wright, R. B., and Converse, S. A. (1992), "Method Bias and Concurrent Verbal Protocol in Software Usability Testing," in *Proceedings of the Human Factors and Ergonomics Society 36th Annual Meeting*, Human Factors and Ergonomics Society, Santa Monica, CA, pp. 1220–1224.

Wright, P. C., and Monk, A. F. (1991), "A Cost-Effective Evaluation Method for Use by Designers," *International Journal of Man–Machine Studies*, Vol. 35, pp. 891–912.

CHAPTER 50

WEB SITE DESIGN AND EVALUATION

Kim-Phuong L. Vu
California State University–Long Beach
Long Beach, California

Robert W. Proctor
Purdue University
West Lafayette, Indiana

1 INTRODUCTION

Virtually all users have encountered examples of good and bad sites in their interactions with the World Wide Web (referred to as the Web). Poor Web designs often lead to user frustration because users cannot easily access the information they are seeking, and this frustration may cause them to abandon the site. Having users abandon a site is undesirable in almost any case, as it defeats the purpose of disseminating information or providing services on the Web. Moreover, abandonment of a site is particularly harmful to e-businesses because it results in the loss of potential customers. Organizations that are successful in developing a good Web site and maintaining it will thus have a competitive edge over their rivals.

The first question to ask when designing a Web site should be: What is the goal of the Web site? Is the site's goal to sell the most widgets that Company XYZ produces? Is it to provide a resource where scientific information relating to a certain topic can be found? Or, is it to be an entertaining site that users can visit in their recreation time? Although the task of posing this question may be simple, the answer is critical for determining the nature of the Web site, as well as driving the design and evaluation process for the site (see, e.g., Zhu et al., 2005). Clearly defining the goals for the Web site helps the designer focus on those aspects that are important to attaining the goal and determining the emphasis that is to be placed on each component.

Several basic types of Web sites are commonly encountered (Irie, 2004): information dissemination, portal, community, search, e-commerce, company information, and entertainment. Each of these types of Web sites is designed for specific purposes or goals and has characteristics unique to achieving those goals:

- News/information dissemination
  - *Goal*: to provide users with news or information
  - *Characteristic*: mostly text-based, with minor graphics; simple and consistent navigation
- Portal
  - *Goal*: to provide links to other Web sites or resources
  - *Characteristic*: mostly uniform resource locator (URL) links, with minor descriptions of the linked resources; usually organized alphabetically, or by a keyword or theme
- Community/communication
  - *Goal*: to provide a medium that promotes the community and provides opportunity for communication and interaction among users
  - *Characteristic*: usually in the form of message boards or chat rooms
- Search
  - *Goal*: to facilitate retrieval of specific information or resources
  - *Characteristic*: usually in the form of a search engine; the returned results page resembles a portal

- E-commerce
  - *Goal*: to allow companies to sell products or services; to allow users to purchase products or services
  - *Characteristic*: usually consists of a searchable catalog of products; must include a mechanism for secure online monetary transactions
- Company/organization/product information
  - *Goal*: to provide specific information about a company, organization, or product
  - *Characteristic*: emphasis on company's or product's image/logo
- Entertainment
  - *Goal*: to provide pleasant interactions or entertainment resources for users
  - *Characteristic*: usually places emphasis on aesthetics; games and videos

Once the goal for the Web site has been established, the next step is to determine what content should be placed in the site. It is important to distinguish content from aesthetics. Content design focuses primarily on the substance or information that is contained in the Web site, whereas aesthetics design focuses primarily on trying to make the site visually pleasant or enjoyable. A good Web site can be informative as well as pretty and creative, which means that designing for content and aesthetics is not mutually exclusive. However, because the goal of many Web sites is to provide some type of service, it is usually more important to design for content than aesthetics.

2 WEB SITE DESIGN

Web sites should be designed with the end users in mind at all times because the sites are intended to support user activities (Proctor and Vu, 2004). The design process for Web interfaces proceeds in a manner similar to traditional usability engineering life cycles, which include a requirements analysis phase, a design, testing, and development phase, and an installation phase (see, e.g., Mayhew, 2005). It is important to note that design, evaluation, and development are iterative processes. That is, the design of the Web site should be evaluated as it is being developed so that usability problems can be identified and fixed. However, for ease of presentation, Web design issues are covered in this section of the chapter, and Web evaluation techniques are covered separately in Section 3.

2.1 Components of a Web Site

There are several major components of a Web site: its content, architecture or structure and organization, the presentation of the content, and the programming logic that is used to integrate the content and its presentation within the site's structure. Content design and presentation can be broken down further into specific components, such as page design, navigation, use of multimedia, search design, and URL design (see, e.g., Nielsen, 2000). One way to determine the process involved in Web site design is to look at current design practices. Newman et al. (2003; see also Newman and Landay, 2000) conducted a study in which 11 expert Web site designers were interviewed about a current or recently completed Web design project to determine the practices in which Web designers customarily engage during the design process. They identified several specific areas to which the expert designers referred when describing the design process of a Web site:

- *Information design* includes identifying and grouping content so that individual components can be integrated and organized into a coherent whole.
- *Navigation design* includes methods for users to move through or access different parts of the Web site.
- *Graphics design* includes how to present individual pieces of information or content visually to the users (e.g., through images).
- *Information architecture* includes how to combine the information and navigation components so that the entire Web site functions as a unified entity.
- *User interface design* refers to designing and evaluating the usability of Web site, including its informational and navigational components.

Although the areas listed above emphasize the major areas of concern for Web design, as identified by several Web design experts, there is some degree of overlap between the different areas. Newman et al. (2003) noted that most of the time, the designers indicated that work in the areas of information design and navigation design preceded work on the graphics design. Because the Web consists of many individual components, designers should use the goal of the Web site to determine which components are more critical than others.

2.2 Content Preparation

The Web provides a medium for exchange of information and services to a global audience. Because of the potential impact that the Web can have on the success of an organization or company, it is important to have an effective content design that promotes the goals of the Web site. Unfortunately, it is often the case that many Web sites are not designed in an effective manner. For example, although the goal of an e-commerce Web site is to sell the products, one study showed that users were not able to find specific items on e-commerce Web sites 36% of the time (Nielsen et al., 2000). If users cannot find an item, they cannot buy it. Thus, e-commerce Web sites that are designed to structure and organize their content in a manner that promotes the ease with which users can locate specific items will have a competitive edge over rival Web sites designed to achieve the same goal.

Content preparation refers to the processes involved in determining the information for the Web site to convey, and how to organize, structure, and present that information so that it can be retrieved easily and efficiently when needed. Proctor et al. (2002b) summarized four major areas that Web designers should emphasize for content preparation (see also Proctor et al., 2003). These areas include:

1. *Knowledge elicitation*: determining what type of information should be conveyed
2. *Structure and organization of information*: determining the best method to structure and organize information
3. *Retrieval of information*: determining the best methods for helping users search and retrieve information
4. *Presentation of information*: determining the medium in which information should be presented to the user

2.3 Knowledge Elicitation

When designing for the Web, knowledge should be elicited from two classes of users: experts and end users (Proctor et al., 2002). Experts in Web design can provide valuable information, including (1) how to organize and present information in a manner that is consistent with human information-processing capabilities, (2) the functions and features that good Web sites should possess, (3) methods for enhancing the Web site's efficiency or effectiveness; and so on. The information that end users provide is usually not expert advice regarding how to design a Web site, but rather, information that is intended to help designers understand (1) the computing skills or level of knowledge of the end users; (2) the users' mental models, or representations, of the content that is contained in the Web site; (3) the specific pieces of information that users need when performing a task; and so on. Another way to characterize the different information provided by experts and end users is that end users reveal the type of information needed by them to achieve their goals, whereas experts determine how to organize and present that information in the most effective manner.

2.3.1 Eliciting Knowledge from Experts

A lot of information can be gained about how to design for the Web from interviewing experts and observing their work. Interviews are a well-known knowledge elicitation technique (e.g., Shadbolt and Burton, 1995) and can be one of the best methods for obtaining in-depth data from experts. In the study by Newman et al. (2003) mentioned earlier, expert designers were interviewed extensively to obtain information about their thought processes and involvement during an entire design process. The experts were asked to walk the interviewer through each phase of the project, showing examples of their "work in progress" if possible. Newman et al. report in detail the steps involved in creating a tutorial for a suite of CAD tools that can be accessed through the intranet within a company or through the Internet remotely. The key steps identified by the expert are as follows:

1. *Discovery phase*: gathering background information about the project to determine the project scope, goals, and timeline for completion
2. *Design exploration phase*: generating sketches of initial variations of the design, including how the content will be structured, the individual pages, navigation, and the interaction sequence
3. *Design refinement phase*: creating high-fidelity mock-ups of the site, including the home page and second-level pages that can be accessed from the home page; choosing a limited set of the mock-up designs for further refinement and development; selecting a single refined mock-up to prototype in HTML
4. *Production phase*: writing up guidelines for the prototype so that a design team can turn the prototype into a working product

As illustrated by Newman et al.'s (2003) study, interviews from experts and observation of artifacts of their work can provide valuable information about the design process. However, there are numerous other methods that can be used to elicit knowledge from experts. Many of these techniques are described by Proctor et al. (2002b), and include the following:

- *Verbal protocol analysis*: analysis of an expert's problem-solving strategies through examining the verbal protocols, obtained by having the experts "think aloud" as they solve problems or reflect on their thought processes when reviewing recordings of their performance during a problem-solving task.
- *Group task analysis*: analysis of a group of experts' joint depiction about how a specific task is represented and processed. Usually, a flowchart is used to depict the individual steps required for performing a specific task.
- *Narratives and scenarios*: analysis of information contained in stories or narratives that experts tell about their activities. They often reveal valuable information about the goals of a particular task and the sequence of actions and events that lead to particular decisions and outcomes.
- *Critical incident reports*: analysis of critical incidents that tested a person's expertise for the insight they provide into the processes involved an expert's decision making and reasoning when an unexpected or unusual event occurs.

When interviewing design experts is not feasible, a designer should refer to published papers and chapters on specific Web design issues of interest written by experts in the field [see Ratner (2003), Bidgoli (2004), and Proctor and Vu (2005) for edited volumes on the Internet and Web design].

When it comes to evaluating expertise, one question that can be asked is: Who is the expert (Flach, 2000)? Should Web sites be designed according to the wishes of the expert designers or are the end users the expert in this case? After all, the end users are the people for whom the Web site is intended. Most Web professionals would agree that the input of the end users is important when designing a Web site and evaluating its effectiveness. However, the end users should not be allowed to direct the design process because the features they desire for the Web site may be hard to implement or may lead to a less than optimal design when all factors are considered.

Below, different techniques for eliciting knowledge from end users are discussed. Many of these methods and techniques are used in the exploratory phases of Web design to understand the end users and their goals, representations of the task, likes and dislikes, and preferences. As a result, the term *understanding the user* is used to refer to this process rather than *knowledge elicitation* (see, e.g., Volk and Wang, 2005).

2.3.2 Understanding the User

The Web affords the opportunity for a site to be accessed anywhere, at anytime, and with various devices (e.g., computer, laptop, personal data assistant, and cellular phone). As a result, in initial or exploratory phases of Web design, it is important to obtain background information about the targeted user population. It is especially important to obtain information relating to the users' cognitive and physical capabilities, the tasks that they are likely to perform, the information needed to perform those tasks, and their roles and responsibilities during exploratory phases (Stanney et al., 1997). For example, if the Web site is targeted for use by older adults, it is recommended that designers avoid using colors from the short-wavelength end of the visual spectrum (i.e., blue and green) and increase the resolution of elements on the screen so that the items can be seen more easily (Bitterman and Shalev, 2004). Moreover, different types of users may need to access different parts of the Web site. For example, an e-commerce merchant may want to access the U.S. Postal Service's Web site to request or print out mailing labels, or to schedule a package pickup, whereas both the merchant and the consumer may access the site to track a package. Because different users have unique goals, the Web site must be designed to accommodate the tasks that the different user groups want and need to perform.

There are many different types of methods that can be employed to understand the users (see Proctor et al., 2002b; Volk and Wang, 2005, for reviews). Most of these methods are aimed at obtaining information regarding what users need to complete their task on the Web site, along with users' preferences for the site's options and features. Below are brief descriptions of the major methods, along with their goals and characteristics.

Interviews As with experts, end users can be interviewed to obtain in-depth knowledge about their characteristics, opinions, and preferences. There are two main types of interviews: structured and unstructured. In a structured interview, the interviewer asks the users questions from a prearranged list. In an unstructured interview, the interviewer allows users to express their thoughts on a topic freely. Structure and unstructured are two ends of a continuum. Most interviews fall somewhere in between these two ends, with the interviewer asking general questions, but the direction of the interview is dependent on the user's answers. Interviews of end users provide a large amount of qualitative data in which the evaluators often organize into topics or perform a content analysis of the data to identify themes and categories of information. The interviewers should try to avoid imposing their beliefs or opinions during the process. Caution must be taken when framing the questions to avoid biasing the users' answers, and at least two different evaluators should code the data to ensure that their interpretations of the answers match to some degree.

Surveys and Questionnaires Surveys and questionnaires consist of questions used to gather information about a user (e.g., How old are you?; How often do you make online purchases?), obtain users' likes and dislikes [e.g., On a scale of 1 (do not like it at all) to 10 (like it a lot), indicate how much you like pop-up windows], and preferences (e.g., Do you prefer a dark font color on a light background or a light font color on a dark background?). Some advantages of using surveys and questionnaires include (1) obtaining data from a large sample in different demographic areas or specified users groups (e.g., previous customers or persons on an e-mail list), (2) the data can be obtained in a relatively short period of time, (3) if there are no open-ended questions, the data can be coded and summarized relatively easily, and (4) data from questionnaires and surveys can also be used to develop user profiles. However, some disadvantages of using surveys and questionnaires include (1) not all targeted users chose to fill them out and submit their answers; (2) it may be difficult to verify the identity of the participants who submit their answers through the Web; (3) if the default code for "no response" on an item is not coded as such, the default value may be mistaken as the user's response rather than no response; (4) the framing of questions may affect the validity of the results; and (5) users' judgments and preferences may not correlate with their actual performance.

Focus Groups Unlike surveys and questionnaires, focus groups consist of a smaller number of users (usually, 5 to 10), but allow the users to interact with one another when discussing and evaluating different aspects or issues of the design. Problems and concerns about the Web site that may not have been identified by a single user can emerge during the group's interactions. A moderator usually directs the group to ensure that everyone participates and stays on the task so that all the topics wanting to be covered will get covered and that no single participant dominates the session. Proctor et al. (2002b) noted that focus groups are best used to attain high-level goals, such as generating a list of functions or features for a product. Focus groups yield large amounts of qualitative data that must be organized and summarized. As with the previous methods, focus groups also suffer from the fact that the design judgments and preferences produced by the group may not correspond to designs that benefit performance. Furthermore, since the tasks and interface are evaluated in a context different from what the user is likely to experience during "real" use of the Web site, issues identified in the focus group may not always be of items of major concern.

Naturalistic Observation Observation and nonparticipatory contextual inquiries reflect more naturalistic techniques for understanding users. These methods rely on observing users' everyday interactions with the Web site in their natural surroundings, and provide researchers with background information about the context in which a product is being used. This can be done by having researchers observe users performing Web tasks from "afar." It is important that the evaluator remain unnoticed, so that users' natural behaviors can be observed. Often, it is difficult for evaluators to remain unobtrusive. However, this can be done through the use of one-way mirrors or video cameras. Connecting rooms with one-way mirrors are often found in usability labs (described later) but do not capture the users in their natural environment unless they access the Web site in public locations (e.g., a library). Video cameras (or Web cameras) can be set in public locations to observe users. However, it is important that these cameras not be visible to users because they may change their interaction pattern if they know that they are being observed. Because data from observational methods are based on the users' actions, the conflict between users' verbal report and their actions is no longer an issue. However, it is often difficult for the observer to remain unnoticed, and the data obtained from the observer may reflect his or her biases and/or interpretation of the observed events.

Ethnographic Studies With ethnographic studies, which emerged from the field of anthropology, the researcher seeks to understand the users by immersing himself or herself in the targeted users' culture or work environment (Millen, 2000). The goal of the researcher is to become a natural member of the group so that he or she will be able to understand the views of the user groups and work with them to design products that meet their needs. Ethnographic studies can yield a customer–partner relationship that results in an effective medium for identifying the users' needs (Volk and Wang, 2005). However, there are many drawbacks to ethnographic methods as well. Ethnographic studies take a long time to conduct, suffer from the same disadvantages of interpretation and bias as other methods based on self-report or observational data, and may not result in general guidelines for product designs since the data are based on a very specific group of people.

User Diary The user diary allows the user to observe and record, in a diary, his or her actions with a product over a period of time. A typical diary study involves asking participants to keep a diary of their daily activities with the product for a few days or weeks. However, Rieman (1993) noted that is often inconvenient for participants to keep a diary for more than two weeks. The diary method is based on a real-time tracking system. Diary logs can reveal qualitative data such as "critical incidents" that have occurred as well as descriptive data such as the time spent on each task. The disadvantage of traditional diary methods is that they rely on the conscientiousness of the participants, which may result in diaries of different qualities. Participants may forget to log the activities, may log information that they think the evaluators want, or provide vague or general summaries. However, with the availability of wireless connections, an online or video diary can be obtained by wearing a wireless video camera or Web camera that records the users' interaction with different Web sites. The video diary circumvents the traditional problems of users waiting until a convenient time to enter the data in their diary.

Web Server Log Files A benefit of collecting data about Web site usage is that the site itself can be used as a data collection tool because the actions of the users are recorded as the users interact with the Web site. Server logs can provide designers valuable information about existing Web sites, such as "who" is visiting the site, the pages within the site that the users visit and the order in which those pages are accessed, how long the user spends looking at a particular page, and what items users search for (Pearrow, 2000). The benefits of evaluating data from log files are that a large amount of data can be obtained from users who access the site and the data collection process does not interfere with the users' interaction with the site. However, because information irrelevant to the design goals is logged, it may be difficult to sort through log files and important variables of interest may not have been logged.

2.3.3 Summary

This section summarized the major methods that are used to determine the processes involved in the Web site's design life cycle and how to extract the content to be conveyed by the Web site from experts and end users. Although these knowledge elicitation methods can yield substantial insight and data regarding the knowledge, expertise, and characteristics of designers and end users, the main drawback is that the methods rely on self-report data. Self-report data are vulnerable because there are well-known biases in self reports (e.g., Isaacs, 1997), not all cognitive processes can be articulated clearly [see Anderson (1982) for a discussion of declarative vs. procedural knowledge], analysis is based on the interpretation of the evaluators (see Proctor et al., 2002b; Volk and Wang, 2005), and users' preferences and judgments may not correlate with their performance (e.g., Bailey, 1993; Nielsen, 2001; Vu and Proctor, 2003).

2.4 Structuring and Organizing the Content

Once designers have identified or elicited the content that needs to be conveyed, they must structure and organize it in a manner that will allow effective presentation and retrieval of that information. It is important for designers to plan carefully how to structure and organize the content to be conveyed by the Web site because poor design of the information architecture leads to poor usability (Nielsen, 2000), making the Web site more difficult to use and less enjoyable to the end user. To create a good Web site, designers should be familiar with the following concepts (e.g., Chou, 2002):

- *Organizational schemes*: how the information is organized (e.g., by name, date, type, association).
- *Organizational structure*: how the Web site is structured (e.g., hierarchical structure) and the programming logic that is used to access, retrieve, and present different pieces of information.
- *Labeling*: how specific items are referred to. Labels should be used consistently throughout the Web site.
- *Navigation*: how to set paths that allow users to find their way through the Web site.

2.4.1 Organizational Schemes

One simple, yet successful method that can be used to determine relationships between fixed sets of categories is concept, or card, sorting. In a concept-sorting task, users are given cards that contain concept words and are asked to organize these words into discrete categories based on the relationships between the words. Users are often also asked to come up with a global label for each pile, or category, and then the categories can be used to determine how the various concepts should be organized (see, e.g., Vaughan et al., 2001). There are many other more sophisticated methods for organizing and structuring content for the Web (see Proctor et al., 2002b), some of which are described briefly below.

Objects/Actions Interface Model (Shneiderman, 1997) This model focuses on decomposing complex information found in Web sites into manageable hierarchies of objects (e.g., networks) and actions (e.g., searching). Specifically, the designers should focus on how the objects and actions are represented in the task and interface. For example, an e-commerce site may consist of individual objects such as pencils and paper. These objects can be aggregated into classes such as school supplies or stationery. Similarly, users can perform individual actions in the site, such as clicking on a hyperlink to get from one page to another, or aggregate actions, such as scanning a list of different types of office supplies to find and link to a specific category of supplies. To link the objects and actions into the Web interface, designers can use common metaphors (e.g., a file cabinet) to organize objects and "handles" (e.g., pull-down menu) or a magnifying glass to represent an action (zooming feature) or type of action that can be performed.

Ecological Interface Design (Vicente and Rasmussen, 1992) This design process emphasizes that the way in which information is represented in a good display depends on the users' knowledge level and the type of behavior in which the users' engage. The ecological interface design is built on two concepts from cognitive engineering: the abstraction hierarchy and the skills–rules–knowledge behavior framework. The abstraction hierarchy is a multilevel knowledge representation framework that can be used to identify the information content and structure of the Web interface. The skills–rules–knowledge framework is used to distinguish the three modes of behaviors of users (Rasmussen, 1986). *Skill-based behavior* arises when users interact with a system on a regular basis and routine commands can be performed "automatically." *Rule-based behavior* occurs when users are confronted with a novel situation but can apply rules to solve them that they have learned previously. *Knowledge-based behavior* arises when a completely unfamiliar event occurs and the user must invoke his or her problem-solving skills to continue the task. By understanding the behavioral mode and constraints placed on the end user, the goal of ecological interface design is to organize and structure the information in the most meaningful manner to display to the user.

Latent Semantic Analysis (Landauer, 1998; Landauer et al., 1998) This analysis provides a valuable tool for organizing and structuring conceptual knowledge based on the relatedness of the items. Latent semantic analysis considers the frequencies with which words occur in various contexts. Each word then occupies a position in the semantic space based on a large corpus of text and is linked with other words through common labels or features. Words that are more highly associated are represented closer in space. Latent semantic analysis can be used as a tool to structure and organize information because it can easily determine similarity relationships between words, sentences, or larger text units.

Extensible Markup Language (XML) XML is a markup language that provides designers with the option of tagging the content with standard markup indicators (e.g., </TITLE>), as well as new indicators that may not have been used before. XML provides a mechanism to impose standards or constraints on how the content is specified so that it can be stored, retrieved, and organized easily.

Use of the Semantic Web The Semantic Web (http://www.w3.org/2001/sw/) provides a framework to transform documents written in natural language into machine-readable form. The machine-readable annotations can be used to for organizing and retrieving Web content. There are several tools available for creating Semantic Web markups (see Golbeck et al., 2005).

2.4.2 Labeling

Labels are used extensively on the Web to represent particular pieces of information or a categories of information. Labels are usually keywords or short descriptors that highlight the type of content that the user will encounter when accessing information categorized by the label. Category labels can be assigned manually by the designer or can be generated automatically or by computer programs. Qin (2004) recommends consulting existing encyclopedias and reference books in the domain in which the Web site is being created so that the designer can be familiar with the classification schemes already being used. Designers can also use key terms or keywords from indexes of books in the domain to familiarize themselves with the vocabulary and ontology that the end users are also familiar with.

Labels are used to identify different components of the Web site, including:

1. *Page titles.* Every page of a Web site has a title. Coming up with a good label and description for the page is critical because "you get 40 to 60 characters to explain what people will find on your page. Unless the title makes it absolutely clear what the page is about, users will never open it" (Nielsen, 2000, p. 123).
2. *Headings.* Heading labels are used to signal to the user what content is forthcoming as he or she scans or scrolls down a Web page. Because many users tend to skim a Web page when looking for information rather than reading the content carefully, a good heading will alert the user to pay attention to the forthcoming material.
3. *Hyperlinks.* Hyperlinks take users to other places inside or outside the Web site. Hyperlinks have brief labels that inform users about the content that they will access by clicking on the link (e.g., if the link is to a file) or information about where the link will take them (e.g., if the link is taking users to another page). Hyperlinks should be made self-explanatory so that users do not have to guess where they will be going by clicking on them (Pearrow, 2000).
4. *Images.* There are many images that can be displayed on a Web site. When using images, ALT (alternative) tags should be used so that a textual label or description of the image can be displayed to the user when the image cannot be seen (e.g., text-based browser). Von Ahn and Dabbish (2004) introduced a method of labeling images with a computer game call ESP. The ESP game has a pair of users' guess labels for images until a match is obtained between the pair. They found that, on average, 3.89 labels are agreed upon each minute by participants in the game. The quality of these labels was validated by independent tests of search precision and comparison with descriptors generated by participants in an experiment. Coming up with good labels for the images can help produce more efficient image search as well as organization of the images.
5. *Icons.* Icons are symbolic representations of items, actions, or concepts. A good icon conveys an appropriate label for the content. In addition to visual icons, earcons (or auditory icons) are also being used on the Web (see Hempel and Altinsoy, 2005). All icons should include a descriptive label so that they can be indexed and referred to. The label should also be displayed to the users when a cursor is placed over the icon in case the user has trouble identifying the meaning of the icon.

It is important that the same item and/or category within a Web site be given the same label consistently throughout the Web site. It may confuse users, for example, to refer to the search feature as "search" in one part of the Web site and "find" in another part of the site. The label should also be representative of the type of information that it is linked to and should be a term that is recognized by the end users. By producing good labels for the individual components of the Web site, the designer should have an easier time organizing and structuring the content. Kantor (2003b) suggests the following guidelines for coming up with good labels:

- Never assume that users are familiar with the acronyms, programs, and jargon that the company or designer may use or that they are familiar with how the site is structured.
- Be consistent.
- Use labels that are familiar to users.

2.4.3 Organizational Structure

To design an information architecture for a Web site successfully, the designer must take into account the goals and purpose of the Web site, the nature of the information that needs to be conveyed, and the type of information that is need to perform a particular task. The organizational structure refers to the physical structuring of information based on its relationship with other components of the site. The most common types of organizational structures are hierarchical, network, linear, and database oriented (Kantor, 2003a).

Hierarchical Structures Most Web sites are organized using some type of hierarchical structure. Basically, this structure consists of a high-level, or global, category and is branched into subcategories underneath it (see Figure 1*a*). The subcategories can be broken down into smaller components until the elemental objects are reached. Sometimes, hierarchical structures are represented as tree diagrams. Shneiderman's (1997) objects–action interface model is based on a hierarchical organization. Hierarchical organizations are good because they represent how users naturally link concepts and categories together. Kantor (2003a) notes that although categories within the hierarchy should be both inclusive (provide all the relevant data) and exclusive (fits into only one category), this is rarely the case in practice because information can be categorized in a variety of ways. When designing hierarchies, the designer must also take into account the issue of breadth (how much should be included in each level)

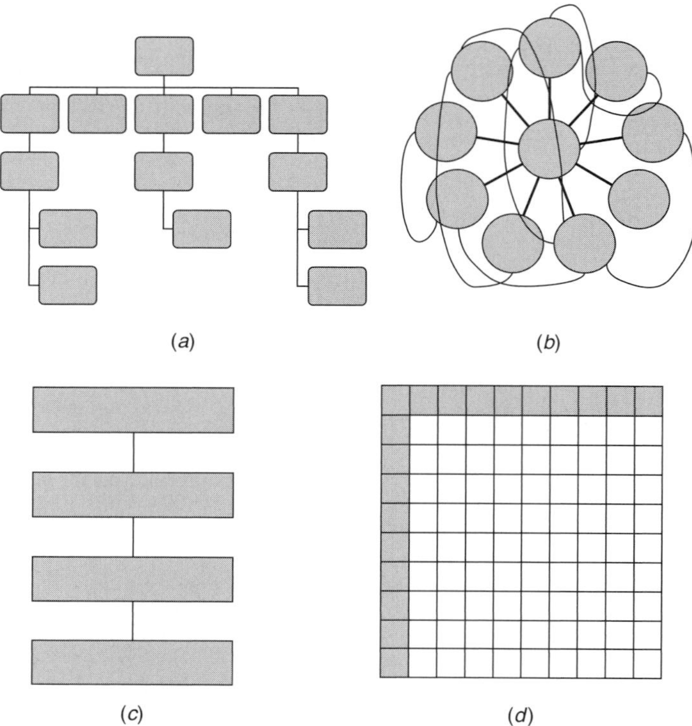

Figure 1 (*a*) Hierarchical, (*b*) networked, (*c*) linear, and (*d*) database structures.

and depth (how deep the levels should go). Tullis et al. (2005) concluded from a review of several studies examining the breadth vs. depth issue that breadth wins over depth in complex or ambiguous situations and that in simple situations, depth wins over breadth.

Linear Structures Linear information structures connect pieces of information serially and do not allow for branching of information as hierarchical structures do (see Figure 1*c*). In essence, linear structures have depth and no breadth. A linear structure is most appropriate for organizing and structuring information that should be accessed in order (e.g., step-by-step instructions or alphabetical listing of items). Linear structures can be embedded into a larger hierarchical structure. For example, the index of an online book can be structured hierarchically according to the first letter (A–Z) of the keywords or concepts, but the concepts themselves can be organized linearly or alphabetically under the first letters.

Networked Structures The Web itself is organized using networked structures. With this organization, users can move around the information architect by clicking on *hyperlinks* (see Figure 1*b*). The hyperlink can transport the user to an area of the Web site that does not have to be adjacent to their starting point. This structure is useful for organizing information that is related, but does not necessarily fall into the same category. For example, when searching for specific items such as dishes or plates on an e-commerce site, the results returned could include links to other kitchenware, such as cutlery sets.

Database Structures Information within a Web site can also be organized and stored in a database structure (see Figure 1*d*). Database structures usually rely heavily on a search feature that allows users to retrieve the information. For example, the catalog Web site for a library may use a database structure in which users search for information within well-defined categories (e.g., call number, author, title, and date).

The type of structure used to design a Web site should reflect the purpose and context of the site as well as the type of tasks the site is designed to support. It is also important to note that Web sites do not have to be organized solely in terms of one type of structure. Hybrid structures can be developed to support the specific needs of the users or for specific tasks that are performed within the Web site.

2.4.4 Navigation

Once the structure of the Web site is determined, designers must focus on the navigation system that allows users to access the content from different areas of the site. Navigation through a Web site is usually implemented through the use of hyperlinks that connect one Web page to another or a search engine that returns results that are relevant to the keyword or terms that a user enters. Because issues relating to information search are discussed in the following section, the primary emphasis of this section will be on navigation. A navigation scheme provides users with a set of paths that allow them to move through a Web site, such as from the homepage to a specific section of the site.

Avoiding Getting Users Lost Because a Web site can contain a huge amount of information that is distributed throughout the site, the content needs to be structured in a simple manner that is intuitive to the users. If users become overwhelmed by the information or if the site is structured poorly, users can get lost in a Web site. Once the user is lost, he or she must either start over from the beginning or abandon the task. The former adds to the time needed to complete the task, and the latter results in failure of the task.

One way to help users navigate through a Web site is to communicate the structure of the site to them (Proctor et al., 2002). If users understand the nature of how the site is structured and organized, they may be able to navigate through it more easily and quickly. Some techniques that can help communicate the structure of the site to users include:

1. *Site maps.* An outline of the Web site illustrates how the various sections and individual pages are organized within the site and where they are located.

2. *Interactive navigation displays.* Interactive navigation displays provide users with a graphical display of their current position within a Web site and provide information about how the users got there, how they can get back to a previously visited page, and where they can go next. For example, navigational breadcrumbs are trails of hyperlinked page titles, usually located at the top of each page that shows the users how they arrived at the current page. Users can click on the hyperlinks of pages visited previously to move back.

3. *Obvious and consistent major navigation controls.* Navigation controls for the major sections should be placed in prominent locations on each page of the site. The navigational controls should also be located in the same place on each page so the user can know where to find them (Najjar, 2001). Usually, navigation controls are placed on the top or left side of the screen (Pearrow, 2000). If the page links to lower levels, navigational controls to access those levels should also be placed on the page.

4. *Avoidance of orphan pages.* Make sure that users are not stranded when they click on a link that transports them to a page that does not link back to the site. If the link is to an external Web page, warn the user that he or she is leaving the site and/or have the linked page open in a separate window.

5. *Mark for links visited.* Separate pages that users have visited from ones that they have not by using different colors for the links. Most Web sites tend to use the color blue to indicate active hyperlinks that have not been visited and purple or burgundy to mark hyperlinks that lead to visited pages. Halverson and Hornof (2004) showed that differentiating visited and unvisited hyperlinks by color can improve search performance by narrowing the search space.

6. *Back button and history.* Users use the back button to leave a page approximately two-thirds of the time (see Dix, 2005), especially as a means to correct mistakes, to avoid being stuck at a "dead end," and to explore or browse the site. Because the back button is used often, users should be allowed the option of returning to a site visited previously by clicking on the back button. The history feature provided by the Web browser is also a means for users to return to a site visited previously. Because history logs can be set to record the navigation paths for long periods of time (i.e., days and months), it is important to give informative page titles so that users can find the page again when scanning the history log.

Presenting Navigation Options Tullis et al. (2005) provided a summary of several different techniques that are being used for presenting navigation options, including:

- *List of static links.* A list of links leads to subsections on sequential pages.
- *Index or table of contents layout.* Links are organized by topics in tables or by columns and rows.
- *Expanding and contracting outlines.* Links can expand when clicked to reveal subsections or contract to hide them.
- *Pull-down menus.* Heading links result in a pull-down menu of subsections either beneath the heading or to its side.

Tullis et al. concluded that users' performance was better when the navigation controls were presented with simple static listing of links or index/table of contents organization rather than with dynamic techniques.

2.4.5 Summary

Many methods can be used to connect or link various parts of a site so that it becomes a unified whole. As aforementioned, the most common technique is the use of hyperlinks. When a hyperlink is activated by clicking on it, the user is moved automatically to a different Web page. Although users can move throughout the site serially by relying on hyperlinks on successive pages, they can take shortcuts or jump from one section of the Web site to another by using the search function of the site and accessing the links returned.

2.5 Retrieval of Information

The search function is one of the most important elements of a home page and should be placed in a location that allows users to find it easily. Why is the search feature so important? The answer is simple. Users like to use the search feature to locate information that is of interest to them. Kobayashi and Takeda (2000) indicated: "About 85% of Web users surveyed claimed to be using search engines or some kind of search tool to find specific information of interest" (p. 146). According to the U.S. Department of Commerce's Economics and Statistics Administration and National Telecommunications and Information Administration (2002), searching for information is one of the most highly ranked online activities, second to e-mailing and instant messaging. Their report stated that one-third of Americans use the Internet to search for products and service information, with 36.2% searching for product–service information, 33.3% searching for information about the news, weather, or sports, and 7.5% searching for job information. A summary of statistics by Greenspan (2004) indicated that in the week of May 9 to 15, 5.5% of all U.S. Internet visits were to the three major search engines: Google.com, Yahoo.com, and MSN.com.

Although there has been much research about search engine designs for the Web (see Fang et al., 2005, for a review), the focus in this chapter is on local search engines designed for searching within a particular Web site. One benefit of designing a search function for a particular Web site is that the content that has to be located and retrieved is much smaller than that on the entire World Wide Web. However, many issues and techniques involved in designing global search engines for searching the Web are applicable to designing search features for an individual Web site.

Users can engage in two types of behavior when searching for information: browsing and keyword search (Chen et al., 1997). *Browsing* refers to the act of scanning, reviewing, or skimming the contents of a Web page to find interesting or relevant information. *Keyword searching* refers to the process of entering a keyword, term, or phrase into a search engine in attempt to find a particular piece of information. Of course, these two searching behaviors are not mutually exclusive. For example, even if users engage in keyword searching, they are likely to browse the returned results page to find a particular link to which they want to go.

2.5.1 Browsing for Information

Browsing is a strategy used by beginners when they do not know exactly what they are looking for. However, even more experienced Web users tend to engage in browsing behavior when they are exploring or investigating a topic. Because there is much information that a Web site can contain, browsing may not be a very efficient way to look for information in the Web site. Moreover, if a Web site is large, users can get sidetracked by the various links that they encounter even when they have a purpose for their search. Fang et al. (2005) summarized several types of browsing behaviors that reflect goal-directed or nondirected search behavior:

1. *Search-oriented browsing*: directed search aimed at accomplishing a specific goal. *Example:* looking for a specific section or link in an e-commerce Web site that contains information about laptop computers.
2. *Reviewing or general-purpose browsing*: scanning through and reviewing information or Web pages related to the users' general goals but not necessarily needed to accomplish a specific task. *Example:* browsing through the electronics section of an e-commerce Web site to determine what different types of computer products are available and reading information about them.
3. *Scanning browsing*: scanning through information without reviewing it. *Example:* scanning the headings on the home page of a Web site to find interesting topics.
4. *Serendipitous browsing*: just looking to see what is in the Web site, without a specific goal but with the possibility that the user may stumble into something of interest.

2.5.2 Keyword Search

Keyword search reflects more goal-directed behavior. That is, users are looking for a particular piece of information at the Web site that is relevant to attaining their goals. Keyword search is conducted through a Web site's search engine, in which users enter a query and results matching the query or relevant to it are returned. Keyword search gives users direct access to Web pages within the site without navigating through it serially. For example, if the user wants to find the Web page for the Department of Psychological Sciences at Purdue University, he or she can enter the department name into Purdue University's search engine and have the URL for the Web page returned, rather than browsing through several layers of Purdue University's Web site to find the page. Keyword search can be more efficient than browsing if the search engine is powerful (i.e., does not return many irrelevant results). In the following sections, users' keyword search behaviors are reviewed.

Finding the Search Function For users to use the search function of the site, they must be able to find it. Thus, the search feature should be represented in a manner that will be recognized by users and located in a place where users expect to find the search feature. The former point is captured in Nielsen et al.'s (2000) study, in which "users told us that when they looked for the search function, they looked for 'one of the little boxes.' Tabs and links to a separate Search page just didn't work for them" (p. 7). Search features are often located at the top or bottom of a Web page and should be placed on every Web page so that users can have the option of performing a search instead of browsing through the contents or returning to the home page.

Simple versus Advanced Search *Simple search* refers to search engines that primarily use keywords to find information, and *advanced search* engines allow additional filters. Many advanced search features include Boolean operators (OR and AND). Eastman and Jansen (2003) noted that 90% of Web searches used extremely simple queries and only 10% used advanced query options. They conducted a study examining the effectiveness of searches with and without advanced operators on global search engines and found that the use of advanced operators did not result in significantly better results. Furthermore, Nielsen (2000) reported that users have trouble with Boolean operators because they often confuse AND with OR, and vice versa. Given that a majority of users have trouble with advanced search, the Web site's default search feature should be a simple search, providing an option for advanced search for users who want to use the feature.

Fast and Accurate Search A good search feature or engine should be fast and accurate. Speed refers to how long a user has to wait before the results returned are displayed. Users tend to rate speed as an important factor affecting their preference for a Web site. For example, users in Lightner's (2003) survey on preferences for e-commerce sites indicated that navigation speed and buying speed were the third and fourth characteristics that determined their overall satisfaction. Thus, the search feature for a Web site should provide fast, high-quality results. The speed at which a search can be performed depends on the size of the search space and the power of the search engine, and the quality of the information retrieved depends on the precision of the engine (see, e.g., Kobayashi and Takeda, 2000). *Precision* can be defined simply as the proportion of the number of relevant documents retrieved by the search over the number of relevant documents in the search space. The use of wide search boxes can improve the quality of the search because it allows more room for users to type more descriptive words that will help result in better matches (Nielsen, 2000).

Dictionaries and Thesauruses Typographical errors (typos) occur when users enter keywords or terms into a search engine. Dictionaries can be used to help correct typos. When dictionaries are used to correct a user's spelling, users should be notified of it (Proctor et al., 2002b). For example, before results are returned, a message can be presented indicating that there were "no matches for onebedroom apartments, did you mean one-bedroom apartments instead?" By notifying the user that the search engine noted a potential typo and corrected it, users will understand why they are receiving the results that are being displayed. Different users may also use different words to describe the same item, so including a thesaurus can help users find items that are stored in the Web site even if it does not exactly match the label that was assigned to the item.

Better Indexing of Terms Queried terms are usually matched against index terms in order to find matches. Inclusion of a thesaurus will help avoid the problem of different people using different terms to describe the same item. Meta tags can also be used to index Web pages within a site. Meta tags allow designers to specify additional keywords for a Web page that is indexed. It is very labor-intensive to create good meta tags for a Web page, but the investment in creating them may be worthwhile because global search engines they can use meta tags to lead users to the Web page. However, not all search engines support meta tags (Sullivan and Sherman, 2001).

Using Relevance to Rank the Presentation of Results Search engines for a site usually return a list of Web pages that "matches" the queried topic. Usually, short descriptions of the Web page are provided underneath or next to the hyperlink that will point the user to the page itself. Depending on the nature of the Web site, many Web pages can match the queried topic. Because users do not typically read or skim all the results that are returned, the linked Web pages should be ordered according to their judged relevance to the queried topic to help users locate the desired information quickly. There are many tools and algorithms that have been developed for determining relevance (see Kobayashi and Takeda, 2000; Fang et al., 2005).

Allowing Users to Return to the Results Page When a user follows a link to a Web page listed on the results page, it may or may not be the particular Web page for which the user is looking. Thus, users should always be allowed to link back to the results page or navigate backward (use the back button of the browser) to return to the results page. When users return to the results page, links visited should be made distinct from new links to show the user where they have been and new Web pages that they have not visited. As noted in Section 2.4.4, blue links should be used to designate unvisited sites and purple or burgundy links to mark sites that have been visited.

2.5.3 Summary

The search feature of a Web site is important because users are prone to use search features to locate information. If users cannot find what they are looking for, they may abandon a Web site. The search feature should be in the form of a "box" and should include algorithms that return relevant results quickly.

2.6 Information Presentation

Successful structuring and organization of the components in a Web site will lead to more efficient navigation and search of information. The next step is to determine the best methods to present to users the information included in a Web site. Effective presentation of information is critical because it allows the user to extract the relevant information that they need to accomplish their goals. Fogg et al. (2003) conducted a survey in which 2684 participants evaluated the credibility of Web sites. They found that the "look" of the Web site was more important than any other factor in determining the credibility of the site. In fact, 46.1% of the comments received were devoted to issues of presentation and design, including the visual design, layout, color schemes, and so on. Because designers of the Web site built it, they have a good understanding of where everything is located and know why the information is presented in the manner that it is. However, some designers may fail to recognize that just because the organization and presentation of information is intuitive to them does not mean that it will be intuitive to the end users. Subsequent sections are devoted to the topic of information presentation at the global level and page level and how to present information in a manner that is accessible to users or different cultures and to users with disabilities.

2.6.1 Global Site Design

Simplicity, Straightforwardness, and Easiness Information presented in a Web site should be made simple and straightforward (Nielsen, 2000). This will help the users locate the information they are looking for and will aid in the ease with which they can use the Web site. Although fancy designs may be aesthetically pleasing, they may distract users or "hide" relevant features or information presented in the site. In the spirit of parsimony, the simplest design that includes all the relevant and important features and information is probably the best.

Browsers and Service Providers It is important to know what types of browsers users will be using to access the site and to present information in a manner that meets the requirements of the browser. If the information is presented in a manner not supported by the browser (e.g., frames and flash), the user will not be able to see it. The three most common browsers that need to be taken into account when designing how to present information are Netscape Navigator/Communicator, Internet Explorer, and text-based browsers (Irie, 2004). Designers need to note that the same Web site may look different when accessed by different browsers (see, e.g., Figure 2). For example, text-based browsers will not display graphics, and the use of ALT tags, describe earlier, can be implemented to designate what the image was supposed to be. Furthermore, because newer versions of these browsers are released periodically, designers should present the information to accommodate the requirements not only of current versions but of back versions as well. Nielsen (2000) analyzed statistics of browser versions and noted that the only about 2% of users per week upgrade or change to a newer version of a browser when moving from version 1 to 2 or 2 to 3, and only 1% from version 3 to 4. Thus, it seems appropriate to keep at least one year behind the latest browser, if not more.

It is also important to keep in mind the service providers that users will use to connect to the Web. For example, Najjar (2001) indicated that e-commerce sites should present their content to fit inside a 625 × 270 pixel window to accommodate AOL users because at that time, AOL users were more than three times more likely to purchase products online than other users. Thus, designing an e-commerce site to accommodate that population of users was critical.

Scrolling and Paging Web designers often struggle with how much information should be presented on a given page of Web site. Information that is presented in large amounts becomes lengthy and takes up a lot of space, and users are likely to have to scroll down the page to read or review the entire content of the page. Users do not like to scroll, so some designers recommend placing important or critical information "above the fold," within the space that can be viewed without scrolling (e.g., Pearrow, 2000). To avoid having to scroll, information can be decomposed into smaller chunks and presented on different pages. Users can access the information on the next page by clicking on a link or forward button. This method does not require users to scroll but does require them to page through the information serially (although some sites allow users to "jump" pages by providing links to numbered pages that represent the chronology of the information).

Baker (2003) found that it took participants 19 seconds longer to read paged text than scrolled text, but there was no significant difference in the level of comprehension between the two groups. He also found that users took 55 seconds longer to search for information in paged text than in scrolled text. Thus, it may be better to place all the information on a single page and require users to scroll than to break up the information across different pages. However, Tullis et al. (2005) noted that home pages and navigation pages should be kept short. One reason why these two types of pages should be kept short is that they have the goal of pointing users to the information that they need rather than being the source of that information.

The scrolling referred to above involves vertical scrolling or moving down a page by scrolling. However, if the Web site is designed to be too wide, users may have to scroll horizontally to see the information at the "edges." In general, it is recommended that horizontal scrolling should be avoided unless the task involves item comparison, in which case horizontal scrolling may facilitate the task (see Najjar, 2001).

Secondary and Pop-up Windows *Secondary windows* are windows that open up with the Web page or information that users request, whereas *pop-up windows* are usually Web pages, advertisements, or information that "pops up" or appears without the users' request. Pop-up windows are usually considered annoying by users and should not be implemented in a Web site. Moreover, Fogg et al. (2003) indicated that, "pop-up ads were widely disliked and typically reduced perceptions of site credibility" (p. 7). Secondary windows, on the other hand, can be used appropriately to present information since users are requesting the information and the information is not being imposed on them. Secondary windows are especially helpful for presenting additional information such as online help or detailed product information (see Tullis et al., 2005). When a second window is being used to present additional information, users should be notified of its use by presenting the secondary window in a smaller size that does not occlude the original screen or by prompting the user that the another window will be opened to display the requested information.

Figure 2 The same Web site displayed in (*a*) Internet Explorer and (*b*) Netscape 7.0.

Frames Frames divide the browser screen into distinct parts, each of which can be navigated and scrolled through independently. Frames can be particularly useful for displaying information that needs to be visible while the user scrolls down a page (see Dix, 2005). For example, providing navigational tools in a separate frame allow users to access them even when the users scrolls "below the fold" of the content material in another frame (see Figure 3). However, frames have drawbacks. First, not all browsers support the use of frames. Frames can also make the Web site or pages from the Web site harder to find by search engines because the page that sets the frame is indexed instead of the content page (see Wisman, 2004). Finally, there can be problems with bookmarking and printing content from Web sites with frames because the page that sets the frame is the one that is often marked or printed, or the last framed accessed is printed.

URL Design URLs are Web addresses that direct the users to the desired Web page. The URL of a Web site starts with the prefix "http://"; however, some Web servers will load the page without the prefix.

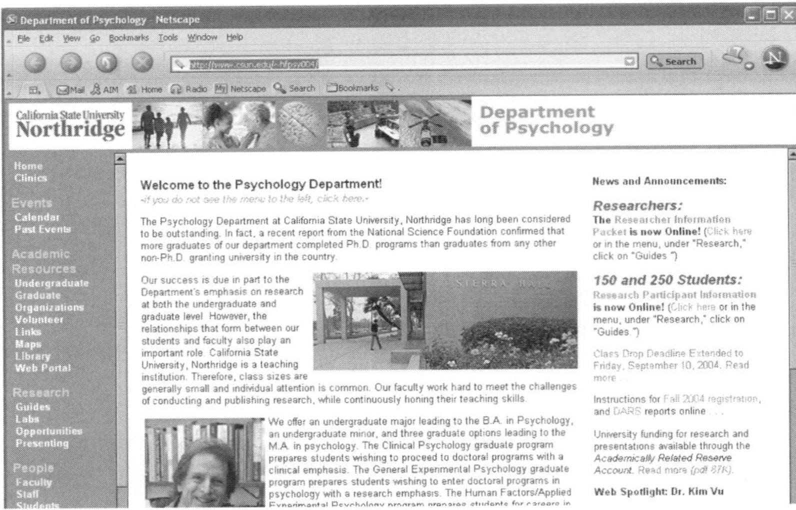

Figure 3 Web site that uses frames for navigational controls.

After the prefix are the letters "www" for World Wide Web, followed by a domain name and specific page titles, if there are any. Nielsen (2000) noted that the domain name is the most important component of the URL because if users can recall or guess the correct domain name, they can get to the home page and find their way to specific information from there. Because URLs must be specified completely accurately, Nielsen recommends that they be made as short as possible, with common words, all lowercase letters, and without any special characters. Because some users will choose to "cut and paste" URLs into browsers (e.g., from a referral in an e-mail) or link to a specific URL, it is important that the URL not be broken (e.g., by inserting a return break at the end to "wrap" the URL). As mentioned earlier, one problem with a frame is that it breaks the URL, making it difficult or impossible for users to bookmark the specific Web page.

Error Messages There are many possible reasons for errors to occur. Some of these errors may be due to the Web browser that the user is using, whereas others can be due to the design of a specific Web site. Lazar and Huang (2002) examined error messages for Web browsers and noted that many error messages are confusing and difficult for users to understand. Although the designer cannot control the error messages produced by the Web browser, care should be taken in error messages returned by the individual Web site. For example, if the Web site asks the user to enter the date in a specific format such as MM\DD\YYYY, but the user enters the date as MM\DD\YY, the error message returned should not be something obscure such as "PAGE ERROR: INVALID REFERENCE, NOT SET TO AN INSTANCE OF THE REFERENCE" but should indicate the error so that the user can fix the problem. A better error message would be: "FORMAT FOR DATE IS NOT VALID, PLEASE ENTER DATE IN THE FORM OF MM\DD\YYYY" (see also Figure 4). This message specifically tells the user that the date field is incorrect rather than making the user guess what is meant by "INVALID REFERENCE," as in the former error message example. Several Web pages have also been designed to mark the field or entry in which the error was made to help users find where the error occurred.

2.6.2 Page Design

Home Page Design The home page is usually, but not always, the first page that a user encounters when entering a Web site. If users access specific pages in the site (e.g., from a hyperlink on a global search engine's results page or from another Web site), they should be allowed to have direct access to the home page through a link or "home" button. The home page should communicate the site's purpose, include information about the site or organization hosting the site, and contain navigation functions to major subsections of the site. Nielsen and Tahir (2002) examined the usability of 50 Web sites' home pages and suggested 113 guidelines for designing usable home pages. Some of the more important guidelines are summarized below.

- Make the home page unique from other pages in the site.
- Present the site's, or company's, name and/or logo in a prominent location on the page.
- Include a statement or tag line that summarizes the purpose of the site, but do not use phrases or language that will be difficult for users to figure out.
- Emphasize high-priority tasks or sections that users will want to access, and label sections in

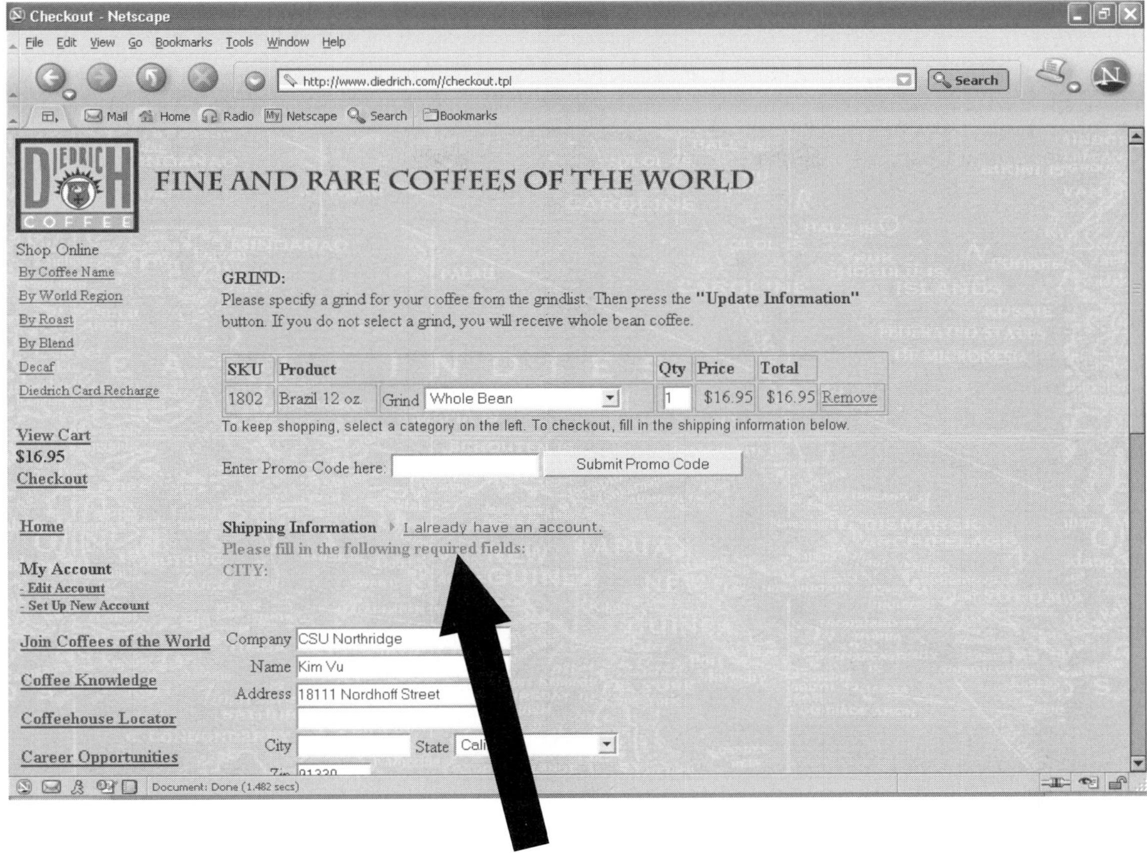

Figure 4 Good error message from a Web site.

a manner that will help users accomplish their goals.

- Differentiate links and make them scannable.
- Provide information about the company or organization hosting the site, including corporate, contact information, and frequently asked questions.
- Avoid redundant content, exclamation marks, and use of all UPPERCASE letters.
- Place a search box on the home page that searches the entire site. Do not offer a feature to search the Web, but limit the search to the site.
- Use graphics to show content and never animate critical elements.
- Do not make users scroll horizontally.

Page Layout The layout of the page should be designed in a manner consistent with users' expectations. For example, the home page link or logo linking to the home page is usually located at the top-left corner of the page, navigational links are located at the left column of the page, and the main content is located in the center. Guidelines and standards have been developed to try to make Web pages more consistent, and user testing has shown some layouts to be more effective than others. Tullis et al. (2005) conducted a thorough review of research relating to page-layout issues and recommended the following based on the literature:

- *Use a fluid layout rather than a fixed layout.* A fixed layout does not change with the size of the browser window, whereas a fluid layout adjusts the elements to the browser window as well as the printed page.
- *Use a medium level of white space.* White space can be used to reduce clutter, although too much white space may increase the time needed for users to locate items if it causes a need to scroll down to see information.
- *Place items in locations where users are likely to expect them.* Because the Web has been around for some time, users have become aware of what a "typical" Web page should look like.

Number 1 / January 1, 1998
Volume 40
Number 12 / December 1, 1997
Number 11 / November 1, 1997
Number 10 / October 1, 1997
Number 9 / September 1, 1997
Number 8 / August 1, 1997
Number 7 / July 1, 1997
Number 6 / June 1, 1997
Number 5 / May 1, 1997
Number 4 / April 1, 1997
Number 3 / March 1, 1997
Number 2 / February 1, 1997
Number 1 / January 1, 1997

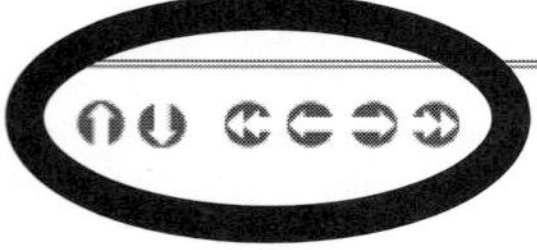

Add Publication to Shopping Cart

(*a*)

$299,900	Encino	2 beds	2 baths

Condo

MARVELOUS UNIT IN ENCINO Quiet complex, big living room w/fireplace. 2BR, 2BA, swimming pool, two car parking.
more info | contact agent | map

$299,900	Lake Elizabeth	4 beds	3 baths

BEAUTIFUL LAKE ELIZABETH HOME, An opportunity to have fresh air with lake and mountain views in this amazing 4BR + 3BA, 3,200 + sq. ft. home. Call...
more info | contact agent | map

$299,000	Northridge	2 beds	1 bath

Very Special! 2+1 feat. hrdwd flrs, new cstm built kitchen & 1 car private gar. Excellent neighborhood! $299,000
more info | contact agent | map

(*b*)

Figure 5 (*a*) Bad image link because the circled images look like navigational buttons but are nonresponsive; (*b*) good image link that directs the user to a page with a larger view of the image.

Links Links are important elements of a Web page because users can use the links to go to other Web pages that are connected to the current one. There are two types of links, text-based links and image/icon links. The mouse cursor changes shape when it is placed over an active text or image link. Text-based links are usually marked by highlighting the text on a Web site in color and underlining it. Visited and unvisited links should be distinguished clearly to help users know where they have been. Using different colors to mark visited and unvisited links helps reduce search time (Halverson and Hornof, 2003), but care should be taken as to which color to use. Halverson and Hornof showed that search time is increased when red-text links are present on a Web page.

Icon or image-based links are graphics or pictures that link to another page when they are clicked. It is important to make links distinctive from nonlinks so that users know by looking at the image or icon what they can and cannot click on to take them to another Web page, rather than having to place their cursor over each item to find which is an active link (see Figure 5*a*). An image link is particularly useful when it provides a visual representation of the information requested or the item or product available from the Web site. Because text-based links usually download faster than image- or icon-based links since they are smaller in size, thumbnail image links can be used to lead users to an enlarged picture or image to reduce page load times when many images are presented on a page (e.g., a returned search page) (see Figure 5*b*).

Portals, Web sites that are intended to provide links to other Web sites or online resources, often display many links on a single page. One question that arises is how many links should be displayed on a page. Bernard et al. (2002) had participants locate specific links on a search engine's result page for which the number of links returned was 100, with the number of links present on each page being 10, 50, or 100. Task completion time was shortest for the page with 50 links and longest for the page with 10 links. Users rated the 10- and 50-link per page formats roughly equal in terms of how easy it was to find information on the page and in terms of being professional looking, but showed a slightly greater preference for the 50 links to a page layout.

Text versus Graphics and Images Text is the most common method used to convey information. This is especially true since there are some Web browsers that do not support graphics, and skilled users may know how to turn off the graphics mode in browsers that support them. Thus, the question of interest is whether graphics should be used with the text and, if so, what combination of text and graphics is optimal. Sears et al. (2000) found that all participants in their study indicated that Web sites that contained graphics were more attractive than those that contained only text. Although the use of graphics and animation can distract users, some types of information can be conveyed better by a "picture" than by words (see, e.g., Plaue et al., 2004). Sears et al. conducted an experiment examining a version of the 1997 Microsoft Web site that was modified to make the site self-contained (i.e., not linked externally) and a simplified version of the site (animations were removed and graphics simplified, graphical links were transformed to text links, etc.). They found that users were able to find desired information more easily with the original site than with the simplified site. Thus, there does not seem to be a problem with including some graphics in a Web site, as long as ALT tags or labels are used to indicate what the graphics are supposed to be in browsers that do not support them.

Graphics should not be used gratuitously, though, because they tend to be large in size, slowing the download time. As described in the next section, slow download times often frustrate users. The use of graphics and images is important to many e-commerce sites because users often want to see the product. Because search results from an e-commerce site tend to return many products that fit the queried item, including detailed images of each product may not be possible. In this case, small thumbnails of the image should be provided, with the option for users to access a more detailed version of the image if desired.

Animation Animation can be applied to graphics to mimic real-life movement. Animation is frequently used as an attention getter in Web pages (Nielsen, 2000; Lazar, 2003). People are particularly sensitive to movement in the peripheral visual field due to the human sensory and perceptual system (see Chapter 3). Nielsen indicated that the use of animation is good for the purposes of (1) demonstrating continuity, transitions, or changes over time; (2) illustrating three-dimensional structures; and (3) attempting to attract attention. However, it has been shown that having animation present on the screen decreases performance on other tasks (e.g., Zhang, 2000). Zhang had participants perform an identification task in which users had to report the number of times the target information (i.e., strings of letters) appeared in an 8×10 table of a Web page. On the page containing the search array, animation was presented randomly to the top, sides, or bottom of the table. Zhang found that performance in the identification task decreased in the animation conditions compared to baseline conditions in which no animation was present. Furthermore, the animation particularly hurt performance when it contained information similar to the targeted information that was irrelevant to the task.

Animated banners are typically used as attention getters for advertisements. However, there is at least some evidence suggesting that users do not recall animated banners more than static banners (Bayles, 2002). Furthermore, advertisements, in general, tend to be viewed negatively, especially pop-up ads (Fogg et al., 2003). Thus, designers should take caution when determining whether to include animation. Animation is not always bad, but its presence can hurt performance on other tasks.

Page Load Times Although advances in technological development for high-speed connectivity have increased the availability of high-speed connections such as digital subscriber line (DSL) and cable modems, many users continue to connect via dial-up modems (U.S. Department of Commerce, 2002). The U.S. Department of Commerce's survey of connection types in 2001 showed that 80% of users reported using dial-up connections. Even though this percentage is likely to decrease in the future due to greater availability and lower costs of higher-speed connections, designers must still take into account that a significant number of users will access the Web via dial-up connections. It is important to take the users' connection speed into account because the connection speed may determine some of the users' behaviors. For example, a user might have no qualms about downloading a large media file when he or she is connected to the Web through a cable connection, but might hesitate to do so when connected through a phone modem.

Slow download times for Web pages is disliked by users (e.g., Lightner et al., 1996). There are many reasons why users may react unfavorably to long download times (e.g., Bouch et al., 2000), including (1) not knowing whether the user made an error, (2) not wanting to wait "forever" to complete a task, (3) thinking that the site is not secure or well designed, and (4) loosing his or her train of thought or task set during the delay.

Jacko et al. (2000) conducted an experiment evaluating the effects of network delay on users' perceptions of the quality of Web documents. They manipulated the download time (mean delay times were 575, 3500, and 6750 ms for the fast, medium, and slow conditions, respectively) of Web pages from a Web site with only text or one with text and graphics. The content of both Web sites were the same. Users judged the text and graphic documents to be of higher quality than the text-only documents at short delays, but text-only documents were judged to be of higher quality at long delays. Jacko et al. noted that the lower-quality judgments for graphic/text pages with long-delays may be due to users attributing the long delay to graphics involved in the document. Sears et al. (2000) also showed that users with slower connections are also less impressed with Web page graphics than those who tend to have access to faster connections.

Nielsen (2000) recommended that Web pages be formatted in a manner that can be downloaded in 10 s or less. However, a recent study by Galletta et al. (2004) suggested that delays should be less than 8 s to promote positive attitudes from users and less than 4 s to encourage users to continue with the task or to revisit the site later. Bouch et al. (2000) found that users' tolerance for delays decreased as the time they spent on the task increased, but if the page loaded incrementally, users were more tolerant of longer delays. If it is not possible for a page to download in the recommended time, users should be notified of this in advance. One way to help minimize the download time of Web pages is to reduce graphics to thumbnails and allow users to "zoom" or access larger images separately (perhaps by using an image link).

2.7 Designing for Accessibility and Universal Access

One benefit of having a Web site is that it can be accessed from anywhere and at any time with a variety of computing devices. For a Web site to be of maximal use, it should be designed proactively to promote accessibility from the early design phases and throughout the development life cycle. Stephanidis and Akoumianakis (2005) consider access to be a contextual issue that is determined by three major parameters: target user, the access terminal or interaction platform, and the task context.

Although a particular Web site can be viewed by anyone who can arrive at its "address," it is likely that the site is not designed to target all users, but rather a subset of users. For example, a Web site devoted to being a portal to refereed publications in cognitive neuroscience is unlikely to be used by elementary school children. Thus, this site does not have to take into account the cognitive capabilities of elementary school children, but rather, it should be designed to promote easy access of research materials for scientists and medical professionals from all over the world. The Web site should also be designed to be usable on different computing devices that the target users are likely to use for accessing the information, such as personal computers, laptops, and personal data assistants. Finally, the design of the Web site should take into account the context in which the site will be used, such as to browse through the contents to see the latest research in the area or to look up a specific piece of information of interest. In the following sections, Web design issues for different classes of users will be discussed.

2.7.1 Cross-Cultural Designs

Cultural differences in the user population relating to time, language, communication style, and social context may affect Web use (Choong et al., 2005). For example, users can be grouped into two general classes in terms of their perception of time: monochronic and polychronic (see, e.g., Rose et al., 2003). Monochronic cultures view time in linear fashion and are very task-oriented. Polychronic cultures view time in a more flexible manner in which several tasks can be worked on concurrently and the users switch back and forth between the various tasks. Rose et al. conducted a study with users from two countries whose cultures are primarily monochronic (United States and Finland) and two whose cultures are primarily polychronic (Egypt and Peru). They found that all users had a negative attitude toward delays, but it was less pronounced for the polychronic cultures than for the monochronic cultures.

Web designers should not only take into account cultural issues such as time perception, but also the resources that the users are likely to have access to. For example, Sears et al. (2000) showed that Swiss users were less impressed by graphical Web pages than American users partly because they had access to slower connections. Thus, Web designers should take into account cross-cultural issues in developing Web sites such as the representation of meaning through icons, symbols, and colors, the use of graphics, and the differences in page layout when translating images and text from different languages [see Choong et al. (2005) and Chapter 7 for reviews on cross-cultural Web design]. Some of the guidelines summarized by Choong et al. include:

- Use unambiguous language.
- Allow extra space for text.
- Accommodate text reproduction methods.
- Do not embed text in icons.
- Use an appropriate method of sequence and order in lists.
- Take linguistic differences into account.
- Take the direction in which text is read into account.
- Be aware that variations exist within the same language.
- Provide natural layout orientation for information to be scanned.

- Provide layout orientation compatible with the language being presented for menu designs.
- Note that icons designed and tested well in one region may not be accepted by people in other regions.
- Provide a combination of text and picture when designing icons.
- Examine the textual component in the graphics on the Web carefully when it is intended for a global audience.

2.7.2 Designing for the Elderly

The population of older adults is large and increasing rapidly. It is estimated that by the year 2050 the number of adults 55 years of age or older will exceed 120 million (U.S. Census Bureau, 2002). Bitterman and Shalev (2004) indicated: "Senior citizens are connected to the Internet for the longest time at a single sitting—more than any other age group" (p. 25). Furthermore, Internet use trends from 1997 to 2001 indicate that people who used the Internet when they were younger will continue to use it as they become older (U.S. Department of Commerce, 2002). Thus, Web sites should be designed to take into account the capabilities of older adults.

Because declines in sensory, cognitive, and motor functioning with age can affect the users' ability to interact with computing devices, it is important to take these considerations into account when designing a Web site. Below are some design guidelines relating to perceptual and cognitive changes with age that Web designers should take into account [based on recommendations from Bitterman and Shalev (2004) and Fisk et al., 2004)].

Perceptual changes:

- Use larger letters and avoid decorative or cursive fonts (e.g., 12-point sans serif letters).
- Avoid style sheets that prevent users from increasing the font size.
- Use bright colors and avoid combinations of colors of short wavelengths (blue–violet–green).
- Maximize contrast (at least 50 : 1 contrast, e.g., black text on white background).
- Increase the resolution of elements on the screen.
- Minimize clutter.
- Avoid moving and scrolling text.
- Provide redundant information (e.g., graphic and text or visual and auditory).
- Avoid high-frequency sounds (above 4000 Hz).

Cognitive changes:

- Use a consistent structure for the site and layout of the Web page.
- Convey the system's status clearly.
- Provide clear feedback for errors.
- Minimize demands on working memory.
- Allow for recognition of information, rather than requiring recall.
- Group information.
- Provide a site map that illustrates where the user is.
- Emphasize important and relevant information.
- Provide a "help" section or "support" information.

2.7.3 Accessibility for Users with Disabilities

In the United States, approximately 8.5% of the population has at least one of the following types of disability: blind or severe visual impairment, deaf or severe hearing impairment, difficulty walking, difficulty typing, or difficulty leaving home (U.S. Department of Commerce, 2002). Web sites should be designed to take this population into account. Designers should not regard having to make the site more accessible to users with disabilities as an extra burden because Web sites that are designed to improve accessibility for disabled populations often produce benefits to people without disabilities as well (Vanderheiden, 2005). The Web Content Accessibility Guidelines 1.0 (Chisholm et al., 2001) recommend the following to promote the development of Web content that will be accessible to people with disabilities:

1. Provide equivalent alternatives to auditory and visual content.
2. Do not rely on color alone.
3. Use markup and style sheets, and do so properly.
4. Clarify natural language use.
5. Create tables that transform gracefully.
6. Ensure that pages featuring new technologies transform gracefully.
7. Ensure user control of time-sensitive content changes.
8. Ensure direct accessibility of embedded user interfaces.
9. Design for device independence.
10. Use interim solutions.
11. Use W3C (World Wide Web Consortium) technologies and guidelines.
12. Provide context and orientation information.
13. Provide clear navigation mechanisms.
14. Ensure that documents are clear and simple.

Web pages designed following these guidelines should be accessible by users with diverse sensory capabilities.

2.8 Security and Privacy

The Web allows users to access information, exchange information, and perform transactions online. However, there are well-founded concerns about privacy and security since personal data collected about users need to be protected. Some users try to avoid Web sites that ask for personal information because they fear misuse of this information. For example, a user may be hesitant to register his e-mail address with a site in order to view its contents because he does not want to deal with the possibility of receiving unsolicited e-mails from the site's organization or one of its affiliates. To date, the most prominent effort for online privacy protection is the World Wide Web Consortium's (W3C) Platform for Privacy Preferences (P3P) project (see http://www.w3.org/P3P/). The goal of the P3P project is to enable Web sites to encode their data-collection and data-use practices in a machine-readable XML format and to provide a simple and automated mechanism for users to specify their privacy preferences through a Preference Exchange Language called APPEL. Through the use of P3P and APPEL, a Web site's privacy policy can be checked against the user's privacy preferences, to determine whether the site's data-collection and data-use practices are acceptable to the user. Although the P3P project is a significant step toward developing privacy protection mechanisms, it is still in development and has its limitations.

In addition to privacy concerns, issues relating to information security have received tremendous attention in recent years because much of our personal information flows through the Web as we receive online services or make online transactions. As a result, it is essential for Web sites to have reliable security systems that protect the sites against information theft, denial of service, and fraud (see, e.g., Schultz, 2005). The most common method used by Web sites to identify and authenticate users is the user name–password combination. However, despite its pervasive implementation, it is well known that the user name–password combination is a relative weak security method because many users fail to adopt crack-resistant passwords (Proctor et al., 2002a). Users also tend to create passwords that are easy to remember (e.g., Riddle et al., 1989; Klein, 1990) or write them down. Furthermore, because different sites have different requirements for acceptable passwords, users have trouble remembering unique passwords for multiple accounts (Vu et al., 2004). One reason why the user name–password method is still popular despite its limitations is that it is relatively easy to implement and is accepted by users (as opposed to more intrusive methods, such as biometrics).

Web sites should also be designed to protect against breaches in security such as hacker attacks (unauthorized access to the information stored by the Web site), denial of service attacks (which cause the network to slow down or Web site to stop working), Web defacements (unauthorized modification of the content of the site), and computer viruses (Schultz, 2005). Techniques to counter security breaches are beyond the scope of this chapter. However, given that the users' trust in the Web site can be obliterated by security breaches and the potential costs and legal ramifications associated with the breaches can be great, it is important to emphasize that care should be taken when designing the Web site to incorporate security mechanisms.

3 EVALUATING WEB USABILITY

Web usability refers to the ease with which a Web site can be used, its efficiency and effectiveness, and the satisfaction it provides to users (e.g., Brink et al., 2002). Web usability is extremely important because it is a major determinant for whether or not a particular Web site will be successful (Nielsen, 2000). Because Web usability is an area within the general domain of computer software usability (see Zhu et al., 2005), many issues relevant to software usability discussed in Chapter 49 also apply to Web usability.

Although the topics of Web design and Web evaluation were covered separately in this chapter for ease of presentation, designing and evaluating Web usability often occur in iterative cycles (Mayhew, 2005). Web evaluation should be implemented early in the development process and throughout its life cycle. It is not adequate to conduct usability evaluations on a final product to catch the "bugs." Moreover, usability problems discovered on a final product may not always be fixable. Thus, it is important to emphasize that usability evaluations should be implemented in early, as well as late, phases of design, where usability problems can be identified and more easily fixed.

Web usability evaluation is dependent on the goal or purpose of a Web site. These goals and purposes were elicited from the organization or users at the start of the design process covered earlier in the chapter and should include consideration of:

1. *Target users.* Information and descriptions of the target users should include demographics, Web and computer experience, core-user tasks, task environment, and preferences. Characteristics of the target users can be obtained by examining user profiles (Fleming, 1998) or personas (Cooper, 1999). *User profiles* are brief summaries of real users' characteristics. *Personas* do not refer to real people but to a representation of a type or category of users. Personas are typically used when designers do not have access to user profiles but have a general idea of the characteristics that actual users might have, and can be based on the designers' knowledge of who the real users are likely to be.

2. *Core user tasks.* Core user tasks are those tasks that are frequently performed by the target users. Core user tasks are highly dependent on the nature of the Web site. Below are examples of some likely core user tasks for each of the major types of Web sites:

- *News/information dissemination*: find out the late-breaking news; check the weather or stock market.
- *Portal*: find Web sites relating to Web design; find electronic resources on the topic of usability.
- *Communication*: chat with other users with a common interest in e-business; e-mail a friend.

- *Search*: find information on a topic such as cancer; find a professor's e-mail contact information in a university's online directory.
- *E-commerce*: purchase Jamaican Blue Mountain coffee; purchase a computer online.
- *Company/product information*: find the location of a store; track delivery status of a product.
- *Entertainment*: watch the a movie trailer, play a game online.

The core set of user tasks can be defined at different levels. For example, higher-level tasks can include browsing, researching, communicating, searching, purchasing, and accessing entertainment. Each of the higher-level tasks can be broken down into subtasks such as locating a product on Web page, "placing" it into the "shopping cart," providing billing information, and providing shipping information. Once the goals of the Web site are defined and the core user tasks are established, one of several usability methods can be used to evaluate whether the Web site is successful in conveying the purpose of the organization and supporting the users in accomplishing their tasks.

3.1 Evaluation Methods

There are several general classes of usability tests, although some of the specific methods under each can cut across categories.

- Interviews, focus groups, and surveys/questionnaires
- Naturalistic observation
- Participatory evaluation (ethnographic methods and diary studies)
- Web-based methods (automated sessions, Web logs, and opinion polls)
- Prototyping (paper prototypes and interactive prototypes)
- Usability inspections (heuristic evaluation, cognitive walkthrough, and alternative viewing tools)
- Usability lab testing (usability and performance testing)

The first four classes of usability tests were introduced earlier as methods to help elicit information about users and core user tasks when determining what content to include in the Web site. However, they are also used as techniques for evaluating usability later in the design process. For example, interviews, focus groups, and surveys/questionnaires can be used to determine the difficulties that users have while interacting with the Web site and their opinions, attitudes, satisfaction, and preferences for the different features, layout, or structure of the Web site. Similarly, naturalistic observation can be used to evaluate users' performance on different tasks and facial or bodily expressions while interacting with a Web site. Participatory evaluations can be used to evaluate the Web site in the context of the users' social and task environment or by having users' log usability problems and concerns as they encounter them in daily use of the Web site.

As mentioned earlier, the Web itself is a tool for performing evaluations on a particular Web site. When a site is accessed, the Web server records information about the files sent to a browser, which can include the date, time, host domain (the Web address of the requesting browser), file name, referrer (the URL of the page that provided the link), and so on. Not only can Web logs reveal information about the users, such as how many users visit a Web site and from which domain they are visiting, but Web logs can also be used to evaluate the site by examining the frequency with which specific Web pages in the site are visited, the amount of time spent on a page or task, the success versus bail-out rate, and navigational paths that users take while interacting with the site. Web log data may also provide some clues for usability problems. For example, if a specific Web page is not visited frequently, designers can conduct further investigation to determine whether the lack of access to the page is due to where it is placed within the site or the content that it conveys.

Instrumented Web browsers can also be used to conduct automated sessions of remote usability tests. Participants recruited for these tests are provided with the instrumented version of the browser, for which they are asked to complete certain tasks by visiting one or more Web sites. Performance measures such as time on task, successes and failures, links followed, and files opened are recorded. Users can also indicate when they have completed a task (either successfully or unsuccessfully) by clicking on a "finish" button and then be prompted to continue to the next task. Automated sessions may be best for summative usability evaluations (Jacques and Savastano, 2001).

Prototyping, usability inspections, and usability lab tests are used more often in later phases of design, after a working prototype has been developed and can be evaluated. It is important to note, though, that any of these usability methods can be used throughout the design process. For example, a questionnaire can be administered to users at the start of the design phase to determine the types of tasks that users are likely to perform on the Web site, in the middle of the development life cycle to determine what users like or dislike about the current version of the Web site and what they would like to see in the final version of it, or near the completion of the site to obtain users' feedback and impressions of the overall functioning of the site.

3.1.1 Prototyping

Prototyping is a particularly useful tool to evaluate the usability of a Web design, especially at earlier phases of development. Alternative designs for the site can quickly be "mocked up" and tested by usability experts or end users. Prototypes can be as basic as drawn images and features on a piece of paper (low fidelity) or more advanced, to mimic the "look and feel" of a real Web site (high fidelity). The main distinction between a prototype Web site and a real one is that the prototype is not fully functioning, but rather, is a representation of the site along with simulations of the functions and features of the site. For example, designers can use a high-fidelity prototype of a Web site to study the presentation of the search results page by having users click on a search button to take them to a predesigned results page. That is, the results page is not returned by the actual search, but is linked to "fake" data.

Low-fidelity prototypes consist of paper mock-ups, storyboards, or paper prototypes. These prototypes are usually hand-drawn, without the detail or polished look of a high-fidelity design. The goal of using low-fidelity prototypes is to convey the conceptual design rather than to test its features. As shown by Newman et al.'s (2003) study with expert Web designers, paper prototypes are still commonly used, even though it may not take much time to implement the same design in HTML to be presented on the Web. The Web environment does, however, make it easier to design a "skeleton" Web site with interactive features (Pearrow, 2000).

- *Paper prototypes.* Paper prototypes are hand-drawn designs of different screens or static printouts of screen shots. They allow designers to focus on the global properties of the Web site and extreme flexibility for changing components of the design. Functional aspects of the design can be illustrated via buttons and links and additions to the design (e.g., to mimic a drop-down menu) can be made on Post-It notes.
- *Interactive prototypes.* As aforementioned, it is easier to build interactive prototypes of a Web site using HTML. Pearrow (2000) made two distinctions for developing interactive prototypes: horizontal and vertical. The emphasis for a horizontal prototype is to enable all the top-level functions, whereas the emphasis for a vertical prototype is to enable the functions along a particular path (e.g., a task to be completed).

3.1.2 Usability Inspection Methods

Usability inspection methods are typically used to evaluate the usability of a system without testing end users (see, e.g., Cockton et al., 2003). Inspection methods cannot replace user testing (see, e.g., Zhu et al., 2005) but can reveal many usability problems quickly, before end users have to be recruited and tested. They are also typically considered to be less expensive than many other methods used to evaluate usability because the expert needs to be able to access the Web site only during his or her evaluation. User-based testing, on the other hand, often requires the cost associated with setting up a usability lab and payment for the usability expert as well as the end users participating in the test. Inspection methods are generally based on design guidelines or recommendations, best practices, and particular findings derived from research studies and theoretical frameworks. The two best known and most frequently used inspection methods are the heuristic evaluation and cognitive walkthrough. However, since the Web can be accessed from a variety of platforms, alternative viewing tools are often used to ensure the usability of the site when accessed by different computing devices.

Heuristic Evaluation With a heuristic evaluation, usability experts examine the functions, features, layout, and content of a Web site and determine whether the site's format, structure, and functions are consistent with established guidelines or design recommendations (e.g., Nielsen, 1993). Some of these heuristic guidelines for Web design include (based on Pearrow, 2000):

- Chunk together related information.
- Use the inverted pyramid style of writing.
- Place important information "above the fold."
- Avoid gratuitous use of features.
- Make the pages scannable.
- Keep download and response times low.

Usually, three to five evaluators are needed to find the majority (at least two-thirds) of the usability problems (Nielsen and Molich, 1990). A single evaluator typically finds only about 35% of usability problems (Nielsen, 1993). Furthermore, the more familiar and experienced the evaluators are with human factors and usability engineering, the more effective they become at identifying usability problems. In addition to evaluating whether the Web site adheres to design principles and recommendations, evaluators can determine whether the site contains common Web design mistakes and remove them before they become usability problems. The top 10 mistakes in Web design posted by J. Nielsen for 2002 and 2003 are listed in Table 1.

Cognitive Walkthrough With the cognitive walkthrough, usability evaluators "walk through" the steps that a user would execute when performing specific tasks on the Web site. The evaluator tries to perform the task from the user's perspective and identifies any problems that users are likely to encounter. The method focuses on how easy the Web site is to use and on how easy the functions in the site are to learn and use (e.g., Polson et al., 1992). For each step of the task, the evaluators are encouraged to ask themselves the following questions (e.g., Wharton et al., 1992):

- Will the user form the right goal for the task?
- Will the user notice that the correct action is available?
- Will the user associate the correct action with the correct control or feature?
- Will the user receive feedback about their progress in the task?

Table 1 Top 10 Web Design Mistakes in 2002 and 2003

2002	2003
1. No prices	1. Unclear statement of purpose
2. Inflexible search engines	2. New URLs for archived content
3. Horizontal scrolling	3. Undated content
4. Fixed font size	4. Small thumbnail images of big, detailed photos
5. Blocks of text	5. Overly detailed ALT text
6. JavaScript in links	6. No "what-if" support
7. Infrequently asked questions in FAQ (frequently asked questions)	7. Long lists that can't be winnowed by attributes
8. Collecting e-mail addresses without a privacy policy	8. Products sorted only by brand
9. URL greater than 75 characters	9. Overly restrictive form entry
10. "Mailto" links in unexpected locations	10. Pages that link to themselves

Source: Based on Jakob Nielsen's Alertbox, available at http://www.useit.com. Readers should visit useit.com to read more details about the mistakes.

Before the evaluators perform the cognitive walkthrough, they should prepare a list of tasks to be performed and questions to be answered when performing each task to guide them through the process. Evaluators need to ensure that the types of tasks they will perform during the evaluation are representative of the tasks that the users typically perform and that a range of difficulty levels be included. The walkthrough itself can sometimes be a tedious and slow process, but there have been some suggested streamlined versions of the cognitive walkthrough that are intended to be more efficient (e.g., Spencer, 2000).

Blackmon et al. (2002, 2003) introduced a variant of the cognitive walkthrough designed especially for the Web. Their Cognitive Walkthrough for the Web (CWW) evaluation process simulates the users' step-by-step interactions with the Web site in a goal-directed task, such as searching for specific information in the site. CWW uses latent semantic analysis to identify usability problems associated with heading/link titles. More specifically, CWW can identify headings or links that are (1) likely to be confusable with other headings/links in the page or site, (2) unfamiliar to the target audience, and (3) likely to lead to a specific goal (i.e., the goal can be classified under competing headings). Blackmon et al. (2003) used CWW to identify usability problems associated with the headings/link titles of an experimental Web site designed to present encyclopedia articles. They showed that the performance on the site improved significantly after the site was repaired for problems identified by CWW.

In general, cognitive walkthroughs are particularly beneficial for helping the design team to think along the lines of users' goals and knowledge. Cognitive walkthroughs, though, are less successful than heuristic evaluations at identifying more serious usability problems (see Cockton et al., 2003).

3.1.3 Evaluations with Alternative Viewing Tools

Because there are many computing devices that allow users to access the Web from any place at any time, it is important that designers also consider how the Web site will be displayed on a variety of browsers (e.g., Internet Explorer vs. Netscape) and platforms (e.g., Web TV, PDAs, mobile phones). Alternative viewing tools are viewers, or software tools, that are intended to show designers how a Web site would look on different devices and platforms. In addition to the platform issue, Web designers should also keep several earlier versions of popular browsers because it may take up to two or three years before users "upgrade" from one browser version to another (Nielsen, 2000).

3.1.4 Usability Lab Tests

Usability lab testing is probably the best method for evaluating the Web site in terms of its effectiveness in promoting successful interactions with the users. Most usability experts would agree that the methods described earlier are useful tools when evaluating Web usability, but they cannot replace usability lab tests. As a result, many of the techniques described above are used in addition to usability lab testing at various phases of the design life cycle. The basic methodology and procedure on how to conduct a usability test have been described in Chapter 49 (see also Pearrow, 2000; Brinck et al., 2002; Dumas, 2003), and are discussed only briefly in this chapter.

A usability lab test for Web sites involves observing and analyzing users' task performance (e.g., successful completion rates and failure rates, task completion times, and errors) while interacting with a Web site, as well as their thought processes during the task (e.g., understandings, misunderstandings, and preferences). Ideally, usability lab tests should occur several times throughout the design phase so that iterative evaluations can be made.

During the test session, the evaluators should be separated from the participant by a one-way mirror, so that the participant can be observed but the evaluator does not "get in the way." Communication with the participant is usually done through an intercom system, and the evaluators should allow time for users to adjust to this communication medium. If the designer wants to evaluate the effectiveness of alternative Web designs or page layouts, the users' performance with the various versions of the Web site can be evaluated. Before bringing users into the lab, designers should determine the goal for the test (e.g., test the navigation paths or the efficiency with which the task can be performed), the population from which the sample users will be recruited, the specific tasks that users are to perform, the procedures that are to be followed, as well as how the data will be coded and analyzed.

1. *Sample of representative users.* The users that are recruited to participate in the usability test should be as representative as possible of the targeted user group. Each test participant is usually asked to perform several tasks and to provide detail information about their interaction with the Web site. Because of the extensive nature of usability testing, relatively few users are selected and tested. Usually four or five users are needed to find about 80% of the usability problems (Brink et al., 2002), but a larger sample size (e.g., 8 to 10 users) can increase the number of usability problems detected as well as providing a more representative sample of the target population.

2. *Sample of representative tasks.* The specific tasks that users are asked to perform should be based on a subset of the core user tasks that were identified earlier in the design process. The task should be representative, both in terms of the type of task (search or navigation) and level of difficulty. If the version of the Web site has not been placed on the Web, the designers should simulate response times for the system to match the connection speed that the end users are likely to encounter.

3. *Procedures to be followed.* Typically, a usability test starts with an introductory session in which participants are given an opportunity to familiarize themselves with the lab setup and ask any questions that they might have. Participants are usually given consent forms to sign, which include information about the nature of the test, duration of the test, type of compensation, confidentiality of their data, and a statement indicating that their participation is voluntary. Some organizations also require participants to sign a nondisclosure agreement. During the test, the experimenter should try to keep the participants on task but should not have too much interaction with the participants so that the experimenter does not bias the outcome of the test. The experimenter should be empathetic when participants get frustrated during the test. Participants should not leave the test thinking that it was their fault that they had difficulties with the tasks. After the usability test is over, questionnaires or interviews can be administered to collect information about the participants' background and/or their opinions about the tasks.

Although the costs of performing usability tests are much higher than those associated with performing usability inspections, usability testing is better than usability inspection methods at finding general, severe, and recurring problems. It is generally agreed that usability lab testing should not be skipped just because the design was tested previously for usability with inspection methods.

4 CONCLUSIONS

Web sites should be designed and evaluated for usability from their conception and throughout the development cycle. The impact of designing for Web usability can be seen by examining case studies in which the return of investment is immense. Below are just a few examples of usability success stories (see also Bias et al., 2003; Bertus and Bertus, 2005).

- Diamond Bullet Design's redesigned version of a state government's portal site resulted in improvements on task success rate and completion time. It was estimated that the redesigned site would result in a savings of $1.2 million per year for residents of the state (Withrow et al., 2000).
- Dell Computer's investment to improve the usability of its e-commerce Web site in fall of 1999 led to a dramatic increase in sales, going from $1 million per day in September 1998 to $34 million per day in March 2000 (Black, 2002).
- Staples.com's investment in usability to improve their registration process resulted in a 53% decrease in drop-off rates and an increase in sales (Roberts-Witts, 2001).

As illustrated by these examples, it is wise to take usability into account when designing Web sites. Web resources can change on an hourly, daily, weekly, monthly, or yearly basis. Thus, it is important that the Web site be maintained so that it provides accurate content and continues to be highly usable. However, maintaining a good Web site is a challenge for designers because new technologies continue to emerge, making it difficult to keep up with the evolving Web.

REFERENCES

Anderson, J. R. (1982), "Acquisition of Cognitive Skill," *Psychological Review*, Vol. 89, pp. 369–406.

Bailey, R. W. (1993), "Performance Versus Preference," in *Proceedings of the Human Factors and Ergonomics Society 37th Annual Meeting*, Human Factors and Ergonomics Society, Santa Monica, CA, pp. 282–286.

Baker, J. R. (2003), "The Impact of Paging vs. Scrolling on Reading Online Text Passages," retrieved July, 5, 2004, from http://psychology.wichita.edu/surl/usabilitynews/51/paging_scrolling.htm.

Bayles, M. E. (2002), "Designing Online Banner Advertisements: Should we Animate?" in *Proceedings of the SIGCHI Conference on Human Factors in Computing Systems: Changing our World, Changing Ourselves*, ACM, Minneapolis, MN, pp. 363–368.

Bernard, M., Baker, R., and Fernandez, M. (2002), "Paging vs. Scrolling: Looking for the Best Way to Present Search Results," *Usability News 4.1*, www.usabilitynews.org.

Bertus, E., and Bertus, M. (2005), "Determining the Value of Human Factors in Web Design," in *Handbook of Human Factors in Web Design*, R. W. Proctor and K.-P. L. Vu, Eds., Lawrence Erlbaum Associates, Mahwah, NJ, pp. 679–687.

Bias, R. G., Mayhew, D. J., and Upmanyu, D. (2003), "Cost Justification," in *The Human–Computer Interaction Handbook: Fundamentals, Evolving Technologies and Emerging Applications*, J. A. Jacko and A. Sears, Eds., Lawrence Erlbaum Associates, Mahwah, NJ, pp. 1202–1212.

Bidgoli, H., Ed. (2004), *The Internet Encyclopedia*, Wiley, New York.

Bitterman, N., and Shalev, I. (2004), "The Silver Surfer: Making the Internet Usable for Seniors," *Ergonomics in Design*, Vol. 12, pp. 24–28.

Black, J. (2002), "Usability Is Next to Profitability," *Business Week Online*, December 4; retrieved July 21, 2004, from http://www.businessweek.com/technology/content/dec2002/tc2002124_2181.htm.

Blackmon, M. H., Polson, P. G., Kitajima, M., and Lewis, C. (2002), "Cognitive Walkthrough for the Web," in *Proceedings of the SIGCHI Conference on Human Factors in Computing Systems: Changing Our World, Changing Ourselves*, ACM, Minneapolis, MN, pp. 463–470.

Blackmon, M. H., Kitajima, M., and Polson, P. G. (2003), "Repairing Usability Problems Identified by the Cognitive Walkthrough for the Web," in *Proceedings of the Conference on Human Factors in Computing Systems, CHI 2003,* ACM, Ft. Lauderdale, FL, pp. 497–504.

Bouch, A., Kuchinsky, A., and Bhatti, N. (2000), "Quality Is in the Eye of the Beholder: Meeting Users' Requirements for Internet Quality of Service," in *Proceedings of CHI 2000*, ACM, Amsterdam.

Brink, T., Gergle, D, and Wood, S. D. (2002), *Usability for the Web: Designing Web Sites That Work*, Morgan Kaufmann, San Francisco.

Chen, B., Wang, H., Proctor, R. W., and Salvendy, G. (1997), "A Human-Centered Approach for Designing World Wide Web Browsers," *Behavior Research Methods, Instruments and Computers*, Vol. 29, pp. 172–179.

Chisholm, W., Vanderheiden, G., and Jacobs, I. (2001), "Web Content Accessibility Guidelines 1.0," *Interactions*, July–August, pp. 35–54.

Choong, Y.-Y., Plocher, T., and Rau, P. L. P. (2005), "Cross-Cultural Web Design," in *Handbook of Human Factors in Web Design*, R. W. Proctor and K.-P. L. Vu, Eds., Lawrence Erlbaum Associates, Mahwah, NJ, pp. 284–300.

Chou, E. (2002), "Redesigning a Large and Complex Website: How to Begin, and a Method for Success, in *Proceedings of the 30th Annual ACM SIGUCCS Conference on User Services*, ACM, Providence, RI, pp. 22–28.

Cockton, G., Lavery, D., and Woolrych, A. (2003), "Inspection-Based Evaluations," in *The Human–Computer Interaction Handbook: Fundamentals, Evolving Technologies and Emerging Applications*, J. A. Jacko and A. Sears, Eds., Lawrence Erlbaum Associates, Mahwah, NJ, pp. 1118–1138.

Cooper, A. (1999), *The Inmates Are Running the Asylum: Why High-Tech Products Drive Us Crazy and How to Restore the Sanity*, Howard W. Sams, Indianapolis, IN.

Dix, A. (2005), "Human–Computer Interaction and Web Design," in *Handbook of Human Factors in Web Design*, R. W. Proctor and K.-P. L. Vu, Eds., Lawrence Erlbaum Associates, Mahwah, NJ, pp. 176–189.

Dumas, J. S. (2003), "User-Base Evaluations," in *The Human–Computer Interaction Handbook: Fundamentals, Evolving Technologies and Emerging Applications*, J. A. Jacko and A. Sears, Eds., Lawrence Erlbaum Associates, Mahwah, NJ, pp. 1093–1117.

Eastman, C. M., and Jansen, B. J. (2003), "Coverage, Relevance, and Ranking: The Impact of Query Operators on Web Search Engine Results," *ACM Transactions on Information Systems*, Vol. 21, pp. 383–411.

Fang, X., Chen, P., and Chen, B. (2005), "User Search Strategies and Search Engine Interface Design," in *Handbook of Human Factors in Web Design*, R. W. Proctor and K.-P. L. Vu, Eds., Lawrence Erlbaum Associates, Mahwah, NJ, pp. 193–210.

Fisk, A. D., Rogers, W. A., Charness, N., Czaja, S. J., and Sharit, J. (2004), *Designing for Older Adults: Principles and Creative Human Factors Approaches*, CRC Press, Boca Raton, FL.

Flach, J. M. (2000), "Discovering Situated Meaning: An Ecological Approach to Task Analysis," in *Cognitive Task Analysis*, J. M. Shraggen, S. F. Chipman, and V. L. Shalin, Eds., Lawrence Erlbaum Associates, Mahwah, NJ, pp. 87–100.

Fleming, J. (1998), *Web Navigation: Designing the User Experience*, O'Reilly & Associates, Sebastopol, CA.

Fogg, B. J., Soohoo, C., Danielson, D. R., Marable, L., Stanford, J., and Tauber, E. R. (2003), "How Do Users Evaluate the Credibility of Web Sites? A Study with over 2,500 Participants," in *Proceedings of the 2003 Conference on Designing for User Experiences*, ACM, New York.

Galletta, D. F., Henry, R., McCoy, S., and Polak, P. (2004), "Web Site Delays: How Tolerant Are Users?" *Journal of the Association for Information Systems*, Vol. 5, pp. 1–28.

Golbeck, J., Alford, A., Alford, R., and Hendler, J. (2005), "Organization and Structure of Information Using Semantic Web Technologies," in *Handbook of Human Factors in Web Design*, R. W. Proctor and K.-P. L. Vu, Eds., Lawrence Erlbaum Associates, Mahwah, NJ, pp. 176–189.

Greenspan, R. (2004), "Google Gains Overall, Competition Builds Niches," retrieved July 5, 2004, from http://www.clickz.com/stats/big_picture/applications/article.php/3362591.

Halverson, T., and Hornof, A. J. (2004), "Link Colors Guide a Search," in *Proceedings of CHI 2004*, ACM, Vienna, pp. 1367–1370.

Hempel, T., and Altinsoy, E. (2005), "Multimodal User Interface: Designing Media for the Auditory and Tactile Channel," in *Handbook of Human Factors in Web Design*, R. W. Proctor and K.-P. L. Vu, Eds., Lawrence Erlbaum Associates, Mahwah, NJ, pp. 134–156.

Irie, R. L. (2004), "Web Site Design," in *The Internet Encyclopedia*, Vol. 3, H. Bidgoli, Ed., Wiley, New York, pp. 768–775.

Isaacs, E. A. (1997), "Interviewing Customers: Discovering What They Can't Tell You," in *Proceedings of the Conference on Human Factors in Computing Systems, CHI 1997*, ACM, Atlanta, GA, March 22–27, pp. 180–181.

Jacko, J., Sears, A., and Borella, M. S. (2000), "The Effect of Network Delay and Media on User Perceptions of Web Resources," *Behaviour and Information Technology*, Vol. 19, pp. 427–439.

Jacques, R., and Savastano, H. (2001), "Remote vs. Local Usability Evaluation of Web Sites," in *Proceedings of IHM-HCI 2001*, Lille, France, September 10–14, pp. 91–92.

Kantor, P. L. (2003a), "Information Structures," retrieved July 4, 2004 from http://academ.hvcc.edu/~kantopet/

site_design/index.php?page=info+structures&parent =organizing+info.

Kantor, P. L. (2003b), "Using Labels," retrieved July 4, 2004 from http://academ.hvcc.edu/~kantopet/site_design/index.php?page=using+labels&parent=labeling+systems.

Klein, D. (1990), "Foiling the Cracker: A Survey of, and Improvements to, Password Security," in *Proceedings of the USENIX Security Workshop*.

Kobayashi, M., and Takeda, K. (2000), "Information Retrieval on the Web," *ACM Computing Surveys*, Vol. 32, pp. 144–173.

Landauer, T. K. (1998), "Learning and Representing Verbal Meaning: The Latent Semantic Analysis Theory," *Current Directions in Psychological Sciences*, Vol. 7, pp. 161–164.

Landauer, T. K., Foltz, P., and Laham, R. D. (1998), "An Introduction to Latent Semantic Analysis," *Discourse Processes*, Vol. 25, pp. 259–284.

Lazar, J. (2003), "The World Wide Web," in *The Human–Computer Interaction Handbook: Fundamentals, Evolving Technologies and Emerging Applications*, J. A. Jacko and A. Sears, Eds., Lawrence Erlbaum Associates, Mahwah, NJ, pp. 714–730.

Lazar, J., and Huang, Y. (2002), "Designing Improved Error Messages for Web Browsers," in *Human Factors and Web Development*, J. Ratner, Ed., Lawrence Erlbaum Associates, Mahwah, NJ, pp. 167–182.

Lightner, N. J. (2003), "What Users Want in E-Commerce Design: Effects of Age, Education, and Income," *Ergonomics*, Vol. 46, pp. 153–168.

Lightner, N. J., Bose, I., and Salvendy, G. (1996), "What Is Wrong with the World Wide Web? A Diagnosis of Some Problems and Prescription of Some Remedies," *Ergonomics*, Vol. 39, pp. 995–1004.

Mayhew, D. J. (2005), "A Design Process for Web Usability," in *Handbook of Human Factors in Web Design*, R. W. Proctor and K.-P. L. Vu, Eds., Lawrence Erlbaum Associates, Mahwah, NJ, pp. 338–356.

Millen, D. (2000), "Rapid Ethnography: Time Deepening Strategies for HCI Field Research," in *Proceedings of the Conference for Designing Interactive Systems: Process, Practices, Methods and Techniques*, ACM, New York, pp. 280–286.

Najjar, L. (2001), "E-Commerce User Interface Design for the Web," in *Usability Evaluation and Interface Design: Cognitive Engineering, Intelligent Agents, and Virtual Reality*, M. J. Smith, G. Salvendy, D. Harris, and R. J. Koubek, Eds., Lawrence Erlbaum Associates, Mahwah, NJ, Vol. 1, pp. 843–847.

Newman, M. W., and Landay, J. A. (2000), "Sitemaps, Storyboards, and Specifications: A Sketch of Web Site Design Practice," in *Proceedings of the Symposium on Designing Interactive Systems, DIS 2000*, ACM, New York, pp. 263–274.

Newman, M. W., Lin, J., Hong, J. I., and Landay, J. A. (2003), "DENIM: An Informal Web Site Design Tool Inspired by Observations of Practice," *Human–Computer Interaction*, Vol. 18, pp. 259–324.

Nielsen, J. (1993), *Usability Engineering*, Academic Press, San Diego, CA.

Nielsen, J. (2000), *Designing Web Usability*, New Riders, Indianapolis, IN.

Nielsen, J. (2001), "First Rule of Usability? Don't Listen to Users," http://www.useit.com/alertbox/20010805.html.

Nielsen, J., and Molich, R. (1990), "Heuristic Evaluation of User Interfaces," in *Proceedings of Conference on Human Factors in Computing Systems, CHI 1990*, ACM, Seattle, WA, April 1–5, pp. 249–256.

Nielsen, J., and Tahir, M. (2002), *Homepage Usability: 50 Websites Deconstructed*, New Riders, Indianapolis, IN.

Nielsen, J., Snyder, C., Molich, R., and Farrell, S. (2000), *E-Commerce User Experience*, Nielsen Norman Group, Fremont, CA.

Pearrow, M. (2000), *Web Site Usability Handbook*, Charles River Media, Hingham, MA.

Plaue, C., Miller, T., and Stasko, J. (2004), "Is a Picture Worth a Thousand Words? An Evaluation of Information Awareness Displays," in *Proceedings of Graphics Interface 2004*, Canadian Human–Computer Communications Society, London, Ontario.

Polson, P. G., Lewis, C., Rieman, J., and Wharton, C. (1992), "Cognitive Walkthroughs: A Method for Theory-Based Evaluation of User Interfaces," *International Journal of Man–Machine Studies*, Vol. 36, pp. 741–773.

Proctor, R. W., and Vu, K.-P. L. (2004), "Human Factors and Ergonomics for the Internet," in *The Internet Encyclopedia*, Vol. 2, H. Bidgoli, Ed., Wiley, Hoboken, NJ, pp. 141–149.

Proctor, R. W., and Vu, K.-P. L., Eds. (2005), *Handbook of Human Factors in Web Design*, Lawrence Erlbaum Associates, Mahwah, NJ.

Proctor, R. W., Lien, M.-C., Vu, K.-P. L., Schultz, E. E., and Salvendy, G. (2002a), "Influence of Restrictions on Password Generation and Recall," *Behavior Research Methods, Instruments, and Computers*, Vol. 34, pp. 163–169.

Proctor, R. W., Vu, K.-P. L., Salvendy, G., and 19 other authors (2002b), "Content Preparation and Management for Web Design: Eliciting, Structuring, Searching, and Displaying Information," *International Journal of Human–Computer Interaction*, Vol. 14, pp. 25–92.

Proctor, R. W., Vu, K.-P. L., Najjar, L., Vaughan, M. W., and Salvendy, G. (2003), "Content Preparation and Management for E-Commerce Websites," *Communications of the ACM*, Vol. 46, No. 12, pp. 289–299.

Qin, J. (2004), "Web Content Management," in *The Internet Encyclopedia*, Vol. 3, H. Bidgoli, Ed., Wiley, Hoboken, NJ, pp. 687–698.

Rasmussen, J. (1986), *Information Processing and Human–Machine Interaction: An Approach to Cognitive Engineering*, North-Holland, Amsterdam.

Ratner, J. (2003), *Human Factors and Web Development*, Lawrence Erlbaum Associates, Mahwah, NJ.

Riddle, B. L., Miron, M. S., and Semo, J. A. (1989), "Passwords in Use in a University Timesharing Environment," *Computers and Security*, Vol. 8, pp. 569–579.

Rieman, J. (1993), "The Diary Study: A Workplace-Oriented Research Tool to Guide Laboratory Efforts Collecting," in *Proceedings of the Conference on Human Factors in Computing Systems, INTERCHI '93*, ACM, Amsterdam, April, 24–29, pp. 321–326.

Roberts-Witts, S. L. (2001), "A Singular Focus," *PC*, September, retrieved July 24, 2004 from http://www.pcmag.com/article2/0,4149,16651,00.asp.

Rose, G. M., Evaristo, R., and Straub, D. (2003), "Culture and Consumer Response to Web Download Time: A Four Continent Study of Mono and Polychronism," *IEEE Transactions on Engineering Management*, Vol. 50, pp. 31–44.

Schultz, E. E. (2005), "Web Security and Privacy," in *Handbook of Human Factors in Web Design*, R. W. Proctor and K.-P. L. Vu, Eds., Lawrence Erlbaum Associates, Mahwah, NJ, pp. 613–625.

Sears, A., Jacko, J. A., and Dubach, E. M. (2000), "International Aspects of World Wide Web Usability and the Role of High-End Graphical Enhancements," *International Journal of Human–Computer Interaction*, Vol. 12, pp. 241–261.

Shadbolt, N., and Burton, M. (1995), "Knowledge Elicitation: A Systematic Approach," in *Evaluation of Human Work*, 2nd ed., J. R. Wilson and E. N. Corlett, Eds., Taylor & Francis, London, pp. 406–440.

Shneiderman, B. (1997), "Designing Information-Abundant Web Sites: Issues and Recommendations," *International Journal of Human–Computer Studies*, Vol. 47, pp. 5–29.

Spencer, R. (2000), "The Streamlined Cognitive Walkthrough Method: Working Around Social Constraints Encountered in a Software Development Company," in *Proceedings of the Conference on Human Factors in Computing Systems, CHI 2000, ACM*, The Hague, The Netherlands, April 1–6, pp. 353–359.

Stanney, K. M., Maxey, J., and Salvendy, G. (1997), "Socially-Centered Design," in *Handbook of Human Factors and Ergonomics*, 2nd ed., G. Salvendy, Ed., Wiley, New York, pp. 637–656.

Stephanidis, C., and Akoumianakis, D. (2005), "A Design Code of Practice for Universal Access: Methods and Techniques," in *Handbook of Human Factors in Web Design*, R. W. Proctor and K.-P. L. Vu, Eds., Lawrence Erlbaum Associates, Mahwah, NJ, pp. 239–250.

Sullivan, D., and Sherman, C. (2001), "Search Engine Features for Webmasters," http://searchendingewatch.com/webmasters/features.html.

Tullis, T. S., Catani, M., Chadwick-Dias, A., and Cianchette, C. (2005), "Presentation of Information," in *Handbook of Human Factors in Web Design*, R. W. Proctor and K.-P. L. Vu, Eds., Lawrence Erlbaum Associates, Mahwah, NJ, pp. 107–133.

U.S. Census Bureau, Population Division (2002), "National Population Projections," retrieved February 22, 2003, from http://www.census.gov/population/www/projections/natsum-T3.html.

U.S. Department of Commerce, Economics and Statistic Administration and National Telecommunications and Information Administration (2002), "A Nation Online: How Americans Are Expanding Their Use of the Internet," retrieved July 5, 2004, from http://www.ntia.doc.gov/ntiahome/dn/anationonline2.pdf.

Vanderheiden, G. C. (2005), "Access to Web Content by Those Operating Under Constrained Conditions," in *Handbook of Human Factors in Web Design*, R. W. Proctor and K.-P. L. Vu, Eds., Lawrence Erlbaum Associates, Mahwah, NJ, pp. 267–283.

Vaughan, M. W., Candland, K. M., and Wichansky, A. M. (2001), "Information Architecture of a Customer Web Application: Blending Content and Transactions," in *Usability Evaluation and Interface Design: Cognitive Engineering, Intelligent Agents, and Virtual Reality*, M. J. Smith, G. Salvendy, D. Harris, and R. J. Koubek, Eds., Lawrence Erlbaum Associates, Mahwah, NJ, Vol. 1, pp. 833–837.

Vicente, K. J., and Rasmussen. J. (1992), "Ecological Interface Design: Theoretical Foundations," *IEEE Transactions on Systems, Man, and Cybernetics*, Vol. 22, pp. 589–606.

Volk, F., and Wang, H. (2005), "Understanding Users: Some Qualitative and Quantitative Methods," in *Handbook of Human Factors in Web Design*, R. W. Proctor and K.-P. L. Vu, Eds., Lawrence Erlbaum Associates, Mahwah, NJ, pp. 303–320.

von Ahn, L., and Dabbish, L. (2004), "Labeling Images with a Computer Game," in *Proceedings of CHI 2004,* ACM, Vienna.

Vu, K.-P. L., and Proctor, R. W. (2003), "Naïve and Experienced Judgments of Stimulus–Response Compatibility: Implications for Interface Design," *Ergonomics*, Vol. 46, pp. 169–187.

Vu, K.-P. L., Tai, B.-L., Bhargav, A., Schultz, E. E., and Proctor, R. W. (2004), "Promoting Memorability and Security of Passwords Through Sentence Generation," in *Proceedings of the 48th Annual Meeting of the Human Factors and Ergonomics Society*, Human Factors and Ergonomics Society, Santa Monica, CA, pp. 1478–1482.

Wharton, C., Bradford, J., Jeffries, R., and Franzke, M. (1992), "Applying Cognitive Walkthroughs to More Complex User Interfaces: Experiences, Issues, and Recommendations," in *Proceedings of the Conference on Human Factors in Computing Systems, CHI 1992, ACM*, Monterey, CA, May, 3–7, pp. 381–388.

Wisman, R. (2004), "Web Search Fundamentals," in *The Internet Encyclopedia*, Vol. 3, H. Bidgoli, Ed., Wiley, Hoboken, NJ, pp. 724–737.

Withrow, J., Brink, T., and Speredelozzi, A. (2000), *Comparative Usability Evaluation for an E-Government Portal*, Diamond Bullet Design Report U1-00-2, Diamond Bullet Design, Ann Arbor, MI.

Zhang, P. (2000), "The Effects of Animation on Information Seeking Performance on the World Wide Web: Securing Attention or Interfering with Primary Tasks?" *Journal of the Association for Information Systems*, Vol. 1, pp. 1–28.

Zhu, W., Vu, K.-P. L., and Proctor, R. W. (2005), "Evaluating Web Usability," in *Handbook of Human Factors in Web Design*, R. W. Proctor and K.-P. L. Vu, Eds., Lawrence Erlbaum Associates, Mahwah, NJ, pp. 321–337.

CHAPTER 51

DESIGN OF E-BUSINESS WEB SITES

Jonathan Lazar
Towson University
Towson, Maryland

Andrew Sears
UMBC
Baltimore, Maryland

1 INTRODUCTION

E-commerce and e-business have become buzzwords during the last 10 years. Thousands of Web sites have been built to help sell products and services or to carry out other types of business. Over time, many of these Web sites have failed and disappeared. During 1999–2001, many e-businesses, even those that were well known and spent millions on advertising, failed. In many cases, these businesses failed because they had poor business models. That is, the businesses did not offer products or services that people wanted, they did not take in sufficient revenue, or they could not deliver on the services that they promised. This chapter does not address the issues involved in developing effective models for e-businesses. Instead, the focus is on designing the store front for these businesses—the Web site.

Although an effective business model is necessary for an e-business to succeed, it is not sufficient by itself. Another important consideration is the design of the organization's Web site. With traditional stores, many factors can result in customers leaving without purchasing the desired product even when it is available, including an unpleasant environment or difficulty finding the product. Similarly, a well-designed e-business's site can lead to increased sales by making products easier to find and guiding users through the purchase. Other potential benefits include an increase in repeat business, more referrals, increased trust in the organization, and an improved public perception of the organization. An e-business site that provides high-quality information that facilitates purchasing decisions can even reduce the number of products returned. Simply making more information, or even the precise information that customers need, available does not necessarily ensure successful purchases. Availability is the first step, but effective design that presents information when it is needed by the customer is also important. By ensuring that customers have the right information at the right time, they should be less surprised by the outcome of their transaction and less likely to return a product, roll back a transaction, or express dissatisfaction with the process or organization.

Ensuring that customer needs are satisfied as an e-business site is designed does take time, and short development time lines only put extra pressure on managers to ignore usability and move a Web site to implementation before it is "ready for prime time." However, in the long term, this is often a mistake. Common but easily avoidable problems relating to poor or confusing design can be avoided with appropriate user involvement and testing. For instance, one Web site that sells books changed their design so that users place their items in a "bag" instead of a shopping cart, and then exit the Web site (Nielsen, 2000b). Users are increasingly familiar with the common metaphor of placing items into a shopping cart and then going to the "checkout," a metaphor that builds on the way most physical stores operate. However, if you place items in a bag and go to the exit, most people would consider that shoplifting! Interestingly, sales dropped when this metaphor was introduced to customers. This is a simple and perhaps entertaining example of what not to do when building an e-business site.

Unfortunately, situations such as this, where decisions are made without adequately consulting users and understanding their needs, occur relatively frequently. Studies on user frustration related with Web sites show that users waste nearly 40 to 50% of their time due to frustrating experiences that could be alleviated by utilizing well-established design practices and applying known guidelines for human–computer interaction (Lazar et al., (2003b); Ceaparu 2004).

There are many considerations when building an effective e-business site. There are also many references on how to build e-business sites (Norris and West, 2001) or how to market and manage an e-business (Turban et al., 2000). Unfortunately, little scientific research has focused explicitly on the relationship between usability and the success of e-businesses. Much of what has been published consists of experiences and opinions rather than hard scientific evidence. This chapter focuses primarily on those issues that are unique to e-business Web sites. We do not address issues that apply equally to both traditional and e-businesses (e.g., developing business models), general interface design (e.g., selecting colors and fonts), or generic Web design (e.g., page layout and navigation). Instead, we focus on important issues that are related directly to the design of effective, customer-centered, e-business sites.

We begin by summarizing some of the case studies described elsewhere to provide additional examples of how improved design can result in tangible benefits for an organization. Next, we review the basic categories of e-business sites. Our purpose is to highlight the differing goals and how these translate into different metrics when evaluating the success of usability activities. Cost justification is discussed since this is a critical activity whenever there is an effort to introduce usability activities into an already tight development time line. Cost justification is addressed in more detail elsewhere, so our coverage focuses on those issues that are unique to e-business sites. The remainder of the chapter focuses on design (e.g., trust and accessibility) and process (e.g., user involvement and evaluation activities) as they apply to the design of effective e-business Web sites.

2 CASE STUDIES

Numerous case studies are available on the Internet. A smaller number have been accepted for presentation at conferences, including the 2003 Designing the User Experience Conference sponsored by ACM and the CHI 2002/AIGA Experience Design Forum. Several of these are summarized below.

2.1 Redesigning an Entertainment Club Web Site

Fletcher and Brookman (2002) discuss a project with a straightforward goal: increasing their conversion rate (the percentage of visitors that become club members) for visitors to the Columbia House Music or DVD club Web site. They considered click-through rates from banner ads acceptable, but conversion rates were described as being "extremely low." Through a combination of user testing and analysis of log files, five issues were identified as contributing to the low conversion rate: (1) pages that loaded too slowly; (2) a confusing process for joining the club; (3) providing customers with too many choices; (4) a checkout process that was too long and cumbersome; and (5) customers receiving too little feedback as they selected products.

A new version of the site was developed with the goal of addressing these problems. Specific goals included developing an interface that made it simple to select multiple products, making it clear which steps were required and in what order, simplifying the checkout process, and speeding downloads such that they would be completed in less than 10 seconds when customers were using slower network connections. The ultimate goal was to accomplish all of this in less than one month and to double conversion rates as a result. Features were identified, prototypes developed, designs evaluated, and the new site was launched in just 20 days. Conversion rates increased by 180%.

2.2 Redesigning a Business School Web Site

Brink et al. (2003) discuss the redesign of a Web site of more than 3000 pages for the University of Michigan Business School. Once again, their goal was clear. The school was working to brand itself as one of the top business schools in the country. Redesigning their Web site was considered an important part of this task, and the goal was to create an attractive and functional site that allowed for "simple and successful user experiences."

The site was large, vast amounts of information were available, but the existing design was inconsistent and did not present the desired image. Three major goals were defined: (1) leveraging existing content and resources (including integration of the public Web site and the school's intranet), (2) using the Internet for marketing and to build relationships with companies and prospective students, and (3) presenting a consistent image of the school's identity and brand. The team employed a user-centered design process that began with defining usability metrics that would be used as criteria for success. Metrics included task completion rates, task completion times, subjective ratings of specific tasks (e.g., difficulty and perceived speed), and subjective ratings of the overall experience (e.g., attractiveness and ease of use).

Strategy and user needs analysis was based on competitive analysis and focus groups with the intent of defining user needs and the goals for the site. Conceptual design involved defining important tasks, developing scenarios, and conducting walkthroughs. Next, prototypes were developed and evaluated using both user testing and expert reviews. Rapid prototyping tools were used to allow designs to be modified quickly. Evaluations compared the original site to competitors' sites, various prototypes of the new site, and a near-final version of the site just before the formal launch.

Content was reorganized, terminology made more consistent, creating clear links to other resources, creating a consistent layout and use of color for pages throughout the site, simplifying navigation, and making the site accessible for people with disabilities. The results included:

- *A 50% increase in task completion rates.* The task completion rate increased from 62% to 92% while competitors' sites resulted in task completion rates of approximately 72 to 76%.
- *A 39% reduction in task completion times.* The time required dropped from 73 seconds to only 44 seconds. Competitors' sites required approximately 100 to 110 seconds.
- *Improved satisfaction ratings.* Task ratings improved from 3.6 to 2.5 on a 7-point Likert scale. Competitors' sites received ratings of 3.6. Overall satisfaction with the site also improved from 4.2 to 2.0, compared with ratings for competitors' sites, which averaged 3.4.

2.3 Understanding and Increasing Credibility

Fogg et al. (2003) report on a study involving over 2600 participants that was designed to provide insights into the factors affecting credibility. The goal was to enhance our understanding of the factors that lead people to believe, or not believe, the information they find online. Clearly, credibility is an important issue for e-commerce sites. Ten Web sites in 10 different content categories were selected, including sites that focused on e-commerce, entertainment, finance, and travel. Comments from participants were coded and analyzed, providing insights into the factors that were viewed as influencing the credibility of the various sites. Differences between categories (e.g., e-commerce vs. sports) were also explored.

Over 46% of the comments mentioned the appearance of the site ("design look") when discussing the credibility of the sites. This was the single issue mentioned most frequently. Other issues mentioned frequently include information design/structure (28.5%), information focus (25.1%), company motive (15.5%), usefulness of information (14.8%), accuracy of information (14.3%), name recognition and reputation (14.1%), advertising (13.8%), and bias of information (11.6%). Of these issues, several are clearly usability/design-related (i.e., design look, information design/structure), while others focus more on the content that is being provided and the organization providing that content. Nine other issues were mentioned in at least 3% of the comments, including several additional usability-related items: functionality of site (8.6%) and readability of the text.

Differences were noted as the various categories of sites were analyzed separately. For example, name recognition and reputation as well as customer service were considered more important for e-commerce sites. In contrast, information design and functionality became more important when evaluating search sites. The authors discuss their results in relation to earlier research as well as prominence-interpretation theory (Fogg, 2002). As an example, they discuss the importance of privacy policies. In earlier studies, participants indicated that Web sites with privacy policies were considered more credible (Fogg et al., 2000, 2001). However, results from the current study indicate that users rarely noticed privacy policies. The question is what benefit to privacy policies really provide if they are not noticed. Several of their concluding suggestions for increasing the credibility of a Web site are related directly to traditional design/usability activities: (1) visual design matters—invest here, (2) make careful decisions about prominence, and (3) ask users about credibility.

3 TYPES OF E-BUSINESS WEB SITES

Web sites can serve a variety of goals as defined by the organization. Some provide information, others sell products or services. Some connect individual consumers to businesses, others connect businesses to each other, and still others connect consumers to each other. These differing goals result in different metrics being applied when assessing the effectiveness of a Web site. We divide e-commerce sites into two broad categories: transaction- and information-oriented. Within each category, subcategories can be defined based on the business models employed and type of consumer that is targeted.

Some Web sites represent pure e-businesses (e.g., amazon.com), where the primary opportunity to interact with the organization is through the Web. Even pure e-businesses often have call centers that can handle purchases in addition to questions, returns, and other inquiries. The key is that customers typically do not have the option of visiting a traditional store to interact with these organizations.

Hybrid businesses provide both online and traditional store fronts (e.g., barnesandnoble.com). Hybrids experience many of the same challenges as pure e-businesses, but they also have a preexisting reputation based on their traditional stores. Hybrids may want customers to complete their transactions on the Web but can also benefit if the Web site drives customers to their traditional stores.

We describe some of the more common types of Web sites below. We begin by discussing the three most common types of transaction-oriented e-business sites. Next, we review three key types of information-oriented e-business sites. However, it is important to note that other types of e-business sites exist and new hybrids are being created constantly.

3.1 Transaction-Oriented Sites

Many e-business sites exist to generate transactions. This includes the sale of both products and services. Transactions may involve both businesses and individual consumers. These sites must provide high-quality, credible information. However, they must also support specific tasks such as the comparison of different products or services. Customers must be able to indicate which products or services they wish to purchase and then complete the transaction. As a result, these sites must gather information about individuals (e.g., addresses) and payments (e.g., credit card numbers), making security an important concern.

For pure e-businesses, if the Web-based transaction fails or is aborted, it is likely that the sale is lost. If the consumer is not comfortable providing payment information, is not confident the organization will deliver the product or service requested, or other concerns about the electronic transaction, the sale may be lost. Conversion rates are critical for pure e-businesses, as these Web-based transactions are their primary (and perhaps only) source of revenue.

Hybrid businesses have the same concerns and also benefit from high conversion rates. The key difference is that hybrid businesses also have traditional stores. Although the goal may be to encourage Web-based transactions, a Web site that drives customers to the organization's traditional stores can also be considered successful. In fact, many consumers now use the Web sites of hybrid businesses to research products that they ultimately intend to purchase at a local retailer. These indirect sales are an important source of revenue but are obviously more difficult to track.

3.1.1 Business-to-Consumer (B2C) Transactions

Many Web sites offer products or services directly to consumers. The process is relatively simple: Consumers (users) find the items that they want to purchase, they complete a transaction in which they request a product or service and provide a method of payment, and the product or service is shipped or provided to them. Many of the best known e-commerce sites operate in this fashion; however, the scope of what is offered can vary greatly. Although a site such as Amazon.com offers almost any type of product that a consumer would want to purchase, other sites are much more limited in scope, offering, for instance, only shoes or home electronic equipment. For these types of sites, ease of use is paramount, because consumers can easily switch and purchase from other vendors. It is important to remember that for some sites, such as food vendors, most consumers have many choices close to their home, and therefore, even a short delay in processing, or a problem in usability, will keep users from becoming customers (Henderson 1998).

3.1.2 Business-to-Business (B2B) Transactions

Instead of targeting individual consumers, some organizations focus on providing products and services to other businesses. Business purchasing is totally different from consumer purchasing. Whereas most consumers purchase online with their credit card or some form of digital cash, most organizations have far more complex rules for purchasing. Businesses may have contracts in place that allow them to purchase certain items only from certain organizations. Some types of purchases (in certain item categories) might require multiple administrative approvals. Purchases over a certain amount might require multiple bids. Purchase orders or personalized bills might also be required. Because of the complexity of business purchasing, having electronic means to do so can certainly help speed up the purchasing process. At the same time, B2B e-business sites must be flexible to do business in the multiple ways required by various organizations. An example of a business to business (B2B) site is for CDWG, a company that provides technology and service primarily to government and businesses (Figure 1). The user base for B2B sites tends to be different from B2C sites. For B2B sites, the person performing the transactions is likely to be an administrator in some form—the person responsible for purchasing, acquisitions, or equipment. This person will probably be familiar with the process and terminology involved in purchasing, but may not be an expert in complex technology.

eBay is one widely recognized Web site that does not focus on selling its own products or services. Instead, eBay serves as a go-between, helping to match buyers with sellers. These businesses typically operate by charging a small fee for each transaction. There are other sites that offer services: referrals, or even services, where the goal is simply to match up people who can offer a product or service with people who are in need of a product or service. These sites do not offer any product themselves but must offer the highest quality of marketplace. For instance, eBay offers various ways to determine the trustworthiness of the seller (see Section 5.1), methods for secure payment, escrow, and means of mediation. These sites must also be vigilant to ensure that illegal products or services are not being offered.

3.2 Information-Oriented Sites

Some Web sites do have the goal of selling products or services. Frequently, these sites exist to provide information about an organization. This may include information about products and services, but customers wishing to complete a transaction must follow up with the organization using more traditional mechanisms. For these sites, providing easy access to the right information is critical. Visitors must be able to locate the necessary information to allow appropriate follow-up, whether it is a request for more information, placing an order or a product or service, or visiting the location of an affiliated store or business. Such sites must be credible, helping visitors build a positive impression of the sponsoring organization.

3.2.1 Purely Informational Sites

Some e-business Web sites do not expect for the site itself to be profitable. Rather, the Web site simply serves as a marketing tool, just like a television commercial or radio advertisement. The Web site simply

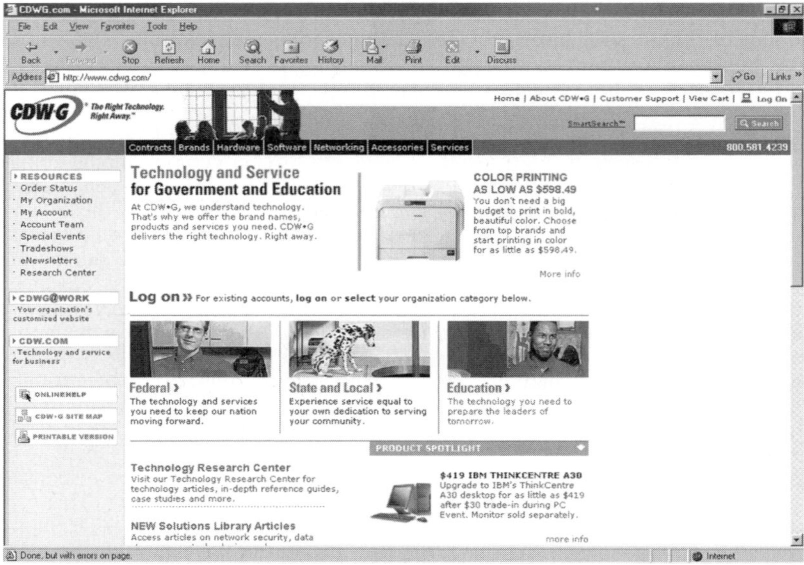

Figure 1 CDWG.com, a business-to-business Web site: consumer to consumer (C2C).

markets the organization, provides information about the organization and the products or services offered, but encourages consumers to follow up with a phone call or visit to the nearest store or company office. An example of a purely informational e-business Web site is Kevin's Candy (Figure 2). It is possible for some companies that they do not have the computer support staff to operate an e-business, but they did want to have information about their products available on the Web. For other products that require in-person testing, it simply might not make sense to sell products online (e.g., easy chairs). The e-business site might provide information about a service that requires a visit (e.g., custom framing). Finally, it might be that the goal of the site is simply to get the consumer to visit a store or business location.

3.2.2 Profits through Advertising

Some sites survive through advertising. These sites may provide information, search facilities, or even match individuals based on their interests or needs. The key is that the Web site does not sell any products or services. Frequently, these sites will record information about the topics that a visitor has viewed in an attempt to place advertising in front of visitors that are deemed more likely to be interested in the specific product or service being advertised. Some entertainment sites work on this model. Some medical information sites (e.g., webmd.com) work on this model.

3.2.3 Profits through Subscriptions

Finally, some Web sites provide information or resources to users in exchange for a fee. Often, such sites make a select subset of their content available for free to entice more people to subscribe. Since subscriptions are typically purchased online, these sites often support limited transactions. The information itself is the service being provided. Some news sites, such as the *Wall Street Journal*, run on the subscription model. Other sites that are subscription-based include alumni sites (such as classmates.com) and dating sites (such as match.com), and game-playing sites.

4 COST JUSTIFICATION

It takes time and money to ensure that a Web site, any type of Web site, is easy to use and meets the needs of users. Further, those who are in charge of monitoring the bottom line are likely to question whether the resulting usability activities (such as user profiling, requirements gathering, prototyping, and usability testing) are worth the time and expense involved. With insufficient planning, user involvement and usability engineering activities that are necessary to ensure effective solutions can lengthen time lines and add significant expense to the development process. However, when carefully planned, in advance, the time required can be minimized and the costs controlled, while providing great benefits. Furthermore, consumers frequently cite ease of use of an e-business site as being more important than product cost (Marcus, 2002).

Although some may argue that usability problems can be fixed later, there are two major flaws in this argument. A poorly designed Web site can result in a variety of negative consequences, such as drawing negative attention to the site and organization and causing users to abandon purchases due to frustration. Importantly, reversing the consequences of these negative experiences is often difficult, if not impossible. Second, it is much more expensive to fix usability flaws

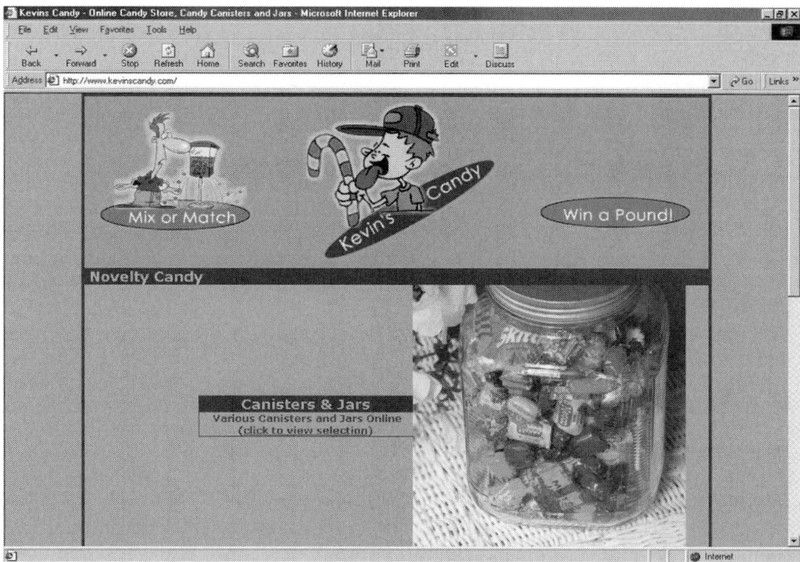

Figure 2 Kevin's Candy, a purely informational e-business site.

after a site has been implemented vs. designing the site correctly in the first place (Marcus, 2002). The later in the design and implementation the usability fixes take place, the more expensive the fixes will be (Marcus, 2002). In addition, unless you invest in usability engineering activities, you won't know what changes should be made during a Web site redesign to improve usability and increase your return on investment (Mayhew and Bias, 2003).

The costs of the usability engineering activities are clear, measurable, and upfront. These costs can include consulting costs for experts, payments to participants, distribution of surveys, and rental of usability equipment. Unfortunately, the benefits of usability engineering can be unclear, difficult to measure, indirect, and in the future. Potential benefits of usability engineering activities might include an improved public perception, increased purchases in the future, or happier customers (that tell other potential customers how positive an experience they had), all of which are very difficult to measure but attribute directly to usability engineering activities and expenses (Marcus, 2002). Increased sales/transactions, increased traffic, and increased retention of customers are easier to measure (Marcus, 2002). Other benefits may include reduced expenses through other communication media (e.g., calls to the phone center, fewer catalogs or brochures that need to be mailed) as well as fewer errors in business transactions, which result in large costs to businesses (Mayhew and Bias, 2003). Bias et al. (2003) highlight a number of additional potential benefits, including reduced costs due to more efficient development practices, fewer calls for customer support, and a reduction in documentation and training materials. They also repeat some standard benefits, including increased sales, more return customers, and additional referrals. Interestingly, they also stress less obvious possibilities, including reduced opportunities for lawsuits and public relations problems as well as additional positive press in trade-related publications.

Understanding the variety of benefits an organization may receive, when they will be experienced, and what dollar value can be associated with each benefit is a significant challenge. As part of this process, it is important to identify which specific benefits are relevant for the current project. This brings us back to the different types of e-commerce sites and their differing goals. Increased revenue from transactions, advertising, or subscriptions are the most obvious measures of success for many sites. Similarly, fewer calls to a call center, printing fewer catalogs, increased conversion rates, higher click-through rates, and a reduction in aborted transactions are easy to measure and appreciate. More pages being visited and longer dwell times may prove important when selling advertising and may even be useful for hybrid businesses (perhaps leading to more indirect sales at their traditional locations, but may be of little value to a pure e-business that is transaction-oriented. More repeat business or an increase in referrals provides clear benefits but can be difficult to measure in the short term. Still other benefits, such as improved public perceptions and happier customers, can be difficult to convert to dollars even if measured.

Since cost justification is not unique to e-businesses, readers are referred to more comprehensive discussions elsewhere. Specifically, Mayhew and Bias (2003) provide a comprehensive discussion of cost justification in the context of Web development activities. Bias et al. (2003) discuss cost justification in the context of more general human–computer interaction activities, but they also argue that cost justification is even more important when developing Web sites. They make this argument, in part, because unlike traditional software, most Web sites can be accessed for free (and therefore there is no cost associated with abandoning the site).

We conclude this section by providing a few additional examples of the benefits usability engineering activities can provide when developing e-business sites. Each example resulted in measurable benefits that can be converted into increased profitability.

IBM's e-commerce Web site was very difficult to use. IBM realized that the two most used features were the "search" function and the "help" button and concluded that users could not understand how to use the site (Tedeschi, 1999). After a redesign effort that cost millions of dollars, sales increased 400% and use of the "help" button decreased 84%. The functionality stayed the same, allowing users to perform the same tasks they could with the previous version, but the redesigned site was easier to use.

Macy's realized that their search engine often failed to provide the information users were looking for. For example, customers could search for a "\$35 tee" and receive only a fraction of the desired results. To their customers, a "\$35 tee" also meant a "T-shirt that costs \$34.95." By redesigning the search engine to better understand their customers goals, their conversion rate went up 150% (Kemp, 2001).

Staples.com provides another example of how improved usability can result in quantifiable benefits. Staples used feedback from users to improve and simplify their online registration pages. After various improvements were implemented, the registration drop-off rate (the number of people who begin registering but fail to complete the registration) decreased by 53% (Roberts-Witt, 2001).

5 DESIGN CONSIDERATIONS

There are many considerations in designing an e-business site that will represent a company successfully. General design guidelines for Web sites also provide useful guidance for e-business sites. For instance, it is well established in the literature that Web sites should download quickly, as the longer the user waits for a Web site to download, the more negative the user's perception of the Web site (Ramsay et al., 1998; Jacko et al., 2000; Sears and Jacko, 2000; Sears et al., 2000; Lazar et al., 2004b). This is often one of the biggest problems with Web sites. The sites take so long to download that the user has either lost interest, moved on to another Web site, given up, perceived an error, or repeatedly hit the "reload" button to the point that an error has actually occurred (Lazar, 2003; Lazar et al., 2004b).

Page design and navigation design provide additional examples of issues that are important for e-business sites. However, like download speeds, these issues are viewed as generic usability concerns regardless of the nature of the Web site. Effective screen layouts can help users find information more quickly, while ineffective solutions (such as multiple cascading menus or pull-down menus) may hide relevant information, emphasize irrelevant information, and frustrate users. Bad combinations of color (e.g., red text on a black background) can hurt users' eyes and make it harder for them to discern the text. Too much information on a page, without some sort or organization, can overwhelm users. Figure 3 displays an e-business

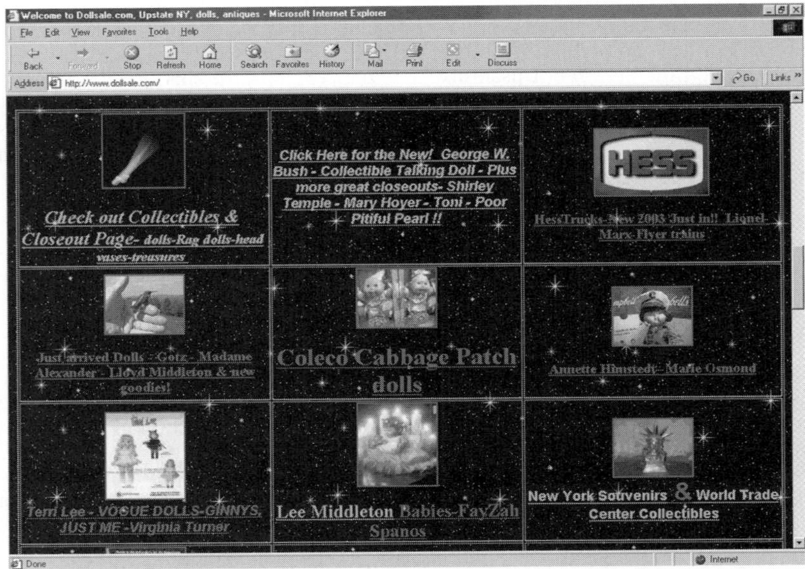

Figure 3 Web site with inappropriate layout and use of color.

Table 1 Customer-Centered Design Guidelines

Home page	The Web page should be clean and not cluttered with text and graphics. The width of a page should be less than the width of the browser window to avoid horizontal scrolling.
Navigation	Text on the links or buttons should be self-explained and descriptive. When linking to another product-related Web site, link to the exact product page instead of the home page of that site.
Categorization	Categorize products in a way that is meaningful to regular customers. The depth of the categories should be no more than three.
Product information	Present accurate, consistent, and detailed descriptions of products. Provide accurate and full pictures of products. Present the size of products in a measurable and comparable way. Present the inventory information of a product in the beginning. Present products in a table with enough information to make a purchasing decision such as prices and features for easy comparison. Present related charges up front and in an accurate way. Same products should be presented in the same page, same position. Products shouldn't be removed from the page because they are out of stock.
Shopping cart	In the shopping cart page, provide a link that directs the customer back to the page that he or she left for continuing the shopping.
Checkout and registration	Only ask for necessary and meaningful information such as name and address; no marketing questions. Allow customers to browse the site without logging in.
Customer service	Provide a 1–800 number for customers to call. State the return policy clearly in a prominent place.

Source: Fang and Salvendy (2003).

site selling dolls, which uses poor color combinations and an overwhelming number of graphics.

Navigation is central to all Web design. Web sites may include hundreds or even thousands of pages. As with any other Web page, navigation is central to the design of effective e-business sites. For transaction-oriented sites, customers must be able to find the desired product or service and navigate through the process of purchasing their selections. For information-oriented sites, customers must be able to find the desired information easily or they are likely to abandon their efforts. Text links, graphical links, and search mechanisms are all important navigation techniques. Especially for e-business sites with hundreds or thousands of pages, a search engine is especially important (Proctor et al., 2002). Not only must the navigation be technically usable, but the terminology used for navigation must also be based on what users already know. Often, the easiest way to improve the usability of an e-business site is simply to change the terminology being used, from terminology that is based on the terminology used within the business, to the terminology actually being used by other businesses and by customers. This may seem counterintuitive—that the most powerful fixes are the simplest to make—but terminology really does make a difference.

General guidelines for designing effective Web sites are plentiful. For additional guidance on these more generic issues, readers may wish to start with Chapter 50 of this book or see Lazar (2003). Multiple books have been written on the topic of good Web design (Fleming, 1998; Spool et al., 1999; Nielsen, 2000a; Lazar, 2001). In addition to these more general guidelines that all Web sites must address, there are additional issues that are particularly important for e-business sites. E-business sites are not simply a new type of information retrieval (Miles et al., 2000). There are many complicating factors for e-commerce sites, such as trust and credibility, security, internationalization, browser compatibility, and accessibility. Some of these concerns are summarized nicely in a set of design guidelines for customer-centered e-commerce design from Fang and Salvendy (2003), and these guidelines are presented in Table 1. Another short set of e-commerce design guidelines from IBM are provided in Table 2. A longer set of e-commerce design guidelines are provided by Rohn (1998). More information on specific concerns for e-business sites are discussed in detail in the following sections.

5.1 Trust and Credibility

Trust and credibility are related but are not necessarily the same. Credibility is a characteristic of the Web site or the organization. Customers will believe the information provided by the site if they consider it credible. Trust determines whether or not a customer will place confidence in the Web site. This is critical if the customer is expected to engage in a transaction or provide information. Customers must be able to trust that the product or service they order will be delivered and that any information they provide will be used properly and protected from theft. These are both important issues for e-business sites. Regardless of what type of e-business site, the site must appear to be credible and to represent a credible organization. For transaction-oriented sites, the site and organization must instill a sense of trust. For a business-to-consumer

Table 2 E-Commerce Design Guidelines

Customer support	Provide contact information on every page. Provide assistance when users have forgotten their passwords. Provide clear and informative error messages. Address user's frequently asked questions. Provide simple definitions and explanations of important terms. Provide product selection assistance. Provide assistance to guide users through multiple-step processes. Provide shipping information. Provide mechanisms that allow users to monitor the status of their orders. Provide an easy means to change orders submitted. State clearly and prominently all terms and conditions related to customer transactions. Provide customizable shopping lists if your users routinely buy the same items. Provide registered customers access to information on their previous purchases.
Trust	Provide access to a privacy policy from every page, and highlight it whenever users give personal information. Explain the benefits users receive from sharing personal information. Provide mechanisms for controlling how personal information is used. Use a secure Web server to collect customer data and complete transactions. Communicate that ordering online is secure. Display endorsements and affiliations that create a feeling of trust and security. Provide background on your company.
Product navigation	Provide easy, fast paths from the storefront to detailed product information. Provide different site paths to facilitate different shopping strategies. Provide links to shopping pages from a variety of other pages and sites. Provide shortcuts to the most popular products. Display products simultaneously to facilitate comparison. Give users control over which products they compare. Provide easily navigable and enticing product lists. Enable users to browse sequentially through product descriptions within categories. Support easy navigation between the order list and other shopping pages.
Product information	Offer a range of products that meet users' expectations. Provide pictures of all physical merchandise. Provide information about availability. Display prices prominently. Provide detailed product information. Disclose the most important product information first. Display unintrusive promotions on key pages. Facilitate cross-selling and up-selling without annoying or distracting users.
Purchase transaction	Provide an order list page that supports reviewing, editing, and submitting an order. Provide at least two forms of confirmation that the order has been received. Provide mechanisms for fast-path purchasing. Make the order form as simple and brief as possible. Provide alternative methods for ordering products. Enable users to change an order at any point prior to submitting it.

Source: IBM (2004).

transaction-oriented site, the consumer must feel that they are purchasing a product from a company that is trustworthy and therefore will deliver the item. Things can be even more complicated for consumer-to-consumer sites because there are multiple layers of trust: The customer must trust the intermediary site (e.g., eBay) and must also trust the person selling the product through the intermediary.

How can Web sites for e-business be designed so that they appear to be trustworthy, which in turn may increase business? It depends on who is providing the service or product. Hybrid organizations which also operate traditional stores may be at an advantage. Consumers often have predefined perceptions of traditional stores, much of which can transfer to the online branch of the organization. Of course, there are new issues that must be dealt with, such as security, but the general perception of the parent organization may alleviate some concerns. For instance, if a business has been in existence way before the advent of the Web, the e-business Web site should state that. See Figure 4, the Web site for Lammes Candies. The home page notes that Lammes has been in business since 1885. It also displays the product very clearly, in a way that the user can easily appreciate. For pure e-businesses, the Web site must make the business

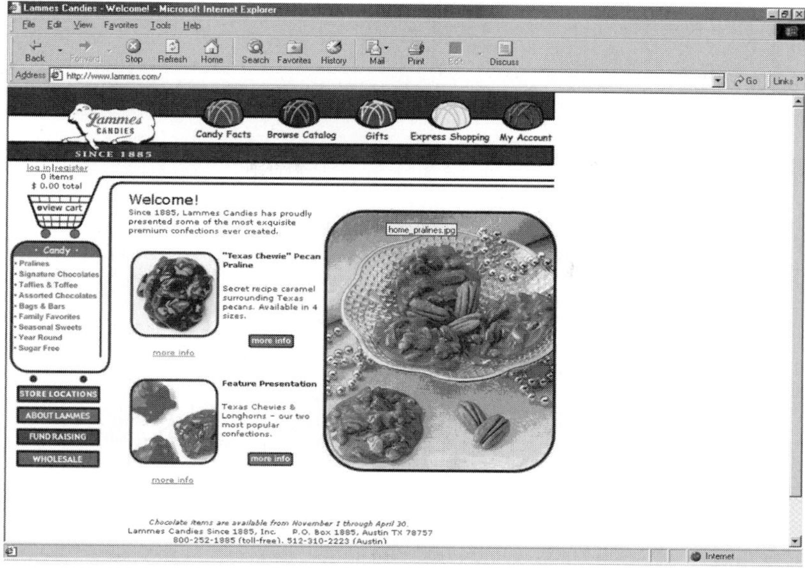

Figure 4 E-business that highlights the long history of the company.

appear trustworthy. If another customer is providing the product or service, the e-business site must provide opportunities to establish the trustworthiness of the product or service provider.

There are many design features that can be added so that customers perceive that a company is one that can be trusted. For instance, one concern with e-businesses is that they can be fly-by-night operations, which will take money and then not provide a service. Highlighting the established nature of the company by discussing the organization's history or physical facilities can help establish the trustworthiness of a company. Third-party endorsements may also allow customers to feel that they can trust the organization behind the Web site. The following guidelines for establishing that a Web site is trustworthy or credible are from Fogg et al. (2001).

1. Design Web sites to convey the real-world aspect of the organization.
2. Make Web sites easy to use.
3. Include markers of expertise.
4. Include markers of trustworthiness.
5. Tailor the user experience.
6. Avoid overly commercial elements.
7. Avoid the pitfalls of amateurism.

For e-businesses where the site provides a link between one customer/business and another customer/business, it is important to distinguish between the trustworthiness of two different parties: the go-between site (e.g., eBay) and the actual product or service provider. For a go-between site, the site simply facilitates the transaction, linking the buyer and seller of the product or service. Therefore, the important design issues related to trust on the go-between site include making sure that the site properly handles the customer information that it collects and that it uses secure payment methods.

There is an entirely separate set of issues related to presenting the trustworthiness of the end product or service provider. For consumers and businesses to engage in transactions with unknown strangers, the intentions and trustworthiness of the strangers (the other party in the transaction) must be established (Shneiderman, 2000), especially since it is unlikely that the parties will ever meet and it is likely that the parties to the transaction are separated geographically. Customers and businesses need assurances that the other party will fulfill their end of the bargain. The best way that this has been done is to provide a mechanism for reputation management. This approach has been used in both social interaction and e-commerce transactions online (Preece, 2000; Chen and Singh, 2001). In a reputation manager, people or businesses that have previously interacted with a person or organization have the opportunity to post information about their experiences as well as the quality of a product or service received (Jensen et al., 2002). This may be as simple as providing a numerical rating or as complex as engaging in an online discussion about one's experiences. These online communities on e-business sites can help establish the trustworthiness of the product or service provider, especially as the number of transactions they engage in increases. Figure 5 displays the reputation manager from eBay.

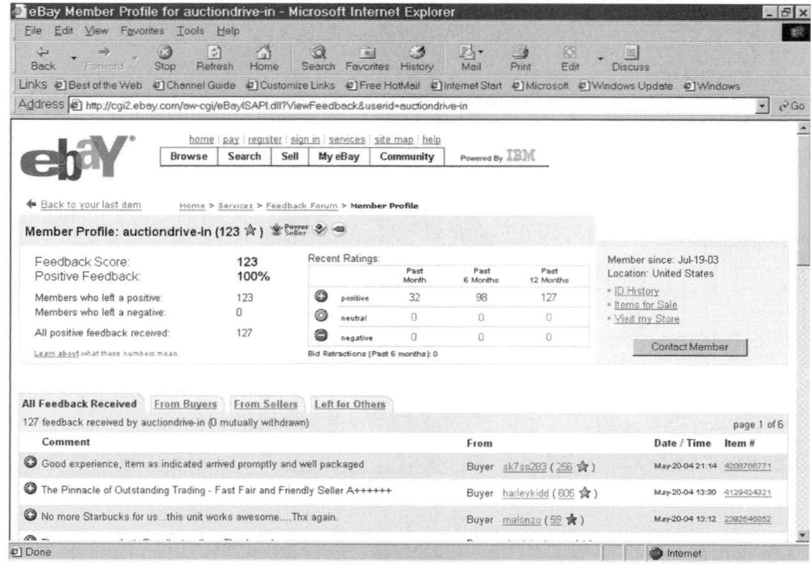

Figure 5 Reputation manager from eBay.

By using the reputation manager, customers and businesses can ensure that other consumers have been more or less pleased by their experiences, providing some level of certainty that the product or service provider is trustworthy. Overall rankings, as well as individual comments, are provided.

As outlined above, Fogg et al. (2003) found that both usability issues and the content of the Web site can contribute to the credibility of a site. Basic interface design was surprisingly important. In addition, the quality, relevance, and accuracy of the information provided by the site were among the most frequently mentioned issues considered when evaluating the credibility of Web sites. Presenting biased information appears to harm credibility, even if the goal is to sell a product or service. The motives of the organization, presence of advertising, name recognition, and existing reputation of the organization were also considered. Interestingly, although privacy statements have been suggested as a useful way to increase the credibility of an organization, Fogg et al. found that users rarely noticed these statements.

5.2 Security

Security is not just a matter of getting information across the Internet without it being incepted by a third party. One must also ensure that the information, once it is received, is used properly and protected from theft. It is still not uncommon for the media to report that private information has been disclosed inappropriately or stolen from an organization's computers. Universities have accidentally made private information about students available on the Internet, and companies have had thousands of credit card numbers stolen from their computers. E-businesses must define and enforce policies regarding the use and storage of information gathered from customers.

The first step in addressing security is arguably the use of the correct network protocols. Secure network connections significantly reduces opportunities for information to be stolen in transit. The second step is to ensure that only those people who need access to this confidential information are granted access. This simple step has been overlooked on more than one occasion, resulting in large quantities of private information being shared throughout an organization or, even worse, with anyone that has Internet access. Organizations must also determine what information needs to be saved and what should be deleted as soon as possible. Many organizations seek to simplify future transactions by saving addresses, credit card information, and customer preferences. If this information is saved, it is important that ever effort be made to secure this information properly. However, one should note that even when an organization has the best of intentions, it is possible that a security hole will be exploited and information will be stolen. In other situations, information can and should be deleted immediately. For example, recently the author received an e-mail from a small organization that runs annual conferences. The e-mail was alerting the author to the potential theft of credit card information, as someone had hacked into one of the organization's computers. This situation was unnecessary since the credit card information in question was associated with the annual events that occurred the preceding year as well as the year before that. Once these earlier events had taken place and the funds had been collected, there was no reason to save this information. If the credit card information had been deleted after it was no longer needed, the entire

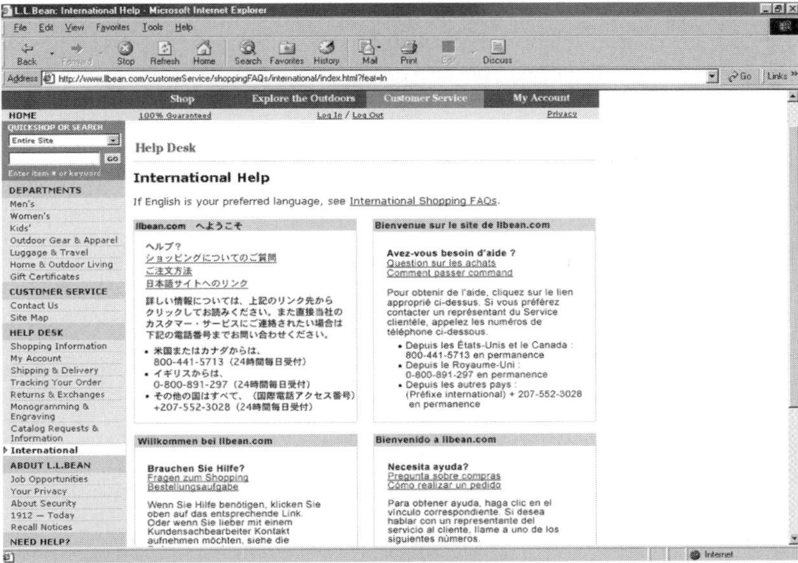

Figure 6 L.L. Bean Web site, offering help in four languages.

problem would have been avoided. See Chapter 48 for additional information about information security.

5.3 Internationalization

It is important to recognize that e-businesses are inherently international, whether or not it is intended. As a result, when an e-business site is built or redesigned, it is important to consider cross-country and cross-cultural considerations. If it is expected that there will be many users outside the country of e-business origin, both of these issues must be addressed. If a company is to do business across national borders, it must consider the laws of all countries involved, relating to tax, documentation, currency, and legality of transactions. International business transactions can be especially hard, because there are so many complicating factors, such as current exchange rates and shipping laws.

In some cases, companies will offer certain information in multiple languages. For instance, L.L. Bean, a clothing and luggage manufacturer in the United States, offers help for international customers in four different languages (Figure 6). Frequently, large companies will simply set up separate sites for business in each country, especially for business-to-consumer transactions. For instance, Amazon has multiple Web sites set up to address the large number of countries in which it does business. Rockport, a comfort shoe provider, provides different Web sites for each of five countries served (United States, Canada, UK/Ireland, Japan, and Sweden). See Figure 7 for a screen shot of the main page of the Rockport web site.

By offering one Web site for each country, this can simplify issues related to pricing (what currency?), shipping (from/to where?), and sizing (metric system or not?). Determining which language to use can be more difficult. For example, Canada requires that every Web site provide information in both French and English. Switzerland could be considered even more complex as there are four national languages, but there are no official requirements as to which language or languages must be used on Web sites. As a result, most large commercial Web sites in Switzerland contain all information in German, French, and Italian. In contrast, Swiss companies targeting international customers often use English as their primary language.

Sears et al. (2000) provide additional guidance for those designing Web sites for international audiences. Although internationalization involves far more than just changing the language used on the site, this is one important step. Issues to consider include the way that text expands or contracts as it is translated between languages. For example, when text is translated from English to other languages, if often gets longer (Belge, 1995). As a result, the layout of the screen and space allocated to lists must be adjusted. Since all concepts do not translate directly, it is important that the resulting text be proofread to ensure that it conveys the meaning intended.

Although most countries use the Arabic number system, formats differ as you move between countries. One simple example is the use of periods and commas when writing numbers in the United States and Europe. Dates are formatted differently and currencies vary. Images, symbols, and icons convey different meanings as you move between countries, while some metaphors may work in one location but not in another. An examples cited frequently is that of the "trash can" on the Macintosh, which is often interpreted by British users as a mailbox. The metaphor of a mailbox is a frequently used interface, but different cultures (and

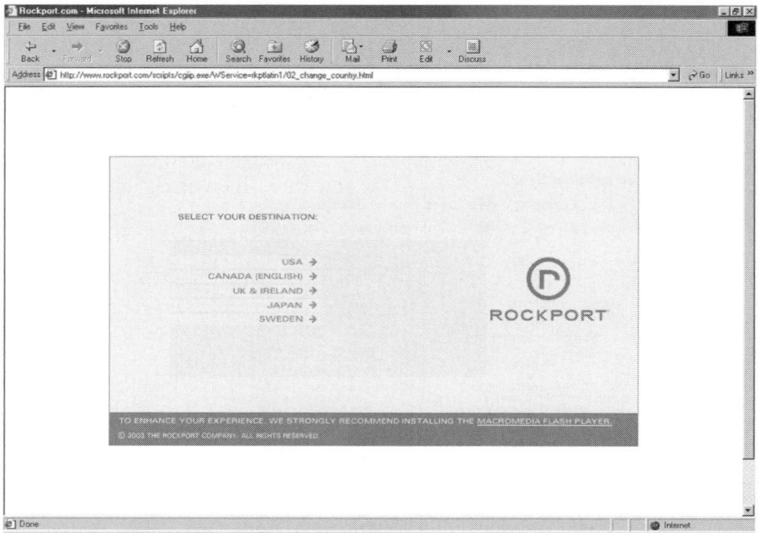

Figure 7 Rockport Web site, showing multiple country options.

countries) have different ideas of what mailboxes look like. Social norms determine what is acceptable as you move from one location to another. As a result, one must take great care when using images of religious symbols, the human body, and hand gestures (Russo and Boor, 1993). Colors also convey different concepts in different cultures: Red is often used to represent danger in the United States, but it represents happiness in China (Salomon, 1990).

Translating the language is a start, but internationalization requires much more. Instead, Web sites should be localized, modifying the site to meet the needs of a specific user population (Alvarez et al., 1998). For instance, user populations from different countries could have different needs related to interface design (Lightner et al., 2002). In one study it was found that Turkish users especially wanted the indications of a secure e-business site clearly present. In addition, Turkish users wanted the opportunity to bid or negotiate the product price, as this is a well-established practice in their culture (Lightner et al., 2002). Usability testing should consider the differences between users and should include representatives from each key region before the site is released to that user population (Nielsen, 1996). Many of the internationalization issues that are relevant to general interface design are also important in the context of the Web. For more information about cross-cultural interface design, see Marcus (2003).

5.4 Browser Compatibility

How a Web page appears can differ depending on the browser through which it is viewed. Although the World Wide Web Consortium (http://www.w3.org) provides standards for HTML and other languages used to present information on the Web, different browsers continue to interpret the same code in different ways (Lazar, 2001). Some Web sites acknowledge this inconsistency, highlighting that their site is "best when viewed with" a particular browser.

A number of factors must be considered when thinking about browser compatibility: the brand, version, and platform. Browser brands include Internet Explorer, Netscape Navigator, Mozilla, and Neoplanet. Internet Explorer has an overwhelming market share, but this does not simplify the process of developing e-business sites. For instance, the different versions of Internet Explorer and different platform versions of Internet Explorer interpret and support HTML, JavaScript, Cascading Style Sheets (CSS), and other Web languages and standards in different ways (Niederst, 1999). The same is true for browsers developed by other companies.

Because of these incompatibilities, it is important to test an e-business Web site with multiple browsers before the site is launched (or to retest the site as changes are made). During the requirements-gathering process, one goal should be to determine which browsers are used most often by the target user population. The prototype site should be tested with those browsers that are popular with the target users. In some instances, developers will go as far as determining what browser is being utilized (e.g., using JavaScript) and then deliver a Web page that is customized for use with that specific browser (Hysell, 1998). Such practices can create significant challenges as content is updated and new versions of browsers are released.

5.5 Accessibility

Accessibility is an increasingly important issue for the Web in general as well as e-business sites. An accessible Web site is a site that can be used effectively by people with disabilities. To illustrate the magnitude of this issue, note that in the United States alone there are more than 50 million people with some type of disability (Paciello, 2000; Slatin and Rush, 2003). This includes over 6 million persons with significant visual impairments that hinder their use of traditional displays (Ghaoiu et al., 2001), nearly 9 million with motor impairments that interfere with the use of traditional keyboard and mouse-based interfaces, 28 million with significant hearing loss, and almost 2 million with impairments that interfere with communicating thoughts and needs to others. To understand the complexity of this task, one must only consider the diversity of challenges experienced by people with disabilities. Sears (2003) provides a more comprehensive discussion of accessibility and the Web.

5.5.1 Visual Impairments

Visual impairments are arguably the single most important category of disabilities to consider when designing for the Web. Visual abilities fall along a continuum. We provide a brief overview of the issues but refer the reader to Jacko et al. (2003) for a more comprehensive discussion. Common measures of visual abilities include visual acuity, contrast sensitivity, field of view, and color perception, all of which are important when interacting with the Web. Visual acuity measures the ability to distinguish fine detail (Kline and Schieber, 1985), while contrast sensitivity assesses the ability to detect patterned stimuli (Wood and Troutbeck, 1994). A person's field of view is the area over which effective vision is maintained (Kline and Schieber, 1985), and color perception addresses a person's ability to discern and identify color (Kraut and McCabe, 1994).

People who are blind cannot make effective use of visual displays and often interact with computers using either text-to-speech (TTS) technologies or braille displays. Issues to consider include difficulty accessing graphical content, careful design of auditory descriptions when they are used, and difficulties with cursor positioning tasks. Further, many of the techniques used to deliver information to blind users are transient, often placing greater demands on the person's working memory.

In addition to the blind, many more users have low vision. Low vision refers to a wide range of impairments, including reduced visual acuity, contrast sensitivity, or field of view as well as color perception difficulties. In the context of the WWW, low vision becomes a significant concern when people can no longer interact effectively with traditional computer interfaces without alteration. Examples include people with extreme near- or farsightedness, blurred vision, tunnel vision, or night blindness. Consequences include difficulty reading text, identifying feedback or small changes on the screen, and making use of graphics or video. Since people with low vision typically prefer to make use of their residual visual capabilities, solutions designed for people who are blind are unlikely to be widely accepted (Edwards, 1995). At the same time, low vision encompasses a wide range of impairments, so the specific difficulties that any person may experience are likely to be unique. Simplifying the visual design, reducing clutter, organizing information carefully, and reducing clutter can all make interactions easier for persons with low vision. It is important to recognize that these users may enlarge the fonts or employ screen magnification software. Increasing the contrast between the foreground and background can also prove beneficial, as can avoiding complex or patterned backgrounds.

One must also be careful when using color since approximately 8% of men and 0.4% of women have difficulty perceiving colors. Most of these people have difficulty distinguishing two colors (e.g., red and green are the most frequently confused pair of colors). Relatively few people cannot distinguish any colors and see only shades of gray. Colors should be selected such that people can still extract the necessary information even when the colors are not perceived as intended. A simple solution is to ensure that color is not the sole means of communicating information. Also, colors that vary in two or more dimensions (i.e., hue, saturation, brightness) are easier to distinguish.

5.5.2 Hearing Impairments

Hearing impairments are becoming increasingly important when designing for the Web as the use of multimedia, video, and audio grows. Users should be able to adjust the volume, pitch, and other characteristics of the acoustic information they receive, but this is typically accomplished using standard or specialized software on their own computer. However, people with significant hearing impairments may require redundant delivery mechanisms, such as visual effects that supplement attention-attracting sounds and either transcripts or closed captioning for audio information. Auditory impairments are discussed in more detail in Jacko et al. (2003).

5.5.3 Cognitive Impairments

Like visual abilities, cognitive capabilities also fall along a continuum. While Newell et al. (2003) provide a comprehensive discussion of information technologies and cognitive impairments, we provide a brief overview of some of the key issues to consider when designing for the Web. The specific challenges that people with cognitive impairments experience vary dramatically. Areas that often result in significant challenges for people with cognitive impairments include language comprehension, memory impairments, and difficulties focusing on a task. Simplifying the vocabulary, avoiding complex syntax, reducing demands on the users' memory, and reducing complexity are the first steps to take. Simplifying the design does not translate into creating an unattractive, boring Web site but does suggest highlighting important information, using a standardized layout and color scheme, simplifying graphics, and recognizing that every visual or auditory element is competing for a user's attention. Navigation can prove particularly challenging since it often involves problem-solving skills (e.g., determining which links to follow) while placing demands on the user's memory (e.g., where have I been and what am I trying to find). Importantly, these recommendations not only make a Web site more effective for people with cognitive impairments, but often prove beneficial for traditional computer users as well.

5.5.4 Physical Impairments

Physical impairments can affect all aspects of life. Our focus is on those impairments that may affect a person's ability to utilize the WWW. For a more comprehensive discussion of physical impairments, refer to Sears and Young (2003). Physical impairments affect muscle power (e.g., paralysis, weakness), movement control (e.g., uncontrolled movements, tremors, coordination difficulties), mobility (e.g., range and ease of motion of the joints), and physical structures (e.g., missing fingers). The cause may be a health condition or an injury. Often, people with physical impairments will use alternative input devices to replace or supplement the traditional keyboard and mouse. Although these devices typically emulate the standard keyboard or mouse, they usually result in slower data entry rates, and higher error rates are not uncommon.

Physical impairments can affect the most basic of tasks, including pressing keys on a keyboard, positioning the cursor, and clicking on objects, all of which are important for Web-based interactions (Trewin and Pain, 1999). In many cases, both the size of the target and the distance the cursor must move negatively affect cursor-positioning activities (e.g., Radwin et al., 1990; Casali, 1992; Sears et al., 2002). The most fundamental steps in addressing physical impairments include avoiding device or platform specific capabilities and allowing users to complete all tasks through keyboard-based interactions.

5.5.5 Assistive Technologies

It is important to note that many persons with perceptual or motor disabilities use assistive technologies (such as eye tracking, screen readers, screen magnification software, refreshable braille displays, alternative keyboards, and alternative pointing devices) when interacting with computers (Alliance for Technology Access, 2000). For these interactions to be successful, Web sites must be designed to be flexible enough to be accessed by all of these various assistive technology devices. Fortunately, this does not necessarily need to change the appearance of the Web site for traditional users. In fact, most of the necessary adjustments are back-end coding changes, which provide equivalents for graphical content, graphical navigation, or interface widgets (such as a submit button) that cannot easily be utilized by a user with disabilities (Slatin and Rush, 2003).

5.5.6 Making a Site Accessible

The first step is awareness, followed by carefully considering the diverse capabilities of the people who are likely to visit the site. Numerous tools and guidelines are available to help in making accessible Web sites. Many Web development tools (such as FrontPage and DreamWeaver) now provide functionality to test for accessibility as sites are being developed (Slatin and Rush, 2003). Basic guidelines were outlined above and in Chapter 50, but more detailed guidelines are available elsewhere (e.g., Sears, 2003).

The World Wide Web Consortium provides a comprehensive set of guidelines and resources through the Web Accessibility Initiative (WAI) to help developers (http://www.w3.org/wai). One set of accessibility guidelines available through the Web Accessibility Initiative are called the Web Content Accessibility Guidelines (WCAG). A number of governments have based their rules for Web accessibility on the WCAG. For instance, the U.S. government's rules related to accessibility, called Section 508 (http://www.section508.gov), have a set of guidelines for Web accessibility that are roughly equivalent to priority level 1 of the WCAG. These guidelines apply only to U.S. government Web sites or contractors of the U.S. government. Other countries have similar laws, which frequently relate to government information, but in some cases they also apply to limited types of businesses. A short list of accessibility heuristics, known as the "quick tips to make accessible Web sites," showcases some of the major accessibility problems for Web sites (http://www.w3.org/wai).

- *Images and animations.* Use the alt attribute to describe the function of each visual.
- *Image maps.* Use the client-side map and text for hotspots.
- *Multimedia.* Provide captioning and transcripts of audio, and descriptions of video.
- *Hypertext links.* Use text that makes sense when read out of context. For example, avoid "click here."
- *Page organization.* Use headings, lists, and consistent structure. Use CSS for layout and style where possible.
- *Graphs and charts.* Summarize or use the longdesc attribute.
- *Scripts, applets, and plug-ins.* Provide alternative content in case active features are inaccessible or unsupported.
- *Frames.* Use the noframes element and meaningful titles.
- *Tables.* Make line-by-line reading sensible. Summarize.
- *Check your work.* Validate. Use tools, checklist, and guidelines at http://www.w3.org/TR/WCAG.

Automated software tools are available to evaluate the accessibility of existing Web pages. These tools typically evaluate Web pages in comparison to the WAI guidelines or government regulations, pointing out accessibility flaws (Lazar et al., 2003a). There are many different types of tools. Some are Web-based and allow limited free use (e.g., Bobby, available at http://bobby.watchfire.com/bobby/html/en/index.jsp).

Others are free, but must be downloaded and installed (e.g., A-Prompt, available at http://aprompt.snow.utoronto.ca/). Still others are expensive but powerful desktop applications (e.g., RAMP, InFocus). All of these tools work, at various levels of detail, in the same manner. A Web page is submitted by typing in a file name or URL. The application then returns a set of accessibility violations found on the page. The applications differ in how specific the violation information is, how much help is provided in fixing the violation, and how many manual checks are required (Lazar et al., 2003a). A manual check is when the developer is instructed that a page could have a violation of a specific guideline and is instructed that human examination is required.

Given the existence of interface guidelines and automated software tools, one may expect that Web accessibility levels would be high. Unfortunately, the opposite is true. A number of studies have examined accessibility levels on the Web. Generally, these studies report that between 75 and 98% of Web sites are considered inaccessible (Sullivan and Matson, 2000; Stowers, 2002; Lazar et al., 2003a). Given the direct relationship between accessibility and profitability, one may expect that e-business sites would have levels of accessibility higher than other categories of Web sites. Surprisingly, one recent study found that only 9 to 16% of e-business sites were considered accessible (depending on which set of guidelines, Section 508, or WAI, was used) (Loiacono and McCoy, 2004). This may be due to several common myths. One such myth is that an accessible Web site must be all text and no graphics (Lazar et al., 2004a). Others include the view that users with disabilities represent only a small number of potential customers and that they don't use computers (Lazar et al., 2004a). Developers and webmasters of e-business sites would be wise to make sure that their sites are accessible, which can help increase customers of the e-business and improve overall profitability.

5.6 Content Preparation

One final area of importance in design for e-business is the topic of content preparation. Content is an important consideration for all types of Web sites, including e-business sites (Fuccella and Pittolato, 1999). Simply put, a Web site must have content that users want, otherwise, users will not continue to return to that site, even if it is 100% usable (Small and Arnone, 2000; Lazar, 2001). It is important to determine from user (customer) involvement exactly what users want in an e-business site (Proctor et al., 2002). More information on involvement methods is given later in the chapter. For e-business sites, it is helpful to understand exactly how people perform business transactions. For instance, what mental model do they use? How are the tasks organized in their mind? How is information structured? (Proctor et al., 2002) For instance, there are many parts of the business environment that might be informal, that might not be known to Web developers, but must be modeled in e-business site development (e.g., special discounts for long-term customers). If organizational employees typically monitor orders from vendors, what information is needed to perform that task? (Proctor et al., 2002) What information must be modeled on the Web to match the information provided by employees in a non-Web environment? Content preparation really consists of two separate areas: determining specifically what content is needed by users, and then determining how that information is structured (Proctor et al., 2002). If information is structured to match the users' models of how that information is structured, it can lower reliance on search engines (Proctor et al., 2002), which most users typically have problems with (Marchionini, 1995). The next section covers the methods needed to determine both the content needed and the structuring of the information.

6 PROCESS CONSIDERATIONS

6.1 User Involvement

There are numerous usability engineering activities that can be employed during development to ensure that e-business sites are effective. These activities can be split into those that take place early in development (those that are more centered on the used tasks and profiles) and those that take place later in development (which are more focused on interface development) (Lazar et al., 2004c). Activities that take place earlier in development include surveys, focus groups, ethnography, interviews, and naturalistic inquiry. These activities can help gain more knowledge about who the users are, their usability needs, and their needs related to content (Lazar, 2001; Proctor et al., 2002; Lazar et al., 2004c). Which specific activities to employ in a given situation depends on the purpose of the e-business site, how well defined the tasks are, and how well defined the user population is as well as the time and resources that are available. For instance, if the tasks are relatively well defined before development, usability might be more of the concern of the user involvement activities. On the other hand, if a new intranet is being built so that two businesses that have an ongoing supplier relationship can constantly communicate to deliver parts in a just-in-time fashion, a large amount of time will need to be spent trying to understand the task and content needs (Lazar et al., 2004c). In a situation like that, an involvement method such as participatory design, where users are involved in all stages of development, as a part of the development team, might be most effective (Schuler and Namioka, 1993; Ellis and Kurniawan, 2000). As content begins to become defined, another helpful user involvement activity is card sorting, where users sort many content pieces into organized categories, which can help developers get a better understanding of the mental models that users have of how information is structured (Proctor et al., 2002). Surveys in multiple formats (paper, phone, Web-based, e-mailed) seem to be the most popular method for user involvement in Web development projects (Lazar and Preece, 2001; Lazar et al., 2003c, 2004c).

Other user involvement activities can take place later in the development process. These activities tend to focus on the interface itself. Whereas the earlier stage activities tend to focus on understanding the users, their usability needs, their task needs, their content needs, the user involvement at later stages focuses on the interface, and making sure that those user needs are reflected in the interface built. Most later-stage user involvement focuses on usability evaluation.

6.2 Usability Evaluation

Usability evaluation is a critical component of any usability engineering process. Regardless of the type of e-business Web site, usability evaluation can prove beneficial by ensuring that tasks are effectively supported and user reactions to their experiences will be appropriate. There are three main approaches to usability evaluation: inspection-based evaluations, user-based testing, and automated testing. Evaluating the usability of e-commerce sites is arguably similar to evaluating any other Web site. However, those issues identified earlier in this chapter (e.g., trust, credibility, security, accessibility, and internationalization) should receive additional attention. Addressing these issues does not mean that different, or additional, evaluation activities are necessary. Instead, one must simply ensure that these issues receive adequate attention during the evaluations that do occur.

6.2.1 Inspection-Based Techniques

Nielsen and Mack (1994) define inspection-based evaluation techniques as a collection of methods that rely on evaluators inspecting usability-related aspects of user interfaces (Nielsen and Mack, 1994). Evaluators may be usability specialists but do not have to be. Inspection-based techniques tend to require less formal training, can often be applied quickly, can be applied throughout the development process, and reduce the need for test users. Inspection-based techniques do not necessarily identify all of the usability problems with a system, but they can identify numerous usability problems and they frequently do allow for the most serious problems to be found. At the same time, inspection-based evaluations identify some issues as problems that would not actually hinder users. Common inspection-based techniques include heuristic evaluations, guidelines reviews, and cognitive walkthroughs (Nielsen, 1994).

Many of the issues involved in conducting effective inspection-based evaluations are discussed in more detail in Chapter 50. Cockton et al. (2003) provide a comprehensive review of inspection-based evaluation techniques, including a brief overview of heuristic evaluations, guideline-based methods, cognitive walkthroughs, and heuristic walkthroughs. Sears (1997) provides a more detailed discussion of heuristic walkthroughs, a hybrid technique that combines the benefits of cognitive walkthroughs and heuristic evaluations.

6.2.2 User-Based Techniques

Perhaps the most popular approach to usability evaluation is user-based testing. Unlike inspection-based techniques, which typically involve just the development team, user-based testing brings representative users into the process. Most often, these test participants are asked to attempt to perform a series of representative tasks. As with inspection-based evaluations, there are a variety of ways to approach user-based testing. The most common are classic usability testing and think-aloud evaluations.

For classic usability testing, users are observed as they focus on interacting with the system. Various types of data can be collected, including task completion times, error rates, how many tasks are completed successfully, and subjective satisfaction ratings, which are typically provided via questionnaire after the participant completes the assigned tasks. Much of this data can be recorded automatically, but video and audio records may also be made to allow for more detailed post-hoc review of the participant's experience. During the think-aloud varient of usability testing, participants are asked to think out loud as they interact with the system. This can provide valuable insights into why the participants experienced difficulties—something that can be difficult to determine through traditional usability testing. User-based usability testing is discussed in detail in Chapter 49 and in Dumas (2003).

6.6.3 Automated Evaluations

Automated evaluations can be accomplished using a variety of software packages, including LIFT (http://www.usablenet.com/), RAMP, (http://www.section508ok.com/), InFocus (http://www.ssbtechnologies.com/), and WebSAT (http://zing.ncsl.nist.gov/WebTools/). These programs review a series of screens and use preexisting interface guidelines to identify flaws that have been known to cause problems for users. Other software tools exist that focus on specific categories of usability problems. For example, Bobby (http://bobby.watchfire.com/) focuses explicitly on accessibility-oriented concerns. Automated evaluations can be effective in identifying some usability concerns but are not sufficient if used in isolation. Model-based evaluations (Kieras, 2003) can also allow for automated evaluations, but significant time and expertise is often required to configure these techniques to evaluate a specific system.

7 CONCLUSIONS

This chapter covers human factor design issues for e-business Web sites. While many of the general human factor issues for Web sites apply to e-business Web sites and are covered elsewhere, this chapter focuses on some of the unique human factor issues for e-business Web sites, such as cost justification, trust and credibility, transaction processing, and content preparation. It is hoped that this information will help designers in making their e-business Web sites more effective by better meeting users' needs and providing improved interfaces.

REFERENCES

Alliance for Technology Access (2000), *Computer and Web Resources for People with Disabilities*, Hunter House, Berkeley, CA.

Alvarez, M., Kasday, L., and Todd, S. (1998), "How We Made the Web Site International and Accessible: A Case Study," presented at the 1998 Human Factors and the Web Conference, http://www.research.att.com/conf/hfweb/proceedings/alvarez/.

Belge, M. (1995), "The Next Step in Software Internationalization," *Interactions*, Vol. 2, No. 1, pp. 21–25.

Bias, R., Mayhew, D., and Upmanyu, D. (2003), "Cost-Justification," in *The Human–Computer Interaction Handbook*, J. Jacko and A. Sears, Eds., Lawrence Erlbaum Associates, Mahwah, NJ, pp. 1202–1212.

Brink, T., Pritula, N., Lock, K., Speredelozzi, A., and Monan, M. (2003), Making an Impact: Redesigning a Business School Web Site Around Performance Metrics," in *Proceedings of the 2003 Conference on Designing for User Experiences*, pp. 1–15.

Casali, S. (1992), "Cursor Control Device Used by Persons with Physical Disabilities: Implications for Hardware and Software Design," in *Proceedings of the Annual Meeting of the Human Factors and Ergonomics Society*, pp. 311–315.

Ceaparu, I., Lazar, J., Bessiere, K., Robinson, J., and Shneiderman, B. (2004), "Determining Causes and Severity of End-User Frustration," *International Journal of Human–Computer Interaction*, Vol. 17, No. 3, pp. 333–356.

Chen, M., and Singh, J. (2001), "*Computing and Using Reputations for Internet Ratings*," in *Proceedings of the ACM Conference on E-Commerce*, pp. 154–162.

Cockton, G., Lavery, D., and Woolrych, A. (2003), "Inspection-Based Evaluations," in *The Handbook of Human–Computer Interaction*, J. Jacko and A. Sears, Eds., Lawrence Erlbaum Associates, Mahwah, NJ, pp. 1118–1138.

Dumas, J. (2003), "User-Based Evaluations," in *The Handbook of Human–Computer Interaction*, J. Jacko and A. Sears, Eds., Lawrence Erlbaum Associates, Mahwah, NJ, pp. 1093–1117.

Edwards, A. (1995), "Computers and People with Disabilities," in *Extra-ordinary Human–Computer Interaction*, A. Edwards, Ed., Cambridge University Press, New York, pp. 19–43.

Ellis, R. D., and Kurniawan, S. (2000), "Increasing the Usability of Online Information for Older Users: A Case Study in Participatory Design," *International Journal of Human–Computer Interaction*, Vol. 12, No. 2, pp. 263–276.

Fang, X., and Salvendy, G. (2003), "Customer-Centered Rules for Design of E-Commerce Web Sites," *Communications of the ACM*, Vol. 46, No. 12, pp. 332–336.

Fleming, J. (1998), *Web Navigation: Designing the User Experience*, O'Reilly & Associates, Sebastopol, CA.

Fletcher, D., and Brookman, A. (2002), "Making Joining Easy: Case of an Entertainment Club Website," in *Proceedings of the Conference on Human Factors in Computing Systems: Case Studies of the CHI2002/AIGA Experience Design Forum*, pp. 1–16.

Fogg, B. (2002), *Prominence-Interpretation Theory: Explaining How People Assess Credibility*, Stanford Persuasive Technology Lab, Palo Alto, CA; www.captology.stanford.edu/PIT.html.

Fogg, B., Marshall, J., Laraki, O., Osipovich, A., Varma, C., Fang, N., Paul, J., Rangnekar, A., Shon, J., Swani, P., and Treinen, M. (2000), "Elements That Affect Web Credibility: Early Results from a Self-Report Study," in *Proceedings of the 2000 ACM Conference on Human Factors in Computing Systems*, pp. 295–296.

Fogg, B., Marshall, J., Laraki, O., Osipovich, A., Varma, C., Fang, N., Paul, J., Rangnekar, A., Shon, J., Swani, P., and Treinen, M. (2001), "What Makes Web Sites Credible? A Report on a Large Quantitative Study," in *Proceedings of the Conference on Human Factors in Computing, CHI 2001*, pp. 61–68.

Fogg, B., Soohoo, C., Danielson, D., Marable, L., Stanford, J., and Tauber, E. (2003), "How Do Users Evaluate the Credibility of Web Sites? A Study with over 2,500 Participants," in *Proceedings of the 2003 Conference on Designing for User Experiences*, pp. 1–15.

Fuccella, J., and Pittolato, J. (1999), "Giving People What They Want: How to Involve Users in Site Design," *IBM DeveloperWorks*, http://www-4.ibm.com/software/developer/library/design-by-feedback/expectations.html.

Ghaoiu, C., Mann, M., and Ng, E. (2001), "Designing a Humane Multimedia Interface for the Visually Impaired," *European Journal of Engineering Education*, Vol. 26, No. 2, pp. 139–149.

Henderson, R., RIckwood, D., and Roberts, P. (1998), "The Beta Test of an Electronic Supermarket," *Interacting with Computers*, Vol. 10, No. 4, pp. 385–399.

Hysell, D. (1998), "Meeting the Needs (and Preferences) of a Diverse World Wide Web Audience," in *Proceedings of the ACM 16th Annual International Conference on Computer Documentation*, pp. 164–172.

IBM (2004), "IBM Web Design Guidelines for E-Commerce," http://www-306.ibm.com/ibm/easy/eou_ext.nsf/publish/611.

Jacko, J., Sears, A., and Borella, M. (2000), "The Effect of Network Delay and Media on User Perceptions of Web Resources," *Behaviour and Information Technology*, Vol. 19, No. 6, pp. 427–439.

Jacko, J., Vitense, H., and Scott, I. (2003), "Perceptual Impairments and Computing Technologies," in *The Handbook of Human–Computer Interaction*, J. Jacko and A. Sears, Eds., Lawrence Erlbaum Associates, Mahwah, NJ, pp. 504–522.

Jensen, C., Davis, J., and Farnham, S. (2002), "*Finding Others Online: Reputation Systems for Social Spaces Online*," in *Proceedings of the ACM Conference on Human Factors in Computing Systems*, pp. 447–454.

Kemp, T. (2001), "Macy's Doubles Conversion Rate," *InternetWeek.com*, November 28, http://www.internetwk.com/story/INW20011128S20010004.

Kieras, D. (2003), "Model-Based Evaluations," in *The Handbook of Human–Computer Interaction*, J. Jacko and A. Sears, Eds., Lawrence Erlbaum Associates, Mahwah, NJ, pp. 1139–1151.

Kline, D., and Schieber, F. (1985), "Vision and Aging," in *Handbook of the Psychology of Aging*, J. Birren and K. Schaie, Eds., Van Nostrand Reinhold, New York, pp. 296–331.

Kraut, J., and McCabe, C. (1994), "The Problem of Low Vision," in *Principles and Practices of Ophthalmology*, D. Albert, F. Jakobiec, and N. Robinson, Eds., Elsevier, Philadelphia, PA, pp. 3664–3683.

Lazar, J. (2001), *User-Centered Web Development*, Jones & Bartlett, Sudbury, MA.

Lazar, J. (2003), "The World Wide Web," in *The Handbook of Human–Computer Interaction*, J. Jacko and A. Sears, Eds., Lawrence Erlbaum Associates, Mahwah, NJ, pp. 714–730.

Lazar, J., and Preece, J. (2001), "Using Electronic Surveys to Evaluate Networked Resources: From Idea to Implementation," in *Evaluating Networked Information Services: Techniques, Policy, and Issues*, C. McClure and J. Bertot, Eds., Information Today, Medford, NJ, pp. 137–154.

Lazar, J., Beere, P., Greenidge, K., and Nagappa, Y. (2003a), "Web Accessibility in the Mid-Atlantic United States: A Study of 50 Web Sites," *Universal Access in the Information Society*, Vol. 2, No. 4, pp. 331–341.

Lazar, J., Bessiere, K., Ceaparu, I., Robinson, J., and Shneiderman, B. (2003b), "Help! I'm Lost: User Frustration in Web Navigation," *IT and Society*, Vol. 1, No. 3, pp. 18–26.

Lazar, J., Jones, A., and Greenidge, K. (2003c), "Web-Star: A Survey Tool for Analyzing User Requirements for Web Sites," in *Proceedings of the Information Resource Management Association 2003 International Conference*, pp. 917–918.

Lazar, J., Dudley-Sponaugle, A., and Greenidge, K. (2004a), "Improving Web Accessibility: A Study of Webmaster Perceptions," *Computers in Human Behavior*, Vol. 20, No. 2, pp. 269–288.

Lazar, J., Meiselwitz, G., and Norcio, A. (2004b), "A Taxonomy of User Perception of Error on the Web," *Universal Access in the Information Society*, Vol. 3, No. 3.

Lazar, J., Ratner, J., Jacko, J., and Sears, A. (2004c), "User Involvement in the Web Development Process: Methods and Cost-Justification," presented at the 10th International Conference on Industry, Engineering and Management Systems.

Lightner, N., Yenisey, M., Ozok, A., and Salvendy, G. (2002), "Shopping Behaviour and Preferences in E-Commerce of Turkish and American University Students: Implications from Cross-Cultural Design," *Behaviour and Information Technology*, Vol. 21, No. 6, pp. 373–385.

Loiacono, E., and McCoy, S. (2004), "Web Site Accessibility: An Online Sector Analysis," *Information Technology and People*, Vol. 17, No. 1, pp. 87–101.

Marchionini, G. (1995), *Information Seeking in Electronic Environments*, Cambridge University Press, Cambridge.

Marcus, A. (2002), "Return on Investment for Usable UI Design," *User Experience*, Vol. 2, No. 1, pp. 25–31.

Marcus, A. (2003), "Global and Intercultural User-Interface Design," in *The Handbook of Human–Computer Interaction*, J. Jacko and A. Sears, Eds., Lawrence Erlbaum Associates, Mahwah, NJ, pp. 441–463.

Mayhew, D., and Bias, R. (2003), "Cost-Justifying Web Usability," in *Human Factors and Web Development*, 2nd ed., J. Ratner, Ed., Lawrence Erlbaum Associates, Mahwah, NJ, pp. 63–87.

Miles, G., Howes, A., and Davies, A. (2000), "A Framework for Understanding Human Factors in Web-Based Electronic Commerce," *International Journal of Human–Computer Studies*, Vol. 52, pp. 131–163.

Newell, A., Carmichael, A., Gregor, P., and Alm, N. (2003), "Information Technology for Cognitive Support," in *The Handbook of Human–Computer Interaction*, J. Jacko and A. Sears, Eds., Lawrence Erlbaum Associates, Mahwah, NJ, pp. 464–481.

Niederst, J. (1999), *Web Design in a Nutshell*, O'Reilly & Associates, Sebastopol, CA.

Nielsen, J. (1994), *Usability Engineering*, Academic Press, San Diego, CA.

Nielsen, J. (1996), "International Usability Engineering," in *International User Interfaces*, E. DelGaldo and J. Nielsen, Eds., Wiley, New York, pp. 1–19.

Nielsen, J. (2000a), *Designing Web Usability: The Practice of Simplicity*, New Riders, Indianapolis, IN.

Nielsen, J. (2000b), "Why Doc Searls Doesn't Sell Any Books," http://www.useit.com.

Nielsen, J., and Mack, R., Eds. (1994), *Usability Inspection Methods*, Wiley, New York.

Norris, M., and West, S. (2001), *E-Business Essentials: Technology and Network Requirements for Mobile and Online Markets*, Wiley, Chichester, West Sussex, England.

Paciello, M. (2000), *Web Accessibility for People with Disabilities*, CMP Books, Lawrence, KS.

Preece, J. (2000), *Online Communities: Designing Usability, Supporting Sociability*, Wiley, New York.

Proctor, R., Vu, K., Salvendy, G., et al. (2002), "Content Preparation and Management for Web Design: Eliciting, Structuring, Searching, and Displaying Information," *International Journal of Human–Computer Interaction*, Vol. 14, No. 1, pp. 25–92.

Radwin, R., Vanderheiden, G., and Lin, M. (1990), "A Method for Evaluating Head-Controlled Computer Input Devices Using Fitts' Law," *Human Factors*, Vol. 32, No. 4, pp. 423–438.

Ramsay, J., Barbesi, A., and Preece, J. (1998), "A Psychological Investigation of Long Retrieval Times on the World Wide Web," *Interacting with Computers*, Vol. 10, No. 1, pp. 77–86.

Roberts-Witt, S. (2001), "A Singular Focus," *PC*, September 25.

Rohn, J. (1998), "Creating Usable E-Commerce Sites," *Standard View*, Vol. 6, No. 3, pp. 110–115.

Russo, P., and Boor, S. (1993), "How Fluent Is Your Interface? Designing for International Users," in *Proceedings of INTERCHI 1993*, pp. 342–347.

Salomon, G. (1990), "New Uses for Color," in *The Art of Human–Computer Interaction*, B. Laurel, Ed., Addison-Wesley, Reading, MA, pp. 269–278.

Schuler, D., and Namioka, A., Eds., (1993), *Participatory Design: Principles and Practices*, Lawrence Erlbaum Associates, Mahwah, NJ.

Sears, A. (1997), "Heuristic Walkthroughs," *International Journal of Human–Computer Interaction*, Vol. 9, No. 3, pp. 213–234.

Sears, A. (2003), "Universal Usability and the Web," in *Human Factors and Web Development*, 2nd ed., J. Ratner, Ed., Lawrence Erlbaum Associates, Mahwah, NJ, pp. 21–46.

Sears, A., and Jacko, J. (2000), "Understanding the Relation Between Network Quality of Service and the Usability of Distributed Multimedia Documents," *Human–Computer Interaction*, Vol. 15, No. 1, pp. 43–68.

Sears, A., and Young, M. (2003), "Physical Disabilities and Computing Technologies: An Analysis of Impairments," in *The Handbook of Human–Computer Interaction*, J. Jacko and A. Sears, Eds., Lawrence Erlbaum Associates, Mahwah, NJ, pp. 482–503.

Sears, A., Jacko, J., and Dubach, E. (2000), "International Aspects of WWW Usability and the Role of High-End Graphical Enhancements," *International Journal of Human–Computer Interaction*, Vol. 12, pp. 243–263.

Sears, A., Lin, M., and Karimullah, A. (2002), "Speech-Based Cursor Control: Understanding the Effects of Target Size, Cursor Speed, and Command Selection,"

Universal Access in the Information Society, Vol. 2, No. 1, pp. 30–43.

Shneiderman, B. (2000), "Designing Trust into Online Experiences," *Communications of the ACM*, Vol. 43, No. 12, pp. 57–59.

Slatin, J., and Rush, S. (2003), *Maximum Accessibility*, Addison-Wesley, Reading, MA.

Small, R., and Arnone, M. (2000), "Evaluating the Effectiveness of Web Sites," in *Human-Centered Methods in Information Systems: Current Research and Practice*, B. Clarke and S. Lehaney, Eds., Idea Group Publishing, Hershey, PA, pp. 91–101.

Spool, J., Scanlon, T., Schroeder, W., Snyder, C., and DeAngelo, T. (1999), *Web Site Usability: A Designer's Guide*, Morgan Kaufmann, San Francisco.

Stowers, G. (2002), "The State of Federal Web Sites: The Pursuit of Excellence," http://endowment.pwcglobal.com/pdfs/StowersReport0802.pdf.

Sullivan, T., and Matson, R. (2000), "Barriers to Use: Usability and Content Accessibility on the Web's Most Popular Sites," in *Proceedings of the ACM Conference on Universal Usability*, pp. 139–144.

Tedeschi, B. (1999), "Good Web Site Design Can Lead to Healthy Sales," *The New York Times*, August 30.

Trewin, S., and Pain, H. (1999), "Keyboard and Mouse Errors Due to Motor Disabilities," *International Journal of Human–Computer Studies*, Vol. 50, No. 2, pp. 109–144.

Turban, E., Lee, J., King, D., and Chung, H. (2000), *Electronic Commerce: A Managerial Perspective*, Prentice Hall, Upper Saddle River, NJ.

Wood, J., and Troutbeck, R. (1994), "Effect of Age and Visual Impairment on Driving and Vision Performance," *Transportation Research Record*, No. 1428, pp. 84–90.

CHAPTER 52

AUGMENTED COGNITION IN HUMAN–SYSTEM INTERACTION

Dylan Schmorrow
Office of Naval Research
Arlington, Virginia

Kay M. Stanney
University of Central Florida
Orlando, Florida

Glenn Wilson
Air Force Research Laboratory
Wright-Patterson Air Force Base, Ohio

Peter Young
Colorado State University
Fort Collins, Colorado

1 INTRODUCTION

> The fig tree is pollinated only by the insect *Blastophaga grossorun*. The larva of the insect lives in the ovary of the fig tree, and there it gets its food. The tree and the insect are thus heavily interdependent: the tree cannot reproduce without the insect; the insect cannot eat without the tree; together, they constitute not only a viable but a productive and thriving partnership. This cooperative "*living together in intimate association, or even close union, of two dissimilar organisms*" is called symbiosis.... "Man-computer symbiosis" is a subclass of man-machine systems. There are many man-machine systems. At present, however, there are no man-computer symbioses.... The hope is that, in not too many years, human brains and computing machines will be coupled together very tightly, and that the resulting partnership will think as no human brain has ever thought and process data in a way not approached by the information-handling machines we know today. (Licklider, 1960, pp. 4–5)

Although this was written over 40 years ago, Licklider's vision fully characterizes the current status of interactive computing and contemporary aspirations for its future. Historically, visionaries such as Licklider (1960) and Engelbart (1963) suggested that human–computer symbiosis should augment human intelligence and extend human cognitive abilities. Yet such intelligence augmentation has so far proved elusive for interactive system developers. There is a burgeoning paradigm shift in interactive computing that has the potential to realize these visionary projections; it is called *augmented cognition*.

Augmented cognition is a constellation of desires, concepts and goals aimed at maximizing human cognitive abilities through the unification of humans and computational systems (Schmorrow and McBride, 2004). As Licklider (1960) suggested, human brains and computing machines should be coupled together very tightly. The essence of augmented cognition is to achieve such coupling by leveraging the latest in powerful imaging techniques that enable mapping of distinct and detailed functions of the brain. Specifically, augmented cognition seeks to revolutionize the way that humans interact with computers by coupling traditional electromechanical interaction devices (e.g., mouse, joystick) with psychophysiological interaction (e.g., eye blinks, respiration, heart rate, electroencephalogram), such that subtle human physiological indicators can be used to direct human–system interaction. Fundamental research and related technology developments in this domain have centered on leveraging data from physiological indicators to alleviate and maximize throughput of human information-processing bottlenecks [e.g., sensory, working memory (WM), attention, executive function (EF)]. The basis for much of this work is grounded in the view that human information-processing capabilities are fundamentally the weak link in the symbiotic relationship between humans and computers. As computational prowess continues to increase, human and computer capabilities are ever more reliant on each other to achieve maximal performance. Demanding conditions, such as those associated with homeland security or military operations, call for expertise not from a specific human or computer system, but from a linked human–machine dyad. A dyad that is functionally a human and their computational system, which through shared experience and insight into how they both function, will jointly deliver solutions at a previously unimagined rate far surpassing that of a solitary entity.

Common within a majority of augmented cognition endeavors is the attempt to understand intrinsically how human information processing works so that augmentation schemes might be developed and effectively exploited to enhance human processing capacity. Thus, the central vision of augmented cognition is to extend human abilities substantially via computational technologies designed explicitly to address human information processing limitations.

1.1 Human Information Processing Limitations

Current understanding of human information processing suggests that information is perceived through multiple sensory processors. This information is then perceptually encoded (i.e., stimulus is identified and recognized), processed by a WM subsystem that is regulated and controlled by attention via the EF, which may be supported by long-term memory (LTM), to arrive at a decision, which in turn triggers a human response (Baddeley, 1986, 1990, 2000; Wickens, 1992). Within human information processing there are thus several "bottlenecks" or points of limited processing capacity, including sensory memory, WM, attention, and executive function.

1.1.1 Sensory Memory Bottleneck

Sensory memory is responsible for encoding information and converting it to a usable mental form (Atkinson and Shiffrin, 1968, 1971). There is a different sensory memory system for each of the human senses, including visual, auditory, tactile (haptic), olfactory, and gustatory. Behavioral studies suggest that human information processing begins with information being perceived on average in about 100 ms (Cheatham and White, 1954; Harter, 1967) by one of the sensory processors. The visual *iconic* sensory memory modality has been suggested to have an average capacity of about 17 items, and this iconic percept is fleeting, decaying completely, on average, in about 200 ms if it does not transfer to WM (Sperling, 1960, 1963; Averbach and Coriell, 1961; Neisser, 1967). Audition, or *echoic* sensory memory, is suggested to have an average capacity of five items and is a bit more persistent, with the "internal echo" lasting an average of about 1.5 seconds (Neisser, 1967; Darwin et al., 1972). *Haptic* sensory memory is very limited in terms of capacity (Watkins and Watkins, 1974; Mahrer and Miles, 2002) and has a decay rate between 2 and 8 seconds (Bliss et al., 1966; Posner and Konick, 1966; Lachman et al., 1979). Little is known about olfactory and gustatory sensory memories. In general, a considerable amount of information can be perceived if it is allocated across *multiple* sensory systems. Thus, *given the limited capacity of sensory memory, augmented cognition seeks to enhance sensory perception by exploiting multiple sensory channels for increased input capacity* (see Table 1). Sensory stimuli that have passed the sensory memory bottleneck and are

Table 1 Tenets of Augmented Cognition

Human Information-Processing Bottleneck	Tenet
Sensory memory	Augmented cognition seeks to enhance sensory perception by exploiting multiple sensory channels for increased input capacity.
Working memory	Augmented cognition seeks to support simultaneous processing of competing tasks by allocating data streams strategically to various multimodal sensory systems while maintaining multimodal information demands within working memory capacity.
Attention	Augmented cognition seeks to equip computers such that they become aware of subtle cues emanating from humans indicating how they are prioritizing incoming information (i.e., directing attention) and capitalize on these cues to enhance human information processing.
Executive function	Augmented cognition seeks to enhance information processing by directing the recall of contextual information that cues the optimal interpretation of incoming information and moderates the effects of modality switching.

rapidly decaying must then compete for the drastically limited resources of WM and attention.

1.1.2 Working Memory Bottleneck

Working memory allows people to maintain and manipulate information that has been perceived by sensory memory and is currently available in a short-term memory store. In general, WM is described as a functional multiple component of cognition "that allows humans to comprehend and mentally represent their immediate environment, to retain information about their immediate past experience, to support the acquisition of new knowledge, to solve problems, and to formulate, relate, and act on current goals" (Baddeley and Logie, 1999, p. 29). It is considered a temporary active storage area where information is manipulated and maintained for executing simple and complex tasks (e.g., serial recall, problem solving). Working memory is divided into separate processes that are required for short-term storage (according to Baddeley and Logie's (1999) model, these include the phonological loop and visuospatial sketchpad) and for allocating attention and coordinating maintained information (i.e., the executive function).

Working memory is still being defined, and recent research has suggested dissociations in both the phonological loop (i.e., phonological store vs. articulatory rehearsal mechanism) (Baddeley and Logie, 1999) and visuospatial sketchpad (visual form and color recognition vs. localization) (Carlesimo et al., 2001; Mendez, 2001; Pickering, 2001). In general, WM is said to have a limited capacity of about seven chunks, a rapid decay rate of about 200 ms, and a recognize-act processing time of 70 ms, on average (Miller, 1956; Card et al., 1983). Recent research suggests, however, that presenting information multimodally can in fact enhance human information processing via an increase in WM capacity, with gains on the order of three times Miller's (1956) "magic number" of seven being realized in one recent study (Samman et al., 2004). These gains could be tempered if the costs for modality switching are high; this is discussed in the next two sections.

Given separable WM components and WM capacity enhancements based on modality, Wickens's (1984) Multiple Resource Theory (MRT) can be expanded to suggest that *modality-based* resources can be utilized strategically at different points in user interaction to streamline a user's cognitive load (Stanney et al., 2004). In such a case, total WM capacity will depend on how dissimilar streams of information are in terms of modality. An expanded MRT would address how to allocate multimodal WM resources, particularly during multitasking, in such a way as to allow attention to be time-shared among various tasks. Thus, *augmented cognition seeks to support simultaneous processing of competing tasks by strategically allocating data streams to various multimodal sensory systems while maintaining multimodal information demands within WM capacity* (see Table 1).

1.1.3 Attention Bottleneck

Three general categories of attention theories can be found in the literature: (1) "cause" theories, in which attention is suggested to modulate information processing (e.g., via a spotlight that functions as a serial scanning mechanism or via limited resource pools); (2) "effect" theories, in which attention is suggested to be a by-product of information processing among multiple systems (e.g., stimulus representations compete for neuronal activation); and (3) hybrids that combine cause-and-effect theories (Fernandez-Duque and Johnson, 2002). In general, attention is suggested to be a selective process via which stimulus representations are transferred between sensory memory and WM and then contributes to the processing of information once in working memory. Attention improves human performance on a wide range of tasks, minimizes distractions, and facilitates access to awareness (i.e., focused attention). In the best case, attention helps to filter out irrelevant multimodal stimuli. In the worst case, critical information is lost due to overload of incoming information, stimulus competition, or distractions. Thus, if one were to try to enhance WM via multimodal interaction, such stimulation would impose a trade-off between the benefits of incorporating additional sensory systems and the costs associated with dividing attention between various sensory modalities. Attention must thus be moderated judiciously to enhance human–computer symbiosis. Augmented cognition seeks to "build systems that sense, and share with users, natural signals about attention to support ... fluid mixed-initiative collaboration with computers ... an assessment of a user's current and future attention (could thus) be employed to triage computational resources" (Horvitz et al., 2003, p. 52). Thus, *with augmented cognition, computers will become aware of subtle cues emanating from humans indicating how they are prioritizing incoming information (i.e., directing attention) and will capitalize on these cues to enhance human information processing* (see Table 1).

1.1.4 Executive Function Bottleneck

The EF system is suggested to be responsible for selection, initiation, and termination of human information processing routines (e.g., encoding, storing, and retrieving) (Matlin, 1998; Baddeley, 2003). It controls (i.e., focuses, divides, and switches) attention, integrates information from WM subcomponents, and connects WM with contextually triggered information from LTM. The EF is thus associated with regulatory processes underlying the control of human information processing and sheds light on operational costs associated with these control activities (Zakay and Block, 2004). The EF is thought to be especially active in handling novel situations (i.e., those with contextual ambiguity), such as those involving planning or decision making, error correction or troubleshooting, novel sequences of actions or responses, danger or technical difficulty, or the need to overcome habitual responses (Norman and Shallice, 1980; Shallice, 1982). When a person faces such contextual ambiguity during human information processing, high-level control functions of the EF become engaged. During such processing, a person will retrieve the multiple interpretations associated with a given uncertain situation, choose the more likely interpretation based on context and frequency of occurrence, discard alternative interpretations, and mark that point in their information representation as a choice point (Zakay and Block, 2004). Reducing contextual ambiguity, and thus effortful EF processing, would involve easing selection among multiple interpretations by increasing the number of contextual cues associated with any given alternative.

As indicated previously, frequent switching between one modality or task and another will incur a cost of switching that will be associated with inhibitions of responses to the previous modality stimuli or task, selection and activation of the response best associated with the new modality or task context, and resequencing of these stimuli. Since more frequent switching may entail greater contextual changes, it is expected to engage effortful EF processing. Thus, it is important during modality switching to consider the cost of such contextual changes. *Augmented cognition seeks to enhance information processing by directing recall of contextual information that cues optimal interpretation of incoming information and moderates the effects of modality switching* (see Table 1).

2 COGNITIVE STATE ASSESSORS

Augmented cognition seeks to enhance human–system interaction substantially by adopting a paradigm shift from primarily passive systems dependent on user input to proactive systems that gauge and detect, via diagnostic psychophysiological sensors, human information processing bottlenecks and then employing augmentation strategies to overcome these limitations. To realize this paradigm shift, one must first be able to characterize *cognitive state* such that the noted bottlenecks can be monitored and regulated appropriately. Research in psychophysiology, principally through brain-imaging techniques, has established a correspondence between cognitive processors and particular brain structures that have an identifiable locus in the brain. This allows use of neural signals from those structures as a diagnostic tool of *cognitive load*, which can be measured in real time while a person is engaged with an interactive system. Such psychophysiological data streams can be used to characterize cognitive state, specifically current load on information processing bottlenecks.

2.1 Psychophysiological Techniques for Capturing a Cognitive State

Many human–system interactive situations do not provide sufficient human performance information that can be used to infer cognitive state or what shall herein be called an operator's functional state (OFS). This is especially true of highly automated systems, which for the most part put the human in a monitoring role (Byrne and Parasuraman, 1996). Because system monitoring does not require overt behavioral responses, it is difficult to assess user state. Thus, a user may not be in an optimal state at all times, and system corrections or malfunctions may not be detected and responded to correctly. A methodology is needed that provides accurate assessment of OFS in the absence of overt performance data and to provide additional information when performance data are available. Psychophysiological measures have been suggested to fill this role.

Psychophysiological signals are always present and can often be collected unobtrusively, thereby providing a source of uninterrupted information about user state (Kramer, 1991; Wilson and Eggemeier, 1991; Scerbo et al., 2001; Wilson, 2002a). Correlations between psychophysiological measures and OFS have been described (Wilson and Schlegel, 2003). Although these correlations do not prove causality, they do suggest that psychophysiological measures can be used to assess OFS and further, that this information can be used to modify system parameters to meet the momentary needs of users (i.e., cognitive augmentation via adaptive aiding). Of the several criteria for implementation of OFS driven adaptive aiding, three crucial ones are that (1) significant and meaningful system performance improvements must be demonstrated; (2) the sensors used must be nonintrusive to a user's primary task, as this would hinder human–system performance; and (3) their use must be acceptable to users.

For widespread adoption, it must be demonstrated that OFS assessment and aiding either (1) improve human performance and enhance job success for work-related applications, or (2) enhance the interactive experience for entertainment-based applications. An example of a successful application of adaptive aiding is the use of antigravity (anti-*g*) suits, which require wearing additional gear that inflates at predetermined *g*-levels. These suits have been proven to save lives because they can prevent *g*-induced loss of consciousness in jet pilots and have therefore met with wide acceptance.

2.1.1 Current Status

In the past, the typical approach when using psychophysiological measures to assess OFS was to collect one or more measures and demonstrate that statistically significant differences exist between at least two levels of task demand or human state such as fatigue. Most of this research has been conducted in the laboratory. However, a growing body of research is expanding into operational environments. Psychophysiological measures have been applied successfully in driving, flight, and other test and evaluation environments (Wilson, 2002a). For example, heart rate has been shown to be increased significantly under high mental workload conditions compared to low mental workload conditions during flight (Hankins and Wilson, 1988; Wilson, 2002b). Electroencephalography (EEG), a physiological measure of the momentary functional state of cerebral structures, provides useful information about both high cognitive workload and inattention (Kramer, 1991; Wilson and Eggemeier, 1991; Gundel and Wilson, 1992; Sterman and

Mann, 1995). For example, theta-band EEG activity has been reported to increase with increased task demands (Gundel and Wilson, 1992; Gevins et al., 1998; Hankins and Wilson, 1998).

2.1.2 Current Technology for Recording Psychophysiological Data

Numerous psychophysiological measures have been shown to provide valuable information concerning OFS in real-world operational environments (Wilson, 2002a; Wilson and Schlegel, 2003). Because of the restrictions of the operational environment, some psychophysiological methods cannot be used. For example, positron emission tomography (PET), functional magnetic resonance imaging (fMRI), and magnetoencephalography (MEG) are not practical OFS gauges because the associated recording equipment is too restrictive, too large, and requires special shielding, among other prohibiting conditions. Even those measures that are less prohibitive have drawbacks. Almost all currently available, operationally useful psychophysiological sensors require contact with a user's body and use some form of electrolyte sensors. This is the case for EEG, electrocardiography (ECG), electrooculography (EOG), and electromyography (EMG). Users typically do not like to wear such sensors and associated equipment. Further, the sensors are usually attached to the skin with some type of adhesive, and repeated application in a day-to-day operational environment may cause skin irritation. There are less invasive options, such as pupillometry and eye point of regard, which are typically recorded with head-mounted sensors. Off-head eye point of regard devices are available but they restrict head movement, which limits their applicability in real-world environments.

2.1.3 New Sensor Technologies

New sensor technologies promise to provide users with more acceptable recording methods and valuable OFS data. Sensors that require only "dry" (no electrolyte or adhesive) contact with the skin have been developed (Kingsley et al., 2002; Trejo et al., 2003). Two approaches that are being explored for dry EEG sensors are capacitive coupled and optical sensors. These technologies can also be used to record ECG, EMG, and EOG. Currently, EEG can be recorded from nonhairy skin areas such as the forehead, but the goal is to be able to record EEG from anywhere on the scalp using these sensors. Eye activity can be recorded using video cameras that image the face from a distance, requiring no actual contact with users. Additionally, new sensor technology has been developed that provides measures of brain activity using blood flow technology. For example, functional near-infrared (fNIR) sensors provide information about brain oxygen levels, cortical blood volume, and neuronal activity.

2.1.4 Functional Near-Infrared Sensors

Using near-infrared light emitters, near-infrared energy can be directed through the scalp and skull and reflected from underlying cerebral tissue. Two types of cerebral information can be obtained from fNIR. The first type is hemodynamic response, reflecting oxyhemoglobin and deoxyhemoglobin concentrations in the brain. The consensus is that increased brain activity results in increased levels of local oxyhemoglobin and decreased levels of deoxyhemoglobin (Gratton and Fabiani, 2001). These responses have been used to investigate cognitive activity (Hock et al., 1997; Villringer and Chance, 1997; Takeuchi, 1999). The second type of information that can be obtained from fNIR is to detect changes in the optical characteristics in brain tissue that are related to neuronal activity (Gratton and Fabiani, 2001). The exact cause of these optical changes is not totally understood. This latter method is said to provide millisecond temporal resolution; the first method is much slower. For either procedure the infrared emitters and sensors have only to touch the scalp rather than being affixed to it (see Figure 1). The emitter–sensor unit can be held in place using a strap or cap arrangement. fNIR systems have been developed that function on hairy areas of the scalp and so are not restricted to the forehead region. This developing technology holds a great deal of promise for advancing our understanding of cognition and may be used more readily in operational environments than sensor technologies that require adhesives.

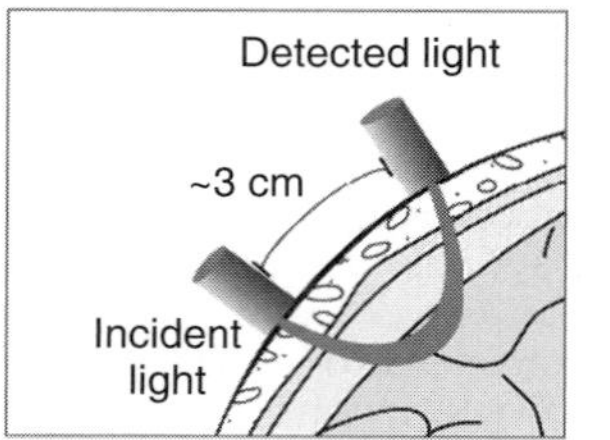

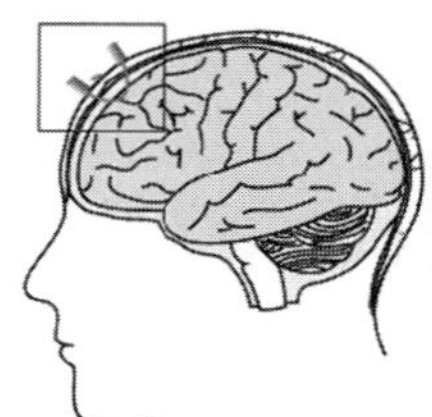

Figure 1 Locations of the infrared emitter and detector area important to ensure that cortical tissue is imaged. (From Downs and Downs, 2004.)

2.2 Transforming Sensors into Cognitive-State Gauges

To be useful, real-time assessment of cognitive activity using psychophysiological measures must be transformed from individual measures to cognitive gauges. Whereas consideration of individual measures provides valuable information, augmented cognition requires gauges that are composite estimates characterizing the functional state of a user (such as those to gauge load on the human information-processing bottlenecks, as well as others, such as Kolmogorov entropy of EEG signals and task load, which are mentioned in Sections 3.1.3 and 3.1.4). Given the complexity inherent to most operational environments, it is not sufficient simply to be aware that statistical changes exist in several measures. Measures or gauges must be able to characterize the functional state of a user such that this information can be used to implement adaptive aiding (i.e., triggering of augmentation strategies) in real time in real-world situations. In 2003, the U.S. Defense Department Defense Advanced Research Project Agency (DARPA) conducted a technology integration experiment (TIE) with various psychophysiological sensors (i.e., EEG, event-related potential, fNIR, pupil dilation, heart rate variability, arousal, galvanic skin response) to demonstrate the feasibility of simultaneous data collection (Morrison et al., 2003). The TIE demonstrated that real-time computation of sensor data to produce online gauge information was feasible, and further confirmed that several sensor technologies could be combined with minimal interference. However, substantial variability between human participants in gauge sensitivity suggested the need for additional research. Additional research also needs to be focused on how to transform sensors to specific OFS gauges, such as gauges to measure the load on the human information processing bottlenecks. Thus, *augmented cognition seeks to leverage a set of psychophysiological gauges that allow for real-time assessment of cognitive state, particularly current load on information processing bottlenecks, which can then be transformed directly into computer control commands for triggering implementation of augmentation strategies.*

3 HUMAN–SYSTEM AUGMENTATION

In Section 1, various human information processing bottlenecks were discussed (i.e., sensory memory, WM, EF, attention). In Section 2, means of gauging the current cognitive load on a person were considered. Augmented cognition seeks to overcome the noted points of limited capacity processing through the utilization of human–system augmentation strategies, which will be triggered by cognitive state gauges. It is suggested that through augmentation strategies, the cost of these bottlenecks (e.g., degraded human performance due to overload, underload, stress, losses in situational awareness, or emotional state) can be overcome.

3.1 Augmentation Strategies

In conventional human–system interaction, an excessive amount of cognitively demanding tasks can be imposed on a user. In such situations, human information processing can break down at any of the bottlenecks. Instead of overloading users, interactive systems should seek to achieve cognitive congeniality (Kirsh, 1996) by (1) *presenting* an optimal level of task-relevant information and ensuring that it is readily perceived, (2) minimizing cognitive load on WM by *sequencing* and *pacing* tasks appropriately, and (3) reducing the number and cost of mental computations required for task success by *delegating* tasks when appropriate. Taken together, these strategies should increase the speed, accuracy, and robustness of human–system interaction. Each of these augmentation strategies (i.e., task presentation, sequencing, pacing, and delegation) is discussed below. It should be noted that other such strategies can and should be identified. Additional augmentation strategies to consider include but are not limited to techniques for supporting information filtering and triage, multitasking, mixed-initiative interaction, and context-sensitive interaction (Horvitz et al., 2003).

3.1.1 Task Presentation

When designing interactive systems, a central question is which information should be conveyed via which modality. Conventional interactive systems present information to users primarily via visual cues, sometimes offering auditory accessories. Yet to optimize sensory processing, thereby relieving the sensory memory bottleneck, one should consider the types of information each modality is particularly suited to display. Table 2 presents theorized suitability of sensory modalities for conveying various information sources. In addition to suitability, one must consider capacity. As aforementioned, Samman et al. (2004) demonstrated that multimodal WM capacity can reach levels nearly three times that of Miller's (1956) magic number seven. Thus, rather than overloading a single modality, by distributing information across multiple modalities the WM bottleneck can be relieved. Table 3 represents the WM capacity of various modalities based on several studies (Bliss et al., 1966; Sullivan and Turvey, 1974; Smyth and Pendleton, 1990; Keller et al., 1995; Livermore and Laing, 1996; Woodin and Heil, 1996; Feyereisen and Van der Linden, 1997; Matsuda, 1998; Jinks and Laing, 1999; Laska and Teubner, 1999; Frenchman et al., 2003). The numbers in Table 3 suggest the upper limit on the number of items that should be presented via each modality, as individual modality capacity tends to decline during multimodal multitasking even though overall capacity increases (Samman et al., 2004). Thus, with knowledge of the information sources constituting a given application, a determination of optimal modalities can be made to direct multimodal task presentation. More specifically, after characterizing a given application's information sources via a task analysis, first a matching to the optimal modality can be determined using Table 2. Then, given the outcome of the related OFS gauges (i.e., current load on sensory and WM bottlenecks), a determination of reserve capacity can be estimated using Table 3 and a selection of the optimal modality made (i.e., the one with the best match from Table 2 and adequate reserve capacity). The applied implication is that in cognitively demanding task environments, not only should information be presented in a modality that is most suitable but also in one that is not currently fully loaded, thereby easing the sensory memory and WM bottlenecks. Thus, the first augmentation strategy is to *identify the optimal modality by which to present information based on consideration of suitability principles as well as current psychophysiological measures of cognitive load* (see Table 4).

3.1.2 Task Sequencing

Once the modality by which to present an information source is determined, the information event can be scheduled. The MRT (Wickens, 1984) suggests that people are more efficient in time-sharing tasks when different resources are utilized in terms of

Table 2 Theorized Suitability of Modalities for Conveying Various Information Sources[a]

Information Source	Sensory Modality					
	Visual	Verbal	Tactile	Kinesthetic	Tonal	Olfactory
Spatial acuity (size, distance, position)	++	□	+	+		−−
2D localization (absolute/relative location in 2D)	++	+	+	□	+	−−
3D localization (absolute/relative location in 3D)		□	+	□	+	−−
Change over time	++	+	+		□	−−
Persistent attention	++	−−	++		−−	
Absolute quantitative parameters	++	+	−−	−−	−−	−−
Temporal (e.g., duration, interval, rhythm)	□		+	+	++	−−
Instructions	+	++				
Rapid cuing (e.g., alerts, warning)	+	+	++		++	
Surface characteristics (e.g., roughness, texture)	+		++			
Hand–eye coordination (e.g., object manipulation)			++			
Memory aid (e.g., recognition of a formerly perceived object)	+	+	−	−	++	++
Affective or ambient information	□	□			+	++

[a] *Key:* ++, best modality; +, next best; □, neutral; −, not well suited, but possible; −−, unsuitable.
Source: Adapted from ETSI (2002).

Table 3 WM Capacity of Various Sensory Modalities

WM Subsystem	Capacity
Visual	2–5
Verbal	4–7
Spatial	5–7
Tactile	3–5
Kinesthetic	3–5
Tonal	4–6
Olfactory	3–4

sensory stimuli modality (e.g., visual, auditory), WM processing codes (e.g., spatial, verbal), and response modality (e.g., vocal, manual). For example, various studies have suggested that a person can recall more in two tasks with different types of materials combined than in a single task, especially if the modalities or types of representation are very different (Klapp and Netick, 1988; Penney, 1989; Baddeley, 1990; Cowan, 2001; Sulzen, 2001). More recent MRT efforts have suggested that task interference can be minimized by leveraging opposite ends of four task dimensions, including processing stages (perception, cognition, response), perceptual (sensory), modality (visual, verbal, spatial, tactile, kinesthetic, tonal, olfactory), visual processing channels (focal, ambient), and WM processing codes (spatial, verbal) (Wickens, 2002). An applied implication of this theory is that time sharing of tasks should be more effective with cross-modal as compared to intramodal information displays. Thus, through systematic sequencing of tasks, simultaneous processing of competing tasks can be allocated strategically across various multimodal sensory systems in an effort to maintain multimodal information demands within WM capacity. Beyond addressing the WM bottleneck, this augmentation strategy can assist in prioritizing incoming information by sequencing cues according to priority, thereby directing attention. When applying this strategy, it is essential to ensure that there is a means to avoid the adaptive state from oscillating too frequently. This can be done through the application of robust controllers (see Section 4). Through systematic control of the adaptive state, this strategy also addresses the EF bottleneck by moderating the effects of modality switching.

To determine task sequencing (i.e., ordering and combining of tasks), a conflict matrix could be calculated following Wickens's (2002) approach, in which the amount of conflict between resource pairs for

Table 4 Augmentation Strategies

Augmentation Strategy	Description	Human Information-Processing Bottleneck Addressed
Task presentation	Identify optimal modality by which to present information based on consideration of suitability principles and current psychophysiological measures of cognitive load	Sensory and working memory
Task sequencing	Assign modalities to information sources and schedule them, considering priority, such that they minimize interference over the performance period while leveraging robust controllers to moderate effects of modality switching	Sensory and working memory, attention, executive function
Task pacing	Provide external pacing of tasks, which could be achieved by monitoring behavioral entropy	Working memory, attention, executive function
Task delegation	Direct assisted explicit task delegation based on psychophysiological indexes of task load	Attention, executive function

task couplings is determined. This calculation factors in both conflict and task difficulty (i.e., resource demands), resulting in a task interference value. This could be done in conjunction with a time-line analysis (Sarno and Wickens, 1995), which calculates resource demand levels of time-shared tasks over the time during which the tasks are to be performed. In allocating resources, these principles could be coupled with a scheme of task priorities (as derived through an a priori task analysis), which taken together could guide task ordering and combining given current resource constraints (i.e., task interference values and OFS gauge outputs from all four bottlenecks). The second augmentation strategy is thus to *assign modalities to information sources and then schedule them, considering priority, such that they minimize interference over the performance period while leveraging robust controllers to moderate the effects of modality switching* (see Table 4). This should help relieve the sensory, WM, attention, and EF bottlenecks.

3.1.3 Task Pacing

Time management is an essential component of many dynamic task situations (and is also critical to feedback stability of closed-loop systems, see Section 4). Yet, in cognitively demanding task environments, pacing skills can decline rapidly, as temporal judgments depend on the amount of attentional resources allocated to a temporal processor (Casini and Macar, 1999). Further, internal (self) pacing has been shown via EEG signals to impose higher human information processing demands compared to externally (e.g., via metronome) paced tasks (Gerloff et al., 1998). Disruption of an orderly rhythm is thought to increase the entropy of the human information-processing system, thereby increasing information content due purely to asynchronous pacing of a task. Such disruption can occur when a person becomes overloaded with information, as this often results in delayed event detection and more corrective responses (Boer, 2001). Interestingly, Boer (2001) developed a simple but highly predictive linear model based on Wickens and Hollands's (2000) MRT, which predicted the effect of various tasks on steering entropy and driver performance. The model demonstrated that steering entropy was affected primarily by loading of spatial tasks, as would be predicted by MRT because driving is a highly spatial task. Thus, to achieve effective time management, a potential augmentation strategy would be to *provide external pacing of tasks, which could be achieved by monitoring behavioral entropy* (see Table 4). Specifically, the Kolmogorov entropy (K-entropy) of EEG signals can be used to assess information flow (Pravitha et al., 2003). K-entropy is proportional to the rate at which information about the state of a dynamical system is lost in the course of time. This entropy index has been shown to fluctuate with changes in the complexity of human information processing, such as that imposed by fatigue (leading to a lesser extent of information flow through particular brain regions) (Rekha et al., 2003) or information overload (King, 1991) while remaining quite stable during performance of demanding cognitive tasks (Pravitha et al., 2003). Thus, using K-entropy of EEG signals to direct task pacing should help relieve the WM, attention, and EF bottlenecks, as it could help minimize pace the processing of incoming information and minimizes disruptions.

3.1.4 Task Delegation

In the context of augmented cognition, the purpose of dynamic task delegation would be to increase information throughput by balancing the utilization of human resources across a network of users. Task delegation allows for distribution of task demands across individuals as well as coordination between humans and automated systems. In task delegation, certain actions required by a particular task performer are delegated to another performer or back to the system itself once task load gets above some threshold (Dearden et al., 2000; Hoc, 2001; Debernard et al., 2002). Such handing off can be implicit (i.e., imposing an allocation based on current OFS load predictions) or explicit, in that it requires an action from the task performer prior to allocation. Although implicit allocation has been shown to lead to better performance than explicit, implicit allocation does not always meet with user acceptance, as humans like to maintain control of dynamic task situations and become anxious when they lose control (Hock et al., 2002). This, in turn, could affect behavioral entropy, thereby affecting system pacing. Taken together, this could affect system stability properties negatively (see Section 4). Assisted explicit allocation is a compromise, where after detecting an overload using an OFS gauge of task load, such as the task engagement index used by Prinzel et al., (2000, 2003), the interactive system would make an allocation proposal which the human would be able to veto but would not be in charge of allocating. This cooperative task allocation strategy generally leads to effective performance while avoiding complacency by requiring the human to cooperate in the allocation process. Thus, a fourth potential augmentation strategy would be to *direct assisted explicit task delegation based on psychophysiological indexes of task load* (see Table 4). This should help relieve the attention and EF bottlenecks, as it eases the need to determine what to attend to.

4 ROBUST CONTROLLERS

Although augmentation strategies have the potential to enhance human performance through reducing the load on human information processing bottlenecks, they could also lead to an adaptive state that oscillates too frequently, thereby destabilizing human–system interaction over time. Thus, there is a need to identify techniques for ensuring that changes requested through the augmentation strategies are implemented so as to maintain system stability and enhance human performance. Mathematical system theory deals with the modeling, analysis, and design of complex dynamic systems. Robust control theory is a discipline of mathematical system theory that is concerned with the analysis and design of feedback controllers for situations where there is only partial or incomplete knowledge of the underlying system dynamics. In the work discussed in this chapter, whereby a user's display/input is adapted based on his or her measured cognitive load, it is important to note that a feedback loop is being *closed* around the human. Moreover, since the underlying system dynamics involve the human, it is certainly true that only partial knowledge concerning a user's state will be available, hence the need for this section on robust control.

4.1 Control System Models

Recent developments in the field of cognitive neuroscience have heralded a great deal of change in what is known about human mental operations (Posner and DiGirolamo, 2000). As has been discussed, these advances have the potential to allow psychophysiological indicators to direct human–system interaction (Farwell and Donchin, 1988). The ability to use sensors to measure the cognitive performance of a user immediately through psychophysiological characteristics, and virtually instantly adapt a system to meet user needs, presents an exciting new paradigm in interactive systems. The introduction of such real-time adaptive aiding offers the prospect of radically altering how humans interact with computer technology. However, one important aspect of such a potential change in the nature of human–system interaction is the inherent difference between open- and closed-loop systems.

Even well-understood, stable open-loop systems will show very different performance under closed-loop operation. A simple example of this effect can be seen when bringing a speaker and a microphone (connected to each other) too close together. A well-known audio feedback effect occurs as the signal from the speaker runs through the microphone, back out of the speaker, back into the microphone, and so on. The resulting feedback loop is (typically) unstable and produces a familiar (and unpleasant) sound. The volume of this sound may grow or decay (corresponding to unstable and stable feedback systems, respectively), depending on the proximity of the microphone to the speaker (which implicitly sets the loop gain in the feedback system). Thus, two perfectly well-behaved open-loop systems (speaker and microphone) may or may not be closed-loop stable, depending on how feedback is applied. A more precise quantitative example of such behavior for an augmented cognition system will be provided later, where it is shown that a stable open-loop system may generate a stable or unstable closed-loop system, depending on how feedback is designed.

Although a great deal about human performance may be understood, the nature of the shift from an open- to a closed-loop system is a unique type of change. As a result, many standard predictable aspects of cognitive and motor performance may operate in drastically different ways in closed-loop systems. A prime candidate for understanding such closed-loop circumstances is through the use of engineering control systems theory. [For a discussion of the pros and cons of various types of models, see Baron et al. (1990).] Control systems theory deals with fundamental properties of systems as described (typically) by mathematical models. It provides a framework and tools for analyzing fundamental system properties, such as performance, noise rejection, and stability, and offers systematic approaches for designing systems with these desired properties.

The idea of applying control theory to humans has some history, with Wiener (1948) widely considered to be the first person to draw parallels between control systems in machines and the organization present within some living systems. However, few attempts have been made to apply control systems theory to human–system interaction (Flach, 1999; Jagacinski and Flach, 2003; Young et al., 2004), and thus this is an exciting area of research where much remains to be done. One notable exception that the current effort draws from is Card et al.'s (1983) Model Human Processor (MHP). The MHP is a human information-processing model consisting of a basic block diagram interconnect model of a human, with an associated estimate of the time taken by each processing stage to process relevant data. For augmented cognition purposes, the three most relevant stages (i.e., blocks) are probably the perceptual, cognitive, and motor processors. This is illustrated in Figure 2, which shows a human operator piloting a vehicle. In this example, information from the operator's system display would first pass through the operator's perceptual (i.e., sensory) processor, being perceived, on average, in about 100 ms (Cheatham and White, 1954; Harter, 1967). Perceived information would then be available to the cognitive processor, which has an average cycle time of 70 ms. The cognitive processor would then make a decision, and that decision would be implemented by the motor processor, which has an average cycle time of 70 ms, with a resulting action on the vehicle controls. Note that these three blocks provide an internal model of the operator's interaction with the external vehicle displays and controls. This block diagram model not only characterizes the flow of information and commands between the vehicle and operator but also enables us to access the internal state of the operator at various stages in the process. This allows modeling of what an augmented cognition system might have access to (internal to the human; e.g., load on human information processing bottlenecks) and how those data might be used to direct closed-loop human–system interaction.

If one considers a control systems model incorporating the flow of human information processing, the time taken by each block adds time delay to the model. However, it does much more than that. As indicated in the early discussion on bottlenecks, it also implies a certain *bandwidth* for the system, both in terms of channel capacity and because signals that vary more rapidly than the time constant of the system (i.e., high-frequency signals) do not pass through it. Hence the processing blocks act as low-pass filters, only allowing through signals that are below the system bandwidth. For example, humans do not generally perceive the flicker on a computer monitor because it typically occurs at a frequency (100 Hz) higher than that of the perceptual processor's bandwidth of only about 10 Hz. As a first attempt at modeling such phenomenon, the effects of time lags in human perceptual, cognitive, and motor processing blocks are considered. This results in a dynamic model of the form shown in Figure 3.

Note that the setup depicted in Figure 3 is a generic dynamic model of any one of the MHP components (perceptual, cognitive, motor) shown in Figure 2 (although the model parameters will be different for each). The dynamic models associated with each MHP component ("first-order lag" and "time delay") of the block model are given, respectively, in the time domain (i.e., convolution representation) as

$$y(t) = \frac{1}{\tau}\int_0^t e^{-(t-\gamma)/\tau} u(\gamma)\, d\gamma$$

$$z(t) = y(t - \tau)$$

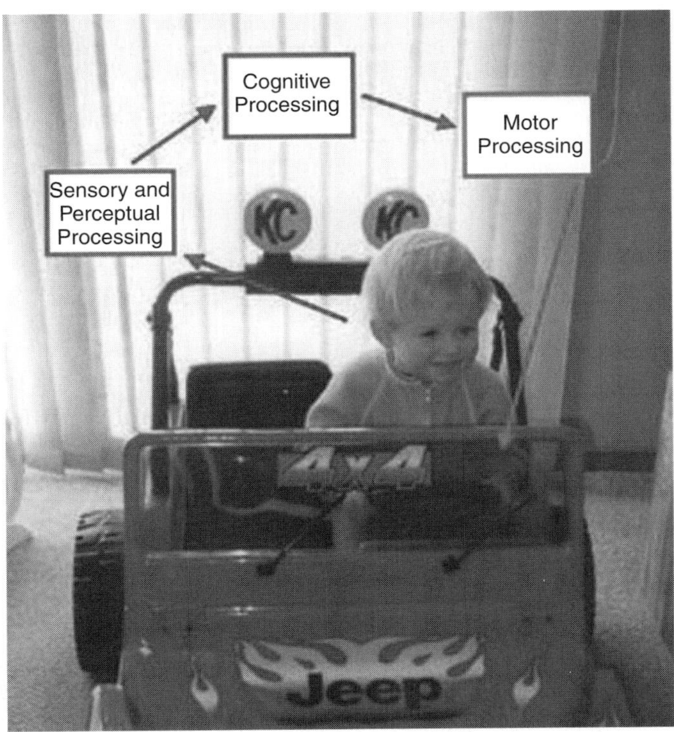

Figure 2 Human information-processing model.

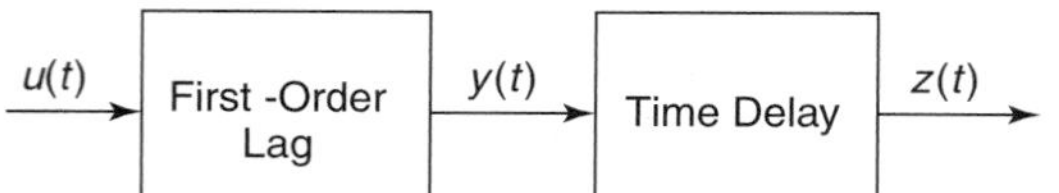

Figure 3 Block model for each component.

for each processing block [with overall input $u(t)$ and output $z(t)$], with the time constant τ taken from the relevant processing time in the MHP model. The first-order lag models the dynamic relationship between input and output signals, which captures the bandwidth effect described earlier. This is most easily seen using the Laplace transform to transform this model from the time domain to an equivalent frequency-domain representation:

$$Y(s) = G(s)U(s)$$

where the function $G(s)$ is given as

$$G(s) = \frac{1}{1 + s\tau}$$

This is known as the *transfer function* of the system. [See Phillips and Parr (1999) for an overview of transform methods for signals and systems; see Ogata (2002) for an overview of the application of these techniques to dynamic systems and feedback control.] A key point is that the time-domain convolution operator has been transformed into a simple multiplication operator in the frequency domain. That multiplication operator, $G(s)$, is both complex-valued and frequency-varying. The function $G(s)$ captures the frequency response of the system in both magnitude and phase.

To see this, one can evaluate the transfer function along the imaginary axis, that is, substitute $s = j\omega$ into the model (equivalent to specializing the Laplace transform to a Fourier transform) to yield

$$G(j\omega) = \frac{1}{1 + j\omega\tau} = \frac{1}{\sqrt{1 + (\omega\tau)^2}} e^{-j \tan^{-1} \omega\tau}$$

which is the frequency response of the system (with ω the real-valued frequency). This has the desired low-pass frequency response. Low-frequency (slowly varying) signals pass through almost unattenuated, but higher-frequency (rapidly varying) signals are more and more attenuated until hardly any of the signal passes through the system at all. This variation of the magnitude response with frequency in the first-order lag block is what accounts for the computer monitor effect (i.e., lack of perceiving flicker) described earlier (one could not account for this effect with a time-delay block alone because the frequency response of a pure time delay is flat, i.e., no variation of magnitude with frequency).

Note that this magnitude response comes with an associated phase response. Low-frequency signals pass though this system with almost undistorted phase. However, as frequency increases, the signals start to incur phase lag, which ultimately reaches 90° at high frequency. Phase lag has a destabilizing effect on closed-loop feedback systems, so understanding the relationship between magnitude and phase of different frequency signals as they pass through the system is of crucial importance in designing any feedback control system.

These various steps have provided the separate pieces necessary to build a model of an entire open-loop system. Since transfer functions operate by multiplication, models for the individual blocks can be cascaded. These are linear models and therefore they commute, so the order of cascade can be changed, and hence time delays can be accumulated into a single block if desired. This now provides a quantitative dynamic model for the human as illustrated in Figure 4.

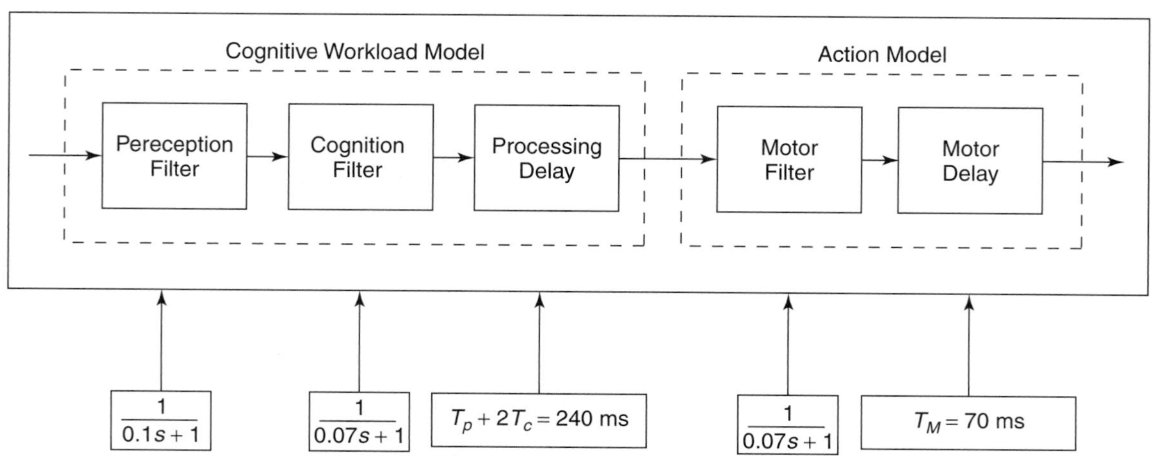

Figure 4 Dynamic control system model of the human.

Note that, as discussed above, this model captures the gain–phase relationship with frequency, which is crucial if the model is to be used in a feedback control loop.

This model should allow accurate predictions of open-loop performance and other properties of the system to be made. However, it is important to note that this control theory–based model is in a form that will also allow for prediction of how performance and properties are modified when transforming to a closed-loop setup, which is described in the following section.

4.1.1 Augmented Cognition Closed-Loop Models

Augmented cognition aims to provide display and information systems that take measurements from OFS gauges, such as those described in Section 2, and use these data to dynamically adapt human–system interaction. The sensor dynamics of any future OFS gauges are still to be determined, so as a starting point such sensors are modeled here as simple first-order lags with a time constant of $\tau = 1$ second. The sensor data would be used to dynamically change inputs to a user by directing instantiation of augmentation strategies, such as those described in Section 3. As an example, consider an application where workload is reduced via the *task delegation* augmentation strategy (Wickens et al., 1998). In such an application, using OFS gauges to detect cognitive overload (e.g., through a EEG-derived index of task engagement) (Prinzel et al., 2003), lower-priority tasks would be offloaded to automated agents, with the goal of maintaining users working at their maximum capacity. Such a closed-loop human–system interaction model was implemented in the Matlab/Simulink simulation environment, which is illustrated in Figure 5.

Various pieces of an augmented cognition system can be seen in the model in Figure 5, including the human perceptual, cognitive, and motor processors. Note both the OFS gauge that detects the state of the human user (i.e., cognitive work overload measurement) and the augmentation strategies (i.e., within the PID controller) that will alter the input to the human. The rest of the model contains task inputs to the system, displayed outputs at various points (e.g., actual vs. measured cognitive workload), and a simple model of performance errors resulting from cognitive overload. The feedback loop being closed is now apparent in this simulation model, which drives the need for a systematic control theory approach.

4.2 Controller Analysis and Design

Even this simple model has already produced some important findings. In particular, one major finding from initial efforts with the model is to show how dynamic instability can result from introducing feedback within the system. That is to say that rapid detection of cognitive state under high workload might result in input being removed, which would reduce workload and hence information would be added, which would once again result in high workload, and the cycle repeats. This simple illustration indicates how users might find their display cycling rapidly through cluttered and decluttered states as a result of changes detected in workload. Control theory offers a means to remove such instability and optimize performance.

Figure 6 shows results from three simulations of a task overload situation. The input to each of these simulations is the same: Initially, the user is fully loaded (and making no errors), and then a step increase in workload is introduced 1 second into the simulation. This results in task overload from that point on, with subsequent performance errors. Note that each of these simulations uses the same system model, so the only difference is how (or if) the feedback control (i.e., augmentation strategy) is applied.

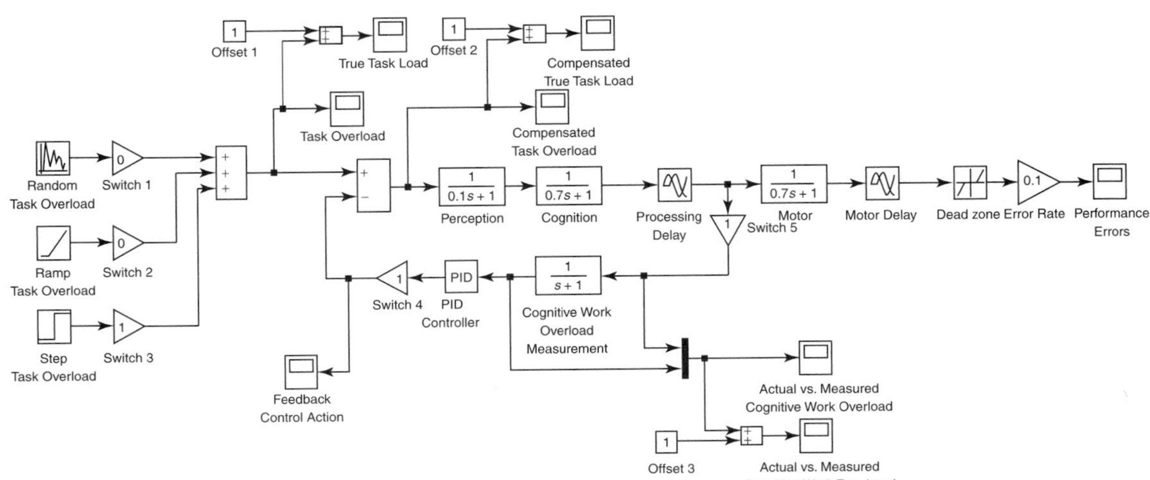

Figure 5 Matlab/Simulink model of closed-loop augmented cognition human–system interaction.

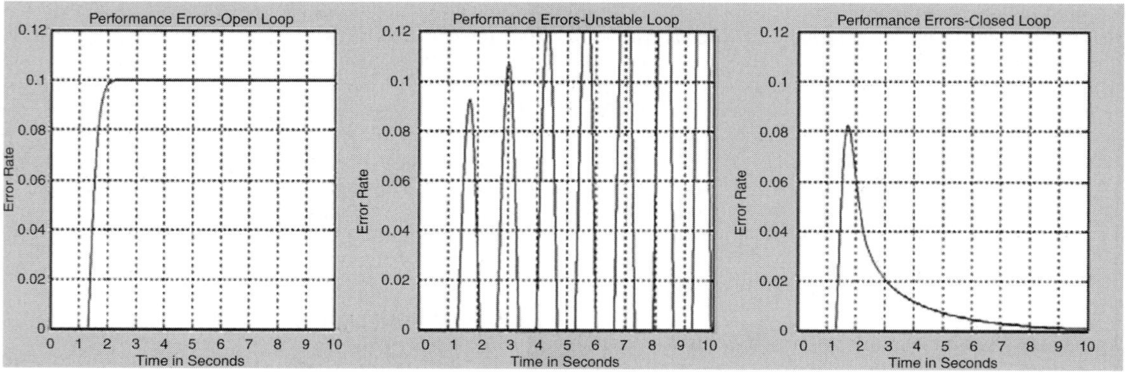

Figure 6 Simulation results for an augmented cognition closed-loop dynamic model.

Starting from the left, the first plot of Figure 6 shows the resulting performance errors for an open-loop simulation (i.e., with the augmented cognition system disabled). As the workload of the task increases, the plot shows how the number of errors quickly rises to a certain level and stays there. The next panel shows a poorly designed augmented cognition system. This system utilizes simple proportional control; that is, the control action $c(t)$ that reduces task workload to the user is just directly proportional to measured overload $m(t)$. Thus, the controller is of the form

$$c(t) = Km(t)$$

and the control designer simply chooses the proportionality constant or controller gain K, which determines when an augmentation strategy (i.e., task delegation in this case) is to be implemented. High-gain controllers, with large K values, use a high-magnitude feedback signal that tries aggressively to drive the control loop to the desired point (for fast or high performance). If K is chosen too aggressively, however, the closed-loop system will approach (or even exceed) stability margins. In this example, the gain K is chosen poorly, resulting in instability of the type described above, with the input being reduced rapidly and then increased, resulting in highly fluctuating performance from the user. Note that the precise values of K that drive the system into instability depend on the specific problem (and can be predicted accurately with control theory methods), but they can certainly occur at plausible real-world values (in this example, $K = 2.8$).

Proportional control is what people often think of when they consider feedback. A simple version is the cruise control in a car, which moves the gas pedal in a manner proportional to the difference between the desired and actual speeds. However, this simple control strategy can deliver only limited performance improvements, even when designed correctly. For instance, one could never get steady-state errors down to zero with this type of control. This approach is limited because it utilizes the same gain for all frequencies (and hence all signals), so one does not have sufficient degrees of freedom to exploit any trade-offs in the design. A very common type of controller used in engineering applications is the proportional–integral–derivative (PID) controller. This generates a corrective action from a measurement of the form

$$c(t) = K_P m(t) + K_I \int_0^t m(\tau)\,d\tau + K_D \frac{dm(t)}{dt}$$

There are now three constants to be chosen (designed), K_P, K_I, and K_D, which correspond, respectively, to the amounts of proportional, integral, and derivative feedback used in the closed loop. Note that the integral action effectively includes memory and thus allows better compensation at low frequency and hence improved steady-state performance. The derivative action essentially includes anticipation, which allows for improved high-frequency performance, resulting in better transient response and improved stability properties. The overall controller has frequency-varying gain, which allows design trade-offs to be exploited more properly. The right panel of Figure 6 shows a functional closed-loop system using a well-designed PID controller to deliver closed-loop stability and good performance. It is clear that even maximum errors never reach the level of the open-loop (automation-free) system and that they quickly drop to minimal levels (asymptotically approaching zero) without any undesirable oscillatory transient response.

4.2.1 Human Dynamics and Achievable Performance

The first benefit of the modeling approach described above is that it provides some *proof of concept* for the augmented cognition concept: namely, to show precisely how an integrated system of OFS gauges, augmentation strategies, and robust controllers can combine to augment performance. The caveat from this work, however, is to note that such systems need to be designed carefully, with a systematic control theory approach rather than simple heuristic tuning, else augmented cognition may fail to fulfill its potential.

Fortunately, systematic modeling can offer assistance in terms of determining the nature of information required and parameters necessary for driving specific OFS gauges. The types of questions that could be addressed by this type of analysis include:

- What time constant/bandwidth is necessary for a particular OFS gauge to have a significant useful effect (i.e., how fast)?
- What resolution is required of the OFS gauge (i.e., how accurate)?
- How much noise can reasonably be tolerated on any given measurement?
- What would additional measurements/gauges offer?
- What performance level could be achieved (given the above)?

These questions should be addressed in future work in the area of modeling and analyses. Note that both qualitative and quantitative analyses can be carried out, and both have their uses (e.g., qualitative analysis might steer one toward a particular technology, whereas quantitative analysis might allow one to design and implement it accurately). Note also that specific scenarios can be carried out in a simulation, which would allow one to test out certain strategies repeatedly and reliably before going to the expense of constructing an experimental setup, including low-probability events that might not occur in an experimental setting. Furthermore, control theory includes powerful analysis tools that go well beyond simple simulation to address fundamental trade-offs and limitations inherent in any feedback loop (Doyle et al., 1992).

Ultimately, the modeling strategies described in this section would aim to predict the impact on human–system performance of various augmentation strategies for changing how information is provided to a user. In addition, they have the potential to highlight areas that would receive particular benefit from such augmentation. Thus, overall, this work can provide the basis for future systematic closed-loop analysis and controller design, bringing to bear powerful tools from engineering control theory. The power of such analysis tools is demonstrated in the next section.

4.2.2 Individual Differences and Robustness Analysis

Preliminary robustness analysis was conducted on the closed-loop system with PID controller modeled in Figure 5. The theory of robust control deals with systems subject to uncertainty such as any closed-loop augmented cognition system would be subject to due to individual differences (among other reasons, as noted above). Control theory provides a means of examining what performance on such a system will be, rather than just an idealized simulation. It also allows for examination of variation between users, since relevant parameters in the model can be varied (e.g., speed of the MHP processors) to determine to what extent a given control scheme is robust against such variations.

The theoretical tools used here to model individual differences were based on the structured singular value (SSV), or μ, and its extensions to handle real parametric uncertainty (Young, 2001). The idea is that one first has to use linear fractional transformations (LFTs) to rearrange the problem into canonical $M-\Delta$ form, as illustrated in Figure 7. Here $M(s)$ collects all the known dynamics of the (closed-loop) system, and Δ is a (block) diagonal structured perturbation, which in the case of individual differences analysis will consist of real parametric uncertainty representing variation in the parameters of the model. Thus, this approach handles LFT (block diagram) perturbations rather than handing perturbed coefficients directly in a (transfer function) model, but this apparent limitation is readily overcome, as we illustrate below.

The individual differences analysis considered variations in two time constants (i.e., speed of the perceptual and cognitive processors). These could arise due to variations among users, but could also be introduced through inaccuracies in the modeling approach. To realize this analysis, these variations were cast as a block diagram perturbation. This can be done by noting the interconnect in Figure 8, which shows an example of rearranging parametric uncertainty as an LFT (block diagram).

Mathematically straightforward block diagram calculations now reveal that the transfer function in Figure 8 is represented by

$$\frac{1}{1+s(\tau+\Delta\tau)}$$

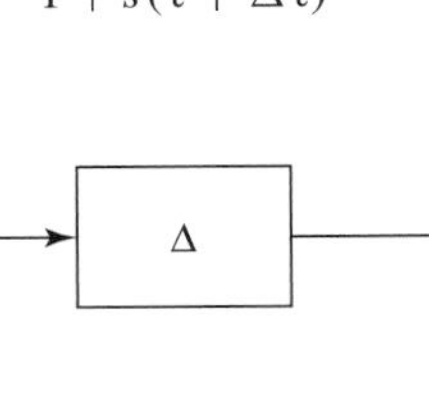

Figure 7 Canonical form for SSV analysis.

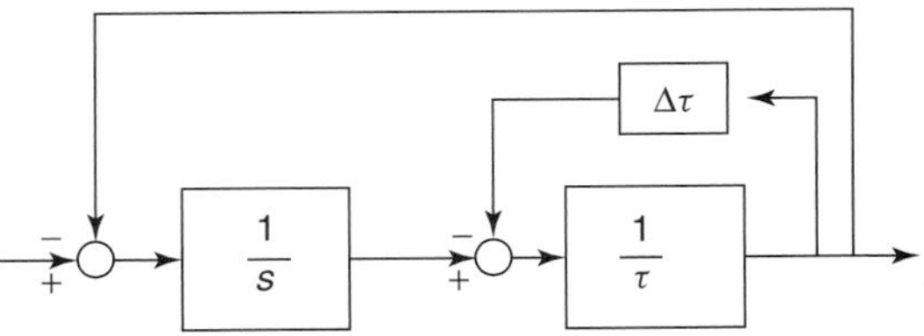

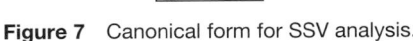

Figure 8 Variation in time constant as an LFT perturbation.

so that the block diagram perturbation in Figure 8 actually becomes a perturbed coefficient in the transfer function model, and to be specific it represents a perturbation in the time constant of the first-order lag model.

This approach was applied to the closed-loop augmented cognition system model represented in Figure 5, considering a parametric variation in the time constants of the first-order lag models of the perceptual and cognitive processor blocks. Note that the motor processor block is not in the feedback loop in this scenario, so it does not affect stability and hence was not included in the robustness analysis presented here. The LFT and μ analysis machinery could then be applied to this block diagram. The mathematics of this approach is quite involved and uses computational complexity theory, complex analysis, and linear algebra among others (Young, 2001). Space constraints prevent going into any kind of detailed explanation here, but the end result of this analysis was to give a parameter range over which (robust) stability is *guaranteed*. This means that no parameter combination in the allowed range can cause instability. For example, in this case one could guarantee that no person with a combination of processing time constants for the perceptual and cognitive processors in the ranges specified would cause the closed-loop augmented cognition system in Figure 5 to go unstable. It is important to note the power of this *guarantee*, because one cannot get such guarantees from any amount of exhaustive simulation or testing (it is always possible that a parameter combination is missed, which causes a problem no matter how many variations are tried).

The results of this analysis showed that both the perceptual and cognitive processor time constants could be reduced to very small numbers (practically all the way down to zero), indicating that faster user response than predicted was no problem. The upper limits for the perceptual and cognitive processor time constants were found to be 3.7 and 2.6 seconds, respectively. Thus, it is possible for the system to go unstable with slower users. However, the degree of robustness afforded by a PID controller is huge. Specifically, in this example the time constants are a factor of more than 37 times greater than those nominally assumed (e.g., the MHP's "slowman" to "fastman" range for the perceptual processor is 150 ms and for the cognitive processor is 145 ms) (Card et al., 1983), meaning that tremendous variability in the perceptual and cognitive processor time constants can be tolerated between users and tasks. These robustness analysis results were also confirmed by a simulation model, which showed stable behavior for all parameter variations in the range allowed but which could be driven unstable by parameter combinations outside these ranges (Young et al., 2004). This individual-differences analysis serves to illustrate what could be done when an augmented cognition system is coupled with a systematic control theoretic approach.

4.2.3 Robust Controller Synthesis

The control theory methods reviewed above can facilitate the design of high-performance closed-loop systems, even for systems whose dynamics are only partially known (Packard and Doyle, 1993; Zhou et al., 1996; Young, 2001). This is not achieved by optimizing nominal performance measures as in classical optimal control techniques such as linear quadratic Gaussian/linear quadratic regulator control (Ogata, 2002). Rather, these new approaches attempt to optimize robust performance measures utilizing techniques such as μ-synthesis (Packard and Doyle, 1993). In this way, systems can be designed which are insensitive (or robust) to variations in the system that are naturally occurring but hard to predict a priori (e.g., differences between users). The mathematical machinery underlying such techniques is quite involved, and the associated optimization problems can be nonconvex and even NP-hard. At first sight, such problems may appear to be intractable, and indeed, global minima usually cannot be guaranteed. Nevertheless, practical computation schemes have been developed using approximation schemes such as upper and lower bounds. These schemes are capable of finding very good approximate solutions in a reasonable amount of time. Moreover, there are numerically efficient implementations available of the associated algorithms, usually in convenient Matlab form (Balas et al., 1991), so such designs can readily be carried out (with the appropriate software) in a reasonable time using current computer hardware.

All this adds up to the fact that developers of augmented cognition systems have at their disposal a number of powerful tools for robust controller analysis and synthesis. These theoretical techniques offer the potential of safely optimizing performance in an augmented cognition system while maintaining guaranteed closed-loop stability.

5 APPLICATION DOMAINS

As with any scientific discovery or technical innovation, there are multiple paths upon which technology components will advance. Identifying specific applications at the dawn of a significant advance in our understanding of a field of study is problematic in that the assumptions on which hypothesized applications are based are very likely to be flawed. The assumptions that must be made (and that are likely to be wrong) include:

- Who will take advantage of emergent technology components?
- What components of the emergent technology will ultimately prove most useful and robust?
- When will the various components of the emergent technology be validated sufficiently for incorporation into real-world systems?
- Where will the emergent technology components be found to be most useful?
- Why will the emergent technology components be seen as beneficial?

- How will the emergent technology components be used?

Given the challenges inherent in answering these questions, a good starting point is to identify potential general application domains and then extract examples from these domains in hopes of describing potential uses of emergent technology components. The general application domains likely to be affected most by augmented cognition technology components include operational domains such as truck driving and power plant operation that would benefit from real-time cognitive readiness and assessment capabilities; educational domains, such as a scenario-based training system that can adapt in real time to trainee performance, as assessed by both overt behavior and cognitive state (as captured by OFS gauges); and clinical domains such as medical applications, where for example, the real-time attention processes of children with attention-deficit hyperactivity disorder (ADHD) are monitored and reinforcement interventions for "paying attention" are employed. The value of augmented cognition to the operational, educational, and clinical application domains should be noted; but the specific examples reviewed below should not be assumed to be predictions of actual application areas.

Another way to attempt to glimpse the future of applications is to examine early prototypes that incorporate the underlying science and technology of interest. The practice of postulating potential futures is common when one has demonstrated technology; fortunately, it is not unknown at the beginnings of basic science or the technology development process either. When he was president and CEO of Bellcore, George Heilmeier (1999), insisted that before starting off on any scientific endeavor or technology development project, the following questions be addressed:

- What are you trying to do? Articulate your objectives using absolutely no jargon.
- How is it done today, and what are the limits of current practice?
- What's new in your approach, and why do you think it will be successful?
- Who cares? If you're successful, what difference will it make?
- What are the risks and payoffs?
- How much will it cost? How long will it take?
- What are the midterm and final "exams" to check for success?

Conveniently, augmented cognition as a field of study was born in part through significant investment by DARPA in the Improving Warfighter Information Intake Under Stress Program, where advanced thought toward eventual application was mandatory. One can look to the applications and prototypes generated from this technology initiative to understand potential applications more clearly and as a guide for attempting to postulate potential futures and to answer the two sets of questions presented at the beginning of this section.

DARPA's Improving Warfighter Information Intake Under Stress Program focused initially on challenges and opportunities associated with real-time monitoring of cognitive states with psychophysiological sensors. The ultimate goal was to demonstrate the use of this underlying technology to increase human information processing substantially in four operational military applications. The applications included applying augmented cognition component technologies to a military driving platform, an unmanned vehicle interface platform, command and control platforms, and a dismounted soldier platform. These operationally focused applications provided proof of concepts of benefits of the emergent component technologies; they also provided testbeds for refinement and testing of new methods of incorporating these technologies (the results of many of these efforts are available in the *Proceedings of the First International Conference on Augmented Cognition*, 2005).

There are also many potential nonmilitary operational applications of augmented cognition. The commercial sector is beginning to consider augmented cognition technologies for incorporation into operational systems. For instance, in 2002, IEEE held their seventh Conference on Human Factors and Power Plants, which focuses on new trends, and dedicated an entire session to reviewing state-of-the-art augmented cognition technologies and assessing the maturity level of these technologies. Components of interest included methods to detect and measure a power plant operator's workload, strategies to facilitate multitasking in multimodal environments, and support for intelligent interruption and context recovery. Furthermore, NASA is supporting the development of personal monitoring capabilities to support both intelligent system automation and human performance aids (Prinzel et al., 2000, 2003). The demanding environment of complex missions and associated dynamic information processing demands require NASA to seek enhanced system capabilities and maximized human performance. Leveraging augmented cognition, NASA hopes to increase the ability of a single human to perform numerous tasks while maintaining the strictest margins of safety.

To date, nonoperational domains such as training and clinical domains have been the least affected by augmented cognition technology. However, it is in the training area that the likelihood of successful application development would seem most promising. Joseph Cohn and Amy Kruse of DARPA's Improving Warfighter Information Intake Under Stress Program have suggested that the development of an augmented cognition system that could turn novices rapidly into experts would revolutionize the training community. Such a system would identify a person's current level of expertise and would allow the person to be guided rapidly to heightened levels of sustained performance in a context-independent fashion. Additionally, a person's cognitive performance during training could be periodically or continuously assessed to ensure that their training was proceeding appropriately. Cohn and Kruse's research seeks to develop a multilevel approach to training that capitalizes on being able to observe patterns at both the overt behavioral level and at a deeper structure neuro-imaging level. They point to research results from the neurosciences which indicate that activation of specific brain regions is correlated to novice and expert behaviors, giving evidence for a neural correlate to observed expert behavior. Additionally, they suggest that changes in these structures can be assessed over time to track progression toward "expert" neural activation. An application that could characterize expert performance, identify where in the novice–expert continuum a trainee's performance lies, and then mold the trainee's patterns to more closely reflect an expert's would revolutionize training.

Finally, within the clinical domain one can imagine that by leveraging augmented cognition technology, clinicians would be better able to diagnose, evaluate, and mitigate cognitive and learning decrements. Developments of emergent augmented cognition technology components have been least aligned with such potential clinical applications. However, successes in the operational and training domains will probably accelerate application developments in the clinical domain. Additionally, continued investment in real-time diagnostic tools by the National Institutes of Health will probably create a marketplace for new medical tools and associated applications that leverage augmented cognition technology.

In summary, applications of augmented cognition are in their infancy. Examples of possible applications of underlying technology components can readily be imagined; however, successful instantiation and usefulness of system applications can only be guessed at. There is significant evidence to suggest that the technology components are ready for insertion into mature applications, and that the operational, educational, and clinical domains have capability gaps that call for technology solutions offered by the field of augmented cognition. Thus, although mature augmented cognition science and technology components have now embarked on the path toward application, the only certainty along this journey is that the applications developed will be like no others that have come before them.

6 CONCLUSIONS

Augmented cognition seeks to achieve Licklider's (1960) vision of human–computer symbiosis, where human brains and computing machines are tightly coupled, thereby achieving a partnership that surpasses the information-handling capacity of either entity alone. Such improvement in human–system capability is clearly a worthy goal, whether the context is clinical restoration of function, educational applications, market-based improvements in worker efficiency, or warfighting superiority. Augmented cognition is an attempt to realize a revolutionary paradigm shift in interactive computing, not by optimizing the *friendliness* of connections between human and computer but by achieving a *symbiotic dyad* of silicon- and carbon-based enterprises.

ACKNOWLEDGMENTS

This material is a based on work supported in part by DARPA's Improving Warfighter Information Intake Under Stress Program. Any opinions, findings, and conclusions or recommendations expressed in this material are those of the authors and do not necessarily reflect the views or the endorsement of DARPA.

REFERENCES

Atkinson, R. C., and Shiffrin, R. M. (1968), "Human Memory: A Proposed System and Its Control Processes," in *The Psychology of Learning and Motivation*, K. W. Spence and J. T. Spence, Eds., Academic Press, New York.

Atkinson, R. C., and Shiffrin, R. M. (1971), "The Control of Short Term Memory," *Scientific American*, Vol. 225, No. 2, pp. 82–90.

Averbach, E., and Coriell, A. S. (1961), "Short-Term Memory in Vision," *Bell System Technical Journal*, Vol. 40, pp. 309–328.

Baddeley, A. (1986), *Working Memory*, Oxford University Press, New York.

Baddeley, A. (1990), *Human Memory: Theory and Practice*, Allyn & Bacon, Boston.

Baddeley, A. (2000), "Short-Term and Working Memory," in *The Oxford Handbook of Memory*, E. Tulving and F. Craik, Eds., Oxford University Press, New York, pp. 77–92.

Baddeley, A. (2003), "Working Memory: Looking Back and Looking Forward," *Nature Reviews: Neuroscience*, Vol. 4, pp. 829–839.

Baddeley, A., and Logie, R. (1999), "Working Memory: The Multiple Component Model," in *Models of Working Memory*, A. Miyake and P. Shah, Eds., Cambridge University Press, New York, pp. 28–61.

Balas, G., Doyle, J., Glover, K., Packard, A., and Smith, R. (1991), *The μ Analysis and Synthesis Toolbox*, MathWorks and MUSYN, Natick, MA.

Baron, S., Kruser, D. S., and Messick, B. (1990), *Quantitative Modeling of Human Performance in Complex, Dynamic Systems*, National Academy Press, Washington, DC.

Bliss, J. C., Crane, H. D., Mansfield, P. K., and Townsend, J. T. (1966), "Information Available in Brief Tactile Presentations," *Perception and Psychophysics*, Vol. 1, pp. 273–283.

Boer, E. R. (2001), "Behavioral Entropy as a Measure of Driving Performance," keynote address delivered at Driving Assessment 2001, Aspen, CO, August 14–17; retrieved February 6, 2004, from http://www.ppc.uiowa.edu/driving-assessment/2001/Summaries/Driving%20Assessment%20Papers/44_boer_edwin.pdf.

Byrne, E. A., and Parasuraman, R. (1996), "Psychophysiology and Adaptive Automation," *Biological Psychology*, Vol. 42, pp. 249–268.

Card, S. K., Moran, T. P., and Newell, A. (1983), *The Psychology of the Human–Computer Interaction*, Lawrence Erlbaum Associates, Mahwah, NJ.

Carlesimo, G. A., Perri, R., Turriziani, P., Tomaiuolo, F., and Caltagirone, C. (2001), "Remembering What but Not Where: Independence of Spatial and Visual Working Memory in the Human Brain," *Cortex*, Vol. 37, pp. 519–537.

Casini, L., and Macar, F. (1999), "Multiple Approaches to Investigate the Existence of an Internal Clock Using

Attentional Resources," *Behavioural Processes*, Vol. 45, No. 1–3, pp. 73–85.

Cheatham, P. G., and White, C. T. (1954), "Temporal Numerosity, III: Auditory Perception of Number," *Journal of Experimental Psychology*, Vol. 47, pp. 425–428.

Cowan, N. (2001), "The Magical Number 4 in Short-Term Memory: A Reconsideration of Mental Storage Capacity," *Behavioral and Brain Sciences*, Vol. 24, pp. 87–114.

Darwin, C. J., Turvey, M. T., and Crowder, R. G. (1972), "An Auditory Analogue of the Sperling Partial Report Procedure: Evidence for Brief Auditory Storage," *Cognitive Psychology*, Vol. 3, pp. 255–267.

Dearden, A., Harrison, M., and Wright, P. (2000), "Allocation of Function: Scenarios, Context and the Economics of Effort," *International Journal of Human–Machine Studies*, Vol. 52, pp. 289–318.

Debernard, S., Cathelain, S., Crevits, I., and Poulain, T. (2002), "AMANDA Project: Delegation of Tasks in the Air Traffic Control Domain," presented at the 5th International Conference on the Design of Cooperative Systems, COOP '02, Saint Raphael, France, June 4–7.

Downs, T., and Downs, H. (2004), "Hairy Situations: fNIR Technology and Novasol," presented at the DARPA PI Meeting, Augmented Cognition: Improving Warfighter Information Under Stress, Orlando, FL, January 6–8.

Doyle, J. C., Francis, B. A., and Tannenbaum, A. R. (1992), *Feedback Control Theory*, Macmillan, New York.

Engelbart, D. C. (1963), "A Conceptual Framework for the Augmentation of Man's Intellect," in *Vistas in Information Handling*, P. W. Howerton, Ed., Spartan Books, Washington, DC, pp. 1–29.

ETSI (2002), *Human factors (HF): Guidelines on the Multimodality of Icons, Symbols, and Pictograms*, Report ETSI EG 202 048 v 1.1.1 (2002–08), European Telecommunications Standards Institute, Sophia Antipolis, France.

Farwell, L. A., and Donchin, E. (1988), "Talking off the Top of Your Head: Toward a Mental Prosthesis Utilizing Event-Related Brain Potentials," *Electroencephalography and Clinical Neurophysiology,* Vol. 70, No. 6, pp. 510–523.

Fernandez-Duque, D., and Johnson, M. L. (2002), "Cause and Effect Theories of Attention: The Role of Conceptual Metaphors," *Review of General Psychology*, Vol. 6, No. 2, pp. 153–165.

Feyereisen, P., and Van der Linden, M. (1997), "Immediate Memory for Different Kinds of Gestures in Younger and Older Adults," *Current Psychology of Cognition*, Vol. 16, pp. 519–533.

Flach, J. M. (1999), "Beyond Error: The Language of Coordination and Stability," in *Human Performance and Ergonomics*, P. A. Hancock, Ed., Academic Press, San Diego, CA, pp. 109–128.

Frenchman, K. A. R., Fox, A. M., and Maybery, M. T. (2003), "The Hand Movement Test as a Tool in Neuropsychological Assessment: Interpretation Within a Working Memory Theoretical Framework," *Journal of the International Neuropsychological Society*, Vol. 9, pp. 633–641.

Gerloff, C., Richard, J., Hadley, J., Schulman, A. E., Honda, M., and Hallett, M. (1998), "Functional Coupling and Regional Activation of Human Cortical Motor Areas During Simple, Internally Paced and Externally Paced Finger Movements," *Brain*, Vol. 121, pp. 1513–1531.

Gevins, A., Smith, M. E., Leong, H., McEvoy, L, Whitfield, S., Du, R., and Rush, G. (1998), "Monitoring Working Memory Load During Computer-Based Tasks with EEG Pattern Recognition Methods," *Human Factors*, Vol. 40, pp. 79–91.

Gratton, G., and Fabiani, M. (2001), "The Event-Related Optical Signal: A New Tool for Studying Brain Function," *International Journal of Psychophysiology*, Vol. 42, pp. 109–121.

Gundel, A., and Wilson, G. F. (1992), "Topographical Changes in the Ongoing EEG Related to the Difficulty of Mental Task," *Brain Topography*, Vol. 5, pp. 17–25.

Hankins, T. C., and Wilson, G. F. (1998), "A Comparison of Heart Rate, Eye Activity, EEG and Subjective Measures of Pilot Mental Workload During Flight," *Aviation, Space and Environmental Medicine*, Vol. 69, pp. 360–367.

Harter, M. R. (1967), "Excitability and Cortical Scanning: A Review of Two Hypotheses of Central Intermittency in Perception," *Psychological Bulletin*, Vol. 68, pp. 47–58.

Heilmeier, G. H. (1999), "1999 Woodruff Distinguished Lecture Transcript: From POTS to PANS.com—Transitions in the World of Telecommunications for the Late 20th Century and Beyond," retrieved September 15, 2004, from http://www.me.gatech.edu/me/publicat/99trans.html.

Hoc, J. M. (2001), "Towards a Cognitive Approach to Human–Machine Cooperation in Dynamic Situations," *International Journal of Human–Computer Studies*, Vol. 54, pp. 509–540.

Hock, C., Villringer, K., Muller-Spahn, F., Wenzel, R., Heekeren, H., Schuh-Hofer, S., et al. (1997), "Decrease in Parietal Cerebral Hemoglobin Oxygenation During Performance of a Verbal Fluency Task: Inpatients with Alzheimer's Disease Monitored by Means of Near-Infrared Spectroscopy (NIRS), Correlation with Simultaneous rCBF-PET Measurements," *Brain Research*, Vol. 755, pp. 293–303.

Horvitz, E., Kadie, C. M., Paek, T., and Hovel, D. (2003), "Models of Attention in Computing and Communications: From Principles to Applications," *Communications of the ACM*, Vol. 46, No. 3, pp. 52–59.

Jagacinski, R. J., and Flach, J. M. (2003), *Control Theory for Humans: Quantitative Approaches to Modeling Performance*, Lawrence Erlbaum Associates, Mahwah, NJ.

Jinks, A., and Laing, D. G. (1999), "Temporal Processing Reveals a Mechanism for Limiting the Capacity of Humans to Analyze Odor Mixtures," *Cognitive Brain Research*, Vol. 8, No. 3, pp. 311–325.

Keller, T. A., Cowan, N., and Saults, J. S. (1995), "Can Auditory Memory for Tone Pitch Be Rehearsed?" *Journal of Experimental Psychology: Learning, Memory, and Cognition*, Vol. 21, pp. 635–645.

King, C. C. (1991), "Fractal and Chaotic Dynamics in Nervous Systems," *Progress in Neurobiology*, Vol. 36, pp. 279–308; retrieved February 11, 2003, from http://www.math.auckland.ac.nz/~king/Preprints/pdf/BrChaos.pdf.

Kingsley, S. A., Sriram, S., Pollick, A., Caldwell, J., Pearce, F., and Sing, H. (2002), "Physiological Monitoring with High-Impedance Optical Electrodes (Photrodes™), presented at the 23rd Annual Army Science Conference, Orlando, FL, December 2–5.

Kirsh, D. (1996), "Adapting the Environment Instead of Oneself," *Adaptive Behavior*, Vol. 4, pp. 415–452.

Klapp, S. T., and Netick, A. (1988), "Multiple Resources for Processing and Storage in Short-Term Working Memory," *Human Factors*, Vol. 30, pp. 617–632.

Kramer, A. F. (1991), "Physiological Measures of Mental Workload: A Review of Recent Progress," in *Multiple Task Performance*, D. Damos, Ed., Taylor & Francis, London, pp. 279–328.

Lachman, R., Lachman, J. L., and Butterfield, E. C. (1979), *Cognitive Psychology and Information Processing: An Introduction*, Lawrence Erlbaum Associates, Mahwah, NJ.

Laska, M., and Teubner, P. (1999), "Olfactory Discrimination Ability of Human Subjects for Ten Pairs of Enantiomers," *Chemical Senses*, Vol. 24, No. 2, pp. 161–170.

Licklider, L. C. R. (1960), "Man–Computer Symbiosis," *IRE Transactions on Human Factors in Electronics*, Vol. 1, pp. 4–11; http://www.memex.org/licklider.pdf.

Livermore, A., and Laing, D. G. (1996), "Influence of Training and Experience on the Perception of Multicomponent Odor Mixtures," *Journal of Experimental Psychology: Human Perception and Performance*, Vol. 22, No. 2, pp. 267–277.

Mahrer, P., and Miles, C. (2002), "Recognition Memory for Tactile Sequences," *Memory*, Vol. 10, No. 1, pp. 7–20.

Matlin, M. W. (1998), *Cognition*, 4th ed., Hartcourt Brace, Fort Worth, TX.

Matsuda, M. (1998), "Visual Span of Detection and Recognition of a Kanji Character Embedded in a Horizontal Row of Random Hiragana Characters," *Japanese Psychological Research*, Vol. 40, pp. 125–133.

Mendez, M. F. (2001), "Visuospatial Deficits with Preserved Reading Ability in a Patient with Posterior Cortical Atrophy," *Cortex*, Vol. 37, pp. 539–547.

Miller, G. A. (1956), "The Magical Number Seven Plus or Minus Two: Some Limits on Our Capacity for Processing Information," *Psychological Review*, Vol. 63, pp. 81–97.

Morrison, J. G., Kobus, D., and St. John, M. (2003), "DARPA Augmented Cognition Technology Integration Experiment (TIE)," retrieved September 3, 2004, from http://www.tadmus.spawar.navy.mil/AugCog_Brief.pdf.

Neisser, U. (1967), *Cognitive Psychology*, Appleton-Century-Crofts, New York.

Norman, D. A., and Shallice, T. (1980), "Attention to Action: Willed and Automatic Control of Behaviour," University of California San Diego CHIP Report 99, reprinted in M. Gazzaniga, Ed., *Cognitive Neuroscience: A Reader*, Blackwell, Oxford, 2000.

Ogata, K. (2002), *Modern Control Engineering*, 4th ed., Prentice Hall, Upper Saddle River, NJ.

Packard, A. K., and Doyle, J. C. (1993), "The Complex Structured Singular Value," *Automatica*, Vol. 29, pp. 71–109.

Penney, C. G. (1989), "Modality Effects and the Structure of Short-Term Verbal Memory," *Memory and Cognition*, Vol. 17, pp. 398–422.

Phillips, C. L., and Parr, J. M. (1999), *Signals, Systems, and Transforms*, 2nd ed., Prentice Hall, Upper Saddle River, NJ.

Pickering, S. J. (2001), "Cognitive Approaches to the Fractionation of Visuospatial Working Memory," *Cortex*, Vol. 37, pp. 457–473.

Posner, M. I., and DiGirolamo, G. J. (2000), "Cognitive Neuroscience: Origins and Promise," *Psychological Bulletin*, Vol. 126, pp. 873–889.

Posner, M. I., and Konick, A. F. (1966), "Short-Term Retention of Visual and Kinesthetic Information," *Organizational Behavior and Human Performance*, Vol. 1, pp. 71–88.

Pravitha, R., Sreenivasan, R., and Nampoori, V. P. N. (2003), "Complexity of Brain Dynamics Inferred from the Sample Entropy Analysis of Electroencephalogram," in *Proceedings of the National Conference on Nonlinear Systems and Dynamics*, Karagpur, West Bengal, India, December 28–30.

Prinzel, L. J., Freeman, F. G., Scerbo, M. W., Mikulka, P. J., and Pope, A. T. (2000), "A Closed-Loop System for Examining Psychophysiological Measures for Adaptive Task Allocation," *International Journal of Aviation Psychology*, Vol. 10, pp. 393–410.

Prinzel, L. J., Parasuraman, R., Freeman, F. G., Scerbo, M. W., Mikulka, P. J., and Pope, A. T. (2003), *Three Experiments Examining the Use of Electroencephalogram, Event-Related Potentials, and Heart Rate Variability for Real-Time Human-Centered Adaptive Automation Design*, NASA TP-2003-212442, NASA STI Program Office, Hanover, MD; retrieved February 9, 2004, from http://www.techreports.larc.nasa.gov/ltrs/PDF/2003/tp/NASA-2003-tp212442.pdf.

Rekha, M., Pravitha, R., Nampoori, V. P. N., and Sreenivasan, R. (2003), "Effect of Fatigue on Mental Performance: A Nonlinear Analysis," in *Proceedings of the National Conference on Nonlinear Systems and Dynamics*, Karagpur, West Bengal, India, December 28–30.

Samman, S. N., Stanney, K. M., Dalton, J., Ahmad, A., Bowers, C., and Sims, V. (2004), "Multimodal Interaction: Multi-capacity Processing Beyond 7 + / − 2," in *Proceedings of the 48th Annual Human Factors and Ergonomics Society Meeting*, New Orleans, LA, September 20–24.

Sarno, K. J., and Wickens, C. D. (1995), "The Role of Multiple Resources in Predicting Time-Sharing Efficiency," *International Journal of Aviation Psychology*, Vol. 5, pp. 107–130.

Scerbo, M. W., Freeman, F. G., Mikulka, P. J., Parasuraman, R., Di Nocero, F., and Prinzel, L. J. (2001), *The Efficacy of Psychophysiological Measures for Implementing Adaptive Technology*, NASA TP-2001-211018, NASA Langley Research Center, Hampton, VA.

Schmorrow, D., and McBride, D. (2004), "Introduction," Special Issue on Augmented Cognition, *International Journal of Human–Computer Interaction*, Vol. 17, No. 2, pp. 127–130.

Shallice, T. (1982), "Specific Impairments of Planning," *Philosophical Transactions of the Royal Society London, Series B*, Vol. 298, pp. 199–209.

Smyth, M. M., and Pendleton, L. R. (1990), "Space and Movement in Working Memory," *Quarterly Journal of Experimental Psychology*, Vol. 42, pp. 291–304.

Sperling, G. (1960), "The Information Available in Brief Visual Presentations," *Psychological Monographs*, Vol. 74, No. 11, Whole No. 498.

Sperling, G. (1963), "A Model of Visual Memory Tasks," *Human Factors*, Vol. 5, No. 1, pp. 19–31.

Stanney, K., Samman, S., Reeves, L., Hale, K., Buff, W., Bowers, C., Goldiez, B., Nicholson, D., and Lackey, S. (2004), "A Paradigm Shift in Interactive Computing: Deriving Multimodal Design Principles from Behavioral and Neurological Foundations," *International Journal of Human–Computer Interaction*, Vol. 17, No. 2, pp. 229–257.

Sterman, M. B., and Mann, C. A. (1995), "Concepts and Applications of EEG Analysis in Aviation Performance Evaluation," *Biological Psychology*, Vol. 40, pp. 115–130.

Sullivan, E. V., and Turvey, M. T. (1974), "On the Short-Term Retention of Serial, Tactile Stimuli," *Memory and Cognition*, Vol. 2, pp. 600–606.

Sulzen, J. (2001), "Modality Based Working Memory," School of Education, Stanford University, retrieved February 5, 2003, from http://ldt.stanford.edu/~jsulzen/james-sulzen-portfolio/classes/PSY205/modality-project/paper/modality-expt-paper.PDF.

Takeuchi, Y. (1999), "Change in Blood Volume in Brain During a Simulated Aircraft Landing Task," *Journal of Occupational Health*, Vol. 42, pp. 60–65.

Trejo, L. J., Wheeler, K. R., Jorgensen, C. C., Rosipal, R., Clanton, S. T., Matthews, B., Hibbs, A. D., Matthews, R., and Krupka, M. (2003), "Multimodal Neuroelectric Interface Development," *IEEE Transactions on Neural Systems and Rehabilitation Engineering*, Vol. 11, No. 2, pp. 199–204.

Villringer, A., and Chance, B. (1997), "Non-invasive Optical Spectroscopy and Imaging of Human Brain Function," *Trends in Neuroscience*, Vol. 20, pp. 435–442.

Watkins, D. H., and Watkins, O. C. (1974), "A Tactile Suffix Effect," *Memory and Cognition*, Vol. 2, pp. 176–180.

Wickens, C. D. (1984), *Engineering Psychology and Human Performance*, HarperCollins, New York.

Wickens, C. D. (1992), *Engineering Psychology and Human Performance*, 2nd ed., HarperCollins, New York.

Wickens, C. D. (2002), "Multiple Resources and Performance Prediction," *Theoretical Issues in Ergonomics Scientific*, Vol. 3, No. 2, pp. 159–177.

Wickens, C. D., and Hollands, J. G. (2000), *Engineering Psychology and Human Performance*, 3rd ed., Prentice Hall, Upper Saddle River, NJ.

Wickens, C. D., Mavon, A. S., Parasuraman, R., and McGee, A. P. (1998), *The Future of Air Traffic Control: Human Operators and Automation*, National Academy Press, Washington, DC.

Wiener, N. (1948), *Cybernetics, or Control and Communication in the Animal and the Machine*, MIT Press, Cambridge, MA.

Wilson, G. F. (2002a), "Psychophysiological Test Methods and Procedures," in *Handbook of Human Factors Testing and Evaluation*, 2nd ed., S. G. Charlton and T. G. O'Brien, Eds., Lawrence Erlbaum Associates, Mahwah, NJ, pp. 127–156.

Wilson, G. F. (2002b), "An Analysis of Mental Workload in Pilots During Flight Using Multiple Psychophysiological Measures," *International Journal of Aviation Psychology*, Vol. 12, pp. 3–18.

Wilson, G. F., and Eggemeier, F. T. (1991), "Physiological Measures of Workload in Multitask Environments," in *Multiple-Task Performance*, D. Damos, Ed., Taylor & Francis, London, pp. 329–360.

Wilson, G. F., and Schlegel, R. E., Eds. (2003), *Operator Functional State Assessment*, Final Report, RTO-TR-HFM-104, NATO, Paris.

Woodin, M. E., and Heil, J. (1996), "Skilled Motor Performance and Working Memory in Rowers: Body Patterns and Spatial Positions," *Quarterly Journal of Experimental Psychology*, Vol. 49, pp. 357–378.

Young, P. M. (2001), "Structured Singular Value Approach for Systems with Parametric Uncertainty," *International Journal of Robust and Nonlinear Control*, Vol. 11, pp. 653–680.

Young, P. M., Clegg, B., and Smith, C. A. P. (2004), "Dynamic Models of Augmented Cognition," *International Journal of Human–Computer Interaction*, Vol. 17, No. 2, pp. 259–273.

Zakay, D., and Block, R. A. (2004), "Prospective and Retrospective Duration Judgments: An Executive Control Perspective," *Acta Neurobiolagiae Experimentalis*, Vol. 64, pp. 319–328.

Zhou, K., Doyle, J. C., and Glover, K. (1996), *Robust and Optimal Control*, Prentice Hall, Upper Saddle River, NJ.

PART 9

DESIGN FOR INDIVIDUAL DIFFERENCES

CHAPTER 53

DESIGN FOR PEOPLE WITH FUNCTIONAL LIMITATIONS

Gregg C. Vanderheiden
University of Wisconsin–Madison
Madison, Wisconsin

1 INTRODUCTION

1.1 Not a Special Population but a Continuum—and an End Game for Most of Us

Often, the topic of design for human disability and aging is thought of as a special topic, vertical market, or special application. Although there are special products or assistive technologies designed specifically for use by people with disabilities, they constitute only a small portion of the total number of products that need to be designed to accommodate persons with functional limitations. In addition to the specially designed tools, everyone, including those with disabilities, needs to access a wide range of technologies found in their everyday lives: at home, at school, on the job, and in the community. It is toward the more accessible design of everyday products that this chapter is directed.

Another common misconception is that the population in question is small. Although there are many different types and degrees of disabilities, some of which represent smaller numbers of people, cumulatively those with disabilities represent around 20% of the population. In addition, a majority of people who live beyond age 75 will experience functional limitations due to disability. Approximately 64% of those who live beyond age 75 will have functional limitations, and 41% of them will have severe functional limitations (Kraus et al., 1996). In addition, many of these people experience multiple functional limitations.

1.2 Multiplier Effect

Designing products for personal use constitutes a significant portion of the market. When designing products to be used by families or within industry, the impact is multiplied. As a family unit consists of three or four people, the percentage of families who have people with disabilities is much higher. When you turn to industry, particularly large industries, you find that the percentage of industries that employ people with disabilities is very high. Thus, if you are designing products and systems for use by larger industries, you will find that almost all of the customer base will have employees with disabilities.

1.3 Who Is Included in the Category "Disabled and Elderly Persons"

In considering product design, it is important to note that there is no clear line between people who are categorized as disabled and those who are not. A performance or ability distribution for a given skill or ability is generally a continuous function rather than bimodal with distinctive able and disabled groups. This distribution includes a small number of people who have exceptionally high ability, a larger number with midrange ability, and another longer tail representing those with little or no ability in a particular area. In looking at such a distribution, it is impossible simply to draw a vertical line and separate able-bodied from disabled persons. It is also important to note that each aspect of ability has a separate distribution. Thus, a person who is poor along an ability distribution in one dimension (e.g., vision) may be at the other end of the distribution (i.e., excellent) with regard to another dimension (e.g., hearing or IQ). Thus, people do not fall at the lower or upper end of the distribution overall, but generally fall into different positions, depending on the particular ability being measured.

1.4 The 95th Percentile Illusion

It should be clear that even if elderly and disabled persons are included in the mainstream design process, it is not possible to design all products and devices so that they are usable by all people. There will always be a "tail" of people who are unable to use a given product. To include a sizable portion of the population in the category "those who can use a product with little or no difficulty," the 95th percentile data are often used. The problem is that there are no 95th percentile data for specific designs—there are only data with regard to individual physical or sensory characteristics. Thus, there are 95th percentile data for height, vision, hearing, and so on. As a result, it is not possible to determine when a product can be used by 95% of people. It is only possible to estimate when a product can be used by 95% of the population along any one dimension. Since people in the 5% tail for any one dimension (e.g., height) are usually not the same people as those in the 5% tail along another dimension (e.g., vision) (Kroemer, 1990), it is possible to design a product using 95th percentile data and end up with a product that can be used by far less than 95% of the population.

To illustrate this phenomenon, imagine a minipopulation of 10 people. Ten percent of them (1 of 10) have one short leg, 10% have a visual impairment, 10% have a missing arm, 10% are short, and 10% cannot hear. Let's assume that we design a product that required 90th percentile ability along each of the dimensions of height, vision, leg use, arm use, and hearing. In this instance we would end up with a product that was in fact usable by only 50% of this population. This occurs because although only 10% of this minipopulation is limited in any single dimension, different people fall into the 10% tail for each dimension, and only 50% of the population is within the 90th percentile for all five areas.

In real life, the effect is not quite this dramatic, and its calculation is not as simple. First, the percentage of people with disabilities is less than 10% along any one dimension. Second, there is often overlap where one person would have more than one disability (elderly persons, for example). On the other hand, there is a much wider range of different individual types of disability. In addition, the data from which the 95th percentiles are calculated often exclude persons with disabilities (Kroemer, 1990), making the percentage who could use the design(s) smaller than one would first calculate.

2 DISABILITY IS A CONSEQUENCE, NOT A CONDITION

> Disability is the inability to accommodate to the world as it is currently designed.

The quote above is a paraphrase of Ralph Caplan's "Disability is the inability to accommodate poor design," with an emphasis on the fact that design can be changed, and thus so can disability. In looking at the impact of disability and its relationship to design, it is often useful to use a model such as that shown in Figure 1, which is similar to the World Health Organization model for disability (WHO, 1980). The model shows the relationship that both impairment and design have in creating disabilities. It also shows how circumstance can create similar reduced abilities in anyone, including those without functional impairment. Combined with poor design, these circumstances can also lead to situations where people experience circumstantial disabilities or inabilities to carry out certain tasks. Thus, in addition to generally making products easier for everyone to use, better or more universal design can make a product usable even when people are under stressed conditions. Take, for example, a mother whose young son just fell and cut his head. She makes the mistake of mentioning the doctor, and is now trying to use the phone while holding her screaming, kicking son in one arm to keep him from running off and hiding. Because of the screaming, she can hear very little and has some of the same functional problems as those of a person with a hearing impairment. Because her son is kicking and thrashing, she has poor motor control and has only one hand available. Because he is bleeding profusely, she is also highly distracted and is able to bring only limited cognitive skills and attention to the task at hand.

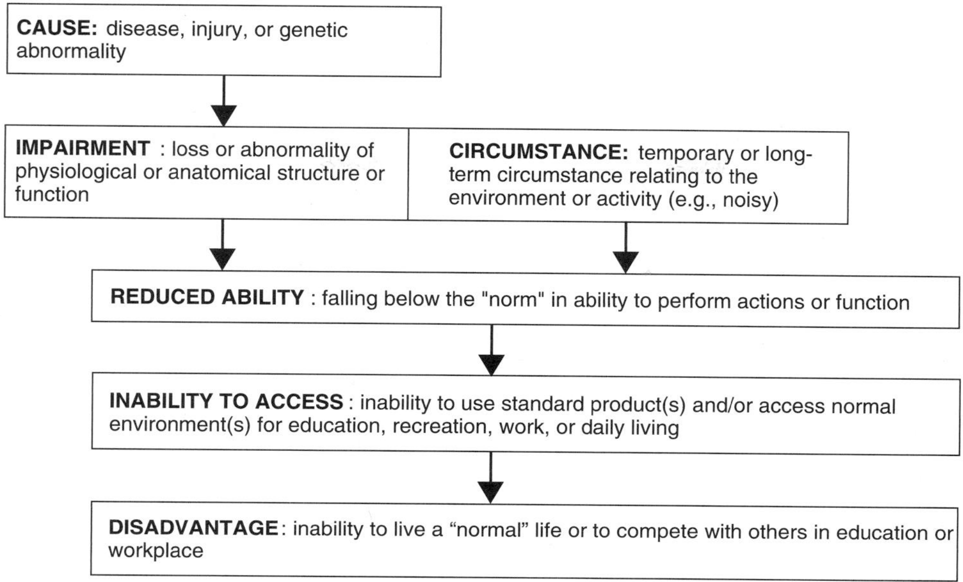

Figure 1 Cause–effect model shows the role that both impairment and design play in disability, as well as the parallel role that conditions or circumstances can play.

2.1 Three Approaches

There are basically three ways to address the problem faced by those who are unable to use the world around them:

1. *Change the person.* This may be accomplished through surgery, education, skill development, skill practice, or by teaching strategies, tricks, or "secrets" for doing things or for doing things more easily.
2. *Provide the person with tools.* This includes prosthetics and orthotics (e.g., eyeglasses, braces, artificial limbs) or assistive technologies [wheelchairs, alternative writing aids, telecommunication devices for the deaf (TDDs/TTs), etc.].
3. *Change the way that the world is designed.* Develop more universal and accessible designs.

Ergonomics is involved in all three of these areas by (1) developing new techniques and strategies that can allow an unaided person to perform better in the workplace, home, or community; (2) developing specialized tools or assistive technologies that can maximize the use of residual skills and abilities and compensate for missing abilities; and, of course, (3) changing the design of the world in general so that it is more usable with a wider range of skills and abilities. The focus of this chapter is on the third approach: designing the world to be more universally usable.

3 UNIVERSAL DESIGN

Universal design is the term that has been given to the practice of designing products or environments which can be used effectively and efficiently by people with a wide range of abilities operating in a wide range of situations. This includes people with no limitations as well as those operating with functional limitations relating to disabilities or simply by circumstance. For example, products developed using universal design principles would be flexible enough to be usable by people with no limitations as well as those:

- Who cannot see the product because they are blind or because their eyes are occupied temporarily (e.g., driving a car).
- Who cannot use their hands well because of aging or a physical disability or because their hands are temporarily full, cold, or gloved.
- Who cannot speak or are in an environment where speech is not practical (library or noisy crowd).
- Who cannot hear the product because they are deaf or because they are in a very noisy environment (e.g., an airplane or a shopping mall at Christmas).
- Who have learning disabilities or who are able to divert only part of their attention to the task at hand.
- Whose primary language is sign language or a foreign language.
- Who are very young or very old (or the like).

An ideal design is one that is attractive, easy to learn, effective, and whose functions can be accessed efficiently and used by everyone across the full range of circumstances that could occur for its intended use. A good design is a commercially practical, mass market design that is usable by and attractive to the maximum possible number and diversity of users, given the best of today's collective knowledge, technologies, and materials.

3.1 Non-Disability-Related Reasons for Universal Design

3.1.1 Benefits for All

A general characteristic of good universal design is that it benefits many more people without disabilities than those with disabilities. This, of course, follows from the design benefiting everyone and the fact that there are more people without disabilities than those with disabilities. The sidewalk curbcut is a prime example of this, as are ramps in general. Although originally designed for users of wheelchairs, they are also used by parents pushing baby carriages, people pulling baggage carriers, bicycle riders, skateboard users, kids on tricycles, and any number of other people. Even people walking can be observed to veer from their path to walk up a curbcut rather than stepping up a curb. In another example, a technique called a "talking fingertip" was used to allow persons who are blind to access and use touchscreen-based kiosks. Once implemented, however, it was found that it was also very useful for persons with low vision as well as those who could not read due to literacy or language problems.

3.1.2 New Insights

Studying the use of products by people with functional limitations can also provide insights into a design that might not otherwise be achieved. For example, it is much easier to determine which elements in a kitchen require greater strength by testing a person who is weak or who has poor grasp than it would be by employing someone with normal or extraordinary strength. Even if such a person were asked which things required more or less effort, the mere fact of having so much strength in reserve would cause the person to use it unconsciously.

3.1.3 Lower-Cost Design

Universal design can also lead to insights that result in lower-cost designs. Although universal designs are usually thought of as being more expensive, this is generally not the case. If one discounts the time it takes to reorient one's thinking and familiarize oneself with the characteristics and constraints of people with functional limitations, the resulting designs can be both easier to use and less expensive.

One example of this is the current design of elevators and their alert bells. In the past, people with disabilities had a problem getting onto elevators when they were arranged in elevator banks. Often, by the time the person using a wheelchair got to the elevator that had opened, the door had closed. New standards were proposed which would require that elevator doors stay open for a longer period of time to allow them to be boarded successfully by wheelchair users. This caused problems, because it increased the number of elevators that needed to be installed in buildings to ensure adequate service to all floors. In some thin, tall buildings, this could result in using up a substantial portion of the building for elevators.

After an injunction was sought to stop the standards, the designers and consumer advocates sat down to study the problem anew. It was determined that the problem was *not* the time it took to board the elevator but the time it took to get in front of the elevator. Since the elevators were computer controlled, and the computers knew where the elevators were going in advance of their arrival, it was quickly determined that lighting the alert light and sounding the bell in advance would allow persons with disabilities to position themselves in front of the elevator door and be able to board as it opened. Testing bore this out, and it was found that people in wheelchairs as well as everyone else could actually begin the boarding process much more quickly and in much less time than the elevators were then staying open. Following the modification in timing of the alert light and bell, designers were able to decrease the time that the doors stayed open, allowing builders either to use fewer elevators or to provide better service to the floors.

Approximately one-third of persons with disabilities who can and would like to work are unemployed. This amounts to approximately 2 million people (Kraus and Stoddard, 1989). Figuring an average annual salary of $20,000, that amounts to $40 billion in lost productivity as well as several billion dollars in lost tax revenues. This is in addition to the large costs in the form of transfer payments made to those who cannot live independently. Total expenditures, public and private, for people with disabilities is estimated at between $200 and $270 billion per year (Figure 2). It was estimated at $355 billion in 1997 (Brandt and Pope, 1997) and to rise to $426 billion in 2002 (Braddock, 2001). What portion of this could be saved if the design of the environment allowed people to live more independently or to stay on their jobs longer?

4 DEMOGRAPHICS

As shown in Figures 3 to 5, the prevalence of the various types of functional limitation (visual, hearing, physical, cognitive) varies significantly as a function of age. In children we see a much higher percentage of mental retardation and language and learning disabilities than of other disabilities (Figure 3). As people age, sensory and physical disabilities become more prevalent (Figure 4). Not evident from these charts is the fact that in older persons we see a much higher incidence of multiple disabilities, including combinations such as hearing and visual impairments, which interfere with many of the adaptive strategies developed for those who have hearing or visual impairments alone. Finally, we can see that the percentage of people who have functional limitations within the population increases sharply as a function of age. In fact, a wide majority of those over the age of 75 (Figure 5) will

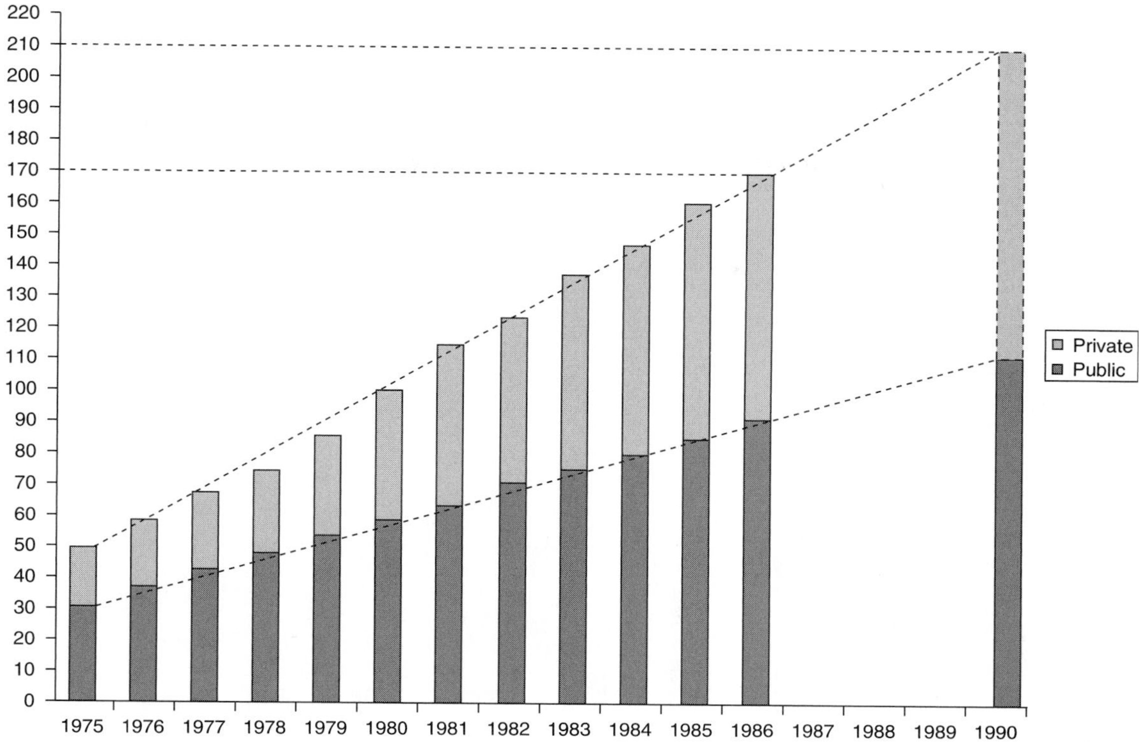

Figure 2 Disability expenditures in the United States in billions of dollars, public and private sectors. (Data from Berkowitz and Greene, 1989.)

have functional limitations, and almost half will have severe functional limitations of one type or another. Thus, over our lifetimes (if we live long enough), most of us will not only benefit from but will require more universal design.

4.1 Characteristics of Users with Functional Limitations

In considering design for people with functional limitations, it is important to examine their abilities both without and with tools and strategies that they normally employ. For example, it is important to look at an amputee's abilities both with and without different types of artificial limbs. These present very different mechanical and manipulative characteristics. Many touch buttons, for example, cannot be activated using different types of artificial arms. Acoustic wave touch screens may be accessible using soft plastic cosmetic arms, but not hooks.

It is also important to consider people without their assistive devices, since many people do not have them, either because they cannot afford them or because they prefer to avoid the stigma (e.g., not wanting to use hearing aids or even very strong glasses). This, of course, adds to the variability and complicates any attempts at comprehensive surveying of needs, abilities, or characteristics. Although a comprehensive survey of the types of assistive technologies used by persons with different disabilities cannot be presented here, a partial listing is provided in Table 1. For a more comprehensive review, readers are referred to Galvin and Scherer (1995) or Cook and Hussey (2002).

5 RESEARCH IN ERGONOMICS AND PEOPLE WITH FUNCTIONAL LIMITATIONS

Most of the research on people with functional limitations is taken not from research on people with "disabilities" but rather from experiments done with "normal" persons operating under stress or adverse conditions (e.g., blinded by smoke, encumbered by a spacesuit). These studies represent much more controlled conditions than those represented by the great diversity of types, combinations, and degrees of disability, but do yield interesting information that can be used by people with disabilities. As noted above, the results of work with persons with disabilities can also be applied to these other environments/locations where people have reduced abilities due to circumstance.

There are major problems in carrying out research in that the variation and range of ability or constraint is so great. Visual impairments, for example, can take a very wide range of forms, and each of these can

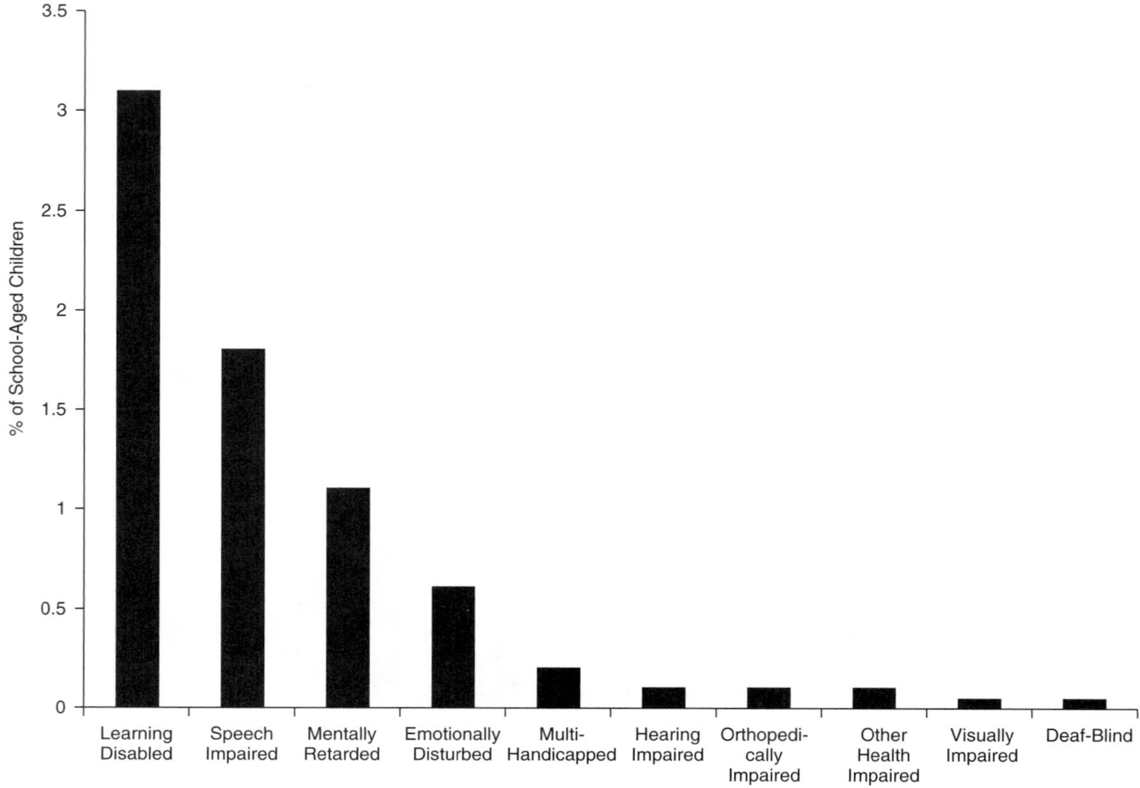

Figure 3 Prevalence of impairments (primary diagnosis) for school-aged children (3–21 years) in the United States. Each child is counted in only one category. (Data from Kraus and Stoddard, 1989, based on reports of the Office of Special Education and Rehabilitation Services, 1988. OSEP state reported date, 1986–1987 school year.)

vary in degree from very mild to severe reduction or total loss. As a result, it is not possible to make blanket statements about these populations. Instead, the research generally tries to characterize the diversity, to quantify numbers of people within particular ranges, and/or to chart the functional characteristics for major groups. For example, people who are experiencing hearing loss due to aging tend to lose hearing at certain frequencies more than others. People with photosensitive epilepsy tend to be much more susceptible to certain frequencies than to others. (These results are reflected in the design guidelines that follow.) The fact that there are no set patterns and that one can find people with just about any type, degree, and combination of disabilities makes developing design guidelines difficult. However, design principles do exist, as well as strategies that can significantly increase the accessibility and usability of products by a much wider range of people.

Note that this chapter refers to persons rather than populations. *Populations* tends to imply somewhat homogeneous groups (although there may be variance within the group). When talking about people with functional limitations, we are talking about something that is a continuum that flows across many dimensions simultaneously. Probably a classic example is people who are older, who may have reductions in visual, hearing, physical, and/or cognitive abilities simultaneously. These abilities will also take different tracks and combinations in different people, and will be progressive over time, making design of environments and products challenging. Clearly, designs must be flexible to accommodate different people, but in these cases they must be flexible to accommodate the same person over time, or sometimes during different periods of the same day.

6 OVERVIEW BY MAJOR DISABILITY GROUP

Although there is a tremendous variety of specific causes, as well as combinations and severity of disabilities, we can most easily relate their basic impact to the use of consumer products by looking at five major categories of impairment: (1) visual impairments, (2) hearing impairments, (3) physical impairments, (4) cognitive/language impairments, and (5) seizure disorders. In addition, we discuss the special case of seizure disorders as well as some of the common situations of multiple impairments.

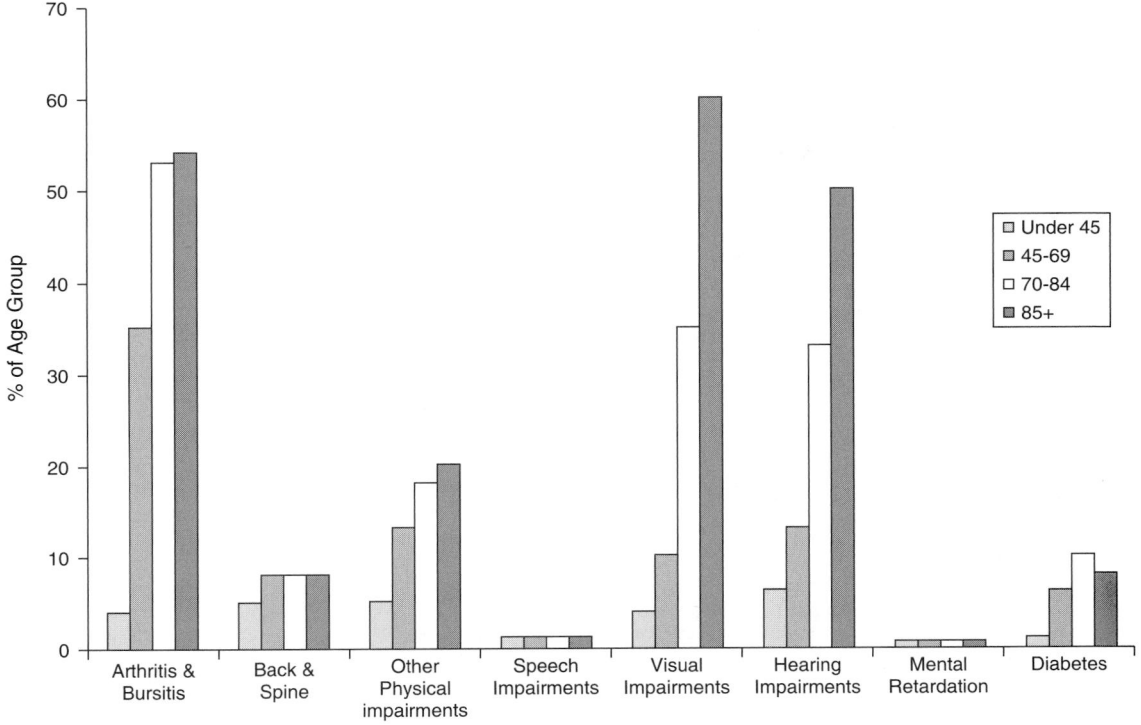

Figure 4 U.S. prevalence of selected impairments within age groups. Data categories are not exclusive. (Data from LaPlante, 1988, based on the National Health Interview Surveys, 1983–1985. Tabulations from public use tapes.)

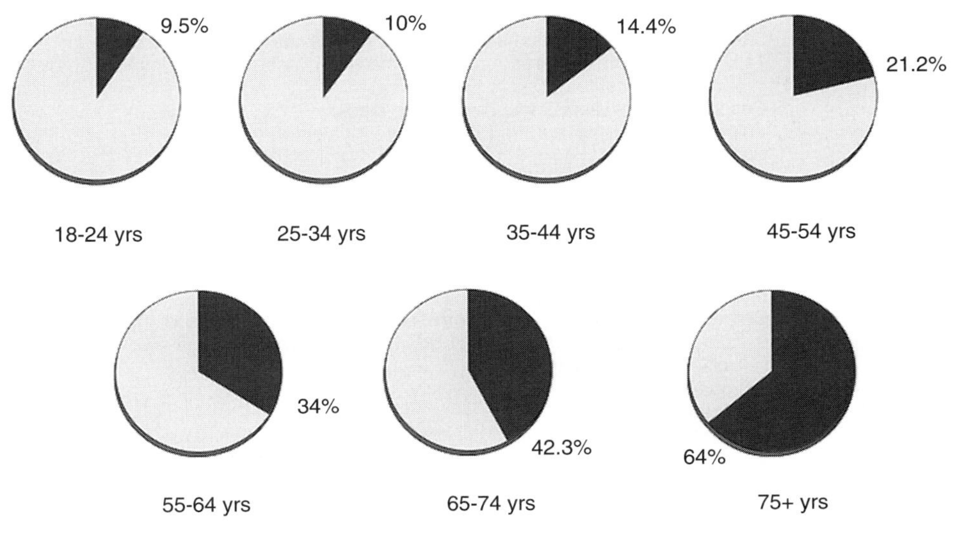

Figure 5 Functional limitations as a function of age. (Data from Kraus and Stoddard, 1989, based on U.S. Bureau of Census Series P-70, No.8, Survey SIPP, 1984.)

Table 1 Partial List of Assistive Technologies

Type of Disability or Functional Limitation	Assistive Technologies and Strategies Used
Hearing impairment	Hearing aids
	Amplifiers
	Assistive listening devices (remote microphone that transmits to a receiver worn by the user)
	Cochlear implants
	Headphones
	Inductive loops (that couple to the hearing aid)
	Direct connection wire (from audio device to the hearing aid)
	T-coils (in hearing aid, to couple to telephone earpiece through induction)
	(See also deafness strategies and technologies)
Deafness	Telecommunication device for the deaf or text telephone (TDD/TT)
	Relay service (a special operator or interpreter with a TDD/TT)
	Video relay service
	Closed captions
	Sign language
	Sign language interpreters
	Lipreading
Low vision	Lights
	Magnifiers
	Telescopes
	Closed-circuit television
	(See also blindness strategies and technologies)
Blindness	Braille (used by approximately 15% of those who are blind)
	Dynamic braille displays (1, 12, 20, or 40 braille cells with pins that move up and down to form the braille characters)
	Tactile symbols and shapes
	Raised line drawings
	Long cane
	Tape recorders
	Synthetic speech
	Synthetic speech or braille-based portable notetakers
	Talking clocks, watches, calculators
	Satellite positioning systems and electronic map databases (emerging)
	Talking signs (infrared broadcasters in the environment that are picked up by small handheld units)
	Descriptive television (audio description track)
	Voice output screen readers (on computer systems)
Physical impairment	Reachers
	Artificial arms, legs, and hands or hooks
	Canes and crutches
	Walkers
	Wheelchairs
	Splints and braces
	Mouthsticks, headsticks
	Communication, writing, and control aids using a wide variety of input techniques, including sip and puff, Morse code, eyegaze, joystick, single-switch scanning, multiswitch encoding, etc.
	Keyguards
	Hand and arm rests
	Universal remote controllers
Speech impairment	Voice amplifiers
	Voice synthesizers
	Artificial larynxes
Cognitive impairment	Memory aids
	Cuing systems
	Calculators
	Text-to-speech aids
	Network-based assistance on demand

6.1 Visual Impairments

Visual impairment represents a continuum from people with very poor vision, to people who can see light but no shapes, to people who have no perception of light at all. However, for general discussion it is useful to think of this population as representing two broad groups: those with low vision and those who are legally blind. Approximately 3.4 million people over age 40 have a visual impairment (including blindness). Blindness alone affects approximately 1 million people over 40 (National Eye Institute, 2002). In the elderly population the percentage of persons with visual impairments is very high.

As established by the American Medical Association in 1934, a person is termed *legally blind* when their visual acuity (sharpness of vision) is 20/200 or worse *after correction*, or when their field of vision is less than 20° in the best eye after correction (Hoover and Bledsoe, 1981). *Low vision* includes problems (after correction) such as dimness of vision, haziness, film over the eye, foggy vision, extreme near- or farsightedness, distortion of vision, spots before the eyes, color distortions, visual field defects, tunnel vision, no peripheral vision, abnormal sensitivity to light or glare, and night blindness. There are approximately 2.4 million people in the United States with severe visual impairments who are not legally blind (National Eye Institute, 2002). Many diseases causing severe visual impairments are common in those who are aging (glaucoma, cataracts, macular degeneration, and diabetic retinopathy). With current demographic trends toward a larger proportion of elderly, the incidence of visual impairments will certainly increase.

6.1.1 Functional Limitations Caused by Visual Impairments

Those who are legally blind may retain some perception of shape and contrast or of light vs. dark (the ability to locate a light source), or they may be totally blind (having no awareness of environmental light). Those with visual impairments have the most difficulty with visual displays and other visual output (e.g., hazard warnings). In addition, there are problems in utilizing controls where labeling or actual operation is dependent on vision (e.g., where eye–hand coordination is required, as with a computer mouse). Written operating instructions and other documentation may be unusable, and there can be difficulties in manipulation (e.g., insertion/placement, assembly).

Because many people with visual impairments still have some visual capability, many of them can read with the assistance of magnifiers, bright lighting, and glare reducers. Many such people with low vision are helped immensely by use of larger lettering, sans serif typefaces, and high-contrast coloring. Those with color blindness may have difficulty differentiating between certain color pairs. This generally does not pose much of a problem except in those instances when information is color coded or where color pairs are chosen that result in poor figure–ground contrast. Key coping strategies for people with more severe visual impairments include the use of braille and large raised lettering. Note, however, that braille is preferred by only 10% of blind people (normally those blind from early in life). Raised lettering must be large and is therefore better for indicating simple labels than for extensive text.

6.2 Hearing Impairments

Hearing impairment is one of the most prevalent chronic disabilities. Approximately 28 million people in the United States have hearing impairments (National Institute on Deafness and Other Communication Disorders, n.d.). Approximately 0.3% of the U.S population have severe to profound impairments (Gallaudet Research Institute, 2002). *Hearing impairment* means any degree and type of auditory disorder; *deafness* means an extreme inability to discriminate conversational speech through the ear. Deaf people, then, are those who cannot use their hearing for communication. People with a lesser degree of hearing impairment are called *hard of hearing* (Schein, 1981). Usually, a person is considered deaf when sound must reach at least 90 decibels (5 to 10 times louder than normal speech) to be heard, and even amplified speech cannot be understood.

Hearing impairments can be found in all age groups, but loss of hearing acuity is part of the natural aging process. Of those aged 55 to 64, 15.4% have hearing impairments, and 29.1% over age 65 have hearing impairments (National Center for Health Statistics, 1994). The number of people with hearing impairments will increase with the increasing age of the population and the increase in the severity of noise exposure.

Hearing impairment may be sensorineural or conductive. *Sensorineural hearing loss* involves damage to the auditory pathways within the central nervous system, beginning with the cochlea and auditory nerve, and including the brain stem and cerebral cortex (this prevents or disrupts interpretation of the auditory signal). *Conductive hearing loss* is damage to the outer or middle ear, which interferes with sound waves reaching the cochlea. Causes of both types of hearing loss include heredity, infections, tumors, accidents, and aging (presbycusis, or "old hearing") (Schein, 1981).

6.2.1 Functional Limitations Caused by Hearing Impairments

The primary difficulty for people with hearing impairment in using standard products is receiving auditory information. This problem can be compensated by presenting auditory information redundantly in visual and/or tactile form. If this is not feasible, an alternative solution to this problem would be to provide a mechanism, such as a jack, which would allow the user to connect alternative output devices. Increasing the volume range and lowering the frequency of products with high-pitched auditory output would be helpful to some less severely impaired persons. (Progressive hearing loss usually occurs in higher frequencies first.)

Although not yet prevalent there is much talk of using voice input on commercial products in the future. This, too, will present a problem for many deaf persons. Whereas many have some residual speech, which they work to maintain, those who are deaf from birth or a very early age often are also nonspeaking or have speech that cannot be recognized using current voice input technology. Thus, alternatives to voice input will be necessary to these people to access products with voice input.

Familiar coping strategies for hearing-impaired people include the use of hearing aids, sign language, lipreading, and TDDs (telecommunication devices for the deaf). Some hearing aids are equipped with a T-coil as well, which provides direct inductive coupling with a second coil (such as in a telephone receiver) to reduce ambient noise. Some other commercial products could make use of this capability. ASL (American Sign Language) is commonly used by people who are deaf. It should be noted, however, that this is a completely different language from English. Thus, deaf people who primarily use ASL may understand English only as a second language, and may therefore not be as proficient with English as are native speakers. Finally, telecommunication devices for the deaf (TDDs) are becoming more common in households and businesses as a means for deaf and hard of hearing people to communicate over the phone. TDDs have always used the Baudot code, but newer ones receive both Baudot and ASCII.

6.3 Physical Impairments

6.3.1 Functional Limitations Caused by Physical Impairments

Problems faced by people with physical impairments include poor muscle control, weakness and fatigue, difficulty walking, talking, seeing, speaking, sensing, or grasping (due to pain or weakness), difficulty reaching things, and difficulty doing complex or compound manipulations (push and turn). People with spinal cord injuries may be unable to use their limbs and may use mouthsticks for most manipulations. Twisting motions may be difficult or impossible for people with many types of physical disabilities (including cerebral palsy, spinal cord injury, arthritis, multiple sclerosis, muscular dystrophy, etc.).

Some people with severe physical disabilities may not be able to operate even well-designed products directly. They usually must rely on assistive devices that take advantage of their specific abilities and on their ability to use these assistive devices with standard products. Commonly used assistive devices include mobility aids (e.g., crutches, wheelchairs), manipulation aids (e.g., prosthetics, orthotics, reachers), communication aids (e.g., single switch–based artificial voice), and computer–device interface aids (e.g., eyegaze-operated keyboard).

6.3.2 Nature and Causes of Physical Impairments

Neuromuscular impairments include:

- Paralysis (total lack of muscular control in part or most of the body)
- Weakness (paresis; lack of muscle strength, nerve enervation, or pain)
- Interference with control, via spasticity (where muscles are tense and contracted), ataxia (problems in accuracy of motor programming and coordination), and athetosis (extra, involuntary, uncontrolled, and purposeless motion)

Skeletal impairments include joint movement limitations (either mechanical or due to pain), small limbs, missing limbs, or abnormal trunk size. Some major causes of these impairments are described next.

Arthritis Arthritis is defined as pain in joints, usually reducing range of motion and causing weakness. Rheumatoid arthritis is a chronic syndrome. Osteoarthritis is a degenerative joint disease. An estimated 43 million people in the United States suffer from rheumatic disease. That number is expected to increase to 60 million by the year 2020 (National Institute on Arthritis and Musculoskeletal and Skin Diseases, 2002).

Cerebral Palsy (CP) Cerebral palsy is defined as damage to the motor areas of the brain prior to brain maturity (most cases of CP occur before, during, or shortly following birth). There are more than 760,000 people (children and adults) with CP in the United States, and 8000 infants are diagnosed with CP annually (United Cerebral Palsy, 2001). CP is a type of injury, not a disease (although it can be caused by a disease) and does not get worse over time; it is also not "curable." Some causes of cerebral palsy are high temperature, lack of oxygen, and injury to the head. The most common types are (1) spastic, where a person moves stiffly and with difficulty; (2) ataxic, characterized by a disturbed sense of balance and depth perception; and (3) athetoid, characterized by involuntary, uncontrolled motion. Most cases are combinations of the three types.

Spinal Cord Injury Spinal cord injury can result in paralysis or paresis (weakening). The extent of paralysis or paresis and the parts of the body affected are determined by how high or low on the spine the damage occurs and the type of damage to the cord. Quadriplegia involves all four limbs and is caused by injury to the cervical (upper) region of the spine; paraplegia involves only the lower extremities and occurs where injury was below the level of the first thoracic vertebra (mid-lower back). There are 222,000 to 285,000 people with spinal cord injuries in the United States, with 11,000 new cases projected annually. Of all spinal cord injuries, 42.6% result in a diagnosis of paraplegia; 56.4%, of tetraplegia. Car accidents are the most frequent cause (50.4%), followed by falls and jumps (23.8%) and gunshot wounds or other violent acts (11.2%) (Spinal Cord Injury Information Network, 2004).

Head Injury (Cerebral Trauma) The term *head injury* is used to describe a wide array of injuries, including concussion, brain stem injury, closed head injury, cerebral hemorrhage, depressed skull fracture, foreign object (e.g., bullet), anoxia, and postoperative infections. Like spinal cord injuries, head injury and stroke often result in paralysis and paresis, but there can be a variety of other effects as well. Annually, about 1.5 million Americans suffer head injuries that include cerebral trauma (Kraus and McArthur, 1996). However, many of these are not disabled permanently or severely.

Stroke (Cerebral Vascular Accident; CVA) The three main causes of stroke are thrombosis (blood clot in a blood vessel blocks blood flow past that point), hemorrhage (resulting in bleeding into the brain tissue; associated with high blood pressure or rupture of an aneurism), and embolism (a large clot breaks off and blocks an artery). The response of brain tissue to injury is similar whether the injury results from direct trauma (as above) or from stroke. In either case, function in the area of the brain affected either stops altogether or is impaired (Anderson, 1981).

Loss of Limbs or Digits (Amputation or Congenital) This may be due to trauma (e.g., explosions, mangling in a machine, severance, burns) or surgery (e.g., due to cancer, peripheral arterial disease, diabetes). Usually, prosthetics are worn, although these do not result in full return of function. There are approximately 1.2 million persons living with the absence of a limb in the United States. New cases of limb loss occur at the rate of approximately 62 per 10,000 persons (National Limb Loss Information Center, 2002).

Parkinson's Disease This is a progressive disease of older adults characterized by muscle rigidity, slowness of movements, and a unique type of tremor. There is no actual paralysis. The usual age of onset is 50 to 70, and the disease is relatively common. It is estimated that 1.5 million Americans have Parkinson's disease, with 60,000 new cases diagnosed each year (National Parkinson Foundation, n.d.).

Multiple Sclerosis (MS) Multiple sclerosis is defined as a progressive disease of the central nervous system characterized by the destruction of the insulating material covering nerve fibers. The problems these patients experience include poor muscle control; weakness and fatigue; difficulty walking, talking, seeing, sensing, or grasping objects; and intolerance of heat. Onset is between the ages of 10 and 40. This is one of the most common neurological diseases, affecting as many as 400,000 people in the United States alone (National Multiple Sclerosis Society, 2004).

ALS (Lou Gehrig's Disease) ALS (amyotrophic lateral sclerosis) is a fatal degenerative disease of the central nervous system characterized by slowly progressive paralysis of the voluntary muscles. The major symptom is progressive muscle weakness involving the limbs, trunk, breathing muscles, throat, and tongue, leading to partial paralysis and severe speech difficulties. This is not a rare disease (five cases per 100,000). It strikes mostly those between ages 40 and 70, and men three times as often as women. Duration from onset to death is about 1 to 10 years, with 50% of those diagnosed living 3 or more years (Robert Packard Center for ALS Research, 2004).

Muscular Dystrophy (MD) Muscular dystrophy is a group of hereditary diseases causing progressive muscular weakness; loss of muscular control; contractions; and difficulty in walking, breathing, reaching, and use of hands involving strength.

6.4 Cognitive/Language Impairments

6.4.1 Functional Limitations Caused by Cognitive/Language Impairments

The type of cognitive impairment can vary widely, from severe retardation to inability to remember, to the absence or impairment of specific cognitive functions (most particularly, language). Therefore, the types of functional limitations that can result also vary widely.

Cognitive impairments are varied but may be categorized as memory, perception, problem solving, and conceptualizing disabilities. Memory problems include difficulty getting information from short-term, long-term, and remote memory. This includes difficulty recognizing and retrieving information. Perception problems include difficulty taking in, attending to, and discriminating sensory information. Difficulties in problem solving include recognizing the problem; identifying, choosing, and implementing solutions; and evaluation of outcome. Conceptual difficulties can include problems in sequencing, generalizing previously learned information, categorizing, cause and effect, abstract concepts, comprehension, and skill development. Language impairments can cause difficulty in comprehension and/or expression of written and/or spoken language.

There are very few assistive devices for people with cognitive impairments. Simple cuing aids or memory aids are sometimes used. As a rule, these people benefit from use of simple displays; low language loading; use of patterns; simple, obvious sequences; and cued sequences.

6.4.2 Types and Causes of Cognitive/Language Impairments

Mental Retardation A person is considered mentally retarded if he or she has an IQ below 70 (average IQ is 100) and if they have difficulty functioning independently. An estimated 6.2 to 7.5 million Americans have mental retardation (Batshaw, 1997). For most, the cause is unknown, although infections, Down syndrome, premature birth, birth trauma, or lack of oxygen may all cause retardation. Those with mild retardation have an IQ between 55 and 69 and achieve the fourth- to seventh-grade levels in education. They usually function well in the community and hold semiskilled and unskilled jobs. People with moderate retardation have an IQ between 40 and 54 and are trainable in educational skills and independence. They can learn to recognize symbols and simple words, achieving approximately a second-grade level. They often live in group homes and work in sheltered workshops.

Language and Learning Disabilities Aphasia, an impairment in the ability to interpret or formulate language symbols as a result of brain damage, is frequently caused by left cerebral vascular accident (stroke) or head injury. Specific learning disabilities are chronic conditions of presumed neurological origin that interfere selectively with the development, integration and/or demonstration of verbal and/or nonverbal abilities. Aside from their specific learning disability, many people with learning disabilities are highly intelligent. Approximately 8% of children age 6 to 11 years have learning disabilities (Pastor and Reuben, 2002).

Age-Related Disease Alzheimer's disease is a degenerative disease that leads to progressive intellectual decline, confusion, and disorientation. Dementia is a brain disease that results in the progressive loss of mental functions, often beginning with memory, learning, attention, and judgment deficits. The underlying cause is obstruction of blood flow to the brain. Some kinds of dementia are curable, whereas others are not.

6.5 Seizure Disorders

A number of injuries or conditions can result in seizure disorders. Epilepsy is a chronic neurological disorder. It is reported that approximately 300,000 people have a first seizure each year, and between 181,000 new cases of epilepsy are diagnosed each year (Epilepsy Foundation, 2003). A seizure consists of an explosive discharge of nervous tissue, which often starts in one area of the brain and spreads through the circuits of the brain like an electrical storm. The seizure discharge activates the circuits in which it is involved, and the function of these circuits will determine the clinical pattern of the seizure. Except at those times when this electrical storm is sweeping through it, the brain is working perfectly well in a person with epilepsy. Seizures can vary from momentary loss of attention to grand mal seizures, which result in the severe loss of motor control and awareness. Seizures can be triggered in people with photosensitive epilepsy by rapidly flashing lights, particularly in the range 10 to 25 Hz (Harding and Jeavons, 1994).

6.6 Multiple Impairments

It is common to find that whatever caused a single type of impairment also caused others. This is particularly true where disease or trauma is severe or in the case of impairments caused by aging. Deaf–blindness is one commonly identified combination. Most of these people are neither profoundly deaf nor legally blind, but are both visual and hearing impaired to the extent that strategies for deafness or blindness alone won't work. People with developmental disabilities may have a combination of mental and physical impairments that result in substantial functional limitations in three or more areas of major life activity. Diabetes, which can cause blindness, also often causes loss of sensation in the fingers. This makes braille or raised lettering impossible to read. Cerebral palsy is often accompanied by visual impairments, by hearing and language disorders, or by cognitive impairments.

7 DESIGN GUIDELINES

To facilitate use by product design teams, this section is organized functionally rather than by disability area. Functional categories are as follows:

- *Output/displays*: includes all means of presenting information to the user
- *Input/controls*: includes keyboards and all other means of communicating to the product
- *Manipulations*: includes all actions that must be performed directly by a person in concert with the product or for routine maintenance (e.g., inserting disk, loading tape, changing ink cartridge)
- *Documentation*: primarily operating instructions
- *Safety*: includes alarms and protection from harm

Each guideline is phrased as an objective, followed by a statement of the problem(s) faced by people with disabilities. The problem statement is accompanied by more specific examples. Next, design options are presented to provide some suggestions as to how the objective could be achieved. The guidelines are stated as generically as possible. Therefore, all, some, or none of the design options and ideas presented may apply in the case of any specific product. The recommended approach is to implement those options that together go the longest way toward achieving the objective of the guideline for your product. It is understood that this is not an ideal world, so it may currently be too expensive to implement all those ideas that would best achieve the objective. It is also anticipated that there will be other ways of meeting accessibility objectives than those discussed here, and such discoveries are encouraged.

7.1 Output/Displays

Maximize the number of people who can/will ...

O-1 hear auditory output clearly enough.
O-2 not miss important information if they cannot hear.
O-3 have a line of sight to visual output and reach printed output.
O-4 see visual output clearly enough.
O-5 not miss important information if they cannot see.
O-6 understand the output (visual, auditory, other).
O-7 view the output display without triggering a seizure.

O-1 Maximize the Number of People Who Can Hear Auditory Output Clearly Enough

Problem Information presented auditorially (e.g., synthesized speech, cuing and warning beeps, buzzers, tones, machine noises) may not be heard effectively.

Example: People who have mildly to moderately impaired hearing may not be able to discern sounds that are too low in volume. People who have mild hearing impairments may be unable to turn the volume up sufficiently in some environments (e.g., libraries, where others would be disturbed, or in noisy environments, where even the highest volume is insufficient). People with moderate hearing impairments are often unable to hear sounds in higher frequencies (above 2000 Hz) (Hunt, 1970). People with hearing aids may have difficulty separating background noise from the desired auditory information. People with cognitive impairments may easily be distracted by too much background noise. Auditory information that is short or not repeated or repeatable (e.g., a short beep or voice message) may be missed or not understood.

Note: Severely hearing impaired (and deaf) people cannot use audio output at all. See section O-2 for how to address this problem.

Design Options and Ideas to Consider:

- Providing a volume adjustment, preferably using a visual volume indicator. Sound should be intelligible (undistorted) throughout the volume range
- Making audio output (or volume range if adjustable) as loud as practical
- Using sounds that have strong middle- to low-frequency components (500 to 3000 Hz)
- Providing alerts and other auditory warnings that include at least two strong middle- to low-frequency components, with recommended ranges of 300 to 750 Hz for one component and 500 to 3000 Hz for the other (Berkowitz and Casali, 1990)
- Providing a headphone jack to enable a person with impaired hearing to listen at high volume without disturbing others, to enable such a person to isolate themselves effectively from background noise, and to facilitate use of neck loops and special amplifiers (Figures 6 and 7; see additional information below)
- Providing a separate volume control for the headphone jack so that people without hearing impairments can listen as well (at standard listening levels)
- When a headphone jack is not possible:
    - Placing the sound source on the front of the device and away from loud mechanisms would facilitate hearing
    - Locating the speaker on the front of the device would also facilitate use of a small microphone and amplifier to pick up and present the information (via speaker, neck loop, or vibrator)

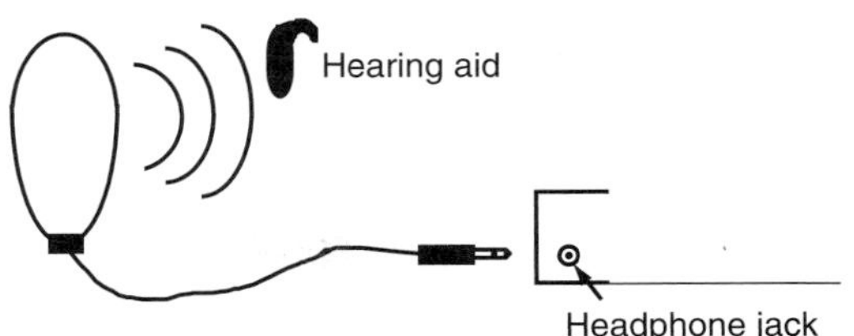

Figure 6 A neck ring or ear loop can be plugged into a headphone jack on an audio source and provide direct inductive coupling between the audio source and a special induction coil on a person's hearing aid. This cuts out background noise that would be picked up by the hearing aid's microphone and provides clearer reception of the audio signal.

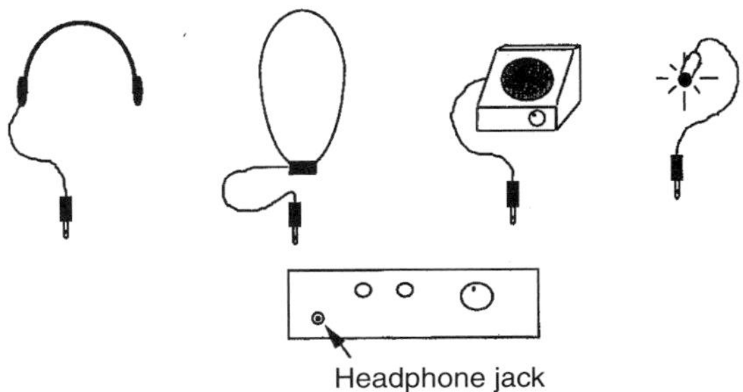

Figure 7 A headphone jack permits the connection of headphones, neck/ear loops, amplifiers, or sound indication lights.

- Facilitating direct use of the telecoil in hearing aids by incorporating a built-in inductive loop in a product (e.g., in telephone receiver's earpiece)
- Reducing the amount of unmeaningful sound produced by the product (i.e., background noise)
- Repeating the message
- Having a warning beep precede the message to allow people to attend
- Using a male (lower) voice for speech synthesis
- Providing control of speech rate or speed
- Providing methods for pausing, rewinding, and repeating speech
- Presenting auditory information continuously or periodically until the desired message is confirmed or acted upon (spoken messages could automatically repeat or have a mechanism for the user to ask for them to be repeated)

O-2 Maximize the Number of People Who Will Not Miss Important Information If They Cannot Hear

Problem Audio output (e.g., synthesized speech, cuing and warning beeps, buzzers, tones) may not be

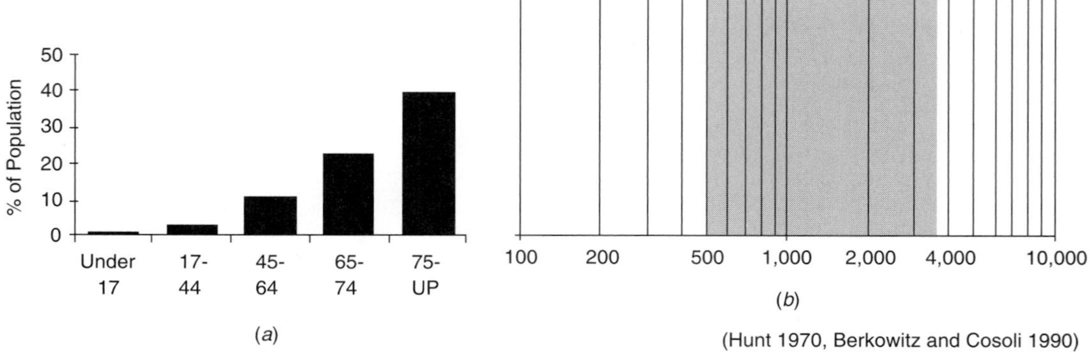

Figure 8 (*a*) Hearing loss as a function of age; (*b*) recommended frequency for alerting devices. [(*a*) From Schow et al., 1978; (*b*) based on Hunt, 1970, and Berkowitz, 1990.]

heard at all or may be insufficient for communicating information effectively.

Example: People who are severely hearing impaired or deaf may not hear audio output, even at high volume and low frequencies (Figure 8). People with language or cognitive impairments may not be able to respond to information given only in auditory form. (This may also be true if the language used is not a person's primary language.) People who are deaf–blind may not hear audio output. People with *standard* hearing must sometimes use products in environments where the sound must be turned off (e.g., libraries) or where the environment is too noisy to hear any sound output reliably.

Design Options and Ideas to Consider:

- Providing all important auditory information in visual form as well (or having it available; includes any speech output as well as auditory cues and warnings)
- Providing a tactile indication of auditory information (e.g., vibrating alarms)
- Facilitating the connection or use of tactile aids
- Providing an optional remote audio/visual or tactile indicator

O-3 Maximize the Number of People Who Will Have Line of Sight to Visual Output and Reach Printed Output

Problem Visual displays or printouts may be unreadable due to their placement.

Example: People who are in a wheelchair or who are extremely short may be unable to read displayed information due to the physical placement or angle of the display screen. People in wheelchairs, with missing or paralyzed arms, or with ability to move limited by cerebral palsy or disease (e.g., severe arthritis, MS, ALS, muscular dystrophy) may be unable to reach printed output (e.g., receipts produced by an automatic teller machine), due to printer placement.

Design Options and Ideas to Consider:

- Locating display screens so they are readable from varying heights, including a wheelchair (see Section I-1 for specific anthropomorphic data and Section O-4 regarding image height)
- Locating printed output within easy reach of those who are in wheelchairs
- Facilitating manipulation of printouts by reaching and grasping aids
- Providing redundant audio output in addition to visual display if the visual display cannot be made physically accessible to a person in a wheelchair (see Section O-5)

O-4 Maximize the Number of People Who Can See Visual Output Clearly Enough

Problem Visual output (e.g., information presented on screens, paper printouts, cuing and warning lights or dials) may not be seen effectively.

Example: People who are visually impaired may not be able to see output that is too small. Those who are visually impaired may have difficulty discerning complex typefaces or graphics. People who are color blind may not be able to differentiate between certain color pairs. People with poor vision have more difficulty seeing letters or pictures against a background of similar hue or intensity (low contrast). People with visual impairments may be much more sensitive to glare (Figure 9). Those who have visual impairments may not be able to see detail in low lighting. Some people with severe lack of head control (e.g., cerebral palsy) may not be able to maintain continuous eye contact with a display and therefore may miss portions of dynamic (i.e., moving, changing) displays.

See Section O-5 for guidelines for people who cannot use visual output at all and Section O-6 for problems in understanding displayed output.

Design Options and Ideas to Consider:

- Making letters and symbols on visual output as large as possible or practical

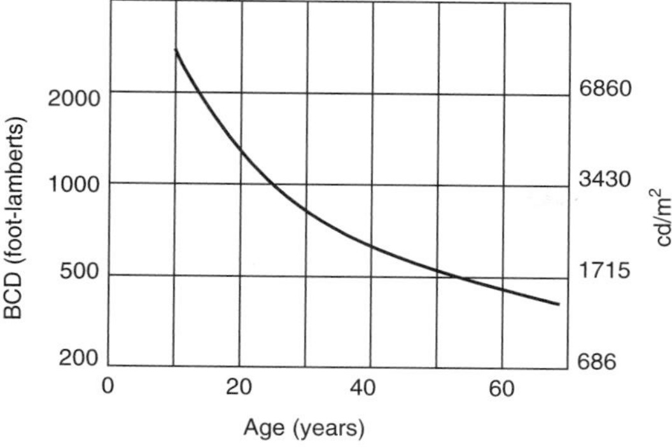

Figure 9 Ability to tolerate glare decreases sharply as a function of age. Data are based on a 1° glare source size and a background luminance of 1.6 foot-lamberts. (From Bennett, 1977a.)

- Using upper- and lowercase type to maximize readability
- Making sure that letter spacing, the space between lines (leading), and the distance between messages are sufficient that the letters and messages stand out distinctly from each other
- Providing adjustable display image size
- Providing a video jack for attaching larger-image displays or utilizing special assistive devices (e.g., electronic magnifiers; see additional information below)
- Using high contrast between text or graphics and background
- Keeping letters and symbols on visual output as simple as possible; using sans serif typefaces for nontext lettering (e.g., labels, dials, displays) (see Section D-1)
- Using only black and white or using colors that vary in intensity so that the color itself carries no information
- Providing adjustable color selection (hue and/or intensity)
- Replacing or supplementing color coding with different shape or relative position coding
- Providing contrast and/or brightness adjustment
- Minimizing glare (e.g., by employing filtering devices on display screens and/or avoiding shiny surfaces and finishes)
- Providing the best possible lighting for displays or areas containing instrumentation (good even illumination without hot spots and brighter than background illumination)
- Providing adjustable speed for dynamic displays (so they can be slowed down for those who lack motor control)
- Avoiding use of light blue color to convey important information (harder for aging eyes to perceive; see below)
- Increasing contrast on LCD displays by allowing user to adjust viewing angle

O-5 Maximize the Number of People Who Will Not Miss Important Information If They Cannot See

Problem Visual output (e.g., information presented on screens, paper printouts, cuing and warning lights, and dials) may not be seen at all by some users.

Example: People who are severely visually impaired or blind may not be able to see visual output, even when magnified and clarified (as recommended in Section O-4). People who cannot read may be unable to use visually presented text. People who are deaf and blind may only be able to perceive tactile output. People who do not have any visual impairment may miss warnings, cues, or other information if it is presented only in visual form while their attention is diverted.

Design Options and Ideas to Consider:

- Providing all important visual information (redundantly) in audio and/or tactile form.
- Accompanying visual cues and warnings by a sound, one component of which is of middle to low frequency (500 to 3000 Hz) (see Section O-1)
- Making information that is displayed visually (both text and graphics) also available electronically at an external connection point (standard or special port) to facilitate the use of special assistive devices (e.g., voice synthesizers, braille printers), preferably in an industry or company standard format (see Figure 17)

O-6 Maximize the Number of People Who Can Understand the Output (Visual, Auditory, Other)

Problem Visual and/or auditory output may be confusing or difficult to understand.

Example: Some people with specific learning disabilities or with reduced or impaired cognitive abilities are easily confused by complex screen layouts (e.g., multiple "windows" of information), have difficulty understanding complex or sophisticated verbal (printed or spoken) output, or have a short attention span, and are easily distracted when reviewing a screen display. For many people who are deaf as well as many other U.S. citizens, English is a second language and not well understood.

Design Options and Ideas to Consider:

- Using simple screen layouts, or providing the user with the option to look at one thing at a time
- Shortening menus
- Hiding (or layering) seldom used commands or information
- Keeping language as simple as possible
- Accompanying words with pictures or icons (Note, however, that the use of graphics may

present more difficulty for people who are blind. See Section O-5.)

- Using Arabic rather than Roman numerals (e.g., use 1, 2, 3 instead of I, II, III)
- Using attention-attracting (e.g., underlining, boldfacing) and grouping techniques (e.g., putting a box around things or color blocking)
- Highlighting key information
- Putting most important information at the beginning of written text (but not spoken announcements, where it might be missed)
- Providing an attention-getting sound or words before audio presentation
- Keeping auditory presentations short
- If providing menu choices, always state the choice first and the action second (e.g., *use for deposits, press 1*; do not use *press 1 for deposits*)
- Having auto-repeat or a means to repeat auditory messages
- Presenting information in as many (redundant) forms as possible/practical (i.e., visual, audio, and tactile) or providing as many display options as possible
- Providing digital readouts for product-generated numbers where the numeric or precise value is important
- Providing dials or bar graphs where qualitative information is more important (e.g., half full, full) (See Sections I-4 and I-6)

O-7 Maximize the Number of People Who Can View the Output Display without Triggering a Seizure

Problem People with seizure sensitivities (e.g., epilepsy) may be affected by screen cursor or display update frequencies, increasing the chance of a seizure while working on or near a display screen.

Design Options and Ideas to Consider:

- Avoiding screen refresh or update flicker or flashing frequencies which are most likely to trigger seizure activity (Figure 10 provides a general overview of the frequencies most likely to trigger a seizure.)
- Avoiding flashing where there are more than three flashes within any 1-second period where the combined area of the flashing would occupy more than 25% of the central vision (central 10°) (Harding and Jeavons, 1994)

For the general flash threshold, a flash is defined as a pair of opposing changes in luminance (i.e., an increase in luminance followed by a decrease, or a decrease followed by an increase) of 20 candelas per rectangle meter (cd/m^2) or more and where the screen luminance of the darker image is below 160 cd/m^2. For the red flash threshold, a flash is defined as any pair of opposing transitions to or from a saturated red at any luminance level.

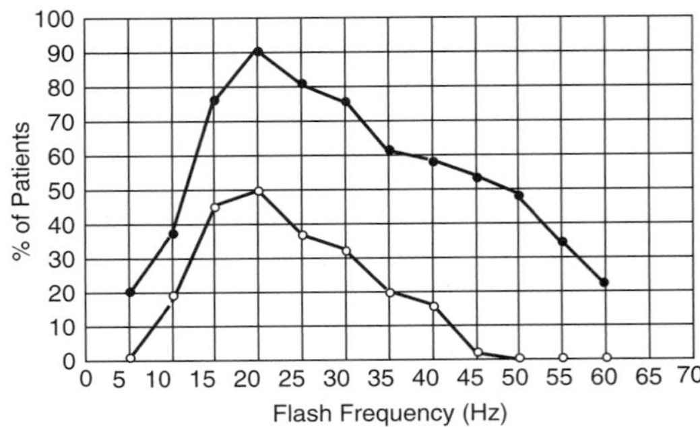

Figure 10 Percentage of photosensitive patients in whom a photoconvulsive response was elicited by a 2-second train of flashes with eyes open and closed. As can be seen, the greatest sensitivity is at 20 Hz, with a steep drop-off at higher and lower frequencies. (From Jeavons and Harding, 1975.)

Note: Video waveform luminance is not a direct measure of display screen brightness. Not all display devices have the same gamma characteristic, but a display with a gamma value of 2.2 may be assumed for the purpose of determining electrical measurements made to check compliance with these guidelines. For the purpose of measurements made to check compliance with these guidelines, pictures are assumed to be displayed in accordance with the home viewing environment described in Recommendation ITU-R BT.500, in which peak white corresponds to a screen illumination of 200 cd/m^2. Specifications are based on "ITC Guidance Note for Licensees on Flashing Images and Regular Patterns in Television" (revised and reissued in July 2001).

7.2 Input/Controls

Maximize the number of people who can ...

I-1 reach the controls.
I-2 find the individual controls/keys if they cannot see them.
I-3 read the labels on the controls/keys.
I-4 determine the status or setting of the controls if they cannot see them.
I-5 physically operate controls and other input mechanisms.
I-6 understand how to operate controls and other input mechanisms.
I-7 connect special alternative input devices.

I-1 Maximize the Number of People Who Can Reach the Controls

Problem Controls, keyboards, and so on may be unreachable or unusable.

Example: People who use a wheelchair, who are very weak, or who are extremely short may be unable to reach some controls, keypads, and so on, well enough to use them. People with poor motor control may be able to reach the controls but may find the knobs, buttons, and so on, too small or close together to operate accurately. People with severe weakness may be able to reach the controls but may find the act of reaching or holding position in order to manipulate the controls too tiring.

Design Options and Ideas to Consider:

- Locating controls, keyboards, and so on, so they are within easy reach of those who are in wheelchairs or have limited reach
- Locating controls so that the user can reach and use them with the least change in body position
- Locating controls that must be used constantly in the closest positions possible and where there is wrist or arm support
- Using keys with smaller radiuses of curvature (with sharper edges)
- Providing a (redundant) speech recognition input option
- Offering remote controls (wired, wireless, or bus operated)

I-2 Maximize the Number of People Who Can Find the Individual Controls/Keys If They Cannot See Them

Problem People with visual impairments may be unable to find controls.

Example: People who are severely visually impaired may be unable to locate controls tactilely because they are on a flat membrane or glass panel (e.g., calculators, microwave ovens) or because they are placed too close together or in a complicated arrangement. People who have diabetes may have both visual impairments and failing sensation in fingertips, making it difficult to locate controls that have only subtle tactile cues.

Design Options and Ideas to Consider:

- Varying the size of controls (also texture or shape), with the most important being larger to facilitate their location and identification
- Providing controls whose shapes are associated with their functions
- Providing sufficient space between controls for easy tactile location and identification as well as easier labeling (large print or braille)
- Using keys with small radius of curvature on edges (sharper edges)
- Locating controls adjacent to what they control
- Making layout of controls logical and easy to understand, to facilitate tactile identification (e.g., stove burner controls in corresponding locations to actual burners)
- Providing a raised lip or ridge around flat (membrane or glass) panel buttons
- Providing a (redundant) speech recognition input option

See Figures 11 and 12.

I-3 Maximize the Number of People Who Can Read the Labels on the Controls/Keys

Problem Labels on controls, keys, and so on, are difficult or impossible to see, due to their size, color, or location.

Example: People with low vision may have difficulty identifying controls or keys on a keyboard because the label lettering is too small and/or because the contrast between letters/graphics and background is poor. People with color blindness may have difficulty distinguishing controls that are color-coded or use certain pairs of colors for labels and background. People with physical impairments may have difficulty reading labels on the sides or backs of objects. People who are blind may not be able to see printed labels at all.

Design Options and Ideas to Consider:

- Making lettering used for labels as large as possible or practical
- Making sure that–letter spacing,–the space between lines (leading), and the distance between labels are sufficient that the letters and labels stand out distinctly from each other
- Placing important labels or instructions on the front or an easily accessible side of large or stationary devices, where they can be read from wheelchairs
- Using sans serif fonts for nontext lettering (e.g., labels, dials)
- Using high contrast between letters/graphics and background
- Providing sufficient illumination of controls and instructions
- Supplementing color coding with use of different button/key shape or letter/graphic labels
- Providing color choices for color-coded buttons
- Providing tactile labels
- Avoiding use of blues, greens, and violets to encode information (since the yellowing of the cornea can cause confusions between some shades of these colors)
- Using easily interchangeable keycaps to allow replacement with special or optional keycaps
- Arranging controls in groupings that facilitate tactile identification (e.g., using small groups of keys that are separated from the other keys, or placing frequently used keys near tactile landmarks such as along the edges of a keyboard)
- Using established layouts for keyboards (e.g., typewriter, adding machine, phone)
- Using voice output to "speak" the names of keys or buttons as they are pressed (This capability would need to be turned on and off as needed.)
- If a flat membrane panel cannot be avoided, providing a stick-on tactile overlay that provides tactile demarcation of the key locations and functions

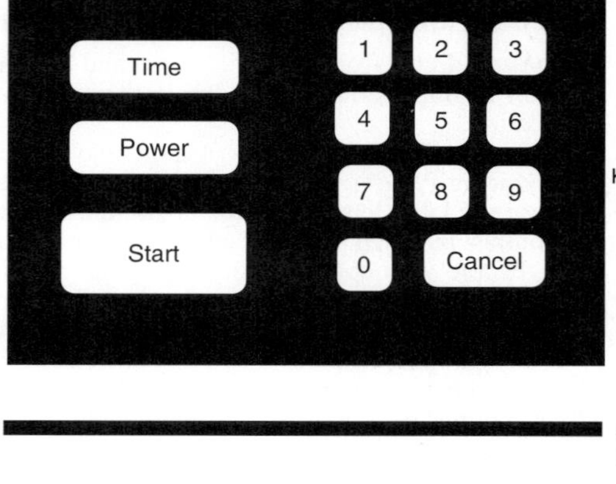

Keypad on which edge views below are based.

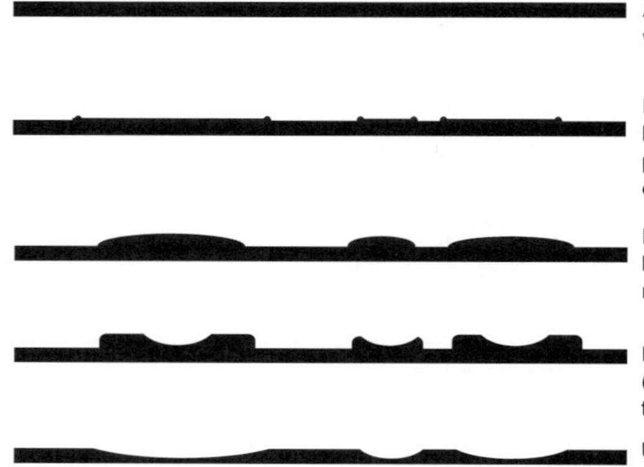

A flat membrane or glass keypad provides no tactile indication as to where the keys are, even if one memorizes the arrangement.

Providing a slight raised lip around the keys allows their location to be discerned easily by touch. The ridge around the key also helps prevents slipping off of the key when using a mouthstick, reacher, etc., to press the keys.

Raised bumps are tactilely discernable but it is harder to press the key without slipping off, particlary if one is using a mouthstick, reacher, or other manipulative aid.

Raised keys with indents provide better feed back then just indents (as in example above) especially if the keys have different shapes or textures that correspond to their function.

Using indentations or hollows on the touchpad most of the advantage of ridges but is easier to clean. Hollows can be the same size as the key orof a consistent small circular size centered on the keys. Shallow edges such as those on the left button are harder to sense with fingers than the sharper curve of the middle button.

Figure 11 The shape of a key or button can have a significant effect on people's ability to (and operate) it accurately locate.

See Sections O-4 and O-6 for related guidelines for output/displays.

I-4 Maximize the Number of People Who Can Determine the Status or Setting of the Controls If They Cannot See Them

Problem: Determination of control status or setting may depend solely on vision.

Example: People with visual impairments may be unable to see a control setting or on/off indicator (e.g., where a dial is set, whether a button is pushed in, whether a light is on, flashing or off, or what a numeric setting on a visual display reads).

Design Options and Ideas to Consider:

- Providing multisensory indication of the separate divisions, positions, and levels of the controls (e.g., use of detents or clicks to indicate center position or increments, raised lines)
- Using absolute reference controls (e.g., pointers) rather than relative controls (e.g., pushbuttons to increase/decrease, or round, unmarked knobs)
- Using moving pointers with stationary scales
- Providing multisensory indications of control status (e.g., in addition to a status light indicating "on," or providing an intermittent audible tone and/or tactilely discernable vibration)
- Using direct keypad input
- Providing speech output to read or confirm the setting

See Sections O-3, O-4, and O-5 for design options covering visual displays.

See Figures 13 and 14.

I-5 Maximize the Number of People Who Can Physically Operate Controls and Other Input Mechanisms

Problem Controls (or other input mechanisms) may be difficult or impossible for those with physical disabilities to operate effectively.

Example: People with severe weakness may be unable to operate controls at all, or may have great difficulty performing constant, uninterrupted input. People with only one arm or without arms (but utilizing

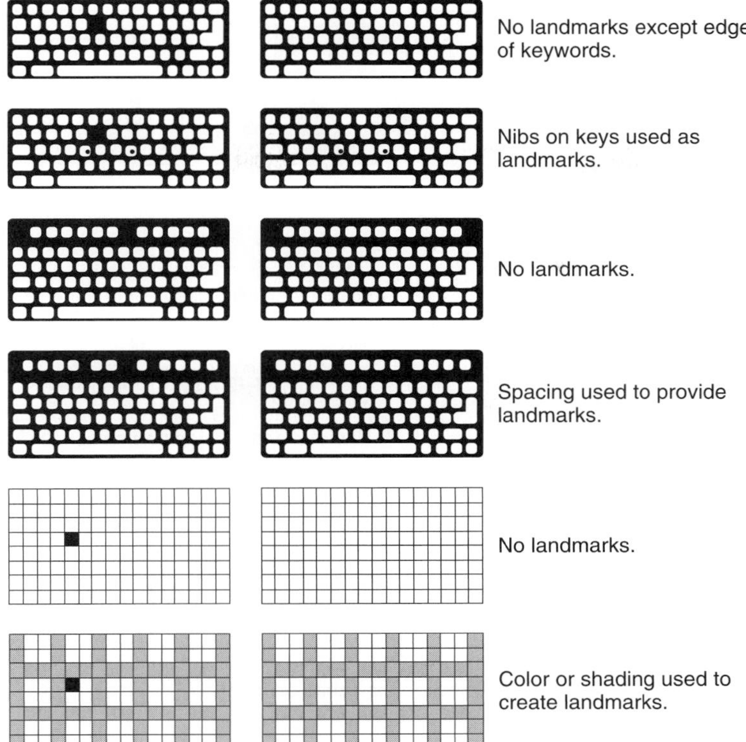
No landmarks except edges of keywords.
Nibs on keys used as landmarks.
No landmarks.
Spacing used to provide landmarks.
No landmarks.
Color or shading used to create landmarks.

Figure 12 Quick self-demonstration of the impact of landmarks on key finding by people who cannot see labels on a key due to blindness or very low vision. *Instructions:* For each keyboard, visually locate the key on the right-hand keyboard that corresponds to the key marked on the left. Note the increase in speed and accuracy when landmarks (nibs or breaks in the key patterns) are provided.

assistive devices such as headsticks or mouthsticks) may not be able to activate multiple controls or keys at the same time. People with artificial hands or reaching aids may have difficulty grasping small knobs or operating knobs or switches which require much force. People with poor coordination or impaired muscular control have slower or irregular reaction times, making time-dependent input unreliable. People lacking fine movement control may be unable to operate controls requiring accuracy (e.g., a mouse or joystick) or twisting or complex motions. People with limited movement control (including tremor, incoordination, or those using headsticks or mouthsticks) can inadvertently bump extra controls on their way to a nearby desired control.

Design Options and Ideas to Consider:

- Minimizing the need for strength by minimizing the force required as much as possible or by providing adjustable force on mechanical controls
- If stiff resistance is provided to prevent accidental activation it could drop off after activation. Other non-strength related safety interlocks could also be considered.
- Spacing the controls to provide a guard space between controls, thus also leaving room for adaptations such as attaching levers to hard-to-turn knobs or room to replace knobs with larger, easier-to-turn knobs or cranks
- Minimizing or providing alternatives to performing constant, uninterrupted actions (e.g., button locks or push on–push off buttons to eliminate the need to press some buttons continuously)
- Where simultaneous actions are required (e.g., pressing shift or control key while typing another key), providing an alternative method to achieve a result that does not require simultaneous actions (e.g., sequential option as in StickyKeys; see below)
- Providing for operation with the left or right hand
- Using concave and/or nonslip buttons, which are easier to use with mouthsticks or headsticks;

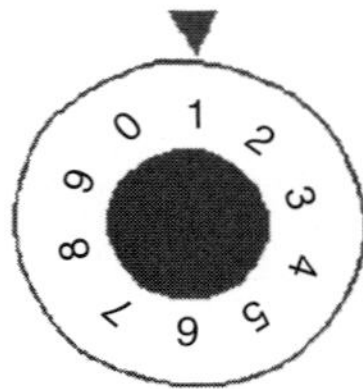

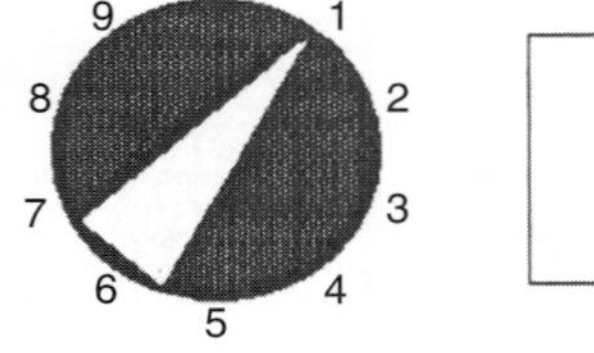

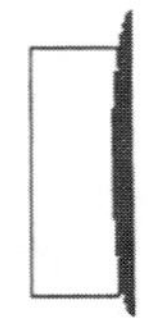

Side view

- No nonvisual indication of setting. If vision blurred, one cannot tell setting.
- Difficult to put large print or braille labels on knob.
- (Also harder to grasp and requires twisting motion.)

- Highly visible raised pointer.
- Instant tactile indication of orientation allows setting to be read even if user is blind.
- Easy to put larger-print or braille labels on back panel.
- Use of detents (large and small) can facilitate internumeral settings.
- Black base disk provides high contrast and helps in control location/orientation on panel.
- (Design is also easy to grasp and can be turned by pushing the point around — no twisting if the knob turns freely enough.)

FOR EXAMPLE: What are the settings of the knobs below?

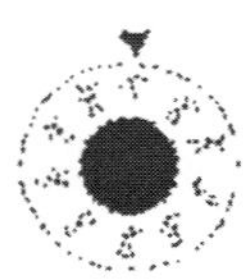

Figure 13 The design of a knob can greatly affect its usability by people with low vision or blindness.

POOR:
Round smooth knob, no tactile orientation cue.

BETTER:
Has tactile orientation cue but user has to feel around to find it.

BETTER:
Orientation cue is less ambiguous. However, the user must still feel the ends to be sure which is the pointer end.

BEST:
Has tactile orientation cue which is unambiguous and can be felt immediately upon grasping knob.

Figure 14 Knob design can have substantial effect on usability by people who are blind.

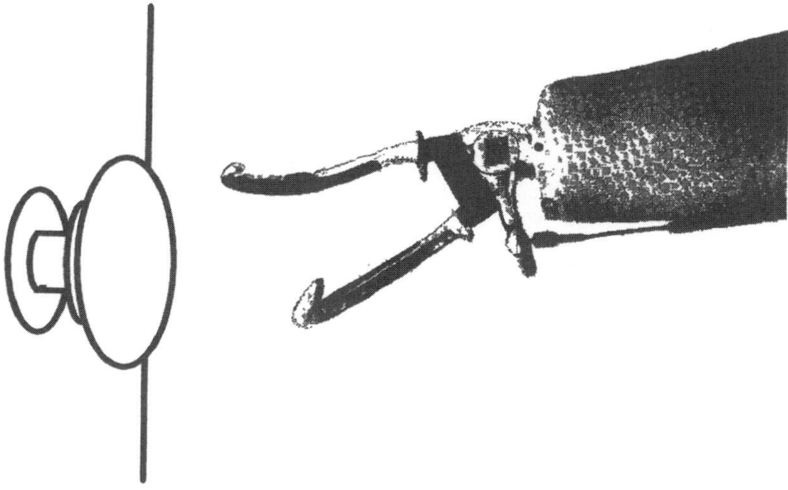

Figure 15 People with arthritis, artificial hands, hooks, disabilities that restrict wrist rotation, or disabilities that cause weakness have difficulty with knobs or controls that require twisting. Also difficult for people with loss of upper body strength, range of motion, and flexibility, as is common with elderly persons. These really should be avoided in bathrooms, where soap and water create slippery environment. (Lever handles, now required in many building codes, facilitate access.)

on flat membrane keypads, providing a ridge around buttons

- If product requires a quick response (i.e., a reaction time of less than 5 seconds, or release of a key or button in less than 1.5 seconds), allowing the user to adjust the time interval or to have a non-time-dependent alternative input method
- If product requires fine motor control, providing an alternative mechanism for achieving the same objectives that does not require fine motor control (e.g., on a mouse-based computer, provide a way to achieve mouse actions from the keyboard)
- Avoiding controls that require twisting or complex motions (e.g., push and turn) (*Note:* There are rotating knobs that do not require twisting)
- Spacing, positioning, and sizing controls to allow manipulation by people with poor motor control or arthritis
- Where many keys must be located in close proximity, providing an option that delays the acceptance of input for a preset, adjustable amount of time (i.e., the key must be held down for the preset amount of time before it is accepted), thus helping some users who would otherwise bump and activate keys on the way to pressing their desired key (*Note:* This option must be difficult to invoke accidentally and be provided on request only, as it can have the effect of making the keyboard appear to be "broken" to naive users)
- Making keyboards adjustable from horizontal (0 to 15 inches is standard) (Grandjean, 1987; Mueller, 1990)
- Providing an optional keyguard or keyguard mounting for keyboards
- Providing optional (redundant) voice control
- Providing textured controls (avoid slippery surfaces/controls)
- Providing means to stabilize body part used to operate the device

See Figure 15.

I-6 Maximize the Number of People Who Can Understand How to Operate Controls and Other Input Mechanisms

Problem The layout, labeling, or method of operating controls and other input mechanisms can be confusing or unclear.

Example: People with reduced or impaired cognitive function may be confused by complex, cluttered control layouts, with many and/or many types of controls; may have difficulty making selections from large sets; may have trouble remembering sequences (see also section M-5); may be confused by dual-purpose controls, or may not relate appropriately to control settings indicated solely by notches/dots or numbers. People with reduced or impaired cognitive function, language impairments, illiteracy, or for whom English is a second language may have difficulty relying solely on textual labels, especially where abbreviations are used, and sometimes have difficulty making associations between label and control, or may have trouble with timed responses involving text.

Design Options and Ideas to Consider:

- *Reducing the number of controls*
  - Limiting the number of choices where practical
  - Using layering of controls where only the most frequent or necessary controls or commands are visible unless you open a door or ask for additional levels of commands (e.g., hiding less frequently used controls, or at least grouping the most frequently used controls together and placing them prominently)
  - Where possible, making products automatic or self-adjusting, thus removing the need for controls (e.g., TV fine tuning and horizontal hold)
- *Simplifying the controls*
  - Minimizing dual-purpose controls
  - Using direct selection techniques where practical (selection techniques where the person need only make a single, simple, non-time-dependent movement to select)
  - Using visual/graphic indications for settings along with, or instead of, numbers or notches/dots (i.e., substitute concrete indications for abstract indications)
  - Reducing or eliminating lag/response times
  - Minimizing ambiguity
  - Providing a busy indicator or, preferably, a progress indicator when a product is busy and cannot take further input or when there is a delay before the requested action is taken
  - Integrating, grouping, and otherwise arranging controls to indicate function or sequence of operation
- *Making labels easy to understand*
  - Placing the label on or, less preferably, immediately adjacent to, the control (does not apply to scales, which should not be on the controls but on the background)
  - Placing a line around the button and label (or from button to label) to show association (should be kept away from any lettering especially if it is raised to avoid tactile confusion with the lettering)
  - Using simple concise language
  - Using redundant labeling (e.g., color code plus label)
  - Avoiding abbreviations in labeling (e.g., PrtScr, FF, C)
  - Leaving space around keys (makes it easier to match labels to keys and easier to add special labels)
  - Using multisensory presentation of feedback information
  - Providing labels on interinterval marks (see additional information below)
- *Reducing, eliminating, or providing cues for sequences*
  - Allowing use of programmable function keys or using a "default" mode
  - Using preprogrammed buttons for common sequences
  - Allowing entry of a short code to program a longer sequence (e.g., new service with *TV Guide* and VCR programming; see below)
  - Simplifying required sequences, limiting the number of steps
  - Arranging controls to indicate sequence of operation
  - Adding memory cues or simple operating instructions on the device where possible
  - Cueing required sequences of action
  - Providing an easy exit that returns the user to the original starting point from any point in the program/sequence (exit should be prominent and clear)
- *Building on users' experiences (make the similarity obvious)*
  - Laying out controls to follow function
  - Making operation of controls follow movement stereotypes (see below)
  - Using common layouts or patterns for controls
  - Using common color-coding conventions in addition to textual or graphic labeling
  - Standardizing by using the same shape/color/icon/label for the same function or action (within and across products and manufacturers)

I-7 Maximize the Number of People Who Can Connect Special Alternative Input Devices

Problem Standard controls (or other input mechanisms) cannot be made accessible for all of those with severe impairments.

Example: People with paralysis of their arms, severe weakness, tremor, or other severe physical impairments may not be able to use controls or input mechanisms that require the use of hands. Blind persons cannot use input devices that require constant eye–hand coordination and visual feedback (e.g., a standard computer mouse, trackball, or touch screen without special accommodation).

Design Options and Ideas to Consider:

- Providing a standard infrared remote control (e.g., VCRs, TV sets, stereos sets)
- Providing alternative means for eye–hand coordination input devices (e.g., mice, trackballs, relative joysticks) or allowing special devices to be substituted by the user, which will achieve as many of the functions as possible
- Providing tactile or auditory cues to allow direct use of touch pads or techniques to allow touch

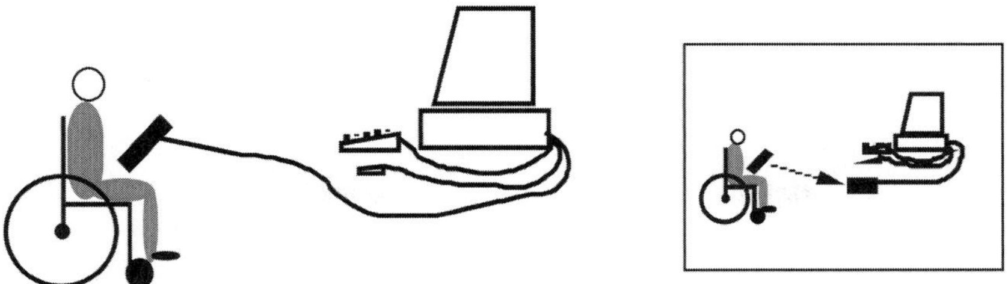

Figure 16 By building a USB option into a computer or other device it is possible for users who cannot use the standard keyboard and mouse to create "authentic" keystrokes and mouse movements by sending signals into the USB port. This would allow these people to access the computer and all of its software.

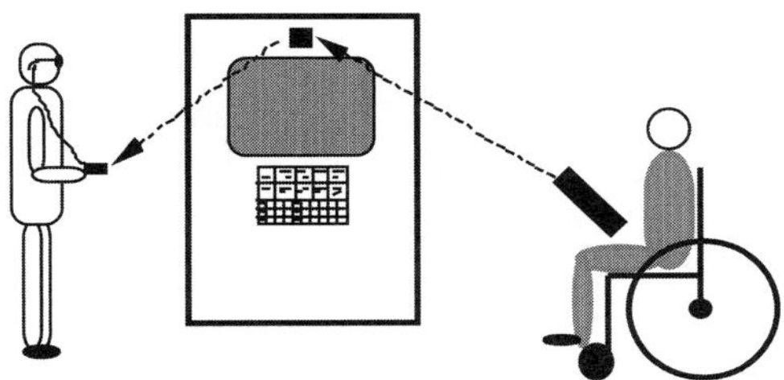

People who are blind or unable to read the displayed information could use an assistive device and have information presented in auditory or tactile (braille) form and to provide input to the terminal.

People who are unable to operate the standard controls could use an assistive device to control the terminal using an input system they can control (eyegaze, sip and puff, single-switch scanning, etc.).

- **Public information terminal**
- **Restaurant and hotel guide at airport**
- **Automated teller machine**
- **Electronic building directory**
- **Point-of-sale terminal**
- **Information or sales kiosk at airport or mall or other public information/atransaction terminal**

Figure 17 A wireless bidirectional link could provide a low-cost environment and vandal-resistant mechanism for connecting assistive devices to information, control, and transaction terminals.

screens to function alternatively as auditory or tactile touch pads

- Providing a standard connection point (connector or infrared link) for special alternative input devices (e.g., eyegaze keyboards, communication aids)

See Figures 16 and 17.

7.3 Manipulations

This includes all actions that must be directly performed by a person in concert with the device or for routine maintenance (e.g., inserting disk, loading tape, changing ink cartridge)

Maximize the number of people who can ...

M-1 physically insert and/or remove objects as required to operate a device.
M-2 physically handle and/or open the product.
M-3 remove, replace, or reposition often-used detachable parts.
M-4 understand how to carry out the manipulations necessary to use the product.

M-1 Maximize the Number of People Who Can Physically Insert and/or Remove Objects as Required to Operate a Device

Problem Insertion and/or removal of objects required to operate some devices (e.g., diskettes, compact disks, cassette tapes, credit cards, keys, coins, currency) may be physically impossible. In addition, damage to the object or device can occur from unsuccessful attempts.

Example: People using mouthsticks or other assistive devices may have difficulty grasping an object and manipulating it as required to insert or retrieve it from the device. People with poor motor control may be unable to place a semifragile object accurately into the device and retrieve without damage (e.g., bending of floppy disk or credit card). People with severe weakness may have difficulty reaching the slot (or positioning the object) for insertion or removal. People who are blind may be unable to determine proper orientation or alignment for insertion (e.g., hotel key card may be held upside down, backward, or at the wrong angle).

Design Options and Ideas to Consider:

- *Facilitating orientation and insertion*
  - Ensuring that objects can be inserted (and removed) with minimal user reach and dexterity
  - Providing a simple funneling system or other self-guidance/orienting mechanism that will position the object properly for insertion
  - Allowing receptacles to be repositioned or reangled to be more reachable
  - Whenever possible, allowing the object to be inserted in several ways (e.g., a six-sided wrench can be positioned on a mating bolt six different ways; two-sided keys can be inserted upside down)
  - Providing visual contrast between the insertion point and the rest of the device (making a more obvious "target")
  - Clearly marking the proper orientation both visually and tactilely
- *Facilitating removal*
  - Providing ample ejection distance to facilitate easy gripping and removal (ejection distance as large as possible while retaining a stable ejection; see Figure 18)
  - Using pushbutton ejection or automatic (motorized) ejection mechanism
- *Facilitating handling*
  - Making objects to be inserted rugged and able to take rough handling
  - Using objects with high friction surfaces for ease in grasping

M-2 Maximize the Number of People Who Can Physically Handle and/or Open the Product

Problem Handles, doorknobs, drawers, trays, and so on, may be impossible for some people to grasp or open.

Example: People using mouthsticks or other assistive devices may be unable to grasp handles, doorknobs, and so on, in order to open or operate the product, and may find it impossible to open doors or drawers without handles (e.g., those using recessed "lips," or those utilizing only side pressure to open).

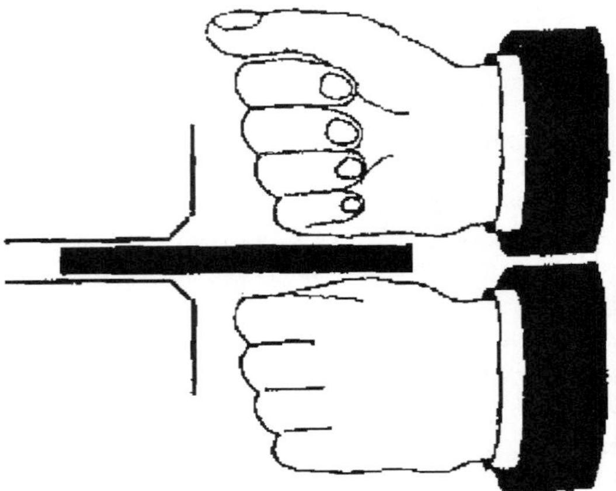

Figure 18 Mechanisms that eject items at least 1 inch and preferably 2 inches facilitate grasping the item with tools, reachers, teeth, or fists for those who cannot use their hands or fingers effectively.

People with limited arm and hand movement (due to arthritis or cerebral palsy, for example) may have problems grasping handles that are in-line (straight). People with only one hand or with poor coordination may have difficulty opening products that require two simultaneous actions (e.g., stabilizing while opening or operating two latches that spring closed).

Design Options and Ideas to Consider:

- Using doors with open handles, levers, or doors that are pushed, then spring open
- Avoiding use of knobs or lips to open products
- Avoiding dual latches that must be operated simultaneously
- Using latches that are operable with a closed fist
- Using bearings for drawers or heavy objects that must be moved
- Providing electric pushbutton or remote control power openers
- Shaping product and door handles, and so on, to minimize the need for bending the wrist or body

See Section I-5 for additional suggestions.

M-3 Maximize the Number of People Who Can Remove, Replace, or Reposition Often-Used Detachable Parts

Problem Covers, lids, and other detachable parts may be difficult to remove, replace, or reposition.

Example: People with poor motor control may be unable to replace a cover or lid once it has been detached because it was dropped to the floor or into an inaccessible part of the product. People with weakness may have difficulty repositioning a keyboard, monitor, or TV set if the resistance to movement is high.

Design Options and Ideas to Consider:

- Employing devices with covers or lids that could be hinged, have sliding covers, or be operated electronically
- Tethering covers and lids with a cord or wire
- Making device components repositionable with a minimum of force
- Eliminating or limiting tasks needed for consumer assembly, installation, or maintenance of product

M-4 Maximize the Number of People Who Can Understand How to Carry Out The Manipulations Necessary to Use the Product

Problem Some people may have difficulty remembering how to operate the product, performing tasks in the correct order or within the required time, making choices, doing required measurements, or problem solving.

Example: Some people (particularly those with learning disabilities or cognitive impairments) have difficulty remembering codes required to operate a device (e.g., PIN number for automated teller machine). They may also be unable to remember which control to push to start or stop the device, have difficulty with serial order recall (the ability to remember items or tasks in sequence), and thus cannot follow complex or numerous steps, and have a slower or delayed reaction time, due to their inability to remember things quickly or to make responses that are dependent on timed input. Some get confused when there is a time lag for a response after they issue a command or when they expect an immediate result, and have trouble in choosing from available selection options (e.g., selecting paper size on a printer, choosing settings on a stereo set). Some cannot understand the concept of measuring or quantifying. Some have significant difficulty finding out what and where the problem is when a device is not functioning properly, and may have difficulty identifying solutions to problems they have identified.

Design Options and Ideas to Consider: Many of the problems in this category are similar to the problems outlined in Section I-6 and many of the same design ideas would apply, including the following:

- Keeping things as simple as possible
- Providing cues or prompts for sequences of actions required
- Writing the instructions directly on the device
- Having programmable keys for commonly used sequences
- Providing an easy way out of any situation
- Eliminating any timed responses (or make the times adjustable)
- Providing feedback to the user when the device is busy or "thinking"
- Hiding seldom-used controls which are not used primarily, to limit available choices

Other design suggestions include:

- Incorporating premeasuring methods whenever a quantifiable amount is required
- Providing prompts to inform users about the source(s) of problems and lead them to action to be taken to solve the problems (e.g., lights and color-coded pictorials used in copying machines)
- Eliminating or simplifying consumer assembly, installation, and maintenance of the product
- Providing a "standard" key or default mode to operate standardized functions (e.g., a key on the copier to give standard-sized copies)
- Providing an automatic mode so that the machine will make self-adjustments

7.4 Documentation

Maximize the number of people who can . . .

D-1 access the documentation.
D-2 understand the documentation.

D-1 Maximize the Number of People Who Can Access the Documentation

Problem Printed documentation (e.g., operating or installation instructions) may not be readable.

Example: People with low vision may not be able to read documentation due to small size or poor format. Poor choice of colors may make diagrams ambiguous for people with color blindness. People who are blind cannot use printed documentation, especially graphics. People with severe physical impairments may find it difficult or impossible to handle printed documentation.

Design Options and Ideas to Consider:

- Providing documentation in alternate formats: electronic, large-print, audio tape, and/or braille
- Using large fonts
- Using sans serif fonts
- Making sure that letter spacing, the space between lines (leading), and the distance between topics are sufficient that the letters and topics stand out from each other distinctly
- Supplementing any information that is presented via color coding so it can be interpreted in some other way which does not rely on color (e.g., bar charts may use various black-and-white patterns under the colors or patterns in the colors)
- Providing a text description of all graphics (this is especially important for use in electronic, taped, and large-print forms)
- Providing basic instructions directly on the device as well as in the documentation
- Making printed documentation "scanner/OCR-friendly" (see below)

D-2 Maximize the Number of People Who Can Understand the Documentation

Problem Printed documentation (e.g., operating or installation instructions) may not be understandable.

Example: People with cognitive impairments may have difficulty following multistep instructions. People with language difficulties or for whom English is a second language (including people with deafness) may have difficulty understanding complex text. People with learning difficulties may have difficulty distinguishing directional terms.

Design Options and Ideas to Consider:

- Providing clear, concise descriptions of the product and its initial setup
- Providing descriptions that do not require pictures (words and numbers used redundantly with pictures and tables), at least for all the basic operations (see below)
- Formatting with plenty of white space used to create small text groupings and bullet points
- Highlighting key information by using large, bold letters and putting it near the front of text
- Providing step-by-step instructions which are numbered, bulleted, or have check boxes
- Using affirmative instead of negative or passive statements
- Keeping sentence structure simple (i.e., one clause)
- Supplying a glossary
- Avoiding directional terms (e.g., left, right, up, down) where possible
- Providing a basic "bare bones" form or section to the documentation that gets you up and running with just the basic features

See also Sections O-6, I-6 and M-4.

7.5 Safety

Maximize the number of people who can . . .

S-1 perceive hazard warnings.
S-2 use the product without injury due to unperceived hazards or the user's lack of motor control.

S-1 Maximize the Number of People Who Can Perceive Hazard Warnings

Problem Hazard warnings (alarms) are missed, due to monosensory presentation or lack of understandability.

Example: People with hearing impairments may not hear auditory alarms that have only a narrow frequency spectrum. People who are deaf may not hear auditory alarms. People with visual impairments may not see visual warnings. People with cognitive impairments may not understand the nature of a warning quickly enough.

Design Options and Ideas to Consider:

- Using a broad-frequency spectrum with at least two frequency components between 500 and 3000 Hz for alarm signals
- Using redundant visual and auditory format for alarms (e.g., flashing lights plus alarm siren)
- Reducing glare on any surfaces containing warning messages
- Using common color-coding conventions and/or symbols along with simple warning messages
- Providing an optional, portable, vibrating module for use by persons who are deaf

S-2 Maximize the Number of People Who Can Use the Product without Injury due to Unperceived Hazards or the User's Lack of Motor Control

Problem Users are injured because they are unaware of an "obvious" hazard or because they lack sufficient motor control to avoid hazards.

Example: People with visual impairments may not see a hazard that is obvious to those with average sight. People with lack of strength or muscle control may inadvertently topple a device while in use so that it injures them. People with incoordination or lack of muscle control may inadvertently put their limbs or fingers in places not intended for contact or other hazardous places (e.g., the cassette tape drive of a stereo contains sharp edges that can cut fingers jammed inside with force). People with cognitive impairments may be unable to remember to shut off devices when not in use.

Design Options and Ideas to Consider:

- Avoiding pinch points on moving parts
- Eliminating or audibly warning of hazards that rely on the user's visual ability to avoid
- Making all surfaces, corners, protrusions, and device entrances free of sharp edges or extreme heat
- Deburring any internal parts accessible by a body part, even if contact with a body part is not normally expected (e.g., inside an open cassette tape door on a stereo set)
- Providing automatic shutoff of devices that would present a hazard if left on (e.g., irons)
- Ensuring that devices have stable, nonslip bases, or the ability to be attached to a stable surface (see below)

7.6 Universal Design Tools

A group of architects, product designers, and human factors engineers have gotten together to develop a common set of universal design principles and guidelines (see Table 2). Members of the team also have developed tools that work with the principles including a Guide to Evaluating the Universal Design Performance of Products For those on the Internet, the most current version of the principles and guidelines

Table 2 Principles of Universal Design

PRINCIPLE ONE: Equitable Use
The design is useful and marketable to people with diverse abilities.
Guidelines:
1a. Provide the same means of use for all users: identical whenever possible; equivalent when not.
1b. Avoid segregating or stigmatizing any users.
1c. Make provisions for privacy, security, and safety equally available to all users.
1d. Make the design appealing to all users.

PRINCIPLE TWO: Flexibility in Use
The design accommodates a wide range of individual preferences and abilities.
Guidelines:
2a. Provide choice in methods of use.
2b. Accommodate right- or left-handed access and use.
2c. Facilitate the user's accuracy and precision.
2d. Provide adaptability to the user's pace.

PRINCIPLE THREE: Simple and Intuitive Use
Use of the design is easy to understand, regardless of the user's experience, knowledge, language skills, or current concentration level.
Guidelines:
3a. Eliminate unnecessary complexity.
3b. Be consistent with user expectations and intuition.
3c. Accommodate a wide range of literacy and language skills.
3d. Arrange information consistent with its importance.
3e. Provide effective prompting and feedback during and after task completion.

PRINCIPLE FOUR: Perceptible Information
The design communicates necessary information effectively to the user, regardless of ambient conditions or the user's sensory abilities.
Guidelines:
4a. Use different modes (pictorial, verbal, tactile) for redundant presentation of essential information.
4b. Maximize "legibility" of essential information.
4c. Differentiate elements in ways that can be described (i.e., make it easy to give instructions or directions).
4d. Provide compatibility with a variety of techniques or devices used by people with sensory limitations.

PRINCIPLE FIVE: Tolerance for Error
The design minimizes hazards and the adverse consequences of accidental or unintended actions.
Guidelines:
5a. Arrange elements to minimize hazards and errors: most used elements, most accessible; hazardous elements eliminated, isolated, or shielded.
5b. Provide warnings of hazards and errors.
5c. Provide fail safe features.
5d. Discourage unconscious action in tasks that require vigilance.

PRINCIPLE SIX: Low Physical Effort
The design can be used efficiently and comfortably and with a minimum of fatigue.
Guidelines:
6a. Allow user to maintain a neutral body position.
6b. Use reasonable operating forces.
6c. Minimize repetitive actions.
6d. Minimize sustained physical effort.

PRINCIPLE SEVEN: Size and Space for Approach and Use
Appropriate size and space is provided for approach, reach, manipulation, and use regardless of user's body size, posture, or mobility.
Guidelines:
7a. Provide a clear line of sight to important elements for any seated or standing user.
7b. Make reach to all components comfortable for any seated or standing user.
7c. Accommodate variations in hand and grip size.
7d. Provide adequate space for the use of assistive devices or personal assistance.

Source: Bettye Rose Connell, Mike Jones, Ron Mace, Jim Mueller, Abir Mullick, Elaine Ostroff, Jon Sanford, Ed Steinfeld, Molly Story, and Gregg Vanderheiden, © 1997 NC State University, The Center for Universal Design.

can be found at the author's Web site, listed at the end of the chapter. The Trace Center at the University of Wisconsin–Madison has also developed an online design tool. The tool facilitates the design of more accessible mainstream products by highlighting aspects that contribute to enhanced and expanded usability. It also provides strategies, techniques, and examples for various product types.

8 CONCLUSIONS

Universal design should not really exist as a separate topic. In fact, it is just an extension of good human factors design today. The fact that it is currently a separate topic is probably an artifact of both the heavy military influence in the early ergonomic design process and the focus on serving the largest and most homogeneous segment of the population. However, legislation and commercial interests, a shifting and aging population, and the high costs of health care are combining to provide increased emphasis on this area. In the computer area, Apple, IBM, Microsoft, Digital Equipment Corporation, Sun, and other computer companies are all expanding the human interface and general design of their products to allow them to accommodate people with a much wider range of skills and abilities. Similarly, homebuilders, household product manufacturers, and others are extending and modifying their lines to serve people with more diverse abilities. There is an acute shortage, however, of people with background and experience in what might be *universal design*. It is to be hoped that over time the term and the field of universal design will fade as it becomes part and parcel of the standard design process.

FOR FURTHER INFORMATION

Further information on topics covered in this chapter, as well as updated versions of design guidelines, resource materials, and references, can be found at the Web site located at trace.wisc.edu.

REFERENCES

Anderson, T. P. (1981), "Stroke and Cerebral Trauma: Medical Aspects," in *Handbook of Severe Disability*, W. C. Stolov and M. R. Clowers, Eds., U.S. Department of Education, Rehabilitation Services Administration, Washington, DC.

Batshaw, M. (1997), *Children with Disabilities*, Paul H. Brookes, Baltimore.

Bennett, C. A. (1977). The demographic variables of discomfort glare. *Lighting Design and Application*, Vol. 7, pp. 22–24.

Berkowitz, J. P., and Casali, S. P. (1990), "Influence of Age on the Ability to Hear Telephone Ringers of Different Spectral Content," *Proceedings of the Human Factors Society 34th Annual Meeting*, Vol. 1, pp. 132–136.

Berkowitz, M. (1990).

Berkowitz, M., and Greene, C. (1989), "Disability Expenditures," *American Rehabilitation*, Vol. 15, No. 1, pp. 7–15, 29.

Braddock, D. L. (2001), "Public Financial Support for Disability at the Dawn of the 21st Century," *American Journal on Mental Retardation*, Vol. 107, No. 6, pp. 478–489.

Brandt, E. N., Jr., and Pope, A. M., Eds. (1997), *Enabling America: Assessing the Role of Rehabilitation Science and Engineering*, National Academies Press, Washington, DC.

Caplan, R. (1992), "Disabled by Design," *Interior Design*, Vol. 63, August, pp. 88–91.

Cook, M., and Hussey, S. M. (1995), *Assistive Technologies: Principles and Practice*, Mosby–Year Book, St. Louis, MO.

Epilepsy Foundation (2003), "Epilepsy and Seizure Statistics," retrieved September 13, 2004, from http://www.epilepsyfoundation.org/answerplace/statistics.cfm.

Gallaudet Research Institute (2002), "Current Estimates: Where to Look for Answers," retrieved September 9, 2004, from http://gri.gallaudet.edu/Demographics/deaf-US.php.

Galvin, J., and Scherer, M. (1996), *Evaluating, Selecting, and Using Appropriate Technology*, Aspen, Gaithersburg, MD.

Grandjean, E., Ed. (1987), *Ergonomics of Computerized Offices*, Taylor & Francis, Bristol, PA.

Harding, G. F. A., and Jeavons, P. M. (1994), *Photosensitive Epilepsy*, MacKeith Press, London.

Hoover, R. E., and Bledsoe, C. W. (1981), "Blindness and Visual Impairments," in *Handbook of Severe Disability*, W. C. Stolov and M. R. Clowers, Eds., U.S. Department of Education, Rehabilitation Services Administration, Washington, DC.

Hunt, R. M. (1970), "Determination of an Effective Tone Ringer Signal," presented at the 38th Convention of the Audio Engineering Society, New York.

Jeavons, P. M., and Harding, G. R. A. (1975), *Photosensitivity Epilepsy*, Heinemann, London.

Kraus, J. F., and MacArthur, D. L. (1996), "Epidemiologic Aspects of Brain Injury," *Neurologic Clinics*, Vol. 14, No. 2, pp. 435–450.

Kraus, L., and Stoddard, S. (1989), *Chartbook on Disability in the United States*, prepared for U.S. Department of Education, National Institute on Disability and Rehabilitation Research, Washington, DC.

Kraus, L., Stoddard, S., and Gilmartin, D. (1996), *Chartbook on Disability in the United States*, prepared for U.S. Department of Education, National Institute on Disability and Rehabilitation Research, Washington, DC.

Kroemer, K. H. E. (1990), *Engineering Physiology Bases of Human Factors/Ergonomics*, Van Nostrand Reinhold, New York.

LaPlante, M. P. (1988), *Data on Disability from the National Health Interview Survey, 1983–85*, U.S. Department of Education, National Institute on Disability and Rehabilitation Research, Washington, DC.

Mueller, J. (1990), *The Workplace Workbook: An Illustrated Guide to Job Accommodation and Assistive Technology*, RESNA Press, Washington, DC.

National Center for Health Statistics (1994), *Data from the National Health Interview Survey*, Seri. 10, No. 188, Table 1.

National Eye Institute (2002), "Vision Problems in the U.S.: Prevalence of Adult Vision Impairment and Age-Related Eye Diseases in America," retrieved September 9, 2004, from http://www.nei.nih.gov/eyedata/pdf/VPUS.pdf.

National Institute on Arthritis and Musculoskeletal and Skin Diseases (2002), "Questions and Answers About Arthritis and Rheumatic Diseases," retrieved September 13, 2004, from http://www.niams.nih.gov/hi/topics/arthritis/ArthRheumdisRP.pdf.

National Institute on Deafness and Other Communication Disorders (n.d.), "Statistics About Hearing Disorders, Ear Infections and Deafness," retrieved September 9, 2004, from http://www.nidcd.nih.gov/health/statistics/hearing.asp.

National Limb Loss Information Center (2002), "Limb Loss in the United States," retrieved September 13, 2004, from http://www.amputee-coalition.org/fact_sheets/limbloss_us.pdf.

National Multiple Sclerosis Society (2004), "About MS," retrieved September 13, 2004, from http://www.nationalmssociety.org/Who%20gets%20 MS.asp.

National Parkinson Foundation (n.d.), "About Parkinson Disease," retrieved September 13, 2004, from http://www.parkinson.org/site/pp.asp?c=9dJFJLPwB&b=71125.

Pastor, P. N., and Reuben, C. A. (2002), "Attention Deficit Disorder and Learning Disability: United States, 1997–98," National Center for Health Statistics, *Vital Health Statistics*, Vol. 10, No. 206.

Robert Packard Center for ALS Research at John Hopkins (2004), "About ALS," retrieved September 13, 2004, from http://www.alscenter.org/about_als/stats.cfm.

Schein, J. D. (1981), "Hearing Impairments and Deafness," in *Handbook of Severe Disability*, W. C. Stolov and M. R. Clowers, Eds., U.S. Department of Education, Rehabilitation Services Administration, Washington, DC.

Schow, R. L., et al. (1978), *Communication Disorders of the Aged: A Guide for Health Professionals*, University Park Press, Baltimore, MD.

Spinal Cord Injury Information Network (2004), "Spinal Cord Injury Facts and Figures at a Glance," retrieved September 13, 2004, from http://images.main.uab.edu/spinalcord/pdffiles/factsfig.pdf.

United Cerebral Palsy (2001), "Cerebral Palsy Facts and Figures," retrieved September 13, 2004, from http://www.ucp.org/ucp_generaldoc.cfm/1/9/37/37-37/447.

WHO (1980), *International Classification of Impairments, Disabilities, and Handicaps: A Manual of Classification Relating to the Consequences of Disease*, World Health Organization, Geneva, Switzerland.

UNIVERSAL DESIGN RESOURCES

The Accessible Housing Design File, 1991, Barrier Free Environments, Inc., Wiley, New York.

Beautiful Barrier-Free: A Visual Guide to Accessibility, 1992, C. Liebrock and S. Behar, Wiley, New York.

Building for a Lifetime: The Design and Construction of Fully Accessible Homes, 1994, M. Wylde, A. Baron-Robbins, and S. Clark, Taunton Press, Newton, CT.

Building Sight: A Handbook of Building and Interior Design Solutions to Include the Needs of Visually Impaired People, 1995, P. Barker, J. Barrick, and R. Wilson, American Foundation for the Blind Press, New York.

A Consumer's Guide to Home Adaptation, 1992, Adaptive Environments Center, Inc., 374 Congress Street, Suite 301, Boston, MA 02210; 617-695-1225 (V/TTY); http://www.adaptenv.org/index.php?option=Resource&articleid=219&topicid=25.

Definitions: Accessible, Adaptable and Universal Design (Fact Sheet), 1991, Center for Universal Design, North Carolina State University, Box 8613, Raleigh, NC 27695-8613; 919-515-3082 (V/TTY); http://www.design.ncsu.edu/cud/pubs/center/fact_sheets/housdef.htm.

Design for Dignity: Studies in Accessibility, 1993, W. Lebovitch, Wiley, New York.

"Designing Consumer Product Displays for the Disabled," in *Proceedings of the Human Factors Society 34th Annual Meeting*, 1990, J. T. Ward, Vol. 1, pp. 448–451.

The Directory of Accessible Building Products 2004 (updated annually), National Association of Home Builders Research Center, 400 Prince George's Boulevard, Upper Marlboro, MD 20772-8731; 301-249-4000; http://www.nahbrc.org/tertiaryR.asp?TrackID=&CategoryID=1652&DocumentID=2591.

Ergonomics at Work, 1987, D. J. Osborne, Wiley, New York.

Human Factors in Engineering and Design, 6th ed., 1987, M. S. Sanders and E. J. McCormick, McGraw-Hill, New York.

Human Factors Research Needs for an Aging Population, 1990, S. J. Czaja, Ed., Panel on Human Factors Research Issues for an Aging Population, Committee on Human Factors, Commission on Behavioral and Social Sciences and Education, National Research Council, National Academies Press, Washington, DC; http://books.nap.edu/catalog/1518.html.

Making Life More Livable: Simple Adaptations for Living at Home After Vision Loss, 2001, I. R. Dickman, American Foundation for the Blind Press, New York.

Play for All Guidelines: Planning, Design and Management of Outdoor Play Settings for All Children, 1992, R. Moore, S. Goltsman, and D. Oacofano, Eds., MIG Communications, Berkeley, CA.

Practicing Universal Design: An Interpretation of the ADA, 1994, W. L. Wilkoff and L. W. Abed, Wiley, New York.

Symbol Sourcebook: An Authoritative Guide to International Graphic Symbols, 1972, H. Dreyfuse, McGraw-Hill, New York.

Tactile Graphics, 1992, P. K. Edman, American Foundation for the Blind Press, New York.

Touch, Representation, and Blindness, 2000, M. A. Heller, Oxford University Press, New York.

Transgenerational Design, 1994, J. J. Pirkle, Van Nostrand Reinhold, New York.

Universal Design File, 1998, M. F. Story, J. L. Mueller, and R. Mace, Center for Universal Design, North Carolina State University, Box 8613, Raleigh, NC 27695-8613; 800-647-6700; http://www.design.ncsu.edu/cud/.

Universal Design Handbook, 2001, W. Preiser, and E. Ostroff, Eds., McGraw-Hill Professional, New York.

Universal Design Newsletter, quarterly, Universal Designers and Consultants, Inc., 6 Grant Avenue, Takoma Park, MD 20912-4324; 301-270-2470 (V/TTY); http://www.universaldesign.com/newsletters.php.

Universal Kitchen and Bathroom Planning: Design that Adapts to People, 1998, M. J. Peterson, McGraw-Hill, New York.

ORGANIZATIONS

AccessIT

National Center on Accessible Information Technology in Education
Box 357920
University of Washington
Seattle, WA 98195-7920
866-968-2223 (V)
866-866-0162 (TTY)
http://www.washington.edu/accessit/index.php

AccessIT, at the University of Washington, promotes the use of electronic and information technology (E&IT) for students and employees with disabilities in educational institutions at all academic levels. The Web site contains a searchable, growing database of questions and answers regarding accessible E&IT.

Funding for AccessIT is provided by the National Institute on Disability and Rehabilitation Research.

Adaptive Environments
374 Congress Street, Suite 301
Boston, MA 02210
617-695-1225 (V/TTY)
http://www.adaptiveenvironments.org

Adaptive Environments is a nonprofit organization that works on consulting, training, and education projects intended to promote the practices of universal design. It is also home to the New England ADA and Accessible Information Technology Center, funded by the National Institute on Disability and Rehabilitation Research.

Center for Inclusive Design and Environmental Research (IDEA)
School of Architecture and Planning, SUNY–Buffalo
378 Hayes Hall
3435 Main Street
Buffalo, NY 14214-3087
716-829-3458 × 329
http://www.ap.buffalo.edu/~idea

The Center for Inclusive Design and Environmental Access (IDEA) at SUNY–Buffalo provides resources and technical expertise in architecture, product design, and facilities management. IDEA is funded as an RERC on Universal Design of the Built Environment by the National Institute on Disability and Rehabilitation Research.

Center for Universal Design
College of Design
North Carolina State University
Brooks Hall, Room 104
50 Pullen Road
Campus Box 8613
Raleigh, NC 27695-8613
800-647-6777
http://www.design.ncsu.edu/cud/

The Center for Universal Design at North Carolina State University is a national research, information, and technical assistance center that evaluates, develops, and promotes universal design in housing, public and commercial facilities, and related products. The CUD is funded as an RERC on Universal Design of the Built Environment by the National Institute on Disability and Rehabilitation Research.

Inclusive Technologies
Temper Complex
37 Miriam Drive
Matawan, NJ 07747
732-441-0831 (V/TTY)
http://www.inclusive.com

Inclusive Technologies provides a full range of consulting services to companies, consumer organizations, researchers, and policymakers on how products can better meet the needs of all users, including users with disabilities and elders.

Information Technology Technical Assistance and Training Center (ITTATC)
Center for Assistive Technology and Environmental Access
Georgia Institute of Technology
490 10th Street NW
Atlanta, GA 30318
866-948-8282 (V/TTY)
http://www.ittatc.org

The ITTATC at the Georgia Institute of Technology provides information related to Section 508, Section 255, and universal design of E&IT products. IIIATC is funded by the National Institute on Disability and Rehabilitation Research.

J.L. Mueller, Inc.
4717 Walney Knoll Court
Chantilly, VA 22021
703-222-5808
http://home.earthlink.net/~jlminc/

J.L. Mueller, Inc. is the consulting practice of Jim Mueller, an industrial designer, who works providing job accommodation assistance, workplace design, and product design research. The Web site includes a wealth of information about the principles of universal design, applicable legislation, and workplace accommodations.

Technology Access Program
Gallaudet University
101 Kendall Hall
800 Florida Avenue, NE
Washington, DC 20002
202-651-5257 (V/TTY)
http://tap.gallaudet.edu

Gallaudet University's Technology Access Program focuses on technologies and services that eliminate communication barriers traditionally faced by deaf and hard-of-hearing people; it is a key resource on TTYs and universal access to all forms of telecommunication.

Trace Research and Development Center
University of Wisconsin–Madison
2107 Engineering Centers Building
1550 Engineering Drive
Madison, WI 53706
608-262-6966 (V)
608-263-5408 (TTY)
http://trace.wisc.edu

The Trace Center at the University of Wisconsin–Madison focuses on the design of mainstream information technology and telecommunications products and systems for us by all people. Trace is also the home of the RERC on Information Technology Access and (in partnership with Gallaudet University) the RERC on Telecommunication Access, both funded by the National Institute on Disability and Rehabilitation Research.

Universal Design Education Online
http://www.udeducation.org

Universal Design Education Online is a resource for those interested in teaching about universal design and accessibility. This new project, funded by the National Institute on Disability and Rehabilitation Research, is compiling educational materials related to universal design for download and use by others.

CHAPTER 54
DESIGN FOR AGING

Timothy A. Nichols, Wendy A. Rogers, and Arthur D. Fisk
Georgia Institute of Technology
Atlanta, Georgia

1 INTRODUCTION

Age is a critical variable relevant to design considerations in human factors research and practice. This conclusion is founded on three primary facts:

1. The number of older adults in developed countries today is higher than ever and is increasing.
2. There are critical age-related differences between younger and older adults which necessitate specific design considerations.
3. The proportion of older adults within the global workforce and of all users of systems and products is increasing steadily (Figure 1).

1.1 Increasing Population over 65

The world's older adult (65+ years) population is increasing by approximately 800,000 each month (Kinsella and Velkoff, 2001). Over the last decade, it has been clearly documented in worldwide population estimates that the average lifespan is increasing, as is the average age of the world's population (U.S. Bureau of the Census, 1996) (Figure 2).

1.2 Design-Critical Age-Related Differences

Why is it important to consider older adult users? The statistics of older adult population rates alone are not cause for the field of human factors to take notice. The meaningful issue is whether older adults require significantly different design considerations than younger adults. This issue is addressed in the present chapter by specifying the cognitive, perceptual, motor, and motivation differences between the two age groups and also considering whether these differences translate into functional differences. For example, age-related response time differences of 1.5 seconds may not be functionally meaningful when searching for information on the Internet, but this age-related difference could be critical in the driving environment or when responding to a medical emergency in the home where delayed responses can lead to serious consequences. Primary goals of this chapter are to describe functionally meaningful changes that occur with age, their

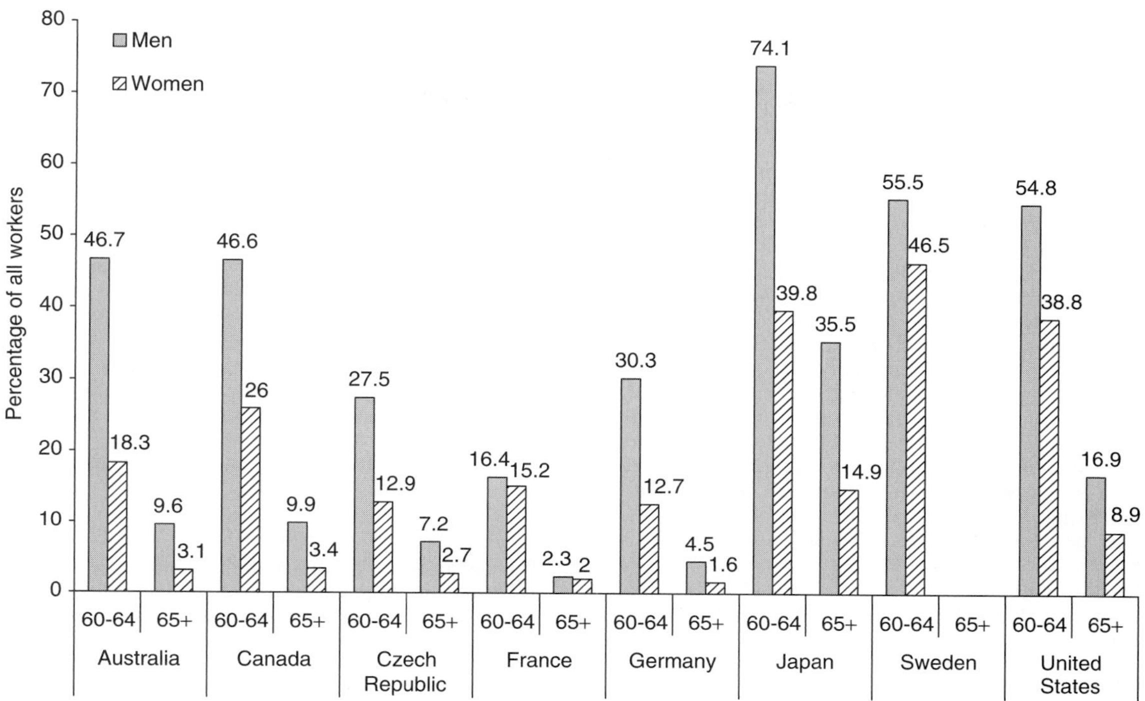

Figure 1 Percentage of men and women between 60 and 64 and 65+ in selected countries who are in the workforce.

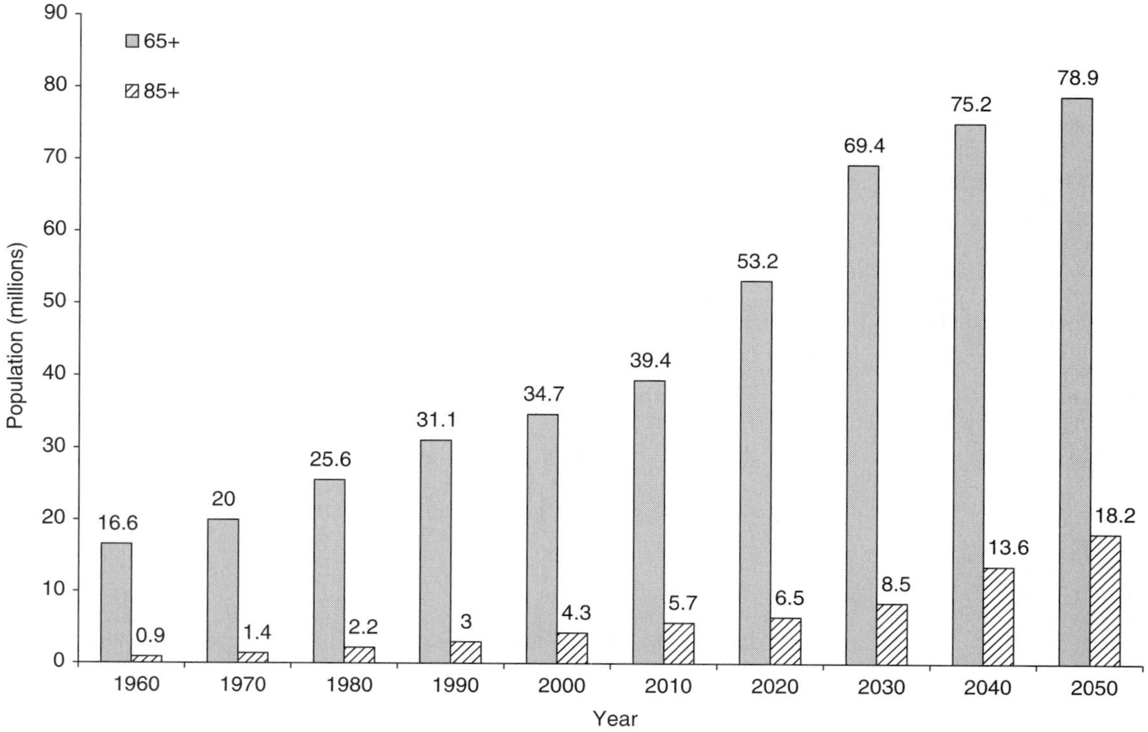

Figure 2 Population estimates for the age groups 65+ and 85+ through the year 2050.

effects on performance, and to recommend design considerations and design principles.

1.3 Older Adults' Use of Technology

Given the need for age-specific design considerations, a critical third issue for human factors researchers and practitioners is to understand the extent to which older adults interact with systems and products. Certainly, if older adults comprise only a very small proportion of users of systems and products, this would temper the degree to which the field must be concerned with the age-related changes and characteristics of older adults. However, this is not the case. In a recent survey of product use (Hancock et al., 2001), adults over age 65 reportedly used a variety of household products as frequently as adults under age 65, and they used health care products more frequently. Furthermore, as discussed in Section 6, many older adults express a desire to learn to use new technologies (e.g., Rogers et al., 1998). In summary, the capabilities and limitations of the growing older adult population must be understood and accounted for in the design process and in human factors research to ensure that this segment of the population can interact with products and systems in a safe, efficient, and effective manner.

1.4 Definitions

1.4.1 Age

The focus of the chapter is on the relevance of "age" to design. However, age is no more than a chronological marker for the myriad changes that people undergo as they age—hence the term age- *related* changes. The changes discussed here are not caused by a person's age, but rather are correlated with age, and there is substantial variability in the degree to which individual older adults show some changes and the degree of change for a particular person. Furthermore, given the continuous nature of this variable, is it proper or even useful to think of age in discrete chunks (e.g., younger vs. older)? In fact, there are several reasons to do so (see Nichols et al., 2003, for a review). For research purposes, we recommend defining "older" adults as 65 to 85 years, "middle-aged" adults as 40 to 55, and "younger" adults as 18 to 30. Subgroups such as young-old (e.g., 56 to 64) and oldest-old (85+), are often used in aging research, and the younger group may at times be extended (e.g., 18 to 39). These age range decisions are often, in part, based on the user population to which the research is intended to generalize, as well as the age-related changes that are being investigated. For example, younger and older groups in a study of air traffic controllers would be based on different age ranges than a study of normal driving. The standard maximum age for an air traffic controller is 56, whereas the maximum age of drivers extends into the 80s and 90s. Thus, what is older in one context differs from what is older in the other context. We recommend that researchers and practitioners follow the above age-range guidelines for the following reasons:

- *Variability in performance.* Age-related variance will be controlled to some degree (as opposed to a study that uses a sample ranging from 40 to 75). As will be highlighted, considerable age-related changes occur across the lifespan, and those changes may have a significant impact on behavior and task performance.
- *Precision and consistency.* By following appropriate age classification guidelines and reporting participants' age, the field of human factors will better provide precise and useful information on the performance of adults of different ages and design will better fit the user.
- *Parsimony.* For many studies, it may be simpler to think of a single variable than the host of variables affected by age. Although age is merely a marker predicting performance, age provides a broad and useful designate for many of the age-related changes discussed in this chapter.

1.4.2 Products and Systems

The design issues discussed in this chapter are intended to be relevant to a wide range of products and systems (i.e., any system or product older adults might use). This is necessarily general, as older adults interact with a wide range of products and systems in their daily activities. Although the guidelines are intended to be general, their applicability will differ as a function of the demands of the task at hand (e.g., a technology that requires fine motor control will be more influenced by age-related changes in movement control). Guidelines specific to particular devices and contexts are available in Fisk et al. (2004).

1.5 How to Use the Recommendations

The recommendations presented in this chapter provide a means of reducing the potential solution space for resolving design-related questions. However, these recommendations do not provide a complete solution. Variability at all levels of behavior in older adults is a hallmark characteristic of aging, making iterative design and user testing with older adults crucial. The design issues discussed here should be beneficial in understanding and predicting older adults' performance, but the chapter is not intended to replace necessary human factors methods such as iterative design and user testing. For the implications and design recommendations, our goal is to provide general design principles relevant to design for older users. However, many of the age-related changes discussed in this chapter occur gradually and improvements in design and training targeted at older adults are likely to benefit younger and middle-aged adults as well (Fisk, 1999).

1.6 Overview of the Chapter

The remainder of this chapter is organized as follows: First we discuss aspects of perception, movement control and biomechanics, cognition, language, and motivation from the perspective of age-related differences. Within each section we present general implications for design. We then present a "case study" of a telemedicine system designed for home use. This case study will be used to illustrate the implications of the age-related changes for system design.

1.6.1 Age-Related Changes

The following categories of age-related changes are reviewed briefly and then discussed in terms of the relevance of such changes to design: (1) perception, (2) movement control and biomechanics, (3) cognitive processes, (4) language, and (5) motivation.

1.6.2 Design Implications and Suggestions

Following each section on age-related changes, the implications of these data will be explored in greater depth in a review of the design implications of the age-related changes. We also present a hypothetical case study of a home-based telemedicine system for older adults. While the discussion accompanying each age-related change will involve specific consequences for design, the goal of the latter case study is to provide robust design recommendations based on consideration of multiple age-related changes.

In providing design suggestions, there is a tension between providing overly general suggestions that are not specific enough to be useful vs. overly specific suggestions that do not generalize across products or systems. Our approach is to focus on implications and recommendations for those systems and products particularly relevant to a given age-related change. Thus, when we present design implications at the end of each section, these will be reasonably broad (i.e., the design implications for the selective attention section will be relevant to the design of automobiles but not to the design of furniture).

1.6.3 Consistent Themes and Guidelines

It is clear from any review of the literature on human factors and aging that there are consistent themes that are relevant to design considerations for an older population (Table 1). These themes are evident in many of the design suggestions and implications in this chapter.

Table 1 Six Major Themes of Designing for Aging

Design Theme	Definition
Provide environmental support through context, cues, and organization	Performance can be supported by placing information in the task environment, which reduces cognitive demands on the user.
Improve the physical stimulus	Age-related perceptual declines can be offset to some degree by improving the physical stimulus and increasing older adults' ability to perceive and recognize the stimulus.
Display information consistently	Some level of consistency is required for learning to occur, and with greater degree of consistency, learning is more efficient.
Provide training	Through appropriate training, older adults' performance can be brought closer in line with that of younger adults.
Capitalize on crystallized knowledge	Fact knowledge is generally unaffected by aging, and design can take advantage of this knowledge.
Recognize knowledge, interest, and motivation	Older adults' motivation and desire to interact with technology are often underestimated.

2 PERCEPTION

Age-related changes in vision and audition are perhaps the most immediately obvious and pervasive changes that humans undergo as they age. With a focus on vision and audition, in the following section we present several age-related changes in perception, accompanied by implications for design where applicable (see Table 2 for a summary of the visual and auditory changes). Most products and systems are designed to provide information via the visual and auditory modalities, and these two sensory systems have been well studied in the aging population. These systems show significant and substantial declines in older adults. For information to affect behavior, the information must first enter the sensory system and then be encoded. If, due to sensory system degradation, information is sensed or perceived incompletely, it may be processed incorrectly.

2.1 Vision

In general, the ability to resolve an image accurately is dependent on the available luminance and the contrast of the scene. The most common age-related causes of visual impairment are age-related macular degeneration (ARMD), cataracts, and glaucoma (Desai et al., 2001) (Figure 3). Due to changes in the structure of the aging eye and visual processing system, older adults are less able to resolve details and are less sensitive to critical environmental characteristics such as luminance, contrast, color, and motion.

2.1.1 Acuity

Although visual impairment can affect people across the lifespan, age is the best predictor of visual decline. Visual acuity, the best known measure of visual ability, is typically measured relative to what a "normal" person can see at 20 feet away (thus, the measure 20/20) (Snellen, 1862, cited in Bennett, 1965). Acuity is affected by various eye pathologies (many of which are more commonly observed with age), as well as deterioration of the brain's visual pathways. In ARMD, the most common cause of vision loss in older adults, people experience a reduced ability to resolve fine detail (Bellman and Holz, 2001). People with ARMD

Table 2 Age-Related Changes in Vision and Audition

	Visual Changes
Visual acuity	The ability to resolve detail decreases.
Visual accommodation	The ability to focus on close objects decreases.
Color vision	The ability to discriminate and perceive shorter wavelength light decreases.
Contrast detection	The ability to detect contrast decreases.
Dark adaptation	The ability to adapt quickly to darker conditions decreases.
Glare	The susceptibility to glare increases.
Illumination	More illumination is required to see adequately.
Motion perception	Motion is not as readily detected and motion estimation is reduced.
Useful field of view	The useful visual field is reduced.
	Auditory Changes
Auditory acuity	The ability to detect sound decreases, particularly at higher frequencies and particularly for males.
Auditory localization	The ability to localize sound decreases, particularly at higher frequencies and when directly in front of or behind the user.
Audition in noise	The ability to perceive speech and complex sounds decreases.

suffer from reduced central vision due to degeneration of the *macula*, an area in the center of the retina. Cataracts also affect acuity, resulting in a clouding of the lens, but these are often treatable, whereas vision loss due to ARMD cannot currently be restored. Glaucoma results from damage to the optic nerve due to an increase in pressure in the eye, as fluid flow through the front of the eye is hindered. The buildup of fluid at the front of the eye results in increased pressure at the back of the eye, causing irreparable degeneration of the optic nerve fibers (Goldstein, 1999). As a result, peripheral vision deteriorates, and if not treated, central vision will deteriorate as well.

Older adults are able to compensate for loss of acuity. For example, despite poorer visual acuity, older adults were shown to perceive blurred text signs better than younger adults (Kline et al., 1999). In this study, the acuity of both younger and older adults was reduced artificially and the size at which blurred text signs could be read was measured. The primary findings were that both age groups were better able to read familiar text signs than novel text signs and that when both age groups' acuity was comparably impaired (e.g., at 20/40), older adults could read text at a smaller size than younger adults. Presumably the nature of optical blur is such that low-vision persons are less affected (Legge et al., 1987a). Thus, older adults' ability to deal with blurred information may be related to their ability to adapt to changes in visual acuity as well as to rely on *top-down processing* (i.e., interpreting information based on well-learned, crystallized knowledge).

Design Implications and Suggestions The loss in acuity has profound effects on the way in which information should be displayed for older adults. Increasing the size, brightness, and contrast of an item will improve older adults' perception of information. For example, text should typically be displayed in a 12-point font size or greater. It is especially important for novel information to be made perceptually salient for older adults. Top-down processing can result in the correct identification of a perceptually indistinct stimulus (Kline et al., 1999), but when processing is primarily bottom-up (as with novel stimuli), clarity of the stimulus is crucial. For example, a driver who has reduced near vision may not be able to perfectly discriminate the points on the odometer, but the consistent spacing and common look of the odometer should provide sufficient information about the vehicle's speed. Thus, it is important to design for such consistent aspects of a system, because for users who rely on these consistent aspects to aid their top-down processing interpretation of displays can be supported (this is a form of environmental support).

If a display cannot be changed to accommodate low-acuity people, it is important to maximize the effect of top-down processing. Contextual cues, another form of environmental support, can be provided to increase the likelihood that a stimulus will be recognized. For example, a driver who has trouble reading traffic control signs from afar may rely on color and shape of the signs, which allows the driver to identify the type of sign. However, acuity is not the only age-related decrement in visual perception that must be considered.

2.1.2 Accommodation

Older adults have difficulty with visual accommodation (termed *presbyopia*), which involves adjusting the curvature of the lens to focus on objects of different depths. In fact, reductions in accommodative ability are primarily responsible for losses in acuity in near vision, typically starting at age 40. By age 65, lens accommodation is so reduced that only objects at a certain distance can be focused on the retina, meaning that information not displayed at this distance cannot be clearly perceived by the person (B. Schneider and Pichora-Fuller, 2000).

Design Implications and Suggestions The need to focus at different distances should be minimized as much as possible. In a system with multiple displays, all of the displays should be placed as close to the optimal reading distance from the user as possible. This will reduce the necessity for head movement to bring information into perfect clarity as the multiple displays are scanned.

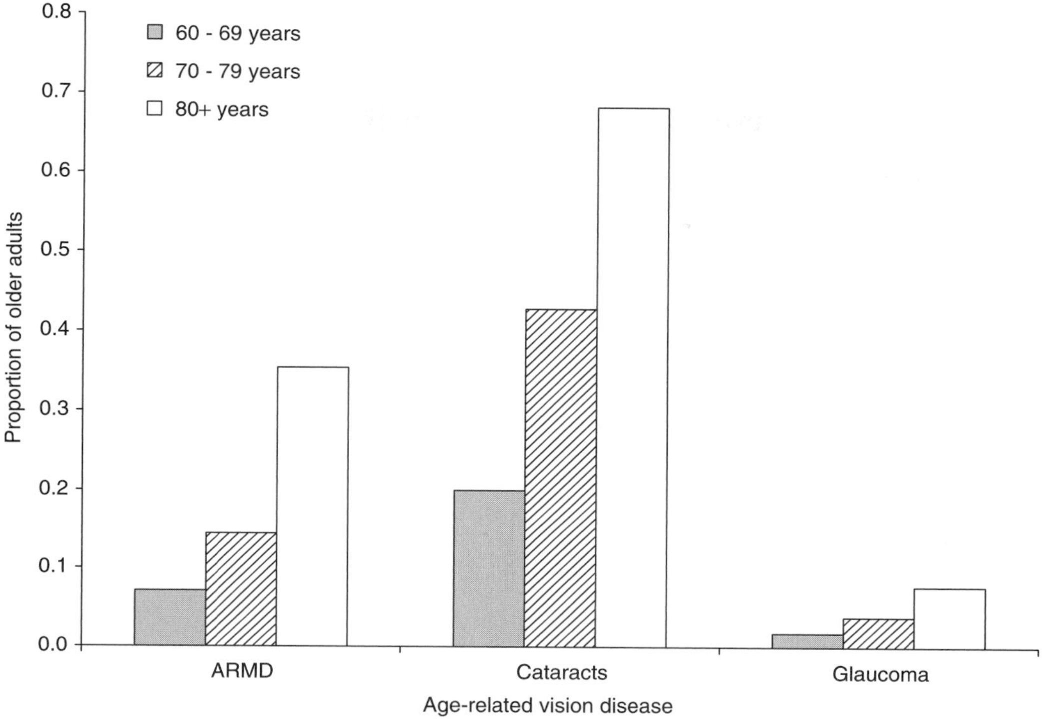

Figure 3 Prevalence of age-related macular degeneration (ARMD), cataracts, and glaucoma across three groups of older adults. (Adapted from Desai et al., 2001.)

2.1.3 Color Vision

Older adults are less able to discriminate shorter-wavelength light, such as blues and greens, due to a yellowing of the lens (Said and Weale, 1959). Furthermore, the ability to discriminate color declines with age (Kraft and Werner, 1999), and age-related differences in color discrimination increase at lower light levels and lower color saturations (Pinckers, 1980; Knoblauch et al., 1987).

Design Implications and Suggestions Color coding is still a feasible option in information visualization (e.g., representing multidimensional data at a single time), but color codes should avoid shorter-wavelength light as a general rule or use only a single blue or green, thus eliminating the need to compare within this range of color. Color coding should not be used when many distinct levels are required, and colors should be well saturated.

2.1.4 Contrast Detection

The importance of high contrast in detecting stimuli and in tasks such as reading is important for adults of all ages, but particularly for older adults. People with poorer acuity are more affected by reductions in contrast whether they are younger adults (Legge et al., 1987b) or older adults (Fozard, 1990). However, even if matched for visual acuity, older adults have reduced contrast sensitivity (Mitzner and Rogers, 2003). Reduced contrast sensitivity is due in part to the scattering of light as it enters the eye, such that light from the image is scattered across the retina, creating a more uniform dispersal of light on the retina (B. Schneider and Pichora-Fuller, 2000).

Design Implications and Suggestions Optimal contrast in which text should be presented is a bright white on black, or vice versa. If colors are used to present information, colors close together on the spectrum should not be used together (e.g., a red icon on an orange background). Furthermore, the deleterious effects of reduced contrast can be alleviated through the use of context, essentially allowing top-down processes to aid the older user in identifying the stimulus correctly (Mitzner and Rogers, 2003).

2.1.5 Illumination, Glare, and Dark/Light Adaptation

In the aging eye, changes in the cornea scatter light before it reaches the retina, the lens absorbs more light, and pupil size is reduced, allowing less light to reach the retina (B. Schneider and Pichora-Fuller, 2000). Even though less light reaches the retina, glare is problematic for older adults. Glare occurs when a person is exposed to levels of light higher than the eye is currently adapted. Thus, glare is a notable problem when driving at night. Similarly, bright sunlight can cause glare from either direct line of sight or from reflected surfaces. Due to the scattering of light in the older eye, glare is more of a problem for older adults, as the excess light is more distributed across the eye, essentially reducing perception for a greater degree of the visual field. The aging eye also is slower to adapt to light and dark. With reductions in the amount of light that reaches the retina, the retina is therefore slower to adapt to the changing light conditions. Furthermore, the chemical processes that cause dark adaptation are slowed with age (Jackson et al., 1999).

Design Implications and Suggestions These age-related changes in the sensation of light cannot be solved via an external perceptual aid, as they are specifically due to deterioration of the cornea. Older adults will benefit from increased illumination in all environments, including driving and activities at home and work. Unfortunately, Charness and Dijkstra (1999) demonstrated that the homes of older adults are not lit optimally, particularly at night (although older adults appeared to compensate for their visual impairments by using significantly more light in their homes than younger adults).

Appropriate lighting is critical in optimizing the perception of information. If possible, increase the level of illumination to at least 100 cd/m^2, as measured by the reflection from reading surfaces (Charness and Dijkstra, 1999). Lighting levels should be even when possible, in roadways and in office and home layouts. To reduce glare, light sources should be diffused and positioned to create ambient light as opposed to direct light. Mirrors and shiny surfaces should be avoided, as the undiffused reflections can cause glare. Multiple light sources serve to reduce harsh shadows and to even out the light in the environment.

These age-related changes in vision are critically relevant in the driving domain, which is one of the most perceptually and cognitively demanding tasks as well as one of the most widely and frequently performed. When driving at night, illumination levels are already extremely low. Glare can occur as the result of passing headlights and streetlights, causing older adults to be blinded temporarily by the dispersed light across their retinas. When driving from daylight into a tunnel, older adults' visual perception will suffer relative to younger adults. In the design of roads and tunnels, lighting should be made as constant as possible, to reduce the negative effects of low illumination, glare, and slower adaptation to dark and light.

2.1.6 Useful Field of View

Useful field of view (UFOV) refers to the size of the visual field that may be perceived in a single glance and is a measure of both processing speed and attention (Owsley et al., 1991; Roenker et al., 2003). That is, UFOV can be thought of as the subset of the total visual field that is available for processing (thus, the similarity with the construct of attention). The UFOV may change within a person, depending on the nature of the task being performed (Owsley et al., 1991). For example, one's UFOV may be larger when driving on a road with no traffic, whereas one may experience the phenomenal sense of constricted vision when driving in the rain and heavy traffic. Research has shown that older adults have a restricted UFOV, which has been linked to driving accidents (Owsley et al., 1991; Roenker et al., 2003).

Design Implications and Suggestions The broad implications for changes in UFOV are that practitioners cannot assume that a user will necessarily notice, use, or respond to information falling within the visual field. Age must be considered, and under some circumstances the UFOV of the user population might need to be assessed directly. With training, as people are able to perform subtasks more efficiently, the UFOV can effectively be increased. Thus, the implications for design are to know the UFOV of the user population in the context of the task, to ensure that stimuli are presented within their UFOV, and if necessary, to provide training to increase UFOV for the users.

2.1.7 Perception of Motion

Older adults are less able to detect motion relative to younger adults. To investigate sensitivity to motion, researchers have created sets of elements, a subset of which move synchronously (Trick and Silverman, 1991; Tran et al., 1998). Older adults appear to have a higher element threshold than younger adults, indicating that they required more movement of elements to detect motion in the array.

Age-related differences in perception of motion have been investigated in a driving scenario (Atchley and Andersen, 1998; Andersen et al., 2000). A three-dimensional display was used to simulate a vehicle windshield and display an environment at one of two velocities followed by a constant deceleration. The deceleration occurred such that the car would either stop short of an obstacle, stop directly at the obstacle, or crash into the obstacle. The display was stopped prior to stopping or crashing, and the participants were required to indicate what the outcome would be. Relative to younger adults, older adults were less sensitive to detecting collisions, indicating that a crash was inevitable when no collision occurred.

Design Implications and Suggestions Motion detection findings suggest that in combination with age-related declines in movement speed and visual perception, older adults' declining perception of potential collision information could be a factor in their ability to safely avoid vehicular collisions. When motion is a critical cue, it must be accentuated for older adults. However, this design implication must be incorporated carefully because, all other things being equal, motion can be more of a distraction for older adults than for younger adults.

2.2 Audition

Auditory information is presented in a wide variety of environments. Museums and exhibition halls may play descriptive recordings at displays, training materials may include a video with a model explaining how to perform some task, computers in the office or home emit alerts and other auditory signals, and many systems rely on auditory stimuli to communicate system status information. Safe and efficient system interaction can depend on the user's ability to hear normally, but audition is another perceptual domain wherein older adults show declines.

2.2.1 Auditory Acuity

In addition to an overall decrement in auditory acuity, age-related losses in hearing occur differentially across frequency ranges, with greater loss occurring for higher frequencies (greater than 8000 Hz) (B. Schneider and Pichora-Fuller, 1998; Fozard and Gordon-Salant, 2001). Furthermore, men have worse high-frequency perception than women (Moscicki et al., 1985).

Design Implications and Suggestions A high-frequency stimulus is sometimes used as an alert or indicator in computer applications. Older adults, particularly males, may not perceive these stimuli. In fact, the age-related changes in auditory acuity suggest that high-frequency sounds should be avoided in any system or product that older adults might use. A high-frequency alert or indicator may be differentiated from background noise better than lower frequencies, but if it is not perceived, it is useless. If auditory stimuli are designed to attract attention when the user's vision is elsewhere, auditory alerts should not exceed 4000 Hz (Fisk et al., 2004). For non-alert-related stimuli, it is important to provide the user with control over the intensity of the stimulus. Because of individual differences in overall thresholds, volume control should be provided so that people can calibrate for themselves. However, often it is not sufficient to provide overall volume control but rather the ability to modulate various frequencies.

2.2.2 Localization

Data suggest that older adults are less adept at localizing sounds in space, specifically being prone to front/back localization errors (Abel et al., 2000). When high-frequency deficits occur, localization is more difficult in the elevation dimension (up vs. down) than in the azimuth (right vs. left) (Noble et al., 1994). Furthermore, higher-frequency stimuli are harder to localize for all ages because high-frequency stimuli reach both ears at the same time (Lorenzi et al., 1999).

Design Implications and Suggestions The reduced ability to localize high-frequency sounds is another reason to avoid high-frequency auditory stimuli. When an auditory stimulus is intended to direct the older user's attention to the source of the stimulus, the stimulus should be presented between 5000 and 8000 Hz. Furthermore, auditory stimuli designed to orient attention should not be presented directly behind or in front of the user. This is especially relevant in workstations or other scenarios where the user is likely to remain in the same space. Sounds that must be localized should be presented for durations long enough for people to turn their heads and localize the sound, thus eliminating the error-prone front/back scenario.

2.2.3 Degraded Stimulus Environment

Noises are not often pure auditory stimuli. Many auditory signals as well as speech occur within a noisy environment—for example, at a workstation with the hum of computer fans and conversing co-workers in the background. Research has shown that older adults have greater difficulty than younger adults perceiving speech in such degraded auditory conditions. There is some debate concerning whether the locus of the difference is primarily cognitive or perceptual (B. Schneider et al., 2000). Regardless of the absolute locus of the effect, noise degrades auditory perception more for older than for younger adults.

Design Implications and Suggestions This difficulty in perception under noisy conditions demonstrates the importance of using cues in the visual modality instead of the auditory when presenting information in a potentially noisy environment. However, auditory cues can be used to augment visual cues, via redundant or dual coding. Dual coding is beneficial even in quiet environments, as users' visual attention may be directed elsewhere when information needs to be communicated to them. Younger adults may perceive an auditory stimulus easily, but older adults will have more difficulty. Speech perception specifically can be hindered in high-noise environments, particularly if the people have poor hearing (B. Schneider et al., 2000). However, given that the locus of the problem is perceptual (as opposed to cognitive), age-related differences are likely to be evident for auditory stimuli other than speech.

For optimal perception, the signal should be presented independent of any noise. For example, in training materials, there should be no sound except for the relevant instructional materials (e.g., no background music). If the auditory signal can be amplified independent of background noise, users should be offered this capability (e.g., headphones at a museum display). If this is not possible (such as in the automated speech in an elevator or subway car), text should be presented to provide redundant information. Compressed speech is more difficult for older adults to perceive [although Sharit et al., (2003) indicate that 10% compression has little effect on young, middle-aged, or older adults]. It is recommended that speech rates not exceed 140 words per minute (Fisk et al., 2004). In public presentation of information, where ambient speech and other noise may be present, provide wireless headphones to amplify the signal, if feasible. Sound-absorbing materials on floors, walls, and ceilings may be used.

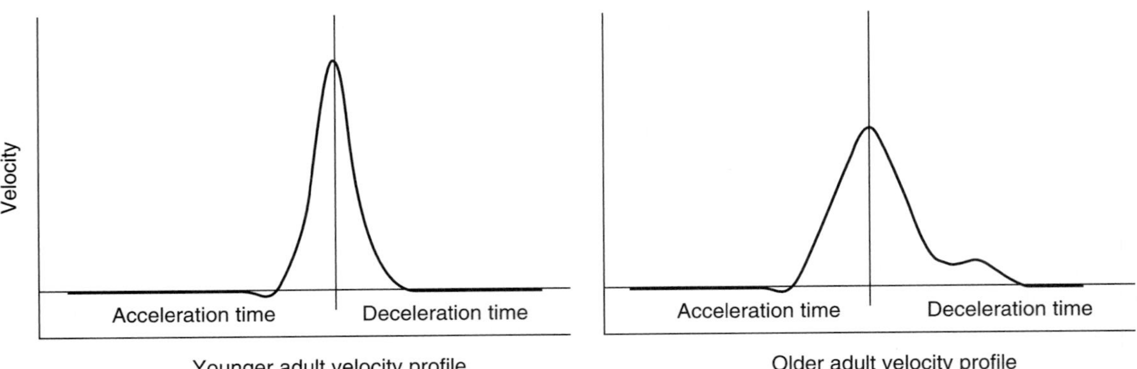

Figure 4 Examples of a younger adult's and an older adult's velocity profiles in a reaching task. Note the longer time to make the movement and greater variability during the deceleration stage in the older adult profile. (Adapted from Ketcham et al., 2002.)

3 MOVEMENT AND BIOMECHANICS

3.1 Movement Speed

Movement speed is the speed with which a person can make a movement after the requisite cognitive processes to start the movement have occurred (Spirduso, 1995). In general, older adults are slower in their movements than younger adults (Stelmach and Nahom, 1992). Figure 4 shows the typical finding of greater overall time to make a reaching movement for older adults relative to younger adults and the slower peak velocity.

Design Implications and Suggestions Reductions in movement speed are relevant to a wide range of activities and scenarios performed by older adults. Traffic light timers should be set to provide adequate walking time for older pedestrians. The required double-click speed for mouse buttons on public computers should be amenable to older adult users. Other examples include the speed of revolving doors, self-operated credit card terminals, and the time required for text entry in cell phones. Any task that requires relatively rapid movement is potentially difficult for many older users.

3.2 Movement Control

In effect, older adults will be slower in tasks that involve grasping (Carnahan et al., 1993), reaching (Seidler-Dobrin and Stelmach, 1998), and continuous movement (Wishart et al., 2000). Their movements involve more submovements and shorter initial primary submovements than those of younger adults (Walker et al., 1997; Seidler-Dobrin and Stelmach, 1998; Smith et al., 1999). This essentially results in slower, more variable movements in older adults. Furthermore, as a movement task becomes more difficult, older adults slow at a greater rate than do younger adults (Ketcham and Stelmach, 2001).

Age-related differences in computer mouse performance have been reported. For example, in four common mouse tasks—pointing, clicking, double-clicking, and dragging—older adults were slower in their movements, produced more submovements, and made more errors, particularly for double-clicking tasks (e.g., moving out of the icon range before it was double-clicked) (Smith et al., 1999; see also Walker et al., 1997). Also, older adults are less able to coordinate multiple movements with multiple body parts as well as younger adults, such as bimanual tasks (e.g., twisting-off a lid) (Stelmach et al., 1988).

Design Implications and Suggestions The variability in movement control and speed contribute to older adults' greater variability in their use of input devices, both within and between people (Rogers et al., in press). It is possible that with increased perceptual feedback, older adults can be trained to lengthen their initial movements, thereby also reducing the required number of submovements. For example, mice could be operated in conjunction with software that provides tactile feedback as the cursor nears an icon or target. "Sticky" icons have been employed that attract the cursor, within some user-defined sensitivity range. Users should be provided with the option to increase icon size, effectively improving the physical stimulus.

Older adults have been shown to have difficulty decelerating as they approach a target and trouble staying at the target once reached (Smith et al., 1999). This is consistent with the common finding that older adults employ more submovements as they home in on a target (Walker et al., 1997). Discrete hand movements are often required in computer tasks, whether with a mouse, touch screen, light pen, or other input device.

In software design, icons should be a reasonable size, as older adults have difficulty navigating to small targets (Walker et al., 1996). Furthermore, older adults are slowed in their navigation more than younger adults by targets embedded with other stimuli (Rogers et al., in press). Thus, in addition to making icons and target stimuli easier to select, creating appropriately sized icons can help to separate those relevant targets perceptually from irrelevant or background stimuli.

Older adults' difficulty with coupled movements and other movement coordination tasks can cause problems for bimanual tasks, such as opening a twist-top pill bottle or reaching and grasping coordinated movements. Thus, products that require a coordination of multiple movements should be redesigned if possible. For example, pill bottles designed for older adults should employ a tabbed top.

3.3 Balance

Falls due to loss of balance are a serious problem for older adults (Horak et al., 1989; Agnew and Suruda, 1993). Postural sway increases with age (Borger et al., 1999; Kristinsdottir et al., 2001), leading to loss of balance and falls. Postural sway is affected by several variables, including poorer vision, which reduces older adults' ability to detect movement cues in the environment that would indicate sway; reduced sensitivity to vibrations in the lower limbs, which reduces the contraction signals sent to muscles (Kristinsdottir et al., 2001); and cognitive demands (Lajoie et al., 1993; Brown et al., 1999; Melzer et al., 2001). It appears that older adults rely more on dynamic visual cues to aid in their balance. When visual cues indicated that they were moving (although they were in fact stationary), older adults were much more likely to exhibit postural sway (Borger et al. 1999). Older adults also tend to have greater postural sway (relative to younger adults) when in a moving visual environment and when flooring was more compliant (e.g., carpet vs. hardwood flooring), age-related differences in postural sway were greater (Redfern et al., 1997).

Balance is affected by cognitive demands. When participants were asked to perform simple subtraction while stationary and while recovering from a small movement of the platform on which they were standing, older and younger adults performed similarly on the subtraction task prior to perturbations of their platform. However, older adults were repeatedly and significantly slower after a perturbation, suggesting that the cognitive demands of maintaining balance interfered with their ability to perform the subtraction task (Brown et al., 1999).

Design Implications and Suggestions Because older adults may be more prone to losing their balance while getting on and off moving sidewalks, warnings should be provided with a long lead time prior to the need to correct posture. Apparent motion and true motion in the environment can also increase the likelihood that older adults will lose their balance due to their reliance on perceptual cues. Examples include wall-sized animated advertisements and moving subway cars. Environmental support may be provided to reduce this problem. For example, in designing platforms in front of moving subway cars, a row of lights can be provided above the railway cars, resulting in a stationary stimulus to counter the perceptual cues of the train. Cues may be beneficial in other contexts as well. On walls next to stairs and pedestrian ramps, arrows can be displayed indicating the direction of the change of slope in the walkway. For any scenario where balance may be an issue, handrails are highly recommended.

3.4 Locomotion

The slower walking speed of older adults is due to several factors: preference/strategy, decreases in strength, joint motion, and endurance. The decrease in locomotion speed is the primary factor in gait changes as well. Gait changes include shorter steps and an increase in the time that both feet are on the ground, increasing support and balance (Spirduso, 1995; Lockhart, 1997). However, when cognitive demands are increased, walking becomes more demanding, presenting functional problems, particularly for older adults. For example, in one study, age-related differences in a memory task were greater when both age groups were walking than when they were stationary (Li et al., 2001). The researchers interpreted the results in terms of older adults selecting the more important walking task over the memory task (presumably staying balanced was considered more important than memorizing words). Cognitive demands can also result in movement decrement as well, as older adults walked at a slower speed when performing a cognitive secondary task (Lajoie et al., 1996).

Design Implications and Suggestions Older adults walk more slowly than younger adults, and dual-task data suggest that older adults may change their movement speed and their cognitive performance. Slower locomotion for older adults has implications for pedestrian crossings, vehicles with automatic doors (such as elevators and subway trains), and other situations in which older adults are expected to keep pace with younger persons. Attendants should be trained to be prepared for these locomotion limitations. Particularly in instances where cognitive demand is high, such as crossing busy streets, walking in the downtown district of a large city, or any novel walking environment, the walking speed and balance of older adults may be significantly compromised. Although several studies have shown that the cognitive task is compromised as opposed to the walking task, this is probably due to the relative insignificance of the cognitive task. When searching for landmarks in an unfamiliar business district, the cognitive task may be more important, and attention to one's walking may be reduced.

3.5 Strength

Muscular strength is maintained through much of adulthood, beginning to fall off around age 60 (Ketcham and Stelmach, 2001). Strength drops off as a result of muscle mass loss, but with appropriate exercise and training regimens, strength and muscle losses can be abated to some degree.

Design Implications and Suggestions As with changes in speed, changes in strength have widespread effects on the functional limitations of older adults. For those systems and products that may have older adults in the user population, designers must consider the reductions in strength for all tasks that require actions such as pushing, pulling, lifting, twisting, and pressing. Some pill bottles are now designed to be secured such that minimal force is required to open them, and other products that require actions such as these must be designed with older adults' reductions in strength in mind. If force requirements for a product cannot be reduced, assistive aids should be provided.

3.6 Force Control

Controlling one's force correctly can be critical in maintaining one's safety (e.g., holding onto a grab bar in the shower), avoiding embarrassment (e.g., accidentally crushing a beverage-filled Styrofoam cup at a party), or simply manipulating an everyday device [e.g., rotating the jog dial on a personal digital assistant (PDA)]. We focus primarily on control of force involving the hand and fingers. Older adults employ a grip force up to twice that of younger adults, and their force beyond that required is also twice that of younger counterparts (Cole, 1991). This indicates both a perceptual component (older adults, particularly over 60 years, do not accurately perceive the friction of the object being gripped) and a strategic component (perhaps aware of their inclination to misgrip objects or misperceive friction characteristics, they overgrip intentionally) (Cole et al., 1999).

Design Implications and Suggestions The ability to control the force employed in a task is decreased significantly as one reaches old age. This has design implications for the structural soundness of products and the design of controls. For example, some computer mice have scroll wheels that can function as a button when clicked. If this button is too sensitive, older adults may have considerably more trouble than younger adults in scrolling without activating the function intended by the button. In designing handles for gripping, extra texture should be provided to offset the perceptual loss experienced by older adults. This can help to reduce the overgripping behavior, which may cause early fatigue when using the product.

4 COGNITION

4.1 Attention

Attention is not a unitary scientific construct. Indeed, there are well-known varieties of attention (James, 1890/1950; Parasuraman and Davies, 1984). Two such categories of attention are selective attention and attentional capacity. Research on selective attention has focused on the ability to focus on and process a restricted set of goal-relevant information while ignoring available information that is not relevant to the goal (Johnston and Dark, 1986). Attentional capacity research has investigated the amount of "mental work" that humans can perform at a given time, often employing dual-task methods, where the trade-off in performance between the two tasks can provide a measure of attentional capacity given to each task. Selective attention and attentional capacity are affected deleteriously by age (see Rogers and Fisk, 2001, for a review), but certain interventions and design provisions have been shown to reduce these decrements to some degree.

4.1.1 Selective Attention

Selectively attending to goal-relevant stimuli and ignoring goal-irrelevant stimuli is required for efficient performance in any task. Selective attention involves purposefully shifting attention to different stimuli and categories of stimuli in the environment. For example, when driving, a person may be actively searching for certain elements or groups of elements, such as other cars, pedestrians, and traffic signs and signals. Irrelevant stimuli such as a car alarm or brightly colored or waving advertising sign can distract the driver momentarily. The degree of susceptibility to distraction and the duration of the distraction can obviously have severe consequences in this task domain and in others. Older adults are susceptible to distracting effects of irrelevant stimuli in the environment (Rogers and Fisk, 2001).

Selective Inhibition Deficits There has been considerable research in the field of cognitive aging, suggesting that older adults have relatively more difficulty inhibiting irrelevant information (e.g., Hasher and Zacks, 1988; Stoltzfus et al., 1993). However, there is not a general deficit of inhibition because certain inhibitory systems appear unaffected by aging (e.g., Connelly and Hasher, 1993; Kramer et al., 1994). For example, younger and older adults were equally able to adjust their focus of attention to include or exclude information (Hartley et al., 1992), suggesting that older adults, in this case, did not have greater difficulty inhibiting irrelevant information. Older adults were also able to inhibit the location of irrelevant stimuli as well as younger adults, although they were impaired relative to younger adults in their ability to suppress the identities of irrelevant stimuli (Connelly and Hasher, 1993).

This reduced efficiency in selectively deploying attention can have deleterious consequences for performance, especially in highly attention-demanding tasks such as driving. Older adults seem to be inefficient in novel searching visual environments, for example, searching in the same area repeatedly (Maltz and Shinar, 1999), suggesting that they are not monitoring where they have searched previously. In this driving-based task, adults were less likely to maintain attention in a search task, less likely to discriminate previously attended areas, and less likely to attend selectively all relevant areas of the display. It is important to note that after extensive consistent training, search performance is still slower for older adults, but the qualitative aspects of the search, such as the learning curve, are similar for younger and older adults (Fisk and Rogers, 1991).

Attending to Multiple Tasks Younger and older adults appear to deploy attention differently across multiple simultaneous tasks after training (Sit and Fisk, 1999). Younger and older participants viewed a screen with a different task in each quadrant and were encouraged (via a points-based system) to attend primarily to one task over the other three (but overall performance was encouraged as well). After training on the multiple-task display, older adults were still performing at a lower level than younger adults, although age-related differences in performance had lessened over training. They were then required to attend more to a different task at transfer—that is, the task demands did not change, but the way in which attention was deployed was changed. Older adults focused primarily on the newly important task, to the exclusion of the rest of the tasks. Younger adults were able to perform effectively on three of the four tasks, including the newly important task. The findings suggested two things: (1) age-related differences in a divided attention task requiring different selective attention allocation strategies are attenuated after training, and (2) older adults are less able than younger adults to strategically change the way in which they perform a previously learned task.

Design Implications and Suggestions Age-related changes in selective attention must be carefully considered in any environment in which multiple stimuli are presented. Tasks and environments with multiple displays and controls, such as driving, cockpits, security and surveillance tasks, medical displays, and industrial control panels, all require the user to attend to a subset of a multitude of auditory and visual stimuli in the environment.

One way to improve older adults' ability to select goal-relevant stimuli from a distraction-filled environment is to increase the perceptual salience of the goal-relevant stimuli (or decrease the salience of distracting stimuli) (Shaw, 1990, 1991). When the physical stimulus is improved, the contrast between goal-relevant and goal-irrelevant stimuli is increased, and the relevant cue can be used more efficiently. This form of environmental support can help guide a user's selective attention to relevant stimuli.

In addition to increasing the perceptual salience of relevant stimuli, whenever possible, these stimuli should be made salient to the user via explicit means as well. For example, older users should be told how best to differentiate and ignore distracting stimuli (i.e., essentially giving older adult users a specific strategy to follow). Consider using an instructional manual to put together a complex product. The manual may include information about safety, rebates, warranty information, and so on, in addition to the specific, sequential assembly information. In designing this manual, the actual assembly steps should be perceptually salient, relative to other information in the manual. For example, assembly steps should be boxed in a particular color or set off with bold header information. This enables older users to better identify the relevant assembly information in a field of irrelevant information (for the assembly task). Furthermore, the designers of the manual should explicitly inform the user of the cues they should be looking for, as opposed to requiring users to recognize the relevance or meaning of different cues on their own.

Older adults' attention is more likely to be drawn to perceptually salient stimuli in the task environment. It is important to minimize the attention-attracting nature of irrelevant stimuli. Attention-attracting perceptual characteristics include flashing, moving, bright, loud, and unexpected stimuli. Through training, older adults can improve their ability to successfully select a subset of relevant stimuli. This critical finding suggests that the decrement in selective attention is, at least in part, a more labile age-related change and that, with training, older adults are capable of selecting important stimuli among distracting stimuli. Hence, in situations requiring selective attention, training should be given to users until the criterion performance level is reached.

4.1.2 Attentional Capacity

Divided Attention Users can attend to and cognitively process a limited amount of information at a time. The hypothetical construct of attentional resources has been used to explain the capacity to process, think about, and cognitively manipulate information at a given time (Wickens, 1984). Older adults are presumed to have a reduction in the processing resources available to them to perform attention-demanding tasks (e.g., Crossley and Hiscock, 1992). For example, older adults have relatively more difficulty maintaining appropriate levels of performance when required to perform multiple tasks at once (Kramer and Larish, 1996), that is, under divided attention conditions. Clearly, older adults experienced a greater decrement in performance transferring from single to dual tasks than did younger adults (see, e.g., McDowd and Craik, 1988). In a study of dual-task performance, testing the efficacy of various vehicular travel aids for different age groups (e.g., automated visual map aids, synthetic speech, paper maps), older drivers had more safety-related errors in this dual-task environment than did younger adults (Dingus et al., 1997). However, when redundant auditory guidance was provided in addition to the automated map aid, older adults performed more safely than without the redundant information.

Visual Clutter Even within what seems to be a single task, attention can be overloaded and performance can suffer. In a search-type task performed on a busy, noisy visual display, people are required to orient and reorient their attention as they scan the display. As stimuli become more similar or increase in number, identifying and comparing stimuli becomes more demanding (Shiffrin, 1988). In general, older adults have more difficulty in higher-clutter environments (e.g., Schieber and Goodspeed, 1997). In an investigation of the effect of clutter in a typical driving scene (i.e., a typical two-lane rural highway, a commercial district in a small city, and a downtown metropolis scene), older adults' speed and accuracy at detecting target signs was considerably poorer than younger adults (Schieber and Goodspeed, 1997). This pattern was observed whenever even a small amount of clutter was present.

Automatic Processing With repeated exposure to consistent stimulus mappings, people can be trained to automatically attend to and respond to stimuli that are consistently meaningful in the task (e.g., targets in a search task). This has been termed an automatic attention response (W. Schneider and Shiffrin, 1977; Shiffrin and Schneider, 1977). Consistent mapping refers to the degree to which stimuli belong to a given category when they appear. In experimental search tasks, consistent mapping occurs when target stimuli never appear as distractor stimuli and distractor stimuli never appear as targets. The concept of consistent mapping (and conversely, varied mapping) extends directly to natural tasks such as driving (brake lights are consistently paired with a slowing of the vehicle) and computing (icons are consistently mapped with the applications they represent, and keyboard keys are consistently mapped with respect to their location and function). For these consistent features in the environment, younger and older users can develop very quick and accurate responses, but older adults' responding will be less efficient than younger adults'.

An automatic attention response can be important in the fast, accurate detection of stimuli in the environment. For example, brake lights may cause a driver automatically to attend to the lights and respond by braking, given enough exposure to consistent instances of brake lights co-occurring with the slowing of the vehicle ahead. Older adults may not develop new automatic attention responses to learned stimuli in visual search tasks, but considerable performance improvements occur under consistent conditions (Fisk et al., 1988; Fisk and Rogers, 1991; Rogers, 1992).

Design Implications and Suggestions

Performance Gains with Training Consider the plight of novice drivers. In this new environment, with new displays and controls, there are a host of stimuli to monitor, to search for, and perhaps to respond to. Each and every scenario is completely new, and even after some experience, novel instances occur with high frequency (e.g., rarer scenarios such as passing a cyclist or avoiding a large pothole). For these novice drivers, speaking to a passenger (let alone speaking on a mobile phone), adjusting the radio, or waving to someone on the sidewalk can be difficult and error prone (either in the main driving task or the secondary task). However, over time, considerable learning occurs. Similar instances occur multiple times and are remembered, allowing faster responses subsequently. The location of the gas, clutch, and brake pedals are quickly located and never confused. Eventually, attention can safely be divided between driving and other tasks. In the same way, through sufficient, appropriately designed training, performance on a task can become considerably more efficient. Older adults' performance can improve greatly through such experience, and this is the goal of training programs. For example, initial age-related differences in a divided-attention task can be attenuated (though still present) after training (e.g., Rogers et al., 1994; Sit and Fisk, 1999).

Part-Task Training To reduce the demands on attention in dual-task conditions, participants can be trained on certain parts of the task at different times before being trained on the whole task—that is, part-task training. For example, participants without computer experience must often undergo mouse training before beginning a study involving computers with mice as the control. Thus, participants are able to devote a majority of their attention to the experimental task instead of dividing their attention between an untrained, novel device and the task. To assess the benefits of part-task training, Kramer et al. (1995) presented a dual-task condition to older adults, with half of the participants required to devote an equal amount of attention to each task and the other half required to pay more attention to one task at certain times and to the other task at other times. Thus, the second group practiced on, essentially, only part of the task at different times throughout training. By the end of training, older adults in the part-task group demonstrated considerably more learning than those in the first group.

Redundant Information When information can be presented redundantly in certain dual-task conditions, older adults benefit from the reduction in task demands (e.g., Dingus et al., 1997). Redundant information has been shown to be important in cluttered environments as well. Thus, providing redundant information may be a means of providing environmental support to compensate for age-related reductions in attentional capacity.

Clutter in the Visual Environment The effects of clutter in a visual environment such as driving can be very detrimental to performance, particularly for older adults. Older adults spend more time than younger adults searching in a cluttered environment and spend more time making decisions about stimuli (Ho et al., 2001). This can obviously be detrimental to driving performance and to one's safety. However, this situation can be ameliorated if the user is provided attentional cues designed to support the user's performance. This sort of environmental support was tested for younger and older participants in a driving simulation task that involved making quick decisions about whether a left-hand turn could be performed safely (Staplin and Fisk, 1991). When normal driving contextual information was present in the task (i.e., clutter stimuli such as lampposts, trees, and houses), both younger and older adults benefited from receiving a cue about the future status of the intersection, thus aiding their left-hand-turn decision.

4.2 Memory

Memory can be divided into three general stages. If information is to be remembered, it must first be encoded, then it must be stored or represented in some way in the brain, and then it must be retrieved from storage. Within this basic framework, researchers have focused on various types of memory, such as memory for information that occurred at a certain time and place, memory for facts, memory for procedures, memory to do something in the future, and memory for the source of information.

Age-related declines in some aspects of memory (such as working memory and episodic memory) are well documented (Zacks et al., 2000). In some cases, these age-related declines can be improved by placing information in the task environment, instead of requiring people to maintain the information in memory (see Table 3 for several memory conclusions). In other cases, there are minimal changes across the lifespan, such as in semantic memory (memory for facts) and procedural memory (memory for how to perform an activity or sequence of actions) (Zacks et al., 2000).

4.2.1 Working Memory

Working memory can be thought of as information that is actively being processed and "used" (Baddeley, 1986). Similar to the concept of capacity limitations on attentional processing, it is typically measured via span tasks, which measure the number of elements that can be kept activated in working memory. Reduced working memory capacity in older adults is a hallmark finding in the cognitive aging literature (see Zacks et al., 2000, for a review), and a meta-analysis of the literature revealed a sizable age-related difference (Verhaeghen et al., 1993).

Table 3 Five Conclusions about Memory and Aging

Conclusion	Examples
Older adults maintain their semantic memory, but their ability to retain episodic memories is decreased.	Older adults can remember the names of friends they would like to e-mail, but remembering a new e-mail password will be difficult.
Maintenance of episodic memories can be particularly hampered when atypical, distracting elements are present when retrieval is attempted.	Older adults are more likely to forget to pick up milk on the way home if traffic is unusually hectic.
The retention of habits, or processes performed automatically, is relatively spared in older adults.	Although learning a new software package may be difficult, older adults retain the learned ability to type.
Working memory capacity is reduced in older adults.	Older adults will need to playback a set of auditory instructions more times than younger counterparts.
When older adults are required to perform self-initiated processes, memory is more difficult. Environmental support can reduce these difficulties.	Older adults will have difficulty recalling the color, make, and model of a hit-and-run vehicle, but if given a lineup, they will probably be able to identify the correct vehicle.

Source: Adapted from Fisk and Rogers (2002).

Design Implications and Suggestions The well-documented age-related decrement in working memory capacity has clear implications for designers. Older users should not be required to keep multiple items in memory. E-commerce Web sites should provide comparison programs that allow customers to compare similar products as opposed to keeping in memory variables such as price and features. A telephone voice menu should have deep as opposed to broad menu structures, so that users do not have to keep too many options in memory before they make a selection. In general, information should be displayed to the user (i.e., putting the information in the environment), as opposed to requiring users to rely on their working memory.

4.2.2 Episodic Memory

Significant age-related differences exist in the ability to recall various events and instances which are referred to as episodic memories. In a typical episodic memory study, participants are shown stimuli and asked to recall them at a later time (see Tulving, 2002, for a review). Age-related deficits in these tasks are commonly found. In this section we discuss two ways that this fundamental age-related difference in memory can be addressed and supported: memory strategies and supportive information.

Memory Strategies When older adults are required to elaborate internally the stimulus to be remembered, it is later recalled with greater accuracy (Park et al., 1990b; Verhaeghen et al., 1993; Dunlosky and Hertzog, 1998). For example, when older adults are required to generate words for later recall as opposed simply to reading a list, they recall more accurately (Hirshman and Bjork, 1988; Johnson et al., 1989). Some studies have demonstrated that older adults may engage in suboptimal encoding strategies, which may explain their relative deficit in remembering (Rogers and Gilbert, 1997; but see Dunlosky and Hertzog, 1998).

In assessing older adults' associative learning ability, Rogers and Gilbert (1997) found that some older adults chose continually to use an inefficient strategy to perform a task, whereas nearly all the younger adults employed the optimal strategy. The task presented a consistent set of word pairs at the top of the screen and a test word pair in the center. Participants were required to indicate whether the test pair was one of the pairs at the top of the screen. The task involved multiple trials, such that it was optimal to attempt to memorize the word pairs above as opposed to searching for a match on each trial. Older adults were less likely than younger adults to adopt optimal strategies spontaneously in this associative learning task, but they were able to use the optimal strategy if encouraged to do so. A larger study of individual differences with this same task replicated the finding that older adults were less likely overall to adopt an optimal strategy (Rogers et al., 2000). However, for those older adults who did adopt the appropriate strategy, the age-related differences in learning were reduced. Taken together, these studies suggest that in certain cases, older adults' performance may be improved by providing optimal strategies for performing a task.

Cognitive Support Memory researchers have conducted numerous studies testing the effects of memory cues and aging. Relatively few studies have shown a greater benefit of memory cues for older adults than for younger adults; however, in general, these studies show that older adults' memory can be improved through the use of memory cues. For example, adults of all ages increase their recall when some aspect of the stimulus is present at recall (e.g., a word fragment vs. freely recalling a word); this is the basis for Craik's environmental support framework (Craik, 1986). Memory retrieval has also been shown to increase when stimuli are studied in a visually distinctive context (Park et al., 1990a). Younger and older participants studied a set of different objects on either a colorfully distinctive background or a plain background. They were later asked to place either labeled note cards or the original items back in the original spatial arrangement. Both age groups benefited equally from the distinctive background condition as well from the use of the original items.

In another study, researchers presented target words (i.e., words to be remembered) to older and younger adults, either integrating the target with an object in the environment (e.g., "The key fit the lock on the file cabinet," where a file cabinet was in the test room) or integrating the target with an object not in the environment (e.g., "The key fit the lock on the car," where no car was present in the test room) (Earles et al., 1996). Younger adults benefited more than older adults from this combination of environmental cues and the target word.

Design Implications and Suggestions

Strategy Suggestions Older adults' episodic recall can be improved through the use of different strategies. One of the more commonly used is a form of encoding elaboration. For example, to remember a pair of words better, it can be helpful to construct a distinctive, imagistic sentence or concept that links the two words (e.g., "dog" and "spoon" → "The dog balanced the spoon on its nose"). Other heuristics can be employed, such as creating acronyms from a series of words to be remembered, repeating a set of instructions, or creating a link between a stimulus and some internal concept. However, many people do not use such memory strategies spontaneously outside the lab. Therefore, a training system should provide strategy suggestions when memory is required.

Although older adults are not as likely as younger adults to adopt optimal strategies spontaneously, they are able to utilize these optimal strategies if encouraged to do so. There are a number of workable methods for encouraging participants to use memory strategies. These include explicitly instructing the participants (Hulicka and Grossman, 1967; Haider and Frensch, 1996; Nichols and Fisk, 2001), pretraining on an orienting task that makes apparent the optimal strategy (Hulicka and Grossman, 1967; Doane et al., 1996; Rogers and Gilbert, 1997), or giving intertask tests that will encourage use of the desired strategy (Rogers and Gilbert, 1997).

Physical Reminders Given older adults' knowledge about their memory capabilities, physical reminders can play an important role in older adults' lives. Older adults reportedly use various physical reminders for medications (Park et al., 1992; Sanchez et al., 2003), and adults over age 71 years benefited most from the physical reminders (Park et al., 1992). Some suggestions for reminders include the following:

1. Physical reminders should be placed in visually salient places, where the person will see them. For example, one can place medications for the following day next to one's toothbrush at night.
2. Given age-related issues in *source monitoring*, the ability to remember the source of an event, the reminder should provide relevant information instead of simply reminding that there is *something* to be remembered.
3. Automated visual and auditory reminders, such as those used in PDA software and Microsoft Outlook, can minimize the need for older adults to search actively for the reminder.

Environmental Support Framework Older adults' memory should benefit from distinctive contexts insofar as the context is present when retrieval occurs. The context serves to provide additional retrieval cues for the person. Thus, if the context is absent, recollection may be harmed. The environmental support framework is based on the notion that older adults appear to have more difficulty with effortful, internally driven processes, such as freely recalling an event or appropriately employing a task strategy (Craik, 1986; McDowd and Shaw, 2000). Because of these difficulties, older adults rely more on contextual or environmental information to aid their performance. Thus, providing useful information in the environment of a task can aid performance. However, because there is inevitably distracting stimuli in the environment in addition to the supportive information, older adults may have difficulty ignoring the irrelevant stimuli.

If memory requirements in a task are offset by the existence of readily accessible information in the environment, attentional and memory processes can be allocated elsewhere, to more demanding aspects of a task. For example, in a typical software application, instead of using function buttons with icons only, at least provide the option for turning on text labels, which eliminates the need for novice users to experiment with and memorize the functions of the buttons.

Environmental support can come in various forms:

1. Provide some characteristic of the stimulus to be recalled to aid, or cue, recollection.

2. Provide an outline or map of the material. For Web browsing tasks, navigation aids should be provided if desired by the user. These can provide a visual history of where the user has been, reducing the reliance on memory for the structure of the Web site.

3. Physical aids constitute a form of environmental support. They serve to remind the user of the previous encoding instance. Specific aids, or cognitive prosthetics, can be used to assist older adults' memory (Morrow, 2003)—for example, "intelligent" PDA software designed to store grocery lists and to recommend items that have been purchased commonly in the past.

4. Structuring text appropriately can benefit recall of the text (discussed below in Section 4.2.5).

5. Search and detection of target stimuli benefit from consistent arrays of irrelevant stimuli (Chun and Jiang, 1998; Jiang and Chun, 2001). That is, attention can be guided to targets by the knowledge gathered from the consistent arrays of distractor stimuli.

4.2.3 Prospective Memory

Prospective memory involves remembering to do something in the future and is essential in planning and completing general daily activities (e.g., fulfilling appointments, performing household chores). It can also be critically important for safety, such as remembering to take medications or turn off the oven. There are two categories of prospective memory—time-based and event-based—which differ in the degree to which the rememberer must rely on self-initiated cues (Einstein et al., 1995; Park et al., 1997). In event-based prospective memory, the person is cued about the to-be-remembered information by some external event or stimulus (e.g., remembering to take a medication when a timer goes off); whereas in time-based prospective memory, the person must remember after some amount of time has passed (e.g., remembering to take a medication at two o'clock in the afternoon). As is typically found in cognitive aging, when self-initiated processing is required, older adults' performance suffers (Craik, 1986), and age-related differences in prospective memory are greater for time-based situations (Einstein et al., 1995).

Although older adults' prospective memory is impaired relative to younger adults (Park et al., 1997; Zacks et al., 2000), the memory phenomenon under heavy task demands may provide a better representation of prospective memory in real tasks. Under these demanding conditions, age-related differences are exacerbated. Einstein and colleagues (Einstein et al., 1997) investigated prospective memory in the context of additional, distracting activities (essentially, a divided-attention manipulation). They found that older adults had a more difficult time remembering to do the to-be-remembered action than younger adults when attentional demands were high.

Design Implications and Suggestions Older adults' difficulty with prospective memory has important implications for the design of memory aids and reminders. Physical reminders such as cognitive prosthetics are critical in reducing the functional effects of the prospective memory age-related difference. Whenever prospective memory is required (such as in remembering to take one's medications at various points throughout the day), time-based prospective memory tasks should be turned into event-based memory tasks. Alerts can be built into cell phones or PDAs or other small, unobtrusive devices, providing users with a memory aid. Essentially, these function as environmental supports, reducing older adults' need to rely on self-initiated memory processes.

4.2.4 Source Errors

As with all human memory, recall for the context in which a memory was first created is not necessarily reliable (Johnson et al., 1993). For example, one might recall that a car owner's manual provided information about an automotive repair but not which section of the manual it occurred in (i.e., external source monitoring). Several studies have shown that older adults are poorer than younger adults at source monitoring (Cohen and Faulkner, 1989; Ferguson et al., 1992; Johnson et al., 1995). For example, when the source of information was distinct (i.e., the gender and appearance of two speakers), older adults' source monitoring was on par with younger adults (Ferguson et al., 1992; Johnson et al., 1995). However, older adults were less able to utilize multiple cues to aid their memory for source (Ferguson et al., 1992), and when additional cognitive processing was performed between source and test, older adults had difficulty retaining the link between the distinctive perceptual source information and other aspects of the source, such as what the source said (Johnson et al., 1995).

Design Implications and Suggestions Older adults' memory for source can benefit from perceptual disambiguation. In the repair manual example above, the chapters may be made more distinctive by using different-colored paper in each chapter. The memory for a given chapter includes the color of the paper, and this memory for color may be cued when the user flips past that color in the manual. Thus, the user does not need to actively retrieve the section of the manual, but instead, can easily recognize a particularly salient feature.

Because memory for source is impaired in older adults, particularly under intervening cognitively demanding conditions, memory aids should be provided to reduce the need for older adults to rely on self-initiated retrieval processes. For example, an older adult may not recall whether medical advice was given to them by their doctor or by a friend. This problem could be reduced if all medical advice given by their physician was also provided to them via a text file or printed transcript. This added information would serve as a redundant cue to the information source and provide the information in the world, rather than requiring the person to rely on memory for the source.

4.2.5 Semantic Memory

Despite the general negative view of aging and memory presented thus far, there are several characteristics of memory and knowledge that remain relatively robust across the lifespan, most notably, semantic memory (memory for facts or general knowledge, sometimes referred to as crystallized knowledge) (Cattell, 1963). Designers should take advantage of older adults' relatively preserved semantic memory.

Semantic memory has been used to improve older adults' memory for events in the future, specifically in the domain of medication instructions and health appointments. For example, using older adults' schemata (or crystallized knowledge structures about some domain) about a given memory task can be used to construct aids to help them in the memory task (Morrow et al., 1998, 2000). Younger and older adults share a schema for how medication reminders should be worded and arranged—specifically, they preferred shorter messages, incorporating, in order, time to take a medication, required dosage, duration one should take the medication, health warnings, and side effects (Morrow et al., 2000). When this schema knowledge was incorporated in an automated phone message application to present medication information in a way that followed their schema of presentation or violated the schema, both younger and older adults benefited from the schema-consistent version, such that they recalled the relevant information more accurately.

Design Implications and Suggestions Older adults' knowledge of existing systems and devices can be used as a tool to design systems that can easily be used and understood by older adults. Knowledge engineering, a technique that facilitates the understanding of how tasks are performed by gathering the knowledge used within a specific process is a critical phase in the design process, particularly with older adults, given their differences in knowledge and experience. Furthermore, older adults' crystallized knowledge may be extended to novel domains, where their knowledge can be transferred to similar novel applications and technology (this idea is discussed in Section 5.2).

5 LANGUAGE

Cognition and language are closely related, but given the importance of language in instruction design, text design, and other issues relevant to human factors, it is discussed within its own section. Language comprehension remains a critical function of people's lives as they age. With advancing age, people must be able efficiently to read labels of new medications, the instruction manuals of novel devices such as wheelchairs and health-related devices, and the warning materials that accompany these and other products and systems.

Often, older adults perform well in comprehending spoken and written language (Wingfield and Stine-Morrow, 2000). For example, they are able to comprehend figurative language as well as younger adults (Szuchman and Erber, 1990), and they are able to create an appropriate mental representation of text (Radvansky et al., 1990). However, there are several factors that can negatively influence older adults' comprehension of spoken and written language, more so than for younger adults. In general, these are related to demands on working memory (see Wingfield and Stine-Morrow, 2000, for a review).

5.1 Sentence Structure

Comprehension can be improved by reducing processing demands on older readers. For example, working memory can be taxed if many words and clauses bisect the subject and verb in a sentence (Norman et al., 1992; Wingfield and Stine-Morrow, 2000). Left-branching sentences contain a particularly difficult clause, as it comes between the subject and verb in the main clause and requires the maintenance of the subject while simultaneously processing the embedded clause. Left-branching sentences are particularly detrimental to older adults' comprehension (Norman et al., 1992).

Design Implications and Suggestions When designing instructions, warnings, and other text-based materials, the limitations of older adults' working memory should be considered. Comprehension will be improved if subjects and predicates are near each other, minimizing the need to keep the subject of the sentence in working memory for long periods. Following the guidelines put forth by Norman et al. (1992) and Kemper (1987) can greatly improve the readability of text, primarily by reducing the demand on working memory (see Table 4). For clearest writing, subject and predicate should be within close proximity.

5.2 Inferencing and Figurative Language

Research suggests that older adults are at a disadvantage relative to younger adults in making appropriate inferences (Hamm and Hasher, 1992). Furthermore, this age-related difficulty may be exacerbated in instances where older adults cannot rely on their crystallized knowledge (Arenberg and Robertson-Tchabo, 1985; Hancock et al., in press). Older adults interpret figurative language well (e.g., metaphors) (Szuchman and Erber, 1990). The use of figurative language taps the rich structure of knowledge that older adults have built across their lifetime and allows them to constrain their inferences with that knowledge.

Design Implications and Suggestions Often, it is necessary to make inferences beyond what is present in the text. For example, if a user manual for a lawn mower tells the user to "disable the starting mechanism before replacing the blade," an important inference one might make would be that one should also disable the starting mechanism before removing debris from the undercarriage of the mower. It is important to minimize the need for inferencing beyond the text. This can be especially critical in

Table 4 Examples of Cognitively Demanding Prose

Example	Problem	Revision
To change the level of the fluid in the round canister above the rotary encoder and the pressure dial, rotate it clockwise.	The reader must identify the referent for the pronoun "it." There are several intervening phrases between the pronoun and referent, causing a load on the reader's working memory.	To change the level of the fluid in the round canister, rotate the canister clockwise. The canister is located above the rotary encoder and the pressure dial.
The screw that is used to secure the tray above the cabinets so that the compartment is accessible is located in the yellow bag.	The subject ("screw") and predicate ("is located") are separated by a long clause, causing a load on the reader's working memory.	Locate the screw in the yellow bag. This screw is used to secure the tray above the cabinets so that the compartment is accessible.

the construction of warning materials (Hancock et al., in press). Older adults' ability to interpret figurative language combined with their high degree of semantic knowledge suggests that considerable information may be communicated through brief figurative text. That is, the figurative text cues older adults' extensive and rich semantic network of knowledge, potentially communicating considerably more information than is solely within the figurative text. This can be especially useful in the design of space-limited text, such as warning labels on pill bottles or cleaners, where space is severely restricted.

6 BELIEFS, ATTITUDES, AND MOTIVATION

Ageism biases include the belief that older adults are reluctant to interact with technology. Focus group research shows that older adults are actually motivated to use products when they are informed about the benefits (Melenhorst et al., 2001). In fact, reduced usage rates among older adults may be the result of a poor understanding of the benefits of the product, reduced income, and difficulty using certain products (Fisk et al., 2004). For example, many older adults appear interested and motivated to use the Web. A Web usage questionnaire of middle-aged (40 to 59 years), young-old (60 to 74 years), and old-old (75 to 92 years) showed usage proportions of 56%, 25%, and 10%, respectively (Morrell et al., 2000). When asked about their desires for Web activities, regardless of whether they currently used the Web, older adults' top three preferred activities were to use e-mail, to obtain information on travel or pleasure, and to obtain health-related information. Of the Web users, 66% reported being online for one year or more and were using the Web for an hour or more several times per week. Of nonusers, middle-aged users expressed the most interest in using the Web, whereas the old-old users expressed the least interest. Relative to other age groups the percentage of older adults who use computers is lower, but their usage rates continue to grow (Kinsella and Velkoff, 2001).

Experience with new technologies may increase older adults' willingness to use them. For example, after completing an experiment with an automated teller machine (ATM) simulator, the number of older adults who expressed interest in interacting with an ATM increased from 28% to 60% (Rogers et al., 1996). The relatively short experiment provided sufficient exposure to and knowledge about the system to motivate participants to use the technology.

Knowledge and anxiety about computers appear to directly influence interest in computers, mediating the role of age in computer interest as evidenced by significant correlations between age and computer knowledge and computer anxiety in a structural equation model (Ellis and Allaire, 1999). The r^2 value for computer interest was 0.49, indicating that nearly half of the variance in interest was captured by the model. The effect of age on computer interest was mediated primarily by computer knowledge and anxiety, although some age-related variance was directly related to computer interest (indicating other possible predictors).

Design Implications and Suggestions Computer knowledge is highly related to computer anxiety (Ellis and Allaire, 1999), suggesting the importance of educating older adults about computers. Understanding the potential benefits of computer technology and reducing anxiety associated with computers may increase older adults' willingness to use computers, and perhaps, technology in general (e.g., Czaja and Sharit, 1998). It is important to understand older adults' attitudes for any system or product they may encounter. Once older adults are made knowledgeable about the system or product (including the knowledge of how to interact with it successfully, the knowledge that they are capable of interacting with it successfully, and the knowledge about how it can benefit them), their interest and motivation to use it will increase, which in turn will result in greater attention to the system or product and lead to more efficient learning and utilization of the technology.

7 CASE STUDY: HOME TELEMEDICINE SYSTEM FOR OLDER ADULTS

We next present a hypothetical case study to illustrate the design implications derived from the age-related changes reviewed in this chapter. We have chosen a telemedicine system because of the increased exposure older adults will have with such systems and because telemedicine systems offer the opportunity to discuss most of the aspects of aging we outlined above.

Telemedicine is the deployment of health care across physical and/or temporal space, and home telemedicine systems are growing in popularity (Kaufman et al., 2003). Simple computer-mediated, home telemedicine systems can include a patient-administered medical device (e.g., a blood glucose meter), a personal computer, and an Internet connection. Patients may upload medical information to their physician's Web site or send information via e-mail. More complex systems may involve multiple patient-administered medical devices and software applications for displaying, uploading, and downloading both personal health data and a physician's advice and recommendations.

Home-based telemedicine applications include telerehabilitation, telecardiology, and general health care for a range of conditions, including such diseases as diabetes. For the purposes of this case study, we present a hypothetical home telemedicine system for monitoring one's blood glucose levels and related factors such as exercise and diet.

7.1 Description of the System

This system involves a self-administered blood glucose meter, a personal computer, a small Web camera, and related software. In this system, the patient interacts with (1) a blood glucose meter (BGM), (2) a heart rate monitor, (3) a PDA, (4) a personal computer with Internet access, (5) a Web camera (i.e., a camera that can transfer live video via the Internet), (6) software, and (7) instructional materials for all hardware and software. We assume a strict regimen that patients must follow, with multiple tasks throughout the day. The goal of this system is to keep careful measure of the patients' health and to teach patients about their condition. Patients keep track of their blood glucose levels, nutritional intake, and heart rate during strategic points throughout the day (using the devices listed above). At the end of the day, the information is uploaded to the patient's computer and sent to the primary care physician via a Web site. The BGM and PDA are kept with the patient at all times. Thus, patients must learn to interact with several different technological products. We will assume that these products are novel to the patients; thus, they must learn how to interact with them efficiently.

Several critical tasks must be performed during this process:

1. The initial task is to learn how to use the hardware and software.
2. Patients measure their blood glucose level with the meter three times per day. This information is stored in the meter during the day.
3. Patients keep track of the approximate sugar intake in their daily diet. This information is kept in a notepad by the patient during the day.
4. These data are entered into the computer each evening. The meter is attached via USB, whereas the other data are physically entered into a spreadsheet-like software application.
5. After the data are entered, they must be uploaded to the physician's computer.
6. Every two weeks, patients discuss their health care with their physician live via the Web camera. Other communication takes place via e-mail.
7. Patients are encouraged to monitor their data on a daily basis, and the data from all the variables can be displayed in different ways, including the number of variables displayed and the length of time to be displayed. For example, blood glucose level and diet can be displayed in combination over the course of a week, allowing the patient to see how the two variables interact.

7.2 Information Presentation

In this hypothetical system, several critical displays provide an interface to the older adult user. These include the computer display, the display windows of the BGM and heart rate monitor, and the PDA display. On many of these displays, instructional materials and other information are displayed in text and icon form.

7.2.1 Text

The type of text that is most commonly used in technological systems (such as the home telemedicine system) is primarily instructional text, informational text, and labels. That is, verbose and complex writing is typically unnecessary, and clarity and precision should be the primary objective. In creating instructional text for a computerized training program or for printed materials, organize the relevant information following standard information display guidelines. Avoid creating an interface with multiple frames, segmentation lines, text boxes, headings, icons, and links. The goal is to present minimal distraction and to present relevant information as clearly as possible.

- Group related elements (such as navigational links).
- Align elements in a list, generally to the left.
- Utilize common symbols to convey meaning efficiently (but avoid novel symbols or highly similar symbols with different meanings).
- Base text on a grid layout. Following this layout throughout an interface creates a common look and feel to the interface.

Perceptual Considerations When presenting textual information on a computer or in print, the text should be presented in high contrast (Figure 5) and in an easily readable font (e.g., sansserif fonts) (Morrell

Low-contrast text is difficult to read.
Serif and script fonts are difficult to read.
Eight-point font is too small.
This test has appropriate font, size (12 pt), and contrast.

Figure 5 Examples of poor and appropriate text presentation. (Adapted from Fisk et al., 2004.)

and Echt, 1997). Text should be divided into sections, with perceptually salient headers or labels (e.g., **Step 2: Replace the item**). Additionally, strategic use of white space will help to separate sections, reducing the need for visual search. Lines of text should not run across the length of a computer screen, and horizontal scrolling should be avoided. Text should be centered on the screen if possible, and in both printed text and text presented via computer, lines should run for 6 to 8 inches, to reduce the need for long visual scanning (Ellis and Kurniawan, 2000). If done consistently, this organization provides support for older adults in scanning to find headings and in reading the text. Text should be presented in 12-point type, and the highest possible contrast should be used. For example, a BGM can include a small black-on-gray LCD display to display blood glucose levels. However, this display will be less readily perceived by older adults than will a larger, lighted, high-contrast display.

Cognitive Considerations Text should not contain left-branching sentences or sentences with many clauses in them, which overload working memory. For adults of all ages, technical text should be written at a sixth-grade level (G. McLaughlin, 1969). Text should be organized via a small number of organizational principles. For example, in an instructional manual, chapters, subsections, and headers should use consistent conventions (e.g., subsection headings can be presented in a unique font size or in bold type).

7.2.2 Icons

Icons and symbols are often used in instructional manuals and software applications to convey meaning in minimum physical space. This can lead to small icons that are difficult for older adults to see and confusing icons that are difficult to interpret. In the hypothetical system, icons may be used in the spreadsheet software to represent functions such as "upload" or "save," they may be presented on the small buttons of the BGM, and they may be used to denote helpful hints and warnings in the instruction manuals.

Perceptual Considerations Besides icons that are too small, a graphically complex icon may be difficult for lower-acuity persons to perceive correctly. Also, if icons and symbols are color coded, care should be taken to avoid multiple colors from the high-frequency end of the color spectrum (e.g., blues, greens) and to use highly distinctive hues to avoid reducing the contrast of the icon.

Cognitive Considerations By serving as graphical representations of concepts and instructions, symbols and icons can be very effective in communicating a large amount of information in a small space. However, evidence suggests that older adults may have more trouble than younger adults in understanding symbols and icons, and usability testing should be conducted with older adults to define all icons (Hancock et al., 2004). When used, icons should always be accompanied by a textual label and description, at least initially. Structurally complex icons can also be problematic, not only in terms of comprehension, but for older adults, in terms of perception.

7.2.3 Information Visualization

The hypothetical software allows for multiple related variables to be presented at a single time, with the intent of providing information about the relationship among variables such as blood glucose levels, physical activity, and time of day. Data visualization can refer to a simple representation of data in a pictorial fashion, such as a pie chart. In this context, data visualization would require a multivariate display for representing the various relevant characteristics of the data in a multidimensional space. The limiting factor in data visualization can often be traced to the person viewing the representation. An understanding of human cognition and visualization constraints are essential to the development and selection of effective display characteristics. However, there has been little specific focus on age-related differences in information-visualization capabilities—thus, the best design approach here would be iterative user testing of the software with older adults. In addition, age-specific training might also have to be provided to enable older adults to capitalize on the functionality of the software (Hickman et al., 2003).

7.2.4 Modality of Communication

Physician–patient communication occurs in two primary ways in this system. Patients hold remote meetings with their physician, and the physician answers e-mailed questions from patients about their treatment and progress. Furthermore, patients can contact technical support if they have questions about the system itself. These three types of communication should be presented in specific modalities.

1. Physician–patient meetings should be done with video and audio, via webcams and computers. This allows both the patient and physician to observe nonverbal cues during the communication, which is important in physician–patient communication and provides the patient with a more personal experience.
2. In response to patients' questions about their treatment, physicians should provide e-mailed (text) responses. Text can be reviewed more quickly and easily than video or audio recordings and saved or printed more easily for future reference.
3. When a patient has a question about how to use a device properly or how to perform a task with the software, the technician should reply with a well-structured video reply. For any sequential task, video provides the optimal presentation modality. (Preferably these are prerecorded and usability-tested support videos, to relieve technicians from creating optimal videos on the fly.) When using speech to communicate information (e.g., a voice-over in a training video for the BGM), uncompressed speech is optimal.

7.3 Auditory Information as Primary and Secondary Cues

In this telemedicine system, auditory cues should be used both as (1) attention-gathering alerts and (2) redundant information accompanying visual cues. In scenarios where the user may be engaged in tasks independent of the system, auditory alerts are optimal given that the devices are kept on the patient's person during the day. For example, auditory stimuli can be used as a reminder that either breakfast has not been eaten or that the user has not yet entered nutritional information that morning. Auditory stimuli can be useful as redundant information when the user is interacting with the system (i.e., already attending to the devices or computer). For example, the computer software may present a unique auditory cue when the patient's daily data have finished uploading to the physician's Web site or if the upload did not function properly. Visual cues (such as a warning dialog box) would serve as the primary cue in this case, but the auditory alert would serve to attract attention in the event that the patient's visual attention is diverted.

7.4 Input Devices

7.4.1 Perceptual Considerations

Given the necessarily small size of buttons on handheld devices, labeling these can be difficult, particularly when considering the reduced acuity of older users. High-contrast symbols are generally necessary as opposed to text, and these should be clearly defined in instructional materials and training programs. Tactile feedback should not be used as the primary feedback for input controls. Older adults are less sensitive than younger adults to tactile feedback (e.g., Thornberry and Mistretta, 1981). Thus, when a button is pressed, visual or auditory feedback should be the primary means of feedback (discussed in Section 7.3).

7.4.2 Motor Considerations

PDAs, BGMs, and other small handheld devices necessarily have small controls—small jog dials, small buttons, and small input devices, such as pens for touch screens. Reductions in force control and movement control are critical considerations in designing these devices.

Jog Dials Jog dials, typically used for rotating through options in small handheld devices, should have sufficiently stiff, discrete stopping points when rotated, given older adults' force control deficiencies. The "teeth" on the dial should provide enough friction to account for older adults' more slippery skin (Cole, 1991).

Buttons Buttons should be far enough apart to minimize accidental activations. Buttons should be firm, so that they are not depressed accidentally when older users rests their finger on them, which may be likely given reductions in force control. Tactile feedback of a button press should be provided as redundant feedback to visual and/or auditory feedback to inform the user that the control has been activated.

Touch Screens Touch screens are not optimal for older adults for small targets, due to high variability in movement control. For example, if selecting one of four $\frac{1}{4}$-inch-wide options on a PDA screen, a jog dial should be used to cycle through the options and make a selection, as opposed to requiring the older user to select the option with a touch pen or with a finger. If touch screens are used in small devices, selection areas should be maximized to increase accuracy. In general, touch screens should be used for ballistic movements (particularly when screen real estate is large), but for precise control, indirect pointing devices such as a rotary device or mouse should be used (Charness et al., 2004; Rogers et al., in press). Older adults have difficulty making accurate, fine motor movements as well as making fast movements. Fine mouse movements should be minimized if possible, and the option to control the double-click speed and gain on the mouse should be available.

7.5 Training and Instructional Materials

Training can cover a multitude of design errors (although it should not be used as a crutch by designers). For example, people are entirely capable of learning the meaning of an obscure icon, provided that it is associated consistently with the same function and is not easily confused with other icons. Older adults may have unique training requirements. For example, older adults may be less familiar with computer technologies, and as a result, may require training for basic features of a system. For example, mouse training, instruction on windowing, or search tool training may be required before other, higher-level aspects of the system are trained. Older adults may also be less confident in their ability to interact successfully with novel technological devices such as PDAs (due in part to less experience).

Inexperienced older users may have an incorrect mental model of the system or no model at all. Often these models are constructed from repeated interaction with the system; hence, novice users should first be trained to form an appropriate model of the system. Older adults may be able to adopt new strategies that are optimal for the task (Rogers and Gilbert, 1997) and develop new mental models successfully (Gilbert and Rogers, 1999). Furthermore, older adults may expect a product or system to "act" like systems or products they have learned in the past. For example, the webcam can be started through the software application on the computer, as opposed to physically depressing a button to initiate the camera. This may not match the older patients' preconceptions of how a camera should work. However, this is clearly not a problem of comprehension, but simply a difference in experience.

Training programs and instructional materials should be developed for the use of the handheld devices and the computer software. In designing training programs and materials for the current hypothetical system, several considerations should be made. Many of these are relevant to training adults of all ages.

7.5.1 Duration of Training

Based on training research with simple and complex stimulus situations, older adults should receive about one and a half to two times the amount of training required by younger adults (Fisk et al., 2004). Extended training (or *overtraining*) can help solidify the process being trained and improve retention (Jones, 1989). Older adults tend to benefit more from overtraining than younger adults, who presumably reach a more stable area of the learning curve earlier (e.g., Sharit et al., 2003). Short breaks should be provided when training sessions run for 30 minutes or more.

7.5.2 Format of Training

Appropriately constructed part-task training can be helpful in training complex tasks. Part-task training involves practice on some subset of a task before practice on the whole task (Kramer et al., 1995). Selecting the appropriate subset is, of course, critical. A thorough task analysis is required before a task is divided into parts, particularly in complex tasks. Artificial divisions of the task (especially in the simplest tasks) will be detrimental. As an example, in a search and detection task with a substantial motor component, the motor aspect of the task could be trained independent of the visual search aspect. When proficiency is reached, the two can be combined and additional training provided. In the present system, a mouse training application should be an optional training program for novice users, to be completed prior to using the computer software. However, if two aspects of a task will need to be closely integrated, such as related motor movements with different hands, dividing the task will be detrimental to older adults' performance (Korteling, 1993).

7.5.3 Flexibility of Training

A flexible training program is important for older users, as the variability in experience, skills, and knowledge can be high in an older population. The training application for the software program should assess older adults' computer knowledge and be flexible enough to provide low-level training to novice users, but to skip these aspects of the training program for users who demonstrate proficiency.

7.5.4 Active Learning

Active training is essential; passive observation leads to little, if any, learning (W. Schneider, 1985). Training programs should make use of the same interface on which users will be performing the actual tasks or as close as possible. Tutorials for using the computer software (e.g., uploading data) can easily be presented using the same graphical user interface as the actual application. This allows the user to interact actively with the interface, as opposed to reading instructional text or viewing pictures of the interface. However, BGM training is not easily administered using the actual device. In this case, several options exist.

- A training video can be provided to present the steps involved in taking a blood sample with the BGM (Mykityshyn et al., 2003). Video and audio are preferable to audio alone in sequential tasks (e.g., A. McLaughlin et al., 2002).
- A graphical representation of the BGM can be used in a software training application. The user can interact with the BGM via a mouse, creating an active learning environment for the user.

7.5.5 Feedback in Training

Feedback should be provided for every interaction with the system (e.g., a button press should result in a corresponding change on the screen as well as an auditory cue), but specifically in a training environment, feedback is critical in making trainees aware of mistakes and in creating an appropriate mental model. Feedback is particularly critical for older adults (for a review, see Kausler, 1991). Given that they may experience a higher degree of anxiety at learning to interact with novel technology (Fisk et al., 2004), this feedback should be communicated as clearly as possible. Consider the graphical user interface (GUI) training application for the BGM.

- *Feedback should be immediate.* If a user performs steps out of order, the training application should provide immediate feedback to prevent this incorrect order of operations from becoming a learned procedure.
- *Feedback should be specific.* Users should be informed of their incorrect action and shown the correct action.
- *Feedback should be succinct.* Removing the user from the training program for an extended period of time to explain a mistake will prevent the user from quickly learning the correct procedure.

7.5.6 Consistency

Learning will not occur for completely inconsistent information, but like younger adults, older adults can learn under partially consistent conditions (Meyer and Fisk, 1998). However, for older adults, consistent relationships between stimuli or aspects of a task or system should be identified explicitly. Given that a goal of the system is educational, older users should be trained to recognize the relationships between their blood glucose levels and multiple critical variables. These relationships can be graphically displayed across time, but the relationships will not be perfectly consistent. However, over repeated exposure to the semiconsistent relationships, they will be learned.

Instructions can explicitly identify consistent relationships among stimuli. Thus, older users will learn these semiconsistent relationships better if they are identified explicitly by the software or by the physician. Consistency should be employed across all aspects of the training and instructional materials. For example, the introduction of an instruction manual should inform users that certain symbols will be used throughout to designate certain points in the text. Consistent aspects can be similarly marked or grouped together to emphasize their relationship. Conversely, if grouped items are not related to each other, this may cause confusion.

7.5.7 Importance of Task Analysis

A device that seems simple to designers often is much more complicated and difficult to use for novice users. For example, a human factors–based task analysis demonstrated that proper use of a BGM required 52 steps as opposed to the three suggested by the manufacturer's advertising materials (Rogers et al., 2001). By performing a comprehensive task analysis, the individual requirements for successful interaction with the system or device will be identified and accounted for in the training and instructional materials. Furthermore, if part-task training is plausible, a careful analysis of all aspects of the task is necessary to segment the task appropriately. Task analysis will result in an understanding of problems and errors that can occur. These issues should be anticipated in instructional manuals: for example, in a "frequently asked questions" section of a manual.

8 CONCLUSIONS

Older adults comprise a significant portion of users of technological products, thus demanding the attention of human factors professionals. We have described a wide array of age-related factors that have functional significance for the performance of older adults, and provided design implications and recommendations based on the existing literature. Human factors professionals will benefit by considering the age of their users and designing appropriately.

Several themes in the design guidelines follow from the review of age-related changes and from the case study. The importance of environmental support, or taking cognitively demanding requirements from the user and putting information in the task environment, is pervasive. Especially for older adults, this supportive information can direct attention to relevant stimuli, cue users' memory, support tasks such as reading, visual search, and even balance, and can free up valuable cognitive resources that can be applied to other aspects of the task. Resources can also be allocated to other tasks if the stimulus itself is improved. Despite pervasive perceptual declines in the aging system, the perceptual stimulus can be greatly improved by increasing the size or intensity of a stimulus or increasing the lighting of an environment. The designer must understand the limitations of older users before even such a simple design adjustment can be made.

For tasks that are not readily performed by older adults after these two guidelines have been followed, older adults may still be capable of attaining proficiency through appropriate training. Older adults are able to achieve skilled performance through training provided that the information to be learned is at least semiconsistent. Older adults follow a power law of learning for consistent information similar to that of younger adults. Thus, older adults are certainly not limited in all aspects relative to their younger counterparts. In fact, designers should attempt to take advantage of those areas where older adults surpass the capabilities of younger adults, specifically in their semantic knowledge. However, none of these design guidelines are useful if older adults choose not to use the system or product. Fortunately, data suggest that older adults are willing, and it is critical that the benefits of new systems and products are communicated to potential older users to increase their motivation to learn to use these new technologies.

Age-related changes in capabilities of older adults have been well studied. An understanding of such differences provides constraints for the design space of new products and new instantiations of existing products. Clearly, iterative design and user testing will always be essential for good design, and older adults must be included in the usability test group. However, our goal in this chapter was to provide a summary of the literature on aging to enable designers to start from an informed position about how systems and products should be designed if they are to be used safely and effectively by older adults.

ACKNOWLEDGMENTS

The authors were supported in part by grant P01 AG17211 from the National Institutes of Health (National Institute on Aging) under the auspices of the Center for Research and Education on Aging and Technology Enhancement (CREATE).

REFERENCES

Abel, S. M., Giguere, C., Consoli, A., and Papsin, B. (2000), "The Effect of Aging on Horizontal Plane Sound Localization," *Journal of the Acoustical Society of America*, Vol. 108, No. 2, pp. 743–752.

Agnew, J., and Suruda, A. (1993), "Age and Fatal Work-Related Falls," *Human Factors*, Vol. 35, pp. 731–736.

Andersen, G. J., Cisneros, J., Saidpour, A., and Atchley, P. (2000), "Age-Related Differences in Collision Detection During Deceleration," *Psychology and Aging*, Vol. 15, pp. 241–252.

Arenberg, D., and Robertson-Tchabo, E. A. (1985), "Adult Age Differences in Memory and Linguistic Integration Revisited," *Experimental Aging Research*, Vol. 11, pp. 187–191.

Atchley, P., and Andersen, G. J. (1998), "The Effects of Age, Retinal Eccentricity, and Speed on the Detection of Optic Flow Components," *Psychology and Aging*, Vol. 13, pp. 297–308.

Baddeley, A. (1986), *Working Memory*, Clarendon Press, Oxford.

Bellmann, C., and Holz, F. G. (2001), "Visual Impairment and Functional Deficits in Age-Related Macular Degeneration," in *On the Special Needs of Blind and Low-Vision Seniors*, H. -W. Wahl and H. -E. Schulze, Eds., IOS Press, Amsterdam, pp. 49–57.

Bennett, A. G. (1965), "Ophthalmic Test Types," *British Journal of Physiological Optics*, Vol. 22, pp. 238–271.

Borger, L. L., Whitney, S. L., Redfern, M. S., and Furman, J. M. (1999), "The Influence of Dynamic Visual Environments on Postural Sway in the Elderly," *Journal of Vestibular Research*, Vol. 9, pp. 197–205.

Brown, L., Shumway-Cook, A., and Woollacott, M. (1999), "Attentional Demands and Postural Recovery: The Effects of Aging," *Journals of Gerontology: Medical Sciences*, Vol. 54A, pp. M165–M171.

Carnahan, H., Goodale, M. A., and Marteniuk, R. G. (1993), "Grasping Versus Pointing and the Differential Use of Visual Feedback," *Human Movement Science*, Vol. 12, pp. 219–234.

Cattell, R. B. (1963), "Theory of Fluid and Crystallized Intelligence: A Critical Experiment," *Journal of Educational Psychology*, Vol. 54, pp. 1–22.

Charness, N., and Dijkstra, K. (1999), "Age, Luminance, and Print Legibility in Homes, Offices, and Public Places," *Human Factors*, Vol. 41, No. 2, pp. 173–193.

Charness, N., Holley, P., Feddon, J., and Jastrzembski, T. (2004), "Light Pen Superiority over a Mouse: Age, Hand, and Practice Effects," *Human Factors*, Vol. 46, pp. 373–384.

Chun, M., and Jiang, Y. (1998), "Contextual Cueing: Implicit Learning and Memory of Visual Context Guides Spatial Attention," *Cognitive Psychology*, Vol. 36, pp. 28–71.

Cohen, G., and Faulkner, D. (1989), "Age Differences in Source Forgetting: Effects on Reality Monitoring and on Eyewitness Testimony," *Psychology and Aging*, Vol. 4, No. 1, pp. 10–17.

Cole, K. J. (1991), "Grasp Force Control in Older Adults," *Journal of Motor Behavior*, Vol. 23, No. 4, pp. 251–258.

Cole, K. J., Rotella, D. L., and Harper, J. G. (1999), "Mechanisms for Age-Related Changes of Fingertip Forces During Precision Gripping and Lifting in an Adult," *Journal of Neuroscience*, Vol. 19, No. 8, pp. 3238–3247.

Connelly, S. L., and Hasher, L. (1993), "Aging and the Inhibition of Spatial Location," *Journal of Experimental Psychology: Human Perception and Performance*, Vol. 19, pp. 1238–1250.

Craik, F. I. M. (1986), "A Functional Account of Age Differences in Memory," in *Human Memory and Cognitive Capabilities: Mechanisms and Performances*, F. Klix and H. Hagendorf, Eds., Elsevier North-Holland, Amsterdam, pp. 409–422.

Crossley, M., and Hiscock, M. (1992), "Age-Related Differences in Concurrent-Task Performance of Normal Adults: Evidence for a Decline in Processing Resources," *Psychology and Aging*, Vol. 7, pp. 499–506.

Czaja, S. J., and Sharit, J. (1998), "The Effect of Age and Experience on the Performance of a Data Entry Task," *Journal of Experimental Psychology: Applied*, Vol. 4, pp. 332–351.

Desai, M., Pratt, L. A., Lentzner, H., and Robinson, K. N. (2001), "Trends in Vision and Hearing Among Older Americans," in *Aging Trends*, Vol. 2, National Center for Health Statistics, Hyattsville, MD.

Dingus, T. A., Hulse, M. C., Mollenhauer, M. A., Fleischman, R. N., McGehee, D. V., and Manakkal, N. (1997), "Effects of Age, System Experience, and Navigation Technique on Driving with an Advanced Traveler Information System," *Human Factors*, Vol. 39, No. 2, pp. 177–199.

Doane, S. M., Alderton, D. L., Sohn, Y. W., and Pellegrino, J. W. (1996), "Acquisition and Transfer of Skilled Performance: Are Visual Discrimination Skills Stimulus Specific?" *Journal of Experimental Psychology: Human Perception and Performance*, Vol. 22, pp. 1218–1248.

Dunlosky, J., and Hertzog, C. (1998), "Aging and Deficits in Associative Memory: What Is the Role of Strategy Use? *Psychology and Aging*, Vol. 13, pp. 597–607.

Earles, J., Smith, A. D., and Park, D. C. (1996), "Adult Age Differences in the Effects of Environmental Context on Memory Performance," *Experimental Aging Research*, Vol. 22, pp. 267–280.

Einstein, G. O., McDaniel, M. A., Richardson, S. L., Guynn, M. J., and Cunfer, A. R. (1995), "Aging and Prospective Memory: Examining the Influences of Self-Initiated Retrieval Processes," *Journal of Experimental Psychology: Learning, Memory, and Cognition*, Vol. 21, No. 4, pp. 996–1007.

Einstein, G. O., Smith, R. E., McDaniel, M. A., and Shaw, P. (1997), "Aging and Prospective Memory: The Influence of Increased Task Demands at Encoding and Retrieval," *Psychology and Aging*, Vol. 12, pp. 479–488.

Ellis, R. D. and Allaire, J. (1999), "Modeling Computer Interest in Older Adults: The Role of Age, Education, Computer Knowledge, and Computer Anxiety," *Human Factors*, Vol. 41, No. 3, pp. 345–355.

Ellis, R. D. and Kurniawan, S. H. (2000), "Increasing the Usability of On-Line Information for Older Users: A Case Study in Participatory Design," *International Journal of Human–Computer Interaction*, Vol. 12, pp. 263–276.

Ferguson, S. A., Hashtroudi, S., and Johnson, M. K. (1992), "Age Differences in Using Source-Relevant Cues," *Psychology and Aging*, Vol. 7, pp. 443–452.

Fisk, A. D. (1999), "Human Factors and the Older Adult," *Ergonomics in Design*, Vol. 7, No. 1, pp. 8–13.

Fisk, A. D., and Rogers, W. A. (1991), "Toward an Understanding of Age-Related Memory and Visual Search Effects," *Journal of Experimental Psychology: General*, Vol. 120, pp. 131–149.

Fisk, A. D., and Rogers, W. A. (2002), "Health Care of Older Adults: The Promise of Human Factors Research," in *Human Factors Interventions for the Health Care of Older Adults*, W. A. Rogers and A. D. Fisk, Eds., Lawrence Erlbaum Associates, Mahwah, NJ, pp. 31–46.

Fisk, A. D., McGee, N. D., and Giambra, L. M. (1988), "The Influence of Age on Consistent and Varied Semantic Category Search Performance," *Psychology and Aging*, Vol. 3, pp. 323–333.

Fisk, A. D., Rogers, W. A., Charness, N., Czaja, S. J., and Sharit, J. (2004), *Designing for Older Adults: Principles and Creative Human Factors Approaches*, CRC Press, Boca Raton, FL.

Fozard, J. L. (1990), "Vision and Hearing in Aging," in *Handbook of the Psychology of Aging*, 3rd ed., J. E. Birren and K. W. Schaie, Eds., Academic Press, San Diego, CA, pp. 150–170.

Fozard, J. L., and Gordon-Salant, S. (2001), "Sensory and Perceptual Changes with Aging," in *Handbook of the Psychology of Aging*, 5th ed., J. E. Birren and K. W. Schaie, Eds., Academic Press, San Diego, CA, pp. 241–266.

Gilbert, D. K., and Rogers, W. A. (1999), "Age-Related Differences in the Acquisition, Utilization, and Extension of a Spatial Mental Model," *Journals of Gerontology: Psychological Sciences*, Vol. 54B, pp. P246–P255.

Goldstein, E. B. (1999), *Sensation and Perception*, Brooks/Cole, Pacific Grove, CA.

Haider, H., and Frensch, P. A. (1996), "The Role of Information Reduction in Skill Acquisition," *Cognitive Psychology*, Vol. 30, pp. 304–337.

Hamm, V. P., and Hasher, L. (1992), "Age and the Availability of Inferences," *Psychology and Aging*, Vol. 7, pp. 56–64.

Hancock, H. E., Fisk, A. D., and Rogers, W. A. (2001), "An Evaluation of Warning Habits and Beliefs Across the Adult Life Span," *Human Factors*, Vol. 43, No. 3, pp. 343–354.

Hancock, H. E., Rogers, W. A., Schroeder, D., and Fisk, A. D. (2004), "Safety Symbol Comprehension: Effects of Symbol Type, Familiarity, and Age," *Human Factors*, Vol. 46, No. 2, pp. 183–195.

Hancock, H. E., Fisk, A. D., and Rogers, W. A. (in press), "Comprehending Product Warning Information: Age-Related Effects of Memory, Inferencing, and Knowledge," *Human Factors*.

Hartley, A. A., Kieley, J., and McKenzie, C. R. M. (1992), "Allocation of Visual Attention in Younger and Older Adults," *Perception and Psychophysics*, Vol. 52, pp. 175–185.

Hasher, L., and Zacks, R. T. (1988), "Working Memory, Comprehension, and Aging: A Review and a New View," in *The Psychology of Learning and Motivation*, G. H. Bower, Ed., Academic Press, New York, Vol. 22, pp. 193–225.

Hickman, J. M., Rogers, W. A., and Fisk, A. D. (2003), "Age-Related Effects of Training on Developing a System Representation," in *Proceedings of the Human Factors and Ergonomics Society 47th Annual Meeting*, Human Factors and Ergonomics Society, Santa Monica, CA.

Hirshman, E., and Bjork, R. (1988), "The Generation Effect: Support for a Two-Factor Theory," *Journal of Experimental Psychology: Learning Memory and Cognition*, Vol. 14, pp. 484–494.

Ho, G., Scialfa, C. T., and Caird, J. K. (2001), "Visual Search for Traffic Signs: The Effects of Clutter, Luminance, and Aging," *Human Factors*, Vol. 43, pp. 194–207.

Horak, F. B., Shupert, C. L., and Mirka, A. (1989), "Components of Postural Dyscontrol in the Elderly: A Review," *Neurobiolology of Aging*, Vol. 10, pp. 727–738.

Hulicka, I. M. and Grossman, J. L. (1967), "Age Group Comparisons for the Use of Mediators in Paired-Associate Learning," *Journal of Gerontology*, Vol. 22, pp. 46–51.

Jackson, G. R., Owsley, C., and McGwin, G. (1999), "Aging and Dark Adaptation," *Vision Research*, Vol. 39, pp. 3975–3982.

James, W. (1890/1950), *The Principles of Psychology*, Vol. 1, Dover, New York.

Jiang, Y., and Chun, M. M. (2001). "Selective Attention Modulates Implicit Learning," *Quarterly Journal of Experimental Psychology*, Vol. 54A, pp. 1105–1124.

Johnson, M. K., Hashtroudi, S., and Lindsay, D. S. (1993), "Source Monitoring," *Psychological Bulletin*, Vol. 114, pp. 3–28.

Johnson, M. K., De Leonardis, D., Hashtroudi, S., and Ferguson, S. A. (1995), "Aging and Single Versus Multiple Cues in Source Monitoring," *Psychology and Aging*, Vol. 10, pp. 507–517.

Johnson, M. M., Schmitt, F. A., and Pietrukowicz, M. (1989), "The Memory Advantages of the Generation Effect: Age and Process Differences," *Journals of Gerontology: Psychological Sciences*, Vol. 44, No. 3, pp. P91–P94.

Johnston, W. A., and Dark, V. J. (1986), "Selective Attention," *Annual Review of Psychology*, Vol. 37, pp. 43–75.

Jones, M. B. (1989), "Individual Differences in Skill Retention," *American Journal of Psychology*, Vol. 102, pp. 183–196.

Kaufman, D. R., Patel, V. L., Hilliman, C., Morin, P. C., Pevzner, J., Weinstock, G. R., Shea, S., and Starren, J. (2003), "Usability in the Real World: Assessing Medical Information Technologies in Patients' Homes," *Journal of Biomedical Informatics*, Vol. 36, pp. 45–60.

Kausler, D. H. (1991), *Experimental Psychology, Cognition, and Human Aging*, Springer-Verlag, New York.

Kemper, S. (1987), "Constraints on Psychological Processes in Discourse Production," in *Psycholinguistic Models of Production*, H. W. Dechert and M. Raupach, Eds., Ablex, Westport, CT, pp. 185–188.

Ketcham, C. J., and Stelmach, G. E. (2001), "Age-Related Declines in Motor Control," in *Handbook of the Psychology of Aging*, 5th ed., J. E. Birren and K. W. Schaie, Eds., Academic Press, San Diego, CA, pp. 313–348.

Ketcham, C. J., Seidler, R. D., van Gemmert, A. W. A., and Stelmach, G. E. (2002), "Age Related Kinematic Differences as Influenced by Task Difficulty, Target-Size, and Movement Amplitude," *Journals of Gerontology: Psychological Sciences and Social Sciences*, Vol. 57B, pp. P54–P64.

Kinsella, K., and Velkoff, V. A. (2001), *An Aging World: 2001*, U.S. Census Bureau, Series P95/01-1, U.S. Government Printing Office, Washington, DC.

Kline, D. W., Buck, K., Sell, Y., Bolan, T., and Dewar, R. E. (1999), "Older Observers' Tolerance of Optical Blur: Age Differences in the Identification of Defocused Text Signs," *Human Factors*, Vol. 41, No. 3, pp. 356–364.

Knoblauch, K., Saunders, F., Kusuda, M., Hynes, R., Podgor, M., Higgens, K. E., and deMosasterio, M. (1987), "Age and Illuminance Effects in the Farnsworth–Munsell 100-Hue Test," *Applied Optics*, Vol. 26, pp. 1441–1448.

Korteling, J. E. (1993), "Effects of Age and Task Similarity on Dual-Task Performance," *Human Factors*, Vol. 35, No. 1, pp. 99–113.

Kraft, J. M., and Werner, J. S. (1999), "Aging and the Saturation of Colors, 1: Colorimetric Purity Discrimination," *Journal of the Optical Society of America A*, Vol. 16, pp. 223–230.

Kramer, A. F., and Larish, J. (1996), "Aging and Dual-Task Performance," in *Aging and Skilled Performance*, W. A. Rogers, A. D. Fisk, and N. Walker, Eds., Lawrence Erlbaum Associates, Mahwah, NJ, pp. 83–112.

Kramer, A. F., Humphrey, D. G., Larish, J. F., Logan, G. D., and Strayer, D. L. (1994), "Aging and Inhibition: Beyond a Unitary View of Inhibitory Processing in Attention," *Psychology and Aging*, Vol. 9, pp. 491–512.

Kramer, A. F., Larish, J. F., and Strayer, D. L. (1995), "Training for Attentional Control in Dual-Task Settings: A Comparison of Young and Old Adults," *Journal of Experimental Psychology: Applied*, Vol. 1, pp. 50–76.

Kristinsdottir, E. K., Fransson, P. -A., and Magnusson, M. (2001), "Changes in Postural Control in Healthy Elderly Are Related to Vibration Sensation, Vision and Vestibular Asymmetry," *Acta Otolaryngologica*, Vol. 121, pp. 700–706.

Lajoie, Y., Teasdale, N., Bard, C., and Fleury, M. (1993), "Attentional Demands for Static and Dynamic Equilibrium," *Experimental Brain Research*, Vol. 97, No. 1, pp. 139–144.

Lajoie, Y., Teasdale, N., Bard, C., and Fleury, M. (1996), "Attentional Demands for Walking: Age-Related Changes," in *Changes in Sensory Motor Behavior in Aging*, A. Ferrandez and N. Teasdale, Eds., Elsevier, Amsterdam, Vol. 114, pp. 235–256.

Legge, G. E., Mullen, K. T., Woo, G. C., and Campbell, F. W. (1987a), "Tolerance to Visual Defocus," *Journal of the Optical Society of America*, Vol. A4, pp. 851–863.

Legge, G. E., Rubin, G. S., and Luebker, A. (1987b), "Psychophysics of Reading, V: The Role of Contrast in Normal Vision," *Vision Research*, Vol. 27, pp. 1165–1171.

Li, K. Z. H., Lindenberger, U., Freund, A. M., and Baltes, P. B. (2001), "Walking While Memorizing: Age-Related Differences in Compensatory Behavior," *Psychological Science*, Vol. 12, pp. 230–237.

Lockhart, T. E. (1997), "Ability of Elderly People to Traverse Slippery Walking Surfaces," in *Proceedings of the Human Factors and Ergonomics Society 41st Annual Meeting*, Human Factors and Ergonomics Society, Santa Monica, CA, pp. 125–129.

Lorenzi, C., Gatehouse, S., and Lever, C. (1999), "Sound Localization in Noise in Hearing-Impaired Listeners," *Journal of the Acoustical Society of America*, Vol. 105, No. 6, pp. 3454–3463.

Maltz, M., and Shinar, D. (1999), "Eye Movements of Younger and Older Drivers," *Human Factors*, Vol. 41, No. 1, pp. 15–25.

McDowd, J. M., and Craik, F. I. M. (1988), "Effects of Aging and Task Difficulty on Divided Attention Performance," *Journal of Experimental Psychology: Human Perception and Performance*, Vol. 14, pp. 267–280.

McDowd, J., and Shaw, R. J. (2000), "Attention and Aging," in *The Handbook of Aging and Cognition*, 2nd ed., F. I. M. Craik and T. A. Salthouse, Eds., Lawrence Erlbaum Associates, Mahwah, NJ, pp. 221–292.

McLaughlin, A. C., Rogers, W. A., and Fisk, A. D. (2002), "Effectiveness of Audio and Visual Training Presentation Modes for Glucometer Calibration," in *Proceedings of the Human Factors and Ergonomics Society 45th Annual Meeting*, Human Factors and Ergonomics Society, Santa Monica, CA.

McLaughlin, G. (1969), "SMOG Grading: A New Readability Formula," *Journal of Reading*, Vol. 12, No. 8, pp. 639–646.

Melenhorst, A. S., Rogers, W. A., and Caylor, E. C. (2001), "The Use of Communication Technologies by Older Adults: Exploring the Benefits from the User's Perspective," in *Proceedings of the Human Factors and Ergonomics Society 45th Annual Meeting*, Human Factors and Ergonomics Society, Santa Monica, CA.

Melzer, I., Benjuya, N., and Kaplanski, J. (2001), "Age Related Changes of Postural Control: The Effect of Cognitive Task," *Gerontology*, Vol. 47, pp. 189–194.

Meyer, B., and Fisk, A. D. (1998), "Toward an Understanding of Age-Related Use of Incidental Consistency," in *Proceedings of the Human Factors and Ergonomics Society 42nd Annual Meeting*, Human Factors and Ergonomics Society, Santa Monica, CA.

Mitzner, T. L., and Rogers, W. A. (2003), "Age-Related Differences in Reading Text Presented with Degraded Contrast," in *Proceedings of the Human Factors and Ergonomics Society 47th Annual Meeting*, Human Factors and Ergonomics Society, Santa Monica, CA.

Morrell, R. W., and Echt, K. V. (1997), "Designing Written Instructions for Older Adults: Learning to Use Computers," in *Handbook of Human Factors and the Older Adult*, A. D. Fisk and W. A. Rogers, Eds., Academic Press, San Diego, CA, pp. 335–361.

Morrell, R. W., Mayhorn, C. B., and Bennett, J. (2000), "A Survey of World Wide Web Use in Middle-Aged and Older Adults," *Human Factors*, Vol. 42, pp. 175–182.

Morrow, D. G. (2003), "Technology as Environmental Support for Older Adults' Daily Activities," in *The Impact of Technology on Successful Aging*, N. Charness and W. Schaie, Eds., Springer-Verlag, New York, pp. 290–305.

Morrow, D. G., Leirer, V. O., Carver, L. M., and Tanke, E. D. (1998), "Older and Younger Adult Memory for Health Appointment Information: Implications for Automated Telephone Messaging Design," *Journal of Experimental Psychology: Applied*, Vol. 4, No. 4, pp. 352–374.

Morrow, D. G., Carver, L. M., Leirer, V. O., and Tanke, E. D. (2000), "Medication Schemas and Memory for Automated Telephone Messages," *Human Factors*, Vol. 42, pp. 523–540.

Moscicki, E. K., Elkins, E. F., Baum, H. M., and McNamara, P. M. (1985), "Hearing Loss in the Elderly: An Epidemiologic Study of the Framingham Study Cohort," *Ear and Hearing*, Vol. 6, No. 8, pp. 184–190.

Mykityshyn, A. L., Fisk, A. D., and Rogers, W. A. (2002), "Learning to Use a Home Medical Device: Mediating Age-Related Differences with Training," *Human Factors*, Vol. 44, pp. 354–364.

Nichols, T. A., and Fisk, A. D. (2001), "Age-Related Differences in Using Environmental Support: Resources or Strategies?" presented at the 109th Annual APA Convention, San Francisco, August.

Nichols, T. A., Rogers, W. A., and Fisk, A. D. (2003), "Do You Know How Old Your Participants Are? Recognizing the Importance of Participant Age Classifications," *Ergonomics in Design*, Vol. 11, pp. 22–26.

Noble, W., Byrne, D., and Lepage, B. (1994), "Effects on Sound Localization of Configuration and Type of Hearing Impairment," *Journal of the Acoustical Society of America*, Vol. 95, No. 2, pp. 992–1005.

Norman, S., Kemper, S., and Kynette, D. (1992), "Adults' Reading Comprehension: Effects of Syntactic Complexity and Working Memory," *Journal of Gerontology*, Vol. 47, No. 4, pp. 258–265.

Owsley, C., Ball, K., Sloane, M., Roenker, D., and Bruni, J. (1991), "Visual/Cognitive Correlates of Vehicle Accidents in Older Drivers," *Psychology and Aging*, Vol. 6, pp. 403–415.

Parasuraman, R., and Davies, D. R., Eds. (1984), *Varieties of Attention*, Academic Press, New York.

Park, D. C., Cherry, K. E., Smith, A. D., and LaFronza, V. N. (1990a), "Effects of Distinctive Context on Memory for Objects and Their Locations in Young and Older Adults," *Psychology and Aging*, Vol. 5, pp. 250–255.

Park, D. C., Smith, A. D., Morrell, R. W., Puglisi, J. T., and Dudley, W. N. (1990b), "Effects of Contextual Integration on Recall of Pictures in Older Adults," *Journals of Gerontology: Psychological Sciences*, Vol. 45, pp. P52–P58.

Park, D. C., Morrell, R. W., Frieske, D., and Kincaid, D. (1992), "Medication Adherence Behaviors in Older Adults: Effects of External Cognitive Supports," *Psychology and Aging*, Vol. 7, pp. 252–256.

Park, D. C., Morrell, R., Hertzog, C., Kidder, D., and Mayhorn, C. (1997), "Effect of Age on Event-Based and Time-Based Prospective Memory," *Psychology and Aging*, Vol. 12, pp. 314–327.

Pinckers, A. (1980), "Color Vision and Age," *Ophthalmologica*, Vol. 181, pp. 23–30.

Radvansky, G. A., Gerard, L. D., Zacks, R. T., and Hasher, L. (1990), "Younger and Older Adults Use of Models as Representations for Text Materials," *Psychology and Aging*, Vol. 5, pp. 209–214.

Redfern, M. S., Moore, P. L., and Yarsky, C. M. (1997), "The Influence of Flooring on Standing Balance Among Older Persons," *Human Factors*, Vol. 39, No. 3, pp. 445–455.

Roenker, D. L., Cissell, G. M., Ball, K. K., Wadley, V. G., and Edwards, J. D. (2003), "Speed-of-Processing and Driving Simulator Training Result in Improved Driving Performance," *Human Factors*, Vol. 45, pp. 218–233.

Rogers, W. A. (1992), "Age Differences in Visual Search: Target and Distractor Learning," *Psychology and Aging*, Vol. 7, pp. 526–535.

Rogers, W. A., and Fisk, A. D. (2001), "Understanding the Role of Attention in Cognitive Aging Research," in *Handbook of the Psychology of Aging*, 5th ed., J. E. Birren and K. W. Schaie, Eds., Academic Press, San Diego, CA, pp. 267–287.

Rogers, W. A., and Gilbert, D. K. (1997), "Do Performance Strategies Mediate Age-Related Differences in Associative Learning?" *Psychology and Aging*, Vol. 12, pp. 620–633.

Rogers, W. A., Bertus, E. L., and Gilbert, D. K. (1994), "A Dual-Task Assessment of Age Differences in Automatic Process Development," *Psychology and Aging*, Vol. 9, pp. 398–413.

Rogers, W. A., Fisk, A. D., Mead, S. E., Walker, N., and Cabrera, E. F. (1996), "Training Older Adults to Use Automatic Teller Machines," *Human Factors*, Vol. 38, pp. 425–433.

Rogers, W. A., Meyer, B., Walker, N., and Fisk, A. D. (1998), "Functional Limitations to Daily Living Tasks in the Aged: A Focus Group Analysis," *Human Factors*, Vol. 40, pp. 111–125.

Rogers, W. A., Hertzog, C., and Fisk, A. D. (2000), "Age-Related Differences in Associative Learning: An Individual Differences Analysis of Ability and Strategy Influences," *Journal of Experimental Psychology: Learning, Memory, and Cognition*, Vol. 26, pp. 359–394.

Rogers, W. A., Mykityshyn, A. L., Campbell, R. H., and Fisk, A. D. (2001), "Analysis of a 'Simple' Medical Device," *Ergonomics in Design*, Vol. 9, pp. 6–14.

Rogers, W. A., Fisk, A. D., McLaughlin, A. C., and Pak, R. (in press), "Touch a Screen or Turn a Knob: Choosing the Best Device for the Job," *Human Factors*.

Said, F. S., and Weale, R. A. (1959), "The Variation with Age of the Spectral Transmissivity of the Living Human Crystalline Lens," *Gerontologica*, Vol. 3, pp. 213–231.

Sanchez, J., Nichols, T. A., Mitzner, T. L., Rogers, W. A., and Fisk, A. D. (2003), "Medication Adherence Strategies in Older Adults," in *Proceedings of the Human Factors and Ergonomics Society 47th Annual Meeting*, Human Factors and Ergonomics Society, Santa Monica, CA.

Schieber, F., and Goodspeed, C. H. (1997), "Nighttime Conspicuity of Highway Signs as a Function of Sign Brightness, Background Complexity and Age of Observer," in *Proceedings of the Human Factors and Ergonomics Society 41st Annual Meeting*, Human Factors and Ergonomics Society, Santa Monica, CA, pp. 1362–1366.

Schneider, B., and Pichora-Fuller, M. K. (2000), "Implications of Sensory Deficits for Cognitive Aging," in *The Handbook of Aging and Cognition*, 2nd ed., F. I. M. Craik and T. Salthouse, Eds., Lawrence Erlbaum Associates, Mahwah, NJ, pp. 155–219.

Schneider, B., Daneman, M., Murphy, D., and See, S. (2000), "Listening to Discourse in Distracting Settings: The Effects of Aging," *Psychology and Aging*, Vol. 15, pp. 110–125.

Schneider, W. (1985), "Toward a Model of Attention and the Development of Automatic Processing," in *Attention and Performance XI*, M. Posner and O. S. Marin, Eds., Lawrence Erlbaum Associates, Mahwah, NJ, pp. 475–492.

Schneider, W., and Shiffrin, R. M. (1977), "Controlled and Automatic Human Information Processing, I: Detection, Search, and Attention," *Psychological Review*, Vol. 84, pp. 1–66.

Seidler-Dobrin, R. D., and Stelmach, G. E. (1998), "Persistence in Visual Feedback Control by the Elderly," *Experimental Brain Research*, Vol. 119, pp. 467–474.

Sharit J., Czaja S. J., Nair S., and Lee C. C. (2003), "Effects of Age, Speech Rate, and Environmental Support in Using Telephone Voice Menu Systems," *Human Factors*, Vol. 45, No. 2, pp. 234–251.

Shaw, R. J. (1990), "Older Adults Sometimes Benefit from Environmental Support: Evidence from Reading Distorted Text," in *Proceedings of the Human Factors Society 34th Annual Meeting*, Human Factors Society, Santa Monica, CA, pp. 168–172.

Shaw, R. J. (1991), "Age-Related Increases in the Effects of Automatic Semantic Activation," *Psychology and Aging*, Vol. 6, pp. 595–604.

Shiffrin, R. M. (1988), "Attention," in *Stevens' Handbook of Experimental Psychology*, 2nd ed., R. C. Atkinson, R. J. Herrnstein, G. Lindzey, and R. D. Luce, Eds., Wiley, New York, pp. 739–811.

Shiffrin, R. M., and Schneider, W. (1977), "Controlled and Automatic Human Information Processing, II: Perceptual Learning, Automatic Attending, and a General Theory," *Psychological Review*, Vol. 84, pp. 127–190.

Sit, R. A., and Fisk, A. D. (1999), "Age-Related Performance in a Multiple-Task Environment," *Human Factors*, Vol. 41, pp. 26–34.

Smith, M. W., Sharit, J., and Czaja, S. J. (1999), "Aging, Motor Control, and the Performance of Computer Mouse Tasks," *Human Factors*, Vol. 41, No. 3, pp. 389–396.

Spirduso, W. W. (1995), *Physical Dimensions of Aging*, Human Kinetics, Champaign, IL.

Staplin, L. K., and Fisk, A. D. (1991), "Left-Turn Intersection Problems: A Cognitive Engineering Approach to Improve the Safety of Young and Old Drivers," *Human Factors*, Vol. 33, pp. 559–571.

Stelmach, G. E., and Nahom, A. (1992), "Cognitive–Motor Abilities of the Elderly Driver," *Human Factors*, Vol. 34, No. 1, pp. 53–65.

Stelmach, G. E., Amrhein, P. C., and Goggin, N. L. (1988), "Age Differences in Bimanual Coordination," *Journals of Gerontology: Psychological Sciences*, Vol. 43, pp. P18–P23.

Stoltzfus, E. R., Hasher, L., Zacks, R. T., Ulivi, M. S. and Goldstein, D. (1993), "Investigations of Inhibition and Interference in Younger and Older Adults," *Journal of Gerontology: Psychological Sciences*, Vol. 48, pp. P179–P188.

Szuchman, L. T., and Erber, J. T. (1990), "Young and Older Adults' Metaphor Interpretation: The Judgments of Professionals and Nonprofessionals," *Experimental Aging Research*, Vol. 16, No. 2, pp. 62–72.

Thornberry, J. M., and Mistretta, C. M. (1981), "Tactile Sensitivity as a Function of Age," *Journal of Gerontology*, Vol. 36, No. 1, pp. 34–39.

Tran, D. B., Silverman, S. E., Zimmerman, K., and Feldon, S. E. (1998), "Age-Related Deterioration of Motion Perception and Detection," *Graefe's Archive of Clinical Experimental Ophthalmology*, Vol. 236, No. 4, pp. 269–273.

Trick, G. L., and Silverman, S. E. (1991), "Visual Sensitivity to Motion: Age-Related Changes and Deficits in Senile Dementia of the Alzheimer Type," *Neurology*, Vol. 41, No. 9, pp. 1437–1440.

Tulving, E. (2002), "Episodic Memory: From Mind to Brain," *Annual Review of Psychology*, Vol. 53, pp. 1–25.

U.S. Bureau of the Census (1996), *Population Projections of the United States by Age, Sex, Race, and Hispanic Origin: 1995–2050*, Current Population Reports P25–1130, U.S. Government Printing Office, Washington, DC.

Verhaeghen, P., Marcoen, A., and Goossens, L. (1993), "Facts and Fiction About Memory Aging: A Quantitative Integration of Research Findings," *Journals of Gerontology: Psychological Sciences*, Vol. 48, pp. P157–P171.

Walker, N., Millians, J., and Worden, A. (1996), "Mouse Accelerations and Performance of Older Computer Users," in *Proceedings of the Human Factors and Ergonomics Society 40th Annual Meeting*, Human Factors and Ergonomics Society, Santa Monica, CA, pp. 151–154.

Walker, N., Philbin, D. A., and Fisk, A. D. (1997), "Age-Related Differences in Movement Control: Adjusting Submovement Structure to Optimize Performance," *Journal of Gerontology: Psychological Sciences*, Vol. 52B, pp. P40–P52.

Wickens, C. D. (1984), "Processing Resources in Attention," in *Varieties of Attention*, R. Parasuraman and D. R. Davies, Eds., Academic Press, Orlando, FL, pp. 63–102.

Wingfield, A., and Stine-Morrow, E. A. L. (2000), "Language and Speech," in *Handbook of Aging and Cognition*, 2nd ed., F. I. M. Craik and T. A. Salthouse, Eds., Lawrence Erlbaum Associates, Mahwah, NJ, pp. 359–416.

Wishart, L. R., Lee, T. D., Murdoch, J. E., and Hodges, N. J. (2000), "Aging and Bimanual Coordination: Effects of Speed and Instructional Set on In-Phase and Anti-phase Patterns," *Journals of Gerontology: Psychological Sciences*, Vol. 55B, pp. P85–P94.

Zacks, R. T., Hasher, L., and Li, K. Z. H. (2000), "Human Memory," in *Handbook of Aging and Cognition*, 2nd ed., T. A. Salthouse and F. I. M. Craik, Eds., Lawrence Erlbaum Associates, Mahwah, NJ, pp. 293–357.

CHAPTER 55
DESIGN FOR CHILDREN

Juan Pablo Hourcade*
U.S. Census Bureau[†]
Washington, D.C.

1 INTRODUCTION

In this chapter we aim to provide readers with basic information on how to design usable, beneficial technologies for children. Included is the most valuable information the author has learned while designing, developing, and evaluating technologies for children at the University of Maryland and the U.S. Census Bureau. We begin by discussing children's evolving needs and abilities and their impact on the design of technologies, including design guidelines and recommendations. To help readers avoid hazardous designs, we then cover various ways in which children can be harmed by technologies. We conclude by sharing advice on activities to conduct during the design process, how to assemble design teams, and how to incorporate children into the design process.

2 UNDERSTANDING CHILDREN'S NEEDS AND ABILITIES

In this section we provide an overview of children's needs and abilities by discussing theories of learning and the evolution of children's cognitive and motor abilities. As theories of learning and development are discussed, their practical consequences in terms of design guidelines and recommendations are highlighted. One aspect of the highlighted consequences is the way in which technologies can play a positive role in children's learning and development. Another aspect involves the interaction styles that best accommodate children's limited abilities. These limited abilities also call for the avoidance of some interaction styles for certain age groups.

While discussing the ideas of some of the most influential experts in child development, this section is organized around the theories of Jean Piaget, perhaps the most influential voice in child development during the twentieth century. He conducted research and published over a span of 70 years, from the early 1900s to the 1970s. Although his ideas and theories have often been contested, expanded, and revised, they still provide a useful starting point for discussing child development.

2.1 How Children Learn

Technology designers cannot effectively support child development without an understanding of how children learn. Piaget's ideas provide a useful context to discuss the design consequences of the processes and factors that affect learning.

2.1.1 Piaget's Adaptation

Piaget believed that children learn by actively constructing knowledge structures or schemata. This construction occurs as children add to and modify their knowledge structures while experiencing the world. Piaget referred to this process as *adaptation*. In Piaget's view, adaptation is composed of assimilation and accommodation. *Assimilation* occurs when a new experience fits within existing knowledge structures. *Accommodation*, in turn, occurs when a new experience does not fit into existing knowledge structures and thus these structures must be modified (Piaget, 1963).

*Present address: University of Iowa, Iowa City, Iowa.

[†]This report is released to inform interested parties of ongoing research and to encourage discussion. The views expressed are those of the author and not necessarily those of the U.S. Census Bureau.

2.1.2 Factors Affecting Intellectual Development

Piaget cited four major factors that influence adaptation and thus affect intellectual development in children: maturation, experience, social interaction, and emotions (Piaget and Inhelder, 1969). In this section we discuss the design implications of these four factors. Although the first factor, maturation, is in the hands of nature, its consequences must be taken into account in the design of user interfaces. The other three factors provide opportunities for technologies to play a positive role. Perhaps Seymour Papert, one of the earliest proponents of the use of computers in children's education, was thinking of these three factors when he proposed that Piagetian learning works best when children build meaningful items that they share with others (Papert, 1991). Building provides experiences, the social nature of the activity enables social interactions, and a meaningful item gives the emotional motivation.

Maturation Children's physical maturation affects and limits what children are able to accomplish at different age levels. As a child matures physically, possibilities appear for learning new concepts. However, Piaget thought learning would not occur because of physical maturation alone. Although maturation may limit what children can do at a certain age, it does not guarantee that development will occur (Piaget and Inhelder, 1969).

One obvious way to take maturation into account is to ensure that technologies are sized appropriately. Maria Montessori, the great Italian educator, was known for advocating child-sized environments in preschools. She thought children could learn better in classrooms that fit their size. Appropriately sized tools and technology not only provide better ergonomics, but also better motivate children to complete tasks by being more comfortable (Montessori, 1964; Jeong and Park, 1990; Mandal 1997).

Children's cognitive abilities also improve with age. Information-processing speed and working memory have both been shown to increase with age. Kail has proposed an exponential model for this improvement, with dramatic gains in early childhood that slow down through adolescence and reach their peak in young adulthood (Kail, 1991; Swanson, 1999). Kail's model predicts greater variability in performance between children of the same age at younger ages. This prediction has been confirmed through a number of studies (Kerr, 1975; Salmoni and McIlwain, 1979; Jones, 1991; Joiner et al., 1998; Hourcade et al., 2004c). Designers need to take this variability into account, since knowing the abilities of an average child may not be enough.

Working memory has an effect on children's ability to process both verbal and visual information and to recall it temporarily. It can have an effect, for example, on the number of steps that children can remember in order to accomplish a task, or on the amount of visual complexity that a graphical user interface should have. Working memory has also been associated with language comprehension, learning, and reasoning tasks (Baddeley, 1992).

The ability to process visual information can also have an impact on some motor skills, such as pointing at a target. This is of particular relevance to the design of point-and-click graphical user interfaces. Numerous experiments by research psychologists have shown a clear pattern of improvement in pointing tasks with age (Kerr, 1975; Wallace et al., 1978; Salmoni and McIlwain, 1979; Sugden, 1980). Part of the reason for these improvements is that older children can process visual information more quickly. This enables them to adjust pointing movements more often than younger children are able to, resulting in smoother motion when approaching a target. A recent study highlighted the nature of children's difficulties using mice in pointing tasks (Hourcade et al., 2004c). Plots of the paths taken to click on targets showed that children had difficulty controlling the mouse (see Figure 1). This poor control led to significant differences in terms of efficiency, accuracy, and target reentry.

When children are faced with icons that are difficult to click on due to their small size, they look for strategies to avoid this problem. A common strategy, observed often by this chapter's author, is to click repeatedly in the area surrounding the target to increase the chances of hitting it. This sometimes leads children to continue clicking on a target even after it has been clicked on. To avoid problems with these types of interactions, programmers should make sure that they ignore any input between the time that a target is

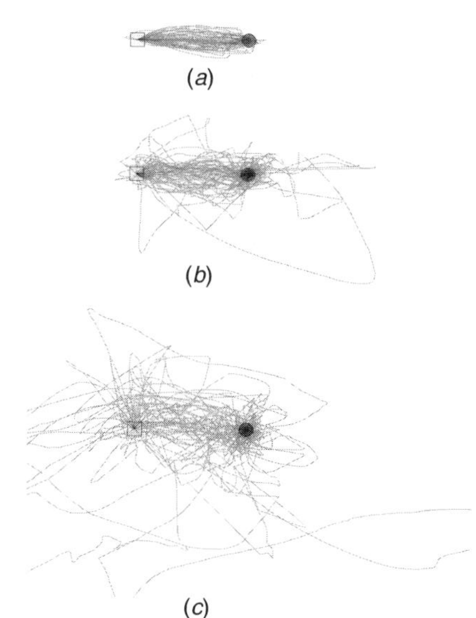

Figure 1 Paths taken by participants to click on a 32-pixel target at a distance of 256 pixels: (*a*) paths taken by all adult participants; (*b*) paths taken by all 5-year-old participants; (*c*) paths taken by all 4-year-old participants. (From Hourcade et al., 2004c.)

clicked on and the time when the program visually updates itself as a consequence of the click.

Another design recommendation for young children's software is to give all mouse buttons the same functionality. Children on average cannot tell left from right with respect to themselves until age 6, and with respect to other objects until age 8 (Kaluger and Kaluger, 1979). A recent study provided some evidence of young children's clicking preferences, suggesting that most preschool children would face difficulty with interfaces that provide different functionality through each mouse button (Hourcade et al., 2004b). An alternative design is to provide functionality only through the left mouse button. This may be useful for children who may be ready developmentally to begin learning about using multiple buttons, as it could provide them with a scaffold to learn these interactions. Designers thinking of multiple button interactions should make sure that they test them with children in the target age group.

A better documented issue in the design of children's interactions with computers is that children have difficulty dragging a mouse. Joiner et al. (1998) and Inkpen (2001) provide clear evidence of the advantages of point-and-click over drag-and-drop techniques. Point-and-click can be adapted to perform tasks similar to those accomplished with drag-and-drop by picking up an object with one click and dropping it with the next click. This technique is sometimes referred to as *click-and-carry*.

Anecdotal evidence also suggests that children require quicker response times than adults do. Long wait times can cause children to lose interest in a technology and may prevent them from making an association between an action and a delayed response.

Experience Piaget believed that natural maturation was not the only factor in children's development. He thought that children's experience in interacting with objects in the environment also contributed to development. Piaget thought that these experiences included interacting with objects to learn about their properties, and acting upon objects to learn about the consequences of actions. Piaget thought the latter type of interaction provided children with ways of creating logical and mathematical knowledge structures (Piaget and Inhelder, 1969).

Maria Montessori agreed with Piaget in the importance of experiences by stressing the need for children to experience the world themselves rather than being told about it. She believed that children need to be allowed to do something themselves in order to learn. That is, for example, why she believed in providing children with real tools (Montessori, 1964).

In a similar vein, Papert has argued for technologies to support children in building their own intellectual structures through experiences otherwise unavailable. His ideas were behind the creation of Logo, a programming language that enabled children to create drawings by instructing a "drawing turtle" to move around the screen (Papert, 1993). Papert's ideas have also influenced a number of technologies that give children the ability to program and even design physical objects and environments (Martin et al., 2000; Montemayor et al., 2002; Eisenberg, 2004).

Children can also augment their experiences by exploring virtual environments that take them to places that are not physically accessible (Roussou, 2004). Technologies can also be used to support explorations in the real world, augmenting children's experiences (Rogers et al., 2004). Similarly, children can learn about places, cultures, people, science, and nature through digital libraries, and data portals.

Social Interaction Piaget thought social interaction was both necessary and essential. He believed that this social dimension is what enables the transmission of knowledge between generations (Piaget and Inhelder, 1969). However, Lev Vygotsky, a still influential Russian psychologist, was the first person to highlight the importance of society in children's development. He believed that human learning was social in nature and that language, signs, and writing played a crucial role. Like Piaget, he developed his ideas through observation and experimentation (Vygotsky, 1978).

Rather than being interested in when children are ready to accomplish a task individually, Vygotsky was interested in the process that takes them there. In observing children, Vygotsky noticed that when encountering difficulty, they seek help from adults or more experienced children. He pointed out that before children can accomplish a task by themselves, they need help from others, often referred to as *scaffolding*. He defined a *zone of proximal development*, denoting what children can accomplish with the assistance of others but not individually. He believed that this zone reveals the areas of development that are in the process of maturation. As such, Vygotsky believed that children's education should concentrate in these areas of potential learning. Vygotsky thought that good learning occurs only before children are individually ready to learn. Once a child internalizes the processes that enable him or her to accomplish tasks with others, he or she learns those processes individually (Vygotsky, 1978).

To support this type of learning, technologies can provide children with new ways of collaborating and sharing. The Internet in particular can enable children to collaborate and share their creations with children and adults anywhere in the world. These collaborations have the potential of providing children with previously unavailable learning opportunities. For example, children may be better able to internalize knowledge about another culture if they can interact with a person from that culture rather than simply reading about it in a book or watching a video.

Design teams should also make an effort to incorporate collaboration into existing technologies, or at least to design technologies such that they will not hamper collaboration. It is particularly difficult, for example, for multiple children to collaborate on a personal computer. Even though children often congregate around a computer, only one child gets to interact with it. One way around this is to provide multiple input devices to one computer (Hourcade et al., 2004a). Another way is to build technologies that are not tied to the personal computer: for example, tangible computing, where computing power is added to physical objects (Eisenberg, 2004; Stringer et al., 2004), and interactive rooms, where physical objects around a room can be programmed to create immersive environments (Montemayor et al., 2002).

Emotions Piaget thought that emotions are at the root of behavior patterns. He thought that the needs to grow, to love and be loved, and to assert oneself to a great degree drive children's quest for learning (Piaget and Inhelder, 1969). Piaget is not alone in highlighting the importance of emotions. John Dewey, an influential American educator, thought teachers should help children learn by taking their interests into account. In addition to Dewey, Vygotsky and Montessori thought that learning activities should be relevant and meaningful to children's lives. Vygotsky also expressed his views on the importance of play as a way for children to use their imagination, act older than they are, and exercise self-control (Dewey, 1959; Montessori, 1964; Vygotsky, 1978).

Technologies can play a role in helping children enjoy learning activities. Through their versatility, they can make it easier to find motivating activities to foster the learning of valuable concepts. One trend in this area is the use of games as part of the learning experience (Malone, 1980; Robertson and Good, 2004).

2.2 Piaget's Developmental Stages

Piaget believed that children do not understand adult thought until they have reconstructed it. To do this they have to build knowledge structures, and in doing so have to go through a series of necessary steps. In this gradual evolution, every step depends and builds on the preceding step. Piaget grouped these steps into stages that every child must go through before achieving adult intellectual abilities. Progressing through the stages involves improvements in both motor and cognitive skills. Children go through the stages in the same order and cannot skip any stages (Piaget and Inhelder, 1969; Piaget, 1973).

As a guideline, Piaget provided some age limits for each stage. These limits refer only to average ages, as there is great variability in the ages at which children will reach each stage. In addition to hereditary differences, sociocultural issues also play a role (Piaget, 1973). This is a crucial point to take into account, as designing only for the average child of a certain age may not take advantage of the abilities of some children and could be too challenging for others. Despite this caveat, Piaget's stages can provide guidelines for the abilities that children can be expected to exhibit at different ages.

Currently, Piaget's ideas on developmental stages are controversial. One problem with this theory is that it considers what children are capable of accomplishing individually. Vygotsky, on the other hand, would argue that it is best to know what children can accomplish with help from others. Piaget's developmental stages have also been criticized because they consider abstract thought the highest form of development. Others have argued that people develop different types of intelligence (e.g., social, musical, athletic, logical) and that abstract thought demonstrates advanced development in only one of those types of intelligence (Singer and Revenson, 1996).

Despite these criticisms, from a very practical point of view, Piaget's developmental stages provide technology designers and developers with a way of identifying possible issues they may find with a particular age group. These issues may not apply to all children in that age group and may go away when children receive some help. But they are likely to have an impact on the design of technologies and should not be ignored. Similarly, designers should be aware of children's skill levels in their target population that may affect their interactions, such as reading, typing, and understanding content. In the remainder of this section we outline children's developmental issues during each stage and discuss the impact of these issues on design.

2.2.1 Sensory-Motor Stage (Birth to Age 2)

At the sensory-motor stage, according to Piaget, children understand the world through perceptions and objects with which they have had direct experience. They learn through their senses and reflexes. Discovered behaviors are repeated and applied to a variety of objects. This explains why a lot of toys for children this age allow them to experiment with their senses by seeing, touching, tasting, listening, and smelling (Piaget, 1963). It is important to provide children in this age group with a safe environment to explore with all their senses. In particular, it is valuable to provide toys that will react to children's actions, thus helping them create new knowledge structures (Garhart Mooney, 2000). Play in this stage should not have rules, though, as children are not interested in winning and get satisfaction instead from imitation, repetition, and experiencing the world through their senses (Piaget, 1962).

2.2.2 Preoperational Stage (Ages 2 to 7)

In Piaget's view, as children acquire language, they reach the preoperational stage. Children at this stage are egocentric, having difficulty understanding objects, people, and situations from someone else's point of view (Piaget, 1995a,b). This is a reason that it is often difficult to have a group of preschool children compromise or understand each other's concerns. This does not mean that children do not enjoy participating in activities with others. In these activities, play usually occurs in parallel and often involves pretending or make-believe activities. As they play, children follow personal sets of rules while imitating other children or adults. At this stage, children obey literally rules from adults or other figures of authority (Piaget, 1962). This should be of concern to designers of technologies or content that may tell children to do something.

Children also show a great level of curiosity at this stage, exploring the world around them. They will learn from their own experiences rather than from what they are told. A great deal of this learning will occur through play (Piaget, 1962). They also have a tendency to overgeneralize based on their experience. This leads them to make up explanations based on limited experiences (Piaget, 1995a). For example, children at this stage often assume that technologies that do not respond to their interactions as expected are broken. These children also are likely to believe that objects such as stuffed animals are living creatures.

Children at this stage can concentrate on only one characteristic of an object at a time. For example, it can be difficult for children to think of their teacher as the parent of other children. Furthermore, they are not capable of understanding hierarchies (Piaget 1995a,c). Thus, interfaces should avoid features that require children to think of more than one characteristic of an object at a time and should avoid the use of hierarchies.

2.2.3 Concrete Operations Stage (Ages 7 to 11)

As children learn to read and write, they also gain the ability to perform mental operations. They can count, add, subtract, multiply, and divide in their heads. They can apply these operations to concrete objects only, not to abstract concepts. They can mentally reverse their actions (Piaget, 1995c). Thus, in searching for a file in a computer, a child may be able to remember the decisions made when deciding where to place it.

Unlike preoperational children, children in this stage are able to think in terms of hierarchies (Piaget, 1995c). They can understand that all poodles are dogs, that all dogs are mammals, and that all mammals are animals. Hence, at this stage, they are able to navigate through hierarchies in user interfaces, such as file dialogues.

Children in the concrete operations stage also become capable of cooperating. They are more likely to understand someone else's point of view (Piaget 1995a,b). This means that they can discuss issues in groups, can play in teams, and can author and create together. By this stage children are also able to arrange items quickly based on quantity. They can also understand how items conserve certain characteristics even when they change their physical appearance. Piaget refers to this last concept as *conservation* (Piaget, 1995c).

2.2.4 Formal Operations Stage (Ages 11 to 16)

As children reach the formal operations stage, they begin to become capable of thinking abstractly. They can formulate hypotheses and use deductive reasoning. They are able to study a problem and logically evaluate a number of possible solutions before making a decision (Piaget, 1995c). Children at this stage also develop a personal value system and a sense of morality. They are capable of agreeing to a set of rules in order to better interact with others and become members of society (Piaget, 1962).

3 AVOIDING HAZARDS

Even if they are designed with adaptation, maturation, experience, social interaction, emotion, and developmental stages in mind, technologies are not always going to have a positive impact on children. They can be poorly designed and poorly used, potentially causing harm to children. Designers of technologies have a responsibility to minimize negative experiences. In this section we give an overview of the many ways in which technologies can be hazardous to children.

3.1 Physical Health

Technologies for children need to follow common sense in their design to avoid physical injuries. The American Academy of Pediatrics recommends that physical products for children not have sharp edges, be made of toxic materials, have parts that could fit in a young child's mouth or otherwise cause choking hazards, have loose string or rope, make loud noises, or have parts that can squeeze children's fingers or other body parts. These products should be easy to clean, especially when they are targeted toward infants and toddlers, who may put them in their mouths (American Academy of Pediatrics, 2003b).

Technology designers should also be aware of the less immediate impact of their products. One such possible negative impact is obesity, which has been linked with children's heavy television watching. Although this impact has been well documented through several studies, the reasons behind it are still unclear. Part of the problem may be due to a more sedentary lifestyle (American Academy of Pediatrics, n.d.). Another reason that has been explored is the exposure to advertisements for unhealthy foods, which affects children's food choices (Henry J. Kaiser Family Foundation, 2004). Regardless of the reason, such a lifestyle, when not including enough physical activity, has been strongly linked with obesity, type 2 diabetes, and cardiovascular disease (American Diabetes Association, 2000; American Academy of Pediatrics, 2003a). Technology designers should be aware of these issues and take steps to avoid children spending excessive amounts of time on sedentary activities. Content providers should also understand the consequences of the types of advertisements and attitudes they promote with respect to eating habits.

3.2 Emotional and Social Health

Use of technologies that are not designed for collaboration or social activities can lead to children having reduced social interactions. The American Academy of Pediatrics has pointed at how heavy television watching can lead to reduced time talking with friends and family and exploring the world (American Academy of Pediatrics, n.d.). A recent study also found that children who are heavy watchers of television spend less time playing outside (Rideout et al., 2003). The media content that children access can also affect emotional health by causing fear, depression, nightmares, and sleep problems (American Academy of Pediatrics, 2001b).

Designers should make technologies that will encourage or at least not hinder collaboration and social interaction. Besides benefiting children in their social growth, such technologies are likely to improve learning opportunities. Designers should also be mindful of not delivering inappropriate content that could cause emotional distress in children.

3.3 Intellectual Development

Early television exposure has been shown to decrease children's attention spans. A study found a correlation between the amount of time watching television between ages 1 and 3, and attention problems at age 7 (Christakis et al., 2004). Problems with concentration and patience could harm children's chances of learning in school settings. There is also evidence that heavy television watching reduces time spent reading by children 6 and under in the United States. In children 4 to 6 years old, heavy television watchers were less likely to know how to read. These children also spend less time playing outside (Rideout et al., 2003).

The type of play available through many media and interactive products can get in the way of the imaginative type of play that is encouraged by developmental psychologists. The key is whether play allows children to use their imagination, create their own rules, and act in a role of their choice, or whether it limits them to predefined scenarios, rules, and roles. Although the latter kind of play may provide a motivation to use the technology, it will not give children the benefits of imaginative, unstructured play.

Design teams need to make an effort to avoid giving children technologies that do not provide a chance for interaction, do not accommodate social activities, and do not allow children to use their imagination. Instead, teams could take Papert's suggestion of providing children with opportunities to build meaningful public objects (Papert, 1991).

3.4 Moral Development

Content in media accessed by children can have a negative impact on their moral development. Viewing of television violence during childhood has been linked with violent and aggressive behavior during childhood and adulthood in both males and females regardless of socioeconomic status, intellectual ability, and parenting factors such as aggression and television habits (Henry J. Kaiser Family Foundation, 2003; Huesmann et al., 2003). Violent video games have also been linked to aggression (Henry J. Kaiser Family Foundation, 2002). To avoid these issues, children's content providers should show the effects of violence, avoid showing violence as a means to resolve conflict, avoid positive portrayals of weapons, eliminate the use of violence in entertaining ways, eliminate gratuitous violence, and avoid rewarding violence in video games (American Academy of Pediatrics, 2001b).

Content can also have negative effects in terms of risky sexual behavior, drinking, and smoking. These activities are often portrayed in the media as something casual, fun, and exciting (American Academy of Pediatrics, n.d., 2001a). Content providers need to either avoid showing these behaviors, or show them together with their negative consequences.

Media content can also have a negative impact in the creation of gender and racial stereotypes. Television, for example, tends to stereotype gender roles (Signorielli, 1998). Video games can also be problematic in terms of both gender and racial stereotypes (Children Now, 2001). Content providers need to avoid showing biases in their selection of characters. A possible solution in interactive technologies is to give children choices as to how they want their characters to look.

4 DESIGNING FOR CHILDREN

Learning about children's needs and abilities and about the possible hazards children may face when interacting with technologies is a must for all designers of children's technologies. However, it is often not enough to develop good-quality products. Designing for children poses specific challenges that are best addressed by multidisciplinary teams following design methodologies specifically adapted for the task. In this section we present advice on how to put together effective multidisciplinary teams and explore design methodologies based on the role that children play in the design process. The advice is based on the author's several years of experience with multidisciplinary teams designing technologies for children (Benford et al., 2000; Druin et al., 2001; Hourcade et al., 2003, 2004a, 2004c).

4.1 Multidisciplinary Teams

Assembling a multidisciplinary team is key to successful designs. The first ingredient is a team leader. Team leaders need to be able to communicate with each of the disciplines represented in their teams. They need to be able to motivate people from different backgrounds and be ready to act as translators and arbitrators between the various disciplines. They should also have experience working with children and designing technologies for them. It is best if the remaining members of the team have had experience working in multidisciplinary teams. The disciplines needed in these multidisciplinary teams can be divided roughly into three categories: builders, user population experts, and domain experts.

4.1.1 Builders

Builders are responsible for being experts at putting together every aspect of a technology, from its look and feel to its code. Depending on the type of technology, builders can be computer scientists, engineers, artists, and so on. These experts will contribute by both planning how to build and actually doing the building. For example, some computer scientists may design the architecture of a system while others will do the actual programming. A visual designer could contribute opinions on the look and feel for a technology, and an artist will actually create the graphics. It is best if those responsible for the look and feel of the product have experience creating designs for children.

4.1.2 User Population Experts

Although builders can put together a product all by themselves, a team would not be complete without at least one member who is an expert on children as a user population. People in this role can be psychologists, educators, teachers, and children. Yes, children. There are no better experts at being children than children themselves. All team members need to contribute their understanding of children's needs and abilities. Children who form part of the team should match as closely as possible the age and background of the target population. Adult members should have experience with children in the targeted age group.

4.1.3 Domain Experts

Domain experts specialize in the subject being addressed by a technology. If the technology creates music, the domain expert will be a musician. If it is a portal to government statistics, the expert will be a government statistician. These experts are key to ensuring that the product being designed provides the best quality and experience within the subject. In a manner similar to that with builders, it helps if the domain experts have experience working with children or making products for them.

4.2 Design Methodologies and the Role of Children

The process of designing technology for children can be approached in many different ways. The choices are often restricted by practical concerns such as budgets and time constraints. Whether the team is following the old waterfall model or the more current iterative and incremental methods, the level of children's participation in the design process can play a crucial role. This prompted Allison Druin, an influential scholar in the field of interaction design and children, to propose the discussion of design methodologies based on the role children play in the design process (Druin, 2002). Druin says children can participate in the design process as users, testers, informants, or partners. Following is an overview of the types of activities and design methods used when children participate in each of their possible roles.

4.2.1 Children as Users

Montessori (1964) affirmed the need to observe children to learn what they need and what prevents them from learning. Vygotsky (1978) highlighted the same need to learn about when they need help to acquire a skill or concept. Technology developers can also learn by observing children. Together with testing before and after using a technology, children can participate in the design process as users by being observed.

Children participate as users at the beginning and end of the design process. Observing them at the beginning of the design process can provide key information for task analysis. Children's activities can be observed to learn where technologies could aid children and to better understand their needs. Children can also be observed using competing technologies or technologies similar to the one being developed. This can yield some information on the features that work, those that do not work, and those that need to be added.

At the end of the design process, children can be tested on a skill or concept before and after using a technology, to assess its impact. Children can also be observed while using the newly developed technology to learn about its positive and negative aspects.

Observation is a very practical and easy way of having children participate in the design process. Observing children in public places is free and does not require permissions from parents, schools, or institutional review boards. Testing, on the other hand, requires more planning and permissions, but often, these tests are easier to develop than technology-based testing environments. Testing for long-term effects, on the other hand, can prove logistically challenging.

Despite its advantages, having children participate only as users greatly limits their role. The most limiting aspect is that these children do not affect the design of the technology during the design process. The lack of direct design input from children has the potential of leading to the development of unappealing technologies that are difficult to use.

4.2.2 Children as Testers

Children as testers move from the endpoints of the design process to the middle of the process. Children can test ideas on paper, as low-fidelity prototypes, or as fully functional technologies at each design iteration. Testing greatly broadens the impact that children can have on the development process.

The value of testing increases if the design team decides to use iterative design methods. By testing paper and other low-tech prototypes, designers and developers can eliminate many design bugs before implementing any of the technology. This can be a valuable time-and-money saver. When testing early ideas and paper prototypes, designers should be careful to stay away from abstract ideas, but present concrete concepts instead. Otherwise, children's developmental issues could get in the way, unless the target population is teenagers.

A useful technique that can bridge the gap between low-tech prototypes and implemented technologies is known as *Wizard of Oz*. This technique, also used with adults, makes a child being tested think that he or she is using a technology that actually works when actually a human being is controlling the responses to the child's input. Höysniemi et al. (2004) provide a literature review of the use of Wizard of Oz techniques together with a useful example applied to a learning technology for children.

Testing in the later stages of development is perhaps the most common way in which children currently participate in the design process. This type of testing is crucial for ensuring that no major issues exist with the technology before it is released. Performing this type of testing alone, though, is not recommended, because it may be too costly to fix basic design problems that could easily have been uncovered with earlier testing.

Testing technologies for children is not as simple as testing those for adults. It is usually more difficult to find children to test technologies than it is to find adults. Parents should be asked for permission to have their children participate in testing. Children should never be forced to participate, even if their parents gave permission. Children should also be able to stop participating in a test if they wish to do so and should be informed that they have this choice. For young children, adults conducting the tests should ensure that the children feel comfortable as they test the technology. If signs of discomfort are observed, the children should be told to stop interacting with the technology. This is necessary with younger children because they may not always voice their discomfort in front of an authority figure. For more detailed guidelines on testing with children, refer to Hanna et al. (1997).

Testing throughout the development process can go a long way toward avoiding poor designs. However, it does not provide children a chance to give their ideas to the design team. The design ideas still come from adults. For projects to succeed when children participate as testers and users, the design team must have a lot of expertise in the design of children's technologies, and the design ideas must be based on sound educational or developmental theories (e.g., Cassell and Ryokai, 2001; Wyeth and Purchase, 2003).

4.2.3 Children as Informants

When children are informants, they actually do get to share ideas and opinions with the design team. Rather than being members of the design team, they act as consultants, making their contributions at key points of the development process. The design team decides when it is best for children to participate as informants.

Working with children during task analysis can provide further information on the challenges and expectations involved in completing tasks that are to be supported by technology. Children can also provide feedback and ideas by trying out existing technologies. As prototypes and design ideas are developed, children can provide feedback when the design team has a number of ideas or questions on how to move forward. Personal interviews, written questionnaires, and focus groups can provide children with an opportunity to have their voices heard (Read et al., 2004).

A way around not being able to work with children more often has been proposed by Antle (2004). While developing an Internet application for children on a short schedule, her team consulted children at key points in the development process. To fill the times in between, her team developed a set of personas that defined a set of representative children who were likely to use the technology. The characteristics of these personas were based on a number of factors, including the perceived characteristics of the children the design team worked with, and relevant child development literature. The personas enabled the design team to question their design decisions from a different point of view. Although certainly not as effective as having real children give their opinions, personas can provide a way for design teams to consider issues from a child's perspective.

For teams designing on the run, which is often the case in industry, having children participate as informants may be the best choice. If design teams are working 80-hour weeks, there will certainly not be an opportunity to have children participate in the design process as equal partners. Instead, children can participate as informants providing feedback, opinions, and ideas at critical points.

4.2.4 Children as Design Partners

When children participate in the design process as design partners, they actually join the design team. They become equal partners in decisions leading to the design and implementation of technology. In this partnership, ideas come from a process of collaboration between adults and children. Children do not tell adults what to do, but they do play a significant role in shaping the outcome of the process.

Druin has pioneered the concept of children as design partners, setting up design team partnerships with elementary school children at the University of New Mexico in the mid-1990s, and since 1998 at the University of Maryland (Druin, 1999, 2002). This chapter's author participated in the teams at the University of Maryland for several years.

Druin's approach is to set up a group of six to eight children to work on a set of projects. Children are recruited through word of mouth, and their parents make a commitment to have their children participate in the design team for one year. Children in the teams need not be particularly smart or technology-savvy. It does help, though, if they are willing to share their opinions and listen to others.

Druin's teams first meet for an intensive two-week camp during the summer when children are introduced to each other, to the adults they will be working with, and to the idea of being designers and inventors. Teams then meet twice a week during the school year. These meetings are referred to as *design sessions*. Rather than working on one project, Druin's teams typically concentrate on several projects at a time, although only one project is usually treated in each design session.

Design sessions use an approach called *cooperative inquiry*, designed for multigenerational teams to work together. Cooperative inquiry combines a number of techniques used previously for adults and tunes them for use with elementary school children. This approach has been used in the development of children's technologies such as KidPad (Hourcade et al., 2004a) and the International Children's Digital Library (Hourcade et al., 2003). It is currently being used in a variety of projects, including the design of children's user interfaces to Census Bureau data. Cooperative inquiry is composed of three techniques: technology immersion, contextual inquiry, and participatory design (Druin, 1999).

Technology immersion is meant to expose nontechnical team members, especially children, to types of technologies they have not experienced. The exposure should be concentrated and guided by the team members experiencing it. The benefit to the children and other nontechnical members is a new awareness of the potential of the technologies. The benefit to the rest of the team is a first look at how children might interact with such technologies. Technology immersion is particularly useful for projects that plan to use innovative technologies, such as novel input and output devices. In such projects, technology immersion sessions are likely to occur toward the beginning of the design process.

Contextual inquiry sessions involve children and adults performing tasks while other adults and children observe them and take notes. The observers and the observed may switch roles. At the end of each session, the team identifies the positive and negative aspects of the interactions and suggests improvements. This is often done with the help of sticky notes filled out as observations occur. Contextual inquiry is useful throughout the development process to evaluate design and technologies and develop new ideas. At the beginning of the development process, team members may be observed using existing technologies that provide experiences similar to those that the team wants to design. Team members may also be observed performing specific tasks without a technology. As designs are developed, teams can use contextual inquiry sessions to evaluate and improve them. The same can be done as prototypes and technologies are implemented. As the project progresses and prototypes are delivered more often, the frequency of contextual inquiry sessions increases. This aids in eliminating design issues and bugs in the technology being developed.

Although technology immersion and contextual inquiry sessions are capable of generating ideas, the technique used for generating most ideas is *participatory design*. In participatory design sessions, children and adults divide into small teams to develop designs to address specific problems. These small teams should ideally be composed of two or three children and at least one adult. When working in these teams, adults and children voice their ideas and elaborate them. Then they use low-tech prototyping materials to sketch their solutions. These materials can include paper, markers, cardboard, crayons, tape, fabric, glue, socks, and so on. At the end of the session, all the teams come together and share design ideas. Participatory design sessions are critical at the beginning of the design process to obtain basic design ideas. As prototypes are developed, participatory design sessions can aid the team in designing interactions for new features to be added.

To organize and conduct design sessions, adult members of the design team have to fulfill certain tasks that go beyond participating in cooperative inquiry activities. These tasks involve facilitating sessions, asking research questions, and documenting. Perhaps the most important task is to facilitate the sessions. A session facilitator leads design sessions, motivating both children and adults to participate, dividing them into groups if necessary, and ensuring that research objectives are met. Motivating and communicating with a 7-year-old and a tenured faculty member at the same time is not an easy task. Thus, an important skill for facilitators is to have the ability to communicate comfortably with both adults and children.

Asking research questions is another task that needs to be fulfilled by adults. The facilitator is often the one to ask the research questions. Stating research questions gives direction to the design session. Asking the questions requires knowledge of where the project is heading and what issues need to be addressed. Although a facilitator need not be involved in the daily activities of the project, the person responsible for asking research questions must be so involved. One can think of facilitating as providing the syntax for a design session, while asking research questions provides the semantics.

During and after a design session, it is important to document the process. Team members can take pictures, record video, and take notes during sessions. After design sessions, adult team members need to meet to discuss the outcomes of the session and make decisions about action items. At this point, note taking is also key to documenting the action items and conclusions reached through the session. Documenting the process ensures that no ideas are lost. It is also a way to keep track of where ideas come from and of documenting the evolution of design ideas. For academics, documenting is key for writing design briefings and sharing lessons learned through research papers and videos.

Druin and her colleagues at the University of Maryland developed the idea of cooperative inquiry for working with a group of elementary school children that meets on a regular basis outside school. This type of arrangement does not meet everyone's needs. Some teams are designing technologies for children of different ages and abilities (Gibson et al., 2002; Knudtzon et al., 2003; Guha et al., 2004). Other teams are not able to meet with children outside school and have to meet groups of children at their schools during the school day (Taxen et al., 2001; Rode et al., 2003). In addition, some researchers have proposed alternative activities (Iversen, 2002; Bekker et al., 2003).

Having children join teams as design partners gives them a greater voice in the design process. Their needs and abilities can be taken into account more easily. Adults can also be more aware of cultural differences between generations. Design decisions will probably include input from children, helping avoid designs that could be difficult to understand or uninteresting for other children.

Despite these advantages, a number of issues make it difficult to work with children as design partners. The first issue is that it takes time to develop a multigenerational design team. Children do not become inventors and designers overnight. They need time to develop the self-confidence necessary to tell an adult researcher that his or her ideas will not work. It also takes time for them to realize that their ideas can actually make it into real products. In most cases, children do not make valuable contributions on a regular basis until they have been part of a design team for about a year. It is crucial then to have at least half of the children return to their design team every year. This leads to the issue of continuity. Not only is it important for some children to return every year, it is also important for the team to meet on a regular basis throughout the school year. Putting together a team that works for a month, then does not meet for six months, will probably not help children develop into valuable contributors. These requirements can make design partnerships with children very difficult to implement for teams that have tight deadlines or short-term projects.

A second issue is logistics. Not all researchers and designers have the ability to meet with children on a regular basis in a suitable space. It is also difficult to recruit children whose parents can reliably bring them to design team meetings on a regular basis. The children should also be able to work together. Children who are socially challenged can disrupt design sessions and distract the team from its goals.

An additional issue with multigenerational design teams is that due to their small size, they are not representative of the entire target population. Hence, the children in the team are likely to bias the design toward their personal needs and abilities, cultural background, socioeconomic status, and likes and dislikes. A strategy to address this issue is to work with a second larger and more representative group of children as informants. They can validate the work of the smaller group at key points in the design process (Druin et al., 2001).

5 CONCLUSIONS

Technologies have the potential to provide children with a wealth of valuable experiences never before available. To design such technologies, design teams need to be aware of what constitutes a positive experience, how to avoid designing negative experiences, and what approaches and techniques to use during the development process.

Positive experiences can occur when children's needs and abilities are taken into account. Such experiences are more likely when children use technologies to learn about the world through motivating social and creative experiences. These technologies should be sized appropriately based on children's physical maturation. Similarly, user interface complexity should be designed, keeping in mind children's limited information-processing speed and working memory. Designers should be mindful of the challenges that children face with input devices. Larger pointing targets, no dragging, and not providing different functionality through each mouse button are recommended. Designers should similarly be aware of how children's intellectual abilities change with age, taking Piaget's developmental stages as a guide. Piaget's stages suggest that designers should, for example, beware of using hierarchies with children younger than 7, or abstract concepts with children younger than 11.

Design teams must also avoid creating technologies that can lead to negative experiences. Sedentary activities and content that promotes unhealthy eating habits can lead to obesity. Technologies can isolate children with activities that do not encourage creativity. Instead, technologies should encourage reading, and imaginative, creative social activities. In addition, content with violence and other risky behaviors should only be made accessible in an age-appropriate manner that at the very least makes clear the negative consequences of such behaviors. Design teams must also be careful of not using stereotypes when designing characters.

Assembling multidisciplinary design teams is a key to creating successful technologies for children. It is also critical to know about the various ways of incorporating children into the design process. Each role that children play in the design process has its own benefits and limitations. If possible, given project constraints, design teams should incorporate children, having them participate of the design process as design partners.

Following these recommendations, technologies designed for children can show marked differences with similar-purpose technologies designed for adults. For example, Figure 2 shows the spiral book reader from the International Children's Digital Library (Hourcade et al., 2003) next to Microsoft Reader 2.1.1 (Microsoft, 2004). The spiral reader was designed for early elementary school children reading digital books that have many pictures. It was developed in a design partnership with children, taking into account children's needs and developing abilities. Microsoft Reader 2.1.1 was designed for adults reading longer digital books that are mostly text. Figure 2 shows the spiral reader in overview mode, which gives children the ability to assess the quality and number of the pictures in a book, something they take into consideration before deciding to read a book. Microsoft Reader does not provide an overview. All pointing targets in the spiral reader are large and very visual, designed to be easy to click on and not require reading. The spiral reader also makes each displayed page a pointing target, making the operation of jumping to another page very concrete. This jumping is animated, giving the effect of flipping through pages. Microsoft Reader 2.1.1, on the other hand, gives users the ability to jump to pages by clicking on one of the bars at the bottom of the screen, which are smaller, harder to click on, and more abstract.

The differences between the spiral reader and Microsoft Reader 2.1.1 illustrate the need for products designed specifically for children. In providing guidelines and recommendations on the design of children's technologies, we have aimed to educate and inspire both novices and seasoned designers to create more technologies designed with children's needs and abilities in mind. It is also the author's wish that this chapter contribute toward an informed, ethical, and active community of designers of children's technologies.

ACKNOWLEDGMENTS

I would like to thank Allison Druin for her guidance during my graduate studies at the University of

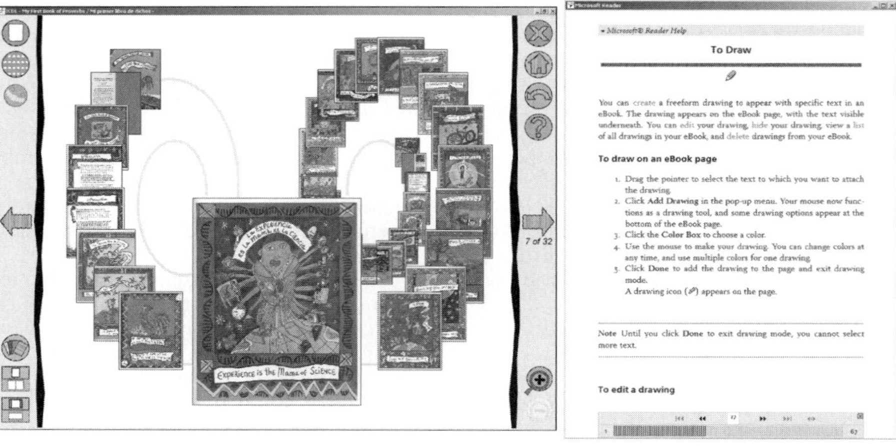

Figure 2 (*a*) International Children's Digital Library's spiral reader; (*b*) Microsoft Reader 2.1.1. [(*a*) From Hourcade et al., 2003; (*b*) from Microsoft, 2004.]

Maryland and for teaching me about the world of interaction design and children. I would also like to acknowledge Ben Bederson, my doctoral advisor at the University of Maryland, for his constant support and insights. Thanks also to Betty Murphy, at the U.S. Census Bureau, who helped edit this chapter. And thank you to all the children who have helped me design technologies over the years.

REFERENCES

American Academy of Pediatrics (n.d.), "Television and the Family," retrieved March 8, 2003, from http://www.aap.org/family/tv1.htm.

American Academy of Pediatrics (2001a), "Children, Adolescents, and Television," *Pediatrics*, Vol. 107, No. 2, pp. 423–426.

American Academy of Pediatrics (2001b), "Media Violence," *Pediatrics*, Vol. 108, No. 5, pp. 1222–1226.

American Academy of Pediatrics (2003a), "Prevention of Pediatric Overweight and Obesity," *Pediatrics,* Vol. 112, No. 2, pp. 424–430.

American Academy of Pediatrics (2003b), "Selecting Appropriate Toys for Young Children: The Pediatrician's Role," *Pediatrics,* Vol. 111, No. 4, pp. 911–913.

American Diabetes Association (2000), "Type 2 Diabetes in Children and Adolescents," *Pediatrics,* Vol. 105, No. 3, pp. 671–680.

Antle, A. (2004), "Supporting Children's Emotional Exploration and Expression in Online Environments," in *Proceedings of Interaction Design and Children 2004*, ACM Press, New York, pp. 97–104.

Baddeley, A. (1992), "Working Memory," *Science*, Vol. 255, No. 5044, pp. 556–559.

Bekker, M., Beusmans, J., Keyson, D., and Lloyd, P. (2003), "KidReporter: A User Requirements Gathering Technique for Designing with Children," *Interacting with Computers*, Vol. 15, No. 3, pp. 187–202.

Benford, S., Bederson, B. B., Akesson, K., Bayon, V., Druin, A., Hansson, P., Hourcade, J. P., Ingram, R., Neale, H., O'Malley, C., Simsarian, K., Stanton, D., Sundblad, Y., and Taxén, G. (2000), "Designing Storytelling Technologies to Encourage Collaboration Between Young Children," in *Proceedings of Human Factors in Computing Systems, CHI 2000*, ACM Press, New York, pp. 556–563.

Cassell, J., and Ryokai, K. (2001), "Making Space for Voice: Technologies to Support Children's Fantasy and Storytelling," *Personal and Ubiquitous Computing*, Vol. 5, pp. 169–190.

Children Now (2001), "Fair Play? Violence, Gender and Race in Video Games," retrieved March 8, 2003, from http://npin.org/library/2002/n00739/n00739.pdf.

Christakis, D. A., Zimmerman, F. J., DiGiuseppe, D. L., and McCarty, C. A. (2004), "Early Television Exposure and Subsequent Attentional Problems in Children," *Pediatrics*, Vol. 113, No. 4, pp. 708–713.

Dewey, J. (1959), *The School and Society*, University of Chicago Press, Chicago.

Druin, A. (1999), "Cooperative Inquiry: Developing New Technologies for Children with Children," in *Proceedings of Human Factors in Computing Systems, CHI '99*, ACM Press, New York, pp. 223–230.

Druin, A. (2002), "The Role of Children in the Design of New Technology," *Behaviour and Information Technology*, Vol. 21, No. 1, pp. 1–25.

Druin, A., Bederson, B. B., Hourcade, J. P., Sherman, L., Revelle, G., Platner, M., and Weng, S. (2001), "Designing a Digital Library for Young Children: An Intergenerational Partnerships," in *Proceedings of the Joint Conference on Digital Libraries, JCDL 2001*, ACM Press, New York, pp. 398–405.

Eisenberg, M. (2004), "Tangible Ideas for Children: Materials Science as the Future of Educational Technology," in *Proceedings of Interaction Design and Children 2004*, ACM Press, New York, pp. 19–26.

Garhart Mooney, C. (2000), *Theories of Childhood: An Introduction to Dewey, Montessori, Erickson, Piaget and Vygotsky*, Redleaf Press, St. Paul, MN.

Gibson, L., Gregor, P., and Milne, S. (2002), "Designing with 'Difficult' Children," in *Proceedings of the*

Interaction Design and Children International Workshop, Shaker Publishing, Maastricht, Netherlands, pp. 42–52.

Guha, M. L., Druin, A., Chipman, G., Fails, J. A., Simms, S., and Farber, A. (2004), "Mixing Ideas: A New Technique for Working with Young Children as Design Partners," in *Proceedings of Interaction Design and Children 2004*, ACM Press, New York, pp. 35–42.

Hanna, L., Risden, K., and Alexander, K. (1997), "Guidelines for Usability Testing with Children," *Interactions*, Vol. 4, No. 5, pp. 9–14.

Henry J. Kaiser Family Foundation (2002), "Key Facts: Children and Video Games," retrieved April 20, 2004, from http://www.kff.org.

Henry J. Kaiser Family Foundation (2003), "Key Facts: TV Violence," retrieved April 20, 2004, from http://www.kff.org.

Henry J. Kaiser Family Foundation (2004), "Issue Brief: The Role of Media in Childhood Obesity," retrieved April 20, 2004, from http://www.kff.org.

Hourcade, J. P., Bederson, B. B., Druin, A., Rose, A., Farber, A., and Takayama, Y. (2003), "The International Children's Digital Library: Viewing Digital Books Online," *Interacting with Computers*, Vol. 15, pp. 151–167.

Hourcade, J. P., Bederson, B. B., and Druin, A. (2004a), "Building KidPad: An Application for Children's Collaborative Storytelling," *Software Practice and Experience*, Vol. 34, DOI 10.1002/spe.598.

Hourcade, J. P., Bederson, B. B., and Druin, A. (2004b), "Preschool Children's Use of Mouse Buttons," in *Extended Abstracts of Human Factors in Computing Systems, CHI 2004*, ACM Press, New York, pp. 1411–1412.

Hourcade, J. P., Bederson, B. B., Druin, A., and Guimbretiere, F. (2004c), "Differences in Pointing Task Performance Between Preschool Children and Adults Using Mice," *ACM Transactions on Computer–Human Interaction*, Vol. 11, No. 4, pp. 357–386.

Höysniemi, J., Hämäläinen, P., and Turkki, L. (2004), "Wizard of Oz Prototyping of Computer Vision Based Action Games for Children," in *Proceedings of Interaction Design and Children 2004*, ACM Press, New York, pp. 27–34.

Huesmann, L. R., Moise-Titus, J., Podolski, C. L., and Eron, L. D. (2003), "Longitudinal Relations Between Children's Exposure to TV Violence and Their Aggressive and Violent Behavior in Young Adulthood: 1977–1992," *Developmental Psychology*, Vol. 39, No. 2, pp. 201–221.

Inkpen, K. M. (2001), "Drag-and-Drop Versus Point-and-Click: Mouse Interaction Styles for Children," *ACM Transactions on Computer–Human Interaction*, Vol. 8, No. 1, pp. 1–33.

Iversen, O. S. (2002), "Designing with Children: The Video Camera as an Instrument of Provocation," in *Proceedings of Interaction Design and Children International Workshop*, Shaker Publishing, Maastricht, Netherlands, pp. 73–81.

Jeong, B. Y., and Park, K. S. (1990), "Sex Differences in Anthropometry for School Furniture Design," *Ergonomics*, Vol. 33, No. 2, pp. 1511–1521.

Joiner, R., Messer, D., Light, P., and Littleton, K. (1998), "It Is Best to Point for Young Children: A Comparison of Children's Pointing and Dragging," *Computers in Human Behavior*, Vol. 14, No. 3, pp. 513–529.

Jones, T. (1991), "An Empirical Study of Children's Use of Computer Pointing Devices," *Journal of Educational Computing Research*, Vol. 7, No. 1, pp. 61–76.

Kail, R. (1991), "Developmental Change in Speed of Processing During Childhood and Adolescence," *Psychological Bulletin*, Vol. 109, No. 3, pp. 490–501.

Kaluger, G., and Kaluger, M. F. (1979), *Human Development: The Span of Life*, 2nd ed., C.V. Mosby. St. Louis, MO.

Kerr, R. (1975), "Movement Control and Maturation in Elementary-Grade Children," *Perceptual and Motor Skills*, Vol. 41, pp. 151–154.

Knudtzon, K., Druin, A., Kaplan, N., Summers, K., Chisik, Y., Kulkarni, R., Moulthrop, S., Weeks, H., and Bederson, B. (2003), "Starting an Intergenerational Technology Design Team: A Case Study," in *Proceedings of Interaction Design and Children 2003*, ACM Press, New York, pp. 51–58.

Malone, T. W. (1980), "What Makes Things Fun to Learn? Heuristics for Designing Instructional Computer Games," in *Proceedings of the 3rd ACM SIGSMALL Symposium and the First SIGPC Symposium on Small Systems*, ACM Press, New York, pp. 162–169.

Mandal, A. C. (1997), "Changing Standards for School Furniture," *Ergonomics in Design*, Vol. 5, pp. 28–31.

Martin, F., Mikhak, B., Resnick, M., Silverman, B., and Berg, R. (2000), "To Mindstorms and Beyond: Evolution of a Construction Kit for Magical Machines," in *Robots for Kids: New Technologies for Learning*, A. Druin and J. Hendler, Eds., Morgan Kaufmann, San Francisco, pp. 9–33.

Microsoft (2004), "Microsoft Reader," retrieved August 3, 2004, from http://www.microsoft.com/reader.

Montemayor, J., Druin, A., Farber, A., Simms, S., Churaman, W., and D'Amour, A. (2002), "Physical Programming: Designing Tools for Children to Create Physical Interactive Environments," in *Proceedings of Human Factors in Computing Systems, CHI 2002*, ACM Press, New York, pp. 299–306.

Montessori, M. (1964), *The Montessori Method*, A. E. George, Trans., Schocken, New York.

Papert, S. (1991), "Situating Constructionism," in *Constructionism*, I. Harel and S. Papert, Eds., Ablex, Norwood, NJ, pp. 1–12.

Papert, S. (1993), *Mindstorms: Children, Computers, and Powerful Ideas*, 2nd ed., Basic Books, New York.

Piaget, J. (1962), *Play, Dreams and Imitation in Childhood*, C. Gattegno and F. M. Hodgson, Trans., W.W. Norton, New York.

Piaget, J. (1963), *The Origins of Intelligence in Children*, M. Cook, Trans., W.W. Norton, New York.

Piaget, J. (1973), *The Child and Reality*, A. Rosin, Trans., Grossman, New York.

Piaget, J. (1995a), "Judgement and Reasoning in the Child," in *The Essential Piaget*, H. E. Gruber and J. J. Voneche, Eds., Jason Aronson, London, pp. 89–117.

Piaget, J. (1995b), "The Language and Thought of the Child," in *The Essential Piaget*, H. E. Gruber and J. J. Voneche, Eds., Jason Aronson, London, pp. 65–88.

Piaget, J. (1995c), "Logic and Psychology," in *The Essential Piaget*, H. E. Gruber and J. J. Voneche, Eds., Jason Aronson, London, pp. 445–477.

Piaget, J., and Inhelder, B. (1969), *The Psychology of the Child*, H. Weaver, Trans., Basic Books, New York.

Read, J. C., MacFarlane, S., and Gregory, P. (2004), "Requirements for the Design of a Handwriting Recognition

Based Writing Interface for Children," in *Proceedings of Interaction Design and Children 2004*, ACM Press, New York, pp. 81–88.

Rideout, V. J., Vandewater, E. A., and Wartella, E. A. (2003), "Zero to Six: Electronic Media in the Lives of Infants, Toddlers, and Preschoolers," The Henry J. Kaiser Family Foundation, retrieved April 20, 2004, from http://www.kff.org.

Robertson, J., and Good, J. (2004), "Children's Narrative Development Through Computer Game Authoring," in *Proceedings of Interaction Design and Children 2004*, ACM Press, New York, pp. 57–64.

Rode, J. A., Stringer, M., Toye, E. F., Simpson, A. R., and Blackwell, A. F. (2003), "Curriculum-Focused Design," in *Proceedings of Interaction Design and Children 2003*, ACM Press, New York, pp. 119–126.

Rogers, Y., Price, S., Fitzpatrick, G., Fleck, R., Harris, E., Smith, H., Randell, C., Muller, H., O'Malley, C., Stanton, D., Thompshon, M., and Weal, M. (2004), "Ambient Wood: Designing New Forms of Digital Augmentation for Learning Outdoors," in *Proceedings of Interaction Design and Children 2004*, ACM Press, New York, pp. 3–10.

Roussou, M. (2004), "Learning by Doing and Learning Through Play: An Exploration of Interactivity in Virtual Environments for Children," *Computers in Entertainment*, Vol. 2, No. 1, pp. 1–23.

Salmoni, A. W., and McIlwain, J. S. (1979), "Fitts' Reciprocal Tapping Task: A Measure of Motor Capacity?" *Perceptual and Motor Skills*, Vol. 49, pp. 403–413.

Signorielli, N. (1998), "Television and the Perpetuation of Gender-Role Stereotypes," *AAP News*, February, pp. 103–104.

Singer, D. G., and Revenson, T. A. (1996), *A Piaget Primer: How a Child Thinks*, rev. ed., Plume, New York.

Stringer, M., Toye, E. F., Rode, J. A., and Blackwell, A. F. (2004), "Teaching Rhetorical Skills with a Tangible User Interface," in *Proceedings of Interaction Design and Children 2004*, ACM Press, New York, pp. 11–18.

Sugden, D. A. (1980), "Movement Speed in Children," *Journal of Motor Behavior*, Vol. 12, pp. 125–132.

Swanson, H. L. (1999), "What Develops in Working Memory? A Life Span Perspective," *Developmental Psychology*, Vol. 35, No. 4, pp. 986–1000.

Taxen, G., Druin, A., Fast, C., and Kjellin, M. (2001), "KidStory: A Technology Design Partnership with Children," *Behaviour and Information Technology*, Vol. 20, No. 2, pp. 119–125.

Vygotsky, L. S. (1978), in *Mind in Society: The Development of Higher Psychological Processes*, M. Cole, V. John-Steiner, S. Scribner, and E. Souberman, Eds., Harvard University Press, Cambridge, MA.

Wallace, S. A., Newell, K. M., and Wade, M. G. (1978), "Decision and Response Times as a Function of Movement Difficulty in Preschool Children," *Child Development*, Vol. 49, pp. 509–512.

Wyeth, P., and Purchase, H. C. (2003), "Using Developmental Theories to Inform the Design of Technology for Children," in *Proceedings of Interaction Design and Children 2003*, ACM Press, New York, pp. 93–100.

CHAPTER 56

DESIGN FOR ALL: COMPUTER-ASSISTED DESIGN OF USER INTERFACE ADAPTATION

Constantine Stephanidis
Foundation for Research and Technology–Hellas (FORTH) and University of Crete
Heraklion, Crete, Greece

Margherita Antona and Anthony Savidis
Foundation for Research and Technology–Hellas (FORTH)
Heraklion, Crete, Greece

1 INTRODUCTION

The increased importance of user interface design methodologies, techniques, and tools in the context of the development and evolution of the information society has been widely recognized in the recent past in the light of the increasing influence of interactive technologies on everyone's life and activities, and of the difficulty of developing usable and attractive interactive services and products (e.g., [Winograd et al., 1996; Winograd, 2001). As the information society develops further, the issue of human–computer interaction (HCI) design becomes even more prominent when considering the notions of *universal access* (Stephanidis et al., 1998, 1999; Stephanidis, 2001a) and *universal usability* (Shneiderman, 2000), aiming at providing access to diverse products and services to anyone, anywhere, and at any time, through a variety of computing platforms and devices. Design for universal access in the information society has often been defined as *design for diversity*, based on consideration of the several dimensions of diversity that emerge from the broad range of user characteristics, the changing nature of human activities, the variety of contexts of use, the increasing availability and diversification of information, the variety of knowledge sources and services, and the proliferation of diverse technological platforms that occur in the information society.

These issues imply an explicit design focus to address diversity systematically, as opposed to afterthoughts or ad hoc approaches, as well as an effort toward reconsidering and redefining the concept of Design for All in the context of HCI (Stephanidis, 2001a). In the emerging information society, therefore, universal access becomes predominantly an issue of design, and the question arises of how it is possible to design systems that take diversity into account and satisfy the variety of implied requirements. Recent work has highlighted a shift of perspective and reinterpretation of HCI design, in the context of universal access, from current artifact-oriented practices toward a deeper and multidisciplinary understanding of the diverse factors shaping interaction with technology, such as users' characteristics and requirements and contexts of use (Stephanidis et al., 1998, 1999; Akoumianakis and Stephanidis, 2001a; Stephanidis, 2001a), and has proposed solutions for methods, techniques, and codes of practice that enable designers to take into account diversity in a proactive manner and address it appropriately in the design of interactive artifacts (Stephanidis, 2001a; Stephanidis and Akoumianakis, 2003). In the framework of such efforts, the concept of Design for All has been reinterpreted and redefined in the domain of HCI. One of the main concepts proposed in such a context as a solution for catering to the needs and requirements of a diverse user population in a variety of context of use is that of automatic user interface adaptation (Stephanidis, 2001a, c). Despite the progress that has been made, however, the practice of designing for diversity remains difficult, due to intrinsic complexity of the task, the current limited expertise of designers and practitioners in designing interfaces capable of automatic adaptation, and the current limited availability of appropriate supporting tools.

The rationale behind this chapter is that the wider practice and adoption of an appropriate design method, supported through appropriate tools, has the potential to contribute to overcoming the difficulties outlined above. Toward this end, after highlighting the main issues involved in the effort of designing for diversity, we describe briefly a design method, the unified user interface design method, which has been developed in recent years to facilitate the design of user interfaces with automatic adaptation behavior (Savidis et al., 2001; Savidis and Stephanidis, 2004a). Subsequently, we discuss a support tool named MENTOR which embodies the unified user interface design process and assists designers in its systematic conduction. Two design cases conducted using MENTOR are also reported briefly. MENTOR is claimed to provide user interface designers with a suitable instrument for practicing user interface design in a Design for All perspective, and for collecting, in the long term, a number of design cases to be studied for further advancing design knowledge, practices, and support in a universal access perspective.

2 APPROACHES, METHODS, AND TECHNIQUES

Universal access implies the accessibility and usability of information society technologies by anyone, anywhere, any time, with the aim to enable equitable access and active participation of potentially all citizens in existing and emerging computer-mediated human activities, by developing universally accessible and usable products and services capable of accommodating individual user requirements in different contexts of use and independent of the location, target machine, or run-time environment. The origins of the concept of universal access are to be identified in early approaches to accessibility, targeted primarily toward providing access to computer-based applications by users with disabilities. Subsequently, accessibility-related methods and techniques have been generalized and extended toward more generic and inclusive approaches. HCI design approaches targeted to support universal access are often grouped under the term Design for All.

2.1 Reactive versus Proactive Strategies

In the past, the term *computer accessibility* was usually associated with access to interactive computer-based systems by people with disabilities. In traditional efforts to improve accessibility, the main direction followed has been to enable disabled users to access interactive applications originally developed for able-bodied users through appropriate adaptations.

Two main technical approaches to adaptation have been followed. The first is to treat each application separately and take all the necessary implementation steps to arrive at an alternative accessible version: *product-level adaptation*. Product-level adaptation often implies redevelopment from scratch. Due to the high costs associated with this strategy, it is considered the least favorable option for providing alternative access. The second alternative is to intervene at the level of the particular interactive application environment (e.g., MS-Windows, X Windowing System), to provide appropriate software and hardware technology so as to make that environment alternatively accessible (*environment-level adaptation*). The latter option extends the scope of accessibility to cover potentially all applications running under the same interactive environment rather than a single application and is therefore acknowledged as a more promising strategy.

The approaches described above have given rise to several methods for addressing accessibility, including techniques for the configuration of input–output at the level of user interface, the provision of alternative access systems, such as screen readers for blind users and scanning techniques for the motor impaired. The majority of efforts in this line of work have focused on the issue of accessibility of graphical environments by blind users (e.g., Mynatt and Weber, 1994).

Despite progress, the prevailing practices aiming to provide alternative access systems, either at the product or environment level, have been criticized for their essentially reactive nature (Stephanidis and Emiliani, 1999). Although the reactive approach to accessibility may be the only feasible solution in certain cases (Vanderheiden, 1998), it suffers from some serious shortcomings, especially when considering the radically changing technological environment—in particular, the emerging information society technologies. The critique is grounded on two lines of argumentation. The first is that reactive solutions typically provide limited and low-quality access. This is evident in the context of nonvisual interaction, where the need to provide genuine nonvisual user interfaces that are not simply adaptations of visual dialogues has been identified (Savidis and Stephanidis, 1995).

The second line of critique concerns the economic feasibility of the reactive approach to accessibility. Reactive approaches, based on a posteriori adaptations, although important in partial solution of some of the accessibility problems of people with disabilities, are not feasible in sectors of the industry characterized by rapid technological change. By the time a particular access problem has been addressed, technology has advanced to a point where the same or a similar problem recurs. The typical example that illustrates this state of affairs is the case of blind people's access to computers. Each generation of technology (e.g., DOS environment, windowing systems, and multimedia) caused a new generation of accessibility problems to blind users, addressed through dedicated techniques such as text-to-speech translation for the DOS environment, off-screen models, and filtering for the windowing systems.

In some cases, adaptations may not be possible without loss of functionality. For example, in early versions of windowing systems, it was impossible for the programmer to obtain access to certain window functions, such as window management. In subsequent versions, this shortcoming was addressed by the vendors of such products, allowing certain adaptations on interaction objects on the screen. Finally, adaptations are programming intensive and therefore are expensive and difficult to implement and maintain. Minor changes in product configuration or user interface may require substantial resources to rebuild the accessibility features. From the above it becomes evident that the reactive paradigm to accessible products and services does not suffice to cope with the rapid technological change and the evolving human requirements. At the same time, the proliferation of interactive products and services in the information society, as well as of technological platforms and access devices, brought about the need to reconsider the issue of access under a proactive perspective, resulting in more generic solutions. This entails an effort to build access features into a product starting from its conception, throughout the entire development life cycle. In the context of the emerging information society, therefore, universal access becomes predominantly an issue of design, and the question arises of how it is possible to design systems that permit systematic and cost-effective approaches to accommodating all users. Toward this end, the concept of Design for All has been revisited in recent years in the context of HCI (Stephanidis et al., 1998, 1999). Design for All is well known in several engineering disciplines, such as civil engineering and architecture, with many applications in interior design and building and road construction (Story, 1998). In the context of universal access, Design for All has a broad and multidisciplinary connotation, abstracting over different perspectives (Bühler and Stephanidis, 2004), such as (1) design of interactive products, services, and applications that are suitable for most potential users without any modification; (2) design of products that have standardized interfaces, capable of being accessed by specialized user interaction devices; and (3) design of products that are easily adaptable to a variety of users (e.g., by incorporating adaptable or customizable user interfaces).

From the above it follows that Design for All either subsumes, or is a synonym of, terms such as *accessible design, inclusive design, barrier-free design,* and *universal design*, each highlighting different aspects of the concept. In this chapter we foster a broad perspective on Design for All in HCI as the conscious and systematic effort, proactively, to apply principles and methods and employ appropriate tools, in order to develop interactive products and services that are accessible and usable by all citizens of the information society, thus avoiding the need for a posteriori adaptations or specialized design (Stephanidis et al., 1998, 1999).

In such a context, it is unrealistic to expect that a single interface design will ensure high-quality interaction for diverse user groups and contexts of use. Consequently, the outcome of the design process in a universal access perspective is not intended to be a "singular" design but a design space populated with appropriate alternatives, together with the rationale underlying each alternative (i.e., the specific user- and usage-context characteristics for which each alternative has been designed).

2.2 Design for All

The advancement of Design for All perspective toward the design of interactive products and services has brought various multidisciplinary approaches. For example, there are lines of work that aim to consolidate existing wisdom on accessibility, in the form of general guidelines or platform- or user-specific recommendations (e.g., for graphical user interfaces or the Web). Guidelines reflect previous experience gained as well as best practice available for designing accessible interactive software (including content). Several collection guidelines are available (e.g., Thoren, 1993; Bergman and Johnson, 1995; HFES/ANSI, 1997; Nicolle and Abascal, 2001) to facilitate access to computer-based equipment and services by people with disabilities. This approach consolidates the large body of knowledge regarding people with disabilities and alternative assistive technology access in an attempt to formulate ergonomic design guidelines that cover a wide range of disabilities. Similarly, there have been proposals of physical accessibility guidelines of consumer product controls (see, e.g., "Accessible Design of Consumer Products"*; see also Rahman and Sprigle, 1997). In recent years there has also been a trend for major software vendors to provide accessibility guidance as part of their mainstream products and services. In fact, all major actors, such as Microsoft, IBM, and Sun, provide documentation and insights as to the accessibility guidelines applicable to their own platforms or products. Moreover, with the advent of the World Wide Web, the issue of its accessibility recurred and was followed up by an effort undertaken in the context of the World Wide Web Consortium (W3C)† to provide a collection of accessibility guidelines for Web-based products and services (W3C–WAI, 1999). The systematic collection, consolidation, and interpretation of guidelines is also pursued in the context of international collaborative and standardization initiatives (e.g., ISO TC 159/SC 4/WG 5).

Despite the usefulness of guidelines, several factors often impede their use in the context of universal access. The first relates to the scope of guidelines currently available (i.e., the type and range of accessibility issues that can be addressed adequately by

*http://trace.wisc.edu/docs/consumer_product_guidelines/consumer.htm.

†http//www.w3c.org/WAI.

available knowledge and the types of solutions that can be generated. The vast majority of existing accessibility guidelines have been formulated on the basis of formative experimentation with people with disabilities, offer disability-oriented recommendations, and necessitate substantial interpretation before they can provide practical support for wider and more proactive accounts of design.

Another relevant line of work is related to user-centered design, which is often claimed to have an important contribution to make in the context of universal access and Design for All (Bevan, 2001; Stephanidis, 2001a), as its human-centered protocols and tight design-evaluation feedback loop replace technocentric practices with a focus on the human aspects of technology use. Design for universal access is in essence user-centered. However, it goes beyond user-centered design by breaking away from the traditional perspective of "typical" users interacting with a desktop machine in a business environment and by identifying and addressing user- and context-related diversity (Stephanidis, 2001a; Stephanidis and Akoumianakis, 2003). Although user-centered design and similar approaches foster maintaining a multidisciplinary and user-involving perspective in systems development, they do not specify how designers can cope with radically different user groups whose requirements are not known a priori. In particular, with the advent of the Internet and the emergence of a highly distributed and collaborative computing paradigm, it is difficult for designers to anticipate who the user may be. For example, Web sites can, in principle, be accessed by anyone possessing an Internet connection and a modem, irrespective of age, gender, educational background, and level of expertise. This requires that the outcomes of designing for diversity not be constrained to single artifacts but rather, to entire design spaces, containing alternative options intended for different user groups or contexts of use. As a result, designers should elicit requirements for all target user groups using methods that are best suited for each case. Additionally, design spaces need to be organized in such a way that they embody the contextual information required to differentiate design alternatives and specify when a particular option should be preferred. Toward this end, the notion of interface adaptation plays a critical role.

2.3 Design for All as the Design of Automatic Adaptation

In light of the above, it appears that single artifact-oriented design approaches offer limited possibilities of addressing the issue raised by universal access. A critical property of interactive artifacts becomes, therefore, their capability for intelligent adaptation and personalization (Akoumianakis et al., 2000; Stephanidis, 2001a,c). In this context, adaptation refers both to the system's capability to tailor aspects of its interactive behavior prior to an interactive session, in anticipation of a user's requirements (adaptability), as well as to run-time dialogue enhancements on the basis of dynamically acquired and maintained knowledge regarding the user (adaptivity). This perspective on adaptation widens the assumption, common in earlier approaches to intelligent interface adaptation (e.g., Dieterich et al., 1993; Hook, 2000), that intelligence entails dynamically enhancing the interaction with a single design artifact, and introduces the concept of adaptation as a context-sensitive processing which encompasses (1) the identification of plausible design alternatives for the variety of users and contexts of use in which an artifact is to be encountered; (2) the unification of alternative concrete design artifacts into abstract, generalized design patterns; (3) a method to allow the mapping of an abstract design pattern into the appropriate concrete/physical artifact; and (4) the capability to enhance interaction dynamically with each of the multiple physical artifacts as the need arises. A user interface complying with the above notion of intelligence is, therefore, able to dynamically undertake the required transformation, prior to and during interaction, so as to provide the appropriate interactive behavior for each user category to accomplish a given task.

From a universal access perspective, adaptation needs to be "designed into" the system rather than being decided upon and implemented a posteriori. This raises several requirements for the design methods and techniques that can be used. A broad view of how tasks are accomplished by different users across different interaction platforms and contexts of use is necessary. For this purpose, methods are needed to allow capturing alternative embodiments of artifacts depicting the diverse contexts of use that may be encountered (i.e., variation in users, platforms, environment), as well as structuring and organizing design alternatives in a manner that can be appropriated by suitable user interface development techniques.

The scope of design for diversity in universal access is broad and complex, since it involves issues pertaining to context-oriented design, diverse user requirements, and adaptable and adaptive interactive behaviors. This complexity arises from the numerous dimensions that are involved, and the multiplicity of aspects in each dimension. In this context, designers should be prepared to cope with large design spaces to accommodate design constraints posed by diversity in the target user population and the emerging contexts of use in the information society (Stephanidis and Akoumianakis, 2003). Moreover, adaptation is likely to predominate as a technique for addressing the compelling requirements for customization, accessibility, and high quality of interaction. Thus, it must be carefully planned, designed, and accommodated into the life cycle of an interactive system, from the early exploratory phases of design, through to evaluation, implementation, and deployment. Additionally, design for diversity is anticipated to be an incremental process, in which designers need to invest effort in anticipating new as well as changing requirements, and accommodating them explicitly in design through continuous updates. The unified user interface design method has been developed in recent years to facilitate the design of user interfaces with automatic adaptation behavior (Savidis et al., 2001; Savidis and Stephanidis, 2004a).

2.4 Toward Design-for-All Practice: Support Tools

In the past, Design for All has often been criticized on the grounds of practicality and cost justification (Bergman and Johnson, 1995). However, universal design in HCI products should not be conceived as an effort to advance a single solution for everybody but as an approach to provide products that can automatically address the possible range of human abilities, skills, requirements, and preferences. As discussed in the preceding two sections, the scope of design for diversity in universal access is broad and complex, and it becomes increasingly obvious that prevailing design practices (e.g., user-centered design), although useful, do not suffice to address universal access goals explicitly (Stephanidis and Akoumianakis, 2003). However, recent work has demonstrated that universal access is neither a utopia nor "wishful thinking," has pointed out the benefits anticipated in the context of the information society, and has developed appropriate methodologies, development frameworks, and applications (Savidis et al., 2001; Savidis and Stephanidis, 2004a). These efforts have also pointed out the compelling need of making available appropriate support tools for the entire development process, including design, as well as the limited value of currently available design support tools in a universal access perspective (Stephanidis et al., 1999).

In such a context, computational environments are considered as having the potential to lead to both cost justification and improved practices in designing for diversity, as they may automate certain tasks, guide designers toward specific targets, and provide extensible support for capturing, consolidating, and reusing previous experience. On these bases, the investigation and development of tools for a range of design tasks have been proposed, including tools for working with guidelines, critiquing tools, tools for capturing and reusing past experience, evaluating designs, capturing design rationale, embedding rationale into designs, providing computational support for metric-based techniques, and generating specifications that meet predetermined usability targets. A related challenge concerns the development of methods and tools capable of making universal access not only technically, but also economically feasible in the long term (Vernardakis et al., 1997, 2001; Stephanidis et al., 1998). Although the field lacks substantial data and comparative assessments as to the costs of designing for the broadest possible population, it has been argued that in the medium to long term, the cost of inaccessible systems is comparatively much higher and is likely to increase even more given the current statistics classifying the demand for accessible products (Bergman and Johnson, 1995). In the past, the availability of tools was an indication of the maturity of a sector and a critical factor for technological diffusion. As an example, graphical user interfaces became popular once tools for constructing them became available, either as libraries of reusable elements (e.g., toolkits), or as higher-level systems (e.g., user interface builders and user interface management systems). As design methods and techniques for addressing diversity are anticipated to involve complex design processes and have a higher entrance barrier with respect to more traditional artifact-oriented methods, it is believed that the provision of appropriate design tools can contribute to overcoming some of the difficulties that hinder the wider adoption of design methods and techniques appropriate for universal access, in terms of both quality and cost, by making the complex design process less resource demanding and better supporting design reuse (Savidis and Stephanidis, 2001c).

Finally, another prominent challenge in the context of universal access has been identified as the need to develop large-scale case study applications that provide instruments for further experimentation and ultimately improving the empirical basis of the field by collecting knowledge on how design for diversity may be practiced. Such case studies should not only aim to demonstrate technical feasibility but also to assess the economic efficiency and efficacy of competing technological options in the longer term (Stephanidis et al., 1998; Stephanidis, 2001c). Clearly, however, large design case studies are difficult to develop and analyzed on paper. It is even more difficult to transmit design outcomes to the implementation phase in a suitable form. The need for appropriate computational tools therefore emerges also in the light of facilitating the entire development of articulated and complex applications, as well as the encoding of concrete design cases for further study and refinement of methods and techniques.

3 UNIFIED USER INTERFACES DEVELOPMENT FRAMEWORK FOR USER INTERFACE ADAPTATION

Designing for universal access means designing for diversity in both end users and contexts of use, and implies making alternative design decisions at various levels of the interaction design, leading to inherent diversity in the final design outcomes. Toward this end, a design method targeted toward the development of a single interface design instance is inappropriate, since it cannot accommodate diversity. Therefore, there is a need for a systematic process in which alternative designs for different design parameters may be supported (Savidis et al., 2001; Savidis and Stephanidis, 2004a). The unified user interfaces development framework introduces a novel approach to intelligent adaptation, the main objective in such a context being to ensure that each end user is provided with the most appropriate interactive experience at run time. Producing and enumerating distinct interface designs through the conduct of multiple design processes would be an impractical solution, since the overall cost for managing such a large number of independent design processes in parallel, and for implementing each interface version separately, would be unacceptable (Savidis and Stephanidis, 2004a). Instead, a design process is required that is capable of leading to a single design outcome that appropriately structures multiple designs and their underlying user- and context-related parameters, therefore facilitating, on the one hand, the mapping of design to a target software system implementation, and on the other hand, the maintenance, updating, and extension of design itself. The unified user interface design method, which is part of the unified user interface development framework, supports such a process.

3.1 Unified User Interfaces

The unified user interface development methodology (Stephanidis, 2001a,b; Savidis and Stephanidis, 2004b) has been proposed as a complete technological solution for supporting universal access of interactive applications and services. Unified user interfaces convey a new perspective in the development of user interfaces, providing a principled and systematic approach toward coping with diversity in the target user requirements, tasks, and environments of use. The notion of a unified user interface originated from research efforts aiming to address the issues of accessibility and interaction quality for people with disabilities (Stephanidis and Emiliani, 1999). The theoretical grounds upon which unified user interface development methodology is based is provided by the concept of user interfaces for all, rooted in the idea of applying universal design in the field of HCI (Stephanidis, 2001a). Subsequently, these principles were extended and adapted to depict a general proposition for HCI design and development, and were extensively tested and validated in the course of real development projects (Stephanidis et al., 2001, 2003).

Unified user interfaces provide an engineering methodology supporting automatic adaptation of user interfaces as a technical path toward universal access. A *unified user interface* comprises a single (unified) interface specification that exhibits the following properties (Savidis and Stephanidis, 2004b):

1. It embeds representation schemes for user- and usage-context parameters and accesses user- and usage-context information resources (e.g., repositories, servers) to extract or update such information.
2. It is equipped with alternative implemented dialogue patterns (i.e., implemented dialogue artifacts) appropriately associated with different combinations of values for user- and usage-context-related parameters. The need for such alternative dialogue patterns is identified during the design process when, given a particular design context, for differing user- and usage-context attribute values, alternative design artifacts are deemed necessary to accomplish optimal interaction.
3. It embeds design logic and decision-making capabilities that support activating at run time the most appropriate dialogue patterns according to particular instances of user- and usage-context parameters and is capable of interaction monitoring to detect changes in parameters.

As a consequence, a unified interface realizes (1) user-adapted behavior (user awareness, i.e., the interface is capable of selecting automatically interaction patterns appropriate to the particular user), and (2) usage-context-adapted behavior (usage context awareness, i.e., the interface is capable of selecting automatically interaction patterns appropriate to the particular physical and technological environment). From a user perspective, a unified user interface can be considered as an interface tailored to personal attributes and to the particular context of use, while from the designer perspective it can be seen as an interface design populated with alternative designs, each alternative addressing specific user- and usage-context parameter values. Finally, from an engineering perspective, a unified user interface is a repository of implemented dialogue artifacts from which the most appropriate according to the specific task context are selected at run time by means of an adaptation logic supporting decision making.

At run time, the adaptations may be of two types: (1) adaptations driven from initial user and context information known prior to the initiation of interaction, and (2) adaptations driven by information acquired through interaction-monitoring analysis. The former behavior is referred to as *adaptability* (i.e., initial automatic adaptation) reflecting the interface's capability initially to tailor itself automatically to each individual end user in a particular context. The latter behavior, referred to as *adaptivity* (i.e., continuous automatic adaptation), characterizes the interface's capability to cope with the dynamically changing or evolving user and context characteristics. Adaptability is crucial to ensure accessibility, since before initiation of interaction, it is essential to provide a fully accessible interface instance to each end user. Furthermore, adaptivity can be applied only on accessible running interface instances (i.e., ones with which the user is capable of performing interaction), since interaction monitoring is required for the identification of changing or emerging decision parameters that may drive dynamic interface enhancements. This combination of adaptation characteristics and behavior makes unified user interfaces suitable and appropriate for supporting universal access (Stephanidis, 2001b).

The concept of unified user interface is supported by a specifically developed architecture (Savidis and Stephanidis, 2001a). The unified user interface development paradigm is general enough so as not to exclude particular design and implementation practices while offering sufficient detail to drive the engineering process. As for any new development paradigm, unified user interface development requires some initial investment to be effectively adopted, assimilated, and applied. However, if the constructed software products are intended to be used by user populations with diverse requirements, operated in different usage contexts, it is argued that the gains will outweigh the overhead of additional resources that need to be invested (Stephanidis, 2001a; Savidis and Stephanidis, 2004b). A particularly important aspect of developing unified user interfaces concerns their design.

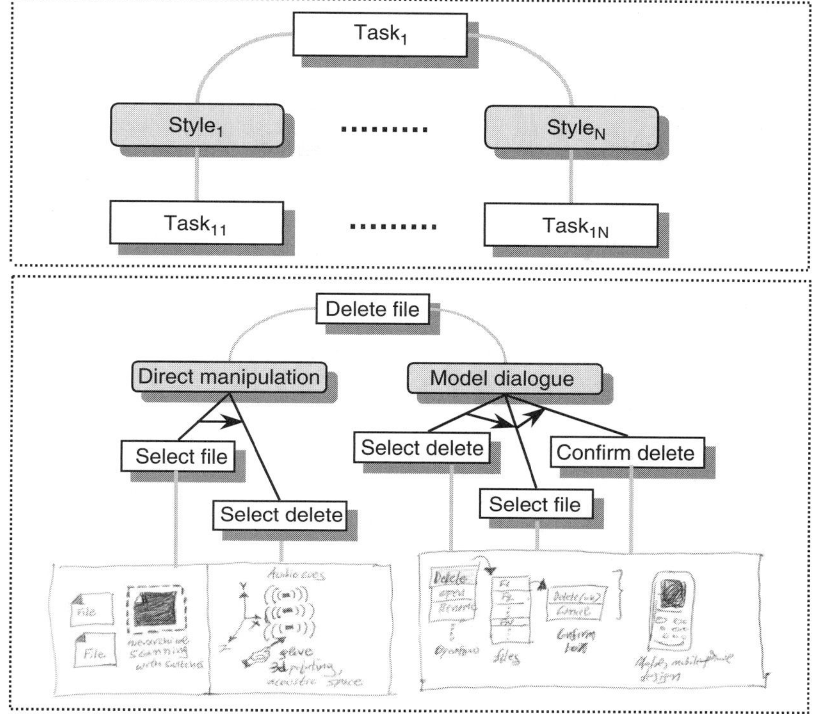

Figure 1 Polymorphic task hierarchy concept. (From Savidis and Stephanidis, 2004a.)

3.2 Unified User Interface Design

The unified user interface design method is a hybrid process-oriented design method enabling the organization of diversity-based design decisions around a single hierarchical structure (Savidis et al., 2001; Savidis and Stephanidis, 2004a). It proposes a specific design process to cater to the management of an evolving design space, in which alternative design artifacts can be associated with variations of the design problem parameters.

The method addresses the following main objectives:

1. Enabling the collection and organization of all design alternatives required for adaptation into a single unified user interface design space that can be produced during a single design phase.

2. Documenting, for each alternative design artifact, a design rationale, in a form that facilitates implementation, including the run-time relationships with the rest of the artifacts within the same design context as well as the specific associated problem parameter values.

3. Supporting design evolution by enabling the effective extension of different design contexts by addressing new (combination of) user- and usage-context attribute values.

The process of designing unified user interfaces does not lead to a single design outcome but to a structured design space. It collects and represents alternative designs appropriately, along with the conditions under which each design should be instantiated at run time (i.e., an adaptation-oriented design rationale). The unified user interface design method encompasses a variety of techniques, such as task analysis, abstract design, design polymorphism, and design rationale (Savidis and Stephanidis, 2004a). Next we discuss representational issues in unified user interface design and provide an overview of the process underlying user interface design and the related outcomes.

3.2.1 Polymorphic Task Hierarchies

The basic representation adopted in unified user interface design, called *polymorphic task hierarchy* (Savidis et al., 2001), combines (1) hierarchical task analysis, (2) design polymorphism [i.e., the possibility of assigning alternative decompositions to the same (sub)task if required based on (combinations of) design parameters], and (3) user-task-oriented operators. Figure 1 depicts an example of polymorphic task hierarchy, illustrating how two alternative dialogue styles for a "delete file" task may be designed. Alternative decomposition styles are depicted in the upper part of the figure, and an exemplary polymorphic decomposition, which includes physical design annotation, appears in the lower part.

The hierarchical decomposition adopts the original properties of hierarchical task analysis (Kirwan and Ainsworth, 1992), enriched with the capability to differentiate and represent design alternatives for the same task, mapping to varying design parameters through polymorphism. Task operators are based on the powerful Communicating Sequential Processes (CSP) language for describing the behavior of reactive systems (Hoare, 1978) and enable the expression of dialogue control flow formulas for task accomplishment. However, the designer is not constrained to use CSP operators exclusively. For example, for describing user actions for device-level interaction (e.g., drawing, drag-and-drop, concurrent input), an event-based representation (e.g., Hartson et al., 1990) may also be used.

In a polymorphic task hierarchy, the root represents design abstractions and leaf nodes represent concrete interaction components. Polymorphic decomposition leads from abstract design pattern to a concrete artifact. Three categories of design artifacts may be subject to polymorphism on the basis of user- and usage-context parameter values:

- *User tasks*, relating to what the user has to do; user tasks are the center of the polymorphic task decomposition process.
- *System tasks*, representing what the system has to do or how it responds to particular user actions (e.g., feedback); in the polymorphic task decomposition process, they are treated in the same manner as user tasks.
- *Physical designs*, which concern the interface components on which user actions are to be performed; physical interface structure may also be subject to polymorphism.

User tasks, and in certain cases, system tasks, are not necessarily related to physical interaction but may represent abstraction on either user or system actions. System tasks and user tasks may be freely combined within task "formulas," defining how sequences of user-initiated actions and system-driven actions interrelate. The physical design, providing the interaction context, is associated with a particular user task and provides the physical dialogue pattern associated with a task-structure definition. Hence, it plays the role of annotating the task hierarchy with physical design information.

Each alternative polymorphic decomposition is called a *decomposition style*, or simply a *style*, and is given a unique name. Alternative task subhierarchies are attached to their respective styles. Polymorphism constitutes a technique for potentially increasing the number of alternative interface instances represented by a typical hierarchical task model. However, the unified user interface design method does not require the designer to follow the polymorphic task decomposition all the way down the user–task hierarchy until primitive actions are met. A nonpolymorphic task can be specialized at any level, following any design method chosen by the interface designer. When polymorphism is applied at the level of top or main tasks (e.g., edit a document, send an e-mail, perform spell checking, construct graphic illustrations), the interface instances designed are likely to be affected by structural differences, resulting in alternative versions of the same interactive environment, such as in the case of multiplatform interfaces (i.e., the effect seems global). Polymorphism on middle hierarchy levels (e.g., dialogue boxes for setting parameters, executing selected operations, editing retrieved items) introduces the effect of overall similarities with localized differences in interactive components and intermediate subdialogues. Finally, polymorphism at the lowest levels of the hierarchy, concerning primitive interactive actions supported by physical artifacts (e.g., pressing a button, moving a slider, defining a stroke with the mouse) causes differences on device-level input syntax and/or on the type of interaction objects in some interface components. If polymorphism is not applied, a task model represents a single interface design instance on which no run-time adaptation is applied. There is therefore a fundamental link between adaptation capability and polymorphism (Savidis and Stephanidis, 2004a).

The unified user interface design emphasizes capturing of the more abstract structures and patterns inherent in the interface design, enabling hierarchical incremental specialization toward the lower physical level of interaction, therefore making it possible to introduce design alternatives as close as possible to physical design. This makes it easier to update and extend the design space, since modifications due to the consideration of additional values of design parameters (e.g., considering new user- and usage-context attribute values) can be applied locally to the lower levels of the design without affecting the rest of the design space.

3.2.2 Design Process

The key elements of the unified user interface design process are (Savidis and Stephanidis, 2004a):

- A hierarchical design discipline building on the notion of task analysis, empowered by the introduction of *task-level polymorphism*
- An iterative design process model, emphasizing *abstract task analysis* with incremental polymorphic physical specialization
- A formalization *of run-time relationships* among the alternative design artifacts associated with the same design context
- Documentation recording the *consolidated design rationale* of each alternative design artifact

Space of User- and Context-Related Design Parameters An essential prerequisite for conducting unified user interface design is the conceptual categorization of diversity aspects in all relevant dimensions (users, context of use, access terminal or platform), and the identification of the target design parameters for each design case (Savidis et al., 2001). There is no predefined or fixed set of attribute categories or values, which are chosen as part of the

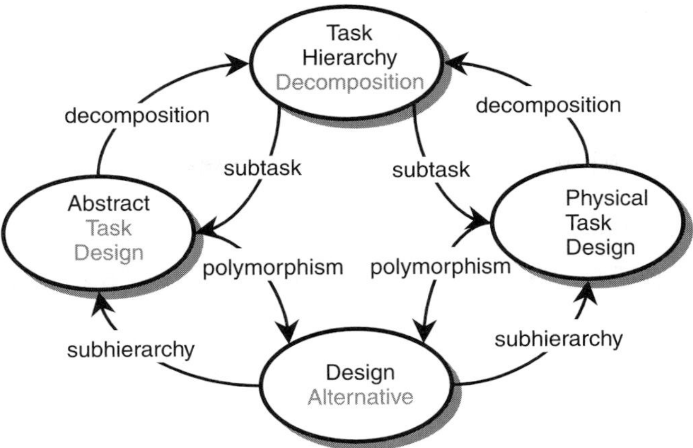

Figure 2 Polymorphic task decomposition process.

design process (e.g., by interface designers or human factors experts). Values do not need to be finite sets. The broader the set of values, the higher the potential for polymorphism (i.e., for alternative designs). For instance, commercial systems realizing a single design for an "average" user have no differentiation capability at all.

Unified user interface design does not involve specific ontologies or predefined models of design parameters. This is based on the consideration that the degree to which a comprehensive taxonomy (or ontology) of design parameters can be achieved based on the current knowledge and wisdom on designing for diversity cannot be determined, and besides, not all parameters in such a taxonomy would be relevant for different design cases. For example, according to the targeted final user groups, different sets of design guidelines may need to be taken into account during design. Therefore, designers should be free to use case-specific design parameters, to experiment with different sets of them, and to create their own (partial) taxonomies according to the type of design cases they address and the target user population. Accumulated experience in designing for diversity is expected to lead progressively to the identification of more commonly valid classifications of design parameters. For example, human abilities relevant for determining alternative choices in a design case are likely to affect similar design cases. Some examples of attribute classes that designers may consider in unified user interface design are general computer-use expertise, domain-specific knowledge, role in an organizational context, motor abilities, sensory abilities, mental abilities, and so on.

Since the unified user interface design method does not pose any restrictions as to the attribute categories considered relevant or the target value domains of such attributes, it needs to provide a suitable framework in which the definition of user- and usage-context attributes constitutes an explicit part of the design process. It is the responsibility of interface designers to choose appropriate attributes and corresponding value ranges as well as to define appropriate design alternatives when necessary. For simplicity, designers may choose to elicit only those attributes from which differentiated design decisions are likely to emerge. The construction of context and platform attributes may follow the same representation approach as users' characteristics. Examples of potential context attributes are acoustic noise and light sources, while examples of potential relevant platform attributes are processor speed, memory, secondary storage, peripheral equipment, resolution, physical screen size, and graphics capabilities.

Combinations of design parameters in the form of triads <User profile, Platform, Context> constitute execution contexts. Each different style should be designed so as to facilitate specific task execution context(s).

Polymorphic Task Decomposition Process The unified user interface design process realizes a hierarchical decomposition of tasks, starting from the abstract level, by specializing incrementally in a polymorphic fashion toward the physical level of interaction (Savidis et al., 2001; Savidis and Stephanidis, 2004a). In this process, different designs are likely to be associated with different (combinations of) user- and usage-context attribute values.

The polymorphic decomposition process is depicted in Figure 2. It starts from abstract- or physical-task design, depending on whether or not top-level user tasks can be defined as being abstract. An abstract task can be decomposed either in a polymorphic fashion, if user- and usage-context attribute values require different dialogue patterns, or following a unimorphic decomposition scheme. In the latter case, the transition is realized via a decomposition action, leading to the task hierarchy decomposition state. Polymorphic decomposition, on the other hand, leads to the design alternative subhierarchies state. Reaching this state means that the required alternative dialogue

Task: Delete File	
Style: Direct Manipulation	**Style:** Modal Dialogue
Users and Contexts: Expert, Frequent, Average	**Users and Contexts:** Casual, Naive
Targets: Speed, naturalness, flexibility	**Targets:** Safety, guided steps
Properties: Object first, function next	**Properties:** Function first, object next
Relationships: Exclusion (with all)	**Relationships:** Exclusion (with all)

Figure 3 Design rationale documentation.

styles have been identified, each initiating a distinct subhierarchy decomposition process. Hence, each such subhierarchy initiates its own instance of polymorphic task decomposition process. While initiating each distinct process, the designer may start from either the abstract task design state or the physical task design state. The former is pursued if the top-level task of the particular subhierarchy is an abstract one, whereas the second occurs when the top-level task involves physical interaction. From this state, the subtasks identified need to be decomposed further. For each subtask at the abstract level, there is a subtask transition to the abstract task design state. Otherwise, if the subtask involves physical interaction means, a subtask transition is performed to the physical task design state.

Physical tasks may be further decomposed in either a unimorphic fashion or a polymorphic fashion. These two alternative design possibilities are indicated by the decomposition and polymorphism transitions, respectively.

In summary, the rules to be applied in polymorphic task decomposition are:

- If a given task does not involve physical interaction, start from abstract task design.
  - Apply polymorphism if decision parameters impose the need for alternative styles on user/system tasks and/or physical structure.
  - Apply decomposition, when alternative designs are needed for the same style.
- If a given task involves physical interaction, start from physical task design and:
  - Apply polymorphism if decision parameters impose the need for alternative styles on user/system tasks and/or physical structure.
  - Apply decomposition when an alternative design is needed to realize the same style.

In unified user interface design, one of the key issues is to decide in which cases the diversity in design parameters leads to different concrete interface artifacts. Designers should take care that every decomposition step satisfies all constraints imposed by the combination of target user- and usage-context attribute values. Polymorphic decomposition is required when different styles are appropriate for different execution contexts based on the designer's decision. Differentiation decisions can be based on consolidated design knowledge, if available (e.g., design guidelines for specific target user groups, target platforms), on the results of surveys or evaluation experiments, for example. There is no automatism in deciding when and how adaptation should be applied in the final interface. This important aspect of the design process is likely to require in-depth experience on the part of designers, who will need to provide the designed unified user interfaces with an adaptation logic suitable for run-time execution.

Adaptation-Oriented Design Rationale In the depicted process, the following primary decisions need to be made: (1) at which points of a task hierarchy polymorphism should be applied, based on the considered (combinations of) user- and usage-context attributes; and (2) how different styles behave at run time; this is performed by assigning to pair(s) of style (groups) design relationships. These decisions need to be documented in a design rationale recorded by capturing, for each subtask in a polymorphic hierarchy, the underlying design logic, which directly associates user-/usage-context parameter values with the artifacts designed.

Such a rationale should document (Savidis and Stephanidis, 2004a): (1) the related task, (2) design targets leading to the introduction of the style, (3) supported execution context, (4) style properties, and (5) design relationships with competing styles.

In Figure 3 an instance of such a documentation record is depicted, adopting a tabular notation. Styles can be evaluated and compared with respect to any design parameter (e.g., performance measures, heuristics, user satisfaction). Evaluation or comparison results can form part of the design rationale as annotations.

Four fundamental relationships among alternative styles (concerning the same polymorphic artifact) have been identified, reflecting the way in which artifacts may be employed during interaction for an individual user in a particular context (Savidis and Stephanidis, 2004a):

1. *Exclusion.* Only one of the alternative styles may be present at run time.
2. *Compatibility.* Any of the alternative styles may be present at run time.
3. *Substitution.* When the second style is activated at run time, the first should be deactivated.
4. *Augmentation.* On the presence of a style at run time, a second style may also be activated.

These relationships in fact express the adaptation run-time behavior of the designed interface, reflecting real-world design scenarios, and are motivated by the observation that different styles are not always mutually exclusive, even if they correspond to different (combinations of) design parameters values, since there are cases in which it is meaningful to make artifacts belonging to alternative styles concurrently available in a single adapted interface instance.

In the context of unified user interface design, the notion of design rationale has a fundamentally different objective with respect to well-known design space analysis methods (e.g., Buckingham Shum, 1996). As already mentioned, previous approaches to design rationale mainly represent argumentation about design alternatives and assessments before reaching final design decisions, whereas in the case of unified user interface design, the rationale records the different user- and usage-context attributes motivating the final design decisions (Savidis and Stephanidis, 2004a). This obviously does not exclude recording for future reference the underlying decision criteria (e.g., guidelines followed, performed experiments).

3.2.3 Design Outcomes

Summarizing, the outcomes of the unified user interface design method include (1) the polymorphic task hierarchy; (2) the design space populated by the produced physical designs; and (3) for each polymorphic artifact in the task hierarchy, a design rationale recording its run-time adaptation logic based on user- and context-related parameters. The unified user interface design method, being general and nonprescriptive, does not require any specific format or notation for delivering the outcomes noted above, and designers are free of using any convenient encoding on a case basis or according to their personal preferences. For example, polymorphic task hierarchies can be visualized through graphlike structures annotated with text, and design rationales can be encoded into tables. Physical design can also be conducted and delivered using any suitable device, such as textual design specifications, or in the case of graphical interfaces, mock-ups and prototypes.

Combined, the outcomes of the unified user interface design process are meant to:

1. Facilitate design reuse and incremental design. This is made possible by the linking of physical design artifacts to their respective tasks and execution contexts, and by the systematic organization and documentation of all design artifacts, facilitating the expansion and modification of existing design cases to cater for different sets of design parameters (e.g., finer-graded individual user attributes).

2. Facilitate the implementation of self-adapting interfaces through the provision of all the necessary knowledge for run-time decision making concerning alternative design artifacts. In fact, a distinctive property of the polymorphic task hierarchy is that it can be mapped into a corresponding set of specifications from which interactive behaviors can be generated. This is an important contribution of the method to HCI design, since it potentially bridges the gap between design and implementation, which has traditionally challenged user interface engineering (Savidis and Stephanidis, 2004a).

The implementation orientedness of the design outcomes has a practical impact on the types of representations that are most suitable to be mapped directly to interface specifications. For example, the representation of end-user characteristics is best developed in terms of attribute-value pairs, using any suitable formalism, to allow the straightforward encoding of the relevant parameters into attribute-value-based user profiles at run time. Concerning the polymorphic task hierarchy itself, it obviously needs to be hierarchically encoded with appropriate reference to the properties of each node in the hierarchy (i.e., subtask or style). This can be performed using a simple relational database or any equivalent encoding form. On the other hand, the design rationale associated with polymorphic task hierarchies is best represented in a directly computable form [such as the Decision Making Specification Language (DMSL) (Savidis and Stephanidis, 2004b; Savidis et al., 2004a), which is directly amenable to implementation in a decision-making component].

3.2.4 Applications and Support Tools

The unified user interface design method has been applied and validated in large-scale applications. These applications include the universally accessible AVANTI Web browser (Stephanidis et al., 2001) as well as the PALIO adaptive hypermedia framework for universally accessible Web services (Stephanidis et al., 2003). The experience acquired in these developments has confirmed that the unified user interface approach is feasible and effective for large applications targeted toward satisfying the needs and requirements of diverse target user groups in diverse contexts of use, and at the same time has confirmed the need for tool support for the unified user interface development method.

A number of tools have been developed in recent years to support the implementation of unified user interfaces, including a high-level language for user

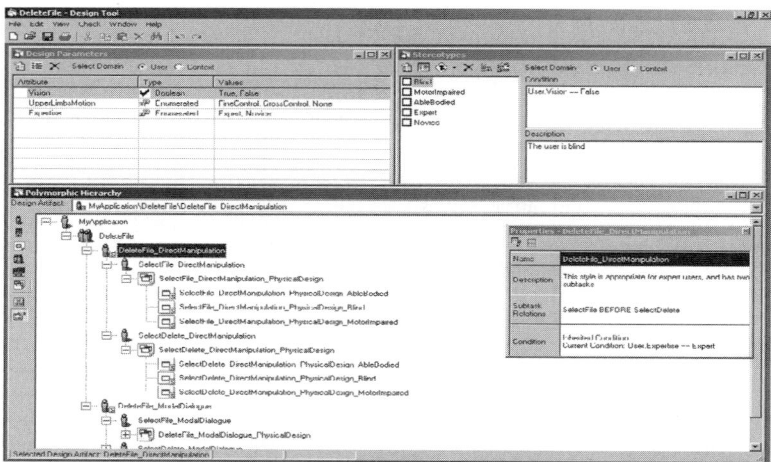

Figure 4 MENTOR interactive environment.

interface specification (Savidis and Stephanidis, 1997) and a user interface management system that generates the implementation automatically from such high-level specifications (Savidis and Stephanidis, 2001b). These are not discussed further in the present context, as they are not directly relevant to the issues addressed in this chapter.

Concerning design, the availability of appropriate support environments has been identified as one of the factors that can, potentially, contribute to mitigate the complexity and consequent difficulties in conducting the method (Savidis and Stephanidis, 2001c). For supporting the design of unified user interfaces, a design-oriented user modeling tool called USE-IT (Akoumianakis and Stephanidis, 2001b) has been developed, targeted toward the derivation of adaptation recommendations for the lexical level of interaction. A different support tool for unified user interface design, named MENTOR, is targeted toward assisting designers in learning and practicing the unified user interface design process, in particular in designing the adaptation behavior of unified user interfaces.

4 COMPUTER-BASED ASSISTANCE FOR CONDUCTING UNIFIED USER INTERFACE DESIGN

MENTOR is a prototype support tool for the process of unified user interface design, which has been developed to provide an instrument for widening and improving the learning and practice of adaptation-based user interface design. MENTOR targets the community of interface designers and does not assume deep knowledge of the unified user interface design method or particular HCI modeling techniques while supporting designers more experienced in adaptation design in performing their work effectively.

MENTOR's functionality includes editing facilities for (1) encoding declarations (signatures) of design parameters attributes and related value spaces, (2) encoding polymorphic task hierarchies, (3) creating stereotypes of adaptation conditions, and (4) attaching information to the artifacts (nodes) in the polymorphic task hierarchy. Additionally, MENTOR provides automated verification mechanisms for the adaptation logic embedded in unified user interface design cases as well as the automated generation of ready-to-implement interface specifications.

Figure 4 depicts the overall interactive environment of MENTOR. Four main editing environments are available: (1) the design parameters editor, (2) the stereotypes editor, (3) the polymorphic task hierarchy editor, and (4) the properties editor. The MENTOR's design parameters editor supports the encoding of design parameter attributes and related value spaces. These constitute the "vocabulary" for defining the *adaptation space* of the unified user interface under design. Parameters can belong to two different domains: the user domain, referring to parameters representing user characteristics, and the context domain, referring to parameters representing characteristics of the context of use and of the interactive platform(s) of the unified user interface under design.

Figure 5 depicts the design parameters editor, which displays parameters attributes, type, and value ranges. Designers can add any number of new design parameters and subsequently, edit the type and range of values. Deletion of a design parameter is also possible as long as such a parameter is not used for the current design case. The tool also offers the possibility of importing design parameters from other design cases, to support the reusability of previous designs. In the latter case, consistency checking between the parameters already declared in the current design case and the parameters selected for importing is performed by the system automatically.

The polymorphic task hierarchy editor allows designers to perform polymorphic task decomposition and encode the results in a hierarchy. Figure 6 depicts

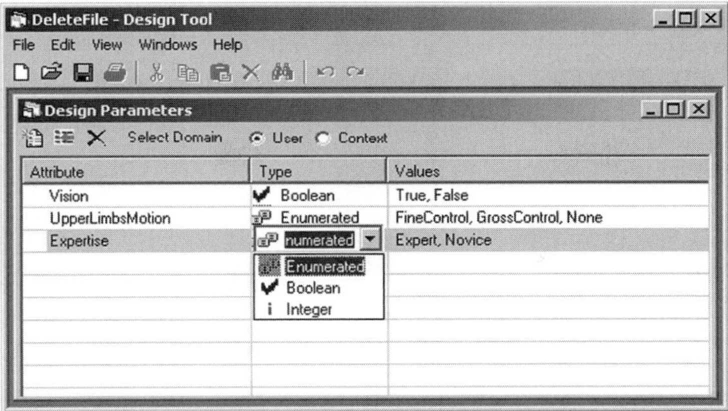

Figure 5 MENTOR design parameters editor. Three types of design parameters values are allowed: (1) enumerated [i.e., values belong to a list of (more than two) strings specified by the designer]; (2) Boolean (i.e., values True or False); and (3) integer, which are specified by supplying the minimum and maximum bounds of the integer range allowed as a value. Value ranges define the space of legal values for a given attribute.

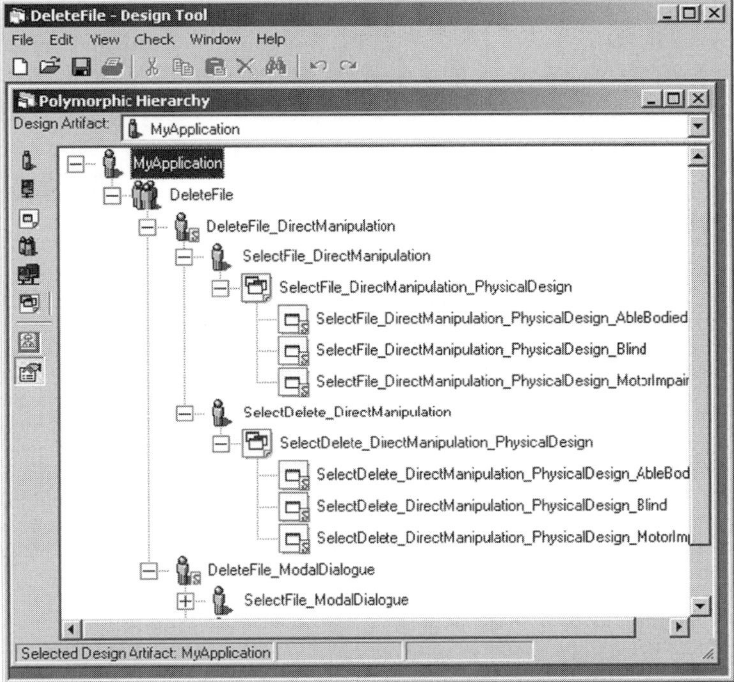

Figure 6 Polymorphic task hierarchy editor. The available decomposition steps for each artifact are reflected both in the docked toolbar and the pop-up menus of the polymorphic task hierarchy editor. When an artifact is deleted, its children are also deleted, to preserve the correctness of the resulting hierarchy (since the children of the deleted task may not be appropriate children of the deleted task father's node). Artifacts can only be pasted as children of nodes that admit the category of the copied artifact as decomposition. Pasting results in producing a copy of the copied artifact at the selected point in the hierarchy, including all its properties if appropriate for the pasting point.

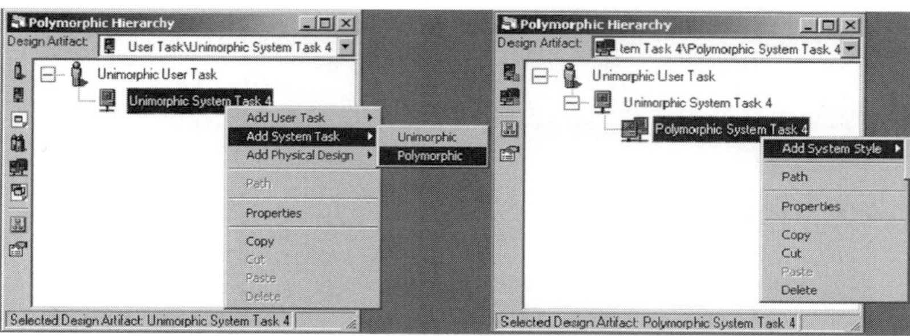

Figure 7 Contextualized decomposition actions in the polymorphic task hierarchy editor.

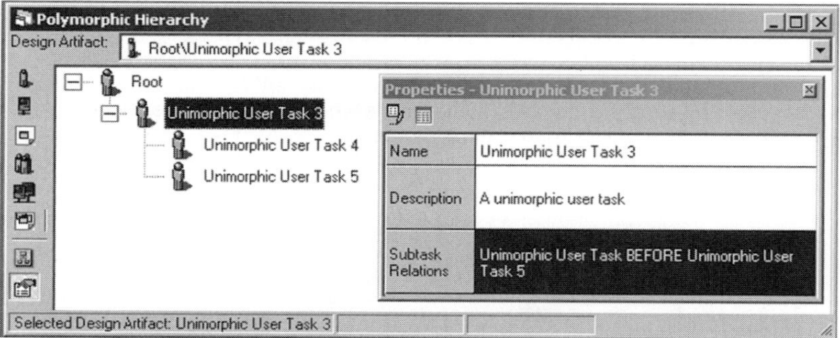

Figure 8 Editing properties of unimorphic tasks. The properties editor appears as a floating window in the overall tool environment, thus ensuring that the polymorphic task hierarchy editor is not hidden while the user is editing properties, and always displays the data related to the currently selected artifact in the polymorphic task hierarchy editor. Properties are shown in a simple tablelike form that can be edited directly in place.

an overview of the editor. The editor's main function is to guide the decomposition process by contextualizing the decomposition actions available according to the category of design artifact selected in the hierarchy. Figure 7 presents an example of such contextualization for unimorphic and polymorphic tasks, respectively. Part of the polymorphic task decomposition process consists in assigning specific properties to the artifacts in the polymorphic hierarchy. Different categories of artifacts involve different properties, some of which, as discussed later in this section, are particularly important for the purposes of the overall adaptation design of the resulting interface. In MENTOR, artifact properties are entered through a properties editor, which displays the contextualized properties of the artifact selected currently in the polymorphic task hierarchy editor. Figure 8 depicts the appearance of the property editor when a unimorphic artifact is selected in the hierarchy.

For more complex properties that need to be set using selection rather than text entry, such as those of polymorphic artifacts or interaction styles, in-place mini-editors are displayed within the editing space of the specific property in the properties editor. Figure 9 depicts the appearance of the properties editor when a style is selected in the polymorphic task hierarchy. The most important piece of information to be attached to styles concerns the user and context parameter instantiations that define the style appropriateness at run time. Style conditions are therefore at the heart of unified user interface design. In MENTOR, they are entered as properties of styles and are formulated in a very simple expression language, specifically the condition fragment of the DMSL (Savidis and Stephanidis, 2004b; Savidis et al., 2004a) for adaptation decision making in unified user interfaces. An example of style condition in DMSL is *User.vision == False*, indicating a certain style (e.g., a nonvisual interaction style) appropriate for blind users.

Semantically, conditions are partial descriptions of users or contexts in the space defined by the current design parameters and refer to sets of users or contexts for which they hold true at run time. In MENTOR, three different ways of entering style conditions are provided to support designers in the difficult task of establishing the adaptation logic of the design interfaces:

1. *By free text typing.*
2. *By using a condition editor* (Figure 9*b*) displayed in the properties editor. The condition editor supports the easy and syntactically correct editing of

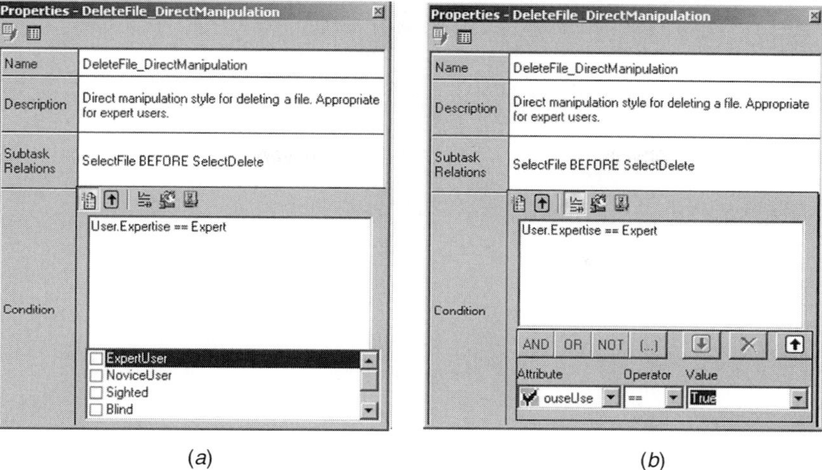

(*a*) (*b*)

Figure 9 Editing properties of styles. (*a*) Available stereotypes are displayed for selection and inclusion in s style condition; (*b*) the condition editor automates text input in formulating conditions, through the use of combo boxes for selections and buttons for conjunction, disjunction, and negation operators, as well as parentheses.

conditions and facilitates beginner users of MENTOR to acquire familiarity with the DMSL expression language, practically guiding designers to compose conditions.

3. *By using stereotypes* (i.e., predefined conditions). In MENTOR, stereotypes can be defined, previously or in parallel with the polymorphic task decomposition, using a stereotypes editor that allows associating a DMSL condition (e.g., *User.vision == False*) with a name (e.g., *blind user*). Figure 10 depicts the appearance of the stereotypes editor. In the properties definition phase, the designer can select among the stereotypes defined in the stereotypes editor in order to compose style conditions (see Figure 9*a*).

Automated verification facilities for DMSL conditions are also included in MENTOR to facilitate designers in the production of a correct and verifiable ready-to-implement adaptation logic. These include verification of the lexical and syntactic correctness, as well as the verifiability of each DMSL expression separately (in both stereotypes and styles). For example, designers need to make sure that a contradictory condition such as *User.vision == False AND User.vision == True* will be assigned to a style and propagated to the user interface implementation phase. Additionally, hierarchical relations among styles in the polymorphic task hierarchy are also checked. Each style in the task hierarchy implicitly inherits the conditions on its ancestor styles, and each polymorphic decomposition applies within the design context (i.e., specified conditions) of a higher-level polymorphic decomposition in the same hierarchy branch, if present. For example, consider a set of conditions specifying that in a nonvisual dialogue, if the user does not have visual abilities, a certain style should apply, whereas if the user has visual abilities, a different style should be displayed. This specification is useless in the first part: No adaptation is necessary to address the needs of a blind user in the specific context, since the dialogue is already nonvisual at such a stage. Additionally, the second part of such a condition specification is "out of context," since it makes little sense to specify a style for a sighted user in the middle of a nonvisual dialogue. Therefore, the verification mechanism of MENTOR checks that:

1. The inherited and local conditions are compatible (i.e., not inconsistent). For example, designers need to make sure that a condition *User.vision == False* is not assigned to a style in a hierarchy branch following a style with the condition *User.vision == True*.
2. The local condition of a style is more specific than the inherited condition (i.e., it restricts the set of user and context parameters combination for which the description holds with respect to the one inherited).

Another important aspect of unified user interface design toward determining the run-time adaptation behavior of a user interface is the assignment of adaptation relations (selecting among incompatibility, compatibility, augmentation, and substitution) between different styles of a polymorphic artifact. In MENTOR, adaptation relations are formulated as properties of polymorphic artifacts. Figure 11 depicts the appearance of the properties editor when a polymorphic artifact is selected in the polymorphic task hierarchy.

MENTOR also supports verifying that the conditions on two styles related through a particular relation are compatible with the type of relation. For example, if two styles are defined as incompatible, their conditions must not be consistent. For instance, if the first style has the condition *User.vision == False*

(*a*)

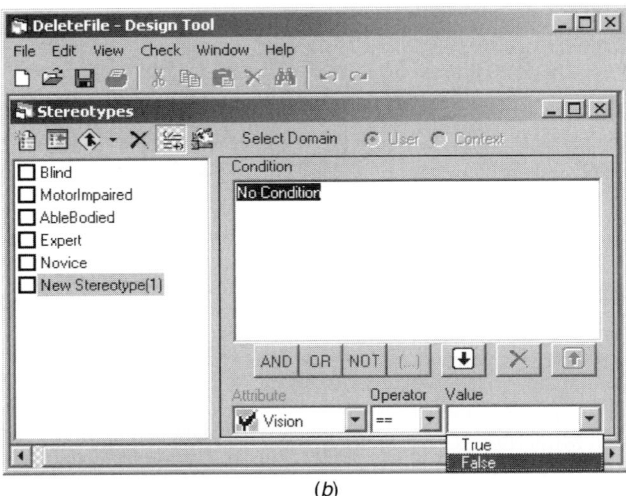

(*b*)

Figure 10 (*a*) Standard appearance of the stereotypes editor; (*b*) condition editor activated in the stereotypes editor.

and the second style has the condition *User.expertise == Expert*, the two styles cannot be incompatible, as they do not refer to mutually exclusive combination of design parameters (the user can be at the same time blind and expert). This checking facility is introduced to ensure that the resulting run-time adaptation logic will be semantically sound on the basis of the declared design parameters and will not contain ambiguities resulting in problems when applying adaptations.

Finally, the documentation of adaptation design is supported in MENTOR through the automated collection of all available information in a design case into a textual design documentation report to be used subsequently for several purposes, such as design reviewing and evaluation, interface documentation, and, most important, implementation. The design report contains the project's design parameters and defined stereotypes, a textual representation of the polymorphic task hierarchy, the properties of each designed artifact, and the designed adaptation logic (i.e., a list of adaptation design decisions) in the form of DMSL rules produced by the tool automatically on the basis of the current style conditions and adaptation relations in polymorphic artifacts of the design case. DMSL rules as produced by MENTOR can be directly embedded in the decision-making component of the unified interface designed. Figure 12 depicts an example of a design report produced by MENTOR.

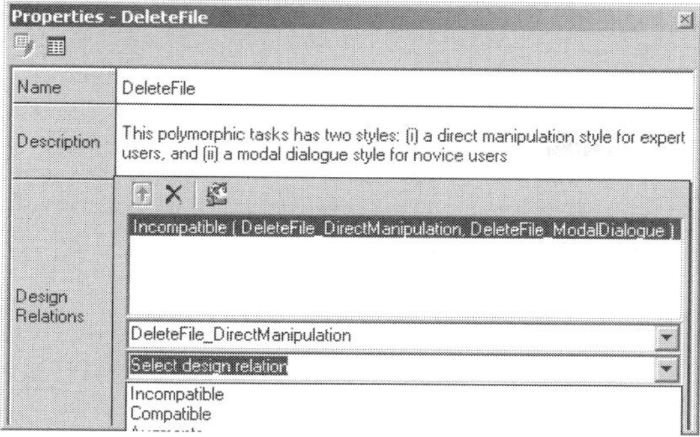

Figure 11 Editing properties of polymorphic artifacts. The dialogue at the bottom of the editor makes it possible to set the adaptation relation that must hold at run time among the children styles of the polymorphic artifact. Binary adaptation relations are constructed by sequential selection of the first relation argument (i.e., one of the styles defined as children of the currently selected polymorphic artifact), the type of the relation (choosing among the four types available in the unified user interface design method: incompatibility, compatibility, augmentation, and substitution), the second relation argument (another style among the ones defined as children of the currently selected polymorphic artifact).

5 PRACTICING UNIFIED USER INTERFACE DESIGN: CASE STUDIES USING MENTOR

One of the aims of MENTOR is to contribute to widening the practice of Design for All through supporting the progressive accumulation of design cases and of the related design experience and knowledge, in particular regarding adaptation, by offering means for extending and reusing (parts of) past design cases. Toward this end, MENTOR has been used and validated in two different middle-scale case studies in different application domains.

5.1 Shopping Cart Case Study

The first validation case study of MENTOR concerns the application of the unified user interface design method to a shopping cart component of a Web portal, to be accessed by first-time and habitual users through standard desktop-based Web browsers as well as through mobile phones with different screen sizes. The shopping cart design case has been conducted by an experienced user interface designer with some basic knowledge but no practical training in unified user interface design.

Figure 13 depicts the design parameters defined in the shopping cart design case. The rationale behind the adoption of the user-related parameters mentioned was that novice users should be provided with enough guidance to be able to complete their shopping tasks successfully, whereas more experienced users would find such a type of guidance too detailed. Additionally, the shopping cart needed to be designed in such a way as to allow nonregistered users to access some of the available functionality while offering full functionality to registered users. For example, a nonregistered user can order a product through the shopping cart, whereas a registered user has the possibility, in addition to performing orders, to constantly check the current content of the shopping cart, to be informed about newly available products, and to receive proposals based on the previous shopping history. The difference in the available functionality for registered and nonregistered users aims at attracting new users to the Web portal while providing incentives for registering. Additionally, for users who visit the Web portal more than once, shopping requirements and preferences may be inferred based on previous navigation and used subsequently in adapting the interface to display products accordingly.

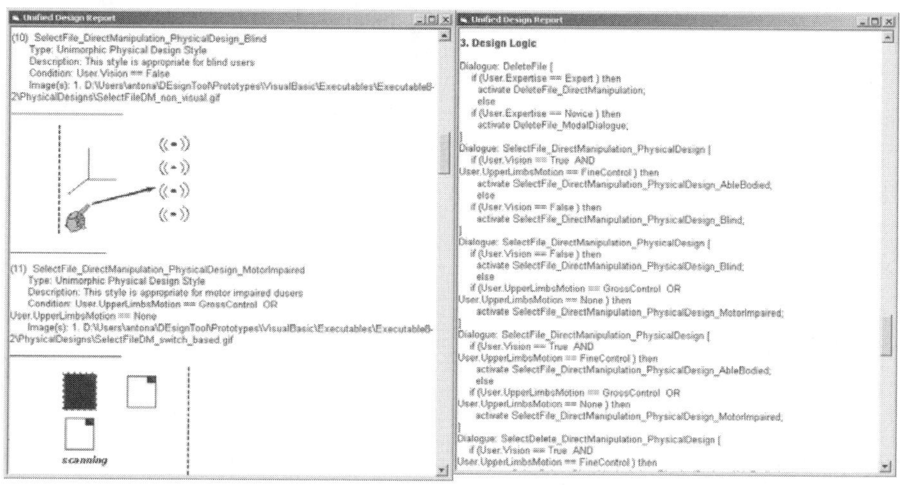

Figure 12 Design report generated by MENTOR.

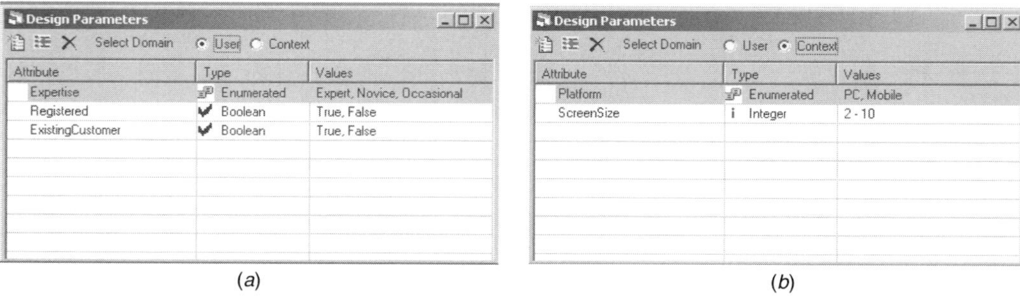

Figure 13 Design parameters of the shopping cart design case: (*a*) user-related parameters; (*b*) context-related parameters.

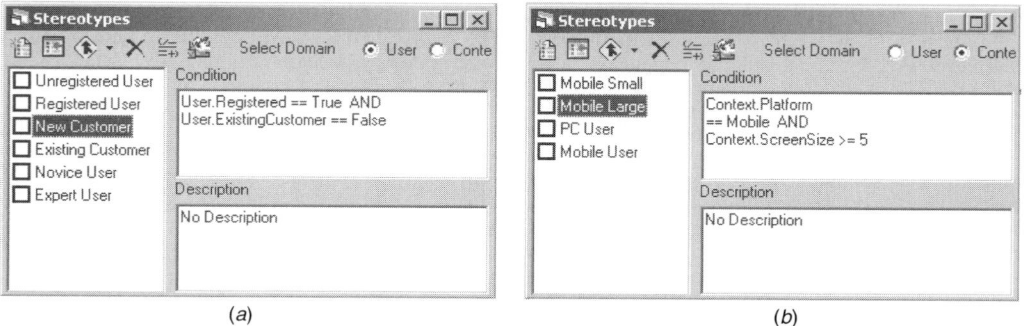

Figure 14 Defined stereotypes in the shopping cart design case: (*a*) defined user; (*b*) context stereotype.

Significant differentiation also occurs in relation to the technological platform used to access the portal. An important characteristic in this respect concerns the available screen size, which is clearly very different in the case of personal computers and mobile phones. Additional considerations that affect the interface design concern differences in input techniques, memory, network connection reliability, and so on.

Finally, an aspect that has been considered as fundamental in the shopping cart design case is the different behavior of users when using personal computers and mobile phones to access the Web. In the first case, users often navigate through available pages independently of a precise goal; in the second case, the navigation behavior is much more targeted to the accomplishment of specific tasks, due to the restrictions that access to the Web through mobile phones present. Therefore, it is important that the mobile phone interface supports effective task accomplishment, while the PC interface should also support pleasant navigation.

Figure 14 depicts the defined user and context stereotypes based on the previously identified design parameters. The polymorphic task hierarchy for the shopping cart design case, depicted in Figure 15, is based on an initial task decomposition of 12 basic tasks: *Registration, Login, Select Shopping Cart from Menu, Add Product to Shopping Cart, View Products in Shopping Cart, Update Products in Shopping Cart, Delete Products from Shopping Cart, Make New Order, View Order Details, Add to Wishing List, Navigation*, and *Error Messages.*

Figure 16 depicts in more detail the portion of the polymorphic task hierarchy concerning the *Login* task, which consists of three unimorphic user tasks and one polymorphic physical design with two incompatible styles, targeted toward the PC and the mobile platforms, respectively. The second style is also polymorphic and decomposes into two incompatible styles targeted toward small and large size mobile phone screens, respectively.

In total the case study involves about 250 design artifacts. Polymorphism occurs mainly at the level of physical design, where completely different styles are defined for interaction through personal computers or mobile phones. An example is illustrated in Figure 17, where the appearance of the product search page on both PCs and mobile phones is displayed. Some cases of design differentiation also occur in dialogue design when the limited screen size of mobile phones requires task simplification (i.e., reduction of functionality) with respect to the desktop interactive environment. Figure 18 shows a part of the design report for the *Login* task, including some of the artifacts' properties (e.g., the images associated with the two mobile phone physical design styles) and the DMSL rules generated.

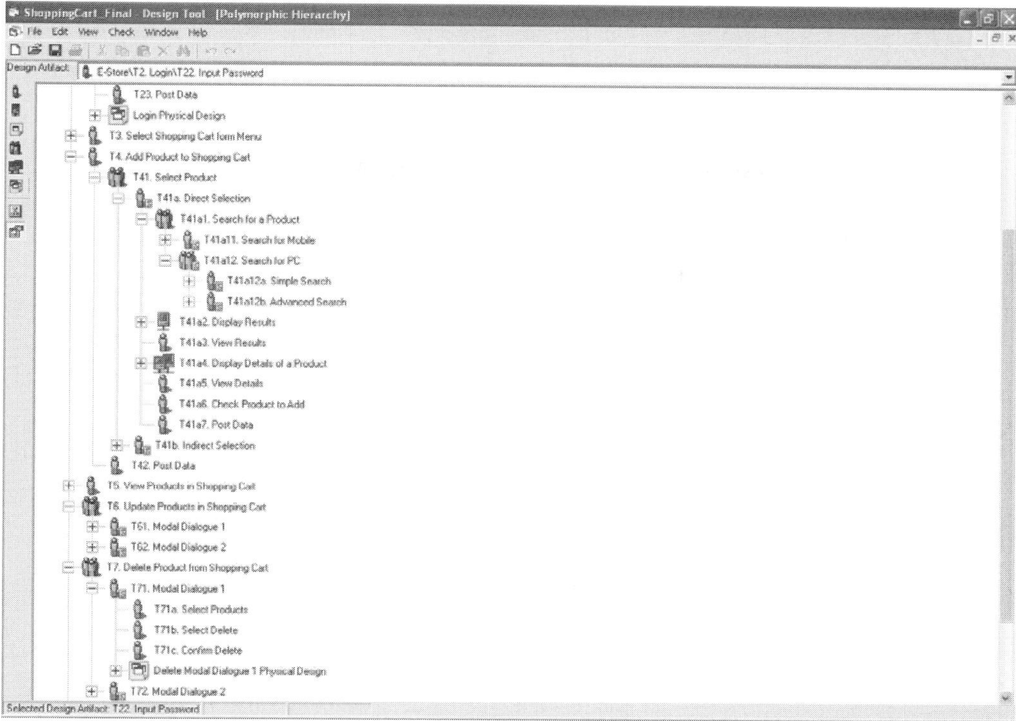

Figure 15 Polymorphic task hierarchy of the shopping cart.

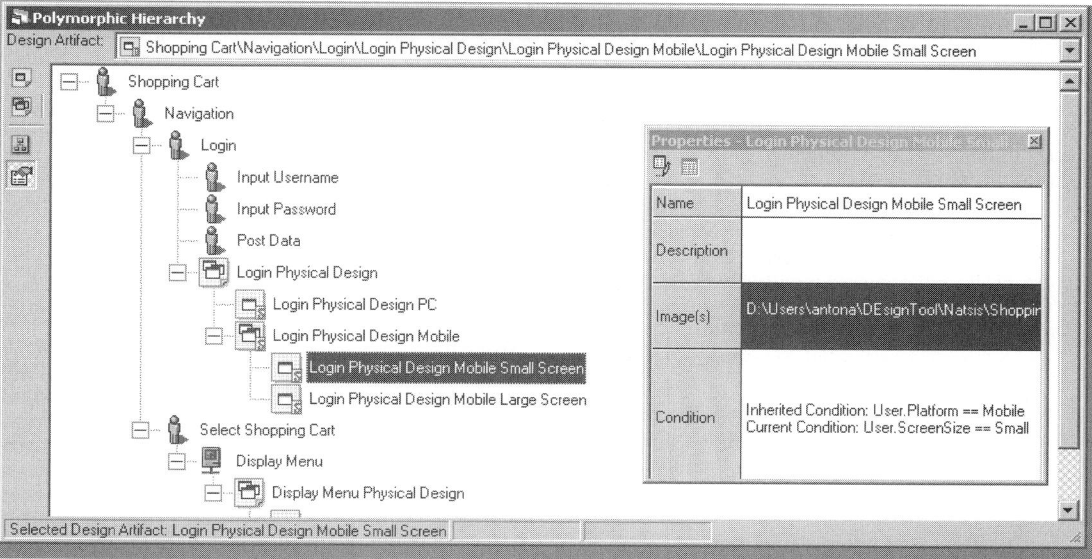

Figure 16 Login task in the shopping cart polymorphic task hierarchy.

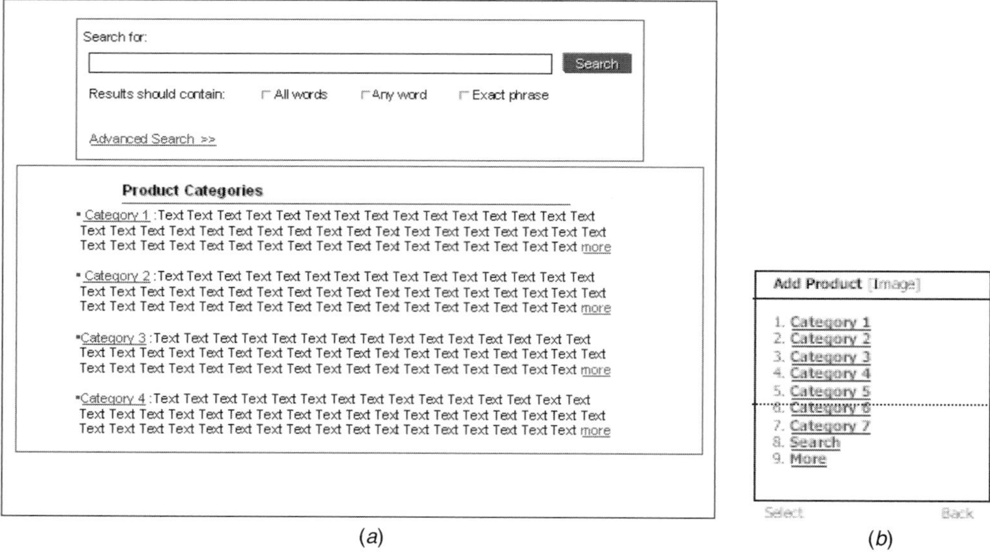

Figure 17 Physical designs of the product searching functionality on (*a*) PCs and (*b*) mobile phones.

5.2 Electronic Health Care Record Case Study

A second case study conducted using MENTOR is presented in Savidis et al. (2004b). In this second experiment, the main objective was to illustrate use of the unified user interface design method in the context of a design scenario in health telematics. The scenario considered exhibits the following dimensions of diversity:

- Doctors can have access to a distributed electronic health records (EHR) system through a PC, PDA, or mobile phone, viewing all types of data for any patient.
- Ambulance paramedic staff is equipped with a PDA or a mobile phone to perform specific tasks and view patient health data which are critical in emergency situations (e.g., allergies, vaccinations).
- Nurses in the hospital can have access to patient administrative and medication data through a PC or a PDA.
- Patients themselves can have access to their data through a PC, PDA, or mobile phone, and can opt to access all data or only the most recent (e.g., last week's blood test results).
- Users of all roles can easily send urgent notifications, including data from the EHR, to other users that may need to be informed (e.g., particular allergy data sent from the ambulance to a doctor in the hospital to obtain advice for emergency treatment, abnormal test results sent from the doctor to the patient, or vice versa).

Reflecting the above, the user and context parameters defined in the design case concern the different user roles in the scenario and the platforms used for accessing the system in the considered situations. Figure 19 depicts the MENTOR design parameters editor with the user and context platform design parameters defined for the case study.

In the polymorphic task decomposition, two main user tasks were identified: *Login* and *View EHR* (see Figure 20). The *Login* task is decomposed into three unimorphic user tasks: *Input User Name, Input User Password*, and *Login Command*. The availability of three different interaction platforms (PC, PDA, and mobile phone) requires the adoption of different interaction styles for each of these tasks as well as for the overall *Login* dialogue. Therefore, the related physical designs are polymorphic and decompose further into three unimorphic physical design styles, one for each available platform (see Figure 21).

The *View EHR* task exhibits different characteristics from the *Login* task. First, it is a polymorphic user task (i.e., it is decomposed into two different styles according to the user role in the EHR system). In this case, polymorphism does not concern the physical design of the task, but its subtask decomposition: professional users (i.e., doctors, paramedics, and nurses) need to select a patient for viewing the related EHR, and patient themselves access their personal data directly after login. As a consequence, the *View EHR* task is decomposed as depicted in Figure 22.

Both styles defined for the *View EHR* polymorphic task are abstract (i.e., they are not associated with a physical design). Instead, they are further decomposed into user and system unimorphic tasks. Figure 23 shows the properties defined for the *View EHR* task

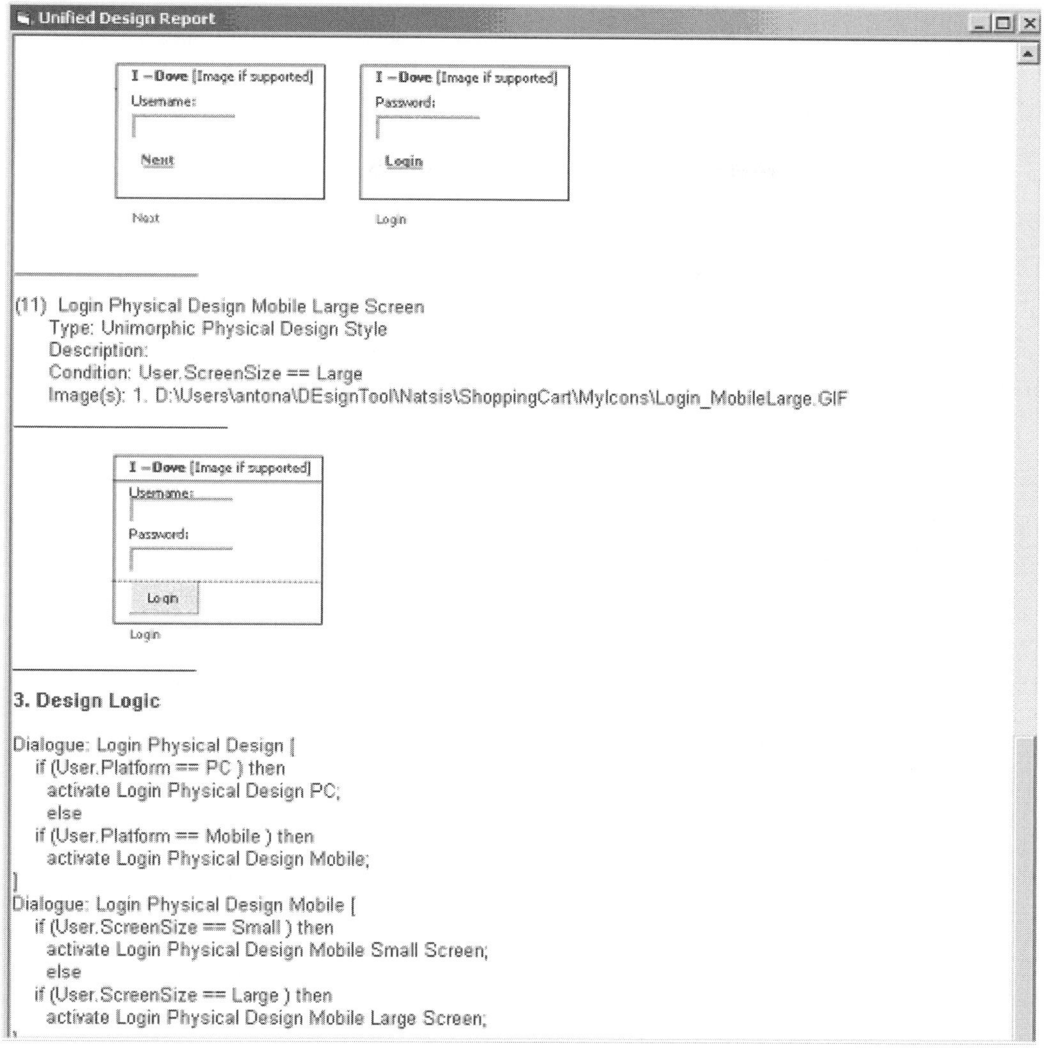

Unified Design Report
I – Dove [Image if supported]
Username:
Next
I – Dove [Image if supported]
Password:
Login
Next
Login
(11) Login Physical Design Mobile Large Screen
Type: Unimorphic Physical Design Style
Description:
Condition: User.ScreenSize == Large
Image(s): 1. D:\Users\antona\DEsignTool\Natsis\ShoppingCart\MyIcons\Login_MobileLarge.GIF
I – Dove [Image if supported]
Username:
Password:
Login
Login
3. Design Logic
Dialogue: Login Physical Design [
if (User.Platform == PC) then
activate Login Physical Design PC;
else
if (User.Platform == Mobile) then
activate Login Physical Design Mobile;
]
Dialogue: Login Physical Design Mobile [
if (User.ScreenSize == Small) then
activate Login Physical Design Mobile Small Screen;
else
if (User.ScreenSize == Large) then
activate Login Physical Design Mobile Large Screen;

Figure 18 Part of the design report for the shopping cart design case.

UUID_Validation_IS4ALL - Design Tool
File Edit View Windows Help
Design Parameters
Select Domain
User
Context
Attribute
Type
Values
Role
Enumerated
Doctor, Nurse, Paramedic, Patient
UUID_Validation_IS4ALL - Design Tool
File Edit View Windows Help
Design Parameters
Select Domain
User
Context
Attribute
Type
Values
Platform
Enumerated
PC, PDA, MobilePhone
Location
Enumerated
Hospital, Ambulance, Home, OnTheMove

(*a*) (*b*)

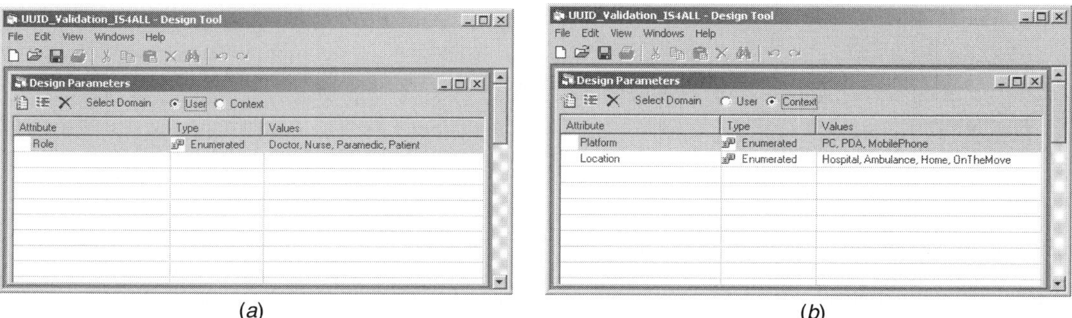

Figure 19 Design parameters of the EHR design case: (*a*) user; (*b*) context.

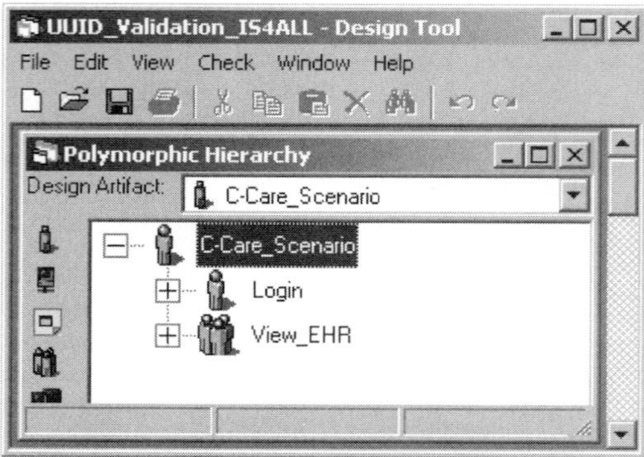

Figure 20 High-level tasks of the EHR design case.

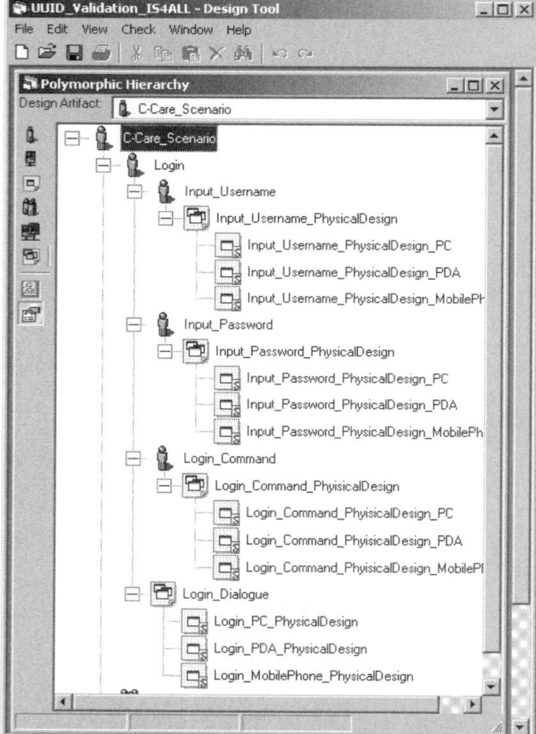

Figure 21 Polymorphic decomposition of the *Login* task in the EHR design case.

Figure 23*a*) and its styles (Figure 23*b* and *c*). 23*b* and *c*. The two styles are defined as incompatible, as they are to be provided to users with different roles and access rights. The condition assigned to the style designed for patients (Figure 23*c*) straightforwardly reflects the specific user role (patient), while the condition assigned to the style defined for professional users is constituted by a disjunction of simple

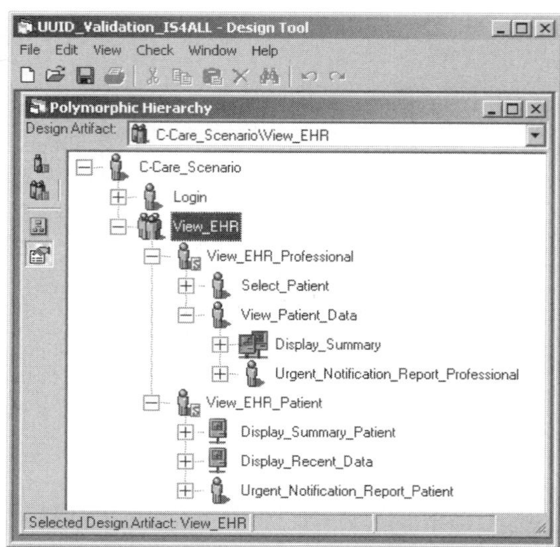

Figure 22 Polymorphic decomposition of the *View EHR* task.

conditions expressing user roles: The user in this case is either a doctor, a paramedic, or a nurse. The adaptation logic generated by MENTOR for the EHR design case appears in Figure 24.

5.3 Designers' Experience

The MENTOR's validation case studies have confirmed its overall usefulness and its advantages compared with conducting unified user interface design without computational support. The designers who conducted the case study were able rapidly to acquire familiarity with the unified user interface design method and with use of the tool itself, and expressed the opinion that the tool appropriately reflects and complements the method and significantly simplifies the conduct of polymorphic task decomposition. In particular, the tool has been found helpful in progressively revising and refining the polymorphic task hierarchy of the case studies and in assigning and revising style conditions and design relations. The systematic organization of design cases as supported by MENTOR has helped designers to articulate in a complete and coherent fashion the polymorphism required by the dimensions of diversity considered in the specific design case. The verification facilities have also been found particularly effective in helping the designer to detect and correct inconsistencies or inaccuracies in the style conditions. Furthermore, the tool has been considered as particularly useful in providing the automatic generation of the unified interface adaptation logic, which in the shopping cart case study, was integrated directly in the prototype implementation of the component. Overall, the design cases have demonstrated that it is possible for designers who are novice in the practice of the unified user interface design method to quickly manage the conduct of the method and the use of MENTOR,

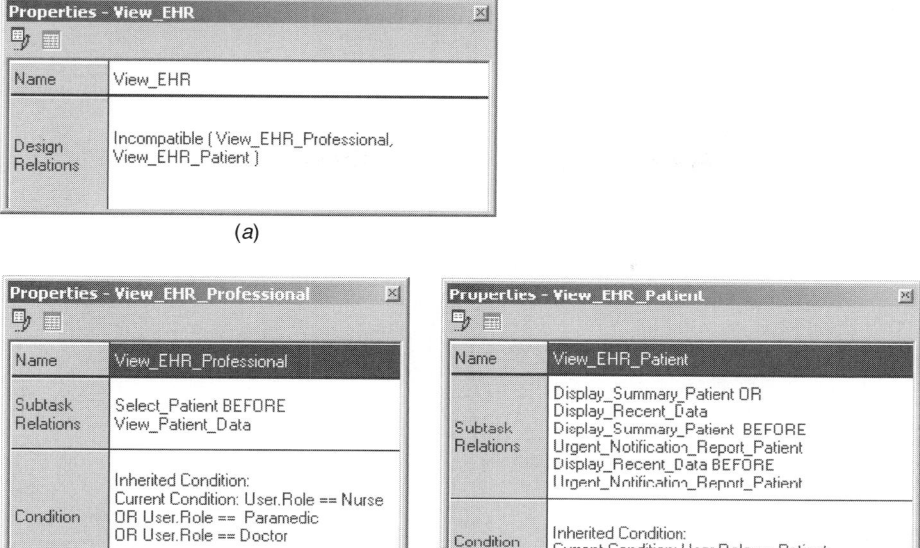

Figure 23 Properties of the *View EHR* polymorphic user task and of its two decomposition styles.

and to produce correct and ready-to-implement design specifications for middle-scale design projects. On the other hand, the main problems encountered in using the tool concerned the need for more extensive support in editing the polymorphic task hierarchy (e.g., extended copy and paste functionality for larger parts of the hierarchy) and in making more visible the current completeness status of the design case (completed tasks vs. pending tasks).

In the case studies, the unified user interface design method has been applied to a design case involving a small set of relevant design parameters and exhibiting characteristics of system access through different platforms and in different contexts as well by users with different roles and access rights. Use of the method has led to the design of different interface instances that are adapted to (1) different user roles, access rights, and levels of expertise; and (2) different platforms and contexts of use.

The validation case studies have involved many, albeit not all, of the conceptual tools of the unified user interface design method. For example, all types of decompositions have been used, and instances of user, system, and physical tasks appear in the polymorphic hierarchy produced. Use of the method in the validation case study, although limited for simplicity to a simple set of design parameters, shows that the method is general enough for systematically capturing and rationalizing diversity in design, and offers an appropriate instrument for proceeding from the conceptualization of diversity at an abstract level to concrete physical designs reflecting the diversification identified. The case studies could easily be extended to consider other potential contexts by identifying the new relevant design parameters and adding to the polymorphic task hierarchy the required new tasks and alternative styles. Therefore, it can be concluded that the method facilitates incremental design, allowing progressive catering for emerging needs and requirements in the field of application (in the specific cases of e-commerce and health telematics). All the steps involved in the design cases have been facilitated by MENTOR, which has also supported automatic derivation of the complete interface adaptation logic. Therefore, it can be concluded that the tool has been useful in supporting use of the unified user interface design method in the specific application domains.

6 SUMMARY AND CONCLUSIONS

Recent progress in the field of universal access and Design for All (i.e., access by anyone, anywhere, and any time to interactive products and services in the information society) has highlighted a shift of perspective and reinterpretation of HCI design from current artifact-oriented practices toward a deeper and multidisciplinary understanding of the diverse factors shaping interaction with technology, such as users' characteristics and requirements and contexts of use, and has proposed solutions for methods, techniques, and codes of practice that make possible proactive and appropriate consideration of diversity in the design of interactive artifacts.

As a consequence, user interface design methodologies, techniques, and tools acquire increased importance in the context of universal access and strive

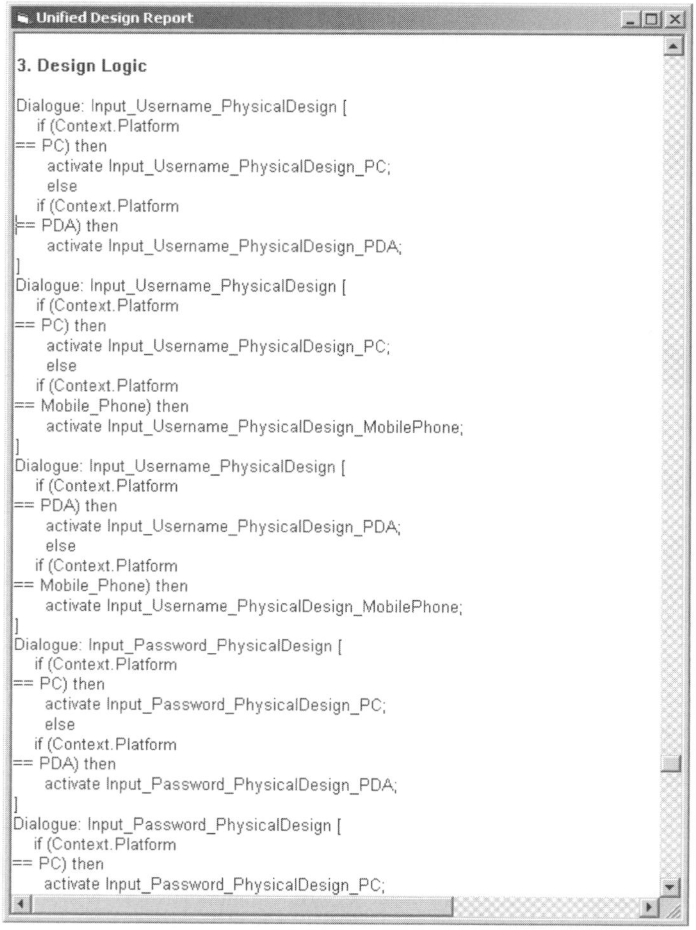

Unified Design Report

3. Design Logic

```
Dialogue: Input_Username_PhysicalDesign [
   if (Context.Platform
== PC) then
     activate Input_Username_PhysicalDesign_PC;
     else
   if (Context.Platform
== PDA) then
     activate Input_Username_PhysicalDesign_PDA;
]
Dialogue: Input_Username_PhysicalDesign [
   if (Context.Platform
== PC) then
     activate Input_Username_PhysicalDesign_PC;
     else
   if (Context.Platform
== Mobile_Phone) then
     activate Input_Username_PhysicalDesign_MobilePhone;
]
Dialogue: Input_Username_PhysicalDesign [
   if (Context.Platform
== PDA) then
     activate Input_Username_PhysicalDesign_PDA;
     else
   if (Context.Platform
== Mobile_Phone) then
     activate Input_Username_PhysicalDesign_MobilePhone;
]
Dialogue: Input_Password_PhysicalDesign [
   if (Context.Platform
== PC) then
     activate Input_Password_PhysicalDesign_PC;
     else
   if (Context.Platform
== PDA) then
     activate Input_Password_PhysicalDesign_PDA;
]
Dialogue: Input_Password_PhysicalDesign [
   if (Context.Platform
== PC) then
     activate Input_Password_PhysicalDesign_PC;
```

Figure 24 Adaptation logic of the EHR design case in DMLS, as generated by MENTOR.

toward approaches that support *design for diversity*, based on consideration of the several dimensions of diversity that emerge from the broad range of user characteristics, the changing nature of human activities, the variety of contexts of use, the increasing availability and diversification of information, the variety of knowledge sources and services, and the proliferation of diverse technological platforms that occur in the information society. Two main dimensions of such a perspective are its user-oriented focus, targeted toward capturing and collecting the requirements of a diversity of users in a diversity of usage context, and its adoption of intelligent interface adaptation as a technological basis, viewing design as the organization and structuring of an entire design space of alternatives to cater for diverse requirements. Under the universal access perspective, adaptation needs to be "designed into" the system rather than decided upon and implemented a posteriori. Unified user interface design, discussed in Section 3, has been proposed in recent years as a method to support the design of user interfaces which automatically adapt to factors that affect their accessibility and usability, such as the abilities and characteristics of different user groups, but also factors related to the context of use and the access technological platforms. Such a method has a strong process-oriented perspective and is characterized by a series of features that make it uniquely suited toward supporting user interface design for universal access. These include, besides process orientation, an open-world view of the design domain (i.e., user- and context-oriented factors relevant for design, interaction platforms, languages and artifacts, and possible types of adaptation); a focus on supporting implementation and fostering design reusability and extendibility; and a limited need for formal design representations and modeling.

Despite recent progress, the practice of designing for diversity remains difficult, due to intrinsic complexity of the task and the current limited expertise of designers and practitioners. Toward overcoming such a difficulty, tool support is required for appropriate structuring of the design process and to help designers in creatively producing and documenting interface adaptations. The interactive design environment MENTOR, discussed in Section 4, is intended to contribute toward overcoming current difficulties in systematically practicing and further studying the process of user interface design from a universal access perspective. MENTOR embodies the unified user interface design method and provides practical integrated support for all its phases, by appropriately guiding the process and structuring the outcomes of design steps through appropriate editing facilities. Additionally, MENTOR provides support for the transition from design to development of unified user interfaces through the availability of automated verification mechanisms for the adaptation logic designed, and through the automated generation of ready-to-implement interface specifications, including the adaptation logic. MENTOR has been used and validated in two middle-sized design cases, reported briefly in Section 5, that have confirmed the usefulness of the tool and its potential contribution toward practical support of designers in applying the unified user interface design method systematically in different application domains.

Support tools such as MENTOR are claimed to have a significant role to play toward widening and improving the practice of Design for All, and ensuring a more effective transition from the design to the implementation phase. Additionally, such tools can provide support for the progressive accumulation of design cases and of the related design experience and knowledge, in particular regarding adaptation, by offering means for extending and reusing (parts of) past design cases.

REFERENCES

Akoumianakis, D., and Stephanidis, C. (2001a), "Rethinking HCI in Terms of Universal Design," in *Universal Access in HCI: Towards an Information Society for All*, Volume 3 of *Proceedings of the 9th International Conference on Human–Computer Interaction*, HCI International 2001, C. Stephanidis, Ed., New Orleans, LA, August 5–10, Lawrence Erlbaum Associates, Mahwah, NJ, pp. 8–12.

Akoumianakis, D., and Stephanidis, C. (2001b), "USE-IT: A Tool for Lexical Design Assistance," in *User Interfaces for All: Concepts, Methods, and Tools* C. Stephanidis, Ed., Lawrence Erlbaum Associates, Mahwah, NJ, pp. 469–487.

Akoumianakis, D., Savidis, A., and Stephanidis, C. (2000), "Encapsulating Intelligent Interactive Behaviour in Unified User Interface Artefacts," *International Journal of Interacting with Computers*, special issue, "The Reality of Intelligent Interface Technology," Vol. 12, No. 4, pp. 383–408.

Bergman, E., and Johnson, E. (1995), "Towards Accessible Human–Computer Interaction," in *Advances in Human–Computer Interaction*, J. Nielsen, Ed., Vol. 5, Ablex Publishing, Stamford, CT.

Bevan, N. (2001), "Quality in Use for All," in *User Interfaces for All: Concepts, Methods, and Tools*, C. Stephanidis, Ed., Lawrence Erlbaum Associates, Mahwah, NJ, pp. 353–368.

Buckingham Shum, S. (1996), "Design Argumentation as Design Rationale," in *The Encyclopedia of Computer Science and Technology*, A. Kent, and J. G. Williams, Eds., Marcel Dekker, New York.

Bühler, C., and Stephanidis, C. (2004), "European Cooperation Activities Promoting Design for All in Information Society Technologies", in *Proceedings of the 9th International Conference on Computers Helping People with Special Needs*, ICCHP 2004, Paris, July 7–9, Springer-Verlag, Berlin, pp. 80–87.

Dieterich, H., Malinowski, U., Kuhme, T., and Schneider-Hufschmidt, M. (1993), "State of the Art in Adaptive User Interfaces," in *Adaptive User Interfaces: Principles and Practice*, M. Schneider-Hufschmidt, T. Kuhme, and U. Malinowski, Eds., Elsevier Science, Amsterdam, pp. 13–48.

Hartson, H. R., Siochi, A. C., and Hix, D. (1990), "The UAN: A User-Oriented Representation for Direct Manipulation Interface Design," *ACM Transactions on Information Systems*, Vol. 8, No. 3, pp. 181–203.

HFES/ANSI (1997), *Draft HFES/ANSI 200 Standard*, Section 5, "Accessibility," Human Factors and Ergonomics Society, Santa Monica, CA.

Hoare, C. A. R. (1978), "Communicating Sequential Processes," *Communications of the ACM*, Vol. 21, No. 8, pp. 666–677.

Hook, K. (2000), "Steps to Take Before Intelligent User Interfaces Become Real," *Interacting with Computers*, Vol. 12, pp. 409–426.

Kirwan, B., and Ainsworth, L. K. (1992), *A Guide to Task Analysis*, Tayler & Francis, London.

Mynatt, W. E., and Weber, G. (1994), "Nonvisual Presentation of Graphical User Interfaces: Contrasting Two Approaches," in *Proceedings of the ACM Conference on Human Factors in Computing Systems*, CHI'94, Boston, pp. 166–172.

Nicolle, C., and Abascal, J., Eds. (2001), *Inclusive Design Guidelines for HCI*, Taylor & Francis, London.

Rahman, M., and Sprigle, S. (1997), "Physical Accessibility Guidelines of Consumer Product Controls," *Assistive Technology Journal*, Vol. 9, pp. 3–14.

Savidis A., and Stephanidis C. (1995), "Developing Dual User Interfaces for Integrating Blind and Sighted Users: The HOMER UIMS," in *Proceedings of the ACM Conference on Human Factors in Computing Systems*, CHI '95, Denver, CO, ACM Press, New York, pp. 106–113.

Savidis, A., and Stephanidis, C. (1997), "Agent Classes for Managing Dialogue Control Specification Complexity: A Declarative Language Framework," in *Design of Computing Systems: Cognitive Considerations, Proceedings of the 7th International Conference on Human–Computer Interaction*, HCI International '97, G. Salvendy, M. J. Smith, and R. J. Koubek, Eds., San Francisco, August 24–29, Elsevier Science, Amsterdam, Vol. 1, pp. 461–464.

Savidis, A., and Stephanidis, C. (2001a), "The Unified User Interface Software Architecture," in *User Interfaces for All: Concepts, Methods, and Tools*, C. Stephanidis, Ed., Lawrence Erlbaum Associates, Mahwah, NJ, pp. 389–415.

Savidis, A., and Stephanidis, C. (2001b), "The I-GET UIMS for Unified User Interface Implementation," in *User Interfaces for All: Concepts, Methods, and Tools*, C. Stephanidis, Ed., Lawrence Erlbaum Associates, Mahwah, NJ, pp. 489–523.

Savidis, A., and Stephanidis, C. (2001c), "Designing with Varying Design Parameters: The Unified Design Process," in *Proceedings of Workshop 14, "Universal Design: Towards Universal Access in the Info Society,"* organized in the context of the ACM Conference on Human Factors in Computing Systems, CHI 2001, Seattle, WA, March 31–April 5.

Savidis, A., and Stephanidis, C. (2004a), "Unified User Interface Design: Designing Universally Accessible Interactions," *International Journal of Interacting with Computers*.

Savidis, A., and Stephanidis, C. (2004b), "Unified User Interface Development: Software Engineering of Universally Accessible Interactions," *Universal Access in the Information Society*, Vol. 3, No. 3.

Savidis, A., Akoumianakis, D., and Stephanidis, C. (2001), "The Unified User Interface Design Method," in *User Interfaces for All: Concepts, Methods, and Tools*, C. Stephanidis, Ed., Lawrence Erlbaum Associates, Mahwah, NJ, pp. 417–440.

Savidis, A., Antona, M., and Stephanidis, C. (2004), "A Decision Making Specification Language for Verifiable Interface Adaptation Logic," to appear in *International Journal of Software Engineering and Knowledge Engineering*.

Savidis A., Antona, M., and Stephanidis, C. (2005b), "Applying the Unified User Interface Design Method in Health Telematics," in *Universal Access Code of Practice in Health Telematics. A Design Code of Practice*, C. Stephanidis, Ed., Springer-Verlag, New York, pp. 115–140.

Shneiderman, B. (2000), "Pushing Human–Computer Interaction Research to Empower Every Citizen: Universal Usability," *Communications of the ACM*, Vol. 43, No. 5, pp. 85–91.

Stephanidis, C. (2001a), "User Interfaces for All: New Perspectives into Human–Computer Interaction," in *User Interfaces for All: Concepts, Methods, and Tools*, C. Stephanidis, Ed., Lawrence Erlbaum Associates, Mahwah, NJ, pp. 3–17.

Stephanidis, C. (2001b), "The Concept of Unified User Interfaces," in *User Interfaces for All: Concepts, Methods, and Tools*, C. Stephanidis, Ed., Lawrence Erlbaum Associates, Mahwah, NJ, pp. 371–388.

Stephanidis, C. (2001c), "Adaptive Techniques for Universal Access," *User Modelling and User Adapted Interaction International Journal*, 10th Anniversary Issue, Vol. 11, No. 1–2, pp. 159–179.

Stephanidis, C., and Akoumianakis, D. (2003), "A Design Code of Practice for Universal Access: Methods and Techniques," in *The Handbook of Human Factors in Web Design*, R. Proctor and K. Vu, Eds., Lawrence Erlbaum Associates, Mahwah, NJ.

Stephanidis, C., and Emiliani, P. L. (1999), "Connecting to the Information Society: A European Perspective," *Technology and Disability Journal*, Vol. 10, No. 1, pp. 21–44.

Stephanidis C. (Ed.), Salvendy, G., Akoumianakis, D., Bevan, N., Brewer, J., Emiliani, P. L., Galetsas, A., Haataja, S., Iakovidis, I., Jacko, J., Jenkins, P., Karshmer, A., Korn, P., Marcus, A., Murphy, H., Stary, C., Vanderheiden, G., Weber, G., and Ziegler, J. (1998), "Toward an Information Society for All: An International R&D Agenda," *International Journal of Human–Computer Interaction*, Vol. 10, No. 2, pp. 107–134.

Stephanidis, C. (Ed.), Salvendy, G., Akoumianakis, D., Arnold, A., Bevan, N., Dardailler, D., Emiliani, P. L., Iakovidis, I., Jenkins, P., Karshmer, A., Korn, P., Marcus, A., Murphy, H., Oppermann, C., Stary, C., Tamura, H., Tscheligi, M., Ueda, H., Weber, G., and Ziegler, J. (1999), "Toward an Information Society for All: HCI Challenges and R&D Recommendations," *International Journal of Human–Computer Interaction*, Vol. 11, No. 1, pp. 1–28.

Stephanidis, C., Paramythis, A., Sfyrakis, M., and Savidis, A. (2001), "A Case Study in Unified User Interface Development: The AVANTI Web Browser," in *User Interfaces for All: Concepts, Methods, and Tools*, C. Stephanidis, Ed., Lawrence Erlbaum Associates, Mahwah, NJ, pp. 525–568.

Stephanidis, C., Paramythis, A., Zarikas, V., and Savidis, A. (2004), "The PALIO Framework for Adaptive Information Services," in *Multiple User Interfaces: Cross-Platform Applications and Context-Aware Interfaces*, A. Seffah and H. Javahery, Eds., Wiley, Chichester, UK, pp. 69–92.

Story, M. F. (1998), "Maximising Usability: The Principles of Universal Design," *Assistive Technology*, Vol. 10, No. 1, pp. 4–12.

Thoren, C. (1993), *Nordic Guidelines for Computer Accessibility*, Nordic Committee on Disability, Swedish Handicap Institute, Stockholm, Sweden.

Vanderheiden, G. (1998), "Universal Design and Assistive Technology in Communication and Information Technologies: Alternatives or Compliments?" *Assistive Technology*, Vol. 10, No. 1, pp. 29–36.

Vernardakis, N., Stephanidis, C., and Akoumianakis, D. (1997), "Transferring Technology Toward the European Assistive Technology Industry: Mechanisms and Implications," *Assistive Technology*, Vol. 9, No. 1, pp. 34–36.

Vernardakis, N., Akoumianakis, D., and Stephanidis, C. (2001), "Economics and Management of Innovation," in *User Interfaces for All: Concepts, Methods, and Tools*, C. Stephanidis, Ed., Lawrence Erlbaum Associates, Mahwah, NJ, pp. 609–632.

W3C–WAI (1999), "Web Content Accessibility Guidelines 1.0," http://www.w3.org/TR/WCAG10/.

Winograd, T. (2001), "From Programming Environments to Environments for Designing," in *User Interfaces for All: Concepts, Methods, and Tools*, C. Stephanidis, Ed., Lawrence Erlbaum Associates, Mahwah, NJ, pp. 165–181.

Winograd, T., Bennett, J., De Young, L., and Hartfield, B., Eds. (1996), *Bringing Design to Software*, Addison-Wesley, Reading, MA.

PART 10

SELECTED APPLICATIONS IN HUMAN FACTORS AND ERGONOMICS

CHAPTER 57

HUMAN FACTORS AND ERGONOMICS STANDARDS

Bohdana Sherehiy, Waldemar Karwowski, and David Rodrick*
University of Louisville
Louisville, Kentucky

1 INTRODUCTION

A *standard* is *a documented agreement containing technical specifications or other precise criteria, to be used consistently as rules, guidelines, or definitions of characteristics, to ensure that materials, products, processes and services are fit for the purpose served by those making reference to the standard* (ISO, 2004). Three main levels of the standardization process can be distinguished: national, regional, and international (see Figure 1). At the highest and broadest level of applicability are the international standards. The basis for worldwide standardization in all areas is provided primarily by three organizations: the International Organization for Standardization (ISO), International Electrotechnical Commission (IEC), and International Telecommunications Union (ITU). Standards related to human factors and ergonomics are developed by the International Organization for Standardization (ISO).

In Europe, there are three standardization bodies: the European Committee for Standardization (CEN), European Committee for Electrotechnical Standardization (CENELEC) and European Telecommunications Standards Institute (ETSI). Their mission is to develop and achieve a coherent set of voluntary standards as a basis for a Single European Market/European Economic Area (Wetting, 2002). At the national level almost every nation has its own national body for standards development. Examples of the national standardization organizations include the American National Standards Institute (ANSI), British Standards Institution (BSI), Deutsches Institut für Normung (DIN), and Association Française de Normalisation (AFNOR). Standards can also be prepared by technical societies, labor organizations, consumer organizations, trade associations, and governmental agencies.

International, regional, and national standards are distinguished by documented standards development procedures. These procedures have been designed to ensure that all interested parties that can be affected by a particular standard will have an opportunity to represent their interest and participate in the standards development process. For example, ISO standards are developed by technical committees, which consist of experts from the industrial, technical, and business sectors that are in need of standards. Many ISO national members apply public review procedures in order to consult draft standards with the interested parties, including representatives of government agencies, industrial and commercial organizations, professional and consumer associations, and the general public (ISO, 2004). The ISO national bodies are expected to take into account any feedback they receive and to present a consensus position to appropriate technical committees.

*Present address: Florida State University, Tallahassee, Florida.

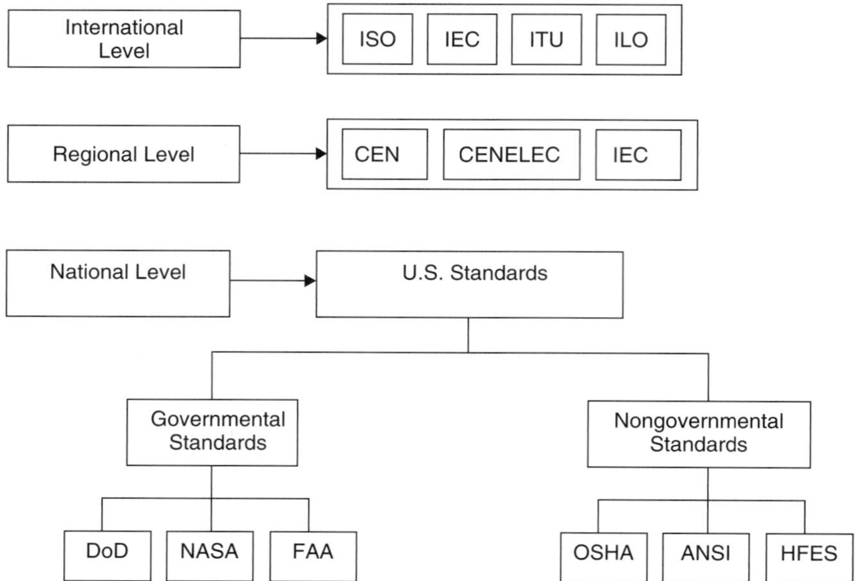

Figure 1 Hierarchy of standards levels.

Standards are necessary to provide quality control and to support legislation and regulations to ensure an equal-opportunity and fairly operating international market. The main purpose of standardization is to achieve uniformity and interchangeability. Standardization limits the diversity of sizes, shapes, or component designs and prevents the generation of unneeded variation of products which do not provide unique services. Standardization is also the means by which society gathers and disseminates technical information (Spivak and Brenner, 2001). Harmonization of standards reduces trade barriers; promotes safety; allows interoperability of products, systems, and services; and promotes common technical understanding (Wetting, 2002).

The need for standardization from the human factors and ergonomics viewpoints can be illustrated by many "horror stories" following World War II. The war required that pilots fly on different types of aircraft which had no standard control arrangement in the cockpits (McDaniel, 1996). Many planes crashed because the pilots used wrong controls based on erroneously applied behavioral patterns. The human factors solutions included standardization of a single arrangement for engine controls and development of distinct shapes for the control handles (McDaniel, 1996). In this chapter we provide an overview of the human factors and ergonomics (HFE) standardization efforts around the world. It should be noted that standards focusing mainly on the safety engineering area are not covered.

2 ISO STANDARDS FOR ERGONOMICS

The International Organization for Standardization (ISO) was created in 1947 to coordinate the development of international standards. ISO is a worldwide federation of national standards bodies from 148 countries, one from each country. ISO is a nongovernmental organization that considers the interests of users, producers, consumers, governments, and the scientific community. The mission of ISO is to promote the development of standardization and related activities in the world in order to facilitate the international exchange of goods and services and to develop cooperation in the spheres of intellectual, scientific, technological, and economic activity (ISO, 2004).

In 1975, the International Organization for Standardization formed technical committee TC 159 to develop standards in the field of ergonomics (Parsons et al., 1995). The scope of the ISO/TC 159 activity has been described as standardization in the field of ergonomics, including terminology, methodology, and human factors data. According to the agreed scope, the ISO Technical Committee 159 (through standardization and coordination of related activities) promotes the adaptation of working and living conditions to human anatomical, psychological, and physiological characteristics in relation to the physical, sociological, and technological environment. Among the main objectives of such efforts are safety, health, well-being, and effectiveness (Parsons, 1995c). It should be noted that because of historical and organizational factors, many standards in the field of ergonomics are not developed by the ISO TC 159.

At present, the ISO TC 159 organizational structure is administrated by the German Standards Association

(DIN). The ergonomics standardization group consists of four subcommittees: SC 1, SC 3, SC 4, and SC 5. The subject areas of subcommittees and their organizational structure are presented in the Table 1. Before any of the ISO documents can become a standard, they have to undergo several stages:

- *Stage 1*: Approved Work Item (AWI); Work Item (WI)
- *Stage 2*: Working Draft (WD)
- *Stage 3*: Committee Draft (CD); Committee Draft Technical Report (CD TR); Committee Draft Technical Specification (CD TS)
- *Stage 4*: Committee Draft for Vote (CDV); Draft International Standard (DIS); Final Committee Draft (FCD); Draft Technical Report (DTR); Draft Technical Specification (DTS)
- *Stage 5*: Final Draft International Standard (FDIS)
- *Stage 6*: International Standard (ISO); Technical Report (ISO TR); Technical Specification (ISO TS)

Table 1 Organizational Structure of ISO TC 159

Committee	Title
TC 159/SC 1: Ergonomic guiding principles	
TC 159/SC 1/WG 1	Principles of the design of work systems
TC 159/SC 1/WG 2	Ergonomic principles related to mental work
TC 159/SC 1/WG 4	Usability of everyday products
TC 159/SC 3: Anthropometry and biomechanics	
TC 159/SC 3/WG 1	Anthropometry
TC 159/SC 3/WG 2	Evaluation of working postures
TC 159/SC 3/WG 4	Human physical strength: manual handling and force limits
TC 159/SC 3/WG 5	Ergonomic procedures for applying anthropometry and biomechanics standards
TC 159/SC 4: Ergonomics of human–system interaction	
TC 159/SC 4/WG 1	Fundamentals of controls and signaling methods
TC 159/SC 4/WG 2	Visual display requirements
TC 159/SC 4/WG 3	Control, workplace, and environmental requirements
TC 159/SC 4/WG 5	Software ergonomics and human–computer dialogues
TC 159/SC 4/WG 6	Human-centered design processes for interactive systems
TC 159/SC 4/WG 8	Ergonomic design of control centers
TC 159/SC 5: Ergonomics of the physical environment	
TC 159/SC 5/WG 1	Thermal environments
TC 159/SC 5/WG 2	Lighting environments
TC 159/SC 5/WG 3	Danger signals and communication in noisy environments

Table 2 ISO Standards for Ergonomic Guiding Principles

Reference Number	Title
ISO 6385:2004	Ergonomic principles in the design of work systems
ISO 10075:1991	Ergonomic principles related to mental workload: General terms and definitions
ISO 10075-2:1996	Ergonomic principles related to mental workload, part 2: Design principles
ISO/FDIS 10075-3	Ergonomic principles related to mental workload, part 3: Principles and requirements concerning methods for measuring and assessing mental workload
ISO/CD 20282-1	Ease of operation of everyday products, part 1: Context of use and user characteristics
ISO/CD TS 20282-2	Ease of operation of everyday products, part 2: Test method

2.1 Ergonomics Guiding Principles

The standards concerned with the ergonomics basic principles are elaborated by the TC 159/SC 1 subcommittee. The list of the published standards and standards in development for the ergonomics guiding principles are provided in Table 2. ISO 6385:2004 is a basic standard that states the objectives of the ergonomics system design and provides definitions of basic terms and concepts in ergonomics (HFE). This standard establishes ergonomics principles of the work system design as basic guidelines. Such guidelines should be applied for the design of optimal working conditions with regard to human well-being, safety, and health, with consideration of technological and economic efficiency (Parsons, 1995a).

The ISO 10075 standard dealing with mental workload is comprised of three parts. The first part presents terminology and main concepts. Part 2 covers guidelines on the design of work systems, including task, equipment, workspace, and work conditions with reference to the mental workload. Part 3, which provides guidelines on measurement and assessment of mental workload, is currently at the stage of a *Final Draft International Standard*. The third part specifies the requirements for the measurement instruments to be met at different levels of precision in measuring mental workload. In these standards it was stated that any human activity, even those that are considered primarily as physical activities, includes a mental workload (Nachreiner, 1995). Therefore, the described standards on mental workload are relevant to all kinds of work design.

2.2 Anthropometry and Biomechanics

The standards related to anthropometry and biomechanics are developed by the TC 159/SC 3 subcommittee. This subcommittee consists of four working groups: Anthropometry (WG 1), Evaluation of

Table 3 Published ISO Standards and Standards under Development for Anthropometry and Biomechanics

Reference Number	Title
ISO 7250:1996	Basic human body measurements for technological design
ISO 11226:2000	Ergonomics: Evaluation of static working postures
ISO 11228-1:2003	Ergonomics: Manual handling, part 1: Lifting and carrying
ISO 14738:2002	Safety of machinery: Anthropometric requirements for the design of workstations at machinery
ISO 14738:2002/Cor 1:2003	Corrigendum
ISO 15534-1:2000	Ergonomic design for the safety of machinery, part 1: Principles for determining the dimensions required for openings for whole-body access into machinery
ISO 15534-2:2000	Ergonomic design for the safety of machinery, part 2: Principles for determining the dimensions required for access openings
ISO 15534-3:2000	Ergonomic design for the safety of machinery, part 3: Anthropometric data
ISO 15535:2003	General requirements for establishing anthropometric databases
ISO/TS 20646-1:2004	Ergonomic procedures for the improvement of local muscular workloads, part 1: Guidelines for reducing local muscular workloads
ISO/CD 11228-2	Ergonomics: Manual handling, part 2: Pushing and pulling
ISO/CD 11228-3	Ergonomics: Manual handling, part 3: Handling of low loads at high frequency
ISO/DIS 15536-1	Ergonomics: Computer manikins and body templates, part 1: General requirements
ISO/CD 15536-2	Ergonomics: Computer manikins and body templates, part 2: Structures and dimensions
ISO/FDIS 15537	Principles for selecting and using test persons for testing anthropometric aspects of industrial products and designs
ISO/DIS 20685	Three-dimensional scanning methodologies for internationally compatible anthropometric databases

working postures (WG 2), Human physical strength (WG 3), and Manual handling and heavy weights (WG 4). The list of the published standards and standards in development for anthropometry and biomechanics are presented in Table 3. The description of anthropometric measurements, which can be used as a basis for definition and comparison of population groups, is provided in the ISO 7250:1996 standard. In addition to the lists of the basic anthropometric measurements, this document contains definitions and measuring conditions.

The three-part standards for the safety of machinery (ISO 15534) provides guidelines for determining the dimensions required for openings for access for machinery. The first part of this standard (ISO 15534-1:2000) presents principles for determining the dimensions for opening for whole-body access to machinery; the second part (ISO 15534-2:2000) specifies dimensions for the access openings. The third part of the safety of machinery standards (ISO 15534-3:2000) provides the requirements for the human body measurements (anthropometric data) that are needed for the calculation of access opening dimensions for machinery specified in the two previous parts of this standard (Parsons, 1995c). The anthropometric data are based on static measurements on nude people and representative of the European population of men and women.

ISO 14738:2002 describes principles for deriving dimensions from anthropometric measurements and applying them to the design of workstations at nonmobile machinery. This standard also specifies the body space requirements for equipment during normal operation in sitting and standing positions. ISO 15535:2003 specifies general requirements for anthropometric databases and their associated reports that contain measurements taken in accordance with ISO 7250. This standard presents such information as characteristics of the user population, sampling methods, and measurement items and statistics, to make international comparison possible among various population segments.

ISO 11228-1:2003 describes limits for manual lifting and carrying with consideration, respectively, the intensity, frequency, and duration of the task. The limits recommended can be used in the assessment of several task variables and the health risk evaluation of the working population (Dickinson, 1995). This standard does not include holding of objects (without walking), pushing or pulling of objects, lifting with one hand, manual handling while seated, and lifting by two or more people. Holding, pushing, and pulling objects are included in other parts of ISO 11228, which are currently at the stage of committee drafts. ISO/TS 20646-1:2004 present guidelines for application of various ergonomics standards related to local muscular workload (LMWL) and specify activities to reduce LMWL in workplaces.

2.3 Ergonomics of Human–System Interaction

The TC 159/SC 4 subcommittee develops the standards related to ergonomics of human–system interaction. The subcommittee are divided into six working groups, which consider the standards related to the following topics: controls and signaling methods, visual display requirements, control, workplace and environmental requirements, software ergonomics and human–computer dialogue, human-centered design processes for interactive systems, and ergonomics design of control centers.

2.3.1 Controls and Signaling Methods

ISO 9355, *Ergonomic Requirements for the Design of Displays and Control Actuators*, provides guidelines for the design of displays and control actuators on work equipment, especially machines (see Table 4). A list of all parts of the standard ISO 9355 is presented in Table 4. Part 1 describes general principles of human interactions with display and controls. The other two parts provides recommendation on the selection, design, and location of information displays (part 2) and control actuators (part 3). Part 4 covers general principles for the location and arrangement of display and actuators.

2.3.2 Visual Display Requirements

The multipart standard ISO 9241, *Ergonomic Requirements for Office Work with Visual Display Terminals (VDTs)*, is believed to be the most important and known standard for ergonomic design (Stewart, 1995; Eibl, 2005). This standard presents general guidance and specific principles that need to be considered in the design of equipment, software, and tasks for office work with VDTs. All parts of the standard ISO 9241, *Ergonomic Requirements for Office Work with Visual Display Terminals (VDTs)*, are presented in Table 5.

ISO 9241 standard describes the basic underlining principles of the user performance approach (part 1). Part 2 describes how task requirements may be identified and specified in organizations and how task requirements can be incorporated into the system design and implementation process. Parts 3 through 9 provide assistance in the procurement and specification of the hardware and environmental components. Three parts presents image quality requirements (performance specification) for different types of displays: white-and-black display (part 3), color displays (part 8), and display with reflections (part 7). Part 4 provides criteria for the keyboard and part 9 for nonkeyboard input devices. Parts 5 and 6 establish ergonomic principles for the appropriate design and procurement of workstation, workstation equipment, and work environment for office work with VDTs (Eibl, 2005). Those two parts includes such issues as technical design of furniture and equipment for the workplace, space organization and workplace layout, physical characteristics of office work environment: lighting, noise, and vibrations. Part 10 presents core ergonomics principles that should be applied to the design of dialogues between humans and information systems. These principles were intended for use in specifications, design, and evaluation of dialogues for office work with visual display terminals (VDTs). Part 11 defines usability and specifies the usability evaluation in terms of the user performance and satisfaction measures (Dzida, 1995). Part 12 provides ergonomic recommendations for information presentation on the text-based displays and graphical user interfaces. Part 13 presents recommendations for different types of user guidance attributes of software interfaces, such as feedback, status, help, and error handling. Parts 14 to 17 deal with particular kinds of dialogue styles: menus, commands, direct manipulation, and form filling.

Table 4 ISO Standards for Controls and Signaling Methods

Reference Number	Title
ISO 9355-1:1999	Ergonomic requirements for the design of displays and control actuators, part 1: Human interactions with displays and control actuators
ISO 9355-2:1999	Ergonomic requirements for the design of displays and control actuators, part 2: Displays
ISO/DIS 9355-3	Safety of machinery: Ergonomic requirements for the design of signals and control actuators, part 3: Control actuators
ISO/DIS 9355-4	Safety of machinery: Ergonomic requirements for the design of displays and control actuators, part 4: Location and arrangement of displays and control actuators

Table 5 ISO 9241: *Ergonomic Requirements for Office Work with Visual Display Terminals*

Reference Number	Title
ISO 9241-1:1997	Part 1: General introduction
ISO 9241-2:1992	Part 2: Guidance on task requirements
ISO 9241-3:1992	Part 3: Visual display requirements
ISO 9241-4:1998	Part 4: Keyboard requirements
ISO 9241-5:1998	Part 5: Workstation layout and postural requirements
ISO 9241-6:1999	Part 6: Guidance on the work environment
ISO 9241-7:1998	Part 7: Requirements for display with reflections
ISO 9241-8:1997	Part 8: Requirements for displayed colors
ISO 9241-9:2000	Part 9: Requirements for nonkeyboard input devices
ISO 9241-10:1996	Part 10: Dialogue principles
ISO 9241-11:199	Part 11: Guidance on usability
ISO 9241-12:1998	Part 12: Presentation of information
ISO 9241-13:1998	Part 13: User guidance
ISO 9241-14:1997	Part 14: Menu dialogues
ISO 9241-15:1997	Part 15: Command dialogues
ISO 9241-16:1999	Part 16: Direct manipulation dialogues
ISO 9241-17:1998	Part 17: Form filling dialogues

The ISO 13406 standard provides recommendations additional to those of ISO 9241 in respect to visual displays based on flat panels. Two parts of this standard cover image quality requirements for the ergonomic design and evaluation of flat panel displays. ISO 14915 provides additional recommendations to ISO 9241 concerning multimedia presentations.

2.3.3 Software Ergonomics

ISO 14915, *Software Ergonomics for Multimedia User Interfaces*, specifies recommendations and principles for the design of interactive multimedia user interfaces that integrate various media, such as static text, graphics, and images, and dynamic media such as: audio, animation, and video. This standard focuses on issues related to integration of different media; hardware issues and multimodal input are not considered. The standard consist of three parts (see Table 6), which address general design principles (part 1), multimedia navigation and control (Part 2), and media selection and combination (part 3). The Committee draft ISO/CD 23973 considers ergonomics design principles for World Wide Web user interfaces.

2.3.4 Ergonomic Design of Control Centers

ISO 11064, *Ergonomic Design of Control Centers*, specifies requirements and presents principles for the ergonomic design of control centers. The list of all parts of ISO 11064 is provided in Table 7. The six parts of this standard are concerned with the following issues: principles for the design of control centers, principles of control suite arrangements, control room and workstation layout and dimensions, displays and controls, environmental requirements, evaluation of control rooms, and ergonomic requirements for specific applications.

Table 6 ISO Standards for Software Ergonomics

Reference Number	Title
ISO 14915-1:2002	Software ergonomics for multimedia user interfaces, part 1: Design principles and framework
ISO 14915-2:2003	Software ergonomics for multimedia user interfaces, part 2: Multimedia navigation and control
ISO 14915-3:2002	Software ergonomics for multimedia user interfaces, part 3: Media selection and combination
ISO/CD 23973	Software ergonomics for World Wide Web user interfaces

Table 7 ISO 11064: *Ergonomic Design of Control Centers*

Reference Number	Title
ISO 11064-1:2000	Part 1: Principles for the design of control centers
ISO 11064-2:2000	Part 2: Principles for the arrangement of control suites
ISO 11064-3:1999	Part 3: Control room layout
ISO 11064-4:2004	Part 4: Layout and dimensions of workstations
ISO/DIS 11064-6	Part 6: Environmental requirements for control centers
ISO/CD 11064-7	Part 7: Principles for the evaluation of control centers

2.3.5 Human–System Interaction

Issues of accessibility in designing usable systems are covered in two ISO Standards. ISO/AWI 16071 provides guidance on accessibility in reference to software, and ISO/TS 16071:2003 in reference to the human–computer interfaces. The guidelines on the human-centered design process throughout the life cycle of computer-based interactive systems are described in ISO 13407:1999 and ISO/TR 18529:2000.

Usability methods supporting human-centered design are described in ISO/TR 16982:2002. Further standards concerned with the human–system interaction address such issues as development and design of icons (ISO 11581), design of typical controls for multimedia functions (ISO 18035), icons for typical WWW browsers (ISO 18036), and definitions and metrics concerning software quality (ISO 9126). Table 8 shows the list of published ISO standards and standards in development for human–system interaction.

2.4 Ergonomics of the Physical Environment

The ISO TC159 SC5 document contains an international standard in the area of the ergonomics of the physical environment. The subcommittee is divided into three working groups: thermal environments (WG 1), lighting (WG 2), and danger signals and communication in noisy environments (WG 3).

2.4.1 Ergonomics of the Thermal Environment

The standards on the ergonomics of thermal environments are concerned with heat stress, cold stress, and thermal comfort as well as with the thermal properties of clothing and metabolic heat production due to activity (Olesen, 1995). Physiological measures, skin reaction to contact with hot, moderate, and cold surfaces, and thermal comfort requirements for people with special requirements are also considered. The list of all standards and standards in development on thermal environment ergonomics are presented in the Tables 9 and 10, respectively.

The main thermal comfort standard, ISO 7730, provides a method for predicting the thermal sensation and the degree of discomfort, which can also be used to specify acceptable environmental conditions for comfort. This method is based on the predicted mean vote (PMV) and predicted percentage of dissatisfied (PPD) thermal comfort indices (Olesen and Parsons, 2002). It also provides methods for the assessment of local discomfort caused by draughts, asymmetric radiation, and temperature gradients. Other thermal environment

Table 8 Published ISO Standards and Standards under Development for Human–System Interaction

Reference Number	Title
ISO 13407:1999	Human-centered design processes for interactive systems
ISO 1503:1977	Geometrical orientation and directions of movements
ISO/AWI 1503	Ergonomic requirements for design on spatial orientation and directions of movements
ISO/AWI 16071	Ergonomics of human–system interaction: Guidance on software accessibility
ISO/TS 16071:2003	Ergonomics of human–system interaction: Guidance on accessibility for human–computer interfaces
ISO/TR 16982:2002	Ergonomics of human–system interaction: Usability methods supporting human–centered design
ISO/PAS 18152:2003	Ergonomics of human–system interaction: Specification for the process assessment of human–system issues
ISO/TR 18529:2000	Ergonomics of human–system interaction: Human-centered lifecycle process descriptions
ISO 13406-1:1999	Ergonomic requirements for work with visual displays based on flat panels, part 1: Introduction
ISO 13406-2:2001	Ergonomic requirements for work with visual displays based on flat panels, part 2: Ergonomic requirements for flat panel displays
ISO/CD 9241-301	Ergonomic requirements and measurement techniques for electronic visual displays, part 301: Introduction
ISO/CD 9241-302	Ergonomic requirements and measurement techniques for electronic visual displays, part 302: Terminology
ISO/CD 9241-303	Ergonomic requirements and measurement techniques for electronic visual displays, part 303: Ergonomic requirements
ISO/AWI 9241-304	Ergonomic requirements and measurement techniques for electronic visual displays, part 304: User performance test method
ISO/CD 9241-305	Ergonomic requirements and measurement techniques for electronic visual displays, part 305: Optical laboratory test methods
ISO/CD 9241-306	Ergonomic requirements and measurement techniques for electronic visual displays, part 306: Field assessment methods
ISO/CD 9241-307	Ergonomic requirements and measurement techniques for electronic visual displays, part 307: Analysis and compliance test methods
ISO/CD 9241-110	Ergonomics of human–system interaction, part 110: Dialogue principles
ISO/CD 9241-400	Physical input devices: Ergonomic principles
ISO/AWI 9241-410	Physical input devices: Design criteria for products
ISO/AWI 9241-420	Physical input devices, part 420: Ergonomic selection procedures
ISO/IEC 11581-1:2000	Information technology: User system interfaces and symbols–Icon symbols and functions, part 1: Icons — general
ISO/IEC 11581-2:2000	Information technology: User system interfaces and symbols–Icon symbols and functions, part 2: Object icons
ISO/IEC 11581-3:2000	Information technology: User system interfaces and symbols–Icon symbols and functions, part 3: Pointer icons
ISO/IEC 11581-5:2004	Information technology: User system interfaces and symbols–Icon symbols and functions, part 5: Tool icons
ISO/IEC 11581-6:1999	Information technology: User system interfaces and symbols–Icon symbols and functions, part 6: Action icons
ISO/IEC 9126-1:2001	Software engineering: Product quality, part 1: Quality model
ISO/IEC TR 9126-2:2003	Software engineering: Product quality, part 2: External metrics
ISO/IEC TR 9126-3:2003	Software engineering: Product quality, part 3: Internal metrics
ISO/IEC TR 9126-4:2004	Software engineering: Product quality, part 4: Quality in use metrics

standards address such issues as thermal comfort for people with special requirements (ISO/TS 14415), responses on contact with surfaces at moderate temperature (ISO 13732, part 2), and thermal comfort in vehicles (ISO 14505, parts 1–4). Standards concerned with thermal comfort assessment specify measuring instruments (ISO 7726), methods for estimation of metabolic heat production (ISO 8996), estimation of clothing properties (ISO 9920), and subjective assessment methods (ISO 10551). ISO 11399:1995 provides information needed for the correct and effective application of international standards concerned with the ergonomics of the thermal environment.

2.4.2 Communication in Noisy Environments

The standards for communication in noisy environments includes warnings, danger signals, and speech. The list of related standards is provided in Table 11. The ISO 7731:1986 document specifies the requirements and test methods for auditory danger signals and gives guidelines for the design of the signals in the public and in workplaces. This document also provides definitions to guide in the use of the standards concerned with noisy environment. Criteria for the perception of the visual danger signals are provided in ISO 11428:1996. This international standard specifies the safety and ergonomic requirements and the corresponding physical measurements.

ISO 11429:1996 specifies a system of danger and information signals in reference to various degrees of urgency. This standard applies to all danger signals that have to be clearly perceived and differentiated, from extreme urgency to "all clear." Guidance on delectability is provided in terms of luminance, illuminance, and contrast, considering both surface and point sources. ISO 9921-1:1996 describes a method for prediction of the effectiveness of speech communication in the presence of noise generated by machinery as well as in any other noisy environment. The following parameters are taken into account in this standard: the ambient noise at the speaker's position, the ambient noise at the listener's position, the distance between the communication partners, and a variety of physical and personal conditions (Parsons, 1995b). ISO/TR 19358:2002 deals with the testing and assessment of speech-related products and services.

2.4.3 Lighting of Indoor Work Systems

ISO 8995 (1989), *Principles of Visual Ergonomics: The Lighting of Indoor Work Systems*, was developed by the ISO 159 SC5 WG2 "Lighting" group in collaboration with the International Commission on Illumination (CIE). This standards describes the principles of the visual ergonomics, identifies factors that influence visual performance, and presents criteria for the achievement of an acceptable visual environment (Parsons, 1995b).

3 CEN STANDARDS FOR ERGONOMICS

In Europe, there are three standardization organizations: the European Committee for Standardisation (Comité Européen de Normalisation, CEN), the European Committee for Electrotechnical Standardisation (CENELEC), and the European Telecommunications Standards Institute (ETSI). Their aim is development and achievement of a coherent set of voluntary standards that can provide a basis for a single European market/European economic area without internal frontiers for goods and services inside Europe. Their work is carried out in conjunction with worldwide bodies and the national standards bodies in Europe (Wetting, 2002). Members of the European Union (EU) and the European Fair Trade Association (EFTA) have agreed to implement CEN standards in their national system and to withdraw conflicting national standards.

In 1987, the Comité Européen de Normalisation (CEN) established CEN/TC 122, "Ergonomics", which is responsible for development of the European ergonomic standards (Dul, et al., 1996). The scope of the CEN/TC 122 is standardization in the field of ergonomics principles and requirements for the design of work systems and work environments, including machinery and personal protective equipment, to promote the health, safety, and well-being of the human operator and the effectiveness of the work (CEN,

Table 9 Published ISO Standards for Ergonomics of the Thermal Environment

Reference Number	Title
ISO 7243:1989	Hot environments: Estimation of the heat stress on working man, based on the WBGT-index (wet bulb globe temperature)
ISO 7726:1998	Ergonomics of the thermal environment: Instruments for measuring physical quantities
ISO 7730:1994	Moderate thermal environments: Determination of the PMV and PPD indices and specification of the conditions for thermal comfort
ISO 7933:1989	Hot environments: Analytical determination and interpretation of thermal stress using calculation of required sweat rate
ISO 8996:1990	Ergonomics: Determination of metabolic heat production
ISO 9886:1992	Evaluation of thermal strain by physiological measurements
ISO 9920:1995	Ergonomics of the thermal environment: Estimation of the thermal insulation and evaporative resistance of a clothing ensemble
ISO 10551:1995	Ergonomics of the thermal environment: Assessment of the influence of the thermal environment using subjective judgment scales
ISO/TR 11079:1993	Evaluation of cold environments: Determination of requisite clothing insulation (IREC)
ISO 11399:1995	Ergonomics of the thermal environment: Principles and application of relevant international standards
ISO 12894:2001	Ergonomics of the thermal environment: Medical supervision of individuals exposed to extreme hot or cold environments
ISO 13731:2001	Ergonomics of the thermal environment: Vocabulary and symbols
ISO/TS 13732-2:2001	Ergonomics of the thermal environment: Methods for the assessment of human responses to contact with surfaces, part 2: Human contact with surfaces at moderate temperature

Table 10 ISO Drafts and Standards under Development for Ergonomics of the Thermal Environment

Reference Number	Title
ISO/DIS 7730	Ergonomics of the thermal environment: Analytical determination and interpretation of thermal comfort using calculation of the PMV and PPD indices and local thermal comfort
ISO/FDIS 7933	Ergonomics of the thermal environment: Analytical determination and interpretation of heat stress using calculation of the predicted heat strain
ISO/FDIS 8996	Ergonomics of the thermal environment: Determination of metabolic rate
ISO/CD 9920	Ergonomics of the thermal environment: Estimation of the thermal insulation and evaporative resistance of a clothing ensemble
ISO/CD 11079	Ergonomics of the thermal environment: Determination and interpretation of cold stress when using required clothing insulation (IREQ) and local cooling effects
ISO/DIS 13732-1	Ergonomics of the thermal environment: Methods for the assessment of human responses to contact with surfaces, part 1: Hot surfaces
ISO/DIS 13732-3	Ergonomics of the thermal environment: Touching of cold surfaces, part 3: Ergonomics data and guidance for application
ISO/CD TS 14415	Ergonomics of the thermal environment: Application of international standards to the disabled, the aged and other handicapped persons
ISO/DIS 14505-1	Ergonomics of the thermal environment: Evaluation of thermal environment in vehicles, part 1: Principles and methods for assessment of thermal stress
ISO/DIS 14505-2	Ergonomics of the thermal environment: Evaluation of thermal environment in vehicles, part 2: Determination of equivalent temperature
ISO/CD 14505-3	Ergonomics of the thermal environment: Thermal environment in vehicles, part 3: Evaluation of thermal comfort using human subjects
ISO 15265	Ergonomics of the thermal environment: Risk assessment strategy for the prevention of stress or discomfort in thermal working conditions
ISO/CD 15743	Ergonomics of the thermal environment: Working practices in cold: Strategy for risk assessment and management

Table 11 ISO Standards for Danger Signals and Communication in Noisy Environments

Reference Number	Title
ISO 7731:1986	Ergonomics: Danger signals for public and work areas — Auditory danger signals
ISO 11428:1996	Ergonomics: Visual danger signals — General requirements, design and testing
ISO 11429:1996	Ergonomics: System of auditory and visual danger and information signals
ISO 9921-1:1996	Ergonomic assessment of speech communication, part 1: Speech interference level and communication distances for persons with normal hearing capacity in direct communication (SIL method)
ISO/TR 19358:2002	Ergonomics: Construction and application of tests for speech technology

Table 12 Organizational Structure of CEN/TC 122

Working Group	Title
CEN/TC 122/WG 1	Anthropometry
CEN/TC 122/WG 2	Ergonomic design principles
CEN/TC 122/WG 3	Surface temperatures
CEN/TC 122/WG 4	Biomechanics
CEN/TC 122/WG 5	Ergonomics of human–computer interaction
CEN/TC 122/WG 6	Signals and controls
CEN/TC 122/WG 8	Danger signals and speech communication in noisy environments
CEN/TC 122/WG 9	Ergonomics of personal protective equipment (PPE)
CEN/TC 122/WG 10	Ergonomic design principles for the operability of mobile machinery
CEN/TC 122/WG 11	Ergonomics of the thermal environment
CEN/TC 122/WG 12	Integrating ergonomic principles for machinery design

2004). The organizational structure of the CEN/TC 122 is presented in Table 12.

The ISO and CEN have signed a formal agreement *Agreement on Technical Cooperation Between ISO and CEN* (the Vienna Agreement) that established close cooperation between these standardization bodies. ISO and CEN decided to harmonize the development of their standards and to cooperate regarding exchange of information and standards drafting. According to this agreement, the ISO standards are adopted by CEN, and vice versa. Table 13 presents published CEN ergonomics standards. Most of the ergonomic standards published by CEN/TC 122 is

Table 13 Published CEN Standards for Ergonomics

CEN Reference	Title	ISO Standard
	Ergonomics Principles	
EN ISO 10075-1:2000	Ergonomic principles related to mental workload, part 1: General terms and definitions	ISO 10075:1991
EN ISO 10075-2:2000	Ergonomic principles related to mental workload, part 2: Design principles	ISO 10075-2:1996
ENV 26385:1990	Ergonomic principles of the design of work systems	ISO 6385: 1981
EN ISO 6385:2004	Ergonomic principles in the design of work systems	ISO 6385:2004
	Anthropometrics and Biomechanics	
EN 1005-1:2001	Safety of machinery: Human physical performance, part 1: Terms and definitions	
EN 1005-2:2003	Safety of machinery: Human physical performance, part 2: Manual handling of machinery and component parts of machinery	
EN 1005-3:2002	Safety of machinery: Human physical performance, part 3: Recommended force limits for machinery operation	
EN 13861:2002	Safety of machinery: Guidance for the application of ergonomics standards in the design of machinery	
EN 547-1:1996	Safety of machinery: Human body measurements, part 1: Principles for determining the dimensions required for openings for whole body access into machinery	
EN 547-2:1996	Safety of machinery: Human body measurements, part 2: Principles for determining the dimensions required for access openings	
EN 547-3:1996	Safety of machinery: Human body measurements, part 3: Anthropometric data	
EN 614-1:1995	Safety of machinery: Ergonomic design principles, part 1: Terminology and general principles	
EN 614-2:2000	Safety of machinery: Ergonomic design principles, part 2: Interactions between the design of machinery and work tasks	
EN ISO 7250:1997	Basic human body measurements for technological design	ISO 7250:1996
EN ISO 14738:2002	Safety of machinery: Anthropometric requirements for the design of workstations at machinery	ISO 14738:2002
EN ISO 15535:2003	General requirements for establishing anthropometric databases	ISO 15535:2003
	Ergonomics Design of Control Centers	
EN ISO 11064-1:2000	Ergonomic design of control centers, part 1: Principles for the design of control centers	ISO 11064-1:2000
EN ISO 11064-2:2000	Ergonomic design of control centers, part 2: Principles for the arrangement of control suites	ISO 11064-2:2000
EN ISO 11064-3:1999	Ergonomic design of control centers, part 3: Control room layout	ISO 11064-3:1999
EN ISO 11064-3:1999/AC:2002	Ergonomic design of control centers, part 3: Control room layout	ISO 11064-3:1999/Cor.1:2002
	Human–System Interaction	
EN ISO 13406-1:1999	Ergonomic requirements for work with visual display based on flat panels, part 1: Introduction	ISO 13406-1:1999
EN ISO 13406-2:2001	Ergonomic requirements for work with visual displays based on flat panels, part 2: Ergonomic requirements for flat panel displays	ISO 13406-2:2001
EN ISO 13407:1999	Human-centered design processes for interactive systems	ISO 13407:1999
EN ISO 13731:2001	Ergonomics of the thermal environment: Vocabulary and symbols	ISO 13731:2001
EN ISO 14915-1:2002	Software ergonomics for multimedia user interfaces, part 1: Design principles and framework	ISO 14915-1:2002
EN ISO 14915-2:2003	Software ergonomics for multimedia user interfaces, part 2: Multimedia navigation and control	ISO 14915-2:2003
EN ISO 14915-3:2002	Software ergonomics for multimedia user interfaces, part 3: Media selection and combination	ISO 14915-3:2002
EN ISO 9921:2003	Ergonomics: Assessment of speech communication	ISO 9921:2003

Table 13 ***(continued)***

CEN Reference	Title	ISO Standard
	Danger Signals	
EN 457:1992	Safety of machinery: Auditory danger signals — General requirements, design and testing	ISO 7731:1986, modified
EN 842:1996	Safety of machinery: Visual danger signals — General requirements, design and testing	
EN 981:1996	Safety of machinery: System of auditory and visual danger and information signals	
	Thermal Environments	
EN 12515:1997	Hot environments: Analytical determination and interpretation of thermal stress using calculation of required sweat rate	ISO 7933:1989 modified
EN 27243:1993	Hot environments: estimation of the heat stress on working man, based on the WBGT-index (wet bulb globe temperature)	ISO 7243:1989
EN 28996:1993	Ergonomics: Determination of metabolic heat production	ISO 8996:1990
EN ISO 10551:2001	Ergonomics of the thermal environment: Assessment of the influence of the thermal environment using subjective judgement scales	ISO 10551:1995
EN ISO 11399:2000	Ergonomics of the thermal environment: Principles and application of relevant international standards	ISO 11399:1995
EN ISO 12894:2001	Ergonomics of the thermal environment: Medical supervision of individuals exposed to extreme hot or cold environments	ISO 12894:2001
EN ISO 7726:2001	Ergonomics of the thermal environment: Instruments for measuring physical quantities	ISO 7726:1998
EN ISO 7730:1995	Moderate thermal environments: Determination of the PMV and PPD indices and specification of the conditions for thermal comfort	ISO 7730:1994
EN ISO 9886:2001	Evaluation of thermal strain by physiological measurements	ISO 9886:1992
EN ISO 9886:2004	Ergonomics: Evaluation of thermal strain by physiological measurements	ISO 9886:2004
EN ISO 9920:2003	Ergonomics of the thermal environment: Estimation of the thermal insulation and evaporative resistance of a clothing ensemble	ISO 9920:1995
ENV ISO 11079:1998	Evaluation of cold environments: Determination of required clothing insulation (REQ)	ISO/TR 11079:1993
EN 13202:2000	Ergonomics of the thermal environment: Temperatures of touchable hot surfaces — Guidance for establishing surface temperature limit values in production standards with the aid of EN 563	
EN 563:1994	Safety of machinery: Temperatures of touchable surfaces — Ergonomics data to establish temperature limit values for hot surfaces	
EN 563:1994/A1:1999	Safety of machinery: Temperatures of touchable surfaces — Ergonomics data to establish temperature limit values for hot surfaces	
EN 563:1994/A1:1999/AC:2000	Safety of machinery: Temperatures of touchable surfaces — Ergonomics data to establish temperature limit values for hot surfaces	
EN 563:1994/AC:1994	Safety of machinery: Temperatures of touchable surfaces — Ergonomics data to establish temperature limit values for hot surfaces	
	Displays and Control Actuators	
EN 894-1:1997	Safety of machinery: Ergonomics requirements for the design of displays and control actuators, part 1: General principles for human interactions with displays and control actuators	
EN 894-2:1997	Safety of machinery: Ergonomics requirements for the design of displays and control actuators, part 2: Displays	
EN 894-3:2000	Safety of machinery: Ergonomics requirements for the design of displays and control actuators, part 3: Control actuators	

Table 14 CEN Standards for Ergonomics under Development

Reference Number	Title	DAV
prEN ISO 13732-3	Ergonomics of the thermal environment: Methods for the assessment of human responses to contact with surfaces, part 3: Cold surfaces (ISO/FDIS 13732-3:2004)	2005-01
prEN 1005-4	Safety of machinery: Human physical performance, part 4: Evaluation of working postures and movements in relation to machinery	2003-09
CEN/TC 122 N 291	Personal protective equipment: Ergonomic principles, part 2: Application of anthropometric measurements in design and specification	2002-09
prEN 13921-3	Personal protective equipment: Ergonomic principles, part 3: Biomechanical characteristics	2003-10
prEN 13921-4	Personal protective equipment: Ergonomic principles, part 4: Thermal characteristics	2003-10
prEN 13921-6	Personal protective equipment: Ergonomic principles, part 6: Sensory factors	2003-10
prEN 14386	Safety of machinery: Ergonomic design principles for the operability of mobile machinery	2005-02
prEN ISO 15537	Principles for selecting and using test persons for testing anthropometric aspects of industrial products and designs (ISO/DIS 15537:2002)	2004-05
	Safeguarding crushing points by means of limitation of the active forces	1998-11
prEN 894-4	Safety of machinery: Ergonomic requirements for the design of displays and control actuators, part 4: Location and arrangement of displays and control actuators	2003-04
EN ISO 11064-4:2004	Ergonomic design of control centres, part 4: Layout and dimensions of workstations (ISO 11064-4:2004)	2004-07
ISO/CD 11064-5	Ergonomic design of control centres, part 5: Displays and controls	2003-10
prEN ISO 11064-6	Ergonomic design of control centres, part 6: Environmental requirements for control centres (ISO/DIS 11064-6:2003)	2003-10
ISO/CD 11064-7	Ergonomic design of control centres, part 7: Principles for the evaluation of control centres	2003-10
prEN 13921-1	Personal protective equipment: Ergonomic principles, part 1: General guidance	2003-10
ISO/NP 12892	Reach envelopes	2003-01
prEN 1005-5	Safety of machinery: Human physical performance, part 5: Risk assessment for repetitive handling at high frequency	2005-04
prEN ISO 15536-1	Ergonomics: Computer manikins and body templates, part 1: General requirements (ISO/DIS 15536-1:2002)	2003-12
prEN ISO 15536-2	Ergonomics: Computer manikins and body templates, part 2: Verification of function and validation of dimensions for computer manikin systems	2002-10
prEN ISO 10075-3	Ergonomic principles related to mental workload, part 3: Principles and requirements concerning methods for measuring and assessing mental workload	2004-10
prEN 614-1 rev	Safety of machinery: Ergonomic design principles, part 1: Terminology and general principles	2004-01
prEN ISO 8996 rev	Ergonomics of the thermal environment: Determination of metabolic rate	2004-10
prEN ISO 7933	Ergonomics of the thermal environment: Analytical determination and interpretation of heat stress using calculation of the predicted heat strain	2004-10
ISO/NP 15743	Ergonomics of the thermal environment: Working practices for cold indoor environments	2003-07
prEN ISO 7730 rev	Ergonomics of the thermal environment: Analytical determination and interpretation of thermal comfort using calculation of the PMV and PPD indices and local thermal comfort (ISO/DIS 7730:2003)	2005-05
prEN ISO 11079	Evaluation of cold environments: Determination of required clothing insulation (IREQ) (will replace ENV ISO 11079:1998)	2003-03
prEN ISO 20685	3D scanning methodologies for internationally compatible anthropometric databases (ISO/DIS 20685:2004)	2005-06
prEN ISO 15265	Ergonomics of the thermal environment: Risk assessment strategy for the prevention of stress or discomfort in thermal working conditions	2004-09
prEN ISO 13732-1	Ergonomics of the thermal environment: Methods for the assessment of human responses to contact with surfaces, part 1: Hot surfaces (ISO/DIS 13732-1:2004)	2006-01
prEN ISO 23973	Software ergonomics for World Wide Web user interfaces	2006-01
prEN ISO 14505-1	Ergonomics of the thermal environment: Evaluation of thermal environment in vehicles, part 1: Principles and methods for assessment of thermal stress	2006-02

Table 14 (*continued*)

Reference Number	Title	DAV
prEN ISO 14505-2	Ergonomics of the thermal environment: Evaluation of thermal environment in vehicles, part 2: Determination of equivalent temperature	2006-02
	Ergonomics of the thermal environment: Application of international standards to the disabled, the aged and other handicapped persons	2004-12
prEN ISO 9920 rev	Ergonomics of the thermal environment: Estimation of the thermal insulation and evaporative resistance of a clothing ensemble	2006-05
	Ergonomics of the thermal environment: Thermal environments in vehicles, part 3: Evaluation of thermal comfort using human subjects (ISO/CD 14505-3)	2007-01
	Ergonomics of human–system interaction: Ergonomic requirements and measurement techniques for electronic visual displays — Introduction	2007-04
	Ergonomics of human–system interaction: Ergonomic requirements and measurement techniques for electronic visual displays — Terminology	2007-04
	Ergonomics of human–system interaction: Ergonomic requirements and measurement techniques for electronic visual displays — Ergonomic requirements	2007-04
	Ergonomics of human–system interaction: Ergonomic requirements and measurement techniques for electronic visual displays — User performance test methods	2007-04
	Ergonomics of human–system interaction: Ergonomic requirements and measurement techniques for electronic visual displays — Optical laboratory test methods	2007-04
	Ergonomics of human–system interaction: Ergonomic requirements and measurement techniques for electronic visual displays — Field assessment methods	2007-04
	Ergonomics of human–system interaction: Ergonomic requirements and measurement techniques for electronic visual displays — Analysis and compliance test methods	2007-04
EN ISO 14738:2002/prAC	Safety of machinery: Anthropometric requirements for the design of workstations at machinery (ISO 14738:2002)	2003-05

adoption, or adaptation, of ISO standards. The CEN ergonomic standards in development are shown in Table 14.

3.1 Other International Standards Related to Ergonomics

For historical and organizational factors, many ISO and CEN standards in the field of ergonomics have not been developed by the technical committees ISO TC 159 and CEN and TC 122. Some ergonomics areas covered by other ISO and CEN technical committees are presented in Table 15. The list of published ISO standards related to the ergonomics area, but developed by groups other than the TC 159 committee, are provided in Table 16.

4 ILO GUIDELINES FOR OCCUPATIONAL SAFETY AND HEALTH MANAGEMENT SYSTEMS

The popularity and success of a systematic and standardized approach to the management systems introduced by the ISO led to the view that this type of approach can also improve the management of occupational safety and health. Following this idea, the International Labor Organization (ILO) developed voluntary guidelines on OSH management systems which reflect ILO values and ensure protection of workers' safety and health (ILO-OSH, 2001). The International Labor Organization (ILO) was founded at the Versailles Congress in 1919 and became a specialized agency of the United Nations (UN) in 1946. The ILO objective is the promotion of social justice and internationally recognized human and labor rights (ILO, 2004). ILO represents the interests of three parties treated equally: employers, employee organizations, and government agencies.

The ILO-OSH (2001) guidelines provide recommendations concerning design and implementation of OSH MS that allows for integration of OSH with the general enterprise management system. The ILO guidelines state that these recommendations are addressed to all who are responsible for the occupational safety and health management. These guidelines are nonmandatory and are not intended to replace national laws and regulations. The ILO-OSH (2001) document distinguished two levels of guideline application: national and organizational. At the national level ILO-OSH (2001) provides recommendations for the establishment of a national framework for occupational safety and health management systems (OSH-MS). The guidelines suggest that this process should be supported by the provision of the relevant national laws and regulations.

Establishment of a national framework for OSH-MS included the following actions (ILO-OSH, 2001): (1) nomination of competent institution(s) for OSH-MS,

Table 15 Ergonomic Areas Covered in Standards Developed by the Other ISO and CEN Technical Committees

Topic	Technical Committee	
	ISO	CEN
Safety of machines	TC 199	TC 114
Vibration and shock	TC 108	TC 211
Noise and acoustics	TC 43	TC 211
Lighting		TC 169
Respiratory protective devices		TC 79
Eye protection		TC 85
Head protection		TC 158
Hearing protection		TC 159
Protection against falls	TC 94	TC 160
Foot and leg protection		TC 161
Protective clothing		TC 162
Radiation protection	TC 85	
Air quality	TC 146	
Assessment and workplace exposure		TC 137
Office machines	TC 95	
Information procession	TC 97	
Road vehicles	TC 22	
Safety color and signs	TC 80	
Graphical symbols	TC 145	

Source: Dul et al. (1996).

(2) formulation of a coherent national policy, and (3) development of national and tailored guidelines. The process of establishment of a national framework for OSH-MS and its components is presented in Figure 2.

At the organizational level ILO-OSH (2001), guidelines establish employer responsibilities regarding occupational safety and health management, and emphasize the importance of compliance with national laws and regulations. ILO-OSH (2001) suggests that OSH management system elements be integrated into overall organizational policy and management strategies actions (ILO-OSH, 2001). The OSH management systems in the organization consist of five main sections: policy, organizing, planning and implementation, evaluation, and action for improvement. These elements correspond to the Demming cycle of plan–do–check–act, internationally accepted as the basis for the systems approach to management. The OSH-MS main sections and their elements are listed in Table 17.

ILO-OSH (2001) guidelines require establishment by the employer of the OSH policy in consultation with workers and their representatives, and define the content of such policy. The ILO-OSH guidelines also indicate the importance of OSH policy integration and compatibility with other management systems in the organization. These guidelines emphasize the necessity of worker participation in the OSH management system in the organization. Therefore, workers should be consulted regarding OSH activities and should be encouraged to participate in OSH-MS, including a safety and health committee. The organizing section of the guidelines underlines the need for allocation of responsibility and accountability for the implementation and performance of the OSH management system to the senior management. This section also includes requirements related to competence and training in the OSH field and defines the necessary documentation and communications activities. The planning and implementation section includes the elements of initial review, system planning, development and implementation, OSH objectives, and hazard prevention. The initial review identifies the actual states of the organization with regards to the OSH and creates the baseline for OSH policy implementation. The evaluation section consists of performance monitoring and measurement, investigation of work-related diseases and incidents, audit, and management review. The guidelines require carrying out internal audits of the OSH-MS according to the policies established. Action for improvement includes the elements of preventive and corrective action and continual improvement. The final section underlines the need for continual improvement of OSH performance through the development of policies, systems, and techniques to prevent and control work-related injuries and diseases.

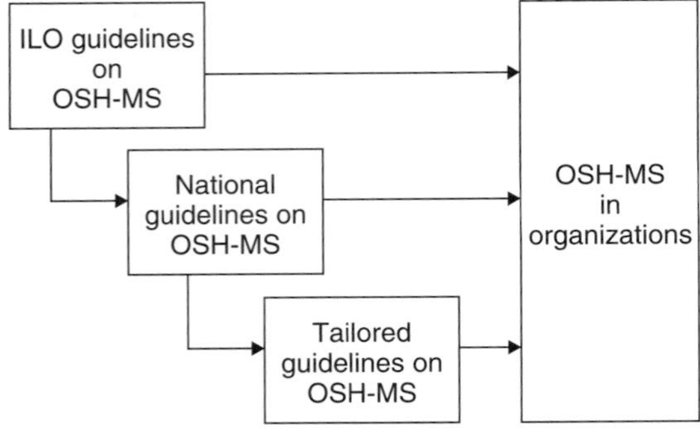

Figure 2 Establishment of a national framework for the OSH-MS. (From ILO-OSH, 2001.)

5 U.S. STANDARDS FOR HUMAN FACTORS AND ERGONOMICS

5.1 U.S. Government Standards

Among the HFE U.S. government standards, two documents are usually mentioned as basic: a military standard providing human engineering design criteria (MIL-STD-1472), and a human–system integration standard (NASA-STD-300) (Chapanis, 1996; McDaniel, 1996). In addition, there are more specific standards that have been developed by such departments as the Department of Defense, Department of Transportation, Department of Energy, and U.S. Nuclear Regulatory Commission. Additionally, a large number of handbooks that contain more detailed and

Table 16 HFE Standards Published by Other Than TC 159 ISO Technical Committees

Reference Number	Title
	CIE: International Commission on Illumination
ISO/CIE 8995:2002	Lighting of indoor workplaces
	JTC 1/SC 6: Telecommunications and Information Exchange between Systems
ISO/IEC 10021-2:2003	Information technology: Message handling systems (MHS) — Overall architecture
	JTC 1/SC 7: Software and System Engineering
ISO/IEC TR 9126-4:2004	Software engineering: Product quality, part 4: Quality in use metrics
ISO/IEC 12119:1994	Information technology: Software packages — Quality requirements and testing
ISO/IEC 12207:1995	Information technology: Software life cycle processes
ISO/IEC 14598-1:1999	Information technology: Software product evaluation, part 1: General overview
ISO/IEC 14598-4:1999	Software engineering: Product evaluation, part 4: Process for acquirers
ISO/IEC 14598-6:2001	Software engineering: Product evaluation, part 6: Documentation of evaluation modules
ISO/IEC 15288:2002	Systems engineering: System life cycle processes
ISO/IEC TR 15504-5:1999	Information technology: Software process assessment, part 5: An assessment model and indicator guidance
ISO/IEC 15910:1999	Information technology: Software user documentation process
ISO/IEC 18019:2004	Software and system engineering: Guidelines for the design and preparation of user documentation for application software
ISO/IEC TR 19760:2003	Systems engineering: A guide for the application of ISO/IEC 15288 (System life cycle processes)
ISO/IEC 20926:2003	Software engineering: IFPUG 4.1 unadjusted functional size measurement method — Counting practices manual
ISO/IEC 20968:2002	Software engineering: Mk II function point analysis — Counting practices manual
	JTC 1/SC 22: Programming Languages, Their Environments and System Software Interfaces
ISO/IEC TR 11017:1998	Information technology: Framework for internationalization
ISO/IEC TR 14252:1996	nformation technology: Guide to the POSIX Open System Environment
ISO/IEC TR 15942:2000	Information technology: Programming languages — Guide for the use of the Ada programming language in high integrity systems
	JTC 1/SC 27: IT Security Techniques
ISO/IEC TR 13335-4:2000	Information technology: Guidelines for the management of IT security, part 4: Selection of safeguards
ISO/IEC 21827:2002	Information technology: Systems security engineering — Capability maturity model
	JTC 1/SC 35: User Interfaces
ISO/IEC 15411:1999	Information technology: Segmented keyboard layouts
ISO/IEC 18035:2003	Information technology: Icon symbols and functions for controlling multimedia software applications
	TC 8/SC 5: Ships' Bridge Layout
ISO 8468:1990	Ship's bridge layout and associated equipment: Requirements and guidelines
ISO 14612:2004	Ships and marine technology: Ship's bridge layout and associated equipment — Additional requirements and guidelines for centralized and integrated bridge functions
	TC 8/SC 6: Navigation
ISO 16273:2003	Ships and marine technology: Night vision equipment for high-speed craft — Operational and performance requirements, methods of testing and required test results
	TC 20: Aircraft and Space Vehicles
ISO/TR 10201:2001	Aerospace: Standards for electronic instruments and systems
	TC 20/SC 1: Aerospace Electrical Requirements
ISO 6858:1982	Aircraft: Ground support electrical supplies — General requirements
	TC 20/SC 14: Space Systems and Operations
ISO 16091:2002	Space systems: Integrated logistic support
ISO 17399:2003	Space systems: Man–systems integration
	TC 21/SC 3: Fire Detection and Alarm Systems
ISO 12239:2003	Fire detection and fire alarm systems: Smoke alarms

Table 16 (continued)

Reference Number	Title
	TC 22/SC 3: Electrical and Electronic Equipment
ISO 11748-2:2001	Road vehicles: Technical documentation of electrical and electronic systems, part 2: Documentation agreement
ISO/TR 15497:2000	Road vehicles: Development guidelines for vehicle based software
	TC 22/SC 13: Ergonomics Applicable to Road Vehicles
ISO 2575:2004	Road vehicles: Symbols for controls, indicators and tell-tales
ISO 3958:1996	Passenger cars: Driver hand-control reach
ISO 4040:2001	Road vehicles: Location of hand controls, indicators and tell-tales in motor vehicles
ISO 6549:1999	Road vehicles: Procedure for H- and R-point determination
ISO/TR 9511:1991	Road vehicles: Driver hand-control reach — In-vehicle checking procedure
ISO/TS 12104:2003	Road vehicles: Gearshift patterns — Manual transmissions with power-assisted gear change and automatic transmissions with manual-gearshift mode
ISO 12214:2002	Road vehicles: Direction-of-motion stereotypes for automotive hand controls
ISO 15005:2002	Road vehicles: Ergonomic aspects of transport information and control systems — Dialogue management principles and compliance procedures
ISO 15007-1:2002	Road vehicles: Measurement of driver visual behaviour with respect to transport information and control systems, part 1: Definitions and parameters
ISO/TS 15007-2:2001	Road vehicles: Measurement of driver visual behaviour with respect to transport information and control systems, part 2: Equipment and procedures
ISO 15008:2003	Road vehicles: Ergonomic aspects of transport information and control systems — Specifications and compliance procedures for in-vehicle visual presentation
ISO/TS 16951:2004	Road vehicles: Ergonomic aspects of transport information and control systems (TICS) — Procedures for determining priority of on-board messages presented to drivers
ISO 17287:2003	Road vehicles: Ergonomic aspects of transport information and control systems — Procedure for assessing suitability for use while driving
	TC 22/SC 17: Visibility
ISO 7397-1:1993	Passenger cars: Verification of driver's direct field of view, part 1: Vehicle positioning for static measurement
ISO 7397-2:1993	Passenger cars: Verification of driver's direct field of view, part 2: Test method
	TC 23/SC 3: Safety and Comfort of the Operator
ISO 4254-1:1989	Tractors and machinery for agriculture and forestry: Technical means for ensuring safety, part 1: General
ISO/TS 15077:2002	Tractors and self-propelled machinery for agriculture and forestry: Operator controls — Actuating forces, displacement, location and method of operation
	TC 23/SC 4: Tractors
ISO 4253:1993	Agricultural tractors: Operator's seating accommodation — Dimensions
ISO 5721:1989	Tractors for agriculture: Operator's field of vision
	TC 23/SC 7: Equipment for Harvesting and Conservation
ISO 8210:1989	Equipment for harvesting: Combine harvesters — Test procedure
	TC 23/SC 14: Operator Controls, Operator Symbols and Other Displays, Operator Manuals
ISO 3767-1:1998	Tractors, machinery for agriculture and forestry, powered lawn and garden equipment: Symbols for operator controls and other displays, part 1: Common symbols
ISO 3767-2:1991	Tractors, machinery for agriculture and forestry, powered lawn and garden equipment: Symbols for operator controls and other displays, part 2: Symbols for agricultural tractors and machinery
ISO 3767-3:1995	Tractors, machinery for agriculture and forestry, powered lawn and garden equipment: Symbols for operator controls and other displays, part 3: Symbols for powered lawn and garden equipment
ISO 3767-5:1992	Tractors, machinery for agriculture and forestry, powered lawn and garden equipment: Symbols for operator controls and other displays, part 5: Symbols for manual portable forestry machinery
	TC 23/SC 15: Machinery for Forestry
ISO 11850:2003	Machinery for forestry: Self-propelled machinery — Safety requirements

Table 16 ***(continued)***

Reference Number	Title
	TC 23/SC 17: Manually Portable Forest Machinery
ISO 8334:1985	Forestry machinery: Portable chain-saws — Determination of balance
ISO 11680-1:2000	Machinery for forestry: Safety requirements and testing for pole-mounted powered pruners, part 1: Units fitted with an integral combustion engine
ISO 11680-2:2000	Machinery for forestry: Safety requirements and testing for pole-mounted powered pruners, part 2: Units for use with a back-pack power source
ISO 11681-1:2004	Machinery for forestry: Portable chain-saw safety requirements and testing, part 1: Chain-saws for forest service
ISO 11681-2:1998	Machinery for forestry: Portable chain-saws Safety requirements and testing, part 2: Chain-saws for tree service
ISO 11806:1997	Agricultural and forestry machinery: Portable hand-held combustion engine driven brush cutters and grass trimmers — Safety
ISO 14740:1998	Forest machinery: Backpack power units for brush-cutters, grass-trimmers, pole-cutters and similar appliances — Safety requirements and testing
	TC 23/SC 18: Irrigation and Drainage Equipment and Systems
ISO/TR 8059:1986	Irrigation equipment: Automatic irrigation systems — Hydraulic control
	TC 38: Textiles
ISO 15831:2004	Clothing: Physiological effects — Measurement of thermal insulation by means of a thermal manikin
	TC 43/SC 1: Noise
ISO 11690-1:1996	Acoustics: Recommended practice for the design of low-noise workplaces containing machinery, part 1: Noise control strategies
ISO 15667:2000	Acoustics: Guidelines for noise control by enclosures and cabins
	TC 46: Information and Documentation
ISO 7220:1996	Information and documentation: Presentation of catalogues of standards
	TC 59/SC 3: Functional/User Requirements and Performance in Building Construction
ISO 6242-1:1992	Building construction: Expression of users' requirements, part 1: Thermal requirements
ISO 6242-2:1992	Building construction: Expression of users' requirements, part 2: Air purity requirements
ISO 6242-3:1992	Building construction: Expression of users' requirements, part 3: Acoustical requirements
	TC 67: Materials, Equipment and Offshore Structures for Petroleum, Petrochemical and Natural Gas Industries
ISO 13879:1999	Petroleum and natural gas industries: Content and drafting of a functional specification
ISO 13880:1999	Petroleum and natural gas industries: Content and drafting of a technical specification
	TC 67/SC 6: Processing Equipment and Systems
ISO 13702:1999	Petroleum and natural gas industries: Control and mitigation of fires and explosions on offshore production installations — Requirements and guidelines
ISO 15544:2000	Petroleum and natural gas industries: Offshore production installations — Requirements and guidelines for emergency response
ISO 17776:2000	Petroleum and natural gas industries: Offshore production installations — Guidelines on tools and techniques for hazard identification and risk assessment
	TC 69: Applications of Statistical Methods
ISO 10725:2000	Acceptance sampling plans and procedures for the inspection of bulk materials
	TC 72/SC 5: Industrial Laundry and Dry-Cleaning Machinery and Accessories
ISO 8230:1997	Safety requirements for dry-cleaning machines using perchloroethylene
ISO 10472-1:1997	Safety requirements for industrial laundry machinery, part 1: Common requirements
	TC 72/SC 8: Safety Requirements for Textile Machinery
ISO 11111:1995	Safety requirements for textile machinery
	TC 85/SC 2: Radiation Protection
ISO 17874-1:2004	Remote handling devices for radioactive materials, part 1: General requirements

(*continued overleaf*)

Table 16 *(continued)*

Reference Number	Title
	TC 92/SC 3: Fire Threat to People and Environment
ISO/TS 13571:2002	Life-threatening components of fire: Guidelines for the estimation of time available for escape using fire data
ISO 10068:1998	Mechanical vibration and shock: Free, mechanical impedance of the human hand–arm system at the driving point
ISO 13090-1:1998	Mechanical vibration and shock: Guidance on safety aspects of tests and experiments with people, part 1: Exposure to whole-body mechanical vibration and repeated shock
	TC 92/SC 4: Fire Safety Engineering
ISO/TR 13387-1:1999	Fire safety engineering, part 1: Application of fire performance concepts to design objectives
	TC 94/SC 4: Personal Equipment for Protection Against Falls
ISO 10333-6:2004	Personal fall-arrest systems, part 6: System performance tests
	TC 94/SC 13: Protective Clothing
ISO 11393-4:2003	Protective Clothing for Users of Hand-Held Chain-Saws, Part 4: Test Methods and Performance Requirements for Protective Gloves
ISO 13688:1998	Protective clothing: General requirements
ISO 16603:2004	Clothing for protection against contact with blood and body fluids: Determination of the resistance of protective clothing materials to penetration by blood and body fluids — Test method using synthetic blood
ISO 16604:2004	Clothing for protection against contact with blood and body fluids: Determination of resistance of protective clothing materials to penetration by blood-borne pathogens — Test method using Phi-X 174 bacteriophage
	TC 101: Continuous Mechanical Handling Equipment
ISO/TR 5045:1979	Continuous mechanical handling equipment: Safety code for belt conveyors — Examples for guarding of nip points
	TC 108/SC 2: Measurement and Evaluation of Mechanical Vibration and Shock as Applied to Machines, Vehicles and Structures
ISO 14964:2000	Mechanical vibration and shock: Vibration of stationary structures — Specific requirements for quality management in measurement and evaluation of vibration
	TC 108/SC 4: Human Exposure to Mechanical Vibration and Shock
ISO 2631-1:1997	Mechanical vibration and shock: Evaluation of human exposure to whole-body vibration, part 1: General requirements
ISO 2631-2:2003	Mechanical vibration and shock: Evaluation of human exposure to whole-body vibration, part 2: Vibration in buildings (1 Hz to 80 Hz)
ISO 2631-4:2001	Mechanical vibration and shock: Evaluation of human exposure to whole-body vibration, part 4: Guidelines for the evaluation of the effects of vibration and rotational motion on passenger and crew comfort in fixed-guideway transport systems
ISO 2631-5:2004	Mechanical vibration and shock: Evaluation of human exposure to whole-body vibration, part 5: Method for evaluation of vibration containing multiple shocks
ISO 5349-1:2001	Mechanical vibration: Measurement and evaluation of human exposure to hand-transmitted vibration, Part 1: General requirements
	TC 108/SC 4: Human Exposure to Mechanical Vibration and Shock
ISO 5982:2001	Mechanical vibration and shock: Range of idealized values to characterize seated-body biodynamic response under vertical vibration
ISO 6897:1984	Guidelines for the evaluation of the response of occupants of fixed structures, especially buildings and off-shore structures, to low-frequency horizontal motion (0,063 to 1 Hz)
ISO 8727:1997	Mechanical vibration and shock: Human exposure — Biodynamic coordinate systems
ISO 9996:1996	Mechanical vibration and shock: Disturbance to human activity and performance — Classification
ISO 13091-1:2001	Mechanical vibration — Vibrotactile perception thresholds for the assessment of nerve dysfunction, part 1: Methods of measurement at the fingertips

Table 16 (*continued*)

Reference Number	Title
ISO 13091-2:2003	Mechanical vibration — Vibrotactile perception thresholds for the assessment of nerve dysfunction, part 2: Analysis and interpretation of measurements at the fingertips
	TC 121/SC 1: Breathing Attachments and Anaesthetic Machines
ISO 7767:1997	Oxygen monitors for monitoring patient breathing mixtures: Safety requirements
	TC 121/SC 3: Lung Ventilators and Related Equipment
ISO 8185:1997	Humidifiers for medical use: General requirements for humidification systems
IEC 60601-1-8:2003	Medical electrical equipment, part 1–8: General requirements for safety — Collateral standard: General requirements, tests and guidance for alarm systems in medical electrical equipment and medical electrical systems
IEC 60601-2-12:2001	Medical electrical equipment, part 2–12: Particular requirements for the safety of lung ventilators — Critical care ventilators
	TC 123/SC 5: Quality Analysis and Assurance
ISO 12307-1:1994	Plain bearings: Wrapped bushes, part 1: Checking the outside diameter
	TC 127/SC 1: Test Methods Relating to Machine Performance
ISO 8813:1992	Earth-moving machinery: Lift capacity of pipelayers and wheeled tractors or loaders equipped with side boom
	TC 127/SC 2: Safety Requirements and Human Factors
ISO 2860:1992	Earth-moving machinery: Minimum access dimensions
ISO 2867:1994	Earth-moving machinery: Access systems
ISO 3164:1995	Earth-moving machinery: Laboratory evaluations of protective structures — Specifications for deflection-limiting volume
ISO 3411:1995	Earth-moving machinery: Human physical dimensions of operators and minimum operator space envelope
ISO 3449:1992	Earth-moving machinery: Falling-object protective structures — Laboratory tests and performance requirements
ISO 3450:1996	Earth-moving machinery: Braking systems of rubber-tyred machines — Systems and performance requirements and test procedures
ISO 3457:2003	Earth-moving machinery: Guards — Definitions and requirements
ISO 3471:1994	Earth-moving machinery: Roll-over protective structures — Laboratory tests and performance requirements
ISO 3471:1994/Amd 1:1997	Laboratory tests and performance requirements
ISO 5006-2:1993	Earth-moving machinery: Operator's field of view, part 2: Evaluation method
ISO 5006-3:1993	Earth-moving machinery: Operator's field of view, part 3: Criteria
ISO 5010:1992	Earth-moving machinery: Rubber-tyred machines — Steering requirements
ISO 5353:1995	Earth-moving machinery, and tractors and machinery for agriculture and forestry: Seat index point
ISO 6682:1986	Earth-moving machinery: Zones of comfort and reach for controls
ISO 7096:2000	Earth-moving machinery: Laboratory evaluation of operator seat vibration
ISO 8643:1997	Earth-moving machinery: Hydraulic excavator and backhoe loader boom-lowering control device — Requirements and tests
ISO 9244:1995	Earth-moving machinery: Safety signs and hazard pictorials — General principles
ISO/TR 9953:1996	Earth-moving machinery: Warning devices for slow-moving machines — Ultrasonic and other systems
ISO 10262:1998	Earth-moving machinery: Hydraulic excavators — Laboratory tests and performance requirements for operator protective guards
ISO 10567:1992	Earth-moving machinery: Hydraulic excavators — Lift capacity
ISO 10570:2004	Earth-moving machinery: Articulated frame lock — Performance requirements
ISO 10263-1:1994	Earth-moving machinery: Operator enclosure environment, part 1: General and definitions
ISO 10263-2:1994	Earth-moving machinery: Operator enclosure environment, part 2: Air filter test
ISO 10263-3:1994	Earth-moving machinery: Operator enclosure environment, part 3: Operator enclosure pressurization test method
ISO 10263-4:1994	Earth-moving machinery: Operator enclosure environment, part 4: Operator enclosure ventilation, heating and/or air-conditioning test method

(*continued overleaf*)

Table 16 (*continued*)

Reference Number	Title
ISO 10263-5:1994	Earth-moving machinery: Operator enclosure environment, part 5: Windscreen defrosting system test method
ISO 10263-6:1994	Earth-moving machinery: Operator enclosure environment, part 6: Determination of effect of solar heating on operator enclosure
ISO 10533:1993	Earth-moving machinery: Lift-arm support devices
ISO 10968:1995	Earth-moving machinery: Operator's controls
ISO 11112:1995	Earth-moving machinery: Operator's seat — Dimensions and requirements
ISO 12117:1997	Earth-moving machinery: Tip-over protection structure (TOPS) for compact excavators — Laboratory tests and performance requirements
ISO 12508:1994	Earth-moving machinery: Operator station and maintenance areas — Bluntness of edges
ISO 13333:1994	Earth-moving machinery: Dumper body support and operator's cab tilt support devices
ISO 13459:1997	Earth-moving machinery: Dumpers — Trainer seat/enclosure
ISO 17063:2003	Earth-moving machinery: Braking systems of pedestrian-controlled machines — Performance requirements and test procedures
	TC 130: Graphic Technology
ISO 12648:2003	Graphic technology: Safety requirements for printing press systems
ISO 12649:2004	Graphic technology: Safety requirements for binding and finishing systems and equipment
	TC 131/SC 9: Installations and Systems
ISO 4413:1998	Hydraulic fluid power: General rules relating to systems
ISO 4414:1998	Pneumatic fluid power: General rules relating to systems
	TC 136: Furniture
ISO 5970:1979	Furniture: Chairs and tables for educational institutions — Functional sizes
	TC 163/SC 2: Calculation Methods
ISO 13790:2004	Thermal performance of buildings: Calculation of energy use for space heating
	TC 171/SC 2: Application Issues
ISO/TR 14105:2001	Electronic imaging: Human and organizational issues for successful electronic image management (EIM) implementation
	TC 172/SC 9: Electro-Optical Systems
ISO 11553:1996	Safety of machinery: Laser processing machines — Safety requirements
	TC 173: Assistive Products for Persons with Disability
ISO 11199-1:1999	Walking aids manipulated by both arms: Requirements and test methods, part 1: Walking frames
ISO 11199-2:1999	Walking aids manipulated by both arms: Requirements and test methods, part 2: Rollators
ISO 11334-1:1994	Walking aids manipulated by one arm: Requirements and test methods, part 1: Elbow crutches
ISO 11334-4:1999	Walking aids manipulated by one arm: Requirements and test methods, part 4: Walking sticks with three or more legs
	TC 173/SC 3: Aids for Ostomy and Incontinence
ISO 15621:1999	Urine-absorbing aids: General guidance on evaluation
	TC 173/SC 6: Hoists for Transfer of Persons
ISO 10535:1998	Hoists for the transfer of disabled persons: Requirements and test methods
	TC 176/SC 1: Concepts and Terminology
ISO 9000:2000	Quality management systems: Fundamentals and vocabulary
	TC 176/SC 2: Quality Systems
ISO 9004:2000	Quality management systems: Guidelines for performance improvements

Table 16 ***(continued)***

Reference Number	Title
	TC 178: Lifts, Escalators and Moving Walks
ISO/TS 14798:2000	Lifts (elevators), escalators and passenger conveyors: Risk analysis methodology
	TC 184: Industrial Automation Systems and Integration
ISO 11161:1994	Industrial automation systems: Safety of integrated manufacturing systems — Basic requirements
	TC 184/SC 4: Industrial Data
ISO 10303-214:2003	Industrial automation systems and integration: Product data representation and exchange, part 214: Application protocol — Core data for automotive mechanical design processes
	TC 184/SC 5: Architecture, Communications and Integration Frameworks
ISO 15704:2000	Industrial automation systems: Requirements for enterprise-reference architectures and methodologies
ISO 16100-1:2002	Industrial automation systems and integration: Manufacturing software capability profiling for interoperability, part 1: Framework
	TC 188: Small Craft
ISO 15027-3:2002	Immersion suits, part 3: Test methods
	TC 199: Safety of Machinery
ISO 12100-2:2003	Safety of machinery: Basic concepts, general principles for design, part 2: Technical principles
ISO 13849-1:1999	Safety of machinery: Safety-related parts of control systems, part 1: General principles for design
ISO 13851:2002	Safety of machinery: Two-hand control devices — Functional aspects and design principles
ISO 13856-1:2001	Safety of machinery: Pressure-sensitive protective devices, part 1: General principles for design and testing of pressure-sensitive mats and pressure-sensitive floors
ISO 14121:1999	Safety of machinery: Principles of risk assessment
ISO 14123-2:1998	Safety of machinery: Reduction of risks to health from hazardous substances emitted by machinery, part 2: Methodology leading to verification procedures
ISO/TR 18569:2004	Safety of machinery: Guidelines for the understanding and use of safety of machinery standards
	TC 204: Intelligent Transport Systems
ISO 15623:2002	Transport information and control systems: Forward vehicle collision warning systems — Performance requirements and test procedures
	TC 210: Quality Management and Corresponding General Aspects for Medical Devices
ISO 14969:1999	Quality systems: Medical devices — Guidance on the application of ISO 13485 and ISO 13488
ISO 14971:2000	Medical devices: Application of risk management to medical devices
	TC 212: Clinical Laboratory Testing and in Vitro Diagnostic Test Systems
ISO 15190:2003	Medical laboratories: Requirements for safety
ISO 15197:2003	In vitro diagnostic test systems: Requirements for blood-glucose monitoring systems for self-testing in managing diabetes mellitus
	TMB: Technical Management Board
IWA 1:2001	Quality management systems: Guidelines for process improvements in health service organizations
ISO/IEC Guide 50:2002	Safety aspects: Guidelines for child safety
ISO/IEC Guide 71:2001	Guidelines for standards developers to address the needs of older persons and persons with disabilities
	CASCO: Committee on Conformity Assessment
ISO/IEC 17025:1999	General requirements for the competence of testing and calibration laboratories

descriptive information concerning human factor and ergonomics guidelines, preferred practices, methodology, and reference data that may be needed during the design of equipment and systems have also been developed. The handbooks provide assistance in the use and application of relevant government standards during the design process.

5.1.1 Military Standards

The set of consensus military standards was developed by human factors engineers from the U.S. military's three services (Army, Navy, and Air Force), industry, and technical societies (McDaniel, 1996). As a result of standardization reform in the late 1990s, most of the single-service standards were canceled and were integrated into a few Department of Defense standards and handbooks. However, the distinction between two main categories of human factors military standards—general (MIL-STD-1472 and related handbooks) and aircraft (JSSG 2010 and related handbooks)—remain unchanged, which reflects the criticality of aircraft design. The list of the main military standards and handbooks are presented in Table 18.

The basic human engineering principles, design criteria, and practices required for integration of humans with systems and facilities are established in MIL-STD-1472F, *Human Engineering Design Criteria for Military Systems, Equipment and Facilities.* This standard document can be applied to the design of all systems, subsystems, equipment, and facilities, not only military but commercial as well. MIL-STD-1472F includes requirements for displays, controls, control-display integration, anthropometry, ground workspace design, environment, design for maintainability, design of equipment for remote handling, small systems and equipment, operational and maintenance ground/shipboard vehicles, hazards and safety, aerospace vehicle compartment design requirements, and human–computer interface. MIL-STD-1472 also includes nongovernmental standards ANSI/HFS 100 on visual display terminal (VDT) workstations. After standardization reform the design data and information part of MIL-STD-1472F was removed and inserted into MIL-HDBK-759.

Another important military standard document is MIL-HDBK-46855, *Human Engineering Requirements for Military Systems Equipment and Facilities.* This handbook presents human engineering program tasks, procedures, and preferred practices. MIL-HDBK-46855 covers such topics as analysis functions, including human performance parameters, equipment capabilities, and task environments design; test and evaluation; workload analysis; dynamic simulation; and data requirements. This handbook also adopted materials from DOD-HDBK-763, *Human Engineering Procedures Guide*, concerned with human engineering methods and tools, which remained stable over time. The newest rapidly evolving automated human engineering tools are not described in MIL-HDBK-46855 but can be found at Directory of Design Support Methods (DSSM) on the MATRIS Web site (http://dtica.dtic.mil/ddsm/).

Table 17 ILO-OSH-MS Main Sections and Their Elements

Section	Elements
Policy	
	3.1. Occupational safety and health policy
	3.2. Worker participation
Organizing	
	3.3. Responsibility and accountability
	3.4. Competence and training
	3.5. OSH management system documentation
	3.5. Communication
Planning and implementation	3.6. Initial review
	3.7. System planning and implementation
	3.8. Occupational safety and health objectives
	3.9. Hazard prevention
Evaluation	3.10. Performance monitoring and measurement
	3.11. Investigation of work-related incidents and their impact on BHP
	3.12. Audit
	3.13. Management review
Action for improvement	3.15. Preventive and corrective action
	3.16. Continual improvement

Source: ILO-OSH (2001).

Other military standards cover such topics as standard practice for conducting system safety (MIL-STD-882D); acoustical noise limits, testing requirements, and measurement techniques (MIL-STD-1474D); physical characteristics of symbols for army systems displays (MIL-STD-1477C); and symbology requirements for aircraft displays (MIL-STD-1787C). The definitions for all human factors standard documents are provided in MIL-HDBK-1908B, *Department of Defense Handbook: Definitions of Human Factors Terms.*

5.1.2 Other Government Standards

The lists of other government standards are provided in Table 19. NASA-STD-3000 provides generic requirements for space facilities and related equipment important for proper human–system integration. This document is integrated with the Web site, which also offers video images from space missions that illustrate human factors design issues. This standard document is not

Table 18 Military Standards and Handbooks for Human Factors and Ergonomics

Document Number	Title	Date	Source
Standards			
MIL-STD-882D	Standard practice for system safety	2000	http://assist.daps.dla.mil/docimages/0001/95/78/std882d.pd8
MIL-STD-1472F	Human engineering	1999	http://assist.daps.dla.mil/docimages/0001/87/31/milstd14.pd1
MIL-STD-1474D	Noise limits	1997	http://assist.daps.dla.mil/docimages/0000/31/59/1474d.pd1
MIL-STD-1477C	Symbols for army systems displays	1996	http://assist.daps.dla.mil/docimages/0000/42/03/69268.pd9
MIL-STD-1787C	Aircraft display symbology	2001	Controlled distribution document
Handbooks			
DOD-HDBK-743A	Anthropometry of U.S. military personnel	1991	http://assist.daps.dla.mil/docimages/0000/40/29/54083.pd0
MIL-HDBK-759C	Human engineering design guidelines	1995	http://assist.daps.dla.mil/docimages/0000/40/04/mh759c.pd8
MIL-HDBK-767	Design guidance for interior noise reduction in light-armored tracked vehicles	1993	http://assist.daps.dla.mil/docimages/0000/13/24/767.pd1
MIL-HDBK-1473A	Color and marking of army materiel	1997	http://assist.daps.dla.mil/docimages/0000/85/40/hdbk1473.pd6
MIL-HDBK-1908B	Definitions of human factors terms	1999	http://assist.daps.dla.mil/docimages/0001/81/33/1908hdbk.pd9
MIL-HDBK-46855	Human engineering requirements for military systems equipment and facilities		

limited to any specific NASA, military, or commercial program and can be applied to almost any type of equipment. NASA-STD-3000 consists of two volumes: Volume I, *Man–Systems Integration Standards*, presents all of the design standards and requirements, and Volume II, *Appendices*, contains the background information related to standards. NASA-STD-3000 covers the following areas of human factors: anthropometry and biomechanics, human performance capabilities, natural and induced environments, health management, workstations, activity centers, hardware and equipment, design for maintainability, and facility management.

Standards of the Federal Aviation Administration (FAA) are concerned with the following topics: human factors design criteria oriented to the FAA mission and systems (HF-STD-001); design and evaluation of air traffic control systems (DOT-VNTSC-FAA-95-3); elements of the human engineering program (FAA-HF-001); evaluation of human factors criteria conformance of equipment that interface with the operator (FAA-HF-002) and with the maintainer (FAA-HF-003).

In their standard DOE-HDBK-1140-2001, the Department of Energy (DOE) provides the system maintainability design criteria for DOE systems, equipment, and facilities. The Federal Highway Administration (FHA) establishes standards concerning the development and operation of traffic management centers (FHWA-JPO-99-042). FHA also describes human factors guidelines and recommendations for design of advanced traveler information systems (ATISs), commercial vehicle operations (CVOs), and accommodation of older drivers and pedestrians. The Nuclear Regulatory Commission provides guidelines of HFE conformance evaluation of the interface design of nuclear power plant systems (NUREG-0700 and NUREG-0711). FED-STD-795, which has been developed for use in federal and federally funded facilities, establishes standards for facility accessibility by physically handicapped persons.

5.2 OSHA Standards

Development of occupational safety and health standards in the United States is mandated by the general duty clause, Section 5(a)(1), of the *Occupational Safety and Health Act of 1970*, which states: "Each employer shall furnish to each of his employees, employment and a place of employment which is free from recognized hazards that are causing or are likely to cause death or serious harm to his employees." In general, penalties related to deficient and unsafe working conditions have been issued under this general duty clause. The general duty clause has also been supplemented by the *Americans with Disabilites Act* (Public Law 101–336, 1990). The disabilities act has an important bearing on ergonomics design of workplaces. The ADA prohibits disability-based discrimination in hiring practices and requires that all employers make reasonable accommodations to

Table 19 U.S. Government Human Factors/Ergonomics Standards

Document Number	Title	Date	Source
	National Aeronautics and Space Administration		
NASA-STD-3000B	Man–Systems Integration Standards	1995	http://msis.jsc.nasa.gov
	Department of Transportation, Federal Aviation Administration		
HF-STD-001	Human Factors Design Standard	2003	http://www.hf.faa.gov/docs/508/docs/wjhtc/hfds.zip
DOT-VNTSC-FAA-95-3	Human Factors in the Design and Evaluation of Air Traffic Control Systems	1995	http://www.hf.faa.gov/docs/volpehndk.zip
FAA-HF-001	Human Engineering Program Plan	1999	http://www.hf.faa.gov/docs/did_001.htm
FAA-HF-002	Human Engineering Design Approach Document — Operator	1999	http://www.hf.faa.gov/docs/did_002.htm
FAA-HF-003	Human Engineering Design Approach Document — Maintainer	1999	http://www.hf.faa.gov/docs/did_003.htm
FAA-HF-004	Critical Task Analysis Report	2000	http://hfetag.dtic.mil/docs-hfs/faa-hf-004_critical_task_analysis_report.doc
FAA-HF-005	Human Engineering Simulation Concept	2000	http://hfetag.dtic.mil/docs-hfs/faa-hf-005_human-engineering_simulation.doc
	Department of Transportation, Federal Highway Agency		
FHWA-JPO-99-042	Preliminary Human Factors Guidelines for Traffic Management Centers	1999	http://plan2op.fhwa.dot.gov/pdfs/pdf2/edl10303.pdf
FHWA-RD-98-057	Human Factors Design Guidelines for Advanced Traveler Information Systems (ATIS) and Commercial Vehicle Operations (CVO)	1998	http://www.fhwa.dot.gov/tfhrc/safety/pubs/atis/index.html
FHWA-RD-01-051	Guidelines and Recommendations to Accommodate Older Drivers and Pedestrians	2001	http://www.tfhrc.gov/humanfac/01105/cover.htm
FHWA-RD-01-103	Highway Design Handbook for Older Drivers and Pedestrians	2001	http://www.tfhrc.gov/humanfac/01103/coverfront.htm
	Department of Energy		
DOE-HDBK-1140-2001	Human Factors/Ergonomics Handbook for the Design for Ease of Maintenance	2001	http://tis.eh.doe.gov/techstds/standard/hdbk1140/hdbk11402001_part1.pdf
	Multiple Departments		
FED-STD-795	Uniform Federal Accessibility Standards	1988	http://assist.daps.dla.mil/docimages/0000/46/05/53835.pd5

working conditions to allow qualified disabled workers to perform their job functions.

In 1990, OSHA issued a set of voluntary guidelines entitled *Ergonomics Program Management Guidelines for Meatpacking Plants* (OSHA 3123), which have been used successfully by many types of industries, including those from outside the food production business. In 2000, the U.S. government proposed the *Ergonomics Program Rule* (*Federal Register*, November 14, 2000, Vol. 65, No. 220). The main elements of the standard included (1) training in basic ergonomics awareness, (2) providing medical management of work-related musculoskeletal disorders, (3) implementing a quick fix or going to a full program, and (4) implementing a full ergonomic program when indicated, including such elements as management leadership, employee participation, job hazard analysis, hazard reduction and control, training, and program evaluation. However, the regulation was repealed in March 2001.

Recently, OSHA has developed a four-pronged comprehensive approach to ergonomics designed to address musculoskeletal disorders (MSDs) in the workplace. The four segments of the OSHA's strategy were stated as follows:

1. *Guidelines*: to develop industry- or task-specific guidelines for industries based on current incidence rates and available information about effective and feasible solutions
2. *Enforcement*: to conduct inspections for ergonomic hazards and issue citations under the general duty clause and to issue ergonomic hazard alert letters where appropriate
3. *Outreach and assistance*: to provide assistance to businesses, particularly small businesses, and help them proactively address ergonomic issues in the workplace
4. *National advisory committee*: to charter an advisory committee that will be authorized to, among other things, identify gaps in research to the application of ergonomics and ergonomic principles in the workplace

Recently, OSHA has also published three voluntary guidelines to assist employers of the specific type of industries in recognizing and controlling hazards: (1) *Nursing Home Guideline* (issued on March 13, 2003), (2) *Draft Guideline for Poultry Processing* (issued on June 3, 2003), and (3) *Guideline for the Retail Grocery Industry* (issued on May 28, 2004). OSHA plans to develop additional voluntary guidelines with the use of a standard protocol (OSHA, 2004). The objective of this protocol is to establish a fair and transparent process for developing industry- and task-specific guidelines that will assist employers and employees in recognizing and controlling potential ergonomic hazards. By using this protocol, each set of guidelines will address a particular industry or task. It is intended that the industry- and task-specific guidelines will generally be presented in three major parts:

1. Program management recommendations for management practices addressing ergonomic hazards in the industry or task
2. Work site analysis recommendations for work site/workstation analysis techniques geared to the specific operations that are present in the industry or task
3. Hazard control recommendations that contain descriptions of specific jobs and that detail the hazards associated with the operation, possible approaches to controlling the hazard, and the effectiveness of each control approach

Since there are many different types of work-related hazards and injuries, and controls vary from industry to industry and task to task, OSHA expects that the scope and content of the guidelines will vary.

5.3 Other Standards for Occupational Safety and Health

In 2000, the National Safety Council (NSC), acting on behalf of the Accredited Standards Committee (ASC), has issued a draft document (known as Z-365) entitled *Management of Work-Related Musculoskeletal Disorders (MSD)*. The draft defines the following areas of importance to preventing work-related injuries: (1) management responsibility, (2) employee involvement, (3) training, (4) surveillance, (5) evaluation and management of work-related MSD cases, (6) job analysis and design, and (7) follow-up.

Independent of the efforts noted above, in 2001 another ANSI committee, *ASC Z-10, Occupational Health Safety Systems*, was formed under the auspices of the American Industrial Hygiene Association (AIHA). The main objective of ASC Z-10 is to develop a standard of management principles and systems for improving the occupational safety and health in companies.

5.4 ANSI Standards

The following HFE-relevant standards have been developed by the American National Standards Institute (ANSI).

5.4.1 Human Factors Engineering of Visual Display Terminals

ANSI/HFS 100–1988 presents ergonomics principles related to visual display terminals. The standard has been updated by BSR/HFES 100 Draft Standard dated 3/31/02.

5.4.2 Human Factors Engineering of Computer Workstations

According to Albin (2004), the BSR/HFES 100 Human Factors Engineering of Computer Workstations (HFES 100) is *a specification of the recommended human factors and ergonomic principles related to the design of the computer workstation, and is intended for fixed, office-type computer workstations for individuals who are moderate to intensive computer users*. This standard is organized into four major chapters: (1) installed systems, (2) input devices, (3) visual displays, and (4) furniture. The installed systems chapter specifies how to arrange all the workstation system components to match the capabilities of the intended user. The input devices chapter focuses on the design of input devices (including the issues of physical size, operation force, handedness, etc). The visual displays chapter discusses the human factors in the design of monochrome and color CRT and flat-panel displays. The furniture chapter provides design specifications for workstation components, including chairs and desks. The major topics described in each of these chapters are listed in Table 20.

5.4.3 Ergonomic Requirements for Software User Interfaces

The HFES/HCI 200 Committee, which operates under the auspices of the Human Factors and Ergonomics Society's Technical Standards Committee, has been working on development of a proposed U.S. national standard for software user interfaces. This standard will provide requirements and recommendations for software interfaces, with a primary focus on business and personal computing applications. The standard is related to the ISO 9241 series of user interface standards. The topics described in each section of the HFES 200 standard are listed in the Table 21.

5.4.4 Ergonomic Guidelines for the Design, Installation, and Use of Machine Tools

ANSI B11, *Technical Report: Ergonomic Guidelines for the Design, Installation and Use of Machine Tools*, is a consensual ergonomic guidelines developed by the Machine Tool Safety Standards Committee (B11) of the American National Standards Institute. The subcommittee responsible for the preparation of these guidelines consisted of representatives from manufacturing, higher education, safety, design, and ergonomics. The document specifies ergonomic guidelines to assist in the design, installation, and use of individual and integrated machine tools and auxiliary components in manufacturing systems.

The guidelines document underlines the importance of three basic ideas for achievement of effective and safe design, installation, and use of machine tools: (1) communication among all persons involved with the machine tools (users, installers, manufacturers, and designers), (2) dissemination of knowledge concerning ergonomics concepts and principles among all persons, and (3) the ability to apply ergonomics concepts and principles effectively to machine tools and auxiliary components. The guidelines document states that the provision of worker safety, work efficiency, and optimization of the entire production system requires consideration of the following ergonomics issues:

- The variation in employee physiological and psychological characteristics such as strength and capacity
- Incorporation of ergonomics concepts and principles into all new project, tool, machine, and work processes at the beginning of the process
- The goal that routine tasks that are to be done precisely, rapidly, and continuously, especially tasks in hazardous environments, should be performed by machines

Table 20 Main Chapters and Topics of the Human Factors Engineering of Computer Workstations: BSR/HFES 100 Draft Standard

Chapter	Topics
Installed systems	Hardware components, noise, thermal comfort, and lighting
Input devices	Keyboards, mouse and puck devices, trackballs, joysticks, styluses and light pens, tablets and overlays, touch-sensitive panels
Visual displays	Monochrome and color CRT, and flat-panel displays (viewing characteristics, contrast, legibility, etc)
Furniture	Specifications for workstation components (chairs, desks, etc.); postures (reference postures, reclined sitting, upright sitting, declined sitting and standing); anthropometry

Table 21 Topics Addressed in the Ergonomic Requirements for Software User Interfaces: HFES 200-1998

Chapter	Topics
Accessibility	Keyboard input; multiple keystrokes Customization; repeat rates; acceptance delays Pointer alternative; accelerators; remapping; navigation *Display fonts:* size, legibility, styles, colors *Audio output:* volume and frequencies, customization, content and alerts, graphics *Color:* palettes, background–foreground, customization, coding *Errors and persistence:* online documentation and help *Customization:* cursor, button presses, click interval, pointer speed, chording *Window appearance and behavior:* navigation and location, window focus, titles *Input focus:* navigation, behavior, order, location
Color	*Color selection:* chromostereopsis, blending and depth effects, use of blue and red, identification and contrast *Color assignments:* conventions, uniqueness and reuse, naming, cultural assignments *General use consideration:* number of colors, highlighting, positioning and separation *Special uses:* warnings, coding, state indications, pointers, area identification
Voice and telephony	*Speech recognition (input):* commands, vocabularies, prompts, consistency, feedback, error handling, dictation *Speech output:* vocabularies message format, speech characteristics, dialogue techniques, physical properties, alerting tones, stereophonic presentation *Nonspeech auditory output:* consistency, tone format, critical messages, frequency, amplitude Interactive voice response
Technical sections	Presentation of information, user guidance, menu dialogues, command dialogues, direct manipulation, dialogue boxes, and form-filling dialogue windows

- The goal that tasks that require judgment and integration of information (i.e., the tasks that humans do best) should be assigned to workers
- The knowledge that a system that does not consider human limits such as information handling, perception, reach, clearance, posture, or strength exertion can predispose to accident or injury

The documents also recommend matching the design of the tool or process with the physical characteristics and capabilities of workers, to ensure accommodation, compatibility, operability, and maintainability of the machine tools and/or auxiliary components.

5.5 State-Mandated Occupational Safety and Health Standards

The states of California (1997) and Washington (2000) have adopted statewide ergonomics standards. However, the Washington ergonomics standard was repealed in 2003. At present, employers in California are required to comply with provisions of the ergonomics standard that focuses on work-related repetitive injuries (for more information, see www.dir.ca.gov/title8/5110/html). In general, states with OSHA-approved occupational safety and health programs may follow OSHA's approach to ergonomics: to adopt ergonomic standards, include ergonomics in standards establishing safety and health program requirements, and utilize the general duty authority for enforcement purposes (Seabrook, 2001; Stuart-Buttle, 2005).

5.6 Other Standardization Efforts

The American Conference of Governmental Industrial Hygienists (ACGIH) (www.acgih.org) established threshold limit values (TLVs) for the following physical categories of work: (1) hand–arm and whole-body vibration, (2) thermal stress, (3) hand activity level ("monotask" jobs, performed for four hours or more), and (4) lifting tasks (load limits based on lift frequency, task duration, horizontal distance, and height at the start of the lift). Other organizations that develop HFE-related standards include the American Society of Mechanical Engineers (ASME), American Society for Testing and Materials (ASTM); Institute of Electrical and Electronics Engineers (IEEE), Society of Automotive Engineers (SAE), and National Institute of Standards and Technology (www.nist.gov).

6 ISO 9000-2000: QUALITY MANAGEMENT STANDARDS

Quality standards can also play an important role in assuring safety and health at the workplace. ISO stipulates that if a quality management system is implemented appropriately utilizing the eight quality management principles (see below) and in accordance with ISO 9004, all of an organization's interested parties should benefit. For example, people in the organization will benefit from (1) improved working conditions, (2) increased job satisfaction, (3) improved health and safety, (4) improved morale, and (5) improved stability of employment, and the society at large will benefit from (1) fulfillment of legal and regulatory requirements, (2) improved health and safety, (3) reduced environmental impact, and (4) increased security.

As discussed by Hoyle (2001), the term ISO 9000 refers to a set of quality management standards. ISO 9000 currently includes three quality standards: ISO 9000:2000, ISO 9001:2000, and ISO 9004:2000. ISO 9001:2000 presents requirements; ISO 9000:2000 and ISO 9004:2000 present guidelines. ISO first published its quality standards in 1987, revised them in 1994, and then republished an updated version in 2000. These new standards are referred to as the ISO 9000:2000 Standards.

It is recommended that the ISO 9001:2000 standard be used if an organization is seeking to establish a management system that provides confidence in the conformance of its product to established requirements. The standard recognizes that the word *product* applies to services, processed material, and hardware and software intended for, or required by, the customer (Hoyle, 2001).

The ISO 9000:2000 standards apply to all types of organizations, including manufacturing, service, government, and education. The standards are based on eight *quality management principles*:

- *Principle 1*: customer focus
- *Principle 2*: leadership
- *Principle 3*: involvement of people
- *Principle 4*: process approach
- *Principle 5*: system approach to management
- *Principle 6*: continual improvement
- *Principle 7*: factual approach to decision making
- *Principle 8*: mutually beneficial supplier relationships

There are five sections in the standard that specify activities that need to be considered when implemented to the quality management system. According to Hoyle (2001), following a description of the activities that are used to supply products, the organization may exclude the parts of the product realization section that are not applicable to its operations. The requirements in the other four sections, such as quality management system, management responsibility, resource management, and measurement analysis and improvement, apply to all organizations, and the organization needs to demonstrate how it applies them to the organization's quality manual or other documentation. These five sections of ISO 9001:2000 define what an organization should do consistently to provide products that meet customer and applicable statutory or regulatory requirements and enhance customer satisfaction by improving its quality management system. ISO 9004:2000 can be used to extend the benefits obtained from ISO 9001:2000 to employees, owners, suppliers, and society in general.

Table 22 ISO 9000 Quality Management Standards and Guidelines

Standard or Guideline	Purpose
ISO 9000:2000, Quality management systems: Fundamentals and vocabulary	Establishes a starting point for understanding the standards and defines the fundamental terms and definitions used in the ISO 9000 family to avoid misunderstandings in their use
ISO 9001:2000, Quality management systems: Requirements	Requirement standard to be used to assess the organization's ability to meet customer and applicable regulatory requirements and thereby address customer satisfaction; now the only standard in the ISO 9000 family against which third-party certification can be carried
ISO 9004:2000, Quality management systems: Guidelines for performance improvements	Provides guidance for continual improvement of an organization's quality management system to benefit all parties through sustained customer satisfaction
ISO 19011, Guidelines on Quality and/or Environmental Management Systems Auditing (currently under development)	Provides an organization with guidelines for verifying the system's ability to achieve defined quality objectives (use internally or for auditing suppliers)
ISO 10005:1995, Quality management: Guidelines for quality plans	Provides guidelines to assist in the preparation, review, acceptance, and revision of quality plans
ISO 10006:1997, Quality management: Guidelines to quality in project management	Guidelines to help the organization to ensure the quality of both project processes and project products
ISO 10007:1995, Quality management: Guidelines for configuration management	Gives an organization guidelines to ensure that a complex product continues to function when components are changed individually
ISO/DIS 10012, Quality assurance requirements for measuring equipment, part 1: Metrological confirmation system for measuring equipment	Gives an organization guidelines on the main features of a calibration system to ensure that measurements are made with the accuracy intended
ISO 10012-2:1997, Quality assurance for measuring equipment, part 2: Guidelines for control of measurement of processes	Provides supplementary guidance on the application of statistical process control when this is appropriate for achieving the objectives of part 1
ISO 10013:1995, Guidelines for developing quality manuals	Provides guidelines for the development, and maintenance of quality manuals tailored to specific needs
ISO/TR 10014:1998, Guidelines for managing the economics of quality	Provides guidance on how to achieve economic benefits from the application of quality management
ISO 10015:1999, Quality management: Guidelines for training	Provides guidance on the development, implementation, maintenance, and improvement of strategies and systems for training that affects the quality of products
ISO/TS 16949:1999, Quality systems: Automotive suppliers — Particular requirements for the application of ISO 9001:1994	Provides sector-specific guidance to the application of ISO 9001 in the automotive industry

ISO 9001:2000 and ISO 9004:2000 are harmonized in structure and terminology to assist an organization to move smoothly from one to the other. Both standards apply a process approach. Processes are recognized as consisting of one or more linked activities that require resources and must be managed to achieve predetermined output. The output of one process may form directly the input to the next process, and the final product is often the result of a network or system of processes. The eight quality management principles stated in ISO 9000:2000 and ISO 9004:2000 provide the basis for the performance improvement outlined in ISO 9004:2000. The ISO 9000 standards cluster also includes other 10000 series standards. Table 22 shows a list of the relevant standards and their purposes.

As discussed by Hoyle (2001), ISO requires that the organization determine what it needs to do to satisfy its customers, establish a system to accomplish its objectives, and measure, review, and continually improve its performance. More specifically, the ISO 9001 and 9004 requirements stipulate that an organization must:

1. Determine the needs and expectations of customers and other interested parties
2. Establish policies, objectives, and a work environment necessary to motivate the organization to satisfy these needs
3. Design, resource, and manage a system of interconnected processes necessary to implement the policy and attain the objectives
4. Measure and analyze the adequacy, efficiency, and effectiveness of each process in fulfilling its purpose and objectives
5. Pursue the continual improvement of the system from an objective evaluation of its performance

ISO identified several potential benefits of using the quality management standards. These benefits may include the connection of quality management systems to organizational processes, encouragement of a natural progression toward improved organizational performance, and consideration of the needs of all interested parties.

7 CONCLUSIONS

Although human factor and ergonomics standards cannot guarantee appropriate workplace design, they can provide clear and well-defined requirements and guidelines, and therefore the basis for good ergonomics design. Standards for workstation design and the work environment can ensure the safety and comfort of working people through establishing requirements for optimal working conditions. By providing consistency in the human–system interface and improving ergonomics quality of the interface components, ergonomics standards can also contribute to the enhanced systems usability and overall system performance. This benefit is based on the general requirement of harmonization across different tools and systems, to support user performance and avoid unnecessary human errors.

One of the most important benefits from standardization efforts is a formal recognition of the significance of ergonomics requirements and guidelines for system design on the national and international levels (Harker, 1995). The consensus procedure applied to standards development demands consultation with a wide range of commercial, professional, and industrial organizations. Therefore, the decision to develop standards and a consensus of diverse organizations concerning the need for standards reflects the formal recognition that there are important human factors and ergonomics issues that need to be taken into account during the design and development of workplaces and systems.

Standards represent the essence of the best available knowledge and practice extracted from a variety of academic sources, presented in the way that is easy to use by professional designers, and to include this knowledge in the design process. The consensus procedure makes the standards under development known and available to interested parties and the general public. Such a procedure also facilitates dissemination and promotion of human factors and ergonomics knowledge across the world of nonexperts.

REFERENCES

Albin, T. J. (2004), Board of Standards Draft Standard.

CEN (2004), European Standardization Committee Web site, http://www.cenorm.be/cenorm/index.htm.

Chapanis, A. (1996), *Human Factors in Systems Engineering*, Wiley, New York.

Dickinson, C. E. (1995), "Proposed Manual Handling International and European Standards," *Applied Ergonomics*, Vol. 26, No. 4, pp. 265–270.

Dul, J., de Vlaming, P. M., and Munnik, M. J. (1996), "A review of ISO and CEN Standards on Ergonomics," *International Journal of Industrial Ergonomics*, Vol. 17, No. 3, pp. 291–297.

Dzida, W. (1995), "Standards for User-Interfaces," *Computer Standards and Interfaces*, Vol. 17, No. 1, pp. 89–97.

Eibl, M. (2005), "International Standards of Interface Design," in *Handbook of Human Factors and Ergonomics Standards* W. Karwowski, Ed., Lawrence Erlbaum Associates, Mahwah, NJ.

Harker, S. (1995), "The Development of Ergonomics Standards for Software," *Applied Ergonomics*, Vol. 26, No. 4, pp. 275–279.

Hoyle, D. (2001), *ISO 9000: Quality Systems Handbook*, Butterworth-Heinemann, Oxford.

Human Factors and Ergonomics Society (2002), Board of Standards Review/Human Factors and Ergonomics Society 100, *Human Factors Engineering of Computer Workstations: Draft Standard for Trial Use*, HFES, Santa Monica, CA.

ILO (2004), International Labor Organization Website, http://www.ilo.org/public/english/index.htm.

ILO-OSH (2001), *Guidelines on Occupational Safety and Health Management Systems*, ILO-OSH 2001, International Labour Office, Geneva.

ISO (2004), International Standardization Organization Website, http://www.iso.org/iso/en/ISOOnline.openerpage.

McDaniel, J. W. (1996), "The Demise of Military Standards May Affect Ergonomics," *International Journal of Industrial Ergonomics*, Vol. 18, No. 5–6, pp. 339–348.

Nachreiner, F. (1995), "Standards for Ergonomics Principles Relating to the Design of Work Systems and to Mental Workload," *Applied Ergonomics*, Vol. 26, No. 4, pp. 259–263.

Olesen, B. W. (1995), "International Standards and the Ergonomics of the Thermal Environment," *Applied Ergonomics*, Vol. 26, No. 4, pp. 293–302.

Olesen, B. W., and Parsons, K. C. (2002), "Introduction to Thermal Comfort Standards and to the Proposed New Version of EN ISO 7730," *Energy and Buildings*, Vol. 34, No. 6, pp. 537–548.

OSHA (2004), Protocol for Developing Industry and Task Specific Ergonomic Guidelines, http://www.osha.gov/SLTC/ergonomics/guidelines_protocol.html.

Parsons, K. (1995a), "Ergonomics and International Standards," *Applied Ergonomics*, Vol. 26, No. 4, pp. 237–238.

Parsons, K. C. (1995b), "Ergonomics of the Physical Environment: International Ergonomics Standards Concerning Speech Communication, Danger Signals, Lighting, Vibration and Surface Temperatures," *Applied Ergonomics*, Vol. 26, No. 4, pp. 281–292.

Parsons, K. C. (1995c), "Ergonomics and International Standards: Introduction, Brief Review of Standards for Anthropometry and Control Room Design and Useful Information," *Applied Ergonomics*, Vol. 26, No. 4, pp. 239–247.

Parsons, K. C., Shackel, B., and Metz, B. (1995), "Ergonomics and International Standards: History, Organizational Structure and Method of Development," *Applied Ergonomics*, Vol. 26, No. 4, pp. 249–258.

Seabrook, K. A. (2001), "International Standards Update: Occupational Safety and Health Management Systems," in *Proceedings of the American Society of Safety Engineers' 2001 Professional Development Conference*, Anaheim, CA.

Spivak, S. M., and Brenner, F. C. (2001), *Standardization Essentials: Principles and Practice*, Marcel Dekker, New York.

Stewart, T. (1995), "Ergonomics Standards Concerning Human–System Interaction: Visual Displays, Controls and Environmental Requirements," *Applied Ergonomics*, Vol. 26, No. 4, pp. 271–274.

Stuart-Buttle, C. (2005), "Overview of International Standards and Guideliness," in *Handbook of Human Factors and Ergonomics Standards and Guidelines*, W. Karwowski, Ed., Lawrence Erlbaum Associates, Mahwah, NJ.

Wetting, J. (2002), "New Developments in Standardization in the Past 15 Years: Product Versus Process Related Standards," *Safety Science*, Vol. 40, No. 1–4, pp. 51–56.

CHAPTER 58

HUMAN FACTORS AND ERGONOMICS IN MEDICINE

Pascale Carayon
University of Wisconsin–Madison
Madison, Wisconsin

Wolfgang Friesdorf
Technische Universitat Berlin
Berlin, Germany

1 CHARACTERISTICS OF THE HEALTH CARE INDUSTRY

It is important to understand the unique characteristics of medicine and the health care industry in order to know how human factors and ergonomics concepts, models, and methods can be applied or need to be modified and adapted.

1.1 Health Services Industry

In 2002, the health services industry was the largest in the United States, providing 12.9 million jobs (Bureau of Labor Statistics, 2004). The industry is comprised of the following segments (Bureau of Labor Statistics, 2004):

- Hospitals, which employ 41% of all workers. Two out of three hospital workers work in institutions with more than 1000 workers.
- Nursing and residential care facilities. Nursing aides provide the majority of direct care.
- Physician offices. Physicians and surgeons have private practices or work in groups of physicians with similar or different specialties.
- Offices of dentists.
- Offices of other health practitioners, such as chiropractors.
- Home health care services. This is one of the fastest-growing sectors of the economy.
- Outpatient care centers, such as dialysis centers and free-standing outpatient surgery centers.
- Other ambulatory health care services.
- Medical and diagnostic laboratories.

The health services industry has 40 occupations and professions categorized as follows: (1) management, business, and financial occupations (4.8% of employment); (2) professional and related occupations (43.5%); (3) service occupations (31.4%); and (4) office and administrative support functions

(18%). The employment in the health services industry is expected to grow by 28% between 2002 and 2012. Some of the fastest-growing occupations include social and human service assistants (growth of 63.7%), medical assistants (63.2%), home health aides (growth of 54.5%), physician assistants (growth of 54.1%), and medical records and health information technicians (growth of 51.9%). A number of factors contribute to the growth in employment in the health services industry (Sultz and Young, 2001). The aging population is a primary factor that increases the demand for workers who provide long-term care, such as nursing home care and home health care. The increasing implementation of medical and nonmedical technology also has implications for the number and skill requirements of the health care workforce (Sultz and Young, 2001). Health care reforms have shifted health care delivery sites from acute care hospitals to ambulatory, home care, and long-term care settings (Sultz and Young, 2001).

The health care industry is a people-intensive industry involving many different types and categories of workers, patients and their families, communities, and society at large. As Van Cott (1994) says: "The health-care system is people-centered and people-driven." Therefore, the discipline of human factors and ergonomics has much to offer to improve the performance, quality, and safety of the health care system. The current health care system is very decentralized and is comprised of a range of subsystems connected with each other and including caregivers and patients (Institute of Medicine, 2001). The subsystems include hospitals, community pharmacies, clinics, laboratories, long-term facilities, and others. Because of the variety of systems and subsystems, different goals, values, beliefs, and norms of behavior are at work in the health care systems (Van Cott, 1994).

1.2 Complexity of Health Care

Different dimensions of system complexity have been identified (Perrow, 1984; Vicente, 1999) (see Table 1), and health care systems possess many of these the characteristics. Health care is composed of many different elements and forces. It has been estimated that medicine has to deal with about 500,000 illnesses and, therefore, diagnoses.

The health care industry represents a major part of the economy of industrially developed countries. In Germany, health care is one of the biggest industries and represents 11% of the gross domestic product. In 2002, more than 4 million people worked in this field (i.e., 10% of all employed persons). In Germany, there are more than 2000 hospitals (Statistisches Bundesamt, 2005). The United States spends a large amount of its GDP on health care. In 2000, health care expenditures represented more than 13% of the GDP (Agency for Healthcare Research and Quality, 2002).

Health care is basically a sociotechnical system in which people have a preponderant role (see Section 2): as providers, patients, families, and purchasers. People are customers and consumers of health care, and people are producers of health care. Effective functioning of health care depends largely on people and on

Table 1 Complexity of Health Care Systems

Dimensions of System Complexity	Examples in Health Care
Large problem spaces	About 500,000 illnesses (health care expenditures account for 13% of GDP in the United States)
Social aspects	People-intensive; people-centered; people-driven
Heterogeneous perspectives	Different goals, beliefs, values, and behavior norms; different cultures and subcultures
Distributed character	Geographical dispersion (e.g., home care); telemedicine
Dynamic character	Changes in medical knowledge and technology; delay between action and outcome (e.g., primary care and preventive medicine)
Potentially high hazards	Patient safety and medical errors
Many coupled subsystems	Both tight and loose coupling
Automation	High levels of automation in certain parts of health care (e.g., radiotherapy)
Uncertain data	Imperfect information; imperfect knowledge; patient factors
Mediated interaction with computers	Medical devices and technologies (e.g., endoscopic technologies)
Disturbances	Unanticipated events

communication and coordination among various health care staff members.

Workers in health care come from different backgrounds and may have different values regarding health care, its delivery, and its quality and safety. For instance, a recent study by Thomas et al. (2003) shows the discrepancy in the attitudes of critical care physicians and nurses with regard to teamwork. A total of 90 physicians and 230 nurses from eight nonsurgical intensive care units in six hospitals were surveyed. Thirty-three percent of the nurses rated the quality of communication and collaboration with physicians as high or very high, whereas 73% of the physicians rated the quality of communication and collaboration with nurses as high or very high. Physicians were more likely than nurses to agree with the statement that "input from ICU nurses about patient care is well received in my unit." Such discrepancy between workers in health care can have numerous consequences, such as dissatisfaction and poor well-being, and may also affect expectations regarding performance and ultimately the quality and safety of care provided to patients.

People involved in the delivery of health care may be located in different places. One type of long-term care in which people are geographically dispersed is home services provided by home health care agencies (Wunderlich and Kohler, 2001). The home is fast becoming the primary site of care for most persons with acute or chronic illnesses. Indeed, with the current trend toward ambulatory procedures, and increasing technologies, one-half of patients once cared for in hospitals now receive their care at home. The home health arena is unique because it has so many workers of various skill levels who are dispersed over large geographical areas. Workers in home care function in a geographically distributed environment in which few workers ever see other members of their team on a day-to-day basis. Home health care workers have very high turnover rates (Wunderlich and Kohler, 2001), which may be due to their poor working conditions and low quality of working life (Feldman, 1993; Wunderlich and Kohler, 2001). Stone and Wiener (2001) highlight the major issues affecting long-term-care front-line workers, including difficulties in recruiting and retaining workers. They also discuss many of the negative effects of turnover, such as poorer quality and/or unsafe care, increased stress and frustration on the workers, reduced opportunities for on-the-job training and learning, and less peer support. Reasons for the turnover problem of home health care workers include poor working conditions, perceptions of inadequate resources, and a suboptimal environment for providing good safe care (Blegen, 1993; Aiken et al., 2001; Simmons et al., 2001). The discipline of human factors can provide the models and tools for improving working conditions and therefore for improving retention and reducing turnover of home care workers.

Telemedicine is one form of organizing care that puts distance between the patient and the health care providers. According to the American Telemedicine Association: "Telemedicine is the use of medical information exchanged from one site to another via electronic communications for the health and education of the patient or healthcare provider and for the purpose of improving patient care" (Linkous, 2001). A recent study of the application of telemedicine for intensive care demonstrates its financial and patient outcome benefits. The remote ICU care (eICU) program supplemented existing on-site care activities, which did not change as a result of the program. Patients were monitored by the eICU staff (an intensivist, a critical care nurse, and a clerical person) from noon to 7 a.m. The eICU staff was the primary contact for the on-site nurses and were responsible for contacting physicians, responding to all emergencies, and initiating interventions (where authorized). There is much discussion about whether telemedicine can achieve the same kind of positive results as in the study by Breslow et al. (2004). Such geographical separation poses unique challenges to communication and coordination between health care providers.

Complex systems are dynamic and rapidly changing. The health care industry has seen lots of changes in medical knowledge and technology that have been precipitated by large investments in biomedical research (Institute of Medicine, 2001). Much of the medical knowledge and information on evidence-based practice comes from randomized controlled trials (RCTs). Chassin (1998) shows that the number of publications from RCTs referenced in Medline has increased exponentially from 1966 to 1995: in 1966, there were about 100 RCT articles, whereas in 1995, there were over 100,000 RCT articles. The volume and complexity of this information poses unique challenges to health care practitioners and managers and adds to the complexity of health care system. Another time factor of system complexity is the time lag between action and response. Some health care action can lead very quickly to a response. For instance, using an electric heart defibrillation leads to an immediate physical reaction on the part of the patient. The administration of certain drugs can also lead to immediate physiological reactions. There can also be long delays between actions done to the patient and their consequences. In addition, the consequences may not be easily linked to specific actions, or may not be visible to the health care provider who performed the actions in the first place. This is particularly the case in primary and ambulatory care. For instance, the health effects of preventive services provided by primary care physicians may not be measured for many years. Operating a complex sociotechnical system can produce various hazards. In health care, the main hazards are those done to the patients, such as medical errors and lack of patient safety (see Section 5).

Perrow (1984) defines the following characteristics of coupling in systems. First, "tightly coupled systems have more time-dependent processes," whereas "in loosely coupled systems, delays are possible." Second, the sequencing of tasks or process steps in tightly coupled systems is somewhat fixed and invariant. Third, in tightly coupled systems, there is typically only one way of designing and performing a process. One characteristic of loosely coupled systems is equifinality (i.e., they have a common objective that can be achieved with different processes and tasks). Fourth, tightly coupled systems have little slack. A health care system can either be tightly coupled or loosely coupled (Cook, 2004). An example of tight coupling is the sequence of specific steps in the performance of a surgical procedure (e.g., the surgery cannot occur before the anesthetic has been administered). An example of loose coupling is the preoperative process once a surgery has been decided: A number of tasks must occur, but not necessarily in a specific tight temporal sequence; for instance, the physical exam that the patient needs to have before the surgery can occur within a certain window of time, typically 1 to 3 weeks.

Complex systems tend to be highly automated. Whereas automation is not widespread over all health care, there are parts of health care with high degree of automation. Radiotherapy is an example of patient care that relies on various forms of automation, such as connection between information from imaging devices and treatment devices. Uncertainty is another characteristic of complex systems and is highly present in health care. The sources and types of uncertainty in health care include, for instance, imperfect information, imperfect knowledge regarding medical treatment, and patient factors (e.g., impact of treatment on a particular patient).

In health care, much of the interaction is mediated by devices and technologies. For instance, it is not possible to measure directly blood pressure of a patient during a surgery: there is a piece of equipment and a display that provide information on the patient's blood pressure. That type of technology-mediated interaction between the worker and the work object highlights the importance of cognitive ergonomics in health care.

In complex systems, disturbances are very present and workers need to deal with unanticipated events. To maintain safety and performance, workers need to adapt to those events quickly. Health care is, of course, filled with unanticipated events to which workers need to react quickly to ensure the safety of patients and to maintain adequate performance.

Weinger et al. (2003) conducted a study of nonroutine events in anesthesia. A *nonroutine event* was defined as "any event that is perceived by clinicians or skilled observers to deviate from ideal care for that specific patient in that specific clinical situation." In-person surveys of anesthesia providers show that 27% of anesthesia cases ($N = 277$) had at least one nonroutine event (total of 98 nonroutine events). Data from QI reporting systems in two hospitals yielded information on 135 events. An analysis of all of these 233 nonroutine events produced information on the factors contributing to those events: patient disease/unexpected response (67%); provider supervision, knowledge, experience, and judgment (33%); surgical issues (26%); logistical or system issues (19%); inadequate preoperative patient preparation (17%); equipment failure or usability (16%); coordination/communication (15%); and patient positioning (9%).

The complexity of the health care system has a strong effect on the practice of human factors and ergonomics. For example, human factors and ergonomics projects in health care require one to spend more time on managing and implementing change than on deciding on the content of change (Hignett, 2003).

1.3 Lack of Integration and Standardization

The report "Crossing the Quality Chasm" published by the Institute of Medicine (2001) emphasizes the need to redesign care processes to meet the six challenges of providing safe, effective, efficient, personalized, timely, and equitable care. This report and an earlier IOM report (Kohn et al., 1999) emphasize the lack of integration and standardization of care processes. The IOM recommends an approach where systems and processes are designed for the usual but the unusual is being recognized and planned for (Institute of Medicine, 2001). Routine, predictable tasks and processes can be simplified and standardized (80% of the work), and for the remaining 20%, contingency plans can be designed appropriately.

When leaders at St. Joseph's Community Hospital were working on the design of a new hospital facility, they looked for ways of designing the hospital and its spaces to ensure maximizing efficiency and the quality and safety of care (Reiling et al., 2004). One of the hospital design principles was the complete standardization of patient rooms, including materials, gases, and head wall design.

In an intensive care unit (ICU), it is not rare to find multiple devices and technologies that are not connected or related to each other (Friesdorf et al., 1990). Such a lack of integration between various devices and technologies can contribute to a health care provider's workload for the following reasons: (1) additional training required to understand the functioning of new devices, (2) increased need for documentation, (3) increased need for information processing from multiple sources and via multiple channels, and (4) difficulty of gathering all data required for a specific situation or problem.

However, it is important to recognize the limitations of standardization. Woods and Cook (2001) emphasize that "the search for simplicity in health care delivery will prove difficult" and that coping with and learning from complexity are more fruitful avenues. In addition, standardization may create work systems that are too restrictive for the users. One solution may therefore be the customization, adaptation, or tailoring of systems to the end users. In the context of the design of medical devices, there is some debate about the effectiveness of tailoring. Randell (2003) conducted an observational study of the customization of medical devices by intensive care nurses. She found that nurses performed the following types of device customization: (1) customization to overcome limitations of the device and provide adequate patient care, (2) use of pen and paper to ease the use of the device (e.g., Post-It notes attached to devices), and (3) change in the procedure for use of the device. A study on anesthesia alarms shows that tailoring can lead to improved performance *if* there is time to tailor the alarm, the means for adapting the alarm are present, and the benefits appear to outweigh the costs (Watson et al., 2004).

2 END USERS IN MEDICINE

The involvement and participation of users is a critical principle of human factors and ergonomics. In this section we discuss some of the challenges to the application of this principle in medicine: definition and determination of users, involvement of laypersons who do not have medical/health care knowledge, and challenges related to participatory ergonomics.

2.1 End-User Involvement in Health Care System Design

An overriding principle of human factors is to center the design process around the user, therefore creating a user-centered design (Norman, 1988; Meister and Enderwick, 2001; Wickens et al., 2004). In the design of health care systems, the variety of potential end users needs to be considered in the design cycle.

There is much controversy about the system design process, its characteristics and components, the sequence of tasks, and so on. However, the system design process can be conceptualized as being organized around four considerations (Meister and Enderwick, 2001): analysis of the design problem, generation of alternative solutions, analysis of alternative solutions, and selection of preferred solution. From a human factors point of view, Wickens et al. (2004) describe major stages of system design in which human factors can provide important and useful information:

1. *Front-end analysis.* This stage includes definition of the users, of the functions to be achieved by the system, of the environmental conditions under which the system will be used, and of the users' preferences or requirements for the system. It will typically include a user analysis as well as a task analysis.
2. *Iterative design and test and system production.* Initial specifications are used to create initial design or prototypes. At this stage, human factors input typically consists of identification of human factors criteria to the list of system requirements (e.g., usability requirements), function allocation, and design of support materials.
3. *Implementation and evaluation.* Various methods for system change implementation use basic human factors principles (e.g., participatory ergonomics). The evaluation should consider human factors variables (human performance, health and safety, and well-being).
4. *System operation, maintenance, and disposal.* Various human factors activities occur at those stages, such as ensuring the reliability and functioning of medical equipment and devices for safe operations, and designing an appropriate system for hazardous materials (e.g., needles).

Human factors engineers and ergonomists focus on the interactions between humans and other elements of systems. In health care systems, the humans are varied: health care providers and clinicians, patients and their families, and other types of workers (e.g., housekeeping, biomedical engineering, purchasing, and administration). This large variety adds to the complexity of health care systems and their design. For example, a single device such as an infusion pump is used by multiple users: The *nurse* programs the pump when administering medication to a patient; the *patient* is connected to the infusion via tubing; and the *biomedical engineer* maintains the pump and ensures its calibration. This example shows that a single device is used by different users performing different tasks. It is also important to note that this variety of end users is often related to a variety of physical and organizational settings. In the infusion pump example, the physical environment in which the pump is used varies from patient rooms to engineering laboratory. From an organizational viewpoint, the various users have probably received different levels and types of training with regard to use of the pump.

Defining the *user* in a human factors and ergonomics project in health care is critical (Hignett, 2003). Every person is a potential user of the health care system in any country, but only a small proportion of a country's population is in direct contact with the health care system. Another complicating factor is the fact that very often, patients are not paying for the health care service directly. All of these factors make the definition of *user* a difficult task for the user-centered designer (Hignett, 2003).

2.2 Involvement of Laypersons

Recently, because of the increasing shift toward patient involvement in the care process, there has been much interest in examining the use of health care technologies by laypersons. For instance, automated defibrillators can be found in a variety of places, such as airports and other public places. Such devices need to be designed for people who do not have medical and clinical training. Callejas et al. (2004) show that naive users and video-trained users were able to use two types of automated defibrillator safely. The HeartCare project, led by Brennan and colleagues, determined the effects of using consumer electronics and the World Wide Web to provide sequenced, tailored information and communication utilities to patients recovering from coronary artery bypass graft surgery (Jones and Brennan, 2002; Brennan and Aronson, 2003). According to preliminary evidence, participants in the HeartCare project had more successful recovery than did patients having access to an audiotaped instructional program. The HeartCare patients experienced less depression and fewer negative mood states, had fewer symptoms, and were less bothered by those symptoms. These findings confirm the positive health benefit of an intervention such as HeartCare and also indicate that the benefits may vary over time.

2.3 Participatory Ergonomics

Participatory ergonomics is a powerful method for involving end users in system design (Wilson, 1995). Participation has been used in a variety of human factors processes, such as implementing ergonomic programs (Wilson and Haines, 1997). According to Noro and Imada (1991), participatory ergonomics is a method in which end users of ergonomics (e.g., workers, nurses, patients) take an active role in the identification and analysis of ergonomic risk factors as well as the design and implementation of ergonomic solutions.

Evanoff and his colleagues have conducted studies on participatory ergonomics in health care (Bohr et al., 1997; Evanoff et al., 1999). In one study they examined the implementation of participatory ergonomics teams in a medical center (Bohr et al., 1997). Three groups participated in the study: a group of orderlies from the dispatch department, a group of intensive care unit (ICU) nurses, and a group of laboratory workers. Overall, the team members for the dispatch and laboratory groups were satisfied with the participatory ergonomics process, and these perceptions seem to improve over time. However, the ICU team members expressed more negative perceptions. The problems encountered by the ICU team seem to be related to the lack of time and the time pressures due to the clinical demands. A more in-depth evaluation of the participatory ergonomics program on orderlies showed substantial improvements in health and safety following implementation of the participatory ergonomics program (Evanoff et al., 1999). The studies by Evanoff and colleagues demonstrate the feasibility of implementing participatory ergonomics in health care but highlight the difficulty of the approach in a high-stress, high-pressure environment such as an intensive care unit, where patient needs are critical and patients need immediate or continuous attention. More research is needed to develop ergonomic methods for implementing participatory ergonomics programs in health care. Those programs should lead to improvements in human and organizational outcomes as well as improved quality and safety of care. This research should consider the high-pace, high-pressure work environment of health care.

3 HUMAN FACTORS SYSTEMS APPROACHES APPLIED TO HEALTH CARE

Human factors experts working in health care agree on the need to adapt and adopt systems approaches in health care systems (Cook and Woods, 1994; Bogner, 2004; Vincent, 2004). In this section we review selected human factors systems approaches that have been applied to health care.

3.1 Work System Model

The work system model developed by Carayon and Smith describes the many different elements of work (Smith and Carayon-Sainfort, 1989; Carayon and Smith, 2000). The work system is comprised of five elements: the *individual* performing different *tasks* with various *tools and technologies* in a *physical environment* under certain *organizational conditions* (see Figure 1). Given the complexity of the health care system (see Section 1), it is important to adopt a systems approach to the analysis of health care systems (Vincent, 2004). The work system model has been applied to the analysis of medical errors, such as wrong-site surgery (Carayon et al., 2004a). An example of the application of the work system model is provided below. See Section 3.1.1 for how the work system model can be applied to the analysis of an intensive care unit nurse.

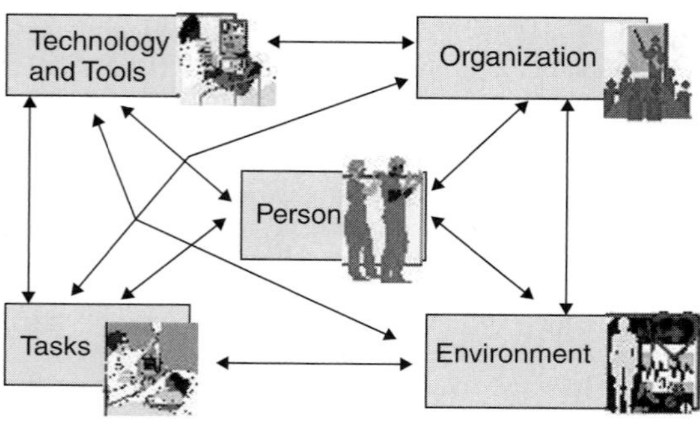

Figure 1 Work system model. (From Smith and Carayon-Sainfort, 1989; Carayon and Smith, 2000.)

3.1.1 Example: Work System Analysis of an ICU Nurse

The following is a brief overview of the various work elements of an ICU nurse's job.

Task The tasks performed by the ICU nurses include (but are not limited to) direct patient care, continuous patient status assessment, carrying out physician orders, medication administration, and family interaction.

Organizational Factors A range of organizational factors are important to understand the job of an ICU nurse. Conflict among nurses and between physicians and nurses has been correlated with high stress and workload in ICUs (Gray-Toft and Anderson, 1981). Studies by Knaus, Rousseau, Shortell, Zimmerman, and colleagues have shown the importance of *caregiver interaction*, which is a composite concept that includes several dimensions, such as communication and coordination (Knaus et al., 1986; Shortell et al., 1994).

Environment Noise and other sensory disruptions abound in the modern ICU setting (Topf, 2000). The physical environment is often crowded and messy, with no one available to help with immediate cleanup of the environment or equipment. The noise, the housekeeping, the level of constant activity, the size of the rooms or physicians' and nurses' personal space (if any), patients/staff coming and going, crowds of people waiting to get a moment of the physician's or nurse's time and attention may all make it more difficult to carry out tasks.

Equipment and Technology The technology, tools, and equipment of the modern ICU have been identified as possible causes of errors and problems (Bracco et al., 2000). The availability of needed supplies, the types of supplies and tools, the technology desired, the working condition of the equipment, and whether or not the new technology is available are but some of the tools and technology issues that can increase workload. Additionally, training and time for acclimation are needed to learn all the new tools and technology.

Establishing Balance in the Work System As an example in the ICU setting, in efforts to reduce workloads and balance the overall work system, physicians and nurses might review how often physical assessments are performed on the patients and who performs them. Typically, both the nurse and the physician perform a patient physical assessment every hour or as needed by a patient's condition. Under this system, both the physician and the nurse perform the patient physical assessment and enter it into the patient's records. The process takes at least several minutes or more, depending on the patient's status, out of every hour. This takes time away from other

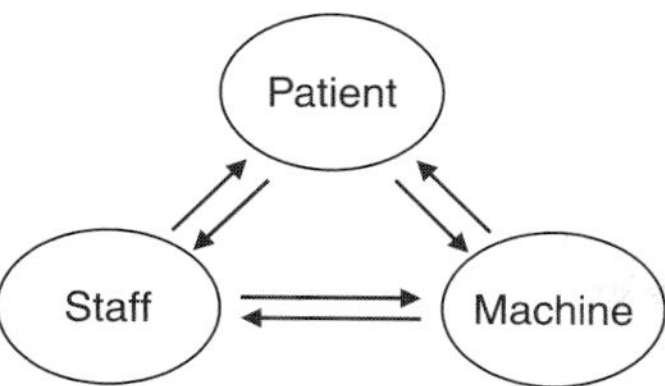

Figure 2 Patient–staff–machine interaction model for health care.

tasks and/or professional activities associated with the patient's care. In addition, others, such as specialty consults may be waiting to review the patient's records. To balance this problem, the ICU physicians and nurses may redesign the patient assessment system based on clinical expertise and cooperation among the physicians and nurses involved.

3.2 Health Care Working System

"Ergonomics (or human factors) is the scientific discipline concerned with the understanding of interactions among humans and other elements of a system ... to optimize human well-being and overall system performance" (International Ergonomics Association, 2000). In industrial work systems this definition leads to a basic person–machine interaction model with a given task objective to produce, maintain, or repair a product. In contrast, the *work object* in health care is the patient to be treated. The patient plays a double role. For a successful treatment a patient is not only a passive work object (e.g., anesthetized patient during surgical intervention), but also an active *coworker* (e.g., a patient taking drugs or cooperating during physiotherapy). The *patient–staff–machine interaction model* takes this double role of the patient into account (Friesdorf et al., 1990) (see Figure 2).

The interactions between staff and machine in a health care working system are similar to those found in industrial work systems. The interactions between patient and machine (e.g., medical devices) include patient monitoring (e.g., measurement of physiologic variables such as ECG and blood pressure) and patient therapy (e.g., pump-driven drug application and artificial ventilation). The interactions between staff and patient typically require some form of communication. Patient, staff, and machine are the core elements of a work system in health care. The interactions between the elements represent the basis of the working processes that are necessary to complete a given task objective (see Figure 3).

The task objective of the patient–staff–machine system is defined by the staff according to the patient's initial status and the anticipated final patient status after task completion. For instance, for a patient who has been diagnosed with gallstones, the task objective could be the surgical removal of the gallbladder. *High quality* is achieved if the task completion corresponds to the task objective. High quality can be achieved only if the working processes are performed faultlessly ("what has to be done?"). *Efficiency* takes the used resources into account. High efficiency can be achieved only by a process organization that is free of process deficits ("how should things be done?").

As soon as what has to be done with a patient is clearly defined, the loss of process quality or process efficiency can have only one cause: the existence of process deficits. These deficits have to be eliminated systematically for safe and efficient patient treatment. What has to be done is defined through the process of medical diagnosis and medical decision making: it can be represented as a task with an objective, a completion outcome, and with quality and efficiency requirements and characteristics.

In high-dependency health care environments such as an operating room and an intensive care unit, the invasive treatment requires several devices and specialists working in parallel with the patient. Dividing a task into subtasks that are performed autonomously by specialists is one strategy for handling complex patient treatment. For instance, in an operating room, the anesthesia intervention is performed in parallel with the

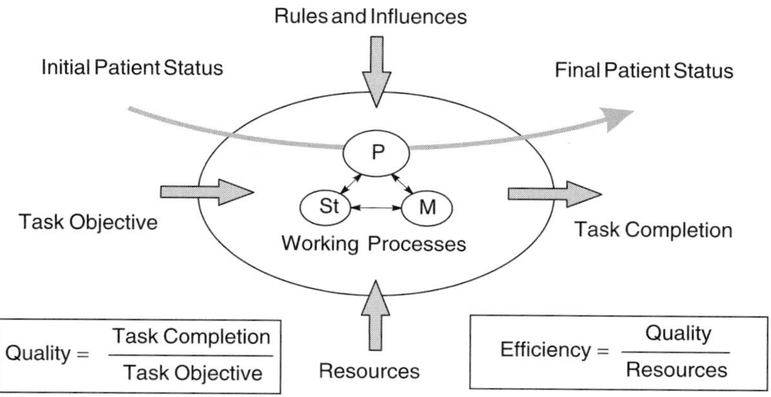

Figure 3 Medical work processes and their context in a work system.

surgical intervention. On the one hand, autonomous task performance by specialists can lead to improvement in treatment quality, but on the other hand, it increases the number of interactions between subsystems, therefore creating the potential for communication errors between those subsystems. The variety of interacting subsystems is also a reason for slowing down change processes (Van Cott, 1994). See Section 4 for the description of a method for analyzing, assessing and optimizing such a complex working system.

3.3 Human Factors Analysis of the Care Process

In medical work systems, three types of work processes can be differentiated (Friesdorf et al., 2002):

- *Primary work processes*: direct patient treatment processes (e.g., a surgical intervention, an intensive care treatment)
- *Secondary work processes*: processes directly supporting the patient treatment processes [e.g., operating room (OR) management, laboratory diagnostics]
- *Tertiary work processes*: processes indirectly supporting the patient treatment processes (e.g., sterilization, cleaning, catering)

Because secondary and tertiary work processes have a rather low level of complexity compared to primary work processes, traditional analysis and optimization strategies for classical work systems (e.g., standardization, statistical process control) can be applied. The complexity of the patient as a work object does not allow an organization to readily apply the flow (or line) production principles of industry to design direct patient treatment processes. Many primary work processes are organized like traditional workshop production. Sometimes the work object (i.e. a patient) has to be moved between locations (e.g., emergency room, OR, ICU, radiology). Sometimes treatment procedures have to be performed on the work object (e.g., a patient who is kept within the same ICU location to get an x-ray).

For the optimization of primary work processes, a new approach is needed. Much is known about medical excellence during decision making [i.e., when assessing a patient's status and deciding on required clinical tasks, and for specific medical procedures (e.g., surgical intervention, induction of anesthesia)]. However, during medical training, physicians are not confronted with issues related to the organization of complex clinical work processes. To optimize primary work processes, additional competencies are required in the following domains: management, process analysis, and systems thinking. The best task definition (what has to be done) can not lead to the desired result if it is not realized with correct work processes (how to do things). With each deficiency (e.g., wrong decision, poor communication) a loss of quality or efficiency is likely to occur. Besides the classical work organization (top-down), the establishment of a sustainable process orientation becomes more important. This can only be achieved by using an additional bottom-up-oriented approach, similar to the idea of business process reengineering for industrial work processes (Hammer and Champy, 1993), but *carefully* adapted to the requirements of medical work systems.

3.4 Levels of Analysis

In an industry such as the automotive industry, extended working processes are decomposed hierarchically down to a level at which the task objective can be defined and working actions can be controlled by a worker or a working team. Thus, the production line of a car can be anticipatorily designed in detail. In health care a similar approach can be attempted. Given a particular diagnosis, the entire treatment can be decomposed at the level of clinical pathways (e.g., admission, diagnostics, preparation, and execution of a surgical intervention, postoperative recovery, and discharge). Only in highly specialized fields such as cataract treatment is decomposition down to a level of working actions possible. In general, the patient's status and the working conditions (see "influences" in Figure 3) are so complex that complete anticipation of all possible actions is not possible (see the discussion on standardization in Section 1.3). Strategic treatment planning adapted to a particular patient is necessary to cope with uncertainty.

In addition to the patient-related uncertainty (e.g., unclear diagnosis), the influences of other system variables have to be considered. What human and technical resources are available to perform the task? How long does it take for these resources to be ready for action? Is the laboratory on duty? For instance, the arrival of a polytraumatized patient into the emergency room may require different decisions depending on the time of the emergency. If the emergency occurs on a Saturday night, we would have to consider the number and characteristics of clinicians on call, the limited availability of laboratory staff, and so on. If the emergency occurs on a Monday morning, other issues, such as the workload of clinicians and the load in the OR, need to be considered. Thus, adaptive decision-making mechanisms on the part of the health care providers are required. When decomposing tasks into subtasks and describing process flows, other system design characteristics need to be considered. The work system model described in Section 3.1 can provide a conceptual framework for identifying all of the relevant system variables.

A similar decomposition of complex tasks is discussed by researchers examining complex decision-making tasks. Rasmussen and Doodstein (1988) describe a mental model with five levels for task subroutines. The "how" question leads to a more detailed level, the "why" question to a higher level, and the "what" question describes the task/subtask at the specific focus level. Figure 4 shows a recursive hierarchical task–process–task model that enables a heuristic decomposition of complex tasks. A task objective is completed by subtasks. A process analysis

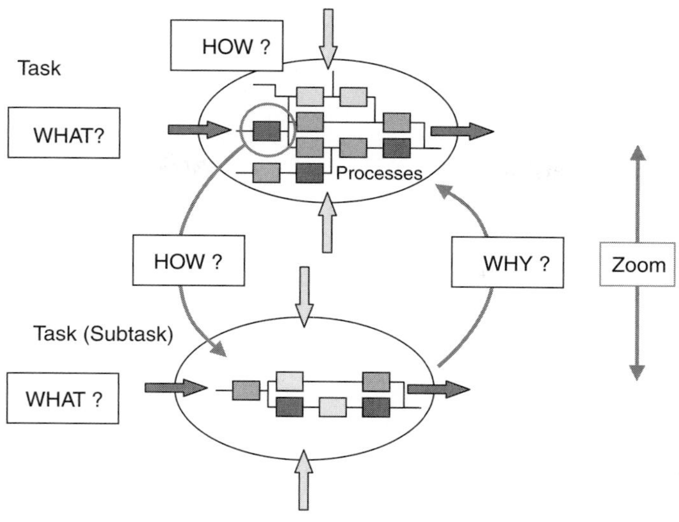

Figure 4 Recursive hierarchical task–process–task model of a complex work system.

answers the "how" question. Critical subtasks can be analyzed on a more detailed level. This can be done recursively until the working processes consist of elementary activities (e.g., adjust an infusion pump, confirm an alarm). The "why" question supports the recomposition of tasks. At each level the definition for quality and efficiency can be used for the assessment of tasks/subtasks.

3.5 Work Process Analysis

An approach to health care process analysis and improvement can be described as a series of seven steps (Marsolek and Friesdorf, 2004):

1. *Project preparation and staff preparation*: definition of the boundaries of the process to be analyzed, kick-off meeting, collection of available data on process, and description of physical layout
2. *Process flow visualization and verification*: staff participation, description and visual representation of the process flow, iterative revisions to the process flow
3. *Hierarchical structuring and process flow quantification*: definition of process modules, listing of all documents used in the process, analysis of the process modules (using the following dimensions: person, machine, measurement, material, methods, milieu, and management)
4. *Identification of process strengths and weaknesses*: participatory identification of process strengths and weaknesses and listing of possible improvements
5. *Development of an optimized process flow*: participatory development and definition of improvement strategy (e.g., "who" has to do "what" "when")
6. *Evaluation of process flow changes and initiation of continuous process control*: definition of goal performance metrics (time, cost, quality, customer satisfaction) and definition of data collection system
7. *Initiation of continuous improvement process*

An example of the application of this process analysis is shown in Figure 5.

3.6 Health Care Team as a System

Teamwork can take different forms (e.g., autonomous work groups vs. manager-led teams) and be implemented in various modes (e.g., temporary vs. permanent teams) (Sainfort et al., 2001b). In health care, people work together in a variety of teams, such as multidisciplinary teams caring for patients with specific clinical conditions (e.g., ICU team) and operating room teams. Members of the teams often come from different disciplines and educational backgrounds (Kohn et al., 1999). The 1999 IOM Report *To Err Is Human: Building a Safer Health System* (Kohn et al., 1999) recommends the establishment of team training programs for personnel in critical care areas, such as the emergency department, intensive care unit, and operating room. The report *Crossing the Quality Chasm*, published by the Institute of Medicine (2001), goes one step further and emphasizes the need to develop effective teams in order to meet the six challenges of providing safe, effective, efficient, personalized, timely, and equitable care.

Human factors research has been done to study teamwork in health care. Guerlain et al. (2002) have developed a system for evaluating performance of

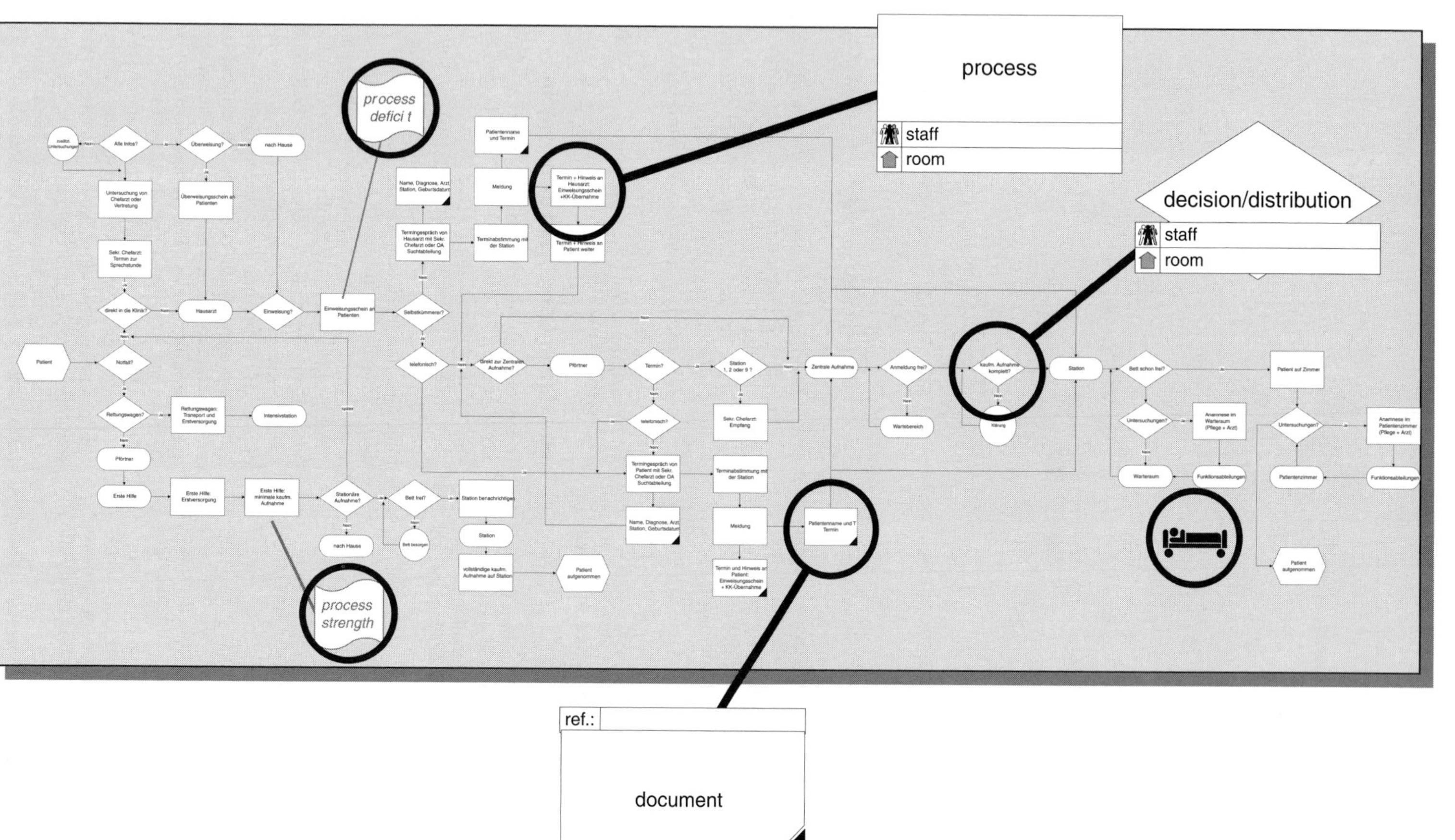

Figure 5 Participatory work process analysis.

surgery teams in the operating room. Helmreich and Schaefer (1994) and Helmreich and Merritt (1998) have examined the performance of OR teams. Their model of OR team performance is comprised of four elements:

- *Team performance input factors*: individual aptitudes, physical condition, team composition, physical environment, professional subcultures, organizational culture, national culture
- *Team performance functions*: team formation and management, surgical and maintenance procedures, communications, decision processes, situational awareness
- *Performance outcomes*: patient well-being, quality and safety of care
- *Individual and organizational outcomes*: attitudes, morale

Research by Salas and colleagues (Tannenbaum et al., 1996; Salas and Cannon-Bowers, 1997; Paris et al., 2000) provides important information on team training that can be used in a health care context: for instance, in the implementation of crew resource management training programs (Salas et al., 1999). Morey et al. (2002) report on the implementation of teamwork in emergency departments and its impact on error reduction and performance improvement. The work redesign had two components: (1) teamwork training and (2) institutionalizing teamwork. The teamwork training curriculum was developed based on crew resource management (CRM) principles (Helmreich and Merritt, 1998; Salas et al., 1999). Physicians, nurses, and technicians in emergency departments of six hospitals were trained in CRM. The training curriculum was organized around five team dimensions: (1) maintain team structure and climate; (2) apply problem-solving strategies; (3) communicate with the team; (4) execute plans and manage workload; and (5) improve team skills. Following the training, a formal teamwork structure was implemented. A team comprised 3 to 10 members, including physicians, nurses, and technicians. The effectiveness of the intervention (training + teamwork implementation) was measured with three types of outcomes: team behaviors, attitudes and opinions, and ED performance (including clinical errors observed). After the teamwork implementation, there were significant improvements in staff attitudes toward teamwork. Data also showed that the number of clinical errors observed went down significantly in the teamwork-trained emergency departments compared to the control group of emergency departments. Another significant finding of this study was that the positive impact of the intervention was maintained to a large extent about eight months after the training. The teamwork intervention was effective in each of the three domains evaluated in the study. The authors of the study emphasize that training physicians, nurses, and technicians was only the first step in the intervention. The training was well received, but implementation of a formal teamwork structure necessitated considerable effort and time on the part of the stakeholders.

4 HUMAN FACTORS IN MEDICAL TECHNOLOGIES

In health care, technologies are often seen as an important solution to improving quality of care and to reduce or eliminate medical errors (Kohn et al., 1999; Bates and Gawande, 2003). These technologies include organizational and work technologies aimed at improving the efficiency and effectiveness of information and communication processes (e.g., computerized order-entry provider systems and electronic medical record systems) and patient care technologies that are involved directly in the care processes (e.g., bar code technology for medication administration). The 1999 IOM report recommends adoption of new technology, such as bar code administration technology, to reduce medication errors (Kohn et al., 1999). However, implementation of new technologies in health care has not been without difficulties or work-arounds. For example, the study of Patterson and colleagues (2002) shows some of the negative side effects of bar code medication administration technology, such as degraded coordination between nurses and physicians. Technologies can change the way work is being performed, and because health care work and processes are complex, negative consequences of new technologies are possible (Cook, 2002).

Efficiency and quality, including patient safety, are primary goals for designing clinical working systems (Carayon, 2003; Carayon et al., 2003). Reducing resources to increase efficiency and improving quality seem to be becoming different goals, even conflicting goals (Reason, 1997). The discipline of human factors and ergonomics might resolve this conflict. However, in the medical–technical industry, human factors knowledge is generally integrated only in a late phase of product development (Jensen, 1999). One aspect of the human factors and ergonomics discipline focuses on the health and safety of health care employees. This is an important goal of human factors and ergonomics, especially given the frequent musculoskeletal disorders experienced by health care workers (Stubbs et al., 1983; Punnett, 1987; Evanoff et al., 1999). This issue is explored further in Chapter 31. The discipline of human factors and ergonomics can also contribute to the development of efficient, safe clinical processes.

4.1 Health Care Technology Design

The human factors characteristics of health care technologies' design should be studied carefully (Battles and Keyes, 2002). An experimental study by Lin et al. (2001) showed the application of human factors engineering principles to the design of the interface of an analgesia device. Results showed that the new interface led to the elimination of drug concentration errors and to a reduction in other errors. A study by Effken et al. (1997) shows the application of a human factors engineering model (i.e., an ecological approach to interface design) to the design of a hemodynamic

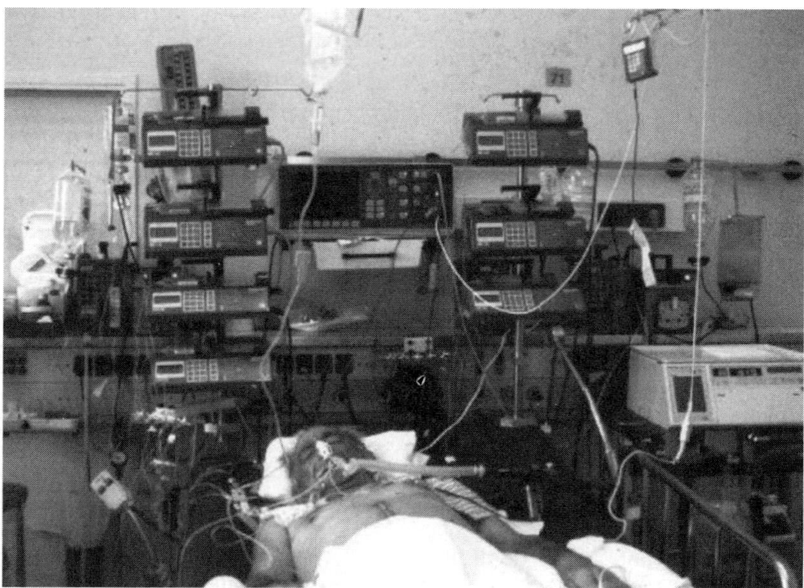

Figure 6 ICU workplace.

monitoring device. New health care technologies may bring their own forms of failure (Reason, 1990; Battles and Keyes, 2002; Cook, 2002). For instance, bar coding technology can prevent patient misidentifications, but the possibility exists that an error during patient registration may be disseminated throughout the information system and may be more difficult to detect and correct than with conventional systems (Wald and Shojania, 2001).

New digital technologies such as surgical navigation or robotic systems are changing clinical working systems dramatically. The risk–benefit assessment of those emerging technologies is difficult. Such an assessment has to consider utilization of the technologies and their impact on task completion (see the concepts of quality and efficiency in Figure 3), as well as the long-term impact, such as a patient's condition five years after intervention. For instance, Cao and Taylor (2004) examined the impact of the introduction of a new robotic technology on performance and communication patterns in an OR team. The new technology, a remote master–slave surgical robot, removes the surgeon from the surgical site. Results of the human factors analysis show large differences in the amount and type of information required by the surgeon to accomplish the procedure with and without the robot. In the robotic work system, the surgeon has additional tasks to perform and additional decisions to make, therefore increasing the cognitive load.

One important health care technology design characteristic is the usability of a medical device. To what extent does the medical device perform in a safe, easy-to-use manner? For a given task and a unified group of users, usability is a measure of the fast (efficient), simple (effective), and satisfying use of a technical device (Ravden and Johnson, 1989; Bevan et al., 1991; Dumas and Redish, 1993; Staggers, 2003). To assess usability, various methods of usability engineering have been developed (Whiteside et al., 1988; Nielsen, 1992, 1993; Wiklund, 1993a; Mayhew, 1999). For the evaluation of the usability of medical technical devices, specific methods have been used, such as user questionnaires, usability inspection methods, and usability tests (Wiklund, 1993b, 1995; Backhaus et al., 2001).

The design of workplaces in ORs and ICUs is determined by a patient's individual treatment requirements. Even though recommendations for the layout of workplaces exist, the number and variety of devices used for any patient lead to an individual arrangement. The patient is integrated in a web of lines, tubes, and cables connected to devices for treatment (e.g., drug delivery pumps and artificial ventilation) and monitoring (e.g., ECG, blood pressure). In an OR or an ICU, it is not rare to find more than 30 connections between a patient and medical devices (see Figure 6). Implementing and removing devices according to the treatment requirements (e.g., an additional drug requires the implementation of an additional pump) includes the risk of errors and mishaps. The concept of an integrated workplace has been proposed, but up to now it has been only partially implemented (Friesdorf et al., 1990).

The concept of an integrated workplace consists of a special and logical structuring (see Figure 7). The ventilator may be placed at an ergonomically adverse location because of the need for remote control via a data manager. The integration and structuring of an

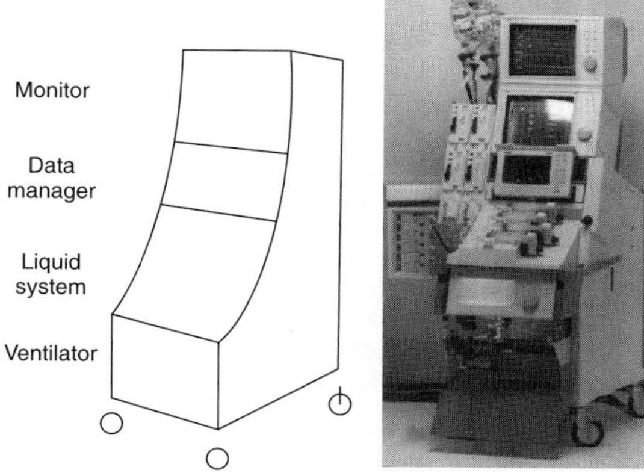

Figure 7 Integrated workplace in an ICU (prototype). (From Friesdorf et al., 1990.)

integrated workplace in an ICU should focus on the following principles (Friesdorf et al., 1990):

- Filtering important information
- Using organ icons and bar graphs for fast orientation
- Separating organ systems vertically: central nervous system, cardiovascular system, ventilation, water–electrolytes
- Separating monitoring variables on the left and related treatment variables on the right
- Employing a touch screen and rotary knob for remote control of all devices with the same control sequence (select by pointing on the touch screen, adjust by turning the knob, and confirm by pushing the knob)

4.2 Health Care Technology Implementation

When looking for solutions to improving the performance, quality, and safety of health care systems, technology may or may not be the only solution. For instance, a study of the implementation of nursing information computer systems in 17 New Jersey hospitals showed many problems experienced by hospitals, such as delays and lack of software customization (Hendrickson et al., 1995). On the other hand, at least initially, nursing staff reported positive perceptions, in particular with regard to documentation (more readable, complete, and timely). However, a more scientific quantitative evaluation of the quality of nursing documentation following the implementation of bedside terminals did not confirm those initial impressions (Marr et al., 1993). The latter result was due to the low use of bedside terminals by nurses. This technology implementation may have ignored the impact of the technology on the tasks performed by the nurses. Nurses may have needed time away from the patient's bedside to organize their thoughts and collaborate with colleagues (Marr et al., 1993). This study demonstrates the need for a human factors systems approach to understand the impact of technology. For instance, instead of using the "leftover" approach to function and task allocation, a human-centered approach to function and task allocation should be used (Hendrick and Kleiner, 2001). This approach considers simultaneous design of the technology and the work system to achieve a balanced work system. One possible outcome of this allocation approach would be to rely on human and organizational characteristics that can foster performance and safety (e.g., autonomy provided at the source of the variance, human capacity for error recovery), instead of trusting the technology completely to achieve high quality and safety of care. It is important to examine for what tasks technology can be useful to provide better, safer care (Hahnel et al., 1992).

The manner in which a new technology is implemented is as critical to its success as its technological capabilities (see, e.g., Eason, 1982; Smith and Carayon, 1995). End-user involvement in the design and implementation of a new technology is a good way to help ensure a successful technological investment. Korunka et al. (1993, 1997) and Korunka and Carayon (1999) have empirically demonstrated the crucial importance of end-user involvement in the implementation of technology to the health and well-being of end users. The implementation of technology in an organization has both positive and negative effects on the job characteristics that ultimately affect individual outcomes (quality of working life, such as job satisfaction and stress; and perceived quality of care delivered or self-rated performance) (Carayon and Haims, 2001). For instance, inadequate planning when introducing a new technology designed to decrease medical errors has led to technology falling short of achieving its patient safety goal (Kaushal and Bates, 2001; Patterson et al., 2002). The most common reason for failure of technology implementations is that the implementation process is treated as a technological problem, and the human and organizational issues are ignored or not recognized (Eason, 1988). When a technology is implemented, several human and organizational issues are important to consider (Carayon-Sainfort, 1992; Smith and Carayon, 1995).

Whenever implementing a technology, one should examine the potential positive *and* negative influences of the technology on the other work system elements (Smith and Carayon-Sainfort, 1989; Kovner et al., 1993; Sheridan and Thompson, 1994; Battles and Keyes, 2002). In a study of the implementation of an electronic medical record (EMR) system in a small family medicine clinic, a number of issues were examined: impact of the EMR technology on work patterns, employee perceptions related to the EMR technology and its potential/actual effect on work, and the EMR implementation process (Carayon and Smith, 2001). Employee questionnaire data showed the following impact of the EMR technology on work. Increased dependence on computers was found, as well as an increase in quantitative workload and a perceived negative influence on performance occurring at least in part from introduction of the EMR (Hundt et al.,

2002). An overview of the implementation of EMR technology in four family practice residency programs highlights some of the benefits of EMR technology and the difficulties associated with their implementation (Swanson et al., 1997). The authors list several benefits: access to clinical information, stimulation of resident research, and greater use of other computer tools. The major barriers were funding, institutional commitment, need for organizational change, and need to create interfaces with outside institutions.

Shortliffe (1999) emphasizes the importance of organizational and management change in successful software implementation in health care systems. He highlights three factors for successful introduction of health care technology: analysis, redesign, and cooperative joint-development efforts. In a similar vein, Blumenthal and Epstein (1996) comment that "the science of behavior modification" has not been applied much in the health care field. It is important to consider the human and organizational aspects that can hinder or foster technological change in health care systems.

The way that change is implemented (i.e., the process of implementation) is central to the successful adaptation of organizations to changes (Korunka et al., 1993; Tannenbaum et al., 1996). A successful technological implementation can be defined by its human and organizational characteristics: reduced/limited negative impact on people (e.g., stress, dissatisfaction) and on the organization (e.g., delays, costs, reduced performance), and increased positive impact on people (e.g., acceptance of change, job control, enhanced individual performance) and on the organization (e.g., efficient implementation process). Various principles for the successful implementation of technological change have been defined in the human factors and ergonomics and business research literature.

Employee participation is a key principle in organizational change (Korunka et al., 1993; Smith and Carayon, 1995; Coyle-Shapiro, 1999). There are research and theory demonstrating the potential benefits of participation in the workplace. Benefits include increased employee motivation and job satisfaction, enhanced performance and employee health, more rapid implementation of technological and organizational change, and more thorough diagnosis and solution formation for ergonomic problems (Gardell, 1977; Lawler, 1986; Noro and Imada, 1991; Wilson and Haines, 1997). The manner in which a new technology is implemented is as critical to its success as its technological capabilities (see, e.g., Eason, 1982; Smith and Carayon, 1995), and end-user participation in the design and implementation of a new technology is a good way to help ensure a successful technological investment. Previous research has made the distinction between *active participation*, where the employees and end users are participating actively in the implementation of the new technology, from *passive participation*, where the employees and end users are *informed about and communicated with* regarding the new technology (Carayon and Smith, 1993).

Technological implementation in health care systems should be considered as an evolving process that requires considerable *learning* and adjustment (Mohrman et al., 1995). The participatory process model developed by Haims and Carayon (1998) specifies the underlying concepts of the learning and adjustment (i.e., action, control, and *feedback*). The importance of feedback in managing the change process is also echoed in the literature on quality management [see, e.g., the plan–do–check–act cycle proposed by Deming (1986)]. Feedback is an important element in order to change behavior (Smith and Smith, 1966) and has been emphasized as an important organizational design element in the health care literature (McDonald et al., 1996; Evans et al., 1998). An aspect of learning in the context of technological change is the type and content of *training* (e.g., Frese et al., 1988; Gattiker, 1992). A questionnaire survey of 244 family practice residents' perceptions regarding the use of EMR showed that those residents who felt that the EMR-related training was adequate were more likely to report benefits due to the EMR, such as decreased time to review past records, increased documentation accuracy, and increased consistency of health maintenance (Aaronson et al., 2001).

Studies have been performed that examine the characteristics of technological change processes that lead to successful implementations in industrial settings. For example, Korunka and Carayon (1999) examined the implementation of information technology in 60 Austrian companies and 18 U.S. companies. Compared with the Austrian implementations, the American implementations were characterized by a higher degree of professionalism (e.g., more use of *project management* tools) and more participation, but at the same time by more negative effects for employees (e.g., personnel reduction).

In summary, the literature on human factors and ergonomics, and change management highlights the following principles for a successful technological change process in health care systems:

- *Employee participation*: extent to which health care staff is involved in various decisions and activities related to the technology implementation
- *Information and communication*: extent to which health care staff is kept informed of the technology implementation through various means of communication
- *Training and learning*: extent and nature of the training provided to the health care staff and extent of learning by the health care staff
- *Feedback*: extent to which feedback is sought after during the technology implementation
- *Project management*: activities related to the organization and management of the technology implementation itself

5 PATIENT SAFETY AND MEDICAL ERRORS

The discipline of human factors and ergonomics has much to offer to the understanding, reduction, and prevention of medical errors, and therefore the improvement of patient safety (as well as employee safety) (Bogner, 1994).

5.1 Patient Safety

The quality and safety of health care is very much discussed across the world [e.g., in Australia (McNeil and Leeder, 1995) and the UK (U.K. Department of Health, 2002)]. In the United States the 1999 publication of a report by the Institute of Medicine has raised the level of public awareness regarding medical errors and patient safety (Kohn et al., 1999).

To what extent do health care systems provide safe high-quality, care? A 2000 report published by the UK Department of Health (2002) provides some data on the extent to which the English health care system fails to provide such care. About 400 people die or are seriously injured in adverse events involving medical devices. About 10,000 people report having experienced serious adverse reactions to drugs. The UK National Health Service pays around £ 400 million a year settlement of clinical negligence claims. Data for the United States indicate that "preventable adverse events are a leading cause of death in the United States" (Kohn et al., 1999). It has been suggested that at least 44,000 and perhaps as many as 98,000 Americans die in hospitals each year as a result of medical errors. Much debate has occurred around the validity of those numbers (Leape, 2000). Whereas there is disagreement regarding the frequency of medical errors in health care, most people agree that system changes need to occur to improve the quality and safety of care (Institute of Medicine, 2001).

Quality of care has been conceptualized and assessed in a variety of manners. First, the performance of a health care practitioner can be evaluated on two dimensions: (1) technical performance, which depends on the knowledge and judgment used to arrive at the diagnostic and strategy of care and the skills in implementing the strategy; and (2) interpersonal performance, which emphasizes the relationship between the practitioner and the patient (Donabedian, 1988). Second, the quality of care can be assessed at various levels: care provided by a practitioner (e.g., physician, nurse) to an individual patient, care provided by a health care institution (e.g., hospital, nursing home), care provided by a health plan, care received by community, and so on (Donabedian, 1988; Brook et al., 1996). Third, quality of care can be evaluated on the basis of structure, process, or outcome (Donabedian, 1988). According to Donabedian (1988), *structure* relates to "the attributes of the settings in which care occurs" and includes material resources, human resources, and organizational structure. *Process* is defined as what is "actually done in giving and receiving care," and *outcome* relates to the "effects of care on the health status of patients and populations." Debate is ongoing as to whether structural, process, or outcome measures of quality should be emphasized (Brook et al., 1996; Clancy and Eisenberg, 1998).

Fourth, quality of care problems have been categorized into misuse (i.e., "occurs when an appropriate service has been selected but a preventable complication occurs and the patient does not receive the full potential benefit of the service"), overuse (i.e., "occurs when a health care service is provided under circumstances in which its potential for harm exceeds the potential benefit"), and underuse (i.e., "failure to provide a health care service when it would have produced a favorable outcome for a patient") (Chassin et al., 1998). According to the 1999 IOM report (Kohn et al., 1999), issues of overuse and underuse should be addressed by changing health care practices and achieving practices consistent with current medical knowledge; and issues of misuse fit the patient safety concerns. However, overuse and underuse can also be related to patient safety, such as too much care provided that put patient safety at risk or too little use of appropriate care that may decrease unnecessary complications (Wakefield, 2001). According to the Agency for Healthcare Research and Quality, "the goal of patient safety is to reduce the risk of injury and harm from preventable medical errors," and according to the IOM, patient safety is freedom from accidental injury (Institute of Medicine, 2001). Patient safety can be considered as one piece of the quality of health care puzzle (Kohn et al., 1999; Institute of Medicine, 2001; Wakefield, 2001).

The different approaches to quality of care and patient safety emphasize the characteristics of the system (or structure) in which care processes occur and which lead to patient outcomes (Bogner, 1994, 2004; Moray, 1994). Therefore, the discipline of human factors and ergonomics has an important role to play in helping in the human-centered design of systems and processes in order to achieve both positive individual and organizational outcomes as well as improved patient outcomes (improved quality and safety of care) (Sainfort et al., 2001a).

5.2 Human Error in Health Care

In the human factors and ergonomics literature, models and approaches of human error have been developed to understand the human mechanisms leading to accidents. One of the most prevalent human error models was defined by Rasmussen (1983) and Reason (1990). This model defines two types of human error: (1) slips and lapses, and (2) mistakes. In turn, mistakes can be categorized as resulting from either rule-based behavior or knowledge-based behavior. This taxonomy of human error has been applied successfully to analyze and evaluate accidents in a range of domains, including the nuclear industry (Rasmussen, 1982; Moray, 1997), aviation (Helmreich and Merritt, 1998) and more recently, in health care (Reason, 2000; Sexton et al., 2000).

Another important distinction brought up by the human error literature is that of active and latent errors (Reason, 1990). *Active errors* have effects that are felt or seen immediately and are associated with the performance of the front-line operators, such as nurses. *Latent errors* are more likely to be related to organizational and management factors that are removed in both time and space from the front-line operations. The distinction between active and latent failures, or between the sharp end and the blunt end (Cook and Woods, 1994), has led to recognition of the importance of organizational, management, and procedural factors in errors and accidents. This had led to the development of a number of models describing the chain of events that can lead to an accident or an adverse outcome. For instance, Vincent and colleagues (1998) proposed an organizational accident model that identifies the following chain of events: latent failures (i.e., management decision, organizational processes) influence conditions of work (i.e., workload, supervision, communication, equipment, knowledge/ability), which in turn can lead to unsafe acts or active failures (i.e., omissions, action slips/failures, cognitive failures, violations) that can lead to accidents or adverse outcomes if the barriers or defense mechanisms are insufficient.

There is increasing recognition in the human error literature of the various levels of factors that can contribute to human error and accidents (Rasmussen, 2000). If the factors are aligned appropriately, like "slices of Swiss cheese," accidents can occur (Reason, 1990). The human error literature has produced a number of lists of factors that can influence clinical practice and affect patient safety. For instance, Vincent et al. (1998) categorize those factors into (1) factors that influence clinical practice (i.e., institutional context, organizational and management factors, work environment, team factors, individual-staff factors, task factors, and patient characteristics), and (2) team factors and their components (i.e., verbal communication, written communication, supervision and seeking help, and structure of team). Human error models and approaches provide much information on how to understand, analyze, and evaluate near misses and accidents in healthcare (van der Schaaf, 1992; Kaplan et al., 1998; Shojania et al., 2002).

6 FUTURE NEEDS

The discipline of human factors and ergonomics has much to offer to improve the performance, quality and safety of health care systems. Given the people-intensive, people-centered, people-driven characteristic of health care, human factors and ergonomics can provide the models, concepts, and methods necessary to consider the people component of health care systems.

Some of the human factors and ergonomics models, concepts, and methods need to be adapted to the characteristics of health care (see Section 1). For instance, a key principle of human factors and ergonomics is user participation and involvement. Two characteristics of health care contribute to the difficulty of implementing this principle. First is the definition of the user (see Section 2). Second, health care is a very dynamic, changing environment with much time pressure. This makes the application of participatory approaches difficult when the users have little time to spare and spend on those human factors and ergonomics activities (see Section 2.3). More work needs to be done to pursue and expand the effort of considering the unique characteristics of health care in human factors and ergonomics work. Hignett (2003) also argues for the discipline of human factors and ergonomics to develop "more context-sensitive methodology" for health care.

Another important issue in human factors and ergonomics in medicine is the necessity of combining human factors technical knowledge with health care knowledge. For instance, in observing medical work such as anesthesia processes, the observer needs to have knowledge of anesthesia to interpret the activities meaningfully (Norros and Klemola, 1999). On the other hand, Weinger et al. (1994) developed a highly structured task analysis that could be used by non-medically trained observers in the observation of an anesthesiologist's work. Carayon et al. (2004b, 2005) discussed criteria to consider when combining human factors and ergonomics and health care knowledge in observations of clinical work. The impact of human factors and ergonomics on medicine will be strengthened by encouraging encounters between human factors and ergonomics and medicine knowledge and the development of collaborations between human factors and ergonomics and health care subject-matter experts.

7 CONCLUSIONS

In this chapter we have reviewed a limited set of human factors and ergonomics issues in medicine. For instance, the issue of working conditions and workload in health care has not been addressed in this chapter, but much has been written on this topic, in particular regarding nursing (Carayon et al., 2003a; Institute of Medicine, 2004). Other important human factors and ergonomics issues in medicine are addressed in Chapters 5, 16, 19, 22, 27, 31, and 50.

REFERENCES

Aaronson, J. W., Murphy-Cullen, C. L., Chop, W. M., and Frey, R. D. (2001), "Electronic Medical Records: The Family Practice Resident Perspective," *Family Medicine*, Vol. 33, No. 2, pp. 128–132.

Agency for Healthcare Research and Quality (2002), *Health Care Costs: Fact Sheet*, Publication 02-P033, AHRQ, Rockville, MD.

Aiken, L. H., Clarke, S. P., Sloane, D. M., Sochalski, J. A., Busse, R., Clarke, H., et al. (2001), "Nurses' Reports on Hospital Care in Five Countries," *Health Affairs*, Vol. 20, No. 3, pp. 43–53.

Backhaus, C., Papanikolaou, M., Kuhnigk, S., and Friesdorf, W. (2001), "Usability-Engineering: Eine Methodenübersicht zur anwendergerechten Gestaltung von Medizinprodukten," *Medizintechnik*, Vol. 121, pp. 133–138.

Bates, D. W., and Gawande, A. A. (2003), "Improving Safety with Information Technology," *New England Journal of Medicine*, Vol. 348, No. 25, pp. 2526–2534.

Battles, J. B., and Keyes, M. A. (2002), "Technology and Patient Safety: A Two-Edged Sword," *Biomedical Instrumentation and Technology*, Vol. 36, No. 2, pp. 84–88.

Bevan, N., Kirakowski, J., and Maissel, J. (1991), "What Is Usability?" in *Human Aspects in Computing: Design and Use*, H. J. Bullinger, Ed., Elsevier, Amsterdam. pp. 651–655.

Blegen, M. A. (1993), "Nurses' Job Satisfaction: A Meta-analysis of Related Variables," *Nursing Research*, Vol. 42, No. 1, pp. 36–41.

Blumenthal, D., and Epstein, A. M. (1996), "Part 6: The Role of Physicians in the Future of Quality Management," *New England Journal of Medicine*, Vol. 335, No. 17, pp. 1328–1331.

Bogner, M. S., Ed. (1994), *Human Error in Medicine*, Lawrence Erlbaum Associates, Mahwah, NJ.

Bogner, M. S. (2004), "Understanding Human Error," in *Misadventures in Health Care: Inside Stories*, M. S. Bogner, Ed., Lawrence Erlbaum Associates, Mahwah, NJ, pp. 1–12.

Bohr, P. C., Evanoff, B. A., and Wolf, L. (1997), "Implementing Participatory Ergonomics Teams Among Health Care Workers," *American Journal of Industrial Medicine*, Vol. 32, pp. 190–196.

Bracco, D., Favre, J. -B., Bissonnette, B., Wasserfallen, J. -B., Revelly, J. -P., Ravussin, P., et al. (2000), "Human Errors in a Multidisciplinary Intensive Care Unit: A 1-Year Prospective Study," *Intensive Care Medicine*, Vol. 27, No. 1, pp. 137–145.

Brennan, P. F., and Aronson, A. R. (2003), "Towards Linking Patients and Clinical Information: Detecting UMLS Concepts in E Mail," *Journal of Biomedical Informatics*, Vol. 36, No. 4–5, pp. 334–341.

Breslow, M. J., Rosenfeld, B. A., Doerfler, M., Burke, G., Yates, G., Stone, D. J., et al. (2004), "Effect of a Multiple-Site Intensive Care Unit Telemedicine Program on Clinical and Economic Outcomes: An Alternative Paradigm for Intensivist Staffing," *Critical Care Medicine*, Vol. 32, No. 1, pp. 31–38.

Brook, R. H., McGlynn, E. A., and Cleary, P. D. (1996), "Quality of Health Care, 2: Measuring Quality of Care," *New England Journal of Medicine*, Vol. 335, No. 13, pp. 966–970.

Bureau of Labor Statistics, U.S. Department of Labor (2004), "Career Guide to Industries, 2004–05 Edition: Health Services," retrieved September 1, 2004, from http://www.bls.gov/oco/cg/cgs035.htm.

Callejas, S., Barry, A., Demertsidis, E., Jorgenson, D., and Becker, L. B. (2004), "Human Factors Impact Successful Lay Person Automated External Defibrillator Use During Simulated Cardiac Arrest," *Critical Care Medicine*, Vol. 32, No. 9.

Cao, C. G. L., and Taylor, H. (2004), "Effects of New Technology on the Operating Room Team," in *Work with Computing Systems, 2004*, H. M. Khalid, M. G. Helander, and A. W. Yeo, Eds., Damai Sciences, Kuala Lumpur, Malaysia.

Carayon, P. (2003), "Macroergonomics in Quality of Care and Patient Safety," in *Human Factors in Organizational Design and Management*, H. Luczak and K. J. Zink, Eds., IEA Press, Santa Monica, CA. pp. 21–35.

Carayon, P., and Haims, M. C. (2001), "Information & Communication Technology and Work Organization: Achieving a Balanced System," In *Humans on the Net-Information & Communication Technology (ICT), Work Organization and Human Beings*, G. Bradley, Ed. Prevent, Stockholm, Sweden, pp. 119–138.

Carayon, P., and Smith, M. J. (1993), "The Balance Theory of Job Design and Stress as a Model for the Management of Technological Change," presented at the 4th International Congress of Industrial Engineering, Marseille, France.

Carayon, P., and Smith, M. J. (2000), "Work Organization and Ergonomics," *Applied Ergonomics*, Vol. 31, pp. 649–662.

Carayon, P., and Smith, P. D. (2001), "Evaluating the Human and Organizational Aspects of Information Technology Implementation in a Small Clinic," in *Systems, Social and Internationalization Design Aspects of Human–Computer Interaction*, M. J. Smith and G. Salvendy, Eds., Lawrence Erlbaum Associates, Mahwah, NJ, pp. 903–907.

Carayon, P., Alvarado, C., and Hundt, A. S. (2003a), *Reducing Workload and Increasing Patient Safety Through Work and Workspace Design*, Institute of Medicine, Washington, DC.

Carayon, P., Alvarado, C. J., Brennan, P., Gurses, A., Hundt, A., Karsh, B. -T., et al. (2003b), "Work System and Patient Safety," in *Human Factors in Organizational Design and Management, VII*, H. Luczak and K. J. Zink, Eds., IEA Press, Santa Monica, CA, pp. 583–589.

Carayon, P., Schultz, K., and Hundt, A. S. (2004a), "Righting Wrong Site Surgery," *Joint Commission Journal on Quality and Safety*, Vol. 30, No. 7, pp. 405–410.

Carayon, P., Wetterneck, T. B., Hundt, A. S., Ozkaynak, M., Ram, P., DeSilvey, J., et al. (2004b), "Assessing Nurse Interaction with Medication Administration Technologies: The Development of Observation Methodologies," in *Work with Computing Systems, 2004*, H. M. Khalid, M. G. Helander, and A. W. Yeo, Eds., Damai Sciences, Kuala Lumpur, Malaysia, pp. 319–324.

Carayon, P., Wetterneck, T. B., Hundt, A. S., Ozkaynak, M., Ram, P., DeSilvey, J., et al. (2005). "Observing Nurse Interaction with Medication Administration Technologies," in *Advances in Patient Safety: From Research to Implementation*, Vol. 2, K. Henriksen, J. B. Battles, E. Marks, and D. I. Lewin, Eds., Agency for Healthcare Research, Rockville, MD, pp. 349–364.

Carayon-Sainfort, P. (1992), "The Use of Computers in Offices: Impact on Task Characteristics and Worker Stress," *International Journal of Human–Computer Interaction*, Vol. 4, No. 3, pp. 245–261.

Chassin, M. R. (1998), "Is Health Care Ready for Six Sigma Quality?" *Milbank Quarterly*, Vol. 76, No. 4, pp. 565–591.

Chassin, M. R., Galvin, R. W., and The National Roundtable on Health Care Quality (1998), "The Urgent Need to Improve Health Care Quality," *Journal of the American Medical Association*, Vol. 280, No. 11, pp. 1000–1005.

Clancy, C. M., and Eisenberg, J. M. (1998), "Outcomes Research: Measuring the End Results of Health Care," *Science*, Vol. 282, No. 5387, pp. 245–246.

Cook, R. (2004), "Observational and Ethnographic Studies: Insights on Medical Error and Patient Safety," presented at the 6th Annual NPSF Patient Safety Congress, Boston.

Cook, R. I. (2002), "Safety Technology: Solutions or Experiments?" *Nursing Economic$*, Vol. 20, No. 2, pp. 80–82.

Cook, R. I., and Woods, D. D. (1994), "Operating at the Sharp End: The Complexity of Human Error," in *Human Error in Medicine*, M. S. Bogner, Ed., Lawrence Erlbaum Associates, Mahwah, NJ, pp. 255–310.

Coyle-Shapiro, J. A. -M. (1999), "Employee Participation and Assessment of an Organizational Change Intervention," *Journal of Applied Behavioral Science*, Vol. 35, No. 4, pp. 439–456.

Deming, W. E. (1986), *Out of the Crisis*, Massachussets Institute of Technology, Cambridge, MA.

Donabedian, A. (1988), "The Quality of Care: How Can It Be Assessed?" *Journal of the American Medical Association*, Vol. 260, No. 12, pp. 1743–1748.

Dumas, J., and Redish, J. C. (1993), *A Practical Guide to Usability Testing*, Ablex, Norwood, NJ.

Eason, K. D. (1982), "The Process of Introducing Information Technology," *Behaviour and Information Technology*, Vol. 1, No. 2, pp. 197–213.

Eason, K. (1988), *Information Technology and Organizational Change*, Taylor & Francis, London.

Effken, J. A., Kim, M. -G., and Shaw, R. E. (1997), "Making the Constraints Visible: Testing the Ecological Approach to Interface Design," *Ergonomics*, Vol. 40, No. 1, pp. 1–27.

Evanoff, V. A., Bohr, P. C., and Wolf, L. (1999), "Effects of a Participatory Ergonomics Team Among Hospital Orderlies," *American Journal of Industrial Medicine*, Vol. 35, pp. 358–365.

Evans, R. S., Pestotnik, S. L., Classen, D. C., Clemmer, T. P., Weaver, L. K., Orme, J. F. J., et al. (1998), "A Computer-Assisted Management Program for Antibiotics and Other Antiinfective Agents," *New England Journal of Medicine*, Vol. 338, No. 4, pp. 232–238.

Feldman, P. H. (1993), "Work Life Improvements for the Home Aide Work Force: Impact and Feasibility," *Gerontologist*, Vol. 33, No. 1, pp. 47–54.

Frese, M., Albrecht, K., Altmann, A., Lang, J., Papstein, P., Peyerl, R., et al. (1988), "The Effects of an Active Development of the Mental Model in the Training Process: Experimental Results in a Word Processing System," *Behaviour and Information Technology*, Vol. 7, No. 3, pp. 295–304.

Friesdorf, W., Schwilk, B., Hahnel, J., Fett, P., and Weideck, H. (1990), "Ergonomics Applied to an Intensive Care Workplace," *Intensive Care World*, Vol. 7, No. 4, pp. 192–198.

Friesdorf, W., Marsolek, I., and Goebel, M. (2002), "Integrative Concepts for the OR," *Journal of Clinical Monitoring and Computing*, Vol. 17, No. 7–8, pp. 489–490.

Gardell, B. (1977), "Autonomy and Participation at Work," *Human Relations*, Vol. 30, pp. 515–533.

Gattiker, U. E. (1992), "Computer Skills Acquisition: A Review and Future Directions for Research," *Journal of Management*, Vol. 18, No. 3, pp. 547–574.

Gray-Toft, P., and Anderson, J. G. (1981), "The Nursing Stress Scale: Development of an Instrument," *Journal of Behavioral Assessment*, Vol. 3, No. 1, pp. 11–23.

Guerlain, S., Shin, T., Guo, H., Adams, R., and Calland, J. F. (2002), "A Team Performance Data Collection and Analysis System," presented at the Human Factors and Ergonomics Society 46th Annual Meeting, Baltimore.

Hahnel, J., Friesdorf, W., Schwilk, B., Marx, T., and Blessing, S. (1992), "Can a Clinician Predict the Technical Equipment a Patient Will Need During Intensive Care Unit Treatment? An Approach to Standardize and Redesign the Intensive Care Unit Workstation," *Journal of Clinical Monitoring*, Vol. 8, No. 1, pp. 1–6.

Haims, M. C., and Carayon, P. (1998), "Theory and Practice for the Implementation of 'In-House,' Continuous Improvement Participatory Ergonomic Programs," *Applied Ergonomics*, Vol. 29, No. 6, pp. 461–472.

Hammer, M., and Champy, C. (1993), *Reengineering the Corporation: A Manifesto for Business Revolution*, Harper Business, New York.

Helmreich, R. L., and Merritt, A. C. (1998), *Culture at Work in Aviation and Medicine*, Ashgate, Aldershot, Hampshire, England.

Helmreich, R. L., and Schaefer, H. (1994), "Team Performance in the Operating Room," in *Human Error in Medicine*, M. S. Bogner, Ed., Lawrence Erlbaum Associates, Mahwah, NJ, pp. 225–253.

Hendrick, H. W., and Kleiner, B. M. (2001), *Macroergonomics: An Introduction to Work System Design*, Human Factors and Ergonomics Society, Santa Monica, CA.

Hendrickson, G., Kovner, C. T., Knickman, J. R., and Finkler, S. A. (1995), "Implementation of a Variety of Computerized Bedside Nursing Information Systems in 17 New Jersey Hospitals," *Computers in Nursing*, Vol. 13, No. 3, pp. 96–102.

Hignett, S. (2003), "Hospital Ergonomics: A Qualitative Study to Explore the Organizational and Cultural Factors," *Ergonomics*, Vol. 46, No. 9, pp. 882–903.

Hundt, A. S., Carayon, P., Smith, P. D., and Kuruchittham, V. (2002), "A Macroergonomic Case Study Assessing Electronic Medical Record Implementation in a Small Clinic," presented at the Human Factors and Ergonomics Society 46th Annual Meeting, Baltimore.

Institute of Medicine, Committee on Quality of Health Care in America (2001), *Crossing the Quality Chasm: A New Health System for the 21st Century*, National Academy Press, Washington, DC.

Institute of Medicine, Committee on the Work Environment for Nurses and Patient Safety (2004), *Keeping Patients Safe: Transforming the Work Environment of Nurses*, National Academy Press, Washington, DC.

International Ergonomics Association (2000), *The Discipline of Ergonomics*, retrieved August 22, 2004, from http://www.iea.cc/ergonomics/.

Jensen, R. C. (1999), "Ergonomics in Health Care Organizations," in *The Occupational Ergonomics Handbook*, W. Karwowski and W. S. Marras, Eds., CRC Press, Boca Raton, FL, pp. 1949–1960.

Jones, J. F., and Brennan, P. F. (2002), "Telehealth Interventions to Improve Clinical Nursing of Elders," *Annual Review of Nursing Research*, Vol. 20, pp. 293–322.

Kaplan, H. S., Battles, J. B., Van der Schaaf, T. W., Shea, C. E., and Mercer, S. Q. (1998), "Identification and Classification of the Causes of Events in Transfusion Medicine" [see comments], *Transfusion*, Vol. 38, No. 11–12, pp. 1071–1081.

Kaushal, R., and Bates, D. W. (2001), "Computerized Physician Order Entry (CPOE) with Clinical Decision Support Systems (CDSSs)," in *Making Health Care Safer: A Critical Analysis of Patient Safety Practices*, K. G. Shojania, B. W. Duncan, K. M. McDonald, and R. M. Wachter, Eds., *Evidence Report/Technology Assessment*, Agency for Healthcare Research and Quality, Rockville, MD, pp. 59–69.

Knaus, W. A., Draper, E. A., Wagner, D. P., and Zimmerman, J. E. (1986), "An Evaluation of Outcome from Intensive Care in Major Medical Centers," *Annals of Internal Medicine*, Vol. 104, pp. 410–418.

Kohn, L. T., Corrigan, J. M., and Donaldson, M. S., Eds. (1999), *To Err Is Human: Building a Safer Health System*, National Academy Press, Washington, DC.

Korunka, C., and Carayon, P. (1999), "Continuous Implementations of Information Technology: The Development of an Interview Guide and a Cross-National Comparison of Austrian and American Organizations," *International Journal of Human Factors in Manufacturing*, Vol. 9, No. 2, pp. 165–183.

Korunka, C., Weiss, A., and Karetta, B. (1993), "Effects of New Technologies with Special Regard for the Implementation Process Per Se," *Journal of Organizational Behavior*, Vol. 14, No. 4, pp. 331–348.

Korunka, C., Zauchner, S., and Weiss, A. (1997), "New Information Technologies, Job Profiles, and External Workload as Predictors of Subjectively Experienced Stress and Dissatisfaction at Work," *International Journal of Human–Computer Interaction*, Vol. 9, No. 4, pp. 407–424.

Kovner, C. T., Hendrickson, G., Knickman, J. R., and Finkler, S. A. (1993), "Changing the Delivery of Nursing Care: Implementation Issues and Qualitative Findings," *Journal of Nursing Administration*, Vol. 23, No. 11, pp. 24–34.

Lawler, E. E., III (1986), *High Involvement Management: Participative Strategies for Improving Organizational Performance*, Jossey-Bass, San Francisco.

Leape, L. L. (2000), "Institute of Medicine Medical Error Figures Are Not Exaggerated," *Journal of the American Medical Association*, Vol. 284, No. 1, pp. 95–97.

Lin, L., Vicente, K. J., and Doyle, D. J. (2001), "Patient Safety, Potential Adverse Drug Events, and Medical Device Design: A Human Factors Engineering Approach," *Journal of Biomedical Informatics*, Vol. 34, No. 4, pp. 274–284.

Linkous, J. D. (2001), "Toward a Rapidly Evolving Definition of Telemedicine," retrieved August 30, 2004, from http://www.atmeda.org/about/aboutata.htm.

Marr, P. B., Duthie, E., Glassman, K. S., Janovas, D. M., Kelly, J. B., Graham, E., et al. (1993), "Bedside Terminals and Quality of Nursing Documentation," *Computers in Nursing*, Vol. 11, No. 4, pp. 176–182.

Marsolek, I., and Friesdorf, W. (2004), "Optimising Clinical Process Flows: Experiences from the Expert Systems," *International Journal of Intensive Care*, Winter Issue, pp. 172–179

Mayhew, D. J. (1999), *The Usability Engineering Lifecycle*, Morgan Kaufmann, San Francisco.

McDonald, C. J., Overhage, J. M., Tierney, W. M., Abernathy, G. R., and Dexter, P. R. (1996), "The Promise of Computerized Feedback Systems for Diabetes Care," *Annals of Internal Medicine*, Vol. 124, No. 1S-II (Suppl.), pp. 170–174.

McNeil, J. J., and Leeder, S. R. (1995), "How Safe Are Australian Hospitals?" *Medical Journal of Australia*, Vol. 163, No. 6, pp. 472–475.

Meister, D., and Enderwick, T. P. (2001), *Human Factors in System Design, Development, and Testing*, Lawrence Erlbaum Associates, Mahwah, NJ.

Mohrman, S. A., Cohen, S. G., and Mohrman, A. M. J. (1995), *Designing Team-Based Organizations: New Forms of Knowledge and Work*, Jossey Bass, San Francisco.

Moray, N. (1994), "Error Reduction as a Systems Problem," in *Human Error in Medicine*, M. S. Bogner, Ed., Lawrence Erlbaum Associates, Mahwah, NJ, pp. 67–91.

Moray, N. (1997), "Human Factors in Process Control," in *Handbook of Human Factors and Ergonomics*, 2nd ed., G. Salvendy, Ed., Wiley, New York, pp. 1944–1971.

Morey, J. C., Simon, R., Jay, G. D., Wears, R. L., Salisbury, M., Dukes, K. A., et al. (2002), "Error Reduction and Performance Improvement in the Emergency Department Through Formal Teamwork Training: Evaluation Results of the MedTeams Project," *Health Service Research*, Vol. 37, No. 6, pp. 1553–1581.

Nielsen, J. (1992), "The Usability-Engineering-Lifecycle," *IEEE Computer*, Vol. 25, No. 3, pp. 12–22.

Nielsen, J. (1993), *Usability Engineering*, Morgan Kaufmann, Amsterdam.

Norman, D. A. (1988), *The Psychology of Everyday Things*, Basic Books, New York.

Noro, K., and Imada, A. (1991), *Participatory Ergonomics*, Taylor & Francis, London.

Norros, L., and Klemola, U. -M. (1999), "Methodological Considerations in Analysing Anaesthetists' Habits of Action in Clinical Situations," *Ergonomics*, Vol. 42, No. 11, pp. 1521–1530.

Paris, C. R., Salas, E., and Cannon-Bowers, J. A. (2000), "Teamwork in Multi-person Systems: A Review and Analysis," *Ergonomics*, Vol. 43, No. 8, pp. 1052–1075.

Patterson, E. S., Cook, R. I., and Render, M. L. (2002), "Improving Patient Safety by Identifying Side Effects from Introducing Bar Coding in Medication Administration," *Journal of the American Medial Informatics Association*, Vol. 9, pp. 540–553.

Perrow, C. (1984), *Normal Accidents: Living with High-Risk Technologies*, Basic Books, New York.

Punnett, L. (1987), "Upper Extremity Musculoskeletal Disorders in Hospital Workers," *Journal of Hand Surgery*, Vol. 12A, No. 5, Pt. 2, pp. 858–862.

Randell, R. (2003), "User Customisation of Medical Devices: The Reality and the Possibilities," *Cognition, Technology and Work*, Vol. 5, pp. 163–170.

Rasmussen, J. (1982), "Human Errors: A Taxonomy for Describing Human Malfunction in Industrial Installations," *Journal of Occupational Accidents*, Vol. 4, pp. 311–333.

Rasmussen, J. (1983), "Skills, Rules, and Knowledge; Signals, Signs, and Symbols, and Other Distinctions in Human Performance Models," *IEEE Transactions on Systems, Man, and Cybernetics*, Vol. 13, No. 3, pp. 257–266.

Rasmussen, J. (2000), "Human Factors in a Dynamic Information Society: Where Are We Heading?" *Ergonomics*, Vol. 43, No. 7, pp. 869–879.

Rasmussen, B., and Doodstein, L. P. (1988), "Information Technology and Work," in *Handbook of Human–Computer Interaction*, M. Helander, Ed., North-Holland, Amsterdam.

Ravden, S., and Johnson, G. (1989), *Evaluating Usability of Human–Computer Interfaces: A Practical Method*, Wiley, New York.

Reason, J. (1990), *Human Error*, Cambridge University Press, Cambridge.

Reason, J. (1997), *Managing the Risks of Organizational Accidents*, Ashgate, Brookfield, Vermont.

Reason, J. (2000), "Human Error: Models and Management," *British Medical Journal*, Vol. 320, No. 7237, pp. 768–770.

Reiling, J. G., Knutzen, B. L., Wallen, T. K., McCullough, S., Miller, R. H., and Chernos, S. (2004), "Enhancing the Traditional Design Process: A Focus on Patient

Safety," *Joint Commission Journal on Quality Improvement*, Vol. 30, No. 3, pp. 115–124.

Sainfort, F., Karsh, B., Booske, B. C., and Smith, M. J. (2001a), "Applying Quality Improvement Principles to Achieve Healthy Work Organizations," *Journal on Quality Improvement*, Vol. 27, No. 9, pp. 469–483.

Sainfort, F., Taveira, A. D., Arora, N., and Smith, M. J. (2001b), "Teams and Team Management and Leadership," in *Handbook of Industrial Engineering*, G. Salvendy, Ed., Wiley, New York, pp. 975–994.

Salas, E., and Cannon-Bowers, J. A. (1997), "Methods, Tools, and Strategies for Team Training," in *Training for a Rapidly Changing Workforce: Applications of Psychological Research*, M. A. Quinones and A. Ehrenstein, Eds., American Psychological Association, Washington, DC, pp. 249–279.

Salas, E., Prince, C., Bowers, C. A., Stout, R. J., Oser, R. L., and Cannon-Bowers, J. A. (1999), "A Methodology for Enhancing Crew Resource Management Training," *Human Factors*, Vol. 41, No. 1, pp. 161–172.

Sexton, J. B., Thomas, E. J., and Helmreich, R. L. (2000), "Error, Stress, and Teamwork in Medicine and Aviation: Cross Sectional Surveys," *British Medical Journal*, Vol. 320, No. 7237, pp. 745–749.

Sheridan, T. B., and Thompson, J. M. (1994), "People Versus Computers in Medicine," in *Human Error in Medicine*, M. S. Bogner, Ed., Lawrence Erlbaum Associates, Mahwah, NJ, pp. 141–158.

Shojania, K. G., Wald, H., and Gross, R. (2002), "Understanding Medical Error and Improving Patient Safety in the Inpatient Setting," *Medical Clinics of North America*, Vol. 86, No. 4, pp. 847–867.

Shortell, S. M., Zimmerman, J. E., Rousseau, D. M., Gillies, R. R., Wagner, D. P., Draper, E. A., et al. (1994), "The Performance of Intensive Care Units: Does Good Management Make a Difference?" *Medical Care*, Vol. 32, No. 5, pp. 508–525.

Shortliffe, E. H. (1999), "The Evolution of Electronic Medical Records," *Academic Medicine*, Vol. 74, No. 4, pp. 414–419.

Simmons, B. L., Nelson, D. L., and Neal, L. (2001), "A Comparison of the Positive and Negative Work Attitudes of Home Health Care and Hospital Nurses," *Health Care Management Review*, Vol. 26, No. 3, pp. 63–74.

Smith, M. J., and Carayon, P. (1995), "New Technology, Automation, and Work Organization: Stress Problems and Improved Technology Implementation Strategies," *International Journal of Human Factors in Manufacturing*, Vol. 5, No. 1, pp. 99–116.

Smith, M. J., and Carayon-Sainfort, P. (1989), "A Balance Theory of Job Design for Stress Reduction," *International Journal of Industrial Ergonomics*, Vol. 4, pp. 67–79.

Smith, K. U., and Smith, M. F. (1966), *Cybernetic Principles of Learning and Educational Design*, Holt, Rhinehart and Winston, New York.

Staggers, N. (2003), "Human Factors: Imperative Concepts for Information Systems in Critical Care," *AACN Clinical Issues*, Vol. 14, No. 3, pp. 310–319.

Statistisches Bundesamt (2005), Gesundheitswesen—Krankenhäuser und Vorsorgeoder Rehabilitationseinrichtungen—Einrichtungen, Betten und Patientenbewegung, http://www.destatis.de/themen/d/thm_gesundheit.php# Krankenhäuser.

Stone, R. I., and Wiener, J. M. (2001), *Who Will Care for Us? Addressing the Long-Term Workforce Crisis*, Urban Institute and American Association of Homes and Services for the Aging, Washington, DC.

Stubbs, D. A., Buckle, P., Hudson, M. P., Rivers, P. M., and Worringham, C. J. (1983), "Back Pain in the Nursing Profession, I: Epidemiology and Pilot Methodology," *Ergonomics*, Vol. 26, pp. 755–765.

Sultz, H. A., and Young, K. M. (2001), *Health Care USA: Understanding Its Organization and Delivery*, Aspen, Gaithersburg, MD.

Swanson, T., Dostal, J., Eichhorst, B., Jernigan, C., Knox, M., and Roper, K. (1997), "Recent Implementations of Electronic Medical Records in Four Family Practice Residency Programs," *Academic Medicine*, Vol. 72, No. 7, pp. 607–612.

Tannenbaum, S. I., Salas, E., and Cannon-Bowers, J. A. (1996), "Promoting Team Effectiveness," in *Handbook of Work Group Psychology*, M. A. West, Ed., Wiley, New York, pp. 503–530.

Thomas, E. J., Sexton, J. B., and Helmreich, R. L. (2003), "Discrepant Attitudes About Teamwork Among Critical Care Nurses and Physicians," *Critical Care Medicine*, Vol. 31, No. 3, pp. 956–959.

Topf, M. (2000), "Hospital Noise Pollution: An Environmental Stress Model to Guide Research and Clinical Interventions," *Journal of Advanced Nursing*, Vol. 31, No. 3, pp. 520–528.

U.K. Department of Health (2002), *An Organisation with a Memory: Report of an Expert Group on Learning from Adverse Events in the NHS*, U.K. Department of Health, London.

Van Cott, H. (1994), "Human Errors: Their Causes and Reduction," in *Human Error in Medicine*, M. S. Bogner, Ed., Lawrence Erlbaum Associates, Mahwah, NJ, pp. 53–65.

van der Schaaf, T. W. (1992), Near Miss Reporting in the Chemical Process Industry, Unpublished Doctoral thesis, Eindhoven University of Technology, Eindhoven, the Netherlands.

Vicente, K. J. (1999), *Cognitive Work Analysis*, Lawrence Erlbaum Associates, Mahwah, NJ.

Vincent, C. A. (2004), "Analysis of Clinical Incidents: A Window on the System Not a Search for Root Causes," *Quality and Safety in Health Care*, Vol. 13, pp. 242–243.

Vincent, C., Taylor-Adams, S., and Stanhope, N. (1998), "Framework for Analysing Risk and Safety in Clinical Medicine," *British Medical Journal*, Vol. 316, No. 7138, pp. 1154–1157.

Wakefield, M. K. (2001), "The Relationship Between Quality and Patient Safety," in *Lessons in Patient Safety*, L. Zipperer and S. Cushman, Eds., National Patient Safety Foundation, Chicago, pp. 15–19.

Wald, H., and Shojania, K. (2001), "Prevention of Misidentifications," in *Making Health Care Safer: A Critical Analysis of Patient Safety Practices*, D. G. Shojania, B. W. Duncan, K. M. McDonald, and R. M. Wachter, Eds., Publication 01-E058, Agency for Healthcare Research and Quality, Washington, DC, pp. 491–503.

Watson, M., Sanderson, P., and Russell, W. J. (2004), "Tailoring Reveals Information Requirements: The Case of Anaesthesia Alarms," *Interacting with Computers*, Vol. 16, No. 2, pp. 271–293.

Weinger, M. B., Herndon, O. W., Zornow, M. H., Paulus, M. P., Gaba, D. M., and Dallen, L. T. (1994), "An Objective Methodology for Task Analysis and Workload

Assessment in Anesthesia Providers," *Anesthesiology*, Vol. 80, No. 1, pp. 77–92.

Weinger, M. B., Slagle, J., Jain, S., and Ordonez, N. (2003), "Retrospective Data Collection and Analytical Techniques for Patient Safety Studies," *Journal of Biomedical Informatics*, Vol. 36, No. 1–2, pp. 106–119.

Whiteside, J., Bennett, J., and Holtzblatt, K. (1988), "Usability Engineering: Our Experience and Evolution," in *Handbook of Human–Computer Interaction*, M. Helander, Ed., North-Holland, Amsterdam, pp. 791–817.

Wickens, C. D., Lee, J. D., Liu, Y., and Becker, S. E. G. (2004), *An Introduction to Human Factors Engineering*, 2nd ed., Prentice Hall, Upper Saddle River, NJ.

Wiklund, M. E. (1993a), "How to Implement Usability Engineering," *Medical Device and Diagnostic Industry*, Vol. 15, No. 9, pp. 68–73.

Wiklund, M. E. (1993b), "Usability Tests of Medical Products as a Prelude to the Clinical Trial," *Medical Device and Diagnostic Industry*, Vol. 3, No. 7, pp. 68–73.

Wiklund, M. E. (1995), *Medical Device and Equipment Design: Usability Engineering and Ergonomics*, Interpharm Press, Buffalo Grove, IL.

Wilson, J. R. (1995), "Ergonomics and Participation," in *Evaluation of Human Work*, 2nd ed., J. R. Wilson and E. N. Corlett, Eds., Taylor & Francis, London, pp. 1071–1096.

Wilson, J. R., and Haines, H. M. (1997), "Participatory Ergonomics," in *Handbook of Human Factors and Ergonomics*, 2nd ed., G. Salvendy, Ed., Wiley, New York, pp. 490–513.

Woods, D., and Cook, R. (2001), "From Counting Failures to Anticipating Risk: Possible Futures for Patient Safety," in *Lessons in Patient Safety*, L. Zipperer and S. Cushman, Eds., National Patient Safety Foundation, Chicago, pp. 89–97.

Wunderlich, G. S., and Kohler, P. O., Eds. (2001), *Improving the Quality of Long-Term Care*, National Academy Press, Washington, DC.

CHAPTER 59

HUMAN FACTORS AND ERGONOMICS IN MOTOR VEHICLE TRANSPORTATION

David W. Eby and Barry H. Kantowitz
University of Michigan Transportation Research Institute
Ann Arbor, Michigan

1 CHAPTER ORGANIZATION

This chapter consists of three parts. The first (Sections 3 to 5) focuses on the three components of motor vehicle transportation: the operator, the vehicle, and the roadway. We have placed a greater emphasis on the operator because if vehicles, technology, and roadways are to be designed and used safely, issues related to the human element, such as impairment and medical conditions, are critical to an understanding. In addition, many of the operator-based issues are relevant to all modes of transportation. In the second part (Section 6) we address emerging technology, called *intelligent transportation systems*, designed specifically to improve the safety and mobility of roadway users. In the final part we address, in detail, two important topics in motor vehicle transportation human factors: commercial vehicle operations (Section 7) and aging drivers (Section 8). Although this chapter is centered entirely on motor vehicle transportation, the information illustrates human factors problems that are applicable to all modes of transportation. Thus, this chapter can provide guidance for research and understanding of rail, marine, pedestrian, and aviation transportation.

2 INTRODUCTION

2.1 Historical Perspective

Informal human factors activity in transportation occurred long before the automobile. For example, safety belts were first developed in the 1880s to prevent people from getting bounced out of horse-drawn buggies. Traditionally, the mission of human factors and ergonomic activity in motor vehicle transportation has been to improve traffic safety and mobility by improving the "fit" between users, vehicles, and the transportation environment. Early work focused primarily on vehicle design. As vehicles became more complicated, additional controls and displays were added. Much early research was conducted to determine how best to design the human–machine interface (HMI). As you will see in this chapter, this research continues. Measurements of the human body have been used to design comfortable seats

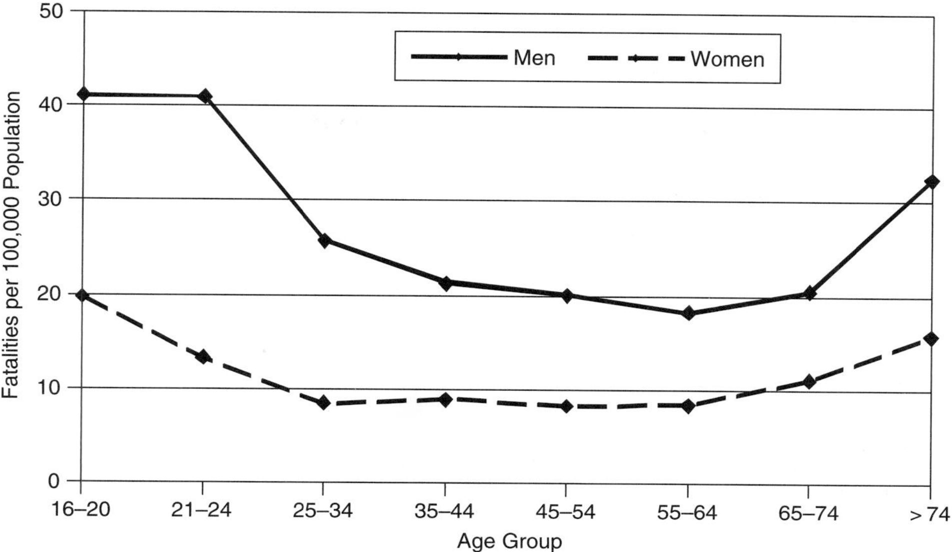

Figure 1 U.S. fatal crash rates per 100,000 population in 2002. (Data from NHTSA, 2003a.)

and vehicle cockpits, where displays and controls are placed at optimal locations and distances. Power steering, power brakes, and automatic transmissions have reduced the demands of the driving task. Early research also helped to develop roadways that are safer and improve mobility. One particularly innovative roadway improvement was the introduction of "mileposts" by ancient Romans in about 250 B.C. (Bryan, 2004). More recently, in the United States, the first centerline pavement markings were painted in 1918 in Detroit, Michigan (Bryan, 2004). Reportedly, the public generally believed that centerlines would be a traffic safety hazard. These innovations, and the public's response to them, continue today.

As computer, communication, and information technologies have advanced, they have been applied to the transportation environment, not always successfully. As discussed in this chapter, the proliferation of cellular phones has improved productivity but may also have decreased safety. The most promising applications of these technologies have come in the form of intelligent transportation systems (ITSs). Because of the complexity of these systems, however, much work is yet to be done to ensure that they enhance safety and mobility while still being acceptable to the public.

People, too, have changed. Advances in nutrition, medicine, and emergency services have increased the life expectancy of people who live in developed countries. In the United States, for example, life expectancy increased from 49 years in 1901 to slightly more than 77 years in 2001. This increased life expectancy, coupled with the "baby boom" following World War II, has led most Western nations to experience a gradual aging of the driving population. In the United States, drivers over the age of 64 are expected to account for about 20% of the total driving population by 2050 (U.S. Department of Commerce, 2001). Coincident with this aging will be a greater proportion of drivers with medical conditions who are taking medications that could impair driving abilities.

2.2 Crashes

The mobility provided by motor vehicle transportation comes at a price. The World Health Organization has estimated that traffic crashes are the third-leading cause of death and injury in the world and focused their 2004 World Health Day on this topic. Traffic crashes in the United States are the leading cause of death for people age 2 through 33 years. In 2003, 42,643 people were killed in over 6 million police-reported crashes, and nearly 3 million were injured (National Highway Traffic Safety Administration, NHTSA, 2004a). NHTSA (2004a) estimates that the economic cost of these crashes was more than \$230 billion. These numbers represent good news, however, in that the United States has continued to reduce the motor vehicle fatality rate. In 1993, 1.75 people died in crashes per 100 million vehicle miles traveled (VMT). This rate reached its historical low point in 2003 (the newest information that is available): 1.48 deaths per 100 million VMT (NHTSA, 2004a). Clearly, there is still much human factors work to be done to continue this downward trend in the fatality rate.

Crashes differ by age and gender. Figure 1 shows that for both men and women, crashes are highest for young drivers, decline up to about 30 years of age, remain about the same until around age 60, and then begin to increase again. For all age groups, men have higher crash rates than women. Current emphasis in human factors research has been to prevent the

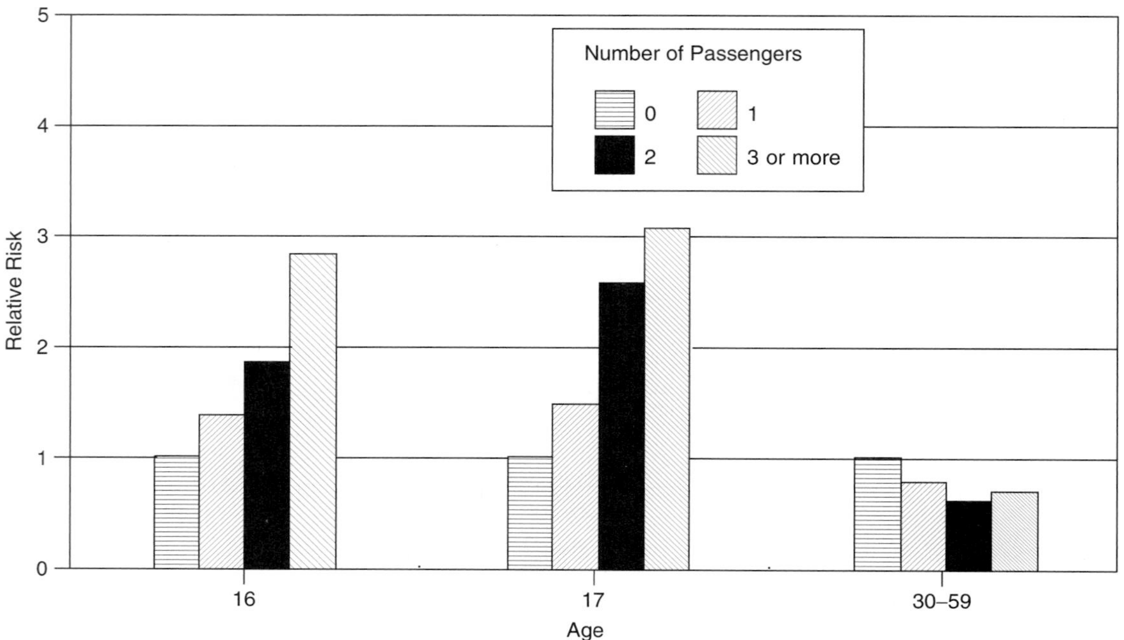

Figure 2 Relative risks for driver death by driver age and the number of passengers in the vehicle at the time of the crash. (Data from Chen et al., 2000.)

elevated crash rate for young people and both delay and flatten the upturn in rates found at the older ages. There are also efforts to shrink the difference in rates between men and women by bringing down the fatal crash rates for men. Unfortunately, little progress has been made in this area.

3 OPERATOR-BASED FACTORS

3.1 Young Drivers

In the United States and many other countries, the acquisition of a driver's license is considered a rite of passage—a salient marker for the transition from adolescence to adulthood. Yet it appears that many young people are not able to handle the responsibilities of operating a motor vehicle. According to the National Center for Health Statistics (2004), motor vehicle crashes are the leading cause of death for 15 to 25 year olds, despite the fact that motor vehicle fatality rates for this age group have declined significantly since 1975. On a per population basis, drivers under age 25 in the United States had the highest rate of involvement in fatal crashes of any age group in 2002, and their fatality rate based on vehicle miles traveled was four times greater than the comparable rate for drivers age 26 to 65 (NHTSA, 2003a). Teenage drivers have by far the highest fatal crash involvement rate of any age group based on number of licensed drivers. Motor vehicle injury rates also show that teenagers continue to have vastly higher injury rates than those of the population in general.

There is also considerable evidence that young people, more frequently than others, speed (e.g., Soliday, 1974; Konecni et al., 1976; Wasielewski, 1984; Jonah, 1986), travel with shorter headways (e.g., Wasielewski, 1984), run yellow lights (e.g., Konecni et al., 1976), and fail to use safety belts (see e.g., Eby et al., 2000; Glassbrenner, 2004). Not surprisingly, these behaviors result in young driver crashes that differ from the crashes of other drivers. Crash data show that young driver crashes are more likely to be at-fault crashes, to involve speeding, to be alcohol-related, to involve a single vehicle, and to occur at night (Williams et al., 1995; Cammisa et al., 1999).

Young drivers seem to be particularly susceptible to the influences of passengers, as evidenced by crash records (Doherty et al., 1998; Preusser et al., 1998). New drivers, in particular, are at higher risk of crash when passengers are present. Chen et al. (2000) compared fatal crash risks of 16- and 17-year-old drivers over a five-year period (1992–1997) with the fatal crash risk of drivers 30 to 59 years of age. Using zero passengers in each age group as the reference group, these researchers found that the relative risk of a fatal crash increased significantly with each additional passenger for the 16- and 17-year-old drivers, whereas the relative fatal crash risk decreased with additional passengers for the 30- to 59-year-old age group. These data are shown in Figure 2. The study also found that fatal crashes for 16- and 17-year-old drivers were more likely when passengers were male, in their teens, and age 20 to 29.

3.1.1 Risky Driving Behaviors

Why are young drivers at such high risk of automobile crash? A commonly suggested reason is that young drivers tend to engage in risky driving behaviors, actions that increase, above some threshold, the objective likelihood of a crash or the severity of injury should a crash occur (Eby and Molnar, 1998). There are at least two main factors underlying this class of behaviors: inexperience with driving and psychological immaturity. Because both of these factors tend to vary with the age of the young driver, they are often studied simultaneously by using the driver age as an independent variable. However, work has suggested that both factors, independently, are associated with elevated crash risk.

In a study in Canada, Cooper et al. (1995) found that between the ages of 15 and 55, those drivers with one year of driving experience tended to have higher crash rates than drivers of the same age with two to three years of experience when at-fault crashes were considered. Interestingly, when not-at-fault crashes were considered, the opposite was found; that is, over all ages studied, drivers with one year of experience had a slight but significantly lower crash rate than that of drivers with two to three years of experience. This study highlights the important role that experience plays in crash likelihood, at least in at-fault crashes. Further, since no difference was found between drivers with two to three years of driving experience, the study showed that the most important experiential aspects of learning to drive are probably acquired during the first year of driving.

Human development is a dynamic process, continuing throughout the life span. As reviewed by Eby and Molnar (1998), several cognitive processes continue to develop through the first 25 years of life. Collectively, these developing processes can lead drivers to engage in risky driving, increasing their chance of a crash. Although a complete review of these processes is beyond the scope of this chapter, we review two of the more important processes. For further information the reader is referred to the COMSIS Corporation/Johns Hopkins University (1995), Jonah (1997), and Eby and Molnar (1998).

Sensation Seeking Past research has documented the negative traffic safety consequences of people who have a need for high levels of arousal. These people are commonly called *sensation seekers* [see, e.g., Jonah (1997) and Zuckerman (1979, 1990) for excellent reviews]. Zuckerman (1994, p. 27) defines *sensation seeking* behavior as "the seeking of varied, novel, complex, and *intense* sensations and experiences, and the willingness to take physical, social, *legal*, and *financial* risks for the sake of such experiences" (italic his). Thus, in terms of driving, a sensation seeker might perform risky driving behaviors simply to experience a situation in which physiological arousal will be elevated. There is good evidence that sensation seeking has a strong biological component (Zuckerman, 1994).

Who are the sensation seekers? Zuckerman and his colleagues (e.g., Zuckerman et al., 1964, 1978; Zuckerman and Link, 1968; Zuckerman, 1971) have developed a test in which behaviors related to sensation seeking are self-reported. This test, called the *sensation seeking scale* (SSS), has been used extensively to define the demographics of sensation seeking and its relationship to unsafe driving behaviors. Several studies using this measure have shown that males score higher than females on the total SSS (e.g., Zuckerman and Neeb, 1980; Perez et al., 1986; Teraski et al., 1987; Björk-Åkesson, 1990; Russo et al. 1991, 1993). Research has also documented that scores on the SSS tend to increase with age up to about 16 to 19 years and then decline gradually through the life span (e.g., Farley and Cox, 1971; Zuckerman et al., 1978; Magaro et al., 1979; Zuckerman and Neeb, 1980; Ball et al., 1984; Giambra et al., 1992; Russo et al., 1993). Figure 3 is a composite of the results from two studies (Zuckerman et al., 1978; Russo et al., 1993) and shows the typical relationship between total SSS scores, age, and gender. As can be seen in this figure, the difference between males and females on the total score on the SSS is consistent through the life span, and total SSS scores vary consistently as a function of age, reaching a peak at around 16 to 19 years of age. As reviewed by Jonah (1997), sensation seeking is related to drinking and driving in the young driver population. In the population of college-age and younger drivers, research has shown that self-reported impaired drivers, drivers convicted multiple times for drunk driving, and those arrested for drunk driving following a collision or violation score significantly higher on the SSS than do those in comparison groups (Lastovicka et al., 1987; Arnett, 1990; McMillen et al., 1992a; Arnett et al., 1997). Scores on the SSS have also been shown to correlate positively with driving speed (e.g., Zuckerman and Neeb, 1980; Clement and Jonah, 1984; Heino et al., 1992; Lajunen and Summala, 1996; Arnett et al., 1997; Jonah et al., 1997) and to be related to nonuse of safety belts (Clement and Jonah, 1984; Wilson and Jonah, 1988; Beirness, 1995; Jonah et al., 1997).

Collectively, the results comparing scores on the SSS with risky driving behaviors show that at least a link exists between the two. When coupled with the fact that sensation seeking has a developmental component, it explains, in part, why risky driving behaviors are more common among young drivers, in particular young males.

Risk Perception Our thoughts about risks and how we assess them have been termed *risk perception* (see, e.g., Fischhoff et al., 1981; DeJoy, 1989, 1990). Because risky driving behaviors are both a public health issue (they increase the risk of injury) and a legal issue (they are illegal), these two types of perceived risk are relevant for traffic safety. For each type of risk, there are two perceived probabilities that are important: the probability of the negative event occurring and the severity of the negative outcome. In the public health domain, the negative event is a crash and the severity of the outcome is the extent of injury. In the legal domain, the negative event is

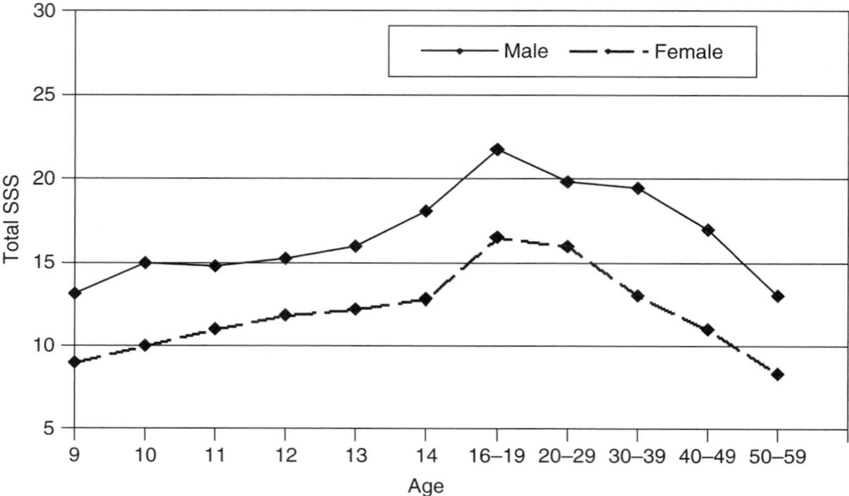

Figure 3 Total sensation seeking score by age and gender. (Data from Russo et al., 1993; Zuckerman et al., 1978.)

getting pulled over by police and receiving a citation or being arrested, and the severity of outcome is the costs associated with the police action (i.e., fines, increased insurance premiums, jail, etc.). Regardless of the domain, the two components interact to influence behavior. For example, if the perceived severity of the outcome is quite small (i.e., low fines), a high perceived chance of receiving a citation will not change behavior. Conversely, if a person thinks the event will never happen (i.e., the person believes that he or she will never crash), a high perceived severity of the outcome will not influence behavior.

A number of studies have investigated perceptions of traffic crash and injury risk by age, and the majority have found that young drivers tend to perceive less risk in specific crash scenarios and general driving than do older drivers (e.g., Finn and Bragg, 1986; Sivak et al., 1989; Tränkle et al., 1990; Groeger and Chapman, 1996). Young drivers also tend to see themselves as less likely to be in a crash than others in their own age group (e.g., Svenson, 1981; Svenson et al., 1985; Finn and Bragg, 1986; Matthews and Moran, 1986; DeJoy, 1989, 1990). Several driving situations are judged as less risky by young drivers compared to older drivers. These situations include tailgating, driving at night, driving on curves, driving on hills, urban driving, driving on bald tires, encountering a slow driver on the road, driving on wet roads, speeding, and drinking and driving (Finn and Bragg, 1986; Matthews and Moran, 1986; Tränkle et al., 1990). Work has also shown that young people tend to perceive less risk of a crash when they are driving than when they are passengers, a result not found with older people (Bragg and Finn, 1985; McKenna, 1993; Greening and Chandler, 1997).

3.2 Drowsy Driving

Sleep is a basic biological need of all humans. Each day, we progress among states of consciousness from waking to daydreaming to sleeping. As we slip from being awake to being asleep, we experience a state of consciousness known as *drowsiness*. Drowsiness is characterized by a strong need to sleep and can impair reaction time, vigilance, attention, and information processing (Haraldsson et al., 1990; Kribbs and Dinges, 1994; Dinges, 1995). Two closely related concepts are fatigue and inattention. *Fatigue* results from prolonged activity and leads to a withdrawal of attention to the activity. Thus, fatigue can lead to the same impairments as drowsiness. *Inattention*, on the other hand, is the broad category of all circumstances that result from a withdrawal of attention from a primary task such as driving. Thus, fatigue, drowsiness, drug or alcohol impairment, and distraction are all forms of inattention.

3.2.1 Causes of Drowsiness

Although alcohol and medications can cause drowsiness, there are three primary causes (NHTSA, 2002). The first is *sleep loss or restriction*, which results in a person getting less sleep than is biologically necessary. Numerous factors can reduce sleep duration, including work schedules, family responsibilities, insomnia, and the choice to engage in activities other than sleeping, such as socializing. The second cause is sleep *fragmentation*, in which the natural progression of sleep is interrupted, leading to a reduction in the quality and duration of sleep. Fragmentation can be caused by untreated sleep disorders such as sleep apnea or narcolepsy or by external factors such as spousal snoring, noisy neighbors, or job-related duties, as with on-call workers. A final primary cause is *disruption of circadian rhythms* (the process governing the sleep and wake cycle each day). The most familiar cause of circadian disruption occurs when traveling rapidly across time zones, resulting in jet lag. Shift work can also disrupt the circadian rhythms by requiring a person to remain awake when the body is ready to sleep.

3.2.2 Drowsy Driving and Crashes

One of the most difficult transportation safety issues to quantify is the relationship between drowsiness and crashes. It is widely recognized that drowsiness is underreported in crash records. Drowsiness is inherently difficult to identify because there are no physical markers that can be tested for after the crash as there are in the case of alcohol, and because the stimulation of crash involvement may mask symptoms of drowsiness. Moreover, police officers are typically not trained to identify drowsiness-related crashes and have numerous other duties at crash scenes. It is therefore likely that only the more pronounced cases of drowsy driving are detected and recorded in crash data.

Accordingly, researchers have to rely on inference to quantify and understand the drowsy-driving crash problem. Police crash reports (e.g., Langlois et al., 1985; Knipling and Wang, 1994; Pack et al., 1995), self-reported drowsiness (e.g., Maycock, 1996; McCartt et al., 1996; Royal, 2003), and surrogate crash measures (e.g., Wang et al., 1996) have all been utilized to help understand drowsiness-related crashes. NHTSA (2004b) has defined a surrogate measure of drowsy-driving crashes as a crash that occurs between the hours of midnight and 6 : 00 a.m., involving a single vehicle and a sober driver traveling alone, with the car leaving the roadway without any attempt to avoid the crash. Based on this definition, NHTSA estimates that drowsy driving causes 100,000 injuries and 1550 fatalities each year.

Recent work, however, suggests that these estimates may be quite low. The Gallup Organization conducted a nationally representative survey of drivers on distracted and drowsy driving (Royal, 2003). This survey found that 11% of drivers reported having nodded off or fallen asleep while driving in the past six months. Of these drivers, only 28% had a drowsy-driving episode between midnight and 6 : 00 a.m. Thus, 72% of the drowsy-driving episodes would not have fallen within NHTSA's measure. Clearly, further research is necessary to fully understand the magnitude of drowsy-driving crashes both at night and during the daytime.

3.2.3 Characteristics of Drowsy-Driving Crashes

Despite the difficulty of determining the incidence of drowsy driving from crash records, studies have addressed the characteristics of drowsy-driving crashes. Collectively, these studies show that drowsy-driving crashes are more common for young male drivers (e.g., Wang et al., 1996), those who work night shifts (e.g., Stutts et al., 2003), drivers who take soporific medications (Ray et al., 1992), and drivers who have untreated sleep disorders (Aldrich, 1989). Interviews of crash-involved drivers have found that when compared to drivers in non-sleep-related crashes, drivers in sleep-related crashes report getting fewer hours of sleep per night, poorer sleep quality, greater daytime sleepiness, and more frequent drowsy-driving incidents (Stutts et al., 2003). These findings are consistent with what is known about the causes of drowsiness.

3.3 Impairment

Driving while impaired by alcohol or illicit drugs is considered by many to be the most preventable traffic safety problem, in that this behavior is entirely voluntary. For decades legislation has been enacted and enforcement programs have been conducted to prevent impaired driving. Nevertheless, impairment continues to be a contributing factor in a large proportion of crashes.

3.3.1 Alcohol

The consumption of alcohol is an important aspect of most Western cultures. Recent comparisons across countries show that Luxembourg has the highest rate of alcohol consumption at 12 liters of pure alcohol per person per year, in the form of beer, wine, and spirits (Kempton, 2004). By comparison, the United States has a rate of about 6.5 liters. Alcohol is a central nervous system depressant and its effects on the body are well known and pervasive: impaired judgment, slowed reaction time, impaired balance, blurred vision, drowsiness, and loss of coordination.

Alcohol has been shown to affect abilities related to driving even at low blood alcohol concentration (BAC) levels. In an excellent review of this literature, Moskowitz and Fiorentino (2000) concluded that some driving skills are affected adversely at small elevations above zero (i.e., less than one full drink for many people); the majority of studies showed impairment of skills at a BAC of 0.05 grams per deciliter (g/dL, a level that is legal in all states); and all drivers experienced impairment in at least some skills at a BAC of 0.08 g/dL (the minimum illegal level in all U.S. states) (Insurance Institute for Highway Safety, 2004). According to Moskowitz et al. (2000), the magnitude of these impairments at various BAC levels was not significantly different by age, gender, or drinking practices, at least for people between 19 and 70 years of age who were at least moderate, but not very heavy drinkers.

Alcohol-Involved Crashes Recent figures from NHTSA (2004a) show that alcohol was involved in 17,013 fatal crashes in 2003, accounting for 40% of fatal crashes in that year. In addition, NHTSA (2004a) estimates that another 275,000 people were injured in alcohol-related crashes. Although these figures are alarming, they represent the latest in a consistent decrease in alcohol-related crash fatalities over the last decade.

The NHTSA (2004a) study showed that alcohol-involved crashes tend to occur during times that people usually consume alcohol. In 2003, alcohol-involved crashes and alcohol-involved fatal crashes were five times and three times more likely during nighttime, respectively. These crashes were also most likely

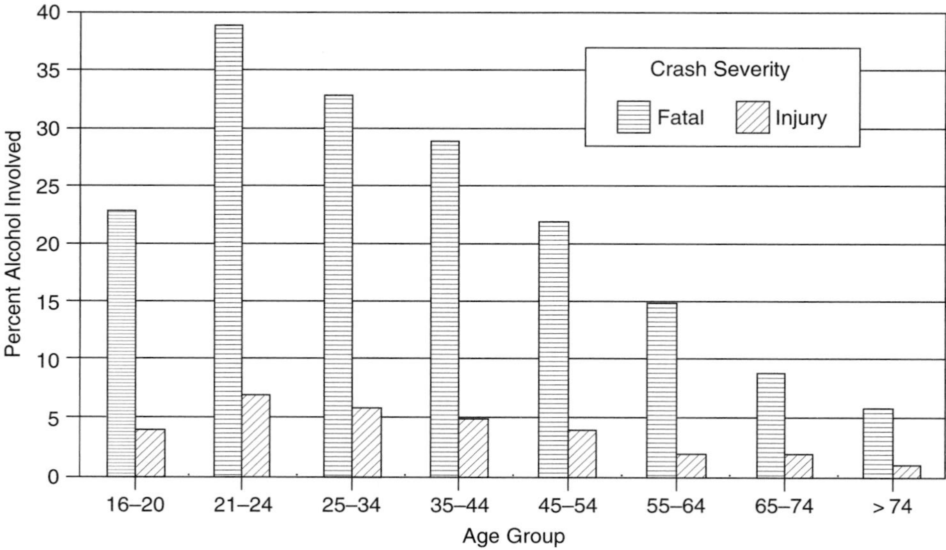

Figure 4 Percent of alcohol-involved crashes (BAC of 0.01 or greater) by age and severity in the United States in 2002. (Data from NHTSA, 2003a.)

to involve a single vehicle. All types of alcohol-involved crashes were more likely during the weekend than during weekdays. Alcohol-involved fatal crashes were also more likely for motorcycle drivers than for drivers of other vehicle types. The NHTSA (2004a) data also identified driver characteristics associated with alcohol-involved crashes. Men were twice as likely to be involved in drunk-driving fatal crashes as women. As shown in Figure 4, of all drivers involved in fatal crashes with a BAC level of 0.08 g/dL or higher, those in the 21- to 24-year-old age group had the highest proportion of involvement, with the proportions decreasing with age. Drivers younger than the legal drinking age of 21 had the same percentage of involvement as drivers in the 45- to 54-year-old age group. Thus, it is clear that once alcohol consumption is legal, fatal crashes for those drivers increase precipitously for the next few years.

Recidivism The NHTSA (2004a) data also showed that a driver with a previous drunk-driving conviction was nine times more likely to be involved in a fatal alcohol-involved crash in 2003. High rates of drunk-driving recidivism are common (see, e.g., Jones and Lacey, 2000; Eby et al., 2002b). Beerman et al. (1988) compared drunk-driving recidivists (three or more drunk-driving convictions in 12 years) with drivers having one or two drunk-driving convictions on several factors. They found that the recidivists had significantly more moving violations, were more likely to be unemployed, were more likely to have a criminal record, were more often arrested for drunk driving on weekdays in the afternoon, were more likely to have been driving on a suspended or revoked license, and were more likely to have refused the breath test. Other work has found that drunk-driving recidivists are more likely to have a history of poor interpersonal relationships, be unmarried, and to be white (Beirness and Simpson, 1997; Abdel-Aty and Abdelwahab, 2000; Jones and Lacey, 2000).

Ignition Interlock Systems One of the most effective countermeasures for the deterrence of drunk-driving recidivism is alcohol-ignition-interlock technology. The ignition-interlock system is designed to be installed in a convicted drunk driver's vehicle as part of his or her sanctioning. In concept, the system requires that the driver pass a BAC test before the vehicle can be started. All systems also include various safeguards to prevent circumvention, such as random retesting while driving, checks for nondriver samples, and removal detection. Several studies have found that the interlock system is successful in reducing drunk driving (e.g., Marques et al., 1999) and repeat drunk-driving convictions (Beck et al., 1997; Weinrath, 1997; Eby et al., 2002b). The long-term effectiveness of this technology after the device has been removed is not known.

3.3.2 Illicit Drugs

In the last several years, attention has turned to the problem of illicit-drug-impaired driving. Unlike alcohol, there are no easily administered tests to determine the presence of an illicit drug in a driver. Thus, there is little information available on the incidence of illicit-drug-impaired driving (Jones et al., 2003). Although the magnitude of the drug-impaired driving problem is difficult to quantify, there are some data addressing the problem. Data from studies on injured or killed drivers show evidence of marijuana use in about 9% of cases (Cimbura et al., 1982; Terhune et al., 1992) and cocaine use in as much as 2% of cases (Terhune et al., 1992). These percentages have, however, been found to be much higher in special populations. Williams et al. (1985), for example, studied a sample of 440 fatally injured young male drivers (15 to 34 years of age) in California and found marijuana and cocaine use in 37 and 11% of cases, respectively. Studies on heavy-truck drivers using both randomly selected paid volunteers (Lund et al., 1988) and fatally injured drivers (Crouch et al., 1993) have found marijuana use ranging from 13 to 16% and cocaine use as frequent as 8%. Waller et al. (1995) studied drivers who presented at an emergency department in Michigan following a motor vehicle crash. After examination, they found that marijuana, cocaine, and/or opiates were present in 14% of the drivers. Thus, based on these studies and others [see Jones et al. (2003) for an excellent review], it appears that marijuana and cocaine are the most commonly used illicit drugs among drivers.

Effects of Marijuana on Driving Performance Skills As described by Couper and Logan (2004), marijuana produces a spectrum of behavioral and physiological effects, with the effects increasing in magnitude with the dose level. In their summary of the research on the effects of marijuana and driving, Couper and Logan (2004) conclude that this drug can impair driving performance for up to three hours. The specific performance skills that are affected include lane keeping, reaction time, time and distance perception, headway keeping, and vigilance. Performance decrements are increased during both high- and low-driving-demand situations.

Effects of Cocaine on Driving Performance Skills Cocaine is both a stimulant and an anesthetic. As with marijuana, this drug produces a variety of behavioral and physiological effects (Couper and Logan, 2004). The effects of cocaine on driving performance skills are somewhat complex. One panel of experts concluded that a single low dose of cocaine may improve driving performance, but only for fatigued or drowsy drivers (Couper and Logan, 2004) The panel also thought that significant decrements in performance would occur in high or multiple doses. Observational, case study, and police report analysis research (summarized by Couper and Logan, 2004) shows that cocaine impairment can lead to poor speed control (speeding), risk taking in driving, inattention, and aggressive driving. The relationship of these performance decrements to crashes is not known.

3.4 Distraction

Safe operation of a motor vehicle requires that a driver focus a substantial portion of his or her attentional resources on driving-related tasks, including monitoring the roadway, anticipating the actions of other drivers, and controlling the vehicle. A driver may also, however, be engaged in other nondriving activities that compete for his or her attentional resources. As these nondriving activities increase, the driver allocates greater attention to them, or the driver's attentional capacity is reduced (e.g., fatigue), and there is a reduction in the attentional resources necessary for safe driving. Driver inattention has been found to be a major factor in traffic crashes, with 20 to 50% of crashes involving some form of inattention (Sussman et al., 1985; Wang et al., 1996; Goodman et al., 1997; Ranney et al., 2001; Stutts et al., 2001).

As described in Section 3.2, one form of inattention is driver distraction, which results from a triggering event (Stutts et al., 2001). A distracted driver has delayed recognition of information necessary for safe driving because an event inside or outside the vehicle has attracted the driver's attention (Stutts et al., 2001). A distracted driver may be less able to respond appropriately to changing road and traffic conditions, leading to an increased likelihood of crash. Driver distraction has been estimated to be a contributing factor in 8 to 13% of tow-away crashes (Wang et al., 1996; Stutts et al., 2001).

Determining the effect of driver distraction on crash risk has proven challenging. Crash reports from which detailed crash databases are derived often lack good information about distraction-related events leading up to a crash, and surrogate measures of distraction-related crashes, such as rear-end crashes, can be overly subjective and inaccurate. In addition, even when crash data contain good distraction-related information, interpretation of these data is difficult because information about the frequency of exposure to the distraction scenario is not available. Scenarios that lead to driver distraction can be either classified as occurring outside or inside the vehicle.

3.4.1 Distractions outside the Vehicle

Exterior Incident This scenario refers to an event outside the vehicle that draws the driver's attention. A wide range of incidents are possible and include, but are not limited to, crashes, police activity, vehicle actions, and pedestrians. Several studies have found that exterior incidents are the most frequent contributor to distraction-related crashes (Wang et al., 1996; General Assembly of the Commonwealth of Pennsylvania, 2001; Stutts et al., 2001; Glaze and Ellis, 2003). Although the frequency with which a driver encounters an exterior incident is unknown, one would think the exposure to this type of potential distractor is quite high (perhaps multiple incidents per trip). In an attempt to further delineate the most common types of exterior incidents involved in distraction-related crashes, Stutts et al. (2001) examined a sample of crash narratives from two years of Crashworthiness Data System (CDS) database files. They found that the most common exterior incidents in distraction-related crashes involved traffic or a vehicle, such as a vehicle swerving or changing lanes, an emergency vehicle, or bright vehicle lights. The next two most common incidents were police activity and an animal in the roadway, followed by, in order of frequency, people or objects in the roadway, sunlight/sunset, crash scene (rubbernecking), and road construction.

Looking at Scenery or Landmarks Another potential distraction outside a vehicle is scenery or landmarks. In a recent study by Virginia Commonwealth University (Glaze and Ellis, 2003), researchers analyzed more than 2800 surveys filled out by police officers at driver-inattention-related crash scenes regarding the main distraction that contributed to the crash. In nearly 10% of cases, looking at scenery/landmarks was reported. This distraction factor was second only to exterior incidents.

3.4.2 Distractions inside the Vehicle

Passengers Travel with a passenger occurs in about one-third of automobile trips in the United States. Given the incredible variety of human interactions, it is not surprising that some of these interactions can be distracting to an automobile driver and may lead to an increased risk of crash. For young drivers in the United States, at least, analyses have shown that the rate of crashes increases with the number of passengers present in the vehicle, and crash risk is increased even further when the passengers themselves are young (Doherty et al., 1998; Chen et al., 2000; Williams, 2003). On the other hand, research on nonteenage drivers has found either no change or a reduction in crash risk when passengers are present (Doherty et al., 1998; Vollrath et al., 2002; Williams, 2003). Thus, it may be that young drivers are more susceptible than older people to the distracting influence of passengers, or that the interactions that young people have with their passengers are qualitatively different. Analyses of distraction-related crash data files have found passenger-related distractions to be a relatively common triggering event for the crash (General Assembly, 2001; Stevens and Minton, 2001; Stutts et al., 2001; Glaze and Ellis, 2003; Royal, 2003). In their analysis of CDS crash narratives, Stutts et al. (2001) found that verbal interaction with the passenger was the most common passenger-related event, followed by tending a child or infant, and the passenger doing something (e.g., yelling, reaching, fighting).

Entertainment System The vast majority of motor vehicles are equipped with entertainment systems that include radios, cassette players, and/or compact disk (CD) players. Operation of these systems usually involves manual manipulation of buttons, knobs, and media, as well as visual input, leading to a potential for physical, cognitive, and visual distraction. Analyses by several researchers have shown that adjusting an entertainment system is one of the leading in-vehicle triggering events for distraction-related tow-away crashes (Wang et al., 1996; Stutts et al., 2001), distraction-related police-reported crashes (Glaze and Ellis, 2003), and distraction-related fatal crashes (Stevens and Minton, 2001).

Cellular Phone Use of cellular (mobile) phones while driving is a growing traffic safety concern. Cellular phone ownership has been increasing rapidly over the last several years and is predicted to rise to more than 80% by 2005 (Telecompetition Inc., 2001). Self-reported data show that about two-thirds of cellular phone use occurs while in a motor vehicle (Insurance Research Council, 1997; Bureau of Transportation Statistics, 2000; Gallup, 2001). Direct observation studies of cellular phone use have found that about 3 to 5% of the driving population are conversing on a handheld cellular phone at any given moment during daylight hours (NHTSA, 2001; Reinfurt et al., 2001; Eby and Vivoda, 2003; Eby et al., 2003a). According to NHTSA (2001) estimates, this use rate equates to about 600,000 drivers using a cellular phone at any given time during daylight hours in the United States.

Evidence obtained from simulated driving (e.g., McKnight and McKnight, 1993; Serafin et al., 1993; Alm and Nilsson, 1995; de Waard et al., 2001; Strayer and Johnston, 2001; Consiglio et al., 2003) and on-the-road driving (e.g., Brookhuis et al., 1991; Tijerina et al., 1995; Hancock et al., 2003) has shown that use of a mobile phone can lead to decrements in tasks required for safe driving. There is general agreement in the literature that the most distracting activities involving cellular phone use are dialing and receiving phone calls (see, e.g., Zwahlen et al., 1988; Brookhuis et al., 1991; Green, 2000; Tijerina et al., 2000; Alm and Nilsson, 2001). In addition, use of handheld phones tends to be associated with greater decrements in driving performance than hands-free phones, but the conversations tend to be equally distracting, especially when the information content is high (see, e.g., McKnight and McKnight, 1993; Strayer and Johnston, 2001; Patten et al., 2004; Consiglio et al., 2003).

Evidence is also mounting, although still far from conclusive, that the use of cellular phones increases crash risk. In their analysis of the CDS data, Stutts et al. (2001) found that cellular phone use or dialing was implicated in about 1.5% of distraction-related crashes. One would expect this percentage to increase as the predicted use of cellular phones increases. More recent work in Virginia has found that about 5% of distraction-related crashes involve cellular phones (Glaze and Ellis, 2003). Utilizing self-reported data on cell phone crash involvement, Royal (2003) estimates that 292,000 U.S. drivers reported cell phone involvement in a crash in the past five years. Results from epidemiological studies in which cellular phone use has been linked with crash records are beginning to support the hypothesis that use of a cellular phone while driving increases crash risk (Violanti and Marshall, 1996; Koushki et al., 1999; Redelmeier and Tibshirani, 1997; Sagberg, 2001; Laberge-Nadeau et al., 2003).

Route-Guidance System One of the most widely available in-vehicle advanced technologies is the route-guidance system. These systems provide the driver with information about a route to a destination supplied by the driver. Because these systems use vehicle location technology, such as global positioning system (GPS) technology, route directions can be timed to correspond with the driver's information needs as he or she drives. There is little information about the incidence of route-guidance systems in vehicles or the frequency with which they are used.

Analysis of the crash databases yielded no instances in which use of a route-guidance system was indicated as a contributing factor in distraction-related crashes (Stevens and Minton, 2001; Stutts et al., 2001). In addition, natural use studies of various route guidance systems have found no adverse effect on traffic safety nor any increase in self-reported distraction (see, e.g., Perez et al., 1996; Eby et al., 1997; Kostyniuk et al., 1997a, b). Despite these results, there is general agreement in the literature that the function of destination entry is quite distracting if it involves visual displays and manual controls [see Tijerina et al. (2000) for an excellent summary of this work]. Although most destination entry would probably occur in a stationary vehicle, Green (1997) has pointed out that there are several scenarios in which a driver might engage in destination entry while driving and in turn be at greater risk for a distraction-related crash: The driver is in a hurry and enters the destination after starting the trip; the driver changes his or her mind about the destination after starting trip; the driver gets other information, such as a radio traffic report, then decides to change the route; the driver entered the wrong destination; or the driver does not know the exact destination prior to departure and enters the actual destination later. Thus, there are several scenarios in which use of a route-guidance system could lead to distraction-related crashes.

Eating or Drinking Many of us would agree that eating and drinking in the car is a common activity for drivers. Certainly, the activity leads to physical distraction, as it requires the driver to hold food or a beverage. Eating and drinking in a vehicle can also result in cognitive and visual distraction as the driver attempts to locate items or prevent them from spilling. Thus, eating and drinking in a vehicle may be a contributing factor in distraction-related crashes. Indeed, Stutts et al. (2001) have found evidence for the presence of this activity in about 2% of distraction-related crashes in the CDS database. In-vehicle eating or drinking has also been indicated in about 5% of police-reported crashes in Pennsylvania (General Assembly of the Commonwealth of Pennsylvania, 2001) and a small number of fatal, distraction-related crashes in the UK (Stevens and Minton, 2001).

Jenness et al. (2002) investigated the distracting effects of eating a cheeseburger during simulated driving. Based on lane keeping, minimum speed violations, and glances-away-from-the-road measures, the researchers concluded that eating a cheeseburger was about as distracting as using a voice-activated dialing system. In-vehicle eating, however, was less distracting than adjusting an entertainment system or reading directions.

Vehicle Controls Motor vehicles have a variety of systems that the driver controls, including lights, safety belts, turn signals, windshield wipers, and heating/ventilation/air conditioning (HVAC). Operation of these systems through steering-wheel or dashboard controls can draw attention away from driving, leading to distraction. Generally, most systems, except for HVAC, are simple controls that require little attention to operate, at least in a familiar vehicle. However, HVAC systems, which generally have at least two controls with multiple settings, can lead to distraction even in a familiar vehicle. Studies that have investigated distraction-related crashes in various databases have found that adjustments of vehicle controls account for about the same frequency of distraction-related crashes as eating and drinking—about 2 to 5% (General Assembly, 2001; Stevens and Minton, 2001; Stutts et al., 2001).

Objects Moving in the Vehicle People often transport objects in their vehicles, such as groceries, packages, purses, laptop computers, and briefcases. If these objects are not secured, the kinematics of normal driving can cause them to slide along the vehicle floor or fall off the seat. These events can draw attention away from the driving task during braking and/or turning, which are critical safety-related maneuvers. People also transport pets, which if not constrained, can move about the vehicle, causing distractions. An object moving in a vehicle does seem to be a factor in distraction-related crashes. Stutts et al. (2001) found that a moving object in the vehicle was the triggering event in about 4% of distraction-related crashes in the CDS database, with the percentage as high as 7.6% in some years. Little is known about the frequency of this distraction-related event. However, anecdotally, one would expect that the majority of people transport objects on nearly every trip. The frequency with which these objects move and whether this movement attracts the driver's attention is unknown.

Smoking The Centers for Disease Control and Prevention (CDC) estimate that about 23% of the adult population are current smokers, with little change in prevalence over the last several years (CDC, 2002). Given that many jurisdictions are banning smoking in public buildings, the vehicle may be one of the few places left, besides at home, where a person can smoke. Thus, smoking while driving is likely to be a frequent activity. Cigarette smoking has been identified as a contributing factor in about 1% of distraction-related crashes in the CDS (Stutts et al., 2001), nearly 5% of distraction-related crashes in Pennsylvania (General Assembly, 2001), and in a small percentage of fatal distraction-related crashes in the UK (Stevens and Minton, 2001). These percentages were similar to those for the involvement of cellular phone use in distraction-related crashes. Analysis of the CDS narratives showed that in order of prevalence, smoking-related distractions were lighting a cigarette, reaching or looking for a cigarette, a cigarette blowing back into the vehicle, and dropping a cigarette (Stutts et al., 2001).

Other Scenarios A number of other distracted driving scenarios have been discussed in the literature but little empirical data were available to assess them.

These in-vehicle scenarios, however, may be serious and may be particularly suitable for human factor interventions. The first is reading. Clearly, driving and reading can lead to visual, cognitive, and physical distraction. Reading printed materials such as a book, newspaper, or mail is considered by 80% of people surveyed nationally to distract drivers enough to make driving more dangerous (Royal, 2003). More than one-half of respondents also considered looking at maps or written directions to be activities that make driving more dangerous. The second is use of wireless technology. Wireless technology is proliferating and includes personal digital assistants (PDAs), wireless e-mail, pagers, and beepers. One would expect that use of these technologies while driving will become more frequent in the future. Royal (2003) found that remote Internet equipment, such as PDAs, was the second most frequently selected distracting activity after reading. About 40% of respondents thought that pagers or beepers were distracting enough to make driving more dangerous. The third scenario is personal grooming. This activity involves a range of behaviors and probably leads to some level of visual, physical, and cognitive distraction. More than 60% of respondents in a nationwide telephone survey thought that personal grooming was one of the most distracting activities for drivers (Royal, 2003).

3.5 Medical Conditions

Perhaps the most controversial topic in traffic safety is determining fitness to drive for people with medical conditions, in particular for older drivers. The issue is complex because the same medical diagnosis can lead to different physical effects and/or different levels of impairment in different people; properly treated medical conditions can improve driving ability; medications can lead to side effects that compromise driving safety; and multiple medical conditions and/or medications can have widely varying effects on driving ability. Furthermore, despite volumes of research attempting to correlate scores on various assessment instruments with traffic crashes, researchers have found few instruments that predict traffic crashes with even moderate precision. Without delving into this controversy, we present what is known about the more common medical conditions and traffic safety.

3.5.1 Heart Disease

Coronary heart disease (CHD) is relatively uncommon for young people but is the leading cause of death among U.S. residents age 65 and over (Kannel et al., 1990). There is at least a 100-fold increase in the risk of cardiac death for a 65-year-old man compared with a 35-year-old man (U.S. Public Health Service, 1990). Although CHD is a chronic, progressive, and disabling disease, 41% of deaths from CHD are sudden (Kannel et al., 1990).

The incidence of death from heart disease while driving is probably very low (Janke, 1994). In the few situations where this has occurred, many of the stricken drivers are able to stop their automobiles or otherwise prevent a crash or injury to others [see Epstein et al. (1996) for a review of cases]. In general, only 0.45 to 1.0 per 1000 motor vehicle crashes are caused by sudden incapacitation due to CHD (Gerber et al., 1966; Herner et al., 1966). Janke (1994) reviewed several studies that showed there is no significant increased crash risk among drivers with chronic heart disease. In fact, in some cases, the crash risk was reduced. It is possible that the reduction in risk is related to changes in lifestyle, such as retirement, and driving compensation, such as avoiding heavy traffic (Janke, 1994). Thus, there does not seem to be a serious problem with heart disease and driving safety among older drivers, because those with heart disease self-restrict their driving activities.

3.5.2 Cardiac Arrhythmia

Arrhythmia is an irregular rhythm of the heart not occurring in the acute phase of myocardial infarction or as a result of drug toxicity or electrolyte imbalance (Canadian Cardiovascular Society, 1996). The presence of arrhythmia may pose a problem for safe driving because of a common treatment for the disease: implantable cardioverter defibrillators (ICDs). ICDs are used to manage arrhythmia by delivering a high-energy shock to the heart. This shock can sometimes result in loss of consciousness (syncope) or a brief impairment of voluntary motor activities (Kou et al., 1991; Epstein et al., 1996). The incidence of syncope after shock has been found to be as high as 9% (Kou et al., 1991). Syncopal episodes caused by ICDs while driving, however, are not common. Beauregard et al. (1995) estimated the risk to be 0.0011% per kilometer driven of syncope caused by ICDs while driving. Currently, assessment of a person's fitness to drive after receiving an ICD is difficult (Gimbel, 2004). To date there are no clinical predictors for syncope related to arrhythmia, and even history of syncope or the absence of it does not predict future occurrences (Kou et al., 1991). Recent research, however, suggests that well-treated arrhythmia should not preclude a person from driving (Bleakley and Akiyama, 2003).

3.5.3 Syncope

Syncope can occur from a variety of causes, including a sudden fall of blood pressure, a neurological pathology, and an increase in blood sugar (Rehm and Ross, 1995). Although syncope can happen at any age, it is most frequent in the older population, with an incidence of about 3% for those 65 years of age or older (Savage et al., 1985; Bonema and Maddens, 1992; Kapoor, 1994). In about 40% of those who experience syncope, no cause for it can be found (Kapoor et al., 1983, 1989; Spudis et al., 1986). The risk of syncope associated with driving is particularly low. The chance that a person who already has experienced one episode of syncope will eventually faint while driving is 0.33% per driver-year. The risk of syncope causing a crash or injury is even lower (Sheldon and Koshman, 1995). There is still little agreement about whether syncopal people should drive. Published guidelines range for complete driving cessation of one to three months after a single episode of syncope to driving cessation for one year to complete cessation after multiple episodes of syncope (Decter et al., 1994; Sheldon and Koshman, 1995; Canadian Cardiovascular Society, 1996).

3.5.4 Stroke

Stroke becomes more likely as a person ages. The prevalence rates for the 65 years of age and older age group are as much as 10 times higher than the overall population prevalence rate (Kurtzke, 1985). The physical effects of a stroke are based largely on the location in the brain where the stroke occurred, but these effects can include partial or incomplete paralysis, impaired spatial abilities, agnosias, aphasia, attention deficits, impaired recognition ability, reduced numerical ability, and emotional disruptions (Lings and Jensen, 1991). As with other medical conditions, there are no universally accepted criteria for assessing fitness to drive after a stroke.

3.5.5 Diabetes Mellitus

Diabetes mellitus causes a variety of vascular problems that can lead to various conditions, including heart attacks, visual deficits, kidney disease, and loss of feeling in the extremities (Davidson, 1991). Diabetes is classified into two types: type 1 (insulin dependent) and type 2 (noninsulin dependent). Five to 10% of all people diagnosed with diabetes have type 1 diabetes and the remaining have type 2 diabetes (National Diabetes Data Group, 1979; CDC, 1997). The prevalence of diabetes in the U.S. population ranges from 2 to 6% (Davidson, 1991; Hu et al., 1993; CDC, 1997) and is more likely in older adulthood (Davidson, 1991; Hansotia, 1993; Hu et al., 1993). Insulin and other medications used to control diabetes may actually increase the risk of traffic crashes because the frequency and severity of hypoglycemia are increased among type 1 diabetes patients treated intensively (DCCT Research Group, 1987).

The studies relating crash risk and diabetes have found inconsistent results. In her review of various disabilities, Janke (1994) noted that studies have shown mixed crash-risk results, and of those that do show increased crash risk, the risk is not excessively inflated. In particular, type 1 patients seem to show the greatest risk, but they comprise only 5 to 10% of the diabetic population.

3.5.6 Epilepsy

Epilepsy is a chronic neurologic condition characterized by abnormal electrical charges to the brain, which result in seizures (Adams and Victor, 1989). Seizures can range from dramatic grand mal seizures to subtle seizures that result in changes in cognition and consciousness (Browne and Feldman, 1983; Doege and Engelbert, 1986). Although the cause of epilepsy is not known, in about 75% of cases, risk factors include vascular disease, head trauma, congenital or perinatal factors, syncope, central nervous system infections, and neoplasms (Hauser and Kurland, 1975). Epilepsy is a common neurologic condition in the United States, with prevalence estimated at 5 to 7 per 1000 persons in the general population and lower than this for the older adult population (Hauser and Kurland, 1975; Haerer et al., 1986; Hauser and Hesdorffer, 1990; CDC, 1994). Effective treatment, usually pharmacology, can prevent seizures in most persons with epilepsy. Depending on the frequency of seizures, epilepsy can have a negative effect on driving safety. Seizures adversely affect driving ability by causing an alteration or loss of consciousness and motor control. This potential for crashes is supported by studies that have found an increased risk of crashes and injury among drivers with epilepsy (Popkin and Waller, 1989; Hansotia and Broste, 1991; Taylor et al., 1996).

3.5.7 Dementia/Alzheimer's

Dementia/Alzheimer's (DA) is recognized as intellectual deterioration in an adult that is severe enough to interfere with occupational or social performance (McKhann et al., 1984). Older people with dementia are affected not only by the physical changes that normally accompany aging, but also by intellectual and cognitive impairment (Kapust and Weintraub, 1992). Cognitive deterioration is the central feature of DA (Smith and Kiloh, 1981). DA can be caused by a variety of medical conditions, including stroke, hypothyroidism, traumatic head injuries, brain tumors, carbon monoxide poisoning, and alcoholism (Haase, 1977; Katzman, 1987). Prevalence estimates are inconsistent due to variations in how AD is diagnosed and range from 4 to 16% of the older adult population (Terry and Katzman, 1983; Evans et al., 1989; Cushman, 1992; Adler et al., 1996). Three severity stages of DA have been indexed by the Clinical Dementia Rating Scale: early, middle, and late (Hughes et al., 1982). Progression usually spans an average of eight years from the time symptoms first appear, although DA has been known to last as long as 25 years.

Despite the disability caused by the DA, many patients continue to drive. Studies show that about 30 to 45% of patients with DA diagnosis were still driving (Lucas-Blaustein et al., 1988; Carr et al., 1990; Logsdon et al., 1992), and 80% of these patients drove alone (Lucas-Blaustein et al., 1988). DA patients who had traffic crashes since their illness onset were equally likely to still be driving as those who did not have crashes (Lucas-Blaustein et al., 1988). As these data suggest, DA has deleterious effects on driving capabilities. Although many of these results have yet to be replicated adequately, studies have shown several driving problems associated with DA, including getting lost while driving, even in familiar areas (Lucas-Blaustein et al., 1988; Underwood, 1992; Adler et al., 1996), vehicle speed control (Odenheimer et al., 1994), particularly consistently driving below posted speed limits (Lucas-Blaustein et al., 1988), signaling lane changes (Hunt et al., 1993; Odenheimer et al., 1994), checking blind spots before lane changes (Hunt et al., 1993), maintaining lateral lane position (Odenheimer et al., 1994), judgment in traffic (Hunt et al., 1993), running stop signs (Cushman, 1992), and recognizing and obeying traffic signs (Hunt, 1991; Cushman, 1992; Hunt et al., 1993; Mitchell et al., 1995; Adler et al., 1996). As DA progresses, errors such as these become more frequent (Fox et al., 1997).

Given the driving problems associated with DA, it is not surprising to find that patients with DA tend to have an elevated crash risk (Reger et al., 2004). In one study, DA subjects had 263.2 crashes per million vehicle miles of travel, compared with 14.3 crashes per million vehicle miles of travel for the older-adult controls and 5.7 crashes for the general driving population under age 56 (Dubinsky et al., 1992). In another study of subjects diagnosed with DA, 47% had at least one crash and 30% had at least one major crash causing personal injury or over $500 of vehicle damage. Only 10% of control subjects had crashes (Friedland et al., 1988). Drachman and Swearer (1993) found that of 83 patients with DA, 26% had crashes while driving after diagnosis. During the same time period, only 8% of 83 matched controls had crashes.

4 VEHICLE-BASED FACTORS

4.1 Vehicle Design

4.1.1 Anthropometry

Anthropometry refers to measures of the body both statically and when in motion. It is important to take into account measurements of the human body when designing a vehicle so that the person can effectively operate controls, easily view displays, see the roadway environment, and travel in the vehicle comfortably. Unfortunately for designers, the dimensions of the human body vary greatly, making it impossible to design vehicles that will fit the extremes of human anthropometry. Therefore, designers often have a goal of accommodating the 95th percentile driver. Achievement of this goal indicates that some feature of the vehicle design will accommodate 95% of drivers, usually excluding the smallest and largest 2.5% (Dewar, 2002). Given the number of vehicle features and widely varying human dimensions, many design features will not accommodate a large number of drivers.

The three critical body measurements used by designers are eye height, leg reach, and arm reach. These measurements are calculated relative to a measurement of the hip location, or H-point. These measurements are also dependent on how the driver sits and his or her head position. Based on these measurements, the proper placement for vehicle controls, displays, and mirrors can be determined using guidelines developed by the Society of Automotive Engineers (SAE).

4.1.2 Controls

Any vehicle contains a number of controls. As discussed by Dewar (2003), there are four features of vehicle control: type (e.g., button, knob), location (e.g., steering column, door), operation (e.g., in/out, rotate), and coding (e.g., color, shape). Proper combination of these features can make them easier to operate, reducing the workload on the driver and the likelihood of a crash.

An important factor in determining how to combine features is driver expectancies or stereotypes about the control (Wierwille and McFarlane, 1993; Rubens et al., 1994). For example, drivers may expect a toggle switch on a vehicle's roof to move toward the back of the vehicle to open the sunroof or a rotary knob to be turned clockwise to increase the radio volume. Controls should be designed with these expectancies in mind, as violating an expectancy can lead to misuse of the control, especially during high-workload situations. Again, there are SAE guidelines for locating controls and lists of driver expectancies for various controls. With the introduction of new technologies into vehicles, such as night vision systems, it may become increasingly difficult to develop controls that do not violate expectancies.

4.1.3 Displays

In addition to controls, vehicles also contain a number of displays. The three features of a display are the information content (e.g., numerical, pictograph), placement (e.g., instrument panel, mirror), and modality (e.g., visual, auditory). With the driver workspace becoming increasingly cluttered with visual and auditory displays, such as navigation systems and crash warning technologies, proper development of displays is essential for minimizing driver distraction.

As with controls, driver expectancies play a critical role in the understanding of displays. This is particularly true when the information content of the displays is developed. Many vehicle displays utilize pictographs (simple meaningful pictures) to identify a control or transmit information to the driver. Developing good pictographs is difficult. Research has shown that some pictographs are understood by most drivers, whereas others are understood by few (Hoffmeister et al., 1995). Sayer and Green (1988) had drivers compare 25 automotive pictographs from the International Standards Organization (ISO) with alternative pictographs for comprehension. They found that several of the alternative symbols were superior to the ones from the ISO. Further research to improve poorly understood pictographs is warranted. In addition, it may be possible to improve pictograph comprehension through an educational or public information campaign.

Display placement is dictated mainly by visibility. The displays should not be occluded, be easily distinguishable from other displays, and minimize driver eye and head movements to be viewed. Dewar (2003) recommends that complex displays be located high on the instrument panel so that less time is spent with eyes off the road. Until recently, the primary modes of display were visual and auditory. With the increasing complexity of ITS technology, other display modalities have been investigated. For example, some advanced crash-warning systems use the tactile mode in the form of a haptic seat, to inform the driver of a hazard (Delphi, 2003).

4.1.4 Windows and Mirrors

The safe operation of a motor vehicle requires that the driver be able to have good visibility of the roadway environment in front of, next to, and behind the vehicle. Good visibility is both a function of a wide field of view and good optical quality. The front and some of the side roadway environment is viewed directly through windows. A window's field of view is related to its size and distance from the driver's eyes. The transmission of light through the glass, however, can be influenced by weather, dirt, angle relative to the driver's eyes, and the level of tinting. Under most normal circumstances these factors do not affect light transmission to the point of being a safety hazard; however, at night and for drivers with reduced visual capacities (such as elderly drivers), these factors may compromise safety.

The roadway environment behind and next to the vehicle is viewed indirectly through the use of mirrors. A mirror's field of view is determined by its size, degree and sign of curvature, and the distance to the driver's eyes. The mirror's field of view can be increased slightly by the driver moving his or her head lateral to the mirror. To increase the field of view and decrease the size of a vehicle's blind spot, convex mirrors are often used as the driver-side mirror. Although the curvature increases the area that can be seen, it produces distortions of size and distance that can make backing a vehicle dangerous.

4.2 Headlights

Many dangerous intersections are lighted during the nighttime hours to improve drivers' ability to see other vehicles and pedestrians. The vast majority of intersections and roadways, however, are not lighted. When traveling these sections of roadway at night, the illumination necessary for safe driving is provided by the vehicle's headlight system.

4.2.1 Nighttime Crashes

According to NHTSA (2004a) data, the deadliest time for crashes is between midnight and 3 : 00 a.m. Fatal crash rates are also elevated at other nighttime hours relative to daytime hours. Pedestrians are particularly at risk for fatal crashes. About 30% of all fatal pedestrian crashes occur between 8 : 00 p.m. and midnight (NHTSA, 2004a). As discussed previously, factors such as drowsiness and alcohol use contribute to these nighttime crashes, and research has shown that nighttime visibility is a major factor in many of these crashes (Owens and Sivak, 1996; Sullivan and Flannagan, 2002).

4.2.2 Role of Target Contrast

It is generally believed that vehicle headlights function by making objects in the roadway environment brighter. As discussed by Olson (2003), however, this view is incorrect. Simply making a target brighter will not necessarily improve its visibility unless it can be made vastly brighter. The important criterion for vehicle headlights is to increase a target's contrast relative to its background. Under the dim-illumination conditions experienced at night, the eye can generally respond only to luminance contrast, that is, the differences in the amount of light returned from a surface (in perceptual terms this corresponds roughly to brightness contrast). In other words, for headlights to be effective, they have to either increase or decrease a target's illumination relative to its background.

4.2.3 Nighttime Visibility versus Glare

Although there are several issues associated with the design of effective headlights, the main difficulty is providing good visibility to the nighttime driver while minimizing the amount of glare for drivers in approaching vehicles. Unfortunately, the ideal location for directing the headlight illumination is about where the other driver's eyes are located. This is where the high-beam headlight system directs light, and all of us have experienced the glare associated with an oncoming vehicle using a high-beam system. Ideal visibility, therefore, cannot be achieved when other vehicles are on the road. Instead, engineers have also designed a low-beam system that directs as much light as possible in front of the vehicle but below the horizon and away from the eyes of approaching drivers. Unfortunately, signage and other important characteristics of the roadway environment are located above the visual horizon so that some light must also be directed above the horizon. Thus, headlight design and visibility factors produce a trade-off between good visibility and the reduction of glare. It remains to be seen whether future technology, such as nighttime vision enhancement systems, can improve on this trade-off.

5 ROADWAY-BASED FACTORS

5.1 Roadway Design

At an earlier time, civil engineers devoted their attention only to concrete, grades, and following design standards. Fortunately, the modern civil engineer is aware that human factors considerations are also important. Europe is the world leader in the application of human factors to highway design, but the United States has a growing interest in this area.

For the last decade, European ergonomic researchers have been studying the *self-organizing road*. A self-organizing road increases the probability that a driver will automatically select the appropriate speed and steering behavior for the roadway without depending on road signs. The geometric features (e.g., road curvature, road width, type of road shoulder, gradient) of the road encourage the desired driver behavior and do not rely on the driver's ability or willingness to read and obey road signs. A perfect self-organizing road would not require speed limit signs and curve advisory warnings.

The roundabout, common in Europe, is a good example of a self-organizing road. The driver must slow down entering the roundabout, and pavement markings are often used to help the driver perceive this lower speed requirement. Roundabouts are a product of human-centered roadway analysis. In the United States, drivers are held responsible for following the rules and driving correctly. In Europe, there is much greater realization that drivers will inevitably make errors, so that the roadway design community must anticipate such errors and mitigate their effects. For example, a common driver error is running through a red traffic light. This may lead to a crash where one vehicle strikes another at a 90° angle, inflicting great damage. Such a crash cannot occur at a roundabout for two reasons: First, roundabouts do not have traffic signals, and second, any crash in a roundabout is at an angle much less than 90°, so that damage is minimized.

Another example of human-centered highway design is the 2 + 1 roadway design used in Sweden, Finland, and Germany (National Cooperative Highway Research Program, 2003). This roadway has three lanes, but none of them is a center lane. Instead, the passing lane alternates in a systematic and predictable manner from one side of the road to the other. Drivers need not, and do not, speed up at the end of a passing lane as is common in the United States because they know, from experience and road signage, that another passing opportunity will occur soon. Although passing lanes on two-lane highways are often used on grades in the United States, this differs from the 2 + 1 design, where a passing lane is always present on one side of the road, regardless of topography. Furthermore, a chain barrier of steel ropes separates the two directions of traffic on a 2 + 1 roadway. This barrier prevents vehicles from crossing to the wrong side of the highway and has resulted in an 85% decrease in highway fatalities. The chain barrier catches the vehicle as opposed to the solid barriers used in the United States, which can deflect vehicles back into the traffic stream. Although the chain barrier inflicts physical damage to the vehicle, it is far better to suffer property damage than loss of life. Indeed, European insurance companies have agreed to cover most of the costs of such physical damage, including repairing the chain barrier.

5.2 Intersection Design

Civil engineers design intersections to move traffic efficiently and safely. These two criteria, however, can be at odds with each other. For example, the protected left-turn lane improves safety but reduces traffic flow through the intersection. In other situations, the two criteria are complementary, such as the roundabout discussed previously. A third important criterion in intersection design is complexity. Examples of complex intersections include five or more legged intersections, roundabouts, and four-legged skewed (or ×-shaped) intersections. Most improvements in the last few decades in intersection safety have come in the form of improved signing, marking, and signalization. Even with these improvements, older drivers have a higher risk of crash involvement for left-hand turns (Federal Highway Administration, FHWA, 2004; Schieber, 2004). The *Older Driver Highway Design Handbook* (FHWA, 1998) offers the following guidelines for improving intersection negotiation safety for elderly drivers:

- Wherever possible, use 12-ft lane widths with 4-ft shoulders.
- Intersecting roadways should meet at 90° angles.
- For left- and right-turn lane treatments, provide raised channelization with sloping curbed medians.
- Use positive offset of opposing left-turn lanes.

The last recommendation is provided because older drivers tend not to position themselves in the intersection prior to making a turn (FHWA, 2004). The offset is designed to provide more room within the intersection for older drivers to move into prior to executing the turn.

5.3 Traffic Control Devices

A critical aspect of driving is to visually gather information about the roadway environment. This information is used for current driving, such as keeping a vehicle in the lane or speed control, as well as for appropriately setting expectations for the upcoming roadway, such as curves or changes in the roadway geometry. In addition, the information is used for helping drivers safely negotiate intersections. Although there are a variety of traffic control devices (TCD), they can be classified as either signs, signals, or pavement markings. Regardless of the type of TCD, it must attract the driver's attention (conspicuity), it must be easily read or interpreted (legibility), it must be understandable (comprehension), and its information must be able to be acquired quickly, within 1 or 2 seconds (Dewar and Olson, 2002).

5.3.1 Signs

Signs are used to convey to the driver information that is critical for the driving task. The information conveyed by signs fall into three categories: those that inform drivers of required upcoming actions (e.g., stop sign), those that warn drivers of possible dangerous situations (e.g., "bridge may be icy" sign), and those that inform drivers of regulations (e.g., speed limit signs). Because of the critical information they convey, signs must be conspicuous, legible, and quickly comprehended under a variety of weather and driver conditions.

To enhance these factors, symbols are often used on signs instead of words. Symbols have the additional benefit of being language independent, so that nonnative drivers can utilize the signage. However, as with the pictograms used for vehicle controls, much of the driving public does not understand what some of the symbols indicate (Dewar et al., 1994). Greater effort should be made to standardize symbols among countries and to educate the public about symbol meaning.

Word signs are by far the most common sign found on roadways, and much research has gone into their development. Early work on sign legibility did not consider the diminished visual capacities of older drivers, and many signs on the road today are difficult for older drivers to read. Work has shown that older drivers need signs that are as much as 30% larger to be read as well as younger drivers can read them (reviewed in Schieber, 2004). Even with good legibility, drivers of all ages sometimes do not understand what the words mean. For example, Hawkins et al. (1993) found that fewer than one-half of 1745 Texas drivers tested understood a sign with the words LIMITED SIGHT DISTANCE. Thus, an important human factors issue in sign development is to determine how to convey complex information in a few words or a symbol.

5.3.2 Signals

Signals are used to convey information to drivers at roadway areas in which vehicles come into conflict, such as crossing paths at an intersection. Because they are placed in the driver's field of view and are lighted, signals are generally conspicuous and legible. Comprehension of certain signals, however, may be poor. The ubiquitous three-light traffic signal is well understood by drivers. Other signals, such as protected left turns and freeway lane control signals, can be difficult for drivers to understand (Hummer et al., 1990; Ullman, 1993). A study of signal comprehension by age found that older drivers (age 60 or above) had poorer comprehension than younger drivers of left-turn signals and emergency flashing modes of traffic lights (Drakopoulos and Lyles, 1997). Signal comprehension should be addressed in educational programs for older drivers.

5.3.3 Pavement Markings

Pavement markers include painted lines, painted curbs, raised/reflective markers, rumble strips, and word or symbol messages. These TCD are used for assisting in lane keeping, channelization, conveying warnings of speed changes, warning drivers about obstacles, and identifying lane-specific characteristics (e.g., turn lane, high-occupancy vehicle lanes). Conspicuity of pavement markings depends on the contrast of the marking and roadway. Contrast can be improved by having bright roadway lighting and by using retroreflective material or paint. Conspicuity and legibility can be reduced due to a number of factors, including poor weather, aging or faded paint, and poor lighting. The study in Texas found surprisingly poor comprehension of pavement markings (Hawkins et al., 1993). For example, nearly one-fourth of drivers in their study did not understand the meaning of a single-broken-yellow-centerline or a solid-white-edge line.

6 INTELLIGENT TRANSPORTATION SYSTEMS

ITSs are now available to perform a bewildering variety of functions that have the potential to aid drivers by increasing safety and improving mobility. These technologies have been classified into three major categories (Wang et al., 2002):

- *Infrastructure-based ITSs*: systems that depend exclusively on roadside equipment
- *Vehicle-based ITSs*: systems that depend exclusively on in-vehicle equipment
- *Cooperative ITSs*: systems that combine in-vehicle and other external equipment

Infrastructure-based ITSs include roadside traffic management systems, roadside traveler information systems, intersection traffic management systems, and pedestrian protection systems. A vehicle-based ITS uses an intelligent vehicle to warn the driver of impending collisions, monitor driver status and performance, enhance driver vision, aid navigation, and notify emergency services in case of collision. Cooperative ITSs are in the planning stage and can improve functions such as navigation by combining routing and congestion information.

Human factors research is needed to ensure that the information provided by these systems aids the driver without creating distractions that direct attention away from the primary task of controlling a vehicle safely on public roads or overloading the driver with too much information. As the amount of in-vehicle information increases, so do the risks associated with processing this information while a vehicle is in motion. For example, there is mounting evidence that talking on any kind of cellular phone while driving increases the risk of a crash.

6.1 In-Vehicle Information Systems

Although early research on driver information systems went under the rubric of advanced traveler information systems (ATISs), the newer term *in-vehicle information system* (IVIS) is used to eliminate ATIS devices, such as travel kiosks, that are not inside a vehicle. In this section we discuss four kinds of IVIS devices: navigation and routing systems, vision enhancement systems, collision warning systems, and safety-belt reminder systems as well as a new device under development, called an *in-vehicle workload manager*, that attempts to control and coordinate several IVIS devices that currently function independently within a vehicle. Current IVIS implementations suffer from what has been called in aviation, *clumsy automation*. The promises of IVISs will not be realized completely and safely until human factors research provides a validated conceptual model of how drivers process cognitive information inside a moving vehicle (Kantowitz, 2000).

6.1.1 Navigation and Routing Systems

The most popular ATIS and IVIS applications are navigation and routing systems. Indeed, for the past several years anyone renting a car from Hertz has been able to obtain such a system. Although first introduced in luxury vehicles, such systems have migrated to midpriced vehicles and will eventually be available at reasonable prices for almost all vehicles as economies of scale are applied to the technology of in-vehicle global positioning systems.

There has been substantial human factors research on these systems. For example, a six-year multimillion dollar project sponsored by FHWA has generated 75 human factors ATIS design guidelines that continue to be widely used by industry (Campbell et al., 1999). These guidelines can be downloaded from the internet by searching the FHWA Turner–Fairbank Highway Research Center human factors report database.

One important human factors issue for IVISs is whether drivers will trust systems that are not 100% reliable. Although highway engineers would like to provide up-to-date accurate information, it would be quite expensive and difficult for this information to be completely error-free. People vary in their reaction to unreliable systems. Most people will not put additional funds into a vending machine that has failed to dispense a product once initial funding was committed. Fortunately, this type of one-trial extinction does not apply to navigation systems. Kantowitz et al. (1997) used a real-time ATIS simulator to study how drivers select routes and react to incorrect information. The simulator displayed live video of traffic conditions that corresponded to the route selected. Drivers could query the system about traffic congestion before selecting route segments. Drivers were paid cash rewards to reach destinations on time; as they were delayed by traffic congestion, their disappearing reward was displayed in real time. This produced considerable frustration and emotional responses. Nevertheless, drivers accepted some unreliability, particularly in an unknown city. Using a simulator allowed two cities, one real and known to drivers and one unknown new city, to be matched exactly on road topography. Drivers demanded higher reliability in the known city but still accepted some error in the system.

Another simulator ATIS study compared a no-ATIS control condition to basic and enhanced ATIS devices (Llaneras and Lerner, 2000). The basic ATIS condition provided descriptive information about traffic incidents and congestion. The enhanced ATIS condition added additional information about alternative routes, incident details, real-time traffic map, and live video traffic images. Route diversions were selected by 42% of the drivers in the enhanced ATIS condition, 13% in the basic ATIS, and 4% in the no-ATIS control condition. These diversions resulted in travel-time savings.

Of course, people drive real cars on real roads, not simulators, so simulator results need to be validated against on-road data. Although simulators offer excellent experimental control, economy, and safety—no driver has yet died in a simulator crash—in-vehicle systems eventually must be evaluated on the road, even though this research is expensive. Eby and Kostyniuk (1999) conducted such a real-world evaluation that compared written instructions to two commercial IVISs: Ali-Scout, with dynamic route information, and TetraStar, with static route information. Few drivers got lost, not even those unfamiliar with the area, using written instructions. Trip duration was shorter for the drivers using ATISs. Dynamic route information did not provide significant advantages, but this was probably due to occasional failures of the beacon system used by Ali-Scout. Modern dynamic systems use GPS and sophisticated algorithms to support navigation when the GPS is not available (e.g., blocked by tall buildings in dense urban areas).

Driver acceptance of ATISs is high and there have been great improvements in the performance and usability of current systems compared to the first generation of navigation devices. Although some of this improvement comes from better signal technology, human factors researchers can take pride in their contribution to the development of a device that improves mobility and has high consumer acceptance. Future human factors research will be required to ensure that only those IVIS features that are consistent with safe control of the vehicle are available when the vehicle is in motion.

6.1.2 Vision Enhancement Systems

Vision enhancement systems (VESs) are designed to provide drivers with roadway information that is either difficult or impossible for the driver to obtain through direct vision. Using technology developed by the military, vision enhancement systems use infrared cameras to detect pedestrians and animals. The thermal imaging display is usually placed above the dashboard so that the driver can map the information from the system onto the visual world that he or she sees in front of the vehicle. Through the display, the driver is able to view the forward scene for up to 700 meters. All warm objects in the forward scene are highlighted in the thermal imaging display with a brightness that is directly related to temperature. Since living beings tend to have a high temperature relative to other objects at night, humans and animals stand out visually in the display. These systems can already be found in luxury vehicles.

Studies of VESs have utilized both simulators and on-the-road studies (e.g., Raytheon Commercial Infrared and ElCAN-Teaxs Optical Technology, 2000; Druid, 2002). Collectively, drivers report that they can intuitively interpret the thermal imaging display; however, this ability seems to be reduced when the display is not positioned above the steering wheel, and display fields of view that are similar to the driver's field of view were preferred. Such findings are a success story for human factors engineers, in that the systems were designed taking into account the driver's mental model of the visual scene.

Whether VESs improve safety is not yet known. Driver comments suggest that VESs do not increase distraction, reduce the debilitating effects of headlight glare, and can easily be incorporated into the normal driving routine. On the other hand, the display can be difficult to read for drivers with bifocals (Raytheon Commercial Infrared and ElCAN-Teaxs Optical Technology, 2000; Druid, 2002). VESs may be particularly useful for improving the mobility of elderly drivers who tend to place nighttime driving restrictions on themselves in response to declining visual abilities at night (see Section 8). However, because of other declining abilities, such as the ability to divide attention, careful research must be conducted to ensure that a VES does not increase distraction or workload to unsafe levels with this age group.

6.1.3 Collision Warning Systems

A current focus of ITS research is the development of systems that enhance driver awareness of potential dangerous driving situations so that the driver can take appropriate evasive actions or, if appropriate, not perform a maneuver. These technologies are known as *collision warning systems* (CWSs). CWSs utilize sensors, such as radars and cameras, positioned around the vehicle to determine the locations of other traffic and lane pavement markings. When a potentially hazardous situation arises, a CWS warns the driver of the hazard. More advanced systems are in development that not only warn the driver but also take over partial control of the vehicle's operation to avoid the hazard. Three near-term systems are under evaluation: collision avoidance, intelligent cruise control, and lane departure warning systems. Although each of these systems has unique characteristics, the following human factors issues are important to the safety and effectiveness of these systems:

- Correctly determining the level of threat
- Indicating effectively the level of threat
- The modality for threat presentation
- How to present threat information during a time when driving task demand is likely to be high (driver can easily be distracted)
- Reducing false alarms and increasing trust of the system

As a driver moves along a roadway, the distances to other vehicles changes. In addition, the driver may intend to make maneuvers, such as changing lanes when a vehicle is in the blind spot. Determining which intervehicle distances and situations are likely to lead to a collision and the probability of collision translate into a level of threat. Accurate identification of threat levels is important for user acceptance. If a driver is constantly warned about situations that are not very hazardous, the system will become a nuisance. On the other hand, high-threat situations should always be detected or the driver will lose trust in the system. One potentially effective means for displaying level of threat has been utilized in a field operational test of an automotive collision avoidance system (NHTSA, 2003b). The display uses a vehicle rear-end icon that both gets larger and changes color the closer a vehicle comes to a rear-end collision. The icon also begins flashing as the collision becomes more eminent. The visual display is accompanied by a tone.

Different levels of threat may also necessitate different means of warning the driver. For low-threat situations, the workload and distraction potential for a driver is low, allowing for greater time to process the warning. In a high-threat situation in which a collision is eminent, however, the system should not increase the driver's workload and level of distraction because he or she will need to respond quickly to the hazard. Researchers have begun to use nonvisual warnings that are linked to the location of the hazard. For example, Delphi Electronics Corporation in the SAfety VEhicle(s) using adaptive Interface Technology (SAVE-IT) Program (Delphi, 2003) has proposed to give high threat collision warnings through a haptic seat. For threats to the right-hand side of the vehicle, the right side of the driver seat vibrates, and similarly for left-hand threats. The benefits of haptic warnings are that they do not produce visual distraction and they direct a driver's attention to the side of the vehicle to which he or she should be attending.

6.1.4 Safety Belt Reminder Systems

The single most effective technology for reducing or preventing injuries from a motor vehicle crash is the safety belt. This system is, however, effective only if it is used. The most recent nationwide survey of safety belt use in the United States (the National Occupant Protection Use Survey, NOPUS) estimated that 80% of front-outboard motor vehicle occupants use their safety belt (Glassbrenner, 2004). Although this is the highest rate ever in the United States, the rate is lower than that in many other developed countries (e.g., Boase et al., 2004) and shows that a significant portion of U.S. travelers do not use safety belts, even though it is mandated in all but one state.

Recent attention has turned to the development of in-vehicle technologies for increasing belt use (NHTSA, 2003c; Transportation Research Board, 2003). One promising technology is the safety belt reminder system. Since 1975, all new vehicles in the United States have been required to display a 4- to 8-second signal if the driver does not use the safety belt after starting the vehicle. Once the belt is fastened, the signal stops. This relatively benign reminder system is easily ignored. Therefore, current research has focused on the development of more effective and acceptable safety belt reminder systems (Harrison et al., 2000; Dahlstedt, 2001; Fildes et al., 2002; Bentley et al., 2003; Williams and Wells, 2003; Eby et al., 2004). The impetus of this research has been to determine qualitatively which signals, signal presentation methods, and systems would be most likely to get a user to buckle up and would be acceptable in a vehicle. Starting with enhanced reminder systems already installed in new vehicles, such as the Ford BeltMinder system (Williams et al., 2002; Williams and Wells, 2003), several ideas have been investigated, ranging from signals that get more intense the faster the vehicle moves to interlocking the entertainment or heating/cooling system to belt use.

Effectiveness and acceptability, however, can be at odds with one another in reminder systems; that is, a highly intrusive system would be so unacceptable, even though the driver would be more likely to use his or her belt to stop the annoyance, that he or she would

Timeline	Start of trip	4–8 sec after ignition on	Car starts moving	Car travels on patrolled roadways
Safety Belt Use Group	Full-time user	Part-time user for cognitive or personal reasons	Part-time user due to low perceived risk of crash or ticket	Full-time nonuser
Type of System Engaged	No system engaged	Reminder system	Annoyance system	Interlock system
System Goals	System invisible to driver	System effectiveness and user acceptibility are maximized	System effectiveness is optimized; acceptibility is minimized	User acceptability is minimized
System Signal Usage	No signal	Signal is selected by user	Continuous buzzer	Radio/ entertainment interlock
Level	I	II	III	IV

Figure 5 Framework proposed for in-vehicle technology designed to promote safety belt use. As the trip progresses without the driver buckling up, more intrusive systems are engaged. (From Eby et al., 2004.)

not want the system in the vehicle. In recognition of this, researchers have proposed adaptive reminder systems that become more intrusive the longer the driver travels without using the safety belt (Fildes et al., 2002; Transportation Research Board, 2003; Eby et al., 2004).

In research sponsored by Toyota Motor Corporation, Eby et al. (2004) conducted a nationwide telephone survey and a series of focus groups with part-time belt users. Based on this qualitative research, they developed the model safety belt use reminder system shown in Figure 5. In this model, the system changes its signal and presentation method to become increasingly intrusive as the trip progresses. The model was developed by first categorizing drivers by their belt use (full-time, part-time, or hardcore nonuser). Part-time users were further categorized by their reason for nonuse, either cognitive (e.g., forgetting) or low perceived risk of a crash or ticket. At the start of each trip, the system assumes that the driver is a full-time belt user and displays nothing. As the trip proceeds, and the driver does not buckle up, the system becomes more intrusive until it finally begins to shut down the entertainment system. For each driver category, the criteria for effectiveness and acceptability differ.

Although the system proposed by Eby et al. (2004) is thought to be effective, it would clearly not be acceptable to some part-time users and to all nonusers. Thus, legislation is likely to be required before such a system is used in vehicles. Recent direct observation results of the Ford BeltMinder (a system that goes up to level II in Figure 5) has found a 5 percentage point higher belt use rate for drivers in vehicles with the system (Williams et al., 2002). Thus, it is worthwhile to pursue legislation to require level IV systems in new vehicles.

6.1.5 Workload Managers

There are now such a variety of in-vehicle devices available, with more coming every day, that a driver could be completely overloaded even when the vehicle is parked. A workload manager is an in-vehicle system that manages allocation of function to prevent telematic devices from taking so much of the driver's attention that the vehicle can no longer be controlled safely on the roadway.

If vehicles are to be operated while drivers are performing simultaneous telematic tasks, a workload manager is needed to minimize decrements in driving performance. A superior workload manager must accomplish several goals:

- Evaluate the workload imposed by in-vehicle devices.
- Evaluate the workload imposed by driving, including traffic and road conditions.
- Evaluate driver capability, including fatigue and the influence of alcohol and drugs.
- Calculate the total driver workload.
- Establish a red-line value for the workload based on driver capability.
- Control telematic devices when workload is excessive, including inhibiting displays.
- Control vehicle functions when workload is excessive, including limiting vehicle speed.

Workload managers are examples of a larger class of human–machine systems called intelligent interfaces (Kantowitz, 1989). An intelligent interface is a closed-loop system that uses feedback to modify system behavior until a goal is realized. Kantowitz

(2001) has explained how intelligent interfaces can be applied to manage driver workload by using dynamic allocation of function (Kantowitz and Sorkin, 1987). The simplest workload manager uses a binary strategy whereby specific subsystems are either on or off. For example, an in-vehicle cell phone could be turned off when driver workload is too high, with messages diverted to a mailbox (Piechulla et al., 2003).

But there are far more sophisticated solutions than simple binary filtering based on a red-line threshold. If we think of intelligent control of a system as a continuum bounded by complete manual control at one end and complete automatic control at the other, an optimal intelligent interface could assume any state along this continuum. This optimal interface could control modes and subtasks for in-vehicle devices and so transition more gracefully and continuously than a simple binary interface that energizes or disconnects entire telematic devices. For example, data entry for a navigation system could be disabled while a vehicle is in motion while other, less intrusive navigation functions could be maintained. Kantowitz (2001) has discussed potential difficulties, such as uncertainty about system mode that has caused serious aviation accidents, in sophisticated intelligent interfaces.

This continuous approach will require cognitive task analyses applied to the specific telematic systems being evaluated. Such analyses require a cognitive model, such as the Kantowitz–Knight hybrid model or other computational model, that relates telematic task components to workload. Indeed, without a computational model, every new telematic system would require a new empirical evaluation which is hardly practical for the automotive industry.

Methodological Considerations There are many ways to measure driver workload, ranging from subjective scales, to physiological measures, to objective secondary tasks, to variations of primary-task performance; it would require an entire chapter just to review methodological issues. But we would be remiss if we did not at least mention three techniques that offer considerable potential.

The peripheral detection task (PDT) uses a spatial array of lights and records reaction time (RT) (Harms and Patten, 2003). Stimuli are usually presented asynchronously (Kantowitz et al., 1987) independent of the driving task. Stimuli are presented frequently and a simple RT task (Donders A task) is used. The diagnostic value of a single stimulus is low, but many stimuli are presented to make up for this deficiency. Increased RT is the index of driver workload. But if the asynchronous secondary task is embedded more frequently than might occur in a naturalistic setting, the secondary task can create an artificial elevation in operator workload. Thus, it is vital to include single-task task control conditions for both driving and secondary tasks to provide an appropriate baseline for workload. A synchronous secondary task minimizes this risk by presenting far fewer stimuli, yoked to critical events in the driving task. Kantowitz and Premkumar (2004) found substantial decrements in performing simple telematic tasks using a synchronous modified PDT task. Furthermore, they compared simple- and choice-reaction (Donders B) secondary tasks and found that the choice-reaction task was more sensitive for higher levels of telematic workload, although both reaction tasks were reliable indices of driver workload.

Steering entropy (Boer, 2001) is also an excellent index of driver workload. It is unobtrusive but suffers from the need to record and calculate it over a large time window. Although Kantowitz and Premkumar (2004) were able to demonstrate strong workload effects of steering entropy using only a 60-second window, this may still be too large to be used in a dynamic workload manager. More research on this metric is needed with shorter time windows.

Visual occlusion is an old task (Senders et al., 1967) that is making a comeback. Driver vision is intentionally eliminated for brief periods. The longer that vision can be occluded without inducing driving error, the less the driver workload (Tsimhoni and Green, 2001).

7 COMMERCIAL VEHICLE OPERATIONS

The ability to move freight and people efficiently and safely is a crucial part of an economy. Although much of this movement of people involves air, water, and rail travel, the largest component of freight movement involves large trucks. In this section we review some of the human factor and ergonomic issues regarding large trucks.

7.1 Crashes

The yearly number of large truck occupant deaths has declined since 1975 in the United States. During this same time period, however, both the U.S. population and the number of truck miles traveled (TMT) have increased. Truck and car occupant deaths resulting from crashes involving trucks as a function of population and TMT is shown in Figure 6. These data show that deaths to passenger vehicle occupants (calculated either by population or TMT) are greater than fatality rates for truck occupants. All rates have generally declined between about 1979 and 1993. After 1993, all rates remain generally the same.

An apparently important factor in large-truck crashes is drowsiness and fatigue. In his analysis of UMTRI trucks involved in fatal accidents (TIFA) and fatality analysis reporting system (FARS) data sets, Campbell (2002) concluded that fatigue or drowsiness was involved in about 2% of medium- and large-truck fatal crashes. For many of the reasons discussed in Section 3.2, Campbell cautions that this percentage underestimates the true incidence of fatal fatigue or drowsy crashes involving trucks. Nevertheless, detailed examination of these cases allowed Campbell (2002) to determine the characteristics of truck-involved fatal fatigue or drowsy crashes. These crashes were common for drivers employed by for-hire carriers and for drivers on long-haul trips, are more common between midnight and 6 a.m., and are more likely during the first few hours of a trip,

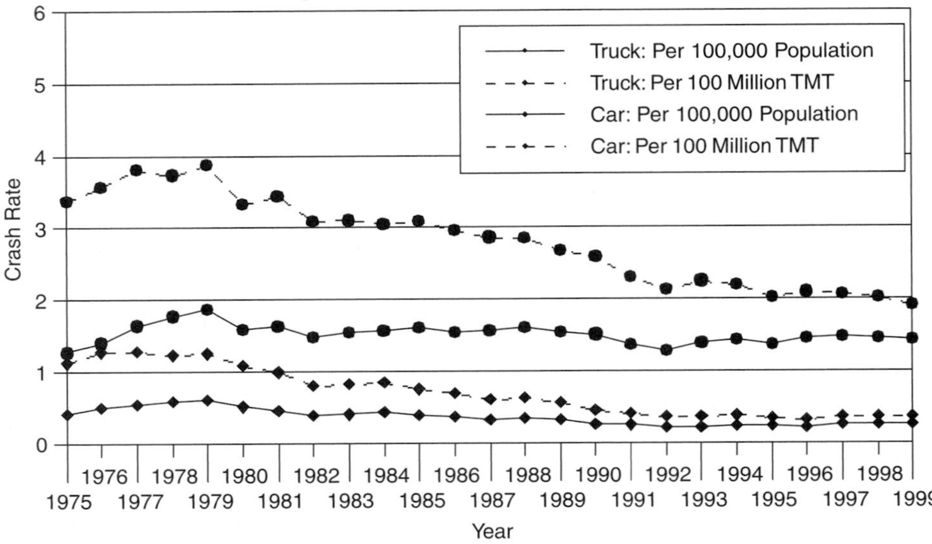

Figure 6 Comparison of crash rates between heavy trucks and passenger cars by year. (Data from Lyman and Braver, 2002.)

suggesting that a driver begins his or her trip fatigued or drowsy.

7.2 Hours of Service

There is an economic motivation for truck owners and some drivers to drive as many hours as possible. Indeed, Williams and Monaco (2001) have found that there is a link between economic pressures on drivers, crashes, and violations of hours of service regulations. To manage the risk of driver fatigue and drowsiness, governments set hours of service regulations. These regulations stipulate the legal lengths of driving, on-duty, and break times. In the United States, these regulations had remained virtually unchanged since 1939. In response to data on the fatigue and drowsiness problem in fatal truck crashes, the U.S. government revised the regulations in 2003 and they were implemented in January 2004 (Keller and Associates, 2003). Among other things, the new U.S. hours of service regulations stipulate the following:

- Provide 10-hour breaks for rest between on-duty times.
- Allow 11 hours of driving time after a 10-hour break.
- After a 10-hour break, a driver may not drive beyond the fourteenth consecutive hour after coming on duty, even if shorter breaks are taken during this time. A 10-hour break resets the 14-hour clock.
- Drivers may drive up to 60 hours in a seven-day schedule or 70 hours in an eight-day schedule, depending on the frequency with which the company operates their trucks. This 60- or 70-hour clock can be reset with 34 consecutive hours off.

The Federal Motor Carrier Safety Administration Division (Keller and Associates, 2003) estimates that these regulations will save up to 75 lives per year and prevent up to 1326 crashes per year due to driver fatigue or drowsiness. It is too early to assess the accuracy of these predictions.

7.3 Human Factors and Ergonomics Concerns

Besides the issue of fatigue and drowsiness, four other safety issues are of critical importance: comfort, driver workload, ingress/egress, and safety belt use. As with passenger vehicles, comfort involves anthropometrically appropriate seats and instrument panels. Heavy trucks, however, have quite different dynamics than passenger vehicles and produce larger-magnitude vertical vibration or bounce. To dampen these motions, advanced seat technology, including air suspended seats with shock absorbers, are utilized. These design features reduce vertical vibration but also produce a decoupling of the driver and the cab motion. These independent motions require the driver to monitor and operate controls that are constantly in motion relative to the driver's eyes and arms. Although this effect has not been researched, it seems likely that these dynamics would increase driver fatigue over a long trip (Woodrooffe, 2004).

The commercial truck driver's driving task demand is greater than for passenger vehicle drivers in that trucks are more difficult to operate, he or she must rely more on mirrors for traffic maneuvers, and passengers vehicles do not drive safely around trucks (Kostyniuk et al., 2002). In addition, the driver may also have to respond to dispatch messages and operate a routing or navigation display. As with passenger vehicles, truck driver workload can be lessened by appropriate control and display placement and labeling. Distraction management systems are also a potentially useful application for improving truck driver safety.

Heavy-truck vehicle cabs are located approximately 3 ft off the ground. Generally, ingress and egress is performed through use of a ladder and can lead to injury. According to work performed in Australia (Road Freight Transport, 2000), the main risk factors in accessing trucks are slippery footholds and handholds, lack of uniformity in step design, too-high first step, unstable first step, poorly design handholds (location, diameter, and coordination with footholds), too-high second and subsequent steps, and lack of step conspicuity. Each of these factors could be addressed through design changes.

Research shows that both commercial light-vehicle (Eby et al., 2002a) and heavy-vehicle occupants (Knoblauch et al., 2003) have surprisingly low safety belt use. In a direct observation study conducted in 12 states, Knoblauch et al. (2003) found that heavy-truck occupants use belts less than one-half the time, a rate that is more than 25 percentage points below the national average for passenger vehicles. These low use rates are also reflected in truck fatal crash data, where only 10% of truck occupants in fatal crashes were found using safety belts in an Australian study (Haworth et al., 1999).

7.4 Technology

Many of the technologies discussed in previous sections of this chapter, such as advanced traveler information systems, collision avoidance, workload managers, and safety belt reminder systems, could be applied effectively to commercial vehicle operations to enhance safety and efficiency. Current efforts, however, have been directed at developing vehicle location technology (VLT) for improving safety and productivity. VLT uses GPS information and telecommunications to broadcast the latitude and longitude to a central controller or dispatcher. Geographic information system software can then place the vehicle on a map and determine the vehicle's speed relative to the roadway's speed limit, whether or not the truck is allowed to travel on that roadway, whether the vehicle needs to be rerouted due to nonrecurrent congestions, and so on. This technology can be particularly useful for heavy trucks that exceed the normal truck design dimensions such as long combination vehicles (LCVs), which consist of a tractor with two or three trailers. LCVs are beneficial because they transport greater amounts of freight at relatively little added operating cost. In Canada, many LCVs operate under permits with strict safety requirements that restrict driving speeds, roadway use, and hours of service.

Woodrooffe (2001) and Woodrooffe and Ash (2001) conducted a safety and economic benefits analysis of LCVs in Alberta, Canada. Woodrooffe found that LCVs had the lowest crash rate of all vehicle classes, including passenger vehicles. Woodrooffe concluded that the Alberta infrastructure, which includes the AVL technology described previously, was "a vital influencing factor in the creation of a safe operating environment" (Woodrooffe, 2001, p. v). The economic analysis showed similar positive results. In summary, the study found that freight carried by LCVs in Alberta has resulted in an annual savings of $42 million CDLs, a 32% reduction in diesel fuel consumed, and about a 40% reduction in pavement wear (Woodrooffe and Ash, 2001). Clearly, application of AVL technology and the proper monitoring and analysis of the location data could provide great safety and economic benefits.

8 AGING DRIVERS

8.1 Crashes

The issue of older driver safety continues to be an important research topic despite the fact that it has received increasing attention in the past decade. It is clear that people are living longer lives than in the past. In the United States, the proportion of people who are 65 years of age or older has grown from less than 10% in 1950 to the current rate of about 13%. By 2050, the percentage of the U.S. population over 65 years of age is projected to reach nearly 20% (U.S. Department of Commerce, 2001). In terms of absolute numbers, those over 65 years of age will increase from about 35 million currently to about 70 million in 30 years (U.S. Department of Commerce, 2001). As described by Hakamies-Bloomqvist (2004), it is less clear whether older drivers are at a higher risk of crash than younger drivers. The basis of this issue is that the typical measures of exposure (population, licensed drivers, and vehicle miles traveled) are either potentially biased or are difficult to determine accurately. On the other hand, there is strong evidence that for a crash of given dimensions, older drivers are more likely to be injured than younger drivers, due, presumably, to increased frailty (see, e.g., Evans, 1991). As such, older drivers are likely to be overrepresented in fatal and serious crashes (Hauer, 1988; Maycock, 1997).

Older drivers as a group tend to be involved in different types of crashes than are younger drivers. For example, compared with younger drivers, drivers over 65 years of age, particularly drivers over 75 years of age, have more vehicle-to-vehicle collisions, more intersection crashes, and fewer alcohol-involved crashes (e.g., Hauer, 1988; Hakamies-Bloomqvist, 1994, 2004; Dulisse, 1997). Such findings are in line with what is known about driver behavior in this age group. Older drivers as a group adjust their driving to reduce the demands of the driving task (Gallo et al., 1999; Kostyniuk et al., 2000a); that is, older drivers tend to travel slower and choose times, roadways, and routes that make them feel safest. Such findings suggest that unlike crashes among younger drivers, older-driver crashes do not result from risk taking or careless driving but from age-related declines in driving abilities.

8.2 Declining Abilities

It is well established that aging can lead to declines in abilities related to safe driving [see Eby et al. (1998) for a review]. These declines result from age-related medical conditions and the medications used to treat the conditions, as well as from the effect that increasing age has on the various systems of the human organism. Declines can be classified into three categories: visual, cognitive, and psychomotor.

8.2.1 Visual Abilities

A host of visual changes can occur with age that may adversely affect a person's ability to drive safely in certain driving conditions. In general, older people have slower and more restricted ability to move the eyes to fixate an object as well as a decline in the ability to accommodate (focus). Thus, older drivers are more likely to miss critical information in the roadway environment or be slower to detect and recognize objects. Older people also have greater difficulty seeing at night, and the time for the visual system to recover from the blinding effects of glare has been found to increase with age. An elderly person's ability to see detail (static visual acuity), such as reading a street sign, is about half that of a teenager's. The decline in ability to see details while they are in motion (dynamic visual acuity) is even worse. Thus, as a person ages, spatial details, such as letters on a sign, need to be increased in size to be seen at the same distance by a young person. The visual field of older people shrinks with age and the useful portion of the remaining field (i.e., the area in which unexpected stimuli can be detected) also shrinks, meaning objects in the visual periphery are more difficult to detect, and greater head movements are required to scan the same visual area than by a young person.

8.2.2 Cognitive Abilities

Several cognitive processes have been found to show age-related declines. Both divided and selective attention abilities have been shown to decrease with age. Thus, older persons are more likely to be distracted by a given distraction scenario and are more likely to have difficulty detecting critical information in the roadway environment. It has also been found that older people process information more slowly than young people do. A significant correlation has been found between "hesitancy in decision making" and crash rates among people over 60 years of age (West et al., 1992). Finally, spatial cognition ability (the ability to think about spatial relationships) tends to decline with increasing age, leading to difficulty remembering routes and navigating.

8.2.3 Psychomotor Abilities

Psychomotor abilities refer to the coordinated and controlled ability to move or orient parts of the body (Kelso, 1982) and are of obvious necessity to operate a motor vehicle. Both simple and choice RTs have been found to increase with age, with a greater decrease found for choice RTs. Thus, as the driving task demands increase, older adults will take longer to respond to something than will younger adults. Flexibility, in particular neck flexibility, has been found to decrease with age. Reduced flexibility results in the driver having to rely more on mirrors and eye movements to detect objects behind and next to his or her vehicle. Coupled with the fact that eye movement ability and the size of the visual field is reduced with age, this compensation to decreased flexibility is often inadequate. Another safety issue associated with decreased flexibility is that safety belts can be difficult to fasten, leading to decreased use. Finally, increasing age can lead to a decreases in strength and stamina for the legs and arms. As such, older drivers have more difficulty steering, braking, and accelerating, especially during long trips. In fact, one good predictor of poor driving performance is how fast a person can walk up and back a 10-ft distance (Eby et al., 2003b).

8.3 Reduction or Cessation of Driving

As described by Eby et al. (2003b), the declines in visual, cognitive, and psychomotor driving abilities, coupled with increasing negative driving experiences and family feedback, lead older drivers to start losing confidence and to feel discomfort about their driving. In response to this discomfort, older people begin to engage in driving compensation. Many compensation strategies have been documented and are either self-imposed restrictions on driving or adaptations to simplify the driving task. Older drivers as a group tend to avoid driving at night and in unfamiliar areas (Benekohal et al., 1994), during times and at locations of heavy traffic (Knoblauch et al., 1995), and during inclement weather (Stewart et al., 1993). These self-restrictive behavioral compensation strategies ultimately lead to a reduction in the independent mobility of older persons. Older drivers employ several strategies to reduce the demands of driving, including choosing less complex routes, driving more slowly, changing lanes less frequently, avoiding unprotected left turns, and utilizing a "copilot" to perform some of the driving tasks, such as reading signage (Lerner et al., 1990; Benekohal et al., 1994; Kostyniuk et al., 1997c; Smiley, 2004). Many of these simplification strategies, however, may increase crash risk because they increase the chances of conflict with drivers not engaging in these behaviors (Smiley, 2004).

The decision to stop driving is a difficult one for nearly all drivers. Loss of driving privileges not only means a loss of independence, but for many older drivers it also represents entry into the final stage of life. It is not surprising, therefore, that older people, particularly men, continue driving past the time when they can safely operate a motor vehicle (Kostyniuk et al., 2000b). Those that do give up driving, either voluntarily or involuntarily, cite several reasons: advice from a physician, anxiety about driving, vision problems, medical conditions, advice from family and friends, movement problems, presence of adequate alternative transportation, high cost of automobile ownership, crash involvement, or license revocation (Persson, 1993).

8.4 Automotive Adaptive Equipment

One means for increasing the length of time for older driver safe mobility is to equip automobiles with adaptive equipment to help maintain driving proficiency in the face of declining abilities. After assessment to determine the types and degree of decline, appropriate adaptive equipment can be installed in a vehicle. According to Koppa (2004), adaptive equipment is available to assist older drivers in the following: vehicle access, occupant protection, manipulating controls that must be operated while the vehicle is moving, manipulating controls that can be operated while the vehicle is at rest, manipulating controls that can be operated when the vehicle is parked, manipulating primary controls (accelerating, braking, and steering), and navigation. Common types of adaptive equipment include hand controls, assist grips, signal switches, wide-angle and multiplanar mirrors, and window tinting. The safe and effective use of adaptive equipment also involves training that should include using the equipment under low-risk situations (Molnar et al., 2003).

8.5 Intelligent Transportation Systems

ITS technology holds promise for maintaining and enhancing the safe mobility of older adults. All of the ITS applications reviewed in this chapter, and many others, would seem to be beneficial to older drivers. Until recently, however, little of this technology was developed to take into account the unique requirements of older drivers. In an excellent review of older drivers and ITSs, Caird (2004) points out that although older drivers are the group most likely to benefit from ITSs, they are also the group most likely to suffer from the effects of ITSs. Poorly designed ITS applications could increase distractions for older users, leading to a higher risk of crash. On the other hand, systems designed for optimal use by older drivers would also be beneficial to drivers of all ages.

It is clear that older drivers use ITS applications differently then younger drivers (Eby and Kostyniuk, 1998; Stamatiadis, 1998; Caird, 2004). For example, in an evaluation of navigation assistance applications, Kostyniuk et al. (1997c) found that older drivers used the system more frequently than young people, entered a greater number of destinations into the system, and utilized the technology with a copilot. Understanding these patterns of use for the various ITS applications that are being developed is crucial for optimizing the benefits of ITSs for all users. Such research is lagging (Caird, 2004).

Studies have also found that older drivers take much longer to learn how to use ITS technology (Kostyniuk et al., 1997c; Caird, 2004). Whether this is a cohort effect of people who did not grow up using computer technologies, or an effect of aging per se, is not known. Nevertheless, it is clear that acceptance of ITS applications by older drivers will be largely dependent on the quality of training received.

ACKNOWLEDGMENTS

The authors are indebted to Lidia P. Kostyniuk, John Woodrooffe, and Lisa J. Molnar for their insights and comments on this chapter. We thank Mary Chico and Renée St. Louis for their editorial assistance.

REFERENCES

Abdel-Aty, M. A., and Abdelwahab, H. T. (2000), "Exploring the Relationship Between Alcohol and Driver Characteristics in Motor Vehicle Accidents," *Accident Analysis and Prevention*, Vol. 32, pp. 473–482.

Adams, R. D., and Victor, M. (1989), *Principles of Neurology*, 4th ed., McGraw-Hill, New York.

Adler, G., Rottunda, S. J., and Dusken, M. W. (1996), "The Driver with Dementia," *American Journal of Geriatric Psychiatry*, Vol. 4, pp. 110–120.

Aldrich, M. S. (1989), "Automobile Accidents in Patients with Sleep Disorders," *Sleep*, Vol. 12, pp. 487–494.

Alm, H., and Nilsson, L. (1995), "The Effects of a Mobile Telephone Task on Driver Behaviour in a Car Following Situation," *Accident Analysis and Prevention*, Vol. 27, pp. 707–715.

Alm, H., and Nilsson, L. (2001), "The Use of Car Phones and Changes in Driver Behaviour," *International Journal of Vehicle Design*, Vol. 26, pp. 4–11.

Arnett, J. (1990), "Drunk Driving, Sensation Seeking, and Egocentrism Among Adolescents," *Personality and Individual Differences*, Vol. 11, pp. 541–546.

Arnett, J., Offer, D., and Fine, M. A. (1997), "Reckless Driving in Adolescence: 'State' and 'Trait' Factors," *Accident Analysis and Prevention*, Vol. 29, pp. 57–63.

Ball, I. L., Farnill, D., and Wangeman, J. F. (1984), "Sex and Age Differences in Sensation Seeking: Some National Comparisons," *British Journal of Psychology*, Vol. 75, pp. 257–265.

Beauregard, L. A., Barnard, P. W., Russo, A. M., and Waxman, H. L. (1995), "Perceived and Actual Risks of Driving in Patients with Arrhythmia Control Devices," *Archives of Internal Medicine*, Vol. 155, pp. 609–613.

Beerman, K. A., Smith, M. M., and Hall, R. L. (1988), "Predictors of Recidivism in DUIIs," *Journal of Studies on Alcohol*, Vol. 49, pp. 443–449.

Beck, K., Rauch, W., Baker, E., and Williams, A. (1997), "Effects of Ignition Interlock License Restrictions on Drivers with Multiple Alcohol Offenses: A Random Trial in Maryland," *American Journal of Public Health*, Vol. 89, pp. 1696–1700.

Beirness, D. J. (1995), "The Relationship Between Lifestyle Factors and Collisions Involving Young Drivers," in *New to the Road: Reducing the Risks for Young Motorists International Symposium*, UCLA Brain Information Service/Brain Research Institute, Los Angeles, CA.

Beirness, D. J., and Simpson, H. M. (1997), *Study of the Profile of High-Risk Drivers*, Report TO-13108 E, Traffic Injury Research Foundation, Ottawa, Ontario, Canada.

Benekohal, R. F., Michaels, R. M., Shim, E., and Resende, P. T. V. (1994), "Effects of Aging on Older Drivers' Travel Characteristics," *Transportation Research Record*, Vol. 1438, pp. 91–98.

Bentley, J. J., Kurrus, R., and Beuse, N. (2003), *Qualitative Research Regarding Attitudes Towards Four Technologies Aimed at Increasing Safety Belt Use*, Report. 2003-1, Equals Three Communication, Bethesda, MD.

Björk-Åkesson, E. (1990), *Measuring Sensation Seeking*, *Göteborg Studies in Educational Sciences* 75, Acta Universitatis Gothoburgenis, Göteborg, Sweden.

Bleakley, J. F., and Akiyama, T. (2003), "Driving and Arrhythmias: Implications of New Data," *Cardiac Electrophysiology Review*, Vol. 7, pp. 77–79.

Boase, P., Jonah, B. A., and Dawson, N. (2004), "Occupant Restraint Use in Canada," *Journal of Safety Research*, Vol. 35, pp. 223–229.

Boer, E. R. (2001), "Behavioral Entropy as a Measure of Driving Performance," in *Proceedings of the First International Driving Symposium on Human Factors in Driver Assessment, Training and Vehicle Design*, Aspen, CO, pp. 225–229.

Bonema, J. D., and Maddens, M. E. (1992), "Syncope in Elderly Patients: Why Their Risk Is Higher," *Postgraduate Medicine*, Vol. 91, pp. 129–144.

Bragg, B. W. E., and Finn, P. (1985), "Influence of Safety Belt Use on Perception of the Risk of an Accident," *Accident Analysis and Prevention*, Vol. 17, pp. 15–23.

Brookhuis, K. A., de Vries, G., and de Waard, D. (1991), "The Effects of Mobile Telephoning on Driver Performance," *Accident Analysis and Prevention*, Vol. 23, pp. 309–316.

Browne, T. R., and Feldman, R. G. (1983), *Epilepsy*, Little, Brown, Boston.

Bryan, S. K. (2004), "Civil Engineering: What, Why, How," retrieved September 2004, from http://www.cityofcanton.com/safetyservice/trafficeng/history.html.

Bureau of Transportation Statistics (2000), *August 2000 Household Survey Results*, U.S. Department of Transportation, Washington, DC.

Caird, J. (2004), "In-Vehicle Intelligent Transportation Systems: Safety and Mobility of Older Drivers," in *Transportation in an Aging Society: A Decade of Experience*, Transportation Research Board, Washington, DC, pp. 236–255.

Cammisa, M. X., Williams, A. F., and Leaf, W. A. (1999), "Vehicles Driven by Teenagers in Four States," *Journal of Safety Research*, Vol. 30, pp. 25–30.

Campbell, K. (2002), "Estimates of the Prevalence and Risks of Fatigue in Fatal Accidents Involving Medium and Heavy Trucks," in *Proceedings of the International Truck and Bus Safety Research and Policy Symposium*, Report E01-2510-001-001-02, University of Tennessee, Knoxville, TN.

Campbell, J. L., Carney, C., and Kantowitz, B. H. (1999), "Developing Effective Human Factors Design Guidelines: A Case Study," *Transportation Human Factors*, Vol. 1, pp. 207–224.

Canadian Cardiovascular Society (1996), "Assessment of the Cardiac Patient for Fitness to Drive: 1996 Update," *Canadian Journal of Cardiology*, Vol. 12, pp. 1164–1170.

Carr, D. B., Jackson, T., and Alguire, P. (1990), "Characteristics of an Elderly Driving Population Referred to a Geriatric Assessment Center," *Journal of the American Geriatric Society*, Vol. 38, pp. 1145–1150.

CDC (Centers for Disease Control and Prevention) (1994), "Prevalence of Self-Reported Epilepsy: United States, 1986–1990," *Morbidity and Mortality Weekly Report*, Vol. 43, pp. 810–811, 817–818.

CDC (1997), *National Diabetes Fact Sheet*, National Center for Chronic Disease Prevention and Health Promotion, Centers for Disease Control and Prevention, Atlanta, GA.

CDC (2002), "Cigarette Smoking Among Adults: United States, 2000," *Morbidity and Mortality Weekly Report*, Vol. 51, No. 29, pp. 642–645.

Chen, L., Baker, S. P., Braver, E. R., and Li, G. (2000), "Carrying Passengers as a Risk Factor for Crashes Fatal to 16- and 17-Year-Old Drivers," *Journal of the American Medical Society*, Vol. 283, pp. 1578–1582.

Cimbura, G., Lucas, D. M., Bennett, R. C., Warren, R. A., and Simpson, H. M. (1982), "Incidence and Toxicological Aspects of Drugs Detected in 484 Fatally Injured Drivers and Pedestrians in Ontario," *Journal of Forensic Sciences*, Vol. 27, pp. 855–867.

Clement, R., and Jonah, B. A. (1984), "Field Dependence, Sensation Seeking and Driving Behaviour," *Personality and Individual Differences*, Vol. 5, pp. 87–93.

COMSIS Corporation/Johns Hopkins University (1995), *Understanding Youthful Risk Taking and Driving*, Report DOT HS-808-318, U.S. Department of Transportation, Washington, DC.

Consiglio, W., Driscoll, P., Witte, M., and Berg, W. P. (2003), "Effect of Cellular Phone Conversations and Other Potential Interference on Reaction Time in a Braking Response," *Accident Analysis and Prevention*. Vol. 35, pp. 495–500.

Cooper, P. J., Pinili, M., and Chen, W. (1995), "An Examination of the Crash Involvement Rates of Novice Drivers Aged 16 to 55," *Accident Analysis and Prevention*, Vol. 27, pp. 89–104.

Couper, F. J., and Logan, B. K. (2004), *Drugs and Human Performance Fact Sheets*, Report DOT-HS-809-725, U.S. Department of Transportation, Washington, DC.

Crouch, D. J., Birky, M. M., Gust, S. W., Rollins, D. E., Walsh, M. J., Moulden, K., and Beckel, R. W. (1993), "The Prevalence of Drugs and Alcohol in Fatally Injured Truck Drivers," *Journal of Forensic Sciences*, Vol. 38, pp. 1342–1353.

Cushman, L. A. (1992), *The Impact of Cognitive Decline and Dementia on Driving in Older Adults*, AAA Foundation for Traffic Safety, Washington, DC.

Dahlstedt, S. (2001), *Perception of Some Seat Belt Reminder Sounds*, Report VTI-77A-2001, Swedish National Road and Transport Research Institute, Stockholm, Sweden.

Davidson, M. B. (1991), *Diabetes Mellitus: Diagnosis and Treatment*, 3rd ed., Churchill Livingstone, New York.

DCCT Research Group (1987), "Diabetes Control and Complications Trial (DCCT): Results of a Feasibility Study," *Diabetes Care*, Vol. 10, pp. 1–19.

Decter, B. M., Goldner, B., and Cohen, T. J. (1994), "Vasovagal Syncope as a Cause of Motor Vehicle Accidents," *American Heart Journal*, Vol. 127, pp. 1619–1621.

DeJoy, D. M. (1989), "The Optimism Bias and Traffic Accident Risk Perception," *Accident Analysis and Prevention*, Vol. 21, pp. 333–340.

DeJoy, D. M. (1990), "Gender Differences in Traffic Accident Risk Perception," in *Proceedings of Human Factors Society 34th Annual Meeting*, Human Factors Society, Santa Monica, CA.

Delphi Electronics Corporation (2003), "Proposal from Delphi Delco Electronics Systems for the Safety Vehicle(s) Using Adaptive Interface Technology (SAVE-IT)," retrieved September 2004, from http://www.volpe.dot.gov/opsad/saveit/docs.html.

de Waard, D., Brookhuis, K., and Hernández-Gress, N. (2001), "The Feasibility of Detecting Phone-Use Related Driver Distraction," *International Journal of Vehicle Design*, Vol. 26, pp. 83–95.

Dewar, R. W. (2002), "Vehicle Design," in *Human Factors in Traffic Safety*, R. W. Dewar and P. L. Olson, Eds., Lawyers and Judges Publishing Company, Tucson, AZ.

Dewar, R. W., and Olson, P. L. (2002), "Traffic Control Devices," in *Human Factors in Traffic Safety*, R. E. Dewar and P. L. Olson, Eds., Lawyers and Judges Publishing Company, Tucson, AZ.

Dewar, R. W., Kline, D. W., and Swanson, H. A. (1994), "Age Differences in Comprehension of Traffic Sign Symbols," *Transportation Research Record*, Vol. 1456, pp. 1–10.

Dinges, D. (1995), "An Overview of Sleepiness and Accidents," *Journal of Sleep Research,* Vol. 4, pp. 4–14.

Doege, T. C., and Engelbert, A. L., Eds. (1986), *Medical Conditions Affecting Drivers*, American Medical Association, Chicago.

Doherty, S. T., Andrey, J. C., and MacGregor, C. (1998), "The Situational Risks of Young Drivers: The Influence of Passengers, Time of Day and Day of Week on Accident Rates," *Accident Analysis and Prevention*, Vol. 30, pp. 45–52.

Drachman, D. A., and Swearer, J. M. (1993), "Driving and Alzheimer's Disease: The Risk of Crashes," *Neurology*, Vol. 43, pp. 2448–2456.

Drakopoulos, A., and Lyles, R. W. (1997), "Driver Age as a Factor in Comprehension of Left-Turn Signals," *Transportation Research Record*, Vol. 1573, pp. 76–85.

Druid, A. (2002), "Vision Enhance System: Does Display Position Matter?" Master's thesis, Linköping University, Vargarda, Sweden.

Dubinsky, R. M., Williamson, A., Gray, C. S., and Glatt, S. L. (1992), "Driving in Alzheimer's Disease," *Journal of the American Geriatric Society*, Vol. 40, pp. 1112–1116.

Dulisse, B. (1997), "Older Drivers and Risk to Other Road Users," *Accident Analysis and Prevention*, Vol. 29, pp. 573–582.

Eby, D. W., and Kostyniuk, L. P. (1998), "Maintaining Older Driver Mobility and Well-Being with Traveler Information Systems," *Transportation Quarterly*, Vol. 52, pp. 45–53.

Eby, D. W., and Kostyniuk, L. P. (1999), "An On-the-Road Comparison of In-Vehicle Navigation Assistance Systems," *Human Factors*, Vol. 41, pp. 295–311.

Eby, D. W., and Molnar, L. J. (1998), *Matching Safety Strategies to Youth Characteristics: A Literature Review of Cognitive Development*, Report DOT-HS-808-927, U.S. Department of Transportation, Washington, DC.

Eby, D. W., and Vivoda, J. M. (2003), "Driver Hand-Held Mobile Phone Use and Safety Belt Use," *Accident Analysis and Prevention*, Vol. 35, pp. 893–895.

Eby, D. W., Kostyniuk, L. P., Streff, F. M., and Hopp, M. L. (1997), *Evaluating the Perceptions and Behaviors of Ali-Scout Users in a Naturalistic Setting*, Report UMTRI-97-08, University of Michigan Transportation Research Institute, Ann Arbor, MI.

Eby, D. W., Trombley, D., Molnar, L. J., and Shope, J. T. (1998), *The Assessment of Older Driver's Capabilities: A Review of the Literature*, Report UMTRI-98-24, University of Michigan Transportation Research Institute, Ann Arbor, MI.

Eby, D. W., Molnar, L. J., and Olk, M. L. (2000), "Trends in Driver and Front-Right Passenger Safety Belt Use in Michigan: 1984 to 1998," *Accident Analysis and Prevention*, Vol. 32, pp. 837–843.

Eby, D. W., Fordyce, T. A., and Vivoda, J. A. (2002a), "A Comparison of Safety Belt Use in Commercial and Noncommercial Vehicles," *Accident Analysis and Prevention*, Vol. 34, pp. 285–291.

Eby, D. W., Kostyniuk, L. P., Spradlin, H., Sudharsan, K., Zakrajsek, J. S., and Miller, L. L. (2002b), *An Evaluation of Michigan's Repeat Alcohol Offender Laws*, Report UMTRI-2002-23, University of Michigan Transportation Research Institute, Ann Arbor, MI.

Eby, D. W. Kostyniuk, L. P., and Vivoda, J. M. (2003a), "Risky Driving: The Relationship Between Cellular Phone and Safety Belt Use," *Transportation Research Record*, No. 1843, pp. 20–23.

Eby, D. W., Molnar, L. J., Shope, J. T., Vivoda, J. M., and Fordyce, T. A. (2003b), "Improving Older Driver Knowledge and Awareness Through Self-Assessment: The Driving Decisions Workbook," *Journal of Safety Research*, Vol. 34, pp. 371–381.

Eby, D. W., Molnar, L. J., Kostyniuk, L. P., and Shope, J. T. (2004), *Developing an Effective and Acceptable Safety Belt Reminder System*, Report UMTRI-2004-29, University of Michigan Transportation Research Institute, Ann Arbor, MI.

Epstein, A. E., Miles, W. M., Benditt, D. G., Camm, A. J., Darling, E. J., Friedman, P. L., Garson, A., Harvey, J. C., Kidwell, G. A., Klein, G. J., Levine, P. A., Marchlinski, F. E., Prystowsky, E. N., and Wilkoff, B. L. (1996), "Personal and Public Safety Issues Related to Arrhythmias That May Affect Consciousness: Implications for Regulation and Physician Recommendations," *Circulation*, Vol. 94, pp. 1147–1166.

Evans, L. (1991), *Traffic Safety and the Driver*, Van Nostrand Reinhold, New York.

Evans, D. A., Funkenstein, H. H., Albert, M. S., Scherr, P. A., Cook, N. R., Chown, M. J., Hebert, L. E., Hennekens, C. H., and Taylor, J. O. (1989), "Prevalence of Alzheimer's Disease in a Community Population of Older Persons: Higher Than Previously Reported," *Journal of the American Medical Association,* Vol. 262, pp. 2551–2556.

Farley, F. H., and Cox, S. O. (1971), "Stimulus-Seeking Motivation in Adolescents as a Function of Age and Sex," *Adolescence*, Vol. 6, pp. 207–218.

FHWA (1998), *Older Driver Highway Design Handbook*, Report FHWA-RD-97-135, Federal Highway Administration, Washington, DC.

FHWA (2004), *Older Drivers at Intersections: 10 Brief Issues*, Federal Highway Administration, Washington, DC.

Fildes, B., Fitzharris, M., Koppel, S., and Vulcan, P. (2002), *Benefits of Seat Belt Reminder Systems*, Monash University Accident Research Center, Melbourne, Victoria, Australia.

Finn, P., and Bragg, B. W. E. (1986), "Perception of the Risk of an Accident by Young and Older Drivers," *Accident Analysis and Prevention*, Vol. 18, pp. 289–298.

Fischhoff, B., Lichtenstein, S., Slovic, P., Derby, S. L., and Keeney, R. L. (1981), *Acceptable Risk*, Cambridge University Press, Cambridge.

Fox, G. K., Bowden, S. C., Bashford, G. M., and Smith, D. S. (1997), "Alzheimer's Disease and Driving: Prediction and Assessment of Driving Performance," *Journal of the American Geriatric Society*, Vol. 45, pp. 949–953.

Friedland, R. P., Koss, E., Kumar, A., Gaine, S., Metzler, D., Haxby, J. V., and Moore, A. (1988), "Motor Vehicle

Crashes in Dementia of the Alzheimer Type," *Annals of Neurology*, Vol. 24, pp. 782–786.

Gallo, J. J., Rebok, G. W., and Lesiker, S. E. (1999), "The Driving Habits of Adults Aged 60 Years and Older," *Journal of the American Geriatrics Society,* Vol. 47, pp. 335–341.

Gallup (2001), "Half of All Americans Own a Cellular Phone: Many Chat Frequently Behind the Wheel," retrieved September 18, 2001, from http://www.gallup.com/poll/releases/pr0000426.asp.

General Assembly of the Commonwealth of Pennsylvania (2001), *Driver Distractions and Traffic Safety*, General Assembly of the Commonwealth of Pennsylvania, Joint State Government Commission, Harrisburg, PA.

Gerber, S. R., Joliet, B. V., and Feegel, J. R. (1966), "Single Vehicle Accidents in Cuyahoga (Ohio), 1958–1963," *Journal of Forensic Sciences*, Vol. 11, pp. 144–151.

Giambra, L. M., Camp, C. J., and Grodsky, A. (1992), "Curiosity and Stimulation Seeking Across the Adult Life Span: Cross-Sectional and Seven-Year Longitudinal Findings," *Psychology and Aging*, Vol. 7, pp. 150–157.

Gimbel, J. R. (2004), "When Should Patients Be Allowed to Drive After ICD Implantation?" *Cleveland Clinic Journal of Medicine*, Vol. 71, pp. 125–128.

Glassbrenner, D. (2004), *Safety Belt Use in 2004: Overall Results*, Report DOT HS 809 708, U.S. Department of Transportation, Washington, DC.

Glaze, A. L., and Ellis, J. M. (2003), *Pilot Study of Distracted Drivers*, Virginia Commonwealth University, Richmond, VA.

Goodman, M., Bents, F., Tijerina, L., Wierwille, W., Lerner, N., and Benel, D. (1997), *An Investigation of the Safety Implications of Wireless Communications in Vehicles*, Report DOT-HS-808-635, U.S. Department of Transportation, Washington, DC.

Green, P. (1997), "Potential Safety Impacts of Automotive Navigation Systems," presented at the 2nd Annual Automotive Land Navigation Conference, Detroit, MI.

Green, P. (2000), "Crashes Induced by Driver Information Systems and What Can be Done to Reduce Them," in *Proceedings of the 2000 International Congress on Transportation Electronics*, Society of Automotive Engineers, Warrendale, PA.

Greening, L., and Chandler, C. C. (1997), "Why It Can't Happen to Me: The Base Rate Matters, but Overestimating Skill Leads to Underestimating Risk," *Journal of Applied Social Psychology*, Vol. 27, pp. 760–780.

Groeger, J. A., and Chapman, P. R. (1996), "Judgement of Traffic Scenes: The Role of Danger and Difficulty," *Applied Cognitive Psychology*, Vol. 10, pp. 349–364.

Haase, G. R. (1977), "Diseases Presenting as Dementia," in *Dementia*, 2nd ed., C. E. Wells, Ed., F.A. Davis, Philadelphia.

Haerer, A. F., Anderson, D. W., and Schoenberg, B. S. (1986), "Prevalence and Clinical Features of Epilepsy in a Biracial United States Population," *Epilepsia*, Vol. 27, pp. 66–75.

Hakamies-Bloomqvist, L. (1994), "Aging and Fatal Accidents in Male and Female Drivers," *Journals of Gerontology: Social Sciences*, Vol. 49, pp. S286–S290.

Hakamies-Bloomqvist, L. (2004), "Safety of Older Persons in Traffic," in *Transportation in an Aging Society: A Decade of Experience*, Transportation Research Board, Washington, DC, pp. 22–35.

Hancock, P. A., Lesch, M., and Simmons, L. (2003), "The Distraction Effects of Phone Use During a Critical Driving Maneuver," *Accident Analysis and Prevention*, Vol. 35, pp. 501–514.

Hansotia, P. (1993), "Seizure Disorders, Diabetes Mellitus, and Cerebrovascular Disease: Considerations for Older Drivers," *Clinical Geriatric Medicine*, Vol. 9, pp. 323–339.

Hansotia, P., and Broste, S. (1991), "The Effect of Epilepsy and Diabetes Mellitus on the Risk of Automobile Accidents," *New England Journal of Medicine*, Vol. 324, pp. 22–26.

Haraldsson, P. O., Carenfelt, C., Laurel, H., and Tornros, J. (1990), "Driving Vigilance Simulator Test," *Acta Otolaryngologica (Stockholm)*, Vol. 110, pp. 136–140.

Harms, L., and Patten, C. (2003), "Peripheral Detection as a Measure of Driver Distraction: A Study of Memory-Based Versus System-Based Navigation in a Built-up Area," *Transportation Research, Part F*, Vol. 6, pp. 23–36.

Harrison, W. A., Senserrick, T. M., and Tingvall, C. (2000), *Development and Trial of a Method to Investigate the Acceptability of Seat Belt Reminder Systems*, Report 170, Monash University Accident Research Centre, Melbourne, Victoria, Australia.

Hauer, E. (1988), "The Safety of Older Persons at Intersections," in *Special Report 218: Transportation in an Aging Society: Improving Mobility and Safety for Older Persons*, Vol. 2, Transportation Research Board, Washington, DC.

Hauser, W. A., and Hesdorffer, D. C. (1990), *Epilepsy: Frequency, Causes and Consequences*, Epilepsy Foundation of America, New York.

Hauser, W. A., and Kurland, L. T. (1975), "The Epidemiology of Epilepsy in Rochester, Minnesota, 1935 through 1967," *Epilepsia*, Vol. 16, pp. 1–66.

Hawkins, H. G., Jr., Womack, K. N., and Mounce, J. M. (1993), "Driver Comprehension of Regulatory Signs, Warning Signs, and Pavement Markings," *Transportation Research Record*, No. 1403, pp. 67–82.

Haworth, N., Bowland, L., and Foddy, B. (1999), *Seat Belts for Truck Drivers*, Report AP-141/99, Monash University, Melbourne, Victoria, Australia.

Heino, A., van den Molen, H. H., and Wilde, G. J. S. (1992), *Risk Homeostasis Process in Car Following Behaviour: Individual Differences in Car Following and Perceived Risk*, Report VK 92-02, Rijksuniversiteit Groningen Haven, Groningen, The Netherlands.

Herner, B., Smedby, B., and Ysander, L. (1966), "Sudden Illness as a Cause of Motor-Vehicle Accidents," *British Journal of Industrial Medicine*, Vol. 23, pp. 37–41.

Hoffmeister, D. H., Arsenault, R., and Crandall, H. E. (1995), "Recognition of Automobile Climate Control Symbols," in *Proceedings of the 39th Annual Meeting of the Human Factors and Ergonomics Society*, HFES, Santa Monica, CA, pp. 1132–1136.

Hu, P. S., Young, J. R., and Lu, A. (1993), *Highway Crash Rates and Age-Related Driver Limitations: Literature Review and Evaluation of Data Bases*, Oak Ridge National Laboratory, Oak Ridge, TN.

Hughes, C. P., Berg, L., Danziger, W. L., Coben, L. A., and Martin, R. L. (1982), "A New Clinical Scale for the Staging of Dementia," *British Journal of Psychiatry*, Vol. 140, pp. 556–572.

Hummer, J. E., Montgomery, R. E., and Sinha, K. C. (1990), "Motorists Understanding of and Preferences for Left-Turn Signals," *Transportation Research Record*, No. 1281, pp. 136–147.

Hunt, L. A. (1991), "Dementia and Road Test Performance," in *Proceedings of Strategic Highway Research Program and Traffic Safety on Two Continents*, Göteborg, Sweden.

Hunt, L. A., Morris, J. C., Edwards, D., and Wilson, B. S. (1993), "Driving Performance in Persons with Mild Senile Dementia of the Alzheimer Type," *Journal of the American Geriatric Society*, Vol. 41, pp. 747–753.

Insurance Institute for Highway Safety (2004), "State DWI/DUI Laws as of July 2004," retrieved September 2004, from http://www.iihs.org/safety_facts/state_laws/dui.htm.

Insurance Research Council (1997), *Public Attitude Monitor*, IRC, Wheaton, IL.

Janke, M. K. (1994), *Age-Related Disabilities That May Impair Driving and Their Assessment: A Literature Review*, Report RSS-94-156, California Department of Motor Vehicles, Sacramento, CA.

Jenness, J. W., Lattanzio, R. J., O'Toole, M., and Taylor, N. (2002), "Voice Activated Dialing or Eating a Cheeseburger: Which Is More Distracting During Simulated Driving?" in *Proceedings of the Human Factors and Ergonomics Society 46th Annual Meeting*, HFES, Santa Monica, CA, pp. 592–596.

Jonah, B. A. (1986), "Accident Risk and Risk-Taking Behavior Among Young Drivers," *Accident Analysis and Prevention*, Vol. 18, pp. 255–271.

Jonah, B. A. (1997), "Sensation Seeking and Risky Driving: A Review and Synthesis of the Literature," *Accident Analysis and Prevention*, Vol. 29, pp. 651–665.

Jonah, B. A., Thiessen, R., Au-Yeung, E., and Vincent, A. (1997), "Sensation Seeking and Risky Driving," presented at the Risk-Taking Behavior and Traffic Safety Symposium, Cape Cod, MA.

Jones, R. K., and Lacey, J. H. (2000), *State of Knowledge of Alcohol-Impaired Driving: Research on Repeat DWI Offenders*, Report DOT-HS-809, U.S. Department of Transportation, Washington, DC.

Jones, R. K., Shinar, D., and Walsh, J. M. (2003), *State of Knowledge of Drug-Impaired Driving*, Report DOT-HS-809-642, U.S. Department of Transportation, Washington, DC.

Kannel, W. B., Gagnon, D. R., and Cupples, L. A. (1990), "Epidemiology of Sudden Coronary Death: Population at Risk," *Canadian Journal of Cardiology*, Vol. 6, pp. 439–444.

Kantowitz, B. H. (1989), "Interfacing Human and Machine Intelligence," in *Intelligent Interfaces: Theory, Research and Design*, Vol. 3, P. A. Hancock and M. H. Chignell, Eds., North-Holland, Amsterdam, pp. 49–67.

Kantowitz, B. H. (2000), "In-Vehicle Information Systems: Premises, Promises and Pitfalls," *Transportation Human Factors*, Vol. 2, pp. 359–379.

Kantowitz, B. H. (2001), "Using Microworlds to Design Interfaces That Minimize Driver Distraction," in *Proceedings of the First International Driving Symposium on Human Factors in Driver Assessment, Training and Vehicle Design*, Aspen, CO, pp. 42–57.

Kantowitz, B. H. and Premkumar, S. (2004), *Safety Vehicles Using Adaptive Interface Technology: Identify Demand Levels of Telematic Tasks*, University of Michigan Transportation Research Institute, Ann Arbor, MI.

Kantowitz, B. H., and Sorkin, R. D. (1987), "Allocation of Functions," in *Handbook of Human Factors*, G. Salvendy, Ed., Wiley, New York.

Kantowitz, B. H., Bortolussi, M. R., and Hart, S. G. (1987), "Measuring Pilot Workload in a Motion Base Simulator, III: Synchronous secondary tasks," *Proceedings of the Human Factors Society*, Vol. 31, pp. 834–837.

Kantowitz, B. H., Hanowski, R. J., and Kantowitz, S. C. (1997), "Driver Acceptance of Unreliable Traffic Information in Familiar and Unfamiliar Settings," *Human Factors*, Vol. 39, pp. 164–176.

Kapoor, W. N. (1994), "Syncope in Older Persons," *Journal of the American Geriatric Society*, Vol. 42, pp. 426–436.

Kapoor, W. N., Karpf, M., Wieand, S., Peterson, J. R., and Levey, G. S. (1983), "A Prospective Evaluation and Follow-up of Patients with Syncope," *New England Journal of Medicine*, Vol. 309, pp. 197–204.

Kapoor, W. N., Hammill, S. C., and Gersh, B. J. (1989), "Diagnosis and Natural History of Syncope and the Role of Invasive Electrophysiologic Testing," *American Journal of Cardiology*, Vol. 63, pp. 730–734.

Kapust, L. R., and Weintraub, S. (1992), "To Drive or Not to Drive: Preliminary Results from Road Testing of Patients with Dementia," *Journal of Geriatric Psychiatry and Neurology*, Vol. 5, pp. 210–216.

Katzman, R. (1987), "Alzheimer's Disease: Advances and Opportunities," *Journal of the American Geriatric Society*, Vol. 35, pp. 69–73.

Keller and Associates (2003), *Supervisor's Guide to Hours of Service: Motor Carrier Safety Report*, J.J. Keller & Associates, Neenah, WI.

Kelso, J. A. S. (1982), *Human Motor Behavior: An Introduction*, Lawrence Erlbaum Associates, Mahwah, NJ.

Kempton, H. (2004), *World Drink Trends 2004*, World Advertising Research Center, Henley on Thames, Oxfordshire, England.

Knipling, R., and Wang, J. (1994), *Crashes and Fatalities Related to Driver Drowsiness/Fatigue*, Research Note, National Highway Traffic Safety Administration, Washington, DC.

Knoblauch, R. L., Nitzburg, M., and Seifurt, R. F. (1995), "Older Driver Freeway Needs and Capabilities," in *Compendium of Technical Papers, Institute of Transportation Engineers 65th Annual Meeting*, ITE, Washington, DC, pp. 629–633.

Knoblauch, R. L., Cotton, R., Nitzburg, M., Seifurt, R. F., Shapiro, G., and Broene, P. (2003), *Safety Belt Usage by Commercial Motor Vehicle Drivers*, Federal Motor Carrier Safety Administration, Washington, DC.

Konecni, V. J., Ebbesen, E. B., and Konecni, D. K. (1976), "Decision Processes and Risk Taking in Traffic: Driver Response to the Onset of Yellow Light," *Journal of Applied Psychology*, Vol. 61, pp. 359–367.

Koppa, R. (2004), "Automotive Adaptive Equipment and Vehicle Modifications," in *Transportation in an Aging Society: A Decade of Experience*, Transportation Research Board, Washington, DC, pp. 227–235.

Kostyniuk, L. P., Eby, D. W., Christoff, C., Hopp, M. L., and Streff, F. M. (1997a), *The FAST-TRAC Natural Use Leased-Car Study: An Evaluation of User Perceptions and Behaviors of Ali-Scout by Age and Gender*, Report UMTRI-97-09, University of Michigan Transportation Research Institute, Ann Arbor, MI.

Kostyniuk, L. P., Eby, D. W., Hopp, M. L., and Christoff, C. (1997b), *Driver Response to the TetraStar Navigation Assistance System by Age and Sex*, Report UMTRI-97-33, University of Michigan Transportation Research Institute, Ann Arbor, MI.

Kostyniuk, L. P., Streff, F. M., and Eby, D. W. (1997c), *"The Older Driver and Navigation Assistance Systems,"*

Report UMTRI-97-47, University of Michigan Transportation Research Institute, Ann Arbor, MI.

Kostyniuk, L. P., Shope, J. T., and Molnar, L. J. (2000a), *Reduction and Cessation of Driving Among Older Drivers in Michigan: Final Report*, Report UMTRI-2000-06, University of Michigan Transportation Research Institute, Ann Arbor, MI.

Kostyniuk, L. P., Trombley, D. A., and Shope, J. T. (2000b), *The Process of Reduction and Cessation of Driving Among Older Drivers: A Review of the Literature*, Report UMTRI-98-23, University of Michigan Transportation Research Institute, Ann Arbor, MI.

Kostyniuk, L. P., Streff, F. M., and Zakrajsek, J. (2002), *Identifying Unsafe Driver Actions That Lead to Fatal Car–Truck Crashes*, AAA Foundation for Traffic Safety, Washington, DC.

Kou, W. H., Calkins, H., Lewis, R. R., Bolling, S. F., Kirsch, M. M., Langberg, J. J., de Buitleir, M., Sousa, J., El-atassi, R., and Morady, F. (1991), "Incidence of Loss of Consciousness During Automatic Implantable Cardioverter-Defibrillator Shocks," *Annals of Internal Medicine*, Vol. 115, pp. 942–945.

Koushki, P. A., Ali, S. Y., and Al-Saleh, O. I. (1999), "Driving and Using Mobile Phones: Impacts on Road Accidents," *Transportation Research Record*, No. 1694, pp. 27–33.

Kribbs, N., and Dinges, D. (1994), "Vigilance Decrement and Sleepiness," in *Sleep Onset Mechanisms*, J. Harsh and R. Ogilvie, Eds., American Psychological Association, Washington, DC, pp. 113–125.

Kurtzke, J. F. (1985), "Epidemiology of Cerebrovascular Disease," in *Cerebrovascular Survey Report for the National Institutes of Neurological and Communicative Disorders and Stroke*, rev. ed., F. McDowell and L. R. Caplan, Eds., Whiting Press, Rochester, MN.

Laberge-Nadeau, C., Maag, U., Bellavance, F., Lapierre, S. D., Desjardins, D., Messier, S., and Saïdi, A. (2003), "Wireless Telephones and the Risk of Road Crashes," *Accident Analysis and Prevention*, Vol. 35, pp. 649–660.

Lajunen, T., and Summala, H. (1996), "Effects of Driving Experience, Personality, Driver's Skill and Safety Orientation on Speed Regulation and Accidents," presented at the International Conference on Traffic and Transport Psychology, Valencia, Spain, May.

Langlois, P. H., Smolensky, M. H., Hsi, B. P., and Weir, F. W. (1985), "Temporal Patterns of Reported Single Vehicle Car and Truck Accidents in Texas, USA during 1980–83," *Chronobiology International*, Vol. 2, pp. 131–140.

Lastovicka, J. L., Murray, J. P., Joachimsthaler, E. A., Bhalla, G., and Scheurich, J. (1987), "A Lifestyle Typology to Model Young Male Drinking and Driving," *Journal of Consumer Research*, Vol. 14, pp. 257–263.

Lerner, N. D., Morrison, M. L., and Ratte, D. J. (1990), *Older Drivers' Perceptions of Problems in Freeway Use*, AAA Foundation for Traffic Safety, Washington, DC.

Lings, S., and Jensen, P. B. (1991), "Driving After Stroke: A Controlled Laboratory Investigation," *International Disability Studies*, Vol. 13, pp. 74–82.

Llaneras, R. E., and Lerner, N. D. (2000), "The Effects of ATIS on Driver Decision Making," *ITS Quarterly*, Summer.

Logsdon, R. G., Teri, L., and Larson, E. B. (1992), "Driving and Alzheimer's Disease," *Journal of General Internal Medicine*, Vol. 7, pp. 583–588.

Lucas-Blaustein, M. J., Filipp, L., Dungan, C., and Tune, L. (1988), "Driving in Patients with Dementia," *Journal of the American Geriatric Society*, Vol. 36, pp. 1087–1092.

Lund, A. K., Preusser, D. F., Blomberg, R. D., and Williams, A. F. (1988), "Drug Use by Tractor-Trailer Drivers," *Journal of Forensic Sciences*, Vol. 33, pp. 648–661.

Lyman, S., and Braver, E. R. (2002), "Occupant Deaths in Large Truck Crashes in the United States: 25 Years of Experience," in *Proceedings of the International Truck and Bus Safety Research and Policy Symposium*, Report E01-2510-001-001-02, University of Tennessee, Knoxville, TN.

Magaro, P. A., Smith, P., Cionini, L., and Velicogna, F. (1979), "Sensation Seeking in Italy and the United States," *Journal of Social Psychology*, Vol. 109, pp. 159–165.

Marques, P. R., Voas, R. B., Tippetts, A. S., and Beirness, D. J. (1999), "Behavioral Monitoring of DUI Offenders with Alcohol Ignition Interlock Recorder," *Addiction*, Vol. 94, pp. 1861–1870.

Matthews, M. L., and Moran, A. R. (1986), "Age Differences in Male Drivers' Perception of Accident Risk: The Role of Perceived Driving Ability," *Accident Analysis and Prevention*, Vol. 18, pp. 299–313.

Maycock, G. (1996), "Sleepiness and Driving: The Experience of UK Car Drivers," *Journal of Sleep Research*, Vol. 5, pp. 220–237.

Maycock, G. (1997), *The Safety of Older Car Users in the European Union*, Foundation for Road Safety Research, Basingstoke, Hampshire, England.

McCartt, A. T., Ribner, S. A., Pack, A. I., and Hammer, M. C. (1996), "The Scope and Nature of the Drowsy Driving Problem in New York State," *Accident Analysis and Prevention*, Vol. 28, pp. 511–517.

McKenna, F. P. (1993), "It Won't Happen to Me: Unrealistic Optimism or Illusion of Control?" *British Journal of Psychology*, Vol. 84, pp. 39–50.

McKhann, G., Drachman, D., Folstein, M. F., Katzman, R., Price, D., and Stadlan, E. M. (1984), "Clinical Diagnosis of Alzheimer's Disease: Report of the NINCDS-ADRDA Work Group," *Neurology*, Vol. 34, pp. 939–944.

McKnight, A. J., and McKnight, A. S. (1993), "The Effects of Cellular Phone Use upon Driver Attention," *Accident Analysis and Prevention*, Vol. 25, pp. 259–265.

McMillen, D. L., Adams, M. S., Wells-Parker, E., Pang, M. G., and Anderson, B. J. (1992a), "Personality Traits and Behaviors of Alcohol Impaired Drivers: A Comparison of First and Multiple Offenders," *Addictive Behaviors*, Vol. 17, pp. 407–414.

Mitchell, R. K., Castleden, C. M., and Fanthome, Y. (1995), "Driving, Alzheimer's Disease and Ageing: A Potential Cognitive Screening Device for All Elderly Drivers," *International Journal of Geriatric Psychiatry*, Vol. 10, pp. 865–869.

Molnar, L. J., Eby, D. W., and Miller, L. L. (2003), *Promising Approaches to Enhancing Elderly Mobility*, Report UMTRI-2003-14, University of Michigan Transportation Research Institute, Ann Arbor, MI.

Moskowitz, H., and Fiorentino, D. (2000), *A Review of the Literature on the Effects of Low Doses of Alcohol on Driving-Related Skills*, Report DOT-HS-809-028, U.S. Department of Transportation, Washington, DC.

Moskowitz, H., Burns, M., Fiorentino, D., Smiley, A., and Zador, P. (2000), *Driver Characteristics and Impairment at Various BACs*, U.S. Department of Transportation, Washington, DC.

National Center for Health Statistics (2004), "Health, United States, 2003," retrieved September 2004, from www.cdc.gov/nchs/products/pubs/pubd/hus/trendtables.htm.

National Cooperative Highway Research Program (2003), *Application of European 2 + 1 Roadway Designs*, NCHRP Report 275, Transportation Research Board, Washington, DC.

National Diabetes Data Group (1979), "Classification and Diagnosis of Diabetes Mellitus and Other Categories of Glucose Intolerance," *Diabetes*, Vol. 28, p. 1039.

NHTSA (National Highway Traffic Safety Administration) (2001), *Passenger Vehicle Driver Cell Phone Use Results from the Fall 2000 National Occupant Protection Use Survey*, Report DOT-HS-809-293, U.S. Department of Transportation, Washington, DC.

NHTSA (2002), *Drowsy Driving and Automobile Crashes*, U.S. Department of Transportation, Washington, DC.

NHTSA (2003a), *Traffic Safety Facts, 2002*, U.S. Department of Transportation, Washington, DC.

NHTSA (2003b), *Automotive Collision Avoidance System Field Operational Test*, Report DOT-HS-809-600, U.S. Department of Transportation, Washington, DC.

NHTSA (2003c), *Initiatives to Address Safety Belt Use*, Report NHTSA-2003-14620, U.S. Department of Transportation, Washington, DC.

NHTSA (2004a), *Traffic Safety Facts, 2003*, U.S. Department of Transportation, Washington, DC.

NHTSA (2004b), *What Is the Mission of the Drowsy Driving Program?* retrieved September 2004, from http://www.nhtsa.dot.gov/people/injury/drowsy_driving1/index.html.

Odenheimer, G. L., Beaudet, M., Gette, A. M., Albert, M. S., Grande, L., and Minaker, K. L. (1994), "Performance-Based Driving Evaluation of the Elderly Driver: Safety, Reliability and Validity," *Journal of Gerontology: Medical Science*, Vol. 49, pp. 153–159.

Olson, P. L. (2003), "Visibility with Motor Vehicle Headlamps," in *Human Factors in Traffic Safety*, R. W. Dewar and P. L. Olson, Eds., Lawers and Judges Publishing Company, Tucson, AZ.

Owens, D. A., and Sivak, M. (1996), "Differentiation of Visibility and Alcohol as Contributing Factors to Twilight Road Fatalities," *Human Factors*, Vol. 38, pp. 680–689.

Pack, A. I., Pack, A. M., Rodgman, E., Cucchaira, A., Dinges, D. F., and Schwab, C. W. (1995), "Characteristics of Crashes Attributed to the Driver Having Fallen Asleep," *Accident Analysis and Prevention*, Vol. 27, pp. 769–775.

Patten, C. J. D., Kircher, A., Östlund, J., and Nilsson, L. (2004), "Using Mobile Telephones: Cognitive Workload and Attention Resource Allocation," *Accident Analysis and Prevention*, Vol. 36, pp. 341–350.

Perez, J., Ortet, G., Pla, S., and Simo, S. (1986), "A Junior Sensation Seeking Scale," *Personality and Individual Differences*, Vol. 7, pp. 915–918.

Perez, W. A., Van Aerde, M., Rakha, H., and Robinson, M. (1996), *TravTek Evaluation Safety Study*, Report FHWA-RD-95-188, U.S. Department of Transportation, Washington, DC.

Persson, D. (1993), "The Elderly Driver: Deciding When to Stop," *Gerontologist*, Vol. 33, pp. 88–91.

Piechulla, W., Mayser, C., Gehrke, H., and König, W. (2003), "Reducing Drivers' Mental Workload by Means of an Adaptive Man–Machine Interface," *Transportation Research, Part F*, Vol. 6, pp. 233–248.

Popkin, C. L., and Waller, P. F. (1989), "Epilepsy and Driving in North Carolina: An Exploratory Study," *Accident Analysis and Prevention*, Vol. 21, pp. 389–393.

Preusser, D. F., Ferguson, S. A., and Williams, A. F. (1998), "The Effect of Teenage Passengers on the Fatal Crash Risk of Teenage Drivers," *Accident Analysis and Prevention*, Vol. 30, pp. 217–222.

Ranney, T. A., Garrott, R., and Goodman, M. J. (2001), "NHTSA Driver Distraction Research: Past, Recent, and Future," in *Proceedings of the 17th International Technical Conference on the Enhanced Safety of Vehicles*, Report 233, CD-ROM, U.S. Department of Transportation, Washington, DC.

Ray, W. A., Fought, R. L., and Decker, M. D. (1992), "Psychoactive Drugs and the Risk of Injurious Motor Vehicle Crashes in Elderly Drivers," *American Journal of Epidemiology*, Vol. 136, pp. 873–883.

Raytheon Commercial Infrared and ElCAN-Teaxs Optical Technology (2000), *NightDriver*™ *Thermal Imaging Camera and HUD Development Program for Collision Avoidance Applications*, Report DOT-HS-809-163, U.S. Department of Transportation, Washington, DC.

Redelmeier, D. A., and Tibshirani, R. J. (1997), "Association Between Cellular-Telephone Calls and Motor Vehicle Collisions," *New England Journal of Medicine*, Vol. 336, No. 7, pp. 453–457.

Reger, M. A., Welsh, R. K., Watson, G. S., Cholerton, B., Baker, L. D., and Craft, S. (2004), "The Relationship Between Neurophysiological Functioning and Driving Ability in Dementia: A Meta-analysis," *Neuropsychology*, Vol. 18, pp. 85–93.

Rehm, C. G., and Ross, S. E. (1995), "Syncope as Etiology of Road Crashes Involving Elderly Drivers," *American Surgeon*, Vol. 61, pp. 1006–1008.

Reinfurt, D. W., Huang, H. F., Feaganes, J. R., and Hunter, W. W. (2001), *Cell Phone Use While Driving in North Carolina*, University of North Carolina Highway Safety Research Center, Chapel Hill, NC.

Road Freight Transport (2000), *Road Freight Transport Health and Safety Guide*, Road Freight Transport, Division of Workplace Health and Safety, Brisbane, Queensland, Australia.

Royal, D. (2003), *National Survey of Distracted and Drowsy Driving Attitudes and Behaviors: 2002*, Vol. 1, *Findings Report*, Gallup Organization, Washington, DC.

Rubens, S., Barber, P., and Bradley, M. (1994), "Expectancies Related to Automobile Controls: Function Specific Effects," in *Contemporary Ergonomics*, S. A. Robertson, Ed., Taylor & Francis, London, pp. 211–216.

Russo, M. F., Lakey, B. B., Christ, M. A. G., Frick, P. J., McBurnett, K., Walker, J. L., Loeber, R., Stouthhamer-Loeber, M., and Green, S. (1991), "Preliminary Development of a Sensation Seeking Scale for Children," *Personality and Individual Differences*, Vol. 12, pp. 399–405.

Russo, M. F., Stokes, G. S., Lakey, B. B., Christ, M. A. G., McBurnett, K., Loeber, R., Stouthhamer-Loeber, M., and Green, S. (1993), "A Sensation Seeking Scale in Children: Further Refinement and Psychometric Development," *Journal of Psychopathology and Behavioral Assessment*, Vol. 15, pp. 69–86.

Sagberg, F. (2001), "Accident Risk of Car Drivers During Mobile Telephone Use," *International Journal of Vehicle Design*, Vol. 26, pp. 57–69.

Savage, D. D., Corwin, L., McGee, D. L., Kannel, W. B., and Wolf, P. A. (1985), "Epidemiological Features of Isolated Syncope: The Framingham Study," *Stroke*, Vol. 16, pp. 626–629.

Sayer, J. R., and Green, P. (1988), *Current ISO Automotive Symbols Versus Alternatives: A Preference Study*, Report SP-752, Society of Automotive Engineers, Detroit, MI.

Schieber, F. (2004), "Highway Research to Enhance Safety and Mobility of Older Persons," in *Transportation in an Aging Society: A Decade of Experience*, Transportation Research Board, Washington, DC, pp. 125–154.

Senders, J. W., Kristofferson, A. B., Levison, W. H., Dietrich, C. W., and Ward, J. L. (1967), *The Attentional Demand of Automobile Driving*, Highway Research Record 195, National Academy of Sciences, Transportation Research Board, Washington, DC, pp. 15–33.

Serafin, C., Wen, C., Paelke, G., and Green, P. (1993), *Development and Human Factors Tests of Car Phones*, Report UMTRI-93-17, University of Michigan Transportation Research Institute, Ann Arbor, MI.

Sheldon, R., and Koshman, M. L. (1995), "Can Patients with Neuromediated Syncope Safely Drive Motor Vehicles?" *American Journal of Cardiology*, Vol. 75, pp. 955–956.

Sivak, M., Soler, J., Tränkle, U., and Spagnhol, J. M. (1989), "Cross-Cultural Differences in Driver Risk-Perception," *Accident Analysis and Prevention*, Vol. 21, pp. 355–362.

Smiley, A. (2004), "Adaptive Strategies of Older Drivers," in *Transportation in an Aging Society: A Decade of Experience*, Transportation Research Board, Washington, DC, pp. 36–43.

Smith, J. S., and Kiloh, L. G. (1981), "The Investigation of Dementia: Results in 200 Consecutive Admissions," *Lancet*, Vol. 1, pp. 824–827.

Soliday, S. M. (1974), "Relationship Between Age and Hazard Perception in Automobile Drivers," *Perceptual and Motor Skills*, Vol. 39, pp. 335–338.

Spudis, E. V., Penry, J. K., and Gibson, P. (1986), "Driving Impairment Caused by Episodic Brain Dysfunction: Restrictions for Epilepsy and Syncope," *Archives of Neurology*, Vol. 43, pp. 558–564.

Stamatiadis, N. (1998), "ITS and Human Factors and the Older Driver: The U.S. Experience," *Transportation Quarterly*, Vol. 52, pp. 91–101.

Stevens, A., and Minton, R. (2001), "In-Vehicle Distraction and Fatal Accidents in England and Wales," *Accident Analysis and Prevention*, Vol. 33, pp. 539–545.

Stewart, R. B., Moore, M. T., Marks, G., May, F. E., and Hale, W. E. (1993), *Driving Cessation and Accidents in the Elderly: An Analysis of Symptoms, Diseases, Cognitive Dysfunction, and Medications*, AAA Foundation for Traffic Safety, Washington, DC.

Strayer, D. L., and Johnston, W. A. (2001), "Driven to Distraction: Dual-Task Studies of Simulated Driving and Conversing on a Cellular Telephone," *Psychological Science*, Vol. 12, pp. 642–466.

Stutts, J. C., Reinfurt, D. W., and Rodgman, E. A. (2001), "The Role of Driver Distraction in Crashes: An Analysis of 1995–1999 Crashworthiness Data System Data," in *45th Annual Proceedings of the Association for the Advancement of Automotive Medicine*, AAAM, Des Plaines, IA, pp. 287–301.

Stutts, J. C., Wilkins, J. W., Osberg, J. S., and Vaughn, B. V. (2003), "Driver Risk Factors for Sleep-Related Crashes," *Accident Analysis and Prevention*, Vol. 35, pp. 321–331.

Sullivan, J. M., and Flannagan, M. J. (2002), "The Role of Ambient Light Level in Fatal Crashes: Inferences from Daylight Saving Time Transitions," *Accident Analysis and Prevention*, Vol. 34, pp. 487–498.

Sussman, E. D., Bishop, H., Madnick, B., and Walter, R. (1985), "Driver Inattention and Highway Safety," *Transportation Research Record*, No. 1047, pp. 40–48.

Svenson, O. (1981), "Are We All Less Risky and More Skillful Than Our Fellow Drivers?" *Acta Psychologica*, Vol. 47, pp. 143–148.

Svenson, O., Fischhoff, B., and MacGregor, D. (1985), "Perceived Driving Safety and Seat Belt Usage," *Accident Analysis and Prevention*, Vol. 17, pp. 119–133.

Taylor, J., Chadwick, D., and Johnson, T. (1996), "Risk of Accidents in Drivers with Epilepsy," *Journal of Neurology, Neurosurgery and Psychiatry*, Vol. 60, pp. 621–627.

Telecompetition Inc. (2001), *U.S. Mobil Phone Subscribers and Penetration Rates*, Telecompetition Inc., San Ramon, CA.

Teraski, M., Shiomi, K., Kishimoto, Y., and Hiraoka, K. (1987), "A Japanese Version of the Sensation Seeking Scale," *Japanese Journal of Psychology*, Vol. 58, pp. 42–48.

Terhune, K. W., Ippolito, C. A., Hendricks, D. L., Michalovic, J. G., Bogema, S. C., Santinga, P., Blomberg, R., and Preusser, D. F. (1992), *The Incidence and Role of Drugs in Fatally Injured Drivers*, Report DOT-HS-808-81, U.S. Department of Transportation, Washington, DC.

Terry, R. D., and Katzman, R. (1983), "Senile Dementia of the Alzheimer Type," *Annals of Neurology*, Vol. 14, pp. 497–506.

Tijerina, L., Kiger, S. M., Rockwell, T. H., and Tornow, C. (1995), "Workload Assessment of In-Cab Text Message System and Cellular Phone Use by Heavy Vehicle Drivers on the Road," in *Proceedings of the Human Factors and Ergonomics Society 39th Annual Meeting*, HFES, Santa Monica, CA, pp. 1117–1121.

Tijerina, L., Johnston, S., Parmer, E., Winterbottom, M. D., and Goodman, M. (2000), *Driver Distraction with Wireless Telecommunications and Route Guidance Systems*, Report DOT-HS-809-069, U.S. Department of Transportation, Washington, DC.

Tränkle, U., Gelau, C., and Metker, T. (1990), "Risk Perception and Age-Specific Accidents of Young Drivers," *Accident Analysis and Prevention*, Vol. 22, pp. 119–125.

Transportation Research Board (2003), *Buckling Up: Technologies to Increase Seat Belt Use*, Transportation Research Board, Washington, DC.

Tsimhoni, O., and Green, P. (2001), "Visual Demand of Driving and the Execution of Display-Intensive In-Vehicle Tasks," in *Proceedings of the Human Factors and Ergonomics Society 45th Annual Meeting*, HFES, Santa Monica, CA, pp. 1586–1590.

Ullman, G. L. (1993), "Motorist Interpretation of MUTCD Freeway Lane Signals," *Transportation Research Record*, No. 1403, pp. 49–56.

Underwood, M. (1992), "The Older Driver: Clinical Assessment and Injury Prevention," *Archives of Internal Medicine*, Vol. 152, pp. 735–740.

U.S. Department of Commerce (2001), *An Aging World: 2001*, U.S. Department of Commerce, Bureau of the Census, Washington, DC.

U.S. Public Health Service (1990), *Vital and Health Statistics*, Ser. 24, No. 4, Report DHHS-PHS 90–1954, Public Health Service, Hyattsville, MD.

Violanti, J. M., and Marshall, J. R. (1996), "Cellular Phones and Traffic Accidents: An Epidemiological Approach," *Accident Analysis and Prevention*, Vol. 28, pp. 265–270.

Vollrath, M., Meilinger, T., and Krüger, H. -P. (2002), "How the Presence of Passengers Influences the Risk of a Collision with Another Vehicle," *Accident Analysis and Prevention*, Vol. 34, pp. 649–654.

Waller, P., Blow, F., Maio, R., Hill, E., Singer, K., and Schaeffer, N. (1995), "Crash Characteristics and Injuries of Drivers Impaired by Alcohol/Drugs," in *Proceedings of the 13th International Conference on Alcohol, Drugs, and Traffic Safety, TS-'95*, National Health and Medical Research Council, Adelaide, Australia.

Wang, J. -S., Knipling, R. R., and Goodman, M. J. (1996), "The Role of Driver Inattention in Crashes: New Statistics from the 1995 Crashworthiness Data System," in *40th Annual Proceedings of the Association for the Advancement of Automotive Medicine*, AAAM, Des Plaines, IA, pp. 377–392.

Wang, U., Hasson, P., and Lister, M. (2002), "Safer Roads Thanks to ITS," *Public Roads*, Vol. 65, No. 6, pp. 14–18.

Wasielewski, P. (1984), "Speed as a Measure of Driver Risk: Observed Speeds Versus Driver and Vehicle Characteristics," *Accident Analysis and Prevention*, Vol. 16, pp. 89–104.

Weinrath, M. (1997), "Ignition Interlock Program for Drunk Drivers: A Multivariate Test," *Crime Delinquency*, Vol. 43, pp. 42–59.

West, R. L., Crook, T. H., and Barron, K. L. (1992), "Everyday Memory Performance Across the Life-Span: Effects of Age and Noncognitive Individual Differences," *Psychology and Aging*, Vol. 7, pp. 72–82.

Wierwille, W. W., and McFarlane, J. (1993), "Role of Expectancy and Supplementary Cues for Control Operation," in *Automotive Ergonomics*, B. Peacock and W. Karwowski, Eds., Taylor & Francis, London, pp. 269–298.

Williams, A. F. (2003), "Teenage Drivers: Patterns of Risk," *Journal of Safety Research*, Vol. 34, pp. 5–15.

Williams, E., and Monaco, K. (2001), "Accidents and Hours-of-Service Violations Among Over-the-Road Drivers," *Journal of the Transportation Research Forum*, Vol. 40, pp. 105–115.

Williams, A. F., and Wells, J. K. (2003), "Drivers' Assessment of Ford's Belt Reminder System," *Traffic Injury Prevention*, Vol. 4, pp. 358–362.

Williams, A. F., Peat, M. A., Crouch, D. J., Wells, J. K., and Finkle, B. S. (1985), "Drugs in Fatally Injured Young Male Drivers," *Public Health Reports*, Vol. 100, pp. 19–25.

Williams, A. F., Preusser, D. F. Ulmer, R. G., and Weinstein, H. B. (1995), "Characteristics of Fatal Crashes of 16-Year-Old Drivers: Implications for Licensure Policies," *Journal of Public Health Policy*, Vol. 16, pp. 347–360.

Williams, A. F., Wells, J. K., and Farmer, C. M. (2002), "Effectiveness of Ford's Belt Reminder System in Increasing Seat Belt Use," *Injury Prevention*, Vol. 8, pp. 293–296.

Wilson, R. J., and Jonah, B. A. (1988), "The Application of Problem Behavior Theory to the Understanding of Risky Driving," *Alcohol, Drugs and Driving*, Vol. 4, pp. 173–191.

Woodrooffe, J. (2001), *Long Combination Vehicle (LCV) Safety Performance in Alberta, 1995 to 1998*, Woodrooffe & Associates, Carleton Place, Untario, Canada.

Woodrooffe, J. (2004), Personal communication, September.

Woodrooffe, J., and Ash, L. (2001), *Economic Efficiency of Long Combination Transport Vehicles in Alberta*, Woodrooffe & Associates, Carleton Place, Untario, Canada.

Zuckerman, M. (1971), "Dimensions of Sensation Seeking," *Journal of Consulting and Clinical Psychology*, Vol. 36, pp. 45–52.

Zuckerman, M. (1979), *Sensation Seeking: Beyond the Optimal Level of Arousal*, Lawrence Erlbaum Associates, Mahwah, NJ.

Zuckerman, M. (1990), "The Psychophysiology of Sensation Seeking," *Journal of Personality*, Vol. 58, pp. 313–345.

Zuckerman, M. (1994), *Behavioral Expressions and Biosocial Bases of Sensation Seeking*, Cambridge University Press, New York.

Zuckerman, M., and Link, K. (1968), "Construct Validity for the Sensation Seeking Scale," *Journal of Consulting and Clinical Psychology*, Vol. 32, pp. 420–426.

Zuckerman, M., and Neeb, M. (1980), "Demographic Influences in Sensation Seeking and Expressions of Sensation Seeking in Religion, Smoking, and Driving Habits," *Personality and Individual Differences*, Vol. 1, pp. 197–206.

Zuckerman, M., Kolin, I., Price, L., and Zoob, I. (1964), "Development of a Sensation Seeking Scale," *Journal of Consulting Psychology*, Vol. 28, pp. 477–482.

Zuckerman, M., Eysenck, S. B. G., and Eysenck, H. J. (1978), "Sensation Seeking in England and America: Cross-Cultural, Age, and Sex Comparisons," *Journal of Consulting and Clinical Psychology*, Vol. 46, pp. 139–149.

Zwahlen, H. T., Adams, C. C., Jr., and Schwartz, P. J. (1988), "Safety Aspects of Cellular Telephones in Automobiles," in *Proceedings of the 18th International Symposium on Automotive Technology and Automation*, Croyden, London.

CHAPTER 60

HUMAN FACTORS AND ERGONOMICS IN AUTOMATION DESIGN

John D. Lee
University of Iowa
Iowa City, Iowa

1 INTRODUCTION

Automation has a long history marked by many successes and equally notable failures. In the early nineteenth century the Luddites in northern England protested against the introduction of automation in the weaving industry by sabotaging the machines. Although the term *luddite* now refers to technophobes, these people correctly foresaw some of the highly damaging changes that the introduction of automation would bring to their lives. Automation and the industrial revolution radically changed the craft-centered culture of the time. More recently, information technology has had an equally important effect on industries as diverse as process control, aviation, and ship navigation.

The Luddites foresaw the threat to their lifestyle. Of greater concern are situations in which people fail to recognize the risks of adopting automation and are surprised by unanticipated effects (Sarter et al., 1997). Automation frequently surprises designers, operators, and managers with unforeseen mishaps. As an example, the cruise ship *Royal Majesty* ran aground because the global positioning system (GPS) signal was lost and the position estimation reverted to position extrapolation based on speed and heading (dead reckoning). For over 24 hours, the crew followed the compelling electronic chart display and did not notice that the GPS signal had been lost or that the position error had been accumulating. The crew failed to heed indications from boats in the area, lights on the shore, and even salient changes in water color that signal shoals. The surprise of the GPS failure was discovered only when the ship ran aground (National Transportation Safety Board, NTSB, 1997; Lutzhoft and Dekker, 2002). This mishap demonstrates the power of technology to either make us smart or surprisingly stupid (Norman, 1993). Automation exemplifies this power.

Automation has been defined as a device or system that performs a function previously performed by a human operator (Parasuraman et al., 2000). However, automation does not simply supplant the person, but enables new activities, creates new roles for the person, and changes activities in unexpected ways (Woods, 1994). As a result, automation often results in surprises at many levels, from the societal, as with the Luddites, to the individual, as with the *Royal Majesty*. For automation to achieve its promise, its design must anticipate these changes. The need to anticipate and avert automation-related surprises is more difficult now than ever. One of the ironies in automation design is that as automation increasingly supplants human control, it becomes increasingly important for designers to consider the contribution of the human operator (Bainbridge, 1983). This chapter draws upon over 30 years of research to identify general automation-related failures and to identify strategies for improving automation design.

2 AUTOMATION PROMISES AND PITFALLS

Automation has many clear benefits. In the case of the control of cargo ships and oil tankers, it has made it possible to operate a vessel with as few as 8 to 12 crew members compared to the 30 to 40 that were required 40 years ago (Grabowski and Hendrick, 1993). In the case of aviation, automation has reduced flight times and increased fuel efficiency (Nagel, 1988). Similarly, automation in the form of decision-support systems has been credited with saving millions of dollars in guiding policy and production decisions (Singh and Singh, 1997). Automation promises greater efficiency, lower workload, and fewer human errors; however, these promises are not always fulfilled.

Many pitfalls plague the introduction of automation. Well-documented failures of information technology show that it seldom provides the promised economic benefits (Landauer, 1995) and often fails to provide promised safety benefits (Perrow, 1984). When automation is introduced to eliminate human error, the result is sometimes new and often more catastrophic errors (Sarter and Woods, 1995). Automation often fails to provide expected benefits because it does not simply replace the human in performing a task, but also transforms the job and introduces a new set of tasks. Operators then often receive inadequate feedback and support in performing these new tasks. Automation also fails because the role of the person performing the task is often underestimated, particularly the ability to compensate for the unexpected. Although automation can handle typical cases it often lacks the flexibility of humans needed to handle unanticipated situations. Although any list of automation-related problems and surprises will be incomplete, the following represent some important instances:

- Out-of-the-loop unfamiliarity
- Clumsy automation
- Automation-induced errors
- Inappropriate trust (misuse, disuse, and complacency)
- Behavioral adaptation
- Inadequate training and skill loss
- Job satisfaction and health
- Eutactic behavior

2.1 Out-of-the-Loop Unfamiliarity

Out-of-the-loop unfamiliarity refers to the diminished ability of people to detect automation failures and to resume manual control (Endsley and Kiris, 1995). Several factors underlie this problem. First, automation may reduce feedback, and this feedback may be qualitatively different than that received when operating under manual control (McFadden et al., 2003). With manual control operators may have both proprioceptive and visual cues, whereas under automatic control they may have only visual cues (Wickens and Kessel, 1981). Automation also reduces feedback because it distances operators from the process. Introducing automation into papermaking plants eliminated the informal feedback associated with vibrations, sounds, and smells that many operators relied upon (Zuboff, 1988). Second, monitoring the performance of automation involves passive observation of changes in system state, which is qualitatively different than the active monitoring associated with manual control (Gibson, 1962; Eprath and Young, 1981). In manual control, perception actively supports control, and control actions guide perception (Flach and Jagacinski, 2002). Monitoring automatic control disrupts this process. Third, automatic control can induce the operator to disengage and direct attention to other activities, further compromising the feedback from the system. The tendency to rely complacently on automation, particularly during multitask situations, may reflect this tendency to disengage from the monitoring task (Parasuraman et al., 1993, 1994; Metzger and Parasuraman, 2001). Finally, the operator's mental model may be inadequate to guide expectations, situation awareness, and control. In particular, the automation may use control algorithms that are at odds with the control strategies and mental model of the person, making it difficult to anticipate the actions and limits of the automation (Goodrich and Boer, 2003). Operators with substantial previous experience and well-developed mental models detect disturbances more rapidly than operators without this experience, but extended periods of monitoring automatic control may undermine this skill and diminish operators' ability to generate expectations of correct behavior (Wickens and Kessel, 1981). This skill loss may also undermine operators' self-confidence, which can make them less inclined to intervene (Lee and Moray, 1994). Overall, out-of-the-loop unfamiliarity stems from disrupted feedback that diminishes situation awareness, the ability to form correct expectations, and the ability to control the system manually.

An example of out-of-the-loop unfamiliarity occurs in driving. Adaptive cruise control (ACC) has the potential to induce out-of-the-loop unfamiliarity, leading to delayed and less effective braking responses in situations in which the ACC is not able to respond to a braking lead vehicle (Stanton and Young, 1998). ACC uses sensors to maintain not only a set speed, as with conventional cruise control, but also a set distance to cars ahead. When drivers engage ACC, they no longer receive the haptic feedback that conveys the degree of braking needed to slow the vehicle in response to the braking behavior of vehicles ahead. Drivers also revert to passive monitoring of other vehicles rather than directing their attention toward the active control of their headway. Most important, ACC may induce drivers to direct their attention to nondriving activities such as cell phone conversations or reading the newspaper (Ward, 2000). Such distractions clearly delay driver response (Lee and Strayer, 2004). More subtly, drivers may have a poor mental model of the ACC control algorithms and so may not be able to anticipate situations that lie beyond the capabilities of the automation. The difficulty of anticipating the behavior of the automation has also been seen in aviation, in which verbal protocol data show that pilots have problems with automation because of poor feedback and difficulties developing expectations regarding the behavior of the automation (Olson and Sarter, 2001).

2.2 Clumsy Automation

Clumsy automation refers to the situation in which automation makes easy tasks easier and hard tasks harder (Wiener, 1989). As Bainbridge (1983) notes, designers often leave the operator with the most difficult tasks—those the designers are unable to automate. Because the easy tasks have been automated, the operator has less experience and an impoverished context for responding to the difficult tasks, as a result of the out-of-the-loop problem mentioned above. In this situation, automation has the effect of both reducing workload during already low-workload periods and increasing it during high-workload periods. For example, a flight management system tends to make the low-workload phases of flight (such as straight and level flight or a routine climb) easier, but high-workload phases (such as the maneuvers in preparation for landing) more difficult, as pilots have to share their time between landing procedures, communication, and programming the flight management system. Such effects are seen not only in aviation but also in the operating room (Cook et al., 1990b; Woods et al., 1991). Clumsy automation also results from the unfortunate tendency of operators to be more willing to delegate tasks to automation during periods of low workload, compared to situations of high workload (Bainbridge, 1983). This observation demonstrates that clumsy automation is not simply a technical problem, but one that depends on operator attitudes such as trust.

Operators do not respond passively to the introduction of clumsy automation, instead they actively tailor their tasks or tailor the automation and adapt to poorly designed automation (Cook et al., 1990a). These strategies can mask the effects of clumsy automation in routine situations and make it appear more effective than it really is. When operators encounter abnormal situations, the problems of clumsy automation may emerge unexpectedly.

An example of potentially clumsy automation in maritime navigation occurs when the GPS is integrated with digital charts to create electronic chart display and information systems (ECDISs). When combined with existing advanced maritime navigation systems (e.g., automatic radar plotting aid) these technological innovations tend to reduce repetitive physical activity while potentially increasing the mental demands made on the crew. The reduction in physical demands implies the possibility of reducing the number of personnel required on the bridge from as many as four people (captain, watch officer, helmsman, and lookout) to one. Recent studies suggest that under proper conditions, workload declines and performance rises with one-person operations (Schuffel et al., 1988); however, this research has addressed only routine performance and has not considered more stressful conditions. Software failures and dense traffic situations combine to increase the workload substantially relative to the traditional system (Lee and Sanquist, 1996). This finding is consistent with poorly designed automation in the aviation and operating room, which reduces workload under routine conditions but increases it during stressful conditions (Wiener, 1989; Woods, 1991).

Clumsy automation can occur at the individual and organizational levels. Automation promises to reduce the need for human labor; during routine circumstances, fewer people are able to control the system effectively. The dramatic reduction in crew members needed to operate large ships testifies to this fact. However, clumsy automation at the macro level can occur when abnormal situations or high-tempo operations challenge the resources of the diminished crew (Lee and Morgan, 1994). Frequently, the wider effects of automation on training and recruitment go unexamined (Strain and Eason, 2000). Clumsy automation at the micro level of the operator and the macro level of the organization represents critical challenges in anticipating the effect of automation.

2.3 Automation-Induced Errors

Automation-induced errors refer to the new forms of human error that sometimes accompany the introduction of automation. Managers and system designers often introduce automation to eliminate human error. Ironically, new and more disastrous errors can sometimes result. Automation often extends the scope of human actions and delays feedback associated with those actions. As a consequence, human errors may be more likely to go undetected and do more damage.

One unexpected outcome of poorly designed automation are *brittle failures*. Brittle failures are typical of human–automation interactions in which novel problems arise or even simple data-entry mistakes are made with systems that completely automate the decision process and leave operators to assess the automation's decision (Roth et al., 1987; Roth and Woods, 1988). Such failures contrast with *graceful degradation*, a common characteristic of time-tested manual processes. Brittle failures are characterized by a sudden and dramatic decline in system performance, whereas graceful degradation is characterized by a more gradual and predictable decline. For example, a flight-planning system for pilots can induce dramatically poor decisions because it assumes that weather forecasts represent reality and lacks the flexibility to consider situations in which the actual weather might deviate from the forecast (Smith et al., 1997). In maritime navigation, electronic charts introduce the potential for brittle failures in position estimation. Electronic charts distance mariners from the process of recording vessel position, leaving them with little insight into the factors that might lead to erroneous position estimates. The manual process of recording a position on a paper chart superimposes at least two position estimates, one based on extrapolation of the previous position and one based on visual bearings or other position information. These complementary position estimates help identify errors in determining position (Hutchins, 1995). Unlike the manual position recording on paper charts, electronic charts show the quality of the position estimation only indirectly, in terms of GPS signal quality; however, many mariners have little understanding of the relevance of these numbers, and gross errors in position can result (Lee and Sanquist, 2000).

Mode errors are perhaps the most pervasive of the automation-induced errors (Woods, 1994; Sarter and Woods, 1995). These arise when operators fail to detect the mode or recognize the consequence of mode transitions in complex automation. Substantial research with cockpit automation demonstrates that flight management systems often surprise pilots with unexpected mode transitions. These complex systems use a combination of the pilots' commands and system coupling to transition between modes. Mode transitions are often not commanded explicitly by the pilot and sometimes go unnoticed (Sarter and Woods, 1995).

Electronic charts in maritime navigation also offer the potential for mode errors. Such charts have several modes for determining a ship's position. One uses GPS data, another uses position extrapolation based on speed and heading (dead reckoning) to estimate the ship's position. If the GPS signal is lost, the electronic chart system changes automatically to the dead reckoning mode. This mode transition is signaled by a short alarm. If the alarm is not detected, however, the mariner may not notice that the GPS signal is no longer the basis for position estimates. Furthermore, many electronic charts do not maintain a continuous visual record of the vessel track. A track line is shown as long as the same chart or scale is used, but if the scale is changed, the track line is lost. The lack of track line continuity further undermines the ability of mariners to detect a transition from GPS to dead reckoning position estimates. If the mariner does not notice this mode transition, the ship can drift many miles from the intended course while the electronic chart continues to display the position as if the vessel were following that course precisely. This is exactly what happened in the grounding of the cruise ship *Royal Majesty*, where the GPS signal was lost and the position estimation transitioned to the dead reckoning mode. The mode transition was noticed only when the ship ran aground (NTSB, 1997).

Automation can also introduce *configuration errors*. Many forms of automation involve complex configurations or setups, and mistakes made during this process can later prove disastrous. For example, with electronic charts that aid maritime navigation, it is possible to configure the system to test the actual position automatically against the intended track using a feature in which an acceptable safety margin can be specified. If the ship deviates beyond this distance, an alarm sounds (provided that the feature was engaged and the GPS is functioning normally). Failing to engage this feature could jeopardize ship safety if mariners have come to rely on the automated warning. Also, because any one of several mariners can configure the system, system configuration and behavior can change in unanticipated ways as different mariners enter different safety margins. The danger of an inappropriate or unanticipated chart configuration is not a failure mode associated with paper charts but represents an automation-induced error that can threaten ship safety.

Configuration and mode errors, together with the possibility of brittle failures, tend to undermine individual reliability and may have even greater detrimental effects on team performance (Skitka et al., 2000b). These new automation-related errors may be particularly troublesome if the automation also undermines effective error-correcting strategies such as feedback and redundancies in the multiperson position-fixing process (Hutchins, 1995). For example, because poor position fixes are visible to all team members, crew members who generate these fixes are corrected quickly. Automation also creates the potential for one team member to induce configuration errors for other team members. These factors all point toward the need to consider automation-induced errors at the level of the team and at the level of the individual.

2.4 Inappropriate Trust: Misuse, Disuse, and Complacency

Misuse refers to the failures that occur when people inadvertently violate critical assumptions and rely on automation inappropriately, whereas *disuse* signifies failures that occur when people reject the capabilities of automation (Parasuraman and Riley, 1997). Another useful distinction in how operators use automation is that of reliance and compliance (Meyer, 2001). *Reliance* refers to the situation in which the operator does not act because the automation has not issued a warning or seems to be performing adequately. In contrast, *compliance* refers to the situation in which the operator acts in response to a warning or command from the automation. Overreliance results in errors of omission, and overcompliance results in errors of commission (Skitka et al., 2000a).

Misuse and disuse of automation may depend on certain attitudes of users, such as trust and self-confidence (Lee and Moray, 1994; Dzindolet et al., 2001). As an example, the difference in operators' trust in a route planning aid and their self-confidence in their own ability was highly predictive of reliance on the aid (de Vries et al., 2003). Many studies have demonstrated that trust is a meaningful concept to describe human–automation interaction, in both naturalistic settings (Zuboff, 1988) and laboratory settings (Halprin et al., 1973; Lee and Moray, 1992; Muir and Moray, 1996; Lewandowsky et al., 2000).

People tend to rely on automation they trust and to reject automation they do not trust. In the context of operator reliance on automation, trust has been defined as an attitude that the automation will help achieve an operator's goals in a situation characterized by uncertainty and vulnerability (Lee and See, 2004).

Inappropriate reliance associated with misuse and disuse depends in part on how well trust matches the true capabilities of the automation. Calibration, resolution, and specificity of trust describe the match between trust and the capabilities of automation. *Calibration* refers to the correspondence between a person's trust in automation and the automation's capabilities (Lee and Moray, 1994; Lee and See, 2004). Definitions of the appropriate calibration of trust parallel those of misuse and disuse in describing appropriate reliance. Overtrust is poor calibration in which trust exceeds system capabilities; with distrust, trust falls short of automation capabilities. Figure 1 shows good calibration as the diagonal line where the level of trust matches automation capabilities. Above this line is overtrust and below is distrust. Overreliance on automation has sometimes been termed *complacency* and can result from trusting the automation more than is warranted.

Resolution refers to how precisely a judgment of trust differentiates levels of automation capability (Cohen et al., 1999). Figure 1 shows that poor resolution occurs when a large range of automation capability maps onto a small range of trust. With low resolution, large changes in automation capability are reflected in small changes in trust. *Specificity* refers to the degree to which trust is associated with a particular component or aspect of the trustee. Functional specificity describes the differentiation of functions, subfunctions, and modes of automation. With high functional specificity, a person's trust reflects capabilities of specific subfunctions and modes. Low functional specificity means that the person's trust reflects the capabilities of the entire system. Specificity can also describe changes in trust as a function of the situation or over time. High temporal specificity means that a person's trust reflects moment-to-moment fluctuations in automation capability, whereas low temporal specificity means that the trust reflects only long-term changes in automation capability. Although temporal specificity implies a generic change over time as the person's trust adjusts to failures in the automation, temporal specificity also addresses adjustments that should occur when the situation or context changes and affects the capability of the automation. High functional and temporal specificity increase the likelihood that the level of trust will match the capabilities of a particular element of the automation at a particular time. Good calibration, high resolution, and high specificity of trust can mitigate misuse and disuse of automation.

The information required to support appropriate trust can be considered in terms of attributional abstraction, which varies from the demonstrations of competence to the intentions of the automation (Lee and See, 2004). A recent review of trust literature concluded that three general levels summarize the bases of trust: ability, integrity, and benevolence (Mayer et al., 1995). Lee and Moray (1992) made similar distinctions in defining the factors that influence trust in automation and identified performance, process, and purpose as the general bases of trust.

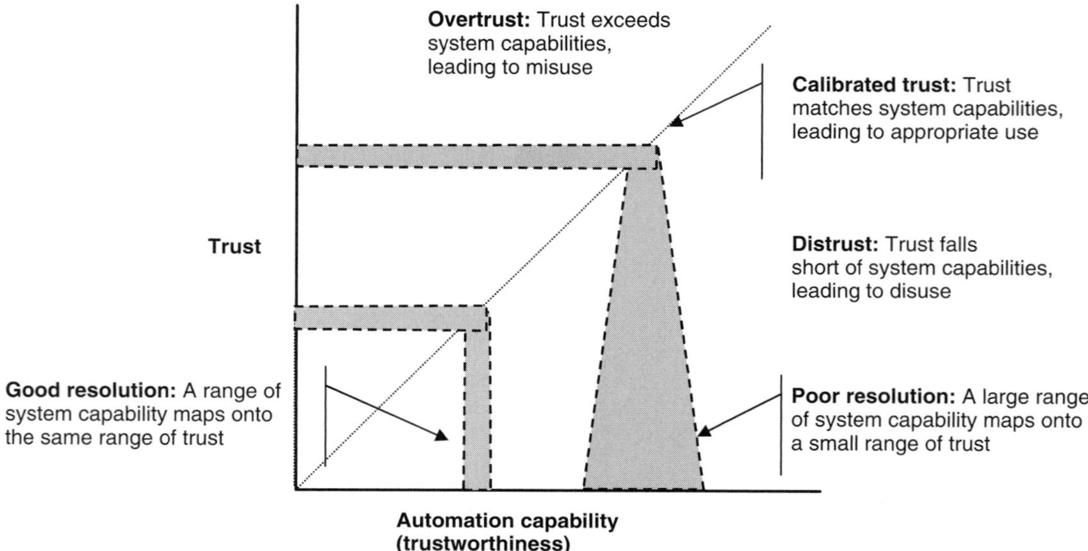

Figure 1 Calibration, resolution, and automation capability define appropriate trust in automation. Overtrust may lead to misuse, and distrust may lead to disuse. (Reprinted with permission from Lee and See, 2004. Copyright 2004 by the Human Factors and Ergonomics Society. All rights reserved.)

Performance refers to the current and historical performance and reliability of the automation. Performance information describes *what* the automation does. More specifically, performance refers to the competency or expertise of the system, as demonstrated by its ability to achieve the operator's goals. Because performance is linked to the ability to achieve specific goals, it demonstrates the task- and situation-dependent nature of trust. This is similar to Sheridan's (1992) concept of robustness. The operator will tend to trust automation that performs in a manner that reliably achieves his or her goals.

Process is the degree to which the algorithms of the automation are appropriate for the situation and able to achieve the operator's goals. Process information describes *how* the automation operates. In interpersonal relationships, this corresponds to the consistency of actions associated with adherence to a set of acceptable principles (Mayer et al., 1995). Process as a basis for trust reflects a shift away from focus on specific behaviors and toward qualities and traits attributed to the automation. With the process dimension, trust is in the automation and not in the specific actions of the automation. As an example, knowing why automation failed increased trust even when it was not warranted (Dzindolet et al., 2003). In contrast, trust tends to drop with any sign of incompetence of the automation, even if the overall system performance is unaffected (Muir and Moray, 1996). Thus, the process basis of trust relies on dispositional attributions and inferences and is similar to Sheridan's (1992) concept of understandability. The operator will tend to trust the automation if its algorithms can be understood and it seems capable of achieving the operator's goals in the current situation.

Purpose refers to the degree to which the automation is being used within the realm of the designer's intent. It addresses the question of *why* the automation was developed. With interpersonal relationships, this depends on the intentions and motives of the trustee. This can take the form of abstract, generalized value congruence (Sitkin and Roth, 1993), which can be described as whether and to what extent the trustee has a motive to lie (Hovland et al., 1953). The purpose basis of trust reflects the attribution of these characteristics to the automation. Frequently, whether or not this attribution takes place will depend on whether the designer's intent has been communicated to the operator. If so, the operator will tend to trust the automation to achieve the goals it was designed to achieve. Often, the complexity, authority, and autonomy of the automation lead to a perceived animacy, in which the automation seems capable of independent and willful action independent of the operator (Sarter and Woods, 1994). In this situation, the intents that the operator infers may have little relationship to the purpose of the design, leading to a serious miscalibration of trust.

Although trust depends heavily on the interactions between an operator and the automation, the team and organizational structure within which they function may have an important effect on the diffusion of trust among co-workers. Communication with co-workers augments direct interaction with the automation and may have a strong influence on trust in the automation. Research has not yet considered the evolution of trust in multiperson groups that share responsibility for managing automation. In this situation, people must not only develop appropriate trust in the automation, but must also develop appropriate trust in the other people who manage the automation (Lee and See, 2004).

One of the ironies of automation is that operators often express a desire for simple and reliable automation, but want the automation to aid them with their most complex tasks (Tenney et al., 1998). Similarly, a highly sensitive warning system that results in many warnings can undermine trust because operators feel that the warnings fail to reflect the danger of the situation accurately (Gupta et al., 2002). These results suggest that understandable and reliable performance on easy tasks may not leave operators willing to rely on the automation to handle more difficult situations. Designing automation to promote appropriate trust may help resolve these conflicts. Ideally, trust in automation guides reliance when the complexity of a system makes complete understanding impractical and when the situation demands adaptive behavior (Lee and See, 2004). However, how to design automation to promote appropriate trust, particularly for complex automation that cannot be fully understood by the operator, is a substantial challenge.

2.5 Behavioral Adaptation

Behavioral adaptation refers to the tendency of operators to adapt to the new capabilities of the automation, particularly to change behavior so that the potential safety benefits of the technology are not realized. Automation intended by designers to enhance safety may instead lead operators to reduce effort and leave safety unaffected or even diminished. Behavioral adaptation occurs at the individual (Wilde, 1988, 1989; Evans, 1991), organizational (Perrow, 1984), and societal levels (Tenner, 1996).

Antilock brake systems (ABSs) for cars show this effect. An ABS modulates brake pressure automatically to maintain maximum brake force without skidding. This automation makes it possible for drivers to maintain control in extreme crash avoidance maneuvers, which should enhance safety. However, ABSs have not produced the intended safety benefits, in part because with them drivers tend to drive less conservatively, adopting higher speeds and shorter following distances (Sagberg et al., 1997). Similarly, vision enhancement systems make it possible for drivers to see more at night, potentially enhancing safety; however, drivers tend to adapt to the systems by increasing their speed (Stanton and Pinto, 2000).

A related form of behavioral adaptation undermines the benefits of automation in that the presence of the automation causes a diffusion of responsibility and a tendency to exert less effort when automation is available (Mosier et al., 1998; Skitka et al., 2000a). As a result, people tend to commit more omission errors (failing to detect events not detected by the automation) and more commission errors (concurring incorrectly with erroneous detection of events by the automation) when they work with automation. Automation can lead people to conserve cognitive effort rather than increase detection performance. This effect parallels the adaptation of people when they work in groups. Diffusion of responsibility leads people to perform more poorly when they are part of a group compared to individually (Skitka et al., 1999). A similar phenomenon is seen with decision-support automation. People often use decision-support systems to reduce effort rather than to enhance decision quality (Todd and Benbasat, 1999, 2000). The strong tendency of people to minimize effort and adapt their behavior to the most salient feedback they receive merits careful consideration in the design and implementation of automation.

2.6 Inadequate Training and Skill Loss

Inadequate training and skill loss refers to the situation in which the introduction of automation leaves the operator without the appropriate skills to accommodate the demands of the job. In situations in which the automation takes on the tasks previously assigned to the operator, the skills of the operator may atrophy as they go unexercised (Endsley and Kiris, 1995). This is a particular concern in aviation, where pilots' aircraft handling skills may degrade when they rely on the autopilot. In response, some pilots disengage the autopilot and fly the aircraft manually to maintain their skills (Billings, 1997).

Automation can also change the task of the operator such that new skills are needed. Sophisticated automation eliminates many physical tasks and leaves complex cognitive tasks that may appear superficially easy, leading to less emphasis on training and a poor understanding of the automation. On ships, misunderstanding of new radar and collision avoidance systems has contributed to accidents (NTSB, 1990). One contribution to such accidents is training and certification that fail to reflect the demands of the automation. An analysis of the exam used by the U.S. Coast Guard to certify radar operators indicated that 75% of the items assess skills that have been automated and are not required by the new technology (Lee and Sanquist, 2000). The new technology makes it possible to monitor a greater number of ships, enhancing the need for interpretive skills such as understanding the rules of the road and the automation. These are the very skills that are underrepresented on the Coast Guard exam. Ironically, while increasing levels of automation may relieve the operator of some tasks, they are likely to create new and more complex tasks that require more, not less, training.

2.7 Job Satisfaction and Health

The issues noted above have addressed primarily the direct performance problems associated with automation. The issue of job satisfaction goes well beyond performance to consider the morale and moral implications of a worker whose job is being changed by automation. Automation that is introduced merely because it increases the profit of the company may not necessarily be well received. Automation often has the effect of deskilling a job, making suddenly obsolete, skills that operators worked for years to perfect. Properly implemented, automation should re-skill workers and make it possible for them to leverage their old skills into new ones that are extended by the automation. Many operators are highly skilled and proud of their craft; automation can thus either empower or demoralize them (Zuboff, 1988). Unhappy operators may fail to capitalize on the potential of an automated system or may even actively sabotage the automation, similar to what the Luddites did.

Automation may also contribute to an environment in which the demands of the work increase but the decision latitude decreases. Such an environment can undermine worker health, leading to problems ranging from increased heart disease to increased incidents of depression (Vicente, 1999). However, if automation extends the capability of the operator, it can enhance both satisfaction and health if operators are given sufficient decision latitude. As an example, night shift operators had greater decision latitude than that of day shift operators who worked under the eye of the managers. The night shift operators used this latitude to learn how to manage the automation more effectively (Zuboff, 1988). These effects demonstrate the need to consider the management and implementation of the automation.

2.8 Eutactic Behavior

Eutactic behavior is behavior that approximates an optimal or satisficing response to the automation (Moray, 2003). As a consequence, eutactic behavior is not an instance of inappropriate reliance on automation, but an instance of appropriate reliance that may be inconsistent with the expectations of the designers or managers. Misuse and disuse may sometimes reflect poorly calibrated trust, automation bias, or complacency. However, misuse and disuse may also reflect eutactic behavior and appropriate reliance once the costs and benefits are assessed completely. Automation that is generally reliable, relieves the operator of substantial mental effort, and fails with modest cost to system performance should be relied upon, even if it fails periodically. Careful monitoring to catch the occasional failure may not be worth the effort. What may appear to be complacent behavior may actually be appropriate given the costs of monitoring. Such behavior is eutactic and not a consequence of overtrust in the automation. Discriminating between complacency and eutactic behavior requires an optimization analysis of the cost function that includes the cost of failing to detect failures *and* the cost of monitoring (Moray, 2003). Similarly, disuse may also be appropriate. Introducing automation to which operators can delegate tasks also introduces new tasks associated with programming, engaging, monitoring, and disengaging the automation (Kirlik, 1993). The burden of managing the automation can outweigh the benefit. In this situation, the aid will go unused by a well-adapted operator (Kirlik, 1993). Such behavior is eutactic and is not a consequence of distrust of the automation.

2.9 Interaction between Automation Problems

Although described independently, the problems of automation often reflect an interacting and dynamic process. One problem may lead to another. Figure 2 summarizes the general problems with automation and identifies some of the important interactions. In many of these relationships, positive feedback reinforces the problem, creating vicious cycles that exacerbate the difficulty. As an example, inadequate training and skill loss may lead the operator to disengage from the monitoring task. This, in turn, will exacerbate the out-of-the-loop unfamiliarity, which will further undermine the operator's skills, and so on. A similar dynamic exists between clumsy automation and automation-induced errors. Clumsy automation produces workload peaks, which increase the chance of mode and configuration errors. Recovering from these errors can further increase workload, and so on. Designing and implementing automation without regard for human capabilities and defining the human role as a by-product has been referred to automation abuse (Parasuraman and Riley, 1997) and is likely to initiate the negative dynamics shown in Figure 2.

3 TYPES OF AUTOMATION

The first step in minimizing the problems and maximizing the benefits of automation is to clarify what is meant by the term *automation*. Automation is not a homogeneous technology. Instead, there are many types of automation and each poses different design challenges. Automation can highlight, alert, filter, interpret, decide, and act for the operator. It can assume different degrees of control and can operate over time scales that range from milliseconds to months. The type of automation, its limits, the operating environment, and human characteristics interact to produce the problems just discussed. Descriptions of automation from different perspectives can reveal the implications of automation for system performance. One such description considers automation in terms of the four stages of human information processing and levels of automation (Parasuraman et al., 2000). Another description considers popular metaphors for automation: tools, prostheses, and agents. Finally, automation can be considered in terms of the scope of the tasks it supports: strategic, tactical, and operational. Any such low-dimensional description of a high-dimensional space will certainly fail to capture important distinctions; nevertheless, these perspectives can make meaningful distinctions that can support design decisions.

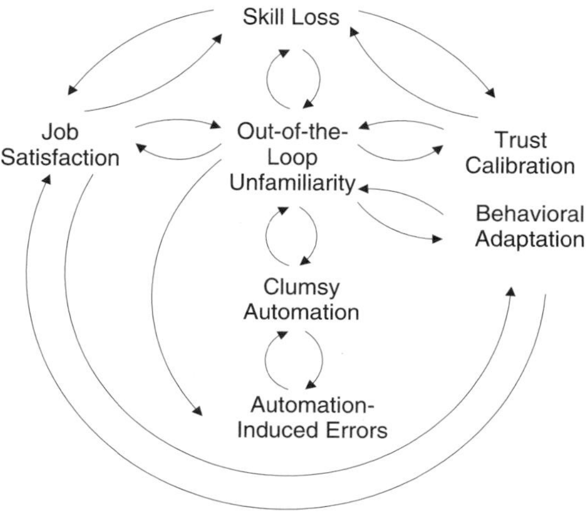

Figure 2 Interactions among the problems with automation.

3.1 Information Processing Stages and Levels of Automation

If automation is considered as technology that replaces the human in performing a function, it is then reasonable to describe automation in terms of the information-processing functions of the person. Although imperfect, the information process model of human cognition provides a useful engineering approximation that has been widely applied to system design (Broadbent, 1958; Rasmussen, 1986). The basic information processing functions—information acquisition, information analysis, action selection, and action implementation—provide simple distinctions that can describe human and automation functions in a common language. A different type of automation corresponds to each stage of information processing. For each of these four functions, different degrees of automation are possible, ranging from full automation to manual control (Sheridan and Verplank, 1978). Information-processing stages and the degree of automation combine to describe a wide array of automation in a way that can guide automation design (Parasuraman et al., 2000).

Information acquisition automation refers to technology that complements the process of human attention. Such automation highlights targets (Yeh and Wickens, 2001; Dzindolet et al., 2002), provides alerts and warnings (Bliss, 1997; Bliss and Acton, 2003), and organizes, prioritizes, and filters information. Highlighting targets exemplifies a low degree of information acquisition automation because it preserves the underlying data and allows operators to guide their attention to the information they believe to be most critical. Filtering exemplifies a high degree of automation, and operators are forced to attend to the information the automation deems relevant. *Information analysis* refers to technology that supplants perception and working memory in the interpretation of a situation. Such automation supports situation assessment and diagnosis. As an example, critiquing a diagnosis generated by the operator represents a low degree of automation, whereas automation that provides a single diagnosis represents a high degree of automation. *Action selection automation* refers to technology that combines information in order to make decisions for the operator. Unlike information acquisition and analysis, action selection automation suggests or decides on actions using assumptions about the state of the world and the costs and values of the possible options (Parasuraman et al., 2000). Providing the operator with a list of suggested options represents a relatively low level of action selection automation. In contrast, automation that commands the operator to respond, as in the verbal "pull up, pull up" command of the ground proximity warning system, represents a high level of action selection automation. *Action implementation automation* supplants the operators' activity in executing a response. Olson and Sarter (2001) describe two degrees of action implementation automation *management by consent*, in which the automation acts only with the consent of the operator, and a greater degree of automation *management by exception*, in which automation initiates activities autonomously.

Each of these four stages of automation combines with the degree of automation to describe how the technology supplants the operator's role in perceiving and responding to the environment. Figure 3 shows two hypothetical systems. System B replaces the operator to a relatively high degree for all information-processing stages. In contrast, system A represents a generally lower level of automation, with only a moderate degree of automation in the information acquisition stage (Parasuraman et al., 2000).

3.2 Tool, Prostheses, and Agents

As automation becomes more complex, considering it simply as a replacement for the information-processing functions of the operator may not differentiate adequately between important types of automation. In many situations, automation is not merely a system that operators engage and disengage. Often, automation consists of a complex array of modes and levels that operators must manage. Interacting with the automation involves coordinating multiple goals and strategies to select a mode of operation that fits the situation (Olson and Sarter, 2000). The simple distinction of engaging manual and automatic control does not capture the complexity of many types of automation. Important design issues emerge as automation evolves from a tool the operator uses to act on the environment, to a prosthesis that replaces a human ability, to an agent that acts on behalf of the operator. The metaphors of automation as a tool, prosthesis, and agent provide complementary perspectives to the information-processing metaphor of automation.

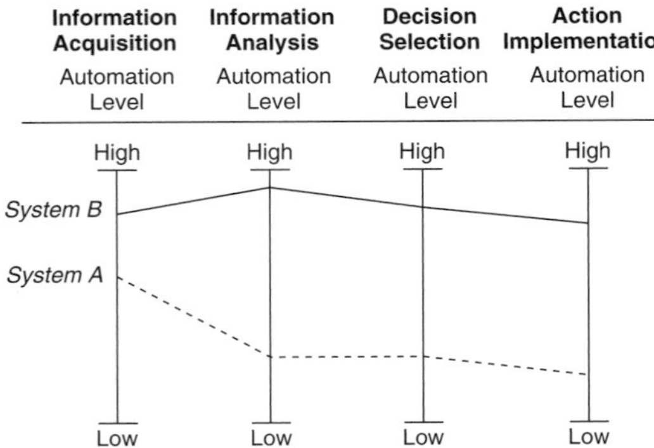

Figure 3 Two examples of automation defined by a profile of the degree of automation over the four information-processing stages. (From Parasuraman et al., 2000, © IEEE 2000.)

Automation, considered as a *cognitive tool,* extends and complements human capabilities. According to the tool metaphor of automation, operators work directly on the environment, but automation augments their interactions. Just as a hammer augments human action in physical tasks, automation can augment operators in cognitive tasks (Woods, 1987). The benefit of automation as a tool is that its influence is clear and its failures are obvious.

Automation, considered as a *cognitive prosthesis,* acts to replace human function with a more capable computer version. Often, designers adopt this approach in an attempt to enhance system performance or safety by eliminating human error. A cognitive prosthesis eliminates a variable or error-prone aspect of human behavior and replaces it with a consistent computer-based process. The cost of this approach is lost flexibility and the ability to adapt to unforeseen situations (Roth et al., 1988). For these reasons, the cognitive prosthesis approach is most appropriate for routine, low-risk situations where decision consistency is more important than adapting to unusual situations. Automation that must accommodate unusual circumstances should adapt a cognitive tool perspective that complements rather than replaces human decision making.

Automation considered as an *agent* acts as a semi-autonomous partner with the operator. According to the agent metaphor, the operator no longer acts directly on the environment but acts through an intermediary agent (Lewis, 1998) or intelligent associate (Jones and Jacobs, 2000). As an agent, automation initiates actions that are not in direct response to operators' commands. The authority, autonomy, and complexity of many advanced automated systems makes them appear to be intentional agents, even if the designers had not intended to adopt this metaphor (Sarter and Woods, 1997). This autonomy and authority can lead to instances of poor coupling and coordination breakdowns because the agents fail to communicate their intentions (Sarter and Woods, 2000; Hoc, 2001). One of the greatest challenges with automated agents is that of mutual intelligibility. Instructing the agent for even simple tasks can be onerous, and agents who try to infer operators' intent and act autonomously can surprise operators, who might lack mental models of agent behavior. One approach is for the agents to learn and adapt to the characteristics of the operator through a process of remembering what they have been being told to do in similar situations (Bocionek, 1995). After the agent completes a task, it can be equally challenging to make the results meaningful to the operator (Lewis, 1998). Because of these characteristics, agents are most useful for highly repetitive and simple activities, where the cost of failure is limited. In high-risk situations, constructing effective management strategies and providing feedback to clarify agent intent and communicate behavior becomes critical (Olson and Sarter, 2000; Sarter, 2000).

The differences between automation as a tool, prosthesis, and agent reflect a shift in the locus of control. With a tool, the operator firmly maintains control, but with an agent, the locus of control is more ambiguous and may pass back and forth between the operator and the automation. Ambiguity in the locus of control introduces important considerations regarding inferred intent and the dynamic coordination of actions (Woods, 1994).

The metaphors of tool, prosthesis, and agent complement the information-processing description of automation in important ways. The information-processing metaphor emphasizes the idea that automation replaces the person in performing a function but that function and system remain unchanged. Other metaphors, such as that of automation as an agent, emphasize the far-reaching changes that automation may induce. Rarely is automation a simple replacement of the human. As one researcher put it: "We have seen that technology change produces a complex set of effects. In other words, automation is a wrapped package—a package that consists of changes on many different dimensions bundled together as a hardware/software system" (Woods, 1994). Just as the information-processing metaphor of automation leverages a long history of experimental psychology research, the agent metaphor may leverage recent developments in distributed cognition and team effectiveness (Seifert and Hutchins, 1992; Hutchins, 1995). Such a shift may lead to a change in the boundaries that define the unit of analysis, from one centered on a single operator and a single element of automation to one that considers multioperator, multiautomation interactions (Hollan et al., 2000; Gao and Lee, 2004). Although agent interactions are most obvious when one considers group work, they also offer a powerful metaphor for describing the emergent behavior that is human cognition (Minsky, 1986).

3.3 Multilevel Control

The scope of automation varies dramatically, from decision-support systems that guide corporate strategies over months and years to antilock brake systems that modulate brake pressure over milliseconds. Substantially different human limitations govern operator interaction with automation at these extremes. A three-level structure that has been used to describe driver behavior seems appropriate for discussing the more general issue of human–automation coordination (Michon, 1985; Ranney, 1994). Figure 4 shows three levels of control that provide a framework for considering issues of coordination and communication of intent. Each level of the figure defines a different level of control that could be supported by a different type of automation. *Strategic automation*

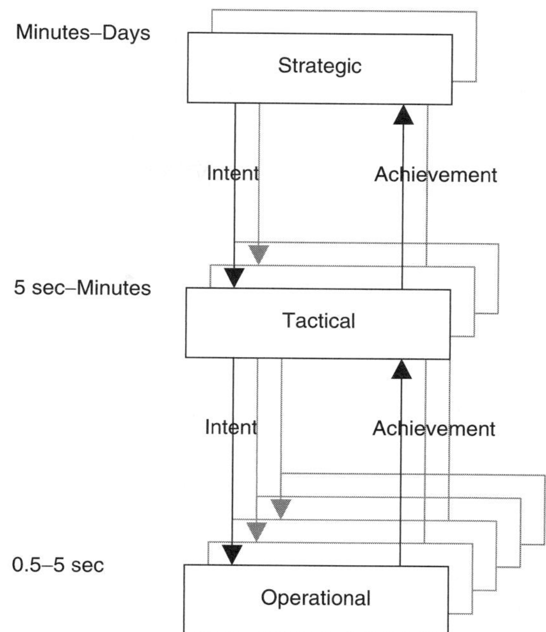

Figure 4 Strategic, tactical, and operational automation describe types of automation that reveal important coordination and feedback requirements.

concerns balancing values and costs as well as defining goals; *tactical automation*, on the other hand, involves priorities and coordination. Finally, *operational automation* has to do with perceptual cues and motor response.

The bottom of Figure 4 shows operational automation, which governs system behavior over the span of approximately 0.5 s to 5 s. Automation at this level concerns the moment-to-moment control of dynamic processes. An example in the driving domain is adaptive cruise control (ACC), which controls the speed of the car and its distance to the vehicle ahead. The middle of Figure 4 shows tactical automation, which governs system response over a time span of seconds to minutes. In driving, this automation would include route guidance systems that notify drivers of upcoming turns. At the top of the figure is strategic automation, which governs behavior from minutes to days. In driving, this automation helps drivers to select routes and plan trips.

The multilevel control perspective shown in Figure 4 identifies design considerations for the different types of automation and the interaction between different elements of automation. First, automation at one level may have unanticipated effects on behavior at another level. For example, automatic control at the lower level might lead people to adopt different behaviors at a higher level, such as when ACC reduces the attention needed in routine car following at the operational level and influence decisions at the tactical level, such as deciding to engage in a cell phone conversation. Second, time constants have a critical effect on monitoring and control behavior. Detection of low-frequency events requires sustained monitoring best suited to time scales at the operational level (0.5 to 5 seconds), but such events often occur on a time scale that is orders of magnitude greater. At the other extreme are systems that demand responses on a time scale so short that it exceeds human capabilities. For these systems, automation may need to assume final authority for actions (Moray et al., 2000; Inagaki, 2003). The three-level structure highlights qualitative differences in the time constants of system control that automation design should consider (Hoc, 1993). Third, this perspective highlights the critical issue of communicating intent and the achievement of intent. Figure 4 highlights this requirement because automation at the operational level must be coordinated to achieve the intent developed at the tactical level. Adequate performance of an element of automation at the operational level does not guarantee success at the tactical level unless it is coordinated properly. Automation at one level of control must be managed to minimize interference between agents that might otherwise jeopardize achieving common tasks at another level of control (Hoc, 2001). Finally, Figure 4 points to the need to consider what some have termed *macrocognition* (Klein et al., 2003). Macrocognitive processes include situation assessment, planning, and coordination. Typical laboratory studies of automation have focused on *microcognition*, associated with operational and tactical levels; however, many critical problems with automation lie at the strategic level and in the interaction between the strategic and tactical levels.

The term *automation* represents a broad array of technology, and no one dimension or framework will capture the many factors that contribute to the problems often encountered with its implementation. Metaphors of information-processing systems, tools, prostheses, agents, and multilevel control all provide complementary perspectives on the nature of automation and how it influences operator performance. These perspectives are not mutually exclusive. A given instance of automation could be described using any or all of the three perspectives. Each provides a different description to guide automation design. Similarly, each perspective provides a partial and distorted description of the true complexities of automation. Although each perspective is limited, each can enhance our understanding of human-automation interaction.

4 STRATEGIES TO ENHANCE HUMAN–AUTOMATION INTERACTION

Defining the problems encountered with automation should instill caution in those who believe that automation can enhance system performance and safety by replacing the human operator. The perspectives on the nature and types of automation reveal the complexity of automation. Neither caution nor perspective, however, are sufficient to develop successful automation. In this section we describe specific strategies for designing effective automation, which include:

- Fitts's list and function allocation
- Dynamic function allocation (adaptable and adaptive automation)
- Matching automation to human performance characteristics
- Representation aiding and multimodal feedback
- Matching automation to mental models
- Formal automation analysis techniques

4.1 Fitts's List and Function Allocation

One approach to automation is to assess each function and determine whether a human or automation would perform that function better (Kantowitz and Sorkin, 1987; Sharit, 2003). Functions better performed by automation are automated and the operator remains responsible for the rest and for recovering during the periodic failures of the automation. Fitts's list provides a heuristic basis for determining the relative performance of humans and automation for each function (Fitts, 1951). Table 1 shows a revised Fitts list for the stages of automation identified earlier. The relative capability of the automation and human depend on the stage of automation (Sheridan, 2000).

Using the heuristics in Table 1 to determine which functions should be automated mitigates skill loss and lack of training by clearly identifying the human role in a system. This approach also enhances job satisfaction by designing a role for the operator that is compatible with human capabilities. Ideally, the function allocation process should not focus on what functions should be allocated to the automation or to the human, but should identify how the human and the automation can complement each other in jointly satisfying the functions required for system success (Hollnagel and Bye, 2000).

Applying the information in Table 1 to determine an appropriate allocation of function has, however, substantial weaknesses. One weakness is that there are many interconnections between functions. Any description of functions is a somewhat arbitrary decomposition of activities that masks complex interdependencies. As a consequence, automating functions as if they were independent has the tendency to fractionate the operator's role, leaving the operator with only those tasks too difficult to automate (Bainbridge, 1983). Automation must be designed to support the job of the operator as an integrated whole. Another weakness with this approach is the situation dependence of the automation and human performance. The same function may require improvisation in some circumstances and precise application of a fixed response in others. Another weakness is that the work and the automation coevolve, with the automation making unanticipated work practices possible and the work leading to unanticipated applications of the automation (Dearden et al., 2000). A final weakness with function allocation using Fitts's list is the diminishing list of situations in which the human abilities

Table 1 Fitts's List: Relative Strengths of Automation and Humans for the Four Information-Processing Stages

Information-Processing Stage	Humans Are Better At:	Automation Is Better At:
Information acquisition	Detecting small amounts of visual, auditory, or chemical signals	Monitoring processes
	Detecting a wide range of stimuli	Detecting signals beyond human capability
Information analysis	Perceiving patterns and making generalizations	Ignoring extraneous factors and making quantitative assessments
	Exercising judgment	Consistent application of precise criteria
	Recall of related information and development of innovative associations between items	Storing information for long periods and recalling specific parts and exact reproduction
Action selection	Improvising and using flexible procedures	Repeating the same procedure in precisely the same manner many times
	Reasoning inductively and correcting errors	Reasoning deductively
Action implementation	Switching between actions as demanded by the situation	Performing many complex operations at once
	Adjusting dynamically to a wide range of conditions	Responding quickly and precisely

exceed those of the automation. Strict adherence to the application of Fitts's list to allocate functions between people and machines has been widely recognized as problematic (Parasuraman et al., 2000; Sheridan, 2000).

Although imperfect, Table 1 contains some general considerations that can improve design. People tend to be effective with complete patterns and less good with highly precise repetition. Human memory tends to organize large amounts of related information in a network of associations that can support effective judgments requiring the consideration of many factors. People also adapt, improvise, and accommodate unexpected variability. For these reasons it is important to leave the "big picture" to the human and the details to the automation (Sheridan, 2002).

4.2 Dynamic Function Allocation: Adaptable and Adaptive Automation

Using Fitts's list or some other method to allocate functions between humans and automation results in static function allocation in which the division of labor is fixed by the designer. Functions once performed by the human are now performed by automation. Static allocation of function contrasts with dynamic allocation of function, in which adaptable and adaptive automation makes it possible to adjust the division of labor between the human and the automation over time (Scerbo, 1996; Sarter and Woods, 1997). Dynamic allocation of function addresses the need to adjust the degree and type of automation according to individual differences, the state of the operator, and the state of the system. Adaptable and adaptive automation is often preferable to automation that is fixed and rigid.

Adaptable automation is that which the operator can engage or disengage as needed. The operator adapts the level and type of automation to the situation. Giving operators the option of manual or automatic control can be more effective than making available only automatic or only manual control (Harris et al., 1995). More generally, adaptable automation gives operators additional degrees of freedom needed to accommodate unanticipated events (Hoc, 2000). The decision to rely on the automation or to intervene with manual control depends on many factors, including perceived risk, workload, trust, and self-confidence (Riley, 1989, 1994; Lee and Moray, 1994). To the extent that operators trust the automation appropriately and have appropriate self-confidence, they tend to rely on the automation appropriately and avoid some of the out-of-the-loop unfamiliarity problems. Allowing operators to transition easily between automatic and manual control can also mitigate clumsy automation. On the other hand, one of the critical deficiencies of adaptable automation is that it gives the operator the additional tasks of engaging and disengaging the automation. If the effort associated with these tasks is great, adaptable automation can increase the workload of demanding situations, and thus become an example of clumsy automation.

Adaptive automation goes a step further than adaptable automation by automatically adjusting the level of automation based on the operator's performance, the operator's state, or the task situation (Rouse, 1988; Byrne and Parasuraman, 1996). Often, adaptive automation focuses on increasing the level of automation when either the operator's workload increases or the operator's capacity decreases. One way to estimate operator workload is through physiological measures such as heart rate and electroencephalograph (EEG) signals (Byrne and Parasuraman, 1996). For example, it is possible to moderate an operator's workload by using closed-loop control algorithms to adjust the level of automation according to the operator's EEG signal (Prinzel et al., 2000). Other estimates of workload depend on models that relate the task situation to expected cognitive load and operator performance. For example, by combining operator performance and task variables it is possible to engage automation and mitigate predictable workload increases (Scallen and Hancock, 2001). Most promising is an approach that combines data from all three sources along with model-based predictions of workload. By engaging higher levels of automation during periods of high workload, adaptive automation promises to solve some of the problems of clumsy automation.

Alleviating overload is often the motive behind the development of adaptive automation. It may be equally important, however, to consider how it can mitigate problems of underload. Both underload and overload stress an operator's ability to respond (Hancock and Warm, 1989), and automation that returns tasks to the operator during underload situations may place operators in a less stressful situation. Similarly, operators who monitor reliable automation for long periods become surprisingly inefficient at detecting automation failures. Adaptive automation can mitigate this automation-induced complacency by returning manual control periodically to the operator (Parasuraman et al., 1996).

Adaptive automation is a sort of meta-automation that can suffer from some of the same problems of automation if implemented improperly. Adaptive automation relieves the operator of the task of engaging and disengaging the automation, but it imposes the additional task of monitoring the adaptive automation, which can also increase workload (Kaber et al., 2001). In addition, adaptive automation faces challenging measurement and control problems. Adaptive automation depends on a precise measure of operator state, which can include physiological variables. If the time constant of these variables is longer than the time constant of the demands of the environment, automation will not adapt quickly enough. Even if operator state can be measured in a precise and timely manner, developing control algorithms that relate the operator state to an appropriate level of automation is difficult. Many of the limits of applying the Fitts list to static allocation of function also make dynamic allocation of function a challenge. Finally, even if an appropriate algorithm for adjusting the automation dynamically can be defined, the operator might respond in unexpected ways. For example, operators may manipulate their physiological state to influence the automation (Byrne and Parasuraman, 1996). Most important, operators may not understand the adaptive automation and so will view the system as behaving erratically. Such dynamic changes also introduce interface inconsistencies and increase the potential for mode errors.

4.3 Matching Automation to Human Performance Characteristics

Another approach to automation design considers how operators respond to different types of imperfect automation (Parasuraman et al., 2000). How well an operator is able to recognize and recover from automation failures often governs overall system performance. As a consequence, an important approach to automation design is to consider how human performance characteristics interact with the type of automation. The objective of this design approach is to minimize the tension that arises from mismatches between human performance characteristics and the type of automation (Sharit, 2003). A specific example of this approach considers the levels of automation and types of automation as defined by the stages of information processing. Primary considerations for automation design include workload, situation awareness, complacency, and skill maintenance (Parasuraman et al., 2000). These considerations do not specify a universally applicable degree of automation for each information-processing stage. Instead, appropriate automation design depends on the reliability of the automation and the consequences of failure as well as on technical and economic considerations (Parasuraman et al., 2000). In the context of air traffic control, human performance characteristics argue for the following upper bounds on the level of automation: information acquisition (high), information interpretation (high), action selection (medium), and action implementation (medium).

As an example, displays that indicate the status of the system (information interpretation automation) are preferable to those that advise the operator on how to respond (action selection automation) (Crocoll and Coury, 1990). Specifically, alerts regarding hazardous road conditions presented as a command (e.g., merge left) led to more dangerous lane changes compared to the same information presented as a notification (e.g., road construction in right lane) (Lee et al., 1999). Similar findings for a decision aid to help pilots make decisions regarding the dangers of aircraft icing suggest that status displays are preferable to command displays in high-risk domains where the automation is imperfect (e.g., space flight, medicine, and process control) (Sarter and Schroeder, 2001). Action implementation automation can be helpful when reliable but even more dangerously compelling when unreliable. Operators benefited more from action implementation automation than from action selection automation, but only when the automation performed normally (Endsley and Kaber, 1999). Although a greater degree of automation enhances performance and reduces workload during routine situation, it can also reduce situation awareness and undermine the ability to respond—when the automation fails, operators perform better with lower levels of automation (Kaber et al., 2000). In addition to the reliability of the automation, time pressure influences the benefit of a greater degree of automation. Pilots preferred management by consent, a relatively low level of automation; however, during periods of high time pressure and high workload, they preferred management by exception, a higher level of automation (Olson and Sarter, 2000).

Expert systems are a form of automation that has frequently failed to meet expectations. Typically, expert systems act as a prosthesis, supposedly replacing flawed and inconsistent human reasoning with more precise computer algorithms. Unfortunately, the level of automation associated with such an approach often conflicts with the range of situations the automation must face: The system gives the wrong answer when confronted with cases for which the automation is not fully competent. In addition, the operator typically plays a passive role of data entry and assessment of the decisions made by the automation and brittle failures result (Roth et al., 1988). A less automated approach, which places the automation in the role of critiquing the operator, has met with much more success. In critiquing, the computer presents alternative interpretations, hypotheses, or choices that complement those of the operator (Guerlain et al., 1999; Sniezek et al., 2002). A specific example is a decision support system for blood typing (Guerlain et al., 1999). Rather than using the expert system as a cognitive prosthesis to identify blood types, the critiquing approach suggests alternative hypotheses regarding possible interpretations of the data. In cases where the automation was fully competent, the operators made correct diagnoses 100% of the time, compared to 33 to 63% for those without the critiquing system. In cases where the critiquing system was not fully competent, performance degraded gracefully and operators still correctly diagnosed 32% more cases than those without the critiquing system. In situations where the automation is imperfect or the cost of failure is high, a lower level of automation, such as that used in the critiquing approach, is less likely to induce errors. Although much of the benefit of a critiquing system stems from the lower degree of automation and the greater involvement of the operator in the decision process, representation aiding plays an important role in supporting efficient operator–automation interaction.

4.4 Representation Aiding and Multimodal Feedback

Even if the type of automation is well matched to the task situation and human capabilities, inadequate feedback can undermine human–automation interaction. Inadequate feedback underlies many of the problems with automation from developing appropriate trust and clumsy automation to the out-of-the-loop phenomenon (Norman, 1990). However, providing sufficient feedback without overwhelming the operator is a critical design challenge. Poorly presented or excessive feedback can increase operator workload and undermine the benefits of the automation (Entin et al., 1996). In addition, without the proper context, abstraction, and integration, information regarding the behavior of complex automation may not be understandable. Representation aiding and multimodal feedback are two approaches that can help people understand how the automation works and how it is performing.

Representation aiding capitalizes on the power of visual perception to convey complex dynamic relationships. For example, graphical representations for pilots can augment the traditional airspeed indicator with target airspeeds and acceleration indicators. Integrating this information into a traditional flight instrument allows pilots to assimilate automation-related information with little extra effort (Hollan et al., 2000). Using a display that combines pitch, roll, altitude, airspeed, and heading can directly specify task-relevant information such as what is "too low" (Flach, 1999). Integrating automation-related information with traditional displays and combining low-level data into meaningful information are two important ways to enhance feedback without overwhelming the operator.

In regard to process control, Guerlain et al. (2002) identified three specific strategies for visual representation of complex process control algorithms. First, create visual forms whose emergent features correspond to higher-order relationships. Emergent features are salient symmetries or patterns that depend on the interaction of the individual data elements. A simple emergent feature is *parallelism*, which can occur with a pair of lines. Higher-order relationships are combinations of the individual data elements that govern system behavior. The boiling point of water is a higher-order relationship that depends on temperature and pressure. Second, use appropriate visual features to represent the dimensional properties of the data. For example, magnitude is a dimensional property that should be displayed using position or size on a visual display, not color or texture. Third, place data in a meaningful context. The meaningful context for any variable depends on what comparisons need to be made. For automation, this includes the allowable ranges relative to the current control variable setting, and the output relative to the desired level. Figure 5 shows some of the principles of representation aiding—use analog rather than digital or text, provide meaningfully integrated rather than raw data, and provide a context to support visual rather than mental comparisons.

Representation aiding makes it more likely that operators will trust automation more appropriately. However, trust also depends on more subtle elements of the interface (Lee and See, 2004). In many cases, trust and credibility depend on surface features of the interface that have no obvious link to the true capabilities of the system (Briggs et al., 1998; Tseng and Fogg, 1999). For example, in an online survey of over 1400 people, Fogg et al. (2001b) found that for Web sites, credibility depends heavily on "real-world feel," which is defined by factors such as response speed, a physical address, and photos of the organization. Similarly, a formal photograph of the author enhanced trustworthiness of a research article, whereas an informal photograph decreased trust (Fogg et al., 2001a). These results show that trust tends to increase when information is displayed in a way that provides concrete details that are consistent and clearly organized.

A similar pattern of results appears in studies of automation for target detection. Increasing image realism increased trust and led to greater reliance of the cueing information (Yeh and Wickens, 2001). Similarly, the tendency of pilots to follow the advice of the system blindly increased when the aid included detailed pictures (Ockerman, 1999). Just as highly realistic images can increase trust, degraded imagery can decrease trust, as was shown in a target cueing situation (MacMillan et al., 1994). Adjusting image quality and adding information to the interface regarding the capability of the automation can promote appropriate trust. In a signal detection task, the reliability of the sources was coded with different levels of luminance, leading participants to weigh reliable sources more

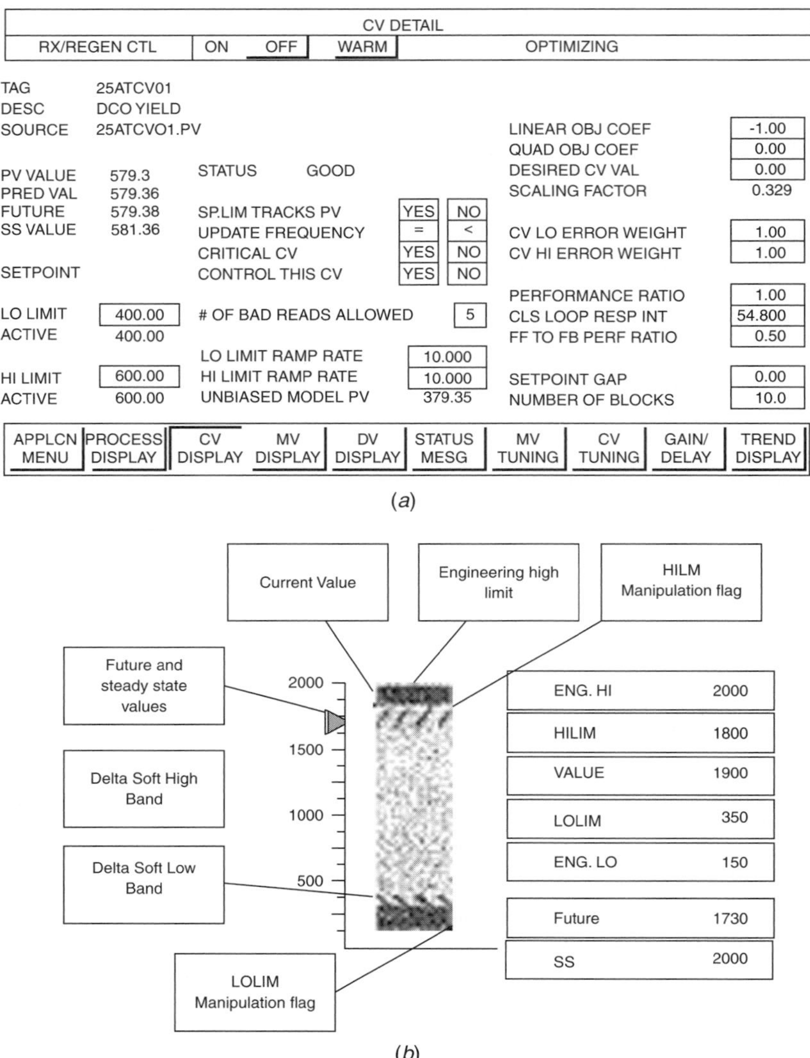

Figure 5 (*a*) Comparison of a traditional interface for automation; (*b*) example of representation to support operator understanding of automation.

than unreliable ones (Montgomery and Sorkin, 1996). These results suggest that the particular interface form can increase the level of trust, particularly the emphasis on concrete realistic representations.

Representation aiding tends to focus on interfaces that require focal as opposed to peripheral vision. Operators already face substantial demands on focal vision, and presenting automation-related information in that channel may overwhelm the operator. *Multimodal feedback* provides operators with information through haptic, tactile, auditory, and peripheral vision to avoid overwhelming the operator. Haptic feedback has proved more effective in alerting pilots to mode changes in cockpit automation compared to visual cues (Sklar and Sarter, 1999). Pilots receiving visual alerts detected 83% of the mode changes; those with haptic warnings detected 100% of the mode changes. Importantly, the haptic warnings did not interfere with the performance of concurrent visual tasks. Similarly, peripheral visual cues also helped pilots detect uncommanded mode transitions and did not interfere with concurrent visual tasks any more than did currently available automation feedback (Nikolic and Sarter, 2001). Haptic warnings may also be less annoying and acceptable compared to auditory warnings (Lee et al., 2004). Although promising, multimodal interfaces lack the resolution of visual interfaces, making it difficult to convey complex relationships and detailed information.

4.5 Matching Automation to Mental Models

The complexity of automation sometimes makes it difficult to convey its behavior using representation aiding or multiple-modal feedback. Sometimes a more effective strategy is to simplify the automation (Riley, 2001) or to match its algorithms to the operators' mental model (Goodrich and Boer, 2003). This is particularly true when a technology-centered approach to automation design has created an overly complex array of modes and features. The out-of-the-loop unfamiliarity problems result partially from the difficulties that operators have in generating correct expectations for the counterintuitive behavior of complex automation. Automation designed to perform in a manner consistent with operators' preferences and expectations can make it easier for operators to recognize failures and intervene.

Adaptive cruise control (ACC) is a specific example of where matching the mental model of the operator may be quite effective. Because drivers must focus their attention on the road, representation aiding could be distracting. Because ACC can apply only moderate levels of braking, drivers must intervene if the car ahead brakes heavily. If drivers must intervene, they must quickly enter the control loop because fractions of a second matter. If the automation behaves in a manner consistent with that of the driver, he or she will be more likely to detect and respond to the operational limits of the automation (Goodrich and Boer, 2003). To design an ACC algorithm consistent with drivers' mental models, driver behavior was partitioned according to perceptually relevant variables of inverse time to collision and time headway. Inverse time to collision (T_c^{-1}) is the relative velocity divided by the distance between the vehicles. Time headway (T_h) is the distance between vehicles divided by the velocity of the driver's vehicle. Using these variables, it is possible to identify the boundary that separates speed regulation and headway maintenance from active braking associated with collision avoidance. Figure 6 shows this boundary in the space defined by time headway and inverse time to collision. This boundary provides a template for designing ACC—the ACC should signal the driver to intervene as the driving situation crosses the boundary.

For situations in which the metaphor for automation is an agent, the mental model that people may adopt to understand the automation is that of a human collaborator. If the template for understandable automation is the operator's mental model, an agent should respond as would a human. Specifically, Miller (2002) suggests that computer etiquette may have an important influence on human–automation interaction. Etiquette may influence trust because category membership associated with adherence to a particular etiquette helps people to infer how the automation will perform. Specific rules for automation etiquette adapted from Miller and Funk (2001) include:

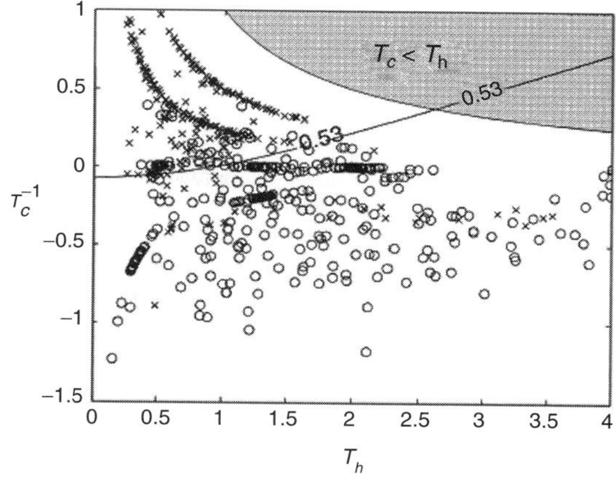

Figure 6 Driver braking behavior, showing a clear boundary between headway maintenance (○) and collision avoidance (×) that could be used to define operational limits of ACC. (From Goodrich and Boer, 2003, © IEEE 2002.)

- Make many correct interactions for every erroneous interaction.
- Make it very easy to override the automation.
- Do not make the same mistake twice—stop a behavior if corrected by the operator.
- Do not enable interaction features just because they are possible.
- Explain what is being done and why.
- Be able to take instruction.
- Do not assume every operator is the same—be sensitive and adapt to individual, contextual, and cultural differences.
- Be aware of what the operator knows and do not repeat unnecessarily.
- Use multiple modalities to communicate.
- Try not to interrupt.
- Be cute only if it furthers specific interaction goals.

Developing automation etiquette could promote appropriate trust, but it could lead to inappropriate trust if people infer inappropriate category memberships and develop distorted expectations regarding the capability of the automation. Even in simple interactions with technology, people often respond as they would to another person (Reeves and Nass, 1996; Nass and Lee, 2001). If anticipated, this tendency could help operators develop appropriate expectations regarding the behavior of the automation; however, unanticipated anthropomorphism could lead to surprising misunderstandings of the automation.

An important prerequisite for designing automation according to the mental model of the operator is the existence of a consistent mental model. Individual differences may be difficult to accommodate. This is particularly true for automation that acts as an agent, in which a mental model–based design must conform to complex social and cultural expectations. In addition, the mental model must be consistent with the physical constraints of the system if the automation is to work properly (Vicente, 1990). Mental models often contain misconceptions, and transferring these to the automation could be counterproductive or deadly. Even if operators have a single mental model that is consistent with the system constraints, automation based on a mental model may not achieve the same benefits of automation based on more sophisticated algorithms. In this case, designers must consider the trade-off between the benefits of a complex control algorithm and the costs of a poorly understood system. Representation aiding can mitigate this trade-off.

4.6 Formal Automation Analysis Techniques

Effective representation aiding depends on identifying the relevant information needed to understand the behavior of the automation. With complex automation, this can be a substantial challenge. One approach to meeting this challenge is to use formal verification techniques (Leveson, 1995; Degani and Heymann, 2002). Specifically, state machines can define the behavior of the automation and the operator's model. The state machine that defines the operator's model is constructed from the training materials and the information available on the interface. State machines provide a formal modeling language to define mismatches between the operator's model of the automation and the automation. These mismatches cause automation-related errors and surprises to occur. More specifically, legal and illegal states are defined by the task constraints that the automation and operator must satisfy. When the automation model enters an illegal state and the operator's model does not, the analysis predicts that the associated ambiguity will surprise operators and lead to errors (Degani and Heymann, 2002). Such ambiguities have been discovered in actual aircraft autopilot systems but have not been attributed correctly to deficiencies in the interface and training material (Degani and Heymann, 2002). Such mismatches between the operator and automation models indicate deficiencies in the operator's mental model that should be addressed by changing the automation, training, or interface. The state machine formalism makes it possible to generate training and interface requirements automatically.

Often, designers overestimate the benefit of automation because of the surprising interactions between the automation, environment, and the operator. Formal analysis that considers these interactions in terms of expected value calculations can reduce the surprise and guide design. In the example of a rear-end collision warning system for cars, a Bayesian approach combined with signal detection theory shows that the posterior probability of a collision situation given a warning is surprisingly low because the base rate of collision situation is so low (Parasuraman et al., 1997). This analysis shows that the selection of a detection threshold should consider the base rate; otherwise, the relatively high rate of false alarms could undermine driver acceptance.

More generally, calculating the expected value of manual and automatic control provides a rigorous means of selecting the best alternative (Sheridan and Parasuraman, 2000). In the simplest case this involves comparing the expected value of the operator and automation response to a binary failure state—a system is operating normally or it has failed. The expected value calculation combines the benefits and costs of four general responses to the system: a true positive, a true negative, a false negative, and a false positive. The expected value of the automation and the expected value of the operator response depend on the costs of being wrong and the benefits of being correct, together with the prior probabilities of the failure and the probabilities of the automation and operator being wrong and correct. If the expected value for automatic control is greater than the expected value for manual control, automation should be implemented. A similar analysis shows that the time-dependent value of automation makes it reasonable to give the automation final authority in some situations, such as in guiding the pilot to make go/no go decisions in aborting a takeoff (Inagaki, 2003). A similar analysis might help designers balance the information-processing demands of feedback regarding automation behavior with the time demands of the situation. Experiments assessing human interaction with automation should consider this calculation in defining experimental conditions, defining the reward structure, and interpreting the participants' behavior (Bettman and Payne, 1990; Payne et al., 1992; Meyer, 2004), otherwise, it is impossible to differentiate automation bias from eutactic behavior.

An expected value analysis provides a way to formalize the cost–benefit analysis that might otherwise be guided by the qualitative Fitts list heuristics. Although it promises to precisely quantify otherwise ambiguous decisions, estimating the numbers required to support the calculations can be a challenge. The costs and probabilities of rare, catastrophic events are notoriously difficult to estimate. More subtly, operator performance may affect the prior probabilities of events such that good operators experience fewer failures than do poor operators. In this situation, the automation will perform more poorly for better operators (Meyer and Bitan, 2002). Although precise probabilities and values may be difficult or impossible to estimate, such an approach is quite useful even if only relative benefits and costs of the automation and operator can be estimated (Sheridan and Parasuraman, 2000).

Simulation can also guide designers to consider the costs and benefits of automation more thoroughly. A simulation of a supervisory control situation shows that well-adapted operators are sensitive to the costs of engaging and disengaging automation (Kirlik, 1993). This simulation analysis identifies how the time costs of engaging the automation interact with the dynamics of the environment to undermine the value of the automation. A generally similar analysis argues that designers must make the normative strategy less effortful than competing strategies if operators are to use the automation effectively (Todd and Benbasat, 2000). More generally, simulation models that capture the human performance consequences of different levels of automation, automation reliability, and the environmental constraints are needed to support design. For example, a connectionist model of complacency provides a strong theoretical basis that accounts for empirical findings (Farrell and Lewandowsky, 2000). Cognitive architectures such as ACT-R also offer a promising approach to modeling human–automation interaction (Anderson and Libiere, 1998). Although ACT-R may not be able to capture the full complexity of this interaction, it may provide a useful tool for approximating the costs and benefits of various automation alternatives (Byrne and Kirlik, 2005).

5 EMERGING CHALLENGES

Substantial progress has been made regarding how to design automation to support people effectively. However, continuous advances in software and hardware development combined with an ever-expanding range of applications make future problems with automation likely. The following section highlights some of these emerging challenges. The first is the demands of managing a new type of automation, *swarm automation*, in which many semiautonomous agents work together. The second is the implication of automation in large interconnected networks of people and other automated elements, where issues of coordination and competition become critical. Automation in this environment requires considerations beyond those of the typical single operator interacting with one or two elements of automation. The third is the introduction of automation into daily life: specifically, automation in the car. These three examples represent some of the challenges associated with new types of automation, new types of human–automation organizations, and new application domains.

5.1 Swarm Automation

Swarm automation is an alternative approach to automation that may make it possible to respond to environmental variability while reducing the chance of system failure. These capabilities have important applications in a wide range of domains, including planetary exploration, unmanned aerial vehicle reconnaissance, landmine neutralization, or even data exploration, where hundreds of simple agents might be more effective than a single complex agent. Biology-inspired roboticists provide a specific example of swarm automation. Instead of the traditional approach of relying on one or two larger robots, they employ swarms of insect robots as an alternative (Brooks et al., 1990; Johnson and Bay, 1995). The swarm robot concept assumes that small machines with simple reactive behaviors can perform important functions more reliably and with lower power and mass requirements than can larger robots (Beni and Wang, 1993; Brooks and Flynn, 1993; Fukuda et al., 1998). Typically, the simple programs running on an insect robot are designed to elicit desirable emergent behaviors in the swarm (Sugihara and Suzuki, 1990; Min and Yin, 1998). For example, a large group of small robots might be programmed to search for concentrations of particular mineral deposits by building on the foraging algorithms of honeybees or ants.

In addition to physical examples of swarm automation, swarm automation has potential in searching large complex data sets for useful information. For example, the pervasive issue of data overload and the difficulties associated with effective information retrieval suggest a particularly useful application of swarm automation. Current approaches to searching large complex data sources, such as the Internet, are limited. People are likely to miss important documents, disregard data that represent a significant departure from initial assumptions, misinterpret data that conflict with an emerging understanding, and disregard more recent data that could revise interpretation (Patterson, 1999). These issues can be summarized as the need to broaden searches to enhance opportunity to discover highly relevant information, promote recognition of unexpected information to avoid premature fixation on a particular viewpoint or hypothesis, and manage data uncertainty to avoid misinterpretation of inaccurate or obsolete data (Woods et al., 1999). These represent important challenges that may require innovative design concepts and significant departures from current tools (Patterson, 1999). Just as swarm automation might help explore physical spaces, it might also help explore information spaces.

Managing swarm automation requires a qualitatively different approach than that of more traditional automation (Lee, 2001). Swarms of bees and ants, in which many simple individuals combine to behave as a single entity, provide some useful insights into the characteristics of swarm behavior and how they might be managed (Bonabeau et al., 1997). A defining characteristic of swarm behavior is that it emerges from parallel interaction between many agents. For example, swarms of bees adjust their foraging behavior to the environment dynamically in a way that does not depend on the performance of any individual. A colony of honeybees functions as a large, diffuse, amoeboid entity that can extend over great distances and simultaneously tap a vast array of food sources (Seeley, 1997). Direct control of this emergent behavior is not possible. Instead, mechanisms influencing individual elements of the swarm indirectly influence swarm behavior. Two particularly important mechanisms are positive feedback and random variation. Positive feedback reinforces existing activities, and random variation generates new activities and encourages adaptation (Resnick, 1991). One way that positive feedback and random variation combine to influence behavior is through *stimergy*, in which communication and control occurs through a dynamically evolving structure. Through stimergy, social insects communicate directly through the products of their work (e.g., the bees' honeycomb and the termites' chambers). A specific example of stimergy is the pheromone trail that guides the self-organizing foraging behavior of ants. Stimergy in foraging behavior involves a trade-off of speed of trail establishment and search thoroughness; a trail that is more quickly established will sacrifice the thoroughness of the search. Stimergy represents a powerful alternative to a static set of instructions that specify a sequence of activity. Parallel interaction between many agents, positive feedback, random variation, and stimergy make it possible for many simple individuals to produce complex group behavior (Bonabeau et al., 1997). However, such control mechanisms may be difficult for operators to understand.

The concept of hortatory control describes some of the challenges of controlling swarm automation. Hortatory control applies in situations where the system being controlled retains a high degree of autonomy and operators must exert indirect rather than direct control (Murray and Liu, 1997). Interacting with swarm automation requires people to consider swarm dynamics independent of the individual agents. In these situations it is most useful for the operator to control parameters affecting group rather than individual agents and for the operators to receive feedback about group rather than individual behavior. Swarm automation has great potential to extend human capabilities, but only if a thorough empirical and analytic investigation identifies the display requirements, feasible control mechanisms, and the range of swarm dynamics that can be comprehended and controlled by humans.

5.2 Management of Complex Networks of Operators and Automation

As automation becomes pervasive, it creates complex networks of increasingly tightly coupled elements. In this situation, the appropriate unit of analysis may shift from a single operator interacting with a single element of automation to that of multiple operators interacting with multiple elements of automation. Important dynamics can only be explained with this more complex unit of analysis. More so than single-operator situations, in these highly coupled systems, poor coordination between operators and inappropriate reliance on automation can degrade the decision-making performance and lead to catastrophes (Woods, 1994). As an example, the largest power grid failure in the nation's history occurred on August 14, 2003. In this failure, the flow of approximately 61,800 megawatts of electricity was disrupted, leaving 50 million customers from Ohio to New York and parts of Canada without power. An important contribution to this event was a lack of cooperation between two regional electrical grid operators that monitor the same region. These operators manage the flow of the electricity from suppliers to distributors. Poor communication and a failure to exchange detailed information on their operations prevented them from understanding and responding to changes in the power grid. Similar failures occur in supply chains as well as petrochemical processes, where people and automation sometimes fail to coordinate their activities.

Supply chains represent an increasingly important example of multioperator multiautomation. A supply chain is composed of a network of suppliers, transporters, and purchasers who work together, usually as a decentralized virtual company, to convert raw materials into products for end users. The growing popularity of supply chains reflects the general trend of companies to move away from vertical integration, where a single company converts raw materials into products for end users. Many manufacturers increasingly rely on supply chains; a typical U.S. company purchases 55% of the value of its products from other companies (Dyer and Singh, 1998). Efficient supply chains play a critical role in maintaining the economic health of the U.S. economy.

However, supply chains suffer from serious problems that erode their promised benefits. One is the *bullwhip effect*, in which small variations in end-item demand induce large-order oscillations, excess inventory, and backorders (Sterman, 1989). This effect can have enormous consequences on a company's efficiency and value. As an example, news reports of supply chain glitches associated with the bullwhip effect resulted in abnormal declines of 10.28% in companies' stock price (Hendricks and Singhal, 2003). Automation that forecasts demands can moderate these oscillations (Lee and Whang, 2000; Zhao and Xie, 2002). However, people must trust and rely on that automation, and substantial cooperation between supply chain members must exist to share such information.

Another major problem facing supply chains is the breakdown in cooperation as relationships between members of a supply chain devolve through an escalating series of conflicts that has been termed a *vicious cycle* (Akkermans and van Helden, 2002). Such conflicts can have dramatic negative consequences for a supply chain. For example, a strategic alliance between Office Max and Ryder International Logistics devolved into a legal fight in which Office Max sued Ryder for \$21.4 million and then Ryder sued Office Max for \$75 million (Handfield and Bechtel, 2002). Beyond the legal costs, these breakdowns can threaten competitiveness and undermine the market value of the company (Dyer and Singh, 1998). Vicious cycles also undermine information sharing, which can exacerbate the bullwhip effect. Companies that develop cooperative relationships can save money through shared resources, such as trucking companies in Chicago that can reduce their fleets by 25% by sharing containers (Smilowitz, in review). Even with the substantial benefits of cooperation, supply chains frequently fall into a vicious cycle in which poor cooperation leads to further poor cooperation. Trust between people plays a critical role in developing and sustaining cooperative relationships. People must trust each other to share information, and this trust can be undermined if poorly managed automation of one supply chain member compromises the success of another.

The bullwhip effect and vicious cycles and other supply chain problems reflect the influence of inappropriate actions at the local level that drive dysfunctional network dynamics. These effects are unique to highly coupled networks and require a unit of analysis that goes beyond the single person interacting with a single element of automation. As exemplified by the bullwhip effect and vicious cycles, the problems of supply chain management reflect generic challenges in using decentralized control to achieve a central objective. Decentralized networks promise efficiency and the capacity to adapt to unexpected perturbations, but their complexity and inefficient information sharing can lead people to respond to local rather than global considerations. Automation can alleviate the tendency for attention to local goals to magnify a small disturbance into a widespread disruption. However, too little or too much trust in automation leads to inappropriate reliance, which can induce dysfunctional dynamics, such as the bullwhip effect and vicious cycles.

Other domains share the general promise and pitfalls of modern supply chain management. For example, power grid management involves a decentralized network that makes it possible to supply the United States efficiently with power, but it can fail catastrophically when cooperation and information sharing breaks down (Zhou et al., 2003). Similarly, datalink-enabled air traffic control makes it possible for pilots to negotiate flight paths efficiently, but it can fail when pilots have trouble anticipating the complex dynamics of the system (Olson and Sarter, 2001; Mulkerin, 2003). Also, grid computing makes its enormous computing power available for use by many independent agents, but it can fail if load balancing and job scheduling do not consider global considerations (Lorch and Kafura, 2002; Chervenak et al., 2003). Overall, technology is creating many highly interconnected networks that have great potential but that also raise important concerns. Resolving these concerns partially depends on designing effective multioperator, multiautomation interactions.

5.3 Driving and Roadway Safety

Much of the existing research on automation has focused on operators of large complex systems for which expensive automation has been practical to develop. As computer and sensor technology become more affordable, automation will become more common in systems encountered in day-to-day life. Automation for cars and trucks is an example of automation that will touch the day-to-day lives of many people. Vehicle automation may touch more peoples' lives and have a greater safety consequence than any other type of automation. In the United States alone, people drive over 2 trillion miles a year in cars and light trucks (Pickrell and Schimek, 1999). The safety consequence is equally impressive. Over 6 million crashes kill approximately 42,000 people each year and result in an economic cost of over $164 billion per year (Wang et al., 1999). Motor vehicle crashes are also the leading cause of workplace injuries, being responsible for 42% of work-related fatalities (Bureau of Labor Statistics, 2003). Automation in cars and trucks has the potential to influence the safety and comfort of many people.

Functions that vehicle automation might support range from routing and navigation to collision avoidance and vehicle control (Lee, 1997). Table 2 shows some of the many examples of current and potential types of vehicle automation. Currently, examples include navigation systems that use GPS data and electronic map databases to give drivers turn-by-turn directions. Also, adaptive cruise control uses sensors and new control algorithms to extend cruise control so that cars slow down automatically and maintain a safe distance from the car ahead. Many vehicles even have a system that uses sensor data (e.g., airbag deployment) to detect a crash, calls for emergency aid, and then transmits the crash location using the car's GPS. The potential of automation to enhance the safety and comfort of drivers is substantial.

Designing automation to support driving confronts many of the same challenges as those found with automation in other domains. Sensor imperfections and complexity of the driving environment make adaptive cruise control and collision warning systems fallible. Recent studies suggest that adaptive cruise control may induce complacency and the potential of over trust. Specifically, many drivers intervene too slowly to prevent a collision when the adaptive cruise

Table 2 Automation for Driving and Other In-Vehicle Technology

General Functions	Specific Examples
Routing and navigation	Trip planning, multimode travel coordination and planning, predrive route and destination selection, dynamic route selection, route guidance, route navigation, automated toll collection, route scheduling, posttrip summary
Motorist services	Broadcast services/attractions, services/attractions directory, destination coordination, delivery-related information
Augmented signage	Guidance sign information, notification sign information, regulatory sign information
Safety and warning	Immediate hazard warning, road condition information, aid request, vehicle condition monitoring, driver monitoring, sensory augmentation
Collision avoidance and vehicle control	Forward object collision avoidance, road departure collision avoidance, lane change and merge collision avoidance, intersection collision avoidance, railroad crossing collision avoidance, backing aid, vehicle control
Driver comfort, communication, and convenience	Real-time communication; asynchronous communication; contact search and history; entertainment and general information; heating, ventilation, air conditioning, and noise

Source: Adapted from Lee and Kantowitz (2005).

control fails to brake (Stanton et al., 1997). Behavioral adaptation also threatens to undermine the safety benefits of automation. Automation aimed to enhance safety, such as an antilock brake system (ABS), has not produced the expected safety benefits because drivers with an ABS tend to change their driving behavior and follow more closely (Sagberg et al., 1997). A similar response may occur with collision warning systems that aim to give drivers advance notice of impending collisions. Such systems may lead some drivers to think they can safely engage in distracting activities, such as reading or watching DVDs, while driving. Understanding how to develop vehicle automation to enhance safety such that behavioral adaptation does not erode its benefits is a critical challenge.

Another challenge that confronts the design of vehicle automation is the potential for driver confusion in the face of many poorly integrated systems. Similar problems of automation coordination and integration have occurred with maritime navigation aids (Lee and Sanquist, 2000), flight management systems (Sarter and Woods, 1995), and medical devices (Cook et al., 1990a). Already, early examples of vehicle automation show the substantial confusion and frustration associated with poorly integrated systems, such as the recent controversy and confusion regarding the 700 features of the BMW iDrive (Norman, 2003). Forward object, road departure, lane change, and intersection collision warning systems may all populate the car of the future, and identifying which warning has been activated may be a challenge for drivers. To avoid such confusion requires a design approach that considers the overall driving ecology and the information needed to negotiate it rather than an approach focused on sensor technology and arbitrarily defined collision types.

Unlike operators of automation in domains such as aviation and process control, drivers do not receive specific training on how to operate particular features of their car. In addition, drivers belong to a very heterogeneous group that spans a wide range of age, experience, and goals for driving. The difficulty of providing systematic training for automotive automation and the diversity of drivers makes it likely that many drivers will misunderstand and misuse vehicle automation. Drivers misunderstand even a simple system, such as an ABS, and benefit from training on how to use it (Mollenhauer et al., 1997). More complex systems such as adaptive cruise control may confuse drivers, particularly as they move from a vehicle they are accustomed to driving to one they are not (e.g., a rental car). Ensuring that all drivers are properly trained is much more difficult than ensuring that process control operators or pilots understand the automation they manage. Automation that affects day-to-day life, such as vehicle automation, faces the particular challenges of being understood and used appropriately by a highly diverse array of potential users.

6 AUTOMATION – DOES IT NEED US?

The Luddites faced the prospect of automation changing their lives, and we face a similar prospect today. Increasingly sophisticated automation makes it possible to replace the human being in many situations, and the situations in which humans outperform automation are diminishing rapidly. Although the need for human adaptability, creativity, and flexibility make complete automation of most systems infeasible, the increasing capability of automation may eliminate even these reasons to include human operators. Soon, automating based on the criterion of whether the human or machine is better suited to perform a task may be irrelevant. This situation requires a deeper consideration of the purpose of technology (Hancock, 1996). Although automation allows people to avoid dangerous and unpleasant situations, unrestrained automation may eliminate activities that provide intrinsic enjoyment and purpose to life (Nickerson, 1999). Ironically, automating everything that is technologically possible or even everything that enhances system efficiency and safety may have the unanticipated effect of diminishing the lives of the people that automation should ultimately serve. Like the Luddites, we may ultimately need to confront the issue of whether automation needs us. "At least we have it in our power to say no to new technology, or do we?"(Sheridan, 2000).

REFERENCES

Akkermans, H., and van Helden, K. (2002), "Vicious and Virtuous Cycles in ERP Implementation: A Case Study of Interrelations Between Critical Success Factors," *European Journal of Information Systems*, Vol. 11, No. 1, pp. 35–46.

Anderson, J. R., and Libiere, C. (1998), *Atomic Components of Thought*, Lawrence Erlbaum Associates, Mahwah, NJ.

Bainbridge, L. (1983), "Ironies of Automation," *Automatica*, Vol. 19, No. 6, pp. 775–779.

Beni, G., and Wang, J. (1993), "Swarm Intelligence in Cellular Robotic Systems," in *Robots and Biological Systems: Towards a New Bionics*, P. Dario, G. Sansani, and P. Aebischer, Eds., Springer-Verlag, Berlin.

Bettman, J. R. J. E. J., and Payne, J. W. (1990), "A Computational Analysis of Cognitive Effort in Choice," *Organizational Behavior and Human Decision Processes*, Vol. 45, pp. 111–139.

Billings, C. E. (1997), *Aviation Automation: The Search for a Human-Centered Approach*, Lawrence Erlbaum Associates, Mahwah, NJ.

Bliss, J. P. (1997), "Alarm Reaction Patterns by Pilots as a Function of Reaction Modality," *International Journal of Aviation Psychology*, Vol. 7, No. 1, pp. 1–14.

Bliss, J., and Acton, S. A. (2003), "Alarm Mistrust in Automobiles: How Collision Alarm Reliability Affects Driving," *Applied Ergonomics*, Vol. 34, pp. 499–509.

Bocionek, S. R. (1995), "Agent Systems That Negotiate and Learn," *International Journal of Human–Computer Studies*, Vol. 42, No. 3, pp. 265–288.

Bonabeau, E., Theraulaz, G., Deneubourg, J. L., Aron, S., and Camazine, S. (1997), "Self-Organization in Social Insects," *Trends in Ecology and Evolution*, Vol. 12, No. 5, pp. 188–193.

Briggs, P., Burford, B., and Dracup, C. (1998), "Modeling Self-Confidence in Users of a Computer-Based System Showing Unrepresentative Design," *International*

Journal of Human–Computer Studies, Vol. 49, No. 5, pp. 717–742.

Broadbent, D. E. (1958), *Perception and Communication*, Pergamon Press, London.

Brooks, R. A., and Flynn, A. M. (1993), "A Robot Being," in *Robots and Biological Systems: Towards a New Bionics*, P. Dario, G. Sansani, and P. Aebischer, Eds., Springer-Verlag, Berlin.

Brooks, R. A., Maes, P., Mataric, M. J., and More, G. (1990), "Lunar Base Construction Robots," in *Proceedings of the 1990 International Workshop on Intelligent Robots and Systems*, pp. 389–392.

Bureau of Labor Statistics (2003), *Census of Fatal Occupational Injuries*, U.S. Department of Labor, Washington, DC.

Byrne, M. D., and Kirlik, A. (2005), "Using Computational Cognitive Modeling to Diagnose Possible Sources of Aviation Error," *International Journal of Aviation Psychology*, Vol. 15, No. 2, pp. 135–155.

Byrne, E. A., and Parasuraman, R. (1996), "Psychophysiology and Adaptive Automation," *Biological Psychology*, Vol. 42, No. 3, pp. 249–268.

Chervenak, A., Deelman, E., Kesselman, C., Allcock, B., Foster, I., Nefedova, V., et al. (2003), "High-Performance Remote Access to Climate Simulation Data: A Challenge Problem for Data Grid Technologies," *Parallel Computing*, Vol. 29, No. 10, pp. 1335–1356.

Cohen, M. S., Parasuraman, R., and Freeman, J. (1999), *Trust in Decision Aids: A Model and Its Training Implications*, Technical Report USAATCOM TR 97-D-4, Cognitive Technologies, Inc., Arlington, VA.

Cook, R. I., Woods, D. D., and Howie, M. B. (1990a), "The Natural History of Introducing New Information Technology into a Dynamic High-Risk Environment," in *Proceedings of the Human Factors Society 34th Annual Meeting*, Human Factors Society, Santa Monica, CA.

Cook, R. I., Woods, D. D., McColligan, E., and Howie, M. B. (1990b), "Cognitive Consequences of 'Clumsy' Automation on High Workload, High Consequence Human Performance," presented at the Space Operations, Applications and Research Symposium, SOAR '90, NASA Johnson Space Center, Houston, TX.

Crocoll, W. M., and Coury, B. G. (1990), "Status or Recommendation: Selecting the Type of Information for Decision Aiding," in *Proceedings of the Human Factors Society 34th Annual Meeting*, Human Factors Society, Santa Monica, CA, Vol. 2, pp. 1524–1528.

Dearden, A., Harrison, M., and Wright, P. (2000), "Allocation of Function: Scenarios, Context and the Economics of Effort," *International Journal of Human–Computer Studies*, Vol. 52, No. 2, pp. 289–318.

Degani, A., and Heymann, M. (2002), "Formal Verification of Human–Automation Interaction," *Human Factors*, Vol. 44, No. 1, pp. 28–43.

Deutsch, M. (1958), "Trust and Suspicion," *Journal of Conflict Resolution*, Vol. 2, No. 4, pp. 265–279.

Deutsch, M. (1960), "The Effect of Motivational Orientation upon Trust and Suspicion," *Human Relations*, Vol. 13, pp. 123–139.

de Vries, P., Midden, C., and Bouwhuis, D. (2003), "The Effects of Errors on System Trust, Self-Confidence, and the Allocation of Control in Route Planning," *International Journal of Human–Computer Studies*, Vol. 58, No. 6, pp. 719–735.

Dyer, J. H., and Singh, H. (1998), "The Relational View: Cooperative Strategy and Sources of Interorganizational Competitive Advantage," *Academy of Management Review*, Vol. 23, No. 4, pp. 660–679.

Dzindolet, M. T., Pierce, L. G., Beck, H. P., Dawe, L. A., and Anderson, B. W. (2001), "Predicting Misuse and Disuse of Combat Identification Systems," *Military Psychology*, Vol. 13, No. 3, pp. 147–164.

Dzindolet, M. T., Pierce, L. G., Beck, H. P., and Dawe, L. A. (2002), "The Perceived Utility of Human and Automated Aids in a Visual Detection Task," *Human Factors*, Vol. 44, No. 1, pp. 79–94.

Dzindolet, M. T., Peterson, S. A., Pomranky, R. A., Pierce, L. G., and Beck, H. P. (2003), "The Role of Trust in Automation Reliance," *International Journal of Human–Computer Studies*, Vol. 58, No. 6, pp. 697–718.

Endsley, M. R., and Kaber, D. B. (1999), "Level of Automation Effects on Performance, Situation Awareness and Workload in a Dynamic Control Task," *Ergonomics*, Vol. 42, No. 3, pp. 462–492.

Endsley, M. R., and Kiris, E. O. (1995), "The Out-of-the-Loop Performance Problem and Level of Control in Automation," *Human Factors*, Vol. 37, No. 2, pp. 381–394.

Entin, E. B., Entin, E. E., and Serfaty, D. (1996), "Optimizing Aided Target-Recognition Performance," in *Proceedings of the Human Factors and Ergonomics Society*, HFES, Santa Monica, CA, Vol. 1, pp. 233–237.

Eprath, A. R., and Young, L. R. (1981), "Monitoring vs. Man-in-the-Loop Detection of Aircraft Control Failures," in *Human Detection and Diagnosis of System Failures*, J. Rasmussen and W. B. Rouse, Eds., Plenum Press, New York, pp. 143–154.

Evans, L. (1991), *Traffic Safety and the Driver*, Van Nostrand Reinhold, New York.

Farrell, S., and Lewandowsky, S. (2000), "A Connectionist Model of Complacency and Adaptive Recovery Under Automation," *Journal of Experimental Psychology: Learning, Memory, and Cognition*, Vol. 26, No. 2, pp. 395–410.

Fitts, P. M. (1951), *Human Engineering for an Effective Air Navigation and Traffic Control System*, National Research Council, Washington, DC.

Flach, J. M. (1999), "Ready, Fire, Aim: A 'Meaning-Processing' Approach to Display Design," in *Attention and Performance XVII, Cognitive Regulation of Performance: Interaction of Theory and Application*, D. Gophen and A. Koriat, Eds., MIT Press, Cambridge, MA, pp. 197–221.

Flach, J. M., and Jagacinski, R. J. (2002), *Control Theory for Humans*, Lawrence Erlbaum Associates, Mahwah, NJ.

Fogg, B., Marshall, J., Kameda, T., Solomon, J., Rangnekar, A., Boyd, J., et al. (2001a), "Web Credibility Research: A Method for Online Experiments and Early Study Results," in *Proceedings of the CHI Conference on Human Factors in Computing Systems*, pp. 293–294.

Fogg, B., Marshall, J., Laraki, O., Osipovich, A., Varma, C., Fang, N., et al. (2001b), "What Makes Web Sites Credible? A Report on a Large Quantitative Study," presented at the CHI Conference on Human Factors in Computing Systems, Seattle, WA.

Fukuda, T., Funato, D., Sekiyama, K., and Arai, F. (1998), "Evaluation on Flexibility of a Swarm Intelligent System, in *Proceedings of the 1998 IEEE International Conference on Robotics and Automation*, pp. 3210–3215.

Gao, J., and Lee, J. D. (2004), "Information Sharing, Trust, and Reliance: A Dynamic Model of Multi-operator Multi-automation Interaction," in *Proceedings of the 5th Conference on Human Performance, Situation Awareness and Automation Technology*, D. A. Vincenzi, M. Mouloua, and P. A. Hancock, Eds., Lawrence Erlbaum Associates, Mahwah, NJ, Vol. 2, pp. 34–39.

Gibson, J. J. (1962), "Observations on Active Touch," *Psychological Review*, Vol. 69, pp. 477–491.

Goodrich, M. A., and Boer, E. R. (2003), "Model-Based Human-Centered Task Automation: A Case Study in ACC System Design," *IEEE Transactions on Systems, Man and Cybernetics, Part A, Systems and Humans*, Vol. 33, No. 3, pp. 325–336.

Grabowski, M. R., and Hendrick, H. (1993), "How Low Can We Go? Validation and Verification of a Decision Support System for Safe Shipboard Manning," *IEEE Transactions on Engineering Management*, Vol. 40, No. 1, pp. 41–53.

Guerlain, S. A., Smith, P. J., Obradovich, J. H., Rudman, S., Strohm, P., Smith, J. W., et al. (1999), "Interactive Critiquing as a Form of Decision Support: An Empirical Evaluation," *Human Factors*, Vol. 41, No. 1, pp. 72–89.

Guerlain, S., Jamieson, G. A., Bullemer, P., and Blair, R. (2002), "The MPC Elucidator: A Case Study in the Design for Human–Automation Interaction," *IEEE Transactions on Systems, Man and Cybernetics, Part A, Systems and Humans*, Vol. 32, No. 1, pp. 25–40.

Gupta, N., Bisantz, A. M., and Singh, T. (2002), "The Effects of Adverse Condition Warning System Characteristics on Driver Performance: An Investigation of Alarm Signal Type and Threshold Level," *Behaviour and Information Technology*, Vol. 21, No. 4, pp. 235–248.

Halprin, S., Johnson, E., and Thornburry, J. (1973), "Cognitive Reliability in Manned Systems," *IEEE Transactions on Reliability*, Vol. 22, pp. 165–169.

Hancock, P. A. (1996), "Teleology of Technology," in *Automation and Human Performance*, R. Parasuraman and M. Mouloua, Eds., Lawrence Erlbaum Associates, Mahwah, NJ, pp. 461–498.

Hancock, P. A., and Warm, J. S. (1989), "A Dynamic Model of Stress and Sustained Attention," *Human Factors*, Vol. 31, No. 5, pp. 519–537.

Handfield, R. B., and Bechtel, C. (2002), "The Role of Trust and Relationship Structure in Improving Supply Chain Responsiveness," *Industrial Marketing Management*, Vol. 31, No. 4, pp. 367–382.

Harris, W. C., Hancock, P. A., Arthur, E. J., and Caird, J. K. (1995), "Performance, Workload, and Fatigue Changes Associated with Automation," *International Journal of Aviation Psychology*, Vol. 5, No. 2, pp. 169–185.

Hendricks, K. B., and Singhal, V. R. (2003), "The Effect of Supply Chain Glitches on Shareholder Wealth," *Journal of Operations Management*, Vol. 21, No. 5, pp. 501–522.

Hoc, J. M. (1993), "Some Dimensions of a Cognitive Typology of Process-Control Situations," *Ergonomics*, Vol. 36, No. 11, pp. 1445–1455.

Hoc, J. M. (2000), "From Human–Machine Interaction to Human–Machine Cooperation," *Ergonomics*, Vol. 43, No. 7, pp. 833–843.

Hoc, J. H. (2001), "Towards a Cognitive Approach to Human–Machine Cooperation in Dynamic Situations," *International Journal of Human–Computer Studies*, Vol. 54, No. 4, pp. 509–540.

Hollan, J., Hutchins, E., and Kirsh, D. (2000), "Distributed Cognition: Toward a New Foundation for Human–Computer Interaction Research," *ACM Transactions on Computer–Human Interaction*, Vol. 7, No. 2, pp. 174–196.

Hollnagel, E., and Bye, A. (2000), "Principles for Modelling Function Allocation," *International Journal of Human–Computer Studies*, Vol. 52, No. 2, pp. 253–265.

Hovland, C. I., Janis, I. L., and Kelly, H. H. (1953), *Communication and Persuasion: Psychological Studies of Opinion Change*, Yale University Press, New Haven, CT.

Hutchins, E. (1995), *Cognition in the Wild*, MIT Press, Cambridge, MA.

Inagaki, T. (2003), "Automation and the Cost of Authority," *International Journal of Industrial Ergonomics*, Vol. 31, No. 3, pp. 169–174.

Johnson, P. J., and Bay, J. S. (1995), "Distributed Control of Simulated Autonomous Mobile Robot Collectives in Payload Transportation," *Autonomous Robots*, Vol. 2, No. 1, pp. 43–63.

Jones, P. M., and Jacobs, J. L. (2000), "Cooperative Problem Solving in Human–Machine Systems: Theory, Models, and Intelligent Associate Systems," *IEEE Transactions on Systems, Man and Cybernetics, Part C, Applications and Reviews*, Vol. 30, No. 4, pp. 397–407.

Kaber, D. B., Onal, E., and Endsley, M. R. (2000), "Design of Automation for Telerobots and the Effect on Performance, Operator Situation Awareness, and Subjective Workload," *Human Factors and Ergonomics in Manufacturing*, Vol. 10, No. 4, pp. 409–430.

Kaber, D. B., Riley, J. M., Tan, K. W., and Endsley, M. R. (2001), "On the Design of Adaptive Automation for Complex Systems," *Internation Journal of Cognitive Ergonomics*, Vol. 5, No. 1, pp. 37–57.

Kantowitz, B. H., and Sorkin, R. D. (1987), "Allocation of Functions," in *Handbook of Human Factors*, G. Salvendy, Ed., Wiley, New York, pp. 355–369.

Kirlik, A. (1993), "Modeling Strategic Behavior in Human–Automation Interaction: Why an 'Aid' Can (and Should) Go Unused," *Human Factors*, Vol. 35, No. 2, pp. 221–242.

Klein, G., Ross, K. G., Moon, B. M., Klein, D. E., Hoffman, R. R., and Hollnagel, E. (2003), "Macrocognition," *IEEE Intelligent Systems*, Vol. 18, No. 3, pp. 81–85.

Landauer, T. K. (1995), *The Trouble with Computers: Usefulness, Usability, and Productivity*, MIT Press, Cambridge, MA.

Lee, J. D. (1997), "A Functional Description of ATIS/CVO Systems to Accommodate Driver Needs and Limits," in *Ergonomics and Safety of Intelligent Driver Interfaces*, Y. I. Noy, Ed., Lawrence Erlbaum Associates, Mahwah, NJ, pp. 63–84.

Lee, J. D. (2001), "Emerging Challenges in Cognitive Ergonomics: Managing Swarms of Self-Organizing Agent-Based Automation," *Theoretical Issues in Ergonomics Science*, Vol. 2, No. 3, pp. 238–250.

Lee, J. D., and Kantowitz, B. K. (2005), "Network Analysis of Information Flows to Integrate In-Vehicle Information Systems," *International Journal of Vehicle Information and Communication Systems*, Vol. 1, No. 1/2, pp. 24–43.

Lee, J. D., and Moray, N. (1992), "Trust, Control Strategies and Allocation of Function in Human–Machine Systems," *Ergonomics*, Vol. 35, No. 10, pp. 1243–1270.

Lee, J. D., and Moray, N. (1994), "Trust, Self-Confidence, and Operators' Adaptation to Automation," *International Journal of Human–Computer Studies*, Vol. 40, pp. 153–184.

Lee, J. D., and Morgan, J. (1994), "Identifying Clumsy Automation at the Macro Level: Development of a Tool to Estimate Ship Staffing Requirements," in *Proceedings of the Human Factors and Ergonomics Society 38th Annual Meeting*, Vol. 2, pp. 878–882.

Lee, J. D., and Sanquist, T. F. (1996), "Maritime Automation," in *Automation and Human Performance*, R. Parasuraman and M. Mouloua, Eds., Lawrence Erlbaum Associates, Mahwah, NJ, pp. 365–384.

Lee, J. D., and Sanquist, T. F. (2000), "Augmenting the Operator Function Model with Cognitive Operations: Assessing the Cognitive Demands of Technological Innovation in Ship Navigation," *IEEE Transactions on Systems, Man and Cybernetics, Part A, Systems and Humans*, Vol. 30, No. 3, pp. 273–285.

Lee, J. D., and See, K. A. (2004), "Trust in Technology: Designing for Appropriate Reliance," *Human Factors*, Vol. 46, No. 1, pp. 50–80.

Lee, J. D., and Strayer, D. L. (2004), Preface to a Special Section on Driver Distraction, *Human Factors*, Vol. 46, pp. 583–586.

Lee, H. L., and Whang, S. J. (2000), "Information Sharing in a Supply Chain," *International Journal of Technology Management*, Vol. 20, No. 3–4, pp. 373–387.

Lee, J. D., Gore, B. F., and Campbell, J. L. (1999), "Display Alternatives for In-Vehicle Warning and Sign Information: Message Style, Location, and Modality," *Transportation Human Factors Journal*, Vol. 1, No. 4, pp. 347–377.

Lee, J. D., Hoffman, J. D., and Hayes, E. (2004), "Collision Warning Design to Mitigate Driver Distraction," in *Proceedings of CHI*, ACM, New York, pp. 65–72.

Leveson, N. G. (1995), *Safeware: System Safety and Computers*, Addison-Wesley, Reading, MA.

Lewandowsky, S., Mundy, M., and Tan, G. (2000), "The Dynamics of Trust: Comparing Humans to Automation," *Journal of Experimental Psychology: Applied*, Vol. 6, No. 2, pp. 104–123.

Lewis, M. (1998), "Designing for Human–Agent Interaction," *AI Magazine*, Vol. 19, No. 2, pp. 67–78.

Lorch, M., and Kafura, D. (2002), "Supporting Secure Ad-Hoc User Collaboration in Grid Environments," in *Grid Computing*, Grid 2002, Vol. 2536, pp. 181–193.

Lutzhoft, M. H., and Dekker, S. W. A. (2002), "On Your Watch: Automation on the Bridge," *Journal of Navigation*, Vol. 55, No. 1, pp. 83–96.

MacMillan, J., Entin, E. B., and Serfaty, D. (1994), "Operator Reliance on Automated Support for Target Acquisition," in *Proceedings of the Human Factors and Ergonomics Society 38th Annual Meeting*, HFES, Santa Monica, CA, Vol. 2, pp. 1285–1289.

Mayer, R. C., Davis, J. H., and Schoorman, F. D. (1995), "An Integrative Model of Organizational Trust," *Academy of Management Review*, Vol. 20, No. 3, pp. 709–734.

McFadden, S., Vimalachandran, A., and Blackmore, E. (2003), "Factors Affecting Performance on a Target Monitoring Task Employing an Automatic Tracker," *Ergonomics*, Vol. 47, No. 3, pp. 257–280.

Metzger, U., and Parasuraman, R. (2001), "The Role of the Air Traffic Controller in Future Air Traffic Management: An Empirical Study of Active Control Versus Passive Monitoring," *Human Factors*, Vol. 43, No. 4, pp. 519–528.

Meyer, J. (2001), "Effects of Warning Validity and Proximity on Responses to Warnings," *Human Factors*, Vol. 43, No. 4, pp. 563–572.

Meyer, J. (2004), "Conceptual Issues in the Study of Dynamic Hazard Warnings," *Human Factors*, Vol. 46, No. 2, pp. 196–204.

Meyer, J., and Bitan, Y. (2002), "Why Better Operators Receive Worse Warnings," *Human Factors*, Vol. 44, No. 3, pp. 343–353.

Michon, J. A. (1985), "A Critical View of Driver Behavior Models: What Do We Know, What Should We Do? in *Human Behavior and Traffic Safety*, L. Evans and R. C. Schwing, Eds., Plenum Press, New York, pp. 485–520.

Miller, C. A. (2002), "Definitions and Dimensions of Etiquette," in *Etiquette for Human–Computer Work*, C. Miller, Ed., Technical Report FS-02-02, American Association for Artificial Intelligence, Menlo Park, CA, pp. 1–7.

Miller, C. A., and Funk, H. B. (2001), "Associates with Etiquette: Communication to Make Human–Automation Interaction More Natural, Productive and Polite," in *Proceedings of the 8th European Conference on Cognitive Science Approaches to Process Control*, Munich, September 24–26, pp. 1–8.

Min, T. W., and Yin, H. K. (1998), "A Decentralized Approach for Cooperative Sweeping by Multiple Mobile Robots," in *Proceedings of the 1998 IEEE/RSJ International Conference on Intelligent Robots and Systems.*

Minsky, M. (1986), *The Society of Mind*, Simon & Schuster, New York.

Mollenhauer, M. A., Dingus, T. A., Carney, C., Hankey, J. M., and Jahns, S. (1997), "Antilock Brake Systems: An Assessment of Training on Driver Effectiveness," *Accident Analysis and Prevention*, Vol. 29, No. 1, pp. 97–108.

Montgomery, D. A., and Sorkin, R. D. (1996), "Observer Sensitivity to Element Reliability in a Multielement Visual Display," *Human Factors*, Vol. 38, No. 3, pp. 484–494.

Moray, N. (2003), "Monitoring, Complacency, Scepticism and Eutactic Behaviour," *International Journal of Industrial Ergonomics*, Vol. 31, No. 3, pp. 175–178.

Moray, N., Inagaki, T., and Itoh, M. (2000), "Adaptive Automation, Trust, and Self-Confidence in Fault Management of Time-Critical Tasks," *Journal of Experimental Psychology: Applied*, Vol. 6, No. 1, pp. 44–58.

Mosier, K. L., Skitka, L. J., Heers, S., and Burdick, M. (1998), "Automation Bias: Decision Making and Performance in High-Tech Cockpits," *International Journal of Aviation Psychology*, Vol. 8, No. 1, pp. 47–63.

Muir, B. M., and Moray, N. (1996), "Trust in Automation, 2: Experimental Studies of Trust and Human Intervention in a Process Control Simulation," *Ergonomics*, Vol. 39, No. 3, pp. 429–460.

Mulkerin, T. (2003), "Free Flight Is in the Future: Large-Scale Controller Pilot Data Link Communications Emulation Testbed," *IEEE Aerospace and Electronic Systems Magazine*, Vol. 18, No. 9, pp. 23–27.

Murray, J., and Liu, Y. (1997), "Hortatory Operations in Highway Traffic Management," *IEEE Transactions on Systems, Man and Cybernetics, Part A, Systems and Humans*, Vol. 27, No. 3, pp. 340–350.

Nagel, D. C. (1988), "Human Error in Aviation Operations," in *Human Factors in Aviation*, E. Weiner and D. Nagel, Eds., Academic Press, New York, pp. 263–303.

Nass, C., and Lee, K. N. (2001), "Does Computer-Synthesized Speech Manifest Personality? Experimental Tests of Recognition, Similarity-Attraction, and Consistency-Attraction," *Journal of Experimental Psychology: Applied*, Vol. 7, No. 3, pp. 171–181.

Nickerson, R. S. (1999), in *Automation, Technology and Human Performance*, M. W. Scerbo and M. Mouloua, Eds., Lawrence Erlbaum Associates, Mahwah, NJ, pp. 11–19.

Nikolic, M. I., and Sarter, N. B. (2001), "Peripheral Visual Feedback: A Powerful Means of Supporting Effective Attention Allocation in Event-Driven, Data-Rich Environments," *Human Factors*, Vol. 43, No. 1, pp. 30–38.

Norman, D. A. (1990), "The 'Problem' with Automation: Inappropriate Feedback and Interaction, Not 'Overautomation,'" *Philosophical Transactions of the Royal Society, London, Series B, Biological Sciences*, Vol. 327, No. 1241, pp. 585–593.

Norman, D. A. (1993), *Things That Make Us Smart*, Addison-Wesley, Reading, MA.

Norman, D. A. (2003), "Interaction Design for Automobile Interiors," retrieved April 8, 2004, from http://www.jnd.org/dn.mss/InteractDsgnAutos.html.

NTSB (1990), *Marine Accident Report: Grounding of the U.S. Tankship* Exxon Valdez *on Bligh Reef, Prince William Sound, Valdez, Alaska, March 24, 1989*, Report NTSB/MAR90/04, National Transportation Safety Board, Washington, DC.

NTSB (1997), *Marine Accident Report: Grounding of the Panamanian Passenger Ship* Royal Majesty *on Rose and Crown Shoal near Nantucket, Massachusetts, June 10, 1995*, Report NTSB/MAR97/01, National Transportation Safety Board, Washington, DC.

Ockerman, J. J. (1999), "Over-Reliance Issues with Task-Guidance Systems," in *Proceedings of the Human Factors and Ergonomics Society 43rd Annual Meeting*, pp. 1192–1196.

Olson, W. A., and Sarter, N. B. (2000), "Automation Management Strategies: Pilot Preferences and Operational Experiences," *International Journal of Aviation Psychology*, Vol. 10, No. 4, pp. 327–341.

Olson, W. A., and Sarter, N. B. (2001), "Management by Consent in Human–Machine Systems: When and Why It Breaks Down," *Human Factors*, Vol. 43, No. 2, pp. 255–266.

Parasuraman, R., and Riley, V. (1997), "Humans and Automation: Use, Misuse, Disuse, Abuse," *Human Factors*, Vol. 39, No. 2, pp. 230–253.

Parasuraman, R., Molloy, R., and Singh, I. (1993), "Performance Consequences of Automation-Induced 'Complacency'" *international Journal of Aviation Psychology*, Vol. 3, No. 1, pp. 1–23.

Parasuraman, R., Mouloua, M., and Molloy, R. (1994), "Monitoring Automation Failures in Human–Machine Systems," in *Human Performance in Automated Systems: Current Research and Trends*, M. Mouloua and R. Parasuraman, Eds., Lawrence Erlbaum Associates, Mahwah, NJ, pp. 45–49.

Parasuraman, R., Mouloua, M., and Molloy, R. (1996), "Effects of Adaptive Task Allocation on Monitoring of Automated Systems," *Human Factors*, Vol. 38, No. 4, pp. 665–679.

Parasuraman, R., Hancock, P. A., and Olofinboba, O. (1997), "Alarm Effectiveness in Driver-Centred Collision-Warning Systems," *Ergonomics*, Vol. 40, No. 3, pp. 390–399.

Parasuraman, R., Sheridan, T. B., and Wickens, C. D. (2000), "A Model for Types and Levels of Human Interaction with Automation," *IEEE Transactions on Systems, Man and Cybernetics, Part A, Systems and Humans*, Vol. 30, No. 3, pp. 286–297.

Patterson, E. S. (1999), "A Simulation Study of Computer-Supported Inferential Analysis Under Data Overload," in *Proceedings of the Human Factors and Ergonomics Society 43rd Annual Meeting*, Vol. 1, pp. 363–368.

Payne, J. W., Bettman, J. R., Coupey, E., and Johnson, E. J. (1992), "A Constructive Process View of Decision-Making: Multiple Strategies in Judgment and Choice," *Acta Psychologica*, Vol. 80, No. 1–3, pp. 107–141.

Perrow, C. (1984), *Normal Accidents*, Basic Books, New York.

Pickrell, D., and Schimek, P. (1999), "Growth in Motor Vehicle Ownership and Use: Evidence from the Nationwide Personal Transportation Survey," *Journal of Transportation and Statistics*, Vol. 2, No. 1, pp. 1–18.

Prinzel, L. J., Freeman, F. C., Scerbo, M. W., Mikulka, P. J., and Pope, A. T. (2000), "A Closed-Loop System for Examining Psychophysiological Measures for Adaptive Task Allocation," *International Journal of Aviation Psychology*, Vol. 10, No. 4, pp. 393–410.

Ranney, T. A. (1994), "Models of Driving Behavior: A Review of Their Evolution," *Accident Analysis and Prevention*, Vol. 26, No. 6, pp. 733–750.

Rasmussen, J. (1986), *Information Processing and Human–Machine Interaction: An Approach to Cognitive Engineering*, North-Holland, New York.

Reeves, B., and Nass, C. (1996), *The Media Equation: How People Treat Computers, Television, and New Media Like Real People and Places*, Cambridge University Press, New York.

Rempel, J. K., Holmes, J. G., and Zanna, M. P. (1985), "Trust in Close Relationships," *Journal of Personality and Social Psychology*, Vol. 49, No. 1, pp. 95–112.

Resnick, M. (1991), *Turtles, Termites, and Traffic Jams: Explorations in Massively Parallel Microworlds*, MIT Press, Cambridge, MA.

Riley, V. (1989), "A General Model of Mixed-Initiative Human–Machine Systems," in *Proceedings of the Human Factors Society 33rd Annual Meeting*, pp. 124–128.

Riley, V. A. (1994), "Human Use of Automation," Ph.D. dissertation, University of Minnesota, Minneapolis, MN.

Riley, V. (2001), "A New Language for Pilot Interfaces," *Ergonomics in Design*, Vol. 9, No. 2, pp. 21–27.

Ross, W., and LaCroix, J. (1996), "Multiple Meanings of Trust in Negotiation Theory and Research: A Literature Review and Integrative Model," *International Journal of Conflict Management*, Vol. 7, No. 4, pp. 314–360.

Roth, E. M., and Woods, D. D. (1988), "Aiding Human Performance, I: From Cognitive Analysis to Support Systems," Vol. 51, No. 1, pp. 39–64.

Roth, E. M., Bennett, K. B., and Woods, D. D. (1987), "Human Interaction with an 'Intelligent' Machine," *International Journal of Man–Machine Studies*, Vol. 27, pp. 479–526.

Roth, E. M., Bennett, K. B., and Woods, D. D. (1988), "Human Interaction with an 'Intelligent' Machine," in *Cognitive Engineering in Complex, Dynamic Worlds*, Academic Press, London, pp. 23–69.

Rotter, J. B. (1967), "A New Scale for the Measurement of Interpersonal Trust," *Journal of Personality*, Vol. 35, No. 4, pp. 651–665.

Rouse, W. B. (1988), "Adaptive Aiding for Human/Computer Control," *Human Factors*, Vol. 30, pp. 431–443.

Sagberg, F., Fosser, S., and Saetermo, I. A. F. (1997), "An Investigation of Behavioural Adaptation to Airbags and Antilock Brakes Among Taxi Drivers," *Accident Analysis and Prevention*, Vol. 29, No. 3, pp. 293–302.

Sarter, N. B. (2000), "The Need for Multisensory Interfaces in Support of Effective Attention Allocation in Highly Dynamic Event-Driven Domains: The Case of Cockpit Automation," *International Journal of Aviation Psychology*, Vol. 10, No. 3, pp. 231–245.

Sarter, N. B., and Schroeder, B. (2001), "Supporting Decision Making and Action Selection Under Time Pressure and Uncertainty: The Case of In-Flight Icing," *Human Factors*, Vol. 43, No. 4, pp. 573–583.

Sarter, N. B., and Woods, D. D. (1994), "Decomposing Automation: Autonomy, Authority, Observability and Perceived Animacy," in *Human Performance in Automated Systems: Current Research and Trends*, M. Mouloua and R. Parasuraman, Eds., Lawrence Erlbaum Associates, Mahwah, NJ, pp. 22–27.

Sarter, N. B., and Woods, D. D. (1995), "How in the World Did We Ever Get in That Mode? Mode Error and Awareness in Supervisory Control," *Human Factors*, Vol. 37, No. 1, pp. 5–19.

Sarter, N. B., and Woods, D. D. (1997), "Team Play with a Powerful and Independent Agent: Operational Experiences and Automation Surprises on the Airbus A-320," *Human Factors*, Vol. 39, No. 4, pp. 553–569.

Sarter, N. B., and Woods, D. D. (2000), "Team Play with a Powerful and Independent Agent: A Full-Mission Simulation Study," *Human Factors*, Vol. 42, No. 3, pp. 390–402.

Sarter, N. B., Woods, D. D., and Billings, C. E. (1997), "Automation Surprises," in *Handbook of Human Factors and Ergonomics*, 2nd ed., G. Salvendy, Ed., Wiley, New York, pp. 1926–1943.

Scallen, S. F., and Hancock, P. A. (2001), "Implementing Adaptive Function Allocation," *International Journal of Aviation Psychology*, Vol. 11, No. 2, pp. 197–221.

Scerbo, M. W. (1996), "Theoretical Perspectives on Adaptive Automation," in *Automation and Human Performance*, R. Parasuraman and M. Mouloua, Eds., Lawrence Erlbaum Associates, Mahwah, NJ, pp. 37–63.

Schuffel, J. J., Boer, P. A., and van Breda, L. (1988), "The Ship's Wheelhouse of the Nineties: The Navigation Performance and Mental Workload of the Officer of the Watch," *Journal of Navigation*, Vol. 42, No. 1, pp. 60–72.

Seeley, T. D. (1997), "Honey Bee Colonies Are Group-Level Adaptive Units," *American Naturalist*, Vol. 150, pp. S22–S41.

Seifert, C. M., and Hutchins, E. L. (1992), "Error as Opportunity: Learning in a Cooperative Task," *Human–Computer Interaction*, Vol. 7, pp. 409–435.

Sharit, J. (2003), "Perspectives on Computer Aiding in Cognitive Work Domains: Toward Predictions of Effectiveness and Use," *Ergonomics*, Vol. 46, No. 1–3, pp. 126–140.

Sheridan, T. B. (1992), *Telerobotics, Automation, and Human Supervisory Control*, MIT Press, Cambridge, MA.

Sheridan, T. B. (2000), "Function Allocation: Algorithm, Alchemy or Apostasy?" *International Journal of Human–Computer Studies*, Vol. 52, No. 2, pp. 203–216.

Sheridan, T. B. (2002), *Humans and Automation*, Wiley, New York.

Sheridan, T. B., and Ferrell, W. R. (1974), *Man–Machine Systems: Information, Control, and Decision Models of Human Performance*, MIT Press, Cambridge, MA.

Sheridan, T. B., and Hennessy, R. T. (1984), *Research and Modeling of Supervisory Control Behavior*, National Academy Press, Washington, DC.

Sheridan, T. B., and Parasuraman, R. (2000), "Human Versus Automation in Responding to Failures: An Expected-Value Analysis," *Human Factors*, Vol. 42, No. 3, pp. 403–407.

Sheridan, T. B., and Verplank, W. L. (1978), *Human and Computer Control of Undersea Teleoperators*, Technical Report, Man–Machine Systems Laboratory, Department of Mechanical Engineering, MIT, Cambridge, MA.

Singh, D. T., and Singh, P. P. (1997), "Aiding DSS Users in the Use of Complex OR Models," *Annals of Operations Research*, Vol. 72, pp. 5–27.

Sitkin, S. B., and Roth, N. L. (1993), "Explaining the Limited Effectiveness of Legalistic 'Remedies' for Trust/Distrust," *Organization Science*, Vol. 4, No. 3, pp. 367–392.

Skitka, L. J., Mosier, K. L., and Burdick, M. (1999), "Does Automation Bias Decision-Making? *International Journal of Human–Computer Studies*, Vol. 51, No. 5, pp. 991–1006.

Skitka, L. J., Mosier, K., and Burdick, M. D. (2000a), "Accountability and Automation Bias," *International Journal of Human–Computer Studies*, Vol. 52, No. 4, pp. 701–717.

Skitka, L. J., Mosier, K. L., Burdick, M., and Rosenblatt, B. (2000b), "Automation Bias and Errors: Are Crews Better Than Individuals?" *International Journal of Aviation Psychology*, Vol. 10, No. 1, pp. 85–97.

Sklar, A. E., and Sarter, N. B. (1999), "Good Vibrations: Tactile Feedback in Support of Attention Allocation and Human–Automation Coordination in Event-Driven Domains," *Human Factors*, Vol. 41, No. 4, pp. 543–552.

Smilowitz, K. (in review), "Multi-resource Routing with Flexible Tasks: An Application in Drayage Operations," *IEMS Working Paper*.

Smith, P. J., McCoy, E., and Layton, C. (1997), "Brittleness in the Design of Cooperative Problem-Solving Systems: The Effects on User Performance," *IEEE Transactions on Systems, Man and Cybernetics, Part A, Systems and Humans*, Vol. 27, No. 3, pp. 360–371.

Sniezek, J. A., Wilkins, D. C., Wadlington, P. L., and Baumann, M. R. (2002), "Training for Crisis Decision-Making: Psychological Issues and Computer-Based Solutions," *Journal of Management Information Systems*, Vol. 18, No. 4, pp. 147–168.

Stanton, N. A., and Pinto, M. (2000), "Behavioural Compensation by Drivers of a Simulator When Using a Vision Enhancement System," *Ergonomics*, Vol. 43, No. 9, pp. 1359–1370.

Stanton, N. A., and Young, M. S. (1998), "Vehicle Automation and Driving Performance," *Ergonomics*, Vol. 41, No. 7, pp. 1014–1028.

Stanton, N. A., Young, M., and McCaulder, B. (1997), "Drive-by-Wire: The Case of Driver Workload and

Reclaiming Control with Adaptive Cruise Control," *Safety Science*, Vol. 27, No. 2–3, pp. 149–159.

Sterman, J. D. (1989), "Modeling Managerial Behavior: Misperceptions of Feedback in a Dynamic Decision-Making Experiment," *Management Science*, Vol. 35, No. 3, pp. 321–339.

Stickland, T. R., Britton, N. F., and Franks, N. R. (1995), "Complex Trails and Simple Algorithms in Ant Foraging," *Proceedings of the Royal Society of London, Series B, Biological Sciences*, Vol. 260, No. 1357, pp. 53–58.

Strain, J., and Eason, K. (2000), "Exploring the Implications of Allocation of Function for Human Resource Management in the Royal Navy," *International Journal of Human–Computer Studies*, Vol. 52, No. 2, pp. 319–334.

Sugihara, K., and Suzuki, I. (1990), "Distributed Motion Coordination of Multiple Mobile Robots," in *Proceedings of the 5th IEEE International Symposium on Intelligent Control*, pp. 138–143.

Tenner, E. (1996), *Why Things Bite Back: Technology and the Revenge of Unanticipated Consequences*, Alfred A. Knopf, New York.

Tenney, Y. J., Rogers, W. H., and Pew, R. W. (1998), "Pilot Opinions on Cockpit Automation Issues," *International Journal of Aviation Psychology*, Vol. 8, No. 2, pp. 103–120.

Todd, P., and Benbasat, I. (1999), "Evaluating the Impact of DSS, Cognitive Effort, and Incentives on Strategy Selection," *Information Systems Research*, Vol. 10, No. 4, pp. 356–374.

Todd, P., and Benbasat, I. (2000), "Inducing Compensatory Information Processing Through Decision Aids That Facilitate Effort Reduction: An Experimental Assessment," *Journal of Behavioral Decision Making*, Vol. 13, No. 1, pp. 91–106.

Tseng, S., and Fogg, B. J. (1999), "Credibility and Computing Technology," *Communications of the ACM*, Vol. 42, No. 5, pp. 39–44.

Vicente, K. J. (1990), "Coherence- and Correspondence-Driven Work Domains: Implications for Systems Design," *Behaviour and Information Technology*, Vol. 9, pp. 493–502.

Vicente, K. J. (1999), *Cognitive Work Analysis: Towards Safe, Productive, and Healthy Computer-Based Work*, Lawrence Erlbaum Associates, Mahwah, NJ.

Wang, J. A., Knipling, R. R., and Blincoe, L. J. (1999), "The Dimensions of Motor Vehicle Crash Risk," *Journal of Transportation and Statistics*, Vol. 2, No. 1, pp. 19–44.

Ward, N. J. (2000), "Automation of Task Processes: An Example of Intelligent Transportation Systems," *Human Factors and Ergonomics in Manufacturing*, Vol. 10, No. 4, pp. 395–408.

Wickens, C. D., and Kessel, C. (1981), "Failure Detection in Dynamic Systems," in *Human Detection and Diagnosis of System Failures*, J. Rasmussen and W. B. Rouse, Eds., Plenum Press, New York, pp. 155–169.

Wiener, E. L. (1989), *Human Factors of Advanced Technology ("Glass Cockpit") Transport Aircraft*, NASA Contractor Report 177528, NASA Ames Research Center, Moffett Field, CA.

Wilde, G. J. S. (1988), "Risk Homeostasis Theory and Traffic Accidents: Propositions, Deductions and Discussion of Dissension in Recent Reactions," *Ergonomics*, Vol. 31, No. 4, pp. 441–468.

Wilde, G. J. S. (1989), "Accident Countermeasures and Behavioral Compensation: The Position of Risk Homeostasis Theory," *Journal of Occupational Accidents*, Vol. 10, No. 4, pp. 267–292.

Woods, D. D. (1987), "Commentary: Cognitive Engineering in Complex and Dynamic Worlds," *International Journal of Man–Machine Studies*, Vol. 27, pp. 571–585.

Woods, D. D. (1991), "Nosocomial Automation: Technology-Induced Complexity and Human Performance," in *Proceedings of the International Conference on Systems, Man, and Cybernetics*, pp. 1279–1282.

Woods, D. D. (1994), "Automation: Apparent Simplicity, Real Complexity," in *Human Performance in Automated Systems: Current Research and Trends*, M. Mouloua and R. Parasuraman, Eds., Lawrence Erlbaum Associates, Mahwah, NJ, pp. 1–7.

Woods, D. D., Potter, S. S., Johannesen, L., and Holloway, M. (1991), *Human Interaction with Intelligent Systems: Trends, Problems, New Directions*, CSEL Report 1991-001, Ohio State University, Columbus, OH.

Woods, D. D., Patterson, E. S., Roth, E. M., and Christoffersen, K. (1999), "Can We Ever Escape from Data Overload? A Cognitive Systems Diagnosis," in *Proceedings of the Human Factors and Ergonomics 43rd Annual Meeting*, Vol. 1, pp. 174–179.

Yeh, M., and Wickens, C. D. (2001), "Display Signaling in Augmented Reality: Effects of Cue Reliability and Image Realism on Attention Allocation and Trust Calibration," *Human Factors*, Vol. 43, pp. 355–365.

Zhao, X. D., and Xie, J. X. (2002), "Forecasting Errors and the Value of Information Sharing in a Supply Chain," *International Journal of Production Research*, Vol. 40, No. 2, pp. 311–335.

Zhou, T. S., Lu, J. H., Chen, L. N., Jing, Z. J., and Tang, Y. (2003), "On the Optimal Solutions for Power Flow Equations," *International Journal of Electrical Power and Energy Systems*, Vol. 25, No. 7, pp. 533–541.

Zuboff, S. (1988), *In the Age of Smart Machines: The Future of Work Technology and Power*, Basic Books, New York.

CHAPTER 61

HUMAN FACTORS AND ERGONOMICS IN MANUFACTURING AND PROCESS CONTROL

Dieter Spath, Martin Braun, and Lorenz Hagenmeyer
Fraunhofer Institute for Industrial Engineering
Stuttgart, Germany

1 MANUFACTURING

1.1 Definitions

Manufacturing is a human-driven transformation process. Using energy and labor, this process creates superposable economic or consumer goods from natural or raw materials (Westkämper and Warnecke, 2002). *Process control* is a method for the quality assurance of industrial process and enterprise systems. A control system usually regulates continuous operations or processes. Process control considers especially quality, risk, reliability, and safety. Most often it refers to systems in which material and energy flows interact with one another and are transformed (Woods and Hanes, 1986). Contrary to manufacturing systems, process control systems are often very large and highly complex. In some cases, time delay and system dynamics are extreme (Moray, 1997).

Associated working characteristics, however, come together as a result of increasing automation and microcomputer-supported control in both manufacturing and process control. As a result, the manufacturing area is moving away farther and farther from force-focused physical activity in favor of cognitive control activity. At this stage, most of the same or at least similar design aspects must be considered for work design. Differences between the two aspects are expected to virtually disappear in the future. In this chapter, production processes are important. In principle, production includes manufacturing-oriented production as well as process-oriented production. Several differences regarding ergonomic work design in manufacturing and in process control are addressed.

Business management considers production as one of the classical functions of an enterprise where production includes all production process tasks, including manufacturing and assembly as well as disposition, logistics, planning, control, preventive maintenance, and operative quality assurance. Moreover, in addition to these factors, human resources management is responsible for integrating, qualifying, and informing employees. A comprehensive approach to production refers not only to technological concerns but also to work's social, cultural, and ethical concepts. The production process can only be optimal once all of these factors have been achieved. Hence, a production system incorporates the entire production process, which is independent of the actual business department where these tasks are fulfilled (Spath, 2003).

1.2 Historical Overview

The historical development of production concepts involves primarily technical development as characterized by three industrial revolutions (see Figure 1). During the *first industrial revolution*, prime movers were developed. Characteristic of this period is the question of power transformation. During the *second industrial revolution*, the technical development of production was at the forefront. As a result of organizational work measures, time became an important production factor. During the *third industrial revolution*, informational

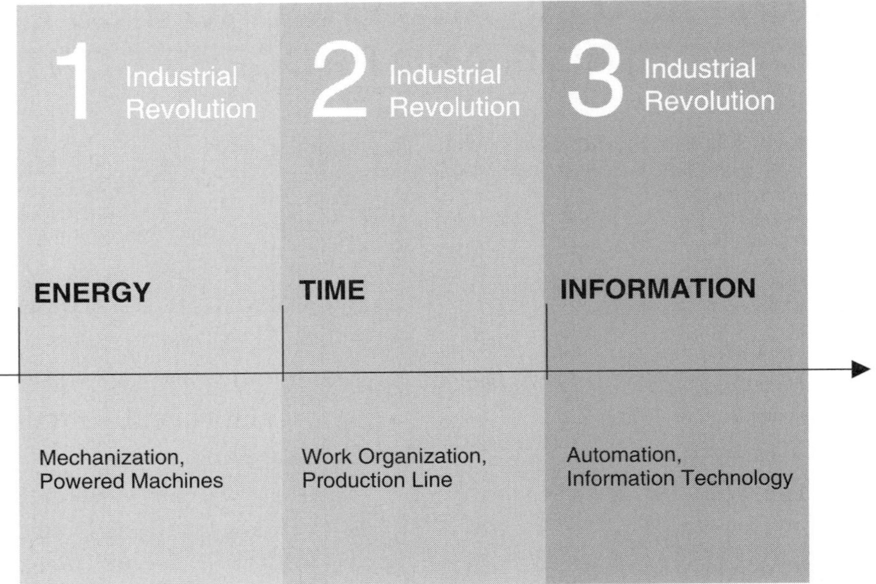

Figure 1 History of industrial development.

technology and automation were highlighted. Characteristic of this era is the availability of knowledge.

1.2.1 Early Industrialization

Business production concepts have been influenced for decades by the *scientific business management* ideas of Taylor (1911). Taylor examined the effects on work performance of monetary incentive systems along with those of workplace and tool design. The basic assumption was that the average worker is motivated to work efficiently primarily by financial considerations. The working human being was considered as an all-purpose machine. This assumption leads to the separation of manual and mental work. In this way, one became independent of the know-how of the skilled worker. Complex tasks divided into sufficiently small subtasks could be done by almost everyone. This division of work called for a strong and responsible hierarchical operational division, which is still an influential characteristic of many businesses. The most important characteristics of the Tayloristic work structure are:

- Separation of planning and implementing tasks
- Individual incentives (e.g., the piece wage)
- Many hierarchical ranks
- Low qualification standards
- Small specified work content amount
- One best way for every work sequence

It was not until the middle of the twentieth century that there was a gradual turn from the great Taylorism division of work structure. In the 1960s, teamwork was used in place of these concepts. Teamwork calls for the worker to strive toward self-realization and autonomy. Work should correspond to personal advancement criteria. Given the complex nature of these challenges, work science was needed to develop appropriate design strategies. Although numerous promising design concepts were developed, it is to be noted that the findings obtained were implemented insufficiently. As a result of economic pressure, it was not until recently that businesses began to use these scientific human-centered concepts.

1.2.2 Current Positions

In the long run, progressive technical development, varying market conditions, and changes in human attitudes toward work significantly influence production concepts. At the same time, technical, scientific, and human factors influencing production concepts are not isolated from each other, but are, rather, to be observed in their alternating dependency stress field. The most important factors influencing the design of production concepts are discussed below.

The Human Over the past few years, the application of flexible production systems has become more widespread. This was a result of market demand to manufacture small lot sizes economically simultaneously with an increase in the variety of products. A high level of mechanization as well as intense informational relationships is characteristic of flexible production systems. Altering stress situations for the working person developed: While physical stress was at the forefront earlier, the human worker today must increasingly face psychological stress (Braun, 2003). Examples of this are:

- Control and preventive maintenance of large equipment of constantly increasing complexity (e.g., excessive demands arising from the need for rapid adjustment, or insufficient demands as a result of automatic operation)
- Attention when working with dangerous materials or during dangerous processes
- Exposure to industrial robots
- Working with computer networking systems
- Innovative information and communication systems
- New forms of work and company organization
- Decentralization of responsibility

Despite the change in working conditions, physical strain resulting from working conditions still plays an important role. Physical strain results from stress caused by the manual handling of loads, by unnatural body movements, and by forced body postures. This physical strain can lead to severe health hazards. In upcoming years, companies will be confronted continuously with the effects of demographic developments (e.g., a somewhat overaged worker population). Production tasks in the future will be carried out predominantly by older employees. Therefore, occupational health care and qualification concepts will become increasingly important (Pack et al., 2000).

The Market In the past few years, a buyer's market has come into existence which places manufacturing businesses at the forefront of new challenges. For example, product life cycles are being shortened, and hence time to make profits from selling products is constantly decreasing. This pressure has an effect on employees, causing them to complain about stress, hecticness, deadline pressure, and excessive demands.

Technology Increasing automation and use of information technology (IT) are particularly important in this area. Simultaneously, a paradigm shift is emerging from technocentric work design to anthropological work design: Factors of human-centered work design, in addition to technical factors, are taken into consideration when planning and operating production systems (Bullinger, 2003). Thus, the importance of technology is not in question; the question that does arise is how work can be designed and organized for human applications amid the use of existing and developing technology.

1.2.3 Future Production Perspectives

The market-driven development and introduction of new products is a crucial task for enterprises in order to obtain competitive advantages. At the same time, increasing efficiency must be promoted. This can be achieved by the strategies discussed below (Westkämper, 1996, 2004).

Time to Market To attain and use innovations successfully with the goal of having a timely advantage over competitors, the process chain must be supported effectively from concept to finished product. Thus, knowledge and tools have to be supplied to decrease the time and cost of development and to reduce production preparation (Spath et al., 2001). The following are regarded as seminal: the synchronization of product and production development, market and objective orientation considering the entire process chain, preventive optimization of product development with intelligence systems, integrated engineering and product management systems and the application of generative production processes for prototype construction.

Versatile Business Structures Proximity to customers and to the market is a strategic success factor. The related fulfillment of customer wishes and supply of products in a shorter time result in significant advantages. Since the market is dynamic, only flexible organization and short-term versatile business structure deliver justifiable economic results. First, the resources (i.e., manufacturing resources, employees, material) must be adjusted constantly according to demand. Currently, reaction times are located in the middle range. A high level of division of working processes, as well as lengthy logistical methods, prevent modification and adaptability.

Short-term production structure modifications can be achieved by production networks, self-organization, self-optimization, flexible work systems, and the use of intelligent methods of production. Slogans such as *agile manufacturing* or *bionic manufacturing* characterize the actual discussion concerning product philosophies to bring about higher levels of action and modification. Integrating these concepts into an enterprise's practice will, however, happen only over the course of time.

Production Facing the Performance–Precision Trade-off In recent years, performance potential has been improved in many types of production technology. The development of materials, sensors, and actuators as well as the recognition of relationships between process parameters and achievable efficiency levels and precision were of utmost importance. In practice, however, businesses are often hesitant to put these scientific concepts into use. One reason for this are increasing demands under existing conditions of process dependability and control. Significant time and cost savings can be attained if a heightened level of performance and precision can be reached simultaneously with the achievement of stable production in operationally unstable departments.

Technical solutions are found to be: using intelligent and/or self-learning process control systems, integrated procedures, integrated process chains, process-integrated quality control, just-in-time information supply, automatic error compensation, and machine and system diagnostics. These techniques can be utilized in almost all aspects of production.

Automation and Humanization The idea of completely automated factories has existed for some time. It has been supported by developments and by enhanced possibilities of automation technology. The workplace's management of supply, waste, and operating materials and informational processing was intended to be completely integrated. These concepts proved, however, to have many shortcomings. Despite this, areas of physical work, particularly heavy work, will decrease further in all operating sectors as machines take on this type of work while contributing to rationalization. Moreover, in the future, work tasks will be supported increasingly by information technologies. Furthermore, this further advancement of rationalization will be a recurring factor in light of work tasks. Routine information work will then be increasingly automated (Bubb, 2000), leading to a shift from muscle work to brain work in the production sector also.

As a result of these experiences, numerous enterprises changed their strategy by placing more value on employee qualifications. In this way, effective and increased short-term performance could be achieved. At the same time, enterprises learned how to simplify operations and better control processes. As a result of increasing product complexity and product operations, some method of simplification becomes a necessity. New work organization concepts are needed to better apply available human resources.

Adaptive Production The vision of adaptive production is, on the one hand, not to be affected by turbulence in the business environment and, on the other, even to use this turbulence to develop a competitive advantage. Adaptive production solution approaches are manifold. The following characteristics serve as the foundation for future production systems and structures:

- Technical systems are growing more and more complex, resulting in more difficult handling and system comprehension.
- Information as a production factor is becoming increasingly important.
- Advancement in economical efficiency is being achieved by system considerations rather than by individual optimization.
- Employee qualification development is becoming increasingly important.
- Interdisciplinary teamwork is being increased.
- Questions relating to resource-adapting production are becoming centralized.

1.3 Holistic Production Systems

Holistic production systems were developed to deal with upcoming operational challenges. They originated as a result of industrial use and combined systemic and holistic approaches.

1.3.1 Definitions and Concepts

A *system* is an integrated structural combination of details, things, or procedures which either exist in nature or are human-made. A system consists of interrelated subsystems and single system elements. It is confined by the system's boundaries. Combinations, connections, and interactions within a system are described in a multidimensional and process-oriented manner.

Systemic perception requires that elements and methods be networked with each another. Thus, important considerations are optimal flow of material and information. The output of one method serves as an input for the next method. By networking the flow of material and information paths, transferred volumes, waiting time, repacking, reformatting, and so on, are reduced to a minimum.

Holism in this context describes integrated consideration of the entire system and its effective combinations. Often, the notion *integrated* is used synonymously with *holistic* within the given context. The human, the organization, and the technology should be taken into consideration particularly when dealing with holistic or integrated production systems (Spath, 2003). As a result, different historically developed management concepts have been integrated (see Figure 2).

A holistic production system covers at least the production process, including planning functions, manufacturing and assembly, control, logistics, and quality control. Enterprises aim more and more for the continuous design of all enterprise processes in order to become an integrated system. This calls for expansion of the order fulfillment process, the acquiring process, the product development process, and the sales and distribution process.

1.3.2 Design Principles

Holistic production systems are adapted specifically to markets, products, technologies, and corporate cultures. Therefore, systematic principles of design are applied to assure that all partial solutions fit together, resulting in a complete system. The following are the main principles of design for holistic production systems (Scholtz et al., 2003).

Partially Autonomous Working Groups Partially autonomous working groups have a comprehensive and challenging work task which they carry out on their own. The team takes on planning tasks such as work distribution in the group, planning the order in which assignments are processed, appointment compliance, acknowledgment of lot size per unit of time, material disposition, and so on. In addition, the team takes on tasks such as machine maintenance and repair, cleaning and transportation tasks, and quality control. Team members take turns being the team leader, therefore also taking on the tasks of superiors. Team members are enabled to fulfill these tasks by means of qualification activities. The concept corresponds to a longtime existing demand for the creation of human working conditions that advance worker capabilities and personal development possibilities. Figure 3 illustrates a working group

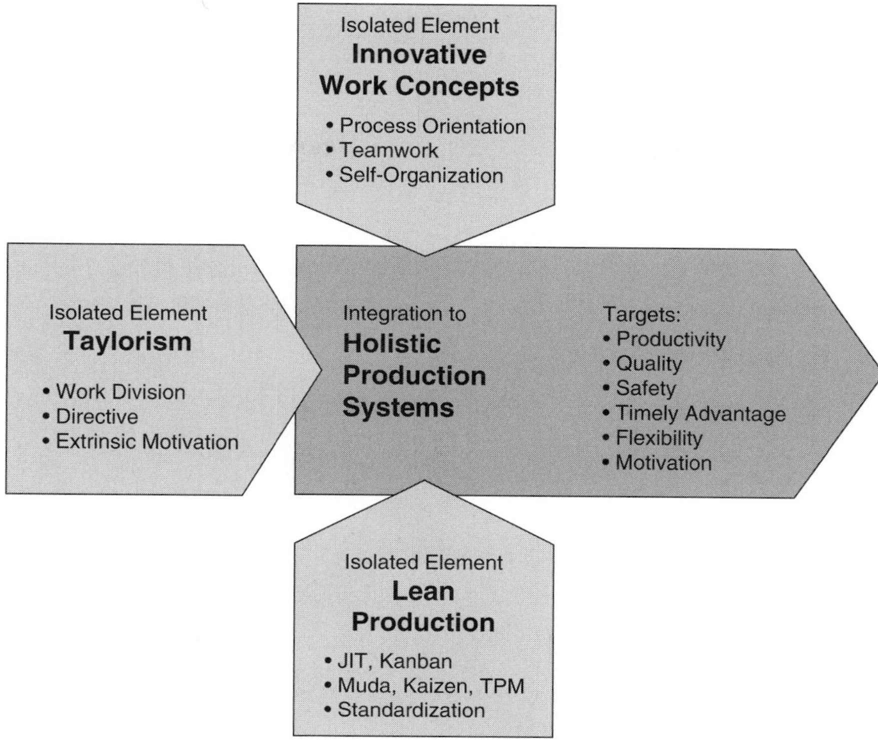

Figure 2 Development of holistic production systems. (From Korge and Scholtz, 2004.)

situation at an ergonomically optimized assembly workstation.

Process Orientation Process orientation is defined as cooperation among production functions, replacing observation of isolated functions. Here the business process ranges from first ideas to final customer supply, including intermediate steps such as the production itself. Process orientation assumes that production reacts flexibly to supply and demand. Extensive electronic data handling accounts for process-oriented informational and logistical tasks.

Just in Time *Just in time* (JIT), the concept of timely supply according to demand, is an essential and integral part of process-oriented work organization. With this concept, only a limited number of warehouses are needed. Therefore, minimal capital commitment will be achieved for both raw materials and supplies. The parts needed for each process step are produced in just the amount needed for a specific time frame or a specific amount of final product. Therefore, JIT means that only the essential parts are produced at the right time in the desired quality and free of waste. It has to be noted that the supply industry is included in the JIT concept. The most important methods and concepts of JIT are one-piece flow (piece production); first in, first out (FIFO); small-sized loading equipment, rapid changeover [single minute exchange of die (SMED)], and Kanban.

Continuous Improvement Process The continuous improvement process is a fundamental element of modern work organization. The goal of this process is to always be on the lookout for possible methods of advancement and better solutions.

The best people to offer advancement suggestions are the workers themselves, who are advised to incorporate this attitude as part of their own work habits. Contrary to the traditional employee suggestion system, it ought not to be a one-time happening but ought to comprise a large number of individual suggestions coming from all employees. Basically, it is an attitude that directs employee attention to a critical consideration of the establishment: analysis and development of problem solutions, and suggestions for improvement. The recommendations are collected and made available for all to see.

The continuous improvement process can be a fixed element of teamwork (e.g., by means of quality circles). Most often, employees are prompted to provide ideas by financial rewards (e.g., an incentive plan). The risk that employees themselves rationalize by means of their own troubleshooting competencies should not go unnoticed. If this is the case, active employee participation in a continuous improvement

Figure 3 Group work assembly operation at an ergonomically optimized workstation. (Courtesy of Fraunhofer IAO.)

process will decrease. Worker participation works only if employees are aware of the fact that no disadvantages can arise from their participation. Therefore, a continuous improvement process should not only aim for cost savings, personnel reduction, or work compression. The continuous improvement process should emphasize that it is long term in nature and aims for improved work organization, higher employee qualifications, more use of ergonomic technology, stress reduction, and the like.

Professional Work Routines To maintain the results of a continuous improvement process and to prevent them from being erased by factors such as worker exchange, professional work routines are established. They define the manner in which a work process is carried out. The goal is to increase process reliability and efficiency. Furthermore, professional work routines aim for simplification. It is wise to reduce varied solutions to as few variances as possible. The necessary degree of flexibility with respect to the customer may not, however, be confined by standardized solutions. Methods regarding professional work routines include standardized working papers, standardized shift change, standardized equipment, standardized quality operations, and specified cycles of preventive maintenance and documentation.

Management by Objectives Here, an agreement is made with individual employees or groups of employees regarding performance during a given time frame, after which it is decided whether or not the goals have been reached. Many factors can be stipulated: for example, the amount of revenue aimed for during a given time frame, the date until which a product is to be manufactured, or when an order must be completed. Similarly, there are stipulations regarding employee performance: for example, to increase operational readiness. The extent and degree of the objectives achieved form the foundation for further personnel decisions, such as promotion prospects, shifting, or termination. Management by objectives is often associated with performance-related recompensation, forming an instrument for performance control. One advantage of management by objectives is that employees decide how they will reach a goal. For this reason, employees have great leeway in carrying out their jobs.

Management by objectives should be developed cooperatively. Cooperation furthers communication, thus creating mutual trust. The employee no longer feels excluded and knows why these goals are to be reached. These goals should be neither inconsistent nor too detailed, but relevant and easy to measure.

Robust Processes Robust processes aim to provide customers with products at a level of performance that is both flawless and reliable (e.g., a zero-error goal). This goal can only be reached, not when the quality is reviewed, but when it is produced. Preventive methods should help to identify and avoid errors before the start of production. Using error-tracking routines, errors are identified so that they can be corrected immediately. By analyzing the source of the error and removing it, errors are prevented from recurring. Well-known robust process concepts and methods include the quality circle, the account force diagram, marginal samples, the quality alarm, the machine breakpoint, total productive maintenance, Poka Yoke, the FMEA-process, and quality agreements.

1.4 Human Role in Production

Contributions by the worker to economic business success strongly influence the significance of human factors.

1.4.1 Business Interests

Businesses have an economic interest in healthy, motivated, and qualified employees because they provide production efficiency. Frustrated, passive, or aggressive employees are less inclined to perform well; moreover, they are susceptible to disease. Disease- and accident-caused failures and early disablement weaken an enterprise's production. Operational surveys show that 1% of absentism influences approximately 1% of personnel costs. Adequate compensation for absent employees in many divisions is not possible without conflict, resulting in loss of flexibility, quality defects, and production interruption—ending ultimately in loss of orders. In the course of flexible and rational operational strategies and a loss of missing personnel compensation buffers, the economic consequences of absentism and insufficient employee engagement are becoming increasingly important (Braun, 2003).

1.4.2 Motives for Performance and Cooperation

Employees are motivated to do good and engaged work if their personal interests agree with those of the company and if their performance is acknowledged. If agreement is lost during the course of changing processes, and/or acknowledgment from superiors is missing, resistance develops. In operational practice, resistance to changes leads to a considerable loss of performance and cooperation. As a result, managers of large enterprises dispose of between 50 and 80% of their work by addressing internal operational resistance (Spath et al., 2003). A survey of approximately 2000 employees taken by Gallup (2002) in Germany showed that about 80% of those surveyed were insufficiently involved in their company; approximately 70% showed lack of engagement, acting rather passively when dealing with their superiors; and 15% of those surveyed were downright displeased and often aggressive toward their superiors, displaying this by poor productivity. Such negative attitudes result in:

- A high number of absent days
- Readiness to leave the company as soon as an opportunity arises
- Nonexistent career intentions with the current employer
- A bad attitude toward others
- A small chance of recommending one's workplace to friends or relatives
- A small chance of recommending one's own product
- The missing element of enjoyment at work

As a result of lower productivity and a high number of absences, the macroeconomic loss resulting from unengaged employees adds up to over 10% of the annual gross national product.

1.4.3 Focusing the Healthy Worker

Always when discussing cooperation, flexible reaction, and innovation, the worker is irreplaceable, due to his or her thinking and communication abilities as well as potential for ideas, creativity, and curiosity. The employees' interest and will to change are essential factors in every increase in productivity. Great enterprises have recognized that healthy, qualified, and efficiency-motivated employees are one of their main assets.

Empirical comparative studies support the fact that technology does not play as meaningful a role in the development of a great enterprise as is often suggested. Actually, technology does not trigger changes, only picks up the pace (Collins and Porras, 1994). During the course of process changes, a successful enterprise must provide its employees with reliable orientation and demanding motivation. Relevant experience from great enterprises shows that problems such as motivation, engagement, and change are solved on their own when human-oriented working conditions are provided. From the factors mentioned above regarding production concepts, it may be concluded that the ongoing automatization and flexibilization of business processes similarly call for a consideration of human-oriented work design.

2 HUMAN-ORIENTED WORK DESIGN

2.1 Objectives

The goal of human-oriented (or ergonomic) work design is to balance the strain on the worker. By doing this, human performance potential for the production of goods and services is promoted and the untimely deterioration of human resources is countered. This goal is sought through the use of technical, medical, psychological, social, and ecological knowledge (Bullinger, 1994). It is to be noted that the concepts presented in this section (i.e., the classical concepts of work science, particularly systematic work design) have been researched and well documented since the 1970s (Helander, 1995; Salvendy, 1997; Karwowski and Salvendy, 1998; Karwowski, 2001). It is therefore assumed that the reader has knowledge of these concepts. However, since their use is still basic, the concepts are presented concisely here and referenced accordingly. Use of these design concepts is a necessary requirement for efficient, competitive, and sustainable production. Moreover, further innovative production system concepts and chances are presented in this chapter.

Human-oriented work design is concerned on the one hand with the worker, the development of his or her skills, capacities, and abilities, and on the other hand with an analysis of variables possibly influencing performance. Further tasks concern the design of technical setups and organizational structures that people use for work. The goal is to reach an optimal customization of setups and structures with respect to abilities and skills that have been detected. This can be achieved by three operational and design approaches: (1) fitting the work to the worker in the design of working conditions, (2) fitting the worker to the work with respect to assignment and qualification, and (3) fitting the workers to each other (this can result only indirectly from organizational and technical work design).

A five-stage system is drawn up for human-oriented work design and assessment. This system has interdependent criteria: The criteria of a lower level must be achieved so that criteria at a higher level can come into play. The criteria are described in increasing order as follows (Luczak et al., 1987):

1. *Damage elimination*. Work must be tolerable and free of damage, and long-term effects must be considered: that is, work must be tolerable and free of damage not only when done once or for only a short period of time but must fulfill these criteria when done multiple times during the course of an entire professional career. Here work time, work intensity, and environmental conditions are particularly significant.

2. *Feasibility*. Work tasks, in particular tool handling, must be feasible. Human tasks regarding individual abilities and skills must not lead to one-sided strain. For this purpose, limits are generally established by human biomechanics or mental capacity.

3. *Reasonability*. The question concerning *reasonability* is a personal one, which can only be answered by the individual worker. Personal experience, is, however, related to the cultural environment and previous know-how. According to this criterion, reasonable work should leave room for maneuver for the individual with respect to work task design and the work environment.

4. *Satisfaction*. Work should be individually satisfying and personality promoting. This can be achieved by the inclusion of psychological know-how and knowledge of the cultural environment in the work design. Acknowledgment, motivation, reward, and superior leadership behavior are all important here.

5. *Social compatibility*. Social compatibility means that employees are involved in the work design based on a cooperative organization. Task-oriented work structures achieve this criterion in particular when employees are involved in the work design process.

2.2 Humanization and Rationalization

Human-oriented work design follows humanization and rationalization objectives. *Rationalization* describes the substitution of inherited procedures with more practical and improved procedures. *Humanization* refers to the design of the work based on human well-being. Work design follows the goals of both humanization and rationalization. For this it is necessary to find compatibility conditions for both objectives, which in practice often position themselves in a difficult manner (Bullinger and Braun, 2001).

In recent years, under the premise of humanization, enterprises have increased their efforts to accommodate the work organization to the growing demands of individuals for larger task variety, self-realization, and self-affirmation. The working person should there by be better motivated and ultimately made more satisfied by his or her work. As a result, enterprises expect better product quality together with a decline in fluctuation and absentism. In conjunction with this, enterprises ultimately expect a better level of quality in their business processes.

2.3 Basic Strategies of Human-Oriented Work Design

Focusing on the working person in the design and assessment of a work system, the following work description criteria and assessment levels of human work can be used:

- *Feasibility*: anthropometrical, psychophysical, and biomechanical threshold values for a short loading time
- *Achievability*: physiological and medical threshold values for a long loading time
- *Reasonability*: sociological, specific group, and individual threshold values for a long loading time
- *Satisfaction*: individual social–psychological threshold values for long and short validity periods.

These explanations make clear that the first two criteria, feasibility and achievability, are achieved by measures of engineering; the other two, reasonability and satisfaction, are achieved by social science knowledge. However, it should not be concluded that the two approaches can be separated.

When considering a manual assembly operation, feasibility limits arise for the worker which result from the motion speed and motion accuracy required. Figure 4 illustrates this situation. At the point where the resulting work task's stress parameters does not lead to a feasible work situation, automation has to take place. Where work is achievable but not tolerable, the work content must be redesigned (i.e., the work must be restructured). Ergonomically optimal conditions are aimed for where tolerable combinations exist.

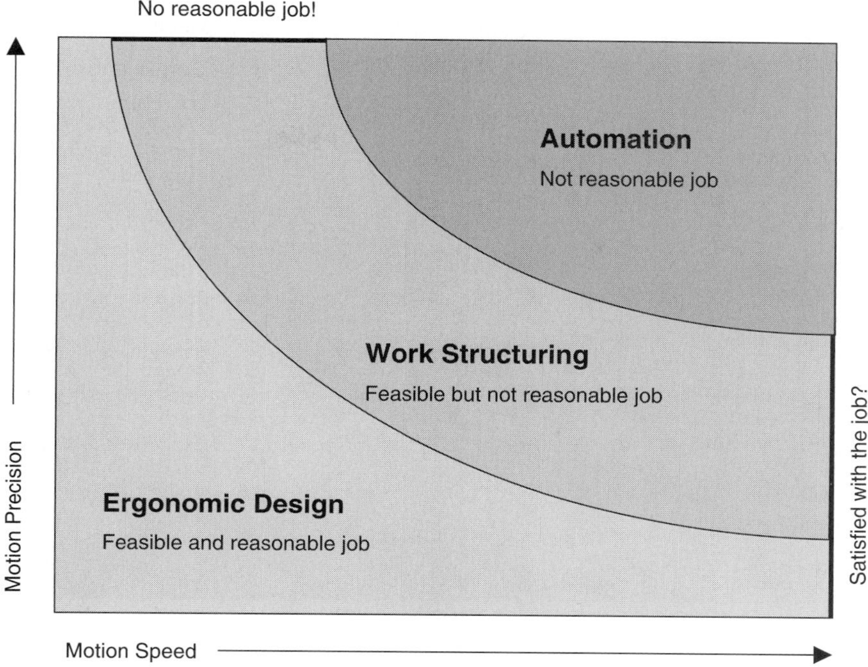

Figure 4 Dimensions of work design strategies.

The three work design approaches differ in the areas of humanization and rationalization. The border area, in which work is in fact in principle reasonable but not achievable, comprises most of the problems. These strategies hold true for informational work as well.

2.3.1 Ergonomics

Ergonomics aims to prevent human impairment, especially by excluding all influences that restrict efficiency or cause physical impairment. So far, good results regarding this task have been achieved. In particular, anthropometric *workplace design* methods are so well engineered that it is surprising in practice if one consistently finds workplaces where important measurement questions remain unconsidered. Computer-aided human models, among others, are installed for workplace design. The human models include anthropometric and biomechanical modules and databanks for comfort estimation (see Figure 5).

A further task is *product ergonomics*, which deals with ergonomic product design with respect to the task to be performed. This is particularly essential for products used daily and when using handheld tools, to allow for optimal performance. Technical function-oriented design should not be done for its own sake. The technically optimal product ultimately is not of much help if people are not able to accomplish what it is possible to achieve (Bullinger, 1997). For this reason, ergonomic work design always advances work efficiency.

Due to multiple requirements, ergonomic work design has becoming increasingly complex. This complexity can be accomplished methodically only by using an integrated procedure (see Figure 6). Integration takes place in three dimensions. In an integration of requirements, all of the relative design influence factors and requirements are registered. In terms of methodical integration, it is essential to define and combine established procedures and methods during the design phases. The operation and flow of information for a design project will ultimately be adjusted in line with organizational integration.

2.3.2 Work Structuring

Work structuring involves the organization of work as well as its requirements so that a preferable gain or increase in work performance is in accordance with the ambitions, and aims of capabilities the individual (Bullinger and Braun, 2001). The interesting thing about this definition is the use of the terms *performance, work performance, ambition*, and *aims*. According to this, the goals of work structuring allow for different interpretations that are dependent on the objectives concerning humanization and rationalization. The following goals pertain to this fact: (1) solving economic problems (such as insufficient flexibility, poor production activity, lack of quality), (2) solving personnel problems (such as high numbers of fluctuation, lack of work morale, signs of dissatisfaction), and (3) technical system redesign and operational organization.

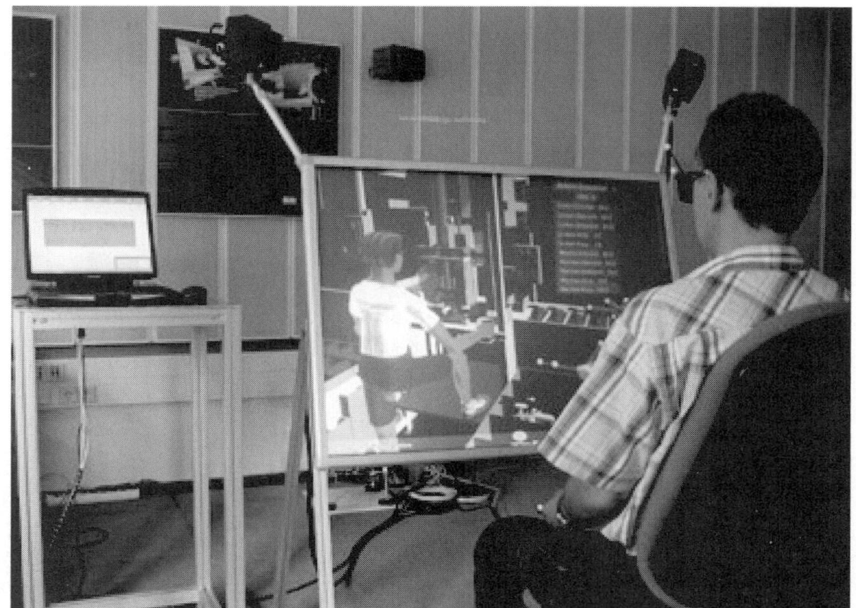

Figure 5 Anthropometric workplace design using the virtual human model ANTHROPOS on a Picasso-three-dimensional-projection system. (Courtesy of Fraunhofer IAO.)

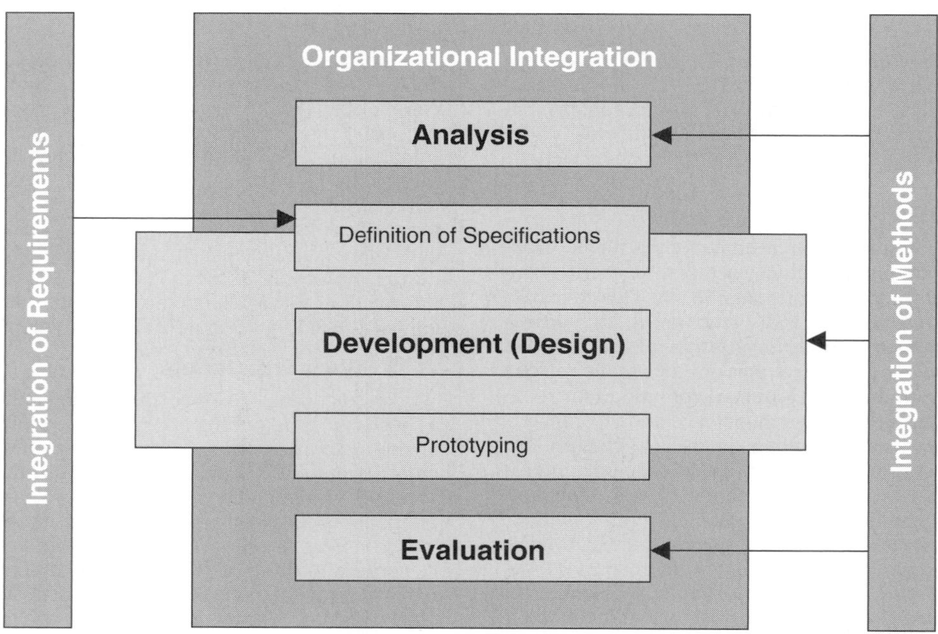

Figure 6 Aspects of integrated ergonomic work design.

At first sight it appears as if economic and human problems are handled in the same way. An exact analysis of human-oriented problems shows, however, that the removal of these problems also enhances performance. For this reason, human-oriented design measures form the basis for economic motives. This does not stop humanization but challenges work design, making suggestions for human-oriented designs that resist reviews of economic efficiency.

2.3.3 Automation

The level of automation is measured by the number of subfunctions performed by humans and machines. Automation aims to shift tasks from humans to machines, thus causing a significant increase in productivity by multiplying human strength. This fact cannot be denied in any discussion of the consequences of automation. However, it accounts for the increasing need to work on solving automation-related problems.

Finding an automation strategy that is acceptable under human and economic aspects is very difficult. The problem becomes clear once it is considered that for economic and technical reasons, automation often takes place such that the human workforce can be omitted entirely, or at most, there is one task, and a few subfunctions, remaining for the human. In many cases only monotonous tasks remain for the workers, since machines may carry out superior work with high speed and precision. If the work is done in a strained work environment, humans expose themselves to health hazards.

Hence, for acceptable work environments, automation considerations must be reinforced by associated work structure considerations. Thus, planning should emphasize the work structure methods where automation is an alternative solution to the problem. This is the only possible way to avoid nonqualified tasks remaining in highly mechanized structures.

Another fundamental requirement is to improve the automation technique in such a way that working humans are relieved of specific tasks, especially when working in a strained environment. Automated handling machines are an example of this. Appropriate use of automation techniques favors the following humanization effects (Bullinger and Braun, 2001): (1) the disappearance of monotonous tasks; (2) the disappearance of heavy physical strain, resulting from unfavorable body position and exertion (e.g., when lifting heavy loads); (3) the disappearance of unfavorable environmental influences (e.g., resulting from heat, filth, and noise); and (4) a decline in accident risk.

2.4 Work System Design

The systemic approach was introduced in Section 1.3.1. To comprehend and design complex systems, the use of systemic models has proven valuable (Spath and Dill, 2002). Basic principles of systemic work design relating to complex production systems are discussed next.

2.4.1 The Work System and Its Elements

The human, the workplace, the work environment, and the work organization shape the production process. They are linked with one another to make up the work system. This systemic model is able to provide systematization for measures of analysis as well as planning, enforcing, and reacting. It can be applied to an individual workplace (first-order work system) as well as a network enterprise (nth-order work system). The work system and its single elements as well as respective design methods are presented later (see Figure 7) as are systemic design approaches.

Fundamental work system elements include the following:

1. A *work task* is an assignment for humans to perform a job, which serves for the achievement of objectives. It illustrates the purpose of the work system. The work task is often referred to as the *targeted work result*.
2. *Tools* include equipment, machines, organizational tools, and so on, which are in any way involved in a work system to accomplish the work task. From a systemic point of view, tools are inactive elements.
3. The *workplace* is a spatial area where one or more people are designated to accomplish a task.
4. Physical, chemical, biological, social, and cultural conditions are identified by the *work environment* which surrounds one's workplace. These conditions influence the system behavior and feature elements.
5. The *human* is the active element of the work system. It is only the human who is able to become active and bring other system elements into action.

Traditional work system design methods are focused primarily on isolated system elements. The previous remarks make it clear, however, that only integrative work design methods in a systemic context can lead to human-oriented and economically effective work systems.

2.4.2 Goals and Methods of Design

Work system design aims at optimal coaction of humans, tools, and work tasks. This goal is to be accomplished with a consideration of the working person's ability as well as his or her individual and social needs. By minimizing impairments, human-oriented work design helps to promote the worker's capability and health continuously. Furthermore, to improve work system effectiveness, work design has the goal of increasing efficiency and reliability by optimizing human–machine interfaces and by influencing work behavior.

In the following, starting with the system model, basic methods for work design are presented. First, isolated work system elements are considered. Subsequently, information for the design of the complete system is given. The interaction of system elements is thus taken into consideration: A change in one element has an influence on all the other elements. Closer

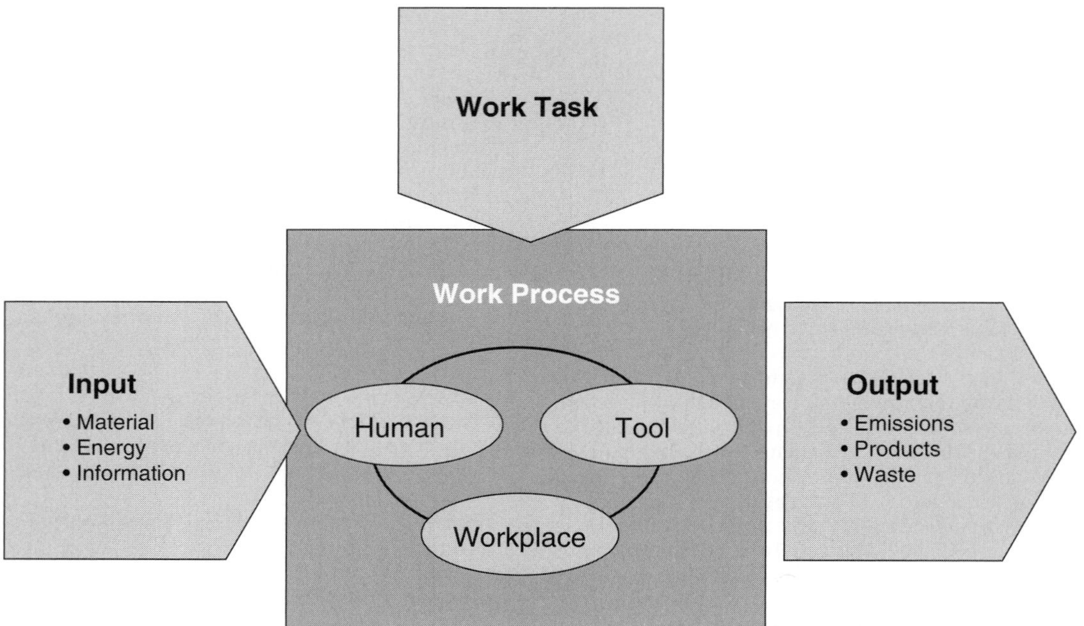

Figure 7 Work system and its elements.

consideration of the methods presented is to be found in the corresponding literature (Helander, 1995; Salvendy, 1997; Luczak, 1998).

2.4.3 The Working Human

The human worker, the active element of the work system, holds particular importance. It is assumed that a long-term effective work system is always a human-oriented system. To design a human-oriented system, the basic characteristics of humans must be identified. These are presented next. The physical and mental dimensions are differentiated here, although the two are closely related.

Stress and Strain The simplified stress–strain model presented in Figure 8 is an appropriate model to use to describe stress (workload), individual capacities, and strain of workers. The term *stress* is defined as all external demands on humans resulting from the workplace, the work process, and all environmental influences. The term *strain* is defined as the reaction of the organism to stress, which is influenced by individual human characteristics. Individual capacity is the factor that combines stress with strain. Human physical and mental capacities are not a constant variable but are subject to change.

The total amount of human stress at work results from the level of stress as well as from the duration of the stress. Human stress is divided into stress that is quantatively and non-quantatively measurable. Quantatively measurable stress can be estimated using physical methods; non-quantatively measurable stress can often only be documented descriptively.

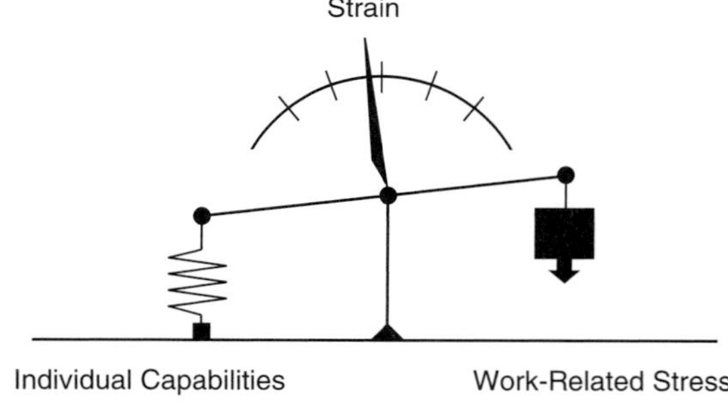

Figure 8 Relationship between stress and strain.

Direct measurement of strain is not possible since every stress can result in different strain for different people. Nevertheless, the analysis and quantification of strain is important in order that (1) work tolerability can be evaluated, (2) capacity limits can be determined, (3) critical strain reactions can be avoided during work design, and (4) physiologically adequate design of rest time is possible. Strains can be measured indirectly by physiological parameters (e.g., frequency of heartbeat, hormone concentration in blood), performance analysis, and subjective techniques (e.g., standardized questionnaires). The goal of human-oriented work design is to create an adequate balance between strain overload and underload.

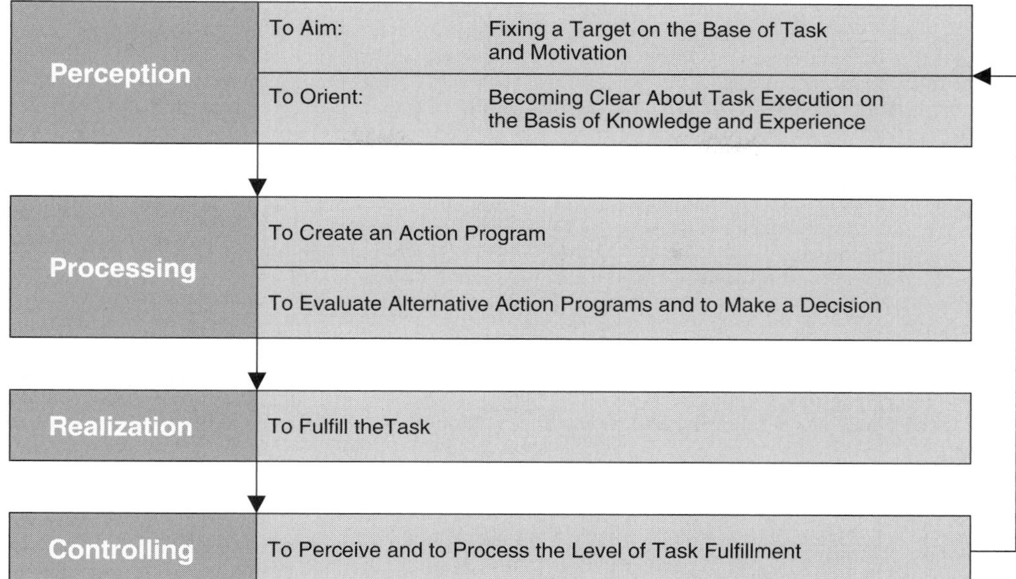

Figure 9 Mental process of action regulation. (Data from Hacker, 1998.)

Capability and Motivation Physiological and psychological capacities affect human capability (Luczak, 1998). Human capability fluctuates interindividually and intraindividually, increasing with training and declining with fatigue. An appropriate recreation phase design maintains ability during work (Spath et al., 2003). Human motivation is no less relevant for efficient production. Even well-educated employees bring little added value when they are not willing to bring performance (i.e., they are unmotivated) (see Section 1.4).

Numerous models exist that describe capability and motivation. At this time, the psychological concept of *action regulation theory* should be mentioned. This theory relates predominantly to work tasks, specifically to observable and conscious processes. Unconscious mental processes (e.g., formation of opinion) remain unconsidered. Action regulation assumes that human action is goal-oriented, corresponds to external matters, involves social connections, and shows process characteristics. Consequently, the theory describes ongoing processes of action, starting with aims and continuing to the point of reaching the goal, and presumes an interaction of mental processes and observable activity. Hereby, two regulation procedures are differentiated (Hacker, 1998):

1. *Stimulus regulation*: It is determined if an action is done. Individual motives, attitudes, and preferences are the basis for these procedures.
2. *Execution regulation*: It is determined how an action is done. Individual subgoals and mean–methods choice and execution control are the basis for these procedures.

Figure 9 presents the structure of the mental process. According to this model, human action is a control circuit that is influenced significantly by motivation, knowledge, and experience.

Action regulation theory assumes that task requirements and stress are independent of one another. Psychological requirements should give workers the opportunity to decide on goals and methods independently. The most important influencing factors are task variety, communication, and cooperation. Mental stress at the interface between the individual and the organization is defined as a restraint. Regulation restraints describe discrepancies between working conditions and goal attainment as well as excessive demands, including time pressure and monotonous jobs.

This theory follows the simple realization that people who perform their work as a result of their own interests and convictions are considerably more productive than those who only feel subject to work. It is clear that *motivation*, the sum of action, behavior, and behavioral tendencies, is a deciding factor in the relationship between humans and work. Contrary to the human biological stimulus, individual motives are learned and incorporated into the socialization process. The motivational encouragement of employees has special importance for work design targeting productivity and humanity.

Motivational theories can always be only partial theories since they explain only parts of human behavior within the stimulus–reaction scheme. As a result of their importance, the theories of Maslow (1987) and Hertzberg et al. (1967) are mentioned as examples. Maslow (1987) establishes a hierarchy of individual requirements. A level of requirement becomes relevant in terms of motivation only once the preceding level of requirement is fulfilled. Thus, according to this theory, it makes little sense, for instance, to motivate an employee to self-realization using a general offer of further training when the security of his or her workplace (e.g., for economic reasons) has not been dealt with.

Hertzberg et al. (1997) differentiate between hygiene factors, whose noncompliance leads to work dissatisfaction, and motivators, whose compliance leads to work satisfaction and motivation. Thus, poor internal company politics lead to intense work dissatisfaction, whereas good internal politics prevent this from happening. This does not, however, lead conversely to work satisfaction: Acknowledgment from superiors has strong motivating effects, but lack of acknowledgment does not necessarily lead to work dissatisfaction.

For business practices, the following design concepts are derived from psychological theories:

1. *Reduction of time constraints.* Using buffer banks between individual workstations, employees have the opportunity to work detached from fixed-time work cycles.
2. *Job rotation.* The work content of individual jobs does not change, but employees exchange their workplaces systematically.
3. *Job enlargement.* Job content is increased. Employees are given more similar assignments at the same qualification level, leading to longer work cycles.
4. *Job enrichment.* The job content is changed so that individual employees have greater task variety, resulting in higher qualification requirements.
5. *Partially autonomous team work.* The work team is given a job assignment. This assignment is separated into subtasks within the team. The team can organize the work itself within certain limits (i.e., time targets, technically marginal conditions). This type of work structuring offers a good chance for individual work design. However, it holds the risk of social conflicts arising within the group.

Consequences of an Inappropriate Workload

Fatigue, monotony, mental saturation, reduced vigilance, and stress are closely related to the problem of strain. These categories describe both a procedural happening and an internal human condition.

Fatigue describes a protecting stoppage of motivation resulting from a continuous task performed over the course of hours up to a day. Considering the nature of the strain, one must differentiate between physical and psychological fatigue. Physical fatigue is attributed to a displacement in the physiological–chemical equilibrium of an organism. Psychological fatigue is an action regulation impairment. It is an aftereffect of psychologically demanding tasks which are characterized by information reception and processing. Monitoring, inspection, and control tasks, which are found especially in process control, are considered to be primarily psychologically demanding tasks. Alertness adaptation and focus on defined task content are typically necessary for such tasks. As a general rule, cognitively overstraining conditions bring about sporadic functional ability decreases in the fields of perception, memory, and thinking. Fatigue-caused capability impairments can be eliminated temporarily by job rotation, environmental influences, or stimulants. They are eliminated completely by sufficient sleep (Richter and Hacker, 1998).

Monotony is similar to a state of fatigue and is generated by a lack of stimulation or by conditions with minimal changes in stimulus structure. Symptoms of monotony include feelings of fatigue, sleepiness, listlessness, and attention decline. The decrease in motivation and reaction ability associated with monotony is reflected in unsteady and decreasing performance. Monotony arises when the fulfillment of a job does not allow a complete solution but at the same time does not offer enough possibilities for a mental analysis of the task (ISO, 1991).

According to ISO (1991), *mental saturation* is a condition of nervousness, restlessness, and intensely affective decline resulting from an undemanding, repetitive task or situation. The concerned person perceives his or her job as being senseless; listlessness and irritation result. Continuation of the task is carried out reluctantly. In the long run, psychological saturation leads to the employee's "internal termination" by denying his or her own initiative and operational readiness. These sentiments are triggered by monotonous tasks that are below the employee's level of qualification. Permanent problems, unfulfilled needs, and unreached personal goals can also lead to mental saturation.

Diminishing vigilance (also called *hypovigilance*) is a condition of reduced mental activation that results from monitoring tasks that lack diversification. It is the consequence of qualitatively undemanding tasks in a highly passive line of work, a low amount of environmental stimulus, and the lack of environmental diversification that results from concentrating on few inputs. The mental tension required for the deliberate balance of functional impairment with reduced vigilance provides an additional source of mental exhaustion. Hypovigilance systems that recognize such conditions and support people accordingly (e.g., by targeted activation) are being researched and developed. Monotony, mental saturation, and diminishing vigilance, which all resemble fatigue, have in common that they result from an "underdemand" on human capability; they can be removed by expanding task variety (Spath et al., 2003).

Stress is a condition of the organism that develops when a person has recognized that his or her well-being or integrity is in danger and that he or she must use all available energy for his or her self-protection and self-defense (Cofer and Appley, 1964). All situations that are experienced as unpleasant or threatening can trigger stress. At the same time, in addition to intensity and exposure time, the amount of stress depends on experiences that the person has had in similar situations, their disposition, and the situational factors.

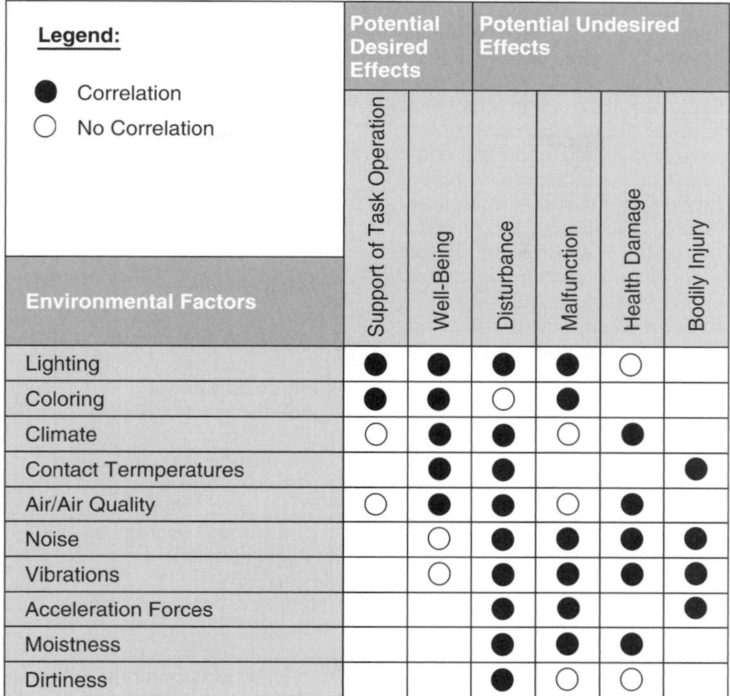

Legend: ● Correlation ○ No Correlation

Environmental Factors	Potential Desired Effects		Potential Undesired Effects			
	Support of Task Operation	Well-Being	Disturbance	Malfunction	Health Damage	Bodily Injury
Lighting	●	●	●	●	○	
Coloring	●	●	○	●		
Climate	○	●	●	○	●	
Contact Termperatures		●	●			●
Air/Air Quality	○	●	●	○	●	
Noise		○	●	●	●	●
Vibrations		○	●	●	●	●
Acceleration Forces			●	●		●
Moistness			●	●	●	
Dirtiness			●	○	○	

Figure 10 Environmental factors at the workplace and their effects. (From Bullinger, 1994.)

The human organism is not able to support long-term stress. The physiological mechanism of stress reaction breaks down at the point when the stress factor cannot be removed by means of coping or avoidance. Long, continuous amounts of work-induced stress lead to excessively increased levels of alertness which do not subside sufficiently at the end of the workday. The consequences are sleep disorders, impulse liability, and internal unrest. Long-term memory is weakened and muscle activation, including speech, becomes aggravated. The responsiveness of perception is reduced. Consequences are erroneous actions, erroneous estimation, and anomie.

2.4.4 The Working Environment

Relevant factors of ergonomic working environment design are lighting, noise, mechanical vibrations, climate, harmful substances, and radiation. The sphere of influence for these factors is found most notably in the direct environment of the workplace as well as in the work tools used. For example, a decrease in machine noise is primarily a constructive problem. Given that the machine is, however, located at the workplace, its sound emission also influences the quality of the workplace.

Some environmental factors are used purposefully in ergonomic work design (see Figure 10); others have undesirable effects. The design goal is to reduce intensity, exposure time, and impact frequency so as to avoid excessive strain. At the same time, one should note that eliminating all environmental influences can have disadvantageous consequences. As an example, psychic problems arise for inhabitants as a result of extreme isolation of apartments protected from outside noise. Completely eliminating noise and vibration emission while using an electrical razor irritates the user, leading him to reject the product. In general, environmental influences are not to be eliminated but to be optimized.

The environmental factors of climate, noise, and lighting are identified most frequently as being problematic during production. These factors are addressed below in more detail.

Lighting Most human sensory perception happens over the visual canal. Appropriate lighting is required for visual information intake. By using appropriate lighting, performance and work safety can be heightened and visual strain can be reduced. Visual perception is thus dependent on light intensity. Therefore, light intensity must be adjusted for the visual task. As a consequence of the aging process, older people require more light to be able to carry out visual tasks precisely. It is to necessary to ensure that light intensity always is sufficiently high.

Moreover, a visual object is recognizable only if it has at least minimum contrast (i.e., a luminous density difference within its environment). The ability to perceive contrasts depends on an object's size, luminous density, perception time, and level of adaptation. The higher the level of lighting, the greater the contrast must be. If the contrast is too strong, glare arises. Direct glare results from glancing directly at a luminous source. It is often found that the absolute light density value is too high (e.g., when looking at the sun). Reflex glare is a result of luminous source reflections on reflective surfaces. Light density differences, which are too large in the visual field (i.e., intense contrasts) lead to relative glare. Furthermore, relative glare causes eyestrain, which results from adaptation.

Contrast depends on the surface condition of the object observed, the angle of light, and the distribution of light density. Good contrast reproduction is achieved by matte material surfaces and by lighting arranged laterally above the workplace. The direction of light at workplaces with visual display units is important so that reflections do not occur on the visual displays.

In practice, the following rules for lighting design are helpful:

- Adequate level of light intensity
- Harmonious distribution of light intensity
- Glare limitation
- High-contrast reproduction
- Proper direction of light
- Accurate amount of shade
- Proper lighting color and appropriate color reproduction
- High degree of energy efficiency

Noise Within the variety of environmental factors acting as strain variables for the working human, noise is the most influential factor. In the last few decades, noise has become a severe problem, due to the high numbers of compensable occupational illnesses resulting from long-lasting noise exposure and emerging noise-induced hearing loss.

A progressive loss of hearing results from lasting noise exposure or in specific cases from short exposure to high levels of noise. Mental reactions such as disturbance and annoyance can be observed at low noise pressure levels. These reactions depend mainly on the position of the person to the origin of noise and his or her momentary disposition (e.g., mood, tension). At approximately 65 dB, a vegetative nervous system reaction is established, such as a change in breathing rate. An irreversible effect of hearing loss is possible when noise exceeds 85 dB. Noise-induced hearing loss makes it difficult to sense acoustic signals and speech. This can lead to higher accident risk. Sound emissions and their effect on people are directly related. The effect of noise on people is assessed by the noise rating level of an eight-hour work shift.

The goal of ergonomic work design is to prevent the development of noise entirely. Basic *primary* measures are preferred to reduce noise. These measures prevent noise from developing (e.g., by using electrical motors instead of pneumatics). Since these measures often have constructive machine and equipment demands, they are considered particularly during production planning and related machine construction; in ongoing production, primary measures are generally very expensive. In this case, *secondary* measures, which prevent sound spread, are aimed for. For instance, noisy machines can be grouped in a separate space, or machines can be enclosed. *Tertiary* measures such as earplugs should be used when all technically possible and economically justifiable efforts concerning noise reduction do not lead to a noise level below 85 dB.

Climate Climate is an important environmental factor at the workplace. Its importance is a result of multiple interactions with the human organism. In fact, the number of workplaces under extreme climatic conditions declines (e.g., foundry). On the other hand, the higher demands in the field of workplace design for a comfortable climate call for ergonomic recommendations and rules.

Climate is not a consistent dimension; it is, rather, a generic term, influenced by air temperature, humidity, air movement, and thermal radiation. Instead of dealing with exact definitions, there is an emphasis on practical design directions. The various climate factors are integrated into one variable by a cumulative climate indicator. The principles of the cumulative climate indicator are based on combinations of four main variables: air temperature, humidity, thermal radiation, and air movement speed. These variables bring about a similarly subjective climatic sensation to which climates are to be compared.

In principle, the human body tries to establish equilibrium between body heat production and external climatic influences. The goal of this regulation is to maintain a normal temperature of approximately 37°C and connect it with a certain level of comfort. Within low environmental temperatures, the lack of warmth is balanced by an increase in body temperature. Within a higher environmental temperature, the body tries to eliminate excess warmth in the environment by transpiration.

In the production environment *temperatures* between −50°C and +50°C are most common [workplaces with more extreme thermal radiation (e.g., as appears in foundry) are not considered]. This climate variety underscores the need for adequate climate design. For climate assessment, subjective influence variables beyond those relating to environmental climate are also considered. Among these are clothing and work intensity as well as the condition and constitution of each person. At the assessment of climate comfort it is to consider, individual differences in climate perception can arise: For example, in a study of easy office work performed by normally clothed people ($n = 1296$), a majority of those surveyed felt a temperature of approximately 21°C to be neither too cold nor too warm. This is also considered to be a state of neutral temperature. The interesting result of the survey was that a scarce 20% of those surveyed found this temperature to be too warm, and approximately 20% found it to be too cold. Consequently, a significant portion of those surveyed were dissatisfied with

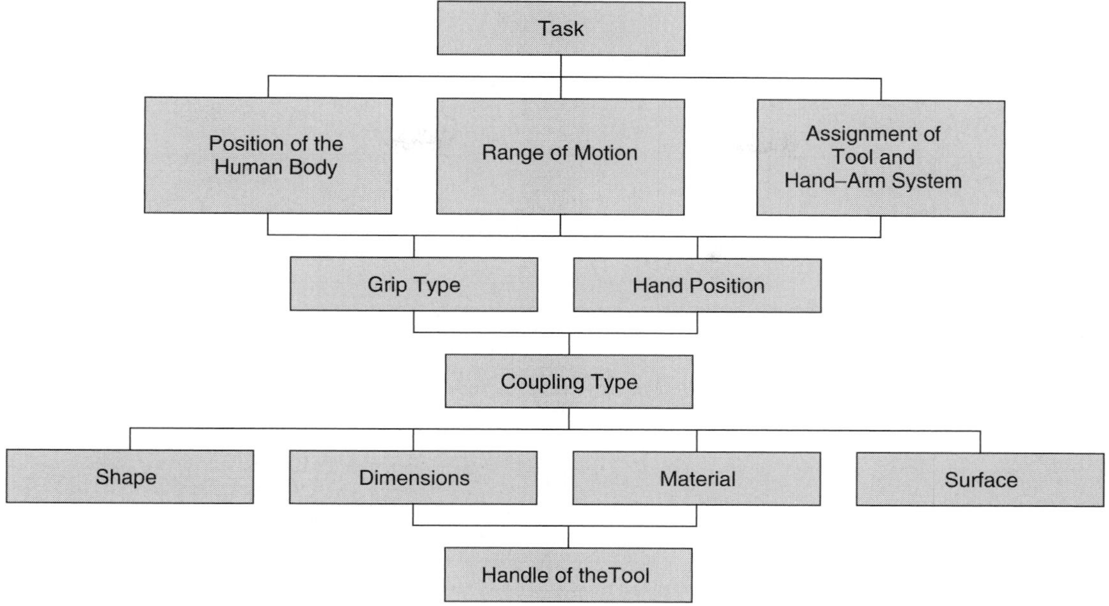

Figure 11 Algorithm for the design of work tools. (From Bullinger, 1994.)

the climate (Fanger, 1972). Thus, climate sensation is very subjective; high average summer temperatures are felt to be more pleasant than those in winter. Ultimately, claims also hold for climate design, which is to create damage-free, better executable, and in general, achievable conditions. In the ideal situation, however, a comfortable climate is reached in which the body heat balance turns out to be neutral.

2.4.5 Tools

In principle, it is differentiated between the hand side and the work side of a tool, ergonomics having the most influence on the hand-side design of work. A deductive approach (from general to specific) works best when designing an ergonomic handle. Inductive approaches that start with decisions about shape are usually doomed to fail or require extensive reworking, which can be time consuming and expensive. Ergonomic tools are created for the human user, taking individual abilities and skills into account, helping to prevent one-sided stress during work and increasing efficiency.

The relevant measured variables on the hand and work sides of a tool affect:

- Body position, posture, and range of motion
- Hand position, grip type, and connection type
- Handle shape, dimensions, material, and texture
- Function direction and force direction
- Accuracy, speed, and resistance

A systematic approach is essential to ensure that these variables are brought into the design process in the right amount and in the right order. It starts with a focus on the task to be performed (i.e., the examination of working conditions). Only after further detailed analysis do design parameters (i.e., shape, dimensions, material, and texture) actually become part of the process (see Figure 11). In this way, tool handle design becomes a creative and analytical process instead of a purely technical and aesthetic task. The procedure conforms to classical scientific engineering methods, which may be extended with usability engineering practices. More information may be found in Chapter 49.

2.4.6 Workplace

The basic requirement for the human and economic application of personnel is an ergonomic workplace. Workplaces arranged inadequately ergonomically not only affect occupational safety and possibly employees' health, but in addition, limit effective application. Due to the fact that skeletal, muscle, and connective tissue diseases cause the longest illness-related absences from work, it is important that special attention be paid to workplace design with respect to dimensions and forces.

The economic advantages of workstations designed ergonomically correctly can be observed both directly and indirectly. On the one hand, processing time is decreased by favorable workstations; on the other hand, absences are reduced. Both human body dimensions and physical forces are at the core of ergonomic work design. Considering human body dimensions is especially important since it is here that minimal values

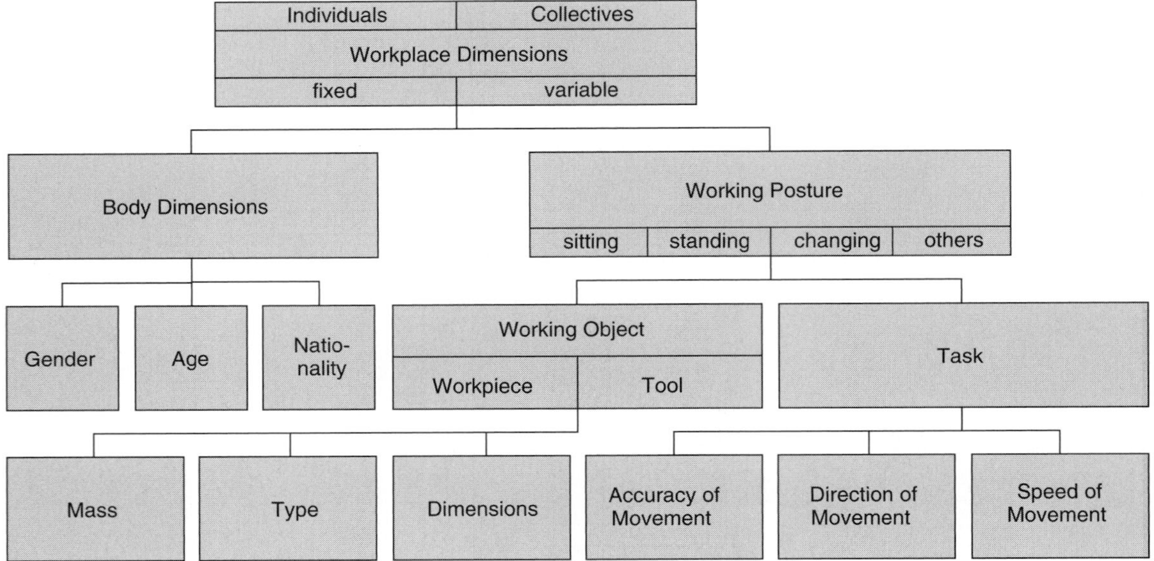

Figure 12 Factors of workplace measure specification. (From Bullinger, 1994.)

are not always crucial, unlike in the case of physical strength.

Anthropometry Anthropometry is the study of the various dimensions of the human body as well as their exact measurement. Since everyone has different anthropometric values, the only way to create an ergonomically correct and efficient working environment is to adjust workplaces to fit individuals. The most important criteria for determining workplace dimensions are depicted in Figure 12.

Physical measurements vary from person to person. Defining an "average human" does not make sense since designs often have to accommodate the smallest or largest physical dimensions among their users. If designs were based on average dimensions, half of the population would be worried about hitting their heads on doorframes while the other half would fear not being able to reach an emergency power switch in the event of an accident. The definition of minimum and maximum for broad subpopulations has proven to be a more workable alternative, leading to the adoption and use of the term *percentile*.

The customary limits for the adjustment range of an object being adapted to the human body are the 5th and 95th percentiles. Since the variance of the residual group of extremes is overproportionally large, 95% of the users can be addressed by only approximately one-fourth of the entire variation range. However, when using anthropometrical data, it should be kept in mind that these data are not valid permanently. The current acceleration phenomenon results in the fact that physical dimensions are increasing slowly and continually. For this reason, several projects examining actual anthropometric dimensions were carried out worldwide from 1999 to 2003. A project named CAESAR, for which representative anthropometric data were collected both in Europe and the United States (CAESAR, 2004), is an example.

The action space for body parts is confined by the anatomically maximal rotary area, the displacement area and bent angles. Since details about optimal, not maximal areas and angles are required for workplace design, this maximal action space plays a secondary role in ergonomics. There are neutral positions between the extreme values in which muscle activity and tendonal and ligamental strain are minimal. Muscle exhaustion is at its lowest level in these positions, which are generally considered to be subjectively comfortable but do not have to inevitably be in the respective centralized position.

The same holds true for visual space. The work process must be controlled and checked visually for nearly all assignments. Besides body position and body posture of workplace action space extremities, visual space parameters are important for design.

Dimensioning of Workplaces The nature of the work task influences primarily the choice of working height and the decision as to whether to establish a sitting, standing, or sitting–standing workplace. In principle, minor motional limitations exist at a standing workplace. For this reason, great force evolvement is possible. The sitting workplace is advantageous for precisional work and reduces posture work. From a physiological standpoint, a sitting workplace is preferred to a standing workplace since work chairs corresponding to ergonomic requirements can reduce the number of continuous muscle contractions. Standing does not, in fact, cause too much muscle

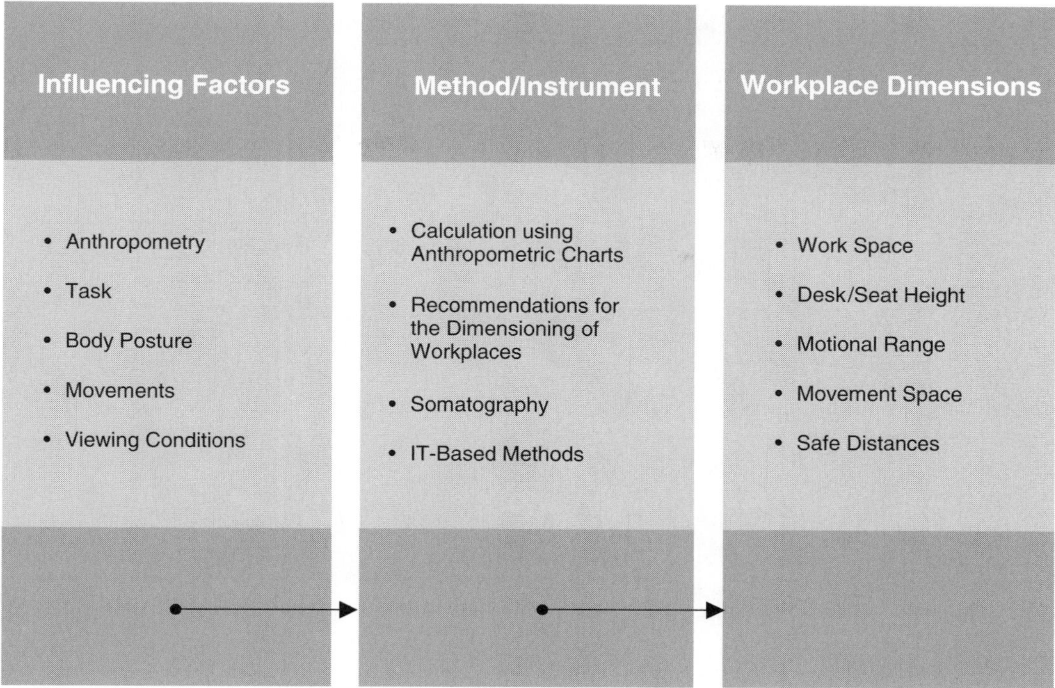

Influencing Factors
Method/Instrument
Workplace Dimensions
• Anthropometry
• Task
• Body Posture
• Movements
• Viewing Conditions
• Calculation using Anthropometric Charts
• Recommendations for the Dimensioning of Workplaces
• Somatography
• IT-Based Methods
• Work Space
• Desk/Seat Height
• Motional Range
• Movement Space
• Safe Distances

Figure 13 Usual methods for workplace design.

activity but does strain the ankles and attached tendons and ligaments. This leads to increased blood pressure in the legs.

It should be noted, however, that there is still no comfortable body posture, nor any type of sitting posture, that can be used for a long period of time. Therefore, a workplace should be set up so that it is possible for the employee to alternate between sitting and standing tasks, thus having balanced movement, which is more favorable for the human body than a forced motionless posture. This fact is expressed by the proverb: "The most ergonomic (sitting) position is the next one." Forced posture results from:

- Not enough free space at the workplace (i.e., foot space, leg space)
- Unfavorable working height (i.e., height of chair, height of table)
- Unfavorable position of work objects (i.e., too large a frontal or side displacement, too high or too low a displacement)
- Arrangement of body displacement from work tools and manual controls
- Unfavorable position of displays
- Limited freedom of movement (e.g., projecting components of machinery)

A workplace adjusted to human dimensions is necessary to provide for natural body posture and motion sequences when carrying out an assignment. Taking into consideration the multiple influencing factors with respect to dimensioning of workplaces, the particular design method takes on a different meaning. Some methods are appropriate for a quick qualitative examination of workplaces, whereas others are complex and hence produce detailed results. An overview of methods and tools for workplace design is presented in Figure 13.

Here, only the two most important methods are presented. More detailed descriptions may be found in Chapter 22.

1. *Calculation using body measurement charts.* Using body measurement charts, specific workstation measurements are defined pertaining to the 5th percentile (e.g., hand and foot operating space), whereas others pertain to the 95th percentile (open space for feet and knees). Briefly, it is necessary that a small person be able to reach all relevant elements and that a large person not hit himself or herself on anything at that same workstation.

2. *IT-based methods.* The common use of computer-aided design (CAD) systems in the area of work design opens up new possibilities for heightened efficiency, especially for workplace design. In particular, variant design and simple and quick changes of particular components of a work system improve the ability of proper ergonomic design for the entire

Figure 14 IT-based workplace design using VirtualANTHROPOS. (Courtesy of Fraunhofer IAO.)

work system. Depending on the nature and capacity of available hardware and software, human models with different degrees of detail and simulation possibilities are available. Two-dimensional models, corresponding in principle to digital templates, are standard. Three-dimensional human models such as JACK (Badler, 1997; Allbeck and Badler, 2002) or ANTHROPOS (Lippmann, 1997) make possible a detailed ergonomic analysis of the entire workplace system. Future tools will integrate the functions of several systems: It is already possible to analyze and optimize an entire work system in a three-dimensional stereo projection room with the help of VirtualANTHROPOS (see Figure 14).

2.4.7 Manual Handling of Loads

Despite the use of handling technology, heavy loads have to been moved often during industrial production. When handling loads manually, one hazard is the constant and long-lasting application of strong physical forces. Here, back stress is so intense that abnormal aftereffects are not to be ruled out: Acute limited functional impairment can develop (such as pulled muscles and blockage of vertebral joints while lifting loads). Furthermore, chronic impairment can develop, along with steadily increasing and continuous medical conditions (e.g., deteriorating intervertebral disks, expansion of ligaments, tenosynovitis, and muscle tension). These conditions cause pain and often limit human flexibility. They can lead to durable disablement.

When constant or heavy lifting is avoided, strains are thus effectively reduced. This is especially true for young people (because of the reduced stress capacity of their spines) and for women (because of the lower average stress capacity of their spines compared to those of men). When this is not possible, appropriate measures have to be taken, on the basis of a work analysis and risk assessment, to keep employee health hazards as low as possible. For risk assessment, a variety of proven methods are available. These methods include operational terms and conditions and corresponding normative guidelines. Resulting measures to be taken can be carried out with the aid of technical, organizational, or personal methods. The following recommendations are for the manual handling of loads:

- A supply of appropriate work tools and utilities (lifting belts, lifting platforms, etc.)
- Load carrying and transportation close to the body
- Favorable load lifting with deposition heights between 70 and 110 cm aboveground
- Adequate motional range for load handling
- Alternation between straining and unstraining tasks
- Adequate relaxation time

Employee information about the correct way to lift and carry loads is necessary so that employees can avoid health impairment. Here, information about straight posture of the spine is important, leading to an even amount of stress allocation. Heightened strain on the spine results from the weight of the load; load allocation is uneven whenever the torso is not straight,

Figure 15 Depiction of a typical process control workstation. (Courtesy of BKB Göppingen.)

at which times the danger of intervertebral disk or vertebral damage increases considerably.

2.4.8 Human–Machine Interface

All components of a work system for functional interaction between humans and a technical system are combined under the term *human–machine interface*. Processes to be supervised and controlled by humans generate a multitude of information, such as that on operating status, task conditions, or tool life. Human receptors take in this information both directly and indirectly. This information is then processed in the brain, and in the form of information or an operation, is referred back to the process.

Considered formally, the function block source, sender, and receiver is defined for the transfer of information. Information is transferred to the technical system in the form of an operation by using either actuator components or control elements (i.e., switches, levers, buttons) or by using complex informational input systems (such as speech input systems, graphic charts, and keyboards). The system can communicate with the user in turn through optical, acoustic, or haptic modes of interaction. Today, more and more automatic announcement and multi-/hypermedia systems are in use, due to highly sophisticated language and image-processing systems as well as highly effective data storage.

When designing modes of interaction, the human, along with his or her different individual attributes, is to be considered: Human expectations are attributed to inherited, or in the technical and social environment, acquired behavioral stereotypes, which are dependent on certain population affiliations (e.g., left-handed people). These behavioral stereotypes are introduced as a part of the person's proficiency requirements with the work system. If one wants to use them during the task or at least respond to them, they must be considered during display and manual control design (i.e., the display and manual controls must be designed compatibly).

Compatibility of informational input and output of work tools is accomplished when certain human expectations regarding statistical, spatial, and mapping aspects of dynamic procedures correspond. In the dynamic case, movement compatibility is of importance. A well-known example is that of steering a vehicle. When turning the steering wheel right, a vehicle is expected to turn right; the same is true of the converse. To design informational input and output elements customized to the work task, it is necessary to analyze the respective assignment. The choice, arrangement, and design of input and output media result from this procedure.

In recent years during a period of increasing automation, the trend has been to move away from manual tasks such as those in classical manufacturing, toward control and monitoring tasks, up to now found mainly in process control. Microcontrollers are now built into most machines. For this reason, human–machine interaction is also changing. Classical machine interfaces, which use mainly discrete operating elements, are found only rarely. They have been replaced by display screens directly on the machine as well as in control centers (see Figure 15). As a result of this, ergonomical software considerations [often referred to as *human–computer interaction* (HCI)] are of great and still growing importance in all types of production, whereas classical human–machine interaction is losing ground, remaining in service only for specific interfaces. Thus, the design of information flow based on cognitive considerations becomes a new challenge for human–machine interaction.

If up to now humans and machines were considered to be two independent partners, the human now moves into focus as the central component of the human–machine system, to which the machine is to be adjusted (Oborne and Arnold, 2001). Since the human changes and studies further through interaction with the machine, however, the best possible mode of interaction is also changing constantly. It is here that the human–machine system is considered to be dynamic. "So the goal is to create supportive dynamic environments that enable individuals to work at their safest and most effective levels; not just to design the environment to fit the individual in some static sense" (Oborne and Arnold, 2001).

The central tool used to accomplish this goal is *usability engineering*. Lin et al. (1997) define *usability* as "the ease with which a ... product can be used to perform its designated task by its users at a specific criterion." According to Bevan (1999), four steps are required to design for good usability: (1) understanding and specifying the context of use, which leads to (2) specifying the user and the organizational requirements, which leads to (3) defining product design solutions, which leads to (4) evaluating the design against the requirements.

For monitoring tasks in constant use in process control, a good interaction system design also implies an important safety aspect. Information must be gathered and processed reliably according to its importance. For this purpose, the human being must not be overextended or unchallenged. In the latter case, a decrease in vigilance can result from monotony, so that the user is not able to act quickly enough in an acutely critical situation. The result is the threat of greater damage to the human and the environment. In addition to ergonomic design of modes of interaction, such as the geometrical arrangement of display elements, vigilance management systems can also bring about an important safety contribution by detecting decreasing vigilance and warning the respective user, correspondingly, by maintaining vigilance through appropriate activation.

3 ORGANIZATIONAL DEVELOPMENT

So far, the work system, its elements, and important methods for humanly adapted work design have been introduced, but these methods relate primarily to isolated elements of a work system. For complex production system design, a limitation on individual elements does not suffice. In fact, integrated systemic consideration is required, in line with organizational design. Design principles and measures of humanly adapted and efficient organizational development in production systems, whose conceptual basic principles have already been described, are presented next.

3.1 Design Principles for Production Systems

3.1.1 From Execution- to Object-Oriented Work Content

Organizational production according to the principles of Taylorism leads to a smaller work content with a greater division of work. As a result, employees identify themselves with their work and work results only minimally. Besides these human problems, the strong division of labor also brings about organizational problems which result from the boundary of the interval units. An example of this is the sometimes troublesome coordination of targets and costs.

Group work, embedded in decentralized enterprise structures, can serve to resolve this problem. Groups are established according to the object principle. An example of this is the complete assembly of all attached parts and aggregates for an instrument panel (object) of a car (product). Later in the production procedure, the completely assembled instrument panel is then installed in the body of the car. This procedure generates small, manageable units whose range of work is such that the division of labor is counteracted. In addition, leeway and range for decision by individual employees is broadened and the work processes can be steered independently and coordinated by the group.

The object-oriented concept has become accepted most notably in the form of production and assembly insulars. The group of employees thus has the task of producing and assembling a spectrum of particles entirely and self-dependently. They also take on tasks from which the individual employee is excluded by the performance principles of Taylorism. The additional work content occurs as a result of (1) job enlargement, by taking over preliminary and downstream functions (e.g., material provision, examination of the particles); and (2) job enrichment, by taking over related production and preparation functions (e.g., setup and maintenance of installation, material disposition).

3.1.2 From One- to Multidimensional Qualifications

In the work system, complex work structures, such as production and assembly insulars, require employee methodical skills in addition to technical skills. These are needed since employees are to handle greater work content in terms of the scope of services and efficiency diversity within these production structures. Cooperation within the group also puts new demands on the cooperational and coordinational abilities of the individual employee. It is obvious that the implementation of innovative production structures with conventional, strongly subject-oriented training concepts on a wider basis becomes a problem since a sufficient number of qualified employees are not available right away. Here, an innovative further training concept is required if companies are to provide technically well-trained employees with methodical and social skills. The existing training methods must be changed or supplemented with new training methods so that they can meet the demands.

3.1.3 From Fixed to Flexible Work Times

The shortening of work times, particularly noticeable in industrial countries, calls for the development of more flexible work models so that the work time factor does not become a locational disadvantage for the enterprise. Such disadvantages are, among others, a result of the pressure to employ capital-intensive production systems most optimally in order to remain competitive in the international market. The requirement for work time flexibility is, however, a decoupled arrangement of working and operating time (i.e., more flexible employee assignment during extended production unit operating times).

In recent years, the following main points of work time design in production have been recognized:

- Flexible work time models
- Seasonalization of work time
- Work time differentiation (maintenance of preferably high work time volume for highly qualified employees, of whom not very many can be found in the job market)
- Part-time work
- Specialized, individualized work time models instead of collective work time models

Flexible work time models must be attractive so that employees will accept them; that is, the employees have to understand the advantages that derive from these models.

3.1.4 From Reactive to Preventive Work Design

By means of ergonomic work design, employees should be protected from work-related stress and strain. Traditionally, this task was done in the business field of occupational health and safety (OHS). In the past, OHS focused predominantly on human protection from hazards. Such a reactive process, does not, however, meet the requirements of a truly human-centered work design. Therefore, it has been recognized that the preventive dimension of OHS has to be strengthened (i.e., the protection of human health is achieved through the configuration and elimination of the causes of hazard before damage occurs).

In practice, preventive methods have not yet been implemented to the degree desirable. This is due especially to the fact that possible sources of hazard must be identified in an early stage of production system planning process before the design measures are realized. Otherwise, practical adjustment of ergonomic design measures might become impossible during the course of the project for economic reasons.

3.2 Organization from a Systemic View

3.2.1 Basic Principles

In systemic organization, the principles of system theory are implemented at the company level. Here a company or business is described as a target-oriented economic and sociotechnical system featuring openness and complexity. Taking economic aspects into consideration, the company aims for an efficient use of resources (Probst, 1997). Thus, companies are complex, open systems which are connected with the environment through a multitude of exchange relationships. Here, *complexity* is defined as the grade of intellectual ability to comprehend and control a system. Complexity implies incomplete descriptiveness and marginal predictability. Increasing complexity results in deterministic management strategies losing more and more of their effectiveness. The following errors often occur (Dörner, 2003):

1. *Inadequate target recognition.* An appointed target is often specified insufficiently since only isolated individual aspects are recognized.
2. *Isolated analysis of the situation.* During system design, only well-known sections are highlighted; as a result of missing structural and organizational principles, a meaningful analysis of the information collected is not reached.
3. *Irreversible concentration.* By focusing on single problems, the consequences of actions in other areas of the system or further problems remain unnoticed.
4. *Unnoticed side effects.* The principles of multicausality and multifinality remain unnoticed in monocausal cause–effect thinking.
5. *Tendency of overmodulation.* If timely delays of cause and effect are not understood in complex situations, system overmodulation can result.

3.2.2 Systemic Project Management

Systemic project management connects enterprise strategy and practical process design. Systemic project management takes into account the following basic principles (Ulrich and Probst, 1988):

1. *Systemic thinking.* Only projects planned as workable systems are able to follow dynamic processes through self-adaptation of their internal structure. For this reason, projects and their elements require autonomous decisional and coordinational instances as well as independent monitoring. Projects require an efficient information system to take in and exchange internal and external information.
2. *Self-organization.* Self-organized systems with adaptation abilities can process considerably more information than a hierarchical structure. This is important primarily when dealing with complex tasks. In addition, self-organized systems can integrate a higher number of external relationships and can respond to system changes quickly and adequately.
3. *Organization of self-supervision.* The most important instrument in project management is the steering of information and communication in relation to time, status, content, and the people concerned.

4. *Networked thinking.* In contrast to linear thinking, networked thinking provides the opportunity to realize a multitude of influencing factors as well as their mutual interference. This is particularly important for interdisciplinary tasks.

5. *Multiple perspectives.* The *single point of view* is replaced by a *varied state of interpretation*, which opens up varied options for operation, with their respective advantages and disadvantages. Instead of there being a single "correct" method, a situation is considered from different perspectives and thought out in light of its consequences.

The effects of systemic principles on practical management show in the comparison below (Ulrich and Probst, 1988).

Evolutionary Target Development in Place of Fixed Objectives In traditional project management, the binding objective is seen as a central requirement, but multiple changes reinforce the questioning of the targets originally defined for the progressing project. Insisting on once-defined targets does not prevent an appropriate adjustment for circumstances, which may have changed. Nevertheless, in systemic project management, targets are also defined and are questioned critically after every cycle. Environmental changes are considered as well as experiences. This leads to a refocusing and a changed position of targets in the project.

Planning as a Means for Communication and Orientation In traditional project management, planning serves to operationalize project targets. A plan should contain reliable information about task structuring as well as the resource and cost situation. Here, a plan is understood to be an exact anticipation of the future. Systemic project management, however, defines as being a means of orientation and communication. Prognoses are limited only as a result of changing environmental conditions. In systemic project management, plans are made in order to diverge from them if necessary. In view of complexity and dynamics, project plans serve to restate beliefs about solution methods. In this way, plans support communication and orientation; they form the source required for learning and change processes.

Project Management for the Adaptation of Networked Structures Traditional project management assigns people to certain jobs. As a general rule, the jobs themselves pertain to the structure of the project plan, by which the classification of project employees is defined in the project hierarchy. In this way, technical experts are isolated in groups working on subprojects and communication is determined. Systemic project management actively changes project organization when necessary: for example, as a result of environmental influences. Interdisciplinary networked structures are thus preferred. Similarly, complexity is reduced by involving customers as an important element of project organization or by reflection though external experts.

Project Control Using a General Framework In traditional project management, project control has a direct influence on target size, cost, and time. Deviations are often corrected by direct intervention with these types of outcomes. In systemic project management, control occurs in principle by setting a general framework. For example, if a technical integration is delayed repeatedly, it is not useful to intervene directly in integration details. Rather, it is appropriate to examine the basic principles of integration: for example, involvement in information flow or in the general process.

Project Control through the Combination of Qualitative Variables Traditionally, examination of the project state is based on the investigation of quantitatively measurable and assessable variables. The complexity of a project, however, cannot be realized exclusively by quantitative variables. In systemic project management, both qualitative and quantitative measures are incorporated into assessment and control. Qualitative assessment permits a variety of interpretations. It is crucially important to secure systematic investigation, to integrate different perspectives, and to interpret and feed back valuations. This also means that no definitive valuation can result because the evolutionary controlling process always allows for more courses of action. However, this does not remove the responsibility for the decision. Practical experience with systemic management shows that a project is ultimately successful if it manages proactively to take in environmental changes and resulting opportunities.

4 HUMAN RESOURCE MANAGEMENT

In this section, essential aspects of human-oriented work design are discussed in detail from human resource management perspectives. One focus of human resource management rests on the accomplishment of demographic change challenges, which are closely related to measures of qualification and occupational health.

4.1 Accomplishing Demographic Change

4.1.1 Initial Situation

Demographical developments in nearly all industrial nations are leading to an increase in the elderly worker population. More dramatic than the decline in the absolute number of working people is the change in age composition as the number of younger workers decreases slowly but continually, and the group of elderly people able to work will be growing continuously in upcoming years. With this in mind, there will be an increase in the average age of staff and more competition in recruiting the junior workforce. Youth-centered enterprise principles, short-term success-oriented personnel policy, and general conditions that favor the retirement of older workers do not measure up to this development (Pack et al., 2000).

Age becomes a problem in professional life particularly when employees remain in stressful tasks for a long duration of time and when the relevant required specific capacity is used up to the point that individual resources no longer satisfy the work requirements. Ongoing physical and psychological stress in poorly designed work environments is partly responsible for physical decline and decreasing mental flexibility. For example, a lack of physical demand resulting from one-sided body posture, such as sitting continuously while working, leads to a reduction in physical efficiency and ultimately to the same result as in capacity overload: that is, musculoskeletal system damage.

The condition of an employee's health is not based primarily on his or her calendar age but by the result of past work conditions. A cutback in older employees' work performance does not generally apply—it always relates to specific tasks and work requirements. For example, a machine operator with damaged intervertebral disks might no longer be able to operate a machine but might be able to work in administration.

4.1.2 Designing Age-Based Work

There is no standard solution for designing age-based jobs, personnel placement, and work. The acceptable method for an enterprise depends on the specific initial conditions of the enterprise. In general, however, a basic sensibility on the topic of age is important. Age-based work design remains an illusion as long as production strategies aim exclusively for short-term economies of scale and profit increases. Indeed, such strategies cause long-term damage to the companies themselves, as misunderstood examples of *lean management* show. Here, the principally correct idea of reducing the number of hierarchical levels causes a number of layoffs, which result in a massive lack of qualified personnel (Pack et al., 2000).

To face the possible risks of demographic change for performance and innovation ability, work design and human resources management must aim both at nurturing employee psychological and physical capacity during their entire professional lifetimes and at developing the specific capabilities of older employees. During work design, it must be considered that with any work, physical and psychological capacity changes through training and learning in a mid- to long-term way depending on requirements. Therefore, work should be designed so that an assorted amount of changing body posture and movements as well as varied mental specifications is required to accomplish assignments.

Effective concepts of aged-based work design consider not only elderly people who already demonstrate performance limitations, but also people at the beginning of their careers, as projected damage to health should be counteracted as early as possible. For this purpose, rethinking is necessary by both employees and managers. Criteria for appropriate personal development processes should thus confirm that through the change of different work specifications, (1) new knowledge is gained, (2) the development of health-damaging strain and stress situations is halted, (3) new social configurations (work groups and the like) are experienced and thus new social key qualifications are learned, and (4) individual willingness and ability to deal with new working situations and requirements are supported.

4.2 Qualification

In modern production systems, knowledge represents an important resource for the creation of goods and services. Knowledge and capability influence the changeability, innovation power, and competitiveness of an enterprise significantly. Thus, it is important that employees be qualified. A superior target for operational qualification measures is the training of comprehensive decision making and responsibility as a sum of technical, method, social, and self-competence. Subject-specialized standard and procedural knowledge is losing importance as a basis for operational decision making and responsibility. In the course of further production procedures, operational decision making and responsibility will be based on prognosis and diagnosis, which allow for the development of professional decision making and responsibility with interventional knowledge.

The required availability of task-specific decision making and responsibility within complex production systems places heightened demands on the qualification process. Qualification processes are therefore especially concerned with the premises of flexible, decentralized organizational concepts:

- Qualification takes place increasingly in knowledge networks which are marked by exchange of knowledge and experience.
- Qualification strategies are based on short-term learning content and on the integration of learning and implementation phases.
- Qualification methods incorporate learners' self-organization, for example, by means of computer-aided self-learning.

Whereas in traditional qualification strategies, knowledge is taught, is supplemented periodically, and is spatially separated, competence is currently more and more job- and application-oriented. Poor predictability of future qualification places increasing demands on flexibility and rapidness of learning. Up-to-date qualification concepts abandon teaching the completeness of knowledge but instead use elements of knowledge for the development of problem solution strategies, enabling the learner to reflect new problems. A dynamic knowledge transfer is requested by the increasing amount of knowledge required during the course of one's professional life and by the short-lived validity of knowledge. Specific professional knowledge becomes old relatively quickly, and life-long learning is becoming more important. On-the-job training as a supplement to vocational training has become the rule.

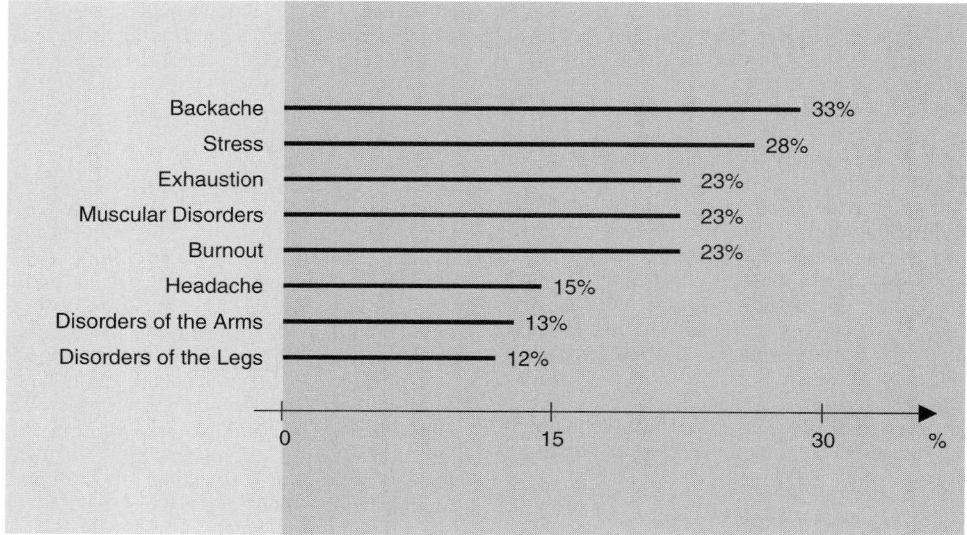

Figure 16 Health problems caused by work: fraction of complaints per 100 employees surveyed in countries within the European Union. (Data from Paoli and Merllié, 2001.)

4.3 Prevention of Occupational Health Problems

4.3.1 Stress Situation

Considering all possible results of innovative work structures, their negative effects on employee health and well-being cannot be overlooked. Increasing requirements on time and on-site flexibility as well as increasing pressure to perform are only a few effects of structural changes. Mental stress and chronic illnesses thus step into the forefront. These factors can lead to underperformance and reduced output.

In recent years, mental health disorders have spread among employees epidemically. More than half of the employees in the countries of the European Union complain about physical health damage caused by work (see Figure 16). In Germany, losses resulting from work-related health disorders are estimated to be over €3 billion a year; stress resulting from work is the cause of 7% of early retirements. The most common causes of underperformance and absences are related to mental disorders (WHO/ILO, 2000). In the working society, the mental health and well-being of employees gain importance as favorable goals that are worthy of being sustained and protected.

4.3.2 Definition of Health

Traditionally, *health* is defined as the absence of illness. According to common understanding, health includes the goals of competence development as well as broad physical, mental, and social well-being (WHO, 1986). Health includes the ability to define and pursue personal lifelong goals, the ability to adapt to changing environmental conditions, and the ability to take part in such changes. Therefore, a healthy person is goal-oriented and acts actively within his or her world to develop his or her abilities further. Health relies on personal and organizational resources. Occupational health is a precondition of productive coping with work challenges as well as a result of adequate working conditions.

4.3.3 Prevention Strategies

For a long time, occupational health prevention was determined primarily by the defense of acute risk hazards and accidents at work as well as by the impact of individual stress and illness factors with a specific known effect on health. In addition to these risks, focus is on the complex correlations of work factors and health effects. These complex correlations do not comply with the prototype of specific cause–effect relationships. They can only be influenced in part by detailed regulations. Furthermore, the health resources concept sets priorities by abandoning a risk-oriented approach (Braun, 2003).

Attempts at preventing occupational health risks include the stabilization and advancement of physical and mental health resources as well as the mobilization of employee capabilities. Physical and mental resources should be developed to cope with health disorders (see Figure 17). In addition to avoiding diseases caused by work, the stabilization of health with respect to potential illness-causing influences, as well as premature weakening processes caused by work, is being sought.

The resource-oriented definition of the health concept focuses on the human ability for competence development as a basis for health. Accordingly, the prevention of health concerns aims for a process that will enable humans to have greater self-responsibility

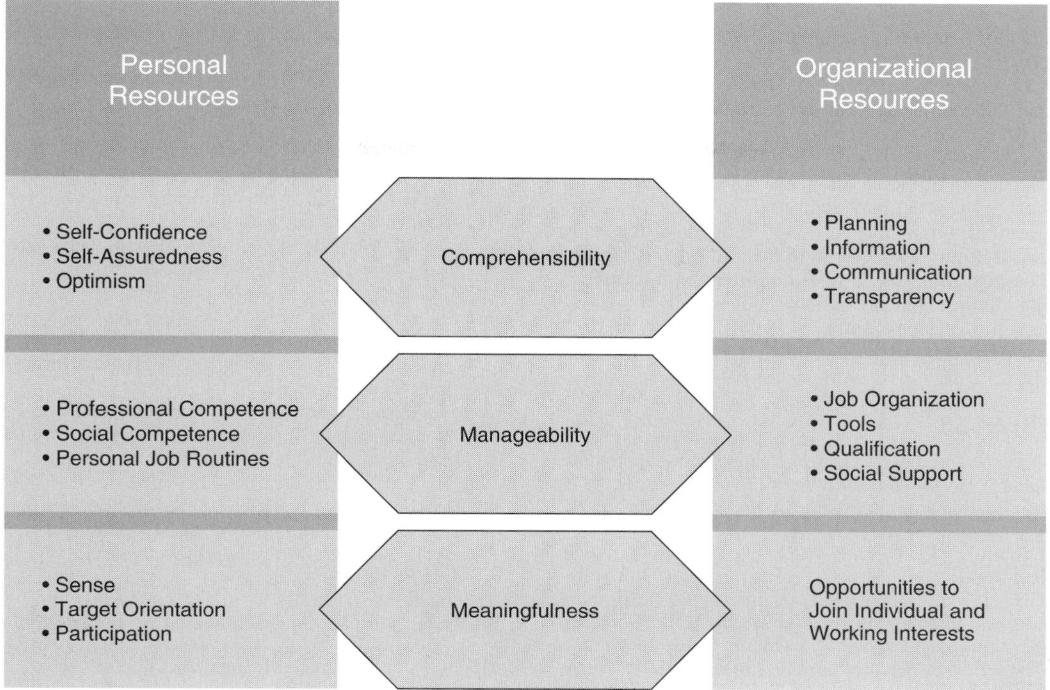

Figure 17 Organizational and personal health resources according to the coherence model by Antonovsky. (From Spath et al., 2003.)

for strengthening their health and, moreover, will enable them to do so. Advancement of work satisfaction and social well-being resulting from cooperation among all parties and a positive setup of the operational environment come into focus.

5 CONCLUSIONS

In recent years, producers have made numerous efforts to increase their performance, efficiency, and flexibility and thus to satisfy market demands. Fundamental characteristics of production concepts are teamwork and business process design. Despite great efforts and undeniable successes, a number of deficiencies still emerge, as a critical look at production processes and results shows: for example, long cycle time and time of delivery.

In view of the obvious limits of technology-oriented strategies, it has recently been recognized that targeted successes in production can be reached only through the holistic coaction of humans, technology, and organization. Holistic production systems integrate into one system elements of the organization of assembly, process, work, quality control, and continued improvement which up to now have been considered merely separated. Design principles, methods, and tools are incorporated into a single integrated production system. The traditional methods of human-oriented work design are thus not to be disregarded. They are to be seen as a necessary requirement for success, but they must be supplemented with innovative concepts.

One aspect that is critical for success is the development of production systems from operational practice. A truly efficient production system can only be realized when considering the specifications of the company or business. Comprehending business culture and human values make up the specific requirements for successful implementation. It is necessary to involve all participants, to identify good practices, and to fit this to the business standards. In this way, the system is comprehensible and practicable for all participants.

The comprehension of human success factors such as qualification, information, and participation is of particular importance. This calls for meaning, commitment, and readiness to take on responsibility and also calls for win–win situations as well as leadership through personal commitment.

In the past, producers have relied primarily on specific established success factors, which were mostly formally documented. They were comprehensible by means of strategy papers, plans, organization charts, job descriptions, operating instructions, and target systems. Currently, a tendency can be noted toward a stronger emphasis on soft success factors, which are geared to the conditions and interests of the working human. These factors include capabilities, values, cultures, and participation. Soft success factors are often impossible to define clearly. Although these factors remain somewhat latent, they affect the performance of businesses crucially, as numerous company examples document: These companies are extremely successful in relation to economic efficiency and competitive ability, customer satisfaction, and employee health.

It is now obvious in view of the variety of business influences that restructuring measures and technological advantage are no longer sufficient to maintain a company's competitive ability. Ultimately, only healthy, motivated, well-qualified human beings generate distinctive and hence successful competitive products and services. Consequently, healthy workers, together with adequate working conditions, are the focus of an innovative work design strategy.

REFERENCES

Allbeck, J., and Badler, N. (2002), "Embodied Autonomous Agents," in *Handbook of Virtual Environments*, K. Stanney, Ed., Lawrence Erlbaum Associates, Mahwah, NJ, pp. 313–332.

Antonovsky, A. (1979), *Health, Stress and Coping*, Wiley, New York.

Badler, N. (1997), "Virtual Humans for Animation, Ergonomics, and Simulation," presented at the IEEE Workshop on Non-rigid and Articulated Motion, Puerto Rico, June.

Bevan, N. (1999), "Quality in Use: Meeting User Needs for Quality," *Journal of System Software*, Vol. 49, pp. 89–96.

Braun, M. (2003), "Gesundheitspräventive Arbeitsgestaltung und Unternehmensentwicklung," *Das Gesundheitswesen*, Vol. 65 No. 12, pp. 698–703.

Bubb, H. (2000), "Wie sehen die Arbeitsplätze der Zukunft aus?" *VDI Ingenium*, Vol. 5.

Bullinger, H.-J. (1994), *Ergonomie: Produkt- und Arbeitsplatzgestaltung*, Teubner, Stuttgart, Germany.

Bullinger, H.-J. (1997), "Mechanische Werkzeuge und Maschinen," in *Handbuch der Arbeitswissenschaft*, H. Luczak and W. Volpert, Eds., Schäffer Poeschel, Stuttgart, Germany, pp. 598–601.

Bullinger, H.-J. (2003), *Früherkennung von Qualifikationserfordernissen in Europa*, Bertelsmann, Bielefeld, Germany.

Bullinger, H.-J., and Braun, M. (2001), "Arbeitswissenschaft in der sich wandelnden Arbeitswelt," in *Erträge der interdisziplinären Technikforschung*, G. Ropohl, Ed., Schmidt, Berlin, Germany, pp. 109–124.

CAESAR (2004), retrieved June 30, 2004, from http://store.sae.org/caesar.

Cofer, C., and Appley, M. (1964), *Motivation: Theory and Research*, Wiley, New York.

Collins, J., and Porras, J. (1994), *Built to Last: Successful Habits of Visionary Companies*, HarperCollins, New York.

Dörner, D. (2003), *Die Logik des Misslingens*, Rowohlt, Reinbek, Germany.

Fanger, P. O. (1972), *Thermal Comfort: Analysis and Applications in Environmental Engineering*, McGraw-Hill, New York.

Gallup (2002), "Nur 16 Prozent der Arbeitnehmer in Deutschland sind engagiert am Arbeitsplatz," retrieved December 10, 2002, from http://www.gallup.de/Mitarbeiterzufriedenheit.htm.

Hacker, W. (1998), *Allgemeine Arbeitspsychologie: Psychische Regulation von Arbeitstätigkeiten*, Huber, Berne, Switzerland.

Helander, M. (1995), *A Guide to the Ergonomics of Manufacturing*, Taylor & Francis, London.

Hertzberg, F., Mausner, B., and Snyderman, B. (1967), *The Motivation to Work*, Wiley, New York.

ISO (1991), *Ergonomic Principles Related to Mental Work-Load*, ISO 10075, International Organization for Standardization, Geneva.

Karwowski, W., Ed. (2001), *International Encyclopedia of Ergonomics and Human Factors*, Taylor & Francis, London.

Karwowski, W., and Salvendy, G., Eds. (1998), *Ergonomics in Manufacturing: Raising Productivity Through Workplace Improvement*, Society of Manufacturing Engineers, Dearborn, MI.

Korge, A., and Scholtz, O. (2004), "Ganzheitliche Produktionssysteme," *Werkstattstechnik Online*, Vol. 94, No. 1–2, pp. 2–6.

Lin, H. X., Chong, Y.-Y., and Salvendy, G. (1997), "A Proposed Index of Usability: A Method for Comparing the Relative Usability of Different Software Systems," *Behaviour and Information Technology*, Vol. 16, pp. 267–278.

Lippmann, R. (1997), "Anthropos, Produkt und Methode zur rechnergestützten Ergonomieanalyse," in *Software-Werkzeuge zur ergonomischen Arbeitsgestaltung*, K. Landau, H. Luczak, and W. Laurig, Eds., REFA (Verbaud für Arbeitsgestaltung, Belriebsorganisation und Unternehmensentwicklung, Association for Work Design, Industrial Organization and Corporate Development), Darmstadt, Germany.

Luczak, H. (1998), *Arbeitswissenschaft*, Springer, Berlin, Germany.

Luczak, H., Volpert, W., Raeithel, A., and Schwier, W. (1987), *Arbeitswissenschaft; Kerndefinition–Gegenstandsbereich–Forschungsgebiete*, Rationalisierungs-Kuratorium, Eschborn, Germany.

Maslow, A. (1987), *Motivation and Personality*, 3rd ed., Harper Collins, New York.

Moray, N. (1997), "Human Factors in Process Control," in *Handbook of Human Factors and Ergonomics*, 2nd ed., G. Salvendy, Ed., Wiley, New York, pp. 1944–1971.

NIOSH (1994), *Applications Manual for the Revised NIOSH Lifting Equation*, Publication 94–110, National Institute for Occupational Safety and Health, Washington, DC.

Oborne, D. J., and Arnold, K. M. (2001), "Human–Machine Interaction: Usability and User Needs of the System," in *Handbook of Industrial, Work and Organizational Psychology*, N. Anderson, D. S. Ones, H. K. Sinangil, and C. Viswesvaran, Eds., Sage Publications, London.

Paoli, P., and Merllié, D. (2001), *Third European Survey on Working Conditions 2000*, European Foundation for the Improvement of Working and Living Conditions, Dublin, Ireland.

Pack, J., Buck, H., Kistler, E., Mendius, H., Morschhäuser, M., and Wolff, H. (2000), *Zukunftsreport demographischer Wandel: Innovationsfähigkeit in einer alternden Gesellschaft*, Bundesministerium für Bildung und Forschung, Bonn, Germany.

Probst, G. (1997), *Selbstorganisation: Ordnungsprozesse in sozialen Systemen aus ganzheitlicher Sicht*, Parey, Berlin.

Richter, P., and Hacker, W. (1998), *Belastung und Beanspruchung: Stress, Ermüdung und Burnout im Arbeitsleben*, Heidelberg, Asanger, Germany.

Salvendy, G., Ed. (1997), *Handbook of Human Factors and Ergonomics*, 2nd ed., Wiley, New York.

Scholtz, O., Korge, A., and Schlauss, S. (2003), "Was ein Produktionssystem ausmacht," in *Ganzheitlich Produzieren*, D. Spath, Ed., Logis, Stuttgart, Germany, pp. 53–84.

Sensation (2004), Web site of the Sensation project, retrieved June 1, 2004, from http://www-sensation-eu.org.

Spath, D. (2003), "Revolution durch Evolution," in D. Spath, Ed., *Ganzheitlich Produzieren*, Logis, Stuttgart, Germany, pp. 15–44.

Spath, D., and Dill, C. (2002), "Ist Flexibilität genug? Turbulenzeny sind nur mit systemischem Denken zu bewältigen," in *Erfolg in Netzwerken*, J. Milberg and G. Schuh, Eds., Springer, Berlin, Germany, pp. 161–175.

Spath, D., Dill, C., and Scharer, M. (2001), *Vom Markt zum Markt: Produktentsehung als zyklischer Prozess*, Logis, Stuttgart, Germany.

Spath, D., Braun, M., and Grunewald, P. (2003), *Gesundheits- und leistungsförderliche Gestaltung geistiger Arbeit*, Schmidt, Berlin.

Taylor, F. W. (1911), *The Principles of Scientific Management*, Harper, New York.

Ulrich, H., and Probst, G. (1988), *Anleitung zum ganzheitlichen Denken und Handeln*, Haupt, Stuttgart, Germany.

Westkämper, E. (1996), "Neue Perspektiven der Fertigungstechnik," *Holz- und Kunststoffverarbeitung*, Vol. 31, No. 5, pp. 74–79.

Westkämper, E. (2004), "Structural Change in Manufacturing: Caused by Turbulent Influencing Factors," in *Proceedings of the International Conference on Competitive Manufacturing* (COMA), D. Dimitrov, Ed., Stellenbosch, South Africa, pp. 21–27.

Westkämper, E., and Warnecke, H.-J. (2002), *Einführung in die Fertigungstechnik*, 5th ed., Teubner, Stuttgart, Germany.

Woods, D., and Hanes, L. (1986), "Human Factors Challenges in Process Control: The Case of the Nuclear Power Plants," in *Handbook of Human Factors*, G. Salvendy, Ed., Wiley, New York.

WHO (1986), *Ottawa Charter for Health Promotion*, World Health Organization, Ottawa, Ontario, Canada.

WHO/ILO International Labour Organization (2000), *Mental Health and Work: Impact, Issues and Good Practices*, World Health Organization, Geneva.

INDEX

UNIVERSITY OF STRATHCLYDE
- 6 JUN 2007
UNIVERSITY LIBRARY